Effects of Drugs on Clinical Laboratory Tests

THIRD EDITION

Donald S. Young, MB, PhD
Director, Division of Laboratory Medicine
Department of Pathology and Laboratory Medicine
University of Pennsylvania
Philadelphia, Pennsylvania

Introduction	i to ii
Laboratory Test Index	1–1 to 1–9
Drug Index	2–1 to 2–18
Laboratory Test Listings	3–1 to 3–381
Drug Listings	4–1 to 4–462
References	5–1 to 5–67

AACC PRESS

2029 K Street, NW
Seventh Floor
Washington, DC 20006

ISBN 0-915274-53-1

Database editing, book assembly and typography by Lexi-Comp Inc, Hudson, Ohio

INTRODUCTION

This third edition of the Effects of Drugs on Clinical Laboratory Tests is a major expansion and revision of the edition published in 1975. Many of the drugs that are most commonly used now were not available when the previous edition was prepared. Many new laboratory tests also have been added reflecting the real progress that has been made over the past 15 years in the field of laboratory medicine. Although several laboratory tests included in the previous editions are rarely performed except in major medical centers or research environments, these tests have been retained since they may still be requested and performed in laboratories of tertiary-care centers or in research environments. Some tests are included that are infrequently performed in this country but still are used elsewhere. The major sources used for developing this compilation were clinical and clinical laboratory journals. However, I have included the entire content of the publication by Dr. O. Sonntag entitled "Arzeneimittel – Interferenzen" published in 1985 by Thieme, Stuttgart, West Germany and containing the results of Dr. Sonntag's personal studies of the effects of many different pure drugs on a wide variety of analytical methods. I am most grateful to Dr. Sonntag for permission to include his work cited as reference 3393 in the text. I am responsible for any errors in the translation from German and misinterpretations.

With the great number of drugs available and the very large number of publications concerned with the effects of drugs on clinical laboratory tests it is inevitable that many effects that should have been included in this compilation have been omitted since it is difficult to review all relevant literature. Furthermore, is often difficult to decide whether a specific reported effect should be included in the text. Some effects undoubtedly pertain to unique situations associated with a particular patient and perhaps with an unusual dose of drug. Some studies reporting an effect of a drug on a analyte measured by a specific method have used a quantity of a drug far greater than likely to occur under normal conditions in a body fluid. In the clinical situation the altered metabolism caused by the patient's disease may mean that effects occurring in an ill patient and a healthy volunteer may differ. Not all documented effects refer to both situations. Most studies reported of the effects of a drug refer to its administration to healthy individuals or to individuals with specific diseases and may not be relevant for the situation in a particular patient. Concurrent administration of other drugs may alter the actions of a drug so that different effects may be observed, even in the same patient, on different occasions.

It is important to note that some effects of drugs on analytical methods are due to metabolites of the drugs and not to the administered drug. For many studies of drugs on analytical methods the study conditions do not mimic in vivo conditions eg, no protein or an inappropriate amount of protein is used in the test system. Also the pure parent compound rather than the reactive metabolite may be studied. Sometimes additives in the tablet or capsule containing the drug may cause effects. Note also that slight variations in the analytical method may cause considerable differences from the reported magnitude of the effects. Many descriptions of the effects of drugs on laboratory tests fail to state the concentration of drug at which the effect was observed or report effects at inappropriate or non physiological concentrations. Other reports fail to indicate whether the effect was sufficiently large to have clinical significance or was only statistically significant but unlikely to influence the management of a patient. The effects contained in this publication are listed as if only one drug was administered at a time which is rarely the case in hospital practice. Thus, the effects listed here may in reality be considerably modified.

It is important to note that the effects of drugs listed here have been reported in one or more publications. It does not mean that an abnormal test result observed in a patient may be attributable to a particular drug that was being administered to that patient. Any effect should be considered in the context of the patient's illness, concomitant treatment and biological factors. This compilation should only serve as a guide to possible causes of abnormal test results. It certainly cannot provide the definitive explanation for abnormal test results when a particular combination of drug and abnormal test result occurs in a patient.

The directory is organized in five sections. The first lists the laboratory tests. The second lists the search terms, in many cases the drugs that are the main focus of the directory. It includes the generic names of all the drugs included in the interactions as well as the proprietary names of these drugs. The compilation is oriented to generic names but this directory may be used to identify the correct generic search term when only a proprietary name is known. Note that this section also includes some entries related to the general class of drugs, eg, anticonvulsants. This may require some cross-referencing to identify additional drug effects. The third and fourth sections are the major components of the compilation. The third is a sort by laboratory test and the fourth a sort by drug or factor altering test results. Each entry comprises the drug name, the body fluid and analyte measured and whether the effect is a physiological or pharmacological (*in vivo*) effect, or an analytical one affecting the measurement procedure (*in vitro* effect). A brief description of the mechanism of the effect follows and the entry is completed with a reference number. To facilitate sorting I have attempted to list all analytes as if they were measured only in either plasma or serum and not necessarily as reported in the original publication. The fifth section lists the references by number in the text, names of authors, title of article and journal source. The format has been designed to be compatible with that used in Friedman RB, and Young, DS: Effects of Disease on Clinical Laboratory Tests (2nd ed), AACC Press, Washington, DC, 1989.

I am grateful to Richard B. Friedman and Steven Entine for their efforts to develop and merge the computer files of the Effects of Drugs and to adapt them for publication. Lucy Marshall, who did much of the background research and data entry here, deserves special thanks. Carol Schwartz, who entered many items into the master database in Wisconsin, contributed much to the project. I also recognize the contribution of David K. Ream who contributed many ideas to the refinement of the files for publication. Jay L. Katzen, Robert D. Kerscher and Leonard L. Lance of Lexi-Comp were responsible for all the painstaking details of composition but, at the same time, provided considerable input to improve the format and utility of the publication. Lenne P. Miller and Janice Saylor of the AACC steered the project through to completion. However, I must assume responsibility for any errors which regretably may be present because of the size and complexity of the project. Please contact me when you identify any misinterpretation or other error so that the database may be updated and corrected.

— Donald S. Young
June, 1990

1 LABORATORY TEST INDEX

A

Absorbance at 450 nm
ACE
    see Angiotensin-Converting
       Enzyme (ACE)
Acenocoumarol
Acetaldehyde
Acetaldehyde Oxidase
Acetaminophen
Acetaminophen Screening Test
Acetate
Acetic Acid
Acetoacetate
    see also
       Ketones
Acetoacetate Decarboxylase
Acetone
    see also
       Ketones
Acetylcholine Receptor Antibodies
Acetylcholinesterase
N-Acetylglucosaminidase
N-Acetylmetanephrine
N-Acetylnormetanephrine
Acetylsalicylic Acid
Acid Alpha-D-Glucoside Glucohydrolase
    see α-Glucosidase
Acid α-D-Glucoside Glucohydrolase
    see α-Glucosidase
Acid Glycoprotein, $\alpha 1$
    see α_1-Acid Glycoprotein
α_1-Acid Glycoprotein
Acid Phosphatase
Acid Phosphatase, Prostatic
Acid Phosphatase, Tartrate Labile
ACP
    see Acid Phosphatase
ACTH
    see Corticotropin
Activated Partial Thromboplastin Time
    see Partial Thromboplastin Time
Acute Phase Proteins
    see α_1-Antitrypsin
Acute Phase Reactant
    see C-Reactive Protein
Acylcholine Acyl Hydrolase
    see Pseudocholinesterase
Adenosine Deaminase
Adenosine Deaminase Binding Protein
Adenosine Monophosphate
    see Cyclic Adenosine
       Monophosphate
Adenosine Triphosphate (ATP)
S-Adenosylmethionine
Adenylate Kinase
Adipate
Adrenaline
    see Epinephrine

Adrenocorticotropic Hormone
    see Corticotropin
AFP
    see α-Fetoprotein
Agglutination Tests
    see also
       Rheumatoid Factor
Agglutinins
AHF-Like Antigen
AHF Procoagulant Activity
ALA
    see Aminolevulinic Acid
Alanine
    see also
       Amino Acids
Alanine Aminopeptidase
Alanine Aminotransferase
Albumin
Alcohol
    see also
       Ethanol
Aldehyde Dehydrogenase
Aldolase
Aldosterone
Alkaline Phosphatase
Alkaline Phosphatase Isoenzymes
Alkaloids
Alloxanthine
Alpha$_1$-Antichymotrypsin
    see α_1-Antichymotrypsin
Alpha$_1$-Lipoprotein
    see also
       Lipoproteins
Alpha-Aminobutyrate
    see α-Amino-N-Butyric Acid
Alpha-Amino-N-Butyric Acid
    see α-Amino-N-Butyric Acid
    see also
       Amino Acids
Alpha-Fetoprotein
    see α-Fetoprotein
Alpha-Glucosidase
    see α-Glucosidase
Alpha-Hydroxybutyrate Dehydrogenase
    see Hydroxybutyrate
       Dehydrogenase
Alpha-Hydroxy-dehydroepiandrosterone
    see 16-α-Hydroxy-
       dehydroepiandrosterone
Alpha-Hydroxy-dehydroepiandrosterone
   Glucuronide
    see also
      15-α-Hydroxy-
       dehydroepiandrosterone
       Glucuronide
      16-α-Hydroxy-
       dehydroepiandrosterone
       Glucuronide

Alpha-Hydroxyprogesterone
    see 16-α-Hydroxyprogesterone
Alpha-Lipoproteins
    see α-Lipoproteins
    see also
       Lipoproteins
ALT
    see Alanine Aminotransferase
Aluminum
Amikacin
Amino-4-Imidazole-5-Carboxamide
   Ribotide (AICAR)
Aminoacid Arylpeptidase
Amino Acids
    see also
       Alanine
       α-Amino-N-Butyric Acid
       β-Amino-Isobutyric Acid
       Arginine
       Asparagine
       Aspartic Acid
       Citrulline
       Cystathionine
       Cysteine
       Cystine
       Glutamic Acid
       Glutamine
       Glycine
       Histidine
       Homocitrulline
       Homocystine
       Hydroxyproline
       Isoleucine
       Leucine
       Lysine
       Methionine
       Methylhistidine
       Ornithine
       Phenylalanine
       Proline
       Sarcosine
       Serine
       Taurine
       Threonine
       Tryptophan
       Tyrosine
       Valine
Aminobenzoic Acid
α-Aminobutyrate
    see α-Amino-N-Butyric Acid
α-Amino-N-Butyric Acid
    see also
       Amino Acids
Amino-Isobutyric Acid
    see β-Amino-Isobutyric Acid
β-Amino-Isobutyric Acid
    see also
       Amino Acids
Aminolevulinic Acid

δ-Aminolevulinic Acid
  see Aminolevulinic Acid
Aminolevulinic Acid Dehydrase
Amino Nitrogen
  see α-Amino-Nitrogen
α-Amino-Nitrogen
Aminopyrine
Aminosalicylic Acid
Amitriptyline
Ammonia
Ammoniacal Silver Nitrate
Ammonium Ions
Amobarbital
Amoxicillin
AMP
  see Cyclic Adenosine
    Monophosphate
Amphetamine
Ampicillin
Amylase
Amylase, Pancreatic
Amylase, Salivary
Androgens
  see also
    Androstenedione
    Androsterone
    Testosterone
5-Androstene-3 β, 17 β-Diol
Androstenedione
  see also
    Androgens
Androstenedione, Free
Androstenetriol Sulfate
Androsterone
  see also
    Androgens
Angiotensin
Angiotensin-Converting Enzyme (ACE)
Angiotensin I
Angiotensin II
Angiotensinogen
Anisindione
Anthranilic Acid
Antibodies to DNA
Antibodies to dsDNA
Antibodies to Histones
Antibodies to Native DNA
  see Antibodies to dsDNA
Antibodies to Ribonuclear Protein
  see Antibodies to RNP
Antibodies to RNP
Antibodies to SCL-70
Antibodies to Sjögren Syndrome A
Antibodies to Sjögren Syndrome B
Antibodies to SmAg
Antibodies to Smith Antigen
  see Antibodies to SmAg
Antibodies to SS-A
  see Antibodies to Sjögren
    Syndrome A
Antibodies to SS-B
  see Antibodies to Sjögren
    Syndrome B
Antibody Evaluation
Antibody Titer
  see also
    Antibodies to dsDNA
    Antibodies to Histones
    Antibodies to RNP
    Antibodies to SCL-70
    Antibodies to Sjögren Syndrome
      A
    Antibodies to Sjögren Syndrome
      B
    Antibodies to SmAg
    Antimitochondrial Antibodies
    Antinuclear Antibodies
    Antithyroglobulin Antibodies
α₁-Antichymotrypsin
Antidiuretic Hormone
Antihemophilic Factor
  see Factor VIII

Antimitochondrial Antibodies
Antinuclear Antibodies
Antinucleoprotein Antibodies
Antipyrine
Antismooth Muscle Antibodies
Antithrombin III
Antithrombin III Antigen
Antithrombin Titer
Antithyroglobulin Antibodies
Antithyroid Antibodies
  see Antithyroglobulin Antibodies
α₁-Antitrypsin
Apolipoprotein A
Apolipoprotein AI
Apolipoprotein AII
Apolipoprotein B
Apolipoprotein B100
Apolipoprotein CII
Apolipoprotein CIII
Apolipoprotein D
Apolipoprotein E
Appearance
Aprindine
APTT
  see Partial Thromboplastin Time
Arabinose
Arginase
Arginine
  see also
    Amino Acids
Argininosuccinate Lyase
Arsenic
Arylsulfatase A
Ascorbic Acid
Asparagine
  see also
    Amino Acids
Aspartate Aminotransferase
Aspartic Acid
  see also
    Amino Acids
Aspirin Esterase
AST
  see Aspartate Aminotransferase
Atenolol
Atrial Natriuretic Peptide

B

Bacteria
Barbital
Barbiturate
Basal Metabolic Rate
Base Deficit
Base Excess
Basophilic Stippling
Basophils
B-Cell Differentiation Factor
B-Cells
  see Lymphocyte B-Cells
BEI
  see Butanol Extractable Iodine
    (BEI)
Bence-Jones Protein
Benzoic Acid
Beta₁C-Globulin
  see Complement C3
Beta₁E-Globulin
  see Complement C4
Beta₂-Glycoprotein
  see β₂-Glycoprotein
Beta₂-Microglobulin
  see β₂-Microglobulin
Beta-Amino-Isobutyric Acid
  see β-Amino-Isobutyric Acid
  see also
    Amino Acids
Beta-Carotene
  see Carotene

Beta-Endorphin
  see β-Endorphin
Beta-Galactosidase
  see β-Galactosidase
Beta-Globulin
  see β-Globulin
Beta-Globulin C/A
  see β₁-Globulin C/A
Beta-Glucosidase
  see β-Glucosidase
Beta-Glucuronidase
  see β-Glucuronidase
Beta-Glusosaminidase
  see β-Glusosaminidase
Beta-Hydroxybutyrate
  see also
    Ketones
Beta-Hydroxybutyric Acid
  see β-Hydroxybutyrate
Beta-Hydroxycortisol
  see 6-β-Hydroxycortisol
Beta-Lipoprotein Cholesterol
  see β-Lipoprotein Cholesterol
Beta-Lipoproteins
  see β-Lipoproteins
Beta-Thromboglobulin
  see β-Thromboglobulin
Bicarbonate
Bile
Bile Acids
  see also
    Chenodeoxycholic Acid
    Cholic Acid
    Deoxycholic Acid
Bile Acids, Conjugated
Bile Salts
Bilirubin
Bilirubin, Conjugated
Bilirubin, Direct
Bilirubin, Indirect
Bilirubin, Neonatal
Bilirubin, Unconjugated
Biotin
Bishydroxycoumarin
Bismuth
Blast Count
Bleeding Time
Blood Urea Nitrogen
  see Urea Nitrogen
B Lymphocyte
  see Lymphocyte B-Cells
Bogen Test
Bradykinin
Bromide
BSP Retention
BUN
  see Urea Nitrogen
Butabarbital
Butanol Extractable Iodine (BEI)
Butyric Acid

C

C₁-Inhibitor
C1q Complement
  see Complement C1q
C3 Complement
  see Complement C3
C4 Complement
  see Complement C4
Cadmium
Cadmium Flocculation
Caffeine
Calcifediol
Calcitonin
Calcitriol
  see 1,25-Dihydroxy Vitamin D₃
Calcium
Calcium, Ultrafiltrable
Calculi

cAMP
  see Cyclic Adenosine
    Monophosphate
Capillary Fragility
Caproic Acid
Carbamazepine
Carbohydrate, Protein Bound
Carbon Dioxide Partial Pressure
  see pCO_2
Carbonic Anhydrase
Carbon Monoxide
  see Carboxyhemoglobin
Carboxyhemoglobin
Carcinoembryonic Antigen
Carnitine
Carotene
β-Carotene
  see Carotene
Carotenoids, Total
Casts
  see also
    Granular Casts
    Hyaline Casts
Catalase
Catecholamines
  see also
    Dopamine
    Epinephrine
    Homovanillic Acid
    Metanephrine
    Metanephrines, Total
    Norepinephrine
    Normetanephrine
Catechol-O-Methyl Transferase
CCK
  see Cholecystokinin (CCK)
CEA
  see Carcinoembryonic Antigen
Cefoperazone
Cefsulodin
Ceftazidime
Ceftriaxone
Cells
Cell Water
Cephalexin
Cephalin
Cephalin Flocculation
Cephalin Time
Cephaloridine
Cephalothin
Cephradine
Ceruloplasmin
Cervical Secretion
Chenodeoxycholic Acid
  see also
    Bile Acids
Chloral Hydrate
Chloramphenicol
Chlordiazepoxide
Chloride
Chlormethiazole
Chloroquine
Chlorothiazide
Chlorphenmetrazine
Chlorphentermine
Chlorpromazine
Chlorpropamide
Chlortetracycline
Chloylglycine
Cholecystokinin (CCK)
Cholesterol
Cholesterol Esters
Cholesterol, Free
Cholesterol, HDL
  see Cholesterol, High Density
    Lipoprotein
Cholesterol, High Density Lipoprotein
Cholesterol, LDL
  see Cholesterol, Low Density
    Lipoprotein
Cholesterol, α-Lipoprotein
Cholesterol, β-Lipoprotein

Cholesterol, Low Density Lipoprotein
  see β-Lipoprotein Cholesterol
Cholesterol, Very Low Density
  Lipoprotein
Cholesterol, VLDL
  see Cholesterol, Very Low Density
    Lipoprotein
Cholic Acid
  see also
    Bile Acids
Cholinesterase
  see Pseudocholinesterase
Christmas Factor
  see Factor IX
Chromium
Chromosomes
Chylomicrons
Chymotrypsin
Citrate
Citrulline
  see also
    Amino Acids
CK
  see Creatine Kinase
Clindamycin
Clonazepam
Clotermine
Clot Lysis
Clot Retraction
Clotting Time
CO_2 Content
  see Bicarbonate
Coagulation Time
Cobalamin
  see Vitamin B_{12}
Cobalt
Cocaine
Cold Agglutinins
Color
Complement
  see Complement CH_{50}
Complement APH_{50}
Complement C1q
Complement C3
Complement C3c
Complement C4
Complement CH_{50}
Complement Factor B
Complement, Total Hemolytic
  see Complement CH_{50}
Compound S
  see 11-Deoxycortisol
Congo Red Test
Conjugated Bilirubin
  see Bilirubin, Direct
Coombs' Test
Coombs' Test, Direct
Coombs' Test, Indirect
Copper
Coproporphyrin
  see also
    Porphyrins
Corticosteroid-Binding Globulin
  see Transcortin
Corticosteroids
Corticosterone
Corticotropin
Cortisol
Cortisol, Free
Cortisol, Protein Bound
Cortisone
Cotinine
C-Peptide
CPK
  see Creatine Kinase
C-Reactive Protein
Creatine
Creatine Kinase
Creatine Kinase Isoenzymes
Creatine Phosphate
Creatine Phosphokinase
  see Creatine Kinase

Creatinine
Creatinine Clearance
  see also
    Glomerular Filtration Rate (GFR)
CRP
  see C-Reactive Protein
Cryofibrinogen
Crystals
Cyanmethemoglobin
Cyanocobalamin
  see Vitamin B_{12}
Cyclic Adenosine-3,5-Monophosphate
  see Cyclic Adenosine
    Monophosphate
Cyclic Adenosine Monophosphate
Cyclic AMP
  see Cyclic Adenosine
    Monophosphate
Cyclosporine
Cystathionine
  see also
    Amino Acids
Cysteine
  see also
    Amino Acids
Cystine
  see also
    Amino Acids
Cystine Aminopeptidase

D

Dapsone
DDT (Chlorophenothane)
Dehydroascorbic Acid
7-Dehydrocholesterol
Dehydroepiandrosterone (DHEA)
Dehydroepiandrosterone Sulfate
  (DHEA-S)
Delta-Aminolevulinic Acid
  see Aminolevulinic Acid
Demethylchlortetracycline
Deoxycholic Acid
  see also
    Bile Acids
Deoxycorticosterone
  see also
    11-Hydroxycorticosteroids
11-Deoxycorticosterone
11-Deoxycortisol
Deoxycytidine
Desmethylcotinine
Desmosterol
Dexamethasone
Dexamethasone Suppression
DHEA
  see Dehydroepiandrosterone
    (DHEA)
DHEA Sulfate
  see Dehydroepiandrosterone
    Sulfate (DHEA-S)
Diagnex Blue Excretion
Dialyzable Free Thyroxine
Diamine Oxidase
Diatrizoate
  see also
    Glomerular Filtration Rate (GFR)
Diatrizoate Clearance
Diazepam
Dicumarol
Diffusible Phosphate
Digitoxin
Digoxin
Digoxin Clearance
Dihydroxyphenylacetic Acid
2,5-Dihydroxyphenylacetic Acid
  see Homogentisic Acid
Dihydroxyphenylalanine
3,4-Dihydroxyphenylethylamine
  see Dopamine

3,4-Dihydroxyphenylethylene Glycol
Dihydroxyphenylglycol
1,25-Dihydroxy Vitamin D$_3$
24,25-Dihydroxy Vitamin D
3,4-Dimethoxyphenylethylamine (3,4-DMPEA)
Diphenylhydantoin
2,3-Diphosphoglycerate
2,3-Diphosphoglycerate Mutase
Direct Bilirubin
    see Bilirubin, Direct
Disopyramide
DNA, Double Stranded Antibodies
    see Antibodies to dsDNA
DOC
    see Deoxycorticosterone
    see 11-Hydroxycorticosteroids
Dopa
    see Dihydroxyphenylalanine
Dopamine
    see also
        Catecholamines
Dopamine Hydroxylase
Dopa Screening Test
Double Stranded DNA Antibodies
    see Antibodies to dsDNA
Doxycycline
dsDNA Antibodies
    see Antibodies to dsDNA

E

EDTA Clearance
    see also
        Glomerular Filtration Rate (GFR)
Effective Renal Plasma Flow
Elastase 2
Electrophoresis
    see Lipoprotein Electrophoresis
    see Protein Electrophoresis
Electrophoretic Index
Encainide
Endoplasmic Reticulum Antibody
β-Endorphin
Enteroglucagon
Eosinophils
Epinephrine
    see also
        Catecholamines
ERPF
    see Effective Renal Plasma Flow
Erythrocytes
Erythrocyte Sedimentation Rate
Erythrocyte Survival
Erythromycin
Erythropoietin
Estradiol
    see also
        Estrogens
Estradiol Binding Globulin
Estradiol, Free
Estriol
    see also
        Estrogens
Estriol-3-Glucuronide
Estrogen Binding Globulin
Estrogens
    see also
        Estradiol
        Estriol
        Estrone
Estrogens (Conjugated)
Estrone
    see also
        Estrogens
Estrone Sulfate
Ethanol
Ethchlorvynol
Ethosuximide
Ethylene Glycol

Etiocholanolone
Euglobulin Clot Lysis Time
Euglobulin Lysis Time
    see Euglobulin Clot Lysis Time
Expiratory Volume
Extracellular Fluid (ECF)
Extractable Nuclear Antigen
    see Antibodies to RNP

F

Factor I
    see Fibrinogen
Factor II
Factor V
Factor VII
Factor VIII
Factor VIII Antigen
Factor IX
Factor X
Factor XI
Factor XII
FANA
    see Antinuclear Antibodies
Fat
Fatty Acids, Free (FFA)
Fatty Acids, Nonesterified (NEFA)
    see Fatty Acids, Free (FFA)
Fatty Acids, Total
Fenfluramine
Ferric Chloride Test
Ferritin
Ferroxidase
α-Fetoprotein
FFA
    see Fatty Acids, Free (FFA)
Fibrin Degradation Products
    see Fibrin Split Products (FSP)
Fibrinogen
Fibrinolysin
    see Plasminogen Antigen
Fibrinolysis
Fibrinolytic Activity
Fibrinolytic Time
Fibrinopeptide A
Fibrin Split Products (FSP)
Fibronectin
FIGLU (N-Formiminoglutamic Acid)
Flecainide
Fluoride
Flurazepam
Flurbiprofen
Folate
Folic Acid
    see Folate
Follicle Stimulating Hormone (FSH)
Formaldehyde
Formic Acid
N-Formiminoglutamic Acid
    see FIGLU (N-Formiminoglutamic Acid)
Fouchet Test
FPN Test
Fractional Sodium Excretion
Free Cholesterol
    see Cholesterol, Free
Free Fatty Acids
    see Fatty Acids, Free (FFA)
Free Reverse Tri-iodothyronine (T$_3$)
    see Tri-iodothyronine (T$_3$), Reverse Free
Free Thyroxine
    see Thyroxine (T$_4$), Free
Free Triiodothyronine
    see Tri-iodothyronine (T$_3$), Free
Free Tryptophan
    see Tryptophan, Free
Free Water Clearance
Fructosamine
Fructose

Fructose-6-Phosphate
FSH
    see Follicle Stimulating Hormone (FSH)
FSH Response to LHRH
FSP
    see Fibrin Split Products (FSP)
FT$_4$
    see Thyroxine (T$_4$), Free
FTI
    see Thyroxine (T$_4$) Index, Free (FTI)
Fucose
α-Fucosidase
Fumarate
Furosemide

G

G-6-PD
    see Glucose-6-Phosphate Dehydrogenase
Galactose
Galactose-1-Phosphate Uridyl Transferase
β-Galactosidase
Gamma-Glutamyltransferase (GGT)
    see γ-Glutamyltransferase (GGT)
Gamma-Glutamyl Transpeptidase
    see γ-Glutamyltransferase (GGT)
Gamma-Interferon
    see γ-Interferon
Gastric Inhibitory Polypeptide
Gastrin
GC Globulin
Gentamicin
GFR
    see Glomerular Filtration Rate (GFR)
GGT
    see γ-Glutamyltransferase (GGT)
GGTP
    see γ-Glutamyltransferase (GGT)
Globulin
α_1-Globulin
α_2-Globulin
β-Globulin
β_1-α-Globulin
β_1C-Globulin
    see Complement C3
β_1E-Globulin
    see Complement C4
β_1-Globulin C/A
γ-Globulin
Globulins, α_1
    see α_1-Globulin
Globulins, β_1A
    see β_1-α-Globulin
Globulins, β_1C
    see Complement C3
Globulins, β_1E
    see Complement C4
Globulins, α_2
    see α_2-Globulin
Globulins, γ
    see γ-Globulin
Glomerular Filtration Rate (GFR)
Glucagon
Glucagon, Pancreatic
Glucaric Acid
Glucocorticoids
Glucose
Glucose-6-Phosphatase
Glucose-6-Phosphate
Glucose-6-Phosphate Dehydrogenase
Glucose Clearance
Glucose Tolerance
α-Glucosidase
β-Glucosidase
Glucuronic Acid
β-Glucuronidase

β-Glusosaminidase
Glutamate Dehydrogenase
Glutamic Acid
    see also
        Amino Acids
Glutamic Oxaloacetic Transaminase
    see Aspartate Aminotransferase
Glutamic Pyruvic Transaminase
    see Alanine Aminotransferase
Glutamine
    see also
        Amino Acids
γ-Glutamyltransferase (GGT)
γ-Glutamyl Transpeptidase
    see γ-Glutamyltransferase (GGT)
Glutarate
Glutathione
Glutathione Reductase
Glutathione S-Transferase
Glutethimide
Glycerol
Glycine
    see also
        Amino Acids
Glycocholic Acid
Glycogen Synthetase
Glycoprotein, α1
    see α₁-Acid Glycoprotein
β₂-Glycoprotein
Glycosylated Hemoglobin
    see Hemoglobin A₁c
Glycosylated Proteins
Gold
Gonadotropin, Chorionic
    see also
        Pregnancy Tests
Gonadotropin, Pituitary
    see also
        Follicle Stimulating Hormone
            (FSH)
        Luteinizing Hormone (LH)
Gonadotropins
GOT
    see Aspartate Aminotransferase
GPT
    see Alanine Aminotransferase
Granular Casts
Granulocytes
Griseofulvin
Growth Hormone
GT
    see γ-Glutamyltransferase (GGT)
GTT
    see Glucose Tolerance
Guaiacols Spot Test
Guanase
Guanidinosuccinic Acid

H

Hageman Factor
    see Factor XII
Haptoglobin
hCG
    see also
        Pregnancy Tests
HCl
    see Hydrochloric Acid
HDL₂-Cholesterol
HDL₂a-Cholesterol
HDL₂b-Cholesterol
HDL₃-Cholesterol
HDL-Cholesterol
    see Cholesterol, High Density
        Lipoprotein
Heinz Body Formation
Hematocrit
Hemoglobin
Hemoglobin A₁c

Hemolytic Complement
    see Complement CH₅₀
Hemopexin
Heparin Sulfate
Hepatic Lipase
Hepatic Triglyceride Lipase
Heroin
Hexokinase
Hexosamine
Hexosaminidase
Hexosaminidase A
    see Hexosaminidase
Hexosaminidase B
    see Hexosaminidase
hGH
    see Growth Hormone
High Density Cholesterol
    see Cholesterol, High Density
        Lipoprotein
Hippuran Clearance
Hippuran Retention
Hippuric Acid
Histamine
Histamine Test
Histidine
    see also
        Amino Acids
Histone Antibodies
    see Antibodies to Histones
HMPG (4-Hydroxy-3-Methoxy-
    Phenylethylene Glycol)
HMPG Sulfate
Homocitrulline
    see also
        Amino Acids
Homocystine
    see also
        Amino Acids
Homogentisic Acid
Homovanillic Acid
    see also
        Catecholamines
α₂-HS Glycoprotein
HVA
    see Homovanillic Acid
Hyaline Casts
Hydantoin-5-Propionic Acid
Hydrazine
Hydrochloric Acid
Hydrochlorothiazide
Hydrocortisone
    see Cortisol
Hydrogen
Hydromorphone
2-Hydroxy-4-Ketoglutarate
11-Hydroxyandrosterone
3-Hydroxyanthranilic Acid
4-Hydroxybenzoic Acid
4-Hydroxybenzylamine
3-Hydroxybutyrate
β-Hydroxybutyrate
    see also
        Ketones
Hydroxybutyrate Dehydrogenase
α-Hydroxybutyrate Dehydrogenase
    see Hydroxybutyrate
        Dehydrogenase
β-Hydroxybutyric Acid
    see β-Hydroxybutyrate
11-Hydroxycorticosteroids
    see also
        Deoxycorticosterone
17-Hydroxycorticosteroids
17-Hydroxycorticosterone
18-Hydroxycorticosterone
6-β-Hydroxycortisol
Hydroxycotinine
16-α-Hydroxy-dehydroepiandrosterone
α-Hydroxy-dehydroepiandrosterone
    see 16-α-Hydroxy-
        dehydroepiandrosterone

15-α-Hydroxy-dehydroepiandrosterone
    Glucuronide
16-α-Hydroxy-dehydroepiandrosterone
    Glucuronide
α-Hydroxy-dehydroepiandrosterone
    Glucuronide
    see also
        15-α-Hydroxy-
            dehydroepiandrosterone
            Glucuronide
        16-α-Hydroxy-
            dehydroepiandrosterone
            Glucuronide
18-Hydroxydeoxycorticosterone
Hydroxy-Diphenylhydantoin
2-Hydroxyestrone
11-Hydroxyetiocholanolone
α-Hydroxyglutarate
5-Hydroxyindoleacetic Acid (5-HIAA)
3-Hydroxykynurenine
Hydroxy-Methoxymandelic Acid
    see also
        Vanillylmandelic Acid
Hydroxynalidixic Acid
    see also
        Vanillylmandelic Acid
17-Hydroxypregnenolone
16-α-Hydroxyprogesterone
17-Hydroxyprogesterone
α-Hydroxyprogesterone
    see 16-α-Hydroxyprogesterone
6-Hydroxyprogesterone Metabolite
Hydroxyproline
    see also
        Amino Acids
5-Hydroxytryptamine Glucuronide
5-Hydroxytryptamine (Serotonin)
5-Hydroxytryptophol
25-Hydroxy Vitamin D₃
25-Hydroxy Vitamin D
    see 25-Hydroxy Vitamin D₃
3-Hydroxyxanthurenic Acid
Hypoxanthine

I

ICD
    see Isocitrate Dehydrogenase
Icteric Index
Iditol Dehydropenase
    see Sorbitol Dehydrogenase
IDL
    see Intermediate density lipoprotein
        (IDL)
IgA
    see Immunoglobulin IgA
IgD
    see Immunoglobulin IgD
IgE
    see Immunoglobulin IgE
IgG
    see Immunoglobulin IgG
IgM
    see Immunoglobulin IgM
IL-1
    see Interleukin-1
IL-2
    see Interleukin-2
IL-2 Receptor Expression
Imidazoleacetic Acid
Imidazolepyruvic Acid
Imipramine
Immune Complexes
Immunoglobulin IgA
    see also
        Immunoglobulins
Immunoglobulin IgD
    see also
        Immunoglobulins

Immunoglobulin IgE
　see also
　　Immunoglobulins
Immunoglobulin IgG
　see also
　　Immunoglobulins
Immunoglobulin IgM
　see also
　　Immunoglobulins
Immunoglobulin Light Chains
　see also
　　Bence-Jones Protein
Immunoglobulins
　see also
　　Immunoglobulin IgA
　　Immunoglobulin IgD
　　Immunoglobulin IgE
　　Immunoglobulin IgG
　　Immunoglobulin IgM
Indican
Indirect Bilirubin
　see Bilirubin, Indirect
Indocyanine Green
Indocyanine Green Clearance
Indoleacetic Acid
Indolylacryloylglycine
Indomethacin
Indoxyl Sulfate
　see Indican
INH-Ketoglutarate
INH-Pyruvate
Insulin
Insulin Tolerance
γ-Interferon
Interleukin-1
Interleukin-2
Intermediate density lipoprotein (IDL)
Intrinsic Factor
Inulin Clearance
Iodide
Iodine, Total
Ionized Calcium
Iron
Iron-Binding Capacity, Total (TIBC)
　see also
　　Transferrin
Iron-Binding Capacity, Unsaturated
　(UIBC)
Iron Saturation
　see also
　　Iron-Binding Capacity, Total
　　(TIBC)
Isocitrate Dehydrogenase
Isohomovanillic Acid
Isoleucine
　see also
　　Amino Acids
Isoniazid
Isopropanol
Isovaleric Acid
Isovanillic Acid
^{131}I Uptake

K

Kallikrein
Kaolin-Cephalin Time
Ketoconazole
11-Ketoetiocholanolone
17-Ketogenic Steroids
α-Ketoglutarate
Ketones
　see also
　　Acetoacetate
　　Acetone
　　β-Hydroxybutyrate
17-Ketosteroids
17-KGS
　see 17-Ketogenic Steroids
Kininogen

17-KS
　see 17-Ketosteroids
Kynurenic Acid
Kynurenine

L

Lactate
Lactate Dehydrogenase
Lactate Dehydrogenase Isoenzymes
Lactose
Lactose Tolerance
LAP
　see Leucine Aminopeptidase
Latex Fixation
　see Rheumatoid Factor
LCAT
　see Lecithin Cholesterol
　　Acyltransferase (LCAT)
LD-1
　see Lactate Dehydrogenase
　　Isoenzymes
LD-2
　see Lactate Dehydrogenase
　　Isoenzymes
LD-3
　see Lactate Dehydrogenase
　　Isoenzymes
LD-4
　see Lactate Dehydrogenase
　　Isoenzymes
LDH
　see Lactate Dehydrogenase
LDL-Cholesterol
　see Cholesterol, Low Density
　　Lipoprotein
LDL + VLDL Cholesterol
　see also
　　Cholesterol, Low Density
　　　Lipoprotein
　　Cholesterol, Very Low Density
　　　Lipoprotein
Lead
LE Cells
Lecithin
　see also
　　Phospholipids, Total
Lecithin Cholesterol Acyltransferase
　(LCAT)
Leucine
　see also
　　Amino Acids
Leucine Aminopeptidase
Leucine Aminopeptidase Isoenzymes
Leukoagglutinins
Leukocyte Esterase
Leukocyte Migration Inhibition
Leukocytes
Levodopa
　see Dihydroxyphenylalanine
Levonorgestrel
LH
　see Luteinizing Hormone (LH)
LH Response to LHRH
Lidocaine
Lincomycin
Linoleate
Linoleic Acid
Lipase
Lipid Glycerol
Lipids
　see also
　　Cholesterol
　　Phospholipids, Total
　　Triglycerides
Lipids, Total
α_1-Lipoprotein
　see also
　　Lipoproteins
Lipoprotein Cholesterol

β-Lipoprotein Cholesterol
Lipoprotein Electrophoresis
Lipoprotein, High Density
　see Cholesterol, High Density
　　Lipoprotein
Lipoprotein Lipase
　see also
　　Postheparin Lipolytic Activity
　　Postheparin Lipoprotein Lipase
Lipoprotein, Low Density
　see Cholesterol, Low Density
　　Lipoprotein
Lipoprotein Lp(a)
Lipoproteins
　see also
　　α-Lipoproteins
　　β-Lipoproteins
　　Lipoproteins, Pre-β
α-Lipoproteins
β-Lipoproteins
Lipoproteins, Pre-β
　see also
　　Lipoproteins
Lipoprotein, Very Low Density
　see Cholesterol, Very Low Density
　　Lipoprotein
Lithium
Lithocholic Acid
Lorazepam
Low Density Lipoprotein Cholesterol
　see Cholesterol, Low Density
　　Lipoprotein
LSD
　see Lysergic Acid Diethylamide
　　(LSD)
L/S Ratio
Lupus Erythematosus Cells
　see LE Cells
Luteinizing Hormone (LH)
Lymphocyte Autoantibodies
Lymphocyte B-Cells
Lymphocyte T-Cells
Lymphocyte T-Helper Cells
Lymphocyte Mitotic Index
Lymphocyte Response
Lymphocytes
Lymphocyte T-Suppressor Cells
Lymphocyte Transformation
Lymphocytic Response to PHA
Lymphocytotoxic Antibodies
Lysergic Acid Diethylamide (LSD)
Lysine
　see also
　　Amino Acids
Lysolecithin
Lysozyme

M

Macroglobulin, α_2
　see α_2-Macroglobulin
α_2-Macroglobulin
Magnesium
Malate
Malate Dehydrogenase
Malondialdehyde
Manganese
Mannose
α-Mannosidase
Mazindol
MB
　see Creatine Kinase
MCH
MCHC
MCV
Mean Corpuscular Hemoglobin
　see MCH
Mean Corpuscular Hemoglobin
　Concentration
　　see MCHC

Mean Corpuscular Volume
    see MCV
Megaloblasts
Megestrol
Melanin
    see Thormählen Test
α-Melanocyte Stimulating Hormone
Melanogen
Melatonin
Meperidine
Meprobamate
6-Mercaptopurine
Mercury
Mescaline
Metanephrine
    see also
        Metanephrines, Total
Metanephrines, Total
    see also
        Catecholamines
Methacycline
Methadone
Methamphetamine
Methane
Methanol
Methapyrilene
Methaqualone
Methemalbumin
Methemoglobin
Methemoglobin Reductase
Methionine
    see also
        Amino Acids
Methionine Sulfoxide
Methotrexate
2-Methoxyestrone
2-Methoxyphenoxy-Lactic Acid
3-Methoxytyrosine
3-O-Methyl-D-Glucose
Methylhistamine
1,4-Methylhistamine
Methylhistidine
    see also
        Amino Acids
1-Methylhistidine
    see Methylhistidine
3-Methylhistidine
    see Methylhistidine
Methylimidazoleacetic Acid
Methylmalonic Acid
Methylnicotinamide
N-Methylnicotinamide
Methylphenidate
Methylprednisolone
Metoprolol
Metyrapone
Metyrapone Test
Mexiletine
Mg
    see Magnesium
Microglobulin, β₂
    see β₂-Microglobulin
β₂-Microglobulin
Microscopy
Microsomal Aminopeptidase
Midazolam
Migration Inhibition Factor
Millons Test
Monoamine Oxidase
Monoamine Oxidase B
Monocytes
Monoglyceride Lipase
Morphine
Motilin
Moxalactam
Mucopolysaccharides
Mucoprotein
Muramidase
    see Lysozyme
Myeloblasts
Myelocytes
Myoglobin

Myokinase

N

Na
    see Sodium
Na/K ATPase
Nalidixic Acid
α-Naphthol
Naphthylamidase Isoenzymes
Narcotics
NBT Test
NEFA
    see Fatty Acids, Free (FFA)
Neurophysin
Neurotensin
Neutral Fats
Neutral Steroids
Neutral Sterols
Neutrophil Elastase
Neutrophils
Nickel
Nicotine
Nifedipine
Nitrite
Nitroblue Tetrazolium Reduction Test
    see NBT Test
Nitrofurantoin
Nitrogen
Nitrogen Balance
Nitroglycerin
Non-esterified Fatty Acids
    see Fatty Acids, Free (FFA)
Nonprotein Nitrogen
Noradrenaline
    see Norepinephrine
Norepinephrine
    see also
        Catecholamines
Normeperidine
Normetanephrine
    see also
        Metanephrines, Total
Nortriptyline
NPN
    see Nonprotein Nitrogen
5'-Nucleotidase

O

O₂ Saturation
    see Oxygen Saturation
Obermayer Test
Occult Blood
Odor
17-OHCS
    see 17-Hydroxycorticosteroids
Oleic Acid
Oncotic Pressure
Organic Acids
Ornithine
    see also
        Amino Acids
Ornithine Carbamoyltransferase (OCT)
Orosomucoid
    see α₁-Acid Glycoprotein
Orotic Acid
Orthophosphate
Orthophosphoric Monoester
  Phosphohydrolase
    see Acid Phosphatase
Osmolality
Osmolar Clearance
Osmotic Fragility
Osteocalcin
Oxalate
Oxazepam
Oxygen-Hemoglobin Affinity

Oxygen Partial Pressure
    see pO₂
Oxygen Saturation
Oxyphenbutazone
Oxytetracycline
Oxytocin
Oxytocinase

P

paCO₂
    see pCO₂
PAH Clearance
Palmitic Acid
Palmitoleic Acid
Pancreatic Polypeptide
paO₂
    see pO₂
Paraldehyde
Para-Nitrophenol
Parathyroid Hormone
Partial Thromboplastin Time
Pb
    see Lead
PBI
    see Protein Bound Iodine (PBI)
PCE
    see Pseudocholinesterase
pCO₂
Penicillamine
Penicillin
Penicillin G
Penicillin V
Pentazocine
Pentobarbital
Pepsin
Pepsinogen
Pepsinogen I
Peripheral Smear
Peroxidase
pH
Phencyclidine
Phendimetrazine
Phenethicillin
Pheneturide
Phenindione
Phenmetrazine
Phenobarbital
Phenol
Phenol Turbidity
Phenothiazine
Phenprocoumon
Phentermine
Phentolamine Test
Phenylacetylglutamine
Phenylalanine
    see also
        Amino Acids
Phenylbutazone
Phenylethylamine
Phenylketones
Phenylpropanolamine
Phenyramidol
Pheochromocytoma Test
Phosphatase, Acid
    see Acid Phosphatase
Phosphatase, Alkaline
    see Alkaline Phosphatase
Phosphate
Phosphate Clearance
Phosphatides, Total
Phosphoenolpyruvate
Phosphoethanolamine
Phosphofructokinase
Phosphoglucomutase
2-Phosphoglycerate
3-Phosphoglycerate
6-Phosphoglycerate Dehydrogenase
Phosphoglycerate Kinase
Phosphohexoseisomerase

Phospholipids, High Density Lipoprotein
Phospholipids, Low Density Lipoprotein
Phospholipids, Total
 see also
  Lecithin
Phosphorus
 see Phosphate
Phosphorus Clearance
 see Phosphate Clearance
Phosphoserine
Phthalates
Phytofluene
Pimelate
Pituitary Gonadotropin
Placental Lactogen
Placental Protein S
Plasma Thromboplastin Antecedent
 see Factor XI
Plasma Thromboplastin Component
 see Factor IX
Plasmin
Plasmin Activity
Plasminogen
Plasminogen Antigen
Platelet Adhesiveness
Platelet Aggregation
Platelet Associated IgG
Platelet Factor 4
Platelets
Platinum
pO₂
Polysaccharide, Protein Bound
Porphobilinogen
Porphyrins
 see also
  Coproporphyrin
  Protoporphyrin
Porter-Silber Chromogens
 see 17-Hydroxycorticosteroids
Postheparin Hepatic Lipase
Postheparin Hepatic Triglyceride Lipase
Postheparin Lipase
Postheparin Lipolytic Activity
Postheparin Lipoprotein Lipase
Potassium
Potassium Clearance
PRA
 see Renin Activity
Prazosin
Prealbumin
Precipitable Iodine
Prednisolone
Pregnancy-Associated Plasma Protein
 A
Pregnancy Protein
Pregnancy Tests
Pregnanediol
Pregnanetriol
Pregnenolone
Prekallikrein
Primidone
Proaccelerin
 see Factor V
Probenecid
Procainamide
Progesterone
Progestins
Proinsulin
Prolactin
Prolactin Response to TRH
Proline
 see also
  Amino Acids
Proline Hydroxylase
Promyeloblasts
Propafenone
Propantheline
Properdin
Propionic Acid
Propoxyphene
Propranolol
Propylthiouracil

Proquazone
Prostaglandin, 6-Keto-F₁α
Prostaglandin A
Prostaglandin E
Prostaglandin E1
Prostaglandin E2
Prostaglandin E, 19-Hydroxy
Prostaglandin F
Prostaglandin F₁α
Prostaglandin F2α
Prostaglandin F, 19-Hydroxy
Prostaglandin G2
Protein
Protein Bound Iodine (PBI)
Protein Electrophoresis
Prothrombin
 see Factor II
Prothrombin Time
Protoporphyrin
 see also
  Porphyrins
Pseudocholinesterase
Pseudoephedrine
Pseudouridine
PSP Excretion
PTA
 see Factor XI
PTC
 see Factor IX
PTH
 see Parathyroid Hormone
Pyridoxal
Pyridoxal Phosphate
Pyridoxamine
Pyridoxamine Phosphate
4-Pyridoxic Acid
Pyridoxine
Pyrophosphate
Pyrrole-2-Carboxylate
Pyruvate
Pyruvate Kinase
Pyruvate Tolerance

Q

Quinidine
Quinine
Quinolinic Acid

R

Ranitidine
Recalcification Time
Red cell-associated Immunoglobulin IgG
Red Cell Count
 see Erythrocytes
Reduced Glutathione
Renal Blood Flow (RBF)
Renin Activity
Renin Substrate
Respiratory Peak Flow
Reticulocytes
Retinol
 see Vitamin A
Retinol Binding Protein
Retinol Esters
Reverse Triiodothyronine
 see Tri-iodothyronine (T₃), Reverse
Rheumatoid Factor
Riboflavin
Ribonuclear Protein Antibodies
 see Antibodies to RNP
Ribonuclease
Ribose
Rifampin
rT₃
 see Tri-iodothyronine (T₃), Reverse
RT₃U
 see T₃ Uptake

S

Salicylate
Sarcosine
 see also
  Amino Acids
SCL-70 Antibodies
 see Antibodies to SCL-70
Scleroderma Antibody
 see Antibodies to SCL-70
Secobarbital
Sedoheptulose
Selenium
Semen Analysis
 see Sperm Count
Serine
 see also
  Amino Acids
Seromucoid
Serotonin
 see 5-Hydroxytryptamine
  (Serotonin)
Sex Hormone Binding Globulin
SGOT
 see Aspartate Aminotransferase
SGPT
 see Alanine Aminotransferase
Short-Chain Fatty Acids
Sialic Acid
Sickle Cells
Siderophilin
 see Transferrin
Silicon
Sjögren Syndrome A Antibodies
 see Antibodies to Sjögren
  Syndrome A
Sjögren Syndrome B Antibodies
 see Antibodies to Sjögren
  Syndrome B
Smith Antigen Antibodies
 see Antibodies to SmAg
Sodium
Somatomedin-C
Somatostatin
Somatotropin
 see Growth Hormone
Sorbitol Dehydrogenase
Specific Gravity
Sperm Count
Sperm Morphology
Sperm Motility
Standard Bicarbonate
Stearic Acid
Sterols
Stuart Factor
 see Factor X
Suberate
Succinate
Succinate Dehydrogenase
Sucrose
Sugar
Sulfa as Sulfanilamide
Sulfadiazine
Sulfamethoxazole
Sulfamethoxine
Sulfamethoxypyridazine
Sulfanilamide
Sulfasymazine
Sulfatase
Sulfate
Sulfathiazole
Sulfhemoglobin
Sulfinpyrazone
Sulfisoxazole
Superoxide Dismutase
Synephrine

T

T_3
   see Tri-iodothyronine (T_3)
T_3 Binding Capacity
T_3 Uptake
T_4
   see Thyroxine (T_4)
Taurine
   see also
      Amino Acids
Terbutaline
Testosterone
   see also
      Androgens
Testosterone Binding
Testosterone Binding Globulin
Testosterone, Free
Tetracycline
Tetrahydroaldosterone
Tetrahydrocortisol
Tetrahydrocortisone
Tetrahydrodeoxycorticosterone
Theophylline
Thiamine
Thiocyanate
Thormählen Test
Threonine
   see also
      Amino Acids
Thrombin Time
β-Thromboglobulin
Thromboplastin Generation
Thromboxane
Thromboxane B_2
Throxine-Binding Prealbumin
   see Prealbumin
Thymidine Incorporation
Thymol Turbidity
Thyro-Binding Index
Thyrocalcitonin
   see Calcitonin
Thyroglobulin
Thyroglobulin Antibodies
   see Antithyroglobulin Antibodies
Thyroid Antibodies
Thyroid Stimulating Hormone (TSH)
Thyrotropin
   see Thyroid Stimulating Hormone
      (TSH)
Thyroxine (T_4)
Thyroxine (T_4) Binding Globulin
Thyroxine (T_4), Free
Thyroxine (T_4) Index, Free (FTI)
Thyroxine (T_4) (Murphy-Pattee)
TIBC
   see Iron-Binding Capacity, Total
      (TIBC)
Tissue Plasminogen Activator Antigen
Titratable Acidity
T Lymphocyte
   see Lymphocyte T-Cells
Tobramycin
Tocopherol
   see Vitamin E (Tocopherol)
α-Tocopherol
Tolbutamide
Toluene
Total Iron-Binding Capacity
   see Iron-Binding Capacity, Total
      (TIBC)
Transcortin
Transferrin
Transferrin Saturation
   see Iron Saturation
Transketolase
Trehalase
Trichlorethanol
Tricyclic Antidepressants
Triglyceride Lipase
Triglycerides
Triglycerides, High Density Lipoprotein

Triglycerides, Low Density Lipoprotein
Triglycerides, Very Low Density
   Lipoprotein
Trihydroxyphenylacetate
Tri-iodothyronine (T_3)
Tri-iodothyronine (T_3), Free
Tri-iodothyronine (T_3) Index, Free
Tri-iodothyronine (T_3), Reverse
Tri-iodothyronine (T_3), Reverse Free
Triiodothyronine Uptake
   see T_3 Uptake
Trimethoprim
Triosephosphate Isomerase
Trypsin
Trypsin Inhibitor
Trypsinogen
Tryptamine
Tryptophan
   see also
      Amino Acids
Tryptophan, Free
TSH
   see Thyroid Stimulating Hormone
      (TSH)
TSH Response to TRH
TTC Test
Tubular Maximum for Phosphate
Tyramine
Tyramine Test
Tyrosine
   see also
      Amino Acids

U

Unconjugated Bilirubin
   see Bilirubin, Indirect
Urea Clearance
Urea Nitrogen
Uric Acid
Uric Acid Clearance
Urobilin
Urobilinogen
Urocanic Acid
Urocanylglycine
Urokinase
Uropepsin
Uropepsinogen
Uroporphyrin
   see also
      Porphyrins

V

Valine
   see also
      Amino Acids
Valproic Acid
Vancomycin
Vanillic Acid
Vanillylamine
Vanillylmandelic Acid
   see also
      Catecholamines
Vasoactive Intestinal Polypeptide
Vasopressin
   see Antidiuretic Hormone
Very Low Density Lipoprotein
   Cholesterol
   see Cholesterol, Very Low Density
      Lipoprotein
Viscosity
Vital Capacity
Vitamin A
Vitamin B_1
   see Thiamine
Vitamin B_2
   see Riboflavin

Vitamin B_6
   see Pyridoxamine
Vitamin B_{12}
Vitamin B_{12} (Unsaturated)
Vitamin C
   see Ascorbic Acid
Vitamin D_3
   see 1,25-Dihydroxy Vitamin D_3
   see 25-Hydroxy Vitamin D_3
Vitamin D Binding Globulin
Vitamin E (Tocopherol)
VLDL-Apoprotein B
VLDL-Cholesterol
   see Cholesterol, Very Low Density
      Lipoprotein
VLDL-Remnant Apoprotein B
Volume

W

Warfarin
Wassermann Reaction
Water
Water Clearance
Water Clearance, Free
WBC Count
   see Leukocytes
WBC Esterase
   see Leukocyte Esterase
White Blood Cells
   see Leukocytes

X

Xanthine
Xanthine Calculi
Xanthochromia
Xanthurenic Acid
Xylose
   see also
      Xylose Excretion
Xylose Excretion

Z

Zimmerman Reaction
   see 17-Ketosteroids
Zinc
Zinc Sulfate Turbidity

2 DRUG INDEX

A

Abbokinase®
  see Urokinase
Accutane®
  see Isotretinoin
Acebutolol
Acenocoumarol
Acetaldehyde
Acetaminophen
Acetanilid
Acetates
Acetazolamide
Acetic Acid
Acetoacetate
Acetohexamide
Acetohydroxamic Acid
Acetone
Acetophenazine
Acetophenone
Acetoxymethylprogesterone
  see Medroxyprogesterone
Acetrizoate
Acetylacetone
N-Acetyl-P-Aminophenol
  see Acetaminophen
Acetylcholine
N-Acetylcysteine
Acetylglucosamine
Acetylhistamine
L-α-Acetylmethadol (LAAM)
Acetylpenicillamine
Acetylphenylhydrazine
Acetylsalicylic Acid
Acetylsulfadiazine
Acetyltryptophan
Achromycin®
  see Tetracycline
Aciclovir
  see Acyclovir
Acidosis
Acids
Acid Urine
Aclovate®
  see Alclometasone
Aconitine
Acridine
Acridine Orange
Acriflavine
ACTH
  see Corticotropin
Acthar®
  see Corticotropin
Actidose-Aqua®
  see Charcoal
Actinomycin
Actinomycin D
  see Dactinomycin

Activated carbon
  see Charcoal
ACV
  see Acyclovir
Acycloguanosine
  see Acyclovir
Acyclovir
Adapin®
  see Doxepin
Adenine
Adenine Deoxyribose
Adenine Diphosphate
Adenine Monophosphate
Adenosine
Adenosine Triphosphate
ADH
  see Vasopressin
Adipex-P®
  see Phentermine
Adiphenine
ADR
  see Doxorubicin
Adrenalin®
  see Epinephrine
β-Adrenergic Drugs
Adriamycin®
  see Doxorubicin
Adsorbonac® Ophthalmic
  see Sodium Chloride
Advil®
  see Ibuprofen
Aeroseb-HC®
  see Hydrocortisone
Aflatoxin
Afrin®
  see Oxymetazoline
Aging
Ajmaline
Akineton®
  see Biperiden
Alanine
β-Alanine
Albumin
Albuminar®-25
  see Albumin
Albumin, Human
  see Albumin
Albumisol®
  see Albumin
Albuterol
Albutoin
Alclofenac
Alclometasone
Alcohols
Alconefrin®
  see Phenylephrine
Alcuronium
Alcuronium Chloride

Aldactone®
  see Spironolactone
Aldatense
Aldomet®
  see Methyldopa
Aldosterone
Aldrin
Aletamine
Alimenazine Tartrate
  see Trimeprazine
Alkalies
Alkaline Antacids
Alkaline Urine
Alkalosis
Alkeran®
  see Melphalan
Alkoxyglycerols
Allantoin
Allopurinol
Alloxan
Allymid
Aloin
L-ALpha-Acetylmethadol (LAAM)
  see L-α-Acetylmethadol (LAAM)
Alpha-Amino-N-Butyric Acid
  see α-Amino-N-Butyric Acid
Alpha-Di-Hydroxyprogesterone
  see 20-α-Di-Hydroxyprogesterone
Alpha-Interferon
  see α-Interferon
Alpha-Ketobutyric Acid
  see α-Ketobutyric Acid
Alpha-Ketoglutarate
  see α-Ketoglutarate
Alphaprodine
Alprazolam
Alprenolol
ALternaGEL®
  see Aluminum Hydroxide
Althesin
Altitude
Alu-Cap®
  see Aluminum Hydroxide
Aluminum Hydroxide
Aluminum Hydroxide and Magnesium Hydroxide
  see Simethicone
Aluminum Nicotinate
Aluminum Salts
Alu-Tab®
  see Aluminum Hydroxide
Amantadine
Ambazone
Ambenonium
Ambulation
Amdinocillin
Americaine®
  see Benzocaine

A-methaPred®
see Methylprednisolone
Amethopterin
see Methotrexate
Amicar®
see Aminocaproic Acid
Amidate®
see Etomidate
Amidopyrine
see Aminopyrine
Amidotrizoic Acid
see Diatrizoic acid
Amikacin
Amikin®
see Amikacin
Amiloride
Aminacrine
Amines
Amino Acids
9-Aminoacridine Hydrochloride
see Aminacrine
Aminoantipyrine
Aminobenzoic Acid
Aminobutyric Acid
α-Amino-N-Butyric Acid
Aminocaproic Acid
Amino-Cerv™ Vaginal Cream
see Urea
Aminoglutethimide
Aminoguanidine
Aminohippuric Acid
Aminoimidazoleacetic Acid
β-Amino-Isobutyric Acid
Aminomel LX 6
Aminomethylcyclohexane
Aminophenazone
Aminophenol
Aminophyllin™
see Aminophylline
Aminophylline
Aminopropionitrile
Aminopterin
Aminopyrimidine
Aminopyrine
Aminosalicylic Acid
Aminothiazole
δ-Amino Valeric Acid
Amiodarone
Amipaque®
see Metrizamide
Amisometradine
Amitriptyline
Ammonia
Ammonium Acetate
Ammonium Chloride
Ammonium Ions
Ammonium Nitrate
Ammonium Oxalate
Ammonium Salts
Amobarbital
Amodiaquine
Amoxapine
Amoxicillin
Amoxil®
see Amoxicillin
Amphenone B
Amphetamine
Amphojel®
see Aluminum Hydroxide
Amphotericin B
Ampicillin
Amrinone
Amygdalin
Amyl Alcohol
Amyl Nitrite
Amytal®
see Amobarbital
Anabolic Steroids
Anacin®
see Aspirin
Anadrol®
see Oxymetholone

Analgesics
Ancef®
see Cefazolin
Ancobon®
see Flucytosine
Ancrod
Androgens
Androstenedione
Androsterone
Anectine® Chloride
see Succinylcholine
Anectine® Flo-Pack®
see Succinylcholine
Anesthetic Agents
Aneurine Hydrochloride
see Thiamine
Angiotensin
Anhydron®
see Cyclothiazide
Anileridine
Aniline
Aniline Dyes
Anisindione
Anspor®
see Cephradine
Antabuse®
see Disulfiram
Antacids
Antazoline
Anthiolimine
Anthranilic Acid
Anthraquinone
Antibiotics
Anticoagulants
Anticonvulsants
Antidiuretic Hormone
see Vasopressin
Antifungal Agents
Antihistamines
Antihypertensive Agents
Antilirium®
see Physostigmine
Antilymphocytic Agents
Antimalarials
Antiminth®
see Pyrantel
Antimony Compounds
Antineoplastic Agents
Antipyretics
Antipyrine
Antiseptics
Antithrombin III
ANTU
Anturane®
see Sulfinpyrazone
Apalcillin
APAP
see Acetaminophen
Apiol
Apomorphine
Apresoline®
see Hydralazine
Aprindine
Apronalide
Aprotinin
AquaMEPHYTON®
see Phytonadione
Aquasol-A®
see Vitamin A
Aquasol® E
see Vitamin E
Arabinose
Ara-C
see Cytarabine
Aralen® Phosphate
see Chloroquine
Aramine®
see Metaraminol
Arginine
Aristocort® Forte
see Triamcinolone

Aristocort®
see Triamcinolone
Aristospan®
see Triamcinolone
Arsenates
Arsenicals
Arsenobenzenes
Artane®
see Trihexyphenidyl
ASA
see Aspirin
A.S.A.®
see Aspirin
Ascorbate-2-Sulfate
Ascorbic Acid
Ascriptin®
see Aspirin
Asendin®
see Amoxapine
see Amoxapine
Asparaginase
Asparagine
Asparagus
Aspartic Acid
Aspidium
Aspirin
Aspoxicillin
Astramorph™ PF
see Morphine
Atabrine®
see Quinacrine
Atarax®
see Hydroxyzine
Atenolol
Ativan®
see Lorazepam
Atromid-S®
see Clofibrate
Atropine
Atropisol®
see Atropine
Attenuvax®
see Measles Vaccine
Augmentin®
Auranofin
Aureomycin®
see Chlortetracycline
Aurothioglucose
Aurothiomalate
Aventyl® Hydrochloride
see Nortriptyline
Avlosulfon®
see Dapsone
Avocados
Azactam®
see Aztreonam
Aza Drugs
Azapropazone
Azaserine
Azathioprine
Azathymine
Azauridine
Azide
Azidothymidine
Aziridine
Azlin®
see Azlocillin
Azlocillin
Azmacort™
see Triamcinolone
Azolid®
see Phenylbutazone
Azosemide
Azostix®
Azosulfamide
Aztreonam
Azulfidine®
see Sulfasalazine
Azuresin

B

Baciguent®
  see Bacitracin
Bacillus Clamette-Guérin
  see BCG Vaccine
Bacitracin
Bacterial Contamination
Bacteriuria
Bactrim™
  see Co-Trimoxazole
Baking Soda
  see Sodium Bicarbonate
BAL in Oil®
  see Dimercaprol
Bamethan
Bananas
Barbital
Barbiturates
Barium
Baro-CAT®
  see Barium
Bayer® Aspirin
  see Aspirin
BCG Vaccine
BCNU
  see Carmustine (BCNU)
Beclamide
Beclomethasone
Beclovent®
  see Beclomethasone
Beconase®
  see Beclomethasone
Bed Rest
Beef NPH Iletin® II
  see Insulin
Beef Regular Iletin® II
  see Insulin
Beets
Belladonna
Benadryl®
  see Diphenhydramine
Bendrofluazide
Benemid®
  see Probenecid
Benfluorex
Benserazide
Benylin® Cough Syrup
  see Diphenhydramine
Benzaldehyde
Benzaprine
Benzazepine
Benzbromaron
  see Benzbromarone
Benzbromarone
Benzene
Benziodarone
Benzocaine
Benzoic Acid
Benzphetamine
Benzquinamide
Benzthiazide
Benztropine
Benzyl Alcohol
Beryllium Salts
Beta-Adrenergic Drugs
  see β-Adrenergic Drugs
Beta-Alanine
  see β-Alanine
Beta-Amino-Isobutyric Acid
  see β-Amino-Isobutyric Acid
Beta,Beta-Dimethylcysteine
  see Penicillamine
Betadine® Douche
  see Povidone-Iodine
Betadine®
  see Povidone-Iodine
Beta-Endorphin
  see β-Endorphin
Betalin®S
  see Thiamine
Betamethasone

Betapen®-VK
  see Penicillin V
Beta-Propiolactone
  see β-Propiolactone
Beta-Sitosterol
  see β-Sitosterol
Betazole
Bethanechol
Bethanidine
Bezafibrate
Bicarbonates
Bicine
BiCNU®
  see Carmustine (BCNU)
Bile
Bile Salts
Bilirubin
Bilirubin Glucuronide
Biliverdin
Biperiden
Biphenabid
  see Probucol
Biphetamine®
  see Dextroamphetamine
Bisacodyl
Bishydroxycoumarin
  see Dicumarol
Bismuth Salts
Bismuth Subnitrate
Bisoprolol
Bisulfite
Blenoxane®
  see Bleomycin
Bleomycin
Bleph®-10
  see Sulfacetamide
Blindness
BLM
  see Bleomycin
Blocadren®
  see Timolol
Blood
Blood Collection Tube
Blood Group
Blood Pressure
Blood Transfusions
Body Habitus
Bombesin
Bopindolol
Boric Acid
Borofax®
  see Boric Acid
Boxidine
Bran
Brethine®
  see Terbutaline
Brevital® Sodium
  see Methohexital
Bricanyl®
  see Terbutaline
Bromate
Bromazepam
Bromelains
Bromides
Bromisovalum
Bromocriptine
Bromodeoxyuridine
Brompheniramine
Bucindol
Bufferin®
  see Aspirin
Buformin
Bufotenine
Bumetanide
Bumex®
  see Bumetanide
Buminate®
  see Albumin
Bunamiodyl
Bunitrolol
Bupropion
Burimamide

Busulfan
Butabarbital
Butacaine
Butanol
Butaperazine
Butazolidin®
  see Phenylbutazone
Butethamine
Butisol Sodium®
  see Butabarbital
Butizide
Butorphanol
Butylgallate

C

Cadmium
Caffeine
Calan®
  see Verapamil
Calciferol™
  see Ergocalciferol
Calcimar®
  see Calcitonin
Calcitonin
Calcium
Calcium and Vitamin D
  see Ergocalciferol
Calcium Bromogalactogluconate
Calcium Carbimide
Calcium Carbonate
Calcium Chloride
Calcium Gluconate
Calcium Ions
Calcium Salts
Calibind®
  see Cellulose Phosphate
Calomel
Campesterol
Cannabis
Canrenoate Potassium
Cantharides
Cantharone®
  see Cantharides
Canthoxanthine
Cantil®
  see Mepenzolate
Capastat® Sulfate
  see Capreomycin
Capoten®
  see Captopril
Capozide®
  see Captopril
Capreomycin
Caproic Acid
Caproxamine
Caprylate
Captopril
Carafate®
  see Sucralfate
Carbacrylamine Resin
Carbamazepine
Carbarsone
Carbenicillin
Carbenoxolone
Carbidopa
Carbimazole
Carbochromen
Carbocromen
  see Chromonar Hydrochloride
Carbohydrate
Carbonates
Carbon Disulfide
Carbon Monoxide
Carboplatin
Carbromal
Carbutamide
Cardene®
  see Nicardipine

Cardio-Green®
see Indocyanine Green
Cardioquin®
see Quinidine
Cardizem®
see Diltiazem
Carfecillin
Carinamide
Carisoprodol
Carmine
Carmol®
see Urea
Carmustine (BCNU)
Carotene
Carotenoid
Carphenazine
Carprazidil
Carrots
Carteolol
Cartrol®
see Carteolol
Cascara
Castor Oil
Catapres®
see Clonidine
Catarase®
see Chymotrypsin
Catechol
Catecholamines
Cathartics
Ceclor®
see Cefaclor
Cedilanid-D®
see Deslanoside
Cefaclor
Cefadyl®
see Cephapirin
Cefamandole
Cefazedone
Cefazolin
Cefizox®
see Ceftizoxime
Cefobid®
see Cefoperazone
Cefodizime
Cefoperazone
Ceforanide
Cefotan®
see Cefotetan
Cefotaxime
Cefotetan
Cefotiam
Cefoxitin
Cefpirome
Cefsulodin
Ceftazidime
Ceftin®
see Cefuroxime
Ceftizoxime
Ceftriaxone
Cefuroxime
Celestone®
see Betamethasone
Celiprolol
Cellobiose
Cellopentose
Cellotriose
Cellulose
Cellulose Acetate
Cellulose Phosphate
Celontin®
see Methsuximide
Cephalexin
Cephaloglycin
Cephaloridine
Cephalosporin
Cephalothin
Cephapirin
Cephradine
Cephulac®
see Lactulose

Cerespan®
see Papaverine
Cerubidine®
see Daunorubicin
Cerulein
C.E.S.
see Estrogens, Conjugated
Cetoxime
Cetoxine
see Cetoxime
Charcoal
Chenodeoxycholic Acid
Chewing Gum
Chiniofon
Chloral Hydrate
Chlorambucil
Chloramine
Chloramphenicol
Chlorate
Chlordane
Chlordiazepoxide
Chlorhexidine
Chloride Salts
Chlorinated Insecticides
Chlorine
Chlormadinone
Chlormerodrin
Chlormethiazole
Chlormethine
Chlormezanone
Chlorobutanol
Chloroethane
see Ethyl Chloride
Chloroform
Chloroguanide
Chloromycetin®
see Chloramphenicol
Chlorophenothane
Chlorophenylalanine
Chlorophyll
Chloroptic®
see Chloramphenicol
Chloroquine
Chlorothiazide
Chlorphenamine
see Chlorpheniramine
Chlorphenesin
Chlorpheniramine
Chlorphentermine
Chlorpromazine
Chlorpropamide
Chlorprothixene
Chlortetracycline
Chlorthalidone
Chlor-Trimeton®
see Chlorpheniramine
Chlorzoxazone
Chocolate
Cholecystokinin
Choleretics
Cholestanol
Cholesterol
Cholesteryl Oleate
Cholestyramine
Cholinergics
Cholografin® Meglumine
see Iodipamide
Choloxin®
see Dextrothyroxine
Chromates
Chromatography
Chromium
Chromonar Hydrochloride
Chronulac®
see Lactulose
Chrysarobin
Chyluria
Chymotrypsin
Cibalith-S®
see Lithium
Cibenzoline
Cimetidine

Cinchophen
Cin-Quin®
see Quinidine
Cipro™
see Ciprofloxacin
Ciprofloxacin
Cisplatin
Citrates
Citruplexina
Claforan®
see Cefotaxime
Clazolam
Cled Medium
Cleocin®
see Clindamycin
Clindamycin
Clinoril®
see Sulindac
Clioquinol
see Iodochlorhydroxyquin
Clofibrate
Clomid®
see Clomiphene
Clomiphene
Clomipramine
Clonazepam
Clonidine
Clopamide
Clopenthixol
Clorazepate
Clorexolone
Clorgiline
Clorprenaline
Clotermine
Clothiapine
Cloxacillin
Cloxapen®
see Cloxacillin
Clozapine
Coactin®
see Amdinocillin
Coal Tar
Cobalt
Cobalt Salts
Coca Cola™
Cocaine
Cocoa
Codeine
Coffee
Cogentin®
see Benztropine
Colace®
see Docusate Sodium
Colchicine
Cold Agglutinins
Cold Exposure
Colestid®
see Colestipol
Colestipol
Colistimethate
Colistin
Collidine
Coly-Mycin® M Parenteral
see Colistimethate
Coly-Mycin® S
see Colistin
Compazine®
see Prochlorperazine
Concentrated Urine
Congo Red
Contact With Clot
Copper
Corbadrin
Cordarone®
see Amiodarone
Corgard®
see Nadolol
Cork Stoppers
Cortef®
see Hydrocortisone
Cortenema®
see Hydrocortisone

Corticosteroids
Corticosterone
Corticotropin
Cortifoam®
    see Hydrocortisone
Cortisol
    see Hydrocortisone
Cortisone
Cortisporin®
    see Bacitracin
Cortone®
    see Cortisone
Cortrophin®
    see Corticotropin
Cortrosyn®
    see Corticotropin
Cosmegen®
    see Dactinomycin
Cotrifamole
Co-Trimoxazole
Coughing
Cough Medicines
Coumadin®
    see Warfarin
Coumaric Acid
Coumarin
Creatine
Creatinine
Creatinine Clearance
Cremomycin
Creosote
Cresol
Cromoglycate
Croton Oil
Cryoglobulin
Crystodigin®
    see Digitoxin
Cuprimine®
    see Penicillamine
CYA
    see Cyclosporine
Cyanates
Cyanide Antidote Kit
    see Amyl Nitrite
Cyanides
Cyclacillin
Cyclamate
Cyclazocine
Cyclic AMP
Cyclobenzaprine
Cyclofenil
Cycloheximide
Cyclohexylamine
Cycloleucine
Cyclopentamine
Cyclopenthiazide
Cyclophosphamide
Cyclopropane
Cycloserine
Cyclosporine
Cyclothiazide
Cyproheptadine
Cyproterone
Cysteine
Cystine
Cystografin®
    see Diatrizoic acid
Cytadren®
    see Aminoglutethimide
Cytarabine
Cytomel®
    see Liothyronine
Cytosar®
    see Cytarabine
Cytosine Arabinosine
    see Cytarabine
Cytoxan®
    see Cyclophosphamide

D

Dactinomycin
Dalmane®
    see Flurazepam
Danazol
Dandruff Medication
Danocrine®
    see Danazol
Dantrium®
    see Dantrolene
Dantrolene
Dapsone
Daranide®
    see Dichlorphenamide
Daraprim®
    see Pyrimethamine
Daricon®
    see Oxyphencyclimine
Darvon®
    see Propoxyphene
Daunorubicin
o'-DDD
    see Mitotane
Debrisoquin
Decaborane
Decadron®
    see Dexamethasone
Deca-Durabolin®
    see Nandrolone
Decamethonium
Decholin®
    see Dehydrocholic Acid
Declomycin®
    see Demeclocycline
Deferoxamine
Dehydration
Dehydrobenzperidole
Dehydrocholic Acid
Dehydroemetine
Dehydroepiandrosterone
Delalande 69276
Delalutin®
    see Hydroxyprogesterone
Delestrogen®
    see Estradiol
Delsym®
    see Dextromethorphan
Delta-Amino Valeric Acid
    see δ-Amino Valeric Acid
Delta-Cortef®
    see Prednisolone
Deltasone®
    see Prednisone
Demecarium
Demeclocycline
Demecolcine
Demerol®
    see Meperidine
Demulen®
    see Ethynodiol
Deoxycholate
11-Deoxycortisol
21-Deoxycortisol
Deoxy-Glucose
Deoxypyridoxine
Depakene®
    see Valproic Acid
Depakote®
    see Valproic Acid
Depo-Medrol®
    see Methylprednisolone
Depo-Provera®
    see Medroxyprogesterone
Deprenyl
Dermoplast®
    see Benzocaine
DES
    see Diethylstilbestrol
Deserpidine
Desferal® Mesylate
    see Deferoxamine

Desiccated Thyroid
Desipramine
Deslanoside
Desmethyldiazepam
Desmosterol
Desogestrel
Desoximetasone
Desoxycorticosterone
Desoxyephedrine Hydrochloride
    see Methamphetamine
Desoxyn®
    see Methamphetamine
Desyrel®
    see Trazodone
Detergents
Dexamethasone
Dexedrine®
    see Amphetamine
    see Dextroamphetamine
Dextran
Dextran 40
Dextran 60
Dextran 70
Dextran 75
Dextroamphetamine
Dextromethorphan
Dextromoramide
Dextropropoxyphene
    see Propoxyphene
Dextrothyroxine
DH-581
    see Probucol
DHE 45®
    see Dihydroergotamine
DHT
    see Dihydrotachysterol
Diabinese®
    see Chlorpropamide
Diaβeta®
    see Glyburide
Dialume®
    see Aluminum Hydroxide
1,2-Diaminopropane
Diamox®
    see Acetazolamide
Diapamide
Diaprim
Diatrizoic acid
Diazepam
Diazo Dyes
Diazoxide
Dibenzyline®
    see Phenoxybenzamine
Dichloralphenazone
Dichlorobenzene
Dichlorphenamide
Diclofenac
Dicloxacillin
Dicumarol
Didrex®
    see Benzphetamine
Didronel®
    see Etidronate
Dieldrin
Dienestrol
Diethazine
Diethylaminoethanol
Diethylpropion
Diethylstilbestrol
Diethylstilbestrol Enseals®
    see Diethylstilbestrol
Diflunisal
Difluorophosphate
Digibind®
    see Digoxin-Specific Fab
Digitalis
Digitonin
Digitoxin
Digoxin
Digoxin Immune Fab
    see Digoxin-Specific Fab
Digoxin-Specific Fab

Dihydralazine
Dihydroergotamine
Dihydromorphine
Dihydrostreptomycin
Dihydrotachysterol
Dihydrotestosterone
17,21-Di-Hydroxy-4-Pregnene-3,20-Dione
Dihydroxyacetone
3,4-Dihydroxybenzoic Acid
Dihydroxycholecalciferol
Dihydroxymandelic Acid
Dihydroxyphenylacetic Acid
3,4-Dihydroxyphenylacetic Acid
20-α-Di-Hydroxyprogesterone
Diiodocaffeine
Diiodohydroxyquin
Diiodoquinoline
Diiodotyrosine
Dilaudid®
    see Hydromorphone
Diltiazem
Dilute Urine
Dimenhydrinate
Dimercaprol
Dimercaptoethane
Dimetane®
    see Brompheniramine
Dimethadione
Dimethindene
Dimethoxyphenyl Penicillin Sodium
    see Methicillin
β,β-Dimethylcysteine
    see Penicillamine
N-Dimethyldiazepam
Dimethyl Ketone
    see Acetone
3,4-Dimethylphenol
Dimethyltryptamine
Dimethyl Tubocurarine
Dimpylate
Dinitrophenol
Dionosil Oily®
    see Propyliodone
Dioxane
Diphenhydramine
2,3-Diphosphoglycerate
Diphosphonate
Diprolene®
    see Betamethasone
Dipropylacetic Acid
    see Valproic Acid
Diprosone®
    see Betamethasone
Diprotrizoate
Dipyridamole
Dipyrone
Disalcid®
    see Salsalate
Disopyramide
Disulfiram
Disulphine Blue
Dithiazanine
Dithionite
Dithiothreitol
Diulo™
    see Metolazone
Diuretics
Diuril®
    see Chlorothiazide
Diurnal Variation
DL-Methionine
    see Methionine
DNR
    see Daunorubicin
Doans® Pills
Dobutamine
Dobutrex®
    see Dobutamine
Docusate Sodium
Dolobid®
    see Diflunisal

Dolophine®
    see Methadone
Domeboro® Otic
    see Acetic Acid
Domperidone
Dopa
    see Levodopa
Dopamine
Dopar®
    see Levodopa
Dopram®
    see Doxapram
Doriden®
    see Glutethimide
Doryx®
    see Doxycycline
Doxapram
Doxazosin
Doxepin
Doxinate®
    see Docusate Sodium
Doxorubicin
Doxychel®
    see Doxycycline
Doxycycline
DPA
    see Valproic Acid
Dramamine®
    see Dimenhydrinate
Drisdol®
    see Ergocalciferol
Driving
Dromostanolone
Dulcolax®
    see Bisacodyl
Duphalac®
    see Lactulose
Durabolin®
    see Nandrolone
Duramorph®
    see Morphine
Durathesia®
    see Procaine
Duvoid®
    see Bethanechol
DV® Cream
    see Dienestrol
Dyazide®
    see Triamterene
Dydrogesterone
Dymelor®
    see Acetohexamide
Dynapen®
    see Dicloxacillin

E

Earlobe Blood
Echothiophate
Econochlor®
    see Chloramphenicol
Econopred® Plus
    see Prednisolone
Econopred®
    see Prednisolone
Ecotrin®
    see Aspirin
Ectylurea
Edecrin®
    see Ethacrynic Acid
Edrophonium
EDTA
E.E.S.®
    see Erythromycin
Efodine®
    see Povidone-Iodine
Efudex®
    see Fluorouracil
Eggplant

EHDP
    see Etidronate
Elase®
    see Fibrinolysin
Elavil®
    see Amitriptyline
Electrocautery
Electroshock
Elspar®
    see Asparaginase
Emete-Con®
    see Benzquinamide
Empirin®
    see Aspirin
E-Mycin®
    see Erythromycin
Enalapril
Enalaprilat
    see Enalapril
Endep®
    see Amitriptyline
β-Endorphin
Endotoxin
Endralazine
Enduron®
    see Methyclothiazide
Enflurane
Enoxacin
Enoximone
Enprostil
Entsufon
    see pHisoHex®
Ephedrine
Epicillin
Epifrin®
    see Epinephrine
Epinal®
    see Epinephrine
Epinephrine
Epinine
EpiPen® Jr
    see Epinephrine
EpiPen®
    see Epinephrine
Epsom Salts
    see Magnesium Sulfate
Equanil®
    see Meprobamate
Erect Posture
Ergocalciferol
Ergothionine
Ergot Preparations
Eryc®
    see Erythromycin
Ery-Tab®
    see Erythromycin
Erythrocin®
    see Erythromycin
Erythrocytes
Erythromycin
Erythrosine
Eserine
    see Physostigmine
Eskalith®
    see Lithium
Estinyl®
    see Ethinyl Estradiol
Estrace®
    see Estradiol
Estraderm®
    see Estradiol
Estradiol
Estradiol-17β
Estriol
Estrogenic Substance Aqueous
    see Estrone
Estrogens
Estrogens, Conjugated
Estrone
Ethacrynate
    see Ethacrynic Acid
Ethacrynic Acid

Ethambutol
Ethamivan
Ethamolin®
    see Ethanolamine
Ethanediol
Ethanoic Acid
    see Acetic Acid
Ethanol
Ethanolamine
Ethaverine
Ethchlorvynol
Ether
Ethinamate
Ethinyl Estradiol
Ethiodized Oil
Ethiodol®
    see Ethiodized Oil
Ethionamide
Ethionine
Ethosuximide
Ethotoin
Ethoxazene
Ethoxycaffeine
Ethoxyethanol
Ethoxzolamide
Ethrane®
    see Enflurane
Ethyl Acetate
Ethyl Alcohol
    see Ethanol
Ethylamine
Ethyl Aminobenzoate
    see Benzocaine
Ethyl Biscoumacetate
Ethyl Chloride
Ethylene
Ethylenediamine
Ethylene Glycol
Ethylestrenol
Ethylmethylketone
Ethylnorepinephrine
Ethylphenacemide
Ethylphenylhydantoin
    see Ethotoin
I-N-Ethyl Sisomicin
    see Netilmicin
Ethynodiol
Etidronate
Etintidine
Etiocholanolone
Etodolac
Etomidate
Etoposide
Etretinate
Etryptamine
Euthroid®
    see Liotrix
Eutonyl®
    see Pargyline
Evans Blue
Exchange Transfusion
Exurate®
    see Benzbromarone

F

Factor IX Complex, Human
Famotidine
Fansidar®
Fasting
Fastin®
    see Phentermine
Fat Emulsions
Fatty Acids
Fatty Foods
Fava Beans
5-FC
    see Flucytosine
Feldene®
    see Piroxicam

Felodipine
Fencamfamin
Fenclofenac
Fenfluramine
Fenofibrate
Fenoprofen
Fenoterol
Fenpipramide
Fentanyl
Feosol®
    see Ferrous Sulfate
Feosol® Spansules®
    see Ferrous Sulfate
Fergon®
    see Ferrous Gluconate
Fer-In-Sol®
    see Ferrous Sulfate
Ferricyanide
Ferrous Ascorbate
Ferrous Gluconate
Ferrous Sulfate
Ferulic Acid
Fever
Fibrin Hydrolysate
Fibrinogen
Fibrinolysin
Filter Paper
Fish
Flagyl®
    see Metronidazole
Flavaspidic Acid
Flavoxate
Flaxedil®
    see Gallamine
Flecainide
Fleet® Babylax®
    see Glycerin
Fleet® Enema
    see Sodium Phosphate
Fleet® Mineral Oil Enema
    see Mineral Oil
Fleet® Phospho®-Soda
    see Sodium Phosphate
Flexeril®
    see Cyclobenzaprine
Florantyrone
Floropryl®
    see Isoflurophate
Floxuridine
Flubiprofen
Flucloxacillin
Flucytosine
Fludrocortisone
Flufenamic Acid
Flumecinolone
Flumethiazide
Flunarizine
Fluocinolone
Fluorescein
Fluorescite®
    see Fluorescein
Fluorides
Fluoroborate
Fluorodeoxyuridine
    see Floxuridine
5-Fluoronicotinic Acid
Fluorophenylalanine
Fluoropyrimidine
Fluorouracil
5-Fluorouracil
    see Fluorouracil
Fluor-I-Strips®
    see Fluorescein
Fluosol-DA
Fluothane®
    see Halothane
Fluoxymesterone
Fluphenazine
Flurazepam
Flurazepam Metabolite
5-Flurocytosine
    see Flucytosine

Fluroxene
Fluspirilene
Folacin
    see Folic Acid
Folate
    see Folic Acid
Folic Acid
Folvite®
    see Folic Acid
Food
Forane®
    see Isoflurane
Formaldehyde
Fortaz®
    see Ceftazidime
Fototar®
    see Coal Tar
Freezing
Fructose
Fructose-1,6-Diphosphate
Fruit
Fruit Sugar
    see Fructose
5-FU
    see Fluorouracil
FUDR®
    see Floxuridine
Fulvicin® P/G
    see Griseofulvin
Fulvicin-U/F®
    see Griseofulvin
Fumagillin
Fungizone®
    see Amphotericin B
Furacin®
    see Nitrofurazone
Furadaltone
Furadantin®
    see Nitrofurantoin
Furapromidium
Furazolidone
Furazolium
Furosemide
Furoxone®
    see Furazolidone
Fusaric Acid
Fuscin
Fusidic Acid

G

Galactosamine
Galactose
Gallamine
Gallium Nitrate
Gallopamil
Gamastan®
    see Immune Serum Globulin
Gamma-Globulin
    see Immune Serum Globulin
Gammar®
    see Immune Serum Globulin
Ganglionic Blocking Agents
Gantanol®
    see Sulfamethoxazole
Gantrisin®
    see Sulfisoxazole
Garamycin®
    see Gentamicin
Gargles
Garlic
Gasoline
Gastrin
Gemfibrozil
Gemonil®
    see Metharbital
Genoptic®
    see Gentamicin
Gentamicin
Gentisic Acid

Geocillin®
see Carbenicillin
Geopen®
see Carbenicillin
Glass Containers
Glassware
Glaucarubin
Glibenclamide
see Glyburide
Glibonuride
Glipizide
Glisoxepide
Globulin
γ-Globulin
see Immune Serum Globulin
Glucagon
Glucocorticoids
Glucosamine
Glucose
Glucose-6-Phosphate
Glucosidase
Glucosulfone
Glucotrol®
see Glipizide
Glucuronic Acid
Glutamate
Glutamic Acid
Glutarimide
Glutathione
Glutethimide
Glybenclamide
see Glyburide
Glybenzcyclamide
see Glyburide
Glyburide
Glyceraldehyde
Glyceric Acid
Glycerin
Glycerol
see Glycerin
Glyceryl Guaiacolate
see Guaifenesin
Glyceryl Trinitrate
see Nitroglycerin
Glycine
Glycocholate
Glycocyamidine
Glycocyamine
Glycopyrrolate
Glycylglycine
Glycylproline
Glycyrrhiza
Glymidine
Gold
Gonadotropin
GPA-1714
Griseofulvin
Growth Hormone
Guaiacol
Guaifenesin
Guanabenz
Guanazole
Guancydine
Guanethidine
Guanfacine
Guanidine
Guanidinoacetic Acid
Guanidinosuccinic Acid
Guanoclor
Guanoxan
Gum Tragacanth

H

H₂O₂
see Hydrogen Peroxide
Haemaccel
Hair Lacquer
Hair Treatments

Halcion®
see Triazolam
Haldol®
see Haloperidol
Haldrone®
see Paramethasone
Halofenate
Haloperidol
Halotestin®
see Fluoxymesterone
Halothane
HCTZ
see Hydrochlorothiazide
Heat
Height
Hematin
Hematoxylin
Hemodialysis
Hemoglobin
Hemolysis
Hemosiderin
Heparin
Heparin Lock Flush
see Heparin
HEPES
Hep-Lock®
see Heparin
Heptabarbital
Heptachlor
Heroin
Heroin Withdrawal
Herplex®
see Idoxuridine
Hetacillin
Hexa-Betalin®
see Pyridoxine
Hexachlorobenzene
Hexachlorphene
see pHisoHex®
Hexamethylmelamine
Hexocyclium
Hexoprenaline
Hexoses
Hibiclens®
see Chlorhexidine
High Carbohydrate Diet
High Sodium Diet
Hippuric Acid
Hiprex®
see Methenamine
Histalog™
see Betazole
Histamine
Histidine
HMM
see Hexamethylmelamine
HMPG
HN₂
see Mechlorethamine
Homocitrulline
Homocysteic Acid
Homocystine
Homogentisic Acid
Homoserine
Homovanillic Acid
Hospitalization
H.P. Acthar® Gel
see Corticotropin
Humatin®
see Paromomycin
Humorsol®
see Demecarium
Humulin® L
see Insulin
Humulin® N
see Insulin
Humulin® R
see Insulin
Humulin® U
see Insulin
Hurricaine®
see Benzocaine

HXM
see Hexamethylmelamine
Hycanthone
Hydantoin Derivatives
Hydeltrasol®
see Prednisolone
Hydralazine
Hydration
Hydrazine Derivatives
Hydrea®
see Hydroxyurea
Hydrochloric Acid
Hydrochlorothiazide
Hydrochlorothiazide and Triamterene
see Triamterene
Hydrocortisone
Hydrocortone®
see Hydrocortisone
Hydrocyanic Acid
HydroDIURIL®
see Hydrochlorothiazide
Hydroflumethiazide
Hydrogen Peroxide
Hydrogen Sulfide
Hydromorphone
Hydroquinone
4-Hydroxy-3-Methoxyphenylethylene glycol
see HMPG
17-Hydroxy-4-Pregnene-3,20-Dione
Hydroxyacetamide
4-Hydroxyacetanilide
Hydroxyaminobutyric Acid
Hydroxyamphetamine
Hydroxyanthranilic Acid
Hydroxybenzoic Acid
Hydroxybutyrate
Hydroxychloroquine
Hydroxycinnamic Acid
Hydroxydione
Hydroxyhexamide
Hydroxyhippuric Acid
5-Hydroxyindoleacetic Acid
Hydroxykynurenine
Hydroxymandelic Acid
5-(p-Hydroxyphenyl)-5-Phenylhydantoin
Hydroxyphenylacetic Acid
2-Hydroxyphenylacetic Acid
3-hydroxyphenylacetic Acid
4-Hydroxyphenylacetic Acid
Hydroxyphenylpyruvic Acid
4-Hydroxyphenylpyruvic Acid
Hydroxyprogesterone
17-Hydroxyprogesterone
Hydroxyproline
Hydroxyquinoline
5-Hydroxytryptamine
5-Hydroxytryptophan
Hydroxyurea
Hydroxyzine
Hygroton®
see Chlorthalidone
Hyoscine-N-Butylbromide
Hyoscine Hydrobromide
see Scopolamine
Hyperstat® I.V.
see Diazoxide
Hyperventilation
Hypochlorites
Hypothermia
Hypoxanthine
Hypoxia
Hyskon®
see Dextran
Hytakerol®
see Dihydrotachysterol
Hytone®
see Hydrocortisone
Hytrin®
see Terazosin

I

Ibidomide Hydrochloride
  see Labetalol
Ibopamine
Ibufenac
Ibuprofen
Ice-Eating
Icterogenin
Idoxuridine
IDU
  see Idoxuridine
Ifex®
  see Ifosfamide
Ifosfamide
IG
  see Immune Serum Globulin
Iletin® I
  see Insulin
Ilosone®
  see Erythromycin
Ilotycin®
  see Erythromycin
Imferon®
  see Iron Dextran
Imidazole
Imidazolepyruvic Acid
I.M. Injections
Imipenem
Imipramine
Immune Serum Globulin
Imodium®
  see Loperamide
Imuran®
  see Azathioprine
Indandione Derivatives
Indapamide
Inderal® LA
  see Propranolol
Inderal®
  see Propranolol
Indican
Indigo Blue
Indigo Carmine
  see Indigotindisulfonate
Indigotin
Indigotindisulfonate
Indocin®
  see Indomethacin
Indocyanine Green
Indole
Indoleacetic Acid
Indolepropionic Acid
Indolepyruvic Acid
Indomethacin
Inflamase®
  see Prednisolone
INH
  see Isoniazid
Inocor®
  see Amrinone
Inositol
Insecticides
Insulatard® NPH Human
  see Insulin
Insulatard® NPH
  see Insulin
Insulin
Intensain®
  see Chromonar Hydrochloride
α-Interferon
Intra-Amniotic Saline
Intro A®
  see α-Interferon
Intropin®
  see Dopamine
Inulin
Inversine®
  see Mecamylamine
Iobenzamic Acid
Iodates
Iodides

Iodinated Glycerol
Iodine
Iodine Containing Drugs
Iodipamide
Iodized Oil
Iodoacetate
Iodoacetic Acid
Iodoalphionic Acid
Iodoamide
Iodoantipyrine
Iodocasein
Iodochlorhydroxyquin
Iodoform
Iodohippurate
Iodophor Detergents
Iodopyracet
Iodoxyl
Iohexol
Ionamin®
  see Phentermine
Ion Exchange Resins
Iopamidol
Iopanoic Acid
Iophendylate
Iophenoxic Acid
Iopydone
Iothalamate
Iothiouracil
Ioxitalamic Acid
Ipecac
Ipodate
Iprindole
Iproniazid
Iproveratril Hydrochloride
  see Verapamil
IPV
  see Poliomyelitis Vaccine
Iron
Iron Dextran
Iron Salts
Iron Sorbitex
ISG
  see Immune Serum Globulin
Ismelin®
  see Guanethidine
Ismotic®
  see Isosorbide
Isocarbazide
Isocarboxamide
Isocarboxazid
Isodine®
  see Povidone-Iodine
d-Isoephedrine Hydrochloride
  see Pseudoephedrine
Isoethadione
  see Paramethadione
Isoflurane
Isoflurophate
Isoleucine
Isometheptene
Isomil®
Isoniazid
Isonicotinic Acid
Isonicotinic Acid Hydrazide
  see Isoniazid
Isoprenaline
Isopropamide
Isopropanol
Isopropyl Alcohol
  see Isopropanol
Isopropyl Dipyrone
Isopropylepinephrine
Isopropylnorepinephrine
Isoproterenol
Isoptin®
  see Verapamil
Isopto® Carpine
  see Pilocarpine
Isopto® Eserine
  see Physostigmine
Isopto® Frin
  see Phenylephrine

Isopto® Hyoscine
  see Scopolamine
Isosorbide
Isotretinoin
Isuprel®
  see Isoproterenol
Itraconazole
Ivadantin®
  see Nitrofurantoin

J

Jaundice

K

Kanamycin
Kantrex®
  see Kanamycin
Kaolin
Kaon-Cl®
  see Potassium Chloride
KCl
  see Potassium Chloride
K-Dur®
  see Potassium Chloride
Keflex®
  see Cephalexin
Keflin®
  see Cephalothin
Kefurox®
  see Cefuroxime
Kefzol®
  see Cefazolin
Kemadrin®
  see Procyclidine
Kenacort®
  see Triamcinolone
Kenalog®
  see Triamcinolone
Kerosene
Ketalar®
  see Ketamine
Ketamine
Ketanserin
Keto Acids
α-Ketobutyric Acid
Ketoconazole
α-Ketoglutarate
Ketones
Ketoprofen
Ketosis
KI
  see Potassium Iodide
Klonopin™
  see Clonazepam
K-Lor™
  see Potassium Chloride
Kōnyne®-HT
  see Factor IX Complex, Human
K-PHOS® M.F.
  see Phosphorus
K-PHOS® Neutral
  see Phosphorus
K-PHOS® Original M.F.
  see Phosphorus
Kwashiorkor
Kynurenine

L

Labetalol
Lactate
Lactate Dehydrogenase
Lactation
Lactinex®
  see Lactobacillus Acidophilus

Lactobacillus Acidophilus
Lactoflavin
   see Riboflavin
Lactose
Lactulose
Lanoxicaps®
   see Digoxin
Lanoxin®
   see Digoxin
Larodopa®
   see Levodopa
Larotid®
   see Amoxicillin
Lasix®
   see Furosemide
Latamoxef
   see Moxalactam
Latamoxef Disodium
   see Moxalactam
Laxatives
L.C.D.
   see Coal Tar
Lead
Lecithin
Lente® Iletin® II (Beef)
   see Insulin
Lente® Iletin® II (Pork)
   see Insulin
Lente® Iletin® I
   see Insulin
Lente®
   see Insulin
Leucine
Leucovorin
Leukeran®
   see Chlorambucil
Leukocytes
Levamisole
Levarterenol
Levodopa
Levoglutamide
Levomepromazine
   see Methotrimeprazine
Levonorgestrel
Levoprome®
   see Methotrimeprazine
Levorphanol
Levothyroxine
Levulose
   see Fructose
Libritabs®
   see Chlordiazepoxide
Librium®
   see Chlordiazepoxide
Lidocaine
Lidoflazine
Light
Lignocaine Hydrochloride
   see Lidocaine
Lincocin®
   see Lincomycin
Lincomycin
Linoleamide
Liothyronine
Liotrix
Lipemia
Lipochrome
Lipomul®
Liposol
Lipuria
Liquaemin®
   see Heparin
Liqui-Char®
   see Charcoal
Liquid antidote
   see Charcoal
Liquid Soap
Lisuride
Lithane®
   see Lithium
Lithium
Lithium Heparin

Lithium Lactate
Lithobid®
   see Lithium
Lithostat®
   see Acetohydroxamic Acid
Lithotabs®
   see Lithium
LMD®
   see Dextran
Local Anesthetics
Loniten®
   see Minoxidil
Loperamide
Lopid®
   see Gemfibrozil
Loprazolam
Lopressor®
   see Metoprolol
Lorazepam
Lorcainide
Lorelco®
   see Probucol
Losec®
   see Omeprazole
Lovastatin
Loxapine
Loxitane®
   see Loxapine
Lozol®
   see Indapamide
Lucanthone
Ludiomil®
   see Maprotiline
Lugol's Iodine
Lugol's Solution
   see Potassium Iodide
Luminal®
   see Phenobarbital
Lutidine
Lynestrenol
Lysergide
Lysine
Lysodren®
   see Mitotane
Lysol®

M

Macrodantin®
   see Nitrofurantoin
Macrodex®
   see Dextran
Mafenide
Magnesia Magma
   see Magnesium Hydroxide
Magnesium
Magnesium Carbonate
Magnesium Hydroxide
Magnesium Salts
Magnesium Sulfate
Magnesium Trisilicate
Malic Acid
Malonic Acid
Maltose
Mandelamine®
   see Methenamine
Mandelic Acid
Mandol®
   see Cefamandole
Manganese
Manganese Dioxide
Manganese Salts
Mannitol
Mannose
MAO Inhibitors
Maolate®
   see Chlorphenesin
Maprotiline
Marital Status
Marophen

Marplan®
   see Isocarboxazid
Matulane®
   see Procarbazine
Mayo Enema
Mazindol
MD-50®
   see Diatrizoic acid
Meals
Measles Vaccine
Meat
Meat Extract
Mebanazine
Mebaral®
   see Mephobarbital
Mebhydrolin
Mebutamate
Mecamylamine
Mechlorethamine
Medazepam
Medihaler-Epi®
   see Epinephrine
Medihaler-Iso®
   see Isoproterenol
Medrol®
   see Methylprednisolone
Medroxyprogesterone
Mefenamic Acid
Mefoxin®
   see Cefoxitin
Mefruside
Megace®
   see Megestrol
Megestrol
Meglumine
Meglumine Ioglycamate
Melanin
Melarsonyl
Melarsoprol
Melphalan
Menadione
Menopause
Menotropins
Menstruation
Mepacrine
Mepacrine Hydrochloride
   see Quinacrine
Mepazine
Mepenzolate
Meperidine
Mephenesin
Mephenesin Carbamate
Mephenoxalone
Mephentermine
Mephenytoin
Mephobarbital
Mephyton®
   see Phytonadione
Meprednisone
Meprobamate
Meptazinol
Meralluride
Merbromin
Mercaptoethane
Mercaptoethanol
Mercaptomerin
Mercaptopurine
6-Mercaptopurine
   see Mercaptopurine
Mercaptopyruvate
D-3-Mercaptovaline
   see Penicillamine
Mercurial Diuretics
Mercury Compounds
Mersalyl
Merthiolate®
   see Thimerosal
Mesantoin®
   see Mephenytoin
   see Mephenytoin
Mescaline

Mesenex®
see Mesna
Mesna
Mesoridazine
Mesterolone
Mestinon® Bromide
see Pyridostigmine
Mestranol
Metahexamide
Metahydrin®
see Trichlormethiazide
Metamucil®
see Psyllium Fibre
Metandren®
see Methyltestosterone
Metanephrine
Metaphosphoric Acid
Metaproterenol
Metaraminol
Metaxalone
Metformin
Methacholine
Methacycline
Methadone
Methadone Metabolite
Methamphetamine
Methandriol
Methandrostenolone
Methanol
Methantheline
Methaphentermine
Methapyrilene
Methaqualone
Metharbital
Methazolamide
Methdilazine
Methemoglobin
Methenamine
Methergoline
Methicillin
Methimazole
Methiodol
Methionine
Methocarbamol
Methohexital
Methoin
see Mephenytoin
Methotrexate
Methotrimeprazine
Methoxamine
Methoxsalen
Methoxyflurane
Methoxyphenamine
4-Methoxyphenol
Methoxypsoralen
see Methoxsalen
Methoxytyramine
Methsuximide
Methyclothiazide
Methylacetoxyprogesterone
see Medroxyprogesterone
Methylamine
Methylaminoantipyrine
Methylamphetamine
Methylbromide
Methyldopa
Methyldopa Hydrazine
Methyldopamine
Methylene Blue
Methylenedioxyamphetamine
Methylephedrine
Methylethinylestradiol
Methylmalonic Acid
Methylphenidate
Methylphenobarbital
Methylphenylethylhydantoin
see Mephenytoin
Methylprednisolone
Methylpromazine
Methylserotonin
Methylsulfonal
Methyltestosterone

Methylthiouracil
Methyltryptophan
Methylurea
Methyprylon
Methysergide
Metiamide
Metiazinic Acid
Meticorten®
see Prednisone
Metoclopramide
Metolazone
Metopirone®
see Metyrapone
Metoprolol
Metrifonate
Metrizamide
Metrizoate
Metronidazole
Metyrapone
Mevacor®
see Lovastatin
Mevinolin
see Lovastatin
Mexate®
see Methotrexate
Mexiletine
Mexitil®
see Mexiletine
Mezlin®
see Mezlocillin
Mezlocillin
Miacalcin®
see Calcitonin
Mianserin
Micatin®
see Miconazole
Miconazole
Micro-K®
see Potassium Chloride
Micronase®
see Glyburide
microNefrin®
see Epinephrine
Microsulfon®
see Sulfadiazine
Midazolam
Milkinol®
see Mineral Oil
Milk of Magnesia
see Magnesium Hydroxide
Milontin®
see Phensuximide
Miltown®
see Meprobamate
Mineral Oil
Minipress®
see Prazosin
Minocin® IV
see Minocycline
Minocin®
see Minocycline
Minocycline
Minoxidil
Mintezol®
see Thiabendazole
Miochol®
see Acetylcholine
Mithramycin
Mitomycin
Mitotane
Mitoxantrone
Mixtard® Human 70/30
see Insulin
Mixtard®
see Insulin
MK-270
Modane® Mild
see Phenolphthalein
Molindone
MOM
see Magnesium Hydroxide

Monacolin K
see Lovastatin
Monistat™
see Miconazole
Monoamine oxidase inhibitors
see MAO Inhibitors
Monofluorosulphonate
Monoiodotyrosine
8-MOP
see Methoxsalen
MOPP
More Attenuated Enders Strain
see Measles Vaccine
Morinamide
Morphine
Morpholine
Motrin®
see Ibuprofen
Mouth Washes
Moxalactam
Moxam®
see Moxalactam
6-MP
see Mercaptopurine
MPCA
MS
see Morphine
MS Contin®
see Morphine
MSIR®
see Morphine
MTX
see Methotrexate
Mumps Vaccine
Mumpsvax®
see Mumps Vaccine
Muro 128® Ophthalmic
see Sodium Chloride
Muscle Massage
Muscular Exercise
Mushroom Poisoning
Mustard
Mustard Gas
Mustargen® Hydrochloride
see Mechlorethamine
Mustine
see Mechlorethamine
Mutamycin®
see Mitomycin
Muzolimine
Myambutol®
see Ethambutol
Mycifradin® Sulfate
see Neomycin
Mycostatin®
see Nystatin
Mykrox®
see Metolazone
Myleran®
see Busulfan
Mylicon®
see Simethicone
Myochrysine®
see Gold
Myoglobin
Mysoline®
see Primidone

N

Nadolol
Nafarelin Acetate
Nafcil™
see Nafcillin
Nafcillin
Nafenopin
Nalfon®
see Fenoprofen
Nalidixic Acid
Nalorphine

Naloxone
Naltrexone
Nandrolone
Naphazoline
Naphthalene
Naphthol
Naphthoxyacetic Acid
Naprosyn®
   see Naproxen
Naproxen
Naqua®
   see Trichlormethiazide
Narcan®
   see Naloxone
Narcotic Antagonists
Narcotics
Nardil®
   see Phenelzine
Navane®
   see Thiothixene
Nebcin®
   see Tobramycin
NegGram®
   see Nalidixic Acid
Nembutal® Sodium
   see Pentobarbital
Neoarsphenamine
Neocinchophen
Neoloid®
   see Castor Oil
Neo-Mull-Soy®
Neomycin
Neosporin®
   see Bacitracin
Neostigmine
Neo-Synephrine®
   see Phenylephrine
Nephrox®
   see Aluminum Hydroxide
Neptazane®
   see Methazolamide
Nessler's Reagent
Netilmicin
Netromycin®
   see Netilmicin
Neuromuscular Relaxants
Neutral Red
Neutra-Phos®
   see Phosphorus
Niacin
Niacinamide
Niagara Sky Blue
Nialamide
Nicardine
Nicardipine
Nicergoline
Nickel
Nicorette®
   see Nicotine
Nicotinamide
   see Niacinamide
Nicotine
Nicotinic Acid
   see Niacin
Nifedipine
Niflumic Acid
Nifurtimox
Nilstat®
   see Nystatin
Nilvadipine
Nipagin
Niridazole
Nisoldipine
Nitrates
Nitrazepam
Nitric Acid
Nitrites
Nitrobenzene
Nitro-Bid®
   see Nitroglycerin
Nitro-Dur®
   see Nitroglycerin

Nitrofurans
Nitrofurantoin
Nitrofurazone
Nitrogen Mustard
   see Mechlorethamine
Nitrogen Oxides
Nitroglycerin
Nitrol®
   see Nitroglycerin
Nitrophenol
Nitrostat®
   see Nitroglycerin
Nitrous Oxide
Nizoral®
   see Ketoconazole
Noctec®
   see Chloral Hydrate
Noise
Noludar®
   see Methyprylon
Nolvadex®
   see Tamoxifen
Nonsteroidal Anti-Inflammatory Agent
   see NSAID (Nonsteroidal Anti-Inflammatory Drugs)
Noramidopyrine
Nordiazepam
Norethandrolone
Norethandrostenolone
Norethindrone
Norethisterone
Norethynodrel
Norfenefrin
Norfloxacin
Norgesic®
   see Orphenadrine
Norgestrel
Norleucine
Norlutin®
   see Norethindrone
Normal Human Serum Albumin
   see Albumin
Normal Saline
   see Sodium Chloride
Normal Serum Albumin (Human)
   see Albumin
Normetanephrine
Normodyne®
   see Labetalol
Normorphine
Noroxin®
   see Norfloxacin
Norpace®
   see Disopyramide
Norpramin®
   see Desipramine
Norpropoxyphene
Norpseudoephedrine
Nortriptyline
Norvaline
Novaminsulfon
Novantrone®
   see Mitoxantrone
Novobiocin
Novocain®
   see Procaine
Novolin® 70/30
   see Insulin
Novolin® L
   see Insulin
Novolin® N
   see Insulin
Novolin® R Penfill
   see Insulin
Novolin® R
   see Insulin
NPH
   see Insulin
NPH Iletin®I
   see Insulin
NPH Purified Pork
   see Insulin

NSAID (Nonsteroidal Anti-Inflammatory Drugs)
NSD 3004
Nucleoproteins
Nuprin®
   see Ibuprofen
Nydrazid®
   see Isoniazid
Nystatin

O

Obesity
Octopamine
Ofloxacin
Ohio 469
Oleandomycin
Oleate
Oleum Ricini
   see Castor Oil
Omeprazole
Omnipaque®
   see Iohexol
Omnipen®
   see Ampicillin
Onions
Ophthalgan®
   see Glycerin
Opiates
Opipramol
Opium Alkaloids
Oral Contraceptives
Oral Resins
Oranges
Orap™
   see Pimozide
Orasone®
   see Prednisone
Oreton®
   see Methyltestosterone
Organidin®
   see Iodinated Glycerol
Organophosphorus Insecticides
Orinase®
   see Tolbutamide
Ornithine
Orotic Acid
Orphenadrine
Ortho® Dienestrol
   see Dienestrol
Orudis®
   see Ketoprofen
Os-Cal® 250
   see Calcium
Os-Cal® 500
   see Calcium
Osmitrol®
   see Mannitol
Osmoglyn®
   see Glycerin
Ouabain
Ovulation
Oxacillin
Oxalacetate
Oxalate
Oxalate/Fluoride
Oxandrolone
Oxaprozin
Oxazepam
Oxedrine
Oxilapine Succinate
   see Loxapine
Oxmetidine
Oxoprogesterone
Oxprenolol
Oxsoralen®
   see Methoxsalen
Oxycodone
Oxymetazoline
Oxymetholone

Oxyphenbutazone
Oxyphencyclimine
Oxyphenisatin
Oxyquinoline
11-Oxysteroids
Oxytetracycline
Oxytocin
Ozone

P

Pain
Palmitic Acid
Pamaquine
Pamelor®
   see Nortriptyline
Pamisyl®
   see Aminosalicylic Acid
L-PAM
   see Melphalan
Pancreozymin
Pancuronium
Panmycin®
   see Tetracycline
Pantopon®
   see Opium Alkaloids
Panwarfin®
   see Warfarin
Papaverine
Paracetamol
   see Acetaminophen
Paracort®
   see Prednisone
Paradione®
   see Paramethadione
Paraflex®
   see Chlorzoxazone
Parafon Forte™ DSC
   see Chlorzoxazone
Paraldehyde
Paramethadione
Paramethasone
Paraplatin®
   see Carboplatin
Parasympathol
Parathiazine
Parathion
Parathyroid Extract
Parathyroid Hormone
Paredrine®
   see Hydroxyamphetamine
Parenogen®
   see Fibrinogen
Pargyline
Parity
Parlodel®
   see Bromocriptine
Parnate®
   see Tranylcypromine
Paromomycin
PAS
   see Aminosalicylic Acid
Pathocil®
   see Dicloxacillin
Pavabid®
   see Papaverine
Pavulon®
   see Pancuronium
PCE®
   see Erythromycin
Pectin
Pedameth®
   see Methionine
Pediapred®
   see Prednisolone
PediaProfen™
   see Ibuprofen
Peganone®
   see Ethotoin
Pempidine

Penbritin®
   see Ampicillin
Penethamate
Penicillamine
D-Penicillamine
   see Penicillamine
Penicillin
Penicillin G
Penicillin V
Pentaerythritol
Pentagastrin
Pentam-300®
   see Pentamidine
Pentamidine
Pentaquine
Pentazocine
Penthrane®
   see Methoxyflurane
Pentobarbital
Pentolinium
Pentoses
Pentothal® Sodium
   see Thiopental
Pentoxifylline
Pentrax®
   see Coal Tar
Pentylenetetrazole
Pen.Vee® K
   see Penicillin V
Pepcid®
   see Famotidine
Peptavlon®
   see Pentagastrin
Perchlorate
Pergolide
Pergonal®
   see Menotropins
Perhexilene
Periactin®
   see Cyproheptadine
Peribedil
Pericyazine
   see Propericiazine
Peridex®
   see Chlorhexidine
Perindopril
Periodate
Peritrate®
   see Pentaerythritol
Peritrate® SA
   see Pentaerythritol
Permanganate
Peroxide
Perphenazine
Persantine®
   see Dipyridamole
Pertofrane®
   see Desipramine
PETN
   see Pentaerythritol
Phazyme®-95
   see Simethicone
Phenacemide
Phenaglycodol
Phenanthroline
Phenantoin
   see Mephenytoin
Phenazocine
Phenazone
Phenazopyridine
Phencyclidine
Phendimetrazine
Phenelzine
Phenergan®
   see Promethazine
Phenethicillin
Phenethylamine
Phenindamine
Phenindione
Pheniprazine
Phenmetrazine
Phenobarbital

Phenobarbitone
   see Phenobarbital
Phenolphthalein
Phenols
Phenolsulfonphthalein
Phenothiazines
Phenoxybenzamine
Phenoxymethicillin
Phenoxypropazine
Phenprocoumon
Phenpromethamine
Phensuximide
Phentermine
Phentolamine
Phenylalanine
Phenylalanine Mustard
   see Melphalan
Phenylbutazone
Phenylenediamine
Phenylephrine
Phenylethylmalonylurea
   see Phenobarbital
Phenylhydrazine
Phenylmercuric Acetate
Phenylpropanolamine
Phenylpyruvic Acid
Phenyl Salicylate
Phenylthiourea
Phenylurea
Phenyramidol
Phenytoin
Phetharbital
pHisoHex®
Phleomycin
Phloridzin
Phosgene
Phosphastrate
Phosphates
Phosphoenolpyruvate
3-Phosphoglyceric Acid
Phospholine Iodide®
   see Echothiophate
Phospholipids
Phosphorus
Phospho-Soda®
Phthalylsulfathiazole
Phylloerythrinogen
Phylloquinone
   see Phytonadione
Physical Stress
Physical Training
Physostigmine
Phytate
Phytofluene
Phytomenadione
   see Phytonadione
Phytonadione
Picoline
Picric Acid
PIDH
Pilocarpine
Pimozide
Pindolol
Pineapples
Pink Capsules
Pipamazine
Pipemidic Acid
Piperacetazine
Piperacillin
Piperacil®
   see Piperacillin
Piperazine
Piperidine
Pipobroman
Pirenzepine
Piretanide
Piroxicam
Pitocin®
   see Oxytocin
Pitressin®
   see Vasopressin

Pizotifen
see Pizotyline
Pizotyline
Placidyl®
see Ethchlorvynol
Plantains
Plaquenil®
see Hydroxychloroquine
Plasma
Plastic
Plastic Tubing
Platinol®
see Cisplatin
Plums
Poison Ivy
Poliomyelitis Vaccine
see Poliomyelitis Vaccine
Polycillin®
see Ampicillin
Polymagma
Polymethylmethacrylate
Polymox®
see Amoxicillin
Polymyxin
Polymyxin E
see Colistin
Polystyrene Sulfonate
Polysucrose
Polytar®
see Coal Tar
Polythiazide
Pondimin®
see Fenfluramine
Ponstel®
see Mefenamic Acid
Pork NPH Iletin® II
see Insulin
Pork Regular Iletin® II
see Insulin
Porphobilinogen
Porphyrins
Posterior Pituitary
Potassium
Potassium Aminobenzoate
Potassium Chlorate
Potassium Chloride
Potassium Iodide
Potassium Iodide Enseals®
see Potassium Iodide
Potassium Oxalate
Potassium Salts
Povidone-Iodine
Practolol
Prasterone
Prazosin
Pred Forte®
see Prednisolone
Pred Mild®
see Prednisolone
Prednisolone
Prednisone
Pregnancy
Pregnanedione
Pregnanetriol
Pregnenolone
Preludin®
see Phenmetrazine
Premarin®
see Estrogens, Conjugated
Prenalterol
Prenylamine
Presamine®
see Imipramine
Prilocaine
Primaquine
Primaxin®
see Imipenem
Primethamine
Primidone
Principen®
see Ampicillin

Priscoline®
see Tolazoline
Pro-Banthine®
see Propantheline
Probenecid
Probucol
Procainamide
Procaine
Procan® SR
see Procainamide
Procarbazine
Procardia®
see Nifedipine
Prochlorperazine
Proctocort™
see Hydrocortisone
Procyclidine
Profenal®
see Suprofen
Profilnine® Heat-Treated
see Factor IX Complex, Human
Proflavine
Progabide
Progesterone
Progestogens
Proguanil
Prolactin
Proline
Prolixin®
see Fluphenazine
Proloprim®
see Trimethoprim
Promazine
Promethazine
Promit®
see Dextran
Pronestyl®
see Procainamide
Propafenone
Propamidine
Propanidid
Propanol
Propantheline
Propericiazine
β-Propiolactone
Propionate
Propionic Acid
Proplex®
see Factor IX Complex, Human
Propoxyphene
Propranolol
Propylene Glycol
Propylhexedrine
Propyliodone
2-Propylpentanoic Acid
see Valproic Acid
Propylthiouracil
2-Propylvaleric Acid
see Valproic Acid
Proscillaridin
ProSobee®
Prostaglandin-α_1
Prostaglandin-α_2
Prostaglandin-15-EPI-A$_2$
Prostaglandin F2α
Prostaglandins
Prostaphlin®
see Oxacillin
Prostate Palpation
Prostigmin®
see Neostigmine
Protamine
Protamine, Zinc and Iletin® II (Beef)
see Insulin
Protamine, Zinc and Iletin® II (Pork)
see Insulin
Protamine, Zinc and Iletin®I
see Insulin
Protein
Protein Deficiency
Protein Hydrolysate
Proteinuria

Protionamide
Protocatechuic Acid
Protriptyline
Proventil®
see Albuterol
Provera®
see Medroxyprogesterone
Provocholine®
see Methacholine
Pseudoephedrine
Pseudomonas
Psyllium Fibre
Pteroylglutamic Acid
see Folic Acid
PTU
see Propylthiouracil
Purge®
see Castor Oil
Purified Pork Insulin
see Insulin
Purinethol®
see Mercaptopurine
Purodigin®
see Digitoxin
Puromycin
Putrescine
Pyrantel
Pyrazinamide
Pyrazinoic Acid
Pyrazinoic Acid Amide
see Pyrazinamide
Pyrazolones
Pyribenzamine®
Pyridamole
Pyridine
Pyridinol Carbamate
Pyridium®
Pyridostigmine
Pyridoxal
Pyridoxamine
Pyridoxine
Pyridyltetrazole
Pyrilamine
Pyrimethamine
Pyrithioxine
Pyritinol
Pyrogallol
Pyrogens
Pyrophosphate
Pyruvate
Pyruvate Kinase
Pyuria
PZI
see Insulin

Q

Quaternary Ammonium Compounds
Quemid®
see Cholestyramine
Questran®
see Cholestyramine
Quinacrine
Quinaglute® Dura-Tabs®
see Quinidine
Quinethazone
Quingestanol
Quinidex® Extentabs®
see Quinidine
Quinidine
Quinine
Quinine Iodobismuthate
Quinocide

R

Race
Racemic Amphetamine Sulfate
see Amphetamine

Radioactive Compounds
Radioactive Iodine
Radiographic Agents
Ragweed
Ramipril
Ranitidine
Rauwolfia
Recumbency
Red Dyed Drugs
Red Dyed Food
Reducing Diet
Regitine®
 see Phentolamine
Reglan®
 see Metoclopramide
Regonol®
 see Pyridostigmine
Regular
 see Insulin
Regular Insulin
 see Insulin
Regular Purified Pork
 see Insulin
Rela®
 see Carisoprodol
Renacidin®
Renal Transplant
Reserpine
Resorcinol
Retin-A™
 see Tretinoin
Retinoic Acid
 see Tretinoin
Retrovir®
 see Azidothymidine
R-Gene® 10
 see Arginine
R-Gene®
 see Arginine
Rhamnose
Rheomacrodex®
 see Dextran
Rhubarb
Riboflavin
Ribose
Ridaura®
 see Auranofin
Rifadin®
 see Rifampin
Rifampicin
 see Rifampin
Rifampin
Rimeterol
Ristocetin
Ritalin®
 see Methylphenidate
Ritodrine
RMS®
 see Morphine
RO4-2137
Robaxin®
 see Methocarbamol
Robaxisal®
 see Methocarbamol
Robinul®
 see Glycopyrrolate
Robitussin®
 see Guaifenesin
Rocephin®
 see Ceftriaxone
Roferon-A®
 see α-Interferon
Rolitetracycline
Room Temperature
Rotenone
Roxanol™
 see Morphine
Roxithromycin
Rubella Virus Vaccine
Rubeola Vaccine
 see Measles Vaccine

S

Saccharated Iron Oxide
Saccharose
Salbutamol
 see Albuterol
Salicylate
Salicylazosulfapyridine
 see Sulfasalazine
Saline
Saliva
Salk
 see Poliomyelitis Vaccine
Salol
Salsalate
Saluron®
 see Hydroflumethiazide
Sandfly Fever
Sandimmune®
 see Cyclosporine
Sansert®
 see Methysergide
Santonin
L-Sarcolysin
 see Melphalan
Sauerkraut
SC-16102
Scopolamine
Scopolamine Bromide
Season
Seclazone
Secobarbital
Seconal™
 see Secobarbital
Secretin
Secretin-Kabi
 see Secretin
Sectral®
 see Acebutolol
Selenium
Selsun®
 see Selenium
Semilente® Iletin® I
 see Insulin
Semilente® Purified Pork
 see Insulin
Senna
Septra®
 see Co-Trimoxazole
Serax®
 see Oxazepam
Serentil®
 see Mesoridazine
Serine
Seromycin® Pulvules®
 see Cycloserine
Serpasil®
 see Reserpine
Serum
Sex Difference
Sexual Activity
Silicones
Silver
Simethicone
Sinequan®
 see Doxepin
Sisomicin
Site of Collection
β-Sitosterol
Sitosterols
Skatole
Skelaxin®
 see Metaxalone
SKF-12185
Skin Lightening Cream
Sleep
Sleep Deprivation
Slo-bid™
 see Theophylline
Slo-Phyllin®
 see Theophylline

Slow-K®
 see Potassium Chloride
Smallpox Vaccine
Smog
Smoking
SMX-TMP
 see Co-Trimoxazole
Sodium
Sodium Acid Carbonate
 see Sodium Bicarbonate
Sodium Azide
Sodium Bicarbonate
Sodium Bisulfite
Sodium Bromide
Sodium Cellulose Phosphate
 see Cellulose Phosphate
Sodium Chloride
Sodium Citrate
Sodium Enibomal
Sodium Etidronate
 see Etidronate
Sodium Fluoride
Sodium Heparin
Sodium Hydrogen Carbonate
 see Sodium Bicarbonate
Sodium Iodate
Sodium Lauryl Sulfate
Sodium Loading
Sodium Methicillin
 see Methicillin
Sodium Nitroprusside
Sodium Oxalate
Sodium Phosphate
Sodium Phosphate ^{32}P
Sodium Phytate
Sodium Pyruvate
Sodium Restriction
Sodium Salicylate
Sodium Salts
Sodium Sulfate
Sodium Taurocholate
Sodium L-Tri-iodothyronine
 see Liothyronine
Soft Water Areas
Solaquin® Forte®
 see Hydroquinone
Solganal®
 see Aurothioglucose
Solu-Cortef®
 see Hydrocortisone
Solu-Medrol®
 see Methylprednisolone
Soma®
 see Carisoprodol
Somatostatin
Somatotropin
Sominex®
Somnos®
 see Chloral Hydrate
Somophyllin®-CRT
 see Theophylline
Somophyllin®-DF
 see Aminophylline
Somophyllin®
 see Aminophylline
Sorbitol
Sorbose
Sotalol
Sparine®
 see Promazine
Spartene
Spectinomycin
Spermatozoa
Spinach
Spinal Anesthesia
Spironolactone
SSKI
 see Potassium Iodide
Stadol®
 see Butorphanol
Standing of Sample
Stanozolol

Staphcillin®
   see Methicillin
Starvation
Stearate
Stearic Acid
Stearylamine
Stelazine®
   see Trifluoperazine
Stibophen
Stigmasterol
Stilbestrol
   see Diethylstilbestrol
Stilphostrol®
   see Diethylstilbestrol
Storage of Sample
Stoxil®
   see Idoxuridine
STP
Streptase®
   see Streptokinase
Streptokinase
Streptomycin
Streptonigrin
Streptozocin
Stress
Stress Cold Exposure
Strong Iodine Solution
   see Potassium Iodide
Strontium
Strychnine
SU-9055
Sublimaze®
   see Fentanyl
Succinic Acid
Succinimide
Succinylcholine
Succinylsulfathiazole
Sucralfate
Sucrose
Sudafed®
   see Pseudoephedrine
Sulamyd® Sodium
   see Sulfacetamide
Sulfacarbamide
Sulfacetamide
Sulfachlorpyridazine
Sulfadiazine
Sulfadimethoxine
Sulfadimidine
Sulfadoxine
Sulfaguanidine
Sulfamerazine
Sulfamethizole
Sulfamethoxazole
Sulfamethoxydiazine
Sulfamethoxypyridazine
Sulfamethoxypyridine
Sulfanilamide
Sulfaphenazole
Sulfapyridine
Sulfarthrol
Sulfasalazine
Sulfates
Sulfathiazole
Sulfhydryl Compounds
Sulfinpyrazone
Sulfisoxazole
Sulfobromophthalein
Sulfomethane
Sulfonamides
Sulfones
Sulfonylureas
Sulforidazine
Sulfoxone
Sulindac
Sulpiride
Sulpiridine
Sulthiame
Sultopride
Sumycin®
   see Tetracycline
Suntan Oil

SuperChar®
   see Charcoal
Suprofen
Suramin
Surgery
Surital®
   see Thiamylal Sodium
Sus-Phrine®
   see Epinephrine
Suxamethonium
   see Succinylcholine
Sweat
Sweet Potatoes
Symmetrel®
   see Amantadine
Synacort®
   see Hydrocortisone
Synalgos®
   see Aspirin
Syntocinon®
   see Oxytocin
Syringes
Syrosingopine

T

T₃ Thyronine Sodium
   see Liothyronine
Tacaryl®
   see Methdilazine
Tagamet®
   see Cimetidine
Talwin® NX
   see Pentazocine
Talwin®
   see Pentazocine
Tambocor®
   see Flecainide
Tamoxifen
Tao®
   see Troleandomycin
Tapazole®
   see Methimazole
Taractan®
   see Chlorprothixene
Tartrates
Taurine
Taurocholate
Tazidime™
   see Ceftazidime
TCN
   see Tetracycline
Tea
Tegison®
   see Etretinate
Tegopen®
   see Cloxacillin
Tegretol®
   see Carbamazepine
Tegrin®
   see Coal Tar
Telepaque®
   see Iopanoic Acid
Temaril®
   see Trimeprazine
Temocillin
Tenex®
   see Guanfacine
Tenoretic®
   see Atenolol
Tenormin®
   see Atenolol
Tensilon®
   see Edrophonium
Tenuate®
   see Diethylpropion
Tepanil®
   see Diethylpropion
Teprotide
Terazosin

Terbutaline
Terramycin®
   see Oxytetracycline
Teslac®
   see Testolactone
TESPA
   see Thiotepa
Testolactone
Testosterone
Tetrabenazine
Tetracaine
Tetrachloroethylene
Tetrachlorothyronine
Tetracosactrin
Tetracycline
Tetracyn®
   see Tetracycline
Tetraethyl-Lead
Tetragastrin
Tetrahydrocannabinol
Tetrahydrocortisol
Tetrahydro-DOC
Tetrahydro Reichstein's S
Tetraiodofluorescein
Tetraiodophthalein
Tetralin
Tetridaz-B
T/Gel®
   see Coal Tar
Thallium
Tham®
   see Tromethamine
Theelin®
   see Estrone
Thenalidine
Theo-24®
   see Theophylline
Theobromine
Theo-Dur®
   see Theophylline
Theolair™
   see Theophylline
Theophylline
Theophylline Ethylenediamine
   see Aminophylline
Thiabendazole
Thiacetazone
Thiamazole
   see Methimazole
Thiamine
Thiamylal Sodium
Thiazides
Thiazolsulfone
Thiethylperazine
Thimerosal
Thiobarbituric Acid
Thiocarlide
Thiocolchicine
Thiocyanate
Thioglycolate
Thioguanine
Thiols
Thioneine
Thiopental
Thiopronine
Thiopropazate
Thioridazine
Thiosemicarbazones
Thiotepa
Thiothixene
Thiouracil
Thiourea
Thorazine®
   see Chlorpromazine
Thorium Dioxide
Threonine
Thrombin
Thymol
Thyroid
Thyrolar®
   see Liotrix

Thyroliberin (TRH)
  see Thyrotropin-Releasing
    Hormone (TRH)
Thyronine
Thyrotropin
Thyrotropin-Releasing Hormone (TRH)
L-Thyroxine
  see Levothyroxine
Thytropar®
  see Thyrotropin
Tiaprofenic Acid
Ticarcillin
Ticar®
  see Ticarcillin
Timegadine
Timolol
Timoptic®
  see Timolol
Tin
TMP
  see Trimethoprim
TMP-SMX
  see Co-Trimoxazole
Tobacco Smoke
Tobramycin
Tobrex®
  see Tobramycin
Tocainide
Tofranil-PM®
  see Imipramine
Tofranil®
  see Imipramine
Tolazamide
Tolazoline
Tolbutamide
Tolectin®
  see Tolmetin
Tolinase®
  see Tolazamide
Tolmetin
Tolonium
Toluene
Toluene Diamine
Tomatoes
Tonocard®
  see Tocainide
Toothpaste
Topicort®
  see Desoximetasone
Torasemide
Tosylate Bretylium
Tourniquet
Tral®
  see Hexocyclium
Trancopal®
  see Chlormezanone
Trandate®
  see Labetalol
Tranquilizers
Transamine Sulphate
  see Tranylcypromine
Transderm-Nitro®
  see Nitroglycerin
Transderm Scop®
  see Scopolamine
Transport of Specimen
Tranxene®
  see Clorazepate
Tranylcypromine
Trazodone
Trecator®-SC
  see Ethionamide
Trental®
  see Pentoxifylline
Tretinoin
Triamcinolone
Triamterene
Triaziquone
Triazolam
Tribromethanol
Trichloracetic Acid
Trichlorethanol

Trichlormethiazide
Trichloroethylene
Tricine
Tridione®
  see Trimethadione
Triethanolamine
Triethylenemelamine
Triflocin
Trifluoperazine
Trifluperidol
Triflupromazine
Trihexyphenidyl
Tri-iodothyroacetic Acid
Tri-iodothyronine
Trilafon®
  see Perphenazine
Trimazosin
Trimeprazine
Trimethadione
Trimethaphan
Trimethobenzamide
Trimethoprim
Trimethoprim and Sulfamethoxazole
  see Co-Trimoxazole
Trimethylpsoralen
  see Trioxsalen
Trimetozine
Trimetrexate
Trimox®
  see Amoxicillin
Trimpex®
  see Trimethoprim
Trinitrotoluene
Triolein
Trioxsalen
Tripamide
Tripelennamine
Triphosadenine
Triprolidine
Tris
Tris(2-Butoxyethyl) Phosphate (TBEP)
Tris Buffer
  see Tromethamine
Tris(hydroxymethyl)aminomethane
  see Tromethamine
Trisoralen®
  see Trioxsalen
Triton X-100
Trizma
Trobicin®
  see Spectinomycin
Trofan®
  see Tryptophan
Trolamine
Troleandomycin
Tromethamine
Troxidone
  see Trimethadione
Trypan Blue
Tryparsamide
Trypsin
Tryptamine
Tryptophan
L-Tryptophan
  see Tryptophan
Tryptophan-Thiol
TSPA
  see Thiotepa
Tuaminoheptane
Tubocurarine
d-Tubocurarine Chloride
  see Tubocurarine
Tums®
  see Calcium Carbonate
Turpentine
Twin Pregnancy
Tybamate
Tylenol®
  see Acetaminophen
Typhoid Vaccine
Tyramine
Tyropanoic Acid

Tyrosine
2-Tyrosine
3-Tyrosine
Tyrothricin

U

Ultralente® Iletin®I
  see Insulin
Ultralente® Purified Beef
  see Insulin
Ultralente®
  see Insulin
Ultraviolet Light
Unipen®
  see Nafcillin
Uracil Mustard
Uranium
Urates
Urea
Uremia
Urethan
Uric Acid
Urine Flow
Urine pH
Urine Turbidity
Urine Volume
Urised®
  see Methenamine
Urispas®
  see Flavoxate
Urobilin
Urokinase
Ursodeoxycholic Acid

V

Vaginal Powder
Valine
Valisone®
  see Betamethasone
Valium®
  see Diazepam
Valmid®
  see Ethinamate
Valproic Acid
Vanadate
Vanadium
Vancenase®
  see Beclomethasone
Vanceril®
  see Beclomethasone
Vancocin®
  see Vancomycin
Vancomycin
Vanilla
Vanillic Acid
Vanillin
Vanillylandelic acid
  see Vanillylmandelic Acid (VMA)
Vanillylmandelic Acid (VMA)
Vaponefrin®
  see Epinephrine
Vaseretic®
  see Enalapril
Vasocon Regular®
  see Naphazoline
Vasopressin
Vasotec®
  see Enalapril
Vasoxyl®
  see Methoxamine
V-Cillin K®
  see Penicillin V
Veetids®
  see Penicillin V
Vegetables
Vegetarian Diet

Velban®
 see Vinblastine
Velosef®
 see Cephradine
Velosulin® Human
 see Insulin
Velosulin®
 see Insulin
Venipuncture
Venoms
Ventolin®
 see Albuterol
VePesid®
 see Etoposide
Veracillin®
 see Dicloxacillin
Verapamil
Vercyte®
 see Pipobroman
Verdoglobin
Versed®
 see Midazolam
Vesprin®
 see Triflupromazine
Vibramycin®
 see Doxycycline
Vibra-Tabs®
 see Doxycycline
Viloxazine
Vinblastine
Vincaleukoblastine
 see Vinblastine
Vincristine
Vinyl Chloride
Vinyl Ether
Vioform®
 see Iodochlorhydroxyquin
Viomycin
Visken®
 see Pindolol
Vistaril®
 see Hydroxyzine
Vitamin A
Vitamin A Acid
 see Tretinoin
Vitamin B$_1$
 see Thiamine
Vitamin B$_2$
 see Riboflavin
Vitamin B$_3$
 see Niacinamide
Vitamin B$_6$
 see Pyridoxamine
Vitamin B$_6$ Depletion
Vitamin B$_{12}$
Vitamin B Complex
Vitamin C
 see Ascorbic Acid
Vitamin D
Vitamin D$_2$
 see Ergocalciferol
Vitamin E
Vitamin G
 see Riboflavin
Vitamin K
Vitamin K$_1$
 see Phytonadione
Vitamin Preparations
Vivactil®
 see Protriptyline
VLB
 see Vinblastine
Voltaren®
 see Diclofenac
VoSol®
 see Acetic Acid
VP-16
 see Etoposide
VP-16-213
 see Etoposide

W

Walnuts
Warfarin
Warfarin Metabolites
Water
Water Load
Weight
Wellbutrin®
 see Bupropion
Westcort® Cream
 see Hydrocortisone
Winstrol®
 see Stanozolol
Wyamine® Sulfate
 see Mephentermine
Wymox®
 see Amoxicillin
Wytensin®
 see Guanabenz

X

Xanax®
 see Alprazolam
Xanthine
Xanthophyll
Xanthurenic Acid
X-Prep® Liquid
 see Senna
X-Ray Therapy
Xylene
Xylitol
Xylocaine®
 see Lidocaine
Xylo-Pfan®
 see Xylose
Xylose
Xylulose

Y

Yeast
Yutopar®
 see Ritodrine

Z

Zanosar®
 see Streptozocin
Zantac®
 see Ranitidine
Zarontin®
 see Ethosuximide
Zaroxolyn®
 see Metolazone
Zero Gravity
Zetar®
 see Coal Tar
Zimeldine
Zinacef®
 see Cefuroxime
Zinc
Zirconium Salts
Zolyse®
 see Chymotrypsin
Zomepirac
ZORprin®
 see Aspirin
Zovirax®
 see Acyclovir
Zoxazolamine
Zyloprim®
 see Allopurinol

3 LABORATORY TEST LISTINGS

Absorbance at 450 nm

Amniotic Fluid   Decrease   Analytical
Dexamethasone   Time-dependent decrease when added to bilirubin containing amniotic fluid *in vitro*. 2453
Prednisolone   Time-dependent reduction in absorbance of bilirubin when added *in vitro*. 2453

Acenocoumarol

Plasma   Increase   Physiological
Amiodarone   Significant potentiation of action so reduced dose of anticoagulant required. 0004

Plasma   Decrease   Physiological
Rifampin   Decreased serum concentration due to hepatic enzyme induction. 0014

Acetaldehyde

Blood   Increase   Physiological
Disulfiram   Ethanol metabolism diverted (10 times normal concentration). 1343
Ethanol   Marked increase in some individuals after 0.4 g alcohol/kg in 56 healthy males. 0015
Sulfonylureas   Mechanism not yet elucidated. 1343

Acetaldehyde Oxidase

Test Conditions   Decrease   Analytical
Disulfiram   Inhibitory effect observed. 2893

Acetaminophen

Serum   No Effect   Analytical
Acetylsalicylic Acid   At 500 mg/L had no effect on HPLC method. 3432
Aspirin   Has no effect on direct acid/ferric reduction method of Liu and Oka. 2195
Caffeine   At concentration of 10 mg/L had no effect on HPLC method. 3432
Chloramphenicol   No effect at therapeutic concentration on o-cresol reaction based method. 0621
Codeine   At concentration of 10 mg/L had no effect on HPLC method. 3432
Diazepam   At concentration of 2 mg/L had no effect on HPLC method. 3432
Disulphine Blue   Using a modified Glynn and Kendal method. 1464

Nitrazepam   At 2 mg/L no effect on HPLC method. 3432
Oxazepam   At 2 mg/L had no effect on HPLC method. 3432
Phenobarbital   At 10 mg/L had no effect on HPLC method. 3432
Salicylate   At 200 and 500 mg/L had no effect on HPLC method. 3432
Streptomycin   No effect at therapeutic concentration on method using o-cresol. 0621
Theobromine   At 10 mg/L had no effect on HPLC method. 3432
Theophylline   At 20 mg/L had no effect on HPLC method. 3432

Serum   Increase   Analytical
Aspirin   With Glynn and Kendal technique even with arithmetic corrections, due to metabolites. 0199   Significant positive effect on unmodified Glynn-Kendal technique. 2934
Salicylate   Increases results with unmodified Glynn and Kendal technique. 0199

Serum   Increase   Physiological
Metoclopramide   Increased rate of absorption with faster gastric emptying. 3794

Serum   Decrease   Analytical
Tetracycline   Slight reduction of up to 18% at just above upper therapeutic concentration on on method using o-cresol reaction. 0621

Serum   Decrease   Physiological
Food   Absorption delayed when taken with food. 3024
Propantheline   Delayed absorption with delayed gastric emptying. 3794

Urine   Increase   Physiological
Phenytoin   Significantly greater proportion excreted as glucuronide in response to coadministration of drug. 0394
Rifampin   Significantly greater proportion excreted as glucuronide in response to co-administration of drug. 0394
Smoking   Significantly greater amount excreted as glucuronide when ingested in smokers. 0394

Acetaminophen Screening Test

Urine   Positive   Analytical
Acetaminophen   Blue with o-cresol at therapeutic concentration. 3326   Red color with 1-naphthol at therapeautic concentration. 3326
Aminophenol   Blue with o-cresol at therapeutic concentration. Red color with 1-naphthol at therapeutic concentrations. 3326
Co-Trimoxazole   Red color with 1-naphthol at therapeutic concentrations. 3326
Phthalylsulfathiazole   Red color with 1-naphthol at therapeutic concentrations. 3326

Acetaminophen Screening Test (continued)

Urine   Positive   Analytical (continued)

Succinylsulfathiazole   Red color with 1-naphthol at therapeutic concentrations. 3326

Sulfadiazine   Red color with 1-naphthol at therapeutic concentrations. 3326

Sulfadimethoxine   Red color with 1-naphthol at therapeutic concentrations. 3326

Sulfadimidine   Red color with 1-naphthol at therapeutic concentrations. 3326

Sulfaguanidine   Red color with 1-naphthol at therapeutic concentrations. 3326

Sulfamethoxypyridine   Red color with 1-naphthol at therapeutic concentrations. 3326

Sulfaphenazole   Red color with 1-naphthol at therapeutic concentrations. 3326

Sulfisoxazole   Red color with 1-naphthol at therapeutic concentrations. 3326

Urine   Negative   Analytical

Acetaminophen   Negative Bratton-Marshall reaction at therapeutic concentrations. 3326

Albuterol   No reaction with o-cresol at therapeutic concentrations. 3326

Aluminum Hydroxide   No reaction with o-cresol at therapeutic concentrations. 3326

Aminophenol   Negative Bratton-Marshall reaction at therapeutic concentrations. 3326

Amitriptyline   No reaction with o-cresol at therapeutic concentrations. 3326

Amobarbital   No reaction with o-cresol at therapeutic concentrations. 3326

Aspirin   No reaction with o-cresol at therapeutic concentrations. 3326

Bendrofluazide   No reaction with o-cresol at therapeutic concentrations. 3326

Bisacodyl   No reaction with o-cresol at therapeutic concentrations. 3326

Chloral Hydrate   No reaction with o-cresol at therapeutic concentrations. 3326

Chlordiazepoxide   No reaction with o-cresol at therapeutic concentrations. 3326

Chlorpromazine   No reaction with o-cresol at therapeutic concentrations. 3326

Co-Trimoxazole   No reaction with o-cresol at therapeutic concentrations. 3326

Diazepam   No reaction with o-cresol at therapeutic concentrations. 3326

Docusate Sodium   No reaction with o-cresol at therapeutic concentrations. 3326

Ferrous Gluconate   No reaction with o-cresol at therapeutic concentrations. 3326

Folic Acid   No reaction with o-cresol at therapeutic concentrations. 3326

Furosemide   No reaction with o-cresol at therapeutic concentrations. 3326

Haloperidol   No reaction with o-cresol at therapeutic concentrations. 3326

Imipramine   No reaction with o-cresol at therapeutic concentrations. 3326

Lithium   No reaction with o-cresol at therapeutic concentrations. 3326

Methsuximide   No reaction with o-cresol at therapeutic concentrations. 3326

Orphenadrine   No reaction with o-cresol at therapeutic concentrations. 3326

Phenobarbital   No reaction with o-cresol at therapeutic concentrations. 3326

Phenytoin   No reaction with o-cresol at therapeutic concentrations. 3326

Phthalylsulfathiazole   No reaction with o-cresol at therapeutic concentrations. 3326

Pimozide   No reaction with o-cresol at therapeutic concentrations. 3326

Potassium Salts   No reaction with o-cresol at therapeutic concentrations. 3326

Primidone   No reaction with o-cresol at therapeutic concentrations. 3326

Prochlorperazine   No reaction with o-cresol at therapeutic concentrations. 3326

Promazine   No reaction with o-cresol at therapeutic concentrations. 3326

Propericiazine   No reaction with o-cresol at therapeutic concentrations. 3326

Senna   No reaction with o-cresol at therapeutic concentrations. 3326

Spironolactone   No reaction with o-cresol at therapeutic concentrations. 3326

Succinylsulfathiazole   No reaction with o-cresol at therapeutic concentrations. 3326

Sulfadiazine   No reaction with o-cresol at therapeutic concentrations. 3326

Sulfadimethoxine   No reaction with o-cresol at therapeutic concentrations. 3326

Sulfadimidine   No reaction with o-cresol at therapeutic concentrations. 3326

Sulfaguanidine   No reaction with o-cresol at therapeutic concentrations. 3326

Sulfamethoxypyridine   No reaction with o-cresol at therapeutic concentrations. 3326

Sulfaphenazole   No reaction with o-cresol at therapeutic concentrations. 3326

Sulfisoxazole   No reaction with o-cresol at therapeutic concentrations. 3326

Sulthiame   No reaction with o-cresol at therapeutic concentrations. 3326

Tetracycline   No reaction with o-cresol at therapeutic concentrations. 3326

Thioridazine   No reaction with o-cresol at therapeutic concentrations. 3326

L-Thyroxine   No reaction with o-cresol at therapeutic concentrations. 3326

Trihexyphenidyl   No reaction with o-cresol at therapeutic concentrations. 3326

Trimethadione   No reaction with o-cresol at therapeutic concentrations. 3326

Vitamin Preparations   No reaction with o-cresol at therapeutic concentrations. 3326

Acetate

Blood   Increase   Physiological

Ethanol   Mean concentration much higher in chronic alcoholics and heavy drinkers than in nonalcoholics and occasional drinkers. 1998

Acetic Acid

Urine   Increase   Physiological

Bacteriuria   Experimental studies — all 11 species produced much acid. 1483

Feces   Increase   Physiological

Ampicillin   Mild effect when given orally to 6 healthy volunteers for 6 d: normalized after 6 weeks. 1664

Clindamycin   Marked proportional increase in 6 healthy individuals fed orally for 6 d. 1664

Acetoacetate

Serum   Increase   Physiological

Aspirin   Due to late metabolic acidosis and renal impairment. 1343

Disulfiram In 1 of 6 volunteers rapid and short lasting effect. *3482*
Ethanol Sustained lipolysis with i.v. administration. *0202*
Starvation Due to metabolic acidosis. *0021*
Streptozocin Temporary effect following infusion. *2532*

Serum   Decrease   Physiological
MPCA Anti-lipolytic effect of drug. *1531*

Urine   No Effect   Analytical
Thymol No effect on Gerhardt procedure. *2981*

Urine   Increase   Analytical
Acetates May interfere with Gerhardt FeCl$_3$ procedure. *0459*
Antipyrine Interferes with ferric chloride test. *2425*
Aspirin Reacts with Gerhardt FeCl$_3$ procedure. *3879*
Cyanates May produce interfering color with Gerhardt FeCl$_3$. *1022*
Phenothiazines Metabolites react with FeCl$_3$. *3879*

Urine   Increase   Physiological
Aspirin Acidotic response especially in children. *1302*
Starvation Due to metabolic acidosis. *0021*

Acetoacetate Decarboxylase

Serum   No Effect   Analytical
Cyanides No inhibition produced. *3684*
Pyruvate No inhibition observed. *3684*

Serum   Increase   Analytical
Riboflavin Activation of enzyme produced. *3684*

Serum   Decrease   Analytical
Heat Inactivation of enzyme observed at 80° C. *3684*
Iodoacetate Inhibition of enzyme produced. *3684*
Mercury Compounds HgCl$_2$ causes denaturation. *3684*
Urea Inhibition of enzyme produced. *3684*

Acetone

Blood   Increase   Physiological
Disulfiram In all of 6 volunteers rapid and short lasting effect. *3482*
Isopropanol Metabolite of isopropanol. *2348*

Urine   No Effect   Analytical
Thymol No effect on Rothera procedure. *2981*

Urine   Increase   Analytical
Cellulose Acetate Gives false positive with Rothera's test. *0251*
Phenolsulfonphthalein Chromogenicity in color reaction (Rothera test). *3879*
Sulfobromophthalein Development of color at alkaline pH (Rothera test). *3879*

Urine   Increase   Physiological
Isopropanol Metabolized partially to acetone. *1488*
Methanol Slight to moderate effect in poisoning. *1343*

Acetylcholine Receptor Antibodies

Serum   No Effect   Physiological
Thiopronine No occurrence of myasthenia gravis or other autoimmune syndromes reported. *1761*

Serum   Increase   Physiological
Penicillamine Myasthenia gravis may occur as autoimmune syndrome, possibly as result of direct binding of drug to receptor. *1761*
Pyrithioxine With dose of 400-600 mg/d; associated with development of immune complex nephritis and other penicillamine-like toxic reactions. *1761*

Acetylcholinesterase

Serum   Decrease   Physiological
Ranitidine Inhibited by therapeutic doses. *1938*

Red Blood Cells   No Effect   Physiological
Smoking No significant difference between smokers and non-smokers. *1548*

Red Blood Cells   Decrease   Physiological
Ethanol Observed in the blood of 36 alcoholics in comparison with 41 healthy volunteers *in vitro* studies showed effect in proportion to concentration of ethanol. *1442* Reduced activity observed in alcoholics: immediate effect observed with in vitro incubation. *1442* Mean activity in cells of alcoholics significantly less than in normals. Same effect observed with *in vitro* studies — inhibition in proportion to ethanol concentration with alcohol concentrations at those found in clinically drunk humans. *1442*

N-Acetylglucosaminidase

Serum   No Effect   Physiological
Gold No significant changes of microsomal enzyme in patients treated with parenteral drug. *2933*

Serum   Increase   Physiological
Pregnancy Three times normal in last trimester. *0987*

Urine   No Effect   Analytical
Ammonium Chloride No effect at 50 mmol/L on 2 colorimetric analytical methods. *1354*
Ascorbic Acid No effect at 100 mmol/L on 2 colorimetric analytical methods. *1354*
Bilirubin No effect at 5 g/L on 2 colorimetric analytical methods. *1354*
Cefoxitin At 2 g/L on 2 colorimetric analytical methods. *1354*
Cisplatin At 1 mg/L had no effect on 2 colorimetric analytical methods. *1354*
Dimercaptoethane At 60 mmol/L on 2 colorimetric analytical methods. *1354*
Doxorubicin At 50 mg/L on 2 colorimetric analytical methods. *1354*
EDTA At 20 mmol/L on 2 colorimetric analytical methods. *1354*
Folic Acid At 50 mmol/L no effect on 2 colorimetric analytical methods. *1354*
Gentamicin No effect at 50 g/L on 2 colorimetric analytical methods. *1354*
Glucose No effect at 100 mmol/L on 2 colorimetric analytical methods. *1354*
Glutathione At 1 mmol/L had no effect on 2 colorimetric analytical methods. *1354*
Hemoglobin At 75 mg/L had no effect on 2 colorimetric analytical methods. *1354*
Heparin At 200,000 U/L had no effect on 2 colorimetric analytical methods. *1354*
Ifosfamide At 1 g/L had no effect on 2 colorimetric analytical methods. *1354*
Leucovorin Had no effect at 10 g/L on 2 colorimetric analytical methods. *1354*
Mercaptoethane At 60 mmol/L on 2 colorimetric analytical methods. *1354*
Methotrexate At 1 mmol/L on 2 colorimetric analytical methods. *1354*

N-Acetylglucosaminidase *(continued)*

Urine   No Effect   Analytical *(continued)*
Penicillamine   At 20 mmol/L on 2 colorimetric analytical methods. *1354*
Pyridium®   At 40 μmol/L on 2 colorimetric analytical methods. *1354*
Sodium Salts   No effect on 2 colorimetric methods with 1 g/L sodium azide, 100 mmol/L sodium bicarbonate or 10 mmol/L sodium fluoride. *1354*
Theophylline   At 3 mg/L had no effect on 2 colorimetric analytical methods. *1354*
Thimerosal   At 20 mg/L had no effect on 2 colorimetric analytical methods. *1354*
Xylose   At 10 mmol/L had no effect on 2 colorimetric analytical methods. *1354*

Urine   No Effect   Physiological
Pipemidic Acid   No significant effect in healthy individuals and those with infections of lower urinary tract but reduction in patients with pyelonephritis when given for 10 d. *0673*
Sisomicin   No significant effect seen in 23 patients given therapeutic amounts for 2 weeks. *2593*

Urine   Increase   Physiological
Aspirin   Marked increase observed even when as little as 3.5 g aspirin ingested daily. *2866*
Cisplatin   Fivefold increase 12 children given drug 100 mg/m^2 for 6 h. Changes observed before alteration in serum creatinine. *1355* Commonly observed increased excretion as a result of nephrotoxicity. *1354*
Diatrizoic acid   Doubled day of bilateral renal arteriography, almost reverted to normal following day. *2594*
Furosemide   Excretion increased by drug induced diuresis. *1492*
Gentamicin   Marked effect especially with treatment for more than 12 d. *2593*
Gold   In 77% in 31 patients treated with parenteral gold. *2933*
Ifosfamide   Manifestation of nephrotoxicity in 12 children given drug 1.6 g/m^2 for 5 d. *1355* Commonly observed: seen in all patients in one study in spite of concomitant. *1354*
Methotrexate   Manifestation of nephrotoxicity in 12 children receiving chemotherapy: up to 5 fold increase in excretion. *1355*
Radiographic Agents   Marked increase within 1 d when used for arteriography: effect persisted for many days: effect less when used for urography. *3259*

N-Acetylmetanephrine

Urine   Increase   Physiological
MAO Inhibitors   Due to inhibition of amine oxidase. *0426*

N-Acetylnormetanephrine

Urine   Increase   Physiological
MAO Inhibitors   Due to inhibition of amine oxidase. *0426*

Acetylsalicylic Acid

Serum   Increase   Physiological
Antacids   Increased rate of absorption due to faster drug release. *3794*

Serum   Decrease   Physiological
Charcoal   Reduced absorption due to adsorption. *3794*
Food   Absorption reduced when taken with food. *3024*

Blood   Increase   Physiological
Aspirin   1 g orally up to 100 mg/L, 12 g = 400 mg/L. *2348*

α_1-Acid Glycoprotein

Serum   No Effect   Physiological
Muscular Exercise   No effect of exercise observed. *2846*
Smoking   No significant difference between smokers and non-smokers. *0809*
Timegadine   In 23 patients with rheumatoid arthritis given 250 to 750 mg/d for 48 weeks. *2345*

Serum   Increase   Physiological
Aging   Some increase with age. *0954*
Anticonvulsants   In 8 epileptic patients receiving phenobarbital with carbamazepine or phenytoin. *2670*
Carbamazepine   Significantly higher in children treated with drug compared with Controls. *2994*
Muscular Exercise   Occurs within 15 minutes, persists for 1 d. *1491*
Oxymetholone   Anabolic metabolic effect. *0227*
Sex Difference   Approximately 10 mg/dL higher in men. *0954*

Serum   Decrease   Physiological
Estrogens   Altered metabolism. *0473*
Norethindrone   Metabolic estrogen effect. *0474*
Oral Contraceptives   Metabolic changes in liver synthesis (estrogen). *2189*
Penicillamine   In 21 patients with rheumatoid arthritis treated for 48 weeks given 250 to 750 mg daily. *2345*
Tamoxifen   Significant change due to mild estrogenic effect of drug in breast cancer patients. *3070*

Acid Phosphatase

Serum   No Effect   Analytical
Contact With Clot   pH change prevented 5 h at room temperature, 24 at 4°. *3217*
Disulphine Blue   With p-nitrophenylphosphate as substrate. *1464*
Hemolysis   If moderate no effect with thymolphthalein phosphate. *0877*
Storage of Sample   If sodium citrate added for 7-8 d at 25, 5, -15°. *0877*   For 115 d at -20°. *3217*   Storage at -20° for over 4 mo. *3873*

Serum   No Effect   Physiological
Smoking   No significant difference between heavy smokers and nonsmokers. *0961*

Serum   Increase   Analytical
Hemolysis   Erythrocytes contain considerable activity. *2981*
Serum   Release of enzyme from erythrocytes. *3879*

Serum   Increase   Physiological
Androgens   Effect of hormone in females. *1022*
Clofibrate   Reported effect (?mechanism). *1022*
Prostate Palpation   Release into bloodstream. *1238*

Serum   Decrease   Analytical
Contact With Clot   Labile at room temperature. *3879*
Copper   Cupric ions inhibit red cell enzyme. *3588*
Detergents   10-37% inhibition at 0.6-0.8 mg/ml. *3217*
Ethanol   Inhibits prostatic component. *3587*
Fluorides   Enzyme activity inhibited by fluoride. *0011*
Formaldehyde   Inhibits erythrocytic component. *0078*
Heparin   Reported observation (significant observation). *0839*
Oxalate   Inhibition of enzyme in laboratory procedures. *3588*
Phosphates   Inhibits reaction if high concentration in substrate. *1022*
Standing of Sample   pH increases, inactivates enzyme (by 10% at 4 h). *3217*
Storage of Sample   28% decrease 6 mo at -10°. *1180*   50% activity lost after 5 h at room temperature. *3873*
Tartrates   Strong inhibitors of prostatic enzyme. *3588*

Serum   Decrease   Physiological
Aging   Steep fall to about age 20 y then almost steady. *1043*
Diurnal Variation   25-50% less between 9 am and 3 am. *3217*
Fluorides   *In vivo* inhibition observed. *1168*

Urine   Decrease   Analytical
Storage of Sample   Destruction of enzyme if frozen (prevented by albumin). *3217*

Prostatic Fluid   Decrease   Physiological
Diethylstilbestrol   Markedly lower in benign hypertrophy. *1917*

Red Blood Cells   Decrease   Analytical
Heparin   Reported observation (significant observation). *0839*

White Blood Cells   No Effect   Analytical
Alloxan   Glycerophosphate enzyme no effect with 45 mmol/L. *0183*
Citrates   Glycerophosphate enzyme no effect of 0.3 mol/L. *0183*
Fluorides   Phenylphosphate enzyme no effect of 15 mmol/L. *0183*

White Blood Cells   Increase   Analytical
Citrates   Phenylphosphate enzyme stimulated by 0.3 mol/L. *0183*
Tartrates   Phenylphosphate enzyme activated. *0183*

White Blood Cells   Decrease   Analytical
Alloxan   Phenylphosphate enzyme inhibited by 45 mmol/L. *0183*
Fluorides   Glycerophosphate enzyme inhibited by 15 mmol/L. *0183*
Heparin   More than 1 unit/mL inhibits by 80%. *0839*
Tartrates   Glycerophosphate enzyme inhibited by 1.5 mmol/L. *0183*

Acid Phosphatase, Prostatic

Serum   Decrease   Physiological
Ketoconazole   In men with prostatic cancer with 400 mg every 8 h and prolonged treatment. *3390*

Acid Phosphatase, Tartrate Labile

Serum   No Effect   Analytical
Hemolysis   No effect with hemoglobin up to 1 g/dL. *3217*

Adenosine Deaminase

Serum   No Effect   Analytical
Storage of Sample   For 30 d at 4-8°, 30 d at -20°. *3217*   No effect 1 week at 4°, 1 mo at -20°. *3873*

Adenosine Deaminase Binding Protein

Urine   No Effect   Physiological
Doxorubicin   No effect in 12 children given drug for 3 d. *1355*
Etoposide   No effect in 12 children given drug for 2 d. *1355*

Urine   Increase   Physiological
Cisplatin   Fivefold increase 12 children given drug 100 mg/m² for 6 h. Changes observed before alteration in serum creatinine. *1355*
Ifosfamide   Manifestation of nephrotoxicity in 12 children given drug 1.6g/m² for 5 d. *1355*
Methotrexate   Manifestation of nephrotoxicity in 12 children receiving chemotherapy; up to 5-fold increase in excretion. *1355*

Adenosine Triphosphate (ATP)

Muscle   Increase   Physiological
Physical Training   Resting concentration in muscle increased by training. *1875*

Red Blood Cells   No Effect   Physiological
Smoking   No significant difference between smokers and non-smokers. *1548*

Red Blood Cells   Decrease   Physiological
Amino Acids   Associated with low serum phosphate. *3619*
Propranolol   Significant effect, also inhibits glucose utilization by 60%. *2770*
Race   Significantly lower in blacks than caucasians. *2523*
Sleep Deprivation   Sleep deprivation for several days causes decrease. *2222*
Smoking   Significantly less in smokers. *3112*

S-Adenosylmethionine

Blood   Decrease   Physiological
Levodopa   O-methylation of catecholamines slowed. *2337*

Adenylate Kinase

Serum   Increase   Physiological
Muscular Exercise   Observed after protracted exercise. *0932*

Adipate

Urine   Increase   Physiological
Lithium   May be considerable effect. *2107*

Agglutinins

Blood   Increase   Physiological
Mushroom Poisoning   Result of poisoning. *0279*

Blood   Positive   Physiological
Croton Oil   May induce formation. *0279*

AHF-Like Antigen

Plasma   Increase   Physiological
Muscular Exercise   Proportional increase in response to extent. *0292*
Pregnancy   Progressive increase throughout pregnancy. *0292*

AHF Procoagulant Activity

Blood   Increase   Physiological
Muscular Exercise   Proportional increase in response to extent and antigen. *0292*
Pregnancy   Progressive increase throughout pregnancy. *0292*

Alanine

Plasma   Increase   Physiological
Glucose   Rise related to gluconeogenesis after load. *1403*
Glutamic Acid   Small increase noted after single 6 g dose. *3444*
Histidine   Interferes with clearance after oral load. *1644*
Tetracosactrin   In healthy volunteers given 1 mg intramuscularly for up to 60 h. *1813*
Valproic Acid   When 1 g given orally to fasting individuals. *3633*

Alanine (continued)

Plasma   Decrease   Physiological

Ethanol   Causes conversion to lactate. *2013*

Oral Contraceptives   Hormonal effect (second part of cycle). *0742*

Starvation   Lowered after 1 d lowest on third day. *0021*

Urine   Increase   Physiological

Lead   Occurs with poisoning. *0987*

Urine   Decrease   Physiological

Ascorbic Acid   In 10 healthy females given 10 g/d. *3624*

Alanine Aminopeptidase

Serum   No Effect   Physiological

Ethanol   Alcohol consumption did not correlate with activity in men or women. *2509*

Urine   No Effect   Physiological

Pipemidic Acid   No significant effect in healthy individuals and those with infections of lower urinary tract but reduction in patients with pyelonephritis when given for 10 d. *0673*

Urine   Increase   Physiological

Ceftazidime   Significant increase in association with reduced GFR. *0051*

Cisplatin   Fivefold increase 12 children given drug 100 mg/m² for 6 h. Changes observed before alteration in serum creatinine. *1355*

Diatrizoic acid   Increased almost 3-fold on day of, and day after, bilateral renal arteriography in patients with hypertension. *2594*

Gentamicin   Marked effect especially with treatment for more than 12 d. *2593*

Gold   In 13% (not significant) increase of microsomal enzyme in 31 patients treated with parenteral gold. *2933*

Ifosfamide   Manifestation of nephrotoxicity in 12 children given drug 1.6 g/m² for 5 d. *1355*

Methotrexate   Manifestation of nephrotoxicity in 12 children receiving chemotherapy: up to 5 fold increase in excretion. *1355*

Radiographic Agents   Marked increase within 1 d when used for arteriography: effect persisted for many days: effect less when used for urography. *3259*

Sisomicin   Approximately 8 fold increase in 23 patients given therapeutic amounts of drug for 2 weeks. *2593*

Urine   Decrease   Physiological

Salicylate   When 0.5 g given orally to 20 healthy adult volunteers and urine studied over next 3 h. *1828*

Alanine Aminotransferase

Serum   No Effect   Analytical

Acetaminophen   At acute overdose concentration (10 mg/dL) on SMAC® method. *3719*   On continuous method at 10 times maximal therapeutic concentration. *1785*   No effect at therapeutic concentration on Reflotron method. *1984*

Acetylsalicylic Acid   At 5 times upper limit of therapeutic range on methods on SMAC®, Abbott-VP, aca, Cobas-Bio and KDA. *2138* At 8300 μmol/L on continuous analytical method. *1785*   No effect at therapeutic concentration on Reflotron method. *1984*

Aminophenazone   On continuous method at 10 times maximal therapeutic concentration. *1785*

Aminosalicylic Acid   At 5 times upper limit of therapeutic range on methods on Abbott-VP, aca, and Cobas-Bio. *2138*

Amitriptyline   At acute overdose concentration (2.5 mg/dL) on SMAC® method. *3719*

Amphetamine   At therapeutic concentration on Reflotron method. *1984*

Ampicillin   No effect on Reflotron method at therapeutic concentration. *1984*

Ascorbic Acid   No effect at therapeutic concentration on Reflotron method. *1984*

Barbital   At acute overdose concentration (20 mg/dL) on Technicon® SMAC® method. *3719*

Bezafibrate   No effect at therapeutic concentration on Reflotron method. *1984*

Bromazepam   At 5 times therapeutic concentration on methods on SMAC®, Abbott-VP, aca, Cobas-Bio and KDA. *2138*

Caffeine   At acute overdose concentration (10 mg/dL) on Technicon® SMAC® method. *3719*   On either continuous or colorimetric methods at 10 times maximal therapeutic methods. *1785* At therapeutic concentration on Reflotron method. *1984*

Carbenicillin   At 5 times upper limit of therapeutic range on methods on SMAC®, Abbott-VP, aca, Cobas-Bio, and KDA. *2138*

Carbochromen   No effect at therapeutic concentration on Reflotron method. *1984*

Cefoxitin   At 5 times upper limit of therapeutic range on methods on SMAC®, Abbott-VP, aca, Cobas-Bio and KDA. *2138*

Cephalothin   At 5 times upper limit of therapeutic range on methods on SMAC®, Abbott-VP, aca, Cobas-Bio and KDA. *2138*

Chloramphenicol   No effect at therapeutic concentration on Reflotron method. *1984*

Chlordiazepoxide   At 5 times upper limit of therapeutic range on methods on SMAC®, Abbott-VP, aca, and Cobas-Bio. *2138*   At acute overdose concentration (5 mg/dL) on Technicon® SMAC® method. *3719*   No effect at therapeutic concentration on Reflotron method. *1984*

Chlorpheniramine   At acute overdose concentration (20 mg/dL) on Technicon® SMAC® method. *3719*

Cocaine   At acute overdose concentration (2.5 mg/dL) on Technicon® SMAC® method. *3719*

Codeine   At acute overdose concentration (2.0 mg/dL) on Technicon® SMAC® method. *3719*

Cyclosporine   At concentration of 41.5 mmol/L (50 μg/L) on methods on Technicon® SMAC® II and Hitachi® 705. *1123*

Cysteine   At 5 times upper limit of therapeutic range on methods on SMAC®, Abbott-VP, aca, Cobas-Bio, and KDA. *2138*

Cystine   At 5 times upper limit of therapeutic range on methods on SMAC®, Abbott-VP, aca, Cobas-Bio and KDA. *2138*

Diazepam   At 5 times upper limit of therapeutic range on methods on SMAC®, Abbott-VP, aca, Cobas-Bio, and KDA. *2138*   At acute overdose concentration (2.5 mg/dL) on Technicon® SMAC® method. *3719*

Diclofenac   On continuous method at 10 times maximal therapeutic concentration On colorimetric method at 10 times maximal therapeutic concentration. *1785*

Diethylpropion   At acute overdose concentration (10 mg/dL) on Technicon® SMAC® method. *3719*

Diphenhydramine   At acute overdose concentration (20 mg/dL) on Technicon® SMAC® method. *3719*

Disulphine Blue   Using recommended Scandinavian method. *1464*

Ethanol   At acute overdose concentration (20 mg/dL) on Technicon® SMAC® methods. *3719*

Ethaverine   No effect at therapeutic concentration on Reflotron method. *1984*

Ethchlorvynol   At acute overdose concentration (20 mg/dL) on Technicon® method. *3719*

Ethinamate   At acute overdose concentration (20 mg/dL) on Technicon® SMAC® method. *3719*

Fluosol-DA   On SMA IIC at concentration of 50%. *2518*

Flurazepam   At 5 times upper limit of therapeutic range on methods on SMAC®, Abbott-VP, aca, Cobas-Bio, and KDA. *2138*   At acute overdose concentration (2.5 mg/dL) on Technicon® SMAC® method. *3719*

Furosemide   No effect at therapeutic concentration on Reflotron method. *1984*

Glibenclamide   No effect at therapeutic concentration on Reflotron method. *1984*

Glutethimide   At acute overdose concentration (5 mg/dL) on Technicon® SMAC® method. *3719*

Ibuprofen   At 5 times upper limit of therapeutic range on methods on SMAC®, Abbott-VP, aca, Cobas-Bio and KDA. *2138*   On continuous method at 10 times maximal therapeutic concentration. *1785*

Indomethacin On continuous method at 10 times maximal therapeutic concentration. On colorimetric method at 10 times maximal therapeutic concentration. *1785* No effect at therapeutic concentration on Reflotron method. *1984*

Ketoprofen On colorimetric method at 10 times maximal therapeutic concentration. *1785*

Lidocaine No effect SMA 12/60 method with 3.5 mg/dL. *2636*

Meperidine At acute overdose concentration (5 mg/dL) on Technicon® SMAC® method. *3719*

Meprobamate At acute overdose concentration (20 mg/dL) on Technicon® SMAC® method. *3719*

Mesoridazine At acute overdose concentration (20 mg/dL) on Technicon® SMAC® method. *3719*

Methadone At acute overdose concentration (2.5 mg/dL) on Technicon® SMAC® method. *3719*

Methanol At acute overdose concentration (20 mg/dL) on Technicon® SMAC® method. *3719*

Methaqualone At acute overdose concentration (2.5 mg/dL) on Technicon® SMAC® method. *3719* No effect at therapeutic concentration on Reflotron method. *1984*

Methicillin At 5 times upper limit of therapeutic range on methods on SMAC®, Abbott-VP, Cobas-Bio, aca, and KDA. *2138*

Methotrexate At 5 times upper limit of therapeutic range on method on Abbott-VP, aca, Cobas-Bio, Hitachi® 705 and KDA. *2138*

Methyldopa No effect at therapeutic concentration on Reflotron method. *1984*

Methylphenidate At acute overdose concentration (20 mg/dL) on Technicon® SMAC® method. *3719*

Methyprylon At acute overdose concentration (10 mg/dL) on Technicon® SMAC® method. *3719*

Morphine At acute overdose concentration (20 mg/dL) on Technicon® SMAC® method. *3719*

Naproxen At 5 times upper limit of therapeutic range on methods on SMAC®, Abbott-VP, aca, Cobas-Bio, and KDA. *2138*

Niacin No effect at therapeutic concentration on Reflotron method. *1984*

Nitrofurantoin On routine methods in use on SMAC®, Ektachem®, Abbott-VP, Cobas-Bio, aca, Hitachi® 705, KDA at 5 times normal upper therapeutic concentration. *2138* No effect at therapeutic concentration on Reflotron method. *1984*

Nortriptyline At acute overdose concentration (20 mg/dL) on Technicon® SMAC® method. *3719*

Oxazepam No effect at therapeutic concentration on Reflotron method. *1984*

Oxytetracycline No effect at therapeutic concentration on Reflotron method. *1984*

Penicillin G At 5 times upper limit of therapeutic range on methods on SMAC®, Abbott-VP, Cobas-Bio, aca, and KDA. *2138*

Pentobarbital At acute overdose concentration (20 mg/dL) on Technicon® SMAC® method. *3719*

Perphenazine At acute overdose concentration (20 mg/dL) on Technicon® SMAC® method. *3719*

Phenazopyridine No effect at therapeutic concentration on Reflotron method. *1984*

Phenobarbital At acute overdose concentration (20 mg/dL) on Technicon® SMAC® method At acute overdose concentration (20 mg/dL) on Technicon® SMAC® method. *3719* On continuous method at 10 times maximal therapeutic concentration. On colorimetric method at 10 times maximal therapeutic concentration. *1785* No effect at therapeutic concentration on Reflotron method. *1984*

Phenprocoumon No effect at therapeutic concentration on Reflotron method. *1984*

Phenytoin At acute overdose concentration (20 mg/dL) on Technicon® SMAC® method. *3719* No effect at therapeutic concentration on Reflotron method. *1984*

Plasma No significant difference as measured on ABA-100. *2228*

Probenecid No effect at therapeutic concentration on Reflotron method. *1984*

Procaine No effect at therapeutic concentration on Reflotron method. *1984*

Promethazine At acute overdose concentration (20 mg/dL) on Technicon® SMAC® method. *3719*

Propoxyphene At acute overdose concentration (2.5 mg/dL) on Technicon® SMAC® method. *3719*

Pyribenzamine® At acute overdose concentration (20 mg/dL) on Technicon® SMAC® methods. *3719*

Pyridamole No effect at therapeutic concentration on Reflotron method. *1984*

Pyritinol No effect at therapeutic concentration on Reflotron method. *1984*

Quinidine At acute overdose concentration (20 mg/dL) on Technicon® SMAC® method. *3719* No effect at therapeutic concentration on Reflotron method. *1984*

Quinine At acute overdose concentration (1.5 mg/dL) on Technicon® SMAC® method. *3719*

Rifampin At 5 times upper limit of therapeutic range on method on Abbott-VP, aca, Cobas-Bio, and KDA. *2138*

Secobarbital At acute overdose concentration (20 mg/dL) on Technicon® SMAC® method. *3719*

Storage of Sample No effect 1 week at 4°, 1 mo at -20°. *1563* For 3 d at 30°, 1 week refrigerated. *0877*

Sulfamethoxazole No effect at therapeutic concentration on Reflotron method. *1984*

Tetracycline At 5 times upper limit of therapeutic range on methods on SMAC®, Abbott-VP, Cobas-Bio, aca, and KDA. *2138*

Theophylline At 5 times upper limit of therapeutic range on methods on SMAC®, Abbott-VP, aca, Cobas-Bio and KDA. *2138* No effect at therapeutic concentration on Reflotron method. *1984*

Thiopental At acute overdose concentration (20 mg/dL) on Technicon® SMAC® method. *3719*

Trimethoprim No effect at therapeutic concentration on Reflotron method. *1984*

Serum   No Effect   Physiological

Cromoglycate No clinical significant change observed. *0484*

Cyclophosphamide No significant perturbation in 41 patients. *2406*

Cyclosporine No effect observed in 23 patients with inflammatory ocular disease treated for for 6 mo. *2713*

Desipramine In 42 children or adolescents treated for up to 24 mo. *1624*

Ethanol No effect over 110 h following ingestion of 0.75 g/kg body weight on 3 consecutive evenings. *2133* No significant difference in specimens from volunteers and values compared with control values for up to 110 h, although component/protein ration decreased in all cases. *2134*

Famotidine In 9 patients with Zollinger-Ellison syndrome when followed for 33 weeks. *1667*

Fenoprofen In 49 patients with juvenile rheumatoid arthritis when some abnormal results in patients with same therapeutic outcome given aspirin. *0465*

Fructose No effect observed after i.v. infusion. *1160*

Glucose No effect observed after i.v. infusion. *1160*

Halothane No effect seen in patients who demonstrated mild increase of GST activity. *1707*

Isotretinoin In 18 patients with severe acne given 0.8 mg/kg daily for 3 mo: changes reverted to normal after treatment stopped. *2242* Significant effect at 6 weeks in 7 patients with severe rosacea treated with 1 mg/kg/d. *2307*

Methadone No toxicity if liver function tests normal initially. *2012*

Methotrexate In 20 patients with psoriasis on long-term therapy. *1556*

Metrifonate No effect reported on hepatic function. *2427*

Muscular Exercise Insignificant effect of 12 minutes on cycle-ergometer. *1231*

Sorbitol Increase by 0.3-0.8 mg/dL after i.v. infusion. *1160*

Tolmetin Observed in single case of 49 y old man who had taken drug as needed for arthritis for 1 y. *3421*

Valproic Acid In chronically treated patients with monotherapy only, but increases when combined with other antiepileptics. *1453*

Xylitol No effect reported after i.v. infusion. *1160*

Serum   Increase   Analytical

Acetaminophen On colorimetric method at 10 times therapeutic concentration. *1785*

Alanine Aminotransferase (continued)

Serum Increase Analytical (continued)

Aminophenazone On continuous method at 10 times maximal therapeutic concentration. *1785*

Aminosalicylic Acid At 5 times upper limit of therapeutic range on methods on KDA and SMAC®. *2138*

Chlordiazepoxide At 5 times upper limit of therapeutic range on kinetic methods on KDA. *2138*

Erythromycin Interferes with colorimetric procedures. *2220*

Hemolysis Effect minimal as only 3-5 times concentration in RBC as in sera. *0877*

Lipemia Turbidity may affect some methods. *0579* With inverse colorimetry: resultant effect of clearing of lipemia and formation of reaction product during SMAC® measurement. *2466*

Serum Increase Physiological

Acetaminophen Hepatic necrosis with dose of 10 g reported. *0438* Manifestation usually of single toxic high dose ingestion in suicide attempt. Liver damage usually of centrizonal hemorrhagic hepatic necrosis. *3948*

Acetohexamide May cause intrahepatic cholestatic jaundice. *2220*

Acetophenazine Cholestasis with intrahepatic obstruction. *2313*

Ajmaline Intrahepatic cholestasis. *0268* Hepatotoxicity and intrahepatic cholestasis reported. *3215* In 4 patients with centrilobular cholestasis and mild hepatocytic lesions: recovery with drug withdrawal. *2729*

Allopurinol Reversible clinical hepatotoxicity reported. *2220* In 6 of 73 patients treated for 6 mo. *3565*

Alprazolam Isolated case, abnormal test value reverted to normal with withdrawal of drug. *3079*

Aluminum Nicotinate Rare hepatotoxic effects. *1678*

Aminobenzoic Acid Hepatotoxicity with centrolobular necrosis. *1948*

Aminoglutethimide Casual association between drug administration and cholestasis in 2 cases. *2779*

Aminopyrine Liver damage up to hepatic necrosis seen. *2427*

Aminosalicylic Acid May cause cytotoxic hepatocellular damage. *0248* Rare side effect. *1292*

Amiodarone Up to 1.5 to 3 times normal values in 55% patients with hepatitis occurring in 4%. *2011* Asymptomatic change in liver function with enzyme activity up to 3 times normal. *2478* 1.5 to 4 fold increase in 15 of 100 patients: effect correlated with drug concentration. *1509* In 5 of 5 patients with toxic and hypersensitivity liver injury. *2987*

Amitriptyline Rare cases of transient cholestasis. *2482* Occasional case of hypersensitivity associated hepatitis. *0098*

Ammonia May produce severe liver damage. *0279*

Amodiaquine Reported hepatotoxicity. *2313* Abnormality usually mild but associated with an immunoallergic hepatotoxicity. *2081*

Amphotericin B Hepatotoxicity reported. *2313* Noted in one patient with acute myelogenous leukemia when treated with high dose. Probable idiosyncratic response. *2443*

Ampicillin Probable effect of i.m. injection. *1966*

Amyl Alcohol May cause liver damage if ingested. *1302*

Anabolic Steroids Cholestatic syndrome (intrahepatic cholestasis). *2220*

Androgens May cause cholestatic syndrome. *2220*

Anesthetic Agents Occurs without permanent liver damage. *3661*

Antifungal Agents Hepatotoxicity may occur. *1237*

Antimony Compounds Hepatotoxic effect. *2220*

Apiol May cause severe liver damage. *0279*

Aprindine Hepatitis is manifestation of chronic toxicity. *3215* Hepatitis in 2 patients within 3 weeks of start of therapy resolved with withdrawal of drug. *1570*

Arsenicals Hepatotoxic effect (cholestasis/cholangiolitis). *2220*

Asparaginase Hepatotoxicity. *1519*

Aspidium May cause hepatic toxicity. *2313*

Aspirin Prolonged use may cause hepatic toxicity. *1022* Noted in some patients at serum concentrations above 25 mg/dL without signs of hypersensitivity reactions but dose related. *3963* In 15% trials of treatment for juvenile rheumatoid arthritis. *0239*

Aspoxicillin In 2 of 5 patients given up to 112 mg/kg intravenously in 3 or 4 divided doses. *1517*

Azapropazone In 9 of 83 patients treated for 6 mo. *3565*

Azaserine May cause hepatotoxicity. *3835*

Azathioprine May cause hepatotoxicity. *3433* Minimal cholestasis and reversible portal fibrosis observed in 8 patients: occasional hepatotoxicity. *0036* Canalicular cholestasis and centrilobular ballooning of hepatocytes in one patient. *0846*

Barbiturates Occurs with poisoning, probable muscle origin. *3901*

Barium Severe liver damage if tannic acid in enema. *0071*

Benzene May cause liver damage. *0279*

Benziodarone May cause hepatic toxicity. *2313*

Bismuth Salts Hepatotoxicity. *2313*

Boric Acid May cause liver damage. *0279*

Butaperazine Cholestatic hepatitis with obstruction. *0071*

Carbamazepine Cholestatic and hepatocellular damage. *2907*

Carbarsone May cause hepatitis/liver necrosis. *0071*

Carbenicillin Elevation reported, ?due to hepatotoxicity. *1680* In several patients as evidence of drug induced hepatotoxicity. *1371* In 7 of 27 patients given 270 mg/kg i.v. for 6 d. *3275*

Carbenoxolone Associated with myopathy. *2458*

Carbimazole Isolated case of jaundice conclusively linked to drug. *0911*

Carbon Disulfide May cause liver damage. *0279*

Carbromal Occurs with poisoning, probable muscle origin. *3901*

Carbutamide Hepatotoxicity. *2313*

Carmustine (BCNU) Usually mild and return to normal over few days. *2406*

Carphenazine Probable cholestatic effect (reversible). *2313*

Cefamandole Occurs in about 4% treated cases. *3134*

Cefoperazone Transient increase in 5% of 450 patients. *0690*

Cefoxitin Reported incidence of 3%. *3134*

Cefuroxime Observed in 24% of 89 patients in one study, but many had pre-existing liver problems. *2615*

Cephaloglycin Slight elevation reported. *1680*

Cephaloridine Transient slight rise ?hepatotoxicity. *1019*

Cephalosporin Observed with all cephalosporins at frequency of from 1 to 7%. *2615*

Cephalothin Single case reported. *2427*

Chenodeoxycholic Acid Slight transient increases may be observed: clinically significant hepatic injury in 3%. *3198* Mild elevation usually less than 3 times normal in up to 30% patients receiving 750 mg/d. *3199*

Chiniofon Large doses may occasionally cause hepatic damage. *1343*

Chloral Hydrate May cause liver damage. *0279*

Chlorambucil Occasional hepatotoxicity reported. *1343*

Chloramphenicol Hepatotoxic-cholestatic effect. *2220*

Chlorate May cause liver damage. *0279*

Chlordane Due to hepatotoxicity. *1302*

Chlordiazepoxide May produce hepatotoxic effect. *0006*

Chlorinated Insecticides May cause liver damage. *0279*

Chlormezanone Occasional reversible cholestasis. *2313*

Chloroform Hepatotoxic effect with necrosis. *2313*

Chlorophenothane May cause liver damage. *0279*

Chlorothiazide May cause cholestatic jaundice. *2313*

Chlorpromazine May be damage of biliary canaliculi. *3835*

Chlorpropamide May cause cytotoxic liver damage. *0248*

Chlorprothixene May cause hepatotoxicity (reversible). *2313*

Chlortetracycline Hepatotoxic effect with centrolobular necrosis. *1237*

Chlorthalidone Single case of severe reversible myopathy noted. *1793*

Chlorzoxazone May cause hepatotoxicity. *2313*

Cholestyramine   Few cases reported, probably not hepatotoxicity. *1680*

Chromates   May cause liver damage. *0279*

Cimetidine   In isolated case treated for 7 mo with 400 mg/d progressed to bridging hepatic necrosis. *3085*

Cinchophen   May cause hepatotoxicity (viral-hepatitis like). *2313*

Ciprofloxacin   Rare cases reported during clinical trials. *0151*

Cisplatin   Slight and transient increases in 45 patients receiving drug. *3698*

Clindamycin   Mild transient rises seen. *2806*   Uncommon side effect. *3864*

Clofibrate   Hepatotoxic effect. *2313*   Single case of granulomatous hepatitis associated with drug administration. *2813*

Clonidine   Single probable case of toxic hepatitis. *1277*

Clopamide   Single case of diuretic associated myopathy. *1793*

Cloxacillin   Increases linked to drug administration observed in some hemodialysis patients. *0309*

Codeine   Rise in intrabiliary pressure especially if liver disease. *3505*

Colchicine   Possible hepatotoxic effect. *2220*

Copper   Toxic effect, sharp rise with hemolytic crisis. *0385*

Cortisone   May contribute to long duration of hepatitis. *2220*

Cough Medicines   One case chronic use terpin and codeine. *1069*

Coumarin   May cause hepatic toxicity. *3835*

Cyclofenil   Slightly and transiently increased in 6 of 19 patients over 4 mo of treatment, especially with high doses. *2786*   In 1 case, but total of 30 cases reported up to 1980; Considered to be metabolic idiosyncrasy and reversible in all cases. *2672*

Cyclophosphamide   May cause hepatotoxicity. *3433*

Cyclopropane   May cause hepatotoxicity (lasts several days). *2313*

Cycloserine   May cause hepatotoxicity. *2313*

Cyclosporine   Increase to 109 U/L on average as manifestation of hepatotoxicity in 18 bone-marrow transplant recipients. *0174*   Both early and late toxicity observed but effects mild in comparison with nephrotoxicity in renal transplant patients. *1959*

Cyclothiazide   Occasional intrahepatic cholestatic jaundice. *1680*

Cyproheptadine   Isolated case of jaundice due to cholestasis: reversed on withdrawal of drug. *1562*

Cytarabine   Hepatotoxic effect. *1343*

Danazol   Significant effect at 4,8, 12 and 16 weeks after 600 mg daily in 18 women with endometriosis. *0591*

Dantrolene   Chronic active hepatitis reported in a few patients. *2260*   In 1.8% of 1044 patients, hepatocellular damage with acute or subacute hepatic disease or chronic active hepatitis. *3656*

Dapsone   Sulfone syndrome in one 16 year old girl given 50 mg daily short-term administration. *3604*

Desipramine   Rare transient cholestasis. *2313*

Diazepam   Isolated case of drug induced hepatitis. *3562*

Dicumarol   May be high enough to simulate myocardial infarction. *0248*

Dienestrol   Reported to cause cholestasis. *1680*

Diethylstilbestrol   Hepatotoxicity with centrolobular necrosis. *1948*

Diiodohydroxyquin   Hepatic damage reported in animals. *1343*

Dinitrophenol   Hepatotoxicity with centrolobular necrosis. *1343*

Disopyramide   Occasional case of cholestatic jaundice reported. *3215*

Disulfiram   One case of questionable cholestasis reported. *1680*   Reversible toxic liver damage in nonalcoholic woman. *2017*   In 6 patients: proved to be drug associated in one by challenge test. *2912*   Observed in one patient and similar results followed challenge test. *2491*

Doxorubicin   May produce idiosyncratic hepatic dysfunction, as observed in 6 patients. *0184*

Ectylurea   May cause cholestasis with cholangiolitis. *2313*

Enflurane   In up to 10 % patients given 2 to 3 administrations of anesthetic agent. *1089*   Evidence of hepatocellular damage; characteristically centrilobular necrosis: mechanism of injury most probably metabolic idiosyncrasy. *2157*

Epinephrine   Following single injection of compound in oil. *1264*

Erythromycin   May cause hepatic toxicity. *1142*   Estolate produces mild hepatotoxicity in about 15% treated individuals. Jaundice in 2% patients on drug for more than 2 weeks. *2792*   Observed to some extent in some patients with different erythromycin salts: usually cholestatic hepatitis. *1895*

Estradiol   Cholestatic effect. *2313*

Estrogens   May cause cholestasis, rare hepatocellular degeneration. *2220*

Estrone   May cause cholestasis. *2313*

Ethacrynic Acid   Cholestasis with hepatocellular damage. *0812*

Ethambutol   Decreased liver function reported. *0071*

Ethanol   Increased in normals after 1 dose 3 g/kg body weight. *1621*   In 43 to 53% of male alcoholics in various studies with higher proportion with abnormal AST. *3124*

Ethchlorvynol   Occurs with poisoning, probable muscle origin. *3901*

Ether   Hepatic disturbance (transient). *2313*

Ethionamide   Intrahepatic bile duct damage, toxic hepatitis. *2034*

Ethotoin   Hepatotoxicity. *2313*

Ethoxazene   Hepatotoxic effect. *2313*

Ethyl Chloride   May cause liver damage. *0071*

Etoposide   Hepatic toxicity reported in 3% treated patients. *0141*

Etretinate   In 2 of 20 patients treated for 1 y. *1866*   In one 74-year old woman treated for severe psoriasis: normalized with withdrawal of therapy and recurred when reinstituted. *3789*

Fansidar®   Biopsy-proven granulomatous hepatitis due to sulfadoxine moiety of drug reported in two individuals receiving drug prophylactically possibly due to sulfonamide hypersensitivity. *2099*

Fava Beans   May cause liver damage. *0279*

Florantyrone   Hepatotoxic effect. *2313*

Floxuridine   Manifestation of toxicity. *1680*

Flucloxacillin   In a single patient on hemodialysis: abnormal results only mild. *2200*

Flucytosine   Reversible hepatotoxicity in 10%. *3443*

Flufenamic Acid   Transient elevation reported. *2427*

Fluoxymesterone   Intrahepatic cholestatic jaundice. *1343*

Fluphenazine   Hypersensitivity response. *2313*   Isolated case of hepatotoxicity with centrilobular cholestasis. *3372*

Flurazepam   May cause hepatic toxicity. *3016*

Fluroxene   Potentially hepatotoxic. *0071*

Fusidic Acid   Reported in 6 patients in UK, although causal relationship could not be established with certainty. *3535*

Gentamicin   May cause hepatotoxicity. *2220*

Glibenclamide   Intrahepatic cholestasis described in 61 y old diabetic. *3894*

Glyburide   Rare and reverts to normal if therapy continued. *0125*

Glycopyrrolate   Hepatotoxicity. *2313*

Gold   Hepatotoxicity with centrolobular necrosis. *2313*   Due to toxic effect on drug on liver. *1674*

GPA-1714   4 of 11 patients increased when dose of approximately 3 g. *3328*

Griseofulvin   May be hepatotoxic. *2220*

Guanethidine   Possible myopathic complication. *2220*

Guanoclor   Possible effect on liver. *2220*

Guanoxan   May cause hepatic toxicity. *2313*

Halofenate   Mild transient effect observed ?origin. *0161*

Haloperidol   May cause hepatocellular changes. *2313*   In isolated case producing cholestatic liver disease: typical frequency of liver disease is 0.2%. *0910*

Halothane   Allergic hepatic hypersensitivity. *3294*   Liver disease from mild focal necrosis and inflammation to massive necrosis. *2260*   Granulomatous hepatitis in single case following second exposure to anesthetic. *3266*   In one nurse occupationally exposed to gas. *1900*   Low frequency but greater number of abnormalities than with enflurane. Occurred most frequently in obese patients. *1089*

Hemolysis   Released from RBCs and platelets. *2313*

Alanine Aminotransferase *(continued)*

Serum Increase Physiological *(continued)*

Heparin   In 59% patients receiving bovine deprived drug. In 27% patients receiving porcine derived drug. *0968*   Mean maximal value 67 U/L in 89% of 46 patients given up to 120,000 IU daily subcutaneously after 8 d. *2454*   On 5th and 10th postoperative day (from 27 to 40 U/L) compared with control group. *2595*   In 2/3 of patients treated with heparin for deep venous thrombosis. *1064*   In patients with cerebrovascular accidents after low-dose heparin treatment. *2671*

Heptachlor   Liver damage occurs as late effect of poisoning. *1302*

Heroin   Normal usually unless alcoholism also. *3471*

Hycanthone   Transient change in liver function. *0071*

Hydantoin Derivatives   Severe delayed hypersensitivity reported. *2427*

Hydralazine   Isolated case with moderate hepatomegaly, abnormal findings resolved when drug discontinued. *1161*   Centrilobular necrosis observed in 3 patients 2 to 4 months after drug therapy for hypertension. *1740*

Hydrazine Derivatives   May cause hepatotoxicity. *2313*

Hydrochlorothiazide   May cause cholestasis or cholangiolitic hepatitis. *1680*

Hydroflumethiazide   May cause intrahepatic cholestasis. *1680*

Hydroxyacetamide   Toxicity effect. *2313*

Ibufenac   Hepatotoxic effect. *2313*

Ibuprofen   Rare hepatotoxicity reported: associated in one case with Stevens-Johnson syndrome. *3460*

Icterogenin   Causes intrahepatic cholestasis. *0153*

Ifosfamide   Manifestation of hepatotoxicity. *3675*

Imipramine   May cause cholestatic jaundice. *1873*   Either due to hepatotoxicity or hypersensitivity in 33 year old woman. *2501*

Indandione Derivatives   Hepatocellular damage with cholestasis. *2313*

Indomethacin   Cytotoxic and cholestatic liver damage. *1908*

α-Interferon   Dose dependent effect in 3 of 81 patients with malignant disease. *3299*

Iopanoic Acid   May cause hepatotoxicity. *2313*

Iprindole   Hepatic cholestasis without inflammation. *0038*

Iproniazid   Prolonged use may cause hepatotoxicity. *3835*

Iron Salts   May cause severe liver damage with poisoning. *0279*

Isocarboxazid   Cholestatic effect. *2313*

Isoniazid   Probable intrahepatic cholestatic jaundice. *2220*   Mild liver injury occurs in approximately 10% patients taking drug, possibly due to conversion of drug to acetylhydrazine or related hepatotoxic derivatives. *2260*   Liver toxicity occurred in 18% patients receiving combined isoniazid and rifampin: effect slight in 14% and severe in 4%. *1409*   Risk of hepatitis caused by or exacerbated by drug over 12 mo 5.2 per 1,000 patients; clinically picture resembles viral hepatitis. *1292*   15 of 89 patients developed significant liver disease, typically hepatitis, although one developed cholestasis, in patients taking drug for at least 2 mo. *0901*   Observed in up to 0.5% patients slightly greater than in placebo treated group. *1725*   In 3.3% children also treated with rifampin: all reactions occurring in first 10 weeks. *2630*

Isopropanol   Transient and mild late toxic effect. *1302*

Isotretinoin   In 10 of 523 patients with mean daily dose of 109 mg for 150 d. *3867*

Itraconazole   Mild reversible increases without serious hepatotoxicity. *3567*

Kanamycin   May cause hepatotoxicity. *2313*

Kerosene   Hepatic damage with large doses. *1343*

Ketamine   In 14 of 34 individuals who had drug as anesthetic for intermediate operations. *0973*

Ketoconazole   Transient abnormalities of liver function observed in 10% patients but true hepatic injury in only 0.1 to 1%. Probably idiosyncrasy involved but may be immune hypersensitivity in some cases. *3390*   Delayed reaction to drug after withdrawal in single case of chronic candidiasis in 61 y old woman. *3513*   In 4 of 36 patients treated with 200 mg daily over 8 mo. *3034*   Rapidly developing liver failure in 67 year old woman taking 200 mg drug daily for 2 mo. *0960*

Levamisole   In 1 of 11 patients receiving postoperative radiation treatment following breast cancer. *2960*

Levodopa   Transient effect returns to normal. *1680*   Increase by 17% for 2 mo although later normalized, in patients with Parkinsons disease. *1417*

Lincomycin   Hepatotoxic-cholestatic effect. *2220*

Lipemia   Associated with alcoholism. *2313*

Lovastatin   Tendency to rise especially with higher doses but not increased above 3 times upper limit of normal. *1530*

Lucanthone   Chronic toxicity may cause liver damage. *1343*

MAO Inhibitors   Intrahepatic cholestasis. *2220*

Maprotiline   Slight reversible rise in approximately 1% patients. *0214*

Mechlorethamine   May cause cytotoxic (hepatocellular) damage. *0248*

Melarsonyl   May produce hepatotoxicity. *0071*

Melarsoprol   May produce hepatotoxicity. *0071*

Mepazine   Cholestatic effect (up to 6 times normal). *1713*

Meperidine   May cause rise in intrabiliary pressure. *3505*

Mephenytoin   Hepatotoxicity. *2313*

Meprobamate   May cause cholestatic (hepatocanalicular) jaundice. *0248*

Mercaptopurine   May cause hepatotoxicity (centrolobular necrosis). *0204*   Intrahepatic cholestasis observed in several patients. *2406*

Metahexamide   Hepatotoxicity (viral hepatitis type). *2313*

Metaxalone   Hepatotoxic effect. *2313*

Methacycline   Possible hepatotoxicity. *0071*

Methandriol   Intrahepatic cholestatic jaundice. *1343*

Methandrostenolone   Up to 3-6 times normal due to cholestasis. *1713*

Methdilazine   Rare case of cholestasis. *1680*

Methenamine   Mild transient effect in some cases. *1680*

Methimazole   May affect liver function (cholestasis). *2313*   Rare case of cholestatic jaundice reported. *1774*

Methotrexate   May cause cytotoxic hepatocellular damage. *0248*   In 152 of 250 courses of treatment but most regressed. *2406*

Methoxsalen   Hepatotoxic effect. *2313*

Methoxyflurane   Hepatic toxicity. *3294*   In 2 pregnant women when used as analgesia for labor: hepatitis observed. *0872*

Methyldopa   Result of hepatocellular damage or cholestasis. *3879*   Disturbances in liver function in 5 to 35% of patients treated for hypertension. Hepatocellular injury may occur after short or long-term exposure. *0162*   Chronic active hepatitis associated with immune hemolytic anemia in one case. *3269*   Drug induced hepatitis with severe, chronic, aggressive inflammation. *0211*

Methyltestosterone   Cholestatic effect (moderate elevation). *2313*

Methylthiouracil   Hepatotoxic effect. *2313*

Mexiletine   Occurs in fewer than 1% treated patients. *2011*

Mithramycin   Hepatocellular damage observed. *0071*   Consistent moderate increase in most cases. *2406*

Mitomycin   Centrilobular stasis in 5 of 6 patients and toxic hepatitis in the other. *2406*

Mitoxantrone   Mild transient effects in 11 of 26 patients. *2703*

Molindone   Single case report (?cause). *1905*

Morphine   May cause rise in intrabiliary pressure. *3505*

Moxalactam   In about 3% patients as reported from several studies. *0592*

Muscle Massage   Significant effect, but not to pathological level 8 h after. *0423*

Mushroom Poisoning   May produce severe liver damage. *0279*

Nafcillin   1 of 32 patients developed abnormal liver function on 3rd day of treatment. *2548*

Nalidixic Acid   May cause cholestatic jaundice. *2808*

Nandrolone   Reported to affect liver function. *2220*

Naproxen   In 6% of patients receiving drug for juvenile rheumatoid arthritis. *0239*

Niacin   Intrahepatic cholestasis observed rarely. *1253*

Niacinamide   Rare probable reversible toxic effect. *3875*

Nitrofurans   Hepatotoxicity. *2313*

Nitrofurantoin May cause cholestatic jaundice. *2808* Moderate increase chronic active hepatitis, much more common in women than men, but still rare. *2444* Moderate increase chronic active hepatitis, much more common in women than men, but still rare. *2444*

Nitrophenol Hypersensitive intrahepatic cholestasis. *1434*

Norethandrolone Usually reversible cholestasis. *1280*

Norethindrone Intrahepatic cholestatic jaundice. *2313*

Norethisterone In 5.6% patients being treated for advanced or recurrent breast cancer. *2073*

Norethynodrel Cholestatic effect. *2313*

Norfloxacin In 2 of 1540 patients in clinical trials. *0731*

Nortriptyline May cause cholestasis. *0071*

Novobiocin Intrahepatic cholestasis may occur. *2313*

Ofloxacin Very infrequent side effect observed during many courses of treatment. *1838*

Oleandomycin May cause hepatotoxicity (cholestatic syndrome). *2313*

Omeprazole Pronounced rise in activity in one patient on eighth day of treatment. *1432*

Oral Contraceptives Cholestatic-hepatotoxic effect. *2313* Significant increase observed within 3 mo usually continuing for several y. *3170*

Organophosphorus Insecticides Hepatotoxic effect. *0405*

Oxacillin Possible hepatotoxic effect. *2220* Typically occurring 3 to 14 d after treatment started, may occur without eosinophilia. *2374* Reversible phenomenon observed in 8 patients given drug i.v. *2664* Drug associated hepatitis in patients given high dose drug i.v. *2675* Asymptomatic hepatic dysfunction in 5 cases following high dose treatment: reversible. *2835* Reversible high-dose associated liver injury. *2431* Abnormal liver function in 1 of 8 treated patients on day 15. *2548*

Oxazepam Possible hepatotoxicity. *2313*

Oxymetholone Cholestatic effect (increased up to 6 times normal). *1713*

Oxyphenbutazone Possible hepatotoxicity. *2313*

Oxyphenisatin May cause hypersensitivity reaction. *2970* Liver damage may present as acute hepatitis or as more advanced chronic disease: usual liver disease is chronic active hepatitis. *2260*

Papaverine Probable hypersensitivity reaction. *3039*

Paraldehyde Possible hepatotoxicity. *2313*

Paramethadione Possible hepatotoxicity. *2313*

Pargyline Possible hepatotoxicity (reversible). *2313*

Penicillamine Possible hepatotoxicity. *2313* In 6 of 99 patients treated for rheumatoid arthritis: evidence of toxic liver necrosis observed in two cases. *3887* Case of acute cholestatic jaundice in patient with systemic lupus erythematosus. *3242*

Phenacemide May affect liver function in about 2% cases. *2313*

Phenazopyridine Hepatotoxic effect. *2313*

Phenelzine May cause hypersensitive hepatitis. *0071*

Phenindione May modify liver function (cholestasis). *2313*

Pheniprazine May cause hepatocellular jaundice. *2313*

Phenobarbital Hepatotoxicity with centrolobular necrosis. *1948*

Phenothiazines May cause cholestatic hepatitis (in up to 4% patients). *2313* Liver damage observed after different phenothiazines administered in succession. *1916*

Phenoxypropazine Produced fatal hepatotoxicity. *2427*

Phenprocoumon Mild effect but recurrently in 2 patients having repeated exposure to drug. *3347*

Phenylbutazone May cause cholestatic and cytotoxic jaundice. *0248* Frequent hypersensitivity reaction but also with hepatotoxic potential: may be hepatocellular injury or systemic vasculitis. *0290*

Phenytoin Intrahepatic cholestatic jaundice. *2313*

Phosphorus Hepatotoxic effect with necrosis. *2313*

Picric Acid May cause acute hepatic damage. *1343*

Pindolol Minor persistent increases reported in about 7% patients but not progressive. *1208*

Piperacetazine Transient reversible effect. *1227*

Piperacillin In 1 of 29 patients given 181 mg/kg i.v. for 6 d. *3275*

Piroxicam Occasional case of liver damage reported. *2824* Some transient increases reported. *2692*

Polythiazide May affect liver function. *2313*

Probenecid Hepatotoxic effect (centrolobular necrosis). *2313*

Procainamide Hepatotoxic effect. *2313*

Prochlorperazine Cholestatic effect. *0071*

Progabide Several fold increase in one patient necessitating stopping treatment. *3188*

Progesterone Hepatotoxicity. *2313*

Promazine May cause cholestatic (hepatocanalicular) jaundice. *0248*

Promethazine May cause cholestasis. *2313*

Propoxyphene Cholestatic effect. *1954*

Propranolol Low incidence drug induced increase. *2427*

Propylthiouracil Hepatotoxic effect (centrolobular necrosis). *2313* Single case of chronic active hepatitis reported. 2 others with bridging necrosis. Recovery with drug withdrawal. *2260* In one 24 year old woman who developed fulminant hepatic failure with lymphocyte sensitization. *2437*

Protionamide Reported to affect liver function. *2583*

Protriptyline Transient reversible cholestasis. *0071*

Pyrazinamide Hepatic toxicity reported (viral-hepatitis like). *3193* Hepatotoxic in 1 to 5% patients (previously reported to be toxic in as many as 25%). *3683* Low toxicity with daily doses of 20 to 30 mg/kg body weight but overall incidence of hepatitis of 0.3%. *0682*

Quinacrine Hepatitis reported (centrolobular necrosis). *0071*

Quinethazone May cause cholestatic jaundice. *2313*

Quinidine Isolated case of hepatic toxicity (granulomatous hepatitis). *0862* In single case of severe quinidine hypersensitivity. *0858* Hypersensitivity reaction reported in one patient. *3350*

Rifampin Hepatic toxicity occurs in up to 7% patients. *1553* Minimal abnormalities of liver function are common: severe liver damage in about 0.6% patients. *3683* Observed in one patient receiving drug who developed porphyria cutanea tarda. *2441* In 61% of 18 children receiving drug in combination with isoniazid. *2186*

Spectinomycin Mechanism not discussed. *0122*

Stanozolol May cause intrahepatic cholestasis. *0757*

Starvation With absolute fast 15 d ?hepatic focal necrosis. *0422*

Stibophen Hepatitis reported. *0071*

Streptokinase Substantial but not as marked as GGT. *3189*

Sulfachlorpyridazine Occasional hepatitis-like reaction. *1680*

Sulfadiazine May cause cholestasis with cholangiolitis. *1948*

Sulfadimethoxine Reversible hypersensitive cholestatic response. *1888*

Sulfamethizole Reversible cholestasis. *1974*

Sulfamethoxazole Cholestatic jaundice may occur. *1974* Occasional case of cholestasis with or without hepatic necrosis. *3452* When given with trimethoprim, fever and malaise: successful response to treatment. *1969*

Sulfamethoxypyridine Reversible cholestasis. *2313*

Sulfanilamide May cause reversible cholestasis. *2220*

Sulfapyridine May cause reversible cholestasis. *2220*

Sulfasalazine Occasional case of drug-induced toxic hepatitis. *1422* Rare hepatotoxicity reported associated with noncaseating granuloma or focal inflammation with or without necrosis. *1127*

Sulfisoxazole May cause cholestasis. *1974*

Sulfonamides May cause cholestasis. *2313* Hepatitis-like reaction in patients with multiple episodes of jaundice after repeated exposure to drug. *1741*

Sulfones May affect liver function. *2313*

Sulfonylureas Moderate elevation with cholangiolitis. *3171*

Sulindac Isolated case of cholestatic jaundice 12 d after therapy started. *2371*

Testosterone May cause cholestatic and cytotoxic liver damage. *0248*

Tetracycline Hepatotoxic especially in pregnant women. *2808*

Thallium Due to hepatic necrosis. *1302*

Thiabendazole Rare cholestasis. *2413*

Thiacetazone May cause cholestatic (hepatocanalicular) jaundice. *0248* May cause hepatitis. *1292*

Thiazides May cause cholestasis. *2313*

Thiethylperazine Few cases of cholestasis reported. *1680*

Thiocarlide Reported effect in 4% subjects. *2427*

Alanine Aminotransferase (continued)

Serum Increase Physiological (continued)

Thiocyanate May cause hepatic necrosis. 1343

Thioguanine May cause cholestasis. 2313

Thiopental 1 case out of 24 3-5 d after anesthesia, more longer period after anesthesia pyrexia and jaundice in one case after repeated anesthesia. 0391

Thioridazine Hepatotoxicity. 1857

Thiosemicarbazones May affect liver function. 2313

Thiothixene Hepatotoxic effect (reversible cholestasis). 2313

Thiouracil May cause cholestasis with cholangiolitis. 1948

Thorium Dioxide Increased about 10% after injection. 3147

Tocainide Occasional abnormality reported associated with reversible hepatitis. 2011

Tolazamide Cholestatic effect. 2313

Tolbutamide May cause cytotoxic liver damage or cholestasis. 0248

Toluene Diamine May cause cholestasis with cholangiolitis. 1948

Tranylcypromine May affect liver function (?hypersensitivity). 2313

Tretinoin In 1 of 11 patients given drug orally. 3867

Trichlormethiazide May rarely cause cholestasis. 1680

Trichloroethylene Potentially hepatotoxic. 0071

Triethylenemelamine Hepatotoxicity (prolonged treatment, large dose). 0071

Trifluoperazine Cholestatic effect. 1713

Trifluperidol Hepatocellular changes observed. 1343

Trimeprazine May cause cholestasis. 0071

Trimethadione May cause hepatotoxicity with necrosis. 2313

Trimetrexate Transient effect in small proportion of patients although reverted to normal with continued treatment. 0058

Trioxsalen May affect liver function. 2313

Troleandomycin In 30% individuals treated with 1 g/d for 3-4 weeks. Jaundice typically occurs after 2 weeks mixed in type with both cholestasis and mild hepatocytic necrosis. 2792

Tryparsamide May cause liver damage. 0071

Uracil Mustard May affect liver function. 2313

Urethan May cause liver damage and necrosis. 0071

Valproic Acid Observed in 4 of 25 patients treated with drug: reversible with reduction of dose or withdrawal of drug. 3847 Dose related increase in 44% of treated patients. 0950 Rise in about 15 to 30% cases transient and usually maximal to 10 to 12 weeks after beginning treatment. 0504 Transient increase in 44% of patients without change of dosage. 1781 In 4 of 25 patients, reversed with withdrawal of treatment. 3847 Fatal case reported in child also given phenytoin. 1626 Acute hepatic centrilobular necrosis with severe fatty change in small number of patients. 3763 In 1 of 9 patients with epilepsy poorly controlled. 3508

Vanadium Mild hepatic dysfunction observed with poisoning. 2217

Verapamil Observed in a single case, reverted to normal as soon as drug withdrawn. 2561

Vinyl Ether Potentially hepatotoxic. 0071

Vitamin A Observed with acute intoxication. 3867

Warfarin Rare cases of intrahepatic cholestasis in patients with previous history of hypersensitivity reported. 2940

Zimeldine 7 of 14 patients treated for more than 1 week demonstrated toxic syndrome ? immunological mechanism or related to high initial dose. 2075 Reported in 2 patients in association with headaches, possibly due to fall in blood concentration of 5-hydroxytryptamine. 3387 In 64% of 21 inpatients with endogenous depression given 225 mg daily for 4 weeks. 2082 In several of 147 patients although 16 had initially high values. 3741

Zoxazolamine May produce viral-hepatitis like syndrome. 1948

Serum Decrease Analytical

Acetylsalicylic Acid At 8300 µmol/L on colorimetric analytical method. 1785

Fluosol-DA Increase by 86% on Du Pont aca III at concentration of 50%. 2518

Ibuprofen On colorimetric method at 10 times maximal therapeutic concentration. 1785

Ketoprofen On continuous method. 1785

Liposol Changed from 41.7 to 8.27 U/L in 26 randomly selected patient sera, before and after addition on Beckman Astra methods. 2054

Methotrexate At 5 times upper limit of therapeutic range on method on SMAC®. 2138

Pyruvate High concentrations may deplete NADH so even normal enzyme activity gives depletion reaction on Technicon® SMAC®. 1511

Rifampin At 5 times upper limit of therapeutic range on method on SMAC®. 2138

Storage of Sample 11% loss in 1 d at -20° 20% decrease in 1 week at -20° if high activity. 3217

Serum Decrease Physiological

Meals 2 h after standard meal in men. 3455 Observed in women. 3319

Muscular Exercise Effect of physical training. 2981

Penicillamine Increase by 45% in 3 patients with Wilson's disease although effect disappeared after 12 mo. 1417

Phenothiazines Observed in treatment of female schizophrenics. 0289

Vitamin B$_6$ Depletion More affected than AST. 0255

Urine Increase Physiological

Proteinuria Effect much less than with AST. 2897

Red Blood Cells No Effect Physiological

Oral Contraceptives No effect, but stimulated in vivo by B$_6$. 3049 No effect even if treated for 3 y. 3048

Red Blood Cells Increase Physiological

Oral Contraceptives Significantly higher than in women not using oral contraceptives. 0955

Red Blood Cells Decrease Physiological

Oral Contraceptives Possibly associated with vitamin B$_6$ deficiency. 0563

Prasterone Effect observed when given orally. 2210

Albumin

Serum No Effect Analytical

Acetaminophen At 5 times therapeutic concentration on BCG methods on SMAC®, Ektachem® 400, Hitachi® 705 and KDA. 2138 At acute overdose concentration (10 mg/dL) on SMAC® method. 3719 At concentration of 1500 mg/L had no effect on BCG method. 3393

Acetazolamide No effect at 12 mg/dL on SMA 12/60 method. 2636 At concentration of 1,000 mg/L had no effect on BCG method. 3393

Acetylsalicylic Acid On bromcresol green method on SMA II at physiological concentration. 1787 At 5 times upper limit of therapeutic range on methods on SMAC®, Ektachem®, Hitachi® 705 and KDA. 2138 At concentration of 2,000 mg/L had no effect on BCG method. 3393

Allopurinol At concentration of 5 mg/L had no effect on BCG method. 3393

Aminopyrine At concentration of 125 mg/L had no effect on BCG method. 3393

Aminosalicylic Acid At 5 times upper limit of therapeutic range on methods on SMAC®, Ektachem®, Hitachi® 705 and KDA. 2138 At concentration of 460 mg/L had no effect on BCG method. 3393

Amiodarone No effect at up to 10.0 mg/L on bromcresol green method on Technicon® SMA II and bromcresol purple on Abbott-VP. Metabolite desethylamiodarone also had no effect. 3318

Amitriptyline At acute overdose concentration (2.5 mg/dL) on SMAC® method. 3719 At concentration of 25 mg/L had no effect on BCG method. 3393

Amobarbital At concentration of 150 mg/L had no effect on BCG method. 3393

Amphotericin B At concentration of 20 mg/L had no effect on BCG method. *3393*

Ampicillin At concentration of 40 mg/L had no effect on BCG method. *3393*

Ascorbic Acid At concentration of 500 mg/L had no effect on BCG method. *3393*

Aspirin No significant effect on BCG method at 600 mg/L. No significant effect on HABA method at 1 g/L. *2625*

Barbital At acute overdose concentration (20 mg/dL) on Technicon® SMAC® method. *3719* At concentration of 500 mg/L had no effect as measured by BCG method. *3393*

Bethanechol No effect at 0.09 mg/dL on SMA 12/60 method. *2636*

Bilirubin No effect if bromocresol green used in method. *2131*

Boric Acid At concentration of 50 mg/L had no effect as measured by BCG method. *3393*

Bromazepam At 5 times therapeutic concentration on BCG methods on SMAC®, Ektachem® 400, Hitachi® 705 and KDA. *2138*

Butabarbital At concentration of 100 mg/L had no effect as measured by BCG method. *3393*

Caffeine At acute overdose concentration (10 mg/dL) on Technicon® SMAC® method. *3719* At concentration of 160 mg/L had no effect on BCG method. *3393*

Carbenicillin At 5 times upper limit of therapeutic range on BCG method on SMAC®, Ektachem® 400 Hitachi® 705 and KDA. *2138*

Cefazolin At concentration of 117 mg/L had no effect on BCG method. *3393*

Cefoxitin At 5 times upper limit of therapeutic range on methods on SMAC®, Ektachem®, Hitachi® 705 and KDA. *2138*

Cephalothin At 5 times upper limit of therapeutic range on methods on SMAC®, Ektachem®, Hitachi® 705 and KDA. *2138* At concentration of 1,000 mg/L had no effect on BCG method. *3393*

Chlordiazepoxide At 5 times upper limit of therapeutic range on methods on SMAC®, Ektachem®, Hitachi® 705 and KDA. *2138* At acute overdose concentration (5 mg/dL) on Technicon® SMAC® method. *3719* At concentration of 50 mg/L had no effect on BCG method. *3393*

Chlormezanone At concentration of 100 mg/L had no effect on BCG method. *3393*

Chlorphenesin At concentration of 150 mg/L had no effect on BCG method. *3393*

Chlorpheniramine At acute overdose concentration (20 mg/dL) on Technicon® SMAC® method. *3719* At concentration of 3 mg/L had no effect on BCG method. *3393*

Chlorpromazine At concentration of 3 mg/L had no effect on BCG method. *3393*

Chlorpropamide At concentration of 150 mg/L had no effect on BCG method. *3393*

Chlorprothixene At concentration of 1 mg/L had no effect on BCG method. *3393*

Cimetidine At concentration of 1 mg/L had no effect on BCG method. *3393*

Cloxacillin At concentration of 5 mg/L had no effect on BCG method. *3393*

Cocaine At acute overdose concentration (2.5 mg/dL) on Technicon® SMAC® method. *3719* At concentration of 25 mg/L had no effect on BCG method. *3393*

Codeine At acute overdose concentration (2.0 mg/dL) on Technicon® SMAC® method. *3719* At concentration of 20 mg/L had no effect on BCG method. *3393*

Contact With Clot If plastic disc used to separate for 48 h. *1863*

Cyclosporine At concentration of 41.5 mmol/L (50 µg/L) on methods on Technicon® SMAC® II and Hitachi® 705. *1123*

Cysteine At 5 times upper limit of therapeutic range on methods on SMAC®, Ektachem®, Hitachi® 705 and KDA. *2138*

Cystine At 5 times upper limit of therapeutic range on methods on SMAC®, Ektachem®, 705 and KDA. *2138*

Desipramine At concentration of 8 mg/L had no effect on BCG method. *3393*

Deslanoside No effect on SMA 12/60 method at 0.06 mg/dL. *2636*

Dextran At concentration of 30,000 mg/L had no effect on BCG method. *3393*

Diazepam At 5 times upper limit of therapeutic range on methods on SMAC®, Ektachem®, Hitachi® 705 and KDA. *2138* At acute overdose concentration (2.5 mg/dL) on Technicon® SMAC® method. *3719* At concentration of 25 mg/L had no effect on BCG method. *3393*

Diclofenac On bromcresol green method on SMA II at physiological concentration. *1787*

Diethylaminoethanol At concentration of 30 mg/L had no effect on BCG method. *3393*

Diethylpropion At acute overdose concentration (10 mg/dL) on Technicon® SMAC® method. *3719* At concentration of 100 mg/L had no effect on BCG method. *3393*

Digitoxin No effect at 21 mg/dL on SMA 12/60 method. *2636* At concentration of 6 mg/L had no effect on BCG method. *3393*

Digoxin No effect at 0.04 mg/dL on SMA 12/60 method. *2636* At concentration of 9 mg/L had no effect on BCG method. *3393*

Dimethadione At concentration of 2,000 mg/L had no effect on BCG method. *3393*

N-Dimethyldiazepam At concentration of 6 mg/L had no effect on BCG method. *3393*

Diphenhydramine At acute overdose concentration (20 mg/dL) on Technicon® SMAC® method. *3719* At concentration of 200 mg/L had no effect on BCG method. *3393*

Disopyramide At concentration of 4 mg/L had no effect on BCG method. *3393*

Disulfiram At concentration of 120 mg/L had no effect on BCG method. *3393*

Ethacrynic Acid At 2.5 mg/dL no effect on SMA 12/60 method. *2636*

Ethanol At acute overdose concentration (20 mg/dL) on Technicon® SMAC® methods. *3719*

Ethchlorvynol At acute overdose concentration (20 mg/dL) on Technicon® method. *3719* At concentration of 400 mg/L had no effect on BCG method. *3393*

Ethinamate At acute overdose concentration (20 mg/dL) on Technicon® SMAC® method. *3719* At concentration of 200 mg/L had no effect on BCG method. *3393*

Ethosuximide At concentration of 390 mg/L had no effect on BCG method. *3393*

Fluosol-DA On SMA IIC at concentration of 50% On Du Pont aca III at concentration of 50%. *2518*

Flurazepam At 5 times upper limit of therapeutic range on methods on SMAC®, Ektachem®, Hitachi® 705 and KDA. *2138* At acute overdose concentration (2.5 mg/dL) on Technicon® SMAC® method. *3719* At concentration of 25 mg/L had no effect on BCG method. *3393*

Furosemide No effect at 1.4 mg/dL on SMA 12/60 method. *2636* At concentration of 4 mg/L had no effect on BCG method. *3393*

Gentamicin At concentration of 150 mg/L had no effect on BCG method. *3393*

Glutethimide At acute overdose concentration (5 mg/dL) on Technicon® SMAC® method. *3719* At concentration of 50 mg/L had no effect on BCG method. *3393*

Heparin No effect on binding of spectrum AB_2. *0583* No significant effect on BCG method on Du Pont aca II or Beckman Astra 8 analyzer. *1494*

Ibuprofen At 5 times upper limit of therapeutic range on methods on SMAC®, Ektachem®, Hitachi® 705 and KDA. *2138* On bromcresol green method on SMA II at physiological concentration. *1787* At concentration of 200 mg/L had no effect on BCG method. *3393*

Indomethacin On bromcresol green method on SMA II at physiological concentration. *1787* At concentration of 13 mg/L had no effect on BCG method. *3393*

Insulin At concentration of 3 mg/L had no effect on BCG method. *3393*

Iproniazid At concentration of 40 mg/L had no effect on BCG method. *3393*

Isoniazid At concentration of 100 mg/L had no effect on BCG method. *3393*

Ketoprofen On bromcresol green method on SMA II at physiological concentration. *1787* At concentration of 60 mg/L had no effect on BCG method. *3393*

Lidocaine No effect SMA 12/60 method with 3.5 mg/dL. *2636*

Albumin (continued)

Serum   No Effect   Analytical   (continued)

Lorazepam   At concentration of 0.05 mg/L had no effect on BCG method. *3393*

Mannitol   No effect on SMA 12/60 method at 445 mg/dL. *2636*

Meperidine   At acute overdose concentration (5 mg/dL) on Technicon® SMAC® method. *3719*   At concentration of 50 mg/L had no effect on BCG method. *3393*

Mephenesin   At concentration of 100 mg/L had no effect on BCG method. *3393*

Meprobamate   At acute overdose concentration (20 mg/dL) on Technicon® SMAC® method. *3719*   At concentration of 200 mg/L had no effect on BCG method. *3393*

Meralluride   No effect at 0.07 mL/dL on SMA 12/60 method. *2636*

Mesoridazine   At acute overdose concentration (20 mg/dL) on Technicon® SMAC® method. *3719*

Methadone   At acute overdose concentration (2.5 mg/dL) on Technicon® SMAC® method. *3719*   At concentration of 25 mg/L had no effect on BCG method. *3393*

Methamphetamine   At concentration of 2 mg/L had no effect on BCG method. *3393*

Methanol   At acute overdose concentration (20 mg/dL) on Technicon® SMAC® method. *3719*

Methapyrilene   At concentration of 13 mg/L had no effect on BCG method. *3393*

Methaqualone   At acute overdose concentration (2.5 mg/dL) on Technicon® SMAC® method. *3719*   At concentration of 25 mg/L had no effect on BCG method. *3393*

Methicillin   At 5 times upper limit of therapeutic range on SMAC®, Abbott-VP, Ektachem®, Hitachi® 705 and KDA. *2138*   At concentration of 900 mg/L had no effect on BCG method. *3393*

Methohexital   At concentration of 50 mg/L had no effect on BCG method. *3393*

Methotrexate   At 5 times upper limit of therapeutic range on method on SMAC®, Ektachem®, Hitachi® 705, and KDA. *2138*

Methsuximide   At concentration of 40 mg/L had no effect on BCG method. *3393*

Methyldopa   At concentration of 80 mg/L had no effect on BCG method. *3393*

Methylphenidate   At acute overdose concentration (20 mg/dL) on Technicon® SMAC® method. *3719*   At concentration of 200 mg/L had no effect on BCG method. *3393*

Methylphenobarbital   At concentration of 150 mg/L had no effect on BCG method. *3393*

Methyprylon   At acute overdose concentration (10 mg/dL) on Technicon® SMAC® method. *3719*   At concentration of 100 mg/L had no effect on BCG method. *3393*

Metoprolol   At concentration of 0.34 mg/L had no effect on BCG method. *3393*

Morphine   At acute overdose concentration (20 mg/dL) on Technicon® SMAC® method. *3719*   At concentration of 200 mg/L had no effect on BCG method. *3393*

Moxalactam   At concentration of 96 mg/L had no effect on BCG method. *3393*

Nafcillin   At concentration of 50 mg/L had no effect on BCG method. *3393*

Naproxen   At 5 times upper limit of therapeutic range on methods on SMAC®, Ektachem®, Hitachi® 705 and KDA. *2138*

Nitrofurantoin   On routine methods in use on SMAC®, Ektachem®, Abbott-VP, Cobas-Bio, aca, Hitachi® 705, KDA at 5 times normal upper therapeutic concentration. *2138*   At concentration of 4 mg/L had no effect on BCG method. *3393*

Nortriptyline   At acute overdose concentration (20 mg/dL) on Technicon® SMAC® method. *3719*   At concentration of 200 mg/L had no effect on BCG method. *3393*

Orphenadrine   At concentration of 9 mg/L had no effect on BCG method. *3393*

Ouabain   At 0.02 mg/dL no effect on SMA 12/60 method. *2636*

Oxazepam   At concentration of 1 mg/L had no effect on BCG method. *3393*

Papaverine   At concentration of 10 mg/L had no effect on BCG method. *3393*

Paraldehyde   At concentration of 2,000 mg/L had no effect on BCG method. *3393*

Penicillin G   At 5 times upper limit of therapeutic range on methods on SMAC®, Ektachem®, Hitachi® 705 and KDA. *2138*   At concentration of 2,000 mg/L had no effect as measured by BCG method. *3393*

Pentobarbital   At acute overdose concentration (20 mg/dL) on Technicon® SMAC® method. *3719*   At concentration of 340 mg/L had no effect on BCG method. *3393*

Perphenazine   At acute overdose concentration (20 mg/dL) on Technicon® SMAC® method. *3719*   At concentration of 200 mg/L had no effect on BCG method. *3393*

Phencyclidine   At concentration of 6 mg/L had no effect on BCG method. *3393*

Phenobarbital   No significant effect on HABA method at 25 mg/L   No significant effect on BCG method at 25 mg/L. *2625*   At acute overdose concentration (20 mg/dL) on Technicon® SMAC® method   At acute overdose concentration (20 mg/dL) on Technicon® SMAC® method. *3719*   At concentration of 250 mg/L had no effect on BCG method. *3393*

Phensuximide   At concentration of 120 mg/L had no effect on BCG method. *3393*

Phenylbutazone   At concentration of 750 mg/L had no effect on BCG method. *3393*

Phenylephrine   At concentration of 4 mg/L had no effect on BCG method. *3393*

Phenytoin   No effect at 1.8 mg/dL on SMA 12/60 method. *2636*   At acute overdose concentration (20 mg/dL) on Technicon® SMAC® method. *3719*   At concentration of 240 mg/L had no effect on BCG method. *3393*

Probenecid   At concentration of 1300 mg/L had no effect on BCG method. *3393*

Procainamide   No effect on SMA 12/60 method at 35 mg/dL. *2636*   At concentration of 50 mg/L had no effect on BCG method. *3393*

Procaine   At concentration of 2 mg/L had no effect on BCG method. *3393*

Prochlorperazine   At concentration of 1 mg/L had no effect on BCG method. *3393*

Promethazine   At acute overdose concentration (20 mg/dL) on Technicon® SMAC® method. *3719*   At concentration of 200 mg/L had no effect on BCG method. *3393*

Propoxyphene   At acute overdose concentration (2.5 mg/dL) on Technicon® SMAC® method. *3719*   At concentration of 25 mg/L had no effect on BCG method. *3393*

Propranolol   No effect at 0.1 mg/dL on SMA 12/60 method. *2636*   At concentration of 0.2 mg/L had no effect on BCG method. *3393*

Pyribenzamine®   At acute overdose concentration (20 mg/dL) on Technicon® SMAC® methods. *3719*

Quinidine   At 26 mg/dL no effect on SMA 12/60 method. *2636*   At acute overdose concentration (20 mg/dL) Technicon® SMAC® method. *3719*   At concentration of 210 mg/L had no effect on BCG method. *3393*

Quinine   At acute overdose concentration (1.5 mg/dL) on Technicon® SMAC® method. *3719*   At concentration of 30 mg/L had no effect on BCG method. *3393*

Reserpine   No effect at 0.02 mg/dL on SMA 12/60 method. *2636*

Rifampin   At 5 times upper limit of therapeutic range on methods on SMAC®, Ektachem®, Hitachi® 705 and KDA. *2138*

Salicylate   No significant effect even at 1 g/L on BCG method. *2625*   At concentration of 600 mg/L had no effect on BCG method. *3393*

Secobarbital   At acute overdose concentration (20 mg/dL) on Technicon® SMAC® method. *3719*   At concentration of 200 mg/L had no effect on BCG method. *3393*

Standing of Sample   If on polystyrene beads 24 h at 25°. *2624*

Storage of Sample   No effect for 10 y stored at -10°. *1180*

Strychnine   At concentration of 12 mg/L had no effect on BCG method. *3393*

Sulfadiazine   At concentration of 1500 mg/L had no effect on BCG method. *3393*

Sulfaguanidine   At concentration of 500 mg/L had no effect on BCG method. *3393*

Sulfanilamide   At concentration of 1,000 mg/L had no effect on BCG method. *3393*

Sulfisoxazole   No significant effect on BCG method at 500 mg/L No significant effect on HABA method at 500 mg/L. *2625*

Tetracycline   At 5 times upper limit of therapeutic range on methods on SMAC®, Ektachem®, Hitachi® 705 and KDA. *2138*   At concentration of 300 mg/L had no effect on BCG method. *3393*

Theobromine   At concentration of 2,000 mg/L had no effect on BCG method. *3393*

Theophylline   At 5 times upper limit of therapeutic range on methods on SMAC®, Ektachem®, Hitachi® 705 and KDA. *2138*   At concentration of 2,000 mg/L had no effect on BCG method. *3393*

Thiamazole   At concentration of 114 mg/L had no effect on BCG method. *3393*

Thiopental   At acute overdose concentration (20 mg/dL) on Technicon® SMAC® method. *3719*   At concentration of 200 mg/L had no effect on BCG method. *3393*

Timolol   At concentration of 0.01 mg/L had no effect on BCG method. *3393*

Tolbutamide   At concentration of 100 mg/L had no effect on BCG method. *3393*

Tribromethanol   At concentration of 90 mg/L had no effect on BCG method. *3393*

Trichlorethanol   At concentration of 1,000 mg/L had no effect on BCG method. *3393*

Trifluoperazine   At concentration of 1 mg/L had no effect on BCG method. *3393*

Tripelennamine   At concentration of 200 mg/L had no effect on BCG method. *3393*

Vitamin Preparations   No effect at expected concentration with SMA 12/60 procedure. *2637*

Warfarin   At concentration of 100 mg/L had no effect on BCG method. *3393*

Serum   No Effect   Physiological

Acetylsalicylic Acid   No effect seen in patients undergoing treatment. *1787*

Amphotericin B   No effect seen in 10 patients with 6 weeks treatment. *0240*

Carbamazepine   In about 20 epileptic patients treated for 2 y. *0873*   Osteomalacia observed in 3 of 31 patients given average of 758 mg/d for average of 20.5 mo. *2635*

Danazol   No change during treatment of women with endometriosis. *2599*   No change in 18 women with endometriosis given 600 mg/d for 6 mo. *0591*

Diclofenac   No effect seen in patients undergoing treatment. *1787*

Halothane   In one nurse occupationally exposed to gas. *1900*

Indomethacin   No effect seen in patients undergoing treatment. *1787*

Isotretinoin   In 18 patients with severe acne given 0.8 mg/kg daily for 3 mo: changes reverted to normal after treatment stopped. *2242*

Ketoprofen   No effect seen in patients undergoing treatment. *1787*

Levonorgestrel   No significant difference over 3 y between implant recipients and IUD users. *0763*

Lithium   No difference between treated manic depressives and controls. *2053*

Meals   No effect after standard breakfast. *0579*

Norethindrone   No effect observed. *0474*

Penicillamine   In 21 patients with rheumatoid arthritis treated for 48 weeks given 250 to 750 mg daily. *2345*

Phenytoin   In about 20 patients treated for 2 y. *0873*

Quinidine   Hypersensitivity observed in 32 of 487 patients who had received drug. *1261*

Saccharated Iron Oxide   In 9 individuals given 40 mg/d i.v. daily for up to 42 d. *2662*

Starvation   Usually unaffected if previously well fed. *3783*

Terbutaline   In 6 normal men treated with therapeutic amounts for 2 weeks. *3177*

Timegadine   In 23 patients with rheumatoid arthritis given 250 to 750 mg/d for 48 weeks. *2345*

Serum   Increase   Analytical

Ampicillin   Produces increase and noisy peaks with automated BCG procedure. *0287*

Caproxamine   At 1.5 mg/dL conventional methods if added to sera. *1758*

Cefotaxime   Slight effect of drug and metabolite on method on Parallel. *0201*

Disulphine Blue   At physiological concentration on bromocresol method for at least 2 d. *1464*

Fluosol-DA   Albumin results very variable but apparent effect. *2518*

Hemoglobin   If no blank with HABA dye, slight effect BCG. *2467*   100 mg/dL equivalent to 100 mg/dL BCG method. *3217*

Heparin   Large amounts may cause HABA binding to fibrinogen. *2228*   Promotes binding of HABA dye to globulins. *3505*

Lipemia   Affects most methods unless blank correction. *3873*   Nonblanked BCG procedure in patient specimens measured on SMAC® in comparison with specimens following ultra-centrifugation. *2466*

Liposol   Changed from 34.7 to 39.3 g/L in 26 randomly selected patient sera, before and after addition on Beckman Astra methods. *2054*

Phenazopyridine   Contributes to absorption binding procedure. *0583*

Serum   Increase   Physiological

Alcuronium   Sensitivity to drug associated with albumin concentration. *3481*

Dehydration   By approximately 11.6% after heat exposure for 2-11 h. *3254*

Diurnal Variation   Follows calcium; increases blood volume. *1829*

Erect Posture   Average 0.42 g/dL 30 minutes after rising. *3217*

Gallamine   Sensitivity to drug correlated with concentration. *3481*

Levonorgestrel   In a group of women using levonorgestrel covered rods versus controls using copper IUD. *0899*

Meals   Affected by meals (most marked in women). *3455*

Muscular Exercise   Significant effect with 12 minutes on cycle-ergometer. *1231*   Approximately 10% increase immediately after, delayed fall. *2846*

Progesterone   Not observed with combined therapy. *0795*

Season   Proportional increase though total protein falls in summer. *2123*

Tourniquet   Affects all protein bound constituents also. *1563*

Serum   Decrease   Analytical

Aspirin   Decreased dye binding capacity. *3505*

Bilirubin   If HABA, methyl orange, or biuret methods used. *2131*   Albumin-bilirubin binding firm with eosin procedure. *1673*

Carbenicillin   At concentrations above 5,000 mg/L lowered the concentration as measured by BCG method. *3393*

Citrates   Dilutional effect when compared against serum or heparin as anticoagulant. *3361*

EDTA   At concentrations above 25,000 mg/L lowered concentration as measured by BCG method. *3393*

Hemolysis   Affects measurement by HABA, biuret, methyl orange. *3873*

Heparin   Significant reduction with bromocresol purple method on aca II: concentration effect less at higher pH values. *1494*   Negative bias progressively greater with increasing amount of heparin on dye-binding procedures using bromocresol purple but not with bromocresol green. *1494*

Penicillin   Competes with HABA for binding (slight effect). *0583*   At concentrations above 18,000 mg/L (normal therapeutic concentration about 12 mg/L) lowered concentration as measured by BCG method. *3393*

Plasma   Increase by 0.12 mg/dL HABA dye on SMA 12/60 method. *2228*

Salicylate   At 400 mg/L decreases HABA method by 10%. *2625*

Sulfonamides   Competes with HABA for binding sites. *0583*

Serum   Decrease   Physiological

Acetaminophen   In severe cases of poisoning (also in some mild). *3470*

Albumin (continued)

Serum   Decrease   Physiological (continued)

Alcuronium Chloride   Negative correlation between sensitivity and albumin concentration. *3481*

Altitude   After 6 weeks at 5400 metres. *2957*

Amiodarone   Approximately 10% reduction compared with controls when patients treated for 6 mo: probable direct effect on synthesis or clearance. *1173*

Anticonvulsants   In 8 epileptic patients receiving phenobarbital with carbamazepine or phenytoin. *2670*

Asparaginase   Hepatotoxicity (observed in 80% patients). *2657* In 71% children and 82% adults reported in various studies. *2406* Depressed in up to 70% of patients treated in different studies: effects usually mild. *0558*

Aspirin   Progressive effect during three days but also noted over longer term study at beginning, eventually reverted towards normal. Effect statistically significant. *3076*

Azathioprine   May cause hepatotoxicity. *3433*

Carbon Disulfide   Effect greater than on total protein. *0818*

Cathartics   Rare protein-losing gastroenteropathy. *1343*

Chlorpropamide   In isolated case with diabetes who developed proliferative glomerulonephritis indicating immunologically mediated reaction. *0146*

Cyclophosphamide   May cause hepatotoxicity. *3433*

Dantrolene   Single observation, ?significance. *0659*

Dapsone   In 2 patients given long-term treatment. *1935* Significant effect after 32 mo treatment in 1 woman with dermatitis herpetiformis. *1164* To below 2.0 g/dL in one woman with dermatitis herpetiformis: possibly associated with increased intravascular catabolism of albumin. *0740* Sulfone syndrome in one 16 year old girl given 50 mg daily short-term administration. *3604*

Dextran   30% after Macrodex®, no effect after Rheomacrodex® *3346*

Estrogens   Metabolic changes in liver synthesis. *2189*

Ethanol   0.5% increase in heavy versus occasional drinkers. *3273*

Ethinyl Estradiol   Metabolic effect. *0473* Metabolic effect. *0473* In post-menopausal women with 50 µg/d for 14 d. *3474*

Halothane   May induce hepatitis. *0593* Liver disease from mild focal necrosis and inflammation to massive necrosis. *2260*

Heroin   Observed in addicts with renal disease. *2914*

Hydralazine   Centrilobular necrosis observed in 3 patients 2 to 4 months after drug therapy for hypertension. *1740*

Ibuprofen   Significantly lower in patients receiving drug than in controls (4.0 g/dL vs 4.2 g/dL); same effect noted before and after treatment Started. *1787*

Lactation   Remains about 1 g/dL below normal. *0987*

Meals   Significant effect in subjects 20-40 y. *3455*

Mestranol   Metabolic effect. *3258*

Methyldopa   Disturbances in liver function in 5 to 35% of patients treated for hypertension. Hepatocellular injury may occur after short- or long-term exposure. *0162*

Niacin   Decreased synthesis due to liver damage. *3196*

Nitrofurantoin   In association with chronic active hepatitis. *3488* In association with chronic active hepatitis. *3488*

Oral Contraceptives   Not significant after 3 mo cessation, significant after 3 y. *3143* 10% fall if used for several months. *1650*

Oxyphenisatin   May cause hypersensitivity reaction. *2970* Observed with chronic active hepatitis induced by drug. *0903*

Prednisone   Fell within 3 d in adult hospitalized patients given 40-50 mg/d orally. concentration paralleled that of zinc. *3790*

Pregnancy   Falls about 1 g/dL in last 2 trimesters. *0987*

Pyrazinamide   Hepatotoxicity (viral-hepatitis like). *0071*

Quinidine   In single case of severe quinidine hypersensitivity. *0858*

Salsalate   Associated with minimal change nephrotic syndrome in one patient. *3664*

Smoking   Approximately 20% reduction in smokers vs controls. *0809*

Thorium Dioxide   May produce severe liver damage with years. *3147*

Valproic Acid   Observed in 4 of 25 patients treated with drug: reversible with reduction of dose or withdrawal of drug. *3847* Reduced to below 35 g/L in 4 of 9 patients. *3508*

Urine   No Effect   Analytical

Bacteriuria   No effect on automated BCG method. *2919*

Urea   On nephelometric method of Killingsworth. *1931*

Urine   No Effect   Physiological

Cimetidine   No change observed during treatment of 13 ulcer patients. *0649*

Urine   Increase   Physiological

Nifedipine   Increase from 14 to 28 µg/min 40 to 60 minutes after single oral dose of 20 mg sublingually. *0650*

Radiographic Agents   Mean increase from 34 mg/d to 1873 mg/d in 37 patients day following arteriography. *2594*

Test Conditions   Decrease   Analytical

Bilirubin   Competes for binding sites ANSA method. *3596*

Oleate   Competes for binding sites ANSA method. *3596*

Alcohol

Breath   Increase   Analytical

Isopropanol   Can produce measurable levels. *1238*

Breath   Increase   Physiological

Ethanol   Presence of ingested alcohol. *1343*

Methanol   Presence of ingested alcohol. *1238*

Aldehyde Dehydrogenase

Red Blood Cells   Increase   Physiological

Ethanol   With acute oral administration in normal subjects but reduction seen in chronic. *1575*

Red Blood Cells   Decrease   Physiological

Ethanol   Mean decrease to 4.98 mIU/mg protein in chronic alcoholics from 8.25 mIU/mg in controls: unrelated to degree of alcohol ingestion or extent of liver damage. *2175*

Aldolase

Serum   No Effect   Analytical

Hemolysis   Slight hemolysis has no effect. *2981*

Storage of Sample   For 2 d room temperature, 21 d at 4-8°. *3217* 5 h at room temperature, 5 d at 4°, 15 at -15°. *0877* No effect 1 week at 4°, 1 mo at -20°. *3873*

Serum   No Effect   Physiological

Muscular Exercise   Physical activity has no effect. *2981*

Serum   Increase   Analytical

Contact With Clot   Due to release from platelets during clotting. *0067*

Hemolysis   High content in erythrocytes and platelets. *0797*

Serum   Increase   Physiological

Aspirin   Experimental effect seen in rabbits with prolonged use. *3092*

Carbenoxolone   Due to hypokalemic myopathy. *0232*

Chlorinated Insecticides   Presumed damage of liver cells. *0405*

Clopamide   Single case of diuretic associated myopathy. *1793*

Corticotropin   Probably due to muscle damage at site of injection. *0275*

Cortisone   In experimental animals ?of muscle origin. *3835*

Epinephrine   Following single injection of compound in oil. *1264*

Ethanol Dramatic increase of LD-1 and LD-2 observed in acute alcoholic myopathy and less marked increase with chronic myopathy. *3124*

Muscular Exercise Effect of physical training. *2981*

Organophosphorus Insecticides Hepatotoxic effect. *0405*

Quinidine If given intramuscularly, possibly due to increase in intracellular calcium. *3215*

Tetracosactrin Probably due to muscle damage at injection site. *0275*

Thiabendazole Probably of muscle origin although taken orally. *2427*

Thorium Dioxide Time related liver damage (eventual tumor). *3147*

Vasopressin Rhabdomyolysis observed in 2 patients following intravenous drug administration. *0029*

Serum   Decrease   Physiological

Phenothiazines In schizophrenics with high initial values. *2427*

Probucol Mechanism obscure. *0806*

Aldosterone

Plasma   No Effect   Analytical

Ranitidine No effect of up to 5 µg/mL on method of Sancho and Haber. *3130*

Plasma   No Effect   Physiological

Arginine No significant changes observed in 6 normal and 4 diabetic subjects after i.v. infusion. *2328*

Carbohydrate No effect when starved patients refed. *1242*

Carprazidil No consistent effect in 15 men with mild to moderate essential hypertension treated for up to 16 weeks. *1268*

Domperidone After 10 mg intravenously in 8 healthy males. *3406* No effect 15 minutes after 1 mg/kg intravenously in healthy volunteers. *3426*

Dopamine Did not change with infusion of up to 3.0 µg/kg/min for 2 h in 6 normal men. *2149*

Enalapril No significant effect in patients receiving drug. *1616*

Gentamicin Effect observed in 2 patients following larger doses of gentamicin. Associated with massive urinary loss of magnesium and potassium. *0224*

Indomethacin Insignificant change with 1 week treatment. *2464*

Naproxen In 10 furosemide treated patients with well controlled congestive heart failure. *1042*

Nifedipine Possibly due to drug's calcium antagonizing action since calcium is required to secrete aldosterone. *1609* Insignificant change from 108 to 121 pg/mL in young hypertensives after 3 h. *1609*

Oral Contraceptives Inconsistent response after 4 weeks. *3785* Normal activity in users in spite of changes in renin substrate concentration. *1333*

Oxaprozin No significant change observed in 7 volunteers treated for up to 1 week. *2464*

Prazosin No significant change with monotherapy for 12 mo in 15 patients. *3531*

Ranitidine Bolus injections have no effect on basal secretion. *2538*

Rimeterol No significant change in 4 healthy men given therapeutic i.v. dose. *2807*

Sulindac In 10 furosemide treated patients with well controlled congestive heart failure given 400 mg/d for 2 d. *1042*

Verapamil No change compared with placebo in 11 patients treated with up to 360 mg daily for 6 weeks. *3396*

Plasma   Increase   Physiological

Amiloride 3 fold increase observed in 6 healthy subjects given 10 mg/d for 8 d. *3737* In 5 normal subjects given 75 mg daily for 7 d. *2440*

Angiotensin Increase of 4 times normal after i.v. for 1 h. *1145*

Azosemide Effect observed in normal male volunteers after 60 mg orally. *2658*

Chlorthalidone At least 50% increase in pre- and post-menopausal women given 100 mg/d for 6 weeks. *0402* At 3 d but not after 3 mo in 8 patients with mild hypertension given 50 mg/d. *3040*

Erect Posture Increases approximately 6 times if normal diet. *0071*

Furosemide Slower response than angiotensin to i.v. injection. *2359*

Heat After week of thermal stress increased by 76%. *0206*

Hemodialysis Mean increase of 8.2 ng/dL during dialysis. *2350*

Hydralazine Significant effect following 20 mg i.v. in 30 to 60 minutes. *1609*

Laxatives Marked increase in chronic abuser with dehydration. *3682*

Menstruation In 3/4 of subjects peak in mid or late luteal phase. *1885*

Metoclopramide Secretion increased but mechanism mediating effect not elucidated but probably through a factor whose concentration is increased in serum. *0569* Observed in normal subjects after 10 mg drug over 2 h; higher response in patients with primary aldosteronism. *3883* Peak 15 minutes after injection: 3 fold higher than control at peak. *0497* Increased by single i.v. bolus of 10 mg: probably related to basal activity of renin angiotensone system. *2543* Increase by 99% 15 minutes after 1 mg/kg intravenously in healthy volunteers. *3426*

Opiates Significant increase within 1 h of i.m. injection of hyoscine, meperidine or omnopon. *1929*

Potassium If K raised by up to 1 meq/L (orally or i.v.). *0914*

Pregnancy Possible compensatory mechanism. *2180*

Sodium Restriction Trend in 2 d, maximum in 5, with decreased sodium. *0495*

Spironolactone Marked increase in supine concentration on constant diet. *1144* Mean increase of 386 pg/mL in 3 men given drug 100 mg bid for 1 week. *3736* In 5 normal subjects given 300 mg daily for 7 d. *2440* Observed with chronic administration of drug. *2836*

Starvation Significant effect in obese after 10 d. *1242*

Triamterene Mean increase of 125 pg/mL in 3 men given drug 50 mg tid for 1 week. *3736*

Verapamil Concentration initially lower in hypertensives and increased gradually up to fourth month. Exact mechanism still to be elucidated. *3956*

Plasma   Decrease   Physiological

Albuterol No significant change in 4 healthy men given i.v. infusion of therapeutic amount. *2807*

Altitude Effect most marked in older subjects. *1836*

Captopril From 1.08 nmol/L to 0.22 nmol/L after 3 d treatment in 1 patient with Bartter's syndrome. *1558* Parallel decline with sodium in congestive heart failure patients. *2592* Fell in conjunction with fall of systemic arterial pressure. *1716* In 11 patients with resistant heart failure reduced mean concentration from 62 ng/dL to 26 ng/dL. *1165* Gradual reduction probably due to longer half-life than angiotensin II. *0177* With doses up to 800 mg/d for 10 d in 23 hypertensive patients. *2836*

Cyclosporine Inappropriately low in several patients receiving drug possibly due to concomitant beta-blockers. *2836*

Desoxycorticosterone Although on low sodium diet. *3263*

Enalapril Gradual reduction following effect on angiotensin II due to longer half-life. *0177* Increase by 1 mo as part of long-term response in 6 responders of 10 treated hypertensives. *1166*

Etomidate Clear suppression when induction compared with induction by thiopentone. *0063*

Furosemide May be marked reduction in patients with congestive cardiac failure when given drug i.v. if initial concentration high, otherwise usually slight increase. *2513*

Glycyrrhiza Hormonal like action of drug. *0232*

Heparin Reversible toxic effect on glomerulosa cells (Probably). *2638* Possibly due to toxic effect on glomerulosa cells: effect readily reversible with withdrawal of drug. *2638*

High Sodium Diet Reverse effect with low sodium diet. *3263*

Indomethacin Marked reversal on withdrawal of drug in one young woman. *3541*

NSAID (Nonsteroidal Anti-Inflammatory Drugs) Associated with inhibition of prostaglandin synthesis. *1561*

Aldosterone *(continued)*

Plasma Decrease Physiological *(continued)*

Perindopril After doses of 4 mg twice daily within 4 h of administration. *0550*

Pregnancy In pre-eclampsia compared with normal pregnancy. *2180*

Ramipril In response to 10 mg and 20 mg on successive days in 9 patients with severe chronic congestive heart failure. *0764*

Ranitidine Significant reduction in plasma aldosterone in both recumbent overnight concentration and after 2 h ambulation with 3-day oral course. *3130*

Recumbency One sixth of standing concentration. *0733*

Saline Short term response if given i.v. in hypertensives. *1911*

Sodium Salts Comparable decreases observed with chloride and citrate salts in 5 men with essential hypertension. *2037*

Verapamil Significant effect in 15 patients with uncomplicated essential hypertension. *0867*

Urine No Effect Analytical

Androsterone No significant effect on RIA procedure of Drewes. *0952*

Corticosterone No significant effect on RIA procedure of Drewes. *0952*

Cortisone No significant effect on RIA procedure of Drewes. *0952*

Desoxycorticosterone No significant effect in RIA procedure of Drewes. *0952*

Estradiol No significant effect in RIA procedure of Drewes. *0952*

Etiocholanolone No significant effect in RIA procedure of Drewes. *0952*

Hydrocortisone No significant effect in RIA procedure of Drewes. *0952*

Prasterone No significant effect in RIA procedure of Drewes. *0952*

Progesterone No significant effect RIA procedure of Drewes. *0952*

Storage of Sample At room temperature for 45 d if acidified. *1180* 1 mo frozen, 7 d room temperature. *0952*

Testosterone No significant effect RIA procedure of Drewes. *0952*

Tetrahydrocortisol No significant effect RIA procedure of Drewes. *0952*

Tetrahydro-DOC No significant effect on RIA procedure of Drewes. *0952*

Tetrahydro Reichstein's S 1 mg/L equivalent to 2 μg/L by method of Drewes. *0952*

Urine No Effect Physiological

Labetalol With treatment for 4 mo in 15 patients with essential hypertension variable effect observed. *3776*

Oxaprozin No significant change observed in 7 volunteers treated for up to 1 week. *2464*

Oxytocin No significant or consistent effect. *1344*

Vasopressin If not overhydrated no effect. *1344*

Urine Increase Analytical

Androsterone 1 mg/L equivalent to 2 μg/L in method of Drewes. *0952*

Copper If in reagents affects RIA procedures. *0952*

Corticosterone 1 mg/L equivalent to 11 μg/L in method of Drewes. *0952*

Cortisone 1 mg/L equivalent to 8 μg/L in method of Drewes. *0952*

Desoxycorticosterone 1 mg/L equivalent to 38 μg/L in method of Drewes. *0952*

17,21-Di-Hydroxy-4-Pregnene-3,20-Dione 1 mg/L equivalent to 6 μg/L by method of Drewes. *0952*

Etiocholanolone 1 mg/L equivalent to 2 μg/L in method of Drewes. *0952*

Hydrocortisone 1 mg/L equivalent to 4 μg/L in method of Drewes. *0952*

17-Hydroxy-4-Pregnene-3,20-Dione 1 mg/L equivalent to 5 μg/L by method of Drewes. *0952*

Prasterone 1 mg/L equivalent to 2 μg/L in method of Drewes. *0952*

Progesterone 1 mg/L equivalent to 11 μg/L by method of Drewes. *0952*

Testosterone 1 mg/L equivalent to 8 μg/L by method of Drewes. *0952*

Tetrahydrocortisol 1 mg/L equivalent to 2 μg/L by method of Drewes. *0952*

Tetrahydro-DOC 1 mg/L equivalent to 28 μg/L by method of Drewes. *0952*

Urine Increase Physiological

Angiotensin Approximate doubling in response to i.v. infusion. *3174*

Azosemide Effect observed in normal male volunteers after 60 mg orally. *2658*

Chlorthalidone Single case of severe reversible myopathy noted. *1793*

Corticotropin Effect less marked than on tetrahydro-DOC. *3174*

Diurnal Variation High 6 am- 3 pm, low at night (postural effect). *3886*

Erect Posture Increased for about 6 h after standing up. *3886*

Felodipine With 20 mg daily in 10 men with essential hypertension over 8 weeks. *1887*

Furosemide Effect observed after 40 mg orally in normal male volunteers. *2658*

Labetalol With treatment for 4 mo in 15 patients with essential hypertension variable effect observed. *3776*

Lithium Occurs after initial fall. *2527*

Nifedipine In 25 patients with mild to moderate primary hypertension given up to 80 mg daily. *2659*

Oral Contraceptives Estrogen effect (in small number of people). *0585*

Pregnancy Possible compensatory mechanism during last trimester. *2180*

Starvation Two fold increase by 2nd week of fasting. *3783*

Urine Decrease Physiological

Altitude Effect observed on acute exposure. *1836*

Aminoglutethimide May inhibit aldosterone production. *2775*

Amphenone B May inhibit aldosterone production. *2775*

Captopril Fell in conjunction with fall of systemic arterial pressure. *1716* Sustained effect: extent related to pretreatment plasma renin activity. *0177* With doses up to 800 mg/d for 10 d in 23 hypertensive patients. *2836*

Chlorophenothane May inhibit synthesis of hormone. *2775*

Clonidine Secondary to effect on renin. *1627*

Desoxycorticosterone Although on low sodium diet. *3263*

Enalapril Sustained effect: extent related to pretreatment plasma renin activity. *0177*

Fludrocortisone Marked reduction following administration. *3174*

Glucocorticoids May inhibit aldosterone production. *3045*

Glycyrrhiza Pseudoaldosteronism effect. *3755*

Heparin Suppressed secretion with protracted treatment. *1680*

High Sodium Diet Reverse effect with low sodium diet. *3263*

Indomethacin Marked reversal on withdrawal of drug in one young woman. *3541* With 1 week treatment fell from 43 to 18 μg/24 h. *2464*

Labetalol With treatment for 4 mo in 15 patients with essential hypertension variable effect observed. *3776*

Metoprolol In 15 patients with essential hypertension treated for 4 weeks, associated with reduction in sympathetic tone and reduced activity of renin-aldosterone system. *1209*

Metyrapone May inhibit aldosterone production. *2775*

Propranolol Change less than with renin. *0523*

SU-9055 May inhibit aldosterone production. *2775*

Alkaline Phosphatase

Serum   No Effect   Analytical

Acetaminophen   At acute overdose concentration (10 mg/dL) on SMAC® method. *3719*

Acetazolamide   No effect at 12 mg/dL on SMA 12/60 method. *2636*

Acetylsalicylic Acid   At 5 times upper limit of therapeutic range on methods on SMAC®, Abbott-VP, aca, Cobas-Bio, Hitachi® 705 and KDA. *2138*   At 8300 µmol/L on continuous method. *1785*

Aminophenazone   On continuous method at 10 times maximal therapeutic concentration. *1785*

Aminosalicylic Acid   At 5 times upper limit of therapeutic range on methods on SMAC®, Abbott-VP, aca, Cobas-Bio, Hitachi® 705 and KDA. *2138*

Amitriptyline   At acute overdose concentration (2.5 mg/dL) on SMAC® method. *3719*

Barbital   At acute overdose concentration (20 mg/dL) on Technicon® SMAC® method. *3719*

Bethanechol   No effect at 0.09 mg/dL on SMA 12/60 method. *2636*

Bilirubin   No effect up to 15 mg/dL method of Prcksch. *2878*

Bromazepam   At 5 times therapeutic concentration on methods on SMAC®, Abbott-VP, aca, Cobas-Bio, Hitachi® 705 and KDA. *2138*

Caffeine   At acute overdose concentration (10 mg/dL) on Technicon® SMAC® method. *3719*   On continuous method at 10 times maximal therapeutic concentration. *1785*

Carbenicillin   At 5 times upper limit of therapeutic range on methods on SMAC®, Abbott-VP, aca, Cobas-Bio, Hitachi® 705 and KDA. *2138*

Cefoxitin   At 5 times upper limit of therapeutic range on methods on SMAC®, Abbott-VP, aca, Cobas-Bio, Hitachi® 705 and KDA. *2138*

Cephalothin   At 5 times upper limit of therapeutic range on methods on SMAC®, Abbott-VP, aca, Cobas-Bio, Hitachi® 705 and KDA. *2138*

Chlordiazepoxide   At 5 times upper limit of therapeutic range on methods on SMAC®, Abbott-VP, aca, Cobas-Bio, Hitachi® 705 and KDA. *2138*   At acute overdose concentration (5 mg/dL) on Technicon® SMAC® method. *3719*

Chlorpheniramine   At acute overdose concentration (20 mg/dL) on Technicon® SMAC® method. *3719*

Cocaine   At acute overdose concentration (2.5 mg/dL) on Technicon® SMAC® method. *3719*

Contact With Clot   If plastic disc used to separate for 48 h. *1863*

Cyclosporine   At concentration of 41.5 mmol/L (50 µg/L) on methods on Technicon® SMAC® II and Hitachi® 705. *1123*

Cysteine   At 5 times upper limit of therapeutic range on methods on KDA and Cobas-Bio. *2138*

Cystine   At 5 times upper limit of therapeutic range on methods on KDA and Cobas-Bio. *2138*

Deslanoside   No effect on SMA 12/60 method at 0.06 mg/dL. *2636*

Diazepam   At 5 times upper limit of therapeutic range on methods on SMAC®, Abbott-VP, aca, Cobas-Bio, Hitachi® 705 and KDA. *2138*   At acute overdose concentration (2.5 mg/dL) on Technicon® SMAC® method. *3719*

Diclofenac   On continuous method at 10 times maximal therapeutic concentration. *1785*

Diethylpropion   At acute overdose concentration (10 mg/dL) on Technicon® SMAC® method. *3719*

Digitoxin   No effect at 21 mg/dL on SMA 12/60 method. *2636*

Digoxin   No effect at 0.04 mg/dL on SMA 12/60 method. *2636*

Diphenhydramine   At acute overdose concentration (20 mg/dL) on Technicon® SMAC® method. *3719*

Disulphine Blue   With p-nitrophenylphosphate as substrate. *1464*

Ethacrynic Acid   At 2.5 mg/dL no effect on SMA 12/60 method. *2636*

Ethanol   At acute overdose concentration (20 mg/dL) on Technicon® SMAC® methods. *3719*

Ethchlorvynol   At acute overdose concentration (20 mg/dL) on Technicon® method. *3719*

Ethinamate   At acute overdose concentration (20 mg/dL) on Technicon® SMAC® method. *3719*

Fluosol-DA   On SMA IIC at concentration of 50%. On Du Pont aca III at concentration of 50%. *2518*

Flurazepam   At 5 times upper limit of therapeutic range on methods on SMAC®, Abbott-VP, Cobas-Bio, aca, Hitachi® 705 and KDA. *2138*   At acute overdose concentration (2.5 mg/dL) on Technicon® SMAC® method. *3719*

Furosemide   No effect at 1.4 mg/dL on SMA 12/60 method. *2636*

Glutethimide   At acute overdose concentration (5 mg/dL) on Technicon® SMAC® method. *3719*

Heparin   Little or no effect observed. *2228*

Ibuprofen   At 5 times upper limit of therapeutic range on methods on Abbott-VP, aca, Cobas-Bio, Hitachi® 705 and KDA. *2138*   On continuous method at 10 times maximal therapeutic concentration. *1785*

Indomethacin   On continuous method. *1785*

Ketoprofen   On continuous method at 10 times maximal therapeutic concentration. *1785*

Lidocaine   No effect SMA 12/60 method with 3.5 mg/dL. *2636*

Lipemia   In patient specimens measured on SMAC® in comparison with specimens following ultra-centrifugation. *2466*

Liposol   Changed from 205.3 to 223.3 U/L in 26 randomly selected patient sera, before and after addition on Beckman Astra methods. *2054*

Mannitol   No effect on SMA 12/60 method at 445 mg/dL. *2636*

Meals   If nitrophenol used as substrate. *3434*

Meperidine   At acute overdose concentration (5 mg/dL) on Technicon® SMAC® method. *3719*

Meprobamate   At acute overdose concentration (20 mg/dL) on Technicon® SMAC® method. *3719*

Meralluride   No effect at 0.07 mL/dL on SMA 12/60 method. *2636*

Mesoridazine   At acute overdose concentration (20 mg/dL) on Technicon® SMAC® method. *3719*

Methadone   At acute overdose concentration (2.5 mg/dL) on Technicon® SMAC® method. *3719*

Methanol   At acute overdose concentration (20 mg/dL) on Technicon® SMAC® method. *3719*

Methaqualone   At acute overdose concentration (2.5 mg/dL) on Technicon® SMAC® method. *3719*

Methicillin   At 5 times upper limit of therapeutic range on methods on SMAC®, Abbott-VP, aca, Cobas-Bio, Hitachi® 705 and KDA. *2138*

Methotrexate   At 5 times upper limit of therapeutic range on method on Abbott-VP, aca, Cobas-Bio, Hitachi® 705 and KDA. *2138*

Methylphenidate   At acute overdose concentration (20 mg/dL) on Technicon® SMAC® method. *3719*

Methyprylon   At acute overdose concentration (10 mg/dL) on Technicon® SMAC® method. *3719*

Morphine   At acute overdose concentration (20 mg/dL) on Technicon® SMAC® method. *3719*

Naproxen   At 5 times upper limit of therapeutic range on methods on Abbott-VP, aca, Cobas-Bio, Hitachi® 705 and KDA. *2138*

Nitrofurantoin   On routine methods in use on Abbott-VP, Cobas-Bio, aca, Hitachi® 705, KDA at 5 times normal therapeutic concentration. *2138*

Nortriptyline   At acute overdose concentration (20 mg/dL) on Technicon® SMAC® method. *3719*

Ouabain   At 0.02 mg/dL no effect on SMA 12/60 method. *2636*

Oxalate   Not inhibited when used as anticoagulant. *3217*

Oxalate/Fluoride   Little or no effect observed. *2228*

Penicillin G   At 5 times upper limit of therapeutic range on methods on SMAC®, Abbott-VP, Cobas-Bio, aca, Hitachi® 705 and KDA. *2138*

Pentobarbital   At acute overdose concentration (20 mg/dL) on Technicon® SMAC® method. *3719*

Perphenazine   At acute overdose concentration (20 mg/dL) on Technicon® SMAC® method. *3719*

Phenobarbital   At acute overdose concentration (20 mg/dL) on Technicon® SMAC® method At acute overdose concentration (20 mg/dL) on Technicon® SMAC® method. *3719*   On continuous method. *1785*

Alkaline Phosphatase *(continued)*

Serum   No Effect   Analytical *(continued)*

Phenytoin   No effect at 1.8 mg/dL on SMA 12/60 method. *2636* At acute overdose concentration (20 mg/dL) on Technicon® SMAC® method. *3719*

Procainamide   No effect on SMA 12/60 method at 35 mg/dL. *2636*

Promethazine   At acute overdose concentration (20 mg/dL) on Technicon® SMAC® method. *3719*

Propoxyphene   At acute overdose concentration (2.5 mg/dL) on Technicon® SMAC® method. *3719*

Propranolol   No effect at 0.1 mg/dL on SMA 12/60 method. *2636*

Pyribenzamine®   At acute overdose concentration (20 mg/dL) on Technicon® SMAC® methods. *3719*

Quinidine   At 26 mg/dL no effect on SMA 12/60 method. *2636* At acute overdose concentration (20 mg/dL) Technicon® SMAC® method. *3719*

Quinine   At acute overdose concentration (1.5 mg/dL) on Technicon® SMAC® method. *3719*

Reserpine   No effect at 0.02 mg/dL on SMA 12/60 method. *2636*

Rifampin   On method involving hydrolysis of p-nitrophenyl phosphate. *0218* At 5 times upper limit of therapeutic range on methods on SMAC®, Cobas-Bio, Abbott-VP, aca, Hitachi® 705 and KDA. *2138*

Secobarbital   At acute overdose concentration (20 mg/dL) on Technicon® SMAC® method. *3719*

Storage of Sample   8 h room temperature, 7 d 4°, 180 d -20°. *3217* No effect 8 h at room temperature, 1 week at -20°. *3873* For 7 d at 25 or 4°. *0877*

Tetracycline   At 5 times upper limit of therapeutic range on methods on SMAC®, Abbott-VP, Cobas-Bio, aca, Hitachi® 705 and KDA. *2138*

Theophylline   At 5 times upper limit of therapeutic range on methods on Abbott-VP and aca. *2138*

Thiopental   At acute overdose concentration (20 mg/dL) on Technicon® SMAC® method. *3719*

Vitamin Preparations   No effect at expected concentration with SMA 12/60 procedure. *2637*

Serum   No Effect   Physiological

Amiodarone   No effect observed in 100 patients. *1509*

Aspirin   When approximately 3 g/d ingested for several weeks. *3076*

Bran   No significant effect with 38 g/d for 5 weeks. *1540*

Carbamazepine   Observed 2 times upper limit of normal in none of 36 adult inpatient epileptics (dose and duration of treatment unknown). *1162*

Chenodeoxycholic Acid   After 3 mo in patients with gallstones. *0274*

Cyclophosphamide   No significant perturbation in 41 patients. *2406*

Danazol   No abnormalities observed in patients taking drug. *3906*

Desipramine   In 42 children and adolescents treated for up to 24 mo. *1624*

Ethanol   No significant difference between heavy and occasional drinkers. *3273* Regardless of amount consumed, acute ingestion had no effect on activity. *0890*

Heparin   No effect of subcutaneous drug after 8 d. *2454* No effect of surgery in this test group or control group. *2595* After low-dose heparin treatment. *2671*

Ketoconazole   In 9 healthy men given up to 1200 mg/d for 1 week. *1300*

Levonorgestrel   No significant difference over 3 y between implant recipients and IUD users. *0763* In a group of women using levonorgestrel covered rods versus controls using copper IUD. *0899*

Light   Sunlight no effect in elderly. *1615*

Methadone   No toxicity if liver function tests normal initially. *2012*

Methotrexate   No effect in long-term treatment of 20 patients with psoriasis noted. *1556* In 20 patients with psoriasis on long-term therapy. *1556*

Muscular Exercise   Insignificant effect with 12 minutes on cycle-ergometer. *1231*

Penicillamine   In rheumatoid arthritis patients when treated for 6 mo. Changes not related to activity of disease. *1794* But decreased in granulocytes related to transient changes in zinc metabolism. *1794*

Phenobarbital   No effect observed in 14 adult epileptic inpatients given drug alone (dose and duration of treatment alone). *1162*

Prednisone   No consistent change noted in 14 hospitalized adult patients when given 40-50 mg/d. *3790*

Serum   Increase   Analytical

Acetaminophen   On continuous method at 10 times maximal therapeutic concentration. *1785*

Bilirubin   Forms diazonium salt with alpha-naphthyl phosphate. *0251*

Cefotaxime   Consistent slight increase on American Monitor Parallel method by metabolite. *0201*

Fluorescein   Slight effect at 200 mg/L on Ektachem® 700 method. *1722*

Hemoglobin   Linear increase of 2 U/L per 0.1 g/dL method of Proksch. *2878*

Hemolysis   RBC concentration 6 x serum so effect minimal. *1563*

Ibuprofen   At 5 times upper limit of therapeutic range on method on SMAC®. *2138*

Magnesium Salts   Activators of enzyme in laboratory procedures. *3588*

Manganese Salts   Activators of enzyme in laboratory procedures. *3588*

Meals   If phenylphosphate used in substrate. *3434*

Methotrexate   At 5 times upper limit of therapeutic range on method on SMAC®. *2138*

Naproxen   At 5 times upper limit of therapeutic range on methods on SMAC®. *2138*

Nitrofurantoin   19% increase on p-nitrophenyl phosphate method on SMAC® at 5 times normal therapeutic concentration. *2138*

Pindolol   At 5 mg/L conventional methods when added to serum. *1758*

Standing of Sample   Significant if on polystyrene beads 24 h at 25°. *2624*

Storage of Sample   Up to 2.7% in 1 d room temperature, 5% at 4°, 5% frozen. *2332* May increase 5-10% after 1 d at 4°. *3873*

Sulfobromophthalein   If final color developed in alkaline solution. *0583* Interferes with Bessey procedure. *0607*

Serum   Increase   Physiological

Acetaminophen   In severe cases of poisoning (also in some mild). *3470* With overdose and centrolobular hepatic necrosis usually associated with high drug concentration. *1771*

Acetohexamide   May cause intrahepatic cholestatic jaundice. *0652*

Acetophenazine   Cholestatic hepatitis with obstruction. *2313*

Aging   Peaks at puberty then steep fall. *0887*

Ajmaline   Intrahepatic cholestasis. *0268* Hepatotoxicity and intrahepatic cholestasis reported. *3215* In 4 patients with centrilobular cholestasis and mild hepatocytic lesions: recovery with drug withdrawal. *2729*

Albumin   Placental alkaline phosphate if from Pitman-Moore or Parke-Davis. *1488*

Allopurinol   Reversible clinical hepatotoxicity noted. *0652* Occasional granulomatous hepatitis as well as massive hepatic necrosis and severe hepatitis. *2389* In 1 of 73 patients treated for 6 mo. *3565*

Aluminum Hydroxide   Mild effect with drug associated osteomalacia. *1054* Observed in one patient secondary to ingestion of large amounts of compound for long time. *1313*

Amantadine   Mild transient increase, other liver function tests normal. *3213*

Ambulation   Following prolonged bed rest (?increase bone formation). *1688*

Aminoglutethimide   Casual association between drug administration and cholestasis in 2 cases. *2779*

Aminosalicylic Acid   Reversible cholestasis may occur. *1713*

Amiodarone Up to 1.5 to 3 times normal values in 55% patients with hepatitis occurring in 4%. *2011* Asymptomatic change in liver function with enzyme activity up to 3 times normal. *2478* In 4 of 5 patients with toxic and hypersensitivity liver injury. *2987*

Amitriptyline Rare cases of cholestasis (transient). *2482* Occasional case of hypersensitivity associated hepatitis. *0098*

Amodiaquine Reported hepatotoxicity. *2313* Abnormality usually mild but associated with an immunoallergic hepatotoxicity. *2081*

Amphotericin B Due to hepatocellular dysfunction. *2313* Noted in one patient with acute myelogenous leukemia when treated with high dose. Probable idiosyncratic response. *2443*

Anabolic Steroids Cholestatic syndrome. *1343*

Androgens May cause cholestatic syndrome. *1022*

Anticonvulsants Occurs in 24% children (90% bony origin). *1695* In 8 epileptic patients receiving phenobarbital with carbamazepine or phenytoin. *2670*

Antifungal Agents Hepatotoxicity may occur. *1237*

Antimony Compounds Hepatotoxic effect. *2220*

Aprindine Hepatitis is manifestation of chronic toxicity. *3215* Hepatitis in 2 patients within 3 weeks of start of therapy resolved with withdrawal of drug. *1570*

Arsenicals Hepatotoxic effect (cholestasis/cholangiolitis). *2220*

Asparaginase May cause hepatotoxicity (frequent). *1519* In 31% children and 47% adults reported in various studies. *2406* Increased in 30-35% of patients treated in different studies: effects usually mild. *0558* Hemorrhagic pancreatitis in fewer than 0.5% treated patients. *1398*

Aspidium May cause hepatic toxicity. *2313*

Aspirin Prolonged use may cause hepatic toxicity. *2111*

Azathioprine May be very high, liver damage or biliary stasis. *1680* Minimal cholestasis and reversible portal fibrosis observed in 8 patients: occasional hepatotoxicity. *0036* Canalicular cholestasis and centrilobular ballooning of hepatocytes in one patient. *0846*

Barbiturates Rare case of hepatotoxicity. *0357*

Benziodarone May cause hepatic toxicity. *2313*

Bismuth Salts Hepatotoxicity. *2313*

Bleomycin Rare mild abnormality reported: reversible. *2406*

Blood Group O secretor subjects — highest intestinal component. *3743*

Bromocriptine In 10 of 45 patients with parkinson's disease increased by up to 25% without other laboratory abnormalities. *2167*

Capreomycin Reported effect on liver. *1705*

Captopril Isolated case of cholestasis in patient receiving 25 mg tid for 1 mo. *2732* Characteristic pattern of hepatocellular jaundice in one patient. But with secondary cholestatic elements. *3688*

Carbamazepine Cholestatic and hepatocellular damage. *2907* 4 of 21 patients with epilepsy (19%) treated with drug only for average of 40 mo. *1625* In about 20 epileptic patients treated for 2 y. *0873* Granulomatous hepatitis observed in small proportion of patients treated for less than 1 mo; resolved within 3 d of cessation of treatment. *2153* Osteomalacia observed in 3 of 31 patients given average of 758 mg/d for average of 20.5 mo. *2635*

Carbenicillin Transient elevations reported. *1343* In 4 of 27 patients given 270 mg/kg i.v. for 6 d. *3275*

Carbimazole Isolated case of jaundice conclusively linked to drug. *0911* Reversible hypersensitivity response to drug. *1790*

Carbutamide Hepatotoxicity. *2313*

Carmustine (BCNU) Usually mild and return to normal over few days. *2406*

Cefamandole Fewer than 3 of 53 patients developed reversible hepatotoxicity. *0969*

Cefazolin Reversible hepatotoxicity noted in some patients. *0969* In 4 of 31 treated cases: transient 2 to 3 fold increase. *2273*

Cefoxitin Transient 2 to 3-fold increase in 3 of 31 patients following colorectal surgery. *2273*

Ceftizoxime In 4 of 110 patients: reversible hepatotoxicity noted. *0969* In 3 of 30 patients following colorectal surgery had transient 2 to 3-fold increase. *2273*

Cephaloridine Mechanism not listed. *1488*

Cephalosporin Observed with all cephalosporins at frequency of from 1 to 7%. *2615*

Chenodeoxycholic Acid Mild transient elevation observed. *0808* In 7-2% in 916 patients given up to 750 mg/d for 2 y. *3199*

Chloramphenicol ?hepatotoxic-cholestatic effect. *2220*

Chlordiazepoxide Infrequent cholestatic effect. *2220*

Chlorinated Insecticides Presumed damage of liver cells. *0405*

Chlormezanone Occasional reversible cholestasis. *2313*

Chloroform Hepatotoxic effect with necrosis. *2313*

Chlorothiazide May cause cholestatic jaundice. *2313*

Chlorpromazine Hepatic sensitivity to drug (in up to 2% patients). *3283* In one patient who developed cholestasis with fever and leukopenia after 260 mg drug. *2840*

Chlorpropamide Infrequent intrahepatic cholestasis. *1713* Cholestatic jaundice observed in one patient in association with red cell aplasia (prevalence of jaundice as high as 0.5%). *1284*

Chlorprothixene Hepatotoxicity (reversible) cholestatic effect. *2313*

Chlortetracycline Hepatotoxic effect with centrolobular necrosis. *1237*

Chlorzoxazone May cause hepatotoxicity. *2313*

Cimetidine Mild increase reversible cholestatic jaundice in a few children. *2172* In isolated case treated for 7 mo with 400 mg/d progressed to bridging hepatic necrosis. *3085*

Cinchophen May cause intrahepatic cholestasis. *2313*

Ciprofloxacin Rare cases reported during clinical trials. *0151*

Clindamycin Transient abnormality noted. *3653* Uncommon side effect. *3864*

Clofibrate Hepatotoxic effect. *2313* Single case of granulomatous hepatitis associated with drug administration. *2813*

Clonidine Single probable case of toxic hepatitis. *1277*

Colchicine Possible hepatotoxic effect. *2220*

Cough Medicines One case chronic use terpin and codeine. *1069*

Coumarin May cause hepatotoxicity. *3835*

Cyclofenil In 1 case, but total of 30 cases reported up to 1980; Considered to be metabolic idiosyncrasy and reversible in all cases. *2672*

Cyclophosphamide May cause hepatotoxicity. *2313*

Cyclopropane May cause hepatotoxicity (lasts several days). *2313*

Cycloserine May cause hepatotoxicity. *2313*

Cyclosporine Persistent increase reported in some renal transplant patients occurring within of start of treatment, possibly due to cholestasis. *0683* Increase to 124 U/L on average as manifestation of hepatotoxicity in 18 bone-marrow transplant recipients. *0174* Persistent increase observed in some renal transplant patients: speculation that it is due to increased sensitivity of bone to action of parathyroid hormone. *2201* Slight increase or normal in association with hepatotoxicity in renal transplant patients. *1959*

Cyproheptadine Isolated case of jaundice due to cholestasis: reversed on withdrawal of drug. *1562*

Cytarabine In two cases with acute leukemia: unlikely to have been due to other drugs co-administered: effects slight. *2827* At 3 g/m² every 12 h for 6 d, some mild and reversible abnormalities in about 75% of 12 cases. *1579* Slight effect and returned to normal while still under treatment. *2406*

Dantrolene Observed effect, cause uncertain. *0659* Acute hepatitis as result of drug ingestion in 4 patients returned to normal after drug stopped. *3837* In 1.8% of 1044 patients, hepatocellular damage with acute or subacute hepatic disease or chronic active hepatitis. *3656*

Dapsone In 3 individuals treated for 3 to 36 weeks. *3478* Associated with hemolytic anemia when occurs as main side effect. *3423* Sulfone syndrome in one 16 year old girl given 50 mg daily short-term administration. *3604*

Desipramine Transient cholestasis (rare). *2313* Significant effect in 46 patients but values did not rise above reference range. *2871*

Diazepam Mild effect reported in one patient. *1878* Isolated case of drug induced hepatitis. *3562*

Disopyramide Occasional case of cholestatic jaundice reported. *3215*

Alkaline Phosphatase *(continued)*

Serum Increase Physiological *(continued)*

Disulfiram One possible case of cholestasis reported. *1680* Reversible toxic liver damage in nonalcoholic woman. *2017* In 6 patients: proved to be drug associated in one by challenge test. *2912* Observed in one patient and similar results followed challenge test. *2491*

Doxorubicin May produce idiosyncratic hepatic dysfunction, as observed in 6 patients. *0184*

Ectylurea May cause cholestasis with cholangiolitis. *2313*

Enflurane Evidence of hepatocellular damage; characteristically centrilobular necrosis: mechanism of injury most probably metabolic idiosyncrasy. *2157*

Erythromycin May cause intrahepatic cholestasis (reversible). *1142* Moderate increase with hepatitis mixed in type with both cholestasis and mild necrosis. Ethylsuccinate and propionate derivatives produce similar jaundice. *2792* Observed to some extent in some patients with different erythromycin salts: usually cholestatic hepatitis. *1895*

Estradiol Occurs in up to 50% cases. *2508*

Estriol Possibly related to impaired hepatic excretion. *1498*

Estrogens Transient increases 2 to 50% reported. *2086*

Estrone Occurs in up to 50% cases. *2508*

Ethanol Ratio of GGT to alkaline phosphatase increased in patients with alcoholic liver disease. *2059* Increased in alcoholic liver disease. *0625*

Ether Hepatic disturbance (transient). *3218*

Ethionamide Hepatotoxic effect. *2313*

Ethotoin Probable idiosyncratic hepatitis. *2313*

Ethoxazene Hepatotoxic effect. *2313*

Ethylphenacemide Frequently occurs. *1343*

Etoposide Hepatic toxicity reported in 3% treated patients. *0141*

Etretinate In one 74-year old woman treated for severe psoriasis: normalized with withdrawal of therapy and recurred when reinstituted. *3789*

Factor IX Complex, Human 62% of cases developed hepatitis after use. *2663*

Fansidar® Biopsy-proven granulomatous hepatitis due to sulfadoxine moiety of drug reported in two individuals receiving drug prophylactically possibly due to sulfonamide hypersensitivity. *2099*

Fat Emulsions Prolonged infusion may impair liver function. *0120*

Fatty Foods Effect most marked 4 h after meal in O secretors. *3743*

Flecainide Mild increases in activity reported. *2011*

Florantyrone Hepatotoxic effect. *2313*

Floxuridine Hepatotoxic manifestation. *1680*

Flucloxacillin In a single patient on hemodialysis: abnormal results only mild. *2200*

Flucytosine Reversible hepatotoxicity in 10%. *3443* Occurs in about 25% treated patients: usually mild but necessitates discontinuation of therapy. *3567*

Fluoxymesterone Intrahepatic cholestatic jaundice. *1343*

Fluphenazine Hypersensitivity response. *2313* Isolated case of hepatotoxicity with centrilobular cholestasis. *3372*

Flurazepam May cause hepatic toxicity. *3016*

Fusidic Acid Reported in 6 patients in UK, although causal relationship could not be established with certainty. *3535* Jaundice developed in 34% of 112 patients given drug, highest when drug given intravenously. Jaundice resolved when drug withdrawn. *1690*

Gentamicin Hepatotoxic effect. *1343*

Glibenclamide Intrahepatic cholestasis described in 61 y old diabetic. *3894*

Glyburide Rare and reverts to normal if therapy continued. *0125*

Glycopyrrolate Hepatotoxicity. *2313*

Gold Rare side effect with liver damage (reversible). *0085* Cholestatic jaundice in 3 patients with rheumatoid arthritis given gold sodium thiomalate: all patients recovered spontaneously. *1084* In 3 patients with rheumatoid arthritis during chrysotherapy. *0993*

Griseofulvin May be hepatotoxic. *2220*

Guanoxan May cause hepatic toxicity. *2313*

Haloperidol May cause hepatocellular changes. *2313* In isolated case producing cholestatic liver disease: typical frequency of liver disease is 0.2%. *0910*

Halothane Reversible depression of liver function. *3294* In one case after repeated exposure to anesthetic: evidence of acute hepatitis with bridging necrosis and piecemeal necrosis. *2027* Granulomatous hepatitis in single case following second exposure to anesthetic. *3266* In one nurse occupationally exposed to gas. *1900* Low frequency but greater number of abnormalities than with enflurane. Occurred most frequently in obese patients. *1089*

Height Positively correlated in children. *0887*

Heroin Normal usually unless alcoholism also. *3471*

Hydantoin Derivatives Severe delayed hypersensitivity reported. *2427*

Hydralazine Isolated case with moderate hepatomegaly, abnormal findings resolved when drug discontinued. *1161* Cholestatic jaundice observed in one patient. *3469* Asymptomatic or symptomatic reversible hepatitis-like reaction or granulomatous hepatitis. *2374* Centrilobular necrosis observed in 3 patients 2 to 4 months after drug therapy for hypertension. *1740*

Hydrazine Derivatives May cause hepatotoxicity. *2313*

Hydroflumethiazide May cause intrahepatic cholestasis. *1680*

Hydroxyacetamide Hepatotoxicity. *1022*

Ibufenac Hepatotoxic effect. *2313*

Ibuprofen Rare case of hepatotoxicity: in one case associated with Stevens-Johnson syndrome. *3460*

Icterogenin Causes intrahepatic cholestasis. *0153*

Idoxuridine Cholestatic jaundice reported in one case. *0837*

Ifosfamide Manifestation of hepatotoxicity. *3675*

Imipramine May cause cholestatic jaundice. *1873* Either due to hepatotoxicity or hypersensitivity in 33 year old woman. *2501*

Indomethacin Cytotoxic and cholestatic liver damage. *1908*

Iprindole Mild elevation with cholestasis. *0038*

Iproniazid May cause cholestatic jaundice. *2313*

Isocarboxazid Cholestatic effect. *2313*

Isoniazid Probable cholestatic effect. *2220* 15 of 89 patients developed significant liver disease, typically hepatitis, although one developed cholestasis, in patients taking drug for at least 2 mo. *0901*

Isotretinoin In 7 patients with severe rosacea treated with 1 mg/kg/d for 12 weeks. Effects possibly due to induction of hepatic microsomal enzymes. *2307* In 10 of 523 patients with mean daily dose of 109 mg for 150 d. *3867*

Kanamycin May cause hepatotoxicity. *2313*

Ketamine In 14 of 34 individuals who had drug as anesthetic for intermediate operations. *0973*

Ketoconazole Delayed reaction to drug after withdrawal in single case of chronic candidiasis in 61 y old woman. *3513* Rapidly developing liver failure in 67 year old woman taking 200 mg drug daily for 2 mo. *0960*

Levodopa Rare elevation reported. *0071*

Lincomycin Hepatotoxic-cholestatic effect. *1022*

Lithium Effect in 20% manic depresses treated for minimum of 20 mo. Markedly increased effect on bone isoenzyme observed in 66% manic depressives treated for minimum of 20 mo. *0488* Significant increase noted in drug treated population and bone isoenzyme increased in 27 of 41 such patients. In 19 patients bone isoenzyme increased although total normal. *0488*

MAO Inhibitors Cholestatic effect. *2220*

Meals Affected by meals (most marked in women). *3455* Fatty meals especially in O secretors. *3873*

Mechlorethamine Hepatotoxic effect. *0248*

Menopause Change of 2.1 King-Armstrong units from 5th to 6th decade. *3830*

Mepazine May alter liver function (cholestasis). *2313*

Mephenytoin Hepatotoxicity. *2313*

Meprobamate May cause cholestatic (hepatocanalicular) jaundice. *0248*

Mercaptopurine Hepatotoxicity (centrolobular necrosis). *2313* Intrahepatic cholestasis observed in several patients. *2406*

Metahexamide Hepatotoxicity (viral hepatitis type). *2313*

Metaxalone Hepatotoxic effect. *2313*

Methandriol Intrahepatic cholestatic jaundice. *1343*

Methandrostenolone   Intrahepatic cholestatic jaundice. *1343*

Methimazole   May affect liver function (cholestasis). *2313*   Rare case of cholestatic jaundice reported. *1774*

Methotrexate   Hepatotoxic effect (seen in 5% cases psoriasis). *2608*   In psoriatic patients receiving drug in comparison with topically treated and controls. *0364*

Methoxsalen   Hepatotoxic effect. *2313*

Methoxyflurane   Hepatic toxicity. *3294*   In 2 pregnant women when used as analgesia for labor: hepatitis observed. *0872*

Methyldopa   Intrahepatic cholestatic jaundice. *1023*   Disturbances in liver function in 5 to 35% of patients treated for hypertension. Hepatocellular injury may occur after short or long-term exposure. *0162*   Drug induced hepatitis with severe, chronic, aggressive inflammation. *0211*

Methyltestosterone   Cholestatic effect. *1538*   In 1 of 60 patients (female transsexuals or impotent males) receiving 150 mg day for long term. *3805*

Methylthiouracil   Hepatotoxic effect. *2313*

Minocycline   Conceivable complication of therapy. *0135*

Mithramycin   Noted in approximately 50% of treated patients. *2406*

Mitotane   Possible hepatotoxic effect (reported in dogs,rats). *1680*

Morphine   Associated with abnormal liver function. *2313*

Moxalactam   In about 3% patients as reported from several studies. *0592*

Nalidixic Acid   May cause cholestatic jaundice. *2808*

Nandrolone   Reported to affect liver function. *2220*

Niacin   May cause impairment of hepatic function. *2726*

Niacinamide   Rare probable reversible toxic effect. *3875*

Nialamide   Probable hypersensitive hepatitis. *0071*

Nitrofurans   Hepatotoxicity. *2313*

Nitrofurantoin   May cause cholestatic jaundice. *2808*   Moderate increase chronic active hepatitis, much more common in women than men, but still rare. *2444*   Moderate increase chronic active hepatitis, much more common in women than men, but still rare. *2444*   May follow acute hepatitis with or without cholestasis or as chronic active hepatitis or chronic granulomatous reaction. *2374*

Nitrophenol   Hypersensitive intrahepatic cholestasis. *1434*

Norethandrolone   Cholestasis produced without cholangiolitis. *1538*

Norethandrostenolone   Intrahepatic cholestatic jaundice. *3197*

Norethindrone   Intrahepatic cholestatic jaundice. *1343*

Norethisterone   In 5.6% patients being treated for advanced or recurrent breast cancer. *2073*

Norethynodrel   Due to cholestasis. *1778*

Nortriptyline   May cause cholestatic jaundice. *0071*

Novobiocin   Intrahepatic cholestatic jaundice can occur. *1343*

Oleandomycin   Cholestatic jaundice reported. *0070*

Oral Contraceptives   May cause cholestasis, rare hepatocellular degeneration. *1713*

Organophosphorus Insecticides   Hepatotoxic effect. *0405*

Oxacillin   Cholestatic jaundice reported in one case. *1022*   Asymptomatic hepatic dysfunction in 5 cases following high dose treatment: reversible. *2835*   Reversible high-dose associated liver injury. *2431*

Oxazepam   Possible hepatotoxicity. *2313*

Oxymetholone   Cholestatic effect. *1713*   Cholestatic hepatitis developed in two patients. *3921*   One case of peliosis hepatitis reported. *2539*

Oxyphenbutazone   Possible hepatotoxicity. *2313*

Oxyphenisatin   Hepatic toxicity if over prolonged period. *2757*   Observed with chronic active hepatitis induced by drug. *0903*

Papaverine   Reversible hepatotoxic effect. *3039*   In 2 patients: fever anorexia and jaundice 4 to 5 weeks after start of treatment due to pericholangitis: reversible. *3371*   In 6 of 14 patients abnormal liver function, possibly due to allergic response and to metabolic aberration in host. *2746*

Paraldehyde   Possible hepatotoxicity. *2313*

Paramethadione   Possible hepatotoxicity. *2313*

Pargyline   Possible hepatotoxicity (reversible). *2313*

Penicillamine   Possible hepatotoxicity. *1022*   In 6 of 99 patients treated for rheumatoid arthritis: evidence of toxic liver necrosis observed in two cases. *3887*   Case of acute cholestatic jaundice in patient with systemic lupus erythematosus. *3242*

Perhexilene   Transient elevation observed. *1619*

Phenacemide   May affect liver function in about 2% cases. *2313*

Phenazopyridine   Hepatotoxic effect. *2313*

Phenindione   May modify liver function (cholestasis). *2313*

Pheniprazine   May cause hepatocellular jaundice. *2313*

Phenobarbital   May cause osteomalacia (?also liver effect). *1396*   Increased activity occurs early in epileptic children. *2162*

Phenothiazines   Intrahepatic cholestatic syndrome. *1563*   Liver damage observed after different phenothiazines administered in succession. *1916*

Phenoxypropazine   Produced fatal hepatotoxicity. *2427*

Phenprocoumon   Mild effect but recurrently in 2 patients having repeated exposure to drug. *3347*

Phenylbutazone   Hepatitis may occur as complication of therapy. *2534*   Observed in 7 patients with side effects due to drug. *3409*   In 2 patients subsided after drug withdrawn associated with granulomas and cholestatic-hepatocellular injury. *1730*   Observed in one patient: clear evidence of intrahepatic cholestasis. *3670*   Frequent hypersensitivity reaction but also with hepatotoxic potential: may be hepatocellular injury or systemic vasculitis. *0290*

Phenytoin   Hepatotoxicity with centrolobular necrosis. *2313*   Observed 2 times upper limit of normal in 6 of 112 adult epileptic in patients. *1162*   In 42% of 60 epileptics treated for 10 y or more. *2493*

Phosphorus   Hepatotoxic effect with necrosis. *2313*

Piperacillin   In 1 of 29 patients given 181 mg/kg i.v. for 6 d. *3275*

Piroxicam   Occasional case of liver damage reported. *2824*

Polythiazide   May affect liver function. *2313*

Practolol   After 4 y treatment 2 cases of primary biliary cirrhosis. *0496*

Pregnancy   2 to 3 times normal activity in third trimester. *3588*

Probenecid   Hepatotoxic effect (centrolobular necrosis). *2313*

Procainamide   Reversible hepatic toxicity reported. *1022*

Prochlorperazine   Cholestatic effect. *0071*

Progesterone   Hepatotoxicity. *2313*

Promazine   May cause cholestatic (hepatocanalicular) jaundice. *0248*

Promethazine   May cause cholestasis. *2313*

Propoxyphene   Hepatic toxicity (cholestatic hepatitis). *1954*   In 3 patients within 10 d of start of drug treatment. *1156*   Hepatotoxic response observed in 2 patients. *2110*

Propranolol   Low incidence drug induced increase. *2427*

Propylthiouracil   Hepatotoxic effect (centrolobular necrosis). *2313*   Temporary effect associated with drug administration regressed with cessation. *2731*   In one 24 year old woman who developed fulminant hepatic failure with lymphocyte sensitization. *2437*   Drug induced liver damage: either acute or chronic active hepatitis. *3787*

Protein Hydrolysate   Transient effect hyperalimentation in infants. *3614*

Protionamide   Temporary side effect. *2427*

Protriptyline   Transient reversible cholestasis. *0071*

Pyrazinamide   Hepatotoxic effect (viral-hepatitis like). *2313*   Hepatotoxic in 1 to 5% patients (previously reported to be toxic in as many as 25%). *3683*

Quinacrine   Hepatotoxic effect. *2313*

Quinethazone   May cause cholestatic jaundice. *2313*

Quinidine   May cause mild hepatic impairment in few patients. *2531*   In single case of severe quinidine hypersensitivity. *0858*   Hypersensitivity observed in 32 of 487 patients who had received drug. *1261*   Isolated case of reversible granulomatous hepatitis from quinidine hypersensitivity and granuloma induction within 3 d of quinidine readministration. *0448*

Rifampin   Can be serious but reversible liver damage. *2806*   Minimal abnormalities of liver function are common: severe liver damage in about 0.6% patients. *3683*   Observed in one patient receiving drug who developed porphyria cutanea tarda. *2441*

Sex Difference   In men up to and including 5th decade. *2130*

Spectinomycin   Mechanism not discussed. *0122*

Alkaline Phosphatase *(continued)*

Serum   Increase   Physiological *(continued)*

Stanozolol   May cause intrahepatic cholestasis. *1680*

Streptokinase   Substantial not as marked as GGT. *3189*   Transient dysfunction of liver, returning to normal when treatment discontinued. *3122*

Sulfadiazine   May cause intrahepatic cholestasis. *1434*

Sulfadimethoxine   Granulomatous reaction in liver. *1049*

Sulfamethizole   Reversible cholestasis. *1974*

Sulfamethoxazole   Cholestatic jaundice may occur. *1974*   Occasional case of cholestasis with or without hepatic necrosis. *3452*   When given with trimethoprim, fever and malaise: successful response to treatment. *1969*

Sulfamethoxypyridine   Reversible cholestasis. *2313*

Sulfanilamide   May cause reversible cholestasis. *2220*

Sulfapyridine   May cause reversible cholestasis. *2220*

Sulfasalazine   Occasional case of drug-induced toxic hepatitis. *1422*   Rare hepatotoxicity reported associated with noncaseating granuloma or focal inflammation with or without necrosis. *1127*   In 2 cases with drug induced hepatotoxicity of hypersensitivity type of reaction: reversed on drug withdrawal. *3364*   As part of systemic hypersensitivity reaction in woman with inflammatory bowel disease. *3402*

Sulfisoxazole   Reversible cholestasis. *2313*

Sulfonamides   Hypersensitivity reaction with cholestasis. *1199*   Hepatitis-like reaction in patients with multiple episodes of jaundice after repeated exposure to drug. *1741*

Sulfones   May affect liver function. *2313*

Sulfonylureas   Elevation with cholangiolitis. *3171*

Sulindac   Isolated case of cholestatic jaundice 12 d after therapy started. *2371*

Testosterone   May cause cholestasis. *2313*

Tetracycline   May occur with cholestasis. *1488*   2 cases of drug-associated fatty liver of pregnancy. *3795*

Thiabendazole   Rare cholestasis. *1778*   One case of drug-induced cholestasis reported. *2962*

Thiacetazone   May cause cholestatic (hepatocanalicular) jaundice. *0248*

Thiazides   Cholestatic hepatitis reported. *1343*

Thioguanine   May cause cholestasis. *2313*   Transient increase in 3 of 29 patients with polycythemia rubra vera. *2450*

Thiopental   1 case out of 24 3-5 d after anesthesia, more longer period after anesthesia pyrexia and jaundice in one case after reported anesthesia. *0391*

Thioridazine   Hypersensitivity reaction. *1857*

Thiosemicarbazones   May affect liver function. *2313*

Thiothixene   Hepatotoxic effect (reversible cholestasis). *1022*

Thiouracil   May cause intrahepatic cholestasis. *1434*

Thorium Dioxide   Very common (related to time since injection). *3147*

Tolazamide   Intrahepatic cholestatic jaundice. *0085*

Tolbutamide   May cause intrahepatic cholestatic syndrome. *0248*   Bone isoenzyme activity increased: inversely influenced by serum 25-hydroxyvitamin D concentration. *3458*

Tolmetin   Observed in single case of 49 y old man who had taken drug as needed for arthritis for 1 yr. *3421*

Tranylcypromine   May affect liver function (?hypersensitivity). *2313*

Tretinoin   In 1 of 11 patients given drug orally. *3867*

Trichlormethiazide   May rarely cause cholestasis. *1680*

Trifluoperazine   Cholestatic effect. *1713*

Trimethadione   Hepatotoxicity with centrolobular necrosis. *2313*

Trimethoprim   Transient increase in the absence of abnormal liver function tests in two patients when drug combined with sulfamethoxazole. *0893*

Trioxsalen   May affect liver function. *2313*

Uracil Mustard   May affect liver function. *2313*

Valproic Acid   Reported in children with coadministration only. *1781*   Acute hepatic centrilobular necrosis with severe fatty change in small number of patients. *3763*   In 1 of 9 patients with epilepsy poorly controlled. *3508*

Vitamin A   Observed with acute intoxication. *3867*

Vitamin D   May be affected in some cases. *0988*

Warfarin   Rare cases of intrahepatic cholestasis in patients with previous history of hypersensitivity reported. *2940*

Xylitol   Dose related hepatocellular damage. *3212*

Zimeldine   7 of 14 patients treated for more than 1 week demonstrated toxic syndrome ? immunological mechanism or related to high initial dose. *2075*   In 45% of 21 inpatients with endogenous depression given 225 mg daily for 4 weeks. *2082*   In several of 147 patients although 16 had initially high values. *3741*

Serum   Decrease   Analytical

Albumin   If prepared from venous blood. *0583*

Arsenicals   Arsenates are inhibitors of enzyme in lab procedures. *3588*

Beryllium Salts   Inhibitors of enzyme in laboratory procedures. *3588*

Citrates   25 mmol/L inhibits 50% activity. *3217*   Inhibitory effect observed. *2228*   Dilutional effect when compared against serum or heparin as anticoagulant. *3361*

Contact With Clot   Reported effect ?temperature or stability. *3879*

Cyanides   Inhibitors of enzyme in laboratory methods. *3588*

Cysteine   At 5 times upper limit of therapeutic range on methods on Abbott-VP, aca, and Hitachi® 705. *2138*

Cystine   At 5 times upper limit of therapeutic range on methods on SMAC®, Abbott-VP, aca, and Hitachi® 705. *2138*

Detergents   17-24% inhibition at 0.6-0.8 mg/ml. *3217*

EDTA   Inhibitory effect observed. *2228*   50 mmol/L is completely inhibitory. *3217*

Fluorides   Inhibitory action on enzyme. *1563*

Glycine   If used as buffer in lab procedure as complexes Mg. *3588*

Oxalate   Inhibition of enzyme in laboratory procedures. *1022*

Phosphates   Inhibits reaction if high concentration in substrate. *1022*

Plasma   Compared with serum on ABA-100, SMA 12/60 methods. *2228*

Storage of Sample   25% decrease after 6 mo at -10°. *1180*

Sulfhydryl Compounds   Inhibit enzyme activity in laboratory methods. *3588*

Theophylline   At 5 times upper limit of therapeutic range on methods on SMAC®, KDA, Cobas-Bio and Hitachi® 705. *2138*   Approximately 10% reduction in activity at drug concentration of 2 mg/dL. *3719*

Zinc   Inhibitors of enzyme in laboratory procedures. *3588*

Serum   Decrease   Physiological

Aldatense   Increase by 80% in a group of hypertensives compared with controls. *1611*

Azathioprine   Improves biliary excretion in biliary cirrhosis. *3063*

Clofibrate   Continuous reduction from 82 U/L to 52 U/L in 26 type IIa hyperlipoproteinemic patients = same response seen in type IV patients. *3169*   Reduction of bone isoenzyme when activity increased in patients on chronic hemodialysis. *1930*   Significant reduction in 27 patients, half of whom had hypertriglyceridemia, after 1 week. *1121*

Estradiol-17β   In post-menopausal women given different amounts of drug. *0654*

Ethinyl Estradiol   In post-menopausal women with 50 μg/d for 14 d. *3474*   Increase by 30% after 2 weeks in 5 patients with primary biliary cirrhosis. *1419*   Significant effect in postmenopausal women: less marked in perimenopausal women. *2311*

Fluorides   *In vivo* inhibition with doses for fluoridation. *1112*

Fluspirilene   Slight but observable trend. *0221*

Meals   Significant effect observed. *3455*

Oral Contraceptives   Slight decrease reported in one study. *2427*   Significant effect observed within 3 mo usually continuing for several y. *3170*   But effects less marked with low estrogen preparations. *1568*

Prednisone   Augmented response in cirrhotics when drug added to therapeutic regime. *0651*

β-Propiolactone   Increase by 8 U/L when plasma from BPL treated blood compared with plasma in 25 specimens. *0217*

Tamoxifen   Observed in nonresponders with breast cancer. *1122*
Trifluoperazine   Not very marked. *0221*

Plasma   No Effect   Analytical
Codeine   At acute overdose concentration (2.0 mg/dL) on Technicon® SMAC® method. *3719*

Urine   No Effect   Analytical
Storage of Sample   No effect 2 d at 4°, 1 week at -20°. *3873*

Urine   Increase   Physiological
Aspirin   Due to nephrotoxic effect of drug. *2897*   Marked increase observed even when as little as 3.5 g aspirin ingested daily. *2866*
Kanamycin   Due to nephrotoxic effect of drug. *2897*
Polymyxin   Due to nephrotoxic effect of drug. *2897*
Proteinuria   In renal disease with necrosis or altered GFR. *2897*
Radiographic Agents   May occur after i.v. pyelography or aortography. *0078*
Streptomycin   Due to nephrotoxic effect of drug. *2897*
Sulfonamides   Due to nephrotoxic effect of drug. *2897*

Urine   Decrease   Analytical
Ethionamide   Interference with determination procedure. *2897*
Nitrofurantoin   Interference with determination method. *2620*

White Blood Cells   No Effect   Analytical
Heparin   No effect observed. *0839*

White Blood Cells   Increase   Physiological
Oral Contraceptives   Observed effect. *1487*
Pregnancy   Occurs after third month. *0987*

White Blood Cells   Decrease   Analytical
EDTA   Reported in presence of low concentrations of metals. *1896*

White Blood Cells   Decrease   Physiological
Penicillamine   In rheumatoid arthritis patients when treated for 6 mo. Changes not related to activity of disease. *1794*

Alkaline Phosphatase Isoenzymes

Serum   Increase   Physiological
Phenytoin   Hepatic isoenzyme relatively greater increase than bone fraction in association with increased total enzyme activity (bone still largest single component). *1449*

Serum   Decrease   Analytical
Phosphastrate   False low intestinal component (phenylalanine inhibitor). *3845*

Alkaloids

Urine   No Effect   Analytical
Acetaminophen   With iodoplatinate of Frings TLC procedure at 10 mg/dL. With ninhydrin in Frings TLC procedure at 10 mg/dL. *1204*
Amitriptyline   With ninhydrin in Frings TLC procedure at 10 mg/dL. *1204*
Amobarbital   With ninhydrin in Frings TLC procedure at 10 mg/dL. With iodoplatinate of Frings TLC procedure at 10 mg/dL. *1204*
Amphetamine   With iodoplatinate of Frings TLC procedure at 5 mg/dL. *1204*
Benztropine   With ninhydrin in Frings TLC procedure at 1 mg/dL. *1204*
Butabarbital   With ninhydrin in Frings TLC procedure at 10 mg/dL. With iodoplatinate of Frings TLC procedure at 10 mg/dL. *1204*

Carisoprodol   With ninhydrin in Frings TLC procedure at 10 mg/dL. With iodoplatinate of Frings TLC procedure at 10 mg/dL. *1204*
Chloral Hydrate   With ninhydrin in Frings TLC procedure at 10 mg/dL. With iodoplatinate of Frings TLC procedure at 10 mg/dL. *1204*
Chlordiazepoxide   With ninhydrin in Frings TLC procedure at 10 mg/dL. *1204*
Chloroquine   With ninhydrin in Frings TLC procedure at 10 mg/dL. *1204*
Chlorpheniramine   With ninhydrin in Frings TLC procedure at 12 mg/dL. *1204*
Chlorpromazine   With ninhydrin in Frings TLC procedure at 10 mg/dL. *1204*
Clorazepate   With ninhydrin in Frings TLC procedure at 10 mg/dL. *1204*
Cocaine   With ninhydrin in Frings TLC procedure at 10 mg/dL. *1204*
Codeine   With ninhydrin in Frings TLC procedure at 10 mg/dL. *1204*
Diazepam   With ninhydrin on Frings TLC procedure at 10 mg/dL. *1204*
Diethylpropion   With ninhydrin on Frings TLC procedure at 25 mg/dL. *1204*
Diphenhydramine   With ninhydrin on Frings TLC procedure at 10 mg/dL. *1204*
Doxepin   With ninhydrin on Frings TLC procedure at 1 mg/dL. *1204*
Ephedrine   With iodoplatinate on Frings TLC procedure at 10 mg/dL. With ninhydrin on Frings TLC procedure at 1 mg/dL. *1204*
Ethchlorvynol   With iodoplatinate on Frings TLC procedure at 10 mg/dL. With ninhydrin on Frings TLC procedure at 10 mg/dL. *1204*
Ethinamate   With iodoplatinate on Frings TLC procedure at 10 mg/dL. With ninhydrin on Frings TLC procedure at 10 mg/dL. *1204*
Fluphenazine   With ninhydrin on Frings TLC procedure at 10 mg/dL. *1204*
Flurazepam   With ninhydrin on Frings TLC procedure at 1 mg/dL. *1204*
Glutethimide   With iodoplatinate on Frings TLC procedure at 10 mg/dL. With ninhydrin on Frings TLC procedure at 10 mg/dL. *1204*
Glycopyrrolate   With ninhydrin on Frings TLC procedure at 2 mg/dL. With iodoplatinate on Frings TLC procedure at 2 mg/dL. *1204*
Haloperidol   With ninhydrin on Frings TLC procedure at 1 mg/dL. *1204*
Hydromorphone   With ninhydrin on Frings TLC procedure at 2 mg/dL. *1204*
Imipramine   With ninhydrin on Frings TLC procedure at 10 mg/dL. *1204*
Mebutamate   With ninhydrin on Frings TLC procedure at 10 mg/dL With iodoplatinate of Frings TLC procedure at 10 mg/dL. *1204*
Meperidine   With ninhydrin on Frings TLC procedure at 10 mg/dL. *1204*
Mephenesin   With iodoplatinate of Frings TLC procedure at 10 mg/dL. With ninhydrin on Frings TLC procedure at 10 mg/dL. *1204*
Mephenesin Carbamate   With ninhydrin on Frings TLC procedure at 10 mg/dL. *1204*
Meprobamate   With ninhydrin on Frings TLC procedure at 10 mg/dL. With iodoplatinate of Frings TLC procedure at 10 mg/dL. *1204*
Mescaline   With iodoplatinate of Frings TLC procedure at 10 mg/dL. *1204*
Methadone   With ninhydrin on Frings TLC procedure at 10 mg/dL. *1204*
Methamphetamine   With iodoplatinate of Frings TLC procedure at 10 mg/dL. *1204*
Methapyrilene   With ninhydrin on Frings TLC procedure at 25 mg/dL. *1204*
Methaqualone   With ninhydrin on Frings TLC procedure at 10 mg/dL. *1204*
Methdilazine   With ninhydrin on Frings TLC procedure at 10 mg/dL. *1204*

Alkaloids (continued)

Urine   No Effect   Analytical (continued)

Methylphenidate   With ninhydrin on Frings TLC procedure at 1 mg/dL. *1204*

Methyprylon   With ninhydrin on Frings TLC procedure at 10 mg/dL. With iodoplatinate of Frings TLC procedure at 10 mg/dL. *1204*

Morphine   With ninhydrin on Frings TLC procedure at 10 mg/dL. *1204*

Oxazepam   With ninhydrin on Frings TLC procedure at 10 mg/dL. With iodoplatinate of Frings TLC procedure at 10 mg/dL. *1204*

Oxycodone   With ninhydrin on Frings TLC procedure at 5 mg/dL. *1204*

Oxymetazoline   With ninhydrin on Frings TLC procedure at 0.5 mg/dL. *1204*

Pentazocine   With ninhydrin on Frings TLC procedure at 10 mg/dL. *1204*

Pentobarbital   With ninhydrin on Frings TLC procedure at 10 mg/dL. With iodoplatinate of Frings TLC procedure at 10 mg/dL. *1204*

Perphenazine   With ninhydrin on Frings TLC procedure at 10 mg/dL. *1204*

Phenobarbital   With iodoplatinate of Frings TLC procedure at 10 mg/dL. With ninhydrin on Frings TLC procedure at 10 mg/dL. *1204*

Phenylephrine   With iodoplatinate of Frings TLC procedure at 10 mg/dL. With ninhydrin on Frings TLC procedure at 1 mg/dL. *1204*

Phenylpropanolamine   With iodoplatinate of Frings TLC procedure at 1 mg/dL. *1204*

Phenytoin   With ninhydrin on Frings TLC procedure at 10 mg/dL. With iodoplatinate on Frings TLC procedure at 10 mg/dL. *1204*

Procainamide   With ninhydrin on Frings TLC procedure at 10 mg/dL. With iodoplatinate of Frings TLC procedure at 10 mg/dL. *1204*

Prochlorperazine   With ninhydrin on Frings TLC procedure at 10 mg/dL. *1204*

Promazine   With ninhydrin on Frings TLC procedure at 10 mg/dL. *1204*

Promethazine   With ninhydrin on Frings TLC procedure at 10 mg/dL. *1204*

Propoxyphene   With ninhydrin on Frings TLC procedure at 10 mg/dL. *1204*

Pyrilamine   With ninhydrin on Frings TLC procedure at 0.5 mg/dL. *1204*

Pyrimethamine   With ninhydrin on Frings TLC procedure at 1 mg/dL. *1204*

Quinidine   With ninhydrin on Frings TLC procedure at 10 mg/dL. *1204*

Quinine   With ninhydrin on Frings TLC procedure at 10 mg/dL. *1204*

Salicylate   With iodoplatinate of Frings TLC procedure at 100 mg/dL. With ninhydrin on Frings TLC procedure at 100 mg/dL. *1204*

Scopolamine   With ninhydrin on Frings TLC procedure at 10 mg/dL. *1204*

Secobarbital   With iodoplatinate of Frings TLC procedure at 10 mg/dL. With ninhydrin on Frings TLC procedure at 10 mg/dL. *1204*

Sodium Bromide   With iodoplatinate of Frings TLC procedure at 100 mg/dL. With ninhydrin on Frings TLC procedure at 100 mg/dL. *1204*

Strychnine   With ninhydrin on Frings TLC procedure at 10 mg/dL. *1204*

Sulfanilamide   With iodoplatinate of Frings TLC procedure at 100 mg/dL. With ninhydrin on Frings TLC procedure at 100 mg/dL. *1204*

Tetracycline   With iodoplatinate of Frings TLC procedure at 2 mg/dL. With ninhydrin on Frings TLC procedure at 2 mg/dL. *1204*

Thiobarbituric Acid   With ninhydrin on Frings TLC procedure at 10 mg/dL. With iodoplatinate of Frings TLC procedure at 10 mg/dL. *1204*

Thiopropazate   With ninhydrin on Frings TLC procedure at 10 mg/dL. *1204*

Thioridazine   With ninhydrin on Frings TLC procedure at 10 mg/dL. *1204*

Thiothixene   With ninhydrin on Frings TLC procedure at 10 mg/dL. *1204*

Trifluoperazine   With ninhydrin on Frings TLC procedure at 10 mg/dL. *1204*

Triflupromazine   With ninhydrin on Frings TLC procedure at 10 mg/dL. *1204*

Trihexyphenidyl   With ninhydrin on Frings TLC procedure at 5 mg/dL. *1204*

Trimethobenzamide   With ninhydrin on Frings TLC procedure at 25 mg/dL. *1204*

Tybamate   With iodoplatinate of Frings TLC procedure at 10 mg/dL. With ninhydrin on Frings TLC procedure at 10 mg/dL. *1204*

Urine   Increase   Analytical

Amitriptyline   Red/purple with iodoplatinate in Frings TLC procedure. *1204*

Amphetamine   Reacts with ninhydrin on TLC method of Frings. *1204*

Benztropine   Purple with iodoplatinate in Frings TLC procedure. *1204*

Chlordiazepoxide   Red/purple with iodoplatinate in Frings TLC procedure. *1204*

Chloroquine   Purple with iodoplatinate in Frings TLC procedure. *1204*

Chlorpheniramine   Purple with iodoplatinate in Frings TLC procedure. *1204*

Chlorpromazine   Purple with iodoplatinate in Frings TLC procedure. *1204*

Clorazepate   Purple with iodoplatinate in Frings TLC procedure. *1204*

Cocaine   Purple with iodoplatinate in Frings TLC procedure. *1204*

Codeine   Purple with iodoplatinate in Frings TLC procedure. *1204*

Diazepam   Purple with iodoplatinate on Frings TLC procedure. *1204*

Diethylpropion   Purple with iodoplatinate on Frings TLC procedure. Reacts with ninhydrin on TLC method of Frings. *1204*

Diphenhydramine   Red/purple with iodoplatinate on Frings TLC procedure. *1204*

Doxepin   Purple with iodoplatinate on Frings TLC procedure. *1204*

Flurazepam   Reacts with ninhydrin on TLC method of Frings Purple with iodoplatinate on Frings TLC procedure. *1204*

Flurazepam Metabolite   Purple with iodoplatinate on Frings TLC procedure. *1204*

Haloperidol   Purple with iodoplatinate on Frings TLC procedure. *1204*

Hydromorphone   Purple with iodoplatinate on Frings TLC procedure. *1204*

Imipramine   Purple with iodoplatinate on Frings TLC procedure. *1204*

Meperidine   Red/purple with iodoplatinate on Frings TLC procedure. *1204*

Mescaline   Reacts with ninhydrin on TLC method of Frings. *1204*

Methadone   Red/purple with iodoplatinate on Frings TLC procedure. *1204*

Methadone Metabolite   Reacts with ninhydrin on TLC method of Frings. *1204*

Methamphetamine   Reacts with ninhydrin on TLC method of Frings. *1204*

Methapyrilene   Purple with iodoplatinate on Frings TLC procedure. *1204*

Methaqualone   Purple with iodoplatinate on Frings TLC procedure. *1204*

Methdilazine   Purple with iodoplatinate on Frings TLC procedure. *1204*

Methylphenidate   Purple with iodoplatinate on Frings TLC procedure. *1204*

Morphine   Blue with iodoplatinate in Frings TLC procedure. *1204*

Nicotine   Blue with iodoplatinate in Frings TLC procedure. *1204*

Oxycodone   Purple with iodoplatinate on Frings TLC procedure. *1204*

Oxymetazoline   Gray with iodoplatinate in Frings TLC procedure. *1204*

Pentazocine   Red/purple with iodoplatinate on Frings TLC procedure. *1204*

Perphenazine   Purple with iodoplatinate on Frings TLC procedure. *1204*

Phenmetrazine   Red/purple with iodoplatinate on Frings TLC procedure. Reacts with ninhydrin on TLC method of Frings. *1204*

Phenylpropanolamine   Reacts with ninhydrin on TLC method of Frings. *1204*

Primaquine   Pink with iodoplatinate in Frings TLC procedure Reacts with ninhydrin on TLC method of Frings. *1204*

Prochlorperazine   Purple with iodoplatinate on Frings TLC procedure. *1204*

Promazine   Purple with iodoplatinate on Frings TLC procedure. *1204*

Promethazine   Purple with iodoplatinate on Frings TLC procedure. *1204*

Propoxyphene   Purple with iodoplatinate on Frings TLC procedure. *1204*

Pyrilamine   Purple with iodoplatinate on Frings TLC procedure. *1204*

Pyrimethamine   Purple with iodoplatinate on Frings TLC procedure. *1204*

Quinidine   Purple with iodoplatinate on Frings TLC procedure. *1204*

Quinine   Purple with iodoplatinate on Frings TLC procedure. *1204*

Scopolamine   Purple with iodoplatinate on Frings TLC procedure. *1204*

Strychnine   Purple with iodoplatinate on Frings TLC procedure. *1204*

Thiopropazate   Purple with iodoplatinate on Frings TLC procedure. *1204*

Thioridazine   Purple with iodoplatinate on Frings TLC procedure. *1204*

Thiothixene   Red/purple with iodoplatinate on Frings TLC procedure. *1204*

Trifluoperazine   Purple with iodoplatinate on Frings TLC procedure. *1204*

Triflupromazine   Purple with iodoplatinate on Frings TLC procedure. *1204*

Trihexyphenidyl   Red/purple with iodoplatinate of Frings TLC procedure. *1204*

Trimethobenzamide   Red/purple with iodoplatinate on Frings TLC procedure. *1204*

Alloxanthine

Serum   Increase   Physiological
Allopurinol   Accumulates with chronic administration. *1343*

Aluminum

Serum   Increase   Physiological
Aluminum Hydroxide   In 7 nondialyzed patients with chronic renal failure (increase from mean 1.7 to to 3.6 μg/dL) with 15-18 g/d. *3533*   Moderate increase after ingestion of aluminum hydroxide as antacid. *3348*

Citrates   When co-administered with aluminum hydroxide ingested as antacid, probably due to formation and absorption of aluminum citrate complexes. *3348*

Sucralfate   Serum concentration in uremic patients comparable to that when patients receiving aluminum hydroxide. *2139*

Valproic Acid   Reported in one woman on chronic hemodialysis. *2587*

Amikacin

Serum   Decrease   Analytical
Heparin   Interferes with biological, radioenzymatic and homogeneous enzyme immunoassay but not RIA techniques due to inhibition of acetyltransferase enzymes. *2021*

Amino-4-Imidazole-5-Carboxamide Ribotide (AICAR)

Urine   Increase   Physiological
Aminosalicylic Acid   Occurs with vitamin B_{12} deficiency. *3773*

Anticonvulsants   Occurs with megaloblastic anemia. *3773*

Barbiturates   Occurs if megaloblastic anemia. *3773*

Colchicine   Megaloblastic anemia with B_{12} deficiency. *3773*

Cycloserine   Occurs with folic acid deficiency. *3773*

Estrogens   Occurs with megaloblastic anemia. *3773*

Ethanol   May occur with folic acid or B_{12} deficiency. *3773*

Isoniazid   Occurs if megaloblastic anemia. *3773*

Metformin   May cause megaloblastic anemia. *3605*

Methotrexate   Occurs with induced folic acid deficiency. *3773*

Neomycin   If megaloblastic anemia develops. *3773*

Oral Contraceptives   Occurs if megaloblastic anemia develops. *3773*

Pentamidine   If megaloblastic anemia develops. *3773*

Phenobarbital   If megaloblastic anemia develops. *3025*

Phenytoin   Occurs with impaired absorption of folic acid. *3773*

Primidone   If megaloblastic anemia occurs. *3025*

Pyrimethamine   If megaloblastic anemia occurs. *3898*

Triamterene   If megaloblastic anemia occurs. *3773*

Aminoacid Arylpeptidase

Serum   Increase   Physiological
Muscular Exercise   Increased by exertion. *1491*

Physical Training   Higher at rest in trained athletes. *1491*

Amino Acids

Serum   No Effect   Analytical
Diatrizoic acid   Concentration too low in serum to affect ninhydrin based procedures. *0379*

Storage of Sample   Heparinized plasma stable for 4 d at 30°. *1712*

Serum   Increase   Analytical
EDTA   May be contaminated with ninhydrin positive constituents. *3230*

Hemolysis   Causes release of intracellular constituents. *3230*

Heparin   May cause hemolysis release ninhydrin positive constituents. *3230*

Standing of Sample   Reported observation. *0355*

Sulfonamides   Reported observation. *0355*

Sweat   If contaminated from analyst. *3230*

Uric Acid   Reported observation. *0355*

Serum   Increase   Physiological
Amino Acids   Marked if intravenously, less if intraduodenal. *2918*

Bismuth Salts   Toxicity effect. *1022*

Glucocorticoids   Due to breakdown of tissue proteins. *0071*

Levarterenol   Catabolic effect. *1022*

11-Oxysteroids   Promote tissue catabolism. *0355*

Serum   Decrease   Physiological
Diethylstilbestrol   Specific amino acids affected in men. *0742*

Epinephrine   Associated with gluconeogenesis. *0355*

Glucose   (some affected only) deposited in muscle. *1403*

Insulin   Metabolic effect. *0355*

Menstruation   Fall 17-22nd day compared with days 2-7. *0743*

Progesterone   Given i.m. to men decreases free and total. *0743*   Specific amino acids affected in men. *0742*

Urine   No Effect   Analytical
Storage of Sample   Acidified to pH 3-5 stable for 4 d at 30°. *1712*

Amino Acids (continued)

Urine No Effect Analytical (continued)
Thymol No effect on ninhydrin. *2981*

Urine No Effect Physiological
Ovulation No significant effect following ovulation. *0743*

Urine Increase Analytical
Aminocaproic Acid Reacts with ninhydrin, measured as amino acid. *3920*

Amphetamine Reacts with ninhydrin; extra spot with TLC, high voltage electrophoresis. *2855*

Ampicillin Presence of drug as additional spot. *2856*

Cephalexin Reacts with ninhydrin; extra spot with TLC, high voltage electrophoresis. *2855*

Cled Medium Artifactual aminoaciduria if in contact. *2756*

Colistin Reacts with ninhydrin; extra spot with TLC, high voltage electrophoresis. *2855*

Dopamine Reacts with ninhydrin; extra spot TLC, high voltage electrophoresis. *2855*

Ephedrine Reacts with ninhydrin; extra spot TLC, high voltage electrophoresis. *2855*

Epinephrine Reacts with ninhydrin; extra spot TLC, high voltage electrophoresis. *2855*

Ethylenediamine Reacts with ninhydrin; extra spot TLC, high voltage electrophoresis. *2855*

Gentamicin Reacts with ninhydrin; extra spot TLC, high voltage electrophoresis. *2855*

Hydroxyaminobutyric Acid False positive ninhydrin reacting spot on TLC Additional spot by high voltage electrophoresis. *3106*

Levarterenol Reacts with ninhydrin; extra spot TLC, high voltage electrophoresis. *2855*

Levodopa Reacts with ninhydrin; extra spot TLC, high voltage electrophoresis. *2855*

Mafenide Reacts with ninhydrin; extra spot TLC, high voltage electrophoresis. *2855*

Metanephrine Reacts with ninhydrin; extra spot TLC, high voltage electrophoresis. *2855*

Methamphetamine Reacts with ninhydrin; extra spot TLC, high voltage electrophoresis. *2855*

Methyldopa Reacts with ninhydrin; extra spot TLC, high voltage electrophoresis. *2855*

Neomycin Reacts with ninhydrin; extra spot TLC, high voltage electrophoresis. *2855*

Normetanephrine Reacts with ninhydrin; extra spot TLC, high voltage electrophoresis. *2855*

Penicillamine Reacts with ninhydrin; extra spot TLC, high voltage electrophoresis. *2855*

Phenylephrine Reacts with ninhydrin; extra spot TLC, high voltage electrophoresis. *2855*

Phenylpropanolamine Reacts with ninhydrin; extra spot TLC, high voltage electrophoresis. *2855*

Polymyxin Reacts with ninhydrin; extra spot TLC, high voltage electrophoresis. *2855*

Urine Increase Physiological
Aspirin Two fold increase after 1.6 g in normals. *1026*

Bismuth Salts Due to proximal tubular dysfunction (Fanconi syndrome). *2911*

Corticotropin Catabolism of body tissues. *3879*

Hydrocortisone Glucocorticoid hormonal action. *1841*

Insulin Metabolic effects. *1022*

Lead Transient observation in poisoning. *0489*

Parathyroid Extract Hormonal action. *1343*

Tetracycline Nephrotoxic effect with degraded tetracycline. *1237*

Triamcinolone Negative nitrogen balance, protein catabolism. *1680*

Urine Decrease Physiological
Epinephrine Metabolic effect associated with gluconeogenesis. *2313*

Insulin Metabolic effects. *1237*

Urine Positive Analytical
Acetaminophen Metabolite reacting with ninhydrin migrates near dihydroxyphenylalanine with 2-dimensional high voltage electrophoresis. *3304*

Ampicillin Produces yellow spots with ninhydrin and thin-layer chromatography. *1278*

Brompheniramine Orange-brown spot with ninhydrin on thin-layer chromatography. *1278*

Carbenicillin Unusual ninhydrin positive spot on TLC. *1596*

Cephalexin Yellow spot with ninhydrin on thin-layer chromatography. *1278*

Cloxacillin 2 orange colored spots with ninhydrin on paper or thin-layer chromatography, not present following peroxide oxidation of specimen but new spot appeared. *1596*

Diatrizoic acid May comigrate with asparagine and glutamine on TLC and with isoleucine on amino acid analyzer. *0379*

DL-Methionine Reddish-pink spot with DL-methionine with ninhydrin on thin-layer chromatography. *1278*

Dopamine Unusual ninhydrin reacting spot on TLC. *1596*

Erythromycin Yellow spot with ninhydrin on thin-layer chromatography. *1278*

Isomil® Reddish-pink spot with DL-methionine with ninhydrin on thin-layer chromatography. *1278*

Kanamycin Reacts with ninhydrin; extra spot TLC, high voltage electrophoresis. *2855* Unusual ninhydrin positive spot observed with TLC. *1596*

Neo-Mull-Soy® Reddish-pink spot with DL-methionine with ninhydrin on thin-layer chromatography. *1278*

Neomycin Purple spot with ninhydrin on thin-layer chromatography when combined with triamcinolone and nystatin in Kenacomb. *1278*

Nystatin Purple spot on thin-layer chromatography with ninhydrin when combined with neomycin and triamcinolone in Kenacomb. *1278*

Penicillamine 2 unusual purple spots with ninhydrin on thin-layer chromatography of urine from patients with Wilson's disease. Spots stain gray with isatin. *1596*

Phenobarbital Unusual ninhydrin positive spot in all systems adjacent to threonine and orange in color. *1596*

Phenylephrine Orange-brown spot with ninhydrin on thin-layer chromatography. *1278*

Phenylpropanolamine Unusual ninhydrin reacting spot observed on TLC. *1596*

Primidone Unusual orange spot with ninhydrin migrating in all TLC systems close to threonine. *1596*

ProSobee® Reddish-pink spot with DL-methionine with ninhydrin on thin-layer chromatography. *1278*

Pseudoephedrine Red spot with ninhydrin on thin-layer chromatography when combined with triprolidine in Actifed®. *1278*

Triamcinolone Purple spot with ninhydrin on thin-layer chromatography when combined with neomycin and nystatin in Kenacomb. *1278*

Triprolidine Red spot with ninhydrin on thin-layer chromatography when combined with pseudoephedrine in Actifed®. *1278*

Hair No Effect Physiological
Race Same in cephalic hair from black, mongol, caucasian. *1321*

Test Conditions Increase Analytical
Aspartic Acid Reacts in spot test with ninhydrin. *3765*

Epinephrine Positive spot test with ninhydrin. *3765*

Insulin Positive spot test with ninhydrin. *3765*

Kynurenine Positive spot test with ninhydrin. *3765*

Methoxytyramine Positive spot test with ninhydrin. *3765*

Normetanephrine Positive spot test with ninhydrin. *3765*

Phenylalanine Positive spot test with ninhydrin. *3765*

Tryptophan Positive spot test with ninhydrin. *3765*

Tyramine Positive spot test with ninhydrin. *3765*

Aminobenzoic Acid

Urine   Increase   Physiological
Procainamide   Up to 10% excreted as this. *1343*

α-Amino-N-Butyric Acid

Serum   Increase   Physiological
Starvation   Occurs with starvation even for 1 d. *0021*

β-Amino-Isobutyric Acid

Plasma   Increase   Physiological
Ethanol   Due to increased hepatic production and release into circulation. *2428*

Urine   Increase   Physiological
Lead   Nephrotoxic effect with lead poisoning. *0987*
X-Ray Therapy   Due to tissue destruction. *2378*

Aminolevulinic Acid

Serum   Increase   Physiological
Amiodarone   Increased in all of 10 patients by average of 79% over 18 mo. *1509*
Anticonvulsants   Highly significant increase when multiple drugs given (146 nmol/L versus 99 nmol/L in Controls. *1348*
Carbamazepine   Modest but highly significant statistical increase in drug treated population (132 nmol/L versus 99 nmol/L). *1348*
Lead   If over 20 μg/dL indicates poisoning. *0489*
Phenobarbital   Slight but not significant increase (112 nmol/L versus 99 nmol/L in controls). *1348*
Phenytoin   Significant increase to 124 nmol/L from 99 nmol/L in controls. *1348*
Valproic Acid   Mean increase to 130 nmol/L from 99 nmol/L in controls. *1348*

Serum   Decrease   Physiological
Hematin   In a patient with acute intermittent porphyria. *0414*

Urine   No Effect   Analytical
Storage of Sample   If acidified in dark 2% decrease in 3 d. *3217*

Urine   Increase   Analytical
Ammonia   If no preliminary separation. *2981*
Glucosamine   If no preliminary separation procedure. *2981*
Glycine   If no preliminary separation. *2981*
Penicillin   Derivative reacts with Ehrlich's reagent. *1997*
Penicillin G   In method of Mauzerall and Granick interacts with acetylacetone which reacts with p-dimethylmenine benzaldehyde. *3640*
Porphobilinogen   Unless removed by column in method of Vincent. *3718*

Urine   Increase   Physiological
Aminopyrine   May precipitate attack of acute porphyria. *1016*
Anticonvulsants   Reported in one child. *2427*
Apronalide   May precipitate acute porphyria. *1322*
Barbiturates   May precipitate acute porphyria. *1322*
Chlordiazepoxide   May precipitate acute porphyria. *1322*
Chlorpropamide   May precipitate cutaneous porphyria. *1322*
Diazepam   May precipitate attack of acute porphyria. *1016*
Dichloralphenazone   May precipitate acute porphyria. *1322*
Ergot Preparations   May precipitate attack of acute porphyria. *1016*
Estrogens   May precipitate porphyria attack. *1322*
Ethanol   May precipitate attack of porphyria. *1322*
Fasting   In normals, if porphyria may precipitate attack. *0938*

Fatty Foods   Effect of high fat diet on hepatic porphyria. *0938*
Glutethimide   May precipitate acute porphyria. *1322*
Glycine   In patients with acute porphyria in remission. *0938*
Gold   Occurs with panmyelopathy. *1343*
Griseofulvin   May precipitate acute porphyria attack. *1322*
Hydantoin Derivatives   May precipitate attack of acute porphyria. *1016*
Isopropyl Dipyrone   May precipitate attack of acute porphyria. *1016*
Lead   Observed in poisoning. *1343*
Meprobamate   May precipitate acute porphyria. *1322*
Methyldopa   May precipitate acute porphyria attack. *1322*
Methylsulfonal   May precipitate acute porphyria. *1322*
Methyprylon   Ala-synthetase stimulated in animals. *1343*
Oral Contraceptives   May precipitate porphyria attack. *1322*
Pentazocine   May precipitate acute porphyria attack. *1322*
Phenytoin   May precipitate acute porphyria. *1322*
Progestogens   May precipitate attack of acute porphyria. *1016*
Succinimide   May precipitate acute porphyria. *1322*
Sulfomethane   May provoke attack of porphyria. *1322*
Sulfonamides   May precipitate acute porphyria. *1322*
Tolbutamide   May precipitate porphyria attack. *1322*

Urine   Decrease   Analytical
Light   Unless preserved with acid in dark. *3718*
Storage of Sample   50% with light and no preservative in 24 h. *3217*

Urine   Decrease   Physiological
Cimetidine   Marked effect in patient with acute intermittent porphyria possibly due to inhibition of hepatic delta-aminolevulinic acid synthetase. *1659*
Hematin   In a patient with elevation due to porphyria. *0414*
Vitamin E   Possible reduction to normal levels in porphyria. *2549*

Aminolevulinic Acid Dehydrase

Serum   Decrease   Physiological
Ethanol   Falls and returns to normal inversely with alcohol concentration when taken acutely; tends to be low in chronic alcoholics. *3124*

Red Blood Cells   No Effect   Analytical
Sodium Oxalate   No effect compared with heparin. *0400*

Red Blood Cells   Decrease   Analytical
EDTA   15% decrease compared with heparin sample. *0400*
Oxalate/Fluoride   15% decrease compared with heparin sample. *0400*
Storage of Sample   85% 24 h at room temperature, up to 25% at 4°. *0400*

Red Blood Cells   Decrease   Physiological
Hospitalization   About 60% of normal individuals. *0400*
Lead   Negative correlation with blood concentration of lead. *0489*
Smoking   13% lower in smokers than nonsmokers. *1548*
Tetraethyl-Lead   Powerful inhibiting action. *0258*

α-Amino-Nitrogen

Plasma   No Effect   Analytical
Hemoglobin   20 g/dL has no effect on Goodwin procedure. *1712*
Storage of Sample   For long period if plasma is frozen Several days in protein free filtrate in refrigerator At refrigerator temperature after 48 h. *1712*

Plasma   No Effect   Physiological
Starvation   Usually unaffected by starvation. *3783*

α-Amino-Nitrogen *(continued)*

Plasma Increase Analytical

Ammonia Unless removed during analytic procedure. *1563*

Contact With Clot Proteolysis occurs even in refrigerator. *1563*

Hemolysis Release of glutathione from erythrocytes. *1563*

Room Temperature Proteolysis occurs even in refrigerator, satisfactory if frozen. *1563*

Sulfamethoxazole Measured in naphthoquinone method of Frame. *2981*

Sulfonamides Reacts with naphthoquinone (method of Frame). *1563*

Uric Acid Reacts in naphthoquinone method of Frame. *1563*

Plasma Increase Physiological

Corticotropin Tissue and protein catabolism. *1237*

Epinephrine Metabolic effect gluconeogenesis. *0355*

Meals Small transient rise after protein meals. *2312*

11-Oxysteroids Promote tissue catabolism. *1022*

Prednisolone After 4 d of oral therapy. *2296*

Protein Rises after meals, falls to normal after 4 h. *0987*

Serum Release from RBC with clotting (up to 40% high). *1563*

Tranylcypromine All amino acids increased except those listed below. *0824*

Plasma Decrease Physiological

Growth Hormone Increased peripheral removal. *2312*

Insulin Increased uptake by tissues. *1237*

Oral Contraceptives Anabolic effect of synthetic steroids (progestogen). *0743*

Progesterone Catabolic effect. *0743*

Urine No Effect Analytical

Creatinine No effect if dinitrofluorobenzene method used. *1346*

Storage of Sample At room temperature for 4 d without preservative. *1712*

Urea No effect if dinitrofluorobenzene method used. *1346*

Urine No Effect Physiological

Progesterone Almost no effect in normal males. *3571*

Urine Increase Analytical

Ammonia Unless removed during analytic procedure. *1563*

Sulfonamides Reacts with naphthoquinone (method of Frame). *1563*

Urine Increase Physiological

Aspirin Inhibition of reabsorption, increased protein catabolism. *1343*

Cadmium Occasionally with long exposure, proteinuria. *2217*

Corticotropin Catabolism of body tissues. *3879*

Cortisone Increased tissue catabolism. *3879*

Lead Occurs with poisoning. *0987*

Meals Significant effect observed after each meal. *3571*

Mercury Compounds Fanconi syndrome with laxative overuse. *2108*

Oxalate Oxalic acid may cause renal damage. *0459*

11-Oxysteroids Promote tissue catabolism. *2313*

Phosphorus Due to nephrotoxicity. *1302*

Pregnancy Renal threshold of amino acids reduced. *0987*

Urine Decrease Physiological

Diurnal Variation Progressive fall (except for meals) from 8 am to 5 am. *3571*

Aminopyrine

Serum Decrease Physiological

Phenobarbital Reported interaction due to alteration of metabolism. *2042*

Aminosalicylic Acid

Serum Increase Physiological

Probenecid Increases by 2-4 fold (inhibits renal excretion). *1487*

Serum Decrease Physiological

Diphenhydramine Delayed absorption due to delayed gastric emptying. *3794*

Amitriptyline

Serum No Effect Physiological

Chlordiazepoxide No effect on serum concentration if given together. *3321*

Diazepam No effect on serum concentration if given together. *3321*

Nitrazepam No effect on serum concentration if given together. *3321*

Oxazepam No effect on serum concentration if given together. *3321*

Serum Increase Analytical

Cyclobenzaprine Reported interference with EMIT, TLC, GC, LC and GC/MS: may also interfere with other antidepressants. *2888*

Serum Increase Physiological

Methylphenidate Inhibits metabolism of tricyclic antidepressants. *1487*

Serum Decrease Physiological

Barbiturates Stimulates metabolism of tricyclic antidepressants. *1487*

Ammonia

Plasma No Effect Analytical

Acetaminophen At concentration of 50 mg/L had no effect on Ektachem® method. *3393*

Acetylsalicylic Acid At concentration of 300 mg/L had no effect on Ektachem® method. *3393*

Allantoin On indophenol reaction with 5,000 nmoles. *1290*

Asparagine On indophenol reaction with 5,000 nmoles. *1290*

Bilirubin No effect on measurement procedure. *1936*

Cysteine On indophenol reaction with 1,000 nmoles. *1290*

Dextran At concentration of 10,000 mg/L had no effect on Ektachem® method. *3393*

Diatrizoic acid At concentration of 5,000 mg/L had no effect on Ektachem® method. *3393*

EDTA Satisfactory as anticoagulant. *0705* At concentration of 800 mg/L had no effect on Ektachem® method. *3393*

Hippuric Acid On indophenol reaction with 5,000 nmoles. *1290*

Levoglutamide On indophenol reaction with 5,000 nmoles. *1290*

Lithium Heparin At concentration of 500 mg/L had no effect on Ektachem® method. *3393*

Meprobamate At concentration of 20 mg/L had no effect on Ektachem® method. *3393*

Mercaptopurine At concentration of 150 mg/L had no effect on Ektachem® method. *3393*

Phenobarbital At concentration of 30 mg/L had no effect on Ektachem® method. *3393*

Phenytoin At concentration of 20 mg/L had no effect on Ektachem® method. *3393*

Potassium Oxalate At concentration of 8,000 mg/L had no effect on Ektachem® method. *3393*

Room Temperature If heparinized and in ice, acid added in 15 minutes. *2877*

Sodium Fluoride At concentration of 10,000 mg/L had no effect on Ektachem® method. *3393*

Sodium Oxalate At concentration of 1540 mg/L had no effect on Ektachem® method. *3393*

Storage of Sample   Stable in iced plasma for several h. *0582*
On blood ammonia for 7 d on dry ice. *0904*
Sulfathiazole   At concentration of 60 mg/L had no effect on
Ektachem® method. *3393*
Urea   On indophenol reaction with 25,000 nmoles. *1290*

Plasma   No Effect   Physiological
Carbamazepine   No striking effect when given to epileptic
patients. *3697*
Clopamide   No significant effects observed. *1487*
Isoniazid   No effect observed in most studies. *1487*
Phenobarbital   No striking abnormality when given to epileptics.
3697
Valproic Acid   No striking abnormality when given to epileptics.
3697   In patients on monotherapy but increases seen with
coadministration of other antiepileptics. *1454*

Plasma   Increase   Analytical
Ammonium Salts   Increased blood levels. *1238*
Contact With Clot   Breakdown of urea if delay in analysis. *1237*
Filter Paper   Especially if acid-washed paper (absorbs NH_3 from
air). *0579*
Fluorides   Cause increase if used as anticoagulants. *0705*
Hemoglobin   Slight interference with color development. *1936*
Hemolysis   Some interference with color development. *1237*
Heparin   Contains variable amounts of ammonium salts. *0579*
Oxalate   Effect if used as anticoagulants. *0705*
Storage of Sample   3 fold increase in 24 h, 8 fold in 7 d. *3217*
Increase by 0.017 μg/mL blood/min at 25°. *0904*
Urea   Breakdown of urea to ammonia. *1022*

Plasma   Increase   Physiological
Acetazolamide   ?diverts NH_3 from kidney to general circulation.
1238
Ammonium Acetate   Maximum in cirrhotics after oral dose. *1289*
Asparaginase   May be marked, associated with abnormal liver
function. *1597*   Hemorrhagic pancreatitis in fewer than 0.5%
treated patients. *1398*
Asparagine   Potential source of ammonia. *1943*
Barbiturates   Impaired metabolism in dogs. *3601*
Blood Transfusions   Especially if stored previously. *3217*
Chlorothiazide   Decreased potassium and alkalosis. *1238*
Chlorthalidone   Partially due to decreased potassium and alkalo-
sis. *0109*
Ethacrynic Acid   ?due to hypokalemia and alkalosis. *1488*
Ethanol   Dose related significant effect. *2924*
Fibrin Hydrolysate   Due to high ammonia content of solution.
3746
Furapromidium   Occasional increase observed. *2427*
Furosemide   Acts like thiazides causes hypokalemia and alkalo-
sis. *1488*
Glucose   May increase as glucose increases in cirrhotics. *1488*
Hydroflumethiazide   May occur especially if pre-existing hepatic
impairment. *1680*
Ion Exchange Resins   Mechanism not reported ?depends on
resin. *1488*
Isoniazid   Due to metabolism of isonicotinic acid. *3951*
Levoglutamide   Potential source of additional ammonia. *1943*
Mercurial Diuretics   Presumed effect as may precipitate hepatic
coma. *1487*
Morphine   Impairs ability of liver to metabolize NH_3 in dogs. *3601*
Muscular Exercise   Tissue catabolism. *1237*
Oral Resins   Exchanged for other ions in gastrointestinal tract.
1022
Tetracycline   If given i.v. in large doses. *1488*
Thiazides   Associated with K depletion and alkalosis. *1238*
Thiopental   Impaired metabolism in liver of dogs. *1487*

Valproic Acid   Occasional hyperammonia when given alone,
more frequent when given with phenytoin when effect also more
marked. *2301*   In 29 of 55 patients receiving drug versus none in
control population on other anticonvulsants; values especially high
when phenytoin also taken. *2528*   About half of patients comedicated with phenobarbital had increase; especially noted when concentrations high. *1454*   Nondose dependent effect in some
patients with both normal and abnormal liver function. *1247*   Probably due to inhibition of enzymes involved in glycine clearance in
liver analogous to ketotic hyperglycinemias. *1759*   Associated with
inhibition of carbamyl phosphate synthetase I and interference with
mitochondrial glycine transport. *0736*

Plasma   Decrease   Analytical
Alanine   With 2500 nmoles produces 8% inhibition indophenol
react. *1290*
Arginine   With 2500 nmoles produces 5% inhibition of indophenol
reaction. *1290*
Aspartic Acid   With 1,000 nmoles produces 12% inhibition of
indophenol react. *1290*
Creatine   With 1,000 nmoles produces 7% inhibition of indophe-
nol react. *1290*
Creatinine   With 500 nmoles produces 14% inhibitor indophenol
reaction. *1290*
Cysteine   With 500 nmoles produces 30% inhibitor indophenol
reaction. *1290*
Glutamic Acid   With 1,000 nmoles produces 9% inhibition of
indophenol reaction. *1290*
Glycine   With 2500 nmoles produces 63% inhibitor indophenol
reaction. *1290*
Guanidine   With 500 nmoles produces 34% inhibitor indophenol
reaction. *1290*
Guanidinoacetic Acid   With 1,000 nmoles produces 22% inhibi-
tion of indophenol reaction. *1290*
Histidine   With 250 nmoles produces 17% inhibitor indophenol
reaction. *1290*
Hydroxyproline   With 500 nmoles produces 17% inhibitor indophenol reaction. *1290*
Isoleucine   With 500 nmoles produces 59% inhibitor indophenol
reaction. *1290*
Leucine   With 500 nmoles produces 9% inhibitor indophenol reaction. *1290*
Levoglutamide   Inhibits indophenol color in Berthelot reaction.
2632
Lysine   With 500 nmoles produces 6% inhibitor indophenol reac-
tion. *1290*
Methionine   With 250 nmoles produces 7% inhibitor indophenol
reaction. *1290*
Phenylalanine   With 250 nmoles produces 14% inhibitor indophenol reaction. *1290*
Phosphates   Inhibits form of indophenol color in Berthelot reaction. *2632*
Proline   With 1,000 nmoles produces 20% inhibition of indophenol
reaction. *1290*
Serine   With 2500 nmoles produces 26% inhibitor indophenol
reaction. *1290*
Threonine   With 1,000 nmoles produces 9% inhibition of indophenol reaction. *1290*
Tromethamine   Indophenol color formation (Berthelot reaction)
inhibition. *2632*
Tryptophan   With 25 nmoles produces 10% inhibitor indophenol
reaction. *1290*
Tyrosine   With 50 nmoles produces 6% inhibitor indophenol reaction. *1290*
Uric Acid   With 500 nmoles produces 18% inhibitor indophenol
reaction. *1290*
Valine   With 500 nmoles produces 47% inhibitor indophenol reaction. *1290*

Plasma   Decrease   Physiological
Acetohydroxamic Acid   Potent urease inhibitor, effect seen in 1
patient. *1487*
Arginine   Capable of reacting with ammonia. *1943*
Diphenhydramine   Reported effect in exogenous NH_3 toxicity.
1942

Ammonia (continued)

Plasma Decrease Physiological (continued)
Glutamic Acid Capable of reacting with ammonia. *1943*
Isocarboxazid Reportedly effective in reducing NH_3 intoxication. *1488*
Kanamycin Impairs NH_3 production by gut bacteria. *1487*
Lactobacillus Acidophilus Causes reduction in hepatic encephalopathy. *2249*
Lactulose In patients with hepatic encephalopathy. *1446*
Levodopa Observed in one case ?unrelated. *1487*
MAO Inhibitors Reported effect in exogenous NH_3 toxicity. *1942*
Neomycin Reduces NH_3 producing bacteria in gastrointestinal tract. *3889*
Potassium K repletion in hepatic coma may reduce NH_3. *2498*
Sodium Salts Reported effect. *2313*
Tetracycline Reduces production by gut bacteria. *1487*

Urine No Effect Analytical
Thymol No effect on Nessler, Berthelot methods. *2981*

Urine No Effect Physiological
Metolazone No effect on carbonic anhydrase. *3358*
Triamterene Does not cause excessive excretion. *1680*

Urine Increase Physiological
Glycine Same extent in gouty and normal subjects. *3929*
Methenamine Hydrolyzed in acid urine (to formaldehyde also). *1678*
Neomycin May occur if hypokalemia induced. *1077*
Pregnancy Output increased in last trimester. *0987*
Protein Due to ingestion of protein. *0987*
Starvation Occurs with metabolic acidosis. *3067*
Triflocin Inhibits carbonic anhydrase. *0033*

Urine Decrease Physiological
Acetazolamide Increased alkalinity. *1343*
Glucose In starved individuals even at 7.5 g/d. *1242*
Mafenide Inhibits carbonic anhydrase if applied topically. *3813*
Secretin Reduced formation with increased alkalinization. *2656*

CSF Decrease Physiological
Lactulose If given as enema has marked effect. *1920*

Feces No Effect Physiological
Lactulose Although increased volume of feces. *0032*

Feces Increase Physiological
Mannitol Modest rise observed. *0032*
Sodium Sulfate Modest rise observed. *0032*
Sorbitol Modest rise observed. *0032*

Ammoniacal Silver Nitrate

Urine Positive Analytical
Homogentisic Acid Rapid darkening of solution. *3588*
Melanin Slow darkening. *3588*

Test Conditions Positive Analytical
Ascorbic Acid Positive spot test with Tollen's reagent. *3765*
Cresol Positive spot test with Tollen's reagent. *3765*
Dopamine Positive spot test with Tollen's reagent. *3765*
Epinephrine Positive Tollen's test for phenols. *3765*
Ferulic Acid Positive spot test with Tollen's reagent. *3765*
Homogentisic Acid Positive spot test with Tollen's reagent. *3765*
Homovanillic Acid Positive spot test with Tollen's reagent. *3765*
Hydroxyanthranilic Acid Positive spot test with Tollen's reagent. *3765*

5-Hydroxyindoleacetic Acid Positive spot test with Tollen's reagent. *3765*
Hydroxykynurenine Positive spot test with Tollen's reagent. *3765*
Hydroxyphenylacetic Acid Positive spot test given by o-hydroxyderivative. *3765*
5-Hydroxytryptamine Positive spot test with Tollen's reagent. *3765*
Levarterenol Positive spot test with Tollen's reagent. *3765*
Metanephrine Positive spot test with Tollen's reagent. *3765*
Methoxytyramine Positive spot test with Tollen's reagent. *3765*
Normetanephrine Positive spot test with Tollen's reagent. *3765*
Uric Acid Positive spot test with Tollen's reagent. *3765*
Vanillic Acid Positive spot test with Tollen's reagent. *3765*
Vanillylmandelic Acid (VMA) Positive spot test with Tollen's reagent. *3765*
Xanthurenic Acid Positive spot test with Tollen's reagent. *3765*

Ammonium Ions

Urine No Effect Physiological
Acetaminophen Without overdose not usually any effect. *2866*

Amobarbital

Blood Increase Physiological
Amobarbital 600 mg orally produces concentration of 9.6 mg/L. *2348*

Urine No Effect Analytical
Chlorphentermine No interference using TLC with ethyl acetate: methanol: water: ammonium hydroxide and modified Dragendorff's reagent for detection. *3868*
Clotermine No interference on TLC using ethyl acetate: methanol: water: ammonium hydroxide and modified Dragendorff's reagent for detection. *3868*
Diethylpropion No interference with TLC using ethyl acetate: methanol: water: ammonium hydroxide and modified Dragendorff's reagent for detection. *3868*
Fenfluramine No interference on TLC using ethyl acetate: methanol: water: ammonium hydroxide and Dragendorff's reagent for detection. *3868*
Mazindol No interference on TLC using ethyl acetate: methanol: water: ammonium hydroxide and modified Dragendorff's reagent for detection. *3868*
Phendimetrazine No interference with TLC using ethyl acetate: methanol: water: ammonium hydroxide and modified Dragendorff's reagent for detection. *3868*
Phenmetrazine No interference on TLC using ethyl acetate: methanol: water: ammonium hydroxide and modified Dragendorff's reagent for detection. *3868*
Phentermine No interference on TLC using ethylacetate: methanol: water: ammonium hydroxide and modified Dragendorff's reagent for detection. *3868*

Amoxicillin

Serum Decrease Physiological
Food Absorption delayed when taken with food Absorption reduced when taken with food. *3024*

Amphetamine

Serum No Effect Physiological
Dextroamphetamine Not detected after 30 mg/d. *2348*

Urine No Effect Analytical
Acetazolamide At 50 mg/L on fluorescent method of Hayes. *1534*
Adiphenine No effect at 100 mg/L on method of Rutter. *3097*

Aletamine   No effect at 100 mg/L on method of Rutter. *3097*

Allantoin   At 50 mg/L on fluorescent method of Hayes. *1534*

Allopurinol   No effect at 100 mg/L on method of Rutter. *3097*

Alphaprodine   No effect at 100 mg/L on method of Rutter. *3097*

Ambazone   No effect at 100 mg/L on method of Rutter. *3097*

Aminacrine   No effect at 100 mg/L on method of Rutter. *3097*

Aminophenol   No effect at 100 mg/L on method of Rutter. *3097*

Aminosalicylic Acid   At 50 mg/L on fluorescent method of Hayes. *1534*

Amisometradine   At 50 mg/L on fluorescent method of Hayes. *1534*

Ampicillin   At 50 mg/L on fluorescent method of Hayes. *1534*

Anileridine   At 50 mg/L on fluorescent method of Hayes. *1534*

Bamethan   No effect at 100 mg/L on method of Rutter. *3097*

Benzocaine   At 50 mg/L on fluorescent method of Hayes. *1534*

Benzphetamine   No effect at 100 mg/L on method of Rutter. *3097*

Betazole   No reaction with NBD chloride procedure of Monforte. *2473*

Bethanidine   No effect at 100 mg/L on method of Rutter. *3097*

Buformin   At 50 mg/L on fluorescent method of Hayes. *1534*

Butacaine   At 50 mg/L on fluorescent method of Hayes. *1534*

Caffeine   At 50 mg/L on fluorescent method of Hayes. *1534*

Cetoxime   No effect at 100 mg/L on method of Rutter. *3097*   At 50 mg/L on fluorescent method of Hayes. *1534*

Chlorhexidine   No effect at 100 mg/L on method of Rutter. *3097*

Chlorphentermine   At 50 mg/L on fluorescent method of Hayes. *1534*   No interference using TLC with ethyl acetate: methanol: water: ammonium hydroxide and modified Dragendorff's reagent for detection. *3868*

Clorprenaline   No effect at 100 mg/L on method of Rutter. *3097*

Clotermine   No interference on TLC using ethyl acetate: methanol: water: ammonium hydroxide and modified Dragendorff's reagent for detection. *3868*

Codeine   No reaction with NBD chloride procedure of Monforte. *2473*

Cyclopentamine   No effect at 100 mg/L on method of Rutter. *3097*

Cycloserine   At 50 mg/L on fluorescent method of Hayes. *1534*

Debrisoquin   No effect at 100 mg/L on method of Rutter. *3097*

Desipramine   No reaction with NBD chloride of Monforte. *2473*   No effect at 100 mg/L on method of Rutter. *3097*

Diethylpropion   At 50 mg/L on fluorescent method of Hayes. *1534*   No interference with TLC using ethyl acetate: methanol: water: ammonium hydroxide and modified Dragendorff's reagent for detection. *3868*

Dihydralazine   No effect at 100 mg/L on method of Rutter. *3097*

Dimethyltryptamine   At 50 mg/L on fluorescent method of Hayes. *1534*

Dopamine   At 50 mg/L on fluorescent method of Hayes. *1534*   No effect at 100 mg/L on method of Rutter. *3097*

Ephedrine   At 50 mg/L on fluorescent method of Hayes. *1534*   No effect at 100 mg/L on method of Rutter. *3097*

Epinephrine   No reaction with NBD chloride of Monforte. *2473*   At 50 mg/L on fluorescent method of Hayes. *1534*   No effect at 100 mg/L on method of Rutter. *3097*

Ethambutol   No effect at 100 mg/L on method of Rutter. *3097*

Ethylnorepinephrine   No effect at 100 mg/L on method of Rutter. *3097*

Fencamfamin   At 50 mg/L on fluorescent method of Hayes. *1534*   No effect at 100 mg/L on method of Rutter. *3097*

Fenfluramine   At 50 mg/L on fluorescent method of Hayes. *1534*   No effect at 100 mg/L on method of Rutter. *3097*   No interference on TLC using ethyl acetate, methanol, water, ammonium hydroxide and modified Dragendorff's reagent for detection. *3868*

Fenpipramide   No effect at 100 mg/L on method of Rutter. *3097*

Guanoclor   No effect at 100 mg/L on method of Rutter. *3097*

Guanoxan   No effect at 100 mg/L on method of Rutter. *3097*

Histamine   No effect at 100 mg/L on method of Rutter. *3097*

Hydralazine   At 50 mg/L on fluorescent method of Hayes. *1534*

Hydroxyamphetamine   No effect at 100 mg/L on method of Rutter. *3097*   At 50 mg/L on fluorescent method of Hayes. *1534*

5-Hydroxytryptamine   No effect at 100 mg/L on method of Rutter. *3097*

Imipramine   At 50 mg/L on fluorescent method of Hayes. *1534*

Iproniazid   No effect at 100 mg/L on method of Rutter. *3097*

Isocarboxazid   No reaction with NBD chloride of Monforte. *2473*

Isometheptene   No effect at 100 mg/L on method of Rutter. *3097*

Isoniazid   At 50 mg/L on fluorescent method of Hayes. *1534*   No reaction with NBD chloride of Monforte. *2473*   No effect at 100 mg/L on method of Rutter. *3097*

Isoprenaline   No effect at 100 mg/L on method of Rutter. *3097*   At 50 mg/L on fluorescent method of Hayes. *1534*

Levarterenol   No reaction with NBD chloride procedure of Monforte. *2473*   No effect at 100 mg/L on method of Rutter. *3097*

Levodopa   No reaction with NBD chloride procedure of Monforte. *2473*

Mazindol   No interference on TLC using ethyl acetate: water: ammonium hydroxide and modified Dragendorff's reagent for detection. *3868*

Meperidine   At 50 mg/L on fluorescent method of Hayes. *1534*

Mephentermine   No effect at 100 mg/L on method of Rutter. *3097*   At 50 mg/L on fluorescent method of Hayes. *1534*

Metaraminol   No reaction with NBD chloride procedure of Monforte. *2473*   No effect at 100 mg/L on method of Rutter. *3097*

Methoxyphenamine   No effect at 100 mg/L on method of Rutter. *3097*

Methylamphetamine   No effect at 100 mg/L on method of Rutter. *3097*

Methyldopa   At 50 mg/L on fluorescent method of Hayes. *1534*

Methylephedrine   At 50 mg/L on fluorescent method of Hayes. *1534*

Methylphenidate   At 50 mg/L on fluorescent method of Hayes. *1534*

Morphine   No reaction with NBD chloride procedure of Monforte. *2473*

Naphazoline   No reaction with NBD chloride procedure of Monforte. *2473*

Niacinamide   No effect at 100 mg/L on method of Rutter. *3097*

Nialamide   No effect at 100 mg/L on method of Rutter. *3097*

Nicotine   No effect at 100 mg/L on method of Rutter. *3097*   At 50 mg/L on fluorescent method of Hayes. *1534*

Nitrazepam   No effect at 100 mg/L on method of Rutter. *3097*

Norpseudoephedrine   At 50 mg/L on fluorescent method of Hayes. *1534*

Orphenadrine   At 50 mg/L on fluorescent method of Hayes. *1534*

Oxedrine   No effect at 100 mg/L on method of Rutter. *3097*

Pargyline   At 50 mg/L on fluorescent method of Hayes. *1534*

Phenazopyridine   No effect at 100 mg/L on method of Rutter. *3097*

Phencyclidine   No effect at 100 mg/L on method of Rutter. *3097*

Phendimetrazine   No interference with TLC using ethyl acetate: methanol: water: ammonium hydroxide and modified Dragendorff's reagent for detection. *3868*

Phenelzine   At 50 mg/L on fluorescent method of Hayes. *1534*

Phenmetrazine   At 50 mg/L on fluorescent method of Hayes. *1534*   No effect at 100 mg/L on method of Rutter. *3097*   No interference on TLC using ethyl acetate: methanol: water: ammonium hydroxide and modified Dragendorff's reagent for detection. *3868*

Phentermine   At 50 mg/L on fluorescent method of Hayes. *1534*   No interference on TLC using ethylacetate: methanol: water: ammonium hydroxide and modified Dragendorff's reagent for detection. *3868*

Phenylephrine   No effect at 100 mg/L on method of Rutter. *3097*

Phenylpropanolamine   At 50 mg/L on fluorescent method of Hayes. *1534*

Piperazine   No effect at 100 mg/L on method of Rutter. *3097*

Prenylamine   No effect at 100 mg/L on method of Rutter. *3097*

Primaquine   At 50 mg/L on fluorescent method of Hayes. *1534*

Proguanil   No effect at 100 mg/L on method of Rutter. *3097*

Propamidine   No effect at 100 mg/L on method of Rutter. *3097*

Propoxyphene   At 50 mg/L on fluorescent method of Hayes. *1534*

Amphetamine *(continued)*

Urine   No Effect   Analytical   (continued)

Propylhexedrine   No effect at 100 mg/L on method of Rutter. *3097*

Pseudoephedrine   At 50 mg/L on fluorescent method of Hayes. *1534*   No effect at 100 mg/L on method of Rutter. *3097*

Putrescine   No effect at 100 mg/L on method of Rutter. *3097*

STP   At 50 mg/L on fluorescent method of Hayes. *1534*

Thioridazine   No reaction with NBD chloride procedure of Monforte. *2473*

Tranylcypromine   No reaction with NBD chloride of Monforte. *2473*   At 50 mg/L on fluorescent method of Hayes. *1534*   No effect at 100 mg/L on method of Rutter. *3097*

Tryptamine   At 50 mg/L on fluorescent method of Hayes. *1534*

Tyramine   At 50 mg/L on fluorescent method of Hayes. *1534*   No effect at 100 mg/L on method of Rutter. *2173*

Urine   Increase   Analytical

Amphetamine   Yields fluorophor with NBD chloride reaction. *2473*

Butethamine   Yields fluorophor with NBD chloride reaction. *2473*

Cyclopentamine   Yields fluorophor with NBD chloride reaction. *2473*

Ephedrine   Reacts with methyl orange in method of Frings. *2474*   Yields fluorophor with NBD chloride reaction. *2473*

Ethionamide   No effect at 100 mg/L on method of Rutter. *3097*

Mephentermine   Yields fluorophor with NBD chloride reaction. *2473*

Methamphetamine   Yields fluorophor with NBD chloride reaction. *2473*

Methaphentermine   Reacts with methyl orange in method of Frings. *2474*

Methoxamine   Yields fluorophor with NBD chloride reaction. *2473*

Methoxyphenamine   Yields fluorophor with NBD chloride reaction. *2473*

Methylphenidate   Yields fluorophor with NBD chloride reaction. *2473*

Nialamide   Yields fluorophor with NBD chloride reaction. *2473*

Nortriptyline   Yields fluorophor with NBD chloride reaction. *2473*

Phenethylamine   False positive fluorescent method of Hayes at 50 mg/L. *1534*

Phenpromethamine   Yields fluorophor with NBD chloride reaction. *2473*

Phenylephrine   Yields fluorophor with NBD chloride reaction. *2473*

Phenylpropanolamine   Yields fluorophor with NBD chloride reaction. *2473*   Reacts with methyl orange method of Frings. *2474*

Prilocaine   Yields fluorophor with NBD chloride reaction. *2473*

Propylhexedrine   Yields fluorophor with NBD chloride reaction. *2473*

Tuaminoheptane   False positive fluorescent method of Hayes at 50 mg/L. *1534*

Urine   Increase   Physiological

Pentazocine   Interferes with colorimetric methyl orange method. *1207*

Urine pH   Excretion increased with alkaline urine. *1636*

Urine   Positive   Analytical

Isometheptene   Reacts as if amphetamine in EMIT screening and confirmatory assays (Note compound is a component of Midrin® used to treat migraine). *2146*

Ampicillin

Serum   Decrease   Physiological

Food   Absorption reduced when taken with food. *3024*

Amylase

Serum   No Effect   Analytical

Acetaminophen   At 5 times therapeutic concentration on maltotriose method on aca, maltotetrose method on Cobas-Bio and amylopectin method on Ektachem®. *2138*

Acetylsalicylic Acid   At 5 times upper limit of therapeutic range on methods on aca, Cobas-Bio, and Ektachem®. *2138*

Aminomel LX 6   At concentration of 16 g/L had no effect on maltotetrose method. *3393*   At concentration of 16 g/L had no effect on p-nitrophenylmaltoheptoside method. *3393*   At concentration of 16 g/L had no effect on p-nitrophenylmaltopentoside/hexoside method. *3393*

Aminosalicylic Acid   At 5 times upper limit of therapeutic range on methods on aca, Cobas-Bio and Ektachem®. *2138*

Ascorbic Acid   Up to 1 mmol/L on method of Rauscher et al. *2926*   At concentration of 1,000 mg/L had no effect on maltotetrose method. *3393*

Bilirubin   No effect on Phadebas procedure. *1726*   At 0.17 mmol/L on method of Rauscher et al. *2926*

Bromazepam   At 5 times therapeutic concentration on methods on aca, Cobas-Bio, Ektachem®. *2138*

Carbenicillin   At 5 times upper limit of therapeutic range on methods on aca, Cobas-Bio and Ektachem® 400. *2138*

Cefoxitin   At 5 times upper limit of therapeutic range on methods on aca, Cobas-Bio and Ektachem®. *2138*

Cephalothin   At 5 times upper limit of therapeutic range on methods on aca, Cobas-Bio and Ektachem®. *2138*

Chlordiazepoxide   At 5 times upper limit of therapeutic range on methods on aca, Cobas-Bio and Ektachem®. *2138*

Cyclosporine   At concentration of 41.5 mmol/L (50 µg/L) on methods on Technicon® SMAC® II and Hitachi® 705. *1123*

Cysteine   At 5 times upper limit of therapeutic range on methods on aca, Cobas-Bio, and Ektachem®. *2138*

Cystine   At 5 times upper limit of therapeutic range on methods on aca, Cobas-Bio and and Ektachem®. *2138*

Dextran 40   At concentration of 80,000 mg/L had no effect on maltotetrose method. *3393*   At concentration of 80,000 mg/L had no effect on p-nitrophenylmaltoheptoside method. *3393*   At concentration of 80,000 mg/L had no effect on p-nitrophenylmaltopentoside/-hexoside method. *3393*

Diazepam   At 5 times upper limit of therapeutic range on methods on aca, Cobas-Bio, and Ektachem®. *2138*

EDTA   At concentration of 5380 mg/L had no effect on maltopentose method on aca. *3393*   At concentration of 1170 mg/L had no effect on p-nitrophenylmaltopentoside/hexaoside method. *3393*

Flurazepam   At 5 times upper limit of therapeutic range on methods on aca, Cobas-Bio, and Ektachem®. *2138*

Glucose   No effect on Phadebas procedure. *1726*   At 100 mmol/L on method of Rauscher et al. *2926*

Haemaccel   At concentration of 7,000 mg/L had no effect on maltotetrose method. *3393*   At concentration of 7,000 mg/L had no effect on p-nitrophenylmaltoheptoside method. *3393*   At concentration of 7,000 mg/L had no effect on p-nitrophenylmaltoheptoside/hexoside method. *3393*

Hemoglobin   No effect on Phadebas procedure. *1726*   With up to 35 µmol/L with method of Rauscher et al. *2926*

Heparin   No effect on amylase activity. *3873*   With sodium salt at 750 mg/L on method of Rauscher et al. *2926*   At concentration of 400 mg/L had no effect on maltopentose method on aca. *3393*   At concentration of 5 mg/L had no effect on maltotetrose method. *3393*   At concentration of 28600 mg/L had no effect on p-nitrophenylmaltopentoside/hexoside method. *3393*

Ibuprofen   At 5 times upper limit of therapeutic range on methods on aca, Ektachem® and Hitachi® 705. *2138*

Levodopa   At concentration of 1,000 mg/L had no effect on maltotetrose method. *3393*

Lipemia   Up to 11.4 mmol/L with method of Rauscher et al. *2926*

Mannitol   At concentration of 40,000 mg/L had no effect on maltotetrose method. *3393*   At concentration of 40,000 mg/L had no effect on p-nitrophenylmaltoheptoside method. *3393*   At concentration of 40,000 mg/L had no effect on p-nitrophenylmaltopentoside/-hexoside method. *3393*

Methicillin   At 5 times upper limit of therapeutic range on methods on aca, Cobas-Bio and Ektachem®. *2138*

Methotrexate At 5 times upper limit of therapeutic range on methods on aca, Cobas-Bio and Ektachem®. *2138*

Naproxen At 5 times upper limit of therapeutic range on methods on aca, Cobas-Bio and Ektachem®. *2138*

Nitrofurantoin On routine methods in use on SMAC®, Ektachem®, Abbott-VP, Cobas-Bio, aca, Hitachi® 705, KDA at 5 times normal upper therapeutic concentration. *2138*

Penicillin G At 5 times upper limit of therapeutic range on methods on aca, Cobas-Bio and Ektachem®. *2138*

Plasma No significant difference from serum with diamyl procedure. *2228*

Rifampin At 5 times upper limit of therapeutic range on methods on aca, Cobas-Bio and Ektachem®. *2138*

Sodium Fluoride At concentration of 15 mg/L had no effect on p-nitrophenylmaltopentoside/-hexoside method. *3393*

Sodium Oxalate At concentration of 3080 mg/L had no effect on p-nitrophenylmaltopentoside/hexoside method. *3393*

Storage of Sample 1 week at room temperature, several months at 4°. *0877* 1 week at 4°, 1 mo at -20 at normal concentration. *1563* 7 d at room temperature and 4°, 150 at -20°. *3217*

Tetracycline At 5 times upper limit of therapeutic range on methods on aca, Cobas-Bio and Ektachem®. *2138*

Theophylline At 5 times upper limit of therapeutic range on aca, Cobas-Bio and Ektachem®. *2138*

Serum   No Effect   Physiological

Aspirin When approximately 3 g/d ingested for several weeks. *3076*

Muscular Exercise No effect even with strenuous exercise. *2845*

Serum   Increase   Analytical

Chloride Salts Chloride enhances enzyme activity. *3295*

Disulphine Blue At physiological concentration for more than 2 d on Phadebas Cibachron blue starch-complex reaction. *1464*

Fluorescein At expected concentration of 1.0 to 100 mg/L interfered with method on Ektachem® -700. *1722*

Fluorides Effect of fluoride, activates amylase. *3588*

Hemolysis Reported effect. *0248*

Lipemia Turbidity may affect some methods. *0579*

Pancreozymin Preparation contains amylase. *1237*

Saliva Saliva contains amylase (may affect pipetting). *1237*

Serum   Increase   Physiological

Aminosalicylic Acid May cause acute pancreatitis. *1343*

Aprotinin One case allergic pancreatitis reported. *2427*

Asparaginase May cause pancreatic toxicity. *1519* Reports vary from incidence of 2.5 to 16% cases of acute pancreatitis: usually mild. *0558* Hemorrhagic pancreatitis in fewer than 0.5% treated patients. *1398*

Aspirin Single case reported. *3879*

Azathioprine Unusual side effect but may cause pancreatitis. *1680* 6.2% of 116 patients receiving only drug demonstrated clinical pancreatitis. *2281*

Benzthiazide Rare, may occur as with other thiazides. *1680*

Bethanechol May cause increased secretion, spasm of sphincter of Oddi. *0816*

Chlorothiazide Infrequent consequence of therapy. *0363* 10 of 20 patients developed 50 to 100% increase in serum amylase shortly after beginning treatment. *2281*

Chlorthalidone May precipitate acute pancreatitis. *1819*

Cholinergics Cause spasm of sphincter of Oddi. *0127*

Cisplatin 10 cycles in 4 patients associated with increased activity: increase mild and transient up to 2 x normal limit. *3698*

Codeine May cause spasm of sphincter of Oddi. *1343*

Copper Pancreatitis with dialysis induced toxicity. *1955*

Corticosteroids Condition like acute idiopathic pancreatitis. *3498*

Corticotropin Increase may be marked (both short and long term). *0622*

Cyclothiazide Occasional case of pancreatitis observed. *1680*

Cyproheptadine Mechanism not listed. *1488*

Dexamethasone May cause pancreatitis as side effect. *1680*

Diatrizoic acid Blockage of pancreatic duct for 6-18 h. *0085*

Ethacrynic Acid Isolated case of acute pancreatitis. *1488*

Ethanol Due to stimulation of pancreatic secretion. *1238*

Fentanyl May cause spasm of sphincter of Oddi. *0071*

Fludrocortisone May cause hemorrhagic pancreatitis. *1680*

Furosemide May induce mild pancreatitis. *3505* Isolated case of acute hemorrhagic pancreatitis. *1820* Not significant increase in total amylase in 12 hypertensives Increase by 16% on average in 12 hypertensives. *2015*

Glucocorticoids Well documented early and late pancreatitis. *2427*

Histamine Subcutaneous injection may cause acute pancreatitis. *3206*

Hydrochlorothiazide Acute pancreatitis may occur. *3572*

Hydroflumethiazide Acute pancreatitis may occur with thiazides. *1680*

Indomethacin Single case reported (?correct implication). *1420*

Isoniazid Reported cause of acute pancreatitis, much doubt. *1488*

Meperidine May cause spasm of sphincter of Oddi. *3316*

Mercaptopurine One case of hemorrhagic pancreatitis. *2427*

Methacholine Stimulates pancreatic secretion, constricts ampulla. *1343* Pancreatic stimulation, constriction of ampulla. *1237*

Methanol Elevation due to pancreatitis. *1238*

Methyclothiazide Pancreatitis may occur with thiazide therapy. *1680*

Methylprednisolone To well above normal in 10 patients given 1 g/d i.v. for 3 d. Maximum effect days 5 to 8. *0804*

Morphine Causes spasm of sphincter of Oddi for 48 h. *3505*

Narcotics Cause spasm of sphincter of Oddi. *0127*

Nitrofurantoin Isolated case confirmed by rechallenge, rapidly resolved on withdrawal of drug. Edema of pancreatic head also caused jaundice. *2571* Isolated case confirmed by rechallenge, rapidly resolved on withdrawal of drug. Edema of pancreatic head also caused jaundice. *2571*

Opium Alkaloids Impaired excretion spasm of sphincter of Oddi. *0652*

Oral Contraceptives In approximately 23%: probably of liver origin. *0024* Isolated case of acute pancreatitis reported. *2519*

Oxyphenbutazone Parotitis is rare complication of therapy. *1412* Occasional case of acute swelling of salivary glands. *0642*

Pancreozymin Effect seen when pancreatic disorders,?in normals. *3334*

Paramethasone May cause pancreatitis occasionally. *1680*

Pentazocine Causes spasm of sphincter of Oddi. *1606*

Phenylbutazone Parotitis may occur as rare complication. *1412* Sialadenitis reported to occur occasionally but in 5 of 7 patients in this study. *3409*

Polythiazide Rare pancreatitis reported with other thiazides. *1680*

Potassium Iodide Parotitis reported as result of treatment. *1680*

Prednisolone May rarely cause pancreatitis. *0019*

Prednisone May cause pancreatitis. *1680*

Pregnancy Approximately 70% higher than in normal women. *0024*

Procyclidine May cause acute parotitis (theoretical effect). *1680*

Radiographic Agents Cholangiography may cause transient increase. *1487*

Sex Difference Women approximately 16% higher than men. *0024*

Smoking Basal activity 100% higher in heavy smokers than non-smokers. *0961*

Sulfachlorpyridazine Occasional cause of pancreatitis. *1680*

Sulfamethizole Case of pancreatitis reported. *0236*

Sulfisoxazole May cause pancreatitis occasionally. *1680*

Tetracycline Toxic effect especially in pregnant women. *1488*

Thiazides May increase up to 200% in a week. *0729*

Triamcinolone May occasionally cause pancreatitis. *1680*

Trichlormethiazide May occasionally cause pancreatitis. *1680*

Amylase *(continued)*

Serum Increase Physiological *(continued)*

Valproic Acid Almost 20% patients had mild increase. *0213* Apparently drug related case of acute pancreatitis. *3152* Associated with dysphagia and epigastric discomfort in one adult patient who subsequently died. *2530* Reported in one woman on chronic hemodialysis. *2587* In 1 of 100 epileptic children when given alone or combined with other anticonvulsants. *0737*

X-Ray Therapy May be up to 20 fold increase at peak with whole body irradiation, mainly of salivary isoenzyme. *1837*

Serum Decrease Analytical

Cefotaxime General effect of both drug and metabolite on Parallel method. *0201*

Citrates Inhibition of enzyme activity. *2364* Sodium salt inhibits by 16% at 5 g/L method of Rauscher et al. *2926*

EDTA 15% inhibition at 1 g/L, 51% inhibition at 5 g/L, 73% inhibition at 10 g/L of method of Rauscher et al using nitrophenylmaltoheptaoside as substrate. *2926*

Fluorides Sodium salt inhibits by 4% at 10 g/L method of Rauscher et al. *2926*

Oxalate Approximately 20% decrease compared with serum/heparinized plasma. *0583* Sodium salt inhibits by 15% at 2 g/L with method of Rauscher et al. *2926*

Storage of Sample At pancreatitis level loss of 30% 18 h at 4 or -20°. *1670*

Serum Decrease Physiological

Anabolic Steroids Mean decrease from 112 to 89 units in patients with chronic pancreatitis after 3 weeks treatment with combination of anabolic steroids. *3617*

Propylthiouracil Reported effect. *0248*

Urine No Effect Analytical

Fluorescein In specimen spiked to 250 mg/L on Ektachem® 700 method. *1722*

Proteinuria No effect on Phadebas procedure. *1726*

Storage of Sample No effect 2 d at 4°, 1 week at -20°. *3873* 1 week at room temperature, months at 4° if no bacteriuria. *0877*

Thymol No effect on starch hydrolysis. *2981*

Urine No Effect Physiological

Oral Contraceptives Although high in serum, suggestive of hepatic origin. *0024*

Urine Increase Physiological

Morphine Observed in some patients after administration. *1397*

Nitrofurantoin Isolated case confirmed by rechallenge, rapidly resolved on withdrawal of drug. Edema of pancreatic head also caused jaundice. *2571* Isolated case confirmed by rechallenge, rapidly resolved on withdrawal of drug. Edema of pancreatic head also caused jaundice. *2571*

Pancreatic Juice No Effect Analytical

Radiographic Agents No effect of Gastrografin®. *0741*

Pancreatic Juice No Effect Physiological

5-Hydroxytryptamine No effect following injection. *3579*

Pancreatic Juice Increase Physiological

Secretin Normal above 14.9 U/kg/80 minutes. *0486*

Amylase, Pancreatic

Serum Increase Physiological

Furosemide Increase by 17% on average in 12 hypertensives. *2015*

Methylprednisolone To well above normal in 10 patients given 1 g/d i.v. for 3 d. Maximum effect days 5 to 8. Possible subclinical damage of pancreatic acinar cell *0804*

Amylase, Salivary

Serum No Effect Physiological

Methylprednisolone No effect in patients given drug i.v. although pancreatic component significantly affected (Amylase). *0804*

Androgens

Plasma No Effect Physiological

Diurnal Variation More effect of exercise than time of day. *0417*

Urine Decrease Physiological

Medroxyprogesterone Inhibition of steroid biosynthesis in adrenals. *3632*

5-Androstene-3 β, 17 β-Diol

Plasma Decrease Physiological

Testosterone In power athletes with 26 weeks on steroid self administration. *3088*

Androstenedione

Plasma No Effect Physiological

Cyclosporine In 16 patients who developed hypertrichosis but also given added cortisone. *3190*

Cyproterone In 27 women treated for total of 194 cycles with 50 μg ethinyl estradiol and 2.0 mg cyproterone acetate. *2280*

Dexamethasone No effect of treatment on initially high concentrations in hirsute women. *0776*

Digoxin No significant effect with long-term administration. *2576*

Ethanol No significant effect on concentration in individuals receiving alcohol over several days: no effect on pattern of secretion in healthy men. *1352*

Plasma Increase Physiological

Clomiphene Liberates LH, anti-steroid hormone effect of drug. *0609*

Cyproterone By up to 450% if acetate derivative used. *3394*

Danazol Increase by 70% in 7 normal subjects given drug 800 mg daily. *2224*

Dehydroepiandrosterone Mean increase from 4.3 to 8.6 nmol/L in 5 normal men given 1600 mg/d orally for 28 d. *2578*

Levonorgestrel After 1 mo in 25 female volunteers when given as subdermal implant. *2687*

Menstruation In follicular phase, highest in mid third of cycle. *1831*

Metyrapone Increase in 3-8 h in men, marked increase in women. *2600*

Smoking Significantly higher mean value in male current smokers than in nonsmokers with apparent dose response relationship. *0238*

Testosterone In power athletes with 26 weeks on steroid self administration. *3088*

Plasma Decrease Physiological

Carbamazepine Within 7 d of starting 400 mg/d treatment in 6 healthy males probably due to induction of hepatic monooxygenase activity. *0707*

Ketoconazole Observed in 2 h after single 200 mg dose in normal men. *3390*

Levonorgestrel In 17 women using Norplant implants. *1799*

Androstenedione, Free

Serum Decrease Physiological

Ketoconazole Observed in 2 h after single 200 mg dose in normal men. *3390*

Androstenetriol Sulfate

Urine   No Effect   Analytical
Ampicillin   No effect on GLC method. *3622*

Urine   Decrease   Physiological
Ampicillin   Increase by 52% after 6 d in pregnant women. *3622*

Androsterone

Plasma   No Effect   Physiological
Obesity   In females but falls with caloric restriction. *3391*

Plasma   Increase   Physiological
Corticotropin   Hormonal effect. *2435*

Urine   No Effect   Physiological
Metoclopramide   No significant influence of drug on excretion in menstruating women. *0096*

Urine   Increase   Physiological
Corticotropin   Normal response to 2 d ACTH test. *1474*
Gonadotropin   150% increase when given to normal men. *3231*

Urine   Decrease   Physiological
Dexamethasone   Suppression of ACTH. *2420*
Oral Contraceptives   Compared with controls — details not discussed. *0527*

Angiotensin

Plasma   Increase   Physiological
Oral Contraceptives   Twofold increase. *0585*
Sleep   Rises during sleep. *1190*

Angiotensin-Converting Enzyme (ACE)

Serum   No Effect   Analytical
Acetylsalicylic Acid   No effect observed even with large amounts. *3038*
Bilirubin   No effect on colorimetric method of Boomsma and Schalekamp. *2814*
Hemoglobin   No effect reported on colorimetric method of Boomsma and Schalekamp. *2814*
Lipemia   No effect reported on colorimetric method of Boomsma and Schalekamp. *2814*
Prednisolone   After 96 h incubation *in vitro* at high doses produced marked reduction: unlikely to be factor *in vivo*. *3074*
Prednisone   No effect on colorimetric method of Boomsma and Schalekamp. *2814*

Serum   No Effect   Physiological
Indomethacin   In 9 uncomplicated essential hypertensives receiving chronic enalapril treatment. *3126*
Nifedipine   No significant effect within 3 h after 10 mg orally. *1609*
Prednisolone   75 mg drug orally had no effect in 10 patients with sarcoidosis on diurnal variation or enzyme activity. *3580*

Serum   Increase   Analytical
Freezing   Freezing with subsequent thawing increases activity by 15%. *3221*
Uric Acid   At concentrations above 600 μmol/L on method of Boomsma and Schalekamp. *2814*

Serum   Increase   Physiological
Tri-iodothyronine   In 7 normal women aged 18-27 y given 25 μg/d three times daily for 14 d. Effect observed on endothelium-associated proteins but not on hepatically synthesized proteins. *1383*

Serum   Decrease   Analytical
Captopril   31% inhibition of method using benzyloxycarbonyl-phenylalanyl-histidyl-leucine as substrate. *3221*   Marked inhibition of enzyme by drug. *3038*
EDTA   40% inhibition reported with method using benzyloxy-carbonyl-phenylalanyl-histidyl-lecuine as substrate. *3221*
Hydroxyquinoline   100% inhibition with method using benzyloxy-carbonyl-phenylalanyl-histidyl-leucine as substrate with 8 hydroxy-quinolone. *3221*
Methylprednisolone   Moderate inhibition by doses generally greater than used clinically. *3038*   After 96 h incubation *in vitro* at high concentrations but no effect under normal measurement conditions. *3074*
Phenanthroline   80% inhibition of method using benzyloxy-carbonyl-phenylalanyl-histidyl-leucine as substrate. *3221*

Serum   Decrease   Physiological
Captopril   May be marked reduction in drug-treated patients. *1855*
Enalapril   Good correlation between serum concentration of drug and inhibition of enzyme in both acute and chronic studies. *1747*   Significant fall even on first day of treatment. *1616*   Marked effect in 4 d as part of long-term response in 6 responders of 10 treated hypertensives. *1166*
Magnesium Sulfate   Decreased in 16 women with pregnancy induced hypertension 1-8 h after treatment then plateaued. *1213*
Prednisone   Marked reduction after 1 week in patients in whom initial value was high, no effect in others. *3038*
Ramipril   In response to 10 mg and 20 mg on successive days in 9 patients with severe chronic congestive heart failure. *0764*
Teprotide   43% inhibition reported with method using benzyloxy-carbonyl-phenylalanyl- histidyl-leucine as substrate. *3221*

Plasma   Decrease   Physiological
Perindopril   With single doses of 8 to 16 mg produced reduction to less than 10% of control in 4 h with lasting effect for 72 h. *0550*

Angiotensin I

Plasma   Increase   Analytical
Protein   A protein normally present may interfere. *2705*

Plasma   Increase   Physiological
Enalapril   Increases as angiotensin II falls after first dose. *1616*
Ethinyl Estradiol   Approximately x 3 in 2 weeks in normals. *2357*
Nifedipine   In young but not old people within 3 h of 10 mg orally. *1609*   From 1923 to 2669 pg/mL in young hypertensives after 3 h. *1609*

Plasma   Decrease   Physiological
Recumbency   Fell by just over half in 3 h. *0733*

Blood   Increase   Physiological
Perindopril   After doses of 4 mg twice daily within 4 h of administration. *0550*

Angiotensin II

Plasma   No Effect   Physiological
Felodipine   With 20 mg daily in 10 men with essential hypertension over 8 weeks. *1887*
Indomethacin   In 12 patients with coronary artery disease, although systolic blood pressure increased and coronary blood flow decreased. *1158*
Verapamil   In 15 patients with uncomplicated essential hypertension. *0867*   No change compared with placebo in 11 patients treated with up to 360 mg daily for 6 weeks. *3396*

Plasma   Increase   Analytical
Contact With Clot   No effect until delay exceeds 5 h. *0977*
Protein   A protein normally present may interfere. *2705*

Angiotensin II (continued)

Plasma   Increase   Physiological
Amiloride   Change varies with renin activity. *2440*
Bendrofluazide   Approximately doubled in 5 or 6 hypertensives given 5 mg daily for 6 weeks. *3866*
Estrogens   Occurs within 5 d, increase may be 3 times normal. *0557*
Furosemide   Rapid response to i.v. injection. *0977*
Metolazone   2 to 3 fold increase after 6 weeks treatment with 5 mg daily in 5 or 6 hypertensives. *3866*
Nifedipine   In young but not old people within 3 h of 10 mg orally. *1609*   From 167 to 215 pg/mL in young hypertensives after 3 h. *1609*
Oral Contraceptives   Elevated to 3 times normal during administration. *0557*
Pregnancy   Increase in 1st trimester due to activity of renin system. *3345*
Sodium Restriction   Trend in 2 d, maximum in 5, with decreased sodium. *0495*
Spironolactone   Change varies with renin activity. *2440*

Plasma   Decrease   Analytical
Hemolysis   Significant effect observed. *0977*

Plasma   Decrease   Physiological
Captopril   Significant effect in 1 h after single dose of 25 mg. *0181*   Parallel decline with sodium in congestive heart failure patients. *2592*   Fell in parallel with reduction of blood pressure. *1716*   Prompt and striking reduction following oral administration over 30 minutes. *0177*
Enalapril   Significant fall after first dose, subsequently remains depressed. *1616*   Prompt and striking reduction following i.v. administration over 30 min: more gradual reduction when given orally. *0177*
Perindopril   After doses of 4 mg twice daily within 4 h of administration. *0550*
Ramipril   In response to 10 mg and 20 mg on successive days in 9 patients with severe chronic congestive heart failure. *0764*

Angiotensinogen

Plasma   Increase   Physiological
Oral Contraceptives   Threefold increase with 0.5 mg mestranol. *0585*

Anisindione

Plasma   Increase   Physiological
Quinidine   Potentiates action. *1487*

Anthranilic Acid

Urine   No Effect   Physiological
Smoking   No effect of smoking for 1 week. *0068*

Urine   Decrease   Physiological
Sleep Deprivation   Noticed after 5 d (?due to relative B_6 deficiency). *2032*

Antibodies to DNA

Serum   Decrease   Physiological
Captopril   In 3 of 78 patients, of IgM class, treated for mean of 11 mo. *1850*

Serum   Positive   Physiological
Penicillamine   In 6 patients with rheumatoid arthritis developed systemic lupus erythematosus-like syndrome. *0623*

Antibodies to dsDNA

Serum   No Effect   Physiological
Hydralazine   Observed in one patient with induced systemic lupus erythematosus. *3784*
Quinidine   Reported in 5 cases of an SLE-like syndrome induced by the drug. *2093*

Serum   Positive   Physiological
Chlorpromazine   In 40% to native-DNA of schizophrenic patients given long-term treatment. *3937*

Antibodies to Histones

Serum   No Effect   Physiological
Quinidine   Reported in 5 cases of an SLE-like syndrome induced by the drug. *2093*

Serum   Positive   Physiological
Hydralazine   Observed in 9 patients developing rapidly progressive glomerulonephritis during treatment. *0366*
Procainamide   Antibodies to histone induced by drug. *2848*

Antibodies to RNP

Serum   No Effect   Physiological
Quinidine   Reported in 5 cases of an SLE-like syndrome induced by the drug. *2093*

Antibodies to SCL-70

Serum   No Effect   Physiological
Quinidine   Reported in 5 cases of an SLE-like syndrome induced by the drug. *2093*

Antibodies to Sjögren Syndrome A

Serum   No Effect   Physiological
Quinidine   Reported in 5 cases of an SLE-like syndrome induced by the drug. *2093*

Antibodies to Sjögren Syndrome B

Serum   No Effect   Physiological
Quinidine   Reported in 5 cases of an SLE-like syndrome induced by the drug. *2093*

Antibodies to SmAg

Serum   No Effect   Physiological
Quinidine   Reported in 5 cases of an SLE-like syndrome induced by the drug. *2093*

Antibody Evaluation

Serum   Decrease   Analytical
Detergents   Increased concentrations cause decreased agglutination on AutoAnalyzer. *1424*

α_1-Antichymotrypsin

Serum   No Effect   Physiological
Smoking   No significant difference between smokers and non-smokers. *0809*

Antidiuretic Hormone

Plasma   No Effect   Analytical
Storage of Sample   No effect probably stored frozen for 45 d. 1180

Plasma   No Effect   Physiological
Chlorpromazine   No effect of intramuscular injection of drug. 2920
Dobutamine   In 10 patients with congestive cardiac failure. 3654
Enoximone   In 10 patients with congestive cardiac failure. 3654
Haloperidol   No effect of intravenous injection. 2920
Naloxone   No effect of 10 mg on basal values. 0176
Ramipril   In response to 10 mg and 20 mg on successive days in 9 patients with severe chronic congestive heart failure. 0764
Verapamil   No change compared with placebo in 11 patients treated with up to 360 mg daily. 3396

Plasma   Increase   Physiological
Azosemide   Effect observed in normal male volunteers after 60 mg orally. 2658
Chlorthalidone   Secreted in response to hyponatremia. 1129
Diurnal Variation   Secretion increased at night. 2222
Erect Posture   In comparison with sitting (3.1 pg/mL versus 1.4 pg/mL). 2650
Ether   Effect in moderate to deep anesthesia. 1343
Furosemide   Effect observed after 40 mg orally in normal male volunteers. 2658
Hydrochlorothiazide   Secreted in response to hyponatremia. 1129
Lithium   Also acts like vasopressin. 1792   Occurs in majority of patients with polyuria consistent with defect in water balance at level of kidney (lithium administration now most common cause of nephrogenic diabetes Insipidus). 3118
Methyclothiazide   Secreted in response to hyponatremia. 1129
Nicotine   In smokers, maximum immediately after smoking. 1703
Polythiazide   Secreted in response to hyponatremia. 1129

Plasma   Decrease   Physiological
Ethanol   Initial fall after ingestion of 75 mL although rose later. 1007
Recumbency   In comparison with sitting (0.4 pg/mL versus 1.4). 3241

Urine   No Effect   Analytical
Storage of Sample   At room temperature for 45 d if acidified. 1180

Urine   No Effect   Physiological
Diurnal Variation   No effect observed on hourly measurements. 1190
Sleep   No effect observed on hourly measurements. 1190
Stress Cold Exposure   Response to cold exposure. 3217

Urine   Increase   Physiological
Chlorpropamide   Stimulates release from neurohypophysis. 1239
Glyburide   Slight (not significant) effect. 2499
Imipramine   Mean excretion of 10.6 mU/h in 7 normal and 10 depressive patients when given 75 mg/d compared to control of 2.6 mU/h. 2889
Pain   Also seen more markedly with emotional stress. 1990   Response to muscle pain. 3217
Stress   With mental stress increase of 50%. 3217

Antimitochondrial Antibodies

Serum   No Effect   Physiological
Cimetidine   No effect noted in 12 patients treated for acid-peptic disease. 2368

Serum   Positive   Physiological
Halothane   Liver disease from mild focal necrosis and inflammation to massive necrosis. 2260
Oxyphenisatin   Very rare observation with chronic hepatitis. 1297

Antinuclear Antibodies

Serum   No Effect   Physiological
Cimetidine   No effect noted in 12 patients treated for acid-peptic disease. 2368
Methyldopa   In a study of 9 hypertensives. 1909
Phenytoin   Same frequency before and after drug therapy. 0228
Propylthiouracil   Incidence no different from controls. 0616

Serum   Increase   Physiological
Captopril   In 13 of 78 patients, mainly of IgM class treated for mean of 11 mo. 1850
Labetalol   In one patient previously treated with methyldopa and atenolol. 1400
Nitrofurantoin   Increased to 1:640 chronic active hepatitis, much more common in women than men, but still rare. 2444   Increased to 1:640 chronic active hepatitis, much more common in women than men, but still rare. 2444
Phenytoin   Occasional increase; possible subclinical collagen-vascular disorder. 1502
Propylthiouracil   Observed in 10 of 53 treated cases without adequate criteria for diagnosis of SLE. 3870
Quinidine   Reported in 5 cases of an SLE-like syndrome induced by the drug. 2093

Serum   Positive   Physiological
Acebutolol   In 3 of 11 patients treated for 48 weeks: readministration after titer became negative resulted in significant rise. 0386   Observed in 18.6% diabetics with drug versus 3.8% in normals and 1.3% in diabetics without drug. 2104   In 20% of 35 men and 44% of 23 women in comparison with 10.2% in men and 12.9% women with other antihypertensive drugs. 0419   Developed in 8 of 9 patients treated for 12 to 24 weeks. 0680
Anticonvulsants   Related to number of drugs, higher in women. 3797
Chlorpromazine   Single case reported associated with rash and fever. 0963   In 63% of schizophrenic patients given long-term treatment. 3937
Ethosuximide   Related to number of drugs, higher in women. 3797
Hydralazine   More common in slow acetylators. 2783   Raised in patients who developed symptoms of lupus, unrelated to dose of drug. 0353
Isoniazid   Up to 78% tuberculous patients develop antibodies. 0043
Methyldopa   More common in females than males. 1692   Observed in one patient with hepatocellular damage of moderate severity due to sensitization by drug. 0874
Mexiletine   Occurs in fewer than 1% treated patients. 2011
Oral Contraceptives   Slight effect in users. 2335
Oxyphenisatin   May cause hypersensitivity reaction. 2970   Liver damage may present as acute hepatitis or as more advanced chronic disease: usual liver disease is chronic active hepatitis. 2260
Penicillamine   In 6 patients with rheumatoid arthritis developed systemic lupus erythematosus-like syndrome. 0623
Phenytoin   Elicited in 25% subjects treated. 0616   Related to number of drugs, higher in women. 3797
Procainamide   Reported to occur in 50% patients. 3505   50-80% patients have antibodies after 3 to 6 mo. 1958
Smoking   If smokers all ages both sexes. 2335
Tocainide   Rare finding in fewer than 0.2% patients. 2011
Trimethadione   Related to number of drugs, higher in women. 3797

Antinucleoprotein Antibodies

Serum   Positive   Physiological
Chlorpromazine   In 58% of schizophrenic patients given long-term treatment. *3937*

Antipyrine

Plasma   No Effect   Physiological
Diphenhydramine   No significant effect on half-life. *3466*
Methaqualone   No significant effect on half-life. *3466*
Nitrazepam   No significant effect on half-life. *3466*

Plasma   Increase   Physiological
Allopurinol   Impairs metabolism. *2641*
Cimetidine   Clearance reduced: probably by inhibition of drug metabolism with increased half-life. *3256*   Plasma clearance reduced from 54 mL/min to 48 mL/min in 7 healthy subjects given 1 g/d. *3004*
Disulfiram   Inhibits hydroxylation, prolongs action. *2641*
Nortriptyline   Impairs metabolism. *2641*

Plasma   Decrease   Physiological
Amobarbital   Significant effect on half-life. *3466*

Antismooth Muscle Antibodies

Serum   No Effect   Physiological
Cimetidine   No effect noted in 12 patients treated for acid-peptic disease. *2368*

Serum   Increase   Physiological
Nitrofurantoin   Increased to 1:640 chronic active hepatitis, much more common in women than men, but still rare. *2444*   Increased to 1:640 chronic active hepatitis, much more common in women than men, but still rare. *2444*

Serum   Positive   Physiological
Methyldopa   Disturbances in liver function in 5 to 35% of patients treated for hypertension. Hepatocellular injury may occur after short- or long-term exposure. *0162*   Chronic active hepatitis associated with immune hemolytic anemia in one case. *3269*
Oxyphenisatin   As component of hypersensitivity response. *1297* Liver damage may present as acute hepatitis or as more advanced chronic disease: usual liver disease is chronic active hepatitis. *2260*

Antithrombin III

Plasma   No Effect   Analytical
Acetylsalicylic Acid   At concentration of 3 mg/L had no effect on aca method. *3393*
Cephalosporin   At concentration of 30 mg/L had no effect on aca method. *3393*
Chloramphenicol   At concentration of 25 mg/L had no effect on aca method. *3393*
Chlorpromazine   At concentration of 3 mg/L had no effect on aca method. *3393*
Dextran 70   At concentration of 1500 mg/L had no effect on aca method. *3393*
Digitoxin   At concentration of 20 mg/L had no effect on aca method. *3393*
Digoxin   At concentration of 10 mg/L had no effect on aca method. *3393*
EDTA   At concentration of 1170 mg/L had no effect on aca method. *3393*
Heparin   At concentration of 2860 mg/L had no effect on aca method. *3393*
Penicillin   At concentration of 100 mg/L had no effect on aca method. *3393*

Protamine   At concentration of 250 mg/L had no effect on aca method. *3393*
Sodium Fluoride   At concentration of 4238 had no effect on aca method. *3393*
Sodium Oxalate   At concentration of 17109 mg/L had no effect on aca method. *3393*
Tetracycline   At concentration of 1,000 mg/L had no effect on aca method. *3393*
Tobramycin   At concentration of 15 mg/L had no effect on aca method. *3393*

Plasma   No Effect   Physiological
Norethisterone   In 75 women who had received up to 24 intragluteal injections of 200 mg every 56 d. *1666*
Oral Contraceptives   No significant changes at various intervals during treatment compared with controls. *1418*
Tri-iodothyronine   In 7 normal women aged 18-27 y given 25 μg/d three times daily for 14 d. Effect observed on endothelium-associated proteins but not on hepatically synthesized proteins. *1383*

Plasma   Decrease   Physiological
Apalcillin   In 21 volunteers with doses up to 225 mg/kg. *1262*
Asparaginase   Mean decrease of 68% (immunological) and 74% (functional assay) in different studies. *0558*
Estrogens   ?predisposing cause of thrombosis. *3961*
Ethinyl Estradiol   In over 20% — not dose related. *3960*
Mestranol   In over 20% — not dose related. *3960*
Oral Contraceptives   ?predisposing cause of thrombosis. *3961* Slight reduction during treatment with drug low in estrogen content: note effect not as marked as reported when high estrogens content preparations used. *1795*
Tamoxifen   Lowered functional activity in 42% treated patients. *1033*

Antithrombin III Antigen

Plasma   Decrease   Physiological
Oral Contraceptives   By activity unchanged in women given low dose ethinyl estradiol and norethindrone. *2623*

Antithrombin Titer

Plasma   Decrease   Physiological
Pregnancy   Occurs with pregnancy. *0987*

Antithyroglobulin Antibodies

Serum   Increase   Physiological
Nitrofurantoin   Mild increase developed lupus-like syndrome associated with pulmonary reaction. *3250*

α_1-Antitrypsin

Serum   No Effect   Analytical
Hemolysis   On AutoAnalyzer immunological method. *2297*
Storage of Sample   For 7 d at 4 or 30°, indefinitely at -70°. *0573*

Serum   No Effect   Physiological
Anticonvulsants   In 8 epileptic patients receiving phenobarbital with carbamazepine or phenytoin. *2670*
Menstruation   No effect observed during menstrual cycle. *2031*
Muscular Exercise   No effect of exercise observed. *2846*
Oral Contraceptives   No significant changes at various intervals during treatment compared with controls. *1418*
Smoking   No significant difference between smokers and non-smokers. *0809*

Tri-iodothyronine In 7 normal women aged 18-27 y given 25 µg/d three times daily for 14 d. Effect observed on endothelium-associated proteins but not on hepatically synthesized proteins. *1383*

Serum Increase Physiological

Aging Very slight increase with age. *0954*

Aminocaproic Acid Inhibits activation of plasminogen, plasmin and trypsin. *1988*

Dextran Possibly associated with underlying disease. *3346*

Estrogens Metabolic effect. *0227*

Ethanol Heterozygotes with alcoholic liver disease have higher concentration than usual mean for healthy PI MZ individuals. *3001*

Muscular Exercise Significant effect at 15 minutes, partial return by 1 d. *1491*

Oral Contraceptives Metabolic changes in liver synthesis (estrogen). *2189* In women given low dose ethinyl estradiol and norethindrone. *2623* But effects less marked with low estrogen preparations. *1568*

Oxymetholone Metabolic effect. *0227*

Physical Training Defense against proteolytic activity in athletes. *1491*

Pregnancy In last trimester compared with controls. *2385*

Sex Difference Approximately 30 mg/dL higher in women. *0954*

Streptokinase Significant effect after infusion in infarct patients. *1356*

Tamoxifen When 10 mg given twice daily to 30 Z homozygous α^1-antitrypsin-deficient subjects for 30 d although change slight. *3808* Significant change due to mild estrogenic effect of drug in breast cancer patients. *3070*

Typhoid Vaccine May be considerable rise after single injection. *2031*

Apolipoprotein A

Serum No Effect Physiological

L-Thyroxine Insignificant change from 2.7 to 2.8 g/L in 11 hypothyroid women treated with 0.1 to 0.2 mg daily. *1201*

Apolipoprotein AI

Serum No Effect Physiological

Acebutolol Increase by 1 mg/dL in 11 patients given 400 mg daily for 3 mo. *2550*

Bezafibrate In 11 hypertriglyceridemic subjects. *3291*

Bunitrolol When 30 mg/d given for 12 weeks in normolipidemic patients with mild essential hypertension. *3151*

Captopril Increase by 1 mg/dL in 15 patients given 75 mg daily for 8 weeks. *2550*

Carbimazole From 2.62 to 2.82 g/L in 12 hyperthyroid women patients treated with 10-30 mg daily. *1201*

Chenodeoxycholic Acid No significant effect of 12 mo treatment in 252 patients. *0046*

Chlorthalidone During diuretic therapy for 6 weeks in 23 subjects. *1307* In 22 premenopausal and 18 postmenopausal women given 100 mg/d for 6 weeks. *0402* Decrease by -1 mg/dL in 10 patients given 100 mg daily for 1 mo. *2550* No effect of monotherapy for 6 weeks. *0089*

Etretinate In 13 patients with hyperkeratotic disorders given 50 mg daily for 2 mo. *3659*

Furosemide Decrease by -2 mg/dL in 12 patients given 40 to 80 mg daily for 3 mo. *2550*

Glucose In response to infusion of 1 g/kg body weight in normal volunteers and non-insulin dependent diabetics. *0476*

Guanabenz Increase by 5 mg/dL in patients given 4 to 8 mg daily for 12 to 16 weeks. *2550*

Hydrochlorothiazide Usual observation when thiazides given to patients. *3780*

Indapamide In 13 hypertensive patients with diabetes treated for 24 weeks. *2686*

Isotretinoin In 12 patients with hyperkeratotic disorders given 40 mg daily for 2 mo. *3659*

Levonorgestrel When 150 µg levonorgestrel given with 30 µg ethinyl estradiol for 3 mo. Results normal within 2 mo after treatment stopped. *0312*

Methyldopa In several patients treated for 2-4 mo. *0088* Decrease by -2 mg/dL in 11 patients given 750 mg daily for 8 weeks. *2550*

Nifedipine Increase by 3 mg/dL in patients given 30 to 60 mg daily for 8 weeks. *2550*

Oral Contraceptives Typically insignificant change over 6 cycles in women with low dose ethinyl estradiol with low dose levonorgestrel or desogestrel. *1249*

Pindolol No significant changes after 12 mo treatment. *2116* No significant changes found during 12 mo treatment. *2117*

Prazosin Increase by 1 mg/dL in 16 patients given 1 to 3 mg daily for 8 weeks. *2550*

Prednisone No change noted after 1 mo but ratio of HDL-C to apo A-I increased, apparent in 48 h. *3949*

Propranolol No change observed in 11 patients. *2476* Decrease by -3 mg/dL in 8 patients given 30 to 60 mg daily for 8 weeks. *2550* No change observed in 11 patients. *2476*

Trichlormethiazide Decrease by -5 mg/dL in 15 patients given 4 mg daily for 3 mo but note marked difference between responders and nonresponder populations. *2550*

Serum Increase Physiological

Aminoglutethimide In 73 patients with advanced breast cancer receiving 500 mg/d with 40 mg hydrocortisone. *0415*

Captopril After 12 weeks treatment in 18 patients with mild essential hypertension. *3150*

Clofibrate Increase by 10 to 20% in 10 hyperlipidemic patients. *2577*

Clopamide No significant change in 17 individuals treated for less than 1 y when used as sole treatment. *3184*

Cyclofenil Increase by 15% after 2 and 4 mo after therapy in 19 patients given 600 mg daily for 4 mo. *2786*

Desogestrel Increase by 20% when 150 µg desogestrel given with 30 µg ethinyl estradiol for 3 mo. Results normal within 2 mo after treatment stopped. *0312*

Ethanol Increase by 6.5% in 78 intemperate drinkers on average. *2882* Significantly higher in drinkers and falls with abstinence, but increases with resumption of moderate drinking. *0565* In normal volunteers after 60-70 g/d for 2 weeks but concentration did not exceed normal range. However turnover substantially increased. *2282* Initially high values at end of alcoholic debauch fell to normal with cessation of drinking. *1011*

Gemfibrozil Increase by 29% in 6 patients with primary familial endogenous hypertriglyceridemia from baseline values. Synthetic rates of apo A-I and apo A-II increased by 27% and 34% respectively. *3117*

Lovastatin Slight effect in clinical trial of 101 patients with heterozygous familial hypercholesterolemia. *1530*

Niacin Increase by 12% in 34 hypercholesterolemic individuals with 1.5 g/d for 1 mo then 3.0 g/d up to 6 mo. *1967*

Nifedipine In 11 patients with 80 mg daily for 6 weeks. *3702* Increase by 5 % in 11 individuals treated for 1.5 mo. *0088*

Phenytoin Mean of 2.1 g/L versus 1.8 g/L in controls in 28 patients treated with 200-300 mg/d for 1 to 35 y. *2598*

Serum Decrease Physiological

Atenolol Reduction by 6% reported from one study. *0088*

Ethanol Significantly lower in alcoholic men than in controls. *1701*

Medroxyprogesterone Increase by 7% in 11 men with sexual deviation syndrome given approximately 1273 mg over a total of approximately 17 d. *0643*

Probucol In 50 diabetics given 500 mg/d for 16 weeks and reduction greatest in highest cholesterol and triglyceride patients. *1523*

Smoking In young nonobese normolipidemic men (same pattern as observed in others with increased risk of atherosclerosis. *0845* Mean 1.09 mg/dL vs 1.29 mg/dL in controls. *0845*

Stanozolol Increase by 41% in 10 normolipidemic postmenopausal osteoporotic women treated for 6 weeks. *3526*

Laboratory Test Listings

Apolipoprotein AII

Serum   No Effect   Physiological

Bezafibrate In 11 hypertriglyceridemic subjects. *3291*

Bunitrolol When 30 mg/d given for 12 weeks in normolipidemic patients with mild essential hypertension. *3151*

Chenodeoxycholic Acid No significant effect of 12 mo treatment in 252 patients. *0046*

Chlorthalidone During diuretic therapy for 6 weeks in 23 subjects. *1307* In 22 premenopausal and 18 postmenopausal women given 100 mg/d for 6 weeks. *0402* No effect of monotherapy for 6 weeks. *0089*

Clopamide No significant change in 17 individuals treated for less than 1 y when used as sole treatment. *3184*

Etretinate In 13 patients with hyperkeratotic disorders given 50 mg daily for 2 mo. *3659*

Glucose In response to infusion of 1 g/kg body weight in normal volunteers and non-insulin dependent diabetics. *0476*

Hydrochlorothiazide Usual observation when thiazides given to patients. *3780*

Isotretinoin In 12 patients with hyperkeratotic disorders given 40 mg daily for 2 mo. *3659*

Methyldopa In several patients treated for 2-4 mo. *0088*

Niacin In 34 hypercholesterolemic individuals with 1.5 g/d for 1 mo, then 3.0 g/d up to 6 mo. *1967*

Phenytoin No significant difference between treated epileptics and controls. *2598*

Pindolol No significant changes after 12 mo treatment. *2116* No significant changes found during 12 mo treatment. *2117*

Prednisone No change noted after 1 mo. *3949*

Smoking Mean 0.38 mg/dL vs 0.36 mg/dL in controls. *0845*

Serum   Increase   Physiological

Captopril After 12 weeks treatment in 18 patients with mild essential hypertension. *3150*

Ethanol Increase by 45% in 78 intemperate drinkers on average. *2882* Significantly higher in drinkers and falls with abstinence, but increases with resumption of moderate drinking. *0565* Significantly higher in alcoholic women than in controls. *1701*

Gemfibrozil Increase by 38% in 6 patients with primary familial endogenous hypertriglyceridemia from from baseline values. Synthetic rates of apo A-I and apo A-II increased by 27% and 34% respectively. *3117*

Lovastatin Slight effect in clinical trial of 101 patients with heterozygous familial hypercholesterolemia. *1530*

Nifedipine In 11 patients with 80 mg daily for 6 weeks. *3702* Increase by 7 % in 11 individuals treated for 1.5 mo. *0088*

Serum   Decrease   Physiological

Stanozolol Increase by 24% in 10 normolipidemic postmenopausal osteoporotic women treated for 6 weeks. *3526*

Apolipoprotein B

Serum   No Effect   Physiological

Acebutolol Decrease by -2 mg/dL in 11 patients given 400 mg daily for 3 mo. *2550*

Bunitrolol When 30 mg/d given for 12 weeks in normolipidemic patients with mild essential hypertension. *3151*

Captopril After 12 weeks treatment in 18 patients with mild essential hypertension. *3150* Decrease by -2 mg/dL in 15 patients given 75 mg daily for 8 weeks. *2550*

Chenodeoxycholic Acid No significant effect of 12 mo treatment in 252 patients. *0046*

Chlorthalidone During diuretic therapy for 6 weeks in 23 subjects. *1307* No effect of monotherapy for 6 weeks. *0089* In 22 premenopausal women given 100 mg/d for 6 weeks. *0402*

Clopamide No significant change in 17 individuals treated for less than 1 y when used as sole treatment. *3184*

Desogestrel (ratio apo B: apo A-I decreased by 17%) when 150 μg desogestrel given with 30 μg ethinyl estradiol for 3 mo. Results normal within 2 mo after treatment stopped. *0312*

Furosemide Increase by 2 mg/dL in 12 patients given 40 to 80 mg daily for 3 mo. *2550*

Guanabenz Increase by 2 mg/dL in patients given 4 to 8 mg daily for 12 to 16 weeks. *2550*

Indapamide In 18 patients treated for 6 weeks. *0089*

Methyldopa In several patients treated for 2-4 mo. *0088* Increase by 1 mg/dL in 11 patients given 750 mg daily for 8 weeks. *2550*

Metoprolol No significant effect in 15 hypertensive patients given 200 mg/d for 10 weeks. *1118*

Nifedipine In 11 patients with 80 mg daily for 6 weeks. *3702* Decrease by -6 mg/dL in patients given 30 to 60 mg daily for 8 weeks. *2550*

Oral Contraceptives Typically insignificant change over 6 cycles in women with low dose ethinyl estradiol with low dose levonorgestrel or desogestrel. *1249*

Prazosin Decrease by -8 mg/dL in 16 patients given 1 to 3 mg daily for 8 weeks. *2550* No significant effect in 15 hypertensives given 4 mg/d for 10 weeks. *1118*

Propranolol Increase by 5 mg/dL in 8 patients given 30 to 60 mg daily for 8 weeks. *2550* No change observed in 11 patients. *2476*

Stanozolol Insignificant increase with 6 weeks treatment reverted to normal by 5 weeks after treatment stopped. *0047*

Trichlormethiazide Increase by 1 mg/dL in 15 patients given 4 mg daily for 3 mo but note marked difference between responders and nonresponder populations. *2550*

Serum   Increase   Physiological

Atenolol Increase by 3% reported from one study. *0088*

Chlorthalidone Increase by 16% in 18 postmenopausal women given 100 mg/d for 6 weeks. *0402* Increase by 11 mg/dL in 10 patients given 100 mg daily for 1 mo. *2550*

Ethanol In normal volunteers after 60-70 g/d for 2 weeks but concentration did not exceed normal range. However turnover substantially increased. *2282*

Etretinate From 133 to 147% in 13 patients with hyperkeratotic disorders given 50 mg daily for 2 mo. *3659*

Isotretinoin From 132 to 157% in 12 patients with hyperkeratotic disorders given 40 mg daily for 2 mo. *3659*

Levonorgestrel Increase by 19% (apo B: apo A-I increased by 18%) when 150 μg levonorgestrel given with 30 μg ethinyl estradiol for 3 mo. Results normal within 2 mo after treatment stopped. *0312*

Smoking Mean 0.94 mg/dL vs 0.76 mg/dL in controls. *0845*

Serum   Decrease   Physiological

Lovastatin Increase by 23% at 40 mg bid in clinical trial of 101 patients with heterozygous familial hypercholesterolemia. *1530*

Medroxyprogesterone Increase by 15% in 11 men with sexual deviation syndrome given approximately 1273 mg over a total of approximately 17 d. *0643*

Apolipoprotein B100

Serum   Decrease   Physiological

Indapamide In 13 hypertensives patients with diabetes treated for 24 weeks. *2686*

Apolipoprotein CII

Serum   No Effect   Physiological

Acebutolol No change in 11 patients given 400 mg daily for 3 mo. *2550*

Bunitrolol When 30 mg/d given for 12 weeks in normolipidemic patients with mild essential hypertension. *3151*

Captopril No change in 15 patients given 75 mg daily for 8 weeks. *2550* After 12 weeks treatment in 18 patients with mild essential hypertension. *3150*

Chlorthalidone Increase by 1 mg/dL in 10 patients given 100 mg daily for 1 mo. *2550*

Furosemide Increase by 1 mg/dL in 12 patients given 40 to 80 mg daily for 3 mo. *2550*

Guanabenz   No change in patients given 4 to 8 mg daily for 12 to 16 weeks. *2550*
Methyldopa   Increase by 2 mg/dL in 11 patients given 750 mg daily for 8 weeks. *2550*
Nifedipine   No change in patients given 30 to 60 mg daily for 8 weeks. *2550*
Prazosin   No change in 16 patients given 1 to 3 mg daily for 8 weeks. *2550*
Propranolol   Increase by 2 mg/dL in 8 patients given 30 to 60 mg daily for 8 weeks. *2550*
Trichlormethiazide   Increase by 2 mg/dL in 15 patients given 4 mg daily for 3 mo but note marked difference between responders and nonresponder populations. *2550*

Serum   Increase   Physiological
Glucose   Significant change in normals in response to infusion of 1 g/kg body weight, although not very marked. *0476*

Serum   Decrease   Physiological
Probucol   In 50 diabetics given 500 mg/d for 16 weeks and reduction greatest in highest cholesterol and triglyceride patients. *1523*

Apolipoprotein CIII

Serum   No Effect   Physiological
Acebutolol   Increase by 1 mg/dL in 11 patients given 400 mg daily for 3 mo. *2550*
Bunitrolol   When 30 mg/d given for 12 weeks in normolipidemic patients with mild essential hypertension. *3151*
Captopril   Increase by 1 mg/dL in 15 patients given 75 mg daily for 8 weeks. *2550*   After 12 weeks treatment in 18 patients with mild essential hypertension. *3150*
Chlorthalidone   Increase by 2 mg/dL in 10 patients given 100 mg daily for 1 mo. *2550*
Furosemide   Increase of 4 mg/dL in 12 patients given 40 to 80 mg daily for 3 mo. *2550*
Guanabenz   No change in patients given 4 to 8 mg daily for 12 to 16 weeks. *2550*
Nifedipine   No change in patients given 30 to 60 mg daily for 8 weeks. *2550*
Prazosin   No change in 16 patients given 1 to 3 mg daily for 8 weeks. *2550*
Smoking   Mean 0.068 mg/dL vs 0.057 mg/dL in controls. *0845*

Serum   Increase   Physiological
Glucose   Significant change in normals in response to infusion of 1 g/kg body weight, although not very marked. *0476*
Methyldopa   Increase by 3 mg/dL in 11 patients given 750 mg daily for 8 weeks. *2550*
Propranolol   Increase by 4 mg/dL in 8 patients given 30 to 60 mg daily for 8 weeks. *2550*
Trichlormethiazide   Increase by 6 mg/dL in 15 patients given 4 mg daily for 3 mo but note marked difference between responders and nonresponder populations. *2550*

Apolipoprotein D

Serum   No Effect   Physiological
Smoking   Mean 0.045 mg/dL 0.046 mg/dL in controls. *0845*

Serum   Decrease   Physiological
Stanozolol   Increase by 23% with 6 weeks treatment reverted to normal by 5 weeks after treatment stopped. *0243*   By 23% with 6 week's treatement: reverted to normal by 5 weeks after treatment stopped. *0047*

Apolipoprotein E

Serum   No Effect   Physiological
Acebutolol   Increase by 1 mg/dL in 11 patients given 400 mg daily for 3 mo. *2550*
Bunitrolol   When 30 mg/d given for 12 weeks in normolipidemic patients with mild essential hypertension. *3151*
Captopril   After 12 weeks treatment in 18 patients with mild essential hypertension. *3150*   No change in 15 patients given 75 mg daily for 8 weeks. *2550*
Methyldopa   Increase by 1 mg/dL in 11 patients given 750 mg daily for 8 weeks. *2550*
Nifedipine   No change in patients given 30 to 60 mg daily for 8 weeks. *2550*
Propranolol   Increase by 2 mg/dL in 11 patients given 30 to 60 mg daily for 8 weeks. *2550*
Trichlormethiazide   Increase by 1 mg/dL in 15 patients given 4 mg daily for 3 mo but note marked difference between responders and nonresponder populations. *2550*

Serum   Increase   Physiological
Smoking   In young nonobese normolipidemic men (same pattern as observed in others with increased risk of atherosclerosis). *0845* Mean 0.060 mg/dL vs 0.050 mg/dL in controls. *0845*

Serum   Decrease   Physiological
Guanabenz   No change in patients given 4 to 8 mg daily for 12 to 16 weeks. *2550*

Plasma   No Effect   Physiological
Prednisone   No change noted after 1 mo. *3949*

Appearance

Urine   Abnormal   Analytical
Carbonates   Cloudy, soluble in dilute acetic acid. *0443*
Chyluria   Milky, soluble in either. *0443*
Phosphates   Cloudy, soluble in dilute acetic acid. *0443*
Urates   Cloudy: dissolve at 60°. *0443*
Uric Acid   Cloudy: dissolve at 60°. *0443*

Urine   Abnormal   Physiological
Bacteriuria   Cloudy, insoluble in dilute acetic acid. *0443*
Erythrocytes   Cloudy, lyse in dilute acetic acid. *0443*
Leukocytes   Cloudy, insoluble in dilute acetic acid. *0443*
Lipuria   Opalescent, milky soluble in either. *0443*
Pyuria   Milky, leukocytes insoluble in dilute acetic acid. *0443*
Radiographic Agents   Cloudy, in acid urine. *0443*
Spermatozoa   Cloudy, insoluble in dilute acetic acid. *0443*
Sulfonamides   Turbid especially, if given with methenamine-due to urates. *1680*
Yeast   Cloudy, insoluble in dilute acetic acid. *0443*

Aprindine

Serum   Increase   Physiological
Amiodarone   Concentration increased so dose can be reduced by 50%. *3950*

Arabinose

Urine   No Effect   Physiological
Diurnal Variation   No evidence of cycle observed. *0273*

Urine   Increase   Physiological
Fruit   Increased by much fruit in diet compared with fasting. *0273*
Vegetables   Increase compared with fasting with high intake. *0273*

Arginase

Serum   No Effect   Analytical
Storage of Sample   1 d at room temperature, 2-3 at 4 d, 60 at -20°. *3217*

Serum   Increase   Analytical
Hemolysis   Ratio of 1,000 to 1 between cells and serum. *0579*

Serum   Increase   Physiological
Copper   Increased when hemolytic crisis occurs. *0385*

Arginine

Plasma   Increase   Analytical
Guanidinoacetic Acid   Measured as analyte in method of Bacchus. *0196*

Plasma   Increase   Physiological
Histidine   Interferes with clearance after oral load. *1644*

Plasma   Decrease   Analytical
Hemolysis   Due to arginase in RBCs. *3230*

Plasma   Decrease   Physiological
Glucose   Probable muscle uptake. *1403*

Urine   Increase   Physiological
Cycloleucine   Reversible marked aminoaciduria. *0499*

Argininosuccinate Lyase

Serum   No Effect   Analytical
Storage of Sample   For 6 d at -15°. *0877*

Serum   Decrease   Analytical
Storage of Sample   10% loss per day at 4°. *0877*

Arsenic

Urine   Increase   Physiological
Dimercaprol   If poisoning due to arsenic. *0071*

Arylsulfatase A

Serum   Increase   Physiological
Ethanol   Activity more than double in acute alcoholism. *1265*

Urine   No Effect   Physiological
Urine Volume   Excretion not dependent on urine flow. *2320*

Urine   Increase   Physiological
Diurnal Variation   Maximum 6-12 am, minimum late afternoon evening. *2320*

Ascorbic Acid

Plasma   No Effect   Analytical
Ascorbate-2-Sulfate   Little conjugate in blood so no effect. *0209*
Storage of Sample   No significant change 21 d at -70°. *0441*

Plasma   No Effect   Physiological
Anticonvulsants   In 146 epileptics with long-term treatment. *2007*
Vitamin B₁₂   No effect if individuals saturated with vitamin C. *2468*

Plasma   Increase   Physiological
Ascorbic Acid   Significant effect in 50 volunteers given 2 g/d for 2 mo. *1039*
Ovulation   Sharp increase most likely at ovulation. *2998*
Sex Difference   Postpubertal women higher than men same diet. *2998*

Plasma   Decrease   Analytical
Metaphosphoric Acid   Increase by 10% if stored at -20° for 21 d. *0441*
Storage of Sample   Immediate decrease of 7% on freezing at -20°. *0441*
Trichloracetic Acid   Increase by 10% if stored at -20° for 21 d. *0441*

Plasma   Decrease   Physiological
Aspirin   Uptake into leukocytes decreased also. *2205*   In children and adults taking large amount of aspirin since it potentiates excretion of vitamin C. *3023*
Estrogens   Effect marked at 2 weeks, < effect in platelets. *1847*
Menstruation   Reverse pattern to dehydroascorbic acid. *2998*
Oral Contraceptives   Maximum effect at 2 weeks, greater effect in platelets. *1847*
Pregnancy   Falls significantly after 20 weeks. *2206*
Smoking   Lower concentration observed than in nonsmokers. *2768*   Significantly lower in smoking adolescent women than in nonsmokers: partly to reduced dietary intake. *1901*

Blood   No Effect   Analytical
Storage of Sample   Stable 6 h at 25°, stable 1 mo 4° if + TCA. *0442*

Blood   Increase   Physiological
Ascorbic Acid   Significant effect in 50 volunteers given 2 g/d for 2 mo. *1039*
Diurnal Variation   At peak in early morning. *2204*

Blood   Decrease   Analytical
Storage of Sample   Stable after loss of 10-20% in 3 d at -70°. *0442*

Blood   Decrease   Physiological
Hemodialysis   Low values observed with long term dialysis. *3885*

Urine   No Effect   Physiological
Menstruation   No change reported. *2998*

Urine   Increase   Analytical
Ascorbate-2-Sulfate   Affects DNPH procedures. *0209*

Urine   Increase   Physiological
Ascorbic Acid   Significant effect in 50 volunteers given 2 g/d for 2 mo. *1039*
Aspirin   Reported effect. *0805*
Corticotropin   Mobilizes vitamin stored in adrenal cortex. *0418*
Diurnal Variation   At maximum in first am urine specimen. *2204*
Ovulation   Excretion correlated with that of LH. *2206*

Urine   Decrease   Analytical
Storage of Sample   50% deterioration per h at 20°. *1506*

Urine   Decrease   Physiological
Estrogens   Significant effect (50%) on comparable diets. *1507*
Oral Contraceptives   Significant effect 50% decrease on comparable diets. *1507*
Ovulation   Sharp decrease reported. *2998*

White Blood Cells   No Effect   Physiological
Vitamin B₁₂   No effect if individuals saturated with vitamin C. *2468*

White Blood Cells   Increase   Physiological
Ascorbic Acid   Significant effect in 50 volunteers given 2 g/d for 2 mo. *1039*

White Blood Cells   Decrease   Physiological
Antibiotics   Observed effect especially in elderly. *3554*
Aspirin   Prolonged administration decreases concentration in buffy coat. *1132*
Ethanol   In 35 patients with alcohol related illness. *0207*
Oral Contraceptives   Mean 19 mg/100 g (control 25.7). *2380*
Smoking   Lower concentration observed in smokers than non-smokers. *2768*
Tetracycline   Other antibiotics also have effect. *3554*

Asparagine

Plasma   Decrease   Analytical
Storage of Sample   Even with prolonged storage at -20°. *3230*

Aspartate Aminotransferase

Serum   No Effect   Analytical
Acetaminophen   At acute overdose concentration (10 mg/dL) on SMAC® method. *3719*   At 10 times maximal therapeutic concentration on colorimetric method. *1785*   No effect at therapeutic concentration on Reflotron method. *1984*
Acetazolamide   No effect at 12 mg/dL on SMA 12/60 method. *2636*
Acetylsalicylic Acid   At 5 times upper limit of therapeutic range on method on SMAC®, Abbott-VP, Cobas-Bio, aca, and Hitachi® 705. *2138*   At 8300 µmol/L on continuous and colorimetric methods. *1785*   No effect at therapeutic concentration on Reflotron method. *1984*
Aluminum Salts   No effect on Karmen procedure. No effect on Babson procedure. *2760*
Aminosalicylic Acid   At 5 times upper limit of therapeutic range on methods on Abbott-VP, aca, Cobas-Bio and Hitachi® 705. *2138*
Amitriptyline   At acute overdose concentration (2.5 mg/dL) on SMAC® method. *3719*
Amphetamine   At therapeutic concentration on Reflotron method. *1984*
Ampicillin   No effect on Reflotron method at therapeutic concentration. *1984*
Ascorbic Acid   No effect at therapeutic concentration on Reflotron method. *1984*
Azathioprine   No effect on Babson procedure. No effect on Karmen procedure. *2760*
Barbital   At acute overdose concentration (20 mg/dL) on Technicon® SMAC® method. *3719*
Bethanechol   No effect at 0.09 mg/dL on SMA 12/60 method. *2636*
Bezafibrate   No effect at therapeutic concentration on Reflotron method. *1984*
Bromazepam   At 5 times therapeutic concentration on methods on SMAC®, Abbott-VP, aca, Cobas-Bio, Hitachi® 705 and KDA. *2138*
Caffeine   At acute overdose concentration (10 mg/dL) on Technicon® SMAC® method. *3719*   On either continuous or colorimetric methods at 10 times maximal therapeutic concentration. *1785*   At therapeutic concentration on Reflotron method. *1984*
Carbenicillin   At 5 times upper limit of therapeutic range on methods on SMAC®, Abbott-VP, aca, Cobas-Bio, Hitachi® 705 and KDA. *2138*
Carbochromen   No effect at therapeutic concentration on Reflotron method. *1984*
Cefoxitin   At 5 times upper limit of therapeutic range on methods on SMAC®, Abbott-VP, aca, Cobas-Bio, Hitachi® 705 and KDA. *2138*
Cephalothin   At 5 times upper limit of therapeutic range on methods on SMAC®, Abbott-VP, aca, Cobas-Bio, Hitachi® 705 and KDA. *2138*

Chloramphenicol   No effect at therapeutic concentration on Reflotron method. *1984*
Chlordiazepoxide   At 5 times upper limit of therapeutic range on methods on SMAC®, Abbott-VP, aca, Cobas-Bio and Hitachi® 705. *2138*   At acute overdose concentration (5 mg/dL) on Technicon® SMAC® method. *0899*   No effect at therapeutic concentration on Reflotron method. *1984*
Chlorothiazide   No effect on Babson procedure. No effect on Karmen procedure. *2760*
Chlorpheniramine   At acute overdose concentration (20 mg/dL) on Technicon® SMAC® method. *3719*
Cocaine   At acute overdose concentration (2.5 mg/dL) on Technicon® SMAC® method. *3719*
Codeine   At acute overdose concentration (2.0 mg/dL) on Technicon® SMAC® method. *3719*
Cyclosporine   At concentration of 41.5 mmol/L (50 µg/L) on methods on Technicon® SMAC® II and Hitachi® 705. *1123*
Cysteine   At 5 times upper limit of therapeutic range on methods on SMAC®, Abbott-VP, aca, Cobas-Bio, Hitachi® 705 and KDA. *2138*
Cystine   At 5 times upper limit of therapeutic range on methods on SMAC®, Abbott-VP, aca, Cobas-Bio, Hitachi® 705 and KDA. *2138*
Deslanoside   No effect on SMA 12/60 method at 0.06 mg/dL. *2636*
Diazepam   At 5 times upper limit of therapeutic range on methods on SMAC®, Abbott-VP, aca, Cobas-Bio, Hitachi® 705 and KDA. *2138*   At acute overdose concentration (2.5 mg/dL) on Technicon® SMAC® method. *3719*
Diclofenac   On colorimetric method at 10 times maximal therapeutic concentration. *1785*
Diethylpropion   At acute overdose concentration (10 mg/dL) on Technicon® SMAC® method. *3719*
Digitoxin   No effect at 21 mg/dL on SMA 12/60 method. *2636*
Digoxin   No effect at 0.04 mg/dL on SMA 12/60 method. *2636*
Diphenhydramine   At acute overdose concentration (20 mg/dL) on Technicon® SMAC® method. *3719*
Disulphine Blue   Using recommended Scandinavian method. *1464*
EDTA   No effect on activity. *3165*
Ethacrynic Acid   At 2.5 mg/dL no effect on SMA 12/60 method. *2636*
Ethanol   At acute overdose concentration (20 mg/dL) on Technicon® SMAC® methods. *3719*
Ethaverine   No effect at therapeutic concentration on Reflotron method. *1984*
Ethchlorvynol   At acute overdose concentration (20 mg/dL) on Technicon® method. *3719*
Ethinamate   At acute overdose concentration (20 mg/dL) on Technicon® SMAC® method. *3719*
Ferrous Sulfate   No effect on Karmen procedure. No effect on Babson procedure. *2760*
Fluosol-DA   On SMA IIC at concentration of 50%. *2518*
Flurazepam   At 5 times upper limit of therapeutic range on methods on SMAC®, Abbott-VP, Cobas-Bio, aca, Hitachi® 705 and KDA. *2138*   At acute overdose concentration (2.5 mg/dL) on Technicon® SMAC® method. *3719*
Furosemide   No effect at 1.4 mg/dL on SMA 12/60 method. *2636*   No effect on Karmen procedure. *2760*   No effect at therapeutic concentration on Reflotron method. *1984*
Glibenclamide   No effect at therapeutic concentration on Reflotron method. *1984*
Glutethimide   At acute overdose concentration (5 mg/dL) on Technicon® SMAC® method. *3719*
Heparin   No effect on activity. *3165*
Ibuprofen   At 5 times upper limit of therapeutic range on methods on SMAC®, Abbott-VP, aca, Cobas-Bio, Hitachi® 705 and KDA. *2138*   On colorimetric method at 10 times maximal therapeutic concentration. *1785*
Indomethacin   On continuous method at 10 times maximal therapeutic concentration. On colorimetric method at 10 times maximal therapeutic concentration. *1785*   No effect at therapeutic concentration on Reflotron method. *1984*

Aspartate Aminotransferase *(continued)*

Serum   No Effect   Analytical *(continued)*

Ketoprofen   On colorimetric method at 10 times maximal therapeutic concentration. *1785*

Mannitol   No effect on SMA 12/60 method at 445 mg/dL. *2636*

Meperidine   No effect on Karmen procedure. No effect on Babson procedure. *2760*   At acute overdose concentration (5 mg/dL) on Technicon® SMAC® method. *3719*

Meprobamate   At acute overdose concentration (20 mg/dL) on Technicon® SMAC® method. *3719*

Meralluride   No effect at 0.07 mL/dL on SMA 12/60 method. *2636*

Mesoridazine   At acute overdose concentration (20 mg/dL) on Technicon® SMAC® method. *3719*

Methadone   At acute overdose concentration (2.5 mg/dL) on Technicon® SMAC® method. *3719*

Methanol   At acute overdose concentration (20 mg/dL) on Technicon® SMAC® method. *3719*

Methaqualone   At acute overdose concentration (2.5 mg/dL) on Technicon® SMAC® method. *3719*   No effect at therapeutic concentration on Reflotron method. *1984*

Methicillin   At 5 times upper limit of therapeutic range on methods on SMAC®, Abbott-VP, aca, Cobas-Bio, Hitachi® 705 and KDA. *2138*

Methotrexate   At 5 times upper limit of therapeutic range on methods on SMAC®, Abbott-VP, aca, Cobas-Bio, Hitachi® 705 and KDA. *2138*

Methyldopa   No effect on Karmen procedure. *2760*   No effect at therapeutic concentration on Reflotron method. *1984*

Methylphenidate   At acute overdose concentration (20 mg/dL) on Technicon® SMAC® method. *3719*

Methyprylon   At acute overdose concentration (10 mg/dL) on Technicon® SMAC® method. *3719*

Metronidazole   No effect noted on coupled NADH procedure on Technicon® SMA II if appropriate blanking used. *1628*   On continuous flow procedure if blank correction incorporated. *1628*

Morphine   At acute overdose concentration (20 mg/dL) on Technicon® SMAC® method. *3719*

Naproxen   At 5 times upper limit of therapeutic range on methods on SMAC®, Abbott-VP, aca, Cobas-Bio, Hitachi® 705 and KDA. *2138*

Niacin   No effect at therapeutic concentration on Reflotron method. *1984*

Nitrofurantoin   On routine methods in use on SMAC®, Abbott-VP, aca, Cobas-Bio, Hitachi® 705 at 5 times normal therapeutic concentration. *2138*   No effect at therapeutic concentration on Reflotron method. *1984*

Nortriptyline   At acute overdose concentration (20 mg/dL) on Technicon® SMAC® method. *3719*

Ouabain   At 0.02 mg/dL no effect on SMA 12/60 method. *2636*

Oxalate   No effect on activity. *3165*

Oxazepam   No effect at therapeutic concentration on Reflotron method. *1984*

Oxytetracycline   No effect at therapeutic concentration on Reflotron method. *1984*

Penicillin G   At 5 times upper limit of therapeutic range on methods on SMAC®, Abbott-VP, Cobas-Bio, aca, Hitachi® 705 and KDA. *2138*

Pentobarbital   At acute overdose concentration (20 mg/dL) on Technicon® SMAC® method. *3719*

Perphenazine   At acute overdose concentration (20 mg/dL) on Technicon® SMAC® method. *3719*

Phenazopyridine   No effect at therapeutic concentration on Reflotron method. *1984*

Phenobarbital   At acute overdose concentration (20 mg/dL) on Technicon® SMAC® method. *3719*   On continuous method at 10 times maximal therapeutic concentration. On colorimetric method at 10 times maximal therapeutic concentration. *1785*   No effect at therapeutic concentration on Reflotron method. *1984*

Phenprocoumon   No effect at therapeutic concentration on Reflotron method. *1984*

Phenytoin   No effect at 1.8 mg/dL on SMA 12/60 method. *2636*   At acute overdose concentration (20 mg/dL) on Technicon® SMAC® method. *3719*   No effect at therapeutic concentration on Reflotron method. *1984*

Plasma   No significant difference from serum on SMA 12/60, LKB 8600. *2228*

Prednisolone   No effect on Karmen procedure. No effect on Babson procedure. *2760*

Probenecid   No effect at therapeutic concentration on Reflotron method. *1984*

Procainamide   No effect on SMA 12/60 method at 35 mg/dL. *2636*

Procaine   No effect at therapeutic concentration on Reflotron method. *1984*

Promethazine   At acute overdose concentration (20 mg/dL) on Technicon® SMAC® method. *3719*

Propoxyphene   At acute overdose concentration (2.5 mg/dL) on Technicon® SMAC® method. *3719*

Propranolol   No effect on Karmen procedure. *2760*   No effect at 0.1 mg/dL on SMA 12/60 method. *2636*

Pyribenzamine®   At acute overdose concentration (20 mg/dL) on Technicon® SMAC® methods. *3719*

Pyridamole   No effect at therapeutic concentration on Reflotron method. *1984*

Pyritinol   No effect at therapeutic concentration on Reflotron method. *1984*

Quinidine   At 26 mg/dL no effect on SMA 12/60 method. *2636*   At acute overdose concentration (20 mg/dL) Technicon® SMAC® method. *3719*   No effect at therapeutic concentration on Reflotron method. *1984*

Quinine   At acute overdose concentration (1.5 mg/dL) on Technicon® SMAC® method. *3719*

Reserpine   No effect at 0.02 mg/dL on SMA 12/60 method. *2636*

Rifampin   At 5 times upper limit of therapeutic range on methods on SMAC®, Cobas-Bio, Abbott-VP, aca, Hitachi® 705 and KDA. *2138*

Secobarbital   At acute overdose concentration (20 mg/dL) on Technicon® SMAC® method. *3719*

Standing of Sample   If on polystyrene beads 24 h at 25°. *2624*

Storage of Sample   2 d room temperature, 14 at 4°, 30 at -20°. *3217*   No effect 1 week at 4°, 1 mo at -20°. *1563*

Sulfamethoxazole   No effect at therapeutic concentration on Reflotron method. *1984*

Tetracycline   At 5 times upper limit of therapeutic range on methods on SMAC®, Abbott-VP, Cobas-Bio, aca, Hitachi® 705 and KDA. *2138*

Theophylline   At 5 times upper limit of therapeutic range on methods on SMAC®, Abbott-VP, aca, Cobas-Bio, Hitachi® 705 and KDA. *2138*   No effect at therapeutic concentration on Reflotron method. *1984*

Thiopental   At acute overdose concentration (20 mg/dL) on Technicon® SMAC® method. *3719*

Thioridazine   No effect on Karmen procedure. No effect on Babson procedure. *2760*

Trimethoprim   No effect at therapeutic concentration on Reflotron method. *1984*

Vitamin Preparations   No effect at expected concentration with SMA 12/60 procedure. *2637*

Serum   No Effect   Physiological

Aspirin   When approximately 3 g/d ingested for several weeks. *3076*

Chenodeoxycholic Acid   After 3 mo in patients with gallstones. *0274*

Cromoglycate   No clinical significant change observed. *0484*

Cyclophosphamide   No significant perturbation in 41 patients. *2406*

Cyclosporine   No effect observed in 23 patients with inflammatory ocular disease treated for for 6 mo. *2713*

Desipramine   Increase observed in 4 of 46 patients but previously increased value decreased in 5 others so probably not dose-dependent toxic effect. *2871*   In 42 children and adolescents treated for up to 24 mo. *1624*

Enflurane In all patients given 2 to 3 administrations of anesthetic agent. *1089*

Ethanol No effect over 110 h following ingestion of 0.75 g/kg body weight on 3 consecutive evenings. *2133* No significant difference in specimens from volunteers and values compared with control values for up to 110 h, although component/protein ratio decreased in all cases. *2134* Regardless of amount consumed, acute ingestion had no effect on activity. *0890*

Famotidine In 9 patients with Zollinger-Ellison syndrome when followed for 33 weeks. *1667*

Fenoprofen In 49 patients with juvenile rheumatoid arthritis when some abnormal results in patients with same therapeutic outcome given aspirin. *0465*

Fructose No effect observed after i.v. infusion. *1160*

Glucose No effect observed after i.v. infusion. *1160*

Height No correlation observed in children. *0887*

Isotretinoin In 18 patients with severe acne given 0.8 mg/kg daily for 3 mo: changes reverted to normal after treatment stopped. *2242*

Levonorgestrel No significant difference over 3 y between implant recipients and IUD users. *0763* In a group of women using levonorgestrel covered rods versus controls using copper IUD. *0899*

Methadone No evidence of toxicity if initially normal liver function test. *2012*

Methotrexate No effect in long-term treatment of 20 patients with psoriasis noted. *1556*

Muscular Exercise Insignificant effect of 12 minutes on cycle-ergometer. *1231* No effect after 2 h march. *3217*

Physical Training No effect noticed following 10 weeks training. *1696*

β-Propiolactone But inhibition when drug added to plasma negated by increase from hemolysis. *0217*

Protein Hydrolysate No effect with hyperalimentation in infants. *3614*

Sorbitol Increase by 0.3-0.8 mg/dL after i.v. infusion. *1160*

Xylitol No effect reported after i.v. infusion. *1160*

Serum Increase Analytical

Acetaminophen Reacts in Morgenstern procedure. *0583* At therapeutic concentration may affect SMA 12/60 method. *3335*

Acetoacetate Interference in diazonium end point reactions. *0077* With non specific diazo procedure on SMA 12/60. *3217*

Acetylsalicylic Acid At 5 times upper limit of therapeutic range on method on KDA. *2138*

Aminophenazone On continuous method at 10 times maximal therapeutic concentration. *1785*

Aminophenol At 1 mmol/L affects SMA 12/60 method. *3335*

Aminosalicylic Acid At 1 mmol/L affects SMA 12/60 method. *3335* With diazonium end point method on AutoAnalyzer. *1310* At 5 times upper limit of therapeutic range on methods on KDA and SMAC®. *2138*

Ascorbic Acid At 1 mmol/L affects SMA 12/60 method. *3335*

Bilirubin Interference in diazonium end point reactions. *0077*

Chlordiazepoxide At 5 times upper limit of therapeutic range on kinetic methods on KDA. *2138*

Diclofenac On continuous method at 10 times maximal therapeutic concentration. *1785*

Epinephrine At 1 mmol/L affects SMA 12/60 method. *3335*

Erythromycin Colorimetric assay if DNPH or diazonium salt used. *3505*

Ethyl Acetate Significant effect when added to serum, diazo reaction. *3217*

Fluorescein Slight effect at 200 mg/L on Ektachem® 700 method. *1722*

Fluorides Increased activity observed with colorimetric method. *3165*

Gentamicin False elevation with Babson procedure. *3107*

Hemoglobin If greater than 25 mg/dL. *3165*

Hemolysis Just visible (50 mg/dL) produces 1-2% increase. *0877*

Hydralazine At 1 mmol/L affects SMA 12/60 method. *3335*

Isoniazid At therapeutic concentration may affect SMA 12/60 method. *3335*

Isoproterenol At 1 mmol/L affects SMA 12/60 method. *3335*

Ketosis With non specific diazo procedure on SMA 12/60. *3217*

Levodopa At 1 mmol/L affects SMA 12/60 method. *3335*

Lipemia Turbidity may affect some methods. *0579* With inverse colorimetry: resultant effect of clearing of lipemia and formation of reaction product during SMAC® measurement. *2466*

Liposol Changed from 49.3 to 56.6 U/L in 26 randomly selected patient sera, before and after addition on Beckman Astra methods. *2054*

Methyldopa Measured as product in AutoAnalyzer Babson method. *2760* At 1 mmol/L affects SMA 12/60 method. *3335*

Nitrofurantoin 36% increase on kinetic MDH procedure on KDA at 5 times normal therapeutic concentration. *2138*

Phenelzine At 1 mmol/L affects SMA 12/60 method. *3335*

Pyridoxal 25 μmol/L produces average increase of 16%. *2953*

Sulfathiazole At 1 mmol/L affects SMA 12/60 method. *3335*

Tolbutamide At 1 mmol/L affects SMA 12/60 method. *3335*

Serum Increase Physiological

Acetaminophen Manifestation usually of single toxic high dose ingestion in suicide attempt. Liver damage usually of centrizonal hemorrhagic hepatic necrosis. *3948* With overdose and centrolobular hepatic necrosis usually associated with high drug concentration. *1771* Hepatic necrosis with dose of 10 g reported. *0438*

Acetohexamide May cause intrahepatic cholestatic jaundice. *2220*

Acetophenazine Cholestasis with intrahepatic obstruction. *2313*

Ajmaline Intrahepatic cholestasis. *0268* Hepatotoxicity and intrahepatic cholestasis reported. *3215* In 4 patients with centrilobular cholestasis and mild hepatocytic lesions: recovery with drug withdrawal. *2729*

Allopurinol Reversible clinical hepatotoxicity reported. *2220* Occasional granulomatous hepatitis as well as massive hepatic necrosis and severe hepatitis. *2389* In 6 of 73 patients treated for 6 mo. *3565*

Alprazolam Isolated case, abnormal test value reverted to normal with withdrawal of drug. *3079*

Aluminum Nicotinate Rare hepatotoxic effects. *1678*

Amantadine Reported effect, no mechanism discussed. *2313*

Aminobenzoic Acid Toxic hepatitis reported. *1343*

Aminocaproic Acid Rare myopathy six weeks after treatment. *0296*

Aminoglutethimide Casual association between drug administration and cholestasis in 2 cases. *2779*

Aminopyrine Liver damage up to hepatic necrosis seen. *2427*

Aminosalicylic Acid May cause cytotoxic hepatocellular damage. *0248* Rare side effect. *1292*

Amiodarone Up to 1.5 to 3 times normal values in 55% patients with hepatitis occurring in 4%. *2011* Asymptomatic change in liver function with enzyme activity up to 3 times normal. *2478* 1.5 to 4 fold increase in 15 of 100 patients: effect correlated with drug concentration. *1509* In 5 of 5 patients with toxic and hypersensitivity liver injury. *2987*

Amitriptyline Rare cases of transient cholestasis. *2482* Occasional case of hypersensitivity associated hepatitis. *0098*

Ammonia May produce severe liver damage. *0279*

Amodiaquine Reported hepatotoxicity. *2313* Abnormality usually mild but associated with an immunoallergic hepatotoxicity. *2081*

Amphotericin B Hepatotoxicity reported. *2313* Noted in one patient with acute myelogenous leukemia when treated with high dose. Probable idiosyncratic response. *2443*

Ampicillin Probable effect of i.m. injection (especially in infants). *1966*

Anabolic Steroids Cholestatic syndrome (intrahepatic cholestasis). *2220*

Androgens May cause cholestatic syndrome. *2220*

Anesthetic Agents Occurs without permanent liver damage. *3661*

Anticonvulsants Observed with chronic administration. *3319*

Antifungal Agents Hepatotoxicity may occur. *1237*

Antimony Compounds Hepatotoxic effect. *2220*

Aspartate Aminotransferase (continued)

Serum Increase Physiological (continued)

Apiol May cause severe liver damage. 0279

Aprindine Hepatitis is manifestation of chronic toxicity. 3215 Hepatitis in 2 patients within 3 weeks of start of therapy resolved with withdrawal of drug. 1570

Arsenicals Hepatotoxic effect (cholestasis/cholangiolitis). 2220

Asparaginase Hepatotoxicity. 1519 In 46% children and 63% adults reported in various studies. 2406 Increased in 35-45% of patients treated in different studies: effects usually mild. 0558 Hemorrhagic pancreatitis in fewer than 0.5% treated patients. 1398

Aspidium May cause hepatic toxicity. 2313

Aspirin Prolonged administration may cause hepatic toxicity. 2111 Noted in some patients at serum concentrations above 25 mg/dL without signs of hypersensitivity reactions, but dose related. 3963 In 15% trials of treatment for juvenile rheumatoid arthritis. 0239

Aspoxicillin In 2 of 5 patients given up to 112 mg/kg intravenously in 3 or 4 divided doses. 1517

Azapropazone In 9 of 83 patients treated for 6 mo. 3565

Azaserine May cause hepatotoxicity. 3835

Azathioprine May cause hepatotoxicity. 3433 Minimal cholestasis and reversible portal fibrosis observed in 8 patients: occasional hepatotoxicity. 0036 Canalicular cholestasis and centrilobular ballooning of hepatocytes in one patient. 0846

Barbiturates Occurs with poisoning, probable muscle origin. 3901

Barium Severe liver damage if tannic acid in enema. 0071

Benzene May cause liver damage. 0279

Benziodarone May cause hepatic toxicity. 2313

Bethanechol Impaired excretion due to spasm of sphincter. 0652

Bismuth Salts Possibly of renal origin due to nephrotoxicity. 2911

Bleomycin Rare mild abnormality reported: reversible. 2406

Boric Acid May cause liver damage. 0279

Butaperazine Cholestatic hepatitis with obstruction. 0071

Capreomycin Reported effect on liver. 1705

Captopril Isolated case of cholestasis in patient receiving 25 mg tid for 1 mo. 2732 Characteristic pattern of hepatocellular jaundice in one patient. But with secondary cholestatic elements. 3688

Carbamazepine Cholestatic and hepatocellular damage. 2907 Granulomatous hepatitis observed in small proportion of patients treated for less than 1 mo; resolved within 3 d of cessation of treatment. 2153

Carbarsone May cause hepatitis/liver necrosis. 0071

Carbenicillin Effect of drug ?hepatotoxic. 1237 In 6 of 27 patients given 270 mg/kg i.v. for 6 d. 3275

Carbenoxolone Associated with myopathy. 2458

Carbimazole Isolated case of jaundice conclusively linked to drug. 0911 Reversible hypersensitivity response to drug. 1790

Carbon Disulfide May cause liver damage. 0279

Carbon Monoxide Associated with hepatomegaly. 1302

Carbromal Occurs with poisoning, probable muscle origin. 3901

Carbutamide Hepatotoxicity. 2313

Carmustine (BCNU) Usually mild and return to normal over few days. 2406

Carphenazine Probable cholestatic effect (reversible). 2313

Cefamandole Occurs in about 4% treated cases. 3134 Fewer than 3 of 53 patients developed reversible hepatotoxicity. 0969

Cefazolin Reversible hepatotoxicity noted in some patients. 0969 In 4 of 31 treated cases: transient 2 to 3 fold increase. 2273

Cefoperazone Transient increase in 5% of 450 patients. 0690

Cefoxitin Reported incidence of 3%. 3134 Transient 2 to 3-fold increase in 3 of 31 patients following colorectal surgery. 2273

Ceftizoxime In 4 of 110 patients: reversible hepatotoxicity noted. 0969 In 3 of 30 patients following colorectal surgery had transient 2 to 3-fold increase. 2273

Cefuroxime Observed in 24% of 89 patients in one study, but many had pre-existing liver problems. 2615

Cephaloglycin Slight elevation reported. 1680

Cephaloridine Transient slight rise ?hepatotoxicity. 1019

Cephalosporin Observed with all cephalosporins at frequency of from 1 to 7%. 2615

Cephalothin Transient rise reported. 1680

Chenodeoxycholic Acid Mild transient elevation observed. 0808 Slight transient increases may be observed: clinically significant hepatic injury in 3%. 3198 Mild elevation usually less than 3 times normal in up to 30% patients receiving 750 mg/d. 3199

Chloral Hydrate May cause liver damage. 0279

Chlorambucil Hepatotoxicity with centrolobular necrosis. 1948

Chloramphenicol Hepatotoxic-cholestatic effect. 2220 Isolated case of hepatitis and pancytopenia reported which resolved with discontinuation of drug. 0602

Chlorate May cause liver damage. 0279

Chlordiazepoxide May produce hepatotoxic effect. 0006

Chlorinated Insecticides May cause liver damage. 0279

Chlormezanone Occasional reversible cholestasis. 2313

Chloroform Hepatotoxic effect with necrosis. 2313

Chlorophenothane May cause liver damage. 0279

Chloroquine Reported effect no mechanism cited. 2313

Chlorothiazide May cause cholestatic jaundice. 2313

Chlorpromazine Hepatic sensitivity to drug (in up to 2% patients). 3283 In one patient who developed cholestasis with fever and leukopenia after 260 mg drug. 2840

Chlorpropamide May cause cytotoxic liver damage. 0248 Cholestatic jaundice observed in one patient in association with red cell aplasia (prevalence of jaundice as high as 0.5%). 1284 In isolated case of hypersensitivity reaction producing hepatic granulomas also in bone marrow. 2988

Chlorprothixene May cause hepatotoxicity (reversible). 2313

Chlortetracycline Hepatotoxic effect with centrolobular necrosis. 1237

Chlorthalidone Single case of severe reversible myopathy noted. 1793

Chlorzoxazone May cause hepatotoxicity. 2313

Cholestyramine Few cases reported, probably not hepatotoxicity. 1680

Cholinergics Impaired excretion due to spasm of sphincter. 0652

Chromates May cause liver damage. 0279

Cimetidine Reversible cholestatic jaundice in a few children. 2172 In isolated case treated for 7 mo with 400 mg/d progressed to bridging hepatic necrosis. 3085

Cinchophen May cause hepatotoxicity (viral-hepatitis like). 2313

Ciprofloxacin Rare cases reported during clinical trials. 0151

Cisplatin Slight and transient increases in 45 patients receiving drug. 3698

Clindamycin Transient abnormality noted. 3653 Uncommon side effect. 3864

Clofibrate Transiently elevated during early therapy. 0318

Clonidine Single probable case of toxic hepatitis. 1277

Clopamide Single case of diuretic associated myopathy. 1793

Cloxacillin Few cases reported, ?due to i.m. injection. 0071

Codeine May cause rise in intrabiliary pressure. 3505

Colchicine Possible hepatotoxic effect. 2220

Copper Toxic effect, sharp rise with hemolytic crisis. 0385

Cortisone May contribute to long duration of hepatitis. 2220

Cough Medicines One case chronic use terpin and codeine. 1069

Coumarin May cause hepatic toxicity. 3835

Cyclofenil Slightly and transiently increased in 6 of 19 patients over 4 mo of treatment, especially with high doses. 2786 In 1 case, but total of 30 cases reported up to 1980; Considered to be metabolic idiosyncrasy and reversible in all cases. 2672

Cyclophosphamide May cause hepatotoxicity. 3433

Cyclopropane May cause hepatotoxicity (lasts several days). 2313

Cycloserine May cause hepatotoxicity. 2313

Cyclosporine Both early and late toxicity observed but effects mild in comparison with nephrotoxicity in renal transplant patients. 1959

Cyclothiazide Occasional intrahepatic cholestatic jaundice. 1680

Cyproheptadine Isolated case of jaundice due to cholestasis: reversed on withdrawal of drug. 1562

Cytarabine In two cases with acute leukemia: unlikely to have been due to other drugs co-administered: effect slight. *2827* At 3 g/m² every 12 h for 6 d, some mild and reversible abnormalities in about 75% of 12 cases. *1579* Slight effect and returned to normal while still under treatment. *2406*

Danazol In 1 of 4 men and 0 in 7 women. *3906*

Dantrolene Observed effect, cause uncertain. *0659* Chronic active hepatitis reported in a few patients. *2260* Acute hepatitis as result of drug ingestion in 4 patients returned to normal after drug stopped. *3837* In 1.8% of 1044 patients, hepatocellular damage with acute or subacute hepatic disease or chronic active hepatitis. *3656*

Dapsone Sulfone syndrome in one 16 year old girl given 50 mg daily short-term administration. *3604*

Dehydroemetine Possibly due to generalized myositis. *2427*

Desipramine Rare transient cholestasis. *2313* Idiosyncratic response observed in 4 of 46 patients. *2871*

Diatrizoic acid Has caused severe muscle spasm and renal failure. *2312*

Diazepam Isolated case of drug induced hepatitis. *3562*

Dicloxacillin Mild hepatic dysfunction observed. *0118*

Dicumarol May simulate myocardial infarct (greater than ALT). *0248*

Dienestrol Reported to cause cholestasis. *1680*

Diethylstilbestrol Hepatotoxicity with centrolobular necrosis. *1948*

Dinitrophenol Hepatotoxicity with centrolobular necrosis. *1343*

Disopyramide Occasional case of cholestatic jaundice reported. *3215*

Disulfiram Reversible toxic liver damage in nonalcoholic woman. *2017* Observed in one patient and similar results followed challenge test. *2491*

Doxorubicin May produce idiosyncratic hepatic dysfunction, as observed in 6 patients. *0184*

Doxycycline Reported effect (?hepatic origin). *0071*

Ectylurea May cause cholestasis with cholangiolitis. *2313*

Enflurane Slight effect but decreased in others. *0999* Evidence of hepatocellular damage; characteristically centrilobular necrosis: mechanism of injury most probably metabolic idiosyncrasy. *2157*

Epinephrine Following single injection of compound in oil. *1264*

Erythromycin Causes hepatic toxicity in some cases. *1142* Estolate produces mild hepatotoxicity in about 15% treated individuals. Jaundice in 2% patients on drug for more than 2 weeks. *2792* Observed to some extent in some patients with different erythromycin salts: usually cholestatic hepatitis. *1895*

Estradiol Cholestatic effect. *2313*

Estrogens May cause cholestasis, rare hepatocellular degeneration. *2220*

Estrone May cause cholestasis. *2313*

Ethacrynic Acid Cholestasis with hepatocellular damage. *0812*

Ethambutol Decreased liver function reported. *0071*

Ethanol Increased in normals after 1 dose 3 g/kg body weight. *1621* 17% increase in heavy versus occasional drinkers. *3273* Observed in different studies in from 18-100% chronic alcoholics or heavy drinkers. *3124* Sensitive indicator of continued alcohol abuse in individuals with known liver disease. *1040* Increased in alcoholic liver disease. *0625*

Ethchlorvynol Occurs with poisoning, probable muscle origin. *3901*

Ether Hepatic disturbance (transient). *3218*

Ethionamide Intrahepatic cholestatic jaundice. *1343*

Ethosuximide Increased in one third cases. *2427*

Ethotoin Hepatotoxicity. *2313*

Ethoxazene Hepatotoxic effect. *2313*

Ethyl Chloride May cause liver damage. *0071*

Etoposide Hepatic toxicity reported in 3% treated patients. *0141*

Etretinate In 2 of 20 patients treated for 1 y. *1866* In one 74-year old woman treated for severe psoriasis: normalized with withdrawal of therapy and recurred when reinstituted. *3789*

Factor IX Complex, Human 62% cases developed hepatitis after use. *2663*

Fansidar® Biopsy-proven granulomatous hepatitis due to sulfadoxine moiety of drug reported in two individuals receiving drug prophylactically possibly due to sulfonamide hypersensitivity. *2099*

Fava Beans May cause liver damage. *0279*

Fibrinolysin Hepatitis as late complication of administration. *0071*

Florantyrone Hepatotoxic effect. *2313*

Floxuridine Manifestation of toxicity. *1680*

Flucloxacillin In a single patient on hemodialysis: abnormal results only mild. *2200*

Flucytosine Reversible hepatotoxicity in 10%. *3443* Occurs in about 25% treated patients: usually mild but necessitates discontinuation of therapy. *3567*

Flufenamic Acid Transient elevation reported. *2427*

Fluorides Due to anoxia with tissue damage. *0010*

Fluoxymesterone Intrahepatic cholestatic effect. *0019*

Fluphenazine Hypersensitivity response. *2313* Isolated case of hepatotoxicity with centrilobular cholestasis. *3372*

Flurazepam May cause hepatic toxicity. *3016*

Fluroxene Potentially hepatotoxic. *0071*

Fusidic Acid Reported in 6 patients in UK, although causal relationship could not be established with certainty. *3535* Jaundice developed in 34% of 112 patients given drug, highest when drug given intravenously. Jaundice resolved when drug withdrawn. *1690*

Gentamicin May cause hepatotoxicity (or due to i.m. injection). *2220*

Glibenclamide Intrahepatic cholestasis described in 61 y old diabetic. *3894*

Glyburide Rare and reverts to normal if therapy continued. *0125*

Glycopyrrolate Hepatotoxicity. *2313*

Gold Hepatotoxicity with centrolobular necrosis. *2313* Cholestatic jaundice in 3 patients with rheumatoid arthritis given gold sodium thiomalate: all patients recovered spontaneously. *1084* In 3 patients with rheumatoid arthritis during chrysotherapy. *0993* Due to toxic effect on drug on liver. *1674*

Griseofulvin May be hepatotoxic. *2220*

Guancydine Transient increases observed in some cases. *0668*

Guanethidine Possible myopathic complication. *2220*

Guanoclor Possible effect on liver. *2220*

Guanoxan May cause hepatic toxicity. *2313*

Halofenate Mild transient effect observed ?origin. *0161*

Haloperidol May cause hepatocellular changes. *2313* In isolated case producing cholestatic liver disease: typical frequency of liver disease is 0.2%. *0910*

Halothane Allergic hepatic hypersensitivity. *3294* Liver disease from mild focal necrosis and inflammation to massive necrosis. *2260* In one case after repeated exposure to anesthetic: evidence of acute hepatitis with bridging necrosis and piecemeal necrosis. *2027* Granulomatous hepatitis in single case following second exposure to anesthetic. *3266* Low frequency but greater number of abnormalities than with enflurane. Occurred most frequently in obese patients. *1089*

Hemolysis Released from RBCs and platelets. *2313*

Heparin In 27% patients receiving porcine derived drug. In 59% patients receiving bovine derived drug. *0968* Mean maximal value 40 U/L in 82% of 46 patients given up to 120,000 IU daily subcutaneously after 8 d. *2454* But significantly different from other individuals in control population. *2595* In 2/3 of patients treated with heparin for deep venous thrombosis. *1064* In patients with cerebrovascular accidents after low-dose heparin treatment. *2671*

Heptachlor Liver damage occurs as late effect of poisoning. *1302*

Heroin Normal usually unless alcoholism also. *3471*

Hetacillin Observed with other penicillins (cause?). *1680*

Hycanthone Transient change in liver function. *0071*

Hydralazine Isolated case with moderate hepatomegaly, abnormal findings resolved when drug discontinued. *1161* Asymptomatic or symptomatic reversible hepatitis-like reaction or granulomatous hepatitis. *2374* Cholestatic jaundice observed in one patient. *3469* Centrilobular necrosis observed in 3 patients 2 to 4 months after drug therapy for hypertension. *1740*

Hydrazine Derivatives May cause hepatotoxicity. *2313*

Aspartate Aminotransferase *(continued)*

Serum Increase Physiological *(continued)*

Hydrochlorothiazide May cause cholestasis or cholangiolitic hepatitis. *1680*

Hydroflumethiazide May cause intrahepatic cholestasis. *1680*

Hydroxyacetamide Toxicity effect. *2313*

Hypothermia Due to hypoxia, acid-base imbalance, hypotension. *2258*

Ibufenac Hepatotoxic effect. *2313*

Ibuprofen Rare hepatotoxicity reported: associated in one case with Stevens-Johnson syndrome. *3460*

Icterogenin Causes intrahepatic cholestasis. *0153*

Idoxuridine Cholestatic jaundice reported in one case. *0837*

Ifosfamide Manifestation of hepatotoxicity. *3675*

I.M. Injections Muscle damage. *3218*

Imipramine May cause cholestatic jaundice. *1873* Either due to hepatotoxicity or hypersensitivity in 33 year old woman. *2501*

Indandione Derivatives Hepatocellular damage with cholestasis. *2313*

Indomethacin Cytotoxic and cholestatic liver damage. *1908*

α-Interferon Dose dependent effect in 3 of 81 patients with malignant disease. *3299*

Iopanoic Acid May cause hepatotoxicity. *2313*

Iprindole Hepatic toxicity. *3505*

Iproniazid Prolonged use may cause hepatotoxicity. *3835*

Iron Salts May cause severe liver damage with poisoning. *0279*

Isocarboxazid Cholestatic effect. *2313*

Isoniazid Probable intrahepatic cholestatic jaundice. *2220* Mild liver injury occurs in approximately 10% patients taking drug, possibly due to conversion of drug to acetylhydrazine or related hepatotoxic derivatives. *2260* In 18.3% versus 6.7% of controls in adult patients receiving drug for prophylaxis. *0555* Liver toxicity occurred in 18% patients receiving combined isoniazid and rifampin: effect slight in 14% and severe in 4%. *1409* Risk of hepatitis caused by or exacerbated by drug over 12 mo 5.2 per 1,000 patients; clinically picture resembles viral hepatitis. *1292* 15 of 89 patients developed significant liver disease, typically hepatitis, although one developed cholestasis, in patients taking drug for at least 2 mo. *0901* Occurs in about 20% of treated patients due to acetylhydrazine metabolite: higher incidence in older and in consumers of excess alcohol. *0053* Observed in up to 0.5% patients slightly greater than in placebo treated group. *1725* In 3.3% children also treated with rifampin: all reactions occurring in first 10 weeks. *2630*

Isotretinoin In 7 patients with severe rosacea treated with 1 mg/kg/d for 12 weeks. Effects possibly due to induction of hepatic microsomal enzymes. *2307* Reversible dose-related effect noted in patients with myelodysplastic syndrome and leukemia at high doses. *1319* Significant effect at 6 weeks in 7 patients with severe rosacea treated with 1 mg/kg/d. *2307*

Itraconazole Mild reversible increases without serious hepatotoxicity. *3567*

Kanamycin May cause hepatotoxicity. *2313*

Ketamine In 14 of 34 individuals who had drug as anesthetic for intermediate operations. *0973*

Ketoconazole Transient abnormalities of liver function observed in 10% patients but true hepatic injury in only 0.1 to 1%. Probably idiosyncrasy involved but may be immune hypersensitivity in some cases. *3390* Delayed reaction to drug after withdrawal in single case of chronic candidiasis in 61 y old woman. *3513* In 4 of 36 patients treated with 200 mg daily over 8 mo. *3034* Rapidly developing liver failure in 67 year old woman taking 200 mg drug daily for 2 mo. *0960*

Levamisole Mild increase in 2 of 11 patients at 2 and 6 mo respectively after start of treatment. Normalized with withdrawal of treatment. *2723* In 1 of 11 patients receiving postoperative radiation treatment following breast cancer. *2960*

Levodopa Transient effect, normalizes despite continuation. *0071* Increase by 30% for 2 mo although later normalized, in patients with Parkinsons disease. *1417*

Lincomycin Hepatotoxic-cholestatic effect. *2220*

Lipemia Associated with alcoholism. *2313*

MAO Inhibitors Intrahepatic cholestasis. *2220*

Maprotiline Slight reversible rise in approximately 1% patients. *0214*

Meals Observed in both sexes. *3319* Affected by meals (most marked in women). *3455*

Mechlorethamine May cause cytotoxic (hepatocellular) damage. *0248*

Melarsonyl May produce hepatotoxicity. *0071*

Melarsoprol May produce hepatotoxicity. *0071*

Mepazine Cholestatic effect (up to 6 times normal). *1713*

Meperidine May cause rise in intrabiliary pressure. *0248*

Mephenytoin Hepatotoxicity. *2313*

Meprobamate May cause cholestatic (hepatocanalicular) jaundice. *0248*

Mercaptopurine May cause hepatotoxicity (centrolobular necrosis). *0204* Picture of cholestasis in 10 of 19 leukemic patients, but drug co-administered with Adriamycin® in all and additional drugs in others. *3021* Intrahepatic cholestasis observed in several patients. *2406*

Mesoridazine Transient effect noted. *0071*

Metahexamide Hepatotoxicity (viral hepatitis type). *2313*

Metaxalone Hepatotoxic effect. *2313*

Methacholine Impairs excretion by spasm of sphincter of Oddi. *0652*

Methacycline Possible hepatotoxicity. *0071*

Methandriol Intrahepatic cholestatic jaundice. *1343*

Methandrostenolone Up to 3-6 times normal due to cholestasis. *1713* In 2 of 6 body-builders taking up to 20 mg/d intermittently for a year or more. *3290*

Methdilazine Rare case of cholestasis. *1680*

Methenamine Mild transient effect in some cases. *1680*

Methimazole May affect liver function (cholestasis). *2313* Rare case of cholestatic jaundice reported. *1774*

Methotrexate Hepatotoxicity (drug induced cirrhosis). *2313* In 152 of 250 courses of treatment but most regressed. *2406*

Methoxsalen Hepatotoxic effect. *2313*

Methoxyflurane Hepatic toxicity. *3294* In 2 pregnant women when used as analgesia for labor: hepatitis observed. *0872*

Methyldopa Result of hepatocellular damage or cholestasis. *3879* Chronic active hepatitis associated with immune hemolytic anemia in one case. *3269* Disturbances in liver function in 5 to 35% of patients treated for hypertension. Hepatocellular injury may occur after short or long-term exposure. *0162* Chronic active hepatitis associated with immune hemolytic anemia in one case. *3269* Drug induced hepatitis with severe, chronic, aggressive inflammation. *0211*

Methylenedioxyamphetamine Reported effect ?of muscle origin. *2975*

Methyltestosterone Cholestatic effect (moderate elevation). *1713* In 1 of 60 patients (female transsexuals or impotent males) receiving 150 mg day for long term. *3805*

Methylthiouracil Hepatotoxic effect. *2313*

Mexiletine Occurs in fewer than 1% treated patients. *2011*

Minocycline Conceivable complication of therapy. *0135*

Mithramycin Hepatocellular damage observed. *0071* Consistent moderate increase in most cases. *2406*

Mitomycin Centrilobular stasis in 5 of 6 patients and toxic hepatitis in the other. *2406*

Mitoxantrone Mild transient effects in 11 of 26 patients. *2703*

Morphine May cause rise in intrabiliary pressure. *3505*

Moxalactam In about 3% patients as reported from several studies. *0592*

Muscular Exercise Marked after exercise (less in trained individuals). *2981*

Mushroom Poisoning May produce severe liver damage. *0279*

Nafcillin Possibly due to trauma of injection. *2313* 1 of 32 patients developed abnormal liver function on 3rd day of treatment. *0052*

Nalidixic Acid May cause cholestatic jaundice. *2808*

Nandrolone Reported to affect liver function. *2220*

Naproxen In 6% of patients receiving drug for juvenile rheumatoid arthritis. *0239*

Narcotics Impaired excretion due to spasm of sphincter. *0652*

Niacin Intrahepatic cholestasis observed rarely. *1253*

Niacinamide Rare probable reversible toxic effect. *3875*

Nialamide Probable hypersensitive hepatitis. *0071*

Nitrofurans Hepatotoxicity. *2313*

Nitrofurantoin May cause cholestatic jaundice. *2808* Moderate increase chronic active hepatitis, much more common in women than men, but still rare. *2444* Moderate increase chronic active hepatitis, much more common in women than Men but still rare. *2444* May follow acute hepatitis with or without cholestasis or as chronic active hepatitis or chronic granulomatous reaction. *2374*

Nitrophenol Hypersensitive intrahepatic cholestasis. *1434*

Norethandrolone Cholestasis produced without cholangiolitis. *3172*

Norethindrone Intrahepatic cholestatic jaundice. *2313*

Norethynodrel Cholestatic effect. *2313*

Norfloxacin In 2 of 1540 patients in clinical trials. *0731*

Nortriptyline May cause cholestasis. *0071*

Novobiocin Intrahepatic cholestasis may occur. *2313*

Ofloxacin Very infrequent side effect observed during many courses of treatment. *1838*

Oleandomycin May cause hepatotoxicity (cholestatic syndrome). *2313*

Opium Alkaloids Possibly due to spasm of sphincter of Oddi. *0652*

Oral Contraceptives Hepatotoxic effect (cholestasis induced). *2086* Significant increase observed within 3 mo usually continuing for several y. *3170*

Organophosphorus Insecticides Hepatotoxic effect. *0405*

Oxacillin Possible hepatotoxic effect. *2220* Typically occurring 3 to 14 d after treatment started, may occur without eosinophilia. *2374* Hypersensitivity reaction observed in one patient with Staphylococcus Aureus endocarditis. *0882* Reversible phenomenon observed in 8 patients given drug i.v. *2664* Drug associated hepatitis in patients given high dose drug i.v. *2675* Asymptomatic hepatic dysfunction in 5 cases following high dose treatment: reversible. *2835* Reversible high-dose associated liver injury. *2431* Abnormal liver function in 1 of 8 treated patients on day 15. *2548*

Oxazepam Possible hepatotoxicity. *2313*

Oxymetholone Cholestatic effect (increased up to 6 times normal). *1713* Cholestatic hepatitis developed in two patients. *3921* One case of peliosis hepatitis reported. *2539*

Oxyphenbutazone Possible hepatotoxicity. *2313*

Oxyphenisatin Hepatic toxicity if over prolonged period. *2757* Liver damage may present as acute hepatitis or as more advanced chronic disease: usual liver disease is chronic active hepatitis. *2260* Observed with chronic active hepatitis induced by drug. *0903*

Papaverine Probable hypersensitivity reaction. *3039* In 2 patients: fever anorexia and jaundice 4 to 5 weeks after start of treatment due to pericholangitis: reversible. *3371* In 6 of 14 patients abnormal liver function, possibly due to allergic response and to metabolic aberration in host. *2746*

Paraldehyde Possible hepatotoxicity. *2313*

Paramethadione Possible hepatotoxicity. *2313*

Pargyline Possible hepatotoxicity (reversible). *2313*

Penicillamine Possible hepatotoxicity. *2313* In 6 of 99 patients treated for rheumatoid arthritis: evidence of toxic liver necrosis observed in two cases. *3887* Case of acute cholestatic jaundice in patient with systemic lupus erythematosus. *3242*

Penicillin Nonspecific hepatitis without cholestasis. *3879*

Perchlorate Single case of acute yellow atrophy of liver. *2427*

Perhexilene Transient elevation observed. *1619*

Phenacemide May affect liver function in about 2% cases. *2313*

Phenazopyridine Hepatotoxic effect. *2313*

Phenelzine May cause hypersensitive hepatitis. *0071*

Phenindione May modify liver function (cholestasis). *2313*

Pheniprazine May cause hepatocellular jaundice. *2313*

Phenobarbital Hepatotoxicity with centrolobular necrosis. *1948*

Phenothiazines May cause cholestatic hepatitis (in up to 4% patients). *2313*

Phenoxypropazine Produced fatal hepatotoxicity. *2427*

Phenprocoumon Mild effect but recurrently in 2 patients having repeated exposure to drug. *3347*

Phenylbutazone May cause cholestatic and cytotoxic jaundice. *0248* Observed in 7 patients with side effects due to drug. *3409* In 2 patients subsided after drug withdrawn associated with granulomas and cholestatic-hepatocellular injury. *1730* Frequent hypersensitivity reaction but also with hepatotoxic potential: may be hepatocellular injury or systemic vasculitis. *0290*

Phenytoin Intrahepatic cholestatic jaundice. *2313*

Phosphorus Hepatotoxic effect with necrosis. *2313*

Pindolol Minor persistent increases reported in about 7% patients but not progressive. *1208*

Piperacetazine Transient reversible increase. *1227*

Piperacillin In 1 of 29 patients given 181 mg/kg i.v. for 6 d. *3275* Transient increase in 1 of 20 treated patients. *3132* In 1 of 59 patients given drug as sole agent with many patients with severe illness. *3874*

Piroxicam Occasional case of liver damage reported. *2824* Some transient increases reported. *2692*

Polythiazide May affect liver function. *2313*

Probenecid Hepatotoxic effect (centrolobular necrosis). *2313*

Procainamide Hepatotoxic effect. *2313*

Prochlorperazine Cholestatic effect. *0071*

Progabide Several fold increase in one patient necessitating stopping treatment. *3188*

Progesterone Hepatotoxicity. *2313*

Promazine May cause cholestatic (hepatocanalicular) jaundice. *0248*

Promethazine May cause cholestasis. *2313*

Propoxyphene Cholestatic effect. *1954* In 3 patients within 10 d of start of drug treatment. *1156* Hepatotoxic response observed in 2 patients. *2110*

Propranolol Occasionally seen, probably not due to hepatotoxicity. *2525*

Propylthiouracil Hepatotoxic effect (centrolobular necrosis). *2313* Single case of chronic active hepatitis reported. 2 others with bridging necrosis. Recovery with drug withdrawal. *2260* Temporary effect associated with drug administration regressed with cessation. *2731* In one 24 year old woman who developed fulminant hepatic failure with lymphocyte sensitization. *2437* Drug induced liver damage: either acute or chronic active hepatitis. *3787*

Protionamide Reported to affect liver function. *2583*

Protriptyline Transient reversible cholestasis. *0071*

Pyrantel Transient increases reported. *2503*

Pyrazinamide Hepatic toxicity reported (viral-hepatitis like). *3193* Hepatotoxic in 1 to 5% patients (previously reported to be toxic in as many as 25%). *3683* Low toxicity with daily doses of 20 to 30 mg/kg body weight but overall incidence of hepatitis of 0.3%. *0682*

Pyridoxine Significant increase in elderly after administration. *2427*

Quinacrine Hepatitis reported (centrolobular necrosis). *0071*

Quinethazone May cause cholestatic jaundice. *2313*

Quinidine May cause mild hepatic impairment in few patients. *2531* In single case of severe quinidine hypersensitivity. *0858* Hypersensitivity observed in 32 of 487 patients who had received drug. *1261* Hypersensitivity reaction reported in one patient. *3350*

Rifampin Hepatic toxicity occurs in up to 7% patients. *1553* Minimal abnormalities of liver function are common: severe liver damage in about 0.6% patients. *3683* In 83% of 18 children receiving drug in combination with isoniazid. *2186*

Salicylate Concentration related effect in 9 of 17 patients with acute rheumatic fever also given phenoxymethyl penicillin: in patients with low albumin effect most marked. *1294*

Sex Difference In men up to and including 5th decade. *2130*

Stanozolol May cause intrahepatic cholestasis. *0757*

Starvation With absolute fast 15 d ?hepatic focal necrosis. *0422*

Stibophen Hepatitis reported. *0071*

Streptokinase Substantial effect not as marked as GGT. *3189* Earlier peak of CK-MB observed in patients on thrombolytic therapy due to reperfusion of ischemic myocardium. *1341* Transient dysfunction of liver, returning to normal when treatment discontinued. *3122*

Sulfadiazine May cause cholestasis with cholangiolitis. *1948*

Sulfadimethoxine Granulomatous reaction in liver. *1049*

Sulfamethizole Reversible cholestasis. *1974*

Aspartate Aminotransferase *(continued)*

Serum Increase Physiological *(continued)*

Sulfamethoxazole Cholestatic jaundice may occur. *1974* Occasional case of cholestasis with or without hepatic necrosis. *3452*

Sulfamethoxypyridine Reversible cholestasis. *2313*

Sulfanilamide May cause reversible cholestasis. *2220*

Sulfapyridine May cause reversible cholestasis. *2220*

Sulfasalazine Occasional case of drug-induced toxic hepatitis. *1422* As part of systemic hypersensitivity reaction in woman with inflammatory bowel disease. *3402* Rare hepatotoxicity reported associated with noncaseating granuloma or focal inflammation with or without necrosis. *1127* Hepatotoxic hypersensitivity reaction like sulfonamide hypersensitivity. *2214* In 2 cases with drug induced hepatotoxicity of hypersensitivity type of reaction: reversed on drug withdrawal. *3364*

Sulfisoxazole Reversible cholestasis. *2313*

Sulfonamides May cause cholestasis. *2313*

Sulfones May affect liver function. *2313*

Sulfonylureas Moderate elevation with cholangiolitis. *3171*

Sulindac Isolated case of cholestatic jaundice 12 d after therapy started. *2371*

Tamoxifen Isolated case of drug induced cholestasis in one patient. *0372*

Testosterone May cause cholestatic and cytotoxic liver damage. *0248*

Tetracycline Hepatotoxic especially in pregnant women. *2808* 2 cases of drug-associated fatty liver of pregnancy. *3795*

Thiabendazole Rare cholestasis and parenchymal liver damage. *2413* One case of drug-induced cholestasis reported. *2962*

Thiacetazone May cause cholestatic (hepatocanalicular) jaundice. *0248* May cause hepatitis. *1292*

Thiazides May cause cholestasis. *2313*

Thiethylperazine Few cases of cholestasis reported. *1680*

Thiocyanate May cause hepatic necrosis. *1343*

Thioguanine May cause cholestasis. *2313* Transient increase in 3 of 29 patients with polycythemia rubra vera. *2450*

Thiopental 1 case out of 24 3-5 d after anesthesia, more longer period after anesthesia pyrexia and jaundice in one case after repeated anesthesia. *0391*

Thioridazine Hepatotoxicity. *1857*

Thiosemicarbazones May affect liver function. *2313*

Thiothixene Hepatotoxic effect (reversible cholestasis). *2313*

Thiouracil May cause cholestasis with cholangiolitis. *1948*

Thorium Dioxide Increased in about 15% after injection. *3147*

Tocainide Occasional abnormality reported associated with reversible hepatitis. *2011*

Tolazamide Cholestasis may occur. *2313*

Tolbutamide May cause cytotoxic liver damage or cholestasis. *0248*

Toluene Diamine May cause cholestasis with cholangiolitis. *1948*

Tranylcypromine May affect liver function (?hypersensitivity). *2313*

Tretinoin In 1 of 11 patients given drug orally. *3867*

Trichlormethiazide May rarely cause cholestasis. *1680*

Trichloroethylene Potentially hepatotoxic. *0071*

Triethylenemelamine Hepatotoxicity (prolonged treatment, large dose). *0071*

Trifluoperazine Cholestatic effect. *1713*

Trimeprazine May cause cholestasis. *0071*

Trimethadione May cause hepatotoxicity with necrosis. *2313*

Trimetrexate Transient effect in small proportion of patients although reverted to normal with continued treatment. *0058*

Trioxsalen May affect liver function. *2313*

Troleandomycin In 30% individuals treated with 1 g/d for 3-4 weeks. Jaundice typically occurs after 2 weeks mixed in type with both cholestasis and mild hepatocytic necrosis. *2792*

Tryparsamide May cause liver damage. *0071*

Uracil Mustard May affect liver function. *2313*

Urethan May cause liver damage and necrosis. *0071*

Valproic Acid Observed in 4 of 25 patients treated with drug: reversible with reduction of dose or withdrawal of drug. *3847* Slight increase in 20% patients on chronic monotherapy: enzyme activity linearly and directly correlated with drug concentration. *1453* Dose related increase in 44% of treated patients. *0950* Rise in about 15 to 30% cases transient and usually maximal to 10 to 12 weeks after beginning treatment. *0504* Transient increase in 44% of patients without change of dosage. *1781* Observed in 3% of 109 patients: reports of fatalities in some studies. *1780* Fatal case reported in child also given phenytoin. *1626* Acute hepatic centrilobular necrosis with severe fatty change in small number of patients. *3763* In 44% of 100 epileptic children when given alone or combined with other anticonvulsants. *0737* In 4 of 9 patients with epilepsy poorly controlled. *3508*

Vasopressin Rhabdomyolysis observed in 2 patients following intravenous drug administration. *0029*

Verapamil Observed in a single case, reverted to normal as soon as drug withdrawn. *2561*

Vinyl Ether Potentially hepatotoxic. *0071*

Vitamin A Observed with acute intoxication. *3867*

Warfarin Rare cases of intrahepatic cholestasis in patients with previous history of hypersensitivity reported. *2940*

X-Ray Therapy Occurs with local irradiation injury. *0987*

Xylitol Dose related hepatocellular damage. *3212*

Zimeldine 7 of 14 patients treated for more than 1 week demonstrated toxic syndrome ? immunological mechanism or related to high initial dose. *2075* Reported in 2 patients in association with headaches, possibly due to fall blood concentration of 5-hydroxytryptamine. *3387* In 64% of 21 inpatients with endogenous depression given 225 mg daily for 4 weeks. *2082* In several of 147 patients although 16 had initially high values. *3741*

Zoxazolamine May produce viral-hepatitis like syndrome. *1948*

Serum Decrease Analytical

Acetaminophen On continuous method at 10 times maximal therapeutic concentration. *1785*

Aminophenazone On continuous method at 10 times maximal therapeutic concentration. *1785*

Clothiapine At 9 mg/L when added to serum and conventional method. *1758*

Detergents 16% inhibition at 0.6-0.8 mg/ml. *3217*

Fluosol-DA Increase by 100% on Du Pont aca III at concentration of 50%. *2518*

Ibuprofen On continuous method at 10 times maximal therapeutic concentration. *1785*

Ketoprofen On continuous method. *1785*

Metronidazole Artifactual depression of activity in almost all patients with NADH-coupled analytical methods. *3056* With continuous flow endpoint reaction method because of drug's high absorbance at 340 nm, although no effect on kinetic methods. *0881* On Kessler method with continuous flow AutoAnalyzers with apparent reduction of activity by as much as 50% at concentration of 8 mg/L. *3592*

Oxalacetate Inhibits colorimetric method. *3835*

Pindolol At 5 mg/L conventional methods when added to serum. *1758*

Pyruvate High concentrations may deplete NADH so even normal enzyme activity gives depletion reaction on Technicon® SMAC®. *1511*

Rifampin At 3.5 mg/L and above decreased activity when methods using absorbance at 340 nm used, since drug absorbs at this wavelength. *0218*

Storage of Sample Stable for less time at -20° if high activity. *3217*

Serum Decrease Physiological

Aging From 47 U/L at 6 y to 35 at 20 y (women). From 45 U/L at 6 y to 29 at 20 y (men). *0887*

Ethinyl Estradiol Increase by 50% after 2 weeks in 5 patients with primary biliary cirrhosis. *1419*

Fluorides *In vivo* inhibition with low doses. *1168*

Fluspirilene Slight but observable trend. *0221*

Hemodialysis Possible loss of pyridoxal phosphate. *3885*

Meals Significant effect observed. *3455*

Penicillamine Increase by 47% in 3 patients with Wilson's disease although effect disappeared after 12 mo. *1417*
Prednisone Augmented response in cirrhotics when drug added to therapeutic regime. *0651*
Progesterone Observed in healthy individuals when only treatment. *0991*
Trifluoperazine Noticeable effect. *0221*
Vitamin B$_6$ Depletion Effect observed at once, less marked then SGPT. *0255*

Plasma   Increase   Physiological
Dapsone Components of typical sulfone or dapsone syndrome. *2026*

Urine   Increase   Physiological
Acetrizoate Sustained increase due to nephrotoxic action. *3538*
Proteinuria Associated with infection and tubular damage. *2897*
Radiographic Agents Transient increase if injected in renal artery. *3538*

CSF   Increase   Analytical
Hemolysis If more than 400 RBC or 200 WBC per cubic mm. *0877*

Red Blood Cells   No Effect   Physiological
Ethanol Activation by pyridoxine in 35 patients with alcohol related illness. *0207*

Red Blood Cells   Increase   Physiological
Oral Contraceptives Compared with normal, B$_6$ stimulation no effect. *3049* Elevated when treated for 6 mo or longer. *3048*

Red Blood Cells   Decrease   Physiological
Anticonvulsants In treated patients compared with controls. *2949*

Aspartic Acid

Plasma   Increase   Analytical
Storage of Sample Even with prolonged storage at -20° Large amounts in platelets and leukocytes. *3230*

Plasma   Increase   Physiological
Glutamic Acid Small increase noted after single 6 g dose. *3444*

Plasma   Decrease   Physiological
Tranylcypromine Observed effect in normal individuals. *0824*

Urine   Increase   Physiological
Ascorbic Acid In 10 healthy females given 10 g/d. *3624*

Aspirin Esterase

Serum   Increase   Physiological
Aspirin In all patients except cirrhotics. *0150*

Atenolol

Serum   No Effect   Physiological
Ranitidine Minimal effect of coadministration on atenolol pharmacokinetics. *1938*

Atrial Natriuretic Peptide

Plasma   No Effect   Physiological
Opiates No effect either with or without anesthesia or surgery. *1929*

Plasma   Increase   Physiological
Nifedipine After 10 mg sublingually increased from mean 19.4 pg/mL to 24.1 pg/mL at 90 minutes. *2917*

Bacteria

Urine   Negative   Analytical
Bilirubin May interfere with color on Microstix. *0129*
Methylene Blue May interfere with color on Microstix. *0129*
Phenazopyridine May interfere with color on Microstix. *0129*

Barbital

Serum   Increase   Physiological
Metharbital Metabolic conversion *in vivo*. *3054*

Barbiturate

Serum   No Effect   Analytical
Digitalis No interference with UV absorption methods. *2981*
Glutarimide No effect on UV absorption methods. *2981*
Morphine No interference with UV absorption methods. *2981*
Oxytetracycline No effect on UV absorption methods. *2981*
Penicillin No effect on UV absorption methods. *2981*
Phenytoin No effect on UV method of Broughton. *1427*
Storage of Sample No effect on UV absorption methods. *2981*

Serum   Increase   Analytical
Acetates Affect colorimetric method with cobalt acetate. *3054*
Antipyrine Extraction properties of free form similar. *3608*
Aspirin May interfere with UV spectrophotometry. *0582*
Atropine False positive screen test with Hg complex formation. *0581*
Chlordiazepoxide Occurs with toxic doses diphenylcarbazone procedure. *0583*
Chlorpheniramine At toxic levels (over 4 mg/dL) only on screen test. *0583*
Creatinine Affect colorimetric method with cobalt acetate. *3054*
Diazepam At toxic levels on diphenylcarbazone procedure. *0583*
Filter Paper Affects diphenylcarbazone procedure if contains tin. *0583* May contain tin, react in Baer's procedure. *2300*
Glutethimide At 5 mg/dL false positive with diphenylcarbazone. *0583*
Meperidine False positive screen test with Hg complex formation. *0581*
Methyprylon Affects diphenylcarbazone above 150 mg/dL only. *0583*
Nitrazepam Forms Hg complex with diphenylcarbazone/Baer's method. *0294*
Phenytoin At 8 mg/dL false positive with diphenylcarbazone. *0583*
Propionate Affect colorimetric method with cobalt acetate. *3054*
Stearate Affect colorimetric method with cobalt acetate. *3054*
Sulfonamides May interfere with UV absorption method. *2981*
Theophylline Identical ultraviolet Spectra. *3608* In certain gas-chromatographic procedures may produce falsely high values. *0852*
Tin Affects diphenylcarbazone procedure if in filter paper. *0583*
Tolbutamide At 10 mg/dL affects diphenylcarbazone procedure. *0583*

Serum   Increase   Physiological
Primidone Phenobarbital is major metabolite. *1225*

Serum   Decrease   Analytical
Aspirin Increased absorbance at acid pH, affects UV methods. *0583*
Filter Paper If siliconized paper used for method of Broughton. *2935*

Barbiturate (continued)

Serum Decrease Analytical (continued)

Glutethimide Negative interference with colorimetric method. *0293*

Phenytoin Negative interference with diphenylcarbazone method. *0293*

Urine No Effect Analytical

Acetaminophen With diphenylcarbazone in Frings TLC procedure at 10 mg/dL. With mercuric SO_4 in Frings TLC procedure at 10 mg/dL. With mercurous NO_3 in Frings TLC procedure at 1 mg/dL. With vanillin on Frings procedure at 10 mg/dL. *1204*

Amitriptyline With diphenylcarbazone in Frings TLC procedure at 10 mg/dL. With mercurous NO_3 in Frings TLC procedure at 10 mg/dL. With vanillin on Frings procedure at 10 mg/dL. With mercuric SO_4 in Frings TLC procedure at 10 mg/dL. *1204*

Amobarbital With vanillin on Frings procedure at 10 mg/dL. *1204*

Amphetamine With diphenylcarbazone in Frings TLC procedure at 10 mg/dL. With mercuric SO_4 in Frings TLC procedure at 10 mg/dL. With mercurous NO_3 in Frings TLC procedure at 10 mg/dL. With vanillin on Frings procedure at 10 mg/dL. *1204*

Benztropine With diphenylcarbazone in Frings TLC procedure at 1 mg/dL. With mercurous NO_3 in Frings TLC procedure at 1 mg/dL. With mercuric SO_4 in Frings TLC procedure at 1 mg/dL. With vanillin on Frings procedure at 1 mg/dL. *1204*

Bromides With mercuric SO_4 in Frings TLC procedure at 100 mg/dL. *1204*

Butabarbital With vanillin on Frings procedure at 10 mg/dL. *1204*

Carisoprodol With mercuric SO_4 in Frings TLC procedure at 10 mg/dL. With mercurous NO_3 in Frings TLC procedure at 10 mg/dL. With vanillin on Frings procedure at 10 mg/dL. With diphenylcarbazone in Frings TLC procedure at 10 mg/dL. *1204*

Chloral Hydrate With mercurous NO_3 in Frings TLC procedure at 10 mg/dL. With mercuric SO_4 in Frings TLC procedure at 10 mg/dL. With vanillin on Frings procedure at 10 mg/dL. With diphenylcarbazone in Frings TLC procedure at 10 mg/dL. *1204*

Chlordiazepoxide With diphenylcarbazone in Frings TLC procedure at 10 mg/dL. With mercuric SO_4 in Frings TLC procedure at 10 mg/dL. With vanillin on Frings procedure at 10 mg/dL. With mercurous NO_3 in Frings TLC procedure at 10 mg/dL. With diphenylcarbazone in Frings TLC procedure at 10 mg/dL. *1204*

Chloroquine With mercuric SO_4 in Frings TLC procedure at 10 mg/dL. With diphenylcarbazone in Frings TLC procedure at 10 mg/dL. With mercurous NO_3 in Frings TLC procedure at 10 mg/dL. With vanillin on Frings procedure at 10 mg/dL. *1204*

Chlorpheniramine With mercurous NO_3 in Frings TLC procedure at 12 mg/dL. With diphenylcarbazone in Frings TLC procedure at 12 mg/dL. With vanillin on Frings procedure at 12 mg/dL. With mercuric SO_4 in Frings TLC procedure at 12 mg/dL. *1204*

Chlorpromazine With diphenylcarbazone in Frings TLC procedure at 10 mg/dL. With mercurous NO_3 in Frings TLC procedure at 10 mg/dL. *1204*

Clorazepate With diphenylcarbazone in Frings TLC procedure at 10 mg/dL. With mercurous NO_3 in Frings TLC procedure at 10 mg/dL. With mercuric SO_4 in Frings TLC procedure at 10 mg/dL. *1204*

Cocaine With mercuric SO_4 in Frings TLC procedure at 10 mg/dL. With vanillin on Frings procedure at 10 mg/dL. With mercurous NO_3 in Frings TLC procedure at 10 mg/dL. With diphenylcarbazone in Frings TLC procedure at 10 mg/dL. *1204*

Codeine With mercurous NO_3 in Frings TLC procedure at 10 mg/dL. With mercuric SO_4 in Frings TLC procedure at 10 mg/dL. With diphenylcarbazone in Frings TLC procedure at 10 mg/dL. *1204*

Diazepam With mercurous NO_3 on Frings TLC procedure at 10 mg/dL. With diphenylcarbazone on Frings TLC procedure at 10 mg/dL. With vanillin on Frings procedure at 10 mg/dL. With mercuric SO_4 on Frings TLC procedure at 10 mg/dL. *1204*

Diethylpropion With mercuric SO_4 on Frings TLC procedure at 25 mg/dL. With mercurous NO_3 on Frings TLC procedure at 25 mg/dL. With vanillin on Frings procedure at 25 mg/dL. With diphenylcarbazone on Frings TLC procedure at 25 mg/dL. *1204*

Diphenhydramine With diphenylcarbazone on Frings TLC procedure at 10 mg/dL. With vanillin on Frings procedure at 10 mg/dL. With mercurous NO_3 on Frings TLC procedure at 10 mg/dL. With mercuric SO_4 on Frings TLC procedure at 10 mg/dL. *1204*

Doxepin With mercurous NO_3 on Frings TLC procedure at 1 mg/dL. With vanillin on Frings procedure at 10 mg/dL. With diphenylcarbazone on Frings TLC procedure at 1 mg/dL. With mercuric SO_4 on Frings TLC procedure at 10 mg/dL. *1204*

Ephedrine With diphenylcarbazone on Frings TLC procedure at 10 mg/dL. With mercuric SO_4 on Frings TLC procedure at 10 mg/dL. With vanillin on Frings procedure at 10 mg/dL. With mercurous NO_3 on Frings TLC procedure at 10 mg/dL. *1204*

Ethchlorvynol With vanillin on Frings procedure at 10 mg/dL. With mercurous NO_3 on Frings TLC procedure at 10 mg/dL. With mercuric SO_4 on Frings TLC procedure at 10 mg/dL. With diphenylcarbazone on Frings TLC procedure at 10 mg/dL. *1204*

Ethinamate With vanillin on Frings procedure at 10 mg/dL. *1204*

Fluphenazine With diphenylcarbazone on Frings TLC procedure at 10 mg/dL. With mercurous NO_3 on Frings TLC procedure at 10 mg/dL. *1204*

Flurazepam With mercuric SO_4 on Frings TLC procedure at 10 mg/dL. With mercurous NO_3 on Frings TLC procedure at 10 mg/dL. With vanillin on Frings procedure at 10 mg/dL. With diphenylcarbazone on Frings TLC procedure at 10 mg/dL. *1204*

Glutethimide With vanillin on Frings procedure at 10 mg/dL. With mercuric SO_4 on Frings TLC procedure at 1 mg/dL. *1204*

Glycopyrrolate With vanillin on Frings procedure at 2 mg/dL. With mercurous NO_3 on Frings TLC procedure at 2 mg/dL. With diphenylcarbazone on Frings TLC procedure at 2 mg/dL. With mercuric SO_4 on Frings TLC procedure at 2 mg/dL. *1204*

Haloperidol With vanillin on Frings procedure at 1 mg/dL. With mercurous NO_3 on Frings TLC procedure at 1 mg/dL. With diphenylcarbazone on Frings TLC procedure at 1 mg/dL. With mercuric SO_4 on Frings TLC procedure at 1 mg/dL. *1204*

Hydromorphone With diphenylcarbazone on Frings TLC procedure at 2 mg/dL. With mercuric SO_4 on Frings TLC procedure at 2 mg/dL. With vanillin on Frings procedure at 2 mg/dL. With mercurous NO_3 on Frings TLC procedure at 2 mg/dL. *1204*

Imipramine With diphenylcarbazone on Frings TLC procedure at 10 mg/dL. With mercurous NO_3 on Frings TLC procedure at 10 mg/dL. *1204*

Mebutamate With mercuric SO_4 on Frings TLC procedure at 10 mg/dL. With mercurous NO_3 on Frings TLC procedure at 10 mg/dL. With diphenylcarbazone on Frings TLC procedure at 10 mg/dL. With vanillin on Frings procedure at 10 mg/dL. *1204*

Meperidine With mercurous NO_3 on Frings TLC procedure at 10 mg/dL. With mercuric SO_4 on Frings TLC procedure at 10 mg/dL. With vanillin on Frings procedure at 10 mg/dL. With diphenylcarbazone on Frings TLC procedure at 10 mg/dL. *1204*

Mephenesin With diphenylcarbazone on Frings TLC procedure at 10 mg/dL. With mercurous NO_3 on Frings TLC procedure at 10 mg/dL. With mercuric SO_4 on Frings TLC procedure at 10 mg/dL. *1204*

Meprobamate With diphenylcarbazone on Frings TLC procedure at 10 mg/dL. With mercuric SO_4 on Frings TLC procedure at 10 mg/dL. With mercurous NO_3 on Frings TLC procedure at 10 mg/dL. *1204*

Mescaline With mercuric SO_4 on Frings TLC procedure at 10 mg/dL. With vanillin on Frings procedure at 10 mg/dL. With diphenylcarbazone on Frings TLC procedure at 10 mg/dL. With mercurous NO_3 on Frings TLC procedure at 10 mg/dL. *1204*

Methadone With vanillin on Frings procedure at 10 mg/dL. With mercurous NO_3 on Frings TLC procedure at 10 mg/dL. With mercuric SO_4 on Frings TLC procedure at 10 mg/dL. With diphenylcarbazone on Frings TLC procedure at 10 mg/dL. *1204*

Methamphetamine With mercurous NO_3 on Frings TLC procedure at 10 mg/dL. With mercuric SO_4 on Frings TLC procedure at 10 mg/dL. With diphenylcarbazone on Frings TLC procedure at 10 mg/dL. With vanillin on Frings procedure at 10 mg/dL. *1204*

Methapyrilene With mercurous NO_3 on Frings TLC procedure at 25 mg/dL. With mercuric SO_4 on Frings TLC procedure at 25 mg/dL. With diphenylcarbazone on Frings TLC procedure at 25 mg/dL. *1204*

Methaqualone With diphenylcarbazone on Frings TLC procedure at 10 mg/dL. With mercurous NO_3 on Frings TLC procedure at 10 mg/dL. With vanillin on Frings procedure at 10 mg/dL. With mercuric SO_4 on Frings TLC procedure at 10 mg/dL. *1204*

Methdilazine With diphenylcarbazone on Frings TLC procedure at 10 mg/dL. With mercurous NO$_3$ on Frings TLC procedure at 10 mg/dL. *1204*

Methylphenidate With vanillin on Frings procedure at 10 mg/dL. With diphenylcarbazone on Frings TLC procedure at 10 mg/dL. With mercuric SO$_4$ on Frings TLC procedure at 10 mg/dL. With mercurous NO$_3$ on Frings TLC procedure at 10 mg/dL. *1204*

Methyprylon With mercuric SO$_4$ on Frings TLC procedure at 10 mg/dL. With diphenylcarbazone on Frings TLC procedure at 10 mg/dL. With mercurous NO$_3$ on Frings TLC procedure at 10 mg/dL. With vanillin on Frings procedure at 10 mg/dL. *1204*

Morphine With diphenylcarbazone on Frings TLC procedure at 10 mg/dL. With mercuric SO$_4$ on Frings TLC procedure at 10 mg/dL. With vanillin on Frings procedure at 10 mg/dL. With mercurous NO$_3$ on Frings TLC procedure at 10 mg/dL. *1204*

Oxazepam With diphenylcarbazone on Frings TLC procedure at 10 mg/dL. With mercuric SO$_4$ on Frings TLC procedure at 10 mg/dL. With vanillin on Frings procedure at 10 mg/dL. With mercurous NO$_3$ on Frings TLC procedure at 10 mg/dL. *1204*

Oxycodone With mercurous NO$_3$ on Frings TLC procedure at 5 mg/dL. With diphenylcarbazone on Frings TLC procedure at 5 mg/dL. With vanillin on Frings procedure at 5 mg/dL. With mercuric SO$_4$ on Frings TLC procedure at 10 mg/dL. *1204*

Oxymetazoline With diphenylcarbazone on Frings TLC procedure at 0.5 mg/dL. With vanillin on Frings procedure at 0.5 mg/dL. With mercuric SO$_4$ on Frings TLC procedure at 0.5 mg/dL. With mercurous NO$_3$ on Frings TLC procedure at 0.5 mg/dL. *1204*

Pentazocine With diphenylcarbazone on Frings TLC procedure at 10 mg/dL. With mercurous NO$_3$ on Frings TLC procedure at 10 mg/dL. With mercuric SO$_4$ on Frings TLC procedure at 10 mg/dL. *1204*

Pentobarbital With vanillin on Frings procedure at 10 mg/dL. *1204*

Perphenazine With mercurous NO$_3$ on Frings TLC procedure at 10 mg/dL. With diphenylcarbazone on Frings TLC procedure at 10 mg/dL. *1204*

Phenmetrazine With mercuric SO$_4$ on Frings TLC procedure at 10 mg/dL. With diphenylcarbazone on Frings TLC procedure at 10 mg/dL. *1204*

Phenobarbital With vanillin on Frings procedure at 10 mg/dL. *1204*

Phenylephrine With mercuric SO$_4$ on Frings TLC procedure at 10 mg/dL. With vanillin on Frings procedure at 10 mg/dL. With diphenylcarbazone on Frings TLC procedure at 10 mg/dL. With mercurous NO$_3$ on Frings TLC procedure at 10 mg/dL. *1204*

Phenylpropanolamine With mercurous NO$_3$ on Frings TLC procedure at 10 mg/dL. With mercuric SO$_4$ on Frings TLC procedure at 10 mg/dL. With diphenylcarbazone on Frings TLC procedure at 10 mg/dL. With vanillin on Frings procedure at 10 mg/dL. *1204*

Phenytoin With vanillin on Frings procedure at 10 mg/dL. *1204*

Primaquine With mercuric SO$_4$ on Frings TLC procedure at 10 mg/dL. With diphenylcarbazone on Frings TLC procedure at 10 mg/dL. *1204*

Primethamine With diphenylcarbazone on Frings TLC procedure at 10 mg/dL. *1204*

Procainamide With vanillin on Frings procedure at 10 mg/dL. With diphenylcarbazone on Frings TLC procedure at 10 mg/dL. With mercuric SO$_4$ on Frings TLC procedure at 10 mg/dL. With mercurous NO$_3$ on Frings TLC procedure at 10 mg/dL. *1204*

Prochlorperazine With mercurous NO$_3$ on Frings TLC procedure at 10 mg/dL. With diphenylcarbazone on Frings TLC procedure at 10 mg/dL. *1204*

Promazine With diphenylcarbazone on Frings TLC procedure at 10 mg/dL. With mercurous NO$_3$ on Frings TLC procedure at 10 mg/dL. *1204*

Promethazine With diphenylcarbazone on Frings TLC procedure at 10 mg/dL. With mercurous NO$_3$ on Frings TLC procedure at 10 mg/dL. *1204*

Propoxyphene With mercurous NO$_3$ on Frings TLC procedure at 10 mg/dL. With vanillin on Frings procedure at 10 mg/dL. With diphenylcarbazone on Frings TLC procedure at 10 mg/dL. With mercuric SO$_4$ on Frings TLC procedure at 10 mg/dL. *1204*

Pyrilamine With mercuric SO$_4$ on Frings TLC procedure at 10 mg/dL. With vanillin on Frings procedure at 10 mg/dL. With diphenylcarbazone on Frings TLC procedure at 10 mg/dL. With mercurous NO$_3$ on Frings TLC procedure at 10 mg/dL. *1204*

Pyrimethamine With vanillin on Frings procedure at 10 mg/dL. With mercuric SO$_4$ on Frings TLC procedure at 10 mg/dL. *1204*

Quinidine With mercuric SO$_4$ on Frings TLC procedure at 10 mg/dL. With vanillin on Frings procedure at 10 mg/dL. With diphenylcarbazone on Frings TLC procedure at 10 mg/dL. With mercurous NO$_3$ on Frings TLC procedure at 10 mg/dL. *1204*

Quinine With mercuric SO$_4$ on Frings TLC procedure at 10 mg/dL. With diphenylcarbazone on Frings TLC procedure at 10 mg/dL. With vanillin on Frings procedure at 10 mg/dL. With mercurous NO$_3$ on Frings TLC procedure at 10 mg/dL. *1204*

Salicylate With diphenylcarbazone on Frings TLC procedure at 100 mg/dL. With vanillin on Frings procedure at 100 mg/dL. With mercurous NO$_3$ on Frings TLC procedure at 10 mg/dL. With mercuric SO$_4$ on Frings TLC procedure at 100 mg/dL. *1204*

Scopolamine With mercuric SO$_4$ on Frings TLC procedure at 10 mg/dL. With diphenylcarbazone on Frings TLC procedure at 10 mg/dL. With mercurous NO$_3$ on Frings TLC procedure at 10 mg/dL. *1204*

Secobarbital With vanillin on Frings procedure at 10 mg/dL. *1204*

Sodium Bromide With diphenylcarbazone on Frings TLC procedure at 100 mg/dL. With vanillin on Frings procedure at 100 mg/dL. With mercurous NO$_3$ on Frings TLC procedure at 100 mg/dL. *1204*

Strychnine With mercuric SO$_4$ on Frings TLC procedure at 10 mg/dL. With mercurous NO$_3$ on Frings TLC procedure at 10 mg/dL. With diphenylcarbazone on Frings TLC procedure at 10 mg/dL. With vanillin on Frings procedure at 10 mg/dL. *1204*

Sulfanilamide With diphenylcarbazone on Frings TLC procedure at 100 mg/dL. With mercuric SO$_4$ on Frings TLC procedure at 100 mg/dL. *1204*

Tetracycline With diphenylcarbazone on Frings TLC procedure at 2 mg/dL. With mercurous NO$_3$ on Frings TLC procedure at 2 mg/dL. With mercuric SO$_4$ on Frings TLC procedure at 2 mg/dL. With vanillin on Frings procedure at 2 mg/dL. *1204*

Thiobarbituric Acid With vanillin on Frings procedure at 10 mg/dL. With mercurous NO$_3$ on Frings TLC procedure at 10 mg/dL. With diphenylcarbazone on Frings TLC procedure at 10 mg/dL. With mercuric SO$_4$ on Frings TLC procedure at 10 mg/dL. *1204*

Thiopropazate With diphenylcarbazone on Frings TLC procedure at 10 mg/dL. With mercurous NO$_3$ on Frings TLC procedure at 10 mg/dL. *1204*

Thioridazine With mercurous NO$_3$ on Frings TLC procedure at 10 mg/dL. With diphenylcarbazone on Frings TLC procedure at 10 mg/dL. *1204*

Thiothixene With mercuric SO$_4$ on Frings TLC procedure at 10 mg/dL. With mercurous NO$_3$ on Frings TLC procedure at 10 mg/dL. With diphenylcarbazone on Frings TLC procedure at 10 mg/dL. *1204*

Trifluoperazine With diphenylcarbazone on Frings TLC procedure at 10 mg/dL. With mercurous NO$_3$ on Frings TLC procedure at 10 mg/dL. *1204*

Triflupromazine With diphenylcarbazone on Frings TLC procedure at 10 mg/dL. With mercurous NO$_3$ on Frings TLC procedure at 10 mg/dL. *1204*

Trihexyphenidyl With mercuric SO$_4$ on Frings TLC procedure at 10 mg/dL. With mercurous NO$_3$ on Frings TLC procedure at 5 mg/dL. With diphenylcarbazone on Frings TLC procedure at 5 mg/dL. With vanillin on Frings procedure at 5 mg/dL. *1204*

Trimethobenzamide With diphenylcarbazone on Frings TLC procedure at 25 mg/dL. With mercuric SO$_4$ on Frings TLC procedure at 25 mg/dL. With mercurous NO$_3$ on Frings TLC procedure at 25 mg/dL. With vanillin on Frings procedure at 25 mg/dL. *1204*

Tybamate With mercuric SO$_4$ on Frings TLC procedure at 10 mg/dL. With diphenylcarbazone on Frings TLC procedure at 10 mg/dL. With mercurous NO$_3$ on Frings TLC procedure at 10 mg/dL. With vanillin on Frings procedure at 10 mg/dL. *1204*

Urine   Increase   Analytical

Amobarbital White spot with mercuric SO$_4$ in Frings TLC procedure. With mercurous NO$_3$ reagent in Frings TLC procedure. Purple with diphenylcarbazone in Frings TLC procedure. *1204*

Butabarbital With mercurous NO$_3$ reagent in Frings TLC procedure. Purple with diphenylcarbazone in Frings TLC procedure. White spot with mercuric SO$_4$ in Frings TLC procedure. *1204*

Chlorpromazine Pink spot with vanillin in Frings TLC procedure. Pink spot with mercuric SO$_4$ in Frings TLC procedure. *1204*

Barbiturate (continued)

Urine Increase Analytical (continued)

Clorazepate Yellow spot with vanillin in Frings TLC procedure. 1204

Codeine Purple spot with vanillin in Frings TLC procedure. 1204

Ethinamate With mercurous NO_3 reagent on Frings TLC procedure. Blue with diphenylcarbazone in Frings TLC procedure. White spot with mercuric SO_4 on Frings TLC procedure. 1204

Fluphenazine Orange spot with vanillin on Frings procedure. Orange spot with mercuric SO_4 in Frings TLC procedure. 1204

Glutethimide With mercurous NO_3 reagent on Frings TLC procedure. Purple with diphenylcarbazone on Frings TLC procedure. 1204

Imipramine Blue/green spot with vanillin in Frings TLC procedure. Blue spot with mercuric SO_4 in Frings TLC procedure. 1204

Mephenesin Pink spot with vanillin in Frings TLC procedure. 1204

Mephenesin Carbamate Pink/orange spot with vanillin in Frings TLC procedure. 1204

Meprobamate Yellow spot with vanillin on Frings TLC procedure. 1204

Methapyrilene Purple spot with vanillin on Frings TLC procedure. 1204

Methdilazine Pink spot with vanillin in Frings TLC procedure. Pink/orange with mercuric SO_4 in Frings TLC procedure. 1204

Pentazocine Pale-purple spot with vanillin in Frings TLC procedure. 1204

Pentobarbital White spot with mercuric SO_4 on Frings TLC procedure. Purple with diphenylcarbazone on Frings TLC procedure. With mercurous NO_3 reagent on Frings TLC procedure. 1204

Perphenazine Pink spot with vanillin in Frings TLC procedure. Pink spot with mercuric SO_4 in Frings TLC procedure. 1204

Phenobarbital White spot with mercuric SO_4 on Frings TLC procedure, Purple with diphenylcarbazone on Frings TLC procedure. With mercurous NO_3 reagent on Frings TLC procedure. 1204

Phenytoin With mercurous NO_3 reagent on Frings TLC procedure. White spot with mercuric SO_4 on Frings TLC procedure, Purple with diphenylcarbazone on Frings TLC procedure. 1204

Primaquine Orange spot with vanillin in Frings TLC procedure. With mercurous NO_3 reagent on Frings TLC procedure. 1204

Prochlorperazine Pink spot with vanillin in Frings TLC procedure. Orange spot with mercuric SO_4 in Frings TLC procedure. 1204

Promazine Orange spot with mercuric SO_4 in Frings TLC procedure. Orange spot with vanillin in Frings TLC procedure. 1204

Promethazine Pink/orange with mercuric SO_4 in Frings TLC procedure. Pink spot with vanillin in Frings TLC procedure. 1204

Pyrimethamine With mercurous NO_3 reagent on Frings TLC procedure. 1204

Secobarbital White spot with mercuric SO_4 on Frings TLC procedure, Purple with diphenylcarbazone on Frings TLC procedure. With mercurous NO_3 reagent on Frings TLC procedure. 1204

Sulfanilamide Yellow spot with vanillin on Frings TLC procedure With mercurous NO_3 reagent on Frings TLC procedure. 1204

Thiopropazate Pink spot with vanillin in Frings TLC procedure. Pink/orange with mercuric SO_4 in Frings TLC procedure. 1204

Thioridazine Blue spot with mercuric SO_4 in Frings TLC procedure. Violet spot with vanillin on Frings TLC procedure. 1204

Thiothixene Orange spot with vanillin in Frings TLC procedure. 1204

Trifluoperazine Orange spot with mercuric SO_4 in Frings TLC procedure. Orange spot with vanillin in Frings TLC procedure. 1204

Triflupromazine Orange spot with mercuric SO_4 in Frings TLC procedure. Orange spot with vanillin in Frings TLC procedure. 1204

Basal Metabolic Rate

Patient No Effect Physiological

Dextrothyroxine No significant effect noted in normal subjects. 0019

Smoking No significant constant change. 0251

Patient Increase Physiological

Amphetamine Metabolic effect of drugs. 2220

Aspirin Reported metabolic effect. 2220

Atropine Temporary effect observed. 0251

Caffeine Metabolic effect of drug (temporary effect). 2220

Epinephrine Temporary effect observed after injection. 0251

Lactation Remains elevated from pregnancy. 0987

Levarterenol Normal metabolic response. 1176

Levothyroxine Metabolic effect of hormone (maximum at 1 week). 2220

Liothyronine Metabolic effect of hormone. 2220

Pain Effect observed if severe pain. 0251

Posterior Pituitary Temporary effect after injection. 0251

Pregnancy Moderately increased especially in last 3 mo. 0987

Thyroid Metabolic effect of hormone. 2220

Patient Decrease Physiological

Barbiturates Decreases rate by about 10%. 0251

Corticotropin Metabolic action of hormone. 2220

Glucocorticoids Metabolic action of hormones. 2220

Menstruation Lowest at time of bleeding, peaks before next period. 0251

Morphine Metabolic effect of drug. 2220

Narcotics Metabolic effect of drugs. 2220

Sleep Slight effect observed. 0251

Base Deficit

Blood Increase Physiological

Hypothermia Due to hypoxia, acid-base imbalance, hypotension. 2258

Blood Decrease Physiological

Starvation Marked metabolic acidosis within 5 d. 3067

Base Excess

Blood No Effect Physiological

Physical Training No significant difference observed. 1491

Basophilic Stippling

Red Blood Cells Abnormal Physiological

Lead Occurs in 60% childhood cases. 0489

Basophils

Blood Increase Physiological

Desipramine Mild effect noted in some patients. 2427

Blood Decrease Physiological

Procainamide Occasional severe reduction noted. 1586

Propanidid Marked fall within 3 minutes of i.v. inject. 2212

Thiopental Significant fall within 3 minutes of i.v. injection. 2212

Bence-Jones Protein

Urine Positive Analytical

Hemoglobin False positive heat test, paraprotein on electrophoresis. 2755

Urine Positive Physiological

Tetracycline Nephrotoxic effect with degraded tetracycline. 1022

Benzoic Acid

Urine   Increase   Physiological
Bacteriuria   8 of 11 species hydrolyzed hippuric acid. *1483*

Bicarbonate

Serum   No Effect   Analytical

Acetaminophen   At concentration of 1500 mg/L had no effect on method using phenolphthalein method. *3393*

Acetazolamide   At concentration of 1,000 mg/L had no effect on phenolphthalein method. *3393*

Allopurinol   At concentration of 5 mg/L had no effect on method using phenolphthalein. *3393*

Aminosalicylic Acid   At concentration of 460 mg/L had no effect on method using phenolphthalein. *3393*

Amitriptyline   At concentration of 20 mg/L had no effect on method using phenolphthalein. *3393*

Amobarbital   At concentration of 150 mg/L had no effect on method using phenolphthalein. *3393*

Amphotericin B   At concentration of 20 mg/L had no effect on method using phenolphthalein. *3393*

Ampicillin   At concentration of 40 mg/L had no effect on method using phenolphthalein. *3393*

Ascorbic Acid   At concentration of 500 mg/L had no effect on method using phenolphthalein. *3393*

Barbital   At concentration of 500 mg/L had no effect on method using phenolphthalein. *3393*

Caffeine   At concentration of 160 mg/L had no effect on method using phenolphthalein. *3393*

Carbenicillin   At concentration of 15,000 mg/L had no effect on method using phenolphthalein. *3393*

Cefazolin   At concentration of 110 mg/L had no effect on method using phenolphthalein. *3393*

Cephalothin   At concentration of 1,000 mg/L had no effect on method using phenolphthalein. *3393*

Chlordiazepoxide   At concentration of 5 mg/L had no effect on method using phenolphthalein. *3393*

Chlormezanone   At concentration of 100 mg/L had no effect on method using phenolphthalein. *3393*

Chlorphenesin   At concentration of 150 mg/L had no effect on method using phenolphthalein. *3393*

Chlorpheniramine   At concentration of 200 mg/L had no effect on method using phenolphthalein. *3393*

Chlorpromazine   At concentration of 3 mg/L had no effect on method using phenolphthalein. *3393*

Chlorpropamide   At concentration of 150 mg/L had no effect on method using phenolphthalein. *3393*

Chlorprothixene   At concentration of 1 mg/L had no effect on method using phenolphthalein. *3393*

Cimetidine   At concentration of 1 mg/L had no effect on method using phenolphthalein. *3393*

Cloxacillin   At concentration of 5 mg/L had no effect on method using phenolphthalein. *3393*

Cocaine   At concentration of 3 mg/L had no effect on method using phenolphthalein. *3393*

Codeine   At concentration of 1 mg/L had no effect on method using phenolphthalein. *3393*

Contact With Clot   If plastic disc used to separate for 48 h. *1863*

Desipramine   At concentration of 8 mg/L had no effect on method using phenolphthalein. *3393*

Diazepam   At concentration of 1.5 mg/L had no effect on method using phenolphthalein. *3393*

Diethylaminoethanol   At concentration of 30 mg/L had no effect on method using phenolphthalein. *3393*

Digitoxin   At concentration of 6 mg/L had no effect on method using phenolphthalein. *3393*

Digoxin   At concentration of 9 mg/L had no effect on method using phenolphthalein. *3393*

Dimethadione   At concentration of 2,000 mg/L had no effect on method using phenolphthalein. *3393*

N-Dimethyldiazepam   At concentration of 6 mg/L had no effect on method using phenolphthalein. *3393*

Diphenhydramine   At concentration of 10 mg/L had no effect on method using phenolphthalein. *3393*

Disopyramide   At concentration of 4 mg/L had no effect on method using phenolphthalein. *3393*

Disulfiram   At concentration of 120 mg/L had no effect on method using phenolphthalein. *3393*

Disulphine Blue   By continuous flow cresol-red reaction. *1464*

Ethchlorvynol   At concentration of 400 mg/L had no effect on method using phenolphthalein. *3393*

Ethosuximide   At concentration of 390 mg/L had no effect on method using phenolphthalein. *3393*

Fluosol-DA   On SMA IIC at concentration of 50%. *2518*   By 18% on Kodak Ektachem® 400 at concentration of 50%. *2518*

Flurazepam   At concentration of 0.1 mg/L had no effect on method using phenolphthalein. *3393*

Furosemide   At concentration of 4 mg/L had no effect on method using phenolphthalein. *3393*

Gentamicin   At concentration of 150 mg/L had no effect on method using phenolphthalein. *3393*

Hemolysis   Calculated error less than 1%/g. *0515*

Hydralazine   At concentration of 6 mg/L had no effect on method using phenolphthalein. *3393*

Indomethacin   At concentration of 13 mg/L had no effect on method using phenolphthalein. *3393*

Insulin   At concentration of 3 mg/L had no effect on method using phenolphthalein. *3393*

Iproniazid   At concentration of 40 mg/L had no effect on method using phenolphthalein. *3393*

Isoniazid   At concentration of 100 mg/L had no effect on method using phenolphthalein. *3393*

Lipemia   In patient specimens measured on SMAC® in comparison with specimens following ultra-centrifugation. *2466*

Lorazepam   At concentration of 0.05 mg/L had no effect on method using phenolphthalein. *3393*

Meperidine   At concentration of 60 mg/L had no effect on method using phenolphthalein. *3393*

Mephenesin   At concentration of 100 mg/L had no effect on method using phenolphthalein. *3393*

Meprobamate   At concentration of 25 mg/L had no effect on method using phenolphthalein. *3393*

Methamphetamine   At concentration of 2 mg/L had no effect on method using phenolphthalein. *3393*

Methapyrilene   At concentration of 13 mg/L had no effect on method using phenolphthalein. *3393*

Methicillin   At concentration of 900 mg/L had no effect on method using phenolphthalein. *3393*

Methohexital   At concentration of 50 mg/L had no effect on method using phenolphthalein. *3393*

Methotrimeprazine   At concentration of 1 mg/L had no effect on method using phenolphthalein. *3393*

Methsuximide   At concentration of 40 mg/L had no effect on method using phenolphthalein. *3393*

Methyldopa   At concentration of 80 mg/L had no effect on method using phenolphthalein. *3393*

Metoprolol   At concentration of 0.34 mg/L had no effect on method using phenolphthalein. *3393*

Morphine   At concentration of 1 mg/L had no effect on method using phenolphthalein. *3393*

Nafcillin   At concentration of 50 mg/L had no effect on method using phenolphthalein. *3393*

Nitrofurantoin   At concentration of 4 mg/L had no effect on method using phenolphthalein. *3393*

Nortriptyline   At concentration of 2 mg/L had no effect on method using phenolphthalein. *3393*

Orphenadrine   At concentration of 9 mg/L had no effect on method using phenolphthalein. *3393*

Oxazepam   At concentration of 1 mg/L had no effect on method using phenolphthalein. *3393*

Papaverine   At concentration of 10 mg/L had no effect on method using phenolphthalein. *3393*

Bicarbonate (continued)

Serum No Effect Analytical (continued)

Paraldehyde At concentration of 2,000 mg/L had no effect on method using phenolphthalein. *3393*

Penicillin G At concentration of 2,000 mg/L had no effect as measured by method using phenolphthalein. *3393*

Pentobarbital At concentration of 340 mg/L had no effect on method using phenolphthalein method. *3393*

Perphenazine At concentration of 1 mg/L had no effect on measurement by phenolphthalein method. *3393*

Phenobarbital At concentration of 250 mg/L had no effect on method using phenolphthalein. *3393*

Phensuximide At concentration of 120 mg/L had no effect on method using phenolphthalein. *3393*

Phenylbutazone At concentration of 750 mg/L had no effect on method using phenolphthalein. *3393*

Phenytoin At concentration of 240 mg/L had no effect on method using phenolphthalein. *3393*

Plasma No significant difference from serum by SMA 12/60. *2228*

Probenecid At concentration of 1300 mg/L had no effect on method using phenolphthalein. *3393*

Procainamide At concentration of 50 mg/L had no effect on method using phenolphthalein. *3393*

Procaine At concentration of 2 mg/L had no effect on method using phenolphthalein. *3393*

Prochlorperazine At concentration of 1 mg/L had no effect on method using phenolphthalein. *3393*

Promethazine At concentration of 200 mg/L had no effect on method using phenolphthalein. *3393*

Propranolol At concentration of 0.2 mg/L had no effect on method using phenolphthalein. *3393*

Quinidine At concentration of 210 mg/L had no effect on method using phenolphthalein. *3393*

Quinine At concentration of 30 mg/L had no effect on method using phenolphthalein. *3393*

Salicylate At concentration of 500 mg/L had no effect on method using phenolphthalein. *3393*

Secobarbital At concentration of 100 mg/L had no effect on method using phenolphthalein. *3393*

Storage of Sample No difference with/without oil if tube full. *3217*

Strychnine At concentration of 12 mg/L had no effect on method using phenolphthalein. *3393*

Sulfadiazine At concentration of 1500 mg/L had no effect on method using phenolphthalein. *3393*

Sulfaguanidine At concentration of 500 mg/L had no effect on method using phenolphthalein. *3393*

Sulfanilamide At concentration of 1,000 mg/L had no effect on method using phenolphthalein. *3393*

Tetracycline At concentration of 4 mg/L had no effect on method using phenolphthalein. *3393*

Theobromine At concentration of 2,000 mg/L had no effect on method using phenolphthalein. *3393*

Theophylline At concentration of 2,000 mg/L had no effect on method using phenolphthalein method. *3393*

Thiopental At concentration of 70 mg/L had no effect on method using phenolphthalein method. *3393*

Timolol At concentration of 0.01 mg/L had no effect on method using phenolphthalein. *3393*

Tolbutamide At concentration of 100 mg/L had no effect on method using phenolphthalein. *3393*

Transport of Specimen By pneumatic tube if specimen protected. *3217*

Tribromethanol At concentration of 90 mg/L had no effect on method using phenolphthalein. *3393*

Trichlorethanol At concentration of 1,000 mg/L had no effect on method using phenolphthalein. *3393*

Trifluoperazine At concentration of 1 mg/L had no effect on method using phenolphthalein. *3393*

Warfarin At concentration of 100 mg/L had no effect on method using phenolphthalein. *3393*

Serum No Effect Physiological

Amiloride No effect seen in 13 men treated for 8 weeks. *0656*

Aspirin When approximately 3 g/d ingested for several weeks. *3076*

Meals No effect after standard breakfast. *0579*

Storage of Sample Insignificant change at 2 h at room temperature. *0931*

Serum Increase Analytical

Bilirubin At 300 mg/L increased concentration by as much as 13 mmol/L when Worthington procedure used on Olympus Demand. *1746*

Naproxen Falsely high value, linearly correlated with serum drug concentration with Technicon® RA-1000 ion-specific electrode. *1512*

Serum Increase Physiological

Aldosterone Metabolic alkalosis. *1022*

Aspirin Later alteration of acid base balance. *0652*

Bendrofluazide Metabolic alkalosis. *1129* By 2 mmol/L approximately in 5 or 6 hypertensives given 5 mg daily for 6 weeks. *3866*

Betamethasone May cause hypokalemic alkalosis. *1680*

Bicarbonates Induces metabolic alkalosis. *0652*

Bumetanide By approximately 1 meq/L. *2666*

Carbenicillin Observed in approximately 8% patients. *1947*

Carbenoxolone Hypokalemic alkalosis. *2458*

Cephalexin If given with gentamicin to leukemics. *3922*

Chlorthalidone Metabolic alkalosis in severe cases. *1129* From 2 to 3 mmol/L at doses from 25 to 200 mg/d in 37 patients over 8 weeks. *3643*

Dexamethasone May cause hypokalemic alkalosis. *1680*

Enflurane Postoperatively moderately high but still normal. *0999*

Ethacrynic Acid Associated with hypochloremic alkalosis. *3214*

Fludrocortisone May cause hypokalemic alkalosis. *1680*

Furosemide In 24% in 204 hospitalized patients receiving the drug. *3419*

Gentamicin If given with cephalexin to leukemics. *3922*

Hydrochlorothiazide May cause metabolic alkalosis. *1129*

Hydrocortisone May cause hypochloremic hypokalemic alkalosis. *1343*

Hydroflumethiazide May cause hypochloremic alkalosis. *1680*

Laxatives If chronic abuse occurs. *3682*

Mayo Enema May cause retention from bicarbonate in enema. *0726*

Meralluride Alkalosis may occur with loss of chloride. *1680*

Methyclothiazide Diuretic action (hypochloremic alkalosis). *1353*

Methylprednisolone May cause hypokalemic alkalosis. *1680*

Metolazone Diuretic action of drug acting on distal tubules. *2816* Increase by about 3-5 mmol/L after 6 weeks treatment with 5 mg daily in 5 or 6 hypertensives. *3866*

Paramethasone May cause hypokalemic alkalosis. *1680*

Polythiazide Metabolic alkalosis with marked diuresis. *1129*

Prednisolone Hypochloremic alkalosis may occur. *0019*

Prednisone May cause hypokalemic alkalosis. *1680*

Probucol Mechanism obscure (effect slight). *0806*

Season Significant effect, maximum in April, minimum in summer. *1172*

Sodium Salts Significant but small effect observed with citrate but not chloride salt in 5 men with essential hypertension. *2037*

Thiazides Over 60% have value over 30 meq/L in long term. *0269*

Tromethamine Correction of respiratory acidosis. *3006*

Viomycin Altered electrolyte balance. *1237*

Serum Decrease Analytical

Acetylsalicylic Acid At concentrations greater than 1,000 mg/L (10 times upper limit of therapeutic range) when using a phenolphthalein indicator method. *3393*

Contact With Clot pH shift in drawn blood allows at least 5% drop. *1563*

EDTA Significant alteration of pH in drawn sample. *1563*

Fluorides Alteration of pH with loss of CO_2. *3879*

Liposol Changed from 23.2 to 18.7 mmol/L in 26 randomly selected patient sera, before and after addition on Beckman Astra methods. *2054*

Oxalate Significant alteration of pH if used as anticoagulant. *0579*

Standing of Sample By 1.5 mmol/L in open AutoAnalyzer cup in 15 minutes. *3217*

Storage of Sample Substantial decrease after 14 d at 4°. *3862* 6 mmol/L in 2.5-4 h with no oil at room temperature. *3217*

Serum   Decrease   Physiological

Acetazolamide Carbonic anhydrase inhibition in renal tubules. *0652*

Amiloride Reduces alkalosis induced by other diuretics. *0142*

Aspirin Initial acidosis with excessive doses. *1022*

Dimercaprol Associated with metabolic acidosis. *1022*

Dimethadione Displacement of bicarbonate by dimethadione. *3882*

EDTA Nephrotoxic effect (especially Calcium EDTA). *1237*

Ethanol Causes lactic acidosis. *2667*

Fructose Metabolic response to large dose fructose. *3113*

Hypothermia Due to hypoxia, acid-base imbalance, hypotension. *2258*

Lipomul® Nephrotoxic effect with azotemia. *1022*

Menstruation Varies in parallel with sodium. *1086*

Mercury Compounds May be depressed in established poisoning. *0987*

Metformin May cause marked acidosis (lactic acidosis). *1022*

Methanol Metabolic acidosis due to organic acid accumulation. *1343*

Methicillin Nephrotoxicity may cause azotemia. *2313*

Methylenedioxyamphetamine Respiratory acidosis. *2975*

Muscular Exercise Vigorous exercise for 30 minutes depresses by 3 meq/L. *1086*

Nitrofurantoin Nephrotoxicity may cause azotemia. *1022*

NSD 3004 Long acting carbonic anhydrase inhibitor. *2232*

Paraldehyde Metabolic acidosis in paraldehyde habitues. *1343*

β-Propiolactone Increase by 14 mmol/L when plasma from BPL treated blood compared with plasma in 25 specimens. *0217*

Starvation Due to metabolic acidosis. *2522*

Streptozocin Associated with low potassium and polyuria. *2532*

Tetracycline May cause acidosis with nephrotoxicity. *0071*

Triamterene Nephrotoxic effect. *1022*

Trimethadione Converted to dimethadione *in vivo* with same effects. *3882*

Xylitol May cause azotemia and acidosis. *3576*

Blood   No Effect   Physiological

Bucindol No effects noted in 8 patients following several weeks on treatment. *1283*

Urine   No Effect   Analytical

Storage of Sample If no bacterial growth room temperature for 45 d. *1180*

Thymol No effect on Van Slyke, Natelson methods. *2981*

Urine   No Effect   Physiological

Metolazone No effect on carbonic anhydrase. *3358*

Urine   Increase   Physiological

Acetazolamide Inhibition of carbonic anhydrase. *1343*

Amiloride Associated with potassium sparing diuretic action. *2618*

Aspirin Response to respiratory alkalosis of early toxicity. *1343*

Lithium Occurs on first day of treatment only. *1792*

Mafenide Inhibits carbonic anhydrase if applied topically. *3813*

Methyclothiazide Slight effect only. *1680*

Parathyroid Extract Inhibits tubular exchange of Na for h. *2611*

Polythiazide But effect minimal. *1680*

SC-16102 Minor increase observed within 2 h. *1804*

Secretin When i.v. infusion given. *2656*

Urine   Decrease   Physiological

Diazoxide Effect not as marked as on sodium excretion. *0124*

Bile   No Effect   Physiological

Secretin In normals, increased in cirrhotics. *0399*

Duodenal Contents   Increase   Physiological

Amino Acids With duodenal infusion, no effect if given i.v. *2918*

Pancreatic Juice   No Effect   Physiological

5-Hydroxytryptamine No effect following injection. *3579*

Pancreatic Juice   Increase   Physiological

Secretin Normal above 90 meq/L. *0486*

Pancreatic Juice   Decrease   Physiological

Ethanol Oral or i.v. affects concentration and output. *2504*

Smoking Chronic effect heavy smokers, less in light. *0554*

Bile

Urine   No Effect   Analytical

Chloroform As preservative has no effect on Stix tests. *2308*

Phenothiazines Ictotest® unaffected. *1487*

Urine   Increase   Analytical

Acriflavine Produces yellow color when urine shaken. *0459*

Aminosalicylic Acid Red color with Fouchet procedure. *0121*

Aspirin Purple color with Fouchet procedure. *0121*

Chlorpromazine Alleged interference with BiliLabstix®. *1488*

Ethoxazene Atypical red color with BiliLabstix® and Ictotest®. *1488*

Flufenamic Acid Probably due to interfering metabolite. *1488*

Formaldehyde May give yellow ring with Gmelin test. *0459*

Mefenamic Acid Reported interference with testing procedure. *1881*

Phenazopyridine False positive with Ictotest®, BiliLabstix®. *2566*

Phenothiazines Alleged interference with BiliLabstix®. *1488*

Thymol Affects determination of bile acids (Hay's test). *2981*

Urine   Increase   Physiological

Acetaminophen Hepatic necrosis may occur with dose of 10 g. *0438*

Acetohexamide May cause intrahepatic cholestatic jaundice. *2313*

Acetophenazine Cholestasis with intrahepatic obstruction. *2313*

Ajmaline Intrahepatic cholestasis. *0268*

Allopurinol Hepatotoxic effect. *2220*

Aminosalicylic Acid Hepatotoxicity. *2313*

Amitriptyline Due to transient cholestasis. *2313*

Amodiaquine Reported hepatotoxicity. *2313*

Amphotericin B Reported hepatotoxicity. *2313*

Anabolic Steroids Intrahepatic cholestatic jaundice. *1343*

Androgens Intrahepatic cholestatic jaundice. *1343*

Antifungal Agents Hepatotoxicity may occur. *1237*

Antimony Compounds Hepatotoxicity. *2313*

Arsenicals Hepatotoxicity (cholestasis/cholangiolitis). *2313*

Aspidium May cause hepatic toxicity. *2313*

Benziodarone May cause hepatic toxicity. *2313*

Bismuth Salts Hepatotoxicity. *2313*

Carbutamide Hepatotoxicity. *2313*

Carphenazine Probable cholestatic effect (reversible). *2313*

Chloramphenicol Hepatotoxicity. *2313*

Chlordiazepoxide Hepatotoxic effect. *2313*

Chlormezanone Occasional reversible cholestasis. *2313*

Chlorothiazide May cause cholestatic jaundice. *2313*

Chlorpropamide Cholestatic jaundice. *0085*

Bile *(continued)*

Urine   Increase   Physiological *(continued)*

Chlorprothixene   May cause hepatotoxicity (reversible). *1237*

Chlortetracycline   Hepatotoxic effect with centrolobular necrosis. *1237*

Chlorzoxazone   May cause hepatotoxicity. *1022*

Cinchophen   May cause hepatotoxicity (viral-hepatitis like). *2313*

Clofibrate   Hepatotoxic effect. *2313*

Colchicine   May cause hepatotoxicity. *2313*

Coumarin   May cause hepatotoxicity. *3835*

Cyclophosphamide   May cause hepatotoxicity. *2313*

Cyclopropane   May cause hepatotoxicity (lasts several days). *2313*

Cycloserine   May cause hepatotoxicity. *2313*

Desipramine   Rare transient cholestasis. *2313*

Dinitrophenol   Due to hepatotoxicity. *1302*

Doxepin   Theoretical possibility due to class of compound. *0071*

Ectylurea   May cause cholestasis with cholangiolitis. *2313*

Erythromycin   Intrahepatic cholestatic jaundice. *0085*

Estradiol   Cholestatic effect. *2313*

Estrogens   Due to cholestasis. *2313*

Estrone   May cause cholestasis. *2313*

Ether   Hepatic disturbance (transient). *2313*

Ethionamide   Hepatotoxic effect. *2313*

Ethotoin   Hepatotoxicity. *2313*

Ethoxazene   Hepatotoxic effect. *2313*

Florantyrone   Hepatotoxic effect. *2313*

Fluoxymesterone   Intrahepatic cholestatic jaundice. *1343*

Fluphenazine   Hypersensitivity response. *2313*

Flurazepam   May cause hepatotoxicity. *3016*

Glycopyrrolate   Hepatotoxicity. *2313*

Gold   Hepatotoxicity with centrolobular necrosis. *2313*

Guanoxan   May cause hepatic toxicity. *2313*

Haloperidol   May cause hepatocellular changes. *2313*

Halothane   Allergic hepatic hypersensitivity. *2313*

Hydrazine Derivatives   May cause hepatotoxicity. *2313*

Hydroxyacetamide   Hepatotoxicity. *2313*

Ibufenac   Hepatotoxic effect. *2313*

Icterogenin   Causes intrahepatic cholestasis. *0153*

Imipramine   May cause cholestatic jaundice. *2313*

Indandione Derivatives   May cause hepatotoxicity. *2313*

Indomethacin   Cytotoxic and cholestatic liver damage. *0085*

Iprindole   Due to hepatic toxicity. *0038*

Iproniazid   Prolonged use may cause hepatotoxicity. *2313*

Isocarboxazid   Cholestatic effect. *2313*

Isoniazid   Intrahepatic cholestatic jaundice. *0085*

Lincomycin   Occurs with hepatotoxicity. *2313*

MAO Inhibitors   Hepatotoxic effect. *2313*

Mechlorethamine   Hepatotoxic effect. *0248*

Mepazine   May alter liver function (cholestasis). *2313*

Mephenytoin   Hepatotoxicity. *2313*

Mercaptopurine   Hepatotoxic effect. *2313*

Metahexamide   Hepatotoxicity (viral hepatitis type). *2313*

Metaxalone   Hepatotoxic effect. *2313*

Methandriol   Intrahepatic cholestatic jaundice. *1343*

Methandrostenolone   Intrahepatic cholestatic jaundice. *1343*

Methimazole   May affect liver function. *2313*

Methotrexate   Hepatotoxicity. *2313*

Methoxsalen   Hepatotoxic effect. *2313*

Methoxyflurane   Hepatotoxic effect. *2313*

Methyldopa   Occurs as result of hepatocellular damage. *3646*

Methyltestosterone   Cholestatic effect. *1713*

Methylthiouracil   Hepatotoxic effect. *2313*

Niacin   May be impaired hepatic function. *2313*

Nialamide   Probable hypersensitive hepatitis. *0071*

Nitrofurans   Hepatotoxicity. *2313*

Nitrofurantoin   May cause cholestatic jaundice. *2313*

Norethandrolone   Cholestatic effect. *3172*

Norethandrostenolone   Intrahepatic cholestatic jaundice. *3197*

Norethindrone   Intrahepatic cholestatic jaundice. *1343*

Norethynodrel   Intrahepatic cholestatic jaundice. *1343*

Nortriptyline   May cause cholestatic jaundice. *0071*

Novobiocin   Intrahepatic cholestatic jaundice can occur. *1343*

Oleandomycin   Hepatotoxic effect. *2313*

Organophosphorus Insecticides   Hepatotoxic effect. *0405*

Oxacillin   Reversible hepatocellular dysfunction. *2313*

Oxazepam   Possible hepatotoxicity. *2313*

Oxyphenbutazone   Possible hepatotoxicity. *2313*

Oxyphenisatin   May cause hypersensitivity reaction. *2970*

Paraldehyde   Possible hepatotoxicity. *2313*

Paramethadione   Possible hepatotoxicity. *2313*

Pargyline   Possible hepatotoxicity (reversible). *2313*

Perphenazine   Low incidence of jaundice reported. *2313*

Phenacemide   May affect liver function in about 2% cases. *2313*

Phenazopyridine   Hepatotoxic effect. *2313*

Phenelzine   May cause hypersensitive hepatitis. *0071*

Phenindione   May modify liver function (cholestasis). *2313*

Pheniprazine   May affect liver function. *2313*

Phenothiazines   May cause cholestatic hepatitis. *2313*

Phenytoin   Hepatotoxicity. *2313*

Phosphorus   Hepatotoxic effect with necrosis. *2313*

Polythiazide   May affect liver function. *2313*

Probenecid   Possible hepatotoxicity. *2313*

Procainamide   Hepatotoxic effect. *2313*

Prochlorperazine   Cholestatic effect. *0071*

Progesterone   Hepatotoxicity. *2313*

Promazine   May cause cholestasis. *2313*

Promethazine   May cause cholestasis. *2313*

Propoxyphene   Cholestatic effect. *1954*

Propylthiouracil   Hepatotoxic effect (centrolobular necrosis). *2313*

Protriptyline   Transient reversible cholestasis. *0071*

Pyrazinamide   Hepatotoxic effect (viral-hepatitis like). *2313*

Quinacrine   Hepatotoxic effect. *2313*

Quinethazone   May cause cholestatic jaundice. *2313*

Sulfadimethoxine   Granulomatous reaction in liver. *1049*

Sulfamethizole   Reversible cholestasis. *1974*

Sulfamethoxazole   Reversible cholestasis. *2313*

Sulfamethoxypyridine   Reversible cholestasis. *2313*

Sulfisoxazole   May cause reversible cholestasis. *2313*

Sulfonamides   May cause cholestasis. *2313*

Sulfones   May affect liver function. *2313*

Testosterone   May cause cholestasis. *2313*

Thiacetazone   May cause cholestasis. *0248*

Thioguanine   May cause cholestasis. *2313*

Thioridazine   Hepatotoxicity. *1857*

Thiosemicarbazones   May affect liver function. *2313*

Thiothixene   Hepatotoxic effect (reversible cholestasis). *2313*

Tolazamide   Cholestatic effect. *2313*

Tolbutamide   Cholestatic jaundice reported. *1488*

Tranylcypromine   May affect liver function (?hypersensitivity). *2313*

Trimeprazine   May cause cholestasis. *0071*

Trimethadione   May cause hepatotoxicity with necrosis. *2313*

Trioxsalen   May affect liver function. *2313*

Uracil Mustard   May affect liver function. *2313*

Urine   Decrease   Analytical

Bacteriuria   Negative due to bacterial metabolism of bili diglucuronide. *2308*

Bile Acids

Serum   Increase   Analytical
Fusidic Acid   In methods using 3-alpha-hydroxysteroid dehydrogenase for which drug functions as substrate. *1485*

Serum   Increase   Physiological
Acetaminophen   Manifestation usually of single toxic high dose ingestion in suicide attempt. Liver damage usually of centrizonal hemorrhagic hepatic necrosis. *3948*   With overdose and centrolobular hepatic necrosis usually associated with high drug concentration. *1771*

Cimetidine   8 to 9 fold increase reversible cholestatic jaundice in a few children. *2172*

Isoniazid   Reported increases when given alone or with rifampin in the absence of other abnormal liver function tests. *0306*   15 of 89 patients developed significant liver disease, typically hepatitis, although one developed cholestasis, in patients taking drug for at least 2 mo. *0901*   In 72% of 61 patients studied for 80 d when treatment combined with rifampin but most other liver function tests including bilirubin could be normal. *0305*

Rifampin   Marked changes may occur when given alone or in combination with isoniazid in the absence of other abnormal liver function tests, possibly due to inhibition of uptake of bile acids into hepatocytes. *0306*   Significant increase possibly due to blocking uptake by plasma membrane of hepatocytes. *1224*

Serum   Decrease   Physiological
Cholestyramine   Binds acids in gastrointestinal tract. *1680*

Bile   Decrease   Physiological
Secretin   Slight in normals, marked in cirrhotics. *0399*

Feces   Increase   Physiological
Cholestyramine   Due to increased binding in gastrointestinal tract. *1680*

Neomycin   Precipitates bile salts in gastrointestinal tract. *1077*

Feces   Decrease   Physiological
Clofibrate   After 2 weeks treatment significant fall. *2463*

High Carbohydrate Diet   Marked if given i.v. *0880*

Bile Acids, Conjugated

Serum   No Effect   Analytical
Bilirubin   At up to 300 mg/L had insignificant effect on Becton-Dickinson RIA method. *2715*

Bile Salts

Bile   No Effect   Physiological
Chenodeoxycholic Acid   No change in patients with gallstones. *0274*

Feces   Decrease   Physiological
Cimetidine   By about 30%; in 17 patients with cystic fibrosis receiving constant concomitant therapy with pancreatic enzymes. *0624*

Bilirubin

Serum   No Effect   Analytical
Acetaminophen   At 5 times therapeutic concentration on routine methods in use on SMAC®, aca, Cobas-Bio, Ektachem®, Hitachi® and KDA. *2138*   At acute overdose concentration (10 mg/dL) on SMAC® method. *3719*   At concentration of 50 mg/L had no effect on Ektachem® method. *3393*   At concentration of 1500 mg/L had no effect on Jendrassik and Grof method. *3393*

Acetazolamide   No effect at 12 mg/dL on SMA 12/60 method. *2636*   At concentration of 1,000 mg/L had no effect on Jendrassik and Grof method. *3393*

Acetylsalicylic Acid   On Jendrassik — Grof, dimethylsulfoxide and spectrophotometric procedures with therapeutic concentrations. *1786*   On diazo method on SMA II at physiological concentration. *1787*   At 5 times upper limit of therapeutic range on methods on SMAC®, aca, Cobas-Bio, Ektachem®, Hitachi® 705 and KDA. *2138*   With 8326 μmol/L on Jendrassik-Grof method. With 8326 μmol/L on method using dimethylsulfoxide. With 8326 μmol/L on direct spectrophotometric method. *1786*   With 8326 μmol/L on direct spectrophotometric method. *1786*   At concentration of 300 mg/L had no effect on Ektachem® method. *3393*   At concentration of 2,000 mg/L had no effect on Jendrassik and Grof method. *3393*

Allopurinol   At concentration of 136 mg/L had no effect on Jendrassik and Grof method. *3393*

Amines   Spectropolarographic procedure of Grahnen. *1374*

Aminophenazone   At therapeutic concentrations no effect on methods of Jendrassik and Grof and those using spectrophotometry. *1786*

Aminophenol   At concentration of 109 mg/L had no effect on Jendrassik and Grof method. *3393*

Aminopyrimidine   At concentration of 57 mg/L had no effect on Ektachem® method. *3393*

Aminopyrine   At concentration of 125 mg/L had no effect on Jendrassik and Grof method. *3393*

Aminosalicylic Acid   At 5 times upper limit of therapeutic range on methods on SMAC®, Ektachem®, Hitachi® 705 and KDA. *2138*   At concentration of 230 mg/L had no effect on Ektachem® method. *3393*   At concentration of 460 mg/L had no effect on Jendrassik and Grof method. *3393*

Aminothiazole   At concentration of 23 mg/L had no effect on Ektachem® method. *3393*

Amitriptyline   At acute overdose concentration (2.5 mg/dL) on SMAC® method. *3719*   At concentration of 25 mg/L had no effect on Jendrassik and Grof method. *3393*

Amobarbital   At concentration of 150 mg/L had no effect on Jendrassik and Grof method. *3393*

Ampicillin   At concentration of 40 mg/L had no effect on Jendrassik and Grof method. *3393*

Ascorbic Acid   On SMAC® method at therapeutic concentration. *3719*   At concentration of 40 mg/L had no effect on Ektachem® method. *3393*

Barbital   At acute overdose concentration (20 mg/dL) on Technicon® SMAC® method. *3719*   At concentration of 500 mg/L had no effect as measured by Jendrassik and Grof method. *3393*

Bethanechol   No effect at 0.09 mg/dL on SMA 12/60 method. *2636*

Boric Acid   At concentration of 50 mg/L had no effect as measured by Jendrassik and Grof method. *3393*

Bromazepam   At 5 times therapeutic concentration on methods on SMAC®, aca, Cobas-Bio, Ektachem®, Hitachi® 705 and KDA. *2138*

Butabarbital   At concentration of 100 mg/L had no effect as measured by Jendrassik and Grof method. *3393*

Caffeine   No effect on Jendrassik-Grof, dimethylsulfoxide and spectrophotometric methods at therapeutic concentrations. *1786*   At acute overdose concentration (10 mg/dL) on Technicon® SMAC® method. *3719*   At concentration of 194 mg/L had no effect on Jendrassik and Grof method. *3393*

Carbenicillin   At 5 times upper limit of therapeutic range on methods on SMAC®, Cobas-Bio, Ektachem® 400, Hitachi® 705 and KDA. *2138*   At concentration of 2,000 mg/L had no effect on Jendrassik and Grof method. *3393*

Cefamandole   At 2.50 mmol/L on method on Eppendorf Epos. *1414*

Cefazedone   On method on Eppendorf Epos at concentration of 2.50 mmol/L. *1414*

Cefazolin   At 2.50 mmol/L on method on Eppendorf Epos. *1414*   At concentration of 117 mg/L had no effect on Jendrassik and Grof method. *3393*

Cefodizime   No effect at 2.50 mmol/L on method on Eppendorf Epos. *1414*

Cefoperazone   At 2.50 mmol/L on method on Eppendorf Epos. *1414*

Cefotaxime   No effect at 2.50 mmol/L on method on Eppendorf Epos. *1414*

Cefotiam   At 2.50 mmol/L on method on Eppendorf Epos. *1414*

Bilirubin (continued)

Serum   No Effect   Analytical (continued)

Cefoxitin At 5 times upper limit of therapeutic range on methods on SMAC®, aca, Cobas-Bio, Ektachem®, Hitachi® 705 and KDA. *2138* At concentrations up to 2.50 mmol/L on methods on Eppendorf Epos. *1414*

Cefpirome At concentration of up to 2.50 mmol/L on methods of Eppendorf Epos. *1414*

Ceftriaxone On method on Eppendorf Epos at concentration of 2.50 mmol/L. *1414*

Cefuroxime On method on Eppendorf Epos at concentration of 2.50 mmol/L. *1414*

Cephaloridine At 2.50 mmol/L on method on Eppendorf Epos. *1414*

Cephalothin At 5 times upper limit of therapeutic range on methods on SMAC®, aca, Cobas-Bio, Ektachem®, Hitachi® 705 and KDA. *2138* No effect a 2.50 mmol/L on method on Eppendorf Epos. *1414* At concentration of 1,000 mg/L had no effect on Jendrassik and Grof method. *3393*

Chlordiazepoxide At 5 times upper limit of therapeutic range on methods on SMAC®, aca, Cobas-Bio, Ektachem®, Hitachi® 705 and KDA. *2138* At acute overdose concentration (5 mg/dL) on Technicon® SMAC® method. *3719* At concentration of 50 mg/L had no effect on Jendrassik and Grof method. *3393*

Chlormezanone At concentration of 100 mg/L had no effect on Jendrassik and Grof method. *3393*

Chlorphenesin At concentration of 150 mg/L had no effect on Jendrassik and Grof method. *3393*

Chlorpheniramine At acute overdose concentration (20 mg/dL) on Technicon® SMAC® method. *3719* At concentration of 200 mg/L had no effect on Jendrassik and Grof method. *3393*

Chlorpromazine At concentration of 3 mg/L had no effect on Jendrassik and Grof method. *3393*

Chlorpropamide At concentration of 150 mg/L had no effect on Jendrassik and Grof method. *3393*

Chlorprothixene At concentration of 1 mg/L had no effect on Jendrassik and Grof method. *3393*

Cimetidine At concentration of 1 mg/L had no effect on Jendrassik and Grof method. *3393*

Cloxacillin At concentration of 5 mg/L had no effect on Jendrassik and Grof method. *3393*

Cocaine At acute overdose concentration (2.5 mg/dL) on Technicon® SMAC® method. *3719* At concentration of 25 mg/L had no effect on Jendrassik and Grof method. *3393*

Codeine At acute overdose concentration (2.0 mg/dL) on Technicon® SMAC® method. *3719* At concentration of 20 mg/L had no effect on Jendrassik and Grof method. *3393*

Cyclosporine At concentration of 41.5 mmol/L (50 µg/L) on methods on Technicon® SMAC® II and Hitachi® 705. *1123*

Cysteine At 5 times upper limit of therapeutic range on methods on SMAC®, aca, Cobas-Bio, Ektachem®, Hitachi® 705 and KDA. *2138*

Cystine At 5 times upper limit of therapeutic range on methods on SMAC®, aca, Cobas-Bio, Ektachem®, Hitachi® 705 and KDA. *2138*

Desipramine At concentration of 8 mg/L had no effect on Jendrassik and Grof method. *3393*

Deslanoside No effect on SMA 12/60 method at 0.06 mg/dL. *2636*

Dextran At concentration of 10,000 mg/L had no effect on Ektachem® method. *3393*

Diatrizoic acid At concentration of 5,000 mg/L had no effect on Ektachem® method. *3393*

Diazepam At 5 times upper limit of therapeutic range on methods on SMAC®, aca, Cobas-Bio, Ektachem®, Hitachi® 705 and KDA. *2138* At concentration of 25 mg/L had no effect on Jendrassik and Grof method. *3393* At acute overdose concentration (2.5 mg/dL) on Technicon SMAC® method. *3719* At 5 times upper limit of therapeutic range on methods on SMAC®, aca, Cobas-Bio, Ektachem®, Hitachi® 705 and KDA. *2138*

Diclofenac No effect at therapeutic concentrations on methods of Jendrassik-Grof and those using spectrophotometry. *1786* On diazo method on SMA II at physiological concentration. *1787*

Diethylaminoethanol At concentration of 30 mg/L had no effect on Jendrassik and Grof method. *3393*

Diethylpropion At acute overdose concentration (10 mg/dL) on Technicon® SMAC® method. *3719* At concentration of 100 mg/L had no effect on Jendrassik and Grof method. *3393*

Digitoxin No effect at 21 mg/dL on SMA 12/60 method. *2636* At concentration of 6 mg/L had no effect on Jendrassik and Grof method. *3393*

Digoxin No effect at 0.04 mg/dL on SMA 12/60 method. *2636* At concentration of 9 mg/L had no effect on Jendrassik and Grof method. *3393*

Dimethadione At concentration of 2,000 mg/L had no effect on Jendrassik and Grof method. *3393*

N-Dimethyldiazepam At concentration of 6 mg/L had no effect on Jendrassik and Grof method. *3393*

Diphenhydramine At acute overdose concentration (20 mg/dL) on Technicon® SMAC® method. *3719* At concentration of 200 mg/L had no effect on Jendrassik and Grof method. *3393*

Disopyramide At concentration of 4 mg/L had no effect on Jendrassik and Grof method. *3393*

Disulfiram At concentration of 120 mg/L had no effect on Jendrassik and Grof method. *3393*

Disulphine Blue With sulfanilic acid/caffeine-benzoate accelerator. *1464*

EDTA At concentration of 8,000 mg/L had no effect on Ektachem® method. *3393*

Epinephrine At concentration of 183 mg/L had no effect on Jendrassik and Grof method. *3393*

Erythromycin At concentration of 73 mg/L had no effect on Jendrassik and Grof method. *3393*

Ethacrynic Acid At 2.5 mg/dL no effect on SMA 12/60 method. *2636*

Ethanol At acute overdose concentration (20 mg/dL) on Technicon® SMAC® methods. *3719*

Ethchlorvynol At acute overdose concentration (20 mg/dL) on Technicon® method. *3719* At concentration of 400 mg/L had no effect on Jendrassik and Grof method. *3393*

Ethinamate At acute overdose concentration (20 mg/dL) on Technicon® SMAC® method. *3719* At concentration of 200 mg/L had no effect on Jendrassik and Grof method. *3393*

Ethosuximide At concentration of 390 mg/L had no effect on Jendrassik and Grof method. *3393*

Fatty Acids Spectropolarographic procedure of Grahnen. *1374*

Fluorescein Minimal interference at expected concentration on Ektachem® 700. *1722*

Fluorouracil At concentration of 130 mg/L had no effect on Jendrassik and Grof method. *3393*

Fluosol-DA On SMA IIC at concentration of 50%. *2518*

Flurazepam At 5 times upper limit of therapeutic range on methods on SMAC®, Cobas-Bio, aca, Ektachem®, Hitachi® 705 and KDA. *2138* At acute overdose concentration (2.5 mg/dL) on Technicon® SMAC® method. *3719* At concentration of 25 mg/L had no effect on Jendrassik and Grof method. *3393*

Furosemide No effect at 1.4 mg/dL on SMA 12/60 method. *2636* At concentration of 4 mg/L had no effect on Jendrassik and Grof method. *3393*

Gentamicin At concentration of 150 mg/L had no effect on Jendrassik and Grof method. *3393*

Glutethimide At acute overdose concentration (5 mg/dL) on Technicon® SMAC® method. *3719* At concentration of 50 mg/L had no effect on Jendrassik and Grof method. *3393*

Hydralazine At concentration of 160 mg/L had no effect on Jendrassik and Grof method. *3393*

Ibuprofen No effect at therapeutic concentrations on Jendrassik-Grof, dimethylsulfoxide and spectrophotometric methods. *1786* At 5 times upper limit of therapeutic range on methods on SMAC®, aca, Cobas-Bio, Ektachem®, Hitachi® 705 and KDA. *2138* On diazo method on SMA II at physiological concentration. *1787* At concentration of 200 mg/L had no effect on Jendrassik and Grof method. *3393*

Indican No effect on Jendrassik-Grof procedures at concentrations up to 0.9 mmol/L. *2844*

Indomethacin No effect at therapeutic concentrations on Jendrassik-Grof, dimethylsulfoxide and spectrophotometric methods. *1786* On diazo method on SMA II at physiological concentration. *1787* At concentration of 13 mg/L had no effect on Jendrassik and Grof method. *3393*

Insulin At concentration of 3 mg/L had no effect on Jendrassik and Grof method. *3393*

Iproniazid At concentration of 40 mg/L had no effect on Jendrassik and Grof method. *3393*

Isoniazid At concentration of 100 mg/L had no effect on Jendrassik and Grof method. *3393*

Ketoprofen No effect at therapeutic concentrations on Jendrassik-Grof, dimethylsulfoxide and spectrophotometric methods. *1786* On diazo method on SMA II at physiological concentration. *1787* At concentration of 60 mg/L had no effect on Jendrassik and Grof method. *3393*

Levodopa At concentration of 6 mg/L had no effect on Ektachem® method. *3393*

Lidocaine No effect SMA 12/60 method with 3.5 mg/dL. *2636*

Lithium Heparin At concentration of 8,000 mg/L had no effect on Ektachem® method. *3393*

Lorazepam At concentration of 0.05 mg/L had no effect on Jendrassik and Grof method. *3393*

Mannitol No effect on SMA 12/60 method at 445 mg/dL. *2636*

Meperidine At acute overdose concentration (5 mg/dL) on Technicon® SMAC® method. *3719* At concentration of 60 mg/L had no effect on Jendrassik and Grof method. *3393*

Mephenesin At concentration of 100 mg/L had no effect on Jendrassik and Grof method. *3393*

Meprobamate At acute overdose concentration (20 mg/dL) on Technicon® SMAC® method. *3719* At concentration of 200 mg/L had no effect on Jendrassik and Grof method. *3393*

Meralluride No effect at 0.07 mg/dL on SMA 12/60 method. *2636*

Mercaptomerin At concentration of 58 mg/L had no effect on Jendrassik and Grof method. *3393*

Mercaptopurine At concentration of 152 mg/L had no effect on Jendrassik and Grof method. *3393*

Mesoridazine At acute overdose concentration (20 mg/dL) on Technicon® SMAC® method. *3719*

Methadone At acute overdose concentration (2.5 mg/dL) on Technicon® SMAC® method. *3719* At concentration of 25 mg/L had no effect on Jendrassik and Grof method. *3393*

Methamphetamine At concentration of 2 mg/L had no effect on Jendrassik and Grof method. *3393*

Methanol At acute overdose concentration (20 mg/dL) on Technicon® SMAC® method. *3719*

Methapyrilene At concentration of 13 mg/L had no effect on Jendrassik and Grof method. *3393*

Methaqualone At acute overdose concentration (2.5 mg/dL) on Technicon® SMAC® method. *3719* At concentration of 25 mg/L had no effect on Jendrassik and Grof method. *3393*

Methicillin At 5 times upper limit of therapeutic range on SMAC®, aca, Cobas-Bio, Ektachem®, Hitachi® 705 and KDA. *2138* At concentration of 900 mg/L had no effect on Jendrassik and Grof method. *3393*

Methohexital At concentration of 50 mg/L had no effect on Jendrassik and Grof method. *3393*

Methotrexate No significant effect at 5 times upper limit of therapeutic range on SMAC®, aca, Cobas-Bio, Hitachi® 705 and KDA. *2138*

Methotrimeprazine At concentration of 1 mg/L had no effect on Jendrassik and Grof method. *3393*

Methsuximide At concentration of 40 mg/L had no effect on Jendrassik and Grof method. *3393*

Methylphenidate At acute overdose concentration (20 mg/dL) on Technicon® SMAC® method. *3719* At concentration of 200 mg/L had no effect on Jendrassik and Grof method. *3393*

Methylphenobarbital At concentration of 150 mg/L had no effect on Jendrassik and Grof method. *3393*

Methyprylon At acute overdose concentration (10 mg/dL) on Technicon® SMAC® method. *3719* At concentration of 100 mg/L had no effect on Jendrassik and Grof method. *3393*

Metoprolol At concentration of 0.34 mg/L had no effect on Jendrassik and Grof method. *3393*

Morphine At acute overdose concentration (20 mg/dL) on Technicon® SMAC® method. *3719* At concentration of 200 mg/L had no effect on Jendrassik and Grof method. *3393*

Moxalactam At 2.50 mmol/L on method on Eppendorf Epos. *1414* At concentration of 96 mg/L had no effect on Jendrassik and Grof method. *3393*

Nafcillin At concentration of 50 mg/L had no effect on Jendrassik and Grof method. *3393*

Naproxen At 5 times upper limit of therapeutic range on methods on SMAC®, aca, Cobas-Bio, Ektachem®, Hitachi® 705, aca, and KDA. *2138*

Nitrofurantoin On routine methods in use on SMAC®, Abbott-VP, Cobas-Bio, aca, Hitachi® 705, KDA at 5 times normal upper therapeutic concentration. *2138* At concentration of 4 mg/L had no effect on Jendrassik and Grof method. *3393*

Nortriptyline At acute overdose concentration (20 mg/dL) on Technicon® SMAC® method. *3719* At concentration of 200 mg/L had no effect on Jendrassik and Grof method. *3393*

Orphenadrine At concentration of 9 mg/L had no effect on Jendrassik and Grof method. *3393*

Ouabain At 0.02 mg/dL no effect on SMA 12/60 method. *2636*

Oxacillin At concentration of 40 mg/L had no effect on Jendrassik and Grof method. *3393*

Oxazepam At concentration of 1 mg/L had no effect on Jendrassik and Grof method. *3393*

Papaverine At concentration of 10 mg/L had no effect on Jendrassik and Grof method. *3393*

Paraldehyde At concentration of 2,000 mg/L had no effect on Jendrassik and Grof method. *3393*

Penicillin G At 5 times upper limit of therapeutic range on methods on SMAC®, aca, Abbott-VP, Ektachem®, Hitachi® 705 and KDA. *2138* At concentration of 2,000 mg/L had no effect as measured by Jendrassik and Grof method. *3393*

Pentobarbital At acute overdose concentration (20 mg/dL) on Technicon® SMAC® method. *3719* At concentration of 340 mg/L had no effect on Jendrassik and Grof method. *3393*

Perphenazine At acute overdose concentration (20 mg/dL) on Technicon® SMAC® method. *3719* At concentration of 200 mg/L had no effect on Jendrassik and Grof method. *3393*

Phencyclidine At concentration of 6 mg/L had no effect on Jendrassik and Grof method. *3393*

Phenelzine At concentration of 136 mg/L had no effect on Jendrassik and Grof method. *3393*

Phenobarbital No effect at therapeutic concentrations on Jendrassik-Grof, dimethylsulfoxide and spectrophotometric methods. *1786* At acute overdose concentration (20 mg/dL) on Technicon® SMAC® method. *3719* At concentration of 250 mg/L had no effect on Jendrassik and Grof method. *3393*

Phenols Spectropolarographic procedure of Grahnen. *1374*

Phensuximide At concentration of 120 mg/L had no effect on Jendrassik and Grof method. *3393*

Phenylbutazone At concentration of 750 mg/L had no effect on Jendrassik and Grof method. *3393*

Phenylephrine At concentration of 4 mg/L had no effect on Jendrassik and Grof method. *3393*

Phenytoin No effect at 1.8 mg/dL on SMA 12/60 method. *2636* At acute overdose concentration (20 mg/dL) on Technicon® SMAC® method. *3719* At concentration of 240 mg/L had no effect on Jendrassik and Grof method. *3393*

Plasma No significant difference from serum on SMA 12/60 method. *2228*

Potassium Oxalate At concentration of 8,000 mg/L had no effect on Ektachem® method. *3393*

Probenecid At concentration of 1300 mg/L had no effect on Jendrassik and Grof method. *3393*

Procainamide No effect on SMA 12/60 method at 35 mg/dL. *2636* At concentration of 350 mg/L had no effect on Ektachem® method. *3393* At concentration of 50 mg/L had no effect on Jendrassik and Grof method. *3393*

Procaine At concentration of 2 mg/L had no effect on Jendrassik and Grof method. *3393*

Prochlorperazine At concentration of 1 mg/L had no effect on Jendrassik and Grof method. *3393*

Bilirubin (continued)

Serum   No Effect   Analytical (continued)

Promethazine   At acute overdose concentration (20 mg/dL) on Technicon® SMAC® method. *3719*   At concentration of 200 mg/L had no effect on Jendrassik and Grof method. *3393*

Propoxyphene   At acute overdose concentration (2.5 mg/dL) on Technicon® SMAC® method. *3719*   At concentration of 25 mg/L had no effect on Jendrassik and Grof method. *3393*

Propranolol   No effect at 0.1 mg/dL on SMA 12/60 method. *2636*   At concentration of 0.2 mg/L had no effect on Jendrassik and Grof method. *3393*

Propylthiouracil   At concentration of 170 mg/L had no effect on Jendrassik and Grof method. *3393*

Pyribenzamine®   At acute overdose concentration (20 mg/dL) on Technicon® SMAC® methods. *3719*

Quinidine   At 26 mg/dL no effect on SMA 12/60 method. *2636*   At acute overdose concentration (20 mg/dL) Technicon® SMAC® method. *3719*   At concentration of 210 mg/L had no effect on Jendrassik and Grof method. *3393*

Quinine   At acute overdose concentration (1.5 mg/dL) on Technicon® SMAC® method. *3719*   At concentration of 30 mg/L had no effect on Jendrassik and Grof method. *3393*

Reserpine   No effect at 0.02 mg/dL on SMA 12/60 method. *2636*

Rifampin   Minimal effect on Jendrassik/Grof methods, although some effect at drug concentration of 15 mg/L. *0218*   At 5 times upper limit of therapeutic concentration on methods on SMAC®, aca, Hitachi® 705 and KDA. *2138*

Salicylate   At concentration of 350 mg/L had no effect on Ektachem® method. *3393*   At concentration of 500 mg/L had no effect on Jendrassik and Grof method. *3393*

Secobarbital   At acute overdose concentration (20 mg/dL) on Technicon® SMAC® method. *3719*   At concentration of 200 mg/L had no effect on Jendrassik and Grof method. *3393*

Sodium Citrate   At concentration of 10500 mg/L had no effect on Ektachem® method. *3393*

Sodium Fluoride   At concentration of 10,000 mg/L had no effect on Ektachem® method. *3393*

Standing of Sample   If on polystyrene beads 24 h at 25°. *2624*

Storage of Sample   No effect for 6 mo at -10°. *1180*

Streptomycin   At concentration of 58 mg/L had no effect on Jendrassik and Grof method. *3393*

Strychnine   At concentration of 12 mg/L had no effect on Jendrassik and Grof method. *3393*

Sulfadiazine   At concentration of 150 mg/L had no effect on Ektachem® method. *3393*   At concentration of 1500 mg/L had no effect on Jendrassik and Grof method. *3393*

Sulfaguanidine   At concentration of 500 mg/L had no effect on Jendrassik and Grof method. *3393*

Sulfamethoxazole   At concentration of 60 mg/L had no effect on Ektachem® method. *3393*

Sulfanilamide   At concentration of 103 mg/L had no effect on Ektachem® method. *3393*   At concentration of 1,000 mg/L had no effect on Jendrassik and Grof method. *3393*

Sulfathiazole   At concentration of 60 mg/L had no effect on Ektachem® method. *3393*   At concentration of 255 mg/L had no effect on Jendrassik and Grof method. *3393*

Sulfisoxazole   At concentration of 60 mg/L had no effect on Ektachem® method. *3393*

Sulfonamides   spectropolarographic procedure of Grahnen. *1374*

Sulforidazine   At concentration of 200 mg/L had no effect on Jendrassik and Grof method. *3393*

Tetracycline   At 5 times upper limit of therapeutic range on methods on SMAC®, Cobas-Bio, aca, Ektachem®, Hitachi® 705 and KDA. *2138*   At concentration of 10 mg/L had no effect on Ektachem® method. *3393*   At concentration of 300 mg/L had no effect on Jendrassik and Grof method. *3393*

Theobromine   At concentration of 2,000 mg/L had no effect on Jendrassik and Grof method. *3393*

Theophylline   At 5 times upper limit of therapeutic range on methods on SMAC®, Cobas-Bio, aca, Ektachem®, Hitachi® 705 and KDA. *2138*   At concentration of 2,000 mg/L had no effect on Jendrassik and Grof method. *3393*

Thiamazole   At concentration of 114 mg/L had no effect on Jendrassik and Grof method. *3393*

Thiopental   At acute overdose concentration (20 mg/dL) on Technicon® SMAC® method. *3719*   At concentration of 200 mg/L had no effect on Jendrassik and Grof method. *3393*

Timolol   At concentration of 0.01 mg/L had no effect on Jendrassik and Grof method. *3393*

Tobramycin   At concentration of 5 mg/L had no effect on Ektachem® method. *3393*

Tolbutamide   At concentration of 270 mg/L had no effect on Jendrassik and Grof method. *3393*

Transport of Specimen   By pneumatic tube if specimen protected. *3217*

Tribromethanol   At concentration of 90 mg/L had no effect on Jendrassik and Grof method. *3393*

Trichlorethanol   At concentration of 1,000 mg/L had no effect on Jendrassik and Grof method. *3393*

Trifluoperazine   At concentration of 1 mg/L had no effect on Jendrassik and Grof method. *3393*

Tripelennamine   At concentration of 200 mg/L had no effect on Jendrassik and Grof method. *3393*

Warfarin   At concentration of 100 mg/L had no effect on Jendrassik and Grof method. *3393*

Serum   No Effect   Physiological

Acetaminophen   Insignificant displacement from protein in neonates. *3748*

Acetazolamide   At pharmacological concentration probably no significant displacement from protein occurs. *3748*

Acetylsalicylic Acid   No effect seen in patients undergoing treatment. *1787*

Aminophylline   No significant displacement from protein in neonates. *3748*

Aminosalicylic Acid   Probable insignificant displacement from protein in neonates. *3748*

Amiodarone   No effect observed in 100 patients. *1509*   In 0 of 5 patients with toxic and hypersensitivity liver injury. *2987*

Amitriptyline   Insignificant protein-displacement effect in neonates. *3748*

Ampicillin   Insignificant displacement from protein in neonates. *3748*

Azlocillin   Probably clinically insignificant displacement from protein in neonates. *3748*

Aztreonam   Probably clinically insignificant displacement from protein in neonates. *3748*

Barbital   Insignificant protein-displacing effect in neonates. *3748*

Bumetanide   Although displacement observed *in vitro*, insignificant effect likely *in vivo* at pharmacological concentration. *3748*

Caffeine   No significant displacement from protein in neonates. *3748*

Carbamazepine   Insignificant displacement from protein in neonates. *3748*

Cephalothin   Clinically insignificant displacement from protein in neonates. *3748*

Chloramphenicol   Clinically insignificant displacement from protein in neonates. *3748*

Chlordiazepoxide   Insignificant protein displacement effect in neonates. *3748*

Chlorothiazide   Displacement from protein observed *in vitro*, but unlikely at pharmacological concentrations *in vivo*. *3748*

Chlortetracycline   Critically insignificant displacement from protein in neonates. *3748*

Ciprofloxacin   Clinically insignificant displacement from protein in neonates. *3748*

Clonazepam   Insignificant displacement from protein in neonates. *3748*

Cyclophosphamide   No significant perturbation in 41 patients. *2406*

Desipramine   No effect observed in 46 treated patients. *2871*   Insignificant protein-displacement effect in neonates. *3748*   In 42 children and adolescents treated for up to 24 mo. *1624*

Diazepam   Insignificant displacement from protein in neonates. *3748*

Diclofenac   No effect seen in patients undergoing treatment. *1787*

Diflunisal At pharmacological concentrations has little or no protein displacement effect. *3748*

Dihydrostreptomycin Clinically insignificant displacement from protein in neonates. *3748*

Enoxacin Clinically insignificant displacement from protein in neonates. *3748*

Ethacrynic Acid Although *in vitro* effect observed no significant effect at pharmacological concentrations. *3748*

Ethanol No significant difference between heavy and occasional drinkers. *3273* Clinically insignificant displacement from protein in neonates. *3748*

Ethosuximide Insignificant displacement of protein in neonates. *3748*

Etretinate In 2 of 20 patients treated for 1 y. *1866*

Fenoprofen At pharmacological concentrations has little or no protein displacement effect. *3748*

Flubiprofen At pharmacological concentrations has little or no effect on displacement from protein. *3748*

Flucloxacillin In a single patient on hemodialysis: abnormal results only mild. *2200*

Furosemide Although *in vitro* effect observed at pharmacological concentrations no significant effect. *3748*

Gentamicin Clinically insignificant displacement from protein in neonates. *3748*

Glutethimide Insignificant protein-displacement effect in neonates. *3748*

Halothane In one nurse occupationally exposed to gas. *1900*

Hydralazine Isolated case with moderate hepatomegaly, abnormal findings resolved when drug discontinued. *1161*

Ibuprofen At pharmacological concentrations has little or no protein displacement effect. *3748* No effect seen in patients undergoing treatment. *1787*

Imipenem Probably clinically insignificant displacement from protein in neonates. *3748*

Imipramine Insignificant protein displacement effect in neonates. *3748*

Indomethacin No effect seen in patients undergoing treatment. *1787* Although tightly bound to protein at pharmacological concentrations has little or no displacement effect. *3748*

Isotretinoin In 18 patients with severe acne given 0.8 mg/kg daily for 3 mo: changes reverted to normal after treatment stopped. *2242*

Kanamycin Clinically insignificant displacement from protein in neonates. *3748*

Ketoprofen No effect seen in patients undergoing treatment. *1787*

Levonorgestrel No significant difference over 3 y between implant recipients and IUD users. *0763* In a group of women using levonorgestrel covered rods versus controls using copper IUD. *0899*

Lincomycin Clinically insignificant displacement from protein in neonates. *3748*

Maprotiline Insignificant displacement from protein in neonates. *3748*

Meprobamate Insignificant displacement from protein in neonates. *3748*

Methadone No toxicity if liver function tests normal initially. *2012*

Methicillin Clinically insignificant displacement from protein in neonates. *3748*

Methohexital Insignificant protein-displacing effect in neonates. *3748*

Metoclopramide No significant displacement from protein observed in neonates. *3748*

Metrizamide Probably nonclinically significant protein displacement effect in neonates. *3748*

Nalidixic Acid No clinically significant displacement from protein likely in neonates. *3748*

Naproxen At pharmacological concentration has little or no protein displacement effect. *3748*

Nisoldipine In 14 mild to moderately hypertensive noninsulin dependent diabetic patients. *2653*

Nitrazepam Insignificant displacement from protein in neonates. *3748*

Nortriptyline Insignificant protein-displacement effect in neonates. *3748*

Oxacillin Asymptomatic hepatic dysfunction in 5 cases following high dose treatment: reversible. *2835*

Oxazepam Insignificant displacement from protein in neonates. *3748*

Oxyphenbutazone Probably little or no effect on displacement from protein. *3748*

Oxytetracycline Clinically insignificant displacement from protein in neonates. *3748*

Penicillin G Insignificant displacement from protein in neonates. *3748*

Phenobarbital Insignificant protein-displacement effect in neonates. *3748*

Phenytoin Insignificant protein displacement effect in neonates. *3748* In about 20 patients treated for 2 y. *0873*

Primidone Insignificant displacement from protein in neonates. *3748*

Rifampin Probably no significant displacement from protein in neonates. *3748*

Streptokinase Transient dysfunction of liver, returning to normal when treatment discontinued. *3122*

Streptomycin Clinically insignificant displacement from protein in neonates. *3748*

Sulfasalazine Characteristically associated with rare cases of hepatotoxicity. *1127*

Tobramycin Clinically insignificant displacement from protein in neonates. *3748*

Trimethoprim No clinically significant displacement from protein likely at pharmacological concentrations in neonates. *3748*

Vancomycin Clinically significant displacement from protein in neonates. *3748*

Zimeldine In 1 study involving 147 patients no significant change seen. *3741*

Serum Increase Analytical

Amino Acids If given i.v. may affect accuracy of estimation. *0120*

Aminophenol At 1 mmol/L affects SMA 12/60 method. *3335*

Aminosalicylic Acid Marked increase at upper limit of therapeutic range on aca method. At 5 times upper limit of therapeutic range on Cobas-Bio method. *2138*

Amphotericin B At concentrations above 96 mg/L (therapeutic concentration about 3.7 mg/L) raised concentration as measured by Jendrassik and Grof method. *3393*

Ascorbic Acid At therapeutic concentration may affect SMA 12/60 method. *3335*

Bilirubin Conjugated bilirubin increases polarographic concentration of unconjugated. *1374*

Carbenicillin At 5 times upper limit of therapeutic range on aca method. *2138*

Carotene Color may interfere with direct methods. *1022*

Carrots Color may be misinterpreted. *1238*

Dextran Causes turbidity with methanol in Evelyn-Malloy method. *3505*

Epinephrine At 1 mmol/L affects SMA 12/60 method. *3335*

Ethoxazene Postulated production of color with Ehrlich's diazo reaction. *0085*

Fat Emulsions If given i.v. may affect accuracy of measurement. *0120*

Fluosol-DA Increase by 500% on Du Pont aca III at concentration of 50%. *2518*

Hemoglobin Percentage of unconjugated decreases with increased Hemoglobin. *2313* At 1 g/L by 1% polarometric procedure of Grahnen. *1374* Cause spurious elevation in direct procedures. *0583*

Histidine Reacts with diazo reagent. *1563*

Indican May produce brown diazo color in uremic serum. *1563* Bilirubin A-Gent Abbott 2,4-dichlorophenyl diazonium procedure increases bilirubin by 50 mg/L for each 1 mmol/L of added indican. Bilirubin C-system (Boehringer-Mannheim) 2,5-dichlorophenyl diazonium procedure increased bilirubin by 33 mg/L per mmol/L of added indican. Slight effect on Harleco micro bilirubin reagent set (Malloy-Evelyn procedure). *2844*

Bilirubin *(continued)*

Serum Increase Analytical *(continued)*

Isoproterenol At 1 mmol/L affects SMA 12/60 method. *3335*

Levodopa At 1 mmol/L affects SMA 12/60 method. *3335* Theoretically reacts with diazo reagent. *3220* At concentrations above 80 mg/L raised concentration as measured by Jendrassik and Grof method. *3393*

Lipemia Interferes with direct spectroscopic methods. *0582*

Lipochrome May be measured as bilirubin in some methods. *1237*

Methanol False value if impure methanol used (Evelyn-Malloy). *0579*

Methotrexate Clinically significant effect at upper limit of therapeutic range on method on Ektachem®. *2138*

Methyldopa At 1 mmol/L affects SMA 12/60 method. *3335* Theoretically reacts with diazo reagent. *3220* At concentrations above 65 mg/L (normal therapeutic concentration 2 mg/L) raised concentration as measured by Jendrassik and Grof method. *3393*

Nitrofurantoin Clinically significant effect at upper limit of normal therapeutic range on Ektachem® 400/700 method. *2138* Clinically significant effect at upper limit of normal therapeutic range on Ektachem® 400/700 method. *2138*

Novobiocin Interference by metabolite (Evelyn-Malloy). *2313* Yellow metabolite affects direct methods. *0583*

Oxytetracycline At concentrations above 50 mg/L raised concentration as measured by Jendrassik and Grof method. *3393*

Phenazopyridine Postulated increased color with diazotization. *0652*

Phenelzine At 1 mmol/L affects SMA 12/60 method slightly. *3335*

Propranolol 4-hydroxypropranolol metabolite increases bilirubin concentration when measured with diazo reaction. Metabolite present as sulfate or glucuronide. *0041*

Reserpine At concentrations above 61 mg/L raised concentration as measured by Jendrassik and Grof method. *3393*

Rifampin Due to yellow coloration of drug. *2820* Marked effect when measured by bilirubinometer. *0218* At 5 times upper limit of therapeutic concentration on methods on Ektachem® and Cobas-Bio. *2138*

Storage of Sample Standards deteriorate at -23°, stable at -70°. *0941*

Tyrosine Reacts with diazo reagent. *1563*

Vitamin A Interferes with analysis. *1238*

Xanthophyll May deepen color of serum (?significance). *1022*

Serum Increase Physiological

Acetaminophen Hepatic damage reported with overdose. *3052* Manifestation usually of single toxic high dose ingestion in suicide attempt. Liver damage usually of centrizonal hemorrhagic hepatic necrosis. *3948* With overdose and centrolobular hepatic necrosis usually associated with high drug concentration. *1771*

Acetanilid Hemolysis with G-6-PD deficiency. *2313*

Acetazolamide Single case of cholestatic jaundice reported. *2018*

Acetohexamide May cause intrahepatic cholestatic jaundice. *0652*

Acetophenazine Cholestasis with intrahepatic obstruction. *1237*

Acetylphenylhydrazine May cause hemolytic anemia. *2313*

Ajmaline Intrahepatic cholestasis. *0268* Hepatotoxicity and intrahepatic cholestasis reported. *3215* In 4 patients with centrilobular cholestasis and mild hepatocytic lesions: recovery with drug withdrawal. *2729*

Allopurinol Reported in renal failure also in other patients. *2165*

Aluminum Nicotinate Rare hepatotoxic effects. *1678*

Aminobenzoic Acid Hepatotoxicity with centrolobular necrosis. *1948*

Aminoglutethimide Casual association between drug administration and cholestasis in 2 cases. *2779*

Aminopyrine Hemolysis in G-6-PD deficiency. *0248*

Aminosalicylic Acid Reversible cholestasis caused by drug. *0277*

Aminothiazole Jaundice with febrile reactions. *1343*

Amiodarone Hepatitis occurs in 4% patients. *2011*

Amitriptyline Rare cases of transient cholestasis. *2482*

Amodiaquine Reported effect. *2313* Abnormality usually mild but associated with an immunoallergic hepatotoxicity. *2081*

Amphotericin B May cause hepatocellular dysfunction. *1343* Noted in one patient with acute myelogenous leukemia when treated with high dose. Probable idiosyncratic response. *2443*

Amyl Nitrite With increased hemolysis. *2426*

Anabolic Steroids Cholestatic syndrome. *1022*

Androgens May cause cholestatic syndrome. *1022*

Anisindione Observed with other indandiones. *1680*

Antifungal Agents Hepatotoxic effect. *1237*

Antimalarials May cause hemolytic anemia. *1902*

Antimony Compounds Hepatotoxic effect. *2220*

Antipyretics Occurs with hemolytic anemia. *1902*

Antipyrine Hemolysis in G-6-PD deficiency. *0248*

Aprindine Hepatitis is manifestation of chronic toxicity. *3215* Hepatitis in 2 patients within 3 weeks of start of therapy resolved with withdrawal of drug. *1570*

Arsenicals Hepatotoxic effect (cholestasis/cholangiolitis). *2220*

Asparaginase Up to 4 mg/dL (dose related effect). *2657* In 29% children and 51% adults reported in various studies. *2406* Increased in 30-60% of patients treated in different studies: effects usually mild. *0558* Hemorrhagic pancreatitis in fewer than 0.5% treated patients. *1398*

Aspidium May cause hepatic toxicity. *1022*

Aspirin Competition for albumin binding. *1516* Clinically significant displacement from protein in neonates. *3748*

Azathioprine May cause hepatotoxicity (not usually very high). *3433* Canalicular cholestasis and centrilobular ballooning of hepatocytes in one patient. *0846* Hypersensitivity reaction without effect on aminotransferases. *0827*

Barbiturates Rare cases of jaundice following use. *0357*

Benziodarone May cause hepatic toxicity. *2313*

Benzthiazide Rare, may occur as with other thiazides. *1680*

Bethanechol Impaired excretion due to spasm of sphincter. *0652*

Bismuth Salts Hepatotoxicity. *2313*

Bleomycin Rare mild abnormality reported: reversible. *2406*

Blindness Significantly higher (average 0.17 mg/dL) than normal. *1637*

Boric Acid Toxicity effect with hepatomegaly. *1302*

Bunamiodyl Competition for hepatic uptake of unconjugated. *0349*

Busulfan Occurs with hemolytic anemia. *1680*

Butaperazine Cholestatic hepatitis with obstruction. *0071*

Cadmium May occur even following respiratory exposure. *1302*

Captopril Isolated case of cholestasis in patient receiving 25 mg tid for 1 mo. *2732* Characteristic pattern of hepatocellular jaundice in one patient. But with secondary cholestatic elements. *3688*

Carbamazepine Cholestatic and hepatocellular damage. *2907* Granulomatous hepatitis observed in small proportion of patients treated for less than 1 mo: resolved within 3 d of cessation of treatment. *2153*

Carbarsone Cholestasis with cholangiolitis may occur. *1948*

Carbenoxolone Possible hepatotoxic effect of drug. *2458*

Carbimazole Isolated case of jaundice conclusively linked to drug. *0911*

Carbutamide Hepatotoxicity. *2313*

Carmustine (BCNU) Usually mild and return to normal over few days. *2406*

Carphenazine Probable cholestatic effect (reversible). *2313*

Cefazolin In 4 of 31 treated cases: transient 2 to 3 fold increase. *2273*

Cefoperazone Significant displacement in neonates but possibly only at concentrations above therapeutic. *3748*

Cefoxitin Transient 2 to 3-fold increase in 3 of 31 patients following colorectal surgery. *2273*

Ceftizoxime In 3 of 30 patients following colorectal surgery had transient 2 to 3-fold increase. *2273*

Cephalothin May cause hemolytic anemia. *1377*

Chenodeoxycholic Acid In 4-9% in 916 patients given up to 750 mg/d for 2 y. *3199*

Chloral Hydrate Probably nonspecific effect on cells. *1500*

Chlorambucil Hepatotoxicity with centrolobular necrosis. *1948*

Chloramphenicol Hepatotoxic-cholestatic effect. *2220*

Chlordane Due to hepatotoxicity. *1302*

Chlordiazepoxide Infrequent cholestatic effect. *1022*

Chlormezanone Occasional reversible cholestasis. *2313*

Chloroform Hepatotoxic effect with necrosis. *2313*

Chloroquine Hemolytic anemia in G-6-PD deficiency. *1680*

Chlorothiazide Reported to cause cholestatic jaundice. *2745*

Chlorpromazine Sensitivity reaction (may cause jaundice in infant). *3283* In one patient who developed cholestasis with fever and leukopenia after 260 mg drug. *2840*

Chlorpropamide Cholestatic jaundice of allergic nature may occur. *1488* Isolated case of drug-induced hemolytic anemia. *3111* Cholestatic jaundice observed in one patient in association with red cell aplasia (prevalence of jaundice as high as 0.5%). *1284*

Chlorprothixene Hepatotoxicity (reversible) cholestatic effect. *2313*

Chlortetracycline Hepatotoxic effect with centrolobular necrosis. *1237*

Chlorzoxazone May cause hepatotoxicity. *2313*

Cholinergics Impaired excretion due to spasm of sphincter. *0652*

Cimetidine Reversible cholestatic jaundice in a few children. *2172* In isolated case treated for 7 mo with 400 mg/d progressed to bridging hepatic necrosis. *3085*

Cinchophen May cause hypersensitive cholestasis. *2313*

Clindamycin Occurs especially if pre-existing liver disease. *2806* Uncommon side effect. *3864*

Clofibrate Hepatotoxic effect. *2313* Single case of granulomatous hepatitis associated with drug administration. *2813*

Clonidine Toxic hepatitis reported. *1277*

Colchicine May cause hepatic toxicity. *2313*

Copper Marked increase with hemolysis of toxicity. *0385*

Coumarin May cause hepatotoxicity. *3835*

Cyclofenil In 1 case, but total of 30 cases reported up to 1980; Considered to be metabolic idiosyncrasy and reversible in all cases. *2672*

Cyclophosphamide May cause hepatotoxicity. *2313*

Cyclopropane May cause hepatotoxicity (lasts several days). *2313*

Cycloserine May cause hepatotoxicity. *2313*

Cyclosporine Observed in 1 of 23 patients with inflammatory ocular disease treated for 6 mo. *2713* Transient increase observed in 20% of renal transplant patients. *0683* In 18 of 21 patients after bone marrow transplantation. *0174* Both early and late toxicity observed but effects mild in comparison with nephrotoxicity in renal transplant patients. *1959* Association in 4 cadaveric renal transplant patients with cyclosporine trough concentration (reversible). *2091*

Cyclothiazide Occasional intrahepatic cholestatic jaundice. *1680*

Cyproheptadine Isolated case of jaundice due to cholestasis: reversed on withdrawal of drug. *1562*

Cytarabine At 3 g/m² every 12 h for 6 d, some mild and reversible abnormalities in about 75% of 12 cases. *1579* Significant elevation in 7 of 42 patients with previously normal liver function. *2406*

Dantrolene Isolated observation in one individual. *0659* Acute hepatitis as result of drug ingestion in 4 patients returned to normal after drug stopped. *3837* In 1.8% of 1044 patients, hepatocellular damage with acute or subacute hepatic disease or chronic active hepatitis. *3656*

Dapsone In 3 individuals treated for 3 to 36 weeks. *3478* Associated with hemolytic anemia when occurs as main side effect. *3423* Sulfone syndrome in one 16 year old girl given 50 mg daily short-term administration. *3604*

Desipramine Rare transient cholestasis. *1488*

Dextrothyroxine One case of possible cholestasis reported. *0019*

Diatrizoic acid Possible displacement from protein because of high circulating concentration achievable. *3748*

Diazepam Presumed hepatic toxic effect. *3879*

Dicloxacillin Possible displacement from protein especially in critically ill neonates. *3748*

Dienestrol May cause cholestasis. *1680*

Diethylstilbestrol Hepatotoxicity with centrolobular necrosis. *1948*

Dimercaprol May cause hemolysis with G-6-PD deficiency. *0248*

Dinitrophenol Due to toxic hepatitis. *1302*

Diphenhydramine May occur with hemolytic anemia. *3602*

Dipyrone May cause hemolytic anemia. *2365*

Disopyramide Occasional case of cholestatic jaundice reported. *3215*

Disulfiram One possible case of cholestasis reported. *1680* In 6 patients proved to be drug associated in one by challenge test. *2912* In 6 patients: proved to be drug associated in one by challenge test. *2491*

Doxepin Theoretical possibility due to class of compound. *0071*

Ectylurea May cause cholestasis with cholangiolitis. *2313*

Enflurane Evidence of hepatocellular damage; characteristically centrilobular necrosis: mechanism of injury most probably metabolic idiosyncrasy. *2157*

Erythromycin Causes cholestasis in approximately 15% patients. *1142* Estolate produces mild hepatotoxicity in about 15% treated individuals. Jaundice in 2% patients on drug for more than 2 weeks. *2792* Observed to some extent in some patients with different erythromycin salts: usually cholestatic hepatitis. *1895*

Estradiol Cholestasis may occur. *2313*

Estrogens May cause cholestasis (may be hepatocellular degeneration). *2313*

Estrone May cause cholestasis. *2313*

Ethacrynic Acid Cholestasis or hepatocellular damage. *3214*

Ethanol May cause centrolobular necrosis of liver. *1948* Increased in alcoholic liver disease. *0625*

Ether Hepatic disturbance (transient). *3218*

Ethionamide Hepatotoxicity in about 2% cases. *1488*

Ethosuximide Rare idiosyncratic hepatitis reported. *0071*

Ethotoin Probable idiosyncratic hepatitis. *2313*

Ethoxazene Hepatotoxic effect. *1022*

Ethyl Chloride May cause liver damage. *0071*

Etoposide Hepatic toxicity reported in 3% treated patients. *0141*

Etretinate No marked effect seen but slight change noted. *3867*

Factor IX Complex, Human 62% cases developed hepatitis after use. *2663*

Fasting Increase of 1.3 g after 24 h and 2.2x after 48 h in normals. *0237*

Fava Beans May cause hemolysis with G-6-PD deficiency. *0248*

Flavaspidic Acid Inhibits UDP-glucuronyl transferase. *2621*

Floxuridine Hepatotoxic manifestation. *1680*

Flucytosine Reversible hepatotoxicity in 10%. *3443*

Fluorouracil Single case reported. *1488*

Fluoxymesterone Intrahepatic cholestatic jaundice. *1343*

Fluphenazine Hypersensitivity response. *2313*

Flurazepam May cause hepatic toxicity. *3016*

Fluroxene Potentially hepatotoxic. *0071*

Fructose Increase by 0.3-0.8 mg/dL after i.v. infusion. *1160*

Furadaltone May cause hemolysis with G-6-PD deficiency. *0248*

Furazolidone May cause hemolysis with G-6-PD deficiency. *0248*

Fusidic Acid Clinically significant displacement from protein in neonates. *3748* Reported in 6 patients in UK, although causal relationship could not be be established with certainty. *3535* Jaundice developed in 34% of 112 patients given drug, highest when drug given intravenously. Jaundice resolved when drug withdrawn. *1690*

Gentamicin Affects liver function. *0071*

Glibenclamide Intrahepatic cholestasis described in 61 y old diabetic. *3894*

Glucose Increase by 0.3-0.8 mg/dL after rapid i.v. infusion. *1160*

Glycopyrrolate Hepatotoxicity. *2313*

Gold Rare side effect with liver damage. *2427* Cholestatic jaundice in 3 patients with rheumatoid arthritis given gold sodium thiomalate: all patients recovered spontaneously. *1084* In 3 patients with rheumatoid arthritis during chrysotherapy. *0993* Due to toxic effect on drug on liver. *1674*

Guanoxan May cause hepatic toxicity. *2313*

Bilirubin *(continued)*

Serum Increase Physiological *(continued)*

Haloperidol May cause hepatocellular changes. *2313* In isolated case producing cholestatic liver disease: typical frequency of liver disease is 0.2% In isolated case producing cholestatic liver disease: typical frequently of liver disease is 0.2%. *0910*

Halothane Reversible depression of liver function. *3294* In one case after repeated exposure to anesthetic: evidence of acute hepatitis with bridging necrosis and piecemeal necrosis. *2027* Granulomatous hepatitis in single case following second exposure to anesthetic. *3266*

Hycanthone Induces form of acute toxic hepatitis. *1079*

Hydantoin Derivatives Severe delayed hypersensitivity reported. *2427*

Hydralazine Observed in a case of obstructive jaundice with pancytopenia. *3469* Asymptomatic or symptomatic reversible hepatitis-like reaction or granulomatous hepatitis. *2374* Centrilobular necrosis observed in 3 patients 2 to 4 months after drug therapy for hypertension. *1740*

Hydrazine Derivatives May cause hepatotoxicity. *2313*

Hydrochlorothiazide Cholestatic jaundice reported effect of thiazides. *2745*

Hydroflumethiazide May cause intrahepatic cholestatic jaundice. *1680*

Hydroxyacetamide Hepatotoxicity. *2313*

Ibufenac Probable effect as bilirubin clearance reduced. *1499*

Ibuprofen Rare hepatotoxicity reported: associated in one case with Stevens-Johnson syndrome. *3460*

Icterogenin Causes intrahepatic cholestasis. *0153*

Idoxuridine Cholestatic jaundice reported in one case. *0837*

Imipramine May cause cholestatic jaundice. *1873*

Indandione Derivatives As result of hypersensitivity. *2427*

Indomethacin Cytotoxic and cholestatic types of liver damage. *1908*

Insecticides May cause hemolytic anemia. *2365*

Iodipamide Clinically significant displacement from protein potentially in neonates. *3748*

Iopanoic Acid May cause hepatotoxicity. *2313* Clinically significant displacement observable in neonates. *3748*

Ipodate Probably due to competition for excretion. *0652*

Iprindole Associated with low serum phosphate. *0038*

Iproniazid May cause cholestatic and cytotoxic jaundice. *0248*

Iron Salts With poisoning 3-4 d after ingestion. *3817*

Isocarboxazid Cholestatic effect. *2313*

Isoniazid Intrahepatic cholestasis. *1022* 15 of 89 patients developed significant liver disease, typically hepatitis, although one developed cholestasis, in patients taking drug for at least 2 mo. *0901*

Isotretinoin Reversible dose-related effect noted in patients with myelodysplastic syndrome and leukemia at high doses. *1319*

Kanamycin May cause hepatotoxicity. *2313*

Ketoconazole Delayed reaction to drug after withdrawal in single case of chronic candidiasis in 61 y old woman. *3513* Rapidly developing liver failure in 67 year old woman taking 200 mg drug daily for 2 mo. *0960*

Lead May cause hemolytic anemia. *0788*

Levodopa Rare elevation reported. *0071*

Lincomycin Hepatotoxic-cholestatic effect. *2220*

Lugol's Iodine Occasional hypersensitive response to iodines. *0071*

MAO Inhibitors Viral hepatitis-like jaundice in some patients. *3265*

Meals Affected by meals (most marked in women). Observed in both sexes. *3455*

Mefenamic Acid May cause autoimmune hemolytic anemia. *1078*

Meglumine Ioglycamate Possible clinically significant displacement from protein in neonates. *3748*

Melphalan May cause hemolytic anemia. *2365*

Mepazine Cholestatic effect. *1713*

Mephenytoin May cause hemolytic anemia. *2365*

Meprobamate May cause cholestatic (hepatocanalicular) jaundice. *0248*

Mercaptopurine May cause increase, especially if prior damage. *3293* Picture of cholestasis in 10 of 19 leukemic patients, but drug co-administered with Adriamycin® in all and additional drugs in others. *3021* Reported in up to 53% leukemic patients treated with drug. *2406*

Mesoridazine Jaundice as manifestation of hepatotoxicity. *1680*

Metahexamide Hepatotoxicity (viral hepatitis type). *2313*

Metaxalone Hepatotoxic effect. *2313*

Methacholine Impairs excretion through biliary tract. *0652*

Methandriol Intrahepatic cholestatic jaundice. *1343*

Methandrostenolone Due to cholestasis. *1713*

Methanol Manifestation of liver damage. *1238*

Methdilazine Rare case of cholestasis. *1680*

Methimazole Toxic effect associated with bone marrow depression. *0262* May affect liver function (cholestasis). *2313* Rare case of cholestatic jaundice reported. *1774*

Methotrexate May cause cytotoxic hepatocellular damage. *0248* In 10 of 250 courses of treatment but most regressed. *2406*

Methotrimeprazine Three cases reported. *1488*

Methoxsalen Hepatotoxic effect. *2313*

Methoxyflurane Hepatic toxicity. *3294* In 2 pregnant women when used as analgesia for labor: hepatitis observed. *0872*

Methsuximide Hepatic damage reported. *0071*

Methyclothiazide Jaundice may occur with thiazide therapy. *1680*

Methyldopa Mild hepatocellular jaundice in about 1% cases. *1023* Chronic active hepatitis associated with immune hemolytic anemia in one case. *3269* Disturbances in liver function in 5 to 35% of patients treated for hypertension. Hepatocellular injury may occur after short or long-term exposure. *0162*

Methylene Blue May cause hemolysis with G-6-PD deficiency. *0248*

Methyltestosterone Cholestatic effect. *0153* In 1 of 60 patients (female transsexuals or impotent males) receiving 150 mg day for long term. *3805*

Methylthiouracil Hepatotoxic effect. *2313*

Metrizoate Theoretically possible effect in neonates because of high circulating concentration. *3748*

Mitoxantrone Mild transient effects in 11 of 26 patients. *2703*

Morinamide Jaundice reported as side effect. *2427*

Morphine Associated with abnormal liver function. *2313*

Moxalactam Clinically significant displacement from protein in neonates. *3748*

Nalidixic Acid May cause cholestatic jaundice. *2808*

Naphthalene May cause hemolysis with G-6-PD deficiency. *0248*

Naproxen Isolated case of hemolytic anemia: one other case reported previously. *1681*

Niacin May cause impairment of hepatic function. *2726*

Niacinamide Rare probable reversible toxic effect. *3875*

Nialamide Probable hypersensitive hepatitis. *0071*

Nitrofurans Hepatotoxicity or due to hemolytic anemia. *2313*

Nitrofurantoin May cause hemolytic anemia or cholestasis. *1974* Chronic active hepatitis much more common in women than men, but still rare. *2444* Chronic active hepatitis much more common in women than men, but still rare. *2444* May follow acute hepatitis with or without cholestasis or as chronic active hepatitis or chronic granulomatous reaction. *2374*

Nitrofurazone May cause hemolysis with G-6-PD deficiency. *0248*

Nitrophenol Hypersensitive intrahepatic cholestasis. *1434*

Norethandrolone Intrahepatic cholestasis produced (up to 20%). *1280*

Norethandrostenolone Intrahepatic cholestatic jaundice. *3197*

Norethindrone Intrahepatic cholestatic jaundice. *1343*

Norethisterone In 5.6% patients being treated for advanced or recurrent breast cancer. *2073*

Norethynodrel Intrahepatic cholestatic jaundice. *1343*

Nortriptyline May cause cholestatic jaundice. *0071*

Novobiocin Especially in newborn: inhibits conjugating mechanism. *0350*

Ohio 469 Preliminary finding of significant increase post operatively. 1647

Oleandomycin May cause intrahepatic cholestatic jaundice. 1237

Oral Contraceptives Interferes with canalicular excretion. 0348

Organophosphorus Insecticides Hepatotoxic effect. 0405

Oxacillin Cholestatic jaundice reported in one case. 2737 Reversible high-dose associated liver injury. 2431

Oxazepam Possible hepatotoxicity. 2313

Oxymetholone Cholestatic effect. 1713 Cholestatic hepatitis developed in two patients. 3921 One case of peliosis hepatitis reported. 2539

Oxyphenbutazone Possible hepatotoxicity. 2313

Oxyphenisatin Hepatic toxicity if over prolonged period. 2757 Liver damage may present as acute hepatitis or as more advanced chronic disease: usual liver disease is chronic active hepatitis. 2260 Observed with chronic active hepatitis induced by drug. 0903

Pamaquine May cause hemolysis in G-6-PD deficiency. 0248

Papaverine Small effect, reversible hypersensitivity reaction. 3267 In 2 patients: fever anorexia and jaundice 4 to 5 weeks after start of treatment due to pericholangitis reversible. 3371 In 6 of 14 patients abnormal liver function, possibly due to allergic response and to metabolic aberration in host. 2746

Paraldehyde Possible hepatotoxicity. 2313

Paramethadione Possible hepatotoxicity. 2313

Pargyline Possible hepatotoxicity (reversible). 2313

Penicillamine Case of acute cholestatic jaundice in patient with systemic lupus erythematosus. 3242

Penicillin Nonspecific hepatitis without cholestasis. 3879

Penicillin V Possible clinically significant displacement from protein especially in critically ill neonates. 3748 Isolated case of hemolytic anemia with IgM antibody. 0360

Pentaquine May cause hemolysis with G-6-PD deficiency. 0248

Perchlorate Single case of acute yellow atrophy of liver. 2427

Perphenazine Low incidence of jaundice reported. 0071

Phenacemide May affect liver function in about 2% cases. 2313

Phenazopyridine Single report of jaundice (?due to hemolysis). 1652

Phenelzine May cause hypersensitive hepatitis. 0071

Phenindione Probable effect as bilirubin clearance reduced. 1499

Pheniprazine Hypersensitive hepatitis. 2313

Phenobarbital Hepatotoxicity with centrolobular necrosis. 1948

Phenothiazines Hypersensitivity cholestatic reaction may occur. 1343 Liver damage observed after different phenothiazines administered in succession. 1916

Phenoxymethicillin Possible clinically significant displacement from protein especially in especially in critically ill neonates. 3748

Phenoxypropazine Produced fatal hepatotoxicity. 2427

Phenprocoumon Mild effect but recurrently in 2 patients having repeated exposure to drug. 3347

Phenylbutazone Granulomatous reaction in liver. 1338 Clinically significant displacement from protein in neonates. 3748 Observed in 7 patients with side effects due to drug. 3409 In 2 patients subsided after drug withdrawn associated with granulomas and cholestatic-hepatocellular injury. 1730 Observed in one patient: clear evidence of intrahepatic cholestasis. 3670 Frequent hypersensitivity reaction but also with hepatotoxic potential: may be hepatocellular injury or systemic vasculitis. 0290

Phenylhydrazine Causes hemolysis. 0348

Phenytoin Rare hypersensitivity reaction. 1488

Phosphorus Hepatotoxic effect with necrosis. 2313

Phytonadione Large dose effect, or with G-6-PD deficiency. 1022

Piperacetazine Rare jaundice (reversible)(like infect hepatitis). 0071

Piperazine May cause hemolytic anemia. 2365

Pipobroman May cause hemolytic anemia. 1022

Piroxicam Occasional case of liver damage reported. 2824

Polythiazide May affect liver function. 2313

Practolol After 4 y treatment 2 cases of primary biliary cirrhosis. 0496

Primaquine May cause hemolysis with G-6-PD deficiency. 0248

Probenecid Hepatotoxic effect (centrolobular necrosis). 2313

Procainamide Reversible hepatic toxicity reported. 1488 In 3% of all patients receiving drug, but in 14% of patients with direct Coombs' test. 1958

Procarbazine Reported effect. ?mechanism. 1488

Prochlorperazine Cholestatic effect. 0071

Progesterone Hepatotoxicity also transient familial increase. 2313

Promazine May cause cholestatic (hepatocanalicular) jaundice. 0248

Promethazine May cause cholestasis. 2313

Propoxyphene Cholestatic effect. 1954 In 3 patients within 10 d of start of drug treatment. 1156 Hepatotoxic response observed in 2 patients. 2110

Propylthiouracil Rare case of hepatotoxicity reported. 1488 Temporary effect associated with drug administration regressed with cessation. 2731 In one 24 year old woman who developed fulminant hepatic failure with lymphocyte sensitization. 2437 Drug induced liver damage: either acute or chronic active hepatitis. 3787

Protein Hydrolysate Hyperalimentation i.v. infants may cause cholestasis. 3614

Protionamide Temporary side effect. 2427

Protriptyline Transient reversible cholestasis. 0071

Pyrazinamide Hepatic toxicity (approximately 3 %) (dose related). 3193 Hepatotoxic in 1 to 5% patients (previously reported to be toxic in as many as 25%). 3683

Quinacrine Hemolysis may occur with G-6-PD deficiency. 0248 Hepatotoxic effect. 2313

Quinethazone May cause cholestatic jaundice. 2745

Quinidine Causes hemolytic anemia. 2365 Hypersensitivity observed in 32 of 487 patients who had received drug. 1261

Quinine May cause hemolytic anemia. 2365

Radiographic Agents Competition for excretion through bile canaliculi. 0127

Rifampin Inhibits hepatic excretion. 0013 Minimal abnormalities of liver function are common: severe liver damage in about 0.6% patients. 3683 Observed in one patient receiving drug who developed porphyria cutanea tarda. 2441 Significant increase in all subjects after single dose of 900 mg. 1224

Salicylate Clinically significant displacement from protein in neonates. 3748

Sex Difference In males at all ages. 2130

Sodium Iodate Clinically significant effect with displacement from protein possible in neonate. 3748

Sorbitol Increase by 0.3-0.8 mg/dL after i.v. infusion. 1160

Stanozolol May cause intrahepatic cholestasis. 1343

Starvation Begins within 8 h, falls within 8 h of refeeding. 2696

Stibophen Produces hemolytic anemia. 2365

Streptomycin May cause hemolytic anemia. 2313

Sulfacetamide May cause hemolysis in G-6-PD deficiency. 0248

Sulfachlorpyridazine Occasional hepatitis-like reaction. 1680

Sulfadiazine May cause intrahepatic cholestasis. 1434 Displacement from protein in neonates. 3748

Sulfadimethoxine Granulomatous reaction in liver. 1049

Sulfadoxine Isolated cases reported when given with pyrimethamine for treatment of malaria. 2669

Sulfamethizole May cause cholestatic jaundice. 1974

Sulfamethoxazole Reversible cholestasis or hemolytic anemia. 2313 Displacement from protein in neonates. 3748 Occasional case of cholestasis with or without hepatic necrosis. 3452

Sulfamethoxypyridine May cause hemolysis with G-6-PD deficiency. 0248

Sulfanilamide May cause hemolysis with G-6-PD deficiency. 0248

Sulfapyridine May cause hemolysis with G-6-PD deficiency. 0248 Displacement from protein in neonates. 3748

Bilirubin (continued)

Serum Increase Physiological (continued)

Sulfasalazine Theoretical displacement from protein in neonates although not observed in practice. 3748 Occasional case of drug-induced toxic hepatitis. 1422 As part of systemic hypersensitivity reaction in woman with inflammatory bowel disease. 3402 Hepatotoxic hypersensitivity reaction like sulfonamide hypersensitivity. 2214 In 2 cases with drug induced hepatotoxicity of hypersensitivity type of reaction: reversed on drug withdrawal. 3364 Occasional case of drug-induced toxic hepatitis. 1422

Sulfinpyrazone Jaundice reported in cases with overdose. 1680

Sulfisoxazole May cause cholestatic jaundice or hemolysis. 1974 Displacement from protein in neonates. 3748

Sulfonamides Hepatic/cholestatic effect, also affects albumin binding. 0348 Hepatitis-like reaction in patients with multiple episodes of jaundice after repeated exposure to drug. 1741

Sulfones May cause hemolysis with G-6-PD deficiency. 0248

Sulfonylureas In some cases probably attributable to drug. 2427

Sulfoxone May cause hemolysis in G-6-PD deficiency. 0248

Sulindac Isolated case of cholestatic jaundice 12 d after therapy started. 2371

Tamoxifen Isolated case of drug induced cholestasis in one patient. 0372

Testosterone May cause cholestatic and cytotoxic liver damage. 0248

Tetracycline Hepatic injury may occur especially if given i.v. 0348 2 cases of drug-associated fatty liver of pregnancy. 3795

Thiabendazole Rare cholestasis. 2413 One case of drug-induced cholestasis reported. 2962

Thiacetazone May cause cholestatic (hepatocanalicular) jaundice. 0248

Thiazides Cholestatic hepatitis reported to occur. 1343

Thiazolsulfone Hemolytic anemia. 2313

Thiethylperazine Few cases of cholestasis reported. 1680

Thiocarlide Rare observation. 2427

Thioguanine May cause cholestasis. 2313

Thiopental 1 case out of 24 3-5 d after anesthesia, more longer period after anesthesia. 0391

Thioridazine Hepatotoxicity (questionable). 1857

Thiosemicarbazones May affect liver function. 2313

Thiothixene Hepatotoxic effect (reversible cholestasis). 2313

Thiouracil May cause intrahepatic cholestasis. 1434

Thorium Dioxide Time related liver damage. 3147

Tolazamide Cholestatic jaundice. 2313

Tolbutamide Cholestatic jaundice reported. 1488

Tolmetin Observed in single case of 49 y old man who had taken drug as needed for arthritis for 1 y. 3421

Toluene Diamine May cause cholestasis with cholangiolitis. 1948

Tranylcypromine May affect liver function (?hypersensitivity). 2313

Trichlormethiazide May rarely cause cholestasis. 1680

Trichloroethylene Potentially hepatotoxic. 0071

Triethylenemelamine May cause hemolytic anemia. 2365

Trifluoperazine Cholestatic effect. 1713

Trifluperidol Elevated in chronic and acute hepatitis. 1022

Triflupromazine Liver dysfunction exceptionally rare. 0071

Trimeprazine May cause cholestasis. 0071

Trimethadione Hepatotoxicity with centrolobular necrosis. 2313

Trimethobenzamide Rare cases of jaundice reported. 1680

Trimetrexate Transient effect in small proportion of patients although reverted to normal with continued treatment. 0058

Trioxsalen May affect liver function. 2313

Troleandomycin In 4% individuals treated with 1 g/d for 3-4 weeks. Jaundice typically occurs after 2 weeks mixed in type with both cholestasis and mild hepatocytic necrosis. 2792

Tyrothricin Hemolysis even when applied topically. 0071

Uracil Mustard May affect liver function. 2313

Urethan Hepatotoxicity with centrolobular necrosis. 0071

Valproic Acid Acute hepatic centrilobular necrosis with severe fatty change in small number of patients. 3763

Verapamil Observed in a single case, reverted to normal as soon as drug withdrawn. 2561

Vinyl Ether Potentially hepatotoxic. 0071

Vitamin K Large doses in neonates or G-6-PD deficiency. 0248

Xylitol Mainly indirect — by 2 to 3 times normal in some. 0925

Zoxazolamine May produce viral-hepatitis like syndrome. 1948

Serum Decrease Analytical

Aminophenazone With 540 μmol/L on dimethylsulfoxide method. 1785

Ascorbic Acid At concentrations above 200 mg/L (normal maximum serum concentration 34 mg/L) lowered concentration as measured by Jendrassik and Grof method. 3393

Caffeine Allegedly lower values when used as coupling agent. 1022

Fluorescein By up to 25% at up to 100 mg/L on Beckman Astra method. 1722

Hemoglobin Competes for nitrite with sulfanilic acid. 0583

Hemolysis Methemoglobin formation affects diazotization. 2362

Light Drawn specimens decrease up to 30 %/h in bright daylight. 0251

Liposol Changed from 25.5 to 3.1 mg/L in 26 randomly selected patient sera, before and after addition on Beckman Astra methods. 2054

Pindolol At 5 mg/L conventional methods when added to serum. 1758

Protein Absence of protein affects diazotization. 2313

Sodium Chloride Inhibition of diazo test reported. 1563

Theophylline Causes depression of color formation. 2313

Urea Brown color in diazo reaction of uremics. 1563

Serum Decrease Physiological

Aspirin Progressive effect during three days but also noted over longer term study at beginning, eventually reverted towards normal. Effect statistically significant. 3076

Barbiturates Induces glucuronyl transferase in newborn infants. 1488

Carbamazepine In about 20 epileptic patients treated for 2 y. 0873 Significant reduction compared with controls in patients treated for average of 20.5 mo due to hepatic microsomal enzyme induction. 2635

Chlorophenothane Effect on congenital non-hemolytic hyperbilirubinemia. 3505

Corticosteroids When elevated as increases bile flow. 2246

Ethanol Decreases in infant if given to pregnant woman. 2160

Flumecinolone Inhibits hyperbilirubinemia of term or premature newborns due to enzyme induction. 2304

Halofenate No reversal of early effect with long treatment. 1708

Heparin In majority of patients after 8 d after low-dose heparin treatment. 2671

Isotretinoin In 7 patients with severe rosacea treated with 1 mg/kg/d for 12 weeks effect possibly due to induction of hepatic microsomal enzymes. 2307 Significant at 12 weeks in 7 patients with severe rosacea treated with 1 mg/kg/d. 2307

Light Breakdown of bilirubin in vivo and in vitro. 3660

Orotic Acid ?induction effect in premature infants only. 3220

Penicillin In newborn combined with sulfisoxazole has effect. 3505

Phenazopyridine Atypical color causes interference. 2313

Phenobarbital Induces hepatic microsomal enzymes especially in pregnant women. 3505

Prednisone Augmented response in cirrhotics when drug added to therapeutic regime. 0651

β-Propiolactone Increase by 1 μmol/L when plasma from BPL treated blood compared with plasma in 25 specimens. 0217

Sulfisoxazole In newborn combined with penicillin has effect. 3505

Sulfonamides Displacement from albumin binding sites. 3505

Thioridazine ?due to effect on bilirubin metabolism. 1629

Urine   No Effect   Analytical
Thymol   No effect on Fouchet procedure. *2981*

Urine   Increase   Physiological
Acetohexamide   May cause intrahepatic cholestatic jaundice. *2313*
Acetophenazine   Cholestasis with intrahepatic obstruction. *2313*
Chlorprothixene   May cause hepatotoxicity (reversible). *2313*
Dapsone   Associated with hemolytic anemia in sulfone syndrome. *3604*
Ethoxazene   Hepatotoxic effect. *1022*
Fluphenazine   Hypersensitivity response. *2313*
Imipramine   May cause cholestatic jaundice. *1873*
Methyldopa   Occurs as result of hepatocellular damage. *3646*
Norethandrolone   Cholestasis produced without cholangiolitis. *3172*
Perphenazine   Low incidence of jaundice reported. *2313*
Phenothiazines   May cause cholestatic hepatitis. *2313*
Sulfadimethoxine   Granulomatous reaction in liver. *1049*
Tolmetin   Observed in single case of 49 y old man who had taken drug as needed for arthritis for 1 y. *3421*

Amniotic Fluid   Decrease   Analytical
Transport of Specimen   Cardboard container cause decrease of 450 nm peak. *0694*

Duodenal Contents   Increase   Physiological
Amino Acids   Increase by 800% with duodenal infusion, no effect if given i.v. *2918*

Bilirubin, Conjugated

Serum   Increase   Analytical
Fluorescein   At expected concentration of 1 to 100 mg/L interfered with method on Ektachem® 700. *1722*
Methotrexate   At concentration of 1,000 μmol/L as much as 38 mg/L increase in concentration on Kodak Ektachem® 400. *3518*

Serum   Increase   Physiological
Cholestyramine   In 19 patients suffering from intrahepatic cholestasis of pregnancy. *1545*
Halothane   Granulomatous hepatitis in single case following second exposure to anesthetic. *3266*

Bilirubin, Direct

Serum   No Effect   Analytical
Cyclosporine   At concentration of 41.5 mmol/L (50 μg/L) on methods on Technicon® SMAC®-II and Hitachi® 705. *1123*
Hemoglobin   At concentrations up to 1.5 g/L on bilirubin oxidase method of Doumas et al. *0940*

Serum   No Effect   Physiological
Streptokinase   Although transient dysfunction of liver observed, which reverts to normal when treatment stopped. *3122*

Serum   Increase   Analytical
Carrots   Color may be misinterpreted. *1238*
Dextran   Turbidity develops with Evelyn-Malloy method. *0652*
Fluorescein   At expected concentration of 1.0 to 100 mg/L interfered with method on Ektachem®-700. *1722*
Hemoglobin   Occurs with direct spectrophotometric method. *0454*
Vitamin A   Interferes with analysis. *1238*

Serum   Increase   Physiological
Acetaminophen   Hepatic necrosis with dose of 10 g reported. *0438*
Acetanilid   Hemolysis with G-6-PD deficiency. *2313*
Acetohexamide   May cause hemolytic anemia. *1019*

Acetylphenylhydrazine   May cause hemolytic anemia. *2313*
Aminopyrine   May cause hemolytic anemia. *2313*
Aminosalicylic Acid   Reversible cholestasis caused by drug. *0277*
Anabolic Steroids   Cholestatic syndrome. *2313*
Androgens   May cause cholestatic syndrome. *2313*
Antimalarials   May cause hemolytic anemia. *1902*
Antipyretics   Occurs with hemolytic anemia. *1902*
Aspidium   Probable inhibition of uptake of bilirubin by liver. *2427*
Aspirin   Occurs with hemolytic anemia. *3505*
Captopril   Characteristic pattern of hepatocellular jaundice in one patient. But with secondary cholestatic elements. *3688*
Cephalothin   May cause hemolytic anemia. *1377*
Chlorpromazine   Sensitivity reaction to drug (in up to 2% patients). *3283*   In one patient who developed Cholestasis with fever and Leukopenia after 260 mg drug. *2840*
Clofibrate   Single case of granulomatous hepatitis associated with drug administration. *2813*
Cytarabine   In two cases with acute leukemia: unlikely to have been due to other drugs co-administered: effects slight. *2827*
Dapsone   Sulfone syndrome in one 16 year old girl given 50 mg daily short-term administration. *3604*
Dipyrone   May cause hemolytic anemia. *2365*
Doxorubicin   May produce idiosyncratic hepatic dysfunction, as observed in 6 patients. *0184*
Erythromycin   Reported effect. *2313*
Fava Beans   May cause hemolysis with G-6-PD deficiency. *0248*
Flavaspidic Acid   Inhibits uptake of bilirubin by liver cells. *0348*
Fluphenazine   Relatively high compared with total. *1343*
Flurazepam   May cause hepatic toxicity. *3016*
Gold   Cholestatic jaundice in 3 patients with rheumatoid arthritis given gold sodium thiomalate; all patients recovered spontaneously. *1084*   In 3 patients with rheumatoid arthritis during chrysotherapy. *0993*
Haloperidol   In isolated case producing cholestatic liver disease; typical frequency of liver disease is 0.2%. *0910*
Hydralazine   Centrilobular necrosis observed in 3 patients 2 to 4 months after drug therapy for hypertension. *1740*
Ibuprofen   Rare hepatotoxicity reported; associated in one case with Stevens-Johnson syndrome. *3460*
Indomethacin   Cytotoxic and cholestatic liver damage. *1908*
Iprindole   Associated with low serum phosphate. *0038*
Iron Salts   With poisoning 3-4 d after ingestion. *3817*
Ketoconazole   Rapidly developing liver failure in 67 year old woman taking 200 mg drug daily for 2 mo. *0960*
Melphalan   May cause hemolytic anemia. *2365*
Mephenytoin   May cause hemolytic anemia. *2365*
Methanol   Manifestation of liver damage. *1238*
Methyldopa   Mild hepatocellular jaundice may occur. *1237*
Nalidixic Acid   Hemolytic anemia especially if G-6-PD deficiency. *3877*
Niacinamide   Rare probable reversible toxic effect. *3875*
Norethandrolone   Reversible cholestasis produced. *1280*
Novobiocin   Competes for conjugation in liver. *0348*
Oral Contraceptives   Hypersensitivity to estrogen component. *1488*
Phenindione   Probable effect as bilirubin conjugation affected. *1499*
Phenothiazines   Relatively large increase compared with total. *1343*
Phenprocoumon   Mild effect but recurrently in 2 patients having repeated exposure to drug. *3347*
Phenylbutazone   In 2 patients subsided after drug withdrawn, associated with granulomas and cholestatic-hepatocellular injury. *1730*   Observed in one patient; clear evidence of intrahepatic cholestasis. *3670*
Phenylhydrazine   Causes hemolysis. *0348*
Propylthiouracil   Temporary effect associated with drug administration regressed with cessation of treatment. *2731*   In one 24 y old woman who developed fulminant hepatic failure with lymphocyte sensitization. *2437*

Bilirubin, Direct (continued)

Serum Increase Physiological (continued)
Protein Hydrolysate Hyperalimentation i.v. infants may cause cholestasis. *3614*
Quinacrine Hemolysis may occur with G-6-PD deficiency. *1237*
Quinidine Causes hemolytic anemia. *2365*
Quinine May cause hemolytic anemia. *2365*
Rifampin Transient effect. *1217*
Stibophen Produces hemolytic anemia. *2365*
Sulfadimethoxine Granulomatous reaction in liver. *1049*
Sulfasalazine Occasional case of drug-induced toxic hepatitis. *1422*
Tetracycline Cholestatic effect. *0945* 2 cases of drug-associated fatty liver of pregnancy. *3795*
Thorium Dioxide Time related liver damage. *3147*
Verapamil Observed in a single case reverted to normal as soon as drug withdrawn. *2561*
Xylitol Increased up to 2 mg/dL following loading test. *3212*

Serum Decrease Analytical
Caffeine Allegedly lower values if used as coupling agent. *1238*
Hemolysis Interference with direct spectrophotometry. *1563*

Serum Decrease Physiological
Desipramine Slight effect observed in 46 treated patients. *2871*
Light Breakdown of bilirubin (less sensitive than indirect). *1237*

Bilirubin, Indirect

Serum Increase Physiological
Procainamide In 3% of all patients receiving drug, but in 14% of patients with direct Coombs' test. *1958*

Bilirubin, Neonatal

Serum Decrease Analytical
Fluorescein Minimal interference at expected concentration on Ektachem® 700. *1722*

Bilirubin, Unconjugated

Serum Decrease Analytical
Fluorescein At expected concentration of 1 to 100 mg/L interfered with method on Ektachem® 700. *1722*
Methotrexate Falsely low values in serum of patients containing large amount of drug following i.v. infusion. *3518*

Biotin

Serum Decrease Physiological
Anticonvulsants Marked effect in 146 epileptics with long-term treatment. *2007*
Carbamazepine Dose related effect observed in long term treated epileptic compared with controls. *2008*
Levodopa Associated with burning feet syndrome. *0824*
Phenobarbital Dose related in long-term treated epileptics compared with controls. *2008*
Phenytoin Dose related in long-term treated epileptics compared with controls. *2008*
Valproic Acid Dose related effect in long term treated epileptics compared with controls effect not as marked as with other anticonvulsants. *2008*

Bishydroxycoumarin

Plasma No Effect Physiological
Aluminum Hydroxide No effect if co-administered. *0079*

Plasma Increase Physiological
Allopurinol Impairs metabolism. *2641* Long term use of drug prolongs half-life. *2927*
Magnesium Hydroxide 75% increase in peak concentration if coadministered. *0079*
Nortriptyline Impairs metabolism. *2641*

Plasma Decrease Physiological
Phenytoin Average decrease from 29-21 μg/mL (DPH dose approximately 1 g/d). *2041*

Bismuth

Serum Increase Physiological
Bismuth Salts Measurable in poisoning. *2427*

Urine Increase Physiological
Bismuth Salts Measurable with poisoning. *2427*

Blast Count

Bone Marrow Decrease Physiological
Isotretinoin Therapeutic response is normalization but effect not usually observed until after 2 weeks. *1319*

Bleeding Time

Blood No Effect Physiological
Ether Anesthesia has no effect. *1343*

Blood Increase Physiological
Ampicillin Dose related prolongation of bleeding time in all 5 volunteers studied. *0492*
Aspirin Also inhibits platelet glycolysis. *1132* Effect slight in normal subjects, but technical variables such as direction of incision may have influence. *2434*
Dextran Observed effect but explanation uncertain. *2313*
Fluroxene Prolonged during anesthesia, normal in 24 h. *0071*
Mithramycin Reversible effect. *0071*
Nifedipine Observed 1 h after ingestion of 20 mg in 20 people. *0796*
Penicillin G Progressive lengthening with dose of 24 million U/d from 3.2 to 6.1 minutes on average, longer with higher dose. *0492*
Streptokinase Dissolves blood clots. *2313*
Warfarin May be prolonged. *1302*

Patient No Effect Physiological
Cyclopropane Anesthesia has no effect. *1343*
Ketoprofen In acute studies but prolonged in subacute in 11 patients given drug intravenously. *1234*
Methicillin Even at 300 mg/kg/d had no effect in volunteers. *0492*

Patient Increase Physiological
Aminocaproic Acid Effect noted in majority of patients after 24 g/d after 9 d treatment. *1387*
Aspirin Significant effect on template bleeding time over 6 d of study. *0895*
Diflunisal Moderate but not significant increase at 1,000 mg bid whereas 250 mg and 500 mg had no effect. Effect reverted to normal in 24 h. *1386*
Ethanol Observed after i.v. ethanol in normals. *1528* Potentiation of aspirin induced increased bleeding time in all patients but to variable extent. *0895*

Halothane   Significant correlation between aggregation *in vitro* and bleeding time *in vivo*. *0800*

Heparin   Doubled after i.v. injection of 100 U/kg in 34 normal subjects: increase unrelated to changes in platelet count. *1544*

Moxalactam   Reversible bleeding diathesis observed in 5 patients. *3792*

Nafcillin   Reportedly 4 times normal in two patients, reverted to normal on withdrawal. *0052*

Piroxicam   Prolongs bleeding time and decreases platelet aggregation. *2692*

Propranolol   Greater than 200% increase observed in 3 of 5 volunteers over 4 day ingestion. *0970*

Valproic Acid   In 3 of 9 patients with poorly controlled epilepsy. *3508*

Bogen Test

Urine   Positive   Analytical
Acetone   Reacts as if ethanol. *1302*
Ether   Reacts as if ethanol. *1302*
Formaldehyde   Reacts as if ethanol. *1302*
Isopropanol   Reacts as if ethanol. *1302*
Methanol   Reacts as if ethanol. *1302*

Bradykinin

Serum   Increase   Physiological
Teprotide   Transient effect with i.v. or subcutaneous drug in one renal transplant patient. *3689*

Bromide

Serum   Increase   Analytical
Halothane   Peak value of from 0.65 mmol/L to 2.25 mmol/L 48 to 72 h after anesthesia in 16 patients. *3597*
Hemolysis   May cause color in protein-free filtrate. *0579*

Serum   Increase   Physiological
Bromides   Concentration measurable with therapeutic doses. *2427*
Bromisovalum   Metabolite (concentration may exceed 25 meq/L). *1343*
Calcium Bromogalactogluconate   In patient taking 12 g drug daily for 3 mo serum concentration of 18.7 mmol/L. *2867*
Carbromal   Metabolite concentration may exceed 25 meq/L. *1343*

Urine   No Effect   Analytical
Acetaminophen   No effect on method of Frings. *1204*
Amitriptyline   No effect on method of Frings. *1204*
Amobarbital   No effect on method of Frings. *1204*
Amphetamine   No effect on method of Frings. *1204*
Benztropine   No effect on method of Frings. *1204*
Butabarbital   No effect on method of Frings. *1204*
Carisoprodol   No effect on method of Frings. *1204*
Chloral Hydrate   No effect on method of Frings. *1204*
Chlordiazepoxide   No effect on method of Frings. *1204*
Chloroquine   No effect on method of Frings. *1204*
Chlorpheniramine   No effect on method of Frings. *1204*
Chlorpromazine   No effect on method of Frings. *1204*
Clorazepate   No effect on method of Frings. *1204*
Cocaine   No effect on method of Frings. *1204*
Codeine   No effect on method of Frings. *1204*
Diazepam   No effect on method of Frings. *1204*
Diethylpropion   No effect on method of Frings. *1204*
Diphenhydramine   No effect on method of Frings. *1204*
Doxepin   No effect on method of Frings. *1204*
Ephedrine   No effect on method of Frings. *1204*
Ethchlorvynol   No effect on method of Frings. *1204*

Ethinamate   No effect on method of Frings. *1204*
Fluphenazine   No effect on method of Frings. *1204*
Flurazepam   No effect on method of Frings. *1204*
Glutethimide   No effect on method of Frings. *1204*
Glycopyrrolate   No effect on method of Frings. *1204*
Haloperidol   No effect on method of Frings. *1204*
Hydromorphone   No effect on method of Frings. *1204*
Imipramine   No effect on method of Frings. *1204*
Mebutamate   No effect on method of Frings. *1204*
Meperidine   No effect on method of Frings. *1204*
Mephenesin   No effect on method of Frings. *1204*
Meprobamate   No effect on method of Frings. *1204*
Mescaline   No effect on method of Frings. *1204*
Methadone   No effect on method of Frings. *1204*
Methamphetamine   No effect on method of Frings. *1204*
Methapyrilene   No effect on method of Frings. *1204*
Methaqualone   No effect on method of Frings. *1204*
Methdilazine   No effect on method of Frings. *1204*
Methylphenidate   No effect on method of Frings. *1204*
Methyprylon   No effect on method of Frings. *1204*
Morphine   No effect on method of Frings. *1204*
Oxazepam   No effect on method of Frings. *1204*
Oxycodone   No effect on method of Frings. *1204*
Oxymetazoline   No effect on method of Frings. *1204*
Pentazocine   No effect on method of Frings. *1204*
Pentobarbital   No effect on method of Frings. *1204*
Perphenazine   No effect on method of Frings. *1204*
Phenmetrazine   No effect on method of Frings. *1204*
Phenobarbital   No effect on method of Frings. *1204*
Phenylephrine   No effect on method of Frings. *1204*
Phenylpropanolamine   No effect on method of Frings. *1204*
Phenytoin   No effect on method of Frings. *1204*
Primaquine   No effect on method of Frings. *1204*
Promazine   No effect on method of Frings. *1204*
Promethazine   No effect on method of Frings. *1204*
Propoxyphene   No effect on method of Frings. *1204*
Pyrilamine   No effect on method of Frings. *1204*
Pyrimethamine   No effect on method of Frings. *1204*
Quinidine   No effect on method of Frings. *1204*
Quinine   No effect on method of Frings. *1204*
Salicylate   No effect on method of Frings. *1204*
Scopolamine   No effect on method of Frings. *1204*
Secobarbital   No effect on method of Frings. *1204*
Sodium Bromide   No effect on method of Frings. *1204*
Strychnine   No effect on method of Frings. *1204*
Sulfanilamide   No effect on method of Frings. *1204*
Tetracycline   No effect on method of Frings. *1204*
Thiobarbituric Acid   No effect on method of Frings. *1204*
Thiopropazate   No effect on method of Frings. *1204*
Thioridazine   No effect on method of Frings. *1204*
Thiothixene   No effect on method of Frings. *1204*
Trifluoperazine   No effect on method of Frings. *1204*
Triflupromazine   No effect on method of Frings. *1204*
Trihexyphenidyl   No effect on method of Frings. *1204*
Trimethobenzamide   No effect on method of Frings. *1204*
Tybamate   No effect on method of Frings. *1204*

Urine   Increase   Physiological
Thiazides   In cases of bromide intoxication. *1343*

Urine   Decrease   Physiological
Sleep   Negligible excretion during night after administration. *3896*

BSP Retention

Serum   No Effect   Physiological

Aging   No change with increasing age noted. *1976*

Chenodeoxycholic Acid   No effect noticed even if ICD increased. *0274*

Niacin   Usual unless liver previously damaged. *1622*

Serum   Increase   Analytical

Ethoxazene   Increases absorbancy in test, falsely high result. *1488*

Hemolysis   Absorbance of Hemoglobin varies with pH (affects some procedures). *1237*

Heparin   Color intensity increased in serum, wavelength shifted. *0721*

Phenazopyridine   Increased spectral absorbancy in colorimetric reading. *3879*

Phenolphthalein   Color development on alkalinization of sample. *3879*

Phenolsulfonphthalein   May increase colorimetric reading. *1488*

Riboflavin   If given i.v. affects method of Gaebler. *2381*

Vitamin Preparations   If given i.v. and contain riboflavin and B vitamins. *2381*

Serum   Increase   Physiological

Acetohexamide   May cause intrahepatic cholestatic jaundice. *0652*

Acetophenazine   Cholestasis with intrahepatic obstruction. *2313*

Allopurinol   Associated with reversible hepatotoxicity. *3226*

Aminosalicylic Acid   Reversible cholestasis caused by drug. *2707*

Amitriptyline   Due to transient cholestasis. *2313*

Amobarbital   Probably nonspecific effect on cells. *3699*

Amodiaquine   Reported hepatotoxicity. *2313*

Amphotericin B   Due to hepatocellular dysfunction. *2313*

Anabolic Steroids   Cholestatic syndrome. *1022*

Androgens   May cause cholestatic syndrome. *1022*

Antifungal Agents   Hepatotoxic effect may impede clearance. *1237*

Antimony Compounds   Hepatotoxic effect. *2220*

Arsenicals   Hepatotoxic effect (cholestasis/cholangiolitis). *2220*

Asparaginase   Hepatotoxicity (usually mild). *1519*   In 57% children and 84% adults reported in various studies. *2406*

Aspidium   Probable inhibition of uptake by liver. *2427*

Aza Drugs   May cause hepatotoxicity. *1237*

Barbiturates   May increase retention if given within 24 h. *1342*

Benziodarone   May cause hepatic toxicity. *2313*

Bethanechol   Impaired excretion due to spasm of sphincter. *0652*

Bismuth Salts   Hepatotoxicity. *2313*

Bunamiodyl   Competes for hepatocellular protein binding sites. *0327*

Capreomycin   Transient increase in 5% subjects. *2427*

Carbamazepine   Cholestatic and hepatocellular damage. *1488*

Carbarsone   Toxic hepatitis may occur. *2575*

Carbenoxolone   Reversible hepatotoxic effect. *2458*

Carbutamide   Hepatotoxicity. *2313*

Carphenazine   Probable cholestatic effect (reversible). *2313*

Chloral Hydrate   Probably nonspecific effect on cells. *1500*

Chloramphenicol   Hepatotoxicity. *2220*

Chlordiazepoxide   Infrequent cholestatic effect. *1713*

Chloroform   Hepatotoxic effect with necrosis. *2313*

Chlorothiazide   May cause cholestatic jaundice. *2313*

Chlorpromazine   Induces transient cholestatic hepatitis (1 case). *0900*

Chlorpropamide   Infrequent cholestatic effect. *1713*

Chlorprothixene   May cause hepatotoxicity (reversible). *2313*

Chlortetracycline   Hepatotoxic effect with centrolobular necrosis. *2425*

Chlorzoxazone   May cause hepatotoxicity. *2313*

Choleretics   Reported effect (?competition for excretion). *1022*

Cholestyramine   Few cases reported, probably not hepatotoxicity. *1680*

Cholinergics   Impaired excretion due to spasm of sphincter. *0652*

Cinchophen   May cause hepatotoxicity (viral-hepatitis like). *2313*

Clindamycin   Probable hepatotoxic effect. *1083*

Clofibrate   Intrahepatic cholestasis reported. *1022*

Clomiphene   May cause hepatotoxicity. *1237*

Clonidine   Single probable case of toxic hepatitis. *1277*

Colchicine   May cause hepatic toxicity. *2313*

Copper   Increased with hemolytic crisis. *0385*

Coumarin   May cause hepatotoxicity. *3835*

Cyclophosphamide   May cause hepatotoxicity. *2313*

Cyclopropane   May cause hepatotoxicity (lasts several days). *2313*

Cycloserine   May cause hepatotoxicity. *2313*

Danazol   Develops in over one third of patients taking drug. *3906*

Dehydrocholic Acid   Hepatic uptake or biliary excretion impaired. *1488*

Desipramine   Rare transient cholestasis. *2313*

Ectylurea   May cause cholestasis with cholangiolitis. *2313*

Endotoxin   In almost 50% patients with induced fever. *0375*

Erythromycin   Intrahepatic cholestatic jaundice. *0085*

Estradiol   Cholestatic effect in almost all subjects. *2508*

Estrogens   Cholestatic effect in most cases (dose related). *2508*

Estrone   Occurs in almost all cases. *2508*

Ethambutol   Few cases reported. *2427*

Ether   Hepatic disturbance (transient). *2313*

Ethinyl Estradiol   Depresses hepatic secretory transport maximum. *1487*

Ethionamide   Hepatotoxic effect. *2313*

Ethotoin   Probable idiosyncratic hepatitis. *2313*

Ethoxazene   Hepatotoxic effect. *2425*

Etiocholanolone   In almost 50% patients with induced fever. *0375*

Fat Emulsions   Long infusion may impair liver function. *0120*

Fever   In almost 50% patients with induced fever. *0375*

Flavaspidic Acid   Competition for hepatocellular binding sites. *2621*

Florantyrone   Hepatotoxic effect. *1022*

Floxuridine   Manifestation of toxicity. *1680*

Fluoxymesterone   ?hepatotoxic, cholestatic effect. *1237*

Fluphenazine   Hypersensitivity response. *2313*

Flurazepam   May cause hepatotoxicity. *3016*

Fluroxene   Potentially hepatotoxic. *0071*

Glycopyrrolate   Hepatotoxicity. *2313*

Gold   Hepatotoxicity with centrolobular necrosis. *2313*

Guanoxan   May cause hepatic toxicity. *1210*

Haloperidol   May cause hepatocellular changes. *2313*

Halothane   Reversible depression of liver function. *3294*

Hycanthone   Induces form of acute toxic hepatitis. *1079*

Hydrazine Derivatives   May cause hepatotoxicity. *2313*

Hydroflumethiazide   May cause intrahepatic cholestasis. *1680*

Hydroxyacetamide   Hepatotoxicity. *2313*

Hydroxyurea   Abnormal retention reported. *1680*

Ibufenac   Hepatotoxic effect. *2313*

Icterogenin   Causes intrahepatic cholestasis. *0153*

Imipramine   May cause cholestatic jaundice. *2447*

Indandione Derivatives   May cause hepatotoxicity. *2313*

Indomethacin   Cytotoxic and cholestatic liver damage. *1908*

Iodipamide   Reported without evidence of liver damage. *2220*

Iodoalphionic Acid   Effect reported without abnormal liver function. *2220*

Iopanoic Acid   Competes for hepatocellular protein binding sites. *3200*

Iproniazid   May cause cholestatic jaundice. *2313*

Isocarboxazid   Intrahepatic cholestasis reported. *0553*

Isoniazid   Intrahepatic cholestatic jaundice. *0085*

Kanamycin   May cause hepatotoxicity. *1022*

Levodopa   Mild and transient effect. *0824*

Lipomul®   Part of fat-overloading syndrome. *2427*

MAO Inhibitors   Intrahepatic cholestasis with possible cell damage. *1237*

Meals   Capacity of liver to eliminate BSP decreased. *0579*

Mechlorethamine   Hepatotoxic effect. *0248*

Mepazine   Cholestatic effect. *1713*

Meperidine   Due to spasm of sphincter of Oddi. *0553*

Mephenytoin   Hepatotoxicity. *2313*

Mercaptopurine   Hepatotoxicity (centrolobular necrosis). *2313*

Mestranol   Depresses hepatic secretory transport maximum. *1487*

Metahexamide   Hepatotoxicity (viral hepatitis type). *2313*

Metaxalone   Hepatotoxic effect. *1022*

Methacholine   Impairs excretion by spasm of sphincter of Oddi. *0652*

Methadone   Hepatotoxic effect or spasm of sphincter of Oddi. *3879*

Methandriol   Intrahepatic cholestatic jaundice. *1343*

Methandrostenolone   Cholestatic phenomenon. *0757* Hepatotoxic effect (common). *2313*

Methimazole   May affect liver function (cholestasis). *2313*

Methotrexate   Hepatotoxicity, may be post-necrotic cirrhosis. *1576*

Methoxsalen   Hepatotoxic effect. *2313*

Methoxyflurane   Hepatotoxicity. *2313*

Methyldopa   Result of hepatocellular damage (cholestasis also). *3879*

Methyltestosterone   Transport and conjugation of BSP impaired. *0153*

Methylthiouracil   Hepatotoxic effect. *2313*

Morphine   Abnormal liver function tests reported. *3879*

Muscular Exercise   During infusion (related to extent of activity). *1171*

Nafcillin   Reported effect (?hepatotoxicity). *0071*

Narcotics   Impaired excretion — spasm of sphincter of Oddi. *0652*

Niacin   Liver function impairment, ?competition for conjugated. *2726*

Nialamide   Hepatotoxic/cholestatic syndromes. *1343*

Nitrofurans   Hepatotoxicity. *2313*

Nitrofurantoin   Intrahepatic cholestatic jaundice. *1045*

Norethandrolone   Transport and conjugation of BSP impaired. *0153*

Norethandrostenolone   Impaired hepatic uptake and excretion. *0652*

Norethindrone   Cholestatic phenomenon (occurs in 20%). *3332*

Norethynodrel   Cholestatic phenomenon. *0757*

Novobiocin   Competition for excretion, may be actual damage. *0350*

Oleandomycin   May cause intrahepatic cholestasis. *1237*

Opium Alkaloids   Impaired excretion spasm of sphincter of Oddi. *0652*

Oral Contraceptives   Occurs in 40%, estrogen depresses secretory mechanism. *2086*

Organophosphorus Insecticides   Hepatotoxic effect. *0405*

Oxacillin   Cholestatic jaundice reported in one case. *1022*

Oxandrolone   Slight increase in one child (other liver function tests normal). *1750*

Oxazepam   Possible hepatotoxicity. *2313*

Oxymetholone   Cholestatic effect. *1713* Cholestatic hepatitis developed in two patients. *3921*

Oxyphenbutazone   Possible hepatotoxicity. *2313*

Oxyphenisatin   May cause hypersensitivity reaction. *2970*

Paraldehyde   Possible hepatotoxicity. *2313*

Paramethadione   Possible hepatotoxicity. *2313*

Pargyline   Possible hepatotoxicity (reversible). *2313*

Phenacemide   May alter liver function in about 2% cases. *2313*

Phenazopyridine   Hepatic damage reported in one case. *1652*

Phenelzine   Hepatotoxic/cholestatic syndromes. *1343*

Phenindione   May modify liver function (cholestasis). *2313*

Pheniprazine   May affect liver function. *2313*

Phenothiazines   May cause cholestatic hepatitis (in up to 4% patients). *2313*

Phenprocoumon   Mild effect but recurrently in 2 patients having repeated exposure to drug. *3347*

Phenylbutazone   May cause cholestasis. *2313*

Phenytoin   Due to hypersensitivity or intrahepatic cholestasis. *1488*

Phosphorus   Hepatotoxic effect with necrosis. *2313*

Polythiazide   May affect liver function. *2313*

Pregnancy   May be moderately abnormal in last month. *0987*

Probenecid   Hepatotoxic effect (centrolobular necrosis). *3879*

Procainamide   Reported case of hepatic toxicity. *1488*

Prochlorperazine   Cholestatic effect. *0071*

Progesterone   Hepatic toxicity occasional occurrence. *2498*

Promazine   May cause cholestasis. *2313*

Promethazine   May cause cholestasis. *2313*

Propoxyphene   Cholestatic effect. *1954*

Propylthiouracil   Hepatotoxic effect (centrolobular necrosis). *2313*

Pyrazinamide   Hepatic toxicity (approximately 4%) (dose related). *3193*

Quinacrine   Hepatotoxic effect. *2313*

Quinethazone   May cause cholestatic jaundice. *2313*

Radiographic Agents   Compete for excretory mechanism. *0553*

Reducing Diet   Impaired hepatic perfusion (up to 40). *3217*

Rifampin   Inhibits hepatic excretion (probably). *1745* Initial increase of serum concentration due to reduced excretion in bile. *0014*

Spinal Anesthesia   Seen postoperatively, related to decreased blood flow. *1343*

Stanozolol   Due to intrahepatic cholestasis. *0757*

Streptokinase   Possible acute hypoxic or toxic liver damage. *3189*

Sulfadimethoxine   Reversible cholestasis. *1049*

Sulfamethizole   Reversible cholestasis. *1974*

Sulfamethoxazole   Reversible cholestasis. *2313*

Sulfamethoxypyridine   Reversible cholestasis. *2313*

Sulfisoxazole   May cause cholestasis. *1974*

Sulfonamides   May cause cholestasis. *2808*

Sulfones   May affect liver function. *2313*

Testosterone   May cause cholestasis. *2313*

Tetracycline   Cholestatic especially in pregnant women. *2808*

Tetraiodophthalein   Competitive uptake and excretion of dye. *1745*

Thiabendazole   Rare cholestasis. *2413*

Thiacetazone   May cause cholestasis. *0248*

Thiazides   Reduced plasma volume and hepatic blood flow. *1488*

Thioguanine   May cause cholestasis. *2313*

Thioridazine   Hepatotoxicity. *1857*

Thiosemicarbazones   May affect liver function. *2313*

Thiothixene   Hepatotoxic effect (reversible cholestasis). *2313*

Thorium Dioxide   May induce liver damage in half cases. *3147*

Tolazamide   Cholestasis may occur. *2313*

Tolbutamide   May cause cytotoxic liver damage or cholestasis. *0248*

Tranylcypromine   Hepatotoxic/cholestatic syndromes. *1343*

Trifluoperazine   Cholestatic effect. *1713*

Trimethadione   Hepatotoxicity with centrolobular necrosis. *2313*

Trioxsalen   May affect liver function. *2313*

Troleandomycin   In 50% individuals treated with 1 g/d for 3-4 weeks. Jaundice typically occurs after 2 weeks mixed in type with both cholestasis and mild hepatocytic necrosis. *2792*

Uracil Mustard   May affect liver function. *2313*

Serum   Decrease   Analytical

Phenazopyridine   High absorbancy in blank if acidified. *0583*

BSP Retention (continued)

Serum   Decrease   Analytical (continued)
Storage of Sample   Decreased up to 4% at room temperature for 16 h. 3873

Serum   Decrease   Physiological
Barbiturates   Increases conjugation with glutathione. 1343
Phenobarbital   Increased clearance by liver in newborns. 3915
Prednisone   Augmented response in cirrhotics when drug added to therapeutic regime. 0651
Proteinuria   False low due to loss of protein bound BSP. 0553

Butabarbital

Blood   Increase   Physiological
Butabarbital   600 mg orally produces concentration of 14 mg/L. 2348

Butanol Extractable Iodine (BEI)

Serum   No Effect   Analytical
Diatrizoic acid   No effect reported with Hypaque®. 0830

Serum   Increase   Physiological
Iodochlorhydroxyquin   Contains organically bound iodine. 2145
Oral Contraceptives   Due to increased thyroxine binding globulin. 0071

Butyric Acid

Urine   Increase   Physiological
Bacteriuria   Some organisms can produce in urine. 1483

C₁-Inhibitor

Serum   Increase   Physiological
Danazol   Approximate doubling in 5 patients with hereditary angioneurotic edema treated for up to 10 mo. 3219

Cadmium

Serum   Increase   Analytical
Sodium Salts   May interfere with atomic absorption measurements. 2155

Urine   Increase   Physiological
EDTA   If poisoning due to cadmium. 0071
Proteinuria   Especially if due to cadmium poisoning. 2155

Cadmium Flocculation

Serum   Increase   Physiological
Thorium Dioxide   May produce severe liver damage after years. 3147

Caffeine

Serum   No Effect   Physiological
Methoxsalen   Peak concentration and time to reach this not affected although mean elimination greatly increased. 2342

Serum   Increase   Physiological
Caffeine   24-hour average concentration correlated poorly with caffeine intake. 2125

Coca Cola™   Maximum (6.3 mg/dL) at 60-120 minutes after 36 fl oz. 2303
Coffee   Maximum (6.7 mg/dL) at 30 minutes after 2 cups. 2303
Tea   Maximum (7.5 mg/dL) at 30 minutes after 3 cups. 2303

Calcifediol

Serum   No Effect   Physiological
Ethanol   In 8 men who had abused alcohol for at least 10 y. 0347

Calcitonin

Serum   No Effect   Physiological
Parathyroid Hormone   In response to infusion of 150 units in 10 normal subjects. 0467

Plasma   No Effect   Physiological
Aluminum Hydroxide   With 15-18 g/d in 7 nondialyzed patients with chronic renal failure. 3533
Epinephrine   Following i.v. infusion for 1 h to produce concentration up to 945 pg/mL. 0401
Pentagastrin   Intravenous infusion had no effect in normals. 0856
Rifampin   No change observed in 8 healthy men taking drug for 2 weeks. 0479

Plasma   Increase   Physiological
Ethinyl Estradiol   Significant effect during and after in 7 healthy postmenopausal women treated for 12 weeks. 3467
Pentagastrin   Slight effect seen in hypocalcemics if given i.v. 0856
Pregnancy   Progressive inc, with decrease of C cell reserves. 1960

Plasma   Decrease   Physiological
Aging   Generally higher in infants at birth. 3127
Phenytoin   In epileptic children in comparison with controls. 2030

Calcium

Serum   No Effect   Analytical
Acetaminophen   At 5 times therapeutic concentration on o-CPC methods on SMAC®, Abbott-VP, aca Hitachi® 705 and KDA and arsenazo III method on Ektachem® 400. 2138   At acute overdose concentration (10 mg/dL) on SMAC® method. 3719   At concentration of 1500 mg/L had no effect on cresolphthalein method. 3393
Acetazolamide   No effect at 12 mg/dL on SMA 12/60 method. 2636   At concentration of 1,000 mg/L had no effect on mercurimetric method. 3393
Acetylsalicylic Acid   On cresolphthalein complexone method on SMA II at physiological concentration. 1787   At 5 times upper limit of therapeutic range on method on SMAC®, Abbott-VP, aca, Ektachem®, Hitachi® 705 and KDA. 2138   At concentration of 2,000 mg/L had no effect on cresolphthalein method. 3393
Albumin   method of Morin at 8 mg/dL. 2485
Allopurinol   At concentration of 136 mg/L had no effect on cresolphthalein method. 3393
Aminopyrine   At concentration of 125 mg/L had no effect on cresolphthalein method. 3393
Aminosalicylic Acid   At 5 times upper limit of therapeutic range on methods on SMAC®, Abbott-VP, aca, Ektachem®, Hitachi® 705 and KDA. 2138   At concentration of 460 mg/L had no effect on cresolphthalein method. 3393
Amitriptyline   At acute overdose concentration (2.5 mg/dL) on SMAC® method. 3719   At concentration of 25 mg/L had no effect on cresolphthalein method. 3393
Amobarbital   At concentration of 150 mg/L had no effect on cresolphthalein method. 3393
Amphotericin B   At concentration of 96 mg/L had no effect on cresolphthalein method. 3393
Ampicillin   At concentration of 40 mg/L had no effect on cresolphthalein method. 3393

Antibiotics No effect on fluorescence of calcein. *2395*

Ascorbic Acid At concentration of 2,000 mg/L had no effect on cresolphthalein method. *3393*

Barbital At acute overdose concentration (20 mg/dL) on Technicon® SMAC® method. *3719* At concentration of 500 mg/L had no effect as measured by cresolphthalein method. *3393*

Barbiturates No effect on fluorescence of calcein. *2395*

Bethanechol No effect at 0.09 mg/dL on SMA 12/60 method. *2636*

Bilirubin With method of Morin at 10 mg/dL. *2485*

Boric Acid At concentration of 50 mg/L had no effect as measured by cresolphthalein method. *3393*

Bromazepam At 5 times therapeutic concentration on methods of SMAC®, Abbott-VP, aca, Ektachem® 400, Hitachi® 705 and KDA. *2138*

Butabarbital At concentration of 100 mg/L had no effect as measured by cresolphthalein method. *3393*

Caffeine At acute overdose concentration (10 mg/dL) on Technicon® SMAC® method. *3719* At concentration of 194 mg/L had no effect on cresolphthalein method. *3393*

Carbenicillin At 5 times upper limit of therapeutic range on methods on SMAC®, Abbott-VP, aca, Ektachem®, Hitachi® 705 and KDA. *2138* At concentration of 15,000 mg/L had no effect on cresolphthalein method. *3393*

Cefazolin At concentration of 110 mg/L had no effect on cresolphthalein method. *3393*

Cefoxitin At 5 times upper limit of therapeutic range on methods on SMAC®, Abbott-VP, aca, Ektachem®, Hitachi® 705 and KDA. *2138*

Cephalothin At 5 times upper limit of therapeutic range on methods on SMAC®, Abbott-VP, aca, Ektachem®, Hitachi® 705 and KDA. *2138* At concentration of 1,000 mg/L had no effect on cresolphthalein method. *3393*

Chlordiazepoxide At 5 times upper limit of therapeutic range on methods on SMAC®, Abbott-VP, aca, Ektachem®, Hitachi® 705 and KDA. *2138* At acute overdose concentration (5 mg/dL) on Technicon® SMAC® method. *3719* At concentration of 50 mg/L had no effect on cresolphthalein method. *3393*

Chlormezanone At concentration of 100 mg/L had no effect on cresolphthalein method. *3393*

Chlorphenesin At concentration of 150 mg/L had no effect on cresolphthalein method. *3393*

Chlorpheniramine At acute overdose concentration (20 mg/dL) on Technicon® SMAC® method. *3719* At concentration of 200 mg/L had no effect on cresolphthalein method. *3393*

Chlorpromazine At concentration of 3 mg/L had no effect on cresolphthalein method. *3393*

Chlorprothixene At concentration of 1 mg/L had no effect on cresolphthalein method. *3393*

Cimetidine At concentration of 1 mg/L had no effect on cresolphthalein method. *3393*

Cloxacillin At concentration of 5 mg/L had no effect on cresolphthalein method. *3393*

Cocaine At acute overdose concentration (2.5 mg/dL) on Technicon® SMAC® method. *3719* At concentration of 25 mg/L had no effect on cresolphthalein method. *3393*

Codeine At acute overdose concentration (2.0 mg/dL) on Technicon® SMAC® method. *3719* At concentration of 20 mg/L had no effect on cresolphthalein method. *3393*

Contact With Clot If plastic disc used to separate for 48 h. *1863*

Cyclosporine At concentration of 41.5 mmol/L (50 µg/L) on methods on Technicon® SMAC® II and Hitachi® 705. *1123*

Cysteine At 5 times upper limit of therapeutic range on methods SMAC®, Abbott-VP, aca, Ektachem®, Hitachi® 705 and KDA. *2138*

Cystine At 5 times upper limit of therapeutic range on methods on SMAC®, Abbott-VP, aca, Ektachem®, Hitachi® 705 and KDA. *2138*

Desipramine At concentration of 8 mg/L had no effect on cresolphthalein method. *3393*

Deslanoside No effect on SMA 12/60 method at 0.06 mg/dL. *2636*

Diazepam At 5 times upper limit of therapeutic range on methods on SMAC®, Abbott-VP, aca, Ektachem®, Hitachi® 705 and KDA. *2138* At acute overdose concentration (2.5 mg/dL) on Technicon® SMAC® method. *3719* At concentration of 25 mg/L had no effect on cresolphthalein method. *3393*

Diclofenac On cresolphthalein complex one method on SMA II at physiological concentration. *1787* At concentration of 23 mg/L had no effect on cresolphthalein method. *3393*

Diethylaminoethanol At concentration of 30 mg/L had no effect on cresolphthalein method. *3393*

Diethylpropion At acute overdose concentration (10 mg/dL) on Technicon® SMAC® method. *3719* At concentration of 100 mg/L had no effect on cresolphthalein method. *3393*

Digitoxin No effect at 21 mg/dL on SMA 12/60 method. *2636* At concentration of 6 mg/L had no effect on cresolphthalein method. *3393*

Digoxin No effect at 0.04 mg/dL on SMA 12/60 method. *2636* At concentration of 9 mg/L had no effect on cresolphthalein method. *3393*

Dimethadione At concentration of 2,000 mg/L had no effect on cresolphthalein method. *3393*

N-Dimethyldiazepam At concentration of 6 mg/L had no effect on cresolphthalein method. *3393*

Diphenhydramine At acute overdose concentration (20 mg/dL) on Technicon® SMAC® method. *3719* At concentration of 200 mg/L had no effect on cresolphthalein method. *3393*

Disopyramide At concentration of 4 mg/L had no effect on cresolphthalein method. *3393*

Disulfiram At concentration of 120 mg/L had no effect on cresolphthalein method. *3393*

Disulphine Blue By continuous flow cresolphthalein complexone reaction. *1464*

EDTA No effect on flame photometric methods. *2220*

Epinephrine At concentration of 183 mg/L had no effect on cresolphthalein method. *3393*

Erythromycin At concentration of 73 mg/L had no effect on cresolphthalein method. *3393*

Ethacrynic Acid At 2.5 mg/dL no effect on SMA 12/60 method. *2636*

Ethanol At acute overdose concentration (20 mg/dL) on Technicon® SMAC® methods. *3719*

Ethchlorvynol At acute overdose concentration (20 mg/dL) on Technicon® method. *3719* At concentration of 400 mg/L had no effect on cresolphthalein method. *3393*

Ethinamate At acute overdose concentration (20 mg/dL) on Technicon® SMAC® method. *3719* At concentration of 200 mg/L had no effect on cresolphthalein method. *3393*

Ethosuximide At concentration of 390 mg/L had no effect on cresolphthalein method. *3393*

Fluorouracil At concentration of 130 mg/L had no effect on cresolphthalein method. *3393*

Fluosol-DA On SMA IIC at concentration of 50%. By 18% on Kodak Ektachem® 400 at concentration of 50%. *2518* On Du Pont aca III at concentration of 50%. *2518*

Flurazepam At 5 times upper limit of therapeutic range on methods on SMAC®, Abbott-VP, aca, Ektachem®, Hitachi® 705 and KDA. *2138* At acute overdose concentration (2.5 mg/dL) on Technicon® SMAC® method. *3719* At concentration of 25 mg/L had no effect on cresolphthalein method. *3393*

Furosemide No effect at 1.4 mg/dL on SMA 12/60 method. *2636* At concentration of 4 mg/L had no effect on cresolphthalein method. *3393*

Gentamicin At concentration of 150 mg/L had no effect on cresolphthalein method. *3393*

Glutethimide At acute overdose concentration (5 mg/dL) on Technicon® SMAC® method. *3719* At concentration of 50 mg/L had no effect on cresolphthalein method. *3393*

Hemoglobin method of Morin at 500 mg/dL. *2485*

Hydrocortisone No effect on fluorescence of calcein. *2395*

Ibuprofen At 5 times upper limit of therapeutic range on methods on SMAC®, Abbott-VP, aca, Ektachem®, Hitachi® 705 and KDA. *2138* On cresolphthalein complexone method on SMA II at physiological concentration. *1787* At concentration of 200 mg/L had no effect on cresolphthalein method. *3393*

Calcium (continued)

Serum   No Effect   Analytical (continued)

Indomethacin   On cresolphthalein complexone method on SMA II at physiological concentration. *1787*   At concentration of 13 mg/L had no effect on cresolphthalein method. *3393*

Insulin   At concentration of 3 mg/L had no effect on cresolphthalein method. *3393*

Iproniazid   At concentration of 40 mg/L had no effect on cresolphthalein method. *3393*

Isoniazid   At concentration of 100 mg/L had no effect on cresolphthalein method. *3393*

Isoprenaline   At concentration of 211 mg/L had no effect on cresolphthalein method. *3393*

Ketoprofen   On cresolphthalein complexone method on SMA II at physiological concentration. *1787*   At concentration of 60 mg/L had no effect on cresolphthalein method. *3393*

Levodopa   At concentration of 197 mg/L had no effect on cresolphthalein method. *3393*

Lidocaine   No effect SMA 12/60 method with 3.5 mg/dL. *2636*

Lipemia   In patient specimens measured on SMAC® in comparison with specimens following ultra-centrifugation. *2466*

Lorazepam   At concentration of 0.05 mg/L had no effect on cresolphthalein method. *3393*

Magnesium   method of Morin at 10 mg/dL. *2485*   On HPE method of American Monitor. *0140*

Mannitol   No effect on SMA 12/60 method at 445 mg/dL. *2636*

Meperidine   At acute overdose concentration (5 mg/dL) on Technicon® SMAC® method. *3719*   At concentration of 60 mg/L had no effect on cresolphthalein method. *3393*

Mephenesin   At concentration of 100 mg/L had no effect on cresolphthalein method. *3393*

Meprobamate   At acute overdose concentration (20 mg/dL) on Technicon® SMAC® method. *3719*   At concentration of 200 mg/L had no effect on cresolphthalein method. *3393*

Meralluride   No effect at 0.07 mL/dL on SMA 12/60 method. *2636*

Mercaptomerin   At concentration of 58 mg/L had no effect on cresolphthalein method. *3393*

Mercaptopurine   At concentration of 152 mg/L had no effect on cresolphthalein method. *3393*

Mesoridazine   At acute overdose concentration (20 mg/dL) on Technicon® SMAC® method. *3719*

Methadone   At acute overdose concentration (2.5 mg/dL) on Technicon® SMAC® method. *3719*   At concentration of 25 mg/L had no effect on cresolphthalein method. *3393*

Methamphetamine   At concentration of 2 mg/L had no effect on cresolphthalein method. *3393*

Methanol   At acute overdose concentration (20 mg/dL) on Technicon® SMAC® method. *3719*

Methapyrilene   At concentration of 13 mg/L had no effect on cresolphthalein method. *3393*

Methaqualone   At acute overdose concentration (2.5 mg/dL) on Technicon® SMAC® method. *3719*   At concentration of 25 mg/L had no effect on cresolphthalein method. *3393*

Methicillin   At 5 times upper limit of therapeutic range on methods on SMAC®, Abbott-VP, aca, *2138*   At concentration of 900 mg/L had no effect on cresolphthalein method. *3393*

Methohexital   At concentration of 50 mg/L had no effect on cresolphthalein method. *3393*

Methotrexate   At 5 times upper limit of therapeutic range on method on SMAC®, Abbott-VP, aca, Ektachem®, Hitachi® 705 and KDA. *2138*

Methotrimeprazine   At concentration of 1 mg/L had no effect on cresolphthalein method. *3393*

Methsuximide   At concentration of 40 mg/L had no effect on cresolphthalein method. *3393*

Methyldopa   At concentration of 80 mg/L had no effect on cresolphthalein method. *3393*

Methylphenidate   At acute overdose concentration (20 mg/dL) on Technicon® SMAC® method. *3719*   At concentration of 200 mg/L had no effect on cresolphthalein method. *3393*

Methylphenobarbital   At concentration of 150 mg/L had no effect on cresolphthalein method. *3393*

Methyprylon   At acute overdose concentration (10 mg/dL) on Technicon® SMAC® method. *3719*   At concentration of 100 mg/L had no effect on cresolphthalein method. *3393*

Metoprolol   At concentration of 0.34 mg/L had no effect on cresolphthalein method. *3393*

Morphine   At acute overdose concentration (20 mg/dL) on Technicon® SMAC® method. *3719*   At concentration of 200 mg/L had no effect on cresolphthalein method. *3393*

Nafcillin   At concentration of 50 mg/L had no effect on cresolphthalein method. *3393*

Naproxen   At 5 times upper limit of therapeutic range on methods on SMAC®, Abbott-VP, aca, Ektachem®, Hitachi® 705 and KDA. *2138*

Nitrofurantoin   On routine methods in use on SMAC®, Ektachem®, Abbott-VP, Cobas-Bio, aca, Hitachi® 705, KDA at 5 times normal upper therapeutic concentration. *2138*   At concentration of 4 mg/L had no effect on cresolphthalein method. *3393*

Nortriptyline   At acute overdose concentration (20 mg/dL) on Technicon® SMAC® method. *3719*   At concentration of 200 mg/L had no effect on cresolphthalein method. *3393*

Orphenadrine   At concentration of 9 mg/L had no effect on cresolphthalein method. *3393*

Ouabain   At 0.02 mg/dL no effect on SMA 12/60 method. *2636*

Oxacillin   At concentration of 40 mg/L had no effect on cresolphthalein method. *3393*

Oxazepam   At concentration of 1 mg/L had no effect on cresolphthalein method. *3393*

Oxytetracycline   At concentration of 50 mg/L had no effect on cresolphthalein method. *3393*

Palmitic Acid   On atomic absorption procedures at 4 mmol/L With dialysis procedures at 2 mmol/L. *2673*

Papaverine   At concentration of 10 mg/L had no effect on cresolphthalein method. *3393*

Paraldehyde   At concentration of 2,000 mg/L had no effect on cresolphthalein method. *3393*

Penicillin G   At 5 times upper limit of therapeutic range on methods on SMAC®, Abbott-VP, aca, Ektachem®, Hitachi® 705 and KDA. *2138*   At concentration of 2,000 mg/L had no effect as measured by cresolphthalein method. *3393*

Pentobarbital   At acute overdose concentration (20 mg/dL) on Technicon® SMAC® method. *3719*   At concentration of 340 mg/L had no effect on cresolphthalein method. *3393*

Perphenazine   At acute overdose concentration (20 mg/dL) on Technicon® SMAC® method. *3719*   At concentration of 200 mg/L had no effect on cresolphthalein method. *3393*

Phencyclidine   At concentration of 6 mg/L had no effect on cresolphthalein method. *3393*

Phenelzine   At concentration of 136 mg/L had no effect on cresolphthalein method. *3393*

Phenobarbital   At acute overdose concentration (20 mg/dL) on Technicon® SMAC® method. *3719*   At concentration of 250 mg/L had no effect on cresolphthalein method. *3393*

Phensuximide   At concentration of 120 mg/L had no effect on cresolphthalein method. *3393*

Phenylbutazone   At concentration of 750 mg/L had no effect on cresolphthalein method. *3393*

Phenylephrine   At concentration of 4 mg/L had no effect on cresolphthalein method. *3393*

Phenytoin   No effect at 1.8 mg/dL on SMA 12/60 method. *2636* At acute overdose concentration (20 mg/dL) on Technicon® SMAC® method. *3719*   At concentration of 240 mg/L had no effect on cresolphthalein method. *3393*

Phosphates   On HPE method of American Monitor. *0140* method of Morin at 20 mg/dL. *2485*

Plasma   No significant difference from serum on SMA 12/60 method. *2228*

Probenecid   At concentration of 1300 mg/L had no effect on cresolphthalein method. *3393*

Procainamide   No effect on SMA 12/60 method at 35 mg/dL. *2636*   At concentration of 50 mg/L had no effect on cresolphthalein method. *3393*

Procaine   At concentration of 2 mg/L had no effect on cresolphthalein method. *3393*

Prochlorperazine   At concentration of 1 mg/L had no effect on cresolphthalein method. *3393*

Promethazine At acute overdose concentration (20 mg/dL) on Technicon® SMAC® method. *3719* At concentration of 200 mg/L had no effect on cresolphthalein method. *3393*

Propoxyphene At acute overdose concentration (2.5 mg/dL) on Technicon® SMAC® method. *3719* At concentration of 25 mg/L had no effect on cresolphthalein method. *3393*

Propranolol No effect at 0.1 mg/dL on SMA 12/60 method. *2636* At concentration of 0.2 mg/L had no effect on cresolphthalein method. *3393*

Propylthiouracil At concentration of 170 mg/L had no effect on cresolphthalein method. *3393*

Pyribenzamine® At acute overdose concentration (20 mg/dL) on Technicon® SMAC® methods. *3719*

Quinidine At 26 mg/dL no effect on SMA 12/60 method. *2636* At acute overdose concentration (20 mg/dL) Technicon® SMAC® method. *3719* At concentration of 210 mg/L had no effect on cresolphthalein method. *3393*

Quinine At acute overdose concentration (1.5 mg/dL) on Technicon® SMAC® method. *3719* At concentration of 30 mg/L had no effect on cresolphthalein method. *3393*

Reserpine No effect at 0.02 mg/dL on SMA 12/60 method. *2636* At concentration of 61 mg/L had no effect on cresolphthalein method. *3393*

Rifampin At 5 times upper limit of therapeutic range on method on SMAC®, Abbott-VP, aca, Ektachem®, Hitachi® 705 and KDA. *2138*

Salicylate At concentration of 500 mg/L had no effect on cresolphthalein method. *3393*

Secobarbital At acute overdose concentration (20 mg/dL) on Technicon® SMAC® method. *3719* At concentration of 200 mg/L had no effect on cresolphthalein method. *3393*

Standing of Sample If on polystyrene beads 24 h at 25°. *2624*

Stearic Acid On atomic absorption procedures at 4 mmol/L With dialysis procedures at 2 mmol/L. *2673*

Storage of Sample Room temperature 8 h, 4° 1 d, frozen 1 y. *3873*

Streptomycin At concentration of 58 mg/L had no effect on cresolphthalein method. *3393*

Strychnine At concentration of 12 mg/L had no effect on cresolphthalein method. *3393*

Sulfadiazine At concentration of 1500 mg/L had no effect on cresolphthalein method. *3393*

Sulfaguanidine At concentration of 500 mg/L had no effect on cresolphthalein method. *3393*

Sulfanilamide At concentration of 1,000 mg/L had no effect on cresolphthalein method. *3393*

Sulfathiazole At concentration of 255 mg/L had no effect on cresolphthalein method. *3393*

Sulforidazine At concentration of 200 mg/L had no effect on cresolphthalein method. *3393*

Tetracycline At 5 times upper limit of therapeutic range on methods on SMAC®, Abbott-VP, aca, Ektachem®, Hitachi® 705 and KDA. *2138* At concentration of 300 mg/L had no effect on cresolphthalein method. *3393*

Theobromine At concentration of 2,000 mg/L had no effect on cresolphthalein method. *3393*

Theophylline At 5 times upper limit of therapeutic range on methods on SMAC®, Abbott-VP, aca, Ektachem®, Hitachi® 705 and KDA. *2138* At concentration of 120 mg/L had no effect on cresolphthalein method. *3393*

Thiamazole At concentration of 114 mg/L had no effect on cresolphthalein method. *3393*

Thiopental At acute overdose concentration (20 mg/dL) on Technicon® SMAC® method. *3719* At concentration of 200 mg/L had no effect on cresolphthalein method. *3393*

Timolol At concentration of 0.01 mg/L had no effect on cresolphthalein method. *3393*

Tolbutamide At concentration of 270 mg/L had no effect on cresolphthalein method. *3393*

Transport of Specimen By pneumatic tube if specimen protected. *3217*

Tribromethanol At concentration of 90 mg/L had no effect on cresolphthalein method. *3393*

Trichlorethanol At concentration of 1,000 mg/L had no effect on cresolphthalein method. *3393*

Trifluoperazine At concentration of 1 mg/L had no effect on cresolphthalein method. *3393*

Tripelennamine At concentration of 200 mg/L had no effect on cresolphthalein method. *3393*

Vitamin Preparations No effect at expected concentration with SMA 12/60 procedure. *2637*

Warfarin At concentration of 100 mg/L had no effect on cresolphthalein method. *3393*

Serum   No Effect   Physiological

Acetylsalicylic Acid No effect seen in patients undergoing treatment. *1787*

Aging No significant change from age 6-20 y. *0887*

Aluminum Hydroxide With 15-18 g/d in 7 nondialyzed patients with chronic renal failure. *3533* Observed in one patient secondary to ingestion of large amounts of compound for long time. *1313*

Amiloride No effect seen in 13 men treated for 8 weeks. *0656*

Bed Rest May be no change in total although ionized increased. *1539*

Bucindol No effects noted in 8 patients following several weeks on treatment. *1283*

Calcium Carbonate No rise after 4 g observed. *3302*

Cisplatin No effect noted in 11 children treated for 0.1 to 3.8 y. *3286*

Diazepam No effect when given in normal therapeutic amounts to elderly. *3925*

Diclofenac No effect seen in patients undergoing treatment. *1787*

Dihydroxycholecalciferol No effect observed. *0469*

Diurnal Variation Slightly higher midday, lowest at night (not significant). *3851*

Ethinyl Estradiol No effect during and after in 7 healthy postmenopausal women treated for 12 weeks. *3467*

Furosemide In 8 normal subjects associated with secondary hyperparathyroidism. *1214*

Ibuprofen No effect seen in patients undergoing treatment. *1787*

Indomethacin No effect seen in patients undergoing treatment. *1787*

Ketoconazole In 9 healthy men given up to 1200 mg/d for 1 week. *1300*

Ketoprofen No effect seen in patients undergoing treatment. *1787*

Levonorgestrel No significant difference over 3 y between implant recipients and IUD users. *0763* In a group of women using levonorgestrel covered rods versus controls using copper IUD. *0899*

Lithium No significant effect observed with chronic treatment. *0488*

Meals Not significantly affected by nonstandardized meals. *3455*

Medroxyprogesterone No difference observed compared with controls. *3329*

Nifedipine No effect of 80 mg/d for 2 d. *3694*

Nitrazepam No effect when given in normal therapeutic amounts to elderly. *3925*

Pentagastrin Intravenous infusion had no effect in normals. *0856*

Pindolol No significant effect even with 6 weeks treatment. *3279*

Prednisone Although given for 1 to 50 mo. *3840*

Propranolol In 340 patients given drug for 10 weeks to reduce diastolic BP to less than 90 mm Hg. *3707*

Saccharated Iron Oxide In 9 individuals given 40 mg/d i.v. daily for up to 42 d. *2662*

Sucralfate No effect in uremic patients or when previous aluminum hydroxide regime replaced. *2139*

Theophylline No effect in response to i.v. infusion. *3935*

Tolbutamide Slight but not significant reduction, associated with increased bone organic matrix turnover. *3458*

Triamterene No significant effect observed. *3745*

Water No discernible difference 2 h after drinking 1 glass. *2123*

Serum   Increase   Analytical

Aminophenol At 0.1 mmol/L affects SMA 12/60 method. *3335* At concentrations above 10 mg/L raised concentration as measured by cresolphthalein method. *3393*

Calcium   (continued)

Serum   Increase   Analytical   (continued)

Bilirubin   If high may confuse EDTA titration endpoint. *1563*

Calcium Salts   Possible contamination of distilled water. *3879*

Cefotaxime   Statistically significant effect of metabolite on Parallel method. *0201*

Chlorpropamide   At concentrations above 7.5 mg/L (normal therapeutic concentration 150 mg/L) raised concentration as measured by cresolphthalein method. *3393*

Copper   Interferes with EDTA titration procedures. *1563*

Cork Stoppers   May be considerable increase with storage. *3217*

Filter Paper   Ordinary paper contains high concentration. *0579*

Heparin   If calcium salt used may affect result. *3873*

Hydralazine   At 1 mmol/L has slight effect on SMA 12/60 method. *3335*   At concentrations above 150 mg/L (normal therapeutic concentration 2.3 mg/L) raised concentration as measured by cresolphthalein method. *3393*

Iron Salts   Interfere with direct EDTA titration. *1563*

Lipemia   Causes turbidity in Ferro-Ham procedure. *1563*

Magnesium   May be up to 15% of total precipitate. *0140*

Magnesium Salts   Measured as Calcium in some EDTA procedures. *1563*

Oxalate   May be coprecipitation of Na, K, Mg and protein. *1563*

Potassium   Affects flame photometry if poor instrument. *1563*

Sodium Salts   Affect flame photometry if poor instrument. *1563*

Standing of Sample   Marked top to bottom if frozen and unmixed. *2674*

Storage of Sample   Significant increase after 12 d at 4°. *3862*

Sulfobromophthalein   Affects methylthymol method of Gindler/King. *1286*   May interfere with colorimetric procedures. *2425*

Zinc   May affect titration with EDTA procedures. *0583*

Serum   Increase   Physiological

Alkaline Antacids   Theoretical possibility with absorption of Calcium salts. *3369*

Aluminum Hydroxide   Small increase to 2.37-2.40 mmol/L during administration of 100 mL gel/d for for 28 d. *0564*   Mild effect with osteomalacia due to inhibition of entry into bone. *1054*

Anabolic Steroids   Positive effect on calcium retention. *0070*

Androgens   Positive effect on calcium retention. *0070*

Antacids   May occur if calcium containing preparations. *2427*

Bed Rest   Increase of approximately 0.1 meq/L. *1688*

Bendrofluazide   Due to increased concentration of protein due to temporary Na depletion. *2727*

Blindness   Slight effect (average 0.03 meq/L) higher. *1637*

Calcium Gluconate   Marked effect noted in newborns. *1550*

Calcium Salts   Direct effect of increased gut absorption. *0071*

Cellulose Phosphate   Without changing calcium balance. *2711*

Chlorothiazide   Impaired excretion with chronic administration. *3505*   Due to increased tubular reabsorption but usually associated with mild primary hyperparathyroidism. *1897*

Chlorthalidone   Occurs in normal and hyperparathyroid. *2122*

Dienestrol   May occur with extended high dosage. *1680*

Diethylstilbestrol   Rapid increase in 24 h in patients with breast cancer. *3519*

Dihydrotachysterol   Weak antirachitic activity. *2313*   Delayed elimination of drug from plasma in hypothyroid state. *2061*

Diurnal Variation   Constant during day but maximum 8 pm, minimum 3 am. *1829*

Dromostanolone   Usually if osteolytic metastases. *0071*

Erect Posture   Standing for 15 minutes causes increase of 6.7%. *3476*

Ergocalciferol   Enhances absorption from gastrointestinal tract. *0652*

Estradiol   May occur with high doses. *1680*

Estrogens   Positive effect on calcium retention. *1343*

Ethanol   0.6% increase in heavy versus occasional drinkers. *3273*

Fluoxymesterone   May occur in immobilized patients or if cancer. *1678*

Hydrochlorothiazide   Impaired excretion (probably also released from bone). *3505*   Increase by 0.3 mg/dL in 343 patients with hypertension given drug for 10 weeks. *3707*   Increased renal tubular reabsorption, usually associated with mild primary hyperparathyroidism. *1054*

Lithium   Mean effect in 130 manic depressives versus controls (about 2% increase). *2053*   Increased above normal in about 13% patients, Reversible. *3118*   Probably due to unmasking of primary hyperparathyroidism. *1054*   Increase observed for next few hours when drug given at 10 p.m. *2401*   8% higher than in controls in 12 patients taking drug for 2 to 13 y. *1175*   In 7 of 97 patients not necessarily attributable to drug. *0699*

Magnesium Sulfate   In one woman given i.v. drug because of Crohn's disease on calcium supplements. *2552*

Meals   Affected by meals (most marked in women). *3455*   Effect of metabolic alkalosis. *3234*

Menopause   Change of 0.06 meq/L from 5th to 6th decade. *3830*

Methandrostenolone   May occur spontaneously, but especially if breast cancer. *1680*

Methyclothiazide   ?related to decreased excretion. *1680*

Methyltestosterone   May occur especially in women on therapy for breast cancer. *1680*

Muscular Exercise   Observed effect with 12 minutes cycle-ergometer. *1231*

Nandrolone   May occur in women with neoplasm of breast. *2681*

Nisoldipine   Change from average of 2.4 to 2.5 mmol/L in 14 mild to moderately hypertensive noninsulin dependent diabetic patients. *2653*

Oral Contraceptives   Increased ingestion would increase concentration. *0652*

Oxymetholone   May occur spontaneously or if carcinoma of breast. *1679*

Parathyroid Extract   Increased calcium mobilization from bone. *0652*

Polystyrene Sulfonate   Increased administered calcium (if in Calcium form). *3505*

Polythiazide   May be increased by up to 0.35 meq/L. *2427*

Progesterone   Probable effect with remission of metastases. *2313*

β-Propiolactone   Increase by 0.1 mmol/L when plasma from BPL treated blood compared with plasma in 25 specimens. *0217*

Propranolol   Increase by 0.34 mg/dL in approximately 120 patients with essential hypertension treated for 1 y. *3708*

Season   Significant effect, sine pattern, maximal in December and May. *1172*

Secretin   Increased by 0.6 meq/L in normals, ?cause. *0470*

Tamoxifen   Especially in people with pre-existing hypercalcemia and with widespread skeletal metastases. *1542*   In patients with previous hypercalcemia and in patients with metastatic bone disease. Probable overall incidence less than 0.1%. *2749*   In 2 patients in whom calcium rose to above 18.0 mg/dL: reverted to normal once drug withdrawn. *2069*

Testolactone   Probable effect during remission of metastases. *0071*

Testosterone   Reported with therapy of breast cancer. *2582*

Theophylline   11 of 60 patients with theophylline toxicity but reverted to normal when drug withdrawn. Also in normals with 400 mg bid for 5 d. *2383*

Thiazides   May increase up to 0.25 meq/L. *2745*

Tourniquet   Increase by 0.11 meq/L over 3 minutes. *1704*   Due to relative increase of plasma proteins. *0884*

Trichlormethiazide   Impaired excretion. *3505*

Vitamin A   Observed with acute intoxication. *3867*

Vitamin D   Effect of increased gastrointestinal tract absorption. *3879*

Serum   Decrease   Analytical

Aspirin   Depresses fluorescence of calcein method. *2395*

Bilirubin   Quenches fluorescence in method of Rushton. *3091*   If calcein titration used. *0055*

Contact With Clot   If prolonged facilitates passage of H_2O from cells. *1563*   Absorption by container. *1237*

EDTA   If determined by oxalate or other precipitation technicians. *3505*

Fluorides May precipitate calcium (forms insoluble salt). *1563*

Hemoglobin Interference with calcein titration. *0055*

Hemolysis Interferes with calcein fluorescence methods. *2395*

Heparin Interferes with EDTA and fluorometric methods. *3879*

Liposol Changed from 90.0 to 84.1 mg/L in 26 randomly selected patient sera, before and after addition on Beckman Astra methods. *2054*

Oxalate Precipitation of calcium oxalate (may be incomplete). *1563* Unless lanthanum added may decrease atomic absorption. *0583*

Palmitic Acid If titrimetric methods used at concentration over 1.5 mmol/L. *2673*

Phosphates Inhibit emission in some flame methods Compete with EDTA for Calcium (slight effect some methods). *1563*

Protein With atomic absorption unless added to standards. *0583*

Stearic Acid If titrimetric methods used at concentration over 1.5 mmol/L. *2673*

Sulfadiazine Depresses fluorescence of calcein method. *2395*

Sulfadimethoxine Depresses fluorescence of calcein method. *2395*

Sulfates With atomic absorption. *0583* Theoretical inhibition of emission in flame methods. *1563*

Sulfisoxazole Depresses fluorescence of calcein method. *2395*

Sulfonamides Depress fluorescence of calcein method. *2395*

Serum Decrease Physiological

Acetazolamide Defective Calcium and PO_4 reabsorption can be induced. *0652*

Albuterol Significant dose-related effect after i.v. infusion of therapeutic doses in 4 healthy male volunteers. *2807*

Ambulation But not to normal in 4 d from high of bed rest. *1539*

Anticonvulsants Found in 30% children on prolonged therapy. *1695*

Asparaginase Observed in 60% (?due to hypoalbuminemia). *2657*

Aspirin Progressive effect during three days but also noted over longer term study at beginning, eventually reverted towards normal. Effect statistically significant. *3076*

Bran 0.15 meq/L with 38 g/d for 5 weeks. *1540*

Calcitonin Decrease of greater than 0.5 meq/L observed. *1439*

Carbamazepine Observed in 3 of 21 (14%) patients whose epilepsy was treated only with drug (mean duration of treatment 40 mo). *1625* In about 20 epileptic patients treated for 2 y. *0873* Osteomalacia observed in 3 of 31 patients given average of 758 mg/d for average of 20.5 mo. *2635*

Carbenoxolone Aldosterone like effect of drug. *2458*

Chlorothiazide Initial response to administration. *0105*

Cisplatin Observed in 5.8% of 17 patients receiving drug, associated with low serum magnesium increased urine excretion and decreased intestinal absorption of Mg. *3468* Marked effect in 15 patients treated with drug. *3964* In 3rd and 4th week of treatment of one girl with malignant ovarian germ cell tumor probably related to renal tubular dysfunction. *2064* Calcium regulation impaired by magnesium deficit. *0368*

Citrates Complexes calcium. *2313*

Corticosteroids Antagonizes action of vitamin D and parathyroid. *0071*

Cortisone If elevated due to sarcoidosis or vitamin D. *0071*

Diatrizoic acid ?due to small amount of EDTA in medium. *2043*

Diuretics Excretion enhanced by most diuretics. *0105*

Dromostanolone May occur with regression. *0071*

EDTA Complexes calcium by chelation. *2005*

Enflurane Slight decrease noted postoperatively. *0999*

Ethanol Secondary hyperparathyroidism. Decreased intestinal absorption and reduced bone mass. *1246* At lower limit of normal in 8 men who had abused alcohol for at least 10 y. *0347* Concentration in chronic alcoholics significantly less than in controls. *0367*

Ethinyl Estradiol Significant fall, associated with decreased albumin. *3923* In post-menopausal women with 50 μg/d for 14 d. *3474* Significant effect in postmenopausal women: less marked in perimenopausal women. *2311*

Ethylene Glycol Characteristic finding, often marked, due to deposition of oxalate and interference with normal homeostasis. *3325*

Etidronate In majority of patients with hypercalcemia of malignancy when drug given i.v. for 3-5 d. *1521*

Fluorides Combines with calcium, accretion in bone crystals. *0010*

Furosemide If increase due to dihyrotachysterol, also if normal. *2944* Diuretic action (different effect hydrochlorothiazide). *0468*

Gallium Nitrate In two patients with hypercalcemia due to parathyroid carcinoma. *3764*

Gastrin Mean decrease of 0.7 mg/dL after 5 h. *2370*

Gentamicin Effect observed in 2 patients following larger doses of gentamicin. Associated with massive urinary loss of magnesium and potassium. *0224*

Glucagon Occurs 10 to 20 minutes after administration if hypoglycemia. *2207*

Glucocorticoids Effective if hypercalcemia due to sarcoid, vitamin D. *0071*

Glucose May fall by 0.5 meq/L during glucose tolerance test. *2207*

Hydrochlorothiazide Increase by 0.13 mg/dL in approximately 175 patients with essential hypertension treated for 1 y. *3708*

Insulin Reported effect. *0355*

Laxatives Excessive use may have effect. *2313*

Light In elderly in absence of sun. *1615*

Magnesium Salts Competes with calcium for gastrointestinal tract absorption. *2005*

Mercurial Diuretics Enhances excretion reducing serum concentration. *0105*

Mercury Compounds Induced with chronic laxative ingestion. *2108*

Mestranol Increased sensitivity to calcitonin in postmenopausal women. *0037*

Methicillin Reported effect (?mechanism). *2313*

Methoxyflurane Slight lowering reported in one case. *2876*

Mithramycin Inhibition of bone resorption of calcium. *3349*

Oral Contraceptives Seen in osteoporosis, ?due to fall in albumin. *3923* Isolated case of acute pancreatitis reported. *2519*

Paramethadione Theoretical effect of type of drug. *1452*

Phenobarbital Metabolic effect with chronic therapy (osteomalacia). *1452* Slight but significant reduction compared with untreated elderly at doses other than for treatment of epilepsy. *3925*

Phenytoin Metabolic effect with chronic therapy. *1452* With or without osteoporosis in chronic users. *2042* May produce osteomalacia with protracted treatment. *0121* In 7% of 60 epileptics treated for 10 y or more. *2493* In 7% of 60 epileptics treated for 10 y or more. *2493* In about 20 patients treated for 2 y. *0873*

Phosphates Transient effect due to colloid retention in liver. *0639*

Phytate Falls initially and remains low. *2947*

Polystyrene Sulfonate Exchanged for potassium, increased fecal loss. *2313*

Prednisone If elevation due to sarcoidosis or vitamin D. *0071*

Pregnancy Due to decreased albumin (loss of protein bound calcium). *3329*

Probucol Mechanism obscure (effect slight). *0806*

Recumbency Decreased by about 6.7 % with 15 minutes lying down. *3476*

Rimeterol Dose related significant change in 4 healthy men given therapeutic i.v. dose. *2807*

Saline Effect of isotonic solution if hypercalcemia. *0071*

Sodium Phytate Decreased gastrointestinal tract absorption. *0071*

Sodium Sulfate If given i.v. may cause hypocalcemia. *0071*

Streptozocin To normal in few cases of following administration. *2087*

Sulfates Brief hypocalcemic effect if infused i.v. *2313*

Tetracycline Observed in pregnant women. *2427*

Trimethadione Theoretical effect of type of drug. *1452*

Viomycin May cause tetany with electrolyte imbalance. *0071*

Calcium *(continued)*

Urine   No Effect   Analytical
Thymol   No effect on flame photometric methods. *2981*

Urine   No Effect   Physiological
Citrates   Insignificant change with sodium citrate in 5 patients with uric acid lithiasis. *3116*

Theophylline   In normals given 400 mg bid for 5 d. *2383*

Urine   Increase   Analytical
Calcium Salts   Possible contamination of distilled water. *3879*

Magnesium   Interferes with cresolphthalein procedures. *0140*

Oxalate   May be coprecipitation of Na, K, Mg. *1563*

Standing of Sample   Marked top to bottom if frozen and unmixed. *2674*

Urine   Increase   Physiological
Acetazolamide   Inhibits tubular reabsorption. *2745*

Aluminum Hydroxide   Observed in one patient secondary to ingestion of large amounts of compound for long time. *1313*

Amiloride   Observed in normals if given for 7 d. *3745*

Ammonium Chloride   During day of administration and following day. *2092*

Asparaginase   Atypical response but observed in some. *2657*

Azosemide   Tended to increase but not marked. *2047*

Bed Rest   Up to 50 meq/d in healthy males. *1688*

Bendrofluazide   Initial diuretic response. *0105*

Bumetanide   After 1 mg maximum within 2 h over by 4 h. *0822*

Cadmium   Occasionally observed with proximal tubular dysfunction. *2217*

Calcitonin   Acts independently of parathyroid. *1439*

Calcium Salts   Increased excretion with large doses. *1343*

Chlorothiazide   Initial diuretic response. *0105*

Cholestyramine   Binds heavy metals. *1022*

Corticosteroids   Metabolic effect. *2220*

Corticotropin   Average or large doses promote excretion. *1680*

Dexamethasone   Metabolic effect. *1680*

Dihydrotachysterol   Hypercalciuric effect. *1022*

Dihydroxycholecalciferol   30-100% increase within 2 d up to 2.7 μg/d. *0469*

Dimercaprol   Effective but less good than edetic acid. *0071*

Diuretics   Excretion enhanced by most diuretics. *0105*

Diurnal Variation   Excretion maximum at night. *2222*

EDTA   Excretion of chelate. *0071*

Ergocalciferol   As result of increased absorption. *0652*

Ethacrynic Acid   Impaired reabsorption. *2745*

Furosemide   Impaired reabsorption (initial effect only). *2745*   In 8 normal subjects associated with secondary hyperparathyroidism. *1214*   Slight increase with administration. *2047*   In 8 normal subjects associated with secondary hyperparathyroidism. *1214*

Glucocorticoids   Promote renal excretion. *0071*

Glucose   Nutrient induced augmentation ?mechanism. *2126*

Growth Hormone   Normal metabolic response. *2414*

Mannitol   Initial diuretic response. *0105*

Meralluride   Reabsorption impaired. *2745*

Mercaptomerin   Reabsorption impaired. *2745*

Mercurial Diuretics   Reabsorption impaired. *2745*

Mersalyl   Reabsorption impaired. *2745*

Metolazone   Maximum diuretic response of 0.4 μEq/min. *3358*

Mithramycin   Inhibition of bone resorption of calcium. *3349*

Nandrolone   Due to hypercalcemia. *2681*

Parathyroid Extract   Due to mobilization from bone. *1841*

Prednisolone   Effect of all corticosteroids. *0019*

Saline   With isotonic saline loading. *3439*   Hypertonic solution has calciuretic effect. *0071*

Season   Up to 50% in July more than January with same diet. *2481*

Sleep   Related to inactivity: max at night. *1139*

Sodium Salts   Observed with chloride salt supplementation in 5 men with essential hypertension. *2037*

Sodium Sulfate   Promotes excretion. *1343*

Spironolactone   Probably artifact as tablets each contain 40 mg calcium. *2863*

Sucrose   Nutrient induced augmentation ?mechanism. *2126*

Sulfates   Due to diuresis and $CaSO_4$ formation. *0071*

Torasemide   Similar effects to that of furosemide. *0919*

Triamcinolone   Typical action of all corticosteroids. *1680*

Triamterene   Impairs reabsorption. *2745*

Viomycin   Promotes urinary loss. *1237*

Vitamin D   Associated with hypercalcemia. *1022*

Urine   Decrease   Analytical
Alkaline Urine   Due to precipitated Calcium salts. *1841*

Oxalate   May be incomplete precipitation. *1563*

Phosphates   Inhibit emission in some flame methods Compete with EDTA for Calcium (more marked than serum). *1563*

Standing of Sample   Precipitation of calcium salts unless acidified. *3879*

Storage of Sample   Unless acidified. *3873*

Sulfates   Inhibit emission in some flame methods. *1563*

Urine   Decrease   Physiological
Aging   In women over 50 y more constant than younger. *2481*

Ambulation   Following several weeks of bed rest. *1688*

Angiotensin   Hemodynamic effect of drug. *2173*

Bendrofluazide   Impaired excretion may occur with thiazides. *2727*

Bicarbonates   Protects skeleton from reabsorption of PO_4. *0243*

Blindness   Significantly lower (average 92 meq/d) than normal. *1637*

Cellulose Phosphate   By impairing absorption decreases output. *2711*   As result of decreased gastrointestinal absorption. *2710*

Chloroquine   In 2 patients with sarcoidosis given 500 mg daily while receiving corticosteroid at same time. *2639*

Chlorothiazide   Impaired excretion with chronic administration. *3505*

Chlorthalidone   Decreased in normals and hyperparathyroid. *2122*

Citrates   From 154 to 99 mg/d on average with potassium citrate in 5 patients with uric acid lithiasis. *3116*   From 154 mg/d to 99 mg/d with 60 meq/d K salt: but not significantly to 139 mg/d from 154 mg/d with sodium salt. *3116*

Ethinyl Estradiol   Associated with decreased serum concentration. *3923*   In post-menopausal women with 50 μg/d for 14 d. *3474*   Significant effect in postmenopausal women: less marked in perimenopausal women. *2311*

Gallium Nitrate   In two patients with hypercalcemia due to parathyroid carcinoma. *3764*

Hydrochlorothiazide   Observed effect. *0451*

Lithium   ?affects Calcium dependent catecholamine system. *2275*   Picture of primary hyperparathyroidism observed in patients given treatment for long time, but mechanism not fully understood. *3118*   Reduced at night in lithium treated patients versus other psychiatric patients and healthy controls. *2401*

Mestranol   Increased sensitivity to calcitonin in postmenopausal women. *0037*

Methyclothiazide   Impairs excretion. *1680*   Significant reduction compared with placebo in normal volunteers. *2052*

Neomycin   Occurs when fecal calcium increased. *1077*

Oral Contraceptives   Occurs with fall in serum concentration. *3923*

Parathyroid Extract   Increases tubular reabsorption. *0071*

Phenytoin   In 60 epileptic patients treated for 10 y or more. *2493*

Phytate   Acts by inhibiting gastrointestinal absorption. *1540*

Polythiazide   Excretion decreased by up to 50%. *2427*

Sodium Phytate   Decreased gastrointestinal tract absorption. *1022*

Sodium Salts   Observed with citrate salt supplementation in 5 men with essential hypertension. *2037*

Starvation   Usually lower than controls. *2382*

Thiazides   By up to 50% (appear to have extra-renal effect). *2745*
Trichlormethiazide   Impaired excretion. *3505*

Duodenal Contents   Increase   Physiological
Calcium   Calcium infusion (10 mg/kg/h) causes increase x 2. *1314*
Ethanol   If given orally but decreased if i.v. *2504*

Duodenal Contents   Decrease   Physiological
Ethanol   If given intravenously only. *2504*

Feces   Increase   Physiological
Bed Rest   Up to 70 meq/d in healthy males. *1688*
Neomycin   Occurs independent of steatorrhea. *1077*
Phosphates   Increased by average of 32 meq/d. *1688*
Polystyrene Sulfonate   Causes impaired absorption. *1022*

Feces   Decrease   Physiological
Ambulation   After several weeks bed rest (gradual effect). *1688*
Chlorothiazide   Accentuates positive calcium balance. *2427*
Ethinyl Estradiol   Decrease less marked than in urine. *3923*
Fluorides   Reduces loss in bone disorders. *1343*
Hydrochlorothiazide   Accentuates positive calcium balance. *2427*
Polythiazide   Accentuates positive calcium balance. *2427*
Vitamin D   Due to excessive absorption. *0987*

Hair   Increase   Analytical
Hair Treatments   Marked with feminine deodorants. *1594*

Hair   Decrease   Analytical
Hair Treatments   Waving and setting lotions cause decrease. *1594*

Pancreatic Juice   Increase   Physiological
Acetazolamide   60 percent increase with infusion in normals. *0984*

Calcium, Ultrafiltrable

Serum   Increase   Physiological
Theophylline   Apparent log dose response correlation with drug concentration in individuals with toxicity. *2383*

Calculi

Urine   Increase   Physiological
Probenecid   May occur if little fluid drunk or excrete much. *0133*
Vitamin D   High incidence in people with self medication. *3559*

Capillary Fragility

Patient   No Effect   Physiological
Cyclopropane   Anesthesia has no effect. *1343*
Ether   Anesthesia has no effect. *1343*

Patient   Increase   Physiological
Rifampin   Reported in patient with macroglobulinemia. *1050*

Caproic Acid

Urine   Increase   Physiological
Bacteriuria   Some organisms can produce in urine. *1483*

Carbamazepine

Serum   No Effect   Analytical
Fluosol-DA   As measured on Du Pont aca III at concentration up to 50%. *2518*

Serum   No Effect   Physiological
Cimetidine   No significant effect on carbamazepine concentration when co-administered. *2147*
Roxithromycin   No effect on pharmacokinetics when added to stable therapeutic regime. *3115*

Serum   Increase   Physiological
Cimetidine   Increased concentration when co-administered, as elimination affected, although urinary excretion increased. *0801*
Danazol   Concentration increased almost 2 fold when drug added for treatment of fibrocystic breast disease. *3946*
Diltiazem   Presumed reduction in elimination of drug: drug dose reduced by 60% to reduce toxicity. *1002*
Erythromycin   Clearance reduced from 0.36 to 0.29 L/kg/d with 1 g/d for 5 d, probably due to effect on metabolism. *3893*   Interferes with liver microsomal metabolism of drug. *1541*
Isoniazid   Due to inhibition of liver enzymes by drug: caused intoxication in 10 of 13 epileptic patients. *3665*
Valproic Acid   Reported increase when valproic acid co-administered. *0119*

Serum   Decrease   Physiological
Carbamazepine   Plasma concentration may decrease by 20 to 30% after 2 to 3 weeks of treatment. *1085*
Charcoal   Reduced absorption due to adsorption. *3794*
Valproic Acid   Slight or no effect in 25 patients studied for 5 to 9 mo. *3829*

Carbohydrate, Protein Bound

Serum   Increase   Physiological
Diurnal Variation   Maximum observed about midday. *2222*

Carbonic Anhydrase

Red Blood Cells   Increase   Physiological
Oral Contraceptives   Measured as B isoenzyme (hormonal effect). *3181*
Pregnancy   Measured as B isoenzyme — progressive increase. *3181*

Red Blood Cells   Decrease   Physiological
Acetazolamide   Inhibitory effect observed. *2893*
Penicillamine   In rheumatoid arthritis patients when treated for 6 mo. Changes not related to activity of disease. *1794*   Related to transient changes in zinc metabolism. *1794*
Sulfonamides   Inhibitory action observed. *2893*
Sulthiame   Metabolic action of drug. *1389*

Carboxyhemoglobin

Blood   Increase   Physiological
Carbon Monoxide   Due to presence of carbon monoxide. *1302*
Driving   Most marked increase in day drivers vs night. *1821*
Smoking   Maximum after regular, minimum after extra-mild. *3094*   Significantly higher in smokers. *3112*   No effect in 7 healthy volunteers after 3 cigarettes. *2352*   Concentration correlated with extent of smoking. *3255*

Carcinoembryonic Antigen

Serum   No Effect   Analytical
Heparin   No activity detected in any heparin preparation when Abbott RIA used. *3903*

Serum   Increase   Analytical
Heparin   Activity present apparently in all heparin preparations when tested by Roche RIA. *3903*

Serum   Positive   Physiological
Smoking   False positives observed in heavy smokers. *0101*

Urine   No Effect   Analytical
Bacteriuria   No effect irrespective of number and type or organisms. *1459*
Glucose   No effect observed with glycosuria. *1459*
Hemoglobin   No effect observed with hematuria. *1459*
Proteinuria   No effect observed with moderate amount. *1459*

Carnitine

Serum   Decrease   Physiological
Valproic Acid   Reversibly reduced, inversely related to drug dose and plasma ammonia concentration. *2660*

Carotene

Serum   No Effect   Analytical
Storage of Sample   No effect at 4° for 6 d. *2734*

Serum   No Effect   Physiological
Anticonvulsants   In 146 epileptics with long-term treatment. *2007*
Pregnancy   No significant effect observed. *1222*

Serum   Increase   Analytical
Canthoxanthine   During administration and for 10 d after may affect certain analytical methods. *3023*
Carotenoid   May interfere with some analytical methods. *3023*

Serum   Increase   Physiological
Pregnancy   Increases with each trimester, maximum at delivery. *0245*
Protein   Rise in parallel with amino acids. *0987*

Serum   Decrease   Analytical
Citrates   Dilutional effect when compared against serum or heparin as anticoagulant. *3361*
Hemolysis   Significant decrease observed with hemolysis. *2734*

Serum   Decrease   Physiological
Aging   Significant correlations especially after 40 y in women increase to age 40. *2880*
Ethinyl Estradiol   Hormonal influence. *1221*
Kanamycin   May induce malabsorption with diarrhea. *1680*
Metformin   Probably associated with malabsorption. *3605*
Mineral Oil   Reduced absorption (may be 50% normal). *1343*   In people on diet high in carotene when drug given before meals. *3023*
Neomycin   Striking effect even if oral supplements. *1077*
Norethindrone   Hormonal influence. *1221*

Carotenoids, Total

Serum   No Effect   Physiological
Clofibrate   Insignificant change in about 26 subjects given 2 g daily. *2875*

Serum   Decrease   Analytical
Citrates   Dilutional effect when compared against serum or heparin as anticoagulant. *3361*

Serum   Decrease   Physiological
Colestipol   Increase by 30% with 30 g daily of drug. *2875*

Casts

Urine   No Effect   Physiological
Chlorthalidone   Normally but augments effects of acidifying agents. *1720*
Hydrochlorothiazide   Normally but augments effects of acidifying agents. *1720*

Urine   Increase   Physiological
Acetaminophen   Renal damage due to hemolysis/anuria. *1343*
Acetanilid   Due to hemolysis and renal damage. *1343*
Acetic Acid   May cause nephrotoxicity. *0279*
Acids   Renal damage following ingestion. *1302*
Alloxan   May cause nephrotoxicity. *0279*
Aloin   May cause nephrotoxicity. *0279*
Amphotericin B   Granular and hyaline casts with toxicity. *0071* Cylindruria may occur. *3567*
Ampicillin   Occurs as result of nephrotoxicity. *3542*
Amyl Nitrite   May cause nephrotoxicity. *0279*
Aniline   Nephrotoxic effect (chronic effect). *0279*
Antimony Compounds   Nephrotoxic effect. *2313*
Antipyrine   May cause nephrotoxicity. *0279*
Apiol   May cause nephrotoxicity. *0279*
Arsenicals   Nephrotoxic effect. *2313*
Aspidium   May cause nephrotoxicity. *0279*
Aspirin   Occurs with poisoning. *1302*
Bacitracin   Nephrotoxic effect (cylindruria may occur). *2313*
Bismuth Salts   Mainly granular due to nephrotoxicity. *2911*
Bromate   Due to renal damage. *1302*
Calcitonin   Possible nephrotoxic effect (increased granular casts). *1490*
Capreomycin   Nephrotoxic effect (usually granular casts). *2313*
Cephaloridine   Nephrotoxicity (hyaline and granular). *1019*
Chloroguanide   Nephrotoxic effect. *2313*
Colistimethate   Nephrotoxic effect. *2313*
Colistin   Cylindruria may occur with nephrotoxicity. *0971*
Cyclosporine   Usually cellular or granular characteristic feature of nephrotoxicity in most common form of renal damage to drug. *0175*
Dinitrophenol   Due to toxic nephritis. *1302*
EDTA   Nephrotoxic effect (especially Calcium EDTA). *2313*
Ethacrynic Acid   Hyaline casts without proteinuria (orosomucoid). *1720*
Furosemide   Hyaline casts without proteinuria (orosomucoid). *1720*
Gentamicin   Observed in association with drug associated nephrotoxicity. *3182*
Griseofulvin   Cylindruria may occur without renal insufficiency. *1343*
Hydrogen Sulfide   Occurs due to nephrotoxicity. *1302*
Ifosfamide   Large number of granular casts in all patients. *3675*
Indomethacin   Granular casts in one patient. *0506*
Iodine Containing Drugs   Hemorrhagic nephritis with toxic doses. *1302*
Isoniazid   Nephrotoxic effect. *2313*
Kanamycin   Nephrotoxic effect (cylindruria and granular casts). *0071*
Kerosene   With severe toxicity following ingestion. *1302*
Lead   Nephrotoxicity with poisoning (cylindruria). *1343*
Melarsonyl   Nephrotoxic effect. *0071*
Melarsoprol   Nephrotoxic effect. *2313*

Mercury Compounds May cause severe nephritis if absorbed. *0071*

Methicillin Nephrotoxic effect (cylindruria observed). *2313*

Methylbromide Due to tubular necrosis. *1302*

Muscular Exercise Hyaline, granular both increased with increased exercise. *1842*

Naphthalene Nephrotoxic effect. *1302*

Neomycin Nephrotoxic effect. *2313*

Paramethadione May have nephrotoxic effect. *2313*

Penicillin Renal cell and other types. *1343*

Phenols Nephrotoxicity with poisoning (usually RBC casts). *1343*

Phosphorus Nephrotoxic effect. *1302*

Polymyxin Cylindruria may occur with daily injection. *1343*

Radiographic Agents Nephrotoxic manifestation. *0652*

Rifampin Rare reported side effect. *1553*

Streptomycin Cylindruria may develop. *1343*

Sulfonamides Hemoglobin casts may cause renal failure. *1343*

Suramin Cylindruria reported. *0071*

Thallium Damage to renal tubular epithelium. *1302*

Tobramycin At dose of 4.5 mg/kg/d for 12 d in 90 patients nephrotoxicity observed in up to 39%, reversible in most. *0677* Manifestation of drug induced nephrotoxicity but tend to decrease as serum creatinine begins to climb. *3182*

Trimethadione May have nephrotoxic effect. *2313*

Vancomycin Occasional evidence of mild nephrotoxicity. *1571*

Viomycin Nephrotoxicity with cylindruria. *0071*

Catalase

Urine   Increase   Physiological

Acetrizoate Sustained increase after renal artery injection. *3538*

Bacteriuria Infected urines have activity. *2897*

Proteinuria Renal damage if no bacteriuria. *2897*

Radiographic Agents Transient increase with slight renal damage if renal artery. *3538*

Catecholamines

Plasma   No Effect   Analytical

Storage of Sample With $NaF-Na_2S_2O_3$ 1 mo frozen for ethylenediamine. *2285*

Plasma   No Effect   Physiological

Indomethacin In 12 patients with coronary artery disease, although systolic blood pressure increased and coronary blood flow decreased. *1158*

Ramipril In response to 10 mg and 20 mg on successive days in 9 patients with severe chronic congestive heart failure. *0764*

Plasma   Increase   Analytical

Ampicillin Concentrated solutions cause striking fluorescence. *0596*

Ascorbic Acid Concentrated solutions cause striking fluorescence. *0596*

Levodopa Measured as epinephrine/norepinephrine by ethylenediamine. *2285*

Methenamine Interference with fluorescence. *0127*

Methyldopa Interference with fluorometric methods. *3146* Reacts like catecholamines, may persist for days. *1343*

Niacin Occurs with large doses, interfering fluorescence. *1488*

Oxytetracycline Interferes with fluorometric methods. *3879*

Protamine Concentrated solutions cause striking fluorescence. *0596*

Quinidine Metabolite causes false positive. *0581*

Quinine Interference by metabolite, no effect *in vitro*. *0579*

Riboflavin Interferes with fluorometric technique. *2425*

Sulfadimidine Striking fluorescence. *0596*

Tea Concentrated solutions cause striking fluorescence. *0596*

Tetracycline Interferes with fluorometric techniques. *0127*

Uremia With fluorescence methods due to interfering substances. *2285*

Vitamin B Complex May be interference with fluorescence. *1488*

Vitamin K May react like epinephrine in Shaw test. *0516*

Plasma   Increase   Physiological

Ajmaline Release of stored norepinephrine. *2220*

Aminophylline Response to i.v. therapeutic dose. *0178*

Chlorpromazine Increased metabolism, decreased organ uptake of norepinephrine. *1237*

Clonidine Increase observed in response to exercise but overall response blunted. *3919*

Cocoa Catechols present in cocoa. *0596*

Cyclopropane Significant increase, may reduce blood to kidneys etc. *0071*

Diazoxide Increased release from tissue. *0116*

Epinephrine Presence of epinephrine. *0758*

Ethanol Slight increase following moderate doses. *0143*

Ether Response to stress. *2869*

Isoproterenol Due to inhalation — effect slight. *0758*

MAO Inhibitors Prevent deamination but not degradation by catechol-o-methyl transferase. *0127*

Methyldopa Pronounced increase observed, reduced by barbiturates. *1846*

Nitroglycerin Effect dosage dependent. *0127*

Perphenazine Increased metabolism, decreased organ uptake of norepinephrine. *1237*

Phenothiazines Increased metabolism, decreased organ uptake of norepinephrine. *0127*

Phentolamine ?due to releasing action or to altered metabolism. *1343*

Promethazine Increased metabolism, decreased organ uptake of norepinephrine. *0596*

Stress Normal metabolic response. *3527*

Plasma   Decrease   Analytical

Cyclopropane Falsely low with ethylene diamine method. *2285*

Ether Falsely low with ethylene diamine method. *2285*

Plasma   Decrease   Physiological

Captopril Mean decrease from 695 ng/L to 476 ng/L but not significant change in patients with heart failure. *1165*

Reserpine Observed normal response to therapy. *2427*

Urine   No Effect   Analytical

Amphetamine No effect observed. *0913*

Aspirin No effect on fluorometric Crout procedure. *0766*

Chlorothiazide No effect on fluorometric method of Crout. *0758*

Digitoxin No effect on fluorometric method of Crout. *0758*

Digoxin No effect observed. *0913*

Diphenhydramine No effect reported on method of Sandhu and Freed. *3135*

Ephedrine No effect observed. *0913*

Hydralazine No effect on fluorometric method of Crout. *0758*

Mecamylamine No effect on fluorometric method of Crout. *0758*

Meprobamate No effect observed. *0913*

Methyldopa Possibly no interference in method Sandhu and Freed. *3135* In volunteers with HPLC methods using electrochemical detection. *2521*

Pentolinium No effect on fluorometric method of Crout. *0758*

Phenoxybenzamine No effect on fluorometric method of Crout. *0758*

Phentolamine No effect on fluorometric method of Crout. *0758*

Propoxyphene No effect on fluorometric Crout procedure. *0766*

Quinidine If Hathaway procedure used. *1487*

Reserpine No effect on fluorometric method of Crout. *0758* No effect reported on method of Sandhu and Freed. *1205*

Storage of Sample No effect at room temperature, several days if acidified. *3873*

Tetracycline No effect reported on method of Sandhu and Freed. *1205*

Catecholamines (continued)

Urine No Effect Analytical (continued)
Triprolidine No effect on method of Sandhu and Freed. *3135*

Urine No Effect Physiological
Acetaminophen No effect with short term ingestion 2.6 g/d. *0766*
Amphetamine No effect observed. *0913*
Aspirin No effect with short term ingestion 2.6 g/d. *0766*
Cannabis Essentially unchanged with moderate dose. *1633*
Chlordiazepoxide No effect of short term ingestion of 30 mg/d. *0766*
Chlorothiazide No effect observed. *0913*
Clopamide Single case of diuretic associated myopathy. *1793*
Diazepam No effect with short term ingestion of 15 mg/d. *0766*
Diphenhydramine No effect with short term ingestion of 150 mg/d. *0766*
Ephedrine No effect observed. *0913*
Glucose No significant response to hyperglycemia. *3373*
Hydralazine No effect observed. *0913*
Mecamylamine No effect observed. *0913*
Meprobamate No effect observed. *0913*
Muscular Exercise After mild or moderate exercise. *3217*
Oral Contraceptives No effect observed with ingestion. *3217*
Phenobarbital No effect short term ingestion of 120 mg/d. *0766*
Phenoxybenzamine No effect observed. *0913*
Phentolamine No effect observed. *0913*

Urine Increase Analytical
Aspirin Interfering fluorescence in many procedures. *2220*
Bananas Catechols in bananas. *0759*
Chloral Hydrate Interferes with fluorometric procedures. *1488*
Chlorpromazine Affects fluorescent procedures. *3217*
Chlortetracycline Interference with fluorometric methods. *2570*
Demeclocycline Produces interfering fluorescence with analysis. *1961*
Dihydroxyphenylacetic Acid Reacts like epinephrine/norepinephrine with ethylenediamine. *2285*
Dopamine Reacts with ethylenediamine as epinephrine, norepinephrine. *2285*
Erythromycin Interferes with fluorometric methods. *0757*
Formaldehyde Interferes with fluorometric method. *1022* Interferes with procedures for analysis. *0757*
Hydralazine Interferes with some fluorometric methods. *2220*
Hydroquinone Reacts like epinephrine with Nelson-Shaw test. *0516*
Isoproterenol Interferes with resin and fluorometric procedures. *1205* Metabolite produces fluorescence in screening. *0583*
Labetalol With fluorometric method of Crout et al. *2430*
Methyldopa Has similar fluorescence. *1841* With trihydroxy-indole fluorometric procedures. *2521*
Niacin May produce interfering fluorescence. *1487*
Oxytetracycline Affects fluorescent procedures. *3217*
Quinidine Affects fluorescent procedures. *3217*
Riboflavin Large doses may produce similar fluorescence. *0583*
Tetracycline Produces interfering fluorescence. *1841*
Vitamin B Complex Large doses may produce similar fluorescence. *0583*

Urine Increase Physiological
Ethanol Produces stress like response. *0143*
Isoproterenol Observed for up to 4 d after drug stopped. *0913*
Muscular Exercise Following vigorous exercise (may be increased 7 fold). *0354*
Nicotine Active component of cigarettes etc: same effect. *1921*
Nitroglycerin 2 fold increase of catecholamines. *1487*
Pain Normal response to stress. *0279*
Prochlorperazine Increased metabolism, decreased organ uptake of norepinephrine. *3879*

Rauwolfia Release of stored norepinephrine. *2220*
Reserpine Release of stored norepinephrine. *2220*
Smoking Increased by 50% with cigar not inhaled Increased x 2 during 4 h with cigarettes inhaled. *1921*
Stress With physical or emotional stress. *3217*
Syrosingopine Release of stored norepinephrine. *2220*
Theophylline Normal response even when given orally. *0178*

Urine Decrease Analytical
Chlorpromazine Affect measurement by Pisano procedure. *3135*
Triamterene When measured with Bio-Rad laboratories catecholamines column test procedure, fluorescence of drug interferes with high blank production. *0246*

Urine Decrease Physiological
Clonidine Dose related effect (primary action of drug). *1627*
Decaborane Markedly decreases output. *3722*
Diurnal Variation Significantly less at night, greater in afternoon that in am. *3616*
Guanethidine Inhibits release of norepinephrine. *3722*
Methyldopa Depletion of tissue stores. *2220*
Ouabain Marked effect but mechanism unexplained. *3166*
Radiographic Agents Competes for excretion after i.v. pyelography. *0913*
Reserpine Decreased norepinephrine synthesis. *0758*
Sleep Significantly less at night. *3616*
Tosylate Bretylium Inhibits release of norepinephrine. *3722*

Catechol-O-Methyl Transferase

Red Blood Cells No Effect Physiological
Aging No difference between ages 16 and 89 y. *1401*
Sex Difference No difference between men and women. *1401*

Red Blood Cells Increase Physiological
Estrogens Observed effect of estrogen contraceptives. *1401*
Oral Contraceptives When high in estrogen content. *1401*

Test Conditions Decrease Physiological
Butylgallate Observed *in vivo* and *in vitro*. *2913*
L-Thyroxine Heavy doses inhibit *in vivo*. *1176*

Cefoperazone

Plasma No Effect Physiological
Probenecid Concomitant administration had no effect on concentration. *0592*

Cefsulodin

Plasma No Effect Physiological
Probenecid Concomitant administration had no effect on concentration. *0592*

Ceftazidime

Plasma No Effect Physiological
Probenecid Concomitant administration had no effect on concentration. *0592*

Ceftriaxone

Plasma No Effect Physiological
Probenecid Concomitant administration had no effect on concentration. *0592*

B-Cell Differentiation Factor

Monocytes   No Effect   Physiological
Methimazole   No effect on production when cells from normal individuals stimulated with mitogens. *3778*

Cells

Urine   Increase   Physiological
Acetaminophen   May be marked increase in renal tubular cells. *1343*   3-6 g for 5 d produces increase in tubular cells but less than with Aspirin or Phenacetin. *2866*
Antipyrine   Renal action of drug. *3824*
Aspirin   Tubular epithelial cells increased initially, may persist. *1343*   Tubular cells increased: occur even with therapeutic doses; effect can be quite marked. *2866*
Bismuth Salts   Mainly tubular due to nephrotoxicity. *2911*
Caffeine   May cause marked increase in renal tubular cells. *1343*
Calcitonin   Possible nephrotoxic effect (increased epithelial cells). *1490*
Castor Oil   Multinucleated giant cells in 50% patients. *2427*
Chloroguanide   May cause transient appearance of epithelial cells. *1343*
Cortisone   Increased number observed with long-term treatment. *2427*
Muscular Exercise   Epithelial cells increase with heavy exercise. *1842*
Thallium   Damage to renal tubular epithelium. *1302*

Cell Water

White Blood Cells   No Effect   Physiological
Diuretics   No change observed in normal individuals. *0997*

Cephalexin

Serum   Increase   Physiological
Probenecid   Reduces renal clearance. *1487*

Serum   Decrease   Physiological
Cholestyramine   Reduced absorption due to adsorption or steatorrhea. *3794*
Food   Absorption delayed when taken with food. *3024*

Cephalin

Serum   Increase   Physiological
Ethinyl Estradiol   Mirrors changes in lipoprotein pattern. *1431*

Cephalin Flocculation

Serum   Increase   Analytical
Ampicillin   Alters time of flocculation. *1966*
Bacterial Contamination   If reagent contaminated turbidity measured. *3879*
Glassware   If heavy metals or strong acids present. *3879*
Heat   As room temperature is increased. *1237*
Light   False positive occur after serum exposed to light. *0244*
Lipemia   Reported effect adding to turbidity. *1563*
Storage of Sample   Reported effect after 2 d at 4°. *1563*

Serum   Increase   Physiological
Acetohexamide   May cause intrahepatic cholestatic jaundice. *2313*
Acetophenazine   Cholestasis with intrahepatic obstruction. *2313*
Aminosalicylic Acid   Reversible cholestasis caused by drug. *2707*

Amitriptyline   Rare cases of transient cholestasis. *2313*
Amodiaquine   Reported hepatotoxicity. *2313*
Amphotericin B   Due to hepatocellular dysfunction. *2313*
Anabolic Steroids   Due to cholestatic syndrome. *2220*
Androgens   Cholestatic phenomenon. *2313*
Antimony Compounds   Hepatotoxic effect. *2220*
Arsenicals   Hepatotoxic effect (cholestasis/cholangiolitis). *2220*
Asparaginase   Hepatotoxicity. *1519*
Aspidium   May cause hepatic toxicity. *2313*
Benziodarone   May cause hepatic toxicity. *2313*
Bismuth Salts   Hepatotoxicity. *2313*
Carphenazine   Probable cholestatic effect (reversible). *2313*
Chloramphenicol   Hepatotoxicity. *2220*
Chlordiazepoxide   Infrequent cholestatic effect. *2313*
Chlormezanone   Occasional reversible cholestasis. *2313*
Chloroform   Hepatotoxic effect with necrosis. *2313*
Chlorothiazide   May cause cholestatic jaundice. *2313*
Chlorpropamide   Hepatocellular damage, affects protein synthesis. *2313*
Chlorprothixene   May cause hepatotoxicity (reversible). *2313*
Chlortetracycline   Hepatotoxic effect with centrolobular necrosis. *1237*
Chlorzoxazone   May cause hepatotoxicity. *2313*
Cinchophen   May cause hepatotoxicity (viral-hepatitis like). *2313*
Clofibrate   Hepatotoxic effect. *2313*
Colchicine   May have hepatotoxic effect. *2313*
Copper   Transient effect in poisoning. *2427*
Coumarin   May cause hepatotoxicity. *3835*
Cyclophosphamide   May cause hepatotoxicity. *2313*
Cyclopropane   May cause hepatotoxicity (lasts several days). *2313*
Cycloserine   May cause hepatotoxicity. *2313*
Desipramine   Rare transient cholestasis. *2313*
Dicloxacillin   Mild hepatic dysfunction noted. *0118*
Ectylurea   May cause cholestasis with cholangiolitis. *2313*
Erythromycin   Weak effect due to cholestasis. *1713*
Estradiol   Cholestatic effect. *2313*
Estrogens   May occur with cholestasis. *2220*
Estrone   May cause cholestasis. *2313*
Ether   Hepatic disturbance (transient). *2313*
Ethionamide   Hepatotoxic effect. *2313*
Ethosuximide   Rare hepatitis reported. *2313*
Ethotoin   Hepatotoxicity. *2313*
Ethoxazene   Hepatotoxic effect. *2313*
Florantyrone   Hepatotoxic effect. *2313*
Flucytosine   Reversible hepatotoxicity in 10%. *3443*
Fluphenazine   Hypersensitivity (maybe normal when other tests not). *2313*
Flurazepam   May cause hepatotoxicity. *3016*
Glycopyrrolate   Hepatotoxicity. *2313*
Gold   Hepatotoxicity with centrolobular necrosis. *2313*
Guanoxan   May cause hepatic toxicity. *2313*
Haloperidol   May cause hepatocellular changes. *2313*
Halothane   Reversible depression of liver function. *2313*
Hydrazine Derivatives   May cause hepatotoxicity. *2313*
Hydroxyacetamide   Hepatotoxicity. *2313*
Ibufenac   Hepatotoxic effect. *2313*
Icterogenin   Causes intrahepatic cholestasis. *0153*
Indandione Derivatives   May cause hepatotoxicity. *2313*
Indomethacin   Cytotoxic and cholestatic liver damage. *1908*
Iopanoic Acid   May cause hepatotoxicity. *2313*
Iproniazid   May cause cholestatic jaundice. *2313*
Isocarboxazid   Hepatotoxic effect. *2313*
Isoniazid   Cytotoxic hepatocellular damage. *0248*
Kanamycin   May cause hepatotoxicity. *2313*
Lincomycin   Hepatotoxic-cholestatic effect. *2313*
MAO Inhibitors   Hepatotoxic effect. *2313*

Cephalin Flocculation *(continued)*

Serum   Increase   Physiological *(continued)*

Mechlorethamine   May cause cytotoxic (hepatocellular) damage. *0248*

Mepazine   May alter liver function (cholestasis). *2313*

Mephenytoin   Hepatotoxicity. *2313*

Mercaptopurine   Hepatotoxicity (centrolobular necrosis). *2313*

Metahexamide   Hepatotoxicity (viral hepatitis type). *2313*

Metaxalone   Hepatotoxic effect. *2313*

Methandrostenolone   Hepatotoxic effect (common). *2313*

Methimazole   Hepatotoxic effect. *2313*

Methotrexate   May cause cytotoxic hepatocellular damage. *0248*

Methoxsalen   Hepatotoxic effect. *2313*

Methoxyflurane   Hepatotoxicity. *2313*

Methsuximide   Hepatic damage reported. *2313*

Methyldopa   Result of hepatocellular damage. *3879*

Methyltestosterone   Cholestatic effect (may not be affected). *1713*

Methylthiouracil   Hepatotoxic effect. *2313*

Nalidixic Acid   May cause cholestatic jaundice. *2808*

Niacin   May impair hepatic function. *2313*

Nitrofurans   Hepatotoxicity. *2313*

Nitrofurantoin   May be cholestatic jaundice. *2313*

Norethandrolone   Cholestasis produced without cholangiolitis. *3172*

Norethindrone   Intrahepatic cholestatic jaundice. *2313*

Norethynodrel   Cholestatic effect. *2313*

Novobiocin   Intrahepatic cholestasis may occur. *2313*

Oleandomycin   Hepatotoxic effect. *2313*

Oral Contraceptives   Observed in many subjects. *2427*

Organophosphorus Insecticides   Hepatotoxic effect. *0405*

Oxacillin   Hepatotoxic effect. *2313*

Oxazepam   Possible hepatotoxicity. *2313*

Oxyphenbutazone   Possible hepatotoxicity. *2313*

Oxyphenisatin   May cause hypersensitivity reaction. *2970*

Paraldehyde   Possible hepatotoxicity. *2313*

Paramethadione   Possible hepatotoxicity. *2313*

Pargyline   Possible hepatotoxicity (reversible). *2313*

Penicillamine   Possible hepatotoxicity. *2313*

Phenacemide   May affect liver function in about 2% cases. *2313*

Phenazopyridine   Hepatotoxic effect. *2313*

Phenindione   May modify liver function (cholestasis). *2313*

Pheniprazine   May affect liver function. *2313*

Phenothiazines   May cause cholestatic hepatitis (in up to 4% patients). *2313*

Phenylbutazone   May cause cholestasis. *2313*

Phenytoin   Hepatotoxicity with centrolobular necrosis. *2313*

Phosphorus   Hepatotoxic effect with necrosis. *2313*

Polythiazide   May affect liver function. *2313*

Probenecid   Hepatotoxic effect (centrolobular necrosis). *2313*

Procainamide   Hepatotoxic effect. *2313*

Prochlorperazine   Cholestatic effect. *0071*

Progesterone   Hepatotoxicity. *2313*

Promazine   May cause cholestasis. *2313*

Promethazine   May cause cholestasis. *2313*

Propoxyphene   Cholestatic effect. *1954*

Propylthiouracil   Hepatotoxic effect (centrolobular necrosis). *2313*

Pyrazinamide   Hepatotoxic effect (viral-hepatitis like). *2313*

Quinacrine   Hepatotoxic effect. *2313*

Quinethazone   May cause cholestatic jaundice. *2313*

Rifampin   Hepatotoxic effect (cholestasis). *2820*

Sulfadimethoxine   Reversible cholestasis. *1049*

Sulfamethizole   Reversible cholestasis. *1974*

Sulfamethoxazole   Reversible cholestasis. *2313*

Sulfamethoxypyridine   Reversible cholestasis. *2313*

Sulfisoxazole   May cause cholestasis. *1974*

Sulfonamides   May cause cholestasis. *2808*

Sulfones   May affect liver function. *2313*

Testosterone   May cause cholestasis. *2313*

Tetracycline   May cause cholestasis. *2313*

Thiabendazole   Rare cholestasis and parenchymal liver damage. *2413*

Thiacetazone   May cause cholestasis. *0248*

Thiazides   May cause cholestasis. *2313*

Thioguanine   May cause cholestasis. *2313*

Thioridazine   Hepatotoxicity. *1857*

Thiosemicarbazones   May affect liver function. *2313*

Thiothixene   Hepatotoxic effect (reversible cholestasis). *2313*

Tolazamide   Cholestasis may occur. *2313*

Tolbutamide   Cholestatic phenomenon. *2313*

Tranylcypromine   May affect liver function (?hypersensitivity). *2313*

Trimethadione   May cause hepatotoxicity with necrosis. *2313*

Trioxsalen   May affect liver function. *2313*

Uracil Mustard   May affect liver function. *2313*

Cephalin Time

Blood   Decrease   Physiological
Oral Contraceptives   Metabolic effect. *2834*

Cephaloridine

Serum   Increase   Physiological
Probenecid   Reduces renal clearance. *1487*

Cephalothin

Serum   Increase   Physiological
Probenecid   Reduces renal clearance. *1487*

Cephradine

Serum   Decrease   Physiological
Food   Absorption delayed when taken with food. *3024*

Ceruloplasmin

Serum   No Effect   Analytical
Bilirubin   No effect observed on Henry PPD procedure. *0877*

Hemoglobin   With up to 200 mg/dL on Henry PPD procedure. *0877*

Storage of Sample   For 2 weeks at 4° or if frozen. *0877*

Serum   No Effect   Physiological
Ascorbic Acid   No effect on protein with 605 mg/d for 3 weeks but up to 21% reduction of oxidase activity. *1752*

Desogestrel   With daily dose of 0.125 mg in 30 healthy female volunteers when given alone. Normal values 30 d after treatment stopped. *3089*

Oxymetholone   No metabolic effect. *0227*

Progesterone   No effect reported over several months. *3505*

Tri-iodothyronine   In 7 normal women aged 18-27 y given 25 μg/d three times daily for 14 d. Effect observed on endothelium-associated proteins but not on hepatically synthesized proteins. *1383*

Serum   Increase   Physiological
Aging   Slight effect in men around 46 y. *0954*

Androgens   Hormonal effect. *0248*

Anticonvulsants   In 8 epileptic patients receiving phenobarbital with carbamazepine or phenytoin. *2670*

Copper   Observed with copper poisoning. *2427*

Estrogens   But no change in ceruloplasmin activity. *2386*

Ethinyl Estradiol   Metabolic effect on liver synthesis (estrogen effect). *1498*

Lactation   Initially high falls to normal. *0987*

Lynestrenol   When 5 mg daily given alone to 30 healthy female volunteers. Normal values 30 d after treatment stopped. *3089*

Muscular Exercise   Occurs within 15 minutes, persists for 1 d. *1491*

Norethindrone   Metabolic estrogen effect. *0474*

Oral Contraceptives   Estrogen effect on liver, no change in activity. *2189*   Increase by 115 to 123% after 3 mo with different preparations. *0311*

Phenobarbital   Observed in both hospitalized and home-living patients with long-term administration. *3803*

Phenytoin   Effect on synthesis of protein in liver. *3505* Observed in both hospitalized and home-living individuals with long-term administration. *3803*

Pregnancy   In last trimester compared with controls. *2385*

Sex Difference   Slightly higher in women (5 mg/dL). *0954*

Tamoxifen   Significant change due to mild estrogenic effect of drug in breast cancer patients. *3070*

Serum   Decrease   Analytical
Hemolysis   May affect linearity of reaction curve. *2981*

Storage of Sample   Unstable at room temperature. *3217*   Oxidase activity decreased at room temperature, no effect for 2 d at 4°. *3873*

Serum   Decrease   Physiological
Asparaginase   Diminished hepatic synthesis. *2657*   Possibly due to inhibition of protein synthesis reported in various studies. *2406*

Levonorgestrel   Slight effect with daily dose of 0.125 mg in 30 healthy female volunteers when given alone. Values normal within 30 d of end of treatment. *3089*

Cervical Secretion

Patient   No Effect   Physiological
Oral Contraceptives   Leucorrhea due to estrogen in 20% women. *2349*

Patient   Increase   Physiological
Estrogens   Leukorrhea in 20% women on oral contraceptives. *2349*

Oral Contraceptives   leukorrhea due to estrogen in 20% women. *2349*

Chenodeoxycholic Acid

Serum   No Effect   Physiological
Oral Contraceptives   No effect observed over 12 mo in 29 women given combination of ethinyl estradiol and norgestrel. *1546*

Phenobarbital   No effect on increased concentration in patients with intrahepatic cholestasis increased of pregnancy. *1545*

Serum   Increase   Physiological
Rifampin   Significant increase possibly due to blocking uptake by plasma membrane of hepatocytes. *1224*

Serum   Decrease   Physiological
Cholestyramine   In 19 patients suffering from intrahepatic cholestasis of pregnancy. *1545*

Feces   Increase   Physiological
Neomycin   Due to altered intestinal flora. *1077*

Chloral Hydrate

Blood   No Effect   Physiological
Chloral Hydrate   Not detectable after 1 g orally. *2348*

Chloramphenicol

Serum   No Effect   Analytical
Acetaminophen   At 100 mg/L no effect on coupled enzymatic procedure. *2490*

Amikacin   No effect at 100 mg/L on coupled enzymatic method. *2490*

Amoxicillin   No effect at 100 mg/L on coupled enzymatic method. *2490*

Ampicillin   At 100 mg/L no effect on coupled enzymatic procedure. *2490*

Augmentin®   No effect at 100 mg/L on coupled enzymatic method. *2490*

Carfecillin   No effect at 100 mg/L on coupled enzymatic method. *2490*

Cephradine   No effect at 100 mg/L on coupled enzymatic method. *2490*

Clindamycin   No effect at 100 mg/L on coupled enzymatic method. *2490*

Colistin   No effect at 100 mg/L on coupled enzymatic method. *2490*

Co-Trimoxazole   No effect at 100 mg/L on coupled enzymatic method. *2490*   No effect at 100 mg/L on coupled enzymatic method. *2490*

Demeclocycline   No effect at 100 mg/L on coupled enzymatic method. *2490*

Doxycycline   No effect at 100 mg/L on coupled enzymatic method. *2490*

Erythromycin   No effect at 100 mg/L on coupled enzymatic method. *2490*

Flucloxacillin   No effect at 100 mg/L on coupled enzymatic method. *2490*

Gentamicin   No effect at 100 mg/L on coupled enzymatic method. *2490*

Lincomycin   No effect at 100 mg/L on coupled enzymatic method. *2490*

Metronidazole   No effect at 100 mg/L on coupled enzymatic method. *2490*

Minocycline   No effect at 100 mg/L on coupled enzymatic method. *2490*

Neomycin   No effect at 100 mg/L on coupled enzymatic method. *2490*

Oxytetracycline   No effect at 100 mg/L on coupled enzymatic method. *2490*

Penicillin V   No effect at 100 mg/L on coupled enzymatic method. *2490*

Rifampin   No effect at 100 mg/L on coupled enzymatic method. *2490*

Salicylate   No effect at 100 mg/L on coupled enzymatic method. *2490*

Streptomycin   No effect at 100 mg/L on coupled enzymatic method. *2490*

Tetracycline   No effect at 100 mg/L on coupled enzymatic method. *2490*

Tobramycin   No effect at 100 mg/L on coupled enzymatic method. *2490*

Trimethoprim   No effect at 100 mg/L on coupled enzymatic method. *2490*

Vancomycin   No effect at 100 mg/L on coupled enzymatic method. *2490*

Chlordiazepoxide

Blood   No Effect   Analytical
Acetaminophen   On method of Riddick at 5 mg/dL. *2983*

Amitriptyline   On method of Riddick at 5 mg/dL. *2983*

Chlordiazepoxide (continued)

Blood   No Effect   Analytical (continued)

Amobarbital   On method of Riddick at 5 mg/dL. *2983*

Amphetamine   On method of Riddick at 5 mg/dL. *2983*

Bromides   On method of Riddick at 5 mg/dL. *2983*

Butabarbital   On method of Riddick at 5 mg/dL. *2983*

Chloral Hydrate   On method of Riddick at 5 mg/dL. *2983*

Chlorpheniramine   On method of Riddick at 5 mg/dL. *2983*

Chlorpromazine   On method of Riddick at 5 mg/dL. *2983*

Cocaine   On method of Riddick at 5 mg/dL. *2983*

Dextroamphetamine   On method of Riddick at 5 mg/dL. *2983*

Diphenhydramine   On method of Riddick at 5 mg/dL. *2983*

Ethchlorvynol   On method of Riddick at 5 mg/dL. *2983*

Flurazepam   On method of Riddick at 5 mg/dL. *2983*

Glutethimide   On method of Riddick at 5 mg/dL. *2983*

Hydromorphone   On method of Riddick at 5 mg/dL. *2983*

Imipramine   On method of Riddick at 5 mg/dL. *2983*

Meprobamate   On method of Riddick at 5 mg/dL. *2983*

Methadone   On method of Riddick at 5 mg/dL. *2983*

Methaqualone   On method of Riddick at 5 mg/dL. *2983*

Methyprylon   On method of Riddick at 5 mg/dL. *2983*

Morphine   On method of Riddick at 5 mg/dL. *2983*

Nalorphine   On method of Riddick at 5 mg/dL. *2983*

Phenobarbital   On method of Riddick at 5 mg/dL. *2983*

Phenytoin   On method of Riddick at 5 mg/dL. *2983*

Prochlorperazine   On method of Riddick at 5 mg/dL. *2983*

Promethazine   On method of Riddick at 5 mg/dL. *2983*

Propoxyphene   On method of Riddick at 5 mg/dL. *2983*

Quinine   On method of Riddick at 5 mg/dL. *2983*

Salicylate   On method of Riddick at 5 mg/dL. *2983*   On method of Riddick at 5 mg/dL. *2983*

Secobarbital   On method of Riddick at 5 mg/dL. *2983*

Thioridazine   On method of Riddick at 5 mg/dL. *2983*

Trifluoperazine   On method of Riddick at 5 mg/dL. *2983*

Trimethobenzamide   On method of Riddick at 5 mg/dL. *2983*

Blood   Increase   Analytical

Chlorpheniramine   May interfere method of Jatlow. *1777*

Diazepam   2 mg/dL equivalent to 1.0 mg/dL by method of Riddick. *2983*

Fluphenazine   2 mg/dL equivalent to 1.0 mg/dL by method of Riddick. *2983*

Methapyrilene   May interfere with method of Jatlow. *1777*

Oxazepam   2 mg/dL equivalent to 3.6 mg/dL by method of Riddick. *2983*   Slight effect only on method of Jatlow. *1777*

Procainamide   2 mg/dL equivalent to 1.1 mg/dL by method of Riddick. *2983*

Sulfadiazine   5 mg/dL equivalent to 1.0 mg/dL by method of Riddick. *2983*

Sulfamerazine   5 mg/dL equivalent to 0.9 mg/dL by method of Riddick. *2983*

Sulfapyridine   5 mg/dL equivalent to 0.9 mg/dL by method of Riddick. *2983*

Sulfisoxazole   5 mg/dL equivalent to 1.0 mg/dL by method of Riddick. *2983*

Tripelennamine   May interfere with method of Jatlow. *1777*

Blood   Increase   Physiological

Chlordiazepoxide   150 mg orally conc=10 mg/L, 600 mg/d = 40 mg/L. *2348*

Cimetidine   Due to impaired clearance and impairment of hydroxylation of desmethyl derivative. *3084*

Urine   Positive   Analytical

Mazindol   Same R$_f$ and color reaction on TLC using ethyl acetate: methanol: water: ammonium hydroxide and modified Dragendorff's reagent. *3868*

Phendimetrazine   Same R$_f$ and color reaction on TLC using ethyl acetate: methanol: water: ammonium hydroxide and modified Dragendorff's reagent. *3868*

Chloride

Serum   No Effect   Analytical

Acetaminophen   At concentration of 1500 mg/L had no effect on mercurimetric method. *3393*

Acetylsalicylic Acid   At concentration of 2,000 mg/L had no effect on mercurimetric method. *3393*

Allopurinol   At concentration of 5 mg/L had no effect on mercurimetric method. *3393*

Aminophenol   At concentration of 5 mg/L had no effect on mercurimetric method. *3393*

Aminosalicylic Acid   At concentration of 460 mg/L had no effect on mercurimetric method. *3393*

Amitriptyline   At concentration of 20 mg/L had no effect on mercurimetric method. *3393*

Amobarbital   At concentration of 150 mg/L had no effect on mercurimetric method. *3393*

Amphotericin B   At concentration of 20 mg/L had no effect on mercurimetric method. *3393*

Ampicillin   At concentration of 40 mg/L had no effect on mercurimetric method. *3393*

Ascorbic Acid   At concentration of 40 mg/L had no effect on Ektachem® method. *3393*   At concentration of 500 mg/L had no effect on mercurimetric method. *3393*

Barbital   At concentration of 500 mg/L had no effect as measured by mercurimetric method. *3393*

Boric Acid   At concentration of 50 mg/L had no effect as measured by mercurimetric method. *3393*

Butabarbital   At concentration of 100 mg/L had no effect as measured by mercurimetric method. *3393*

Caffeine   At concentration of 160 mg/L had no effect on mercurimetric method. *3393*

Carbenicillin   At concentration of 15,000 mg/L had no effect on mercurimetric method. *3393*

Cefazolin   At concentration of 110 mg/L had no effect on mercurimetric method. *3393*

Cephalothin   At concentration of 1,000 mg/L had no effect on mercurimetric method. *3393*

Chlordiazepoxide   At concentration of 5 mg/L had no effect on mercurimetric method. *3393*

Chlormezanone   At concentration of 100 mg/L had no effect on mercurimetric method. *3393*

Chlorphenesin   At concentration of 150 mg/L had no effect on mercurimetric method. *3393*

Chlorpheniramine   At concentration of 2 mg/L had no effect on mercurimetric method. *3393*

Chlorpromazine   At concentration of 3 mg/L had no effect on mercurimetric method. *3393*

Chlorpropamide   At concentration of 150 mg/L had no effect on mercurimetric method. *3393*

Chlorprothixene   At concentration of 1 mg/L had no effect on mercurimetric method. *3393*

Cimetidine   At concentration of 1 mg/L had no effect on mercurimetric method. *3393*

Cloxacillin   At concentration of 5 mg/L had no effect on mercurimetric method. *3393*

Cocaine   At concentration of 3 mg/L had no effect on mercurimetric method. *3393*

Codeine   At concentration of 1 mg/L had no effect on mercurimetric method. *3393*

Contact With Clot   If plastic disc used to separate for 48 h. *1863*

Cyclosporine   At concentration of 41.5 mmol/L (50 μg/L) on methods on Technicon® SMAC® II and Hitachi® 705. *1123*

Desipramine   At concentration of 8 mg/L had no effect on mercurimetric method. *3393*

Diazepam   At concentration of 1.5 mg/L had no effect on mercurimetric method. *3393*

Diethylaminoethanol   At concentration of 30 mg/L had no effect on mercurimetric method. *3393*

Digitoxin At concentration of 6 mg/L had no effect on mercurimetric method. *3393*

Digoxin At concentration of 9 mg/L had no effect on mercurimetric method. *3393*

Dimethadione At concentration of 2,000 mg/L had no effect on mercurimetric method. *3393*

N-Dimethyldiazepam At concentration of 6 mg/L had no effect on mercurimetric method. *3393*

Diphenhydramine At concentration of 10 mg/L had no effect on mercurimetric method. *3393*

Disopyramide At concentration of 4 mg/L had no effect on mercurimetric method. *3393*

Disulfiram At concentration of 120 mg/L had no effect on mercurimetric method. *3393*

Ethchlorvynol At concentration of 400 mg/L had no effect on mercurimetric method. *3393*

Ethosuximide At concentration of 390 mg/L had no effect on mercurimetric method. *3393*

Flurazepam At concentration of 0.1 mg/L had no effect on mercurimetric method. *3393*

Furosemide At concentration of 4 mg/L had no effect on mercurimetric method. *3393*

Gentamicin At concentration of 150 mg/L had no effect on mercurimetric method. *3393*

Hydralazine At concentration of 6 mg/L had no effect on mercurimetric method. *3393*

Indomethacin At concentration of 13 mg/L had no effect on mercurimetric method. *3393*

Insulin At concentration of 3 mg/L had no effect on mercurimetric method. *3393*

Iproniazid At concentration of 40 mg/L had no effect on mercurimetric method. *3393*

Isoniazid At concentration of 100 mg/L had no effect on mercurimetric method. *3393*

Levodopa At concentration of 80 mg/L had no effect on Ektachem® method. *3393*

Lorazepam At concentration of 0.05 mg/L had no effect on mercurimetric method. *3393*

Meperidine At concentration of 60 mg/L had no effect on mercurimetric method. *3393*

Mephenesin At concentration of 100 mg/L had no effect on mercurimetric method. *3393*

Meprobamate At concentration of 25 mg/L had no effect on mercurimetric method. *3393*

Mercaptopurine At concentration of 15 mg/L had no effect on Ektachem® method. *3393*

Methadone At concentration of 7 mg/L had no effect on mercurimetric method. *3393*

Methamphetamine At concentration of 2 mg/L had no effect on mercurimetric method. *3393*

Methapyrilene At concentration of 13 mg/L had no effect on mercurimetric method. *3393*

Methicillin At concentration of 900 mg/L had no effect on mercurimetric method. *3393*

Methohexital At concentration of 50 mg/L had no effect on mercurimetric method. *3393*

Methotrimeprazine At concentration of 1 mg/L had no effect on mercurimetric method. *3393*

Methsuximide At concentration of 40 mg/L had no effect on Ektachem® method. *3393*

Methyldopa At concentration of 80 mg/L had no effect on mercurimetric method. *3393*

Methylphenobarbital At concentration of 150 mg/L had no effect on mercurimetric method. *3393*

Metoprolol At concentration of 0.34 mg/L had no effect on mercurimetric method. *3393*

Morphine At concentration of 1 mg/L had no effect on mercurimetric method. *3393*

Nafcillin At concentration of 50 mg/L had no effect on mercurimetric method. *3393*

Nitrofurantoin At concentration of 4 mg/L had no effect on mercurimetric method. *3393*

Nortriptyline At concentration of 2 mg/L had no effect on mercurimetric method. *3393*

Orphenadrine At concentration of 9 mg/L had no effect on mercurimetric method. *3393*

Oxazepam At concentration of 1 mg/L had no effect on mercurimetric method. *3393*

Papaverine At concentration of 10 mg/L had no effect on mercurimetric method. *3393*

Paraldehyde At concentration of 2,000 mg/L had no effect on mercurimetric method. *3393*

Penicillin G At concentration of 2,000 mg/L had no effect as measured by mercurimetric method. *3393*

Pentobarbital At concentration of 340 mg/L had no effect on mercurimetric method. *3393*

Perphenazine At concentration of 1 mg/L had no effect on mercurimetric method. *3393*

Phencyclidine At concentration of 6 mg/L had no effect on mercurimetric method. *3393*

Phenobarbital At concentration of 250 mg/L had no effect on mercurimetric method. *3393*

Phensuximide At concentration of 120 mg/L had no effect on mercurimetric method. *3393*

Phenylbutazone At concentration of 750 mg/L had no effect on mercurimetric method. *3393*

Phenylephrine At concentration of 4 mg/L had no effect on mercurimetric method. *3393*

Phenytoin At concentration of 240 mg/L had no effect on mercurimetric method. *3393*

Phosphates No effect at 20 mg/dL method of Fingerhut. *1135*

Plasma No significant difference from plasma by chloridometer. *2228*

Probenecid At concentration of 1300 mg/L had no effect on mercurimetric method. *3393*

Procainamide At concentration of 50 mg/L had no effect on mercurimetric method. *3393*

Procaine At concentration of 2 mg/L had no effect on mercurimetric method. *3393*

Prochlorperazine At concentration of 1 mg/L had no effect on mercurimetric method. *3393*

Promethazine At concentration of 1 mg/L had no effect on mercurimetric method. *3393*

Propranolol At concentration of 0.2 mg/L had no effect on mercurimetric method. *3393*

Quinidine At concentration of 210 mg/L had no effect on mercurimetric method. *3393*

Quinine At concentration of 30 mg/L had no effect on mercurimetric method. *3393*

Salicylate At concentration of 500 mg/L had no effect on mercurimetric method. *3393*

Secobarbital At concentration of 100 mg/L had no effect on mercurimetric method. *3393*

Sodium Bromide On Beckman Astra 8 and Corning/EEL 920 chloride meter: chloride meter produced results with 5 mmol/L bromide. *2972*

Storage of Sample No effect after 1 week at room temperature, months at -10. *1563*

Strychnine At concentration of 12 mg/L had no effect on mercurimetric method. *3393*

Sulfadiazine At concentration of 1500 mg/L had no effect on mercurimetric method. *3393*

Sulfaguanidine At concentration of 500 mg/L had no effect on mercurimetric method. *3393*

Sulfanilamide At concentration of 1,000 mg/L had no effect on mercurimetric method. *3393*

Sulfates No effect at 20 mg/dL on method of Fingerhut. *1135*

Tetracycline At concentration of 300 mg/L had no effect on mercurimetric method. *3393*

Theobromine At concentration of 2,000 mg/L had no effect on mercurimetric method. *3393*

Theophylline At concentration of 2,000 mg/L had no effect on mercurimetric method. *3393*

Thiopental At concentration of 200 mg/L had no effect on mercurimetric method. *3393*

Timolol At concentration of 0.01 mg/L had no effect on mercurimetric method. *3393*

Chloride (continued)

Serum   No Effect   Analytical (continued)

Tolbutamide   At concentration of 100 mg/L had no effect on mercurimetric method. 3393

Transport of Specimen   By pneumatic tube if specimen protected. 3217

Tribromethanol   At concentration of 90 mg/L had no effect on mercurimetric method. 3393

Trichlorethanol   At concentration of 1,000 mg/L had no effect on mercurimetric method. 3393

Trifluoperazine   At concentration of 1 mg/L had no effect on mercurimetric method. 3393

Warfarin   At concentration of 100 mg/L had no effect on mercurimetric method. 3393

Serum   No Effect   Physiological

Amiloride   No effect seen in 13 men treated for 8 weeks. 0656

Bopindolol   In 10 hypertensive patients treated for either 1 or 21 d. 1063

Bucindol   No effects noted in 8 patients following several weeks on treatment. 1283

Meals   No effect after standard breakfast. 0579

Starvation   No effect observed usually. 1242

Serum   Increase   Analytical

Bromides   0.03 meq/L at 20 meq/L method of Fingerhut. 1135 Measured as Cl mercurimetric, electrometric methods. 1563

Bromisovalum   Metabolite may be measured as chloride. 1343

Calcium Bromogalactogluconate   Relatively small error by coulometric method as used on Beckman Astra but significant increase at clinical concentrations with thiocyanate methods on Technicon® RA-1000 or SMAC®. 2867

Carbromal   Bromide as metabolite may be measured as chloride. 1343

Cefotaxime   Statistically significant effect of drug on Parallel method. 0201

Fluosol-DA   Unpredictable variation in concentrations as measured on SMA IIC, Beckman Astra 8 and Ektachem® 400. 2518

Glassware   If chloride present. 3879

Iodides   0.51 meq/L at 20 meq/L in method of Fingerhut. 1135

Protein   Failure to precipitate affects Schales method. 1563

Sodium Bromide   As measured by Technicon® C800 instrument in one patient receiving bromide therapy. 2972

Serum   Increase   Physiological

Acetazolamide   Loss of HCO₃ by carbonic anhydrase inhibition. 1022

Ammonium Chloride   Added chloride and metabolic acidosis. 2313

Androgens   May cause salt and water retention. 1343

Aspirin   Progressive effect during three days but also noted over longer term study at beginning. Eventually reverted towards normal. Effect statistically significant. 3076

Azapropazone   In 1 of 83 patients treated for 6 mo. 3565

Azosemide   Delayed effect observed in normal male volunteers after 60 mg orally. 2658

Boric Acid   Toxicity effect (with acute tubular necrosis). 2313

Cannabis   Reported effect. 2220

Chloride Salts   Increased absorption with increased serum levels. 0652

Chlorothiazide   With prolonged therapy hyperchloremic alkalosis. 2220

Cholestyramine   Is chloride salt of basic anion exchange resin. 3879   Severe hyperchloremia up to 128 mmol/L in 2 pediatric patients receiving drug for some weeks. 2872

Corticosteroids   May cause retention. 2313

Cyclosporine   Hyperchloremic acidosis in 7 of 43 renal allograft patients. 0027   Hyperchloremic metabolic acidosis out of proportion to reduction in GFR. 2836

Diazoxide   May cause salt retention. 1343

Estrogens   May cause salt and water retention. 1343

Guanethidine   Salt retention ?due to tubular effect. 1343

Hydrochlorothiazide   Hyperchloremic alkalosis with prolonged therapy. 2220

Hydrocortisone   May cause retention and edema. 0071

Intra-Amniotic Saline   Increased by approximately 3 meq/L after 1 h. 3892

Ion Exchange Resins   Mechanism not cited — ?depends on resin. 0652

Menstruation   Approximately 1.7 meq/L higher premenstrually. 1086

Methoxyflurane   Toxic nephropathy reported. 2980

Methyldopa   May cause salt retention and edema. 1343

Muscular Exercise   Vigorous exercise for 30 minutes increases by 2 meq/L. 1086

Oxyphenbutazone   May cause marked salt retention. 2313

Phenylbutazone   May cause salt retention (?tubular dysfunction). 3687

Potassium Chloride   Added chloride. 2313

Saline   Added chloride. 2313

Triamterene   Nephrotoxic and azotemic effect. 2313

Serum   Decrease   Analytical

Cefotaxime   Statistically significant effect of metabolite on Parallel method. 0201

Contact With Clot   If exposed to air water shift from erythrocytes. 1563

Fluorides   0.05 meq/L at 20 meq/L in method of Fingerhut. 1135

Hemolysis   Dilutional effect. 2313

Lipemia   Increase by 1.5% in patient specimens measured on SMAC® in comparison with specimens ultra-centrifugation. 2466

Liposol   Changed from 102.5 to 96.3 mmol/L in 26 randomly selected patient sera, before and after addition on Beckman Astra methods. 2054

Oxalate   If used as anticoagulant may be passage into cells. 1563

Plasma   Slightly less in plasma by SMA 6/60 method. 2228

Potassium Oxalate   If used as anticoagulant may be shift into cells. 1563

Serum   Decrease   Physiological

Aldosterone   Hypokalemic hypochloremic alkalosis may occur. 1343

Azosemide   Slight effect noted in 2 volunteers. 2047

Bicarbonates   Induces metabolic alkalosis. 0652

Blindness   Significantly lower (average 7 meq/L) than normal. 1637

Bromides   Reversible halide dysequilibration. 0652

Bumetanide   By approximately 3 meq/L. 2666

Carbamazepine   Isolated case of dilutional hyponatremia with water intoxication: 7 previous cases reported. 1882

Carbenoxolone   Average decrease of 3 meq/L, associated with alkalosis. 1179

Chlorpropamide   Drug-induced syndrome of inappropriate ADH secretion. 2601

Chlorthalidone   Single case of severe reversible myopathy noted. 1793   Progressive effect with dose: from 3 mmol/L with 25 mg/d to 6 mmol/L at 200 mg/d in 37 patients treated for 8 weeks. 3643

Clopamide   Single case of diuretic associated myopathy. 1793

Corticosteroids   Metabolic alkalosis with reduced Cl reabsorption. 2448

Corticotropin   May cause hypochloremic alkalosis. 2312

Cortisone   May cause hypochloremic alkalosis. 3961

Dapsone   Sulfone syndrome in one 16 year old girl given 50 mg daily short-term administration. 3604

Diuretics   Diuretic action if excessive. 2313

Ethacrynic Acid   Diuretic action (inhibits tubular reabsorption). 2745

Furosemide   Diuretic action (inhibits tubular reabsorption). 0070 In 36% of 204 hospitalized patients receiving the drug. 3419 Observed after drug administration. 2047   Reduced often in association with hyponatremia. 1391

Glucose   Dilutional effect when infused. 0652

Hydrochlorothiazide May cause marked reduction. *1129*

Hydrocortisone May cause hypochloremic hypokalemic alkalosis. *1343*

Hydroflumethiazide Diuretic action. *1680*

Laxatives If chronic abuse occurs. *3682*

Mannitol Effect if marked diuresis. *0071*

Meralluride Diuretic action. *3879*

Mercurial Diuretics May cause hypochloremic alkalosis and diuresis. *2745*

Mersalyl Consequence of diuretic action. *2745*

Methyclothiazide Diuretic action (hypochloremic alkalosis). *1353*

Metolazone Diuretic action of drug acting on distal tubules. *2816*

Polythiazide Diuretic effect. *1129*

Prednisolone Hypochloremic alkalosis may occur. *0019*

β-Propiolactone Increase by 10 mmol/L when plasma from BPL treated blood compared with plasma in 25 specimens. *0217*

Silver Observed after silver nitrate antisepsis. *2427*

Sodium Salts Significant effect with citrate salt not observed with chloride in 5 men with essential hypertension. *2037*

Thiazides Diuretic action (impaired tubular reabsorption). *2313*

Triamterene Diuretic action with impaired tubular reabsorption. *0652*

Urine   No Effect   Analytical

Storage of Sample At room temperature for 45 d, without preservative. *1180*

Thymol No effect on Schales and Schales method. *2981*

Urine   No Effect   Physiological

Bopindolol In 10 hypertensive patients treated for either 1 or 21 d. *1063*

Sulindac In 10 furosemide treated patients with well controlled congestive heart failure given 400 mg/d for 2 d. *1042*

Urine   Increase   Analytical

Bromides Bromide measured as chloride. *1237*

Glassware If chloride present. *3879*

Urine   Increase   Physiological

Amiloride Diuretic action. *1343*

Ammonium Chloride Diuretic action. *1343*

Azosemide Effect observed in normal male volunteers after 60 mg orally. *2658* Significant effect then gradually decreased. *2047*

Benzthiazide Therapeutic diuretic intent. *1680*

Bumetanide Marked dose related effect. *2666*

Chlorthalidone May produce marked diuresis. *2427*

Clopamide Diuretic action. *2900*

Cyclothiazide Intended diuretic action. *1680*

Diapamide Diuretic action. *2256*

Digitalis Diuretic action in cardiac failure. *1343*

Diurnal Variation Maximum around noon 2pm, minimum during night. *3217*

Ethacrynic Acid Present as major anion. *1343*

Furosemide Diuretic action. *1617* Marked effect: decreased sharply when administration stopped. *2047*

Hydrochlorothiazide Significant effect in 5 h after 150 mg orally. *1189*

Hydroflumethiazide Diuretic action. *1680*

Intra-Amniotic Saline Significant effect for 24 h. *3892*

Isosorbide Similar to natriuretic effect. *2605*

Mefruside Diuretic action. *0180*

Menstruation Post-menstrual diuresis (decrease premenstrually). *0987*

Mercurial Diuretics Therapeutic intent (dominant urinary anion). *2745*

Methyclothiazide Intended diuretic action (maximum in 6 h). *1680*

Metolazone Diuretic action of drug. *1425*

Parathyroid Extract Hormonal action. *1343*

SC-16102 Duration of diuretic action short. *1804*

Spironolactone Diuretic action. *1343*

Thiazides Diuretic action. *1343*

Torasemide Similar effects to that of furosemide. *0919*

Triamterene Diuretic action. *1343*

Viomycin Promotes urinary loss. *0071*

Xanthine Diuretic action. *1343*

Urine   Decrease   Physiological

Acetazolamide Significant effect in 5 h after 500 mg orally. *1189*

Blindness Significantly lower (average 49 meq/d) than normal. *1637*

Corticosteroids Promotes retention (mineralocorticoid effect). *0071*

Cortisone Causes retention. *0071*

Diazoxide Effect over 2 h of 4 mg/kg given i.v. or orally. *1805*

Epinephrine With increased filtration fraction. *1343*

Etodolac Both in normal subjects but also to greater extent in individuals with renal insufficiency. *0453*

Guancydine Reduced renal blood flow causes retention. *2132*

Levarterenol Increased tubular resistance and reabsorption. *1176*

Mafenide Selective retention. *0123*

Naproxen Increase by 26% in 10 furosemide treated patients with well controlled congestive heart failure. *1042*

Sleep Reduced excretion compared with day. *1139*

Spinal Anesthesia Observed in normal males, pregnant women. *1343*

Bile   Increase   Physiological

Secretin Slight in normals, marked decrease in cirrhotics. *0399*

Pancreatic Juice   Increase   Physiological

Acetazolamide 30 percent increase with infusion in normals. *0984*

Chlormethiazole

Serum   Increase   Physiological

Ranitidine At least in one patient prolongation of elimination half-life. *1938*

Chloroquine

Serum   Increase   Physiological

Cimetidine Significant reduction in oral clearance from 0.49 L/d/kg to 0.23 L/d/kg and elimination half-life increased from 3.11 d to 4.62 d in test group. *1052*

Chlorothiazide

Serum   Increase   Physiological

Propantheline Absorption window increased with absorption window or dissolution. *3794*

Serum   Decrease   Physiological

Colestipol Reduced absorption due to binding. *3794*

Metoclopramide Decreased absorption due to absorption window or dissolution. *3794*

Chlorphenmetrazine

Urine   Positive   Analytical

Phentermine Similar R_f and color reaction on TLC using ethyl acetate: methanol: water: ammonium hydroxide and modified Dragendorff's reagent. *3868*

Chlorphentermine

Urine   Positive   Analytical
Phenmetrazine Similar Rf and color reaction on TLC using ethyl acetate: methanol: water: ammonium hydroxide and modified Dragendorff's reagent. *3868*

Chlorpromazine

Serum   Decrease   Physiological
Antacids Absorption decreased by 10 to 45% due to adsorption by antacid. *3794*
Phenobarbital Induces hepatic microsomal enzymes. *1487*

Urine   Increase   Physiological
Phenobarbital Induces hepatic microsomal enzymes. *1487*

Chlorpropamide

Serum   Increase   Physiological
Aspirin Displaces from plasma proteins. *1487*
Co-Trimoxazole Effects of compound may be potentiated. *0679* Effects of compound may be potentiated. *0679*
Dicumarol Increases half life, augments hypoglycemia. *1487*
Trimethoprim May potentiate action by effects on metabolism. *0678*

Blood   Increase   Physiological
Chloramphenicol Inhibits hepatic microsomal metabolizing enzymes thus prolonging half-life. *3864*

Chlortetracycline

Serum   Decrease   Physiological
Antacids Absorption of tetracyclines decreased by up to 80% due to adsorption by antacids. *3794*

Chloylglycine

Serum   Increase   Physiological
Vinyl Chloride Markedly different in individuals with chemical liver injury compared with people with nonchemical liver disease or normals. *2190*

Cholecystokinin (CCK)

Plasma   No Effect   Analytical
Gastrin No significant cross reactivity in RIA procedure of Harvey. *1518*
Glucagon No significant cross reactivity in RIA procedure of Harvey. *1518*
Secretin No significant cross reactivity with RIA procedure of Harvey. *1518*

Cholesterol

Serum   No Effect   Analytical
Acetaminophen At 5 times therapeutic concentration on Liebermann-Burchard method on SMAC® and enzymatic methods on Abbott-VP, Cobas-Bio, Ektachem®, Hitachi® 705 and KDA. *2138* At acute overdose concentration (10 mg/dL) on SMAC® method. *3719* No effect at therapeutic concentration on Reflotron method. *1984* At concentration of 6 mg/L had no effect on CHOD-PAP method. *3393* At concentration of 1500 mg/L had no effect on Liebermann-Burchard method. *3393*
Acetazolamide No effect at 12 mg/dL on SMA 12/60 method. *2636* At concentration of 1,000 mg/L had no effect on Liebermann-Burchard method. *3393*

Acetylsalicylic Acid At therapeutic concentrations on enzymatic and Liebermann — Burchard methods. *1786* On enzymatic method on SMA II at physiological concentration. *1787* At 5 times upper limit of therapeutic range on method on SMAC®, Abbott-VP, Cobas-Bio, Ektachem®, Hitachi® 705 and KDA. *2138* With 8326 μmol/L on cholesterol esterase/oxidase method. *1786* No effect at therapeutic concentration on Reflotron method. *1984* At concentration of 750 mg/L had no effect on CHOD-Iodide method. *3393* At concentration of 750 mg/L had no effect on CHOD-PAP method. *3393* At concentration of 750 mg/L had no effect on catalase-AIDH method. *3393* At concentration of 750 mg/L had no effect on method using catalase-Hantzsch reaction. *3393* At concentration of 2,000 mg/L had no effect on Liebermann-Burchard method. *3393*
Allopurinol At concentration of 250 mg/L had no effect on CHOD-Iodide method. *3393* At concentration of 250 mg/L had no effect on CHOD-PAP method. *3393* At concentration of 250 mg/L had no effect on catalase-AIDH method. *3393* At concentration of 250 mg/L had no effect on method using catalase-Hantzsch reaction. *3393* At concentration of 250 mg/L had no effect on Liebermann-Burchard method. *3393*
Aminophenazone No effect at therapeutic concentration on enzymatic and Liebermann-Burchard methods. *1786*
Aminopyrine At concentration of 125 mg/L had no effect on CHOD-PAP method. *3393*
Aminosalicylic Acid At 5 times upper limit of therapeutic range on methods on SMAC®, Abbott-VP, Cobas-Bio, Ektachem®, Hitachi® 705 and KDA. *2138* At concentration of 460 mg/L had no effect on Liebermann-Burchard method. *3393*
Amitriptyline At acute overdose concentration (2.5 mg/dL) on SMAC® method. *3719* At concentration of 0.1 mg/L had no effect on CHOD-PAP method. *3393* At concentration of 25 mg/L had no effect on Liebermann-Burchard method. *3393*
Amobarbital At concentration of 5 mg/L had no effect on CHOD-PAP method. *3393* At concentration of 150 mg/L had no effect on Liebermann-Burchard method. *3393*
Amphetamine At therapeutic concentration on Reflotron method. *1984*
Ampicillin No effect on Reflotron method at therapeutic concentration. *1984* At concentration of 900 mg/L had no effect on CHOD-PAP method. *3393* At concentration of 900 mg/L had no effect on catalase-AIDH method. *3393* At concentration of 900 mg/L had no effect on catalase-Hantzsch method. *3393* At concentration of 900 mg/L had no effect on Liebermann-Burchard method. *3393*
Ascorbic Acid No effect at therapeutic concentration on Reflotron method. *1984* At concentration of 400 mg/L had no effect on catalase-AIDH method. *3393* At concentration of 1,000 mg/L had no effect on method using catalase-Hantzsch reaction. *3393* At concentration of 500 mg/L had no effect on Liebermann-Burchard method. *3393*
Azapropazone At concentration of 250 mg/L had no effect as measured by CHOD-Iodide method. *3393* At concentration of 250 mg/L had no effect as measured by CHOD-PAP method. *3393* At concentration of 250 mg/L had no effect as measured by catalase-AIDH method. *3393* At concentration of 250 mg/L had no effect as measured by catalase-Hantzsch method. *3393* At concentration of 250 mg/L had no effect as measured by Liebermann-Burchard method. *3393*
Barbital At acute overdose concentration (20 mg/dL) on Technicon® SMAC® method. *3719* At concentration of 500 mg/L had no effect as measured by Liebermann-Burchard method. *3393*
Benzbromarone At concentration of 60 mg/L had no effect as measured by CHOD-Iodide method. *3393* At concentration of 60 mg/L had no effect as measured by CHOD-PAP method. *3393* At concentration of 60 mg/L had no effect as measured by catalase-AIDH method. *3393* At concentration of 60 mg/L had no effect as measured by method using catalase-Hantzsch reaction. *3393* At concentration of 60 mg/L had no effect as measured by Liebermann-Burchard method. *3393*
Bethanechol No effect at 0.09 mg/dL on SMA 12/60 method. *2636*
Bezafibrate No effect at therapeutic concentration on Reflotron method. *1984*

Bilirubin When measured enzymatically with oxidase. *3027* Up to 0.5 mmol/L on Leffler's procedure. *2854* At concentration of 40.0 mg/dL on Ektachem® method. *3407* At 40.0 mg/dL on method recommended on I.L. Multistat III. *3407* At concentration of 40.0 mg/dL on Du Pont aca II. *3407*

Bisacodyl At concentration of 40 mg/L had no effect as measured by CHOD-Iodide method. *3393* At concentration of 40 mg/L had no effect as measured by CHOD-PAP method. *3393* At concentration of 40 mg/L had no effect as measured by catalase-AIDH method. *3393* At concentration of 40 mg/L had no effect as measured by method using catalase-Hantzsch reaction. *3393* At concentration of 40 mg/L had no effect as measured by Liebermann-Burchard method. *3393*

Boric Acid At concentration of 50 mg/L had no effect as measured by Liebermann-Burchard method. *3393*

Bromazepam At 5 times therapeutic concentration on methods on SMAC®, Abbott-VP, Cobas-Bio, Ektachem®, Hitachi® and KDA. *2138*

Bromides Insignificant effect on Leffler's procedure. *1743*

Butabarbital At concentration of 100 mg/L had no effect as measured by Liebermann-Burchard method. *3393*

Butizide At concentration of 440 mg/L had no effect as measured by CHOD-Iodide method. *3393* At concentration of 440 mg/L had no effect as measured by CHOD-PAP method. *3393* At concentration of 440 mg/L had no effect as measured by catalase-AIDH method. *3393* At concentration of 440 mg/L had no effect as measured by method using catalase-Hantzsch reaction. *3393* At concentration of 440 mg/L had no effect as measured by Liebermann-Burchard method. *3393*

Caffeine No effect on enzymatic and Liebermann-Burchard methods at therapeutic concentrations. *1786* At acute overdose concentration (10 mg/dL) on Technicon® SMAC® method. *3719* At therapeutic concentration on Reflotron method. *1984* At concentration of 194 mg/L had no effect on Liebermann-Burchard method. *3393*

Campesterol On GLC procedure of MacGee. *2252*

Carbenicillin At 5 times upper limit of therapeutic range on methods on SMAC®, Abbott-VP, Cobas-Bio, Ektachem® 400, Hitachi® 705 and KDA. *2138* At concentration of 15,000 mg/L had no effect on Liebermann-Burchard method. *3393*

Carbochromen No effect at therapeutic concentration on Reflotron method. *1984*

Cefamandole At 2.50 mmol/L on method on Eppendorf Epos. *1414*

Cefazedone On method on Eppendorf Epos at concentration of 2.50 mmol/L. *1414*

Cefazolin At 2.50 mmol/L on method on Eppendorf Epos. *1414* At concentration of 110 mg/L had no effect on CHOD-PAP method. *3393*

Cefodizime No effect at 2.50 mmol/L on method on Eppendorf Epos. *1414*

Cefoperazone At 2.50 mmol/L on method on Eppendorf Epos. *1414*

Cefotaxime No effect at 2.50 mmol/L on method on Eppendorf Epos. *1414*

Cefotiam At 2.50 mmol/L on method on Eppendorf Epos. *1414*

Cefoxitin At 5 times upper limit of therapeutic range on methods on SMAC®, Abbott-VP, Cobas-Bio, Ektachem®, Hitachi® 705, and KDA. *2138* At concentrations up to 2.50 mmol/L on methods on Eppendorf Epos. *1414*

Cefpirome At concentration of up to 2.50 mmol/L on methods of Eppendorf Epos. *1414*

Ceftriaxone On method on Eppendorf Epos at concentration of 2.50 mmol/L. *1414*

Cefuroxime On method on Eppendorf Epos at concentration of 2.50 mmol/L. *1414*

Cephaloridine At 2.50 mmol/L on method on Eppendorf Epos. *1414*

Cephalothin At 5 times upper limit of therapeutic range on methods on SMAC®, Abbott-VP, Cobas-Bio, Ektachem®, Hitachi® 705 and KDA. *2138* No effect a 2.50 mmol/L on method on Eppendorf Epos. *1414* At concentration of 1,000 mg/L had no effect on CHOD-PAP method. *3393*

Chloramphenicol No effect at therapeutic concentration on Reflotron method. *1984* At concentration of 600 mg/L had no effect on CHOD-Iodide method. *3393* At concentration of 600 mg/L had no effect on CHOD-PAP method. *3393* At concentration of 600 mg/L had no effect on catalase-AIDH method. *3393* At concentration of 600 mg/L had no effect on method using catalase-Hantzsch reaction. *3393* At concentration of 600 mg/L had no effect on Liebermann-Burchard method. *3393*

Chlordiazepoxide At 5 times upper limit of therapeutic range on methods on SMAC®, Abbott-VP, Cobas-Bio, Ektachem®, Hitachi® 705 and KDA. *2138* At acute overdose concentration (5 mg/dL) on Technicon® SMAC® method. *3719* No effect at therapeutic concentration on Reflotron method. *1984* At concentration of 160 mg/L had no effect on CHOD-Iodide method. *3393* At concentration of 160 mg/L had no effect on CHOD-PAP method. *3393* At concentration of 160 mg/L had no effect on catalase-AIDH method. *3393* At concentration of 160 mg/L had no effect on method using catalase-Hantzsch reaction. *3393* At concentration of 160 mg/L had no effect on Liebermann-Burchard method. *3393*

Chlormezanone At concentration of 100 mg/L had no effect on Liebermann-Burchard method. *3393*

Chloroquine At concentration of 60 mg/L had no effect on CHOD-Iodide method. *3393* At concentration of 200 mg/L had no effect on CHOD-PAP method. *3393* At concentration of 60 mg/L had no effect on catalase-AIDH method. *3393* At concentration of 60 mg/L had no effect on method using catalase-Hantzsch reaction. *3393* At concentration of 60 mg/L had no effect on Liebermann-Burchard method. *3393*

Chlorphenesin At concentration of 150 mg/L had no effect on Liebermann-Burchard method. *3393*

Chlorpheniramine At acute overdose concentration (20 mg/dL) on Technicon® SMAC® method. *3719* At concentration of 200 mg/L had no effect on Liebermann-Burchard method. *3393*

Chlorpromazine At concentration of 3 mg/L had no effect on CHOD-PAP method. *3393*

Chlorpropamide At concentration of 150 mg/L had no effect on CHOD-PAP method. *3393*

Chlorprothixene At concentration of 1 mg/L had no effect on Liebermann-Burchard method. *3393*

Chromonar Hydrochloride At concentration of 900 mg/L had no effect on CHOD-Iodide method. *3393* At concentration of 900 mg/L had no effect on CHOD-PAP method. *3393* At concentration of 900 mg/L had no effect on catalase-AIDH method. *3393* At concentration of 900 mg/L had no effect on method using catalase-Hantzsch reaction. *3393* At concentration of 900 mg/L had no effect on Liebermann-Burchard method. *3393*

Cimetidine At concentration of 1 mg/L had no effect on CHOD-PAP method. *3393*

Clofibrate At concentration of 1,000 mg/L had no effect on CHOD-PAP method. *3393* At concentration of 1400 mg/L had no effect on method using catalase-Hantzsch reaction. *3393*

Cloxacillin At concentration of 5 mg/L had no effect on CHOD-PAP method. *3393*

Cocaine At acute overdose concentration (2.5 mg/dL) on Technicon® SMAC® method. *3719* At concentration of 25 mg/L had no effect on Liebermann-Burchard method. *3393*

Codeine At acute overdose concentration (2.0 mg/dL) on Technicon® SMAC® method. *3719* At concentration of 0.1 mg/L had no effect on CHOD-PAP method. *3393* At concentration of 20 mg/L had no effect on Liebermann-Burchard method. *3393*

Corbadrin At concentration of 200 mg/L had no effect on CHOD-PAP method. *3393* At concentration of 200 mg/L had no effect on method using catalase-Hantzsch reaction. *3393* At concentration of 200 mg/L had no effect on Kageyama-Hantzsch method. *3393*

Cyclophosphamide At concentration of 240 mg/L had no effect on CHOD-Iodide method. *3393* At concentration of 240 mg/L had no effect on CHOD-PAP method. *3393* At concentration of 240 mg/L had no effect on catalase-AIDH method. *3393* At concentration of 240 mg/L had no effect on method using catalase-Hantzsch reaction. *3393* At concentration of 240 mg/L had no effect on Liebermann-Burchard method. *3393*

Cyclosporine At concentration of 41.5 mmol/L (50 µg/L) on methods on Technicon® SMAC® II and Hitachi® 705. *1123*

Cysteine At 5 times upper limit of therapeutic range on methods SMAC®, Abbott-VP, Cobas-Bio, Ektachem®, Hitachi® 705 and KDA. *2138*

Cholesterol (continued)

Serum   No Effect   Analytical (continued)

Cystine   At 5 times upper limit of therapeutic range on methods on SMAC®, Abbott-VP, Cobas-Bio, Ektachem®, Hitachi® 705 and KDA. 2138

Desipramine   At concentration of 8 mg/L had no effect on Liebermann-Burchard method. 3393

Deslanoside   No effect on SMA 12/60 method at 0.06 mg/dL. 2636

Desmosterol   On GLC procedure of MacGee. 2252

Dextran 60   At concentration of 6,000 mg/L had no effect on CHOD-Iodide method. 3393   At concentration of 6,000 mg/L had no effect on CHOD-PAP method. 3393   At concentration of 6,000 mg/L had no effect on catalase-AIDH method. 3393   At concentration of 6,000 mg/L had no effect on method using catalase-Hantzsch reaction. 3393   At concentration of 6,000 mg/L had no effect on Liebermann-Burchard method. 3393

Diatrizoic acid   At concentration of 2600 mg/L had no effect on CHOD-Iodide method. 3393   At concentration of 2600 mg/L had no effect on CHOD-PAP method. 3393   At concentration of 2600 mg/L had no effect on catalase-AIDH method. 3393   At concentration of 2600 mg/L had no effect on method using catalase-Hantzsch reaction. 3393   At concentration of 2600 mg/L had no effect on Liebermann-Burchard method. 3393

Diazepam   At 5 times upper limit of therapeutic range on methods on SMAC®, Abbott-VP, Cobas-Bio, Ektachem®, Hitachi® 705 and KDA. 2138   At acute overdose concentration (2.5 mg/dL) on Technicon® SMAC® method. 3719   At concentration of 1.5 mg/L had no effect on CHOD-PAP method. 3393   At concentration of 25 mg/L had no effect on Liebermann-Burchard method. 3393

Diclofenac   No effect at therapeutic concentrations on enzymatic and Liebermann-Burchard methods. 1786   On enzymatic method on SMA II at physiological concentration. 1787   At concentration of 23 mg/L had no effect on CHOD-PAP method. 3393

Diethylaminoethanol   At concentration of 30 mg/L had no effect on Liebermann-Burchard method. 3393

Diethylpropion   At acute overdose concentration (10 mg/dL) on Technicon® SMAC® method. 3719   At concentration of 100 mg/L had no effect on Liebermann-Burchard method. 3393

Digitoxin   No effect at 21 mg/dL on SMA 12/60 method. 2636   At concentration of 6 mg/L had no effect on Liebermann-Burchard method. 3393

Digoxin   No effect at 0.04 mg/dL on SMA 12/60 method. 2636   At concentration of 100 mg/L had no effect on CHOD-PAP method. 3393   At concentration of 100 mg/L had no effect on method using catalase-Hantzsch reaction. 3393   At concentration of 9 mg/L had no effect on Liebermann-Burchard method. 3393

Dimethadione   At concentration of 2,000 mg/L had no effect on Liebermann-Burchard method. 3393

N-Dimethyldiazepam   At concentration of 6 mg/L had no effect on Liebermann-Burchard method. 3393

Diphenhydramine   At acute overdose concentration (20 mg/dL) on Technicon® SMAC® method. 3719   At concentration of 200 mg/L had no effect on Liebermann-Burchard method. 3393

Disopyramide   At concentration of 4 mg/L had no effect on CHOD-PAP method. 3393

Disulfiram   At concentration of 120 mg/L had no effect on Liebermann-Burchard method. 3393

Disulphine Blue   By cholesterol oxidase endpoint method. 1464

Doxepin   At concentration of 150 mg/L had no effect on CHOD-Iodide method. 3393   At concentration of 150 mg/L had no effect on CHOD-PAP method. 3393   At concentration of 150 mg/L had no effect on catalase-AIDH method. 3393   At concentration of 150 mg/L had no effect on method using catalase-Hantzsch reaction. 3393   At concentration of 150 mg/L had no effect on Liebermann-Burchard method. 3393

EDTA   At concentration of 1,000 mg/L had no effect on CHOD-Iodide method. 3393   At concentration of 1,000 mg/L had no effect on CHOD-PAP method. 3393   At concentration of 1,000 mg/L had no effect on catalase-AIDH method. 3393   At concentration of 10,000 mg/L had no effect on method using catalase-Hantzsch reaction. 3393   At concentration of 1,000 mg/L had no effect on Liebermann-Burchard method. 3393

Ethacrynic Acid   At 2.5 mg/dL no effect on SMA 12/60 method. 2636

Ethanol   At acute overdose concentration (20 mg/dL) on Technicon® SMAC® methods. 3719

Ethaverine   No effect at therapeutic concentration on Reflotron method. 1984

Ethchlorvynol   At acute overdose concentration (20 mg/dL) on Technicon® method. 3719   At concentration of 400 mg/L had no effect on Liebermann-Burchard method. 3393

Ethinamate   At acute overdose concentration (20 mg/dL) on Technicon® SMAC® method. 3719   At concentration of 200 mg/L had no effect on Liebermann-Burchard method. 3393

Ethosuximide   At concentration of 390 mg/L had no effect on Liebermann-Burchard method. 3393

Fluorescein   No effect at 200 mg/L on Ektachem® 700 method. 1722

Fluosol-DA   On Du Pont aca III at concentration of 50%. 2518

Flurazepam   At 5 times upper limit of therapeutic range on methods on SMAC®, Abbott-VP, Cobas-Bio, Ektachem®, Hitachi® 705 and KDA. 2138   At acute overdose concentration (2.5 mg/dL) on Technicon® SMAC® method. 3719   At concentration of 0.1 mg/L had no effect on CHOD-PAP method. 3393   At concentration of 25 mg/L had no effect on Liebermann-Burchard method. 3393

Furosemide   No effect at 1.4 mg/dL on SMA 12/60 method. 2636   No effect at therapeutic concentration on Reflotron method. 1984   At concentration of 60 mg/L had no effect on CHOD-Iodide method. 3393   At concentration of 60 mg/L had no effect on CHOD-PAP method. 3393   At concentration of 60 mg/L had no effect on catalase-AIDH method. 3393   At concentration of 60 mg/L had no effect on method using catalase-Hantzsch reaction. 3393   At concentration of 60 mg/L had no effect on Liebermann-Burchard method. 3393

Gentamicin   At concentration of 6 mg/L had no effect on CHOD-PAP method. 3393   At concentration of 150 mg/L had no effect on Liebermann-Burchard method. 3393

Glibenclamide   No effect at therapeutic concentration on Reflotron method. 1984   At concentration of 32 mg/L had no effect on CHOD-Iodide method. 3393   At concentration of 32 mg/L had no effect on CHOD-PAP method. 3393   At concentration of 32 mg/L had no effect on catalase-AIDH method. 3393   At concentration of 32 mg/L had no effect on method using catalase-Hantzsch reaction. 3393   At concentration of 32 mg/L had no effect on Liebermann-Burchard method. 3393

Glutethimide   At acute overdose concentration (5 mg/dL) on Technicon® SMAC® method. 3719   At concentration of 50 mg/L had no effect on Liebermann-Burchard method. 3393

Hemoglobin   On Leffler's procedure at 20 g/L. 1743   With up to 2 g/L on enzymatic procedure. 3027

Hemolysis   Calculated error less than 1 %/g after extract. 0515   When measured enzymatically with oxidase. 3027

Hydralazine   At concentration of 0.5 mg/L had no effect on CHOD-PAP method. 3393   At concentration of 160 mg/L had no effect on Liebermann-Burchard method. 3393

Hyoscine-N-Butylbromide   At concentration of 300 mg/L had no effect on CHOD-Iodide method. 3393   At concentration of 300 mg/L had no effect on CHOD-PAP method. 3393   At concentration of 300 mg/L had no effect on catalase-AIDH method. 3393   At concentration of 300 mg/L had no effect on method using catalase-Hantzsch reaction method. 3393   At concentration of 300 mg/L had no effect on Liebermann-Burchard method. 3393

Ibuprofen   No effect at therapeutic concentrations on Liebermann-Burchard and enzymatic methods. 1786   On enzymatic method on SMA II at physiological concentration. 1787   At 5 times upper limit of therapeutic range on methods on SMAC®, Abbott-VP, Cobas-Bio, Ektachem®, Hitachi® 705 and KDA. 2138   At concentration of 200 mg/L had no effect on CHOD-PAP method. 3393

Indomethacin   No effect at therapeutic concentrations on enzymatic and Liebermann-Burchard methods. 1786   On enzymatic method on SMA II at physiological concentration. 1787   No effect at therapeutic concentration on Reflotron method. 1984   At concentration of 30 mg/L had no effect on CHOD-Iodide method. 3393   At concentration of 30 mg/L had no effect on CHOD-PAP method. 3393   At concentration of 30 mg/L had no effect on catalase-AIDH method. 3393   At concentration of 30 mg/L had no effect on method using catalase-Hantzsch reaction. 3393   At concentration of 30 mg/L had no effect on Liebermann-Burchard method. 3393

Insulin   At concentration of 3 mg/L had no effect on Liebermann-Burchard method. 3393

Iproniazid   At concentration of 40 mg/L had no effect on Liebermann-Burchard method. *3393*

Isoniazid   At concentration of 100 mg/L had no effect on Liebermann-Burchard method. *3393*

Ketoprofen   No effect on Liebermann-Burchard and enzymatic methods at therapeutic concentrations. *1786*   On enzymatic method on SMA II at physiological concentration. *1787*   At concentration of 60 mg/L had no effect on Liebermann-Burchard method. *3393*

Lidocaine   No effect SMA 12/60 method with 3.5 mg/dL. *2636*

Lorazepam   At concentration of 0.05 mg/L had no effect on Liebermann-Burchard method. *3393*

Mannitol   No effect on SMA 12/60 method at 445 mg/dL. *2636*

Meglumine   At concentration of 1200 mg/L had no effect on CHOD-Iodide method. *3393*   At concentration of 1200 mg/L had no effect on CHOD-PAP method. *3393*   At concentration of 1200 mg/L had no effect on catalase-AIDH method. *3393*   At concentration of 1200 mg/L had no effect on method using catalase-Hantzsch reaction. *3393*   At concentration of 1200 mg/L had no effect on Liebermann-Burchard method. *3393*

Meperidine   At acute overdose concentration (5 mg/dL) on Technicon® SMAC® method. *3719*   At concentration of 60 mg/L had no effect on Liebermann-Burchard method. *3393*

Mephenesin   At concentration of 100 mg/L had no effect on Liebermann-Burchard method. *3393*

Meprobamate   At acute overdose concentration (20 mg/dL) on Technicon® SMAC® method. *3719*   At concentration of 25 mg/L had no effect on CHOD-PAP method. *3393*   At concentration of 200 mg/L had no effect on Liebermann-Burchard method. *3393*

Meralluride   No effect at 0.07 mL/dL on SMA 12/60 method. *2636*

Mesoridazine   At acute overdose concentration (20 mg/dL) on Technicon® SMAC® method. *3719*

Methadone   At acute overdose concentration (2.5 mg/dL) on Technicon® SMAC® method. *3719*   At concentration of 25 mg/L had no effect on Liebermann-Burchard method. *3393*

Methamphetamine   At concentration of 2 mg/L had no effect on CHOD-PAP method. *3393*   At concentration of 2 mg/L had no effect on Liebermann-Burchard method. *3393*

Methanol   At acute overdose concentration (20 mg/dL) on Technicon® SMAC® method. *3719*

Methapyrilene   At concentration of 13 mg/L had no effect on Liebermann-Burchard method. *3393*

Methaqualone   At acute overdose concentration (2.5 mg/dL) on Technicon® SMAC® method. *3719*   No effect at therapeutic concentration on Reflotron method. *1984*   At concentration of 25 mg/L had no effect on Liebermann-Burchard method. *3393*

Methicillin   At 5 times upper limit of therapeutic range on methods on SMAC®, Abbott-VP, Cobas-Bio, Ektachem®, Hitachi® 705 and KDA. *2138*   At concentration of 20 mg/L had no effect on CHOD-PAP method. *3393*   At concentration of 900 mg/L had no effect on Liebermann-Burchard method. *3393*

Methohexital   At concentration of 50 mg/L had no effect on Liebermann-Burchard method. *3393*

Methotrexate   At 5 times upper limit of therapeutic range on methods on SMAC®, Abbott-VP, Cobas-Bio, Ektachem® and Hitachi® 705. *2138*   At concentration of 500 mg/L had no effect on CHOD-Iodide method. *3393*   At concentration of 500 mg/L had no effect on CHOD-PAP method. *3393*   At concentration of 500 mg/L had no effect on catalase-AIDH method. *3393*   At concentration of 500 mg/L had no effect on method using catalase-Hantzsch reaction. *3393*   At concentration of 500 mg/L had no effect on Liebermann-Burchard method. *3393*

Methotrimeprazine   At concentration of 1 mg/L had no effect on CHOD-PAP method. *3393*

Methyldopa   No effect at therapeutic concentration on Reflotron method. *1984*   At concentration of 400 mg/L had no effect on catalase-AIDH method. *3393*   At concentration of 400 mg/L had no effect on method using catalase-Hantzsch reaction. *3393*   At concentration of 400 mg/L had no effect on Liebermann-Burchard method. *3393*

Methylphenidate   At acute overdose concentration (20 mg/dL) on Technicon® SMAC® method. *3719*   At concentration of 200 mg/L had no effect on Liebermann-Burchard method. *3393*

Methylphenobarbital   At concentration of 150 mg/L had no effect on Liebermann-Burchard method. *3393*

Methyprylon   At acute overdose concentration (10 mg/dL) on Technicon® SMAC® method. *3719*   At concentration of 100 mg/L had no effect on Liebermann-Burchard method. *3393*

Metoprolol   At concentration of 0.34 mg/L had no effect on CHOD-PAP method. *3393*

Morphine   At acute overdose concentration (20 mg/dL) on Technicon® SMAC® method. *3719*   At concentration of 200 mg/L had no effect on Liebermann-Burchard method. *3393*

Moxalactam   At 2.50 mmol/L on method on Eppendorf Epos. *1414*

Nafcillin   At concentration of 50 mg/L had no effect on Liebermann-Burchard method. *3393*

Naproxen   At 5 times upper limit of therapeutic range on methods on SMAC®, Abbott-VP, Cobas-Bio, Ektachem®, Hitachi® 705 and KDA. *2138*

Niacin   No effect at therapeutic concentration on Reflotron method. *1984*

Niacinamide   At concentration of 40 mg/L had no effect on CHOD-Iodide method. *3393*   At concentration of 40 mg/L had no effect on CHOD-PAP method. *3393*   At concentration of 40 mg/L had no effect on catalase-AIDH method. *3393*   At concentration of 400 mg/L had no effect on catalase-Hantzsch reaction. *3393*   At concentration of 40 mg/L had no effect on Liebermann-Burchard method. *3393*

Niflumic Acid   At concentration of 200 mg/L had no effect on CHOD-PAP method. *3393*

Nitrofurantoin   On routine methods in use on SMAC®, Ektachem®, Abbott-VP, Cobas-Bio, aca, Hitachi® 705, KDA at 5 times normal upper therapeutic concentration. *2138*   No effect at therapeutic concentration on Reflotron method. *1984*   At concentration of 98 mg/L had no effect on CHOD-Iodide method. *3393*   At concentration of 98 mg/L had no effect on CHOD-PAP method. *3393*   At concentration of 98 mg/L had no effect on catalase-AIDH method. *3393*   At concentration of 98 mg/L had no effect on method using catalase-Hantzsch reaction. *3393*   At concentration of 98 mg/L had no effect on Liebermann-Burchard method. *3393*

Norfenefrin   At concentration of 2.4 mg/L had no effect on CHOD-Iodide method. *3393*   At concentration of 2.4 mg/L had no effect on CHOD-PAP method. *3393*   At concentration of 2.4 mg/L had no effect on catalase-AIDH method. *3393*   At concentration of 2.4 mg/L had no effect on method using catalase-Hantzsch reaction. *3393*   At concentration of 2.4 mg/L had no effect on Liebermann-Burchard method. *3393*

Nortriptyline   At acute overdose concentration (20 mg/dL) on Technicon® SMAC® method. *3719*   At concentration of 200 mg/L had no effect on Liebermann-Burchard method. *3393*

Novaminsulfon   At concentration of 900 mg/L had no effect on CHOD-Iodide method. *3393*   At concentration of 900 mg/L had no effect on catalase-AIDH method. *3393*   At concentration of 1160 mg/L had no effect on catalase-Hantzsch method. *3393*   At concentration of 900 mg/L had no effect on Liebermann-Burchard method. *3393*

Orphenadrine   At concentration of 9 mg/L had no effect on Liebermann-Burchard method. *3393*

Ouabain   At 0.02 mg/dL no effect on SMA 12/60 method. *2636*

Oxazepam   No effect at therapeutic concentration on Reflotron method. *1984*   At concentration of 1 mg/L had no effect on Liebermann-Burchard method. *3393*

Oxyphenbutazone   At concentration of 600 mg/L had no effect on CHOD-Iodide method. *3393*   At concentration of 600 mg/L had no effect on CHOD-PAP method. *3393*   At concentration of 600 mg/L had no effect on catalase-AIDH method. *3393*   At concentration of 600 mg/L had no effect on method using catalase-Hantzsch reaction. *3393*   At concentration of 600 mg/L had no effect on Liebermann-Burchard method. *3393*

Oxytetracycline   No effect at therapeutic concentration on Reflotron method. *1984*

Papaverine   At concentration of 10 mg/L had no effect on Liebermann-Burchard method. *3393*

Paraldehyde   At concentration of 2,000 mg/L had no effect on Liebermann-Burchard method. *3393*

Penicillamine   At concentration of 960 mg/L had no effect on CHOD-PAP method. *3393*   At concentration of 960 mg/L had no effect on catalase-AIDH method. *3393*   At concentration of 960 mg/L had no effect on method using catalase-Hantzsch reaction. *3393*   At concentration of 960 mg/L had no effect on Liebermann-Burchard method. *3393*

Cholesterol (continued)

Serum No Effect Analytical (continued)

Penicillin G At 5 times upper limit of therapeutic range on methods on SMAC®, Abbott-VP, Cobas-Bio, Ektachem®, Hitachi® 705 and KDA. *2138* At concentration of 2,000 mg/L had no effect as measured by Liebermann-Burchard method. *3393*

Pentobarbital At acute overdose concentration (20 mg/dL) on Technicon® SMAC® method. *3719* At concentration of 340 mg/L had no effect on Liebermann-Burchard method. *3393*

Perphenazine At acute overdose concentration (20 mg/dL) on Technicon® SMAC® method. *3719* At concentration of 1 mg/L had no effect on CHOD-PAP method. *3393* At concentration of 200 mg/L had no effect on Liebermann-Burchard method. *3393*

Phenazopyridine No effect at therapeutic concentration on Reflotron method. *1984*

Phencyclidine At concentration of 6 mg/L had no effect on Liebermann-Burchard method. *3393*

Phenobarbital No effect on enzymatic and Liebermann-Burchard methods at therapeutic concentrations. *1786* At acute overdose concentration (20 mg/dL) on Technicon® SMAC® method. *3719* No effect at therapeutic concentration on Reflotron method. *1984* At concentration of 352 mg/L had no effect on CHOD-Iodide method. *3393* At concentration of 650 mg/L had no effect on CHOD-PAP method. *3393* At concentration of 352 mg/L had no effect on catalase-AIDH method. *3393* At concentration of 520 mg/L had no effect on method using catalase-Hantzsch reaction. *3393* At concentration of 352 mg/L had no effect on Liebermann-Burchard method. *3393*

Phenothiazines At concentration of 150 mg/L had no effect on CHOD-Iodide method. *3393* At concentration of 150 mg/L had no effect on CHOD-PAP method. *3393* At concentration of 150 mg/L had no effect on catalase-AIDH method. *3393* At concentration of 150 mg/L had no effect on method using catalase-Hantzsch reaction. *3393* At concentration of 150 mg/L had no effect on Liebermann-Burchard method. *3393*

Phenprocoumon No effect at therapeutic concentration on Reflotron method. *1984* At concentration of 80 mg/L had no effect on CHOD-Iodide method. *3393* At concentration of 80 mg/L had no effect on CHOD-PAP method. *3393* At concentration of 80 mg/L had no effect on catalase-AIDH method. *3393* At concentration of 80 mg/L had no effect on method using catalase-Hantzsch reaction. *3393* At concentration of 80 mg/L had no effect on Liebermann-Burchard method. *3393*

Phensuximide At concentration of 120 mg/L had no effect on Liebermann-Burchard method. *3393*

Phenylbutazone At concentration of 280 mg/L had no effect on CHOD-Iodide method. *3393* At concentration of 280 mg/L had no effect on CHOD-PAP method. *3393* At concentration of 280 mg/L had no effect on catalase-AIDH method. *3393* At concentration of 280 mg/L had no effect on method using catalase-Hantzsch reaction. *3393* At concentration of 750 mg/L had no effect on Liebermann-Burchard method. *3393*

Phenylephrine At concentration of 4 mg/L had no effect on Liebermann-Burchard method. *3393*

Phenytoin No effect at 1.8 mg/dL on SMA 12/60 method. *2636* At acute overdose concentration (20 mg/dL) on Technicon® SMAC® method. *3719* No effect at therapeutic concentration on Reflotron method. *1984* At concentration of 20 mg/L had no effect on CHOD-PAP method. *3393* At concentration of 240 mg/L had no effect on Liebermann-Burchard method. *3393*

Prednisolone At concentration of 8 mg/L had no effect on CHOD-PAP method. *3393* At concentration of 8 mg/L had no effect on method using catalase-Hantzsch reaction. *3393*

Probenecid No effect at therapeutic concentration on Reflotron method. *1984* At concentration of 260 mg/L had no effect on CHOD-Iodide method. *3393* At concentration of 280 mg/L had no effect on CHOD-PAP method. *3393* At concentration of 260 mg/L had no effect on catalase-AIDH method. *3393* At concentration of 260 mg/L had no effect on method using catalase-Hantzsch reaction. *3393* At concentration of 1300 mg/L had no effect on Liebermann-Burchard method. *3393*

Procainamide No effect on SMA 12/60 method at 35 mg/dL. *2636*

Procaine No effect at therapeutic concentration on Reflotron method. *1984* At concentration of 2 mg/L had no effect on Liebermann-Burchard method. *3393*

Prochlorperazine At concentration of 1 mg/L had no effect on CHOD-PAP method. *3393*

Promethazine At acute overdose concentration (20 mg/dL) on Technicon® SMAC® method. *3719* At concentration of 1 mg/L had no effect on CHOD-PAP method. *3393* At concentration of 200 mg/L had no effect on Liebermann-Burchard method. *3393*

Propoxyphene At acute overdose concentration (2.5 mg/dL) on Technicon® SMAC® method. *3719* At concentration of 25 mg/L had no effect on Liebermann-Burchard method. *3393*

Propranolol No effect at 0.1 mg/dL on SMA 12/60 method. *2636* At concentration of 0.2 mg/L had no effect on CHOD-PAP method. *3393*

Protein When measured enzymatically with oxidase. *3027*

Pyribenzamine® At acute overdose concentration (20 mg/dL) on Technicon® SMAC® methods. *3719*

Pyridamole No effect at therapeutic concentration on Reflotron method. *1984*

Pyritinol No effect at therapeutic concentration on Reflotron method. *1984*

Quinidine At 26 mg/dL no effect on SMA 12/60 method. *2636* At acute overdose concentration (20 mg/dL) Technicon® SMAC® method. *3719* No effect at therapeutic concentration on Reflotron method. *1984* At concentration of 210 mg/L had no effect on Liebermann-Burchard method. *3393*

Quinine At acute overdose concentration (1.5 mg/dL) on Technicon® SMAC® method. *3719* At concentration of 30 mg/L had no effect on Liebermann-Burchard method. *3393*

Reserpine No effect at 0.02 mg/dL on SMA 12/60 method. *2636*

Rifampin At 5 times upper limit of therapeutic concentration on methods on SMAC®, Abbott-VP, aca, Ektachem®, Hitachi® 705 and KDA. *2138*

Salicylate At concentration of 500 mg/L had no effect on method using catalase-Hantzsch reaction. *3393* At concentration of 500 mg/L had no effect on Liebermann-Burchard method. *3393*

Secobarbital At acute overdose concentration (20 mg/dL) on Technicon® SMAC® method. *3719* At concentration of 1 mg/L had no effect on CHOD-PAP method. *3393* At concentration of 200 mg/L had no effect on Liebermann-Burchard method. *3393*

Serum If heparin used for plasma. *1563*

β-Sitosterol On GLC procedure of MacGee. *2252*

Sitosterols On GLC procedure of MacGee. *2252*

Sodium Citrate At concentration of 5,000 mg/L had no effect on CHOD-Iodide method. *3393* At concentration of 5,000 mg/L had no effect on CHOD-PAP method. *3393* At concentration of 5,000 mg/L had no effect on catalase-AIDH method. *3393* At concentration of 30,000 mg/L had no effect on method using catalase-Hantzsch reaction. *3393* At concentration of 5,000 mg/L had no effect on Liebermann-Burchard method. *3393*

Sodium Fluoride At concentration of 2,000 mg/L had no effect on CHOD-Iodide method. *3393* At concentration of 2,000 mg/L had no effect on CHOD-PAP method. *3393* At concentration of 2,000 mg/L had no effect on catalase-AIDH method. *3393* At concentration of 30,000 mg/L had no effect on method using catalase-Hantzsch reaction. *3393* At concentration of 2,000 mg/L had no effect on Liebermann-Burchard method. *3393*

Sodium Heparin At concentration of 750 mg/L had no effect on CHOD-Iodide method. *3393* At concentration of 750 mg/L had no effect on CHOD-PAP method. *3393* At concentration of 750 mg/L had no effect on catalase-AIDH method. *3393* At concentration of 750 mg/L had no effect on Katalase-Hantzsch method. *3393* At concentration of 750 mg/L had no effect on Liebermann-Burchard method. *3393*

Sodium Oxalate At concentration of 2,000 mg/L had no effect on CHOD-Iodide method. *3393* At concentration of 2,000 mg/L had no effect on CHOD-PAP method. *3393* At concentration of 2,000 mg/L had no effect on catalase-AIDH method. *3393* At concentration of 30,000 mg/L had no effect on method using catalase-Hantzsch reaction. *3393* At concentration of 2,000 mg/L had no effect on Liebermann-Burchard method. *3393*

Spironolactone At concentration of 20 mg/L had no effect on CHOD-Iodide method. *3393* At concentration of 20 mg/L had no effect on CHOD-PAP method. *3393* At concentration of 20 mg/L had no effect on catalase-AIDH method. *3393* At concentration of 20 mg/L had no effect on method using catalase-Hantzsch reaction. *3393* At concentration of 20 mg/L had no effect on Liebermann-Burchard method. *3393*

Standing of Sample If on polystyrene beads 24 h at 25°. *2624*

Stigmasterol   On GLC procedure of MacGee. *2252*

Storage of Sample   No effect 2 d at 4°, 5 y at -20°. *3873*

Strychnine   At concentration of 12 mg/L had no effect on Liebermann-Burchard method. *3393*

Sulfadiazine   At concentration of 1500 mg/L had no effect on Liebermann-Burchard method. *3393*

Sulfaguanidine   At concentration of 500 mg/L had no effect on Liebermann-Burchard method. *3393*

Sulfamethoxazole   No effect at therapeutic concentration on Reflotron method. *1984*

Sulfamethoxydiazine   At concentration of 231 mg/L had no effect on CHOD-PAP method. *3393*   At concentration of 231 mg/L had no effect on catalase-AIDH method. *3393*   At concentration of 231 mg/L had no effect on method using catalase-Hantzsch reaction. *3393*   At concentration of 231 mg/L had no effect on Liebermann-Burchard method. *3393*

Sulfanilamide   At concentration of 1,000 mg/L had no effect on Liebermann-Burchard method. *3393*

Sulforidazine   At concentration of 200 mg/L had no effect on Liebermann-Burchard method. *3393*

Tetracycline   At 5 times upper limit of therapeutic range on methods on SMAC®, Abbott-VP, Cobas-Bio, Ektachem®, Hitachi® 705 and KDA. *2138*   At concentration of 200 mg/L had no effect on CHOD-PAP method. *3393*   At concentration of 200 mg/L had no effect on catalase-AIDH method. *3393*   At concentration of 200 mg/L had no effect on method using catalase-Hantzsch reaction. *3393*   At concentration of 300 mg/L had no effect on Liebermann-Burchard method. *3393*

Theobromine   At concentration of 2,000 mg/L had no effect on Liebermann-Burchard method. *3393*

Theophylline   At 5 times upper limit of therapeutic range on methods on SMAC®, Abbott-VP, Cobas-Bio, Ektachem®, Hitachi® 705 and KDA. *2138*   No effect at therapeutic concentration on Reflotron method. *1984*   At concentration of 20 mg/L had no effect on CHOD-PAP method. *3393*   At concentration of 2,000 mg/L had no effect on Liebermann-Burchard method. *3393*

Thiopental   At acute overdose concentration (20 mg/dL) on Technicon® SMAC® method. *3719*   At concentration of 200 mg/L had no effect on Liebermann-Burchard method. *3393*

Timolol   At concentration of 0.01 mg/L had no effect on CHOD-PAP method. *3393*

Tolbutamide   At concentration of 480 mg/L had no effect on CHOD-Iodide method. *3393*   At concentration of 480 mg/L had no effect on CHOD-PAP method. *3393*   At concentration of 480 mg/L had no effect on catalase-AIDH method. *3393*   At concentration of 480 mg/L had no effect on method using catalase-Hantzsch reaction. *3393*   At concentration of 480 mg/L had no effect on Liebermann-Burchard method. *3393*

Tribromethanol   At concentration of 90 mg/L had no effect on Liebermann-Burchard method. *3393*

Trichlorethanol   At concentration of 12 mg/L had no effect on CHOD-PAP method. *3393*   At concentration of 1,000 mg/L had no effect on Liebermann-Burchard method. *3393*

Trifluoperazine   At concentration of 1 mg/L had no effect on Liebermann-Burchard method. *3393*

Trimethoprim   No effect at therapeutic concentration on Reflotron method. *1984*

Tripelennamine   At concentration of 200 mg/L had no effect on Liebermann-Burchard method. *3393*

Tryptophan   Insignificant effect on Leffler's procedure. *1743*

Vitamin A   Unlikely that affects FeCl3 retention. *1487*

Vitamin B Complex   At concentration of 12.9 mg/L had no effect on CHOD-Iodide method. *3393*   At concentration of 12.9 mg/L had no effect on CHOD-PAP method. *3393*   At concentration of 12.9 mg/L had no effect on catalase-AIDH method. *3393*   At concentration of 12.9 mg/L had no effect on method using catalase-Hantzsch reaction. *3393*   At concentration of 12.9 mg/L had no effect on Liebermann-Burchard method. *3393*

Vitamin Preparations   No effect at expected concentration with SMA 12/60 procedure. *2637*

Warfarin   At concentration of 1.5 mg/L had no effect on CHOD-PAP method. *3393*   At concentration of 100 mg/L had no effect on Liebermann-Burchard method. *3393*

Serum   No Effect   Physiological

Acebutolol   In small numbers of patients when treated for 1-12 mo. *0088*   Decrease by -3 mg/dL in 11 patients given 400 mg daily for 3 mo. *2550*   Insignificant change on average in one 6 mo long study. *2116*   In several studies of about 15 patients treated for 1 to 12 mo. *0089*

Aging   No significant change from age 6-20 y. *0887*

Ascorbic Acid   No influence on concentration of dietary intake in elderly. *1757*

Atenolol   Insignificant change after 6 mo treatment of 14 hypertensives. *0836*   No significant change in 15 hypertensives treated for 8 mo. *1021*   Insignificant change in 3 studies of from 1 to 3 mo. *2116*   In 20 patients with 50 mg/d compared with pre-treatment values after 3 mo. *3071*   Typically no significant change from several studies. *0088*

Bendrofluazide   No significant change in 15 individuals when given as sole treatment for less than 1 y. *3826*   No effect with treatment for several years in many patients. *0089*

Bopindolol   In 24 hypertensives treated with drug for 3 mo. *3667*

Bran   No significant effect with 38 g/d for 5 weeks. *1540*

Bunitrolol   When 30 mg/d given for 12 weeks in normolipidemic patients with mild essential hypertension. *3151*

Captopril   No significant change in patients in 2 studies treated for 2 and 24 mo. *0088*   After 12 weeks treatment in 18 patients with mild essential hypertension. *3150*   No significant change in 7,000 hypertensives treated for 3 y. *1407*   Decrease by -2 mg/dL in 15 patients given 75 mg daily for 8 weeks. *2550*

Carprazidil   No significant effect in 12 subjects treated for 4 mo. *0088*   In one study involving 12 patients treated for 4 mo. *0089*

Celiprolol   In patients with hyperlipoproteinemia types IIa, IIb or IV with 300 mg drug for 4 weeks. *2873*

Chenodeoxycholic Acid   No significant change over 3 mo treatment of gallstones. *0274*   No significant change with 150 mg four times daily in patients with endogenous hypertriglyceridemias. *0601*

Chloramphenicol   No effect seen even when administration orally. *1487*

Chlormadinone   No significant effect observed. *1309*

Chlorthalidone   In 22 premenopausal women without decrease in blood pressure treated for less than 1 y. *0402*   In 22 premenopausal women given 100 mg/d for 6 weeks. *0402*

Cimetidine   In healthy individuals and in subjects with high serum glucose concentration. *1653*   In 25 patients pretreatment 174 mg/dL, after 181 mg/dL over 5 weeks. *3568*

Clopamide   No significant change in 17 individuals treated for less than 1 y when used as sole treatment. *3184*

Danazol   Inconsistent changes observed in 62 patients with endometriosis treated with 600 mg daily for up to 24 weeks. *3222*   Insignificant change in 9 women with endometriosis treated for 6 mo. *0060*   No significant change over 24 weeks in 12 women with endometriosis given 600 mg daily for 24 weeks. *1066*   No effect after 6 mo of 600 mg/d. *3906*

Desogestrel   When 150 μg desogestrel given with 30 μg ethinyl estradiol for 3 mo. Results normal within 2 mo after treatment stopped. *0312*

Diltiazem   No significant change in 31 subjects treated for 6 mo. *0088*

Doxazosin   No significant change in 44 patients treated for 3 mo. *0088*   Compared with controls over 10-12 weeks, although 9% increase in HDL/total cholesterol ratio. *2842*

Enalapril   In 53 patients given up to 160 mg daily or when combined with hydrochlorothiazide. *2277*

Ethanol   No significant difference between heavy and occasional drinkers. *3273*   Usually remain unchanged in response to acute ingestion. *2428*   No effect of ingestion of 1.5 g/kg at night on concentration measured next morning. *3547*

Furosemide   Increase by 2 mg/dL in 12 patients given 40 to 80 mg daily for 3 mo. *2550*

Glibonuride   No significant effect observed after 4 weeks. *0301*

Glucose   In response to infusion of 1 g/kg body weight in normal volunteers and non-insulin dependent diabetics. *0476*

Gum Tragacanth   In 5 male volunteers consuming 9.9 g daily for 21 d. *0990*

Halofenate   No significant effect observed in hyperlipemics. *0161*

Haloperidol   No detectable effect observed in man. *3327*

Cholesterol *(continued)*

Serum No Effect Physiological *(continued)*

Hydrochlorothiazide No effect with 25 or 50 mg/d in hypertensive patients. *1980* In approximately 175 patients with essential hypertension treated for 1 y. *3708* In long term involving many patients. *0089*

Indapamide In 27 subjects treated for less than 1 y with this drug only. *0692* In various studies involving more than 30 patients when used as sole drug for less than 1 y. *0089*

Indomethacin No effect seen in patients undergoing treatment. *1787*

Ketanserin In 50 hypertensive patients given 80 mg/daily for 3 mo. *2171*

Ketoprofen No effect seen in patients undergoing treatment. *1787*

Labetalol No significant change in several studies with patients treated for 1-12 mo. *0088* In 8 patients given 600-1200 mg daily for 4 mo. *1806* In several studies from 1 to 12 mo. *0089*

Levonorgestrel When 150 μg levonorgestrel given with 30 μg ethinyl estradiol for 3 mo. Results normal within 2 mo after treatment stopped. *0312* Insignificant decrease in 11 normolipoproteinemic women given 250 μg/d for 2 weeks. *3593* In a group of women using levonorgestrel covered rods versus controls using copper IUD. *0899*

Meals Not significantly affected by nonstandardized meals. *3455* No effect after standard breakfast. *0579*

Metformin No effect although phenformin causes decrease. *1487*

Methyldopa In several patients treated for 2-4 mo. *0088* Increase by 2 mg/dL in 11 patients given 750 mg daily for 8 weeks. *2550* In 3 studies of 7 to 17 patients for up to 3 mo. *0089*

Metoprolol In several studies for up to 3 mo. *0088* In 20 hypertensive diabetic patients. *3900* No effect in 1 or 3 mo study. *2116* In 20 hypertensives given 200 mg/d for 3 mo. *3071*

MPCA No significant change after 1 mo treatment (270 mg tid). *1423*

Muscular Exercise Observed effect with 12 minutes cycle-ergometer. *1231* But relative reduction of unsaturated acid. *1702*

Nadolol In 94 patients treated for 3 mo. *3706* In 13 patients given 50-200 mg daily for 10 weeks. *1806*

Niacinamide No effect observed in normals. *1622*

Nifedipine No effect observed in 14 patients with essential hypertension treated over 2 mo with 20 mg twice daily. *2119* Decrease by -5 mg/dL in patients given 30 to 60 mg daily for 8 weeks. *2550*

Nisoldipine In 14 mild to moderately hypertensive noninsulin dependent diabetic patients. *2653*

Norethindrone No significant effect with 0.4 mg/d. *1309*

Norethisterone In 75 women who had received up to 24 intragluteal injections of 200 mg every 56 d. *1666*

Oral Contraceptives No effect observed usually. *2427* No difference between treated women and others. *1303* May occur due to opposing effects of estrogen and progestogen components. *0018* In 10 women receiving ethinyl estradiol with norgestrel. *2009* Typically insignificant change over 6 cycles in women with low dose ethinyl estradiol with low dose levonorgestrel or desogestrel. *1249*

Oxprenolol Insignificant effect in 12 patients receiving 160 mg/d for 2 weeks. *1806* In 20 hypertensive men given 160 mg/d for 5 weeks. *2135* In several studies treated for more than 1 mo. *0088*

Oxytetracycline No effect seen even when administered orally. *1487*

Phenytoin In women in 27 patients with transient brain ischemia. *1879* No significant difference between epileptics given 200-300 mg/d for 1 to 35 y and controls. *2598*

Physical Training No difference between trained athletes and others. *1702*

Pindolol No significant change typically seen with treatment for 1 mo or more. *0088* No effect in short-term study of 10 hypertensive men with 15 mg/d. *2135* No significant change in four studies of 1 to 12 mo. *2116*

Practolol Insignificant changes in 2 studies of 2 weeks and 6 mo. *1806*

Prazosin In 22 mild/moderate male hypertensives treated for 8 weeks versus controls. Effect not significant though slight decrease. *1693* Decrease by 12 mg/dL in 16 patients given 1 to 3 mg daily for 8 weeks. *2550* No significant change with monotherapy for 12 mo in 15 patients. *3531* Effect observed with doses as low as 1 to 3 mg/d. *3148* In 16 hypertensive patients given 2 mg daily for 3 mo. *2266*

Prednisolone In men receiving drug for mean of 3.1 y. *1783*

Propranolol In 22 mild/moderate hypertensive males treated for 8 weeks: insignificant increase noted. *1693* General effect observed in multiple studies. *0088* In 23 hypertensive men given up to 160 mg/d for 8 weeks. *2135* Increase by 1.0-1.5 mmol/L in 20 hypertensive diabetic patients. *3900* Insignificant change after 6 mo treatment in 16 hypertensives. *0836* In 53 hypertensives given 80 mg twice daily for 3 mo. *0835* In 340 patients given drug for 10 weeks to reduce diastolic BP to less than 90 mm Hg. *3707* Increase by 2 mg/dL in 8 patients given 30 to 60 mg daily for 8 weeks. *2550* In 23 hypertensive men aged 47-55 y treated for 8 weeks. *1551* In 20 hypertensive diabetic patients. *3900* In 50 volunteers given 160 mg daily for 3 mo. *2171* Range from -2 to 9% change but mostly no change in many studies. *0089*

Quingestanol No significant effect with 300 μg/d 6 mo. *1309*

Ranitidine In 25 patients pretreatment 171 mg/dL, after 175 mg/dL over 5 weeks. *3568*

Reserpine No significant change in small number of patients treated for up to 2.5 mo. *0088*

Smoking No acute short term effects observed. *1927*

Sotalol General effect in 2 studies but overall progressive deterioration in lipid profile. *0088*

Spironolactone No significant change in 17 subjects treated with drug for less than 1 y. *1405*

Stanozolol In 10 normolipidemic postmenopausal osteoporotic women treated for 6 weeks. *3526*

Sulfonylureas Usual effect reported. *0301*

Terbutaline No significant effect observed after 2 weeks in 15 subjects. *1654*

Tetracycline No effect seen even when administered orally. *1487*

Timolol In 15 patients treated for 1 mo. *3826*

Tourniquet No effect for 1 minute. *3540*

Trichlormethiazide Decrease by -17 mg/dL in 15 patients given 4 mg daily for 3 mo but note marked difference between responders and nonresponder populations. *2550*

Trimazosin In 2 studies with 13 and 48 subjects treated for 2.5 to 12 mo. *0088*

Ursodeoxycholic Acid No significant effect with 600 mg daily in patients with endogenous hypertriglyceridemia. *0601*

Verapamil In 64 patients in post-myocardial infarction comparison against placebo. *3489*

Vitamin B$_6$ Depletion No effect observed with 25 d poor diet. *0255*

Vitamin E No significant effect observed. *2158*

Serum Increase Analytical

Aminopyrine Interferes with measurement procedure. *1563*

Amphotericin B At concentrations above 96 mg/L (therapeutic concentration about 3.7 mg/L) raised concentration as measured by Liebermann-Burchard method. *3393*

Aspirin Alleged effect (but also no effect at 30 mg/dL). *1563*

Bilirubin Liebermann-Burchard chromogen 5-9 x cholesterol. Moderate effect with ferric chloride 1 mg equivalent to 4.5 mg/dL with AutoAnalyzer procedure. *3217* Interferes with direct procedures. *1237* Esters increased absorption at 0.2 mmol/L (can be extracted). *2854* At 24.0 mg/dL and above on recommended method on RA-1000. At 5.0 mg/dL and above on methods recommended on BMD-8700 but with ferrocyanide. *3407* At 24.0 mg/dL and above on method recommended on Hitachi® 705 with ferrocyanide. *3407*

Bromides Interferes with Zlatkis-Zak method (up to 10%). *2973*

Cefotaxime Generally significant increase of drug and metabolite on parallel method. *0201*

Chlorpromazine 50 μg in reaction mixture produces color. *0583*

Cholestanol Same retention time procedure of MacGee. *2252* Yields yellow color with Zlatkis-Zak method. *1563*

Corticosteroids Many steroids react with FeCl$_3$ reagent. *1563*

Creatinine   Increase by 0.5% at 11 mg/dL on enzymatic procedure. *0057*

Digitonin   If p-toluenesulfonic acid reaction used. *1563*   Affects L-B and Zlatkis-Zak procedures. *1834*

Fluosol-DA   Increase by 20% on SMA IIC at concentration of 50%. *2518*

Hemoglobin   Increase by 2.1% at 10 mg/dL on enzymatic procedure. *0057*   Interference with Zlatkis-Zak and direct methods. *1237*

Hemolysis   Values high by 10% if extraction not used. *1563*

Iodates   Interference with Zlatkis-Zak reaction. *3505*

Iodides   Interference with Zlatkis-Zak reaction. *3505*

Lipemia   Turbidity if extraction not used. *1563*   To large and variable extent on Technicon® SMAC®. *2466*

Lipochrome   Absorb strongly in blue region may affect L-B method. *1563*

Methandrostenolone   Interferes with Zimmermann reaction. *2220*

Methotrexate   At 5 times upper limit of therapeutic range on enzymatic method on KDA At 5 times upper limit of therapeutic range on enzymatic method on KDA. *2138*

Plasma   Compared with serum if unextracted (SMA 12/60). *2228*

Protein   Tryptophan in protein may react. *2313*

Tetracycline   At concentrations above 50 mg/L (normal therapeutic concentration 8 mg/L) raised concentration as measured by CHOD-Iodide method. *3393*

Thiouracil   Interferes with Zlatkis-Zak reaction. *3505*

Tryptophan   Affects direct reactions using acetic acid. *1563*

Urea   Increase by 0.5% at 60 mg/dL on enzymatic procedure. *0057*

Uric Acid   Increase by 1.6% at 50 mg/dL on enzymatic procedure. *0057*

Viomycin   Interferes with Zlatkis-Zak reaction. *3505*

Vitamin A   Interferes with Zlatkis-Zak reaction. *3505*

Vitamin D   Interferes with Zlatkis-Zak reaction. *3505*

Serum   Increase   Physiological

Acetohexamide   May cause intrahepatic cholestatic jaundice. *2220*

Acetophenazine   Cholestasis with intrahepatic obstruction. *2946*

Acetylsalicylic Acid   Gradual increase during course of treatment with drug. *1787*

Ambulation   30 mg/dL higher in outpatients than inpatients. *2978*

Aminoglutethimide   In 73 patients with advanced breast cancer receiving 500 mg/d with 40 mg hydrocortisone. *0415*

Amiodarone   Observed effect in 3 individuals but without effect on thyroid function. *2833*

Anabolic Steroids   Cholestatic phenomenon. *2220*

Androgens   Cholestatic phenomenon. *2220*

Arsenicals   Hepatotoxic effect (may be very high). *2220*

Ascorbic Acid   When atherosclerotic, ?mobilization from arteries. *3420*

Asparaginase   Unusual response in some patients. *2657*

Atenolol   Increase by 49% after 100 mg/d for 5 weeks in 20 hypertensive men. *2135*   Slight increase (8%) in one 6 mo study. *2116*

Bendrofluazide   5% increase in 66 subjects when treated for less than 1 y when used as only drug. *3706*   5% increase in short term studies. *0089*

Bile Salts   Augment cholesterol pool. *0652*

Blindness   Significantly higher (average 51 mg/dL) than normal. *1637*

Blood Group   Slightly higher level reported in group A males. *0219*

Blood Pressure   Strongly correlated with increased systolic pressure. *2123*

Caffeine   Positive relationship between coffee consumption and lipid concentration: association possibly more with coffee than caffeine. *1447*

Carbimazole   From 4.4 to 5.4 mmol/L in 12 hyperthyroid women patients treated with 10-30 mg daily. *1201*

Carbon Disulfide   With other liver function tests normal on chronic exposure. *0256*

Chenodeoxycholic Acid   Elevation of about 10 mg/dL attributable to drug. *3198*   Mean increase of 20 mg/dL in 252 individuals treated for 12 mo. *0046*   Increase by 10% or more in 82% in 916 patients given up to 750 mg/d for 2 y. *3199*

Chlorpromazine   Associated with hepatocanalicular cholestatic jaundice. *0248*

Chlorpropamide   Infrequent cholestatic effect. *1713*

Chlorthalidone   5% change during monotherapy in 302 subjects for 1 y. *1334*   9% change in 39 subjects treated with drug only for less than 1 y. *1404*   Increase by 13% in 18 postmenopausal women given 100 mg/d for 6 weeks. *0402*   Increase by 10 mg/dL in 1,000 men and women with mild hypertension treated for 1 y. *1334*   Increase by 26 mg/dL in 10 patients given 100 mg daily for 1 mo. *2550*   5% in long term, 9% in short term in several studies with monotherapy. *0089*

Cinchophen   May cause intrahepatic cholestasis. *1434*

Clofibrate   Paradoxical effect in patients with biliary cirrhosis. *2857*

Clonidine   Increase by 6% in 59 patients with primary hypertension given up to 300 mg daily for 6 mo. *1874*

Corticosteroids   Effect of prolonged hormone action. *0697*

Corticotropin   Effect of hormone action after initial fall. *2220*

Cortisone   Increase about 20% with vigorous treatment. *3961*

Cyclofenil   Slight effect in 19 patients given 600 mg daily for 4 mo. *2786*

Cyclophosphamide   Single case of drug induced myxedema. *0681*

Cyproterone   When given in combination with ethinyl estradiol. *2177*

Dantrolene   Isolated observation in one individual. *0659*

Dapsone   In 3 individuals treated for 3 to 36 weeks. *3478*

Diclofenac   Significantly higher (5.16 vs 4.60 mmol/L) than in controls. *1787*

Disulfiram   500 mg/d raised mean concentration from 193 mg/dL to 227 mg/dL after 3 weeks and 264 mg/dL after 6 weeks in alcoholic subjects. No fall in cholesterol with abstinence. *2272*

Epinephrine   Metabolic effect (indirectly through ACTH stimulation). *1488*

Erect Posture   As bound to lipoproteins increases while standing. *2319*

Ethanol   Significant increase in heavy drinkers vs nondrinking or occasional drinkers. *0059*   Significantly higher in alcoholic men than in controls. *1701*

Ether   Reportedly may cause hypercholesterolemia. *1487*

Etretinate   From 5.75 to 6.05 mmol/L in 13 patients with hyperkeratotic disorders given 50 mg daily for 2 mo. *3659*   In 4 of 11 patients who showed hypertriglyceridemia: effect slight but measured over 1 y therapy. *1866*

Fluoxymesterone   Observed effect. *1680*

Furosemide   5% change in 12 normotensive men when treated with drug alone for less than 1 y. *1826*   6% change in 16 subjects treated for less than 1 y with drug only. *1306*

Glutethimide   Significant effect from about 15 d of treatment with 500 mg drug/d in 6 volunteers. *0407*

Gold   May cause hypersensitive cholestasis. *1434*

Heparin   Rebound effect of cessation of treatment. *1680*

Heroin   Observed in addicts with renal disease. *2914*

Hydrochlorothiazide   Observed with 100 mg/d. *3663*   7% change in individuals given drug alone for less than 1 y. *1404*   Increase by 7 mg/dL in 343 patients with hypertension given drug for 10 weeks. *3707*   4-7% in short term study. *0089*

Ibuprofen   Significantly higher in patients receiving drug than in controls (5.26 mmol/L vs 4.60 mmol/L). *1787*

Imipramine   Possible cholestatic effect. *2946*

Indapamide   Slight and insignificant in 13 hypertensive diabetics treated for 24 weeks. *2686*   Increase by 17 mg/dL in 17 patients with essential hypertension treated with 2.5 mg daily for 3 mo. *2010*

Cholesterol (continued)

Serum Increase Physiological (continued)

Isotretinoin In 7 patients with severe rosacea treated with 1 mg/kg/d for 12 weeks. Effects possibly due to induction of hepatic microsomal enzymes. 2307 In both men and women with 1 mg/kg/d for 20 weeks when given for nodulocystic acne. 0326 From 0.9 to 2.2 mmol/L in 18 patients with severe acne given 0.8 mg/kg daily for 3 mo: changes reverted to normal after treatment stopped. 2242 From 0.8 to 1.6 mmol/L at 6 weeks in 7 patients with severe rosacea treated with mg/kg/d. 2307 During treatment but not to abnormal levels. 1818 From 5.75 to 6.49 mmol/L in 12 patients with hyperkeratotic disorders given 40 mg daily for 2 mo. 3659 Increase in both men and women with 1 mg/kg/d for 20 weeks when given for nodulocystic acne. 0326

Levarterenol Reported effect ?mechanism. 1487

Lithium Reported to induce myxedema. 2221

Meals Reported up to 3% increase after meals. 1563 Variable increase of 40 mg/dL in some people. 3217 Affected by meals (most marked in women). 3455

Menopause Change of 27 mg/dL from 5th to 6th decade. 3830

Menstruation Increase immediately before menstruation. 1563

Mepazine Cholestatic effect. 1713

Meprobamate May cause cholestatic (hepatocanalicular) jaundice. 0248

Methandrostenolone Due to cholestasis. 1713

Methimazole Cholestatic effect. 1713

Methyltestosterone May cause hypersensitive cholestasis. 1434

Muscular Exercise Occasional response to exercise. 0582

Nandrolone Due to action on liver. 2681

Nitrophenol Hypersensitive intrahepatic cholestasis. 1434

Norethandrolone Due to cholestasis. 1713

Oral Contraceptives If initially low (no effect if about 200 mg/dL). 0585 Increase by 4 to 15% depending on preparation used. 2860 Extent of effect varies with exact composition of oral contraceptive. 1458 May result with oral contraceptive containing more than 75 µg estrogen. 0018 Slight effect with high-dose combination drugs. 1248 In 10 women receiving ethinyl estradiol with norethindrone. 2009

Oxprenolol Mild effect as possessed relatively weak intrinsic sympathomimetic activity. 2116

Oxymetholone Cholestatic effect. 1713

Paramethadione Possible liver damage. 2313

Penicillamine Single case reported. 1488

Phenothiazines Frequently reported effect, ?mechanism. 2946

Phenylbutazone Observed in one patient: clear evidence of intrahepatic cholestasis. 3670

Phenytoin Hepatotoxicity with centrolobular necrosis. 3678 In men in 27 patients with transient brain ischemia. 1879 Possibly due to subclinical hypothyroidism caused by drug or due to hepatic synthesis stimulation with increase of pool size of bile acids (increase of 6 to 48% in 11 patients). 2767

Pindolol Increase by 5.5% in one study of 4 weeks with 15-30 mg/d. 1806

Piretanide Effect observed with 12 mg/d. 3663

Polythiazide 4% change in 20 people when treated only with drug for less tan 1 y. 1808

Prednisolone Rose from 4.81 to 6.58 mmol/L in women receiving drug for mean of 3.1 y. 1783

Prednisone 17% increase in group of men and women during 1 mo. 3949 Augmented response in cirrhotics when drug added to therapeutic regime. 0651

Pregnancy Increases from 8th week (maximum by 30th). 0987

Prochlorperazine Cholestatic effect. 0071

Promazine May cause cholestatic (hepatocanalicular) jaundice. 0248

Propranolol Increase by 9 mg/dL in approximately 120 patients with essential hypertension treated for 1 y. 3708 Mean increase from 213 to 222 mg/dL in 16 hypertensives given 80 mg daily for up to 12 mo. 2266

Smoking Significantly higher in heavy smokers (?dietary preferences). 1317

Sotalol Significant increase of 16% reported from one long term study (12 mo). 2116 Increased from mean of 5.49 mmol/L to 6.37 mmol/L at 12 mo in group of essential hypertensives. 2120

Spironolactone Maximum average increase of 4% in 3 studies. 0087

Stress Mental stress such as examinations. 3217

Sucrose Effect of sustained high sucrose diet. 3931 Marked increase when sugar substituted for starch. 2036

Sulfadiazine May cause intrahepatic cholestasis. 1434

Sulfonamides Cholestatic effect. 1713

Testosterone Cholestatic effect. 1488

Thiabendazole Cholestatic effect. 2413

Thiacetazone May cause cholestatic (hepatocanalicular) jaundice. 0248

Thiazides Infrequent cholestatic effect. 1713

Thiouracil May cause intrahepatic cholestasis. 1434

Tourniquet 5-20% increase with tourniquet on for 5 minutes. 3476

Trifluoperazine Increase of up to 35 mg/dL reported. 2946

Trimethadione May affect liver function (hepatitis). 2313

Vitamin D In men by 25 mg/dL in ages 35-54 y old. 0793

Serum Decrease Analytical

Acetylsalicylic Acid Increase by 9% with 8326 µmol/L on Liebermann-Burchard method. 1786

Aminoantipyrine At concentrations above 400 mg/L lowered concentration as measured by CHOD-PAP method. 3393

Ampicillin At concentrations above 100 mg/L (therapeutic concentration about 320 mg/L) lowered concentration as measured by CHOD-Iodide method. 3393

Ascorbic Acid Increase by 2.1% at 5 mg/dL on enzymatic procedure. 0057 At concentrations above 50 mg/L (maximum serum concentration 34 mg/L) lowered concentration as measured by CHOD-Iodide method. 3393 At concentrations above 50 mg/L (maximum serum concentration 34 mg/L) lowered concentration as measured by CHOD-PAP method. 3393

Bilirubin Increase by 1.0% at 15 mg/dL on enzymatic procedure. 0057 At concentrations as low as 2.5 mg/dL and higher on BMD-8700 using manufacturer recommended method. At 8.0 mg/dL and above on methods recommended on Hitachi® 705. 3407 At 5.0 mg/dL and above on method recommended on Centrifichem. 3407 From less than 2.5 mg/dL on method recommended on Cobas-Bio. 3407

Bromides Increase by 0.5% at 10 mg/dL (in serum) on enzymatic procedure. 0057

Citrates If used as anticoagulant — causes water shift. 2319 Dilutional effect when compared against serum or heparin as anticoagulant. 3361

EDTA If used as anticoagulant value low by 5-15 mg/dL. 2319

Fluorides Value lower by 30-50 mg/dL. 2319

Globulin Gamma globulin increases viscosity in AutoAnalyzer methods. 1645

Glucose Increase by 1.0% at 500 mg/dL on enzymatic procedure. 0057

Heparin If used as anticoagulant value low by 5-15 mg/dL. 2319

Liposol Changed from 1774 to 0 mg/L in 26 randomly selected patient sera, before and after addition on Beckman Astra methods. 2054

Methylaminoantipyrine At concentrations above 120 mg/L lowered concentration as measured by CHOD-PAP method. 3393

Methyldopa At concentrations above 200 mg/L (normal therapeutic concentration 2 mg/L) lowered concentration as measured by CHOD-Iodide method. 3393 At concentrations above 50 mg/L (normal therapeutic concentration 2 mg/L) lowered concentration as measured by CHOD-PAP method. 3393

Nitrates Interfere with Zlatkis-Zak reaction. 3505

Nitrites 10 µg in reaction mixture decreases by 30%. 0583

Novaminsulfon At concentrations above 150 mg/L (normal therapeutic concentration 15 mg/L) lowered concentration as measured by CHOD-PAP method. 3393

Oxalate If used as anticoagulant (causes water shift). 1563

Penicillamine At concentrations above 10 mg/L (normal therapeutic concentration 11 mg/L) lowered concentration as measured by CHOD-Iodide method. 3393

Plasma Compared with serum if extracted (AA2 procedure). *2228*

Rifampin At 5 times upper limit of therapeutic concentration on method on Cobas-Bio. *2138*

Sulpiride At 15 mg/dL conventional methods when added to serum. *1758*

Thimerosal Marked reduction of results with methods using cholesterol oxidase and p-Aminophenazone. *0471*

Thiouracil At 13 mg/dL decreased by 30-40 mg/dL. *0583*

Serum Decrease Physiological

Allopurinol Hepatotoxic effect. *2220*

Aluminum Nicotinate Therapeutic intent through nicotinic acid release. *0126*

Amiloride Significant fall with treatment in 13 men treated with drug: concentration rose when drug discontinued. *0656*

Aminosalicylic Acid As effective as neomycin, mechanism obscure. *2148*

Amiodarone From 214 to 194 mg/dL in 24 patients in female patients but no effect in men after 90 d. *3392*

Androgens Decreased synthesis. *3879*

Androsterone Therapeutic effect. *1237*

Antimony Compounds Hepatotoxic effect. *2220*

Ascorbic Acid Tends to fall in people under 25 when 1 g/d given. *3420* 16% decrease on average when 1 g/d given to healthy approximately 29 year olds within 2 mo. 14% fall in 58 year olds but required 12 mo. Administration abolished normal rise observed in winter. *0917*

Asparaginase Hepatotoxicity (effect marked). *1519* In 82-85% of patients reported in various studies. *2406*

Aspirin Doses over 5 g reported to have effect. *1343* Progressive effect during three days but also noted over longer term study at beginning, eventually reverted towards normal. Effect statistically significant. *3076*

Azathioprine Improves biliary excretion in biliary cirrhosis. *3063*

Benfluorex Increase by 17% in 12 hypertriglyceridemic type 2 diabetic patients given 150 mg tid after 1 mo treatment. *3385*

Boxidine Inhibits transformation of 7-dehydrocholesterol. *1220*

Bran Possible effect with very large amounts. *1540*

Buformin Probably inhibits synthesis in liver. *0878*

Carbon Disulfide In people with exposure to 37 ppm for 5.7 y. *0818*

Carbutamide Hepatotoxicity. *2313*

Cellulose Observed in young girls with 100 g/d. *1540*

Chloroform Hepatotoxic effect with necrosis. *2313*

Chlorpropamide May inhibit hepatic synthesis (?also absorption). *2220* With 8 weeks treatment of 8 C-peptide negative insulin dependent diabetics. *3260*

Chlortetracycline Hepatotoxic effect with centrolobular necrosis. *1237* Not as effective as neomycin reacts with bile acids. *1487*

Chlorthalidone Statistically significant decrease of 3 mg/dL in 12 men receiving drug in conjunction with diet to reduce cholesterol. *0090*

Cholestyramine Therapeutic goal (?increased binding of bile salts in gut). *1669*

Clofibrate Therapeutic goal (probably diminished synthesis). *0456* By 40% in first year became normal after this in patients with hyperlipoproteinemia and impaired glucose tolerance given diet plus 2 g drug/d. 14 patients monitored for 5 y. *2925* Marked reduction within 6 weeks in 10 hyperlipidemic patients. *2577* Significant reduction in 27 patients, half of whom had hypertriglyceridemia, after 1 week. *1121* In both normals with hypertriglyceridemia and diabetics. *1120* By 40% in first year but normalized later in 14 patients with primary hyperlipoproteinemia given 2.0 g daily over 5 y. *2925*

Clomiphene Possible interference with synthesis. *0071*

Clonidine In 16 patients treated for 2 mo (by 8%). *0088*

Colchicine May have hepatotoxic effect. *3082*

Colestipol Therapeutic intent. *0939*

Corticotropin Ester concentration reduced by stimulation of adrenal. *0652*

Cyproterone In oophorectomized women when given alone. In oophorectomized women (slight effect) when given alone. *2177*

Dehydroepiandrosterone Mean fall from 4.82 to 4.48 mmol/L in 5 men given 1600 mg/d orally for 28 d. *2578*

Dextrothyroxine Therapeutic intent (enhances excretion). *1488*

Doxazosin In 38 treated hypertensives given 1 to 16 mg/d over 10 weeks. *2562*

EDTA Reported to occur if given i.v. *1488*

Erythromycin Hepatotoxic effect. *2220*

Estradiol In 8 post-menopausal, post-oophorectomy women after 100 mg implant in subcutaneous tissue. *3274* By about 5% with 2 mg estradiol valerate/d in 20 normolipoproteinemic women over 3 mo. *3593*

Estradiol-17β Almost 10% reduction in 38 healthy post-menopausal women given 2-4 mg orally or 3 mg cutaneously over 6 mo. *1067*

Estrogens Reduces by up to 18%: used therapeutically. *2508* Increase by 10-13% in post-menopausal women treated for 3 y. *0653*

Estrone To almost normal values towards end of treatment in 20 women at perimenopause. *2058*

Ethanol Occurs when cirrhosis develops in alcoholism. *1621* Low values associated with debauch gradually increased to normal. *1011*

Ethinyl Estradiol Decreases by up to 50% (decreased low density lipoproteins). *1343*

Fenofibrate Increase by 18% over 24 weeks in 92 type IIa given 100 mg 3 times per day with similar responses in type IIb patients when compared with baseline values. *0503*

Fluoxymesterone Observed effect. *1680*

Garlic Significant decrease although given with fatty meal. *0421*

Gemfibrozil Increase by 11% in extensive 5 year double-blinded trial with 2,000 men receiving 600 mg drug twice daily. *1193*

Glucagon Reported effect, mechanism not listed. *0802*

Glyburide Fall by over 50 mg/dL in treated diabetics. *0125*

Guanabenz Mean concentration in 480 patients treated for up to 2 y decreased by 10 mg/dL. Decrease maintained throughout subsequent therapy. *1865* Decrease by -10 mg/dL in patients given 4 to 8 mg daily for 12 to 16 weeks. *2550*

Guanfacine Increase by 14% in 30 patients treated for 2 y. *0088*

Halofenate Irregular effect, mean decrease up to 9%. *0317*

Haloperidol Inhibits cholesterol biosynthesis. *2313*

High Carbohydrate Diet Increase by 40% but maximum if given i.v. *0880*

Hydralazine Increase by 12% in 7 individuals treated for 4 mo. *0088* 12% reduction in 7 patients treated for 4 mo. *0089*

Hydrochlorothiazide In MRFIT study when given with chlorthalidone involving many patients. *0089*

Insulin Therapeutic goal. *3505*

Isoniazid Probable hepatotoxic effect. *2220*

Kanamycin Forms salts with bile acids in gut. *1061*

Ketoconazole After high dose treatment in patients with advanced prostatic cancer; appears to be dose related. *3390*

Levonorgestrel Significant reduction in patients over 3 y compared with patients with IUDs for 2 1/2 y. *0763*

Levothyroxine Often therapeutic intent. *3505*

Lincomycin Hepatotoxic effect. *2220*

Linoleamide Inhibits sterol absorption. *0939*

Liotrix In hypothyroids falls to within normal range. *1680*

Lovastatin Increase by 14 to 34% in clinical trial of 101 patients with heterozygous familial hypercholesterolemia. *1530*

MAO Inhibitors Hepatotoxic effect. *2220*

Medroxyprogesterone Increase by 14% effects observed in 15 postmenopausal women with endometrial cancer after 2 weeks treatment. *2118* Increase by 12% in 11 men with sexual deviation syndrome given approximately 1273 mg over a total of approximately 17 d. *0643*

Menotropins Marked fall in type 2 hyperlipoproteinemia. *2235*

Methandrostenolone Reported effect (may increase as alternative). *1680*

Methyldopa In 17 hypertensive patients given drug for 3 mo. *0967*

Nafenopin More effective than clofibrate. *0939*

Nandrolone Due to action on liver. *2681*

Cholesterol (continued)

Serum   Decrease   Physiological (continued)

Neomycin   Forms salts with bile acids in gut. *3505*   In 20 subjects with type II hyperlipoproteinemia given 2 g/d for 9 mo caused 15% decline in cholesterol. *1618*   Marked effect over diet in 20 patients with type II hyperlipoproteinemia treated over several months. *1618*

Niacin   Therapeutic goal (rebound increase when discontinued). *3505*

Nifedipine   In 23 patients over 60 y old with essential mild to moderate hypertension. *3192*

Ovulation   Decrease at middle of cycle reported. *2319*

Oxandrolone   Anabolic effect. *1308*

Oxymetholone   May cause decrease with therapy. *1679*

Paromomycin   Reduction up to 18%. ?mechanism. *3129*

Pectin   5% decrease observed on adding to diet. *3217*

Pentylenetetrazole   Maximal effect seen after 2 weeks. *0332*

Phenyramidol   Probable inhibition of hepatic microsomal enzymes. *3207*

Pindolol   Lower after 6 mo of therapy than after one. *2117*

Prazosin   In 15 hypertensive patients given 4 mg/d for 10 weeks with mean reduction from 202 to 188 mg/dL. *1118*   Increase by 9% in 23 healthy hypertensives treated for 8 weeks. *2136*   From 5 to 12% in several studies involving 25 to 50 patients for up to 6 mo. *0089*

Probucol   Lowered by more than 20 mg/dL in most patients. *0806*   In 50 diabetics given 500 mg/d for 16 weeks and reduction greatest in highest cholesterol and triglyceride patients. *1523*

Progesterone   Slight effect when only treatment. *0991*

Pyridyltetrazole   Effect greater than with nicotinic acid. *2774*

Race   In blacks significantly < comparable caucasians after age 40. *0286*

Recumbency   10% drop in 30 minutes in normal. *3540*   Decrease of 16% max on lying down (as bound to protein). *3539*

Sitosterols   Inhibits absorption of endogenous and exogenous compound. *1336*

Spironolactone   Increase by 24 mg/dL in 11 men simultaneously with starting diet to reduce cholesterol. *0090*

Tamoxifen   Effect on increased concentration in breast cancer patients. *3070*

Terazosin   Increase by 5.4 mg/dL with up to 20 mg daily over 4 weeks in patients with moderate hypertension. *0857*

Tetracycline   Hepatotoxicity may occur. *2808*

Thyroid   Physiological effect. *1488*

L-Thyroxine   From 7.8 to 6.1 mmol/L in 11 hypothyroid women treated with 0.1 to 0.2 mg daily. *1201*

Tolbutamide   Inhibits hepatic synthesis (?also absorption). *0878*

Trifluperidol   Inhibits biosynthesis in liver. *1343*

Tri-iodothyronine   Physiological consequence of hormone. *3505*

Verapamil   In 12 patients when angina or hypertension treated for 6 weeks. Where change occurred it was of order of 10%. *3756*

Bile   Decrease   Physiological

Chenodeoxycholic Acid   In patients with gallstones, not in normals. *0274*

Cholesterol Esters

Serum   No Effect   Physiological

Meals   No effect after standard breakfast. *0579*

Serum   Increase   Analytical

Contact With Clot   Free decreases at same time due to esterase action. *1563*

Serum   Increase   Physiological

Corticotropin   Initial fall then rise about 10%. *2312*

Cortisone   Increase about 20% with vigorous treatment. *3961*

Oral Contraceptives   Slight effect in primates. *3035*

Serum   Decrease   Physiological

Asparaginase   Marked fall maximal at 4 d after single injection. *3957*

Carbon Disulfide   While cholesterol was increased. *0256*

Jaundice   Obstruction decrease bile salts for esterification. *3947*

Cholesterol, Free

Serum   No Effect   Analytical

Storage of Sample   At room temperature, 4° for 7 d -20 for 28 d. *2512*

Serum   No Effect   Physiological

Cyproterone   When given in combination with Ethinyl Estradiol. *2177*

Serum   Decrease   Physiological

Cyproterone   In oophorectomized women when given alone. *2177*

Cholesterol, High Density Lipoprotein

Serum   No Effect   Physiological

Acebutolol   In small numbers of patients when treated for 1-12 mo. *0088*   Decrease by -2 mg/dL in 11 patients given 400 mg daily for 3 mo. *2550*   Insignificant change on average in one 6 mo long study. *2116*   In several studies of about 15 patients treated for 1 to 12 mo. *0089*

Amiodarone   In 24 patients given drug for 30-90 d. *3392*

Atenolol   Typically no significant change but may be slight reduction. *0088*   No significant change in 15 hypertensives treated for 8 mo. *1021*   After 100 mg/d for 5 weeks in 20 hypertensive men. *2135*   No significant change in one 3 mo long study. *2116*   In 20 patients with 50 mg/d compared with pre-treatment values after 3 mo. *3071*

Bendrofluazide   No significant change in 15 individuals when given as sole treatment for less than 1 y. *3826*

Benfluorex   In 12 hypertriglyceridemic type 2 diabetic patients given 150 mg tid after 1 mo treatment. *3385*

Bopindolol   In 24 hypertensives treated with drug for 3 mo. *3667*

Captopril   After 12 weeks treatment in 18 patients with mild essential hypertension. *3150*   Increase by 1 mg/dL in 15 patients given 75 mg daily for 8 weeks. *2550*

Carprazidil   In one study involving 12 patients treated for 4 mo. *0089*

Chenodeoxycholic Acid   No significant effect of 12 mo treatment in 252 patients. *0046*

Chlorthalidone   Nonsignificant change during monotherapy in 302 subjects for 1 y. *1334*   Nonsignificant change in 39 subjects treated with drug only for less than 1 y. *1404*   In 22 premenopausal women without decrease in blood pressure treated for less than 1 y. *0402*   In 22 premenopausal and 18 postmenopausal women given 100 mg/d for 6 weeks. *0402*   In 1,000 men and women with mild hypertension treated for 1 y. *1334*   Increase by 1 mg/dL in 10 patients given 100 mg daily for 1 mo. *2550*   In long or short term in several studies with monotherapy. *0089*

Cimetidine   In healthy individuals and in subjects with high serum glucose concentration. *1653*

Clonidine   In 59 patients with primary hypertension given up to 300 mg daily for 6 mo. *1874*

Clopamide   No significant change in 17 individuals treated for less than 1 y when used as sole treatment. *3184*

Dehydroepiandrosterone   In 5 normal men given 1600 mg/d orally for 28 d. *2578*

Doxazosin   Compared with controls over 10-12 weeks, although 9% increase in HDL/total cholesterol ratio. *2842*

Estradiol   Insignificant increase with 2 mg estradiol valerate/d in 20 normolipoproteinemic women over 3 mo. *3593*

Ethanol   Not significantly different in alcoholics from nonalcoholic controls. *1701*   No effect of ingestion of 1.5 g/kg at night on concentration measured next morning. morning. *3547*

Etretinate   In 13 patients with hyperkeratotic disorders given 50 mg daily for 2 mo. *3659*

Furosemide Not significant change in 12 normotensive men when treated with drug alone for less than 1 y. *1826* No significant change in 16 subjects treated for less than 1 y with drug only. *1306* Reduced by -4 mg/dL in 12 patients given 40 to 80 mg daily for 3 mo. *2550*

Glucose In response to infusion of 1 g/kg body weight in normal volunteers and non-insulin dependent diabetics. *0476*

Guanabenz No effect in 39 hypertensives treated for 2 y. *1865* Increase by 1 mg/dL in patients given 4 to 8 mg daily for 12 to 16 weeks. *2550*

Hydrochlorothiazide Not significant change in 39 individuals given drug alone for less than 1 y. *1404* No effect with 25 or 50 mg/d in hypertensive patients. *1980* In long term over 1 y in many patients. *0089*

Indapamide In 43 subjects treated with this drug for less than 1 y. *3780* In various studies involving more than 30 patients when used as sole drug for less than 1 y. *0089*

Labetalol No significant change in several studies with patients treated for 1-12 mo. *0088* Although values reported from -12 to +23% in several studies from 1 to 12 mo. *0089*

Levonorgestrel When 150 μg levonorgestrel given with 30 μg ethinyl estradiol for 3 mo. Results normal within 2 mo after treatment stopped. *0312* No difference between implant recipients and iud users. *0763*

Medroxyprogesterone No significant effect in 11 men with sexual deviation syndrome given approximately 1273 mg over a total of approximately 17 d. *0643*

Methyldopa In several patients treated for 2-4 mo. *0088* In 17 hypertensive patients given drug for 3 mo. *0967*

Metoprolol Typically although slight reduction in 1 study. *0088*

Neomycin No significant effect in 20 type II hyperlipoproteinemic subjects given 2 g/d for 9 mo. *1618* No significant effect in 20 patients with type II hyperlipoproteinemia treated over several months. *1618*

Nifedipine In 11 patients with 80 mg daily for 6 weeks. *3702* No effect observed in 14 patients with essential hypertension treated over 2 mo with 20 mg twice daily. *2119* Increase by 1 mg/dL in patients given 30 to 60 mg daily for 8 weeks. *2550*

Oral Contraceptives No difference between treated women and others. *1303* With high dose drugs containing norethisterone or ethynodiol. *1248* In 10 women receiving ethinyl estradiol with norgestrel. *2009* Typically insignificant change over 6 cycles in women with low dose ethinyl estradiol with low dose levonorgestrel or desogestrel. *1249*

Oxprenolol In several studies of 1-4 mo. *0088* In 20 hypertensive men given 160 mg/d for 5 weeks. *2135*

Pindolol No significant change typically seen with treatment for 1 mo or more. *0088* No effect in short-term study of 10 hypertensive men with 15 mg/d. *2135* No significant change in 2 studies of 2 to 12 mo. *2116*

Polythiazide Not significant change in 20 people when treated only with drug for less than 1 year. *1808*

Prazosin In 22 mild/moderate male hypertensives treated for 8 weeks versus controls insignificant reduction noted. *1693* In 23 healthy hypertensives treated for 8 weeks. *2136* No change in 16 patients given 1 to 3 mg daily for 8 weeks. *2550* In several studies involving 25 to 50 patients for up to 6 mo. *0089* In 16 hypertensive patients given 2 mg daily for 3 mo. *2266*

Prednisolone In men receiving drug for mean of 3.1 y. *1783*

Propranolol General effect observed in multiple studies. *0088* In 17 patients with hypertension followed for 3 mo. *0967* In 50 volunteers given 160 mg daily for 3 mo. *2171*

Ranitidine In 25 patients pretreatment 40 mg/dL, after 38 mg/dL over 5 weeks. *3568* In 8 ulcer patients given 300 mg/d for 1 mo. *3858*

Spironolactone No significant change in 17 subjects treated with drug for less than 1 y. *1405* No significant change in 3 studies. *0087*

Terazosin No significant effect with up to 20 mg daily over 4 weeks in patients with moderate hypertension. *0857*

Trimazosin In 2 studies with 13 and 48 subjects treated for 2.5 to 12 mo. *0088*

Ursodeoxycholic Acid No significant effect with 600 mg daily in patients with endogenous hypertriglyceridemia. *0601* In 8 normolipemic patients receiving 1,000 mg drug daily. *2121*

Verapamil In 64 patients in post-myocardial infarction comparison against placebo. *3489*

Serum Increase Physiological

Albuterol After 2 weeks of treatment average increase of 6.9% when receiving 8 mg twice daily. *0640*

Aminoglutethimide In 73 patients with advanced breast cancer receiving 500 mg/d with 40 mg hydrocortisone. *0415*

Ascorbic Acid Significant correlation between drug intake and analyte concentration in elderly. *1757*

Bezafibrate Increase by 21% in 7 hypertriglyceridemic type 2 diabetic patients given 200 mg tid after 1 mo treatment. *3385* Increase by 13% in 11 hypertriglyceridemic subjects. *3291*

Carbimazole From 1.3 to 1.6 mmol/L in 12 hyperthyroid women patients treated with 10-30 mg daily. *1201*

Carprazidil In 15 men with mild to moderate essential hypertension by 26% over 8 weeks. *1268*

Celiprolol From 42 to 54 mg/dL in patients with hyperlipoproteinemia types IIa, IIb or IV with 300 mg drug for 4 weeks. *2873*

Chenodeoxycholic Acid Significant effect with 150 mg four times daily in patients with endogenous hypertriglyceridemias. *0601*

Cimetidine Significant increase in 25 patients over 5 weeks from 37 mg/dL to 42 mg/dL. *3568*

Clofibrate Increase by 10 to 20% in 10 hyperlipidemic patients. *2577*

Cyclofenil By approximately 15% with 2 and 4 mo of therapy in 19 patients given 600 mg daily for 4 mo. *2786*

Desogestrel Increase by 12% (% HDL-cholesterol increased by 15%) when 150 μg desogestrel given with 30 μg ethinyl estradiol for 3 mo. Results normal within 2 mo after treatment stopped. *0312*

Diltiazem Increase by 15% in 31 subjects treated for 6 mo. *0088*

Estradiol In 8 post-menopausal, post-oophorectomy women after 100 mg implant in subcutaneous tissue. *3274* With nonalkylated estrogens, eg valerate salt, in treatment of postmenopausal hormone deficiency. *1037*

Estradiol-17β In 14 oophorectomized women over 6 mo when 50 mg drug given as implant. *1080* By up to 20% in 38 healthy postmenopausal women given 2-4 mg orally or 3 mg cutaneously over 6 mo. *1067*

Estrone Continuous increase, became significant at 12 mo in 20 women at perimenopause. *2058*

Ethanol Increase by 25% in first 2 weeks of ingestion of 30 g alcohol daily, although gradually reverted to normal after another 2 weeks (effect observed in previously non-alcohol drinking healthy young males). *0328* 17% increase in heavy versus occasional drinkers. *3273* Significant increase in heavy drinkers versus non-drinkers or occasional drinker. *0059* Increase by 21% in 78 intemperate drinkers on average. *2882* Possibly due to induction of microsomal enzymes. *2428* Concentration doubled immediately after debauch compared with 10 d later or control subjects. *1011*

Fenofibrate Increase by 11% over 24 weeks in 92 type IIa given 100 mg 3 times per day with similar responses in type IIb patients when compared with baseline values. *0503* Increase by 16% in 9 hypertriglyceridemic type 2 diabetic patients given 100 mg tid after 1 mo treatment. *3385*

Gemfibrozil Increase by 36% in 6 patients with primary familial endogenous hypertriglyceridemia from baseline values. Synthetic rates of apo A-I and apo A-II increased by 27% and 34% respectively. *3117* By more than 10% in extensive 5 year double-blinded trial with 2,000 men receiving 600 mg drug twice daily. *1193* Increase by 50% in 18 patients with chronic renal failure treated with 1200 mg/d for 28 weeks. Simultaneous activation of postheparin plasma lipoprotein and hepatic lipases. Effects reversed when drug discontinued. *2741*

Glutethimide Significant effect from about 15 d of treatment with 500 mg drug/d in 6 volunteers. *0407*

Ketanserin In 50 hypertensive patients given 80 mg/daily for 3 mo. *2171*

Lovastatin Slight effect in clinical trial of 101 patients with heterozygous familial hypercholesterolemia. *1530*

Minoxidil Slight effect (order of 10%) after 3 or 6 mo treatment. *1807*

Niacin Increase by 26% in 34 hypercholesterolemic individuals with 1.5 g/d for 1 mo then 3.0 g/d up to 6 mo. *1967*

Cholesterol, High Density Lipoprotein
(continued)

Serum   Increase   Physiological (continued)

Nifedipine   In 100 patients in a double-blind randomized trial. 3965

Oral Contraceptives   Depending on preparation used. Increase by 17 to 81% depending on preparation used. 2860   If oral contraceptive contains more than 80 µg estrogen. 0018   In women taking high estrogen/low progestin combination. 3733   Increase by 35% with high-dose oral contraceptives that are overly estrogenic. 1248   In 10 women receiving ethinyl estradiol with norethindrone. 2009

Phenytoin   In both men and women in 27 patients with transient brain ischemia. 1879   In 43% of 28 patients treated with 200-300 mg/d for 1 to 35 y (1.87 mmol/L vs 1.51 mmol/L in controls). vs 1.51 mmol/L in controls). 2598

Pindolol   Significant increase during first month of therapy. 2117   20% increase in one 3 mo study. 2116

Prazosin   After 6 mg/d for 12 mo average increase of 17% in 15 patients. 3531   Effect observed with doses as low as 1 to 3 mg/d. 3148   Mean increase from 36 to 40.5 mg/dL in 15 hypertensives given 4 mg/d for weeks. 1118

Prednisone   68% average effect in both men and women. 3949

Terbutaline   10% increase with 2 weeks treatment in 15 subjects. 1654

Serum   Decrease   Analytical

Ascorbic Acid   Decreases of from 0.3 to 10% in 6 methods to determine compound with drug concentration of 2.0 mg/dL. 2500

Serum   Decrease   Physiological

Acebutolol   Slight effect, but no change in ratio of HDL/total cholesterol. 2116   But insignificantly in 18 patients treated for 6 mo. 2115

Atenolol   Reduction by 7 to 12% in 2 studies of 3 to 6 mo. 2116   From 1.21 to 1.13 mmol/L with 100 mg drug/d. 2192

Bisoprolol   From 1.22 to 1.10 mmol/L with 20 mg drug/d. 2192

Bunitrolol   When 30 mg/d given for 12 weeks in normolipidemic patients with mild essential hypertension. 3151

Chenodeoxycholic Acid   Increase by 46% in 8 normolipemic patients receiving 16 g drug daily. 2121

Chlorpropamide   With 8 weeks treatment of 8 C-peptide negative insulin dependent diabetics. 3260

Danazol   After 2 weeks with 600 mg/d in 12 women with endometriosis reduced by 49% and 59% after 6 weeks; returned to normal in 8 weeks after treatment stopped. 1066   By about 45% during first 2 mo in 62 patients with endometriosis treated with 600 mg daily for up to 24 weeks. 3222   Increase by 50% in 6 patients with endometriosis given 600 mg daily for 3 to 6 mo. 2279   In 9 subjects studied for 6 mo significant reduction: returned to normal at end of therapy. 0060   Increase by 49% after 2 weeks and 59% after 8 weeks in 12 women with endometriosis given 600 mg daily for 24 weeks. 1066

Desogestrel   With daily dose of 0.125 mg in 30 healthy female volunteers when given alone. Normal values 30 d after treatment stopped. 3089

Indapamide   Slight effect in 13 hypertensive diabetics treated for 24 weeks. 2686

Isotretinoin   In 7 patients with severe rosacea treated with 1 mg/kg/d for 12 weeks. Effects possibly due to induction of hepatic microsomal enzymes. 2307   In both men and women with 1 mg/kg/d for 20 weeks when given for nodulocystic acne. 0326   From 1.28 to 1.14 mmol/L in 12 patients with hyperkeratotic disorders given 40 mg daily for 2 mo. 3659   From 1.1 to 0.9 mmol/L in 18 patients with severe acne given 0.8 mg/kg daily for 3 mo: changes reverted to normal after treatment stopped. 2242   From 1.30 to 1.04 mmol/L at 6 weeks in 7 patients with severe rosacea treated with 1 mg/kg/d. 2307   Increase in both men and women with 1 mg/kg/d for 20 weeks when given for nodulocystic acne. 0326

Levonorgestrel   With daily dose of 0.125 mg in 30 healthy female volunteers when given alone. Values normal within 30 d of end of treatment. 3089   Increase by 35% in 11 normolipoproteinemic women given 250 µg/d for 2 weeks. 3593

Lynestrenol   When 5 mg daily given alone to 30 healthy female volunteers. Normal values 30 d after treatment stopped. 3089   lcrease by 32% due to progestational activity in 6 women with endometriosis given 5 to 10 mg daily for 6 mo. 2279

Medroxyprogesterone   Increase by 33% effects observed in 15 postmenopausal women with endomentrial cancer after 2 weeks treatment. 2118   Increase by 8% at 2 weeks and more after longer treatment: dose-dependent correlation with results. 1068

Methyldopa   Increase by 10% in 32 middle-aged hypertensive males treated for 6 weeks. 2129   Decrease by -4 mg/dL in 11 patients given 750 mg daily for 8 weeks. 2550   By up to 15% in 3 studies of 7 to 17 patients for up to 3 mo. 0089

Metoprolol   Mean decrease from 37 to 31 mg/dL in 15 hypertensives patients given 200 mg/d for 10 weeks. 1118   From 43 to 35 mg/dL in 18 hypertriglyceridemic hypertensives after 12 weeks. 0341   Increase by 6 to 13% in several studies of 2 to 4 mo with up to 400 mg drug/d. 1806   Increase by 13% after 1 mo in one study and 8% (nonsignificant) in another. 2116   Increase by 6 to 13% in several studies of 2 to 4 mo with up to 400 mg drug/d. 1806   Fell from 1.42 to 1.31 mmol/L in 20 hypertensives given 200 mg/d for 3 mo. 3071

Nadolol   In hypertensive patients while fasting and during and after a meal and an exercise test. ?secondary to reduction of lipoprotein lipase. 2761   Increase by 3% in 13 patients given 50-200 mg daily for 10 weeks. 1806   Significant reduction in fasting concentration and after breakfast. 2761

Norethisterone   In 75 women who had received up to 24 intragluteal injections of 200 mg every 56 d. 1666

Oral Contraceptives   Depending on preparation used. 2860   If oral contraceptive contains more than 50 µg estrogen. 0018   In women taking estrogen/high or low progestin combination. 3733   Increase by 20% with high dose compound containing 500 µg norgestrel. 1248

Oxprenolol   Significant effect in 53 patients given 80 mg/twice daily for 3 mo. 0835   Mild effect as possessed relatively weak intrinsic sympathomimetic activity. 2116

Prednisolone   Fell from 1.99 to 1.10 mmol/L in women receiving drug for mean of 3.1 y. 1783

Probucol   In 50 diabetics given 500 mg/d for 16 weeks and reduction greatest in highest cholesterol and triglyceride patients. 1523

Propranolol   Approximately 10% decrease noted in 22 mild/moderate hypertension treated for 8 weeks: inverse relationship to dose given. 1693   Mean decrease from 49 to 45 mg/dL in approximately 120 hypertensives after 16 weeks therapy. 2331   From 42 to 36 mg/dL in 15 hypertriglyceridemic hypertensives after 12 weeks. 0341   Increase by 13% in 23 hypertensive men given uo to 160 mg/d for 8 weeks. 2135   Marked effect in 53 hypertensives given 80 mg twice daily for 3 mo. 0835   Decrease by -3 mg/dL in 8 patients given 30 to 60 mg daily for 8 weeks. 2550   Increase by 13% in 23 hypertensive men aged 47-55 y treated for 8 weeks. 1551   From no change to 29% reduction in several studies. 0089   Mean decrease from 53 to 48 mg/dL in 16 hypertensives given 80 mg daily for up to 12 mo. 2266

Smoking   Observed in heavy smokers, although rises when smoking given up. 3629

Sotalol   No change in one study: general effect in 2 studies but overall progressive deterioration in lipid profile. 0088   Significant average decrease of 28% reported from one 12 mo study. 2116   Marked reduction and ratio of HDL-cholesterol to total cholesterol in essential hypertensives after 12 mo. 2120

Spironolactone   Average fell from 1.5 to 1.1 mmol/L at 6 mo, and to 1.0 mmol/L in 15 patient with primary hypertension given 100 mg/d. 1072

Stanozolol   Increase by 53% in 10 normolipidemic postmenopausal osteoporotic women treated for 6 weeks. 3526

L-Thyroxine   From 1.6 to 1.4 mmol/L in 11 hypothyroid women treated with 0.1 to 0.2 mg daily. 1201

Trichlormethiazide   Decrease by -11 mg/dL in 15 patients given 4 mg daily for 3 mo but note marked difference between responders and nonresponder populations 2550

Verapamil   In 12 patients with angina or hypertension treated for 6 weeks. Where change occurred it was of order of 10%. 3756

Cholesterol, α-Lipoprotein

Serum   Decrease   Physiological
Oral Contraceptives   Increase by 19 mg/dL in white girl drug users vs controls. *3723*
Smoking   Effect observed in fasting white children aged 8 to 17 y. *3723*

Cholesterol, β-Lipoprotein

Serum   No Effect   Physiological
Smoking   Effect observed in fasting white children aged 8 to 17 y. *3723*

Cholesterol, Low Density Lipoprotein

Serum   No Effect   Physiological
Acebutolol   In small numbers of patients when treated for 1-12 mo. *0088*   Insignificant change on average in one 6 mo long study. *2116*   In several studies of about 15 patients treated for 1 to 12 mo. *0089*
Atenolol   No significant change in 4 studies of from 3 to 6 mo. *2116*   In 20 patients with 50 mg/d compared with pre-treatment values after 3 mo. *3071*   No effect of treatment with 100 mg/d. *2192*
Bisoprolol   No effect with 10 or 20 mg/d. *2192*
Bopindolol   In 24 hypertensives treated with drug for 3 mo. *3667*
Bunitrolol   When 30 mg/d given for 12 weeks in normolipidemic patients with mild essential hypertension. *3151*
Captopril   After 12 weeks treatment in 18 patients with mild essential hypertension. *3150*
Carprazidil   No consistent effect in 15 men with mild to moderate essential hypertension treated for up to 16 weeks. *1268*   In one study involving 12 patients treated for 4 mo. *0089*
Chenodeoxycholic Acid   By 26% in 8 normolipemic patients receiving 16 g drug daily. *2121*
Chlorthalidone   In 22 premenopausal women without decrease in blood pressure treated for less than 1 y. *0402*   In 22 premenopausal and 18 postmenopausal women given 100 mg/d for 6 weeks. *0402*
Cimetidine   In healthy individuals and in subjects with high serum glucose concentration. *1653*
Clonidine   In 59 patients with primary hypertension given up to 300 mg daily for 6 mo. *1874*
Cyclofenil   Nonsignificant tendency to increase in 19 patients given 600 mg daily for 4 mo. *2786*
Diltiazem   No significant change in 31 subjects treated for 6 mo. *0088*
Estradiol   Insignificant decrease with 2 mg estradiol valerate/d in 20 normolipoproteinemic women over 3 mo. *3593*
Ethanol   No effect of ingestion of 1.5 g/kg at night on concentration measured next morning. morning. *3547*
Furosemide   Not significant change in 12 normotensive men when treated with drugs alone for for less than 1 y. *1826*
Hydrochlorothiazide   Not significant change in 39 individuals given drug alone for less than 1 y. *1404*   In short term study involving many patients. *0089*
Indapamide   In 27 subjects treated for less than 1 y with this drug only. *0692*   In various studies involving more than 30 patients when used as sole drug for less than 1 y. *0089*
Labetalol   No significant change in several studies with patients treated for 1-12 mo. *0088*
Levonorgestrel   Insignificant increase in 11 normolipoproteinemic women given 250 μg/d for 2 weeks. *3593*
Methyldopa   In several patients treated for 2-4 mo. *0088*
Metoprolol   Although slight reduction in one study. *0088*   No significant change after 12 weeks in 18 hypertriglyceridemic hypertensives. *0341*   Insignificant change after 1 or 3 mo studies. *2116*   In 20 hypertensives given 200 mg for 3 mo. *3071*
Nadolol   In 13 patients given 50-200 mg daily for 10 weeks. *1806*   No difference between treatment with drug and placebo. *2761*

Nifedipine   No effect observed in 14 patients with essential hypertension treated over 2 mo with 20 mg twice daily. *2119*
Norethisterone   In 75 women who had received up to 24 intragluteal injections of 200 mg every 56 d. *1666*
Oral Contraceptives   No difference between treated women and others. *1303*   No change with high-dose oral contraceptives that are overly estrogenic. *1248*   In 10 women receiving ethinyl estradiol with norethindrone. In 10 women receiving ethinyl estradiol with norgestrel. *2009*   Typically insignificant change over 6 cycles in women with low dose ethinyl estradiol with low dose levonorgestrel or desogestrel. *1249*
Oxprenolol   In several studies of 1-4 mo. *0088*
Phenytoin   In both men and women in 27 patients with transient brain ischemia in 27 patients with transient brain ischemia. *1879*
Pindolol   No significant change typically seen with treatment for 1 mo or more. *0088*   No significant change in 3 studies of 2 to 12 mo. *2116*
Polythiazide   Not significant change in 20 people when treated only with drug for less than 1 year. *1808*
Prazosin   In 16 hypertensive patients given 2 mg daily for 3 mo. *2266*
Prednisone   Insignificant increase (11%) in both men and women after 1 mo. *3949*
Propranolol   General effect observed in multiple studies. *0088*   Or slight reduction in 53 hypertensives given 80 mg twice daily for 3 mo. *0835*   In 17 patients with hypertension followed for 3 mo. *0967*   In 50 volunteers given 160 mg daily for 3 mo. *2171*   In several studies. *0089*
Spironolactone   No significant change in 17 subjects treated with drug for less than 1 y. *1405*
Terbutaline   No significant effect observed after 2 weeks in 15 subjects. *1654*
Ursodeoxycholic Acid   In 8 normolipemic patients receiving 1,000 mg drug daily. *2121*
Verapamil   In 64 patients in postmyocardial infarction in comparison against placebo. *3489*

Serum   Increase   Physiological
Aminoglutethimide   In 73 patients with advanced breast cancer receiving 500 mg/d with 40 mg hydrocortisone. *0415*
Atenolol   Slight increase reported from one study. *0088*   Increase by 5.9% after 100 mg/d for 5 weeks in 20 hypertensive men. *2135*
Bezafibrate   Increase by 16% in 7 hypertriglyceridemic type 2 diabetic patients given 200 mg tid after 1 mo treatment. *3385*
Caffeine   Positive relationship between coffee consumption and lipid concentration: association possibly more with coffee than caffeine. *1447*
Carbimazole   From 2.7 to 3.5 mmol/L in 12 hyperthyroid women patients treated with 10-30 mg daily. *1201*
Celiprolol   Effect slight in patients with hyperlipoproteinemia types IIa, IIb, or IV with 300 mg drug for 4 weeks. *2873*
Chenodeoxycholic Acid   Increase by 12.2 mg/dL in 916 patients given up to 750 mg/d for 2 y. *3199*
Chlorthalidone   10% change during monotherapy in 302 subjects for 1 y. *1334*   10% change in 39 subjects treated with drug only for less than 1 y. *1404*   Increase by 21% in 18 postmenopausal women given 100 mg/d for 6 weeks. *0402*   Increase by 13 mg/dL in 1,000 men and women with mild hypertension treated for 1 y. *1334*   10% in long term, 10% in short term in several studies with monotherapy. *0089*
Clopamide   13% change in 17 individuals treated for less than 1 y when used as sole treatment. *3184*
Danazol   After 2 weeks with 600 mg/d in 12 women with endometriosis increased by 14% and and 34% after 8 weeks, normal in 8 weeks after treatment stopped. *1066*   Constant but significant effect in 62 patients with endometriosis treated with 600 mg daily for up to 24 weeks. *3222*   Increase by 51% in 6 patients with endometriosis given 600 mg daily for 3 to 6 mo. *2279*   Increase by 14% after 2 weeks and 34% after 8 weeks in 12 women with endometriosis given 600 mg daily for 24 weeks. *1066*
Etretinate   From 3.82 to 4.16 mmol/L in 13 patients with hyperkeratotic disorders given 50 mg daily for 2 mo. *3659*
Fenofibrate   Increase by 22% in 9 hypertriglyceridemic type 2 diabetic patients given 100 mg tid after 1 mo treatment. *3385*

Cholesterol, Low Density Lipoprotein
(continued)

Serum Increase Physiological (continued)

Furosemide 15% change in 16 subjects treated for less than 1 y with drug only. *1306*

Glutethimide Significant effect from about 15 d of treatment with 500 mg drug/d in 6 volunteers. *0407*

Hydrochlorothiazide Observed with 100 mg/d. *3663*

Indapamide Slight and insignificant in 13 hypertensive diabetics treated for 24 weeks. *2686*

Isotretinoin In 7 patients with severe rosacea treated with 1 mg/kg/d for 12 weeks. Effects possibly due to induction of hepatic microsomal enzymes. *2307* Increase in both men and women with 1 mg/kg/d for 20 weeks when given for nodulocystic acne. *0326* From 3.92 to 5.33 mmol/L at 6 weeks in 7 patients with severe rosacea treated with 1 mg/kg/d. *2307* From 3.67 to 4.37 mmol/L in 12 patients with hyperkeratotic disorders given 40 mg daily for 2 mo. *3659* Increase in both men and women with 1 mg/kg/d for 20 weeks when given for nodulocystic acne. *0326*

Lynestrenol Increase by 19% due to progestational activity in 6 women with endometriosis given 5 to 10 mg daily for 6 mo. *2279*

Oral Contraceptives Mainly estrogen response. *3525* Increase by 8 to 15% depending on preparation used. *2860* 24% higher median concentration using combination with relatively low estrogen and medium or high progestin. *3733* Increase of 24% when high progestogen compounds ingested. *1248*

Oxprenolol Mild effect as possessed relatively weak intrinsic sympathomimetic activity. *2116*

Pindolol Slight tendency to rise during treatment. *2117*

Piretanide Effect observed with 12 mg/d. *3663*

Propranolol Mean increase from 126 to 134 mg/dL in 16 hypertensives given 80 mg daily for up to 12 mo. *2266*

Sotalol No change or up to 30%: general effect in 2 studies but overall progressive deterioration in lipid profile. *0088* Significant increase of 32% reported from one 12 mo study. *2116*

Spironolactone Maximum average increase of 5% in 3 studies. *0087*

Stanozolol Increase by 21% in 10 normolipidemic postmenopausal osteoporotic women treated for 6 weeks. *3526*

Serum Decrease Physiological

Chlorpropamide With 8 weeks treatment of 8 C-peptide negative insulin dependent diabetics. *3260*

Dehydroepiandrosterone Increase by 7.5% in 5 normal men given 1600 mg/d orally for 28 d. *2578*

Estradiol In 8 post-menopausal, post-oophorectomy women after 100 mg implant in subcutaneous tissue. *3274* With nonalkylated estrogens, eg valerate salt, in treatment of postmenopausal hormone deficiency. *1037*

Estradiol-17β In 14 oophorectomized women over 6 mo when 50 mg drug given as implant. *1080* Marked reduction in 38 healthy post-menopausal women given 2-4 mg orally or 3 mg cutaneously over 6 mo. *1067*

Estrogens In post-menopausal women 11-19% reduction compared with women not taking drug. *3733* In post-menopausal women treated for 3 y. *0653*

Ethanol Associated with alcohol consumption in fasting subjects. *2428* Low values associated with debauch gradually increased to normal. *1011*

Fenofibrate Increase by 20% over 24 weeks in 92 type IIa given 100 mg 3 times per day with similar responses in type IIb patients when compared with baseline values. *0503*

Gemfibrozil Increase by 10% in extensive 5 year double-blinded trial with 2,000 men receiving 600 mg drug twice daily. *1193*

Guanabenz In 39 patients treated for 2 y: approximately 23 mg/dL decrease. *1865*

Ketanserin In 50 hypertensive patients given 80 mg/daily for 3 mo. *2171*

Levonorgestrel Significant reduction in patients over 3 y compared with patients with IUDs for 2 1/2 y. *0763*

Lovastatin Increase by 17 to 39% in clinical trial of 101 patients with heterozygous familial hypercholesterolemia. *1530*

Medroxyprogesterone Increase by 13% in 11 men with sexual deviation syndrome given approximately 1273 mg over a total of approximately 17 d. *0643*

Methyldopa In 17 hypertensive patients given drug for 3 mo. *0967*

Metoprolol Increase by 7 to 8.5% in several studies of 2 to 4 mo with up to 400 mg drug/d. *1806*

Minoxidil Approximately 10% reduction after 3 mo and 20% after 6 mo. *1807*

Neomycin In 20 subjects with type II hyperlipoproteinemia given 2 g/d for 9 mo caused 16% decrease. *1618* Marked effect over diet in 20 patients with type II hyperlipoproteinemia treated over several months. *1618*

Niacin Increase by 21% in 34 hypercholesterolemic individuals with 1.5 g/d for 1 mo then 3.0 g/d up to 6 mo. *1967*

Propranolol Appreciable decrease (130 to 111 mg/dL) after 12 weeks in 15 hypertriglyceridemic hypertensives. *0341*

Tamoxifen Observed in one elderly woman treated for breast cancer. Possibly due to reduction of activities of postheparin plasma lipoprotein lipase and hepatic triglyceride lipase. *0508* From 5.11 to 4.10 mmol/L accountable for much of total change. *3070*

L-Thyroxine From 5.5 to 4.1 mmol/L in 11 hypothyroid women treated with 0.1 to 0.2 mg daily. *1201*

Cholesterol, Very Low Density Lipoprotein

Serum No Effect Physiological

Acebutolol In small numbers of patients when treated for 1-12 mo. *0088* In several studies of about 15 patients treated for 1 to 12 mo. *0089*

Atenolol In 20 patients with 50 mg/d compared with pre-treatment values after 3 mo. *3071*

Bunitrolol When 30 mg/d given for 12 weeks in normolipidemic patients with mild essential hypertension. *3151*

Captopril After 12 weeks treatment in 18 patients with mild essential hypertension. *3150*

Carprazidil No consistent effect in 15 men with mild to moderate essential hypertension treated for up to 16 weeks. *1268* In one study involving 12 patients treated for 4 mo. *0089*

Chenodeoxycholic Acid By 26% in 8 normolipemic patients receiving 16 g drug daily. *2121*

Chlorthalidone In 22 premenopausal women without decrease in blood pressure treated for less than 1 y. *0402* In 22 premenopausal and 18 postmenopausal women given 100 mg/d for 6 weeks. *0402*

Cimetidine In healthy individuals and in subjects with high serum glucose concentration. *1653*

Clonidine In 59 patients with primary hypertension given up to 300 mg daily for 6 mo. *1874*

Clopamide No significant change in 17 individuals treated for less than 1 y when used as sole treatment. *3184*

Dehydroepiandrosterone In 5 normal men given 1600 mg/d orally for 28 d. *2578*

Diltiazem No significant change in 31 subjects treated for 6 mo. *0088*

Estradiol In 8 post-menopausal, post-oophorectomy women after 100 mg implant in subcutaneous tissue. *3274* With nonalkylated estrogens, eg valerate salt, in treatment of postmenopausal hormone deficiency. *1037*

Estradiol-17β Insignificant change in 38 healthy post-menopausal women given 2-4 mg orally or 3 mg cutaneously over 6 mo. *1067*

Ethanol No effect of ingestion of 1.5 g/kg at night on concentration measured next morning. *3547*

Etretinate In 13 patients with hyperkeratotic disorders given 50 mg daily for 2 mo. *3659*

Furosemide No significant change in 16 subjects treated for less than 1 y with drug only. *1306*

Indapamide In 27 subjects treated for less than 1 y with this drug only. *0692* In various studies involving more than 30 patients when used as sole drug for less than 1 y. *0089*

Labetalol No significant change in several studies with patients treated for 1-12 mo. *0088*

Metoprolol   Nonstatistically significant increase from 61 to 67 mg/dL after 12 weeks in 18 hypertriglyceridemic hypertensives. *0341*

Neomycin   No significant effect in 20 patients with type II hyperlipoproteinemia treated over several months. *1618*

Nifedipine   No effect observed in 14 patients with essential hypertension treated over 2 mo with 20 mg twice daily. *2119*

Norethisterone   In 75 women who had received up to 24 intragluteal injections of 200 mg every 56 d. *1666*

Oral Contraceptives   No difference between treated women and others. *1303*

Oxprenolol   In several studies of 1-4 mo. *0088*   In 20 hypertensive men given 160 mg/d for 5 weeks. *2135*

Phenytoin   In both men and women in 27 patients with transient brain ischemia In 27 patients with transient brain Ischemia. *1879*

Pindolol   No effect in short-term study of 10 hypertensive men with 15 mg/d. *2135*

Polythiazide   Not significant change in 20 people when treated only with drug for less than 1 year. *1808*

Prazosin   In 16 hypertensive patients given 2 mg daily for 3 mo. *2266*

Propranolol   General effect observed in multiple studies. *0088* In 17 patients with hypertension followed for 3 mo. *0967*   In 23 hypertensive men given up to 160 mg/d for 8 weeks. *2135*   In several studies. *0089*

Sotalol   No change in one study: general effect in 2 studies but overall progressive deterioration in lipid profile. *0088*

Spironolactone   No significant change in 17 subjects treated with drug for less than 1 y. *1405*   No significant change in 3 studies. *0087*

Ursodeoxycholic Acid   In 8 normolipemic patients receiving 1,000 mg drug daily. *2121*

Serum   Increase   Physiological

Atenolol   Effect observed in 2 studies, but not significant in one other. *0088*   From 1.21 to 1.62 mmol/L in 15 hypertensives treated for 8 mo. *1021*   From 0.90 to 1.14 mmol/L with 100 mg drug/d. *2192*   Increase by 5.9% after 100 mg/d for 5 weeks in 20 hypertensive men. *2135*

Chlorthalidone   7% change in 39 subjects treated with drug only for less than 1 y. *1404*   7% in short term in several studies with monotherapy. *0089*

Furosemide   56% change in 12 normotensive men when treated with drug alone for less than 1 y. *1826*   6% change in 16 subjects treated for less than 1 y with drug only. *1306*

Glutethimide   Significant effect from about 15 d of treatment with 500 mg drug/d in 6 volunteers. *0407*

Hydrochlorothiazide   13% change in individuals given drug alone for less than 1 y. *1404*   13% in a short term involving many patients. *0089*

Isotretinoin   In 7 patients with severe rosacea treated with 1 mg/kg/d for 12 weeks. Effects possibly due to induction of hepatic microsomal enzymes. *2307*   From 0.17 to 0.39 mmol/L at 6 weeks in 7 patients with severe rosacea treated with 1 mg/kg/d. *2307* From 0.69 to 0.93 mmol/L in 12 patients with hyperkeratotic disorders given 40 mg daily for 2 mo. *3659*

Metoprolol   Increase by 30% in one study. *0088*   Rose from 1.00 to 1.29 mmol/L in 20 hypertensives given 200 mg/d for 3 mo. *3071*

Nadolol   Increase by 29% in 13 patients given 50-200 mg daily for 10 weeks. *1806*

Oral Contraceptives   Increase by 16 to 40% depending on preparation used. *2860*

Propranolol   Significant increase from 58 to 80 mg/dL after 12 weeks in 15 hypertriglyceridemic hypertensives. *0341*   In several studies. *0089*   Mean increase from 34 to 41 mg/dL in 16 hypertensives given 80 mg daily for up to 12 mo. *2266*

Tamoxifen   Up to 241 mg/dL observed in one elderly woman treated for breast cancer. Possibly due to reduction of activities of postheparin plasma lipoprotein lipase and hepatic triglyceride lipase. *0508*

Serum   Decrease   Physiological

Atenolol   Increase by 5.9% after 100 mg/d for 5 weeks in 20 hypertensive men. *2135*

Benfluorex   Increase by 20% in 12 hypertriglyceridemic type 2 diabetic patients given 150 mg tid after 1 mo treatment. *3385*

Doxazosin   In 38 treated hypertensives given 1 to 16 mg/d over 10 weeks. *2562*

Estradiol   By more than 50% with 2 mg estradiol valerate/d in 20 normolipoproteinemic women over 3 mo. *3593*

Fenofibrate   Increase by 38% over 24 weeks in 92 type IIa given 100 mg 3 times per day with similar responses in type IIb patients when compared with baseline values. *0503*

Gemfibrozil   Increase by 50% in 18 patients with chronic renal failure treated with 1200 mg/d for 28 weeks. Simultaneous activation of postheparin plasma lipoprotein and hepatic lipases. Effects reversed when drug discontinued. *2741*

Levonorgestrel   By more than 50% in 11 normolipoproteinemic women given 250 μg/d for 2 weeks. *3593*

Methyldopa   In 17 hypertensive patients given drug for 3 mo. *0967*

Metoprolol   Increase by 6 to 7% in several studies of 2 to 4 mo with up to 400 mg drug/d. *1806*

Pindolol   35% reduction in concentration after 2 mo in 11 hypertensive patients but after 16 mo treatment had reverted to normal. *1399*

Prazosin   Increase by 10% in 23 healthy hypertensives treated for 8 weeks. *2136*

Cholic Acid

Serum   No Effect   Physiological

Cholestyramine   In 19 patients suffering from intrahepatic cholestasis of pregnancy. *1545*

Oral Contraceptives   No effect observed over 12 mo in 29 women given combination of ethinyl estradiol and norgestrel. *1546*

Phenobarbital   No effect on increased concentration in patients with intrahepatic cholestasis increased of pregnancy. *1545*

Serum   Increase   Physiological

Rifampin   Significant increase possibly due to blocking uptake by plasma membrane of hepatocytes. *1224*

Feces   Increase   Physiological

Neomycin   Due to altered intestinal flora. *1077*

Chromium

Urine   Increase   Physiological

EDTA   If poisoning due to chromium. *0071*

Chromosomes

Test Conditions   No Effect   Physiological

Chloramphenicol   Not clastogenic in human cells. *3282*

Lead   No effect of chronic occupational exposure. *2648*

Methadone   No effect human leucocytes at concentrations 1/6-3x normal. *1073*

Morphine   No effect human leucocytes at concentrations 1/6-3x normal. *1073*

Penicillin   Not clastogenic in human cells. *3282*

Quinine   No effect human leucocytes at concentrations 1/6-3x normal. *1073*

Streptomycin   Not clastogenic in human cells. *3282*

Tetracycline   Not clastogenic in human cells. *3282*

Test Conditions   Abnormal   Physiological

Acridine   Clastogenic in human diploid fibroblasts *in vitro*. *3282*

Acridine Orange   Clastogenic in human diploid fibroblasts *in vitro*. *3282*

Actinomycin   Clastogenic in human cells. *3282*

Adenine   Clastogenic in human cells. *3282*

Adenine Deoxyribose   Clastogenic in human cells. *3282*

Aflatoxin   Probably clastogenic in human cells. *3282*

Aminopterin   Clastogenic in human lymphocytes in culture. *3282*

Arsenates   Clastogenic in human cells. *3282*

Chromosomes (continued)

Test Conditions   Abnormal   Physiological (continued)

Azathioprine   Clastogenic to bone marrow and leukocytes. *3282*
Aziridine   Clastogenic in human cells. *3282*
Bromodeoxyuridine   Clastogenic in human cells. *3282*
Caffeine   Clastogenic in human cells. *3282*
Chlorambucil   In lymphocytes with chronic lympholeukemia. *2094*
Chloramine   Clastogenic in human lymphocyte cultures. *3282*
Chlormethine   Clastogenic in human cells. *3282*
Chlorpromazine   Clastogenic in human lymphocytes *in vitro.* *3282*
Colchicine   Inhibits mitosis at metaphase in human cells. *3282*
Cyclamate   Potent clastogen in human cells. *3282*
Cyclohexylamine   Potent clastogen in human cells. *3282*
Cyclophosphamide   Clastogenic in human cells. *3282*
Cytarabine   Clastogenic in human cells. *3282*
Daunorubicin   Clastogenic in human cells. *3282*
Ethoxycaffeine   Clastogenic in human cells. *3282*
Floxuridine   Clastogenic in human cells. *3282*
Fluorophenylalanine   Clastogenic in human lymphocyte cultures. *3282*
Lead   Clastogenic in human cells in chronic poisoning. *3282*
Lysergide   Clastogenic in human cells *in vitro* (*?in vitro*). *3282*
Meprobamate   Clastogenic in human lymphocytes *in vitro.* *3282*
Mercaptoethanol   Clastogenic in human cells. *3282*
Mercaptopurine   Clastogenic in human cells. *3282*
Mercaptopyruvate   Clastogenic in human cells. *3282*
Mescaline   Teratogenic in experimental animals. *3282*
Methotrexate   Clastogenic in human lymphocytes in culture. *3282*
Mitomycin   Clastogenic in human cells. *3282*
Nalidixic Acid   Clastogenic in human cells. *3282*
Neutral Red   Clastogenic in human cells. *3282*
Ozone   Clastogenic in human cell cultures. *3282*
Phenytoin   One case abnormal y chromosome. *2154*
Phleomycin   Clastogenic in human cells. *3282*
Piperazine   Clastogenic to bone marrow and leukocytes. *3282*
Proflavine   Clastogenic in human hela cultures. *3282*
Puromycin   Clastogenic in human cells. *3282*
Scopolamine   Clastogenic in human cells. *3282*
Streptonigrin   Clastogenic in human cells *in vitro* for up to 1 mo. *3282*
Tetrahydrocannabinol   Clastogenic in human cells. *3282*
Theobromine   Clastogenic in human cells. *3282*
Theophylline   Clastogenic in human cells. *3282*
Thioguanine   Clastogenic in human cells. *3282*
Thorium Dioxide   Clastogenic in human cells. *3282*
Triaziquone   Clastogenic in human cells. *3282*
Triethylenemelamine   Clastogenic in human cells. *3282*
Ultraviolet Light   Clastogenic in human cells. *3282*
X-Ray Therapy   Clastogenic in human cells in hela culture. *3282*

Chylomicrons

Serum   Increase   Analytical
Bacterial Contamination   If contaminated with phospholipase producing bacteria. *3223*

Serum   Increase   Physiological
Asparaginase   Observed in unusual hyperlipidemic response. *2657*
Ethanol   Due to specific disturbance of the clearing mechanism of triglyceride-rich lipoproteins. *3194*

Chymotrypsin

Duodenal Contents   Increase   Physiological
Calcium   Calcium infusion (10 mg/kg/h) causes increase x 2. *1314*

Pancreatic Juice   Decrease   Physiological
Ethanol   Oral or i.v. cause direct inhibition. *2504*

Citrate

Serum   Increase   Physiological
Blood Transfusions   If citrated blood used. *2220*
Fructose   Rises more rapidly than after glucose. *1709*

Urine   No Effect   Physiological
Bed Rest   Not significantly affected by immobilization. *0863*
Lithium   No consistent effect observed. *2107*   No effect observed after administration. *0409*

Urine   Increase   Physiological
Citrates   From 398 to 856 mg/d on average with potassium citrate in 5 patients with uric acid lithiasis. From 398 to 799 mg/d with sodium citrate in 5 patients with uric acid lithiasis. *3116* Restored to normal levels in 89 patients with hypocitraturic calcium nephrolithiasis treated for up to 4 y. *2708*
Parathyroid Extract   Hormonal action. *1343*

Urine   Decrease   Physiological
Acetazolamide   Alteration of acid base status, diuresis. *2745*
Bendrofluazide   Up to 30% decrease observed. *2427*
Cellulose Phosphate   Observed in 3 of 5 cases. *2710*
Chlorothiazide   Up to 30% decrease observed. *2427*
Hydrochlorothiazide   By up to 30% reported. *3245*
Polythiazide   Excretion decreased by up to 30%. *2427*
Thiazides   By up to 30%. *2745*

Citrulline

Plasma   Increase   Physiological
Histidine   Interferes with clearance after oral load. *1644*

Plasma   Decrease   Analytical
Storage of Sample   If picric acid used for deproteinization. *3230*

Plasma   Decrease   Physiological
Tranylcypromine   Observed effect in normal individuals. *0824*

Urine   Decrease   Analytical
Light   Colored complex with diacetylmonoxime unstable. *2776*

Test Conditions   Increase   Analytical
Fructose   Enhances color development in Archibald procedure. *3424*

Test Conditions   Decrease   Analytical
Cysteine   Inhibits color development in Archibald procedure. *3424*
Glutathione   Inhibits color development in Archibald procedure. *3424*

Clindamycin

Serum   Decrease   Physiological
Kaolin   With pectin causes delayed absorption. *3794*

Clonazepam

Serum   No Effect   Physiological
Valproic Acid   No definite effect in 25 patients studied for 5 to 9 mo. *3829*

Clotermine

Urine   Positive   Analytical
Fenfluramine   Similar R_f and color reaction on TLC using ethyl acetate: methanol: water: ammonium hydroxide and modified Dragendorff's reagent. *3868*

Clot Lysis

Blood   No Effect   Physiological
Sucrose   Effect of sustained high sucrose diet. *3931*

Clot Retraction

Blood   No Effect   Physiological
Ampicillin   In doses as high as 300 mg/kg/d in volunteers. *0492*
Methicillin   No effect with doses as high as 300 mg/kg/d in volunteers. *0492*
Penicillin G   No effect with doses as high as 48 million U/d. *0492*

Blood   Decrease   Physiological
Apronalide   Immunological effect. *1436*
Mithramycin   Reversible poor retraction. *0071*

Clotting Time

Blood   No Effect   Physiological
Cyclopropane   Anesthesia has no effect. *1343*
Ether   No effect of anesthesia. *1343*

Blood   Increase   Physiological
Carbenicillin   Reported effect. *3734*
Dicumarol   Slight effect in glass, greater in silicone. *1343*
Heparin   Concentration related effect. *2649*
Mithramycin   Reversible effect. *0071*
Warfarin   May be prolonged. *1302*

Blood   Decrease   Physiological
Oral Contraceptives   (silicone clotting time) associated with clot problems. *0760*

Coagulation Time

Blood   Increase   Physiological
Anticoagulants   Therapeutic intent. *2313*
Phosphorus   Due to toxicity. *1343*
Tetracycline   Delayed coagulation reported. *2313*

Blood   Decrease   Physiological
Aminophylline   Reported effect. *1343*
Epinephrine   Probably due to increased activity of factor V. *2313*
Garlic   Significant decrease although given with fatty meal. *0421*

Cobalt

Serum   No Effect   Physiological
Piretanide   No significant effect with up to 12 mg/d for 3 mo. *3696*

Cocaine

Urine   Positive   Analytical
Diethylpropion   Similar R_f and color reaction on TLC using ethyl acetate: methanol: water: ammonium hydroxide and modified Dragendorff's reagent. *3868*

Cold Agglutinins

Blood   Positive   Physiological
Pregnancy   Occasional response. *0987*

Color

Serum   Increase   Analytical
Carotenoid   Color serum orange. *3023*

Serum   Increase   Physiological
Oral Contraceptives   Ceruloplasmin may be so high blue-green color. *3505*

Blood   Increase   Physiological
Acetanilid   May cause chocolate colored blood. *0279*
Aniline   May produce chocolate color. *0279*
Carbon Monoxide   May cause cherry-red color (carboxyhemoglobin). *0279*
Chlorate   May cause chocolate color. *0279*
Cyanides   May cause cherry-red color. *0279*

Urine   Increase   Analytical
Acriflavine   Greenish fluorescence. *2448*
Aloin   red-brown/yellow pink (alkaline),yellow brown (acid). *2313*
Aminopyrine   Red brown. *2313*
Aminosalicylic Acid   Abnormal color (not distinctive). *0382*
Amitriptyline   Greenish-blue color. *1022*   Green urine associated with ingestion of drug. *2613*
Aniline Dyes   Red from foods and candy. *0443*
Anisindione   Orange (alkaline), pink-red-brown (acid). *2313*
Anthraquinone   Pink, red, purple, orange and rust. *2425*
Antipyrine   Red brown (green in reflected light). *2313*
Azuresin   Blue or green for a few days after test. *1488*
Beets   Red due to anthocyanins and other pigments. *2313*
Benzene   Red brown. *2313*
Bile   Green color with oxidized bile on standing. *0459*
Bilirubin   Yellowish foam. *2448*
Biliverdin   Greenish foam. *2448*
Carrots   Yellow color soluble in petroleum ether. *2313*
Cascara   Brown (acid),yellow-pink(alkaline)black on standing. *1022*
Chloroquine   Brown color. *1022*
Chlorzoxazone   Orange to purple-red. *1022*
Chrysarobin   Oxidation product colors alkaline urine red. *1343*
Cinchophen   Red brown. *2313*
Copper   Blue diapers (alkaline urine on copper fastenings). *0526*
Creosote   Dark green. *2313*
Cresol   Dark brown on standing. *2313*
Deferoxamine   Forms iron chelate with reddish color. *1488*
Dithiazanine   Blue. *2313*
Doans® Pills   Greenish blue. *2313*
Ethoxazene   red, pink, orange and rust colors. *2425*
Evans Blue   Blue color due to presence of dye. *0459*
Fluorescein   Intravenous administration may cause yellow-orange color. *1488*
Furazolidone   Metabolites may produce brown color. *1488*
Furazolium   Red, pink, purple, orange and rust colors. *1237*
Fuscin   Red from foods and candy. *0443*
Guaiacol   May produce green color. *1187*
Hemoglobin   Clear red to reddish brown. *2448*

Color (continued)

Urine Increase Analytical (continued)

Homogentisic Acid Black (occurs with alkaptonuria). 2448

Ibuprofen Pink, red, purple and rust color. 1237

Indican Occurs if increased and oxidized on standing. 0459

Indigo Blue May produce blue color. 1187

Indigotindisulfonate Color used to measure kidney function. 0071

Iron Salts Iron Sorbitex can cause brown urine (Fe sulfide). 1237

Levodopa Red-tinged on voiding, blackens on standing. 3138

Merbromin Fluorescent pink staining of cells. 0459

Methylene Blue Blue color. 1342

Metronidazole Brown color probably due to metabolite. 1237

Naphthol Dark color on standing. 2313

Niridazole Urine becomes dark. 1343

Nitrobenzene Dark color on standing. 2313

Nitrofurans Brown, green, blue color. 1237

Nitrofurantoin Brown, yellow color. 1022

Pamaquine Brown color. 1022

Phenazopyridine Yellow orange increases with HCl. 0443

Phenindione Red-orange color produced in alkaline urine. 0720

Phenolphthalein Pink, red, purple (alkaline), orange, rust (acid). 0382

Phenols Dark green to brownish black on standing. 2313

Phenolsulfonphthalein Purple red or pink in alkaline urine. 1187

Phenothiazines Pink, red, purple, orange, rust color. 2425

Phensuximide Pink, red, purple, orange and rust color. 2425

Phenyl Salicylate Dark green. 2313

Phenytoin pink, red or red-brown color may occur. 1488

Picric Acid Yellow to red brown. 2313

Primaquine Rusty yellow or brown color. 0382

Pyrogallol Brown to black, darkens on standing. 2313

Quinacrine Deep yellow color on acidification. 1343

Quinine Brown color. 1022

Resorcinol Dark green to greenish blue darkens on standing. 2313

Rhubarb Yellow-brown (acid), yellow-pink (alkaline) darkens. 2313

Riboflavin May produce yellow color with large doses. 1488

Rifampin Red-orange due to drug and metabolites. 1578 Causes orange-pink color in saliva, tears, urine, sweat. 3683

Salol Dark color on standing. 2313

Santonin Bright yellow (NaOH changes to pink, scarlet). 2313

Senna Red, orange or rust. 1187

Sulfamethoxazole Brown color observed. 1022

Sulfobromophthalein Red in alkaline urine. 3362

Sulfonamides Brown color with some sulfonamides. 1022

Tetralin Greenish blue. 2313

Thiazolsulfone Red, pink, purple, orange and rust color. 1237

Thymol Greenish blue. 2313

Tolonium Green and blue color. 1022

Triamterene Green, blue with blue fluorescence. 1022

Urobilin Excess may cause yellow to amber color. 2448

Urine Increase Physiological

Aniline Brown color due to intravascular hemolysis. 1302 May produce red color. 1187

Bacteriuria Blue diapers in babies with P. Aeruginosa (pyocyanin). 0524

Blood Smoky red to brown due to hematuria. 2448

Concentrated Urine Yellow orange with dehydration. 0443

Congo Red Dye not taken up by amyloid. 0071

Dinitrophenol Red brown due to hematuria. 2313

Indandione Derivatives May be orange to red in color. 2425

Indigotindisulfonate Blue-green, decolorized with alkali. 0443

Indomethacin Indirect result of hepatic toxicity, green urine. 1109

Iron Dextran Black urine on standing reported. 1678

Iron Sorbitex Complex with citrate produce black urine. 1187

Lead Red brown (?due to porphyrins and hemoglobin). 2313

Lysol® Brown-black with poisoning. 1187

Mannose Deep orange after i.v. infusion. 0358

Melanin Brown to black. 2448

Mercury Compounds Red brown due to hematuria. 2313

Methemoglobin May produce red-brown urine. 0443 With oxy-hemoglobin produces port-wine color. 1187

Methyldopa May produce brown-black on standing. 0443

Methylsulfonal Red (may provoke porphyria). 2313

Myoglobin Produces red color in urine. 0459

Naphthalene Brown or black due to blood and hemoglobin. 1302

Phenylhydrazine May produce dark brown urine. 1187

Porphyrins Burgundy red, darkens on standing. 2313

Pseudomonas Blue-green color may occur. 0443

Quinacrine Produces yellow coloration. 0443

Sulfomethane Red-brown (may provoke porphyria). 2313

Trinitrotoluene Red brown due to hemoglobin. 2313

Verdoglobin May cause orange-red-brown color. 1187

Urine Decrease Analytical

Ethanol Diuresis induced reduces color. 1488

Methocarbamol Brown, green, blue or black on standing. 0382

Urine Decrease Physiological

Dilute Urine Due to dilution of pigments. 0443

Urine Positive Analytical

Acetaminophen Dark brown urine observed in some patients with overdose, probably due to p-Aminophenol. 0669

Methocarbamol Green color observed. 2613

Feces Increase Analytical

Aluminum Hydroxide White discoloration/speckling if taken orally. 1187

Anthraquinone Brownish staining. 2313

Barium White discoloration/speckling if taken orally. 1187

Bismuth Salts Blackens or discolors stool with 5 g orally. 2313

Calomel Green color observed. 1187

Carmine Produces red color. 1187

Charcoal Black due to ingested material. 1187

Dithiazanine Green to blue. 2313

Hematoxylin Reddish-brown with 1 gram. 0902

Iron Salts Black (gray-black) darkens in air with about 70 mg. 2313

Manganese Dioxide Dark brown to black with 130-140 mg. 0902

Mercury Compounds Green with about 130 mg of calomel. 2313

Phenazopyridine Orange red. 2313

Phenolphthalein Imparts red color. 0071

Santonin Deep yellow with 65-70 mg. 2313

Senna Yellow to brown. 2313

Sulfobromophthalein Alkaline stools may appear bloody. 0583

Tetracycline Red if glucosamine potentiated syrup form. 2313

Feces Increase Physiological

Anticoagulants Red to black due to internal bleeding. 2313

Aspirin Red or black due to gastrointestinal bleeding. 2313

Methylene Blue Blue especially on exposure to air with 130-140 mg. 0902

Rhubarb Yellow with 2 ml extract. 0902

Rifampin Orange-red color due to drug and metabolites. 0019

Salicylate Pink to red black due to gastrointestinal bleeding. 1187

Feces Decrease Analytical

Alkaline Antacids White discoloration or speckling. 2313

Aluminum Salts White discoloration or speckling. 2313

Saliva   Increase   Analytical
Levodopa   Brown reported with treatment. *0798*

Complement APH$_{50}$

Serum   No Effect   Physiological
Oral Contraceptives   Compared with nonusers in first year of use; largely influence of progestogen. *0866*

Complement C1q

Serum   Increase   Physiological
Chlorpropamide   In isolated case with diabetes who developed proliferative glomerulonephritis indicating immunologically mediated reaction. *0146*

Serum   Decrease   Physiological
Hydralazine   Observed in one patient with induced systemic lupus erythematosus. *3784*

Complement C3

Serum   No Effect   Physiological
Anticonvulsants   In 8 epileptic patients receiving phenobarbital with carbamazepine or phenytoin. *2670*
Phenytoin   In about 118 treated epileptics. *0228*

Serum   Increase   Physiological
Oral Contraceptives   Compared with nonusers in first year of use: largely influence of progestogen. *0866*

Serum   Decrease   Analytical
Storage of Sample   Degrades to β_1-globulin, no effect 2 d at room temperature. *0828*

Serum   Decrease   Physiological
Hydralazine   Observed in one patient with induced systemic lupus erythematosus. *3784*
Methyldopa   Observed in one patient with hepatocellular damage of moderate severity due to sensitization by drug. *0874*
Phenytoin   Observed in 50% patients. *0989*

Complement C3c

Serum   Decrease   Physiological
Phenytoin   Observed in 50% patients. *0989*

Complement C4

Serum   No Effect   Physiological
Anticonvulsants   In 8 epileptic patients receiving phenobarbital with carbamazepine or phenytoin. *2670*
Hydralazine   Observed in one patient with induced systemic lupus erythematosus. *3784*
Phenytoin   In about 118 treated epileptics. *0228*

Serum   Increase   Physiological
Danazol   Up to 8 fold increase in 5 patients with hereditary angioneurotic edema treated for up to 10 mo. *3219*
Oral Contraceptives   Compared with nonusers in first year of use: largely influence of progestogen. *0866*

Serum   Decrease   Physiological
Dextran   Complex formation or increased consumption. *3346*
Methyldopa   Observed in one patient with hepatocellular damage of moderate severity due to sensitization by drug. *0874*
Penicillamine   In 6 patients with rheumatoid arthritis developed systemic lupus erythematosus-like syndrome. *0623*

Complement CH$_{50}$

Serum   No Effect   Physiological
Phenytoin   No effect of treatment of epileptics. *0228*

Serum   Increase   Physiological
Chlorpropamide   In isolated case with diabetes who developed proliferative glomerulonephritis indicating immunologically mediated reaction. *0146*
Oral Contraceptives   Compared with nonusers in first year of use: largely influence of progestogen. *0866*

Serum   Decrease   Physiological
Hydralazine   Observed in one patient with induced systemic lupus erythematosus. *3784*

Complement Factor B

Serum   Increase   Physiological
Oral Contraceptives   Compared with nonusers in first year of use: largely influence of progestogen. *0866*

Serum   Decrease   Physiological
Hydralazine   Observed in one patient with induced systemic lupus erythematosus. *3784*

Congo Red Test

Serum   Decrease   Physiological
Proteinuria   May cause urinary loss, false indication of amyloid. *0279*

Coombs' Test

Serum   Increase   Physiological
Moxalactam   Observed in some cases with all cephalosporins. *2615*

Serum   Positive   Physiological
Cefoxitin   6 of 77 patients developed rapidly reversible direct Coombs' test without hemolysis. *2615*
Cephaloglycin   Rare observed side effect. *0131*
Chlorpromazine   Immunological response to drug. *1486*
Chlorpropamide   Immunological response to drug. *1486*
Cyclopenthiazide   Reported observation. *2583*
Dipyrone   May produce immune hemolytic anemia. *0359*
Ethosuximide   Mechanism obscure. *1486*
Hydralazine   Mechanism obscure. *1486*
Isoniazid   Immunological response to drug. *1486*
Levodopa   Autoimmune phenomenon (occurs after several months). *1486*
Mefenamic Acid   Autoimmune phenomenon (occurs after several months). *1486*
Melphalan   Immunological response to drug (gamma antibody). *1486*
Oxyphenisatin   May cause hypersensitivity reaction. *2970*
Penicillamine   In 6 patients with rheumatoid arthritis developed systemic lupus erythematosus-like syndrome. *0623*
Phenylbutazone   Immunological response to drug. *1486*
Phenytoin   Immunological response to drug. *1486*
Procainamide   Mechanism obscure. *1486*
Quinidine   Immunological response to drug (gamma antibody). *1486*
Quinine   Immunological response to drug. *1486*
Streptomycin   Mechanism obscure. *1486*
Sulfonamides   Immunological response to drug. *1486*
Sulfonylureas   May cause immune hemolytic anemia. *0359*
Tetracycline   Mechanism obscure. *1486*

Coombs' Test, Direct

Serum   No Effect   Physiological
Oxacillin   0 of 10 patients demonstrated hypersensitivity reaction. *1963*

Serum   Increase   Physiological
Chlorpropamide   Isolated case of drug-induced hemolytic anemia. *3111*
Tolmetin   Strongly positive: Observed in single case of 49 y old man who had taken drug as needed for arthritis for 1 y. *3421*

Serum   Positive   Physiological
Aminosalicylic Acid   Immunological response to drug. *1486*
Amphotericin B   Single case reported with hemolytic anemia. *1487*
Ampicillin   Hypersensitivity reaction in 39% of 36 patients receiving drug. *1963*
Carbromal   Unusual cause of hemolytic anemia. *2365*
Cefamandole   In about 1% of treated patients (dose related). *3134*
Cephalexin   Incidence low. *1060*
Cephaloridine   No immunological response (complex binds to cell in 8% cases). *1486*
Cephalosporin   Approximately 3% incidence, immunologic effect. *1487*
Cephalothin   Nonimmunologic phenomenon, complex binds to cell. *1486*
Chlorpropamide   Single case with hemolytic anemia. *1487*
Cisplatin   Observed in 2 individuals receiving drug. *1272*
Cyclophosphamide   May cause hemolytic anemia. *2365*
Ethosuximide   3 cases reported with SLE and anemia. *1487*
Hydralazine   May occur if SLE induced. *1487*   Observed in one patient receiving low dose of drug for hypertension. Mechanism involved in producing erythrocyte-reacting antibody not known. *2680*
Insecticides   May cause hemolytic anemia. *2365*
Isoniazid   Complement type of positive response. *1487*
Levodopa   Possible dose related without hemolysis. *1487*
Mefenamic Acid   Drug induces autoimmune phenomenon. *1487*
Mephenytoin   Mechanism obscure. *1486*
Methicillin   In 31% of 45 patients receiving drug (hypersensitivity reaction). *1963*
Methyldopa   Autoimmune phenomenon (occurs with weeks of treatment). *1486*   Observed in 5 of 9 patients typically seen in 20% of treated patients. Probably due to impaired F_c-dependent reticuloendothelial function. *1909*
Methysergide   Single case with retroperitoneal fibrosis. *1487*
Minoxidil   Occurred without hemolysis in one patient. *2174*
Nalidixic Acid   May cause hemolytic anemia with normal G-6-PD. *1281*
Oxyphenisatin   Single case with positive LE cell test. *1487*
Penicillin   Combines to RBC, immunoglobulins develop to drug. *1486*
Penicillin G   44% of 39 patients demonstrated hypersensitivity reaction. *1963*
Penicillin V   Isolated case of hemolytic anemia with IgM antibody. *0360*
Phenylbutazone   Single case after 2 weeks therapy. *1487*
Phenytoin   Few cases with hemolytic anemia. *1487*
Piperacillin   In 1 of 59 patients given drug as sole agent with many patients with severe illness. *3874*
Procainamide   Doubled incidence in individuals receiving drug compared with control with production of red cell autoantibody. *1958*
Quinidine   Associated with hemolytic anemia. *1487*   Occasional hemolytic anemia and hypersensitivity observed in 32 of 487 patients who had received drug. *1261*
Quinine   Associated with hemolytic anemia. *1487*
Rifampin   Weak response in 8% patients. *2843*
Stibophen   Produces hemolytic anemia. *2365*

Tolbutamide   Probably due to autoantibody complex on RBC surface. *0359*   Not definitive proof, hemolytic anemia. *1487*
Triethylenemelamine   May cause hemolytic anemia. *2365*

Plasma   Positive   Physiological
Cefoxitin   In about 2% of treated patients (dose related). *3134*

Blood   Positive   Physiological
Nitrofurantoin   Developed lupus-like syndrome. Syndrome associated with pulmonary reaction. *3250*

Red Blood Cells   Increase   Physiological
Zomepirac   Case reported of immune hemolysis. *3208*

Coombs' Test, Indirect

Serum   Positive   Physiological
Asparaginase   In 4 children who demonstrated hemolytic anemia (direct Coombs' test negative). *0558*
Carbromal   Unusual cause of hemolytic anemia. *2365*
Clonidine   Mechanism not discussed. *3923*
Levodopa   Observed in fewer than 1% of patients. *0824*
Mephenytoin   Mechanism obscure. *1486*
Methyldopa   Autoimmune phenomenon (occurs with weeks of treatment). *1486*
Moxalactam   In up to 0.5% patients as reported from several studies. *0592*
Penicillin   Combines to RBC, immunoglobulins develop to drug. *1486*
Rifampin   Positive response in 33% patients after 3 mo. *2843*

Copper

Serum   No Effect   Physiological
Ascorbic Acid   Not affected by variations in drug intake up to 605 mg/d for 3 weeks. *1752*
Cellulose Phosphate   No significant effect observed. *2710*
Cisplatin   Insignificant change in 15 patients treated with drug. *3964*
Piretanide   No significant effect with up to 12 mg/d for 3 mo. *3696*
Progesterone   No effect reported over several months. *3505*

Serum   Increase   Analytical
Standing of Sample   Marked top to bottom if frozen and unmixed. *2674*

Serum   Increase   Physiological
Carbamazepine   As result of increased ceruloplasmin synthesis. *1275*
Copper   May be normal in toxic cases with hemolysis. *0385*
Estrogens   Associated with increased ceruloplasmin. *1847*
Lactation   Initially high falls to normal. *0987*
Oral Contraceptives   Estrogens increase concentration of binding protein (maybe x 2). *3548*   In 22 women ingesting combination type contraceptives. *1604*
Phenobarbital   Observed in both hospitalized and home-living patients with long-term administration. *3803*
Phenytoin   Observed in both hospitalized and home-living individuals with long-term administration. *3803*
Pregnancy   Metabolic effect maximal in last trimester. *0987*

Serum   Decrease   Analytical
Citrates   Dilutional effect when compared against serum or heparin as anticoagulant. *3361*

Serum   Decrease   Physiological
Acetylpenicillamine   Elimination of heavy metals in cases of mercury poisoning. *1871*

Urine   No Effect   Physiological
Cellulose Phosphate   No significant effect observed. *2710*
Chlordiazepoxide   No significant effect observed. *3728*

Urine   Increase   Physiological
Cisplatin   Marked effect in 15 patients treated with drug. *3964*
Copper   Observed with poisoning. *2427*
Dimercaprol   If cause of poisoning (penicillamine better). *0071*
EDTA   If poisoning due to copper. *0071*
Penicillamine   If poisoning due to copper (also in normals). *0071*
Proteinuria   Loss of protein bound material. *1563*

Hair   No Effect   Physiological
Oral Contraceptives   No significant effect in young women studied for at least 3 mo. *3720*

Hair   Decrease   Analytical
Hair Treatments   Lowering seen with hair tonics. *1594*

Red Blood Cells   No Effect   Physiological
Oral Contraceptives   In 22 women ingesting combination type contraceptives. *1604*

Red Blood Cells   Increase   Physiological
Copper   Concentration gradient from erythrocytes to plasma. *1955*
Lead   Occurs with poisoning. *0987*

White Blood Cells   No Effect   Physiological
Oral Contraceptives   In 22 women ingesting combination type contraceptives. *1604*

Coproporphyrin

Blood   Increase   Physiological
Barbiturates   May precipitate acute cutaneous porphyria. *1322*
Chlorpropamide   May precipitate cutaneous porphyria. *1322*
Estrogens   May precipitate porphyria attack. *1322*
Ethanol   May precipitate attack of porphyria. *1322*
Oral Contraceptives   May precipitate porphyria attack. *1322*
Sulfomethane   May provoke attack of porphyria. *1322*
Tolbutamide   May precipitate cutaneous porphyria. *1322*

Urine   No Effect   Physiological
Cimetidine   No measurable change in one patient with acute intermittent porphyria. *1659*

Urine   Increase   Physiological
Apronalide   May precipitate acute porphyria. *1322*
Barbiturates   May precipitate acute porphyria. *1322*
Chlordiazepoxide   May precipitate acute porphyria. *1322*
Dichloralphenazone   May precipitate acute porphyria. *1322*
Diethylstilbestrol   May induce porphyria cutanea tarda. *3769*
Estrogens   May induce porphyria cutanea tarda. *3769*
Estrogens, Conjugated   May induce porphyria cutanea tarda. *3821*
Ethanol   May precipitate attack of porphyria. *1322*
Fasting   In normals, if porphyria may precipitate attack. *0938*
Glutethimide   May precipitate acute porphyria. *1322*
Gold   Occurs with panmyelopathy. *1343*
Griseofulvin   May precipitate acute porphyria attack. *1322*
Lead   Observed in poisoning (may occur with 6 $\mu g/m^3$ in air). *1343*
Meprobamate   May precipitate acute porphyria. *1322*
Methyldopa   May precipitate acute porphyria attack. *1322*
Methylsulfonal   May precipitate acute porphyria. *1322*
Oral Contraceptives   May induce porphyria cutanea tarda. *3029*
Pentazocine   May precipitate acute porphyria attack. *1322*
Phenytoin   May precipitate acute porphyria. *1322*

Succinimide   May precipitate acute porphyria. *1322*
Sulfomethane   May provoke attack of porphyria. *1322*
Sulfonamides   May precipitate acute porphyria. *1322*
Tolbutamide   May precipitate porphyria attack. *1322*

Urine   Decrease   Physiological
EDTA   If lead poisoning present. *1679*

Feces   Increase   Physiological
Apronalide   May precipitate acute porphyria. *1322*
Barbiturates   May precipitate acute porphyria. *1322*
Chlordiazepoxide   May precipitate acute porphyria. *1322*
Chlorpropamide   May precipitate cutaneous porphyria. *1322*
Dichloralphenazone   May precipitate acute porphyria. *3895*
Diethylstilbestrol   May induce porphyria cutanea tarda. *3769*
Estrogens   May induce porphyria cutanea tarda. *3769*
Estrogens, Conjugated   May induce porphyria cutanea tarda. *3821*
Ethanol   May precipitate attack of porphyria. *1322*
Ethinyl Estradiol   May induce porphyria cutanea tarda. *3769*
Glutethimide   May precipitate acute porphyria. *1322*
Griseofulvin   May precipitate acute porphyria attack. *1322*
Meprobamate   May precipitate acute porphyria. *1322*
Methyldopa   May precipitate acute porphyria attack. *1322*
Methylsulfonal   May precipitate acute porphyria. *1322*
Oral Contraceptives   May precipitate porphyria attack. *1322*
Pentazocine   May precipitate acute porphyria attack. *1322*
Phenytoin   May precipitate acute porphyria. *1322*
Protein   Related to the meat content of diet. *0987*
Succinimide   May precipitate acute porphyria. *1322*
Sulfomethane   May provoke attack of porphyria. *1322*
Sulfonamides   May precipitate acute porphyria. *1322*
Tolbutamide   May precipitate cutaneous porphyria. *1322*

Red Blood Cells   Increase   Physiological
Lead   Occurs with poisoning. *0987*

Corticosteroid-Binding Globulin

Serum   Increase   Physiological
Diethylstilbestrol   Can be doubled in males with treatment. *3571*
Estrogens   Metabolic effect. *0227*
Oral Contraceptives   Due to estrogenic component. *0071*

Serum   Decrease   Physiological
Testosterone   Metabolic effect. *0227*

Corticosteroids

Plasma   Increase   Analytical
Aldosterone   10% (cortisol 100%) competitive protein binding method of Ficher. *1128*
Corticosterone   73% (cortisol 100%) competitive protein binding method of Ficher. *1128*
Cortisone   Cross-react 19% (cortisol 100%) competitive protein binding procedure of Ficher. *1128*
11-Deoxycortisol   100% = cortisol competitive protein binding method of Ficher. *1128*
Heparin   If contaminated by impurities. *1912*
17-Hydroxyprogesterone   84% (cortisol 100%) competitive protein binding method of Ficher. *1128*
Progesterone   28% (cortisol 100%) competitive protein binding method of Ficher. *1128*
Spironolactone   Marked effect on fluorometric procedure. *1487*
Testosterone   13% (cortisol 100%) competitive protein binding method of Ficher. *1128*

Corticosteroids (continued)

Plasma   Increase   Physiological
Corticotropin   Maximum response seen after 4 h. *0168*

Dextroamphetamine   Effect most marked in evenings. *0331*

Insulin   Large effect at 60 minutes after i.v. injection. *3435*

Methamphetamine   Effect most marked in am when given i.v. *1488*

Noise   Reported max at 10,000 Hz. *3351*

Oral Contraceptives   With increased total plasma cortisol. *3525*

Phenytoin   Alters steroid metabolism. *3700*

Pyrogens   Effect at 2 h maximum after 4 h after i.v. *3435*

Tetracosactrin   Therapeutic intent has prolonged action. *1899*

Vasopressin   Significant effect 30-45 minutes after i.v. in normals. *3435*

Plasma   Decrease   Physiological
Albuterol   Significant dose-related effect after i.v. infusion of therapeutic doses in 4 healthy male volunteers. *2807*

Androgens   May cause cholestatic syndrome. *1237*

Dexamethasone   Effect seen following morning if given in evening. *3164*

Oxymetazoline   Clear suppression when induction compared with induction by thiopentone. *0063*

Rifampin   Decreased serum concentration due to hepatic enzyme induction. *0014*

Rimeterol   Dose related significant change in 4 healthy men given therapeutic i.v. dose. *2807*

Urine   No Effect   Analytical
Meprobamate   If Allen correction and Porter-Silber procedure. *0022*

Methenamine   If Allen correction and Porter-Silber procedure. *0022*

Urine   Increase   Analytical
Ascorbic Acid   Butanol extract/no hydrolysis Reddy/Porter-Silber reaction. *0022*

Chloral Hydrate   Butanol extract/no hydrolysis Reddy/Porter-Silber reaction. *0022*

Chlordiazepoxide   Increased absorption at 410 nm in Porter-Silber reaction. *0022*

Colchicine   Increased absorption at 410 nm in Porter-Silber reaction. Butanol extract/no hydrolysis Reddy/Porter-Silber reaction. *0022*

Dexamethasone   Butanol extract/no hydrolysis Reddy/Porter-Silber reaction. *0022*

Fructose   Increased absorption at 410 nm in Porter-Silber reaction. *0022*

Hydroxyzine   Increased absorption at 410 nm in Porter-Silber reaction. *0022*

Menadione   Butanol extract/no hydrolysis Reddy/Porter-Silber reaction. *0022*

Paraldehyde   Butanol extract/no hydrolysis Reddy/Porter-Silber reaction. Increased absorption at 410 nm in Porter-Silber reaction. *0022*

Potassium Iodide   Butanol extract/no hydrolysis Reddy/Porter-Silber reaction. *0022*

Quinidine   Butanol extract/no hydrolysis Reddy/Porter-Silber reaction. *0022*

Quinine   Butanol extract/no hydrolysis Reddy/Porter-Silber reaction. Increased absorption at 410 nm in Porter-Silber reaction. *0022*

Reserpine   Increased absorption at 410 nm in Porter-Silber reaction. *0022*

Spironolactone   Increased absorption at 410 nm in Porter-Silber reaction. *0022*

Sulfamerazine   Butanol extract/no hydrolysis Reddy/Porter-Silber reaction. *0022*

Troleandomycin   Increased absorption at 410 nm in Porter-Silber reaction. *0022*

Urine   Decrease   Analytical
Chlorpromazine   Decreased absorption at 410 nm in Porter-Silber procedure. *0022*

Prochlorperazine   Decreased absorption at 410 nm in Porter-Silber procedure. *0022*

Promethazine   Decreased absorption at 410 nm in Porter-Silber procedure. *0022*

Corticosterone

Plasma   No Effect   Physiological
Amiloride   In 5 normal subjects given 75 mg daily for 7 d. *2440*

Heparin   No effect on components of aldosterone biosynthetic pathway. *2638* No effect noted in spite of effect on aldosterone. *2638*

Metoclopramide   No change observed in normal individuals or in patients with primary aldosteronism after 10 mg drug. *3883* No effect even though other constituents affected by i.v. bolus of 10 mg. *2543*

Oral Contraceptives   Inconsistent response after 4 weeks. *3785*

Spironolactone   In 5 normal subjects given 300 mg daily for 7 d. *2440*

Plasma   Decrease   Physiological
Etomidate   Clear suppression when induction compared with induction by thiopentone. *0063*

Urine   No Effect   Physiological
Corticotropin   No effect observed in normals. *1982*

Metoprolol   In 15 patients with essential hypertension treated for 4 weeks, associated with reduction in sympathetic tone and reduced activity of renin-aldosterone system. *1209*

Corticotropin

Plasma   No Effect   Analytical
Storage of Sample   No effect probably stored frozen for 45 d. *1180*

Plasma   No Effect   Physiological
Domperidone   No effect 15 minutes after 1 mg/kg intravenously in healthy volunteers. *3426*

Valproic Acid   Normal concentration in maternal and umbilical cord blood in one pregnant woman. *1522* In patients with Addison's and Cushing's diseases. *0082*

Plasma   Increase   Physiological
Diurnal Variation   Maximum in early hours of am low in evening. *0926*

Etomidate   Nonsignificant increase vs thiopentone controls in 7 men. *0063* Apparently direct suppressive effect on adrenal cortical function. *0641*

Insulin   Significant effect after i.v. in normals 45-90 minutes. *3435* Response to stress. *2442*

Levodopa   Magnitude variable, stress effect. *3324*

Metoclopramide   Increase by 55% 15 minutes after 1 mg/kg intravenously in healthy volunteers. *3426*

Metyrapone   Response to stress. *2067*

Pyrogens   Large effect maximum at 2-3 h after i.v. *3435*

Sleep   Maximum secretion at night, especially early am. *3217*

Vasopressin   Small but significant effect after i.v. in normals. *3435*

Plasma   Decrease   Physiological
Clonidine   Reduced in response to single oral dose in all of 6 healthy adults studied. *2070*

Dexamethasone   Effect measured after 9 h. *0168*

Fluocinolone   About 23% reduced when 30 g/d drug applied as topical ointment over large area. *0069*

Methylprednisolone   Suppression for 3 weeks following single injection in 12 patients. *1754*

Cortisol

Serum   Decrease   Physiological

Desoximetasone   Rapid and sustained suppression in 5 patients with psoriasis and topical application of glucocorticoid. *1240*

Plasma   No Effect   Analytical

Alclometasone   In 10 volunteers given 30 g over 80% of body surface twice daily. *3586*   With twice daily topical application in 39 children over 3 weeks. *0755*

Aldosterone   Reactivity of less than 1% possible with RIA. *1081*

Blood Collection Tube   No effect on concentration even with storage for 6 d in gel-barrier tubes. *1592*

11-Deoxycortisol   If cross-reactivity of less than 25%. *1081*

21-Deoxycortisol   No effect on phenylhydrazine (Porter-Silber) method. *0719*

Dexamethasone   Reactivity of less than 1% possible with RIA. *1081*

Diflunisal   No effect on FETI methods of Syva® advance. *1687*

Phenaglycodol   No physiological effect observed. *1487*

Triamcinolone   Reactivity of less than 1% possible with RIA. *1081*

Plasma   No Effect   Physiological

Amiloride   In 5 normal subjects given 75 mg daily for 7 d. *2440*

Anticonvulsants   No difference in concentrations in men on long-term treatment. *0257*

Butorphanol   No significant effect in 6 healthy male volunteers given 2 mg i.m. *3033*

Caffeine   No effect of 250 mg in men or women but slight increase with 500 mg after 90 minutes in men. *3418*   In men no significant effect of single dose of 500 mg. In women no significant effect of single dose of 500 mg. *3418*

Cannabis   Essentially unchanged with moderate dose. *1633*

Carbidopa   In 10 normal volunteers given 300 mg daily for 1 week. *0494*

Cephalothin   No physiological effect. *1779*

Chlorophenothane   Occupational exposure no effect. *0674*

Clofibrate   No effect of drug after 1 week in healthy individuals with hypertriglyceridemia. *1120*

Clonidine   No change greater than with normal diurnal variation. *1028*

Cyclosporine   In 16 patients who developed hypertrichosis but also given added cortisone. *3190*

Desipramine   In group of depressed patients. *0560*

Domperidone   After 10 mg intravenously in 8 healthy males. *3406* No effect 15 minutes after 1 mg/kg intravenously in healthy volunteers. *3426*

Dopamine   Did not change with infusion of up to 3.0 µg/kg/min for 2 h in 6 normal men. *2149*

β-Endorphin   No effect with i.v. infusion in depressed patients or methadone treated addicts. *0610*

Enflurane   No difference between pre-induction and values during anesthesia. *3456*

Ethacrynic Acid   No effect observed with therapy. *2427*

Ethanol   No effect in men or women at 100 mg/dL. *3611*

Hemodialysis   No significant effect observed during dialysis. *2350*

Ketoconazole   Observed in 2 h after single 200 mg dose in normal men. *3390*

Levonorgestrel   No effect observed in group of women 20 and 65 mo after levonorgestrel treated rods inserted in uterus. *0899*   In a group of women using levonorgestrel versus control group with copper IUD. *3167*

Lovastatin   No significant change even in patients with familial hypercholesterolemia receiving 40 mg bid. *1530*

Metoclopramide   No change observed in normal individuals or in patients with primary aldosteronism after 10 mg drug. *3883*   No significant change following i.v. injection. *0497*   No effect even though other constituents affected by i.v. bolus of 10 mg. *2543*

Muscular Exercise   Usual response in most subjects. *2105*

Naltrexone   No change in unstimulated values observed. *0176*

Niacin   No effect either if fed or starved. *0661*

Omeprazole   No effect observed in 8 volunteers given 30 mg/d for 28 d. *2516*

Oxytocin   No significant change observed. *1344*

Phenytoin   Although increases hepatic turnover but serum concentration remains in normal range due to increased secretion. *1020*

Potassium   If diurnal effect isolated no change. *0914*

Progestogens   No effect with progestogen — only pill. *0537*

Ramipril   In response to 10 mg and 20 mg on successive days in 9 patients with severe chronic congestive heart failure. *0764*

Ranitidine   No effect with i.v. bolus of much as 300 mg. *0871*

Sexual Activity   Probably no significant effect, may be slight decline after. *1735*

Site of Collection   Independent of site. *3639*

Smoking   No effect on diurnal rhythm. *3626*

Sodium Bromide   No effect in 10 men, 10 women receiving 1 mg/kg/d during 8 weeks or 2 full menstrual cycles. *3141*

Spironolactone   In 5 normal subjects given 300 mg daily for 7 d. *2440*

Sultopride   When given 300-600 mg/d for 5 weeks in 5 schizophrenic women. *2465*

Theophylline   No effect in response to i.v. infusion of aminophylline. *0608*   No effect on hepatic metabolism so concentration unaffected. *1830*

Thyrotropin-Releasing Hormone (TRH)   No effect in normal subjects after i.v. administration. *0100*

Valproic Acid   Normal concentrations in maternal and umbilical cord blood in one pregnant woman. *1522*

Verapamil   With 80 mg 3-4/d orally no significant effect on resting concentration or after ACTH stimulation in mild hypertensives. *3956*

Zimeldine   In group of depressed patients. *0560*

Plasma   Increase   Analytical

Benzyl Alcohol   As contaminant may affect fluorometric procedures. *0583*

Carbamazepine   When determined by Mattingly method. *2901*

Corticosterone   If nonselective extraction and competitive protein binding used. *0512*

Danazol   Probably invalid results because of high cross-reactivity with protein in competitive binding assays. *1479*

11-Deoxycortisol   If nonselective extraction and competitive protein binding used. *0512*

21-Deoxycortisol   Gives 27% of value fluorometrically Gives 87% of value with competitive prot bind techniques. *0719*

Fenfluramine   Fluoresces with sulfuric acid. *0121*

Fusidic Acid   Interfering fluorescence. *1772*

Hemolysis   Produces interference with fluorometric method. *1772*

Hydrocortisone   If non selective extraction and competitive protein binding used. *0512*

17-Hydroxyprogesterone   If nonselective extract and competitive protein binding used. *0512*

Prednisolone   High and equal cross reactivity with RIA and competitive protein binding. *1081*

Prednisone   High and equal cross reactivity with RIA and competitive protein binding. *1081*

Progesterone   If nonselective extraction and competitive protein binding used. *0512*

Quinacrine   Produces interfering fluorescence. *1772*

Spironolactone   Fluorometric methods may be affected. *2237*

Plasma   Increase   Physiological

Altitude   Probably related to hypoxia. *1836*

Anticonvulsants   400 nmol/L versus 260 nmol/L in women on long-term treatment. *0257*

Corticotropin   Therapeutic intent. *1680*

Cortisone   Effect lasts for 24 h at least. *1772*

Cyclic AMP   Hormonal action. *1102*

Diurnal Variation   Maximum observed about 6 am in morning. *2222*

Estrogens   Increases concentration of binding globulin. *1488*

Cortisol (continued)

Plasma Increase Physiological (continued)

Ethanol Effect seen if high doses given i.v. *1789* 12.1 mg/dL in current abusers, 11.0 mg/dL in previous abusers vs 7.7 mg/dL in controls. *3771*

Ether Effect observed in moderate to deep anesthesia. *1343*

Fasting Associated also with increased number of secretory episodes and plasma half-life. *1130*

Glyburide Occasional slightly higher, no significant change after glucose. *0125*

Hydrocortisone Effect lasts for at least 24 h. *1772*

Insulin Marked effect in insulin induced hypoglycemia. *3570*

Lithium Observed in some patients. *2829*

Methadone Significant response to cold not seen in controls. *2956*

Metoclopramide Increase by 75% 15 minutes after 1 mg/kg intravenously in healthy volunteers. *3426*

Naloxone Peak at 60 minutes after start of infusion remained high for duration. *0868*

Opiates Significant increase within 1 h of i.m. injection of hyoscine, meperidine or omnopon. *1929*

Oral Contraceptives Decreases cortisol clearance (estrogen effect). *3332* On combined therapy at 9 am compared with female controls and men. *1668*

Pregnancy Higher especially at night than in normals. *1232*

Prostaglandin F2α Slight effect at 2 μg/kg/min i.v. (only in men). *0735*

Pyrogens After delay of 1 h following injection. *0254*

Smoking Basal levels same in smokers and nonsmokers, but increased significantly on smoking: due to direct effect of nicotine or vasopressin release. *1360*

Stress Marked response to emotional stress. *3217*

Tetracosactrin Rise of 21 μg/dL in 6 h after i.m. injections. *0275* 3 fold increase in healthy volunteers given 1 mg intramuscularly for up to 60 h. *1813*

Vasopressin May be rise of up to 6 μg/dL or more. *3625*

Plasma Decrease Analytical

Danazol Concentrations low: in case of cortisol and testosterone displacement from plasma proteins occurs and protein binding assays are probably invalid. *1479*

Plasma Decrease Physiological

Barbiturates If used preoperatively lower concentration. *3657*

Beclomethasone Significant effect observed in asthmatic children given drug by inhalation reflecting compromizing of pituitary-adrenal axis. *3693*

Betamethasone Very low values observed in children in whom cream applied topically. *0777*

Blindness Significantly lower (average 4.4 μg/dL) than normal. *1637*

Clonidine Lower level in growth hormone deficient children: did not change with 0.1 mg/m² daily for 60 d. *2821* In normal subjects reduced by 50% with 0.15 mg/d but no effect in opiate addicts. *1279*

Danazol Affect binding in women treated with drug so that proportion of free drug high. *1479*

Desoxycorticosterone Rapid and sustained suppression in 5 patients with psoriasis and topical application of glucocorticoid. *1240*

Dexamethasone Marked effect for 4 d after i.v. injection. *0783* Normal response not always observed in psychiatric patients. *2404*

Ephedrine Accelerated steroid clearance due to increased hepatic blood flow and induction of enzymes in liver. *1830*

Etomidate Clear suppression when induction compared with induction by thiopentone. *0063* Apparently direct suppressive effect on adrenal cortical function. *0641* Similar effect to that of thiopental but reduction with induction of anesthesia. *3235* Attributable to direct antisteroidogenic effects on adrenal gland. *0709*

Fluocinolone About 43% suppression in 4 d when 30 g/d drug applied as topical ointment over large area. *0069*

Glucose Progressive fall when serum glucose increased. *3373*

Ketoconazole Significant reduction in patients receiving 800 mg or more daily. *2839*

Levodopa Probably diminished ACTH secretion. *1390*

Lithium Lower in morning and less diurnal variation. *1466* Significant decrease in a.m. cortisol after 1 y in 48 depressed patients; p.m. values also affected in individuals with greatest change in response to treatment. *3357* Typically reduced in patients treated for 1 y as seen in study of 53 patients. *3357*

Methylprednisolone Suppression for 2 weeks following single lumbar extradural injection in 12 patients. *1754*

Metyrapone Normal response to test. *2396*

Morphine In 14 volunteer subjects given 5 mg intravenously. *3953*

Muscular Exercise Slight effect (not significant). *1171*

Nifedipine Slight but progressive decrease over 3 h after oral administration of 10 mg in young people but not in old. *1609*

Norethindrone Slight effect compared with controls. *0261*

Phenytoin Chronic administration effect in Cushings syndrome. *2220* In 10 women receiving anticonvulsants, mainly phenytoin versus 10 controls, drug taken on average for 15 y. *0257*

Ranitidine Compared to control state after 3 d 150 mg/12 h cortisol level significantly lower at rest and after 30 minutes ambulation. *3130*

Triamcinolone Marked decrease in 2 children given injections of compound with suppression of hypothalamic-pituitary axis. *0777*

Plasma Negative Analytical

Fluorescein No result obtainable at concentration of 1.0 mg/L on Abbott TDx and carryover effect into subsequent specimen. *1722*

Urine No Effect Analytical

Glucose On fluorometric method of Ratliff and Hall. *2923*

Hexoses On fluorometric method of Ratliff and Hall. *2923*

Ketones On fluorometric method of Ratliff and Hall. *2923*

Urine No Effect Physiological

Metoprolol In 15 patients with essential hypertension treated for 4 weeks, associated with reduction in sympathetic tone and reduced activity of renin-aldosterone system. *1209*

Oral Contraceptives No significant difference when referenced to creatinine in women on combined therapy versus controls. *1668*

Smoking No effect usual smoking in smokers. *3626*

Urine Increase Analytical

Carbamazepine When determined by Mattingly method. *2901*

Quinacrine Increased unconjugated by fluorometric procedure. *0022*

Spironolactone Fluorescence affects method of Ratliff. *2923*

Urine Increase Physiological

Altitude Probably related to acute hypoxia. *1836*

Corticotropin Progressive increase with repeated injection. *1982*

Oral Contraceptives Significant hormonal effect. *2181*

Pregnancy Significant hormonal effect. *2181*

Saline Reported effect. *1000*

Sodium Chloride Significant effect with up to 16 g/d for 10 d. *1487*

Urine Decrease Analytical

Diazoxide Decreased unconjugated by fluorometric procedure. *0022*

Ethacrynic Acid Decreases unconjugated by fluorometric procedure. *0022*

Hydrochlorothiazide Decreased unconjugated by fluorometric procedure. *0022*

Norethindrone Decreased unconjugated by fluorometric procedure. *0022*

Urine Decrease Physiological

Ethacrynic Acid ?due to changed secretion or renal handling. *1488*

Hydrochlorothiazide Possibly altered cortisol secretion. *1487*

Ketoconazole   Blocks adrenal response to corticotropin and related to serum concentration of drug. *2839*   Significant reduction in 6 patients receiving 1.2 g daily for prostatic cancer, also blunted plasma cortisol response to Synacthen. *3812*
Loprazolam   Observed in 9 poor sleepers given up to 1 mg/d for 3 weeks: overnight urinary cortisol measured. Rebound increase on withdrawal. *0016*
Norethindrone   When results compared with normal menstrual cycle. *0261*
Oral Contraceptives   When results during therapy compared with normal. *0261*
Thiazides   ?changed secretion or renal handling. *1000*
Triazolam   Observed in 9 poor sleepers given 0.5 mg/d for 3 weeks: overnight urinary cortisol measured. Immediate rebound increase on withdrawal. *0016*

Saliva   Increase   Physiological
Oral Contraceptives   Significantly higher on combined therapy versus control women and men. *1668*

Cortisol, Free

Plasma   Increase   Physiological
Danazol   Proportion high due to displacement of hormone from protein. *1479*
Estrogens   Effect like glucocorticoids (?physiological sign). *1849*
Mestranol   Slight effect if over 0.1 mg. *0537*
Oral Contraceptives   Significant hormonal effect. *2181* Estrogen effect. *0585*
Pregnancy   Higher especially at night than in normals (hormonal effect). *1232*

Urine   No Effect   Analytical
Alclometasone   In 10 volunteers given 30 g over 80% of body surface twice daily. *3586*

Urine   No Effect   Physiological
Estrogens   Unaffected if low dose pills. *0537*

Urine   Increase   Physiological
Estrogens   If high dose for cancer, affects diurnal rhythm. *0537*
Oral Contraceptives   50% increase with oral contraceptives. *3217*
Sandfly Fever   3 fold increase over basal with fever. *2929*

Urine   Decrease   Physiological
Ketoconazole   Approximate 50% reduction in patients receiving 800 mg daily. *2839*
Lead   In some men occupationally exposed to large quantities of lead. *0771*

Cortisol, Protein Bound

Serum   Increase   Physiological
Oral Contraceptives   When results on therapy compared with normal. *0261*

Plasma   No Effect   Physiological
Progestogens   No effect with progestogen — only pill. *0537*

Plasma   Increase   Physiological
Estrogens   Hormonal action. *1488*

Cortisone

Serum   Decrease   Physiological
Rifampin   Metabolism enhanced with reduction of circulating drug concentration: cortisol production rate increased. *1830*

Cotinine

Serum   Increase   Physiological
Smoking   Concentration correlated with extent of smoking. *3255*

Urine   Positive   Physiological
Smoking   Normal nicotine metabolite 5-20 cigarettes/d. *0324*

C-Peptide

Plasma   No Effect   Physiological
Cyproterone   Unchanged following oral glucose when given with ethinyl estradiol combination causes insulin resistance. *3239*
Labetalol   In response to i.v. infusion of 100 mg over 10 minutes. *0226*
Omeprazole   No effect observed in 8 volunteers given 30 mg/d for 28 d. *2516*

Plasma   Increase   Physiological
Indapamide   Both mean fasting and stimulated in 13 hypertensive diabetics. *2686*
Piretanide   61% higher than pretreatment level after 8 weeks in 12 male patients with mild hypertension (6 mg bid). *1503*
Rifampin   Increased rate of secretion after oral administration of 100 g glucose. *3534*

Plasma   Decrease   Physiological
Atenolol   Reduced activity at 2 and 3 h during glucose tolerance test versus control. *1977*
Calcitonin   Response to glucose significantly reduced. *2738*

Urine   No Effect   Physiological
Hydrochlorothiazide   In 14 hypertensives men with type 2 diabetes treated for 2 weeks with or without propranolol. *0935*
Propranolol   In 14 hypertensives with type 2 diabetes treated for 3 weeks, with or without added hydrochlorothiazide. *0935*

C-Reactive Protein

Serum   Increase   Physiological
Estrogens   Altered liver metabolism. *3332*
Oral Contraceptives   Estrogen effect. *3332*
Smoking   Up to 11 fold increase in men and 6 fold increase in women. *0809*

Serum   Decrease   Physiological
Oral Contraceptives   Progestogen effect. *3505*

Creatine

Serum   No Effect   Analytical
Dextran   No effect on method of Heinegard and Tiderstrom. *1547*

Serum   Increase   Analytical
Hemolysis   Even if slight may increase by 100-200%. *0904*
Pyruvate   10 mg/L = 0.02 mg/dL by method of Heinegard. *1547*
Sulfobromophthalein   Presence of interfering color. *2313*
Urea   10 g/L=0.03 mg/dL by method of Heinegard. *1547*

Serum   Increase   Physiological
Arginine   Incorporated into urea cycle. *0071*
Diurnal Variation   7 pm value 160% of 7 am concentration. *3217*
Protein   Especially if meat eaten is raw. *0987*

Urine   No Effect   Analytical
Storage of Sample   Stable indefinitely if frozen Without preservative at 30° for 4 h. *0904*
Thymol   No effect on Jaffé, Van Pilsum methods. *2981*

Creatine (continued)

Urine   Increase   Analytical
Phenolsulfonphthalein   Chromogenicity in color reaction. 2313
Storage of Sample   Significant increase 1 d at room temperature. 1563

Urine   Increase   Physiological
Aminopyrine   Reported effect. 2220
Caffeine   Acts on intermediary metabolism. 1343
Cimetidine   Significant effect at 3 weeks in 9 patients given 1.6 g daily. 0978
Fluoxymesterone   For up to 2 weeks when treatment stopped. 1680
Methandrostenolone   May persist up to 2 weeks after treatment. 0019
Methyltestosterone   Metabolic effect ?mechanism. 1343
Oxymetholone   Anabolic effect (possible for 2 weeks after stop). 1679
Pregnancy   General effect with pregnancy. 0987
Protein   Especially if meat eaten is raw. 0987
Starvation   Begins to rise from first day. 2382
Vegetarian Diet   Increased in most vegetarians studied. 3727
X-Ray Therapy   Occurs with tissue destruction. 0987

Urine   Decrease   Physiological
Anabolic Steroids   Anabolic effect. 1237
Androgens   Anabolic effect. 1237
Nandrolone   Anabolic effect. 2681
Thiazides   Clearance tests decreased by 10-20%. 2426

Red Blood Cells   Decrease   Physiological
Pregnancy   Decreased at 3 mo, but increased at 8 mo. 3041

Creatine Kinase

Serum   No Effect   Analytical
Acetaminophen   At acute overdose concentration (10 mg/dL) on SMAC® method. 3719   On continuous method at 10 times maximal therapeutic concentration. 1785
Acetylsalicylic Acid   At 5 times upper limit of therapeutic range on methods on SMAC®, Abbott-VP, aca, Cobas-Bio, and Hitachi® 705. 2138
Aminophenazone   On continuous method at 10 times maximal therapeutic concentration. 1785
Aminosalicylic Acid   At 5 times upper limit of therapeutic range on methods on SMAC®, Abbott-VP, aca, Cobas-Bio and Hitachi® 705. 2138
Amitriptyline   At acute overdose concentration (2.5 mg/dL) on SMAC® method. 3719
Barbital   At acute overdose concentration (20 mg/dL) on Technicon® SMAC® method. 3719
Bromazepam   At 5 times therapeutic concentration on methods on SMAC®, Abbott-VP, aca, Cobas-Bio and and Hitachi® 705. 2138
Caffeine   At acute overdose concentration (10 mg/dL) on Technicon® SMAC® method. 3719   On continuous method at 10 times maximal therapeutic concentration. 1785
Carbenicillin   At 5 times upper limit of therapeutic range on methods on SMAC®, Abbott-VP, aca, Cobas-Bio and Hitachi® 705. 2138
Cefoxitin   At 5 times upper limit of therapeutic range on methods on SMAC®, Abbott-VP, aca, Cobas-Bio, and Hitachi® 705. 2138
Cephalothin   At 5 times upper limit of therapeutic range on methods on SMAC®, Abbott-VP, aca, Cobas-Bio and Hitachi® 705. 2138
Chlordiazepoxide   At 5 times upper limit of therapeutic range on methods on SMAC®, Abbott-VP, aca, Cobas-Bio and Hitachi® 705. 2138   At acute overdose concentration (5 mg/dL) on Technicon® SMAC® method. 3719
Chlorpheniramine   At acute overdose concentration (20 mg/dL) on Technicon® SMAC® method. 3719
Cocaine   At acute overdose concentration (2.5 mg/dL) on Technicon® SMAC® cmethod. 3719

Codeine   At acute overdose concentration (2.0 mg/dL) on Technicon® SMAC® method. 3719
Cyclosporine   At concentration of 41.5 mmol/L (50 μg/L) on methods on Technicon® SMAC® II and Hitachi® 705. 1123
Cysteine   At 5 times upper limit of therapeutic range on methods on Abbott-VP, Cobas-Bio, and Hitachi® 705. 2138
Cystine   At 5 times upper limit of therapeutic range on methods on Abbott-VP, Cobas-Bio, and Hitachi® 705. 2138
Diazepam   At 5 times upper limit of therapeutic range on methods on SMAC®, Abbott-VP, aca, Cobas-Bio, and Hitachi® 705. 2138   At acute overdose concentration (2.5 mg/dL) on Technicon® SMAC® method. 3719
Diclofenac   On continuous method at 10 times maximal therapeutic concentration. 1785
Diethylpropion   At acute overdose concentration (10 mg/dL) on Technicon® SMAC® method. 3719
Diphenhydramine   At acute overdose concentration (20 mg/dL) on Technicon® SMAC® method. 3719
Ethanol   At acute overdose concentration (20 mg/dL) on Technicon® SMAC® methods. 3719
Ethchlorvynol   At acute overdose concentration (20 mg/dL) on Technicon® method. 3719
Ethinamate   At acute overdose concentration (20 mg/dL) on Technicon® SMAC® method. 3719
Flurazepam   At 5 times upper limit of therapeutic range on methods on SMAC®, Abbott-VP, aca, Cobas-Bio, and Hitachi® 705. 2138   At acute overdose concentration (2.5 mg/dL) on Technicon® SMAC® method. 3719
Glutethimide   At acute overdose concentration (5 mg/dL) on Technicon® SMAC® method. 3719
Hemoglobin   If AMP added to inhibit adenylate kinase. 3217
Hemolysis   Slight hemolysis has no effect. 2981
Ibuprofen   At 5 times upper limit of therapeutic range on methods on SMAC®, Abbott-VP, aca, and Hitachi® 705. 2138   On continuous method at 10 times maximal therapeutic concentration. 1785
Indomethacin   On continuous method. 1785
Ketoprofen   On continuous method at 10 times maximal therapeutic concentration. 1785
Liposol   Changed from 322.6 to 344.3 U/L in 26 randomly selected patient sera, before and after addition on Beckman Astra methods. 2054
Meperidine   At acute overdose concentration (5 mg/dL) on Technicon® SMAC® method. 3719
Meprobamate   At acute overdose concentration (20 mg/dL) on Technicon® SMAC® method. 3719
Mesoridazine   At acute overdose concentration (20 mg/dL) on Technicon® SMAC® method. 3719
Methadone   At acute overdose concentration (2.5 mg/dL) on Technicon® SMAC® method. 3719
Methanol   At acute overdose concentration (20 mg/dL) on Technicon® SMAC® method. 3719
Methaqualone   At acute overdose concentration (2.5 mg/dL) on Technicon® SMAC® method. 3719
Methicillin   At 5 times upper limit of therapeutic range on methods on SMAC®, Abbott-VP, Cobas-Bio, aca and Hitachi® 705. 2138
Methotrexate   At 5 times upper limit of therapeutic range on methods on SMAC®, Abbott-VP, aca, Cobas-Bio, and Hitachi® 705. 2138
Methylphenidate   At acute overdose concentration (20 mg/dL) on Technicon® SMAC® method. 3719
Methyprylon   At acute overdose concentration (10 mg/dL) on Technicon® SMAC® method. 3719
Morphine   At acute overdose concentration (20 mg/dL) on Technicon® SMAC® method. 3719
Naproxen   At 5 times upper limit of therapeutic range on methods on SMAC®, Abbott-VP, aca, Cobas-Bio and Hitachi® 705. 2138
Nitrofurantoin   On routine methods in use on SMAC®, Ektachem®, Abbott-VP, Cobas-Bio, aca, Hitachi® 705, KDA at 5 times normal upper therapeutic concentration. 2138
Nortriptyline   At acute overdose concentration (20 mg/dL) on Technicon® SMAC® method. 3719
Penicillin G   At 5 times upper limit of therapeutic range on methods on SMAC®, Abbott-VP, aca, Cobas-Bio and Hitachi® 705. 2138

Pentobarbital At acute overdose concentration (20 mg/dL) on Technicon® SMAC® method. *3719*

Perphenazine At acute overdose concentration (20 mg/dL) on Technicon® SMAC® method. *3719*

Phenobarbital At acute overdose concentration (20 mg/dL) on Technicon® SMAC® method. *3719* On continuous method at 10 times maximal therapeutic concentration. *1785*

Phenytoin At acute overdose concentration (20 mg/dL) on Technicon® SMAC® method. *3719*

Plasma No significant difference from serum by LKB 8600. *2228*

Promethazine At acute overdose concentration (20 mg/dL) on Technicon® SMAC® method. *3719*

Propoxyphene At acute overdose concentration (2.5 mg/dL) on Technicon® SMAC® method. *3719*

Pyribenzamine® At acute overdose concentration (20 mg/dL) on Technicon® SMAC® methods. *3719*

Quinidine At acute overdose concentration (20 mg/dL) on Technicon® SMAC® method. *3719*

Quinine At acute overdose concentration (1.5 mg/dL) on Technicon® SMAC® method. *3719*

Rifampin At 5 times upper limit of therapeutic range on methods on SMAC®, Abbott-VP, aca, Cobas-Bio, and Hitachi® 705. *2138*

Secobarbital At acute overdose concentration (20 mg/dL) on Technicon® SMAC® method. *3719*

Storage of Sample If stored at -20° stable for 2 weeks. *0877* 4 h room temperature, 10 d 4°, 60 at 20°. *3217* 48 h at room temperature, 7 d at 4°, 1 mo at -18. *0877*

Tetracycline At 5 times upper limit of therapeutic range on methods on SMAC®, Abbott-VP, Cobas-Bio, aca and Hitachi® 705. *2138*

Theophylline At 5 times upper limit of therapeutic range on methods on SMAC®, Abbott-VP, Cobas-Bio, aca and Hitachi® 705. *2138*

Thiopental At acute overdose concentration (20 mg/dL) on Technicon® SMAC® method. *3719*

Serum   No Effect   Physiological

Cimetidine No significant effect observed in 9 patients given 1.6 g daily. *0978*

Ethanol No effect over 110 h following ingestion of 0.75 g/kg body weight on 3 consecutive evenings. *2133* No significant difference in specimens from volunteers and values compared with control values for up to 110 h, although component/protein ratio decreases in all cases. *2134*

Heparin After low-dose heparin treatment. *2671*

Meperidine No reported effect. *2427*

Morphine No change, although other enzymes increased. *2427*

Streptokinase Although marked effect on liver. *3189*

Serum   Increase   Analytical

Cefotaxime Statistically significant effect of metabolite on Parallel method. *0201*

Cysteine Stimulates activity of enzyme especially if been stored. *0718* At 5 times upper limit of therapeutic range on methods SMAC® and aca. *2138*

Cystine At 5 times upper limit of therapeutic range on methods on SMAC® and aca. *2138*

Hemoglobin Marked effect on UV coupled procedures with 1 g Hemoglobin. *3217*

Hemolysis 5 U/L in hemolyzed samples. *2744*

Standing of Sample 4 U if on cells for 1 h before separation. *2744*

Storage of Sample 4 U/L if more than 100 h between venesection and assay. *2744*

Thiols Three to tenfold enhancement. *0249*

Serum   Increase   Physiological

Aging 5 U/L in normal adult women age 40 y and above. *2744*

Aminocaproic Acid Rare myopathy 6 weeks after treatment. *0296*

Amphotericin B Rhabdomyolysis caused by severe hypokalemia. *0958*

Ampicillin Probable effect of i.m. injection. *1966*

Analgesics May cause effect if injected i.m. *0249*

Anesthetic Agents Occurs if combined with suxamethonium. *1724*

Barbiturates Occurs with poisoning, probable muscle origin. *3901*

Bezafibrate In four patients with poor renal function produced myolysis as result of overdose due to renal dysfunction. *3086*

Bucindol Effect noted in 3 of 6 patients studied with increased activity originating from skeletal muscle. *1283*

Carbenicillin Probably due to trauma of i.m. injection. *1966*

Carbenoxolone May cause hypokalemic myopathy. *1179*

Carbromal Occurs with poisoning, probable muscle origin. *3901*

Carteolol Significant effect in 10 of 15 patients with essential hypertension. MM isoenzyme most affected. *3149*

Chlorpromazine May be due to injection only (occurs in 20%). *2403*

Chlorthalidone Single case of severe reversible myopathy noted. *1793*

Clindamycin Probably due to muscle damage (common with i.m.). *0841*

Clofibrate Originates from skeletal muscle (in up to 15%). *3505*

Clonidine Temporary effect of unknown significance. *1277*

Clopamide Single case of diuretic associated myopathy. *1793*

Copper Observed in one case of dialysis toxicity. *2217*

Coughing May be doubling with severe coughing. *3179*

Dantrolene Isolated observation, uncertain relevance. *0659*

Digoxin 15 to 17 x increase after i.m. injection (increase for 8 d). *1392*

Diuretics May occur as result of i.m. injections. *0249*

Diurnal Variation Slight increase (3U/L) observed in women from 9 am to 5pm. *2402*

Electrocautery Effect of muscle damage. *1237*

Ethanol Effect noticed in alcoholics after alcohol. *1651*

Ethchlorvynol Occurs with poisoning, probable muscle origin. *3901*

Halofenate Mild transient effect observed ?origin. *0161*

Halothane Marked effect on CK if administered during anesthesia. *2809* Marked effect if given with suxamethonium. *1724*

Hypothermia Due to hypoxia, acid-base imbalance, hypotension. *2258*

I.M. Injections Muscle damage. *3218*

Insulin Is an activator of enzyme. *0688*

Labetalol In 3 of 9 patients with essential hypertension. Note MM isoenzyme most affected. *3149*

Lithium May be high sustained increase although disease controlled. *1357*

Marital Status 4 U in normal married women versus single women. *2744*

Meperidine 2x after single i.m. injection of 50 mg. *3053*

Miconazole Observed as isolated finding in patients given drug i.v. *3462*

Morphine Response to frequent i.m. injections. *0249*

Muscle Massage Significant effect but not to pathological level 1 h after. *0423*

Muscular Exercise Increase by 7% after 12 minutes on cycle-ergometer. *3319* In women increase of 4 U/L if within past 24 h. *2744* Maximum effect observed following day. *1491* Effect of physical training, increased with exercise. *2981*

Narcotics Response to i.m. injections. *0249*

Oral Contraceptives Slight effect only (2 u/L). *2744*

Parity 5 U/L in parous versus other women. *2744*

Penicillamine Marked increase observed in one patient with rheumatoid arthritis after 1 mo treatment. *0947*

Penicillin Frequent injections may cause increase up to 5 times. *1580*

Phenothiazines Probable effect of i.m. injection. *1488*

Phenytoin Myopathy observed in one case as part of hypersensitivity reaction. *1502*

Pindolol In 20 of 25 patients with essential hypertension with increase of 20 to 760% compared with pretreatment values. MM isoenzyme most affected, but 8 of 25 showed slight increase of MB. *3149*

Creatine Kinase *(continued)*

Serum Increase Physiological (continued)

Pregnancy Activity increases during last trimester. *2981*

Propranolol In 4 of 27 patients with essential hypertension, effect mainly in MM isoenzyme: effect not very marked. *3149*

Quinidine If given intramuscularly, possibly due to increase in intracellular calcium. *3215*

Race Significantly higher in blacks than caucasians (for each sex). *2402*

Saline 40% increase after 1 ml saline i.m. *3053*

Streptokinase Earlier peak of CK-MB observed in patients on thrombolytic therapy due to reperfusion of ischemic myocardium. *1341*

Stress Cold Exposure Response to cold exposure. *3217*

Succinylcholine Significant if given with anesthesia (effect of injection). *1724* Occurs if given with halothane. *3628* Marked effect on CK if administered during anesthesia. *2809*

Surgery Significant effect observed 1 d postoperatively. *2809*

Tubocurarine Due to i.m. injections or histamine release. *0688*

Vasopressin Rhabdomyolysis observed in 2 patients following intravenous drug administration. *0029*

Serum Decrease Analytical

Acetylsalicylic Acid With 8300 μmol/L on kinetic method. *1785*

Clothiapine At 9 mg/L when added to serum and conventional method. *1758*

EDTA Inhibits if used as anticoagulant. *3217*

Hemoglobin At 2.5 g/dL inhibits by 12%. *3217*

Hemolysis -3%/g hemoglobin with Oliver procedure. *0515*

Heparin Reported effect. *0248*

Pindolol At 5 mg/L conventional methods when added to serum. *1758*

Storage of Sample Possible loss up to 50% with thawing once when frozen. *0718*

Serum Decrease Physiological

Ethanol 6 U/L in alcohol drinking versus other women. *2744*

Menstruation Decrease of 4 U/L from 12 to 26th day. *2744*

Phenothiazines In schizophrenics with high initial values. *2427*

Physical Training Slight effect noted after 10 weeks training. *1696*

Prednisone Low activities below normal range, observed in several patients some of whom were receiving other drugs but not observed in all patients. *1602*

Pregnancy Significant decrease from 8-20 weeks, max decrease 12-13 weeks. *1933*

Starvation Low values observed with malnutrition. *2382*

Urine Increase Physiological

Acetrizoate Sustained increase after renal artery injection. *3538*

Radiographic Agents Transient increase if injected in renal artery. *3538*

Creatine Kinase Isoenzymes

Serum No Effect Analytical

Cyclosporine At concentration of 41.5 mmol/L (50 μg/L) on methods on Technicon® SMAC® II and Hitachi® 705. *1123*

Fluosol-DA No effect on electrophoresis although added fluorescent band. *2518*

Serum Increase Physiological

Phenytoin Myopathy observed in one case as part of hypersensitivity reaction. Increase of M33 isoenzyem. *1502*

Streptokinase Earlier peak of CK-MB observed in patients on thrombolytic therapy due to reperfusion of ischemic myocardium. *1341*

Creatine Phosphate

Muscle No Effect Physiological

Physical Training No effect of physical training. *1875*

Creatinine

Serum No Effect Analytical

Acebutolol At 100 mg/L on reversed phase LC procedure of Zhiri et al. *3942*

Acetaminophen At 200 mg/L on reversed phase LC procedure of Zhiri et al. *3942* At 5 times upper limit of therapeutic concentration on routine methods in use on SMAC®, Abbott-VP, aca, Cobas-Bio, Ektachem® 400, Hitachi® 705, KDA. *2138* At acute overdose concentration (10 mg/dL) on SMAC® method. *3719* At concentration of 50 mg/L had no effect on Ektachem® method. *3393* At concentration of 180 mg/L had no effect on creatinine iminohydrolase method. *3393* At concentration of 1500 mg/L had no effect on AutoAnalyzer Jaffé method. *3393*

Acetazolamide No effect at 12 mg/dL on SMA 12/60 method. *2636* At concentration of 1,000 mg/L had no effect on AutoAnalyzer Jaffé method. *3393*

Acetoacetate No effect on ion-exchange method of Mitchell. *2459* No effect on method of Polar and Metcoff. *1975* No effect on enzymatic procedure on Ektachem®. *1267* No effect even at 20 mmol/L on creatininase/creatinase/sarcosine/peroxidase BM-BCL method. *0765*

Acetone No effect on method of Heinegard and Tiderstrom. *1547*

Acetylsalicylic Acid On alkaline picrate procedures and Slot method at therapeutic concentrations. *1786* On alkaline picrate method on SMA II at physiological concentration. *1787* At 5 times upper limit of therapeutic range on method on SMAC®, Abbott-VP, Cobas-Bio, aca, Ektachem®, Hitachi® 705 and KDA. *2138* With 8326 μmol/L on AutoAnalyzer Jaffé method. With 8326 μmol/L on picrate method with deproteinization. With 8326 μmol/L on Slot procedure. *1786* At concentration of 300 mg/L had no effect on Ektachem® method. *3393* At concentration of 600 mg/L had no effect on creatinine iminohydrolase method. *3393* At concentration of 2,000 mg/L had no effect on AutoAnalyzer Jaffé method. *3393* At concentration of 600 mg/L had no effect on Jaffé-Fading-Fraction method. *3393* At concentration of 600 mg/L had no effect on Jaffé-Fuller's earth method. *3393* At concentration of 500 mg/L had no effect on kinetic Jaffé method on BKA-2. *3393*

Allopurinol at 200 mg/L on reversed phase LC procedure of Zhiri et al. *3942* At concentration of 60 mg/L had no effect on AutoAnalyzer Jaffé method. *3393* At concentration of 80 mg/L had no effect on Jaffé-Fading-Fraction method. *3393* At concentration of 80 mg/L had no effect on Jaffé-Fuller's earth method. *3393* At concentration of 10 mg/L had no effect on kinetic Jaffé method on BKA-2. *3393*

Aminophenazone At therapeutic concentrations no effect on alkaline picrate and Slot methods. *1786*

Aminophylline At 400 mg/L on reversed phase LC procedure of Zhiri et al. *3942*

Aminopyrine At concentration of 125 mg/L had no effect on AutoAnalyzer Jaffé method. *3393*

Aminosalicylic Acid At 5 times upper limit of therapeutic range on methods on SMAC®, Abbott-VP, aca, Cobas-Bio, Ektachem®, Hitachi® 705 and KDA. *2138* At concentration of 460 mg/L had no effect on AutoAnalyzer method. *3393*

Amiodarone At 50 mg/L on reversed phase LC procedure of Zhiri et al. *3942*

Amitriptyline At 10 mg/L on reversed phase LC procedure of Zhiri et al. *3942* At acute overdose concentration (2.5 mg/dL) on SMAC® method. *3719* At concentration of 0.4 mg/L had no effect on creatinine iminohydrolase method. *3393* At concentration of 25 mg/L had no effect on AutoAnalyzer Jaffé method. *3393*

Amobarbital At concentration of 150 mg/L had no effect on AutoAnalyzer Jaffé method. *3393*

Amphotericin B At concentration of 20 mg/L had no effect on AutoAnalyzer Jaffé method. *3393*

Ampicillin   At concentration of 12 mg/L had no effect on creatinine iminohydrolase method. *3393*   At concentration of 40 mg/L had no effect on AutoAnalyzer Jaffé method. *3393*   At concentration of 600 mg/L had no effect on Jaffé-Fading-Fraction method. *3393*   At concentration of 600 mg/L had no effect on Jaffé-Fuller's earth method. *3393*   At concentration of 1600 mg/L had no effect on kinetic Jaffé method on BKA-2. *3393*

Antazoline   At concentration of 160 mg/L had no effect on Jaffé-Fading-Fraction method. *3393*   At concentration of 160 mg/L had no effect on Jaffé-Fuller's earth method. *3393*

Ascorbic Acid   At 360 mg/dL on ion-exchange method of Mitchell. No effect on Lloyd's procedure. *2459*   No effect on method of Polar and Metcoff. *1975*   At 500 mg/L on reversed phase LC procedure of Zhiri et al. *3942*   On SMAC® method at therapeutic concentration. *3719*   At concentration of 105 mg/L had no effect on creatinine iminohydrolase method. *3393*   At concentration of 400 mg/L had no effect on Jaffé-Fading-Fraction method. *3393*   At concentration of 400 mg/L had no effect on Jaffé-Fuller's earth method. *3393*   At concentration of 100 mg/L had no effect on Jaffé-Heinegard and Tiderstrom method. *3393*   At concentration of 1,000 mg/L had no effect on kinetic Jaffé method on BKA-2. *3393*

Aspirin   At 1.0 g/L on reversed phase LC procedure of Zhiri et al. *3942*   No effect on Jaffé procedure on AutoAnalyzer at concentration of 20 mg/dL. *2433*

Azapropazone   At concentration of 360 mg/L had no effect as measured by Jaffé-Fading-Fraction method. *3393*   At concentration of 360 mg/L had no effect as measured by Jaffé-Fuller's earth method. *3393*   At concentration of 625 mg/L had no effect as measured by kinetic Jaffé method on BKA-2. *3393*

Barbital   At acute overdose concentration (20 mg/dL) on Technicon® SMAC® method. *3719*   At concentration of 500 mg/L had no effect as measured by AutoAnalyzer Jaffé method. *3393*

Benzbromarone   At concentration of 20 mg/L had no effect as measured by Jaffé-Fading-Fraction method. *3393*   At concentration of 20 mg/L had no effect as measured by Jaffé-Fuller's earth method. *3393*   At concentration of 47 mg/L had no effect as measured by kinetic Jaffé method on BKA-2. *3393*

Bethanechol   No effect at 0.09 mg/dL on SMA 12/60 method. *2636*

Bilirubin   By up to 200 mg/L on enzymatic procedure. *2524*

Bisacodyl   At concentration of 4 mg/L had no effect as measured by Jaffé-Fading-Fraction method. *3393*   At concentration of 4 mg/L had no effect as measured by Jaffé-Fuller's earth method. *3393*   At concentration of 2 mg/L had no effect as measured by kinetic Jaffé method on BKA-2. *3393*

Boric Acid   At concentration of 50 mg/L had no effect as measured by AutoAnalyzer Jaffé method. *3393*

Bromazepam   At 5 times therapeutic concentration on methods of SMAC®, Abbott-VP, aca, Cobas-Bio, Hitachi® 705 and KDA. *2138*

Butabarbital   At concentration of 100 mg/L had no effect as measured by AutoAnalyzer Jaffé method. *3393*

Butizide   At concentration of 2.6 mg/L had no effect as measured by Jaffé-Fading-Fraction method. *3393*   At concentration of 2.6 mg/L had no effect as measured by Jaffé-Fuller's earth method. *3393*   At concentration of 2 mg/L had no effect as measured by kinetic Jaffé method on BKA-2. *3393*

Caffeine   At 500 mg/L on reversed phase LC method of Zhiri et al. *3942*   No effect on alkaline picrate and Slot methods at therapeutic concentrations. *1786*   At concentration of 160 mg/L had no effect on AutoAnalyzer Jaffé method. *3393*   At acute overdose concentration (10 mg/dL) on Technicon SMAC® method. *3719*

Carbamazepine   At 100 mg/L on reversed phase LC method of Zhiri et al. *3942*

Carbenicillin   At 5 times upper limit of therapeutic range on methods on SMAC®, Abbott-VP, aca, Ektachem® 400, Cobas-Bio, Hitachi® 705 and KDA. *2138*

Cefamandole   At concentration of 1500 mg/L had no effect on AutoAnalyzer Jaffé method. *3393*   At concentration of 1500 mg/L had no effect on kinetic Jaffé method on aca. *3393*   At concentration of 1500 mg/L had no effect on kinetic method on BKA-2. *3393*

Cefazedone   On Jaffé method at concentration of 2.50 mmol/L. *1414*

Cefazolin   No effect at up to 250 mg/L on Kodak DT-60. *2554*   At concentration of 2,000 mg/L had no effect on AutoAnalyzer Jaffé method. *3393*   At concentration of 2,000 mg/L had no effect on kinetic Jaffé method on aca. *3393*

Cefodizime   No effect on Jaffé method at concentration up to 2.50 mmol/L. *1414*

Cefoperazone   No effect on Jaffé methods. *2024*   At up to 1,000 μg/mL on Technicon® SMAC® Jaffé method. *3405*

Ceforanide   At up to 1,000 μg/mL on Technicon® SMAC® Jaffé method. *3405*

Cefotaxime   No effect on Jaffé methods. *2024*   No effect at up to 250 mg/L on Kodak DT-60. No effect at up to 250 mg/L on Abbott Vision. No effect at up to 250 mg/L on Ames Seralyzer. *2554*   At up to 1,000 μg/mL on Technicon® SMAC® Jaffé method. *3405*   No effect observed on Jaffé procedure at up to 2.50 mmol/L. *1414*   At concentration of 500 mg/L had no effect on AutoAnalyzer Jaffé method. *3393*   At concentration of 500 mg/L had no effect on Jaffé-Fuller's earth method. *3393*   At concentration of 500 mg/L had no effect on kinetic Jaffé method on aca. *3393*

Cefoxitin   At 5 times upper limit of therapeutic range on enzymatic method on Ektachem® 400. *2138*   No effect at up to 250 mg/L on Ektachem® DT-60. *2554*   No effect of up to 160 mg/L on Boehringer creatinase enzymatic kit. *3794*   At concentration of 160 mg/L had no effect on creatinine amidohydrolase method. *3393*   At concentration of 1 mg/L had no effect on creatinine iminohydrolase method. *3393*   At concentration of 500 mg/L had no effect as measured by Jaffé-Fuller's earth method. *3393*   In patients at 2 h after 2 g given i.v. *0976*

Cefsulodin   At up to 1,000 μg/mL on Technicon® SMAC® Jaffé method. *3405*

Ceftazidime   No effect on Jaffé methods. *2024*   No effect at concentrations up to 250 mg/L on Kodak DT-60. No effect at concentrations up to 250 mg/L on Ames Seralyzer. *2554*   No effect at concentrations up to 250 mg/L on Abbott vision. *2554*

Ceftriaxone   On Jaffé method at concentration of 2.50 mmol/L. *1414*

Cefuroxime   On Jaffé method at concentration of 2.50 mmol/L. *1414*

Cephalothin   At 5 times upper limit of therapeutic range on methods on aca, Ektachem® and KDA. *2138*   At concentration of 1,000 mg/L had no effect on kinetic Jaffé method on aca. *3393*

Cephapirin   At concentration of 1,000 mg/L had no effect on AutoAnalyzer Jaffé method. *3393*   At concentration of 1,000 mg/L had no effect on kinetic Jaffé method on aca. *3393*   At concentration of 100 mg/L had no effect on kinetic Jaffé method on BKA-2. *3393*

Chloramphenicol   At concentration of 20 mg/L had no effect on creatinine iminohydrolase method. *3393*   At concentration of 600 mg/L had no effect on Jaffé-Fading-Fraction method. *3393*   At concentration of 600 mg/L had no effect on Jaffé-Fuller's earth method. *3393*   At concentration of 100 mg/L had no effect on kinetic Jaffé method on BKA-2. *3393*

Chlordiazepoxide   At 16 mg/L on reversed phase LC procedure of Zhiri et al. *3942*   At 5 times upper limit of therapeutic range on methods on SMAC®, Abbott-VP, aca, *2138*   At acute overdose concentration (5 mg/dL) on Technicon® SMAC® method. *3719*   At concentration of 20 mg/L had no effect on creatinine iminohydrolase method. *3393*   At concentration of 50 mg/L had no effect on AutoAnalyzer Jaffé method. *3393*   At concentration of 20 mg/L had no effect on Jaffé-Fading-Fraction method. *3393*   At concentration of 20 mg/L had no effect on Jaffé-Fuller's earth method. *3393*   At concentration of 20 mg/L had no effect on kinetic Jaffé method on BKA-2. *3393*

Chlormezanone   At concentration of 100 mg/L had no effect on AutoAnalyzer Jaffé method. *3393*

Chloroquine   At concentration of 0.5 mg/L had no effect on creatinine iminohydrolase method. *3393*   At concentration of 100 mg/L had no effect on Jaffé-Fading-Fraction method. *3393*   At concentration of 100 mg/L had no effect on Jaffé-Fuller's earth method. *3393*   At concentration of 50 mg/L had no effect on kinetic Jaffé method on BKA-2. *3393*

Chlorphenesin   At concentration of 150 mg/L had no effect on AutoAnalyzer Jaffé method. *3393*

Chlorpheniramine   At concentration of 200 mg/L had no effect on AutoAnalyzer Jaffé method. *3393*   At acute overdose concentration (20 mg/dL) on Technicon SMAC® method. *3719*

Chlorpromazine   At concentration of 3 mg/L had no effect on AutoAnalyzer Jaffé method. *3393*

Creatinine (continued)

Serum   No Effect   Analytical (continued)

Chlorpropamide At 5 times therapeutic concentration on Technicon® SMAC®, Beckman Astra and Du Pont aca methods. *2999* At concentration of 150 mg/L had no effect on AutoAnalyzer Jaffé method. *3393*

Chromonar Hydrochloride At concentration of 90 mg/L had no effect on Jaffé-Fading-Fraction method. *3393* At concentration of 90 mg/L had no effect on Jaffé-Fuller's earth method. *3393* At concentration of 180 mg/L had no effect on kinetic Jaffé method on BKA-2. *3393*

Cimetidine At 300 mg/L on reversed phase LC procedure of Zhiri et al. *3942* At concentration of 4 mg/L had no effect on creatinine iminohydrolase method. *3393* At concentration of 1 mg/L had no effect on AutoAnalyzer Jaffé method. *3393*

Clofibrate At concentration of 500 mg/L had no effect on creatinine iminohydrolase method. *3393* At concentration of 400 mg/L had no effect on Jaffé-Fading-Fraction method. *3393* At concentration of 400 mg/L had no effect on Jaffé-Fuller's earth method. *3393*

Clonidine At 10 mg/L on reversed phase LC procedure of Zhiri et al. *3942*

Cloxacillin At concentration of 5 mg/L had no effect on AutoAnalyzer Jaffé method. *3393*

Cocaine At acute overdose concentration (2.5 mg/dL) on Technicon® SMAC® method. *3719* At concentration of 25 mg/L had no effect on AutoAnalyzer method. *3393*

Codeine At acute overdose concentration (2 mg/dL) on Technicon® SMAC® method. *3719* At concentration of 20 mg/L had no effect on AutoAnalyzer Jaffé method. *3393*

Contact With Clot If plastic disc used to separate for 48 h. *1863*

Cyclophosphamide At concentration of 40 mg/L had no effect on Jaffé-Fading-Fraction method. *3393* At concentration of 40 mg/L had no effect on Jaffé-Fuller's earth method. *3393* At concentration of 452 mg/L had no effect on kinetic Jaffé method on BKA-2. *3393*

Cyclosporine At concentration of 41.5 mmol/L (50 μg/L) on methods on Technicon® SMAC® II and Hitachi® 705. *1123*

Cysteine At 5 times upper limit of therapeutic range on methods SMAC®, Abbott-VP, Cobas-Bio, Ektachem®, Hitachi® 705 and KDA. *2138*

Cystine At 5 times upper limit of therapeutic range on methods on SMAC®, Abbott-VP, aca, Cobas-Bio, Ektachem®, Hitachi® 705 and KDA. *2138*

Desipramine At concentration of 8 mg/L had no effect on AutoAnalyzer Jaffé method. *3393*

Deslanoside No effect on SMA 12/60 method at 0.06 mg/dL. *2636*

Dextran At concentration of 10,000 mg/L had no effect on Ektachem® method. *3393*

Dextran 60 At concentration of 6,000 mg/L had no effect on Jaffé-Fading-Fraction method. *3393* At concentration of 6,000 mg/L had no effect on Jaffé-Fuller's earth method. *3393* At concentration of 81250 mg/L had no effect on kinetic Jaffé method on BKA-2. *3393*

Diatrizoic acid At concentration of 5,000 mg/L had no effect on Ektachem® method. *3393* At concentration of 2600 mg/L had no effect on Jaffé-Fading-Fraction method. *3393* At concentration of 2600 mg/L had no effect on Jaffé-Fuller's earth method. *3393* At concentration of 6100 mg/L had no effect on kinetic Jaffé method on BKA-2. *3393*

Diazepam At 50 mg/L on reversed phase LC procedure of Zhiri et al. *3942* At 5 times upper limit of therapeutic range on methods on SMAC®, Abbott-VP, aca, Cobas-Bio, Ektachem®, Hitachi® 705 and KDA. *2138* At acute overdose concentration (2.5 mg/dL) on Technicon® SMAC® method. *3719* At concentration of 25 mg/L had no effect on AutoAnalyzer Jaffé method. *3393*

Diclofenac No effect at therapeutic concentrations on alkaline picrate and Slot methods. *1786* On alkaline picrate method on SMA II at physiological concentration. *1787* At concentration of 23 mg/L had no effect on AutoAnalyzer Jaffé method. *3393*

Diethylaminoethanol At concentration of 30 mg/L had no effect on AutoAnalyzer method. *3393*

Diethylpropion At acute overdose concentration (10 mg/dL) on Technicon® SMAC® method. *3719* At concentration of 100 mg/L had no effect on AutoAnalyzer Jaffé method. *3393*

Digitoxin No effect at 21 mg/dL on SMA 12/60 method. *2636* At concentration of 6 mg/L had no effect on AutoAnalyzer Jaffé method. *3393*

Digoxin No effect at 0.04 mg/dL on SMA 12/60 method. *2636* At 10 mg/L on reversed phase LC procedure of Zhiri et al. *3942* At concentration of 9 mg/L had no effect on AutoAnalyzer Jaffé method. *3393* At concentration of 0.15 mg/L had no effect on Jaffé-Fading-Fraction method. *3393* At concentration of 0.15 mg/L had no effect on Jaffé-Fuller's earth method. *3393* At concentration of 0.15 mg/L had no effect on kinetic Jaffé method on BKA-2. *3393*

Dihydroergotamine At 10 mg/L on reversed phase LC procedure of Zhiri et al. *3942*

N-Dimethyldiazepam At concentration of 6 mg/L had no effect on AutoAnalyzer Jaffé method. *3393*

Diphenhydramine At acute overdose concentration (20 mg/dL) on Technicon® SMAC® method. *3719* At concentration of 200 mg/L had no effect on AutoAnalyzer Jaffé method. *3393*

Disopyramide At 200 mg/L on reversed phase LC procedure of Zhiri et al. *3942* At concentration of 4 mg/L had no effect on AutoAnalyzer Jaffé method. *3393*

Disulfiram At concentration of 120 mg/L had no effect on AutoAnalyzer Jaffé method. *3393*

Disulphine Blue By discrete kinetic alkaline-picrate reaction. *1464*

Doxepin At 1.2 g/L on reversed phase LC procedure of Zhiri et al. *3942* At concentration of 0.3 mg/L had no effect on creatinine iminohydrolase method. *3393* At concentration of 30 mg/L had no effect on Jaffé-Fading-Fraction method. *3393* At concentration of 30 mg/L had no effect on Jaffé-Fuller's earth method. *3393* At concentration of 10 mg/L had no effect on kinetic Jaffé method on BKA-2. *3393*

EDTA No effect on analytical methods reported. *1563* At concentration of 3360 mg/L had no effect on creatinine iminohydrolase method. *3393* At concentration of 1,000 mg/L had no effect on Jaffé-Fading-Fraction method. *3393* At concentration of 1,000 mg/L had no effect on Jaffé-Fuller's earth method. *3393* At concentration of 1,000 mg/L had no effect on kinetic Jaffé method on BKA-2. *3393*

Ethacrynic Acid At 2.5 mg/dL no effect on SMA 12/60 method. *2636*

Ethanol At acute overdose concentration (20 mg/dL) on Technicon® SMAC® method. *1817*

Ethchlorvynol At acute overdose concentration (20 mg/dL) on Technicon® method. *3719* At concentration of 400 mg/L had no effect on AutoAnalyzer Jaffé method. *3393*

Ethinamate At acute overdose concentration (20 mg/dL) on Technicon® SMAC® method. *3719* At concentration of 200 mg/L had no effect on AutoAnalyzer Jaffé method. *3393*

Ethinyl Estradiol At 10 mg/L on reversed phase LC procedure of Zhiri et al. *3942*

Ethosuximide At concentration of 390 mg/L had no effect on AutoAnalyzer Jaffé method. *3393*

Fasting No effect with 96 h fasting when measured by enzymatic method. *2323*

Flucytosine At 100 μg/mL on Jaffé methods on Technicon® SMAC® and Du Pont aca. *3404*

Fluosol-DA On SMA IIC at concentration of 50%. *2518*

Flurazepam At 5 times upper limit of therapeutic range on methods on SMAC®, Abbott-VP, aca, Cobas-Bio, Ektachem®, Hitachi® 705, and KDA. *2138* At acute overdose concentration (2.5 mg/dL) on Technicon® SMAC® method. *3719* At concentration of 25 mg/L had no effect on AutoAnalyzer Jaffé method. *3393*

Furosemide No effect at 1.4 mg/dL on SMA 12/60 method. *2636* At concentration of 5 mg/L had no effect on creatinine iminohydrolase method. *3393* At concentration of 4 mg/L had no effect on AutoAnalyzer Jaffé method. *3393* At concentration of 20 mg/L had no effect on Jaffé-Fading-Fraction method. *3393* At concentration of 20 mg/L had no effect on Jaffé-Fuller's earth method. *3393* At concentration of 30 mg/L had no effect on kinetic Jaffé method on BKA-2. *3393*

Gentamicin At concentration of 150 mg/L had no effect on AutoAnalyzer Jaffé method. *3393* At concentration of 6 mg/L had no effect on Jaffé-Fading-Fraction method. *3393* At concentration of 6 mg/L had no effect on Jaffé-Fuller's earth method. *3393*

Glibenclamide At 10 mg/L on reversed phase LC procedure of Zhiri et al. *3942* At concentration of 3 mg/L had no effect on Jaffé-Fading-Fraction method. *3393* At concentration of 3 mg/L had no effect on Jaffé-Fuller's earth method. *3393* At concentration of 3 mg/L had no effect on kinetic Jaffé method on BKA-2. *3393*

Glutathione No effect on method of Polar and Metcoff. *1975*

Glutethimide At acute overdose concentration (5 mg/dL) on Technicon® SMAC® method. *3719* At concentration of 50 mg/L had no effect on AutoAnalyzer Jaffé method. *3393*

Glyburide At 5 times expected therapeutic concentration had no effect on Du Pont aca, Beckman Astra and Technicon® SMAC® methods. *2999*

Hemolysis Moderate amount no effect on method of Heinegard. *1547* Moderate hemolysis has no effect. *3505*

Heparin At concentration of 1,000 mg/L had no effect on creatinine amidohydrolase method. *3393*

Hydralazine At concentration of 6 mg/L had no effect on AutoAnalyzer Jaffé method. *3393*

Hydrochlorothiazide At concentration of 1.4 mg/L had no effect on creatinine iminohydrolase method. *3393*

Hyoscine-N-Butylbromide At concentration of 12 mg/L had no effect on Jaffé-Fading-Fraction method. *3393* At concentration of 12 mg/L had no effect on Jaffé-Fuller's earth method. *3393*

Ibuprofen No effect at therapeutic concentrations on alkaline picrate and Slot methods. *1786* On alkaline picrate method on SMA II at physiological concentration. *1787* At 5 times upper limit of therapeutic range on methods on SMAC®, Abbott-VP, aca, Cobas-Bio, Ektachem®, Hitachi® 705 and KDA. *2138* At concentration of 200 mg/L had no effect on AutoAnalyzer method. *3393*

Indapamide At 10 mg/L on reversed phase LC procedure of Zhiri et al. *3942*

Indomethacin No effect at therapeutic concentrations on alkaline picrate and Slot methods. *1786* On alkaline picrate method on SMA II at physiological concentration. *1787* At concentration of 6 mg/L had no effect on creatinine iminohydrolase method. *3393* At concentration of 13 mg/L had no effect on AutoAnalyzer Jaffé method. *3393* At concentration of 30 mg/L had no effect on Jaffé-Fading-Fraction method. *3393* At concentration of 30 mg/L had no effect on Jaffé-Fuller's earth method. *3393* At concentration of 40 mg/L had no effect on kinetic Jaffé method on BKA-2. *3393*

Insulin At concentration of 3 mg/L had no effect on AutoAnalyzer Jaffé method. *3393*

Iproniazid At concentration of 40 mg/L had no effect on AutoAnalyzer Jaffé method. *3393*

Isoniazid At 20 mg/L on reversed phase LC procedure of Zhiri et al. *3942* At concentration of 100 mg/L had no effect on AutoAnalyzer method. *3393*

α-Ketoglutarate No effect ion-exchange method of Mitchell. *2459*

Ketoprofen No effect at therapeutic concentrations on alkaline picrate and Slot methods. *1786* On alkaline picrate method on SMA II at physiological concentration. *1787* At concentration of 60 mg/L had no effect on AutoAnalyzer Jaffé method. *3393*

Lactate No effect on method of Polar and Metcoff. *1975*

Levodopa At concentration of 6 mg/L had no effect on creatinine iminohydrolase method. *3393*

Lidocaine No effect SMA 12/60 method with 3.5 mg/dL. *2636*

Lipemia Patient specimens measured on SMAC® in comparison with specimens following ultra- centrifugation. *2466*

Lithium Heparin At concentration of 500 mg/L had no effect on Ektachem® method. *3393*

Lorazepam At 10 mg/L on reversed phase LC procedure of Zhiri et al. *3942* At concentration of 0.05 mg/L had no effect on AutoAnalyzer Jaffé method. *3393*

Mannitol No effect on SMA 12/60 method at 445 mg/dL. *2636*

Meglumine At concentration of 1200 mg/L had no effect on Jaffé-Fading-Fraction method. *3393* At concentration of 1200 mg/L had no effect on Jaffé-Fuller's earth method. *3393* At concentration of 6,000 mg/L had no effect on kinetic Jaffé method on BKA-2. *3393*

Meperidine At acute overdose concentration (5 mg/dL) on Technicon® SMAC® method. *3719* At concentration of 60 mg/L had no effect on AutoAnalyzer method. *3393*

Mephenesin At concentration of 100 mg/L had no effect on AutoAnalyzer Jaffé method. *3393*

Meprobamate At 200 mg/L on reversed phase LC procedure of Zhiri et al. *3942* At acute overdose concentration (20 mg/dL) on Technicon® SMAC® method. *3719* At concentration of 20 mg/L had no effect on Ektachem® method. *3393* At concentration of 200 mg/L had no effect on AutoAnalyzer method. *3393*

Meralluride No effect at 0.07 mL/dL on SMA 12/60 method. *2636*

Mercaptopurine At concentration of 150 mg/L had no effect on Ektachem® method. *3393*

Mesoridazine At acute overdose concentration (20 mg/dL) on Technicon® SMAC® method. *3719*

Metformin At 2 g/L on reversed phase LC procedure of Zhiri et al. *3942*

Methadone At acute overdose concentration (2.5 mg/dL) on Technicon® SMAC® method. *3719* At concentration of 25 mg/L had no effect on AutoAnalyzer method. *3393*

Methamphetamine At concentration of 2 mg/L had no effect on AutoAnalyzer Jaffé method. *3393*

Methanol At acute overdose concentration (20 mg/dL) on Technicon® SMAC® method. *3719*

Methapyrilene At concentration of 13 mg/L had no effect on AutoAnalyzer Jaffé method. *3393*

Methaqualone At acute overdose concentration (2.5 mg/dL) on Technicon® SMAC® method. *3719* At concentration of 25 mg/L had no effect on AutoAnalyzer Jaffé method. *3393*

Methicillin At 5 times upper limit of therapeutic range on SMAC®, Abbott-VP, Cobas-Bio, aca, Ektachem®, Hitachi® 705 and KDA. *2138* At concentration of 900 mg/L had no effect on AutoAnalyzer Jaffé method. *3393*

Methohexital At concentration of 50 mg/L had no effect on AutoAnalyzer Jaffé method. *3393*

Methotrexate At 5 times upper limit of therapeutic range on method on SMAC®, Abbott-VP, Cobas-Bio, Ektachem®, Hitachi® 705 and KDA. *2138* At concentration of 1 mg/L had no effect on Jaffé-Fading-Fraction method. *3393* At concentration of 1 mg/L had no effect on Jaffé-Fuller's earth method. *3393* At concentration of 55 mg/L had no effect on kinetic Jaffé method on BKA-2. *3393*

Methotrimeprazine At concentration of 1 mg/L had no effect on AutoAnalyzer Jaffé method. *3393*

Methsuximide At concentration of 40 mg/L had no effect on Ektachem® method. *3393*

Methyldopa At 30 mg/L on reversed phase LC procedure of Zhiri et al. *3942* At concentration of 20 mg/L had no effect on creatinine iminohydrolase method. *3393* At concentration of 320 mg/L had no effect on Jaffé-Fading-Fraction method. *3393* At concentrations above 50 mg/L (normal therapeutic concentration 2 mg/L) raised concentration as measured by kinetic Jaffé method on BKA-2. *3393*

Methylphenidate At acute overdose concentration (20 mg/dL) on Technicon® SMAC® method. *3719* At concentration of 200 mg/L had no effect on AutoAnalyzer Jaffé method. *3393*

Methylphenobarbital At concentration of 150 mg/L had no effect on AutoAnalyzer Jaffé method. *3393*

Methyprylon At acute overdose concentration (10 mg/dL) on Technicon® SMAC® method. *3719* At concentration of 100 mg/L had no effect on AutoAnalyzer Jaffé method. *3393*

Metoprolol At concentration of 0.34 mg/L had no effect on AutoAnalyzer Jaffé method. *3393*

Morphine At acute overdose concentration (20 mg/dL) on Technicon® SMAC® method. *3719* At concentration of 200 mg/L had no effect on AutoAnalyzer Jaffé method. *3393*

Moxalactam No effect on Jaffé methods. *2024* At up to 1,000 μg/mL on Technicon® SMAC® Jaffé procedure. *3405* At concentration of 400 mg/L had no effect on AutoAnalyzer Jaffé method. *3393* At concentration of 400 mg/L had no effect on kinetic Jaffé method on aca. *3393*

Nafcillin At concentration of 50 mg/L had no effect on AutoAnalyzer Jaffé method. *3393*

Nalidixic Acid At concentration of 60 mg/L had no effect on creatinine iminohydrolase method. *3393*

Creatinine (continued)

Serum No Effect Analytical (continued)

Naproxen At 5 times upper limit of therapeutic range on methods on SMAC®, Abbott-VP, aca, Cobas-Bio, Ektachem®, Hitachi® 705 and KDA. 2138

Niacinamide At concentration of 40 mg/L had no effect on Jaffé-Fading-Fraction method. 3393 At concentration of 40 mg/L had no effect on Jaffé-Fuller's earth method. 3393 At concentration of 120 mg/L had no effect on kinetic Jaffé method on BKA-2. 3393

Nicergoline At 10 mg/L on reversed phase LC procedure of Zhiri et al. 3942

Niflumic Acid At 200 mg/L on reversed phase LC procedure of Zhiri et al. 3942 At concentration of 150 mg/L had no effect on Jaffé-Fading-Fraction method. 3393 At concentration of 150 mg/L had no effect on Jaffé-Fuller's earth method. 3393

Nitrofurantoin On routine methods in use on SMAC®, Ektachem®, Abbott-VP, Cobas-Bio, aca, Hitachi® 705, KDA at 5 times normal upper therapeutic concentration. 2138 At concentration of 4 mg/L had no effect on AutoAnalyzer Jaffé method. 3393

Norfenefrin At concentration of 6 mg/L had no effect on Jaffé-Fading-Fraction method. 3393 At concentration of 6 mg/L had no effect on Jaffé-Fuller's earth method. 3393

Norgestrel At 10 mg/L on reversed phase LC procedure of Zhiri et al. 3942

Nortriptyline At acute overdose concentration (20 mg/dL) on Technicon® SMAC® method. 3719 At concentration of 200 mg/L had no effect on AutoAnalyzer Jaffé method. 3393

Novaminsulfon At concentration of 100 mg/L had no effect on creatinine iminohydrolase method. 3393 At concentration of 8,000 mg/L had no effect on Jaffé-Fading-Fraction method. 3393

Orphenadrine At concentration of 9 mg/L had no effect on AutoAnalyzer Jaffé method. 3393

Ouabain At 0.02 mg/dL no effect on SMA 12/60 method. 2636

Oxalacetate No effect on ion-exchange method of Mitchell. 2459

Oxazepam At 10 mg/L on reversed phase LC procedure of Zhiri et al. 3942 At concentration of 1 mg/L had no effect on AutoAnalyzer method. 3393

Oxyphenbutazone At concentration of 220 mg/L had no effect on creatinine iminohydrolase method. 3393 At concentration of 120 mg/L had no effect on Jaffé-Fading-Fraction method. 3393 At concentration of 120 mg/L had no effect on Jaffé-Fuller's earth method. 3393 At concentration of 600 mg/L had no effect on kinetic Jaffé method on BKA-2. 3393

Papaverine At concentration of 10 mg/L had no effect on AutoAnalyzer Jaffé method. 3393

Paraldehyde At concentration of 2,000 mg/L had no effect on AutoAnalyzer Jaffé method. 3393

Penicillamine At concentration of 480 mg/L had no effect on Jaffé-Fading-Fraction method. 3393 At concentration of 480 mg/L had no effect on Jaffé-Fuller's earth method. 3393 At concentration of 55 mg/L had no effect on kinetic Jaffé method on BKA-2. 3393

Penicillin G At 5 times upper limit of therapeutic range on methods on SMAC®, Abbott-VP, Cobas-Bio, aca, Ektachem®, Hitachi® 705 and KDA. 2138 At concentration of 2,000 mg/L had no effect as measured by AutoAnalyzer Jaffé method. 3393

Pentobarbital At acute overdose concentration (20 mg/dL) on Technicon® SMAC® method. 3719 At concentration of 340 mg/L had no effect on AutoAnalyzer method. 3393

Perphenazine At acute overdose concentration (20 mg/dL) on Technicon® SMAC® method. 3719 At concentration of 200 mg/L had no effect on AutoAnalyzer method. 3393

Phencyclidine At concentration of 6 mg/L had no effect on AutoAnalyzer Jaffé method. 3393

Phenobarbital At 50 mg/L on reversed phase LC procedure of Zhiri et al. 3942 No effect on alkaline picrate and Slot methods at therapeutic concentrations. 1786 At acute overdose concentration (20 mg/dL) on Technicon® SMAC® method. 3719 At concentration of 30 mg/L had no effect on Ektachem® method. 3393 At concentration of 80 mg/L had no effect on creatinine iminohydrolase method. 3393 At concentration of 250 mg/L had no effect on AutoAnalyzer Jaffé method. 3393 At concentration of 80 mg/L had no effect on Jaffé-Fading-Fraction method. 3393 At concentration of 80 mg/L had no effect on Jaffé-Fuller's earth method. 3393 At concentration of 60 mg/L had no effect on kinetic Jaffé method on BKA-2. 3393

Phenothiazines At concentration of 30 mg/L had no effect on Jaffé-Fading-Fraction method. 3393 At concentration of 30 mg/L had no effect on Jaffé-Fuller's earth method. 3393 At concentration of 200 mg/L had no effect on kinetic Jaffé method on BKA-2. 3393

Phenprocoumon At concentration of 6 mg/L had no effect on Jaffé-Fading-Fraction method. 3393 At concentration of 6 mg/L had no effect on Jaffé-Fuller's earth method. 3393 At concentration of 26 mg/L had no effect on kinetic Jaffé method on BKA-2. 3393

Phensuximide At concentration of 120 mg/L had no effect on AutoAnalyzer Jaffé method. 3393

Phenylbutazone At concentration of 750 mg/L had no effect on AutoAnalyzer Jaffé method. 3393 At concentration of 120 mg/L had no effect on Jaffé-Fading-Fraction method. 3393 At concentration of 120 mg/L had no effect on Jaffé-Fuller's earth method. 3393 At concentration of 400 mg/L had no effect on kinetic Jaffé method on BKA-2. 3393

Phenylephrine At concentration of 4 mg/L had no effect on AutoAnalyzer Jaffé method. 3393

Phenytoin No effect at 1.8 mg/L on SMA 12/60 method. 2636 At acute overdose concentration (20 mg/dL) on Technicon® SMAC® method. 3719 At concentration of 20 mg/L had no effect on Ektachem® method. 3393 At concentration of 240 mg/L had no effect on AutoAnalyzer method. 3393

Plasma No significant difference from serum by AutoAnalyzer. 2228

Prednisolone At concentration of 0.23 mg/L had no effect on creatinine iminohydrolase method. 3393 At concentration of 200 mg/L had no effect on Jaffé-Fading-Fraction method. 3393 At concentration of 200 mg/L had no effect on Jaffé-Fuller's earth method. 3393

Probenecid At concentration of 1300 mg/L had no effect on AutoAnalyzer Jaffé method. 3393 At concentration of 200 mg/L had no effect on Jaffé-Fading-Fraction method. 3393 At concentration of 200 mg/L had no effect on Jaffé-Fuller's earth method. 3393 At concentration of 1,000 mg/L had no effect on kinetic Jaffé method on BKA-2. 3393

Procainamide No effect on SMA 12/60 method at 35 mg/dL. 2636 At concentration of 50 mg/L had no effect on AutoAnalyzer Jaffé method. 3393

Procaine At concentration of 2 mg/L had no effect on AutoAnalyzer Jaffé method. 3393

Prochlorperazine At concentration of 1 mg/L had no effect on AutoAnalyzer Jaffé method. 3393

Promethazine At acute overdose concentration (20 mg/dL) on Technicon® SMAC® method. 3719 At concentration of 200 mg/L had no effect on AutoAnalyzer Jaffé method. 3393

Propoxyphene At acute overdose concentration (2.5 mg/dL) on Technicon® SMAC® method. 3719 At concentration of 25 mg/L had no effect on AutoAnalyzer Jaffé method. 3393

Propranolol No effect at 0.1 mg/dL on SMA 12/60 method. 2636 At 100 mg/L on reversed phase LC procedure of Zhiri et al. 3942 At concentration of 0.16 mg/L had no effect on creatinine iminohydrolase method. 3393 At concentration of 0.2 mg/L had no effect on AutoAnalyzer method. 3393

Pyribenzamine® At acute overdose concentration (20 mg/dL) on Technicon® SMAC® methods. 3719

Quinidine At 26 mg/dL no effect on SMA 12/60 method. 2636 At acute overdose concentration (20 mg/dL) Technicon® SMAC® method. 3719 At concentration of 210 mg/L had no effect on AutoAnalyzer Jaffé method. 3393

Quinine At acute overdose concentration (1.5 mg/dL) on Technicon® SMAC® method. 3719 At concentration of 30 mg/L had no effect on AutoAnalyzer Jaffé method. 3393

Reserpine No effect at 0.02 mg/dL on SMA 12/60 method. 2636

Salicylate At concentration of 150 mg/L had no effect on creatinine iminohydrolase method. *3393* At concentration of 500 mg/L had no effect on AutoAnalyzer Jaffé method. *3393*

Scopolamine Bromide At concentration of 20 mg/L had no effect on creatinine iminohydrolase method. *3393* At concentration of 20 mg/L had no effect on kinetic Jaffé method on BKA-2. *3393*

Secobarbital At acute overdose concentration (20 mg/dL) on Technicon® SMAC® method. *3719* At concentration of 200 mg/L had no effect on AutoAnalyzer Jaffé method. *3393*

Sodium Citrate At concentration of 5,000 mg/L had no effect on Jaffé-Fading-Fraction method. *3393* At concentration of 5,000 mg/L had no effect on Jaffé-Fuller's earth method. *3393* At concentration of 5,000 mg/L had no effect on kinetic Jaffé method on BKA-2. *3393*

Sodium Fluoride At concentration of 2,000 mg/L had no effect on Jaffé-Fading-Fraction method. *3393* At concentration of 2,000 mg/L had no effect on Jaffé-Fuller's earth method. *3393* At concentration of 2,000 mg/L had no effect on kinetic Jaffé method on BKA-2. *3393*

Sodium Heparin At concentration of 750 mg/L had no effect on Jaffé-Fading-fraction method. *3393* At concentration of 750 mg/L had no effect on Jaffé-Fuller's earth method. *3393* At concentration of 750 mg/L had no effect on kinetic Jaffé method on BKA-2. *3393*

Sodium Oxalate At concentration of 3,000 mg/L had no effect on Jaffé-Fading-Fraction method. *3393* At concentration of 3,000 mg/L had no effect on Jaffé-Fuller's earth method. *3393*

Sodium Pyruvate No effect ion-exchange method of Mitchell. *2459*

Spironolactone At concentration of 1 mg/L had no effect on creatinine iminohydrolase method. *3393* At concentration of 20 mg/L had no effect on Jaffé-Fading-Fraction method. *3393* At concentration of 20 mg/L had no effect on Jaffé-Fuller's earth method. *3393* At concentration of 50 mg/L had no effect on kinetic Jaffé method on BKA-2. *3393*

Storage of Sample No effect for 6 mo at -10°. *1180*

Strychnine At concentration of 12 mg/L had no effect on AutoAnalyzer Jaffé method. *3393*

Sulfadiazine At concentration of 1500 mg/L had no effect on AutoAnalyzer Jaffé method. *3393*

Sulfaguanidine At concentration of 500 mg/L had no effect on AutoAnalyzer Jaffé method. *3393*

Sulfamethoxazole No effect on alkaline picrate (Jaffé) procedure. *0313* At concentration of 200 mg/L had no effect on AutoAnalyzer Jaffé method. *3393*

Sulfamethoxydiazine At concentration of 200 mg/L had no effect on AutoAnalyzer Jaffé method. *3393* At concentration of 200 mg/L had no effect on Jaffé-Fading-Fraction method. *3393* At concentration of 200 mg/L had no effect on Jaffé-Fuller's earth method. *3393* At concentration of 500 mg/L had no effect on kinetic Jaffé method on BKA-2. *3393*

Sulfamethoxypyridazine At concentration of 70 mg/L had no effect on creatinine iminohydrolase method. *3393*

Sulfanilamide At concentration of 1,000 mg/L had no effect on AutoAnalyzer Jaffé method. *3393*

Sulfathiazole At concentration of 60 mg/L had no effect on Ektachem® method. *3393*

Sulforidazine At concentration of 200 mg/L had no effect on AutoAnalyzer Jaffé method. *3393*

Tetracycline At 5 times upper limit of therapeutic range on methods on SMAC®, Abbott-VP, aca, Cobas-Bio, Ektachem®, Hitachi® 705 and KDA. *2138* At concentration of 60 mg/L had no effect on creatinine iminohydrolase method. *3393* At concentration of 300 mg/L had no effect on AutoAnalyzer Jaffé method. *3393* At concentration of 200 mg/L had no effect on Jaffé-Fading-Fraction method. *3393* At concentration of 200 mg/L had no effect on Jaffé-Fuller's earth method. *3393* At concentration of 40 mg/L had no effect on kinetic Jaffé method on BKA-2. *3393*

Theobromine At concentration of 2,000 mg/L had no effect on AutoAnalyzer method. *3393*

Theophylline At 1.0 g/L on reversed phase LC procedure of Zhiri et al. *3942* At 5 times upper limit of therapeutic range on methods on SMAC®, Abbott-VP, aca, Cobas-Bio, Ektachem®, Hitachi® 705 and KDA. *2138* At concentration of 2,000 mg/L had no effect on AutoAnalyzer Jaffé method. *3393*

Thiamine At 500 mg/L on reversed phase LC procedure of Zhiri et al. *3942*

Thiopental At acute overdose concentration (20 mg/dL) on Technicon® SMAC® method. *3719* At concentration of 200 mg/L had no effect on AutoAnalyzer Jaffé method. *3393*

Timolol At concentration of 0.01 mg/L had no effect on AutoAnalyzer Jaffé method. *3393*

Tolazamide At 5 times therapeutic concentration on Du Pont aca, Beckman Astra and Technicon® SMAC® method. *2999*

Tolbutamide At 5 times therapeutic concentration on Du Pont aca, Beckman Astra and Technicon® SMAC® method. *2999* At concentration of 500 mg/L had no effect on creatinine iminohydrolase method. *3393* At concentration of 100 mg/L had no effect on AutoAnalyzer Jaffé method. *3393* At concentration of 400 mg/L had no effect on Jaffé-Fading-Fraction method. *3393* At concentration of 400 mg/L had no effect on Jaffé-Fuller's earth method. *3393* At concentration of 500 mg/L had no effect on kinetic Jaffé method on BKA-2. *3393*

Transport of Specimen By pneumatic tube if specimen protected. *3217*

Tribromethanol At concentration of 90 mg/L had no effect on AutoAnalyzer Jaffé method. *3393*

Trichlorethanol At concentration of 1,000 mg/L had no effect on AutoAnalyzer Jaffé method. *3393*

Trifluoperazine At concentration of 1 mg/L had no effect on AutoAnalyzer method. *3393*

Trimethoprim On Jaffé reaction with alkaline picrate. *0313* At concentration of 5 mg/L had no effect on AutoAnalyzer Jaffé method. *3393* At concentration of 5 mg/L had no effect on kinetic Jaffé method. *3393*

Tripelennamine At concentration of 200 mg/L had no effect on AutoAnalyzer Jaffé method. *3393*

Troleandomycin At 200 mg/L on reversed phase LC procedure of Zhiri et al. *3942*

Uric Acid No effect on method of Polar and Metcoff. *1975*

Valproic Acid At 1.0 g/L on reversed phase LC procedure of Zhiri et al. *3942* At concentration of 100 mg/L had no effect on creatinine iminohydrolase method. *3393*

Vitamin B Complex At concentration of 12.9 mg/L had no effect on Jaffé-Fading-Fraction method. *3393* At concentration of 12.9 mg/L had no effect on Jaffé-Fuller's earth method. *3393* At concentration of 800 mg/L had no effect on kinetic Jaffé method on BKA-2. *3393*

Vitamin Preparations No effect at expected concentration with SMA 12/60 procedure. *2637*

Warfarin At concentration of 100 mg/L had no effect on AutoAnalyzer Jaffé method. *3393*

Serum   No Effect   Physiological

Acetaminophen Without overdose not usually any effect. *2866*

Acetylsalicylic Acid No effect seen in patients undergoing treatment. *1787*

Alanine Unaffected by alanine administration to fasting obese. *1263*

Amiloride No effect seen in 13 men treated for 8 weeks. *0656*

Aspirin When approximately 3 g/d ingested for several weeks. *3076*

Bopindolol In 10 hypertensive patients treated for either 1 or 21 d. *1063*

Bumetanide No significant effect observed. *2666*

Captopril No significant change in 7,000 hypertensives treated for 3 y. *1407*

Ceftazidime Unchanged in most patients with GFR above 30 mL/min. *0051*

Chenodeoxycholic Acid No significant effect with up to 750 mg/d in 916 patients treated for 2 y. *3199*

Chlorprothixene Constant concentration regardless of dose of drug co-administered and duration. *3268*

Cimetidine No effect in volunteers receiving 1.2 g daily for 12 d. *0300*

Diclofenac No effect seen in patients undergoing treatment. *1787*

Etodolac No cumulative effect observed over 24 h in individuals with normal renal function or with renal insufficiency when given 500 mg bid for 4 d. *0453*

Creatinine (continued)

Serum No Effect Physiological (continued)

Famotidine No significant change from pretreatment values in 9 patients with Zollinger-Ellison syndrome over 33 weeks. *1667*

Furosemide In 8 normal subjects associated with secondary hyperparathyroidism. *1214*

Glyburide No significant effect observed. *0125*

Hydrochlorothiazide In approximately 175 patients with essential hypertension treated for 1 y. *3708*

Ibuprofen No effect during coadministration with digoxin. *2896*

Indomethacin No effect seen in patients undergoing treatment. *1787*

Ketoprofen No effect seen in patients undergoing treatment. *1787*

Labetalol No significant effect in 15 patients with essential hypertension treated for 1 mo. *3776*

Meals No effect after standard breakfast. *0579*

Muscular Exercise Observed effect with 12 minutes cycle-ergometer. *1231*

Nisoldipine In 14 mild to moderately hypertensive noninsulin dependent diabetic patients. *2653*

Piroxicam No effect in spite of effect on urea nitrogen. *2692*

Propranolol No significant effect on renal function in 15 patients with essential hypertension over 1 mo. *3776* In 340 patients given drug for 10 weeks to reduce diastolic BP to less than 90 mm Hg. *3707* In approximately 120 patients with essential hypertension treated for 1 y. *3708*

Radiographic Agents No significant change observed when agents given for either urography or arteriography. *3259*

Ranitidine No effect reported although other drugs with same overall action blocks tubular secretion of creatinine. *3389*

Saccharated Iron Oxide In 9 individuals given 40 mg/d i.v. daily for up to 42 d. *2662*

Starvation Little effect over first few days. *2382*

Sulfamethoxazole Insignificant reduction in 5 volunteers after 7 d. *3078*

Sulindac No significant change in patients with chronic glomerular disease. *0660*

Timegadine In 23 patients with rheumatoid arthritis given 250 to 750 mg/d for 48 weeks. *2345*

Warfarin No effect on renal function noted with continuing administration. *2411*

Serum Increase Analytical

Acetaminophen Mild acute renal failure in two patients following therapeutic ingestion of drug. *1219*

Acetoacetate Slight effect on Lloyd's procedure. Marked effect direct methods. *2459* 1 mmol/L = 0.11 mg/dL method of Heinegard. *1547* Interferes with Grafnetter and AutoAnalyzer methods. *1975* Linear response when added to specimens for analysis by picrate method on Astra™ and SMAC®. Sensitivity of Astra method twice that of SMAC®. Effect less on stored specimens (maybe 5% medical specimens affected). *1267* Can be marked effect on Jaffé based reactions, more marked on Technicon® AutoAnalyzer than Beckman creatinine analyzer. *2470*

Acetohexamide Significant effect on Jaffé methods on Beckman Astra and Du Pont aca by as much as 2.2 or 3.3 mg/dL respectively at therapeutic concentrations and as little as 0.3 mg/dL on SMA II. *3890*

Acetone Interferes with Jaffé method. *1975*

Albumin 50 g/L equivalent to 0.02 mg/dL with method of Heinegard. *1547*

Aminohippuric Acid Chromogenicity in color reaction. *3505*

Arginine If method of Voges-Proskauer used. *1563*

Ascorbic Acid Chromogenicity in color reaction (as reducing agent). *1022* 100 mg/L = 0.01 mg/dL method of Heinegard. *1547* Marked effect direct methods. *2459* At concentrations above 250 mg/L (maximum serum concentration 34 mg/L) raised concentration as measured by AutoAnalyzer Jaffé method. *3393* At concentrations above 25 mg/L (maximum serum concentration 34 mg/L) raised concentration as measured by kinetic Jaffe method. *3393*

Cefamandole Slow and slight reaction in Jaffé methods. *1414*

Cefazolin Increase of 10-20 μmol/L for every 20 mg/L of drug on Abbott Vision. *2554* Slow and slight reaction in Jaffé methods. *1414* At concentrations above 2,000 mg/L (normal therapeutic concentration 150 mg/L) kinetic Jaffé method on BKA-2. *3393* Increase of 10-20 μmol/L for every 20 mg/L of drug on Ames Seralyzer. *2554*

Cefoperazone Slow and slight reaction in Jaffé methods. *1414*

Cefotiam Slow and slight reaction in Jaffé methods. *1414*

Cefoxitin Profound interference with alkaline picrate based reactions. *1414* At 5 times upper limit of therapeutic range on methods on SMAC®, Abbott-VP, aca, Cobas-Bio, Hitachi® 705 and KDA. *2138* Increase of 50-80 μmol/L for every 100 mg/L of drug on Abbott Vision. *2554* When used with Jaffé method on Greiner selective analyzer: concentration related effect. *0061* Concentration dependent increase with kinetic Jaffé reaction on Beckman Astra or Eppendorf system with Merck reagents. *3449* Effect noted on Technicon® SMAC® less than with other commercial systems due to lesser dialysis of drug than creatinine. *2022* Linear correlation between drug concentration and that of analyte as measured by Jaffé reaction. *1414* At concentrations above 80 mg/L (normal therapeutic concentration 150 mg/L) raised the concentration as measured by AutoAnalyzer Jaffé method. *3393* At concentrations above 70 mg/L (normal therapeutic concentration 150 mg/L) raised the concentration as measured by kinetic Jaffé method. *3393* At concentrations above 25 mg/L (normal therapeutic concentration 150 mg/L) raised the concentration as measured by kinetic Jaffé method on aca. *3393* At concentrations above 50 mg/L (normal therapeutic concentration 150 mg/L) raised the concentration as measured by kinetic Jaffé method on BKA-2. *3393* Increase of 50-80 μmol/L for every 100 mg/L of drug on Ames Seraltzer. *2554*

Cefpirome Profound interference with Jaffé reaction. *1414* Linear correlation between drug concentration and apparent analyte concentration. *1414*

Cephaloridine Pronounced interference with alkaline picrate based reactions. *1414* Linear correlation between drug concentration and analyte concentration as measured by Jaffé method. *1414*

Cephalothin Pronounced interference with alkaline picrate based reactions. *1414* At 5 times upper limit of therapeutic range on methods on SMAC®, Abbott-VP, Cobas-Bio and Hitachi® 705. *2138* Dose related interference with Jaffé procedure on Greiner selective analyzer. *0061* Concentration related increase when Jaffé method used. *1414* At concentrations above 500 mg/L (normal therapeutic concentration 17 mg/L) raised concentration as measured by AutoAnalyzer Jaffé method. *3393* At concentrations above 100 mg/L (normal therapeutic concentration 17 mg/L) raised concentration as measured by kinetic Jaffé method on BKA-2. *3393*

EDTA At concentrations above 800 mg/L raised concentration as measured by Ektachem® method. *3393*

Fasting Observed in 5 healthy volunteers when measured by Jaffé method: probably due to accumulation and measurement of acetoacetate (increase by 0.7 mg/dL after 96 h). *2323*

Flucytosine At 10 μg/mL and above on Ektachem® 700 2 slide procedure: change in proportion to concentration. *3404*

Fructose Interference with Jaffé reaction. *3879*

Glucose 7 g/L =0.06 mg/dL method of Heinegard. *1547* Interferes with Jaffé reaction. *2313*

Glycerin 100 mg/L = 0.03 mg/dL method of Heinegard. *1547*

Glycocyamidine Reacts to give false increase with Jaffé reagent. *3505*

Glycocyamine If reaction of Voges-Proskauer used. *1563*

Guanidine If reaction of Voges-Proskauer used. *1563*

Hemolysis At 500 mg/L hemoglobin on enzymatic procedure. *2524* Calculated error 1.8 %/g with alkaline picrate procedure. *0515*

α-Ketoglutarate Slight effect on Lloyd's procedure. Marked effect direct methods. *2459*

Levodopa Acts as reducing agent (probable effect). *0127*

Lidocaine Some interference in some specimens with Gen02 slides for Ektachem® system, but rarely more than 3 mg/L. *1358*

Lipemia At 20 g/L total lipids on enzymatic procedure. *2524*

Lithium Lactate 5 mmol/L = 0.01 mg/dL method of Heinegard. *1547*

Methyldopa Readily oxidized and affects alkaline picrate method. *2313* Above 2 mg/mL even affects Fuller's earth procedures. *2259* At concentrations above 200 mg/L (normal therapeutic concentration 2 mg/L) raised concentration as measured by AutoAnalyzer Jaffé method. *3393* At concentrations above 2,000 mg/L (normal therapeutic concentration 2 mg/L) raised concentration as measured by Jaffé-Fuller's earth method. *3393*

Moxalactam Slow and slight reaction in Jaffé methods. *1414*

Nitrofurantoin At concentrations above 5 mg/L (normal therapeutic concentration 5.5 mg/L) raised concentration as measured by Jaffé-Fading-Fraction method. *3393* At concentrations above 5 mg/L (normal therapeutic concentration 5.5 mg/L) raised concentration as measured by Jaffé-Fuller's earth method. *3393* At concentrations above 18 mg/L (normal therapeutic concentration 5.5 mg/L) raised concentration as measured by kinetic Jaffé method on BKA-2. *3393*

Novaminsulfon At concentrations above 8,000 mg/L (normal therapeutic concentration 15 mg/L) raised concentration as measured by Jaffé-Fuller's earth method. *3393*

Oxalacetate Marked effect direct methods. Moderate effect on Lloyd's procedure. *2459*

Phenolsulfonphthalein Chromogenicity in color reaction. *3505*

Potassium Oxalate At concentrations above 8,000 mg/L when combined with sodium fluoride raised concentration as measured by Ektachem® method. *3393*

Protein 5 g/dL produces color equivalent to 2.2 mg/dL. *0583*

Pyruvate Interferes -AutoAnalyzer method (200 mg/dL = 7.4 mg/dL). *1975*

Resorcinol Falsely high values if Jaffé reaction used. *1563*

Sodium Fluoride At concentrations above 10,000 mg/L when combined with potassium oxalate raised concentration as measured by Ektachem® method. *3393*

Sodium Pyruvate Marked effect direct methods. Moderate effect on Lloyd's procedure. *2459*

Storage of Sample Significant effect after 15 d at 4°. *3862*

Sulfamethoxazole At concentrations above 200 mg/L raised concentration as measured by kinetic Jaffé method. *3393*

Sulfasalazine At concentrations above 500 mg/L (normal therapeutic concentration 70 mg/L) raised concentration as measured by kinetic Jaffé method on BKA-2. *3393*

Sulfobromophthalein Presence of interfering color. *2313*

Uremia Higher concentration of Jaffé chromogens. *2445*

Serum Increase Physiological

Acetaminophen Reversible tubular necrosis reported. *0438*

Acyclovir Reversible increase, especially with doses greater than 5 mg/kg i.v. every 8 h. Major adverse effect occurring in as many as 50% patients. *0936*

Aging From 0.5 mg/dL at 6 y to 0.95 at 20 y (men). From 0.5 mg/dL at 6 y to 0.75 at 16 y (women). *0887*

Aldatense Increase by 10-20% in a group of hypertensives compared with controls. *1611*

Alkaline Antacids May cause milk-alkali syndrome. *2220*

Altitude But less than 0.1 mg/dL. *2957*

Amikacin Nephrotoxicity observed in 7 of 54 patients whose drug concentrations were monitored. *1250*

Amiloride May occur with prolonged therapy. *3711*

Amphotericin B Nephrotoxic effect. *2313* Increased to average of 1.4 mg/dL from 1.0 mg/dL after 2 weeks in 10 patients; further increase with time. *0240* Nephrotoxicity usually develops after a few weeks of treatment, usually reversible unless more than total of 4 g drug given. *3567*

Arginine Incorporated into urea cycle. *0071*

Arsenicals Nephrotoxicity (common with therapeutic doses). *1237*

Aspirin Average increase of 38% in patients and healthy individuals. *0545* Anti-inflammatory doses produced substantial effect in 13 of 23 patients with systemic lupus erythematosus. *2866*

Azapropazone Initial 10% increase observed in 83 treated patients but thereafter no further increase *3565*

Barbiturates Shock and renal failure in intoxication. *2425*

Bethanidine Effect reported in one patient. *3479*

Bezafibrate In four patients with poor renal function produced myolysis as result of overdose due to renal dysfunction. *3086*

Blindness Significantly higher (average 0.2 mg/dL) than normal. *1637*

Bopindolol Minimal effect observed in large population of hypertensives over 12 weeks study period. *3779*

Capreomycin Nephrotoxic effect. *2313*

Captopril Acute reversible renal failure may occur: transient increases common. *3712* Occasional reversible azotemia, either due to hypotension or direct renal damage. *0283* Severe reversible azotemia in few patients with peripheral vascular disease two probably associated with GFR reduction. *0655* Eosinophilic interstitial nephritis and membranous glomerulopathy reported. Cases of nephrotic syndrome also reported. *0929*

Carbutamide Nephrotoxic effect. *2313*

Ceftazidime Effect noted in 3 patients given high doses in relation to their renal function. *0050*

Cefuroxime Mean change from 112 to 137 μmol/L in 3 patients with chronic osteomyelitis over over 14 d when lysine salt used. *3620*

Cephaloridine Nephrotoxic especially if combined with diuretic. *2808*

Cephalothin High rate of nephrotoxicity observed with combined with aminoglycosides, more than when other drugs given with aminoglycosides. *3731*

Chlorpropamide In isolated case with diabetes who developed proliferative glomerulonephritis indicating immunologically mediated reaction. *0146*

Chlorthalidone Nephrotoxic effect. *2220* Effect observed at dose of 100 mg/d and upwards in 37 patients treated for 8 weeks. *3643* Increased in hypertensives; most marked in those individuals whose initial values were lower. *2904*

Cimetidine Rises because of competitive inhibition of creatinine secretion following i.v. bolus of 300 mg. *0535* Single case of reversible renal failure probably attributable to drug. *3243* In 13 ulcer patients average increase of 22% which fell on cessation on therapy. *0649* Small but detectable increases compared with placebo, but rarely exceeding 2 mg/ dl. *2368* Significant effect from first day of treatment and at 3 weeks but not after 12. *0978*

Ciprofloxacin Mild and transient cases reported although 1 case each of acute renal failure, interstitial nephritis and nonspecific nephritis reported in population of 2829 patients. *0151*

Cisplatin Nephrotoxicity evidenced by increase in 12 children receiving chemotherapy. *1355* Observed in one patient, but major side effect renal toxicity. *1194* In 5% of 96 cancer patients given drug for 5 d repeated at 4-6 week intervals. *3119*

Clofibrate Possibly derived from muscle damage (in 15%). *2220*

Clonidine Approximately 0.1 mg/dL. *3923*

Colistimethate Nephrotoxic effect (usually reversible). *2808*

Colistin Nephrotoxic effect (reversible renal damage). *2313*

Co-Trimoxazole Usually reversible effect (in many patients). *1852* By 0.2 mg/dL in 21 patients: probably due to competitive inhibition of tubular secretion of trimethoprim. *0313* Change of 0.12 mg/dL in 5 volunteers after 7 days. *3078*

Cyclosporine Mean change from 1.0 to 1.5 mg/dL in 8 of 23 patients with 2 to 6 mo treatment (patients with inflammatory ocular disease). 30% increase in more than half patients. *2713* In 4 patients with primary biliary cirrhosis in whom creatinine and drug concentration rose as bleeding occurred. *3014* In 62 renal allograft recipients from 28 to 90 d after transplant: effect greater than with azathioprine and prednisolone: reversible. *0635* In 8 of 23 patients with ocular inflammatory disease treated for 1 mo (mean change from 1.0 to 1.5 mg/dL). *2713* Characteristic feature of nephrotoxicity in most common form of renal damage to drug. *0175* In patients with rheumatoid arthritis: significant effect unrelated to initial. *0905*

Danazol Slight effect in 18 women with endometriosis given 600 mg/d for 6 mo. *0591*

Demeclocycline Dose related nephrotoxicity. *2106*

Dextran Blocks tubules causing renal failure. *0647*

Diuretics May be associated with acute sodium depletion. *0105*

Diurnal Variation Maximum about 2 am minimum at 8 am. *3428* 7 pm value 130% of 7 am concentration. *3217*

Doxycycline Nephrotoxic effect. *2802*

Creatinine *(continued)*

Serum   Increase   Physiological *(continued)*

Enalapril   Reversible renal insufficiency reported in some patients without evidence of renal artery stenosis. *0929*   Associated with selective glomerular efferent arteriolar dilatation and possible interference with autoregulatory capacity of kidney in response to severe renovascular hypertension. *0284*

Enflurane   In one patient with initially normal renal function following comparatively long exposure to anesthetic. Normal renal function eventually returned. *1001*

Ethambutol   Rare case of renal damage reported. *0131*

Ethylene Glycol   Due to tubular necrosis caused by toxic metabolites and possibly also due to deposition of calcium oxalate crystals in kidney. *3325*

Etidronate   Transient mild effect in patients with hypercalcemia of malignancy given drug i.v. for 3-5 d. *1521*

Fenoprofen   In 2 isolated cases of patients with arthritis in the absence of hypertension. *0466*

Flucytosine   May have nephrotoxic effect. *1679*

Food   Average increase of 52% 1.5 to 3.5 h after ingestion of 225 g meat. *2996*

Gentamicin   Nephrotoxic effect. *2220*   10.2% incidence in 49 patients given drug by McHenry method versus 8% in 50 patients given drug by Sawchuk/Zaske method. *2339*   Nephrotoxicity observed in 26% patients treated with drug for sepsis. Mean increase for all population studied 0.4 mg/dL. *3359*   24% incidence of nephrotoxicity but unrelated to initial renal function or prior use of drug, drug concentration, amount given, duration of treatment or concurrent treatment with other drugs. *3182*

Griseofulvin   Rare renal damage reported. *0131*

Hydrochlorothiazide   But by less than 0.1 mg/dL in 33 patients with hypertension given drug for 10 weeks to reduce diastolic BP to less than 90 mm Hg. *3707*

Hydroxyurea   ?related to impaired tubular function. *1680*

Ibuprofen   In 2 patients with systemic lupus erythematosus. *1155*   Increase by 40% in patients with chronic glomerular disease. *0660*

Imipramine   Observed in one psychiatric case. *3156*

Indomethacin   Oliguric renal failure developed in one man during treatment: reversible. *1245*   Associated with increasing risk of renal insufficiency in cirrhosis, nephrotic syndrome, decompensated congestive heart failure and chronic renal disease. *0373*   Associated with decreased secretion of renin and aldosterone, decreased sodium delivery to distal tubule and reduction of urinary flow rate. *1228*

Iohexol   Transient increase at 3rd day after i.v. administration for angiography. *1053*

Iopamidol   Transient increase at 3rd day after i.v. administration for angiography. *1053*

Ipodate   Nephrotoxic effect. *0838*

Kanamycin   Nephrotoxic effect (common but slight). *2220*

Lipomul®   Nephrotoxic effect. *2313*

Lithium   Average increase from 0.94 to 1.08 mg/dL in 237 patients on long term treatment. *3705*   Observed in a few patients but risk of renal insufficiency is remote even in patients given drug for many years. *3203*   In 3 of 97 patients not necessarily attributable to drug. *0699*   In lithium treated patients than healthy subjects. *0519*

Mannitol   Due to dehydration. *2313*

Meals   Affected by meals (most marked in women). *3455*

Methicillin   Nephrotoxic effect. *2313*

Methoxyflurane   Impaired renal tubular function. *2355*

Mithramycin   Nephrotoxic effect. *0071*

Mitomycin   Nephrotoxic effect. *2194*

Moxalactam   In about 2% patients as reported from several studies. *0592*

Nalidixic Acid   May cause nitrogen retention. *2808*

Naproxen   In isolated case of patient with arthritis in the absence of hypertension. *0466*

Neomycin   Nephrotoxic effect. *2808*

Netilmicin   44% increase on average in elderly population with initial clearance of 81 mL/min (increase significant). *1203*

Nifedipine   Acute reversible deterioration of renal function in 4 patients with chronic renal failure. *0898*

Nitrofurantoin   Nephrotoxic effect. *2220*   In 8 of 56 acute and 3 of 22 chronic drug induced pulmonary reactions. *1639*   In 8 of 56 acute and 3 of 22 chronic drug induced pulmonary reactions. *1639*

Norfloxacin   Rare complication of treatment. *3750*

NSAID (Nonsteroidal Anti-Inflammatory Drugs)   Reversible acute renal failure associated with many drugs. *1561*

Ofloxacin   Very infrequent side effect observed during many courses of treatment. *1838*

Oxacillin   Transient azotemia with large doses. *0071*

Oxyphenbutazone   Numerous reports of kidney damage up to acute renal failure following therapeutic use of drug. *2866*

Paraldehyde   Possible nephrotoxicity. *1443*

Paramethadione   Possible nephrotoxicity. *2313*

Paromomycin   Frequently observed renal damage. *0131*

Penicillamine   Possible nephrotoxicity. *1237*

Penicillin   Hypersensitivity reaction or nephropathy. *1343*

Pentamidine   Renal toxicity observed in about 25% patients. *3317*   Renal toxicity possibly due to formation of insoluble precipitates of pentamidine with nucleic acids, also associated with hypovolemia. *3445*

Phenylbutazone   May increase especially if coexisting renal damage. *2427*

Phosphorus   Nephrotoxic effect with necrosis. *1237*

Piperacillin   In 1 of 59 patients given drug as sole agent with many patients with severe illness. *3874*

Piroxicam   Occasional drug induced nephrotoxicity, with isolated azotemia, acute interstitial nephritis or nephrotic syndrome. *2824*   Possible increase due to drug in some patients. *3819*

Polymyxin   Nephrotoxic effect. *2808*

Protein   Occurs after large meat intake. *2313*

Radiographic Agents   Occasional effect following aortography. *0065*   Frequency of renal impairment following CT brain scan with infusion 2.1% compared with 1.3% in control group. *0746*   In normal people 2%: in patients with renal dysfunction 30% subclinical damage following non-renal angiography. *0784*

Ramipril   In response to 10 mg and 20 mg on successive days in 9 patients with severe chronic congestive heart failure. *0764*

Reducing Diet   By up to 2 mg/dL (functional impairment). *3217*

Salsalate   Associated with minimal change nephrotic syndrome in one patient. *3664*

Sex Difference   In males at all ages. *2130*

Sisomicin   38% increase in elderly patients with average initial creatinine clearance of 66 mL/min. *1203*

Streptokinase   Nephrotoxic effect (with tubular damage). *2313*

Streptomycin   Nephrotoxicity may occur in 2%. *2808*   Occasional nephrotoxicity, although less than with other aminoglycosides. *3683*

Suprofen   Occasional renal failure observed but mechanism not known. *0009*

Tetracycline   Nephrotoxicity may cause Fanconi like syndrome. *2808*

Thiazides   Nephrotoxic effect with large doses. *2220*

Tobramycin   In 12% of patients with sepsis. Mean increase of only 0.1 mg/dL in all patients studied. *3359*   12% incidence of nephrotoxicity but unrelated to initial renal function or prior use of aminoglycosides, drug concentration, amount given duration of treatment or concurrent treatment with other drugs. *3182*   18.4% incidence of nephrotoxicity in 49 patients given drug by McHenry method versus 16.7% in 48 patients given drug by Sawchuk/Zaske method. *2339*   In 4 of 59 patients whose drug concentrations were monitored. *1250*   At dose of 4.5 mg/kg/d for 12 d in 90 patients nephrotoxicity observed in up to 39%, reversible in most. *0677*

Triamterene   Nephrotoxic effect (causes reduced GFR). *2313*

Trimethoprim   Increase by 0.2 mg/dL in 21 patients when given with sulfamethoxazole; also when given by itself; probably due to competitive inhibition of tubular secretion mechanism. *0313*   In elderly increased by more than 50% and by 20% in young males. Tubular secretion affected. *1880*   Due to co-trimoxazole component, but effect is slight. *3078*

Trimetrexate   Nephrotoxicity reported in some patients following treatment, possibly associated with prior reduced renal function. *1613*

Vancomycin Occasional renal damage (usually reversible). *0131* Occasional evidence of mild nephrotoxicity. *1571* Nephrotoxicity in 5% of 60 patients given drug alone but much higher incidence when given with aminoglycosides. *2654*

Vasopressin Progressive deterioration in renal function observed in one patient given drug i.v. *0029*

Viomycin Nephrotoxic may cause nitrogen retention. *2313*

Vitamin D Manifestation of hypervitaminosis D. *0071*

Weight Positively correlated in children. *0887*

Zimeldine 23 of approximately 147 patients. *2082*

Serum Decrease Analytical

Ascorbic Acid At concentrations above 1,000 mg/L (maximum serum concentration 34 mg/L) lowered concentration as measured by creatinine amidohydrolase method. *3393*

Bilirubin On kinetic Jaffé methods: related to bilirubin concentration, method and temperature dependent, caused by both conjugated and unconjugated bilirubin, unrelated to creatinine concentration: effects due to oxidation of bilirubin by alkali. *1964*

Cephalothin As measured by Du Pont aca due to decrease of absorbance of product with picrate with time. *2025* Positive interference with kinetic Jaffé reactions in first 45 s, thereafter negative effect with reduced concentration. *2025*

Citrates At concentrations above 1,000 mg/L lowered concentration as measured by creatinine amidohydrolase method. *3393*

EDTA At concentrations above 935 mg/L lowered concentration as measured by creatinine amidohydrolase method. *3393*

Liposol Changed from 22.9 to 21.5 mg/L in 26 randomly selected patient sera, before and after addition on Beckman Astra methods. *2054*

Norfenefrin At concentrations above 4 mg/L (normal therapeutic concentration 0.4 mg/L) lowered concentration as measured by kinetic Jaffé method on BKA-2. *3393*

Phenacemide Positive interference with kinetic Jaffé reaction within 21 s but negative result thereafter. *2025* As measured by Du Pont aca due to decrease of absorbance of product with picrate with time. *2025*

Sodium Oxalate At concentrations above 150 mg/L lowered concentration as measured by kinetic Jaffé method on BKA-2. *3393*

Trizma Present in some quality control materials trizma carbonate interferes with single slide creatinine method on Kodak Ektachem®. *0200*

Serum Decrease Physiological

Cannabis Reported effect. *2220*

Ethanol 2.6% decrease in heavy versus occasional drinkers. *3273*

Ibuprofen Significant reduction in patients receiving medication. *1787*

Meals Significant effect in subjects 20-40 y. *3455*

Urine No Effect Analytical

Storage of Sample Stable indefinitely if frozen. *0904* No effect for 4 to 7 d at room temperature. *1563* For 24 h at room temperature if thymol or toluene added Without preservative at 30° for 4 h. *0904*

Thymol No effect on Jaffé, Van Pilsum methods. *2981*

Urine No Effect Physiological

Alanine Unaffected by alanine administration to fasting obese. *1263*

Amiloride No effect seen in 13 men treated for 8 weeks. *0656*

Azosemide No clear pattern observed. *2047*

Cimetidine No significant effect observed in 9 patients given 1.6 g daily. *0978*

Furosemide No clear pattern observed with drug administration. *2047*

Urine Increase Analytical

Acetone 1 mg equivalent to 0.07 mg in automated Jaffé procedure. *2558*

Ascorbic Acid Acts as reducing agent. *0127*

Asparagine 1 mg equivalent to 0.05 mg creatinine. *2558*

Cefoxitin Falsely high values with Jaffé methods. *0976* When used with Jaffé method on Greiner selective analyzer: concentration related effect. *0061*

Cephalothin Dose related interference with Jaffé procedure on Greiner selective analyzer. *0061*

Fructose 1 mg equivalent to 0.07 mg with automated Jaffé. *2558*

Glycocyamidine 1 mg equivalent to 0.30 mg with automated Jaffé. *2558*

Hippuric Acid 1 mg equivalent to 0.08 mg with automated Jaffé. *2558*

Histidine 1 mg equivalent to 0.03 mg with automated Jaffé. *2558*

Indole 1 mg equivalent to 0.1 mg with automated Jaffé. *2558*

Levodopa Probable action as reducing agent. *0127*

Methyldopa Acts as reducing agent with alkaline picrate. *0757*

Nitrofurans React with color reagent. *2220*

Nitrofurazone React with color reagent. *1237*

Phenolsulfonphthalein Interference with Jaffé procedure. *0652*

Pyruvate 1 mg equivalent to 0.1 mg with automated Jaffé procedure. *2558*

Reserpine No effect at 0.02 mg/dL on SMA 12/60 method. *2636*

Urea 1 mg equivalent to 0.05 mg with automated Jaffé procedure. *2558*

Urine Increase Physiological

Corticosteroids Associated with negative nitrogen balance. *2313*

Diurnal Variation Noted in day after meals lowest at night. *0242*

Fluoxymesterone For up to 2 weeks when treatment stopped. *1680*

Food Increased during day of meat consumption in 4 of 7 healthy volunteers. *2580* Average increase of 19% during 24 h after ingestion of 225 g meat. *2996*

Methandrostenolone May persist up to 2 weeks after treatment. *0019*

Methotrexate Manifestation of nephrotoxicity. *1355*

Oxymetholone Anabolic effect (possible for 2 weeks after stop). *1679*

Proteinuria Reported effect (but reverse may occur). *0298*

Urine Decrease Physiological

Altitude Associated with decreased creatinine clearance. *2957*

Anabolic Steroids Anabolic effect. *1237*

Androgens Anabolic effect. *1022*

Nandrolone Anabolic effect. *2681*

Prednisone Excretion less than anticipated in comparison with controls. *1656*

Thiazides Clearance tests decreased by 10-20%. *2426*

Vegetarian Diet Finding in adults protracted vegetarianism. *3727*

Test Conditions Increase Analytical

Creatinine Positive spot test with Jaffé reagent. *3765*

Ethylmethylketone Positive spot test with Jaffé reagent. *3765*

Homogentisic Acid Positive spot test with Jaffé reagent. *3765*

5-Hydroxytryptamine Positive spot test with Jaffé reagent. *3765*

Levarterenol Positive spot test with Jaffé reagent. *3765*

Creatinine Clearance

Serum Decrease Physiological

Prednisone In comparison with controls probably due to decreased muscle mass in drug treated patients in relation to total body weight. *1656*

Urine No Effect Physiological

Acetaminophen Without overdose not usually any effect. *2866*

Cellulose Phosphate No significant effect observed. *2710*

Cyclophosphamide In majority of patients although decreased urine volume. With decreased sodium excretion in one patient. *0854*

Disopyramide No effect when drug given to total of 300 or 600 mg daily. *2992*

Creatinine Clearance *(continued)*

Urine No Effect Physiological *(continued)*

Doxazosin No effect on clearance or on renal blood flow in 24 patients with 6 weeks treatment. *3852*

Enalapril In 53 patients given up to 160 mg daily or when combined with hydrochlorothiazide. *2277*

Etodolac No cumulative effect observed over 24 h in individuals with normal renal function or with renal insufficiency when given 500 mg bid for 4 d. *0453*

Food No effect in individuals in response to ingestion of a cooked meat meal. *2996*

Gallium Nitrate In two patients with hypercalcemia due to parathyroid carcinoma. *3764*

Iron Dextran No effect if prior normal renal function. *1678*

Methoxyflurane No effect observed with high or low concentrations. *1380*

Muzolimine Not changed in 10 hypertensive patients given 30 mg daily for 16 weeks. *1229*

Naproxen When given for 14 d to patients with rheumatoid arthritis and heart failure. *3514*

SC-16102 No effect noted during study. *1804*

Sodium Etidronate No effect of drug alone. *2931*

Sulindac When given for 14 d to patients with rheumatoid arthritis and heart failure. Marked effect. *3514* No significant change in patients with chronic glomerular disease. *0660*

Urine Increase Analytical

Cefoxitin Artifactually increased for more than 4 h after 2 g drug given i.v. *0976*

Levodopa Reducing properties affect Jaffé method. *1136*

Urine Increase Physiological

Furosemide If given intravenously especially. *2818*

Hydration In patients with proteinuria. *2818*

Isosorbide ?due to decreased tubular reabsorption. *2605*

Methylprednisolone Varies with inulin clearance. *3774*

Muscular Exercise 40% decrease with severe exercise Mild, walking at 5.6 km/h produces 20% increase. *3217*

Nifedipine After 10 mg sublingually in next 2 h. *2917*

Proteinuria Compared with inulin clearance (but reverse may occur). *0298*

Sleep Significant effect observed. *1190*

Torasemide Similar effects to that of furosemide. *0919*

Urine Decrease Analytical

Cefoxitin If measured with Jaffé procedure at peak drug concentration, otherwise value will be low, especially near trough concentration. *0061*

Cephalothin If Jaffé procedure used and plasma concentration of drug is near peak; values low if drug concentration low. *0061*

Urine Decrease Physiological

Altitude Possibly due to altered renal blood flow. *2957*

Amphotericin B Nephrotoxicity effect (decrease up to 36%). *0551* Fell to 69 mL/min from 94 mL/min in 10 patients after 2 weeks. Remained at this value with continued treatment. *0240*

Aspirin As a consequence of increased serum concentrations. *0545* Significant effect correlated with plasma salicylate concentration. *2535* Observed even with therapeutic doses. *2866*

Bunamiodyl At dose of 4.5 g; occurred without liver effect. *3796*

Cannabis Temporary decrease noted. *1633*

Carbenoxolone Due to nephropathy. *2458*

Chlorothiazide Reported effect. *2220*

Chlorpropamide In isolated case with diabetes who developed proliferative glomerulonephritis indicating immunologically mediated reaction. *0146*

Chlorthalidone Significantly reduced in 10 hypertensive patients given 25 mg daily for 16 weeks. *1229*

Cimetidine By at least 20% in patients with renal failure maximal after 2 to 3 d. *2085* In 13 ulcer patients average fall of 28 ml/min (26%). *0649* Significant effect within 6 h, normalized after several weeks. *0978*

Cisplatin Observed in one patient, but major side effect renal toxicity. *1194*

Co-Trimoxazole Usually reversible effect (in many patients). *1852*

Dehydration Following overnight dehydration. *2865*

1,2-Diaminopropane In 11 patients who showed glomerular injury including minimal change nephrotic syndrome with a membranous pattern of immune complex deposition as well as other patterns of deposition. *1170*

Diazoxide Effect over 2 h of 4 mg/kg given i.v. *1805*

Diurnal Variation Significant effect with fall at night. *1190*

Ethambutol Rare case of renal damage reported. *0131*

Fenoprofen Rare case of acute tubulointerstitial nephritis with acute renal failure (probable association with drug administration). *3582*

Gold In 11 patients who showed glomerular injury, including minimal change nephrotic syndrome with a membranous pattern of immune complex deposition as well as other patterns of deposition. *1170*

Griseofulvin Rare renal damage reported. *0131*

Guancydine Due to decreased renal blood flow. *2132*

Heroin Observed in some severe cases. *2914*

Hydrochlorothiazide Reported effect. *2220*

Ibuprofen Increase by 28% in patients with chronic glomerular disease but no effect in healthy people. *0660*

Indomethacin Reduced by about 50% in states of diminished circulatory blood volume. *3261*

Iodoalphionic Acid Reported to cause renal failure. *0065*

Iopanoic Acid Reported cause of acute renal failure. *0065*

Lithium Slight decrease: significant negative regression on serum lithium concentration with long-term treatment. *3705*

Lysergide Temporary effect observed. *1633*

Mitomycin Due to nephrotoxicity. *2194*

Muscular Exercise Decrease with heavy exercise. *1842*

Ofloxacin Very infrequent side effect observed during many courses of treatment. *1838*

Oxyphenbutazone Numerous reports of kidney damage up to acute renal failure following therapeutic use of drug. *2866*

Paromomycin Frequently observed renal damage. *0131*

Spectinomycin Mechanism not discussed. *0122*

Starvation Falls by 15% in first 4 d. *1242*

Sulfamethoxazole When given with trimethoprim, fever and malaise: successful response to treatment. *1969*

Sulindac Small but significant effect with treatment for 9 d in 9 patients with stable renal insufficiency. *2457*

Thiazides May cause decrease by up to 20%. *2426*

Triamterene Probably reduced renal blood flow. *3745*

Trimetrexate Nephrotoxicity reported in some patients following treatment, possibly associated with prior reduced renal function. *1613*

Vancomycin Occasional renal damage (usually reversible). *0131*

Viomycin Occasional renal damage observed. *0131*

Cryofibrinogen

Plasma Increase Physiological

Oral Contraceptives Incidence much higher than in controls. *2385*

Crystals

Urine Increase Analytical

Acetazolamide Presence of drug. *1022*

Ampicillin Presence of drug maximal at pH 5. *2856*

Formaldehyde Urea precipitation. *2313*

Methotrexate Crystals of unknown identity (probably drug). *0979*

Thiabendazole Low solubility of drug, especially at neutral pH. *2413*

Urine   Increase   Physiological

Acetylsulfadiazine   Wheatsheaves eccentric binding (not in alkaline urine). *0443*

Acyclovir   Especially with high-dose bolus infusion and with dehydration and pre-existing renal insufficiency. *0936*

Ascorbic Acid   Acidification may precipitate oxalates, urates, cystine. *2063*

Azauridine   May occur due to response to tissue destruction. *0278*

Bilirubin   Red-brown amorphous needles in acid pH. *0443*

Calcium Carbonate   Small colorless dumbbells/spheres (not in acid). *0443*

Cholesterol   Flat colorless plates with notch (acid, neutral pH). *0443*

Ciprofloxacin   Observed in 4 of 2829 patients without change in renal function. *0151*

Co-Trimoxazole   Single case reported. *1852*   May occur with high doses of drug, particularly in patients with severe renal insufficiency. *0679*

Cystine   Colorless, hexagonal, flat (in acid). *0443*

Diatrizoic acid   Colorless, thin, rhombic, may be notch (in acid). *0443*

Ethylene Glycol   Oxalate crystals observed in poisoning. *3325*

Hematin   Small, biconcave, whetstone (in acid). *0443*

Hemosiderin   Clumps golden brown granules (in acid and neutral). *0443*

Hippuric Acid   Colorless needles, rhombic plates, prisms (all pH). *0443*

Indigotin   Amorphous blue or small crystals (all pH). *0443*

Leucine   Yellow spheroids with radial striation (in acid). *0443*

Mannose   Massive uric acid crystalluria after infusion. *0358*

Mercaptopurine   Direct renal damage with doses over 750 mg/sq m. *0980*

Methenamine   Mandelate may occasionally cause crystalluria. *1343*

Norfloxacin   May occur with high doses of drug. *3750*

Oxalate   Oxalate crystals present in urine in poisoning. *1302*

Oxyphenbutazone   May cause crystallization of uric acid. *0019*

Phosphates   Ammonium magnesium or Calcium hydrogen (soluble in acid). *0443*

Primidone   In acute poisoning case crystals = primidone. *0205*

Radiographic Agents   Diatrizoate may produce crystals in acid urine. *0443*

Rhubarb   Calcium oxalate may be deposited. *0152*

Sulfadiazine   Low solubility in acid urine. *2808*

Sulfamerazine   Presence of drug. *0071*

Sulfamethoxazole   Low solubility in acid urine. *2808*

Sulfisoxazole   May occur particularly in acidic urine. *3832*

Sulfonamides   Presence of drug. *2313*

Sulthiame   In poisoning identified as pure drug. *2427*

Triamterene   Either free or as part on conglomerations and also in large round brown bodies; common at pH <6.0. *1070*

Triethylenemelamine   Due to crystalization out of drug. *1680*

Tyrosine   Colorless or yellow needles in sheaves, rosettes. *0443*

Urates   Ca, Mg, K=yellow, NH_4 = brown, sodium=colorless. *0443*

Uric Acid   Yellow, red-brown, many types in acid. *0443*

Xanthine   Rare, colorless, rhombic plates (not in alkaline urine). *0443*

Urine   Positive   Physiological

Trimethoprim   May occur with high doses especially if severe renal insufficiency. *0678*

Cyanmethemoglobin

Blood   Increase   Physiological

Hydrocyanic Acid   Reacts with methemoglobin. *1343*

Cyclic Adenosine Monophosphate

Plasma   No Effect   Physiological

Epinephrine   No effect unless beta-blocking agent also given. *0216*

Furosemide   In 8 normal subjects associated with secondary hyperparathyroidism. *1214*

Plasma   Increase   Physiological

Caffeine   Inhibits phosphodiesterase increases lipolysis. *1529*

Epinephrine   Normal response after subcutaneous injection. *0321*   Response to i.v. infusion in normals. *0216*

Ethanol   At upper limit of normal in 8 men who had abused alcohol for 10 y. *0347*

Glucagon   Normal response after subcutaneous injection. *0321*

Isoproterenol   No effect unless beta-blocking agent also given   Response to i.v. infusion in normals. *0216*

Levarterenol   Response to i.v. infusion in normals No effect unless β-blocking agent also given. *0216*

Probenecid   In 8 healthy young men. Probable effect on carrier mediated process to clear plasma cyclic AMP. *1318*

Urine   No Effect   Physiological

Aluminum Hydroxide   Observed in one patient secondary to ingestion of large amounts of compound for long time. *1313*

Diurnal Variation   No diurnal rhythm observed. *1992*

Ethinyl Estradiol   In post-menopausal women with 50 µg/d for 14 d. *3474*

Probenecid   In 8 healthy young men. *1318*

Recumbency   Posture has no effect. *1992*

Sodium Etidronate   No effect of drug alone. *2931*

Theophylline   In normals given 400 mg bid for 5 d. *2383*

Urine Volume   No significant effect with change in volume. *2691*

Urine   Increase   Physiological

Electroshock   Mean change of from 4.2 µmol/24 h to 14.2 µmol/24 h. *1468*

Epinephrine   2 fold increase in clearance produced. *0321*   Effect less marked than in blood. *0216*

Furosemide   In 8 normal subjects associated with secondary hyperparathyroidism. *1214*

Glucagon   4 fold increase in clearance produced. *0321*

Isoproterenol   Effect less marked than in blood. *0216*

Levarterenol   Effect less marked than in blood. *0216*

Muscular Exercise   Modest increase in normal people. *0747*

Phenytoin   In epileptic children in comparison with controls. *2030*

Urine   Decrease   Physiological

Erect Posture   In normals but increased in hypertensives. *1471*

Isoproterenol   In normals but increased in hypertensives. *1471*

Platelets   Decrease   Physiological

Cimetidine   Probable activation of endogenous cyclic AMP phosphodiesterase involved in favoring action of platelet aggregating agents. *3938*

Cyclosporine

Serum   No Effect   Physiological

Itraconazole   No effect reported on metabolism. *3567*

Nifedipine   No effect on concentration when drug added (20-30 mg/d) to therapeutic regime. *3732*

Serum   Increase   Analytical

Plastic   Spurious peak near cyclosporines A and D when HPLC methods used: observed with Sarstedt Monovette tubes. *2269*

Serum   Increase   Physiological

Cimetidine   Increased concentration presumably due to effect on liver metabolism. *0683*

Cyclosporine (continued)

Serum Increase Physiological (continued)

Diltiazem Probable interference with demethylation and binding to cytochrome P-450. *3732*

Erythromycin One case reported in which marked increase in serum concentration occurred after administration of erythromycin possibly due to inhibition of hepatic clearance. *2881* Because of competition for protein binding sites in serum. *3864*

Ketoconazole Prolongs half-life, probably by competing for metabolizing enzymes. *3809*

Methylprednisolone High doses produce effect probably due to effect on hepatic enzymes. *0683*

Nicardine Doubling of concentration (trough levels) when added to therapeutic regime. *3732*

Sulfamethoxazole Reversible deterioration in renal function with effect on tubular function and nephrotoxicity. *3732*

Trimethoprim Reversible deterioration of renal function with effect on tubular function and nephrotoxicity. *3732*

Serum Decrease Physiological

Carbamazepine Increases hepatic metabolism with rate of hydroxylation and elimination. *3732*

Isoniazid Enhances metabolism by hepatic enzyme induction. *3732*

Phenytoin Marked lowering of serum or blood concentration, presumably due to hepatic enzyme induction. *0683*

Rifampin Marked reduction of concentration due to hepatic enzyme induction. *0683*

Sulfadimidine Reduction when given i.v.; not seen if given orally: mechanism not clear. *3732*

Trimethoprim Marked reduction when drug given with sulfadimidine i.v. due to effect on hepatic metabolism. *0683*

Blood Increase Physiological

Ketoconazole Inhibits function of cytochrome P-450 hepatic enzymes inhibiting clearance of cyclosporine. *1113*

Methyltestosterone One case reported in which administration caused marked increase in blood cyclosporine possibly by inhibition of cytochrome P-450 hepatic enzymes. *2471*

Metoclopramide Coadministered drug causes increased absorption, area under curve and blood concentration. *3732*

Blood Decrease Physiological

Phenobarbital Induces cytochrome P-450 hepatic enzymes thereby increasing clearance of cyclosporine. *0533*

Phenytoin Induces cytochrome P-450 hepatic enzymes thereby increasing clearance of cyclosporine. *1186*

Rifampin Induces cytochrome P-450 hepatic enzymes thereby increasing clearance of cyclosporine. *3668*

Cystathionine

Urine Increase Physiological

Vitamin B$_6$ Depletion Direct correlation between increased excretion and decreased diet. *0148*

Cysteine

Plasma Increase Analytical

Iodoacetate Effect of oxidizing agent (added *in vitro*). *0384*

Cystine

Plasma Decrease Analytical

Hemolysis Due to binding with protein or dilution. *3230*

Urine No Effect Analytical

Storage of Sample 1 week room temperature with HCl, indefinite time if frozen With chloroform added, in refrigerator for 3 mo. *1712*

Urine Increase Physiological

Cycloleucine Reversible marked aminoaciduria. *0499*

Histidine Increased clearance, plasma concentration unchanged. *1644*

Penicillamine If cystinuria. *0071*

Urine Decrease Physiological

Ascorbic Acid In 10 healthy females given 10 g/d. *3624*

Cystine Aminopeptidase

Serum Increase Physiological

Pregnancy Steady increase from 18 to 40 weeks. *0778*

Dapsone

Serum No Effect Analytical

Amitriptyline At 10 mg/dL on colorimetric procedure of Higgins. *1590*

Amobarbital At 5 mg/dL on colorimetric procedure of Higgins. *1590*

Amphetamine At 10 mg/dL on colorimetric procedure of Higgins. *1590*

Ascorbic Acid At 5 mg/dL on colorimetric procedure of Higgins. *1590*

Chlordiazepoxide At 5 mg/dL on colorimetric procedure of Higgins. *1590*

Diazepam At 10 mg/dL on colorimetric procedure of Higgins. *1590*

Ethchlorvynol At 10 mg/dL on colorimetric procedure of Higgins. *1590*

Glutethimide At 10 mg/dL on colorimetric procedure of Higgins. *1590*

Imipramine At 10 mg/dL on colorimetric procedure of Higgins. *1590*

Meprobamate At 10 mg/dL on colorimetric procedure of Higgins. *1590*

Methaqualone At 5 mg/dL on colorimetric procedure of Higgins. *1590*

Pentazocine At 10 mg/dL on colorimetric procedure of Higgins. *1590*

Phenindamine At 10 mg/dL on colorimetric procedure of Higgins. *1590*

Phenmetrazine At 10 mg/dL on colorimetric procedure of Higgins. *1590*

Phenobarbital At 10 mg/dL on colorimetric procedure of Higgins. *1590*

Phenylephrine At 10 mg/dL on colorimetric procedure of Higgins. *1590*

Phenytoin At 10 mg/dL on colorimetric procedure of Higgins. *1590*

Salicylate At 30 mg/dL on colorimetric procedure of Higgins. *1590*

Sulfapyridine At 5 mg/dL on colorimetric procedure of Higgins. *1590*

Trifluoperazine At 10 mg/dL on colorimetric procedure of Higgins. *1590*

Serum Increase Analytical

Procainamide Develop color in procedure of Higgins. *1590*

Sulfanilamide Develops color in procedure of Higgins. *1590*

Serum Increase Physiological

Probenecid 50% increase after 4 h, inhibits excretion. *1487*

Serum Decrease Physiological

Rifampin Increased plasma clearance, but interaction may not be clinically significant. *0014*

DDT (Chlorophenothane)

Serum   Decrease   Physiological
Phenytoin Reported interaction due to alteration of metabolism. *2042*

Dehydroascorbic Acid

Serum   Increase   Physiological
Menstruation At mid cycle, low at ends. *2998*

7-Dehydrocholesterol

Serum   Increase   Physiological
Boxidine Inhibits transformation to cholesterol. *1220*

Dehydroepiandrosterone (DHEA)

Plasma   No Effect   Analytical
Storage of Sample No effect at -20° for 1 y. *0107*

Plasma   No Effect   Physiological
Cyproterone In 27 women treated for total of 194 cycles with 50 μg ethinyl estradiol and 2.0 mg cyproterone acetate. *2280*
Digoxin No significant effect with long-term administration. *2576*
Obesity In females but falls with caloric restriction. *3391*
Omeprazole No effect observed in 8 volunteers given 30 mg/d for 28 d. *2516*

Plasma   Increase   Physiological
Clomiphene Liberates LH, anti-steroid hormone effect of drug. *0609*
Corticotropin Hormonal effect. *2435*

Plasma   Decrease   Physiological
Danazol Significant effect at 8,12 weeks after 600 mg daily in 18 women with endometriosis. *0591*
Testosterone In power athletes with 26 weeks on steroid self administration. *3088*

Urine   No Effect   Analytical
Ampicillin No effect on GLC method. *3622*

Urine   Increase   Physiological
Corticotropin Normal response to 2 d ACTH test. *1474*

Urine   Decrease   Physiological
Ampicillin Increase by 75% after 6 d administration in pregnant women. *3622*
Dexamethasone Suppression of ACTH. *2420*
Oral Contraceptives Compared with controls — details not discussed. *0527*

Dehydroepiandrosterone Sulfate (DHEA-S)

Plasma   No Effect   Physiological
Cyclosporine In 16 patients who developed hypertrichosis but also given added cortisone. *3190*
Danazol No consistent change noted in patients with prostatic cancer. *0693*
Metoclopramide No significant differences in either cycling or post-menopausal woman receiving drug. *0096*
Oxytetracycline Despite decreased urine excretion and increased fecal excretion of estrogens due to decreased hydrolysis by beta-glucuronidase in gastrointestinal tract. *1469*

Plasma   Increase   Physiological
Clomiphene Liberates LH, anti-steroid hormone effect of drug. *0609*

Danazol Significant effect at 2,4 and 8 weeks after 600 mg daily in 18 women with endometriosis. *0591* Increase by 40% in 7 normal subjects given drug 800 mg daily. *2224*
Dehydroepiandrosterone Mean increase of 2.5 to 3.5 fold in 5 normal men given 1600 mg/d orally for 28 d. *2578*

Plasma   Decrease   Physiological
Aging In women from 160 μg/dL at 25 y to 60 at 55 y. *0462*
Carbamazepine Significant reduction in women treated with drug compared with control untreated epileptics. Same effect observed when given in combination with phenytoin. *2141* Within 7 d of starting 400 mg/d treatment in 6 healthy males probably due to induction of hepatic monooxygenase activity. *0707*
Cyproterone In 27 women treated for total of 194 cycles with 50 μg ethinyl estradiol and 2.0 mg cyproterone acetate. *2280*
Dexamethasone Suppressed by treatment in hirsute women from mean of 8.6 to 3.4 μmol/L. *0776*
Oral Contraceptives Usually reduced in response to most regimes. *2280*
Phenytoin Significant reduction in both men and women compared with untreated epileptics. Same effect observed when combined with carbamazepine. *2141*

Urine   No Effect   Physiological
Metoclopramide No significant influence of drug on excretion in menstruating women. *0096*

Demethylchlortetracycline

Serum   Decrease   Physiological
Food Absorption reduced when taken with food. *3024*

Deoxycholic Acid

Serum   No Effect   Physiological
Oral Contraceptives No effect observed over 12 mo in 29 women given combination of ethinyl estradiol and norgestrel. *3445*
Phenobarbital No effect on increased concentration in patients with intrahepatic cholestasis increased of pregnancy. *1545*

Serum   Increase   Physiological
Rifampin Significant increase possibly due to blocking uptake by plasma membrane of hepatocytes. *1224*

Feces   Decrease   Physiological
Neomycin Due to alteration of intestinal flora. *1077*

Deoxycorticosterone

Plasma   No Effect   Physiological
Amiloride In 5 normal subjects given 75 mg daily for 7 d. *2440*
Metoclopramide No effect even though other constituents affected by i.v. bolus of 10 mg. *2543*

Plasma   Increase   Physiological
Pregnancy Increase from 23 weeks, maximum at term. *0498*
Spironolactone In 5 normal subjects given 300 mg daily for 7 d. *2440*

Urine   No Effect   Physiological
Metoprolol In 15 patients with essential hypertension treated for 4 weeks, associated with reduction in sympathetic tone and reduced activity of renin-aldosterone system. *1209*

11-Deoxycorticosterone

Plasma   No Effect   Physiological
Heparin No effect on components of aldosterone biosynthetic pathway. *2638* No effect noted in spite of effect on aldosterone. *2638*

11-Deoxycorticosterone *(continued)*

Plasma   Increase   Physiological
Etomidate   Clear effect 3.5 h after induction demonstrating inhibition of 11β-hydroxylation of glucocorticoid and mineralocorticoid intermediates. *0063*

11-Deoxycortisol

Plasma   Increase   Physiological
Etomidate   Clear effect 3.5 h after induction demonstrating inhibition of 11β-hydroxylation of glucocorticoid and mineralocorticoid intermediates. *0063*
Metyrapone   Normal response to test. *0512*

Plasma   Decrease   Physiological
Megestrol   In 18 postmenopausal women with breast cancer. *0056*

Urine   Increase   Physiological
Corticotropin   Progressive increase with repeated injection. *1982*

Deoxycytidine

Urine   Increase   Physiological
X-Ray Therapy   Due to tissue destruction. *2378*

Desmethylcotinine

Urine   Positive   Physiological
Smoking   Normal nicotine metabolite 5-20 cigarettes/d. *0324*

Desmosterol

Serum   Increase   Physiological
Haloperidol   Further metabolism inhibited so accumulates. *1343*
Trifluperidol   Further metabolism inhibited so accumulates. *1343*

Dexamethasone

Serum   No Effect   Physiological
Theophylline   No effect on hepatic metabolism so concentration unaffected. *1830*

Serum   Increase   Analytical
Estradiol   Very slight cross reactivity procedure of Hichens. *1585*
Hydrocortisone   Very slight cross react procedure of Hichens. *1585*
Testosterone   Very slight cross reactivity procedure of Hichens. *1585*

Serum   Decrease   Physiological
Ephedrine   Metabolic clearance enhanced and conjugated fraction in urine increased. *1830*
Phenobarbital   88% increase in metabolic clearance rate with substantial reduction in half-life. *1830*
Phenytoin   Due to accelerated hepatic clearance of steroid: may given false impression of Cushing's syndrome. *1830*

Dexamethasone Suppression

Patient   No Effect   Physiological
Chlorophenothane   Normal in subjects with occupational exposure. *0674*

Patient   Abnormal   Physiological
Phenytoin   Alters steroid metabolism. *1488*

Diagnex Blue Excretion

Urine   No Effect   Analytical
Storage of Sample   Stable 48 h at room temperature, 10 d at 4°. *1679*

Urine   Increase   Analytical
Methylene Blue   Detection of methylene blue. *3879*
Phenazopyridine   Orange color produces interference. *3879*
Potassium Salts   Displacement of diagnex blue from resin. *3879*
Quinacrine   Release of dye from resin. *1238*
Quinidine   Release of dye from resin. *1238*
Quinine   Release of dye from resin. *2425*
Riboflavin   Interfering color. *3879*
Sodium Salts   Displacement of diagnex blue from resin. *3879*
Triamterene   Increased dye release from resin. *1237*
Vitamin B Complex   Release of dye from resin. *1238*

Urine   Increase   Physiological
Aluminum Salts   Heavy metal displacement of diagnex blue. *3879*
Barium   Heavy metal displacement of diagnex blue. *3879*
Calcium Salts   Heavy metal displacement from resin. *2313*
Iron Salts   Heavy metal displacement of diagnex blue. *3879*
Kaolin   Displacement of diagnex blue from resin. *3879*
Magnesium Salts   Heavy metal displacement of diagnex blue. *3879*
Niacin   Displaces diagnex blue from resin. *3879*
Sulfisoxazole   Displacement of diagnex blue from resin. *3879*

Urine   Decrease   Physiological
Caffeine   As Na benzoate salt; low gastric acidity. *1022*

Dialyzable Free Thyroxine

Serum   Increase   Physiological
Halofenate   Lowers binding to thyroxine binding globulin and albumin. *0831*

Diamine Oxidase

Serum   Increase   Physiological
Pregnancy   Reaches maximum at 24 weeks, then slight fall only. *0595*

Diatrizoate Clearance

Urine   Decrease   Physiological
Aspirin   Observed even with therapeutic doses. *2866*

Diazepam

Serum   No Effect   Analytical
Amitriptyline   On cathode-ray Polarographic method of Berry. *0323*
Chlorpromazine   On cathode-ray Polarographic method of Berry. *0323*
Dichloralphenazone   On cathode-ray Polarographic method of Berry. *0323*
Glutethimide   On cathode-ray Polarographic method of Berry. *0323*
Imipramine   On cathode-ray Polarographic method of Berry. *0323*
Meprobamate   On cathode-ray Polarographic method of Berry. *0323*
Methaqualone   On cathode-ray Polarographic method of Berry. *0323*
Thioridazine   On cathode-ray Polarographic method of Berry. *0323*

Serum   No Effect   Physiological
Ranitidine   Steady-state plasma concentration, clearance and elimination half-life not affected. *1938*

Serum   Increase   Physiological
Cimetidine   40-50% increase due to reduction of total body clearance. *1962* Impaired clearance, increasing half-life by 40%. *3084*
Valproic Acid   Displaces diazepam from plasma protein binding sites and inhibits its metabolism. *0896*

Serum   Decrease   Physiological
Antacids   Delayed absorption due to adsorption. *3794*

Urine   Positive   Analytical
Diethylpropion   Similar R_f and color reaction on TLC using ethyl acetate: methanol: water: ammonium hydroxide and modified Dragendorff's reagent. *3868*

Dicumarol

Plasma   Increase   Physiological
Chloramphenicol   Inhibits hepatic microsomal metabolizing enzymes thus prolonging half-life. *3864*
Magnesium Hydroxide   Increased absorption due to chelation. *3794*

Plasma   Decrease   Physiological
Phenytoin   With long-term therapy: in one study decreased from 20 to 5 μg/mL. *3556*

Diffusible Phosphate

Serum   Increase   Physiological
Sodium Etidronate   Increases tubular reabsorption, no effect on PTH action. *2931*

Digitoxin

Serum   No Effect   Physiological
Ampicillin   Insignificant increase in elimination half-life. *2223*
Spironolactone   May depress level but not to subtherapeutic value due to induction of mixed function oxidases. *2794*

Serum   Increase   Physiological
Cimetidine   Reported to increase concentration, although effect on metabolism uncertain. *0548*

Serum   Decrease   Physiological
Cholestyramine   Reduces bioavailability in gastrointestinal tract due to binding of drug. *2794* Increased elimination rate with interrupted enterohepatic circulation. *3794*
Colestipol   Reduced bioavailability due to binding in gastrointestinal tract. *2794*
Phenobarbital   Stimulates metabolism, induces hepatic microsomal enzymes. *1487* May depress to subtherapeutic value due to induction of mixed function oxidases. *2794*
Phenylbutazone   Due to induction of mixed function oxidases which may reduce to subtherapeutic level. *2794*
Phenytoin   Average decrease from 25 to 10 μg/mL (DPH dose 900 mg/d). *2041* May be reduced to subtherapeutic concentration due to induction of mixed function oxidases. *2794*
Rifampin   May be reduced to subtherapeutic concentration by induction of mixed function oxidases. *2794* Decreased serum concentration due to hepatic enzyme induction. *0014*

Digoxin

Serum   No Effect   Analytical
Amrinone   No effect on Du Pont aca method. *2777*
Diflunisal   No effect on FETI methods of Syva® advance. *1687*

Fluosol-DA   As measured by Abbott TD_x at concentration up to 50%. *2518*
Hemolysis   If dioxane based cocktail used in assay. *3823*

Serum   No Effect   Physiological
Ajmaline   No effect on serum concentration reported. *0344*
Disopyramide   No significant effect after 3 100 mg doses per day. Slight increase noted when daily dose doubled. *2992*
Exchange Transfusion   No effect observed. *0700*
Imipramine   No effect on serum concentration reported. *0344*
Lidoflazine   Reportedly no effect on serum concentration. *0344*
Mexiletine   No significant effect when drug coadministered. *2101*
Procainamide   No significant effect when drug coadministered. *2101*
Spartene   0.8 g/d had no effect on serum concentration. *0282*

Serum   Increase   Analytical
Amiloride   At 50 ng/mL equals 0.6 ng/mL by RIA. *3940*
Amrinone   Positive interference with TD_x method. *2777*
Bilirubin   May affect radioimmunoassay if beta count. *0615*
Canrenoate Potassium   At normal concentrations in serum if no preincubation. *2804*
Digitoxin   At normal concentrations in serum if no preincubation (RIA). *2804* Due to cross-reactivity (RIA) if given i.m. recently. *2033*
Hemolysis   May affect radioimmunoassay with beta count. *0615*
Methylprednisolone   At 50 ng/mL equals 0.2 ng/mL by RIA. *3940*
Ouabain   Reported to affect RIA methods. *3940*
Phenytoin   At 10 mg/mL equals 0.2 ng/mL by RIA. *3940*
Prednisolone   At 50 ng/mL equals 0.5 ng/mL by RIA. *3940* At normal concentrations in serum if no preincubation. *2804*
Prednisone   At normal concentrations in serum if no preincubation. *2804* At 50 ng/mL equals 0.2 ng/mL by RIA. *3940*
Proscillaridin   Reported to affect RIA methods. *3940*
Spironolactone   At normal concentrations in serum if no preincubation. *2804* At 250 ng/mL equals 1.4 ng/mL by RIA. *3940* Cross-reactivity possibly with metabolites with antibodies observed with several radioimmunoassay kits. *3577*
Triamterene   At 500 ng/mL equals 0.3 ng/mL by RIA. *3940*
Uremia   May be affected if beta counting used. *0615*

Serum   Increase   Physiological
Amiodarone   Approximate doubling of effect due to inhibition of metabolism. *2011* Average increase of 69% with possible toxicity but mechanism not yet established. *2794* Serum concentration increased 68 to 800% in presence of normal serum creatinine and urea N, possibly due to inhibited tubular secretion of drug. *1993* Occurs with concomitant administration: mean increase of 280% in four patients, allowing 50%. *1394* Concentration increased within 24 h of dosing: magnitude of interaction dose related. *3950* Following treatment with amiodarone for 1 week in 6 volunteers: serum maximum and area under curve increased. *2293*
Cyclosporine   Higher concentration after digoxin administered for 4 d possibly due to diminished renal clearance. *3732*
Diltiazem   Increased mean trough concentration from 1.11 ng/mL to 1.54 ng/mL after 3 d coadministration in 11 patients with congestive cardiac failure. *0108* Increased plasma concentration after single oral dose, or for 1 week in 6 healthy subjects. Renal clearance decreased. *3917*
Disopyramide   Mean change for 1.3 to 1.5 nmol/L but clinically unimportant: disopyramide concentrations above therapeutic range when this effect noted. *2291*
Erythromycin   Two fold increase noted in some individuals when antibiotic given orally. In 10% patients but bacteria convert digoxin to cardioinactive reduced metabolites. *2179*
Flecainide   Increases serum concentration. *2011*
Gallopamil   16% increase when concomitantly administered with drug. *0283*
Ibuprofen   Significant increase after 7 d treatment with average effect of 59%. *2896*
Nifedipine   45% increase in volunteers given both drugs compared with digoxin alone, but mechanism uncertain. *0281* Increase by 15% in patients to whom 20 mg bid were added to a stable digoxin regime. *1957*

Digoxin (continued)

Serum Increase Physiological (continued)
Nisoldipine Increased trough values by 15%. *1939*

Propafenone 37% increase when drugs coadministered due to decreased renal clearance. *0283*

Propantheline Improves absorption, decreases gastrointestinal motility activity. *2290* With tablets of low bioavailability due to reduction of bowel motility. *2794* Increased absorption with augmented dissolution. *3794*

Quinidine Clearance reduced by coadministration of quinidine: effect further augmented by addition of spironolactone to regime. *1110* Observed in 7 of 9 patients with mean concentration changing from 1.43 to 2.61 nmol/L, probably due to displacement of quinidine from binding sites in tissues and reduced renal clearance. *2291* 20 to 330% increase after 3 d of quinidine in 17 patients. *0790* When coadministered with quinidine: so effect of 0.2 mg with quinidine was comparable to that of 0.4 mg without. *0282* Approximate 2-fold increase due to decreased total body clearance but renal negligible and not affected by chronic renal failure. *1111* Absorption rate constant increased, with decreased lag time and peak time. Systemic availability of digoxin increased from 68% to 79%, but no effect on biotransformation. *2763* Volume of distribution and nonrenal clearance increased, half-time of elimination greatly increased but total clearance and renal clearance greatly decreased. *0344* 2.5 fold increase with more than 50% decrease in renal clearance. *0920*

Quinine Interaction may produce 50% increase in concentration. *0548* Stepwise increase with increasing quinine dose due to impairment of extrarenal digoxin clearance. *2764* Total body clearance reduced with increase of renal elimination half-life. Urine excretion increased. *3758*

Spironolactone Increased concentration when given with spironolactone than when given alone. *3736* Clearance reduced by coadministration of spironolactone: effect more marked when quinidine also administered: renal tubular secretion of digoxin inhibited. *1110* Drug-induced decrease in renal clearance observed. *3577*

Triamterene Increased concentration when given in association with triamterene than when given alone. *3736*

Verapamil Average increase from 0.96 to 1.63 ng/mL in 41 patients when given 240 mg/d. *1952* Coadministration caused decrease of total body clearance and increased plasma half-life from 33.5 to 41.3 h. *2765*

Serum Decrease Analytical
Hemoglobin Estimated by radioimmunoassay unless bleached. *2805*

Storage of Sample 5% 2 d room temperature, 2% 5 d at 4°, ok frozen 1 week. *0615*

Serum Decrease Physiological
Amiloride Mean clearance increased due to increased tubular secretion of digoxin. *3737*

Antacids Reduced absorption due to adsorption and faster gastric emptying. *3794*

Charcoal Reduced absorption due to adsorption. *3794*

Cholestyramine Reduces bioavailability in gastrointestinal tract due to binding of drug. *2794*

Colestipol In one patient drug given for 2 d reduced half-life of digoxin by approximately 50% (from 4 to 2 d). *0548*

Digoxin-Specific Fab Used to remove drug in overdose situations. *1699*

Food Absorption delayed when taken with food. *3024*

Metoclopramide Impairs absorption, increases intestinal activity. *2290* With tablets of low bioavailability due to stimulation of bowel motility. *2794* Reduced absorption with limited dissolution. *3794*

Neomycin At doses of 3 g/d reduces bioavailability. *2794* Reduced absorption due to sprue-like syndrome. *3794* At doses of 1-3 g decreased serum concentration and urinary excretion, also prolonged time to peak concentration. *2178*

Rifampin Due to induction of detoxifying enzymes. *3683* Observed in patients receiving antituberculous treatment in addition to drug for cardiac irregularity. *0549* Decreased serum concentration due to hepatic enzyme induction. *0014*

Sulfasalazine Reduces bioavailability in gastrointestinal tract. *2794*

Serum Negative Analytical
Fluorescein No result obtainable at concentration of 1.0 mg/L on Abbott TD$_x$ and carryover effect into subsequent specimen. *1722*

Urine Increase Physiological
Amiodarone Significant correlation between increases in plasma and urine concentrations. Possible displacement of digoxin from its binding sites. *0942*

Digoxin Clearance

Urine No Effect Physiological
Disopyramide No effect when drug given to total of 300 or 600 mg daily. *2992*

Dihydroxyphenylacetic Acid

Plasma Decrease Physiological
Deprenyl From mean of 730 to 370 ng/mL in 12 depressed or Alzheimer's disease patients given 60 mg drug daily for at least 3 weeks. *1766*

Tranylcypromine From mean of 710 to 63 ng/L in 6 patients with depression or Alzheimer's disease treated with up to 40 mg daily for at least 3 weeks. *1006*

Dihydroxyphenylalanine

Serum Increase Physiological
Methyldopa Hydrazine 5 fold potentiation with pretreatment. *2306*

Plasma No Effect Physiological
Deprenyl In 12 depressed or Alzheimer's disease patients given 60 mg drug daily for at least 3 weeks. *1006*

Tranylcypromine In 6 patients with depression or Alzheimer's disease treated with up to 40 mg daily for at least 3 weeks. *1006*

Plasma Increase Physiological
Carbidopa Concentration doubled in 6 men over 5 h after administration, especially if protein also given. *3842*

Test Conditions Increase Analytical
Levodopa 85% of fluorescence method of Waldmeier. *3735*

Methoxytyramine 24% of fluorescence method of Waldmeier. *3735*

Methyldopa 11.5% of fluorescence method of Waldmeier. *3735*

Methyldopamine 15% of fluorescence method of Waldmeier. *3735*

3,4-Dihydroxyphenylethylene Glycol

Plasma Decrease Physiological
Ethanol Fell immediately after ingestion and remained low for 6 h. *1671*

Dihydroxyphenylglycol

Plasma Decrease Physiological
Deprenyl From mean of 820 to 240 ng/mL in 12 depressed or Alzheimer's disease patients given 60 mg drug daily for at least 3 weeks. *1006*

Tranylcypromine From mean of 850 to 210 ng/L in 6 patients with depression or Alzheimer's disease treated with up to 40 mg daily for at least 3 weeks. *1006*

1,25-Dihydroxy Vitamin D₃

Serum   No Effect   Physiological
Ethanol   In 8 men who had abused alcohol for at least 10 y. *0347*
Nifedipine   No effect of 80 mg/d for 2 d. *3694*
Probenecid   In 8 healthy young men. *1318*
Rifampin   No change observed in 8 healthy men taking drug for 2 weeks. *0479*
Sodium Salts   No significant effect in 5 men with essential hypertension given either chloride or citrate supplement. *2037*

Serum   Increase   Physiological
Aluminum Hydroxide   Observed in one patient secondary to ingestion of large amounts of compound for long time. *1313*
Ethinyl Estradiol   In post-menopausal women with 50 μg/d for 14 d. *3474*   Small effect during and after in 7 healthy postmenopausal women treated for 12 weeks. *3467*

Serum   Decrease   Physiological
Aluminum Hydroxide   Low compared with normals in 7 patients (nondialyzed) with chronic renal failure. Decreased significantly (19.4 to 11.4 pg/mL) with treatment (15-18 g/d). *3533*
Bendrofluazide   In 19 healthy early menopausal women given 5 mg/d with calcium supplement due to primary effect on renal tubules and secondary change on vitamin D metabolism. *2989*
Chloroquine   In 2 patients with sarcoidosis given 500 mg daily while receiving corticosteroid at same time. *2639*
Ketoconazole   Dose dependent effect after administration to normal volunteers. *3390*   In 9 healthy men given up to 1200 mg/d for 1 week. *1300*
Prednisone   In children being treated for renal disease: dose dependent. *0646*

24,25-Dihydroxy Vitamin D

Serum   Increase   Physiological
Bendrofluazide   In 19 healthy early menopausal women given 5 mg/d with calcium supplement due to primary effect on renal tubules and secondary change on vitamin D metabolism. *2989*

3,4-Dimethoxyphenylethylamine (3,4-DMPEA)

Urine   Positive   Physiological
Tea   (pink spot) of dietary origin only. *3425*

Diphenylhydantoin

Serum   No Effect   Analytical
Acetazolamide   On GLC procedure of Papadopoulos if added *in vitro*. *2722*
Amobarbital   On GLC procedure of Papadopoulos when added *in vitro*. *2722*
Beclamide   On GLC procedure of Papadopoulos at *in vivo* concentration. *2722*
Carbamazepine   On GLC procedure of Papadopoulos at *in vivo* concentration. *2722*
Chlordiazepoxide   On GLC procedure of Papadopoulos at *in vivo* concentration. *2722*
Chlorpromazine   No effect on TLC method of Simon, Jatlow. *3323*   On GLC procedure of Papadopoulos at *in vivo* concentration. *2722*
Cyproheptadine   On GLC procedure of Papadopoulos at *in vivo* concentration. *2722*
Diazepam   On GLC procedure of Papadopoulos at *in vivo* concentration. *2722*
Ethosuximide   On GLC procedure of Papadopoulos at *in vivo* concentration. *2722*
Ethotoin   No detectable influence at 40 μg/mL. *3591*   On GLC procedure of Papadopoulos at *in vivo* concentration. *2722*

Fluosol-DA   As measured on Du Pont aca III at concentration up to 50%. *2518*
Haloperidol   On GLC procedure of Papadopoulos at *in vivo* concentration. *2722*
Hemolysis   When determined by gas chromatographic methods. *3828*
Imipramine   On GLC procedure of Papadopoulos at *in vivo* concentration. *2722*
Mephenytoin   On GLC procedure of Papadopoulos at *in vivo* concentration. *2722*
Methsuximide   On GLC procedure of Papadopoulos at *in vivo* concentration. *2722*
Methylphenobarbital   On GLC procedure of Papadopoulos when added *in vitro*. *2722*
Nitrazepam   On GLC procedure of Papadopoulos at *in vivo* concentration. *2722*
Oxytetracycline   On GLC procedure of Papadopoulos when added *in vitro*. *2722*
Phenobarbital   No effect on fluorometric method of Dill. *0906*   No detectable inhibition at 125 μg/ml. *3591*
Primidone   No effect on TLC method of Simon, Jatlow. *3323*
Salicylate   No effect on TLC method of Simon, Jatlow. *3323*
Secobarbital   On GLC procedure of Papadopoulos at *in vivo* concentration. *2722*
Sulthiame   On GLC procedure of Papadopoulos at *in vivo* concentration. *2722*
Thioridazine   On GLC procedure of Papadopoulos at *in vivo* concentration. *2722*
Trifluoperazine   On GLC procedure of Papadopoulos at *in vivo* concentration. *2722*

Serum   No Effect   Physiological
Acenocoumarol   Interactions not been reported (and unlikely to occur due to different routes of elimination). *3556*
Ethanol   No apparent influence when drug coadministered. *0844*
Methylphenidate   No effect observed in 11 patients in one study. *1487*
Phenindione   No effect on metabolism due to different chemical configuration from other coumarins. *3556*
Ranitidine   No significant effect observed on drug metabolism. *3772*
Smoking   No apparent effect when drug coadministered. *0844*
Warfarin   With concurrent therapy typically had no effect. *3556*

Serum   Increase   Analytical
Phenobarbital   Affects titrimetric procedure of Kozelka. *3054*

Serum   Increase   Physiological
Aging   Concentration significantly lower in children at same dose. *3054*
Amiodarone   Pharmacokinetic interaction with clinical significance. *2327*   Up to 3 fold increase when drugs coadministered, due to inhibition of hepatic metabolism. *2366*
Aspirin   May displace from plasma protein. *1487*
Azapropazone   Probably due to decreased clearance of phenytoin. *1257*
Carbamazepine   Significant increase (36%) after drug added to regime. *3945*
Chloramphenicol   May increase from 2.5 to 9 μg/mL (at 2 g/d). *2041*   Inhibition of metabolism. *2042*   Inhibits hepatic microsomal metabolizing enzymes thus prolonging half-life. *3864*
Chlordiazepoxide   Inhibition of metabolism (by 60%). *2042*
Chlorpromazine   Reported impairment of metabolism. *2042*
Cimetidine   Significant increase due to inhibition of metabolism probably by reversibly binding to hepatic microsomal cytochrome P-450. *2147*   Significant increases when coadministered, due to effect on hepatic metabolism. *3120*
Co-Trimoxazole   Effects of compound may be potentiated. *0679*
Diazepam   From 20 to 40 μg/mL (unknown dose) inhibits metabolism. *2422*

Diphenylhydantoin (continued)

Serum Increase Physiological (continued)

Dicumarol May increase concentration 5 to 15 µg/mL (at dose for prothrombin time=30%). *2041* Increase by 38 to 250% in six volunteers probably due to inhibition of para-hydroxylation in liver. *3556*

Disulfiram Inhibits hydroxylation, prolongs action. *2641* May increase from 7/15 to 25/39 µg/mL (at 400-800 mg/d). *2041* Metabolism inhibited, with half-life increased and decrease in mean metabolic clearance rate. *3555*

Estrogens Reported to cause increased plasma levels. *2041*

Ethanol Reported interaction due to alteration of metabolism. *2042*

Ethosuximide May be increased plasma levels (impaired metabolism). *2422*

Halothane Reported impairment of metabolism. *2042*

Isoniazid Impairs metabolism in approximately 10%. *2042* From 12 to 42 µg/mL (with dose of 300 mg). *2041* Increases blood concentration and toxicity. *3683*

Methylphenidate May increase from 9 to 28 µg/mL (at 20-40 mg/d). *2041*

Metronidazole Increased concentration reported in several patients. *2810*

Miconazole Isolated case reported with phenytoin concentration markedly increased after miconazole given i.v., probably due to inhibition of hepatic cytochrome P-450. *3032* Inhibits hepatic metabolism thereby increasing concentration. *3032*

Nitrazepam Probably inhibits metabolism in liver. *1487*

Oral Contraceptives Higher values related to drug dose compared with controls. *0844*

Phenobarbital Competitive inhibition of metabolism. *1487*

Phenprocoumon Half-life increased by 40% in three patients. *3556*

Phenylbutazone Reported impairment of metabolism. *2042*

Phenyramidol May increase concentration from 7 to 12 µg/mL (dose of 1.2 g/d). *2041*

Phenytoin 600 mg orally produces concentration of 10 mg/L. *2348* Therapeutic plasma concentrations obtained within 24 h in most patients: peaking of concentration occurred between 48 and 96 h after loading. *3827*

Prochlorperazine Reported impairment of metabolism. *2042*

Progabide Increased concentration in patients who had a therapeutic response. *3188*

Propoxyphene Reported impairment of metabolism, increased plasma concentration. *2042*

Ranitidine Mean concentration increased from 36.1 µmol/L to 39.3 µmol/L after coadministration. *1938*

Sulfaphenazole Reported impairment of metabolism. *2042*

Sulfisoxazole Displaces from protein *in vitro*. *1487*

Sulthiame Increase from 10 to 20 µg/mL (at dose of 200-800 mg/d). *2041* Probably acts on liver enzymes. *2979*

Thioridazine Two cases reported of increased plasma concentration due to competition for metabolism by cytochrome P-450. *3716* Significant increase in 15% patients with combined therapy, reduction in 7%, No change in others. *3140*

Trimethoprim May potentiate action by effects on metabolism. *0678*

Valproic Acid Caused increased concentration which led to hepatic damage. *2714*

Serum Decrease Analytical

5-(p-Hydroxyphenyl)-5-Phenylhydantoin Reacts quantitatively in radioimmunoassay. *3591*

Mephenytoin Slight effect only at 4 µg/ml. *3591*

Serum Decrease Physiological

Aspirin Statistically significant increase of free fraction (0.13 to 0.16) but lower total concentration. *1178*

Carbamazepine Half life reduced from 10.6 to 6.4 h. *2041*

Charcoal Reduced absorption due to adsorption. *3794*

Chlorophenothane Decrease from 9 to 1 parts per billion (with dose of 300-400 mg). *2041*

Ethanol Half life reduced from 23.5 to 16.3 h. *2041*

Folic Acid If patients folate deficient, stimulate metabolism. *1487* Occurs with pharmacologic doses; lowers plasma concentration. *3024* Increase by 8 to 48% in 4 male patients associated with increase in drug oxidative metabolism. *0307* Due to increased clearance: with doubling of amount of dose excreted in urine. *0308*

Glutamate Occurs with pharmacologic doses: lowers plasma concentrations. *3024*

Heparin Slight effect when high concentrations of nonesterified fatty acids occur due to heparin due to reduced protein binding. *3211* Decreases by 20% with post-heparin increase of free fatty acids. *3210*

Phenothiazines Decreased by more than 40% with start of drug or increased dose. *1455*

Pyridoxine Occurs with pharmacologic doses: lower plasma concentration. *3024*

Theophylline Mean value decreased by 21% when theophylline coadministered; rebound when drug discontinued. *3557*

Valproic Acid Bound concentration falls, although free concentration unchanged. *2475* Proportion of free concentration increased from 9.1% to 15.8% as serum concentration fell from 19.7 to 15.3 µg/mL. *1197* Clearance markedly increased when phenytoin given i.v. *1200* Due to displacement from albumin in 25 patients studied for 5 to 9 mo. *3829* Due to displacement from protein and reduction in total serum concentration but free concentration unchanged. *3144* Displaces from protein-binding sites and inhibited metabolism in some patients. *0510*

CSF Increase Physiological

Phenytoin Concentration same as unbound concentration in plasma. *3054*

2,3-Diphosphoglycerate

Red Blood Cells No Effect Analytical

Phosphates High serum concentration no effect on method of Luisada-Offer. *2226*

Red Blood Cells No Effect Physiological

Hemodialysis Chronic hemodialysis has no effect. *2740*

Methylprednisolone No significant effect following i.v. infusion of drug in patients with myocardial infarction compared with controls. *1560*

Smoking No significant difference between smokers and non-smokers. *1548*

Red Blood Cells Increase Physiological

Androgens Metabolic effect. *1091*

Blood Transfusions Especially if stored blood used. *2226*

Pregnancy Increase by about 30% in normal. *3041*

Testosterone Increase by 2/3 in patients on hemodialysis for 12 weeks. *2618*

Red Blood Cells Decrease Physiological

Amino Acids Associated with low serum phosphate. *3619*

Propranolol Shifts hemoglobin-O_2 dissociation curve to right. *2770*

2,3-Diphosphoglycerate Mutase

Blood No Effect Physiological

L-Thyroxine No effect on amount of 2,3-DPG produced. *3612*

Tri-iodothyronine No effect on amount of 2,3-DPG produced. *3612*

Blood Increase Analytical

Lactate Dehydrogenase Reported to prevent pyruvate inhibition. *3612*

Disopyramide

Serum   No Effect   Physiological
Atenolol   Significant reduction of clearance while half-life, volume of distribution remain unchanged. *0413*
Phenobarbital   Concentration unaffected at level used to induce hepatic enzymes. *1861*
Smoking   Similar half-lives in both smokers and nonsmokers. *1861*

Serum   Increase   Physiological
Amiodarone   Increased concentration with possible adverse effects on EKG. *2298*
Phenytoin   Serum concentration at peak unchanged but area under curve affected suggesting effect on hepatic metabolism. *2597*
Quinidine   When concurrent quinidine given: effect small but significant no significant change in elimination half-life. *0208*

Dopamine

Plasma   Increase   Analytical
Tris   Markedly increased concentration when progressively increasingly diluted with tris and radio enzymatic method used for quantitation. *0768*

Plasma   Increase   Physiological
MAO Inhibitors   Effect observed after single large dose. *1343*

Urine   No Effect   Analytical
Labetalol   On HPLC method with electrochemical detection. *0434*

Urine   No Effect   Physiological
Amantadine   In normals and Parkinson patients. *1817*
Barbiturates   No effect with addiction or withdrawal. *3634*
Ethanol   No change after ingestion after 0.4 g alcohol/kg in 56 healthy males. *0015*
Methyldopa   No effect when methyldopa administered and HPLC used to measure catecholamines. *2521*

Urine   Increase   Physiological
Caffeine   After 250 mg increased for 2 h then returned to baseline in next 2 h. *2751*
Dopamine   Effect noted with infusion of as little as 0.03 µg/kg/min in 6 normal men for 2 h. *2149*
Furosemide   Significant effect in 15 minutes following 30 mg intravenously. *1784*
Levodopa   Response to therapy in Parkinsonism. *2724*
Prochlorperazine   Up to 79% increase after 30 mg in controls. *1607*
Protein   Significant effect of diet on excretion. *1176*

Urine   Decrease   Physiological
Carbidopa   70% reduction in excretion noted over 5 h. *3842*
Glucagon   Marked initial effect after i.v. in patients with congestive cardiac failure. *1607*

Dopamine Hydroxylase

Serum   No Effect   Analytical
Storage of Sample   1 d at 22°, days at -20°. *3807* No effect at -20° for 4 weeks. *1339*

Serum   No Effect   Physiological
Diurnal Variation   No change, or with activity. *3807*
Electroshock   No effect observed. *3807*
Muscular Exercise   No effect observed. *3807*
Prostaglandin F2α   No effect seen with i.v. infusion in pregnant women. *2510*

Serum   Increase   Physiological
Aging   Low or not detectable in newborns. *3807*

Serum   Decrease   Physiological
Disulfiram   Observed in animals. *2913*
Fusaric Acid   Most effective if given with levodopa. *2913* Total inhibitor *in vivo* possible for up to 122 h after 300 mg. *2545*

Dopa Screening Test

Urine   Positive   Analytical
Aspirin   Light amber color produced. *3075*
Diazepam   Very slight purple color produced. *3075*
Phenothiazines   May produce false color (usually buff/amber). *3075*

Doxycycline

Serum   Decrease   Physiological
Ferrous Sulfate   Reduced absorption due to chelation which also affects elimination. *3794*
Food   Absorption reduced when taken with food. *3024*

EDTA Clearance

Urine   No Effect   Physiological
Aspirin   Although creatinine clearance apparently affected. *0545*
Trimethoprim   True glomerular filtration rate not affected by drug. *1880*

Urine   Decrease   Physiological
Cimetidine   Significant effect within 6 h, normalized after several weeks. *0978*
Cisplatin   From 108 to 90 mL/min/1.73 m²; irreversible. *3397*

Effective Renal Plasma Flow

Patient   No Effect   Physiological
Estradiol   No effect observed in humans. *2590*
Metolazone   No effect observed. *3358*
Sulindac   In 9 individuals with stable renal insufficiency when treated for 9 d. *2457*

Patient   Increase   Physiological
Chlorpromazine   Slight increase in renal blood flow. *1343*
Clonidine   Sustained rise after initial drop with i.v. injection. *0477*
Dopamine   Intravenous infusion caused increase from 507 to 798 ml/min. *1326*
Methyldopa   Slight increase or normal in normo- or hypertensives. *1343*
Pregnancy   Normal response in first 8 mo. *0987*

Patient   Decrease   Physiological
Aging   Approximately 45% linear decrease from age 45 to 85 y. *2617*
Amphotericin B   Occurs in high percentage of patients. *0071*
Angiotensin   Marked decrease following administration. *2427*
Diazoxide   After i.v. injection immediate reduction noted. *0124*
Enalapril   Associated with selective glomerular efferent arteriolar dilatation and possible interference with autoregulatory capacity of kidney in response to severe renovascular hypertension. *0284*
Epinephrine   May decrease up to 40% but no effect on blood pressure. *1343*
Ether   Probably due to renal vasoconstriction. *1343*
Ganglionic Blocking Agents   Returns to normal usually within 2 h. *1343*
Guancydine   Due to hypotensive action of drug. *2132*
Halothane   Up to 38% decrease reported in normals. *2427*

Effective Renal Plasma Flow (continued)

Patient   Decrease   Physiological (continued)
Histamine   After 0.3-0.5 mg subcutaneously in humans. *1176*
5-Hydroxytryptamine   30% decrease for 45 minutes after 1 mg i.v. *1176*
Levarterenol   Blood flow reduced, filtration rate unchanged. *1343* Slight fall after i.v. infusion. *2427*
Methoxyflurane   Decrease noted during normal anesthesia. *2427*
Pentazocine   Observed in normal individuals. *1343*
Propranolol   May occur with decreased cardiac output. *3384*
Spinal Anesthesia   In relation to degree of hypotension. *1343*

Elastase 2

Serum   Increase   Physiological
Smoking   Basal activity of immunoreactive pancreatic enzyme. *0961*

Electrophoretic Index

Serum   Increase   Physiological
Estrogens   Increases binding capacity of thyroxine binding globulin. *3871*

Encainide

Serum   Increase   Physiological
Cimetidine   Concomitant administration increases drug concentration by 30 to 40%. *2011*

Endoplasmic Reticulum Antibody

Serum   Positive   Physiological
Methyldopa   Observed in one patient with chronic active hepatitis and cirrhosis. *0259*

β-Endorphin

Plasma   No Effect   Physiological
Clonidine   In normal subjects with 0.15 mg/d but raised to normal values in opiate addicts. *1279*
Dehydrobenzperidole   No reduction of increased values induced by stress when given alone but significant decrease when given with fentanyl also. *2077*

Plasma   Increase   Physiological
Caffeine   Caused increase with peak at 60 minutes after 500 mg in both men and women but with no effect when 250 mg given (increase of order of 25 to 50%). *3418*   500 mg increased concentration in both men and women. *3418*
Ethanol   With severe alcohol abuse in patients with cirrhosis or pancreatitis to 25 pg/mL from 2.5 pg/mL in controls. *3771*

Plasma   Decrease   Physiological
Dexamethasone   Postoperative induced secretion of compound reduced by all amounts of drug given but amount of pain increased. *1497*

Enteroglucagon

Plasma   Decrease   Physiological
Morphine   Postprandial secretion abolished in 6 volunteers after drug given i.v. *0626*

Eosinophils

Blood   No Effect   Physiological
Cromoglycate   No clinical significant change observed. *0484*
Halothane   No significant change noted even with multiple exposures. *1089*
Oxacillin   0 of 10 patients demonstrated hypersensitivity reaction. *1963*

Blood   Increase   Physiological
Acetophenazine   Allergic manifestation. *1680*
Ajmaline   In 4 patients with centrilobular cholestasis and mild hepatocytic lesions: recovery with drug withdrawal. *2729*
Allopurinol   May cause severe sensitivity reaction. *2451* Frequently associated with drug-induced hepatitis when it occurs. *2906*
Aminosalicylic Acid   Hypersensitivity reaction. *1513*
Amitriptyline   Occasional allergic response. *1680* Occasional case of hypersensitivity associated hepatitis. *0098*
Ampicillin   Allergic reaction. *1513* Hypersensitivity reaction in 39% of 36 patients receiving drug. *1963*
Arsenicals   Up to 50% observed in one case, others 10-20%. *1478*
Aspoxicillin   In 1 patient given up to 112 mg/kg intravenously in 3 or 4 divided doses. *1517*
Auranofin   Infrequent reversible side effect (occurring in up to 13% patients). *0617*
Azapropazone   In 1 of 83 patients treated for 6 mo. *3565*
Butaperazine   Allergic manifestation. *1680*
Capreomycin   Allergic reaction (may be up to 35%). *0505*
Captopril   Several cases of rash and eosinophilia reported. *0929*
Carbamazepine   Eosinophilia observed occasionally. *0019* May intensify eosinophilia of filarial infection. *1343* Isolated hypersensitivity reaction reported. Associated with fever, rash, lymphadenopathy, hepatosplenomegaly and asthma. *2156*
Carbenicillin   Manifestation of allergic response. *1680* In 2 of 27 patients given 270 mg/kg i.v. for 6 d. *3275*
Carbon Disulfide   Occasional response. *1302*
Carisoprodol   Allergic manifestation. *1680*
Cefamandole   In 5% cases. *3134*
Cefoperazone   In 8% of 450 patients: effect mild and reversible. *0690*
Cefoxitin   Reported incidence of 3%. *3134*
Cephalexin   Allergic response. *2808* In 16% of 74 patients all children given 25-150 mg/kg/d for 5 to 15 d. *2505*
Cephaloglycin   Allergic response. *1680*
Cephaloridine   May be up to 10% in 1% people. *1019*
Cephalosporin   In about 4% patients treated with drugs. *2615*
Cephalothin   Modest increase with prolonged high dose i.v. administration. *3133* Allergic response. *2416*
Cephapirin   Modest increase with prolonged high dose i.v. administration. *3133*
Cephradine   In 28% of 86 children given 25-110 mg/kg daily orally for 5 to 15 d. *2505*
Chloramphenicol   Allergic reaction. *1513* Occasional hypersensitivity reaction occurs. *3864*
Chlorophenylalanine   Observed in isolated case. *3686*
Chlorpromazine   Often precursor of jaundice. *2427*
Chlorpropamide   Allergic response. *1513* In isolated case with diabetes who developed proliferative glomerulonephritis indicating immunologically mediated reaction. *0146*
Chlorprothixene   Allergic manifestation. *1680*
Chlortetracycline   Probably allergic response. *1680*
Ciprofloxacin   Observed in 16 of 2829 patients and not serious. *0151*
Clindamycin   Occasional allergic response. *2806* Common hypersensitivity reaction. *3864*
Cloxacillin   Hypersensitivity response. *1513*
Dantrolene   Isolated instance reported. *0659*
Dapsone   Observed in 66% (2 of 3 patients) treated with drug who had rheumatoid arthritis. *3360* Components of typical sulfone or dapsone syndrome. *2026*

Demeclocycline   Rare, but reported to occur. *1680*

Desipramine   Allergic reaction. *3857*   In 4 of 46 patients as manifestation of idiosyncratic response but no value exceeded 10% of total white cell count: uncorrelated with drug plasma concentration. *2871*

Dicloxacillin   Mild allergic response. *0118*

Digitalis   Allergic response may be large with toxicity. *1513*

Diurnal Variation   During night (at midnight) minimum in afternoon. *1638*

Doxycycline   Allergic response reported. *2802*

Enalapril   Reversible and associated with rash in one patient. *0929*

Epicillin   Marked increase up to 18% in some cases. *0066*

Erythromycin   Associated with hypersensitivity reaction. *2324* Hypersensitivity reaction in 45% individuals. *2792*

Ethosuximide   Rare eosinophilia reported. *0071*

Etretinate   In one 74-year old woman treated for severe psoriasis: normalized with withdrawal of therapy and recurred when reinstituted. *3789*

Fansidar®   Biopsy-proven granulomatous hepatitis due to sulfadoxine moiety of drug reported in two individuals receiving drug prophylactically possibly due to sulfonamide hypersensitivity. *2099*

Flavoxate   Single reversible case reported. *1680*

Florantyrone   In patients with pre-existing liver disease. *2313*

Fluorides   Eosinophilia reported as allergic reaction. *2427*

Fluphenazine   Allergic response with phenothiazines. *1679*

Gold   Transient allergic response. *0114*   Occurred in 21% patients taking gold sodium thiomalate and in 13% taking auranofin. Occurred in 24% and 30% respectively with toxicity. *0992* Observed in 11% of 64 treated patients. *3360*

Haloperidol   In isolated case producing cholestatic liver disease: typical frequency of liver disease is 0.2%. *0910*

Halothane   Granulomatous hepatitis in single case following second exposure to anesthetic. *3266*

Hetacillin   Probably allergic response. *1680*

Imipramine   Allergic response (may produce Löffler's syndrome). *3280*

Indandione Derivatives   With other signs of hypersensitivity. *2427*

Indomethacin   Oliguric renal failure developed in one man during treatment: reversible. *1245*

Iodides   Allergic response. *2313*

Iprindole   Allergic response reported. *0038*

Isoniazid   Allergic phenomenon. *1513*

Kanamycin   Allergic reaction. *1513*

Levodopa   Occasionally observed without symptoms. *0824*

Lugol's Iodine   Rare allergic response to iodines. *0071*

Marophen   Low incidence reported. *2427*

Mefenamic Acid   Allergic reactions noted. *1680*

Mephenoxalone   Rare side effect. *0071*

Mephenytoin   Rare cases described. *1680*

Meprobamate   Rare allergic manifestation. *1680*

Mesoridazine   Manifestation of allergic reaction. *1680*

Methacycline   May cause allergic response. *1680*

Methicillin   Hypersensitivity reaction. *0212*   In 31% of 45 patients receiving drug (hypersensitivity reaction). *1963*   In 3 of 28 children within 5 to 8 d. *2548*

Methotrimeprazine   Noted with other phenothiazines. *1680*

Methsuximide   Reported effect. *1680*

Methyldopa   Allergic response. *2313*

Methysergide   Transient effect up to 36 h after i.m. injection. *1274*

Minocycline   Allergic response. *1705*

Moxalactam   In about 2.5% patients as reported from several studies. *0592*

Nafcillin   3 of 32 children developed eosinophilia within 1-4 d. *2548*

Nalidixic Acid   Allergic response. *1513*

Naproxen   Hypersensitivity reaction in 3 women with pulmonary infiltrates: resolved when drug discontinued. *0547*

Niridazole   Quite common allergic response. *2427*

Nitrofurans   May be serious anaphylactoid reaction. *0071*

Nitrofurantoin   Allergic response (greater than 1%). *1974*   Mild increase developed lupus-like syndrome associated with pulmonary reaction. *3250*   In 158 of 191 acute and 14 of 32 chronic drug induced pulmonary reactions. *1639*   Mild increase developed lupus-like syndrome associated with pulmonary reaction. *3250*   In 158 of 191 acute and 14 of 32 chronic drug induced pulmonary reactions. *1639*

Norfloxacin   In 2 of 1540 patients in clinical trials. *0731*

Nortriptyline   Presumed allergic response. *1680*

Novobiocin   Allergic reaction. *1513*

NSAID (Nonsteroidal Anti-Inflammatory Drugs)   Observed in 9% of 56 patients treated with these drugs alone for rheumatoid arthritis. *1111*

Nystatin   May cause allergic reaction. *2808*

Oleandomycin   Associated with allergic cholestasis. *1679*

Oxacillin   Reported effect (?allergic). *0071*   Suggestive of hypersensitivity reaction. *2374*   Reversible phenomenon observed in 8 patients given drug i.v. *2664*

Oxazepam   Rare allergic response. *0071*

Papaverine   Allergic response reported to occur. *1680*

Penicillamine   May occur with rash. *0071*   In 14% patients treated with up to 750 mg daily (63 patients studied who had taken penicillamine). *3360*

Penicillin   Allergic reaction (may be up to 20% of all WBC). *0996*

Penicillin G   44% of 39 patients demonstrated hypersensitivity reaction. *1963*

Phenethicillin   Few cases reported only (minor effect). *1680*

Phenindione   Allergic response after 15 d in one patient. *1574*

Phenothiazines   Allergic response. *2313*

Phenylbutazone   In 2 patients subsided after drug withdrawn associated with granulomas and cholestatic-hepatocellular injury. *1730*

Phenytoin   Rare hypersensitivity reaction. *1488*   In 89% patients who developed hepatotoxicity. *2517*   General feature of hepatotoxicity (although this occurs rarely), hypersensitivity usually responsible. *0950*   Associated with several cases of hepatotoxicity due to hypersensitivity reaction. *1994*

Piperacetazine   Occasional allergic response. *1680*

Piperacillin   In 1 of 29 patients given 181 mg/kg i.v. for 6 d. *3275* In 5 of 59 patients given drug as sole agent with many patients with severe illness. *3874*

Potassium Iodide   Allergic response. *1553*

Primidone   May be allergic type of response. *0416*

Probucol   Hypersensitivity response (less than 10%). *0806*

Procainamide   Evidence of allergic response to drug. *3215*

Protriptyline   Rare allergic response. *1678*

Quinethazone   Isolated case report. *2427*

Ranitidine   Isolated case reported within 1 mo of treatment being started, reverted to normal 2 weeks after drug stopped. Probable hypersensitive or idiosyncratic reaction. *3303*

Rifampin   Allergic reaction. *1553*   Isolated case of eosinophilia attributable to drug only. *2520*   Eosinophilia in one patient clearly linked to drug only since declined once drug withdrawn. *2109*

Ristocetin   Allergic response. *2313*

Spironolactone   Relative lymphocytosis and 15% eosinophilia observed in a single patient during period of agranulocytosis. *3487*

Streptokinase   Allergic response. *2313*

Streptomycin   Hematopoietic reaction (occurs in 50% cases). *0071*

Sulfamethoxazole   Allergic response. *1974*   When given with trimethoprim, fever and malaise: successful response to treatment. *1969*

Sulfamethoxypyridine   Observed with hypersensitivity hepatitis. *1987*

Sulfasalazine   As part of systemic hypersensitivity reaction in woman with inflammatory bowel disease. *3402*   Characteristically associated with rare cases of hepatotoxicity. *1127*   Hepatotoxic hypersensitivity reaction like sulfonamide hypersensitivity. *2214* Observed in 11% of 18 patients treated with drug who had rheumatoid arthritis. *3360*

Sulfisoxazole   Allergic response. *1974*

Eosinophils (continued)

Blood Increase Physiological (continued)

Sulfonamides Hypersensitivity response. 1513 Associated with hepatitis like attack in patient with prior episodes of jaundice. 1741

Sulfonylureas May occur in association with cholangiolitis. 3171

Tetracycline Allergic response. 1513

Thiocolchicine May cause eosinophilia. 2143

Thiothixene Reported with other phenothiazines. 1680

Triamterene Allergic response. 1513

Trifluperidol Allergic response. 2313

Vancomycin Allergic response. 1513 Occasional hypersensitivity reaction noted. 1571

Viomycin Allergic response. 1513

X-Ray Therapy Occurs after repeated irradiation. 0251

Blood Decrease Physiological

Amphotericin B Allergic response. 1680

Aspirin May cause aplastic anemia or pancytopenia. 3585

Captopril Isolated cases of pancytopenia, usually with pre-existing renal disease. 1255

Corticotropin Striking response in normals. 1343

Cortisone Normal physiological response to injection. 2222

Desipramine Agranulocytosis. 0748

Diurnal Variation Low between 12 noon and 9 pm, high about 6 am. 2222

Epinephrine May cause eosinopenia (direct action). 1513

Glucocorticoids Reduced inflammatory response. 0071 Characteristic finding as a result of redistribution of cells rather than cell lysis. Effect usually transient lasting less than 24 h. 0410

Hydrocortisone Due to hormonal action on adrenals. 0242

Indomethacin Occasional effect. 0392

Light Sharp decrease (up to 30%) in response to bright light. 1638

Methysergide Starts to occur within 1 h (up to 100% dec). 2427

Niacin By up to 60% in 2 h, increased after 24 h. 1622

Niacinamide No change after 2 h, marked at 4 h. 1622

Prednisone Striking decrease in absolute count with 2 weeks therapy. 3309

Procainamide Occasional severe reduction noted. 1586

Epinephrine

Plasma No Effect Physiological

Azosemide Effect observed in normal male volunteers after 60 mg orally. 2658

Carprazidil No consistent effect in 15 men with mild to moderate essential hypertension treated for up to 16 weeks. 1268

Chlorthalidone In 22 premenopausal and 18 postmenopausal women given 100 mg/d for 6 weeks. 0402

Deprenyl In 12 depressed or Alzheimer's disease patients given 60 mg drug daily for at least 3 weeks. 1006

Desipramine Observed with long term, high or low dose, treatment under experimental conditions. 3066

Disulfiram No effect observed after 1 week. 1845

Erect Posture No effect seen in normals. 1176

Furosemide Effect observed after 40 mg orally in normal male volunteers. 2658

Labetalol No significant effect of acute i.v. administration (as measured by HPLC). 2974

Metoprolol No consistent effect but effect generally high when drug concentration high in 11 healthy young men studied under variety of conditions. 0908 No consistent response in individuals after 50 mg orally following different stresses. 0908

Prostaglandin F2α No effect seen with i.v. infusion in pregnant women. 2510

Site of Collection No difference between antecubital and superior vena cava blood. 3639

Terbutaline In 6 normal men treated with therapeutic amounts for 2 weeks. 3177

Tranylcypromine In 6 patients with depression or Alzheimer's disease treated with up to 40 mg daily for at least 3 weeks. 1006

Plasma Increase Analytical

Labetalol Elutes simultaneously with epinephrine with HPLC and electrochemical detection detection using method of Krstulovic et al. 0433

Phenols Affects trihydroxyindole method. 1176

Tris Markedly increased concentration when progressively increasingly diluted with tris and radio enzymatic method used for quantitation. 0768

Plasma Increase Physiological

Adenosine Increase by 213% after i.v. infusion of 10 to 140 μg/kg/min in 7 normal subjects. 0338

Altitude By third day at 12,000 feet. 2001

Ethanol In individuals with unusual aldehyde dehydrogenase after 0.4 g alcohol/kg in 56 healthy males. 0015

Ether Response to stress. 2869

Fentanyl Mechanism obscure. 1744

Histamine Due to release from adrenals. 1176

Insulin Stimulation of adrenal medulla, ?by hypoglycemia. 2442 Marked effect in insulin induced hypoglycemia. 3570

MAO Inhibitors Effect observed after single large dose. 1343

Meptazinol Almost 2 fold increase in 20 minutes with up to 1.4 mg/kg i.v. 2289

Morphine Mechanism obscure also involved in glucose release. 1744

Muscular Exercise Significant effect after physical stress. 1176

Nicotine Large effect observed. 1176

Opiates Significant increase within 1 h of i.m. injection of hyoscine, meperidine or omnopon. 1929

Pancuronium During halothane anesthesia if also given. 3410

Pentazocine Significant effect after 0.6 mg/kg i.v. in 20 minutes. 2289

Physical Stress Significant effect can be observed. 1176

Propranolol Decreased clearance by 80% in hypertensives and similar effect in normals. 3944

Thyrotropin-Releasing Hormone (TRH) Mean increase of 28% in 2nd to 4th minute after i.v. injection regardless of whether patient was initially hypo- or hyperthyroid. 2921

Plasma Decrease Physiological

Angiotensin Decreased slightly less than norepinephrine increased. 3967

Dextromoramide Mechanism obscure. 1744

Site of Collection Pulmonary artery and right atrium less than arm vein. 3639

Verapamil Significant effect in 15 patients with uncomplicated essential hypertension. 0867

Urine No Effect Analytical

EDTA Stabilizes compounds prevents light destruction. 1868

Storage of Sample No effect at room temperature, several days if acidified. 3873

Urine No Effect Physiological

Acetaminophen No effect with short term ingestion 2.6 g/d. 0766

Amantadine In normals and Parkinson patients. 1817

Ammonium Chloride No significant effect observed. 3616

Aspirin No effect with short term ingestion 2.6 g/d. 0766

Barbiturates No effect with addiction or withdrawal. 3634

Carbidopa No effect over 5 h after ingestion. 3842

Chlordiazepoxide No effect of short term ingestion of 30 mg/d. 0766

Corticotropin No effect observed in normals or rheumatoids. 0279

Diazepam No effect with short term ingestion of 15 mg/d. 0766

Diphenhydramine No effect with short term ingestion of 150 mg/d. 0766

Diuretics No significant effect during diuresis. 3616

Erect Posture   No effect seen in normals. *1176*
Felodipine   With 20 mg daily in 10 men with essential hypertension over 8 weeks. *1887*
Glucagon   No significant effects with i.v. in patients with congestive cardiac failure. *1607*
Glucose   No significant response to hyperglycemia. *3373*
Methacholine   No effect observed. *0279*
Methyldopa   No effect when methyldopa administered and HPLC used to measure catecholamines. *2521*
Phenobarbital   No effect short term ingestion of 120 mg/d. *0766*
Prochlorperazine   No significant effect with up to 30 mg in controls. *1607*
Reserpine   Not affected (unlike norepinephrine). *2702*
Stress Cold Exposure   Response to cold exposure. *3217*

Urine   Increase   Analytical
Labetalol   On HPLC method with electrochemical detection. *0434*

Urine   Increase   Physiological
Altitude   From first day at 12,000 feet. *2001*
Aminophylline   Threefold increase in response to i.v. dose. *0178*
Amphetamine   5.8 fold increase noticed with psychosis. *1825*
Caffeine   After 250 mg increased for 2 h then returned to baseline in next 2 h. *2751*
Dextroamphetamine   For 3 h after administration. *1487*
Diurnal Variation   Maximum excretion about 3 pm. *2222*   Significantly higher during day than at night. *1176*
Dopamine   Effect noted with infusion of 0.3 μg/kg/min and more in 6 normal men for 2 h. *2149*
Ephedrine   For 3 h after administration (slight). *1487*
Ethanol   Slight increase following ingestion. *0143*   Large effect observed. *1176*   In individuals with unusual aldehyde dehydrogenase after 0.4 g alcohol/kg in 56 healthy males. *0015*
Guanethidine   Slight increased output reported. *3083*
Histamine   Due to release from adrenals. *1176*
Insulin   Hypoglycemia produces up to tenfold increase. *0279*
Isoproterenol   Probably small effect with usual doses. *0758*
Methamphetamine   For 3 h after administration (slight). *1487*
Methylphenidate   For 3 h after administration (slight). *1487*
Muscular Exercise   If strenuous may be increased tenfold. *0279*
Nitroglycerin   ?due to adrenergic stimulation of hypotension. *3713*
Noise   Usual drop in afternoon did not occur. *3351*
Pain   Also seen more markedly with emotional stress. *1990*
Phenmetrazine   For 3 h after administration (slight). *1487*
Physical Stress   Significant effect can be observed. *1176*
Stress   With physical or emotional stress. *3217*
Tetrahydrocannabinol   Initial marked increase compared with placebo. *3786*
Urine Flow   Positive correlation between excretion and flow. *1636*
Urine pH   Excretion increased with alkaline urine. *1636*

Urine   Decrease   Physiological
Clonidine   Primary action of drug. *1627*
Cortisone   Decreased output with 200 mg/d. *0279*
Diurnal Variation   Reduction in afternoon compared with morning. *1636*
Recumbency   Significantly less than when ambulant. *3616*

Test Conditions   No Effect   Analytical
Dihydroxymandelic Acid   On fluorescent procedure of Peyrin. *2800*
Dihydroxyphenylacetic Acid   On fluorescent procedure of Peyrin. *2800*
Dopamine   On fluorescent procedure of Peyrin. *2800*
Epinine   On fluorescent procedure of Peyrin. *2800*
HMPG   On fluorescent procedure of Peyrin. *2800*
Homovanillic Acid   On fluorescent procedure of Peyrin. *2800*
Levodopa   On fluorescent procedure of Peyrin. *2800*
Metanephrine   On fluorescent procedure of Peyrin. *2800*

Normetanephrine   On fluorescent procedure of Peyrin. *2800*
Octopamine   On fluorescent procedure of Peyrin. *2800*
Parasympathol   On fluorescent procedure of Peyrin. *2800*
Tyramine   On fluorescent procedure of Peyrin. *2800*
Vanillylmandelic Acid (VMA)   On fluorescent procedure of Peyrin. *2800*

Test Conditions   Increase   Analytical
Epinephrine   34% of fluorescence method of Waldmeier. *3735*
Isopropylepinephrine   25% of fluorescence method of Waldmeier. *3735*
Isopropylnorepinephrine   58% fluorescence of epinephrine in procedure of Peyrin. *2800*
Levarterenol   3.4% fluorescence of epinephrine in procedure of Peyrin. *2800*
Normetanephrine   11.5% of fluorescence method of Waldmeier. *3735*

Erythrocytes

Blood   No Effect   Physiological
Chenodeoxycholic Acid   No significant effect with up to 750 mg/d over 2 y in 916 patients. *3199*
Enalapril   In 53 patients given up to 160 mg daily or when combined with hydrochlorothiazide. *2277*
Fenfluramine   No adverse reaction reported. *0538*
Penicillamine   In 21 patients with rheumatoid arthritis treated for 48 weeks given 250 to 750 mg daily. *2345*
Smoking   No significant difference between smokers and non-smokers. *1548*
Timegadine   In 23 patients with rheumatoid arthritis given 250 to 750 mg/d for 48 weeks. *2345*
Zimeldine   In 1 study involving 147 patients no significant change seen. *3741*

Blood   Increase   Physiological
Androgens   May produce erythrocytemia (increases well-being). *0071*
Cobalt   Reported to cause increased red cell production. *2217*
Corticotropin   Especially marked if given to anemics. *1343*
Dromostanolone   Response to androgens. *0071*
Epinephrine   Due to hemoconcentration. *1343*
Glucocorticoids   Stimulate erythropoiesis. *0071*
Pilocarpine   Probably due to contraction of spleen. *1343*
Smoking   Significantly higher in smokers. *3112*
Vitamin B$_{12}$ Successful treatment may cause mild polycythemia. *2427*

Blood   Decrease   Analytical
Cold Agglutinins   Spurious macrocytosis due to agglutination. *2797*
Storage of Sample   Lysis produced by freezing But only by 2-3% loss if frozen in liquid N_2. *1180*

Blood   Decrease   Physiological
Acetaminophen   May cause hemolytic anemia. *2313*
Acetanilid   Hemolytic anemia/agranulocytosis. *2429*
Acetazolamide   May cause pancytopenia/agranulocytosis. *2313*
Acetohexamide   May cause hemolytic anemia. *1019*
Acetophenazine   Possible aplastic anemia. *2313*
Acetylphenylhydrazine   May cause hemolytic anemia. *2313*
Acyclovir   One case megaloblastic anemia documented. *0936*
Allopurinol   Rare case of anemia may occur. *0071*
Aminobenzoic Acid   Induces malabsorption and megaloblastic anemia. *2716*
Aminopyrine   Hemolytic anemia in G-6-PD deficient persons. *0788*
Aminosalicylic Acid   Hemolytic anemia/megaloblastic anemia. *0788*
Amitriptyline   Rare transient agranulocytosis. *0071*

Erythrocytes *(continued)*

Blood Decrease Physiological *(continued)*

Amodiaquine Occasional aplastic anemia reported. *3717*

Amphetamine Hemolytic anemia (depends on prior sensitivity). *0788*

Amphotericin B Bone marrow depression with hemolytic anemia. *3645* Frequently develops after several weeks of therapy due to interference with production of erythrocytes. *3567*

Ampicillin Reversible anemia may occur. *1679*

Amyl Nitrite Hemolytic anemia (slight or marked effect). *2426*

Aniline May cause marked hemolytic anemia. *0279*

Antimalarials May cause hemolytic anemia. *1902*

Antimony Compounds pancytopenia/may cause hemolytic anemia. *2313*

Antineoplastic Agents May cause aplastic anemia. *2313*

Antipyretics May cause hemolytic anemia. *1902*

Antipyrine Hemolytic anemia in G-6-PD deficient persons. *0788*

Arsenicals Megaloblastic anemia/pancytopenia. *0788*

Aspirin Hemolysis/G-6-PD deficient/gastrointestinal hemorrhage/direct bone marrow depression. *3227*

Auranofin Infrequent reversible side effect (occurring in up to 13% patients). *0617*

Azapropazone In 1 of 83 patients treated for 6 mo. *3565*

Azathioprine Bone marrow depression with anemia. *0071* Principal manifestation of bone marrow depression. *0036* 18 cases of pancytopenia out of 79 cases with hematological problems out of 328 reports of severe adverse effects. *2097*

Barbiturates Aplastic or megaloblastic anemia. *2313*

Benzene Mild macrocytic anemia/aplastic anemia. *2429*

Benzocaine May cause hemolysis. *0380*

Benzthiazide May occur as with other thiazides. *1680*

Bismuth Salts May cause aplastic anemia. *0279*

Bromate May cause hemolysis. *1343*

Brompheniramine Occasionally observed with antihistamines. *1679*

Busulfan pancytopenia/hemolytic anemia. *1680*

Butaperazine Transitory anemia. *0071*

Captopril Isolated case of pancytopenia reported. *1012* Observed in 9 of 12 hypertensive patients on maintenance hemodialysis, maximum effect achieved after about 11 mo. *1608*

Carbamazepine Pancytopenia. *0928*

Carbenicillin Hemolytic anemia reported. *1680*

Carbimazole Occasional aplastic anemia reported. *3717*

Cephaloridine May cause anemia. *0071*

Cephalothin Hemolytic anemia. *1378*

Chlorambucil Dose dependent bone marrow depression. *0071*

Chloramphenicol Dose related response usual/aplastic anemia. *0907* Occurs either as dose-related or idiosyncratic bone marrow suppression: dose- related usually occurs 5-7 d after start of therapy: idiosyncratic is rare occurring 1 case per 40,000 courses of therapy. *3864* Aplastic anemia reported to occur in from 1 to 10,000 or 40,000 cases. *3717*

Chlorate Destroys cells on absorption. *1343*

Chlordane Pancytopenia (AMA Blood dyscrasias). *2429*

Chlordiazepoxide Aplastic anemia occasionally occurs. *3717*

Chlorophenothane Pancytopenia (AMA blood dyscrasias). *2429*

Chloroquine Hemolytic anemia in G-6-PD deficient persons. *3522* Aplastic anemia reported to occur occasionally. *3717*

Chlorothiazide Pancytopenia/aplastic anemia. *0073* Aplastic anemia reported to occur occasionally. *3717*

Chlorpheniramine Reported to cause hemolytic anemia. *1678*

Chlorpromazine Hemolytic anemia. *0725* Occasional aplastic anemia reported. *3717*

Chlorpropamide Mild anemia, rare aplastic anemia. *2801* Isolated case of pure red blood cell aplasia. *2828* Occasional aplastic anemia reported. *3717*

Chlortetracycline Pancytopenia (AMA Blood dyscrasias). *2429*

Chlorthalidone Aplastic anemia may occur. *1680*

Coal Tar May cause hemolytic anemia. *0279*

Colchicine Megaloblastic anemia (impaired absorption of vitamin B_{12}). *3773*

Corticosteroids May cause gastrointestinal bleeding with low iron stores. *0333*

Cresol May cause hemolytic anemia. *0279*

Cyclophosphamide May cause anemia usually reversible. *1680*

Cycloserine May cause megaloblastic anemia. *3773*

Cyclosporine In 6 of 23 patients normochromic normocytic anemia in first 6 mo treatment (patients with inflammatory ocular disease). *2713*

Cyclothiazide Occasional response to thiazides. *1680*

Cytarabine Megaloblastic anemia (relatively infrequent). *2313* Aplasia of bone marrow developed in nearly all of 57 leukemics. *1579*

Dactinomycin Anemia/pancytopenia. *0071*

Dapsone May cause hemolytic anemia. *1612*

Demeclocycline Rare, but reported to occur. *1680*

Desipramine Transient agranulocytosis (up to 25% reported). *0071*

Dicumarol May cause anemia (AMA Blood dyscrasias committee). *2429*

Diethylpropion May cause bone marrow depression. *0134*

Digitalis Aplastic anemia/pancytopenia. *2313*

Diiodohydroxyquin May cause anemia. *2313*

Dimercaprol Hemolytic anemia in G-6-PD deficient persons. *0788*

Dimethindene Rare hemolytic anemia with antihistamines. *1678*

Dinitrophenol May cause anemia/aplastic anemia. *1343*

Diphenhydramine Hemolytic anemia. *3602*

Dipyrone Hemolytic anemia. *2098*

Doxapram Noted postoperatively in a few patients. *2313*

Doxorubicin Impaired but not as marked as WBC decrease. *0387*

EDTA Transient bone marrow depression. *0071*

Estrogens Megaloblastic anemia (impairs absorption of folate). *3773*

Ethanol May affect folic acid absorption and usage. *3773* Increase by 2% decrease in heavy versus occasional drinker. *3273*

Ethosuximide Aplastic anemia/pancytopenia. *2313*

Ethotoin May cause marrow depress or megaloblastic anemia. *0071*

Etretinate No marked effect seen but slight change noted. *3867*

Fava Beans May cause hemolysis with G-6-PD deficiency. *0248*

Fenoprofen In 6.5% in patients treated for juvenile rheumatoid arthritis. *0239*

Floxuridine Bone marrow depression. *1680*

Flucytosine May depress bone marrow function. *1679* Usually mild, occurs in 8 to 13% patients: occasionally pancytopenia occurs. *3567*

Fluorides May cause aplastic anemia. *0279*

Fluorouracil May cause bone marrow depression. *0071*

Fluphenazine Rare pancytopenia with phenothiazines. *1679*

Furadaltone May cause hemolysis with G-6-PD deficiency. *0248*

Furazolidone May cause hemolysis with G-6-PD deficiency. *0248*

Furosemide May cause anemia. *1617*

Gentamicin May cause anemia. *0071*

Glucosulfone Hemolytic anemia. *2429*

Glutethimide May cause aplastic/megaloblastic anemia. *0071*

Gold Aplastic anemia. *0114* Blood dyscrasias major side effect as reported to uk committee on safety of medicines. *0782* Occasional case of aplastic anemia reported. *3717* 7 cases out of 246 treated cases. *0829*

Haloperidol May cause anemia. *2313*

Hetacillin Occasional anemia reported with penicillins. *1680*

Hexachlorobenzene Pancytopenia (AMA Blood dyscrasias). *2429*

Hydralazine Pancytopenia may occur. *0044* Observed in 9 patients developing rapidly progressive glomerulonephritis during treatment. *0366*

Hydrochlorothiazide Aplastic anemia may occur. *1680* Occasional case of aplastic anemia reported. *3717*

Hydroflumethiazide Pancytopenia, aplastic anemia. *3844*

Hydroxychloroquine Anemia. *2313*

Hydroxyurea Anemia (may be transient megaloblastic). *0071*

Ibuprofen In 3% of patients with juvenile rheumatoid arthritis. *0239*

Indomethacin Secondary to gastrointestinal bleed/agranulocytosis/pancytopenia. *0381* Occasional case of aplastic anemia reported. *3717*

Iproniazid Pancytopenia (AMA Blood dyscrasias). *2429*

Isocarboxazid Agranulocytosis/anemia — rare. *2313*

Isoniazid Hemolytic anemia (rare complication). *2313*

Isotretinoin In 10 of 523 patients with mean daily dose of 109 mg for 150 d. *3867*

Lead Hemolytic anemia (with basophilic stippling). *0788*

Levodopa One case of hemolytic anemia reported. *1680*

Lipomul® May be progressive anemia with excess. *2427*

Local Anesthetics Bone marrow depression reported. *2427*

Lysol® May cause intravascular hemolysis. *0629*

MAO Inhibitors Anemia may occasionally occur. *2313*

Mechlorethamine Mild effect. *0071*

Mefenamic Acid Pancytopenia. *2313*

Melarsonyl May cause hemolytic anemia in G-6-PD deficiency. *0071*

Melphalan May cause bone marrow depression (dose related). *0543*

Mepacrine Occasional case of aplastic anemia reported. *3717*

Mepazine Pancytopenia/aplastic anemia. *2313*

Mephenoxalone May cause anemia. *2313*

Mephenytoin Hemolytic anemia. *3008* Occasional case of aplastic anemia reported. *3717*

Meprobamate Aplastic anemia/erythroid hypoplasia. *2429*

Mercaptopurine May occur with bone marrow depression. *2427*

Mercurial Diuretics Anemia may occur. *2313*

Mesoridazine Anemia/aplastic anemia/pancytopenia. *1680*

Methacycline May cause hemolytic anemia. *1680*

Methapyrilene Anemia (AMA Blood dyscrasias). *2429*

Methaqualone One possible case of aplastic anemia reported. *2313*

Methicillin May rarely cause bone marrow depression. *2808*

Methimazole Aplastic anemia. *2313* Occasional case of aplastic anemia reported. *3870*

Methotrexate May = megaloblastic anemia (folic acid antagonist). *3773*

Methsuximide Rare aplastic anemia reported. *2313*

Methyclothiazide Aplastic anemia may occur with thiazides. *1680*

Methyldopa Autoimmune hemolytic anemia/aplastic anemia. *0597* On rare occasion may produce hemolysis. Great majority of patients do not show this. *1909*

Methylene Blue May cause hemolysis with G-6-PD deficiency. *0248*

Methylphenobarbital Megaloblastic anemia. *2313*

Methylthiouracil Occasional case of aplastic anemia reported. *3717*

Mitomycin Pancytopenia. *2194*

Mitoxantrone Erythrocytes not acutely affected but mild anemia in most patients with successive courses. *3289*

Mushroom Poisoning May cause severe hemolytic anemia. *0279*

Mustard Gas May cause aplastic anemia. *2429*

Nalidixic Acid Hemolytic anemia with G-6-PD deficiency. *0279*

Naphthalene May cause hemolysis with G-6-PD deficiency. *0248*

Naproxen In 4% of patients receiving drug for juvenile rheumatoid arthritis. *0239*

Neomycin Megaloblastic anemia (impaired B_{12} absorption). *3773*

Niridazole May cause hemolysis if G-6-PD deficiency. *1343*

Nitrites May cause hemolytic anemia. *2313*

Nitrobenzene Occurs with hemolysis. *2429*

Nitrofurantoin Hypersensitivity (G-6-PD)/megaloblastic anemia. *2358*

Nitrofurazone May cause hemolysis with G-6-PD deficiency. *0248*

Nitrous Oxide Bone marrow depress (of no significance normally). *1343*

Norfloxacin Rare complication of treatment. *3750*

Novobiocin Hemolytic anemia — mild. *0713*

Oleandomycin Probably due to hemolytic anemia. *2313*

Oral Contraceptives May cause megaloblastic anemia. *3773* In 46 oral contraceptive users compared with controls studied over at least 2 y. *1182*

Orphenadrine 2 cases aplastic anemia reported. *1680*

Oxacillin Large doses parenteral penicillin may have effect. *1680*

Oxyphenbutazone May cause blood dyscrasias. *2313* Occasional case of aplastic anemia reported. *3717* Aplastic anemia with death in 38 per 100,000 users. *1723*

Pamaquine May cause hemolytic anemia. *2313* Hemolytic anemia in G-6-PD deficient persons. *0788*

Paramethadione Aplastic anemia may occur rarely. *2313*

Penicillamine Hypochromic anemia in child, menstruating woman. *0074* Rare blood dyscrasias as reported to UK committee on safety of medicines. *0782*

Penicillin Hemolytic anemia due to binding to erythrocytes. *0996*

Pentamidine Megaloblastic anemia — inhibits dihydrofolate reductase. *3773*

Pentaquine May cause hemolysis with G-6-PD deficiency. *0248*

Perchlorate May cause fatal aplastic anemia. *0071* Occasional case of aplastic anemia reported with potassium salt. *3717*

Phenacemide Aplastic anemia/agranulocytosis. *2313*

Phenazopyridine Hemolytic anemia (sensitivity dependent). *2426*

Phenethicillin Theoretically may cause hemolytic anemia. *1680*

Phenobarbital Megaloblastic anemia secondary to disturbance in folic acid metabolism. *1926*

Phenothiazines Hemolytic anemia (sensitivity dependent). *0788*

Phenylbutazone Bone marrow depression/secondary to Na and H_2O retention. *0566* Occasional case of aplastic anemia reported. *3717* Fatal aplastic anemia in 2.2 per 100,000. *1723*

Phenylhydrazine Hemolytic anemia. *2427*

Phenytoin megaloblastic/hemolytic/aplastic anemia/pancytopenia. *3331* Single case of marked aplasia, apparently mediated through an IgG inhibitor requiring the presence of drug to suppress erythroid colony formation *in vitro* inhibitor appears to exert effect on erythroid progenitors at or beyond stage of differentiation of CFU-E. But not on erythroblasts. *0889* Occasional case of aplastic anemia reported. *3717*

Phosphorus Occurs with chronic poisoning. *1302*

Phytonadione Hemolysis may occur with G-6-PD deficiency. *2313*

Picric Acid Hemolytic effect. *1343*

Pipamazine Pancytopenia (AMA Blood dyscrasias). *2429*

Piperacetazine Rare hemolytic anemia/pancytopenia. *0019*

Piperazine May cause hemolytic anemia. *2365*

Pipobroman May cause anemia (bone marrow depression). *0071*

Pregnancy By up to 10-15% due to expanded plasma volume. *0987*

Prilocaine Hemolysis with doses greater than 400 mg. *0071*

Primaquine Hemolytic anemia in G-6-PD deficient persons. *0788*

Primidone Megaloblastic anemia secondary to disturbance in folic acid metabolism. *2968*

Probenecid Hemolytic anemia (sensitivity dependent). *0788*

Procainamide May cause hemolytic anemia. *2583* Pancytopenia may occur with/without lupus-like syndrome. *3215*

Procarbazine Anemia and bone marrow depression. *0071*

Propylthiouracil Occasional immunologically associated anemia. *3870* Occasional case of aplastic anemia reported. *3717*

Pyrazinoic Acid Sideroblastic type of anemia may occur. *0333*

Pyrazolones Aplastic/hemolytic anemia. *2313*

Pyrimethamine Megaloblastic anemia. *3898*

Quinacrine May produce aplastic anemia. *2313* May cause hemolysis with G-6-PD deficiency. *0248*

Quinethazone Theoretical bone marrow depression. *1680*

Erythrocytes *(continued)*

Blood Decrease Physiological *(continued)*

Quinidine Hemolytic anemia (sensitivity dependent) G-6-PD deficiency. *3673* Immune mediated hemolytic anemia associated with high titers IgG antibodies. *3215* Occasional hemolytic anemia and hypersensitivity observed in 32 of 487 patients who had received drug. *1261*

Quinine Hemolytic anemia (sensitivity dependent) G-6-PD deficiency. *0272*

Quinocide May cause hemolysis with G-6-PD deficiency. *1902*

Radioactive Compounds Aplastic anemia. *2313*

Resorcinol Injection or excess absorption may cause hemolysis. *0071*

Rifampin Possible antibody-mediated immune reaction. *3683*

Sodium Phosphate ^{32}P Anemia if excess used. *0071*

Stibophen Produces hemolytic anemia. *2365*

Streptomycin Aplastic/hemolytic anemia (sensitivity dependent). *2313* Occasional case of aplastic anemia reported. *1292*

Sulfacetamide May cause hemolysis with G-6-PD deficiency. *1902*

Sulfachlorpyridazine May be aplastic or hemolytic anemia. *1680*

Sulfadiazine May cause hemolytic anemia. *2313*

Sulfamethizole Hemolytic anemia/aplastic anemia. *2868*

Sulfamethoxazole May cause hemolytic anemia. *1974*

Sulfamethoxypyridine Anemia (AMA Blood dyscrasias). *2429*

Sulfanilamide May cause hemolytic anemia. *0248*

Sulfapyridine May cause hemolytic anemia. *0248*

Sulfasalazine Pancytopenia reported plus reductions of individual series of cells. *2887*

Sulfinpyrazone Rare blood dyscrasia. *0071*

Sulfisoxazole Agranulocytosis/aplastic anemia. *2868* As result of bone marrow depression or hemolysis (in patients with glucose-6-phosphate dehydrogenase deficiency. *3832*

Sulfonamides May cause Heinz-body hemolytic anemia. *2313* Several cases of aplastic anemia reported. *3717*

Sulfones Hemolysis may occur with G-6-PD deficiency. *2313*

Sulfonylureas Pancytopenia. *2313* Occasional case of aplastic anemia reported. *3717*

Sulfoxone May cause hemolytic anemia. *0248*

Suramin Hemolytic anemia reported. *0071*

Tetracycline May cause hemolytic anemia. *2313*

Thiacetazone Rare Stevens-Johnson syndrome induced. *2427*

Thiazides Rare hypersensitive depression of bone marrow. *2313*

Thiazolsulfone May cause hemolytic anemia. *2313*

Thiocolchicine May cause anemia (rare). *2143*

Thiocyanate May cause anemia. *2313* Several cases of aplastic anemia reported. *3717*

Thioridazine Occasionally seen with phenothiazines. *1680*

Thiosemicarbazones Hydrazone complex formed with pyridoxal PO₄. *3025*

Thiotepa May cause bone marrow depression. *0071* 10 of 25 patients given drug intravesically had at least one incident of acute myelosuppression; 5 of 29 patients had chronic myelosuppression. *1632*

Thiothixene Reported with other phenothiazines. *1680*

Tocainide Blood dyscrasias occur in fewer than 0.2% patients. *2011*

Tolazamide May cause hemolytic anemia. *1343*

Tolazoline Pancytopenia (AMA Blood dyscrasias). *2429*

Tolbutamide Agranulocytosis/aplastic anemia. *0073*

Tolmetin In 2% of patients in trials for treatment of juvenile rheumatoid arthritis. *0239*

Tolonium May cause hemolysis with G-6-PD deficiency. *0333*

Triamterene May cause megaloblastic anemia (folic acid antagonist). *3773*

Trichlormethiazide May cause bone marrow depression. *1680*

Triethylenemelamine May cause hemolytic anemia. *2313* May cause bone marrow depression. *0071*

Trifluoperazine Pancytopenia (AMA Blood dyscrasias). *2429*

Trimethadione Aplastic anemia. *2429*

Trimethoprim Folic acid antagonist ?megaloblastic anemia. *3773* Isolated case when drug given alone, but clear evidence of pancytopenia. Megaloblastic anemia may occur with prolonged use. *3284*

Trinitrotoluene May cause hemolysis. *2429* May cause hemolysis with G-6-PD deficiency. *1413*

Tripelennamine Hemolytic anemia. *2427*

Tyrothricin Hemolysis even when applied topically. *0071*

Uracil Mustard May cause bone marrow depression. *0071*

Urethan Bone marrow aplasia/pancytopenia. *2313*

Vinblastine Anemia (secondary to leukopenia). *0071*

Vitamin A With excessive doses and use. *2313*

Vitamin K K₃ and K₄ only, especially with G-6-PD deficiency. *2102*

X-Ray Therapy Occurs with onset of aplastic anemia. *0987*

Urine No Effect Physiological

Chenodeoxycholic Acid No significant effect with up to 750 mg/d in 916 patients treated for 2 y. *3199*

Cyclosporine No effect observed in 23 patients with inflammatory ocular disease treated for 6 mo. *2713*

Timegadine In 23 patients with rheumatoid arthritis given 250 to 750 mg/d for 48 weeks. *2345*

Urine Increase Analytical

pHisoHex® Red globules that look like red blood cells. *1022*

Urine Increase Physiological

Acetaminophen Renal damage due to hemolysis. *1343*

Acetanilid Due to hemolysis and renal damage. *1343*

Acetazolamide Occasionally observed adverse reaction. *1680*

Allopurinol Associated with severe sensitivity reaction. *2451*

Aminopyrine Rare nephrotoxicity reported. *2427*

Aminosalicylic Acid Bleeding caused by drug. *1237*

Amphotericin B Nephrotoxicity. *1022*

Ampicillin Necrosis of tubules due to nephrotoxicity. *3542*

Arsenicals Nephrotoxicity with tubular necrosis. *1343*

Aspirin Initial effect always, may persist. *1343* Observed effect with long term low doses for secondary prevention of coronary heart disease. *2866*

Bacitracin Actual bleeding caused by drug. *1022*

Bismuth Salts May cause severe renal damage. *2427*

Bromate Due to renal damage. *1302*

Caffeine Daily administration causes moderate increase. *2427*

Capreomycin May cause nephrotoxicity. *1680*

Carbamazepine Associated with bleeding tendency. *0115*

Carbon Disulfide Manifestation of renal damage. *0818*

Chloroguanide Drug produces actual bleeding. *1022*

Chlorothiazide occurred in one case after i.v. administration. *1680*

Colchicine May cause bleeding. *2313*

Colistimethate Common nephrotoxic effect. *2808*

Colistin Hematuria may occur with nephrotoxicity. *0971*

Coumarin Actual bleeding may occur with high doses. *1022*

Cyclophosphamide Actual bleeding may be caused by drug. *1022*

Diatrizoic acid Has caused acute renal failure with myelography. *2312*

Dinitrophenol Due to toxic nephritis. *1302*

EDTA Rare side effect, especially with Calcium edetic acid. *1679*

Enflurane In one patient with initially normal renal function following comparatively long exposures to anesthetic. Normal renal function eventually returned. *1001*

Ethacrynic Acid Rare case of hematuria reported. *1680*

Ethylene Glycol Due to nephrotoxicity. *1302*

Fenoprofen Rare case of acute tubulointerstitial nephritis with acute renal failure (probable association with drug administration). *3582*

Formaldehyde Ingestion may cause hematuria and renal damage. *1302*

Gold Produces actual bleeding. *1022*

Heroin   Occasional microscopic evidence in addicts. *2914*

Hydralazine   Reported finding. *2427*

Hydrogen Sulfide   Occurs with nephrotoxicity (usually marked). *1302*

Ibuprofen   In 2 patients with systemic lupus erythematosus. *1155*

Ifosfamide   Occurs in one third patients within 2 d. *3675*

Indandione Derivatives   May cause hematuria — manifestation of overdose. *1237*

Indomethacin   May cause actual bleeding. *1022*   Oliguric renal failure developed in one man during treatment: reversible. *1245*

Iodine Containing Drugs   Hemorrhagic nephritis with toxic doses. *1302*

Ipodate   Nephrotoxic effect. *0838*

Iron Dextran   Reversible after chronic administration in one patient. *1678*

Iron Salts   Hematuria reported after chronic administration Fe sorbitol. *1680*

Kanamycin   Actual bleeding may occur. *1022*

Kerosene   With severe toxicity following ingestion. *1302*

Lead   Nephrotoxicity with poisoning. *1343*

Levodopa   Occasional report of hematuria. *0071*

Lipomul®   May cause actual bleeding. *1022*

Mandelic Acid   Actual bleeding may be caused by drug. *1022*

Mefenamic Acid   Actual bleeding caused by drug. *2313*

Mephenesin   Actual bleeding may be caused by drug. *1022*

Mercaptopurine   Direct renal damage with doses over 750 mg/sq m. *0980*

Mercury Compounds   May cause severe nephritis if absorbed. *0071*

Mersalyl   Actual bleeding may be caused by drug. *1022*

Methenamine   May cause actual bleeding. *1022*

Methicillin   Hypersensitivity reaction, nephrotoxicity. *0212*

Methocarbamol   May cause intravascular hemolysis. *1343*

Methylbromide   Due to tubular necrosis. *1302*

Mitotane   May occasionally produce hematuria. *1680*

Moxalactam   In about 2% patients as reported from several studies. *0592*

Muscular Exercise   Increasing with increasing rates of exercise. *1842*

Naphthalene   Occurs occasionally following inhalation of vapor. *1302*

Oxacillin   Nephrotoxicity with hematuria. *0071*

Oxalate   Nephrotoxic effect if ingested. *1302*

Oxyphenbutazone   May cause actual bleeding. *1022*   Numerous reports of kidney damage up to acute renal failure following therapeutic use of drug. *2866*

Penicillamine   In 1 of 21 patients with rheumatoid arthritis treated for 48 weeks given 250 to 750 mg daily. *2345*

Penicillin   Hypersensitivity reaction, nephrotoxicity. *0212*

Pentamidine   Gross hematuria observed in a single patient. *3317*

Phenindione   May cause actual bleeding. *1022*

Phenolphthalein   May cause acute nephrosis (K deficiency). *2427*

Phenols   Occur with renal damage of poisoning. *1302*

Phensuximide   Hematuria and renal damage reported. *1680*

Phenylbutazone   Actual bleeding caused by drug. *1022*   Observed effect, may occasionally be marked with oliguria or renal failure. *2866*

Phosphorus   Nephrotoxic effect. *1022*

Phytonadione   Actual bleeding caused by drug. *2313*

Piroxicam   Occasional drug induced nephrotoxicity, with isolated azotemia, acute interstitial nephritis or nephrotic syndrome. *2824*

Polymyxin   Actual bleeding may be caused by drug. *1237*

Probenecid   ?due to sensitivity, or toxicity. *0404*

Proguanil   Actual bleeding may be caused by drug. *1022*

Propylthiouracil   In one 24 y old woman who developed fulminant hepatic failure with lymphocyte sensitization. *2437*

Pyrazolones   May cause actual bleeding. *1022*

Radiographic Agents   Nephrotoxic manifestation. *0652*

Rhubarb   Crystalluria may cause renal damage. *0152*

Rifampin   Rare reported side effect. *1553*

Sulfadiazine   Associated with crystalluria and oliguria. *2808*

Sulfamethoxazole   May cause hematuria with crystalluria. *2808*

Sulfonamides   May cause actual bleeding. *1022*

Sulfones   May have nephrotoxic effect. *2313*

Suprofen   Occasional renal failure observed but mechanism not known. *0009*

Suramin   May cause actual bleeding. *1022*

Thiazides   May cause actual bleeding. *1022*

Triethylenemelamine   Due to nephrotoxicity. *1680*

Trimethadione   Hematuria may occur especially in children. *2220*

Trinitrotoluene   May cause hemolysis. *2313*

Trypsin   Occasional reported occurrence. *1022*

Viomycin   May cause actual bleeding. *1237*

Warfarin   Excessive doses may cause hematuria. *0948*

Zimeldine   3 of 14 patients treated for more than 1 week presented mild abnormality. *2075*

Erythrocyte Sedimentation Rate

Blood   No Effect   Physiological

Cromoglycate   No clinical significant change observed. *0484*

Dicumarol   Unaltered by coumarins. *1343*

Diurnal Variation   No effect observed. *0424*

Meals   No effect observed. *0424*

Timegadine   In 23 patients with rheumatoid arthritis given 250 to 750 mg/d for 48 weeks. *2345*

Zimeldine   In 1 study involving 147 patients no significant change seen. *3741*

Blood   Increase   Analytical

Cholesterol   Probably by increasing viscosity. *0459*

Fibrinogen   Increased plasma concentration accelerates rate. *0251*

Blood   Increase   Physiological

Aging   Steady to puberty then rise in men, fall in women. *0887*

Anticonvulsants   Observed with SLE-like syndrome. *0933*

Aspirin   Occurs in some patients (reversible). *1343*

Carbamazepine   Associated with rare cases of granulomatous hepatitis. *3375*

Cephalothin   With high doses and prolonged i.v. administration. *3133*

Cephapirin   With high doses and prolonged i.v. administration. *3133*

Cyclosporine   Mean increase from 35 to 52 mm/1h in 16 patients with inflammatory ocular disease with 3 to 6 mo treatment. *2713*

Dextran   Due to cell aggregating properties. *2313*

Diurnal Variation   Maximum observed in mid afternoon. *2222*

Floxuridine   Manifestation of toxicity. *1680*

Globulin   High molecular weight proteins contribute. *2313*

Glymidine   Slight rise in mean rate reported. *2427*

Hydralazine   Observed in 9 patients developing rapidly progressive glomerulonephritis during treatment. *0366*

Hydrazine Derivatives   Augmentation with SLE-like syndrome. *0933*

Indomethacin   Oliguric renal failure developed in one man during treatment: reversible. *1245*

Isotretinoin   In 50 of 523 patients with mean daily dose of 109 mg for 150 d. *3867*

Lipemia   Large molecular weight components have effect. *2313*

Nitrofurantoin   Developed lupus-like syndrome associated with pulmonary reaction. *3250*   Mild increase developed lupus-like syndrome associated with pulmonary reaction. *3250*

Oral Contraceptives   Associated with increased fibrinogen. *3525*

Phenylbutazone   Observed in 7 patients with side effects due to drug. *3409*

Pregnancy   Normal associated with increased fibrinogen after third month. *0987*

Procainamide   Associated with SLE-like syndrome. *0933*

Erythrocyte Sedimentation Rate (continued)

Blood   Increase   Physiological (continued)

Quinidine   Reported in 5 cases of an SLE-like syndrome induced by the drug. 2093

Sulfamethoxazole   When given with trimethoprim fever and malaise; successful response to treatment. 1969

Tourniquet   Hemoconcentration increases rate. 0459

Vitamin A   Observed effect but explanation unknown. 2427

Blood   Decrease   Analytical

EDTA   Retards rate when compared with heparin. 0251

Fluorides   Retards rate when compared with heparin. 0459

Glucose   High blood sugar lowers sedimentation rate. 2313

Oxalate   Retards rate when compared with heparin. 0251

Quinine   At therapeutic concentration, maximum at 200 minutes with 2 mg/dL. 3000

Blood   Decrease   Physiological

Aspirin   If elevated, reduces toward normal value. 1343

Corticotropin   Particularly in rheumatoid patients. 0251

Cortisone   Particularly in rheumatoid patients. 0251

Gold   With 200 mg aurothioglucose at 4 weeks intervals mean ESR fell from 46 to 26 mm/h in 30 patients. 2619

Penicillamine   In 21 patients with rheumatoid arthritis treated for 48 weeks given 250 to 750 mg daily. 2345

Phospholipids   Possibly by increasing viscosity. 0251

Prednisone   Augmented response in cirrhotics when drug added to therapeutic regime. 0651

Tamoxifen   Observed in nonresponders with breast cancer. 1122

Trimethoprim   With sulfa caused marked decrease in rheumatoids. 1851

Erythrocyte Survival

Blood   No Effect   Physiological

Hydroxyurea   Although reduces rate of iron utilization. 1680

Blood   Decrease   Physiological

Acetanilid   Due to hemolysis. 1343

Aspirin   Large doses increase destruction. 1343

Lead   Due to hemolysis. 0987

Erythromycin

Serum   No Effect   Physiological

Theophylline   No effect on area under curve although appeared to be increased elimination. 1593

Erythropoietin

Serum   Increase   Physiological

Fluoxymesterone   Observed if anemia of renal failure. 1048

Plasma   Increase   Physiological

Anabolic Steroids   Metabolic effect. 0227

Urine   Increase   Physiological

Oxymetholone   Enhances production and excretion if anemia. 1679

Estradiol

Plasma   No Effect   Analytical

Blood Collection Tube   No effect on concentration even with storage for 6 d in gel-barrier tubes. 1592

Plasma   No Effect   Physiological

Anticonvulsants   Insignificantly higher in women on long-term therapy. 0257

Chlorthalidone   In 22 premenopausal and 18 postmenopausal women given 100 mg/d for 6 weeks. 0402

Cimetidine   In 25 men treated for duodenal ulcer or duodenitis. 2368

Cyclosporine   In 16 patients who developed hypertrichosis but also given added cortisone. 3190

Danazol   No consistent change noted in patients with prostatic cancer. 0693   In 7 normal women given 800 mg daily for 2 mo amenorrheic state induced drug. 2224

Dehydroepiandrosterone   In 5 men given 1600 mg/d orally for 28 d. 2578

Dexamethasone   No effect of treatment on initially high concentrations in hirsute women. 0776

Ethanol   No significant effect on concentration in individuals receiving alcohol over several days: no effect on pattern of secretion in healthy men. 1352

Famotidine   No effect of short-term or long-term treatment in male patients with duodenal ulcers. 0727

Ketoconazole   In men with prostatic cancer with 400 mg every 8 h and prolonged treatment. 3390

Levonorgestrel   No effect observed in group of women 20 and 65 mo after levonorgestrel treated rods inserted in uterus. 0899

Metoclopramide   No significant baseline differences in cycling women before and after drug. 0096

Naltrexone   With 50 to 100 mg administered daily to obese subjects over 8 weeks. 0176

Omeprazole   No effect observed in 8 volunteers given 30 mg/d for 28 d. 2516

Oxytetracycline   Despite decreased urine excretion and increased fecal excretion of estrogens due to decreased hydrolysis by beta-glucuronidase in gastrointestinal tract. 1469

Phenytoin   In 10 women receiving anticonvulsants, mainly phenytoin versus 10 controls, drug taken on average for 15 y. 0257

Prostaglandin F2α   When infused into pregnant women. 0023

Ranitidine   After 4 weeks or 6 mo treatment (300 mg and 150 mg daily respectively) in male patients with duodenal ulcer. 0727

Sexual Activity   No effect in women. 3437

Sodium Bromide   No effect in 10 men, 10 women receiving 1 mg/kg/d during 8 weeks or 2 full menstrual cycles. 3141

Spironolactone   No significant effect in males for 2 weeks. 2773

Sulpiride   In 11 healthy women between 6 and 9 weeks of pregnancy given 150 mg daily for 2 weeks. 3916

Sultopride   When given 300-600 mg/d for 5 weeks in 5 schizophrenic women. 2465

Plasma   Increase   Physiological

Clomiphene   Maximum with induced ovulation. 1295   Increase in both follicular and luteal phases over normal. 3005

Diazepam   Observed in 5 of 5 men with gynecomastia. 0316

Estradiol   Marked effect in 8 post-menopausal, post-oophorectomy. 3274

Menstruation   Observed 1-3 d before LH surge Second peak 30-200 pg/mL in luteal phase. 1295

Ovulation   Max of 150-500 pg/mL pre-ovulation. 1295

Pregnancy   Significant increase from 23rd to 41st week. 1143

Smoking   Significantly higher mean value in male current smokers than in nonsmokers with apparent dose response relationship. 0238

Tamoxifen   2 to 8 fold increase in 6 healthy volunteers given 20 mg/d for 5 to 10 d. In postmenopausal breast cancer women who did not respond to drug. 1542

Testosterone   3 fold increase on days 2 to 7 in 11 hypogonadal men given 200 mg cypionate salt intramuscularly. 2555

Plasma   Decrease   Physiological

Aminoglutethimide   By approximately 50% in women with advanced breast cancer (Postmenopausal). 0946

Cimetidine   Significantly reduced only in midproliferative phase of cycle. 0271

Cyproterone In 27 women treated for total of 194 cycles with 50 µg ethinyl estradiol and 2.0 mg cyproterone acetate. *2280*

Danazol But normal mid-follicular values during treatment in 62 patients with endometriosis treated with 600 mg daily for up to 24 weeks. *3222*

Ketoconazole Mild effect in 4 volunteer males with 600 mg doses. Bound and free ratio unchanged in 5 males receiving high doses for long time effect variable but estradiol testosterone ratio persistently increased. *2838*

Levonorgestrel In a group of women using levonorgestrel versus control group with copper IUD. *0899*

Megestrol In 18 postmenopausal women with breast cancer. *0056*

Oral Contraceptives Inhibits physiological rise. *1944*

Urine   No Effect   Physiological
Cyclofenil Normal amounts observed with GC/MS. *0023*

Urine   Increase   Analytical
Ethinyl Estradiol 10% color intensity Kober reaction. Up to 30% if method of Brown used. *0023*
Glycyrrhiza Affects methods of Brown and Beling. *0023*
Levodopa With method of Adlercreutz in one patient. *0023*
Phenolphthalein Affects unmodified Brown/Kober procedure. *0583*
Senna Elevated values method of Brown. *0023*

Urine   Increase   Physiological
Aging Maximal about 30-35 y, smooth fall thereafter. *1304*
Menstruation Peak 1 d before LH peak. *1762*
Spironolactone In males for 2 weeks from 2.6 to 3.5 µg/24 h. *2773*

Urine   Decrease   Analytical
Cascara Affects unmodified Brown/Kober procedure. *0583*
Corticotropin Large amounts affect method of Brown. *0023*
Cortisone Affects unmodified Brown/Kober procedure. *0583*
Cyclofenil Marked effect on colorimetric procedures. *0023*
Diethylstilbestrol Affects unmodified Brown/Kober procedure. *0583*
Vitamin B Complex Affects method of Beling. *0023*

Urine   Decrease   Physiological
Oral Contraceptives Hormonal effect. *2211*
Probenecid Tubular excretion may be blocked. *0023*

Estradiol Binding Globulin

Serum   Increase   Physiological
Oral Contraceptives Metabolic changes in liver synthesis. *2189*

Estradiol, Free

Plasma   Decrease   Physiological
Aminoglutethimide Marked reduction in small number of postmenopausal women with severe breast cancer although no difference in proportion (%) of total estradiol. *0946*

Estriol

Plasma   No Effect   Physiological
Dexamethasone No effect of treatment on initially high concentrations in hirsute women. *0776*
Spironolactone No significant effect in males for 2 weeks. *2773*

Plasma   Increase   Physiological
Pregnancy Average change 2 ng/mL at 20 weeks, 11 at term. *3836*

Plasma   Decrease   Physiological
Albuterol Significant reduction when given to women in premature labor. *1525*
Ampicillin May inhibit synthesis by fetoplacental unit. *0406*   In pregnant women due to alteration of glut flora. *2885*
Hexoprenaline Significant effect when given to women in premature labor. *1525*
Penicillin In pregnant women due to alteration of gut flora. *2885*
Prostaglandin F2α Marked with successful abortion. *3797*

Urine   No Effect   Analytical
Acetazolamide No effect observed on acid hydrolysis. *0023*
Acetone No effect observed on hydrolysis. *0406*
Albumin No effect observed on hydrolysis. *0406*
Amobarbital No interference with GLC. *1163*
Aspirin No effect of 188 mg/L on GLC method. *1163*
Butabarbital No effect on GLC method. *1163*
Chloral Hydrate No effect on GLC method. *1163*
Chlordiazepoxide No effect on GLC method. *1163*
Chlorothiazide No interference with hydrolysis observed. *3060*
Chlorpromazine No effect on GLC method. *1163*
Digoxin No effect on GLC method. *1163*
Ethchlorvynol No effect on GLC method. *1163*
Glucose With 80 mg glucose, enzyme hydrolysis and Kober procedure. *0762*
Glutethimide No effect on GLC method. *1163*
Hydrochlorothiazide At 4 x normal concentration with enzyme hydrolysis and Kober procedure. *0762*
Insulin No effect on GLC method. *1163*
Meprobamate No effect on GLC method. *1163*   If pretreated with alkali in method of Brown. *0023*
Methaqualone No effect on GLC method. *1163*
Methenamine At 1 g/dL with enzyme hydrolysis and Oakey procedure. *0762*
Methyldopa No effect on GLC method. *1163*
Methyprylon No effect on GLC method with 125 mg/L. *1163*
Pentaerythritol No effect on GLC method. *1163*
Pentazocine No effect with 12 mg/L on GLC method. *1163*
Phenobarbital No effect of 50 mg/L on GLC method. *1163*
Phenytoin No effect on GLC method. *1163*
Procainamide No effect of 12 mg/L on GLC method. *1163*
Prochlorperazine No effect of 25 mg/L on GLC method. *1163*
Promazine No effect of 25 mg/L on GLC method. *1163*
Propoxyphene No effect of 12 mg/L on GLC method. *1163*
Secobarbital No effect of 50 mg/L on GLC method. *1163*
Sulfanilamide No effect of 125 mg/L on GLC method. *1163*
Tetracycline No effect on GLC method. *1163*
Thioridazine No effect of 62 mg/L on GLC method. *1163*
Trifluoperazine No effect of 25 mg/L on GLC method. *1163*

Urine   No Effect   Physiological
Bed Rest No significant influence during pregnancy. *0966*
Diurnal Variation Variable changes during day in pregnancy. *0966*

Urine   Increase   Analytical
Glycyrrhiza Affects methods of Brown and Beling. *0023*
Levodopa With method of Adlercreutz in one patient. *0023*
Meprobamate Affects unmodified Brown/Kober procedure. *0583*

Urine   Increase   Physiological
Aging Maximal about 30-35 y, smooth fall thereafter. *1304*
Barbiturates Theoretical due to increased hydroxylation. *0023*
Corticotropin In pregnant increased production of adrenal precursors. *0023*
Menstruation Peak with ovulation, moderately high thereafter. *1762*
Prasterone SO$_4$ comp also affects if i.v. or oral in pregnant women. *0023*

Estriol (continued)

Urine   Increase   Physiological (continued)
Pregnancy   Progressive increase to 36th weeks, then steady. *3201*   Progressive increase to 40th week. *1315*
Spironolactone   In males for 2 weeks from 8.1 to 11.8 µg/24 h. *2773*

Urine   Decrease   Analytical
Aspirin   May affect enzyme hydrolysis. *0023*
Corticotropin   Large amounts affect method of Brown. *0023*
Cortisone   Sometimes low values produced method of Brown. *0023*
Cyclofenil   Less intense effect on colorimetric procedures. *0023*
Diethylstilbestrol   Affects unmodified Brown/Kober procedure. *0583*
Formaldehyde   Interferes with acid hydrolysis stage. *0406*
Fructose   Probably destroys estriol during acid hydrolysis. *0023*
Galactose   Probably destroys estriol during acid hydrolysis. *0023*   Interferes with GLC method. *3186*
Glucose   Interference with GLC method. *3186*   Probably destroys estriol during acid hydrolysis. *0023*
Hydrochlorothiazide   Interferes in hydrolysis of conjugates stage. *3060*   May have slight effect on enzyme hydrolysis. *0023*   Destroys estriol during acid hydrolysis. *0583*
Inulin   Probably destroys estriol during acid hydrolysis. *0023*
Lactose   Probably destroys estriol during acid hydrolysis. *0023*   Interference with GLC method. *3186*
Laxatives   If contain phenolphthalein reduce hydrolysis. *0406*
Mandelic Acid   If acid hydrolysis used with hexamine mandelate. *2427*
Mannose   Interference with GLC method. *3186*
Methenamine   Interferes with hydrolysis stage of methods. *1038*
Phenolphthalein   Interferes with acidic and enzyme hydrolysis. *0406*
Sucrose   Probably destroys estriol during acid hydrolysis. *0023*   Affects hydrolysis of conjugates. *0406*
Vitamin B Complex   Affects method of Beling. *0023*

Urine   Decrease   Physiological
Altitude   At 4200 meters mean 10.7 mg/24 h vs 23.0 mg/24 h at sea level. *0406*
Ampicillin   May diminish synthesis by fetoplacental unit. *0406*
Neomycin   Possibly affects integrity of intestinal microflora. *0023*
Oral Contraceptives   Hormonal effect. *2211*
Oxytetracycline   Probably due to decreased hydrolysis by beta-glucuronidase of estrogen conjugates in intestinal tract, with increased fecal loss. *1469*
Phthalylsulfathiazole   Possibly affects integrity of intestinal microflora. *0023*
Probenecid   Tubular excretion may be blocked. *0023*
L-Thyroxine   Decreased formation due to metabolic action. *0023*

Estriol-3-Glucuronide

Urine   Decrease   Physiological
Neomycin   Possibly affects integrity of intestinal microflora. *0023*
Oxytetracycline   Probably due to decreased hydrolysis by beta-glucuronidase of estrogen conjugates in intestinal tract with increased fecal loss. *1469*
Phthalylsulfathiazole   Possibly affects integrity of intestinal microflora. *0023*

Estrogen Binding Globulin

Serum   Increase   Physiological
Estrogens   Altered liver metabolism. *3332*
Oral Contraceptives   Metabolic effect. *2385*

Estrogens

Plasma   No Effect   Physiological
Prostaglandin F2α   Little effect in normals if given i.v. *1600*

Plasma   Increase   Physiological
Digoxin   2x in men, 1.5 x in postmenopausal women with chronic administration. *3475*
Oral Contraceptives   Often related to nausea. *0585*

Plasma   Decrease   Physiological
Bed Rest   Reported observation. *3217*

Urine   No Effect   Analytical
Aldosterone   At 50 mg/L on fluorescent method of Corns. *0730*
Ampicillin   No effect on method of Brown et al. *3622*
Androstenedione   At 50 mg/L on fluorescent method of Corns. *0730*
Androsterone   At 50 mg/L on fluorescent method of Corns. *0730*
Chlorothiazide   No significant effect Kober reaction and Allen correction. *0023*
Corticosterone   At 50 mg/L on fluorescent method of Corns. *0730*
Dihydrotestosterone   At 50 mg/L on fluorescent method of Corns. *0730*
Etiocholanolone   At 50 mg/L on fluorescent method of Corns. *0730*
Galactose   If sodium borohydride used with fluorescence procedures. *3899*
Glucose   If sodium borohydride used with fluorescence procedures. *3899*
Hydrocortisone   At 50 mg/L on fluorescent method of Corns. *0730*
Hydroxyprogesterone   At 50 mg/L on fluorescent method of Corns. *0730*
Lactose   If sodium borohydride used with fluorescence procedures. *3899*
Prasterone   At 50 mg/L on fluorescent method of Corns. *0730*
Pregnanetriol   At 50 mg/L on fluorescent method of Corns. *0730*
Progesterone   At 50 mg/L on fluorescent method of Corns. *0730*
Sorbitol   On fluorometric procedure at concentration of 1 g/dL. *3899*
Testosterone   At 50 mg/L on fluorescent method of Corns. *0730*

Urine   Increase   Analytical
Anthraquinone   Bright yellow colors with fluorometry. *0023*
Cascara   Reacts with Kober procedure Reacts in Brown's procedure. *0279*
Chlortetracycline   Possible effect (interference not defined). *0022*
Cortisone   Interfere with Brown's method Interferes with Kober procedure. *0279*
Diethylstilbestrol   Interferes in Kober procedure. Reacts in Brown's procedure. *0279*
Ethinyl Estradiol   Reacts in Brown's procedure. Interferes with Kober procedure. *0279*
Glycyrrhiza   Affects colorimetric and fluorometric procedures. *0022*
Levodopa   Affects colorimetric and fluorometric procedures. *0022*
Meprobamate   May react in Brown's and Kober's procedures. *0279*
Methenamine   Interferes in Kober procedure. *0279*
Nandrolone   Normal route of metabolism. *0022*
Naphthol   If applied topically affects colorimetry. *0023*
Penicillin   Activates enzyme used for hydrolysis. *0022*
Phenolphthalein   Affects colorimetric and fluorometric procedures. *0022*   Reacts in Brown's procedure. *0279*
Senna   Interferes in Kober procedure. *0279*
Tetracycline   Possible effect (interference not defined). *0022*

Urine   Increase   Physiological
Acetazolamide   In pregnant but mechanism unknown. *0023*

Chlorpromazine Blocks ovulation, maintains decidual reaction. *1343*

Clomiphene Due to action on hypothalamic-pituitary axis. *2585*

Corticotropin Hormonal effect in men and women. *2211*

Gonadotropin Hormonal action. *2211*

Hydrochlorothiazide In pregnant but mechanism unknown. *0023*

Lactation Excretion 5-10 μg/24 h until normal cycle resumes. *0987*

Ovulation Rises to peak at ovulation falls after. *0987*

Pregnancy Increase from sixth month until term (up to 100 μg/24 h). *0987*

Testosterone Probably due to conversion to estrogen. *3804*

Thymol Marked interference with method of Hainsworth. *1457*

Urine   Decrease   Analytical

Acetazolamide Affects hydrolysis of estrogen conjugates. *0022*

Cascara Affects colorimetric/fluorometric procedures. *0022*

Corticotropin Affects colorimetric/fluorometric procedures. *0022*

Cyclofenil Affects colorimetric/fluorometric procedures. *0022*

Diethylstilbestrol Affects colorimetric/fluorometric procedures. *0022*

Ethinyl Estradiol Affects colorimetric/fluorometric procedures. *0022*

Fructose Affects hydrolysis of estrogen conjugates. *0022*

Galactose Affects fluorometric procedure unless removed. *3899*

Glucose 10% decrease if fluorometric method used. *1457* Affects hydrolysis of estrogen conjugates. *0022*

Hydrochlorothiazide Affects colorimetric/fluorometric procedures. *0022* Affects Kober reaction if Allen correction. *0023*

Inulin Affects hydrolysis of estrogen conjugates. *0022*

Lactose Affects fluorometric procedures unless removed. *3899*

Methenamine Affects hydrolysis of estrogen conjugates. *0022*

Naphthol Affects colorimetric/fluorometric procedures. *0022*

Phenolphthalein Competes for enzyme used for hydrolysis. *0022*

Salicylate Competes for enzyme used for hydrolysis. *0022*

Sucrose Affects hydrolysis of estrogen conjugates. *0022*

Vitamin B Complex Affects colorimetric/fluorometric procedures. *0022*

Urine   Decrease   Physiological

Ampicillin Increase by 46% after 6 d in pregnant, altered gastrointestinal flora. During luteal phase in normal women. During ovulatory phase in normals. *3622*

Dexamethasone Decreased conversion of neutral steroids. *0023*

Menopause Normal response. *0987*

Menstruation Lowest 2-3 d after onset. *0987*

Neomycin In pregnant women due to alteration of gut flora. *2885*

Oral Contraceptives Hormonal effect (decreased by 40%). *3217*

Penicillin In pregnant women due to alteration of gut flora. *2885*

Phenothiazines Block ovulation, inhibit decidual reaction. *1022*

Estrogens (Conjugated)

Urine   No Effect   Analytical

Ampicillin No effect on Ittrich/Kober reaction. *3622*

Urine   Decrease   Physiological

Ampicillin Increase by 50% after 6 d in pregnant women. *3622*

Estrone

Plasma   No Effect   Physiological

Danazol In 7 normal women given 800 mg daily for 2 mo amenorrheic state induced by drug. *2224*

Dehydroepiandrosterone In 5 men given 1600 mg/d orally for 28 d. *2578*

Oxytetracycline Despite decreased urine excretion and increased fecal excretion of estrogens due to decreased hydrolysis by beta-glucuronidase in gastrointestinal tract. *1469*

Prostaglandin F2α When infused into pregnant woman. *0023*

Spironolactone No significant effect in males for 2 weeks. *2773*

Plasma   Increase   Physiological

Digoxin Major estrogen with chronic administration. *3475*

Pregnancy Probably slight increase from 24th to 41st weeks. *1143*

Smoking Significantly higher mean value in male current smokers than in nonsmokers with apparent dose response relationship. *0238*

Plasma   Decrease   Physiological

Cyproterone In 27 women treated for total of 194 cycles with 50 μg ethinyl estradiol and 2.0 mg cyproterone acetate. *2280*

Urine   No Effect   Physiological

Spironolactone No significant effect in men. *2773*

Urine   Increase   Analytical

Ethinyl Estradiol Effect of hot alkali treatment. *0023*

Urine   Increase   Physiological

Aging Maximal about 30-35 y, smooth fall thereafter. *1304*

Menstruation Peak 1 d before LH peak. *1762*

Progesterone Not significant change however. *0991*

Urine   Decrease   Analytical

Diethylstilbestrol method of Brown if very large doses used. *0023*

Senna Affects unmodified Brown/Kober procedure. *0583*

Urine   Decrease   Physiological

Probenecid Tubular excretion may be blocked. *0023*

Estrone Sulfate

Plasma   Increase   Physiological

Pregnancy Progressive increase with time (2x estradiol). *2213*

Ethanol

Serum   No Effect   Physiological

Ranitidine No effect on peak concentration or area under the curve with pretreatment with drug. *3246* Peak plasma concentration and area under curve unaffected. *1938*

Serum   Increase   Physiological

Cimetidine Prior administration of drug caused increased concentration probably due to enhanced absorption rather than effect on metabolism. *3246*

Serum   Decrease   Physiological

Food Absorption slowed when taken with food. *3024*

Blood   No Effect   Analytical

Azide Over 16 weeks at 20° at concentration of 0.5% w/v. *3355*

Bisulfite Over 16 weeks at 20° at concentration of 0.5% w/v. *3355*

Dithionite Over 16 weeks at 20° at concentration of 0.5% w/v. *3355*

Nitrites Over 16 weeks at 20° at concentration of 0.5% w/v. *3355*

Blood   Increase   Physiological

Ethanol 2 g/L = intoxication, stupor at 3 g/L. *2348* Steady fall from 30 minutes after ingestion after 0.4 g alcohol/kg in 56 healthy males. *0015*

Metoclopramide Increased rate of absorption with faster gastric emptying. *3794*

Blood   Decrease   Analytical

Storage of Sample 0.23 mg/dL/h at 20° if no preservative. *3355*

Ethanol *(continued)*

Blood   Decrease   Physiological
Atropine   Reduced rate of absorption due to delayed gastric emptying. *3794*

Propantheline   Reduced rate of absorption with delayed gastric emptying. *3794*

Protein Deficiency   Causes reduced alcohol dehydrogenase activity. *0398*

Urine   Increase   Physiological
Ethanol   At equilibrium approximately 130% blood concentration. *1343*

CSF   Increase   Physiological
Ethanol   Usually at lower concentration than in blood. *1343*

Red Blood Cells   Increase   Physiological
Ethanol   But lower in RBC than in plasma. *1343*

Ethchlorvynol

Serum   No Effect   Analytical
Acetaminophen   At 100 mg/L on GLC procedure of Evenson. *1058*

Aminophylline   At 80 mg/L on GLC procedure of Evenson. *1058*

Amobarbital   At 10 µg/mL on colorimetric method of Wallace. *3753*   At 600 mg/L on GLC procedure of Evenson. *1058*

Amphetamine   At 40 mg/L on GLC procedure of Evenson. *1058*

Aspirin   At 10 µg/mL on colorimetric method of Wallace. *3753*

Caffeine   At 130 mg/L on GLC procedure of Evenson. *1058*

Carbromal   At 10 µg/mL on colorimetric method of Wallace. *3753*

Chlordiazepoxide   At 10 µg/mL on colorimetric method of Wallace. *3753*   At 90 mg/L on GLC procedure of Evenson. *1058*

Chlorpromazine   At 100 mg/L on GLC procedure of Evenson. *1058*

Cocaine   At 8 mg/L on GLC procedure of Evenson. *1058*

Codeine   At 8 mg/L on GLC procedure of Evenson. *1058*

Desipramine   At 100 mg/L on GLC procedure of Evenson. *1058*

Diazepam   At 10 µg/mL on colorimetric method of Wallace. *3753*   At 600 mg/L on GLC procedure of Evenson. *1058*

Diphenhydramine   At 80 mg/L on GLC procedure of Evenson. *1058*

Ethinamate   At 10 µg/mL on colorimetric method of Wallace. *3753*

Glutethimide   At 600 mg/L on GLC procedure of Evenson. *1058*   At 10 µg/mL on colorimetric method of Wallace. *3753*   At 10 mg/L on GLC procedure of Evenson. *1058*

Imipramine   At 100 mg/L on GLC procedure of Evenson. *1058*

Meperidine   At 80 mg/L on GLC procedure of Evenson. *1058*

Meprobamate   At 10 mg/L on GLC procedure of Evenson. *1058*   At 10 µg/mL on colorimetric method of Wallace. *3753*

Methadone   At 8 mg/L on GLC procedure of Evenson. *1058*

Methamphetamine   At 40 mg/L on GLC procedure of Evenson. *1058*

Methapyrilene   At 10 mg/L on GLC procedure of Evenson. *1058*

Methaqualone   At 10 mg/L on GLC procedure of Evenson. *1058*

Methyprylon   At 10 µg/mL on colorimetric method of Wallace. *3753*   At 10 mg/L on GLC procedure of Evenson. *1058*

Morphine   At 8 mg/L on GLC procedure of Evenson. *1058*

Nordiazepam   At 600 mg/L on GLC procedure of Evenson. *1058*

Nortriptyline   At 100 mg/L on GLC procedure of Evenson. *1058*

Pentazocine   At 150 mg/L on GLC procedure of Evenson. *1058*

Pentobarbital   At 600 mg/L on GLC procedure of Evenson. *1058*

Phenacemide   At 10 µg/mL on colorimetric method of Wallace. *3753*

Phenaglycodol   At 10 µg/mL on colorimetric method of Wallace. *3753*

Phenazopyridine   At 250 mg/L on GLC procedure of Evenson. *1058*

Phenobarbital   At 600 mg/L on GLC procedure of Evenson. *1058*   At 10 µg/mL on colorimetric method of Wallace. *3753*

Phenytoin   At 10 µg/mL on colorimetric method of Wallace. *3753*   At 600 mg/L on GLC procedure of Evenson. *1058*

Primidone   At 600 mg/L on GLC procedure of Evenson. *1058*

Prochlorperazine   At 100 mg/L on GLC procedure of Evenson. *1058*

Promazine   At 100 mg/L on GLC procedure of Evenson. *1058*

Propoxyphene   At 70 mg/L on GLC procedure of Evenson. *1058*

Quinidine   At 10 mg/L on GLC procedure of Evenson. *1058*

Salicylate   At 180 mg/L on GLC procedure of Evenson. *1058*   At 180 mg/L on GLC procedure of Evenson. *1058*

Secobarbital   At 10 µg/mL on colorimetric method of Wallace. *3753*   At 600 mg/L on GLC procedure of Evenson. *1058*

Thiopental   At 10 µg/mL on colorimetric method of Wallace. *3753*

Thioridazine   At 100 mg/L on GLC procedure of Evenson. *1058*

Trifluoperazine   At 100 mg/L on GLC procedure of Evenson. *1058*

Serum   Increase   Analytical
Phenazopyridine   False increase with method of Frings. *0576*

Serum   Increase   Physiological
Ethchlorvynol   200 mg orally may produce concentration of 2 mg/L in blood. *2348*

Urine   No Effect   Analytical
Acetaminophen   No effect on method of Frings and Cohen at 10 mg/dL. *1205*

Amitriptyline   No effect on method of Frings and Cohen at 10 mg/dL. *1205*

Amobarbital   No effect on method of Frings and Cohen at 10 mg/dL. *1205*

Amphetamine   No effect on method of Frings and Cohen at 10 mg/dL. *1205*

Aspirin   No effect on Frings and Cohen method at 150 mg/dL. *1205*

Benztropine   No effect on method of Frings and Cohen. *1204*

Bromides   No effect on method of Frings and Cohen at 250 mg/dL. *1205*

Butabarbital   No effect on method of Frings and Cohen at 10 mg/dL. *1205*

Carisoprodol   No effect on method of Frings and Cohen. *1204*

Chloral Hydrate   No effect on method of Frings and Cohen at 10 mg/dL. *1205*

Chlordiazepoxide   No effect on method of Frings and Cohen at 10 mg/dL. *1205*

Chloroquine   No effect on method of Frings and Cohen. *1204*

Chlorpheniramine   No effect on method of Frings and Cohen at 12 mg/dL. *1205*

Chlorpromazine   No effect on method of Frings and Cohen with 20 mg/dL. *1205*

Clorazepate   No effect on method of Frings and Cohen. *1204*

Cocaine   No effect on method of Frings and Cohen at 2 mg/dL. *1205*

Codeine   No effect on method of Frings and Cohen at 2 mg/dL. *1205*

Diazepam   No effect on method of Frings and Cohen at 10 mg/dL. *1205*

Diethylpropion   No effect on method of Frings and Cohen. *1204*

Diphenhydramine   No effect on method of Frings, Cohen at 10 mg/dL. *1205*

Doxepin   No effect on method of Frings and Cohen. *1204*

Ephedrine   No effect on method of Frings and Cohen. *1204*

Ethchlorvynol   No effect on method of Frings and Cohen. *1204*

Ethinamate   No effect on method of Frings and Cohen. *1204*

Flurazepam   No effect on method of Frings and Cohen at 10 mg/dL. *1205*

Glutethimide   No effect on method of Frings and Cohen at 10 mg/dL. *1205*

Glycopyrrolate   No effect on method of Frings and Cohen. *1204*

Haloperidol   No effect on method of Frings and Cohen. *1204*

Hydromorphone   No effect on method of Frings and Cohen at 2 mg/dL. *1205*

Imipramine   No effect on method of Frings and Cohen at 10 mg/dL. *1205*

Mebutamate   No effect on method of Frings and Cohen. *1204*

Meperidine   No effect on method of Frings and Cohen at 10 mg/dL. *1205*

Meprobamate   No effect on method of Frings and Cohen at 20 mg/dL. *1205*

Mescaline   No effect on method of Frings and Cohen. *1204*

Methadone   No effect on method Frings and Cohen at 2 mg/dL. *1205*

Methamphetamine   No effect on method of Frings and Cohen. *1204*

Methapyrilene   No effect on method of Frings and Cohen. *1204*

Methaqualone   No effect on method of Frings and Cohen at 10 mg/dL. *1205*

Methdilazine   No effect on method of Frings and Cohen. *1204* No effect on method Frings and Cohen at 10 mg/dL. *1205*

Methylphenidate   No effect on method of Frings and Cohen. *1204*

Methyprylon   No effect on method Frings and Cohen at 20 mg/ml. *1205*

Morphine   No effect on method Frings and Cohen at 2 mg/dL. *1205*

Nalorphine   No effect on method of Frings and Cohen at 10 mg/dL. *1205*

Oxazepam   No effect on method of Frings and Cohen. *1204*

Oxycodone   No effect on method of Frings and Cohen. *1205*

Oxymetazoline   No effect on method of Frings and Cohen. *1204*

Pentazocine   No effect on method of Frings and Cohen at 10 mg/dL. *1205*

Perphenazine   No effect on method of Frings and Cohen at 10 mg/dL. *1205*

Phenmetrazine   No effect on method of Frings and Cohen. *1204*

Phenobarbital   No effect on method Frings and Cohen at 20 mg/dL. *1205*

Phenylephrine   No effect on method of Frings and Cohen. *1204*

Phenylpropanolamine   No effect on method of Frings and Cohen. *1204*

Phenytoin   No effect on method of Frings and Cohen at 10 mg/dL. *1205*

Primaquine   No effect on method of Frings and Cohen. *1204*

Procainamide   No effect on method of Frings and Cohen at 10 mg/dL. *1205*

Prochlorperazine   No effect on method of Frings and Cohen at 20 mg/dL. *1205*

Promazine   No effect on method of Frings and Cohen at 10 mg/dL. *1205*

Promethazine   No effect on method of Frings and Cohen at 10 mg/dL. *1205*

Propoxyphene   No effect on method of Frings and Cohen at 10 mg/dL. *1205*

Pyrilamine   No effect on method of Frings and Cohen. *1204*

Pyrimethamine   No effect on method of Frings and Cohen. *1204*

Quinidine   No effect on method of Frings and Cohen. *1204*

Quinine   No effect on method of Frings and Cohen. *1204*

Salicylate   No effect on method of Frings and Cohen. *1204*

Scopolamine   No effect on method of Frings and Cohen. *1204*

Secobarbital   No effect on method of Frings and Cohen at 10 mg/dL. *1205*

Sodium Bromide   No effect on method of Frings and Cohen. *1204*

Sominex®   No effect on method of Frings and Cohen. *1205*

Strychnine   No effect on method of Frings and Cohen at 10 mg/dL. *1205*

Sulfanilamide   No effect on method of Frings and Cohen at 100 mg/dL. *1205*

Tetracycline   No effect on method of Frings and Cohen. *1204*

Thiobarbituric Acid   No effect on method of Frings and Cohen. *1204*

Thiopropazate   No effect on method of Frings and Cohen at 10 mg/dL. *1205*

Thioridazine   No effect on method of Frings and Cohen at 10 mg/dL. *1205*

Thiothixene   No effect on method of Frings and Cohen. *1204*

Trifluoperazine   No effect on method of Frings and Cohen at 20 mg/dL. *1205*

Triflupromazine   No effect on method of Frings and Cohen. *1204*

Trihexyphenidyl   No effect on method of Frings and Cohen. *1204*

Trimethobenzamide   No effect on method Frings and Cohen at 25 mg/dL. *1205*

Tybamate   No effect on method of Frings and Cohen. *1204*

Urine   Increase   Analytical
Phenazopyridine   Reacts in method of Frings and Cohen. *1205*

Urine   Increase   Physiological
Ethchlorvynol   200 mg orally may produce concentration of 1 mg/L. *2348*

Ethosuximide

Serum   No Effect   Physiological
Valproic Acid   No definite effect in 25 patients studied for 5 to 9 mo. *3829*

Serum   Increase   Physiological
Valproic Acid   Reported increase when valproic acid co-administered. *0119*

Ethylene Glycol

Urine   Increase   Physiological
Ethylene Glycol   Readily detectable in urine with poisoning. *3325*

Etiocholanolone

Urine   No Effect   Physiological
Metoclopramide   No significant influence of drug on excretion in menstruating women. *0096*

Urine   Increase   Physiological
Corticotropin   Normal response to 2 d ACTH test. *1474*
Gonadotropin   150% increase when given to normal men. *3231*

Urine   Decrease   Physiological
Aging   In women from 2.7 mg/24 h at 25 to 1.3 at 55 y. *0462*
Dexamethasone   Suppression of ACTH. *2420*
Oral Contraceptives   Compared with controls — details not discussed. *0527*

Euglobulin Clot Lysis Time

Blood   Decrease   Physiological
Asparaginase   Occasionally observed with toxicity. *2864*
Clofibrate   Reported observation. *2427*
Dextran   Marked decrease noted. *0320*
Muscular Exercise   Observed effect in exercise. *2847*
Streptokinase   Significant effect in patients with myocardial infarct. *1356*

Expiratory Volume

Patient   No Effect   Physiological
Oral Contraceptives   No significant effect observed. *1185*

Extracellular Fluid (ECF)

Plasma Decrease Physiological
Reducing Diet May constrict by 30% with 300-600 Cal/d. *3217*

Factor I

Plasma No Effect Physiological
Oral Contraceptives Usually no effect observed. *0339*

Factor II

Plasma Increase Physiological
Anticonvulsants In 8 epileptic patients receiving phenobarbital with carbamazepine or phenytoin. *2670*
Estrogens Reported effect. *0429*
Fluoxymesterone Metabolic effect. *0019*
Methandrostenolone Metabolic effect. *0019*
Muscular Exercise Observed effect in exercise. *2847*
Nandrolone Metabolic effect. *2681*
Oral Contraceptives Reported effect of estrogens. *0429*
Oxymetholone Anabolic effect increasing factors. *1679*
Pregnancy Normal response in late pregnancy. *0987*

Plasma Decrease Physiological
Moxalactam Depression in 40 preoperative surgical patients: effects mild. *3121*
Warfarin Therapeutic action. *1680*
Warfarin Metabolites Slight effect with some compounds only. *2159*

Factor V

Plasma No Effect Physiological
Fluroxene No significant effect noted. *1814*
Muscular Exercise Observed effect in exercise. *2847*
Oral Contraceptives Usually no effect observed. *0339*

Plasma Increase Physiological
Anabolic Steroids Metabolic effect. *0227*
Epinephrine Transient effect. *2220*
Estrogens, Conjugated Small increases reported. *0083*
Fluoxymesterone Metabolic effect. *0019*
Methandrostenolone Metabolic effect. *0019*
Nandrolone Metabolic effect. *2681*
Oral Contraceptives Significant effect of combined oral contraceptives. *2456*
Oxymetholone Anabolic effect increasing factors. *1679*
Xanthine Effect of methylxanthines. *1343*

Plasma Decrease Analytical
EDTA Unstable ?affects prothrombin determination. *3217*
Oxalate Unstable ?affects prothrombin determination. *3217*

Plasma Decrease Physiological
Asparaginase Diminished hepatic synthesis. *0578*
Dextran Slight effect (more than hemodilution). *2427*
Heparin Concentration related effect. *1343*
Methyltestosterone Suppresses activity. *2427*

Factor VII

Plasma No Effect Physiological
Ceftazidime In 30 patients with serious infections. *0031*

Plasma Increase Physiological
Anabolic Steroids Metabolic effect. *0227*

Anticonvulsants In 8 epileptic patients receiving phenobarbital with carbamazepine or phenytoin. *2670*
Estrogens Altered protein metabolism. *3332*
Fluoxymesterone Metabolic effect. *0019*
Methandrostenolone Metabolic effect. *0019*
Nandrolone Metabolic effect. *2681*
Oral Contraceptives Estrogen effect (higher than in pregnancy). *3332*
Oxymetholone Anabolic effect increasing factors. *1679*
Pregnancy Normal response in late pregnancy. *0987*

Plasma Decrease Physiological
Asparaginase Diminished hepatic synthesis. *0578*
Aspirin Acts like bishydroxycoumarin. *1343*
Cefamandole Effect on activity in 30 patients with serious infections. Possibly associated with sulfhydryl group. *0031*
Ceftriaxone In 30 patients with serious infections possibly due to presence of sulfhydryl group. *0031*
Dextran Decrease by 50% in normals. *2427*
Dextrothyroxine Reported effect. *0019*
Dicumarol Dose related effect. *1343*
Methyltestosterone Suppresses activity. *2427*
Moxalactam Depression in 40 preoperative surgical patients: effects mild. *3121*
Oral Contraceptives Metabolic effect. *2834*
Phenytoin May cross placenta (vitamin K dependent). *3378*
Vitamin E In 2 patients given 2.3 $g/m^2/d$ intravenously for 4 or more days (effect abrogated by prior administration of menadiol sodium diphosphate). *1554*
Warfarin Therapeutic action. *1680*
Warfarin Metabolites Usually marked effect all compounds. *2159*

Factor VIII

Plasma No Effect Physiological
Danazol In 21 hemophiliacs given drug for 2 weeks. *2627*

Plasma Increase Physiological
Chlormadinone Significant effect observed by 3 mo. *2456*
Epinephrine Occurs following epinephrine i.v. *0883*
Muscular Exercise Biological activity augmented. *0883* Observed effect in exercise. *2847*
Oral Contraceptives Slight effect observed. *1462*
Pregnancy Increased 100% above normal in late pregnancy. *0987*
Stress With post operative stress. *0883*

Plasma Decrease Physiological
Asparaginase Diminished hepatic synthesis. *0578*
Dextran Marked decrease noted in single case. *0320*
Dextrothyroxine Reported effect. *0019*

Factor VIII Antigen

Plasma Increase Physiological
Epinephrine Occurs following epinephrine i.v. *0883*
Muscular Exercise Antigen increased following exercise. *0883*
Stress Occurs with post operative stress. *0883*
Tri-iodothyronine In 7 normal women aged 18-27 y given 25 μg/d three times daily for 14 d. Effect observed on endothelium-associated proteins but not on hepatically synthesized proteins. *1383*

Factor IX

Plasma No Effect Physiological
Danazol In 21 hemophiliacs given drug for 2 weeks. *2627*

Plasma   Increase   Physiological
Chlormadinone   Significant effect observed by 3 mo. *2456*
Estrogens   Altered protein metabolism. *3332*
Estrogens, Conjugated   Small increases reported. *0083*
Oral Contraceptives   Estrogen effect. *3332*
Pregnancy   Normal response in late pregnancy. *0459*

Plasma   Decrease   Physiological
Asparaginase   In 75 to 100% in different studies. *0558*
Dextran   Slight effect (more than hemodilution). *2427*
Dextrothyroxine   Reported effect. *0019*
Dicumarol   Dose related effect. *1343*
Heparin   Reported effect. *1343*
Vitamin E   In 2 patients given 2.3 g/m²/d intravenously for 4 or more days (effect abrogated by prior administration of menadiol sodium diphosphate). *1554*
Warfarin   Therapeutic action. *1680*
Warfarin Metabolites   Usually marked effect all compounds. *2159*

Factor X

Plasma   No Effect   Physiological
Norethisterone   In 75 women who had received up to 24 intragluteal injections of 200 mg every 56 d. *1666*

Plasma   Increase   Physiological
Anabolic Steroids   Metabolic effect. *0227*
Anticonvulsants   In 8 epileptic patients receiving phenobarbital with carbamazepine or phenytoin. *2670*
Estrogens   Slight effect observed. *0585*
Fluoxymesterone   Metabolic effect. *0019*
Methandrostenolone   Metabolic effect. *0019*
Nandrolone   Metabolic effect. *2681*
Oral Contraceptives   Slight effect observed. *1462*
Oxymetholone   Anabolic effect increasing factors. *1679*
Pregnancy   Normal response in late pregnancy. *0987*

Plasma   Decrease   Physiological
Chlormadinone   Significant effect observed by 3 mo. *2456*
Dicumarol   Dose related effect. *1343*
Methyltestosterone   Suppresses activity. *2427*
Oral Contraceptives   Metabolic effect. *2834*
Vitamin E   In 2 patients given 2.3 g/m²/d intravenously for 4 or more days (effect abrogated by prior administration of menadiol sodium diphosphate). *1554*
Warfarin   Therapeutic action. *1680*
Warfarin Metabolites   Usually marked effect all compounds. *2159*

Factor XI

Plasma   Decrease   Physiological
Asparaginase   In 75 to 100% in different studies. *0558*
Captopril   High value of essential hypertension significantly reduced. *2748*
Dextran   Marked reduction in one case. *0320*
Heparin   Concentration related effect. *1343*
Pregnancy   Normal response in late pregnancy. *0987*

Factor XII

Plasma   Increase   Physiological
Oral Contraceptives   Reported effect. *2385*

Plasma   Decrease   Physiological
Captopril   High value of essential hypertension significantly reduced. *2748*

Fat

Feces   Increase   Physiological
Aminosalicylic Acid   Produces steatorrhea possibly because of bile acid chelation. *3023*
Azathioprine   May occasionally cause steatorrhea. *1680*
Bisacodyl   May cause steatorrhea if protracted ingestion. *3855* Observed in laxative abusers. *3023*
Cholestyramine   Occurs with doses over 15 g/d. *1680* Dose-dependent: forms nonabsorbable complexes with bile salts which are lost in feces. *3023*
Colchicine   May cause villus damage and impaired regeneration of the epithelial cells of small intestine. *3023*
Colestipol   Bile acid sequestrant may cause steatorrhea with conventional dosage schedules. *3023*
Gum Tragacanth   In 5 male volunteers consuming 9.9 g daily for 21 d. *0990*
Kanamycin   May induce malabsorption with diarrhea. *1680* Steatorrhea produced probably by causing mucosal damage. *3023*
Lincomycin   Steatorrhea may result from mucosal damage. *3023*
Methotrexate   Probably due to mucosal damage with impaired regeneration of epithelial cells of small intestine. *3023*
Neomycin   Alters intestinal villi, inhibits triglyceride hydrolysis. *3855* Steatorrhea produced because of mucosal damage and rendering bile salts less available for fat absorption. *3023*
Phenolphthalein   Steatorrhea observed in some laxative abusers. *3023*
Polymyxin   Probably because of mucosal damage. *3023*

Feces   Decrease   Physiological
Cimetidine   By about 30%; in 17 patients with cystic fibrosis receiving constant concomitant therapy with pancreatic enzymes. *0624*

Fatty Acids, Free (FFA)

Serum   No Effect   Analytical
Bilirubin   Even at 30 mg/dL on method of Mikac-Devic. *2438* No effect on method of Soloni. *3381* No effect on method of Pinelli. *2819*
Caproic Acid   No effect on method of Pinelli. *2819*
Caprylate   No effect on method of Pinelli. *2819*
Cholesterol   At 5.2 mmol/L — method of Noma. *2607*
EDTA   No effect of anticoagulant on method of Soloni. *3381*
Glucose   Up to 480 mg/dL on method of Soloni. *3381*
Hemolysis   No effect on method of Soloni. *3381*
Heparin   No effect as anticoagulant on method of Soloni. *3381*
Lactate   No effect on method of Pinelli. *2819* At 50 mg/dL method of Soloni. *3381*
Lecithin   No effect on method of Pinelli. *2819*
Niacin   No effect on method of Pinelli. *2819*
Oxalate   No effect as anticoagulant on method of Soloni. *3381*
Phospholipids   NaCl used to eliminate in method of Mikac-Devic. *2438*
Pyruvate   At 10 mg/dL on method of Soloni. *3381*
Salicylate   No effect on method of Pinelli. *2819*
Triolein   At 2.3 mmol/L — method of Noma. *2607*
Uric Acid   At 25 mg/dL on method of Soloni. *3381*

Serum   No Effect   Physiological
Cannabis   No effect observed. *1633*
Chlorthalidone   In 22 premenopausal and 18 postmenopausal women given 100 mg/d for 6 weeks. *0402*
Labetalol   In response to i.v. infusion of 100 mg over 10 minutes. *0226*
Physical Training   No difference between trained athletes and others. *1702*
Pindolol   No significant changes after 12 mo treatment. *2116* Remained constant during 12 mo therapy. *2117*
Pizotyline   No effect after 2 mg i.v. *3411*

Fatty Acids, Free (FFA) *(continued)*

Serum No Effect Physiological *(continued)*

Propranolol In 5 hyperthyroid patients given 10 mg every 8 h for 4 d. *0335*

Site of Collection Independent of site. *3639*

Smoking If no inhalation with cigarettes With half cigar and no inhalation. *1921*

Terbutaline In 6 normal men treated with therapeutic amounts for 2 weeks. *3177*

Tetracosactrin In healthy volunteers given 1 mg intramuscularly for up to 60 h. *1813*

Thyrotropin-Releasing Hormone (TRH) After single dose of 1.0 mg synthetic TRH. *3831*

Timolol In 5 hyperthyroid patients given 10 mg every 8 h for 4 d. *0335*

Serum Increase Analytical

Bilirubin In methods using diethyldithiocarbamate. *2438* Extracted in $CHCl_3$ in procedure of Itaya and Ui. *2079*

Cholesteryl Oleate At 5.2 mmol/L slight effect on method of Noma. *2607*

Heparin Some interference observed if plasma used. *2438* Marked increase in some patients when drug given i.v. Effect probably *in vitro* artifact as can be abolished by protamine inhibition of lipoprotein lipase. *2408*

Lecithin At 1.6 mmol/L = 100 βEq/L method of Noma. *2607*

Storage of Sample 218% at room temperature for 3 d 43% at 4° for 7 d due to lipoprotein lipase. *2512* Of 45% if few h between drawing and analysis. *2438*

Valproic Acid With colorimetric method of Duncombe (overestimation by about 40% at plasma concentration. *0045*

Serum Increase Physiological

Amphetamine But does not modify carbohydrate utilization. *1343*

Benzquinamide 136% increase following single injection i.m. of 300 mg. *1634*

Blood Pressure Strongly correlated with increased systolic pressure. *2123*

Caffeine 220 mg raised nonesterified fatty acids by approximately 30%. *1894* Concentration doubled 1 h after 250 mg and remained increased for 4 h. *2751*

Carbutamide Reported to cause hyperlipemia. *2220*

Chlorpromazine 51% increase following single i.m. injection of 50 mg. *1634*

Coffee Possibly due to stress effect. *2427*

Desipramine 41% increase following single injection i.m. of 50 mg. *1634*

Diazoxide Significant rise observed after oral or i.v. administration. *0124*

Endotoxin Initial rise occurs, normal later. *1230*

Epinephrine Transient metabolic effect. *1343*

Ethanol Altered metabolism with decreased fatty acid oxidation. *1343* But only after high doses. *2428*

Glucose Delayed rise 3 h after administration. *1697*

Growth Hormone Induces lipolysis (?at adenyl cyclase step). *0819*

Heparin Also occurs with situational stresses. *3873* Mean concentration increased 4-fold in 19 patients. *0253* From mean of 0.55 to 2.20 mmol/L following administration in 19 patients: extent correlated with pre-heparin concentration of triglycerides. *0253*

Hypothermia Due to hypoxia, acid-base imbalance, hypotension. *2258*

Isoproterenol As effective as epinephrine. *1343*

Levarterenol Marked increase observed after i.v. infusion. *3114* Metabolic response. *1176*

Levodopa Significant increase if levodopa high in serum (i.v. greater effect). *2997*

Lysergide 73% after 1.5 µg/kg orally. *1634*

Mescaline 115% increase after 5 mg/kg body weight orally. *1634*

Molindone Sustained and significant rise. *1635*

Muscular Exercise Marked increase after strenuous exercise. *1702*

Nicotine 30% increase during smoking of 3 cigarettes. *1921*

Oral Contraceptives If given for 3 mo. *3505*

Prazosin With 2 mg in 12 hypertensives (6 with normal and 6 with abnormal glucose tolerance). *0225*

Reserpine 78% increase after 2.5 mg injected i.m. *1634*

Sandfly Fever Increased 2 fold from baseline. *2929*

Smoking 30% increase during smoking of 3 cigarettes. *1921*

Starvation Due to release and degradation of fatty acids. *2240*

Stress Normal metabolic response. *3527*

Theophylline Rapid pronounced and prolonged rise associated with aminophylline to produce therapeutic concentration of theophylline. *0608* 123% increase after 4 d of treatment in 10 healthy volunteers. *3757*

Tolbutamide Slight initial rise then fall to 25% at 90 minutes. *1526*

Trichlormethiazide Significant increase in fasting state (0.37 vs 0.31 meq/L) and 60 minutes (0.33 vs 0.26 meq/L) after 75 g glucose orally. *2550*

Xanthine Reported effect. *1343*

Serum Decrease Analytical

Citrates As anticoagulant or *in vivo* method of Soloni. *3381*

Storage of Sample Significant decrease at -40° for 28 d. *2512*

Serum Decrease Physiological

Acebutolol Significant effect after one month's treatment and then plateaued. *2116* Significant reduction in 18 patients treated for 6 mo. *2115*

Acetoacetate Effect maximal 40 minutes after load (not marked). *1811*

Amino Acids More marked if intraduodenal than i.v. *2918*

Asparaginase Common hepatotoxic response. *2864*

Aspirin Increased fatty acid oxidation, decreased lipogenesis. *1343* Considerable reduction in concentration in both normals and diabetics and lesser response to oral glucose. *2433*

Atenolol Significant reduction after 3 mo treatment in 14 hypertensives. *0836* Marked effect with 3 mo treatment given 100 mg/d. *0835* In fasting state and after 30 minutes during oral glucose tolerance Test. *1977*

Clofibrate Displacement from albumin. *1343* Reduction observed in diabetics after 1 week treatment. *1120*

Enflurane Fall during anesthesia, but increases in recovery. *2699*

5-Fluoronicotinic Acid Max effect of 50% at 2 h, starts in 15 minutes. *3077*

Glibonuride Small immediate rise then fall to 25% at 90 minutes. *1526*

Glisoxepide Small immediate rise then fall to 25% at 90 minutes. *1526*

Glucose Effect greater in normals than in diabetics. *2837*

Glyburide Protracted fall maximal at 4.5 h. *1526*

Insulin Effect similar in normals and diabetics. *2837*

Levothyroxine Correction of hypothyroid state. *3505*

Meals After mixed meal falls for 2 h below fasting. *1230*

Metoprolol Observed with pharmacological doses in nondiabetic hypertensives. *1410* Significant effect over 3 mo in 53 patients given 100 mg twice daily. *0835*

MPCA Anti-lipolytic effect of drug (acts for 4 h). *1531*

Muscular Exercise Approximately 15% decrease with bicycle pedalling. *1385*

Neomycin Probably related to altered fat absorption. *1077*

Niacin Marked fall then progressive secondary rise. *1727*

Oxprenolol Marked effect in 53 patients given 80 mg/twice daily for 3 mo. *0835*

Propranolol Especially during and after exercise. *0062* Significant reduction of basal concentration after 3 mo treatment in 16 hypertensives, but close to normal after 6 mo. *0836* Marked effect in 53 hypertensives given 80 mg twice daily for 3 mo. *0835* But not significant in hyperthyroid patients given 40 mg every 6 h in 6 patients. *0335* Slight effect with pharmacological doses. *1410*

Propylthiouracil Marked effect in 5 hyperthyroid individuals given 10 mg every 8 h for 4 d. *0335*

Pyridyltetrazole Reduction of up to 200 βEq/L, then rebound increase. *2774*
Sleep Occurs with sleep onset (but in fasting only). *2730*
Sotalol At 1, 3, 6 and 12 mo in group of essential hypertensives given drug orally for 12 mo. *2120*
Streptozocin Mechanism not established yet. *2532*
Sucrose Mean decrease of 400 mg/dL after 1 g/kg/h orally. *0408*

Fatty Acids, Total

Serum   Increase   Physiological
Oral Contraceptives Slight to large effect in primates. *3035*

Fenfluramine

Urine   Positive   Analytical
Clotermine Similar Rfs and color reaction on TLC using ethyl acetate: methanol: water: ammonium hydroxide and modified Dragendorff's reagent. *3868*

Ferric Chloride Test

Urine   Positive   Analytical
Acetoacetate May produce red or red-brown. *0443*
Aminosalicylic Acid Blue-purple color. *0121*
Antipyrine May produce red color. *0443*
Aspirin Red-purple color may mask true color. *0583*
Bilirubin May produce blue-green. *0443*
Chlorpromazine Purple color. *0121*
Cyanates May produce red color. *0443*
Homogentisic Acid Transient blue color. *3588*
Hydroxyphenylacetic Acid 2-hydroxy derivative: may produce mauve. *0443*
Hydroxyphenylpyruvic Acid 4-hydroxy derivative: green fades in seconds. 2-hydroxy derivative: red-brown to green-blue to mauve. *0443*
Imidazolepyruvic Acid Green or blue-green. *0443*
Iodochlorhydroxyquin If on diaper may give false positive. *1680*
α-Ketobutyric Acid Purple: fades to red brown. *0443*
Levodopa When 1-5 g ingested/d=black/brown. *0583*
Melanin Brownish-black color. *3588*
Methotrimeprazine Positive if more than 100 mg/d for 6 d. *2427*
Methyldopa At 0.2 mg/mL in alkaline urine. *2071*
Phenols Derivatives may produce violet color. *0443*
Phenothiazines pink/purple color from metabolites. *0583*
Phenylpyruvic Acid Transient blue-green color formed. *3588*
Pyruvate Deep gold-yellow or green. *0443*
Salicylate May produce stable purple color. *0443*
Xanthurenic Acid Deep green, later brown. *0443*

Ferritin

Serum   No Effect   Physiological
Anticonvulsants In 8 epileptic patients receiving phenobarbital with carbamazepine or phenytoin. *2670*
Oral Contraceptives Smaller proportion of contraceptive users had low values than controls. *1368*

Serum   Increase   Physiological
Ethanol Increases and decreases in parallel with GGT in 9 alcoholics during drinking and withdrawal. *2423*
Iron With 200 mg iron daily, most patients serum ferritin increased as well as bone marrow stainable iron. *1501*
Iron Dextran Peak values 7 to 9 d after i.v. infusion, thereafter declined. *0390*
Oral Contraceptives Mean of 40 ng/mL vs 25 ng/mL in control population followed for at least 2 y. *1182*

X-Ray Therapy Occurs within 2 h of deep x-ray therapy. *0987*

Serum   Decrease   Physiological
Ascorbic Acid During 14 week study of varying intakes of drug probably attributable to phlebotomy only. *1751*

Ferroxidase

Serum   Increase   Physiological
Phenobarbital Significant effect in 40 adult epileptics receiving long term treatment. *3641*

α-Fetoprotein

Serum   Increase   Physiological
Pregnancy Progressive increase until 39 weeks at 200 ng/ml. *1733* Above 250 ng/mL in fetal morbidity or death. *0685*

Urine   Increase   Physiological
Twin Pregnancy Usually twice as high as in single pregnancy. *1732*

Fibrinogen

Plasma   No Effect   Analytical
Acetylsalicylic Acid At concentration of 500 mg/L had no effect on aca method. *3393*
Antithrombin III On clottable protein assay procedure. *3463*
Ascorbic Acid At concentration of 500 mg/L had no effect on aca method. *3393*
Dextran 70 At concentration of 1500 mg/L had no effect on aca method. *3393*
Heparin On potassium mercuric thiocyanate procedure of Roberts. *3003* On clottable protein assay procedure. *3463*
Penicillin At concentration of 100 mg/L had no effect on aca method. *3393*
Protamine At concentration of 250 mg/L had no effect on aca method. *3393*
Pyrantel On potassium mercuric thiocyanate procedure of Roberts. *3003*
Sodium Citrate At concentration of 3794 mg/L had no effect on aca method. *3393*
Storage of Sample For 7 d at room temperature, 4 weeks at 4°. *0573*
Thrombin On potassium mercuric thiocyanate procedure of Roberts. *3003*
Transport of Specimen By pneumatic tube if specimen protected. *3217*

Plasma   No Effect   Physiological
Ampicillin In doses as high as 300 mg/kg/d in volunteers. *0492*
Apalcillin In 21 volunteers with doses up to 225 mg/kg. *1262*
Cyclopropane Anesthesia has no effect. *1343*
Halothane No significant effect noted. *1814*
Methicillin No effect with doses as high as 300 mg/kg/d in volunteers. *0492*
Penicillin G No effect with doses as high as 48 million U/d. *0492*

Plasma   Increase   Analytical
Heparin At concentrations above 2,000 mg/L (normal therapeutic concentration 20 mg/L) raised concentration as measured by aca method. *3393*
Lipemia Maximum effect on turbidimetric procedure. *3463*
Sodium Oxalate At concentrations above 1710 mg/L raised concentration as measured by aca method. *3393*

Plasma   Increase   Physiological
Aspirin Associated with increased sedimentation rate. *1343*
Estrogens Altered liver metabolism. *3332*

Fibrinogen (continued)

Plasma   Increase   Physiological (continued)

Estrogens, Conjugated   Effect in normal females after i.v. administration. 0083

Menstruation   Normal response. 0987

Norethandrolone   Metabolic effect. 0227

Oral Contraceptives   Metabolic changes in liver synthesis (estrogen). 2189

Oxandrolone   Metabolic effect. 0227

Oxymetholone   Metabolic effect. 0227

Pregnancy   Moderate increase by 16th week (plus 33% by term). 0987

Pyrazinamide   Reported effect. 0071

Xanthine   Reported effect. 1343

X-Ray Therapy   Indicative of tissue damage. 0987

Plasma   Decrease   Analytical

Antithrombin III   Maximum effect on turbidimetric procedures. 3463

EDTA   At concentrations above 1170 mg/L lowered concentration as measured by aca method. 3393

Heparin   Maximum effect on turbidimetric procedures. 3463

Plasma   Decrease   Physiological

Anabolic Steroids   Metabolic effect. 0227

Ancrod   Converts to fibrin shreds. 2823

Asparaginase   Marked effect in almost all patients. 2657   In 32 of 33 patients reported in various studies. 2406   In 50 to 100% in different studies. 0558

Bromelains   One drug related case reported. 1680

Cefamandole   Effect observed in some cases: associated with vitamin K associated hypoprothrombinemia. 3103

Clofibrate   Reported observation in some cases. 0620

Danazol   Significant effect in 21 hemophiliacs given drug for 2 weeks. 2627

Dextran   Complex formation or increased consumption. 3346

Estrogens, Conjugated   Small but significant effect after i.v. in normal males. 0083

Fluroxene   From 283 mg/dL to 257 mg/dL postoperatively. 1814

Garlic   Significant decrease although given with fatty meal. 0421

Iron Salts   One case reported with poisoning. 3817

Kanamycin   May occur at beginning of therapy. 1343

Muscular Exercise   Observed effect in exercise. 2847

Oral Contraceptives   When combined estrogen and progestogen. 2427

Oxytocin   Single case of possible reduction observed. 1680

Phosphorus   Probable hepatotoxic effect. 1302

Streptokinase   Significant effect in infarct patients. 1356

Sucrose   Effect of sustained high sucrose diet. 3931

Testosterone   Metabolic effect. 0227

Valproic Acid   Reduced to 0.9 to 1.6 g/L in 9 patients with epilepsy poorly controlled. 3508

Fibrinolysin

Plasma   Increase   Physiological

Muscular Exercise   Activity increased by exercise. 0987

Fibrinolysis

Plasma   Increase   Physiological

Muscular Exercise   Observed effect in exercise. 2847

Fibrinolytic Activity

Plasma   No Effect   Physiological

Aminosalicylic Acid   No effect of 1.5 g/d or after 250 mg i.v. in 6 patients. 3876

Fibrinolytic Time

Plasma   Increase   Physiological

Garlic   Significant effect although given with fatty meal. 0421

Niacin   Significant effect if given parenterally. 0421

Streptokinase   Significant effect if given parenterally. 0421

Urokinase   Significant effect if given parenterally. 0421

Fibrinopeptide A

Plasma   Increase   Physiological

Ethinyl Estradiol   Significant effect when 30 μg of drug combined with 2 different progestins. 2399

Plasma   Decrease   Physiological

Heparin   High values in acute nonlymphocytic leukemia reduced to normal in 14 of 17 patients after intravenous bolus of heparin. 1421

Fibrin Split Products (FSP)

Urine   Decrease   Physiological

Aspirin   Occurs in 2/3 patients with proliferative glomerulonephritis. 0671

Indomethacin   In 2/3 patients with proliferative glomerulonephritis. 0671

Prednisone   In 2-3 d in 2/3 patients with proliferative glomerulonephritis. 0671

Fibronectin

Serum   Increase   Analytical

Serum   Serum concentration 20% higher than all plasma specimens with EDTA heparin, citrate using Boehringer-Mannheim kit and nephelometer. 0770

Serum   Increase   Physiological

Tri-iodothyronine   In 7 normal women aged 18-27 y given 25 μg three times daily for 14 d. Effect observed on endothelium-associated proteins but not on hepatically synthesized proteins. 1383

FIGLU (N-Formiminoglutamic Acid)

Urine   No Effect   Analytical

Storage of Sample   If acidified for 2 weeks at 28°, 1 mo at -20. 1712

Urine   Increase   Physiological

Aminosalicylic Acid   Occurs with vitamin B_{12} deficiency. 3773

Anticonvulsants   Occurs with megaloblastic anemia. 3773

Barbiturates   Occurs if megaloblastic anemia. 3773

Colchicine   Megaloblastic anemia with B_{12} deficiency. 3773

Cycloserine   Occurs with folic acid deficiency. 3773

Estrogens   Occurs with megalobiastic anemia. 3773

Ethanol   May occur with folic acid or B_{12} deficiency. 3773

Isoniazid   Occurs if megaloblastic anemia. 3773

Metformin   May cause megaloblastic anemia. 3605

Methotrexate   Large effect of treatment observed. 3681

Neomycin   If megaloblastic anemia develops. 3773

Oral Contraceptives   Response to histidine tolerance test. 3311 Mild effect in oral contraceptive users. 3310

Pentamidine   If megaloblastic anemia develops. *3773*
Phenobarbital   If megaloblastic anemia develops. *3025*
Phenytoin   Occurs with impaired absorption of folic acid. *3773*
Primidone   If megaloblastic anemia occurs. *3025*
Pyrimethamine   If megaloblastic anemia occurs. *3898*
Tetracycline   Isolated case of folic acid deficiency reported. *1815*
Triamterene   If megaloblastic anemia occurs. *3773*

Flecainide

Serum   Increase   Physiological
Amiodarone   Pharmacokinetic interaction with clinical signifi-cance. *2327*
Cimetidine   Increases elimination half-life and decreases clear-ance by 13 to 27%. *2011*

Fluoride

Serum   Increase   Physiological
Anesthetic Agents   Slight effect following general anesthetic. *0916*
Fluorides   Due to absorbed material in poisoning. *0010*
Halothane   Slight increase following operation. *0916*
Methoxyflurane   Metabolic degradation product of anesthetic. *1177*

Serum   Decrease   Physiological
Aluminum Hydroxide   When aluminum hydroxide coadministered to osteoporotic patients receiving 50 mg fluoride/d. *3415*

Urine   Increase   Physiological
Methoxyflurane   Persisting high concentration many days with renal impairment. *3551*

Feces   Increase   Physiological
Aluminum Hydroxide   In osteoporotic patients receiving fluoride (50 mg/d) with net decreased absorption of fluoride by 57% when aluminum hydroxide co-administered. *3415*

Flurazepam

Urine   Positive   Analytical
Mazindol   Same R_f and color reaction on TLC using ethyl ace-tate: methanol: water: ammonium hydroxide and modified Dragendorff's reagent. *3868*

Flurbiprofen

Serum   Increase   Physiological
Cimetidine   Although peak concentration not increased area under curve increased. *3496*

Folate

Serum   No Effect   Analytical
Ampicillin   If chromatographic procedure of Landon used. *2068* Allegedly no effect on autoclave method. *1815*
Butabarbital   If chromatographic procedure of Landon used. *2068*
Cephaloridine   If chromatographic procedure of Landon used. *2068*
Cloxacillin   If chromatographic procedure of Landon used. *2068*
Erythromycin   If chromatographic procedure of Landon used. *2068*
Gentamicin   If chromatographic procedure of Landon used. *2068*
Lincomycin   If chromatographic procedure of Landon used. *2068*
Methaqualone   If chromatographic procedure of Landon used. *2068*
Nitrazepam   If chromatographic procedure of Landon used. *2068*

Oxytetracycline   If chromatographic procedure of Landon used. *2068*
Penicillin   If chromatographic procedure of Landon used. *2068* Allegedly no effect on autoclave method. *1815*
Pentobarbital   If chromatographic procedure of Landon used. *2068*
Promazine   If chromatographic procedure of Landon used. *2068*
Promethazine   If chromatographic procedure of Landon used. *2068*
Storage of Sample   At -16 or -76° for 1 week radiometric assay. *2455*
Tetracycline   If chromatographic procedure of Landon used. *2068*   Allegedly no effect on autoclave method. *1815*

Serum   No Effect   Physiological
Blood Transfusions   No effect with intraoperative transfusion. *1985*
Carbamazepine   In about 20 epileptic patients treated for 2 y. *0873*
Cyclosporine   No significant effect observed in 23 patients with inflammatory ocular disease treated for 6 mo. *2713*
Menstruation   No effect noticed during cycle. *3459*
Sulfasalazine   No significant difference from controls in 45 out-patients taking drug orally for ulcerative colitis. *2208*

Serum   Increase   Analytical
Folic Acid   Increase by 20% if chloramphenicol method of O'Broin used. *2631*

Serum   Increase   Physiological
Metformin   High in patients if B_{12} malabsorption. *3605*
Nitrous Oxide   If intraoperative exposure of patient is greater than 6 h, but unchanged if exposure for less than 1 h. *1985*
Vegetarian Diet   Average of 6.6 ng/mL versus 4.8 (higher in smokers). *0811*

Serum   Decrease   Analytical
Ampicillin   Inhibits growth of *L. Casei*. *3504*
Chloramphenicol   Inhibits growth of L. Casei. *3504*
Erythromycin   Inhibits growth of L. Casei. *3504*
Lincomycin   Inhibits growth of L. Casei. *3504*
Penicillin   Inhibits growth of L. Casei. *3505*
Plasma   Marked effect in some specimens observed. *2455*
Storage of Sample   Room temperature 24 h without preservative procedure of Mincey. *2455*
Sulfisoxazole   Affects standard autoclave method. *1815*
Tetracycline   Inhibits growth of L. Casei. *3505*

Serum   Decrease   Physiological
Aminopterin   Inhibits dihydrofolate reductase by combining irre-versibly with it. *2062*
Aminosalicylic Acid   May occur with protracted therapy. *2427*
Antacids   Associated with malabsorption. *3023*
Anticonvulsants   May cause megaloblastic anemia. *3025* Marked effect in 146 epileptics with long-term treatment. *2007*
Arsenicals   Megaloblastic anemia after Fowler's solution. *2427*
Aspirin   In 1 study subject brisk, significant but reversible fall in total and bound serum folate. Aspirin *in vitro* also displaced signifi-cant amounts of bound serum folate. *2095*
Barbiturates   May cause megaloblastic anemia (impairs absorp-tion). *3773*
Chloroguanide   Antagonizes folic acid. *0071*
Cycloserine   May cause megaloblastic anemia. *3773*   Low serum folate in half patients treated with cycloserine plus isoniazid in contrast to 2 of 55 treated with isoniazid alone, but mechanism unknown. *2062*
Estrogens   May impair absorption. *3773*
Ethanol   Affects absorption if severe alcoholism. *3773*   In some alcoholics ?decreased myelopoiesis. *2196*   More than half patients with alcoholic cirrhosis had low liver and serum folate, also acutely accelerates deficiency in people with relatively normal folate stores. *2062*

Folate *(continued)*

Serum Decrease Physiological *(continued)*

Glutethimide Characteristic with anticonvulsant intake, possibly due to enhanced catabolism. *3023*

Hemodialysis Repeated may even lead to megaloblastic anemia. *3885*

Isoniazid Low incidence of impaired absorption. *3773*

Levodopa Associated with burning feet syndrome. *0824*

Metformin Due to decreased absorption of dietary folate. *0302*

Methotrexate Inhibits folate reductase. *3025* Significantly higher in patients with psoriasis reflecting decreased dihydrofolate reductase activity. Both oxidized forms pteroylglutamate and dihydrofolate affected. *1556*

Nitrofurans May induce folate deficiency- megaloblastic anemia. *2427*

Nitrofurantoin Inhibits intestinal conjugase. *3058*

Oral Contraceptives Interferes with gastrointestinal absorption. *3802* Greater proportion of contraceptive users had low values than controls. *1368* Mild effect in oral contraceptive users. *3310*

Pentamidine Inhibits dihydrofolate reductase. *3773*

Phenobarbital May cause megaloblastic anemia. *3025* Low serum folate in from 27 to 91% of treated epileptics in different studies. *2062*

Phenytoin May cause megaloblastic anemia (impairs absorption). *3773* Low folate in from 27 to 91% of treated epileptics in different studies. *2062*

Primidone May cause megaloblastic anemia (impairs absorption). *3025* Low serum folate in from 27 to 91% of treated epileptics in different studies. *2062*

Pyrimethamine Inhibits folate reductase, megaloblastic anemia. *3025* Inhibits dihydrofolate reductase. *2062*

Sulfasalazine Folate absorption from gastrointestinal tract may be inhibited to a minor degree: may lead to megaloblastic anemia if severe nutritional deficiency or celiac disease. *2062* Due to malabsorption with prolonged use. *3023*

Tetracycline Isolated case of impaired absorption. *1815*

Triamterene Inhibits dihydrofolate reductase. *3773* Identified as folate antagonist. *3023*

Trimethoprim Usually with sulfa — no hematological abnormality. *2808* Reported association with megaloblastic anemia by inhibiting dihydrofolate reductase although other reports dispute drug as cause of megaloblastic anemia. *2062*

Urine No Effect Physiological

Aspirin Small but insignificant rise due to displacement from binding protein in serum. *2095*

Urine Increase Physiological

Chlortetracycline With dose of 3 g/d. *1343*

Oxytetracycline With dose of 2.5 g/d. *1343*

CSF Decrease Physiological

Anticonvulsants Occurs in many long-treated epileptics. *2967*

Phenobarbital Occurs in many long-treated epileptics. *2967* Low serum folate in from 27 to 91% of treated epileptics in different studies. *2062*

Phenytoin Occurs in many long-treated epileptics. *2967* Low folate in from 27 to 91% of treated epileptics in different studies. *2062*

Primidone Occurs in many long-treated epileptics. *2967* Low folate in from 27 to 91% of treated epileptics in different studies. *2062*

Red Blood Cells No Effect Physiological

Carbamazepine In about 20 epileptic patients treated for 2 y. *0873*

Red Blood Cells Decrease Physiological

Aminopterin Inhibits dihydrofolate reductase by combining irreversibly with it. *2062*

Antacids Associated with malabsorption. *3023*

Anticonvulsants In 39% epileptics on chronic therapy. *3560*

Glutethimide Characteristic with anticonvulsant intake, possibly due to enhanced catabolism. *3023*

Methotrexate Significantly lower in patients with psoriasis with long term treatment reflecting low polyglutamate storage. *1556* In long-term treated group erythrocyte folate significantly reduced reflecting low polyglutamate storage. *1556*

Oral Contraceptives Impaired metabolism due to hormonal factors. *3311* Mild effect in oral contraceptive users. *3310*

Phenobarbital Inverse correlation with drug concentration. *2966* Low serum folate in from 27 to 91% of treated epileptics in different studies. *2062*

Phenytoin Impaired deconjugation of polyglutamates in gut. *3504* Low folate in from 27 to 91% of treated epileptics in different studies. *2062*

Pregnancy Maximum decrease in 2nd or 3rd trimester. *1472*

Primidone Impaired deconjugation of polyglutamates in gut. *0987* Low serum folate in from 27 to 91% of treated epileptics in different studies. *2062*

Pyrimethamine Inhibits dihydrofolate reductase. *2062*

Sulfasalazine Folate absorption from gastrointestinal tract may be inhibited to a minor degree: may lead to megaloblastic anemia if severe nutritional deficiency or celiac disease. *2062* Due to malabsorption with prolonged use. *3023* Inversely correlated with drug dosage in 45 outpatients taking drug orally for ulcerative colitis. *2208*

Trimethoprim Reported association with megaloblastic anemia by inhibiting dihydrofolate reductase although other reports dispute drug as cause of megaloblastic anemia. *2062*

Test Conditions No Effect Analytical

Phenobarbital No effect on *L. casei* or *S. fecalis*. *2427*

Primidone No effect on *L. casei* or *S. fecalis*. *2427*

Trimethadione No effect on *L. casei* or *S. fecalis*. *2427*

Test Conditions Decrease Analytical

Phenytoin Mild depressant effect on *L. casei*. *2427*

Folic Acid

Serum No Effect Analytical

Co-Trimoxazole Drug has no effect on vitamin concentration if measured by radioimmunoassay. *0247*

Serum Decrease Analytical

Co-Trimoxazole Due to effect on Lactobacillus Casei if organism used to measure amount of vitamin. *0247*

Follicle Stimulating Hormone (FSH)

Plasma No Effect Physiological

Butorphanol No significant effect in 6 healthy male volunteers given 2 mg i.m. *3033*

Cimetidine In 3 studied men with duodenal ulcer or duodenitis. *2368* In 6 patients treated for 1 mo with 1 g/d. *3491*

Clonidine No effect seen in 12 healthy adults after single oral dose. *2070*

Cyproterone No change following oral administration. *3394*

Danazol In 7 normal women given 800 mg daily for 2 mo amenorrheic state induced by drug. *2224*

Diurnal Variation No effect observed with 1/2 h intervals in men. *2014*

Dopamine Did not change with infusion of up to 3.0 µg/kg/min for 2 h in 6 normal men. *2149*

Ethanol No consistent effect in healthy men with ingestion over several days. *1352*

Famotidine No effect of short-term or long-term treatment in male patients with duodenal ulcers. *0727*

Flunarizine After 90 d in 8 women. *2265*

Isotretinoin In 7 patients with severe rosacea treated with 1 mg/kg/d for 12 weeks. Effects possibly due to induction of hepatic microsomal enzymes. *2307* With 12 weeks treatment with 1 mg/kg/d. *2307*

Lead In a population of male battery workers compared with control group in cement industry. *0172* In some men occupationally exposed to large quantities of lead. *0771*

Methyldopa No effect in patient on long term treatment who had drug-induced hyperprolactinemia. *0166*

Methylprednisolone No apparent effect, unlike that on testosterone, in 67 y old males with chronic pulmonary disease taking drug for more than 1 mo. *2245*

Metoclopramide No significant baseline differences in cycling women before and after drug. *0096*

Naloxone Infusion did not affect concentration of hormone at concentrations varying from 0.02 to 0.5 mg/kg. *1970*

Naltrexone With 50 to 100 mg administered daily to obese subjects over 8 weeks. *0176*

Prednisone No apparent effect, unlike that on testosterone, in 67 y males with chronic pulmonary disease taking drug for at least 1 mo. *2245*

Prostaglandin F2α When i.v. up to 2 μg/kg/min in men. *0735*

Ranitidine With up to 450 mg/d in 20 males with chronic duodenal ulcer. *2925* No effect with i.v. bolus of much as 300 mg. *0871* After 4 weeks or 6 mo treatment (300 mg and 150 mg daily respectively) in male patients with duodenal ulcer. *0727*

Sexual Activity No effect in either sex. *3437*

Sodium Bromide No effect in 10 men, 10 women receiving 1 mg/kg/d during 8 weeks or 2 full menstrual cycles. *3141*

Sultopride When given 300-600 mg/d for 5 weeks in 5 schizophrenic women. *2465*

Tamoxifen After oral administration of 20 mg/d for 5 or 10 d to 6 healthy women during follicular phase of menstrual cycle. In 10 anovulatory women given up to 40 mg/d for 5 d. *1542*

Thyrotropin-Releasing Hormone (TRH) No effect in normal subjects after i.v. administration. *0100*

Plasma   Increase   Physiological

Bombesin After LHRH after 5 ng/nk/min x 2.5 h in healthy men. *2841*

Cimetidine With 1.2 g/d caused significant increase in periovulatory period. *0271* In 11 male subjects with chronic duodenal ulcer given drug 1 g/d for 3 mo. *3759*

Clomiphene Maximum increase of 350% of control in males. *3145*

Diurnal Variation Cycling throughout day up to 50%, asynchronous with LH. *2542*

Ketoconazole Increase by 63% approximately in normal men with dose effect maximal at 900 mg/d. Due to stimulatory effect of dose dependent fall of testosterone. *1299*

Levodopa Possible increase over fluctuations in controls. *2778*

Menstruation Shorter and lower peak than LH peak at mid cycle. *1762*

Naloxone Significant effect within 60 minutes of start of i.v. infusion. *0868*

Ovulation Midcycle peak twice the baseline. *0112*

Phenytoin Mean 3.7 units/L versus 1.9 in controls in approximately 24 male patients given phenytoin alone or with primidone or phenobarbital. *3609*

Plasma   Decrease   Physiological

Anticonvulsants In 33 male epileptics taking at least one drug for long time. *3020*

Danazol Effect noted in 6 orchidectomized patients with prostatic cancer. *0693*

Diethylstilbestrol At doses affecting pituitary-gonadal axis. *1917*

Estradiol Relatively slower than LH in response to i.v. *3914* Marked effect in 8 post-menopausal, post-oophorectomy women after 100 mg implant in subcutaneous tissue. *3274*

Estrogens Hormonal effect. *3237*

Ethinyl Estradiol In hypogonadal women in 1 week. *3913*

Megestrol In 18 postmenopausal women with breast cancer. *0056*

Mestranol Hormonal effect (inhibitory action of estrogen). *3237*

Oral Contraceptives Over years depressed to 70% of control values. *1340*

Phenothiazines Stimulation effect of gonadotropins inhibited. *2220*

Pimozide Statistically significant decline when given to acutely psychotic males although still within normal range. *3340*

Stanozolol From 2.7 to 1.8 units/L after 1 week in 9 healthy men given 10 mg/d for 14 d. *3354*

Tamoxifen In postmenopausal women with breast cancer treated for 2 weeks. *1542*

Urine   Increase   Physiological

Clomiphene Due to action on hypothalamic-pituitary axis. *2585*

Urine   Decrease   Physiological

Estrogens, Conjugated Slight decrease only even with large doses. *3724*

Oral Contraceptives Marked depression in normal subjects. *3464*

Formaldehyde

Urine   Increase   Physiological
Methanol Metabolite of oxidation. *1343*

Formic Acid

Serum   Increase   Physiological
Methanol Metabolite (with formaldehyde) of oxidation. *1343*

Fouchet Test

Urine   Positive   Analytical
Aspirin Produces purple color. *0459*

FPN Test

Urine   No Effect   Analytical

Acetaminophen No effect at 10 mg/dL on method of Frings. *1204*

Amitriptyline No effect at 10 mg/dL on method of Frings. *1204*

Amobarbital No effect at 10 mg/dL on method of Frings. *1204*

Amphetamine No effect at 10 mg/dL on method of Frings. *1204*

Benztropine No effect at 1 mg/dL on method of Frings. *1204*

Butabarbital No effect at 10 mg/dL on method of Frings. *1204*

Carisoprodol No effect at 10 mg/dL on method of Frings. *1204*

Chloral Hydrate No effect at 10 mg/dL on method of Frings. *1204*

Chlordiazepoxide No effect at 10 mg/dL on method of Frings. *1204*

Chloroquine No effect at 10 mg/dL on method of Frings. *1204*

Chlorpheniramine No effect at 12 mg/dL on method of Frings. *1204*

Chlorpromazine Negative FPN test with 0.4 mg/dL concentration in Frings TLC procedure. *1204*

Clorazepate No effect at 10 mg/dL on method of Frings. *1204*

Cocaine No effect at 10 mg/dL on method of Frings. *1204*

Codeine No effect at 10 mg/dL on method of Frings. *1204*

Diazepam No effect at 10 mg/dL on method of Frings. *1204*

Diethylpropion No effect at 25 mg/dL on method of Frings. *1204*

Diphenhydramine No effect at 10 mg/dL on method of Frings. *1204*

Doxepin No effect at 10 mg/dL on method of Frings. *1204*

Ephedrine No effect at 10 mg/dL on method of Frings. *1204*

Ethchlorvynol No effect at 10 mg/dL on method of Frings. *1204*

Ethinamate No effect at 10 mg/dL on method of Frings. *1204*

Fluphenazine Negative FPN test with 1 mg/dL concentration in Frings TLC procedure. *1204*

Flurazepam No effect at 10 mg/dL on method of Frings. *1204*

Glutethimide No effect at 10 mg/dL on method of Frings. *1204*

Glycopyrrolate No effect at 2 mg/dL on method of Frings. *1204*

FPN Test (continued)

Urine No Effect Analytical (continued)

Haloperidol No effect at 1 mg/dL on method of Frings. *1204*

Hydromorphone No effect at 2 mg/dL on method of Frings. *1204*

Mebutamate No effect at 10 mg/dL on method of Frings. *1204*

Meperidine No effect at 10 mg/dL on method of Frings. *1204*

Mephenesin No effect at 10 mg/dL on method of Frings. *1204*

Meprobamate No effect at 10 mg/dL on method of Frings. *1204*

Mescaline No effect at 10 mg/dL on method of Frings. *1204*

Methadone No effect at 10 mg/dL on method of Frings. *1204*

Methamphetamine No effect at 10 mg/dL on method of Frings. *1204*

Methaqualone No effect at 10 mg/dL on method of Frings. *1204*

Methdilazine Negative FPN test with 0.4 mg/dL concentration in Frings TLC procedure. *1204*

Methylphenidate No effect at 10 mg/dL on method of Frings. *1204*

Methyprylon No effect at 10 mg/dL on method of Frings. *1204*

Morphine No effect at 10 mg/dL on method of Frings. *1204*

Oxazepam No effect at 10 mg/dL on method of Frings. *1204*

Oxycodone No effect at 5 mg/dL on method of Frings. *1204*

Oxymetazoline No effect at 0.5 mg/dL on method of Frings. *1204*

Pentazocine No effect at 10 mg/dL on method of Frings. *1204*

Pentobarbital No effect at 10 mg/dL on method of Frings. *1204*

Perphenazine Negative FPN test with 0.1 mg/dL concentration in Frings TLC procedure. *1204*

Phenmetrazine No effect at 10 mg/dL on method of Frings. *1204*

Phenobarbital No effect at 10 mg/dL on method of Frings. *1204*

Phenylephrine No effect at 10 mg/dL on method of Frings. *1204*

Phenylpropanolamine No effect at 10 mg/dL on method of Frings. *1204*

Phenytoin No effect at 10 mg/dL on method of Frings. *1204*

Primaquine No effect at 10 mg/dL on method of Frings. *1204*

Procainamide No effect at 10 mg/dL on method of Frings. *1204*

Prochlorperazine Negative FPN test with 0.4 mg/dL concentration in Frings TLC procedure. *1204*

Promazine Negative FPN test with 0.4 mg/dL concentration in Frings TLC procedure. *1204*

Promethazine Negative FPN test with 0.1 mg/dL concentration in Frings TLC procedure. Negative FPN test with 0.5 mg/dL concentration in Frings TLC procedure. *1204*

Propoxyphene No effect at 10 mg/dL on method of Frings. *1204*

Pyrilamine No effect at 10 mg/dL on method of Frings. *1204*

Pyrimethamine No effect at 10 mg/dL on method of Frings. *1204*

Quinidine No effect at 10 mg/dL on method of Frings. *1204*

Quinine No effect at 10 mg/dL on method of Frings. *1204*

Salicylate No effect at 100 mg/dL on method of Frings. *1204*

Scopolamine No effect at 10 mg/dL on method of Frings. *1204*

Secobarbital No effect at 10 mg/dL on method of Frings. *1204*

Sodium Bromide No effect at 100 mg/dL on method of Frings. *1204*

Strychnine No effect at 10 mg/dL on method of Frings. *1204*

Sulfanilamide No effect at 100 mg/dL on method of Frings. *1204*

Tetracycline No effect at 10 mg/dL on method of Frings. *1204*

Thiobarbituric Acid No effect at 10 mg/dL on method of Frings. *1204*

Thiopropazate Negative FPN test with 0.4 mg/dL concentration in Frings TLC procedure. *1204*

Thioridazine Negative FPN test with 0.4 mg/dL concentration in Frings TLC procedure. *1204*

Thiothixene No effect at 10 mg/dL on method of Frings. *1204*

Trifluoperazine Negative FPN test with 0.5 mg/dL concentration in Frings TLC procedure. *1204*

Triflupromazine Negative FPN test with 1 mg/dL concentration in Frings TLC procedure. *1204*

Trihexyphenidyl No effect at 5 mg/dL on method of Frings. *1204*

Trimethobenzamide No effect at 25 mg/dL on method of Frings. *1204*

Tybamate No effect at 10 mg/dL on method of Frings. *1204*

Urine Increase Analytical

Aminosalicylic Acid Reported false positive in patients with liver dysfunction. *1204*

Chlorpromazine Pink color observed in method of Frings. Purple with patients also receiving trifluoperazine. Pink FPN test with 0.5 mg/dL concentration in Frings TLC procedure. *1204*

Estrogens Reported false positive in patients with liver dysfunction. *1204*

Methdilazine Pink FPN test with 0.5 mg/dL in Frings TLC procedure. *1204*

Perphenazine Pink FPN test with 0.4 mg/dL concentration in Frings TLC procedure Pink color observed in method of Frings. *1204*

Phenylpyruvic Acid Reported false positive in patients with liver dysfunction. *1204*

Prochlorperazine Pink color observed in method of Frings. Pink FPN test with 0.5 mg/dL concentration in Frings TLC procedure. *1204*

Promazine Orange FPN test with 0.5 mg/dL concentration on Frings procedure Orange color observed in method of Frings. *1204*

Promethazine Pink FPN test with 1 mg/dL concentration in Frings TLC procedure Pink color observed in method of Frings. *1204*

Thiopropazate Pink color observed in method of Frings. Pink FPN test with 0.5 mg/dL concentration in Frings TLC procedure. *1204*

Thioridazine Blue color observed in method of Frings Blue FPN test with 0.5 mg/dL concentration in Frings TLC procedure. *1204*

Trifluoperazine Orange color observed in method of Frings Orange FPN test with 1 mg/dL concentration in Frings TLC procedure Purple with patients also receiving chlorpromazine. *1204*

Triflupromazine Pink FPN test with 2 mg/dL concentration in Frings TLC procedure Pink color observed in method of Frings. *1204*

Urine Positive Analytical

Fluphenazine Pink FPN test with 2 mg/dL concentration in Frings TLC procedure. *1204*

Fractional Sodium Excretion

Urine No Effect Physiological

Bopindolol In 10 hypertensive patients treated for either 1 or 21 d. *1063*

Free Water Clearance

Urine No Effect Physiological

Bopindolol In 10 hypertensive patients treated for either 1 or 21 d. *1063*

Fructosamine

Serum Decrease Analytical

Heparin Significantly lower concentrations than in serum when measured by Roche kit. *1698*

Fructose

Urine Increase Physiological

Sucrose Maximum at 30 minutes, continues for several h. *2551*

Fructose-6-Phosphate

Red Blood Cells   Decrease   Physiological
Amino Acids   Associated with low serum phosphate. *3619*

FSH Response to LHRH

Serum   No Effect   Physiological
Isotretinoin   In 7 patients with severe rosacea treated with 1 mg/kg/d for 12 weeks. Effects possibly due to induction of hepatic microsomal enzymes. *2307*

Fucose

Serum   No Effect   Physiological
Glucose   No change observed during glucose tolerance test in all subjects. *1665*

Serum   Increase   Analytical
Protein   Interferes with method of Dische and Settles. *1665*

Urine   No Effect   Physiological
Diurnal Variation   No evidence of cycle observed. *0273*
Fruit   No effect compared with fasting. *0273*
Vegetables   No effect with high intake compared with controls. *0273*

α-Fucosidase

Serum   Increase   Physiological
Aging   Steep rise to age 25 y then little change. *1043*

Fumarate

Urine   Increase   Physiological
Lithium   May be considerable effect. *2107*

Furosemide

Serum   Decrease   Physiological
Food   Absorption delayed when taken with food. *3024*

Galactose

Urine   Increase   Physiological
Lactose   Maximum at 90 minutes in normals after oral load. *2551*

Galactose-1-Phosphate Uridyl Transferase

Red Blood Cells   No Effect   Analytical
Storage of Sample   14 d at room temperature, for 4 weeks at 4°. *0877*

β-Galactosidase

Serum   Decrease   Physiological
Aging   Steep fall to age 25 y then less steep. *1043*

Urine   No Effect   Physiological
Urine Volume   Excretion not dependent on urine flow. *2320*

Urine   Increase   Physiological
Diurnal Variation   Maximum 6-12 am, minimum late pm, not marked. *2320*

Gastric Inhibitory Polypeptide

Plasma   No Effect   Physiological
Naloxone   In 6 healthy volunteers after ingestion of test meal and intravenous administration of drug. *0626*

Plasma   Decrease   Physiological
Morphine   Reduction in secretion following test meal and drug in 6 healthy volunteers. *0626*

Gastrin

Serum   No Effect   Physiological
Carbohydrate   No effect on endogenous secretion if given orally. *1236*
Cimetidine   No effect reported in spite of long-term treatment. *1667*   No effect noted after short-term treatment and no effect on nocturnal serum concentration. *2368*
Famotidine   No effect in 9 patients with Zollinger-Ellison syndrome treated for 33 weeks. *1667*
Naloxone   In 6 healthy volunteers after ingestion of test meal and intravenous administration of drug. *0626*
Prazosin   With 2 mg in 12 hypertensives (6 with normal and 6 with abnormal glucose tolerance). *0225*
Ranitidine   No effect reported in spite of long term treatment. *1667*

Serum   Increase   Physiological
Amino Acids   Oral feeding produces up to tenfold increase. *1995*
Calcium Carbonate   2 fold increase, maximum 30 to 75 minutes after. *3302*
Calcium Chloride   If given intraluminally 13% increase over NaCl. *1643*
Catecholamines   High concentration associated with pheochromocytoma and on injection. *1533*
Cimetidine   Increase higher in people ingesting drug in response to food. *2368*
Coffee   2-5 fold increase after 350 ml. *3302*
Diurnal Variation   10% change only but maximum at midnight, min early am. *1235*
Epinephrine   High concentrations if pheochromocytoma, also if infused. *3472*
Gastrin   From average fasting level of 87.3 pg/mL to 245 pg/mL/3 h. *2370*
Insulin   Similar response (fairly marked). *3904*
Meals   Similar response (fairly marked). *3904*
Meat   Maximum effect in normals seen in 30 minutes. *2369*
Meat Extract   Similar response (fairly marked). *3904*
Morphine   Secretion prolonged following test meal and drug i.v. in 6 healthy volunteers. *0626*
Omeprazole   Raised to 80.9 pg/mL from 55.5 pg/mL in 8 volunteers after 29 d but reversible. *2516*   Significantly increased after 7 and 14 d treatment but not after a single dose. *1126*
Protein   Oral feeding may produce up to tenfold increase. *1995*

Serum   Decrease   Physiological
Atropine   Significant effect if given i.m. *1236*
Chloroguanide   Significant effect (basal and histamine stimulated) in rats. *0909*
Glucagon   Suppresses response to meals. *0263*
Hydrochloric Acid   Significant effect if given orally. *1236*
Secretin   Basal secretion after injection in normals. *1484*
Streptozocin   Gastric hypersecretion reduced. *2532*

GC Globulin

Serum   Increase   Physiological
Estrogens   Altered metabolism. *0227*
Pregnancy   Observed with steroid administration (50% increase in late pregnancy). *3472*

Gentamicin

Serum   No Effect   Analytical
Bilirubin   No difference between plate-diffusion and tube dilution. *1266*

Cephalosporin   If enzymatic procedure of Daigneault used. *0791*
Chloramphenicol   If enzymatic procedure of Daigneault used. *0791*
Penicillin   If enzymatic procedure of Daigneault used. *0791*
Sodium Taurocholate   No difference between plate-diffusion and tube dilution. *1266*

Serum   Increase   Analytical
Urea   Affects measurement by pH method. *1212*

Serum   Decrease   Analytical
Heparin   If over 50 U/mL affects agar diffusion. *3927*   Interferes with biological, radioenzymatic and homogeneous enzyme immunoassay but RIA techniques due to inhibition of acetyltransferase enzymes. *2021*

Globulin

Serum   Increase   Analytical
Indoleacetic Acid   Reacts as tryptophan in method of Goldenberg. *1329*

Serum   Increase   Physiological
Ethanol   4% increase in heavy versus occasional drinkers. *3273*
Oxyphenisatin   May cause hypersensitivity reaction. *2970*

Serum   Decrease   Physiological
Progesterone   Observed in healthy women when only treatment. *0991*
Pyrazinamide   Reported effect (viral-hepatitis like). *0071*

CSF   Increase   Analytical
Procaine   Gives false positive Pandy test. *0251*

α₁-Globulin

Serum   No Effect   Physiological
Oral Contraceptives   No significant effect at 3 mo cessation or 3 y. *3143*

Serum   Increase   Physiological
Oral Contraceptives   Metabolic change with combined contraceptive. *2312*
Pregnancy   Marked increase in last 2 trimesters. *0987*
X-Ray Therapy   Increase due to tissue damage. *0987*

α₂-Globulin

Serum   Increase   Physiological
Mestranol   Metabolic effect. *3258*
Oral Contraceptives   May be increased by as much as 8 times. *0585*   After 3 y, not after 3 mo cessation. *3143*
Phenytoin   Related to duration of therapy. *2427*
Pregnancy   Occurs in last 2 trimesters. *0987*
Thorium Dioxide   May produce severe liver damage after years. *3147*
X-Ray Therapy   Increased rapidly due to tissue damage. *0987*

Serum   Decrease   Physiological
Asparaginase   Decreased to 70% of control at 2 weeks. *2657*

β-Globulin

Serum   Increase   Physiological
Altitude   43% increase after 6 weeks at 5400 metres. *2957*
Mestranol   Metabolic effect. *3258*
Oral Contraceptives   After 3 y but not after 3 mo cessation. *3143*   Metabolic changes in liver synthesis. *3258*
Pregnancy   Slight increase in second trimester. *0987*
Thorium Dioxide   May produce severe liver damage after years. *3147*

Serum   Decrease   Physiological
Asparaginase   Decreased to 70% of control at 2 weeks. *2657*
Season   Falls during summer. *2123*

β₁-α-Globulin

Serum   Increase   Physiological
Muscular Exercise   Approximately 14% increase immediately after. *2846*
Pregnancy   Response to steroids (increased in late pregnancy). *3472*

Serum   Decrease   Analytical
Storage of Sample   Variable loss of activity after 5 y at -15°. *1180*

Serum   Decrease   Physiological
Asparaginase   Diminished hepatic synthesis. *2657*
Benzene   Significant effect in people occupationally exposed. *3370*
Dextran   Complex formation or increased consumption. *3346*
Phenytoin   Observed in 50% patients. *0989*
Toluene   Significant effect in people occupationally exposed. *3370*
Xylene   Significant effect in people occupationally exposed. *3370*

β₁-Globulin C/A

Serum   Decrease   Physiological
Phenytoin   Probable drug induced immunological effect. *1406*

γ-Globulin

Serum   Increase   Physiological
Altitude   72% increase after 6 weeks at 5400 metres. *2957*
Aminopyrine   Specific antibodies to drug may develop. *2161*
Anticonvulsants   Associated with SLE-like syndrome. *0933*
Asparaginase   Increased continuously to 170% of mean at 4 weeks. *2657*
Hydantoin Derivatives   Observed in cases of hypersensitivity. *2427*
Hydralazine   Reported finding. *2427*
Methimazole   Polyclonal hypergammaglobulinemia reported possibly as result of production of nonspecific drug-stimulated polyclonal antibodies. *3870*
Nitrofurantoin   In 3 of 9 acute and 16 of 20 chronic pulmonary reactions. *1640*   In 3 of 9 acute and 16 of 20 chronic pulmonary reactions. *1640*
Oral Contraceptives   After 3 y, not after 3 mo cessation. *3143*
Phenylbutazone   Observed in 2 cases (?due to antibodies). *2161*
Progesterone   Not observed with combined therapy. *0795*
Propylthiouracil   Possibly related to production of nonspecific drug-stimulated polyclonal antibodies. *3870*
Season   Eskimos- 1.36 g/dL in winter, 2 g/dL in summer. *3742*
Thorium Dioxide   May produce severe liver damage after years. *3147*
Tolazamide   Transient increases reported. *2427*

Tubocurarine   Positive correlation between sensitivity and concentration. *3481*
X-Ray Therapy   May fall in some cases (tissue damage). *0987*
Zirconium Salts   Observed with zirconium granulomas. *2726*

Serum   Decrease   Analytical
Caproxamine   At 1.5 mg/dL conventional methods when added to sera. *1758*
Clothiapine   At 9 mg/L when added to serum and conventional method. *1758*
Storage of Sample   With prolonged storage at -10°. *1180*

Serum   Decrease   Physiological
BCG Vaccine   Associated with severe reaction. *2427*
Gold   Single case reported. *2427*
Methotrexate   Possible immunosuppressive response. *1680*
Prednisone   Augmented response in cirrhotics when drug added to therapeutic regime. *0651*
Pregnancy   May fall slightly in last 3 mo. *0987*
Season   Falls during summer. *2123*

Glomerular Filtration Rate (GFR)

Urine   No Effect   Physiological
Cyclophosphamide   In majority of patients although decreased urine volume. *0854*
Estradiol   No effect observed in humans. *2590*
Levarterenol   After i.v. but increased filtration fraction. *1176*
Methoxyflurane   No effect observed with high or low concentrations. *1380*
Metolazone   No effect observed. *3358*
Muscular Exercise   If exercise moderate. *1842*
Oxaprozin   No significant change observed in 7 volunteers treated for up to 1 week. *2464*
Sleep   No effect observed with sleep only. *1190*
Verapamil   No change compared with placebo in 11 patients treated with up to 360 mg daily for 6 weeks. *3396*

Urine   Increase   Physiological
Carbon Monoxide   Occurs 12 to 24 h after exposure. *3454*
Clonidine   Sustained rise after initial fall with i.v. injection. *0477*
Digitalis   Improvement with relief of edema. *1343*
Methyldopa   Slight increase or normal in normo-, hypertensives. *1343*
Muscular Exercise   If exercise mild. *1842*
Nicardipine   Increase by 35% after 0.5 mg i.v. in 7 patients with mild to moderate hypertension. *0195*
Pregnancy   Normal during first 8 mo but increases up to 50%. *0987*
Protein   Increased in proportion to amount of protein. *0987*

Urine   Decrease   Physiological
Acetrizoate   When 70% injected for aortography. *0065*
Amphotericin B   Occurs in high percentage of patients. *0071*
Angiotensin   Vasoconstrictive effect in kidney. *1343*
Aspirin   Nephrotoxicity of drug occurring acutely. *0267*
Ceftazidime   Decreased by mean of 10 mL/min with 4 g drug/d in 16 patients: initial GFR in patients 30 to 110 mL/min. *0051* Slight but significant reduction after 3 g daily i.v. in 15 patients for 4 to 9 d. *2616*
Cortisone   Probably does this by decreased secretion of creatinine. *0913*
Cyclosporine   51 versus 93 mL/min in 17 cardiac transplant recipients compared with those on azathioprine. *2536*
Diazoxide   Effect over 2 h of 4 mg/kg given i.v. *1805*
Diuretics   Effect on clearances. *2745*
Enalapril   Associated with selective glomerular efferent arteriolar dilatation and possible interference with autoregulatory capacity of kidney in response to severe renovascular hypertension. *0284*
Epinephrine   Slight fall after i.m. injection. *2427*

Ether   Due to renal vasoconstriction. *1343*
Furosemide   Excessive diuresis may cause effect. *2427*
Ganglionic Blocking Agents   Returns to normal usually within 2 h. *1343*
Gentamicin   Using chromium labeled EDTA even with subnormal amounts of drug. Noticeable before effect on serum creatinine. *3621*
Halothane   Up to 19% decrease reported in normals. *2427*
Heroin   Observed in some severe cases. *2914*
Histamine   After 0.3-0.5 mg subcutaneously in humans. *1176*
5-Hydroxytryptamine   After 10 μg/kg i.v. infusion. *1176*
Indomethacin   Transient reduction in individuals during sustained diuresis (fell from 114 mL/minutes to 100 mL/min). *2464*
Isoproterenol   Frequently observed to be diminished. *3205*
Levarterenol   Slight fall after i.v. infusion. *2427*
Methoxyflurane   Decrease noted during normal anesthesia. *2427*
Metolazone   Diuretic action. *3494*
Muscular Exercise   Observed with heavy treadmill exercise. *1842*
Oxprenolol   Observed with acute experiments. *1714*
Polymyxin   Nephrotoxicity increases with continued treatment. *1343*
Propranolol   13% decrease on average in hypertensives. *1714*
Radiographic Agents   If concentrated solutions used for aortography. *0065*
Spinal Anesthesia   Reported effect, related to degree of hypotension. *1343*
Thiazides   Especially if administered i.v. *1343*
Triamterene   Causes reduced glomerular filtration rate. *0652*

Glucagon

Plasma   No Effect   Physiological
Atenolol   No change in basal concentration or after glucose in 14 hypertensives. *0836*
Chlorpropamide   No effect i.v. if given rapidly or slowly. *2766*
Clonidine   No effect when given i.v. to 6 healthy volunteers. *2060* No effect observed on normal response to moderate or heavy exercise. *1797*
Glucocorticoids   Intravenous injection failed to modify concentration. *2296*
Ibopamine   300 mg drug had no effect over period of 90 minutes. *3398*
Naloxone   With infusion in normal and obese subjects. *0176* In 6 healthy volunteers after ingestion of test meal and intravenous administration of drug. *0626*
Omeprazole   No effect observed in 8 volunteers given 30 mg/d for 28 d. *2516*
Pancreozymin   No effect of 1 unit/kg body weight. *2676*
Pirenzepine   Treatment for 1 week had no effect on basal concentrations. *3933*
Prednisolone   Intravenous injection failed to modify concentration. *2296*
Propranolol   In hyperthyroid patients given 40 mg every 6 h in 6 patients. *0335* In 5 hyperthyroid patients given 10 mg every 8 h for 4 d. *0335* No change in basal concentration or after glucose in 16 hypertensives. *0836*
Propylthiouracil   In 5 hyperthyroid individuals given 10 mg every 8 h for 4 d. *0335*
Secretin   No effect observed if injected i.v. *2676*
Tetragastrin   No effect observed if given i.v. *2676*
Theophylline   No effect in response to i.v. infusion of aminophylline. *0608*
Timolol   In 5 hyperthyroid patients given 10 mg every 8 h for 4 d. *0335*
Tolbutamide   No effect i.v. if given slowly or rapidly. *2766*
Verapamil   No effect observed even in individuals fasted for 36 h when infused at rate of 5 mg/h or 3 h. *0103*

Glucagon *(continued)*

Plasma   Increase   Physiological

Arginine   Large effect (by 40 pg/mL) after prednisolone. *2296* Moderate increase after infusion of 30 g. *2676*

Aspirin   Observed in humans and animals but mechanism for this not known. *2433*

Calcitonin   Inhibitory action of oral glucose on glucagon secretion partially prevented in comparison with control. *2738*

Cerulein   After i.v. infusion in normals. *1076*

Danazol   Significant reduction in one woman with systemic lupus erythematosus on withdrawal of drug. *0815*   Concentration above 50 pmol/L observed in 7 women treated with up to 600 mg/d for up to 24 weeks: slight fall with glucose tolerance test. *3839*

Galactose   Much smaller response than after glucose. *3306*

Glucocorticoids   After 4 d fasting level increased 24 pg/ml. *2296* Increase basal and stimulated concentrations and promote gluconeogenesis. *0410*

Glucose   Of gastrointestinal tract origin after glucose load. *2676* Slight increase maximum at 30 minutes after oral dose. *3306*

Guanabenz   In 30 hypertensives: effect most marked in individuals with higher doses. No change observed on withdrawal of drug. *1018*

Hydrochlorothiazide   Higher in 15 hypertensives during treatment with 50 mg twice daily than before treatment and after withdrawal. *1018*

Insulin   Marked increase at 45 minutes after injection. *2676*

Muscular Exercise   May increase 20%, facilitates hepatic glycogenolysis. *1107*

Nifedipine   Significant effect in normal subjects (0.045 vs 0.034 nmol/L). *0637*

Prednisolone   After 4 d fasting level increased 24 pg/ml. *2296*

Plasma   Decrease   Physiological

Atenolol   In 18 patients with mild essential hypertension treated with chlorothiazide concomitantly over 4 weeks. *1035*

Metoprolol   Lower in nondiabetic hypertensives when receiving drug compared with placebo. *1410*

Pindolol   In 18 patients with mild essential hypertension treated with chlorothiazide concomitantly over 4 weeks. *1035*

Propranolol   In 18 patients with mild essential hypertension treated with chlorothiazide concomitantly over 4 weeks. *1035* Slight effect with pharmacological doses. *1410*

Verapamil   Significant change during tolerance test when drug and glucose coadministered. *1115*

Glucagon, Pancreatic

Plasma   No Effect   Physiological

Morphine   No effect observed in 6 healthy volunteers after test meal and drug i.v. *0626*

Naloxone   In 6 healthy volunteers after ingestion of test meal and intravenous administration of drug. *0626*

Glucaric Acid

Urine   Increase   Physiological

Aminopyrine   As result of hepatic enzyme induction. *2220*

Anticonvulsants   Occurs in 94% children (hepatic enzyme induction). *1695*

Antipyrine   Reported induction of hepatic enzymes. *1694*

Barbiturates   As result of hepatic enzyme induction. *2220*

Carbamazepine   Dose dependent correlation: marked effect. *2788*

Ethosuximide   Induces hepatic enzymes. *1694*

Ethylphenacemide   Induces hepatic enzymes (more potent than phenobarbital). *1694*

Glutethimide   Manifestation of hepatic enzyme induction when 500 mg daily ingested for 21 d by 6 volunteers. *0407*

Oral Contraceptives   Reported induction of hepatic enzymes. *1694*

Paramethadione   Probable effect as reported to induce hepatic enzymes. *3882*

Phenobarbital   Due to induction of hepatic enzymes. *2220*   In 98% of patients with epilepsy receiving long term treatment. *3641* Dose dependent effect, greater than other anticonvulsants. *2788*

Phenylbutazone   May induce hepatic enzymes. *0002*

Phenytoin   Induces hepatic enzymes. *1694*   Equipotent with phenobarbital. *2089*   Dose dependent effect next to phenobarbital in potency. *2788*

Phetharbital   Potent enzyme-inducing agent. *1694*

Primidone   More potent inducer of hepatic enzymes than phenobarbital. *2089*   Dose-dependent effect but less potent than carbamazepine. *2788*

Progesterone   Reported induction of hepatic enzymes. *1694*

Rifampin   Manifestation of hepatic enzyme induction. *0014*

Spironolactone   Weak hepatic microsomal enzyme inducer. *3558*

Thiopental   Enhanced excretion with anesthesia. *0003*

Valproic Acid   Selective induction of hepatic metabolizing enzymes. *0848*   Slightly higher but not statistically significant. *2788*

Glucocorticoids

Plasma   Increase   Physiological

Oral Contraceptives   Expected effect observed. *0820*

Glucose

Serum   No Effect   Analytical

Acetaminophen   At 5 times upper limit of therapeutic concentration on routine methods in use on SMAC®, Abbott-VP, aca, Cobas-Bio, Ektachem® 400, Hitachi® 705, KDA. *2138*   No effect observed with colorimetric glucose oxidase procedure on AutoAnalyzer using Boehringer-Mannheim kit. No effect on Beckman glucose analyzer with oxygen produced measured by specific electrode. *3018*   At acute overdose concentration (10 mg/dL) on SMAC® method. *3719* No effect at therapeutic concentration on Reflotron method. *1984* At concentration of 57 mg/L had no effect on GOD/POD-PAP method. *3393*   At concentration of 600 mg/L had no effect on hexokinase/G-6-PDH method. *3393*

Acetazolamide   No effect at 12 mg/dL on SMA 12/60 method. *2636*   At concentration of 200 mg/L had no effect on GOD/POD-PAP method. *3393*

Acetohexamide   No effect on Boehringer GOD-PERID method. *3277*   At concentration of 200 mg/L had no effect on GOD/POD-PAP method. *3393*

Acetylsalicylic Acid   At therapeutic concentrations on hexokinase, glucose dehydrogenase 2,4-dichlorophenol, ABTS, and o-toluidine methods. *1786*   On hexokinase method on SMA II at physiological concentration. *1787*   At 5 times upper limit of therapeutic range on method on SMAC®, Abbott-VP, Cobas-Bio, aca, Ektachem®, Hitachi® 705 and KDA. *2138*   With 8326 µmol/L on hexokinase/G-6-PDH method. With 8326 µmol/L on glucose dehydrogenase method. With 8326 µmol/L on o-toluidine method. *1786*   No effect at therapeutic concentration on Reflotron method. *1984*   At concentration of 100 mg/L had no effect on Ektachem® method. *3393*   At concentration of 1400 mg/L had no effect on hexokinase/G-6-PDH method. *3393*   At concentration of 500 mg/L had no effect on Seralyzer method. *3393*

Ajmaline   At concentration of 3.3 mg/L had no effect on GOD/POD-PAP method. *3393*

Allopurinol   At concentration of 18 mg/L had no effect on Ektachem® method. *3393*   At concentration of 100 mg/L had no effect on GOD/POD-PAP method. *3393*

Aminophenazone   At therapeutic concentrations no effects on hexokinase, glucose dehydrogenase 2.4-dichlorophenol and o-toluidine methods. *1786*   With 540 µmol/L on hexokinase method. With 540 µmol/L on glucose dehydrogenase method. With 540 µmol/L on GOD-Perid/2,4-dichlorophenol method. With 540 µmol/L on o-toluidine method. *1785*

Aminophenol   At concentration of 50 mg/L had no effect on Ektachem® method. *3393*   At concentration of 200 mg/L had no effect on hexokinase/G-6-PDH method. *3393*

Aminopyrine At concentration of 125 mg/L had no effect on hexokinase/G-6-PDH method. *3393*

Aminosalicylic Acid At 1 g/dL on glucose-oxidase methods. At 1 g/dL on p-HBAH procedure of Lever. At 1 g/dL on o-toluidine procedure. *2140* At 5 times upper limit of therapeutic range on methods on SMAC®, Abbott-VP, aca, Cobas-Bio, Ektachem®, Hitachi® 705 and KDA. *2138*

Amitriptyline At acute overdose concentration (2.5 mg/dL) on SMAC® method. *3719* At concentration of 6.3 mg/L had no effect on Ektachem® method. *3393* At concentration of 40 mg/L had no effect on hexokinase/G-6-PDH method. *3393*

Amobarbital At concentration of 5 mg/L had no effect on GOD/POD-PAP method. *3393*

Amphetamine At therapeutic concentration on Reflotron method. *1984*

Ampicillin No effect on hexokinase, glucose-oxidase methods. *2568* No effect on Reflotron method at therapeutic concentration. *1984* At concentration of 180 mg/L had no effect on Ektachem® method. *3393* At concentration of 5 mg/L had no effect on GOD/POD-PAP method. *3393* At concentration of 1200 mg/L had no effect on hexokinase/G-6-PDH method. *3393*

Aprotinin At concentration of 150 kU/L had no effect on GOD/POD-PAP method. *3393*

Ascorbic Acid On p-HBAH procedure of Lever at 1 g/dL. *2140* No effect on Trinder glucose-oxidase method. *2771* Insignificant effect at 5 mg/dL on MBTH procedure of Neeley. *2569* No effect at therapeutic concentration on Reflotron method. *1984* At concentration of 1000 mg/L had no effect on hexokinase/G-6-PDH method. *3393* At concentration of 250 mg/L had no effect on hexokinase/G-6-PDH method on aca. *3393*

Aspirin At 10 mg/dL no effect on glucose oxidase procedure of Gochman. At 10 mg/dL no effect on alkaline ferricyanide procedure. *1311* At 1 g/dL on p-HBAH procedure of Lever. At 1 g/dL on o-toluidine procedure. *2140*

Azapropazone At concentration of 36 mg/L had no effect as measured by Ektachem® method. *3393*

Barbital At acute overdose concentration (20 mg/dL) on Technicon® SMAC® method. *3719*

Benzbromarone At concentration of 8 mg/L had no effect as measured by Ektachem® method. *3393*

Benzoic Acid If saturated on o-toluidine methods. If saturated on alkaline ferricyanide procedures. If saturated on glucose-oxidase methods. If saturated on p-HBAH procedure of Lever. *2140*

Bethanechol No effect at 0.09 mg/dL on SMA 12/60 method. *2636*

Bezafibrate No effect at therapeutic concentration on Reflotron method. *1984*

Bilirubin No effect on hexokinase method of Yee. *3912* At 20 mg/dL on MBTH procedure of Neeley. *2569* At 15 mg/dL on hexokinase method of Coburn. *0676*

Bisacodyl At concentration of 0.2 mg/L had no effect as measured by Ektachem® method. *3393* At concentration of 4,000 mg/L had no effect as measured by hexokinase/G-6-PDH method. *3393*

Bromazepam At 5 times therapeutic concentration on methods of SMAC®, Abbott-VP, aca, Cobas-Bio, Ektachem® 400, KDA and Hitachi® 705. *2138*

Butizide At concentration of 0.2 mg/L had no effect as measured by Ektachem® method. *3393*

Caffeine No effect at therapeutic concentrations on hexokinase, glucose dehydrogenase 2,4-dichlorophenol, ABTS and o-toluidine methods. *1786* At acute overdose concentration (10 mg/dL) on Technicon® SMAC® method. *3719* At therapeutic concentration on Reflotron method. *1984* At concentration of 300 mg/L had no effect on hexokinase/G-6-PDH method. *3393*

Carbenicillin At 5 times upper limit of therapeutic range on methods on SMAC®, Abbott-VP, aca, Ektachem® 400, Cobas-Bio, Hitachi® 705 and KDA. *2138*

Carbochromen No effect at therapeutic concentration on Reflotron method. *1984*

Carbon Monoxide No effect on o-toluidine procedure. *2717*

Cefazolin At concentration of 110 mg/L had no effect on GOD/POD-PAP method. *3393*

Cefoxitin At 5 times upper limit of therapeutic range on methods on SMAC®, Abbott-VP, aca, Cobas-Bio, Ektachem®, Hitachi® 705 and KDA. *2138* At concentration of 1,000 mg/L had no effect on GOD/POD-PAP method. *3393*

Cephalothin At 5 times upper limit of therapeutic range on methods on SMAC®, Abbott-VP, aca, Cobas-Bio, Ektachem®, Hitachi® 705 and KDA. *2138* At concentration of 1046 mg/L had no effect on glucose dehydrogenase method. *3393* At concentration of 1,000 mg/L had no effect on GOD/POD-PAP method. *3393*

Chloramphenicol No effect at therapeutic concentration on Reflotron method. *1984* At concentration of 100 mg/L had no effect on Ektachem® method. *3393* At concentration of 3,000 mg/L had no effect on GOD/POD-PAP method. *3393* At concentration of 400 mg/L had no effect on hexokinase/G-6-PDH method. *3393*

Chlordiazepoxide At 5 times upper limit of therapeutic range on methods on SMAC®, Abbott-VP, aca, Cobas-Bio, Ektachem®, Hitachi® 705 and KDA. *2138* At acute overdose concentration (5 mg/dL) on Technicon® SMAC® method. *3719* No effect at therapeutic concentration on Reflotron method. *1984* At concentration of 1.8 mg/L had no effect on Ektachem® method. *3393* At concentration of 50 mg/L had no effect on GOD/POD-PAP method. *3393* At concentration of 6 mg/L had no effect on hexokinase/G-6-PDH method. *3393*

Chloroquine At concentration of 5 mg/L had no effect on Ektachem® method. *3393* At concentration of 250 mg/L had no effect on GOD/POD-PAP method. *3393*

Chlorpheniramine At acute overdose concentration (20 mg/dL) on Technicon® SMAC® method. *3719*

Chlorpromazine At concentration of 0.1 mg/L had no effect on Ektachem® method. *3393* At concentration of 3 mg/L had no effect on GOD/POD-PAP method. *3393*

Chlorpropamide No effect on Trinder glucose-oxidase procedure. *2771* No effect glucose oxidase (GOD-PERID) method of Boehringer. *3277* On Warner Glucomatic glucose-oxidase method. *2771* At concentration of 1,000 mg/L had no effect on GOD/POD-PAP method. *3393*

Chromonar Hydrochloride At concentration of 18 mg/L had no effect on Ektachem® method. *3393* At concentration of 180 mg/L had no effect on hexokinase/G-6-PDH method. *3393*

Cimetidine At concentration of 4 mg/L had no effect on GOD/POD-PAP method. *3393*

Citrates At 5 g/dL on p-HBAH procedure of Lever. At 5 g/dL on o-toluidine procedure. At 5 g/dL on glucose-oxidase procedures. At 5 g/dL on alkaline ferricyanide procedures. *2140* At concentration of 2,000 mg/L had no effect on hexokinase/G-6-PDH method. *3393*

Clofibrate At concentration of 500 mg/L had no effect on GOD/POD-PAP method. *3393*

Cloxacillin At concentration of 5 mg/L had no effect on GOD/POD-PAP method. *3393*

Cocaine At acute overdose concentration (2.5 mg/dL) on Technicon® SMAC® method. *3719*

Codeine At acute overdose concentration (2.0 mg/dL) on Technicon® SMAC® method. *3719* At concentration of 0.1 mg/L had no effect on GOD/POD-PAP method. *3393*

Contact With Clot If plastic disc used to separate for 48 h. *1863*

Creatinine No effect at 100 mg/dL on glucose oxidase procedure of Gochman. *1311* No effect on Warner Glucomatic method. *2771* At 15 mg/dL on MBTH procedure of Neeley. *2569* No effect on Trinder glucose-oxidase method. *2771*

Cyclophosphamide At concentration of 8 mg/L had no effect on Ektachem® method. *3393* At concentration of 40 mg/L had no effect on GOD/POD-PAP method. *3393*

Cyclosporine At concentration of 41.5 mmol/L (50 µg/L) on methods on Technicon® SMAC® II and Hitachi® 705. *1123*

Cysteine At 10 mg/dL no effect on glucose oxidase procedure of Gochman. *1311* At 5 times upper limit of therapeutic range on methods on SMAC®, Abbott-VP aca, Cobas-Bio, Hitachi® 705 and KDA. *2138*

Cystine At 5 times upper limit of therapeutic range on methods on SMAC®, Abbott-VP, aca, Cobas-Bio, Hitachi® 705 and KDA. *2138*

Deslanoside No effect on SMA 12/60 method at 0.06 mg/dL. *2636*

Dextran At 1 g/dL on MBTH procedure of Neeley. *2569* No effect on hexokinase, glucose oxidase methods. *2568* At concentration of 500 mg/L had no effect on Seralyzer method. *3393*

Dextran 40 At concentration of 2500 mg/L had no effect on hexokinase/G-6-PDH method on aca. *3393*

Glucose (continued)

Serum   No Effect   Analytical (continued)

Dextran 60   At concentration of 1800 mg/L had no effect on Ektachem® method. *3393*   At concentration of 6,000 mg/L had no effect on hexokinase/G-6-PDH method. *3393*

Dextran 70   At concentration of 1,000 mg/L had no effect on Ektachem® method. *3393*

Dextran 75   At concentration of 1500 mg/L had no effect on hexokinase/G-6-PDH method. *3393*

Diatrizoic acid   At concentration of 2600 mg/L had no effect on hexokinase/G-6-PDH method. *3393*

Diazepam   At 5 times upper limit of therapeutic range on methods on SMAC®, Abbott-VP, aca, Cobas-Bio, Ektachem®, Hitachi® 705 and KDA. *2138*   At acute overdose concentration (2.5 mg/dL) on Technicon® SMAC® method. *3719*   At concentration of 4 mg/L had no effect on Ektachem® method. *3393*   At concentration of 1.5 mg/L had no effect on GOD/POD-PAP method. *3393*

Diclofenac   No effect a therapeutic concentrations on hexokinase, glucose dehydrogenase, 2,4-dichloroaience, and o-toluidine methods. *1786*   On hexokinase method on SMA II at physiological concentration. *1787*   At concentration of 23 mg/L had no effect on hexokinase/G-6-PDH method. *3393*

Diethylpropion   At acute overdose concentration (10 mg/dL) on Technicon® SMAC® method. *3719*

Digitoxin   No effect at 21 mg/dL on SMA 12/60 method. *2636*

Digoxin   No effect at 0.04 mg/dL on SMA 12/60 method. *2636* At concentration of 0.015 mg/L had no effect on Ektachem® method. *3393*   At concentration of 0.25 mg/L had no effect on GOD/POD-PAP method. *3393*

Diphenhydramine   At acute overdose concentration (20 mg/dL) on Technicon® SMAC® method. *3719*

Dipyridamole   At concentration of 40 mg/L had no effect on hexokinase/G-6-PDH method. *3393*

Disopyramide   At concentration of 4 mg/L had no effect on GOD/POD-PAP method. *3393*

Disulphine Blue   By continuous-flow glucose oxidase/4-aminophenazone reaction. *1464*

Doxepin   At concentration of 6 mg/L had no effect on Ektachem® method. *3393*

EDTA   At anticoagulant dose on hexokinase method of Coburn. *0676*   At 1.2 g/dL on glucose oxidase procedures   At 1.2 g/dL on o-toluidine procedure. *2140*   At 200 mg/dL on MBTH procedure of Neeley. *2569*   At 1.2 g/dL on p-HBAH procedure of Lever. *2140* No effect on hexokinase method if anticoagulant. *2982*   At 1.2 g/dL on alkaline ferricyanide procedure. *2140*   At concentration of 2,000 mg/L had no effect on hexokinase/G-6-PDH method. *3393*

Epinephrine   At 10 mg/dL no effect on glucose oxidase procedure of Gochman. *1311*

Ergothionine   At 10 mg/dL no effect on glucose oxidase procedure of Gochman. *1311*

Ethacrynic Acid   At 2.5 mg/dL no effect on SMA 12/60 method. *2636*

Ethanol   At 1 g/dL on p-HBAH procedure of Lever.   At 1 g/dL on glucose oxidase procedures.   At 1 g/dL on alkaline ferricyanide procedure.   At 1 g/dL on o-toluidine procedure. *2140*   At acute overdose concentration (20 mg/dL) on Technicon® SMAC® methods. *3719*

Ethaverine   No effect at therapeutic concentration on Reflotron method. *1984*   At concentration of 400 mg/L had no effect on hexokinase/G-6-PDH method. *3393*

Ethchlorvynol   At acute overdose concentration (20 mg/dL) on Technicon® method. *3719*

Ethinamate   At acute overdose concentration (20 mg/dL) on Technicon® SMAC® method. *3719*

Fluorescein   No effect at 200 mg/L on Ektachem® 700 method. *1722*

Fluorides   No effect on hexokinase procedure. *2982*   At 2.5 g/dL on alkaline ferricyanide procedure.   At 2.5 g/dL on p-HBAH procedure of Lever.   At 2.5 g/dL on o-toluidine procedure.   At 2.5 g/dL on glucose oxidase procedures. *2140*

Fluosol-DA   On SMA IIC at concentration of 50%. *2518*   By 18% on Kodak Ektachem® 400 at concentration of 50%. *2518*

Flurazepam   At 5 times upper limit of therapeutic range on methods on SMAC®, Abbott-VP, aca, Cobas-Bio, Ektachem®, Hitachi® 705, and KDA. *2138*   At acute overdose concentration (2.5 mg/dL) on Technicon® SMAC® method. *3719*   At concentration of 0.1 mg/L had no effect on GOD/POD-PAP method. *3393*

Fructose   No effect on GOD-PERID procedure. *2771*   No effect if measured by hexokinase, glucose oxidase. *2568*   At 200 mg/dL on hexokinase method of Coburn. *0676*   No effect on Trinder glucose oxidase procedure   No effect on o-toluidine procedure. *2771*

Furosemide   No effect at 1.4 mg/dL on SMA 12/60 method. *2636* No effect at therapeutic concentration on Reflotron method. *1984* At concentration of 40 mg/L had no effect on Ektachem® method. *3393*   At concentration of 5 mg/L had no effect on GOD/POD-PAP method. *3393*   At concentration of 100 mg/L had no effect on hexokinase/G-6-PDH method. *3393*

Galactose   No effect on Trinder glucose oxidase procedure. *2771* At 200 mg/dL on hexokinase method of Coburn. *0676*   No effect if measured by hexokinase or glucose oxidase. *2568*   No effect on GOD-PERID procedure. *2771*

Gentamicin   At concentration of 10 mg/L had no effect on GOD/POD-PAP method. *3393*

Gentisic Acid   Has no effect on glucose oxidase method of Gochman. *1311*

Glibenclamide   No effect at therapeutic concentration on Reflotron method. *1984*   At concentration of 0.3 mg/L had no effect on Ektachem® method. *3393*

Glutathione   No effect on Warner Glucomatic method.   No effect on Trinder glucose oxidase method. *2771*   At 10 mg/dL no effect on glucose oxidase procedure of Gochman. *1311*

Glutethimide   At acute overdose concentration (5 mg/dL) on Technicon® SMAC® method. *3719*   At concentration of 17.5 mg/L had no effect on Ektachem® method. *3393*

Glyburide   No effect on glucose oxidase method of Boehringer. *3277*

Hemoglobin   At 200 mg/dL on MBTH procedure of Neeley. *2569* No effect on hexokinase method of Yee. *3912*

Heparin   On glucose oxidase procedures.   On o-toluidine procedure.   On alkaline ferricyanide procedures.   On p-HBAH procedure of Lever. *2140*   No effect on hexokinase procedure. *2982*   At concentration of 15,000 mg/L had no effect on GOD/POD-PAP method. *3393*   At concentration of 2,000 mg/L had no effect on hexokinase/G-6-PDH method. *3393*

Hydralazine   Affects Boehringer GOD-PERID method. *3277*   At concentration of 0.5 mg/L had no effect on GOD/POD-PAP method. *3393*

Hydrochlorothiazide   At concentration of 7 mg/L had no effect on GOD/POD-PAP method. *3393*

Hyoscine-N-Butylbromide   At concentration of 2 mg/L had no effect on Ektachem® method. *3393*

Ibuprofen   No effect at therapeutic concentrations on hexokinase, glucose dehydrogenase ABTS, 2,4-dichlorophenol or o-toluidine methods. *1786*   On hexokinase method on SMA II at physiological concentration. *1787*   At 5 times upper limit of therapeutic range on methods on SMAC®, Abbott-VP, aca, Cobas-Bio, Ektachem®, Hitachi® 705 and KDA. *2138*   At concentration of 200 mg/L had no effect on hexokinase/G-6-PDH method. *3393*

Indomethacin   No effect at therapeutic concentrations on hexokinase, glucose dehydrogenase 2,4-dichlorophenol, ABTS and o-toluidine methods. *1786*   On hexokinase method on SMA II at physiological concentration. *1787*   No effect at therapeutic concentration on Reflotron method. *1984*   At concentration of 4 mg/L had no effect on Ektachem® method. *3393*   At concentration of 25 mg/L had no effect on GOD/POD-PAP method. *3393*   At concentration of 40 mg/L had no effect on hexokinase/G-6-PDH method. *3393*

Iodoacetate   At 0.5 g/dL on o-toluidine procedure.   At 0.5 g/dL on alkaline ferricyanide methods.   At 0.5 g/dL on p-HBAH procedure of Lever.   At 0.5 g/dL on glucose oxidase methods. *2140*

Iodoacetic Acid   At 0.5 g/dL on alkaline ferricyanide procedure. At 0.5 g/dL on o-toluidine procedure.   At 0.5 g/dL on p-HBAH procedure of Lever.   At 0.5 g/dL on glucose oxidase procedures. *2140*

Iproniazid   Affects Boehringer GOD-PERID method. *3277*

Isocarbazide   At concentration of 55 mg/L had no effect on GOD-Perid method. *3393*

Isocarboxazid   Affects Boehringer GOD-PERID method. *3277*

Isoniazid   Affects Boehringer GOD-PERID method. *3277*

Isonicotinic Acid At concentration of 5 mg/L had no effect on Ektachem® method. *3393*

Ketoprofen No effect at therapeutic concentrations on hexokinase, glucose dehydrogenase 2,4-dichlorophenol, ABTS and o-toluidine methods. *1786* On hexokinase method on SMA II at physiological concentration. *1787*

Lactose No effect on GOD-PERID procedure. *2771* At 200 mg/dL on hexokinase method of Coburn. *0676* No effect on Trinder glucose oxidase procedure. *2771*

Levarterenol At 10 mg/dL no effect on glucose oxidase procedure of Gochman. *1311*

Levodopa At 10 mg/dL no effect on glucose oxidase procedure of Gochman. *1311* At 10 mg/dL on MBTH procedure of Neeley. *2569* No effect on hexokinase method. *2568*

Lidocaine No effect SMA 12/60 method with 3.5 mg/dL. *2636* At concentration of 3.2 mg/L had no effect on Ektachem® method. *3393*

Lipemia No effect on o-toluidine of Feteris or Cooper. *1206*

Liposol Changed from 1295 to 1274 mg/L (insignificant) in 26 randomly selected patient sera, before and after addition on Beckman Astra methods. *2054*

Lorazepam At concentration of 0.05 mg/L had no effect on GOD/POD-PAP method. *3393*

Maltose No effect on o-toluidine procedure. No effect on GOD-PERID procedure. *2771*

Mannitol No effect on SMA 12/60 method at 445 mg/dL. *2636*

Mannose No effect on hexokinase, glucose oxidase methods. *2568* At 200 mg/dL on hexokinase method of Coburn. *0676*

Meglumine At concentration of 200 mg/L had no effect on Ektachem® method. *3393* At concentration of 2,000 mg/L had no effect on hexokinase/G-6-PDH method. *3393*

Meperidine At acute overdose concentration (5 mg/dL) on Technicon® SMAC® method. *3719*

Meprobamate At acute overdose concentration (20 mg/dL) on Technicon® SMAC® method. *3719* At concentration of 484 mg/L had no effect on Ektachem® method. *3393* At concentration of 25 mg/L had no effect on GOD/POD-PAP method. *3393*

Meralluride No effect at 0.07 mL/dL on SMA 12/60 method. *2636*

Mercaptopurine At concentration of 20 mg/L had no effect on Ektachem® method. *3393*

Mesoridazine At acute overdose concentration (20 mg/dL) on Technicon® SMAC® method. *3719*

Methadone At acute overdose concentration (2.5 mg/dL) on Technicon® SMAC® method. *3719*

Methanol At acute overdose concentration (20 mg/dL) on Technicon® SMAC® method. *3719*

Methaqualone At acute overdose concentration (2.5 mg/dL) on Technicon® SMAC® method. *3719* No effect at therapeutic concentration on Reflotron method. *1984* At concentration of 80 mg/L had no effect on hexokinase/G-6-PDH method. *3393*

Methicillin At 5 times upper limit of therapeutic range on SMAC®, Abbott-VP, Cobas-Bio, aca, Ektachem®, Hitachi® 705 and KDA. *2138* At concentration of 20 mg/L had no effect on GOD/POD-PAP method. *3393*

Methotrexate At 5 times upper limit of therapeutic range on method on SMAC®, Abbott-VP, Cobas-Bio, Ektachem®, Hitachi® 705 and KDA. *2138* At concentration of 1200 mg/L had no effect on Ektachem® method. *3393* At concentration of 25 mg/L had no effect on GOD/POD-PAP method. *3393*

Methotrimeprazine At concentration of 1 mg/L had no effect on GOD/POD-PAP method. *3393*

Methsuximide At concentration of 40 mg/L had no effect on Ektachem® method. *3393*

Methyldopa No effect at therapeutic concentration on Reflotron method. *1984* At concentration of 400 mg/L had no effect on hexokinase/G-6-PDH method. *3393*

Methylphenidate At acute overdose concentration (20 mg/dL) on Technicon® SMAC® method. *3719*

Methylphenobarbital At concentration of 41 mg/L had no effect on Ektachem® method. *3393*

Methyprylon At acute overdose concentration (10 mg/dL) on Technicon® SMAC® method. *3719*

Metoprolol At concentration of 3 mg/L had no effect on GOD/POD-PAP method. *3393*

Metronidazole On glucose oxidase method on Beckman Astra at therapeutic concentration. *0937*

Morphine At acute overdose concentration (20 mg/dL) on Technicon® SMAC® method. *3719*

Nalidixic Acid At concentration of 50 mg/L had no effect on GOD/POD-PAP method. *3393*

Naproxen At 5 times upper limit of therapeutic range on methods on SMAC®, Abbott-VP, aca, Cobas-Bio, Ektachem®, Hitachi® 705 and KDA. *2138*

Niacin No effect at therapeutic concentration on Reflotron method. *1984*

Niacinamide At concentration of 12 mg/L had no effect on Ektachem® method. *3393*

Nipagin At 2 g/dL on glucose oxidase procedure. At 2 g/dL on p-HBAH procedure of Lever. At 2 g/dL on alkaline ferricyanide procedure. At 2 g/dL on o-toluidine procedure. *2140*

Nitrofurantoin No effect at therapeutic concentration on Reflotron method. *1984* On routine methods in use on SMAC®, Ektachem®, Abbott-VP, Cobas-Bio, aca, Hitachi® 705, KDA at 5 times normal upper therapeutic concentration. *2138* At concentration of 5 mg/L had no effect on Ektachem® method. *3393* At concentration of 2.5 mg/L had no effect on GOD/POD-PAP method. *3393* At concentration of 60 mg/L had no effect on hexokinase/G-6-PDH method. *3393*

Noramidopyrine No effect on hexokinase procedures. *3524*

Norfenefrin At concentration of 0.48 mg/L had no effect on Ektachem® method. *3393*

Normetanephrine No effect on glucose oxidase method of Boehringer. *3277*

Nortriptyline At acute overdose concentration (20 mg/dL) on Technicon® SMAC® method. *3719*

Novaminsulfon At concentration of 80 mg/L had no effect on Ektachem® method. *3393* At concentration of 800 mg/L had no effect on hexokinase/G-6-PDH method. *3393*

Opipramol No effect on glucose oxidase method of Boehringer. *3277*

Ouabain At 0.02 mg/dL no effect on SMA 12/60 method. *2636*

Oxalate At 2.5 g/dL on p-HBAH procedure of Lever. At 2.5 g/dL on alkaline ferricyanide procedure. At 2.5 g/dL on glucose oxidase procedure. At 2.5 g/dL on o-toluidine procedure. *2140*

Oxazepam No effect on glucose oxidase method of Boehringer. *3277* No effect at therapeutic concentration on Reflotron method. *1984* At concentration of 30 mg/L had no effect on hexokinase/G-6-PDH method. *3393*

Oxedrine At concentration of 60 mg/L had no effect on GOD/POD-PAP method. *3393*

Oxyphenbutazone At concentration of 12 mg/L had no effect on Ektachem® method. *3393* At concentration of 200 mg/L had no effect on hexokinase/G-6-PDH method. *3393*

Oxytetracycline No effect at therapeutic concentration on Reflotron method. *1984* At concentration of 600 mg/L had no effect on hexokinase/G-6-PDH method. *3393*

Penicillamine At concentration of 36 mg/L had no effect on Ektachem® method. *3393*

Penicillin G At 5 times upper limit of therapeutic range on methods on SMAC®, Abbott-VP, Cobas-Bio, aca, Ektachem®, Hitachi® 705 and KDA. *2138* At concentration of 90 mg/L had no effect as measured by GOD/POD-PAP method. *3393*

Pentobarbital At acute overdose concentration (20 mg/dL) on Technicon® SMAC® method. *3719*

Perphenazine At acute overdose concentration (20 mg/dL) on Technicon® SMAC® method. *3719* At concentration of 1 mg/L had no effect on GOD/POD-PAP method. *3393*

Phenazopyridine No effect at therapeutic concentration on Reflotron method. *1984* At concentration of 120 mg/L had no effect on hexokinase/G-6-PDH method. *3393*

Phenobarbital No effect at therapeutic concentrations on hexokinase, glucose dehydrogenase 2,4-dichlorophenol, ABTS and o-toluidine methods. *1786* At acute overdose concentration (20 mg/dL) on Technicon® SMAC® method. *3719* No effect at therapeutic concentration on Reflotron method. *1984* At concentration of 30 mg/L had no effect on Ektachem® method. *3393* At concentration of 100 mg/L had no effect on GOD/POD-PAP method. *3393* At concentration of 60 mg/L had no effect on hexokinase/G-6-PDH method. *3393*

Glucose *(continued)*

Serum   No Effect   Analytical *(continued)*

Phenols   At 10 mg/dL no effect on glucose oxidase procedure of Gochman. *1311*   Minimal effect at 10 mg/dL glucose oxidase dianisidine procedure. *3037*   At 10 mg/dL on alkaline ferricyanide procedure. *1311*

Phenothiazines   At concentration of 20 mg/L had no effect on Ektachem® method. *3393*

Phenprocoumon   No effect at therapeutic concentration on Reflotron method. *1984*   At concentration of 0.36 mg/L had no effect on Ektachem® method. *3393*   At concentration of 2 mg/L had no effect on hexokinase/G-6-PDH method. *3393*

Phenylbutazone   At concentration of 12 mg/L had no effect on Ektachem® method. *3393*   At concentration of 200 mg/L had no effect on GOD-Perid method. *3393*   At concentration of 200 mg/L had no effect on GOD/POD-PAP method. *3393*   At concentration of 200 mg/L had no effect on hexokinase/G-6-PDH method. *3393*

Phenytoin   No effect at 1.8 mg/dL on SMA 12/60 method. *2636*   At acute overdose concentration (20 mg/dL) on Technicon® SMAC® method. *3719*   No effect at therapeutic concentration on Reflotron method. *1984*   At concentration of 36 mg/L had no effect on Ektachem® method. *3393*   At concentration of 20 mg/L had no effect on GOD/POD-PAP method. *3393*   At concentration of 160 mg/L had no effect on hexokinase/G-6-PDH method. *3393*

Plasma   No significant difference from serum by AutoAnalyzer. *2228*

Potassium Oxalate   At 200 mg/dL on MBTH procedure of Neeley. *2569*

Povidone-Iodine   When used as skin cleansing agent and fingerstick measurement done by Dextrostix®. *2330*

Prednisolone   At concentration of 25 mg/L had no effect on GOD/POD-PAP method. *3393*

Primidone   At concentration of 2.4 mg/L had no effect on Ektachem® method. *3393*

Probenecid   No effect at therapeutic concentration on Reflotron method. *1984*   At concentration of 40 mg/L had no effect on Ektachem® method. *3393*   At concentration of 200 mg/L had no effect on hexokinase/G-6-PDH method. *3393*

Procainamide   No effect on SMA 12/60 method at 35 mg/dL. *2636*

Procaine   No effect at therapeutic concentration on Reflotron method. *1984*   At concentration of 40 mg/L had no effect on hexokinase/G-6-PDH method. *3393*

Prochlorperazine   At concentration of 1 mg/L had no effect on GOD/POD-PAP method. *3393*

Promethazine   At acute overdose concentration (20 mg/dL) on Technicon® SMAC® method. *3719*   At concentration of 1 mg/L had no effect on GOD/POD-PAP method. *3393*

Propoxyphene   At acute overdose concentration (2.5 mg/dL) on Technicon® SMAC® method. *3719*   At concentration of 1.8 mg/L had no effect on Ektachem® method. *3393*

Propranolol   No effect at 0.1 mg/dL on SMA 12/60 method. *2636*   At concentration of 80 mg/L had no effect on GOD/POD-PAP method. *3393*

Pyribenzamine®   At acute overdose concentration (20 mg/dL) on Technicon® SMAC® methods. *3719*

Pyridamole   No effect at therapeutic concentration on Reflotron method. *1984*

Pyritinol   No effect at therapeutic concentration on Reflotron method. *1984*

Quinidine   At 26 mg/dL no effect on SMA 12/60 method. *2636*   At acute overdose concentration (20 mg/dL) Technicon® SMAC® method. *3719*   No effect at therapeutic concentration on Reflotron method. *1984*   At concentration of 5.6 mg/L had no effect on Ektachem® method. *3393*

Quinine   At acute overdose concentration (1.5 mg/dL) on Technicon® SMAC® method. *3719*

Reserpine   No effect at 0.02 mg/dL on SMA 12/60 method. *2636*

Rifampin   At 5 times upper limit of therapeutic concentration on methods on SMAC®, Abbott-VP, Cobas-Bio, Ektachem®, Hitachi® 705 and KDA. *2138*   At concentration of 150 mg/L had no effect on hexokinase/G-6-PDH method. *3393*

Rolitetracycline   At concentration of 4 mg/L had no effect on GOD/POD-PAP method. *3393*

Salicylate   At concentration of 133 mg/L had no effect on Ektachem® method. *3393*

Scopolamine Bromide   At concentration of 20 mg/L had no effect on GOD/POD-PAP method. *3393*

Secobarbital   At acute overdose concentration (20 mg/dL) on Technicon® SMAC® method. *3719*   At concentration of 1 mg/L had no effect on GOD/POD-PAP method. *3393*

Sodium Citrate   At concentration of 500 mg/L had no effect on Ektachem® method. *3393*   At concentration of 20,000 mg/L had no effect on GOD/POD-PAP method. *3393*

Sodium Fluoride   No effect on alkaline ferricyanide method No effect on o-toluidine method. *2771*   At anticoagulant dose on hexokinase method of Coburn. *0676*   No effect on GOD-PERID glucose oxidase method No effect on Trinder glucose oxidase method. *2771*   At 200 mg/dL on MBTH procedure of Neeley. *2569*   No effect on Neocuproin method No effect on Warner Glucomatic method. *2771*   At concentration of 200 mg/L had no effect on Ektachem® method. *3393*   At concentration of 7,000 mg/L had no effect on GOD-Perid method. *3393*   At concentration of 40,000 mg/L had no effect on GOD/POD-PAP method. *3393*

Sodium Heparin   At concentration of 75 mg/L had no effect on Ektachem® method. *3393*

Sodium Oxalate   At concentration of 300 mg/L had no effect on Ektachem® method. *3393*   At concentration of 8,000 mg/L had no effect on GOD/POD-PAP method. *3393*

Sodium Salicylate   At concentration of 350 mg/L had no effect on Ektachem® method. *3393*   At concentration of 1,000 mg/L had no effect on GOD/POD-PAP method. *3393*

Sorbitol   On glucose oxidase procedure at 1 g/dL On p-HBAH procedure of Lever at 1 g/dL On alkaline ferricyanide procedure at 1 g/dL On o-toluidine procedure at 1 g/dL. *2140*

Spironolactone   At concentration of 8 mg/L had no effect on Ektachem® method. *3393*   At concentration of 1 mg/L had no effect on GOD/POD-PAP method. *3393*

Sucrose   At 200 mg/dL on hexokinase method of Coburn. *0676*

Sulfadiazine   No effect on glucose oxidase method of Boehringer. *3277*

Sulfaguanidine   No effect on Boehringer GOD-PERID glucose oxidase method. *3277*

Sulfamethizole   No effect on glucose oxidase method of Boehringer. *3277*

Sulfamethoxazole   No effect at therapeutic concentration on Reflotron method. *1984*   At concentration of 320 mg/L had no effect on hexokinase/G-6-PDH method. *3393*

Sulfamethoxydiazine   At concentration of 20 mg/L had no effect on Ektachem® method. *3393*

Sulfamethoxypyridazine   At concentration of 70 mg/L had no effect on GOD/POD-PAP method. *3393*

Sulfanilamide   At concentration of 300 mg/L had no effect on Seralyzer method. *3393*

Sulfathiazole   At concentration of 50 mg/L had no effect on Ektachem® method. *3393*

Sulfisoxazole   No effect on Boehringer GOD-PERID glucose oxidase method. *3277*

Tetracycline   No effect on glucose oxidase method. *2568*   At 5 times upper limit of therapeutic range on methods on SMAC®, Abbott-VP, aca, Cobas-Bio, Ektachem®, Hitachi® 705 and KDA. *2138*   At concentration of 20 mg/L had no effect on Ektachem® method. *3393*   At concentration of 500 mg/L had no effect on Seralyzer method. *3393*

Theophylline   At 5 times upper limit of therapeutic range on methods on SMAC®, Abbott-VP, aca, Cobas-Bio, Ektachem®, Hitachi® 705 and KDA. *2138*   No effect at therapeutic concentration on Reflotron method. *1984*   At concentration of 23 mg/L had no effect on Ektachem® method. *3393*   At concentration of 20 mg/L had no effect on GOD/POD-PAP method. *3393*

Thimerosal   At 2 g/dL on p-HBAH procedure of Lever. At 2 g/dL on alkaline ferricyanide procedures. At 2 g/dL on glucose oxidase procedures. At 2 g/dL on o-toluidine procedure. *2140*

Thiopental   At acute overdose concentration (20 mg/dL) on Technicon® SMAC® method. *3719*

Thymol   At 0.7 g/dL o-toluidine procedure At 0.7 g/dL on alkaline ferricyanide procedures At 0.7 g/dL on glucose oxidase procedures. *2140*

Timolol   At concentration of 0.01 mg/L had no effect on GOD/POD-PAP method. *3393*

Tolazamide No effect on hexokinase, o-toluidine methods. *2568* If guiacum or phenolaminophenazone with glucose oxidase. *3277* At concentration of 200 mg/L had no effect on GOD/POD-PAP method. *3393*

Tolbutamide No effect on Boehringer GOD-PERID method. *3277* On Warner Glucomatic glucose oxidase method. No effect on Trinder glucose oxidase procedure. *2771* No effect glucose oxidase, o-toluidine, hexokinase procedures. *2568* At concentration of 220 mg/L had no effect on Ektachem® method. *3393* At concentration of 1,000 mg/L had no effect on GOD/POD-PAP method. *3393*

Transport of Specimen By pneumatic tube if specimen protected. *3217*

Trazodone At concentration of 50 mg/L had no effect on GOD/POD-PAP method. *3393*

Trichlorethanol At concentration of 12 mg/L had no effect on GOD/POD-PAP method. *3393*

Trifluoperazine At concentration of 1 mg/L had no effect on GOD/POD-PAP method. *3393*

Trimethoprim No effect at therapeutic concentration on Reflotron method. *1984*

Uric Acid At 10 mg/dL on MBTH procedure of Neeley. *2569* In vivo by 10 mg/L per 10 mg/L GOD-ABTS procedure. *3524* No effect on Warner Glucomatic method. *2771* No effect at 100 mg/dL glucose oxidase method of Gochman. *1311* No effect on Trinder glucose oxidase method. *2771*

Valproic Acid At concentration of 100 mg/L had no effect on GOD/POD-PAP method. *3393*

Vanillin At 10 mg/dL no effect alkaline ferricyanide procedure At 10 mg/dL no effect on glucose oxidase procedure of Gochman. *1311*

Vanillylmandelic Acid (VMA) At 10 mg/dL no effect on glucose oxidase procedure of Gochman. At 10 mg/dL no effect on alkaline ferricyanide procedure. *1311*

Vitamin B Complex At concentration of 2.3 mg/L had no effect on Ektachem® method. *3393*

Vitamin Preparations No effect at expected concentration with SMA 12/60 procedure. *2637*

Warfarin At concentration of 1.5 mg/L had no effect on GOD/POD-PAP method. *3393*

Xylose No effect on glucose oxidase, hexokinase methods. *1856* At 200 mg/dL on hexokinase method of Coburn. *0676*

Serum   No Effect   Physiological

Acetazolamide Little effect observed in normals. *1487*

Acetylsalicylic Acid No effect seen in patients undergoing treatment. *1787*

Aging No significant change from age 6-20 y. *0887*

Amiloride No effect seen in 13 men treated for 8 weeks. *0656*

Atenolol No significant change after 3 or 6 mo treatment in 14 hypertensives. *0836* In 18 patients with mild essential hypertension treated with chlorothiazide concomitantly over 4 weeks. *1035* Compared with controls during oral glucose tolerance test. *1977*

Bendrofluazide After 12 mo treatment of 53 previously untreated hypertensives. *0315*

Cannabis No effect observed. *1633*

Captopril No significant change in 7,000 hypertensives treated for 3 y. *1407*

Chlorthalidone In 22 premenopausal and 18 postmenopausal women given 100 mg/d for 6 weeks. *0402*

Cimetidine Oral drug for 48 h had little effect in normal subjects. *3491*

Clonidine In children with growth hormone deficiency 0.1 mg/m^2 daily for 60 d. *2821*

Cyproheptadine Reported to have no effect. *1487*

Diclofenac No effect seen in patients undergoing treatment. *1787*

Enalapril In 53 patients given up to 160 mg daily or when combined with hydrochlorothiazide. *2277*

Ethanol No change in fasting or postprandial. *1288*

Ethynodiol With doses of 0.25, 0.35, 0.5 mg/d. *1335*

5-Fluoronicotinic Acid No effect seen in normals. *3077*

Glycerin No effect of ingestion of glycerol. *3615*

Hydrochlorothiazide No effect in hypertensives given potassium supplement if demonstrated no effect in response to hydrochlorothiazide alone. *2360*

Ibuprofen No effect seen in patients undergoing treatment. *1787*

Indomethacin No effect seen in patients undergoing treatment. *1787*

Ketoprofen No effect seen in patients undergoing treatment. *1787*

Levodopa No effect observed although increased plasma insulin. *1860*

Levonorgestrel In a group of women using levonorgestrel covered rods versus controls using copper IUD. *0899*

Meals Not significantly affected by nonstandardized meals. *3455*

Molindone Relatively unaffected over short term. *1635*

Muscular Exercise Observed effect with 12 minutes cycle-ergometer. *1231*

Naloxone In 6 healthy volunteers after ingestion of test meal and intravenous administration of drug. *0626*

Naltrexone No effect of long-term administration to obese individuals. *0176*

Nifedipine In 11 patients with 80 mg daily for 6 weeks. *3702*

Nisoldipine In 14 mild to moderately hypertensive noninsulin dependent diabetic patients. *2653*

Obesity Usual unless diabetic. *1106*

Ouabain No effect with i.v. infusion for 1 h. *3166*

Pancreozymin No change after 1 unit/kg i.v. *2676*

Penicillamine In 21 patients with rheumatoid arthritis treated for 48 weeks given 250 to 750 mg daily. *2345*

Phenytoin No significant difference between drug treated epileptics and controls. *2785*

Pindolol In 18 patients with mild essential hypertension treated with chlorthiazide concomitantly over 4 weeks. *1035* No significant changes found during 12 mo treatment. *2117*

Propranolol No effect observed with i.v. infusion. *3114* No effect on short term administration. *1635* After 12 mo treatment of 53 previously untreated hypertensives. *0315* Insignificantly reduced after 3 or 6 mo treatment in 16 nondiabetic hypertensives, but lower after glucose ingestion. *0836* In 5 hyperthyroid patients given 10 mg every 8 h for 4 d. *0335* In 18 patients with mild essential hypertension treated with chlorothiazide concomitantly over 4 weeks. *0041*

Propylthiouracil In 5 hyperthyroid individuals given 10 mg every 8 h for 4 d. *0335*

Recumbency Change of posture had no effect. *3539*

Secretin No change after 1 unit/kg i.v. *2676*

Spironolactone In 15 patients with primary hypertension treated with 100 mg daily for 1 y In 15 patients with primary hypertension treated with 100 mg daily for 1 y. *1072*

Terbutaline In 6 normal men treated with therapeutic amounts for 2 weeks. *3177*

Tetragastrin No change after 4 μg/kg i.v. *2676*

Thiopental No effect except during surgery when causes increase. *0972*

Thyrotropin-Releasing Hormone (TRH) After single dose of 1.0 mg synthetic TRH. *3831*

Triamterene In normals but increased in diabetics. *3744*

Tripamide No effect in hypertensives with or without diabetes. *0711*

Serum   Increase   Analytical

Acetaminophen At 1 mmol/L affects SMA 12/60 method. *3335* In YSI glucose analyzer with potentiometric measurement of hydrogen peroxide produced. Effect can be quite marked. *3018*

Acetylsalicylic Acid Increase by 11% with 8326 μmol/L on glucose-peroxidase method with 2,4-Dichlorophenol. Increase by 3% with 8326 μmol/L on glucose-peroxidase method with ABTS. *1786*

Aminophenol At 1 mmol/L affects SMA 12/60 method. *3335*

Aminosalicylic Acid At 1 g/dL on alkaline ferricyanide=1.7 mmol/L. *2140* At 1 mmol/L affects SMA 12/60 method. *3335* At concentrations above 100 mg/L raised concentration as measured by Ektachem® method. *3393*

Glucose *(continued)*

Serum Increase Analytical *(continued)*

Ascorbic Acid At 1 mmol/L affects SMA 12/60 method. *3335* Increases sensitivity of o-toluidine procedures 1 g/dL equivalent to 3.3 mmol/L with alkaline ferricyanide. *2140* Affects Neocuproin procedure. *2771* At concentrations above 60 mg/L (maximum serum concentration 34 mg/L) raised concentration as measured by glucose dehydrogenase method. *3393*

Bilirubin May have marked effect on o-toluidine method. *3902*

Cefotaxime Statistically significant effect of metabolite on Parallel method. *0201*

Citrates At concentrations above 7591 mg/L raised concentration as measured by Ektachem® method. *3393*

Creatinine Affects Neocuproin procedure. *2771* If measured by alkaline ferricyanide reduction. *2857*

Cysteine At 10 mg/dL affects alkaline ferricyanide procedure. *1311*

Dextran 10 g/dL equivalent to 0.3 mmol/L alkaline ferricyanide. 10 g/dL equivalent to 0.7 mmol/L p-HBAH procedure. *2140* Affects o-toluidine procedure: turbidity at acid pH. *3505* 10 g/dL equivalent to 6.7 mmol/L o-toluidine. 10 g/dL equivalent to 0.3 mmol/L glucose oxidase procedures. *2140*

Dextran 40 At concentrations above 10,000 mg/L raised concentration as measured by Ektachem® method. *3393*

Diclofenac Increase at therapeutic concentration with glucose oxidase/peroxidase method with ABTS. *1786*

EDTA If greater than 1 mg/mL affects o-toluidine procedures. *0724* At concentrations above 100 mg/L raised concentration as measured by Ektachem® method. *3393*

Epinephrine At 1 mmol/L affects SMA 12/60 method. *3335* At 10 mg/dL affects alkaline ferricyanide procedure. *1311*

Ergothionine At 10 mg/dL affects alkaline ferricyanide procedure. *1311*

Fluorides If greater than 5 mg/mL affects o-toluidine procedure. *0724*

Fluosol-DA Increase by 14% on Du Pont aca III at concentration of 50%. *2518*

Fructose Mole for mole effect alkaline ferricyanide. Mole for mole effect p-HBAH procedure of Lever. *2140* Affects Neocuproin procedures. *2771* 1 g/dL equivalent to 0.8 mmol/L o-toluidine procedure. *2140* Non specificity of o-toluidine, Neocuproin, FeCN procedures. *2857*

Galactose Affects o-toluidine procedure. Affects Neocuproin procedures. Affects alkaline ferricyanide method. *2771* Mg for mg effect in MBTH procedure of Neeley. *2569*

Gentisic Acid Falsely high with alkaline ferricyanide procedure. *1311*

Glucuronic Acid Interferes with some methods. *2313*

Glutathione Affects Neocuproin procedure. Affects alkaline ferricyanide method. *2771*

Hemoglobin May have marked effect with o-toluidine. *3902*

Hydralazine At 1 mmol/L affects SMA 12/60 method. *3335*

Hypochlorites Causes direct oxidation of glucose oxidase chromogen. *0583*

Iron Dextran At 5 g Fe/dL on glucose oxidase (slight) procedure. 5 g Fe/dL equivalent to 15.5 mmol/L o-toluidine procedure. 5 g Fe/dL equivalent to 3.5 mmol/L p-HBAH procedure. Slight effect at 5 g Fe/dL on alkaline ferricyanide procedure. *2140*

Iron Sorbitex 5 g Fe/dL equivalent to 7.1 mmol/L o-toluidine. 5 g Fe/dL equivalent to 1.9 mmol/L p-HBAH procedure. 5 g Fe/dL equivalent to 1.0 mmol/L alkaline ferricyanide. 5 g Fe/dL equivalent to 1/6 mmol/L glucose oxidase. *2140*

Isoproterenol At 1 mmol/L affects SMA 12/60 method. *3335*

Lactose 1 g/dL equivalent to 0.3 mmol/L glucose oxidase procedure. *2140* Affects Neocuproin procedures. *2771* 1 g/dL equivalent to 3.0 mmol/L with alkaline ferricyanide procedure. 1 g/dL equivalent to 1.3 mmol/L p-HBAH procedure of Lever. *2140* Produces 33 % color of glucose, o-toluidine method. *0583*

Levarterenol At 10 mg/dL affects alkaline ferricyanide procedure. *1311*

Levodopa At 1 mmol/L affects SMA 12/60 method. *3335* At 10 mg/dL affects alkaline ferricyanide procedure. *1311*

Lipemia Observed effect on "Trucose" o-toluidine method. *1206*

Maltose Affects Neocuproin procedures. *2771* Produces 5 % color of glucose, o-toluidine method. *0583* Affects alkaline ferricyanide method. Affects Trinder procedure (maltase in Fermcozyme). *2771*

Mannose Nonspecificity of FeCN, o-toluidine, Neocuproin. *2857* Mg for mg effect MBTH procedure of Neeley. *2569*

Mercaptopurine At 1 mmol/L affects SMA 12/60 method. *3335*

Methimazole At 1 mmol/L affects SMA 12/60 method. *3335*

Methyldopa At 1 mmol/L affects SMA 12/60 method. *3335*

Metronidazole Increased values observed with Technicon® SMAC® and Du Pont aca hexokinase methods. Normal drug concentration increased glucose by about 60 mg/L. *0936* On hexokinase method on Technicon® SMAC® at therapeutic concentration. *0937*

Nalidixic Acid Copper reduction methods affected. *3505*

Oxazepam Filler affects o-toluidine, Neocuproin methods. *2419*

Povidone-Iodine When povidone-iodine swab used with finger-stick and measured by Chemstrip bG® and Visidex. *1094*

Propylthiouracil At 1 mmol/L affects SMA 12/60 method. *3335*

Ribose Nonspecificity of ferricyanide, Neocuproin methods. *3902*

Rifampin At 5 times upper limit of therapeutic concentration on method on aca. *2138*

Sucrose Probably due to acid hydrolysis in o-toluidine method. *0583* Affects GOD-PERID procedure (sucrase in enzyme). *2771*

Tetracycline Slight effect hexokinase, o-toluidine methods. *2568* Mg for mg (approximately) MBTH procedure of Neeley. *2569*

Tolbutamide False increase (little effect at normal concentration) with glucose oxidase. *3278*

Uric Acid Affects Neocuproin procedure. *2771* If measured by alkaline ferricyanide reduction. *2857*

Xylose Nonspecificity of o-toluidine, FeCN, Neocuproin. *3902*

Serum Increase Physiological

Acetazolamide In prediabetics and if hypoglycemic agents used. *2039*

β-Adrenergic Drugs Masks spurious hypoglycemia from insulin overdose. *1237*

Alanine Slight effect in normals, greater in diabetics. *3880* After prolonged fast in obese. *1263*

Albuterol Significant dose-related effect after i.v. infusion of therapeutic doses in 4 healthy male volunteers. *2807*

Althesin 30 mg/dL increase observed 1 h after. *0291*

Aluminum Nicotinate Produces carbohydrate intolerance over long term. *1979*

Amino Acids Significant if i.v. infusion, not if intraduodenal. *2918*

Aminosalicylic Acid Hyperglycemia reported with protracted therapy. *2495*

Amiodarone Observed effect in 3 individuals but without effect on thyroid function. *2833*

Amitriptyline Reported observation. *1680*

Androgens Reported effect. *0117*

Anesthetic Agents Presumed response to stress. *0251*

Antipyrine Severe hyperglycemia unresponsive to insulin. *2427*

Arginine Slight, more in diabetics and if prednisolone. *3408* Positive correlation in diabetics given i.v. infusion of 0.5 g/kg. *2328*

Asparaginase May be hyperosmotic nonketotic hyperglycemia. *0578* Observed in 9.7% children, some within 1 week of start of treatment, observed most commonly in older children. *2883* Reported in 9.7% children, although hypoglycemia reported occasionally. *0558*

Aspirin Increased absorption and steroid release inhibits TCA cycle. *3505*

Benzthiazide May occur as with other thiazides. *1680*

Benzyl Alcohol Used as saline preservative, effect in mice ?humans. *0210*

Blood Pressure Strongly correlated with increased systolic pressure. *2123*

Caffeine Effect in order of 10 mg/dL only. *0644*

Cannabis Reported effect. *0117*

Cerulein Slight effect after i.v. infusion. *1076*

Chlorothiazide Diabetogenic properties of drug affect glucose tolerance test. *3505*

Chlorpromazine Abnormally high with repeated doses. *1595*

Chlorprothixene Due to altered endocrine function. *1680*

Chlorthalidone Diabetogenic-like action affects pre- or diabetic. *0102* Progressive effect with dose: by 5 mg/dL with 25 mg/d to 14 mg/dL with 200 mg/d in 37 patients treated for 8 weeks. *3643* Increase by 15 mg/dL in 39 mildly hypertensives treated for 6 weeks. *1404*

Clonidine When given i.v. over 10 minutes increase of about 15 mg/dL preceded other changes in normal volunteers. *2060* Increase observed in response to exercise but overall response blunted. *3919*

Clopamide Mild effect ?due to induced hypokalemia. *1487*

Clopenthixol Hyperglycemic effect reported. *2427*

Clorexolone Diabetogenic action. *2313*

Coffee Metabolic action of stimulant. *0117*

Corticosteroids Tends to be high (as in Cushing's syndrome). *1343*

Corticotropin Gluconeogenesis, insulin antagonism. *3879*

Cortisone Gluconeogenesis. *2220*

Cyclic AMP Hormonal action. *1102*

Cyclopropane Moderate with depletion of liver glycogen. *0071*

Desipramine Mechanism not understood. *1680*

Dexamethasone Hormonal action. *2525*

Dextroamphetamine Metabolic effect. *2313*

Dextrothyroxine Effect seen in diabetics. *1488*

Diapamide Alters glucose tolerance. *2256*

Diazoxide Inhibits insulin release, peripheral glucose utilization. *0636* Predictable effect due to direct inhibition of insulin secretion. *0451*

Diltiazem Mean increase from 98 to 105 mg/dL after 16 weeks in approximately 120 hypertensives. *2331*

Dimercaprol Initial response to toxic doses. *2313*

Dopamine Inhibits peripheral utilization, increased glycogenolysis. *0481* 5-7 µg/kg/min i.v. causes small increase. *1176*

Enflurane Effect of anesthesia and surgery. *0999*

Ephedrine Less effective than epinephrine. *1343*

Epinephrine Characteristic action of hormone. *1488*

Estrogens Metabolic effect. *1680*

Ethacrynic Acid Diabetogenic properties. *3505*

Ethanol Transient hyperglycemia in developing intoxication. *1343* 2.2% increase in heavy versus occasional drinkers. *3273*

Ether Metabolic effect (transient effect). *3918*

Ethionamide Hyperglycemia reported. *2220*

Ethylene After prolonged use may be moderate increase. *0071*

Fludrocortisone Endocrine response. *1680*

Fluoxymesterone Endocrine response. *1680*

Furosemide Diabetogenic-like action of drug affects glucose tolerance test. *3505* In 6% in 204 hospitalized patients receiving the drug. *3419*

Glucagon Counteracts hypoglycemia by glycogenolysis. *1019*

Glucocorticoids After i.v. 6 mg/dL at 150 minutes to 15 at 4 h. *2296* Diabetogenic action (increased gluconeogenesis). *2313*

Glucose Normal metabolic response usually maximum at 60 minutes. *3306* Normal response in normals and diabetics to ingestion or infusion. *0476*

Haloperidol Observed endocrinological disorder. *1680*

Halothane Response to stress of surgery. *2701*

Hemodialysis Usual response. *2547*

Heparin Single report of rise of 30 mg/dL. *0721*

Hydrochlorothiazide Diabetogenic-like action of drug. *3505* In 14 hypertensive men with type 2 diabetes by 31% over 3 weeks: effect augmented by coadministered propranolol. *0935* To extent of coma in 2 diabetics given hydrochlorothiazide with propranolol: exact mechanism not known. *2559* Increase by 6 mg/dL in 343 patients with hypertension given drug for 10 weeks. *3707* Increase by 4.7 mg/dL in approximately 175 patients with essential hypertension treated for 1 y. *3708*

Hydroflumethiazide May occur (similar action to other thiazides). *1680*

Hydroxydione Significant increase 1 h after anesthesia. *0291*

Ibopamine Peaked after 45 minutes in all subjects: normalized in 90 minutes. *3398*

Imipramine Reported effect. *0019*

Indapamide And after 75 g load in 13 hypertensive diabetics. *2686* Increase by 7 mg/dL in 17 patients with essential hypertension treated with 2.5 mg daily for 3 mo. *2010*

Indomethacin Rare side effect. *1488* In one patient with psoriatic arthritis, reverted to normal with drug withdrawal. *3600*

Isoniazid Large doses cause hyperglycemia by glycogenolysis. *1488*

Isoproterenol Not as marked as with epinephrine. *1343*

Labetalol Probably in response to norepinephrine release in response to i.v. infusion of 100 mg over 10 minutes. *0226*

Levarterenol Slight increase observed after i.v. infusion. *3114*

Levodopa Probably converted to dopamine which acts. *0481*

Levonorgestrel In implant users in comparison with iud users (average difference of 5 mg/dL). *0763*

Lithium Hyperglycemia been reported after use (transient). *3672* In two patients glycosuria and hyperglycemia occurred when lithium started: tests previously normal. *1812*

Lysergide Reported effect. *0117*

Maltose Marked rise observed with oral load. *2551*

Meals In effect acts as glucose tolerance test. *0579* 2 h after standardized meal in both sexes. *3455*

Medroxyprogesterone Metabolic effect observed after 1 y. *3414*

Meperidine Central effect also involves epinephrine release. *1343*

Meprednisone Alteration of carbohydrate metabolism. *1680*

Metanephrine Intravenous infusion 5 µg/min causes increase of 30%. *1176*

Methyclothiazide Diabetogenic action of thiazides. *1680*

Metolazone Diabetic-like action of diuretics. *2816*

Metoprolol Increase by 1.0 to 1.5 mmol/L in 20 hypertensive diabetic patients. *3900* Small but significant increase in nondiabetic hypertensives. *1410*

Morphine Minor, clinically insignificant increase. *1343*

Muscular Exercise Rise due to adrenal activity. *0987*

Narcotics May be rare insignificant increases. *1487*

Niacin Mechanism not discussed. *1343*

Nicotine Due to adrenal response in poisoning. *1302*

Nifedipine Fasting concentration increased by 10% in normal subjects. *0637*

Nortriptyline Endocrine response. *1680*

Obesity Proportion of subjects with high values increased. *2123*

Ohio 469 10-30% higher in postoperative period. *1647*

Oral Contraceptives Does not affect fasting glucose but alters glucose tolerance test. *3505*

Oxyphenbutazone metabolic/endocrine response. *1680*

Pancreozymin Intravenous infusion causes increase. *1775*

Paraldehyde Has caused transient hyperglycemia. *2220*

Pentamidine Paradoxical effect observed. *1343*

Perphenazine May cause hyperglycemia. *1680*

Phenelzine Decreases glucose tolerance. *3679*

Phenolphthalein Impaired glucose tolerance may occur due to K loss. *0706*

Phenothiazines Probable effect, adrenergic response. *1487*

Phenylbutazone metabolic/endocrine response. *1680*

Phenylephrine Reported to increase sugar. *2726*

Phenytoin Inhibitory effect on insulin secretion. *0789* Metabolic effect of drug. *0121* Small number of cases with hyperglycemia; occasional convulsions and Coma reported. Probably due to decreased insulin secretion. *0598*

Piperacetazine Disordered endocrine response. *0019*

Polythiazide Diabetogenic action. *2313*

Prazosin With 2 mg in 12 hypertensives (6 with normal and 6 with abnormal glucose tolerance). *0225*

Prednisolone After 4 d of oral therapy After i.v. 6 mg/dL at 150 minutes to 15 at 4 h. *2296*

Prednisone Glucocorticoid effect. *2525*

Pregnancy Reported effect. *0279*

Glucose *(continued)*

Serum Increase Physiological *(continued)*

Propranolol Mean increase from 101 to 108 mg/dL after 16 weeks treatment in approximately 120 hypertensives. *2331* Overt diabetes developed or after glucose challenge in 40 hypertensives. *2469* Increase by 1.0-1.5 mmol/L in 20 hypertensive diabetic patients. *3900* To extent of diabetic coma in 2 diabetics when propranolol given with hydrochlorothiazide: exact mechanism not known. *2559* Increase by 2 mg/dL in 340 patients given drug for 10 weeks to reduce diastolic BP to less than 90 mm Hg. *3707* Increase by 8 mg/dL in approximately 120 patients with essential hypertension treated for 1 y. *3708* In 14 hypertensive men with type 2 diabetes when treated for 3 weeks, marked increase when propranolol coadministered. *0935* Slight effect with pharmacological doses. *1410*

Protriptyline Reported effect. *1678*

Quinethazone May precipitate latent diabetes or aggravate exist. *2313*

Recumbency Occurs with reduced physical activity. *0279*

Reserpine Hyperglycemia may follow administration. *2309*

Rifampin Early phase hyperglycemia after oral administration of 100 g glucose. *3534*

Rimeterol Dose related significant change in 4 healthy men given therapeutic i.v. dose. *2807*

Ritodrine Moderate effect when given intravenously for premature labor. *0575*

Season Slight increase during summer. *2123*

Secretin Intravenous infusion causes increase. *1775*

Serum Approximately 13% higher than in whole blood. *3217*

Sex Difference In males except in 6th decade. *2130*

Smoking Effect of stimulant. *0117* No effect in 7 healthy volunteers after 3 cigarettes. *2352*

Starvation Metabolic response (decreased in advanced state). *0279*

Stress Normal metabolic response. *3527*

Sucrose Mean increase of 40 mg/dL after 1 g/kg/h orally. *0408* With tolerance test from 80 to 180 mg/dL. *2551*

Terbutaline Observed following infusion of 0.25 mg or 5 mg 3 times/d on first day of treatment. *0288*

Tetracosactrin From 5.2 to 7.2 mmol/L healthy volunteers given 1 mg intramuscularly for up to 60 h. *1813*

Theophylline Follows overdose, but decreased glucose tolerance also observed in infants and others. *3422* Effect small in response to i.v. infusion of aminophylline. *0608*

Thiabendazole Rare case of hyperglycemia reported. *1488*

Thiazides Decreased glucose tolerance. *0070*

Thiothixene Observed with some phenothiazines. *1678*

Thyroid Metabolic action of hormone. *0652*

Triamcinolone Glucocorticoid effect. *2525*

Triamterene Effect less common than with thiazides. *2029*

Trichlormethiazide Diabetogenic like action of drug: affects glucose tolerance test. *3505* At 30 and 60 minutes in 6 patients given orally 75 g glucose values versus controls were 158 and 170 versus 136 and 156 mg/dL respectively. *2550*

Tyramine Observed with high doses. *1176*

Vinyl Ether Transient effect (less than with ether). *0071*

Serum Decrease Analytical

Acetaminophen At concentrations above 25 mg/L (normal therapeutic concentration 20 mg/L) lowered concentration as measured by GOD-Perid method. *3393*

Aminophenazone Slight reduction at therapeutic concentration on ABTS method. *1786* With 540 µmol/L on GOD-Perid/ABTS method. *1785*

Aminophenol At concentrations above 25 mg/L lowered concentration as measured by GOD-Perid method. *3393*

Ascorbic Acid Negative peaks GOD-PERID procedure. *2771* 17% decrease at 10 mg/dL glucose oxidase dianisidine procedure. *3037* Slight effect with coupled glucose-oxidase method. *1311* At very high concentrations on Glucomatic method. *2771* But reduction insignificant (each 1 mg/dL lowers concentration as measured by 0.65 mg/dL) with glucose oxidase method on SMAC®. *3719* At concentrations above 100 mg/L (maximum serum concentration 34 mg/L) lowered concentration as measured by Ektachem® method. *3393* At concentrations above 125 mg/L (maximum serum concentration 34 mg/L) lowered concentration as measured by GOD-Perid method. *3393* At concentrations above 150 mg/L (maximum serum concentration 34 mg/L) lowered concentration as measured by GOD/POD-PAP method. *3393* At concentrations above 25 mg/L (maximum serum concentration 34 mg/L) lowered concentration as measured by Seralyzer. *3393*

Bilirubin Slight effect on hexokinase procedures. *3217* Effect can be avoided by using Somogyi filtrate. *1563* Very slight effect with automated alkaline ferricyanide. *3217*

Bilirubin Glucuronide Inhibits coupled glucose-oxidase reaction. *0583*

Carbon Monoxide Saturation may cause decrease of 15% with glucose oxidase methods. *2717*

Catechol May inhibit glucose-oxidase procedures. *1563*

Chlorpropamide Negative peaks with Boehringer GOD-PERID method. *2771* At concentrations above 500 mg/L (normal therapeutic concentration 150 mg/L) lowered concentration as measured by GOD-Perid method. *3393*

Contact With Clot Glucose metabolism by cells. *0579* At 4° changes 1-3 mg/dL/h. *1563*

Creatinine Negative peaks GOD-PERID procedure. *2771*

Cysteine May affect some glucose-oxidase procedures. *1563* At 5 times upper limit of therapeutic range on glucose oxidase method on Ektachem®. *2138*

Cystine At 5 times upper limit of therapeutic range on glucose oxidase method on Ektachem®. *2138*

Deoxy-Glucose At 4.0 mmol/L if hexokinase from Leuconostoc Mesent. *0411*

Diatrizoic acid At concentrations above 5140 mg/L (probably below circulating concentration) lowered concentration as measured by Ektachem® method. *3393*

Fructose At 3.7 mmol/L if hexokinase from Leuconostoc Mesent. *0411*

Gentisic Acid 29% decrease at 10 mg/dL glucose oxidase dianisidine procedures. *3037*

Glutathione May affect some glucose oxidase methods. *1563* Negative peaks GOD-PERID procedure. *2771*

Guanoclor Affects glucose oxidase method of Boehringer. *3277*

Hemolysis Affects glucose oxidase procedure, avoidable by Somogyi filtrate. *1563*

Hydralazine Affects glucose oxidase method of Boehringer. *3277* At concentrations above 115 mg/L (normal therapeutic concentration 2.3mg/L) lowered concentration as measured by GOD-Perid method. *3393*

Iproniazid Depresses glucose oxidase method of Boehringer. *3277* At concentrations above 110 mg/L lowered concentration as measured by GOD-Perid method. *3393*

Isocarboxazid Affects glucose oxidase method of Boehringer. *3277*

Isoniazid Affects glucose oxidase method of Boehringer. *3277* At concentrations above 55 mg/L (normal therapeutic concentration 10 mg/L) lowered concentration as measured by GOD-Perid method. *3393* At concentrations above 500 mg/L (normal therapeutic concentration 10 mg/L) lowered concentration as measured by Seralyzer method. *3393*

Levodopa 51% decrease at 10 mg/dL glucose oxidase dianisidine procedure. *3037* May cause marked decrease with glucose oxidase method. *2568* At concentrations above 100 mg/L lowered concentration as measured by Ektachem® method. *3393* At concentrations above 3 mg/L lowered concentration as measured by GOD-Perid method. *3393* At concentrations above 300 mg/L lowered concentration as measured by Seralyzer method. *3393*

Lipemia Increase by 1.9% patient specimens measured on SMAC® in comparison with specimens following ultra-centrifugation. *2466*

Mannose Above 0.28 mmol/L if hexokinase from Leuconostoc Mesent. *0411*

Methyldopa At concentrations above 100 mg/L (normal therapeutic concentration 2 mg/L) lowered concentration as measured by Ektachem® method. *3393* At concentrations above 200 mg/L (normal therapeutic concentration 2 mg/L) lowered concentration as measured by GOD-Perid method. *3393* At concentrations above 7 mg/L (normal therapeutic concentration 2 mg/L) lowered concentration as measured by GOD/POD-PAP method. *3393*

Metronidazole At concentrations above 100 mg/L (normal therapeutic concentration 47.5 mg/L) lowered concentration as measured by hexokinase/G-6-PDH method. *3393* At concentrations above 200 mg/L (normal therapeutic concentration 47.5 mg/L) lowered concentration as measured by aca method. *3393*

Nitrazepam Slight effect glucose oxidase method of Boehringer. *3277*

Noramidopyrine A-V decrease 50 mg/L at *in vivo* concentration GOD-ABTS procedure. *3524*

Novaminsulfon At concentrations above 400 mg/L (normal therapeutic concentration 15 mg/L) lowered concentration as measured by GOD-Perid method. *3393* At concentrations above 200 mg/L (normal therapeutic concentration 15 mg/L) lowered concentration as measured by GOD-PAP method. *3393*

Oxyphenbutazone At concentrations above 25 mg/L (normal therapeutic concentration 20 mg/L) lowered concentration as measured by GOD-Perid method. *3393*

Phenazopyridine Delays coupled glucose oxidase reaction. *0583*

Potassium Oxalate Ineffective as antiglycolytic agent. *2313*

Serum By up to 24% if no antiglycolytic agent used. *2690*

Standing of Sample Significant if on polystyrene beads 24 h at 25°. *2624*

Storage of Sample Significant decrease after 36 d in frozen state. *3862* Significant effect after 6 mo at -10°. *1180*

Sulpiride At 15 mg/dL conventional methods when added to serum. *1758*

Tetracycline Ascorbic acid in prep may affect glucose oxidase procedure. *0583* At concentrations above 20 mg/L (normal therapeutic concentration 8 mg/L) lowered concentration as measured by GOD/POD-PAP me thod. *3393*

Thymol 1 mg/mL inhibits color = 10 mg/dL by o-toluidine. *0724*

Tolazamide Inhibits oxidation chromogen Boehringer glucose oxidase method. *3277* At concentrations above 10 mg/L lowered concentration as measured by GOD-Perid method. *3393*

Tolbutamide Negative peaks with Boehringer GOD-PERID method. *2771* At concentrations above 500 mg/L (normal therapeutic concentration 110 mg/L) lowered concentration as measured by GOD-Perid method. *3393*

Trypan Blue Affects glucose oxidase method of Boehringer. *3277*

Uric Acid If measured by o-toluidine. *2857* Negative peaks GOD-PERID procedure. *2771* If high competes with chromogen in glucose oxidase procedure. *0583*

Serum Decrease Physiological

Acetaminophen Reported effect of metabolite. *0817*

Acetoacetate Effect maximal 40 minutes after loading. *1811*

Acetohexamide Sulfonylurea derivative promotes insulin secretion. *0652*

Alanine After short term fast in obese. *1263*

Allopurinol Hepatotoxic effect. *2220*

Ambulation 1/2 mile walk in normal causes decrease of 5 mg/dL. *1946*

Amino Acids Fall may occur after intraduodenal infusion. *2918*

Aminobenzoic Acid Rare case of hypoglycemia observed. *3251*

Aminosalicylic Acid May cause lowering in diabetics. *2220*

Amitriptyline Reported observation. *1680*

Amphetamine Dextroamphetamine may produce slight effect. *1487*

Anabolic Steroids Anabolic effect in fasting state. *2220*

Androgens Anabolic effect in fasting state. *2220*

Antihistamines May occur in susceptible individuals especially children. *0851*

Antimony Compounds Hepatotoxic effect. *2220*

Arsenicals Hepatotoxic effect. *2220*

Ascorbic Acid Significant reduction in fasting concentration of diabetics treated for 15 d. *3136*

Aspirin In diabetics and if toxic doses ingested. *1022* Also decreased response to oral glucose in normal subjects and diabetics. *2433*

Atropine Possible slight fall if given as premedication. *2394*

Barbiturates Reported effect. *0117*

Bed Rest Observed effect with prolonged inactivity. *0117*

Benzene Toxic effect. *2313*

Blindness Significantly lower (average 14 mg/dL) than normal. *1637*

Buformin Therapeutic intent. *2427*

Calcium Gluconate Slight effect observed in newborns. *1550*

Cannabis Hypoglycemic effect approximately 4 h after use. *0912*

Carbutamide Hepatotoxicity. *2313*

Chloramphenicol Reported effect. *0117*

Chloroform Hepatotoxic effect with necrosis. *2313*

Chlorpropamide Sulfonylurea derivative promotes insulin secretion. *2801*

Cimetidine After 100 mg/h for 4 h caused decrease of 15% at 150 minutes in normal subjects. Mean value fell from 5.4 to 4.8 mmol/L on average in 6 patients given 1 g/d for 1 mo. *3491*

Clofibrate If previously abnormal. *1487* After 1 week in both normals and diabetics but more marked in latter. *1120*

Cyproheptadine Small decrease (10%), often no change. *1488*

Cyproterone When given with ethinyl estradiol: combination causes insulin resistance. *3239*

Desipramine Mechanism not understood. *1680*

Dextroamphetamine Reported to produce decrease of fasting glucose. *1487*

Dicumarol Reported effect. *0117*

Dimercaprol After initial increase. *2313*

Disopyramide To below 10 mg/dL observed in 2 patients both receiving other drugs and occurring with hypotension. *2557*

Erythromycin Hepatotoxic effect. *2220*

Ethacrynic Acid Symptomatic hypoglycemia reported. *3214*

Ethanol Reduced to below 50 mg/dL: effect for up to 24 h. *0586* In individuals who have depleted glycogen stores after 72 h of fasting. *2428*

Fenfluramine Direct effect, increased glucose uptake by muscle. *3638*

Fever Observed effect. *0117*

Fluoxymesterone Metabolic action. *0019*

Fructose Marked fall may occur 1 h after i.v. fructose. *3113*

Glibonuride Therapeutic intent. *1526*

Glipizide At dose of 2.5-5.0 mg produces rapid decrease. *2735*

Glyburide Long acting sulfonylurea = insulin secretion. *2216*

Glymidine Plasma half-life approximately 4 h. *2563*

Guanethidine Antidiabetic activity (rise once stop therapy). *1426*

Haloperidol Insulin like action of drug reported in one case. *1488*

Hydrazine Derivatives May potentiate action of insulin in diabetics. *2220*

Hydroxyhexamide Mild effect (stimulates release of insulin). *3928*

Imipramine Reported effect. *0019*

Insulin Natural action of hormone. *2313*

Isocarboxazid MAO inhibitors have slight effect. *1488*

Lincomycin May occur with hepatotoxicity. *2220*

MAO Inhibitors Mechanism not clear (possible hepatotoxicity). *0025*

Mebanazine Appears to potentiate insulin in diabetics. *2427*

Megestrol In 18 postmenopausal women with breast cancer. *0056*

Metformin Mode of action uncertain (occurs with overdose). *1184*

Methandrostenolone Anabolic effect. *0019*

MK-270 Slight nonsignificant decrease in fasting value. *3099*

Muscular Exercise Occurs with strenuous exercise. *0117*

Nandrolone Anabolic effect. *2681*

Nialamide May prolong action of insulin in diabetics. *2427*

Nifedipine From 102 to 95 mg/dL in 15 hypertensive patients undergoing hemodialysis after 3 weeks treatment. *2986*

Glucose *(continued)*

Serum Decrease Physiological *(continued)*

Nifurtimox   May cause decline. *0071*

Nortriptyline   Endocrine response. *1680*

Onions   As fried onions has effect on diabetes. *1768*

Oral Contraceptives   Estrogen effect of combined oral contraceptive. *0018*   Significantly lower in drug users. *1248*

Oxandrolone   Anabolic effect. *1308*

Oxymetholone   Anabolic effect (possible for 2 weeks after stop). *1679*

Oxytetracycline   Mild hypoglycemic effect observed in diabetics. *3253*

Pain   Response to physical stress. *0117*

Pargyline   Possible hepatotoxic effect. *2313*

Pentamidine   Hypoglycemia — possible effect. *1343*   Hypoglycemia occurred in 10 to 30% patients with pneumocystis carinii pneumonia, usually 5 to 13 d after treatment, but fatal case described after 2 weeks. *3157*   Four patients with pneumocystis carinii developed severe fasting hypoglycemia. *0431*

Phentolamine   Toxic doses over long period of time. *1343*

Phenyramidol   Reported effect. *0117*

Phosphorus   Toxic effect. *2313*

PIDH   Significant reduction after glucose load. *3100*

Piperacetazine   Disordered endocrine response. *0019*

Pizotyline   10% reduction after 3-5 h with 2 mg i.v. *3411*

Potassium Aminobenzoate   Reported effect. *2313*

Potassium Chloride   Drawn into cells with potassium. *2313*

Pregnancy   May be low in occasional cases. *0987*

Probenecid   Reported effect. *0117*

Progesterone   Slight insignificant effect (contrast with estrogens). *0991*

Promethazine   If given i.v. or i.m. *0851*

Propoxyphene   Hypoglycemia allegedly occurred in one case. *3820*

Propranolol   Has slight effect like that of prolonging insulin. *3505*   6 episodes of hypoglycemia observed in 5 nondiabetic patients on chronic hemodialysis, due to beta-adrenergic blockage. *1376*   Rare cases due to inhibition of glycogenolysis in nondiabetics. *2633*

Protriptyline   Reported effect. *1678*

Quinine   Intravenous drug reduced concentration from 88 to 68 mg/dL in normal volunteers. *3814*

Rotenone   Severe hypoglycemia reported with poisoning. *1302*

Sulfaphenazole   Reported effect. *2313*

Sulfonamides   Reported effect. *0117*

Sulfonylureas   May cause severe hypoglycemia. *2427*

Thiabendazole   Asymptomatic lowering observed. *0071*

Thiocarlide   Hypoglycemic reactions observed. *2427*

Thiothixene   Observed with some phenothiazines. *1678*

Thiouracil   Reported effect. *0117*

Tolazamide   Sulfonylurea derivative stimulates insulin secretion. *3278*

Tolbutamide   Therapeutic intent (promotes insulin secretion). *1152*

Tranylcypromine   In animals probably stimulates insulin secretion. *1487*

Tripelennamine   Reported effect. *0117*

Tromethamine   Hypoglycemia especially if i.v., also transient if oral. *3006*

Verapamil   When infused i.v. at rate of 5 mg/h for 3 h in prolonged fasted individuals but not in overnight fasted subjects. *0103*

Plasma   No Effect   Physiological

Timolol   In 5 hyperthyroid patients given 10 mg every 8 h for 4 d. *0335*

Urine   No Effect   Analytical

Acetoacetate   At 5 g/L on Boehringer — Mannheim BM 33071 glucose pad. *3356*

Amdinocillin   On Clinitest® at physiological concentration. On Diastix® at physiological concentration. *2733*   On TesTape® at physiological concentration. *0191*

Amikacin   No effect at up to 2 mg/mL on Clinitest® method and Diastix® and TesTape® procedures. *2250*

Ampicillin   No effect observed with TesTape®. *1100*   No effect at concentration up to 10 mg/mL on Diastix® and TesTape® methods. *2250*   At concentration of 4,000 mg/L had no effect on Diabur-test. *3393*

Ascorbic Acid   With DNSA method at usual concentrations. *0076*   No effect observed with TesTape®. *3505*

Aspirin   With DNSA method at usual concentrations. *0076*

Azlocillin   Concentrations accurately measured by Diastix®. Concentrations accurately measured by TesTape®. *2100*

Aztreonam   Dark green-black color with Clinitest®, but no effect or slight reduction with glucose concentrations of 1% and up. Concentrations measured accurately by Diastix®. Concentrations measured accurately by TesTape®. *2100*   Concentrations measured accurately by TesTape®. *0086*

Bisacodyl   No effect observed with TesTape®. *1100*

Carbenicillin   No influence of drug at up to 20 mg/mL on Diastix® and TesTape® procedures. *2250*

Cefoperazone   Dark green-black color with Clinitest® but no effect or slight reduction with glucose concentrations of 1% and up. Concentration measured accurately by Diastix®. Concentrations measured accurately by TesTape®. *2100*

Cefotaxime   Dark green-black color with Clinitest® but no effect or slight reduction with glucose concentrations of 1% and up. Concentrations measured accurately by Diastix®. Concentrations measured accurately by TesTape®. *2100*

Ceftazidime   Dark green-black color with Clinitest® but no effect or slight reduction with glucose concentrations of 1% and up. Concentrations measured accurately by Diastix®. Concentrations measured accurately by TesTape®. *2100*

Ceftizoxime   Dark green-black color with Clinitest® but no effect or slight reduction with glucose concentrations of 1% and up. Concentrations measured accurately by Diastix®. Concentrations measured accurately by TesTape®. *2100*

Ceftriaxone   Dark green-black color with Clinitest® but no effect or slight reduction with glucose concentrations of 1% and up. Concentrations measured accurately by Diastix®. Concentrations measured accurately by TesTape®. *2100*

Cephalexin   No effect on TesTape®. *1705*   At concentration of 10,000 mg/L had no effect on Diabur-test. *3393*

Cephaloglycin   No effect on TesTape®. *1680*

Cephaloridine   No effect observed with TesTape®. *1019*

Chloral Hydrate   No effect observed with TesTape®. *1100*

Chloramphenicol   At concentration of 400 mg/L had no effect on Diabur-test. *3393*

Chloroform   As preservative has no effect on Stix tests. *2308*

Creatinine   With DNSA method at usual concentrations. *0076*

Cyclacillin   No effect on Diastix® at physiological concentration. No effect on TesTape® at physiological concentration. *2733*

Diazepam   No effect observed on TesTape®. *1100*

Digoxin   No effect observed with TesTape®. *1100*

Ferrous Sulfate   No effect observed with TesTape®. *1100*

Fluorescein   At 250 mg/L in spiked specimen run on Ames Clinitek 200. *1722*

Flurazepam   No effect observed with TesTape®. *1100*

Furosemide   No effect observed with TesTape®. *1100*

Gentamicin   At up to 250 µg/mL had no effect on measured glucose concentrations using Clinitest®, Diastix® and TesTape®. *2250*   At concentration of 200 mg/L had no effect on Diabur-test. *3393*

Glucuronic Acid   No effect on Clinistix®, Labstix® methods. No effect on TesTape®. *3877*

Hemoglobin   At 0.5 g/L on Boehringer-Mannheim BM33071 glucose pads. *3356*

Hydroxybutyrate   Beta-hydroxybutyrate at 10.0 g/L on Boehringer-Mannheim BM33071 glucose pad. *3356*

Levodopa   No effect on TesTape®. *1099*

Metaxalone   No effect on glucose oxidase methods. *1487*

Methyldopa   No effect on glucose oxidase methods. *1099*

Mezlocillin   Concentrations measured accurately by Diastix®. Concentration measured accurately by TesTape®. *2100*

Moxalactam No effect on copper reduction procedures such as Clinitest®. *2615*

Nalidixic Acid No effect with Clinistix®, TesTape®. *3877*

Netilmicin On Clinitest® at physiological concentration. On Diastix® at physiological concentration. On TesTape® at physiological concentration. *2733*

Nitrofurantoin At concentration of 500 mg/L had no effect on Diabur-test. *3393*

Oxalate With DNSA method at usual concentration. *0076*

Penicillin No effect observed with TesTape® No effect on Clinistix® or TesTape®. *2525*

Penicillin G No effect at up to 16.2 mg/mL on any glucose concentration as measured by Diastix® or TesTape®. *2250* At concentration of 3.6 mg/L had no effect as measured by Diabur-test. *3393*

Phenobarbital No effect observed on TesTape®. *1100*

Phenols With DNSA method at usual concentrations. *0076*

Piperacillin Concentrations accurately measured by Diastix®. Concentrations accurately measured by TesTape®. *2100*

Prednisone No effect observed with TesTape®. *1100*

Probenecid No effect on glucose oxidase methods. *1487*

Propoxyphene No effect on TesTape®. *1100*

Salicylate At concentration of 5,000 mg/L had no effect on Diabur-test. *3393*

Secobarbital No effect observed with TesTape®. *1100*

Sisomicin On TesTape® at physiological concentration. On Diastix® at physiological concentration. On Clinitest® at physiological concentration. *2733*

Streptomycin No effect on glucose oxidase methods. *1487* No effect at up to 1 mg/mL on any glucose concentration as measured by Clinitest® Diastix® and TesTape®. *2250*

Sulfacarbamide At concentration of 1600 mg/L had no effect on Diabur-test. *3393*

Sulfamethoxazole At concentration of 1300 mg/L had no effect on Diabur-test. *3393*

Thymol No effect on glucose-oxidase methods. *2981*

Ticarcillin Concentrations measured accurately with Diastix® Concentrations measured accurately with TesTape®. *2100*

Tobramycin No effect at up to 250 μg/mL on any glucose concentration as measured by Clinitest®, Diastix® or TesTape®. *2250*

Trimethoprim At concentration of 500 mg/L had no effect on Diabur-test. *3393*

Uric Acid With DNSA method at usual concentrations. *0076*

Vitamin Preparations No effect observed with TesTape®. *1100*

Urine No Effect Physiological

Chenodeoxycholic Acid No significant effect with up to 750 mg/d in 916 patients treated for 2 y. *3199*

Penicillamine In 21 patients with rheumatoid arthritis treated for 48 weeks given 250 to 750 mg daily. *2345*

Timegadine In 23 patients with rheumatoid arthritis given 250 to 750 mg/d for 48 weeks. *2345*

Urine Increase Analytical

Ascorbic Acid Effect on Clinitest® causes normal to be read as trace at low concentrations: less effect at high concentrations. *0786*

Azlocillin Falsely elevated values with Clinitest®. *2100*

Cyclacillin Falsely elevated value of 0.25% approximately at physiological concentration. *2733*

Furazolidone Due to presence of glucose in vaginal powder. *1237*

Hydrogen Peroxide Oxidises chromogen in glucose oxidase test. *1022*

Hypochlorites Oxidation of chromogen (o-tolidine). *1022* False positive with glucose oxidase procedures. *0443*

Mezlocillin Falsely elevated values with Clinitest®. *2100*

Pentoses Produce orange color with o-toluidine. *0583*

Peroxide May produce false positive with Clinistix®. *1563*

Phenazopyridine False positive reported with TesTape®. *2566*

Piperacillin Falsely elevated values with Clinitest®. *2100*

Ticarcillin Falsely elevated values with Clinitest®. *2100*

Trimetozine False positive with glucose oxidase (Combistix®) reported. *0666*

Vaginal Powder Powders often contain glucose. *2220* Powders may contain glucose. *1488*

Urine Increase Physiological

Acetazolamide Occasionally observed adverse reaction. *1680*

Aminosalicylic Acid Glycosuria reported with protracted therapy. *2495* Glycosuria reported with long therapy. *1487*

Amyl Alcohol Associated with renal damage. *1302*

Antipyrine Renal action of drug. *3824*

Aspirin Inhibits liver and muscle glycogen synthesis due to hyperglycemia. *1343*

Benzthiazide Rare, may occur as with other thiazides. *1680*

Bismuth Salts Due to proximal tubular dysfunction (Fanconi syndrome). *2911*

Cadmium Occasionally observed with proteinuria. *2217*

Captopril Reversible glycosuria reported in one boy with abdominal aortitis and resistant hypertension. *0929*

Carbamazepine Single case of glycosuria reported. *1488*

Chlorothiazide Diabetogenic-like action of drug affects glucose tolerance test. *3505*

Chlorpromazine Due to hyperglycemia. *1680*

Chlorprothixene Due to altered endocrine function. *1680*

Chlorthalidone Diabetogenic-like action of drug affects glucose tolerance test. *3505* Diabetogenic-like action of drug affects glucose tolerance test. *3504*

Corticosteroids As result of hyperglycemia. *2313*

Corticotropin Increased blood glucose, reduced TMG. *3879*

Cyclothiazide Diabetogenic action of thiazides. *1680*

Dexamethasone Associated with hyperglycemia. *1680*

Dextroamphetamine Will occur if hyperglycemia. *2313*

Dextrothyroxine Occurs due to elevation of blood sugar Occurs due to hyperglycemia. *1488*

Diapamide Result of marked hyperglycemia. *2256*

Diazoxide If hyperglycemia occurs. *0636*

EDTA Reported effect. ?mechanism May cause tubular damage. *1343*

Enalapril Reported in one patient with mild uncomplicated essential hypertension. *0929*

Ephedrine May cause glycosuria Increased excretion of glucose. *1022*

Ethacrynic Acid Diabetogenic properties. *3505*

Ether Due to hyperglycemia *2313*

Ethionamide Due to hyperglycemia. *2220*

Fludrocortisone Endocrine response. *1680*

Furosemide Diabetogenic-like action of drug. *3505* Observed in 0.2% of 2580 medical inpatients. *2218*

Glucagon Mobilizes hepatic glycogen, causes glycosuria. *2313* Mobilizes enough glycogen to cause glycosuria. *0071*

Glucocorticoids Due to hyperglycemia. *2313*

Growth Hormone May cause hyperglycemia and glycosuria *2313*

Hydrochlorothiazide Diabetogenic-like action of drug due to hyperglycemia if produced. *3505*

Hydroflumethiazide May occur as consequence of hyperglycemia. *1680*

Ibufenac Augmentation of diabetic glycosuria seen. *2427*

Indomethacin Rare side effect. As result of rare hyperglycemia. *1488*

Isoniazid Glycosuria may follow induced hyperglycemia. *1488* Due to hyperglycemia. *2220*

Lactose Maximum at 90 minutes in normals after oral load. *2551*

Lead Nephrotoxic effect with lead poisoning. *0987* Nephrotoxicity with poisoning. *1343*

Lithium Reported effect of lithium therapy in some cases consequence of hyperglycemia. *1488* In two patients glycosuria and hyperglycemia occurred when lithium started: tests previously normal. *1812*

Meprednisone Alteration of carbohydrate metabolism. *1680*

Mercury Compounds May cause Fanconi syndrome. *2108*

Methyclothiazide May occur as result of hyperglycemia. *1680*

Glucose (continued)

Urine Increase Physiological (continued)

Niacin As result of hyperglycemia. *1343* Due to hyperglycemia. *1253*

Phenothiazines Effect seen especially in long term therapy of diabetics. *2497* Long term effect in some patients. *1487*

Phenylbutazone Observed effect, may occasionally be marked with oliguria or renal failure. *2866*

Phenytoin Occurs with inhibition of insulin secretion Occurs with hyperglycemia. *0789*

Phloridzin Decreases glucose reabsorption. *3343* Increases clearance. *2313*

Piperacetazine Due to induced hyperglycemia. *0019*

Polythiazide Consequence of hyperglycemia. *2313*

Pregnancy May occur with decreased tolerance. *0987*

Quinethazone Occurs as consequence of hyperglycemia. *2313*

Reserpine Occurs as consequence of hyperglycemia. *2309*

Streptozocin Rarely exceeded 25 mg/dL Rarely exceeds 25 mg/dL. *2532*

Sucrose Maximum at 60 minutes, continues for several h. *2551*

Tetracycline Degraded drug may cause Fanconi like syndrome. *2808* Degraded material may cause Fanconi syndrome. *2427*

Theophylline Follows overdose, but decreased glucose tolerance also observed in infants and others. *3422*

Thiazides May occur in prediabetics especially May occur especially in prediabetics. *0460*

Thiothixene Observed with some phenothiazines. *1678*

Triamcinolone Consequence of hyperglycemia. *2525*

Trichlormethiazide Diabetogenic like action of drug Diabetogenic-like action of drug. *3505*

Trimetozine Hyperglycemia reported in one patient. *1487*

Turpentine Manifestation of nephrotoxicity of poisoning Due to ingestion — nephrotoxic effect. *1302*

Urine Decrease Analytical

Acetoacetate At 5 g/L with Hema-Combistix® caused complete inhibition. At 0.75 g/L cause some inhibition. *3356*

Ampicillin Low with Clinistix®, Diastix®. *1100*

Ascorbic Acid Impaired color develop of chromogen in glucose oxidase method. *1022* May inhibit TesTape® and Clinistix®. *3505* At 0.4 g/L interfered with BM33071 in 20% urines containing 1.0 g glucose/L but not with urines containing 5.0 g glucose/L. No effect of ascorbic acid at 0.1 to 0.2 g/L. At 0.4 g/L interfered with Hema-Combistix® in 90% urines containing glucose at 5.0 g/L. No effect of ascorbic acid at 0.1 to 0.2 g/L. *3356* But only very slight effect with BM33071 procedure from Boehringer-Mannheim compared with other dipsticks. *0787* At physiological urine amounts, marked reduction of results with especially Ecur-test, Diabur-test 5,000 and Rapignost basis screen but also TesTape®. *0786* At concentration of about 100 mg/dL made 0.1 g/dL react negatively with Chemstrip® 7, 1-lema-Combistix® and give trace reacftion with Chemstrip® UG. *3968* Significant effect with Redia-test, L-Combur-5-test, Labstix®, Rapignost, Meditest but less effect with BM33.071. *0303*

Aspirin Glucose-oxidase methods inhibited by gentisic acid. *1099*

Bacteriuria False negative through metabolic actions. *2308*

Bilirubin May produce false negative with Clinistix®. *2544*

Bisacodyl Low with Clinistix®, Diastix®. *1100*

Chloral Hydrate Low with Clinistix®, Diastix®. *1100*

Diazepam Low with Clinistix®, Diastix®. *1100*

Digoxin Low with Clinistix®, Diastix®. *1100*

Dipyrone May cause false negative with enzyme tests. *1563*

Epinephrine Inhibits peroxidase reaction of Clinistix®. *2544*

Ferrous Sulfate Low with Clinistix®, Diastix®. *1100*

Flurazepam Low with Clinistix®, Diastix®. *1100*

Formaldehyde Produces orange color (interferes) with o-toluidine. Inhibits reaction in coupled glucose oxidase procedure. *0583*

Furosemide Low with Clinistix®, Diastix®. *1100*

Gentisic Acid Affects Clinistix® at 0.05 mg/ml. *1924*

Homogentisic Acid Reducing substance affects coupled glucose oxidase procedure. *0583* Inhibits peroxidase reaction of Clinistix®. *2544*

Hydroquinone Reported to inhibit glucose oxidase stix reactions. *0076* May inhibit peroxidase reaction on Clinistix®. *2544*

5-Hydroxyindoleacetic Acid Low with Clinistix®, Diastix®. *1100*

Levodopa False negative, inhibition of glucose oxidase method False negative if Clinistix® used (no effect on TesTape®). *1099*

Meralluride Interferes with glucose oxidase method. *2425*

Mercurial Diuretics May cause false negative results with glucose oxidase methods. *1488* May produce false negative with glucose oxidase based Stix. *2308*

Oxytetracycline Affects dipsticks if buffered with ascorbic acid. *1563*

Phenazopyridine False negative with glucose oxidase methods. *2566*

Phenobarbital Low with Clinistix®, Diastix®. *1100*

Prednisone Low with Clinistix®, Diastix®. *1100*

Propoxyphene Low with Clinistix®, Diastix®. *1100*

Secobarbital Low with Clinistix®, Diastix®. *1100*

Tetracycline Prevents oxidation of chromogen in glucose oxidase methods. *2425* False negative dipstick test if buffered with ascorbic. *1103*

Uric Acid Reported to inhibit glucose oxidase in Stix reactions. *0076*

Vitamin Preparations Low with Clinistix®, Diastix®. *1100*

Urine Decrease Physiological

Aspirin May reduce hyperglycemia, glycosuria in diabetes. *1343*

Urine Positive Analytical

Ampicillin At drug concentration of 4 and 10 mg/mL on Clinitest® when no glucose present, but some reduction with 0.5% glucose and higher concentrations. *2250*

Carbenicillin At 10 and 20 mg/mL gave false positive with negative urine using Clinitest® but with positive glucose specimens gave falsely low results. *2250*

Penicillin G At 16.2 mg/mL gave false positive with negative urine but at higher concentrations gave occasional falsely low value. *2250*

Urine Negative Analytical

Ascorbic Acid At concentration of 100 mg/L (normal concentration in urine up to 1290 mg/L) produced false negative result with Diabur-test. *3393*

Red Blood Cells No Effect Analytical

Fructose No effect in RBC hemolysates if hexokinase used. *0785*

Test Conditions Increase Analytical

Cellobiose If glucose-oxidase contaminated with glucosidase. *2803*

Cellopentose If glucose-oxidase contaminated with glucosidase. *2803*

Cellotriose If glucose-oxidase contaminated with glucosidase. *2803*

Glucosidase May be contaminant of glucose oxidase kit. *2803*

Glucose-6-Phosphatase

Serum No Effect Physiological

Pregnancy No significant difference from controls. *1017*

Glucose-6-Phosphate

Red Blood Cells Decrease Physiological

Amino Acids Associated with low serum phosphate. *3619*

Glucose-6-Phosphate Dehydrogenase

Red Blood Cells   No Effect   Analytical
Storage of Sample   If Alsever's solution added for 3.5 weeks at 30°. *0877*

Red Blood Cells   No Effect   Physiological
Smoking   No significant difference between smokers and non-smokers. *1548*

Red Blood Cells   Increase   Analytical
Heparin   Several additional isoenzymes from complexes. *3217*

Red Blood Cells   Decrease   Analytical
Hemoglobin   Minimal quenching of fluorescence may occur. *2264*

Red Blood Cells   Decrease   Physiological
Copper   Strongly inhibited (no activity at 100 μmol/L) *in vitro.* *0432*
Sulfates   Sensitive to sulfate ion at concentration of 5 mmol/L *in vitro.* *0432*

Glucose Clearance

Urine   Increase   Physiological
Phloridzin   Due to decreased reabsorption. *3343*

Glucose Tolerance

Serum   No Effect   Physiological
Aspirin   No significant effect usually observed. *1487*
Captopril   No significant deterioration with long-term treatment in diabetic hypertensives. *3307*
Chlorpromazine   Low dose effect in normal subjects. *3002*
Cyproheptadine   Reported to have no effect. *1487*
Gum Tragacanth   In 5 male volunteers consuming 9.9 g daily for 21 d. *0990*
Nifedipine   In 11 patients with 80 mg daily for 6 weeks. *3702*   In 15 hypertensive patients undergoing hemodialysis after 3 weeks treatment. *2986*
Oral Contraceptives   Estrogen effect of combined oral contraceptive. *0018*
Phenytoin   No consistent change seen in patients on long-term Treatment. *2767*   No significant difference between drug treated epileptics and controls. *2785*
Piretanide   In 12 male patients with mild hypertension (6 mg bid). *1503*
Tripamide   No effect in hypertensives with or without diabetes. *0711*

Serum   Increase   Physiological
Caffeine   Decrease of glucose at 30 and 60 minutes. *1093*
Clofibrate   Improvement relative to degrees of abnormal triglycerides. *0318*   Improved in most patients in 14 patients with primary hyperlipoproteinemia given 2.0 g daily over 5 y. *2925*
Fenfluramine   Significant improvement (mean decrease of 25 mg/dL). *0377*
Fluoxymesterone   Metabolic effect. *0019*
Glibonuride   Reported but not universal effect. *0301*
Glyburide   Marked improvement noted. *2216*
Guanethidine   Antidiabetic activity (change at end of treatment). *1426*
MAO Inhibitors   Seen in diabetics treated with insulin. *2220*
Metformin   Mode of action uncertain. *1184*
Methandrostenolone   Anabolic effect. *0019*
MK-270   Produces flattening of curve. *3099*
Nandrolone   Anabolic effect. *2681*
Niacin   Increases glucose disappearance after i.v. glucose tolerance test. *0819*

Norethisterone   In 75 women who had received up to 24 intragluteal injections of 200 mg every 56 d. *1666*
Pargyline   Flat curve may be produced. *2220*
Phenytoin   Lower in chronic patients than in controls. *0606*

Serum   Decrease   Physiological
Acebutolol   Increased glucose at 60 and 120 minutes without impaired release of insulin. *2115*
Aluminum Nicotinate   Rare hepatotoxic effects. *1678*
Bendrofluazide   In men and women in 3 y study of patients with mild hypertension. *2506*
Caffeine   Reduced pancreatic insulin release in normals. *0049*
Calcitonin   Rise in plasma glucose exaggerated after oral sugar. *2738*
Cannabis   Increased concentrations noted in some subjects at 1/2 to 1 h. *2830*
Chlorothiazide   Diabetogenic-like action of drug. *2745*
Chlorpromazine   Abnormal curves in 40% patients. *2427*   Acute high dose effect in normal subjects. *1044*
Chlorthalidone   Diabetogenic-like action of drug. *2745*   Significant reduction in 10 hypertensive patients given 25 mg daily for 16 weeks. *1229*
Clofibrate   After 1 week in both normals and diabetics but more marked in latter. *1120*
Clopamide   Possible effect of hypokalemia. *3881*
Corticotropin   Corticosteroid impairment of insulin secretion. *0203*
Cortisone   Gluconeogenesis and anti-insulin effects. *2220*
Cyproterone   When given with ethinyl estradiol: combination causes insulin resistance. *3239*
Danazol   In patients receiving drug, although baseline concentration normal. *3839*   In response to oral and i.v. glucose tolerance associated with insulin resistance. *3906*
Desoximetasone   In 2 of 5 patients with psoriasis and topical application of glucocorticoid. *1240*
Desoxycorticosterone   In 2 of 5 patients with psoriasis and topical application of glucocorticoid. *1240*
Dexamethasone   May impair carbohydrate tolerance. *1680* Observed in normal subjects and associated with enhanced insulin, C-peptide and glucagon responses to a test meal and blunted gastric inhibitory polypeptide response. *1411*
Diapamide   Modification of glucose tolerance. *2256*
Diethylstilbestrol   May provoke mild to moderate deterioration. *1849*
Estrogens   Effect variable depending on strength. *1488*   In postmenopausal women on hormone replacement therapy, statistically significant at 30, 60, 90 minutes, but not when fasting and at 120 minutes abnormal in 19% individuals. *3492*
Estrogens, Conjugated   May provoke mild to moderate deterioration. *1849*
Ethacrynic Acid   Diabetogenic-like action of drug. *2745*
Ethanol   Possible diminished tissue uptake. *0934*   In individuals fasted for 12 h in response to acute dose due to decreased peripheral use of glucose with increased glycogenolysis also. *2428*
Ethinyl Estradiol   Hormonal effect. *2525*   In group of women taking drug as anti-androgen therapy. *3239*
Ferrous Ascorbate   Mechanism not reported. *1488*
Fludrocortisone   Endocrine response. *1680*
Furosemide   Diabetogenic-like action of drug. *2745*
Glucagon   Decreased insulin sensitivity. *1271*
Glucocorticoids   In majority of nondiabetic patients although adaptive response usually occurs with time and glucose concentrations revert to pretreatment values. *0410*
Growth Hormone   Reduces disappearance rate after i.v. glucose tolerance test. *0819*
Hydrochlorothiazide   Diabetogenic-like action of drug. *2745*
Hydroflumethiazide   Diabetogenic like action of drug. *1680*
Imipramine   Preliminary observations only reported. *0158*
Iron Salts   As Fe ascorbate before glucose tolerance test. *1487*
Lithium   Associated with hyperglycemia (?mechanism). *1680*
Medroxyprogesterone   Abnormal in 15% may not return to normal in 1 y. *3414*

Glucose Tolerance *(continued)*

Serum   Decrease   Physiological *(continued)*

Mefenamic Acid   Effect noted in a diabetic. *1680*

Mestranol   May provoke mild to moderate deterioration. *1849*

Methandrostenolone   Alters curve in diabetic direction. *2427*

Methylprednisolone   Decreased carbohydrate tolerance. *1680*

Metoprolol   During early part of i.v. glucose tolerance test. *1410*

Muzolimine   Significant reduction in 10 hypertensive patients given 30 mg daily for 16 weeks. *1229*

Niacin   Reduced tolerance observed in diabetics. *2220*

Nifedipine   Both in normal subjects and those with already impaired glucose tolerance but in normals improved tolerance after 60 minutes. *1296*

Nitrofurantoin   Single case reported. *1463*

Norethindrone   After 6 mo treatment compared with control. *3413*

Norethynodrel   Gluconeogenetic effect of steroids. *1237*

Oral Contraceptives   Mainly estrogen effect (reversible in 3 out 4). *1462*   Glucose higher by 11 mg/dL at 1 h in drug users. *1248*

Oxytetracycline   Effect observed in animals. *0220*

Paramethasone   Decreased carbohydrate tolerance. *1680*

Perphenazine   Abnormal curves in 35% subjects. *2427*

Phenolphthalein   Result of hypokalemia. *0706*

Phenothiazines   May produce diabetic type of curve in normals. *2220*

Phenytoin   Decreases insulin excretion. *2220*

Pindolol   Slight effect after 60 minutes, significant effect at 120 minutes during oral glucose tolerance test. *2117*

Polythiazide   Impaired tolerance may occur. *1680*

Prazosin   Decreased glucose response to intravenous glucose tolerance test associated with decreased early insulin response mediated by increased circulating catecholamines. *2191*

Prednisolone   Metabolic effect. *0019*

Prednisone   Endocrine action. *1680*

Pregnancy   Decreased tolerance in last trimester. *0987*   Intravenous tolerances decreases in 2nd to 3rd trimester. *3509*

Propranolol   Overt diabetes developed or after glucose challenge in 40 hypertensives. *2469*   Raised concentration during later part of i.v. glucose tolerance test. *1410*

Quinethazone   Similar effect to thiazides. *1680*

Recumbency   Bed rest associated with increased peripheral insulin resistance. *2187*

Sandfly Fever   With fever decreased by half. *2929*

Spironolactone   Transient effect in 15 primary hypertension patients at 6 mo after 100 mg drug daily. *1072*

Thiazides   Diabetogenic-like action of drug. *2745*

Triamcinolone   Latent diabetes manifestation, endocrine effect. *1680*

Triamterene   Increase up to 250 mg/dL seen in normals even. *3744*

Verapamil   Significant impairment when glucose and drug co-administered. *1115*

α-Glucosidase

Urine   No Effect   Physiological

Urine Volume   Excretion not dependent on urine flow. *2320*

Urine   Increase   Physiological

Diurnal Variation   Maximum 6-12 am, minimum late pm, not marked. *2320*

Proteinuria   Possibly contributed from WBC. *2897*

β-Glucosidase

Serum   No Effect   Physiological

Aging   Little change throughout life. *1043*

Glucuronic Acid

Urine   No Effect   Analytical

Arabinose   Minimal effect on carbazole procedure. On m-hydroxydiphenyl procedure. *0389*

Glucose   On orcinol procedure. On m-hydroxydiphenyl procedure. *0389*

Urine   Increase   Analytical

Arabinose   Large effect on orcinol procedure. *0389*

Glucose   Significant effect on carbazole reaction. *0389*

Urine   Increase   Physiological

Oral Contraceptives   Extent of effect varies with exact composition of oral contraceptive. *1458*

Phenytoin   Due to hepatic enzyme induction in liver. *0848*

β-Glucuronidase

Serum   No Effect   Physiological

Smoking   No significant difference between heavy smokers and nonsmokers. *0961*

Serum   Increase   Physiological

Aging   Little change to age 45 y then sharp increase. *1043*

Anabolic Steroids   Metabolic effect. *0227*

Chlorpromazine   Result of toxic hepatitis. *3835*

Estrogens   Metabolic effect. *0227*

Ethanol   Marked effect in acute alcoholism. *1265*

Fluoxymesterone   Metabolic effect. *0227*

Iproniazid   Effect of toxic hepatitis. *3835*

Methandrostenolone   Metabolic effect. *0227*

Methyltestosterone   Metabolic effect. *0227*

Norethandrolone   Metabolic effect. *0227*

Oral Contraceptives   Altered metabolism (estrogen effect). *0227* Significant effect observed within 3 mo usually continuing for several y. *3170*

Oxymetholone   Metabolic effect. *0227*

Pregnancy   Slight drop initially then large increase towards term. *1119*

Stanozolol   Metabolic effect. *0227*

Urine   Increase   Physiological

Aspirin   Marked increase observed even when as little as 3.5 g aspirin ingested daily. *2866*

Bacteriuria   High activity possible with bacterial contamination. *2897*

Proteinuria   With all acute or inflammatory renal diseases. *2897*

White Blood Cells   Decrease   Analytical

Heparin   Significant effect above 1 unit/tube. *0839*

β-Glusosaminidase

Serum   Increase   Physiological

Aging   Steady increase from puberty onwards. *1043*

Glutamate Dehydrogenase

Serum   No Effect   Analytical

Hemolysis   Slight hemolysis has no effect. *2981*

Muscular Exercise   Physical activity has no effect. *2981*

Storage of Sample   7 d at room temperature and 4°, 14 at -20°. *3217*

Serum   Increase   Physiological

Ethanol   Reflects alcohol induced liver cell necrosis better than AST, ALT and GGT. GDH also increased by recent alcohol consumption reverts rapidly to normal. *3124*

Oral Contraceptives Significant effect observed within 3 mo usually continuing for several y. *3170*

Physical Training Significantly higher than in control subjects. *1491*

Streptokinase Disproportionate increase but not as marked as GGT. *3189*

Serum   Decrease   Analytical

Detergents 100% inhibition at 0.6-0.8 mg/ml. *3217*

Muscular Exercise Observed effect with 12 minutes cycle-ergometer. *1231*

Glutamic Acid

Plasma   Increase   Analytical

Storage of Sample Large amounts in platelets, leukocytes Even with prolonged storage at -20°. *3230*

Plasma   Increase   Physiological

Glutamic Acid Small increase noted after single 6 g dose. *3444*

Plasma   Decrease   Physiological

Oral Contraceptives Hormonal effect (second part of cycle). *0742*

Venipuncture Effect lasts for up to 1 h. *3230*

Glutamine

Serum   Increase   Physiological

Phenytoin Significant increase in children possibly due to inhibition of carbamoylphosphate synthase. *3642*

Plasma   No Effect   Physiological

Carbamazepine No striking effect when given to epileptic patients. *3697*

Phenobarbital No striking abnormality when given to epileptics. *3697*

Valproic Acid No striking abnormality when given to epileptics. *3697*

Plasma   Increase   Physiological

Phenobarbital Significant effect noted in children, possibly due to inhibition of carbamoylphosphate synthase. *3642*

Plasma   Decrease   Analytical

Storage of Sample Even with prolonged storage at -20°. *3230*

γ-Glutamyltransferase (GGT)

Serum   No Effect   Analytical

Acetaminophen No effect at therapeutic concentration on Reflotron method. *1984*

Acetylsalicylic Acid At 5 times upper limit of therapeutic range on methods on SMAC®, Abbott-VP and Hitachi® 705. *2138* No effect at therapeutic concentration on Reflotron method. *1984*

Aminosalicylic Acid At 5 times upper limit of therapeutic range on methods on SMAC®, Abbott-VP and Hitachi® 705. *2138*

Amphetamine At therapeutic concentration on Reflotron method. *1984*

Ampicillin No effect on Reflotron method at therapeutic concentration. *1984*

Ascorbic Acid No effect at therapeutic concentration on Reflotron method. *1984*

Bezafibrate No effect at therapeutic concentration on Reflotron method. *1984*

Bromazepam At 5 times therapeutic concentration on SMAC®, Abbott-VP, and Hitachi® 705. *2138*

Caffeine At therapeutic concentration on Reflotron method. *1984*

Carbenicillin At 5 times upper limit of therapeutic range on methods on SMAC®, Abbott-VP, and Hitachi® 705. *2138*

Carbochromen No effect at therapeutic concentration on Reflotron method. *1984*

Cefoxitin At 5 times upper limit of therapeutic range on methods on SMAC®, Abbott-VP, and Hitachi® 705. *2138*

Cephalothin At 5 times upper limit of therapeutic range on methods on SMAC®, Abbott-VP, and Hitachi® 705. *2138*

Chloramphenicol No effect at therapeutic concentration on Reflotron method. *1984*

Chlordiazepoxide At 5 times upper limit of therapeutic range on methods on SMAC®, Abbott-VP, and Hitachi® 705. *2138* No effect at therapeutic concentration on Reflotron method. *1984*

Cyclosporine At concentration of 41.5 mmol/L (50 μg/L) on methods on Technicon® SMAC® II and Hitachi® 705. *1123*

Cysteine At 5 times upper limit of therapeutic range on methods on SMAC®, Abbott-VP, and Hitachi® 705. *2138*

Cystine At 5 times upper limit of therapeutic range on methods on SMAC®, Abbott-VP, and Hitachi® 705. *2138*

Diazepam At 5 times upper limit of therapeutic range on methods on SMAC®, Abbott-VP, and Hitachi® 705. *2138*

Disulphine Blue Using gamma-glutamyl-3-carboxy-4-nitroanilide as substrate. *1464*

Ethaverine No effect at therapeutic concentration on Reflotron method. *1984*

Flurazepam At 5 times upper limit of therapeutic range on methods on SMAC®, Abbott-VP, and Hitachi® 705. *2138*

Furosemide No effect at therapeutic concentration on Reflotron method. *1984*

Glibenclamide No effect at therapeutic concentration on Reflotron method. *1984*

Ibuprofen At 5 times upper limit of therapeutic range on methods on SMAC®, Abbott-VP, and Hitachi® 705. *2138*

Indomethacin No effect at therapeutic concentration on Reflotron method. *1984*

Methaqualone No effect at therapeutic concentration on Reflotron method. *1984*

Methicillin At 5 times upper limit of therapeutic range on methods on SMAC®, Abbott-VP and Hitachi® 705. *2138*

Methotrexate At 5 times upper limit of therapeutic range on methods on SMAC®, Abbott-VP, and Hitachi® 705. *2138*

Methyldopa No effect at therapeutic concentration on Reflotron method. *1984*

Naproxen At 5 times upper limit of therapeutic range on methods on SMAC®, Abbott-VP, and Hitachi® 705. *2138*

Niacin No effect at therapeutic concentration on Reflotron method. *1984*

Nitrofurantoin On routine methods in use on SMAC®, Ektachem®, Abbott-VP, Cobas-Bio, aca, Hitachi® 705, KDA at 5 times normal upper therapeutic concentration. *2138* No effect at therapeutic concentration on Reflotron method. *1984*

Oxazepam No effect at therapeutic concentration on Reflotron method. *1984*

Oxytetracycline No effect at therapeutic concentration on Reflotron method. *1984*

Penicillin G At 5 times upper limit of therapeutic range on methods on SMAC®, Abbott-VP and Hitachi® 705. *2138*

Phenazopyridine No effect at therapeutic concentration on Reflotron method. *1984*

Phenobarbital No effect at therapeutic concentration on Reflotron method. *1984*

Phenprocoumon No effect at therapeutic concentration on Reflotron method. *1984*

Phenytoin No effect at therapeutic concentration on Reflotron method. *1984*

Probenecid No effect at therapeutic concentration on Reflotron method. *1984*

Procaine No effect at therapeutic concentration on Reflotron method. *1984*

Pyridamole No effect at therapeutic concentration on Reflotron method. *1984*

Pyritinol No effect at therapeutic concentration on Reflotron method. *1984*

Quinidine No effect at therapeutic concentration on Reflotron method. *1984*

γ-Glutamyltransferase (GGT) (continued)

Serum   No Effect   Analytical (continued)

Rifampin On method using hydrolysis of gamma-glutamyl p-nitroanilide with absorbance of product at 410 nm. *0218* At 5 times upper limit of therapeutic range on methods on SMAC®, Abbott-VP, and Hitachi® 705. *2138*

Sulfamethoxazole No effect at therapeutic concentration on Reflotron method. *1984*

Tetracycline At 5 times upper limit of therapeutic range on methods on SMAC®, Abbott-VP and Hitachi® 705. *2138*

Theophylline At 5 times upper limit of therapeutic range on methods on SMAC®, Abbott-VP,and Hitachi® 705. *2138* No effect at therapeutic concentration on Reflotron method. *1984*

Trimethoprim No effect at therapeutic concentration on Reflotron method. *1984*

Serum   No Effect   Physiological

Estrogens No metabolic effect. *0227*

Ethanol Acute consumption has no effect in healthy volunteers or in patients with alcoholic liver disease. *3124* Regardless of amount consumed, acute ingestion had no effect on activity. *0890*

Gold No significant changes of microsomal enzyme in patients treated with parenteral drug. *2933*

Physical Training Not systematically affected by training. *1491*

Valproic Acid In spite of induction of some hepatic enzymes. *0848*

Serum   Increase   Analytical

Glutamate Free glutamate effect 100% in normal range. *0412*

Serum   Increase   Physiological

Acetaminophen In severe cases of poisoning (also in some mild). *3470*

Allopurinol In 3 of 73 patients treated for 6 mo. *3565*

Aminoglutethimide Increased in 26 of 45 patients with breast cancer. *2546*

Amiodarone But effect minimal in 100 treated patients. *1509*

Amodiaquine Abnormality usually mild but associated with an immunoallergic hepatotoxicity. *2081*

Anticonvulsants In 8 epileptic patients receiving phenobarbital with carbamazepine or phenytoin. *2670*

Barbiturates Possibly due to enzyme induction. *3815*

Captopril Characteristic pattern of hepatocellular jaundice in one patient. But with secondary cholestatic elements. *3688*

Carbamazepine Observed 2 times upper limit of normal in 2 of 35 adult inpatient epileptics (dose and duration of treatment unknown). *1162* Associated with rare cases of granulomatous hepatitis. *3375*

Carbimazole Reversible hypersensitivity response to drug. *1790*

Chloramphenicol Isolated case of Hepatitis and Pancytopenia reported, which resolved with discontinuation of drug. *0602*

Chlorpropamide Cholestatic jaundice observed in one patient in association with red cell aplasia (prevalence of jaundice as high as 0.5%). *1284*

Cimetidine In isolated case treated for 7 mo with 400 mg/d progressed to bridging hepatic necrosis. *3085*

Cisplatin Slight and transient increases in 45 patients receiving drug. *3698*

Disulfiram Reversible toxic liver damage in nonalcoholic woman. *2017*

Enflurane In about 10% patients given 2 to 3 administrations of anesthetic agent. *1089*

Ethanol Seen with other liver functions tests normal in chronic alcoholism. *3043* Observed also in moderate or heavy drinkers. *3042* 68% increase in heavy drinkers vs occasional. *3273* Significant increase in heavy drinkers versus nondrinkers or occasional drinker. *0059* Ratio of GGT to alkaline phosphatase increased in patients with alcoholic liver disease. *2059* Sensitive indicator of continued alcohol abuse in individuals with known liver disease. *1040* Increased in alcoholic liver disease. *0625*

Etretinate In 2 of 20 patients treated for 1 y. *1866*

Haloperidol In isolated case producing cholestatic liver disease: typical frequency of liver disease is 0.2%. *0910*

Halothane Granulomatous hepatitis in single case following second exposure to anesthetic. *3266* Low frequency but greater number of abnormalities than with enflurane. Occurred most frequently in obese patients. *1089*

Heparin Increase occurred in 37% of 46 patients given up to 120,000 IU daily subcutaneously after 8 d. *2454* But significantly different from other individuals in control population. *2595*

Ibuprofen Rare case of hepatotoxicity: in one case associated with Stevens-Johnson syndrome. *3460*

Isotretinoin In 7 patients with severe rosacea treated with 1 mg/kg/d for 12 weeks. Effects possibly due to induction of hepatic microsomal enzymes. *2307* From 13.0 to 21.1 U/L in 18 patients with severe acne given 0.8 mg/kg daily for 3 mo: changes reverted to normal after treatment stopped. *2242* Significant effect at 6 weeks in 7 patients with severe rosacea treated with 1 mg/kg/d. *2307* Slight increase with high dose treatment. *1818*

Ketamine In 14 of 34 individuals who had drug as anesthetic for intermediate operations. *0973*

Lidoflazine Positive correlation with clinical state (normal liver function tests). *0322* ?due to changes in vasculature. *1059*

Methotrexate In psoriatic patients receiving drug in comparison with topically treated and controls. *0364*

Methyldopa Drug induced hepatitis with severe, chronic, aggressive inflammation. *0211*

Metrizoate Effect observed in subjects with liver tumors. *0064*

Ofloxacin Very infrequent side effect observed during many courses of treatment. *1838*

Oral Contraceptives Significant effect observed within 3 mo usually continuing for several y. *3170* Extent of effect varies with exact composition of oral contraceptive. *1458* Positive association between activity and oral contraceptive use (average 28% Increase). *0157* But effects less marked with low estrogen preparations. *1568*

Papaverine In 6 of 14 patients abnormal liver function, possibly due to allergic response and to metabolic aberration in host. *2746*

Phenobarbital ?induction or damage of hepatic microsomes. *3044* Observed 2 times upper limit of normal in 2 of 15 adult epileptic inpatients (dose and duration of treatment unknown). *1162*

Phenothiazines Liver damage observed after different phenothiazines administered in succession. *1916*

Phenprocoumon Mild effect but recurrently in 2 patients having repeated exposure to drug. *3347*

Phenytoin Possibly due to enzyme induction. *3815* Observed 2 times upper limit of normal in 58 of 125 adult epileptic inpatients and 5 times upper limit in 9 of 125 (dose and duration of therapy not known). *1162* Mean 3 fold increase in 90% patients after 6 mo treatment — not influenced by age or sex or additional anticonvulsant therapy, accentuated by regular consumption of alcohol. *1897*

Polymethylmethacrylate 11 of 90 total hip arthroplasty and 7 of 23 knee arthroplasty patients had abnormal GGT increases 5-10 d post-surgery: not observed in controls. *2993*

Propoxyphene Hepatotoxic response observed in 2 patients. *2110*

Quinidine Hypersensitivity reaction reported in one patient. *3350* Isolated case of reversible granulomatous hepatitis from quinidine hypersensitivity and granuloma induction within 3 d of quinidine readmination. *0448*

Streptokinase Occurring in approximately 25% increase by 4x. *3189* Transient dysfunction of liver, returning to normal when treatment discontinued. *3122*

Sulfasalazine Occasional case of drug-induced toxic hepatitis. *1422* Rare hepatotoxicity reported associated with noncaseating granuloma or focal inflammation with or without necrosis. *1127*

Sulindac Isolated case of cholestatic jaundice 12 d after therapy started. *2371*

Thiopental 1 case out of 24 3-5 d after anesthesia, more longer period after anesthesia. *0391*

Tolmetin Observed in single case of 49 y old man who had taken drug as needed for arthritis for 1 y. *3421*

Valproic Acid Transient increase in 44% of patients without change of dosage. *1781*

Warfarin Rare cases of intrahepatic cholestasis in patients with previous history of. *2940*

Zimeldine Reported in 2 patients in association with headaches, possibly due to fall in blood concentration of 5-hydroxytryptamine. *3387*

Serum   Decrease   Analytical
Cefotaxime 10% reduction in specimens from patients with liver disease measured on American Monitor Parallel. *0201*
Heparin Activity lower in heparinized specimen than in serum: effect can be overcome by using preincubation period of more than 5 minutes, or by addition of 50 mmol/L sodium chloride to reaction mixture. *3775*

Serum   Decrease   Physiological
Aldatense Increase by 80% in a group of hypertensives compared with controls. *1611*
Clofibrate Significant reduction in 27 patients, half of whom had hypertriglyceridemia, after 1 week. *1121*
Ethinyl Estradiol Increase by 50% after 2 weeks in 5 patients with primary biliary cirrhosis. *1419*

Urine   No Effect   Physiological
Enalapril In 53 patients given up to 160 mg daily or when combined with hydrochlorothiazide. *2277*

Urine   Increase   Physiological
Gold In 6.5% (not significant) in 31 patients treated with parenteral gold. *2933*

Urine   Decrease   Physiological
Salicylate When 0.5 g given orally to 20 healthy adult volunteers and urine studied over next 3 h. *1828*

Glutarate

Urine   Increase   Physiological
Lithium Reversible inhibition of renal transport. *0409*

Glutathione

Plasma   No Effect   Analytical
Storage of Sample In oxalate reduced form stable for 24 h at room temperature. *1712*

Plasma   Increase   Analytical
Hemolysis Reduced and oxidized in RBCs. *3230*
Storage of Sample At refrigerator temperature prone to hemolysis. *1712*

Red Blood Cells   No Effect   Physiological
Smoking No significant difference between smokers and non-smokers. *1548*

Glutathione Reductase

Serum   No Effect   Analytical
Storage of Sample 3 d at room temperature and 4°, 7 at -20°. *3217* No effect 1 week at 4°, 1 mo at -20°. *3873*

Red Blood Cells   No Effect   Physiological
Smoking No significant difference between smokers and non-smokers. *1548*

Red Blood Cells   Decrease   Physiological
Ethanol Activation by riboflavin in 35 patients with alcohol related illness. *0207*
Oral Contraceptives Manifestation of poor riboflavin status in women of low socioeconomic status. *3024*

Glutathione S-Transferase

Serum   Increase   Physiological
Halothane 16 of 20 patients who received halothane anesthesia for minor urological problems showed small transient rise 1 to 3 h after surgery. *1707*

Glutethimide

Blood   Increase   Physiological
Glutethimide 1 g orally may produce concentration of 7 mg/L. *2348*

Glycerol

Serum   No Effect   Physiological
Propranolol In 5 hyperthyroid patients given 10 mg every 8 h for 4 d. *0335*
Tetracosactrin In healthy volunteers given 1 mg intramuscularly for up to 60 h. *1813*

Serum   Increase   Physiological
Ethanol Sustained lipolysis with i.v. administration. *0202*
Glycerin Rose from normal 0.51 mmol/L to 20 mmol/L. *3615*
Muscular Exercise Approximately 10-30% increase with bicycle pedalling. *1385*
Starvation Occurs soon after food withdrawal. *0021*
Valproic Acid When 1 g given orally to fasting individuals. *3633*

Serum   Decrease   Physiological
Amino Acids More marked if intraduodenal than i.v. *2918*
Glibonuride Response similar to that of free fatty acids. *1526*
Glisoxepide Response similar to that of free fatty acids. *1526*
Glyburide Response similar to that of free fatty acids. *1526*
Propranolol In hyperthyroid patients given 40 mg every 6 h in 6 patients. *0335*
Tolbutamide Response similar to that of free fatty acids. *1526*

Blood   No Effect   Physiological
Timolol In 5 hyperthyroid patients given 10 mg every 8 h for 4 d. *0335*

Blood   Decrease   Physiological
Propylthiouracil Marked effect in 5 hyperthyroid individuals given 10 mg every 8 h for 4 d. *0335*

Glycine

Serum   Increase   Physiological
Valproic Acid Associated with inhibition of carbamyl phosphate synthetase I and interference. *0736*

Plasma   No Effect   Physiological
Carbamazepine No striking effect when given to epileptic patients. *3697*
Phenobarbital No striking abnormality when given to epileptics. *3697*
Starvation No effect over 6 d, but increased if protein-free diet. *0021*
Valproic Acid No striking abnormality when given to epileptics. *3697*

Plasma   Increase   Physiological
Glycine Similar marked increase in gouty and normals. *3929*
Histidine Interferes with clearance after oral load. *1644*
Valproic Acid Probably due to inhibition of enzymes involved in glycine clearance in liver analogous to ketotic hyperglycinemias. *1759*

Glycine *(continued)*

Plasma   Decrease   Physiological
Obesity   Characteristic finding correlated with insulin. *1106*
Oral Contraceptives   Hormonal effect (second part of cycle). *0742*

Urine   Increase   Physiological
Glycine   Similar marked increase in gouty and normals. *3929*
Lead   May be transient increase in poisoning. *0489*
Niacinamide   Increased up to 80% with 1 g at 4 h. *1622*

Urine   Decrease   Physiological
Ascorbic Acid   In 10 healthy females given 10 g/d. *3624*
Niacin   2 g causes 30% decrease at 2 h. *1622*

Glycocholic Acid

Serum   Increase   Physiological
Propylthiouracil   In one 24 y old woman who developed fulminant hepatic failure with lymphocyte sensitization. *2437*

Glycogen Synthetase

Muscle   Increase   Physiological
Physical Training   Level of activity directly related to fitness. *3553*

β_2-Glycoprotein

Serum   Decrease   Physiological
Asparaginase   Diminished hepatic synthesis. *2657*

Glycosylated Proteins

Serum   No Effect   Physiological
Desogestrel   When 150 µg desogestrel given with 30 µg ethinyl estradiol for 3 mo. Results normal within 2 mo after treatment stopped. *0312*
Levonorgestrel   When 150 µg levonorgestrel given with 30 µg ethinyl estradiol for 3 mo. Results normal within 2 mo after treatment stopped. *0312*

Gold

Blood   No Effect   Analytical
EDTA   Atomic absorption method of Harth et al. *1515*
Heparin   Atomic absorption method of Harth et al. *1515*
Oxalate   Atomic absorption method of Harth et al. *1515*

Blood   No Effect   Physiological
Penicillamine   No effect in patients with rheumatoid arthritis previously treated with gold. *1324*

Urine   Increase   Physiological
N-Acetylcysteine   Excretion doubled in response to i.v. drug in patients previously given gold. *1312*
Dimercaprol   If poisoning due to gold. *0071*
Penicillamine   Slight effect when given to patients with rheumatoid arthritis previously treated with gold. *1324*

Red Blood Cells   Increase   Physiological
Gold   Up to 45% of blood concentration in RBC. *3366*

Gonadotropins

Plasma   No Effect   Physiological
Clonidine   In children with growth hormone deficiency 0.1 mg/m² daily for 60 d. *2821*

Plasma   Increase   Physiological
Levodopa   Reported metabolic effect. *1390*
Menopause   Normal response. *0987*
Pregnancy   Increased early to yield positive pregnancy test. *0459*
Sleep   Maximum at night, especially in early am. *3217*

Urine   Increase   Physiological
Ovulation   May be temporary increase. *0987*
Pregnancy   HCg increased from 2 to 12 weeks, falls later. *0987*

Urine   Decrease   Physiological
Chlorpromazine   Blocks ovulation, maintains decidual reaction. *1343*
Oral Contraceptives   Hormonal effect. *2211*
Phenothiazines   Associated with other endocrinological changes. *2427*

Granular Casts

Urine   Increase   Physiological
Ibuprofen   In 2 patients with systemic lupus erythematosus. *1155*
Piroxicam   Few: occasional drug induced nephrotoxicity, with isolated azotemia, acute interstitial nephritis or nephrotic syndrome. *2824*

Granulocytes

Blood   No Effect   Physiological
Cimetidine   No decrease in bone marrow granulocyte reserves. *3448*

Blood   Increase   Physiological
Smoking   Approximately 30% — mechanism obscure. *0732*

Blood   Decrease   Physiological
Amoxapine   Observed in one 35 year old woman after receiving 18 g over 57 d. Granulocytes reappeared in peripheral blood on 15th day after treatment stopped. *3238*
Asparaginase   Slight reduction (not dose dependent). *2657*
Carbamazepine   May cause agranulocytosis. *0019*
Fluspirilene   Decreased values noted (not marked). *0221*
Phenylbutazone   Indicative of toxicity developing. *1680*
Ranitidine   Marked reduction in one elderly woman following 2 weeks of treatment with 300 mg/ daily: normalized when treatment stopped. *0463*
Spironolactone   Complete agranulocytosis observed in a single patient. *3487*
Streptomycin   Rare agranulocytosis reported. *1292*
Trifluoperazine   Decreased values noted (not marked). *0221*
Trimeprazine   Absence reported in one case. *0440*

Griseofulvin

Serum   Increase   Physiological
Food   Better absorbed with meal containing fat. *3024*

Serum   Decrease   Physiological
Phenobarbital   Impairs gastrointestinal absorption. *1487*

Growth Hormone

Plasma   No Effect   Analytical

Heparin   With double-antibody procedure and 20-80 I.U./mL. *0761*

Plasma   No Effect   Physiological

Amantadine   No effect observed. *2329*

Amitriptyline   No effect observed with acute or chronic treatment in depressed patients. *1366*

Atenolol   No change in basal concentration or after glucose in 14 hypertensives. *0836*

Butorphanol   No significant effect in 6 healthy male volunteers given 2 mg i.m. *3033*

Caffeine   No significant effect of single dose of 250 mg in men or women. *3418*   In men no significant effect of single dose of 500 mg. *3418*

Carbidopa   In 10 normal volunteers given 300 mg daily for 1 week. *0494*

Chlorpromazine   No significant difference in patients receiving drug on continuing basis versus controls. *3860*

Cimetidine   In 6 patients treated for 1 mo with 1 g/d. *3491*

Clofibrate   No effect of drug after 1 week in healthy individuals with hypertriglyceridemia. *1120*

Clonidine   No effect observed on normal response to moderate or heavy exercise. *1797*

Domperidone   No significant change in concentration after 0.17 mg/kg bolus i.v. in 10 children. *2330*

β-Endorphin   No effect with i.v. infusion in depressed patients or methadone treated addicts. *0610*

Enflurane   No effect unless given with propranolol. *2329*   No effect of anesthesia alone. *2699*

Epinephrine   No effect observed in normals. *0371*

Ethanol   No effect in men or women at 100 mg/dL. *3611*

Glyburide   No effect observed. *0125*

Guanabenz   In 30 patients treated with twice daily doses of from 4 to 32 mg. *1018*

Hydrochlorothiazide   In 15 hypertensives treated with 50 mg twice daily. *1018*

Labetalol   In response to i.v. infusion of 100 mg over 10 minutes. *0226*

Medroxyprogesterone   No effect observed after 1 y. *3414*

Megestrol   In 18 postmenopausal women with breast cancer. *0056*

Metoclopramide   With pharmacological doses in normal men, but increased in hypogonadal men. *0648*   In normal subjects after single i.v. injection of 10 mg. *3721*

Metyrapone   No significant effect with 750 mg orally. *2176*

Naloxone   No significant effect with i.v. infusion. *0868*

Phenylephrine   No effect with/without propranolol. *2329*

Propranolol   No effect unless with epinephrine then sharp increase. *2329*   No effect on concentration at 3 or 6 mo treatment. *0836*

Ranitidine   No effect with i.v. bolus of much as 300 mg. *0871*

Smoking   Mean basal values the same in smokers and non-smokers. *1360*

Spironolactone   In 15 patients with primary hypertension treated with 100 mg daily for 1 y. *1072*

Theophylline   No effect in response to i.v. infusion of aminophylline. *0608*

Thioridazine   No significant difference in patients receiving drug on continuing basis versus controls. *3860*

Thyrotropin-Releasing Hormone (TRH)   No effect in normal subjects after i.v. administration. *0100*

Plasma   Increase   Physiological

Acetoacetate   Effect slight after 40 minutes. *1811*

Alanine   Infusion caused increase by at least 40 pg/mL. *3880*

Amino Acids   Slight effect, mode of infusion makes no difference. *2918*

Apomorphine   Effect greater with 5 μg/kg in 8 depressed individuals than with 1.3 μg/kg of clonidine. *0728*   In control healthy subjects i.v. injection increased mean concentration from 1.8 to 28.3 ng/mL, but weak effect in schizophrenic patients only. *3178*   7 fold increase at 30-45 minutes in 6 children after 12 mg/kg subcutaneously. *2330*

Arginine   Effect also shown by protein meals. *2898*

Clomipramine   In depressive patients: increase in 5 of 8 patients with acute treatment: no effect observed after 28 d. *1366*

Clonidine   In children with growth hormone deficiency 0.1 mg/m² daily for 60 d. *2821*   Peak of 6.4 to 30 ng/mL when given i.v. to 6 healthy normals. *2060*

Cold Exposure   Observed effect in normals. *0371*

Corticotropin   Up to 1 mg i.v. has marked effect. *1898*   Increase in 50%, moderate effect only. *2176*

Cyclic AMP   Hormonal action. *1102*

Desipramine   Substantial effect in some patients in group of depressed patients. *0560*

Diurnal Variation   Maximum 2 h after commencement of sleep. *2743*

Dopamine   Stimulates release in normal men. *1101*

Estrogens   Modulators of high secretion in normals. *2414*

Ethanol   Transient increase with hypoglycemia and decreased insulin. *0202*

Fasting   Values low and also blunted response to TRH. *1130*

Fenfluramine   Action on brain stimulating release. *2160*

Glucagon   Hormonal action (potent stimulant). *1271*

Glucose   Delayed rise after initial slight fall. *1697*

Halothane   ?metabolic response to surgical stress. *2701*

Histamine   Observed effect in normals. *0371*

5-Hydroxytryptamine   Probable effect. *1101*

Insulin   Occurs only when glucose down to 10 mg/dL. *3859*   Increase of 22 ng/mL after 0.1 U/kg. *2176*   Marked effect in insulin induced hypoglycemia. *3570*   In 12 healthy male volunteers significant increase at 30 to 60 minutes following injection. *2322*

Levodopa   In normals single dose cause increase in 1-2 h. *1860*   After i.v. infusion mean rose to 15.5 ng/ml. *1721*

Methamphetamine   Significant rise. *0331*

Methyldopa Hydrazine   2 fold increase when given with Dopa. *2306*

Metoclopramide   Increased from 30 to 60 minutes after single i.v. injection in 6 normal women. *0648*   5 fold increase at 30-45 minutes after 0.17 mg/kg bolus i.v. in 10 children. *2330*   In 5 of 9 cirrhotic male patients after single injection of 10 mg. *3721*

Muscular Exercise   Effect more marked in untrained than trained. *1811*

Niacin   Produced by fall in free fatty acids. *1727*

Oral Contraceptives   During first year of use (may be increased 3 fold). *0071*

Oxprenolol   Observed with chronic treatment of 5 active hypertensive patients. *1090*

Pentagastrin   In 3 of 9 women to 4.87 ng/mL from 1.96 ng/mL, but increase not significant. *3648*

Propranolol   One out of six produce response. *0371*   Marked increase observed in 11 hypertensives also treated with diuretics. Also caused marked effect in 3 of 4 acromegalics. *1090*

Prostaglandin F2α   Slight effect at 2 μg/kg/min i.v. (only in men). *0735*

Pyrogens   After delay of 1 h following injection. *0254*

Sandfly Fever   Increased 8 fold at fasting, 18 fold at 30 minutes. *2929*

Sleep   Maximum 2 h after start of sleep (even with naps). *2743*

Starvation   Increased fifteen-fold to peak on second day. *0021*

Stress   Observed effect in normals. *0371*

Tetracosactrin   Associated with change in cortisol. *0275*

Vasopressin   Observed effect in normals. *0371*

Plasma   Decrease   Physiological

Chlorpromazine   Probably inhibits secretion of pituitary growth hormone. *3297*

Corticosteroids   Suppresses secretion of hormone. *2427*

Corticotropin   Reduces maximum of hormone during sleep. *1301*

Growth Hormone (continued)

Plasma   Decrease   Physiological (continued)

Hemodialysis   Normal response to increased glucose. 2547

Hydrocortisone   Marked effect of 100 mg i.v. by 4 h. 2936

Methyldopa   Slight effect on basal concentration after many mo of treatment for hypertension. Concentration increased to greater extent in response to Insulin in short-term treated individuals than in controls or long-term treated. 3453

Obesity   Little secreted at night compared with normals. 2225

Probucol   Mechanism obscure. 0806

Sultopride   After 1 week, but normal after 3-5 weeks. 2465

Guaiacols Spot Test

Urine   Positive   Analytical

Aspirin   False reaction with screening test of Rogers. 3031

Catechol   Action on procedure of Rogers. 3031

Chlorpromazine   False reaction with screening test of Rogers. 3031

Guaifenesin   False reaction with screening test of Rogers. 3031

HMPG   Action on procedure of Rogers. 3031

MAO Inhibitors   False reaction with screening test of Rogers. 3031

Metanephrine   Action on procedure of Rogers. 3031

Methyldopa   False reaction with screening test of Rogers. 3031

Morphine   False reaction with screening test of Rogers. 3031

Normetanephrine   Action on procedure of Rogers. 3031

Reserpine   False reaction with screening test of Rogers. 3031

Thiopental   False reaction with screening test of Rogers. 3031

Vanillylmandelic Acid (VMA)   Action on procedure of Rogers. 3031

Urine   Negative   Analytical

Dopamine   Action on procedure of Rogers. 3031

Epinephrine   Action on procedure of Rogers. 3031

Homovanillic Acid   Action on procedure of Rogers. 3031

Levarterenol   Action on procedure of Rogers. 3031

Levodopa   Action on procedure of Rogers. 3031

Methoxytyramine   Action on procedure of Rogers. 3031

Guanase

Serum   No Effect   Analytical

Bilirubin   Up to 150 mg/L on method of Nishikawa. 2602

Hemoglobin   Up to 5 g/L on method of Nishikawa. 2602

Storage of Sample   3 d at room temperature, 2 weeks at 4°, 10 mo at -15. 0877

Xanthine   Up to 1.0 mmol/L on method of Nishikawa. 2602

Serum   Increase   Physiological

Aminosalicylic Acid   May cause cytotoxic hepatocellular damage. 0248

Anabolic Steroids   Due to cholestatic syndrome. 2220

Androgens   Cholestatic phenomenon. 2313

Antimony Compounds   Hepatotoxic effect. 2220

Arsenicals   Hepatotoxic effect. 2220

Chlorpropamide   May cause cytotoxic liver damage. 0248

Iproniazid   May cause cholestatic and cytotoxic jaundice. 0248

Isoniazid   Cytotoxic hepatocellular damage. 0248

Mechlorethamine   May cause cytotoxic (hepatocellular) damage. 0248

Methotrexate   May cause cytotoxic hepatocellular damage. 0248

Phenylbutazone   May cause cholestatic and cytotoxic jaundice. 0248

Testosterone   May cause cholestatic and cytotoxic liver damage. 0248

Guanidinosuccinic Acid

Serum   No Effect   Analytical

Arginine   No effect on method of Kamoun, Pleau, Man. 1856

Creatinine   No effect on method of Kamoun, Pleau, Man. 1856

Haptoglobin

Serum   No Effect   Physiological

Hemodialysis   Unless intravascular hemolysis occurs. 3613

Pregnancy   No effect observed. 1498

Sex Difference   No effect observed. 0954

Smoking   No significant difference between smokers and non-smokers. 0809

Tri-iodothyronine   In 7 normal women aged 18-27 y given 25 μg/d three times daily for 14 d. Effect observed on endothelium-associated proteins but not on hepatically synthesized proteins. 1383

Serum   Increase   Analytical

Hemoglobin   If radial immunoassay used. 0455

Serum   Increase   Physiological

Aging   Mean 120 mg/dL at 20 y, 250 at 70 y. 0954

Anabolic Steroids   Metabolic effect. 0227

Anticonvulsants   In 8 epileptic patients receiving phenobarbital with carbamazepine or phenytoin. 2670

Ethylestrenol   Metabolic effect. 0227

Fluoxymesterone   Metabolic effect. 0227

Methandrostenolone   Metabolic effect. 0227

Methyltestosterone   Metabolic effect. 0227

Muscular Exercise   Approximately 17% increase immediately after. 2846

Norethandrolone   Metabolic effect. 0227

Norethindrone   With high dose, prolonged treatment. 0474

Oxandrolone   Metabolic effect. 0227

Oxymetholone   Metabolic effect. 0227

Physical Training   27% increase after 4 weeks training. 1491

Stanozolol   Metabolic effect. 0227

Testosterone   Metabolic effect (if aqueous solution i.m.). 0227

Serum   Decrease   Analytical

Hemoglobin   If peroxidase activation method used. 0455

Serum   Decrease   Physiological

Acetanilid   Hemolysis with G-6-PD deficiency. 2313

Acetohexamide   May cause hemolytic anemia. 1019

Acetylphenylhydrazine   May cause hemolytic anemia. 2313

Aminopyrine   May cause hemolytic anemia. 2313

Aminosalicylic Acid   Effect of hemolytic anemia. 0788

Antimony Compounds   May cause hemolytic anemia. 0279

Asparaginase   Possibly due to inhibition of protein synthesis reported in various studies. 2406

Benzene   May cause hemolytic anemia. 0279

Chlorate   May cause hemolytic anemia. 0279

Chlorpromazine   Hemolytic anemia. 2313

Coal Tar   May cause hemolytic anemia. 0279

Copper   May be hemolysis with poisoning. 2427

Dapsone   Associated with hemolytic anemia when occurs as main side effect. 3423

Dextran   Complex formation or increased consumption. 3346

Dimercaprol   May cause hemolysis. 0248

Diphenhydramine   Consequence of hemolytic anemia. 3602

Dipyrone   May cause hemolytic anemia. 2365

Estrogens   Altered liver metabolism. 3332

Ethinyl Estradiol   Metabolic effect. 0473

Furadaltone   Due to hemolysis. 0248

Furazolidone   Due to hemolysis. 0248

Indomethacin   May cause hemolytic anemia. 0071

Isoniazid   Hemolytic anemia. *2313*
Mephenytoin   May cause hemolytic anemia. *2365*
Methicillin   One doubtful case of hemolytic anemia. *0071*
Methyldopa   May cause hemolytic anemia. *2313*
Methylene Blue   May cause hemolysis. *2313*
Muscular Exercise   Mean effect of 18 mg/dL from pre-exercise. *3217*
Nalidixic Acid   May cause hemolytic anemia especially if G-6-PD deficiency. *3877*
Naphthalene   May cause hemolysis. *0248*
Nitrofurantoin   Hemolytic anemia. *2313*
Nitrofurazone   May cause hemolysis. *0248*
Norethindrone   Metabolic estrogen effect. *0474*
Oral Contraceptives   Metabolic changes in liver synthesis (estrogen). *2189*
Pamaquine   May cause hemolytic anemia. *0788*
Penicillin   May cause hemolytic anemia. *2313*
Pentaquine   May cause hemolysis. *0248*
Phenylhydrazine   May cause hemolysis (hemolytic anemia). *2427*
Pregnancy   Variable response (may be slight increase in absolute mass). *3472*
Primaquine   May cause hemolytic anemia. *2313*
Procainamide   In 3% of all patients receiving drug, but in 14% of patients with direct Coombs' test. *1958*
Quinidine   Causes hemolytic anemia. *2365*
Quinine   May cause hemolytic anemia. *2365*
Resorcinol   Occurs with hemolytic anemia. *2427*
Stibophen   Produces hemolytic anemia. *2365*
Streptomycin   May cause hemolytic anemia. *2313*
Sulfacetamide   May cause hemolysis if G-6-PD deficiency. *0248*
Sulfadiazine   May cause hemolytic anemia. *2313*
Sulfamethizole   May cause hemolytic anemia. *2868*
Sulfamethoxazole   May cause hemolytic anemia. *1974*
Sulfamethoxypyridine   May cause hemolytic anemia. *2429*
Sulfanilamide   May cause hemolytic anemia. *0248*
Sulfapyridine   May cause hemolytic anemia. *0248*
Sulfisoxazole   May cause hemolytic anemia. *1974*
Sulfonamides   May cause hemolytic anemia if G-6-PD deficiency. *0071*
Sulfones   May cause hemolytic anemia. *0248*
Sulfoxone   May cause hemolytic anemia. *0248*
Tamoxifen   Significant change due to mild estrogenic effect of drug in breast cancer patients. *3070*
Thiazolsulfone   Hemolytic anemia. *2313*
Tolmetin   Observed in single case of 49 y old man who had taken drug as needed for arthritis for 1 y. *3421*
Tripelennamine   Hemolytic anemia. *0788*

HDL$_2$-Cholesterol

Serum   No Effect   Physiological
Ranitidine   In 8 ulcer patients given 300 mg/d for 1 mo. *3858*

Serum   Increase   Physiological
Cimetidine   In 8 individuals with peptic ulcer given 1 g/d for 1 mo. *3858*   Increased proportion of HDL-cholesterol when 600 mg given bid to 6 males for 1 week. *2677*
Ethanol   Most of increased HDL-concentration attributable to this protein. *1011*
Niacin   Increase by 36% in 34 hypercholesterolemic individuals with 1.5 g/d for 1 mo then 3.0 g/d up to 6 mo. *1967*

Serum   Decrease   Physiological
Danazol   Increase by 73% after 2 weeks in 12 women with endometriosis given 600 mg daily for 24 weeks. *1066*
Medroxyprogesterone   Increase by 35% effects observed in 15 postmenopausal women with endometrial cancer after 2 weeks treatment. *2118*   Increase by 15% at 2 weeks and more after longer treatment: dose-dependent correlation with results. *1068*

Stanozolol   Increase by 85% in 10 normolipidemic postmenopausal osteoporotic women treated for 6 weeks. *3526*

HDL$_{2a}$-Cholesterol

Serum   No Effect   Physiological
Oral Contraceptives   In 10 women receiving ethinyl estradiol with norethindrone. In 10 women receiving ethinyl estradiol with norgestrel. *2009*

HDL$_{2b}$-Cholesterol

Serum   Increase   Physiological
Oral Contraceptives   In 10 women receiving ethinyl estradiol with norethindrone. *2009*

Serum   Decrease   Physiological
Oral Contraceptives   In 10 women receiving ethinyl estradiol with norethindrone. *2009*

HDL$_3$-Cholesterol

Serum   No Effect   Physiological
Cimetidine   In 8 individuals with peptic ulcer given 1 g/d for 1 mo. *3858*
Medroxyprogesterone   No effect although changes in other fractions. *1068*
Oral Contraceptives   In 10 women receiving ethinyl estradiol with norgestrel. *2009*
Ranitidine   In 8 ulcer patients given 300 mg/d for 1 mo. *3858*

Serum   Increase   Physiological
Niacin   Increase by 35% in 34 hypercholesterolemic individuals with 1.5 g/d for 1 mo then 3.0 g/d up to 6 mo. *1967*
Oral Contraceptives   In 10 women receiving ethinyl estradiol with norethindrone. *2009*

Serum   Decrease   Physiological
Chlorpropamide   With 8 weeks treatment of 8 C-Peptide negative insulin dependent diabetics. *3260*
Danazol   Increase by 29% after 2 weeks in 12 women with endometriosis given 600 mg daily for 24 weeks. *1066*
Medroxyprogesterone   Increase by 15% effects observed in 15 postmenopausal women with endometrial cancer after 2 weeks treatment. *2118*
Stanozolol   Increase by 35% in 10 normolipidemic postmenopausal osteoporotic women treated for 6 weeks. *3526*

Heinz Body Formation

Blood   Positive   Physiological
Acetanilid   Occurs initially prior to overt hemolysis. *1902*
Aminopyrine   Occurs initially before overt hemolysis. *1902*
Antimalarials   Occurs prior to overt hemolysis. *1902*
Antipyretics   Occurs prior to overt hemolysis. *1902*
Antipyrine   Occurs initially prior to hemolysis. *1902*
Arsenicals   Some may cause hemolytic anemia. *0333*
Aspirin   Occur initially but disappear with hemolysis. *1902*
Dapsone   Early stage of hemolytic anemia. *0333*
Furadaltone   Early stage of hemolytic anemia. *0333*
Furazolidone   Early stage of hemolytic anemia. *0333*
Lysol®   May cause intravascular hemolysis. *0629*
Methylene Blue   Early stage of hemolytic anemia. *0333*
Naphthalene   Early stage of hemolytic anemia. *0333*
Nitrofurans   Occurs initially with hemolysis. *1902*
Nitrofurantoin   May cause hemolytic anemia. *0333*
Nitrofurazone   May occur in early stages of hemolytic anemia. *0333*

Heinz Body Formation *(continued)*

Blood   Positive   Physiological *(continued)*

Pamaquine   May occur in early stages of hemolysis. *0333*

Pentaquine   May cause hemolysis with G-6-PD deficiency. *0333*

Phenazopyridine   Associated with hemolytic anemia, methemoglobinemia. *2427*

Phenylhydrazine   May occur in early stages of hemolysis. *0333*

Primaquine   May occur in early stages of hemolysis. *0333*

Procarbazine   May occur with hemolysis. *1680*

Sulfanilamide   May occur in early stages of hemolysis. *0333*

Sulfonamides   Occurs with marked hemolytic anemia. *1902*

Sulfones   Reported effect in first few days. *1343*

Tolonium   May cause hemolysis with G-6-PD deficiency. *0333*

Hematocrit

Blood   No Effect   Physiological

Altitude   No effect if plasma volume not contracted. *2957*

Amiloride   No effect seen in 13 men treated for 8 weeks. *0656*

Ascorbic Acid   No effect with varying degrees of supplementation over 14 week study period in young men. *1751*

Chenodeoxycholic Acid   No significant effect with up to 750 mg/d over 2 y in 916 patients. *3199*

Enalapril   In 53 patients given up to 160 mg daily or when combined with hydrochlorothiazide. *2277*

Fenfluramine   No adverse reaction reported. *0538*

Oral Contraceptives   No significant difference between oral contraceptive users and control adolescents. *1368*   No significant effect in oral contraceptive users. *3310*

Race   No difference between blacks and caucasians. *1869*

Smoking   No significant difference between smokers and nonsmokers. *1548*

Tourniquet   No effect for 1 minute. *3540*

Triamterene   Usually no effect in normals. *3744*

Blood   Increase   Physiological

β-Adrenergic Drugs   But less than expected with change in plasma volume. *1832*

Altitude   At 4500 feet 7-8% higher than at sea level. *3742*

Androgens   Associated with increased well being. *0071*

Atropine   But not to same extent as plasma volume. *1832*

Blood Pressure   Strongly correlated with increased systolic pressure. *2123*

Chlorthalidone   Increase by 1.4% in 39 mildly hypertensives treated for 6 weeks. *1404*

Cobalt   Polycythemia observed with poisoning. *2217*

Corticotropin   Especially in anemics. *0275*

Co-Trimoxazole   May cause megaloblastic anemia. *0745*

Cyclopropane   Possible effect due to decreased plasma volume. *1343*

Diurnal Variation   Maximum observed in morning minimum during night. *2222*

Dromostanolone   Response to androgens. *0071*

Earlobe Blood   Increased up to 15% above fingerstick blood. *0638*

Erect Posture   Standing for 15 minutes causes increase of 12.9%. *3476*

Ethanol   Increase by 1.5% increase in heavy versus occasional drinker. *3273*

Fluoxymesterone   Observed if anemia of renal failure. *1048*

Histamine   Due to increased capillary permeation and hemoconcentration. *1176*

Muscular Exercise   6% increase immediately after exercise normal in 30 minutes. *2846*

Nandrolone   Due to decreased plasma volume. *0330*

Oral Contraceptives   Probably progestogen effect. *3548*

Saccharated Iron Oxide   In 9 individuals given 40 mg/d i.v. daily for up to 42 d. *2662*

Smoking   Highest in smokers who inhale. *2123*   Up to 6% in men (nonsmokers) immediately after 6 cigarettes. *1004*

Spironolactone   When compared with results when patients treated with warfarin alone: augmented effect when spironolactone coadministered probably due to hemoconcentration due to diuresis. *2643*

Vitamin B$_{12}$ Successful treat may cause mild polycythemia. *2427*

Blood   Decrease   Analytical

Cold Agglutinins   Spurious macrocytosis due to agglutination. *2797*

Oxalate   Shrinks RBC, plasma volume increased up to 13%. *1563* With K salt may be 8-13% less than with heparin. *0067*

Blood   Decrease   Physiological

Acetanilid   Hemolytic anemia. *2426*

Acetohexamide   May cause anemia and hemolysis if G-6-PD deficiency. *1019*

Acetylphenylhydrazine   May cause hemolytic anemia. *2313*

Allopurinol   Rare case of anemia reported. *0071*

Aminobenzoic Acid   Induces malabsorption and megaloblastic anemia. *2716*

Aminopyrine   Hemolytic anemia in G-6-PD deficient persons. *0788*

Aminosalicylic Acid   Hemolytic anemia/megaloblastic anemia. *0788*

Amphetamine   Hemolytic anemia (depends on prior sensitivity). *0788*

Amphotericin B   Bone marrow depression with hemolytic anemia. *3645*

Anticonvulsants   May cause megaloblastic/aplastic anemia. *3025*

Antimalarials   May cause hemolytic anemia. *1902*

Antipyretics   May cause hemolytic anemia. *1902*

Antipyrine   Hemolytic anemia in G-6-PD deficient persons. *0788*

Arsenicals   Pancytopenia. *0788*

Asparaginase   May cause anemia. *1519*

Aspirin   Depresses bone marrow, gastrointestinal bleeding, hemolytic anemia. *0788*

Auranofin   Infrequent reversible side effect (occurring in up to 13% patients). *0617*

Azathioprine   In 6 cases of anemia out of 79 cases with hematological problems out of 328 reports of severe adverse effects. *2097*

Barbiturates   May cause megaloblastic anemia. *3773*

Benzazepine   Slight effect observed. *1738*

Benzocaine   May cause hemolysis. *0380*

Bopindolol   After 1 d but reverted to normal in 10 hypertensives treated for 21 d. *1063*

Brompheniramine   Occasionally observed with antihistamines. *1679*

Busulfan   pancytopenia/hemolytic anemia. *1680*

Butaperazine   Transitory anemia. *0071*

Captopril   Observed in 9 of 12 hypertensive patients on maintenance hemodialysis, maximum effect achieved after about 11 mo. *1608*

Carbamazepine   Pancytopenia. *0928*

Carbenoxolone   By approximately 4% in adrenalectomized patients. *3801*

Carbon Disulfide   Hypochromic anemia common. *0818*

Cephaloridine   May cause anemia. *0071*

Cephalothin   Hemolytic anemia. *1378*

Chloramphenicol   Normally slight response/may be pancytopenia. *0907*

Chloroquine   G-6-PD hemolytic anemia/pancytopenia. *3522*

Chlorothiazide   Pancytopenia. *0073*

Chlorpheniramine   Case of aplastic anemia after 10 y of low dose treatment. *1859*

Chlorpromazine   Hemolytic anemia. *0725*

Chlorpropamide   Mild anemia, rare aplastic anemia. *2801*   Isolated case of drug-induced hemolytic anemia. *3111*

Colchicine   Megaloblastic anemia (impaired absorption of vitamin B$_{12}$). *3773*

Copper   With hemolysis of copper toxicity. *0385*

Corticosteroids May cause gastrointestinal bleeding with low iron stores. *0333*

Cycloserine May cause megaloblastic anemia. *3773*

Dantrolene Rare effect of no consequence. *0659*

Dapsone Hemolysis (hemolytic anemia). *1612*

Dehydration Slight hemodilution with low protein fluid. *3254*

Dextran Macrodex® increases blood vol, no effect Rheomacrodex®. *3346*

Dimercaprol Hemolytic anemia in G-6-PD deficient persons. *0788*

Diphenhydramine Hemolytic anemia. *3602*

Dipyrone Hemolytic anemia. *2098*

Doxapram Noted postoperatively in a few patients. *0034*

EDTA May occur with prolonged therapy. *0071*

Estrogens May impair absorption of folate. *3773*

Ethanol May affect folic acid absorption and usage. *3773*

Ethosuximide Aplastic anemia. *0071*

Fava Beans May cause hemolysis with G-6-PD deficiency. *0248*

Floxuridine Bone marrow depression. *1680*

Furadaltone May cause hemolysis with G-6-PD deficiency. *0248*

Furazolidone May cause hemolysis with G-6-PD deficiency. *0248*

Furosemide May cause anemia. *1617*

Glucosulfone Hemolytic anemia. *0788*

Glycerin Hemolysis may occur after i.v. administration. *1448* Maximum decrease about 80 minutes after ingestion. *3615*

Gold Aplastic anemia. *0114*

Hydralazine May cause hemolytic anemia. *0044*

Hydroflumethiazide Pancytopenia. *3844*

Indomethacin Secondary to gastrointestinal bleed/agranulocytosis/pancytopenia. *0381*

Isocarboxazid May occasionally produce anemia. *2427*

Isoniazid Hemolytic anemia/rare megaloblastic anemia. *2313*

Lead Hemolytic anemia. *0788*

Levodopa Mild not related to hemolysis. *0071*

Lysol® May cause intravascular hemolysis. *0629*

MAO Inhibitors Occasional anemia may occur. *2427*

Mefenamic Acid May cause autoimmune hemolytic anemia. *1078*

Melarsonyl May cause hemolytic anemia in G-6-PD deficiency. *0071*

Melphalan Anemia may occur rarely. *2427*

Mephenytoin Hemolytic/aplastic/megaloblastic anemia. *3008*

Meprobamate May cause aplastic anemia. *0333*

Mercaptopurine May occur with bone marrow depression. *2427*

Metformin May be associated with megaloblastic anemia. *3605*

Methimazole Occasional case of aplastic anemia reported. *3870*

Methotrexate May = megaloblastic anemia (folic acid antagonist). *3773*

Methyldopa Autoimmune hemolytic anemia. *0597*

Methylene Blue May cause hemolysis with G-6-PD deficiency. *0248*

Methylphenobarbital May cause megaloblastic anemia. *2313*

Miconazole By more than 4% in 44% treated patients. *3462*

Minocycline May cause hemolytic anemia (rare). *1680*

Nalidixic Acid Hemolytic anemia with G-6-PD deficiency. *0279*

Naphthalene May cause hemolysis with G-6-PD deficiency. *0248*

Neomycin Megaloblastic anemia (impaired B_{12} absorption). *3773*

Niridazole Occurs with marked hemolysis. *1343*

Nitrites May cause hemolytic anemia. *2313*

Nitrobenzene Occurs with hemolysis. *2429*

Nitrofurans May cause hemolytic anemia if G-6-PD deficiency. *0071*

Nitrofurantoin Megaloblastic anemia/hypersensitivity (G-6-PD). *2358*

Nitrofurazone May cause hemolysis with G-6-PD deficiency. *0248*

Novobiocin Hemolytic anemia — mild. *0713*

Oral Contraceptives May cause megaloblastic anemia. *3025* In 46 oral contraceptive users compared with controls studied over at least 2 y. *1182*

Oxyphenbutazone Dilutional effect of water retention. *1343*

Pamaquine Hemolytic anemia in G-6-PD deficient persons. *0788* May cause hemolytic anemia. *2313*

Penicillamine Hypochromic anemia in child, menstruating woman. *0074*

Penicillin Hemolytic anemia due to binding to erythrocytes. *0996*

Pentamidine Megaloblastic anemia — inhibits dihydrofolate reductase. *3773*

Pentaquine May cause hemolysis with G-6-PD deficiency. *0248*

Pentoxifylline In 2 patients receiving other drugs concomitantly but for considerable time previously without ill effects. *2244*

Phenacemide Aplastic anemia. *0071*

Phenazopyridine Hemolytic anemia (sensitivity dependent). *0788*

Phenobarbital Megaloblastic anemia secondary to disturbance in folic acid metabolism. *2968*

Phenothiazines Hemolytic anemia (sensitivity dependent). *0788*

Phenylbutazone Bone marrow depression/secondary to Na and H_2O retention. *0566*

Phenylhydrazine Hemolytic anemia. *0788*

Phenytoin megaloblastic/hemolytic/aplastic anemia,pancytopenia. *3331*

Phytonadione Hemolysis may occur with G-6-PD deficiency. *2313*

Piperacetazine Mild transient decrease with hypotension. *1227*

Piperazine May cause hemolytic anemia. *2365*

Pipobroman Intended effect when polycythemia present. *1680*

Piroxicam Possibly related to drug administration or concomitant therapy in patients with osteoarthrosis. *3564* Reported, unassociated with obvious gastrointestinal bleeding. *2692*

Pregnancy In relation to decreased hemoglobin. *0987*

Prilocaine Hemolysis with doses greater than 400 mg. *0071*

Primaquine Hemolytic anemia in G-6-PD deficient persons. *0788*

Primidone Megaloblastic anemia secondary to disturbabce in folic acid metabolism. *2968*

Probenecid Hemolytic anemia (sensitivity dependent). *0788*

Procainamide Observed with SLE-like syndrome, hemolytic anemia. *0933*

Propranolol Associated with altered morphology of red cells. *2770*

Propylthiouracil Occasional immunologically associated anemia. *3870*

Pyrazinamide Rare anemia reported. *2427*

Pyrazolones Aplastic/hemolytic anemia. *2313*

Pyrimethamine Megaloblastic anemia. *3898*

Quinacrine May cause hemolysis with G-6-PD deficiency. *0248*

Quinidine Hemolytic anemia (sensitivity dependent) G-6-PD deficiency. *3673*

Quinine Hemolytic anemia (sensitivity dependent) G-6-PD deficiency. *0272*

Quinocide May cause hemolysis with G-6-PD deficiency. *1902*

Radiographic Agents Transient fall following rapid i.v. injection. *0611*

Recumbency Change of 7% with lying, 4% with sitting. *3539* 6% drop at 20 minutes in normal. *3540*

Resorcinol Injection or excess absorption may cause hemolysis. *0071*

Spectinomycin Mechanism not discussed. *0122*

Stibophen May produce hemolytic anemia. *2365*

Streptomycin Hemolytic anemia (sensitivity dependent). *0788*

Sulfacetamide May cause hemolysis if G-6-PD deficiency. *0248*

Sulfadiazine May cause hemolytic anemia. *2313*

Sulfamethizole Hemolytic anemia. *2868*

Sulfamethoxazole May cause hemolytic anemia. *1974*

Sulfamethoxypyridine Hemolytic/aplastic anemia. *2313*

Sulfanilamide May cause hemolytic anemia. *0248*

Sulfapyridine May cause hemolytic anemia. *0248*

Sulfisoxazole Agranulocytosis/aplastic anemia. *2868*

Sulfonamides May cause Heinz-body hemolytic anemia. *2808*

Sulfones May cause hemolysis. *0071*

Hematocrit *(continued)*

Blood   Decrease   Physiological *(continued)*

Sulfoxone   May cause hemolytic anemia. *0248*

Suramin   Hemolytic anemia reported. *0071*

Tetracosactrin   Maximum effect observed after 24 h. *0275*

Tetracycline   May cause hemolytic anemia. *2313*

Thiazolsulfone   May cause hemolytic anemia. *2313*

Thioguanine   In 29 patients with polycythemia rubra vera; effect usually apparent in 2 weeks. *2450*

Thioridazine   Significant decrease reported. *1629*

Thiosemicarbazones   Large dose effect. *1343*

Thiotepa   May be rapid decrease. *2427*

Tolbutamide   Agranulocytosis/aplastic anemia. *0073*

Tolmetin   Observed in single case of 49 y old man who had taken drug as needed for arthritis for 1 y. *3421*

Tolonium   May cause hemolysis with G-6-PD deficiency. *0333*

Triamterene   May cause megaloblastic anemia (folic acid antagonist). *3773*

Trimethadione   Aplastic anemia. *3008*

Trimethoprim   Folic acid antagonist ?megaloblastic anemia. *3773*

Trinitrotoluene   May cause hemolysis. *2429*

Tripelennamine   Hemolytic anemia. *0788*

Tyrothricin   May cause hemolysis even if applied topically. *0071*

Urethan   May occur with bone marrow depression. *2427*

Vitamin A   Anemia observed. *2427*

Vitamin K K_3 and K_4 only, especially with G-6-PD deficiency. *2102*

Hemoglobin

Plasma   No Effect   Analytical

Bilirubin   No effect on method of de Mendonca. *0672*

Citrates   No effect of citrated blood on method of de Mendonca. *0672*

Plasma   No Effect   Physiological

Epinephrine   No effect observed with increased blood flow. *1755*

Lipemia   No effect observed with induced lipemia. *1755*

Menstruation   No effect observed. *1755*

Muscular Exercise   No effect with normal activity. *1755*

Niacin   No effect observed with increased blood flow. *1755*

Tourniquet   No effect observed with local circulatory stasis. *1755*

Plasma   Increase   Analytical

Serum   During clotting may be 100 fold increase. *1755*

Transport of Specimen   7.9 mg/dL in pneumatic tube system. *3217*

Plasma   Increase   Physiological

Acetanilid   Occurs with marked hemolysis. *1902*

Aminopyrine   May occur with marked hemolysis. *1902*

Amyl Nitrite   If intravascular hemolysis occurs. *2426*

Antimalarials   Occurs with marked hemolysis. *1902*

Antipyretics   May occur with marked hemolysis. *1902*

Aspirin   Occurs with hemolytic anemia. *1902*

Copper   May occur with hemolysis of toxicity. *0385*

Dimercaprol   Occurs with intravascular hemolysis. *0788*

Fentanyl   In a single patient with low haptoglobin following high dose administration dose related effect. *1719*

Glycerin   Hemolysis may occur after i.v. administration. *1448*

Muscular Exercise   Light activity causes increase x 3-5, heavy increase x 10-30. *2981*

Nitrofurans   Occurs with intravascular hemolysis. *2426*

β-Propiolactone   When plasma from BPL treated blood compared with plasma in 25 specimens. *0217*

Sulfonamides   May occur with intravascular hemolysis. *2426*

Sulfones   May occur with hemolysis. *2426*

Urea   Rapid infusion of concentration solution may cause hemolysis. *1022*

Blood   No Effect   Analytical

Storage of Sample   Stable at room temperature for at least 1 week. *1563*

Blood   No Effect   Physiological

Ascorbic Acid   No effect with varying degrees of supplementation over 14 week study period in young men. *1751*

Bucindol   No effects noted in 8 patients following several weeks on treatment. *1283*

Chenodeoxycholic Acid   No significant effect with up to 750 mg/d over 2 y in 916 patients. *3199*

Cromoglycate   No clinical significant change observed. *0484*

Methotrexate   No effect in long-term treatment of 20 patients with psoriasis noted. *1556*   In 20 patients with psoriasis on long-term therapy. *1556*

Oral Contraceptives   No significant difference between oral contraceptive users and control adolescents. *1368*   No significant effect in oral contraceptive users. *3310*   In 46 oral contraceptive users compared with controls studied over at least 2 y. *1182*

Penicillamine   In 21 patients with rheumatoid arthritis treated for 48 weeks given 250 to 750 mg daily. *2345*

Race   No difference between blacks and caucasians. *1869*

Smoking   No significant difference between smokers and non-smokers. *1548*

Timegadine   In 23 patients with rheumatoid arthritis given 250 to 750 mg/d for 48 weeks. *2345*

Blood   Increase   Analytical

Amino Acids   If i.v. reported to affect accuracy of measurement. *0120*

Bilirubin   In high concentrations affects oxyhemoglobin methods. *2981*

Fat Emulsions   If given i.v. may affect accuracy of measurement. *0120*

Lipemia   May cause elevation by up to 3 g/dL. *0579*

Blood   Increase   Physiological

Aging   13.5 g/dL at 6 to 16 y, at 16 y 15.7, at 20 y 16.3 (men). 13.5 g/dL at 6 y to 14.3 at puberty then steady (women). *0887*

Altitude   At 4500 feet 7-8% higher than at sea level. *3742*

Dromostanolone   Response to androgens. *0071*

Earlobe Blood   Increased up to 15% above fingerstick blood. *0638*

Erect Posture   Due to redistribution of body water. *3879*

Ethanol   1.6% increase in heavy versus occasional drinkers. *3273*

Height   Positively correlated in children. *0887*

Iron Dextran   Typically rises after i.v. infusion. *0390*

Isotretinoin   Therapeutic response is normalization but effect not usually observed until after 3 weeks. *1319*

Meprobamate   May cause aplastic anemia. *0333*

Muscular Exercise   Mild exercise causes transient decrease in blood volume. *1563*

Oral Contraceptives   Increase after 12 mo with low base combined pill regime. *2996*

Saccharated Iron Oxide   In 9 individuals given 40 mg/d i.v. daily for up to 42 d. *2662*

Smoking   Slight effect in all age/class groups. *1195*

Stanozolol   Effective in some patients with aplastic anemia. *1679*

Tourniquet   Venous stasis may cause significant increase. *3879*

Weight   Positively correlated in children. *0887*

Blood   Decrease   Analytical

Storage of Sample   2-3 % loss if frozen with liquid nitrogen. *1180*

Blood   Decrease   Physiological

Acetaminophen   Anemia/pancytopenia. *0071*

Acetanilid   Hemolytic anemia. *2426*

Acetazolamide   Anemia due to pancytopenia/agranulocytosis. *2313*

Acetohexamide   May cause anemia and hemolysis if G-6-PD deficiency. *1019*

Acetylphenylhydrazine   May cause hemolytic anemia. *2313*

Acyclovir   One case megaloblastic anemia documented. *0936*

Allopurinol   Rare case of anemia reported. *0071*

Aminobenzoic Acid   Induces malabsorption and megaloblastic anemia. *2716*

Aminopyrine   Hemolytic anemia in G-6-PD deficient persons. *0788*

Aminosalicylic Acid   Hemolytic anemia/megaloblastic anemia. *0788*   Rare hemolytic anemia. *1292*

Amphetamine   Hemolytic anemia (depends on prior sensitivity). *0788*

Amphotericin B   Bone marrow depression with hemolytic anemia. *3645*   Majority of patients receiving drug have fall of 18-35% in concentration (normocytic normochromic anemia). *2262*

Ampicillin   Reversible hypersensitivity reaction. *1680*

Aniline   May cause marked hemolytic anemia. *0279*

Anticonvulsants   May cause megaloblastic/aplastic anemia. *3025*

Antimalarials   May cause hemolytic anemia. *1902*

Antimony Compounds   May cause hemolytic anemia. *0279*

Antipyretics   May cause hemolytic anemia. *1902*

Antipyrine   Hemolytic anemia in G-6-PD deficient persons. *0788*

Arsenicals   Megaloblastic anemia/pancytopenia. *0788*

Asparaginase   May cause anemia. *1519*

Aspirin   Depresses bone marrow, gastrointestinal bleeding, hemolytic anemia. *0788*

Auranofin   Infrequent reversible side effect (occurring in up to 13% patients). *0617*

Azathioprine   Principal manifestation of bone marrow depression. *0036*   In 6 cases of anemia out of 79 cases with hematological problems out of 328 reports of severe adverse effects. *2097*

Azidothymidine   Profound effect observed in one male homosexual with AIDS unlikely related to other factors. *1157*

Barbiturates   May cause megaloblastic anemia. *3773*

Benzazepine   Slight effect observed. *1738*

Benzene   Due to aplastic anemia. *1343*

Benzocaine   May cause hemolysis. *0380*

Bromate   May cause hemolysis. *1343*

Brompheniramine   Occasionally observed with antihistamines. *1679*

Busulfan   pancytopenia/hemolytic anemia. *1680*

Butaperazine   Transitory anemia reported. *0071*

Captopril   Observed in 9 of 12 hypertensive patients on maintenance hemodialysis, maximum effect achieved after about 11 mo. *1608*

Carbamazepine   Pancytopenia. *0928*   In about 20 epileptic patients treated for 2 y. *0873*

Carbenicillin   Hemolytic anemia reported. *1680*

Carbon Disulfide   Hypochromic anemia common. *0818*

Cefamandole   Rare hemolytic anemia. *3134*

Cefoxitin   Rare case of hemolytic anemia reported. *3134*

Cephaloridine   May cause anemia. *0071*

Cephalothin   Coombs' positive hemolytic anemia. *1378*

Chloramphenicol   Normally slight response/may be pancytopenia. *0907*   Isolated cases of hepatitis and pancytopenia reported which resolved with discontinuation of drug. *0602*

Chlorate   Causes hemolysis. *1343*

Chloroquine   Hemolytic anemia in G-6-PD deficient persons. *3522*

Chlorothiazide   Pancytopenia. *0073*

Chlorpheniramine   Reported to cause hemolytic anemia. *1678* Case of aplastic anemia after 10 y of low dose treatment. *1859* One case of aplastic anemia after 1 mo therapy, reversible with discontinuation of drug discontinuation of drug. *0886*

Chlorpromazine   Hemolytic anemia. *0725*

Chlorpropamide   Mild anemia, rare aplastic anemia. *2801*   Isolated case of drug-induced hemolytic anemia. *3111*

Chlortetracycline   Hemolytic anemia may occur. *1680*

Cholestyramine   Impairs absorption of iron. *1965*

Cimetidine   Reported in one case in which drug given intravenously. *1770*

Cisplatin   By more than 2 g/dL in 2 of 20 patients. *1384* Observed in 2 individuals receiving drug. *1272*   In 23% of 74 evaluable cancer patients given drug for 5 d repeated at 4-6 weeks intervals. *3119*

Coal Tar   May cause hemolytic anemia. *0279*

Colchicine   Megaloblastic anemia (impaired absorption of vitamin B_{12}). *3773*

Copper   May be marked decrease. *2427*

Corticosteroids   May cause gastrointestinal bleeding with low iron stores. *0333*

Co-Trimoxazole   May cause hemolysis if G-6-PD deficiency. *2694*

Cresol   May cause hemolytic anemia. *0279*

Cyclophosphamide   May cause megaloblastic anemia. *2583*

Cycloserine   May cause megaloblastic anemia. *3773*

Cyclosporine   Reduction of 2- 4 g/dL over 6 mo treatment in 6 of 23 patients with inflammatory ocular disease. *2713*   In 4 patients with primary biliary cirrhosis in whom creatinine and drug concentration rose as bleeding occurred. *3014*   Increase by 2 to 4 g/dL associated with normochromic normocytic anemia in 6 of 23 patients with ocular inflammatory disease over 6 mo. *2713*

Cytarabine   May cause anemia. *2313*

Dactinomycin   May cause anemia. *0071*

Dantrolene   Rare effect of no consequence. *0659*

Dapsone   May cause hemolytic anemia. *1612*   Associated with hemolytic anemia when occurs as main side effect. *3423*   Anemia developed in 61% of 51 patients with leprosy. *2410*

Dehydrocholic Acid   Mild hemolytic action if injected. *0071*

Dihydrotachysterol   Adverse effect with severe hypercalcemia. *1680*

Dimercaprol   Hemolytic anemia in G-6-PD deficient persons. *0788*

Dimethindene   Rare hemolytic anemia with antihistamines. *1678*

Dinitrophenol   May cause anemia/aplastic anemia. *1343*

Diphenhydramine   Hemolytic anemia. *3602*

Dipyrone   Hemolytic anemia. *2098*

Diurnal Variation   Falls at night minimum about midnight. *3428*

Doxapram   Noted postoperatively in a few patients. *0034*

Doxycycline   Observed with other tetracyclines. *1680*

EDTA   May occur with prolonged therapy. *0071*

Estrogens   Impairs absorption of folate. *3773*

Ethanol   May affect folic acid absorption and usage. *3773*

Ethosuximide   Aplastic anemia. *0071*

Fava Beans   May cause hemolysis with G-6-PD deficiency. *0248*

Fibrin Hydrolysate   ?due to septicemia or hypophosphatemia. *3746*

Floxuridine   Bone marrow depression. *1680*

Flucytosine   May depress bone marrow function. *1679*

Fluorides   May cause aplastic anemia. *0279*

Furadaltone   May cause hemolysis with G-6-PD deficiency. *0248*

Furazolidone   May cause hemolysis with G-6-PD deficiency. *0248*

Furosemide   May cause anemia. *1617*

Glucosulfone   Hemolytic anemia. *0788*

Glycerin   Hemolysis may occur after i.v. administration. *1448*

Gold   Aplastic anemia. *0114*

Hydralazine   May cause hemolytic anemia. *0044*   Observed in one patient receiving low dose of drug for hypertension. Mechanism involved in producing erythrocyte-reacting antibody not known. *2680*   Observed in a case of obstructive jaundice with pancytopenia. *3469*   Observed in 9 patients developing rapidly progressive glomerulonephritis during treatment. *0366*

Hydroflumethiazide   Pancytopenia. *3844*

Hydroxyurea   Decreases of from 0.5 to 6.0 g/dL common. *2581*

Ice-Eating   Low values observed (probably effect not cause). *0501*

Ifosfamide   Occurs in 32% patients after 150 mg/kg. *3675*

Indomethacin   Secondary to gastrointestinal bleed/agranulocytosis/pancytopenia. *0381*

Isocarboxazid   May occasionally produce anemia. *2427*

Hemoglobin (continued)

Blood Decrease Physiological (continued)

Isoniazid Hemolytic anemia/rare megaloblastic anemia. *2313* Very rare cases of pure red cell aplasia, other cases of hemolytic anemia pyridoxine-responsive sideroblastic anemia also reported. *0667* Rare hemolytic anemia in patients with glucose-6-phosphate deficiency. *1292*

Lead Hemolytic anemia. *0788*

Levodopa Mild not related to hemolysis. *0071*

Lipomul® May be severe hemolytic anemia. *2427*

Lysol® May cause intravascular hemolysis. *0629*

MAO Inhibitors Occasional anemia may develop. *2427*

Mefenamic Acid May cause autoimmune hemolytic anemia. *1078*

Melarsonyl May cause hemolytic anemia in G-6-PD deficiency. *0071*

Melphalan Anemia may occur rarely. *2427*

Mephenytoin Hemolytic/aplastic/megaloblastic anemia. *3008*

Mercaptopurine May occur with bone marrow depression. *2427*

Metformin Associated with impaired B_{12} absorption. *3605*

Methazolamide Some cases of aplastic anemia, and leukopenia reported when used in treatment for glaucoma. *3799*

Methimazole Occasional case of aplastic anemia reported. *3870*

Methotrexate May = megaloblastic anemia (folic acid antagonist). *3773*

Methyldopa Autoimmune hemolytic anemia. *0597* On rare occasion may produce hemolysis. Great majority of patients do not show this. *1909*

Methylene Blue May cause hemolysis with G-6-PD deficiency. *0248*

Methylphenobarbital May cause megaloblastic anemia. *2313*

Minocycline Reported cases of hemolytic anemia. *1680*

Mithramycin Reversible effect. *0071*

Mushroom Poisoning May cause severe hemolytic anemia. *0279*

Nalidixic Acid Hemolytic anemia with G-6-PD deficiency. *0279*

Naphthalene May cause hemolysis with G-6-PD deficiency. *0248*

Naproxen Isolated case of hemolytic anemia: one other case reported previously. *1681*

Neoarsphenamine May cause hemolysis. *1902*

Neomycin Megaloblastic anemia (impaired B_{12} absorption). *3773*

Niridazole Occurs with marked hemolysis. *1343*

Nisoldipine From 14.7 to 14.0 g/dL in 14 mild to moderately hypertensive noninsulin dependent diabetic patients. *2653*

Nitrites May cause hemolytic anemia. *2313*

Nitrobenzene Occurs with hemolysis. *2429*

Nitrofurans May cause hemolytic anemia if G-6-PD deficiency. *0071*

Nitrofurantoin Megaloblastic anemia/hypersensitivity (G-6-PD). *2358*

Nitrofurazone May cause hemolysis with G-6-PD deficiency. *0248*

Novobiocin Hemolytic anemia — mild. *0713*

Oral Contraceptives May cause megaloblastic anemia. *3025*

Oxyphenbutazone Dilutional effect of water retention. *1343*

Pamaquine Hemolytic anemia in G-6-PD deficient persons. *0788* May cause hemolytic anemia. *2313*

Penicillamine Hypochromic anemia in child, menstruating woman. *0074*

Penicillin Hemolytic anemia due to binding to erythrocytes. *0996*

Penicillin V Isolated case of hemolytic anemia with IgM antibody. *0360*

Pentamidine Megaloblastic anemia — inhibits dihydrofolate reductase. *3773*

Pentaquine May cause hemolysis with G-6-PD deficiency. *0248*

Phenacemide Aplastic anemia. *0071*

Phenazopyridine Hemolytic anemia (sensitivity dependent). *0788*

Phenethicillin Theoretical effect of penicillins. *1680*

Phenobarbital Megaloblastic anemia secondary to disturbance in folic acid metabolism. *1926*

Phenothiazines Hemolytic anemia (sensitivity dependent). *0788*

Phenylbutazone Bone marrow depression/secondary to Na and H_2O retention. *0566*

Phenylhydrazine Hemolytic anemia. *0788*

Phenytoin megaloblastic/hemolytic/aplastic anemia, pancytopenia. *3331*

Phytonadione Hemolysis may occur with G-6-PD deficiency. *2313*

Piperacetazine Mild transient decrease with hypotension. *1227*

Piperazine May cause hemolytic anemia. *2365*

Pipobroman Hemolytic anemia has been described. *0070*

Piroxicam Possibly related to drug administration or concomitant therapy in patients with osteoarthrosis. *3564* Reported, unassociated with obvious gastrointestinal bleeding. *2692*

Pregnancy Normally slight reduction (not below 10 mg/dL). *0987*

Prilocaine Hemolysis with doses greater than 400 mg. *0071*

Primaquine Hemolytic anemia in G-6-PD deficient persons. *0788*

Primidone Megaloblastic anemia secondary to disturbance in folic acid metabolism. *2968*

Probenecid Hemolytic anemia (sensitivity dependent). *0788*

Procainamide Observed with SLE-like syndrome, hemolytic anemia. *0933*

Propylthiouracil Occasional immunologically associated anemia. *3870*

Pyrazinamide Rare anemia reported. *2427*

Pyrazolones Aplastic/hemolytic anemia. *2313*

Pyrimethamine Megaloblastic anemia. *3898*

Quinacrine May cause hemolysis with G-6-PD deficiency. *0248*

Quinidine Hemolytic anemia (sensitivity dependent) G-6-PD deficiency. *3673*

Quinine Hemolytic anemia (sensitivity dependent) G-6-PD deficiency. *0272*

Quinocide May cause hemolytic anemia. *1902*

Recumbency Average maximum decrease of 5% occurs in 20 minutes. *1010*

Resorcinol Injection or excess absorption may cause hemolysis. *0071*

Rifampin Observed effect. *1705* Rarely acute hemolytic anemia occurs associated with circulating drug dependent antibodies in high titer. *1292*

Spectinomycin Mechanism not discussed. *0122*

Stibophen May produce hemolytic anemia. *2365*

Streptomycin Hemolytic anemia (sensitivity dependent). *0788*

Sulfacetamide May cause hemolysis if G-6-PD deficiency. *0248*

Sulfadiazine May cause hemolytic anemia. *2313*

Sulfamethizole Hemolytic anemia. *2868*

Sulfamethoxazole May cause hemolytic anemia. *1974*

Sulfamethoxypyridine Hemolytic/aplastic anemia. *2313*

Sulfanilamide May cause hemolytic anemia. *0248*

Sulfapyridine May cause hemolytic anemia. *0248*

Sulfasalazine Hemolysis quite common, but frank hemolytic anemia rare. *2887*

Sulfisoxazole Agranulocytosis/aplastic anemia. *2868*

Sulfonamides Agranulocytosis/hemolytic anemia/sen. dependent (G-6-PD). *2427*

Sulfones May cause hemolysis. *0071*

Sulfoxone May cause hemolytic anemia. *0248*

Suramin Hemolytic anemia reported. *0071*

Tamoxifen In a small proportion of patients with associated reductions of platelet and white cell counts. *1542*

Tetracycline May cause hemolytic anemia. *2313*

Thiacetazone May cause hemolytic anemia; may fall without frank anemia. *1292*

Thiazolsulfone May cause hemolytic anemia. *2313*

Thioguanine In 29 patients with polycythemia rubra vera; effect usually apparent in 2 weeks. *2450*

Thioridazine Significant effect reported. *1629*

Thiosemicarbazones Large dose effect. *1343*

Thiotepa May be rapid decrease. *2427* 10 of 25 patients given drug intravesically had at least one incident of acute myelosuppression; 5 of 29 patients had chronic myelosuppression. *1632*

Tolbutamide Agranulocytosis/aplastic anemia. *0073*

Tolonium May cause hemolysis with G-6-PD deficiency. *0333*

Triamterene May cause megaloblastic anemia (folic acid antagonist). *3773*

Triethylenemelamine May cause hemolytic anemia. *2313*

Trimethadione Aplastic anemia. *3008*

Trimethoprim Folic acid antagonist ?megaloblastic anemia. *3773* Isolated case when drug given alone, but clear evidence of pancytopenia. Megaloblastic anemia may occur with prolonged use. *3284*

Trinitrotoluene May cause hemolysis with G-6-PD deficiency. *1413*

Tripelennamine Hemolytic anemia. *0788*

Tyrothricin May cause hemolysis even if applied topically. *0071*

Urea Rapid infusion of concentration solution may cause hemolysis. *1022*

Urethan May occur with bone marrow depression. *2427*

Vitamin A Anemia observed. *2427*

Vitamin K K_3 and K_4 only, especially with G-6-PD deficiency. *2102*

Zomepirac Case reported of immune hemolysis. *3208*

Urine   No Effect   Analytical

Chloroform As preservative has no effect on Stix tests. *2308*

Fluorescein At 250 mg/L in spiked specimen run on Ames Clinitek 200. *1722*

Urine   No Effect   Physiological

Chenodeoxycholic Acid No significant effect with up to 750 mg/d in 916 patients treated for 2 y. *3199*

Cromoglycate No clinical significant change observed. *0484*

Fluroxene No significant effect noted. *1814*

Halothane No significant effect noted. *1814*

Urine   Increase   Analytical

Bacteriuria False positive with Hematest®. *3879*

Bromides Interferes with guaiac and benzidine tests. *3879*

Copper False positive with guaiac and benzidine tests. *2313*

Ferricyanide Interferes with benzidine test. *1022*

Filter Paper Interferes with benzidine test. *0459*

Formaldehyde Reaction with benzidine. *2313*

Iodides Interferes with guaiac and benzidine tests. *1237*

Nitric Acid Affects benzidine test. *2313*

Permanganate Interferes with benzidine test. *1022*

Urine   Increase   Physiological

Acetanilid Occurs with hemolysis. *1902*

Aminopyrine May occur with marked hemolysis. *1902*

Aminosalicylic Acid Actual bleeding caused by drug. *1022*

Amphotericin B Nephrotoxicity, bleeding actually caused by drug. *1022*

Ampicillin Occurs as result of nephrotoxicity. *3542*

Anisindione Overdose manifestation. *1680*

Antimalarials May occur with marked hemolysis. *1902*

Antipyretics Occurs with marked hemolysis. *1902*

Antipyrine Occurs with marked hemolysis. *1902*

Arsenicals May be marked hematuria. *1302*

Aspirin Occurs with severe hemolytic anemia. *1902*

Auranofin Occasional drug-associated hematuria: may be partially responsible for anemia. *0617*

Azapropazone In 1 of 83 patients treated for 6 mo. *3565*

Bacitracin Actual bleeding caused by drug. *1022*

Bismuth Salts May cause severe renal damage. *2427*

Blood Transfusions May occur especially with frozen glycerolized blood. *2220*

Carbamazepine Associated with bleeding tendency. *0115*

Carbon Disulfide Manifestation of renal damage. *0818*

Chloroguanide Drug produces actual bleeding. *1022*

Chlorpropamide Isolated case of drug-induced hemolytic anemia. *3111*

Colchicine May cause bleeding. *2313*

Colistimethate Common nephrotoxic effect. *2808*

Copper May occur with copper toxicity. *0385*

Coumarin Actual bleeding may occur with high doses. *1237*

Cyclophosphamide Actual bleeding caused by drug. *1022*

Cyclosporine In approximately 5% patients given drug versus 8% given azathioprine in over 200 renal transplant patients. *0572*

Ethosuximide Rare side effect observed. *2042*

Ethylene Glycol Due to nephrotoxicity. *1302*

Fenoprofen In 6.5% in patients treated for juvenile rheumatoid arthritis. *0239* Rare case of acute tubulointerstitial nephritis with acute renal failure (probable association with drug administration). *3582*

Fentanyl In a single patient with low haptoglobin following high dose administration. dose related effect. *1719*

Glycerin Rare transient effect after i.v. infusion. *3615* Isolated cases after 20% solution i.v. *1504*

Gold Produces actual bleeding. *1022*

Hydralazine Reported finding. *2427* Microscopic hematuria observed in 9 patients developing rapidly progressive glomerulonephritis during treatment. *0366*

Hydrogen Sulfide Occurs due to nephrotoxicity. *1302*

Ibuprofen In 3% of patients with juvenile rheumatoid arthritis. *0239*

Ifosfamide Occurs in one third patients within 2 d. *3675*

Indandione Derivatives May cause hematuria — manifestation of overdose. *1237*

Indomethacin Produces actual bleeding. *1022*

Ipodate Nephrotoxic effect. *0838*

Iron Dextran Reversible after chronic administration in one patient. *1678*

Iron Salts Hematuria reported after iron Sorbitex. *2427*

Kanamycin Actual bleeding occurs. *1022*

Lead Occurs with acute hemolytic crisis. *1302*

Levodopa Occasional report of hematuria. *0071*

Lipomul® Produces actual bleeding. *1022*

Lysol® Occurs with massive hemolysis. *0629*

Mandelic Acid Actual bleeding caused by the drug. *1022*

Mefenamic Acid Actual bleeding caused by the drug. *1022*

Mephenesin Actual bleeding caused by drug. *1022*

Mercaptopurine Direct renal damage with doses over 750 mg/sq m. *0980*

Mercury Compounds May cause severe nephritis if absorbed. *0071*

Mersalyl Actual bleeding caused by drug. *1022*

Methenamine Actual bleeding produced by drug. *1022*

Methicillin Hypersensitivity reaction, nephrotoxicity. *0212*

Methocarbamol May cause intravascular hemolysis. *1343*

Methotrexate May cause hematuria. *3920*

Mitotane May occasionally produce hematuria. *1680*

Muscular Exercise May occur after severe exercise. *0987*

Naphthalene Due to G-6-PD related hemolysis or poisoning. *1302*

Naproxen In 2% of patients receiving drug for juvenile rheumatoid arthritis. *0239*

Nitrofurans May occur with severe hemolytic anemia. *1902*

Oxacillin Nephrotoxicity with hematuria. *0071*

Oxyphenbutazone Actual bleeding caused by drug. *1022*

Penicillin Hypersensitivity reaction, nephrotoxicity. *0212*

Pentamidine Gross hematuria observed in a single patient. *3317*

Phenindione Actual bleeding caused by drug. *1022*

Phenolphthalein May cause acute nephrosis (K deficiency). *2427*

Phenols Nephrotoxicity with poisoning. *1343*

Phenylbutazone Actual bleeding caused by drug. *1022*

Phosphorus Actual bleeding caused by drug. *1022*

Phytonadione Actual bleeding caused by drug. *1022*

Hemoglobin (continued)

Urine Increase Physiological (continued)

Piroxicam Occasional drug induced nephrotoxicity, with isolated azotemia, acute interstitial nephritis or nephrotic syndrome. 2824

Polymyxin Actual bleeding caused by drug. 1022

Probenecid Actual bleeding caused by drug. 1022

Proguanil Actual bleeding caused by drug. 1022

Pyrazolones Actual bleeding caused by drug. 1022

Quinine Possible contributing factor. 0071

Sulfadiazine Associated with crystalluria and hematuria. 2808

Sulfamethoxazole May cause hematuria with oliguria. 2808

Sulfonamides Actual bleeding caused by drug (due to hemolysis). 1022

Sulfones May have nephrotoxic effect. 1022

Suprofen Occasional renal failure observed but mechanism not known. 0009

Suramin Actual bleeding caused by drug. 1237

Thiazides Actual bleeding caused by the drug. 1022

Tobramycin At dose of 4.5 mg/kg/d for 12 d in 90 patients nephrotoxicity observed in up to 39%, reversible in most. 0677

Triethylenemelamine Due to nephrotoxicity. 1680

Trimethadione Hematuria may occur especially in children. 2220

Trinitrotoluene May cause hemolysis. 2313

Trypsin Occasional reported occurrence. 1022

Turpentine Nephrotoxic effect following ingestion. 1302

Urea Hemoglobinuria may occur especially if hypothermia. 1022

Vancomycin Occasional evidence of mild nephrotoxicity. 1571

Viomycin Actual bleeding caused by drug. 1237

Vitamin K Effect of treatment of hemorrhagic states in children. 0071

Warfarin Due to overdosage in some cases. 1302

Urine Decrease Analytical

Alkaline Urine May interfere with hemagglutination test. 0017

Ascorbic Acid In large amounts inhibits guaiac test. 1022 At up to 140 mg/dL made Chemstrip® 7 react negatively to 250 mg/L hemoglobin, also with Hema-Combistix® and some reactions were negative at approximately same concentration with Chemstrip® 9. 3968

Tetracycline In large amounts if buffered with ascorbate. 0583

Hemoglobin A₁c

Blood No Effect Physiological

Piretanide In 12 male patients with mild hypertension (6 mg bid). 1503

Blood Increase Analytical

Aspirin Acetylation of hemoglobin simulates glycosylation when measured by HPLC or electrophoresis but no increase observed with isoelectric focusing and colorimetric techniques. 2564

Blood Increase Physiological

Hydrochlorothiazide Increase by 6% in 14 hypertensive type 2 diabetic men over 3 weeks: effect augmented by coadministered propranolol. 0935

Indapamide After 24 weeks in 13 hypertensive diabetics. 2686

Propranolol In 14 hypertensive men with type 2 diabetes when treated for 3 weeks, marked increase when propranolol coadministered. 0935

Blood Decrease Physiological

Benfluorex Significant effect in hypertriglyceridemic type 2 diabetic patients given 150 mg tid after 1 mo treatment. 3385

Nisoldipine From 14.7 to 14.0 g/dL in 14 mild to moderately hypertensive noninsulin dependent diabetic patients. 2653

Hemopexin

Serum Increase Physiological

Phenobarbital Significant effect in 40 adult epileptics receiving long term treatment. 3641

Physical Training 6.6 increase after 4 weeks training. 1491

Pregnancy In effect no change with pregnancy. 3472

Heparin Sulfate

Urine Increase Physiological

Gentamicin Reacts with heparin to form precipitate. 0897

Hepatic Lipase

Serum No Effect Physiological

Ethanol With acute administration after overnight fast. 2428

Serum Decrease Physiological

Ethanol With acute administration to fed individuals. 2428

Hepatic Triglyceride Lipase

Serum No Effect Physiological

Isotretinoin Increase in both men and women with 1 mg/kg/d for 20 weeks when given for nodulocystic acne. 0326

Heroin

Urine Increase Analytical

Quinine Fluoresces, producing interference. 3869

Hexokinase

Red Blood Cells Decrease Physiological

Copper Very sensitive to inhibition by copper in vitro. 0432

Hexosamine

Serum No Effect Analytical

Storage of Sample For 7 d at 30°. 0573

Serum Increase Physiological

Diurnal Variation Maximum observed about midday. 2222

Hexosaminidase

Serum Increase Physiological

Ethanol Observed with chronic alcohol consumption. 3124

Hippuran Clearance

Urine Decrease Physiological

Cimetidine Significant effect within 6 h, normalized after several weeks. 3199

Hippuran Retention

Serum Increase Physiological

Muscular Exercise During infusion (depends on extent of activity). 1171

Hippuric Acid

Urine   Increase   Analytical
Aspirin   Salicyluric acid measured by method of Tomokuni. *3607*
Xylene   Metabolites measured as hippurate (method of Tomokuni). *3607*

Urine   Increase   Physiological
Methenamine   If given as hippurate salt. *1678*
Toluene   Metabolites also increased on exposure. *3607*

Histamine

Plasma   Increase   Analytical
Ammonia   May affect fluorometric method of Shore. *0081*

Plasma   Increase   Physiological
Atropine   Associated with dose given associated with anesthesia. *2212*
Gallamine   Observed with injection for anesthesia. *2212*
Ioxitalamic Acid   Major increase in arterial and mixed venous blood in 30 s following injection for translumbar arterial aortography in 16 patients. *1364*
Meperidine   Associated with dose associated with anesthesia. *2212*
Morphine   Observed with injection associated with anesthesia. *2212*
Propanidid   Immediate rise after i.v. inject without anaphylaxis. *2212*
Radiographic Agents   Observed in some patients if administered i.v. *0449*
Succinylcholine   Observed with injection for anesthesia. *2212*
Thiopental   Normal response even up by 350% at 5 minutes. *2212*
Trimethaphan   Observed with injection associated with anesthesia. *2212*
Tubocurarine   Observed with administration for anesthesia. *2212*

Urine   No Effect   Physiological
Amantadine   In normals and Parkinson patients. *1817*
Aminoguanidine   No effect observed on standard diet. *1381*

Urine   Increase   Physiological
Ethanol   All gastric stimulants produce effect. *0814*
Histamine   Threefold increase in 1 d after injection. *1382*
Meat   All gastric stimulants produce effect. *0814*
Pentagastrin   All gastric stimulants produce effect. *0814*

Histamine Test

Patient   Increase   Physiological
MAO Inhibitors   Enhanced responsiveness. *1325*

Patient   Decrease   Physiological
Thiazides   False negative due to hypotensive effect of drug. *1488*

Histidine

Plasma   No Effect   Analytical
Ammonia   No significant effect fluorometric method of Ambrose. *0081*
Storage of Sample   No effect -20° for 6 mo, decreased at room temperature 3 d. *0081*

Plasma   Increase   Analytical
Aminobutyric Acid   May affect fluorometric method of Ambrose. *0081*
Arginine   May affect fluorometric method of Ambrose. *0081*

Asparagine   May affect fluorometric method of Ambrose. *0081*
Cystine   May affect fluorometric method of Ambrose. *0081*
Hemolysis   Released from red cells. *0081*
Histamine   May produce marked effect on method of Ambrose. *0081*
Kynurenine   May produce slight effect method of Ambrose. *0081*
Tyrosine   May produce moderate effect on method of Ambrose. *0081*

Urine   Increase   Physiological
Pregnancy   Progressive increase observed. *3507*
Protein   Occurs after high meat diet. *0987*
Tetracycline   Unexplained mechanism. *1343*

HMPG (4-Hydroxy-3-Methoxy-Phenylethylene Glycol)

Urine   No Effect   Analytical
Normetanephrine   100 µg = 0-2 µg method of Bigelow. *0343*
Storage of Sample   No effect 6 y at -20° (method of Borud). *0426*
Vanillylmandelic Acid (VMA)   100 µg = 0-2 µg method of Bigelow. *0343*

Urine   No Effect   Physiological
Antibiotics   Orally administered has no effect on excretion. *0426*

Urine   Increase   Physiological
Ajmaline   Release of stored norepinephrine. *2220*
Disulfiram   Inhibition of aldehyde dehydrogenase. *2220*
Ethanol   Following ingestion after 0.4 g alcohol/kg in 56 healthy males. *0015*
Rauwolfia   Release of stored norepinephrine. *2220*
Reserpine   Release of stored norepinephrine. *2220*
Syrosingopine   Release of stored norepinephrine. *2220*

Urine   Decrease   Physiological
MAO Inhibitors   Inhibition of amine oxidase. *0426*
Methyldopa   By more than VMA, ?affects aldehyde reductase Inhibition of Dopa decarboxylase, ?also aldehyde reductase. *0426*
Reserpine   Long term administration produces decrease. *0426*
Verapamil   In 7 chronically ill schizophrenic patients when administered for 5 weeks. *2811*

CSF   Increase   Physiological
Probenecid   60% increase on average if 100 mg/kg over 18 h. *1351*

CSF   Decrease   Physiological
Lithium   Significant reduction with treatment of manic patients. Initial high values correlates with severity of disease. *3517*

HMPG Sulfate

CSF   Increase   Physiological
Probenecid   Small but not significant increase after 9 h. *1351*

Homocitrulline

Plasma   Decrease   Analytical
Storage of Sample   If picric acid used for deproteinization. *3230*

Homocystine

Plasma   Decrease   Analytical
Standing of Sample   Binds to protein with time. *3230*
Storage of Sample   Binds to protein with time. *3230*

Homogentisic Acid

Plasma   Decrease   Analytical
Storage of Sample   25% loss in frozen plasma in 1 week. *1712*

Urine   No Effect   Analytical
Ascorbic Acid   On TLC method of Feldman and Bowman. *1096*
3,4-Dihydroxyphenylacetic Acid   On TLC method of Feldman and Bowman. *1096*
Gentisic Acid   On TLC method of Feldman and Bowman. *1096*
Levodopa   On TLC method of Feldman and Bowman. *1096*
Storage of Sample   If acidified and frozen stable for 1 week. *1712*

Urine   Increase   Analytical
Ascorbic Acid   If method of Briggs used. *1096*
Aspirin   Interferes with measurement procedure. *1238*
Catechol   If method of Nuberger used. *1096*
Gentisic Acid   If method of Nuberger used. *1096*
Hydroquinone   If method of Nuberger used. *1096*
4-Hydroxyphenylpyruvic Acid   If method of Briggs used. *1096*
Levodopa   If method of Briggs used If method of Nuberger used. *1096*
Uric Acid   If method of Walkow, Baumann used. *1096*

Homovanillic Acid

Plasma   No Effect   Physiological
Apomorphine   No effect when given i.v. to either healthy controls or schizophrenics. *3178*

Plasma   Increase   Physiological
Verapamil   In 7 chronically ill schizophrenic patients when administered for 5 weeks. *2811*

Plasma   Decrease   Physiological
Methyldopa Hydrazine   65% reduction when given with Dopa. *2306*

Urine   No Effect   Analytical
Dihydroxymandelic Acid   At high concentrations with method of Kahane. *1843*
Dihydroxyphenylacetic Acid   At high concentrations with method of Kahane. *1843*
Vanillylmandelic Acid (VMA)   At high concentrations with method of Kahane. *1843*

Urine   No Effect   Physiological
Amantadine   In normals and Parkinson patients. *1817*
Barbiturates   No effect with addiction or withdrawal. *3634*

Urine   Increase   Analytical
Aspirin   Affects colorimetric methods of Sandler, Ruthven. *1620* May produce interfering fluorescence. *0583*

Urine   Increase   Physiological
Disulfiram   Probably due to inhibition of dopamine hydroxylase. *3703*
Levodopa   In Parkinsonian patients is response to therapy. *0562*
Pyridoxine   When given to patients on L-Dopa. *1675*
Reserpine   Maximum during second day of treatment. *3838*

Urine   Decrease   Analytical
Aspirin   High blank in fluorometric method of Sato. *1620*

CSF   No Effect   Physiological
Amphetamine   No change in CSF with psychotic dose. *1825*
Bupropion   Insignificant effect in approximately 40 patients with depression or Alzheimer's disease after chronic treatment. *2991*
Chlorophenylalanine   After 1 g/d for 4 d or after single dose. *3342*

Desipramine   In 43 patients with depression or Alzheimer's disease chronically treated with drug. *2991*
Thiopropazate   No significant effect in Huntington's chorea. *2379*

CSF   Increase   Analytical
Aspirin   Interferes with fluorometric method even if 5 d before. *1801*

CSF   Increase   Physiological
Levodopa   ?metabolic response (variable between individuals). *2725*
Probenecid   Approximately 6 fold at 9 h, 9 fold at 18 h. *1351*
Tetrabenazine   Significant effect in chorea after oral administration. *2379*
Verapamil   In 7 chronically ill schizophrenic patients when administered for 5 weeks. *2811*
Zimeldine   Insignificant effect in 43 chronically treated patients with depression or Alzheimer's disease. *2991*

CSF   Decrease   Physiological
Clorgiline   Significant reduction in 43 patients with depression or Alzheimer's disease chronically treated with drug. *2991*
Deprenyl   In 43 patients with depression or Alzheimer's disease chronically treated with drug. *2991*
Diazepam   Probably due to decreased turnover of dopamine. *2725*

α_2-HS Glycoprotein

Serum   Increase   Physiological
Muscular Exercise   Significant effect at 15 minutes, partial return by 1 d. *1491*

Hyaline Casts

Urine   Increase   Physiological
Piroxicam   Many: occasional drug induced nephrotoxicity, with isolated azotemia, acute interstitial nephritis or nephrotic syndrome. *2824*

Hydantoin-5-Propionic Acid

Urine   Increase   Physiological
Methotrexate   Large effect of treatment observed. *3681*

Hydrazine

Plasma   No Effect   Analytical
Dopamine   On fluorometric method of Vickers at 1 μg/mL. *3710*
Phenylhydrazine   On fluorometric method of Vickers at 1 μg/mL. *3710*
Tryptamine   On fluorometric method of Vickers at 10 μg/mL. *3710*
Tyrosine   On fluorometric method of Vickers at 10 μg/mL. *3710*

Plasma   Increase   Analytical
Carbidopa   On fluorometric method of Vickers. *3710*

Hydrochloric Acid

Gastric Material   No Effect   Physiological
Cholecystokinin   May occasionally cause increase. *2049*
Smoking   No effect in 7 healthy volunteers after 3 cigarettes. *2352*

Gastric Material   Increase   Physiological
Acetylcholine   Hypoglycemia is powerful stimulant. *0814*
Ajmaline   Strong stimulant action. *2220*

Betazole   Used to stimulate secretion. *0071*

Caffeine   Effect of i.v. infusion, also if given orally. *0689*

Calcium   Strong response to infusion of Calcium gluconate. *2739*

Calcium Chloride   Intraluminally (3 x increase over sodium chloride). *1643*

Calcium Salts   Rebound increase at night if $CaCO_3$ ingested early in day. *3264*

Chewing Gum   But insignificantly in normals. *1480*

Corticotropin   Effect of protracted therapy. *1680*

Ethanol   Psychically, reflexly and through histamine/gastrin. *1343*

Famotidine   Therapeutic effect as drug is long-acting histamine H_2-receptor antagonist. *1667*

Gastrin   Hormonal action. *2049*

Histamine   Used diagnostically. *0071*

Insulin   Hypoglycemia is powerful stimulant. *0814*

Meat Extract   But less than with insulin. *3904*

Meprednisone   Steroid effect. *1680*

Methysergide   Affect basal juice and after histamine. *2220*

Morinamide   Hyperacidity reported. *2427*

Niacin   Increase up to 230% in patients after 0.5 g orally. *0104*

Pentagastrin   If infused i.v. *0689*

Pilocarpine   Produces secretion like that of vagal stimulation. *1343*

Propanidid   Stimulation paralleled plasma histamine concentration. *2212*

Rauwolfia   Stimulates gastric secretion. *2220*

Reserpine   Excess secretion may activate peptic ulcers. *1343*

Smoking   Effect of 4 to 6 cigarettes in 1 h. *0295*   Most subjects, especially in normal initially, show increase. *0139*

Syrosingopine   Excessive amounts may be released. *2220*

Tetragastrin   Effect similar to pentagastrin. *1482*

Thiopental   Stimulation of secretion parallels plasma histamine. *2212*

Tolazoline   Also enhances histamine stimulation. *1343*

Gastric Material   Decrease   Physiological

Acetazolamide   In large doses reversible effect. *0814*

Atropine   Volume also reduced. *1343*

Barbiturates   Secretion slightly depressed. *1343*

Burimamide   (up to 60%) mainly due to decreased volume. *3905*

Carbon Disulfide   Frequent;also high incidence chronic gastritis. *0818*

Diazepam   Presumed central action lasts for 5 h after 10 mg. *0362*

Diurnal Variation   Minimal at 2 am (almost alkaline). *0814*

Ganglionic Blocking Agents   Volume and acidity generally reduced. *1343*

Glucagon   Intravenously 0.015 mg/kg/h affects for hours. *0814*

Hexocyclium   Effective anticholinergic action. *1679*

5-Hydroxytryptamine   But mucus secretion increased. *3579*

Insulin   Absolute amount decreased by injection or infusion. *0599*

Loperamide   Effect on both basal and submaximal pentagastrin-stimulated gastric acid secretion. Effect dose related. *0559*

Morphine   Slight decrease in secretion of acid. *1343*

Naloxone   Reduces basal and meal stimulated concentration. *0176*

Omeprazole   When stimulated acid output reduced from 27.4 mmol H^+/h to 7.8 mmol H^+/h after 29 d in 8 volunteers but reversible. *2516*

Oxyphencyclimine   Absolute decrease by 60%, but not concentration. *2537*

Pregnancy   May be hyposecretion. *0987*

Propranolol   Affects basal and histamine stimulated. *1273*

Prostaglandin-α_2   Transient effect in all subjects. *0337*

Ranitidine   Reduced to 10% of normal in healthy volunteers. *3388*

Secretin   Pancreatic secretion also at peak. *1989*

Pancreatic Juice   Decrease   Physiological

Secretin   Reciprocal relationship to bicarbonate. *0194*

Hydrochlorothiazide

Serum   Increase   Physiological

Propantheline   Delayed but increased absorption with delayed gastric emptying. *3794*

Hydrocortisone

Serum   Decrease   Physiological

Phenytoin   25% increase in metabolic clearance rate with much reduced half-life. *1830*

Rifampin   Metabolism enhanced with reduction of circulating drug concentration: cortisol production rate increased. *1830*

Hydrogen

Breath   No Effect   Physiological

Gum Tragacanth   In 5 male volunteers consuming 9.9 g daily for 21 d. *0990*

Hydromorphone

Urine   No Effect   Analytical

Chlorphentermine   No interference using TLC with ethyl acetate: methanol: water: ammonium hydroxide and modified Dragendorff's reagent for detection. *3868*

Diethylpropion   No interference with TLC using ethyl acetate: methanol: water: ammonium hydroxide and modified Dragendorff's reagent for detection. *3868*

Fenfluramine   No interference on TLC using ethyl acetate: methanol: water: ammonium hydroxide and modified Dragendorff's reagent for detection. *3868*

Mazindol   No interference on TLC using ethyl acetate: methanol: water: ammonium hydroxide and modified Dragendorff's reagent for detection. *3868*

Phendimetrazine   No interference with TLC using ethyl acetate: methanol: water: ammonium hydroxide and modified Dragendorff's reagent for detection. *3868*

Phenmetrazine   No interference with TLC using ethyl acetate: methanol: water: ammonium and modified Dragendorff's reagent for detection. *3868*

Phentermine   No interference on TLC using ethylacetate: methanol: water: ammonium hydroxide and modified Dragendorff's reagent for detection. *3868*

Urine   Positive   Analytical

Clotermine   No interference on TLC using ethyl acetate: methanol: water: ammonium hydroxide and modified Dragendorff's reagent for detection. *3868*

2-Hydroxy-4-Ketoglutarate

Urine   Increase   Physiological

Lithium   May be considerable effect. *2107*

11-Hydroxyandrosterone

Urine   Increase   Physiological

Corticotropin   Normal response to 2 d ACTH test. *1474*

3-Hydroxyanthranilic Acid

Urine   No Effect   Physiological

Smoking   Probably no significant effect with tryptophan load. *3183*

Urine   Increase   Physiological

Deoxypyridoxine   After 2 g tryptophan load. *3050*

3-Hydroxyanthranilic Acid (continued)

Urine   Increase   Physiological (continued)
Estrogens   Induce tryptophan pyrrolase. *3045*
Glucocorticoids   Induce tryptophan pyrrolase. *3045*
Hydrocortisone   Causes induction of tryptophan pyrrolase. *3047*
Oral Contraceptives   Estrogen component induces tryptophan pyrrolase. *3045*   After 2 g tryptophan load. *3050*
Pregnancy   Causes induction of tryptophan pyrrolase. *3045*

Urine   Decrease   Physiological
Anabolic Steroids   Effect of synthetic androgens given to males. *3045*
Androgens   Effect of synthetic androgens given to males. *3045*

4-Hydroxybenzoic Acid

Urine   Increase   Physiological
Bananas   Possible effect of bacterial metabolism. *3281*
Coffee   Possible effect of bacterial metabolism. *3281*

4-Hydroxybenzylamine

Urine   Increase   Physiological
Mustard   Excreted after mustard eaten. *2784*

3-Hydroxybutyrate

Plasma   No Effect   Physiological
Lactate   No variability with method of Zivin. *3955*

Plasma   Decrease   Physiological
Valproic Acid   78% reduction when 1 g given orally to fasting individuals. *3633*

β-Hydroxybutyrate

Serum   No Effect   Physiological
Albuterol   No significant change in 4 healthy men given i.v. infusion of therapeutic amount. *2807*
Rimeterol   No significant change in 4 healthy men given therapeutic i.v. dose. *2807*

Serum   Increase   Physiological
Ethanol   Associated with increased lactate also. *1532*
Starvation   Gradual increase, due to metabolism of fat. *2240*

Serum   Decrease   Physiological
MPCA   Anti-lipolytic effect of drug. *1531*

Blood   Increase   Physiological
Disulfiram   In 1 of 6 volunteers rapid and short lasting effect. *3482*

Blood   Decrease   Physiological
Alanine   Marked fall when given orally to fasting obese. *1263*

Urine   Increase   Physiological
Starvation   Due to metabolic acidosis. *0021*

Urine   Decrease   Physiological
Alanine   Falls in parallel with blood with chronic administration. *1263*

Hydroxybutyrate Dehydrogenase

Serum   No Effect   Analytical
Acetaminophen   On continuous method at 10 times maximal therapeutic concentration. *1785*
Aminophenazone   On continuous method at 10 times maximal therapeutic concentration. *1785*
Caffeine   On continuous method at 10 times maximal therapeutic concentration. *1785*
Citrates   No effect on activity. *2981*
Diclofenac   On continuous method at 10 times maximal therapeutic concentration. *1785*
EDTA   No effect on activity. *2981*
Ibuprofen   On continuous method at 10 times maximal therapeutic concentration. *1785*
Indomethacin   On continuous method at 10 times maximal therapeutic concentration. *1785*
Ketoprofen   On continuous method at 10 times maximal therapeutic concentration. *1785*
Storage of Sample   5 d at 30 or 4°, at -20° for 10 d. *0877*

Serum   Increase   Analytical
Hemoglobin   Large effect on UV procedures per g Hemoglobin. *3217*
Hemolysis   Erythrocytes contain large amounts. *2981*

Serum   Increase   Physiological
Anticonvulsants   In 4% epileptics on chronic therapy. *3560*
Hypothermia   Due to hypoxia, acid-base imbalance, hypotension. *2258*
Meperidine   May cause spasm of sphincter of Oddi. *2427*
Morphine   Probably due to spasm of sphincter of Oddi. *2427*

Serum   Decrease   Analytical
Acetylsalicylic Acid   With 8300 μmol/L on kinetic method. *1785*
Heparin   Significant inactivation. *2981*
Oxalate   Almost complete inactivation of enzyme. *2981*
Phenobarbital   On continuous method at 10 times maximal therapeutic concentration. *1785*

Red Blood Cells   Increase   Physiological
Anticonvulsants   In 21% epileptics on chronic therapy. *3560*

11-Hydroxycorticosteroids

Plasma   No Effect   Analytical
Carbamazepine   No effect on fluorometric Mattingly procedure at 1 mg/ml. *2977*
Dexamethasone   Presumed effect on fluorescent method of Mejer. *2396*
Phenobarbital   No effect on Mattingly procedure at 200 mg/L. *2977*
Phenytoin   No effect on Mattingly procedure at 1 mg/ml. *2977*
Prednisolone   Presumed effect on fluorometric method of Mejer. *2396*
Prednisone   Presumed effect on fluorometric method of Mejer. *2396*

Plasma   No Effect   Physiological
Ethanol   No effect seen in chronic alcoholics. *1487*
Obesity   In females but falls with caloric restriction. *3391*

Plasma   Increase   Analytical
Carbamazepine   When determined by Mattingly method. *2901*
Cholesterol   Due to nonspecific fluorescence. *1369*
Corticotropin   Increased cholesterol with nonspecific fluorescent procedure. *1369*
Fusidic Acid   When determined by Mattingly method. *2901*
Quinacrine   When determined by Mattingly method. *2901*
Spironolactone   When determined by Mattingly method. *2901*

Plasma   Increase   Physiological

Calcium Gluconate   Transient effect maximum at 15 minutes after i.v. *1487*

Carbenoxolone   Transient effect after 100 mg orally. *1487*

Ethanol   In normals after 1.5 ml/kg. *1487*

Nicotine   Up to 80% increase after heavy smoking. *1922*

Smoking   Up to 77% increase after heavy smoking. *1487*

Sucrose   Increase of 300-400% on high sucrose diet. *3931*

Urine   No Effect   Analytical

Aspirin   No effect on fluorometric Mattingly method. *0766*

Urine   No Effect   Physiological

Acetaminophen   No effect with short term ingestion 2.6 g/d. *0766*

Aspirin   No effect with short term ingestion 2.6 g/d. *0766*

Chlordiazepoxide   No effect of short term ingestion of 30 mg/d. *0766*

Diazepam   No effect with short term ingestion of 15 mg/d. *0766*

Diphenhydramine   No effect with short term ingestion of 150 mg/d. *0766*

Glucose   No significant response to hyperglycemia. *3373*

Noise   Excretion unaffected by 80 decibels for 2 h. *3351*

Phenobarbital   No effect short term ingestion of 120 mg/d. *0766*

Smoking   No effect usual smoking in smokers. *3626*

Urine   Increase   Physiological

Noise   Maximum effect at 10,000 Hz. *3351*

Urine   Decrease   Physiological

Pentazocine   ?due to depression of adrenocortical secretion. *0766*

Propoxyphene   Slight effect only (probably physiological action). *0766*

17-Hydroxycorticosteroids

Plasma   Increase   Physiological

Aspirin   Large doses stimulate adrenocortical activity. *3365*

Betamethasone   Slight effect, absorbed through skin. *0071*

Diurnal Variation   Maximum observed in morning (same as in urine). *2222*

Estrogens   Causes stimulation of pituitary adrenal axis. *3752*

Metyrapone   Indirectly stimulates ACTH production. *2211*

Plasma   Decrease   Physiological

Cromoglycate   Significant effect observed if given prior to exercise. *1641*

Morphine   Inhibits ACTH and pituitary gonadotropin release. *1343*

Urine   No Effect   Analytical

Alclometasone   In 10 volunteers given 30 g over 80% of body surface twice daily. *3586*

Amphetamine   No effect on Glenn-Nelson procedure. *0427*

Aspirin   No effect on Porter-Silber procedure. *0766*

Barbiturates   No interference with Porter-Silber procedure. *0766*

Chloramphenicol   No effect modified Reddy procedure. *1487*

Chlordiazepoxide   No effect Porter-Silber reaction if added *in vitro. 0766*

Dextroamphetamine   No effect modified Glenn-Nelson procedure. *1487*

Nalidixic Acid   No effect with Porter-Silber reaction. *3877*

Penicillin   No effect modified Glenn-Nelson. *1487*

Phenobarbital   No effect on Zimmermann procedure at 200 mg/L. *2977*

Phenytoin   No effect on Zimmermann procedure at 200 mg/L. *2977*

Propoxyphene   Added *in vitro* no effect on Porter-Silber procedure. *0766*

Quinine   No effect modified Porter-Silber reaction. *1487*

Secobarbital   No effect on Glenn-Nelson procedure when added *in vitro. 0427*

Storage of Sample   At room temperature for 45 d if acidified. *1180*

Urine   No Effect   Physiological

Acetaminophen   No effect with short term ingestion 2.6 g/d. *0766*

Amobarbital   No significant effect (?small decrease). *3466*

Aspirin   No effect with short term ingestion 2.6 g/d. *0766*

Barbiturates   No effect with short term ingestion. *3879*

Bed Rest   No effect observed on excretion. *3217*

Chlordiazepoxide   No effect of short term ingestion of 30 mg/d. *0766*

Clopamide   Single case of diuretic associated myopathy. *1793*

Cromoglycate   No proportional increase in relation to blood change. *1641*

Diazepam   No effect with short term ingestion of 15 mg/d. *0766*

Diphenhydramine   No effect with short term ingestion of 150 mg/d. *0766*

Iothalamate   No physiological effect observed. *2573*

Nitrazepam   No effect although 6-hydroxycortisol increased. *3466*

Phenobarbital   No effect short term ingestion of 120 mg/d. *0766* No significant change in response to chronic treatment. *3943*

Smoking   No effect usual smoking in smokers. *3626*

Starvation   No effect observed in normal or obese. *2382*

Urine   Increase   Analytical

Acetazolamide   *In vitro* interference with Glenn-Nelson method. *1022*

Acetone   Reported to interfere with Porter-Silber procedure. *0583*

Antihypertensive Agents   Some may be measured as analytes. *0279*

Ascorbic Acid   Interferes with method of Reddy. *1488*

Carbamazepine   Purple color so impossible to quantify with Silber and Porter method at physiological amounts. *3943*

Cefoxitin   Methodological interference with Porter-Silber reaction not eliminated by sodium bisulfite. *1062* Increased from 3 to 10-fold in urine from patients when Amberlite XAD-2 Clini-Skreen used for measurement. *2023* Substantial effect (up to 3 times actual concentration) when Porter-Silber reaction used on specimen from patients (*in vitro* 5 mg/L reacted as if 14-4 mg/L. *1062*

Cephalothin   5 mg/L reacted as if 7.2 mg/L: effect eliminated if Allen correction used. *1062*

Chloral Hydrate   Interferes with Porter-Silber reaction. *3505*

Chlordiazepoxide   Interferes with Porter-Silber reaction (vitro, vivo). *0427*

Chlormerodrin   Possible interference with Porter-Silber reaction. *0427*

Chlorothiazide   Interferes with Porter-Silber reaction *in vitro. 3879*

Chlorpromazine   Interferes with Porter-Silber reaction. *0427*

Colchicine   Interferes with Porter-Silber reaction. *3505*

Dexamethasone   Measured as endogenous steroids by Reddy method. *2420*

Digitoxin   Moderate effect with *in vitro* test. *1488*

Digoxin   Moderate effect with *in vitro* test. *1488*

Erythromycin   Reported interference with measuring procedure. *1488*

Ethinamate   Minimum effect with Glenn-Nelson method. *1488*

Etryptamine   Glucuronide interferes with Zimmermann reaction. *0427*

Fructose   Interferes with Porter-Silber reaction. *3505*

Glutethimide   Interferes with moderate Glenn-Nelson procedure *in vitro. 0427*

Hydroxyzine   Affects modified Glenn-Nelson method. *0427* Interferes with Porter-Silber reaction. *2425*

Iodides   Interferes with Porter-Silber reaction. *3505*

Meprobamate   Small effect on modified Glenn-Nelson method. *3879* Glucuronide interferes with Porter-Silber reaction. *2220*

Methenamine   Affects Porter-Silber and Reddy methods. *0457*

Methyprylon   Reported effect on Glenn-Nelson procedure. *0427*

Oleandomycin   Interferes with Porter-Silber reaction. *2628*

17-Hydroxycorticosteroids (continued)

Urine   Increase   Analytical (continued)

Paraldehyde   Interferes with Porter-Silber reaction. 3505

Penicillin   ?interferes with Porter-Silber reaction. 2220

Perphenazine   Abnormal color with Glenn-Nelson procedure. 0427

Phenazopyridine   Interferes with modified Glenn-Nelson method. 0427

Phenothiazines   Slight increased absorbance modified Glenn-Nelson method. 0427

Piperidine   Interferes with Porter-Silber reaction *in vitro*. 0427

Potassium Iodide   Affects Porter-Silber reaction. 3217

Prochlorperazine   Interference with Porter-Silber reaction. 0427

Promazine   *In vitro* effect at least on Glenn-Nelson method. 0427

Quinidine   Metabolite interferes with Zimmermann reaction. 1238

Quinine   Reddy method affected, not Porter-Silber. 3505

Spironolactone   Metabolite interferes with Porter-Silber reaction. 3505

Sulfamerazine   Alleged effect on method of Reddy. 1488

Testosterone   Common keto group involved in color reaction. 0757

Tranquilizers   Some artifactually increase value. 0279

Vitamin K   Alleged *in vitro* interference with Reddy method. 1488

Urine   Increase   Physiological

Betamethasone   Slight increase, may be absorbed through skin. 0071

Chlorthalidone   Reported effect. 2220

Corticotropin   Marked response to i.v. infusion. 0361

Cortisone   Measuring excretory products of cortisone. 3879

Diethylstilbestrol   Can be doubled in males with treatment. 3571

Diurnal Variation   Maximum excretion in morning up to midday, minimum at night. 2222

Ethinyl Estradiol   Temporary enhanced excretion on stopping drug. 1685

Gonadotropin   Hormonal action. 3225

Histamine   Due to release from adrenals. 1176

Metyrapone   Normal response to injection is 2-4 times increase. 2211

Muscular Exercise   Response to stress of exercise. 0582

Sleep Deprivation   73% increase between first and second day. 2032

Urine   Decrease   Analytical

Aspirin   Conjugate inhibits β-glucuronidase, dose > 4.8 g/d. 0766

Carbamazepine   Forms colored compound at 430 nm in Zimmermann procedure. 2977

Estrogens   Interferes with Zimmermann reactions. 2425

Hydralazine   *In vitro* effect reported modified Glenn-Nelson. 0427

Phenytoin   Inhibit beta-glucuronidase during hydrolysis. 0022

Prochlorperazine   Abnormal yellow-pink color, blank not adequate. 0583

Promethazine   Interference with Porter-Silber reaction. 3505

Reserpine   Interference with Porter-Silber reaction. 3505

Salicylate   Inhibit beta-glucuronidase during hydrolysis. 0022

Urine   Decrease   Physiological

Aminoglutethimide   Inhibits steroid biosynthesis. 1252

Barbiturates   Chronic ingestion metabolism diverted to 6-β-hydroxy-cortisol. 0766

Blindness   Significantly lower (average 5.1 mg/d) than normal. 1637

Calcium Gluconate   Reduced value reported in a single case. 0531

Carbon Disulfide   Reduction in relation to time of exposure. 0818

Chlorpromazine   Inhibition of hypothalamus and decreased ACTH secretion. 1252

Corticosteroids   If given orally, inhaled or topically. 1487

Corticotropin   Reduction even if given orally or topically. 0446

Dexamethasone   Pituitary feedback with suppression of ACTH. 2164

Estrogens   With chronic ingestion cortisol= 6-beta-hydroxy cortisol. 1884

Ethinyl Estradiol   Inhibits response to metyrapone. Decreased excretion of cortisol metabolites. 0022

Lead   In some men occupationally exposed to large quantities of lead. 0771

Levodopa   Possible inhibition of ACTH secretion. 0022

MAO Inhibitors   Probably due to depressed central synthesis. 0022

Medroxyprogesterone   Inhibition of steroid biosynthesis in adrenals. 3632

Meperidine   Probable effect (inhibits ACTH and PGH release). 1343

Methandrostenolone   Inhibits response to metyrapone. 0022

Methylethinylestradiol   Decreased excretion of cortisol metabolites. 0022

Metyrapone   Direct effect on adrenal steroidogenesis. 0022

Mitotane   Stimulates extra-adrenal hydroxylation of cortisol. 0710

Morphine   Inhibits ACTH and pituitary gonadotropin release. 1343

Norethynodrel   Decreased excretion of cortisol metabolites. 0022

Oral Contraceptives   Probably due to estrogen decreasing cortisol secretion. 2596

Pentazocine   ?due to depression of adrenocortical secretion. 0766

Perphenazine   Acts on hypothalamus to depress ACTH secretion. 0427

Phenobarbital   With chronic ingestion cortisol= 6-beta-hydroxy cortisol. 0546

Phenothiazines   Inhibit release of steroid hormones. 2220

Phenylbutazone   Cortisol metabolism diverted to 6-beta-hydroxy cortisol. 2035

Phenytoin   With chronic ingestion cortisol= 6-beta-hydroxy cortisol. 3800

Progesterone   Exact mechanism not known. 0022

Promazine   Acts on hypothalamus to decrease ACTH secretion. 0427

Propoxyphene   Probable action on hypothalamic pituitary. ACTH secretion. 0766

Rauwolfia   Probably due to depressed central synthesis. 0022

Reserpine   Probably due to depressed central synthesis. 0022

SKF-12185   Direct effect on adrenal steroidogenesis. 0022

17-Hydroxycorticosterone

Plasma   No Effect   Physiological

Levodopa   No effect observed on injection into dogs. 1440

18-Hydroxycorticosterone

Plasma   No Effect   Physiological

Domperidone   After 10 mg intravenously in 8 healthy males. 3406

Metoclopramide   No change observed in normal individuals or in patients with primary aldosteronism after 10 mg drug. 3883

Plasma   Increase   Physiological

Amiloride   In 5 normal subjects given 75 mg daily for 7 d. 2440

Metoclopramide   Observed in normal subjects after 10 mg drug over 2 h; higher response in patients with primary aldosteronism. 3883   Increased by single i.v. bolus of 10 mg: probably related to basal activity of renin angiotensin aldosterone system. 2543

Spironolactone   In 5 normal subjects given 300 mg daily for 7 d. 2440

Urine   No Effect   Physiological

Metoprolol   In 15 patients with essential hypertension treated for 4 weeks, associated with reduction in sympathetic tone and reduced activity of renin-aldosterone system. 1209

6-β-Hydroxycortisol

Urine   No Effect   Physiological
Nitrazepam   No significant effect. *3466*

Urine   Increase   Physiological
Amobarbital   Significant increase (by approximately 50%). *3466*
Chlordiazepoxide   Observed in 2 out of 5 patients. *2682*
Diethylstilbestrol   Increased conversion of cortisol produced. *0022*
Estrogens   Alteration of steroid excretory pattern (long term). *0766*
Mitotane   Altered cortisol metabolism induced by drug. *2492*
Phenobarbital   Alteration of steroid excretory pattern (long term). *0766*   Increased conversion of cortisol produced. *0022*   Chronic treatment leads to substantially increased excretion compared with controls (approximately 9 fold increase). *3943*
Phenylbutazone   Long term change in steroid excretory pattern. *2035*
Phenytoin   Alters steroid metabolism. *3800*   Marked increase reflecting hepatic enzyme induction. *0394*   Manifestation of drug induced 6-beta-hydroxylase activity. *1020*
Phetharbital   Increased to over 400 μg/d. *3292*
Rifampin   Marked increase reflecting hepatic enzyme induction. *0394*
Smoking   Greater excretion in response to hepatic enzyme induction. *0394*

Hydroxycotinine

Urine   Positive   Physiological
Smoking   Normal nicotine metabolite 5-20 cigarettes/d. *0324*

16-α-Hydroxy-dehydroepiandrosterone

Urine   Decrease   Physiological
Ampicillin   Increase by 72% after 6 d administration in pregnant women. *3622*

15-α-Hydroxy-dehydroepiandrosterone Glucuronide

Urine   No Effect   Analytical
Ampicillin   No effect on GLC method. *3622*

16-α-Hydroxy-dehydroepiandrosterone Glucuronide

Urine   Decrease   Physiological
Ampicillin   Increase by 45% after 6 d in pregnant women. *3622*

18-Hydroxydeoxycorticosterone

Plasma   No Effect   Physiological
Spironolactone   In 5 normal subjects given 300 mg daily for 7 d. *2440*

Plasma   Increase   Physiological
Amiloride   In 5 normal subjects given 75 mg daily for 7 d. *2440*

Urine   No Effect   Physiological
Metoprolol   In 15 patients with essential hypertension treated for 4 weeks, associated with reduction in sympathetic tone and reduced activity of renin-aldosterone system. *1209*

Hydroxy-Diphenylhydantoin

Urine   Increase   Physiological
Phenytoin   Normal metabolite (absolute amount variable in individuals). *1269*

2-Hydroxyestrone

Urine   Increase   Physiological
L-Thyroxine   Metabolic effect of hormone administration. *0023*

11-Hydroxyetiocholanolone

Urine   Increase   Physiological
Corticotropin   Normal response to 2 d ACTH test. *1474*

α-Hydroxyglutarate

Urine   Increase   Physiological
Lithium   May be considerable effect. *2107*

5-Hydroxyindoleacetic Acid (5-HIAA)

Urine   No Effect   Analytical
Cresol   No effect 10 mg/dL method of Goldenberg. *1327*
3,4-Dimethylphenol   No effect with 5 mg/dL on method of Goldenberg. *1327*
Guaifenesin   No effect on TLC method of McGregor. *2367*   No effect *in vivo* dose on method of Goldenberg. *1327*
Homovanillic Acid   No effect at 10 mg/dL method of Goldenberg. *1327*   No significant decrease with procedure of Udenfriend. *1097*
4-Hydroxyacetanilide   No effect 10 mg/dL method of Goldenberg. *1327*
4-Hydroxyphenylacetic Acid   No effect 10 mg/dL method of Goldenberg. *1327*
5-Hydroxytryptamine   No effect with 5 mg/dL on method of Goldenberg. *1327*
Indoleacetic Acid   No effect *in vivo* dose on method of Goldenberg. *1327*
Mephenesin   No effect of *in vivo* dose on method of Goldenberg. *1327*
4-Methoxyphenol   No effect with 5 mg/dL on method of Goldenberg. *1327*
Oxazepam   Slight effect when added *in vitro* in method of Udenfriend. *0670*
Standing of Sample   Stable by screening test several days at room temperature. *1027*

Urine   No Effect   Physiological
Amantadine   In normals but increased in Parkinsonism. *1817*
Barbiturates   No effect with addiction or withdrawal. *3634*
Food   No significant change in hourly excretion rate after a meal. *0099*

Urine   Increase   Analytical
Acetaminophen   May cause false high colorimetric results. *0913*   Affects nitrosonaphthol procedures. *1487*
Acetanilid   False positive with nitrosonaphthol, no effect on quantitative test. *2812*
Bananas   Indoles in bananas. *0759*
Chlordiazepoxide   Slight effect observed *in vitro* in method of Udenfriend. *0670*
Coumaric Acid   May interfere with colorimetric methods. *0913*
Diazepam   With method of Udenfriend et al. due to reaction of nitrosonaphthol on the reactive fused benzene ring of the benzodiazepine nucleus. Effect unlikely to produce clinical misinterpretation. Effect also seen with N-desmethyldiazepam. *0670*
Ephedrine   May cause false increase in color. *0913*

5-Hydroxyindoleacetic Acid (5-HIAA)
(continued)

Urine   Increase   Analytical   (continued)

Flurazepam   Slight effect when added in vitro in method of Udenfriend. 0670

Guaifenesin   Interferes with nitrosonaphthol method. 1237 Affects quantitative method of Udenfriend. 0583

Medazepam   Slight effect when added in vitro in method of Udenfriend. 0670

Mephenesin   Metabolite reacts in quantitative Udenfriend procedure. 0583   Interferes with nitrosonaphthol reaction. 3879

Methocarbamol   Affects quantitative method of Udenfriend. 0583 Metabolite allegedly reacts with nitrosonaphthol. 1649

Naproxen   Due to metabolite desmethylnaproxen on spectrophotometric assays but compound is thermolabile and can be destroyed by heat. 3749

Nitrazepam   Slight effect when added in vitro in method of Udenfriend. 0670

Oxprenolol   Interferes with screening tests with nitrosonaphthol. 1803

Phenobarbital   May cause false high colorimetric values. 0913

Phentolamine   May cause falsely high colorimetric values. 0913

Urine   Increase   Physiological

Avocados   May contain large amounts of serotonin. 1487

Caffeine   Alleged effect. 1487

Eggplant   As result of high content in food. 1488

Fluorouracil   In patients with carcinoid, due to cell destruction. 1488

Melphalan   Probably due to tissue destruction if carcinoid. 2215

Methamphetamine   Single instance reported. 2142

Nicotine   Releases serotonin. 1176

Phenmetrazine   Reported effect. 2142

Pineapples   Rich in serotonin. 1167

Plums   Rich in serotonin. 1488

Pregnancy   Moderate increase observed. 0987

Rauwolfia   Result of release of 5-HT from brain, tissues. 1343

Reserpine   Release of 5-HT from brain and tissues. 2425

Sleep Deprivation   Immediate response associated with decreased ATP levels. 2032

Smoking   In habitues and normally nonsmokers. 1176

Walnuts   Rich in serotonin. 1488

Urine   Decrease   Analytical

Alkaline Urine   Unstable at room temperature at alkaline pH, ok if acidified. 0913

Aspirin   Affects procedure of Udenfriend (modest effect). 1097 Reported to affect fluorometric method. 1487

Chlorpromazine   Interferes with method of Goldenberg. 1328 Inhibits color development. 0913

Dihydroxyphenylacetic Acid   Significant decrease with procedure of Udenfriend. 1097

Formaldehyde   Inhibits color develop with nitrosonaphthol. 0583

Gentisic Acid   Significant decrease with procedure of Udenfriend. 1097

Homogentisic Acid   Significant effect procedure of Udenfriend Significant effect procedure of Mustala. 1097

Imipramine   May inhibit color development in reaction. 0913

Keto Acids   Color formation with nitrosonaphthol inhibited. 0583

Levodopa   Moderate effect on procedure of Udenfriend. 1097

Methenamine   Slight false negative effect with nitrosonaphthol. 3305

Phenothiazines   False decrease if nitrosonaphthol used. 3879

Prochlorperazine   Interference with nitrosonaphthol methods. 3065

Promazine   Interferes with method of Goldenberg. 1328

Promethazine   Interference with nitrosonaphthol methods. 3065

Proteinuria   If greater than 2 g/L cause low and variable results. 0913

Urine   Decrease   Physiological

Aging   Lower in elderly compared with 20-60 y. 1176

Chlorophenylalanine   Inhibits biosynthesis of serotonin. 0754

Corticotropin   Mechanism not described. 1488

Ethanol   Serotonin metabolism diverted to 5 hydroxy tryptophol. 0833

Heparin   Reduction in single case carcinoid syndrome. 1742

Hydrazine Derivatives   May potentiate action of drugs on CNS. 2220

Imipramine   May decrease up to 50%: decrease cell permeability to 5-HT. 1520

Isocarboxazid   Due to inhibition of conversion of 5-HT to 5-HIAA. 1488

Isoniazid   Causes decarboxylase inhibition with reduced 5-HT. 0832

Levodopa   In Parkinson's disease ?increased tryptophan pyrrolase activity. 0509

MAO Inhibitors   Inhibition of conversion of 5-HT to 5-HIAA. 1937

Methyldopa   Inhibition of aromatic amino acid decarboxylation. 0832

Streptozocin   If carcinoid treated. 1100

CSF   No Effect   Physiological

Amphetamine   No change in CSF with psychotic dose. 1825

Bupropion   Insignificant effect in approximately 40 patients with depression or Alzheimer's disease after chronic treatment. 2991

Chlorophenylalanine   After 1 g/d for 4 d or after single dose. 3342

Tetrabenazine   No significant effect in Huntington's chorea. 2379

Thiopropazate   No significant effect in Huntington's chorea. 2379

CSF   Increase   Physiological

Probenecid   Approximately 4 fold at 9 h, 5 fold at 18 h. 1351

CSF   Decrease   Physiological

Anticonvulsants   Reduction to 18.3 ng/mL from 25.1 ng/mL in lumbar CSF of treated epileptics. 3926

Clorgiline   Significant reduction in 43 patients with depression or Alzheimer's disease chronically treated with drug. 2991

Deprenyl   In 43 patients with depression or Alzheimer's disease chronically treated with drug. 2991

Desipramine   In 43 patients with depression or Alzheimer's disease chronically treated with drug. 2991

Levodopa   Inhibition of 5-hydroxy-tryptophan hydroxylase. 2725

Nortriptyline   Increase by 4.8 ng/mL in depressed patients. 0167

Zimeldine   Significant effect in 43 chronically treated patients with depression or Alzheimer's disease. 2991

3-Hydroxykynurenine

Urine   No Effect   Physiological

Smoking   Probably no significant effect with tryptophan load. 3183

Urine   Increase   Physiological

Deoxypyridoxine   After 2 g tryptophan load. 2042

Estrogens   Induce tryptophan pyrrolase. 3045

Glucocorticoids   Induce tryptophan pyrrolase. 3045

Hydrocortisone   Causes induction of tryptophan pyrrolase. 3047

Isoniazid   Induces pyridoxal PO_4 deficiency. 3045

Oral Contraceptives   After 2 g tryptophan load. 3050   Estrogen component induces tryptophan pyrrolase. 3045

Pregnancy   Causes induction of tryptophan pyrrolase. 3045

Urine   Decrease   Physiological

Anabolic Steroids   Effect of synthetic androgens given to males. 3045

Androgens   Effect of synthetic androgens given to males. 3045

Hydroxy-Methoxymandelic Acid

Urine   Increase   Analytical
Levodopa   In a patient given Sinemet® (levodopa/carbidopa) using Pisano method. Note high blank. *0698*

Hydroxynalidixic Acid

Urine   Increase   Physiological
Nalidixic Acid   Major metabolites with glucuronide conjugates. *1678*

17-Hydroxypregnenolone

Plasma   Decrease   Physiological
Testosterone   In power athletes with 26 weeks on steroid self administration. *3088*

16-α-Hydroxyprogesterone

Plasma   Increase   Analytical
Desoxycorticosterone   Up to 25% cross reactivity. *0008*
20-α-Di-Hydroxyprogesterone   Up to 25% cross reactivity. *0008*
Progesterone   Up to 25% cross reactivity. *0008*

17-Hydroxyprogesterone

Plasma   No Effect   Physiological
Cyclosporine   In 16 patients who developed hypertrichosis but also given added cortisone. *3190*
Etomidate   No significant difference in 7 men given drug in induction dose versus thiopentone. *0063*

Plasma   Increase   Physiological
Ketoconazole   Observed in normal men due to blockade of 17,20-desmolase by drug, with inconsistent effect on serum estradiol. *1299*

Plasma   Decrease   Physiological
Testosterone   In power athletes with 26 weeks on steroid self administration. *3088*

6-Hydroxyprogesterone Metabolite

Urine   No Effect   Analytical
Ampicillin   No effect on method of James. *3622*

Urine   Decrease   Physiological
Ampicillin   Increase by 29% after 6 d in pregnant women. *3622*

Hydroxyproline

Urine   No Effect   Analytical
Anthranilic Acid   At 53 mg/L on method of Goverde. *1365*
Aspartic Acid   At 300 mg/L on method of Seymour. *3262*
Citruplexina   At 53 mg/L on method of Goverde. *1365*
Glucose   At 10 g/L on method of Goverde. *1365*
Histidine   At 46 mg/L on method of Goverde. *1365*
5-Hydroxyindoleacetic Acid   At 100 mg/L on method of Seymour. *3262*
Indican   At 500 mg/L on method of Seymour. *3262*
Mannitol   At 10 g/L on method of Goverde. *1365*
Storage of Sample   At 30° for 5 d if pH reduced to 1-2. *1712* At 0° under toluene. *0561*
Tryptophan   At 400 mg/L on method of Seymour. *3262*   At 68 mg/L on method of Goverde. *1365*

Tyrosine   At 45 mg/L on method of Goverde. *1365*
Urea   At 10 g/L on method of Goverde. *1365*

Urine   No Effect   Physiological
Propylthiouracil   In hyperthyroid patients and normal controls. *0334*
Timolol   No effect in hyperthyroid patients. *0334*

Urine   Increase   Physiological
Aminopropionitrile   Observed in experimental studies. *3644*
Bed Rest   Maximal during second month bed rest. *1688*
Corticosteroids   Stimulate growth in children. *2427*
Growth Hormone   Metabolic effect. *1343*
Parathyroid Hormone   Due to catabolic action. *3644*
Phenobarbital   Increased excretion occurred early in epileptic children without obvious bone changes suggestive of rickets. *2162*
Phenytoin   In epileptic children in comparison with controls. *2030*
Pregnancy   In last trimester, usually maximum 6-8 d postpartum. *2209*
Thyroid   Due to catabolic action. *3644*
Tolbutamide   Associated with turnover of bone organic matrix. *3458*
Vitamin D   In vitamin-sensitive rickets. *3644*

Urine   Decrease   Physiological
Ambulation   Falls below normal after high of bed rest. *0927*
Antineoplastic Agents   Catabolic action of cytostatics. *3644*
Ascorbic Acid   Slight effect in osteogenesis imperfecta. *3644*
Aspirin   At 100 mg/kg in children has significant effect. *2163*
Calcitonin   Due to anticatabolic action. *3644*
Corticosteroids   Alleged normal metabolic effect. *3644*
Diphosphonate   Normal response in Paget's disease. *3644*
Estradiol   Due to anticatabolic action. *3644*
Estriol   Due to anticatabolic action. *3644*
Ethinyl Estradiol   In post-menopausal women with 50 μg/d for 14 d. *3474*   significant effect in postmenopausal women: less marked in perimenopausal women. *2311*
Gallium Nitrate   In two patients with hypercalcemia due to parathyroid carcinoma. *3764*
Glucocorticoids   In rheumatoids under treatment. *2919*
Mithramycin   Inhibition of bone resorption of calcium. *3349*
Propranolol   In hyperthyroid patients possible effect due to membrane-stabilizing property of drug. *0334*

5-Hydroxytryptamine Glucuronide

Urine   Increase   Physiological
MAO Inhibitors   Due to inhibition of conversion of 5-HT to 5-HIAA. *1343*

5-Hydroxytryptamine (Serotonin)

Plasma   Increase   Physiological
Diurnal Variation   Maximum observed about noon. *2222*
MAO Inhibitors   Effect observed after single large dose. *1343*

Plasma   Decrease   Physiological
Methysergide   Noted in migraine subjects. *0774*
Reserpine   Observed normal response to therapy. *2427*

Blood   No Effect   Physiological
Food   No significant increase for 1 h after a meal. *0099*

Urine   No Effect   Analytical
Storage of Sample   At room temperature for 45 d if acidified. *1180*

Urine   Increase   Analytical
Plantains   5-hydroxy tryptamine contained in plants. *1237*

5-Hydroxytryptophol

Urine   Increase   Physiological
Ethanol   Raises excretion from 2% to 50%. *1176*

25-Hydroxy Vitamin D₃

Serum   No Effect   Physiological
Aluminum Hydroxide   Observed in one patient secondary to ingestion of large amounts of compound for long time. *1313*
Anticonvulsants   No significant difference, although slightly less, in men and women on long-term treatment. *0257*
Bendrofluazide   In 19 healthy early menopausal women given 5 mg/d with calcium supplement due to primary effect on renal tubules and secondary change on vitamin D metabolism. *2989*
Chloroquine   Possibly inhibits conversion of 25-hydroxyvitamin D to 1,25-dihydroxyvitamin D. *2639*
Ketoconazole   After administration to normal volunteers. *3390*
In 9 healthy men given up to 1200 mg/d for 1 week. *1300*
Phenytoin   In 10 women receiving anticonvulsants mainly phenytoin versus 10 controls, drug taken on average for 15 y. *0257*
Probenecid   In 8 healthy young men. *1318*

Serum   Decrease   Physiological
Anticonvulsants   Probably due to increased metabolism of vitamin D. *1451*   Marked effect in 146 epileptics with long-term treatment. *2007*
Carbamazepine   Significantly lower (11.1 ng/mL vs 17.6 ng/mL) in 21 patients treated with drug only for average of 40 mo. *1625*
Corticosteroids   Observed effect in some children. *3023*
Ethanol   Of chronic alcoholics 58% had concentration below normal. *0367*
Ethinyl Estradiol   Small but insignificant effect during and after in 7 healthy postmenopausal women treated for 12 weeks. *3467*
Paramethadione   Theoretical effect of type of drug. *1452*
Phenobarbital   Positive correlation with calcium level. *1452*
Observed effect in some children. *3023*
Phenytoin   Positive correlation with calcium level. *1452*
Reduced by approximately 50% in long term treatment in children: most marked with combination therapy. *1449*   Observed effect in some children. *3023*
Rifampin   Secondary to induction of hepatic microsomal enzymes: decrease of 70% with short course in 8 healthy men. *0479*
Stanozolol   From 64 to 54.5 nmol/L after 1 week in 9 healthy men given 10 mg/d for 14 d. *3354*
Tolbutamide   Associated with turnover of bone organic matrix. *3458*
Trimethadione   Theoretical effect of type of drug. *1452*

Plasma   Decrease   Physiological
Ethanol   Associated with reduced exposure to sunshine plus inadequate supply of vitamin D. Compound formed in liver where vitamin D binding protein also synthesized. Finding sometimes reported in chronic alcoholics. Normal concentration observed typically in well nourished alcoholics. *1246*
Season   Significant decrease both sexes in winter. *3429*

3-Hydroxyxanthurenic Acid

Urine   Increase   Physiological
Oral Contraceptives   After 2 g tryptophan load. *3050*

Hypoxanthine

Serum   Increase   Physiological
Allopurinol   Inhibits xanthine oxidase (slight increase only). *0071*

Urine   Increase   Physiological
Allopurinol   Inhibits xanthine oxidase. *0071*

Icteric Index

Serum   Increase   Analytical
Carotene   Interfering background color. *1237*
Carrots   Color may be misinterpreted. *1237*
Hemoglobin   Contributes background color. *0582*
Hemolysis   Presence of hemoglobin produces red color. *1565*
Lipemia   Turbidity of serum. *1237*
Sweet Potatoes   Due to color. *1238*

Serum   Increase   Physiological
Acetaminophen   May cause hepatic toxicity. *2313*
Novobiocin   Competition for conjugation mechanism. *2425*
Phosphorus   Due to nephrotoxicity. *1302*
Quinacrine   Hemolysis may occur with G-6-PD deficiency. *1237*

IL-2 Receptor Expression

Monocytes   No Effect   Physiological
Methimazole   No effect of the drug on mitogen stimulated mononuclear cells from normal individuals. *3778*

Imidazoleacetic Acid

Urine   Decrease   Physiological
Aminoguanidine   Inhibits diamine oxidase. *1382*

Imidazolepyruvic Acid

Urine   Increase   Physiological
Oral Contraceptives   Mechanism not yet established. *3681*

Imipramine

Serum   Increase   Analytical
Benzaprine   In liquid-chromatographic method described has similar retention time. *3195*

Serum   Increase   Physiological
Cimetidine   Although peak concentration not different: clearance significantly reduced so higher concentration overall. *1555*

Blood   Increase   Physiological
Imipramine   After 150-300 mg/d concentration = 0.1-0.6 mg/L. *2348*

Immune Complexes

Serum   No Effect   Physiological
Methyldopa   In a study of 9 hypertensives. *1909*

Immunoglobulin IgA

Serum   No Effect   Analytical
Hemolysis   On AutoAnalyzer immunological method. *2297*
Storage of Sample   Stable at 4 and 30° for 7 d. *0573*

Serum   No Effect   Physiological
Anticonvulsants   In 8 epileptic patients receiving phenobarbital with carbamazepine or phenytoin. *2670*
Muscular Exercise   No observed effect 15 minutes or 1 d after. *1491*

Serum   Increase   Physiological
Asparaginase   Increased hepatic synthesis. *2657*
Ethanol   In alcoholics continuing to drink. *2065*

Methyldopa Observed in one patient with hepatocellular damage of moderate severity due to sensitization by drug. *0874*
Muscular Exercise Approximately 14% increase immediately after. *2846*
Nitrofurantoin Chronic active hepatitis, much more common in women than men, but still rare. *2444* Chronic active hepatitis much more common in women than men, but still rare. *2444*
Oxyphenisatin As component of hypersensitivity response. *1297*

Serum   Decrease   Physiological
Aurothioglucose Reduced only at 3 mo in 25 patients with rheumatoid arthritis. *3680*
Benzene Significant effect in people occupationally exposed. *3370*
Carbamazepine Significant effect within 1 mo, remained low over next 30 mo. *1282*
Dextran Complex formation or increased consumption. *3346*
Ethanol In alcoholics after 1 y abstinence. *2065*
Glucocorticoids Less marked effect than with IgG. *0410*
Gold Substantial lowering at 3 mo in 25 patients with rheumatoid arthritis treated with gold. *3680*
Methylprednisolone Significant effect in 43% individuals. *0552*
Oral Contraceptives Estrogen effect. *3505*
Penicillamine But not significantly in 21 patients with rheumatoid arthritis treated for 48 weeks given 250 to 750 mg daily. *2345*
Phenytoin Observed in 21% (mechanism not elucidated). *3400* Further decrease below low value in epileptics. *0228* Further decrease from typical low values of epilepsy with drug treatment. *0228* Further decrease below low value in epileptics. *0228*
Pregnancy Also absolute decrease in late pregnancy. *3472*
L-Thyroxine In all 5 children with infantile hypothyroidism soon after start of treatment in 4 concentration returned to normal. *3232*
Timegadine But not significantly in 23 patients with rheumatoid arthritis given 250 to 750 mg/d for 48 weeks. *2345*
Toluene Significant effect in people occupationally exposed. *3370*
Xylene Significant effect in people occupationally exposed. *3370*

Immunoglobulin IgD

Serum   No Effect   Analytical
Storage of Sample Stable at 4 and 30° for 7 d. *0573*

Serum   No Effect   Physiological
Phenytoin In about 118 treated epileptics. *0228* No effect of treatment of epileptics. *0228*

Serum   Increase   Physiological
Pregnancy At labor 0.085 mg/mL versus 0.033 in controls. *1429*

Immunoglobulin IgE

Serum   Increase   Physiological
Penicillin G Present in some patients with drug-induced acute interstitial nephritis. *0145*
Ragweed In allergic patients increase with pollen season. *3932*

Serum   Decrease   Physiological
Phenytoin In about 118 treated epileptics. *0228* Significant decreases in patients with different types of epilepsy. *0228* In about 118 treated epileptics. *0228*

Immunoglobulin IgG

Serum   No Effect   Analytical
Hemolysis On AutoAnalyzer immunological method. *2297*
Storage of Sample Stable at 4 and 30° for 7 d. *0573*

Serum   No Effect   Physiological
Anticonvulsants In 8 epileptic patients receiving phenobarbital with carbamazepine or phenytoin. *2670*
Auranofin No effect of 6 mg/d for 1 y. *0617*
Carbamazepine No effect regardless of duration of treatment. *1282*
Muscular Exercise No observed effect 15 minutes or 1 d after. *1491*
Prednisone Shortens half life but increased synthesis (no net change). *1402*
L-Thyroxine No significant effect in children with infantile hypothyroidism. *3232*

Serum   Increase   Physiological
Asparaginase Increased hepatic synthesis. *2657*
Auranofin 26.5% reduction after 12 weeks treatment in patients with rheumatoid arthritis. *0617*
Methadone Commonly seen in response to treatment. *0781*
Methyldopa Observed in one patient with hepatocellular damage of moderate severity due to sensitization by drug. *0874*
Muscular Exercise Approximately 10% increase immediately after. *2846*
Narcotics Often elevated in addicts (?liver problem). *2012*
Nitrofurantoin Chronic active hepatitis, much more common in women than men, but still rare. *2444* Chronic active hepatitis, much more common in women than men, but still rare. *2444*
Oxyphenisatin Observed with chronic active hepatitis induced by drug. *0903*

Serum   Decrease   Physiological
Aurothioglucose Reduced only at 12 mo in 25 patients with rheumatoid arthritis. *3680*
Benzene Significant effect in people occupationally exposed. *3370*
Dextran Complex formation or increased consumption. *3346*
Diazoxide Effect may be persistent ?mechanism. *0124*
Glucocorticoids May be reduced by 50% for up to 3 mo after 1 week treatment. *0410*
Gold Substantial lowering at 3 mo in 25 patients with rheumatoid arthritis treated with gold. *3680*
Methylprednisolone After 96 mg/d for 5 d. *0552*
Penicillamine But not significantly in 21 patients with rheumatoid arthritis treated for 48 weeks given 250 to 750 mg daily. *2345*
Phenytoin Significant effect due to immunosuppressive action. *2257* Minor decrease noted with treatment of epilepsy. *0228* Minor effect in about 118 treated epileptics. *0228*
Pregnancy Also absolute decrease in late pregnancy. *3472*
Timegadine But not significantly in 23 patients with rheumatoid arthritis given 250 to 750 mg/d for 48 weeks. *2345*
Toluene Significant effect in people occupationally exposed. *3370*
Xylene Significant effect in people occupationally exposed. *3370*

Urine   Increase   Physiological
Diatrizoic acid Marked increased excretion (20 to 60 times baseline) day of bilateral renal arteriography in 23 patients with hypertension, reverted to normal on following day. *2594*
Radiographic Agents Mean increase from 6.1 mg/d to 206.3 mg/d in 37 patients day following arteriography. *2594*

Immunoglobulin IgM

Serum   No Effect   Analytical
Storage of Sample Stable at 4 and 30° for 7 d. *0573*

Serum   No Effect   Physiological
Anticonvulsants In 8 epileptic patients receiving phenobarbital with carbamazepine or phenytoin. *2670*
Benzene No effect in people occupationally exposed. *3370*
Glucocorticoids No effect seen, although other immunoglobulins affected. *0410*

Immunoglobulin IgM *(continued)*

Serum   No Effect   Physiological *(continued)*
Muscular Exercise   No observed effect 15 minutes or 1 d after. *1491*   No effect of exercise observed. *2846*
L-Thyroxine   No significant effect in children with infantile hypothyroidism. *3232*
Toluene   No effect in people occupationally exposed. *3370*
Xylene   No effect in people occupationally exposed. *3370*

Serum   Increase   Physiological
Asparaginase   Increased hepatic synthesis. *2657*
Chlorpromazine   Significant correlation with dose and duration of treatment in schizophrenic patients. *3937*
Methyldopa   Observed in one patient with hepatocellular damage of moderate severity due to sensitization by drug. *0874*
Narcotics   Frequently elevated in addicts (?liver problem). *2012*
Nitrofurantoin   Chronic active hepatitis, much more common in women than men, but still rare. *2444*   Chronic active hepatitis, much more common in women than men, but still rare. *2444*

Serum   Decrease   Analytical
Storage of Sample   Significant effect at -20° for 50 d (though starts immediately). *1367*

Serum   Decrease   Physiological
Aurothioglucose   Substantial reduction at 3 mo in 25 patients with rheumatoid arthritis. *3680*
Azathioprine   Reduced when biliary cirrhosis treated. *3063*
Carbamazepine   Significant effect within 1 mo, slight rebound with continuation of treatment for 3 mo. *1282*
Dextran   Complex formation or increased consumption. *3346*
Gold   Substantial lowering at 12 mo in 25 patients with rheumatoid arthritis treated with gold. *3680*
Methylprednisolone   Noted in 14% individuals. *0552*
Penicillamine   But not significantly in 21 patients with rheumatoid arthritis treated for 48 weeks given 250 to 750 mg daily. *2345*
Phenytoin   Minor effect in about 118 treated epileptics. *0228*   Minor decrease noted with treatment of epilepsy. *0228*
Pregnancy   Also absolute decrease in late pregnancy. *3472*
Timegadine   But not significantly in 23 patients with rheumatoid arthritis given 250 to 750 mg/d for 48 weeks. *2345*

Immunoglobulin Light Chains

Urine   Increase   Physiological
Gentamicin   Marked effect especially with treatment for more than 12 d. *2593*
Sisomicin   Seen in all patients given drug in therapeutic amounts for 2 weeks. *2593*

Immunoglobulins

Serum   No Effect   Physiological
Estrogens   No metabolic effect. *0227*

Serum   Increase   Analytical
Thimerosal   As preservative increases immunodiffusion precipitin rings. *0573*

Serum   Increase   Physiological
Oral Contraceptives   Metabolic changes in liver synthesis. *2189*

Serum   Decrease   Analytical
Thimerosal   Affects immunodiffusion if added as bactericidal. *2579*

Serum   Decrease   Physiological
Estrogens   Altered liver metabolism. *3332*
Oral Contraceptives   Estrogen effect. *3332*

Valproic Acid   Deficiency occurred in 29% of 41 epileptic patients on anticonvulsant therapy. users of drug in general had lower concentrations than in controls. *1827*

Indican

Urine   No Effect   Analytical
Thymol   No effect on Obermayer and Jaffé methods. *2981*

Urine   Increase   Physiological
Tryptophan   Direct correlation in normals with amount ingested. *3606*

Urine   Decrease   Analytical
Formaldehyde   Prevents Obermeyer test reaction. *0459*

Indocyanine Green

Serum   No Effect   Analytical
Bilirubin   No effect at high plasma concentrations. *2398*

Serum   Increase   Physiological
Cyclopropane   Hepatic extraction impaired after injection. *1343*
Pregnancy   Observed in normals greater in pre-eclampsia. *2336*
Rifampin   Delayed elimination following i.v. injection. *2427*

Serum   Decrease   Analytical
Bisulfite   As contaminant of heparin may reduce peak. *0071*

Serum   Decrease   Physiological
Anticonvulsants   Mechanism not yet established. *2398*
Haloperidol   Mechanism not yet established. *2398*
Heroin   Observed in small series with normal liver function test. *2398*
Meperidine   Observed in small series, normal liver function test. *2398*
Methadone   Observed in small series, normal liver function test. *2398*
Morphine   Observed in small series, normal liver function test. *2398*
Nitrofurantoin   Mechanism not yet established. *2398*
Opium Alkaloids   Mechanism not yet established. *2398*
Phenobarbital   Mechanism not yet established. *2398*
Phenylbutazone   Mechanism not yet established. *2398*

Indocyanine Green Clearance

Serum   Decrease   Physiological
Methotrexate   In psoriatic patients receiving drug in comparison with topically treated and controls. *0364*
Vinyl Chloride   Markedly different in individuals with chemical liver injury compared with people with nonchemical liver disease or normals. *2190*

Indoleacetic Acid

CSF   No Effect   Physiological
Anticonvulsants   Insignificant reduction to 2.60 ng/mL in lumbar CSF from 3.74 ng/mL in untreated epileptics. *3926*

CSF   Decrease   Physiological
Nortriptyline   Increase by 2 ng/mL in depressed patients. *0167*

Indolylacryloylglycine

Urine   Increase   Physiological
Light   Probably due to high intensity sunlight. *2302*
Season   Large increase observed with summer. *2302*

Indomethacin

Serum   No Effect   Analytical

Acetaminophen   No effect on HPLC method of Roberts and Smith. *3002*

Allopurinol   No effect on HPLC method of Roberts and Smith. *3002*

Amiodarone   No effect on HPLC method of Roberts and Smith. *3002*

Aspirin   No effect on HPLC method of Roberts and Smith. *3002*

Atenolol   No effect on HPLC method of Roberts and Smith. *3002*

Caffeine   No effect on HPLC method of Roberts and Smith. *3002*

Carbamazepine   No effect on HPLC method of Roberts and Smith. *3002*

Chlormethiazole   No effect on HPLC method of Roberts and Smith. *3002*

Digoxin   No effect on HPLC method of Roberts and Smith. *3002*

Erythromycin   No effect on HPLC method of Roberts and Smith. *3002*

Ethosuximide   No effect on HPLC method of Roberts and Smith. *3002*

Furosemide   No effect on HPLC method of Roberts and Smith. *3002*

Glibenclamide   No effect on HPLC method of Roberts and Smith. *3002*

Lorazepam   No effect on HPLC method of Roberts and Smith. *3002*

Methyldopa   No effect on HPLC method of Roberts and Smith. *3002*

Metronidazole   No effect on HPLC method of Roberts and Smith. *3002*

Nifedipine   No effect on HPLC method of Roberts and Smith. *3002*

Phenobarbital   No effect on HPLC method of Roberts and Smith. *3002*

Phenytoin   No effect on HPLC method of Roberts and Smith. *3002*

Primidone   No effect on HPLC method of Roberts and Smith. *3002*

Theophylline   No effect on HPLC method of Roberts and Smith. *3002*

Valproic Acid   No effect on HPLC method of Roberts and Smith. *3002*

Serum   No Effect   Physiological

Isosorbide   No effect on HPLC method of Roberts and Smith. *3002*

Serum   Increase   Physiological

Probenecid   Inhibits renal tubular secretion. *1487*

INH-Ketoglutarate

Urine   Increase   Physiological

Isoniazid   Major metabolite in urine. *3093*

INH-Pyruvate

Urine   Increase   Physiological

Isoniazid   Major metabolite in urine. *3093*

Insulin

Serum   Increase   Physiological

Desoximetasone   2 to 3 fold increase in 5 patients with psoriasis and topical application of glucocorticoid. *1240*

Plasma   No Effect   Analytical

EDTA   Results same as when serum used. *0112*

Storage of Sample   No change at room temperature for 4 h. *1098*

Plasma   No Effect   Physiological

Atenolol   No change in basal concentration or after glucose in 14 hypertensives. *0836*   In 18 patients with mild essential hypertension treated with chlorothiazide concomitantly over 4 weeks. *1035*   Compared with controls during glucose tolerance test. *1977*

Bendrofluazide   After 12 mo treatment of 53 previously untreated hypertensives. *0315*

Captopril   No decrease during treatment nor effect on glucose tolerance. *2902*

Chlorthalidone   In 22 premenopausal and 18 postmenopausal women given 100 mg/d for 6 weeks. *0402*

Clonidine   No effect observed on normal response to moderate or heavy exercise. *1797*

Danazol   But rose to higher extent than in controls when given oral glucose tolerance test. *3839*

Enflurane   No effect during anesthesia, slight increase after. *2699*

Ethanol   No change in fasting or postprandial. *1288*

Ethinyl Estradiol   In group of women taking drug as anti-androgen therapy. *3239*

Ethynodiol   With doses of 0.25, 0.35, 0.5 mg/d. *1335*

Fenfluramine   Variable response (some increased, some decreased). *0377*

Galactose   No effect observed in peripheral blood. *0310*

Glucocorticoids   Intravenous injection failed to modify concentration. *2296*

Guanabenz   In 30 patients treated with twice daily doses of from 4 to 32 mg. *1018*

Hydrochlorothiazide   In 15 hypertensives treated with 50 mg twice daily. *1018*

Metoprolol   In 20 hypertensive diabetic patients. *3900*   Insignificant change although plasma concentration of glucose changed. *1410*

Muscular Exercise   No effect if of short or moderate duration. *1171*

Naloxone   With infusion in normal and obese subjects. *0176*   In 6 healthy volunteers after ingestion of test meal and intravenous administration of drug. *0626*

Naltrexone   No effect of long-term administration to obese individuals. *0176*

Nifedipine   In 11 patients with 80 mg daily for 6 weeks. *3702*

Omeprazole   No effect observed in 8 volunteers given 30 mg/d for 28 d. *2516*

Ouabain   No effect with i.v. infusion for 1 h. *3166*

Pancreozymin   No effect of 1 unit/kg body weight. *2676*

Phenytoin   No significant difference from controls during glucose tolerance test. *0606*

Pindolol   In 18 patients with mild essential hypertension treated with chlorothiazide. *1035*   No significant change in concentration during glucose tolerance test after drug. *2117*

Pirenzepine   Treatment for 1 week had no effect on basal concentrations. *3933*

Piretanide   In 12 male patients with mild hypertension (6 mg bid). *1503*

Pizotyline   No effect after 2 mg i.v. *3411*

Prednisolone   Intravenous injection failed to modify concentration. *2296*

Propranolol   After 12 mo treatment of 53 previously untreated hypertensives. *0315*   In hyperthyroid patients given 40 mg every 6 h in 6 patients. *0335*   In 18 patients with mild essential hypertension treated with chlorothiazide concomitantly over 4 weeks. *1035*   In 5 hyperthyroid patients given 10 mg every 8 h for 4 d. *0335*   No significant effect although drug affects glucose concentration. *1410*   In 20 hypertensive diabetic patients. *3900*   No change in basal concentrations or after glucose in 16 hypertensives. *0836*

Propylthiouracil   In 5 hyperthyroid individuals given 10 mg every 8 h for 4 d. *0335*

Recumbency   Change of posture had no effect. *3539*

Site of Collection   Independent of site. *3639*

Sultopride   When given 300-600 mg/d for 5 weeks in 5 schizophrenic women. *2465*

Insulin (continued)

Plasma   No Effect   Physiological (continued)

Terbutaline   In 6 normal men treated with therapeutic amounts for 2 weeks. 3177

Theophylline   No effect in response to i.v. infusion of aminophylline. 0608

Thyrotropin-Releasing Hormone (TRH)   After single dose of 1.0 mg synthetic TRH. 3831

Timolol   In 5 hyperthyroid patients given 10 mg every 8 h for 4 d. 0335

Tripamide   No effect in hypertensives with or without diabetes. 0711

Verapamil   No effect observed even in individuals fasted for 36 h when infused at rate of 5 mg/h for 3 h. 0103

Xylitol   Although increased in portal venous blood. 0310

Plasma   Increase   Analytical

Heparin   Spuriously high values reported for immunoassay. 0191

Plasma   Increase   Physiological

Acetoacetate   Maximum observed at 40 minutes after loading dose. 1811

Acetohexamide   Usual effect observed (at 2 mo). 0301

Alanine   Marked increase after fasting in obese. 1263

Albuterol   Significant dose-related effect after i.v. infusion of therapeutic doses in 4 healthy male volunteers. 2807

Amiloride   Marked effect of i.v. injection in rat. 0190

Amino Acids   ?due to action on beta cells. 1775   Max higher if intraduodenal than if i.v. 2918

Arginine   Slight effect and metabolism by tissues. 1071   Large effect (by 60 uU/mL) after prednisolone. 2296

Aspirin   Increased in response to decreased serum glucose. 2433

Calcium Gluconate   Marked effect noted in newborns. 1550

Cannabis   Responsible for hypoglycemia. 0912

Chlorpropamide   Effect observed during tolerance test. 2236 Observed in most patients (especially if low initially). 0301

Cyclic AMP   Hormonal action. 1102

Cyproterone   Fasting concentration when given with ethinyl estradiol: combination causes insulin resistance. 3239

Desoxycorticosterone   2 to 3 fold increase in 5 patients with psoriasis and topical application of glucocorticoid. 1240

Ethanol   In response to 1.5 g/kg at night significantly increased values measured next morning. morning. 3547

Glibonuride   Immediate sharp increase, lasting for 20 minutes. 1526

Glipizide   Stimulated beta cells in pancreas. 2735

Glisoxepide   Immediate sharp increase lasting for 20 minutes. 1526

Glucagon   Stimulates beta cells of pancreas. 3128

Glucose   50 g orally caused increase from 28 to 39 pmol/L. 2939 Marked rise immediately after oral or i.v. administration. 3909 Marked effect (maximum at 60 minutes) for 2 h. 0310   Rises in parallel with glucose in normals or in noninsulin dependent diabetics. 0476

Glyburide   Significant increase reported. 2216

Growth Hormone   Small postabsorptive rise. 0138

Hemodialysis   Normal physiological response to increased glucose. 2547

Ibopamine   Peaked after 45 minutes in all subjects: normalized in 90 minutes. 3398

Insulin   In 12 healthy male volunteers significant increase at 30 to 60 minutes following injection. 2322

Leucine   Facilitates uptake of amino acids by tissues. 1071

Levodopa   During therapy of Parkinsonism. 1860

Lysine   Slight effect, AIDS metabolism of amino acids. 1071

Medroxyprogesterone   Metabolic effect (?glucocorticoid). 3414

Megestrol   In 18 postmenopausal women with breast cancer. 0056

Methionine   Slight effect, AIDS metabolism of amino acids. 1071

Niacin   ?response to increased glucose output. 1253

Norethindrone   During glucose tolerance test after 6 mo treat compared with control. 3413

Obesity   Increase observed in fasting and after glucose. 1015

Oral Contraceptives   In women in whom insulin was initially normal. 3412

Pancreozymin   Intravenous infusion causes increase. 1775

Phenylalanine   Slight effect, AIDS metabolism of amino acids. 1071

Prazosin   With 2 mg in 12 hypertensives (6 with normal and 6 with abnormal glucose tolerance). 0225

Prednisolone   After 4 d of oral therapy. 2296

Protein   Stimulation of beta cells by amino acids. 1775

Quinine   Intravenous drug in normal volunteers increased concentration from 8.9 to 17.1 mU/L. 3814

Rifampin   Increased rate of secretion after oral administration of 100 g glucose. 3534

Rimeterol   Dose related significant change in 4 healthy men given therapeutic i.v. dose. 2807

Sandfly Fever   2 fold at 30 minutes, 3 fold at 60 minutes. 2929

Secretin   More than 2 fold increased with 15 U pulse. 2137   Only from 3 to 6 minutes after i.v. injection. 2676

Spironolactone   Average rose from 16 to 29 mU/L in 15 primary hypertension patients at 6 mo after 100 mg drug daily. 1072

Streptozocin   ?due to release from damaged beta cells. 2532

Sucrose   By about 30% on high sucrose diet. 3931

Terbutaline   Observed following infusion of 0.25 mg or 5 mg 3 times/d on first day of treatment. 0288

Tetracosactrin   From 5.2 to 13.1 mU/L healthy volunteers given 1 mg intramuscularly for up to 60 h. 1813

Tetragastrin   Slight rise but not significant. 2676

Tolazamide   Usual effect observed. 0301

Tolbutamide   Marked rise associated with hypoglycemia. 1152 Slight effect max in 15 minutes. 0310

Trichlormethiazide   Significant increase versus controls in 6 patients given 75 g glucose orally at 30 minutes (79 vs 54 mU/mL) and 60 minutes (99 vs 76 mU/mL). 2550

Valine   Slight effect, AIDS metabolism of amino acids. 1071

Verapamil   Significant change during tolerance test when drug and glucose coadministered. 1115

Xylitol   ?same mechanism as glucose to release in dogs. 2048

Plasma   Decrease   Analytical

Hemolysis   Causes destruction of hormone for radioimmunoassay. 0478

Heparin   Effect in heparinized plasma and serum. 2684

Oxalate   Compared with lithium heparin plasma or serum. 3217

Serum   Consistently higher in plasma than serum. 0112

Storage of Sample   Increase by 75% (serum, plasma) at -20° for 28 mo. 1098

Plasma   Decrease   Physiological

Acetohexamide   Effect observed after 3 mo therapy. 0301

Asparaginase   ?due to decreased production with decreased protein synthesis. 0578

Calcitonin   Response to glucose significantly reduced. 2738

Chlorpropamide   Observed when initial level high. 0301

Cimetidine   After 100 mg/h for 4 h caused decrease of 34% at 150 minutes in normal subjects. 3491

Clofibrate   After 1 week in both normals and diabetics but more marked in latter. 1120

Ethacrynic Acid   Reduction in fasting state noted. 2427

Ethanol   Intravenously caused 65% decrease for 10 mg/dL glucose dec. 3637   Delayed increase seen in all subjects. 3762

Ether   Due to release of epinephrine causing inhibition. 3918

Furosemide   Intravenous njection effect, little on blood sugar. 2427

Hydrochlorothiazide   Intravenous injection has effect with little on sugar. 2427

5-Hydroxytryptamine   Probable action as inhibits secretion. 1101

Metformin   Slight, all hyperlipoproteinemias, marked type i.v. 1430

Morphine Reduction in secretion following test meal and drug in 6 healthy volunteers. *0626*

Muscular Exercise If exercise strenuous. *1171*

Nifedipine Response to glucose challenge significantly reduced in subjects taking drug. *1296* Basal insulin concentration reduced by 26% in normals. *0637* From 20 to 14 uU/mL in 15 hypertensive patients undergoing hemodialysis after 3 weeks treatment. *2986*

Phenytoin Reduces insulin response to glucose challenge. *2276* Lower values in drug treated group than in controls reached significant level at 90 minutes after glucose. *2785*

PIDH Reduces insulin response to glucose load. *3100*

Propranolol Significantly less in nondiabetic treated hypertensives than in diabetic- treated patients. *2469*

Starvation 40% decrease occurs after 1 d, then plateaus. *0021*

Tolazamide Return to normal usually if high initially. *0301*

Tolbutamide Effect observed in some patients. *0301*

Urine   Increase   Physiological

Glucose Occurs 3 h after glucose load. *3080*

Insulin Tolerance

Plasma   Increase   Physiological

Propranolol Decreases glucose rebound at end of test. *3292*

γ-Interferon

Monocytes   No Effect   Physiological

Methimazole No effect on production when cells from normal individuals stimulated with mitogens. *3778*

Interleukin-1

Monocytes   No Effect   Physiological

Methimazole No effect on production when cells from normal individuals stimulated with mitogens. *3778*

Interleukin-2

Monocytes   Increase   Physiological

Methimazole Increased activity in culture supernatants when mononuclear cells from normals stimulated with mitogens. Effect apparent between 24 h and 60 h. *3778*

Intermediate density lipoprotein (IDL)

Serum   No Effect   Physiological

Oral Contraceptives In 10 women receiving ethinyl estradiol with norethindrone. In 10 women receiving ethinyl estradiol with norgestrel. *2009*

Intrinsic Factor

Gastric Material   Decrease   Physiological

Cimetidine Marked effect on basal and stimulated concentrations. *3451*

Inulin Clearance

Urine   No Effect   Physiological

Amiloride No effect on glomerular filtration rate or effective renal plasma flow. *0656* Unchanged or may be slight increase. *3711*

Etodolac No effect with acute or chronic treatment in individuals with normal renal function but transient reduction in people with renal insufficiency. *0453*

Urine   Increase   Analytical

Dextran Interferes with analytical proce[...]

Urine   Increase   Physiological

Dopamine Intravenous infusion caused increas[...] mL/min. *1326*

Levodopa ?secondary to renal vasodilatation or dire[...] tubules. *1136*

Methylprednisolone At 24 h after 1 g i.v. in normals. *377[...]*

Urine   Decrease   Physiological

Aspirin Significant effect correlated with plasma salicylate concentration. *2535* Observed even with therapeutic doses. *2866*

Diazoxide Effect over 2 h of 4 mg/kg given i.v. *1805*

Phloridzin Decrease of up to 30%. *3343*

Iodide

Serum   Increase   Physiological

Erythrosine Significant dose related increases in 30 men receiving 20 to 200 mg/d for 15 d. *1241*

Urine   Increase   Physiological

Erythrosine Significant effect of daily doses of 60 mg and higher in 30 men treated for 15 d. *1241*

Hydrochlorothiazide Significant effect in 5 h after 150 mg orally. *1189*

Urine   Decrease   Physiological

Acetazolamide Significant effect in 5 h after 500 mg orally. *1189*

Iodine, Total

Serum   Increase   Physiological

Iodine Containing Drugs Iodine contamination. *1237*

Oral Contraceptives Altered metabolism. *1462*

Ionized Calcium

Serum   No Effect   Analytical

Potassium With changes in physiological range. *2056*

Storage of Sample If frozen for up to 3 d. *3493*

Serum   No Effect   Physiological

Cisplatin No effect noted in 11 children treated for 0.1 to 3.8 y. *3286*

Epinephrine Following i.v. infusion for 1 h to produce concentration up to 945 pg/mL. *0401*

Furosemide In 8 normal subjects associated with secondary hyperparathyroidism. *1214*

Tourniquet No effect noted for up to 5 minutes. *2056* No effect if used for less than 2 minutes. *3493*

Serum   Increase   Analytical

Collidine At concentrations > 0.1 mmol/L on Calcium specific electrode. *0540*

Glucose At concentrations above 1 g/dL ion specific electrode. *0540*

Imidazole At low concentrations but decrease at high, Calcium specific electrode. *0540*

Iodides Observed interference ion specific electrode. *0540*

Lutidine At concentrations > 0.1 mmol/L on Calcium specific electrode. *0540*

Perchlorate Observed interference ion specific electrode. *0540*

Picoline At concentrations > 0.1 mmol/L on Calcium specific electrode. *0540*

Pyridine At concentrations > 0.1 mmol/L on Calcium specific electrode. *0540*

Sodium As concentration changes from 140 to 168 meq/L. *2056*

(continued)

...0 mmol/L on Cal-

...ion specific elec-

...2.9 meq/L also in patients.

...ed for up to 2 weeks after drug ...d effect. *0451*

...an in controls in 12 patients taking drug for

...e On average 0.025 mmol/L higher in men. *2056*

...lts Slight but not significant increase in 5 men with ...al hypertension given citrate salt supplement. *2037*

...ccinylcholine Rise not marked (i.v. administration). *1047*

Serum   Decrease   Analytical

Amyl Alcohol At 0.1 mmol/L to 0.1 mol/L with Calcium specific electrode. *0540*

Aniline At 0.1 mmol/L to 0.1 mol/L with Calcium specific electrode. *0540*

Benzaldehyde At 0.1 mmol/L to 0.1 mol/L with Calcium specific electrode. *0540*

Benzene At 0.1 mmol/L to 0.1 mol/L with Calcium specific electrode. *0540*

Benzyl Alcohol At 0.1 mmol/L to 0.1 mol/L with Calcium specific electrode. *0540*

Butanol At 0.1 mmol/L to 0.1 mol/L with Calcium specific electrode. *0540*

Ethanediol At 0.1-1.0 mol/L with Calcium specific electrode. *0540*

Ethanol At 0.1 mmol/L to 0.1 mol/L with Calcium specific electrode. *0540*

Ethanolamine At concentrations > 0.1 mmol/L on Calcium specific electrode. *0540*

Glycerin At 0.1-1.0 mol/L with Calcium specific electrode. *0540*

Heat 2-3% at 37° compared with room temperature. *2056*

Heparin Up to 0.03 mmol/L in Vacutainers™. *2056*

Magnesium As concentration increases. *2056*

Methanol At 0.1 mmol/L to 0.1 mol/L on Calcium specific electrode. *0540*

Methylamine At concentrations > 0.1 mmol/L on Calcium specific electrode. *0540*

Morpholine At concentrations > 0.1 mmol/L on Calcium specific electrode. *0540*

Phenols At 0.1 mmol/L to 0.1 mol/L on Calcium specific electrode. *0540*

Piperidine At concentrations > 0.1 mmol/L on Calcium specific electrode. *0540*

Propanol At 0.1 mmol/L to 0.1 mol/L on Calcium specific electrode. *0540*

Storage of Sample 0.04 meq/L in 2 d, 0.08 meq/L in 7 d (?temperature). *1539*

Trolamine At concentration more than 0.1 mmol/L on Calcium specific electrode. *0540* Up to 12% if added to standards. *2056*

Tromethamine At concentrations > 0.1 mmol/L on Calcium specific electrode. *0540*

Trypsin Up to 2.5% if added to standards. *2056*

Serum   Decrease   Physiological

Ambulation But not to normal in 4 d from high of bed rest. *1539*

Betazole Increase by 0.062 mmol/L 1 h after 1.5 mg/kg s.c. *1682*

Citrates Complexes calcium (effect of transfusions). *2825*

EDTA Chelates calcium. *0071*

Ethinyl Estradiol Significant effect in postmenopausal women: less marked in perimenopausal women. *2311*

Gallium Nitrate Change paralleled that of total calcium in two patients with hypercalcemia due to parathyroid carcinoma. *3764*

Gastrin Increase by 0.052 mmol/L 1 h after 30 µg i.v. *1682*

Hyperventilation Effect of respiratory alkalosis. *3234*

Meals Increase by 0.053 mmol/L 1 h after steak meal. *1682* Effect more marked than with respiratory alkalosis. *3234*

Phosphates Slight effect abolished by Calcium infusion. *2951*

Iron

Serum   No Effect   Analytical

Acetaminophen At 5 times therapeutic concentration on Ferrozine method on SMAC®. *2138* At acute overdose concentration (10 mg/dL) on SMAC® method. *3719* At concentration of 100 mg/L had no effect on Ferrozine method. *3393*

Acetylsalicylic Acid On Ramsay and bathophenanthroline methods at therapeutic concentrations. *1786* On Ferrozine method on SMA II at physiological concentration. *1787* At 5 times upper limit of therapeutic range on Ferrozine method of SMAC®. *2138* With 8326 µmol/L on bathophenanthroline method. With 8326 µmol/L on Ramsay method. *1786* At concentration of 1500 mg/L had no effect on Ferrozine method. *3393*

Aminophenazone At therapeutic concentration on Ramsay and bathophenanthroline methods. *1786*

Aminopyrine At concentration of 125 mg/L had no effect on Ferrozine method. *3393*

Aminosalicylic Acid At 5 times upper limit of therapeutic range on Ferrozine method on SMAC®. *2138*

Amitriptyline At acute overdose concentration (2.5 mg/dL) on SMAC® method. *3719* At concentration of 25 mg/L had no effect on Ferrozine method. *3393*

Barbital At acute overdose concentration (20 mg/dL) on Technicon® SMAC® method. *3719* At concentration of 200 mg/L had no effect as measured by Ferrozine method. *3393*

Bilirubin No effect on iron measurement. *2981*

Bromazepam At 5 times therapeutic concentration on Ferrozine method on SMAC®. *2138*

Caffeine No effect of therapeutic concentrations on Ramsay and bathophenanthroline methods. *1786* At acute overdose concentration (10 mg/dL) on Technicon® SMAC® method. *3719* At concentration of 150 mg/L had no effect on Ferrozine method. *3393*

Carbenicillin At 5 times upper limit of therapeutic range on Ferrozine method on SMAC®. *2138*

Cefoxitin At 5 times upper limit of therapeutic range on Ferrozine method on SMAC®. *2138*

Cephalothin At 5 times upper limit of therapeutic range on methods on Ferrozine method on SMAC®. *2138*

Chloramphenicol At concentration of 5,000 mg/L had no effect on Ferrozine method. *3393*

Chlordiazepoxide At 5 times upper limit of therapeutic range on Ferrozine method of SMAC®. *2138* At acute overdose concentration (5 mg/dL) on Technicon® SMAC® method. *3719* At concentration of 50 mg/L had no effect on Ferrozine method. *3393*

Chlorpheniramine At acute overdose concentration (20 mg/dL) on Technicon® SMAC® method. *3719*

Cocaine At acute overdose concentration (2.5 mg/dL) on Technicon® SMAC® method. *3719* At concentration of 25 mg/L had no effect on Ferrozine method. *3393*

Codeine At acute overdose concentration (2.0 mg/dL) on Technicon® SMAC® method. *3719* At concentration of 20 mg/L had no effect on Ferrozine method. *3393*

Contact With Clot If plastic disc used to separate for 24 h. *1863*

Cyclosporine At concentration of 41.5 mmol/L (50 µg/L) on methods on Technicon® SMAC® II and Hitachi® 705. *1123*

Cysteine At 5 times upper limit of therapeutic range on Ferrozine method on SMAC®. *2138*

Cystine At 5 times upper limit of therapeutic range on methods on Ferrozine method on on SMAC®. *2138*

Dextran At concentration of 10,000 mg/L had no effect on Ferrozine method. *3393*

Diazepam At 5 times upper limit of therapeutic range on Ferrozine method on SMAC®. *2138* At acute overdose concentration (2.5 mg/dL) on Technicon® SMAC® method. *3719* At concentration of 25 mg/L had no effect on Ferrozine method. *3393*

Diclofenac No effect at therapeutic concentrations on Ramsay and bathophenanthroline methods. *1786* On Ferrozine method on SMA II at physiological concentration. *1787* At concentration of 23 mg/L had no effect on Ferrozine method. *3393*

Diethylpropion At acute overdose concentration (10 mg/dL) on Technicon® SMAC® method. *3719* At concentration of 100 mg/L had no effect on Ferrozine method. *3393*

Diphenhydramine At acute overdose concentration (20 mg/dL) on Technicon® SMAC® method. *3719* At concentration of 200 mg/L had no effect on Ferrozine method. *3393*

Ethanol At acute overdose concentration (20 mg/dL) on Technicon® SMAC® methods. *3719*

Ethchlorvynol At acute overdose concentration (20 mg/dL) on Technicon® method. *3719* At concentration of 400 mg/L had no effect on Ferrozine method. *3393*

Ethinamate At acute overdose concentration (20 mg/dL) on Technicon® SMAC® method. *3719* At concentration of 200 mg/L had no effect on Ferrozine method. *3393*

Flurazepam At 5 times upper limit of therapeutic range on methods on Ferrozine method on SMAC®. *2138* At acute overdose concentration (2.5 mg/dL) on Technicon® SMAC® method. *3719* At concentration of 25 mg/L had no effect on Ferrozine method. *3393*

Glutethimide At acute overdose concentration (5 mg/dL) on Technicon® SMAC® method. *3719* At concentration of 50 mg/L had no effect on Ferrozine method. *3393*

Hemoglobin With up to 20 mg/dL method of Megraw. *2391* No effect unless concentration greater than 500 mg/dL. *2981*

Heparin At concentration of 1,000 mg/L had no effect on Ferrozine method. *3393*

Ibuprofen No effect at therapeutic concentrations on Ramsay and bathophenanthroline methods. *1786* At 5 times upper limit of therapeutic range on Ferrozine method on SMAC®. *2138* On Ferrozine method on SMA II at physiological concentration. *1787* At concentration of 200 mg/L had no effect on Ferrozine method. *3393*

Indomethacin No effect at therapeutic concentrations on bathophenanthroline and Ramsay methods. *1786* On Ferrozine method on SMA II at physiological concentration. *1787*

Ketoprofen No effect at therapeutic concentrations on Ramsay and bathophenanthroline methods. *1786* On Ferrozine method on SMA II at physiological concentration. *1787* At concentration of 60 mg/L had no effect on Ferrozine method. *3393*

Lipemia If slight or moderate method of Megraw. *2391*

Meperidine At acute overdose concentration (5 mg/dL) on Technicon® SMAC® method. *3719*

Meprobamate At acute overdose concentration (20 mg/dL) on Technicon® SMAC® method. *3719* At concentration of 200 mg/L had no effect on Ferrozine method. *3393*

Mesoridazine At acute overdose concentration (20 mg/dL) on Technicon® SMAC® method. *3719*

Methadone At acute overdose concentration (2.5 mg/dL) on Technicon® SMAC® method. *3719* At concentration of 25 mg/L had no effect on Ferrozine method. *3393*

Methanol At acute overdose concentration (20 mg/dL) on Technicon® SMAC® method. *3719*

Methaqualone At acute overdose concentration (2.5 mg/dL) on Technicon® SMAC® method. *3719* At concentration of 25 mg/L had no effect on Ferrozine method. *3393*

Methicillin At 5 times upper limit of therapeutic range on Ferrozine method on SMAC®. *2138*

Methotrexate At 5 times upper limit of therapeutic range on Ferrozine method on SMAC®. *2138*

Methylphenidate At acute overdose concentration (20 mg/dL) on Technicon® SMAC® method. *3719* At concentration of 200 mg/L had no effect on Ferrozine method. *3393*

Methyprylon At acute overdose concentration (10 mg/dL) on Technicon® SMAC® method. *3719* At concentration of 100 mg/L had no effect on Ferrozine method. *3393*

Morphine At acute overdose concentration (20 mg/dL) on Technicon® SMAC® method. *3719* At concentration of 200 mg/L had no effect on Ferrozine method. *3393*

Naproxen At 5 times upper limit of therapeutic range on Ferrozine method of SMAC®. *2138*

Nitrofurantoin On routine methods in use on SMAC®, Ektachem®, Abbott-VP, Cobas-Bio, aca, Hitachi® 705, KDA at 5 times normal upper therapeutic concentration. *2138*

Nortriptyline At acute overdose concentration (20 mg/dL) on Technicon® SMAC® method. *3719* At concentration of 200 mg/L had no effect on Ferrozine method. *3393*

Penicillin G At 5 times upper limit of therapeutic range on methods on SMAC®. *2138*

Pentobarbital At acute overdose concentration (20 mg/dL) on Technicon® SMAC® method. *3719* At concentration of 340 mg/L had no effect on Ferrozine method. *3393*

Perphenazine At acute overdose concentration (20 mg/dL) on Technicon® SMAC® method. *3719* At concentration of 200 mg/L had no effect on Ferrozine method. *3393*

Phenobarbital No effect at therapeutic concentrations on bathophenanthroline and Ramsay methods. *1786* At concentration of 200 mg/L had no effect on Ferrozine method. *3393*

Phenytoin At acute overdose concentration (20 mg/dL) on Technicon® SMAC® method. *3719* At concentration of 200 mg/L had no effect on Ferrozine method. *3393*

Promethazine At acute overdose concentration (20 mg/dL) on Technicon® SMAC® method. *3719* At concentration of 200 mg/L had no effect on Ferrozine method. *3393*

Propoxyphene At acute overdose concentration (2.5 mg/dL) on Technicon® SMAC® method. *3719* At concentration of 25 mg/L had no effect on Ferrozine method. *3393*

Pyribenzamine® At acute overdose concentration (20 mg/dL) on Technicon® SMAC® methods. *3719*

Quinidine At acute overdose concentration (20 mg/dL) Technicon® SMAC® method. *3719* At concentration of 200 mg/L had no effect on Ferrozine method. *3393*

Quinine At acute overdose concentration (1.5 mg/dL) on Technicon® SMAC® method. *3719* At concentration of 15 mg/L had no effect on Ferrozine method. *3393*

Rifampin At 5 times upper limit of therapeutic range on Ferrozine method on SMAC®. *2138*

Secobarbital At acute overdose concentration (20 mg/dL) on Technicon® SMAC® method. *3719* At concentration of 200 mg/L had no effect on Ferrozine method. *3393*

Sodium Citrate At concentration of 10,000 mg/L had no effect on Ferrozine method. *3393*

Sodium Fluoride At concentration of 4,000 mg/L had no effect on Ferrozine method. *3393*

Storage of Sample No effect 4 d room temperature or 7 d at 4°. *1563*

Sulforidazine At concentration of 200 mg/L had no effect on Ferrozine method. *3393*

Tetracycline At 5 times upper limit of therapeutic range on Ferrozine method on SMAC®. *2138*

Theophylline At 5 times upper limit of therapeutic range on Ferrozine method on SMAC®. *2138*

Thiopental At acute overdose concentration (20 mg/dL) on Technicon® SMAC® method. *3719* At concentration of 200 mg/L had no effect on Ferrozine method. *3393*

Tripelennamine At concentration of 200 mg/L had no effect on Ferrozine method. *3393*

Vitamin B Complex At concentration of 1.4 mg/L had no effect on Ferrozine method. *3393*

Serum   No Effect   Physiological

Captopril In 9 of 12 hypertensive patients although other hematological effects observed. *1608*

Carbenicillin No effect although marrow depressed. *2964*

Cyclosporine No significant effect observed in 23 patients with inflammatory ocular disease treated for 6 mo. *2713*

Diclofenac No effect seen in patients undergoing treatment. *1787*

Ibuprofen No effect seen in patients undergoing treatment. *1787*

Indomethacin No effect seen in patients undergoing treatment. *1787*

Ketoprofen No effect seen in patients undergoing treatment. *1787*

Iron (continued)

Serum   Increase   Analytical
Cefotaxime   General effect of both drug and metabolite on Parallel method. *0201*

Copper   500 μg/dL = 50 μg/dL Ferrozine procedure of White. *3810*

Dextran   Causes turbidity with method of Young and Hicks. *2981*

Disulphine Blue   At physiological concentration for more than 2 d on continuous flow Ferrozine reaction. *1464*

Hemoglobin   Slight effect of 1 g on Tripyridyl-s-triazine procedure. *3217*

Hemolysis   Interferes with colorimetric procedure. *1564*

Trichloracetic Acid   Volume reduction effect (1.5%) with protein loss. *1131*

Serum   Increase   Physiological
Acetylsalicylic Acid   Significant effect correlated with duration of treatment (17.95 μmol/L vs 14.98 μmol/L). *1787*

Blood Transfusions   Siderosis may occur with multiple transfusions. *3742*

Chloramphenicol   Reversible toxic reaction. *2808*   Occurs either as dose-related or idiosyncratic bone marrow suppression: dose-related usually occurs 5-7 d after start of therapy: idiosyncratic is rare occurring 1 case per 40,000 courses of therapy. *3864*

Cisplatin   Mean increase from 67 to 128 μmol/L observed in 14 of 20 patients. Normalized few months after therapy. *1384*

Diurnal Variation   Maximum at 4 pm, minimum at 4 am. *3865*

Estrogens   Usually cause effect. *0121*

Ethanol   Enhances absorption from gastrointestinal tract in alcoholics. *1532*   From 16 mmol/L to 30 mmol/L at 15 h after 100 mg alcohol in 4 healthy male volunteers. *1223*

Iron Dextran   Increased iron stores. *2313*   Rose to exceedingly high values (up to 8,000 μmol/L) immediately after intravenous infusion. *0390*

Iron Salts   Effect of i.m. iron. *1237*

Lead   Reported effect. *0987*

Methicillin   If erythrocyte maturation depressed. *2964*

Methotrexate   Sharp increase 8-12 h after end of cycle. Maximum value of 295% at 48-60 h; after 108 h had returned to normal in only 50%. *3173*

Oral Contraceptives   Increase in available binding protein (plus 20% increase). *3548*   In 46 oral contraceptive users compared with controls studied over at least 2 y. *1182*   Increase after 12 mo with low base combined pill regime. *2996*

Serum   Decrease   Analytical
Deferoxamine   Marked reduction in AutoAnalyzer and other methods. *3180*   At concentrations above 140 mg/L lowered concentration as measured by Ferrozine method. *3393*

EDTA   Interferes with method of Young and Hicks. *2981*   Although no effect with atomic absorption methods low results obtained with colorimetric methods including Ferrozine and bathophenanthroline. *1991*   At concentrations above 100 mg/L lowered concentration as measured by Ferrozine method. *3393*

Lipemia   If marked due to high blank (procedure of Megraw). *2391*

Sodium Oxalate   At concentrations above 300 mg/L lowered concentration as measured by Ferrozine method. *3393*

Serum   Decrease   Physiological
Allopurinol   40% reduction in 1 week (accumulates in liver). *1672*

Aspirin   May be markedly reduced with large doses. *1343*

Cholestyramine   Impairs absorption of iron. *1965*

Corticotropin   Decrease in iron binding globulin. *3879*

Cortisone   Reduced synthesis of transferrin. *3879*

Deferoxamine   Therapeutic intent. *3180*

Diurnal Variation   Towards late afternoon or evening. *3873*

Epinephrine   1 mL 1% solution effect lasts for 6 h. *0336*

Metformin   Associated with impaired B_{12} absorption. *3605*

Muscular Exercise   Response to stress of exercise. *0582*

Oxymetholone   Iron deficiency anemia may occur. *1679*

Phytate   Early effect but increased if continued. *2947*

Piretanide   Slight drop with 12 mg/d for 3 mo. *3696*

Pregnancy   Falls from midterm onwards. *0987*

Urine   Increase   Physiological
Deferoxamine   Primary affinity for trivalent iron. *0071*

Iron Dextran   Approximately 30% Fe Sorbitex in urine in 24 h. *1678*

Penicillamine   If poisoning due to iron. *0071*

Iron-Binding Capacity, Total (TIBC)

Serum   No Effect   Analytical
Bilirubin   No effect on measurement. *2981*

Contact With Clot   If plastic disc used to separate for 24 h. *1863*

Hemoglobin   No effect on measurement. *2981*

Hemolysis   No effect observed with 1 g/dL. *3217*

Serum   No Effect   Physiological
Captopril   In 9 of 12 hypertensive patients although other hematological effects observed. *1608*

Carbenicillin   No effect although marrow depressed. *2964*

Diurnal Variation   As iron and UIBC vary reciprocally. *3217*

Ethinyl Estradiol   No effect if administered alone. *0473*

Serum   Increase   Analytical
Disulphine Blue   At physiological concentration for more than 2 d on magnesium carbonate/ continuous flow Ferrozine reaction. *1464*

Magnesium Carbonate   Variable effect method Young/Hicks if not dry. *0280*

Perchlorate   When Fe salt used to saturate in method Young/Hicks. *1055*

Serum   Increase   Physiological
Estrogens   Usual effect due to increased carrier protein. *0121*

Iron Salts   Effect of i.m. iron. *1237*

Mestranol   20% rise on average. *1660*

Oral Contraceptives   Estrogen or progestogen effect (usually plus 20%). *3548*   In 46 oral contraceptive users compared with controls studied over at least 2 y. *1182*

Pregnancy   With fall of serum iron after midterm. *0987*

Serum   Decrease   Physiological
Chloramphenicol   Decreased uptake by erythroid tissue. *2220*

Corticotropin   Decrease in iron binding globulin. *3879*

Cortisone   Reduced synthesis of transferrin. *3879*

Iron-Binding Capacity, Unsaturated (UIBC)

Serum   Increase   Physiological
Oxymetholone   Decreases percentage saturation of transferrin. *1679*

Serum   Decrease   Physiological
Corticotropin   Decrease in iron binding globulin. *3879*

Ethanol   Up to 90% increase in saturation in alcoholics. *1532*

Iron Dextran   Due to increased availability of iron. *2313*

Iron Salts   Effect of i.m. iron. *1237*

Methicillin   If erythrocyte maturation depressed. *2964*

Iron Saturation

Serum   Increase   Analytical
Cefotaxime   Slight but significant increase with both drug and metabolite on method on American Monitor Parallel. *0201*

Isocitrate Dehydrogenase

Serum   No Effect   Analytical
Storage of Sample No effect 1 week at 4°, 1 mo at -20°. *3873*
For 7 d at 30° or if frozen. *0877* 6 h at room temperature, many days at 4°. *3217*

Serum   Increase   Analytical
Hemolysis Activity high in erythrocytes. *0279*

Serum   Increase   Physiological
Allopurinol Reversible clinical hepatotoxicity reported. *2220*
Aminosalicylic Acid May cause cytotoxic hepatocellular damage. *0248*
Amodiaquine Reported hepatotoxicity. *2313*
Amphotericin B Due to hepatocellular dysfunction. *2313*
Anabolic Steroids Due to cholestatic syndrome. *2220*
Androgens Cholestatic phenomenon. *2313*
Anesthetic Agents Occurs without permanent liver damage. *3661*
Antimony Compounds Hepatotoxic effect. *2220*
Arsenicals Hepatotoxic effect. *3835*
Chenodeoxycholic Acid Slight effect in almost 30% cases. *0274*
Chlordiazepoxide Infrequent cholestatic effect. *2313*
Chlorpromazine May be hypersensitive reaction. *3835*
Clindamycin Mild transient increase (transient). *1083*
Ethanol Occurs within 4 h in normal subjects. *1323*
Iproniazid May cause cholestatic and cytotoxic jaundice. *0248*
Isoniazid Cytotoxic hepatocellular jaundice. *0248*
Mechlorethamine May cause cytotoxic (hepatocellular) damage. *0248*
Methotrexate May cause cytotoxic hepatocellular damage. *0248*
Phenylbutazone May cause cholestatic and cytotoxic jaundice. *0248*
Pregnancy Placental origin in last trimester. *0987*
Pyridinol Carbamate Noticeable effect, cause not discussed. *2560*
Testosterone May cause cholestatic and cytotoxic liver damage. *0248*

Serum   Decrease   Analytical
Storage of Sample Freezing reported to cause 10-25% loss of activity. *0877*

Isohomovanillic Acid

Urine   Increase   Physiological
Levodopa Response to therapy in Parkinson patients. *0703*

Isoleucine

Plasma   Increase   Physiological
Ethanol Due to increased hepatic production and release in to circulation. *2428*
Obesity Characteristic finding correlated with insulin. *1106*
Starvation Marked effect even after 1 d. *0021*

Plasma   Decrease   Physiological
Alanine Decrease of about 40% with 1 week administration. *1263*
Glucose Probable muscle uptake. *1403*
Histidine Probably due to flow into gut after oral load. *1644*
Oral Contraceptives Hormonal effect (second part of cycle). *0742*

Urine   Increase   Physiological
Ascorbic Acid In 10 healthy females given 10 g/d. *3624*
Starvation Observed effect due to tissue catabolism. *0021*

Isoniazid

Serum   Increase   Physiological
Aminosalicylic Acid Inhibits acetylation, increases serum concentration. *1487*

Serum   Decrease   Physiological
Antacids Delayed and reduced absorption due to adsorption and first-pass metabolism. *3794*
Food Absorption reduced when taken with food. *3024*

Isopropanol

Blood   Increase   Physiological
Isopropanol Fatal poisoning at 1 g/L. *2348*

Isovaleric Acid

Urine   Increase   Physiological
Bacteriuria Some organisms can produce in urine. *1483*

Isovanillic Acid

Urine   Increase   Physiological
3,4-Dihydroxybenzoic Acid Significant increase with glycine conjugate. *2870*

[131]I Uptake

Serum   No Effect   Physiological
Acetazolamide No effect on test. *0583*
Anabolic Steroids No effect observed. *1915*
Androgens No effect on uptake reported. *2220*
Aspirin No effect observed. *1915*
Calcium Loading and withdrawal no effect. *0583*
Carbimazole No effect on uptake by thyroid. *2220*
Chlorpheniramine 16 mg/d reduced uptake by 48%. *0583*
Corticosteroids No effect observed. *1915*
Dextrothyroxine No effect reported. *2220*
Diazepam No effect after i.v. administration. *0583*
Digitoxin With 0.01 mg/kg/week orally no effect. *0583*
Estrogens Have no effect on uptake. *1915*
Gold No effect observed. *1915*
Hydrochlorothiazide No change observed. *0583*
Mephenytoin No effect reported. *2220*
Meralluride No effect on thyroid function. *2220*
Mercurial Diuretics No effect in euthyroid subjects. *0583*
Mercury Compounds No effect observed. *1915*
Methimazole No effect on uptake by thyroid. *2220*
Methylthiouracil No effect on uptake by thyroid. *2220*
Metronidazole 1200 mg/d for 1 week no effect. *0583*
Oral Contraceptives Thyroid function unaffected. *0071*
Perchlorate No effect on uptake. *2220*
Perphenazine No effect observed in euthyroid subjects. *2220*
Phenothiazines No effect observed even with prolonged use. *1915*
Phenytoin No effect observed. *1915*
Propranolol No effect in euthyroid patients. *0356*
Propylthiouracil No effect reported. *2220*
Silver No effect reported. *1915*
Sulfobromophthalein No effect on uptake reported. *2220*
Tetracycline No effect with 2 g/d for 11 d. *0583*
Thiazides No effect observed in most patients. *1487*
Thiocyanate Does not affect uptake by thyroid. *2220*

[131]I Uptake (continued)

Serum Increase Physiological

Barbiturates ?due to enzyme induction. *2220*

Chlorpromazine With procyclidine decreased renal clearance (45 to 27 mL/min). *0583*

Estrogens Increased uptake reported. *2220*

Lithium Mechanism unclear. *0544*

Oxymetholone Anabolic effect (possible for 2 weeks after stop). *1679*

Phenothiazines Reported effect in hyperthyroidism. *1488*

Thyrotropin Results vary with thyroid status. *2220*

Serum Decrease Physiological

Acetrizoate Due to iodine component of material. *2220*

Aminobenzoic Acid May impair uptake. *2220*

Aminoglutethimide Reported effect. *2220*

Aminosalicylic Acid May cause goitrous hypothyroidism. *2808*

Amobarbital Amytal®, tuinal contain tetraiodofluorescein. *2652*

Amphenone B Reported effect. *0830*

Ampicillin Omnipen® contains tetraiodofluorescein. *2652*

Antihistamines Uncommon reported effect. *2220*

Aspirin With large doses and chronic administration. *1444*

Barium Theoretically if contaminated with I₂. *0583*

Benziodarone Due to iodine component of drug. *2220*

Bromides Theoretically if contaminated with I₂. *0583*

Brompheniramine Effect observed in some patients. *0830*

Calcium Gluconate Wafer of Upjohn contains tetraiodofluorescein. *2652*

Carbarsone Lilly compound contains tetraiodofluorescein. *2652*

Carbutamide Substantial effect observed in elderly. *0830*

Cascara Lilly compound contains tetraiodofluorescein. *2652*

Chlordiazepoxide ?antithyroid effect. *0234*

Chlorpheniramine Ornade®, Teldrin contain tetraiodofluorescein. *2652*

Clofibrate Up to 2.5 g/d produces effect for up to 4 mo. *0583*

Cobalt Due to impaired synthesis of thyroxine. *1915*

Cobalt Salts Reported effect. *2220*

Corticosteroids Observed effect may last 8 d. *1487*

Corticotropin ?effect on TSH — lasts up to 8 d. *1444*

Cortisone Probably diminishes TSH secretion. *2427*

Cremomycin Merck compound contains tetraiodofluorescein. *2652*

Cyclophosphamide Single case of drug induced myxedema. *0681*

Cycloserine Seromycin® contains tetraiodofluorescein. *2652*

Cyclothiazide Anhydron contains tetraiodofluorescein. *2652*

Demeclocycline Declomycin® contains tetraiodofluorescein. *2652*

Dextrothyroxine Marked effect observed even in normals. *0019*

Diatrizoic acid Due to iodine component of material. *2220*

Diazepam Conflicting reports ?no effect. *1237*

Digitalis Lilly product contains tetraiodofluorescein. *2652*

Digitoxin Purodigin, Crystodigin® contain tetraiodofluorescein. *2652*

Diiodohydroxyquin Effect lasts for several weeks. *1488*

Dimercaprol Elemental iodine trapped in thyroid. *2313*

Diphenhydramine Benadryl® contains tetraiodofluorescein. *2652*

Diprotrizoate Due to iodine component of drug. *2220*

Disulfiram Uncommon reported effect. *2220*

Dithiazanine Due to iodine component of drug. *2220*

Ephedrine Lilly, P-D products contain tetraiodofluorescein. *2652*

Erythromycin Tetraiodofluorescein in pedimycin. *2652*

Estrogens SK-estrogens contain tetraiodofluorescein. *2652*

Ethinyl Estradiol Feminone contains tetraiodofluorescein. *2652*

Fluoxymesterone Metabolic effect. *0019*

Folic Acid Filibon, Iberet® contain tetraiodofluorescein. *2652*

Gallamine Due to iodine component of drug. *2220*

Glucocorticoids Associated with reduced BMR. *2220*

Imipramine Tofranil® contains tetraiodofluorescein. *2652*

Indocyanine Green Contains iodine, inhibits further uptake. *2220*

Iodides Massive doses increase pool. *0583*

Iodinated Glycerol Daily use may give up to 200 mg/d. *0583*

Iodine Containing Drugs Small doses may affect but not PBI. *0583*

Iodipamide Due to iodine component of material. *2220*

Iodized Oil Interferes with uptake. *0071*

Iodoalphionic Acid Organic iodine contamination. *1237*

Iodochlorhydroxyquin Contains organically bound iodine. *2145*

Iodoform Organically bound iodine, inhibits further uptake. *2220*

Iodopyracet Due to iodine component of material. *2220*

Iopanoic Acid Due to iodine component of material. *2220*

Iophenoxic Acid Due to iodine component of material. *2220*

Iopydone Due to iodine component of material. *2220*

Iothalamate Interferes with uptake. *0071*

Iothiouracil Drug consists of organically bound iodine. *0830*

Ipodate Interferes with uptake. *0071*

Isoniazid Reduces uptake. *2220*

Isopropamide Contains iodine — reduces further uptake. *1488*

Levodopa Larodopa® contains tetraiodofluorescein. *2652*

Levothyroxine Due to metabolic effect of drug. *2220*

Lincomycin Lincocin® contains tetraiodofluorescein. *2652*

Liothyronine Except in hyperthyroidism. *2220*

Lithium Lithionate contains tetraiodofluorescein. *2652*

Mephenytoin Mesantoin® contains tetraiodofluorescein. *2652*

Mesoridazine Serentil® contains tetraiodofluorescein. *2652*

Methandrostenolone Also modifies binding of thyroid hormones. *2427*

Methantheline Reported to decrease results. *0583*

Methaqualone Parest® contains tetraiodofluorescein. *2652*

Methenamine Mandelamine® contains tetraiodofluorescein. *2652*

Methimazole Effect may last from 2 to 8 d. *1444*

Nandrolone Anabolic effect. *2681*

Normetanephrine Anhydron contains tetraiodofluorescein. *2652*

Novobiocin Albamycin contains tetraiodofluorescein. *2652*

Oleandomycin Cyclamycin contains tetraiodofluorescein. *2652*

Oxazepam Serax® contains tetraiodofluorescein. *2652*

Oxyphenbutazone Impaired synthesis of thyroxine. *1915*

Oxyphenisatin Dialose contains tetraiodofluorescein. *2652*

Oxytetracycline Terramycin® contains tetraiodofluorescein. *2652*

Penicillin If hydriodide salt given decreases further uptake. *2220* V-cillin contains tetraiodofluorescein. *2652*

Pentobarbital Lilly product contains tetraiodofluorescein. *2652*

Perphenazine Depresses uptake. *0583*

Phenazopyridine Donnasep contains tetraiodofluorescein. *2652*

Phenindione Uncommon reported effect. *2220*

Phenmetrazine Preludin® contains tetraiodofluorescein. *2652*

Phenolphthalein Phenolax contains tetraiodofluorescein. *2652*

Phensuximide Milontin® contains tetraiodofluorescein. *2652*

Phenylbutazone May last up to 2 weeks. *3209*

Polymagma Contains tetraiodofluorescein. *2652*

Promazine Sparine® contains tetraiodofluorescein. *2652*

Promethazine Phenergan®, Mepergan® contain tetraiodofluorescein. *2652*

Propoxyphene Darvon® contains tetraiodofluorescein. *2652*

Propylthiouracil Effect lasts up to 8 d. *1488*

Quinidine Lilly product contains tetraiodofluorescein. *2652*

Resorcinol Reported effect of treatment of varicose ulcers. *1343*

Secobarbital Seconal™, tuinal contain tetraiodofluorescein. *2652*

Sodium Nitroprusside Single case reported (?due to thiocyanate). *2626*

Sulfadiazine Dulcet contains tetraiodofluorescein. *2652*

Sulfonamides Effect lasts about 7 d. *1444*

Sulfonylureas Due to impaired synthesis of thyroxine. *1915*

Tetracycline Panmycin® contains tetraiodofluorescein. *2652*

Tetraiodofluorescein Constituent of capsules of many pharmaceuticals. *2652*

Thyroid Consequence of treatment. *1488*

Tolbutamide Uncommon reported effect. *2220*

Tranylcypromine Parnate contains tetraiodofluorescein. *2652*

Tri-iodothyronine Marked decrease in obese with 150 μg/d. *0458*

Vitamin A Significant effect when administered for 3 weeks. *0830*

Vitamin Preparations If preparations contain iodine. *2220* Daylets contain tetraiodofluorescein. *2652*

Warfarin Panwarfin® contains tetraiodofluorescein. *2652*

Water At 1 mg/L in drinking water. *0583*

Kallikrein

Urine   Decrease   Physiological

Metoprolol In 15 patients with essential hypertension treated for 4 weeks, associated with reduction in sympathetic tone and reduced activity of renin-aldosterone system. *1209*

Kaolin-Cephalin Time

Blood   Decrease   Physiological

Oral Contraceptives Metabolic effect. *2834*

Ketoconazole

Serum   Decrease   Physiological

Rifampin Plasma concentration reduced by about 33% with concomitant administration. *0450*

11-Ketoetiocholanolone

Urine   Increase   Physiological

Corticotropin Normal response to 2 d ACTH test. *1474*

17-Ketogenic Steroids

Urine   No Effect   Analytical

Ampicillin No effect on method of Gray. *3622*

Carbamazepine No effect on method of Normyberski. *2901*

Chlorothiazide Probably minimal interference with Zimmermann reaction. *3505*

Glutethimide Probably minimum interference with Zimmermann reaction. *3505*

Urine   No Effect   Physiological

Carbamazepine No physiological effect observed. *2901*

Iothalamate No physiological effect observed. *2573*

Oxytocin No significant change observed. *1344*

Urine   Increase   Analytical

Acetazolamide Interferes with Zimmermann reaction. *0427*

Acetone Brown color with Zimmermann reaction. *0583*

Acetophenone Interferes in measurement by Zimmermann method. *3505*

Cephaloridine On day of administration — method of Wilson and Lipsett. *1779*

Cephalothin Non specificity of method of Wilson and Lipsett. *1779*

Chlordiazepoxide Interferes with Zimmermann reaction. *2573*

Chlorpromazine Interferes with Zimmermann reaction. *3505*

Digitoxin Interferes with Zimmermann reaction *in vitro*. *0427*

Etryptamine Interferes with Zimmermann reaction. *0427*

Hydralazine Glucuronide interferes with Zimmermann reaction. *3505*

Hydroxyzine Interferes with Zimmermann reaction. *2573*

Meprobamate Glucuronide interferes with Zimmermann reaction. *3505*

Methyprylon Interferes with Zimmermann reaction. *3505*

Nalidixic Acid Interferes with Zimmermann reaction. *2197*

Oleandomycin Interferes with Zimmermann reaction. *3505*

Paraldehyde Reported to affect Zimmermann procedure. *0583*

Penicillin Interferes with Zimmermann (Norymberski) reaction. *3505*

Phenaglycodol Interferes with Zimmermann reaction. *3505*

Phenazopyridine Interferes with Zimmermann reaction. *3505*

Phenothiazines Yield similar color with Zimmermann reaction. *2220*

Quinine Reported to affect Zimmermann reaction. *0583*

Spironolactone Metabolite interferes with Zimmermann reaction. *3505*

Urine   Increase   Physiological

Ampicillin After 14 d in 1 postmenopausal woman. *3622*

Cortisone Metabolic products. *2220*

Metyrapone Normal response to injection is doubling of output. *1680*

Urine   Decrease   Analytical

Chlordiazepoxide Interferes with Zimmermann reaction. *3505*

Glucose Acts as reducing agent in Zimmermann reaction. *0583* Interferes with Norymberski reaction. *3505*

Iodipamide Interferes with reaction. *2573*

Iothalamate Acts as reducing substance (Rutherford and Nelson method). *2573*

Meprobamate Interferes with Zimmermann reaction (Holtorff-Koch). *3125*

Metyrapone Interferes with Zimmermann reaction. *3505*

Radiographic Agents Output halved with Zimmermann procedure. *0583*

Urine   Decrease   Physiological

Ampicillin Increase by 41% after 7 d administration in pregnant women. *3622*

Dexamethasone Suppression of ACTH. *2420*

Oral Contraceptives Probably due to estrogen decreasing cortisol secretion. *2596*

α-Ketoglutarate

Serum   Increase   Physiological

Fructose Rises more rapidly than after glucose. *1709*

Urine   Increase   Physiological

Lithium Reversible inhibition of renal transport. *0409*

Ketones

Serum   No Effect   Physiological

Propranolol In 5 hyperthyroid patients given 10 mg every 8 h for 4 d. *0335*

Tetracosactrin In healthy volunteers given 1 mg intramuscularly for up to 60 h. *1813*

Serum   Increase   Physiological

Acetoacetate Maximum 60 minutes after tolerance load given. *1811*

Albuterol Significant dose-related effect after i.v. infusion of therapeutic doses in 4 healthy male volunteers. *2807*

Aspirin Due to induced acidosis. *1302*

Ethanol May cause marked ketoacidosis. *1291*

Fenfluramine (average increase of 57%). *2753*

Growth Hormone Small postabsorptive rise. *0138*

Methanol Moderate effect in comparison with extent of acidosis. *1343*

Ketones *(continued)*

Serum Increase Physiological *(continued)*
Muscular Exercise Effect marked in untrained individuals only. *1811*

Nifedipine In 15 hypertensive patients undergoing hemodialysis after 3 weeks treatment. *2986*

Paraldehyde Transient hyperglycemia and ketosis. *2220*

Physical Training Increase in untrained individuals after exercise only. *1811*

Rimeterol Dose related significant change in 4 healthy men given therapeutic i.v. dose. *2807*

Starvation Begins increase during 2nd or 3rd day. *3783*

Serum Decrease Physiological
Aspirin Increased oxidation of ketone bodies in diabetics. *1343*

Propranolol In hyperthyroid patients given 40 mg every 6 h in 6 patients. *0335*

Plasma Decrease Physiological
Valproic Acid 60% reduction when 1 g given orally to fasting individuals. *3633*

Blood No Effect Physiological
Timolol In 5 hyperthyroid patients given 10 mg every 8 h for 4 d. *0335*

Blood Decrease Physiological
Propylthiouracil Marked effect in 5 hyperthyroid individuals given 10 mg every 8 h for 4 d. *0335*

Urine No Effect Analytical
Chloroform As preservative has no effect on Stix tests. *2308*

Fluorescein At 250 mg/L in spiked specimen run on Ames Clinitek 200. *1722*

Urine Increase Analytical
Acetaldehyde Affects alkaline nitroprusside procedure. *0583*

Aspirin Reddish color with Gerhardt's test. *1022*

Inositol Possible reported effect. *1841*

Levodopa Affects alkaline nitroprusside procedure. *0583* Intermittent false positive if Ketostix® or Phenistix®. *3884*

Mesna Common if not invariable in patients given i.v. mesna. *0577* False-positive with Multistix® and Chemstrip® but red color can be discharged with glacial acetic acid. *0767*

Methyldopa Affects alkaline nitroprusside procedure. *0583*

Paraldehyde Affects alkaline nitroprusside procedure. *0583* False positive when drug combined with ethanol. *1443*

Phenazopyridine False positive with Ketostix® or FeCl₃. *2566*

Phenolphthalein Presumed false positive from color. *1487* Pink with Rothera, Ketostix® and Acetest® tests. *0121*

Phenolsulfonphthalein Red-purple color with alkaline nitroprusside. *0583* Presumed false positive from color. *1487*

Phenothiazines Pink or purple with Gerhardt's test. *0583*

Phenylpyruvic Acid Reacts with alkaline nitroprusside. *0583* False positive with Ketostix®. *2308*

Pyrazinamide Pink-brown with Ketostix®, Acetest® and Rothera. *0121*

Sulfobromophthalein Produces purple color with alkaline nitroprusside. *0583* False positive color react with BiliLabstix® in acid urine. *1487*

Valproic Acid Single drug eliminated as ketones may give false positive test. *0504*

Urine Increase Physiological
Aspirin Acidotic response especially in children. *1302*

Ethanol Mainly due to β-hydroxy-butyrate, little acetoacetic. *1291*

Ether May follow anesthesia. *1487*

Growth Hormone Small postabsorptive rise. *0138*

Insulin Occurs especially if low liver glycogen stores. *1342*

Isoniazid Mechanism not listed. *1488*

Isopropanol In intoxication acetone is normal metabolite. *1487*

Metformin Associated with lactic acidosis. *1022*

Niacin ?due to hepatic mobilization of ketogenic amino acids. *1253*

Starvation Occurs within 4 d of commencement of fast. *3783*

Urine Decrease Analytical
Phenazopyridine Nitroprusside reaction masked by color. *0583*

Urine Decrease Physiological
Aspirin Increased oxidation of ketone bodies in diabetics *1343*

Glucose In starved individuals even at 7.5 g/d. *1242*

Urine Positive Analytical
Captopril False positive at concentration of 25 mmol/L on Ames Keto-diastix®, also affected Boehringer Combur Test. *1373* Trace to 3 + reactions in 9 patients with both Diastix® and Chemstrip-6®. *3766*

17-Ketosteroids

Plasma Increase Analytical
Cephalothin Marked effect on Zimmermann procedure with 6 g dose. *1932*

Plasma Decrease Physiological
Aging In women from 160 μg/dL at 25 y to 80 μg/dL at 45 y. *0462*

Morphine Inhibits ACTH and pituitary gonadotropin release. *1343*

Urine No Effect Analytical
Acetazolamide Minimal interference with Zimmermann reaction in vitro. *0427*

Amphetamine No effect on Zimmermann reaction. *0427*

Ampicillin No effect on method of Gray. *3622*

Aspirin No effect on Zimmermann procedure. *0766*

Barbiturates No interference with Zimmermann procedure. *0766*

Chlordiazepoxide No effect on Zimmermann reaction if added in vitro. *0766*

Chlorothiazide Minimal interference with Zimmermann reaction in vitro. *0427*

Dextroamphetamine No effect on modified Holtorff-Koch procedure. *1487*

Digitoxin Minimal interference with Zimmermann reaction. *0427*

Gentamicin With normal dose on Zimmermann procedure. *1932*

Glutethimide Minimal interference with Zimmermann reaction in vitro. *0427*

Hydralazine No significant effect with Zimmermann reaction. *0427*

Hydroxyzine No effect reported on Zimmermann reaction. *0427*

Perphenazine Minimal effect on Zimmermann reaction. *0427*

Phenobarbital No effect on Zimmermann procedure at 200 mg/L. *2977*

Phenytoin No effect on Zimmermann procedure at 200 mg/L. *2977*

Prochlorperazine No significant effect with Zimmermann reaction. *0427*

Propoxyphene Added in vitro no effect on Zimmermann reaction. *0766*

Testosterone Measured as a ketosteroid. *1487*

Urine No Effect Physiological
Acetaminophen No effect with short term ingestion 2.6 g/d. *0766*

Aspirin No effect with short term ingestion 2.6 g/d. *0766*

Barbiturates Short term ingestion produces no effect. *0766*

Carbamazepine No physiological effect observed. *2901*

Chlordiazepoxide No effect of short term ingestion of 30 mg/d. *0766*

Clopamide Single case of diuretic associated myopathy. *1793*

Diazepam No effect with short term ingestion of 15 mg/d. *0766*

Diphenhydramine No effect with short term ingestion of 150 mg/d. *0766*

Iothalamate No physiological effect observed. *2573*

Lead In a population of male battery workers compared with control group in cement industry. *0172*

Phenobarbital No effect short term ingestion of 120 mg/d. *0766*

Urine Increase Analytical

Acetone Has absorption at 520 nm in Zimmermann procedure. *0022*

Acetophenone Interferes in measurement by Zimmermann method. *3505*

Antihypertensive Agents Some may be measured as analytes. *0279*

Ascorbic Acid Due to chemical structure affects Zimmermann procedure. *3217*

Cephaloridine On day of administration — Zimmermann reaction. *1779*

Cephalothin Non specificity of Zimmermann reaction on day of administration. *1779* Marked effect on Zimmermann procedure with 6 g dose. *1932*

Chloramphenicol Affects some methods but not modified Reddy method. *1488*

Chlorothiazide Possible interference *in vitro. 0583*

Chlorpromazine Interferes with Zimmermann reaction. *3505*

Cloxacillin Interferes with Zimmermann reaction. *3505*

Dexamethasone Affects method of Reddy. *1686*

Erythromycin Interference with measuring procedure. *1488*

Ethinamate Interferes with Zimmermann reaction. *0427*

Etryptamine Slight increase reported with *in vivo* studies. *0583*

Meprobamate Glucuronide interferes with Zimmermann reaction. *3125*

Methicillin Absorption at 520 nm in Zimmermann procedure. *0022*

Methyprylon Reported with Holtorff-Koch modification of Zimmermann. *0427*

Morphine Due to chemical structure affects Zimmermann procedure. *3217*

Nalidixic Acid Interferes with Zimmermann reaction. *2197*

Oleandomycin Interferes with Zimmermann reaction. *2628*

Oxacillin Absorption at 520 nm in Zimmermann procedure. *0022*

Penicillin Interferes with Zimmermann reaction. *3505*

Phenaglycodol Interferes with Zimmermann reaction. *3505*

Phenazopyridine Interferes with Zimmermann reaction. *3505*

Phenothiazines Increased absorbance abnormal color Zimmermann procedure. *0427*

Piperidine Interference with Zimmermann reaction *in vitro. 0427*

Quinidine Metabolite interferes with Zimmermann reaction. *1022*

Reserpine Due to chemical structure affects Zimmermann procedure. *3217*

Secobarbital Metabolite interferes with Zimmermann reaction. *1022*

Spironolactone Metabolite interferes with Zimmermann reaction. *3505*

Tranquilizers Some artifactually increase value. *0279*

Urine Increase Physiological

Aging Progressive increase to puberty, continues to increase after in men. *0887*

Ampicillin After 14 d in 1 postmenopausal woman. *3622*

Chlorpromazine Alters steroid metabolism. *0427*

Chlorthalidone Alleged increased excretion. *1488*

Corticotropin Stimulation of adrenal. *1488*

Danazol 80% increase in 7 women given 800 mg daily for 2 mo. *2224*

Diurnal Variation Maximum in morning, lowest at night. *2222*

Ethinyl Estradiol Temporary enhanced excretion on stopping drug. *1685*

Gonadotropin Metabolic response. *3225*

Muscular Exercise Effect most marked in well trained individuals. *0185*

Nandrolone Anabolic effect. *2681*

Pregnancy Upper limit of normal at term. *0987*

Testolactone Reported effect. *1488*

Urine Decrease Analytical

Carbamazepine Forms colored compound at 430 nm in Zimmermann procedure. *2977*

Chlordiazepoxide Interferes with Zimmermann reaction (*in vitro, vivo*). *0427*

Chlormerodrin Possible interference with Zimmermann reaction. *0427*

Digoxin Slight effect on Zimmermann reaction *in vitro. 0427*

Estrogens Interferes with Zimmermann reactions. *1237*

Glucose Interferes with Zimmermann reaction. *3505*

Meprobamate Interferes with Zimmermann react (after Allen correct). *3125*

Metyrapone Interferes with Zimmermann reaction. *3505*

Promazine *In vitro* effect at least on Zimmermann reaction. *0427*

Propoxyphene Affects Zimmermann procedure after Allen correction. *0022*

Reserpine Interference with Zimmermann procedure. *0427*

Secobarbital Negative interference on Zimmermann reaction *in vitro. 0427*

Spironolactone Affects Zimmermann procedure after Allen correction. *0022*

Urine Decrease Physiological

Aminoglutethimide Inhibits steroid biosynthesis. *1252*

Ampicillin Increase by 42% after 7 d administration in pregnant women. *3622*

Betamethasone Feedback pituitary suppression of ACTH. *1238*

Blindness Significantly lower (average 3.5 mg/d) than normal. *1637*

Carbon Disulfide Reduction in relation to time of exposure. *0818*

Chlorpromazine Inhibition of hypothalamus and decreased ACTH secretion. *1252*

Corticosteroids If given orally, inhaled or topically. *1487*

Corticotropin Reduction even if orally or topically given. *0446*

Cortisone Pituitary suppression of ACTH. *3879*

Fluoxymesterone Endogenous hormone suppression. *1680*

Menopause Normal response. *0987*

Meperidine Probable effect (inhibits ACTH and PGH release). *1343*

Methandrostenolone Metabolic action of drug. *0019*

Metronidazole If previously elevated, ?depresses adrenal cortex. *1343*

Morphine Inhibits ACTH and pituitary gonadotropin release. *1343*

Noise Reported on exposure to 130 decibels. *3351*

Oral Contraceptives Probable decrease in cortisol secretion. *2596*

Oxymetholone Anabolic effect (possible for 2 weeks after stop). *1679*

Phenothiazines Inhibit release of steroid hormones. *2220*

Phenytoin Alters steroid metabolism. *1488*

Probenecid Decrease of up to 50% reported. *1198*

Propoxyphene Probable action on hypothalamic pituitary. ACTH secretion. *0766*

Pyrazinamide Temperature decreased excretion reported. *1343*

Spironolactone Decreased in 5 of 7 men over 2 weeks. *2773*

Kininogen

Plasma Decrease Physiological

Cyclophosphamide Maximum effect with onset of leukopenia. *0779*

Fluorouracil Maximum effect with onset of leukopenia. *0779*

Kynurenic Acid

Urine   Increase   Physiological
Stress Cold Exposure   Possibly due to activation of tryptophan oxygenase. *1169*

Kynurenine

Urine   No Effect   Physiological
Smoking   Probably no significant effect with tryptophan load. *3183*   No effect of smoking for 1 week. *0068*

Urine   Increase   Physiological
Estrogens   Induce tryptophan pyrrolase. *3045*
Glucocorticoids   Induce tryptophan pyrrolase. *3045*
Hydrocortisone   Causes induction of tryptophan pyrrolase. *3047*
Isoniazid   Induces pyridoxal PO_4 deficiency. *3045*
Oral Contraceptives   Estrogen component induces tryptophan pyrrolase. *3045*
Penicillamine   Induces pyridoxine antagonism. *1343*
Pregnancy   Causes induction of tryptophan pyrrolase. *3045*

Urine   Decrease   Physiological
Anabolic Steroids   Effect of synthetic androgens given to males. *3045*
Androgens   Effect of synthetic androgens given to males. *3045*

Lactate

Serum   No Effect   Analytical
Acetylsalicylic Acid   At concentration of 1400 mg/L had no effect on enzyme method. *3393*
Ampicillin   At concentration of 1200 mg/L had no effect on enzymatic method. *3393*
Ascorbic Acid   At concentration of 600 mg/L had no effect on enzymatic method. *3393*
Bisacodyl   At concentration of 4,000 mg/L had no effect as measured by enzymatic method. *3393*
Caffeine   At concentration of 300 mg/L had no effect on enzymatic method. *3393*
Chloramphenicol   At concentration of 400 mg/L had no effect on enzymatic method. *3393*
Chlordiazepoxide   At concentration of 6 mg/L had no effect on enzymatic method. *3393*
Chromonar Hydrochloride   At concentration of 180 mg/L had no effect on enzymatic method. *3393*
Dextran 60   At concentration of 6,000 mg/L had no effect on enzymatic method. *3393*
Diatrizoic acid   At concentration of 2600 mg/L had no effect on enzymatic method. *3393*
Dipyridamole   At concentration of 40 mg/L had no effect on enzymatic method. *3393*
Ethaverine   At concentration of 400 mg/L had no effect on enzymatic method. *3393*
Furosemide   At concentration of 100 mg/L had no effect on enzymatic method. *3393*
Heparin   No effect observed on enzymatic method. *2268*
Indomethacin   At concentration of 40 mg/L had no effect on enzymatic method. *3393*
Meglumine   At concentration of 1200 mg/L had no effect on enzyme method. *3393*
Methaqualone   At concentration of 25 mg/L had no effect on enzymatic method. *3393*
Methyldopa   At concentration of 400 mg/L had no effect on enzymatic method. *3393*
Nitrofurantoin   At concentration of 60 mg/L had no effect on enzymatic method. *3393*
Novaminsulfon   At concentration of 800 mg/L had no effect on enzymatic method. *3393*
Oxazepam   At concentration of 30 mg/L had no effect on enzymatic method. *3393*

Oxyphenbutazone   At concentration of 120 mg/L had no effect on enzymatic method. *3393*
Oxytetracycline   At concentration of 600 mg/L had no effect on enzymatic method. *3393*
Phenazopyridine   At concentration of 120 mg/L had no effect on enzymatic method. *3393*
Phenobarbital   At concentration of 60 mg/L had no effect on enzymatic method. *3393*
Phenprocoumon   At concentration of 2 mg/L had no effect on enzymatic method. *3393*
Phenytoin   At concentration of 160 mg/L had no effect on enzymatic method. *3393*
Probenecid   At concentration of 200 mg/L had no effect on enzyme method. *3393*
Procaine   At concentration of 40 mg/L had no effect on enzymatic method. *3393*
Rifampin   At concentration of 150 mg/L had no effect on enzymatic method. *3393*
Sulfamethoxazole   At concentration of 320 mg/L had no effect on enzymatic method. *3393*

Serum   No Effect   Physiological
Nifedipine   In 15 hypertensive patients undergoing hemodialysis after 3 weeks treatment. *2986*

Serum   Increase   Analytical
Serum   If no antiglycolytic agent (arises from glucose). *2690*
Standing of Sample   May be significantly increased in heparinized sample if over 15 minutes. *2268*

Serum   Increase   Physiological
Albuterol   Significant dose-related effect after i.v. infusion of therapeutic doses in 4 healthy male volunteers. *2807*
Bicarbonates   Small effect after i.v. Na bicarbonate. *1487*
Buformin   Rises but not usually to level of acidosis. *2427*
Dimercaprol   Associated with metabolic acidosis. *0071*
Epinephrine   Marked effect like beta-adrenergic compounds. *1343*
Ethanol   Causes lactic acidosis. *0586*   Produced pyruvate with increased NADH/NAD ratio. *2428*
Ether   Metabolic effect. *3918*
Ethylene Glycol   Due to NADH formation from oxidation of ethylene glycol and inhibition of tricarboxylic acid cycle by toxic metabolites. *3325*
Fructose   Metabolic response to i.v. or oral fructose. *3113*
Glucose   Increased concentration observed after i.v. infusion. *1160*
Isoniazid   If overdose with inhibition of NAD activity. *1487*
Lactose   Maximum at 1 h, persists for 2 h. *1907*
Metformin   Possibly always with predisposing condition. *2103*
Muscular Exercise   Arterial lactate increased from 5.5 to 20 μmol/ml. *1924*   Considerable effect of exercise. *3873*
Oral Contraceptives   Alteration in carbohydrate metabolism. *2385*
Rimeterol   Dose related significant change in 4 healthy men given therapeutic i.v. dose. *2807*
Ritodrine   Moderate effect when given intravenously for premature labor. *0575*
Sorbitol   In response to i.v. infusion. *1160*
Streptozocin   Temporary effect following infusion. *2532*
Sucrose   Increase maximum 1 h after, persists for 2 h. *1907*
Tetracosactrin   In healthy volunteers given 1 mg intramuscularly for up to 60 h. *1813*
Xylitol   4 times control after loading test. *3212*

Serum   Decrease   Analytical
Hemoglobin   At concentrations where no visible hemolysis: effect greater on Du Pont aca method than Abbott TD_x. *3154*
Malic Acid   Interferes with enzyme methods. *1150*
Pyruvate   High concentrations interfere with enzymatic method. *1150*

Serum   Decrease   Physiological

Morphine   Intravenous administration of 0.33 mg/kg caused 50% decrease. *1487*

Plasma   No Effect   Analytical

Acetaminophen   At concentration of 600 mg/L had no effect on enzymatic method. *3393*

Plasma   No Effect   Physiological

Isoniazid   Probable effect of normal dose and renal function. *1487*

Plasma   Increase   Analytical

Hemolysis   Due to metabolic activity of cells. *3873*

Plasma   Increase   Physiological

Aspirin   Due to late metabolic acidosis and renal impairment. *1343*

Carbamazepine   Dose related effect observed in long term treated epileptic compared with controls. *2008*

Phenobarbital   Dose related in long-term treated epileptics compared with controls. *2008*

Phenytoin   Dose related in long-term treated epileptics compared with controls. *2008*

Valproic Acid   Dose related effect in long term treated epileptics compared with controls effect not as marked as with other anticonvulsants. *2008*   When 1 g given orally to fasting individuals. *3633*

Plasma   Decrease   Analytical

Glyceric Acid   High concentrations interfere with enzymatic method. *1150*

Plasma   Decrease   Physiological

Methylene Blue   Variable response when lactic acidosis. *1487*

Blood   Increase   Physiological

Disopyramide   Marked effect observed in 2 patients both receiving other drugs and occurring with hypotension. *2557*

Methylprednisolone   Significant effect in 10 of 13 patients with myocardial infarction. Effect observed in 1 h, max at 3 h, persisted for 24 h. *1560*

Propylene Glycol   Statistically significant correlation with serum concentration of drug when used as vehicle for other i.v. drugs. *1906*

Urine   No Effect   Physiological

Lithium   No effect observed after administration. *0409*

Saliva   Increase   Physiological

Fructose   Occurs at 1 h, as with blood. *1907*

Lactose   Maximum at 1 h, persists for 2 h. *1907*

Sucrose   Increase maximum 1 h after, persists for 2 h. *1907*

Lactate Dehydrogenase

Serum   No Effect   Analytical

Acetaminophen   At acute overdose concentration (10 mg/dL) on SMAC® method. *3719*   On continuous methods using either pyruvate or lactate as substrate at 10 times maximal therapeutic concentration. *1785*

Acetazolamide   No effect at 12 mg/dL on SMA 12/60 method. *2636*

Acetylsalicylic Acid   At 5 times upper limit of therapeutic range on method on SMAC®, Abbott-VP, Cobas-Bio, and Hitachi® 705. *2138*   At 8300 µmol/L on continuous method with lactate as substrate. At 8300 µmol/L on colorimetric method. *1785*

Aminophenazone   On continuous method with pyruvate as substrate at 10 times maximal therapeutic concentration. On continuous method with lactate as substrate at 10 times maximal therapeutic concentration. On colorimetric method at 10 times maximal therapeutic concentration. *1785*

Aminosalicylic Acid   At 5 times upper limit of therapeutic range on methods on SMAC®, Abbott-VP, Cobas-Bio, Hitachi® 705 and KDA. *2138*

Amitriptyline   At acute overdose concentration (2.5 mg/dL) on SMAC® method. *3719*

Barbital   At acute overdose concentration (20 mg/dL) on Technicon® SMAC® method. *3719*

Bethanechol   No effect at 0.09 mg/dL on SMA 12/60 method. *2636*

Bromazepam   At 5 times therapeutic concentration on SMAC®, Abbott-VP, Cobas-Bio, Hitachi®, KDA. *2138*

Caffeine   At acute overdose concentration (10 mg/dL) on Technicon® SMAC® method. *3719*   On continuous method with pyruvate as substrate at 10 times maximal therapeutic concentration. On colorimetric method at 10 times maximal therapeutic concentration. *1785*

Carbenicillin   At 5 times upper limit of therapeutic range on methods on SMAC®, Abbott-VP, Cobas-Bio, Hitachi® 705 and KDA. *2138*

Cefoxitin   At 5 times upper limit of therapeutic range on methods on SMAC®, Abbott-VP, Cobas-Bio, Hitachi® 705 and KDA. *2138*

Cephalothin   At 5 times upper limit of therapeutic range on methods on SMAC®, Abbott-VP, Cobas-Bio, Hitachi® 705 and KDA. *2138*

Chlordiazepoxide   At 5 times upper limit of therapeutic range on methods on SMAC®, Abbott-VP, Cobas-Bio, Hitachi® 705 and KDA. *2138*   At acute overdose concentration (5 mg/dL) on Technicon® SMAC® method. *3719*

Chlorpheniramine   At acute overdose concentration (20 mg/dL) on Technicon® SMAC® method. *3719*

Cocaine   At acute overdose concentration (2.5 mg/dL) on Technicon® SMAC® method. *3719*

Codeine   At acute overdose concentration (2.0 mg/dL) on Technicon® SMAC® method. *3719*

Cyclosporine   At concentration of 41.5 mmol/L (50 µg/L) on methods on Technicon® SMAC® II and. *1123*

Cysteine   At 5 times upper limit of therapeutic range on methods on SMAC®, Abbott-VP and Hitachi® 705. *2138*

Cystine   At 5 times upper limit of therapeutic range on methods on SMAC®, Abbott-VP, and Hitachi® 705. *2138*

Deslanoside   No effect on SMA 12/60 method at 0.06 mg/dL. *2636*

Diazepam   At 5 times upper limit of therapeutic range on methods on SMAC®, Abbott-VP, Cobas-Bio, Hitachi® 705 and KDA. *2138*   At acute overdose concentration (2.5 mg/dL) on Technicon® SMAC® method. *3719*

Diclofenac   On continuous method with lactate or pyruvate as substrate at 10 times maximal therapeutic concentration. On colorimetric method at 10 times maximal therapeutic concentration. *1785*

Diethylpropion   At acute overdose concentration (10 mg/dL) on Technicon® SMAC® method. *3719*

Digitoxin   No effect at 21 mg/dL on SMA 12/60 method. *2636*

Digoxin   No effect at 0.04 mg/dL on SMA I1/60 method. *2636*

Diphenhydramine   At acute overdose concentration (20 mg/dL) on Technicon® SMAC® method. *3719*

Disulphine Blue   Using recommended Scandinavian method. *1464*

Ethacrynic Acid   At 2.5 mg/dL no effect on SMA 12/60 method. *2636*

Ethanol   At acute overdose concentration (20 mg/dL) on Technicon® SMAC® methods. *3719*

Ethchlorvynol   At acute overdose concentration (20 mg/dL) on Technicon® method. *3719*

Ethinamate   At acute overdose concentration (20 mg/dL) on Technicon® SMAC® method. *3719*

Flurazepam   At 5 times upper limit of therapeutic range on methods on SMAC®, Abbott-VP, aca, Cobas-Bio, Hitachi® 705 and KDA. *2138*   At acute overdose concentration (2.5 mg/dL) on Technicon® SMAC® method. *3719*

Furosemide   No effect at 1.4 mg/dL on SMA 12/60 method. *2636*

Glutethimide   At acute overdose concentration (5 mg/dL) on Technicon® SMAC® method. *3719*

Lactate Dehydrogenase *(continued)*

Serum   No Effect   Analytical *(continued)*

Ibuprofen   At 5 times upper limit of therapeutic range on methods on SMAC®, Abbott-VP, Cobas-Bio, Hitachi® 705 and KDA. *2138*   On continuous method with lactate or pyruvate as substrate at 10 times maximal therapeutic concentration. On colorimetric method at 10 times maximal therapeutic concentration. *1785*

Indomethacin   With lactate or pyruvate as substrate on continuous method. *1785*   On colorimetric method at 10 times maximal therapeutic concentration. *1785*

Ketoprofen   With pyruvate as substrate on continuous method. *1785*

Lidocaine   No effect SMA 12/60 method with 3.5 mg/dL. *2636*

Mannitol   No effect on SMA 12/60 method at 445 mg/dL. *2636*

Meperidine   At acute overdose concentration (5 mg/dL) on Technicon® SMAC® method. *3719*

Meprobamate   At acute overdose concentration (20 mg/dL) on Technicon® SMAC® method. *3719*

Meralluride   No effect at 0.07 mL/dL on SMA 12/60 method. *2636*

Mesoridazine   At acute overdose concentration (20 mg/dL) on Technicon® SMAC® method. *3719*

Methadone   At acute overdose concentration (2.5 mg/dL) on Technicon® SMAC® method. *3719*

Methanol   At acute overdose concentration (20 mg/dL) on Technicon® SMAC® method. *3719*

Methaqualone   At acute overdose concentration (2.5 mg/dL) on Technicon® SMAC® method. *3719*

Methicillin   At 5 times upper therapeutic range on methods on SMAC®, Abbott-VP, Cobas-Bio, Hitachi® 705 and KDA. *2138*

Methotrexate   At 5 times upper limit of therapeutic range on method on Abbott-VP, Cobas-Bio, Hitachi® 705 and KDA. *2138*

Methylphenidate   At acute overdose concentration (20 mg/dL) on Technicon® SMAC® method. *3719*

Methyprylon   At acute overdose concentration (10 mg/dL) on Technicon® SMAC® method. *3719*

Morphine   At acute overdose concentration (20 mg/dL) on Technicon® SMAC® method. *3719*

Naproxen   At 5 times upper limit of therapeutic range on methods on SMAC®, Abbott-VP, Cobas-Bio, Hitachi® 705 and KDA. *2138*

Nitrofurantoin   On routine methods in use on SMAC®, Ektachem®, Abbott-VP, Cobas-Bio, aca, Hitachi® 705, KDA at 5 times normal upper therapeutic concentration. *2138*

Nortriptyline   At acute overdose concentration (20 mg/dL) on Technicon® SMAC® method. *3719*

Ouabain   At 0.02 mg/dL no effect on SMA 12/60 method. *2636*

Penicillin G   At 5 times upper limit of therapeutic range on methods on SMAC®, Abbott-VP, Cobas-Bio, Hitachi® 705 and KDA. *2138*

Pentobarbital   At acute overdose concentration (20 mg/dL) on Technicon® SMAC® method. *3719*

Perphenazine   At acute overdose concentration (20 mg/dL) on Technicon® SMAC® method. *3719*

Phenobarbital   At acute overdose concentration (20 mg/dL) on Technicon® SMAC® method. *3719*   With pyruvate as substrate on continuous method. On colorimetric method at 10 times maximal therapeutic concentration. *1785*

Phenytoin   No effect at 1.8 mg/dL on SMA 12/60 method. *2636*   At acute overdose concentration (20 mg/dL) on Technicon® SMAC® method. *3719*

Plasma   When kinetic method on LKB 8600. *2228*

Procainamide   No effect on SMA 12/60 method at 35 mg/dL. *2636*

Promethazine   At acute overdose concentration (20 mg/dL) on Technicon® SMAC® method. *3719*

Propoxyphene   At acute overdose concentration (2.5 mg/dL) on Technicon® SMAC® method. *3719*

Propranolol   No effect at 0.1 mg/dL on SMA 12/60 method. *2636*

Pyribenzamine®   At acute overdose concentration (20 mg/dL) on Technicon® SMAC® methods. *3719*

Quinidine   At 26 mg/dL no effect on SMA 12/60 method. *2636*   At acute overdose concentration (20 mg/dL) Technicon® SMAC® method. *3719*

Quinine   At acute overdose concentration (1.5 mg/dL) on Technicon® SMAC® method. *3719*

Reserpine   No effect at 0.02 mg/dL on SMA 12/60 method. *2636*

Rifampin   At 5 times upper limit of therapeutic range on method on SMAC®, Abbott-VP, Cobas-Bio and Hitachi® 705. *2138*

Secobarbital   At acute overdose concentration (20 mg/dL) on Technicon® SMAC® method. *3719*

Standing of Sample   If on polystyrene beads 24 h at 25°. *2624*

Storage of Sample   8 h at room temperature, 4 d at 4°, 20 d at -20°. *3217*   No effect 1 week at 4°, 1 mo at -20°. *1563*

Tetracycline   At 5 times upper limit of therapeutic range on methods on SMAC®, Abbott-VP, Cobas-Bio, Hitachi® 705 and KDA. *2138*

Theophylline   At 5 times upper limit of therapeutic range on methods on SMAC®, Abbott-VP, Cobas-Bio, Hitachi® 705 and KDA. *2138*

Thiopental   At acute overdose concentration (20 mg/dL) on Technicon® SMAC® method. *3719*

Triamterene   At 10 mmol/L on UV methods. *0618*

Vitamin Preparations   No effect at expected concentration with SMA 12/60 procedure. *2637*

Serum   No Effect   Physiological

Anticonvulsants   No effect of chronic therapy. *3560*

Aspirin   When approximately 3 g/d ingested for several weeks. *3076*

Flucloxacillin   In a single patient on hemodialysis: abnormal results only mild. *2200*

Fructose   No effect observed with ingestion. *1907*

Height   No correlation in children. *0887*

Hemodialysis   Unless intravascular hemolysis occurs. *3613*

Heparin   No effect of subcutaneous drug after 8 d. *2454*

Lactose   No effect on activity observed. *1907*

Levonorgestrel   No significant difference over 3 y between implant recipients and IUD users. *0763*   In a group of women using levonorgestrel covered rods versus controls using copper IUD. *0899*

Oral Contraceptives   No significant effect observed. *3319*

Streptokinase   Although marked effect on liver. *3189*

Sucrose   Unaffected by ingestion. *1907*

Serum   Increase   Analytical

Acetaminophen   On colorimetric method at 10 times maximal therapeutic concentration. *1785*

Caffeine   On continuous method with lactate as substrate at 10 times maximal therapeutic concentration. *1785*

Contact With Clot   High intracellular concentration of LDH activity. *1563*

Fluosol-DA   Increase by 37% on SMA IIC at concentration of 50%. Increase by 18% on Du Pont aca III at concentration of 50%. *2518*

Hemoglobin   Marked effect on UV procedures. *3217*

Hemolysis   Released from RBCs and platelets. *1237*

Phenobarbital   With lactate as substrate on continuous method. *1785*

Plasma   Plasma high colorimetrically on SMA 12/60. *2228*

Rifampin   At 5 times upper limit of therapeutic range on method on KDA. *2138*

Serum   Up to 40% higher than plasma. *0435*

Standing of Sample   25% increase reported if in contact with clot for 1 h. *0877*

Transport of Specimen   Increase by 50 units in pneumatic tube. *3217*

Triamterene   At 100 μmol/L on fluorometric procedure. *0618*

Serum   Increase   Physiological

Amiodarone   Up to 1.5 to 3 times normal values in 55% patients with hepatitis occurring in 4%. *2011*

Amphotericin B   Noted in one patient with acute myelogenous leukemia when treated with high dose. Probable idiosyncratic response. *2443*

Anabolic Steroids   ?part of cholestatic syndrome. *0583*

Anesthetic Agents Response occurs even with premedication. *3661*

Aspirin Experimental effect seen in rabbits with prolonged use. *3092*

Captopril Isolated case of cholestasis in patient receiving 25 mg tid for 1 mo. *2732* Characteristic pattern of hepatocellular jaundice in one patient. But with secondary cholestatic elements. *3688*

Carbenicillin Transient elevations reported. *1343*

Chloramphenicol Isolated case of hepatitis and pancytopenia reported which resolved with discontinuation of drug. *0602*

Chlorpromazine In one patient who developed cholestasis with fever and leukopenia after 260 mg drug. *2840*

Chlorpropamide Isolated case of drug-induced hemolytic anemia. *3111* In isolated case of hypersensitivity reaction producing hepatic granulomas also in bone marrow. *2988*

Chlorthalidone Single case of severe reversible myopathy noted. *1793*

Cimetidine In isolated case of drug associated bridging hepatic necrosis. *0864* In isolated case treated for 7 mo with 400 mg/d progressed to bridging hepatic necrosis. *3085*

Ciprofloxacin Rare cases reported during clinical trials. *0151*

Clindamycin ?hepatic or muscle origin (transient). *1083*

Clofibrate May cause muscle fiber atrophy in many patients. *1883*

Clopamide Single case of diuretic associated myopathy. *1793*

Codeine May cause rise in intrabiliary pressure. *3505*

Copper Toxic effect, sharp increase with hemolytic crisis. *0385*

Dapsone In 3 individuals treated for 3 to 36 weeks. *3478* Associated with hemolytic anemia when occurs as main side effect. *3423* Sulfone syndrome in one 16 year old girl given 50 mg daily short-term administration. *3604*

Dicumarol May be high enough to simulate myocardial infarction. *0248*

Enflurane In about 10% patients given 2 to 3 administrations of anesthetic agent. *1089*

Ethanol Increased in normals after 1 dose 3 g/kg body weight. *1621* Dramatic increase observed in acute alcoholic myopathy and less marked increase with chronic myopathy. *3124* Slight effect at 15 and 110 h after 0.75 g/kg ingestion. *2134*

Etretinate In one 74-year old woman treated for severe psoriasis: normalized with withdrawal of therapy and recurred when reinstituted. *3789*

Floxuridine Manifestation of toxicity. *1680*

Fluorides Due to anoxia with tissue damage. *0010*

Fluphenazine Isolated case of hepatotoxicity with centrilobular cholestasis. *3372*

Gold Cholestatic jaundice in 3 patients with rheumatoid arthritis given gold sodium thiomalate: all patients recovered spontaneously. *1084*

Halothane Hepatotoxic in rare cases, ?idiosyncrasy. *3294* Low frequency but greater number of abnormalities than with enflurane. Occurred most frequently in obese patients. *1089*

Hemodialysis Variable response of total LDH. *2990*

Heparin Abnormal in 36% patients receiving either porcine or bovine derived drug. *0968* In 1 of 13 patients with cerebrovascular accident after low-dose heparin treatment. *2671*

Hydralazine Isolated case with moderate hepatomegaly, abnormal findings resolved when drug discontinued. *1161* Asymptomatic or symptomatic reversible hepatitis-like reaction or granulomatous hepatitis. *2374*

Ibuprofen Rare case of hepatotoxicity: in one case associated with Stevens-Johnson syndrome. *3460*

Imipramine May cause cholestatic jaundice. *1873*

Levodopa Rare instance of elevation, ?origin. *0071*

Meals Significant effect observed Affected by meals (most marked in women). *3455*

Meperidine May cause rise in intrabiliary pressure. *0248*

Methotrexate Effect in 40% cases of psoriasis (reversible). *2608*

Methyldopa Chronic active hepatitis associated with immune hemolytic anemia in one case. *3269*

Methyltestosterone Infrequent rise observed. *0563*

Mithramycin Hepatocellular damage observed. *0071*

Morphine May cause rise in intrabiliary pressure. *3505*

Muscle Massage Significant effect but not to pathological level 8 h after. *0423*

Muscular Exercise Maximum effect observed following day. *1491* Marked increase with exercise. *2981*

Nitrofurantoin May cause hemolytic anemia. *1974* In 6 of 17 acute and 12 of 19 chronic drug induced pulmonary reactions. *1639* In 16 of 1756 acute and 12 of 19 chronic drug induced pulmonary reactions. *1639* May follow acute hepatitis with or without cholestasis or as chronic active hepatitis or chronic granulomatous reaction. *2374*

Norethandrolone ?part of cholestatic syndrome. *0583*

Norfloxacin In 2 of 1540 patients in clinical trials. *0731*

Ohio 469 Preliminary finding of significant increase postoperatively. *1647*

Oxacillin Reversible high-dose associated liver injury. *2431*

Oxyphenisatin Increase observed with active hepatitis. *0066*

Penicillamine In 6 of 99 patients treated for rheumatoid arthritis: evidence of toxic liver necrosis observed in two cases. *3887*

Phenprocoumon Mild effect but recurrently in 2 patients having repeated exposure to drug. *3347*

Phenylbutazone In 2 patients subsided after drug withdrawn associated with granulomas and cholestatic-hepatocellular injury. *1730*

Physical Training Tends to be higher in trained. *1491* Resting level increased, response to exercise dec. *1696*

Piperacillin In 1 of 59 patients given drug as sole agent with many patients with severe illness. *3874*

Pregnancy Physiological effect observed. *0718*

β-Propiolactone Increase by 75 U/L when plasma from BPL treated blood compared with plasma in 25 specimens. *0217*

Propoxyphene Cholestatic effect. *1954*

Propylthiouracil Temporary effect associated with drug administration regressed with cessation. *2731* In one 24 year old woman who developed fulminant hepatic failure with lymphocyte sensitization. *2437*

Prostate Palpation Release into bloodstream. *1731*

Quinidine May cause mild hepatic impairment in few patients. *2531* If given intramuscularly, possibly due to increase in intracellular calcium. *3215* In single case of severe quinidine hypersensitivity. *0858* Isolated case of reversible granulomatous hepatitis from quinidine hypersensitivity and granuloma induction within 3 d of quinidine readministration. *0448*

Season High in summer, low in winter. *3872*

Starvation With absolute fast 15 d ?hepatic focal necrosis. *0422*

Streptokinase Earlier peak of CK-MB observed in patients on thrombolytic therapy due to reperfusion of ischemic myocardium. *1341*

Sulfamethoxazole May cause hemolytic anemia. *1974*

Sulfamethoxypyridine May occur with preponderance of LD-5. *1987*

Sulfasalazine Rare hepatotoxicity reported associated with non-caseating granuloma or focal inflammation with or without necrosis. *1127* In 2 cases with drug induced hepatotoxicity of hypersensitivity type of reaction: reversed on drug withdrawal. *3364*

Sulfisoxazole May cause hemolytic anemia. *1974*

Sulindac Observed with other laboratory evidence of hepatitis in 1 child. *1889*

Tetracycline 2 cases of drug-associated fatty liver of pregnancy. *3795*

Thiopental pyrexia and jaundice in one case after repeated anesthesia. *0391*

Tolmetin Observed in single case of 49 y old man who had taken drug as needed for arthritis for 1 y. *3421*

Valproic Acid Acute hepatic centrilobular necrosis with severe fatty change in small number of patients. *3763* In 2 of 9 patients with epilepsy poorly controlled. *3508*

Vasopressin Rhabdomyolysis observed in 2 patients following intravenous drug administration. *0029*

Verapamil Observed in a single case, reverted to normal as soon as drug withdrawn. *2561*

Xylitol Dose related hepatocellular damage. *3212*

Lactate Dehydrogenase (continued)

Serum   Decrease   Analytical
Acetylsalicylic Acid   At 5 times upper limit of therapeutic range on method on KDA. *2138*   With 8300 μmol/L on continuous method with pyruvate substrate. *1785*

Ascorbic Acid   At therapeutic concentration may depress SMA 12/60 value. *3335*

Cefotaxime   Generally significant decrease of drug and metabolite on parallel method. *0201*

Detergents   30% inhibition at 0.6-0.8 mg/ml. *3217*

Ketoprofen   With lactate as substrate on continuous method. Marked effect on colorimetric method at 10 times maximal therapeutic concentration. *1785*

Lipemia   With direct colorimetry: resultant effect of clearing of lipemia and formation of reaction product during SMAC® measurement. *2466*

Liposol   Changed from 217.2 to 134.4 U/L in 26 randomly selected patient sera, before and after addition on Beckman Astra methods. *2054*

Methotrexate   At 5 times upper limit of therapeutic range on method on SMAC®. *2138*

Oxalate   Inhibition of enzyme activity. *3588*

Plasma   Probably due to absence of platelet component. *3319*

Storage of Sample   Minimal stability observed at 0° 10% loss even with quick freeze and thaw 10% decrease after 24 h at room temperature. *0877*

Theophylline   Effect small and can be ignored at therapeutic concentration. *3719*

Serum   Decrease   Physiological
Aging   Slow fall to puberty then steep to age 20 y. *0887*

Aldrin   *In vitro* 4.5% decrease at 1 x 10⁻⁴ mol/L. *1270*

Benzene   *In vitro* 6.1% decrease at 1 x 10⁻⁴ mol/L. *1270*

Clofibrate   Possibly derived from muscle damage. *2220*

Dieldrin   *In vitro* 8.0% decrease at 1 x 10⁻⁴ mol/L. *1270*

Fluorides   *In vivo* inhibition with low doses. *1168*

Hemodialysis   Variable response of total LDH. *2990*

Urine   Increase   Physiological
Acetrizoate   Sustained increase due to nephrotoxic action. *3538*

Aspirin   Renal irritation and desquamation of epithelial cells. *2897* Marked increase observed even when as little as 3.5 g aspirin ingested daily. *2866*

Proteinuria   In large number and variety renal diseases. *2897*

Radiographic Agents   Transient increase if injected in renal artery. *3538*

Urine   Decrease   Analytical
Ethionamide   Interference with determination procedure. *2897*

Nitrofurantoin   Interference with determination method. *2620*

Storage of Sample   Loses activity unless analyzed immediately. *3873*

Red Blood Cells   Increase   Physiological
Anticonvulsants   In 38% epileptics on chronic therapy. *3560*

Lactate Dehydrogenase Isoenzymes

Serum   Increase   Physiological
Bismuth Salts   Especially of LDH 1 and 2 — probably of renal origin. *2911*

Ethanol   Dramatic increase of LD-1 and LD-2 observed in acute alcoholic myopathy and less marked increase with chronic myopathy. *3124*

Fluorides   Mainly fraction 4 and 5 in poisoning. *0010*

Hemodialysis   LD-5 of renal origin if acute tubular necrosis. *2990*

Hypoxia   Increase of fraction 5 associated with shock. *3485*

Morphine   Hepatic fraction increase ?due to spasm of sphincter. *2427*

Muscular Exercise   3,4,5 increased but no change in 1,2. *3051*

Physical Training   Heart component increased. *1491*

Starvation   With absolute fast 15 d ?liver fractions increase. *0422*

Xylitol   Mainly due to liver component from liver damage. *3576*

Serum   Decrease   Analytical
Heat   Inactivates liver components to large extent. *0718*

Storage of Sample   LD-5 most heat and cold labile, may decrease with storage. *0877*

Urea   Inactivates cardiac components (LDH 1, LDH 2). *0718*

Serum   Positive   Analytical
Streptokinase   Broadening of band between LD-3 and LD-4 and absent LD-5. Complex formed between LD-3 and IgA. Unusual band disappeared. *3666*

Lactose

Serum   No Effect   Physiological
Lactose   No change with tolerance test in infants. *2551*

Serum   Increase   Physiological
Lactose   With tolerance test from 0.2 to 0.6 mg/dL. *2551*

Urine   Increase   Analytical
Cled Medium   Artifactual presence if in contact. *2756*

Urine   Increase   Physiological
Lactation   Quite common, especially in afternoon. *0987*

Lactose   Less than 1% of oral load. *2551*

Pregnancy   May occur in last trimester especially in afternoon. *0987*

Lactose Tolerance

Serum   Decrease   Physiological
Neomycin   Significantly lowered glucose response noted. *1077*

Latex Fixation

Serum   Positive   Physiological
Procainamide   Associated with drug induced lupus. *0930*

LDL + VLDL Cholesterol

Serum   Decrease   Physiological
Terazosin   Increase by 6.1 mg/dL with up to 20 mg daily over 4 weeks in patients with moderate hypertension. *0857*

Lead

Blood   No Effect   Physiological
Smoking   No difference between smokers and nonsmokers. *0489*

Blood   Increase   Physiological
Ethanol   30% increase in heavy versus occasional drinkers. *3273*

Heat   Increased temperature may cause mobilization of fixed. *0489*

Lead   May be increased much above normal of 50 μg/dL. *0987* In a population of male battery workers compared with control group in cement industry. *0172*

Tetraethyl-Lead   Doubled normal concentration observed. *0258*

Blood   Decrease   Physiological
EDTA   If lead poisoning present. *1679*

Urine   Increase   Analytical
Calcium   75 meq/L equivalent to 110 μg/L by atomic absorption. *3240*

Creatinine 100 mg/dL equivalent to 50 μg/L by atomic absorption. *3240*

Potassium 100 meq/L equivalent to 10 μg/L by atomic absorption. *3240*

Sodium 200 meq/L equivalent to 50 μg/L by atomic absorption. *3240*

Urea 1500 mg/dL equivalent to 60 μg/L by atomic absorption. *3240*

Urine   Increase   Physiological

Dimercaprol If poisoning due to lead. *0071*

EDTA If poisoning due to lead. *0071*

Lead Due to increased body load (in poisoning 80 μg/dL). *1302* In a population of male battery workers compared with control group in cement industry. *0172*

Penicillamine If poisoning due to lead. *0071*

Smoking Non smokers 27.1 μg/L, cigarette 28.6, cigar 29.0 μg/L. *2377*

Tetraethyl-Lead Excretion further increased by chelating agents. *0258*

Urine   Decrease   Analytical

Magnesium 25 meq/L equivalent to 5 μg/L by atomic absorption. *3240*

Semen   Increase   Physiological

Lead In a population of male battery workers compared with control group in cement industry. *0172*

LE Cells

Serum   Positive   Physiological

Oxyphenisatin Liver damage may present as acute hepatitis or as more advanced chronic disease: usual liver disease is chronic active hepatitis. *2260*

Blood   Positive   Physiological

Acebutolol Observed in several patients treated for 12 to 24 weeks. *0680*

Acetazolamide May produce LE-like syndrome. *3505*

Aminosalicylic Acid May produce LE-like syndrome. *2325*

Anthiolimine Observed effect in some cases. *0933*

Anticonvulsants May induce lupus like syndrome in some cases. *1574*

Chlorothiazide May produce LE-like syndrome. *3504*

Chlorpromazine Single case reported associated with rash and fever. *0963*

Chlorprothixene May produce LE-like syndrome. *3504*

Clazolam SLE may occur, usually normalized when stopped. *3059*

Corticosteroids Implicated as activators of SLE. *0962*

Demeclocycline Rare exacerbation may occur. *1680*

Digitalis SLE may occur, usually normalizes when stopped. *3059*

Doxycycline May produce exacerbation of SLE. *1680*

Ethosuximide Rare immune response reported. *0518*

Gold SLE may occur, usually normalizes when stopped. *3059*

Griseofulvin May produce LE-like syndrome. *3504*

Guanoxan SLE may occur, usually normalizes when stopped. *3059*

Hydantoin Derivatives May cause appearance of LE cells. *2220*

Hydralazine More common in slow acetylators. *2783* Observed in one patient with induced systemic lupus erythematosus. *3784* 3% patients developed lupus-like syndrome all of whom were slow acetylators and receiving less than 200 mg drug/d. *0353*

Hydrazine Derivatives May activate lupus erythematosus. *0616*

Isoniazid More common in slow acetylators. *2325*

Mephenytoin SLE may occur, usually normalized when stopped. *3059*

Methimazole Lupus-like syndrome reported. *1680*

Methsuximide Cause of SLE reported. *1680*

Methyldopa Autoimmune phenomenon. *3296*

Methylthiouracil May produce LE-like syndrome. *3505*

Methysergide Observed effect in some cases. *0933*

Minocycline May cause exacerbation of SLE. *1680*

Oral Contraceptives May precipitate or exaggerate LE-like syndrome. *3505*

Oxyphenisatin May cause lupoid hepatitis. *2970*

Paramethadione Rare idiosyncratic response. *2042*

Penicillamine Observed effect in some cases. *0933* In 6 patients with rheumatoid arthritis developed systemic lupus erythematosus-like syndrome. *0623*

Penicillin Allergic response with urticaria in one case. *1343*

Phenobarbital Rare idiosyncratic response. *2042*

Phenolphthalein Associated with hypersensitivity. *2427*

Phenylbutazone May precipitate or exaggerate LE-like syndrome. *3505*

Phenytoin May activate lupus erythematosus. *0616*

Pindolol Drug induced systemic lupus erythematosus reported in one case. *0299*

Primidone Less frequent than with many anticonvulsants. *0416*

Procainamide Reported effect. *2055* 10-20% patients eventually develop drug-induced lupus syndrome. *1958*

Propylthiouracil May produce LE-like syndrome. *3505*

Quinethazone Theoretical possibility of this type of drug. *1680*

Reserpine SLE may occur, usually normalizes when stopped. *3059*

Streptomycin May produce LE-like syndrome. *3505*

Sulfachlorpyridazine Observed occasionally. *1680*

Sulfadimethoxine Implicated as activator of SLE. *0962*

Sulfamethoxypyridine Implicated as activator of SLE. *0962*

Sulfisoxazole May cause LE phenomenon (rare). *1680*

Sulfonamides May produce LE-like syndrome. *3505*

Tetracycline May produce LE-like syndrome. *3505*

Thiazides SLE may occur, usually normalizes when stopped. *3059*

Tocainide Rare finding in fewer than 0.2% patients. *2011*

Trimethadione Rare immune response observed. *1226*

Lecithin

Serum   Increase   Physiological

Ethanol Moderate increase compared with controls in alcoholics especially in younger adults. *0430*

Plasma   Increase   Physiological

Ethinyl Estradiol Mirrors changes in lipoprotein pattern. *1431*

Lecithin Cholesterol Acyltransferase (LCAT)

Serum   Decrease   Physiological

Medroxyprogesterone Significant reduction after treatment. Effects observed in 15 postmenopausal women with endometrial cancer after 2 weeks treatment. *2118*

Plasma   Decrease   Physiological

Stanozolol Increase by 30% with 6 weeks treatment: reverted to normal by 5 weeks after treatment stopped. *0047*

Leucine

Plasma   Increase   Physiological

Obesity Characteristic finding correlated with insulin. *1106*

Starvation Marked effect even after 1 d. *0021*

Plasma   Decrease   Physiological

Alanine Decrease of about 40% with 1 week administration. *1263*

Glucose Probable muscle uptake. *1403*

Histidine Probably due to flow into gut after oral load. *1644*

Leucine *(continued)*

Plasma Decrease Physiological *(continued)*
Oral Contraceptives Hormonal effect (second part of cycle). *0742*

Urine Increase Physiological
Ascorbic Acid In 10 healthy females given 10 g/d. *3624*
Starvation Observed effect due to tissue catabolism. *0021*

Leucine Aminopeptidase

Serum No Effect Analytical
Storage of Sample 7 d at 30°, stable frozen. *0877* No effect 1 week at 4°, 1 mo at -20°. *1563* 1 d room temperature, 21 at 4°, 60 at -20°. *3217*

Serum No Effect Physiological
Estrogens No effect observed. *1498*
Progestogens No effect observed. *1498*

Serum Increase Physiological
Estrogens Altered metabolism. *0227*
Morphine Possibly due to spasm of sphincter of Oddi. *2427*
Oral Contraceptives Possible liver damage. *1462*
Pregnancy Moderately increased throughout pregnancy. *0987*
Quinidine Isolated case of hepatic toxicity (granulomatous hepatitis). *0862*
Thorium Dioxide Common finding (related to time since injection). *3147*

Serum Decrease Analytical
EDTA Inhibits if used as anticoagulant. *3217*

Urine No Effect Analytical
Storage of Sample No effect 1 d at 4°, 1 week at -20°. *3873*

Urine Increase Physiological
Aspirin Marked increase observed even when as little as 3.5 g aspirin ingested daily. *2866*
Colistin Probably associated with proximal renal tubular injury. *0078*
Kanamycin Associated with proximal renal tubular injury. *0078*
Phenolsulfonphthalein Facilitates permeation of enzyme into tubules. *2897*
Polymyxin Facilitates permeation of enzyme into tubules. *2897*
Proteinuria With toxic renal damage. *2897*
Radiographic Agents Facilitates permeation of enzyme into tubules. *2897*
Streptokinase Activates peptidases by plasminogen. *2897*
Streptomycin Facilitates permeation of enzyme into tubules. *2897*
Sulfonamides Facilitate permeation of enzyme into tubules. *2897*
Vasopressin Activates peptidases by release of plasminogen. *2897*
X-Ray Therapy Toxic damage due to released metabolites. *2897*

Urine Decrease Physiological
Aminocaproic Acid Due to antifibrinolytic action. *2897*
Aminomethylcyclohexane Due to antifibrinolytic action. *2897*
Aspirin Due to antifibrinolytic action. *2897*
Salicylate When 0.5 g given orally to 20 healthy adult volunteers and urine studied over next 3 h. *1828*

Leucine Aminopeptidase Isoenzymes

Serum No Effect Physiological
Phenobarbital No effect observed. *3087*

Serum Increase Physiological
Anticonvulsants Increase slower running components. *3087*
Oral Contraceptives Increased slow component ?liver involvement. *3087*
Phenytoin Increased slower running components. *3087*
Primidone Increased slower running components. *3087*

Leukoagglutinins

Serum Increase Physiological
Cephradine In a single patient either due to a toxic effect on bone marrow or immune mediated mechanism, probably immune basis. *2096*

Serum Positive Physiological
Methyldopa Observed in one patient with hepatocellular damage of moderate severity due to sensitization by drug. *0874*
Sulfapyridine Reported observation. *0810*

Leukocyte Esterase

Urine No Effect Analytical
Fluorescein At 250 mg/L in spiked specimen run on Ames Clinitek 200. *1722*

Leukocyte Migration Inhibition

Blood No Effect Physiological
Cimetidine No effect noted in 12 patients treated for acid-peptic disease. *2368*

Leukocytes

Blood No Effect Analytical
Storage of Sample No lysis produced by freezing with liquid N_2. *1180*

Blood No Effect Physiological
Bucindol No effects noted in 8 patients following several weeks on treatment. *1283*
Chenodeoxycholic Acid No significant effect with up to 750 mg/d over 2 y in 916 patients. *3199*
Cromoglycate No clinical significant change observed. *0484*
Enalapril In 53 patients given up to 160 mg daily or when combined with hydrochlorothiazide. *2277*
Famotidine No significant change from pretreatment values in 9 patients with Zollinger-Ellison syndrome over 33 weeks. *1667*
Penicillamine In 21 patients with rheumatoid arthritis treated for 48 weeks given 250 to 750 mg daily. *2345*
Thiopronine Toxicity similar to penicillamine and other sulfhydryl drugs. *1761*
Timegadine In 23 patients with rheumatoid arthritis given 250 to 750 mg/d for 48 weeks. *2345*
Zimeldine In 1 study involving 147 patients no significant change seen. *3741*

Blood Increase Analytical
Cryoglobulin ?particle formation with Coulter model S. *1032*

Blood Increase Physiological
Acids Response to ingestion. *1302*
Alkoxyglycerols Protect against decrease caused by X-rays. *0483*
Allopurinol Hypersensitivity reaction occurs with fever. *2313*
Aminosalicylic Acid Mainly due to eosinophilia due to hypersensitivity. *2313*
Amphetamine Myeloblastic leukemia. *0325*
Amphotericin B Leukocytosis occurs. *1680*
Ampicillin Mainly due to eosinophilia of hypersensitivity. *2313*

Atropine   May cause leukocytosis (effect in children). *0071*

Belladonna   May cause leukocytosis. *0071*

Benzene   Lymphocytic or polymorphonuclear early. *1343*

Bumetanide   By up to 6,000/uL over 3 mo. *2666*

Capreomycin   Due to eosinophilia of hypersensitivity. *2313*

Carbamazepine   Often marked leukocytosis maximal on fourth day. *1343*   Leukocytosis occasionally observed. *0019*   Granulomatous hepatitis observed in small proportion of patients treated for less than 1 mo; results within 3 d of cessation of treatment. *2153*

Cephalothin   Due to eosinophilia of hypersensitivity. *2313*

Chloramphenicol   Leukemia may occur as sequel to marrow depression. *0071*

Chloroform   Normal response to anesthesia. *0251*

Chlorpropamide   Due to eosinophilia of hypersensitivity. *2313*

Cloxacillin   Due to eosinophilia of hypersensitivity. *2313*

Colchicine   Leukocytosis follows initial leukopenia. *1343*

Cold Agglutinins   Spurious high count until blood warmed to 37°. *2515*

Cold Exposure   Physiological response to prolonged cold baths. *0251*

Copper   Marked leukocytosis with dialysis induced toxicity. *1955*   Leukocytosis observed with poisoning. *2427*

Corticotropin   Significant neutrophilic granulocytosis. *2427*

Desipramine   Due to eosinophilia of hypersensitivity. *2313*

Desoximetasone   2 to 3 fold increase in 5 patients with psoriasis and topical application of glucocorticoid. *1240*

Desoxycorticosterone   In 5 patients with psoriasis and topical application of glucocorticoid. *1240*

Dieldrin   May occur in acute response. *1302*

Digitalis   Rare leukocytosis. *2313*

Diurnal Variation   Rises to maximum during day, minimum in early am. *2222*

Dydrogesterone   Mild leukocytosis reported. *0071*

Enflurane   May also partly reflect surgery. *0999*

Epinephrine   Initially due to lymphocytosis later neutrophilia. *3438*

Erythromycin   Leukocytosis observed. *2324*

Ethanol   3.9% increase in heavy versus occasional drinkers. *3273*

Ether   Normal response to anesthesia. *0251*

Etiocholanolone   Pyrogen — increases number of granulocytes. *2427*

Florantyrone   In patients with pre-existing liver disease. *2313*

Fluroxene   Normal response to anesthesia. *3465*

Fluspirilene   Initial effect observed. *0221*

Gold   Due to eosinophilia of hypersensitivity. *2313*

Haloperidol   Rarely reported leukocytosis. *0071*

Halothane   Allergic hypersensitive response. *1343*

Imipramine   (transient 1-6 h) (possibly Löffler's syndrome). *3280*

Iodides   Due to eosinophilia. *2313*

Iron Dextran   Leukemoid reaction reported. *2427*

Isoflurane   Normal response to anesthesia. *3465*

Isoniazid   Due to eosinophilia. *2313*

Isotretinoin   Therapeutic response is normalization but effect not usually observed until after 3 weeks. *1319*

Kanamycin   Due to eosinophilia. *2313*

Levodopa   Unassociated with fever or infection reported. *0824*

Light   Normal response to sun or UV light. *0251*

Lithium   ?drug associated endocrine effect (may double). *2341*   After 4 weeks treatment increase in 21 patients on average from 6.3 to 8.6 thousand/uL. *1798*   Mean increase of 2.2 thousand/uL with treatment. *2078*

Melphalan   4 cases of acute leukemia in 474 patients with ovarian carcinoma: all cases in patients who had received at least 300 mg for 3 y. *1003*

Menstruation   Physiological response. *0251*

Mephenytoin   Rare cases described. *1680*

Mercury Compounds   May induce leukocytosis. *0251*

Methicillin   Due to eosinophilia/leukocytosis. *2313*

Methyldopa   Due to eosinophilia. *2313*

Methysergide   Due to neutrophilia. *2313*

Moxalactam   In up to 0.5% patients as reported from several studies. *0592*

Muscular Exercise   Mainly due to neutrophilia after exercise. *0987*

Nalidixic Acid   Due to eosinophilia. *2313*

Naphthalene   Leukocytosis may occur following ingestion. *1302*

Niacinamide   Marked elevation noted at 24 h. *1622*

Nickel   Occurs with industrial exposure. *2359*

Nitrofurantoin   In 80 of 153 acute and 5 of 33 chronic drug induced pulmonary reactions. *1639*   In 80 of 153 acute and 5 of 33 chronic drug induced pulmonary reactions. *1639*

Novobiocin   Due to eosinophilia. *2313*

Ohio 469   Significant rise postoperatively especially of segmented neutrophils. *1647*

Oleandomycin   Occasional leukocytosis after 2 weeks. *1679*

Oral Contraceptives   ?stimulating effect of steroids on bone marrow. *1141*

Oxyphenbutazone   May cause leukemia type of reaction. *0019*

Paraldehyde   Leukocytosis in severe acute or chronic poisoning. *1343*

Phenindione   Occasional leukocytosis may occur. *1343*

Phenothiazines   Due to eosinophilia or generalized leukocytosis. *2313*

Phenylbutazone   Leukemia drug induced. *2427*

Phenytoin   Due to eosinophilia. *2313*   General feature of hepatotoxicity (although this occurs rarely), hypersensitivity usually responsible. *0950*

Phosphorus   Reported leukocytosis with poisoning. *1302*

Pilocarpine   Probably due to contraction of spleen. *1343*

Potassium Iodide   Due to eosinophilia. *2313*

Prednisolone   leukocytosis observed occasionally. *2313*

Prednisone   Observed in 2 of 676 cases, no obvious cause. *0428*   Significant effect with increase of 6,000/uL by second week. *3309*   Augmented response in cirrhotics when drug added to therapeutic regime. *0651*

Pregnancy   In late pregnancy and at labor. *0987*

Quinidine   Marked leukocytosis in 2 patients in association with quinidine fever, normalized after drug discontinued. *0266*

Quinine   Primary inc, especially lymphocytes (splenic contractions). *1343*

Radioactive Iodine   Incidence of acute leukemia higher. *0071*

Ristocetin   Due to eosinophilia. *2313*

SC-16102   Mechanism not explained. *1804*

Smoking   Higher than in nonsmokers. *2123*   Higher in inhalers than other smokers. *1195*   Significantly higher than in nonsmokers but still in normal range. *1548*   Correlated well with carboxyhemoglobin saturation. Average leukocyte count at upper limit of normal in nonsmokers. *3691*

Streptokinase   Due to eosinophilia. *2313*

Strychnine   Probably due to release of epinephrine from adrenal. *1302*

Sulfamethoxazole   When given with trimethoprim, fever and malaise: successful response to treatment. *1969*

Sulfasalazine   Characteristically associated with rare cases of hepatotoxicity. *1127*   Hepatotoxic hypersensitivity reaction like sulfonamide hypersensitivity. *2214*

Sulfonamides   Associated with hemolysis (especially with long-acting drugs). *2313*

Tetracosactrin   Increase (mainly NPL) of 7,000 seen in 1 d. *3010*

Tetracycline   Leukocytosis with atypical lymphocytes. *1343*

Thiothixene   Transient leukocytosis may occur. *1678*

Triamterene   Due to eosinophilia. *2313*

Trifluperidol   Due to eosinophilia. *2313*

Vancomycin   Due to eosinophilia. *2313*

Viomycin   Due to eosinophilia. *2313*

Blood   Decrease   Analytical

Azathioprine   Low count by Coulter S, ?due to fragile cells. *2227*

Prednisone   Low count by Coulter S, ?due to fragile cells. *2227*

Leukocytes (continued)

Blood Decrease Physiological

Acetaminophen May affect bone marrow function/pancytopenia. *2198*

Acetanilid Leukopenia. *2004*

Acetazolamide Leukopenia/agranulocytosis. *2759*

Acetohexamide Leukopenia and aplastic anemia reported. *2313*

Acetophenazine Transitory leukopenia or agranulocytosis. *0071*

Aging Fall from age 20 to 80 y in non smokers. *1195*

Allopurinol Hypersensitivity reaction (often transient). *2313* In 1 of 73 patients treated for 6 mo. *3565* Noted in 3 individuals undergoing therapeutic starvation. *3224*

Aminoglutethimide May cause leukopenia. *2313*

Aminopyrine Agranulocytosis occurs within minutes. *3941*

Aminosalicylic Acid Leukopenia but no cases of agranulocytosis. *2429*

Amitriptyline Leukopenia/agranulocytosis may occur. *0187*

Amodiaquine Reported effect (AMA — blood dyscrasia committee). *2313* 20 cases associated with agranulocytosis in Switzerland, 12 of which associated with hepatitis. *1610* Occasional aplastic anemia reported. *3717*

Amphotericin B Leukopenia or agranulocytosis. *1680* Occasional complication of treatment. *3567*

Ampicillin Leukopenia. *1370*

Anisindione Observed with other indandiones. *1680*

Antazoline Leukopenia. *3602*

Anticonvulsants May occur with severe megaloblastic anemia. *3773*

Antineoplastic Agents Often therapeutic intent. *2313*

Antipyrine Leukopenia/agranulocytosis. *2427*

Arsenicals Pancytopenia. *0788*

Asparaginase Mild effect in up to 25% patients. *2657*

Aspirin Depresses leukocytosis of acute rheumatic fever. *3825*

Auranofin Infrequent reversible side effect (occurring in up to 13% patients). *0617*

Azathioprine Drug related leukopenia (probable effect). *3287* Principal manifestation of bone marrow depression. *0036* In 30 of 79 cases with hematological problem out of 328 reports of severe adverse effects. *2097*

Azidothymidine Profound effect observed in one male homosexual with AIDS unlikely related to other factors. *1157*

Barbiturates Leukopenia. *3669*

Benzazepine Slight effect observed. *1738*

Benzene Probable myelotoxic effect (usually late). *2429*

Benzthiazide May occur as with other thiazides. *1680*

Bismuth Salts May cause aplastic anemia/agranulocytosis. *2313*

Brompheniramine Occasional leukopenia/agranulocytosis. *1679*

Busulfan Pancytopenia/leukopenia. *2427*

Butaperazine Transitory leukopenia. *0071*

Capreomycin May cause leukopenia or leukocytosis. *1680*

Captopril Isolated reports of neutropenia and agranulocytosis when first introduced and given in high doses. *1761* Agranulocytosis observed in several cases. *3712* Isolated case of pancytopenia reported. *1012* Isolated cases of pancytopenia, usually with pre-existing renal disease. *1255*

Carbamazepine Pancytopenia (leukopenia in 15% patients). *0928* Many patients have drop to 3,000/uL but returns to normal with continued treatment. *1085*

Carbenicillin Leukopenia may occur infrequently. *1680*

Carbimazole Agranulocytosis. *0541* Occasional aplastic anemia reported. *3717* Two case reports with considerable reduction in white cell count and neutrophil count. *1473*

Carbutamide Leukopenia/agranulocytosis. *3878*

Carisoprodol Possible consequence/not marked. *1680*

Cefamandole In 2 patients given 6-9 g daily intravenously. *1646*

Cefoperazone In 2% of 450 patients: resolved on withdrawal of drug. *0690*

Cefotaxime Observed in 0.5% treated patients. *2615*

Cefoxitin Isolated case of drug-induced leukopenia. *1839* Reported in fewer than 0.1% treated patients. *2615*

Ceftazidime Observed in 0.6% of patients receiving drug. *2615*

Ceftriaxone Reported to occur in fewer than 0.1% patients. *2615*

Cefuroxime In one case who developed drug-dependent neutrophil antibodies. *2529*

Cephalexin In 12% of 74 patients all children given 25-150 mg/kg/d for 5 to 15 d. *2505*

Cephaloridine May cause neutropenia/leukopenia. *1019*

Cephalothin Marked decrease 1 case prolonged high i.v. doses. *3133* Rare response to therapy. *2313* In 2 patients given 12 g daily intravenously. *1646*

Cephapirin Marked decrease 1 case prolonged high i.v. doses. *3133*

Cephradine In 12% of 86 children given 25-110 mg/kg daily orally for 5 to 15 d. *2505*

Chlorambucil Neutropenia (dose related). *0543*

Chloramphenicol Pancytopenia/aplastic anemia. *0907* Occurs either as dose-related or idiosyncratic bone marrow suppression: dose- related usually occurs 5-7 d after start of therapy: idiosyncratic is rare occurring 1 case per 40,000 courses of therapy. *3864* Aplastic anemia reported to occur in from 1 to 10,000 or 40,000 cases. *3717* Isolated case of hepatitis and pancytopenia reported which resolved with discontinuation of drug. *0602*

Chlordane Pancytopenia (AMA Blood dyscrasias). *2429*

Chlordiazepoxide Agranulocytosis reported (leukopenia more common). *2313* Aplastic anemia occasionally occurs. *3717*

Chlorophenothane Pancytopenia (AMA blood dyscrasias). *2429*

Chloroquine Agranulocytosis/pancytopenia. *3522* Aplastic anemia reported to occur occasionally. *3717*

Chlorothiazide Leukopenia/pancytopenia. *0073* Aplastic anemia reported to occur occasionally. *3717*

Chlorphenesin Rare leukopenia/agranulocytosis/pancytopenia. *1680*

Chlorpheniramine Leukopenia (AMA Blood dyscrasias). *2429* Case of aplastic anemia after 10 y of low dose treatment. *1859* One case of aplastic anemia after 1 mo therapy, reversible with discontinuation of drug. *0886*

Chlorpromazine Agranulocytosis/leukopenia/granulocytopenia. *2238* Occasional aplastic anemia reported. *3717* Isolated case of agranulocytosis reported in elderly woman (general incidence of 1 in 1300 in psychiatric population dose related). *2299* In one patient who developed cholestasis with fever and leukopenia after 260 mg drug. *2840*

Chlorpropamide Agranulocytosis. *3811* Isolated case of agranulocytosis observed. *3627* Occasional aplastic anemia reported. *3717*

Chlorprothixene Possible hematological disorder. *1343*

Chlortetracycline Pancytopenia (AMA Blood dyscrasias). *2429*

Chlorthalidone Agranulocytosis. *1953*

Cholestyramine Possible relationship to drug administration. *1680*

Cimetidine Proved case of drug induced agranulocytopenia. *0587* Reported in one case in which drug given intravenously. *1770* Significant reduction observed in one patient. *0696*

Cinchophen Leukopenia or agranulocytosis may occur. *2429*

Ciprofloxacin Seen in 7 of 2829 patients: in all cases mild and not clearly drug related. *0151*

Clindamycin May cause agranulocytosis. *2808*

Clofibrate May cause leukopenia. *3008*

Cloxacillin May cause leukopenia. *2313*

Colchicine Leukopenia. *2426*

Colistimethate Neutropenia/leukopenia/granulocytopenia. *0071*

Colistin May cause leukopenia/granulocytopenia. *2313*

Corticosteroids May cause leukopenia. *2313*

Co-Trimoxazole May cause leukopenia. *0745* Isolated cases reported. *0665*

Cycloheximide Leukopenia (AMA Blood dyscrasias). *2429*

Cyclophosphamide Bone marrow depression (reversible leukopenia). *0072*

Cycloserine May occur with megaloblastic anemia. *3773*

Cyclosporine In approximately 1% patients given drug versus 12% given azathioprine in over 200 renal transplant patients. *0572*

Cyclothiazide Occasional response to thiazides. *1680*

Cytarabine Agranulocytosis/leukopenia. *2313* Aplasia of bone marrow developed in nearly all of 57 leukemics. *1579*

Dactinomycin Leukopenia/pancytopenia. *0071*

Dapsone Agranulocytosis. *1612* After 7-9 weeks of treatment with drug plus pyrimethamine in 7 patients taking drug as prophylaxis against malaria. *1202*

Desipramine May cause agranulocytosis. *2313*

1,2-Diaminopropane In 3% of 90 patients with rheumatoid arthritis. *1461*

Diaprim Leukopenia. *2426*

Diazepam Leukopenia. *3017*

Diazoxide May cause leukopenia. *3740*

Dichlorphenamide May cause leukopenia. *2313*

Dicumarol Reported effect (AMA Blood dyscrasias committee). *2429*

Diethazine May cause leukopenia. *2313*

Diethylpropion Marrow depression with leukopenia/agranulocytosis. *2415*

Digitalis Agranulocytosis/leukopenia. *2313*

Diiodohydroxyquin May cause leukopenia. *2313*

Dimethindene Rare reported effect of antihistamines. *1680*

Dinitrophenol May cause agranulocytosis/aplastic anemia. *1343*

Dipyrone Leukopenia or agranulocytosis. *1684*

Doxapram Further decrease noted in patient with leukopenia. *0034*

Doxepin Theoretical possibility due to class of compound. *0071*

Doxorubicin In 70% patients with 60 mg/sq m for 21 d. *0387*

Doxycycline Neutropenia reported. *2802*

EDTA Transient bone marrow depression. *0071*

Erythromycin Leukopenia or neutropenia may occur. *2324*

Ethacrynic Acid Agranulocytosis. *3747*

Ethanol In some alcoholics ?decreases myelopoiesis. *2196*

Ethosuximide Aplastic anemia/pancytopenia/leukopenia in 10%. *0071*

Ethotoin Possible bone marrow depression. *0071*

Ethoxzolamide Leukopenia. *2313*

Ethylphenacemide Leukopenia reported. *1343*

Etoposide Dose-limiting noncumulative leukopenia in 60-90% patients. *0141* 12 of 37 patients had transient leukopenia for 3-4 d. *3891* Occurs 7-10 d after therapy started: recovery occurs by day 20 to 24. *3339*

Flavoxate Single reversible case reported. *1680*

Floxuridine Early indication of toxicity on bone marrow. *1680*

Flucytosine Reported effect (with agranulocytosis). *3443* Usually mild, occurs in 8 to 13% patients: occasionally pancytopenia occurs. *3567*

Fluorides May cause aplastic anemia. *0279*

Fluorouracil May cause bone marrow depression. *0071*

Fluoxymesterone Leukopenia may occur. *0019*

Fluphenazine Agranulocytosis/leukopenia rarely. *1679*

Fumagillin Leukopenia (AMA Blood dyscrasias). *2429*

Furosemide May cause leukopenia or aplastic anemia. *2313*

Glaucarubin Leukopenia. *2313*

Glucosulfone Leukopenia. *2313*

Glutethimide May cause aplastic anemia. *0071*

Gold Aplastic anemia/leukopenia/agranulocytosis. *0114* Blood dyscrasias major side effect as reported to uk committee on safety of medicines. *0782* Occasional case of aplastic anemia reported. *3717* 7 cases out of 246 treated cases. *0829* Reported in 6 patients: brief self-limiting process. *1362* In 3% of 90 patients with rheumatoid arthritis. *1461*

Griseofulvin May cause leukopenia/neutropenia. *2427*

Guanazole Observed in approximately 80% patients. *1581*

Hair Lacquer Some may cause leukopenia (AMA Blood dyscrasias). *2429*

Haloperidol May cause anemia. *2313* Several cases reported but none of agranulocytosis. *0250*

Hetacillin Probably allergic response (rare). *1680*

Hexachlorobenzene Pancytopenia (AMA Blood dyscrasias). *2429*

Hexamethylmelamine Moderate effect in half of patients. *3863*

Hydralazine Agranulocytosis/pancytopenia/leukopenia. *0044* Observed in a case of obstructive jaundice with pancytopenia. *3469*

Hydrochlorothiazide May cause leukopenia/agranulocytosis. *0847* Occasional case of aplastic anemia reported. *3717*

Hydroflumethiazide Pancytopenia, agranulocytosis or aplastic anemia. *3844*

Hydroxychloroquine Leukopenia. *2313*

Hydroxyurea Leukopenia may occur. *2427*

Ibuprofen Rare adverse reactions reported by physicians to UK committee on safety of medicines. *0782* In one elderly man associated with a complement-dependent IgG antibody: reversible with cessation of treatment. *1054*

Idoxuridine Effect of high concentration only. *2313*

Ifosfamide Occurs in 80% patients after 150 mg/kg. *3675*

Imipramine May cause leukopenia/agranulocytosis. *3280*

Indandione Derivatives With other evidence of hypersensitivity. *2313*

Indomethacin Agranulocytosis/pancytopenia. *3012* Rare side effect reported to UK committee on safety of medicines. *0782* Occasional case of aplastic anemia reported. *3717*

α-Interferon Dose dependent leukopenia in 3 of 81 patients with malignant disease. *3299*

Iothiouracil Agranulocytosis. *2313*

Iproniazid Pancytopenia (AMA Blood dyscrasias). *2429*

Isocarboxazid Agranulocytosis/leukopenia — rare. *1612*

Isoniazid Agranulocytosis — rare. *0222* Occasional case of agranulocytosis reported. *3683*

Isotretinoin In 10 of 523 patients with mean daily dose of 109 mg for 150 d. *3867*

Labetalol In one patient previously treated with methyldopa and atenolol. *1400*

Lead Pancytopenia (AMA Blood dyscrasias). *2429*

Levamisole In 2 of 60 patients: sufficiently severe to warrant withdrawal from treatment. *2665* In 16% of 201 patients treated for rheumatoid arthritis had to be withdrawn from study. Occurred after mean treatment time of 7.4 mo. *3709* In 4 of 11 patients with breast cancer and radiation treatment. *2960*

Levodopa Transitory depression in a few patients. *0071*

Lincomycin Agranulocytosis/leukopenia/neutropenia. *0075*

Local Anesthetics Bone marrow depression and agranulocytosis reported. *2427*

Mafenide Probable effect observed in one child. *0123*

MAO Inhibitors Occasional leukopenia/agranulocytosis. *2313*

Mefenamic Acid Pancytopenia (temporary depression may occur often). *2313*

Melphalan May cause bone marrow depression (dose related). *0543*

Mepacrine Occasional case of aplastic anemia reported. *3717*

Mepazine Agranulocytosis. *1138*

Mephenesin May cause leukopenia. *2313*

Mephenoxalone Mild and transitory. *0071*

Mephenytoin Agranulocytosis/aplastic anemia. *3008* Occasional case of aplastic anemia reported. *3717*

Meprobamate Pancytopenia (may be agranulocytosis). *0072*

Meralluride May cause bone marrow depression. *1680*

Mercaptopurine Agranulocytosis. *2426*

Mercurial Diuretics neutropenia/agranulocytosis reported. *2000*

Mesoridazine Transient agranulocytosis reported. *0071*

Methaqualone 1 case of pancytopenia reported. *1680*

Methazolamide Probable effect as like acetazolamide. *2313* Some cases of aplastic anemia, and leukopenia reported when used in treatment for glaucoma. *3799*

Methicillin May cause bone marrow depression. *2313*

Methimazole Rare agranulocytosis or pancytopenia. *2427* Agranulocytosis. *0072* Occasional case of aplastic anemia reported. *3870*

Methocarbamol May cause leukopenia. *2313*

Methotrexate Agranulocytosis/lymphocytopenia. *2426*

Methotrimeprazine Agranulocytosis with long-term high-dose use. *1680*

Leukocytes *(continued)*

Blood Decrease Physiological *(continued)*

Methsuximide Rare aplastic anemia or reversible leukopenia. *2313*

Methyclothiazide Agranulocytosis or aplastic anemia may occur. *1680*

Methyldopa Very rare, may cause granulocytopenia. *1940*

Methylpromazine Leukopenia (AMA Blood dyscrasias). *2429*

Methylthiouracil Occasional leukopenia or agranulocytosis. *2313* Occasional case of aplastic anemia reported. *3717*

Methyprylon Due to toxic action of metabolite. *2313*

Methysergide Neutropenia reported. *2313*

Metronidazole Leukopenia and reduction of polymorphs. *2313*

Mexiletine Occurs in fewer than 1% treated patients. *2011*

Mianserin Decreased concentration reported in 4 individuals who were also receiving other drugs. *0028*

Miconazole Observed as isolated finding in patients given drug i.v. *3462*

Minocycline Reported as side effects. *1680*

Mithramycin Reversible effect. *0071*

Mitomycin Pancytopenia. *2194*

Mitoxantrone Granulocytopenia is dose-limiting toxicity. In phase II studies nadir between 8 and 15 d. Less than 5% patients had nadir below 1,000/uL. Fall with each treatment until 5th or 6th course. *3289*

Moxalactam Reported in fewer than 0.5% treated patients. *2615* In up to 0.5% patients as reported from several studies. *0592*

Neomycin If severe megaloblastic anemia. *3773*

Nialamide Leukopenia reported. *0071*

Nitrofurans May occasionally cause agranulocytosis. *2427*

Nitrofurantoin Leukopenia/agranulocytosis may occur. *0994*

Norfloxacin Reduction but not to below 1,000/uL in 6 of 1540 patients, usually plateaued between 3,000-4,000/uL. *0731* Occurs in about 1% treated cases probably via an immunologic mechanism. *2747*

Nortriptyline May cause agranulocytosis. *0071*

Novobiocin May induce blood dyscrasias. *2313*

Orphenadrine 2 cases aplastic anemia reported. *1680*

Oxacillin May cause bone marrow depression. *2313* In 2 patients receiving 12-15 g/d intravenously. *1646* Observed in one man, recovery began within 2 d of withdrawal. *1844* In two cases in one of which abnormalities developed within 48 h and in other in 17 d. *0485* Single case of drug-related agranulocytosis in patient with prior history of penicillin sensitivity. *3167*

Oxazepam Leukopenia occurs rarely. *2313*

Oxyphenbutazone Agranulocytosis/leukopenia. *2313* Rare side effect reported to UK committee on safety of medicines. *0782* Occasional case of aplastic anemia reported. *3717*

Paramethadione Aplastic anemia/neutropenia. *0071*

Parathiazine May occur with prolonged use. *2313*

Parathion Leukopenia (AMA Blood dyscrasias). *2429*

Penicillamine Leukopenia may occur with rash. *0074* Most serious and potentially life-threatening effect of drug. Either idiosyncratic response seen in first year of treatment. Independent of dosage or more commonly dose related with gradual onset. *1761* Rare blood dyscrasias as reported to UK committee on safety of medicines. *0782* In 3% of 90 patients with rheumatoid arthritis previously treated with gold. *1461* Toxicity observed in some patients with rheumatoid disease. *1891*

Penicillin Agranulocytosis/leukopenia. *2695*

Penicillin G Occasional case of drug associated leukopenia. *1839* In 2 patients receiving 140-150 mg/kg/d intravenously. *1646*

Pentazocine Effect observed in 3 patients although taking other drugs, but effect recurred with reinstitution of treatment. *1655*

Pentoxifylline In 2 patients receiving other drugs concomitantly but for considerable time previously without ill effects. *2244*

Perchlorate May cause fatal aplastic anemia. *2313* Occasional case of aplastic anemia reported with potassium salt. *3717*

Perphenazine Suspected of causing agranulocytosis. *2822*

Phenacemide Aplastic anemia (leukopenia reported most often). *0071*

Phenelzine May cause leukopenia. *0071*

Phenethicillin Theoretically leukopenia may occur. *1680*

Phenindione Agranulocytosis/leukopenia. *2313*

Phenobarbital Pancytopenia (AMA Blood dyscrasias). *2429*

Phenothiazines Agranulocytosis/leukopenia especially if low initially. *2113*

Phensuximide Possible agranulocytosis (very rare). *0071*

Phenylbutazone Agranulocytosis/aplastic anemia/leukopenia. *2930* Rare side effects reported to UK committee on safety of medicines. *0782* Occasional case of aplastic anemia reported. *3717*

Phenytoin megaloblastic/hemolytic/aplastic anemia/pancytopenia. *0072* Occasional case of aplastic anemia reported. *3717*

Phosphorus Reported effect with poisoning. *1302*

Pipamazine Pancytopenia (AMA Blood dyscrasias). *2429*

Piperacetazine Rare leukopenia/agranulocytosis (reversible). *0071*

Piperacillin 6 of 20 patients had small drop in leukocyte count. *3132*

Pipobroman Bone marrow depression may occur after 4 weeks. *1680*

Pirenzepine In isolated cases even though other drugs also being ingested. *3486*

Polythiazide Rare reported side effect. *1680*

Prednisolone Leukopenia. *2313*

Primaquine Agranulocytosis/leukopenia. *2313*

Primidone Pancytopenia (AMA Blood dyscrasias). *2429*

Probenecid Occasional aplastic anemia seen. *1679*

Procainamide Agranulocytosis. *3537* Significant association observed with therapy with sustained release form of drug at normal therapeutic concentration of drug. *1029* Pancytopenia may occur with/without lupus-like syndrome. *3215*

Procarbazine Leukopenia and bone marrow depression. *0071*

Prochlorperazine Agranulocytosis due to interference in development. *0072*

Promazine Agranulocytosis. *1138*

Promethazine Agranulocytosis/leukopenia. *3602*

Propafenone Decreased from mean of 6800 to 5900/uL in 45 patients treated over 1 y. *1477*

Propylthiouracil Agranulocytosis (incidence 1 in 200). *0701* Occasional case of aplastic anemia reported. *3717* In one 24 year old woman who developed fulminant hepatic failure with lymphocyte sensitization. *2437*

Protionamide Single case reported. *2427*

Protriptyline Transient agranulocytosis or leukopenia. *0071*

Pyrazolones Aplastic/hemolytic anemia (may be agranulocytosis). *2313*

Pyrimethamine May occur with severe megaloblastic anemia. *3898*

Quinacrine Leukopenia/agranulocytosis/aplastic anemia. *2426*

Quinethazone Theoretical bone marrow depression. *1680*

Quinidine Agranulocytosis/leukopenia. *2313* Agranulocytosis after 8 weeks of drug treatment. *1008*

Quinine Leukopenia (AMA Blood dyscrasias). *2429*

Race Significantly lower in blacks (applies to both sexes). *1869*

Radioactive Compounds May cause marrow depression. *2313*

Ranitidine Isolated case reported within 1 mo of treatment being started, reverted to normal 2 weeks after drug stopped. Probable hypersensitive or idiosyncratic reaction. *3303* Marked reduction in one elderly woman following 2 weeks of treatment with 300 mg/daily: normalized when treatment stopped. *0463*

Rifampin Toxic or allergic response (usually transient). *2820*

Ristocetin Leukopenia or agranulocytopenia. *2313*

Sodium Phosphate ^{32}**P** Leukopenia if excess used. *0071*

Spironolactone Documented single case after 5 weeks of 100 mg/d when no other treatment given in 70 year old woman. Probable immunoallergic mechanism. *3487*

Streptomycin Agranulocytosis/leukopenia/neutropenia. *2313* Occasional case of aplastic anemia reported. *1292*

Sulfachlorpyridazine Leukopenia, aplastic anemia may occur. *1680*

Sulfadiazine   Agranulocytosis/leukopenia. *2313*

Sulfamethizole   Agranulocytosis. *2868*

Sulfamethoxazole   Toxic reaction to drug. *1974*

Sulfamethoxypyridine   Leukopenia (AMA Blood dyscrasias). *2429*

Sulfasalazine   Pancytopenia reported plus reductions of individual series of cells. *2887*

Sulfisoxazole   Agranulocytosis/aplastic anemia. *2868*

Sulfonamides   Agranulocytosis/aplastic anemia. *2427*   Several cases of aplastic anemia reported. *3717*

Sulfones   Agranulocytosis/leukopenia. *0071*

Sulfonylureas   Pancytopenia/agranulocytosis. *2313*   Occasional case of aplastic anemia reported. *3717*

Sulthiame   leukopenia/also increases half life of diphenylhydantoin. *1343*

Suramin   Agranulocytosis reported. *0071*

Tamoxifen   In up to 20% of treated patients. *1542*

Tetracycline   Leukopenia (AMA Blood dyscrasias). *2429*

Thenalidine   Leukopenia (AMA Blood dyscrasias). *2429*

Thiabendazole   Agranulocytosis/leukopenia. *2413*

Thiacetazone   Drug related agranulocytosis (rare). *2427*

Thiazides   Reported to cause agranulocytopenia. *2745*

Thiocarlide   Associated with relative lymphocytosis and monocytosis. *2427*

Thiocyanate   Several cases of aplastic anemia reported. *3717*

Thioglycolate   Reported effect. *2429*

Thioguanine   May cause marrow depression. *0543*   In 29 patients with polycythemia rubra vera; effect usually apparent in 2 weeks. *2450*   Observed in 7 patients with chronic granulocytic leukemia. *3416*

Thiopropazate   Rare leukopenia. *0071*

Thioridazine   Agranulocytosis due to inhibition of development. *2822*

Thiosemicarbazones   leukopenia/agranulocytosis in 0.5%. *1343*

Thiotepa   May cause bone marrow depression. *0071*   10 of 25 patients given drug intravesically had at least one incident of acute myelosuppression; 5 of 29 patients had chronic myelosuppression. *1632*

Thiothixene   Transitory leukopenia. *2313*

Thiouracil   Agranulocytosis (in about 1% cases). *3618*

Tocainide   Blood dyscrasias occur in fewer than 0.2% patients. *2011*

Tolazamide   Leukopenia/agranulocytosis. *1343*

Tolazoline   Pancytopenia (AMA Blood dyscrasias). *2429*

Tolbutamide   Agranulocytosis/aplastic anemia. *0634*

Tranylcypromine   Theoretical effect of this type of drug. *0071*

Trichlormethiazide   May cause bone marrow depression. *1680*

Triethylenemelamine   May cause bone marrow depression. *0071*

Trifluoperazine   Pancytopenia (AMA Blood dyscrasias). *2427*

Trifluperidol   Rare leukopenia/agranulocytosis. *1343*

Triflupromazine   Agranulocytosis due to inhibition of development. *2822*

Trimeprazine   May cause leukopenia or agranulocytosis. *0071*

Trimethadione   Aplastic anemia. *3008*

Trimethoprim   Isolated case when drug given alone, but clear evidence of pancytopenia. Megaloblastic anemia may occur with prolonged use. *3284*

Tripelennamine   Agranulocytosis/leukopenia (rare). *3602*

Uracil Mustard   May cause bone marrow depression. *0071*

Urethan   Bone marrow aplasia/pancytopenia. *0071*

Valproic Acid   In 27% of 100 epileptic children when given alone or combined with other anticonvulsants. *0737*

Vancomycin   Favorable response of antibiotic-associated colitis to drug. *3561*

Vinblastine   Leukopenia. *0071*

Vincristine   Usually reversible leukopenia. *0071*

Vitamin A   Leukopenia with hypoplastic anemia reported. *3908*

Vitamin K   Pancytopenia reported after K_3 and K_4. *2427*

X-Ray Therapy   Cell destruction in leukemics. *3598*

Zimeldine   7 of 14 patients treated for more than 1 week demonstrated toxic syndrome ? immunological mechanism or related to high initial dose. *2075*

Urine   No Effect   Analytical

Methyldopa   At concentration of 1,000 mg/L had no effect on Cytur-Test. *3393*

Salicylate   At concentration of 1,000 mg/L had no effect on Cytur-Test. *3393*

Urine   No Effect   Physiological

Chenodeoxycholic Acid   No significant effect with up to 750 mg/d in 916 patients treated for 2 y. *3199*

Cyclosporine   No effect observed in 23 patients with inflammatory ocular disease treated for 6 mo. *2713*

Urine   Increase   Physiological

Allopurinol   Associated with severe sensitivity reaction. *2451*

Dantrolene   Possible effect, perhaps urinary infection. *0659*

Fenoprofen   Rare case of acute tubulointerstitial nephritis with acute renal failure (probable association with drug administration). *3582*

Indomethacin   Oliguric renal failure developed in one man during treatment: reversible. *1245*

Isotretinoin   In 10 of 523 patients with mean daily dose of 109 mg for 150 d. *3867*

Moxalactam   In about 2% patients as reported from several studies. *0592*

Urine   Decrease   Analytical

Ascorbic Acid   At concentrations above 2,000 mg/L (maximum concentration in urine up to 1290 mg/L) caused false negative result with Cytur-Test. *3393*

CSF   Increase   Physiological

Ibuprofen   Few cases of aseptic meningitis described. *1276*

Levodopa

Serum   Increase   Physiological

Antacids   Increased absorption with faster gastric emptying. *3794*

Metoclopramide   Increased rate of absorption with faster gastric emptying. *3794*

Serum   Decrease   Physiological

Food   Absorption reduced when taken with food. *3024*

Levonorgestrel

Serum   Decrease   Physiological

Phenytoin   Markedly lessened concentrations when drug co-administered due to enhanced metabolism. *2655*

LH Response to LHRH

Plasma   No Effect   Physiological

Isotretinoin   In 7 patients with severe rosacea treated with 1 mg/kg/d for 12 weeks. Effects possibly due to induction of hepatic microsomal enzymes. *2307*

Lidocaine

Serum   No Effect   Physiological

Ranitidine   Insignificant on drug kinetics when co-administered. *1938*

Serum   Decrease   Analytical

Tris(2-Butoxyethyl) Phosphate (TBEP)   By up to 17% at 12.7 µmol/L when used as plasticizer in evaluated blood tubes. *0892*

Lincomycin

Serum   Decrease   Physiological
Cyclamate   Inhibits gastrointestinal absorption. *1487*
Kaolin   Inhibits gastrointestinal absorption. *1487*

Linoleate

Serum   Increase   Physiological
Muscular Exercise   Both absolutely and relatively. *1702*

Linoleic Acid

Serum   Increase   Physiological
Caffeine   Significant increase within 1 h, still elevated after 4 h, when 250 mg ingested. *2751*

Lipase

Serum   No Effect   Analytical
Calcium   At 5 mmol/L no effect on method of Tietz, Repique. *3589*
Calcium Ions   Up to 5 mmol/L on method of Tietz. *3589*
Hemoglobin   At 5 g/L no effect on method of Tietz, Repique. *3589*
Storage of Sample   No effect 1 week at 4°, 1 mo at -20°. *1563* 7 d at room temperature, 4° and -20°. *3217*

Serum   Increase   Analytical
Albumin   By preventing inactivation of enzyme. *3589*
Bilirubin   Turbidimetric but not titrimetric methods. *0583*
Deoxycholate   Sodium salts prevent inactivation of enzyme. *3589*
Glycocholate   Sodium salts prevent inactivation of enzyme. *3589*
Pancreozymin   Preparation contains lipase. *1237*
Taurocholate   Sodium salts prevent inactivation of enzyme. *3589*

Serum   Increase   Physiological
Asparaginase   Reports vary from incidence of 2.5 to 16% cases of acute pancreatitis: usually mild. *0558*
Azathioprine   6.2% of 116 patients receiving only drug demonstrated clinical pancreatitis. *2281*
Bethanechol   May cause increased secretion, spasm of sphincter of Oddi. *0816*
Cholinergics   Impaired excretion spasm of sphincter of Oddi. *1343*
Codeine   Causes spasm of sphincter of Oddi. *3777*
Ethanol   Chemical or physical pancreatitis. *1237*
Furosemide   Isolated case of acute hemorrhagic pancreatitis. *1820*
Heparin   Increase of 150% 10 minutes after injection. *3026*
Indomethacin   Associated with cholestatic liver damage. *1237*
Meperidine   May cause spasm of sphincter of Oddi. *3316*
Methacholine   Constricts sphincter of Oddi. *0652*
Methylprednisolone   To well above normal in 10 patients given 1 g/d i.v. for 3 d. Maximum effect days 5 to 8, possible subclinical damage of pancreatic acinar cell. *0804*
Morphine   Causes spasm of sphincter of Oddi. *3505*
Narcotics   Impaired excretion — spasm of sphincter of Oddi. *0652*
Oral Contraceptives   Isolated case of acute pancreatitis reported. *2519*
Pentazocine   Causes spasm of sphincter of Oddi. *2425*
Secretin   May cause spasm of sphincter of Oddi. *3777*
Sulfisoxazole   Rare effect on pancreas, salivary glands. *2427*
X-Ray Therapy   May be tripling with total body irradiation. *1837*

Serum   Decrease   Analytical
Calcium Ions   Above 5 mmol/L on method of Tietz. *3589*

Hemoglobin   Inhibits lipase activity. *1936*
Hemolysis   Inhibition of lipase activity. *1566*

Serum   Decrease   Physiological
Protamine   Inhibition occurs whether heparinized or not. *3026*
Saline   At molar concentrations (whether heparinized or not). *3026*

Duodenal Contents   Increase   Physiological
Calcium   Calcium infusion (10 mg/kg/h) causes doubling. *1314*

Pancreatic Juice   No Effect   Analytical
Radiographic Agents   No effect of Gastrografin®. *0741*

Pancreatic Juice   Decrease   Physiological
Ethanol   Oral or i.v. causes direct inhibition. *2504*

Lipid Glycerol

Serum   Increase   Analytical
Stearylamine   Possible effect in method of Horney. *2683*

Serum   Decrease   Analytical
Citrates   Inhibits phospholipase C method of Horney. *2683*
Fluorides   Inhibits phospholipase C method of Horney. *2683*
Phosphates   Inhibits phospholipase with method of Horney. *2683*
Sodium Lauryl Sulfate   Inhibits phospholipase C method of Horney. *2683*

Lipids, Total

Serum   No Effect   Physiological
Niacinamide   No lowering effect observed. *1343*

Serum   Increase   Analytical
Filter Paper   If low fat paper not used in some methods. *0579*

Serum   Increase   Physiological
Asparaginase   Rare response after initial hypolipidemia. *2657*
Chlorpromazine   Xanthomatous biliary cirrhosis may occur. *1978*
Cobalt   Profound hyperlipemia, xanthomatosis in one case. *2217*
Coffee   Significant correlation with intake in coronary disease. *2193*
Endotoxin   Observed response. *1230*
Menopause   Change of 0.2 g/dL from 5th to 6th decade. *3830*
Oral Contraceptives   155% higher in pill-users than controls. *2688*
Protamine   Mechanism not established. *1343*

Serum   Decrease   Physiological
Asparaginase   Parallel decrease in cholesterol. *2657*
Cholestyramine   Lowers bile acids by ionic binding. *2313*
Clofibrate   Normal response (may increase in diabetics). *2427*
Dextrothyroxine   Therapeutic effect. *0019*
EDTA   Occurs if elevated lipids and given i.v. *1488*
Ethanol   Occurs when cirrhosis develops in alcoholism. *1621*
Glucagon   Transfer of lipids to platelets. *0584*
Menotropins   Marked fall in type 2 hyperlipidemia. *2235*
Niacin   Prompt and sustained hypolipemic action. *1680*

CSF   Increase   Physiological
Oral Contraceptives   Altered metabolism. *1601*

Lipoprotein Cholesterol

Serum   No Effect   Physiological
Oral Contraceptives   No difference between treated women and others. *1195*

β-Lipoprotein Cholesterol

Serum   Increase   Physiological
Oral Contraceptives   Icrease by 34 mg/dL in white girl drug users vs controls. *3723*

Lipoprotein Electrophoresis

Serum   No Effect   Analytical
Fluosol-DA   Not affected by presence of compound in specimen. *2518*

Serum   No Effect   Physiological
Quingestanol   No significant effect with 300 μg/d 6 mo. *1309*

Serum   Positive   Analytical
Heparin   Alters electrophoretic pattern. *1951*

Lipoprotein, High Density

Serum   No Effect   Physiological
Oral Contraceptives   In 10 women receiving ethinyl estradiol with norgestrel. *2009*

Serum   Increase   Physiological
Labetalol   In 15 patients with essential hypertension, also decreased total cholesterol: HDL ratio. *3776*
Oral Contraceptives   In 10 women receiving ethinyl estradiol with norethindrone. *2009*

Lipoprotein Lipase

Serum   No Effect   Physiological
Isotretinoin   Increase in both men and women with 1 mg/kg/d for 20 weeks when given for nodulocystic acne. *0326*   Increase in both men and women with 1 mg/kg/d for 20 weeks when given for nodulocystic acne. *0326*
Metoprolol   No significant effect in 15 hypertensive patients given 200 mg/d for 10 weeks. *1118*

Serum   Increase   Physiological
Heparin   Release of tissue lipase into plasma. *1343*
Prazosin   Mean increase from 28.4 to 37.7 μmol/L per minute given 4 mg/d for 10 weeks. *1118*

Serum   Decrease   Physiological
Aging   Mean 0.13 meq/L at age 26 y, 0.06 at 86 y. *0480*
Asparaginase   Observed in unusual hyperlipidemic response. *2657*
Estrogens   Metabolic effect. *0227*
Oral Contraceptives   Reduced response to heparin injection. *0585*
Protamine   Inhibition occurs whether heparinized or not. *3026*
Saline   At molar concentrations (whether heparinized or not). *3026*

Lipoprotein, Low Density

Serum   No Effect   Physiological
Oral Contraceptives   In 10 women receiving ethinyl estradiol with norgestrel. *2009*

Serum   Increase   Physiological
Oral Contraceptives   In 10 women receiving ethinyl estradiol with norethindrone. *2009*

Lipoprotein Lp(a)

Serum   Decrease   Physiological
Stanozolol   By average of 65% with 6 weeks treatment reverted to normal by 5 weeks after treatment stopped. *0047*

Lipoproteins

Serum   No Effect   Analytical
Storage of Sample   For electrophoresis 3 d room temperature, 14 at -20°. *3217*

Serum   No Effect   Physiological
Glibonuride   No significant effect observed after 4 weeks. *0301*

Serum   Increase   Physiological
Coffee   Significant correlation with intake in coronary disease. *2193*
Epinephrine   Effect on low density components. *1343*
Estrogens   If very low density. *3332*
Oral Contraceptives   50% increase after 6 mo use. *3505*

Serum   Decrease   Physiological
Asparaginase   Possible depressed synthesis. *1519*
Meals   Affected by meals (most marked in women). *3455*

α-Lipoproteins

Serum   No Effect   Physiological
Pregnancy   No significant change observed. *3472*

Serum   Increase   Physiological
Carprazidil   In 15 men with mild to moderate essential hypertension by 26% over 8 weeks. *1268*
Estrogens   Metabolic effect (almost doubled in some studies). *0227*
Ethinyl Estradiol   May almost double in some studies. *1343*
Oral Contraceptives   Responsible for increased phospholipids. *0585*

Serum   Decrease   Analytical
Bacterial Contamination   If contaminated with phospholipase producing bacteria. *3223*

β-Lipoproteins

Serum   No Effect   Physiological
Chenodeoxycholic Acid   No effect observed during treatment of gallstones. *0808*
Muscular Exercise   No effect of exercise observed. *2846*

Serum   Increase   Physiological
Oral Contraceptives   Affects cholesterol (approximately 20% increase after 6 mo). *0585*   Extent of effect varies with exact composition of oral contraceptive. *1458*
Pregnancy   Increase by 24% (absolute amount greater) in late pregnancy. *3472*
Smoking   Significantly higher (?due to diet preferences) heavy smokers. *1317*

Serum   Decrease   Analytical
Bacterial Contamination   If contaminated with phospholipase producing bacteria. *3223*
Storage of Sample   Marked with 7 d at -20°, mild 28 d at 0°. *1951*

Serum   Decrease   Physiological
Asparaginase   Observed in unusual hyperlipidemic response. *2657*

β-Lipoproteins *(continued)*

Serum   Decrease   Physiological *(continued)*
Cholestyramine   But does not affect very low density lipoproteins. *1343*
Clofibrate   Therapeutic effect (mechanism disputed). *0071*
Dextrothyroxine   Therapeutic effect. *0071*
Estrogens   Metabolic effect (decrease of 30%). *0227*
Ethinyl Estradiol   Decreases by up to 30%. *1343*
Fenfluramine   Small but significant reduction. *0377*
Light   Fall of 30% reported on short exposure to UV. *2893*
Niacin   Chronic administration has slight effect. *1343*
Sitosterols   Therapeutic intent. *1680*

Lipoproteins, Pre-β

Serum   Increase   Physiological
Asparaginase   Observed in unusual hyperlipidemic response. *2657*
Estrogens   Reported effect. *1343*
Oral Contraceptives   Probably due to increased apoprotein synthesis. *2385*

Serum   Decrease   Analytical
Bacterial Contamination   If contaminated with phospholipase producing bacteria. *3223*

Serum   Decrease   Physiological
Clofibrate   Therapeutic intent. *0165*
Heparin   Nonsustained response to small i.v. injection. *1343*
Niacin   Therapeutic effect. *1343*

Lipoprotein, Very Low Density

Serum   No Effect   Physiological
Oral Contraceptives   In 10 women receiving ethinyl estradiol with norethindrone. *2009*

Serum   Increase   Physiological
Oral Contraceptives   In 10 women receiving ethinyl estradiol with norgestrel. *2009*

Lithium

Serum   No Effect   Analytical
Contact With Clot   No effect observed over 24 h. *2891*
Serum   No difference observed. *2891*
Storage of Sample   No effect 1 week at 4°, frozen for months. *2891*

Serum   No Effect   Physiological
Furosemide   No effect observed in normal volunteers given drug over 2 weeks (40 mg/d). *1782*
Ibuprofen   Inconsistent change when drug added to therapeutic regime in 3 patients. *2903*

Serum   Increase   Analytical
Oxalate   If lithium salt of oxalates. *1563*

Serum   Increase   Physiological
Bendrofluazide   Substantial increase in one patient receiving lithium to whose regime was subsequently added. *1919*
Bumetanide   Substantial increase in one patient receiving lithium to whose regime diuretic was subsequently added. *1919*
Diclofenac   Decreased renal clearance and increased plasma concentration by 26% in 5 normal women. *2943*

Hydrochlorothiazide   Significant increase with 50 mg/d over 2 weeks possible effect on reabsorption in loop of Henle. *1782*   Reported in 2 cases taking lithium: drug also given with triamterene: due to reduced clearance of lithium. *2393*
Ibuprofen   Mean concentration increased by 15% when ibuprofen co-administered with increased RBC to plasma ratio. *2019*
Indomethacin   Concentration increased from 0.9 to 1.4 meq/L within 6 d of indomethacin administration in 3 patients. *2903*
Lithium   Therapeutic level between 0.5 and 1.0 meq/L. *1343*
Metoclopramide   Increased rate of absorption with faster gastric emptying. *3794*
Triamterene   Reported in 2 cases taking lithium: drug also given with hydrochlorothiazide: due to reduced clearance of lithium. *2393*

Serum   Decrease   Physiological
Propantheline   Delayed absorption with delayed gastric emptying. *3794*

Urine   Increase   Physiological
Lithium   Excretion proportional to plasma concentration. *1680*

Urine   Decrease   Physiological
Ibuprofen   Total body and renal clearance significantly reduced during co-administration. *2019*

CSF   Increase   Physiological
Lithium   Concentration about half in serum. *1343*

Red Blood Cells   Increase   Physiological
Lithium   But only to max of 50% in plasma. *1441*

Lithocholic Acid

Feces   Decrease   Physiological
Neomycin   Due to alteration of intestinal flora. *1077*

Lorazepam

Serum   No Effect   Physiological
Cimetidine   Not subject to N-dealkylation or hydroxylation by cytochrome P-450 so not affected by drug. *3084*
Ranitidine   No affected: normally conjugated in liver. *1938*

L/S Ratio

Amniotic Fluid   Increase   Analytical
Chromatography   Higher on TLC if H_2SO_4 spray. Higher on TLC if silica gel h used. *1258*

Amniotic Fluid   Decrease   Analytical
Hemoglobin   Variable effect seen if more than 10%. *0513*

Luteinizing Hormone (LH)

Plasma   No Effect   Physiological
L-α-Acetylmethadol (LAAM)   In 9 male heroin addicts maintained on drug and in 2 weeks following abrupt withdrawal. *2409*
Bromocriptine   No effect observed. *1736*
Butorphanol   No significant effect in 6 healthy male volunteers given 2 mg i.m. *3033*
Carbamazepine   Insignificant fall in 6 healthy males at 14 d after 400 mg/d treatment. *0707*
Cimetidine   In 6 patients treated for 1 mo with 1 g/d. *3491*   In 3 studied men with duodenal ulcer or duodenitis. *2368*   No significant effect of drug during menstrual cycle. *0271*   In 11 male subjects with chronic duodenal ulcer given drug 1 g/d for 3 mo. *3759*
Clonidine   No effect seen in 12 healthy adults after single oral dose. *2070*
Cyproterone   No change following oral administration. *3394*

Danazol No significant change in 5 patients with hereditary angioneurotic edema treated for up to 10 mo. *3219* In 7 normal women given 800 mg daily for 2 mo amenorrheic state induced by drug. *2224*

Desipramine In group of depressed patients. *0560*

Diurnal Variation With 1/2 hourly measurements over 48 h. *2014*

Ethanol No effect in men or women at 100 mg/dL. *3611*

Famotidine No effect of short-term or long-term treatment in male patients with duodenal ulcers. *0727*

Flunarizine After 90 d in 8 women. *2265*

Isotretinoin In 7 patients with severe rosacea treated with 1 mg/kg/d for 12 weeks. Effects possibly due to induction of hepatic microsomal enzymes. *2307* With 12 weeks treatment with 1 mg/kg/d. *2307*

Lead In a population of male battery workers compared with control group in cement industry. *0172* In some men occupationally exposed to large quantities of lead. *0771*

Levodopa No effect after 2 weeks in males. *3338*

Lovastatin No significant change even in patients with familial hypercholesterolemia receiving 40 mg bid. *1530*

Methyldopa No effect in patient on long term treatment who had drug-induced hyperprolactinemia. *0166*

Methylprednisolone No apparent effect, unlike that on testosterone, in 67 y old males with chronic pulmonary disease taking drug for more than 1 mo. *2245*

Metoclopramide No significant baseline differences in cycling women before and after drug. *0096*

Naltrexone With 50 to 100 mg administered daily to obese subjects over 8 weeks. *0176*

Oxytetracycline Despite decreased urine excretion and increased fecal excretion of estrogens due decreased hydrolysis by beta-glucuronidase in gastrointestinal tract. *1469*

Prednisone No apparent effect, unlike that on testosterone, in 67 y males with chronic pulmonary disease taking drug for at least 1 mo. *2245*

Prostaglandin F2α When i.v. up to 2 μg/kg/min in men. *0735* Little effect in normals when given i.v. *1600*

Ranitidine With up to 450 mg/d in 20 males with chronic duodenal ulcer. *2925* No effect with i.v. bolus of much as 300 mg. *0871* After 4 weeks or 6 mo treatment (300 mg and 150 mg daily respectively) in male patients with duodenal ulcer. *0727*

Sexual Activity No effect in either sex. *3437*

Sodium Bromide No effect in 10 men, 10 women receiving 1 mg/kg/d during 8 weeks or 2 full menstrual cycles. *3141*

Sultopride When given 300-600 mg/d for 5 weeks in 5 schizophrenic women. *2465*

Tamoxifen After oral administration of 20 mg/d for 5 or 10 d to 6 healthy women during follicular phase of menstrual cycle. *1542*

Thyrotropin-Releasing Hormone (TRH) No effect seen with 1 mg i.v. in normals. *3822* No effect in normal subjects. *0100*

Zimeldine In group of depressed patients. *0560*

Plasma Increase Physiological

Bombesin After LHRH after 5 ng/nk/min x 2.5 h in healthy men. *2841*

Clomiphene Up to 700% in normal males for first 21 d. *3145* Maximum with induced ovulation. *1295*

Diurnal Variation Rapid cycling throughout day of 100-300%. *2542*

Ethanol Inconsistent change with ingestion over several days in healthy men. *1352*

Ketoconazole Increase by 127% approximately in normal men with dose effect maximal at 900 mg/d. Due to stimulatory effect of dose dependent fall of testosterone. *1299*

Menstruation Sharp single peak about mid-cycle. *1762* Max with ovulation. *1295*

Mestranol Estrogen exerts stimulatory action. *3914*

Naloxone Significant effect within 30 minutes of start of i.v. infusion. *0868* After 0.08 mg/kg body weight i.v. in girls and boys at most advanced stage of gonadal maturation: ineffective in prepubertal and early pubertal children. *2796* Effect observed regardless of amount infused. *1970*

Phenytoin Mean 10.0 units/L versus 4.6 in controls in approximately 24 male patients given phenytoin alone or with primidone or phenobarbital. *3609*

Spironolactone In males for 2 weeks. *2773*

Tamoxifen In 10 anovulatory women given up to 40 mg/d for 5 d. *1542*

Plasma Decrease Physiological

Anticonvulsants In 33 male epileptics taking at least one drug for long time. *3020*

Danazol Effect noted in 6 orchidectomized patients with prostatic cancer. *0693*

Diethylstilbestrol At doses affecting pituitary-gonadal axis. *1917*

Digoxin 50% in men, 40% decrease in postmenopausal women. *3475*

Dopamine Effect noted with infusion of 0.3 μg/kg/min and more in 6 normal men for 2 h. *2149*

Estradiol In response to i.v. infusion for short duration. *3914*

Ethanol Inconsistent change with ingestion over several days in healthy men. *1352*

Ethinyl Estradiol In hypogonadal women in 1 week. *3913*

Megestrol Suppresses LH peak. *3237* In 18 postmenopausal women with breast cancer. *0056*

Methandrostenolone In 4 of 6 body-builders taking up to 20 mg/d intermittently for a year or more. *3290*

Norethindrone Apparent suppression of production or release. *2479*

Oral Contraceptives Combination type pill lowered value to 20% control. *1340*

Phenothiazines Stimulation effect of gonadotropins inhibited. *2220*

Pimozide Statistically significant decline when given to acutely psychotic males although. *3340*

Progesterone Suppresses LH peak. *3237*

Stanozolol From 6.5 to 4.5 units/L after 1 week in 9 healthy men given 10 mg/d for 14 d. *3354*

Tamoxifen In postmenopausal women with breast cancer treated for 2 weeks. *1542*

Thioridazine Significantly less in 42 male schizophrenics than when they ingested other neuroleptic agents. *0500*

Urine Increase Physiological

Clomiphene Due to action on hypothalamic-pituitary axis. *2585*

Estrogens, Conjugated Marked initial effect, falls off with large dose. *3724*

Ovulation Increased by more than 3 times at time of ovulation. *2206*

Urine Decrease Physiological

Chlormadinone Usual effect in normal women. *3464*

Oral Contraceptives Marked depression in normal subjects. *3464*

Lymphocyte Autoantibodies

Blood Decrease Physiological

Carbimazole Impaired thyroid microsomal or thyroglobulin antibody secretion due to effect on lymphocytes within thyroid. *2376*

Lymphocyte B-Cells

Blood Decrease Physiological

Prednisone Reduced but not to same extent as T-Lymphocytes. *3309*

Lymphocyte T-Cells

Blood No Effect Physiological

Ethanol Not changed by previous or current abuse. *3771*

Lymphocyte T-Cells *(continued)*

Blood   Decrease   Physiological
Chlorpromazine   In 13 of 41 patients with schizophrenia with long term treatment. *3937*
Cyclosporine   In renal transplant patients with normal graft function compared with normal controls, higher values seen in individuals with acute cyclosporine nephrotoxicity. *3288*
Prednisone   Proportional greatest reduction with steroids. *3309*

Lymphocyte T-Helper Cells

Blood   Increase   Physiological
Ethanol   But only in patients with alcoholic liver disease or pancreatitis. *3771*

Lymphocyte Mitotic Index

Test Conditions   Decrease   Physiological
Oral Contraceptives   Hormonal action. *1146*

Lymphocyte Response

Blood   No Effect   Physiological
Acyclovir   No effect on blastogenic response to mitogens-phytohemagglutin, pokeweed, and concanavilin A *in vitro. 0147*

Lymphocytes

Blood   No Effect   Physiological
Ascorbic Acid   No effect with megadose supplementation. *1347*
Cefamandole   In 2 patients given 6-9 g daily intravenously. *1646*
Cephalothin   In 2 patients given 12 g daily intravenously. *1646*
Gold   No effect observed with megadose supplementation. *1347*
Lithium   No change observed although WBC count changed. *2078*
Oxacillin   In 2 patients receiving 12-15 g/d intravenously. *1646*
Penicillin G   In 2 patients receiving 140-150 mg/kg/d intravenously. *1646*
Race   No difference between blacks and caucasians. *1870*
Riboflavin   No effect with megadose supplementation. *1347*
Vitamin A   No apparent effect of megadose supplementation. *1347*
Vitamin B$_{12}$ No effect with megadose supplementation. *1347*

Blood   Increase   Physiological
Aminosalicylic Acid   May produce syndrome like infectious mononucleosis. *1343*
Carbon Disulfide   Effect usually slight in poisoning. *1302*
Chlorpropamide   Mild, of no clinical significance. *1680*
Griseofulvin   (relative lymphocytosis). *2427*
Haloperidol   Slight effect may occur. *1680*
Levodopa   Observed with hemolytic anemia. *0824*
Narcotics   Frequent absolute and relative increase in addicts. *2012*
Niacinamide   Up by 25% at 4 h 40% at 24 h. *1622*
Propylthiouracil   In one 24 year old woman who developed fulminant hepatic failure with lymphocyte sensitization. *2437*
Smoking   Approximately 30% — mechanism obscure. *0732* Unrelated to amount of carboxyhemoglobin. *3691*
Spironolactone   Relative lymphocytosis and 15% eosinophilia observed in a single patient during period of agranulocytosis. *3487*
Thiocolchicine   May cause real increase. *2143*

Blood   Decrease   Physiological
Asparaginase   Slight reduction (not dose dependent). *2657*
Chlorambucil   Marked decrease may occur. *1680*
Corticotropin   Marked drop occurs within 2 h. *1343*

Cyclosporine   In renal transplant patients with normal graft function compared with normal controls, higher values seen in individuals with acute cyclosporine nephrotoxicity. *3288*
Diurnal Variation   Fall during day, maximum in early am. *2222*
Folic Acid   Significant effect with megadose supplementation. *1347*
Glucocorticoids   Also decrease in lymphoid tissue. *0071* Characteristic finding as a result of redistribution of cells rather than cell lysis. Effect usually transient lasting less than 24 h. *0410*
Glutamate   Significant effect with megadose supplementation. *1347*
Hydrazine Derivatives   Significant effect observed with megadose supplementation. *1347*
Hydrocortisone   Due to hormonal action on adrenals. *0242*
Ibuprofen   Observed in one child with juvenile rheumatoid arthritis when dose of drug increased. At same time altered liver function occurred. Resolved with cessation of treatment. *3457*
Lithium   ?drug associated endocrine effect. *3313*
Mechlorethamine   Occurs within 24 h. *1343*
Methysergide   Average decrease of 29% noted. *2427*
Niacin   By up to 20% in 2 h, ?normal at 24 h. *1622* Significant effect with megadose supplementation. *1347*
Phenytoin   Dose related associated with decreased DNA synthesis. *2257*
Prednisone   Maximal change in third week of therapy due to redistribution of cells out of circulation. *3309*
Pyridoxine   Significant effect with megadose supplementation. *1347*
Race   In Bantu much below cape colored or whites. *3767*
Thiamine   Significant effect with megadose supplementation. *1347*
Vitamin E   Significant effect with megadose supplementation. *1347*
X-Ray Therapy   Due to cell destruction. *0987*

Lymphocytes   Decrease   Physiological
Furosemide   Significantly reduced concentration in congestive heart failure patients treated with drug. *3101*

Lymphocyte T-Suppressor Cells

Blood   Decrease   Physiological
Ethanol   But only in patients with alcoholic cirrhosis or pancreatitis. *3771*

Lymphocyte Transformation

Blood   Increase   Physiological
Cimetidine   Serum from patients treated with drug enhanced lymphocyte response to phytohemagglutinin. *2368*

Lymphocytic Response to PHA

Blood   Increase   Physiological
Flunarizine   Enhanced response suggesting differential sensitivity of lymphocytes to calcium-entry blockers. *0482*

Blood   Decrease   Physiological
Chlorambucil   Significant reduction with chemotherapy observed. *0445*
Cyclophosphamide   Significant reduction with chemotherapy observed. *0445*
Oral Contraceptives   Hormonal action. *1146*
Pregnancy   Hormonal action. *1146*

Lymphocytotoxic Antibodies

Serum   Positive   Physiological
Procainamide 50-80% patients have antibodies after 3 to 6 mo. *1958*

Lysergic Acid Diethylamide (LSD)

Gastric Material   Increase   Analytical
Ipecac Fluorescent spectrum may interfere. *2511*

Lysine

Plasma   Increase   Physiological
Histidine Interferes with clearance after oral load. *1644*

Urine   Increase   Physiological
Cycloleucine Reversible marked aminoaciduria. *0499*

Lysolecithin

Serum   Increase   Physiological
Ethanol Moderate increase compares with controls in alcoholics especially in younger adults. *3330*

Plasma   Decrease   Physiological
Ethinyl Estradiol Mirrors change in lipoprotein pattern. *1431*
Oral Contraceptives If administered for long period. *2591*

Lysozyme

Serum   No Effect   Analytical
Storage of Sample Stable at -20°. *0877*

Serum   No Effect   Physiological
Ethanol In some alcoholics ?decreases myelopoiesis. *2196*
Gentamicin Remained in normal range in 26 patients given course of treatment. *2593*
Muscular Exercise No effect even with strenuous exercise. *2845*

Serum   Increase   Physiological
Muscular Exercise After protracted exertion. *1491*

Serum   Decrease   Analytical
Heparin Significant inhibition may occur. *3217*

Urine   No Effect   Analytical
Storage of Sample 4 d at room temperature if pH 4.5-6.3. *0877*

Urine   No Effect   Physiological
Diatrizoic acid No significant effect observed in 23 patients with hypertension and bilateral renal arteriography performed. *2594*
Enalapril In 53 patients given up to 160 mg daily or when combined with hydrochlorothiazide. *2277*
Radiographic Agents No change from 3.3 mg/d to 3.3 mg/d in 37 patients day following arteriography. *2594*
Sisomicin No significant effect seen in 23 patients given therapeutic amounts for 2 weeks. *2593*

Urine   Increase   Physiological
Benzene With chronic exposure-related to monocytic leukemia. *0040*
Gentamicin Marked effect especially with treatment for more than 12 d. *2593*
Muscular Exercise Very high clearance: proximal tubular function affected. *2845*
Proteinuria Increased in small percentage of cases. *2897*

Radiographic Agents Marked increase within 1 d when used for arteriography: effect persisted for many days: effect less when used for urography. *3259*
X-Ray Therapy Associated with cell destruction in leukemics Effect observed for more than 45 d after therapy. *3598*

White Blood Cells   No Effect   Analytical
Heparin No effect observed. *0839*

α_2-Macroglobulin

Serum   No Effect   Physiological
Oral Contraceptives No significant changes at various intervals during treatment compared with controls. *1418*
Prednisone No consistent change noted in 14 hospitalized adult patients when given 40-50 mg/d. *3790*

Serum   Increase   Physiological
Estrogens Metabolic effect. *0227*
Mestranol Maximum effect 1 week after treatment. *1660*
Muscular Exercise Approximately 5% increase immediately after. *2846* Significant effect at 15 minutes, partial return by 1 d. *1491*
Oral Contraceptives Metabolic changes in liver synthesis. *2189*
Pregnancy 20% increase in late pregnancy (other report no change). *3472*

Serum   Decrease   Physiological
Dextran Complex formation or increased consumption. *3346*
Streptokinase Significant effect remained low for 2 weeks after infusion. *1356*

Urine   No Effect   Analytical
Urea On nephelometric method of Killingsworth. *1931*

Magnesium

Serum   No Effect   Analytical
Calcium At concentration of 2.5 to 40 mg/dL did not significantly interfere with measurement of magnesium on Du Pont aca-III. *0949*
Citrates Has no effect on atomic absorption procedures. *0556*
Contact With Clot If plastic disc used to separate for 48 h. *1863*
Disulphine Blue By colorimetric reaction of Mann and Yoe. *1464*
Fluosol-DA No effect at 50% on atomic absorption measurement. *2518*
Glucuronic Acid Calcium gluconate has no effect on dihydroxyazobenzene procedure. *3501*
Protein Serum protein no effect on dihyroxyazobenzene procedure. *3501*
Storage of Sample Room temperature 8 h, 4° 1 d, frozen 1 y. *3873*

Serum   No Effect   Physiological
Aging No significant change from age 6-20 y. *0887*
Ammonium Salts No effect observed in spite of urinary change. *2315*
Bumetanide No significant effect observed. *2666*
Calcium Gluconate No effect on dihydroxyazobenzene method. *3501*
Hydrochlorothiazide No significant difference in hypertensive patients with 50 mg/d. *1980*
Muscular Exercise Observed effect with 12 minutes cycle-ergometer. *1231*
Oral Contraceptives No difference from normal observed. *0532*
Pindolol No significant effect even with 6 weeks treatment. *3279*
Spironolactone 100 mg did not increase serum concentration. *3101*
Theophylline No effect in response to i.v. infusion. *3935*
Tourniquet Insignificant over 3 minutes. *1704*

Magnesium *(continued)*

Serum   Increase   Analytical

Cefotaxime   Effect of metabolite on method on American Monitor Parallel. *0201*

Hemoglobin   Slight effect of 1 g on atomic absorption. *3217*

Hemolysis   Calculated error of 7.6%/g. *0515*

Trichloracetic Acid   Enhances absorption in atomic absorption methods. *1769*

Serum   Increase   Physiological

Alkaline Antacids   Theoretical possibility. *3369*

Amiloride   20 mg/d produces significant effect sustained for duration of treatment. *3101*

Aspirin   Prolonged therapy likely to cause elevation. *0251*

Lithium   ?affects membrane transport systems. *3505*   Picture of primary hyperparathyroidism observed in patients given treatment for long term, but mechanism not fully understood. *3118*   Increased for 24 h after drug administered. *2401*

Magnesium Salts   Absorbed from gastrointestinal tract from antacids etc. *1488*

Medroxyprogesterone   Estrogen type of response. *3329*

Menstruation   Significant increase with menstruation (by 0.1-0.2 meq/L). *1172*

Parathyroid Extract   Due to decreased renal excretion. *1343*

Progesterone   Significantly higher than in controls. *0794*

Renacidin®   One case reported of child who developed severe hypermagnesemia. *3853*

Season   Significant effect, max in Feb, min in summer. *1172*

Triamterene   Slight increased effect observed in normals. *3745*

Serum   Decrease   Analytical

Calcium Gluconate   False decrease if measured by titan-yellow. *0091*

Cefotaxime   Marked effect of drug and metabolite on specimens containing tobramycin on American Monitor method on Parallel. *0201*

Citrates   Affects fluorometric 8-hydroxyquinoline procedure. Affects titan yellow procedures. *0556*

Glucuronic Acid   Falsely low with titan yellow method. *3501*

Serum   Decrease   Physiological

Albuterol   Significant dose-related effect after i.v. infusion of therapeutic doses in 4 healthy male volunteers. *2807*

Amphotericin B   (occasional) associated with toxic effect of drug. *0128*   Significant reduction by 2 weeks from 2.35 mg/dL to 2.0 mg/dL and to 1.6 mg/dL by 4 weeks in 10 patients. *0240*

Bendrofluazide   Significant effect noted with prolonged therapy. *3285*

Calcium Salts   Competes for absorption from gut and tubules. *0091*

Carbenoxolone   By approximately 0.1 meq/L in adrenalectomized patients. *3801*

Chlorthalidone   Significant effect with prolonged treatment. *3285*

Cisplatin   In 76% of 50 patients receiving low dose cisplatin in combination with 4 other drugs every 4 weeks. Lower incidence when chemotherapy less frequently. *0520*   Incidence and severity dose-dependent in patients receiving multiple drugs: extent of effect may be related to interaction with another drug. *0520*   Marked effect in 15 patients treated with drug. *3964*   In 3rd and 4th week of treatment of one girl with malignant ovarian germ cell tumor probably related to renal tubular dysfunction. *2064*   Typical response largely avoided by prophylactic magnesium. *2247*   As result of renal tubular defect induced by drug may be severe enough to cause tetany and grand mal fits. *2247*   In 4 % of 140 cycles in 96 cancer patients given drug for 5 d repeated at 4-6 weeks intervals. *3119*   In 41 of 69 gynecologic oncology patients. *0169*   In 22 of 29 of patients with tumors of testis. *3185*   In 76% of 50 patients receiving 5-drug combination including cisplatin: dose dependent effect. *0520*

Citrates   Complexes Mg (may occur with blood transfusions). *0529*

Cyclosporine   Mean of 1.06 meq/L versus 1.33 meq/L in methotrexate treated patients after 3 mo treatment of patients with bone marrow transplant. *1833*

Diatrizoic acid   ?due to small amount of EDTA in medium. *2043*

Digoxin   Important factor in digitalis toxicity, possibly associated with prior diuretic use. *1151*

Ethanol   Following ethanol induced urinary excretion. *3495*   At lower limit of normal in 8 men who had abused alcohol for at least 10 y. *0347*

Furosemide   Observed in several patients on long term therapy or with short term vigorous treatment. *3285*   Significant reduction may occur with prolonged treatment. *3285*

Gentamicin   Effect observed in 2 patients following larger doses of gentamicin. Associated with massive urinary loss of magnesium and potassium. *0224*

Glucagon   Significant effect at 120 minutes of i.v. infusion. *0715*

Hydrochlorothiazide   In 10% patients receiving diuretics, but majority had diseases likely also to hypomagnesemia. *2050*

Insulin   Effect seen in treatment of diabetic coma. *3730*   Significant effect observed at 180 and 210 minutes of glucose tolerance test and when incubated with insulin *in vitro* due to shift of magnesium from plasma to erythrocytes. *2720*

Meralluride   Hypomagnesemia especially if NH_4Cl also given. *2314*

Mercurial Diuretics   Effect most marked if NH_4Cl also given. *2314*

Mercury Compounds   Induced with chronic laxative ingestion. *2108*

Metolazone   In isolated case as result of marked diuresis. *3285*

Neomycin   May be loss in stools due to steatorrhea. *1343*

Oral Contraceptives   0.15 meq/L decrease reported (estrogen effect). *1337*   Generally significant reduction in all women taking oral contraceptives versus age-matched controls. *3430*

Pregnancy   Significant lowering (?related to decreased albumin). *0148*

Rimeterol   Dose related significant change in 4 healthy men given therapeutic i.v. dose. *2807*

Sodium Sulfate   May be excreted combined with sulfate. *0071*

Thiazides   Consequence of diuresis. *1748*

Urine   No Effect   Analytical

Storage of Sample   Allegedly satisfactory for 45 d, no preservative. *1180*

Urine   No Effect   Physiological

Cisplatin   In 11 children given drug for 0.1-3.8 y although serum concentration reduced. *3286*

Hydrochlorothiazide   No significant effect on clearance noted with i.v. infusion of 50 mg, although long-term oral studies suggest some increased loss and reduction of plasma concentration. *3101*

Spironolactone   100 mg/d for 6 mo had sparing properties but may be related to aldosterone status. *3101*

Triamterene   Spared renal excretion with 37.5 mg/d. *3101*

Urine   Increase   Analytical

Barium   Measured by fluorometric method of Schachter. *3573*

Cadmium   Measured in fluorometric method of Schachter. *3573*

Strontium   Measured by fluorometric method of Schachter. *3573*

Tin   Measured by fluorometric method of Schachter. *3573*

Zinc   Measured by fluorometric method of Schachter. *3573*

Urine   Increase   Physiological

Acetazolamide   Reported to cause minor increase in excretion. *3101*

Aldosterone   Magnesium excretion increased to same extent as Potassium. *3730*

Ammonium Chloride   Positive correlation with calcium excretion. *2092*

Ammonium Salts   Diuretic action observed with acidosis. *2315*

Amphotericin B   Following i.v. infusion for 2 h. *2427*   Significant effect at 4 weeks, but not earlier. Fractional magnesium excretion increased. *0240*

Bed Rest   Slight effect only. *1688*

Bumetanide   After 1 mg maximum within 2 h over by 4 h. *0822*
Calcitonin   Acts independently of parathyroid. *1439*
Chlorothiazide   1.5 g/d for 2 d produces 33% increase. *1487*
Cisplatin   Renal loss of magnesium responsible for low serum concentration. *3468*   Urine wasting common manifestation. *0368*   Marked effect in 15 patients treated with drug. *3964*
Cyclosporine   Renal loss due to nephrotoxicity observed in many patients with bone marrow transplantation. *1833*
Diurnal Variation   Excretion maximum at night. *2222*
Ethacrynic Acid   Increase up to seven times reported. *1488*
Ethanol   Increased excretion seen in chronic alcoholics. *3495*
Fructose   Observed after test meals. *2422*
Furosemide   Diuretic action on divalent cations. *3745*   Renal wasting of magnesium reported with loop-blocking diuretics. *3101*
Gentamicin   Effect observed in 2 patients following larger doses of gentamicin. Associated with massive urinary loss of magnesium and potassium. *0224*
Glucose   Nutrient induced augmentation ?mechanism. *2126*
Hydrochlorothiazide   60% increase after 200 mg in 1 dose orally. *1487*
Lithium   Following administration of therapy. *2275*   Increased during day in lithium treated patients versus other psychiatric patients and healthy controls. *2401*
Mercaptomerin   Excretion increased by up to 30%. *3368*
Mercurial Diuretics   Excretion increased by up to 30%. *3368*
Methyclothiazide   Small but significant effect in normal volunteers. *2052*
Metolazone   Maximum diuretic response of 0.4 µEq/min. *3358*
Saline   With isotonic saline loading. *3439*
Sleep   ?related to inactivity: maximum at night. *1139*
Sucrose   Nutrient induced augmentation ?mechanism. *2126*
Thiazides   Increase of 33% reported. *3368*
Torasemide   Similar effects to that of furosemide. *0919*
Triamterene   Increased clearance observed. *3745*

Urine   Decrease   Analytical
Calcium Gluconate   False decrease if measured by titan-yellow. *0091*
Glucuronic Acid   Falsely low with titan yellow method. *3501*

Urine   Decrease   Physiological
Acetazolamide   Reported effect. *3745*   Following acute i.v. administration of 500 mg. *3101*
Amiloride   5 or 10 mg when administered alone, and also blocked enhanced excretion caused by hydrochlorothiazide. *3101*
Angiotensin   Hemodynamic effect of drug. *2173*
Oral Contraceptives   Associated with fall in serum concentration. *1337*
Parathyroid Extract   Decreased clearance. *1439*
Phosphates   Decreases increased excretion of bed rest. *1688*

CSF   Decrease   Physiological
Lithium   Significant effect (reverse of serum). *1441*

Gastric Material   Increase   Physiological
Histamine   Significant increased output although concentration decrease in first 20 minutes. *3252*

Gastric Material   Decrease   Physiological
Salicylate   When 0.5 g given orally to 20 healthy adult volunteers and urine studied over next 3 h. *1828*

Hair   Increase   Analytical
Hair Treatments   Marked with feminine deodorants. *1594*

Hair   Decrease   Analytical
Hair Treatments   With setting permanent waving lotions. *1594*

Lymphocytes   Decrease   Physiological
Furosemide   Significantly reduced concentration in congestive heart failure patients treated with drug. *3101*

Pancreatic Juice   Increase   Physiological
Acetazolamide   Approximately 70% increase with infusion in normals. *0984*

Red Blood Cells   No Effect   Physiological
Cisplatin   In 11 children given drug for 0.1-3.8 y although serum concentration reduced. *3286*

Red Blood Cells   Increase   Physiological
Insulin   Significant effect observed at 180 and 210 minutes of glucose tolerance test and when incubated with insulin *in vitro* due to shift of magnesium from plasma to erythrocytes. *2720*

Malate

Urine   Increase   Physiological
Lithium   May be considerable effect. *2107*

Malate Dehydrogenase

Serum   No Effect   Physiological
Physical Training   Not systematically affected by training. *1491*

Serum   Increase   Physiological
Amiodarone   In 10-20% of 36 patients. *1509*
Muscular Exercise   Significant effect with exercise. *1231*   Significant effect after 2 h march. *3217*

Serum   Decrease   Analytical
Detergents   45% inhibition at 0.6-0.8 mg/ml. *3217*

Malondialdehyde

Serum   No Effect   Physiological
Chlorpromazine   Following treatment in schizophrenic patients significantly lower values. *0351*

Platelets   Decrease   Physiological
Diflunisal   Production reduced with 1,000 mg bid whereas no effect observed at lesser amounts. Effects observed only for 24 h after administration for 1 d. *1386*

Manganese

Serum   No Effect   Physiological
Piretanide   No significant effect with up to 12 mg/d for 3 mo. *3696*

Urine   No Effect   Analytical
Storage of Sample   No effect 45 d room temperature, no preservative. *1180*

Urine   Increase   Physiological
EDTA   If poisoning due to manganese. *0071*

Mannose

Serum   Increase   Physiological
Mannose   Twice normal level in diabetics (after i.v.). *0358*

α-Mannosidase

Serum   Increase   Physiological
Aging   Very slight increase after puberty. *1043*

Mazindol

Urine   Positive   Analytical
Phendimetrazine   Same R_f and color reaction on TLC using ethyl acetate: methanol: water: ammonium hydroxide and modified Dragendorff's reagent. *3868*

MCH

Blood   No Effect   Physiological
Smoking   No significant difference between smokers and non-smokers. *1548*

Blood   Increase   Physiological
Ethanol   3.5% increase in heavy versus occasional drinkers. *3273*
Oral Contraceptives   Slight increase with continuing use. *1141* In 46 oral contraceptive users compared with controls studied over at least 2 y. *1182*

MCHC

Blood   No Effect   Physiological
Captopril   In 9 of 12 hypertensive patients although other hematological effects observed. *1608*
Ethanol   No significant difference between heavy and occasional drinkers. *3273*
Oral Contraceptives   No significant difference between oral contraceptive users and control adolescents. *1368*
Smoking   No significant difference between smokers and non-smokers. *1548*

Blood   Increase   Analytical
Cold Agglutinins   Spurious macrocytosis due to agglutination. *2797*

Blood   Increase   Physiological
Oral Contraceptives   Significant effect in users for less than 5 y. *1141* In 46 oral contraceptive users compared with controls studied over at least 2 y. *1182*

Blood   Decrease   Physiological
Erect Posture   Approximately 1% decrease in vertical but not significant. *1087*
Lead   Hemolytic anemia. *0788*
Penicillamine   Hypochromic anemia in child, menstruating woman. *0074*

MCV

Blood   No Effect   Physiological
Captopril   In 9 of 12 hypertensive patients although other hematological effects observed. *1608*
Methotrexate   No effect in long-term treatment of 20 patients with psoriasis noted. *1556* In 20 patients with psoriasis on long-term therapy. *1556*
Oral Contraceptives   No significant difference between oral contraceptive users and control adolescents. *1368* In 46 oral contraceptive users compared with controls studied over at least 2 y. *1182*
Smoking   No significant difference between smokers and non-smokers. *1548*

Blood   Increase   Analytical
Cold Agglutinins   Spurious macrocytosis due to agglutination. *2797*

Blood   Increase   Physiological
Aminobenzoic Acid   Induces malabsorption and megaloblastic anemia. *2716*
Aminosalicylic Acid   If megaloblastic anemia occurs. *3773*
Anticonvulsants   May cause megaloblastic/aplastic anemia. *3773*

Azathioprine   Macrocytosis in two-thirds of renal transplant patients. *0036*
Barbiturates   May cause megaloblastic anemia. *3773*
Carbamazepine   In about 20 epileptic patients treated for 2 y. *0873*
Colchicine   Megaloblastic anemia with B_{12} deficiency. *3773*
Cycloserine   May cause megaloblastic anemia. *3773*
Estrogens   Megaloblastic anemia. *3773*
Ethanol   3.5% increase in heavy versus occasional drinkers. *3273* By about 1.7 fL for every 10 g alcohol taken daily. *0995* Sensitive indicator of continued alcohol abuse in individuals with known liver disease. *1040* Increased in alcoholic liver disease. *0625*
Ethotoin   Theoretical effect on folic acid metabolism. *1680*
Glutethimide   May cause megaloblastic anemia. *1343*
Isoniazid   If megaloblastic anemia occurs. *2313*
Lead   Rare increase with poisoning. *0987*
Mefenamic Acid   Megaloblastic anemia reported. *0071*
Mephenytoin   May cause megaloblastic anemia. *0071*
Metformin   Occurs if megaloblastic anemia. *3605*
Methotrexate   Occurs with megaloblastic anemia. *3773*
Methylphenobarbital   May cause megaloblastic anemia. *2313*
MOPP   Reflection of bone marrow reaction to cytotoxic therapy in people with malignant disease. *0840*
Neomycin   If megaloblastic anemia develops. *3773*
Nitrofurans   May cause megaloblastic anemia. *2427*
Oral Contraceptives   Occurs if megaloblastic anemia. *3025*
Pentamidine   Megaloblastic anemia. *3773*
Phenobarbital   Megaloblastic anemia secondary to disturbance in folic acid metabolism. *1926*
Phenytoin   May occur with megaloblastic anemia. *3331* In about 20 patients treated for 2 y. *0873*
Primidone   Megaloblastic anemia secondary to disturbance in folic acid metabolism. *2968*
Pyrimethamine   Megaloblastic anemia. *3898*
Triamterene   Megaloblastic anemia. *3773*
Trimethoprim   Occurs with megaloblastic anemia. *3773*

Blood   Decrease   Physiological
Lead   Hemolytic anemia. *0788*
Nitrofurantoin   Megaloblastic anemia/hypersensitivity (G-6-PD). *2358*
Warfarin   May be secondary microcytic hypochromic anemia. *1302*

Megaloblasts

Blood   Increase   Physiological
Glycine   May provoke or accentuate megaloblastosis. *3025*
Methionine   Can aggravate vitamin B_{12} deficiency. *3025*
Serine   Can aggravate vitamin B_{12} deficiency. *3025*

Megestrol

Serum   Increase   Physiological
Megestrol   Progressive increase in concentration with time, regardless of dose given. *0056*

α-Melanocyte Stimulating Hormone

Plasma   No Effect   Physiological
Fluocinolone   When 30 g/d drug applied as topical ointment over large area. *0069*

Melanogen

Urine   Positive   Analytical
Methyldopa   At 0.2 mg/mL in alkaline urine. *2071*

Melatonin

Plasma   Decrease   Physiological
Estradiol   In one male given valerate derivative 10 μg/kg daily. Episodic pattern decreased. *2772*

Meperidine

Serum   Increase   Physiological
Meperidine   100 mg i.m. produces 1 mg/L. *2348*

Serum   Decrease   Analytical
Blood Collection Tube   Reduced concentration when blood drawn into some Vacutainers™. *3834*

Urine   Positive   Physiological
Meperidine   Main excretion product in neonates and pregnant. *2427*

Meprobamate

Blood   Increase   Physiological
Meprobamate   0.4 to 1.2 g orally produces concentrations of 5-15 mg/L. *2348*

6-Mercaptopurine

Plasma   Increase   Physiological
Allopurinol   Impairs metabolism. *2641*

Mercury

Serum   Increase   Physiological
Acetylpenicillamine   Mobilization of mercury in case of poisoning. *1871*
Fish   High concentrations found in some fish-eaters. *3270*

Urine   Increase   Physiological
Acetylpenicillamine   Elimination of heavy metals in cases of mercury poisoning. *1871*
Dimercaprol   If poisoning due to mercury. *0071*
Glassware   Risk of exposure if Van Slyke etc apparatus used. *1504*
Mercury Compounds   Due to ingestion of compound and if poisoning. *3515*
Penicillamine   Related to drug dosage in toxicity cases. *3515*
Penicillin   Acts as chelating agent in acrodynia at least. *2090*
Skin Lightening Cream   If contain Hg but no obvious nephrotoxicity. *0235*
Thimerosal   May occur with overdosage. *1680*

Mescaline

Urine   No Effect   Analytical
Chlorphentermine   No interference using TLC with ethyl acetate: methanol: water: ammonium hydroxide and modified Dragendorff's reagent for detection. *3868*
Clotermine   No interference on TLC using ethylacetate: methanol: water: ammonium hydroxide and modified Dragendorff's reagent for detection. *3868*
Diethylpropion   No interference with TLC using ethyl acetate: methanol: water: ammonium hydroxide and modified Dragendorff's reagent for detection. *3868*
Fenfluramine   No interference on TLC using ethyl acetate: methanol: water: ammonium hydroxide and modified Dragendorff's reagent for detection. *3868*

Mazindol   No interference on TLC using ethyl acetate: methanol: water: ammonium hydroxide and modified Dragendorff's reagent for detection. *3868*
Phendimetrazine   No interference with TLC using ethyl acetate: methanol: water: ammonium hydroxide and modified Dragendorff's reagent for detection. *3868*
Phenmetrazine   No interference with TLC using ethyl acetate: methanol: water: ammonium and modified Dragendorff's reagent for detection. *3868*
Phentermine   No interference on TLC using ethylacetate: methanol: water: ammonium hydroxide and modified Dragendorff's reagent for detection. *3868*

Metanephrine

Urine   No Effect   Analytical
Bromazepam   No effect at 2 mg/L on HPLC method. *0342*
Guanfacine   No effect at 2 mg/L on HPLC method. *0342*
Sulpiride   At 2 mg/L on HPLC method. *0342*
Viloxazine   No influence on liquid chromatographic measurement as drug elutes at different time. *0342*

Metanephrines, Total

Urine   No Effect   Analytical
Ascorbic Acid   At 5 g/L on modified Pisano procedure. *1428*
Chlorpromazine   At 15 mg/L on modified Pisano procedure. *1428*
Diatrizoic acid   No effect if Hypaque® administered within 2 d. *1809*
Ephedrine   At 5 mg/L on modified Pisano procedure. *1428*
Epinephrine   At 8 mg/L on modified Pisano procedure. *1428*
Glucose   At 5 g/L on modified Pisano procedure. *1428*
Imipramine   At 15 mg/L on modified Pisano procedure. *1428*
Labetalol   On method of Bigelow and Weil-Malherbe. *2430*
Levarterenol   At 50 mg/L on modified Pisano procedure. *1428*
Phenylephrine   At 5 mg/L on modified Pisano procedure. *1428*
Vanillin   At 40 mg/L on modified Pisano procedure. *1428*
Vanillylmandelic Acid (VMA)   At 50 mg/L on modified Pisano procedure. *1428*

Urine   Increase   Analytical
Chlorpromazine   Interference in Pisano procedure. *0388*
HMPG   At 2 mg/L modified Pisano procedure. *1428*
Imipramine   Interference with Pisano method. *0388*
Labetalol   With photometric method of Pisano et al and method of Crout et al. *2430* Especially when combined with other antihypertensive drugs and HPLC method unless toluene extraction used. *0751*
Methyldopa   Questionable interference fluorometric methods. *3879*
Oxprenolol   Especially when combined with other antihypertensive drugs and HPLC method used unless toluene extraction used. *0751*
Oxytetracycline   Interferes with fluorometric methods. *3879*
Phenothiazines   Interference in Pisano procedure. *0388*
Phenylephrine   At concentration possibly 10 times higher than would be encountered when measured by Pisano method. *3417*
Sotalol   Cause shift in absorbance peak with pisano method (may be large enough to double apparent concentration). *3161*

Urine   Increase   Physiological
Hydrazine Derivatives   May potentiate action of drugs on CNS. *2220*
MAO Inhibitors   Prevent deamination. *1343*
Prochlorperazine   Increased metabolism, decreased organ uptake of norepinephrine. *3879*

Urine   Decrease   Analytical
Diatrizoic acid   False negative if Renovist or Renografin® within 2 d. *1809*

Metanephrines, Total *(continued)*

Urine   Decrease   Analytical *(continued)*
Meglumine   Inhibits oxidation to vanillin in Pisano procedure. *0583*

Urine   Decrease   Physiological
Levodopa   ?dopamine as neurotransmitter suppresses normetanephrine. *1700*

Urine   Positive   Analytical
Acetaminophen   Interference with unmodified ion-exchange chromatographic method of Shoup and Kissinger. *3861*

Methacycline

Serum   Decrease   Physiological
Ferrous Sulfate   Reduced absorption due to chelation which also affects elimination. *3794*
Food   Absorption reduced when taken with food. *3024*

Methadone

Serum   Decrease   Physiological
Rifampin   Decreased serum concentration due to hepatic enzyme induction. *0014*

Urine   Increase   Analytical
Diphenhydramine   At concentrations above 100 mg/L positive results obtained with Syva® EMIT-ASSAY for drugs of abuse. *0039*

Urine   Decrease   Analytical
Liquid Soap   Progressively more negative values with increasing amounts of several preparations of liquid soap. *0964*

Urine   Positive   Analytical
Mazindol   Same R_f and color reaction on TLC using ethyl acetate: methanol: water: ammonium hydroxide and modified Dragendorff's reagent. *3868*
Phendimetrazine   Same R_f and color reaction on TLC using ethyl acetate: methanol: water: ammonium hydroxide and modified Dragendorff's reagent. *3868*

Methamphetamine

Urine   No Effect   Analytical
Chlorphentermine   No interference using TLC with ethyl acetate: methanol: water: ammonium hydroxide and modified Dragendorff's reagent for detection. *3868*
Clotermine   No interference on TLC using ethyl acetate: methanol: water: ammonium hydroxide and modified Dragendorff's reagent for detection. *3868*
Diethylpropion   No interference with TLC using ethyl acetate: methanol: water: ammonium hydroxide and modified Dragendorff's reagent for detection. *3868*
Fenfluramine   No interference on TLC using ethyl acetate: methanol: water: ammonium hydroxide and modified Dragendorff's reagent for detection. *3868*
Mazindol   No interference on TLC using ethyl acetate: methanol: water: ammonium hydroxide and modified Dragendorff's reagent for detection. *3868*
Phendimetrazine   No interference with TLC using ethyl acetate: methanol: water: ammonium hydroxide and modified Dragendorff's reagent for detection. *3868*
Phenmetrazine   No interference on TLC using ethyl acetate: methanol: water: ammonium hydroxide and modified Dragendorff's reagent for detection. *3868*
Phentermine   No interference on TLC using ethylacetate: methanol: water: ammonium hydroxide and modified Dragendorff's reagent for detection. *3868*

Urine   Increase   Physiological
Urine pH   Excretion increased with alkaline urine. *1636*

Methane

Breath   No Effect   Physiological
Gum Tragacanth   In 5 male volunteers consuming 9.9 g daily for 21 d. *0990*

Methanol

Blood   Increase   Physiological
Ethanol   Competitive inhibition of alcohol dehydrogenase. *2271*
Methanol   Presence of ingested alcohol (fatal at 1 g/L usually). *1343*

CSF   Increase   Physiological
Methanol   Higher concentration than in blood. *1343*

Methapyrilene

Urine   Positive   Analytical
Mazindol   Same R_f and color reaction on TLC using ethyl acetate: methanol: water: ammonium hydroxide and modified Dragendorff's reagent. *3868*
Phendimetrazine   Same R_f and color reaction on TLC using ethyl acetate: methanol: water: ammonium hydroxide and modified Dragendorff's reagent. *3868*

Methaqualone

Serum   No Effect   Analytical
Acetaminophen   At 100 mg/L on GLC procedure of Evenson. *1057*
Aminophylline   At 80 mg/L on GLC procedure of Evenson. *1057*
Amitriptyline   At 100 mg/L on GLC procedure of Evenson. *1057*
Amobarbital   At 20 mg/L on GLC procedure of Evenson. *1057*
Amphetamine   At 5 mg/L on GLC procedure of Evenson. *1057*
Caffeine   At 130 mg/L on GLC procedure of Evenson. *1057*
Chlordiazepoxide   At 90 mg/L on GLC procedure of Evenson. *1057*
Chlorpromazine   At 100 mg/L on GLC procedure of Evenson. *1057*
Codeine   At 8 mg/L on GLC procedure of Evenson. *1057*
Desipramine   At 100 mg/L on GLC procedure of Evenson. *1057*
Desmethyldiazepam   At 50 mg/L on GLC procedure of Evenson. *1057*
Diazepam   At 50 mg/L on GLC procedure of Evenson. *1057*
Diphenhydramine   At 80 mg/L on GLC procedure of Evenson. *1057*
Glutethimide   At 20 mg/L on GLC procedure of Evenson. *1057*
Imipramine   At 100 mg/L on GLC procedure of Evenson. *1057*
Meperidine   At 80 mg/L on GLC procedure of Evenson. *1057*
Meprobamate   At 20 mg/L on GLC procedure of Evenson. *1057*
Methadone   At 8 mg/L on GLC procedure of Evenson. *1057*
Methamphetamine   At 5 mg/L on GLC procedure of Evenson. *1057*
Methapyrilene   At 10 mg/L on GLC procedure of Evenson. *1057*
Methyprylon   At 20 mg/L on GLC procedure of Evenson. *1057*
Morphine   At 20 mg/L on GLC procedure of Evenson. *1057*
Nortriptyline   At 100 mg/L on GLC procedure of Evenson. *1057*
Pentobarbital   At 20 mg/L on GLC procedure of Evenson. *1057*
Phenobarbital   At 20 mg/L on GLC procedure of Evenson. *1057*
Phenytoin   At 4 mg/L on GLC procedure of Evenson. *1057*
Primidone   At 3 mg/L on GLC procedure of Evenson. *1057*
Prochlorperazine   At 100 mg/L on GLC procedure of Evenson. *1057*

Promazine   At 100 mg/L on GLC procedure of Evenson. *1057*
Propoxyphene   At 70 mg/L on GLC procedure of Evenson. *1057*
Quinidine   At 10 mg/L on GLC procedure of Evenson. *1057*
Quinine   At 10 mg/L on GLC procedure of Evenson. *1057*
Salicylate   At 180 mg/L on GLC procedure of Evenson. *1057*   At 180 mg/L on GLC procedure of Evenson. *1058*
Secobarbital   At 20 mg/L on GLC procedure of Evenson. *1057*
Thioridazine   At 100 mg/L on GLC procedure of Evenson. *1057*
Trifluoperazine   At 100 mg/L on GLC procedure of Evenson. *1057*

Serum   Increase   Physiological
Methaqualone   250 mg orally produced 2 mg/L in 30 minutes. *2348*

Serum   Decrease   Analytical
Pentazocine   Insignificant at therapeutic concentration procedure of Evenson. *1057*

Urine   Positive   Analytical
Diethylpropion   Similar R_f and color reaction on TLC using ethyl acetate: methanol: water: ammonium hydroxide and modified Dragendorff's reagent. *3868*

Methemalbumin

Plasma   Increase   Physiological
Antipyrine   Occurs with intravascular hemolysis. *0788*
Copper   Occurs with acute intravascular hemolysis. *0385*
Dimercaprol   Occurs with intravascular hemolysis. *0788*
Dipyrone   Occurs with hemolytic anemia. *2098*
Furadaltone   May occur with hemolysis. *0248*
Furazolidone   May occur with hemolysis. *0248*
Quinine   If given concurrently with pamaquine. *1343*

Plasma   Positive   Physiological
Penicillin V   Isolated case of hemolytic anemia with IgM antibody. *0360*

Methemoglobin

Blood   Increase   Analytical
Lipemia   Produces turbidity. *1456*

Blood   Increase   Physiological
Acetaminophen   May rarely cause hemolysis. *1343*
Acetanilid   Intravascular hemolysis may occur. *2426*
Aminosalicylic Acid   May cause hemolytic anemia. *0071*
Ammonium Nitrate   May cause hemolytic anemia. *2429*
Amyl Alcohol   May cause hemolysis following ingestion. *1302*
Amyl Nitrite   Hemolytic anemia. *2426*
Analgesics   May cause intravascular hemolysis. *3504*
Aniline   Occurs as result of intravascular hemolysis. *1302*
Antimalarials   May cause hemolytic anemia. *1902*
Antipyrine   Increase seldom occurs. *1343*
Aspirin   May cause hemolysis with G-6-PD deficiency. *3581*
Benzocaine   Hemolysis. *0380*
Bismuth Subnitrate   May cause hemolytic anemia. *2429*
Bromate   May cause hemolysis. *1343*
Carrots   With fresh or juice if contaminated with nitrites. *1893*
Chloramphenicol   May cause hemolysis in G-6-PD deficiency. *3581*
Chlorate   May cause hemolysis. *1343*
Copper   May occur with copper toxicity. *0385*
Co-Trimoxazole   Possible effect of drug or sulfonamides. *1678*
Dapsone   May cause hemolytic anemia. *1612*
Dimercaprol   May cause hemolysis in G-6-PD deficiency. *3581*
Dinitrophenol   May cause hemolysis/aplastic anemia. *2429*

Furazolidone   May cause hemolysis with G-6-PD deficiency. *3581*
Glucosulfone   Hemolytic anemia. *0788*
Glutethimide   Bleeding associated with overdosage. *2427*
Hydantoin Derivatives   Reported effect. *2220*
Isoniazid   Reported effect. *1343*
Isosorbide   Significant increase in angina patients but probably not of routine significance, but may be important in anemics or in patients with coronary insufficiency (difference = 1.13 vs 0.99 in controls). *2096*   Commonly used nitrates at regular doses capable of causing usually clinically insignificant increases. *0164*
Local Anesthetics   Reported effect. *2427*
Lysol®   May cause intravascular hemolysis. *0629*
Methicillin   One doubtful case of hemolytic anemia. *0071*
Methylene Blue   May cause hemolysis (also used as treatment). *0397*
Methylsulfonal   May cause hemolytic anemia. *2429*
Naphthalene   May cause hemolysis with G-6-PD deficiency. *3581*
Niridazole   May cause hemolysis if G-6-PD deficiency. *1343*
Nitrates   May cause hemolysis. *2426*
Nitrites   Effect of organic nitrites less than amyl nitrite. *1343*
Nitrobenzene   May cause hemolysis. *2429*
Nitrofurans   May cause hemolytic anemia if G-6-PD deficiency. *0071*
Nitrofurantoin   May cause hemolysis with G-6-PD deficiency. *3581*
Nitrofurazone   May cause hemolytic anemia. *0333*
Nitrogen Oxides   Mild elevation may occur after nitrous oxidase inhalation. *1302*
Nitroglycerin   May cause hemolysis. *2426*   Reported occasional complication of i.v. drug. *1598*
Pamaquine   May cause hemolysis in G-6-PD deficiency. *3581*   May cause hemolytic anemia. *2313*
Pentaquine   May cause hemolysis in G-6-PD deficiency. *3581*
Phenazopyridine   May cause hemolysis. *2426*
Phenylenediamine   May cause hemolysis. *2427*
Phenylhydrazine   May cause hemolysis. *2427*
Phenytoin   Occurs with hemolytic anemia. *2220*
Potassium Chlorate   May cause hemolytic anemia. *2429*
Prilocaine   O-toluidine produced as metabolite causes hemolysis. *0752*
Primaquine   May cause hemolysis in G-6-PD deficiency. *0071*
Probenecid   May cause hemolysis with G-6-PD deficiency. *3581*
Quinacrine   May cause hemolysis with G-6-PD deficiency. *3581*
Quinidine   May cause hemolysis in G-6-PD deficiency. *3581*
Quinine   May cause hemolysis in G-6-PD deficiency. *3581*
Resorcinol   Occurs with hemolytic anemia. *2427*   Injection or excess absorption may cause hemolysis. *0071*
Spinach   If contaminated with nitrites. *1893*
Sulfacetamide   May cause hemolysis if G-6-PD deficiency. *3581*
Sulfachlorpyridazine   Due to hemolysis. *1680*
Sulfamethizole   May cause hemolytic anemia. *2868*
Sulfamethoxypyridine   May cause hemolytic anemia. *2429*
Sulfanilamide   May cause hemolytic anemia. *2429*
Sulfapyridine   May cause hemolytic anemia. *2429*
Sulfisoxazole   May cause hemolytic anemia. *3015*
Sulfomethane   May cause hemolytic anemia. *2429*
Sulfonamides   Agranulocytosis/aplastic or hemolytic anemia. *2427*
Sulfones   Hemolysis may occur with G-6-PD deficiency. *0071*
Sulfoxone   May cause hemolysis in G-6-PD deficiency. *3581*
Thiazolsulfone   Hemolytic anemia. *2313*
Trinitrotoluene   Associated with hemolysis. *2429*
Vitamin A   May cause hemolysis with G-6-PD deficiency. *3581*

Urine   Increase   Physiological
Acetaminophen   Renal damage due to hemolysis. *1343*
Acetanilid   Due to hemolysis and renal damage. *1343*
Aniline   Due to intravascular hemolysis. *1302*

Methemoglobin Reductase

Red Blood Cells   No Effect   Physiological
Smoking   No significant difference between smokers and non-smokers. *1548*

Methionine

Plasma   Increase   Physiological
Starvation   Slight effect after 4 d. *0021*

Plasma   Decrease   Analytical
Iodoacetate   Effect of oxidizing agent (added *in vitro*). *0384*

Urine   Increase   Analytical
Pedameth®   Used for diaper rash in infants. *1749*

Urine   Increase   Physiological
Amobarbital   Contained in large amount in infant formula. *1749*
Neo-Mull-Soy®   Contained in large amount in infant formula. *1749*
ProSobee®   Contained in large amount in infant formula. *1749*

CSF   Decrease   Physiological
Levodopa   In Parkinson patients significant effect after 2 weeks. *1496*

Methionine Sulfoxide

Plasma   Increase   Analytical
Iodoacetate   Effect of oxidizing agent (added *in vitro*). *0384*

Methotrexate

Serum   Increase   Physiological
Aminobenzoic Acid   Displaces from plasma protein binding. *1487*
Aspirin   Displaces from plasma protein binding, if present. *1487*   Clearance by kidneys may be halved by large doses of drug. *2941*
Phenytoin   Displaces from plasma protein binding. *1487*
Sulfisoxazole   Displaces from plasma protein binding. *1487*

2-Methoxyestrone

Urine   Increase   Physiological
L-Thyroxine   Metabolic effect on hormone administration. *0023*

2-Methoxyphenoxy-Lactic Acid

Urine   Increase   Physiological
Guaifenesin   Major metabolite (44% of 1 g in 3 h). *3690*

3-Methoxytyrosine

Serum   Increase   Physiological
Levodopa   Observed after 1 week when given levodopa. *1108*

3-O-Methyl-D-Glucose

Urine   Increase   Analytical
Glucose   Several methods without removal of glucose. *1183*

Methylhistamine

Urine   Increase   Physiological
Aminoguanidine   Methylation increased when diamine oxidase inhibited. *1382*
Menstruation   Highest value observed at this time. *1381*

1,4-Methylhistamine

Urine   No Effect   Physiological
Amantadine   In normals but increased in Parkinsonism. *1817*

Methylhistidine

Urine   Increase   Physiological
Protein   Occurs with high meat diet. *0987*

3-Methylhistidine

Urine   Increase   Physiological
Ascorbic Acid   In 10 healthy females given 10 g/d. *3624*
Food   Quantitatively excreted in 2 d in 7 healthy people. *2580*

Methylimidazoleacetic Acid

Urine   Increase   Physiological
Aminoguanidine   As result of metabolism of increased methyl-histamine. *1382*
Histidine   Doubling in days after injection. *1382*
Smoking   With standard diet higher values than nonsmokers. *1381*

Methylmalonic Acid

Serum   Decrease   Analytical
Propionic Acid   MMA may be converted to propionic in GLC. *0975*

Urine   No Effect   Analytical
Chloral Hydrate   Drug has no effect on GC/MS assay which can be used to assess vitamin B_{12} deficiency. *2614*
Storage of Sample   Stable 7 d room temperature, 4 mo at -20°. *1712*

Urine   No Effect   Physiological
Lithium   No consistent effect observed. *2107*

Urine   Increase   Analytical
Malonic Acid   Measured as analyte. *1256*

Urine   Increase   Physiological
Aminosalicylic Acid   Occurs with vitamin B_{12} deficiency. *3773*
Colchicine   Occurs with impaired absorption of B_{12}. *3773*
Ethanol   May occur if B_{12} deficiency. *3773*

Urine   Decrease   Analytical
Propionic Acid   MMA may be converted to propionic in GLC. *0975*

Methylnicotinamide

Urine   No Effect   Physiological
Smoking   No effect of smoking for 1 week. *0068*   Probably no significant effect with tryptophan load. *3183*

N-Methylnicotinamide

Urine   Increase   Physiological
Chlortetracycline With dose of 3 g/d. *1343*
Oxytetracycline With dose of 2.5 g/d. *1343*

Methylphenidate

Urine   Positive   Analytical
Mazindol Same R_f and color reaction on TLC using ethyl acetate: methanol: water: ammonium hydroxide and modified Dragendorff's reagent. *3868*

Methylprednisolone

Serum   Decrease   Physiological
Phenobarbital 90% increase in metabolic clearance rate with substantial reduction in half-life. *1830*
Phenytoin 130% increase in metabolic clearance rate with much reduced half-life. *1830*

Metoprolol

Serum   Increase   Physiological
Ranitidine Significantly increased area under curve and peak plasma concentration. *1938*

Metyrapone

Serum   Decrease   Physiological
Phenytoin Average decrease from 48 to 7 μg/dL (with 3.5 g DPH). *2041*

Metyrapone Test

Plasma   No Effect   Physiological
11-Deoxycortisol Not measured by Clark/Rubin procedure. *0512*

Patient   No Effect   Physiological
Chlorpromazine May interfere with response to test. *0913*

Patient   Positive   Physiological
Fluoxymesterone At concentration of 23 mg/L had no effect on Ektachem® method. *0019*
Methandrostenolone Anabolic effect. *2681*
Nandrolone Anabolic effect. *2681*
Oxymetholone Anabolic effect (possible for 2 weeks after stop). *1679*

Mexiletine

Serum   Increase   Physiological
Amiodarone Increased concentration with possible adverse effects on EKG. *2298*
Chloramphenicol Decreases clearance of drug and prolongs half-life. *2011*
Cimetidine Decreases clearance of drug and prolongs half-life. *2011*
Dicumarol Decreases clearance and prolongs half-life. *2011*
Disulfiram Decreases clearance and prolongs half-life. *2011*
Isoniazid Decreases clearance and prolongs half-life. *2011*
Methylphenidate Decreases clearance and prolongs half-life. *2011*

Serum   Decrease   Physiological
Phenobarbital Induces hepatic enzymes: may reduce elimination half-life by 50%. *2011*

Phenytoin Hepatic enzyme induction may reduce elimination half-life by 50%. *2011*
Primidone Hepatic enzyme induction may decrease elimination half-life by 50%. *2011*
Rifampin Hepatic enzyme induction may reduce elimination half-life by 50%. *2011*

β_2-Microglobulin

Serum   No Effect   Physiological
Ceftazidime Unchanged in most patients with GFR above 30 mL/min. *0051*
Gentamicin No difference during or following treatment in 26 patients given drug. *2593*
Sisomicin In 23 patients given drug for 2 weeks in therapeutic amounts. *2593*

Serum   Increase   Physiological
Cefuroxime Mean change from 4,000 to 6,000 μg/L in 3 patients with chronic osteomyelitis 14 d when lysine salt used. *3620*
Gentamicin Significant effect observed in the absence of change in the serum creatinine. *3621*

Urine   No Effect   Physiological
Ceftazidime Unchanged in most patients with GFR above 30 mL/min. *0051* No effect on proximal tubular function in 15 patients given 3 g daily for 4-9 d. *2616*
Cimetidine No change observed during treatment of 13 ulcer patients. *0649*
Diatrizoic acid No significant effect observed in 23 patients with hypertension and bilateral renal arteriography performed. *2594*
Radiographic Agents Mean increase from 2.9 mg/d to 5.4 mg/d in 37 patients day following arteriography. *2594*
Tobramycin Up to i.v. dose of 2 mg/kg does not affect renal excretion. *3726*

Urine   Increase   Physiological
Cisplatin Transient 2 to 5-fold increase during treatment. *3397*
Gentamicin Competitively inhibits reabsorption of compound when drug excretion rates exceed 150 mg/min. *3739* Significant effect observed in the absence of change in the serum creatinine. *3621* Marked effect especially with treatment for more than 12 d. *2593*
Nifedipine Increase from 0.12 to 0.74 μg/min 40 to 60 minutes after single oral dose of 20 mg sublingually. *0650*
Sisomicin Increased up to 2 times in 23 patients given therapeutic amounts for 2 weeks. *2593*
Tobramycin Manifestation of drug induced nephrotoxicity but tend to decrease as serum creatinine begins to climb. *3182*

Microscopy

Urine   No Effect   Physiological
Cimetidine Unchanged although there may be slight increases in creatinine, not progressive; decreases with continued treatment. *2368*

Microsomal Aminopeptidase

Serum   No Effect   Physiological
Gold No significant changes in patients treated with parenteral drug. *2933*

Midazolam

Serum   Increase   Physiological
Ranitidine Bioavailability significantly increased. *1938*

Migration Inhibition Factor

Blood   Positive   Physiological
Ajmaline   In 6 patients with as high frequency as with Quinidine, Procainamide and Nifedipine. *3073*

Millons Test

Test Conditions   Positive   Analytical
Aminoimidazoleacetic Acid   Positive spot test for phenols. *3765*
Cresol   Positive spot test for phenols. *3765*
Epinephrine   Positive spot test for phenols. *3765*
Ferulic Acid   Positive spot test for phenols. *3765*
Homovanillic Acid   Positive spot test for phenols. *3765*
Hydroxybenzoic Acid   Positive spot test for phenols. *3765*
Hydroxycinnamic Acid   Positive spot test for phenols. *3765*
Hydroxyhippuric Acid   Positive spot test for phenols. *3765*
Hydroxyphenylacetic Acid   Positive spot test for phenols. *3765*
Methoxytyramine   Positive spot test for phenols. *3765*
Normetanephrine   Positive spot test for phenols. *3765*
Phenols   Positive spot test for phenols. *3765*
Tyramine   Positive spot test for phenols. *3765*
Tyrosine   Positive spot test for phenols. *3765*
Vanillic Acid   Positive spot test for phenols. *3765*
Vanillylmandelic Acid (VMA)   Positive spot test for phenols. *3765*

Monoamine Oxidase

Serum   Increase   Physiological
Levodopa   Increased activity after 2-3 mo therapy. *3623*

Serum   Decrease   Physiological
L-Thyroxine   Observed *in vitro* and *in vivo*. *1176*

Platelets   Decrease   Physiological
Amitriptyline   Mean decrease of 50% in 8 subjects with primary or secondary depression when given 100 to 300 mg/d. *2963*
Haloperidol   Significant reduction in both acute and chronic schizophrenics. Effect seen after 14 d and results did not correlate with response to treatment. *2270*
Smoking   Low enzyme activity identified in platelets from cigarette smokers. *2679*

Monoamine Oxidase B

Platelets   Decrease   Physiological
Ethanol   Marked reduction in alcoholics versus controls (by approximately 38%). *1285*

Monocytes

Blood   Increase   Physiological
Ampicillin   Associated with agranulocytosis. *1370*
Carbenicillin   In 1 of 27 patients given 270 mg/kg i.v. for 6 d. *3275*
Carbon Disulfide   Occurs in 25% cases regardless of duration. *0818*
Chlorpromazine   Occasionally before agranulocytosis. *2238*
Diurnal Variation   Rise during evening fall during early am. *2222*
Griseofulvin   (relative monocytosis). *2427*
Haloperidol   Slight effect may be observed. *1680*
Mephenytoin   Rare cases described. *1680*
Methsuximide   Reported observation. *1680*
Phosphorus   Reported effect of poisoning. *1302*
Piperacillin   In 1 of 29 patients given 181 mg/kg i.v. for 6 d. *3275*
Prednisone   Changes parallel those of neutrophils. *3309*

Propylthiouracil   In one 24 y old woman who developed fulminant hepatic failure with lymphocyte sensitization. *2437*
Smoking   Approximately 30% — mechanism obscure. *0732*

Blood   Decrease   Physiological
Azapropazone   In 1 of 83 patients treated for 6 mo. *3565*
Glucocorticoids   Characteristic finding as a result of redistribution of cells rather than cell lysis. Effect usually transient lasting less than 24 h. *0410*

Monoglyceride Lipase

Serum   No Effect   Physiological
Quingestanol   No significant effect with 300 μg/d 6 mo. *1309*

Morphine

Serum   Increase   Physiological
Morphine   After 10 mg i.v. concentration is 0.1 mg/L in 1 h. *2348*

Serum   Decrease   Physiological
Rifampin   Decreased serum concentration due to hepatic enzyme induction. *0014*

Urine   No Effect   Analytical
Apomorphine   Insignificant cross reactivity with RIA procedures. *2514*
Chlorphentermine   No interference using TLC with ethyl acetate: methanol: water: ammonium hydroxide and modified Dragendorff's reagent for detection. *3868*
Clotermine   No interference on TLC using ethyl acetate: methanol: water: ammonium hydroxide and modified Dragendorff's reagent for detection. *3868*
Cyclazocine   Insignificant cross reactivity with RIA procedures Insignificant cross reactivity with EMIT procedure for opiates. *2514*
Dextromethorphan   Insignificant cross reactivity with RIA procedures Insignificant cross react with hemagglutination inhibition Insignificant cross react with EMIT procedure for opiates. *2514*
Diethylpropion   No interference with TLC using ethyl acetate: methanol: water: ammonium hydroxide and modified Dragendorff's reagent for detection. *3868*
Fenfluramine   No interference on TLC using ethyl acetate: methanol: water: ammonium hydroxide and modified Dragendorff's reagent for detection. *3868*
Levorphanol   Insignificant cross react with hemagglutination inhibition. *2514*
Mazindol   No interference on TLC using ethyl acetate: methanol: water: ammonium hydroxide and modified Dragendorff's reagent for detection. *3868*
Methadone   Insignificant cross reactivity with RIA procedures Insignificant cross reactivity with EMIT procedure for opiates. *2514*
Nalorphine   Insignificant cross react with EMIT procedure for opiates Insignificant cross react with hemagglutination inhibition. *2514*
Naloxone   Insignificant cross reactivity with RIA procedures Insignificant cross react with EMIT procedure for opiates. *2514*
Normorphine   Insignificant cross react with hemagglutination inhibition Insignificant cross react with EMIT procedure for opiates. *2514*
Phendimetrazine   No interference with TLC using ethyl acetate: methanol: water: ammonium hydroxide and modified Dragendorff's reagent for detection. *3868*
Phenmetrazine   No interference on TLC using ethyl acetate: methanol: water: ammonium hydroxide and modified Dragendorff's reagent for detection. *3868*
Phentermine   No interference on TLC using ethylacetate: methanol: water: ammonium hydroxide and modified Dragendorff's reagent for detection. *3868*
Propoxyphene   Insignificant cross reactivity with EMIT procedure for opiates. Insignificant cross reactivity with hemagglutination inhibition. Insignificant cross reactivity with RIA procedures. *2514*

Urine   Increase   Analytical
Codeine Cross react equally (or more) with RIA procedures. Greater reaction than morphine EMIT procedure. Substantial cross reaction hemagglutination inhibition procedure. *2514*
Dihydromorphine Cross reactivity equal (or more) with RIA procedures. *2514*
Heroin Cross reactivity equal (or more) with RIA procedures. Substantial cross reactivity with hemagglutination inhibition. *2514*
Hydromorphone Cross reactivity equal (or more) with RIA procedures. Substantial cross reactivity with hemagglutination inhibition. *2514*
Levorphanol Cross react equally (or more) with RIA procedures. *2514*

Motilin

Plasma   No Effect   Physiological
Naloxone In 6 healthy volunteers after ingestion of test meal and intravenous administration of drug. *0626*

Plasma   Decrease   Physiological
Morphine Postprandial secretion abolished in 6 volunteers after drug given i.v. *0626*

Moxalactam

Plasma   No Effect   Physiological
Probenecid Concomitant administration had no effect on concentration. *0592*

Mucopolysaccharides

Urine   Increase   Analytical
Heparin False positive cetrimide and toluene blue tests — given i.v. *0525*

Mucoprotein

Urine   Increase   Physiological
Muscular Exercise Concentration of Tamm-Horsfall protein increased with decreased volume. *1720*

Myeloblasts

Blood   Increase   Physiological
Amphetamine Myeloblastic leukemia. *0325*
Benzene Chronic exposure may cause acute leukemia. *0040*

Myelocytes

Blood   Increase   Physiological
Chloroguanide Up to 10% in patients with overt malaria. *1343*

Myoglobin

Serum   Increase   Physiological
Ambulation Follows physical exertion. *1657*
Bezafibrate In four patients with poor renal function produced myolysis as result of overdose due to renal dysfunction. *3086*
Copper Observed in one case of dialysis toxicity. *2217*
Streptokinase Earlier peak of CK-MB observed in patients on thrombolytic therapy due to reperfusion of ischemic myocardium. *1341*
Succinylcholine Occasional result of i.v. injection in children. *3098*

Urine   No Effect   Physiological
Thiabendazole Not observed although serum aldolase increased. *2427*

Urine   Increase   Physiological
Ambulation Follows physical exertion. *1657*
Amphotericin B Caused by rhabdomyolysis. *0958*
Barbiturates May be increased in barbiturate poisoning. *0017*
Bezafibrate In four patients with poor renal function produced myolysis as result of overdose due to renal dysfunction. *3086*
Carbenoxolone Myopathy following hypokalemia. *2458*
Carbon Monoxide May be observed in intoxication. *0017*
Chlorthalidone Single case of severe reversible myopathy noted. *1793*
Clopamide Single case of diuretic associated myopathy. *1793*
Electroshock May occur with electrical shock. *0017*
Ethanol May be observed in alcoholism. *0017*
Glycyrrhiza May follow hypokalemia. *2458*
Succinylcholine Occasional result of i.v. injection in children. *3098*
Vasopressin Rhabdomyolysis observed in 2 patients following intravenous drug administration. *0029*

Urine   Decrease   Analytical
Alkaline Urine May interfere with hemagglutination test. *0017*

Myokinase

Serum   Increase   Physiological
Muscle Massage Significant effect 8 h after (may exceed normal limits). *0423*

Na/K ATPase

Red Blood Cells   Increase   Physiological
Enalapril Effect of drug on erythrocyte membrane so intracellular sodium reduced and potassium increased. *2277*

Nalidixic Acid

Serum   Increase   Physiological
Nalidixic Acid After 1 g orally 25-35 μg/mL in 2 h. *1678*

Urine   Increase   Physiological
Nalidixic Acid Major metabolites with glucuronide conjugates. *1678*

α-Naphthol

Urine   Positive   Physiological
Naphthalene Present as metabolite. *1302*

Naphthylamidase Isoenzymes

Serum   Increase   Physiological
Estrogens Metabolic effect. *0227*

Narcotics

Urine   Positive   Analytical
Smoking Nicotine spot in TLC method of Berry, Grove. *0324*

NBT Test

Blood   No Effect   Analytical
EDTA If Ficoll used with method of Gordon. *1349*

Blood   No Effect   Physiological
Estrogens When administered alone no effect observed. *0163*
Estrogens, Conjugated No effect observed. *0163*
Prednisolone No effect observed in uremic patients. *3888*
Progestogens When administered alone no effect observed. *0163*

Blood   Increase   Analytical
Glass Containers When glass used instead of plastic. *2338*

Blood   Increase   Physiological
Indomethacin Mechanism not discussed. *0943*
Typhoid Vaccine False positive after typhoid/paratyphoid vaccination. *2589*

Blood   Decrease   Analytical
EDTA False low result as requires complement. *3507*

Blood   Decrease   Physiological
Antibiotics Index drops in 4-6 h if therapy satisfactory. *0943*
Aspirin Mechanism not discussed. *0943*
Methylprednisolone False negative in one patient on 1 g/d. *2589*
Phenylbutazone Mechanism not discussed. *0943*

Blood   Positive   Physiological
Megestrol With 0.5 mg/d 6-9% with positive levels. *0163*
Oral Contraceptives Observed in 4 out of 6 subjects (?mechanism). *2610*
Pregnancy Occurs at all stages (in approximately 30%). *2908*

Neurophysin

Plasma   Increase   Physiological
Nicotine In smokers, maximum immediately after smoking. *1703*

Neurotensin

Plasma   No Effect   Physiological
Naloxone In 6 healthy volunteers after ingestion of test meal and intravenous administration of drug. *0626*

Plasma   Decrease   Physiological
Morphine Reduction in secretion following test meal and drug in 6 healthy volunteers. *0626*

Neutral Fats

Serum   Increase   Physiological
Pregnancy Observed throughout pregnancy. *0987*

Neutral Steroids

Feces   Decrease   Physiological
Clofibrate After 2 weeks treatment fall observed. *2463*

Neutral Sterols

Feces   Decrease   Physiological
High Carbohydrate Diet Marked if given i.v. *0880*

Neutrophil Elastase

Plasma   No Effect   Physiological
Smoking Between smokers and nonsmokers but significant rise in smokers after 8 h of abstinence and then intense smoking, probably due to *in vivo* release of neutrophil elastase. *0005*

White Blood Cells   Increase   Physiological
Smoking Mean concentration of elastase-derived fibrinopeptide A-alpha-1-21 was 5 fold higher in 10 cigarette smokers than in 20 healthy nonsmokers. Acute effects observed after smoking 3 cigarettes. *3793*

Neutrophils

Serum   Decrease   Physiological
Cephradine In a single patient either due to toxic effect on bone marrow or immune-mediated. *2096*

Blood   No Effect   Physiological
Ascorbic Acid No effect with megadose supplementation. *1347*
Cephalexin In 74 patients all children given 25-150 mg/kg/d for 5 to 15 d. *2505*
Cyclosporine No significant effect observed in 23 patients with inflammatory ocular disease treated for 6 mo. *2713*
Folic Acid No effect with megadose supplementation. *1347*
Glutamate No effect with megadose supplementation. *1347*
Gold No effect observed with megadose supplementation. *1347*
Hydrazine Derivatives No effect observed with megadose supplementation. *1347*
Niacin No effect with megadose supplementation. *1347*
Pyridoxine No effect with megadose supplementation. *1347*
Riboflavin No effect with megadose supplementation. *1347*
Thiamine No effect with megadose supplementation. *1347*
Vitamin A No apparent effect of megadose supplementation. *1347*
Vitamin B_{12} No effect with megadose supplementation. *1347*
Vitamin E No effect with megadose supplementation. *1347*

Blood   Increase   Physiological
Azapropazone In 1 of 83 patients treated for 6 mo. *3565*
Clindamycin Uncommon side effect. *3864*
Cortisone Significant granulocytosis observed. *1343*
Diurnal Variation Maximum between 12 noon and 6 pm, minimum about 6 am. *0242*
Glucocorticoids Characteristic finding as a result of redistribution of cells rather than cell lysis. Effect usually transient lasting less than 24 h. *0410*
Hydrocortisone Due to hormonal action on adrenals. *0242*
Iodochlorhydroxyquin Toxic effect reported. *3569*
Lithium ?drug associated endocrine effect. *3313* Increase largely responsible for increase in WBC count. *2078*
Methysergide Average increase of 23% noted. *2427*
Niacin By up to 100% in 2 h, high at 24 h. *1622*
Niacinamide Up by 40% after 2 g at 4 h. *1622*
Prednisone Maximum reached in second week of therapy, thereafter falls. *3309*
Smoking Correlated well with carboxyhemoglobin saturation. Average leukocyte count at upper limit of normal in nonsmokers. *3691*

Blood   Decrease   Physiological
Acetaminophen May cause neutropenia/pancytopenia. *1343*
Acetazolamide Occasional agranulocytosis reported. *3717*
Ajmaline Isolated reversible cases of agranulocytosis. *3215*
Allopurinol Noted in 3 individuals undergoing therapeutic starvation. *3224* Occasional agranulocytosis reported. *3717*
Aminosalicylic Acid If severe megaloblastic anemia occurs. *3773* Occasional case of agranulocytosis reported. *3717*
Amitriptyline May cause agranulocytosis/neutropenia. *2583*
Amodiaquine Occasional case of agranulocytosis reported. *3717*

Amoxicillin   Absolute count of less than 1500/uL in 54% of 41 children treated for 10 d. *1104*

Ampicillin   Reported observation. *2583*

Anticonvulsants   May occur without effect on white cell count. *1680*

Antipyrine   Myelotoxic action of drugs. *2427*

Aprindine   Agranulocytosis may occur between 4th and 16th week of treatment: may be quite severe: usually reversible. Occurs with frequency of 0.1 to 1.0%. *3215*

Azathioprine   Drug related leukopenia (cured by cessation). *1508* In 6 of 79 cases with hematological problem out of 328 reports of severe adverse effects. *2097*

Brompheniramine   Occasional case of agranulocytosis reported. *3717*

Bumetanide   Occasional case of agranulocytosis reported. *3717*

Captopril   In approximately 0.3% patients: develops within first 3 to 12 weeks of treatment associated with myeloid hypoplasia of bone marrow. *3712*   Isolated cases of pancytopenia, usually with pre-existing renal disease. *1255*   Agranulocytosis reported to occur in 1 of 250 treated patients. *3717*

Carbamazepine   Reported to cause neutropenia/agranulocytosis. *2583*   Neutropenia associated with decrease of WBC in small proportion of patients. *3090*   Occasional case of agranulocytosis reported. *3717*

Carbenicillin   Neutropenia reported occasionally. *1680*

Carbimazole   May cause agranulocytosis/neutropenia. *2583*   Occasional agranulocytosis reported. *3717*   Two case reports with considerable reduction in white cell count and neutrophil count. *1473*   Occasional case of drug-induced neutropenia. *0155*

Cefamandole   Occasional count below 500/uL observed. *2615*

Cefotaxime   Occasional count of less than 500/uL observed. *2615*

Cefoxitin   Isolated case of drug-induced leukopenia. *1839*

Cefuroxime   In one case who developed drug-dependent neutrophil antibodies. *2529*

Cephalexin   Agranulocytosis reported to occur occasionally. *3717*

Cephaloglycin   Neutropenia reported. *1680*

Cephalothin   Neutropenia or leukopenia rare. *0070*   Occasional count below 500/uL observed. *2615*

Cephapirin   Occasional count of below 500/uL reported. *2615*

Cephradine   In 1 of 86 children given 25-110 mg/kg daily orally for 5 to 15 d. *2505*

Chlorambucil   Severe neutropenia may be produced. *1680*

Chloramphenicol   Common toxic reaction/aplastic anemia. *2427* Occurs either as dose-related or idiosyncratic bone marrow suppression: dose- related usually occurs 5-7 d after start of therapy: idiosyncratic is rare occurring 1 case per 40,000 courses of therapy. *3864*   Agranulocytosis reported to occur occasionally. *3717*

Chlordiazepoxide   Occasional case of agranulocytosis reported. *3717*

Chlorothiazide   Occasionally observed. *2427*   Occasional case of agranulocytosis reported. *3717*

Chlorpromazine   Reported to cause agranulocytosis/neutropenia. *2583*   Occasional case of agranulocytosis reported. *3717*   Isolated case of agranulocytosis reported in elderly woman (general incidence of 1 in 1300 in psychiatric population dose related). *2299*

Chlorpropamide   Isolated case of agranulocytosis observed. *3627*   Occasional aplastic anemia reported. *3717*

Chlorthalidone   Few cases reported. *2427*   Occasional case of agranulocytosis reported. *3717*

Cimetidine   Proved case of drug induced agranulocytopenia. *0587*   Occasional case of agranulocytosis reported. *3717*

Ciprofloxacin   Reversible rare effect, often in association with administration of other drugs. *0189*

Clindamycin   Transient neutropenia may occur. *1679*

Clomipramine   Occasional case of agranulocytosis reported. *3717*

Clozapine   Occasional case of agranulocytosis reported. *3717*

Colchicine   Occasional case of agranulocytosis reported. *3717*

Co-Trimoxazole   Also reduced survival of transfused platelets due to drug associated antibodies. *0665*

Cyclopenthiazide   Reported to cause neutropenia/agranulocytosis. *2583*

Cyclophosphamide   Reported observation. *2583*

Dapsone   Occasional case of agranulocytosis reported. *3717* Isolated case of agranulocytoisis reported. *1140*   After 7-9 weeks of treatment with drug plus pyrimethamine in 7 patients taking drug as prophylaxis against malaria. *1202*

Demeclocycline   Rare, but reported to occur. *1680*

Desipramine   Occasional case of agranulocytosis reported. *3717*

Diatrizoic acid   May occur after angiography. *1680*

Diazepam   Transitory neutropenia reported. *0071*   Occasional case of agranulocytosis reported. *3717*

Diazoxide   Occasional case of agranulocytosis reported. *3717*

Digoxin   Reported observation. *2583*

Dipyrone   Rate of agranulocytosis 23.7 times higher than in non-users. *1105*

Disopyramide   Isolated case of agranulocytosis reported. *3215* Occasional case of agranulocytosis reported. *3717*

Doxycycline   Observed with other tetracyclines. *1680*   Occasional case of agranulocytosis reported. *3717*

Ethacrynic Acid   Occasional neutropenia or agranulocytosis. *2427*   Occasional case of agranulocytosis reported. *3717*

Ethanol   In some alcoholics ?decreased myelopoiesis. *2196*

Ethosuximide   Occasional case of agranulocytosis reported. *3717*

Etoposide   Nadir at 7 to 14 d after administration. *0141*

Fenoprofen   Occasional case of agranulocytosis reported. *3717*

Flucytosine   Occasional case of agranualocytosis reported. *3717*

Fluphenazine   Occasional case of agranulocytosis reported. *3717*

Furosemide   Reported to cause neutropenia. *2583*

Gentamicin   Occasional case of agranulocytosis reported. *3717*

Gold   Occasional case of sodium aurothiomalate induced neutropenia. *0155*   Reported in 6 patients: brief self-limiting process. *1362*

Griseofulvin   May cause decrease by up to 20% with fall in total count. *2353*   Occasional case of agranulocytosis reported. *3717*

Hydralazine   Occasional case of agranulocytosis reported. *3717*

Hydrochlorothiazide   Occasionally observed. *2427*   Occasional case of agranulocytosis reported. *3717*

Hydroxychloroquine   Occasional case of agranulocytosis reported. *3717*

Hydroxyurea   May cause agranulocytosis. *2583*

Ibuprofen   In one elderly man associated with a complement-dependent IgG antibody: reversible with cessation of treatment. *1054*   Occasional case of agranulocytosis reported. *3717*

Imipramine   Occasional case of agranulocytosis reported. *3717*

Indomethacin   Rare neutropenia/may also cause aplastic anemia. *1343*   Occasional case of drug-induced neutropenia. *0155*   Occasional case of agranulocytosis reported. *3717*

α-Interferon   Dose dependent leukopenia in 3 of 81 patients with malignant disease. *3299*

Isoniazid   May cause neutropenia. *2583*   Occasional case of agranulocytosis reported. *3717*

Levamisole   Occasional case of agranulocytosis reported. *3717* Causally related to presence of autoantibodies in serum. Granulocytoxins found in 6 of 20 patients. *0951*   In 16% of 201 patients treated for rheumatoid arthritis had to be withdrawn from study. Occurred after mean treatment time of 7.4 mo. *3709*   In 35% of 60 patients treated for rheumatoid arthritis, reversed with withdrawal of drug. *3841*   In 4 of 11 patients with breast cancer and radiation treatment. *2960*   In one patient with acute lymphoblastic leukemia receiving drug with methotrexate. *3849*   Observed in 17 of 174 patients with breast cancer. *3563*   In one patient with herpes simplex but also receiving other drugs. *1864*

Levodopa   Occasional case of agranulocytosis reported. *3717*

Levothyroxine   May cause neutropenia. *2583*

Lincomycin   Occasional case of agranulocytosis reported. *3717*

Mebhydrolin   Occasional case of agranulocytoisis reported. *3717*

Mepazine   Occasional case of agranulocytosis reported. *3717*

Mephenytoin   May occur with/without pancytopenia. *1680* Occasional case of agranulocytosis reported. *3717*

Meprobamate   occasional case of agranulocytosis reported. *3717*

Meralluride   May cause bone marrow depression. *1680*

Methacycline   Neutropenia reported. *1680*

Neutrophils (continued)

Blood Decrease Physiological (continued)

Methazolamide Occasional case of agranulocytosis reported. *3717*

Methicillin Neutropenia with granulocytopenia may occur. *1680*

Methimazole Occasional case of drug-induced neutropenia. *0155* Occasional case of drug-induced neutropenia. *0155*

Methotrimeprazine Occasional case of drug-induced neutropenia. *0155*

Methyldopa Occasional case of agranulocytosis reported. *3717*

Methylpromazine Occasional case of agranulocytosis reported. *3717*

Methylthiouracil Occasional case of agranulocytosis reported. *3717*

Methysergide Reported side effect. *1680*

Metiamide Occasional case of agranulocytosis reported. *3717*

Metronidazole Transient neutropenia may occur. *2427* Occasional case of agranulocytosis reported. *3717*

Mexiletine Occurs in fewer than 1% treated patients. *2011*

Mianserin Decreased concentration reported in 4 individuals who were also receiving other drugs. *0028*

Minocycline May cause hypersensitivity reaction. *1705*

Moxalactam Counts below 500/uL reported in some cases. *2615*

Nafcillin 2 of 32 children developed neutropenia within 4 to 13 d. *2548*

Nitrofurantoin Occasional case of agranulocytosis reported. *3717*

Norfloxacin Occurs in about 1% treated cases probably via an immunologic mechanism. *2747*

Oxacillin Observed in 5 patients given high doses intravenously: reversible with cessation of therapy. *0035* Granulocytopenia in 2 cases within 2 d of treatment being started. *1074* Observed in one man, recovery began within 2 d of withdrawal. *1844* Case observed after i.v. drug in 1 y old child: postulated that mechanism due to toxic effect on maturation of cells. *0657* 5 patients developed neutropenia during high-dose i.v. therapy. *0035* In two cases in one of which abnormalities developed within 48 h and in other in 17 d. *0485* Abnormal liver function in 1 of 8 treated patients on day 15. *2548* Single case of drug-related agranulocytosis in patient with prior history of penicillin sensitivity. *3167*

Oxyphenbutazone May cause neutropenia. *2583* Occasional case of agranulocytosis reported. *3717* Occasional case of drug-induced neutropenia. *0155*

Paramethadione May occur without overall effect on WBC. *1680*

Penicillamine Occasional case of drug-induced neutropenia. *0155* In 14 of 84 patients occurring typically in first 6 mo of treatment. *1892*

Penicillin Agranulocytosis/leukopenia. *2695*

Penicillin G Occasional case of drug associated leukopenia. *1839*

Pentazocine Occasional case of agranulocytosis reported. *3717*

Perchlorate Occasional case of agranulocytosis reported with potassium salt. *3717*

Perphenazine May cause neutropenia/agranulocytosis. *2583*

Phenylbutazone Occasional case of aplastic anemia reported. *3717*

Phenytoin megaloblastic/hemolytic/aplastic anemia,pancytopenia. *0072* Rare cases of neutropenia reported. *0155* Occasional case of aplastic anemia reported. *3717*

Pindolol Occasional case of agranulocytosis reported. *3717*

Piperacillin In 1 of 59 patients given drug as sole agent with many patients with severe illness. *3874*

Pirenzepine In isolated cases even though other drugs also being ingested. *3486*

Polythiazide Rare reported side effect. *1680*

Primidone Occasional case of agranulocytosis reported. *3717*

Procainamide Occasional severe reduction noted. *1586* Occasional case of agranulocytosis reported. *3717*

Prochlorperazine Occasional case of agranulocytosis reported. *3717*

Promazine Occasional case of agranulocytosis reported. *3717*

Promethazine Occasional case of agranulocytosis reported. *3717*

Propranolol Occasional case of agranulocytosis reported. *3717*

Propylthiouracil Occasional reported case of drug-induced neutropenia. *0155* Reported in about 4% of treated individuals. *3870* Occasional case of agranulocytosis reported. *3717*

Pyrazolones Myelotoxic effect of drugs. *2427*

Pyrimethamine Occasional case of agranulocytosis reported. *3717*

Quinidine Occasional agranulocytosis reported. *3215*

Quinine Occasional case of agranulocytosis reported. *3717*

Race Significantly lower in blacks (applies to both sexes). *1870*

Ranitidine Isolated case reported within 1 mo of treatment being started, reverted to normal 2 weeks after drug stopped. Probable hypersensitive or idiosyncratic reaction. *3303*

Rifampin Neutropenia may occur. *3202* Occasional case of agranulocytosis reported. *3717*

Spironolactone Occasional case of agranulocytosis reported. *3798*

Streptomycin Occasional case of agranulocytosis reported. *3717*

Sulfadoxine Isolated cases reported when given with pyrimethamine for treatment of malaria. *2669*

Sulfasalazine Rare cases of neutropenia reported. *0155* Hemolysis quite common, but frank hemolytic anemia rare. *2887*

Sulfonamides Several cases of drug induced neutropenia in patients taking only one drug. *0156*

Tetracycline Neutropenia may occur occasionally. *1680*

Thenalidine Occasional case of drug-induced neutropenia. *0155*

Thiacetazone May cause agranulocytosis. *1292*

Thiamazole Occasional case of drug-induced neutropenia. *0155*

Thiocolchicine May be reduced although lymphocytes increased. *2143*

Thioguanine Observed in 7 patients with chronic granulocytic leukemia. *3416*

Thioridazine Occasional case of agranulocytosis reported. *3717*

Tocainide Blood dyscrasias occur in fewer than 0.2% patients. *2011* Estimated incidence of 0.18%. *3019*

Tolbutamide Occasionally observed. *2427* Occasional case of drug-induced neutropenia. *0155* Occasional case of agranulocytosis reported. *3717*

Trimeprazine Occasional case of agranulocytosis reported. *3717*

Trimethadione Moderate neutropenia quite common. *1343* Occasional agranulocytosis reported. *3717*

Trimethoprim Occasional cases of neutropenia reported. *0155* Observed at least once in 57% of 49 children given trimethoprim-sulfamethoxazole over 10 day treatment period. *1104*

Trimetrexate Neutropenia in from 6 to 30% patients given different regimes. *0058*

Vancomycin Occasional reversible side effect. *1571* Observed in one patient with renal failure after 3 g drug. *0026*

Vitamin A Leukopenia with hypoplastic anemia reported. *1791*

Warfarin May cause neutropenia. *2583*

X-Ray Therapy Also toxic granulation in cells. *0987*

Nickel

Urine Increase Physiological

EDTA If poisoning due to nickel. *0071*

Test Conditions Increase Analytical

Chromium Possible interference with atomic absorption. *3499*

Cobalt Possible interference with atomic absorption. *3499*

Copper Possible interference with atomic absorption. *3499*

Iron If ferrous and atomic absorption used. *3499*

Manganese Possible interference with atomic absorption. *3499*

Zinc Possible interference with atomic absorption. *3499*

Nicotine

Plasma Increase Physiological
Smoking Sharp rise up to about 40 ng/mL with 1 cigarette. *1728*

Blood Increase Physiological
Smoking Concentration less than 0.3 mg/L. *2348*

Urine Increase Physiological
Smoking significantly greater with inhalation than non inhalation. *1921* Normal nicotine metabolite 5-20 cigarettes/d. *0324*

Urine Positive Analytical
Clotermine Similar Rfs and color reaction on TLC using ethyl acetate: methanol: water: ammonium hydroxide and modified Dragendorff's reagent. *3868*
Fenfluramine Similar R_f and color reaction on TLC using ethyl acetate: methanol: water: ammonium hydroxide and modified Dragendorff's reagent. *3868*
Phendimetrazine Same R_f and color reaction on TLC using ethyl acetate: methanol: water: ammonium hydroxide and modified Dragendorff's reagent. *3868*

Nifedipine

Serum Increase Physiological
Ranitidine 25% in mean plasma concentration and area under curve. *1938*

Nitrite

Urine No Effect Analytical
Fluorescein At 250 mg/L in spiked specimen run on Ames Clinitek 200. *1722*

Nitrofurantoin

Serum Decrease Physiological
Magnesium Trisilicate Absorption reduced due to adsorption. *3794*

Urine No Effect Analytical
Chloramphenicol No effect on method of Conklin and Hollifield. *0704*
Colistimethate No effect on method of Conklin and Hollifield. *0704*
Erythromycin No effect on method of Conklin and Hollifield. *0704*
Kanamycin No effect on method of Conklin and Hollifield. *0704*
Methenamine No effect on method of Conklin and Hollifield. *0704*
Nalidixic Acid No effect on method of Conklin and Hollifield. *0704*
Neomycin No effect on method of Conklin and Hollifield. *0704*
Oxacillin No effect on method of Conklin and Hollifield. *0704*
Oxytetracycline At 50 μg on method of Conklin/Hollifield. *0704*
Penicillin No effect on method of Conklin and Hollifield. *0704*
Polymyxin No effect on method of Conklin and Hollifield. *0704*
Streptomycin No effect of 50 μg on method of Conklin/Hollifield. *0704*
Sulfisoxazole No effect with 50 μg method of Conklin/Hollifield. *0704*
Tetracycline No effect on method of Conklin and Hollifield. *0704*

Nitrogen

Urine No Effect Analytical
Storage of Sample No effect 45 d room temperature, no preservative. *1180*

Urine Increase Physiological
Aminopyrine Reported effect. *2220*
Asparaginase Effect greatest in responders to therapy. *2657*
Aspirin Effect observed in adults. *3365*
Bed Rest Negative nitrogen balance (over 1 g/d). *1688*
Chlortetracycline Effect observed in malnourished. *1343*
Ethanol Result of chronic ingestion with negative nitrogen balance. *2428*
Oxytetracycline Observed in malnourished. *1343*
Prednisolone Negative nitrogen balance due to protein catabolism. *0019*
Starvation Marked increase if previously well fed. *3783*
Tetracycline Due to antianabolic effect. *0071*

Urine Decrease Physiological
Neomycin Due to reduced absorption of amino acids etc. *1077*

Feces Increase Physiological
Neomycin Alters intestinal villi, inhibits triglyceride hydrolysis. *3855*

Feces Decrease Physiological
Cimetidine By about 30%; in 17 patients with cystic fibrosis receiving constant concomitant therapy with pancreatic enzymes. *0624*

Nitrogen Balance

Patient No Effect Physiological
Phytate No effect of 2.5 g/d in healthy. *2947*

Patient Positive Physiological
Indolepyruvic Acid Able to replace tryptophan in diet. *2976*

Patient Negative Physiological
Dexamethasone Due to protein catabolism. *1680*
Fludrocortisone Due to protein catabolism. *1680*
Meprednisone Due to protein catabolism. *1680*
Methylprednisolone Due to protein catabolism. *1680*
Prednisone Due to protein catabolism. *1680*
Triamcinolone Due to protein catabolism. *1680*

Nitroglycerin

Serum Decrease Physiological
Nitroglycerin Enhanced metabolism when phenobarbital coadministered. *1598*
Phenobarbital When coadministered with nitroglycerin because of enhanced metabolism. *1598*

Nonprotein Nitrogen

Serum No Effect Analytical
Storage of Sample No effect 1 d at room temperature, several at 4°. *1563*

Serum Increase Analytical
EDTA Theoretical increase of 7 mg/dL as anticoagulant. *1563*
Filter Paper If contains ammonia affects some methods. *1563*
Oxalate Balanced oxalate mixture contains NH_4. *1563*

Serum Increase Physiological
Amphotericin B Nephrotoxic effect. *2313*
Arsenicals Nephrotoxicity (common with therapeutic doses). *1237*
Blindness Significantly higher (average 6 mg/dL) than normal. *1637*
Capreomycin Nephrotoxic effect. *2313*
Carbutamide Nephrotoxic effect. *2313*

Nonprotein Nitrogen (continued)

Serum   Increase   Physiological (continued)

Cephaloridine   Nephrotoxic especially if combined with diuretic. 2808

Chlortetracycline   Effect observed in malnourished. 1343

Colistimethate   Nephrotoxic effect (usually reversible). 2808

Colistin   Nephrotoxic effect (reversible renal damage). 2313

EDTA   Nephrotoxic effect (especially Calcium EDTA). 2313

Kanamycin   Nephrotoxic effect. 2313

Lipomul®   Nephrotoxic effect. 2313

Methicillin   Nephrotoxic effect. 2313

Nalidixic Acid   May cause nitrogen retention. 2808

Neomycin   Nephrotoxic effect. 2313

Nitrofurantoin   Nephrotoxic effect. 2313

Oxalate   Acute renal failure with Calcium oxalate deposition. 1343

Phosphorus   Due to nephrotoxicity. 1302

Polymyxin   Nephrotoxic effect. 2313

Streptomycin   Nephrotoxicity may occur in 2%. 2808

Sulfonamides   Occasional uremia with/without crystalluria. 2313

Tetracycline   Due to nephrotoxicity or antianabolic effect. 2313

Vitamin D   Manifestation of hypervitaminosis D. 2313

Serum   Decrease   Analytical

Storage of Sample   Significant effect at -10° after 6 mo. 3873

Serum   Decrease   Physiological

Pregnancy   Decreased 20-25% (with relatively greater decrease of urea nitrogen). 0459

Norepinephrine

Plasma   No Effect   Physiological

Captopril   No effect of drug and concentration responds appropriately to postural changes. 3712

Carprazidil   No consistent effect in 15 men with mild to moderate essential hypertension treated for up to 16 weeks. 1268

Chlorthalidone   In 22 premenopausal and 18 postmenopausal women given 100 mg/d for 6 weeks. 0402

Deprenyl   In 12 depressed or Alzheimer's disease patients given 60 mg drug daily for at. 1006

Enalapril   Nonsignificant increase as part of long-term response in 6 responders of 10 treated hypertensives. 1166

Enoximone   In 10 patients with congestive cardiac failure. 3654

Erect Posture   No effect seen in normals. 1176

Labetalol   No significant effect of acute i.v. administration (as measured by HPLC). 2974

Site of Collection   Independent of site. 3639

Terbutaline   In 6 normal men treated with therapeutic amounts for 2 weeks. 3177

Tranylcypromine   In 6 patients with depression or Alzheimer's disease treated with up to 40 mg daily for at least 3 weeks. 1006

Plasma   Increase   Analytical

Phenols   Affects trihydroxyindole method. 1176

Tris   Markedly increased concentration when progressively increasingly diluted with tris and radio enzymatic method used for quantitation. 0768

Plasma   Increase   Physiological

Adenosine   Increase by 44% after i.v. infusion of 10 to 140 μg/kg/min in 7 normal subjects. 0338

Altitude   By third day at 12,000 feet. 2001

Angiotensin   Of same magnitude as if norepinephrine given. 3967

Azosemide   Effect observed in normal male volunteers after 60 mg orally. 2658

Chlorpromazine   Increases metabolism, decreases organ uptake. 1343

Desipramine   Observed with long term, high or low dose, treatment under experimental conditions. 3066

Dopamine   Effect noted with infusion of 3.0 μg/kg/min in 6 normal men for 2 h. 2149

Doxazosin   After 6 weeks in 24 patients treated for 6 weeks. 3852

Ethanol   In individuals with unusual aldehyde dehydrogenase after 0.4 g alcohol/kg in 56 healthy males. 0015   Begins about 30 minutes after drinking and lasts for 4 h. 1671

Ether   Response to stress. 2869

Felodipine   When infused in 10 healthy normotensive volunteers. 3352

Furosemide   Effect observed after 40 mg orally in normal male volunteers. 2658

Histamine   Due to release from adrenals. 1176

Insulin   Marked effect in insulin induced hypoglycemia. 3570

Labetalol   In response to i.v. infusion of 100 mg over 10 minutes. 0226

Levarterenol   After i.v. infusion. 3967

MAO Inhibitors   Effect observed after single large dose. 1343

Meptazinol   Almost 2 fold increase in 20 minutes with up to 1.4 mg/kg i.v. 2289

Metoprolol   Significant effect in 11 healthy young men studied under variety of conditions. 0908   General greater increase in subjects after 50 mg drug in response to most stresses. 0908

Muscular Exercise   Significant effect after physical stress. 1176

Nicotine   Significant but smaller effect seen. 1176

Nilvadipine   Slight effect only in 10 individuals with mild hypertension. 3530

Opiates   Significant increase within 1 h of i.m. injection of hyoscine, meperidine or omnopon. 1929

Pancuronium   During halothane anesthesia if also given. 3410

Pentazocine   After up to 0.6 mg/kg i.v. almost 2-fold increase in 10-20 minutes. 2289

Physical Stress   Significant effect can be observed. 1176

Propranolol   Decreased clearance by 20% in hypertensives and similar effect in normals. 3944

Prostaglandin F2α   Significant increase if infused i.v. in pregnant women. 2510

Sex Difference   Significant increase in adult men compared with women. 1176

Thyrotropin-Releasing Hormone (TRH)   Mean increase of 21% in 2nd to 4th minute after i.v. injection regardless of whether patient was initially hypo- or hyperthyroid. 2921

Verapamil   Insignificant change in 15 patients with uncomplicated essential hypertension. 0867

Plasma   Decrease   Physiological

Bromocriptine   Significant effect of 5 mg/d for 5 d in healthy women but not in hyperprolactinemic women. 2283

Dextromoramide   Mechanism obscure. 1744

Disulfiram   Significant decrease after 1 week therapy. 1845

Dobutamine   In 10 patients with congestive cardiac failure. 3654

Fentanyl   Mechanism obscure. 1744

Morphine   Mechanism obscure. 1744

L-Thyroxine   From high pretreatment values to normal range with 30-60 d treatment of 7 hypothyroid women with dry thyroid extract. 2288

Urine   No Effect   Analytical

Labetalol   On HPLC method with electrochemical detection. 0434

Storage of Sample   No effect at room temperature, several days if acidified. 3873

Urine   No Effect   Physiological

Acetaminophen   No effect with short term ingestion 2.6 g/d. 0766

Amantadine   In normals and Parkinson patients. 1817

Ammonium Chloride   No significant effect observed. 3616

Aspirin   No effect with short term ingestion 2.6 g/d. 0766

Barbiturates   No effect with addiction or withdrawal. 3634

Caffeine   Effect not significant over 2 h after ingestion of 250 mg. *2751*

Carbidopa   No effect over 5 h after ingestion. *3842*

Chlordiazepoxide   No effect of short term ingestion of 30 mg/d. *0766*

Dextroamphetamine   No effect observed. *1487*

Diazepam   No effect with short term ingestion of 15 mg/d. *0766*

Diphenhydramine   No effect with short term ingestion of 150 mg/d. *0766*

Diuretics   No significant effect during diuresis. *3616*

Ephedrine   No effect observed. *1487*

Erect Posture   No effect seen in normals. *1176*

Glucose   No significant response to hyperglycemia. *3373*

Insulin   Not appreciably affected by hypoglycemia. *0279*

Methamphetamine   No effect observed. *1487*

Methyldopa   No effect when methyldopa administered and HPLC used to measure catecholamines. *2521*

Methylphenidate   No effect observed. *1487*

Pain   Response to muscle pain. *3217*

Phenmetrazine   No effect observed. *1487*

Phenobarbital   No effect short term ingestion of 120 mg/d. *0766*

Prochlorperazine   No significant effect with up to 30 mg in controls. *1607*

Urine   Increase   Physiological

Altitude   On third day at 12,000 feet. *2001*

Aminophylline   Twofold increase in response to i.v. dose. *0178*

Amphetamine   2.2 fold increase noted with psychosis. *1825*

Azosemide   Effect observed in normal male volunteers after 60 mg orally. *2658*

Chlorthalidone   After 3 d and 3 mo in 8 patients with mild hypertension given 50 mg/d. *3040*

Creatinine Clearance   Positive correlation with clearance. *1636*

Diurnal Variation   Significantly higher during day than at night. *1176* Maximum excretion about 3 pm. *2222*

Erect Posture   3 fold increase compared with lying. *3217* 30 ng/min in upright whereas 10 ng/min lying. *3351*

Ethanol   Slight increase (but less than epinephrine). *0143* In individuals with unusual aldehyde dehydrogenase after 0.4 g alcohol/kg in 56 healthy males. *0015*

Felodipine   With 20 mg daily in 10 men with essential hypertension over 8 weeks. *1887*

Furosemide   Effect observed after 40 mg orally in normal male volunteers. *2658*

Fusaric Acid   May cause increase release early, later increased metabolism. *2702*

Glucagon   Mainly conjugated, after i.v. in patients with congestive cardiac failure. *1607*

Levodopa   No effect on epinephrine excretion. *3146*

Methacholine   Slight increase observed. *0279*

Muscular Exercise   May rise up to 200-300 ng/min. *3351* If strenuous may be increased tenfold. *0279*

Nitroglycerin   ?due to adrenergic stimulation of hypotension. *3713*

Noise   Usual drop in afternoon less marked. *3351*

Pain   Also seen but less markedly with emotional stress. *1990*

Physical Stress   Significant effect can be observed. *1176*

Stress   With physical or emotional stress. *3217*

Stress Cold Exposure   Response to cold exposure. *3217*

Tetrahydrocannabinol   Possible slight increase initially compared with placebo. *3786*

Urine pH   Probable effect with alkaline urine. *1636*

Urine   Decrease   Analytical

Alkaline Urine   May be almost total destruction unless kept at pH 4-5. *1117*

Urine   Decrease   Physiological

Bethanidine   Reported effect. *1488*

Clonidine   Dose related effect (primary action of drug). *1627*

Corticotropin   Observed effect in normals and rheumatoids. *0279*

Guanethidine   Non significant decrease reported. *3083*

Guanoxan   Decreased output reported. *3083*

Lithium   Significant reduction with treatment of manic patients. Initial high values correlated with severity of disease. *3517*

Recumbency   Significantly less than when ambulant. *3616*

Reserpine   Contributes to fall of total catecholamines. *2702*

Zero Gravity   Significant effect observed. *1176*

Test Conditions   No Effect   Analytical

Dihydroxymandelic Acid   On fluorescent procedure of Peyrin. *2800*

Dihydroxyphenylacetic Acid   On fluorescent procedure of Peyrin. *2800*

Epinephrine   On automated fluorometric procedure of Peyrin. *2800*

HMPG   On fluorescent procedure of Peyrin. *2800*

Homovanillic Acid   On fluorescent procedure of Peyrin. *2800*

Isopropylnorepinephrine   On fluorescent procedure of Peyrin. *2800*

Metanephrine   On fluorescent procedure of Peyrin. *2800*

Normetanephrine   On fluorescent procedure of Peyrin. *2800*

Octopamine   On fluorescent procedure of Peyrin. *2800*

Parasympathol   On fluorescent procedure of Peyrin. *2800*

Tyramine   On fluorescent procedure of Peyrin. *2800*

Vanillylmandelic Acid (VMA)   On fluorescent procedure of Peyrin. *2800*

Test Conditions   Increase   Analytical

Dopamine   2.9% fluorescence of norepinephrine (Peyrin procedure). *2800*

Epinine   3.8% fluorescence of norepinephrine (Peyrin procedure). *2800*

Levodopa   2.9% fluorescence of norepinephrine (Peyrin procedure). *2800*

Normeperidine

Urine   Positive   Physiological

Meperidine   Main metabolite in normals. *2427*

Normetanephrine

Urine   No Effect   Analytical

Bromazepam   No effect at 2 mg/L on HPLC method. *0342*

Guanfacine   No effect at 2 mg/L on HPLC method. *0342*

Sulpiride   At 2 mg/L on HPLC method. *0342*

Urine   Increase   Analytical

Viloxazine   Substantial effect on liquid chromatographic measurement as drug elutes at same time. *0342*

Urine   Increase   Physiological

Hydrazine Derivatives   May potentiate action of drugs on CNS. *2220*

MAO Inhibitors   Prevent deamination. *1343*

Urine   Decrease   Physiological

Levodopa   Sharp decrease in normal after 3 g dose. *2634*

Nortriptyline

Serum   No Effect   Physiological

Chlordiazepoxide   No effect on serum concentration if given together. *3321*

Diazepam   No effect on serum concentration if given together. *3321*

Nitrazepam   No effect on serum concentration if given together. *3321*

Nortriptyline (continued)

Serum   No Effect   Physiological (continued)
Oxazepam   No effect on serum concentration if given together. *3321*

Serum   Increase   Physiological
Cimetidine   Area under curve increased, most noticeably for its 10-hydroxy metabolite. *1555*
Nortriptyline   30-160 μg/L after 75-225 mg for 4 d. *2348*

5'-Nucleotidase

Serum   No Effect   Analytical
Hemolysis   No effect hemoglobin up to 1 g/dL. *3217*
Storage of Sample   For 4 d at room temperature or 4°, 4.5 mo if frozen. *0877*

Serum   No Effect   Physiological
Carbamazepine   Observed 2 times upper limit of normal in none of 34 adult inpatient epileptics given drug alone (dose and duration of treatment unknown). *1162*
Phenobarbital   No effect observed in 12 adult epileptic in patients given drug alone (dose and duration of treatment unknown). *1162*

Serum   Increase   Physiological
Acetaminophen   In severe cases of poisoning (also in some mild). *3470*
Acetohexamide   Due to cholestasis. *1778*
Anabolic Steroids   Due to cholestasis. *1778*
Androgens   Due to cholestasis. *1778*
Asparaginase   Observed in up to 25% patients. *2657*   In 15% children and 26% adults reported in various studies. *2406*
Aspirin   Reversible hepatotoxicity with prolonged administration. *3092*
Carbamazepine   Due to cholestasis. *1778*
Carbenoxolone   Possible direct hepatotoxicity of drug. *2458*
Chloramphenicol   Due to cholestasis. *2220*
Chlorothiazide   May cause cholestatic jaundice. *2313*
Chlorpropamide   Due to cholestasis. *1778*
Estradiol   Cholestatic effect. *2313*
Estrogens   May cause cholestasis. *2313*
Ethinyl Estradiol   Slight effect?impaired excretion. *1498*
Imipramine   Due to cholestasis. *1778*
Indomethacin   Due to cholestasis. *1778*
Lincomycin   Due to cholestasis. *1778*
Methyltestosterone   Due to cholestasis. *1778*
Nalidixic Acid   Due to cholestasis. *1778*
Nitrofurantoin   Due to cholestasis. *1778*
Norethandrolone   Due to cholestasis. *1778*
Norethynodrel   Due to cholestasis. *1778*
Oxacillin   Due to cholestasis. *1778*
Phenothiazines   Due to cholestasis. *1778*
Phenytoin   Observed 2 times upper limit of normal in 10 of 127 adult epileptic inpatients. *1162*
Propoxyphene   Due to cholestasis. *1778*
Sulfonamides   May cause cholestasis. *1778*
Thiabendazole   Rare cholestasis. *1778*
Thiazides   Due to cholestasis. *1778*
Tolazamide   Due to cholestasis. *1778*
Tolbutamide   Due to cholestasis. *1778*
Trimethoprim   Transient increase in the absence of abnormal liver function tests in two patients when drug combined with sulfamethoxazole. *0893*

Obermayer Test

Urine   Positive   Analytical
Indican   Produces blue color (indigo blue)in chloroform. *3588*

Urine   Negative   Analytical
Formaldehyde   Inhibits test for indican. *0459*

Occult Blood

Feces   No Effect   Analytical
Bisacodyl   No effect noted on Hemoquant method. *3671*
Cholestyramine   No effect noted on Hemoquant procedure. *3671*
Cimetidine   No effect noted on Hemoquant method. *3671*
Iron   No effect on Hemoquant procedure. *3671*
Iron Salts   If 3,3-dimethylnaphthidine used as chromogen. *0849*
Lactulose   No effect on Hemoquant procedure. *3671*
Metoclopramide   No effect on Hemoquant method. *3671*
Psyllium Fibre   No effect on Hemoquant procedure. *3671*
Ranitidine   No effect on Hemoquant method. *3671*
Sulfobromophthalein   No effect on tests for occult blood. *0583*

Feces   No Effect   Physiological
Fenoprofen   In 49 patients with juvenile rheumatoid arthritis when some abnormal results in patients with same therapeutic outcome given aspirin. *0465*

Feces   Increase   Physiological
Cyclosporine   In approximately 3% patients given drug versus 6% given azathioprine in over 200 renal transplant patients. *0572*
Piroxicam   Gastrointestinal bleeding most common severe problem. *2051*

Feces   Decrease   Analytical
Ascorbic Acid   False reduction at physiological amounts (if taking added vitamin C) on Hemoccult® and other tests with pseudoperoxidase principle eg guaiac, benzidine, or other diamino compound as indicator. *1763*

Feces   Positive   Analytical
Bromides   In vitro reaction, ?high enough concentration in vivo. *0583*
Cimetidine   When added to Hemoccult® test paper as pure chemical applied at pH of gastric juice. *1527*   At concentrations above 1500 mg/L on Hemoccult® method. *3393*
Ferrous Gluconate   50% false positive with Hemoccult®; 65% false positive with Hematest® in 10 male volunteers taking 300 mg 3 times per day for 1 week. *2170*
Ferrous Sulfate   65% false positive with Hemoccult®; 25% false positive with Hematest® in 10 male volunteers taking 300 mg 3 times per day for 1 week. *2170*
Fish   Affect most methods if not withheld for 3 d. *0849*
Iodides   In vitro reaction, ?high enough concentration in vivo. *0583*   Interferes with benzidine test. *0579*
Iron Salts   Interferes with guaiac test (?benzidine). *3584*
Lead   Lead sulfide may simulate melena. *1302*
Povidone-Iodine   With Hemoccult® reaction probably due to oxidation of alpha-guaiaconic acid impregnated test paper. Same effect with Lugol's iodine. *2678*   Occurs with as little as 0.005 mL of a 1 to 1,000 dilution due to iodine component. *0376*

Feces   Positive   Physiological
Acetanilid   Poisoning may cause many gastrointestinal symptoms. *0279*
Acetazolamide   Initial response noted. *1680*
Acids   If ingested may cause bleeding and gastrointestinal symptoms. *0279*
Aconitine   If ingested may cause many gastrointestinal symptoms. *0279*
Alcohols   May cause many gastrointestinal symptoms including bleeding. *0279*

Alkalies   If ingested may cause many gastrointestinal symptoms. *0279*

Aminophylline   Hematemesis may occur early with poisoning. *2427*

Aminopyrine   Possibly due to impaired platelet aggregation. *0605*

Aminosalicylic Acid   Reversible gastritis caused by drug. *1343*

Amphetamine   Reported to cause ulcers of gastrointestinal tract. *0612*

Amphotericin B   Melena and hemorrhagic gastroenteritis. *0071*

Amyl Alcohol   May cause gastrointestinal hemorrhage. *1302*

Anisindione   May indicate toxicity. *1680*

Arsenicals   Observed in several cases with poisoning. *1478*

Aspirin   In over 70% patients when more than 3 g/d given. *3761* In 1.5% trials of treatment for juvenile rheumatoid arthritis. *0239*

Barium   May cause severe gastrointestinal tract hemorrhage. *1343*

Benzene   May cause gastrointestinal bleeding and other symptoms. *0279*

Betamethasone   May activate peptic ulcer. *1680*

Boric Acid   Toxicity effect. *2313*

Cefamandole   Observed in 3 of 37 patients receiving intravenous nutrition. *3134*

Cefoperazone   Associated with vitamin K correctable hypoprothrombinemia but mechanism unclear. *2685*

Chloramphenicol   May be gastrointestinal hemorrhage, with low prothrombin. *0071*   Occasional case of pseudomembranous colitis reported with mucus or bloody diarrhea. *0447*

Chlorophenothane   May cause gastrointestinal tract bleeding and other symptoms. *0279*

Chlorphenesin   Two cases gastrointestinal bleeding reported. *1680*

Chlorpropamide   Associated with severe diarrhea and bleeding. *2801*

Chlortetracycline   Occasional case of pseudomembranous colitis reported with mucus and bloody diarrhea. *0447*

Cinchophen   May cause severe ulceration of gastrointestinal tract. *0612*

Clindamycin   Occasionally bloody mucus, especially in elderly. *1679*   Reported frequency of 1 case of pseudomembranous colitis in 10 treated cases associated with mucus or bloody diarrhea. *0447*

Colchicine   Toxicity effect. *2313*

Copper   May occur with severe poisoning. *2217*

Corticosteroids   May increase incidence of gastric ulcers. *0333*

Cortisone   May cause hemorrhage or ulceration of gastrointestinal tract. *0612*

Cyclophosphamide   Hemorrhagic colitis reported. *2427*

Cytarabine   Gastrointestinal hemorrhage in fewer than 2%. *1680*

Dexamethasone   May aggravate peptic ulcer and cause bleeding. *1680*

Dicumarol   May cause intramural hemorrhage even if no ulcer. *3498*

Diflunisal   Significant increase when 1,000 mg given bid whereas no effect with 250 mg or 500 mg bid. *1386*   Two reported cases of hematemesis attributable to drug, although other drugs also ingested. *2326*

Digitalis   May occasionally cause gastrointestinal hemorrhagic necrosis. *0658*

Dipyrone   May cause gastrointestinal bleeding. *0071*

Ergot Preparations   May cause gastrointestinal bleeding with overdose. *0279*

Ethacrynic Acid   One case reported (26% if given i.v. possibly). *0363*

Fenoprofen   Significant effect but less than with aspirin. *2985* In 9% in patients treated for juvenile rheumatoid arthritis. *0239*

Floxuridine   May cause gastrointestinal bleeding. *3850*

Fludrocortisone   May activate peptic ulcer. *1680*

Fluorides   Due to hemorrhagic gastroenteritis if ingested. *0010*

Fluorouracil   Manifestation of toxicity. *1680*

Fluroxene   In 1 of 8 patients (slight only). *1814*

Formaldehyde   Ingestion may cause hematemesis. *1302*

Glucocorticoids   May cause lower intestinal ulceration, perforation and hemorrhage. *0410*

Gold   Associated with thrombocytopenia. *2427*   Rare side effect as reported to uk committee on safety of. *0782*

Halothane   In 3 of 8 patients (slight only). *1814*

Hetacillin   Caused bleeding with dose of 4 g. *0853*

Histamine   May cause gastric mucosal bleeding if given subcutaneously. *3498*

Hydralazine   Rare gastrointestinal hemorrhage. *1343*

Hydrocortisone   May cause hemorrhage or ulceration of gastrointestinal tract. *0612*

Hypochlorites   Corrosive action on gastrointestinal tract if swallowed. *1302*

Ibufenac   May cause gastrointestinal tract bleeding. *2313*

Ibuprofen   May produce gastric irritation and activate ulcer. *1630* Rare adverse reactions reported by physicians to UK committee on safety of medicines. *0782*

Indomethacin   May cause ulceration of stomach, duodenum, gut. *2412*   Rare side effect reported to UK committee on safety of medicines. *0782*

Iodides   Chronic poisoning may cause bloody diarrhea. *1343*

Iodine Containing Drugs   May occur with toxicological doses. *1302*

Iron Salts   With poisoning may cause gastrointestinal hemorrhage. *3817*

Isopropanol   May cause hematemesis and gastroenteritis. *1302*

Kerosene   Toxic effect if ingested. *1343*

Ketoprofen   Gastrointestinal irritation is probably major side effect as reported to UK committee on safety of medicines. *0782*

Lead   May be bloody diarrhea with poisoning. *1302*

Levarterenol   Diffuse hemorrhagic enteritis with vasoconstriction. *3498*

Levodopa   Single case of gastritis with melena. *2984*

Lincomycin   May cause severe enterocolitis. *3316*   Occasional case of pseudomembranous colitis reported with bloody or mucus diarrhea. *0447*

Lipomul®   May be severe gastrointestinal tract bleeding. *2427*

Meat   Myoglobin in meat, no effect if none for 3 d. *0849*

Medroxyprogesterone   Ischemic colitis reported in one case. *1260*

Mefenamic Acid   Occurs less frequently than with aspirin. *0071*

Melphalan   May cause gastrointestinal hemorrhage. *0071*

Meprednisone   Activation or complication of ulcer. *1680*

Mercury Compounds   Bloody diarrhea occurs with poisoning. *1343*

Methotrexate   May cause hemorrhagic enteritis. *0136*

Nalidixic Acid   May cause bleeding from gastrointestinal tract. *0071*

Naproxen   Several cases of gastrointestinal bleeding reported to UK committee on safety of medicines. *0782*

Nitrites   May cause bloody diarrhea if ingested. *1302*

Novobiocin   Intestinal hemorrhage may occur. *1343*

Oxalate   Occurs after ingestion. *1302*

Oxyphenbutazone   May cause gastrointestinal bleeding. *0071* Rare side effect reported to UK committee on safety of medicines. *0782*

Paraldehyde   Bleeding gastritis in acute or chronic poisoning. *1343*

Paramethadione   Gastrointestinal bleeding may occur (can affect many organs). *1680*

Paramethasone   May cause peptic ulcer with hemorrhage. *1680*

Penicillamine   Occasional side effect as reported to UK committee on safety of medicines. *0782*

Phenolphthalein   Protracted treatment may cause ulceration. *0612*

Phenylbutazone   May cause gastrointestinal tract bleeding. *2313* Rare side effects reported to UK committee on safety of medicines. *0782*

Phenylephrine   May = hemorrhagic enteritis with vasoconstriction. *3498*

Phosphorus   Bloody diarrhea may occur with poisoning. *1343*

Occult Blood *(continued)*

Feces   Positive   Physiological *(continued)*

Potassium Chloride May cause gastrointestinal tract ulceration. *0612*

Potassium Salts May cause ulceration and hemorrhage in gastrointestinal tract. *1680*

Prednisolone May activate peptic ulcer with hemorrhage. *0019*

Prednisone May cause hemorrhage and ulceration of gastrointestinal tract. *0612*

Procarbazine Melena and gastrointestinal tract bleeding. *0071*

Propylthiouracil In one 24 year old woman who developed fulminant hepatic failure with lymphocyte sensitization. *2437*

Pyrazolones May cause gastrointestinal tract bleeding. *2313*

Reserpine May activate peptic ulcers and cause bleeding. *3856*

Silver May cause hemorrhagic gastroenteritis. *1343*

Sulfonamides Hematemesis and melena may occur. *2427*

Sulthiame One case reported with poisoning. *2427*

Tetracycline May occur with hematemesis and melena. *0945*

Thallium Due to ingestion of toxic dose. *1302*

Theophylline In large doses may cause gastric hemorrhage. *1680*

Thiotepa May be ulceration of gastrointestinal tract. *2427*

Triamcinolone May activate peptic ulcer or cause perforation. *1680*

Trimethadione Gastrointestinal bleeding reported (may affect many organs). *1022*

Warfarin May cause intramural hemorrhage even if no ulcer. *3498*

Feces   Negative   Analytical

Ascorbic Acid Interferes with analytic methods. *2313* At concentrations above 30 mg/L produced false negative result as measured by Hemoccult®. *3393*

Odor

Urine   Increase   Analytical

Thiabendazole Odor similar to that after ingestion of asparagine. *2413*

Urine   Increase   Physiological

Asparagus Characteristic after ingestion. *0443* Observed in 43% of 800 volunteers. Reproducible effect inherited as autosomal dominant trait. *2461*

Thymol Characteristic after ingestion. *0443*

Turpentine Odor resembles that of violets. *1302*

Breath   Increase   Physiological

Selenium Garlic-like odor due to dimethyl selenide. *2217*

Oleic Acid

Serum   Increase   Physiological

Caffeine Significant increase within 1 h, still elevated after 4 h, when 250 mg ingested. *2751*

Oncotic Pressure

Plasma   Increase   Physiological

Ethylene Glycol Also observed in patients who ingested several different alcohols. *3325*

Organic Acids

Urine   Increase   Physiological

Carbamazepine Dose related effect observed in long-term treated epileptics compared with controls. *2008*

Phenobarbital Dose related in long-term treated epileptics compared with controls. *2008*

Phenytoin Dose related in long-term treated epileptics compared with controls. *2008*

Valproic Acid Dose related effect in long term treated epileptics compared with controls. Effect not as marked as with other anticonvulsants. *2008*

Ornithine

Plasma   No Effect   Physiological

Carbamazepine No striking effect when given to epileptic patients. *3697*

Phenobarbital No striking abnormality when given to epileptics. *3697*

Tranylcypromine No effect observed. *0824*

Valproic Acid No striking abnormality when given to epileptics. *3697*

Plasma   Increase   Analytical

Hemolysis Due to arginase in RBCs. *3230*

Plasma   Decrease   Physiological

Histidine Probably due to flow into gut after oral load. *1644*

Ornithine Carbamoyltransferase (OCT)

Serum   No Effect   Analytical

Hemolysis Slight has no effect, massive produces slight effect. *0877*

Storage of Sample 1 d in blood at room temperature, 1 week at 4° in serum. *0877* 1 y at -20°. *3217*

Serum   No Effect   Physiological

Surgery No significant effect observed. *2809*

Serum   Increase   Physiological

Aminosalicylic Acid May cause cytotoxic hepatocellular damage. *0248*

Anesthetic Agents Occurs without permanent liver damage. *3661*

Chloroform Hepatotoxic effect with necrosis. *3835*

Chlorpropamide May cause cytotoxic liver damage. *0248*

Clindamycin Mild increase noted (transient). *1083*

Ethanol Occurs at 15 h in normal subjects. *1323* Activity correlates with aminotransferases: may be almost as sensitive as AST in detection of alcoholic liver injury. *3124*

Iproniazid May cause cholestatic and cytotoxic jaundice. *0248*

Isoniazid Cytotoxic hepatocellular damage. *0248*

Mechlorethamine May cause cytotoxic (hepatocellular) damage. *0248*

Methotrexate May cause cytotoxic hepatocellular damage. *0248*

Muscular Exercise Observed to increase following exercise. *1231* Maximum effect observed 7 d after exercise. *1491*

Oral Contraceptives Often raised during first month of treatment. *1498*

Oxyphenisatin Rare increase with chronic active hepatitis. *1297* Observed when combined with iron preparation. *1298*

Phenylbutazone May cause cholestatic and cytotoxic jaundice. *0248*

Pregnancy Physiological and metabolic response. *1498*

Sulfamethoxypyridine Reversible cholestasis may occur. *1986*

Testosterone May cause cholestatic and cytotoxic liver damage. *0248*

X-Ray Therapy Due to breakdown of tissue proteins. *0483*

Serum   Decrease   Analytical

Storage of Sample Freezing and thawing once causes 10% decrease. *0877*

Serum   Decrease   Physiological
Alkoxyglycerols   Protects against elevation caused by X-rays. 0483

Orotic Acid

Urine   Increase   Physiological
Azauridine   Metabolic effect may cause crystalluria. 0278

Orthophosphate

Test Conditions   No Effect   Analytical
Glucose   No effect up to 1 mol/L. 1658
Magnesium Sulfate   No effect up to 1 mol/L. 1658
Pyrophosphate   Due to molybdate catalyzed hydrolysis. 1658
Urea   No effect up to 1 mol/L. 1658

Test Conditions   Decrease   Analytical
Acetates   Interference above 0.7 mol/L on method of Horder. 1658
Boric Acid   Slight interference reported only. 1658
Chloride Salts   85% inhibition at 0.4 mol/L on method of Horder. 1658
Potassium Iodide   Total inhibition at 0.1 mol/L on method of Horder. 1658
Sodium Citrate   Total inhibition at 0.1 mol/L on method of Horder. 1658
Tromethamine   Total inhibition at 0.1 mol/L on method of Horder. 1658

Osmolality

Serum   No Effect   Analytical
Storage of Sample   No effect for 3 h at room temperature, 10 h at 4°. 1810

Serum   No Effect   Physiological
Azosemide   Effect observed in normal male volunteers after 60 mg orally. 2658
Bopindolol   In 10 hypertensive patients treated for either 1 or 21 d. 1063
Bumetanide   No significant effect observed. 2666
Methylprednisolone   No significant effect following i.v. infusion of drug in patients with myocardial infarction compared with controls. 1560

Serum   Increase   Analytical
Citrates   Significant effect when added as anticoagulant. 3361
Glucose   Osmotically active constituent in samples. 1810
Mannitol   Concentration related increase: 6% at 3.1 g/L and 12.3% at 6.2 g/L. 2066
Urea   Osmotically active constituent in samples. 1810

Serum   Increase   Physiological
Arginine   Slight but not significant increase (about 8 mosmol/L) after i.v. infusion in diabetics. 2328
Corticosteroids   Associated with sodium and chloride retention. 1343
Ethanol   Osmotic activity of alcohol-minimum from other sources. 3009   Rise precedes rise in vasopressin following ingestion of 75 mL alcohol. 1007
Ethylene Glycol   Greater if measured by freezing point depression than calculated value. 3325
Glycerin   Rose an average of 19 mosm/kg. 3615
Inulin   Massive doses have marked effect. 0071
Ioxitalamic Acid   Major increase in arterial osmolality within 30 s following injection of 77 ± 16 mL Telebrix 38 for translumbar arterial aortography in 16 patients. 1364
Mannitol   May cause marked dehydration. 2220
Methoxyflurane   Impaired renal tubular function. 0749

Prolactin   Effect not marked. 1662
Sleep   Reversal with diuresis of sleep. 1190

Serum   Decrease   Physiological
Altitude   Significant effect at 3800 m compared with sea level. 3939
Bendrofluazide   Due to hyponatremia. 1129
Carbamazepine   Mean concentration reduced in carbamazepine treated patients, possibly due to stimulation of release of ADH. 2787   Isolated case of dilutional hyponatremia with water intoxication: 7 previous cases reported. 1882
Chlorthalidone   With ADH secretion in response to diuresis. 1129
Cyclophosphamide   Average 15 mosm/kg for 1 d (due to metabolites). 0855
Cyclothiazide   Syndrome of inappropriate ADH secretion seen. 1661
Diurnal Variation   Associated with water retention at night. 1190
Hydrochlorothiazide   Due to hyponatremia of diuretic action. 1129
Ketoconazole   To 248 mosmol/kg in 73 year old man with prostatic cancer treated with 600 mg daily for 2 1/2 mo. 2817
Lorcainide   Significant effect in 16 of 33 patients with organic heart disease and ventricular arrhythmias. Effect observed after single i.v. dose. 3383
Methyclothiazide   With ADH secretion due to hyponatremia. 1129
Polythiazide   ADH secretion with hyponatremia. 1129

Urine   No Effect   Physiological
Bopindolol   In 10 hypertensive patients treated for either 1 or 21 d. 1063

Urine   Increase   Analytical
Glucose   Osmotically active constituent in samples. 1810
Urea   Osmotically active constituent in samples. 1810

Urine   Increase   Physiological
Anesthetic Agents   Induction of anesthesia produces marked effect. 0916
Chlorpropamide   Normal diuretic response also if diabetes insipidus. 2499
Cyclophosphamide   Impaired water excretion 500 mosm/kg increase. 0855
Metolazone   Diuretic action. 3494
Phloridzin   Due to increased excretion of glucose etc. 3343
Vincristine   May produce clinically impaired water excretion. 0854

Urine   Decrease   Physiological
Acetohexamide   Normal slight diuretic response. 2499
Demeclocycline   Maximum osmolality in response to dehydration. 3333
Glyburide   Normal diuretic response. 2499
Lithium   Reduced concentrating ability in lithium treated patients than healthy subjects. 0519
Methoxyflurane   Nephrotoxic effect of drug (dose dependent). 2436
Muscular Exercise   Effect most marked with light exercise. 1842
Starvation   Observed response to stress. 2382
Tolazamide   Normal diuretic response. 2499

Osmolar Clearance

Urine   Increase   Physiological
Bumetanide   Marked dose related effect. 2666
Isosorbide   Diuretic action alone, potentiates others. 2605
Metolazone   Diuretic action of drug. 3494

Urine   Decrease   Physiological
Anesthetic Agents   Significant effect with general anesthesia. 0916
Diazoxide   Effect over 2 h of 4 mg/kg given i.v. 1805

Osmotic Fragility

Red Blood Cells   Increase   Physiological
Dapsone   In 20 of 51 patients with leprosy, most of whom were receiving 51 to 100 mg drug daily. *2410*
Pregnancy   May be occasional increase. *0987*

Osteocalcin

Plasma   Decrease   Physiological
Ethinyl Estradiol   In post-menopausal women with 50 µg/d for 14 d. *3474*

Oxalate

Serum   Increase   Physiological
Methoxyflurane   Metabolic degradation product of drug. *2876*

Urine   No Effect   Analytical
Ascorbic Acid   When boric acid used as diluent for low chromatographic measurement procedures. *3007*   Minimal effect on gas-chromatographic procedure with alkalinization of urine. *2344*

Urine   No Effect   Physiological
Ascorbic Acid   No effect in 50 volunteers given 2 g/d for 2 mo. *1039*
Citrates   In 89 patients with hypocitraturic calcium nephrolithiasis treated for up to 4 y. *2708*

Urine   Increase   Analytical
Ascorbic Acid   With Sigma procedure with diluted or undiluted urine, but pretreatment with ferric chloride causes loss of oxalate. *1305*
Homogentisic Acid   When urine from alkaptonuric measured by Sigma oxalate oxidase method. *0345*

Urine   Increase   Physiological
Ascorbic Acid   Normal metabolite excreted. *2063*
Ethylene Glycol   Increased excretion following ingestion. *3325*
Methoxyflurane   Metabolic degradation product of drug. *2876*
Oxalate   Effect greatest with low calcium diet. *2310*

Urine   Decrease   Analytical
Ascorbic Acid   Reduced concentration observed with methods involving oxalate decarboxylase. *3030*
Calcium Carbimide   Inhibits enzymatic procedure for measurement. *3936*

Urine   Decrease   Physiological
Pyridoxine   In patients with oxalate renal calculi. *1343*

Oxazepam

Serum   No Effect   Physiological
Cimetidine   Not subject to N-dealkylation or hydroxylation by cytochrome P-450 so not affected by drug. *3084*

Oxygen-Hemoglobin Affinity

Blood   Decrease   Physiological
Smoking   Significantly less in smokers. *3112*

Oxygen Saturation

Blood   Increase   Analytical
Lipemia   Slight effect on method using CO-Oximeter™. *0842*

Oxyphenbutazone

Serum   Increase   Physiological
Methandrostenolone   Possibly due to inhibition of metabolism. *1487*

Oxytetracycline

Serum   Increase   Physiological
Bilirubin   Affects fluorometric method of Murthy. *2533*
Hemolysis   Affects fluorometric method of Murthy. *2533*

Serum   Decrease   Physiological
Ferrous Sulfate   Reduced absorption due to chelation which also affects elimination. *3794*
Food   Absorption reduced when taken with food. *3024*

Oxytocin

Plasma   No Effect   Physiological
Naloxone   No effect observed for 120 minutes in healthy male volunteers after intravenous injection of 10 mg. *1648*

Plasma   Increase   Physiological
Oral Contraceptives   Significant effect in women taking oral contraceptives. *3320*

Urine   No Effect   Physiological
Pregnancy   No effect of pregnancy or labor. *0437*

Oxytocinase

Serum   Increase   Physiological
Pregnancy   Rise from tenth week, falls with labor. *0094*   In 3rd trimester (55 U/L versus 2.8 normal). *3677*

PAH Clearance

Serum   No Effect   Physiological
Diatrizoic acid   No significant effect of bilateral renal arteriography in 23 patients with hypertension. *2594*

Urine   No Effect   Physiological
Amiloride   No effect on glomerular filtration rate or effective renal plasma flow. *0656*
Aspirin   Insignificant reduction observed. *2535*
Etodolac   No effect with acute or chronic treatment in individuals with normal renal function but transient reduction in people with renal insufficiency. *0453*
Radiographic Agents   No effect of arteriography. *2594*
Sulindac   No significant change in patients with chronic glomerular disease. *0660*

Urine   Increase   Analytical
Aminobenzoic Acid   Measured as if PAH. *2981*
Aminosalicylic Acid   Very slight effect, measured as if PAH. *2981*
Azosulfamide   Slight effect only, measured as if PAH. *2981*
Sulfonamides   React as if PAH with Bratton-Marshall method. *2981*

Urine   Increase   Physiological
Clonidine   Nonsignificant effect in hypertensives. *0393*
Hydrocortisone   Increases GFR. *1343*
Levodopa   ?secondary to renal vasodilatation or direct action on tubules. *1136*
Methylprednisolone   At 24 h after 1 g i.v. in normals. *3774*

Nicardipine   After 0.5 mg i.v. increased renal blood flow in 7 patients with mild to moderate hypertension by average of 27%. *0195*

Urine   Decrease   Physiological
Acetrizoate   When 70% injected for aortography. *0065*
Aging   Approximately 60% linear decrease from age 40 to 85 y. *2617*
Aspirin   Observed even with therapeutic doses. *2866*
Chlorothiazide   Competitive inhibition of secretion. *1433*
Diazoxide   Effect over 2 h of 4 mg/kg given i.v. *1805*
Ibuprofen   Increase by 35% in patients with chronic glomerular disease. *0660*
Probenecid   Renal clearance impaired. *2375*
Radiographic Agents   If concentrated solutions used for aortography. *0065*
Sulfinpyrazone   Inhibits tubular transport. *0071*

Palmitic Acid

Serum   Increase   Physiological
Caffeine   Significant increase within 1 h, still elevated after 4 h, when 250 mg ingested. *2751*

Palmitoleic Acid

Serum   Increase   Physiological
Caffeine   Significant increase within 1 h, still elevated after 4 h, when 250 mg ingested. *2751*

Pancreatic Polypeptide

Plasma   No Effect   Physiological
Naloxone   In 6 healthy volunteers after ingestion of test meal and intravenous administration of drug. *0626*

Plasma   Decrease   Physiological
Morphine   Postprandial secretion abolished in 6 volunteers after drug given i.v. *0626*
Pirenzepine   Borderline significant reduction from basal mean value of 37 ng/L to 26.2 ng/L after treatment with 100 mg/d for 7 d. *3933*

Paraldehyde

Blood   Increase   Physiological
Paraldehyde   Usually fatal at 500 mg/L. *2348*

Para-Nitrophenol

Urine   Increase   Physiological
Parathion   Roughly increased in proportion to inhibitor of cholinesterase. *3850*

Parathyroid Hormone

Plasma   No Effect   Physiological
Cellulose Phosphate   In patients with absorptive hypercalciuria. *2711*
Chlorthalidone   But probable enhancement of peripheral action. *2122*
Epinephrine   Following i.v. infusion for 1 h to produce concentration up to 945 pg/mL. *0401*
Ethylene Glycol   Although normal homeostasis interfered with. *3325*
Hydrochlorothiazide   No effect (?mechanism of effect on calcium). *3480*

Ketoconazole   After administration to normal volunteers. *3390*
In 9 healthy men given up to 1200 mg/d for 1 week. *1300*
Omeprazole   No effect observed in 8 volunteers given 30 mg/d for 28 d. *2516*
Probenecid   In 8 healthy young men. *1318*
Propranolol   In hyperthyroid patients possible effect due to membrane-stabilizing property of drug. *0334*
Propylthiouracil   In hyperthyroid patients and normal controls. *0334*
Rifampin   No change observed in 8 healthy men taking drug for 2 weeks. *0479*
Sodium Salts   No significant effect in 5 men with essential hypertension given either chloride or citrate supplement. *2037*
Theophylline   In patients with drug toxicity. *2383*
Timolol   No effect in hyperthyroid patients. *0334*

Plasma   Increase   Physiological
Diurnal Variation   Rises after 8 pm, maximum 3 am, normal by 8 am. *1829*
Ethanol   At upper limit of normal in 8 men who had abused alcohol for 10 y. *0347*
Ethinyl Estradiol   Significant increase after in 7 healthy postmenopausal women treated for 12 weeks. *3467*
Furosemide   10% increase in normal individuals after 8 d administration. *1214*
Hydrocortisone   If given i.v. stimulates secretion. *3840*
Lithium   Observed in 21% psychiatric patients given compound, positively correlated with serum lithium concentration. *3118*   38% higher than in controls in 12 patients taking drug for 2 to 13 y. *1175*
Phenytoin   In epileptic children in comparison with controls. *2030*
Phosphates   Up to 125% increase at 1 h after 1 g orally. *2951*
Prednisone   Significant effect although Calcium normal. *3840*

Plasma   Decrease   Physiological
Aluminum Hydroxide   Fell approximately 40% in 7 nondialyzed patients with chronic renal failure with 15-18 g/d. *3533*
Cimetidine   Affect C-terminal component only: observed in normals and patients with renal failure: slight increase of N-terminal component. *1137*
Ethinyl Estradiol   Average decrease of about 25% with RIAs directed against mid-molecule in 10 post-menopausal women treated for 14 d with 50 μg/d. *3474*
Gallium Nitrate   In two patients with hypercalcemia due to parathyroid carcinoma. *3764*
Gentamicin   Effect observed in 2 patients following larger doses of gentamicin. Associated with massive urinary loss of magnesium and potassium. *0224*
Pindolol   Observed within 3 h of treatment: significant effect over 6 weeks. *3279*

Partial Thromboplastin Time

Plasma   No Effect   Physiological
Ampicillin   In doses as high as 300 mg/kg/d in volunteers. *0492*
Apalcillin   In 21 volunteers with doses up to 225 mg/kg. *1262*
Fluroxene   No significant effect noted. *1814*
Ketoprofen   In 11 patients given drug intravenously. *1234*
Methicillin   No effect with doses as high as 300 mg/kg/d in volunteers. *0492*
Moxalactam   Insignificant change noted in preoperative patients. *3121*
Penicillin G   No effect with doses as high as 48 million U/d. *0492*

Plasma   Increase   Analytical
Saline   20% dilution produces significant effect in lab. *1244*

Plasma   Increase   Physiological
Asparaginase   Observed frequently with toxicity. *3358*
Cefamandole   In 7 cases developed vitamin K deficient hypoprothrombinemia. *3103*

Partial Thromboplastin Time (continued)

Plasma Increase Physiological (continued)

Cefoperazone Reported in previous study to occur in about 4% treated patients. Mechanism unclear but possibly due to inhibition of prothrombin production. Effect correctable with vitamin K. *2685*

Chlorpromazine Caused by circulating inhibitor resembling that seen in systemic lupus in schizophrenic patients. *3937*

Dextran 15% increase noted in one case. *0320*

Gold In one 3 y old with rheumatoid arthritis. *1674*

Heparin Concentration related effect. *2649*

Metronidazole In one patient stabilized on warfarin to whose therapeutic regime metronidazole was added. *0850*

Naloxone With multiple doses effect observed (no bleeding). *0019*

Phenytoin May cross placenta (vitamin K dependent). *3378*

Radiographic Agents ?transient inactivation of coagulation factors. *1487*

Tolmetin In 1% of patients in trials for treatment of juvenile rheumatoid arthritis. *0239*

Plasma Decrease Analytical

Hemolysis Significant effect in normal subjects (but not abnormals). *1244*

Oxalate If used as anticoagulant compared with citrate. *3382*

Plasma Decrease Physiological

Estrogens, Conjugated Shortening of time in males after i.v. administration. *0083*

Muscular Exercise Observed effect in exercise. *2847*

Oral Contraceptives Associated with disordered clotting. *0760* In women given low dose ethinyl estradiol and norethindrone. *2623*

Blood No Effect Physiological

Smoking No significant difference between smokers and non-smokers. *1547*

Blood Increase Physiological

Vitamin E In 2 patients given 2.3 g/m²/d intravenously for 4 or more days (effect abrogated by prior administration of menadiol sodium diphosphate). *1554*

pCO₂

Blood No Effect Analytical

Transport of Specimen No effect in pneumatic tube system. *3217*

Blood No Effect Physiological

Smoking No effect observed. *2028*

Triamterene No effect in normals (decreased in diabetics). *3744*

Blood Increase Physiological

Althesin Post-induction measurement in arterial blood. *3160*

Apomorphine May depress respiration. *1343*

Barbiturates Respiratory depressant. *1343*

Codeine May cause respiratory depression. *1343*

Dextromethorphan High doses may produce respiratory depression. *1343*

Diazepam In healthy volunteers but change of short duration. *1041*

Heroin If pulmonary edema occurs with decreased respiration. *3059*

Hydromorphone 10 times as potent as morphine. *0493*

Hypothermia Due to hypoxia, acid-base imbalance, hypotension. *2258*

Meals Slight effect alkaline tide. *0814*

Meperidine Depresses responsiveness of respiratory center to CO₂. *1343*

Methadone May cause diminished pulmonary ventilation. *1343*

Methylenedioxyamphetamine Respiratory acidosis. *2975*

Midazolam In healthy volunteers but of short duration. *1041*

Morphine Diminishes ventilation, causes hypercapnia. *0071*

Nitrogen Oxides Retention of CO₂ occurs. *1302*

Pentazocine High doses produce marked respiratory depression. *1343*

Phenazocine May produce respiratory depression. *1343*

Propoxyphene Large doses may produce respiratory depression. *1343*

Storage of Sample Slight increase observed at 2 h at room temperature. *0931*

Blood Decrease Analytical

Heparin Dilution effect if dead space filled with liquid heparin. *0444*

Storage of Sample By approximately 0.6 mm Hg per h at 0-4°. *3217*

Blood Decrease Physiological

Acetazolamide Usual effect in bronchitics. *2321*

Aspirin In toxicity with increased respiratory rate and pulmonary ventilation. *1343*

Diurnal Variation On waking, decreased suppression of respiratory center. *0814*

Ethamivan Increased depth of respiration and improved pulmonary ventilation. *0071*

Ether May be slight effect during anesthesia. *1343*

Mafenide Inhibits carbonic anhydrase if applied topically. *3813*

Neuromuscular Relaxants Secondary to hyperventilation postoperatively. *1343*

NSD 3004 Arterial blood long acting carbonic anhydrase inhibition. *2232*

Starvation Marked metabolic acidosis within 5 d. *3067*

Tromethamine Correction of respiratory acidosis. *3006*

Xylitol May cause metabolic acidosis. *3576*

Penicillamine

Serum Decrease Physiological

Antacids Reduced absorption due to adsorption and chelation. *3794*

Ferrous Sulfate Reduced absorption due to chelation which also affects elimination. *3794*

Penicillin

Serum Decrease Physiological

Neomycin Decrease by 50% when given orally. *1487* Reduced absorption due to sprue-like syndrome. *3794*

Penicillin G

Serum Decrease Physiological

Food Absorption reduced when taken with food. *3024*

Penicillin V

Serum Decrease Physiological

Food Absorption reduced when taken with food. *3024*

Pentazocine

Blood Increase Physiological

Pentazocine Concentration of 150 µg/L achieved. *2348*

Pentobarbital

Blood   Increase   Physiological
Pentobarbital   600 mg orally produces concentration of 3.3 mg/L. *2348*

Urine   No Effect   Analytical
Chlorphentermine   No interference using TLC with ethyl acetate: methanol: water: ammonium hydroxide and modified Dragendorff's reagent for detection. *3868*
Clotermine   No interference on TLC using ethyl acetate: methanol: water: ammonium hydroxide and modified Dragendorff's reagent for detection. *3868*
Diethylpropion   No interference with TLC using ethyl acetate: methanol: water: ammonium hydroxide and modified Dragendorff's reagent for detection. *3868*
Fenfluramine   No interference on TLC using ethyl acetate: methanol: water: ammonium hydroxide and modified Dragendorff's reagent for detection. *3868*
Mazindol   No interference on TLC using ethyl acetate: methanol: water: ammonium hydroxide and modified Dragendorff's reagent for detection. *3868*
Phendimetrazine   No interference with TLC using ethyl acetate: methanol: water: ammonium hydroxide and modified Dragendorff's reagent for detection. *3868*
Phenmetrazine   No interference on TLC using ethyl acetate: methanol: water: ammonium hydroxide and modified Dragendorff's reagent for detection. *3868*
Phentermine   No interference on TLC using ethylacetate: methanol: water: ammonium hydroxide and modified Dragendorff's reagent for detection. *3868*

Pepsin

Gastric Material   Increase   Physiological
Acetylcholine   Hypoglycemia is powerful stimulant. *0814*
Caffeine   Effect of i.v. infusion, also if given orally. *0689*
Calcium   Marked increase if serum concentration increased above 6 meq/L. *2689*
Histamine   Effect occurs with secretion of acid. *1343*
Insulin   Hypoglycemia is powerful stimulant. *0814*
Pentagastrin   If infused i.v. *0689*
Pilocarpine   Produces secretion like that of vagal stimulation. *1343*
Reserpine   Greatly augments secretion. *0612*
Secretin   Rises to 2.5 times basal, falls in 30 minutes. *1677*
Tolazoline   Also enhances histamine stimulation. *1343*

Gastric Material   Decrease   Physiological
Atropine   Antagonizes cholinergic stimulation. *0814*
Barbiturates   Secretion slightly depressed. *1343*
Cimetidine   Marked effect on basal and stimulated concentrations. *3451*
Ethanol   Unless major psychic component. *1343*
Oxyphencyclimine   Absolute decrease by 60%, but not concentration. *2537*
Pregnancy   May be hyposecretion. *0987*

Pepsinogen

Serum   No Effect   Analytical
Storage of Sample   Stable 4 d at room temperature. *0877*

Pepsinogen I

Serum   Increase   Physiological
Omeprazole   Significant effect after 7 and 14 d therapy (more than doubling concentration). *1126*

Peripheral Smear

Blood   No Effect   Analytical
Storage of Sample   Unstained blood smear usually satisfactory for 45 d. *1180*

Blood   Abnormal   Physiological
Aminosalicylic Acid   Atypical lymphocytosis with eosinophilia. *2427*
Arsenicals   May produce stippling of red cells. *0279*
Gold   May cause stippling of cells. *1205*
Pargyline   Poikilocytosis and inisocytosis common. *2427*

Peroxidase

Urine   Increase   Physiological
Proteinuria   Indicative of presence of blood cells. *2897*

pH

Serum   No Effect   Physiological
Serum   Clotting has no effect. *1233*

Serum   Increase   Analytical
Oxalate   *In vitro* addition of K salt to plasma/serum. *0106*

Serum   Decrease   Analytical
Citrates   *In vitro* addition of sodium salt to plasma/serum. *0106*
EDTA   *In vitro* addition of sodium salt to plasma/serum. *0106*

Blood   No Effect   Analytical
Heparin   Even if syringe dead space contains liquid heparin. *0444*
Transport of Specimen   No effect in pneumatic tube system. *3217*

Blood   No Effect   Physiological
Arginine   No change following i.v. infusion in diabetics. *2328*
Ascorbic Acid   Even when 8 g/m²/d ingested. *0233*
Meperidine   Insignificant effect observed. *2427*
Methylprednisolone   No significant effect following i.v. infusion of drug in patients with myocardial infarction compared with controls. *1560*
Smoking   No effect observed. *2028*
Triamterene   No effect in normals when given 100 mg bid. *3744*

Blood   Increase   Analytical
Heat   Increased 0.015 per degree (may also affect buffers). *0579*
Oxalate   Occurs with Na or K oxalates as anticoagulants. *0579*
Storage of Sample   If sample exposed to air. *3879*

Blood   Increase   Physiological
Acetates   Alkalinizing action due to rapid metabolism. *1343*
Antacids   May cause occasional metabolic alkalosis. *0071*
Aspirin   Initial respiratory alkalosis. *3687*
Carbenicillin   High pH observed in most patients. *1947*
Carbenoxolone   Alkalosis occurs in approximately 15% patients. *2849*
Citrates   Restores bicarbonate reserve, may cause alkalosis. *0071*
Ethacrynic Acid   Hypochloremic alkalosis may occur. *3214*
Glutamic Acid   Sodium salt may cause alkalosis. *0071*
Glycyrrhiza   May cause alkalosis. *2220*
Hyperventilation   Due to respiratory alkalosis. *3234*
Lactate   Used in treat of metabolic acidosis. *0071*
Laxatives   If chronic abuse may cause metabolic alkalosis. *3682*
Mafenide   Usual finding with respiratory alkalosis. *0123*
Mayo Enema   Due to bicarbonate in enema. *0726*
Meals   Effect of metabolic alkalosis. *3234*
Meralluride   Alkalosis may occur with massive diuresis. *1680*

pH (continued)

Blood Increase Physiological (continued)

Mercurial Diuretics May cause systemic alkalosis especially if hypochloremia. *1343*

Phenylbutazone May cause respiratory or metabolic alkalosis. *1680*

Silver Observed after silver nitrate antisepsis. *2427*

Sodium Bicarbonate Affects acid-base balance *in vivo*. *2313*

Sodium Salts Significant but small effect observed with citrate but not chloride salt in 5 men with essential hypertension. *2037*

Triamcinolone May cause hypokalemic alkalosis. *1680*

Tromethamine Can correct metabolic or hypercapnic acidosis. *3006*

Tubocurarine Respiratory alkalosis with low doses. *2220*

Blood Decrease Analytical

Ammonium Oxalate Alters acid-base balance if used as anticoagulant. *2313*

Citrates Significant effect of sodium citrate as anticoagulant. *0579*

EDTA Affects acid-base balance if used as anticoagulant. *2313*

Oxalate Significant effect of NH_4 oxalate as anticoagulant. *0579*

Storage of Sample By approximately 0.006 per h at 0-4°. *3217* If anaerobic storage at 37° or 2 h at room temperature. *3879*

Blood Decrease Physiological

Acetazolamide Due to inhibition of carbonic anhydrase. *2313*

Acetone If ingested may cause acidosis. *0279*

Aminobenzoic Acid Acidosis from use of free acid reported. *1343*

Aminosalicylic Acid Moderate strong acid, loss of fixed cation = acidosis. *1343*

Ammonium Chloride Following administration (metabolic acidosis). *2313*

Arginine Chloride salt tends to cause acidosis. *0071*

Aspirin May cause acidosis later (respiratory and metabolic). *2313* Systemic acidosis common in poisoning, more frequent in chronic situation, in patients with severe manifestations and dehydration. *1251*

Blood Transfusions If massive transfusions given. *2220*

Calcium Chloride Is an acidifying salt. *2313*

Citrates Significant acid shift produced. *1233*

Cyclopropane May cause metabolic acidosis. *1343*

Dimercaprol May induce metabolic acidosis. *0071*

Dimethadione Extracellular acidosis induced. *3882*

Earlobe Blood Significantly less in first sample than in subsequent. *0931*

Ethanol Causes lactic acidosis. *0568*

Ether Metabolic acidosis especially in children. *0071*

Ethoxzolamide May cause metabolic acidosis. *2220*

Ethylene Glycol May cause marked acidosis with poisoning. *0279*

Fluorides Combined respiratory and metabolic acidosis in poisoning. *0010*

Fructose Associated with lactic acidosis. *0744*

Hypothermia Due to hypoxia, acid-base imbalance, hypotension. *2258*

Isoniazid Large doses may produce severe acidosis. *1343*

Mafenide If respiratory impairment as reduced renal buffering. *0123*

Mercury Compounds Induced with chronic laxative ingestion. *2108*

Methanol Causes acidosis. *2313*

Methylenedioxyamphetamine Respiratory acidosis. *2975*

Minocycline May occur if impaired renal function. *1680*

Paraldehyde Acidotic action (decomposes to acetic acid). *2313*

Pentazocine Slight fall (average 0.05) 15 minutes after i.v. *2446*

Spironolactone May cause mild acidosis. *1680*

Starvation Sharp decrease observed sometimes after 2nd day. *3067*

Tetracycline May cause acidosis with renal impairment. *2220*

Trimethadione Converted to dimethadione *in vivo* with same effects. *3882*

Tubocurarine Large dose effect with prolonged recovery. *2220*

Xylitol Pronounced metabolic acidosis in many patients. *3575*

Urine No Effect Analytical

Chloroform As preservative has no effect on Stix tests. *2308*

Fluorescein At 250 mg/L in spiked specimen run on Ames Clinitek 200. *1722*

Urine No Effect Physiological

Methyclothiazide No significant effect usually observed. *1680*

Theophylline In healthy volunteers given drug intravenously. *1593*

Urine Increase Physiological

Acetazolamide Inhibition of carbonic anhydrase. *1343*

Aldosterone H and K ions exchanged for Na ion. *1343*

Amiloride Moderate increase sometimes observed. *1343*

Amphotericin B In absence of acid load indicates pending decrease of GFR. *0536*

Bed Rest Average increase of 0.1 to 0.2 with immobilization. *0863*

Carbenoxolone Impaired acidification with hypokalemia. *2458*

Cimetidine Mean increase of 0.4 in healthy volunteers. *2076*

Citrates From average of 5.35 to 6.68 for potassium citrate and to 6.73 for sodium citrate in 5 patients with uric acid lithiasis. *3116* In 89 patients with hypocitraturic calcium nephrolithiasis treated for up to 4 y. *2708* 60 meq/d increased pH on average by 1.3 with both sodium and potassium salts. *3116*

Epinephrine Metabolic effect. *1622*

Mafenide Inhibits carbonic anhydrase if applied topically. *3813*

Meals Due to alkaline tide. *0814*

Muscular Exercise Effect noted after mild exercise. *1842*

Niacinamide Reduces acidity at 2-4 h. *1622*

Parathyroid Extract Inhibits tubular exchange of Na for h. *2611*

Prolactin Individual response — may be marked. *1662*

Ranitidine Mean increase of 0.4 in healthy volunteers. *3389*

Sodium Bicarbonate Used to alkalinize urine. *0071*

Triamterene Slight alkalinization (mechanism not known). *1343*

Urine Decrease Physiological

Ammonium Chloride To below 5.3 if acidification normal. *0913*

Ascorbic Acid Acidifies urine when ingested in large amounts. *2063* Some effect with ingestion of 3-6 g/d. *0233*

Corticotropin Metabolic effect. *1622*

Diazoxide Due to decreased bicarbonate excretion. *0124*

Diurnal Variation Minimum after midday in afternoon. *2222*

Glucose Associated with increased net acid excretion. *2126*

Methenamine Mandelate is an acidifying agent. *1343*

Methionine Used to acidify urine. *0443*

Metolazone Associated with diuretic response. *3358*

Muscular Exercise At all rates of exercise (acid metabolites). *1842*

Niacin Increases acidity at 4 h. *1622*

Sucrose Associated with increased acid excretion. *2126*

Feces Decrease Physiological

Lactulose When given as retention enema. *1920*

Mannitol No effect observed. *0032*

Sodium Sulfate No effect observed. *0032*

Sorbitol No effect observed. *0032*

Gastric Material No Effect Physiological

Prostaglandin-α_1 No consistent effect noted. *0337*

Prostaglandin-15-EPI-A_2 No consistent effect noted. *0337*

Gastric Material   Increase   Physiological

Cimetidine   Effectively reduces gastric acidity both before and after meals. *2351*

Oxmetidine   pH 5 reached 80 minutes after 200 mg drug. *2578*

Prostaglandin-α_2   Transient inhibition with pH to above 6. *0337*

Ranitidine   pH 5 reached approximately 1 h after 50 mg. *2578*

Gastric Material   Decrease   Physiological

Flucytosine   Effect on both basal and submaximal pentagastrin simulated gastric acid secretion. Effect dose related. *0559*

Smoking   Effect of 4-6 cigarettes in 1 h. *0295*

Muscle   Decrease   Physiological

Muscular Exercise   Fell from normal 6.93 to 6.40 after exercise. *1572*

Phencyclidine

Urine   Positive   Analytical

Diethylpropion   Similar R_f and color reaction on TLC using ethyl acetate: methanol: water: ammonium hydroxide and modified Dragendorff's reagent. *3868*

Phendimetrazine

Urine   Positive   Analytical

Mazindol   Same R_f and color reaction on TLC using ethyl acetate: methanol: water: ammonium hydroxide and modified Dragendorff's reagent. *3868*

Phenethicillin

Serum   Decrease   Physiological

Food   Absorption reduced when taken with food. *3024*

Pheneturide

Serum   No Effect   Analytical

Acetazolamide   On GLC procedure of Papadopoulos if added *in vitro*. *2722*

Amobarbital   On GLC procedure of Papadopoulos when added *in vitro*. *2722*

Beclamide   On GLC procedure of Papadopoulos at *in vivo* concentration. *2722*

Carbamazepine   On GLC procedure of Papadopoulos at *in vivo* concentration. *2722*

Chlordiazepoxide   On GLC procedure of Papadopoulos at *in vivo* concentration. *2722*

Chlorpromazine   On GLC procedure of Papadopoulos at *in vivo* concentration. *2722*

Cyproheptadine   On GLC procedure of Papadopoulos at *in vivo* concentration. *2722*

Diazepam   On GLC procedure of Papadopoulos at *in vivo* concentration. *2722*

Ethosuximide   On GLC procedure of Papadopoulos at *in vivo* concentration. *2722*

Ethotoin   On GLC procedure of Papadopoulos at *in vivo* concentration. *2722*

Haloperidol   On GLC procedure of Papadopoulos at *in vivo* concentration. *2722*

Imipramine   On GLC procedure of Papadopoulos at *in vivo* concentration. *2722*

Mephenytoin   On GLC procedure of Papadopoulos at *in vivo* concentration. *2722*

Methsuximide   On GLC procedure of Papadopoulos at *in vivo* concentration. *2722*

Methylphenobarbital   On GLC procedure of Papadopoulos when added *in vitro*. *2722*

Nitrazepam   On GLC procedure of Papadopoulos at *in vivo* concentration. *2722*

Oxytetracycline   On GLC procedure of Papadopoulos when added *in vitro*. *2722*

Secobarbital   On GLC procedure of Papadopoulos at *in vivo* concentration. *2722*

Sulthiame   On GLC procedure of Papadopoulos at *in vivo* concentration. *2722*

Thioridazine   On GLC procedure of Papadopoulos at *in vivo* concentration. *2722*

Trifluoperazine   On GLC procedure of Papadopoulos at *in vivo* concentration. *2722*

Phenindione

Plasma   Increase   Physiological

Quinidine   Potentiates action. *1487*

Phenmetrazine

Urine   Increase   Physiological

Acid Urine   With overdose, acidification increases excretion. *1678*

Urine   Positive   Analytical

Chlorphentermine   Similar R_f and color reaction on TLC using ethyl acetate: methanol: water: ammonium hydroxide and modified Dragendorff's reagent. *3868*

Phentermine   Similar R_f and color reaction on TLC using ethyl acetate: methanol: water: ammonium hydroxide and modified Dragendorff's reagent. *3868*

Phenobarbital

Serum   No Effect   Analytical

Acetazolamide   On GLC procedure of Papadopoulos if added *in vitro*. *2722*

Amobarbital   On GLC procedure of Papadopoulos when added *in vitro*. *2722*

Beclamide   On GLC procedure of Papadopoulos at *in vivo* concentration. *2722*

Carbamazepine   On GLC procedure of Papadopoulos at *in vivo* concentration. *2722*

Chlordiazepoxide   On GLC procedure of Papadopoulos at *in vivo* concentration. *2722*

Chlorpromazine   On GLC procedure of Papadopoulos at *in vivo* concentration. *2722*

Cyproheptadine   On GLC procedure of Papadopoulos at *in vivo* concentration. *2722*

Diazepam   On GLC procedure of Papadopoulos at *in vivo* concentration. *2722*

Ethosuximide   On GLC procedure of Papadopoulos at *in vivo* concentration. *2722*

Ethotoin   On GLC procedure of Papadopoulos at *in vivo* concentration. *2722*

Fluosol-DA   As measured on Du Pont aca III at concentration up to 50%. *2518*

Haloperidol   On GLC procedure of Papadopoulos at *in vivo* concentration. *2722*

Imipramine   On GLC procedure of Papadopoulos at *in vivo* concentration. *2722*

Mephenytoin   On GLC procedure of Papadopoulos at *in vivo* concentration. *2722*

Mephobarbital   On GLC procedure of Papadopoulos when added *in vitro*. *2722*

Methsuximide   On GLC procedure of Papadopoulos at *in vivo* concentration. *2722*

Methylphenobarbital   On GLC procedure of Papadopoulos when added *in vitro*. *2722*

Nitrazepam   On GLC procedure of Papadopoulos at *in vivo* concentration. *2722*

Phenobarbital (continued)

Serum   No Effect   Analytical (continued)

Oxytetracycline   On GLC procedure of Papadopoulos when added in vitro. 2722

Secobarbital   On GLC procedure of Papadopoulos at in vivo concentration. 2722

Sulthiame   On GLC procedure of Papadopoulos at in vivo concentration. 2722

Thioridazine   On GLC procedure of Papadopoulos at in vivo concentration. 2722

Trifluoperazine   On GLC procedure of Papadopoulos at in vivo concentration. 2722

Serum   Increase   Physiological

Mephobarbital   Metabolic conversion in vivo. 3054

Phenobarbital   600 mg orally produces concentration of 23 mg/L in blood. 2348

Phenytoin   Average increase from 22 to 48 μg/mL (DPH dose 900 mg/d). 2041

Primidone   Metabolic conversion in vivo. 3054

Progabide   Increased concentration in patients who had a therapeutic response. 3188

Valproic Acid   In patients taking primidone. 0504   Reported increase when valproic acid coadministered. 0119   In 25 patients studied for 5 to 9 mo. 3829   Increased half-life, relaxed plasma clearance, and other effects suggesting inhibition of metabolism. 2742   In 4 patients decreased conversion to hydroxyphenyl-phenobarbital. 0511

Serum   Decrease   Physiological

Charcoal   Reduced absorption and increased elimination rate with adsorption reducing availability. 3794

Chlorophenothane   Reported interaction due to alteration of metabolism. 2042

Folic Acid   Occurs with pharmacologic doses; lowers plasma concentration. 3024

Food   Absorption reduced when taken with food. 3024

Glutamate   Occurs with pharmacologic doses: lowers plasma concentrations. 3024

Phenothiazines   But not to quite same extent as for phenytoin. 1455

Pyridoxine   Occurs with pharmacologic doses: lower plasma concentration. 3024

Urine   No Effect   Analytical

Chlorphentermine   No interference using TLC with ethyl acetate: methanol: water: ammonium hydroxide and modified Dragendorff's reagent for detection. 3868

Clotermine   No interference on TLC using ethyl acetate: methanol: water: ammonium hydroxide and modified Dragendorff's reagent for detection. 3868

Diethylpropion   No interference with TLC using ethyl acetate: methanol: water: ammonium hydroxide and modified Dragendorff's reagent for detection. 3868

Fenfluramine   No interference on TLC using ethyl acetate: methanol: water: ammonium hydroxide and modified Dragendorff's reagent for detection. 3868

Mazindol   No interference on TLC using ethyl acetate: methanol: water: ammonium hydroxide and modified Dragendorff's reagent for detection. 3868

Phendimetrazine   No interference with TLC using ethyl acetate: methanol: water: ammonium hydroxide and modified Dragendorff's reagent for detection. 3868

Phenmetrazine   No interference on TLC using ethyl acetate: methanol: water: ammonium hydroxide and modified Dragendorff's reagent for detection. 3868

Phentermine   No interference on TLC using ethylacetate: methanol: water: ammonium hydroxide and modified Dragendorff's reagent for detection. 3868

CSF   Increase   Physiological

Phenobarbital   Concentration about half serum level. 3054

Phenol

Urine   Increase   Physiological

Benzene   Relationship between exposure to benzene vapor. 3301

Phenols   Quantitative relationship between exposure and urine concentration. 2661

Phenol Turbidity

Serum   Increase   Analytical

Heat   Increases greatly with change of temperature. 0579

Phenothiazine

Urine   Positive   Physiological

Phenothiazines   Some detectable up to 18 mo after therapy. 1343

Phenprocoumon

Plasma   No Effect   Physiological

Sulfinpyrazone   No effect on plasma concentration. 2642

Plasma   Increase   Physiological

Allopurinol   Accumulates for several weeks when drugs coadministered. 1765

Phenylbutazone   Increased concentration of anticoagulant displaced from protein with increase of free component. 2642

Phentermine

Urine   Positive   Analytical

Chlorphentermine   Similar R_f and color reaction on TLC using ethyl acetate: methanol: water: ammonium hydroxide and modified Dragendorff's reagent. 3868

Phenmetrazine   Similar R_f and color reaction on TLC using ethyl acetate: methanol: water: ammonium hydroxide and modified Dragendorff's reagent. 3868

Phentolamine Test

Pancreatic Juice   Increase   Physiological
Phenylpropanolamine   Single case reported. 0981

Patient   Increase   Physiological
MAO Inhibitors   Enhanced responsiveness. 1325

Patient   Decrease   Physiological
Thiazides   False negative due to hypotensive effect of drug. 1488

Patient   Positive   Physiological
Ethambutol   False positive, mechanism unknown. 1218

Uremia   Probably related to impaired excretion of drug. 1218

Phenylacetylglutamine

Urine   Increase   Physiological
Protein   From bacterial proteinolysis of unabsorbed. 3233

Phenylalanine

Plasma   No Effect   Analytical
Storage of Sample   In diluted serum in refrigerator for 3 d 4 d at room temperature, 7 d if fluoride added Dried blood spots at room temperature for 11 weeks. *1712*

Plasma   No Effect   Physiological
Levodopa   In patients after 1 week given levodopa. *1108*

Plasma   Increase   Physiological
Co-Trimoxazole   4 h after 0.1 g/kg orally. *1036*
Obesity   Characteristic finding correlated with insulin. *1106*

Plasma   Decrease   Physiological
Ascorbic Acid   Reduces elevated level of premature infants. *1343*
Glucose   Probable muscle uptake. *1403*
Histidine   Probably due to flow into gut after oral load. *1644*

Urine   Increase   Physiological
Ascorbic Acid   In 10 healthy females given 10 g/d. *3624*

CSF   No Effect   Physiological
Probenecid   No effect on concentration observed. *2373*

Phenylbutazone

Serum   No Effect   Physiological
Allopurinol   No significant effect observed on plasma concentrations. *2927*
Amobarbital   No significant effect on half-life. *3466*
Diphenhydramine   No significant effect on half-life. *3466*
Methandrostenolone   Unaffected by concomitant administration. *1487*
Methaqualone   No significant effect on half-life. *3466*
Nitrazepam   No significant effect on half-life. *3466*

Serum   Decrease   Physiological
Charcoal   Reduced absorption and increased elimination rate with adsorption reducing availability. *3794*
Desipramine   Inhibits gastrointestinal absorption. *1487*
Phenobarbital   Reported interaction due to alteration of metabolism. *2042*

Phenylethylamine

Urine   Increase   Physiological
MAO Inhibitors   Prevent deamination. *1343*

Phenylketones

Urine   No Effect   Analytical
Homogentisic Acid   No effect with Phenistix®. *0775*

Urine   Positive   Analytical
Acetoacetate   red/red-brown with $FeCl_3$, no effect with Phenistix®. *0775*
Aminobenzoic Acid   Green with $FeCl_3$. *0775*
Aminosalicylic Acid   Red-brown with $FeCl_3$, pink to purple Phenistix®. *0775*
Antipyrine   Red fading with $FeCl_3$, pink to red with Phenistix®. *0775*
Aspirin   Purple with $FeCl_3$, purple with Phenistix®. *1022*
Bilirubin   Blue-green with $FeCl_3$. *0775*
Chlorpromazine   Light purple with $FeCl_3$, same with Phenistix®. *0775*
Epinephrine   Green in high concentrations with $FeCl_3$. *0775*

Homogentisic Acid   Blue/green fading with $FeCl_3$, nil with Phenistix®. *0775*
Iodochlorhydroxyquin   Green with $FeCl_3$. *0775*
Melanin   Gray to black with $FeCl_3$. *0775*
Methotrimeprazine   Phenistix® positive if more than 100 mg/d for 6 d. *2427*
Nialamide   Fading green with $FeCl_3$, green with Phenistix®. *0775*
Phenols   Violet with $FeCl_3$, nil with Phenistix®. *0775*
Phenothiazines   pink/red-purple with $FeCl_3$, same with Phenistix®. *0775*
Prochlorperazine   Light purple with $FeCl_3$, also with Phenistix®. *0775*
Pyruvate   Deep yellow/green with $FeCl_3$, nil Phenistix®. *0775*
Vanillic Acid   red-violet/brown with $FeCl_3$, brown Phenistix®. *0775*

Urine   Negative   Analytical
Detergents   False negative results (brown color) with $FeCl_3$ test. *3187*

Phenylpropanolamine

Urine   No Effect   Analytical
Chlorphentermine   No interference using TLC with ethyl acetate: methanol: water: ammonium hydroxide and modified Dragendorff's reagent for detection. *3868*
Clotermine   No interference on TLC using ethylacetate: methanol: water: ammonium hydroxide and modified Dragendorff's reagent for detection. *3868*
Diethylpropion   No interference with TLC using ethyl acetate: methanol: water: ammonium hydroxide and modified Dragendorff's reagent for detection. *3868*
Fenfluramine   No interference on TLC using ethyl acetate: methanol: water: ammonium hydroxide and modified Dragendorff's reagent for detection. *3868*
Mazindol   No interference on TLC using ethyl acetate: methanol: water: ammonium hydroxide and modified Dragendorff's reagent for detection. *3868*
Phendimetrazine   No interference with TLC using ethyl acetate: methanol: water: ammonium hydroxide and modified Dragendorff's reagent for detection. *3868*
Phenmetrazine   No interference on TLC using ethyl acetate: methanol: water: ammonium hydroxide and modified Dragendorff's reagent for detection. *3868*
Phentermine   No interference on TLC using ethylacetate: methanol: water: ammonium hydroxide and modified Dragendorff's reagent for detection. *3868*

Phenyramidol

Plasma   Increase   Physiological
Warfarin   Impairs metabolism. *2641*

Pheochromocytoma Test

Patient   Positive   Physiological
Methotrimeprazine   May cause false test as produces hypotension. *1488*

Phosphate

Serum   No Effect   Analytical
Acetaminophen   At acute overdose concentration (10 mg/dL) on SMAC® method. *3719* At concentration of 1500 mg/L had no effect on phosphomolybdate method. *3393* At 5 times therapeutic concentration on molybdate procedures on SMAC®, aca, Hitachi® 705 and KDA. *2138*
Acetazolamide   No effect at 12 mg/dL on SMA 12/60 method. *2636* At concentration of 1,000 mg/L had no effect on phosphomolybdate method. *3393*

Phosphate (continued)

Serum No Effect Analytical (continued)

Acetylsalicylic Acid At 5 times upper limit of therapeutic range on method on SMAC®, aca, Hitachi® 705, and KDA. 2138 At concentration of 2,000 mg/L had no effect on phosphomolybdate method. 3393

Allopurinol At concentration of 136 mg/L had no effect on phosphomolybdate reduction method. 3393

Aminophenol At concentration of 109 mg/L had no effect on phosphomolybdate method. 3393

Aminopyrine At concentration of 125 mg/L had no effect on phosphomolybdate method. 3393

Aminosalicylic Acid On nonprecipitation method of Peynet. 2799 At 5 times upper limit of therapeutic range on methods on SMAC® and KDA. 2138 At concentration of 460 mg/L had no effect on phosphomolybdate method. 3393

Amitriptyline At acute overdose concentration (2.5 mg/dL) on SMAC® method. 3719 At concentration of 25 mg/L had no effect on phosphomolybdate method. 3393

Amobarbital At concentration of 150 mg/L had no effect on phosphomolybdate method. 3393

Amphotericin B At concentration of 96 mg/L had no effect on phosphomolybdate method. 3393

Ampicillin At concentration of 40 mg/L had no effect on phosphomolybdate method. 3393

Ascorbic Acid At concentration of 150 mg/L had no effect as measured by aca method. 3393 At concentration of 2,000 mg/L had no effect as measured by phosphomolybdate method. 3393

Aspirin On nonprecipitation method of Peynet. 2799

Barbital At acute overdose concentration (20 mg/dL) on Technicon® SMAC® method. 3719 At concentration of 500 mg/L had no effect as measured by phosphomolybdate method. 3393

Bethanechol No effect at 0.09 mg/dL on SMA 12/60 method. 2636

Bilirubin If free, on method of Peynet. 2799 On enzymatic procedure with centrifugal analyzer. 2790

Boric Acid At concentration of 50 mg/L had no effect as measured by phosphomolybdate method. 3393

Bromazepam At 5 times therapeutic concentration on methods of SMAC®, aca, Hitachi® 705, KDA. 2138

Butabarbital At concentration of 100 mg/L had no effect as measured by phosphomolybdate method. 3393

Caffeine At acute overdose concentration (10 mg/dL) on Technicon® SMAC® method. 3719 At concentration of 194 mg/L had no effect on phosphomolybdate method. 3393

Carbenicillin At 5 times upper limit of therapeutic range on methods on SMAC®, aca, Hitachi® 705 and KDA. 2138 At concentration of 15,000 mg/L had no effect on phosphomolybdate method. 3393

Cefazolin At concentration of 110 mg/L had no effect on phosphomolybdate method. 3393

Cefoxitin At 5 times upper limit of therapeutic range on methods on SMAC®, aca, Hitachi® 705 and KDA. 2138

Cephalothin At 5 times upper limit of therapeutic range on methods on SMAC®, aca, Hitachi® 705 and KDA. 2138 At concentration of 1,000 mg/L had no effect on phosphomolybdate method. 3393

Chlordiazepoxide At 5 times upper limit of therapeutic range on methods on SMAC®, aca, Hitachi® 705 and KDA. 2138 At acute overdose concentration (5 mg/dL) on Technicon® SMAC® method. 3719 At concentration of 50 mg/L had no effect on phosphomolybdate method. 3393

Chlormezanone At concentration of 100 mg/L had no effect on phosphotungstate reduction method. 3393

Chlorphenesin At concentration of 150 mg/L had no effect on phosphomolybdate method. 3393

Chlorpheniramine At acute overdose concentration (20 mg/dL) on Technicon® SMAC® method. 3719 At concentration of 200 mg/L had no effect on phosphomolybdate method. 3393

Chlorpromazine At concentration of 3 mg/L had no effect on phosphomolybdate method. 3393

Chlorpropamide At concentration of 150 mg/L had no effect on phosphomolybdate method. 3393

Chlorprothixene At concentration of 1 mg/L had no effect on phosphomolybdate method. 3393

Cimetidine At concentration of 1 mg/L had no effect on phosphomolybdate method. 3393

Cloxacillin At concentration of 5 mg/L had no effect on phosphomolybdate method. 3393

Cocaine At acute overdose concentration (2.5 mg/dL) on Technicon® SMAC® method. 3719 At concentration of 25 mg/L had no effect on phosphomolybdate method. 3393

Codeine At acute overdose concentration (2.0 mg/dL) on Technicon® SMAC® method. 3719 At concentration of 20 mg/L had no effect on phosphomolybdate method. 3393

Colchicine On nonprecipitation method of Peynet. 2799

Cysteine At 5 times upper limit of therapeutic range on methods on SMAC®, aca, Hitachi® 705 and KDA. 2138

Cystine At 5 times upper limit of therapeutic range on methods on SMAC®, aca, Hitachi® 705 and KDA. 2138

Desipramine At concentration of 8 mg/L had no effect on phosphomolybdate method. 3393

Deslanoside No effect on SMA 12/60 method at 0.06 mg/dL. 2636

Diazepam At 5 times upper limit of therapeutic range on methods on SMAC®, aca, Hitachi® 705 and KDA. 2138 At acute overdose concentration (2.5 mg/dL) on Technicon® SMAC® method. 3719 At concentration of 25 mg/L had no effect on phosphomolybdate method. 3393

Diethylaminoethanol At concentration of 30 mg/L had no effect on phosphomolybdate method. 3393

Diethylpropion At acute overdose concentration (10 mg/dL) on Technicon® SMAC® method. 3719 At concentration of 100 mg/L had no effect on phosphomolybdate method. 3393

Digitoxin No effect at 21 mg/dL on SMA 12/60 method. 2636 At concentration of 6 mg/L had no effect on phosphomolybdate method. 3393

Digoxin No effect at 0.04 mg/dL on SMA 12/60 method. 2636 At concentration of 9 mg/L had no effect on phosphomolybdate method. 3393

Dimethadione At concentration of 2,000 mg/L had no effect on phosphomolybdate method. 3393

N-Dimethyldiazepam At concentration of 6 mg/L had no effect on phosphomolybdate method. 3393

Diphenhydramine At acute overdose concentration (20 mg/dL) on Technicon® SMAC® method. 3719 At concentration of 200 mg/L had no effect on phosphomolybdate method. 3393

Disopyramide At concentration of 4 mg/L had no effect on phosphomolybdate method. 3393

Disulfiram At concentration of 120 mg/L had no effect on phosphomolybdate method. 3393

Disulphine Blue By continuous-flow phosphomolybdate/stannous chloride hydrazine method. 1464

Epinephrine At concentration of 183 mg/L had no effect on phosphomolybdate method. 3393

Erythromycin At concentration of 73 mg/L had no effect on phosphomolybdate method. 3393

Ethacrynic Acid At 2.5 mg/dL no effect on SMA 12/60 method. 2636

Ethambutol On nonprecipitation method of Peynet. 2799

Ethanol At acute overdose concentration (20 mg/dL) on Technicon® SMAC® methods. 3719

Ethchlorvynol At acute overdose concentration (20 mg/dL) on Technicon® method. 3719 At concentration of 400 mg/L had no effect on phosphomolybdate method. 3393

Ethinamate At acute overdose concentration (20 mg/dL) on Technicon® SMAC® method. 3719 At concentration of 200 mg/L had no effect on phosphomolybdate method. 3393

Ethionamide On nonprecipitation method of Peynet. 2799

Ethosuximide At concentration of 390 mg/L had no effect on phosphotungstate reduction method. 3393

Fluorouracil At concentration of 130 mg/L had no effect on phosphomolybdate method. 3393

Flurazepam At 5 times upper limit of therapeutic range on methods on SMAC®, aca, Hitachi® 705 and KDA. 2138 At acute overdose concentration (2.5 mg/dL) on Technicon® SMAC® method. 3719 At concentration of 25 mg/L had no effect on phosphomolybdate method. 3393

Furosemide   No effect at 1.4 mg/dL on SMA 12/60 method. *2636* At concentration of 4 mg/L had no effect on phosphomolybdate method. *3393*

Gentamicin   At concentration of 150 mg/L had no effect on phosphomolybdate method. *3393*

Glutethimide   At acute overdose concentration (5 mg/dL) on Technicon® SMAC® method. *3719* At concentration of 50 mg/L had no effect on phosphomolybdate method. *3393*

Hydralazine   At concentration of 6 mg/L had no effect on phosphomolybdate method. *3393*

Ibuprofen   At 5 times upper limit of therapeutic range on methods on SMAC®, aca, Hitachi® 705 and KDA. *2138* At concentration of 200 mg/L had no effect on phosphomolybdate method. *3393*

Indomethacin   On nonprecipitation method of Peynet. *2799* At concentration of 13 mg/L had no effect on phosphomolybdate method. *3393*

Insulin   At concentration of 3 mg/L had no effect on phosphomolybdate method. *3393*

Iproniazid   At concentration of 40 mg/L had no effect on phosphomolybdate method. *3393*

Isoniazid   On nonprecipitation method of Peynet. *2799* At concentration of 100 mg/L had no effect on phosphomolybdate method. *3393*

Isoprenaline   At concentration of 211 mg/L had no effect on phosphomolybdate method. *3393*

Ketoprofen   At concentration of 60 mg/L had no effect on phosphomolybdate method. *3393*

Levodopa   At concentration of 197 mg/L had no effect on phosphomolybdate method. *3393*

Lidocaine   No effect SMA 12/60 method with 3.5 mg/dL. *2636*

Lipemia   In patient specimens measured on SMAC® in comparison with specimens following ultra-centrifugation. *2466*

Liposol   Changed from 40.7 to 41.5 mg/L in 26 randomly selected patient sera, before and after addition on Beckman Astra methods. *2054*

Lorazepam   At concentration of 0.05 mg/L had no effect on phosphomolybdate method. *3393*

Mannitol   Usually not high enough concentration to interfere. *0583* No effect on SMA 12/60 method at 445 mg/dL. *2636* No effect on method on AutoAnalyzer even at 15.5 g/L. *2066*

Meperidine   At acute overdose concentration (5 mg/dL) on Technicon® SMAC® method. *3719* At concentration of 60 mg/L had no effect on phosphomolybdate method. *3393*

Mephenesin   At concentration of 100 mg/L had no effect on phosphomolybdate method. *3393*

Meprobamate   At acute overdose concentration (20 mg/dL) on Technicon® SMAC® method. *3719* At concentration of 200 mg/L had no effect on phosphomolybdate method. *3393*

Meralluride   No effect at 0.07 mL/dL on SMA 12/60 method. *2636*

Mercaptomerin   At concentration of 58 mg/L had no effect on phosphomolybdate method. *3393*

Mercaptopurine   At concentration of 152 mg/L had no effect on phosphomolybdate method. *3393*

Mesoridazine   At acute overdose concentration (20 mg/dL) on Technicon® SMAC® method. *3719*

Methadone   At acute overdose concentration (2.5 mg/dL) on Technicon® SMAC® method. *3719* At concentration of 25 mg/L had no effect on phosphomolybdate method. *3393*

Methamphetamine   At concentration of 2 mg/L had no effect on phosphomolybdate method. *3393*

Methanol   At acute overdose concentration (20 mg/dL) on Technicon® SMAC® method. *3719*

Methapyrilene   At concentration of 13 mg/L had no effect on phosphomolybdate method. *3393*

Methaqualone   At acute overdose concentration (2.5 mg/dL) on Technicon® SMAC® method. *3719* At concentration of 25 mg/L had no effect on phosphomolybdate method. *3393*

Methicillin   At 5 times upper limit of therapeutic range on method on KDA. *2138*

Methohexital   At concentration of 50 mg/L had no effect on phosphomolybdate method. *3393*

Methotrexate   At 5 times upper limit of therapeutic range on KDA method. *2138*

Methsuximide   At concentration of 40 mg/L had no effect on phosphomolybdate method. *3393*

Methyldopa   At concentration of 211 mg/L had no effect on phosphomolybdate method. *3393*

Methylphenidate   At acute overdose concentration (20 mg/dL) on Technicon® SMAC® method. *3719* At concentration of 200 mg/L had no effect on phosphomolybdate method. *3393*

Methylphenobarbital   At concentration of 150 mg/L had no effect on phosphomolybdate method. *3393*

Methyprylon   At acute overdose concentration (10 mg/dL) on Technicon® SMAC® method. *3719* At concentration of 100 mg/L had no effect on phosphomolybdate method. *3393*

Metoprolol   At concentration of 0.34 mg/L had no effect on phosphomolybdate method. *3393*

Morphine   At acute overdose concentration (20 mg/dL) on Technicon® SMAC® method. *3719* At concentration of 200 mg/L had no effect on phosphomolybdate method. *3393*

Nafcillin   At concentration of 50 mg/L had no effect on phosphomolybdate method. *3393*

Naproxen   At 5 times upper limit of therapeutic range on methods on SMAC® and KDA. *2138*

Nitrofurantoin   On routine methods in use on SMAC®, Ektachem®, Abbott-VP, Cobas-Bio, aca, Hitachi® 705, KDA at 5 times normal upper therapeutic concentration. *2138* At concentration of 4 mg/L had no effect on phosphomolybdate method. *3393*

Nortriptyline   At acute overdose concentration (20 mg/dL) on Technicon® SMAC® method. *3719* At concentration of 200 mg/L had no effect on phosphomolybdate method. *3393*

Orphenadrine   At concentration of 9 mg/L had no effect on phosphomolybdate method. *3393*

Ouabain   At 0.02 mg/dL no effect on SMA 12/60 method. *2636*

Oxacillin   At concentration of 40 mg/L had no effect on phosphomolybdate method. *3393*

Oxazepam   At concentration of 1 mg/L had no effect on phosphomolybdate method. *3393*

Oxytetracycline   At concentration of 50 mg/L had no effect on phosphomolybdate method. *3393*

Papaverine   At concentration of 10 mg/L had no effect on phosphomolybdate method. *3393*

Paraldehyde   At concentration of 2,000 mg/L had no effect on phosphomolybdate method. *3393*

Penicillin G   At 5 times upper limit of therapeutic range on methods on SMAC®, aca, Hitachi® 705, and KDA. *2138* At concentration of 2,000 mg/L had no effect as measured by phosphomolybdate method. *3393*

Pentobarbital   At acute overdose concentration (20 mg/dL) on Technicon® SMAC® method. *3719* At concentration of 340 mg/L had no effect on phosphomolybdate method. *3393*

Perphenazine   At acute overdose concentration (20 mg/dL) on Technicon® SMAC® method. *3719* At concentration of 200 mg/L had no effect on phosphomolybdate method. *3393*

Phencyclidine   At concentration of 6 mg/L had no effect on phosphomolybdate method. *3393*

Phenelzine   At concentration of 136 mg/L had no effect on phosphomolybdate method. *3393*

Phenobarbital   At acute overdose concentration (20 mg/dL) on Technicon® SMAC® method. *3719* At concentration of 250 mg/L had no effect on phosphomolybdate method. *3393*

Phensuximide   At concentration of 120 mg/L had no effect on phosphomolybdate method. *3393*

Phenylbutazone   On nonprecipitation method of Peynet. *2799* At concentration of 750 mg/L had no effect on phosphomolybdate method. *3393*

Phenylephrine   At concentration of 4 mg/L had no effect on phosphomolybdate method. *3393*

Phenytoin   No effect at 1.8 mg/dL on SMA 12/60 method. *2636* At acute overdose concentration (20 mg/dL) on Technicon® SMAC® method. *3719* At concentration of 240 mg/L had no effect on phosphomolybdate method. *3393*

Probenecid   At concentration of 1300 mg/L had no effect on phosphomolybdate method. *3393*

Procainamide   No effect on SMA 12/60 method at 35 mg/dL. *2636* At concentration of 50 mg/L had no effect on phosphomolybdate method. *3393*

Phosphate *(continued)*

Serum   No Effect   Analytical *(continued)*

Procaine   At concentration of 2 mg/L had no effect on phosphomolybdate method. *3393*

Prochlorperazine   At concentration of 1 mg/L had no effect on phosphomolybdate method. *3393*

Promethazine   At acute overdose concentration (20 mg/dL) on Technicon® SMAC® method. *3719*   At concentration of 200 mg/L had no effect on phosphomolybdate method. *3393*

Propoxyphene   At acute overdose concentration (2.5 mg/dL) on Technicon® SMAC® method. *3719*   At concentration of 25 mg/L had no effect on phosphomolybdate method. *3393*

Propranolol   No effect at 0.1 mg/dL on SMA 12/60 method. *2636*   At concentration of 0.2 mg/L had no effect on phosphomolybdate method. *3393*

Propylthiouracil   At concentration of 170 mg/L had no effect on phosphomolybdate method. *3393*

Pyribenzamine®   At acute overdose concentration (20 mg/dL) on Technicon® SMAC® methods. *3719*

Quinidine   At 26 mg/dL no effect on SMA 12/60 method. *2636*   At acute overdose concentration (20 mg/dL) Technicon® SMAC® method. *3719*   At concentration of 210 mg/L had no effect on phosphomolybdate method. *3393*

Quinine   At acute overdose concentration (1.5 mg/dL) on Technicon® SMAC® method. *3719*   At concentration of 30 mg/L had no effect on phosphomolybdate method. *3393*

Reserpine   No effect at 0.02 mg/dL on SMA 12/60 method. *2636*   At concentration of 61 mg/L had no effect on phosphomolybdate method. *3393*

Rifampin   On nonprecipitation method of Peynet. *2799*   At 5 times upper limit of therapeutic concentration on methods on SMAC® and KDA. *2138*

Salicylate   At concentration of 500 mg/L had no effect on phosphomolybdate method. *3393*

Secobarbital   At acute overdose concentration (20 mg/dL) on Technicon® SMAC® method. *3719*   At concentration of 200 mg/L had no effect on phosphomolybdate method. *3393*

Sodium Citrate   At concentration of 5400 mg/L had no effect on aca method. *3393*

Sodium Heparin   At concentration of 400 mg/L had no effect on aca method. *3393*

Standing of Sample   If on polystyrene beads 24 h at 25°. *2624*

Storage of Sample   At room temperature 8 h, 4° 1 d, frozen 1 y. *3873*

Streptomycin   On nonprecipitation method of Peynet. *2799*   At concentration of 58 mg/L had no effect on phosphomolybdate method. *3393*

Strychnine   At concentration of 12 mg/L had no effect on phosphomolybdate method. *3393*

Sulfadiazine   At concentration of 1500 mg/L had no effect on phosphomolybdate method. *3393*

Sulfaguanidine   At concentration of 500 mg/L had no effect on phosphomolybdate method. *3393*

Sulfanilamide   At concentration of 1,000 mg/L had no effect on phosphomolybdate method. *3393*

Sulfathiazole   At concentration of 255 mg/L had no effect on phosphomolybdate method. *3393*

Sulforidazine   At concentration of 200 mg/L had no effect on phosphomolybdate method. *3393*

Tetracycline   At 5 times upper limit of therapeutic range on methods on SMAC®, aca, Hitachi® 705 and KDA. *2138*   At concentration of 300 mg/L had no effect on phosphomolybdate method. *3393*

Theobromine   At concentration of 2,000 mg/L had no effect on phosphomolybdate method. *3393*

Theophylline   At 5 times upper limit of therapeutic range on methods on SMAC®, aca, Hitachi® 705 and KDA. *2138*   At concentration of 2,000 mg/L had no effect on phosphomolybdate method. *3393*

Thiamazole   At concentration of 114 mg/L had no effect on phosphomolybdate method. *3393*

Thiopental   At acute overdose concentration (20 mg/dL) on Technicon® SMAC® method. *3719*   At concentration of 200 mg/L had no effect on phosphomolybdate method. *3393*

Timolol   At concentration of 0.01 mg/L had no effect on phosphomolybdate method. *3393*

Tolbutamide   At concentration of 270 mg/L had no effect on phosphomolybdate method. *3393*

Transport of Specimen   By pneumatic tube if specimen protected. *3217*

Tribromethanol   At concentration of 90 mg/L had no effect on phosphomolybdate method. *3393*

Trichlorethanol   At concentration of 1,000 mg/L had no effect on phosphomolybdate method. *3393*

Trifluoperazine   At concentration of 1 mg/L had no effect on phosphomolybdate method. *3393*

Tripelennamine   At concentration of 200 mg/L had no effect on phosphomolybdate method. *3393*

Vitamin Preparations   No effect at expected concentration with SMA 12/60 procedure. *2637*

Warfarin   At concentration of 100 mg/L had no effect on phosphomolybdate method. *3393*

Serum   No Effect   Physiological

Aspirin   When approximately 3 g/d ingested for several weeks. *3076*

Bran   No significant effect with 38 g/d for 5 weeks. *1540*

Carbamazepine   Osteomalacia observed in 3 of 31 patients given average of 758 mg/d for average of 20.5 mo. *2635*

Cisplatin   No effect noted in 11 children treated for 0.1 to 3.8 y. *3286*

Diclofenac   No effect seen in patients undergoing treatment. *1787*

Dihydroxycholecalciferol   No effect observed. *0469*

Ethanol   In 8 men who had abused alcohol for at least 10 y. *0347*

Ibuprofen   No effect seen in patients undergoing treatment. *1787*

Ketoconazole   In 9 healthy men given up to 1200 mg/d for 1 week. *1300*

Levonorgestrel   In a group of women using levonorgestrel covered rods versus controls using copper IUD. *0899*

Lithium   No significant effect observed with chronic treatment. *0488*

Nifedipine   No effect of 80 mg/d for 2 d. *3694*

Phenytoin   In 60 epileptic patients treated for 10 y or more. *2493*

Pindolol   No significant effect even with 6 weeks treatment. *3279*

Tourniquet   Insignificant over 3 minutes. *1704*

Triamterene   No significant effect observed. *3745*

Serum   Increase   Analytical

Aminosalicylic Acid   At 5 times upper limit of therapeutic range on methods on aca and Hitachi® 705. *2138*

Bilirubin   With UV molybdate procedure on centrifugal analyzer. *2789*

Cefotaxime   Marked increase with both drug and metabolite on specimens containing gentamicin and tobramycin on American Monitor Parallel method. *0201*

Contact With Clot   Release from erythrocytes and platelets. *1237*

Detergents   Contaminated glassware etc. *2313*

Fluosol-DA   Increase by 16% on SMA IIC at concentration of 50%. Increase by 82% on Du Pont aca III at concentration of 50%. *2518*

Hemoglobin   Slight effect phosphomolybdate reductions. *3217*

Heparin   Phosphate contamination of heparin reported. *0378*

Methicillin   At 5 times upper limit of therapeutic range on methods on SMAC®, aca, and Hitachi® 705. *2138*   At concentrations above 500 mg/L (normal therapeutic concentration 21 mg/L) raised concentration as measured by phosphomolybdate method. *3393*

Methotrexate   At upper limit of therapeutic range on methods on aca and Hitachi® 705 and at 5 times this on method on SMAC®. *2138*

Naproxen   At 5 times upper limit of therapeutic range on methods on aca and Hitachi® 705. *2138*

Rifampin   At 5 times upper limit of therapeutic concentration on methods on aca and Hitachi® 705. *2138*

Serum   Due to release from cells with clotting. *2229*

Standing of Sample   May be doubling in 1 h at 37°. *2228*

Storage of Sample From second day at 4 or 37°. *3862*

Serum Increase Physiological

Alanine Slight effect of chronic administration in fasting. *1263*

Aldatense Increase by 10-20% in a group of hypertensives compared with controls. *1611*

Aluminum Hydroxide In 5 normal subjects given 100 mL/d for 28 d (rose to 1.32 mmol/L from 1.18 mmol/L). *0564* Mild effect with osteomalacia due to inhibition of entry into bone. *1054*

Anabolic Steroids Augments phosphate retention. *0652*

Androgens Augments phosphate balance. *0652*

Bed Rest Increased up to 0.5 mg/dL. *1688*

Diurnal Variation Maximum 2-4 am, minimum 8-10 am. *1829*

Ergocalciferol Better absorption and utilization. *0652*

Ethanol 2.8% increase in heavy versus occasional drinkers. *3273*

Furosemide Temporary increase when fluid losses continuously replaced. *2944*

Growth Hormone Gradual slight rise — metabolic effect. *2414*

Height Positively correlated in children. *0887*

Hemolysis Released from RBCs and platelets. *2334*

Hydrochlorothiazide Altered parathyroid metabolism. *0468*

Lipomul® May occur with azotemia. *2313*

Meals Affected by meals (most marked in women). Observed in men after meals. *3455*

Medroxyprogesterone Observed effect (unlike oral contraceptives). *3329*

Menopause Change of 0.22 mg/dL from 5th to 6th decade. *3830*

Methicillin Occurs with nephrotoxicity. *2313*

Minocycline If impaired renal function. *1680*

Muscular Exercise Observed effect with 12 minutes cycle-ergometer. *1231* Effect of muscular exercise. *0279*

Oral Contraceptives 18% increase reported in some patients. *2427*

Parathyroid Extract Due to increased excretion in urine. *2313*

Phosphates Increase of greater than 2 mg/dL observed. *1439*

Phospho-Soda® High concentration of phosphate absorbed from gut. *3505*

Pregnancy Observed effect ?associated with hypocalcemia. *3329*

Season Significant effect, max in May and June, min in winter. *1172*

Sodium Etidronate Increases tubular reabsorption, no effect on PTH action. *2931*

Tetracycline May occur with nephrotoxicity. *2313*

Vitamin D Effect of increased gastrointestinal tract and renal absorption. *3879*

Xylitol Mechanism not discussed. *3212*

Serum Decrease Analytical

Adenosine Triphosphate At concentrations above 0.56 mmol/L had linear decreasing effect on method of Daly and Ertingshausen. *0628*

Cefotaxime Effect of drug on most specimens (also of metabolite) on method on American Monitor Parallel. *0201*

Citrates Complexes with molybdate decrease color develop. *0583*

Mannitol Inhibition of color development. *0582* 19% reduction at 6.2 g/L on normal phosphate concentration: 23% reduction at 3.1 g/L on increased phosphate specimen on Du Pont aca method affecting also other methods in which Elon used as reducing agent. *2066* Concentration related reduction, possibly related to nature of reducing agent, on Dade® Paramax and Du Pont aca. *1005* At concentrations above 3100 mg/L lowered concentration as measured by aca method. *3393*

Oxalate Complexes molybdate, decreases color development. *0583* Excess oxalate interferes with develop of color. *3502*

Phenothiazines Affect methods using phosphomolybdate. *1013*

Plasma By average of 0.25 mg/dL. *2228*

Promethazine Turbidity produced, PO_4 concentration decrease method of Fiske. *1013*

Tartrates Complexes molybdate, decreased color develop. *0583*

Serum Decrease Physiological

Acetazolamide Defective Calcium and PO_4 reabsorption can be induced. *0652*

Aging Approximately 4.6 mg/dL 6 y to puberty then fall to 3.5 at 20 y. *0887*

Albuterol Significant dose-related effect after i.v. infusion of therapeutic doses in 4 healthy male volunteers. *2807*

Alkaline Antacids Theoretical possibility. *3369*

Aluminum Hydroxide In 7 nondialyzed patients with chronic renal failure: fell from average 6.3 mg/dL to 3.7 mg/dL with 15-18 g/d. *3533* Observed in one patient secondary to ingestion of large amounts of compound for long time. *1313*

Aluminum Salts Binding of phosphate in gastrointestinal tract. *3879*

Amino Acids If given i.v. hyperalimentation. *3619*

Anesthetic Agents Observed after anesthesia. *1237*

Anticonvulsants Disturbance of vitamin D metabolism or hepatic enzyme induction. *3401*

Arginine Fall from 3.3 to 2.2 mg/dL in normals and 3.6 to 3.2 mg/dL in diabetics after i.v. infusion of 0.5 g/kg in 30 minutes. *2328*

Blindness Significantly lower (average 0.2 mg/dL) than normal. *1637*

Calcitonin Due to urinary loss. *1343*

Carbamazepine Hypophosphatemia observed in 1 of 21 patients (5%) treated for average of 40 mo. *1625* In about 20 epileptic patients treated for 2 y. *0873*

Chloroform Follows most forms of anesthesia. *0279*

Diurnal Variation Minimum 11 am, maximum during night about 8 am. *3428*

Epinephrine Increased gluconeogenesis. *0085* Dose-dependent response to i.v. infusion: maximum decrease of 0.6 mg/dL. *0401*

Ether Observed after most types of anesthesia. *1343*

Ethinyl Estradiol Decrease of approximately 2 mg/dL. *2886* In post-menopausal women with 50 μg/d for 14 d. *3474* Significant effect in postmenopausal women: less marked in perimenopausal women. *2311* Small but insignificant effect during and after in 7 healthy postmenopausal women treated for 12 weeks. *3467*

Ethylene Observed after most forms of anesthesia. *0279*

Fibrin Hydrolysate Intracellular transfer. *3746*

Fructose Phosphorylation occurs after i.v. injection. *1709*

Glucose During glucose tolerance test, less marked and longer than calcium. *2207*

Hydrochlorothiazide Observed in some cases prolonged treatment. *1680*

Hyperventilation Effect of respiratory alkalosis. *3234*

Insulin Increased phosphorylation of glucose. *3879*

Levonorgestrel Significant slight reduction in implant recipients compared with IUD users over 3 y. *0763*

Light In elderly in absence of sun. *1615*

Lithium Picture of primary hyperparathyroidism observed in patients given treatment for long time, but mechanism not fully understood. *3118* When given at 10 p.m. drug caused decrease for next few hours. *2401*

Meals Observed in women after meals. *3455* Phosphorylation of glucose and metabolism. *1237*

Menstruation Significant decrease with menstruation (by about 0.3 mg/dL). *1172*

Mestranol Increased sensitivity to calcitonin in postmenopausal women. *0037*

Mithramycin Inhibition of bone resorption of calcium. *3349*

Oral Contraceptives Reduction of approximately 2 mg/dL. *2886*

Phenobarbital Increases clearance (secondary hyperparathyroidism). *1396*

Phenytoin In about 20 patients treated for 2 y. *0873*

Phytate Falls initially and remains low. *2947*

Rimeterol Dose related significant change in 4 healthy men given therapeutic i.v. dose. *2807*

Saccharated Iron Oxide Stepwise significant reduction with 40 mg i.v. daily for up to 42 d. *2662*

Sucralfate Comparable effect to that of aluminum hydroxide in patients with uremia. *2139*

Tetracycline May occur with Fanconi syndrome. *0071*

Phosphate *(continued)*

Serum   Decrease   Physiological *(continued)*
Theophylline   Response to i.v. infusion, returned to baseline in 4 h. *3935*

Urine   No Effect   Analytical
Ascorbic Acid   At 20 mg/dL 1% increase with method of Jung/Parekh. *1835*
Aspirin   No effect at 20 mg/dL on method of Jung/Parekh. *1835*
Citrates   No effect at 100 mg/dL on method of Jung/Parekh. *1835*
Glucose   No effect at 20 mg/dL method of Jung/Parekh. *1835*
Histidine   No effect at 20 mg/dL method of Jung/Parekh. *1835*
Inositol   No effect at 20 mg/dL method of Jung/Parekh. *1835*
Oxalate   No effect at 50 mg/dL method of Jung/Parekh. *1835*
Thymol   No effect on Fiske and Subbarow method. *2981*

Urine   No Effect   Physiological
Citrates   In 89 patients with hypocitraturic calcium nephrolithiasis treated for up to 4 y. *2708*
Dihydroxycholecalciferol   No effect observed. *0469*
Ethacrynic Acid   No effect observed on excretion. *3745*
Lithium   No effect observed in lithium treated patients versus other psychiatric patients and healthy controls. *2401*
Saccharated Iron Oxide   In 9 individuals given 40 mg/d i.v. daily for up to 42 d. *2662*
Sodium Etidronate   Increases tubular reabsorption, no effect on PTH action. *2931*
Theophylline   In normals given 400 mg bid for 5 d. *2383*

Urine   Increase   Physiological
Acetazolamide   Inhibits tubular reabsorption. *0652*
Acetoacetate   Inhibits tubular reabsorption. *0957*
Alanine   Inhibits tubular reabsorption. *0957*
Aminopyrine   Reported effect. *2220*
Asparaginase   Response in all treated patients. *2657*
Aspirin   Inhibits tubular reabsorption. *1343*
Azosemide   Tended to increase but not marked. *2047*
Bed Rest   Increase of approximately 200 mg/d. *1688*
Bicarbonates   Possible competition for same excretory mechanism. *0957*
Bismuth Salts   Fanconi syndrome with poisoning. *2427*
Cadmium   Occasionally observed with proximal tubular dysfunction. *2217*
Calcitonin   Acts independently of parathyroid. *1439*
Cellulose Phosphate   With decrease of activity product ratio. *2710*
Corticosteroids   Metabolic effect. *2220*
Dihydrotachysterol   Diuresis almost as great as with vitamin D_3. *0071*
Furosemide   Slight increase with administration. *2047*
Glycine   Inhibits tubular reabsorption. *0957*
Hydrochlorothiazide   Altered parathyroid metabolism. *0468*
Lead   Occurs with poisoning. *0987*
Mercurial Diuretics   Reported effect. *3204*
Mercury Compounds   Occurs with Fanconi syndrome. *2108*
Mestranol   Increased response to calcitonin. *0037*
Metolazone   Maximum diuretic response up to 8 µEq/min. *3358*
Parathyroid Extract   Increased clearance. *1439*
Phosphates   Doubled excretion compared with bed rest. *1688*
Starvation   Usually increase in response to stress. *2382*
Tetracycline   Degraded material may cause Fanconi syndrome. *2427*
Tryptophan   Inhibits tubular reabsorption. *0957*
Valine   Inhibits tubular reabsorption. *0957*
Vitamin D   May be normal in many cases. *3755*

Urine   Decrease   Analytical
Mannitol   Complexes molybdate, decreases color develop. *0583*
Storage of Sample   Unless acidified. *3873*

Urine   Decrease   Physiological
Alanine   Marked effect when given to fasting obese. *1263*
Aluminum Salts   Decreased absorption and excretion. *0071*
Blindness   Significantly lower (average 176 mg/d) than normal. *1637*
Diurnal Variation   Lowest excretion just after waking. *2222*
Phloridzin   Probably due to increased reabsorption. *3343*

Feces   Increase   Physiological
Bed Rest   Slight effect only. *1688*
Phosphates   Doubled excretion compared with bed rest. *1688*

Feces   Decrease   Physiological
Vitamin D   Due to excessive absorption. *0987*

Gastric Material   No Effect   Analytical
Creatinine   No effect at 20 mg/dL on method of Jung/Parekh. *1835*

Test Conditions   Increase   Analytical
Adenine Diphosphate   Affects Fiske-Subbarow unless Seddon procedure used. *3236*
Adenine Monophosphate   Affects Fiske-Subbarow unless Seddon procedure used. *3236*
Pyrophosphate   Affects Fiske-Subbarow unless Seddon procedure used. *3236*
Triphosadenine   Affects Fiske-Subbarow unless Seddon procedure used. *3236*

Test Conditions   Decrease   Analytical
Vanadate   Interferes with Fiske-Subbarow procedures and thus indirectly with reactions yielding phosphate and measured this way. *1051*

Phosphate Clearance

Urine   Decrease   Physiological
Cannabis   Temporary decrease noted. *1633*
Diatrizoic acid   Reduction from average of 12.9 mL/min/1.73 m^2 to 7.1 mL/min/1.73 m^2 day of bilateral renal arteriography in patients with hypertension. *2594*
Lysergide   Temporary effect observed. *1633*
Phloridzin   Due to reduced excretion. *3343*
Radiographic Agents   Mean decrease from 12.9 mL/min/1.73 m^2 to 7.1 mL/min/1.73 m^2 day following arteriography in 37 patients. *2594*
Sodium Etidronate   Increases tubular reabsorption, no effect on PTH action. *2931*

Phosphatides, Total

Serum   Increase   Physiological
Oral Contraceptives   Altered metabolism. *1601*

Phosphoenolpyruvate

Red Blood Cells   Decrease   Physiological
Amino Acids   Associated with low serum phosphate. *3619*

Phosphoethanolamine

Urine   Increase   Physiological
Ascorbic Acid   In 10 healthy females given 10 g/d. *3624*

Phosphofructokinase

Red Blood Cells   Decrease   Physiological
Copper   several affected *in vitro*. *0432*

Phosphoglucomutase

Serum   No Effect   Analytical
Storage of Sample 8 h room temperature, 2 d at 4°, 4-7 d at -20°. *3217*

Serum   Decrease   Analytical
Storage of Sample After 8 h at room temperature, 2 d at 4°. *3873*

2-Phosphoglycerate

Red Blood Cells   Decrease   Physiological
Amino Acids Associated with low serum phosphate. *3619*

3-Phosphoglycerate

Red Blood Cells   Decrease   Physiological
Amino Acids Associated with low serum phosphate. *3619*

6-Phosphoglycerate Dehydrogenase

Serum   Decrease   Analytical
Storage of Sample 50% in 4 d at 4°. *3217*

Red Blood Cells   Decrease   Analytical
Hemoglobin May cause up to 7% quenching of fluorescence. *2219*

Red Blood Cells   Decrease   Physiological
Copper Strongly inhibited by addition of copper. *0432*

Phosphoglycerate Kinase

Red Blood Cells   Decrease   Physiological
Copper Less marked inhibition *in vitro*. *0432*

Phosphohexoseisomerase

Serum   No Effect   Analytical
Hemolysis Small degree has no effect. *3873*
Storage of Sample 8-12 h at room temperature, 21 d at 4°, 1 y at -20°. *3217* No effect 1 week at 4°, 1 mo at -20°. *3873*

Phospholipids, High Density Lipoprotein

Serum   Increase   Physiological
Estradiol-17β By about 10% in 38 healthy postmenopausal women given 2-4 mg orally or 3 mg subcutaneously over 6 mo. *1067*
Ethanol Increase by 16% in 78 intemperate drinkers on average. *2882*

Phospholipids, Low Density Lipoprotein

Serum   Decrease   Physiological
Estradiol-17β By about 12% in 38 healthy postmenopausal women given 2-4 mg orally or 3 mg subcutaneously over 6 mo. *1067*

Phospholipids, Total

Serum   No Effect   Analytical
Storage of Sample At room temperature or 4° up to 7 d, 28 d at -20°. *2512*

Serum   No Effect   Physiological
Chlorthalidone In 22 premenopausal and 18 postmenopausal women given 100 mg/d for 6 weeks. *0402*
Estradiol-17β Insignificant change in 38 healthy post-menopausal women given 2-4 mg orally or 3 mg cutaneously over 6 mo. *1067*
Ethanol Usually remain unchanged in response to acute ingestion. *2428*
Gum Tragacanth In 5 male volunteers consuming 9.9 g daily for 21 d. *0990*
Muscular Exercise But relative reduction of unsaturated acid. *1702*
Oral Contraceptives Typically insignificant change over 6 cycles in women with low dose ethinyl estradiol with low dose levonorgestrel or desogestrel. *2216*
Physical Training No difference between trained athletes and others. *1702*

Serum   Increase   Physiological
Asparaginase Rare response after initial hypolipidemia. *2657*
Chlorpromazine May cause xanthomatous biliary cirrhosis. *1434*
Cyproterone When given in combination with ethinyl estradiol. *2177*
Epinephrine Metabolic effect. *1343*
Estrogens Altered metabolism. *3332*
Ethanol Moderate increase compared with controls in alcoholics especially in younger adults. *0430*
Ethinyl Estradiol Due to high and very high density lipoproteins. *1431*
Oral Contraceptives About 20% higher in pill-users after 6 mo. *0585*
Pregnancy Increased from 8 weeks. *0987*
Sucrose Marked increase when sugar substituted for starch. *2036*

Serum   Decrease   Analytical
Bacterial Contamination If contaminated with phospholipase producing bacteria. *3223*
Hemolysis Organic phosphate hydrolysis to inorganic. *3873*

Serum   Decrease   Physiological
Asparaginase Parallel decrease in cholesterol. *2657*
Aspirin Increased fatty acid oxidation, decreased lipogenesis. *1343*
Cholestyramine Therapeutic effect. *2220*
Clofibrate Effect less marked than with triglycerides. *0456*
Cyproterone In oophorectomized women (slight effect) when given alone. *2177*
Dextrothyroxine Therapeutic effect. *0071*
Ethanol Occurs when cirrhosis develops in alcoholism. *1621*
Levothyroxine Correction of hypothyroid state. *3505*
Menotropins Marked fall in type 2 hyperlipidemia. *2235*
Niacin Prompt and sustained hypolipemic action. *1680*

Bile   No Effect   Physiological
Chenodeoxycholic Acid No significant effect in patients with gallstones. *0274*

Phosphoserine

Urine   Decrease   Physiological
Ascorbic Acid In 10 healthy females given 10 g/d. *3624*

Phthalates

Plasma   Positive   Physiological
Blood Transfusions When blood stored for few days in plastic pack. *2295*

Phytofluene

Serum   No Effect   Analytical
Vitamin A   No effect on method of Bubb-Murphy. *0517*

Serum   Increase   Physiological
Tomatoes   Dramatic rise within 30 minutes of ingestion. *0517*

Pimelate

Urine   No Effect   Physiological
Lithium   No consistent effect observed. *2107*

Pituitary Gonadotropin

Urine   Decrease   Physiological
Estradiol   In oophorectomized or postmenopausal women. *3603*
Estrogens, Conjugated   Inhibits pituitary gonadotrophic function. *3603*
Estrone   In oophorectomized or postmenopausal women. *3603*
Thioridazine   Total gonadotrophic activity reduced. *3550*

Placental Lactogen

Plasma   No Effect   Physiological
Diurnal Variation   No significant cyclical effect observed. *3201*
Sulpiride   In 11 healthy women between 6 and 9 weeks of pregnancy given 150 mg daily for 2 weeks. *3916*

Placental Protein S

Serum   Increase   Physiological
Heparin   Up to 40 times increase above basal level: possibly due to direct effect of heparin on placenta. *2405*

Plasmin

Plasma   Increase   Physiological
Muscular Exercise   Observed effect in exercise. *2847*
Oral Contraceptives   Associated also with increased plasminogen (common effect). *0339*
Streptokinase   Stimulates conversion of plasminogen to plasmin by forming an activator complex. *3095*
Urokinase   Acts directly in converting plasminogen to plasmin. *3095*

Plasmin Activity

Blood   Decrease   Physiological
Aminocaproic Acid   Therapeutic intent. *1680*

Urine   Increase   Physiological
Muscular Exercise   Associated with proteinuria. *1842*

Plasminogen

Plasma   No Effect   Analytical
Acetylsalicylic Acid   At concentration of 300 mg/L had no effect on aca method. *3393*
Cephalosporin   At concentration of 30 mg/L had no effect on aca method. *3393*
Chloramphenicol   At concentration of 25 mg/L had no effect on aca method. *3393*
Chlorpromazine   At concentration of 5 mg/L had no effect on aca method. *3393*

Dextran 70   At concentration of 1500 mg/L had no effect on aca method. *3393*
Digitoxin   At concentration of 20 mg/L had no effect on aca method. *3393*
Digoxin   At concentration of 2 mg/L had no effect on aca method. *3393*
EDTA   At concentration of 1170 mg/L had no effect on aca method. *3393*
Heparin   At concentration of 2860 mg/L had no effect on aca method. *3393*
Penicillin   At concentration of 100 mg/L had no effect on aca method. *3393*
Protamine   At concentration of 250 mg/L had no effect on aca method. *3393*
Sodium Citrate   At concentration of 3794 mg/L had no effect on aca method. *3393*
Tetracycline   At concentration of 1,000 mg/L had no effect on aca method. *3393*
Tobramycin   At concentration of 15 mg/L had no effect on aca method. *3393*

Plasma   Increase   Physiological
Anabolic Steroids   Metabolic effect. *0227*
Danazol   Significant effect in 21 hemophiliacs given drug for 2 weeks. *2627*
Estrogens   Altered metabolism. *3332*
Ethylestrenol   Metabolic effect. *0227*
Fluoxymesterone   Metabolic effect. *0227*
Methandrostenolone   Metabolic effect. *0227*
Methyltestosterone   Metabolic effect. *0227*
Muscular Exercise   Observed effect in exercise. *2847*
Norethandrolone   Metabolic effect. *0227*
Oral Contraceptives   Metabolic changes in liver synthesis (estrogen). *2189*
Oxandrolone   Metabolic effect. *0227*
Oxymetholone   Metabolic effect. *0227*
Stanozolol   Metabolic effect. *0227*

Plasma   Decrease   Physiological
Asparaginase   Mean decrease of 59% (immunological) and 62% (functional assay) in different studies. *0558*
Dextran   Marked reduction. *0320*
Streptokinase   Effect observed after myocardial infarct. *1356*

Plasminogen Antigen

Plasma   Increase   Physiological
Oral Contraceptives   In women given low dose ethinyl estradiol and norethindrone. *2623*

Platelet Adhesiveness

Blood   Decrease   Physiological
Dicumarol   Related to dose. *1343*
Sucrose   Effect of sustained high sucrose diet. *3931*

Platelet Aggregation

Blood   No Effect   Physiological
Aminosalicylic Acid   No effect on aggregation or fibrinolytic activity in 6 patients with chronic inflammatory bowel disease. *3876*   No effect on epinephrine or ADP threshold value. *3876*
Ampicillin   No effect on collagen or epinephrine induced aggregation. *0492*
Creatinine   No effect observed with concentrations associated with uremia. *0826*
Ethanol   No effect observed when ethanol given alone. *0895*
Guanidinosuccinic Acid   No effect of concentrations occurring in uremia. *0826*

Mesna Not impaired after stimulation with epinephrine, ADP or arachidonic acid. *1567*

Metoprolol No effect observed in response ADP compared with placebo in healthy volunteers. *3714*

Propranolol No effect on ADP induced aggregation. In healthy volunteers after single dose or after 1 week treatment. *3714*

Blood   Increase   Physiological

Dicumarol In response to ADP. *1343*

Estrogens, Conjugated Small increases reported. *0083*

Oral Contraceptives In response to ADP. *0585*

Smoking Following *in vitro* stimulation greater in older smokers than younger, and in those who smoked for longer than shorter time. *2995* Response to thrombin, ADP, collagen and epinephrine increased with cigarettes of higher nicotine content. *2955*

Blood   Decrease   Physiological

Aminopyrine Observed *in vitro*, might cause gastrointestinal bleeding etc. *0605*

Ampicillin Defective platelet aggregation induced by ADP in 4 of 5 volunteers. *0492*

Aspirin Inhibits release of ADP from platelets. *1343* Inhibits collagen induced. *3068*

Bromelains Reduced sensitivity to ADP induced aggregation. *1549*

Dipyridamole Weak inhibition of adenosine deaminase. *1132*

Estrogens Reported to be modified by administration. *0619*

Ethanol Observed after i.v. ethanol in normals. *1528* Highly significant reduction with ADP in 2 h after 100 mg alcohol in 4 healthy volunteers: maximal at 3 h. *1223*

Ethyl Biscoumacetate Inhibits if due to ATP, collagen, epinephrine, thrombin. *0614*

Furosemide Inhibits primary ADP-induced agglutination. Effect seen *in vitro* but not *in vivo*. *3069*

Guaifenesin Reported effect. *3322*

Halothane Significant correlation between aggregation *in vitro* and bleeding time *in vivo*. *0800*

Hydrocortisone Inhibits streptokinase induced aggregation. *3104*

Ibufenac Observed *in vitro*, may cause gastrointestinal bleeding etc. *0605*

Indomethacin Observed *in vitro*, might cause gastrointestinal bleeding etc. *0605*

Ketoprofen In 11 patients given drug intravenously. *1234*

Mechlorethamine Observed *in vitro*, might cause gastrointestinal bleeding etc. *0605*

Mefenamic Acid Observed *in vitro*, might cause gastrointestinal bleeding etc. *0605*

Methicillin In response to ADP in 1 of 5 volunteers receiving up to 300 mg/kg/d. *0492*

Moxalactam Reduced in response to ADP in some patients. *2615* Reduced response to ADP demonstrated *in vitro*. *3792*

Nafcillin Reduced response to ADP collagen and epinephrine. *0052*

Nifedipine Significant reduction of collagen-induced and ADP-induced aggregation probably by inhibiting increase of intra-cytoplasmic Calcium by blocking Calcium channel through platelet membrane: also inhibits platelet aggregability induced by exercise. *3532* Maximal rate in response to ADP reduced by 20-26% and by 23% in response to collagen. *0796*

Nitrofurantoin Significant effect on ADP induced aggregation. *3068*

Penicillin G Defective aggregation in response to ADP at 24 million u/d. *0492*

Phenylbutazone Lack second phase ADP agglutination. *0834*

Prazosin High value in essential hypertension in response to ADP normalized with treatment. *1715*

Propoxyphene Observed *in vitro*, may cause gastrointestinal bleeding etc. *0605*

Sulfinpyrazone Inhibits collagen induced aggregation. *3068*

Urea At concentrations possibly occurring in uremia. *0826*

Warfarin If caused by ADP, thrombin, collagen, epinephrine. *0614*

Platelet Associated IgG

Serum   No Effect   Physiological

Methyldopa In a study of 9 hypertensives. *1909*

Platelet Factor 4

Plasma   No Effect   Physiological

Smoking No significant difference between smokers and non-smokers. *2995*

Platelets

Blood   No Effect   Analytical

Storage of Sample No lysis produced by freezing with liquid N_2. *1180*

Blood   No Effect   Physiological

Ampicillin In doses as high as 300 mg/kg/d in volunteers. *0492*

Bucindol No effects noted in 8 patients following several weeks on treatment. *1283*

Cefamandole Not known as complication. *3134*

Cefoxitin No report of thrombocytopenia in response to therapy. *3134*

Chenodeoxycholic Acid No significant effect with up to 750 mg/d in 916 patients treated for 2 y. *3199*

Cyclopropane Anesthesia has no effect. *1343*

Cyclosporine No significant effect observed in 23 patients with inflammatory ocular disease treated for 6 mo. *2713*

Dapsone After 7-9 weeks of treatment with drug plus pyrimethamine in 7 patients taking drug as prophylaxis against malaria. *1202*

Enalapril In 53 patients given up to 160 mg daily or when combined with hydrochlorothiazide. *2277*

Ethanol No effect observed when ethanol given alone. *0895*

Ether Anesthesia has no effect. *1343*

Etoposide In all of 37 patients given drug orally. *3891*

Haloperidol No significant reduction, in contrast to phenothiazines, in psychiatric patients treated for more than 1 y. *1642*

Halothane No significant effect noted. *1814*

Methicillin No effect with doses as high as 300 mg/kg/d in volunteers. *0492*

Moxalactam Insignificant changes in preoperative surgical patients. *3121*

Nafcillin Unchanged in spite of abnormal bleeding time and aggregation. *0052*

Nifedipine No effect in 20 people following ingestion of 20 mg. *0796*

Penicillin G No effect with doses as high as 48 million U/d. *0492*

Smoking No observed effect. *1004* No significant difference between 5 smokers and nonsmokers. *2995*

Timegadine In 23 patients with rheumatoid arthritis given 250 to 750 mg/d for 48 weeks. *2345*

Zimeldine In 1 study involving 147 patients no significant change seen. *3741*

Blood   Increase   Physiological

Alkoxyglycerols Protect against decrease caused by X-rays. *0483*

Amoxapine Observed in one 35 year old woman after receiving 18 g over 57 d. Marked thrombocytosis on 5th day after cessation of treatment. May be early sign of recovery of bone marrow in drug-associated toxic agranulocytosis. *3238*

Auranofin Infrequent reversible side effect (occurring in up to 13% patients). *0617*

Clindamycin Uncommon side effect. *3864*

Dipyridamole Possibly due to alteration of turnover. *1132*

Epinephrine Normal physiological response to injection. *0251*

Fluroxene Slight from 217 m/cmm to 251 m postoperatively. *1814*

Platelets (continued)

Blood Increase Physiological (continued)

Glucocorticoids Stimulate production of platelets. 0071

Isotretinoin Therapeutic response is normalization but effect not usually observed until after 3 weeks. 1319 In 10 of 523 patients with mean daily dose of 109 mg for 150 d. 3867

Lithium After 4 weeks treatment increase in 21 patients on average from 302 to 342 thousand/uL. 1798

Menstruation Effect observed with menstruation. 0251

Metoprolol When given twice per day for 1 week in healthy volunteers, also after single dose. 3714

Miconazole Occurred in 31% of courses followed. 3462

Moxalactam In up to 0.5% patients as reported from several studies. 0592

Muscular Exercise Effect of sudden exercise. 0987

Oral Contraceptives Reported effect. 0071 Slight effect during treatment period. 1795

Ovulation Increases from 8th to 14th day of cycle. 0987

Phosphorus Reported effect with poisoning. 1302

Pregnancy Normal physiological response. 0251

Propranolol Significant increase in healthy volunteers after single dose or after 1 week treatment. 3714

Sulfarthrol Mechanism of action unknown. 1092

Blood Decrease Analytical

Storage of Sample Destroyed by freezing. 1180

Blood Decrease Physiological

Acetaminophen Single case of antibodies to SO_4 conjugate of drug. 1009

Acetazolamide May cause pancytopenia with aplastic anemia. 2074

Acetohexamide Thrombocytopenia and aplastic anemia reported. 2313

Acetophenazine Agranulocytosis or aplastic anemia. 0071

Albuterol In 5 healthy volunteers given drug i.v.: significant reduction occurred over 6 minutes. 2045

Allopurinol Rare potentially dangerous complication. 1332 In 4 of 73 patients treated for 6 mo. 3565

Allymid Immunologically induced thrombocytopenia. 1877

Alprenolol Marked reduction on two occasions in one patient due to increased destruction. 2267 In one woman with essential hypertension after 1.5 y treatment: disappeared with withdrawal of drug. 0613

Aminobenzoic Acid Induces malabsorption and megaloblastic anemia. 3602

Aminopyrine May cause hemolytic or aplastic anemia. 3941

Aminosalicylic Acid May occur with severe megaloblastic anemia. 3773 Several cases of immune mediated thrombocytopenia reported. 2502 Rare reaction. 1292

Amiodarone Observed in 2 patients with rechallenge probably due to delayed hypersensitivity reaction: occurs early during administration of drug. 3781

Amitriptyline Five cases reported (probably immune response). 2603

Amodiaquine Occasional aplastic anemia reported. 3717

Amphotericin B Bone marrow depression with hemolytic anemia. 3645 Occasional complication of treatment. 3567

Ampicillin May be associated with purpura. 1680 Isolated case of platelet-associated IgG and thrombocytopenia. 1910

Antazoline Thrombocytopenia (immunologically induced). 3602

Anticonvulsants May occur with severe megaloblastic anemia. 3773

Antilymphocytic Agents Observed less commonly with other immunosuppressants. 3433

Antimony Compounds Thrombocytopenia. 2313

Antineoplastic Agents May cause aplastic anemia. 2313

Antipyrine Associated with hemolytic anemia. 2427

Apronalide Thrombocytopenia (immunologically — induced). 2313 Immunological mechanism. 2427

Arsenicals Pancytopenia. 0788

Arsenobenzenes Thrombocytopenia. 2313

Asparaginase May cause bone marrow depression. 2657

Aspirin Decreased platelet survival time, may be purpura. 3566 Several cases of immune-mediated thrombocytopenia reported. 2502

Aurothiomalate In patient with rheumatoid arthritis with shortened platelet survival and platelet phagocytosis. 2144

Azapropazone In 3 of 83 patients treated for 6 mo. 3565

Azathioprine May cause bone marrow depression. 3287 Principal manifestation of bone marrow depression. 0036 In 13 of 79 cases with hematological problem out of 328 reports of severe adverse effects. 2097

Azidothymidine Profound effect observed in one male homosexual with AIDS unlikely related to other factors. 1157

Barbiturates Thrombocytopenia. 2313

Benzene Probable myelotoxic effect (common). 2429

Benzthiazide May occur as with other thiazides. 1680

Bethanidine Mild thrombocytopenia reported. 3856

Bismuth Salts May cause aplastic anemia. 0279

Bleomycin May cause bone marrow aplasia. 1436

Busulfan Pancytopenia/thrombocytopenia. 1680

Butaperazine purpura/pancytopenia observed. 1680

Captopril Isolated case of pancytopenia reported. 1012

Carbamazepine Thrombocytopenia may occur after 1 y (immunologic). 0928 Within 2 weeks of commencement of treatment in a single patient. Associated with petechiae. 3096 Marked effect in one patient with strongly positive migration inhibition (MIF) test. 3308 In one case drug-dependent IgG antibodies identified. 1996

Carbenicillin Occasionally thrombocytopenia may occur. 1680

Carbimazole Occasional aplastic anemia reported. 3717

Carbutamide May cause aplastic anemia or thrombocytopenia. 3878

Cefaclor Rare immune-mediated toxicity. 2615

Cefamandole Rare immune-mediated toxicity reported. 2615

Cefazolin Rare immune-mediated toxicity reported. 2615

Cefoxitin Rare immune-mediated toxicity reported. 2615

Cephalothin Increased resistance to osmotic fragility. 2426

Chlorambucil Dose dependent bone marrow depression. 0071

Chloramphenicol Pancytopenia/aplastic anemia. 0907 Occurs either as dose-related or idiosyncratic bone marrow suppression: dose- related usually occurs 5-7 d after start of therapy: idiosyncratic is rare occurring 1 case per 40,000 courses of therapy. 3864 Aplastic anemia reported to occur in from 1 to 10,000 or 40,000 cases. 3717 Isolated case of hepatitis and pancytopenia reported which resolved with discontinuation of drug. 0602

Chlordane Pancytopenia (AMA Blood dyscrasias). 2429

Chlordiazepoxide May rarely cause bone marrow depression. 1436 Aplastic anemia occasionally occurs. 3717

Chlorophenothane Pancytopenia (AMA blood dyscrasias). 2429

Chloroquine Pancytopenia. 3522 Aplastic anemia reported to occur occasionally. 3717

Chlorothiazide Pancytopenia due to aplastic anemia may occur. 0179 Aplastic anemia reported to occur occasionally. 3717

Chlorphenesin Rare thrombocytopenia/pancytopenia. 1680

Chlorpheniramine Thrombocytopenia reported. 1680 Case of aplastic anemia after 10 y of low dose treatment. 1859 One case of aplastic anemia after 1 mo therapy, reversible with discontinuation of drug. 0886

Chlorpromazine Associated with purpura and pancytopenia. 0071 Occasional aplastic anemia reported. 3717 Significant reduction, although still in normal range in 17 psychiatric patients treated for more than 1 y. 1642

Chlorpropamide Thrombocytopenia or aplastic anemia may occur. 0073 Occasional aplastic anemia reported. 3717

Chlorprothixene Purpura or pancytopenia may occur. 1680

Chlortetracycline Pancytopenia (AMA Blood dyscrasias). 2429

Chlorthalidone Associated with rare agranulocytosis. 2427

Cimetidine Isolated case of drug effect together with psoriasis. 3911 Isolated case of platelet-associated IgG and thrombocytopenia. 1910 Reported in one case in which drug given intravenously. 1770 Significant reduction observed in one patient. 0696 Observed in one patient with cancer in absence of leukopenia: fell again with rechallenge by drug. 2286

Clindamycin   Some cases reported. *0137*

Codeine   Thrombocytopenia reported to occur. *1436*

Colchicine   Selective thrombocytopenia or aplastic anemia. *2426*

Co-Trimoxazole   May cause thrombocytopenia. *0745*   Highly significant effect in warfarin treated patients but no effect on warfarin half-life. *0665*   Effects of compound may be potentiated. *1910*

Cyclopenthiazide   Thrombocytopenia reported. *2583*

Cyclophosphamide   May cause bone marrow depression. *0072*

Cycloserine   May occur with megaloblastic anemia. *3773*

Cyclothiazide   Occasional response to thiazides. *1680*

Cytarabine   Due to bone marrow depression. *2313*   Aplasia of bone marrow developed in nearly all of 57 leukemics. *1579*

Dactinomycin   Thrombocytopenia/pancytopenia. *0071*

Demeclocycline   Rare, but reported to occur. *1680*

Deserpidine   Possible effect as related compounds cause this. *1680*

Desipramine   Immune response after some weeks. *1436*

Dextroamphetamine   Thrombocytopenia reported to occur. *1436*

Dextromethorphan   Thrombocytopenia (AMA Blood dyscrasias). *2429*

1,2-Diaminopropane   In 3% of 90 patients with rheumatoid arthritis. *1461*

Diazoxide   Immunologic response occurs from days to months. *3740*

Dichlorphenamide   Probable effect (like acetazolamide). *0071*

Diethylpropion   Isolated cases of bone marrow depression. *1679*

Diethylstilbestrol   Thrombocytopenia (AMA Blood dyscrasias). *2429*

Digitalis   (rare) pancytopenia/thrombocytopenia. *3924*

Digitoxin   Rare thrombocytopenia (due to immune mechanism). *3924*   Several cases of immune-mediated thrombocytopenia reported. *2502*

Dipyrone   Thrombocytopenic purpura reported. *1680*

Doxepin   Single case observed (probably immune response). *2603*

Doxorubicin   Impaired but not as marked as WBC decrease, possible bone marrow aplasia. *0387*

Doxycycline   Thrombocytopenia reported. *2802*

EDTA   Transient bone marrow depression. *0071*

Erythromycin   Thrombocytopenia reported to occur. *1680*

Estrogens   Thrombocytopenia may occur after some weeks. *1436*

Ethacrynic Acid   Thrombocytopenia reported. *2745*

Ethanol   Toxic marrow suppression, due to chronic alcoholism. *0568*   Decreased lifespan and production in alcoholics. *0739*

Ethinamate   Thrombocytopenia (AMA Blood dyscrasias). *2429*

Ethosuximide   Aplastic anemia/thrombocytopenia/pancytopenia. *0071*

Ethoxzolamide   Thrombocytopenia. *2426*

Etoposide   Nadir at 9 to 16 d after administration. *0141*   Occurs 9-13 d after therapy started. *3339*

Fenoprofen   Single case with approximately 1 g/d for 6 to 8 weeks, although patient also taking niacin. *3330*

Floxuridine   Toxic effect on bone marrow. *3850*

Flucytosine   Occasional thrombocytopenia observed. *1679*   Usually mild, occurs in 8 to 13% patients: occasionally pancytopenia occurs. *3567*

Fluorides   May cause aplastic anemia. *0279*

Fluorouracil   May cause bone marrow depression. *0071*

Fluphenazine   Rare thrombocytopenia with/without purpura. *1679*   Significant reduction observed in 18 psychiatric patients treated for at least 1 y although mean value still in normal range. *1642*

Furosemide   May be associated with purpura. *1617*   Observed in 0.2% of 2580 medical inpatient. *2218*

Glutethimide   Thrombocytopenia or aplastic anemia. *0071*

Glyburide   Rare thrombocytopenia reported. *0125*

Gold   Aplastic anemia (thrombocytopenia). *0114*   Blood dyscrasias major side effect as reported to uk committee on safety of medicines. *0782*   Several cases of platelet-associated IgG and thrombocytopenia. *1910*   Occasional case of aplastic anemia reported. *3717*   Severe thrombocytopenia observed 18 mo after end of gold therapy for rheumatoid arthritis. *3427*   7 cases out of 246 treated cases. *0829*   In 23 patients treated for 25 y: apparently associated with HLA-DR3 alloantigen. *0675*   In 3% of 90 patients with rheumatoid arthritis. *1461*

Guanazole   Observed in almost all patients. *1581*

Heparin   Reported effect following i.v. infusions. *2426*   Count fell in 86% patients with cerebral infarction etc. In 15% count fell by 40%. Note significant association between poor outcome and platelet drop. *0535*   Several cases of immune-mediated thrombocytopenia reported. *2502*   Occurs in about 5% of all patients receiving drug *in vivo*, usually occurs 6-12 d after start of therapy. *1934*   Average 60% decrease in 7 of 137 patients treated for 4 to 24 d. Decreased count observed in 118 patients. *2905*   Frequency of 8% in patients with bovine heparin in this study versus 0% in patients porcine heparin. *2862*   Apparently associated with increased amounts of platelet IgG and C3. *0663*   Greater effect with bovine lung preparations than those from bovine intestinal mucosa. *0276*   In 5.2% of 211 patients the majority of whom had received beef lung heparin. Plasma shown to have a heparin-sensitive antiplatelet antibody. *0664*   In 9 of 37 patients in a coronary care unit, transient and mild. *2574*   Two cases reported: mechanisms not clarified. *2932*   Mild reduction observed in some patients immediately following i.v. infusion. Possibly due to reversible aggregation, margination and sequestration due to direct effect of heparin. Usually subsides spontaneously. More severe fall to 2,000 to 90,000 IU/L after several days. Resolves only when heparin discontinued, possibly due to heparin-dependent platelet membrane antibody. *1573*

Heroin   Several cases of immune mediated thrombocytopenia reported. *2502*

Hetacillin   Thrombocytopenia may occur occasionally. *1680*

Hexachlorobenzene   Pancytopenia (AMA Blood dyscrasias). *2429*

Hexamethylmelamine   Moderate effect in a third of patients. *3863*

Hydantoin Derivatives   Thrombocytopenia (immunologically induced). *2313*

Hydralazine   Pancytopenia may occur with purpura. *0044*   Rare finding in patients treated for hypertension: also observed in neonates. *3818*

Hydrochlorothiazide   Thrombocytopenia reported. *2745*   Several cases of immune mediated thrombocytopenia reported. *2502*   Occasional case of aplastic anemia reported. *3717*

Hydroflumethiazide   Pancytopenia or thrombocytopenia with purpura. *3844*

Hydroxychloroquine   Thrombocytopenia. *2879*

Hydroxyurea   Thrombocytopenia. *0071*

Ibuprofen   Rare adverse reactions reported by physicians to UK committee on safety of medicines. *0782*

Ifosfamide   Occurs in 13% patients after 150 mg/kg. *3675*

Imipramine   May be marrow depression and purpura. *1680*

Immune Serum Globulin   Fall observed several days after injection. *2427*

Indomethacin   Rare agranulocytosis/pancytopenia/aplastic anemia. *3012*   Rare side effect reported to UK committee on safety of medicines. *0782*   Occasional case of aplastic anemia reported. *3717*

α-Interferon   Low value occurred in one patient heavily pretreated with nitrosoureas. *3299*

Iopanoic Acid   Few cases transient thrombocytopenia noted. *2292*   3 episodes in one patient occurring from 8 to 40 d after drug ingestion, resolution occurred in 4 to 8 d. *1711*

Iproniazid   Pancytopenia (AMA Blood dyscrasias). *2429*

Isoniazid   May rarely cause bone marrow aplasia. *0222*

Lead   Pancytopenia (AMA Blood dyscrasias). *2429*

Levamisole   Marked reduction in one patient, recovered on withdrawal of drug, but fell with rechallange. *1014*

Levodopa   Slight effect observed with hemolytic anemia. *0824*

Lincomycin   Rare reversible thrombocytopenia. *1343*

Lipomul®   Thrombocytopenia with fat-overloading. *2313*

Lugol's Iodine   Rare possible response with purpura. *0071*

Platelets (continued)

Blood Decrease Physiological (continued)

Measles Vaccine May cause thrombocytopenic purpura. *2427*

Mebutamate Rare thrombocytopenic purpura. *1680*

Mechlorethamine Bone marrow depression. *0071*

Mefenamic Acid Pancytopenia. *0788*

Melphalan May cause bone marrow depression (dose related). *0543*

Menstruation Increase by 50-70 %, rises to normal by 4th day. *0987*

Mepacrine Occasional case of aplastic anemia reported. *3717*

Mepazine Pancytopenia. *2313*

Mephenytoin Secondary to aplastic anemia/pancytopenia. *3008* Occasional case of aplastic anemia reported. *3717*

Meprobamate Rare bone marrow aplasia, thrombocytopenia. *0072*

Mercaptopurine May cause bone marrow depression. *2426*

Mercurial Diuretics Thrombocytopenia reported (sensitization). *1343*

Mesoridazine Thrombocytopenia may occur. *1680*

Methacycline Thrombocytopenia may occur. *1680*

Methaqualone 1 case of pancytopenia reported. *1680*

Methazolamide Probable effect as like acetazolamide. *2313* Some cases of aplastic anemia, and leukopenia reported when used in treatment for glaucoma. *3799*

Methicillin May rarely cause bone marrow depression. *2808*

Methimazole Thrombocytopenia. *2426* Occasional case of aplastic anemia reported. *3870*

Methotrexate May occur with megaloblastic anemia. *3773*

Methotrimeprazine Noted with other phenothiazines. *1680*

Methsuximide May rarely cause bone marrow aplasia. *1436*

Methyclothiazide Thrombocytopenia with purpura may occur. *1680*

Methyldopa Very rare occurs within days or months. *1940*

Methylphenidate Occasional thrombocytopenic purpura reported. *1678*

Methylthiouracil Occasional case of aplastic anemia reported. *3717*

Methyprylon Isolated reports (?actually responsible). *1680*

Mexiletine Occurs in fewer than 1% treated patients. *2011*

Minocycline Reported as side effects. *1680*

Mithramycin Thrombocytopenia. *0071*

Mitomycin Pancytopenia. *2194*

Mitoxantrone Less than 10% patients develop count of less than 100,000/uL. Count usually falls to 5th to 6th course. *3289*

Moxalactam In up to 0.5% patients as reported from several studies. *0592*

Mumps Vaccine May cause thrombocytopenic purpura. *1680*

Nalidixic Acid Thrombocytopenia. *3877* Thrombocytopenia occurred in 6 cases in Netherlands characteristically within 10-15 d after 4 g daily. Recovered rapidly once drug stopped but reaction could be severe. *2421*

Naphthoxyacetic Acid Reported effect (AMA Blood dyscrasias committee). *2429*

Neomycin If severe megaloblastic anemia. *3773*

Nitrazepam May cause thrombocytopenia. *2583*

Nitrofurantoin Thrombocytopenia. *0994*

Nitroglycerin Immunologic response occurs after 5 mo. *1436*

Nortriptyline Purpura and thrombocytopenia observed. *1680*

Novobiocin Thrombocytopenia (immunologically induced). *1343*

Nystatin Thrombocytopenia (AMA Blood dyscrasias). *2429*

Orphenadrine 2 cases aplastic anemia reported. *1680*

Oxacillin Usually only associated with large parenteral doses. *1680*

Oxprenolol Observed in one patient after slow release drug normalized within 7 d of withdrawal of drug. *0918* Observed with regular preparation in one individual. *1495*

Oxyphenbutazone Thrombocytopenia may occur after 1 week therapy. *2313* Rare side effect reported to UK committee on safety of medicines. *0782* Occasional case of aplastic anemia reported. *3717*

Oxytetracycline Thrombocytopenia reported to occur. *1436*

Paramethadione Aplastic anemia may occur rarely. *0071*

Penicillamine Thrombocytopenia may occur with rash. *0074* Most serious and potentially life-threatening effect of drug. Either idiosyncratic response seen in first year of treatment. Independent of dosage or more commonly dose related with gradual onset. *1761* Rare blood dyscrasias as reported to UK committee on safety of medicines. *0782* Significant effect from 36th week in 21 patients with rheumatoid arthritis treated for 48 weeks given 250 to 750 mg daily. *2345* In 3% of 90 patients with rheumatoid arthritis previously treated with gold. *1461* In 14 of 84 patients occurring typically in first 6 mo of treatment. *1892* Apparently some bone marrow depression in some patients but normal lifespan of cells. *3574* Toxicity observed in some patients with rheumatoid disease. *1891*

Penicillin Agranulocytosis/thrombocytopenia. *0996*

Penicillin G Isolated case of platelet-associated IgG and thrombocytopenia. *1910*

Pentoxifylline In 2 patients receiving other drugs concomitantly but for considerable time previously without ill effects. *2244*

Perchlorate May cause fatal aplastic anemia. *0071* Occasional case of aplastic anemia reported with potassium salt. *3717*

Phenacemide Aplastic anemia may occur rarely. *0071*

Phenobarbital Thrombocytopenia may occur after some time. *2429*

Phenolphthalein Immunologically induced thrombocytopenia. *1877*

Phenothiazines Hemolytic anemia (sensitivity dependent). *0788*

Phenylbutazone May cause aplastic anemia or thrombocytopenia. *0982* Rare side effects reported to UK committee on safety of medicines. *0782* Several cases of immune-mediated thrombocytopenia reported. *2502* Occasional case of aplastic anemia reported. *3717*

Phenytoin megaloblastic/hemolytic/aplastic anemia/pancytopenia. *0072* Several cases of immune-mediated thrombocytopenia reported. *2502* Occasional case of aplastic anemia reported. *3717*

Pipamazine Pancytopenia (AMA Blood dyscrasias). *2429*

Piperacetazine Rare thrombocytopenia (reversible). *0071*

Pipobroman May cause bone marrow depression. *1680*

Pirenzepine In isolated cases even though other drugs also being ingested. *3486*

Poliomyelitis Vaccine May rarely cause thrombocytopenia. *2427*

Polythiazide with/without purpura may occur. *1680*

Potassium Iodide Thrombocytopenia reported to occur. *1436*

Prednisone Immunologic response occurring in months. *1436*

Primidone Pancytopenia (AMA Blood dyscrasias). *2429*

Probenecid Occasional aplastic anemia seen. *1679*

Procainamide Pancytopenia may occur with/without lupus-like syndrome. *3215*

Procarbazine Thrombocytopenia and bone marrow depression. *0071*

Promazine May rarely cause bone marrow aplasia. *1436*

Propranolol Purpura probably reflects allergic response. *1505*

Propylthiouracil Associated with platelet associated IgG. *3870* Occasional case of aplastic anemia reported. *3717* In one 24 year old woman who developed fulminant hepatic failure with lymphocyte sensitization. *2437*

Protriptyline Rare allergic response. *1678*

Pyrazinamide Thrombocytopenia reported to occur. *1436*

Pyrazolones Thrombocytopenia. *2313*

Pyrimethamine May occur with severe megaloblastic anemia. *3898*

Pyrithioxine With dose of 400-600 mg/d: associated with development of immune complex nephritis and other penicillamine-like toxic reactions. *1761*

Quinacrine Thrombocytopenia or aplastic anemia may occur. *2313*

Quinethazone Theoretical bone marrow depression. *1680*

Quinidine Allergic reaction/pancytopenia/purpura. *3673* Several cases of platelet-associated IgG and thrombocytopenia. *1910* Isolated case of reversible granulomatous hepatitis from quinidine hypersensitivity and granuloma induction within 3 d of quinidine readministration. *0448*

Quinine Immunological mechanism. *0756* Several cases of immune thrombocytopenia reported. *2502* Quinine dependent antibody caused platelet lysis. *0662*

Reserpine Thrombocytopenia with purpura may occur. *2429*

Rifampin Occurs with antibody production. *1050* Possible antibody-mediated immune reaction. *3683* Several cases of immune mediated thrombocytopenia reported. *2502* Rare thrombocytopenia reported: rapid fall after dose given. *1292* Isolated case of severe reduction in platelet count after one dose in patient receiving treatment 4 mo previously. *1445*

Ristocetin Thrombocytopenia may occur (toxic to platelets). *2313*

Rubella Virus Vaccine Self limiting occurring within 10 d or purpura. *0241*

Smallpox Vaccine May occasionally cause thrombocytopenic purpura. *2313*

Sodium Enibomal Immunologically induced thrombocytopenic purpura. *1877*

Sodium Phosphate ^{32}P Thrombocytopenia if excess used. *0071*

Spironolactone Immunologically induced thrombocytopenia. *1877*

Stibophen Immunological mechanism (often with purpura). *0756*

Streptomycin Pancytopenia/thrombocytopenia. *0788* Occasional case of aplastic anemia reported. *1292*

Sucrose Thrombocytopenia may occur in h to days. *1436*

Sulfachlorpyridazine Purpura/thrombocytopenia. *1680*

Sulfadimethoxine Aplastic anemia/thrombocytopenia with sulfonamides. *2313*

Sulfadimidine Immunologically-induced thrombocytopenic purpura. *1877*

Sulfamethizole Agranulocytosis/aplastic anemia/thrombocytopenia. *2868*

Sulfamethoxazole Thrombocytopenia. *1974*

Sulfamethoxypyridine Thrombocytopenia may occur after days to weeks. *2429*

Sulfasalazine Several cases of immune-mediated thrombocytopenia reported. *2502* Pancytopenia reported plus reductions of individual series of cells. *2887*

Sulfisoxazole Agranulocytosis/aplastic anemia (after days). *2868* Isolated case of platelet-associated IgG and thrombocytopenia. *1910*

Sulfonamides Agranulocytosis/aplastic anemia. *2427* Several cases of aplastic anemia reported. *3717* Several cases of immune-mediated thrombocytopenia reported. *2502*

Sulfonylureas Pancytopenia. *2313* Occasional case of aplastic anemia reported. *3717*

Sulindac After 3 d of 400 mg/d in one patient reversible but mechanism not understood. No other effects on bone marrow noted. *1867*

Syrosingopine Purpura due to thrombocytopenia may occur. *0019*

Tamoxifen In up to 20% of treated patients given 400 mg/d for 2 d. *1542*

Tetracycline Thrombocytopenia/pancytopenia. *2313*

Thiabendazole Thrombocytopenia. *2413*

Thiazides Reported to cause thrombocytopenia. *2745*

Thiocolchicine May cause thrombocytopenia occasionally. *2143*

Thiocyanate Several cases of aplastic anemia reported. *3717*

Thioguanine Immunologically induced thrombocytopenia. *0543* In 29 patients with polycythemia rubra vera; effect usually apparent in 2 weeks. *2450* Observed in 7 patients with chronic granulocytic leukemia. *3416*

Thiopronine Toxicity similar to penicillamine and other sulfhydryl drugs. *1761*

Thioridazine Occasionally seen with phenothiazines. *1680*

Thiotepa May cause bone marrow depression. *0071* 10 of 25 patients given drug intravesically had at least one incident of acute myelosuppression; 5 of 25 patients had chronic myelosuppression. *1632*

Thiothixene Reported with other phenothiazines. *1680*

Thiouracil Thrombocytopenia. *2604*

Tocainide Blood dyscrasias occur in fewer than 0.2% patients. *2011*

Tolazamide Thrombocytopenia/pancytopenia. *1343*

Tolazoline Pancytopenia (AMA Blood dyscrasias). *2429*

Tolbutamide Agranulocytosis/aplastic anemia. *0634*

Trichlormethiazide May occur with/without purpura, bone marrow depression. *1680*

Triethylenemelamine May cause bone marrow depression. *0071*

Trifluoperazine Pancytopenia (AMA Blood dyscrasias). *2429*

Trimethadione Aplastic anemia may occur rarely. *3008*

Trimethoprim Most common serious toxic effect. *2937* Isolated case when drug given alone, but clear evidence of pancytopenia. Megaloblastic anemia may occur with prolonged use. *3284*

Tripelennamine Selective thrombocytopenia or aplastic anemia. *1436*

Uracil Mustard May cause bone marrow depression. *0071*

Urea Reported effect (AMA Blood dyscrasias committee). *2429*

Urethan Bone marrow aplasia/pancytopenia. *0071*

Valproic Acid Platelet-bound antibody found in 4 of 31 patients and serum antiplatelet antibody found in 1 patient. *3139* In 1 of 100 epileptic children when given alone or combined with other anticonvulsants. *0737* In 4 of 9 patients with epilepsy poorly controlled. *3508*

Vinblastine May cause bone marrow depression. *0071*

Vincristine May cause bone marrow depression. *1343*

Vitamin K Pancytopenia reported after K_3 and K_4. *2427*

X-Ray Therapy Thrombocytopenia. *0987*

Zimeldine 7 of 14 patients treated for more than 1 week demonstrated toxic syndrome ? immunological mechanism or related to high initial dose. *2075*

Urine   Increase   Physiological
Aspirin Loss through damaged glomeruli may occur. *3510*

Platinum

Serum   Increase   Physiological
Cisplatin Marked effect in 15 patients treated with drug. *3964*

Urine   Increase   Physiological
Cisplatin Marked effect in 15 patients treated with drug. *3964*

Red Blood Cells   Increase   Physiological
Carboplatin 1 h after bolus injection 2% of drug bound to erythrocytes, stabilizes after 3 h. *3286*

pO₂

Blood   No Effect   Analytical
Heparin Even if syringe dead space contains liquid heparin. *0444*

Blood   No Effect   Physiological
Smoking No effect observed. *2028*

Blood   Increase   Analytical
Transport of Specimen Increase by 0.5 to 4.5 mm Hg in pneumatic tube. *3217*

Blood   Increase   Physiological
Diazepam In healthy volunteers but changes of short duration. *1041*
Urokinase In patients with pulmonary embolism. *3298*

Blood   Decrease   Analytical
Syringes Occurs through leakage of air in plastic syringes. *3229*

Blood   Decrease   Physiological
Althesin Post-induction measurement in arterial blood. *3160*

pO₂ *(continued)*

Blood   Decrease   Physiological *(continued)*

Barbiturates   Slight decrease during sleep with hypnotic dose. *1343*

Diazepam   In healthy volunteers but changes of short duration. *1041*

Heroin   If pulmonary edema occurs with decreased respiration. *3059*

Hypothermia   Due to hypoxia, acid-base imbalance, hypotension. *2258*

Isoproterenol   By approximately 10 mm Hg in chronic lung disease. *2427*

Meperidine   Significant reduction in arterial oxygen pressure. *2427*

Methylprednisolone   Initial reduction in myocardial infarction patients compared with placebo treated controls (112 mm Hg to 88 mm Hg on average). *1560*

Midazolam   In healthy volunteers but of short duration. *1041*

Polysaccharide, Protein Bound

Serum   Increase   Physiological
Pregnancy   Throughout pregnancy. *0987*

Porphobilinogen

Serum   Decrease   Physiological
Hematin   In a patient with acute intermittent porphyria. *0414*

Urine   No Effect   Analytical
Thymol   No effect on Waldenström method. *2981*

Urine   No Effect   Physiological
Lead   Normal in lead porphyria. *1563*

Urine   Increase   Analytical
Acetylacetone   Reacts with Ehrlich's reagent unless removed. *3155*

Aminosalicylic Acid   Reacts with Ehrlich's reagent. *2425*

Apronalide   Produces red color with Ehrlich's reagent. *1563*

Cascara   Color extractable into chloroform. *1563*

Chlorpromazine   Reacts with Ehrlich's aldehyde reagent. *3879*

Phenothiazines   May react with Ehrlich's aldehyde reagent. *2950*

Procaine   Interferes with Ehrlich's aldehyde reaction. *3879*

Urine   Increase   Physiological
Anticonvulsants   Drug related effect reported in one child. *2427*

Apronalide   May precipitate acute porphyria. *1322*

Barbiturates   May precipitate acute porphyria. *1322*

Chlordiazepoxide   May precipitate acute porphyria. *1322*

Chlorpropamide   May aggravate cutaneous porphyria. *1322*

Dichloralphenazone   May precipitate acute porphyria. *1322*

Estrogens   May precipitate porphyria attack. *1322*

Ethanol   May precipitate attack of porphyria. *1322*

Glutethimide   May precipitate acute porphyria. *1322*

Glycine   In patients with acute porphyria in remission. *0938*

Griseofulvin   May precipitate acute porphyria attack. *1322*

Meprobamate   May precipitate acute porphyria. *1322*

Methyldopa   May precipitate acute porphyria attack. *1322*

Methylsulfonal   May precipitate acute porphyria. *1322*

Oral Contraceptives   May precipitate porphyria attack. *1322*

Pentazocine   May precipitate porphyria attack. *1322*

Phenytoin   May precipitate acute porphyria. *1322*

Succinimide   May precipitate acute porphyria. *1322*

Sulfomethane   May provoke attack of porphyria. *1322*

Sulfonamides   May precipitate acute porphyria. *1322*

Tolbutamide   May precipitate porphyria attack. *1322*

Urine   Decrease   Analytical
Ascorbic Acid   Inhibition of color develop if no prior separation. *1116*

Thiols   Unless prior separation is employed. *1116*

Urea   Inhibits color development unless prior separation. *1116*

Urine   Decrease   Physiological
Actinomycin   Suppressed induction of ALA synthetase in porphyria. *0938*

Cimetidine   Slight decrease in patient with acute intermittent porphyria. *1659*

Glucose   Suppresses ALA synthetase in acute porphyria. *0938*

Hematin   In a patient with elevation due to porphyria. *0414*

Vitamin E   Possible reduction to normal levels in porphyria. *2549*

Porphyrins

Blood   Increase   Physiological
Griseofulvin   May precipitate acute porphyria attack. *2808*

Urine   Increase   Analytical
Acriflavine   Produce fluorescence. *3879*

Ethoxazene   False positive with fluorescent methods. *3879*

Hemoglobin   Affects UV absorption screening methods. *2981*

Oxytetracycline   Produces interfering fluorescence. *1238*

Phenazopyridine   Interference with fluorescence (in screening test). *3879*

Sulfamethoxazole   May cause false positive with fluorescent methods. *1022*

Tetracycline   Interfering fluorescence. *1238*

Urine   Increase   Physiological
Aminopyrine   Stimulates formation of ALA-synthetase. *1343*

Antipyretics   Reported effect. *1237*

Apronalide   May precipitate attack of acute porphyria. *2427*

Barbiturates   May precipitate attack of acute porphyria. *1016*

Carbromal   May precipitate attack of acute porphyria. *2220*

Chloral Hydrate   May precipitate attack of acute porphyria. *2220*

Chlordiazepoxide   May precipitate attack of acute porphyria. *2220*

Chlorpropamide   May precipitate attack of acute porphyria. *1016*

Diazepam   May precipitate attack of acute porphyria. *1016*

Dichloralphenazone   May precipitate attack of acute porphyria. *1016*

Ergot Preparations   May precipitate attack of acute porphyria. *1016*

Estrogens   May precipitate attack of acute porphyria. *1016*

Ethanol   Increased synthesis of porphyrins. *3879*

Ethoxazene   Hepatotoxic effect. *1022*

Glutethimide   May precipitate attack of acute porphyria. *2220*

Griseofulvin   Stimulates formation of ALA-synthetase. *1343*

Hexachlorobenzene   Stimulates formation of ALA-synthetase. *1343*

Hydantoin Derivatives   May precipitate attack of acute porphyria. *1016*

Isopropyl Dipyrone   May precipitate attack of acute porphyria. *1016*

Menstruation   May precipitate episode of hepatic porphyria. *2427*

Meprobamate   May precipitate attack of acute porphyria. *2220*

Methyldopa   May precipitate attack of acute porphyria. *1016*

Methyprylon   May induce porphyria in animals, ?in humans. *1343*

Metyrapone   Reported to precipitate attack of acute porphyria. *2220*

Pentazocine   May precipitate attack of acute porphyria. *1016*

Phenylhydrazine   May occasionally precipitate attack of porphyria. *3879*

Pregnancy   May precipitate episode of hepatic porphyria. *2427*

Progestogens   May precipitate attack of acute porphyria. *1016*

Succinimide   May precipitate attack of acute porphyria. *1016*

Sulfomethane Stimulates formation of ALA-synthetase. *1343*
Sulfonamides May precipitate acute porphyria. *2220*
Vitamin K Reported side effect. *2427*

Urine   Decrease   Analytical
Light Photosensitive. *2981*

Urine   Decrease   Physiological
Oral Contraceptives May occur in patients with established disease. *2427*

Feces   No Effect   Physiological
Cimetidine No measurable change in one patient with acute intermittent porphyria. *1659*

Postheparin Hepatic Lipase

Plasma   No Effect   Physiological
Estradiol No significant change after 2 mg estradiol valerate/d in 20 normolipoproteinemic women over 3 mo. *3593*

Plasma   Increase   Physiological
Levonorgestrel Increase by 64% in 11 normolipoproteinemic women given 250 μg/d for 2 weeks. *3593*

Postheparin Hepatic Triglyceride Lipase

Plasma   Decrease   Physiological
Oral Contraceptives Reduced by 46% in oral contraceptive treated women. *1303*

Postheparin Lipase

Plasma   Decrease   Physiological
Trichlormethiazide Significant effect during 3 mo drug administration (2.3 vs 3.3 μmol free fatty acid/mL/h respectively). *2550*

Postheparin Lipolytic Activity

Plasma   No Effect   Physiological
Quingestanol No significant effect with 300 μg/d 6 mo. *1309*

Postheparin Lipoprotein Lipase

Plasma   No Effect   Physiological
Ethanol With acute administration to fed individuals. *2428*
Levonorgestrel Insignificant increase in 11 normolipoproteinemic women given 250 μg/d for 2 weeks. *3593*

Plasma   Increase   Physiological
Ethanol Fell by 40% 5 d after cessation of debauch to normal concentration. *1011*

Plasma   Decrease   Physiological
Estradiol Increase by 25% with 2 mg estradiol valerate/d in 20 normolipoproteinemic women over 3 mo. *3593*
Ethanol With acute administration after overnight fast. *2428*

Potassium

Serum   No Effect   Analytical
Acetaminophen At concentration of 6 mg/L had no effect on flame-photometric method. *3393* At concentration of 1500 mg/L had no effect on measurement by ISE with predilution. *3393*
Acetazolamide At concentration of 1,000 mg/L had no effect on flame-photometric method. *3393*

Acetylsalicylic Acid At concentration of 100 mg/L had no effect on flame-photometric method. *3393* At concentration of 3,000 mg/L had no effect on ISE measurement with predilution. *3393* At concentration of 1,000 mg/L had no effect on ISE measurement without predilution. *3393*
Allopurinol At concentration of 5 mg/L had no effect on flame-photometric method. *3393* At concentration of 180 mg/L had no effect on ISE measurement without predilution. *3393*
Aminosalicylic Acid At concentration of 460 mg/L had no effect on ISE measurement with predilution. *3393*
Amitriptyline At concentration of 0.1 mg/L had no effect on flame-photometric method. *3393* At concentration of 20 mg/L had no effect on ISE measurement with predilution. *3393*
Amobarbital At concentration of 5 mg/L had no effect on flame-photometric method. *3393* At concentration of 65 mg/L had no effect on ISE measurement with predilution. *3393*
Ampicillin At concentration of 5 mg/L had no effect on flame-photometric method. *3393* At concentration of 1800 mg/L had no effect on ISE measurement without predilution. *3393*
Ascorbic Acid At concentration of 60 mg/L had no effect as measured by flame-photometric method. *3393* At concentration of 40,000 mg/L had no effect on ISE measurement with predilution. *3393* At concentration of 800 mg/L had no effect on ISE measurement without predilution. *3393*
Azapropazone At concentration of 360 mg/L had no effect on ISE measurement without predilution. *3393*
Barbital At concentration of 500 mg/L had no effect on measurement by ISE with predilution. *3393*
Benzbromarone At concentration of 80 mg/L had no effect on ISE measurement without predilution. *3393*
Bisacodyl At concentration of 2 mg/L had no effect on ISE measurement without predilution. *3393*
Butizide At concentration of 1.98 mg/L had no effect on ISE measurement without predilution. *3393*
Carbamazepine At concentration of 20 mg/L had no effect on ISE measurement with predilution. *3393*
Cefazolin At concentration of 110 mg/L had no effect on flame-photometric method. *3393*
Cephalosporin At concentration of 100 mg/L had no effect on ISE measurement with predilution. *3393*
Cephalothin At concentration of 1,000 mg/L had no effect on flame-photometric method. *3393*
Chloral Hydrate At concentration of 200 mg/L had no effect on measurement by ISE with predilution. *3393*
Chloramphenicol At concentration of 6,000 mg/L had no effect on measurement by ISE without predilution. *3393*
Chlordiazepoxide At concentration of 5 mg/L had no effect on flame-photometric method. *3393* At concentration of 20 mg/L had no effect on measurement by ISE with predilution. *3393* At concentration of 18 mg/L had no effect on measurement by ISE without predilution. *3393*
Chlormezanone At concentration of 100 mg/L had no effect on measurement by ISE with predilution. *3393*
Chloroquine At concentration of 50 mg/L had no effect on measurement by ISE without predilution. *3393*
Chlorphenesin At concentration of 150 mg/L had no effect on measurement by ISE with predilution. *3393*
Chlorpromazine At concentration of 3 mg/L had no effect on flame-photometric method. *3393* At concentration of 3 mg/L had no effect on measurement by ISE with predilution. *3393*
Chlorpropamide At concentration of 150 mg/L had no effect on flame-photometric method. *3393*
Chromonar Hydrochloride At concentration of 180 mg/L had no effect on ISE measurement without predilution. *3393*
Cimetidine At concentration of 1 mg/L had no effect on flame-photometric method. *3393*
Clindamycin At concentration of 150 mg/L had no effect on measurement by ISE with predilution. *3393*
Cloxacillin At concentration of 5 mg/L had no effect on flame-photometric method. *3393*
Codeine At concentration of 0.1 mg/L had no effect on flame-photometric method. *3393*
Colistin At concentration of 150 mg/L had no effect on measurement by ISE with predilution. *3393*
Contact With Clot If plastic disc used to separate for 48 h. *1863*

Potassium (continued)

Serum   No Effect   Analytical (continued)

Cyclophosphamide   At concentration of 80 mg/L had no effect on measurement by ISE without predilution. 3393

Cyclosporine   At concentration of 41.5 mmol/L (50 μg/L) on methods on Technicon® SMAC® II and Hitachi® 705. 1123

Dexamethasone   At concentration of 1.4 mg/L had no effect on measurement by ISE with predilution. 3393

Dextran   At concentration of 30,000 mg/L had no effect on Ektachem® ISE method. 3393

Dextran 40   At concentration of 9100 mg/L had no effect on measurement by ISE with predilution. 3393

Dextran 60   At concentration of 18,000 mg/L had no effect on measurement by ISE without predilution. 3393

Dextran 70   At concentration of 5400 mg/L had no effect on measurement by ISE with predilution. 3393

Diatrizoic acid   At concentration of 6080 mg/L had no effect on ISE measurement without predilution. 3393

Diazepam   At concentration of 1.5 mg/L had no effect on flame-photometric method. 3393   At concentration of 10 mg/L had no effect on measurement by ISE with predilution. 3393

Digitoxin   At concentration of 6 mg/L had no effect on measurement by ISE with predilution. 3393

Digoxin   At concentration of 0.002 mg/L had no effect on flame-photometric method. 3393   At concentration of 9 mg/L had no effect on measurement by ISE with predilution. 3393   At concentration of 0.15 mg/L had no effect on measurement by ISE without predilution. 3393

Diphenhydramine   At concentration of 23 mg/L had no effect on measurement by ISE with predilution. 3393

Disopyramide   At concentration of 4 mg/L had no effect on flame photometric method. 3393

Doxepin   At concentration of 60 mg/L had no effect on measurement by ISE without predilution. 3393

EDTA   At concentration of 1,000 mg/L had no effect on measurement by ISE without predilution. 3393

Erythromycin   At concentration of 10 mg/L had no effect on measurement by ISE with predilution. 3393

Ethchlorvynol   At concentration of 400 mg/L had no effect on measurement by ISE with predilution. 3393

Ethosuximide   At concentration of 390 mg/L had no effect on measurement by ISE with predilution. 3393

Fluosol-DA   By 18% on Kodak Ektachem® 400 at concentration of 50%. 2518

Flurazepam   At concentration of 0.1 mg/L had no effect on flame-photometric method. 3393

Furosemide   At concentration of 4 mg/L had no effect on flame-photometric method. 3393   At concentration of 6,000 mg/L had no effect on measurement by ISE with predilution. 3393   At concentration of 400 mg/L had no effect on measurement by ISE without predilution. 3393

Gentamicin   At concentration of 6 mg/L had no effect on flame-photometric method. 3393   At concentration of 14 mg/L had no effect on measurement by ISE with predilution. 3393

Glibenclamide   At concentration of 3 mg/L had no effect on measurement by ISE with predilution. 3393

Glutethimide   At concentration of 10 mg/L had no effect on measurement by ISE with predilution. 3393

Hydralazine   At concentration of 0.5 mg/L had no effect on flame-photometric method. 3393   At concentration of 6 mg/L had no effect on measurement by ISE with predilution. 3393

Hyoscine-N-Butylbromide   At concentration of 20 mg/L had no effect on measurement by ISE without predilution. 3393

Imipramine   At concentration of 30 mg/L had no effect on measurement by ISE with predilution. 3393

Indomethacin   At concentration of 13 mg/L had no effect on measurement by ISE with predilution. 3393   At concentration of 40 mg/L had no effect on measurement by ISE without predilution. 3393

Insulin   At concentration of 3 mg/L had no effect on measurement by ISE with predilution. 3393

Iproniazid   At concentration of 40 mg/L had no effect on measurement by ISE with predilution. 3393

Kanamycin   At concentration of 10 mg/L had no effect on measurement by ISE with predilution. 3393

Levodopa   At concentration of 200 mg/L had no effect on measurement by ISE with predilution. 3393

Lidocaine   At concentration of 0.5 mg/L had no effect on measurement by ISE with predilution. 3393

Lipemia   In patient specimens measured on SMAC® in comparison with specimens following ultra-centrifugation. 2466

Lithium Heparin   At concentration of 3,000 mg/L had no effect on Ektachem® method. 3393

Lorazepam   At concentration of 0.05 mg/L had no effect on flame-photometric method. 3393

Meglumine   At concentration of 2,000 mg/L had no effect on ISE measurement without predilution. 3393

Mephenesin   At concentration of 100 mg/L had no effect on measurement by ISE with predilution. 3393

Mephenytoin   At concentration of 20 mg/L had no effect on measurement by ISE with predilution. 3393

Meprobamate   At concentration of 25 mg/L had no effect on flame-photometric method. 3393   At concentration of 160 mg/L had no effect on measurement by ISE with predilution. 3393

Methapyrilene   At concentration of 13 mg/L had no effect on measurement by ISE with predilution. 3393

Methaqualone   At concentration of 6,000 mg/L had no effect on measurement by ISE with predilution. 3393

Methicillin   At concentration of 20 mg/L had no effect on flame-photometric method. 3393

Methotrexate   At concentration of 80 mg/L had no effect on measurement by ISE without predilution. 3393

Methotrimeprazine   At concentration of 1 mg/L had no effect on flame-photometric method. 3393

Methyldopa   At concentration of 7 mg/L had no effect on flame-photometric method. 3393   At concentration of 100 mg/L had no effect on measurement by ISE with predilution. 3393   At concentration of 800 mg/L had no effect on measurement by ISE without predilution. 3393

Methylprednisolone   At concentration of 100 mg/L had no effect on measurement by ISE with predilution. 3393

Metoprolol   At concentration of 0.34 mg/L had no effect on flame-photometric method. 3393

Minocycline   At concentration of 6 mg/L had no effect on measurement by ISE with predilution. 3393

Nalidixic Acid   At concentration of 10 mg/L had no effect on measurement by ISE with predilution. 3393

Niacinamide   At concentration of 120 mg/L had no effect on measurement by ISE without predilution. 3393

Nitrofurantoin   At concentration of 2.5 mg/L had no effect on flame-photometric method. 3393   At concentration of 14 mg/L had no effect on measurement by ISE with predilution. 3393   At concentration of 50 mg/L had no effect on measurement by ISE without predilution. 3393

Norfenefrin   At concentration of 3 mg/L had no effect on measurement by ISE without predilution. 3393

Novaminsulfon   At concentration of 800 mg/L had no effect on measurement by ISE without predilution. 3393

Orphenadrine   At concentration of 9 mg/L had no effect on measurement by ISE with predilution. 3393

Oxyphenbutazone   At concentration of 120 mg/L had no effect on measurement by ISE without predilution. 3393

Papaverine   At concentration of 10 mg/L had no effect on measurement by ISE with predilution. 3393

Penicillamine   At concentration of 360 mg/L had no effect on measurement by ISE without predilution. 3393

Penicillin   At concentration of 15,000 mg/L had no effect on measurement by ISE with predilution. 3393

Penicillin G   At concentration of 18 mg/L had no effect as measured by flame-photometric method. 3393

Perphenazine   At concentration of 1 mg/L had no effect on flame-photometric method. 3393

Phenazopyridine   At concentration of 20 mg/L had no effect on measurement by ISE with predilution. 3393

Phenobarbital   At concentration of 10 mg/L had no effect on flame-photometric method. *3393*   At concentration of 250 mg/L had no effect on measurement by ISE with predilution. *3393*   At concentration of 60 mg/L had no effect on measurement by ISE without predilution. *3393*

Phenothiazines   At concentration of 200 mg/L had no effect on measurement by ISE without predilution. *3393*

Phenprocoumon   At concentration of 3.6 mg/L had no effect on measurement by ISE without predilution. *3393*

Phensuximide   At concentration of 120 mg/L had no effect on measurement by ISE with predilution. *3393*

Phenylbutazone   At concentration of 750 mg/L had no effect on measurement by ISE with predilution. *3393*   At concentration of 120 mg/L had no effect on measurement by ISE without predilution. *3393*

Phenytoin   At concentration of 20 mg/L had no effect on flame-photometric method. *3393*   At concentration of 240 mg/L had no effect on measurement by ISE with predilution. *3393*

Primidone   At concentration of 10 mg/L had no effect on measurement by ISE with predilution. *3393*

Probenecid   At concentration of 1300 mg/L had no effect on measurement by ISE with predilution. *3393*   At concentration of 400 mg/L had no effect on measurement by ISE without predilution. *3393*

Procaine   At concentration of 2 mg/L had no effect on measurement by ISE with predilution. *3393*

Prochlorperazine   At concentration of 1 mg/L had no effect on flame-photometric method. *3393*

Promethazine   At concentration of 1 mg/L had no effect on flame-photometric method. *3393*

Propoxyphene   At concentration of 5 mg/L had no effect on measurement by ISE with predilution. *3393*

Propranolol   At concentration of 0.2 mg/L had no effect on flame-photometric method. *3393*

Protamine   At concentration of 10 mg/L had no effect on measurement by ISE with predilution. *3393*

Quinidine   At concentration of 210 mg/L had no effect on ISE measurement without predilution. *3393*

Quinine   At concentration of 30 mg/L had no effect on measurement by ISE with predilution. *3393*

Salicylate   At concentration of 100 mg/L had no effect on measurement by ISE with predilution. *3393*   At concentration of 350 mg/L had no effect on measurement by ISE on Ektachem®. *3393*   At concentration of 30,000 mg/L had no effect on measurement by ISE with predilution. *3393*

Secobarbital   At concentration of 1 mg/L had no effect on flame-photometric method. *3393*   At concentration of 100 mg/L had no effect on measurement by ISE with predilution. *3393*

Sodium Fluoride   At concentration of 2,000 mg/L had no effect on measurement by ISE without predilution. *3393*

Sodium Heparin   At concentration of 750 mg/L had no effect on measurement by ISE without predilution. *3393*

Sodium Oxalate   At concentration of 3,000 mg/L had no effect on measurement by ISE without predilution. *3393*

Spironolactone   At concentration of 80 mg/L had no effect on measurement by ISE without predilution. *3393*

Standing of Sample   If on polystyrene beads 24 h at 25°. *2624*

Storage of Sample   No effect for 2 weeks at room temperature or 4°. *1563*

Streptomycin   At concentration of 400 mg/L had no effect on measurement by ISE with predilution. *3393*

Strychnine   At concentration of 12 mg/L had no effect on measurement by ISE with predilution. *3393*

Sulfadiazine   At concentration of 1500 mg/L had no effect on measurement by ISE with predilution. *3393*

Sulfaguanidine   At concentration of 500 mg/L had no effect on measurement by ISE with predilution. *3393*

Sulfamethoxydiazine   At concentration of 200 mg/L had no effect on measurement by ISE with predilution. *3393*   At concentration of 200 mg/L had no effect on measurement by ISE without predilution. *3393*

Sulfanilamide   At concentration of 1,000 mg/L had no effect on measurement by ISE with predilution. *3393*

Sulfisoxazole   At concentration of 80,000 mg/L had no effect on measurement by ISE with predilution. *3393*

Tetracycline   At concentration of 4 mg/L had no effect on flame-photometric method. *3393*   At concentration of 20,000 mg/L had no effect on measurement by ISE with predilution. *3393*   At concentration of 400 mg/L had no effect on measurement by ISE without predilution. *3393*

Theophylline   At concentration of 20 mg/L had no effect on flame-photometric method. *3393*   At concentration of 1,000 mg/L had no effect on measurement by ISE with predilution. *3393*

Timolol   At concentration of 0.01 mg/L had no effect on flame-photometric method. *3393*

Tolbutamide   At concentration of 100 mg/L had no effect on flame-photometric method. *3393*   At concentration of 2,000 mg/L had no effect on measurement by ISE with predilution. *3393*   At concentration of 400 mg/L had no effect on measurement by ISE without predilution. *3393*

Trichlorethanol   At concentration of 12 mg/L had no effect on flame-photometric method. *3393*   At concentration of 1,000 mg/L had no effect on measurement by ISE with predilution. *3393*

Triethanolamine   At concentration of 200 mg/L had no effect on measurement by ISE with predilution. *3393*

Trifluoperazine   At concentration of 1 mg/L had no effect on flame-photometric method. *3393*

Vitamin B Complex   At concentration of 23.3 mg/L had no effect on measurement by ISE without predilution. *3393*

Warfarin   At concentration of 1.5 mg/L had no effect on flame-photometric method. *3393*   At concentration of 100 mg/L had no effect on measurement by ISE with predilution. *3393*

Serum   No Effect   Physiological

Aging   No significant change from age 6-20 y. *0887*

Aspirin   When approximately 3 g/d ingested for several weeks. *3076*

Azosemide   Effect observed in normal male volunteers after 60 mg orally. *2658*

Bendrofluazide   After 12 mo treatment of 53 previously untreated hypertensives. *0315*

Bopindolol   In 10 hypertensive patients treated for either 1 or 21 d. *1063*

Bucindol   No effects noted in 8 patients following several weeks on treatment. *1283*

Captopril   No significant change in 7,000 hypertensives treated for 3 y. *1407*

Cellulose Phosphate   No significant effect observed. *2710*

Etodolac   No cumulative effect observed over 24 h in individuals with normal renal function or with renal insufficiency when given 500 mg bid for 4 d. *0453*

Hydrochlorothiazide   If potassium supplement coadministered with hydrochlorothiazide in hypertensive women. *2360*

Lithium   No significant effect observed with chronic treatment. *0488*   Isolated case of hyperkalemia reported but usually no effect reported. *2836*

Meals   No effect after standard breakfast. *0579*

Metoclopramide   No significant change following i.v. injection. *0497*

Muzolimine   No significant change in 10 hypertensive patients given 30 mg daily for 16 weeks. *1229*

Oxaprozin   No significant change observed in 7 volunteers treated for up to 1 week. *2464*

Pindolol   No significant effect even with 6 weeks treatment. *3279*

Propranolol   After 12 mo treatment of 53 previously untreated hypertensives. *0315*

Ranitidine   No effect of 3-d of 150 mg/12 h. *3130*

Saccharated Iron Oxide   In 9 individuals given 40 mg/d i.v. daily for up to 42 d. *2662*

Spironolactone   In 15 patients with primary hypertension treated with 100 mg daily for 1 y. *1072*

Starvation   Usually no effect, may fall to low normal. *3783*

Terbutaline   No reduction after oral treatment for 13 d. *0288*   In 6 normal men treated with therapeutic amounts for 2 weeks. *3177*

Serum   Increase   Analytical

Calcium   Emission spectrum may interfere. *2313*

Cefotaxime   By up to 0.2 mmol/L with both drug and metabolite on method on American Monitor Parallel. *0201*

Potassium *(continued)*

Serum Increase Analytical *(continued)*

Contact With Clot Release from erythrocytes and platelets. *3879*

Fluorides With inhibition of phosphorylation. *1563*

Fluosol-DA Icrease by 33% on SMA IIC at concentration of 50%. Increase by 23% on SMA IIC at concentration of 50%. *2518*

Liposol Changed from 4.45 to 6.59 mmol/L in 26 randomly selected patient sera, before and after addition on Beckman Astra methods. *2054*

Oxalate If potassium salt of oxalate. *1563*

Procainamide At concentrations above 8 mg/L (normal therapeutic concentration 10 mg/L) raised concentration as measured by ISE with predilution. *3393*

Protein Presence raises temperature, increases emission. *2313*

Sodium Variable concentration affects most flame methods. *0582*

Tobacco Smoke Interferes with flame photometry. *1238*

Transport of Specimen 0.1 mmol/L in pneumatic tube system. *3217*

Serum Increase Physiological

Alprenolol Slight increase in patients treated with moderate doses of drug. *2836*

Amiloride Also inhibits kaliuresis caused by thiazides. *2969* Significant rise in 3 patients with creatinines from 0.2 to 0.70 mmol/L when given with hydrochlorothiazide. *3816* Increase by 0.9 mmol/L over 7 d in 5 volunteers with 75 mg daily. *2440* Binds reversibly to luminal membrane, blocking reabsorption of filtered sodium and inhibiting passive potassium and hydrogen ion secretion. *2836*

Aminocaproic Acid Reported effect especially if renal function impaired. *0130*

Amphotericin B May occur with renal toxicity. *0551*

Arginine Up to 5.6-6.5 mmol/L in diabetics (less in normals) after 0.5 g/kg i.v. infusion in 30 minutes (maximum increase 1.4 mmol/L). *2328* Causes shift of potassium to extracellular compartment, concentration correlated with aminoacid concentration. *2836*

Azathioprine Observed in only 1 of 13 renal transplant patients after 1 mo treatment also with prednisone. *1154*

Blindness Significantly higher (average 0.4 meq/L) than normal. *1637*

Blood Transfusions May occur with release from RBC. *2220* Plasma concentration affected by preservative, storage temperature and length of storage. *2836*

Bopindolol Minimal effect observed in large population of hypertensives over 12 weeks study period. *3779*

Boric Acid Toxicity effect with acute tubular necrosis. *0652*

Cannabis Reported effect. *2220*

Captopril From 2.9 to 3.3 mmol/L after 1 mo in patient with Bartter's syndrome due to inhibited aldosterone production. *1558* Resulting from decreased secretion of aldosterone. *1716* Rise less than 1.0 mmol/L, but greatest in patients with high baseline renin activity. *2836*

Carbacrylamine Resin Part of resin is in form of potassium salt. *2313*

Cephaloridine May occur with nephrotoxicity. *2313*

Cyclosporine Probably occurs as manifestation of nephrotoxicity with tubular dysfunction. *0683* Reported in 7 of 43 patients with good renal function following kidney transplantation: single case reported here which responded to fludrocortisone. *2795* Sustained hyperkalemia (6.0 to 7.1 mmol/L) inappropriate for renal function in in 7 of 43 renal allograft patients. *0027* Observed in 13 of 50 renal transplant patients after 1 mo. *1154* Hyperkalemia out of proportion to reduction in GFR. *2836*

Danazol Slight effect in 18 women with endometriosis given 600 mg/d for 6 mo. *0591*

Digitalis But only with doses 20 to 40 times therapeutic. *2836*

Enalapril Produces mild potassium retention and hyperkalemia. *2836*

Epinephrine Initial rise accompanies glucose mobilization. *1343* Rise within 1 minute of i.v. injection and fall after 4 minutes, eventually dropping below control values. *2836*

Ethanol 3% increase in heavy versus occasional drinkers. *3273*

Glucagon Immediate increase in dogs then fall to normal or less. *0691*

Hemolysis Released from RBCs and platelets. *0579*

Heparin Decreased renal excretion. *1343* Very rare hyperkalemia reported although consistent impairment of adrenocortical synthesis of aldosterone: in cases of hyperkalemia reported selective hypoaldosteronism present. *2836*

Heroin Associated with severe rhabdomyolysis. *2836*

Histamine Marked effect of i.v. injection (?involves gastrointestinal tract). *1343*

Hydrochlorothiazide Marked effect when given with amiloride to 3 patients with renal failure (creatinine 0.2-0.7 mmol/L). *3816*

Indomethacin Marked reversal on withdrawal of drug in one young woman. *3541* Associated with increasing risk of renal insufficiency in cirrhosis, nephrotic syndrome, decompensated congestive heart failure and chronic renal disease. *0373* Rise from 4.3 to 4.6 mmol/L with treatment for 1 week. *2464* Associated with decreased secretion of renin and aldosterone, decreased sodium delivery to distal tubule and reduction of urinary flow rate. *1228* Observed in 3 patients with gouty arthritis who developed renal insufficiency. *1133*

Intra-Amniotic Saline No change until 8 h after then by 0.6 meq/L. *3892*

Isoniazid Reported effect of overdose. *1488*

Lipomul® Nephrotoxic effect. *2313*

Lithium ?by displacing from cells. *2275*

Mannitol Mechanism not discussed. *2480*

Methicillin Possible result of nephrotoxicity. *2313*

Nifedipine In 23 patients over 60 y old with essential mild to moderate hypertension. *3192* At dose of 40 mg/d caused increase of 0.3 mmol/L when also treated with propranolol. *2836*

NSAID (Nonsteroidal Anti-Inflammatory Drugs) Associated with inhibition of prostaglandin synthesis. *1561*

Oxprenolol Slight rise observed in patients treated with moderate doses. *2836*

Penicillin May occur if K salt given i.v. also alkalosis. *3505*

Pindolol Slight rise in patients treated with moderate doses of drug. *2836*

Potassium Chloride Over correction of hypokalemia. *0071* Severe hyperkalemia can develop following single oral dose of 30 to 45 meq in patients either with renal potassium excretion or internal disposal disorders. Potassium chloride supplementation was single major cause of fatal drug reactions. *2836*

β-Propiolactone Significant effect by 0.7 mmol/L when plasma from BPL treated blood compared with plasma in 25 specimens. *0217*

Propranolol Usual effect if aldosterone decreased. *0523* In 15 patients with essential hypertension when treated for 1 mo. *3776* Moderate increase observed in several trials due to redistribution from intracellular to extracellular compartments, not due to retention in body. *2234* Increase by 0.2 mmol/L in 340 patients given drug for 10 weeks to reduce diastolic BP to less than 90 mm Hg. *3707* Increase by 0.17 mmol/L in approximately 120 patients with essential hypertension treated for 1 y. *3708* Approximately 5% increase in both men and women with treatment for 3 y. *2506* Pronounced and prolonged increase in patients who had acute myocardial infarction. *2612* Increases of 0.3 to 0.5 mmol/L when doses of 160 to 640 mg/d given, although never to value above 4.5 mmol/L. *2836*

Ramipril In response to 10 mg and 20 mg on successive days in 9 patients with severe chronic congestive heart failure. *0764*

Saline Effect of hypertonic solution. *2480*

Serum Due to release from erythrocytes with clotting. *3879*

Spironolactone Inhibits Na/K exchange in renal tubules. *1343* Mean increase of 0.4 mmol/L in 3 men given drug for 1 week. *3736* Increase by 0.4 mmol/L after 7 d in 5 volunteers. *2440* Binds to cytoplasmic hormone receptor on pericapillary side of distal tubular cells, blunting normal response to aldosterone. *2836*

Succinylcholine If injected i.v. *3505* Caused transient hyperkalemia in patients undergoing general anesthesia. *2836* Due to increased chemosensitivity of muscle membrane due to development of receptor sites in extrajunctional areas. *1408*

Terbutaline Due to β₂-adrenergic mediated uptake of potassium in skeletal muscle and other tissues (decrease of about 0.8 mmol/L). *1823*

Tetracycline Occurs with azotemia. *0071*

Timolol Slight increase in patients treated with moderate doses of drug. *2836* Pronounced and prolonged increase in patients who had acute myocardial infarction. *2612*

Tourniquet Especially if forearm muscles exercised also. *1563*

Triamterene Potassium sparing action (affects Na/K exchange). *2313* Mean increase of 0.3 mmol/L in 3 men given drug for 1 week. *3736*

Tromethamine Reported effect. *3006*

Serum Decrease Analytical

EDTA Reacts with potassium. *2313*

Plasma Mean 0.4 meq/L higher in serum. *2228*

Sodium Citrate At concentrations above 2500 mg/L lowered concentration as measured by ISE with predilution. *3393*

Sulfasalazine At concentrations above 1,000 mg/L (normal therapeutic concentration 70 mg/L) lowered concentration as measured by ISE without predilution. *3393*

Serum Decrease Physiological

Acetazolamide Diuretic action, carbonic anhydrase inhibition. *0652*

Alanine Significant effect with chronic administration in fasting. *1263*

Albuterol Significant dose-related effect after i.v. infusion of therapeutic doses in 4 healthy male volunteers. *2807*

Aldosterone Hypokalemic hypochloremic alkalosis may occur. *1343*

Aminosalicylic Acid Due to action on renal tubules or vomiting. *2372* Rare side effect. *1292*

Ammonium Chloride Perpetuates potassium deficiency, cation loss. *2313*

Amphotericin B (frequent) associated with renal damage. *0128* Nephrotoxicity usually develops after a few weeks of treatment, usually reversible unless more than total of 4 g drug given. *3567*

Antibiotics May occur with multiple regime, ?redistribution. *3549*

Aspirin Diuretic action, respiratory alkalosis. *0652*

Azosemide Slight effect noted in 2 volunteers. *2047*

Bendrofluazide Marked diuretic response. *1129* In men and women in 3 y study of patients with mild hypertension. *2506* Mean fall from 4.3 to 3.9 mmol/L in 7 essential hypertensives treated for 16 weeks. *0352* From 0.2 to 0.6 mmol/L in 5 or 6 hypertensives given 5 mg daily for 6 weeks. *3866*

Benzthiazide May occur as with other thiazides. *1680*

Betamethasone Occurs infrequently. *0071*

Bicarbonates Induces metabolic alkalosis. *0652*

Bisacodyl Associated with steatorrhea if used in excess. *3855*

Bumetanide By approximately 0.3 meq/L. *2666*

Capreomycin Observed occasionally with therapy. *1679*

Carbenicillin Observed effect ?redistribution or increased excretion. *3549* Nonreabsorbable anion increases electrical negativity of lumen of distal nephron with enhanced potassium and hydrogen excretion. *0645* In one child due to drug having impermeant anion effect on renal tubule. *3431*

Carbenoxolone By approximately 0.5 meq/L in adrenalectomized patients. *3801* Aldosterone like effect. *1159*

Cathartics Excessive use may cause hypokalemia. *1343*

Cephalexin If given with gentamicin to leukemics. *3922*

Chloroquine Marked prolonged reduction in two cases of acute massive intoxication. *2203* Marked prolonged reduction in two cases of acute massive intoxication. *2203*

Chlorothiazide Diuretic action. *3505*

Chlorthalidone Diuretic induced depletion. *1237* 43% patients with hypokalemia given long term 25-100 mg/d. *0314* Significant effect in pre- and postmenopausal women given 100 mg/d for 6 weeks. *0402* Progressive effect with dose: from 0.4 mmol/L with 25 mg/d to 1.0 mmol/L at 200 mg/d in 37 patients treated for 8 weeks. *3643* 43% patients with hypokalemia given long term 25-100 mg/d. *0314* Significant fall from 8 weeks in 10 hypertensive patients given 25 mg daily for for 16 weeks. *1229* 0.5 mmol/L reduction in drug-treated versus placebo-treated patients. *3367* Hyperaldosteronism major causal factor in diuretic induced hypokalemia. *1438* In majority of children during course of treatment. *0197*

Cisplatin Marked effect in 15 patients treated with drug. *3964* Associated with magnesium deficiency. *0368*

Clopamide Result of diuretic action. *3881* Single case of diuretic associated myopathy. *1793*

Clorexolone Diuretic action. *2427*

Corticosteroids Mineralocorticoid effect with increase renal excretion. *0070*

Corticotropin Increased urinary excretion (mineralocorticoid effect). *3879*

Cortisone Mineralocorticoid effect. *3879*

Cyclothiazide Same degree of hypokalemia as hydrochlorothiazide. *0859*

Desoxycorticosterone Occurs as result of increased excretion. *0071*

Dexamethasone May promote increased urinary loss. *1680*

Diapamide Diuretic action. *2256*

Dichlorphenamide Diuretic action (inhibits carbonic anhydrase). *2313*

Diuretics Loss in urine. *2313*

Diurnal Variation Minimum about 10 pm maximum at 8 am. *3428*

Enflurane Slight decrease noted postoperatively. *0999*

Epinephrine Intravenous injection causes sharp rise then fall after 4 minutes. *3064* Profound hypokalemia produced by steady-state infusion at 3-5 nmol/L. *3490*

Ethacrynic Acid Diuretic action. *3505*

Ethanol Inappropriate secretion of ADH in beer drinkers. *0875* Increase by 0.2 mmol/L 2 h after 0.75 g/kg ingestion. *2134*

Ethoxzolamide Probable effect (like acetazolamide). *0071*

Fenoterol Reduction of up to 0.9 mmol/L in normal subjects with inhalation of drug. *1437*

Fludrocortisone May cause increased urinary excretion. *0071*

Food Absorption delayed when taken with food. *3024*

Furosemide Diuretic action. *3505* Clinically important in about 3.6% patients. *1391* In 25% of 204 hospitalized patients receiving the drug. *3419* 12% patients with hypokalemia given 40-80 mg/d over long term. *0314* Observed after drug administration. *2047* In 12% patients with hypokalemia given 40-80 mg/d over long term. *0314*

Gentamicin If given with cephalexin to leukemics. *3922* Effect observed in 2 patients following larger doses of gentamicin. Associated with massive urinary loss of magnesium and potassium. *0224*

Glucocorticoids May cause potassium loss. *0071*

Glucose Shift intracellularly when infusion given. *3362*

Glycyrrhiza Aldosterone like action. *2458*

Hydrochlorothiazide Diuretic action. *3505* During 3 mo treatment in hypertensive patients with either 25 or 50 mg/d, but only significant statistically with 50 mg. *1980* Increase by 0.6 mmol/L in 343 patients with hypertension given drug for 10 weeks. *3707* Increase by 0.6 mmol/L in approximately 175 patients with essential hypertension treated for 1 y. *3708* 28% patients with hypokalemia given long term 25-100 mg/d. *0314*

Hydrocortisone Promotes urinary elimination. *0071*

Hydroflumethiazide Diuretic action. *1680*

Indapamide In 3 of 13 hypertensive patients with diabetes treated for 24 weeks. *2686* Increase by 0.4 mmol/L in 17 patients with essential hypertension treated with 2.5 mg daily for 3 mo. *2010* Mean fall from 4.2 to 3.4 mmol/L in 8 hypertensives treated for 16 weeks. *0352*

Insulin Therapeutic effect, causes intracellular shift. *1342*

Laxatives Excessive use may have effect. *2313*

Levodopa Reduction of up to 0.9 mmol/L in normal subjects with inhalation of drug. *1437*

Lithium Slight decreases observed only. *0530*

Mefruside Diuretic action less effective than thiazides. *3854*

Meralluride Diuretic action. *3505*

Mercurial Diuretics May induce hypokalemia in some cases. *1188*

Mercury Compounds Induced with chronic laxative ingestion. *2108*

Methazolamide With prolonged use (carbonic anhydrase inhibition). *2313*

Methyclothiazide Diuretic action. *1353*

Potassium *(continued)*

Serum Decrease Physiological *(continued)*

Methylprednisolone May cause potassium loss. *1680*

Metoclopramide At dose of 10 mg causes decrease of about 0.3 mmol/L. *2836*

Metolazone Potassium loss in urine. *1425* Significant reduction after 6 weeks treatment of 5 hypertensives given 5 mg daily. *3866* Mean decrease of 0.5-0.6 mmol/L in group of patients as hypertension controlled. *2494*

Mithramycin Depression of level reported. *0071*

Moxalactam Common when coadministered with amikacin. *0592*

Neomycin May occur with neomycin induced malabsorption. *1077*

Neuromuscular Relaxants If K deficiency potentiates effect. *1343*

Nifedipine Single case of drug induced hypokalemia reported. *3599*

Paramethasone May cause potassium loss. *1680*

Penicillin If i.v. sodium penicillin infused. *3505*

Penicillin G Nonreabsorbable anion in distal nephron promoting potassium loss especially with high doses. Effect quite marked in leukemia. *0645*

Phenolphthalein If chronic laxative abuse and aldosteronism. *1148*

Phosphates Drawn into cells with gluconeogenesis. *3879*

Polymyxin Reported to occur in leukemia, maybe steroid effect. *0958* Probably due to direct toxic effect of drug on renal tubular cells. *0645*

Polystyrene Sulfonate Exchanges for sodium (if Na form used). *2313*

Polythiazide Diuretic action. *1129*

Prednisolone Slight mineralocorticoid effect. *0071*

Prednisone Slight mineralocorticoid effect only. *0071*

Quinethazone Diuretic action. *2427*

Rimeterol Dose related significant change in 4 healthy men given therapeutic i.v. dose. *2807*

SC-16102 Duration of diuretic action short. *1804*

Sodium Bicarbonate Causes potassium to shift into cells. *0071*

Sodium Salts Significant effect with citrate salt not observed with chloride in 5 men with essential hypertension. *2037*

Sodium Sulfate May be excreted combined with sulfate. *3300*

Spironolactone Diuretic action (not marked). *3505*

Starvation Probably due to impaired homeostasis. *2522*

Streptozocin Associated with polyuria. *2532*

Sulfates May be excreted combined with sulfate. *2313*

Terbutaline Observed following infusion of 0.25 mg or 5 mg 3 times/d on first day of treatment. *0288*

Tetracycline Fanconi like syndrome may occur with degraded comp. *1488*

Theophylline Response to i.v. infusion, returned to baseline in 4 h. *3935*

Thiazides Consequence of diuresis. *1343*

Triamterene Diuretic action. *3505*

Trichlormethiazide Diuretic action. *3505*

Urea May cause severe depletion with diuresis. *1022*

Viomycin May occur with nephrotoxicity. *2313*

Urine No Effect Analytical

Storage of Sample No change unprocessed for 45 d at room temperature. *1180*

Thymol No effect on flame photometric methods. *2981*

Urine No Effect Physiological

Amiloride No effect seen in 13 men treated for 8 weeks. *0656*

Bopindolol In 10 hypertensive patients treated for either 1 or 21 d. *1063*

Dopamine Did not change with infusion of up to 3.0 μg/kg/min for 2 h in 6 normal men. *2149*

Hydrochlorothiazide In patients who were normokalemic after receiving drug for treatment for uncomplicated systemic hypertension. *2721*

Nifedipine No effect following single 20 mg sublingual dose. *0650*

Ranitidine No effect of 3-d of 150 mg/12 h. *3130*

Spironolactone No significant change in 5 volunteers over 7 d. *2440*

Urine Increase Physiological

Acetazolamide Diuretic action. *1343*

Aldosterone Conservation of Na with increased K loss. *3730*

Ammonium Chloride Diuretic action and hyperchloremic acidosis. *1343*

Antibiotics Some increase often with multiple regimes. *3549*

Aspirin Direct effect on renal tubules. *1343*

Azosemide Slight effect observed in normal male volunteers after 60 mg orally. *2658* Tended to increase but not marked. *2047*

Bed Rest Increased loss of up to 14 g with immobilization. *0863*

Betamethasone Occurs infrequently. *0182*

Bumetanide After 1 mg maximum within 2 h over by 4 h. *0822*

Calcitonin Acts independently of parathyroid. *1439*

Carbenicillin Possible increased distal tubular excretion in some patients. *3549*

Carbenoxolone Aldosterone like effect. *2458*

Cathartics Secondary aldosteronism if blood volume reduced. *1343*

Chlorthalidone May produce marked diuresis. *2427*

Citrates In 89 patients with hypocitraturic calcium nephrolithiasis treated for up to 4 y. *2708*

Clopamide Diuretic action. *2900*

Corticosteroids Promotes excretion (mineralocorticoid effect). *0071*

Corticotropin Mobilization of potassium from tissues. *1343*

Cortisone Promotes excretion. *0071*

Desoxycorticosterone Promotes increased urinary elimination. *0071*

Dexamethasone Metabolic effect. *1680*

Diapamide Diuretic action. *2256*

Diuretics Diuretic action. *2313*

Diurnal Variation Maximum around midday and afternoon. *2222*

Dopamine Directly inhibits tubular solute reabsorption. *1591*

EDTA Occurs particularly if given i.v. *1488*

Ethacrynic Acid Marked diuretic response may occur. *1488*

Fludrocortisone Increases urinary elimination. *0071*

Furosemide Diuretic action. *1617* Slight increase with administration. *2047*

Gentamicin Effect observed in 2 patients following larger doses of gentamicin. Associated with massive urinary loss of magnesium and potassium. *0224*

Glycyrrhiza Aldosterone like action. *2458*

Hydrochlorothiazide Significant effect in 5 h after 150 mg orally. *1189* In hypokalemic patients receiving drug over initial period of therapy: Increase of up to 41 mmol/d. *2721*

Hydrocortisone Increases elimination in urine. *0071*

Intra-Amniotic Saline Significant effect for 24 h. *3892*

Isosorbide Slight effect only if given alone. *2605*

Levodopa ?secondary to renal vasodilatation or direct action on tubules. *1136*

Mafenide Inhibits carbonic anhydrase if applied topically. *3813*

Mefruside Diuretic action, acts up to 20 h. *3854*

Methyclothiazide Small but significant effect in normal volunteers. *2052*

Metolazone Slight diuretic response. *3358*

Niacinamide 1 g causes marked excretion. *1622*

Parathyroid Extract Increased clearance. *1439*

Penicillin If i.v. sodium penicillin infused. *3505*

Prednisolone Slight mineralocorticoid effect. *0071*

Prednisone Slight mineralocorticoid effect only. *0071*

Quinethazone Relatively small increase. *1680*

SC-16102 Duration of diuretic action short. *1804*
Starvation Initially increase then falls to 10-15 meq/d. *3783*
Streptozocin In spite of low serum concentration. *2532*
Sulfates Increased excretion combines with sulfate. *0071*
Thiazides Diuretic action. *1343*
Torasemide Similar effects to that of furosemide. *0919*
Triamcinolone Only in exceedingly large doses. *0071*
Triflocin Diuretic action. *0033*
Viomycin Promotes urinary loss. *0071*
Xanthine Slight diuretic effect only. *1343*

Urine   Decrease   Physiological
Alanine Marked effect when given to fasting obese. *1263*
Amiloride May occur even if marked sodium loss. *1343*
Decreased significantly with start of treatment. *2440*   20 mg/d produces significant effect sustained for duration of treatment. *3101*
Anesthetic Agents Significant effect with general anesthesia. *0916*
Blindness sodium/potassium ratio higher than normal. *1637*
Carbamazepine Isolated case of dilutional hyponatremia with water intoxication: 7 previous cases reported. *1882*
Diazoxide Effect over 2 h of 4 mg/kg given i.v. *1805*
Epinephrine With increased filtration fraction. *1343*
Glucose After oral glucose and fall in serum concentration. *1623*
Growth Hormone Causes positive balance. *0138*
Levarterenol Increased tubular resistance and reabsorption. *1176*
Niacin With 1 g, no change with 2 g. *1622*
Prolactin Reduces renal excretion, not as marked as for sodium. *1662*
Ramipril In response to 10 mg and 20 mg on successive days in 9 patients with severe chronic congestive heart failure. *0764*
Sleep Reduced excretion compared with day. *1139*

Feces   Increase   Physiological
Neomycin Induced by steatorrhea. *1077*

Gastric Material   Increase   Physiological
Histamine Concentration and output increased after subcutaneous injection. *3252*

Red Blood Cells   No Effect   Physiological
Muzolimine Not appreciably affected in 10 hypertensive patients given 30 mg daily for 16 weeks. *1229*

Red Blood Cells   Increase   Physiological
Diurnal Variation Maximum concentration occurs in evening (7-8 pm). *2222*

Red Blood Cells   Decrease   Physiological
Chlorthalidone Not significant but tendency to fall in 10 hypertensive patients given 25 mg for 16 weeks. *3341*
Digoxin 6% drop within 2 d, affects membrane ATP-ase. *1923*
Race Significantly lower in blacks than caucasians. *2523*

Saliva   Increase   Physiological
Clonidine Mechanism not discussed. *0393*

Saliva   Decrease   Physiological
Pilocarpine Produces secretion like plasma ultrafiltrate. *1343*

White Blood Cells   No Effect   Physiological
Diuretics No change observed in normal individuals. *0997*

Potassium Clearance

Urine   Increase   Physiological
Amphotericin B Dose related inverse relationship to GFR. *0536*

Prazosin

Serum   Decrease   Physiological
Heparin Slight effect when high concentrations of nonesterified fatty acids occur due to heparin due to reduced protein binding. *3211*

Prealbumin

Serum   No Effect   Analytical
Storage of Sample No effect for several days at room temperature. *1180*

Serum   No Effect   Physiological
Anticonvulsants In 8 epileptic patients receiving phenobarbital with carbamazepine or phenytoin. *2670*
Halofenate No effect observed after 3 weeks. *2483*
Terbutaline In 6 normal men treated with therapeutic amounts for 2 weeks. *3177*
Tri-iodothyronine In 7 normal women aged 18-27 y given 25 μg/d three times daily for 14 d. Effect observed on endothelium-associated proteins but not on hepatically synthesized proteins. *1383*

Serum   Increase   Physiological
Anabolic Steroids Metabolic effect. *0227*
Norethandrolone Decreases thyroxine binding globulin however. *0042*
Oral Contraceptives Metabolic changes in liver synthesis. *2189*
Prednisone Occurs with decreased thyroxine binding globulin. *0042*
Propranolol In euthyroid patients due to inhibition of peripheral deiodination of thyroxine and on binding protein metabolism. *1174*

Serum   Decrease   Physiological
Amiodarone Approximately 20% reduction compared with controls when patients treated for 6 mo: probable direct effect on clearance or synthesis. *1173*
Estrogens Metabolic effect. *0227*
Pregnancy Depressed by up to 27%. *3472*

Precipitable Iodine

Serum   Increase   Analytical
Tetraiodophthalein Iodine contamination. *1237*

Prednisolone

Serum   Increase   Physiological
Cyclosporine Increase of area under curve and half-life with reduction of body clearance by 22%. *3732*

Serum   Decrease   Physiological
Phenytoin 77% increase in metabolic clearance rate with much reduced half-life. *1830*
Rifampin Significant increase in systemic clearance and plasma concentration and area under curve due to induction of liver enzymes. *2347*

Pregnancy-Associated Plasma Protein A

Serum   Increase   Analytical
EDTA Values higher than serum if used as anticoagulant and test uses crossed immunoelectrophoresis. *3610*
Heparin Values higher than serum when measured by crossed immunoelectrophoresis. *3610*

Pregnancy Protein

Serum   Increase   Physiological
Estrogens   Metabolic effect. *0227*

Pregnancy Tests

Urine   No Effect   Analytical
Boric Acid   Even in gelatin on Planotest, Pregnosticon®. *1075*
Chlordiazepoxide   No effect on UCG test. *1487*

Urine   Decrease   Physiological
Carbamazepine   False negative or inconclusive value with Prepurex, Predictor, Gonavislide, Pregnosticon®. *2182*

Urine   Positive   Analytical
Aspirin   Large dose effect on mouse, rabbit tests. *1487*
Boric Acid   If in gelatin affects Gravindex™. *1075*
Chlordiazepoxide   False positive with Gravindex™. *1487*
Chlorpromazine   Gives false positive with frog, rabbit and immunological test. *3505*
Fluphenazine   False reactions with phenothiazines. *1679*
Methadone   Highest incidence with Gravindex™. *1663*
Pentylenetetrazole   In 1 patient affected hCG and Pregslide tests. *1487*
Phenothiazines   False react with frog, rabbit, immunological tests. *2719*
Promethazine   False positive with Gravindex™. *3529*
Proteinuria   False positive tests if proteinuria present. *3505*
Thioridazine   With Prognosticon and other tests. *1487*

Urine   Positive   Physiological
Butaperazine   May occur with delayed menstruation and ovulation. *1680*
Chlorprothixene   Endocrine abnormality. *1680*
Piperacetazine   Associated with endocrine disorders. *0019*
Thiothixene   Observed with some phenothiazines. *1678*

Urine   Negative   Analytical
Promethazine   False negative with Prepuerin or Pap-test. *3529*

Pregnanediol

Urine   No Effect   Analytical
Ampicillin   No effect on GLC method of Vela. *3622*

Urine   Increase   Analytical
Phenazopyridine   Mechanism unknown. *2220*

Urine   Increase   Physiological
Aging   In women progressive increase to 20 y, slight decrease after. In males progressive increase to 20 y, slight decrease after. *1304*
Corticotropin   Hormonal effect. *2211*
Gonadotropin   Hormonal action. *2211*
Menstruation   Maximum midway between ovulation and menstruation. *1762*
Ovulation   Slight effect 2 d after ovulation. *0987*

Urine   Decrease   Physiological
Ampicillin   Increase by 53% after 5 d administration in pregnant women. During luteal phase in normal women. *3622*
Medroxyprogesterone   Induces anovulatory state. *3632*
Oral Contraceptives   Hormonal effect (may be absent excretion). *2211*
Phenothiazines   Reported effect. *2220*
Progesterone   Significant decrease with treatment (from 3.5 to 2.0 mg/24 h). *0991*

Pregnanetriol

Urine   Increase   Physiological
Aging   In women progressive increase to 20 y, slight decrease after. In men progressive increase to 20 y, slight decrease after. *1304*
Gonadotropin   Variable change when given to normal men. *3231*
Muscular Exercise   Effect most marked in well trained individuals. *0185*

Pregnenolone

Plasma   No Effect   Physiological
Heparin   No effect on components of aldosterone biosynthetic pathway. *2638*   No effect noted in spite of effect on aldosterone. *2638*

Plasma   Decrease   Physiological
Testosterone   In power athletes with 26 weeks on steroid self administration. *3088*

Urine   No Effect   Analytical
Ampicillin   No effect on GLC method of Vela. *3622*

Urine   Decrease   Physiological
Ampicillin   Increase by 44% after 6 d administration in pregnant women. *3622*

Prekallikrein

Plasma   Decrease   Physiological
Captopril   Rapid decrease following institution of therapy. *2748*

Primidone

Serum   No Effect   Analytical
Acetazolamide   On GLC procedure of Papadopoulos when added *in vitro*. *2722*
Amobarbital   On GLC procedure of Papadopoulos when added *in vitro*. *2722*
Beclamide   On GLC procedure of Papadopoulos at *in vivo* concentration. *2722*
Carbamazepine   On GLC procedure of Papadopoulos at *in vivo* concentration. *2722*
Chlordiazepoxide   On GLC procedure of Papadopoulos at *in vivo* concentration. *2722*
Chlorpromazine   On GLC procedure of Papadopoulos at *in vivo* concentration. *2722*
Cyproheptadine   On GLC procedure of Papadopoulos at *in vivo* concentration. *2722*
Diazepam   On GLC procedure of Papadopoulos at *in vivo* concentration. *2722*
Ethosuximide   On GLC procedure of Papadopoulos at *in vivo* concentration. *2722*
Ethotoin   On GLC procedure of Papadopoulos at *in vivo* concentration. *2722*
Haloperidol   On GLC procedure of Papadopoulos at *in vivo* concentration. *2722*
Imipramine   On GLC procedure of Papadopoulos at *in vivo* concentration. *2722*
Mephenytoin   On GLC procedure of Papadopoulos at *in vivo* concentration. *2722*
Methsuximide   On GLC procedure of Papadopoulos at *in vivo* concentration. *2722*
Methylphenobarbital   On GLC procedure of Papadopoulos when added *in vitro*. *2722*
Nitrazepam   On GLC procedure of Papadopoulos at *in vivo* concentration. *2722*
Oxytetracycline   On GLC procedure of Papadopoulos when added *in vitro*. *2722*

Secobarbital   On GLC procedure of Papadopoulos at *in vivo* concentration. *2722*

Sulthiame   On GLC procedure of Papadopoulos at *in vivo* concentration. *2722*

Thioridazine   On GLC procedure of Papadopoulos at *in vivo* concentration. *2722*

Trifluoperazine   On GLC procedure of Papadopoulos at *in vivo* concentration. *2722*

Serum   No Effect   Physiological

Phenothiazines   No significant effect on serum concentration. *1455*

Serum   Increase   Physiological

Valproic Acid   In patients taking primidone. *0504*

Probenecid

CSF   No Effect   Analytical

Phenylalanine   No effect on method of Korf, Van Praag. *2373*

Procainamide

Serum   No Effect   Physiological

Ranitidine   No effect on serum concentration or steady-state pharmacokinetics. *3022*

Serum   Increase   Physiological

Amiodarone   Approximate doubling of effect due to inhibition of metabolism. *2011*   In 11 of 12 treated patients, probably due to an effect on tissue binding or decrease of clearance. *3105*   Average 57% in 12 patients allowing 20% reduction in dosage. *1394*

Cimetidine   NAPA metabolite also increased not due to effect on hepatic blood flow or liver cytochrome enzymes; interaction occurs at renal tubular excretion level. *1588*   Area under curve increased by 43%, renal clearance decreased 36% and decreased ratio of clearance to bioavailability. *3022*   Decreases renal secretion adn that of N-acetylprocainamide. *3389*

Ranitidine   Significantly increased area under curve and reduced renal clearance. *1938*

Plasma   Increase   Physiological

Ranitidine   Renal clearance reduced without change in half-life. Increased area under curve for both procainamide and NAPA. Also slightly reduced gastrointestinal absorption: probably blocks tubular secretion of drug. *3388*

Progesterone

Plasma   No Effect   Analytical

Ampicillin   No effect on GLC method of Willman. *3622*

Corticosterone   1% or less cross reactivity with RIA. *0567*

Desoxycorticosterone   1% or less cross reactivity with RIA. *0567*

Hydrocortisone   1% or less cross reactivity with RIA. *0567*

Hydroxyprogesterone   1% or less cross reactivity with RIA. *0567*

Pregnenolone   1% or less cross reactivity with RIA. *0567*

Plasma   No Effect   Physiological

Cimetidine   No significant effect of drug during menstrual cycle. *0271*

Etomidate   No significant difference in 7 men given drug in induction dose versus thiopentone. *0063*

Famotidine   No effect of short-term or long-term treatment in male patients with duodenal ulcers. *0727*

Heparin   No effect on components of aldosterone biosynthetic pathway. *2638*

Metoclopramide   No significant baseline differences in cycling women before and after drug. *0096*

Prostaglandin F2α   In early luteal phase with i.v. in normals. *1600*

Ranitidine   After 4 weeks or 6 mo treatment (300 mg and 150 mg daily respectively) in male patients with duodenal ulcer. *0727*

Sexual Activity   No effect in women. *3437*

Sodium Bromide   No effect in 10 men, 10 women receiving 1 mg/kg/d during 8 weeks or 2 full menstrual cycles. *3141*

Sulpiride   In 11 healthy women between 6 and 9 weeks of pregnancy given 150 mg daily for 2 weeks. *3916*

Tamoxifen   After oral administration of 20 mg/d for 5 or 10 d to 6 healthy women during follicular phase of menstrual cycle. *1542*

Plasma   Increase   Analytical

Hydroxyprogesterone   46% cross reactivity with alpha compound, 16% with beta. *0567*

Oxoprogesterone   22% cross reactivity with RIA method of Cameron. *0567*

Pregnanedione   11-24% cross reactivity RIA procedure of Cameron. *0567*

Plasma   Increase   Physiological

Clomiphene   Increased over normal in luteal phase. *3005*

Ketoconazole   In men with prostatic cancer with 400 mg every 8 h and prolonged treatment. *3390*

Menstruation   Maximum during mid-point luteal phase. *0876*

Pregnancy   May increase by from 10 to 100 times. *2180*

Plasma   Decrease   Analytical

Blood Collection Tube   When gel-barrier tubes used to collect specimen: effect progressively increases with time. *1592*

Plasma   Decrease   Physiological

Ampicillin   Probably decreases synthesis. *3622*

Cyproterone   In 27 women treated for total of 194 cycles with 50 μg ethinyl estradiol and 2.0 mg cyproterone acetate. *2280*

Danazol   Often to nondetectable level in 62 patients with endometriosis treated with 600 mg daily for up to 24 weeks. *3222*

Ethinyl Estradiol   Luteal phase also shortened. *1800*

Heparin   No effect noted in spite of effect on aldosterone. *2638*

Oral Contraceptives   Less than 100 ng/dL throughout cycle. *0876*

Prostaglandin F2α   Marked with successful abortion. *3797*   In normals with i.v. in late luteal phase. *1600*

Testosterone   In power athletes with 26 weeks on steroid self administration. *3088*

Urine   Decrease   Physiological

Phenothiazines   Associated with other endocrinological changes. *2427*

Progestins

Urine   Decrease   Physiological

Chlorpromazine   Blocks ovulation, maintains decidual reaction. *1343*

Proinsulin

Plasma   No Effect   Physiological

Glyburide   No significant effect after 6 mo. *0965*

Prolactin

Plasma   No Effect   Analytical

Promethazine   No significant response to 25 mg i.m. *1416*

Plasma   No Effect   Physiological

Atropine   No effect on basal concentration in morning or evening but evening response to hypoglycemia significantly inhibited. *2565*

Benztropine   No significant response to 1.0 mg given i.m. *1416*

Caffeine   No significant effect of single dose of 250 mg in men or women. *3418*   In men no significant effect of single dose of 500 mg. In women no significant effect of single dose of 500 mg. *3418*

Prolactin *(continued)*

Plasma   No Effect   Physiological *(continued)*

Chloral Hydrate   No significant effect in response to up to 9 g/d. *1416*

Cimetidine   In 6 patients treated for 1 mo with 1 g/d. *3491*   In 2 studied men with duodenal ulcer or duodenitis. *2368*

Clonidine   In children with growth hormone deficiency 0.1 mg/m^2 daily for 60 d. *2821*   No effect seen in 12 healthy adults after single oral dose. *2070*   No effect observed on normal response to moderate or heavy exercise. *1797*

Clozapine   No response to 12.5 mg given orally. *1416*

Cyclosporine   In 16 patients who developed hypertrichosis but also given added cortisone. *3190*

Cyproheptadine   No effect on basal concentration or response to hypoglycemia in morning or evening. *2565*

Cyproterone   In 27 women treated for total of 194 cycles with 50 μg ethinyl estradiol and 2.0 mg cyproterone acetate 2.0 mg cyproterone acetate. *2280*

Desipramine   No significant effect with 50 mg orally. *1416*

Diazepam   No effect with 20 mg given i.m. *1416*

Diphenhydramine   No significant effect with 50 mg given i.m. *1416*

Ethanol   No effect in men or women at 100 mg/dL. *3611*

Ethinyl Estradiol   At the low doses (20-30 μg) in combined oral contraceptives. *0825*

Glucagon   On already elevated values in newborn. *1435*

Guanabenz   In 30 patients treated with twice daily doses of from 4 to 32 mg. *1018*

Hemodialysis   Elevated value not decreased by dialysis. *2547*

Hydrochlorothiazide   In 15 hypertensives treated with 50 mg twice daily. *1018*

Ibopamine   300 mg drug had no effect over period of 90 minutes. *3398*

Imipramine   Like other tricyclic antidepressants no effect. *0723*

Indomethacin   In up to 35 healthy men given 100 mg/d for 1 week in response to TRH stimulation. *2904*

Insulin   No significant effect with hypoglycemia. *1435*

Ketoconazole   In men with prostatic cancer with 400 mg every 8 h and prolonged treatment. *3390*

Lead   In a population of male battery workers compared with control group in cement industry. *0172*

Lithium   No significant effect with up to 8 g daily orally. *1416*

Naloxone   No significant effect with i.v. infusion. *0868*   Infusion did not affect concentration of hormone at concentrations varying from 0.02 to 0.5 mg/kg. *1970*   Basal and stimulated levels not affected. *0176*

Nifedipine   No effect of 80 mg/d for 2 d. *3694*

Omeprazole   No effect observed in 8 volunteers given 30 mg/d for 28 d. *2516*

Oral Contraceptives   No significant effect of meals or oral contraceptives. *3320*

Oxmetidine   No significant effect during treatment. *2578*

Prazosin   With 2 mg in 12 hypertensives (6 with normal and 6 with abnormal glucose tolerance). *0225*

Promazine   No significant change in response to 25 mg i.m. *1416*

Ranitidine   No difference in TRH-stimulated concentration compared with control in 10 ulcer patients or in prestimulation concentration. *1603*   No significant effect during treatment. *2578*

Sodium Bromide   No effect in 10 men, 10 women receiving 1 mg/kg/d during 8 weeks or 2 full menstrual cycles. *3141*

Plasma   Increase   Physiological

Amitriptyline   In depressive patients: on first day of administration in 6 of 11 patients: nonsignificant decrease after 28 d treatment. *1366*   Marked increase in men and women psychiatric patients treated for up to 4 weeks. *3630*

Amoxapine   Significant effect in both men and women possibly by blocking dopamine receptors. *0723*

Arginine   2 fold increase at 30 minutes with infusion. *1435*

Azosemide   Effect observed in normal male volunteers after 60 mg orally. *2658*

Benserazide   Significant increase in 5 patients with primary hypothyroidism; possibly due to reduction in circulating dopamine. *0869*

Butaperazine   Significant response to 5 mg given orally. *1416*

Butorphanol   Significant rise in 6 healthy male volunteers given 2 mg i.m. *3033*

Carbidopa   In 10 normal volunteers given 300 mg daily for 1 week. *0494*

Chlorpromazine   Marked increase in normals in 2 h. *3631*   Significant increase within 5 minutes of ingestion of 50 mg orally and 3 to 27 times baseline at 2 h. Functions as potent dopamine antagonist in the tuberoinfundibular system. Effect dose related. *3108*   Marked increase in male and female psychiatric patients treated for up to 4 weeks. *3630*   Normal response to intravenous TRH. *0588*

Cimetidine   Following either i.v. or oral drug, may be associated with larger doses of drug, possible effect on dopamine receptors in anterior pituitary, or on inhibition of uptake in peripheral tissues. *2368*   Sustained effect throughout luteal phase of menstrual cycle. *0271*   Significant increase in basal values observed. *2780*   In 5 of 11 male subjects with chronic duodenal ulcer given drug 1 g/d for 3 mo. *3759*

Clomipramine   In depressive patients: temporary increase during first day with lag after drug peak in 6 out of 11 patients. Significant effect after 28 d. *1366*

Danazol   Effect noted in 6 orchidectomized patients with prostatic cancer. *0693*

Desipramine   Acute response in group of depressed patients. *0560*

Diethylstilbestrol   Over 1 week in normal males. *3913*

Diurnal Variation   Maximum in women 1 am to 5 am, in men at 5 am. *2606*

Domperidone   Quick and marked effect when given i.m. to 12 normal subjects and to a group of patients with subclinical hypothyroidism. *0870*   After 10 mg intravenously in 8 healthy males. *3406*

β-Endorphin   Prompt 2-4 fold increase in 4 depressed psychiatric patients with i.v. infusion. *0610*

Enflurane   In both men and women when used to induce anesthesia with peak at 30 minutes. *3456*

Estrogens   Chronically in men with carcinoma of prostate. *3913*

Ethinyl Estradiol   Increase by 1350% in women if on perphenazine. *0522*   In hypogonadal women within 1 week of therapy. *3913*

Flunarizine   Significant increase of baseline value in men and women. *2265*

Fluphenazine   Typical dose-related response to i.m. administered drug due to antidopaminergic action. *2072*   Marked increase in male and female psychiatric patients treated for up to 4 weeks. *3630*

Food   Abrupt increase within 45 minutes of starting a meal. *2895*

Furosemide   Effect observed after 40 mg orally in normal male volunteers. *2658*

Haloperidol   Repeated administration of 1 mg i.m. to 19 normal men produced Reproducible dose-response curve. *2072*

Imipramine   Marked increase in male and female psychiatric patients treated for up to 4 weeks. *3630*

Insulin   In post vagotomy pts marked if i.v. or single injection. *3859*   Marked effect in insulin induced hypoglycemia. *3570*

Labetalol   Marked in women, less so in men: possibly due to antidopaminergic effect of drug in response to i.v. infusion of 100 mg over 10 minutes. *0226*

Lactation   Large response ten to forty days postpartum. *3647*

Loxapine   Significant change in response to 10 mg orally. *1416*

Megestrol   Affects both basal and TRH-stimulated concentration in 18 postmenopausal women with breast cancer. *0056*

Menstruation   Significant higher in luteal than in follicular phase. *1435*

Mestranol   In 31 of 88 oophorectomized women treated for 3 to 11 y. *0230*

Methyldopa Increased concentration after single doses of drug during long-term treatment. *0166* After single doses of 750 or 1,000 mg peak reached 4 to 6 h after administration. With long term administration 3 to 4 fold increase over normal noted. *3453* Marked effect in male and female hypertensives treated for up to 6 weeks. *3630*

Metoclopramide Significant increase in some menstruating women compared with control cycle. *0096* Observed in normal subjects after 10 mg drug over 2 h; higher response in patients with primary aldosteronism. *3883* 11 fold increase after i.v. injection (0.04 mg/kg). *0497* Increased by single i.v. bolus of 10 mg: probably related to basal activity of renin angiotensin aldosterone system. *2543* Response smaller in thyrotoxic than euthyroid patients with long term treatment. *3520* In both hyperthyroid and euthyroid subjects in response to oral drug. *3163*

Molindone Significant response to 5 mg given orally. *1416*

Morphine In 14 volunteer subjects given 5 mg intravenously. *3953* 10 mg i.v. produced prompt and significant increase in 7 hypothyroid and 5 healthy individuals. *0891*

Oral Contraceptives Mean change from 8.9 ng/mL to 10.2 ng/mL at 3 mo in 120 women with Low estrogen pills. *1710* In 30% women to varying extent, not correlated with dose of estrogen or duration of treatment. *2965*

Parathyroid Hormone From 5.1 ng/mL to 14.9 ng/mL on average in 10 normal subjects after infusion of 150 units. *0467*

Pentagastrin Increased in 63% of 9 women: mean concentration increased from 9.67 ng/mL to 16.08 ng/mL in 5 to 10 minutes. *3648*

Perphenazine By approximately 800% in women after 8 mg. *0522* Typical dose-related response to i.m. administered drug due to antidopaminergic action. *2072* Marked effect in male and female psychiatric patients treated for up to 4 weeks. *3630*

Phenytoin Mean 312 mUnits/L versus 207 in controls in approximately 24 male patients given phenytoin alone or with primidone or phenobarbital. *3609*

Pimozide General effect observed. *1736* In acutely psychotic males. *3340*

Pregnancy Rises throughout gestation. *3647*

Prochlorperazine Typical dose-related response to i.m. administered drug due to antidopaminergic actions. *2072*

Promazine Marked effect in male and female psychiatric patients treated for up to 4 weeks. *3630*

Ranitidine No effect up to 100 mg i.v., but mean increase of 1.3 ng/mL in 5 subjects. *0871* Significant increase in basal value. *2780*

Reserpine Dose-related at doses greater than 0.25 mg/d. *0451* Marked effect in male and female hypertensives treated for up to 6 weeks. *3630*

Sex Difference In woman mean 7.9 ng/mL, in men 5.2 ng/ml. *1678*

Sexual Activity 8 fold increase in 2 of 6 women (?breast stimulated). *3437*

Sleep Peak between 5 and 7 am, falls with rising. *3217*

Sulpiride General effect observed. *1736* After 1 and 2 weeks in 11 healthy women between 6 and 9 weeks of pregnancy given 150 mg daily for 2 weeks. *3916*

Sultopride After 1 d, maximum at 1 week increase throughout treatment, probably by blocking pituitary dopamine receptors. *2465*

Thiethylperazine Significant response to drug given i.m. or orally. *1416*

Thioridazine Significant response to drug given orally (50 mg). *1416* Significant increase within 45 minutes of ingestion of 50 mg orally and 8 to 19 times baseline at 2 h. Functions as potent dopamine antagonist in the tuberoinfundibular system. Effect dose related. *3108*

Thiothixene Typical dose-related response to i.m. administered drug due to antidopaminergic action. *2072*

Thyrotropin-Releasing Hormone (TRH) Response within 5 minutes to i.v. injection. *1753* Maximum effect 15-30 minutes after i.v. *1435*

Trifluoperazine Typical response to i.m. administered drug: effect dose related. *2072*

Verapamil In 7 chronically ill schizophrenic patients when administered for 5 weeks. *2811*

Zimeldine Response only after chronic pretreatment in group of depressed patients. *0560*

Plasma Decrease Physiological

Aging Maximum in premature, falls to 5 ng/mL ages 2-12 y. *1435*

Anticonvulsants In 33 male epileptics taking at least one drug for long time. *3020*

Apomorphine In control subjects reduced by 57% when given i.v. *3178*

Bombesin After TRH after 5 ng/nk/min x 2.5 h in healthy men. *2841*

Bromocriptine 70% decrease in healthy women and those with prolactin secreting tumors when given 5 mg/d for 5 d. *2283*

Calcitonin In 9 healthy subjects and 4 patients with hyperprolactinemia after i.v. infusion of salmon preparation. *1729*

Dexamethasone Normal response not always observed in psychiatric patients. *2404*

Dopamine Effect noted with infusion of as little as 0.03 μg/kg/min in 6 normal men for 2 h. *2149*

Levodopa Fall to 8.7% of baseline after 2 h in normals. *0521* Transient effect in nonpuerperal galactorrhea. *2274* Completely suppressed for 1-4 h after 250 mg orally. *1435*

Pergolide Single dose reduced concentration for 24 h in normal subjects, multiple doses reduced concentration by 80%. *2127* Marked reduction in patients with Parkinson's disease to low or undetectable levels. *1956*

Ranitidine Compared with control state after 3 d 150 mg/12 h lower during ambulation and at rest. *3130*

Rifampin For at least 12 h in normals. *0865*

Smoking Significant reduction in women who smoked compared with those who did not. *2609* Mean value less in smokers than nonsmokers but no acute effect of smoking. *1360*

Tamoxifen Significant effect in 6 healthy volunteers given 20 mg/d for 5 to 10 d. *1542*

Valproic Acid Significant effect in women 30-180 minutes after 400 mg orally. *2400* Significant effect in both normal women and in hyperprolactinemic women. *2400*

Water Load Fall in 2 h to 6.9% of baseline in normals. *0521*

Prolactin Response to TRH

Plasma No Effect Physiological

Aspirin In up to 35 healthy volunteers given 3.6 g daily for 1 week. *2904*

Proline

Plasma No Effect Analytical

Cystine No effect on method of Goodwin. *1345*

Histidine No effect on method of Goodwin. *0117*

Lysine No effect on method of Goodwin. *1345*

Phenylalanine No effect on method of Goodwin. *1345*

Plasma No Effect Physiological

Tranylcypromine No effect observed. *0824*

Plasma Increase Analytical

Arginine Slight effect on method of Goodwin. *1345*

Glycylproline Equivalent effect on method of Goodwin. *1345*

Hydroxyproline Slight effect on method of Goodwin. *1345*

Ornithine Slight effect on method of Goodwin. *0117*

Plasma Increase Physiological

Levodopa In Parkinson patients significant effect after 2 weeks. *1496*

Plasma Decrease Physiological

Oral Contraceptives Hormonal effect (second part of cycle). *0742*

Proline Hydroxylase

Serum   Increase   Physiological
Methoxyflurane   Marked elevation observed due to liver cell injury. *3447*

Promyeloblasts

Blood   Increase   Physiological
Amphetamine   Myeloblastic leukemia. *0325*

Propafenone

Plasma   Increase   Physiological
Amiodarone   Increased concentration with probable adverse effects on EKG. *2298*
Food   Maximum plasma drug concentration higher and reached earlier. *0186*

Propantheline

Serum   Decrease   Physiological
Food   Absorption reduced when taken with food. *3024*

Properdin

Serum   Decrease   Physiological
X-Ray Therapy   Due to tissue destruction. *0987*

Propionic Acid

Serum   Increase   Analytical
Methylmalonic Acid   May be formed from methylmalonic acid in GLC. *0975*

Serum   Increase   Physiological
Valproic Acid   Associated with inhibition of carbamyl phosphate synthetase I and interference with mitochondrial glycine transport. *0736*

Urine   Increase   Analytical
Methylmalonic Acid   May be formed from methylmalonic acid in GLC. *0975*

Urine   Increase   Physiological
Bacteriuria   In experimental studies 8 of 11 species produced acid. *1483*

Propoxyphene

Serum   No Effect   Analytical
Amitriptyline   At 10 mg/L on method of Evenson. *1056*
Amobarbital   At 25 mg/L on method of Evenson. *1056*
Amphetamine   At 25 mg/L on method of Evenson. *1056*
Aspirin   At 25 mg/L on method of Evenson. *1056*
Chlordiazepoxide   At 50 mg/L on method of Evenson. *1056*
Diazepam   At 25 mg/L on method of Evenson. *1056*
Doxepin   At 10 mg/L on method of Evenson. *1056*
Ethchlorvynol   At 5 mg/L on method of Evenson. *1056*
Glutethimide   At 25 mg/L on method of Evenson. *1056*
Meprobamate   At 25 mg/L on method of Evenson. *1056*
Methamphetamine   At 40 mg/L on method of Evenson. *1056*
Methyprylon   At 25 mg/L on method of Evenson. *1056*
Oxazepam   At 140 mg/L on method of Evenson. *1056*
Pentobarbital   At 25 mg/L on method of Evenson. *1056*
Phenobarbital   At 25 mg/L on method of Evenson. *1056*
Phenytoin   At 25 mg/L on method of Evenson. *1056*
Primidone   At 25 mg/L on method of Evenson. *1056*
Secobarbital   At 25 mg/L on method of Evenson. *1056*

Serum   Increase   Physiological
Propoxyphene   0.2 mg/L after 200 mg orally, 0.3 mg/L after 50 mg i.v. *2348*

Urine   No Effect   Analytical
Betazole   Less than 1% fluorescence in procedure of Valentour. *3662*
Codeine   Less than 1% fluorescence in procedure of Valentour. *3662*
Desipramine   Less than 1% fluorescence in procedure of Valentour. *3662*
Epinephrine   Less than 1% fluorescence in procedure of Valentour. *3662*
Isocarboxamide   Less than 1% fluorescence in procedure of Valentour. *3662*
Levarterenol   Less than 1% fluorescence in procedure of Valentour. *3662*
Methadone   Less than 1% fluorescence in procedure of Valentour. *3662*
Methoxamine   Less than 1% fluorescence in procedure of Valentour. *3662*
Morphine   Less than 1% fluorescence in procedure of Valentour. *3662*
Naphazoline   Less than 1% fluorescence in procedure of Valentour. *3662*
Phenylpropanolamine   Less than 1% fluorescence in procedure of Valentour. *3662*
Propylhexedrine   Less than 1% fluorescence in procedure of Valentour. *3662*
Thioridazine   Less than 1% fluorescence in procedure of Valentour. *3662*

Urine   Increase   Analytical
Amphetamine   1% fluorescence in procedure of Valentour. *3662*
Butethamine   1% fluorescence in procedure of Valentour. *3662*
Ephedrine   1% fluorescence in procedure of Valentour. *3662*
Methamphetamine   3% fluorescence in procedure of Valentour. *3662*
Methoxyphenamine   1% fluorescence in procedure of Valentour. *3662*
Norpropoxyphene   5% fluorescence in procedure of Valentour. *3662*
Phenylephrine   1% fluorescence in procedure of Valentour. *3662*

Propranolol

Serum   No Effect   Physiological
Enprostil   No effect on propranolol elimination or hepatic metabolism. *2942*
Ranitidine   No significant effect on plasma concentration, area under curve or elimination half-life. *1938*

Serum   Increase   Physiological
Cimetidine   Reduces oral clearance by as much as 50%. *2942*
Etintidine   Significantly increased area under curve and prolonged elimination half-life: also protracted elimination of 4-hydroxypropranolol, an active metabolite. *1676*
Flecainide   Increases hypotensive and negative inotropic effects as well as concentration. *2011*

Serum   Decrease   Analytical
Blood Collection Tube   Spuriously low values when specimen drawn into certain Vacutainers™. *0734*

Serum   Decrease   Physiological
Cholestyramine   Significant reduction of peak plasma concentration and area under curve due to binding in gastrointestinal tract. *1584*

Colestipol Significant reduction of peak concentration and area under curve due to binding in gastrointestinal tract. *1584*

Propylthiouracil

Serum   No Effect   Analytical
Hemolysis No effect with 2,6-DQC (procedure of Ratliff). *2922*
Lipemia No effect with 2,6-DQC (procedure of Ratliff). *2922*

Serum   Increase   Analytical
Propylthiouracil Reacts with 2,6-DQC (procedure of Ratliff). *2922*
Thiourea Reacts with 2,6-DQC (procedure of Ratliff). *2922*

Serum   Decrease   Analytical
Aspirin Produces negative interference (procedure of Ratliff). *2922*

Proquazone

Serum   Decrease   Physiological
Antacids Delayed absorption noted. *3794*

Prostaglandin, 6-Keto-F$_1\alpha$

Plasma   No Effect   Physiological
Prednisone No significant change with up to 25 mg drug for up to 14 d: no effect on production during clotting. *2890*

Plasma   Decrease   Physiological
Cimetidine Probable activation of endogenous cyclic AMP phosphodiesterase involved in favoring action of platelet aggregating agents. *3938*

Urine   No Effect   Physiological
Cyclosporine No difference in excretion between azathioprine and cyclosporine renal- transplant patients(6-keto). *0223*
Sulindac No significant change in patients with chronic glomerular disease. *0660*

Urine   Decrease   Physiological
Etodolac 6-keto- less than 30% in people with normal renal function but about 60% in in individuals with renal insufficiency. *0453*
Ibuprofen Reduced by 80% in patients with chronic glomerular disease and in healthy people. *0660*
Indomethacin In 9 uncomplicated essential hypertensives receiving chronic enalapril treatment. *3126*
Naproxen Increase by 76% in 10 furosemide treated patients with well controlled congestive heart heart failure. *1042*
Smoking In heavy smokers following smoking of 4 cigarettes, but no change in nonsmokers. *2541*

Prostaglandin A

Plasma   Increase   Physiological
Sodium Restriction Increase by 36% to 2.14 ng/mL in normals. *3966*

Plasma   Decrease   Physiological
Sodium Loading 49% to 0.82 ng/mL in normals. *3966*

Prostaglandin E

Plasma   No Effect   Physiological
Sodium Loading No significant effect with variation in intake. *3966*
Sodium Restriction No significant effect with variation in intake. *3966*

Plasma   Decrease   Physiological
Aspirin In up to 35 healthy volunteers given 3.6 g daily for 1 week. *2904*
Indomethacin Decrease by 67% in rats at least. *1760* In up to 35 healthy men given 100 mg/d for 1 week. *2904*

Urine   Increase   Physiological
Angiotensin 2.3 fold increase after 3 h in women. *1211*

Urine   Decrease   Physiological
Aspirin Clearance-reduced increased: clearance with sodium restriction. *2535*

Semen   Decrease   Physiological
Naproxen Significant effect but returned to prior level 1 week after treatment stopped. *0285*

Prostaglandin E1

Plasma   Increase   Physiological
Food Probably due to ingestion of fats. *1385*
Muscular Exercise Increase of 340% (mean after exercise). *1385*

Plasma   Decrease   Physiological
Fasting Mean decrease of 4.5 ng/mL compared with postprandial. *1385*
Tourniquet Mean decrease of 7.1 ng/mL after 5 minutes. *1385*

Prostaglandin E2

Plasma   Increase   Physiological
Diltiazem Increase by 63 pg/mL in 20 patients with essential hypertension with 14 weeks treatment. *3521*

Urine   No Effect   Physiological
Cyclosporine No difference in excretion between azathioprine and cyclosporine renal- transplant patients. *0223*
Naproxen In 11 volunteers in randomized trial. *0452*
Sulindac No significant change in patients with chronic glomerular disease. *0660*

Urine   Decrease   Physiological
Etodolac About 40% reduction in people with normal renal function but 70% in people with renal insufficiency. *0453*
Ibuprofen Reduced by 80% in patients with chronic glomerular disease and in healthy people. *0660*
Indomethacin Marked reversal on withdrawal of drug in one young woman. *3541*
Naproxen When given for 14 d days to patients with rheumatoid arthritis and heart failure. Marked effect. *3514*
Sulindac When given for 14 d to patients with rheumatoid arthritis and heart failure. Marked effect. *3514* In 11 normal volunteers contrary to concept of sparing renal but inhibiting systemic prostaglandins. *0452*

Gastric Material   Increase   Physiological
Smoking In 7 volunteers after 3 cigarettes: reduced from mean 22.8 ng/15 minutes to 12.2 ng/15 minutes. *2352*

Prostaglandin E, 19-Hydroxy

Semen   Decrease   Physiological
Naproxen Significant effect but returned to prior level 1 week after treatment stopped. *0285*

Prostaglandin F

Plasma   No Effect   Physiological
Sodium Loading No significant effect with variation in intake. *3966*

Prostaglandin F (continued)

Plasma   No Effect   Physiological (continued)
Sodium Restriction No significant effect with variation in intake. *3966*

Plasma   Decrease   Physiological
Aspirin In up to 35 healthy volunteers given 3.6 g daily for 1 week. *2904*
Indomethacin In up to 35 healthy men given 100 mg/d for 1 week. *2904*

Urine   No Effect   Physiological
Angiotensin No effect observed after i.v. infusion. *1211*

Semen   Decrease   Physiological
Naproxen Significant effect but returned to prior level 1 week after treatment stopped. *0285*

Prostaglandin $F_{1\alpha}$

Urine   No Effect   Physiological
Sulindac In 10 furosemide treated patients with well controlled congestive heart failure. *1042*

Prostaglandin $F2\alpha$

Plasma   Decrease   Physiological
Pregnancy Lowest values in 2nd and 3rd trimester. *0507*

Urine   Decrease   Physiological
Naproxen When given for 14 d to patients with rheumatoid arthritis and heart failure. Marked effect. *3514*
Sulindac When given for 14 d to patients with rheumatoid arthritis and heart failure. Marked effect. *3514*

Prostaglandin F, 19-Hydroxy

Semen   Decrease   Physiological
Naproxen Significant effect but returned to prior level 1 week after treatment stopped. *0285*

Prostaglandin G2

Urine   Decrease   Physiological
Ibuprofen In 11 healthy volunteers in randomized trial. *0452*

Protein

Serum   No Effect   Analytical
Acetaminophen At 5 times therapeutic concentration on biuret procedures on SMAC®, Abbott-VP, Ektachem® 400, Hitachi® 705 and KDA. *2138* At acute overdose concentration (10 mg/dL) on SMAC® method. *3719* At concentration of 1500 mg/L had no effect on biuret method with blank correction. *3393*

Acetazolamide No effect at 12 mg/dL on SMA 12/60 method. *2636* At concentration of 1,000 mg/L had no effect on biuret method with blank correction. *3393*

Acetylsalicylic Acid At therapeutic concentrations on biuret and spectrophotometric methods. *1786* On biuret method on SMA II at physiological concentration. *1787* At 5 times upper limit of therapeutic range on methods on SMAC®, Abbott-VP, Hitachi® 705 and KDA. *2138* At concentration of 2,000 mg/L had no effect on biuret method with blank correction. *3393*

Allopurinol At concentration of 136 mg/L had no effect on biuret method with blank correction. *3393*

Aminophenazone No effect on biuret and spectrophotometric methods at therapeutic concentration. *1786*

Aminophenol At concentration of 109 mg/L had no effect on biuret method with blank correction. *3393*

Aminopyrine At concentration of 125 mg/L had no effect on biuret method with blank correction. *3393*

Aminosalicylic Acid At 5 times upper limit of therapeutic range on methods on SMAC®, Abbott-VP, Ektachem®, Hitachi® 705 and KDA. *2138* At concentration of 460 mg/L had no effect on biuret method with blank correction. *3393*

Amitriptyline At acute overdose concentration (2.5 mg/dL) on SMAC® method. *3719* At concentration of 25 mg/L had no effect on biuret method with blank correction. *3393*

Amobarbital At concentration of 150 mg/L had no effect on biuret method with blank correction. *3393*

Amphotericin B At concentration of 96 mg/L had no effect on biuret method with blank correction. *3393*

Ascorbic Acid At concentration of 2,000 mg/L had no effect as measured by biuret method with blank correction. *3393*

Barbital At acute overdose concentration (20 mg/dL) on Technicon® SMAC® method. *3719* At concentration of 500 mg/L had no effect as measured by biuret method with blank correction. *3393*

Bethanechol No effect at 0.09 mg/dL on SMA 12/60 method. *2636*

Bilirubin On ACA using dual wavelength biuret procedure. *2782*

Boric Acid At concentration of 50 mg/L had no effect as measured by method using biuret with blank correction. *3393*

Bromazepam At 5 times therapeutic concentration on methods of SMAC®, Abbott-VP, Ektachem® 400, Hitachi® 705 and KDA. *2138*

Butabarbital At concentration of 100 mg/L had no effect on method using biuret with blank correction. *3393*

Caffeine No effect on biuret and spectrophotometric methods at therapeutic concentrations. *1786* At acute overdose concentration (10 mg/dL) on Technicon® SMAC® method. *3719* At concentration of 194 mg/L had no effect on biuret method with blank correction. *3393*

Carbenicillin At 5 times upper limit of therapeutic range on biuret method on KDA. *2138* At concentrations above 500 mg/L raised the concentration as measured by biuret method with blank correction. *3393*

Cefamandole At 2.50 mmol/L on method on Eppendorf Epos. *1414*

Cefazedone On method on Eppendorf Epos at concentration of 2.50 mmol/L. *1414*

Cefazolin At 2.50 mmol/L on method on Eppendorf Epos. *1414* At concentration of 117 mg/L had no effect on biuret method with blank correction. *3393*

Cefodizime No effect at 2.50 mmol/L on method on Eppendorf Epos. *1414*

Cefoperazone At 2.50 mmol/L on method on Eppendorf Epos. *1414*

Cefotaxime No effect at 2.50 mmol/L on method on Eppendorf Epos. *1414*

Cefotiam At 2.50 mmol/L on method on Eppendorf Epos. *1414*

Cefoxitin At 5 times upper limit of therapeutic range on methods on SMAC®, Abbott-VP, Ektachem®, Hitachi® 705 and KDA. *2138* At concentrations up to 2.50 mmol/L on methods on Eppendorf Epos. *1414*

Cefpirome At concentration of up to 2.50 mmol/L on methods of Eppendorf Epos. *1414*

Ceftriaxone On method on Eppendorf Epos at concentration of 2.50 mmol/L. *1414*

Cefuroxime On method on Eppendorf Epos at concentration of 2.50 mmol/L. *1414*

Cephaloridine At 2.50 mmol/L on method on Eppendorf Epos. *1414*

Cephalothin At 5 times upper limit of therapeutic range on methods on SMAC®, Abbott-VP, Ektachem®, Hitachi® 705 and KDA. *2138* No effect a 2.50 mmol/L on method on Eppendorf Epos. *1414*

Chlordiazepoxide At 5 times upper limit of therapeutic range on methods on SMAC®, Abbott-VP, Ektachem®, Hitachi® 705 and KDA Hitachi® 705 and KDA. *2138* At acute overdose concentration (5 mg/dL) on Technicon® SMAC® method. *3719* At concentration of 50 mg/L had no effect on biuret method with blank correction. *3393*

Chlormezanone At concentration of 100 mg/L had no effect on biuret method with blank correction. *3393*

Chlorphenesin At concentration of 150 mg/L had no effect on biuret method with blank correction. *3393*

Chlorpheniramine At acute overdose concentration (20 mg/dL) on Technicon® SMAC® method. *3719* At concentration of 200 mg/L had no effect on biuret method with blank correction. *3393*

Chlorpromazine At concentration of 3 mg/L had no effect on biuret method with blank correction. *3393*

Chlorpropamide At concentration of 150 mg/L had no effect on biuret method with blank correction. *3393*

Chlorprothixene At concentration of 1 mg/L had no effect on biuret method with blank correction. *3393*

Cimetidine At concentration of 1 mg/L had no effect on biuret method with blank correction. *3393*

Cloxacillin At concentration of 5 mg/L had no effect on biuret method with blank correction. *3393*

Cocaine At acute overdose concentration (2.5 mg/dL) on Technicon® SMAC® method. *3719* At concentration of 25 mg/L had no effect on biuret method with blank correction. *3393*

Codeine At acute overdose concentration (2.0 mg/dL) on Technicon® SMAC® method. *3719* At concentration of 20 mg/L had no effect on biuret method with blank correction. *3393*

Contact With Clot If plastic disc used to separate for 48 h. *1863*

Cyclosporine At concentration of 41.5 mmol/L (50 μg/L) on methods on Technicon® SMAC® II and Hitachi® 705. *1123*

Cysteine At 5 times upper limit of therapeutic range on methods on SMAC®, Abbott-VP, Ektachem®, Hitachi® 705 and KDA. *2138*

Cystine At 5 times upper limit of therapeutic range on methods on SMAC®, Abbott-VP, Ektachem®, Hitachi® 705 and KDA. *2138*

Desipramine At concentration of 8 mg/L had no effect on biuret method with blank correction. *3393*

Deslanoside No effect on SMA 12/60 method at 0.06 mg/dL. *2636*

Dextran If dextranase used to remove. *2567* No effect on modified biuret method of Moore. *2477*

Dextran 40 Since 1987 no interference up to 120 g/L on American Monitor Parallel method. *2651* At concentrations up to 30 g/L no effect on Roche Cobas-Bio, Ektachem® 400 and Beckman Astra 8. *0231*

Diazepam At 5 times upper limit of therapeutic range on methods on SMAC®, Abbott-VP, Ektachem®, Hitachi® 705 and KDA. *2138* At acute overdose concentration (2.5 mg/dL) on Technicon® SMAC® method. *3719* At concentration of 25 mg/L had no effect on biuret method with blank correction. *3393*

Diclofenac On biuret method on SMA II at physiological concentration. *1787* At concentration of 23 mg/L had no effect on biuret method with blank correction. *3393* No effect at therapeutic concentrations on Biuret and Spectrophotometric methods. *1786*

Diethylaminoethanol At concentration of 30 mg/L had no effect on biuret method with blank correction. *3393*

Diethylpropion At acute overdose concentration (10 mg/dL) on Technicon® SMAC® method. *3719* At concentration of 100 mg/L had no effect on biuret method with blank correction. *3393*

Digitoxin No effect at 21 mg/dL on SMA 12/60 method. *2636* At concentration of 6 mg/L had no effect on biuret method with blank correction. *3393*

Digoxin No effect at 0.04 mg/dL on SMA 12/60 method. *2636* At concentration of 9 mg/L had no effect on biuret method with blank correction. *3393*

Dimethadione At concentration of 2,000 mg/L had no effect on biuret method with blank correction. *3393*

N-Dimethyldiazepam At concentration of 6 mg/L had no effect on biuret method with blank correction. *3393*

Diphenhydramine At acute overdose concentration (20 mg/dL) on Technicon® SMAC® method. *3719* At concentration of 200 mg/L had no effect on biuret method with blank correction. *3393*

Disopyramide At concentration of 4 mg/L had no effect on biuret method with blank correction. *3393*

Disulfiram At concentration of 120 mg/L had no effect on biuret method with blank correction. *3393*

Epinephrine At concentration of 183 mg/L had no effect on biuret method with blank correction. *3393*

Erythromycin At concentration of 73 mg/L had no effect on biuret method with blank correction. *3393*

Ethacrynic Acid At 2.5 mg/dL no effect on SMA 12/60 method. *2636*

Ethanol At acute overdose concentration (20 mg/dL) on Technicon® SMAC® methods. *3719*

Ethchlorvynol At acute overdose concentration (20 mg/dL) on Technicon® SMAC® method. *3719* At concentration of 400 mg/L had no effect on biuret method with blank correction. *3393*

Ethinamate At acute overdose concentration (20 mg/dL) on Technicon® SMAC® method. *3719* At concentration of 200 mg/L had no effect on biuret method with blank correction. *3393*

Ethosuximide At concentration of 390 mg/L had no effect on biuret method with blank correction. *3393*

Fluorescein No effect at 200 mg/L on Ektachem® 700 method. *1722*

Fluorouracil At concentration of 130 mg/L had no effect on biuret method with blank correction. *3393*

Fluosol-DA On SMA IIC at concentration of 50%. *2518*

Flurazepam At 5 times upper limit of therapeutic range on methods on SMAC®, Abbott-VP, Ektachem®, Hitachi® 705 and KDA. *2138* At acute overdose concentration (2.5 mg/dL) on Technicon® SMAC® method. *3719* At concentration of 25 mg/L had no effect on biuret method with blank correction. *3393*

Furosemide No effect at 1.4 mg/dL on SMA 12/60 method. *2636* At concentration of 4 mg/L had no effect on biuret method with blank correction. *3393*

Gentamicin At concentration of 150 mg/L had no effect on biuret method with blank correction. *3393*

Glutethimide At acute overdose concentration (5 mg/dL) on Technicon® SMAC® method. *3719* At concentration of 50 mg/L had no effect on biuret method with blank correction. *3393*

Hydralazine At concentration of 6 mg/L had no effect on biuret method with blank correction. *3393*

Ibuprofen No effect at therapeutic concentrations on biuret and spectrophotometric methods. *1786* On biuret method on SMA II at physiological concentration. *1787* At 5 times upper limit of therapeutic range on methods on SMAC®, Abbott-VP, Ektachem®, Hitachi® 705 and KDA. *2138* At concentration of 200 mg/L had no effect on biuret method with blank correction. *3393*

Indomethacin No effect at therapeutic concentrations on biuret and spectrophotometric methods. *1786* On biuret method on SMA II at physiological concentration. *1787* At concentration of 13 mg/L had no effect on biuret method with blank correction. *3393*

Insulin At concentration of 3 mg/L had no effect on biuret method with blank correction. *3393*

Iproniazid At concentration of 40 mg/L had no effect on biuret method with blank correction. *3393*

Isoniazid At concentration of 100 mg/L had no effect on biuret method with blank correction. *3393*

Isoprenaline At concentration of 211 mg/L had no effect on biuret method with blank correction. *3393*

Ketoprofen No effect at therapeutic concentrations on biuret and spectrophotometric methods. *1786* On biuret method on SMA II at physiological concentration. *1787* At concentration of 60 mg/L had no effect on biuret method with blank correction. *3393*

Levodopa At concentration of 197 mg/L had no effect on biuret method with blank correction. *3393*

Lidocaine No effect SMA 12/60 method with 3.5 mg/dL. *2636*

Lorazepam At concentration of 0.05 mg/L had no effect on biuret method with blank correction. *3393*

Mannitol No effect on SMA 12/60 method at 445 mg/dL. *2636*

Meperidine At acute overdose concentration (5 mg/dL) on Technicon® SMAC® method. *3719* At concentration of 60 mg/L had no effect on biuret method with blank correction. *3393*

Mephenesin At concentration of 100 mg/L had no effect on biuret method with blank correction. *3393*

Protein (continued)

Serum No Effect Analytical (continued)

Meprobamate At acute overdose concentration (20 mg/dL) on Technicon® SMAC® method. *3719* At concentration of 200 mg/L had no effect on biuret method with blank correction. *3393*

Meralluride No effect at 0.07 mL/dL on SMA 12/60 method. *2636*

Mercaptomerin At concentration of 58 mg/L had no effect on biuret method with blank correction. *3393*

Mercaptopurine At concentration of 152 mg/L had no effect on biuret method with blank correction. *3393*

Mesoridazine At acute overdose concentration (20 mg/dL) on Technicon® SMAC® method. *3719*

Methadone At acute overdose concentration (2.5 mg/dL) on Technicon® SMAC® method. *3719* At concentration of 25 mg/L had no effect on biuret method with blank correction. *3393*

Methamphetamine At concentration of 2 mg/L had no effect on biuret method with blank correction. *3393*

Methanol At acute overdose concentration (20 mg/dL) on Technicon® SMAC® method. *3719*

Methapyrilene At concentration of 13 mg/L had no effect on biuret method with blank correction. *3393*

Methaqualone At acute overdose concentration (2.5 mg/dL) on Technicon® SMAC® method. *3719*

Methicillin At 5 times upper limit of therapeutic range on methods on SMAC®, Abbott-VP, Ektachem®, Hitachi® 705 and KDA. *2138*

Methohexital At concentration of 50 mg/L had no effect on biuret method with blank correction. *3393*

Methotrexate At 5 times upper limit of therapeutic range on methods on SMAC®, Abbott-VP, Hitachi® 705 and KDA. *2138*

Methotrimeprazine At concentration of 1 mg/L had no effect on biuret method with blank correction. *3393*

Methsuximide At concentration of 40 mg/L had no effect on biuret method with blank correction. *3393*

Methyldopa At concentration of 211 mg/L had no effect on biuret method with blank correction. *3393*

Methylphenidate At acute overdose concentration (20 mg/dL) on Technicon® SMAC® method. *3719* At concentration of 200 mg/L had no effect on biuret method with blank correction. *3393*

Methylphenobarbital At concentration of 150 mg/L had no effect on biuret method with blank correction. *3393*

Methyprylon At acute overdose concentration (10 mg/dL) on Technicon® SMAC® method. *3719* At concentration of 100 mg/L had no effect on biuret method with blank correction. *3393*

Metoprolol At concentration of 0.34 mg/L had no effect on biuret method with blank correction. *3393*

Morphine At acute overdose concentration (20 mg/dL) on Technicon® SMAC® method. *3719* At concentration of 200 mg/L had no effect on biuret method with blank correction. *3393*

Moxalactam At 2.50 mmol/L on method on Eppendorf Epos. *1414* At concentration of 96 mg/L had no effect on biuret method with blank correction. *3393*

Naproxen At 5 times upper limit of therapeutic range on methods on SMAC®, Abbott-VP, Ektachem®, Hitachi® 705 and KDA. *2138*

Nitrofurantoin On routine methods in use on SMAC®, Ektachem®, Abbott-VP, Cobas-Bio, aca, Hitachi® 705, KDA at 5 times normal upper therapeutic concentration. *2138* At concentration of 4 mg/L had no effect on biuret method with blank correction. *3393*

Nortriptyline At acute overdose concentration (20 mg/dL) on Technicon® SMAC® method. *3719* At concentration of 200 mg/L had no effect on biuret method with blank correction. *3393*

Orphenadrine At concentration of 9 mg/L had no effect on biuret method with blank correction. *3393*

Ouabain At 0.02 mg/dL no effect on SMA 12/60 method. *2636*

Oxazepam At concentration of 1 mg/L had no effect on biuret method with blank correction. *3393*

Oxytetracycline At concentration of 50 mg/L had no effect on biuret method with blank correction. *3393*

Papaverine At concentration of 10 mg/L had no effect on biuret method with blank correction. *3393*

Paraldehyde At concentration of 2,000 mg/L had no effect on biuret method with blank correction. *3393*

Penicillin G At 5 times upper limit of therapeutic range on methods on SMAC®, Abbott-VP, Ektachem®, Hitachi® 705 and KDA. *2138*

Pentobarbital At acute overdose concentration (20 mg/dL) on Technicon® SMAC® method. *3719* At concentration of 340 mg/L had no effect on biuret method with blank correction. *3393*

Perphenazine At acute overdose concentration (20 mg/dL) on Technicon® SMAC® method. *3719* At concentration of 200 mg/L had no effect on biuret method with blank correction. *3393*

Phencyclidine At concentration of 6 mg/L had no effect on biuret method with blank correction. *3393*

Phenelzine At concentration of 136 mg/L had no effect on biuret method with blank correction. *3393*

Phenobarbital At acute overdose concentration (20 mg/dL) on Technicon® SMAC® method. *3719* No effect at therapeutic concentrations on biuret and spectrophotometric method. *1786* At concentration of 250 mg/L had no effect on biuret method with blank correction. *3393*

Phensuximide At concentration of 120 mg/L had no effect on biuret method with blank correction. *3393*

Phenylbutazone At concentration of 750 mg/L had no effect on biuret method with blank correction. *3393*

Phenylephrine At concentration of 4 mg/L had no effect on biuret method with blank correction. *3393*

Phenytoin No effect at 1.8 mg/dL on SMA 12/60 method. *2636* At acute overdose concentration (20 mg/dL) on Technicon® SMAC® method. *3719* At concentration of 240 mg/L had no effect on biuret method with blank correction. *3393*

Probenecid At concentration of 1300 mg/L had no effect on biuret method with blank correction. *3393*

Procainamide No effect on SMA 12/60 method at 35 mg/dL. *2636* At concentration of 50 mg/L had no effect on biuret method with blank correction. *3393*

Procaine At concentration of 2 mg/L had no effect on biuret method with blank correction. *3393*

Prochlorperazine At concentration of 1 mg/L had no effect on biuret method with blank correction. *3393*

Promethazine At acute overdose concentration (20 mg/dL) on Technicon® SMAC® method. *3719* At concentration of 200 mg/L had no effect on biuret method with blank correction. *3393*

Propoxyphene At acute overdose concentration (2.5 mg/dL) on Technicon® SMAC® method. *3719* At concentration of 25 mg/L had no effect on biuret method with blank correction. *3393*

Propranolol No effect at 0.1 mg/dL on SMA 12/60 method. *2636* At concentration of 0.2 mg/L had no effect on biuret method with blank correction. *3393*

Propylthiouracil At concentration of 170 mg/L had no effect on biuret method with blank correction. *3393*

Pyribenzamine® At acute overdose concentration (20 mg/dL) on Technicon® SMAC® methods. *3719*

Quinidine At 26 mg/dL no effect on SMA 12/60 method. *2636* At acute overdose concentration (20 mg/dL) Technicon® SMAC® method. *3719* At concentration of 210 mg/L had no effect on biuret method with blank correction. *3393*

Quinine At acute overdose concentration (1.5 mg/dL) on Technicon® SMAC® method. *3719* At concentration of 30 mg/L had no effect on biuret method with blank correction. *3393*

Reserpine No effect at 0.02 mg/dL on SMA 12/60 method. *2636* At concentration of 61 mg/L had no effect on biuret method with blank correction. *3393*

Rifampin At 5 times upper limit of therapeutic concentration on methods on SMAC®, Abbott-VP, Hitachi® 705 and KDA. *2138*

Salicylate At concentration of 500 mg/L had no effect on biuret method with blank correction. *3393*

Secobarbital At acute overdose concentration (20 mg/dL) on Technicon® SMAC® method. *3719* At concentration of 200 mg/L had no effect on biuret method with blank correction. *3393*

Storage of Sample No effect for 1 week at room temperature, 1 mo at 4°. *1563*

Streptomycin At concentration of 58 mg/L had no effect on biuret method with blank correction. *3393*

Strychnine At concentration of 12 mg/L had no effect on biuret method with blank correction. *3393*

Sulfadiazine At concentration of 1500 mg/L had no effect on biuret method with blank correction. *3393*

Sulfaguanidine At concentration of 500 mg/L had no effect on biuret method with blank correction. *3393*

Sulfanilamide At concentration of 1,000 mg/L had no effect on biuret method with blank correction. *3393*

Sulfapyridine At concentration of 80 mg/L had no effect on biuret method with blank correction. *3393*

Sulfasalazine No effect when Du Pont aca turbidimetric method is used with 100 fold dilution of sperm. *2484* At concentration of 50 mg/L had no effect on biuret method with blank correction. *3393*

Sulfathiazole At concentration of 255 mg/L had no effect on biuret method with blank correction. *3393*

Sulforidazine At concentration of 200 mg/L had no effect on biuret method with blank correction. *3393*

Tetracycline At 5 times upper limit of therapeutic range on methods on SMAC®, Abbott-VP, Ektachem®, Hitachi® 705 and KDA. *2138*

Theobromine At concentration of 2,000 mg/L had no effect on biuret method with blank correction. *3393*

Theophylline At 5 times upper limit of therapeutic range on methods on SMAC®, Abbott-VP, Ektachem®, Hitachi® 705 and KDA. *2138* At concentration of 2,000 mg/L had no effect on biuret method with blank correction. *3393*

Thiamazole At concentration of 114 mg/L had no effect on biuret method with blank correction. *3393*

Thiopental At acute overdose concentration (20 mg/dL) on Technicon® SMAC® method. *3719* At concentration of 200 mg/L had no effect on biuret method with blank correction. *3393*

Timolol At concentration of 0.01 mg/L had no effect on biuret method with blank correction. *3393*

Tolbutamide At concentration of 270 mg/L had no effect on biuret method with blank correction. *3393*

Transport of Specimen By pneumatic tube if specimen protected. *3217*

Tribromethanol At concentration of 90 mg/L had no effect on biuret method with blank correction. *3393*

Trichlorethanol At concentration of 1,000 mg/L had no effect on biuret method with blank correction. *3393*

Trifluoperazine At concentration of 1 mg/L had no effect on biuret method with blank correction. *3393*

Tripelennamine At concentration of 200 mg/L had no effect on biuret method with blank correction. *3393*

Vitamin Preparations No effect at expected concentration with SMA 12/60 procedure. *2637*

Warfarin At concentration of 100 mg/L had no effect on biuret method with blank correction. *3393*

Serum   No Effect   Physiological

Acetylsalicylic Acid No effect seen in patients undergoing treatment. *1787*

Altitude No effect at 5400 meters unless dehydrated. *2957*

Amphotericin B No effect seen in 10 patients with 6 weeks treatment. *0240*

Bopindolol In 10 hypertensive patients treated for either 1 or 21 d. *1063*

Captopril In 9 of 12 hypertensive patients although other hematological effects observed. *1608*

Diclofenac No effect seen in patients undergoing treatment. *1787*

Enflurane No effect observed. *0999*

Ibuprofen No effect seen in patients undergoing treatment. *1787*

Indomethacin No effect seen in patients undergoing treatment. *1787*

Isotretinoin In 18 patients with severe acne given 0.8 mg/kg daily for 3 mo: changes reverted to normal after treatment stopped. *2242*

Ketoprofen No effect seen in patients undergoing treatment. *1787*

Levonorgestrel No significant difference over 3 y between implant recipients and IUD users. *0763*

Meals No effect after standard breakfast. *0579*

Phenytoin In about 20 patients treated for 2 y. *0873*

Phosphorus Decreased synthesis due to liver damage. *0279*

Starvation Usually unaffected if previously well fed. *3783*

Terbutaline In 6 normal men treated with therapeutic amounts for 2 weeks. *3177*

Serum   Increase   Analytical

Acetylsalicylic Acid Increase by 9% with 8326 μmol/L with direct spectrophotometry at 280 nm. *1786*

Amino Acids If given i.v. may affect accuracy of measurement. *0120*

Ampicillin At concentrations above 500 mg/L raised concentration as measured by biuret method with blank correction. *3393*

Bilirubin No effect up to 29 mg/dL. *1563* Above 5 mg/dL (20 mg/dL = 0.2 g/L). *3217*

Carbenicillin At 5 times upper limit of therapeutic range on methods on SMAC®, Abbott-VP, Ektachem® 400 and Hitachi® 705. *2138*

Cephalothin At concentrations above 500 mg/L (normal therapeutic concentration 17 mg/L) raised concentration as measured by biuret method with blank correction. *3393*

Chloramphenicol At concentrations above 500 mg/L (normal therapeutic concentration 10 mg/L) raised concentration as measured by biuret method with blank correction. *3393*

Dextran Turbidity effect with biuret reaction. *3505* At concentrations greater than 5,000 mg/L (normal therapeutic concentration 14500 mg/L) raised concentration as measured by biuret method with blank correction. *3393*

Disulphine Blue At physiological concentration on an endpoint biuret reaction for at least 2 d. *1464*

Fat Emulsions If given i.v. may affect accuracy of measurement. *0120*

Fluosol-DA By 18% on Kodak Ektachem® 400 at concentration of 50%. *2518*

Hemoglobin 1 mg measured as equivalent to 1.9 mg protein. *1563*

Hemolysis Added protein measured. *2313*

Lipemia Turbidity of serum. *2313*

Liposol Changed from 56.2 to 61.6 g/L in 26 randomly selected patient sera, before and after addition on Beckman Astra methods. *2054*

Methicillin At concentrations above 500 mg/L (normal therapeutic concentration 21 mg/L) raised concentration as measured by biuret method with blank correction. *3393*

Nafcillin At concentrations above 500 mg/L raised concentration as measured by biuret method with blank correction. *3393*

Oxacillin At concentrations above 500 mg/L (normal therapeutic concentration 6 mg/L) raised concentration as measured by biuret method with blank correction. *3393*

Penicillin G At concentrations above 500 mg/L (normal therapeutic concentration 12 mg/L) raised concentration as measured by biuret method with blank correction. *3393*

Phenazopyridine Orange-brown color affects absorbance. *0583*

Plasma Mean 0.24 g/dL due to fibrinogen. *2228*

Radiographic Agents May produce interfering turbidity. *2313*

Rifampin At 5 times upper limit of therapeutic concentration on methods on biuret method on Ektachem®. *2138*

Sulfobromophthalein Color augmentation with alkaline biuret. *0579*

Serum   Increase   Physiological

Anabolic Steroids Associated with increased protein synthesis. *0697*

Androgens Associated with increased protein synthesis. *0697*

Angiotensin Hemoconcentration effect. *1343*

Azosemide Slight and delayed effect observed in normal male volunteers after 60 mg orally. *2658*

Blood Pressure Strongly correlated with increased systolic pressure. *2123*

Bumetanide By up to 0.6 g/dL over 3 mo. *2666*

Clofibrate Reported effect, mechanism not discussed. *2525*

Corticosteroids Physiological doses promote protein synthesis. *0697*

Protein (continued)

Serum Increase Physiological (continued)

Corticotropin Physiological doses promote protein synthesis. *0697*

Dehydration By approximately 15.7% after heat exposure for 2-11 h. *3254*

Digitalis Improved hepatic function and decreased hypovolemia. *1343*

Diurnal Variation Maximum observed after lunch (and during night). *2222* Maximum at 4-8 pm, minimum at 8 am slight effect. *3865*

Epinephrine Due to hemoconcentration. *1343*

Erect Posture Increase by 0.75 g/dL compared with rest and recumbency. *3873*

Growth Hormone Associated with increased protein synthesis. *0697*

Insulin Associated with increased protein synthesis. *0697*

Levonorgestrel In a group of women using levonorgestrel covered rods versus controls using copper IUD. *0899*

Meals Affected by meals (most marked in women). *3455*

Muscular Exercise Significant effect with 12 minutes on cycle-ergometer. *1231* 9% increase immediately after exercise, normal in 30 minutes. *2846*

Oral Contraceptives Significant increase 3 y after administration, after 3 mo cessation. *3143*

Progesterone Metabolic effect. *0795*

β-Propiolactone Increase by 3.5 g/L when plasma from BPL treated blood compared with plasma in 25 specimens. *0217*

Season Significant effect maximal in Nov, Dec, minimal in Jan, June. *1172*

Thyroid Physiological effect exerts anabolic effect. *0697*

Tourniquet Affects all protein bound constituents also. *1563* Increase by 0.5 g/dL over 3 minutes. *1704*

Serum Decrease Analytical

Acetylsalicylic Acid Increase by 1.5% with 8326 μmol/L on biuret method. *1786*

Ammonium Ions Cupric ammonium complex formed (Cu availability decreased). *1563*

Caproxamine At 1.5 mg/dL conventional methods when added to sera. *1758*

Cefotaxime Effect of drug and metabolite on specimens containing gentamicin and tobramycin on American Monitor Parallel. *0201*

Citrates Dilutional effect when compared against serum or heparin as anticoagulant. *3361*

Dextran 40 At concentrations above 30 g/L causes falsely low results with Du Pont aca, although results could be corrected if ethylene glycol previously injected into packs. *0231*

Fluosol-DA Increase by 20% on Du Pont aca III at concentration of 50%. *2518*

Lipemia By less than 1.5% in patient specimens measured on SMAC® in comparison with specimens following ultra-centrifugation. *2466*

Serum Due to lack of fibrinogen present in plasma. *2229*

Storage of Sample Decrease observed after 26 d at 4°. *3862*

Sulfasalazine Spuriously low values with biuret methods on Du Pont aca related to overblanking. *2484*

Serum Decrease Physiological

Arginine Slight transient reduction (by up to 0.7 g/dL) after i.v. infusion in diabetics. *2328*

Aspirin Progressive effect during three days but also noted over longer term study at beginning, eventually reverted towards normal. Effect statistically significant. *3076*

Benzene Reduced synthesis due to liver damage. *0279*

Blindness Significantly lower (average 0.57 g/dL) than normal. *1637*

Carbamazepine In about 20 epileptic patients treated for 2 y. *0873*

Carbon Disulfide Effect slight. *0818*

Cathartics Rare protein losing gastroenteropathy. *1343*

Dantrolene Single observation, ?significance. *0659*

Dapsone Sulfone syndrome in one 16 year old girl given 50 mg daily short-term administration. *3604*

Dextran 30% after Macrodex®, no effect after Rheomacrodex®. *3346*

Diurnal Variation 7.0 g/dL at 8:30 am but 6.6 g/dL at 10 am. *2123*

Estrogens Altered metabolism. *3332*

Floxuridine Manifestation of toxicity. *1680*

Laxatives May occur with continued use. *2220*

Meals Significant effect in subjects 20-40 y. *3455*

Mercury Compounds Due to albuminuria and starvation. *2313*

Oral Contraceptives Estrogen effect. *3332*

Phosgene Decreased synthesis due to liver damage. *0279*

Pyrazinamide Part of hepatotoxicity (viral-hepatitis like). *1343*

Recumbency Average decrease of 1.0 g/dL after 40 minutes. *1010*

Rifampin Due to impaired hepatic metabolism. *2820*

Season Slight decrease during summer. *2123*

Trimethadione Reversible effect due to urinary loss. *2220*

Urine No Effect Analytical

Aminosalicylic Acid No effect on Labstix®, heat with acetic acid tests. *1487*

Ampicillin No effect in patients receiving up to 8 g daily on sulfosalicylic acid, trichloracetic acid or Ponceau S dye methods. *3919*

Bilirubin At 4 mg/dL on TCA dye method of Pesce and Strande. *2791*

Cefamandole No effect on sulfosalicylic acid method and Albustix method at concentration of 1 g/L. *2388*

Cefazolin No effect in 9 patients receiving up to 4 g daily on sulfosalicylic acid, trichloracetic acid and Ponceau S dye methods. *3919*

Cefoperazone No effect at 1 g/L on sulfosalicylic acid and Albustix methods. *2388*

Ceforanide No effect of 1 g/L on sulfosalicylic acid and Albustix methods. *2388*

Cefotaxime No effect of 1 g/L on sulfosalicylic acid and Albustix methods. *2388*

Cefoxitin No effect of 1 g/L on sulfosalicylic acid and Albustix methods. *2388* No effect in 4 patients receiving up to 8 g daily on trichloracetic acid, sulfosalicylic acid and Ponceau S dye methods. *3919*

Ceftazidime No effect of 1 g/L on sulfosalicylic acid and Albustix methods. *2388*

Ceftizoxime No effect of 1 g/L on sulfosalicylic acid and Albustix methods. *2388*

Ceftriaxone No effect at 1 g/L on sulfosalicylic acid and Albustix methods. *2388*

Cephalothin No effect of 1 g/L on sulfosalicylic acid and Albustix methods. *2388*

Chloroform As preservative has no effect on Stix tests. *2308*

Cibenzoline Negative reaction in 23 of 53 patients when sulfosalicylic acid or acetic acid/sodium acetate heat coagulation used to measure protein. *2002*

Co-Trimoxazole No effect of up to 4 tablets daily in 4 patients on sulfosalicylic acid, trichloroacetic acid and Ponceau S Dye Methods. *3919*

Diatrizoic acid No effect on Albustix, Labstix® No effect on heat acetic acid test. *1487*

Fluorescein At 250 mg/L in spiked specimen run on Ames Clinitek 200. *1722*

Gentamicin No difference between sulfosalicylic acid and trichloracetic acid methods in patients receiving therapeutic doses. *3760*

Glibenclamide No significant difference in 12 patients receiving up to 15 mg daily with sulfosalicylic acid and trichloracetic methods. *3919*

Iopanoic Acid No effect on heat acetic acid test No effect on Albustix, Labstix®. *1487*

Moxalactam No effect of 1 g/L on sulfosalicylic acid and Albustix methods. *2388*

Penicillin No effect on Albustix even with massive doses. *0459*

Phenazopyridine No effect sulfosalicylic, heat tests. *1487*

Quaternary Ammonium Compounds With SSA and acetic acid tests. *0443*

Radiographic Agents On Combistix®, Urostix, Albustix etc. *1872*

Salicylate At 100 mg/dL on TCA dye method of Pesce. *2791*

Sulfadiazine At 100 mg/dL on TCA dye method of Pesce. *2791*

Sulfisoxazole At 100 mg/dL on TCA dye method of Pesce. *2791* No effect on Albustix. *0459*

Thymol No effect on biuret, boiling tests. *2981*

Tolbutamide No effect on Albustix. *0459*

Urine Turbidity On Stix tests (bromphenol reagent strip). *0443*

Urine   No Effect   Physiological

Bopindolol In 10 hypertensive patients treated for either 1 or 21 d. *1063*

Chenodeoxycholic Acid No significant effect with up to 750 mg/d in 916 patients treated for 2 y. *3199*

Cimetidine Although drug cleared principally by glomerular filtration and partial reabsorption in proximal renal tubules. *2368*

Cromoglycate No clinical significant change observed. *0484*

Cyclosporine No effect observed in 23 patients with inflammatory ocular disease treated for for 6 mo. *2713*

Timegadine In 23 patients with rheumatoid arthritis given 250 to 750 mg/d for 48 weeks. *2345*

Urine   Increase   Analytical

Acetazolamide Makes urine highly alkaline, causes false positive. *0111* When measured by Ponceau S dye method in comparison with sulfosalicylic acid or trichloracetic acid methods. *3919*

Alkaline Urine Possible false positive with Albustix, Combistix® etc. *1872* Highly buffered urine may affect Combistix®. *2313*

Aminosalicylic Acid Affects acid turbidimetric procedures. *0583* With SSA and heat with acetic acid tests. *0443* Reacts with Folin-Ciocalteu of Lowry procedure. *1237* May cause false positive with sulfosalicylic acid. *1872*

Aspirin Interference with Folin-Ciocalteu reaction. *3879*

Bacteriuria If turbidimetric methods used. *2919*

Bicarbonates Highly alkaline urine causes false positive. *1487*

Bunamiodyl Affects turbidity tests for up to 3 d. *0065*

Carinamide Produces precipitate with acid tests. *0459*

Cephaloridine Affects acid turbidimetric procedures. *0583* Precipitate occurs with sulfosalicylic acid Weak positive with Albustix at high concentrations. *2152*

Cephalothin Affects acid turbidimetric procedures. *0583* Precipitate occurs with sulfosalicylic acid Weak positive with Albustix at high concentrations. *2152*

Chlorhexidine False positive with Stix alkalinization. *0121*

Chlorpromazine Affects turbidity tests for up to 3 d. *0065* In 2 patients receiving therapeutic doses on Ponceau S dye method in comparison with sulfosalicylic acid and trichloracetic acid method. *3919*

Cibenzoline Positive reaction in 23 of 53 patients when bromphenol reagent strips used to measure protein. *2002*

Copper Affects biuret part of Doetsch procedure. *0921*

Diatrizoic acid Interferes with sulfosalicylic, nitric acid tests. *1700* Affects biuret part of Doetsch procedure. *0921* False positive with sulfosalicylic acid. *1487*

Dihydrotachysterol Increased glomerular permeability. *1022*

Dithiazanine Reacts with Folin-Ciocalteu of Lowry procedure. *2425*

Gentamicin On Ponceau S dye method in comparison with sulfosalicylic acid method in 7 patients receiving therapeutic doses. *3760*

Glibenclamide Significant effect with Ponceau S dye method in comparison with sulfosalicylic method in 12 patients receiving up to 15 mg daily. *3919*

Iodoalphionic Acid Affects acid precipitation methods. *0065*

Iodopyracet If acid precipitation methods used. *2958* Gives false positive with turbidity tests. *0757*

Iopanoic Acid Gives turbidity if acid precipitation tests used. *2958* Causes false positive with turbidity tests. *0065*

Iophenoxic Acid Affects turbidity tests for up to 3 d. *0065*

Metahexamide Interference by drug metabolite. *3729* Interference by drug metabolite. *1022*

Penicillin Causes turbidity if sulfosalicylic acid used. *3505* Massive doses may produce turbidity with acid. *0459* With SSA and acetic acid tests. *0443*

Penicillin G When measured by Ponceau S dye method in comparison with sulfosalicylic acid or trichloracetic acid methods in 5 patients receiving therapeutic doses. *3919*

Phenazopyridine False positive with Labstix® etc. *2566*

Promazine Affects turbidity tests for up to 3 d. *0065*

Quaternary Ammonium Compounds Gives false positive by changing pH (stix tests). *3505*

Radiographic Agents Turbidity if acid procedures used. *1700* Affects biuret part of Doetsch procedure. *0921* Affects turbidimetric methods for some days. *3505* Affects sulfosalicylic, heat and acetic acid tests. *1872*

Sodium Bicarbonate False positive with Labstix® due to high pH. *1488*

Streptomycin Reacts as phenol if Folin-Ciocalteu reaction used. *3505*

Sulfamethoxazole May interfere with sulfosalicylic acid methods. *1022*

Sulfisoxazole Causes turbidity if sulfosalicylic acid used. *3505* False positive with Exton's reagent No effect with acetic acid and heat. *0443* May produce turbidity with acid methods. *0459*

Sulfonamides Reacts as phenol with Folin-Ciocalteu reagent. *3505* May cause false positive with sulfosalicylic acid. *1488*

Thymol Excess may give false positive with turbidity procedures. *0459*

Tolbutamide Causes turbidity if sulfosalicylic acid used. *3226* Affects heat and acetic acid test. *1872* Interferes with sulfosalicylic acid method. *3226*

Trifluoperazine In 2 patients receiving up to 30 mg daily on Ponceau S dye method in comparison with sulfosalicylic acid and trichloracetic acid methods. *3919*

Urine   Increase   Physiological

Acetaminophen Nephrotoxic effect of drug. *0438*

Acetanilid Hemolysis may cause renal damage. *1343*

Acids Ingestion may cause severe renal damage. *1302*

Altitude Associated with decreased creatinine clearance. *2957*

Aminophylline Increase in renal disease (occurs early with poisoning). *1022*

Aminopyrine Nephrotoxicity reported. *2427*

Aminosalicylic Acid May occur as result of nephrotoxicity. *2496* May cause nephrotoxicity. *1237*

Amphotericin B Nephrotoxic effect. Nephrotoxicity. *1022* Nephrotoxicity usually develops after a few weeks of treatment, usually reversible unless more than total of 4 g drug given. *3567*

Ampicillin Necrosis of tubules due to nephrotoxicity. *3542*

Antimony Compounds Nephrotoxic effect. *1022*

Antipyrine Renal action of drug. *3824*

Arsenicals Nephrotoxic effect. *1022*

Asparaginase Transient slight effect for few days. *2657* Infrequent transient proteinuria. *0558*

Aspirin May cause nephrotoxicity. *1488* Observed effect with long term low doses for secondary prevention of coronary heart disease. *2866*

Auranofin In 3% of patients receiving drug, but proteinuria did not persist beyond 12 mo: effect less than when gold given. *1886*

Bacitracin May cause nephrotoxicity. *1237* Nephrotoxic effect. *1022*

Beryllium Salts May cause nephrotoxicity. *3204*

Bismuth Salts Nephrotoxic effect. *1022* Nephrotoxic effect reported. *3204*

Boric Acid Due to renal damage. *1302*

Bromate Due to renal damage. *1302*

Bunamiodyl May cause nephrotoxicity. *3204*

Cadmium Generalized impairment of proximal tubular function. *0019*

Cantharides May cause nephrotoxicity. *3204*

Capreomycin Nephrotoxic effect (transient). *1022*

Protein *(continued)*

Urine   Increase   Physiological *(continued)*

Captopril Isolated reports of immune complex glomerulopathy when first introduced and high doses given. *1761* Effect observed in small number of patients with excellent control of hypertension. *0181* Greater than 1.0 g/d occurs in about 1.2% patients may subside with continuing treatment. *3712* Occurs in approximately 1% of 7100 hypertensives most often who had pre-existing renal disease and receiving high doses of drug. *1407* In patients with pre-existing renal dysfunction with proteinuria. *0181* Some patients develop heavy proteinuria during use of drug. Reversible with discontinuation. *2263* Eosinophilic interstitial nephritis and membranous glomerulopathy reported. Cases of nephrotic syndrome also reported. *0929*

Carbamazepine Manifestation of renal damage. *0019*

Carbarsone Nephrotoxic effect. *1022*

Carbenoxolone Due to myoglobinuria and nephropathy. *2458*

Carbon Disulfide Manifestation of renal damage. *0818*

Carbon Monoxide May cause nephrotoxicity. *3204* Associated with oliguria and renal damage. *1302*

Carbutamide Nephrotoxic effect. *3729*

Castor Oil May cause renal damage. *0279*

Cephaloglycin Of cephalosporins one of most likely to cause nephrotoxicity. *2615*

Cephaloridine May be result of nephrotoxicity. *1605* Of cephalosporins one of most likely to produce nephrotoxicity. *2615*

Cephalothin Rare effect following i.v. infusion. *2427*

Chloral Hydrate Renal damage with high concentration Parenchymatous renal injury in chronic intoxication. *1343*

Chlorate May cause renal damage. *0279*

Chlormerodrin May produce nephrotic syndrome. *1188*

Chloroform May cause renal damage. *0279*

Chlorpropamide Nephrotoxic effect. *3729* In isolated case with diabetes who developed proliferative glomerulonephritis indicating immunologically mediated reaction. *0146* Isolated case of drug-induced hemolytic anemia. *3111*

Chlorthalidone Nephrotoxic effect. *1237*

Chromates May cause renal damage. *0279*

Chrysarobin Kidney irritation by metabolite. *0071*

Codeine May cause nephropathy. *2583*

Colistimethate Common nephrotoxic effect. *2808* Nephrotoxic effect. *2313*

Colistin Nephrotoxic effect (reversible renal damage). *2495*

Copper May cause nephrotoxicity. *3204*

Corticosteroids Nephrotoxic especially in children with chronic disease. *1487*

Corticotropin May be nephrotoxic in chronic disease. *1583*

Cresol May cause renal damage. *0279*

Cyclosporine Characteristic feature of nephrotoxicity in most common form of renal damage to drug. *0175*

Dantrolene Possible effect, perhaps urinary infection. *0659*

Demeclocycline Nephrotoxic effect. Nephrotoxicity. *2106*

1,2-Diaminopropane Good correlation between amount of protein and duration of proteinuria moderate doses does not preclude reinstitution of therapy once protein has cleared. *2586* In 16% of 90 patients with rheumatoid arthritis. *1461* HLA DRB positive patients had 11 times risk of this side effect than those antigen. *1379*

Diatrizoic acid Marked increased excretion (20 to 60 times baseline) day of bilateral renal arteriography in 23 patients with hypertension, reverted to normal on following day. *2594*

Dichlorobenzene May cause renal damage. *0279*

Dihydrotachysterol May induce increased glomerular permeability. *1022*

Dinitrophenol Due to nephrotoxicity. *1302*

Dioxane May cause renal damage. *0279*

Dithiazanine Transient proteinuria may occur. *2427*

Doxapram May have nephrotoxic effect ?nephrotoxic effect. *1022*

Doxycycline Nephrotoxic effect. *2802*

EDTA Nephrotoxic effect (especially Calcium EDTA). *0071* Nephrotoxic effect (especially Calcium EDTA). *1237*

Electroshock May cause nephrotoxicity. *3204*

Enflurane In one patient with initially normal renal function following comparatively long exposure to anesthetic. Normal renal function eventually returned. *1001*

Ergot Preparations May cause renal damage with poisoning. *0279*

Ether May cause nephrotoxicity. *3204*

Ethosuximide Possible reversible nephropathy. *1022*

Ethoxyethanol May cause nephrotoxicity. *3204*

Ethylene Glycol May cause nephrotoxicity. *3204* Due to nephrotoxicity. *1302*

Fava Beans May cause renal damage. *0279*

Fenoprofen In 4% in patients treated for juvenile rheumatoid arthritis. *0239* In 2 isolated cases of patients with arthritis in the absence of hypertension. *0466*

Fluorides May cause renal damage. *0279*

Furosemide If pre-existing proteinuria. *2818*

Gasoline May cause renal damage if ingested. *0279*

Gentamicin Nephrotoxic effect. *1022* Manifestation of drug-induced nephrotoxicity. *3182*

Glycerin Hemolysis may occur after i.v. administration. *1448*

Gold May cause nephrotoxicity. *2427* Nephrotoxic effect (at least in 50% cases). *1022* Observed in 16% of 90 patients with rheumatoid arthritis in one study. *0782* Prevalence of gold nephropathy about 1 in 500. Nephropathy not necessarily related to other side effects. *0829* Observed in 3% of 1283 patients ranging from mild to heavy: did not persist beyond 12 mo. Some patients showed membranous glomerulonephritis. *1886* In 11 patients who showed glomerular injury, including minimal change nephrotic syndrome with a membranous pattern of immune complex deposition as well as other patterns of deposition. *1170* Good correlation between amount of protein and duration of proteinuria. Moderate proteinuria does not preclude reinstitution of therapy once protein has cleared. *2586* In 16% of 90 patients with rheumatoid arthritis. *1461* HLA DR3 positive patients had 3 times risk of this side effect than those without antigen. *1379*

Griseofulvin May cause nephrotoxicity. *0633* ?nephrotoxic effect (transient and reversible). *1022*

Heroin Massive proteinuria may occur in addicts. *2914*

Hydralazine May cause nephrotoxicity. *3204*

Hydration In patients with proteinuria. *2818*

Hydrogen Sulfide Occurs due to nephrotoxicity. *1302*

Ibuprofen In 2 patients with systemic lupus erythematosus. *1155*

Indomethacin Reported in one patient. *0506* Oliguric renal failure developed in one man during treatment: reversible. *1245*

Insecticides May cause nephrotoxicity. *3204*

α-Interferon 1 to 2 g/24 h in 2 patients receiving relatively large amount of drug: normalized 1 to 2 weeks after treatment stopped. *3299*

Iodine Containing Drugs Result of hemorrhagic nephritis at toxic doses. *1302*

Ipodate Nephrotoxic effect. *0838*

Iron Dextran Reversible after chronic administration in one patient. *1678*

Iron Salts May cause nephrotoxicity. *1488*

Isoniazid May have nephrotoxic effect. *1488* Nephrotoxic effect. *1022*

Isopropanol Transient and mild late toxic effect. *1302*

Kanamycin Nephrotoxic effect. *1022*

Kerosene Renal damage with large doses. *1343*

Lead Nephrotoxic effect with poisoning Nephrotoxicity with poisoning. *1343*

Lipomul® Nephrotoxic effect. *1022*

Lithium May have slight nephrotoxic effect. *1488*

Lucanthone Chronic toxicity may cause renal damage. *1343*

Mefenamic Acid Nephrotoxic effect. *1237* Nephrotoxic effect. *1022*

Melarsonyl Nephrotoxic effect. *0071*

Melarsoprol Nephrotoxic effect. *2313*

Menstruation Premenstrual urine may contain protein. *0987*

Mercurial Diuretics   May produce nephrotic syndrome. *1188*

Mercury Compounds   Nephrotoxic effect. *1237*

Metaxalone   May have nephrotoxic effect Nephrotoxic effect. *1022*

Methanol   Nephrotoxic effect with poisoning Occurs with poisoning. *1343*

Methenamine   Nephrotoxic in large doses. *1343*   Nephrotoxic in large doses. *1022*

Methicillin   Nephrotoxicity may occur. *0071*   Nephrotoxic effect. *2496*

Methsuximide   Renal damage reported. *1237*

Methylbromide   Due to tubular necrosis. *1302*

Mithramycin   Nephrotoxic effect. *0071*

Mitomycin   Nephrotoxic effect. *2194*

Mitotane   May rarely produce hematuria, renal damage. *1680*

Moxalactam   In about 2% patients as reported from several studies. *0592*

Muscular Exercise   More common with heavy exercise than mild. *1842*

Mushroom Poisoning   May cause nephrotoxicity. *3204*

Naphthalene   Nephrotoxic effect. *1302*

Naproxen   In isolated case of patient with arthritis in the absence of hypertension. *0466*

Neomycin   Nephrotoxic effect. *1022*

Nifedipine   Acute reversible deterioration of renal function in 4 patients with chronic renal failure. *0898*

NSAID (Nonsteroidal Anti-Inflammatory Drugs)   Most of drugs have been associated with reversible clinical syndrome of heavy proteinuria and renal insufficiency. *1561*

Oxacillin   Nephrotoxic effect. *0071*

Oxalate   If ingested may cause nephrotoxicity. *1302*

Oxyphenbutazone   May cause renal damage. *0019*   May be nephrotoxicity. *1680*

Paraldehyde   Nephrotoxic effect (nephrosis with poisoning). *1443*

Paramethadione   May have nephrotoxic effect. *2496*

Paromomycin   Potentially nephrotoxic if given parenterally. *0071*

Penicillamine   Nephrotoxic effect. *1237*   May be minimal and asymptomatic or lead to nephrotic syndrome due to immune complex glomerulitis. *1761*   Observed in 16% of 90 patients with rheumatoid arthritis in one study. *1461*   In 3 of 21 patients with rheumatoid arthritis treated for 48 weeks given 250 to 750 mg daily. *2345*   In 16% of 90 patients with rheumatoid arthritis previously treated with gold. *1461*   In 15 of 84 patients occurring most often in 6 to 12 mo after treatment started. *1892*   Toxicity observed in some patients with rheumatoid disease. *1891*   In 30% patients with rheumatoid arthritis or cystinuria and 4% in patients with Wilson's disease. *0113*   In 10 to 20% of patients with rheumatoid arthritis in first year, of which one third may proceed to nephrotic syndrome. *0753*

Penicillin   Nephrotoxicity may occur with large doses. *1343*

Perchlorate   Reported cases of nephrotic syndrome. *0071*

Phenacemide   Occasional nephropathy. *0071*   Occasional nephropathy. *1022*

Phenindione   Nephrotoxicity may occur. *2496*

Phenolphthalein   May cause acute nephrosis (K deficiency). *2427*

Phenols   Nephrotoxicity with poisoning. *1343*

Phensuximide   Reversible nephropathy (especially in children). *0071*

Phenylbutazone   Reported nephrotoxic effect. *1488*   Observed effect, may occasionally be marked with oliguria or renal failure. *2866*

Phosphorus   Renal toxic effect. *1237*

Picric Acid   Nephrotoxicity. *1343*

Piroxicam   Occasional drug induced nephrotoxicity, with isolated azotemia, acute acute interstitial nephritis or nephrotic syndrome. *2824*

Poison Ivy   May cause nephrotoxicity. *3204*

Polymyxin   Nephrotoxic effect. *1022*   Nephrotoxic effect. *1237*

Pregnancy   Moderate proteinuria common Moderate proteinuria may occur commonly. *0987*

Probenecid   Nephrotoxic effect. *3879*   Nephrotoxic effect. *1488*

Promazine   Affects turbidity tests for up to 3 d. *0065*

Proteinuria   Actual effect. *2897*

Puromycin   May cause nephrotoxicity. *3204*

Pyrazolones   Nephrotoxic effect. *1237*   Nephrotoxic effect. *1022*

Pyrithioxine   With dose of 400-600 mg/d: associated with development of immune complex nephritis and other penicillamine-like toxic reactions. *1761*

Quinine   Rare renal damage May rarely cause renal damage. *1343*

Radiographic Agents   May occur following aortography. *0065*   Observed in small number of patients receiving agents for arteriography. *3259*   Mean increase from 129 mg/d to 2760 mg/d in 37 patients day following arteriography. *2594*

Rhubarb   Crystalluria may cause renal damage. *0152*

Rifampin   Attributed to drug in 2 cases. *2427*

Rolitetracycline   May cause nephropathy. *2583*

Salsalate   Associated with minimal change nephrotic syndrome in one patient. *3664*

Silver   May cause nephrotoxicity. *3204*

Skin Lightening Cream   Trace observed in some cases. *0235*

Stibophen   Renal irritation reported. *0071*

Streptokinase   Renal damage reported. *2313*

Streptomycin   May cause nephrotoxicity. *1343*   Occasional nephrotoxicity, although less than with other aminoglycosides. *3683*

Sulfadiazine   Due to crystalluria and hematuria Associated with crystalluria and hematuria. *2808*

Sulfamethoxazole   May cause hematuria with crystalluria. *2808*

Sulfonamides   May cause nephrotoxicity. *1488*   May cause nephrotoxicity. *2313*

Sulfones   May have nephrotoxic effect. *1022*

Suprofen   Occasional renal failure observed but mechanism not known. *0009*

Suramin   Usual effect during treatment of acute stage. *0071*   Nephrotoxic effect. *1022*

Tartrates   If absorbed may cause renal damage. *1343*

Tetrachloroethylene   May cause nephrotoxicity. *3204*

Tetracycline   Nephrotoxic effect with degraded tetracycline. *1237*

Thallium   Damage to renal tubular epithelium. *1302*

Theophylline   When high doses sodium glycinate salt given. *2313*

Thiabendazole   May cause nephrotoxicity. *1488*

Thiocyanate   May cause nephrosis. *1343*

Thiopronine   Toxicity similar to penicillamine and other sulfhydryl drugs. *1761*

Thiosemicarbazones   Nephrotoxic effect. *1237*   Nephrotoxic effect. *1022*

Tobramycin   Manifestation of drug induced nephrotoxicity but tend to decrease as serum creatinine begins to climb. *3182*   At dose of 4.5 mg/kg/d for 12 d in 90 patients nephrotoxicity observed in up to 39%, reversible in most. *0677*

Tolmetin   In 1% of patients in trials for treatment of juvenile rheumatoid arthritis. *0239*

Triethylenemelamine   Due to nephrotoxicity. *1680*

Trimethadione   May have nephrotoxic effect. *1488*   May have nephrotoxic effect. *1582*

Trypsin   Isolated cases reported. *1680*

Turpentine   May have nephrotoxic effect. *1022*   Nephrotoxic effect. *1237*

Uranium   May cause nephrotoxicity. *3204*

Vancomycin   May cause nephrotoxicity. *3204*   Occasional evidence of mild nephrotoxicity. *1571*

Venoms   May cause nephrotoxicity. *3204*

Viomycin   May have nephrotoxic effect Nephrotoxic effect. *1237*

Vitamin D   Nephrotoxic effect with hypercalcemia. *1022*

Vitamin K   Reported side effect. *2427*

X-Ray Therapy   May cause nephrotoxicity. *3204*

Zimeldine   3 of 14 patients treated for more than 1 week preserved mild abnormality. *2075*

Zoxazolamine   Has nephrotoxic effect May have nephrotoxic effect. *1488*

Protein (continued)

Urine   Decrease   Analytical

Alkaline Urine   Possible false negative with sulfosalicylic acid. Possible false negative with heat and acetic acid. *1872*

Proteinuria   Unable to detect paraproteins with BiliLabstix®. *0490*

Urine   Decrease   Physiological

Captopril   Significant reduction in patients with advanced diabetic nephropathy. *3528*

Enalapril   Often reduction and rarely increased during treatment. *2361*

CSF   No Effect   Analytical

Ascorbic Acid   No effect on Folin-Ciocalteu procedure. *3958*

Bilirubin   At 4 mg/dL with TCA dye method of Pesce and Strande. *2791*

Erythromycin   No effect on Folin-Ciocalteu procedure. *3958*

Salicylate   At 100 mg/dL on TCA dye method of Pesce. *2791*

Sulfadiazine   At 100 mg/dL on TCA dye method of Pesce. *2791*

Sulfisoxazole   At 100 mg/dL on TCA dye method of Pesce. *2791*

CSF   Increase   Analytical

Aminobenzoic Acid   1.0 mg = 1.8 mg in Folin-Ciocalteu procedure. *0583*

Aminosalicylic Acid   Reacts as phenol if Folin-Ciocalteu reaction used. *3505*

Aspirin   False positive with Folin-Ciocalteu reagent. *3958*

Bilirubin   May affect turbidimetric procedures. *0583*

Chloramphenicol   Slight effect on Folin-Ciocalteu procedure. *3958*

Chlorpromazine   Reacts as if phenol with Folin-Ciocalteu reagent. *3505*

Epinephrine   1.0 mg = 1.5 mg in Folin-Ciocalteu procedure. *0583*

Globulin   Color equivalent > albumin with Folin-Ciocalteu. *1196*

5-Hydroxyindoleacetic Acid   1.0 mg = 3.2 mg in Folin-Ciocalteu procedure. *0583*

5-Hydroxytryptamine   1.0 mg = 3.1 mg in Folin-Ciocalteu procedure. *0583*

5-Hydroxytryptophan   1.0 mg = 3.9 mg in Folin-Ciocalteu procedure. *0583*

Imipramine   1.0 mg = 3.5 mg in Folin-Ciocalteu procedure. *0583*

Lidocaine   Reacts with Folin-Ciocalteu reagent. *1237*

Methotrexate   High absorbance with turbidimetric methods. Reacts as if phenol with Folin-Ciocalteu. *3969*   Marked effect on Du Pont aca method (up to 40 times actual concentration) even at therapeutic concentration; if concentration low enough not to produce yellow color protein concentration not significantly affected. *3701*

Oxytetracycline   Reacts as if phenol with Folin-Ciocalteu procedure. *3958*

Penicillin   Reacts as if phenol with Folin-Ciocalteu procedure. *3958*   Causes turbidity if sulfosalicylic acid used. *3505*

Phenothiazines   False positive with Folin-Ciocalteu reagent. *1237*

Procaine   Interferes with Folin-Ciocalteu reagent. *1237*

Radiographic Agents   Causes turbidity if sulfosalicylic acid used. *3505*

Streptomycin   Reacts as phenol if Folin-Ciocalteu reaction used. *3505*

Sulfadiazine   Reacts as if phenol with Folin-Ciocalteu procedure. *3958*

Sulfaguanidine   Reacts as if phenol with Folin-Ciocalteu procedure. *3958*

Sulfamerazine   Reacts as if phenol with Folin-Ciocalteu procedure. *3958*

Sulfanilamide   Reacts as if phenol with Folin-Ciocalteu method. *3958*

Sulfisoxazole   Causes turbidity if sulfosalicylic acid used. *3505*

Sulfonamides   Reacts as phenol with Folin-Ciocalteu reagent. *3505*

Tetracaine   Interferes with Folin-Ciocalteu reagent. *1237*

Tetracycline   Reacts as if phenol with Folin-Ciocalteu procedure. *3958*

Thyronine   1.0 mg = 2.9 mg in Folin-Ciocalteu procedure. *0583*

Tolbutamide   Causes turbidity if sulfosalicylic acid used. *3505*

Tryptophan   Interferes with Folin-Ciocalteu reagent. *3958*

Tyramine   1.0 mg = 2.7 mg in Folin-Ciocalteu procedure. *0583*

Tyrosine   Reacts as if phenol with Folin-Ciocalteu procedure. *3958*

CSF   Increase   Physiological

Ibuprofen   Few cases of aseptic meningitis described. *1276*

Lead   Occurs with lead encephalopathy, encephalitis. *1302*

Mercury Compounds   May produce Guaillain-Barré like syndrome. *3515*

Methotrimeprazine   May occur if cerebral edema etc. *1680*

Perphenazine   Altered proteins reported. *1680*

Piperacetazine   Abnormality produced. *0019*

Poliomyelitis Vaccine   May rarely cause Guaillain-Barré syndrome. *2427*

CSF   Decrease   Analytical

Albumin   Turbidity < globulins with sulfosalicylic acid. *2313*

Cytarabine   False low value with Folin-Ciocalteu reagent False low with turbidimetric method on standing. *3969*

Pancreatic Juice   Increase   Physiological

Acetazolamide   Approximately 40% increase with infusion in normals. *0984*

Pancreozymin   Increase in response to challenge. *0486*

Test Conditions   No Effect   Analytical

Sodium Lauryl Sulfate   Lowry procedure when referenced with chemical only used. *1796*

Sodium Phosphate   Lowry procedure when referenced with chemical only used. *1796*

Sucrose   Lowry procedure when referenced with chemical only used. *1796*

Urea   Lowry procedure when referenced with chemical only used. *1796*

Test Conditions   Increase   Analytical

Acetylglucosamine   Slight reaction with Folin-Ciocalteu reagent. *1569*

Ascorbic Acid   Reacts with Folin-Ciocalteu of Lowry procedure. *0702*

Chloride Salts   If glyoxylic acid used to measure tryptophan. *0093*

Cyanides   Interferes with Folin-Ciocalteu of Lowry method. *2202*

Cysteine   Reacts with Folin-Ciocalteu of Lowry procedure. *0702*

Dithiothreitol   Reacts with Folin-Ciocalteu of Lowry procedure. *0702*

Epinephrine   Reacts with Folin-Ciocalteu of Lowry procedure. *0702*

Fructose   Interferes with Folin-Ciocalteu method of Lowry. *3954*

Galactosamine   Reacts with Folin-Ciocalteu of Lowry method. *1569*

Glucosamine   Hexosamines, N-acetyl-derivatives affect Lowry method. *0304*   Reacts with Folin-Ciocalteu of Lowry method. *1569*

Glucose   Interferes with Folin-Ciocalteu method of Lowry. *3636*

Glycerin   Interferes with Lowry and biuret methods. *3954*

Hematin   Even 1 μg has significant effect on 150 μg protein. *0396*

HEPES   Interferes with Folin-Ciocalteu of Lowry. *3636*

Indole   Reacts with Folin-Ciocalteu of Lowry method. *1569*

Mannitol   Possible measurement by biuret reaction. *3376*

Mannose   Interferes with Folin-Ciocalteu method of Lowry. *3636*

Morphine   Reacts with Folin-Ciocalteu of Lowry method. *0702*

Nalorphine   Reacts with Folin-Ciocalteu of Lowry method. *0702*

Penicillin   Massive doses may cause turbidity with acid tests. *2313*

Phenylthiourea   Reacts with Folin-Ciocalteu of Lowry method. *0702*

Polysucrose Affects Lowry, Folin-Ciocalteu procedures. *2199* Possible measurement by biuret reaction. *3376*

Rhamnose Interferes with Folin-Ciocalteu method of Lowry. *3636*

Saccharose Possible measurement by biuret reaction. *3376*

Sodium Bisulfite At concentrations higher than 0.01 % affects Folin-Ciocalteu. *0702*

Sorbose Interferes with Folin-Ciocalteu method of Lowry. *3636*

Thiamylal Sodium Reacts with Folin-Ciocalteu of Lowry method. *0702*

Thioguanine Reacts with Folin-Ciocalteu of Lowry method. *1569*

Thiourea Reacts with Folin-Ciocalteu of Lowry method. *0702*

Trolamine Possible measurement by biuret reaction. *3376*

Tromethamine Possible measurement by biuret reaction. *3376*

Uric Acid Reacts with Folin-Ciocalteu of Lowry method. *1569*

Xanthine Reacts with Folin-Ciocalteu of Lowry method. *1569*

Xylose Interferes with Folin-Ciocalteu method of Lowry. *3636*

Test Conditions   Decrease   Analytical

Bicine Lowry procedure ?non linear absorption. *1796*

EDTA Variable but significant effect on Lowry procedure. *1796*

Glycine Lowry procedure due to chelation of chemicals. *1796*

Glycylglycine Lowry procedure due to chelation of chemicals. *1796*

Sodium Citrate Lowry procedure due to chelation of chemicals. *1796*

Succinic Acid Lowry procedure due to chelation of chemicals. *1796*

Tricine Lowry procedure, non linear absorption. *1796*

Triton X-100 Lowry procedure, non linear absorption. *1796*

Tromethamine Lowry procedure, non linear absorption. *1796*

Protein Bound Iodine (PBI)

Serum   No Effect   Analytical

Mercury Compounds No effect Barker dry ash procedure. *0354*

Storage of Sample No effect 5 weeks at room temperature. *1563*

Serum   No Effect   Physiological

Acetazolamide Reported to have no clinical effect. *0830*

Antibiotics No effect observed. *0830*

Anticoagulants No effect observed. *0830*

Antihistamines Probably exert no clinically significant effect. *0830*

Busulfan Normal levels observed even over long time. *0583*

Calcium No effect with calcium loading. *0583*

Chlorpromazine No effect observed with normal doses. *0830*

Demeclocycline Although discoloration of thyroid occurs. *1680*

Dextroamphetamine Appears to have no effect. *0830*

Diazepam Conflicting reports ?no effect. *1488*

Digitalis No effect observed. *0830*

Digitoxin With 0.1 mg/kg/week no effect observed. *0583*

Fluorides At concentrations used to cause fluoridation of water. *0830*

Fluorouracil No effect observed. *0830*

Gonadotropin No significant effect observed with chorionic hormone. *0583*

Hydrochlorothiazide No effect observed in one study. *0583*

Hydroxyzine No effect with normal doses. *0830*

Imipramine No effect reported with normal doses. *0830*

Indocyanine Green Not usually affected. *1487*

Iodophor Detergents No effect with occupational exposure. *0583*

Isocarboxazid No effect reported with normal doses. *0830*

Isopropamide No definite effect observed. *0583*

Meals No effect of meals reported. *0830*

Meprobamate No effect with normal doses. *0830*

Methotrexate Reported to have no effect. *0830*

Metronidazole 1200 mg/d no effect over 1 week. *0583*

Minocycline Although thyroid tissue may be stained. *1680*

Muscular Exercise No effect observed. *0830*

Phenylbutazone Probably usual effect. *1487*

Promazine No effect with normal doses. *0830*

Reserpine No effect observed with oral, parenteral administration. *0830*

Sulfonylureas Probably no effect in most patients. *1487*

Testosterone No effect short term administration in men. *0830*

Tetracycline No effect with 2 g/d for 11 d. *0583*

Thiamylal Sodium No effect on PBI observed. *0583*

Thiazides Reported to have no effect. *0830*

Thiopental No effect on PBI in humans. *0583*

Thioridazine No effect on PBI observed. *0583*

Trifluoperazine No effect with normal doses. *0830*

Vincristine Reported to have no effect. *0830*

Vitamin A No effect observed when given for 3 weeks. *0830*

Warfarin No effect with 5 mg/d. *0583*

Serum   Increase   Analytical

Acetrizoate But does not affect T_4 (effect lasts 2 to 4 weeks). *3218*

Antiseptics If contain iodine and used to clean skin. *0830*

Barium Occasionally contaminated with inorganic iodine. *0830*

Benziodarone Contains 46% iodine. *0012*

Bromides May be contaminated with iodine. *1237*

Chiniofon Contains 27.5% iodine. *0012*

Chloride Salts Laboratory contamination may cause significant effect. *0830*

Dandruff Medication Iodine contamination (in some preparations). *1237*

Dextrothyroxine Reacts like levo compound. *0583*

Diatrizoic acid But does not affect T_4, effect lasts 2-6 d. *3218*

Diiodohydroxyquin Contains iodine, duration of effect uncertain. *0354* Contains 65% iodine, effect lasts for some weeks. *0012*

Diiodoquinoline Contains 67% iodine. *0012*

Diprotrizoate But does not affect T_4 (effect for 1-2 weeks). *3218*

Dithiazanine Contains 24.5% iodine. *0012*

Ethiodized Oil Contains iodine. *0071*

Gallamine Organic iodine contamination. *2220*

Gargles Some may contain organic and inorganic iodine. *1237*

Indocyanine Green Organic iodine contamination. *1237*

Iodides Iodide salts if more than 1 g/d lasts 28 d. *0012*

Iodipamide But does not affect T_4 effect lasts 3-4 mo. *3218*

Iodoalphionic Acid Contains iodine (effect lasts 2-12 mo). *1563*

Iodoantipyrine Contains iodine. *2425*

Iodocasein Organic iodine contamination. *1237*

Iodochlorhydroxyquin Contains iodine (effect lasts 2-3 mo). *3652*

Iodoform Contains iodine. *2220*

Iodohippurate But does not affect T_4. *3218*

Iodopyracet Contains iodine effect lasts 2 weeks. *3505*

Iopanoic Acid Organic iodine affects test, effect lasts 1-4 mo. *3218*

Iophendylate Contains iodine effect lasts up to 5 y. *0354*

Iophenoxic Acid Contains iodine effect lasts up to 30 y. *0354*

Iopydone Contains iodine. *2220*

Iothalamate Contains iodine. *2220*

Iothiouracil Contains organic iodine (effect for several months). *3505*

Ioxitalamic Acid May take up to 8 d to become normal. *0149*

Ipodate Contains organic iodine. *1563*

Isopropamide Contains iodine. *0012*

Lugol's Iodine Contains iodine, iodide: effect for up to 3 weeks. *3218*

Meprobamate Contains iodine. *3505*

Mercurial Diuretics Interferes with digestion technique. *0012*

Methiodol But does not affect T_4 (effect lasts 1-2 weeks). *3218*

Mouth Washes Some may contain iodine. *1237*

Protein Bound Iodine (PBI) *(continued)*

Serum   Increase   Analytical   *(continued)*

Oxyquinoline   In vaginal suppositories (often iodinated). *3505*

Penethamate   Contains iodine. *3505*

Penicillin   If given as hydriodide salt. *2220*

Propyliodone   Effect lasts 1-5 mo. *3505*

Radiographic Agents   Most radiopaque media have effect. *3217*

Red Dyed Drugs   Presence of erythrocrine dye. *0097*

Red Dyed Food   Presence of erythrocrine dye. *1237*

Sulfobromophthalein   Organic iodide contamination of some BSP solutions. *2815*

Suntan Oil   Many preparations contain iodine. *2313*

Tetridaz-B   Iodine contamination. *1237*

Vitamin Preparations   Iodine contamination. *1237*

Serum   Increase   Physiological

Albumin   Causes effect if iodinated. *3505*

Bismuth Salts   Many bismuth salts contain iodine. *0012*

Bunamiodyl   Effect up to 2 mo (affects T_4 at high concentrations). *0604*

Chlormadinone   Due to increased binding globulin. *2313*

Clofibrate   Up to 2.5 g caused increase of 1.5 μg/dL. *0583*

Cough Medicines   Some may contain iodides. *1237*

Decamethonium   Contains 49.5% iodine. *0012*

Desiccated Thyroid   Increases by 1-2 μg/dL/d per grain given. *0354*

Dextrothyroxine   Contains 65% iodine (may cause increase to 10-25 μg/dL). *0012*

Diethylstilbestrol   Increases concentration of circulating thyroxine binding globulin. *0012*

Diiodocaffeine   Contains 66.5% iodine. *0012*

Dimethyl Tubocurarine   Contains 35% iodine. *0012*

Diurnal Variation   Small statistically significant effect. *0830*

Erect Posture   Change of up to 0.8 μg/dL observed. *3217*

Erythrosine   Significant dose related increases in 30 men receiving 20 to 200 mg/d for 15 d. *1241*

Estradiol   Increases concentration of circulating thyroxine binding globulin. *0012*

Estrogens   Increases binding capacity of thyroxine binding globulin. *3871*

Estrone   Increases concentration of circulating thyroxine binding globulin. *0012*

Ether   Causes mobilization of PBI stores. *0012*

Ethinyl Estradiol   Increases concentration of circulating thyroxine binding globulin. *0012*

Fluoropyrimidine   Mechanism obscure. *0012*

Insulin   Mechanism obscure. *0012*

Iodinated Glycerol   Effect lasts up to 2 d. *0354*

Iodine Containing Drugs   Iodine contamination. *1237*

Iodized Oil   Does not affect T_4 (effect 1-5 y). *3218*

Iodoxyl   Contains iodine, effect lasts for 1-2 weeks. *0354*

Jaundice   If biliary regurgitation as normal excretion in bile. *3947*

Levodopa   ?effect due to tetraiodofluorescein in capsules. *0420*

Levothyroxine   In patients on thyroid maintenance therapy. *0012*

Lysergide   Stimulates TSH release, ?effect on I 131 uptake. *0012*

Methysergide   Mechanism not determined. *0583*

Nialamide   Variable effect on I 131 uptake in rats. *0012*

Norethindrone   Reported effect but mechanism unknown. *0757*

Norethynodrel   Increases concentration of circulating thyroxine binding globulin. *0012*

Oral Contraceptives   Estrogens increase circulating thyroxine binding globulin. *3548*

Ovulation   Small increase in ovulatory and luteal phases. *0830*

Perphenazine   May contain iodinated contaminants. *1481*

Phenothiazines   Due to increased thyroxine binding globulin with prolonged use. *1915*

Pink Capsules   May contain iodine in dye. *1237*

Plastic Tubing   Bard Intracath tubing contains I_2. *0583*

Potassium Iodide   Small doses elevate in normal range, others above. *0830*

Povidone-Iodine   Found by some investigators only. *2892*

Pregnancy   Observed throughout pregnancy. *0987*

Progesterone   Increased production of thyroxine binding globulin. *3879*

Propylthiouracil   Reduces metabolism of exogenous thyroxine. *0012*

Pyrazinamide   Occurs in first week and after 1 mo. *0830*

Quinine Iodobismuthate   Contains 57% iodine. *0012*

Radioactive Iodine   May cause hypothyroidism (increased by 2-3% per year). *0071*

Reserpine   Increased metabolism by hepatic microsomes. *1237*

Sex Difference   Higher in women at all age levels. *2130*

Stress   Large effect with psychological stress. *3217*

Testosterone   Androgen effect seen in some patients. *1680*

Tetraiodofluorescein   Red dye used to color many pharmaceuticals. *0012*

Thymol   Affects PBI if given as thymol iodide. *0012*

Thyroid   Increases by 1-2 μg/dL/d/1 grain given. *1488*

Thyrotropin   Starts within 15 h, may increase by up to 5.5 μg/dL. *0830*

Toothpaste   Some may contain enough I_2 to cause effect. *0830*

Tourniquet   Occlusion for 5 minutes may cause increase up to 2 μg/dL. *0830*

Vasopressin   Stimulates thyroidal I 131 release. *0012*

Vitamin A   When given in cod liver oil. *2313*

Vitamin D   When given in cod liver oil. *2313*

Serum   Decrease   Analytical

Chlormerodrin   Formation of mercurial iodide during analysis. *0757*

Copper   As contaminant of water may affect analysis. *2387*

Gold   Interferes with chloric acid, Barker methods. *0354*

Hemolysis   5% decrease if at 700 nmol/L. *3257*

Meralluride   Interferes with ceric arsenious acid. *3934*

Mercaptomerin   Formation of mercurial iodide during analysis. *0757*

Mercurial Diuretics   Interfere with acid distill, chloric acid methods. *1488*

Mercury Compounds   Interfere with chloric acid digest procedures. *0012*

Nessler's Reagent   Catalytic effect of iodine inhibited by K+, Hg++. *1563*

Silver   Interferes with ashing procedures. *1915*

Thimerosal   Inhibition of reaction caused by contaminated skin. *3218*

Serum   Decrease   Physiological

Acetazolamide   Inhibits iodination of tyrosine in thyroxine binding globulin. *0012*

Albutoin   Side effects less marked than with other hydantoins. *2449*

Aminobenzoic Acid   Inhibits iodination of tyrosine in thyroxine binding globulin. *0012*

Aminoglutethimide   Inhibits iodination of tyrosine in thyroxine binding globulin. *0012*

Aminosalicylic Acid   Inhibits iodination of tyrosine in thyroxine binding globulin. *3244*

Aminothiazole   Inhibits iodination of tyrosine in thyroxine binding globulin. *0012*

Amphenone B   Inhibits iodination of tyrosine in thyroxine binding globulin. *0012*

Anabolic Steroids   Reduces concentration of circulating thyroxine binding globulin. *0012*

Androgens   Reduced synthesis of thyroxine binding globulin. *2220*

Aspirin   Competes for thyroxine binding prealbumin also uncouples phosphorylation. *1022*

Barbiturates   Competes with T_4 for thyroxine binding prealbumin binding sites. *0583*

Brompheniramine Inhibits iodination of tyrosine in thyroxine binding globulin. *0012*

Carbimazole Inhibits iodination of tyrosine in thyroxine binding globulin. *0012*

Carbutamide Inhibits iodination of tyrosine in thyroxine binding globulin. *0012*

Chlorate Interferes with trapping of iodide by thyroid. *0012*

Chlorpromazine Mechanism obscure (observed if over 600 mg given). *0012*

Chlorpropamide Questionable effect (may be slight in diabetics). *0012*

Cobalt Salts Cause decrease, possibly to extent of goiter. *1343*

Corticosteroids If chronic administration in small or large doses. *0830* Large dose effect with decreased thyroxine binding globulin. *1915*

Corticotropin Reduced thyro-binding globulin. *3879*

Cortisone Inhibits iodination of tyrosine, also hemodilution. *0012*

Cyanides Inhibit iodination of tyrosine in thyroxine binding globulin. *0012*

Desiccated Thyroid Observed with some batches (T_4 converted to T_3). *0757*

Desoxycorticosterone Inhibits iodination of tyrosine in thyroxine binding globulin. *0012*

Diazo Dyes Compete for binding sites. *0583*

Difluorophosphate Interferes with trapping of iodide by thyroid. *0012*

Dinitrophenol Competes with thyroxine for thyroxine binding globulin. *0012*

Disulfiram Uncommon reported effect. *2220*

Epinephrine Causes release of PBI from thyroid. *0012*

Ethionamide Antithyroid effect after several weeks. *1488*

Evans Blue Mechanism obscure. *0012*

Fish May be associated with high Hg content of some fish. *2935*

Fluorides Large doses interfere with I_2 trapping by thyroid. *0012*

Fluoroborate Interferes with trapping of iodide by thyroid. *0012*

Fluoxymesterone Due to decreased binding capacity (androgen effect). *1680*

Gentisic Acid Competes for binding with thyroxine binding prealbumin. *0012*

Glucocorticoids Reduced amount of thyroxine binding globulin. *1237*

Halofenate Inhibits binding of T_4 to thyroxine binding globulin. *2483*

Hemolysis If marked, RBCs contain no I_2 (dilution effect). *0354*

Hydrochlorothiazide Mechanism obscure. *0012*

Hydrocortisone Inhibits iodination of tyrosine in thyroxine binding globulin. *0012*

Hypochlorites Interferes with trapping of iodide by thyroid. *0012*

Iodates Interferes with trapping of iodide by thyroid. *0012*

Isoniazid Reduces thyroid synthesis. *2220*

Liothyronine Remains below normal even with full replacement. *1680*

Liotrix Therapeutic response. *0071*

Lithium Inhibits iodination of tyrosine in thyroxine binding globulin. *0071*

Menstruation Slight fall after menstruation observed. *0830*

Mephenytoin Competes with thyroxine for binding sites. *2220*

Methandrostenolone Metabolic effect. *0019*

Methimazole Inhibits iodination of tyrosine in thyroxine binding globulin. *1444*

Methyclothiazide But no signs of thyroid disturbance. *1680*

Methyltestosterone Lowers concentration of circulating thyroxine binding globulin. *0012*

Methylthiouracil Inhibits iodination of tyrosine in thyroxine binding globulin. *0012*

Mitotane Rare reported side effect. *1680*

Monofluorosulphonate Interferes with trapping of iodide by thyroid. *0012*

Nandrolone Anabolic effect. *2681*

Niagara Sky Blue Mechanism obscure. *0012*

Norethandrolone Concentration of circulating thyroxine binding globulin lowered. *0012*

Norethindrone When treatment results compared with controls. *0260*

Oxymetholone Reduces concentration of circulating thyroxine binding globulin. *0012*

Oxyphenbutazone Inhibits iodination of tyrosine in thyroxine binding globulin. *0012*

Penicillin Competes for thyroxine binding prealbumin binding sites. *0583*

Perchlorate Interferes with trapping of iodine by thyroid. *0012*

Periodate Interferes with trapping of iodine by thyroid. *0012*

Phenindione Inhibits iodination of tyrosine in thyroxine binding globulin. *0012*

Phenothiazines Antithyroid effect of large doses. *0830*

Phenylbutazone Competes with thyroxine for binding on thyroxine binding globulin. *0830*

Phenytoin Competes with thyroxine for binding sites. *1488* Metabolic effect of drug. *0121*

Polythiazide Without clinical effect on thyroid function. *1680*

Prednisolone Inhibits iodination of tyrosine in thyroxine binding globulin. *0012*

Prednisone Inhibits iodination of tyrosine in thyroxine binding globulin. *0012*

Progesterone Inhibits iodination of tyrosine in thyroxine binding globulin. *0012*

Propylthiouracil Thyroxine in circulation decreased. *1488*

Quinidine Occasional effect (usually none). *0830*

Reserpine Rare depression reported (given orally for 1 mo). *2220*

Resorcinol Reported effect of treatment of varicose ulcers. *1343*

Sodium Nitroprusside Single case reported (?due to thiocyanate). *2626*

Stanozolol Observed with other anabolic steroids. *1680*

Sulfamethoxazole Small effect with therapeutic doses. *1488*

Sulfonamides Inhibits iodination of tyrosine in thyroxine binding globulin. *0012*

Sulfonylureas Due to impaired synthesis of thyroxine. *1915*

Testosterone Decreased thyroxine binding globulin synthesis, and affects binding. *3879*

Thiazides Increased excretion with prolonged therapy may occur. *1343*

Thiocarlide Myxedema reported in one case. *2427*

Thiocyanate Blockage of thyroxine biosynthesis. *1238*

Thiopental Inhibits iodination of tyrosine in thyroxine binding globulin. *0012*

Thiouracil Inhibits iodination of tyrosine residues in thyroxine binding globulin. *0012*

Tolbutamide Inhibits iodination of tyrosine residues in thyroxine binding globulin. *0012*

Trichlormethiazide May occur in absence of thyroid hypofunction. *1680*

Tri-iodothyronine Pituitary suppression of TSH. *0012*

Trypan Blue Mechanism obscure. *0012*

Vitamin A Inhibits iodination of tyrosine residues in thyroxine binding globulin. *0012*

Protein Electrophoresis

Serum   No Effect   Analytical

Allopurinol At concentration of 180 mg/L had no effect on automated Olympus-Hite method but with slight displacement of fractions. *3393*

Ampicillin At concentration of 1800 mg/L had no effect on automated Olympus-Hite method. *3393*

Ascorbic Acid At concentration of 800 mg/L had no effect on automated Olympus-Hite method but with slight displacement of fractions. *3393*

Azapropazone At concentration of 360 mg/L had no effect as measured by automated Olympus-Hite method except for slight displacement of fractions. *3393*

Protein Electrophoresis *(continued)*

Serum   No Effect   Analytical *(continued)*

Benzbromarone   At concentration of 80 mg/L had no effect as measured by automated Olympus-Hite method. *3393*

Bisacodyl   At concentration of 2 mg/L had no effect as measured by automated Olympus-Hite method. *3393*

Butizide   At concentration of 1.98 mg/L had no effect as measured by automated Olympus-Hite method. *3393*

Chloramphenicol   At concentration of 6,000 mg/L had no effect on automated Olympus-Hite method but with slight displacement of fractions. *3393*

Chlordiazepoxide   At concentration of 18 mg/L had no effect on automated Olympus-Hite method except for slight displacement of fractions. *3393*

Chloroquine   At concentration of 50 mg/L had no effect on automated Olympus-Hite method. *3393*

Chromonar Hydrochloride   At concentration of 180 mg/L had no effect on automated Olympus-Hite method. *3393*

Cyclophosphamide   At concentration of 80 mg/L had no effect on automated Olympus-Hite method. *3393*

Dextran 60   At concentration of 18,000 mg/L had no effect on automated Olympus-Hite method except for slight displacement of fractions. *3393*

Diatrizoic acid   At concentration of 6080 mg/L had no effect on automated Olympus-Hite method but with displacement of fractions. *3393*

Digoxin   At concentration of 0.15 mg/L had no effect on automated Olympus-Hite method. *3393*

Doxepin   At concentration of 60 mg/L had no effect on automated Olympus-Hite method. *3393*

EDTA   At concentration of 1,000 mg/L had no effect on automated Olympus-Hite method. *3393*

Fluosol-DA   Not affected by presence of compound in specimen. *2518*

Furosemide   At concentration of 400 mg/L had no effect on automated Olympus-Hite method. *3393*

Glibenclamide   At concentration of 3 mg/L had no effect on automated Olympus-Hite method. *3393*

Hyoscine-N-Butylbromide   At concentration of 20 mg/L had no effect on automated Olympus-Hite method. *3393*

Indomethacin   At concentration of 40 mg/L had no effect on automated Olympus-Hite method except for slight displacement of fractions. *3393*

Meglumine   At concentration of 2,000 mg/L had no effect on automated Olympus-Hite method. *3393*

Melphalan   At concentration of 1.5 mg/L had no effect on automated Olympus-Hite method but with slight displacement of fractions. *3393*

Methotrexate   At concentration of 80 mg/L had no effect on automated Olympus-Hite method. *3393*

Methyldopa   At concentration of 800 mg/L had no effect on automated Olympus-Hite method. *3393*

Niacinamide   At concentration of 120 mg/L had no effect on automated Olympus-Hite method. *3393*

Nitrofurantoin   At concentration of 50 mg/L had no effect on automated Olympus-Hite method. *3393*

Norfenefrin   At concentration of 4.8 mg/L had no effect on automated Olympus-Hite method. *3393*

Novaminsulfon   At concentration of 800 mg/L had no effect on automated Olympus-Hite method. *3393*

Oxyphenbutazone   At concentration of 120 mg/L had no effect on automated Olympus-Hite method. *3393*

Penicillamine   At concentration of 360 mg/L had no effect on automated Olympus-Hite method. *3393*

Phenobarbital   At concentration of 60 mg/L had no effect on automated Olympus-Hite method. *3393*

Phenothiazines   At concentration of 200 mg/L had no effect on automated Olympus-Hite method. *3393*

Phenprocoumon   At concentration of 3.6 mg/L had no effect on automated Olympus-Hite method. *3393*

Phenylbutazone   At concentration of 120 mg/L had no effect on automated Olympus-Hite method. *3393*

Prednisolone   At concentration of 7.2 mg/L had no effect on automated Olympus-Hite method except for slight displacement of fractions. *3393*

Probenecid   At concentration of 400 mg/L had no effect on automated Olympus-Hite method. *3393*

Sodium Citrate   At concentration of 5,000 mg/L had no effect on automated Olympus-Hite method. *3393*

Sodium Fluoride   At concentration of 2,000 mg/L had no effect on automated Olympus-Hite method but with slight displacement of fractions. *3393*

Sodium Heparin   At concentration of 750 mg/L had no effect on automated Olympus-Hite method. *3393*

Sodium Oxalate   At concentration of 3,000 mg/L had no effect on automated Olympus-Hite method but with slight displacement of fractions. *3393*

Spironolactone   At concentration of 80 mg/L had no effect on automated Olympus-Hite method. *3393*

Storage of Sample   No effect 3 d at room temperature or 1 mo at 4°. *1563*   No effect at -20° for 6 mo. *0573*

Sulfamethoxydiazine   At concentration of 200 mg/L had no effect on automated Olympus-Hite method. *3393*

Sulfasalazine   At concentration of 1600 mg/L had no effect on automated Olympus-Hite method except for slight displacement of fractions. *3393*

Tetracycline   At concentration of 400 mg/L had no effect on automated Olympus-Hite method. *3393*

Tolbutamide   At concentration of 400 mg/L had no effect on automated Olympus-Hite method. *3393*

Vitamin B Complex   At concentration of 23.3 mg/L had no effect on automated Olympus-Hite method. *3393*

Serum   No Effect   Physiological

Skin Lightening Cream   No evidence of nephrotic syndrome. *0235*

Serum   Positive   Analytical

Hemolysis   May augment beta, between alpha$_2$ and beta. *1563*
Penicillin   Causes bisalbuminemia. *3505*
Radiographic Agents   Produces uninterpretable pattern. *3879*

Plasma   No Effect   Analytical

Acetylsalicylic Acid   At concentration of 1,000 mg/L had no effect on automated Olympus-Hite method. *3393*

Prothrombin Time

Plasma   No Effect   Analytical

Hemolysis   Has negligible effect. *3873*
Storage of Sample   Stable at room temperature for 18 h in oxalated plasma. *3873*

Plasma   No Effect   Physiological

Acetaminophen   No effect observed in volunteers. *3292*

Amiodarone   In 0 of 5 patients with toxic and hypersensitivity liver injury. *2987*

Ampicillin   In doses as high as 300 mg/kg/d in volunteers. *0492*

Anticonvulsants   In 8 epileptic patients receiving phenobarbital with carbamazepine or phenytoin. *2670*

Apalcillin   In 21 volunteers with doses up to 225 mg/kg. *1262*

Ascorbic Acid   No effect on Thrombotest with dose of 1 g/d. *1691*   No effect seen in most patients. *1487*

Aspirin   No effect observed in volunteers. *3292*

Ceftazidime   In 30 patients with serious infections. *0031*

Chloral Hydrate   No significant effect noted in 10 patients on warfarin when receiving 0.5 g daily. *3649*

Chlordiazepoxide   Does not affect action of administered coumarins. *3011*

Chlorothiazide   Does not affect action of administered coumarins. *3011*

Cyclopropane   Anesthesia has no effect. *1343*

Diazepam   No effect when given 15 mg/d. *2682*

Diphenhydramine   Probably no effect on response to anticoagulant. *1487*

Ether   Anesthesia has no effect. *1343*

Fluroxene   No significant effect noted. *1814*

Halothane   In one nurse occupationally exposed to gas. *1900*

Hydroxyzine   Reported to have no effect. *1487*

Ketoprofen   In 11 patients given drug intravenously. *1234*

Meprobamate   Probably no significant effect clinically. *1487*

Methicillin   No effect with doses as high as 300 mg/kg/d in volunteers. *0492*

Nitrazepam   No effect observed with phenprocoumon. *0340*

Penicillin G   No effect with doses as high as 48 million U/d. *0492*

Quinidine   No effect observed in volunteers. *3292*   Hypersensitivity observed in 32 of 487 patients who had received drug. *2913*

Silicones   If in cooking oil — no effect on anticoagulants absorption. *3536*

Temocillin   4 g intravenously 12 hourly had no effect in 8 weeks. *2629*

Plasma   Increase   Analytical

EDTA   When used as anticoagulant. *3962*

Storage of Sample   Oxalated plasma loses activity at 4°. *3873*   Loss of activity observed after 5 y at -15°. *1180*

Plasma   Increase   Physiological

Acenocoumarol   Lowers prothrombin of blood: intended action. *0019*

Acetaminophen   Depresses clotting factor synthesis. *0438*   Manifestation usually of single toxic high dose ingestion in suicide attempt. Liver damage usually of centrizonal hemorrhagic hepatic necrosis. *3948*   Significant effect in patients receiving warfarin, possibly due to interference with synthesis of factors II, VII, IX, and X. *0403*

Acetohexamide   Occurs if failure to excrete bile salts. *0071*

Allopurinol   Patients on coumarins (unconfirmed clinically). *3704*

Aminopyrine   Observed in man and experimental animals. *2313*

Aminosalicylic Acid   Drug suppresses formation of prothrombin. *2916*   Rare side effect. *1292*

Amiodarone   May double when warfarin coadministered. *2011*   Probably by lowering concentrations of vitamin K-dependent coagulation factors. *1470*   Prolongation sufficiently great that 43% reduction in dose of Coumadin® required in 11 patients. *1394*

Amodiaquine   Abnormality usually mild but associated with an immunoallergic hepatotoxicity. *2081*

Anabolic Steroids   Exaggerated response to anticoagulants reported. *1680*

Androgens   Mechanism not established. *0132*

Anesthetic Agents   Observed effect. *1680*

Anisindione   Intended effect. *1680*

Antibiotics   Decreased synthesis of vitamin K by gastrointestinal tract. *3879*

Anticoagulants   Therapeutic intent. *2313*

Antipyrine   Patients on coumarins. *2427*

Asparaginase   Diminished synthesis of prothrombin by liver. *0578*   In 18 patients investigated reported in various studies. *2406*

Aspirin   Large dose effect (decreased synthesis of clot factors). *0894*

Azapropazone   Drug has ulcerogenic potential by itself, but enhanced with warfarin. *0095*

Barbiturates   Theoretically if cholestasis occurs. *2313*

Bromelains   Effect observed over 4 h period in animals. *0071*

Carbenicillin   Effect observed especially in uremics. *1680*

Cathartics   Accelerated passage and decreased absorption of vitamin K. *1487*

Cefamandole   In 3 of 37 patients receiving intravenous nutrition. *3134*   In 7 cases developed vitamin K deficient hypoprothrombinemia. *3103*   4 of 31 patients receiving drug for more than 48 h developed response; reversible but significant bleeding occurred in some. *0329*

Cefoperazone   Reported in previous study to occur in about 4% treated patients. Mechanism unclear but possibly due to inhibition of prothrombin production. Effect correctable with vitamin K. *2685*   Reported effect in some patients with occasional bleeding. *2615*   Associated with vitamin K correctable hypoprothrombinemia but mechanism unclear. *2685*   In 4% of 450 patients but correctable with vitamin K. *0690*

Cefotetan   Reported prolongation and occasional bleeding. *2615*

Ceftriaxone   In 30 patients with serious infections possibly due to presence of sulfhydryl group. *0031*

Cephaloridine   Reported effect. *0071*

Chloral Hydrate   Displaces anticoagulants from albumin. *3379*

Chloramphenicol   May cause lowered prothrombin and hemorrhage. *0894*

Chlordiazepoxide   Associated with failure of excretion of bile salts. *1713*

Chlormadinone   Significant effect observed by 3 mo. *0798*

Chlorpromazine   Associated with failure of excretion of bile salts. *1713*

Chlorpropamide   Associated with failure of excretion of bile salts. *1713*

Chlortetracycline   Reported effect due to action in gastrointestinal tract. *0019*

Chlorthalidone   Patients on coumarins. *2426*

Cholestyramine   Combines with bile acids (vitamin K not absorbed). *0894*

Chymotrypsin   May prolong action of anticoagulants. *1679*

Cimetidine   In 6 patients mean prolongation of time by 12.6 s in patients receiving warfarin, nicoumalone and phenindone. *3256*

Cinchophen   May prolong action of anticoagulants. *1679*

Clofibrate   Displaces anticoagulants from binding protein. *2668*

Corticosteroids   Alleged potentiation of effect of anticoagulants. *1680*

Corticotropin   Prolongs action of anticoagulants. *2313*

Co-Trimoxazole   Several cases of platelet-associated IgG and thrombocytopenia. *2646*

Cremomycin   May cause hypoprothrombinemia. *0071*

Cyclophosphamide   May cause hypoprothrombinemia. *3433*

Dantrolene   Acute hepatitis as result of drug ingestion in 4 patients returned to normal after drug stopped. *3837*

Demeclocycline   Depresses plasma prothrombin activity. *1680*

Dextran   May prolong action of anticoagulants. *1679*

Dextrothyroxine   Increases receptor site affinity for anticoagulants. *3380*

Diazoxide   Displaces anticoagulants from albumin. *3248*

Dicumarol   Dose related effect. *1343*

Dipyrone   May aggravate prothrombin deficiency. *1680*

Disulfiram   Inhibits metabolism of coumarins. *3072*   Augmented S- warfarin hypoprothrombinemia but not that of R- warfarin. *2644*

Diuretics   May prolong action of anticoagulants. *1679*

Doxycycline   Depresses prothrombin activity. *1680*

Erythromycin   Associated with impaired availability of bile salts. *2220*   Increase by 40% in a patient within 3 d of adding drug to a regime stabilized with warfarin: possible effect on metabolism. *1706*

Estrogens, Conjugated   In normal women after i.v. administration. *0083*

Ethacrynic Acid   Displaces coumarins from albumin. *3248*

Ethanol   With large quantities and alcoholism. *2313*

Floxuridine   Manifestation of toxicity. *1680*

Fluorides   Mechanism not discussed causes bleeding. *0010*

Fluoxymesterone   Increased sensitivity to anticoagulant reported. *0019*

Glucagon   Patients on coumarins. *1971*

Glutethimide   Effect observed with overdosage. *2427*

Gold   In one 3 year old with rheumatoid arthritis. *1674*

Guanethidine   Enhances anticoagulant activity. *1343*

Halothane   May induce hepatitis. *0593*   Granulomatous hepatitis in single case following second exposure to anesthetic. *3266*

Heat   Prolonged hot weather may have effect. *1679*

Heparin   Concentration related effect. *2649*

Prothrombin Time (continued)

Plasma Increase Physiological (continued)

Hydroxyzine Prolongs action of anticoagulants. 3879

Indomethacin Displaces anticoagulants from binding protein. 2540

Kanamycin May decrease vitamin K synthesis by gut bacteria. 1487

Ketoconazole Rapidly developing liver failure in 67 year old woman taking 200 mg drug daily for 2 mo. 0960

Laxatives Accelerated gastrointestinal pass and decrease absorption of vitamin K. 1487

Levothyroxine May potentiate action of anticoagulants. 1680

Liotrix May potentiate action of oral anticoagulants. 1680

Magnesium Hydroxide Theoretically if bishydroxycoumarin co-administered. 0079

MAO Inhibitors May prolong action of anticoagulants. 1679

Mefenamic Acid Displaces coumarins from albumin. 3248

Mepazine Associated with failure of excretion of bile salts. 1713

Mercaptopurine Depresses clotting factor synthesis. 3379

Methandrostenolone Prolongs action of anticoagulants. 2313 Also seen with other 17-alkyl substituted steroids. 0998

Methotrexate May occur with hepatic malfunction. 1487

Methoxyflurane In 2 pregnant women when used as analgesia for labor: hepatitis observed. 0872

Methyldopa Enhances anticoagulant activity. 3879

Methylphenidate Inhibits metabolism of coumarins and phytonadione. 1243

Methyltestosterone Associated with failure of excretion of bile salts. 1713

Methylthiouracil Also exaggerated response to anticoagulants. 1973

Metronidazole Potentiates effect on warfarin. 1890 In one patient stabilized on warfarin to whose therapeutic regime metronidazole was added. 0850 Significant increase in concentration of warfarin and prolongation of prothrombin time. Effect noted with S (-) — warfarin but not R (+) — warfarin. 2645 Marked effect reported in a single patient previously stabilized on warfarin but to whose regime metronidazole was added. 1890

Miconazole In patients receiving warfarin, probably due to displacement from plasma protein. 3770 In patients receiving warfarin whose hepatic metabolism is inhibited. 3461

Mineral Oil Inhibits absorption of vitamin K. 2313

Minocycline May prolong action of anticoagulants. 1705

Mithramycin Reversible decreased prothrombin level. 0071

Moxalactam Persistent mild increase of 0.7s in 40 preoperative patients. 3121 Observed in some patients especially malnourished. 2615 Bleeding tendency may occur especially in debilitated patients, also impaired platelet function may occur. 3158 In up to 0.5% patients as reported from several studies. 0592

Muscular Exercise Observed effect in exercise. 2847

Nalidixic Acid Displaces coumarins from albumin. 3248

Nandrolone May increase sensitivity to anticoagulants. 2681

Narcotic Antagonists Reported enhancement of anticoagulant activity. 1487

Narcotics Prolonged use may have effect. 1679

Neomycin Reduces availability of vitamin K. 3650

Niacin Due to hepatocellular and obstructive liver damage. 3196

Niacinamide Rare probable reversible toxic effect. 3875

Norethandrolone Also seen with other 17-alkyl substituted steroids. 0998

Nortriptyline Patients on coumarins (unconfirmed clinically). 3704

Oral Contraceptives Associated with failure of excretion of bile salts. 1713

Oxymetholone Also seen with other 17-alkyl substituted steroids. 0998 May increase sensitivity to anticoagulants. 1679

Oxyphenbutazone Displaces anticoagulants from albumin. 2313

Oxyphenisatin May cause hypersensitivity reaction. 2970

Penicillin Occasional effect. 0019

Phenolphthalein Patients on coumarins. 2427

Phenylbutazone Potentiates action of anticoagulants. 0030 Increased concentration of anticoagulant displaced from protein with increase of free component. 2642 Displaces warfarin from protein binding sites. 2929

Phenyramidol Inhibits metabolism of bishydroxycoumarin. 0600

Phenytoin Patients on coumarins which inhibit its metabolism. 2220 Significant increase when drug added to regime of warfarin on which previously stabilized. 2556

Phosphorus Toxicity effect (hypoprothrombinemia). 2313

Probenecid Probably displaces anticoagulant from binding protein. 1487

Prochlorperazine Associated with impaired excretion of bile salts. 0071

Promazine Associated with failure of excretion of bile salts. 1713

Propylthiouracil Deficiency of prothrombin and proconvertin. 0071 Observed occasionally in absence of abnormal liver function. 3870 In one 24 year old woman who developed fulminant hepatic failure with lymphocyte sensitization. 2437

Pyrazinamide Reduces concentration of prothrombin. 0071

Quinidine If administration with indandiones, coumarin. 1487 Depresses prothrombin formation in liver. 1972

Quinine Depresses prothrombin formation in liver. 0894

Radioactive Compounds Also exaggerated response to anticoagulants. 1343

Radiographic Agents ?transient inactivation of coagulation factors. 1487

Reserpine Long term treatment markedly enhances anticoagulant. 1343

Rifampin May prolong action of anticoagulants. 1679

Streptomycin May decrease vitamin K synthesis by gut bacteria. 1487

Succinylsulfathiazole May cause vitamin K deficiency. 0071

Sulfachlorpyridazine Due to decreased prothrombin. 1680

Sulfamethizole Reaction may occur with all sulfonamides. 1680

Sulfamethoxazole Possibly due to drug-induced vitamin K deficiency. 1974 Displaces warfarin from its binding sites on albumin. 3595

Sulfinpyrazone May prolong action of anticoagulants. 1679 When free warfarin concentration increased due to displacement from protein. 2642

Sulfisoxazole Occurs with failure of excretion of bile salts. 0894

Sulfonamides Potentiate action of administered coumarins. 1973

Tetracycline Associated with cholestasis, and reduced activity. 0945

Thiazides Associated with failure of excretion of bile salts. 1713

Thyroid Prolongs action of anticoagulants. 2313

Tolazamide Occurs with failure to excrete bile salts. 2313

Tolbutamide Occurs with failure to excrete bile salts. 0894

Tosylate Bretylium Enhances anticoagulant activity. 1343

Trifluoperazine Associated with failure of excretion of bile salts. 1713

Valproic Acid Observed in 4 of 25 patients treated with drug: reversible with reduction of dose or withdrawal of drug. 3847 Associated with other effects on coagulation. 0504 In 3 of 9 volunteers with epilepsy poorly controlled. 3508

Vitamin A Hemorrhagic trend especially if vitamin K restricted. 2313

Warfarin Therapeutic intent (inhibits prothrombin formation). 1343

X-Ray Therapy Also exaggerated response to anticoagulants. 1343

Plasma Decrease Physiological

Amobarbital Antagonizes action of coumarins (enzyme induction). 3011

Anabolic Steroids Increased prothrombin as metabolic effect. 2526

Antacids May shorten action of anticoagulants. 1679

Antihistamines Accelerate metabolism of anticoagulants. 2313

Antipyrine Observed in man and experimental animals. 2427

Ascorbic Acid May shorten action of anticoagulants. 1679 Case reported when compound given to patient whose time was increased by anticoagulant. 0233

Aspirin Small dose effect. *2313*

Azathioprine May cause hepatotoxicity. *3433*

Barbital Antagonizes effect of bishydroxycoumarin. *3292*

Barbiturates Metabolism of coumarins enhanced (enzyme induction in liver). *1359*

Butabarbital Decreased response to anticoagulants (enzyme induction). *3379*

Caffeine Observed in patients receiving anticoagulants. *2313*

Carbamazepine Enhances warfarin metabolism by enzyme induction. *1487*

Chloral Hydrate Accelerates rate of inactivation of coumarins. *2313*

Chlordane Induces hepatic metabolism of anticoagulants. *1343*

Chlordiazepoxide May shorten action of anticoagulants. *1679*

Cholestyramine May shorten action of anticoagulants. *1679*

Colchicine Patients on coumarins. *2426*

Corticosteroids Accelerate metabolism of anticoagulants. *2313*

Corticotropin Patients on coumarins (may induce enzymes). *2426*

Cortisone Patients on coumarins. *2407*

Dichloralphenazone Antagonizes action of administered coumarins. *3011*

Digitalis Animal experiments suggest increased coagulability. *2313*

Diuretics In patients receiving anticoagulants. *2313*

EDTA Transient fall during administration (within 12 h). *2313*

Ethanol May shorten action of anticoagulants. *1679*

Ethchlorvynol Decreased potency of coumarin anticoagulants (enzyme induction). *1802*

Glutethimide Increases rate of degradation of administered coumarins. *3011* In 10 patients receiving 0.5 g/d when on warfarin due to hepatic microsomal enzyme Induction. *3649*

Griseofulvin Induces hepatic metabolism of anticoagulants. *0772*

Haloperidol May shorten action of anticoagulants. *1679*

Heptabarbital Induces hepatic metabolism of anticoagulants. *2220*

Kanamycin Effect noted at beginning of therapy. *2427*

Meprobamate Induces hepatic metabolism of anticoagulants. *1343*

Methaqualone In 10 patients receiving 0.3 g/d also on warfarin: effect mild and not significant. Weak inducer of hepatic microsomal enzymes. *3649*

Metharbital Theoretical possibility due to enzyme induction. *1680*

Nicotine May partially counteract action of heparin. *1679*

Oral Contraceptives Decreases response to oral anticoagulants. *0894*

Orphenadrine May enhance metabolism of anticoagulants. *1679*

Paraldehyde May shorten action of anticoagulants. *1679*

Phenobarbital Metabolism enhanced by barbiturates. *0071* With 0.1 g/d in 10 patients receiving warfarin due to induction of hepatic microsomal enzymes. *3649*

Phytonadione Stimulates synthesis of clotting factors. *3379*

Primidone May shorten action of anticoagulants. *1679*

Reserpine Short term treatment blocks action of anticoagulant. *1343*

Rifampin Occurs in some patients, especially if on anticoagulants. *1375* Due to induction of enzymes metabolizing warfarin when coadministered. *3683*

Secobarbital Induces metabolism of administered coumarins. *3011* With 0.1 g/d in 10 patients receiving warfarin due to hepatic microsomal enzyme induction. *3649* Increase in plasma clearance of warfarin. *2647*

Simethicone Impairs absorption of oral anticoagulant. *1487*

Spironolactone When compared with results when patients treated with warfarin alone: augmented effect when spironolactone coadministered probably due to hemoconcentration due to diuresis. *2643*

Tetracycline May partially counteract action of heparin. *1679*

Thiothixene Rarely reported side effect. *0071*

Tolbutamide Stimulates metabolism of anticoagulants. *1487*

Vitamin K Therapeutic intent, affects action of warfarin. *2313*

Xanthine Antagonizes effects of coumarins. *2313*

Blood   No Effect   Physiological

Moxalactam 4 g intravenously 12 hourly had no effect in 6 subjects. *2629*

Blood   Increase   Physiological

Methyldopa Decreased synthesis of prothrombin associated with hepatitis in some cases. *2261*

Vitamin E In 2 patients given 2.3 g/m²/d intravenously for 4 or more days (effect abrogated by prior administration of menadiol sodium diphosphate). *1554*

Protoporphyrin

Blood   Increase   Physiological

Barbiturates May precipitate acute cutaneous porphyria. *1322*

Chlorpropamide May precipitate cutaneous porphyria. *1322*

Estrogens May precipitate porphyria attack. *1322*

Ethanol May precipitate attack of porphyria. *1322*

Oral Contraceptives May precipitate porphyria attack. *1322*

Sulfomethane May provoke attack of porphyria. *1322*

Tolbutamide May precipitate cutaneous porphyria. *1322*

Urine   Increase   Physiological

Chlorpropamide May precipitate cutaneous porphyria. *1322*

Lead In a population of male battery workers compared with control group in cement industry. *0172*

Tetraethyl-Lead Erythrocyte protoporphyrin excretion slight increase. *0258*

Feces   Increase   Physiological

Apronalide May precipitate acute porphyria. *1322*

Barbiturates May precipitate acute porphyria. *1322*

Chlordiazepoxide May precipitate acute porphyria. *1322*

Chlorpropamide May precipitate cutaneous porphyria. *1322*

Dichloralphenazone May precipitate acute porphyria. *1322*

Diethylstilbestrol May induce porphyria cutanea tarda. *3769*

Estrogens May induce porphyria cutanea tarda. *3769*

Estrogens, Conjugated May induce porphyria cutanea tarda. *3821*

Ethanol May precipitate attack of porphyria. *1322*

Ethinyl Estradiol May induce porphyria cutanea tarda. *3769*

Glutethimide May precipitate acute porphyria. *1322*

Griseofulvin May precipitate acute porphyria attack. *1322*

Meprobamate May precipitate acute porphyria. *1322*

Methylsulfonal May precipitate acute porphyria. *1322*

Oral Contraceptives May precipitate porphyria attack. *1322*

Pentazocine May precipitate acute porphyria attack. *1322*

Phenytoin May precipitate acute porphyria. *1322*

Succinimide May precipitate acute porphyria. *1322*

Sulfomethane May provoke attack of porphyria. *1322*

Sulfonamides May precipitate acute porphyria. *1322*

Tolbutamide May precipitate cutaneous porphyria. *1322*

Red Blood Cells   Increase   Physiological

Lead Occurs with poisoning. *0987*

Pseudocholinesterase

Serum   No Effect   Analytical

Freezing Stable for up to 12 mo at -20° C and unaffected by freezing and thawing with benzoyl choline or butyrylthiocholine as substrate. *3635*

Hemolysis If minor does not affect Acholest method. *1680*

Storage of Sample 2 d room temperature, 14 at 4°, months at -20°. *3217* 4 d room temperature, 6 mo at -10°. *1180* No effect 1 week at 4°, 1 mo at -20°. *3873* For 17 d at 23 or 4°, 3 mo at -20°. *0877*

Pseudocholinesterase *(continued)*

Serum   Increase   Analytical
Glutathione   If used as standard reacts with ACH in Levine method. *1024*

Serum   Decrease   Analytical
Fluorides   Total inhibition occurs. *3651*

Serum   Decrease   Physiological
Acetaminophen   In severe cases of poisoning (also in some mild). *3470*
Aldrin   *In vitro* 6.0% decrease at 1×10^{-4} mol/L. *1270*
Ambenonium   Direct action of drug. *1147*
Asparaginase   Induces impairment of hepatic synthesis. *0769*
Barbiturates   May inhibit activity. *1680*
Benzene   *In vitro* 7.1% decrease at 1×10^{-4} mol/L. *1270*
Carbon Disulfide   Not usually severe enough to produce symptoms. *1302*
Cyclophosphamide   Causes inhibition of activity. *3959*
Demecarium   Therapeutic intent. *0071*
Dieldrin   *In vitro* 7.2% decrease at 1×10^{-4} mol/L. *1270*
Dimpylate   *In vitro* 8.2% decrease at 1×10^{-4} mol/L. *1270*
Echothiophate   After few weeks of eyedrop therapy. *1678*
Edrophonium   Direct action of drug. *1147*
Estrogens   Altered liver metabolism. *3332*
Halothane   Low frequency but greater number of abnormalities than with enflurane. Occurred most frequently in obese patients. *1089*
Insecticides   Direct effect of drug. *3523*
Iodipamide   Fairly powerful inhibitor. *2088*
Iopanoic Acid   Potent inhibitor at 0.06 mmol/L. *2088*
Isoflurophate   Therapeutic action of drug. *0071*
Mechlorethamine   Observed activity *in vitro*, probable *in vivo*. *3959*
Metrifonate   Powerful inhibitor of enzyme. *2427*
Neostigmine   Direct effect of drug. *3523*
Neuromuscular Relaxants   If enzyme deficiency may cause toxicity. *1343*
Opium Alkaloids   May cause inhibition of activity. *1680*
Oral Contraceptives   Estrogen effect. *3505* Significant effect observed within 3 mo usually continuing for several y. *3170* Significant reduction by about 12% in drug users versus controls. *2133*
Organophosphorus Insecticides   Hepatotoxic effect. *0405*
Pancuronium   ?*in vivo*, but *in vitro* 25 μmol causes 80% inhibition. *3410*
Parathion   Inhibitory action of drug with poisoning. *1302*
Phenelzine   May cause hypersensitive hepatitis. *3505*
Phenothiazines   Also affects erythrocyte enzyme. *1487*
Physostigmine   Direct effect of drug. *3523*
Procainamide   At therapeutic concentration *in vitro* inhibition of enzyme by 15 to 30%. *1853*
Pyridostigmine   Direct action of drug. *1147*
Streptokinase   Possible acute hypoxic or toxic liver damage. *3189*
Testosterone   Metabolic effect. *0227*
Thiotepa   Mild depressive effects reported. *1487*
Triethylenemelamine   Observed *in vitro*, probable effect *in vivo*. *3959*
Tryptamine   Observed effect in humans. *1176*

Red Blood Cells   Increase   Physiological
Chlorinated Insecticides   Powerful inhibitor of enzyme. *0405*

Red Blood Cells   Decrease   Physiological
Echothiophate   Direct effect of drug (specific inhibitor of enzyme). *3523*
Parathion   Specific inhibitor of enzyme. *3850*
Tryptamine   Observed effect in humans. *1176*

Test Conditions   Decrease   Analytical
Neostigmine   Inhibitory effect observed. *2893*
Quaternary Ammonium Compounds   Inhibitory effect observed. *2893*
Tubocurarine   Inhibitory effect observed. *2893*

Pseudoephedrine

Serum   Increase   Physiological
Aluminum Hydroxide   Increased rate of absorption due to raised gastric pH. *3794*

Serum   Decrease   Physiological
Kaolin   Delayed absorption due to adsorption. *3794*

Pseudouridine

Urine   Increase   Physiological
X-Ray Therapy   Due to tissue destruction. *2378*

PSP Excretion

Urine   Increase   Analytical
Anthraquinone   Interference by color of compound. *1999*
Bile   Interfering color. *2313*
Cascara   Converted to anthraquinone-gastrointestinal tract (red color). *0757*
Ethoxazene   Produces interfering background color. *1022*
Formaldehyde   Produces interfering color reaction. *0757*
Hemoglobin   Interfering color. *2313*
Methenamine   Produces interfering color in urine. *0757*
Phenazopyridine   False color reaction at alkaline pH. *3879*
Phenolphthalein   Color development on alkalinization of urine. *3879*
Sulfobromophthalein   Development of color at alkaline pH. *3879*

Urine   Increase   Physiological
Levodopa   Increased plasma flow. *1136*
Prostaglandin-α_2   If given i.v. as 0.4 μg/min (slight effect). *1739*
Radiographic Agents   Affect renal excretion. *1237*
Sulfamethoxazole   Possible large dose effect. *1022*

Urine   Decrease   Physiological
Acetazolamide   Increased urine flow. *1022*
Aspirin   Competes with PSP for excretion. *3879*
Atropine   Interferes with secretion by tubules. *2981*
Capreomycin   Observed with other signs of decreased renal function. *1680*
Carinamide   Inhibits secretion. *1022*
Diuretics   Blocking of secretory mechanism. *1022*
Iodopyracet   May compete for excretion through renal tubules. *0974*
Novobiocin   ?nephrotoxic effect. *1022*
Penicillin   Interference with PSP excretion. *3879*
Probenecid   Inhibits renal transport of PSP. *3879*
Sulfamethoxazole   More common effect. *1022*
Sulfinpyrazone   Competition for excretion. *0974*
Sulfonamides   Reduces renal excretion of PSP. *3879*
Thiazides   Competition for excretion by tubules. *0974*

Pyridoxal

Serum   No Effect   Analytical
Menstruation   No variation during cycle. *0716*

Serum   Increase   Physiological
Pyridoxal   In response to loading dose. *0716*

Serum   Decrease   Physiological
Iproniazid   Observed in patient with toxicity. *0738*

Urine   No Effect   Physiological
Menstruation   No change observed during cycle. *0716*

Urine   Increase   Physiological
Pyridoxal   In response to loading dose. *0716*

Pyridoxal Phosphate

Serum   No Effect   Physiological
Anticonvulsants   In 146 epileptics with long-term treatment. *2007*
Menstruation   No variation during cycle. *0716*

Serum   Increase   Physiological
Pyridoxal   In response to loading dose. *0716*
Pyridoxine   Increased after administration in normal subjects. *2427*

Serum   Decrease   Physiological
Anticonvulsants   Increase by 20% at 4 weeks, 60% at 12 weeks. *2948*
Ethanol   Progressive decrease observed in alcoholics. *1532* Excess consumption affects formation *in vivo*. *2231* In alcoholic patients with or without liver dysfunction. *2428*
Phenytoin   Increase by 20% at 4 weeks, 60% at 12 weeks. *2948*
Pregnancy   Progressive decrease to term. *1472*
Succinimide   Increase by 20% at 4 weeks, 60% at 12 weeks. *2948*

Plasma   Decrease   Physiological
Vitamin B6 Depletion   No change for 15 d then marked fall. *0255*

Urine   Increase   Physiological
Pyridoxal   In response to loading dose. *0716*

Pyridoxamine

Serum   No Effect   Physiological
Menstruation   No variation during cycle. *0716*

Urine   No Effect   Physiological
Menstruation   No change observed during cycle. *0716*

Pyridoxamine Phosphate

Serum   No Effect   Physiological
Menstruation   No variation during cycle. *0716*

4-Pyridoxic Acid

Serum   No Effect   Physiological
Menstruation   No variation during cycle. *0716*

Serum   Increase   Physiological
Pyridoxal   In response to loading dose. *0716*

Urine   No Effect   Physiological
Menstruation   No change observed during cycle. *0716*

Urine   Increase   Physiological
Pyridoxal   In response to loading dose. *0716*

Urine   Decrease   Physiological
Vitamin B6 Depletion   Zero detectable after 25 d deprivation. *0255*

Pyridoxine

Serum   Decrease   Physiological
Iproniazid   Observed in patient with toxicity. *0738*
Isoniazid   Observed in 13% of 38 children when measured with protozoan procedure: also reported in 2% of adults after 6 mo. *2769*
Levodopa   Associated with burning feet syndrome. *0824*
Vitamin B6 Depletion   20% control value after 5 d, zero after 25 d. *0255*

Urine   No Effect   Physiological
Ascorbic Acid   No effect with varying degrees of supplementation over 14 week study period in young men. *1751*

Urine   Increase   Physiological
Isoniazid   Dose related effect. *3025* Particularly prevalent with use of large doses of drug or in nutritionally deficient persons such as alcoholics. *3683*

Urine   Decrease   Physiological
Starvation   Falls to level below recommended in 3 d. *0714*
Vitamin B6 Depletion   Marked decrease within few days. *0255*

Pyrophosphate

Urine   Increase   Physiological
Bed Rest   Mild elevation with bed rest, fall with reambulation. *0927*
Hydrochlorothiazide   Associated with increased orthophosphate also. *2709*

Urine   Decrease   Physiological
Ambulation   Falls below normal after high of bed rest. *0927*

Pyrrole-2-Carboxylate

Urine   Increase   Physiological
Hydroxyproline   Action of alpha-amino acid oxidases (both d- and l-). *1536*
Proline   Exact mechanism obscure. *1536*

Pyruvate

Serum   No Effect   Physiological
Ritodrine   Little effect noted in patients given drug intravenously for premature labor, with consequent increase in lactate/pyruvate ratio. *0575*

Serum   Increase   Analytical
Hemolysis   Due to metabolic activity of cells. *3873*

Serum   Increase   Physiological
Aspirin   Due to late metabolic acidosis and renal impairment. *1343*
Estrogens   Action like glucocorticoids. *1849*
Ether   Metabolic effect. *3918*
Fructose   Rises more rapidly than after glucose. *1709*
Glucose   Increased concentration observed after i.v. infusion. *1160*
Muscular Exercise   After exercise to exhaustion 1.7 times control. *2846*
Nifedipine   In 15 hypertensive patients undergoing hemodialysis after 3 weeks treatment. *2986*
Oral Contraceptives   Greater increase than normal during glucose tolerance test. *0585* Estrogen effect of combined oral contraceptive. *0018*
Streptozocin   Temporary effect following infusion. *2532*
Sucrose   Increase maximum at 1 h, lactate/pyruvate ratio inc. *1907*

Pyruvate *(continued)*

Serum   Increase   Physiological *(continued)*
Tetracosactrin   In healthy volunteers given 1 mg intramuscularly for up to 60 h. *1813*
Valproic Acid   When 1 g given orally to fasting individuals. *3633*

Serum   Decrease   Analytical
Pyruvate Kinase   If LDH used is contaminated with pyruvate kinase. *1557*

Serum   Decrease   Physiological
Muscular Exercise   Slight fall within 1 h and then steep drop. *0987*
Sorbitol   In response to i.v. infusion (effect slight). *1160*
Xylitol   Slight fall after i.v. infusion reported. *1160*

Urine   No Effect   Physiological
Lithium   No effect observed after administration. *0409*

Saliva   Increase   Physiological
Fructose   Occurs at 1 h, as with blood. *1907*
Sucrose   Increase maximum at 1 h, lactate/pyruvate ratio inc. *1907*

Pyruvate Kinase

Platelets   Increase   Analytical
EDTA   5 fold increase if added to assay mixture. *3543*

Red Blood Cells   No Effect   Analytical
Storage of Sample   If heparin added stable for 8 d at 30° If Alsever's solution added for 3.5 weeks at 30°. *0877*

Red Blood Cells   Increase   Analytical
2,3-Diphosphoglycerate   Activating effect on enzyme. *0370*
EDTA   1.5 fold increase if added to assay mixture. *3543*
Fructose-1,6-Diphosphate   Activating effect on enzyme. *0370*
Glucose-6-Phosphate   Activating effect on enzyme. *0370*
Phosphates   Activating effect on enzyme. *0370*
Phosphoenolpyruvate   Activating effect on enzyme. *0370*
3-Phosphoglyceric Acid   Activating effect on enzyme. *0370*

Red Blood Cells   Decrease   Analytical
Acetates   Observed with 0.3 mmol/L PEP in reaction mixture. *0370*
Alanine   Most marked with larger side chain. *0370*
Bromides   Observed with 0.3 mmol/L PEP in reaction mixture. *0370*
Chloride Salts   Observed with 0.3 mmol/L PEP in reaction mixture. *0370*
1,2-Diaminopropane   Most marked with larger side chain. *0370*
Ethylamine   Most marked with larger side chain. *0370*
Glycine   Most marked with larger side chain. *0370*
Isoleucine   Most marked with larger side chain. *0370*
Leucine   Most marked with larger side chain. *0370*
Methylamine   Most marked with larger side chain. *0370*
Phenylalanine   Most marked with larger side chain. *0370*
Proline   Most marked with larger side chain. *0370*
Propionate   Most marked with larger side chain. *0370*
Valine   Most marked with larger side chain. *0370*

Red Blood Cells   Decrease   Physiological
Copper   several affected *in vitro.* *0432*

White Blood Cells   Increase   Analytical
EDTA   15 fold increase if added to assay mixture. *3543*

Pyruvate Tolerance

Blood   Decrease   Physiological
Ethanol   In 35 patients with alcohol related illness. *0207*

Quinidine

Serum   Increase   Analytical
Triamterene   Interfering fluorescence. *1237*

Serum   Increase   Physiological
Acetazolamide   Alkalinizes urine, increases reabsorption. *1487*
Alkaline Urine   Marked effect if given when pH over 7. *1487*
Amiodarone   Drug reported to produce 50% increase in concentration. *0548*   Rapid effect possibly due to change in protein or receptor binding. *3546*   Average 32% in 11 patients allowing 37% reduction in dosage. *1394*
Chlorothiazide   Alkalinizes urine, increases reabsorption. *1487*
Cimetidine   55% increase in mean half-life: protein binding and urinary excretion unchanged effect studied in 6 healthy volunteers. *1493*   Increased area under curve and prolonged half-life and decreased total body clearance. *1983*
Diuretics   If administration concomitantly and alkalinize urine. *1487*
Hydrochlorothiazide   Alkalinizes urine, increases reabsorption. *1487*
Nifedipine   Marked effect observed once drug discontinued, probably hemodynamically induced interaction. *1082*
Rifampin   Concentration of metabolites increased with probable maintenance of therapeutic effects. *0548*
Sodium Bicarbonate   Alkalinizes urine, increases reabsorption. *1487*
Thiazides   Alkalinize urine, increase reabsorption. *1487*

Serum   Decrease   Analytical
Blood Collection Tube   Observed 4 to 8% reduction in plasma concentration when some Vacutainers™ used compared with glass. *3271*
Tris(2-Butoxyethyl) Phosphate (TBEP)   By up to 32% at 4.4 μmol/L when used as plasticizer in evaluated blood tubes. *0892*

Serum   Decrease   Physiological
Disopyramide   Small effect noted, but elimination half-life not significantly affected. *0208*
Nifedipine   Marked decrease when drugs coadministered but mechanism not clearly understood. *1388*
Phenytoin   Reduces half-life by more than 50% due to induction of hepatic cytochrome P-450 activity. *2020*
Rifampin   Observed in patients receiving antituberculous treatment in addition to drug for cardiac irregularity. *0549*   Reduction in peak concentration reported in one patient. *0014*

Quinine

Serum   Increase   Physiological
Pyrimethamine   May displace from plasma protein binding. *1487*

Urine   Positive   Analytical
Chlorphentermine   Similar R_f and color reaction on TLC using ethyl acetate: methanol: water: ammonium hydroxide and modified Dragendorff's reagent. *3868*
Phenmetrazine   Similar R_f and color reaction on TLC using ethyl acetate: methanol: water: ammonium hydroxide and modified Dragendorff's reagent. *3868*
Phentermine   Similar R_f and color reaction on TLC using ethyl acetate: methanol: water: ammonium hydroxide and modified Dragendorff's reagent. *3868*

Quinolinic Acid

Urine   Increase   Physiological
Deoxypyridoxine   After 2 g tryptophan load. *3050*
Vitamin B₆ Depletion   Observed with experimental dietary deficiency. *3045*

Ranitidine

Serum   Increase   Physiological
Propantheline   Significantly reduced time to peak concentration but area under curve increased. *1938*

Recalcification Time

Blood   Increase   Physiological
Dicumarol   Effect on citrated plasma. *1343*

Red cell-associated Immunoglobulin IgG

Blood   Positive   Physiological
Methyldopa   Observed in 5 of 9 patients typically seen in 20% of treated patients. Probably due to impaired F_c-dependent reticuloendothelial function. *1909*

Reduced Glutathione

Red Blood Cells   Decrease   Physiological
Acetanilid   Sharp fall before overt hemolysis. *1902*
Aminopyrine   Sharp fall before overt hemolysis. *1902*
Antimalarials   Occurs prior to overt hemolysis. *1902*
Antipyretics   Sharp fall before overt hemolysis. *1902*
Aspirin   Occurs initially before overt hemolysis. *1902*
Copper   Marked decrease with hemolysis of toxicity. *0385*
Lysol®   Marked effect with hemolysis. *0629*
Nitrofurans   Occurs with hemolysis. *1902*
Primaquine   Associated with hemolysis and G-6-PD deficiency. *3171*
Sulfonamides   Occurs early before overt hemolysis. *1902*
Sulfones   Falls sharply preceding hemolysis. *1902*

Renal Blood Flow (RBF)

Patient   No Effect   Physiological
Verapamil   No change compared with placebo in 11 patients treated with up to 360 mg daily for 6 weeks. *3396*

Patient   Increase   Physiological
Enalapril   May increase without increase in GFR in patients with normal renal function, but GFR may increase if initial GFR below 80 mL/min/m². *0823*
Levarterenol   After i.v. but increased filtration fraction. *1176*

Patient   Decrease   Physiological
Cyclosporine   320 mL/min versus 480 mL/min in 17 cardiac transplant recipients compared with those on azathioprine. *2536*
Indomethacin   Associated with decreased secretion of renin and aldosterone, decreased sodium delivery to distal tubule and reduction of urinary flow rate. *1228*
Metoclopramide   In 20 patients given 1 to 2.5 mg/kg (from mean 443 mL/min to 387 mL/min). *1737*
Propranolol   13% decrease if continued orally or after i.v. injection. *1714*

Renin Activity

Plasma   No Effect   Physiological
Adenosine   After i.v. infusion of 10 to 140 µg/kg/min in 7 normal subjects. *0338*
Arginine   No significant changes observed in 6 normal and 4 diabetic subjects after i.v. infusion. *2328*
Aspirin   In patients in whom it had been increased with sodium restriction. *2535*
Carbohydrate   No effect when starved patients refed. *1242*
Carprazidil   No consistent effect in 15 men with mild to moderate essential hypertension treated for up to 16 weeks. *1268*
Chlorthalidone   Insignificant change in 10 hypertensives patients given 25 mg daily for 16 weeks. *1229*
Domperidone   After 10 mg intravenously in 8 healthy males. *3406*
Dopamine   Did not change with infusion of up to 3.0 µg/kg/min for 2 h in 6 normal men. *2149*
Gentamicin   Effect observed in 2 patients following larger doses of gentamicin. Associated with massive urinary loss of magnesium and potassium. *0224*
Hemodialysis   Not affected by period requiring dialysis. *0750*
Indomethacin   In 12 patients with coronary artery disease, although systolic blood pressure increased and coronary blood flow decreased. *1158*   Insignificant change with 1 week treatment. *2464*
Labetalol   With treatment for 4 mo in 15 patients with essential hypertension variable effect observed. *3776*
Metoclopramide   No change observed in normal individuals or in patients with primary aldosteronism after 10 mg drug. *3883*   No significant change following i.v. injection. *0497*   No effect even though other constituents affected by i.v. bolus of 10 mg. *2543*
Minoxidil   No significant effect when given alone or with added diuretics at 3 or 6 mo. *1807*
Naproxen   In 10 furosemide treated patients with well controlled congestive heart failure. *1042*
Nilvadipine   In 10 patients with mild essential hypertension. *3530*
Oral Contraceptives   Normal activity in users in spite of changes in renin substrate concentration. *1333*
Oxaprozin   No significant change observed in 7 volunteers treated for up to 1 week. *2464*
Oxytocin   No significant or consistent effect. *1344*
Potassium   No effect observed with infusion. *0914*
Prazosin   No significant change with monotherapy for 12 mo in 15 patients. *3531*
Prenalterol   Not significantly different from control state in 23 patients with stage III heart disease over 1 mo under controlled conditions. *3768*
Ranitidine   No effect of 3-d of 150 mg/12 h. *3130*
Smoking   No difference between smokers and nonsmokers and not affected by acute smoking. *1361*
Sulindac   In 9 individuals with stable renal insufficiency when treated for 9 d. *2457*   No significant change in patients with chronic glomerular disease. *0660*   In 10 furosemide treated patients with well controlled congestive heart failure given 400 mg/d for 2 d. *1042*
Trimazosin   No significant change observed. *0861*
Vasopressin   No demonstrable effect if not overhydrated. *1344*
Verapamil   In 15 patients with uncomplicated essential hypertension. *0867*

Plasma   Increase   Physiological
Aging   Much higher in neonates, than children, than adults. *2006*
Albuterol   Significant dose-related effect after i.v. infusion of therapeutic doses in 4 healthy male volunteers. *2807*
Ambulation   Activity increased by moderate to severe exercise. *0182*
Amiloride   13.6 fold increase, maximal after 5 d treatment. *2440*
Azosemide   Effect observed in normal male volunteers after 60 mg orally. *2658*
Bendrofluazide   Mean increase from 3.3 to 5.6 mmol/L in 7 essential hypertensives treated for 16 weeks. *0352*   Approximately doubled in 5 or 6 hypertensives given 5 mg daily for 6 weeks. *3866*

Renin Activity (continued)

Plasma Increase Physiological (continued)

Captopril From 10.0 to 34.7 pmol/L/s in patient with Bartter's syndrome after 1 mo. *1558* Significant effect in 1 h after single dose of 25 mg. *0181* Marked effect within 30 minutes on active and total renin with reduction of inactive renin. *2748* Gradually increases over 1 h due to negative feedback on renin activity. *0177*

Chlorthalidone At least 3 fold increase in pre- and post-menopausal women given 100 mg/d for 6 weeks. *0402* At 3 d and after 3 mo in 8 patients with mild hypertension given 50 mg/d. *3040*

Diazoxide Effect observed in some hypertensives. *1343*

Diltiazem Increase by 0.8 ng/mL/h in 20 patients with essential hypertension with 14 weeks treatment. *3521*

Dobutamine In 10 patients with congestive cardiac failure from 11.3 to 17.8 ng/mL/h on average. *3654*

Doxazosin Acute increase in 24 patients treated for 6 weeks. *3852*

Enalapril Steady increase from administration of first dose. *1616* Gradually increases over 1 h (due to negative feedback on renin secretion). *0177* Within 4 d as part of long-term response in 6 responders of 10 treated hypertensives. *1166*

Endralazine Significant effect in 20 hypertensive patients (WHO phase I and II). *1524*

Enoximone In 10 patients with congestive cardiac failure from 13.6 to 16.6 ng/mL/h on average. *3654*

Erect Posture Increases approximately 2 times if normal diet. *0071*

Estrogens Due to increased substrate. *0557*

Ethinyl Estradiol Men and women, independent of diet and posture. *2427* Usual but not constant in 2 weeks. *2357*

Fasting During total fasting in obese individuals. *3695*

Felodipine When infused in 10 healthy normotensive volunteers. *3352* With 20 mg daily in 10 men with essential hypertension over 8 weeks. *1887*

Furosemide Most often slight increase in patients with chronic congestive heart failure in response to i.v. drug but not consistent pattern of response. *2513* Effect observed after 40 mg orally in normal male volunteers. *2658*

Heat After week of thermal stress increased by 174%. *0206*

Hydralazine 7 fold increase at 1 mg/kg in rats. *2798* Significant effect following 20 mg i.v. in 30 to 120 minutes. *1609*

Indapamide But not significant in 8 hypertensives treated for 16 weeks. *0352*

Labetalol With treatment for 4 mo in 15 patients with essential hypertension variable effect observed. *3776*

Laxatives Marked increase in chronic abuser with dehydration. *3682*

Menstruation Follows aldosterone in half cases. *1885*

Methyclothiazide Stimulates plasma renin activity. *1353*

Metolazone Approximately doubled after 6 weeks treatment with 5 mg daily in 5 or 6 hypertensives. *3866*

Muzolimine Statistically significant increase in 10 hypertensive patients given 30 mg daily. *1229*

Nifedipine Observed within 3 h in response to 10 mg drug. *1609* From 2.26 to 3.72 ng/mL/h in young hypertensives after 3 h. *1609* In 25 patients with mild to moderate primary hypertension given up to 80 mg daily. *2659*

Opiates Significant increase within 1 h of i.m. injection of hyoscine, meperidine or omnopon. *1929*

Oral Contraceptives Estrogen effect. *3332* Activity increased although concentration lowered. *0557*

Perindopril After doses of 4 mg twice daily within 4 h of administration. *0550*

Pregnancy Higher in first half. *3345* Increased in normal pregnancy, decreased with pre-eclampsia. *1981*

Ramipril In response to 10 mg and 20 mg on successive days in 9 patients with severe chronic congestive heart failure. *0764*

Rimeterol Dose related significant change in 4 healthy men given therapeutic i.v. dose. *2807*

Spironolactone Usual response observed. *0684* Marked increase in supine activity on constant diet. *1144* 9.5 fold increase after 5 d in 5 volunteers. *2440*

Starvation Significant effect in obese after 10 d. *1242*

Plasma Decrease Physiological

Altitude At 12,000 feet (both supine and standing). *2001*

Angiotensin Causes decrease in normals when given i.v. *1145*

Aspirin Mean reduction from 2.94 ng/mL/h to 1.41 ng/mL/h when treatment started, but rose following treatment before returning to baseline in 10 weeks. *0487*

Bopindolol In 19 patients with essential hypertension reduced mean pretreatment concentration of 2.6 μg/L/h to 1.6 μg/L/h with chronic treatment. *1689*

Captopril In 6 nonazotemic patients with cirrhosis and ascites due to increased renal vasodilatation. *2728*

Carbenoxolone Significant decrease after adrenalectomy (standing and lying). *3801*

Clonidine Dose related effect (secondary to action of catecholamines). *1627*

Cyclosporine Significantly lower than in azathioprine treated patients or in those with renal insufficiency, possibly due to expansion of extracellular fluid. *0223* Inappropriately low in several patients receiving drug possibly due to concomitant beta-blockers concentration. *2836*

Desoxycorticosterone Although on low sodium diet. *3263*

Estrogens Inhibition of circulating level by angiotensin. *0557*

Ethinyl Estradiol Usual but not constant in 2 weeks. *2357*

Glycyrrhiza Pseudoaldosteronism effect. *3755*

High Sodium Diet Reverse effect with low sodium diet. *3263*

Hydralazine May be suppressed activity. *2397*

Ibuprofen 59% reduction in patients being treated. *0660*

Indomethacin In 9 uncomplicated essential hypertensives receiving chronic enalapril treatment. *3126* Marked reversal on withdrawal of drug in one young woman. *3541*

Labetalol With treatment for 4 mo in 15 patients with essential hypertension variable effect observed. *3776*

Metoprolol In 15 patients with essential hypertension treated for 4 weeks, associated with reduction in sympathetic tone and reduced activity of renin-aldosterone system. *1209*

Naproxen When given for 14 d to patients with rheumatoid arthritis and heart failure. Marked effect. *3514*

NSAID (Nonsteroidal Anti-Inflammatory Drugs) Associated with inhibition of prostaglandin synthesis. *1561*

Oral Contraceptives Metabolic changes in liver synthesis. *2189* From 30 ng/ml/h to 13 at 1 week onwards. *0265*

Oxprenolol In 10 patients with essential hypertension treated over 5 weeks: significant effect with plasma concentration of about 1,000 ng/mL. *1114*

Pregnancy If pre-eclampsia compared with normal pregnancy. *2180*

Propranolol Observed in all patients, maximum if high initially. *0523* Significant effect over 5 weeks in 10 patients with essential hypertension with mean plasma concentration of drug about 244 ng/mL. *1114*

Recumbency Fell by just over half in 3 h. *0733*

Sodium Salts Comparable decreases observed with chloride and citrate salts in 5 men with essential hypertension. *2037*

Sulindac When given for 14 d to patients with rheumatoid arthritis and heart failure. Marked effect. *3514*

Renin Substrate

Plasma No Effect Physiological

Captopril No significant effect although marked changes in PRA and angiotensin II. *0181*

Menstruation No change observed. *1885*

Plasma Increase Physiological

Estrogens Hormonal action. *0557*

Oral Contraceptives Metabolic changes in liver synthesis. *2189* From 1.3 μg/mL to 4 at 1 week, 6 at 3 mo. *0265* Hormonal effect. *0071*

Pregnancy Higher in second half. *3345*

Plasma   Decrease   Physiological
Oral Contraceptives   Almost 4 fold increase in oral contraceptive users compared with nonusers probably due to stimulated hepatic synthesis. *1333*

Respiratory Peak Flow

Patient   No Effect   Physiological
Oral Contraceptives   No significant effect observed. *1185*

Reticulocytes

Blood   No Effect   Physiological
Captopril   In 9 of 12 hypertensive patients although other hematological effects observed. *1608*
Chenodeoxycholic Acid   No significant effect with up to 750 mg/d over 2 y in 916 patients. *3199*

Blood   Increase   Physiological
Acetanilid   Occurs during recovery from hemolysis. *1902*
Allopurinol   Mild (may be associated with other drugs). *0542*
Aminopyrine   May occur during recovery from hemolysis. *1902*
Amyl Nitrite   Hemolytic anemia (up to 15%). *2426*
Antimalarials   Occurs during recovery from hemolysis. *1902*
Antipyretics   Occurs during recovery from hemolysis. *1902*
Antipyrine   Occurs with hemolytic anemia (recovery). *0788*
Arsenicals   Values observed ranging up to 18%. *1478*
Aspirin   Response during recovery from hemolysis. *1902*
Corticotropin   Especially if given to anemics. *1343*
Dimercaprol   Hemolytic anemia in G-6-PD deficient persons. *0788*
Furadaltone   Occurs with hemolysis. *0248*
Furazolidone   Occurs with hemolysis. *0248*
Hydralazine   Observed in one patient receiving low dose of drug for hypertension. Mechanism involved in producing erythrocyte-reacting antibody not known. *2680*
Lead   Due to stimulation of hemolysis. *1343*
Levodopa   Hemolytic anemia when decarboxylase inhibitor also. *0481*
Methyldopa   Hemolytic anemia of autoimmune type occurring in fewer than 1% of treated patient. *0461*
Naproxen   Isolated case of hemolytic anemia: one other case reported previously. *1681*
Nitrofurans   Occurs during recovery after hemolysis. *1902*
Penicillin   Consequence of hemolytic anemia. *0996*
Pipobroman   Hemolytic anemia has been described. *0070*
Procainamide   In 3% of all patients receiving drug, but in 14% of patients with direct Coombs' test. *1958*
Quinidine   Occasional hemolytic anemia and hypersensitivity observed in 32 of 487 patients who had received drug. *1261*
Sulfonamides   In response to hemolysis occurs with recovery. *1343*
Sulfones   Occurs during recovery phase. *1902*
Tolmetin   Observed in single case of 49 y old man who had taken drug as needed for arthritis for 1 y. *3421*

Blood   Decrease   Physiological
Azathioprine   Gradual reduction observed. *1904*
Chloramphenicol   Reversible toxic reaction or pancytopenia. *2426*
Chlorpropamide   Isolated case of pure red blood cell aplasia. *2828*
Cytarabine   In fewer than 2%. *1680*
Dactinomycin   Reticulocytopenia/pancytopenia. *0071*
Methotrexate   Affects hematopoiesis. *1343*
Sulfonamides   Agranulocytosis with aplastic or hemolytic anemia. *1788*
Vinblastine   Absence from peripheral blood observed. *1343*

Retinol

Plasma   No Effect   Physiological
Ascorbic Acid   No effect with varying degrees of supplementation over 14 week study period in young men. *1751*

Retinol Binding Protein

Serum   No Effect   Physiological
Anticonvulsants   In 146 epileptics with long-term treatment. *2007*
Ethanol   No different in chronic alcoholics and controls. *0367*
Gentamicin   No difference during or following treatment in 26 patients given drug. *2593*
Sisomicin   In 23 patients given drug for 2 weeks in therapeutic amounts. *2593*

Serum   Increase   Physiological
Anticonvulsants   In 8 epileptic patients receiving phenobarbital with carbamazepine or phenytoin. *2670*
Oral Contraceptives   Increased to about 150% of controls in women on a variety of oral contraceptives. *1303*   Values doubled in 6 mo in 8 women taking estrogenic combination but diminished to some extent after 4 y. *0773*   Values almost doubled within 6 mo but diminished somewhat after 4 y. *0773*   Due to direct effect on synthesis of protein. *3023*
Phenobarbital   In comparison with untreated cognitively delayed children 4.0 mg/dL on average versus 3.3 mg/dL. *2003*
Phenytoin   In comparison with untreated cognitively delayed children 4.0 mg/dL versus 3.3 mg/dL. *2003*

Serum   Decrease   Analytical
Citrates   Dilutional effect when compared against serum or heparin as anticoagulant. *3361*

Urine   No Effect   Physiological
Sisomicin   No significant effect seen in 23 patients given therapeutic amounts for 2 weeks. *2593*

Urine   Increase   Physiological
Gentamicin   Marked effect especially with treatment for more than 12 d. *2593*

Retinol Esters

Serum   Increase   Physiological
Oral Contraceptives   Minimally increased fasting concentration in treated women. *1303*

Rheumatoid Factor

Serum   No Effect   Physiological
Penicillamine   In 21 patients with rheumatoid arthritis treated for 48 weeks given 250 to 750 mg daily. *2345*
Timegadine   In 23 patients with rheumatoid arthritis given 250 to 750 mg/d for 48 weeks. *2345*

Serum   Positive   Physiological
Aging   In older nonsmokers. *2335*
Methyldopa   Reported effect. *1343*
Oral Contraceptives   Slight effect in users. *2335*
Oxyphenisatin   Liver damage may present as acute hepatitis or as more advanced chronic disease: usual liver disease is chronic active hepatitis. *2260*
Smoking   In young smokers. *2335*

Riboflavin

Urine   Increase   Physiological
Chlortetracycline   With dose of 3 g/d. *1343*

Riboflavin (continued)

Urine   Increase   Physiological (continued)
Oxytetracycline   With dose of 2.5 g/d. *1343*

Urine   Decrease   Physiological
Starvation   Falls to still acceptable level after 10 d. *0714*

Ribonuclease

Serum   Increase   Physiological
Aging   In males over 30 y, but not females. *1854*
Pregnancy   Significant increase, gradual increase with duration. *0365*
Sex Difference   Higher in nonpregnant women than men. *0365*

Ribose

Urine   No Effect   Physiological
Diurnal Variation   No evidence of cycle observed. *0273*
Fruit   No effect compared with fasting. *0273*
Vegetables   No effect with high intake compared with controls. *0273*

Rifampin

Serum   Increase   Physiological
Furosemide   Effect observed after 40 mg orally in normal male volunteers. *2658*
Probenecid   Decreased hepatic uptake, serum concentration increase by 86%. *1918*

Serum   Decrease   Physiological
Aminosalicylic Acid   May impair gastrointestinal absorption. *1487* Delayed and reduced absorption due to adsorption to bentonite in PAS granules. *3794*
Food   Absorption reduced when taken with food. *3024*

Salicylate

Serum   No Effect   Analytical
Bilirubin   Up to 20 mg/dL no effect on Trinder procedure. *2251*
Disulphine Blue   Using a modified Trinder method. *1464*
Glucose   Up to 1,000 mg/dL no effect on Trinder method. *2251*
Heparin   No effect on Trinder procedure. *2251*
Phenols   With 25 mg/mL no effect on Trinder procedure. *2251*
Urea   Up to 1,000 mg/dL no effect on Trinder method. *2251*

Serum   No Effect   Physiological
Ascorbic Acid   No effect on serum concentration with several days treatment. *1489*

Serum   Increase   Analytical
Acetoacetate   50 mg/dL give equivalent of 1 mg/dL by Trinder. *2251*
Chlorpromazine   At 100 mg/L had significant positive interference on colorimetric methods of Heller (and modified version) and Trinder. *1776*
Diflunisal   At 100 mg/L on colorimetric methods of Keller (and modified version) and Trinder -effect approximately half of that of salicylate. *1776* 2 to 3 times higher concentration than salicylate measured by TDX method. *0799* Half to two thirds concentration of salicylate when measured by Trinder and aca methods. *0799*
Oxalate   250 mg/dL give equivalent of 0.3 mg/dL by Trinder. *2251*
Phenazone   At 100 mg/L has slight effect (8%) on method of Keller but 22% on that of Trinder, also observed in one patient who had taken overdose of drug. *1776*

Serum   Increase   Physiological
Aminobenzoic Acid   By competing for glycine for conjugation. *1343*
Aspirin   Due to ingestion of compound. *1302*   When approximately 3 g/d ingested for several weeks. *3076*
Cimetidine   Influence on serum concentration in drug treated group compared with enteric-coated drug administered group (180 μg/mL vs 161 μg/mL). *3848*
Furosemide   If given concomitantly, competes for excretion. *1678*

Serum   Decrease   Physiological
Aluminum Hydroxide   Significant reduction from 19.8 mg/dL to 15.8 mg/dL with several days antacid, due to increased alkalinity of urine. *1489*
Corticosteroids   Increases glomerular filtration, decreases tubular reabsorption. *1487*   Concomitant administration causes accelerated metabolism, most marked in males. *1830*   Levels did not rise above 150 μg/mL in patients given corticosteroids and 4.8 g aspirin/d. *1372*

Urine   No Effect   Analytical
Acetaminophen   At 10 mg/dL no effect on Trinder method. *1204*
Amitriptyline   At 10 mg/dL no effect on Trinder method. *1204*
Amobarbital   At 10 mg/dL no effect on Trinder method. *1204*
Amphetamine   At 10 mg/dL no effect on Trinder method. *1204*
Benztropine   At 1 mg/dL no effect on Trinder method. *1204*
Bromides   At 100 mg/dL no effect on Trinder method. *1204*
Butabarbital   At 10 mg/dL no effect on Trinder method. *1204*
Carisoprodol   At 10 mg/dL no effect on Trinder method. *1204*
Chloral Hydrate   At 10 mg/dL no effect on Trinder method. *1204*
Chlordiazepoxide   At 10 mg/dL no effect on Trinder method. *1204*
Chloroquine   At 10 mg/dL no effect on Trinder method. *1204*
Chlorpheniramine   At 12 mg/dL no effect on Trinder method. *1204*
Chlorpromazine   At 1 mg/dL no effect on Trinder procedure. *1204*
Clorazepate   At 10 mg/dL no effect on Trinder method. *1204*
Cocaine   At 10 mg/dL no effect on Trinder method. *1204*
Codeine   At 10 mg/dL no effect on Trinder method. *1204*
Diazepam   At 10 mg/dL no effect on Trinder method. *1204*
Diethylpropion   At 25 mg/dL no effect on Trinder method. *1204*
Diphenhydramine   At 10 mg/dL no effect on Trinder method. *1204*
Doxepin   At 10 mg/dL no effect on Trinder method. *1204*
Ephedrine   At 10 mg/dL no effect on Trinder method. *1204*
Ethchlorvynol   At 10 mg/dL no effect on Trinder method. *1204*
Ethinamate   At 10 mg/dL no effect on Trinder method. *1204*
Fluphenazine   Up to 10 mg/dL no effect on Trinder procedure. *1204*
Flurazepam   At 10 mg/dL no effect on Trinder method. *1204*
Glutethimide   At 10 mg/dL no effect on Trinder method. *1204*
Glycopyrrolate   At 2 mg/dL no effect on Trinder method. *1204*
Haloperidol   At 1 mg/dL no effect on Trinder method. *1204*
Hydromorphone   At 2 mg/dL no effect on Trinder method. *1204*
Mebutamate   At 10 mg/dL no effect on Trinder method. *1204*
Meperidine   At 10 mg/dL no effect on Trinder method. *1204*
Mephenesin   At 10 mg/dL no effect on Trinder method. *1204*
Meprobamate   At 10 mg/dL no effect on Trinder method. *1204*
Mescaline   At 10 mg/dL no effect on Trinder method. *1204*
Methadone   At 10 mg/dL no effect on Trinder method. *1204*
Methamphetamine   At 10 mg/dL no effect on Trinder method. *1204*
Methaqualone   At 10 mg/dL no effect on Trinder method. *1204*
Methdilazine   Up to 5 mg/dL no effect on Trinder procedure. *1204*
Methylphenidate   At 10 mg/dL no effect on Trinder method. *1204*
Methyprylon   At 10 mg/dL no effect on Trinder method. *1204*
Morphine   At 10 mg/dL no effect on Trinder method. *1204*

Oxazepam   At 10 mg/dL no effect on Trinder method. *1204*
Oxycodone   At 5 mg/dL no effect on Trinder method. *1204*
Oxymetazoline   At 0.5 mg/dL no effect on Trinder method. *1204*
Pentazocine   At 10 mg/dL no effect on Trinder method. *1204*
Pentobarbital   At 10 mg/dL no effect on Trinder method. *1204*
Phenmetrazine   At 10 mg/dL no effect on Trinder method. *1204*
Phenobarbital   At 10 mg/dL no effect on Trinder method. *1204*
Phenylephrine   At 10 mg/dL no effect on Trinder method. *1204*
Phenylpropanolamine   At 10 mg/dL no effect on Trinder method. *1204*
Phenytoin   At 10 mg/dL no effect on Trinder method. *1204*
Primaquine   At 10 mg/dL no effect on Trinder method. *1204*
Procainamide   At 10 mg/dL no effect on Trinder method. *1204*
Prochlorperazine   Up to 2 mg/dL no effect or Trinder procedure. *1204*
Promazine   Up to 2 mg/dL no effect on Trinder procedure. *1204*
Promethazine   Up to 5 mg/dL no effect on Trinder procedure. *1204*
Propoxyphene   At 10 mg/dL no effect on Trinder method. *1204*
Pyrilamine   At 10 mg/dL has no effect on Trinder method. *1204*
Pyrimethamine   At 10 mg/dL no effect on Trinder method. *1204*
Quinidine   At 10 mg/dL no effect on Trinder method. *1204*
Quinine   At 10 mg/dL no effect on Trinder method. *1204*
Scopolamine   At 10 mg/dL no effect on Trinder method. *1204*
Secobarbital   At 10 mg/dL no effect on Trinder method. *1204*
Strychnine   At 10 mg/dL no effect on Trinder method. *1204*
Sulfanilamide   At 100 mg/dL no effect on Trinder method. *1204*
Tetracycline   At 2 mg/dL no effect on Trinder method. *1204*
Thiobarbituric Acid   At 10 mg/dL no effect on Trinder method. *1204*
Thiopropazate   Up to 5 mg/dL has no effect on Trinder procedure. *1204*
Thioridazine   Up 1 mg/dL no effect on Trinder procedure. *1204*
Thiothixene   At 10 mg/dL no effect on Trinder method. *1204*
Trifluoperazine   Up to 10 mg/dL no effect on Trinder procedure. *1204*
Trihexyphenidyl   At 10 mg/dL no effect on Trinder method. *1204*
Trimethobenzamide   At 25 mg/dL no effect on Trinder method. *1204*
Tybamate   At 10 mg/dL no effect on Trinder method. *1204*

Urine   Increase   Analytical
Chlorpromazine   At 2 mg/dL produces pink color on Trinder procedure. *1204*
Methdilazine   At 10 mg/dL produces pink color on Trinder procedure. *1204*
Prochlorperazine   4 mg/dL produces pink color on Trinder procedure. *1204*
Promazine   4 mg/dL produces orange color on Trinder procedure. *1204*
Promethazine   At 10 mg/dL produces pink color on Trinder procedure. *1204*
Sodium Azide   Produces orange-red color in Trinders procedure. *2894*
Thiopropazate   At 10 mg/dL produces pink color on Trinder procedure. *1204*
Thioridazine   At 2 mg/dL produces green color on Trinder procedure. *1204*

Sarcosine

Urine   Decrease   Physiological
Ascorbic Acid   In 10 healthy females given 10 g/d. *3624*

Secobarbital

Blood   Increase   Physiological
Secobarbital   600 mg orally produces concentration of 4.8 mg/L. *2348*

Urine   No Effect   Analytical
Chlorphentermine   No interference using TLC with ethyl acetate: methanol: water: ammonium hydroxide and modified Dragendorff's reagent for detection. *3868*
Clotermine   No interference on TLC using ethyl acetate: methanol: water: ammonium hydroxide and modified Dragendorff's reagent for detection. *3868*
Diethylpropion   No interference with TLC using ethyl acetate: methanol: water: ammonium hydroxide and modified Dragendorff's reagent for detection. *3868*
Fenfluramine   No interference on TLC using ethyl acetate: methanol: water: ammonium hydroxide and modified Dragendorff's reagent for detection. *3868*
Mazindol   No interference on TLC using ethyl acetate: methanol: water: ammonium hydroxide and modified Dragendorff's reagent for detection. *3868*
Phendimetrazine   No interference with TLC using ethyl acetate: methanol: water: ammonium hydroxide and modified Dragendorff's reagent for detection. *3868*
Phenmetrazine   No interference on TLC using ethyl acetate: methanol: water: ammonium hydroxide and modified Dragendorff's reagent for detection. *3868*
Phentermine   No interference on TLC using ethylacetate: methanol: water: ammonium hydroxide and modified Dragendorff's reagent for detection. *3868*

Sedoheptulose

Urine   Decrease   Analytical
Iron Salts   Ferric iron inhibits cysteine-H_2SO_4 reaction. *2346*

Selenium

Serum   Decrease   Analytical
Citrates   Dilutional effect when compared against serum or heparin as anticoagulant. *3361*

Serum   Decrease   Physiological
Ethanol   Mean reduction to 0.065 μg/mL from 0.100 μg/mL in chronic heavy alcohol ingestion versus controls. *0983*

Urine   Increase   Physiological
Selenium   Ingested in diet as well as in poisoning. *2217*

Serine

Plasma   Increase   Physiological
Glycine   Similar marked increase in gouty and normals. *3929*
Histidine   Interferes with clearance after oral load. *1644*

Urine   Increase   Physiological
Glycine   Similar marked increase in gouty and normals. *3929*

Urine   Decrease   Physiological
Ascorbic Acid   In 10 healthy females given 10 g/d. *3624*

Seromucoid

Serum   No Effect   Analytical
Storage of Sample   For 2-7 d at 30°. *0573*

Serum   Increase   Analytical
Standing of Sample   Alleged if stands on clot for more than 2 h. *0573*

Sex Hormone Binding Globulin

Serum   No Effect   Physiological
Anticonvulsants   In 8 epileptic patients receiving phenobarbital with carbamazepine or phenytoin. *2670*
Cyclosporine   In 16 patients who developed hypertrichosis but also given added cortisone. *3190*
Dehydroepiandrosterone   In 5 men given 1600 mg/d orally for 28 d. *2578*
Smoking   No difference between male smokers and nonsmokers. *0238*
Tetracycline   In 9 men with acne given compound for 3 d. *2884* No effect reported. *1736*

Serum   Increase   Physiological
Anticonvulsants   In 37 male epileptics mean of 49.0 nmol/L versus 23.3 nmol/L in controls receiving chronic therapy. *0803*   83 nmol/L versus 54 nmol/L in women and 32 nmol/L versus 22 nmol/L in men on long-term treatment. *0257*
Carbamazepine   Within 7 d of starting 400 mg/d treatment in 6 healthy males probably due to induction of hepatic monoxygenase activity. *0707*
Dexamethasone   Mean increase from 35.2 to 47.7 nmol/L in 14 women hirsute women. *0776*
Diethylstilbestrol   Observed when given to men with androgen-dependent prostatic cancer. *1736*
Estrogens   Metabolic effect. *0227*
Oral Contraceptives   Increase by 80 to 213% after 3 mo with different preparations. *0311*   Highly significant increase induced by ethinylestradiol. *2280*
Phenytoin   In 10 women receiving anticonvulsants, mainly phenytoin versus 10 controls, drug taken on average for 15 y. *0257* Mean 8.3 mol/L versus 5.0 in controls in approximately 24 male patients given phenytoin alone or with primidone or phenobarbital. *3609*

Serum   Decrease   Physiological
Danazol   Reduction of binding capacity by 80% during treatment of women with endometriosis. *2599*   When used for treatment of endometriosis can cause increase of up to 20%. *1736*   Reduced up to 5 fold in 5 patients with hereditary angioneurotic edema treated for up to 10 mo. *3219*   Marked reduction in women over 3 mo. *3906*
Desogestrel   With daily dose of 0.125 mg in 30 healthy female volunteers when given alone. Normal values 30 d after treatment stopped. *3089*
Levonorgestrel   With daily dose of 0.125 mg in 30 healthy female volunteers when given alone. Values normal within 30 d of end of treatment. *3089*   In 17 women using Norplant implants. *1799*
Lynestrenol   When 5 mg daily given alone to 30 healthy female volunteers. Normal values 30 d after treatment stopped. *3089*
Megestrol   In 18 postmenopausal women with breast cancer. *0056*
Stanozolol   From 20.7 to 12.2 nmol/L after 1 week in 9 healthy men given 10 mg/d for 14 d. *3354*
Testosterone   Increase by 80-90% in power athletes with 26 weeks on steroid self administration. *3088*   Drop below 80% of basal concentration in prepubertal subjects: useful as test of androgen sensitivity. *0270*

Short-Chain Fatty Acids

Feces   No Effect   Physiological
Metronidazole   No significant effect when given orally for 6 d. *1664*

Feces   Decrease   Physiological
Ampicillin   Mild effect when given orally to 6 healthy volunteers for 6 d: normalized after 6 weeks. *1664*
Clindamycin   Marked effect in 6 healthy individuals fed orally for 6 d. *1664*

Sialic Acid

Serum   No Effect   Physiological
Glucose   No change observed during glucose tolerance test in all subjects. *1665*

Serum   Increase   Physiological
Diurnal Variation   Maximum observed about midday. *2222*
Ethylestrenol   Metabolic effect. *0227*
Methandrostenolone   Metabolic effect. *0227*
Oxandrolone   Metabolic effect. *0227*
Oxymetholone   Metabolic effect. *0227*
Stanozolol   Metabolic effect. *0227*
Testosterone   Metabolic effect (if aqueous solution i.m.). *0227*

Serum   Decrease   Physiological
Estrogens   Altered metabolism. *0227*
Oral Contraceptives   Metabolic alteration (estrogen effect). *0227*

Sickle Cells

Blood   Positive   Physiological
Prostaglandins   Shown to cause sickling *in vitro*. *3846*

Silicon

Test Conditions   Increase   Analytical
Phosphorus   Up to 60 mg/L affects Jolles/Neurath method. *1773*

Test Conditions   Decrease   Analytical
Phosphorus   Above 60 mg/L affects Jolles/Neurath method. *1773*

Sodium

Serum   No Effect   Analytical
Acetaminophen   At concentration of 6 mg/L had no effect on flame-photometric method. *3393*   At concentration of 1500 mg/L had no effect on measurement by ISE with predilution. *3393*
Acetazolamide   At concentration of 1,000 mg/L had no effect on flame-photometric method. *3393*
Acetylsalicylic Acid   At concentration of 200 mg/L had no effect on flame-photometric method. *3393*   At concentration of 3,000 mg/L had no effect on ISE measurement with predilution. *3393*   At concentration of 1,000 mg/L had no effect on ISE measurement without predilution. *3393*
Allopurinol   At concentration of 5 mg/L had no effect on flame-photometric method. *3393*   At concentration of 180 mg/L had no effect on ISE measurement without predilution. *3393*
Aminosalicylic Acid   At concentration of 460 mg/L had no effect on ISE measurement with predilution. *3393*
Amitriptyline   At concentration of 0.1 mg/L had no effect on flame-photometric method. *3393*   At concentration of 20 mg/L had no effect on ISE measurement with predilution. *3393*
Amobarbital   At concentration of 5 mg/L had no effect on flame-photometric method. *3393*   At concentration of 65 mg/L had no effect on ISE measurement with predilution. *3393*
Ampicillin   At concentration of 5 mg/L had no effect on flame-photometric method. *3393*
Ascorbic Acid   At concentration of 60 mg/L had no effect as measured by flame-photometric method. *3393*   At concentration of 40,000 mg/L had no effect on ISE measurement with predilution. *3393*   At concentration of 800 mg/L had no effect on ISE measurement without predilution. *3393*
Azapropazone   At concentration of 360 mg/L had no effect on ISE measurement without predilution. *3393*
Barbital   At concentration of 500 mg/L had no effect on measurement by ISE with predilution. *3393*
Benzbromarone   At concentration of 80 mg/L had no effect on ISE measurement without predilution. *3393*

Bisacodyl   At concentration of 2 mg/L had no effect on ISE measurement without predilution. *3393*

Butizide   At concentration of 1.98 mg/L had no effect on ISE measurement without predilution. *3393*

Carbamazepine   At concentration of 20 mg/L had no effect on ISE measurement with predilution. *3393*

Cefazolin   At concentration of 110 mg/L had no effect on flame-photometric method. *3393*

Cephalosporin   At concentration of 100 mg/L had no effect on ISE measurement with predilution. *3393*

Cephalothin   At concentration of 1,000 mg/L had no effect on flame-photometric method. *3393*

Chloral Hydrate   At concentration of 200 mg/L had no effect on measurement by ISE with predilution. *3393*

Chloramphenicol   At concentration of 6,000 mg/L had no effect on measurement by ISE without predilution. *3393*

Chlordiazepoxide   At concentration of 5 mg/L had no effect on flame-photometric method. *3393*   At concentration of 20 mg/L had no effect on measurement by ISE with predilution. *3393*   At concentration of 18 mg/L had no effect on measurement by ISE without predilution. *3393*

Chlormezanone   At concentration of 100 mg/L had no effect on measurement by ISE with predilution. *3393*

Chloroquine   At concentration of 50 mg/L had no effect on measurement by ISE without predilution. *3393*

Chlorphenesin   At concentration of 150 mg/L had no effect on measurement by ISE with predilution. *3393*

Chlorpromazine   At concentration of 3 mg/L had no effect on flame-photometric method. *3393*   At concentration of 3 mg/L had no effect on measurement by ISE with predilution. *3393*

Chlorpropamide   At concentration of 150 mg/L had no effect on flame-photometric method. *3393*

Chromonar Hydrochloride   At concentration of 180 mg/L had no effect on ISE measurement without predilution. *3393*

Cimetidine   At concentration of 1 mg/L had no effect on flame-photometric method. *3393*

Clindamycin   At concentration of 150 mg/L had no effect on measurement by ISE with predilution. *3393*

Cloxacillin   At concentration of 5 mg/L had no effect on flame-photometric method. *3393*

Codeine   At concentration of 0.1 mg/L had no effect on flame-photometric method. *3393*

Colistin   At concentration of 150 mg/L had no effect on measurement by ISE with predilution. *3393*

Contact With Clot   If plastic disc used to separate for 48 h. *1863*

Cyclophosphamide   At concentration of 80 mg/L had no effect on measurement by ISE without predilution. *3393*

Cyclosporine   At concentration of 41.5 mmol/L (50 µg/L) on methods on Technicon® SMAC® II and Hitachi® 705. *1123*

Dexamethasone   At concentration of 1.4 mg/L had no effect on measurement by ISE with predilution. *3393*

Dextran 40   At concentration of 9100 mg/L had no effect on measurement by ISE with predilution. *3393*

Dextran 60   At concentration of 18,000 mg/L had no effect on measurement by ISE without predilution. *3393*

Dextran 70   At concentration of 5400 mg/L had no effect on measurement by ISE with predilution. *3393*

Diatrizoic acid   At concentration of 6080 mg/L had no effect on ISE measurement without predilution. *3393*

Diazepam   At concentration of 1.5 mg/L had no effect on flame-photometric method. *3393*   At concentration of 10 mg/L had no effect on measurement by ISE with predilution. *3393*

Digitoxin   At concentration of 6 mg/L had no effect on measurement by ISE with predilution. *3393*

Digoxin   At concentration of 0.002 mg/L had no effect on flame-photometric method. *3393*   At concentration of 9 mg/L had no effect on measurement by ISE with predilution. *3393*   At concentration of 0.15 mg/L had no effect on measurement by ISE without predilution. *3393*

Diphenhydramine   At concentration of 23 mg/L had no effect on measurement by ISE with predilution. *3393*

Disopyramide   At concentration of 4 mg/L had no effect on flame photometric method. *3393*

Doxepin   At concentration of 60 mg/L had no effect on measurement by ISE without predilution. *3393*

Erythromycin   At concentration of 10 mg/L had no effect on measurement by ISE with predilution. *3393*

Ethchlorvynol   At concentration of 400 mg/L had no effect on measurement by ISE with predilution. *3393*

Ethosuximide   At concentration of 390 mg/L had no effect on measurement by ISE with predilution. *3393*

Flurazepam   At concentration of 0.1 mg/L had no effect on flame-photometric method. *3393*

Furosemide   At concentration of 4 mg/L had no effect on flame-photometric method. *3393*   At concentration of 6,000 mg/L had no effect on measurement by ISE with predilution. *3393*   At concentration of 400 mg/L had no effect on measurement by ISE without predilution. *3393*

Gentamicin   At concentration of 6 mg/L had no effect on flame-photometric method. *3393*   At concentration of 14 mg/L had no effect on measurement by ISE with predilution. *3393*

Glibenclamide   At concentration of 3 mg/L had no effect on measurement by ISE with predilution. *3393*

Glutethimide   At concentration of 10 mg/L had no effect on measurement by ISE with predilution. *3393*

Hydralazine   At concentration of 0.5 mg/L had no effect on flame-photometric method. *3393*   At concentration of 6 mg/L had no effect on measurement by ISE with predilution. *3393*

Imipramine   At concentration of 30 mg/L had no effect on measurement by ISE with predilution. *3393*

Indomethacin   At concentration of 13 mg/L had no effect on measurement by ISE with predilution. *3393*   At concentration of 40 mg/L had no effect on measurement by ISE without predilution. *3393*

Insulin   At concentration of 3 mg/L had no effect on measurement by ISE with predilution. *3393*

Iproniazid   At concentration of 40 mg/L had no effect on measurement by ISE with predilution. *3393*

Kanamycin   At concentration of 10 mg/L had no effect on measurement by ISE with predilution. *3393*

Levodopa   At concentration of 200 mg/L had no effect on measurement by ISE with predilution. *3393*

Lidocaine   At concentration of 0.5 mg/L had no effect on measurement by ISE with predilution. *3393*

Lorazepam   At concentration of 0.05 mg/L had no effect on flame-photometric method. *3393*

Meglumine   At concentration of 1200 mg/L had no effect on ISE measurement without predilution. *3393*

Mephenesin   At concentration of 100 mg/L had no effect on measurement by ISE with predilution. *3393*

Mephenytoin   At concentration of 20 mg/L had no effect on measurement by ISE with predilution. *3393*

Meprobamate   At concentration of 25 mg/L had no effect on flame-photometric method. *3393*   At concentration of 160 mg/L had no effect on measurement by ISE with predilution. *3393*

Methapyrilene   At concentration of 13 mg/L had no effect on measurement by ISE with predilution. *3393*

Methaqualone   At concentration of 6,000 mg/L had no effect on measurement by ISE with predilution. *3393*

Methicillin   At concentration of 20 mg/L had no effect on flame-photometric method. *3393*

Methotrexate   At concentration of 80 mg/L had no effect on measurement by ISE without predilution. *3393*

Methotrimeprazine   At concentration of 1 mg/L had no effect on flame-photometric method. *3393*

Methyldopa   At concentration of 7 mg/L had no effect on flame-photometric method. *3393*   At concentration of 100 mg/L had no effect on measurement by ISE with predilution. *3393*   At concentration of 800 mg/L had no effect on measurement by ISE without predilution. *3393*

Methylprednisolone   At concentration of 100 mg/L had no effect on measurement by ISE with predilution. *3393*

Metoprolol   At concentration of 0.34 mg/L had no effect on flame-photometric method. *3393*

Minocycline   At concentration of 6 mg/L had no effect on measurement by ISE with predilution. *3393*

Nalidixic Acid   At concentration of 10 mg/L had no effect on measurement by ISE with predilution. *3393*

Sodium (continued)

Serum   No Effect   Analytical (continued)

Niacinamide   At concentration of 120 mg/L had no effect on measurement by ISE without predilution. 3393

Nitrofurantoin   At concentration of 2.5 mg/L had no effect on flame-photometric method. 3393   At concentration of 14 mg/L had no effect on method using ISE with predilution. 3393

Novaminsulfon   At concentration of 800 mg/L had no effect on measurement by ISE without predilution. 3393

Oxyphenbutazone   At concentration of 120 mg/L had no effect on measurement by ISE without predilution. 3393

Papaverine   At concentration of 10 mg/L had no effect on measurement by ISE with predilution. 3393

Penicillamine   At concentration of 360 mg/L had no effect on measurement by ISE without predilution. 3393

Penicillin   At concentration of 15,000 mg/L had no effect on measurement by ISE with predilution. 3393

Penicillin G   At concentration of 18 mg/L had no effect as measured by flame-photometric method. 3393

Perphenazine   At concentration of 1 mg/L had no effect on flame-photometric method. 3393

Phenazopyridine   At concentration of 20 mg/L had no effect on measurement by ISE with predilution. 3393

Phenobarbital   At concentration of 10 mg/L had no effect on flame-photometric method. 3393   At concentration of 250 mg/L had no effect on measurement by ISE with predilution. 3393   At concentration of 60 mg/L had no effect on measurement by ISE without predilution. 3393

Phenothiazines   At concentration of 200 mg/L had no effect on measurement by ISE without predilution. 3393

Phenprocoumon   At concentration of 3.6 mg/L had no effect on measurement by ISE without predilution. 3393

Phensuximide   At concentration of 120 mg/L had no effect on measurement by ISE with predilution. 3393

Phenylbutazone   At concentration of 750 mg/L had no effect on measurement by ISE with predilution. 3393   At concentration of 120 mg/L had no effect on measurement by ISE without predilution. 3393

Phenytoin   At concentration of 20 mg/L had no effect on flame-photometric method. 3393   At concentration of 240 mg/L had no effect on measurement by ISE with predilution. 3393

Primidone   At concentration of 10 mg/L had no effect on measurement by ISE with predilution. 3393

Probenecid   At concentration of 1300 mg/L had no effect on measurement by ISE with predilution. 3393   At concentration of 400 mg/L had no effect on measurement by ISE without predilution. 3393

Procainamide   At concentration of 50 mg/L had no effect on measurement by ISE with predilution. 3393

Procaine   At concentration of 2 mg/L had no effect on measurement by ISE with predilution. 3393

Prochlorperazine   At concentration of 1 mg/L had no effect on flame-photometric method. 3393

Promethazine   At concentration of 1 mg/L had no effect on flame-photometric method. 3393

Propoxyphene   At concentration of 5 mg/L had no effect on measurement by ISE with predilution. 3393

Propranolol   At concentration of 0.2 mg/L had no effect on flame-photometric method. 3393

Protamine   At concentration of 10 mg/L had no effect on measurement by ISE with predilution. 3393

Quinidine   At concentration of 210 mg/L had no effect on ISE measurement with predilution. 3393

Quinine   At concentration of 30 mg/L had no effect on measurement by ISE with predilution. 3393

Salicylate   At concentration of 100 mg/L had no effect on measurement by ISE with predilution. 3393   At concentration of 30,000 mg/L had no effect on measurement by ISE with predilution. 3393

Secobarbital   At concentration of 1 mg/L had no effect on flame-photometric method. 3393   At concentration of 100 mg/L had no effect on measurement by ISE with predilution. 3393

Sodium Heparin   At concentration of 750 mg/L had no effect on measurement by ISE without predilution. 3393

Spironolactone   At concentration of 80 mg/L had no effect on measurement by ISE without predilution. 3393

Standing of Sample   If on polystyrene beads 24 h at 25°. 2624

Storage of Sample   No effect for 2 weeks at room temperature or 4°. 1563

Streptomycin   At concentration of 400 mg/L had no effect on measurement by ISE with predilution. 3393

Strychnine   At concentration of 12 mg/L had no effect on measurement by ISE with predilution. 3393

Sulfadiazine   At concentration of 1500 mg/L had no effect on measurement by ISE with predilution. 3393

Sulfaguanidine   At concentration of 500 mg/L had no effect on measurement by ISE with predilution. 3393

Sulfamethoxydiazine   At concentration of 200 mg/L had no effect on measurement by ISE with predilution. 3393   At concentration of 200 mg/L had no effect on measurement by ISE without predilution. 3393

Sulfanilamide   At concentration of 1,000 mg/L had no effect on measurement by ISE with predilution. 3393

Sulfasalazine   At concentration of 1600 mg/L had no effect on measurement by ISE without predilution. 3393

Sulfisoxazole   At concentration of 80,000 mg/L had no effect on measurement by ISE with predilution. 3393

Tetracycline   At concentration of 4 mg/L had no effect on flame-photometric method. 3393   At concentration of 20,000 mg/L had no effect on measurement by ISE with predilution. 3393   At concentration of 400 mg/L had no effect on measurement by ISE without predilution. 3393

Theophylline   At concentration of 20 mg/L had no effect on flame-photometric method. 3393   At concentration of 1,000 mg/L had no effect on measurement by ISE with predilution. 3393

Timolol   At concentration of 0.01 mg/L had no effect on flame-photometric method. 3393

Tolbutamide   At concentration of 100 mg/L had no effect on flame-photometric method. 3393   At concentration of 2,000 mg/L had no effect on measurement by ISE with predilution. 3393   At concentration of 400 mg/L had no effect on measurement by ISE without predilution. 3393

Transport of Specimen   By pneumatic tube if specimen protected. 3217

Trichlorethanol   At concentration of 12 mg/L had no effect on flame-photometric method. 3393   At concentration of 1,000 mg/L had no effect on measurement by ISE with predilution. 3393

Triethanolamine   At concentration of 200 mg/L had no effect on measurement by ISE with predilution. 3393

Trifluoperazine   At concentration of 1 mg/L had no effect on flame-photometric method. 3393

Vitamin B Complex   At concentration of 23.3 mg/L had no effect on measurement by ISE without predilution. 3393

Warfarin   At concentration of 1.5 mg/L had no effect on flame-photometric method. 3393   At concentration of 100 mg/L had no effect on measurement by ISE with predilution. 3393

Serum   No Effect   Physiological

Aging   No significant change from age 6-20 y. 0887

Amiloride   No effect seen in 13 men treated for 8 weeks. 0656

Aspirin   When approximately 3 g/d ingested for several weeks. 3076

Azosemide   Effect observed in normal male volunteers after 60 mg orally. 2658

Bendrofluazide   In 7 essential hypertensives treated for 16 weeks. 0352

Bopindolol   In 10 hypertensive patients treated for either 1 or 21 d. 1063

Bucindol   No effects noted in 8 patients following several weeks on treatment. 1283

Bumetanide   No significant effect observed. 2666

Cellulose Phosphate   With decrease of activity product ratio. 2710

Chlorthalidone   No significant change in 10 hypertensive patients given 25 mg daily for 16 weeks. 1229

Danazol   No change in 18 women with endometriosis given 600 mg/d for 6 mo. 0591

Diuretics   May appear normal if hypovolemia occurs. 0105

Indapamide   In 8 hypertensives treated for 16 weeks. *0352*
Meals   Not significantly affected by nonstandardized meals. *3455* No effect after standard breakfast. *0579*
Muzolimine   No significant change in 10 hypertensive patients given 30 mg daily for 16 weeks. *1229*
Pindolol   No significant effect even with 6 weeks treatment. *3279*
Ranitidine   No effect of 3-d of 150 mg/12 h. *3130*
Starvation   No effect observed usually. *1242*

Serum   Increase   Analytical

Ampicillin   At concentrations above 1800 mg/L (therapeutic concentration about 320 mg/L) raised concentration as measured by ISE without predilution. *3393*
Calcium   Emission spectrum may interfere. *2313*
Cefotaxime   By up to 5 mmol/L with both drug and metabolite on method on American Monitor Parallel. *0201*
Copper   May interfere with flame photometry. *2313*
Detergents   Contaminated glassware etc. *2313*
EDTA   At concentrations above 500 mg/L raised concentration as measured by ISE without predilution. *3393*
Fluorides   If salt of fluoride. *1563*
Fluosol-DA   Unpredictable variation in concentrations as measured on SMA IIC, Beckman Astra 8 and Ektachem® 400. *2518*
Heparin   If sodium salt used may affect result. *3873*
Nitrofurantoin   At concentrations above 40 mg/L (normal therapeutic concentration 5.5 mg/L) raised concentration as measured by ISE without predilution. *3393*
Norfenefrin   At concentrations above 4.8 mg/L (normal therapeutic concentration 0.4 mg/L) raised concentration as measured by ISE without predilution. *3393*
Oxalate   If sodium salt of oxalate. *1563*
Potassium   Affects flame photometry if poor instrument. *1563*
Protein   Presence raises temperature, increases emission. *2313*
Sodium Citrate   At concentrations above 250 mg/L raised concentration as measured by ISE without predilution. *3393*
Sodium Fluoride   At concentrations above 250 mg/L raised concentration as measured by ISE without predilution. *3393*
Sodium Oxalate   At concentrations above 250 mg/L raised concentration as measured by ISE without predilution. *3393*
Standing of Sample   Marked top to bottom if frozen and unmixed. *2674*
Trichloracetic Acid   Volume reduction effect with protein precipitation. *1131*

Serum   Increase   Physiological

Amiloride   Significant effect in 13 men given drug alone: remained high throughout treatment. *0656*
Amino Acids   If Aminosol given i.v. may cause sodium retention. *0120*
Anabolic Steroids   Mineralocorticoid effect with retention. *2313*
Androgens   Mineralocorticoid effect with increased retention. *0070*
Angiotensin   Due to salt retaining action of aldosterone. *1343*
Betamethasone   Occurs infrequently with edema. *0071*
Bicarbonates   Induces metabolic alkalosis. *0652*
Boric Acid   Toxicity effect. *2313*
Cannabis   Reported effect. *2220*
Carbenoxolone   Significant increase (approximately 5 meq/L) in adrenalectomized patients. *3801*
Chlorthalidone   Single case of severe reversible myopathy noted. *1793*
Cholestyramine   Severe hypernatremia (sodium 175 mmol/L) in 2 pediatric patients receiving drug for some weeks. *2872*
Clonidine   Probable direct action on renal tubules. *1343*
Clopamide   Single case of diuretic associated myopathy. *1793*
Corticosteroids   May cause retention. *2313*
Corticotropin   Causes retention with edema. *0071*
Cortisone   Mineralocorticoid effect. *3879*
Desoxycorticosterone   May cause retention and edema. *0071*
Diazoxide   May cause salt retention. *1343*
Diurnal Variation   Maximum observed at night in early am. *2222*

Estrogens   May cause salt and water retention. *1343*
Ethanol   Slightly higher after heavy beer drinking. *0922*   Increase by 2 mmol/L 2 h after ethanol ingestion. *2134*
Fludrocortisone   May cause sodium retention and edema. *0071*
Glucocorticoids   May cause sodium retention. *0071*
Guanethidine   Salt retention ?due to tubular effect. *1343*
Heparin   When 1,000 IU/mL added increased concentration when direct ISE methods used. *3516*
Hydrocortisone   May cause retention and edema. *0071*
Intra-Amniotic Saline   Increased by approximately 4 meq/L after 1 h. *3892*
Isosorbide   Dehydration with overdosage. *3315*
Lactulose   In 20 of 75 courses of treatment of hepatic failure concentration exceeded 145 mmol/L. *2572*
Mannitol   May cause marked dehydration. *2220*
Mayo Enema   High sodium content — may be retained. *0726*
Menopause   Change of 0.7 meq/L from 5th to 6th decade. *3830*
Methandrostenolone   May be affected but water also retained. *0019*
Methoxyflurane   Impaired renal tubular function. *0749*
Methyldopa   May cause salt retention and edema. *2313*
Oral Contraceptives   May cause sodium retention. *2313*
Oxyphenbutazone   May cause marked salt retention. *2313*
Phenelzine   Rare hypernatremia reported. *1680*
Phenylbutazone   May cause salt retention. *3687*
Prednisolone   Very slight mineralocorticoid effect. *0071*
Prednisone   Slight effect only. *0071*
Progesterone   May cause sodium retention. *2220*
Prolactin   Renal retention effect when given i.m. *1662*
β-Propiolactone   Significant effect by 7 mmol/L when plasma from BPL treated blood compared with plasma in 25 specimens. *0217*
Ramipril   In response to 10 mg and 20 mg on successive days in 9 patients with severe chronic congestive heart failure. *0764*
Rauwolfia   May cause electrolyte retention or edema. *2313*
Saline   May occur especially if impaired cardiac/renal function. *0071*
Season   Significant effect, maximum in summer. *1172*
Sodium Bicarbonate   May cause sodium retention. *0071*
Sodium Salts   Slight effect with citrate salt compared with chloride in 5 men with essential hypertension. *2037*
Sodium Sulfate   If given i.v. may cause fluid retention and coma. *3300*
Tetracycline   May cause hypernatremia with renal impairment. *2220*

Serum   Decrease   Analytical

Heparin   At concentration of 300 U/L sodium concentration reduced by up to 5 mmol/L with direct measurement on Corning 902 analyzer. *2287*
Hyoscine-N-Butylbromide   At concentrations above 20 mg/L lowered concentration as measured by ISE without predilution. *3393*
Lipemia   By less than 1.5% in patient specimens measured on SMAC® in comparison with with specimens following ultra-centrifugation. *2466*
Liposol   Changed from 138.8 to 131.6 mmol/L in 26 randomly selected patient sera, before and after addition on Beckman Astra methods. *2054*
Plasma   2.6 meq/L by 12/60, 0.6 meq/L by I.L. *2228*

Serum   Decrease   Physiological

Amiloride   In 3 women when drug given with hydrochlorothiazide probably due to direct effect on distal nephrons. *3545*   Increase by 10 mmol/L over 7 d in 5 volunteers with 75 mg daily. *2440*
Ammonium Chloride   Cation loss as excess chloride excreted. *0652*
Amphotericin B   Significant effect even in normal subjects. *0551*
Arginine   Slight transient reduction (by 4 mmol/L) after i.v. infusion in diabetics. *2328*
Azosemide   Slight effect noted in 2 volunteers. *2047*

Sodium *(continued)*

Serum   Decrease   Physiological *(continued)*

Bendrofluazide   May cause hyponatremia due to potassium depletion. *1129*

Blindness   Significantly lower (average 5 meq/L) than normal. *1637*

Captopril   In five men with congestive heart failure with fall of sodium by 7 mmol/L on 3rd to 4th day. *2592*

Carbacrylamine Resin   Cation-exchange with reduced gastrointestinal tract absorption. *2313*

Carbamazepine   Although mean concentration in population of epileptics not affected, significant reduction in 5 of 80 patients. *2787*   Isolated case of dilutional hyponatremia with water intoxication: 7 previous cases reported. *1882*   In 28 of 674 epileptic patients most often when drug combined with barbiturates. *1848*

Cathartics   Excessive use may cause sodium depletion. *1343*

Chlorothiazide   May cause hyponatremia. *2220*

Chlorpropamide   Induces inappropriate ADH secretion. *3791* Drug-induced syndrome of inappropriate ADH secretion. *2601*

Chlorthalidone   Severe hyponatremia with potassium depletion. *1129*

Cisplatin   In single case in 3rd and 4th treatment course, identified with renal tubular dysfunction. *2064*   In 3rd and 4th week of treatment of one girl with malignant ovarian germ cell tumor probably related to renal tubular dysfunction. *2064*

Cyclophosphamide   Average 8.5 meq/L for 1 (H_2O retention by metabolites). *0855*

Cyclothiazide   Syndrome of inappropriate ADH secretion seen. *1661*

Dapsone   To 117 meq/L in one woman with dermatitis herpetiformis: possibly associated increased intravascular catabolism of albumin. *0740*   Sulfone syndrome in one 16 year old girl given 50 mg daily short-term administration. *3604*

Dichlorphenamide   Cation loss as excess chloride is excreted. *0652*

Diuretics   Diuretic action. *2745*

Ethacrynic Acid   Diuretic action. *2313*

Ethanol   Inappropriate secretion of ADH in beer drinkers. *0875* 0.3% decrease in heavy versus occasional drinkers. *3273*

Furosemide   Diuretic action with sodium depletion. *0070*   In 25% of 204 hospitalized patients receiving the drug. *3419*   Observed after drug administration. *2047*

Glycerin   Maximum effect of less than 5%. *3615*

Hemolysis   Dilution effect. *2313*

Heparin   Increased excretion due to aldosterone suppression. *2313*   When 5,000 IU/mL added reduction of sodium concentration when measured by direct ISE. *3516*

Hydrochlorothiazide   May cause hyponatremia. *1129*   Severe hyponatremia observed in several cases. *0170*

Hydroflumethiazide   Diuretic action. *1680*

Indomethacin   Marked reduction in some neonates (to below 130 mmol/L) within 48 h of administration to close patent ductus arteriosus. *1476*

Ketoconazole   To 121 mmol/L in 73 year old man with prostatic cancer treated with 600 mg daily for 2 1/2 mo. *2817*

Laxatives   Excessive use may have effect. *2313*

Lithium   Due to initial natriuresis and diuresis. *2275*

Lorcainide   Increase by 8 meq/L significant effect in 16 of 33 patients with organic heart disease and ventricular arrhythmias. Effect observed after single i.v. dose. *3383*

Mannitol   Effect if marked diuresis. *0071*

Menstruation   Decrease by about 2.2 meq/L compared with 16 d after. *1086*

Meralluride   Diuretic action. *3879*

Mercurial Diuretics   Diuretic action with sodium depletion. *2313*

Mercury Compounds   May occur with established Hg poisoning. *0987*

Methyclothiazide   Diuretic action. *1353*

Metolazone   Diuretic action of drug acting on distal tubules. *2816*

Miconazole   Occurred in 46% of courses followed, with average decrease of 10 mmol/L. *3462*

Nifedipine   In 23 patients over 60 y old with essential mild to moderate hypertension. *3192*

Nisoldipine   Fell by average of 3 mmol/L in 14 mild to moderately hypertensive noninsulin dependent diabetic patients. *2653*

NSAID (Nonsteroidal Anti-Inflammatory Drugs)   May be associated with water retention. *1561*

Oxytocin   In mother during induction of labor due both to oxytocin and glucose in infusion: also observed in infant. *3337*

Phenoxybenzamine   Isolated case. Presumed to be due to drug-induced inappropriate release of vasopressin. *0159*

Polythiazide   Diuretic action, with K deficiency. *1129*   Severe hyponatremia noted in one patient with mild hypertension. *0170*

Quinethazone   Diuretic action with sodium depletion. *2313*

Silver   Observed after silver nitrate antisepsis. *2427*

Somatostatin   Water intoxication observed in 2 patients given drug i.v., although creatinine remained constant. *1465*

Spironolactone   Aldosterone antagonism with consequent diuresis. *2313*   Reduction by 4 mmol/L after 5 d in 5 volunteers. *2440*

Sulfates   May be excreted combined with sulfate. *2313*

Sulfonylureas   Potentiates vasopressin, causes water retention. *1134*

Theophylline   Significant effect 8 h after i.v. infusion. *3935*

Thiazides   Diuretic action with sodium depletion. *0070*

Triamterene   Diuretic action. *2313*

Trimethoprim   Impairs free water clearance: effect noted when combined with diuretic. *0986*

Urea   May cause severe depletion with diuresis. *1022*

Vasopressin   May occur with water retention. *0071*

Vincristine   May be inappropriate ADH secretion. *1680*

Urine   No Effect   Analytical

Storage of Sample   No change unprocessed for 45 d at room temperature. *1180*

Thymol   No effect on flame photometric methods. *2981*

Urine   No Effect   Physiological

Amiloride   No effect seen in 13 men treated for 8 weeks. *0656*

Bopindolol   In 10 hypertensive patients treated for either 1 or 21 d. *1063*

Chlorthalidone   After 3 d and 3 mo in 8 patients with mild hypertension given 50 mg/d. *3040*

Cimetidine   Although drug cleared principally by glomerular filtration and partial reabsorption in proximal renal tubules. *2368*

Citrates   In 89 patients with hypocitraturic calcium nephrolithiasis treated for up to 4 y. *2708*

Etodolac   No cumulative effect observed over 24 h in individuals with normal renal function or with renal insufficiency when given 500 mg bid for 4 d. *0453*

Oxytocin   No significant or consistent effect. *1344*

Ranitidine   No effect of 3-d of 150 mg/12 h. *3130*

Sulindac   In 9 individuals with stable renal insufficiency when treated for 9 d. *2457*   In 10 furosemide treated patients with well controlled congestive heart failure given 400 mg/d for 2 d. *1042*

Trimazosin   No significant effect noted. *0861*

Urine   Increase   Analytical

Standing of Sample   Marked top to bottom if frozen and unmixed. *2674*

Urine   Increase   Physiological

Acetazolamide   Diuretic action. *1343*

Amiloride   Diuretic action of drug. *2969*   Marked natriuresis observed with start of therapy. *2440*

Ammonium Chloride   Diuretic action. *1343*

Aspirin   Response to respiratory alkalosis of early toxicity. *1343*

Azosemide   Effect observed in normal male volunteers after 60 mg orally. *2658*   Significant effect then gradually decreased. *2047*

Bed Rest   Slight increase during immobilization. *0863*

Benzthiazide   Therapeutic intent (maximum effect 4-6 h). *1680*

Bumetanide   Marked dose related effect. *2666*   After 1 mg maximum within 2 h over 4 h. *0966*

Calcitonin   Acts independently of parathyroid. *0346*

Chlorothiazide   Diuretic action. *2220*
Chlorthalidone   May produce marked diuresis. *2427*
Cisplatin   In single case in 3rd and 4th treatment course, identified with renal tubular dysfunction. *2064*
Clopamide   Diuretic action (rapid effect). *2900*
Cyclothiazide   Intended diuretic action. *1680*
Dexamethasone   Mild diuresis with sodium loss may occur initially. *0071*
Diapamide   Diuretic action. *2256*
Digitalis   Diuretic action in cardiac failure. *1343*
Diurnal Variation   Maximum around midday and afternoon. *2222*
Dopamine   Intravenous infusion in normals (increase = 171 to 571 μEq/min). *1326*   Effect noted with infusion of 3.0 μg/kg/min in 6 normal men for 2 h. *2149*   Directly inhibits tubular solute reabsorption. *1591*
Ethacrynic Acid   Therapeutic intent. *3214*
Furosemide   Diuretic action. *1617*   Therapeutic intent of drug administration. *1214*   Marked effect: decreased sharply when administration stopped. *2047*
Heparin   Due to aldosterone suppression. *0652*
Hydrochlorothiazide   Diuretic action of drug. *2220*
Hydrocortisone   Enhances excretion especially if prior loading. *1343*
Hydroflumethiazide   Diuretic action. *1680*
Intra-Amniotic Saline   Significant effect for 24 h. *3892*
Isosorbide   Diuretic action alone, potentiates others. *2605*
Levodopa   ?secondary to renal vasodilatation or direct action on tubules. *1136*
Lithium   Impairs reabsorption by tubules. *1680*
Mannitol   Slight increase occurs only. *2220*
Mefruside   Diuretic action, less effective than thiazides. *3854*
Menstruation   Post-menstrual diuresis (decreased premenstrually). *0987*
Mercurial Diuretics   Therapeutic intent. *1343*
Methyclothiazide   Intended diuretic action (maximum in 6 h). *1680*
Metolazone   Diuretic action of drug. *1425*
Niacin   20% increase at 4 h after 2 g. *1622*
Niacinamide   1 g causes up to 40% increase at 4 h. *1622*
Nicardipine   Increase by 56% after 0.5 mg i.v. in 7 patients with mild to moderate hypertension. *0195*
Nifedipine   Significant effect following 20 mg sublingually. *0650*   After 10 mg sublingually in next 2 h. *2917*
Paramethasone   May be excreted although edema may occur. *0071*
Parathyroid Extract   Increased clearance. *1439*
Polythiazide   Therapeutic intent (peak effect at 6 h). *1680*
Progesterone   May occur with high doses. *2427*
Quinethazone   Intended effect, high Na/K ratio. *1680*
Recumbency   Metabolic effect. *3578*
Saline   With isotonic saline loading. *3439*
SC-16102   Duration of diuretic action short. *1804*
Secretin   Interferes with reabsorption, exchange for H^+. *2656*
Soft Water Areas   Greater than elsewhere due to added dietary salt. *0813*
Spironolactone   Diuretic action. *1343*   Marked natriuresis with start of treatment. *2440*
Starvation   Initially increase then falls to 1-15 meq/d. *3783*
Sulfates   Increased excretion combines with sulfate. *0071*
Tetracycline   Natriuresis and diuresis effects of drug. *0491*
Thiazides   Diuretic action. *1343*
Torasemide   Similar effects to that of furosemide. *0919*
Triamcinolone   Mild diuresis with sodium loss in first few days. *0071*
Triamterene   Diuretic action (increased clearance). *1343*   Reduces availability of potassium within distal tubular cells for secretion into distal lumen. *2836*
Trichlormethiazide   Therapeutic intent — diuretic action. *1680*
Triflocin   Diuretic action. *0033*

Trimethoprim   Impairs free water clearance: effect noted when combined with diuretic. *0986*
Verapamil   Marked enhancement in 15 patients with uncomplicated essential hypertension. *0867*
Vincristine   May be inappropriate ADH secretion. *1680*
Xanthine   Diuretic action. *1343*

Urine   Decrease   Physiological
Aldosterone   Due to hormonal action. *3730*
Anesthetic Agents   Significant effect with general anesthesia. *0916*
Angiotensin   Due to effect on renal tubules. *1343*
Carbamazepine   Isolated case of dilutional hyponatremia with water intoxication: 7 previous cases reported. *1882*
Carbohydrate   Marked effect when starved patients refed. *1242*
Corticosteroids   Promotes retention (mineralocorticoid effect). *0071*
Cortisone   Causes retention. *0071*
Desoxycorticosterone   Exchanged with potassium ion. *1343*
Diazoxide   Effect over 2 h of 4 mg/kg given i.v. *1805*
Epinephrine   With increased filtration fraction. *1343*
Etodolac   Both in normal subjects but also to greater extent in individuals with renal insufficiency. *0453*
Guancydine   Probable action on renal tubules. *3715*
Indomethacin   Caused by increased tubular absorption with marked oliguria as a result. *3261*
Insulin   Antinatriuresis with/without acidosis may be marked. *3131*
Levarterenol   Increased tubular resistance and reabsorption. *1176*
Lithium   Due to action of aldosterone (initial increase). *2527*
Naproxen   Increase by 26% in 10 furosemide treated patients with well controlled congestive heart failure. *1042*
Prolactin   Reduces renal excretion at tubular level. *1662*
Propranolol   Impaired in normals and patients. *2427*
Ramipril   In response to 10 mg and 20 mg on successive days in 9 patients with severe chronic congestive heart failure. *0764*
Sleep   Reduced excretion compared with day. *1139*   Observed without effect on GFR. *1190*
Spinal Anesthesia   Observed in normal males, pregnant women. *1343*

Feces   Increase   Physiological
Neomycin   Induced by steatorrhea. *1077*

Gastric Material   No Effect   Physiological
Histamine   Output unchanged although concentration decreased. *3252*

Red Blood Cells   No Effect   Physiological
Verapamil   No effect when drug given alone. *2762*

Red Blood Cells   Increase   Physiological
Digoxin   16% increase, due to effect on membrane ATP-ase. *1923*
Race   Significantly higher in blacks than caucasians. *2523*
Verapamil   When given with digoxin increased concentration more than with controls with digoxin only. *2762*

Red Blood Cells   Decrease   Physiological
Pregnancy   Maternal intracellular Na decrease towards term. *0987*

White Blood Cells   No Effect   Physiological
Diuretics   No change observed in normal individuals. *0997*

Test Conditions   Increase   Analytical
Magnesium   By neutron activation 0.65 μg = 1 mg. *3336*

Somatomedin-C

Plasma   Increase   Physiological
Clonidine   In children with growth hormone deficiency 0.1 mg/m² daily for 60 d. *2821*

Somatostatin

Plasma   No Effect   Physiological
Morphine   No effect observed in 6 healthy volunteers after test meal and drug i.v. *0626*
Naloxone   In 6 healthy volunteers after ingestion of test meal and intravenous administration of drug. *0626*

Plasma   Increase   Analytical
Theophylline   With a variety of different antibodies for RIA measurement inhibits tracer binding at concentrations normally achieved therapeutically. *3341*

Plasma   Decrease   Analytical
Theophylline   Interferes with binding in radioimmunoassays with 3 different antisera. *3341*

CSF   Decrease   Physiological
Carbamazepine   Appears to affect compound whereas other drugs do not. *3081*

Sorbitol Dehydrogenase

Serum   No Effect   Analytical
Hemolysis   Slight hemolysis has no effect. *2981*

Serum   Increase   Physiological
Muscular Exercise   Significant effect after 2 h march. *3217*

Serum   Decrease   Analytical
EDTA   Inhibitory effect. *2981*

Serum   Decrease   Physiological
Phenothiazines   Observed with therapy of female schizophrenics. *0289*

Specific Gravity

Urine   Increase   Analytical
Albumin   Each 1 g/dL increases S.G. by 0.003. *0251*
Dextran   High molecular weight. *1022*
Diatrizoic acid   Presence of high molecular weight compound. *1700*
Formaldehyde   Causes slight increase only. *0251*
Glucose   Each 1% increases by 0.004. *0251*
Radiographic Agents   Presence of high molecular weight substance. *1700*   Presence of high molecular weight substance. *3505*

Urine   Increase   Physiological
Diurnal Variation   Urine concentration increased at night. *2222*

Urine   Decrease   Physiological
Carbenoxolone   Impaired concentration with hypokalemia Impaired ability to concentrate with low potassium. *2458*
Colistin   Concentrating ability may be impaired. *0971*
Lithium   Large proportion of patients with polyuria with long term treatment. *3705*
Methoxyflurane   Impaired renal tubular function (dose dependent) Impaired renal tubular function. *0749*
Muscular Exercise   Apparent reduced ability to concentrate at all rates. *1842*
Starvation   Observed response to stress. *2382*

Sperm Count

Semen   No Effect   Physiological
Naproxen   No change with treatment although prostaglandin concentration reduced. *0285*
Ranitidine   With up to 450 mg/d in 20 males with chronic duodenal ulcer. *2925*

Semen   Increase   Physiological
Clomiphene   In men with oligozoospermia but pregnancy rate unchanged. *0953*
Tamoxifen   Increased in oligozoospermic men if FSH concentration low or normal. *0953*

Semen   Decrease   Physiological
Azathioprine   Not uncommon in males on drug. *3061*
Cimetidine   In 7 patients treated for 9 weeks with 1200 mg/d compared with their pretreatment values. *2368*   In 11 male subjects with chronic duodenal ulcer given drug 1 g/d for 3 mo. *3759*
Colchicine   Potent antispermatogenic action observed. *2831*
Cyproterone   May completely inhibit spermatogenesis with high doses. *0953*
Demecolcine   Potent antispermatogenic action observed. *2831*
Fluoxymesterone   After prolonged administration or excess dosage. *1678*
Ketoconazole   With prolonged treatment with 800-1200 mg/d. *3390* Azospermia common in patients receiving drug. *2839*
Lead   In a population of male battery workers compared with control group in cement industry. *0172*   In some men occupationally exposed to large quantities of lead. *0771*
Levonorgestrel   When combined with testosterone enanthate may cause oligospermia. *0953*
Methotrexate   Oligospermia occurs without altering plasma hormone concentrations. *0953*
Methyltestosterone   May occur after prolonged dose or excess administration. *1678*
Sulfasalazine   Often with abnormal sperm in high proportion of men: effect reversible. *0953*

Sperm Morphology

Semen   No Effect   Physiological
Cimetidine   In 11 male subjects with chronic duodenal ulcer given drug 1 g/d for 3 mo. *3759*
Ranitidine   With up to 450 mg/d in 20 males with chronic duodenal ulcer. *2925*

Sperm Motility

Semen   No Effect   Physiological
Cimetidine   In 11 male subjects with chronic duodenal ulcer given drug 1 g/d for 3 mo. *3759*
Naproxen   No change with treatment although prostaglandin concentration reduced. *0285*

Semen   Decrease   Physiological
Lead   In some men occupationally exposed to large quantities of lead. *0771*

Standard Bicarbonate

Serum   Decrease   Physiological
NSD 3004   Arterial blood long acting carbonic anhydrase inhibition. *2232*

Blood   No Effect   Physiological
Physical Training   No significant difference observed. *1491*

Blood   Decrease   Physiological
Muscular Exercise   Decreased to 11 meq/L with intermittent exercise. *1924*

Stearic Acid

Serum   Increase   Physiological
Caffeine   Significant increase within 1 h, still elevated after 4 h, when 250 mg ingested. *2751*

Sterols

Feces   Increase   Physiological
Neomycin   Precipitates bile salts in gastrointestinal tract. *1077*

Suberate

Urine   No Effect   Physiological
Lithium   No consistent effect observed. *2107*

Succinate

Blood   Increase   Physiological
Hypoxia   (arterial blood) reversal of path to oxaloacetate. *1717*

Urine   Increase   Physiological
Lithium   May be considerable effect. *2107*

Succinate Dehydrogenase

Test Conditions   Decrease   Physiological
Malonic Acid   Inhibitory effect marked. *2893*

Sucrose

Serum   Increase   Physiological
Sucrose   With tolerance test from 0.04 to 0.1 mg/dL. *2551*

Urine   Increase   Physiological
Sucrose   But small compared with maltose in tolerance test. *2551*

Sugar

Urine   No Effect   Analytical
Methyldopa   No effect reported at 0.2 mg/mL. *2071*
Thymol   No effect on reducing methods. *2981*

Urine   Increase   Analytical
Acetanilid   Acts as reducing substance with Benedict's reagent. *2425*   Acts as reducing agent with nonspecific tests. *1237*
Amino Acids   False positive caused by some with Benedict's. *1022*
Aminopyrine   Glucuronide metabolism affects Benedict's, Clinitest®. *0583*
Aminosalicylic Acid   Acts as reducing agent with Benedict's reagent. *0110*   Acts as reducing agent with nonspecific methods. *1237*
Amygdalin   Measured as reducing agents with Benedict's. *2313*
Antipyrine   Acts as reducing agent. *2425*
ANTU   May reduce Fehling's and Benedict's solutions. *0279*
Arabinose   False positive with Benedict's. *1563*
Ascorbic Acid   False positive with Benedict's and Clinitest®. *1022*   Positive with Benedict's, Clinitest® at 50 mg/dL. *0583*
Aspidium   Acts as reducing agent. *1022*

Aspirin   False positive with Clinitest® or Benedict's. *1022*   Conjugate may react with Benedict's. *1563*
Bismuth Salts   Interferes with Benedict's reaction. *1022*
Carinamide   False positive with Benedict's. *1563*
Cephalexin   False positive with Benedict's, Fehlings, Clinitest®. *1705*
Cephaloglycin   Affects Benedict's, Fehlings, Clinitest®. *1680*
Cephaloridine   Abnormal dark color with Benedict's and Clinitest®. *1099*
Cephalothin   False positive with copper reduction procedures. *3505*   Brown-black color with Clinitest® (false positive). *0110*
Chloral Hydrate   Excreted as glucuronide false positive with Benedict's. *1022*
Chloramphenicol   False positive with copper reduction procedures. *3505*
Chloride Salts   If anthrone method used. *0093*
Chlortetracycline   Acts as reducing substance. *1238*
Cinchophen   Acts as a reducing substance. *2425*
Creatine   Yellow with Benedict's. *3507*
Creatinine   Acts like reducing substance with Benedict's. *2313*
Cresol   Excreted as glucuronide, reacts with Benedict's. *0629*
Diatrizoic acid   Affects copper reduction techniques due to I_2. *2425*
EDTA   Reduces Benedict's reagent. *1237*
Ethinamate   Theoretical effect as excreted as glucuronide. *1343*
Flurazepam   Excreted as glucuronide, sulfate — affect Benedict's. *3216*
Formaldehyde   May reduce Fehlings and Clinitest®. *1563*
Fructose   False positive with Benedict's, Clinitest®. *1563*
Furazolidone   Metabolites may give false positive with Benedict's. *1488*
Galactose   Gives positive with Benedict's, Clinitest®. *1563*
Gentisic Acid   Reducing substance affects Benedict's, Clinitest®. *0583*
Glucosamine   Positive with Benedict's, Clinitest®. *0583*
Glucose   Positive with Clinitest®, Benedict's, Fehling's. *2308*
Glucuronic Acid   Interferes with Benedict's, Fehlings, Clinitest®. *3877*
Hippuric Acid   Reduces Benedict's solution. *1022*
Histidine   High concentrations cause yellowing of Benedict's color. *3507*
Homogentisic Acid   Reduces Benedict's solution, may affect Clinitest®. *1022*
Hypochlorites   May cause false positive with Clinistix®. *1563*
Indican   False positive with Benedict's. *2448*
Isoniazid   False positive with Benedict's and Clinitest®. *1099*
Ketones   False positive with Benedict's. *1563*
Lactose   False positive with Benedict's, Clinitest®. *1563*
Levodopa   False positive with Clinitest® Produces trace positive if Clinitest® used. *1099*
Lysol®   Excreted glucuronide reacts with Benedict's. *0629*
Maltose   False positive with Benedict's, Clinitest®. *1563*
Melanin   In large quantity may reduce Benedict's. *3588*
Metaproterenol   Interferes with Benedict's reagent. *1022*
Metaxalone   False positive with copper reduction procedures. False positive with Benedict's, Fehling's reactions. *3505*
Methenamine   False positive with Benedict's reagent. *2150*
Methyldopa   False positive with Clinitest®, no effect glucose oxidase method. *1099*
Methyprylon   Excreted as glucuronide, acts as reducing substance. *1343*
Morphine   Interferes with copper reduction method. *2425*
Nalidixic Acid   False positive with Fehlings, Benedict's, Clinitest®. *3877*
Neocinchophen   Affect Benedict's, Clinitest®. *0583*
Niacin   Interferes with Benedict's reagent. *1022*
Nitrofurans   Metabolites reduce Benedict's reagent. *2220*
Nitrofurantoin   Metabolites may reduce Benedict's, yield false positive. *2220*

Sugar (continued)

Urine   Increase   Analytical   (continued)

Nitrofurazone   Metabolites may reduce Benedict's reagent Reducing action of metabolites. *1237*

Nucleoproteins   Interferes with Benedict's-?reducing sugar. *1022*

Oxalate   False positive with Benedict's. *1563*

Oxazepam   Alleged to reduce copper in Somogyi procedure. *0583*

Oxytetracycline   Acts as reducing agent. *3879*

Penicillin   False positive with copper reduction procedures. *3505* Drug and metabolites act as red substances in high concentrations. *0583*

Pentoses   Positive with Clinitest®, Benedict's Fehling's. *2308*

Phenols   Interferes with Benedict's reagent. *1022*

Probenecid   False positive with Benedict's or Clinitest®. *3505* Reducing substances with Benedict's, Clinitest®. *0583*

Protein   False positive with Benedict's. *2313*

Pyrazolones   Interferes with copper reducing methods. *1022*

Quinethazone   False positive with Benedict's. *2313*

Radiographic Agents   Green-black reaction with reducing procedures. *0583*

Rhamnose   Reducing substance affects Benedict's, Clinitest®. *0583*

Rhubarb   May cause false positive with Benedict's. *2448*

Ribose   False positive with Benedict's, Clinitest®. *1563*

Santonin   False positive with Benedict's. *2448*

Sauerkraut   Metabolite may affect Galatest. *1563*

Streptomycin   False positive with copper reduction procedures. *3505* Acts as reducing agent affects Benedict's, Galatest. *3879*

Sulfamethoxazole   May cause positive with fluorescent methods. *1022*

Sulfanilamide   False positive with Benedict's. *1563*

Sulfathiazole   Yellow-orange with Benedict's. *0583*

Sulfonamides   Affects Benedict's and Clinitest®. *1488*

Tetracycline   False positive copper reduction procedure. *3505* False positive with Benedict's and Clinitest®. *1488*

Thiazides   Interferes with Benedict's reagent. *1022*

Trimetozine   Interferes with Benedict's reagent. *1022*

Uric Acid   Acts as reducing agent with Benedict's. *2313*

Xylose   False positive with Benedict's, Clinitest®. *1563*

Xylulose   False positive with Benedict's, Clinitest®. *1563*

Urine   Increase   Physiological

Chloroform   Positive with Benedict's, Fehling's solutions. *3879*

Epinephrine   Positive with Fehling's and Benedict's solutions. *0279*

Ether   May cause positive Benedict's and Fehling's tests. *0279*

Urine   Decrease   Analytical

Sulfonamides   Forms colorless Cu complex with Benedict's. *0583*

Test Conditions   Increase   Analytical

Aminoimidazoleacetic Acid   Positive spot test with Benedict's reagent. *3765*

Epinephrine   Positive spot test with Benedict's reagent. *3765*

Sulfa as Sulfanilamide

Serum   Increase   Analytical

Aminobenzoic Acid   Measured as if sulfonamide. *1343*

Aminohippuric Acid   Measured as if sulfonamide. *2981*

Aminosalicylic Acid   Very slight effect, measured as if sulfa. *2981*

Aniline   Reacts in Bratton-Marshall procedure. *0279*

Azosulfamide   Slight effect, measured as if sulfonamide. *2981*

Benzocaine   Diazotizes and interferes. *0251*

Cresol   Reacts in Bratton-Marshall procedure. *0279*

Indole   Reacts with diazotization. *0251*

Phenols   Reacts in Bratton-Marshall procedure. *0279*

Procaine   Metabolized to PABA which interferes. *1343*

Sulfonamides   Reacts in Bratton-Marshall reaction. *0279*

Tetracaine   Yields diazotization reaction. *0251*

Tryptophan   Yields positive Bratton-Marshall reaction. *0279*

Serum   Increase   Physiological

Sulfachlorpyridazine   After 4 g rises to 22 mg/dL at 3 h. *1680*

Urine   Increase   Analytical

Phenobarbital   Positive reaction but delayed with usual amount present. *0251*

Sulfadiazine

Serum   Increase   Physiological

Sulfinpyrazone   Displaces from protein binding. *1487*

Serum   Decrease   Physiological

Food   Absorption delayed when taken with food. *3024*

Sulfamethoxazole

Serum   Decrease   Physiological

Cholestyramine   Reduced absorption due to adsorption or steatorrhea. *3794*

Urine   Increase   Physiological

Alkaline Urine   If co-trimoxazole administered. *0745*

Sulfamethoxine

Serum   Decrease   Physiological

Food   Absorption delayed when taken with food. *3024*

Sulfamethoxypyridazine

Serum   Decrease   Physiological

Food   Absorption delayed when taken with food. *3024*

Sulfanilamide

Serum   Decrease   Physiological

Food   Absorption delayed when taken with food. *3024*

Sulfasymazine

Serum   Decrease   Physiological

Food   Absorption delayed when taken with food. *3024*

Sulfatase

Urine   Increase   Physiological

Proteinuria   Increased in renal diseases. *2897*

Sulfate

Urine   No Effect   Analytical

Storage of Sample   No effect at 4° or -20 for 30 d. *2243* Satisfactory for 24 h without preservative at room temperature. *1180*

Urine   Increase   Physiological

Parathyroid Extract   Hormonal action. *1343*

Protein   Derived from dietary protein. 0948

Urine   Decrease   Physiological
Acetaminophen   Reduces increased output of rheumatoids. 3102

Sulfathiazole

Serum   Increase   Physiological
Antacids   Due to faster dissolution rate. 3794

Sulfhemoglobin

Blood   Increase   Physiological
Acetanilid   Hemolytic anemia in acute poisoning. 1343
Glucosulfone   Hemolytic anemia. 2429
Phenazopyridine   Positive with severe oxidative hemolysis. 1216
Sulfamethizole   May occur with hemolytic anemia. 2868
Trinitrotoluene   Associated with methemoglobinemia and hemolysis. 2429

Sulfinpyrazone

Serum   Increase   Physiological
Probenecid   Inhibits renal tubular secretion. 1487

Sulfisoxazole

Serum   Increase   Physiological
Sulfinpyrazone   Displaces from protein binding. 1487

Serum   Decrease   Physiological
Food   Absorption delayed when taken with food. 3024

Superoxide Dismutase

Plasma   Increase   Physiological
Heparin   Marked increase when injected i.v. 200 IU/kg body weight. Enzyme probably arises from endothelial cells. 1876

Synephrine

Urine   Increase   Physiological
Oranges   Due to presence in fruit. 2784

T$_3$ Binding Capacity

Serum   Decrease   Physiological
Prednisone   Lowered with 1-2 mg/kg/d for 2-4 weeks in 10 children. 3353

T$_3$ Uptake

Serum   No Effect   Physiological
Aspirin   When approximately 3 g/d ingested for several weeks. 3076
Clomiphene   100 mg/d no effect on test. 0583
Diazepam   No effect after i.v. administration. 0583
Erythrosine   In 30 men receiving up to 200 mg daily for 15 d. 1241
Gold   No effect observed with resin. 1915
Heparin   No effect in fasting people, or when added *in vitro*. 2622
Iodine Containing Drugs   No effect on resin test observed. 1915
Ipodate   Significant increase unrelated to I$_2$ content. 0583
Levodopa   No effect observed in chronic treatment. 1860

Lithium   Observed in one patient who developed thyrotoxicosis without pre-existing goiter. 2418
Mercury Compounds   No effect on resin uptake. 1915
Metronidazole   1200 mg/d for 1 week producing no effect. 0583
Norethindrone   When treatment results compared with controls. 0260
Phenytoin   No effect 300 mg/d for 1 week. 0583   With long-term treatment due to accelerated thyroxine clearance via stimulation of hepatic microsomal enzymes. 2168
Povidone-Iodine   No effect observed. 0583
Primidone   In 5 epileptic patients treated on average for 6 mo. 2168
Silver   No effect on resin procedures. 1915
Smoking   No significant difference between heavy smokers and controls. 3255
Sulfonylureas   Probably no effect in most patients. 1487
Tetracycline   No effect with 2 g/d for 11 d. 0583
Thiamylal Sodium   No effect on test. 0583
Thiopental   No effect on test. 0583
Tri-iodothyronine   No significant effect with 150 μg/d in obese. 0458
Valproic Acid   In 10 epileptic patients treated average of 8 mo. 2168
Warfarin   No effect at dose of 5 mg/d. 0583

Serum   Increase   Analytical
Estrogens   As measured by Thyopac method. 0632
Pregnancy   As measured by Thyopac method. 0632

Serum   Increase   Physiological
Aminoglutethimide   Competes for binding sites. 0583
Aminosalicylic Acid   Resin uptake increase due to impaired synthesis of T$_4$. 1915
Anabolic Steroids   Decrease level of thyroxine binding globulin. 0583
Androgens   Decrease level of thyroxine binding globulin. 0583
Aspirin   Red cell uptake affected, also affects resin test. 1022
Barbiturates   Competes for thyroxine binding prealbumin sites. 0583
Chlorpropamide   Competes for thyroxine binding globulin sites. 0583
Cobalt   Resin uptake increased with impaired synthesis of T$_4$. 1915
Colestipol   Observed in euthyroid patients when drug given together with niacin. 0603
Corticosteroids   Resin uptake increased due to decreased thyroxine binding globulin. 1915
Coumarin   Competes with T$_3$ for thyroxine binding albumin sites. 0583
Dextrothyroxine   Binds to usual T$_4$ sites, decreases thyroxine binding globulin. 0583
Diazo Dyes   Compete for binding sites. 0583
Dicumarol   Red cell uptake affected. 0582
Dinitrophenol   Competes for sites on thyroxine binding prealbumin. 0583
Fluoxymesterone   Affects RBC and resin uptake. 0019
Furosemide   Diminished protein binding at time of peak drug concentration. 3473   Changes observed in opposite direction to total T$_4$ concentration. 2584   Transient significant effect seen 2-5 h after ingestion of various amounts of drug chronically in 34 patients with congestive cardiac failure. 2584
Halofenate   Resin uptake increased by 20% after 3 weeks. 2483
Heparin   Red cell uptake increased, resin unaffected. 3176
Liotrix   Increases but remains within normal range. 1680
Methandrostenolone   Affects RBC and resin uptakes. 0019   In 6 body-builders taking up to 20 mg/d intermittently for a year or more. 3290
Mitotane   Competes for sites on thyroxine binding globulin. 0583
Nandrolone   Anabolic effect. 2681
Oxymetholone   Anabolic effect (possible for 2 weeks after stop). 1679
Oxyphenbutazone   Impaired synthesis of thyroxine. 1915

T₃ Uptake (continued)

Serum   Increase   Physiological (continued)
Penicillin   Competes for thyroxine binding prealbumin sites. *0583*
Phenylbutazone   Resin and red cell uptake affected. *3209*
Phenytoin   Affects both resin and red cell uptake. *1488*
Sulfonamides   Some compete for binding sites. *0583*
Sulfonylureas   Resin uptake increase due to impaired synthesis of T₄. *1915*
Thyroid   Consequence of treatment. *1488*
L-Thyroxine   From 0.59 to 0.98 arbitrary units in 11 hypothyroid women treated with 0.1 to 0.2 mg daily. *1201*
Tolbutamide   Increase by 5-10% with 1 g i.v. *0583*

Serum   Decrease   Analytical
Diflunisal   Same interference as with thyroxine but to lesser extent. *1687*
Pregnancy   As measured by resin sponge using Triosorb. *0632*

Serum   Decrease   Physiological
Amiodarone   From 0.98 to 0.96 in 13 patients treated for average of 17 mo. *0425*
Carbimazole   From 1.33 to 0.90 arbitrary units in 12 hyperthyroid women patients treated with 10-30 mg daily. *1201*
Chlordiazepoxide   ?antithyroid effect (resin test affected). *0234*
Clofibrate   Due to increased thyroxine binding globulin. *2858*
Diazepam   Resin test affected. *2220*
Estrogens   Increases binding capacity of thyroxine binding globulin. *3871*
Iothiouracil   Depresses thyroid function. *0583*
Lithium   Reported to induce myxedema. *2221*   Irreversible myxedema observed in 2 patients treated for 1 y. *2773*
Methimazole   Therapeutic result. *1444*
Oral Contraceptives   Falls from 80-100% to 55-90% (resin test). *0585*   Similar values to those of first trimester of pregnancy. *3403*
Perphenazine   Occurs with prolonged use. *2313*
Phenothiazines   Due to increased thyroxine binding globulin with prolonged use. *1915*
Pregnancy   Increased level of thyroxine binding globulin. *0583*
Propylthiouracil   Action of drug. *1488*
Thiazides   Slight effect observed only. *2392*

Taurine

Plasma   Increase   Analytical
Iodoacetate   Effect of oxidizing agent (added *in vitro*). *0384*
Storage of Sample   Large amounts in platelets, leukocytes. *3230*

Plasma   Increase   Physiological
Histidine   Interferes with clearance after oral load. *1644*

Plasma   Decrease   Physiological
Tranylcypromine   Observed effect in normal individuals. *0824*
Venipuncture   Effect lasts for up to 1 h. *3230*

Urine   Increase   Physiological
X-Ray Therapy   Due to tissue destruction. *2378*

Urine   Decrease   Physiological
Aspirin   Reduces elevated concentration in rheumatoid patients. *3102*
Phenylbutazone   Reduces increased output in rheumatoids. *3102*

Terbutaline

Serum   Increase   Physiological
Oxprenolol   Decreased clearance and increased area under curve when co-administered. *1823*

Testosterone

Serum   No Effect   Analytical
Storage of Sample   For 2 to 4 mo at -20°. *0888*
Tetracycline   No effect given at 100 µg/mL on radioimmunoassay. *2884*

Serum   No Effect   Physiological
L-α-Acetylmethadol (LAAM)   In 9 male heroin addicts maintained on drug and in 2 weeks following abrupt withdrawal. *2409*
Anticonvulsants   In 33 male epileptics taking at least one drug for long time. *3020*   Insignificant increase in 37 men versus control population on chronic therapy. *0803*   No significant difference in men on long-term therapy. *0257*
Carbamazepine   Usually no effect but may be slight increase. *1736*
Chlorthalidone   No significant effect observed although higher incidence of sexual dysfunction than in control population. *1259*   Although higher incidence of sexual dysfunction. *1259*
Cimetidine   In 6 patients treated for 1 mo with 1 g/d. *3491*   In 25 men treated for duodenal ulcer or duodenitis. *2368*
Cyclosporine   In 16 patients who developed hypertrichosis but also given added cortisone. *3190*
Cyproterone   In 27 women treated for total of 194 cycles with 50 µg ethinyl estradiol and 2.0 mg cyproterone acetate. *2280*
Danazol   No consistent change noted in patients with prostatic cancer. *0693*   No change during treatment of women with endometriosis. *2599*
Dehydroepiandrosterone   In 5 men given 1600 mg/d orally for 28 d. *2578*
Dexamethasone   No effect of treatment on initially high concentrations in hirsute women. *0776*
Ethanol   No effect in men or women at 100 mg/dL. *3611*
Famotidine   No effect of short-term or long-term treatment in male patients with duodenal ulcers. *0727*
Haloperidol   Has no effect, unlike some other neuroleptics. *1736*
Heroin   In treated/untreated male addicts. *0780*
Lead   In a population of male battery workers compared with control group in cement industry. *0172*
Levodopa   No significant effect observed after 2 weeks in males. *3338*
Levonorgestrel   No effect observed in group of women 20 and 65 mo after levonorgestrel treated rods inserted in uterus. *0899*
Lovastatin   No significant change even in patients with familial hypercholesterolemia receiving 40 mg bid. *1530*
Methadone   In treated/untreated male addicts. *0780*
Naltrexone   With 50 to 100 mg administered daily to obese subjects over 8 weeks. *0176*
Omeprazole   No effect observed in 8 volunteers given 30 mg/d for 28 d. *2516*
Oxytetracycline   Despite decreased urine excretion and increased fecal excretion of estrogens due to decreased hydrolysis by beta-glucuronidase in gastrointestinal tract. *1469*
Phenytoin   Usual effect. *1736*
Pimozide   General effect observed. *1736*   In acutely psychotic males. *3340*
Primidone   Usual effect but may be slight increase. *1736*
Ranitidine   With up to 450 mg/d in 20 males with chronic duodenal ulcer. *2925*   After 4 weeks and 6 mo treatment (300 mg and 150 mg daily respectively) in male patients with duodenal ulcer. *0727*
Sexual Activity   No effect in men. *3437*
Sodium Bromide   No effect in 10 men, 10 women receiving 1 mg/kg/d during 8 weeks or 2 full menstrual cycles. *3141*
Sulpiride   General effect observed. *1736*   In 11 healthy women between 6 and 9 weeks of pregnancy given 150 mg daily for 2 weeks. *3916*
Valproic Acid   Usual effect although slight increase may occur. *1736*

Serum   Increase   Analytical
Cholesterol   In amounts of 1-20 µg affected competitive protein binding method. *0888*

Contact With Clot 10-25% increase in serum of females when allowed to stand in contact with clot for 5 d at room temperature. *1475*

Danazol May be cross-reactivity in certain radioimmunoassays. *0693* Probably invalid results because of high cross-reactivity with protein in competitive binding assays. *1479* Nonspecificity in RIA kits produced by diagnostic products corporation, Farmos Diagnostica and Bio-RIA. *3276*

Dihydrotestosterone 5-alpha compound affects competitive protein binding method. *2750*

Serum Increase Physiological

Aging In women due to decreased metabolic clearance rate. *1941*

Barbiturates Due to decreased metabolic clearance rate (in women). *1941*

Bromocriptine Causes increase of about 20%. *1736*

Cimetidine In 11 male subjects with chronic duodenal ulcer given drug 1 g/d for 3 mo. *3759* Up to 20% increase on chronic administration and displaces DHT from its binding sites at pituitary and hypothalamic level. *1736*

Clomiphene Liberates LH, anti-steroid hormone effect of drug. *0609*

Danazol Increase by 70% in 7 normal subjects given drug 800 mg daily for 2 mo. *2224*

Diurnal Variation Maximum (160%) at 6 am, minimal (60%) at 10 pm. *3201* Maximum observed about 1 pm. *2222*

Erect Posture In women due to decreased metabolic clearance rate. *1941*

Estrogens Probable effect in women — decreased clearance, increased binding. *1941*

Gonadotropin Twofold increase observed in males. *3338*

Levonorgestrel Approximately 24% increase after 6 mo use of subdermal implant. *2687*

Menstruation Random variations but highest in mid third. *1831*

Oral Contraceptives Metabolic changes in liver synthesis. *2189*

Phenytoin Mean 30.7 nmol/L versus 17.8 in controls in approximately 24 male patients given phenytoin alone or with primidone or phenobarbital. *3609*

Pregnancy Possible effect as increased plasma binding. *1941*

Rifampin Probably due to increase in microsomal activity and increased biosynthesis. *1736*

Sexual Activity Reported effect in males. *1735*

Smoking Marginal effect in male current smokers compared with nonsmokers. *0238*

Testosterone In 11 hypogonadal men given 200 mg cypionate salt intramuscularly, 3 fold rise maximum day 2 to 5. *2555* In power athletes with 26 weeks on steroid self administration. *3088*

Serum Decrease Analytical

Danazol Concentrations low: in case of cortisol and testosterone displacement from plasma proteins occurs and protein binding assays are probably invalid. *1479*

Serum Decrease Physiological

Carbamazepine Within 7 d of starting 400 mg/d treatment in 6 healthy males probably due to induction of hepatic monooxygenase activity. *0707*

Cimetidine Observed in one 66 y old man, together with increased serum gonadotropin possibly due to reversible defect in 17-beta-hydroxysteroid dehydrogenase. *2080*

Cyclophosphamide Due to induced testicular atrophy. *1736*

Cyproterone Increase by 20-40% (antiandrogenic effect). *3394* Decreases hormone in adult men but does not affect concentration in children with precocious puberty. Decreases concentration in hirsute women. *1736*

Danazol Throughout treatment period after 600 mg daily in 18 women with endometriosis. *0591* Affect binding in women treated with drug so that proportion of free drug high. *1479* Marked reduction in some patients in 5 patients with hereditary angioneurotic edema treated for up to 10 mo. *3219*

Dexamethasone Decreased metabolic clearance and binding in women. *1941*

Diethylstilbestrol At doses affecting pituitary-gonadal axis. *1917* When given to men with androgen-dependent prostatic cancer reduced concentration to normal female range. *1736*

Digoxin 30% decrease in men with chronic administration. *3475*

Ethanol Observed in male alcoholics. *0780* Reduced number of episodic bursts of secretion and steady decline in with several days of ingestion in healthy men. *1352*

Glucose Mean maximum fall of 34% normal in 150 minutes. *3751*

Halothane Mechanism unclear ?inhibits release of gonadotropin. *2697*

Ketoconazole Marked effect in 4 volunteer males with 600 mg doses. With long term high dose treatment same effect observed. *2838* Transiently blocks testosterone synthesis. Doses above 800 mg/d may cause more prolonged blockade. *2839*

Levonorgestrel In 17 women using Norplant implants. *1799* In a group of women using levonorgestrel versus control group with copper IUD. *0899*

Medroxyprogesterone Significant reduction after treatment. Effects observed in 15 postmenopausal women with endometrial cancer after 2 weeks treatment. *2118*

Methandrostenolone In 4 of 6 body-builders taking up to 20 mg/d intermittently for a year or more. *3290*

Methylprednisolone Reduced by more than 50% in 14 of 16 67 y old men with chronic pulmonary disease receiving drug for more than 1 mo. Testosterone concentration inversely correlated with dose of drug. Serum protein-binding unaffected. Effect probably due to suppression of secretion of gonadotropin releasing hormone by hypothalamus. *2245*

Metyrapone Fall in 2 h in men, minor increase in women. *2600*

Nafarelin Acetate In all 9 patients with benign prostatic hypertrophy: due to reduced production in testes by inhibition of pituitary release of gonadotropins. Concentration reversibly reduced from normal to castration level. *2793*

Prednisone Reduced by more than 50% in 14 of 16 men aged 67 y with chronic pulmonary disease taking drug for at least 1 mo. Testosterone concentration inversely related to dose of drug. Serum protein binding unaffected. Effect probably due to suppression of secretion of gonadotropin releasing hormone by hypothalamus. *2245*

Spironolactone In males for 2 weeks from 729 ng/mL to 634 mg/ml. *2773* Inhibits biosynthesis in the testis. *0953* Also displaces DHT from its cytosolic receptors. *1736*

Stanozolol Increase by 55% (from 22.1 to 10.6 nmol/L) after 1 week in 9 healthy men given 10 mg/d for 14 d. *3354*

Tetracycline By about 10-20%. *1736* Fell from 21 to 17 nmol/L in 9 men after 3 d treatment: mechanism not established. *2884*

Thioridazine Slight effect in patients on long-term treatment. *1736* Significantly less in 42 male schizophrenics than when they ingested other neuroleptic agents. *0500*

Urine Increase Physiological

Clomiphene Liberates LH, anti-steroid hormone effect of drug. *0229*

Corticotropin Moderate effect in men. *3062*

Gonadotropin Marked effect in normal men. *3062*

Urine Decrease Physiological

Dexamethasone Slight effect in men. *3062*

Testosterone Binding

Plasma Increase Physiological

Twin Pregnancy Greater than with single fetus, less than with triplets. *0888*

Testosterone Binding Globulin

Serum Increase Physiological

Estrogens Metabolic response. *1498*

Oral Contraceptives Metabolic effect. *2385*

Pregnancy Increased during pregnancy, maximum in 3rd trimester. *1498*

Testosterone, Free

Serum   No Effect   Physiological
Dehydroepiandrosterone   In 5 men given 1600 mg/d orally for 28 d. *2578*
Levonorgestrel   In 17 women using Norplant implants. *1799*
Oxytetracycline   Despite decreased urine excretion and increased fecal excretion of estrogens due to decreased hydrolysis by beta-glucuronidase in gastrointestinal tract. *1689*

Serum   Increase   Physiological
Danazol   Causes displacement of hormone from shbg and may result in increase of up to 90%. *1736*   Marked rise observed in women with endometriosis. Due to displacement from sex hormone binding globulin. *2599*   Proportion high due to displacement of hormone from protein. *1479*
Testosterone   4.5 fold increase at days 2-3 in 11 hypogonadal men given 200 mg cypionate salt intramuscularly. *2555*

Serum   Decrease   Physiological
Anticonvulsants   (free) 0.35 nmol/L versus 0.55 nmol/L in 37 men on chronic therapy. *0803*
Carbamazepine   Observed effect. *1736*   Within 7 d of starting 400 mg/d treatment in 6 healthy males probably due to induction of hepatic monooxygenase activity. *0707*
Cyproterone   In 27 women treated for total of 194 cycles with 50 μg ethinyl estradiol and 2.0 mg cyproterone. *2280*
Diethylstilbestrol   In men with androgen-dependent prostatic cancer and reduction of total hormone concentration with SHBG. *1736*
Ketoconazole   Observed in 2 h after single 200 mg dose in normal men. *3390*   Equally reduced with bound testosterone. *2839*
Oral Contraceptives   As a result of markedly increased sex hormone binding globulin. *2280*
Phenytoin   Observed effect. *1736*
Primidone   Observed effect. *1736*
Stanozolol   From 398 to 226 pmol/L after 1 week in 9 healthy men given 10 mg/d for 14 d. *3354*
Tetracycline   By about 10-20%. *1736*
Valproic Acid   Observed effect. *1736*

Tetracycline

Serum   Decrease   Physiological
Ferrous Sulfate   Reduced absorption due to chelation which also affects elimination. *3794*
Food   Absorption reduced when taken with food. *3024*
Magnesium Sulfate   Reduced absorption noted. *3794*

Tetrahydroaldosterone

Urine   Increase   Physiological
Ovulation   Higher in luteal phase. *2850*
Pregnancy   Increases with period of gestation. *2850*

Tetrahydrocortisol

Urine   Increase   Physiological
Diurnal Variation   Maximum excretion in morning (about 10 am). *2222*

Tetrahydrocortisone

Urine   Increase   Physiological
Diurnal Variation   Maximum excretion in early am (about 7 am). *2222*

Tetrahydrodeoxycorticosterone

Urine   No Effect   Physiological
Angiotensin   No significant effect when infused in normals. *3174*

Urine   Increase   Physiological
Corticotropin   Fivefold increase in response to infusion in normals. *3174*

Urine   Decrease   Physiological
Fludrocortisone   Slight effect following administration. *3174*

Theophylline

Serum   No Effect   Analytical
Fluosol-DA   As measured by Syva® EMIT at concentration up to 50%. *2518*

Serum   No Effect   Physiological
Cefaclor   No effect on steady state kinetics in healthy young adults. *1824*
Ranitidine   No significant effect on ranitidine pharmacokinetics. *2859*
Roxithromycin   Little effect on pharmacokinetics when added to stable therapeutic regime. *3115*

Serum   Increase   Analytical
Caffeine   When present in tea, coffee or cola may raise apparent concentration in certain fluoroimmunoassays. *1614*
Dimenhydrinate   When Clinical Assays RIA kit used: almost 3 fold increase observed. *1450*

Serum   Increase   Physiological
Allopurinol   Increased half-life and decreased clearance probably due to effect on hepatic metabolizing enzymes. *2284*
Cimetidine   Plasma clearance reduced from 71 mL/min to 56 mL/min in 7 healthy subjects given 1 g/d. *3004*   Half-life increased by 60% with coadministration. *3782*   Single case reported in which serum concentration doubled: effect reversible: effect possibly due to direct inhibition of hepatic metabolism. *0571*   In group of young adults and elderly with increased area under curve, increased half-life and reduced clearance. *0687*
Ciprofloxacin   In one case clearance fell from 2.3 L/h to 0.8 L/h when drug added to treatment regime. *3583*   Approximate doubling of concentration in 33 hospitalized patients with respiratory tract infections. In 20 patients increase into toxic range. Effect probably due to inhibition of hepatic microsomal enzymes by ciprofloxacin. *2915*
Erythromycin   Clearance significantly reduced. Plasma concentration increased by 28% in 12 patients. *2952*   40% increase in concentration due to 26% reduction in clearance after 1 week therapy. *2057*
α-Interferon   1 d after i.m. injection in 5 patients with hepatitis and 4 healthy volunteers significantly reduced clearance and significant increase of elimination half- life. *3843*
Metoprolol   Clearance reduced by 30-50% in healthy individuals given 200 mg or more per day: possibly due to a metabolite binding to cytochrome P-450. *0712*
Propranolol   Clearance markedly reduced in healthy individuals given drug orally 40 mg/6 h, possibly due to a metabolite binding to cytochrome P-450. *0712*
Ranitidine   Similar effect to cimetidine in reducing plasma clearance. *1938*

Serum   Decrease   Physiological
Carbamazepine   Half-life reduced from 5.25 h to 2.75 h during treatment probably due to enzyme interaction. *3055*
Furosemide   Significantly lower when coadministered with drug. *0594*
Phenytoin   Elimination half-life reduced; total body clearance almost doubled: activity of cytochrome metabolizing enzymes in liver stimulated. *2305*

Rifampin 25% reduction in area under curve when coadministered and reduction of half-life from 7.0 to 4.8 h. *0436* Reduced area under concentration curve and increased metabolic clearance and volume of distribution. *2861* Significantly increased mean oral clearance of drug with elimination rate constant increased by mean of 25%. *3484*

Terbutaline 0.075 mg 3 times per day caused average reduction of serum concentration from 13.8 to 10.8 μg/mL when sustained release drug given. *0807*

Thiamine

Urine   Decrease   Physiological
Starvation Falls to level below recommended in 3 d. *0714*

Thiocyanate

Serum   Increase   Physiological
Smoking In smokers versus nonsmokers (from cyanide metabolites). *2363* 3-6 fold higher in smokers than nonsmokers. *2278* Concentration correlated with extent of smoking. *3255*

Urine   Increase   Physiological
Smoking Doubled excretion in smokers (metabolite of cyanide). *2363* 3-6 fold higher in smokers than nonsmokers. *2278*

Thormählen Test

Urine   Positive   Analytical
Creatinine Large amounts give brown color. *3588*
Melanin Green-blue to blue-black (normal = olive-brown). *3588*
Methyldopa At 0.2 mg/mL in alkaline urine. *2071*

Threonine

Plasma   Increase   Physiological
Histidine Interferes with clearance after oral load. *1644*

Plasma   Decrease   Physiological
Glucose Probable muscle uptake. *1403*
Starvation 33% decrease by third day returns to normal at 6th. *0021*

Urine   Increase   Physiological
Ascorbic Acid In 10 healthy females given 10 g/d. *3624*
Tetracycline Unexplained mechanism. *1343*

Thrombin Time

Plasma   No Effect   Physiological
Ampicillin In doses as high as 300 mg/kg/d in volunteers. *0492*
Apalcillin In 21 volunteers with doses up to 225 mg/kg. *1262*
Cyclopropane Anesthesia has no effect. *1343*
Fluroxene No significant effect noted. *1814*
Methicillin No effect with doses as high as 300 mg/kg/d in volunteers. *0492*
Penicillin G No effect with doses as high as 48 million U/d. *0492*

Plasma   Increase   Physiological
Asparaginase Occurs with decreased fibrinogen concentration. *2657*
Heparin Related to concentration of circulating heparin. *2649*
Streptokinase Reflects extent of fibrinogen breakdown. *3298*
Urokinase In patients with pulmonary embolism. *3298*

Plasma   Decrease   Physiological
Dextran Slight reduction noted in one case. *0320*

β-Thromboglobulin

Serum   Decrease   Physiological
Prazosin 30% higher value in essential hypertensives normalized with treatment. *1715*

Plasma   No Effect   Physiological
Aminosalicylic Acid No effect of 1.5 g/d or after 250 mg i.v. in 6 patients. *3876*
Smoking No significant difference between smokers and non-smokers. *2995*

Thromboplastin Generation

Blood   Increase   Physiological
Oral Contraceptives Significant effect observed. *2456*

Blood   Decrease   Physiological
Heparin Abnormal response (inhibition). *1343*
Protamine Possesses anticoagulant action. *1343*
Tetracycline Impaired rate of regeneration observed. *1343*

Thromboxane

Plasma   No Effect   Physiological
Dexamethasone No effect of up to 100 μmol/L on formation during clotting. *2890*
Hydrocortisone No effect of up to 100 μmol/L on formation during clotting. *2890*
Prednisolone No effect of up to 100 μmol/L on formation during blood clotting. *2890*
Prednisone No significant change with up to 25 mg drug for up to 14 d: no effect on production during clotting. *2890*

Thromboxane B$_2$

Serum   Decrease   Physiological
Aspirin Reduction by about 99% over 24 h, gradually reverted towards normal. *2046*
Ibuprofen Significant reduction in patients during treatment. *0660*
Indomethacin In 9 uncomplicated essential hypertensives receiving chronic enalapril treatment. *3126*
Sulindac 85% reduction in patients with chronic glomerular disease. *0660*

Plasma   Increase   Physiological
Cimetidine Probable activation of endogenous cyclic AMP phosphodiesterase involved in favoring action of platelet aggregating agents. *3938*

Plasma   Decrease   Physiological
Indomethacin Marked fall from 191 to 1.4 ng/mL in 12 patients with coronary artery disease, although systolic blood pressure increased and coronary blood flow decreased. *1158*

Urine   Decrease   Physiological
Ibuprofen In 11 healthy volunteers in randomized trial. *0452*
Naproxen In 11 volunteers in randomized trial. *0452*
Sulindac In 11 normal volunteers contrary to concept of sparing renal but inhibiting systemic prostaglandins. *0790*

Thymidine Incorporation

Lymphocytes   Increase   Physiological
Halothane Liver disease from mild focal necrosis and inflammation to massive necrosis. *2260*

Thymol Turbidity

Serum   No Effect   Analytical

Bilirubin   No effect up to 14 mg/dL. *1563*

Hemoglobin   No effect up to concentration of 14 mg/dL. *1563*

Storage of Sample   Usually stable at 4° for 1 week. *1563*

Serum   Increase   Analytical

Albumin   If high. *2313*

Fatty Foods   Lipemia causes false high values due to turbidity. *1238*

Heparin   Affects physico-chemical properties altering turbidity. *1563*

Lipemia   Due to inherent turbidity. *2313*

Meals   Cause mild lipemia which is measured. *0582*

Serum   Increase   Physiological

Acetohexamide   May cause intrahepatic cholestatic jaundice. *2313*

Acetophenazine   Cholestasis with intrahepatic obstruction. *2313*

Ajmaline   Intrahepatic cholestasis. *0268*

Aminosalicylic Acid   Liver damage may occur with 10% mortality. *2808*

Amitriptyline   Rare cases of transient cholestasis. *2313*

Amodiaquine   Reported hepatotoxicity. *2313*

Amphotericin B   Due to hepatocellular dysfunction. *2313*

Anabolic Steroids   Due to cholestatic syndrome. *2220*

Androgens   Cholestatic phenomenon. *2313*

Antimony Compounds   Hepatotoxic effect. *2220*

Arsenicals   Hepatotoxic effect. *2220*

Asparaginase   Hepatotoxicity. *1519*

Aspidium   May cause hepatic toxicity. *2313*

Benziodarone   May cause hepatic toxicity. *2313*

Bismuth Salts   Hepatotoxicity. *2313*

Carphenazine   Probable cholestatic effect (reversible). *2313*

Cephalothin   Reported in alcoholic patient with cirrhosis. *2313*

Chloramphenicol   Hepatotoxicity. *2220*

Chlordiazepoxide   Infrequent cholestatic effect. *2313*

Chlormezanone   Occasional reversible cholestasis. *2313*

Chloroform   Hepatotoxic effect with necrosis. *2313*

Chlorothiazide   May cause cholestatic jaundice. *2313*

Chlorpropamide   May cause cytotoxic liver damage. *2313*

Chlorprothixene   May cause hepatotoxicity (reversible). *2313*

Chlortetracycline   Hepatotoxic effect with centrolobular necrosis. *1237*

Chlorzoxazone   May cause hepatotoxicity. *2313*

Cinchophen   May cause hepatotoxicity (viral-hepatitis like). *2313*

Clofibrate   Hepatotoxic effect. *2313*

Clorazepate   Mild elevation in 10% patients reported. *1878*

Colchicine   May have hepatotoxic effect. *2313*

Corticosteroids   Physiological doses increase protein synthesis. *0697*

Corticotropin   Physiological dose increase protein synthesis. *0697*

Coumarin   May cause hepatotoxicity. *3835*

Cyclophosphamide   May cause hepatotoxicity. *2313*

Cyclopropane   May cause hepatotoxicity (lasts several days). *2313*

Cycloserine   May cause hepatotoxicity. *2313*

Desipramine   Rare transient cholestasis. *2313*

Diazepam   Mild effect reported in one patient. *1878*

Ectylurea   May cause cholestasis with cholangiolitis. *2313*

Erythromycin   Weak effect due to cholestasis. *1713*

Estradiol   Cholestatic effect. *2313*

Estrogens   Occurs with cholestasis. *2220*

Estrone   May cause cholestasis. *2313*

Ether   Hepatic disturbance (transient). *2313*

Ethionamide   Intrahepatic cholestasis. *2313*

Ethotoin   Hepatotoxicity. *2313*

Ethoxazene   Hepatotoxic effect. *2313*

Florantyrone   Hepatotoxic effect. *2313*

Flucytosine   Reversible hepatotoxicity in 10%. *3443*

Fluphenazine   Hypersensitivity response. *2313*

Flurazepam   May cause hepatotoxicity. *3016*

Glycopyrrolate   Hepatotoxicity. *2313*

Gold   Hepatotoxicity with centrolobular necrosis. *2313*

Growth Hormone   Increased protein synthesis. *0697*

Guanoxan   May cause hepatic toxicity. *2313*

Haloperidol   May cause hepatocellular changes. *2313*

Halothane   Allergic hepatic hypersensitivity. *2313*

Hydrazine Derivatives   May cause hepatotoxicity. *2313*

Hydroxyacetamide   Hepatotoxicity. *2313*

Ibufenac   Hepatotoxic effect. *2313*

Icterogenin   Causes intrahepatic cholestasis. *0153*

Indandione Derivatives   May cause hepatotoxicity. *2313*

Indomethacin   Cytotoxic and cholestatic liver damage. *1908*

Insulin   Associated with increased protein synthesis. *0697*

Iopanoic Acid   May cause hepatotoxicity. *2313*

Iproniazid   May cause cholestatic jaundice. *2313*

Isocarboxazid   Cholestatic effect. *2313*

Kanamycin   May cause hepatotoxicity. *2313*

Lincomycin   Hepatotoxic-cholestatic effect. *2313*

MAO Inhibitors   Intrahepatic cholestasis. *2313*

Mechlorethamine   Hepatotoxic effect. *0248*

Mepazine   May alter liver function (cholestasis). *2313*

Mephenytoin   Hepatotoxicity. *2313*

Mercaptopurine   Hepatotoxicity (centrolobular necrosis). *2313*

Metahexamide   Hepatotoxicity (viral hepatitis type). *2313*

Metaxalone   Hepatotoxic effect. *2313*

Methandrostenolone   Hepatotoxic effect. *2313*

Methimazole   May affect liver function. *2313*

Methotrexate   Hepatotoxic effect. *2313*

Methoxsalen   Hepatotoxic effect. *2313*

Methoxyflurane   Hepatotoxicity. *2313*

Methyldopa   Hepatotoxic effect. *0597*

Methyltestosterone   Cholestatic effect (may not be affected). *1713*

Methylthiouracil   Hepatotoxic effect. *2313*

Nalidixic Acid   May cause cholestatic jaundice. *2425*

Niacin   Impaired hepatic function. *2313*

Nitrofurans   Hepatotoxicity. *2313*

Nitrofurantoin   May be cholestatic jaundice. *2313*

Norethandrolone   Cholestasis produced without cholangiolitis. *3172*

Norethindrone   Intrahepatic cholestatic jaundice. *2313*

Norethynodrel   Cholestatic effect. *2313*

Novobiocin   Intrahepatic cholestasis may occur. *2313*

Oleandomycin   May cause hepatotoxicity. *2313*

Organophosphorus Insecticides   Hepatotoxic effect. *0405*

Oxacillin   Reversible hepatocellular dysfunction. *2313*

Oxazepam   Possible hepatotoxicity. *2313*

Oxyphenbutazone   Possible hepatotoxicity. *2313*

Oxyphenisatin   May cause hypersensitivity reaction. *2970*

Paraldehyde   Possible hepatotoxicity. *2313*

Paramethadione   Possible hepatotoxicity. *2313*

Pargyline   Possible hepatotoxicity (reversible). *2313*

Penicillamine   Possible hepatotoxicity. *2313*

Phenacemide   May affect liver function in about 2% cases. *2313*

Phenazopyridine   Hepatotoxic effect. *2313*

Phenindione   May modify liver function (cholestasis). *2313*

Pheniprazine   May affect liver function. *2313*

Phenothiazines   May cause cholestatic hepatitis (in up to 4% patients). *2313*

Phenylbutazone   May cause cholestasis. *2313*

Phenytoin   Intrahepatic cholestatic jaundice. *2313*

Phosphorus   Hepatotoxic effect with necrosis. *2313*

Polythiazide May affect liver function. *2313*
Probenecid Hepatotoxic effect (centrolobular necrosis). *2313*
Procainamide Hepatotoxic effect. *2313*
Prochlorperazine Cholestatic effect. *0071*
Progesterone Hepatotoxicity. *2313*
Promazine May cause cholestasis. *2313*
Promethazine May cause cholestasis. *2313*
Propoxyphene Cholestatic effect. *1954*
Propylthiouracil Hepatotoxic effect (centrolobular necrosis). *2313*
Pyrazinamide Hepatotoxic effect (viral-hepatitis like). *2313*
Quinacrine Hepatotoxic effect. *2313*
Quinethazone May cause cholestatic jaundice. *2313*
Rifampin Hepatotoxic effect (cholestasis). *2820*
Sulfadimethoxine Reversible cholestasis. *1049*
Sulfamethizole Reversible cholestasis. *1974*
Sulfamethoxazole Reversible cholestasis. *2313*
Sulfamethoxypyridine Reversible cholestasis. *2313*
Sulfisoxazole Reversible cholestasis. *2313*
Sulfonamides May cause cholestasis. *2808*
Sulfones May affect liver function. *2313*
Testosterone May cause cholestasis. *2313*
Tetracycline May cause cholestasis. *2313*
Thiabendazole Rare cholestasis. *2413*
Thiacetazone May cause cholestasis. *0248*
Thiazides May cause cholestasis. *2313*
Thioguanine May cause cholestasis. *2313*
Thioridazine Hepatotoxicity. *1857*
Thiosemicarbazones May affect liver function. *2313*
Thiothixene Hepatotoxic effect (reversible cholestasis). *2313*
Thorium Dioxide May produce severe liver damage after years. *3147*
Thyroid Increase protein metabolism. *0652*
Tolazamide Cholestasis may occur. *2313*
Tolbutamide May cause cytotoxic liver damage or cholestasis. *2313*
Tranylcypromine May affect liver function (?hypersensitivity). *2313*
Trimethadione May cause hepatotoxicity with necrosis. *2313*
Trioxsalen May affect liver function. *2313*
Uracil Mustard May affect liver function. *2313*

Serum   Decrease   Analytical
Albumin High serum concentration causes decrease. *0355*
Clothiapine At 9 mg/L when added to serum and conventional method. *1758*
Heat As room temperature is increased. *0579*
Heparin Concentration related effect. *0583*
Oxalate Decreased turbidity developed. *0583*

Thyro-Binding Index

Serum   Increase   Physiological
Oral Contraceptives For up to 1 mo after stopping treatment. *0913*
Pregnancy Rises from 3 weeks, normal 8 weeks after birth. *0913*

Serum   Decrease   Physiological
Androgens May occur for up to 1 mo after treatment. *0913*
Aspirin For up to 1 week after treatment. *0913*
Corticosteroids May occur for up to 1 week after therapy. *0913*
Corticotropin Occurs for up to 1 week after treatment. *0913*
Estrogens Increases binding capacity of thyroxine binding globulin for up to 1 mo. *3871*
Heparin Effect observed during anticoagulant therapy. *0251*
Phenylbutazone For up to 3 weeks after therapy. *0913*
Phenytoin For up to 3 weeks after therapy. *0913*

Thyroglobulin

Serum   No Effect   Physiological
Carbamazepine No significant effect with 400 mg/d for 21 d. *0708*

Serum   Decrease   Physiological
Neomycin Mean decrease from 17.3 to 11.7 ng/mL when given 2.0 g/d orally for 7 d. *0959*

Thyroid Antibodies

Serum   Positive   Physiological
Lithium In two thirds women with hypothyroidism treated for 2 y. *2185*

Thyroid Stimulating Hormone (TSH)

Serum   No Effect   Analytical
Carbamazepine At therapeutic and toxic concentrations on RIA methods. *0374* No effect observed at both therapeutic and toxic concentrations when added to serum *in vitro*. *0374*

Serum   No Effect   Physiological
Amiodarone Insignificant change in 13 patients treated for average of 17 mo. *0425*
Amitriptyline No effect observed with acute or chronic treatment in depressed patients. *1366*
Anesthetic Agents No effect observed. *2700*
Apomorphine In euthyroids but can lower basal and TRH-stimulated values. *3798*
Arginine No effect observed. *2700*
Asparaginase No effect during treatment within 3 weeks in 14 children. *1543*
Butorphanol No significant effect in 6 healthy male volunteers given 2 mg i.m. *3033*
Caffeine No significant effect of single dose of 250 mg in men or women. *3418* In men no significant effect of single dose of 500 mg. *3418*
Carbamazepine No effect of 400 mg/d on basal or stimulated TSH. *0708* Observed in 9 hypothyroid patients given substitution therapy due to increased extra-thyroidal metabolism of thyroid hormones. *0001* No change in TSH or in its response to TRH. *3798* No significant change in patients on long-term treatment. *2168*
Cimetidine In 6 patients treated for 1 mo with 1 g/d. *3491* No significant changes caused by drug. *2780* No change after 1 g daily for 28 d, although different findings by other group. *3798*
Clomipramine No effect with acute or chronic treatment in depressive patients. *1366*
Clonidine In children with growth hormone deficiency 0.1 mg/m² daily for 60 d. *2821*
Cotrifamole No change in men or women over 10 d. *0686*
Co-Trimoxazole In men or women with treatment for 10 days. *0686*
Danazol No effect after 2 weeks with 800 mg/d but slight decrease after 4 weeks. *0985*
Dopamine Did not change with infusion of up to 3.0 µg/kg/min for 2 h in 6 normal men. *2149*
Endotoxin No effect observed. *2700*
Estrogens High doses did not augment TRH-mediated response in women: in men slight effect only with high doses without influence on basal TSH. *3798*
Famotidine No effect of short-term or long-term treatment in male patients with duodenal ulcers. *0727*
Fenclofenac None of 8 patients receiving 1.2 g/d had altered values. *2971* May be short-term reduction (beginning of therapy) at hypophyseal level. *3798*
Fenoprofen No effect noted physiologically or on methods. *3594*
Flunarizine Basal concentration not affected but increase of domperidone blunted in women. *2265*
Furosemide No effect typically observed in spite of changes in concentration of peripheral hormones. *2584*

Thyroid Stimulating Hormone (TSH)
(continued)

Serum   No Effect   Physiological (continued)

Halothane No appreciable effect observed. 2698

Heroin No significant effect observed. 0193

Heroin Withdrawal No significant effect observed. 0193

Hypothermia No effect observed. 2700

Indomethacin No effect on TSH secretion. 3798

Insulin No effect observed. 2700

Isotretinoin In 7 patients with severe rosacea treated with 1 mg/kg/d for 12 weeks. Effects possibly due to induction of hepatic microsomal enzymes. 2307 Or no significant change in 24 healthy men given 1 mg/kg/d for 16 weeks. 2640 In 7 patients given 1 mg/kg/d for 12 weeks. 2307 In 18 patients with severe acne given 0.8 mg/kg daily for 3 mo: changes reverted to normal after treatment stopped. 2242 Significant at 12 weeks in 7 patients with severe rosacea treated with 1 mg/kg/d. 2307

Lead In some men occupationally exposed to large quantities of lead. 0771

Levodopa No effect observed. 3324

Levonorgestrel No effect observed in group of women 20 and 65 mo after levonorgestrel treated rods inserted in uterus. 0899 In a group of women using levonorgestrel versus control group with copper IUD. 0899

Megestrol In 18 postmenopausal women with breast cancer. 0056

Mestranol Insignificant difference in 19 post-menopausal women receiving mean of 24 μg/d versus controls. 0007

Methadone No significant effect observed. 0193

Methoxyflurane No effect observed with anesthesia. 2698

Metoclopramide Basal values did not change as result of long-term treatment. 3520 In either hyperthyroid or euthyroid subjects in response to oral drug. 3163

Nadolol No effect over 2 weeks in 10 healthy volunteers. 2938

Naloxone No significant effect with i.v. infusion. 0868

Omeprazole No effect observed in 8 volunteers given 30 mg/d for 28 d. 2516

Oral Contraceptives No significant difference from nonpregnant values. 3403 Hypophyseal mediated effect; results generally contradictory. 3798

Parathyroid Hormone In response to infusion of 150 units in 10 normal subjects. 0467

Phenytoin With long-term treatment due to accelerated thyroxine clearance via stimulation of hepatic microsomal enzymes. 2168 No change or in response to TRH: probably central regulation involved. 3798

Primidone In 5 epileptic patients treated on average for 6 mo. 2168

Propranolol In 8 healthy volunteers given 80 mg bid and followed serially, in spite of reduction of free T_4. 3833 In 15 euthyroid hypertensives treated for 30 d given 80 to 480 mg/d. 2016 No significant effect in euthyroid patients. 1174 In 10 healthy volunteers treated for 2 weeks. 2938

Prostaglandin F2α When i.v. up to 2 μg/kg/min in men. 0735

Prostaglandins No effect on TSH secretion. 3798

Ranitidine In 10 peptic ulcer patients no effect of 150 mg twice daily for 28 d. 1603 No effect with i.v. bolus of much as 300 mg. 0871 No effect of treatment on values. 2780 After 4 weeks or 6 mo treatment (300 mg and 150 mg daily respectively) in male patients with duodenal ulcer. 0727

Smoking No significant difference between heavy smokers and controls. 3255

Sodium Bromide No effect in 10 men, 10 women receiving 1 mg/kg/d during 8 weeks or 2 full menstrual cycles. 3141

Spironolactone Increased response to TRH with action upon hypophyseal T_3 receptors. 3798

Stanozolol 1.95 mU/L before treatment, 2.31 after 2 weeks in 9 healthy men given 10 mg/d for 14 d. 3354

Sultopride When given 300-600 mg/d for 5 weeks in 5 schizophrenic women. 2465

Surgery No effect observed. 2700

Terbutaline In 6 normal men treated with therapeutic amounts for 2 weeks. 3177

Theophylline Increased response to TRH: due to increased intracellular ATP which exerts action on TRH. 3798

Valproic Acid In 10 epileptic patients treated average of 8 mo. 2168

Serum   Increase   Physiological

Aminoglutethimide Observed in one patient in whom tested in association with reduced T_4. 3159

Amiodarone Median TSH values in individuals treated for 6 mo increased compared with controls but values still within normal range, probable anterior pituitary effect. 1173 Initial increase returns to normal in a few months possibly due to inhibition of intracellular binding of T_3 to its nuclear receptor by drug. 2327

Benserazide In hypothyroidism: is a dopamine antagonist. 3798 Significant increase in 5 patients with primary hypothyroidism; possibly due to reduction in circulating dopamine. 0869

Chlorpromazine In hypothyroidism although no change in euthyroidism, increases TRH-induced TSH response. 3798

Clomiphene Small but significant effect with basal and TRH-stimulated value: hypophyseal effect. 3798

Diurnal Variation Maximum between 4 and 6 am (men), 9 pm to 9 am (women). 0417 Significant increase maximum between 2 am and 4 am, minimum 6 pm to 8 pm. 2743

Domperidone Small but significant effect when given i.m. to 12 normal subjects and to a group of patients with subclinical hypothyroidism. 0870 After 10 mg intravenously in 8 healthy males. 3406 In euthyroid and hypothyroid patients: drug is a dopamine antagonist. 3798

Erythrosine Significant increase from 1.7 to 2.2 μU/mL after 15 d on up to 200 mg/d in 30 men. 1241

Flunarizine Significant increase in men noted. 2265

Furosemide Increased after drug treatment stopped. 3473

Hemodialysis Increased to abnormal levels in some subjects. 2547

Iobenzamic Acid Blocks intrahypophyseal conversion of T_4 to T_3. 3798

Iodides Also increased response to TRH related to decreased T_4 and T_3 concentrations. 3798

Iopanoic Acid Still increased 1 week after course of agent. 3511 Blocks intrahypophyseal conversion of T_4 to T_3. 3798

Ipodate Affects intrahypophyseal conversion of T_4 to T_3. 3798

Lithium Observed in long-term therapy in manic depressives. 2184 Mechanism unclear. 0544 Compensatory response to reduced peripheral hormone concentrations. Observed in about 15% patients given long-term treatment. TSH response to TRH is increased. 3118 Hypophyseal influence responsible for augmented TRH-induced values. 3798 In 10 of 237 patients with smaller increases in others treated for more than 6 mo: normalized in most cases on withdrawal. 0084 Irreversible myxedema observed in 2 patients treated for 1 y. 2781

Metoclopramide Following 10 mg orally marked increase within 1 h in euthyroid subjects: effect most marked in patients with primary hypothyroidism. 3168 Increased in euthyroidism and hyperthyroidism; maximum effect 3 to 6 h after administration. 3798

Monoiodotyrosine Marked increase within 1 h after 10 mg orally in euthyroid subjects acting through inhibition of tyrosine hydroxylase: effect more marked in hypothyroid subjects. 3168

Morphine 10 mg i.v. produced prompt and significant increase in 7 hypothyroid and 5 healthy individuals. 0891

Phenytoin Slight effect observed in 50% patients at dose of 3 mg/kg/d with normal serum concentration. 1293

Potassium Iodide Increase by 2.3 uU/mL over 11 d in normals. 3658

Prazosin In 19 hypertensives treated for 12 weeks caused change from 3.63 uU/mL to 4.83 uU/mL. 0451

Prednisone Increased about twice with 1-2 mg/kg/d for 2-4 weeks in 10 children. 3353 Basal value increased two fold in 10 children given 1-2 mg/kg/h for 2-4 weeks but TSH response to TRH unchanged. 3353

Radiographic Agents Substantial increase 3 d after oral cholecystography. 2084

Rifampin In one patient with primary hypothyroidism receiving constant replacement dose of L-thyroxine: tri-iodothyronine clearance retarded. 1734

Sulpiridine As a dopamine blocking substance increases basal and TRH-mediated values. *3798*

Thyrotropin-Releasing Hormone (TRH) Dose related when given i.v.-effect 20 mg orally. *1065* Threefold increase in 30 minutes. *0100*

Tyropanoic Acid Intrahypophyseal conversion of T_4 to T_3 involved. *3798*

Serum Decrease Physiological

Apomorphine In hypothyroidism but can lower basal and TRH-stimulated values. *3798*

Aspirin Decreased release after administration. *2427*

Bombesin After TRH after 5 ng/nk/min x 2.5 h in healthy men. *2841*

Carbamazepine Marked reduction in 26 patients on long-term therapy. *2169*

Clofibrate In hypothyroidism but no change in euthyroid patients presumed to have direct action at hypothalamic or pituitary level. *3798*

Corticosteroids Usual response in patients on steroids. *2220*

Danazol Not initially, but after 4 weeks in 12 healthy female postmenopausal volunteers given up to 800 mg daily. *0985*

Dopamine Antagonizes TRH at hypophyseal level: reduces concentration in both euthyroid and hypothyroid individuals. Diminishes TSH response to TRH. *3798*

Fasting Values low and also blunted response to TRH. *1130*

Fusaric Acid Inhibition of dopamine degradation in hypothyroidism. *3798*

Glucocorticoids In euthyroidism and hypothyroidism with diminished response to TRH due to hypophyseal inhibition. *3798*

Hemodialysis Probably effect of heparin. *1816*

Heparin Interferes with thyroxine binding to protein. *1816*

Iodoamide In 17 euthyroid men following arteriography: maximum effect of more than 50% and at 10-12 weeks. *1030*

Levodopa In hypothyroidism but no change in euthyroid — is dopamine precursor. Diminishes response to TRH in hypothyroidism. *3798*

Lisuride Can lower basal and TRH-stimulated values. *3798*

Methergoline In hypothyroidism — is a serotonin antagonist. *3798*

Peribedil Effect noted in hypothyroid subjects. *3798*

Pimozide But effect contrary to assumed mechanism of a dopamine blocker. *3798*

Pyridoxine Decreases concentration in normal and hypothyroid subjects after i.v. injection. *3798*

Somatostatin In hypothyroid and euthyroid patients and reduced response to TRH: drug is inhibitor of TSH release. *3798*

L-Thyroxine In hypothyroidism in response to T_4 analog. *3798*

Tri-iodothyronine Effect marked in euthyroid, hypothyroid subjects. *3655*

Blood Increase Physiological

Lithium In 20% of women, but no men among 53 psychiatric patients treated for 2 y. *2185*

Thyroxine (T_4)

Serum No Effect Analytical

Bilirubin At 10 mg/dL on competitive protein binding procedure of Alexander. *0054*

Blood Collection Tube No effect on concentration even with storage for 6 d in gel-barrier tubes. *1592*

Carbamazepine At therapeutic and toxic concentrations on an equilibrium dialysis method. *0374* No effect observed at both therapeutic and toxic concentrations when added to serum *in vitro*. *0374*

Hemoglobin At 1 g/dL on competitive protein binding procedure of Alexander. *0054*

Hemolysis If minimal on competitive protein binding methods. *3257*

Ioxitalamic Acid No effect observed after i.v. injection. *0149*

Lugol's Iodine No effect on methods. *3218*

Phenylbutazone No effect on competitive protein binding procedure of Alexander. *0054*

Phenytoin No effect on competitive protein binding procedure of Alexander. *0054*

Radiographic Agents Most agents in low concentration. *3217* No effect on competitive protein binding procedure of Alexander. *0054*

Storage of Sample On competitive protein binding at 4 or -20° for 2 weeks. *3257* No effect for several days at room temperature. *1180*

Tri-iodothyronine No effect on competitive protein binding procedure of Alexander. *0054*

Serum No Effect Physiological

Benziodarone Relatively constant in 9 normal volunteers receiving 100 mg three times daily for 14 d due to diversion of peripheral metabolism of T_4 to rT_3 rather than T_3. *3907*

Bromocriptine In euthyroid patients although response to TRH shows no significant change. *3798*

Cimetidine No significant difference in drug treated healthy volunteers: no difference in response to TRH. *1683* In 6 patients treated for 1 mo with 1 g/d. *3491*

Clomiphene Probably direct thyroid inhibition. *3798*

Colestipol Reduced reabsorption of thyroid hormones from gastrointestinal tract. *3798*

Cotrifamole Possible intrathyroidal inhibition of conversion of iodine to organic form. *3798* No change in men or women over 10 d. *0686*

Diazepam Conflicting reports ?no effect. *1488*

Erythrosine In 30 men receiving up to 200 mg daily for 15 d. *1241*

Famotidine No effect of short-term or long-term treatment in male patients with duodenal ulcers. *0727*

Fenoprofen No effect noted physiologically or on methods. *3594*

Glucocorticoids Commonly observed typically with inhibition of conversion. *3798*

Indomethacin In up to 35 healthy men given 100 mg/d for 1 week. *2904*

Insulin No change during insulin induced hypoglycemia. *3570*

Iodoamide In 17 euthyroid men following arteriography for up to 14 weeks. *1030*

Levodopa No effect observed. *0824*

Levonorgestrel No effect observed in group of women 20 and 65 mo after levonorgestrel treated rods inserted in uterus. *0899* In a group of women using levonorgestrel versus control group with copper IUD. *0899*

Methandrostenolone In 6 body-builders taking up to 20 mg/d intermittently for a year or more. *3290*

Methoxyflurane No effect observed with anesthesia. *2698*

Nadolol No effect over 2 weeks in 10 healthy volunteers. *2938*

Neomycin When given 2.0 g/d orally for 7 d. *0959*

Netilmicin No effect when given i.v. with cloxacillin. *0959*

Omeprazole No effect observed in 8 volunteers given 30 mg/d for 28 d. *2516*

Penicillamine In 21 patients with rheumatoid arthritis treated for 48 weeks given 250 to 750 mg daily. *2345*

Prednisone Insignificant change in 10 children treated for 2-4 weeks. *3353*

Primidone In 5 epileptic patients treated on average for 6 mo. *2168*

Season No seasonal variation observed. *2853*

Sodium Bromide No effect in 10 men, 10 women receiving 1 mg/kg/d during 8 weeks or 2 full menstrual cycles. *3141*

Tetraiodofluorescein No effect observed. *3725*

Timegadine In 23 patients with rheumatoid arthritis given 250 to 750 mg/d for 48 weeks. *2345*

Tobramycin When given i.v. with cloxacillin to 13 patients. *0959*

Valproic Acid In 10 epileptic patients treated average of 8 mo. *2168*

Serum Increase Analytical

Cellulose Acetate Due to phthalates as plasticizer in containers. *3153*

Thyroxine (T₄) *(continued)*

Serum Increase Analytical *(continued)*

Delalande 69276 At 37.5 mg/dL usual methods when added to serum. *1758*

Iodine Some organic iodine may elute (odd pattern). *0583*

Iodoalphionic Acid Interferes with direct bromination. *0583*

Iodohippurate Interferes with measurement by column. *0251*

Iopanoic Acid Contaminating iodine affects test. *3218*

Iothiouracil Organic iodine compound (satisfactory with Murphy-Pattee). *0354*

Propyliodone At very high concentrations only. *0354*

Radiographic Agents Orabilix and Dionosil® affect at 1 mg. *3217*

Sulfamethoxazole Erratic elution from column = false result. *0581*

Serum Increase Physiological

Amiodarone Significant increase (mean change from 100 to 155 nmol/L, i.e. 55%) after 2 mo probably impaired peripheral conversion of T₄ to T₃. *1316* 2 of 92 patients developed hyperthyroidism, whereas 11 developed hypothyroidism in study of up to 4 y. *2851* After slight decrease in some patients initially in 10 patients over 18 mo. *1509* Mean concentration of 131 nmol/L in 38 patients receiving mean 1420 mg weekly for more than 9 mo. *3142* From 7.4 to 9.7 μg/dL in 13 patients treated for average of 17 mo. *0425* In 9 men treated for 28 d caused insignificant increase of 1.4 μg/dL. *0534* Slight effect in 15 euthyroid volunteers given 400-600 mg/daily for 2 weeks. *0092*

Clofibrate Insignificant increase due to increased binding capacity (augmentation of transport proteins). *3798*

Desiccated Thyroid Increases by 1-2 μg/dL/d per grain given. *0354*

Dextrothyroxine Mechanism not discussed. *1488*

Eggplant Increased binding capacity (augmentation of transport proteins). *3798*

Estrogens Increases binding capacity of thyroxine binding globulin for up to 1 mo. *0354* Increased binding capacity (augmentation of transport proteins). *3798*

Ether Effect observed in moderate to deep anesthesia. *1343*

Fluorouracil Increased binding capacity (augmentation of transport proteins) *3798*

Glucocorticoids In Hashimoto's disease typically with inhibition of conversion. *3798*

Halofenate Associated with competition for transport proteins. *3798*

Halothane Increase of almost 2 μg/dL observed. *2698* Probably due to release from liver. *3798*

Heroin Significant increase in addicts observed. *0193* Increased binding capacity (augmentation of transport proteins). *3798*

Heroin Withdrawal Significant increase in addicts observed. *0193*

Insulin Probable enhanced release from liver. *3798*

Iobenzamic Acid Impaired conversion of T₄ to T₃. *3798*

Iopanoic Acid Inhibition of conversion of thyroxine to tri-iodothyronine. *3798*

Ipodate Inhibition of T₄ to T₃ conversion. *3798*

Jaundice If biliary obstruction as normal excretion in bile. *3947*

Levarterenol Metabolic response. *1176*

Levodopa During therapy of Parkinsonism. *1860*

Levothyroxine Endogenous hormone suppressed, exogenous measured. *0354*

Lithium Observed in one patient who developed thyrotoxicosis without pre-existing goiter. *2418*

Mestranol From 100 to 133 nmol/L in 19 post-menopausal women receiving mean of 24 μg/d versus controls. *0007*

Methadone Slight but not significant increase. *0193* Increased binding capacity (augmentation of transport proteins). *3798*

Obesity Characteristic finding correlated with insulin. *1106*

Oral Contraceptives Increased binding protein available. *3505* Similar values to those in late first trimester. *3403* Increased binding capacity (augmentation of transport proteins). *3798*

Orphenadrine Reported in two studies. *3798*

Phenytoin In 10 women receiving anticonvulsants, mainly phenytoin versus 10 controls, drug taken on average for 15 y. *0257*

Prazosin In 19 hypertensives treated for 12 weeks caused change from 10.03 μg/dL to 10.85 μg/dL. *0451*

Pregnancy Rises from 3 weeks, normal 8 weeks after birth. *0913*

Propranolol In 12 hyperthyroid patients given 160 mg/d for 2 weeks due to effect on peripheral metabolism. *1514* Increase by 16% in 15 euthyroid hypertensives treated for 30 d given 80 to 480 mg/d. *2016* In 6 patients on large daily doses of drug (480 ± 155 mg) without clinical evidence of hyperthyroidism, due to drug-induced blockage of iodothyronine deiodination. *0722* Slight tendency in 10 healthy volunteers treated for 2 weeks. *2938*

Propylthiouracil In one 24 year old woman who developed fulminant hepatic failure with lymphocyte sensitization. *2437*

Prostaglandin F2α Peak within 10-48 h up to 2 times original. *1718*

Prostaglandins Probably direct thyroid effect. *3798*

Radiographic Agents In first few days following oral cholecystographic agents. *2084*

Tamoxifen In 10 of 50 postmenopausal patients with breast cancer. Also rose in other patients with start of treatment. Effect not observed in normal volunteers. *1350*

Thyroid Increased available thyroxine. *2220*

Thyrotropin Several h after i.m. injection, less than on T₃. *1631*

Thyrotropin-Releasing Hormone (TRH) But no change in % free T₄ (variable response). *1631*

L-Thyroxine Raised but still less than controls with 30-60 d treatment of 7 hypothyroid women with dry thyroid extract. *2288* From 24 to 124 nmol/L 11 hypothyroid women treated with 0.1 to 0.2 mg daily. *1201*

Tyropanoic Acid Impaired conversion of T₄ to T₃. *3798*

Serum Decrease Analytical

Danazol Concentrations low: in case of cortisol and testosterone displacement from plasma proteins occurs and protein binding assays are probably invalid. *1479*

Diflunisal Using FETI in Syva® advance system in a patient receiving 250 mg drug twice per day although mechanism not clear. *1687*

Hemolysis 5% decreased if at 700 nmol/L. *3257* When severe, decreased recovery in competitive protein binding procedures. *0054*

Iothiouracil Depresses thyroid function. *0583*

Oxyphenbutazone Yellow eluate persists in reaction mixture. *0583*

Serum Decrease Physiological

Ambulation Transient due to increased cellular utilization. *0879*

Aminoglutethimide Competes for T₄ binding sites. *0583* On average 1.6 μg/dL decrease in concentration after 3 mo treatment. *3159*

Aminosalicylic Acid May cause goitrous hypothyroidism. *2808* Prolonged administration may lead to hypothyroidism. *1292*

Amiodarone Single case of drug induced myxedema coma and death. *2343*

Anabolic Steroids Decrease amount of thyroxine binding globulin. *0583*

Androgens For up to 1 mo after treatment (decreased thyroxine binding globulin). *0913* Decreased binding capacity (diminution of transport proteins). *3798*

Anticonvulsants 79 nmol/L versus 99 nmol/L in women and 82 nmol/L versus 100 nmol/L in men on long-term treatment. *0257*

Asparaginase From 10.7 to 2.9 μg/dL within 3 weeks in 14 children. *1543*

Aspirin Displaces thyroxine from binding sites (thyroxine binding prealbumin). *3505* Progressive effect during three days but also noted over longer term study at beginning, eventually reverted towards normal. Effect statistically significant. *3076* In up to 35 healthy volunteers given 3.6 g daily for 1 week. *2904*

Barbiturates Compete with T₄ for thyroxine binding prealbumin. *0583*

Bromocriptine In hypothyroid patients although response to TRH shows no significant change. *3798*

Carbamazepine Fall from mean of 82 to 75 nmol/L with 400 mg/d carbamazepine after 14 d treatment in 10 healthy males, secondary to hepatic enzyme induction with accelerated nondeiodinative hepatic hormone disposal. *0708* Observed in 9 hypothyroid patients given substitution therapy due to increased extra-thyroidal metabolism of thyroid hormones. *0001*

Carbimazole From 184 to 101 nmol/L in 12 hyperthyroid women patients treated with 10-30 mg daily. *1201*

Chlorpromazine Increased metabolism by hepatic microsomes. *3505*

Chlorpropamide Displaces thyroxine from thyroxine binding globulin. *3505*

Cholestyramine Decreased intestinal absorption of thyroxine. *3505* Reduced absorption from gastrointestinal tract due to adsorption. *3794*

Clofibrate Due to increased binding capacity of drug. *2858*

Cobalt Impaired synthesis from tyrosine. *1915*

Colestipol Observed in euthyroid patients when drug given together with niacin. *0603*

Corticosteroids May occur for up to 1 week after therapy. *0913*

Corticotropin Decreased thyro-binding globulin (occurs for up to 1 week). *3879*

Cortisone Decreased synthesis. *3505*

Co-Trimoxazole Marked decrease possibly due to intrathyroidal inhibition of conversion of iodine to organic form. *3798* From 97.1 to 83.0 in men and 92.6 to 87.2 nmol/L in women with 10 days treatment. *0686*

Danazol In healthy postmenopausal women after 2 weeks treatment with 400 to 800 mg daily through direct effect on thyroxine binding globulin production at cellular level. *0985* Affect binding in women treated with drug so that proportion of free drug high. *1479* In 12 healthy female post-menopausal volunteers given up to 800 mg daily. *0985* Fell by 26% in 9 individuals due to competition for binding sites or androgen-like effect. *2718*

Dexamethasone 8 mg/d for 2 d reduced concentration in normal individuals due to altered secretion and peripheral metabolism. *0860*

Diazepam Due to competition for transport proteins. *3798*

Diazo Dyes Compete for binding sites. *0583*

Dinitrophenol Displaced from binding sites on thyroxine binding globulin. *3505*

Ethionamide Antithyroid effect after several weeks. *1488*

Evans Blue Compete for binding sites. *0583*

Fenclofenac In 7 out of 8 individuals receiving 1.2 g/d values below normal range. T_4 displaced from binding protein. *2971* From 173 nmol/L to 70 nmol/L in 4 women with thyrotoxicosis given 1.2 g/d for 7 d. *2758*

Furosemide In 4 patients high concentrations of drug inhibit T_4 binding in plasma. *3473* Probably due to enhanced clearance due to displacement from serum protein binding sites. *2584* Associated with high dose i.v. administration at peak drug concentration. *3473* Transient significant effect seen 2-5 h after ingestion of various amounts of drug chronically in 34 patients with congestive cardiac failure. *2584*

Gentisic Acid Competes with T_4 for thyroxine binding prealbumin but not thyroxine binding globulin. *0583*

Glucocorticoids Or no change typically with inhibition of conversion. *3798*

Growth Hormone Observed in most patients initially euthyroid given drug possibly due to inhibition of TSH response to TRH. *1181*

Halofenate Interferes with binding of T_4 to thyroxine binding globulin. *2483*

Heparin Probable modification of thyroxine binding. *3110* Due to competition for transport proteins. *3798*

Hydroxyphenylpyruvic Acid Decreased binding capacity (diminution of transport proteins). *3798*

Iodides Decrease synthesis if diagnostic or therapeutic I 131. *3505* Associated with inhibition of conversion. *3798*

Isotretinoin In 7 patients with severe rosacea treated with 1 mg/kg/d for 12 weeks. Effects possibly due to induction of hepatic microsomal enzymes. *2307* From 98 to 85 nmol/L in 18 patients with severe acne given 0.8 mg/kg daily for 3 mo: changes reverted to normal after treatment stopped. *2242* Significant at 12 weeks in 7 patients with severe rosacea treated with 1 mg/kg/d. *2307*

Lead In some men occupationally exposed to large quantities of lead. *0771*

Liothyronine Depression of endogenous hormone. *2220*

Lithium Reported to induce myxedema. *2221* Reduces thyroidal iodine uptake, iodination of tyrosine, release of T_4, to T_3 and hepatic metabolism of T_4 to T_3. *3118* In 10 of 237 patients with smaller increases in others treated for more than 6 mo: normalized in most cases on withdrawal. *0084* In 20% of women, but no men among 53 psychiatric patients treated for 2 y. *2185* Irreversible myxedema observed in 2 patients treated for 1 y. *2781* Hypothyroidism observed in 1-2% of 2590 psychiatric patients treated with lithium. *2826*

Methimazole Therapeutic intent (stops iodination of tyrosine). *1444*

Methylthiouracil Inhibits synthesis, stops iodination of tyrosine. *0071*

Mitotane Competes with T_4 for thyroxine binding globulin. *0583*

Muscular Exercise Significant effect with strenuous exercise. *3217*

Norethindrone When treatment results compared with controls. *0260*

Oxyphenbutazone Impaired synthesis may occur. *1915*

Penicillin Competes for thyroxine binding prealbumin binding sites. *0583*

Phenobarbital In comparison with untreated cognitively delayed children 7.2 μg/dL on average versus 8.4 μg/dL. *2003* Both significant and nonsignificant changes noted, associated with enzyme induction. *3798*

Phenothiazines Observed with long-term treatment but cause not clear. *3798*

Phenylbutazone Due to impaired synthesis, competes for thyroxine binding albumin. *1915* Reported in two studies due to competition for transport proteins. *3798*

Phenytoin Displaces thyroxine from binding sites. *3505* Slight effect observed in 50% patients at dose of 3 mg/kg/d with normal serum concentration. *1293* 5.5 μg/dL in treated epileptics vs 8.1 μg/dL in controls. *3512* With long-term treatment due to accelerated thyroxine clearance via stimulation of hepatic microsomal enzymes. *2168* In comparison with untreated cognitively delayed children 7.1 μg/dL versus 8.4 μg/dL. *2003*

Potassium Iodide In normals by 1 μg/dL over 11 d. *3658*

Primidone Associated with enzyme induction. *3798*

Propylthiouracil Inhibits synthesis, stops iodination of tyrosine. *1488*

Ranitidine Small reduction noted in 10 ulcer patients after 150 mg twice daily for 28 d ratio: total T_4/total T_3 fell. *1603* Slight reduction after 4 weeks of 300 mg daily, but returned to normal after 6 mo of 150 mg daily in men with duodenal ulcers. *0727*

Reserpine Increased metabolism by hepatic microsomes. *3505*

Rifampin In one patient with primary hypothyroidism receiving constant replacement dose of L-thyroxine: tri-iodothyronine clearance retarded. *1734*

Salicylate Due to competition for transport proteins. *3798*

Smoking Significant reduction in heavy smokers (6.4 μg/dL versus 7.3 μg/dL in controls). *3255*

Sodium Nitroprusside Due to thyroidal inhibition especially in patients with some renal impairment. *3798*

Somatostatin With infusion probable inhibition of TSH release. *3798*

Somatotropin Or no change associated with accelerated conversion. *3798*

Stanozolol From 106.2 to 90.4 nmol/L after 1 week in 9 healthy men given 10 mg/d for 14 d. *3354*

Sulfonamides Acts like thiourea on thyroid gland. *0354*

Sulfonylureas Due to impaired synthesis. *1915* Due to competition for transport proteins. *3798*

Terbutaline From 7.2 to 6.7 μg/dL on average in 6 normal men after 2 weeks treatment. *3177* In 6 normal men after 15 mg/d for 2 weeks with nonsignificant change from 6.7 to 7.2 μg/dL. *3177*

Tetrachlorothyronine Displaces thyroxine from thyroxine binding globulin. *3505*

Tolbutamide Displaces from thyroxine binding globulin. *3505*

Thyroxine (T₄) *(continued)*

Serum   Decrease   Physiological *(continued)*

Tri-iodothyronine   Marked decrease in obese with 150 μg/d. *0458*   Diminished synthesis. *3505*   In 7 normal women aged 18-27 y given 25 μg/d three times daily for 14 d. Effect observed on endothelium-associated proteins but not on hepatically synthesized proteins. *1383*

Urine   No Effect   Physiological

Estrogens   No effect observed but increase on cessation. *0630*

Urine   Increase   Physiological

Ethinyl Estradiol   On withdrawal only in men after 20 μg/d. *0630*
Menstruation   Transient effect during menstruation. *0630*
Phenytoin   Increases during therapy (as binds to thyroxine binding globulin). *0630*
Thyrotropin   Less marked effect than on T₃. *0518*

Thyroxine (T₄) Binding Globulin

Serum   No Effect   Analytical

Carbamazepine   At therapeutic and toxic concentrations on Corning radioimmunometric kit method. *0374*   No effect observed at both therapeutic and toxic concentrations when added to serum *in vitro. 0374*
Diiodotyrosine   On radioimmunoassay procedure of Van Herle. *3676*
Monoiodotyrosine   On radioimmunoassay procedure of Van Herle. *3676*
L-Thyroxine   On radioimmunoassay procedure of Van Herle. *3676*
Tri-iodothyronine   On radioimmunoassay procedure of Van Herle. *3676*

Serum   No Effect   Physiological

Amiodarone   No significant changes after 2 mo. *1316*
Anticonvulsants   In 8 epileptic patients receiving phenobarbital with carbamazepine or phenytoin. *2670*   No difference in men or women on long-term therapy. *0257*
Fenclofenac   In 7 out of 8 patients receiving 1.2 g/d no change observed. *2971*
Furosemide   At same time as thyroxine concentration reduced. *3473*
Halofenate   No effect observed after 3 weeks. *2483*
Omeprazole   No effect observed in 8 volunteers given 30 mg/d for 28 d. *2516*
Phenytoin   In 10 women receiving anticonvulsants, mainly phenytoin versus 10 controls, drug taken on average for 15 y. *0257*
Ranitidine   No difference in thyroid hormone binding protein in 10 ulcer patients after 150 mg twice daily for 28 d. *1603*
Sodium Bromide   No effect in 10 men, 10 women receiving 1 mg/kg/d during 8 weeks or 2 Full menstrual cycles. *3141*
Terbutaline   In 6 normal men treated with therapeutic amounts for 2 weeks. *3177*
Tri-iodothyronine   In 7 normal women aged 18-27 y given 25 μg/d three times daily for 14 d. Effect observed on endothelium-associated proteins but not on hepatically synthesized proteins. *1383*

Serum   Increase   Physiological

Carbamazepine   Observed in 9 hypothyroid patients given substitution therapy due to increased extra-thyroidal metabolism of thyroid hormones. *0001*
Chlormadinone   Therapeutic effect. *2313*
Clofibrate   Increases concentration (small dose effect). *3505*
Diethylstilbestrol   Direct effect of drug. *0012*
Epinephrine   Binding capacity increased after infusion. *1860*
Estradiol   Direct effect of drug. *0012*
Estrogens   Increases binding capacity of thyroxine binding globulin. *3871*
Estrone   Direct effect of drug. *0012*
Ethinyl Estradiol   Direct effect of drug. *0012*

Mestranol   From 24 to 47 mg/L in 19 post-menopausal women receiving mean of 24 μg/d versus controls. *0007*
Norethynodrel   Direct effect of drug. *0012*
Oral Contraceptives   Estrogen activation of carrier protein. *0395*   In response to amount of circulating estrogen. *0585*
Perphenazine   Direct effect of drug. *0012*
Phenothiazines   Occurs with prolonged use. *1915*
Pregnancy   Increased binding capacity observed. *1498*
Progesterone   Increased synthesis. *3879*
Tamoxifen   Raised in all 6 patients with breast cancer studied. *1350*

Serum   Decrease   Physiological

Anabolic Steroids   Direct effect of drug. *0012*
Androgens   Direct effect of drugs. *0012*
Asparaginase   From 29.4 to 8.0 μg/mL within 3 weeks in 14 children. *1543*
Colestipol   Observed in euthyroid patients when drug given together with niacin. *0603*
Corticosteroids   Large dose response. *1915*
Corticotropin   Reduced synthesis. *3879*
Cortisone   Reduced synthesis. *3879*
Danazol   In healthy postmenopausal women after 2 weeks treatment with 400 to 800 mg daily through direct effect on thyroxine binding globulin production at cellular level. *0985*   Approximately halved in 5 patients with hereditary angioneurotic edema treated for up to 10 mo. *3219*   Direct effect on production in 12 healthy female post-menopausal volunteers given up to 800 mg daily. *0985*   Mean fall of 19% in 10 people due to competition for binding sites or androgen-like effect. *2718*
Fluoxymesterone   Metabolic effect. *1680*
Methandrostenolone   Metabolic effect. *0019*
Methyltestosterone   Direct effect of drug. *0012*
Nandrolone   Anabolic effect. *2681*
Norethandrolone   Decreases concentration of thyroxine binding globulin, increases thyroxine binding prealbumin. *3505*
Oxymetholone   Direct effect of drug. *0012*
Phenytoin   19.6 μg/mL in treated epileptics vs 23.4 μg/mL in controls. *3512*
Prednisone   Decreases concentration of thyroxine binding globulin, increases thyroxine binding prealbumin. *3505*
Propranolol   In 8 healthy volunteers given 80 mg bid and followed serially, fell by average of 1.2 mg/L. *3833*   In euthyroid patients due to peripheral inhibition of peripheral deiodination of thyroxine and on binding protein metabolism. *1174*
Stanozolol   From 22.5 to 17.0 mg/L after 1 week in 9 healthy men given 10 mg/d for 14 d. *3354*
Testosterone   Direct effect of drug. *0012*

Thyroxine (T₄), Free

Serum   No Effect   Analytical

Carbamazepine   At therapeutic and toxic concentrations on RIA method. *0374*   No effect observed at both therapeutic and toxic concentrations when added to serum *in vitro. 0374*

Serum   No Effect   Physiological

Amiodarone   In 5 healthy subjects given 600 mg daily for 2 weeks. *1764*
Danazol   No effect after 2 weeks with 800 mg/d but slight increase after 4 weeks. *0985*
Fluoxymesterone   No change observed. *0019*
Isotretinoin   No significant change in 24 healthy men given 1 mg/kg/d for 16 weeks. *2640*
Methandrostenolone   No change observed. *0019*
Oral Contraceptives   No effect observed although PBI may be increased. *2427*   No significant difference from nonpregnant values. *3403*
Oxymetholone   Anabolic effect (possible for 2 weeks after stop). *1679*
Sodium Bromide   No effect in 10 men, 10 women receiving 1 mg/kg/d during 8 weeks or 2 full menstrual cycles. *3141*

Stanozolol   18.5 pmol/L before treatment, 19.0 pmol/L after 2 weeks in 9 healthy men given 10 mg/ day for 14 d. *3354*

Terbutaline   In 6 normal men treated with therapeutic amounts for 2 weeks. *3177*

Serum   Increase   Analytical

Carbamazepine   As measured by addition of drug to control sera and measured by analog radioimmunoassay. *2169*

Heparin   Marked increase in some patients given drug i.v. Effect probably *in vitro* artifact. *2408*   Increased concentration in patients given drug i.v. when measured by equilibrium dialysis and ultrafiltration. *3760*   Observed with equilibrium dialysis technique after i.v. administration. *2183*

Phenytoin   As measured by addition of drug to control sera and measured by analog radioimmunoassay. *2169*

Serum   Increase   Physiological

Amiodarone   Change from mean of 22 to 32 pmol/L after 2 mo. *1316*   In group treated for 6 mo 9 out of 28 had values above normal range. *1173*

Aspirin   Interferes with binding to thyroxine binding globulin and thyroxine binding prealbumin. *2083*

Danazol   Proportion high due to displacement of hormone from protein. *1479*   Slight effect after 4 weeks in 12 healthy female post-menopausal volunteers. *0985*

Estrogens   Increases binding capacity of thyroxine binding globulin. *3871*

Furosemide   Effect similar to that of fT_4 index with analog tracer assay. *2584*   Transient significant effect seen 2-5 h after ingestion of various amounts of drug chronically in 34 patients with congestive cardiac failure. *2584*

Halofenate   Interferes with binding of T_4 to thyroxine binding globulin. *2483*   Associated with competition for transport proteins. *3798*

Hemodialysis   Probably effect of heparin. *1816*

Heparin   Probable modification of thyroxine binding. *2424*   Mean concentration increased by 50% in 19 patients. *0253*   From 23 to 39 pmol/L by equilibrium dialysis and from 17 to 24 pmol/L by Gamma-Coat RIA probably due to activation of lipases by heparin. *0253*

Iopanoic Acid   Inhibition of conversion of thyroxine to tri-iodothyronine. *3798*

Phenylbutazone   Insignificant short-term increase associated with competition for transport proteins. *3798*

Phenytoin   In 10 women receiving anticonvulsants, mainly phenytoin versus 10 controls, drug taken on average for 15 y. *0257*

Propranolol   In 8 healthy volunteers given 80 mg bid and followed serially, increased by average of 3.3 pmol/L. *3833*   Increase by 18% in 15 euthyroid hypertensives treated for 30 d given 80 to 480 mg/d. *2016*   In 6 patients on large daily doses of drug (480 ± 155 mg) without clinical evidence of hyperthyroidism, due to drug-induced blockade of iodothyronine deiodination. *0722*   In euthyroid patients due to peripheral inhibition of peripheral deiodination of thyroxine and on binding protein metabolism. *1174*

Tamoxifen   Observed in one patient with persistently high thyroxine of 50 postmenopausal patients with breast cancer. *1350*

L-Thyroxine   Raised but still less than controls with 30-60 d treatment of 7 hypothyroid women with dry thyroid extract. *2288*

Serum   Decrease   Analytical

Heparin   Marked decrease with Amersham analog method within 1 h of i.v. administration. *2183*

Serum   Decrease   Physiological

Anticonvulsants   16 pmol/L versus 23 pmol/L in women and 16 pmol/L versus 19 pmol/L in men on long-term treatment. *0257*

Asparaginase   From 1.77 to 0.94 ng/dL within 3 weeks in 14 children. *1543*

Carbamazepine   Fall from mean of 16.0 to 14.2 pmol/L with 400 mg/d in healthy males. *0708*   As measured by equilibrium dialysis and analog-type radioimmunoassays; not due to displacement from protein. Therapy. *2169*   Observed in 9 hypothyroid patients given substitution therapy due to increased extra-thyroidal metabolism of thyroid hormones. *0001*

Clofibrate   Percentage of free decreased (large doses). *2858*

Fenclofenac   Nonsignificant decrease reported in several studies. *3798*   Significant decrease but not to normal in 4 women with thyrotoxicosis given 1.2 g/d for 7 d. *2758*

Furosemide   At same time as thyroxine concentration reduced. *3473*

Heroin   Slightly lowered percent in addicts. *0193*

Heroin Withdrawal   Significant lower percentage. *0193*

Isotretinoin   In 7 patients given 1 mg/kg/d for 12 weeks. *2307*

Lead   In some men occupationally exposed to large quantities of lead. *0771*

Levothyroxine   Falls with therapy of hypothyroid state. *1680*

Lithium   Mechanism unclear, non toxic goiter may occur. *3312*

Mestranol   From 19.2 to 15.7 pmol/L in 19 post-menopausal women receiving mean of 24 μg/d versus controls. *0007*

Methadone   Significantly lower percentage. *0193*

Norethindrone   Slight effect when compared with controls. *0260*

Oral Contraceptives   When treatment results compared with controls. *0260*

Phenylbutazone   Long-term effect due to thyroidal inhibition. *3798*

Phenytoin   Competes with T_4 for thyroxine binding globulin, increased liver degradation. *0012*   As measured by equilibrium dialysis and analog-type radioimmunoassays: not due to displacement from binding proteins. *2169*   1.2 ng/dL in treated epileptics vs 1.5 μg/dL in controls. *3512*

Ranitidine   Similar small effect noted as for total hormone concentration: ratio fT_4/fT_3 fell significantly. *1603*

Thyroxine (T_4) Index, Free (FTI)

Serum   No Effect   Physiological

Amiodarone   In 5 healthy subjects given 600 mg daily for 2 weeks. *1764*

Cotrifamole   No change in men or women over 10 d. *0686*

Danazol   No consistent change due to competition for binding sites or androgen-like effect. *2718*

Oral Contraceptives   Thyroid function unaffected. *0071*

Primidone   In 5 epileptic patients treated on average for 6 mo. *2168*

Propranolol   In hyperthyroid patients given 40 mg every 6 h in 6 patients. *0335*   In 5 hyperthyroid patients given 10 mg every 8 h for 4 d. *0335*

Propylthiouracil   In 5 hyperthyroid individuals given 10 mg every 8 h for 4 d. *0335*

Smoking   No significant difference between heavy smokers and controls. *3255*

Timolol   In 5 hyperthyroid patients given 10 mg every 8 h for 4 d. *0335*

Valproic Acid   In 10 epileptic patients treated average of 8 mo. *2168*

Serum   Increase   Analytical

Delalande 69276   At 37.5 mg/dL usual methods when added to serum. *1758*

Serum   Increase   Physiological

Amiodarone   From 7.1 to 9.3 in 13 patients treated for average of 17 mo. *0425*

Amphetamine   In 7.5% of patients admitted to a psychiatric hospital with many having ingested amphetamine. *2489*

Estrogens   Increased binding capacity of thyroxine binding globulin for up to 1 mo. *3871*

Furosemide   occurring 2-5 h after single morning dose of 80 to 250 mg. *2584*   Transient significant effect seen 2-5 h after ingestion of various amounts of drug chronically in 34 patients with congestive cardiac failure. *2584*

Halofenate   Associated with competition for transport proteins. *3798*

Halothane   Probably due to release from liver. *3798*

Oral Contraceptives   May be slightly increased even 1 mo after treatment. *0913*

Orphenadrine   Reported in two studies. *3798*

Thyroxine (T₄) Index, Free (FTI) *(continued)*

Serum   Decrease   Physiological

Androgens Low normal or decreased for 1 mo after therapy. *0913*

Aspirin Low normal or decreased for 1 week after therapy. *0913*

Carbamazepine No significant change in patients on long-term treatment. *2168*

Clomiphene Probably direct thyroid inhibition. *3798*

Corticosteroids Low normal or slight decrease for 1 week after treat. *0913*

Corticotropin Low normal or slight decrease 1 week after therapy. *0913*

Co-Trimoxazole Marked decrease possibly due to intrathyroidal inhibition of conversion of iodine to organic form. *3798* From 95.2 to 82.6 in men and 90.1 to 86.4 in women with 10 days treatment. *0686*

Fenclofenac In 7 out of 8 individuals receiving 1.2 g/d values below normal range. Thyroxine displaced from binding protein. *2971*

Isotretinoin From 91 to 78 in 18 patients with severe acne given 0.8 mg/kg daily for 3 mo: changes reverted to normal after treatment stopped. *2242* Significant at 12 weeks in 7 patients with severe rosacea treated with 1 mg/kg/d. *2307*

Methimazole From 23.8 to 17.0 after 4 weeks treatment with 30 mg daily. *0192*

Phenobarbital Both significant and nonsignificant changes noted, associated with enzyme induction. *3798*

Phenylbutazone Low normal or slight decrease for 3 weeks after therapy. *0913*

Phenytoin Low normal or decrease for up to 3 weeks after therapy. *0913* With long-term treatment due to accelerated thyroxine clearance via stimulation of hepatic microsomal enzymes. *2168*

Primidone Associated with enzyme induction. *3798*

Thyroxine (T₄) (Murphy-Pattee)

Serum   No Effect   Analytical

Iothiouracil No effect on method. *2220*

Serum   No Effect   Physiological

Primidone In 5 epileptic patients treated on average for 6 mo. *2168*

Serum   Increase   Analytical

Dextrothyroxine Reacts as if levo compound. *0583*

Phenytoin Extractable with ethanol, false increase. *0583*

Serum   Increase   Physiological

Estrogens Increases binding capacity of thyroxine binding globulin. *3871*

Halofenate Significant increase observed. *2483*

Oral Contraceptives Increased binding protein available (effect 2-4 weeks). *3175*

Pregnancy Increases amount of thyroxine binding globulin. *0583*

Serum   Decrease   Physiological

Anabolic Steroids Decreases circulating thyroxine binding globulin. *0583*

Androgens Decreases circulating thyroxine binding globulin. *0583*

Aspirin Binds to thyroxine binding prealbumin but not thyroxine binding globulin. *0583*

Chlorpropamide May be competition for thyroxine binding globulin. *0583*

Dinitrophenol Competes with T₄ for thyroxine binding globulin. *0583*

Gentisic Acid Decreases T₄, does not bind to thyroxine binding globulin. *0583*

Iothiouracil Depresses thyroid function. *0583*

Mitotane Competes with T₄ for thyroxine binding globulin. *0583*

Penicillin Binds with thyroxine binding prealbumin but not thyroxine binding globulin. *0583*

Phenylbutazone Binds to thyroxine binding albumin, has antithyroid activity. *0583*

Tissue Plasminogen Activator Antigen

Plasma   Increase   Physiological

Tri-iodothyronine In 7 normal women aged 18-27 y given 25 μg/d three times daily for 14 d. Effect observed on endothelium-associated proteins but not on hepatically synthesized proteins. *1383*

Titratable Acidity

Urine   No Effect   Analytical

Storage of Sample No significant effect if stored at -20° for 30 d. *0627*

Urine   No Effect   Physiological

Acetaminophen Without overdose not usually any effect. *2866*

Cimetidine Although drug cleared principally by glomerular filtration and partial reabsorption in proximal renal tubules. *2368*

Triamterene Does not cause excessive excretion. *1680*

Urine   Increase   Physiological

Metolazone Associated with diuretic response. *3358*

Protein Due to ingestion of protein. *0987*

Starvation Occurs with metabolic acidosis. *3067*

Triflocin Inhibits carbonic anhydrase. *0033*

Urine   Decrease   Physiological

Acetazolamide Increased alkalinity. *1343*

Amphotericin B Development of renal tubular acidosis. *0536*

Lithium Falls on first day of treatment. *1792*

Secretin Reduced formation with increased alkalinization. *2656*

Tobramycin

Serum   Decrease   Analytical

Heparin Interferes with biological, radioenzymatic and homogeneous enzyme immunoassay but not RIA techniques due to inhibition of acetyltransferase enzymes. *2021*

α-Tocopherol

Serum   No Effect   Analytical

Storage of Sample No effect up to 2 weeks frozen. *3692*

Serum   No Effect   Physiological

Ascorbic Acid No effect with varying degrees of supplementation over 14 week study period in young men. *1751*

Serum   Increase   Physiological

Aging From 0.8-1.0 mg/dL at 5 y to 1.2 at 50 y. *2158*

Vitamin E Increase by 50-60% but normal in 4 d. *2158*

Serum   Decrease   Analytical

Citrates Dilutional effect when compared against serum or heparin as anticoagulant. *3361*

Serum   Decrease   Physiological

Cholestyramine ?inhibits synthesis of carrier lipoprotein. *3788*

Clofibrate ?inhibits synthesis of carrier lipoprotein. *3788*

Tolbutamide

Serum   No Effect   Physiological
Probenecid   Probably has no effect on plasma concentration. *1487*
Sulfadiazine   Appears to have no effect on plasma concentration. *1487*
Sulfadimethoxine   Appears to have no effect on plasma concentration. *1487*

Serum   Increase   Physiological
Aspirin   Displaces from plasma proteins if present. *1487*
Chloramphenicol   May cause 3 fold increase in half life. *1487* Inhibits hepatic microsomal metabolizing enzymes thus prolonging half-life. *3864*
Co-Trimoxazole   Effects of compound may be potentiated. *0679*
Dicumarol   Increases half life, augments hypoglycemia. *1487*
Phenylbutazone   Displaces from protein inhibits carboxylation. *1487*
Phenyramidol   Probably impairs metabolism of drug. *1487*
Sulfaphenazole   Displaces from protein, inhibits carboxylation. *1487*
Sulfisoxazole   Displaces from protein. *1487*
Trimethoprim   May potentiate action by effects on metabolism. *0678*

Serum   Decrease   Physiological
Rifampin   Decreased serum concentration due to hepatic enzyme induction. *0014*

Toluene

Blood   Increase   Physiological
Ethanol   During exposure if alcohol ingested significant increase in concentration due to competition for alcohol dehydrogenase. *3738*

Transcortin

Serum   No Effect   Physiological
Danazol   No significant change in 5 patients with hereditary angioneurotic edema treated for up to 10 mo. *3219*
Desogestrel   With daily dose of 0.125 mg in 30 healthy female volunteers when given alone. Normal values 30 d after treatment stopped. *3089*
Levonorgestrel   With daily dose of 0.125 mg in 30 healthy female volunteers when given alone. Values normal within 30 d of end of treatment. *3089*
Progestogens   No effect with progestogen — only pill. *0537*
Testosterone   In power athletes with 26 weeks on steroid self administration. *3088*

Serum   Increase   Physiological
Anabolic Steroids   Metabolic effect. *0227*
Anticonvulsants   470 nmol/L versus 320 nmol/L in women and 420 nmol/L versus 320 nmol/L in men on long-term therapy. *0257* In 8 epileptic patients receiving phenobarbitol with carbmazepine or phenytoin. *2670*
Estrogens   Metabolic response (dose related effect). *1498*
Lynestrenol   When 5 mg daily given alone to 30 healthy female volunteers. Normal values 30 d after treatment stopped. *3089*
Oral Contraceptives   Metabolic changes in liver synthesis. *2189* Increase by 115 to 140% after 3 mo with different preparations. *0311*
Phenytoin   In 10 women receiving anticonvulsants, mainly phenytoin versus 10 controls, drug taken on average for 15 y. *0257*
Pregnancy   Increased binding capacity observed. *1498*

Transferrin

Serum   No Effect   Physiological
Estrogens   No effect observed. *1498*
Progesterone   No significant effect after 6-9 mo. *3505*
Smoking   No significant difference between smokers and non-smokers. *0809*
Tri-iodothyronine   In 7 normal women aged 18-27 y given 25 μg/d three times daily for 14 d. Effect observed on endothelium-associated proteins but not on hepatically synthesized proteins. *1383*

Serum   Increase   Physiological
Diurnal Variation   Maximum at 4-8 pm, minimum at 8 am slight effect. *3865*
Estrogens   Altered liver synthesis. *2386*
Mestranol   Maximum effect 1 week after treatment. *1660*
Muscular Exercise   Raised at 15 minutes, normal within 1 d. *1491* Approximately 10% increase immediately after. *2846*
Oral Contraceptives   Metabolic changes in liver synthesis (estrogen). *2189* In 46 oral contraceptive users compared with controls studied over at least 2 y. *1182*
Physical Training   Significantly higher in athletes. *1491*
Pregnancy   In last trimester compared with controls. *2385*
Sex Difference   Approximately 60 mg/dL higher in women. *0954*

Serum   Decrease   Physiological
Aging   Slight effect with aging. *0954*
Altitude   After rapid ascent to 5400 meters some depletion. *2957*
Asparaginase   Reported effect. *2657* Possibly due to inhibition of protein synthesis reported in various studies. *2406*
Cortisone   Reduced synthesis. *3879*
Dextran   30% after Macrodex®, no effect after Rheomacrodex®. *3346*
Proteinuria   Fall related to permeability of glomerulus. *1945*
Testosterone   Metabolic effect. *0227*

Urine   No Effect   Analytical
Urea   On nephelometric method of Killingsworth. *1931*

Urine   No Effect   Physiological
Gentamicin   No effect observed in 26 patients given drug. *2593*
Sisomicin   No significant change in 23 patients given therapeutic amounts of drug for 2 weeks. *2593*

Urine   Increase   Physiological
Diatrizoic acid   Marked increased excretion (20 to 60 times baseline) day of bilateral renal arteriography in 23 patients with hypertension, reverted to normal on following day. *2594*
Proteinuria   Clearance related to permeability to protein. *1945*
Radiographic Agents   Mean increase from 2.8 mg/d to 147.1 mg/d in 37 patients day following arteriography. *2594*

Transketolase

Red Blood Cells   Decrease   Physiological
Ethanol   Activation by thiamine in 35 patients with alcohol related illness. *0207*

Trehalase

Serum   Increase   Physiological
Muscular Exercise   Observed to increase with exercise. *1231*

Urine   Increase   Physiological
Muscular Exercise   Observed to increase with exercise. *1231*

Trichlorethanol

Blood Increase Physiological
Chloral Hydrate 1 g orally produces concentration of 1.5 mg/L.
2348

Tricyclic Antidepressants

Serum Positive Analytical
Diphenhydramine False positive observed with EMIT-ST test for tricyclic antidepressants. *3399*

Triglyceride Lipase

Serum No Effect Physiological
Quingestanol No significant effect with 300 μg/d 6 mo. *1309*

Serum Increase Physiological
Norethindrone In normal and hyperlipemic women. *1309*

Triglycerides

Serum No Effect Analytical
Acetaminophen At 5 times upper limit of therapeutic concentration on routine methods in use on SMAC®, Abbott-VP, Cobas-Bio, Ektachem® 400, Hitachi® 705, KDA. *2138* At acute overdose concentration (10 mg/dL) on SMAC® method. *3719* No effect at therapeutic concentration on Reflotron method. *1984* At concentration of 300 mg/L had no effect on GPO-PAP method. *3393* At concentration of 1500 mg/L had no effect on lipase/esterase method. *3393*

Acetazolamide At concentration of 1,000 mg/L had no effect on lipase/esterase method. *3393*

Acetylsalicylic Acid On enzymatic methods at therapeutic concentrations. *1786* On enzymatic method on SMA II at physiological concentration. *1787* At 5 times upper limit of therapeutic range on method on SMAC®, Abbott-VP, Ektachem® and Hitachi® 705. *2138* With 8326 μmol/L on enzymatic procedure with lipase. *1786* No effect at therapeutic concentration on Reflotron method. *1984* At concentration of 2,000 mg/L had no effect on lipase/esterase method. *3393*

Ajmaline At concentration of 3.3 mg/L had no effect on GPO-PAP method. *3393*

Allopurinol At concentration of 100 mg/L had no effect on GPO-PAP method. *3393* At concentration of 5 mg/L had no effect on lipase/esterase method. *3393*

Aminophenazone At therapeutic concentration on enzymatic methods. *1786*

Aminosalicylic Acid At 5 times upper limit of therapeutic range on methods on SMAC®, Abbott-VP, Ektachem®, Hitachi® 705 and KDA. *2138* At concentration of 460 mg/L had no effect on lipase/esterase method. *3393*

Amitriptyline At concentration of 132 mg/L had no effect on GPO-PAP method. *3393* At concentration of 25 mg/L had no effect on lipase/esterase method. *3393* At acute overdose concentration (2.5 mg/dL) on SMAC® method. *3719*

Amobarbital At concentration of 5 mg/L had no effect on lipase/esterase method. *3393*

Amphetamine At therapeutic concentration on Reflotron method. *1984*

Amphotericin B At concentration of 20 mg/L had no effect on lipase/esterase method. *3393*

Ampicillin No effect on Reflotron method at therapeutic concentration. *1984* At concentration of 11 mg/L had no effect on GPO-PAP method. *3393* At concentration of 40 mg/L had no effect on lipase/esterase method. *3393*

Aprotinin At concentration of 150 kU/L had no effect on GPO-PAP method. *3393*

Ascorbic Acid No effect at therapeutic concentration on Reflotron method. *1984* At concentration of 500 mg/L had no effect as measured by lipase/esterase method. *3393*

Barbital At acute overdose concentration (20 mg/dL) on Technicon® SMAC® method. *3719* At concentration of 500 mg/L had no effect as measured by lipase/esterase method. *3393*

Bezafibrate No effect at therapeutic concentration on Reflotron method. *1984*

Bilirubin No effect at 17 mg/dL on method of Levy. *2151* At concentration of 40.0 mg/dL on Ektachem® method. At 40.0 mg/dL on recommended methods on Centrifichem. At 40.0 mg/dL on recommended method on I.L. Multistat III. At 40.0 mg/dL on method recommended on Hitachi® 705 with ferrocyanide. *3407* At concentration of 40.0 mg/dL on Du Pont aca II. *3407*

Boric Acid At concentration of 50 mg/L had no effect as measured by lipase/esterase method. *3393*

Bromazepam At 5 times therapeutic concentration on methods on SMAC®, Abbott-VP, Ektachem® 400, Hitachi® 705 and KDA. *2138*

Butabarbital At concentration of 100 mg/L had no effect as measured by lipase/esterase method. *3393*

Caffeine At therapeutic concentration on Reflotron method. *1984* At concentration of 160 mg/L had no effect on lipase/esterase method. *3393*

Carbenicillin At 5 times upper limit of therapeutic range on methods on SMAC®, Ektachem® 400, Hitachi® 705 and KDA. *2138* At concentration of 15,000 mg/L had no effect on lipase/esterase method. *3393*

Carbochromen No effect at therapeutic concentration on Reflotron method. *1984*

Cefazolin At concentration of 110 mg/L had no effect on lipase/esterase method. *3393*

Cefoxitin At 5 times upper limit of therapeutic range on methods on SMAC®, Abbott-VP, Ektachem®, Hitachi® 705 and KDA. *2138* At concentration of 1025 mg/L had no effect on GPO-PAP method. *3393*

Cephalothin At 5 times upper limit of therapeutic range on methods on SMAC®, Abbott-VP, Ektachem®, Hitachi® 705 and KDA. *2138* At concentration of 1,000 mg/L had no effect on lipase/esterase method. *3393*

Chloramphenicol No effect at therapeutic concentration on Reflotron method. *1984* At concentration of 3,000 mg/L had no effect on GPO-PAP method. *3393*

Chlordiazepoxide At 5 times upper limit of therapeutic range on methods on SMAC®, Abbott-VP, Ektachem®, Hitachi® 705 and KDA. *2138* At acute overdose concentration (5 mg/dL) on Technicon® SMAC® method. *3719* No effect at therapeutic concentration on Reflotron method. *1984* At concentration of 51 mg/L had no effect on GPO-PAP method. *3393* At concentration of 50 mg/L had no effect on lipase/esterase method. *3393*

Chlormezanone At concentration of 100 mg/L had no effect on lipase/esterase method. *3393*

Chloroquine At concentration of 250 mg/L had no effect on GPO-PAP method. *3393*

Chlorphenesin At concentration of 150 mg/L had no effect on lipase/esterase method. *3393*

Chlorpheniramine At acute overdose concentration (20 mg/dL) on Technicon® SMAC® method. *3719* At concentration of 200 mg/L had no effect on lipase/esterase method. *3393*

Chlorpromazine At concentration of 3 mg/L had no effect on lipase/esterase method. *3393*

Chlorpropamide At concentration of 150 mg/L had no effect on lipase/esterase method. *3393*

Chlorprothixene At concentration of 1 mg/L had no effect on lipase/esterase method. *3393*

Cimetidine At concentration of 4 mg/L had no effect on GPO-PAP method. *3393* At concentration of 1 mg/L had no effect on lipase/esterase method. *3393*

Clofibrate At concentration of 500 mg/L had no effect on GPO-PAP method. *3393*

Cloxacillin At concentration of 5 mg/L had no effect on lipase/esterase method. *3393*

Cocaine At acute overdose concentration (2.5 mg/dL) on Technicon® SMAC® method. *3719* At concentration of 25 mg/L had no effect on lipase/esterase method. *3393*

Codeine At acute overdose concentration (2.0 mg/dL) on Technicon® SMAC® method. *3719* At concentration of 20 mg/L had no effect on lipase/esterase method. *3393*

Cyclophosphamide   At concentration of 40 mg/L had no effect on GPO-PAP method. *3393*

Cysteine   At 5 times upper limit of therapeutic range on methods on SMAC®, Abbott-VP, Ektachem® and Hitachi® 705. *2138*

Cystine   At 5 times upper limit of therapeutic range on methods on SMAC®, Abbott-VP, Ektachem® and Hitachi® 705. *2138*

Desipramine   At concentration of 8 mg/L had no effect on lipase/esterase method. *3393*

Diazepam   At 5 times upper limit of therapeutic range on methods on SMAC®, Abbott-VP Ektachem®, Hitachi® 705 and KDA. *2138*   At acute overdose concentration (2.5 mg/dL) on Technicon® SMAC® method. *3719*   At concentration of 25 mg/L had no effect on lipase/esterase method. *3393*

Diclofenac   No effect on enzymatic methods at therapeutic concentrations. *1786*   On enzymatic method on SMA II at physiological concentration. *1787*

Diethylaminoethanol   At concentration of 30 mg/L had no effect on lipase/esterase method. *3393*

Diethylpropion   At acute overdose concentration (10 mg/dL) on Technicon® SMAC® method. *3719*   At concentration of 100 mg/L had no effect on lipase/esterase method. *3393*

Digitoxin   At concentration of 6 mg/L had no effect on lipase/esterase method. *3393*

Digoxin   At concentration of 0.25 mg/L had no effect on GPO-PAP method. *3393*   At concentration of 9 mg/L had no effect on lipase/esterase method. *3393*

N-Dimethyldiazepam   At concentration of 6 mg/L had no effect on lipase/esterase method. *3393*

Diphenhydramine   At concentration of 200 mg/L had no effect on lipase/esterase method. *3393*   At acute overdose concentration (20 mg/dL) on Technicon SMAC® method. *3719*

Disopyramide   At concentration of 4 mg/L had no effect on lipase/esterase method. *3393*

Disulfiram   At concentration of 120 mg/L had no effect on lipase/esterase method. *3393*

Disulphine Blue   By kinetic lipase/glycerol kinase method. *1464*

EDTA   At concentration of 2922 mg/L had no effect on GPO-PAP method. *3393*

Ethanol   At acute overdose concentration (20 mg/dL) on Technicon® SMAC® methods. *3719*

Ethaverine   No effect at therapeutic concentration on Reflotron method. *1984*

Ethchlorvynol   At acute overdose concentration (20 mg/dL) on Technicon® method. *3719*   At concentration of 400 mg/L had no effect on lipase/esterase method. *3393*

Ethinamate   At acute overdose concentration (20 mg/dL) on Technicon® SMAC® method. *3719*   At concentration of 200 mg/L had no effect on lipase/esterase method. *3393*

Ethosuximide   At concentration of 390 mg/L had no effect on lipase/esterase method. *3393*

Fluorescein   No effect at 200 mg/L on Ektachem® 700 method. *1722*

Flurazepam   At 5 times upper limit of therapeutic range on methods on SMAC®, Abbott-VP, Ektachem®, Hitachi® 705 and KDA. *2138*   At acute overdose concentration (2.5 mg/dL) on Technicon® SMAC® method. *3719*   At concentration of 25 mg/L had no effect on lipase/esterase method. *3393*

Furosemide   No effect at therapeutic concentration on Reflotron method. *1984*   At concentration of 5 mg/L had no effect on GPO-PAP method. *3393*   At concentration of 4 mg/L had no effect on lipase/esterase method. *3393*

Gentamicin   At concentration of 10 mg/L had no effect on GPO-PAP method. *3393*   At concentration of 150 mg/L had no effect on lipase/esterase method. *3393*

Glibenclamide   No effect at therapeutic concentration on Reflotron method. *1984*

Glucose   No effect at 500 mg/dL on method of Levy. *2151*

Glutethimide   At acute overdose concentration (5 mg/dL) on Technicon® SMAC® method. *3719*   At concentration of 50 mg/L had no effect on lipase/esterase method. *3393*

Hydralazine   At concentration of 6 mg/L had no effect on lipase/esterase method. *3393*

Hydrochlorothiazide   At concentration of 7 mg/L had no effect on GPO-PAP method. *3393*

Ibuprofen   No effect at therapeutic concentrations on enzymatic methods. *1786*   On enzymatic method on SMA II at physiological concentration. *1787*   At 5 times upper limit of therapeutic range on methods on SMAC®, Abbott-VP, Ektachem®, Hitachi® 705 and KDA. *2138*

Indomethacin   No effect on enzymatic methods at therapeutic concentrations. *1786*   On enzymatic method on SMA II at physiological concentration. *1787*   No effect at therapeutic concentration on Reflotron method. *1984*   At concentration of 13 mg/L had no effect on lipase/esterase method. *3393*

Insulin   At concentration of 3 mg/L had no effect on lipase/esterase method. *3393*

Iproniazid   At concentration of 40 mg/L had no effect on lipase/esterase method. *3393*

Isoniazid   At concentration of 100 mg/L had no effect on lipase/esterase method. *3393*

Ketoprofen   No effect at therapeutic concentrations on enzymatic methods. *1786*   On enzymatic method on SMA II at physiological concentration. *1787*

Lorazepam   At concentration of 0.05 mg/L had no effect on lipase/esterase method. *3393*

Meperidine   At acute overdose concentration (5 mg/dL) on Technicon® SMAC® method. *3719*   At concentration of 60 mg/L had no effect on lipase/esterase method. *3393*

Mephenesin   At concentration of 100 mg/L had no effect on lipase/esterase method. *3393*

Meprobamate   At acute overdose concentration (20 mg/dL) on Technicon® SMAC® method. *3719*   At concentration of 200 mg/L had no effect on lipase/esterase method. *3393*

Mesoridazine   At acute overdose concentration (20 mg/dL) on Technicon® SMAC® method. *3719*

Methadone   At acute overdose concentration (2.5 mg/dL) on Technicon® SMAC® method. *3719*   At concentration of 25 mg/L had no effect on lipase/esterase method. *3393*

Methamphetamine   At concentration of 2 mg/L had no effect on lipase/esterase method. *3393*

Methanol   At acute overdose concentration (20 mg/dL) on Technicon® SMAC® method. *3719*

Methapyrilene   At concentration of 13 mg/L had no effect on lipase/esterase method. *3393*

Methaqualone   At acute overdose concentration (2.5 mg/dL) on Technicon® SMAC® method. *3719*   No effect at therapeutic concentration on Reflotron method. *1984*   At concentration of 25 mg/L had no effect on lipase/esterase method. *3393*

Methicillin   At 5 times upper limit of therapeutic range on SMAC®, Ektachem®, Hitachi® 705 and KDA. *2138*   At concentration of 900 mg/L had no effect on lipase/esterase method. *3393*

Methohexital   At concentration of 50 mg/L had no effect on lipase/esterase method. *3393*

Methotrexate   At 5 times upper limit of therapeutic range on methods on Abbott-VP, Ektachem®, Hitachi® 705 and KDA. *2138*   At concentration of 25 mg/L had no effect on GPO-PAP method. *3393*

Methotrimeprazine   At concentration of 1 mg/L had no effect on lipase/esterase method. *3393*

Methsuximide   At concentration of 40 mg/L had no effect on lipase/esterase method. *3393*

Methyldopa   No effect at therapeutic concentration on Reflotron method. *1984*   At concentration of 80 mg/L had no effect on lipase/esterase method. *3393*

Methylphenidate   At acute overdose concentration (20 mg/dL) on Technicon® SMAC® method. *3719*   At concentration of 200 mg/L had no effect on lipase/esterase method. *3393*

Methyprylon   At acute overdose concentration (10 mg/dL) on Technicon® SMAC® method. *3719*   At concentration of 100 mg/L had no effect on lipase/esterase method. *3393*

Metoprolol   At concentration of 3 mg/L had no effect on GPO-PAP method. *3393*   At concentration of 0.34 mg/L had no effect on lipase/esterase method. *3393*

Morphine   At acute overdose concentration (20 mg/dL) on Technicon® SMAC® method. *3719*   At concentration of 200 mg/L had no effect on lipase/esterase method. *3393*

Nafcillin   At concentration of 50 mg/L had no effect on lipase/esterase method. *3393*

Nalidixic Acid   At concentration of 50 mg/L had no effect on GPO-PAP method. *3393*

Triglycerides *(continued)*

Serum   No Effect   Analytical *(continued)*

Naproxen At 5 times upper limit of therapeutic range on methods on Abbott-VP, Ektachem®, Hitachi® 705 and KDA. *2138*

Niacin No effect at therapeutic concentration on Reflotron method. *1984*

Nitrofurantoin On routine methods in use on SMAC®, Ektachem®, Abbott-VP, Cobas-Bio, aca, Hitachi® 705, KDA at 5 times normal upper therapeutic concentration. *2138* No effect at therapeutic concentration on Reflotron method. *1984* At concentration of 0.5 mg/L had no effect on GPO-PAP method. *3393* At concentration of 4 mg/L had no effect on lipase/esterase method. *3393*

Nitroglycerin No effect on method using glycerol kinase and glycerol-3-phosphate on RA-1000 and Ektachem® DT-60. *2588*

Nortriptyline At acute overdose concentration (20 mg/dL) on Technicon® SMAC® method. *3719* At concentration of 200 mg/L had no effect on lipase/esterase method. *3393*

Orphenadrine At concentration of 9 mg/L had no effect on lipase/esterase method. *3393*

Oxazepam No effect at therapeutic concentration on Reflotron method. *1984* At concentration of 1 mg/L had no effect on lipase/esterase method. *3393*

Oxedrine At concentration of 60 mg/L had no effect on GPO-PAP method. *3393*

Oxyphenbutazone At concentration of 120 mg/L had no effect on GPO-PAP method. *3393*

Oxytetracycline No effect at therapeutic concentration on Reflotron method. *1984*

Papaverine At concentration of 10 mg/L had no effect on lipase/esterase method. *3393*

Paraldehyde At concentration of 2,000 mg/L had no effect on lipase/esterase method. *3393*

Penicillin G At 5 times upper limit of therapeutic range on methods on SMAC®, Abbott-VP, Ektachem®, Hitachi® 705 and KDA. *2138* At concentration of 90 mg/L had no effect as measured by GPO-PAP method. *3393* At concentration of 2,000 mg/L had no effect as measured by lipase/esterase method. *3393*

Pentobarbital At acute overdose concentration (20 mg/dL) on Technicon® SMAC® method. *3719* At concentration of 340 mg/L had no effect on lipase/esterase method. *3393*

Perphenazine At acute overdose concentration (20 mg/dL) on Technicon® SMAC® method. *3719* At concentration of 200 mg/L had no effect on lipase/esterase method. *3393*

Phenazopyridine No effect at therapeutic concentration on Reflotron method. *1984*

Phenobarbital No effect at therapeutic concentrations on enzymatic methods. *1786* At acute overdose concentration (20 mg/dL) on Technicon® SMAC® method. *3719* No effect at therapeutic concentration on Reflotron method. *1984* At concentration of 100 mg/L had no effect on GPO-PAP method. *3393* At concentration of 250 mg/L had no effect on lipase/esterase method. *3393*

Phenprocoumon No effect at therapeutic concentration on Reflotron method. *1984*

Phensuximide At concentration of 120 mg/L had no effect on lipase/esterase method. *3393*

Phenylbutazone At concentration of 200 mg/L had no effect on GPO-PAP method. *3393* At concentration of 750 mg/L had no effect on lipase/esterase method. *3393*

Phenytoin At acute overdose concentration (20 mg/dL) on Technicon® SMAC® method. *3719* No effect at therapeutic concentration on Reflotron method. *1984* At concentration of 240 mg/L had no effect on lipase/esterase method. *3393*

Prednisolone At concentration of 230 mg/L had no effect on GPO-PAP method. *3393*

Probenecid No effect at therapeutic concentration on Reflotron method. *1984* At concentration of 1300 mg/L had no effect on lipase/esterase method. *3393*

Procainamide At concentration of 50 mg/L had no effect on lipase/esterase method. *3393*

Procaine No effect at therapeutic concentration on Reflotron method. *1984* At concentration of 2 mg/L had no effect on lipase/esterase method. *3393*

Prochlorperazine At concentration of 1 mg/L had no effect on lipase/esterase method. *3393*

Promethazine At acute overdose concentration (20 mg/dL) on Technicon® SMAC® method. *3719* At concentration of 200 mg/L had no effect on lipase/esterase method. *3393*

Propoxyphene At acute overdose concentration (2.5 mg/dL) on Technicon® SMAC® method. *3719* At concentration of 25 mg/L had no effect on lipase/esterase method. *3393*

Propranolol At concentration of 92 mg/L had no effect on GPO-PAP method. *3393* At concentration of 0.2 mg/L had no effect on lipase/esterase method. *3393*

Pyribenzamine® At acute overdose concentration (20 mg/dL) on Technicon® SMAC® methods. *3719*

Pyridamole No effect at therapeutic concentration on Reflotron method. *1984*

Pyritinol No effect at therapeutic concentration on Reflotron method. *1984*

Quinidine At acute overdose concentration (20 mg/dL) on Technicon® SMAC® method. *3719* No effect at therapeutic concentration on Reflotron method. *1984* At concentration of 210 mg/L had no effect on lipase/esterase method. *3393*

Quinine At acute overdose concentration (1.5 mg/dL) on Technicon® SMAC® method. *3719* At concentration of 30 mg/L had no effect on lipase/esterase method. *3393*

Rifampin At 5 times upper limit of therapeutic concentration on method on SMAC®, Abbott-VP Ektachem® and Hitachi® 705. *2138*

Rolitetracycline At concentration of 4 mg/L had no effect on GPO-PAP method. *3393*

Salicylate At concentration of 594 mg/L had no effect on GPO-PAP method. *3393* At concentration of 500 mg/L had no effect on lipase/esterase method. *3393*

Scopolamine Bromide At concentration of 20 mg/L had no effect on GPO-PAP method. *3393*

Secobarbital At acute overdose concentration (20 mg/dL) on Technicon® SMAC® method. *3719* At concentration of 100 mg/L had no effect on lipase/esterase method. *3393*

Sodium Citrate At concentration of 11764 mg/L had no effect on GPO-PAP method. *3393*

Sodium Fluoride At concentration of 5879 mg/L had no effect on GPO-PAP method. *3393*

Sodium Heparin At concentration of 4,000 mg/L had no effect on GPO-PAP method. *3393*

Sodium Oxalate At concentration of 6160 mg/L had no effect on GPO-PAP method. *3393*

Spironolactone At concentration of 1.3 mg/L had no effect on GPO-PAP method. *3393*

Strychnine At concentration of 12 mg/L had no effect on lipase/esterase method. *3393*

Sulfadiazine At concentration of 1500 mg/L had no effect on lipase/esterase method. *3393*

Sulfaguanidine At concentration of 500 mg/L had no effect on lipase/esterase method. *3393*

Sulfamethoxazole No effect at therapeutic concentration on Reflotron method. *1984*

Sulfamethoxypyridazine At concentration of 70 mg/L had no effect on GPO-PAP method. *3393*

Sulfanilamide At concentration of 1,000 mg/L had no effect on lipase/esterase method. *3393*

Sulforidazine At concentration of 200 mg/L had no effect on lipase/esterase method. *3393*

Tetracycline At 5 times upper limit of therapeutic range on methods on SMAC®, Abbott-VP, Ektachem®, Hitachi® 705 and KDA. *2138* At concentration of 60 mg/L had no effect on GPO-PAP method. *3393* At concentration of 300 mg/L had no effect on lipase/esterase method. *3393*

Theobromine At concentration of 2,000 mg/L had no effect on lipase/esterase method. *3393*

Theophylline At 5 times upper limit of therapeutic range on methods on SMAC®, Abbott-VP, Ektachem®, Hitachi® 705 and KDA. *2138* No effect at therapeutic concentration on Reflotron method. *1984* At concentration of 2,000 mg/L had no effect on lipase/esterase method. *3393*

Thiopental At acute overdose concentration (20 mg/dL) on Technicon® SMAC® method. *3719* At concentration of 200 mg/L had no effect on lipase/esterase method. *3393*

Timolol At concentration of 0.01 mg/L had no effect on lipase/esterase method. *3393*

Tolbutamide At concentration of 540 mg/L had no effect on GPO-PAP method. *3393* At concentration of 100 mg/L had no effect on lipase/esterase method. *3393*

Trazodone At concentration of 50 mg/L had no effect on GPO-PAP method. *3393*

Tribromethanol At concentration of 90 mg/L had no effect on lipase/esterase method. *3393*

Trichlorethanol At concentration of 1,000 mg/L had no effect on lipase/esterase method. *3393*

Trifluoperazine At concentration of 1 mg/L had no effect on lipase/esterase method. *3393*

Trimethoprim No effect at therapeutic concentration on Reflotron method. *1984*

Tripelennamine At concentration of 200 mg/L had no effect on lipase/esterase method. *3393*

Valproic Acid At concentration of 1010 mg/L had no effect on GPO-PAP method. *3393*

Warfarin At concentration of 100 mg/L had no effect on lipase/esterase method. *3393*

Serum   No Effect   Physiological

Acebutolol In small numbers of patients when treated for 1-12 mo. *0088* Increase by 1 mg/dL in 11 patients given 400 mg daily for 3 mo. *2550* Insignificant change on average in one 6 mo long study. *2116* In several studies of about 15 patients treated for 1 to 12 mo. *0089*

Atenolol In 20 patients with 50 mg/d compared with pre-treatment values after 3 mo. *3071*

Bendrofluazide No significant change in 15 individuals when given as sole treatment for less than 1 y. *3826* After 12 mo treatment of 53 previously untreated hypertensives. *0315*

Bunitrolol When 30 mg/d given for 12 weeks in normolipidemic patients with mild essential hypertension. *3151*

Captopril No significant change in patients in 2 studies treated for 2 and 24 mo. *0088* After 12 weeks treatment in 18 patients with mild essential hypertension. *3150* Decrease by -3 mg/dL in 15 patients given 75 mg daily for 8 weeks. *2550*

Carprazidil No consistent effect in 15 men with mild to moderate essential hypertension treated for up to 16 weeks. *1268* In one study involving 12 patients treated for 4 mo. *0089*

Chlormadinone No significant effect observed. *1309*

Chlorthalidone Nonsignificant change during monotherapy in 302 subjects for 1 y. *1334* In 22 premenopausal women without decrease in blood pressure treated for less than 1 y. *0402* In 22 premenopausal and 18 postmenopausal women given 100 mg/d for 6 weeks. *0402* Increase by 6 mg/dL in 10 patients given 100 mg daily for 1 mo. *2550*

Cimetidine In healthy individuals and in subjects with high serum glucose concentration. *1653* In 25 patients pretreatment 139 mg/dL, after 146 mg/dL over 5 weeks. *3568*

Clopamide No significant change in 17 individuals treated for less than 1 y when used as sole treatment. *3184*

Cyproterone In oophorectomized women when given alone. *2177*

Danazol Inconsistent changes observed in 62 patients with endometriosis treated with 600 mg daily for up to 24 weeks. *3222* No significant effect in 9 women with endometriosis treated for 6 mo. *0060* No effect after 6 mo of 600 mg/d. *3906*

Dehydroepiandrosterone In 5 normal men given 1600 mg/d orally for 28 d. *2578*

Diclofenac No effect seen in patients undergoing treatment. *1787*

Diltiazem No significant change in 31 subjects treated for 6 mo. *0088*

Doxazosin No significant change compared with controls. *2842*

Enalapril In 53 patients given up to 160 mg daily or when combined with hydrochlorothiazide. *2277*

Estradiol In 8 post-menopausal, post-oophorectomy women after 100 mg implant in subcutaneous tissue. *3274* With nonalkylated estrogens, eg valerate salt, in treatment of postmenopausal hormone deficiency. *1037*

Estrogens In post-menopausal women treated for 3 y. *0653*

Etretinate In 13 patients with hyperkeratotic disorders given 50 mg daily for 2 mo. *3659*

Furosemide No significant change in 16 subjects treated for less than 1 y with drug only. *1306*

Glibonuride No significant effect observed after 4 weeks. *0301*

Glutethimide No significant effect with 21 d of 500 mg daily in 6 volunteers. *0407*

Guanabenz No effect in 39 hypertensives treated for 2 y. *1865* Decrease by -13 mg/dL in patients given 4 to 8 mg daily for 12 to 16 weeks. *2550*

Gum Tragacanth In 5 male volunteers consuming 9.9 g daily for 21 d. *0990*

Hydrochlorothiazide No effect with 25 or 50 mg/d in hypertensive patients. *1980* Nonsignificant change by 7 mg/dL in 33 patients with hypertension given drug for 10 weeks to reduce diastolic BP to less than 90 mm Hg. *3707* In approximately 175 patients with essential hypertension treated for 1 y. *3708* In short term, and long term involving many patients. *0089*

Ibuprofen No effect seen in patients undergoing treatment. *1787*

Indapamide In 27 subjects treated for less than 1 y with this drug only. *0692* In various studies involving more than 30 patients when used as sole drug for less than 1 y. *0089*

Indomethacin No effect seen in patients undergoing treatment. *1787*

Isotretinoin From 2.12 to 2.72 mmol/L in 12 patients with hyperkeratotic disorders given 40 mg. *3659*

Ketoprofen No effect seen in patients undergoing treatment. *1787*

Labetalol No significant change in several studies with patients treated for 1-12 mo. *0088* In several studies from 1 to 12 mo. *0089*

Methyldopa In several patients treated for 2-4 mo. *0088* In 3 studies of 7 to 17 patients for up to 3 mo. *0089*

Metoprolol Insignificant increase after 12 weeks treatment in hypertriglyceridemic hypertensives. *0341*

MPCA No significant change when given in fed state. *1423*

Muscular Exercise But relative reduction of unsaturated acid. *1702*

Muzolimine Insignificant change in 10 hypertensive patients given 30 mg daily for 16 weeks. *1229*

Neomycin No effect in 20 patients with type II hyperlipoproteinemia treated over several months. *1618*

Nifedipine No effect observed in 14 patients with essential hypertension treated over 2 mo with 20 mg twice daily. *2119* Increase by 4 mg/dL in patients given 30 to 60 mg daily for 8 weeks. *2550*

Norethisterone In 75 women who had received up to 24 intragluteal injections of 200 mg every 56 d. *1666*

Oral Contraceptives Extent of effect varies with exact composition of oral contraceptive. *1458* In 10 women receiving ethinyl estradiol with norethindrone. *2009* Typically insignificant change over 6 cycles in women with low dose ethinyl estradiol with low dose levonorgestrel or desogestrel. *1249*

Oxprenolol In several studies treated for more than 1 mo. *0088* Insignificant effect in 12 patients receiving 160 mg/d for 2 weeks. *1806*

Phenytoin In both men and women in 27 patients with transient brain ischemia. *1879* No overall significant difference between treated epileptics and controls but higher proportion of males had high concentrations than controls. *2598* No consistent change seen in patients on long-term Treatment. *2767*

Pindolol No significant change typically seen with treatment for 1 mo or more. *0088* No effect in short-term study of 10 hypertensive men with 15 mg/d. *2135* Insignificant change in 3 studies of 2 to 12 mo. *2116* Remained constant during 12 mo therapy. *2117*

Prazosin No significant effect in 15 hypertensives given 4 mg/d for 10 weeks. *1118* No significant change with monotherapy for 12 mo in 15 patients. *3531* In 16 hypertensive patients given 2 mg daily for 3 mo. *2266*

Prednisolone In men receiving drug for mean of 3.1 y. *1783*

Prednisone No effect in men during 1 mo. *3949*

Propranolol In 22 mild/moderate hypertensive males treated for 8 weeks: insignificant increase noted. *1693* General effect observed in multiple studies. *0088* After 12 mo treatment of 53 previously untreated hypertensives. *0315*

Triglycerides (continued)

Serum No Effect Physiological (continued)

Quingestanol At 300 μg/d usually no effect. *1309*

Ranitidine In 25 patients pretreatment 121 mg/dL, after 137 mg/dL over 5 weeks. *3568*

Reserpine No significant change in small number of patients treated for up to 2.5 mo. *0088*

Sitosterols No effect observed. *2112*

Sotalol General effect in 2 studies but overall progressive deterioration in lipid profile. *0088*

Spironolactone No significant change in 17 subjects treated with drug for less than 1 y. *1405*

Stanozolol In 10 normolipidemic postmenopausal osteoporotic women treated for 6 weeks. *3526*

Terazosin No significant effect with up to 20 mg daily over 4 weeks in patients with moderate hypertension. *0857*

Terbutaline No significant effect observed after 2 weeks in 15 subjects. *1654*

L-Thyroxine At about 1.3 in 11 hypothyroid women treated with 0.1 to 0.2 mg daily. *1201*

Timolol In 15 patients treated for 1 mo. *3826*

Tourniquet No effect for 1 minute. *3540*

Trimazosin In 2 studies with 13 and 48 subjects treated for 2.5 to 12 mo. *0088*

Ursodeoxycholic Acid No significant effect with 600 mg daily in patients with endogenous hypertriglyceridemia. *0601* In 8 normolipemic patients receiving 1,000 mg drug daily. *2121*

Verapamil In 12 patients when angina or hypertension treated for 6 weeks. Where change occurred it was of order of 10%. *3756* In 64 patients in post-myocardial infarction comparison against placebo. *3489*

Serum Increase Analytical

Acetylsalicylic Acid At 5 times upper limit of therapeutic range on method of KDA. *2138*

Bacterial Contamination If contaminated with phospholipase producing bacteria. *3223*

Bilirubin At 24.0 mg/dL and above on methods recommended on BMD-8700 but with ferrocyanide. At 5.0 mg/dL and above on method recommended on Hitachi® 705 with ferrocyanide. From below 2.5 mg/dL and above on recommended methods on Centrifichem. At 40.0 mg/dL on recommended method on I.L. Multistat III. *3407*

Carbenicillin At 5 times upper limit of therapeutic range on enzymatic method on Abbott-VP. *2138*

Cysteine At 5 times upper limit of therapeutic range on enzymatic method on KDA. *2138*

Cystine At 5 times upper limit of therapeutic range on enzymatic methods on KDA. *2138*

Dihydroxyacetone Measured as glycerol with enzymatic methods. *2981*

Fluosol-DA Directly proportional to Fluosinal concentration since emulsion contains glycerol. *2518*

Glassware Vacutainers™ contaminated with glycerol affect enzyme procedure. *3590*

Glyceraldehyde Measured as glycerol with enzymatic procedures. *2981*

Glycerin When enzymatic procedure used is end product of reaction. *3590*

Lipemia To large and variable extent in patient specimens measured on SMAC® in comparison with specimens following ultra-centrifugation. *2466*

Methicillin At 5 times upper limit of therapeutic range on enzymatic method on Abbott-VP. *2138*

Nitroglycerin In method on Technicon® RA-1000 using glycerol oxidase. *2588*

Propylene Glycol Unless blank or Folch extraction. *2128*

Serum Increase Physiological

Acetylsalicylic Acid Significantly higher in patients receiving drug than in controls (1.69 mmol/L vs 1.47 mmol/L). Also seen in same patients before and after treatment. *1787*

Aging Positive correlation in sedentary but not athletes. *1702*

Aldatense Increase by 30% in a group of hypertensives compared with controls. *1611*

Amiodarone Observed effect in 3 individuals but without effect on thyroid function. *2833*

Asparaginase Rare response after initial hypolipidemia. *2657*

Atenolol Up to 60% increase in several studies although no significant chance in some. *0088* Moderate increase after 6 mo treatment of 14 hypertensives. *0836* Marked effect in 18 patients with mild essential hypertension treated with chlorothiazide concomitantly over 14 weeks. *1035* From 1.94 to 2.44 mmol/L in 15 hypertensives treated for 8 mo. *1021* Increase by 28% after 100 mg/d for 5 weeks in 20 hypertensive men. *2135* Increase by 24 to 62% in 4 studies of from 1 to 6 mo. *2116*

Bendrofluazide 20% increase in 66 subjects when treated for less than 1 y when used as only drug. *3706* 20% increase in short term studies. *0089*

Blood Pressure Strongly correlated with increased systolic pressure. *2123*

Bopindolol After 4-8 weeks by as much as 41% but disappeared after 12 weeks of therapy in 24 hypertensives. *3667*

Carbimazole From 0.77 to 0.89 mmol/L in 12 hyperthyroid women patients treated with 10-30 mg daily. *1201*

Chlorthalidone 15% change in 39 subjects treated with drug only for less than 1 y. *1404* Increase by 10 mg/dL in 1,000 men and women with mild hypertension treated for 1 y. *1334* But insignificant change in 10 hypertensive patients given 25 mg daily for 16 week. *1229* 15% in short term, no effect in long term in several studies with monotherapy. *0089*

Cholestyramine Mechanism obscure (may occur in diabetics). *1822*

Colestipol Increase of 7% noted. *2874*

Cyclofenil Slight increase over 4 mo in 19 patients given 600 mg daily for therapy. *2786*

Cyproterone When given in combination with ethinyl estradiol. *2177*

Danazol Increase by 20% after 2 weeks and 14% after 8 weeks in 12 women with endometriosis given 600 mg daily for 24 weeks. *1066* In 12 women with endometriosis treated with 600 mg/d increased by 20% after 2 weeks. *1066*

Desogestrel Increase by 35% when 150 μg desogestrel given with 30 μg ethinyl estradiol for 3 mo. Results normal within 2 mo after treatment stopped. *0312*

Endotoxin Occurs after free fatty acid rise. *1230*

Estradiol By about 8% (insignificant) with 2 mg estradiol valerate/d in 20 normolipoproteinemic women over 3 mo. *3593*

Estradiol-17β Slight increase in 38 healthy post-menopausal women given 2-4 mg orally or 3 mg cutaneously over 6 mo. *1067*

Estrogens Associated with increased very low density lipoproteins. *1535*

Estrone After 3 mo in 20 women in the perimenopause, then gradual decline to almost. *2058*

Ethanol In fasting hypertriglyceride patients (not normals). *1288* Infusion causes increased hepatic synthesis. *1766* In post-prandial normal and hypertriglyceride patients. *1288* 6.5% increase in heavy versus occasional drinkers. *3273* Significant increase in heavy drinkers versus nondrinkers or occasional drinker. *0059* Usual consequence, but especially with fatty meal. *2428* Significantly higher in alcoholic men than in controls. *1701* In response to 1.5 g/kg at night significantly increased values measured next morning. *3547*

Ethynodiol Decreased removal from circulation. *3505*

Etretinate In 11 of 20 patients but not to more than 2 times normal over 1 y. *1866*

Fructose In experimental animals, contradictory in humans. *1709*

Furosemide 37% change in 12 normotensive men when treated with drug alone for less than 1 year. *1826* Increase by 29 mg/dL in 12 patients given 40 to 80 mg daily for 3 mo. *2550*

Glucocorticoids Mean 16% increased observed with low dose. *0915*

Glucose Significant change in normals in response to infusion of 1 g/kg body weight, although not very marked. *0476*

High Carbohydrate Diet Increase by 60% if oral no effect if i.v. *0880*

Hydrochlorothiazide Observed with 100 mg/d. *3663* 17% change in 39 individuals given drug alone for less than 1 y. *1404* 17% in short term, and increase in MRFIT study when given with chlorthalidone. *0089*

Indapamide Slight and insignificant in 13 hypertensive diabetics treated for 24 weeks. *2686*

Isotretinoin In 7 patients with severe rosacea treated with 1 mg/kg/d for 12 weeks. Effects possibly due to induction of hepatic microsomal enzymes. *2307* Increase in both men and women with 1 mg/kg/d for 20 weeks when given for nodulocystic acne. *0326* Reversible dose-related effect noted in patients with myelodysplastic syndrome and leukemia at high doses. *1319* From 5.1 to 6.1 mmol/L in 18 patients with severe acne given 0.8 mg/kg daily for 3 mo: changes reverted to normal after treatment stopped. *2242* From 5.4 to 6.8 mmol/L at 6 weeks in 7 patients with severe rosacea treated with 1 mg/kg/d. *2307* During treatment but not to abnormal levels. *1818* In 25 of 523 patients with mean daily dose of 109 mg for 150 d. *3867* Increase in both men and women with 1 mg/kg/d for 20 weeks when given for nodulocystic acne. *0326*

Labetalol Increase by 27% in 8 patients given 600-1200 mg daily for 4 mo. *1806*

Levonorgestrel Increase by 48% when 150 μg levonorgestrel given with 30 μg ethinyl estradiol for 3 mo. Results normal within 2 mo after treatment stopped. *0312*

Levothyroxine Effect observed in hypothyroid patients. *2462*

Meals Increase to maximum at 4 h after, minimum 10 h after. *3162*

Menotropins May be slight increase initially. *2235*

Mestranol ?impaired triglyceride removal from circulation. *3505*

Methyldopa Increase by 28% in 32 middle-aged hypertensive males treated for 6 weeks. *2129* Increase by 20 mg/dL in 11 patients given 750 mg daily for 8 weeks. *2550*

Metoprolol In 15 hypertensive patients given 200 mg/d for 10 weeks caused mean increase from 122 to 142 mg/dL. *1118* Up to 34% in some studies but no effect in almost as many. *0088* Insignificant in 20 hypertensive diabetic patients. *3900* Significant effect of 33% and 14% for 1 mo and 3 mo studies respectively. *2116* Rose from 2.51 to 3.41 mmol/L in 20 hypertensives given 200 mg/d for 3 mo. *3071*

Nadolol In hypertensive patients though not significantly, further increased post- prandially and after exercise. *2761* Increase by 15% in 13 patients given 50-200 mg daily for 10 weeks. *1806* Higher in fasting state in patients on treatment, but not significantly higher, maybe secondary to reduction in lipoprotein lipase activity. *2761* In 94 patients treated for 3 mo. *3706*

Obesity Proportion of subjects with high values increases. *2123*

Oral Contraceptives If given for 3 mo impaired removal (estrogen). *3505* Increased in response to a variety of different preparations. *1303* Increase by 25 to 62% depending on preparation used. *2860* Increase by 40 mg/dL in white girl drug users vs controls. *3723* Estrogen effect of combined oral contraceptive. *0018* Positive correlation with serum gamma-glutamyltransferase activity suggesting involvement of hepatic enzyme induction. *2317* With high-dose combination drugs. With drug with high progestogen component induce greater than 50% increase on average. *1248* In 10 women receiving ethinyl estradiol with norgestrel. *2009* Increase by 80% with high dose oral contraceptives that are overly estrogenic. *1248*

Oxprenolol Increase by 22-27% in several studies treated for more than 1 mo. *0088* Marked effect in 53 patients given 80 mg/twice daily for 3 mo. *0835* Mild effect as possessed relatively weak intrinsic sympathomimetic activity. *2116* Increase by 22% in 20 hypertensive men given 160 mg/d for 5 weeks. *2135*

Pindolol In 18 patients with mild essential hypertension treated with chlorothiazide concomitantly over 4 weeks. *1035* Increase by 28% in one study of 4 weeks with 15-30 mg/d. *1806* 28% increase in one 1 mo study. *2116*

Piretanide Effect observed with 12 mg/d. *3663*

Polythiazide 14% change in 20 people when treated only with drug for less than 1 y. *1808*

Practolol Increase by 8 to 16% in 2 studies of 2 weeks to 6 mo respectively. *1806*

Prednisolone Rose from 0.91 to 1.84 mmol/L in women receiving drug for mean of 3.1 y. *1783*

Prednisone In all 6 women slight effect in 1 mo. *3949*

Pregnancy May increase up to 2 1/2 times in normal. *1968*

Propranolol Mean increase from 151 to 197 mg/dL after 16 weeks treatment in approximately 120 hypertensives. *2331* From 256 to 369 mg/dL after 12 weeks treatment in 15 hypertriglyceridemic- hypertensive subjects. *0341* Increase by 24% in 23 hypertensive men given up to 160 mg/d for 8 weeks. *2135* In 20 hypertensive diabetic patients. *3900* Marked increase after 6 mo treatment in 16 patients. *0836* Marked effect in 53 hypertensives given 80 mg twice daily for 3 mo. *0835* Increase by 25 mg/dL in 340 patients given drug for 10 weeks to reduce diastolic BP to less than 90 mm Hg. *3707* Increase by 42 mg/dL in approximately 120 patients with essential hypertension treated for 1 y. *3708* Increase by 15 mg/dL in 8 patients given 30 to 60 mg daily for 8 weeks. *2550* Increase by 24% in 23 hypertensive men aged 47-55 y treated for 8 weeks. *1551* From no change to 65% increase in many studies. *0089* Mean increase from 146 to 211 mg/dL in 16 hypertensives given 80 mg daily for up to 12 mo. *2266*

Smoking Sustained increase with/without inhalation. *1921* Effect observed in fasting white children aged 8 to 17 y. *3723*

Sotalol Significant increase of 66% reported from one long-term study (12 mo). *2116* Simultaneous increase with other lipid changes; change from 1.14 to 1.89 mmol/L over 12 mo. *2120*

Spironolactone Maximum average increase of 19% in 3 studies. *0087*

Starvation Due to release and degradation of fatty acids. *2240*

Sucrose Marked increase when sugar substituted for starch. *2036* Effect of sustained high sucrose diet. *3931*

Tamoxifen Up to 2.8 g/dL observed in one elderly woman treated for breast cancer. Possibly due to reduction of activities of postheparin plasma lipoprotein lipase and hepatic triglyceride lipase. *0508* On slightly higher concentration in breast cancer patients. *3070*

Trichlormethiazide Increase by 44 mg/dL in 15 patients given 4 mg daily for 3 mo but note marked difference between responders and nonresponder populations. *2550*

Serum Decrease Analytical

Acetylsalicylic Acid At concentrations greater than 75 mg/L (within therapeutic range) lowered analyte concentration as measured by GPO-PAP method. *3393*

Ascorbic Acid At concentrations above 30 mg/L (maximum serum concentration 34 mg/L) lowered concentration as measured by GPO-PAP method. *3393*

Bilirubin From less than 2.5 mg/dL on recommended method on RA-1000. At concentrations as low as 2.5 mg/dL and higher on BMD-8700 using manufacturer recommended method. *3407* At 8.0 mg/dL and above on method recommended on Hitachi® 705. *3407* From below 2.5 mg/dL on method recommended on Centrifichem. *3407* At 24.0 mg/dL and above on method recommended on Cobas-Bio. *3407*

Citrates Dilutional effect when compared against serum or heparin as anticoagulant. *3361*

Hydroxyurea Inhibits action of glycerol oxidase included as part of Technicon® RA-1000 method but no effect on procedure used on Beckman Astra-8. *2384*

Levodopa At concentrations above 6 mg/L lowered concentration as measured by GPO-PAP method. *3393*

Liposol Changed from 1752 to 139 mg/L in 26 randomly selected patient sera, before and after addition on Beckman Astra methods. *2054*

Methotrexate At 5 times upper limit of therapeutic range on SMAC® method. *2138*

Methyldopa At concentrations above 8 mg/L (normal therapeutic concentration 2 mg/L) lowered concentration as measured by GPO-PAP method. *3393*

Naproxen At 5 times upper limit of therapeutic range on enzymatic method on SMAC® At 5 times upper limit of therapeutic range on methods on enzymatic method on SMAC®. *2138*

Novaminsulfon At concentrations above 100 mg/L (normal therapeutic concentration 15 mg/L) lowered concentration as measured by GPO-PAP method. *3393*

Plasma Average 4 mg/dL higher in serum. *2228*

Rifampin At 5 times upper limit of therapeutic concentration on method on KDA. *2138*

Storage of Sample Significant decrease 3 d at room temperature. *2512*

Thiamazole At concentrations above 100 mg/L lowered concentration as measured by GPO-PAP method. *3393*

Triglycerides (continued)

Serum   Decrease   Physiological

Amiodarone   From 140 to 123 mg/dL in 24 patients over 90 d treatment. *3392*

Ascorbic Acid   Effect observed in atherosclerotic patients. *3377*

Asparaginase   Noted after first week in some patients. *2657*

Benfluorex   Increase by 33% in 12 hypertriglyceridemic type 2 diabetic patients given 150 mg tid after 1 mo treatment. *3385*

Bezafibrate   Increase by 44% in 7 hypertriglyceridemic type 2 diabetic patients given 200 mg tid after 1 mo treatment. *3385* Increase by 54% in 7 hypertriglyceridemic type 2 diabetic patients given 200 mg tid after 1 mo treatment. *3385* By average of 58% in 11 hypertriglyceridemic subjects. *3291*

Bran   Unprocessed bran for 5 weeks caused decreased of 18 mg/dL. *1540*

Celiprolol   In patients with hyperlipoproteinemia types IIa, IIb or IV with 300 mg drug for 4 weeks. *2873*

Chenodeoxycholic Acid   Mean fall by 20 mg/dL over 3 mo. *0274* Significant effect with 150 mg four times daily in patients with endogenous hypertriglyceridemias. *0601* By 26% in 8 normolipemic patients receiving 16 g drug daily. *2121* By up to 20% in patients given 10 to 25 mg/kg body weight for 6 mo. Increase by 18.6% in 916 patients given up to 750 mg/d for 2 y. *3199*

Chlorthalidone   Statistically significant decrease of 10 mg/dL in 12 men receiving drug in conjunction with diet to reduce cholesterol. *0090*

Cholestyramine   Therapeutic effect. *2220*

Clofibrate   Inhibits transfer from liver to plasma. *0456* Increase by 80% in first year in patients with hyperlipoproteinemia and impaired glucose tolerance given diet plus 2 g drug/d monitored for 5 y. *2925* Marked reduction within 6 weeks in 10 hyperlipidemic patients. *2577* Significant reduction in 27 patients, half of whom had hypertriglyceridemia, after 1 week. *1121* In both normals with hypertriglyceridemia and diabetics. *1120*

Colestipol   Therapeutic intent. *0939*

Dextrothyroxine   Therapeutic effect. *0071*

Doxazosin   Increase by 13% in 44 patients treated for 3 mo. *0088* In 38 treated hypertensives given 1 to 16 mg/d over 10 weeks. *2562*

Ethanol   Initially low immediately after alcoholic debauch. *1011*

Fenfluramine   (not constant finding). *3505*

Fenofibrate   Increase by 38% over 24 weeks in 92 type IIa given 100 mg 3 times per day with similar responses in type IIb patients when compared with baseline values. *0503* Increase by 50% in 9 hypertriglyceridemic type 2 diabetic patients given 100 mg tid after 1 mo treatment. *3385*

Gemfibrozil   Increase by 54% in 6 patients with primary familial endogenous hypertriglyceridemia from baseline values. Synthetic rates of apo A-I and apo A-II increased by 27% and 34% respectively. *3117* Increase by 43% in extensive 5 year double-blinded trial with 2,000 men receiving 600 mg drug twice daily. *1193*

Glucagon   Transfer of lipids to platelets. *0584*

Glyburide   Fall by 50% in treated diabetics. *0125*

Guanfacine   Increase by 15% in 30 patients treated for 2 y. *0088*

Halofenate   Reduced by up to 50%. *2483*

Heparin   Prompt decrease when small dose given i.v. *1343*

Levonorgestrel   Increase by 32 % in 11 normolipoproteinemic women given 250 µg/d for 2 weeks. *3593* Significant reduction in patients over 3 y compared with patients with IUDs for 2 1/2 y. *0763*

Lovastatin   Increase by 14% at 40 mg bid in clinical trial of 101 patients with heterozygous familial hypercholesterolemia. *1530*

Medroxyprogesterone   Significant reduction. Effects observed in 15 postmenopausal women with endometrial cancer after 2 weeks treatment. *2118* Increase by 24% in 11 men with sexual deviation syndrome given approximately 1273 mg over a total of approximately 17 d. *0643*

Menotropins   Marked fall in type 2 hyperlipidemia. *2235*

Metformin   Average of 26% in type i.v. hyperlipoproteinemia. *1430*

Methandrostenolone   Metabolic (anabolic) effect. *3477*

Nafenopin   More effective than clofibrate. *0939*

Neomycin   Consistent but insignificant decrease in concentration in 20 subjects with type ii hyperlipoproteinemia given 2 g/d for 9 mo. *1618*

Niacin   Therapeutic intent. *2736* Increase by 27% in 34 hypercholesterolemic individuals with 1.5 g/d for 1 mo then 3.0 g/d up to 6 mo. *1967*

Nifedipine   In 23 patients over 60 y old with essential mild to moderate hypertension. *3192* In 100 patients in a double-blind randomized trial. *3965*

Nisoldipine   From 1.9 to 1.6 mmol/L in 14 mild to moderately hypertensive noninsulin dependent diabetic patients. *2653*

Norethindrone   In normal and hyperlipemic women. *1309*

Oxandrolone   Increases triglyceride hydrolysis peripherally. *1308*

Oxymetholone   Anabolic effect. *0319*

Physical Training   56 mg/dL versus 85 in untrained (on average). *1702*

Prazosin   In 22 mild/moderate hypertensives treated for 8 weeks versus controls (average effect 9.5%). *1693* Increase by 16% in 23 healthy hypertensives treated for 8 weeks. *2136* Increase by 20 mg/dL in 16 patients given 1 to 3 mg daily for 8 weeks. *2550* Effect observed with doses as low as 1 to 3 mg/d. *3148* From 10 to 22% in several studies involving 25 to 50 patients for up to 6 mo. *0089*

Probucol   In 50 diabetics given 500 mg/d for 16 weeks and reduction greatest in highest cholesterol and triglyceride patients. *1523*

Quingestanol   May be slight effect with 300 µg/d. *1309*

Race   Significantly less in blacks at all ages. *0286*

Recumbency   Decrease of 12% maximum on lying down. *3539* 11% drop at 20 minimum in normal. *3540*

Spironolactone   Increase by 58 mg/dL in 11 men simultaneously with starting diet to reduce cholesterol. *0090* Average fell from 2.4 to 2.0 mmol/L in 15 primary hypertension patients at 6 mo after 100 mg drug daily. *1072*

Sulfonylureas   Effect reported for some compounds. *0301*

Plasma   No Effect   Analytical

Caffeine   No effect on enzymatic methods at therapeutic concentrations. *1786*

Triglycerides, High Density Lipoprotein

Serum   No Effect   Physiological

Atenolol   In 20 patients with 50 mg/d compared with pre-treatment values after 3 mo. *3071*

Chlorthalidone   In 22 premenopausal and 18 postmenopausal women given 100 mg/d for 6 weeks. *0402*

Ethanol   No effect of ingestion of 1.5 g/kg at night on concentration measured next morning. *2545*

Etretinate   In 13 patients with hyperkeratotic disorders given 50 mg daily for 2 mo. *3659*

Glutethimide   No significant effect with 21 d of 500 mg daily in 6 volunteers. *0407*

Isotretinoin   From 0.28 to 0.26 mmol/L in 12 patients with hyperkeratotic disorders given 40 mg daily for 2 mo. *3659*

Levonorgestrel   Insignificant increase in 11 normolipoproteinemic women given 250 µg/d for 2 weeks. *3593*

Metoprolol   In 18 hypertriglyceridemic-hypertensives after 12 weeks. *0341* In 20 hypertensives given 200 mg/d for 3 mo. *3071*

Ursodeoxycholic Acid   In 8 normolipemic patients receiving 1,000 mg drug daily. *2121*

Serum   Increase   Physiological

Atenolol   34% increase reported from one study. *0088*

Estradiol   By about 25% with 2 mg estradiol valerate/d in 20 normolipoproteinemic women over 3 mo. *3593*

Isotretinoin   From 0.14 to 0.31 mmol/L at 6 weeks in 7 patients with severe rosacea treated with 1 mg/kg/d. *2307* In 7 patients with severe rosacea treated with 1 mg/kg/d for 12 weeks. Effects possibly due to induction of hepatic microsomal enzymes. *2307*

Oral Contraceptives   Increase by 17 to 81% depending on preparation used. *2860*

Phenytoin In both men and women in 27 patients with transient brain ischemia. *1879*
Propranolol From 21 to 32 mg/dL in 15 hypertriglyceridemic hypertensives after 12 weeks. *0341*

Serum Decrease Physiological
Chenodeoxycholic Acid In 8 normolipemic patients receiving 1 g drug daily. *2121*
Nifedipine In 11 patients with 80 mg daily for 6 weeks. *3702*

Triglycerides, Low Density Lipoprotein

Serum No Effect Physiological
Atenolol In 20 patients with 50 mg/d compared with pre-treatment values after 3 mo. *3071*
Chlorthalidone In 22 premenopausal and 18 postmenopausal women given 100 mg/d for 6 weeks. *0402*
Estradiol Insignificant increase with 2 mg estradiol valerate/d in 20 normolipoproteinemic women over 3 mo. *3593*
Glutethimide No significant effect with 21 d of 500 mg daily in 6 volunteers. *0407*
Metoprolol Rose from 1 to 1.29 mmol/L in 20 hypertensives given 200 mg/d for 3 mo. *3071*
Phenytoin In both men and women in 27 patients with transient brain ischemia. *1879*
Propranolol No change in 15 hypertriglyceridemic hypertensives after 12 weeks. *0341*
Ursodeoxycholic Acid In 8 normolipemic patients receiving 1,000 mg drug daily. *2121*

Serum Increase Physiological
Atenolol From 0.46 to 0.51 mmol/L in 15 hypertensives treated for 8 mo. *1021* Slight increase reported from one study. *0088*
Ethanol In response to 1.5 g/kg at night significantly increases values measured next morning. *3547*
Isotretinoin From 0.45 to 0.61 mmol/L in 12 patients with hyperkeratotic disorders given 40 mg daily for 2 mo. *3659* In 7 patients with severe rosacea treated with 1 mg/kg/d for 12 weeks. Effects possibly due to induction of hepatic microsomal enzymes. *2307* From 0.26 to 0.56 mmol/L at 6 weeks in 7 patients with severe rosacea treated with 1 mg/kg/d. *2307*
Metoprolol From 24 to 32 mg/dL after 12 wek in 18 hypertriglyceridemic hypertensives. *0341*
Oral Contraceptives Increase by 42 to 60% depending on preparation used. *2860*
Tamoxifen From 0.46 to 0.56 mmol/L accountable for much of total change. *3070*

Serum Decrease Physiological
Chenodeoxycholic Acid In 8 normolipemic patients receiving 1 g drug daily. *2121*
Clonidine Small effect in 59 patients with primary hypertension given up to 300 mg daily for 6 mo. *1874*
Levonorgestrel By about 14 % in 11 normolipoproteinemic women given 250 μg/d for 2 weeks. *3593*

Triglycerides, Very Low Density Lipoprotein

Serum No Effect Physiological
Atenolol In 20 patients with 50 mg/d compared with pre-treatment values after 3 mo. *3071*
Chenodeoxycholic Acid In 8 normolipemic patients receiving 1 g drug daily. *2121*
Chlorthalidone In 22 premenopausal women, without decrease in blood pressure treated for less than 1 y. *0402* In 22 premenopausal and 18 postmenopausal women given 100 mg/d for 6 weeks. *0402*
Cimetidine In healthy individuals and in subjects with high serum glucose concentration. *1653*
Estradiol Insignificant increase with 2 mg estradiol valerate/d in 20 normolipoproteinemic women over 3 mo. *3593*

Estradiol-17β Insignificant change in 38 healthy postmenopausal women given 2-4 mg orally or 3 mg cutaneously over 6 mo. *1067*
Ethanol No effect of ingestion of 1.5 g/kg at night on concentration measured next morning. *3547*
Etretinate In 13 patients with hyperkeratotic disorders given 50 mg daily for 2 mo. *3659*
Glutethimide No significant effect with 21 d of 500 mg daily in 6 volunteers. *0407*
Metoprolol Non-statistically significant increase from 203 to 232 mg/dL after 12 weeks in 18 hypertriglyceridemic hypertensives. *0341* In 20 hypertensives given 200 mg/d for 3 mo. *3071*
Phenytoin In both men and women in 27 patients with transient brain ischemia. *1879*
Spironolactone No significant change in 17 subjects treated with drug for less than 1 y. *1405*
Ursodeoxycholic Acid In 8 normolipemic patients receiving 1,000 mg drug daily. *2121*
Verapamil In 12 patients with angina or hypertension treated for 6 weeks. Where change occurred it was of order of 10%. *3756*

Serum Increase Physiological
Atenolol 48% effect observed in one study. *0088* From 1.21 to 1.62 mmol/L in 15 hypertensives treated for 8 mo. *1021* From 0.90 to 1.14 mmol/L with 100 mg drug/d. *2192*
Bisoprolol Significant effect with 10 mg/d (from 1.04 to 1.31 mmol/L). *2192*
Ethanol Main fraction affected by acute ingestion. *2428*
Furosemide 56% change in 12 normotensive men when treated with drug alone for less than 1 y. *1826* 6% change in 16 subjects treated for less than 1 y with drug only. *1306*
Hydrochlorothiazide Observed with 100 mg/d. *3663*
Isotretinoin From 0.43 to 0.77 mmol/L at 6 weeks in 7 patients with severe rosacea treated with 1 mg/kg/d. *2307* From 1.34 to 1.81 mmol/L in 12 patients with hyperkeratotic disorders given 40 mg daily for 2 mo. *3659* In 7 patients with severe rosacea treated with 1 mg/kg/d for 12 weeks. Effects possibly due to induction of hepatic microsomal enzymes. *2307*
Metoprolol 36% in one study. *0088*
Oral Contraceptives Increase by 16 to 52% depending on preparation used. *2860*
Oxprenolol In several studies of 1-4 mo. *0088* Significant effect in 53 patients given 80 mg/twice daily for 3 mo. *0835*
Piretanide Effect observed with 12 mg/d. *3663*
Propranolol Significant effect in 53 hypertensives given 80 mg twice daily for 3 mo. *0835* Significant increase from 184 mg/dL to 308 mg/dL after 12 weeks in 15 hypertriglyceridemic hypertensives. *0341*

Serum Decrease Physiological
Benfluorex Increase by 36% in 12 hypertriglyceridemic type 2 diabetic patients given 150 mg tid after 1 mo treatment. *3385*
Ethanol Initially low immediately after alcoholic debauch. *1011*
Fenofibrate Increase by 60% in 9 hypertriglyceridemic type 2 diabetic patients given 100 mg tid after 1 mo treatment. *3385*
Gemfibrozil Increase by 50% in 18 patients with chronic renal failure treated with 1200 mg/d for 28 weeks. Simultaneous activation of postheparin plasma lipoprotein and hepatic lipases. Effects reversed when drug discontinued. *2741*
Levonorgestrel Increase by 45% in 11 normolipoproteinemic women given 250 μg/d for 2 weeks. *3593*
Pindolol 35% reduction in concentration after 2 mo in 11 hypertensive patients but after 16 mo treatment had reverted to normal. *1399*

Trihydroxyphenylacetate

Urine Increase Physiological
Levodopa Observed in treatment of Parkinsonian patients. *3137*

Tri-iodothyronine (T₃)

Serum   No Effect   Analytical

Blood Collection Tube   No effect on concentration even with storage for 6 d in gel-barrier tubes. *1592*

Carbamazepine   At therapeutic and toxic concentrations on RIA methods. *0374*   No effect observed at both therapeutic and toxic concentrations when added to serum *in vitro*. *0374*

Diphenhydramine   At acute overdose concentration (20 mg/dL) on Technicon® SMAC® method. *3719*

Storage of Sample   Negligible effect frozen for 2 y. *2853*

Serum   No Effect   Physiological

Amiodarone   Nonsignificant reduction after 2 mo to lower normal range. *1316*   In 5 healthy subjects given 600 mg daily for 2 weeks. *1764*

Caffeine   No significant effect of single dose of 250 mg in men or women. *3418*   In women no significant effect of single dose of 500 mg. *3418*

Cimetidine   In 6 patients treated for 1 mo with 1 g/d. *3491*   No significant changes caused by drug. *2780*

Erythrosine   In 30 men receiving up to 200 mg daily for 15 d. *1241*

Famotidine   No effect of short-term or long-term treatment in male patients with duodenal ulcers. *0727*

Indomethacin   In up to 35 healthy men given 100 mg/d for 1 week. *2904*

Iodoamide   In 17 euthyroid men following arteriography for up to 14 weeks, although initial slight reduction. *1030*

Isotretinoin   No significant change in 24 healthy men given 1 mg/kg/d for 16 weeks. *2640*

Lead   In some men occupationally exposed to large quantities of lead. *0771*

Levonorgestrel   No effect observed in group of women 20 and 65 mo after levonorgestrel treated rods inserted in uterus. *0899*   In a group of women using levonorgestrel versus control group with copper IUD. *0899*

Nadolol   No effect over 2 weeks in 10 healthy volunteers. *2938*

Omeprazole   No effect observed in 8 volunteers given 30 mg/d for 28 d. *2516*

Orphenadrine   Reported in two studies. *3798*

Primidone   In 5 epileptic patients treated on average for 6 mo. *2168*   Associated with enzyme induction. *3798*

Propranolol   In 15 euthyroid hypertensives treated for 30 d given 80 to 480 mg/d. *2016*   In 6 patients on large daily doses of drug (480 ± 155 mg) without clinical evidence of hyperthyroidism, due to drug-induced blockage of iodothyronine deiodination. *0722*   In 5 hyperthyroid patients given 10 mg every 8 h for 4 d. *0335*

Ranitidine   No effect of treatment on values. *2780*   After 4 weeks or 6 mo treatment (300 mg and 150 mg daily respectively) in male patients with duodenal ulcer. *0727*

Season   No seasonal variation observed. *2853*

Sodium Bromide   No effect in 10 men, 10 women receiving 1 mg/kg/d during 8 weeks or 2 full menstrual cycles. *3141*

Timolol   No effect in hyperthyroid patients. *0334*   In 5 hyperthyroid patients given 10 mg every 8 h for 4 d. *0335*

Tobramycin   When given i.v. with cloxacillin to 13 patients. *0959*

Tri-iodothyronine   In 7 normal women aged 18-27 y given 25 µg/d three times daily for 14 d. Effect observed on endothelium-associated proteins but not on hepatically synthesized proteins. *1383*

Valproic Acid   In 10 epileptic patients treated average of 8 mo. *2168*

Serum   Increase   Analytical

Cellulose Acetate   Inhibits binding of T₄ to thyroxine binding globulin. *3153*

Fenoprofen   Doubled in 2 volunteers over 2 weeks as also observed in some patients due to metabolite cross-reacting with antisera from Amersham and to lesser extent with that from Corning. *3594*

L-Thyroxine   33% cross reactivity if product from Cyclo. *3247*

Tri-iodothyroacetic Acid   29% cross reactivity method of Sedadde. *3247*

Serum   Increase   Physiological

Amiodarone   From 120 to 138 ng/dL in 13 patients treated for average of 17 mo. *0425*

Amphetamine   In 17% of patients admitted to a psychiatric hospital with many having ingested amphetamine. *2489*

Benziodarone   Significant effect in 9 normal volunteers receiving 100 mg three times daily for 14 d due to diversion of peripheral metabolism of T₄ to rT₃ rather than T₃. *3907*

Eggplant   Increased binding capacity (augmentation of transport proteins). *3798*

Estrogens   Increased binding capacity (augmentation of transport proteins). *3798*

Fluorouracil   Increased binding capacity (augmentation of transport proteins) *3798*

Halofenate   Associated with competition for transport proteins. *3798*

Hemodialysis   Probably effect of heparin. *1816*

Heparin   Interferes with thyroxine binding to protein. *1816*

Heroin   Slight (not significant) increase in addicts. *0193*   Increased binding capacity (augmentation of transport proteins). *3798*

Heroin Withdrawal   Significant increase observed compared with normals. *0193*

Insulin   Mean increase from 1.86 to 2.51 µg/L at 45 minutes after injection. *3570*

Mestranol   From 2.22 to 2.72 nmol/L in 19 post-menopausal women receiving mean of 24 µg/d versus controls. *0007*

Methadone   Significant increase observed compared with normals. *0193*   Increased binding capacity (augmentation of transport proteins). *3798*

Oral Contraceptives   Similar values to those in late first trimester. *3403*   Increased binding capacity (augmentation of transport proteins). *3798*

Phenytoin   In 10 women receiving anticonvulsants, mainly phenytoin versus 10 controls, drug taken on average for 15 y. *0257*

Prostaglandin F2α   Peak within 10-48 h up to 6 times original. *1718*

Prostaglandins   Probably direct thyroid effect. *3798*

Ranitidine   Small increase noted in ulcer patients after 150 mg twice daily for 28 d. *1603*

Rifampin   In one patient with primary hypothyroidism receiving constant replacement dose of L-thyroxine: tri-iodothyronine clearance retarded. *1734*

Somatotropin   Or no change associated with accelerated conversion. *3798*

Tamoxifen   In 2 of 10 postmenopausal patients with breast cancer with increased thyroxine concentration. *1350*

Terbutaline   From 136 to 160 ng/dL on average in 6 normal men after 2 weeks treatment. *3177*   In 6 normal men after 15 mg/d for 2 weeks with average of 160 ng/dL versus 136 ng/dL in controls. *3177*

Thyrotropin-Releasing Hormone (TRH)   But no change in percent free T₃. *1631*

L-Thyroxine   From 0.7 to 17 nmol/L 11 hypothyroid women treated with 0.1 to 0.2 mg daily. *1201*

Tri-iodothyronine   Marked increase in obese with 150 µg/d. *0458*

Serum   Decrease   Analytical

Kwashiorkor   Effect on Wien lab test possible related to decrease protein. *3674*

Serum   Decrease   Physiological

Amiodarone   Reduced peripheral conversion from thyroxine in 5% patients. *2011*   Mean concentration of 1.89 nmol/L in 38 patients receiving mean 1420 mg weekly for more than 9 mo. *3142*   In 9 men treated for 28 d caused mean decrease of 28 ng/dL. *0534*   Suggestive decrease in 15 euthyroid volunteers given 400-600 mg/daily for 2 weeks. *0092*

Androgens   Decreased binding capacity (diminution of transport proteins). *3798*

Anticonvulsants   1.6 nmol/L versus 1.9 nmol/L in women and 1.7 nmol/L versus 2.0 nmol/L in men on long-term treatment. *0257*

Asparaginase   From 0.99 to 0.35 ng/mL within 3 weeks in 14 children. *1543*

Aspirin In up to 35 healthy volunteers given 3.6 g daily for 1 week. *2904*

Carbamazepine Fall from mean of 1.6 to 1.4 nmol/L with 400 mg/d in healthy males. *0708* Observed in 9 hypothyroid patients given substitution therapy due to increased extra-thyroidal metabolism of thyroid hormones. *0001*

Carbimazole From 4.55 to 1.68 nmol/L in 12 hyperthyroid women patients treated with 10-30 mg daily. *1201*

Cimetidine From average of 1.49 to 1.25 nmol/L after 1.2 g for 5 d in 8 healthy male volunteers reduces response to TRH. *1683*

Clomiphene Probably direct thyroid inhibition. *3798*

Colestipol Reduced reabsorption of thyroid hormones from gastrointestinal tract. *3798*

Cotrifamole Possible intrathyroidal inhibition of conversion of iodine to organic form. *3798* From 2.50 to 2.17 nmol/L in men and 1.90 to 1.65 nmol/L in women for 10 d. *0686*

Co-Trimoxazole Marked decrease possibly due to intrathyroidal inhibition of conversion of iodine to organic form. *2022* From 2.33 to 1.96 nmol/L in men and 1.76 to 1.53 nmol/L in women after 10 days treatemtn. *0686*

Danazol In healthy postmenopausal women after 2 weeks treatment with 400 to 800 mg daily through direct effect on thyroxine binding globulin production at cellular level. *0985* In 12 healthy female post-menopausal volunteers given up to 800 mg daily. *0985*

Dexamethasone 8 mg/d for 2 d reduced concentration in normal and hypothyroid patients. *0860*

Fenclofenac In 7 out of 8 individuals receiving 1.2 g/d values below normal range. Thyroxine displaced from binding protein. *2971* From 6.2 nmol/L to 3.8 nmol/L in 4 women with thyrotoxicosis given 1.2 g/d for 7 d. *2758*

Furosemide At same time as thyroxine concentration reduced. *3473*

Glucocorticoids Typically with inhibition of conversion. *3798*

Iobenzamic Acid Impaired conversion of T_4 to T_3. *3798*

Iodides Associated with inhibition of conversion. *3798*

Iopanoic Acid Inhibition of conversion of thyroxine to tri-iodothyronine. *3798*

Ipodate Inhibition of T_4 to T_3 conversion. *3798*

Isotretinoin In 7 patients with severe rosacea treated with 1 mg/kg/d for 12 weeks. Effects possibly due to induction of hepatic microsomal enzymes. *2307* Significant at 12 weeks in 7 patients with severe rosacea treated with 1 mg/kg/d. *2307*

Lithium Reduces thyroidal iodine uptake, iodination of tyrosine, release of T_4, to T_3 and hepatic metabolism of T_4 to T_3. *3118* In 10 of 237 patients with smaller increases in others treated for more than 6 mo: normalized in most cases on withdrawal. *0084*

Neomycin Mean decrease from 104 to 92 ng/dL when given 2.0 g/d for 7 d orally. *0959*

Netilmicin Mean decrease from 114 ng/dL to 75 ng/dL when combined with cloxacillin given i.v., probably because of increased clearance. Free T_3 proportion increased. *0959*

Oral Contraceptives Due to increase in thyroxine binding globulin. *3175*

Phenytoin Slight effect observed in 50% patients at dose of 3 mg/kg/d with normal serum concentration. *1293* With long-term treatment due to accelerated thyroxine clearance via stimulation of hepatic microsomal enzymes. *2168*

Potassium Iodide In normals by 15 ng/dL over 11 d. *3658*

Prednisone Significant decrease with 1-2 mg/kg/d for 2-4 weeks in 10 children. *3353* Significant effect in 10 children given 1-2 mg/kg/h for 2-4 weeks. *3353*

Propranolol In hyperthyroid patients possible effect due to membrane-stabilizing property of drug. *0334* In 12 hyperthyroid patients given 160 mg/d for 2 weeks due to effect on peripheral metabolism. *1514* In hyperthyroid patients given 40 mg every 6 h in 6 patients. *0335* In thyrotoxic patients due to effect of drug on peripheral conversion of thyroxine to tri-iodothyronine. *1174* In 10 healthy volunteers treated for 2 weeks. *2938*

Propylthiouracil In hyperthyroid patients and normal controls. *0334* Significant reduction in 5 hyperthyroid individuals given 10 mg every 8 h for 4 d. *0335*

Radiographic Agents In first few days following oral cholecystographic agents. *2084*

Salicylate Due to competition for transport proteins. *3798*

Smoking Significant reduction in heavy smokers (181.0 ng/dL versus 204.0 ng/dL in Controls). *3255*

Somatostatin With infusion probable inhibition of TSH release. *3798*

Stanozolol From 2.21 to 1.52 nmol/L after 1 week in 9 healthy men given 10 mg/d for 14 d. *3354*

Tyropanoic Acid Impaired conversion of T_4 to T_3. *3798*

Urine   Increase   Physiological
Thyrotropin Pronounced increase with single i.m. injection. *0631*

Tri-iodothyronine (T_3), Free

Serum   No Effect   Physiological
Fenclofenac Values at lower end of normal range in individuals receiving 1.2 g/d. *2971*

Propranolol In 15 euthyroid hypertensives treated for 30 d given 80 to 480 mg/d. *2016*

Terbutaline In 6 normal men treated with therapeutic amounts for 2 weeks. *3177*

Serum   Increase   Analytical
Carbamazepine As measured by addition of drug to control sera and measured by analog radioimmunoassay. *2169*

Fenoprofen Tripled in 2 volunteers over 2 weeks as also observed in some patients due to metabolite cross-reacting with antisera from Amersham and to lesser extent with that from Corning. *3594*

Phenytoin As measured by addition of drug to control sera and measured by analog radioimmunoassay. *2169*

Serum   Increase   Physiological
Aspirin Interferes with binding to thyroxine binding globulin. *2083*

Heparin Probably acts on thyroxine binding globulin to reduce binding affinity. *1577*

Ranitidine Similar small effect noted as for total hormone concentration:. *1603*

Serum   Decrease   Physiological
Amiodarone In group treated for 6 mo 10 out of 28 had values below normal range. *1173* Common but not invariable response. *2327*

Carbamazepine As measured by equilibrium dialysis and analog-type radioimmunoassays; not due to displacement from protein. *2169* Observed in 9 hypothyroid patients given substitution therapy due to increased extra-thyroidal metabolism of thyroid hormones. *0001*

Phenytoin As measured by equilibrium dialysis and analog-type radioimmunoassays: not due to displacement from binding proteins. *2169*

Propranolol In 8 healthy volunteers given 80 mg bid and followed serially, fell by average of 1.2 pmol/L. *3833* Slight effect in euthyroid patients due to peripheral inhibition of peripheral deiodination of thyroxine and on binding protein metabolism. *1174*

Somatostatin With infusion probable inhibition of TSH release. *3798*

Tri-iodothyronine (T_3) Index, Free

Serum   No Effect   Physiological
Smoking No significant difference between heavy smokers and controls. *3255*

Serum   Increase   Physiological
Amphetamine In 14% of patients admitted to a psychiatric hospital with many having ingested amphetamine. *2489*

Serum   Decrease   Physiological
Fenclofenac In 7 out of 8 individuals receiving 1.2 g/d values below normal range. Thyroxine displaced from binding protein. *2971*

Methimazole From 512 to 368 after 4 weeks treatment with 30 mg daily. *0192*

Tri-iodothyronine (T$_3$), Reverse

Serum   No Effect   Physiological
Carbamazepine   No significant effect with 400 mg/d for 21 d. *0708*   No significant change in patients on long-term treatment. *2168*
Cimetidine   No significant difference in drug treated healthy volunteers: no difference in response to TRH. *1683*
Erythrosine   In 30 men receiving up to 200 mg daily for 15 d. *1241*
Iodoamide   In 17 euthyroid men following arteriography for up to 14 weeks. *1030*
Nadolol   No effect over 2 weeks in 10 healthy volunteers. *2938*
Neomycin   When given 2.0 g/d orally for 7 d. *0959*
Netilmicin   No effect when given i.v. with cloxacillin. *0959*
Phenytoin   With long-term treatment due to accelerated thyroxine clearance via stimulation of hepatic microsomal enzymes. *2168*
Primidone   In 5 epileptic patients treated on average for 6 mo. *2168*
Terbutaline   In 6 normal men treated with therapeutic amounts for 2 weeks. *3177*
Tobramycin   When given i.v. with cloxacillin to 13 patients. *0959*
Valproic Acid   In 10 epileptic patients treated average of 8 mo. *2168*

Serum   Increase   Physiological
Amiodarone   Exceeded normal range after 2 mo in all subjects. *1316*   Significant rise in mean from 0.59 to 1.47 nmol/L in 5 healthy subjects given 600 mg daily for 2 weeks. *1764*   Mean concentration of 0.85 nmol/L in 38 patients receiving mean 1420 mg weekly for more than 9 mo. *3142*   In 9 men treated for 28 d caused increase of 83 ng/dL. *0534*   Significant effect in 15 euthyroid volunteers given 400-600 mg/daily for 2 weeks. *0092*
Benziodarone   Significant effect in 9 normal volunteers receiving 100 mg three times daily for 14 d due to diversion of peripheral metabolism of T$_4$ to rT$_3$ rather than T$_3$. *3907*
Cimetidine   Basal concentration increased by treatment due to drug effect on hepatic metabolism. *2780*
Prednisone   Considerable rise with 1-2 mg/kg/d for 2-4 weeks in 10 children. *3353*
Propranolol   In 6 patients on large daily doses of drug (480 ± 155 mg) without clinical evidence of hyperthyroidism, due to drug-induced blockage of iodothyronine deiodination. *0722*   In euthyroid patients due to peripheral inhibition of peripheral deiodination of thyroxine and on binding protein metabolism. *1174*   In 10 healthy volunteers treated for 2 weeks. *2938*
Radiographic Agents   In first few days following oral cholecystographic agents. *2084*

Serum   Decrease   Physiological
Fenclofenac   From 0.63 nmol/L to 0.52 nmol/L in 4 women with thyrotoxicosis given 1.2 g/d for 7 d. *2758*
Insulin   Fell from mean of 0.184 to 0.171 µg/L but not significant after i.v. injection. *3570*
Prednisone   Significant effect in 10 children given 1-2 mg/kg/h for 2-4 weeks. *3353*

Tri-iodothyronine (T$_3$), Reverse Free

Serum   Increase   Physiological
Propranolol   In 8 healthy volunteers given 80 mg bid and followed serially increased by average of 0.16 nmol/L. *3833*

Trimethoprim

Serum   Decrease   Physiological
Cholestyramine   Delayed absorption due to adsorption. *3794*

Urine   Increase   Physiological
Acid Urine   If co-trimoxazole administered. *0745*

Triosephosphate Isomerase

Red Blood Cells   No Effect   Analytical
Storage of Sample   With Alsever's solution for 3-5 weeks at 30°. *0877*

Trypsin

Serum   Increase   Physiological
Methylprednisolone   Increase by 35-109% in 10 patients given 1 g/d i.v. for 3 d. Maximum effect days 5 to 8 possible subclinical damage of pancreatic acinar cell. *0804*
X-Ray Therapy   May be up to 4 fold increase with total body irradiation. *1837*

Serum   Decrease   Physiological
Anabolic Steroids   Mean decrease from 7.5 to 5.3 mU/dL in patients with chronic pancreatitis after weeks treatment with combination of anabolic steroids. *3617*

Duodenal Contents   Increase   Physiological
Amino Acids   Increase by 200% with duodenal infusion, no effect if given i.v. *2918*
Calcium   Calcium infusion (10 mg/kg/h) causes increase x 2. *1314*

Feces   No Effect   Analytical
Storage of Sample   For 8 d at room temperature, though better stored at 4°. *0877*

Feces   Decrease   Analytical
Diatrizoic acid   Spectrophotometric assay -inhibits enzyme and absorbs. *0741*
Radiographic Agents   Inhibition of esterase by Gastrografin®. *0741*

Trypsin Inhibitor

Serum   Increase   Physiological
Anabolic Steroids   Metabolic effect. *0227*   Mean increase from 401 to 526 mU/dL in patients with chronic pancreatitis after 3 weeks treatment with combination of anabolic steroids. *3617*
Estrogens   Metabolic effect. *0227*

Trypsinogen

Serum   No Effect   Physiological
Smoking   No significant difference in basal immunoreactive cationic trypsinogen between smokers and nonsmokers. *0961*

Tryptamine

Urine   Increase   Physiological
MAO Inhibitors   Due to utilization of alternative pathways. *1343*

Urine   Decrease   Physiological
Carbidopa   Marked effect due to decarboxylase inhibition. *0494*

Tryptophan

Plasma   No Effect   Physiological
Food   No significant increase for 1 h after a meal. *0099*

Plasma   Decrease   Analytical
Light   Affects fluorometric method of Dencla and Dewey. *2114*
Storage of Sample   If picric acid used for deproteinization. *3230*

Plasma   Decrease   Physiological
Alclofenac   (total measured) in rheumatoids on therapy. *0188*

Aspirin (total measured) in rheumatoids on therapy. *0188*
Glucose After 75 g nonprotein bound fell by 35% in 3 h. *2188*
Indomethacin (total measured) in rheumatoids on therapy. *0188*
Stress Cold Exposure Although total amino acids unaffected. *1169*

Urine Increase Physiological
Tetracycline Unexplained mechanism. *1343*

Urine Decrease Analytical
Light Affects fluorometric method of Dencla and Dewey. *2114*

CSF No Effect Physiological
Anticonvulsants Insignificant reduction to 325 ng/mL from 379 ng/mL in untreated epileptics. *3926*

Test Conditions Increase Analytical
Acetyltryptophan Measured as same by Spies/Chambers method. *1415*
Indoleacetic Acid Measured as same as Spies/Chambers method. *1415*
Indolepropionic Acid Measured as same as Spies/Chambers method. *1415*
Methyltryptophan Measured as same in Spies/Chambers method. *1415*
Skatole Measured as same in Spies/Chambers method. *1415*
Tryptamine Measured as same in Spies/Chambers method. *1415*
Tryptophan-Thiol Measured as same in Spies/Chambers method. *1415*

Tryptophan, Free

Serum Increase Physiological
Aspirin In rheumatoids on therapy. *0188*
Indomethacin In rheumatoids on therapy. *0188*

Plasma Increase Physiological
Alclofenac In rheumatoids on therapy. *0188*

TSH Response to TRH

Serum No Effect Physiological
Ethanol No effect of acute alcohol intake. *3798*
Indomethacin In up to 35 healthy men given 100 mg/d for 1 week. *2904*
Insulin Hypoglycemia had no effect on TRH-stimulated TSH release. *3798*
Isotretinoin In 7 patients with severe rosacea treated with 1 mg/kg/d for 12 weeks. Effects possibly due to induction of hepatic microsomal enzymes. *2307*
Prednisone No significant effect with 1-2 mg/kg/d for 2-4 weeks in 10 children. *3353*

Serum Increase Physiological
Amiodarone Maximal increment during treatment 32 mU/mL compared with 20 mU/mL pretreatment. *0534*
Erythrosine Increase from peak increment of 6.3 to 10.5 μU/mL after 15 d on 200 mg/d in 30 men but no significant effect at lower doses. *1241*
Indomethacin In up to 35 healthy men given 100 mg/d for 1 week. *2904*
Phenytoin Major stimulation when patients given 3 mg/kg/d with normal serum concentration. *1293*

Serum Decrease Physiological
Amiodarone In 16 of 44 patients, majority of whom were clinically hyperthyroid: normalized with withdrawal of drug. *3436*
Aspirin In up to 35 healthy volunteers given 3.6 g daily for 1 week. *2904*
Growth Hormone Attributable to T_3 increment through acceleration of T_4 conversion. *3798*

Ranitidine In response to i.v. drug: may cause decrease of basal concentration in hypothyroidism. *1938*

TTC Test

Urine Positive Analytical
Ascorbic Acid Oral ingestion causes reduction of TTC = positive test. *3910*

Tubular Maximum for Phosphate

Urine No Effect Physiological
Ethinyl Estradiol In postmenopausal women with 50 μg/d for 14 d. *3474*

Tyramine

Urine Increase Physiological
MAO Inhibitors Prevent deamination. *1343*

Tyramine Test

Patient Increase Physiological
Guanethidine Increased response to tyramine due to MAO inhibition. *1325*
Hydralazine Increased responsiveness. *1488*
MAO Inhibitors Enhanced responsiveness. *1325*
Methyldopa Enhanced responsiveness reported. *1488*

Patient Decrease Physiological
Reserpine Inhibits responsiveness (produces false negative). *1325*
Thiazides May attenuate response to tyramine. *1034*

Tyrosine

Plasma No Effect Analytical
Alanine On fluorometric procedure of Ambrose. *0080*
β-Alanine On fluorometric procedure of Ambrose. *0080*
α-Amino-N-Butyric Acid On fluorometric procedure of Ambrose. *0080*
β-Amino-Isobutyric Acid On fluorometric procedure of Ambrose. *0080*
δ-Amino Valeric Acid On fluorometric procedure of Ambrose. *0080*
Arginine On fluorometric procedure of Ambrose. *0080*
Asparagine On fluorometric procedure of Ambrose. *0080*
Aspartic Acid On fluorometric procedure of Ambrose. *0080*
Citruplexina On fluorometric procedure of Ambrose. *0080*
Creatine On fluorometric procedure of Ambrose. *0080*
Creatinine On fluorometric procedure of Ambrose. *0080*
Cysteine On fluorometric procedure of Ambrose. *0080*
Diiodotyrosine On fluorometric procedure of Ambrose. *0080*
Ethionine On fluorometric procedure of Ambrose. *0080*
Glutamic Acid On fluorometric procedure of Ambrose. *0080*
Glycine On fluorometric procedure of Ambrose. *0080*
Histamine On fluorometric procedure of Ambrose. *0080*
Histidine On fluorometric procedure of Ambrose. *0080*
Homocitrulline On fluorometric procedure of Ambrose. *0080*
Homocysteic Acid On fluorometric procedure of Ambrose. *0080*
Homocystine On fluorometric procedure of Ambrose. *0080*
Homoserine On fluorometric procedure of Ambrose. *0080*
Isoleucine On fluorometric procedure of Ambrose. *0080*
Kynurenine On fluorometric procedure of Ambrose. *0080*
Leucine On fluorometric procedure of Ambrose. *0080*
Levoglutamide On fluorometric procedure of Ambrose. *0080*

Tyrosine (continued)

Plasma   No Effect   Analytical (continued)
Lysine On fluorometric procedure of Ambrose. *0080*
Methionine On fluorometric procedure of Ambrose. *0080*
Norleucine On fluorometric procedure of Ambrose. *0080*
Norvaline On fluorometric procedure of Ambrose. *0080*
Ornithine On fluorometric procedure of Ambrose. *0080*
Phenylalanine On fluorometric procedure of Ambrose. *0080*
Proline On fluorometric procedure of Ambrose. *0080*
Serine On fluorometric procedure of Ambrose. *0080*
Storage of Sample At -20° for 6 mo. *0080* With fluoride added stable 7 d at 30° 4 d at 30°, stable indefinitely frozen. *1712*
Taurine On fluorometric procedure of Ambrose. *0080*
Threonine On fluorometric procedure of Ambrose. *0080*
Tryptamine On fluorometric procedure of Ambrose. *0080*
Valine On fluorometric procedure of Ambrose. *0080*

Plasma   No Effect   Physiological
Levodopa In patients after 1 week given levodopa. *1108*

Plasma   Increase   Analytical
Aspirin Interferes with fluorometric method of Scott. *3228*
Hemolysis Substantially increased values occur. *0080*
5-Hydroxyindoleacetic Acid 21% fluorescence in procedure of Ambrose. *0080*
2-Hydroxyphenylacetic Acid 8% fluorescence in procedure of Ambrose. *0080*
3-hydroxyphenylacetic Acid 10% fluorescence in procedure of Ambrose. *0080*
4-Hydroxyphenylacetic Acid 170% fluorescence in procedure of Ambrose. *0080*
4-Hydroxyphenylpyruvic Acid 20% fluorescence in procedure of Ambrose. *0080*
5-Hydroxytryptophan 6% fluorescence in procedure of Ambrose. *0080*
Storage of Sample Above 3°, ?hydrolysis of protein. *0080*
Tryptophan 2% fluorescence in procedure of Ambrose. *0080*
Tyramine 190% fluorescence in procedure of Ambrose. *0080*
2-Tyrosine 10% fluorescence in procedure of Ambrose. *0080*
3-Tyrosine 16% fluorescence in procedure of Ambrose. *0080*

Plasma   Increase   Physiological
Diurnal Variation Maximum at about 10 am, minimum about 2-4 am. *2222*
Obesity Metabolic effect (?mechanism). *0843*
Tri-iodothyronine Slight increase in obese with 150 μg/d. *0458*

Plasma   Decrease   Physiological
Androgens In 24 h in boys with delayed puberty. *0843*
Ascorbic Acid Reduces elevated level in premature infants. *1343*
Diethylstilbestrol Observed with 5 mg daily for 5 d. *0843*
Epinephrine Hormonal effect (?exact mechanism). *0843*
Ethinyl Estradiol Induction of tyrosine transaminase by 0.1 mg/d steroid. *3046*
Glucagon Hormonal effect (?exact mechanism). *0843*
Glucose Hormonal effect (?exact mechanism). *0843*
Histidine Probably due to flow into gut after oral load. *1644*
Hydrocortisone Hormonal effect observed in 2 weeks. *0843*
Mesterolone Hormonal effect (?exact mechanism). *0843*
Mestranol Effect of both estrogen and progestogen. *0843*
Oral Contraceptives Hormonal effect (second part of cycle). *0742*
Pregnancy Lowest values (normally higher in women than men). *0843*
Sleep 18% decrease during sleep. *3952*
Stress Cold Exposure Although total amino acids unaffected. *1169*
Testosterone Fall observed in 2 weeks. *0843*

Urea Clearance

Urine   No Effect   Physiological
Meals No significant effect of meals observed. *0459*
Metrifonate No effect reported on renal function. *2427*

Urine   Increase   Physiological
Diuretics May be observed after use of diuretics. *0459*

Urine   Decrease   Physiological
Diurnal Variation Clearance least in morning. *0459*
Erect Posture Decrease observed after rising. *0459*
Muscular Exercise Response to stress of exercise. *0582*

Urea Nitrogen

Serum   No Effect   Analytical
Acetaminophen as routine methods in use on SMAC®, Abbott-VP, Cobas-Bio, Ektachem® 400, Hitachi® 705, KDA. *2138* At acute overdose concentration (10 mg/dL) on SMAC® method. *3719* No effect at therapeutic concentration on Reflotron method. *1984* At concentration of 1500 mg/L had no effect on diacetylmonoxime method. *3393*
Acetazolamide At concentration of 1,000 mg/L had no effect on diacetylmonoxime method. *3393*
Acetylsalicylic Acid At therapeutic concentrations on glutamate dehydrogenase, phenol-hypochlorite and diacetyl monoxime methods. *1786* On diacetyl monoxime method on SMA II at physiological concentration. *1787* At 5 times upper limit of therapeutic range on method on SMAC®, Abbott-VP, Cobas-Bio, Ektachem®, Hitachi® 705 and KDA. *2138* With 8326 μmol/L on method using glutamate dehydrogenase. With 8326 μmol/L on method using phenol hypochlorite. With 8326 μmol/L on method using diacetylmonoxime. *1786* No effect at therapeutic concentration on Reflotron method. *1984* With 8326 μmol/L on method using diacetylmonoxime. *1786* At concentration of 2,000 mg/L had no effect on diacetylmonoxime method. *3393* At concentration of 100 mg/L had no effect on Ektachem® method. *3393* At concentration of 50 mg/L had no effect on urease-Berthelot method. *3393*
Alanine No effect on Berthelot reaction. *1862*
Allopurinol At concentration of 136 mg/L had no effect on diacetylmonoxime method. *3393* At concentration of 18 mg/L had no effect on Ektachem® method. *3393*
Aminophenazone At therapeutic concentrations no effects on glutamate dehydrogenase, phenol-hypochlorite, diacetylmonoxime methods. *1786* With 540 μmol/L on glutamate dehydrogenase method. With 540 μmol/L on phenol-hypochlorite method. With 540 μmol/L on diacetyl monoxime method. *1785*
Aminophenol At concentration of 109 mg/L had no effect on diacetylmonoxime method. *3393*
Aminopyrine At concentration of 125 mg/L had no effect on diacetylmonoxime method. *3393*
Aminosalicylic Acid At 5 times upper limit of therapeutic range on methods on SMAC®, Abbott-VP, Cobas-Bio, Ektachem® and Hitachi® 705. *2138* At concentration of 460 mg/L had no effect on diacetylmonoxime method. *3393*
Amitriptyline At acute overdose concentration (2.5 mg/dL) on SMAC® method. *3719* At concentration of 25 mg/L had no effect on diacetylmonoxime method. *3393* At concentration of 6.3 mg/L had no effect on Ektachem® method. *3393*
Amobarbital At concentration of 150 mg/L had no effect on diacetylmonoxime method. *3393*
Amphetamine At therapeutic concentration on Reflotron method. *1984*
Amphotericin B At concentration of 96 mg/L had no effect on diacetylmonoxime method. *3393*
Ampicillin No effect on Reflotron method at therapeutic concentration. *1984* At concentration of 40 mg/L had no effect on diacetylmonoxime method. *3393* At concentration of 180 mg/L had no effect on Ektachem® method. *3393*
Arginine No effect on Berthelot reaction. *1862*
Ascorbic Acid No effect at therapeutic concentration on Reflotron method. *1984* At concentration of 2,000 mg/L had no effect on diacetylmonoxime method. *3393*

Aspirin   No effect on Berthelot procedure. *1862*

Azapropazone   At concentration of 36 mg/L had no effect as measured by Ektachem® method. *3393*

Barbital   At acute overdose concentration (20 mg/dL) on Technicon® SMAC® method. *3719*   At concentration of 500 mg/L had no effect as measured by diacetylmonoxime method. *3393*

Benzbromarone   At concentration of 8 mg/L had no effect as measured by Ektachem® method. *3393*

Bezafibrate   No effect at therapeutic concentration on Reflotron method. *1984*

Bilirubin   No effect on Berthelot reaction. *1862*   With DMAB procedure of Morin below 3 mg/dL. *2487*

Bisacodyl   At concentration of 0.2 mg/L had no effect as measured by Ektachem® method. *3393*

Boric Acid   At concentration of 50 mg/L had no effect as measured by diacetylmonoxime method. *3393*

Bromazepam   At 5 times therapeutic concentration on methods of SMAC®, Abbott-VP, Cobas-Bio, Ektachem® 400, Hitachi® 705 and KDA. *2138*

Butabarbital   At concentration of 100 mg/L had no effect as measured by diacetylmonoxime method. *3393*

Butizide   At concentration of 0.2 mg/L had no effect as measured by Ektachem® method. *3393*

Caffeine   No effect at therapeutic concentrations on glutamate dehydrogenase, phenol- hypochlorite, diacetylmonoxime methods. *1786*   At acute overdose concentration (10 mg/dL) on Technicon® SMAC® method. *3719*   At therapeutic concentration on Reflotron method. *1984*   At concentration of 194 mg/L had no effect on diacetylmonoxime method. *3393*

Carbenicillin   At 5 times upper limit of therapeutic range on methods on SMAC®, Abbott-VP, Cobas-Bio, Ektachem® 400, Hitachi® 705 and KDA. *2138*   At concentration of 15,000 mg/L had no effect on diacetylmonoxime method. *3393*

Carbochromen   No effect at therapeutic concentration on Reflotron method. *1984*

Cefamandole   At 2.50 mmol/L on method on Eppendorf Epos. *1414*

Cefazedone   On method on Eppendorf Epos at concentration of 2.50 mmol/L. *1414*

Cefazolin   At 2.50 mmol/L on method on Eppendorf Epos. *1414*   At concentration of 117 mg/L had no effect on diacetylmonoxime method. *3393*

Cefodizime   No effect at 2.50 mmol/L on method on Eppendorf Epos. *1414*

Cefoperazone   At 2.50 mmol/L on method on Eppendorf Epos. *1414*

Cefotaxime   No effect at 2.50 mmol/L on method on Eppendorf Epos. *1414*

Cefotiam   At 2.50 mmol/L on method on Eppendorf Epos. *1414*

Cefoxitin   At 5 times upper limit of therapeutic range on methods on SMAC®, Abbott-VP, Cobas-Bio, Ektachem®, Hitachi® 705, and KDA. *2138*   At concentrations up to 2.50 mmol/L on methods on Eppendorf Epos. *1414*

Cefpirome   At concentration of up to 2.50 mmol/L on methods of Eppendorf Epos. *1414*

Ceftriaxone   On method on Eppendorf Epos at concentration of 2.50 mmol/L. *1414*

Cefuroxime   On method on Eppendorf Epos at concentration of 2.50 mmol/L. *1414*

Cephaloridine   At 2.50 mmol/L on method on Eppendorf Epos. *1414*

Cephalothin   At 5 times upper limit of therapeutic range on methods on SMAC®, Abbott-VP, Cobas-Bio, Ektachem®, Hitachi® 705 and KDA. *2138*   No effect a 2.50 mmol/L on method on Eppendorf Epos. *1414*   At concentration of 1,000 mg/L had no effect on diacetylmonoxime method. *3393*

Chloramphenicol   No effect at therapeutic concentration on Reflotron method. *1984*   At concentration of 600 mg/L had no effect on Ektachem® method. *3393*   At concentration of 2500 mg/L had no effect on urease/Berthelot method. *3393*

Chlordiazepoxide   At 5 times upper limit of therapeutic range on methods on SMAC®, Abbott-VP, Cobas-Bio, Ektachem®, Hitachi® 705 and KDA. *2138*   At acute overdose concentration (5 mg/dL) on Technicon® SMAC® method. *3719*   No effect at therapeutic concentration on Reflotron method. *1984*   At concentration of 50 mg/L had no effect on diacetylmonoxime method. *3393*   At concentration of 1.8 mg/L had no effect on Ektachem® method. *3393*

Chlormezanone   At concentration of 100 mg/L had no effect on diacetylmonoxime method. *3393*

Chloroquine   At concentration of 5 mg/L had no effect on Ektachem® method. *3393*

Chlorphenesin   At concentration of 150 mg/L had no effect on diacetylmonoxime method. *3393*

Chlorpheniramine   At acute overdose concentration (20 mg/dL) on Technicon® SMAC® method. *3719*   At concentration of 200 mg/L had no effect on diacetylmonoxime method. *3393*

Chlorpromazine   At concentration of 3 mg/L had no effect on diacetylmonoxime method. *3393*   At concentration of 0.1 mg/L had no effect on Ektachem® method. *3393*

Chlorpropamide   At concentration of 150 mg/L had no effect on diacetylmonoxime method. *3393*

Chlorprothixene   At concentration of 1 mg/L had no effect on diacetylmonoxime method. *3393*

Chromonar Hydrochloride   At concentration of 18 mg/L had no effect on Ektachem® method. *3393*

Cimetidine   At concentration of 1 mg/L had no effect on diacetylmonoxime method. *3393*

Cloxacillin   At concentration of 5 mg/L had no effect on diacetylmonoxime method. *3393*

Cocaine   At acute overdose concentration (2.5 mg/dL) on Technicon® SMAC® method. *3719*   At concentration of 25 mg/L had no effect on diacetylmonoxime method. *3393*

Codeine   At acute overdose concentration (2.0 mg/dL) on Technicon® SMAC® method. *3719*   At concentration of 20 mg/L had no effect on diacetylmonoxime method. *3393*

Contact With Clot   If plastic disc used to separate for 48 h. *1863*

Creatinine   No effect on Berthelot procedure. *1862*

Cyclophosphamide   At concentration of 8 mg/L had no effect on Ektachem® method. *3393*

Cyclosporine   At concentration of 41.5 mmol/L (50 µg/L) on methods on Technicon® SMAC® II and Hitachi® 705. *1123*

Cysteine   At 5 times upper limit of therapeutic range on methods SMAC®, Abbott-VP, Cobas-Bio, Ektachem®, Hitachi® 705 and KDA. *2138*

Cystine   At 5 times upper limit of therapeutic range on methods on SMAC®, Abbott-VP, Cobas-Bio, Ektachem®, Hitachi® 705 and KDA. *2138*

Desipramine   At concentration of 8 mg/L had no effect on diacetylmonoxime method. *3393*

Dextran   Not precipitated in DMAB procedure of Morin. *2487*

Dextran 40   At concentration of 10,000 mg/L had no effect on Ektachem® method. *3393*

Dextran 60   At concentration of 1800 mg/L had no effect on Ektachem® method. *3393*

Diatrizoic acid   At concentration of 1300 mg/L had no effect on Ektachem® method. *3393*

Diazepam   At 5 times upper limit of therapeutic range on methods on SMAC®, Abbott-VP, Cobas-Bio, Ektachem®, Hitachi® 705 and KDA. *2138*   At acute overdose concentration (2.5 mg/dL) on Technicon® SMAC® method. *3719*   At concentration of 25 mg/L had no effect on diacetylmonoxime method. *3393*   At concentration of 4 mg/L had no effect on Ektachem® method. *3393*

Diclofenac   No effect at therapeutic concentrations with glutamate dehydrogenase, phenol hypochlorite and diacetylmonoxime procedures. *1786*   On diacetyl monoxime complexone method on SMA II at physiological concentration. *1787*

Diethylaminoethanol   At concentration of 30 mg/L had no effect on diacetylmonoxime method. *3393*

Diethylpropion   At acute overdose concentration (10 mg/dL) on Technicon® SMAC® method. *3719*   At concentration of 100 mg/L had no effect on diacetylmonoxime method. *3393*

Digitoxin   At concentration of 6 mg/L had no effect on diacetylmonoxime method. *3393*

Urea Nitrogen (continued)

Serum   No Effect   Analytical (continued)

Digoxin   At concentration of 9 mg/L had no effect on diacetylmonoxime method. *3393*   At concentration of 0.015 mg/L had no effect on Ektachem® method. *3393*

Dimethadione   At concentration of 2,000 mg/L had no effect on diacetylmonoxime method. *3393*

N-Dimethyldiazepam   At concentration of 6 mg/L had no effect on diacetylmonoxime method. *3393*

Diphenhydramine   At acute overdose concentration (20 mg/dL) on Technicon® SMAC® method. *3719*   At concentration of 200 mg/L had no effect on diacetylmonoxime method. *3393*

Disopyramide   At concentration of 4 mg/L had no effect on diacetylmonoxime method. *3393*

Disulfiram   At concentration of 120 mg/L had no effect on diacetylmonoxime method. *3393*

Disulphine Blue   By continuous-flow diacetyl monoxime method. *1464*

Doxepin   At concentration of 6 mg/L had no effect on Ektachem® method. *3393*

EDTA   No effect on analytical methods reported. *1563*   At concentration of 100 mg/L had no effect on Ektachem® method. *3393*

Epinephrine   At concentration of 183 mg/L had no effect on diacetylmonoxime method. *3393*

Erythromycin   At concentration of 73 mg/L had no effect on diacetylmonoxime method. *3393*

Ethanol   At acute overdose concentration (20 mg/dL) on Technicon® SMAC® methods. *3719*

Ethaverine   No effect at therapeutic concentration on Reflotron method. *1984*

Ethchlorvynol   At acute overdose concentration (20 mg/dL) on Technicon® method. *3719*   At concentration of 400 mg/L had no effect on diacetylmonoxime method. *3393*

Ethinamate   At acute overdose concentration (20 mg/dL) on Technicon® SMAC® method. *3719*   At concentration of 200 mg/L had no effect on diacetylmonoxime method. *3393*

Ethosuximide   At concentration of 390 mg/L had no effect on diacetylmonoxime method. *3393*

Fasting   No effect with 96 h fasting in 5 healthy volunteers. *2323*

Fluorides   No effect on urease if concentration less than 2 mg/ml. *0279*

Fluorouracil   At concentration of 130 mg/L had no effect on diacetylmonoxime method. *3393*

Fluosol-DA   On SMA IIC at concentration of 50%. *2518*   By 18% on Kodak Ektachem® 400 at concentration of 50%. *2518*

Flurazepam   At 5 times upper limit of therapeutic range on methods on SMAC®, Abbott-VP, Cobas-Bio, Ektachem®, Hitachi® 705 and KDA. *2138*   At acute overdose concentration (2.5 mg/dL) on Technicon® SMAC® method. *3719*   At concentration of 25 mg/L had no effect on diacetylmonoxime method. *3393*

Furosemide   No effect at therapeutic concentration on Reflotron method. *1984*   At concentration of 4 mg/L had no effect on diacetylmonoxime method. *3393*   At concentration of 40 mg/L had no effect on Ektachem® method. *3393*

Gentamicin   At concentration of 150 mg/L had no effect on diacetylmonoxime method. *3393*

Glibenclamide   No effect at therapeutic concentration on Reflotron method. *1984*   At concentration of 0.3 mg/L had no effect on Ektachem® method. *3393*

Glucosamine   No effect on Berthelot reaction. *1862*

Glutethimide   At acute overdose concentration (5 mg/dL) on Technicon® SMAC® method. *3719*   At concentration of 50 mg/L had no effect on diacetylmonoxime method. *3393*

Guanidine   No effect on Berthelot reaction. *1862*

Hemolysis   DMAB procedure of Morin, Hemoglobin below 100 mg/dL. *2487*   Calculated error less than 1%/g. *0515*

Histidine   No effect on Berthelot reaction. *1862*

Hydralazine   At concentration of 160 mg/L had no effect on diacetylmonoxime method. *3393*

Hyoscine-N-Butylbromide   At concentration of 2 mg/L had no effect on Ektachem® method. *3393*

Ibuprofen   No effect at therapeutic concentrations on glutamate, dehydrogenase, phenol- hypochlorite, and diacetylmonoxime methods. *1786*   On diacetyl monoxime method on SMA II at physiological concentration. *1787*   At concentration of 200 mg/L had no effect on diacetylmonoxime method. *3393*   No effect at therapeutic concentrations on glutamate dehydrogenase, phenol- *1786*   On diacetylmonoxime method on SMA II at physiological concentration. *1787*   At concentration of 200 mg/L had no effect on diacetylmonoxime method. *3393*

Indomethacin   No effect at therapeutic concentrations on glutamate dehydrogenase, phenol- hypochlorite and diacetylmonoxime methods. *1786*   On diacetyl monoxime method on SMA II at physiological concentration. *1787*   No effect at therapeutic concentration on Reflotron method. *1984*   At concentration of 13 mg/L had no effect on diacetylmonoxime method. *3393*   At concentration of 4 mg/L had no effect on Ektachem® method. *3393*

Insulin   At concentration of 3 mg/L had no effect on diacetylmonoxime method. *3393*

Iproniazid   At concentration of 40 mg/L had no effect on diacetylmonoxime method. *3393*

Isoniazid   At concentration of 100 mg/L had no effect on diacetylmonoxime method. *3393*

Isoprenaline   At concentration of 211 mg/L had no effect on diacetylmonoxime method. *3393*

Ketoprofen   No effect at therapeutic concentrations on glutamate dehydrogenase, phenol- hypochlorite and diacetylmonoxime methods. *1786*   On diacetyl monoxime method on SMA II at physiological concentration. *1787*   At concentration of 60 mg/L had no effect on diacetylmonoxime method. *3393*

Levodopa   At concentration of 197 mg/L had no effect on diacetylmonoxime method. *3393*

Lidocaine   At concentration of 3.2 mg/L had no effect on Ektachem® method. *3393*

Liposol   Changed from 262 to 257 mg/L in 26 randomly selected patient sera, before and after addition on Beckman Astra methods. *2054*

Lorazepam   At concentration of 0.05 mg/L had no effect on diacetylmonoxime method. *3393*

Lysine   No effect on Berthelot reaction. *1862*

Meglumine   At concentration of 200 mg/L had no effect on Ektachem® method. *3393*

Meperidine   At acute overdose concentration (5 mg/dL) on Technicon® SMAC® method. *3719*   At concentration of 60 mg/L had no effect on diacetylmonoxime method. *3393*

Mephenesin   At concentration of 100 mg/L had no effect on diacetylmonoxime method. *3393*

Meprobamate   At acute overdose concentration (20 mg/dL) on Technicon® SMAC® method. *3719*   At concentration of 200 mg/L had no effect on diacetylmonoxime method. *3393*   At concentration of 484 mg/L had no effect on Ektachem® method. *3393*

Mercaptomerin   At concentration of 58 mg/L had no effect on diacetylmonoxime method. *3393*

Mercaptopurine   At concentration of 152 mg/L had no effect on diacetylmonoxime method. *3393*   At concentration of 15 mg/L had no effect on Ektachem® method. *3393*

Mesoridazine   At acute overdose concentration (20 mg/dL) on Technicon® SMAC® method. *3719*

Methadone   At acute overdose concentration (2.5 mg/dL) on Technicon® SMAC® method. *3719*   At concentration of 25 mg/L had no effect on diacetylmonoxime method. *3393*

Methamphetamine   At concentration of 2 mg/L had no effect on diacetylmonoxime method. *3393*

Methanol   At acute overdose concentration (20 mg/dL) on Technicon® SMAC® method. *3719*

Methapyrilene   At concentration of 13 mg/L had no effect on diacetylmonoxime method. *3393*

Methaqualone   At acute overdose concentration (2.5 mg/dL) on Technicon® SMAC® method. *3719*   No effect at therapeutic concentration on Reflotron method. *1984*   At concentration of 25 mg/L had no effect on diacetylmonoxime method. *3393*

Methicillin   At 5 times upper limit of therapeutic range on methods on SMAC®, Abbott-VP, Cobas-Bio, Ektachem®, Hitachi® 705 and KDA. *2138*   At concentration of 900 mg/L had no effect on diacetylmonoxime method. *3393*

Methohexital   At concentration of 50 mg/L had no effect on diacetylmonoxime method. *3393*

Methotrexate At 5 times upper limit of therapeutic range on method on SMAC®, Abbott-VP, Cobas-Bio, Ektachem®, Hitachi® 705 and KDA. *2138* At concentration of 1100 mg/L had no effect on Ektachem® method. *3393*

Methotrimeprazine At concentration of 1 mg/L had no effect on diacetylmonoxime method. *3393*

Methsuximide At concentration of 40 mg/L had no effect on diacetylmonoxime method. *3393*

Methyldopa No effect at therapeutic concentration on Reflotron method. *1984* At concentration of 211 mg/L had no effect on diacetylmonoxime method. *3393* At concentration of 80 mg/L had no effect on Ektachem® method. *3393*

Methylphenidate At acute overdose concentration (20 mg/dL) on Technicon® SMAC® method. *3719* At concentration of 200 mg/L had no effect on diacetylmonoxime method. *3393*

Methylphenobarbital At concentration of 41 mg/L had no effect on Ektachem® method. *3393*

Methyprylon At acute overdose concentration (10 mg/dL) on Technicon® SMAC® method. *3719* At concentration of 100 mg/L had no effect on diacetylmonoxime method. *3393*

Metoprolol At concentration of 0.34 mg/L had no effect on diacetylmonoxime method. *3393*

Morphine At acute overdose concentration (20 mg/dL) on Technicon® SMAC® method. *3719* At concentration of 200 mg/L had no effect on diacetylmonoxime method. *3393*

Moxalactam At 2.50 mmol/L on method on Eppendorf Epos. *1414* At concentration of 96 mg/L had no effect on diacetylmonoxime method. *3393*

Nafcillin At concentration of 50 mg/L had no effect on diacetylmonoxime method. *3393*

Naproxen At 5 times upper limit of therapeutic range on methods on SMAC®, Abbott-VP, Cobas-Bio, Ektachem®, Hitachi® 705 and KDA. *2138*

Niacin No effect at therapeutic concentration on Reflotron method. *1984*

Niacinamide At concentration of 12 mg/L had no effect on Ektachem® method. *3393*

Nitrofurantoin No effect at therapeutic concentration on Reflotron method. *1984* On routine methods in use on SMAC®, Ektachem®, Abbott-VP, Cobas-Bio, aca, Hitachi® 705, KDA at 5 times normal upper therapeutic concentration. *2138* At concentration of 4 mg/L had no effect on diacetylmonoxime method. *3393* At concentration of 5 mg/L had no effect on Ektachem® method. *3393*

Norfenefrin At concentration of 0.48 mg/L had no effect on Ektachem® method. *3393*

Nortriptyline At acute overdose concentration (20 mg/dL) on Technicon® SMAC® method. *3719* At concentration of 200 mg/L had no effect on diacetylmonoxime method. *3393*

Novaminsulfon At concentration of 80 mg/L had no effect on Ektachem® method. *3393*

Orphenadrine At concentration of 9 mg/L had no effect on diacetylmonoxime method. *3393*

Oxacillin At concentration of 40 mg/L had no effect on diacetylmonoxime method. *3393*

Oxazepam No effect at therapeutic concentration on Reflotron method. *1984* At concentration of 1 mg/L had no effect on diacetylmonoxime method. *3393*

Oxyphenbutazone At concentration of 12 mg/L had no effect on Ektachem® method. *3393*

Oxytetracycline No effect at therapeutic concentration on Reflotron method. *1984* At concentration of 50 mg/L had no effect on diacetylmonoxime method. *3393*

Papaverine At concentration of 10 mg/L had no effect on diacetylmonoxime method. *3393*

Paraldehyde At concentration of 2,000 mg/L had no effect on diacetylmonoxime method. *3393*

Penicillamine At concentration of 36 mg/L had no effect on Ektachem® method. *3393*

Penicillin G At 5 times upper limit of therapeutic range on methods on SMAC®, Abbott-VP, Cobas-Bio, Ektachem®, Hitachi® 705 and KDA. *2138* At concentration of 2,000 mg/L had no effect as measured by diacetylmonoxime method. *3393*

Pentobarbital At acute overdose concentration (20 mg/dL) on Technicon® SMAC® method. *3719* At concentration of 340 mg/L had no effect on diacetylmonoxime method. *3393*

Perphenazine At acute overdose concentration (20 mg/dL) on Technicon® SMAC® method. *3719* At concentration of 200 mg/L had no effect on diacetylmonoxime method. *3393*

Phenazopyridine No effect at therapeutic concentration on Reflotron method. *1984*

Phencyclidine At concentration of 6 mg/L had no effect on diacetylmonoxime method. *3393*

Phenelzine At concentration of 136 mg/L had no effect on diacetylmonoxime method. *3393*

Phenobarbital No effect at therapeutic concentrations on glutamate dehydrogenase, phenol- hypochlorite and diacetylmonoxime methods. *1786* At acute overdose concentration (20 mg/dL) on Technicon® SMAC® method. *3719* No effect at therapeutic concentration on Reflotron method. *1984* At concentration of 250 mg/L had no effect on diacetylmonoxime method. *3393* At concentration of 30 mg/L had no effect on Ektachem® method. *3393*

Phenothiazines At concentration of 20 mg/L had no effect on Ektachem® method. *3393*

Phenprocoumon No effect at therapeutic concentration on Reflotron method. *1984* At concentration of 0.36 mg/L had no effect on Ektachem® method. *3393*

Phensuximide At concentration of 120 mg/L had no effect on diacetylmonoxime method. *3393*

Phenylalanine No effect on Berthelot reaction. *1862*

Phenylbutazone At concentration of 750 mg/L had no effect on diacetylmonoxime method. *3393* At concentration of 12 mg/L had no effect on Ektachem® method. *3393*

Phenylephrine At concentration of 4 mg/L had no effect on diacetylmonoxime method. *3393*

Phenytoin At acute overdose concentration (20 mg/dL) on Technicon® SMAC® method. *3719* No effect at therapeutic concentration on Reflotron method. *1984* At concentration of 240 mg/L had no effect on diacetylmonoxime method. *3393* At concentration of 36 mg/L had no effect on Ektachem® method. *3393*

Plasma No significant difference from serum by SMA 12/60. *2228*

Primidone At concentration of 2.4 mg/L had no effect on Ektachem® method. *3393*

Probenecid No effect at therapeutic concentration on Reflotron method. *1984* At concentration of 1300 mg/L had no effect on diacetylmonoxime method. *3393* At concentration of 40 mg/L had no effect on Ektachem® method. *3393*

Procainamide At concentration of 50 mg/L had no effect on diacetylmonoxime method. *3393* At concentration of 11.4 mg/L had no effect on Ektachem® method. *3393*

Procaine No effect at therapeutic concentration on Reflotron method. *1984* At concentration of 2 mg/L had no effect on diacetylmonoxime method. *3393*

Prochlorperazine At concentration of 1 mg/L had no effect on diacetylmonoxime method. *3393*

Promethazine At acute overdose concentration (20 mg/dL) on Technicon® SMAC® method. *3719* At concentration of 200 mg/L had no effect on diacetylmonoxime method. *3393*

Propoxyphene At acute overdose concentration (2.5 mg/dL) on Technicon® SMAC® method. *3719* At concentration of 25 mg/L had no effect on diacetylmonoxime method. *3393* At concentration of 1.8 mg/L had no effect on Ektachem® method. *3393*

Propranolol At concentration of 0.2 mg/L had no effect on diacetylmonoxime method. *3393*

Propylthiouracil At concentration of 170 mg/L had no effect on diacetylmonoxime method. *3393*

Pyribenzamine® At acute overdose concentration (20 mg/dL) on Technicon® SMAC® methods. *3719*

Pyridamole No effect at therapeutic concentration on Reflotron method. *1984*

Pyritinol No effect at therapeutic concentration on Reflotron method. *1984*

Quinidine At acute overdose concentration (20 mg/dL) Technicon® SMAC® method. *3719* No effect at therapeutic concentration on Reflotron method. *1984* At concentration of 210 mg/L had no effect on diacetylmonoxime method. *3393* At concentration of 5.6 mg/L had no effect on Ektachem® method. *3393*

Quinine At acute overdose concentration (1.5 mg/dL) on Technicon® SMAC® method. *3719* At concentration of 30 mg/L had no effect on diacetylmonoxime method. *3393*

Reserpine At concentration of 61 mg/L had no effect on diacetylmonoxime method. *3393*

Urea Nitrogen *(continued)*

Serum   No Effect   Analytical *(continued)*

Resorcinol   At concentration of 500 mg/L had no effect on Seralyzer method. *3393*

Rifampin   At 5 times upper limit of therapeutic range on method on SMAC®, Abbott-VP, Cobas-Bio, Ektachem®, Hitachi® 705 and KDA. *2138*

Salicylate   At concentration of 500 mg/L had no effect on diacetylmonoxime method. *3393*   At concentration of 133 mg/L had no effect on Ektachem® method. *3393*   At concentration of 300 mg/L had no effect on Seralyzer method. *3393*

Secobarbital   At acute overdose concentration (20 mg/dL) on Technicon® SMAC® method. *3719*   At concentration of 200 mg/L had no effect on diacetylmonoxime method. *3393*

Sodium Citrate   At concentration of 500 mg/L had no effect on Ektachem® method. *3393*

Sodium Heparin   At concentration of 75 mg/L had no effect on Ektachem® method. *3393*

Sodium Oxalate   At concentration of 300 mg/L had no effect on Ektachem® method. *3393*

Sodium Salicylate   At concentration of 350 mg/L had no effect on Ektachem® method. *3393*

Spironolactone   At concentration of 8 mg/L had no effect on Ektachem® method. *3393*

Standing of Sample   If on polystyrene beads 24 h at 25°. *2624*

Storage of Sample   3-5 day room temperature, prolonged storage at -10°. *1180*

Streptomycin   At concentration of 58 mg/L had no effect on diacetylmonoxime method. *3393*   At concentration of 1450 mg/L had no effect on urease/Berthelot method. *3393*

Strychnine   At concentration of 12 mg/L had no effect on diacetylmonoxime method. *3393*

Sulfadiazine   No effect on Berthelot procedure. *1862*   At concentration of 1500 mg/L had no effect on diacetylmonoxime method. *3393*   At concentration of 80 mg/L had no effect on urease/Berthelot method. *3393*

Sulfaguanidine   At concentration of 500 mg/L had no effect on diacetylmonoxime method. *3393*

Sulfamethoxazole   No effect at therapeutic concentration on Reflotron method. *1984*

Sulfamethoxydiazine   At concentration of 20 mg/L had no effect on Ektachem® method. *3393*

Sulfanilamide   At concentration of 1,000 mg/L had no effect on diacetylmonoxime method. *3393*

Sulfathiazole   At concentration of 255 mg/L had no effect on diacetylmonoxime method. *3393*   At concentration of 50 mg/L had no effect on Ektachem® method. *3393*

Sulforidazine   At concentration of 200 mg/L had no effect on diacetylmonoxime method. *3393*

Tetracycline   At 5 times upper limit of therapeutic range on methods on SMAC®, Abbott-VP, Cobas-Bio, Ektachem®, Hitachi® 705 and KDA. *2138*   At concentration of 300 mg/L had no effect on diacetylmonoxime method. *3393*

Theobromine   At concentration of 2,000 mg/L had no effect on diacetylmonoxime method. *3393*

Theophylline   At 5 times upper limit of therapeutic range on methods on SMAC®, Abbott-VP, Cobas-Bio, Ektachem®, Hitachi® 705 and KDA. *2138*   No effect at therapeutic concentration on Reflotron method. *1984*   At concentration of 2,000 mg/L had no effect on diacetylmonoxime method. *3393*   At concentration of 23 mg/L had no effect on Ektachem® method. *3393*

Thiamazole   At concentration of 114 mg/L had no effect on diacetylmonoxime method. *3393*

Thiopental   At acute overdose concentration (20 mg/dL) on Technicon® SMAC® method. *3719*   At concentration of 200 mg/L had no effect on diacetylmonoxime method. *3393*

Timolol   At concentration of 0.01 mg/L had no effect on diacetylmonoxime method. *3393*

Tolbutamide   At concentration of 270 mg/L had no effect on diacetylmonoxime method. *3393*   At concentration of 220 mg/L had no effect on Ektachem® method. *3393*

Transport of Specimen   By pneumatic tube if specimen protected. *3217*

Tribromethanol   At concentration of 90 mg/L had no effect on diacetylmonoxime method. *3393*

Trichlorethanol   At concentration of 1,000 mg/L had no effect on diacetylmonoxime method. *3393*

Trifluoperazine   At concentration of 1 mg/L had no effect on diacetylmonoxime method. *3393*

Trimethoprim   No effect at therapeutic concentration on Reflotron method. *1984*

Tripelennamine   At concentration of 200 mg/L had no effect on diacetylmonoxime method. *3393*

Vitamin B Complex   At concentration of 2.3 mg/L had no effect on Ektachem® method. *3393*

Vitamin Preparations   No effect at expected concentration with SMA 12/60 procedure. *2637*

Warfarin   At concentration of 100 mg/L had no effect on diacetylmonoxime method. *3393*

Serum   No Effect   Physiological

Acetaminophen   Without overdose not usually any effect. *2866*

Acetylsalicylic Acid   No effect seen in patients undergoing treatment. *1787*

Aging   No significant change from age 6-20 y. *0887*

Amiloride   No effect seen in 13 men treated for 8 weeks. *0656*

Aspirin   When approximately 3 g/d ingested for several weeks. *3076*   No effect although creatinine affected. *0545*

Bucindol   No effects noted in 8 patients following several weeks on treatment. *1283*

Chlorprothixene   Constant concentration regardless of dose of drug co-administered and duration. *3268*

Cimetidine   Unchanged although there may be slight increases in creatinine, not progressive: decreases with continued treatment. *2368*   No significant effect observed in 9 patients given 1.6 g daily. *0978*

Cromoglycate   No clinical significant change observed. *0484*

Etodolac   Fewer than 2% patients showed pattern of deviant renal function tests. *3272*

Glyburide   No significant effect observed. *0125*

Indapamide   In 8 hypertensives treated for 16 weeks. *0352*

Indomethacin   No effect seen in patients undergoing treatment. *1787*

Ketoprofen   No effect seen in patients undergoing treatment. *1787*

Labetalol   No significant effect in 15 patients with essential hypertension treated for 1 mo. *3776*

Levonorgestrel   No significant difference over 3 y between implant recipients and IUD users. *0763*   In a group of women using levonorgestrel covered rods versus controls using copper IUD. *0899*

Meals   No effect reported after standard breakfast. *0579*

Nisoldipine   In 14 mild to moderately hypertensive noninsulin dependent diabetic patients. *2653*

Propranolol   No significant effect on renal function in 15 patients with essential hypertension over 1 mo. *3776*   In approximately 120 patients with essential hypertension treated for 1 y. *3708*

Warfarin   No effect on renal function noted with continuing administration. *2411*

Serum   Increase   Analytical

Acetohexamide   Chemical structure produces reaction with diacetyl. *0652*

Acetone   Turbidity in Nesslerization procedures. *1563*

Alkalosis   With Azostix® compared to diacetyl. *1460*

Allantoin   Reacts with xanthydrol in method of Fosse. *1563*

Amino Acids   May react with Berthelot if urease not used. *2981*

Aminophenol   May react in Berthelot procedure. *1862*

Aminosalicylic Acid   Effect marked on DMAB method of Morin. *2487*   Reacts as urea with dimethylaminobenzaldehyde. *3505*   At upper limit of therapeutic range on o-phthaldehyde methoxyquinoline method of KDA. *2138*

Ammonium Salts   If urease used and ammonia measured. *1563*

Asparagine   Blue color with Berthelot's reagent. *1936*

Bilirubin   With DMAB method of Morin above 3 mg/dL. *2487*

Chloral Hydrate   Reacts with Nessler reaction. *3505*

Chloramphenicol   Reacts with Nessler reagent. *3505*

Chlorobutanol   False increase with direct Nesslerization. *1563*

Citruplexina   Absorbance relative to urea = 1.5 diacetyl procedure. *0583*

Creatine   May react with Berthelot if urease not used. *2981*

Creatinine   Increases concentrations affect Nesslerization methods slightly. *1563*

Dextran   Reacts as if urea with dimethylaminobenzaldehyde. *3505*

Guanethidine   Due to chemical similarity to urea. *0085*

Hemoglobin   May affect color of urease/Berthelot procedure. *2981*

Hemolysis   If Berthelot procedure, and wavelength less than 600 nm. *3314*   With method of Morin with hemoglobin above 100 mg/dL. *2487*   Color produced in protein-free filtrates. *0579*

Hydantoin Derivatives   Absorbance relative to urea = 0.3 diacetyl procedure. *0583*

Lipemia   May affect Berthelot reaction. *1862*

Methylurea   Absorbance relative to urea = 1.5 diacetyl procedure. *0583*

Oxalate   Balanced oxalate mixture contains NH_4. *1563*

Phenylurea   Absorbance relative to urea = 2.2 diacetyl procedure. *0583*

Sodium Fluoride   At concentrations above 160 mg/L raised concentration as measured by Ektachem® method. *3393*

Sulfamethoxazole   Effect marked DMAB method of Morin. *2487*

Sulfisoxazole   Effect marked DMAB method of Morin. *2487*

Sulfonamides   Reacts like urea with DMAB and Berthelot reactions. *3505*

Sulfonylureas   Theoretical effect of ureide group diacetyl method. *0583*

Tetracycline   At concentrations above 40 mg/L (normal therapeutic concentration 8 mg/L) raised concentration as measured by Ektachem® method. *3393*

Thiourea   Reacts with xanthydrol in method of Fosse. *1563*

Uric Acid   May react with Berthelot if urease not used. *2981*

Serum   Increase   Physiological

Acetaminophen   Reversible tubular necrosis reported. *0438*   Mild acute renal failure in two patients following therapeutic ingestion of drug. *1219*

Acetazolamide   May occur with prolonged treatment. *2220*

Acetic Acid   May cause nephrotoxicity. *0279*

Acyclovir   Transient rises reported in some patients. *0147*

Alanine   Significant effect of chronic administration in fasting. *1263*

Aldatense   Increase by 10-20% in a group of hypertensives compared with controls. *1611*

Alkaline Antacids   Prolonged use may cause nephrotoxicity. *2220*

Allopurinol   Azotemia as sensitivity reaction in 2 cases. *2451*   In 2 of 73 patients treated for 6 mo. *3565*

Alloxan   If ingested may cause nephrotoxicity. *0279*

Aloin   May cause nephrotoxicity. *0279*

Amantadine   Mild transient elevations reported. *3213*

Amiloride   May occur with excessive saluresis. *2427*

Amino Acids   As result of i.v. infusion. *1215*

Aminopyrine   Rare suggestion of nephrotoxic action. *2427*

Amphotericin B   Nephrotoxic effect. *2313*

Amyl Nitrite   May cause nephrotoxicity. *0279*

Anabolic Steroids   Improves nitrogen balance. *0652*

Androgens   Augments nitrogen balance. *0652*

Aniline   May cause nephrotoxicity (chronic effect). *0279*

Antimony Compounds   Nephrotoxic effect (common at therapeutic doses). *1022*

Antipyrine   May cause nephrotoxicity. *0279*

Apiol   May cause nephrotoxicity. *0279*

Arginine   Incorporated into urea cycle. *0071*

Arsenicals   Nephrotoxic effect (common with therapeutic doses). *1343*

Asparaginase   Occurs in 50% subjects (prerenal origin). *1519*   Increased in from 7.5 to 50% in different studies: usually mild, possibly due to increased availability of ammonia. *0558*

Aspidium   May cause nephrotoxicity. *0279*

Aspirin   May have nephrotoxic effect. *3678*   Anti-inflammatory doses produced substantial effect in 13 of 23 patients with systemic lupus erythematosus. Observed effect with long term low doses for secondary prevention of coronary heart disease. *2866*

Azapropazone   Initial 10% increase observed in 83 treated patients but thereafter no further increase *3565*

Bacitracin   Renal toxicity (especially if given i.v.). *1343*

Bendrofluazide   In men and women in 3 y study of patients with mild hypertension. *2506*

Benzthiazide   May occur as with other thiazides. *1680*

Beryllium Salts   May cause nephrotoxicity. *3204*

Bismuth Salts   Nephrotoxic effect reported, and even renal failure. *3204*

Blood Transfusions   Catabolism of infused protein. *1022*

Boric Acid   Toxicity effect with acute tubular necrosis. *0652*

Boxidine   Possible relationship to tumor destruction. *1220*

Bromate   Anuria and azotemia may occur within hours. *1302*

Bunamiodyl   May cause nephrotoxicity. *3204*

Busulfan   Renal damage may occur from urate deposition. *1343*

Cadmium   May cause nephrotoxicity. *3204*

Calcium Salts   Azotemia with milk-alkali syndrome. *1343*

Cannabis   Reported effect. *2220*

Cantharides   May cause nephrotoxicity. *3204*

Capreomycin   Nephrotoxic effect (in 6% subjects). *1343*

Captopril   Occasional reversible azotemia, either due to hypotension or direct renal damage. *0655*   Severe reversible azotemia in few patients with peripheral vascular disease two weeks after start of therapy, probably associated with GFR reduction. *0655*

Carbamazepine   May cause kidney dysfunction. *0071*

Carbenoxolone   Due to nephropathy with hypokalemia. *0232*

Carbon Disulfide   Renal impairment observed. *0818*

Carbon Monoxide   May cause nephrotoxicity. *3204*

Carbutamide   Nephrotoxic effect. *2313*

Castor Oil   May cause renal damage. *0279*

Cephaloridine   High doses, nephrotoxicity reported. *0085*

Cephalothin   Rare elevation noted (possible nephrotoxicity). *1019*

Chlorate   May cause renal damage. *0279*

Chlormerodrin   May produce nephrotic syndrome. *1188*

Chloroform   May cause renal damage. *0279*

Chlortetracycline   Dose related antianabolic action. *1680*

Chlorthalidone   Nephrotoxic effect. *0070*

Chromates   May cause renal damage. *0279*

Cimetidine   Single case of reversible renal failure probably attributable to drug. *3243*

Ciprofloxacin   Mild and transient cases reported although 1 case each of acute renal failure, interstitial nephritis and nonspecific nephritis reported in population of 2829 patients. *0151*

Cisplatin   Observed in one patient, but major side effect renal toxicity. *1194*   In 11 of 18 cases with doses of less than 50 mg/m^2 surface area: extent of change equivalent to 50% reduction in renal function. *0368*

Clonidine   Associated with decreased GFR at start of therapy. *2961*

Clorazepate   Mild elevation in 10% patients reported. *1878*

Codeine   May cause nephropathy. *2583*

Colistimethate   Nephrotoxic effect (usually reversible). *2313*

Colistin   Nephrotoxic effect/may occur after large doses. *2313*

Copper   May cause nephrotoxicity. *3204*

Co-Trimoxazole   Usually reversible effect (in many patients). *1852*   Uremia occurs if anemia, urine infection G-6-PD deficiency. *2694*

Cremomycin   Dehydration may simulate renal disease. *0071*

Cresol   May cause renal damage. *0279*

Urea Nitrogen (continued)

Serum Increase Physiological (continued)

Cyclosporine Mean change from 13 to 27 mg/dL in 8 of 23 patients with 2 to 6 mo treatment (patients with inflammatory ocular disease). *2713* Increased from 13 to 27 mg/dL in 8 of 23 patients with ocular inflammatory disease. *2713* Characteristic feature of nephrotoxicity in most common form of renal damage to drug. *0175*

Cyclothiazide May occur with decreased renal flow. *1680*

Demeclocycline Dose related nephrotoxicity. *2106*

Dextran Blocks tubules causing renal failure. *0647*

Diatrizoic acid Has caused acute renal failure with myelography. *2312*

Diazepam Slight elevation in one hypertensive. *1878*

Dichlorobenzene May cause renal damage. *0279*

Diclofenac Significantly higher in treated patients (6.87 mmol/L vs 5.32 mmol/L) than in controls. *1787*

Dinitrophenol Reported to cause renal damage. *1343*

Dioxane May cause renal damage. *0279*

Dipyrone Reported to cause anuria. *0071*

Diuretics Reduced clearance. *2745*

Diurnal Variation Maximum after midnight, minimum about 5 pm. *3428*

Doxapram ?nephrotoxic effect. *1022*

Doxycycline Nephrotoxic effect. *2802*

EDTA Nephrotoxicity (especially Calcium EDTA). *1237*

Electroshock May cause nephrotoxicity. *3204*

Enflurane In one patient with initially normal renal function following comparatively long exposure to anesthetic. Normal renal function eventually returned. *1001*

Ergot Preparations May cause renal damage with poisoning. *0279*

Ethacrynic Acid May cause deterioration if renal function impaired. *0085*

Ethambutol Rare renal damage reported. *0131*

Ether May cause nephrotoxicity. *3204*

Ethosuximide Possible reversible nephropathy. *0071*

Ethoxyethanol May cause nephrotoxicity. *3204*

Ethylene Glycol May cause nephrotoxicity. *3204* Due to tubular necrosis caused by toxic metabolites and possibly also due to deposition of calcium oxalate crystals in kidney. *3325*

Fava Beans May cause renal damage. *0279*

Fibrinolysin Renal failure reported (rare complication). *0071*

Flucytosine May have nephrotoxic effect. *1679*

Fluorides May cause renal damage. *0279*

Formaldehyde May cause renal damage if ingested. *0279*

Furosemide ?nephrotoxic effect (reversible) usually dehydration. *0085* In 9 patients with chronic failure when treated for 6 d average plasma urea rose from 18.7 mmol/L to 28.8 mmol/L secondary to reduced urea excretion which occurred in spite of increased urea filtration. *0792* Largely as consequence of volume depletion: effect most marked in patients receiving additional diuretics. *1391* In patients with chronic renal failure: associated with decreased renal clearance and increased tubular reabsorption of urea. *0792* In 8% of 204 hospitalized patients receiving the drug. *3419*

Gasoline May cause renal damage if ingested. *0279*

Gentamicin Nephrotoxic effect with large doses. *2220*

Glycyrrhiza May cause nephropathy. *2458*

Gold May cause nephrotoxicity. *3204*

Griseofulvin Rare renal damage reported. *0131*

Guanethidine Associated with blood pressure reduction, decreased blood flow. *1680*

Guanoclor May decrease renal blood flow. *1022*

Guanoxan Due to decreased blood flow. *2220*

Halothane Observed in postoperative period with norm renal function. *0369*

Hydralazine May cause nephrotoxicity. *3204*

Hydrochlorothiazide May occur with prolonged therapy. *2220* Increase by 2.7 mg/dL in 343 patients with hypertension given drug for 10 weeks to reduce diastolic BP to less than 90 mm Hg. *3707* Increase by 2.5 mg/dL in approximately 175 patients with essential hypertension treated for 1 y. *3708*

Hydroflumethiazide May occur especially if pre-existing renal disease. *1680*

Hydroxyurea Probably associated with tissue destruction. *2581* ?related to impaired tubular function. *1680*

Ibuprofen Significant effect in group of patients with rheumatoid arthritis, although most values remained within normal range. *0885* In 2 patients with systemic lupus erythematosus. *1155*

Ifosfamide Common toxic manifestation. *3675*

Imipramine Observed in one psychiatric case. *3156*

Immune Serum Globulin Nephritis may occur with serum sickness. *2427*

Indomethacin Occasional increase usually to upper normal limit. *2432* Oliguric renal failure developed in one man during treatment: reversible. *1245* Associated with increasing risk of renal insufficiency in cirrhosis, nephrotic syndrome, decompensated congestive heart failure and chronic renal disease. *0373* Associated with decreased secretion of renin and aldosterone, decreased sodium delivery to distal tubule and reduction of urinary flow rate. *1228*

Insecticides May cause nephrotoxicity. *3204*

Iodoalphionic Acid Reported to cause renal failure. *0065*

Ipodate Nephrotoxic effect. *0838*

Iron Salts May cause nephrotoxicity. *3204*

Isosorbide Dehydration with overdosage. *3315*

Kanamycin Nephrotoxic effect (common slight elevation). *0070*

Kerosene Renal damage with large doses. *1343*

Lead May cause nephrotoxicity. *3204*

Levodopa Affects hepatic enzymes, probably not dehydration. *2318*

Lipomul® Possible nephrotoxic effect. *0652*

Meals Affected by meals (most marked in women). *3455*

Mefenamic Acid Reported in one study with normal volunteers. *0071*

Melphalan Azotemia may occur. *0071*

Menopause Change of 1 mg/dL from 5th to 6th decade. *3830*

Mephenesin Fatal anuria with intravascular hemolysis. *1343*

Meralluride May cause transient neutropenia. *1680*

Mercurial Diuretics May produce renal failure or nephrotic syndrome. *1188*

Methacycline Possible nephrotoxicity. *0071*

Methanol May cause nephrotoxicity. *3204*

Methicillin May cause azotemia with nephrotoxicity. *0071*

Methotrexate May cause severe nephropathy, azotemia. *1680*

Methoxyflurane Impaired renal tubular function. *0749*

Methsuximide Renal damage reported. *0071*

Methyclothiazide Slight change only. *1353*

Methylbromide Late effect with tubular necrosis of poisoning. *1302*

Methyldopa ?due to decreased renal blood flow. *3879*

Methysergide Possible decrease in already impaired renal function. *0085*

Metolazone Diuretic action of drug acting on distal tubules. *2816*

Metoprolol Insignificant in 20 hypertensive diabetic patients. *3900*

Minocycline Antianabolic action of class of drugs. *1680*

Mithramycin Nephrotoxic effect. *0071*

Mitomycin Nephrotoxic effect. *2194*

Moxalactam In about 2% patients as reported from several studies. *0592*

Muscular Exercise Raised at 15 minutes, partial return to normal in 24 h. *1491*

Mushroom Poisoning May cause nephrotoxicity. *3204*

Nalidixic Acid May cause nitrogen retention. *2808*

Neomycin Nephrotoxic effect. *1022*

Netilmicin About 2 mg/dL increase in elderly patients with initial poor renal function. *1203*

Nifedipine Acute reversible deterioration of renal function in 4 patients with chronic renal failure. *0898*

Nitrofurantoin Possible decrease in already impaired renal function. *0085*

Norfloxacin Rare complication of treatment. *3750*

Oxacillin Transient azotemia with large doses. *0071*

Oxyphenbutazone May occur with renal damage. *2220* Numerous reports of kidney damage up to acute renal failure following therapeutic use of drug. *2866*

Paraldehyde Possible nephrotoxicity in poisoning. *1443*

Paramethadione Possible nephrotoxicity. *2313*

Pargyline Possible decrease in already impaired renal function. *0085*

Paromomycin Potentially nephrotoxic if given parenterally. *0071*

Penicillamine Possible nephrotoxicity. *1237*

Penicillin Rare reaction to large doses given parenterally. *1680*

Pentamidine Reversible renal dysfunction reported. *1343* Renal toxicity possibly due to formation of insoluble precipitates of pentamidine with nucleic acids, also associated with hypovolemia. *3445* Renal toxicity observed in about 25% patients. *3317*

Perchlorate Reported cases of nephrotic syndrome. *0071*

Phenacemide Occasional nephropathy. *0071*

Phenazopyridine Transient acute renal failure reported. *0071*

Phenindione Nephrotoxicity may occur with tubular necrosis. *1343*

Phensuximide Reversible nephropathy (especially in children). *0071*

Phenylbutazone Reported nephrotoxic effect. *2313*

Phosphorus Nephrotoxic effect. *2313*

Piroxicam Possible increase due to drug in some patients. *3819* Occasional increase noted not progressive. *2692*

Poison Ivy May cause nephrotoxicity. *3204*

Polymyxin Nephrotoxic effect. *1022*

Probenecid Possible nephrotoxicity. *1488*

Propranolol May occur secondary to decreased ERPF. *3384* Increase by 1.1 mg/dL in 340 patients given drug for 10 weeks to reduce diastolic BP to less than 90 mm Hg. *3707* Approximately 10% increase in women without significant change in men when treated over 3 y. *2506* Slight but significant in 20 hypertensive diabetic patients. *3900*

Propylthiouracil Anaphylactic nephritis reported. *0071*

Protein 19, 39, 46 mg/dL on 0.5, 1,5, 2,5 g/kg body weight. *0020*

Puromycin May cause nephrotoxicity. *3204*

Pyrazolones May cause nephrotoxicity. *1237*

Quinethazone Observed effect (?due to dehydration). *2427*

Quinine May cause renal damage (rare). *1343*

Radiographic Agents May produce azotemia or renal failure. *0071*

Rifampin Temporary renal failure reported. *0944*

RO4-2137 Slight increase up to 70 mg/dL in some patients. *1925*

Rolitetracycline May cause nephropathy. *2583*

Season Slight increase during summer. *2123*

Sex Difference In males at all ages except 7th decade. *2130*

Silver May cause nephrotoxicity. *3204*

Spectinomycin Mechanism not discussed. *0122*

Spironolactone Especially if increased at start of therapy. *1680*

Streptokinase Renal tubular damage in one case. *0652*

Streptomycin Nephrotoxicity may occur in 2%. *2808* Occasional nephrotoxicity, although less than with other aminoglycosides. *3683*

Sulfonamides Rare uremia may occur with/without crystalluria. *1343*

Suprofen Occasional renal failure observed but mechanism not known. *0009*

Tartrates If absorbed may cause renal damage. *1343*

Tetrachloroethylene May cause nephrotoxicity. *3204*

Tetracycline Anti-anabolic action, amino acids degraded to urea. *0491* May be dose related renal toxicity. *1680*

Thallium May cause nephrotoxicity. *3204*

Thiazides Nephrotoxic effect in large doses. *1022*

Thiocyanate May cause nephrosis. *1343*

Triamterene Nephrotoxic effect (with excessive diuresis). *1287*

Triethylenemelamine Nephrotoxicity with nitrogen retention. *1680*

Trimethadione Nephropathy reported. *0071*

Trimetrexate Nephrotoxicity reported in some patients following treatment, possibly associated with prior reduced renal function. *1613*

Uranium May cause nephrotoxicity. *3204*

Urea When massive infusion for sickle cell disease. *2354*

Vancomycin Nephrotoxic effect (may even be fatal). *1237* Occasional evidence of mild nephrotoxicity. *1571* Nephrotoxicity in 5% of 60 patients given drug alone but much higher incidence when given with aminoglycosides. *2654*

Vasopressin Progressive deterioration in renal function observed in one patient given drug. *0029*

Venoms May cause nephrotoxicity. *3204*

Vinyl Ether Potentially nephrotoxic. *0071*

Viomycin Frequent complication. *1343*

Vitamin D Manifestation of hypervitaminosis D. *0071*

X-Ray Therapy May cause nephrotoxicity. *3204*

Xylitol May cause azotemia. *3576*

Serum Decrease Analytical

Acidosis Much lower Azostix® than diacetyl value. *1460*

Ascorbic Acid At concentrations above 300 mg/L (maximum serum concentration 34 mg/L) lowered concentration as measured by Ektachem® method. *3393* At concentrations above 20 mg/L (maximum serum concentration 34 mg/L) lowered concentration as measured by Seralyzer. *3393*

Azostix® Result low compared with diacetyl monoxime value. *1460*

Caproxamine At 1.5 mg/dL conventional methods when added to sera. *1758*

Cefotaxime Effect of drug on specimens containing tobramycin on Parallel method. *0201*

Chloramphenicol Inhibits Berthelot reaction. *3505*

Fluorides Urease activity inhibited by fluoride. *0579*

Glassware If mercury present affects Nesslerization. *3879*

Hemolysis Red cell concentration less than serum. *0579*

Levodopa At concentrations above 100 mg/L lowered concentration as measured by Seralyzer method. *3393*

Lipemia Increase by 1.6% in patient specimens measured on SMAC® in comparison with specimens following ultra-centrifugation. *2466*

Mercury Compounds Inhibits urease (note present in Nessler's reagent). *1563*

Sodium Azide At 100 mg/dL causes decrease of 60% diacetyl procedure. *0583*

Streptomycin Inhibits Berthelot reaction. *3505*

Thymol Inhibits urease affecting results. *1237*

Serum Decrease Physiological

Acetohydroxamic Acid Inhibition of urease (limited *in vivo* effect). *3497*

Bendrofluazide Mean increase from 5.6 to 6.5 mmol/L in 7 essential hypertensives treated for 16 weeks. *0352*

Blood Pressure Strongly correlated with increased systolic pressure. *2123*

Enflurane Partly due to surgery possibly. *0999*

Ethanol Increase by 9% decrease in heavy versus occasional drinker. *3273*

Glucose Protein sparing effect of glucose and hemodilution. *3879*

Growth Hormone Metabolic effect. *2414*

Meals Significant effect in subjects 20-40 y. *3455*

Muscular Exercise Observed effect with 12 minutes cycle-ergometer. *1231*

Paramethasone Negative nitrogen balance, protein catabolism. *1680*

Urea Nitrogen *(continued)*

Serum   Decrease   Physiological *(continued)*
Phenothiazines   Decreased urea production if hepatic cirrhosis occurs. *0652*

Pregnancy   Marked anabolic state especially in first 6 mo. *3879*

Smoking   Less than in nonsmokers, lowest if inhale. *2123*

Water   2 h after drinking 1 glass, fell by 1 mg/dL. *2123*

Urine   No Effect   Analytical
Storage of Sample   For several days in refrigerator if pH less than 4. *0904*

Thymol   No effect on Berthelot method. *2981*

Urine   No Effect   Physiological
Azosemide   No clear pattern observed. *2047*

Furosemide   No clear pattern observed with drug administration. *2047*

Urine   Increase   Physiological
Alanine   Significant effect of chronic administration in fasting. *1263*

Protein   Increased in proportion to protein intake. *0987*

Starvation   Begins to rise from first day. *2382*

Urine   Decrease   Analytical
Bacterial Contamination   Rapid evolution of NH_3 changes pH, loss of NH_3. *1563*

Formaldehyde   Inhibits urease (can not be used as preservative). *1563*

Glassware   If mercury present affects Nesslerization. *3879*

Urine   Decrease   Physiological
Blindness   Significantly lower (average 4.3 g/d) than normal. *1637*

Diurnal Variation   Excretion reduced at night. *2222*

Furosemide   Occurs when diuretic given to patients with chronic failure, due to increased tubular reabsorption of urea, presumably in distal part of nephron, secondary to extracellular fluid volume depletion. *0792*

Pregnancy   Especially in first 6 mo. *0987*

Test Conditions   Increase   Analytical
Chlorpromazine   Produces turbidity with Berthelot's reagent. *1936*

Imipramine   Produces turbidity with Berthelot's reagent. *1936*

Levarterenol   Brown with Berthelot's reagent. *1936*

Levoglutamide   Blue color with Berthelot's reagent. *1936*

Nialamide   Bilious yellow-green with Berthelot's reagent. *1936*

Uric Acid

Serum   No Effect   Analytical
Acebutolol   At 100 mg/L on reversed phase LC procedure of Zhiri et al. *3942*

Acetaminophen   At 200 mg/L on reversed phase LC procedure of Zhiri et al. *3942*   At 5 times upper limit of therapeutic concentration on uricase procedures on Abbott-VP, aca, Cobas-Bio, Ektachem® 400, Hitachi® 705. *2138*   At acute overdose concentration (10 mg/dL) on SMAC® method. *3719*   No effect at therapeutic concentration on Reflotron method. *1984*   At concentration of 810 mg/L had no effect on uricase-PAP method. *3393*

Acetazolamide   No effect at 12 mg/dL on SMA 12/60 method. *2636*   At concentration of 1,000 mg/L had no effect on phosphotungstate reduction method. *3393*

Acetylsalicylic Acid   At therapeutic concentrations on uricase-catalase aldehyde dehydrogenase, direct UV-test and phosphotungstate methods. *1786*   On phosphotungstate method on SMA II at physiological concentration. *1787*   At 5 times upper limit of therapeutic range on method on SMAC®, Abbott-VP, Cobas-Bio, aca, Ektachem®, Hitachi® 705 and KDA. *2138*   With 8326 μmol/L on uricase-catalase method with aldehyde dehydrogenase. With 8326 μmol/L on direct UV method with uricase. With 8326 μmol/L on phosphotungstate reduction method. *1786*   No effect at therapeutic concentration on Reflotron method. *1984*   With 8326 μmol/L on phosphotungstate reduction method. *1786*   At concentration of 600 mg/L had no effect on Kageyama-Hantzsch method. *3393*   At concentration of 600 mg/L had no effect on catalase-AIDH method. *3393*   At concentration of 2,000 mg/L had no effect on phosphotungstate reduction method. *3393*   At concentration of 900 mg/L had no effect on uricase-PAP method. *3393*

Ajmaline   At concentration of 3.3 mg/L had no effect on uricase-PAP method. *3393*

Allopurinol   At 200 mg/L on reversed phase LC procedure of Zhiri et al. *3942*   At concentration of 250 mg/L had no effect on Kageyama-Hantzsch method. *3393*   At concentration of 80 mg/L had no effect on catalase-AIDH method. *3393*   At concentration of 5 mg/L had no effect on phosphotungstate reduction method. *3393*   At concentration of 100 mg/L had no effect on uricase method on aca. *3393*   At concentration of 204 mg/L had no effect on uricase-PAP method. *3393*

Aminophenazone   At therapeutic concentrations no effects on methods using uricase-catalase, aldehyde dehydrogenase, direct UV-test phosphotungstate methods. *1786*

Aminophylline   At 400 mg/L on reversed phase LC procedure of Zhiri et al. *3942*

Aminopyrine   At concentration of 125 mg/L had no effect on phosphotungstate reduction method. *3393*

Aminosalicylic Acid   At 5 times upper limit of therapeutic range on methods on SMAC®, Abbott-VP, aca, Cobas-Bio, Ektachem®, Hitachi® 705 and KDA. *2138*   At concentration of 460 mg/L had no effect on phosphotungstate reduction method. *3393*   At concentration of 100 mg/L had no effect on Seralyzer method. *3393*

Amiodarone   At 50 mg/L on reversed phase LC procedure of Zhiri et al. *3942*

Amitriptyline   At 10 mg/L on reversed phase LC procedure of Zhiri et al. *3942*   At acute overdose concentration (2.5 mg/dL) on SMAC® method. *3719*   At concentration of 25 mg/L had no effect on phosphotungstate reduction method. *3393*   At concentration of 40 mg/L had no effect on uricase-PAP method. *3393*

Amobarbital   At concentration of 150 mg/L had no effect on phosphotungstate reduction method. *3393*

Amphetamine   At therapeutic concentration on Reflotron method. *1984*

Amphotericin B   At concentration of 96 mg/L had no effect on phosphotungstate reduction method. *3393*

Ampicillin   At 20 mg/dL on Nishi phosphotungstate procedure. *2230*   No effect on Reflotron method at therapeutic concentration. *1984*   At concentration of 600 mg/L had no effect on Kageyama-Hantzsch method. *3393*   At concentration of 300 mg/L had no effect on catalase-AIDH method. *3393*   At concentration of 40 mg/L had no effect on phosphotungstate reduction method. *3393*

Antazoline   At concentration of 160 mg/L had no effect on Kageyama-Hantzsch method. *3393*   At concentration of 160 mg/L had no effect on catalase-AIDH method. *3393*

Aprotinin   At concentration of 150 kU/L had no effect on uricase-PAP method. *3393*

Ascorbic Acid   At 200 mg/L on uricase procedure of Kabasak. *1840*   No effect on Tripyridyl-s-triazine method of Morin. *2486*   At 500 mg/L on reversed phase LC procedure of Zhiri et al. *3942*   No effect at therapeutic concentration on Reflotron method. *1984*   At concentration of 1,000 mg/L had no effect on Kageyama-Hantzsch method. *3393*   At concentration of 5,000 mg/L had no effect on catalase-AIDH method. *3393*   At concentration of 100 mg/L had no effect on uricase method on aca. *3393*   At concentration of 250 mg/L had no effect on uricase-PAP method. *3393*

Aspirin   No effect on methods using uricase. *2954*   At 1 g/L on uricase procedure of Kabasak. *1840*   At 1.0 g/L on reversed phase LC procedure of Zhiri et al. *3942*

Azapropazone   At concentration of 360 mg/L had no effect as measured by Kageyama-Hantzsch method. *3393*   At concentration of 360 mg/L had no effect as measured by catalase-AlDH method. *3393*

Barbital   At acute overdose concentration (20 mg/dL) on Technicon® SMAC® method. *3719*   At concentration of 500 mg/L had no effect as measured by phosphotungstate reduction method. *3393*

Benzbromarone   At concentration of 20 mg/L had no effect as measured by Kageyama-Hantzsch method. *3393*   At concentration of 20 mg/L had no effect as measured by catalase-AlDH method. *3393*

Bethanechol   No effect at 0.09 mg/dL on SMA 12/60 method. *2636*

Bezafibrate   No effect at therapeutic concentration on Reflotron method. *1984*

Bilirubin   At 200 mg/L on uricase procedure of Kabasak. *1840* With phosphotungstate procedures up to 6.6 mg/dL. *0904*   At concentration of 40.0 mg/dL on Ektachem® method. *3407*   At 40.0 mg/dL on method recommended on Centrifichem. *3407*   At 40.0 mg/dL on method recommended on Cobas-Bio. *3407*   At concentration of 40.0 mg/dL on Du Pont aca II. *3407*

Bisacodyl   At concentration of 20 mg/L had no effect as measured by Kageyama-Hantzsch method. *3393*   At concentration of 4 mg/L had no effect as measured by catalase-AlDH method. *3393*

Boric Acid   At concentration of 50 mg/L had no effect as measured by phosphotungstate reduction method. *3393*

Bromazepam   At 5 times therapeutic concentration on methods of SMAC®, Abbott-VP, aca, Cobas-Bio, Hitachi® 705, Ektachem® 400 and KDA. *2138*

Butabarbital   At concentration of 100 mg/L had no effect as measured by phosphotungstate reduction method. *3393*

Butizide   At concentration of 260 mg/L had no effect as measured by Kageyama-Hantzsch method. *3393*   At concentration of 2.6 mg/L had no effect as measured by catalase-AlDH method. *3393*

Caffeine   At 500 mg/L on reversed phase LC method of Zhiri et al. *3942*   No effect at therapeutic concentrations on uricase-catalase, aldehyde dehydrogenase direct UV-test, and phosphotungstate methods. *1786*   At acute overdose concentration (10 mg/dL) on Technicon® SMAC® method. *3719*   At therapeutic concentration on Reflotron method. *1984*   At concentration of 194 mg/L had no effect on phosphotungstate reduction method. *3393*

Carbamazepine   At 100 mg/L on reversed phase LC method of Zhiri et al. *3942*

Carbenicillin   At 40 mg/dL on Nishi phosphotungstate procedure. *2230*   At 5 times upper limit of therapeutic range on methods on SMAC®, Abbott-VP, aca, Ektachem® 400, Cobas-Bio, Hitachi® 705 and KDA. *2138*   At concentration of 15,000 mg/L had no effect on phosphotungstate reduction method. *3393*

Carbochromen   No effect at therapeutic concentration on Reflotron method. *1984*

Cefamandole   At 2.50 mmol/L on method on Eppendorf Epos. *1414*

Cefazedone   On method on Eppendorf Epos at concentration of 2.50 mmol/L. *1414*

Cefazolin   At 2.50 mmol/L on method on Eppendorf Epos. *1414* At concentration of 117 mg/L had no effect on phosphotungstate reduction method. *3393*

Cefodizime   No effect at 2.50 mmol/L on method on Eppendorf Epos. *1414*

Cefoperazone   At 2.50 mmol/L on method on Eppendorf Epos. *1414*

Cefotaxime   No effect at 2.50 mmol/L on method on Eppendorf Epos. *1414*

Cefotiam   At 2.50 mmol/L on method on Eppendorf Epos. *1414*

Cefoxitin   At 5 times upper limit of therapeutic range on methods on SMAC®, Abbott-VP, aca, Cobas-Bio, Ektachem® 705 and KDA. *2138*   At concentrations up to 2.50 mmol/L on methods on Eppendorf Epos. *1414*   At concentration of 1,000 mg/L had no effect on uricase-PAP method. *3393*

Cefpirome   At concentration of up to 2.50 mmol/L on methods of Eppendorf Epos. *1414*

Ceftriaxone   On method on Eppendorf Epos at concentration of 2.50 mmol/L. *1414*

Cefuroxime   On method on Eppendorf Epos at concentration of 2.50 mmol/L. *1414*

Cephaloridine   At 2.50 mmol/L on method on Eppendorf Epos. *1414*

Cephalothin   At 5 times upper limit of therapeutic range on methods on SMAC®, Abbott-VP, aca, Cobas-Bio, Ektachem®, Hitachi® 705 and KDA. *2138*   No effect a 2.50 mmol/L on method on Eppendorf Epos. *1414*   At concentration of 1,000 mg/L had no effect on phosphotungstate reduction method. *3393*

Chloramphenicol   No effect at therapeutic concentration on Reflotron method. *1984*   At concentration of 600 mg/L had no effect on catalase-AlDH method. *3393*   At concentration of 3,000 mg/L had no effect on uricase-PAP method. *3393*

Chlordiazepoxide   At 16 mg/L on reversed phase LC procedure of Zhiri et al. *3942*   At 5 times upper limit of therapeutic range on methods on SMAC®, Abbott-VP, aca, *2138*   At acute overdose concentration (5 mg/dL) on Technicon® SMAC® method. *3719*   No effect at therapeutic concentration on Reflotron method. *1984*   At concentration of 20 mg/L had no effect on Kageyama-Hantzsch method. *3393*   At concentration of 20 mg/L had no effect on catalase-AlDH method. *3393*   At concentration of 50 mg/L had no effect on phosphotungstate reduction method. *3393*   At concentration of 50 mg/L had no effect on uricase-PAP method. *3393*

Chlormezanone   At concentration of 100 mg/L had no effect on phosphotungstate reduction method. *3393*

Chloroquine   At concentration of 100 mg/L had no effect on Kageyama-Hantzsch method. *3393*   At concentration of 100 mg/L had no effect on catalase-AlDH method. *3393*   At concentration of 250 mg/L had no effect on uricase-PAP method. *3393*

Chlorphenesin   At concentration of 150 mg/L had no effect on phosphotungstate reduction method. *3393*

Chlorpheniramine   At acute overdose concentration (20 mg/dL) on Technicon® SMAC® method. *3719*   At concentration of 200 mg/L had no effect on phosphotungstate reduction method. *3393*

Chlorpromazine   At concentration of 3 mg/L had no effect on phosphotungstate reduction method. *3393*

Chlorpropamide   At concentration of 150 mg/L had no effect on phosphotungstate reduction method. *3393*

Chlorprothixene   At concentration of 1 mg/L had no effect on phosphotungstate reduction method. *3393*

Chromonar Hydrochloride   At concentration of 200 mg/L had no effect on Kageyama-Hantzsch method. *3393*   At concentration of 90 mg/L had no effect on catalase-AlDH method. *3393*

Cimetidine   At 300 mg/L on reversed phase LC procedure of Zhiri et al. *3942*   At concentration of 1 mg/L had no effect on phosphotungstate reduction method. *3393*   At concentration of 4 mg/L had no effect on uricase-PAP method. *3393*

Clofibrate   At concentration of 400 mg/L had no effect on Kageyama-Hantzsch method. *3393*   At concentration of 400 mg/L had no effect on catalase-AlDH method. *3393*   At concentration of 500 mg/L had no effect on uricase-PAP method. *3393*

Clonidine   At 10 mg/L on reversed phase LC procedure of Zhiri et al. *3942*

Cloxacillin   At concentration of 5 mg/L had no effect on phosphotungstate reduction method. *3393*

Cocaine   At acute overdose concentration (2.5 mg/dL) on Technicon® SMAC® method. *3719*   At concentration of 25 mg/L had no effect on phosphotungstate reduction method. *3393*

Codeine   At acute overdose concentration (2.0 mg/dL) on Technicon® SMAC® method. *3719*   At concentration of 20 mg/L had no effect on phosphotungstate reduction method. *3393*

Contact With Clot   No effect 8 h at room temperature if uricase used. *2954*

Creatinine   At 10 mg/dL on copper chelate procedure. At 10 mg/dL on Nishi phosphotungstate procedure. *2230*

Cyclophosphamide   At concentration of 60 mg/L had no effect on Kageyama-Hantzsch method. *3393*   At concentration of 40 mg/L had no effect on catalase-AlDH method. *3393*   At concentration of 40 mg/L had no effect on uricase-PAP method. *3393*

Cyclosporine   At concentration of 41.5 mmol/L (50 µg/L) on methods on Technicon® SMAC® II and Hitachi® 705. *1123*

Cysteine   At 200 mg/L on uricase procedure of Kabasak. *1840* At 5 times upper limit of therapeutic range on uricase methods on Abbott-VP. aca, Cobas-Bio, Ektachem® and Hitachi® 705. *2138*

Uric Acid (continued)

Serum   No Effect   Analytical (continued)

Cystine   At 5 mg/dL on copper chelate procedure. *2230*   At 5 times upper limit of therapeutic range on uricase methods on Abbott-VP, aca, Cobas-Bio, Ektachem® and Hitachi® 705. *2138*

Desipramine   At concentration of 8 mg/L had no effect on phosphotungstate reduction method. *3393*

Deslanoside   No effect on SMA 12/60 method at 0.06 mg/dL. *2636*

Dextran 60   At concentration of 6,000 mg/L had no effect on Kageyama-Hantzsch method. *3393*   At concentration of 6,000 mg/L had no effect on catalase-AIDH method. *3393*

Dextran 70   At concentration of 1,000 mg/L had no effect on Ektachem® method. *3393*

Diatrizoic acid   At concentration of 2600 mg/L had no effect on Kageyama-Hantzsch method. *3393*   At concentration of 2600 mg/L had no effect on catalase-AIDH method. *3393*

Diazepam   At 50 mg/L on reversed phase LC procedure of Zhiri et al. *3942*   At 5 times upper limit of therapeutic range on methods on SMAC®, Abbott-VP, aca, Cobas-Bio, Ektachem®, Hitachi® 705 and KDA. *2138*   At acute overdose concentration (2.5 mg/dL) on Technicon® SMAC® method. *3719*   At concentration of 25 mg/L had no effect on phosphotungstate reduction method. *3393*

Diclofenac   On phosphotungstate method on SMA II at physiological concentration. *1787*   At concentration of 23 mg/L had no effect on phosphotungstate reduction method. *3393*   No effect at therapeutic concentrations on uricase-catalase, aldehyde dehydrogenase, direct UV-test and phosphotungstate methods. *1786*

Diethylaminoethanol   At concentration of 30 mg/L had no effect on phosphotungstate reduction method. *3393*

Diethylpropion   At acute overdose concentration (10 mg/dL) on Technicon® SMAC® method. *3719*   At concentration of 100 mg/L had no effect on phosphotungstate reduction method. *3393*

Digitoxin   No effect at 21 mg/dL on SMA 12/60 method. *2636*   At concentration of 6 mg/L had no effect on phosphotungstate reduction method. *3393*

Digoxin   No effect at 0.04 mg/dL on SMA 12/60 method. *2636*   At 10 mg/L on reversed phase LC procedure of Zhiri et al. *3942*   At concentration of 0.15 mg/L had no effect on Kageyama-Hantzsch method. *3393*   At concentration of 0.15 mg/L had no effect on catalase-AIDH method. *3393*   At concentration of 9 mg/L had no effect on phosphotungstate reduction method. *3393*   At concentration of 0.25 mg/L had no effect on uricase-PAP method. *3393*

Dihydroergotamine   At 10 mg/L on reversed phase LC procedure of Zhiri et al. *3942*

Dimethadione   At concentration of 2,000 mg/L had no effect on phosphotungstate reduction method. *3393*

N-Dimethyldiazepam   At concentration of 6 mg/L had no effect on phosphotungstate reduction method. *3393*

Diphenhydramine   At acute overdose concentration (20 mg/dL) on Technicon® SMAC® method. *3719*   At concentration of 200 mg/L had no effect on phosphotungstate reduction method. *3393*

Disopyramide   At 200 mg/L on reversed phase LC procedure of Zhiri et al. *3942*   At concentration of 4 mg/L had no effect on phosphotungstate reduction method. *3393*

Disulfiram   At concentration of 120 mg/L had no effect on phosphotungstate reduction method. *3393*

Disulphine Blue   By continuous-flow phosphotungstate reduction. *1464*

Doxepin   At 1.2 g/L on reversed phase LC procedure of Zhiri et al. *3942*   At concentration of 30 mg/L had no effect on Kageyama-Hantzsch method. *3393*   At concentration of 30 mg/L had no effect on catalase-AIDH method. *3393*

EDTA   At concentration of 1,000 mg/L had no effect on Kageyama-Hantzsch method. *3393*   At concentration of 1,000 mg/L had no effect on catalase-AIDH method. *3393*   At concentration of 4,000 mg/L had no effect on uricase method on aca. *3393*   At concentration of 2922 mg/L had no effect on uricase-PAP method. *3393*

Erythromycin   At concentration of 73 mg/L had no effect on phosphotungstate reduction method. *3393*

Ethacrynic Acid   At 2.5 mg/dL no effect on SMA 12/60 method. *2636*

Ethanol   At acute overdose concentration (20 mg/dL) on Technicon® SMAC® methods. *3719*

Ethaverine   No effect at therapeutic concentration on Reflotron method. *1984*

Ethchlorvynol   At acute overdose concentration (20 mg/dL) on Technicon® method. *3719*   At concentration of 400 mg/L had no effect on phosphotungstate reduction method. *3393*

Ethinamate   At acute overdose concentration (20 mg/dL) on Technicon® SMAC® method. *3719*   At concentration of 200 mg/L had no effect on phosphotungstate reduction method. *3393*

Ethinyl Estradiol   At 10 mg/L on reversed phase LC procedure of Zhiri et al. *3942*

Ethosuximide   At concentration of 390 mg/L had no effect on phosphotungstate reduction method. *3393*

Etoposide   No effect on uricase method on Du Pont aca at therapeutic concentration. *2333*

Fluorouracil   At concentration of 130 mg/L had no effect on phosphotungstate reduction method. *3393*

Fluosol-DA   On SMA IIC at concentration of 50%. On Du Pont aca III at concentration of 50%. *2518*

Flurazepam   At 5 times upper limit of therapeutic range on methods on SMAC®, Abbott-VP, aca, Cobas-Bio, Ektachem®, Hitachi® 705, and KDA. *2138*   At acute overdose concentration (2.5 mg/dL) on Technicon® SMAC® method. *3719*   At concentration of 25 mg/L had no effect on phosphomolybdate reduction method. *3393*

Furosemide   No effect at 1.4 mg/dL on SMA 12/60 method. *2636*   No effect at therapeutic concentration on Reflotron method. *1984*   At concentration of 20 mg/L had no effect on Kageyama-Hantzsch method. *3393*   At concentration of 20 mg/L had no effect on catalase-AIDH method. *3393*   At concentration of 4 mg/L had no effect on phosphotungstate reduction method. *3393*   At concentration of 5 mg/L had no effect on uricase-PAP method. *3393*

Gentamicin   At concentration of 6 mg/L had no effect on Kageyama-Hantzsch method. *3393*   At concentration of 6 mg/L had no effect on catalase-AIDH method. *3393*   At concentration of 150 mg/L had no effect on phosphotungstate reduction method. *3393*   At concentration of 10 mg/L had no effect on uricase-PAP method. *3393*

Glibenclamide   At 10 mg/L on reversed phase LC procedure of Zhiri et al. *3942*   No effect at therapeutic concentration on Reflotron method. *1984*   At concentration of 3 mg/L had no effect on Kageyama-Hantzsch method. *3393*   At concentration of 3 mg/L had no effect on catalase-AIDH method. *3393*

Glucose   At 5 g/L on uricase procedure of Kabasak. *1840*

Glutathione   If blank at 300 mg/L procedure of Kabasak. *1840*   No effect on Tripyridyl-s-triazine method of Morin. *2486*

Glutethimide   At acute overdose concentration (5 mg/dL) on Technicon® SMAC® method. *3719*   At concentration of 17.5 mg/L had no effect on Kageyama-Hantzsch method. *3393*   At concentration of 50 mg/L had no effect on phosphotungstate reduction method. *3393*

Guanidinoacetic Acid   At 10 mg/dL on Nishi phosphotungstate procedure   At 10 mg/dL on copper chelate procedure. *2230*

Guanidinosuccinic Acid   At 3 mg/dL on Nishi phosphotungstate procedure   At 3 mg/dL on copper chelate procedure. *2230*

Hemolysis   No effect if measured by uricase and is moderate. *2954*

Heparin   At concentration of 1200 mg/L had no effect on uricase method on aca. *3393*

Hydrochlorothiazide   At concentration of 7 mg/L had no effect on uricase-PAP method. *3393*

Hyoscine-N-Butylbromide   At concentration of 60 mg/L had no effect on Kageyama-Hantzsch method. *3393*   At concentration of 12 mg/L had no effect on catalase-AIDH method. *3393*

Hypoxanthine   At 10 mg/dL on Nishi phosphotungstate procedure   At 10 mg/dL on copper chelate procedure. *2230*

Ibuprofen   No effect at therapeutic concentrations on uricase-catalase, aldehyde dehydrogenase direct UV-test and phosphotungstate methods. *1786*   On phosphotungstate method on SMA II at physiological concentration. *1787*   At 5 times upper limit of therapeutic range on methods on SMAC®, Abbott-VP, aca, Cobas-Bio, Ektachem®, Hitachi® 705 and KDA. *2138*   At concentration of 200 mg/L had no effect on phosphotungstate reduction method. *3393*

Indapamide   At 10 mg/L on reversed phase LC procedure of Zhiri et al. *3942*

Indomethacin No effect at therapeutic concentrations on uricase-catalase, aldehyde dehydrogenase, direct UV-test, phosphotungstate methods. *1786* On phosphotungstate method on SMA II at physiological concentration. *1787* No effect at therapeutic concentration on Reflotron method. *1984* At concentration of 30 mg/L had no effect on Kageyama-Hantzsch method. *3393* At concentration of 30 mg/L had no effect on catalase-AIDH method. *3393* At concentration of 13 mg/L had no effect on phosphotungstate reduction method. *3393* At concentration of 25 mg/L had no effect on uricase-PAP method. *3393*

Insulin At concentration of 3 mg/L had no effect on phosphotungstate reduction method. *3393*

Iproniazid At concentration of 40 mg/L had no effect on phosphotungstate reduction method. *3393*

Isoniazid At 20 mg/L on reversed phase LC procedure of Zhiri et al. *3942*

Isoprenaline At concentration of 211 mg/L had no effect on phosphotungstate reduction method. *3393*

Ketoprofen No effect at therapeutic concentrations on uricase-catalase, aldehyde dehydrogenase, direct UV-test and phosphotungstate methods. *1786* On phosphotungstate method on SMA II at physiological concentration. *1787* At concentration of 60 mg/L had no effect on phosphotungstate reduction method. *3393*

Levodopa No increase reported when uricase used. *0071* At concentration of 3 mg/L had no effect on uricase method on aca. *3393*

Lidocaine No effect SMA 12/60 method with 3.5 mg/dL. *2636*

Lipemia In patient specimens measured on SMAC® in comparison with specimen following ultra-centrifugation. *2466*

Liposol Changed from 43.4 to 45.5 mg/L in 26 randomly selected patient sera, before and after addition on Beckman Astra methods. *2054*

Lorazepam At 10 mg/L on reversed phase LC procedure of Zhiri et al. *3942* At concentration of 0.05 mg/L had no effect on phosphotungstate reduction method. *3393*

Mannitol No effect on SMA 12/60 method at 445 mg/dL. *2636*

Meglumine At concentration of 2,000 mg/L had no effect on Kageyama-Hantzsch method. *3393* At concentration of 1200 mg/L had no effect on catalase-AIDH method. *3393*

Meperidine At acute overdose concentration (5 mg/dL) on Technicon® SMAC® method. *3719* At concentration of 60 mg/L had no effect on phosphotungstate reduction method. *3393*

Mephenesin At concentration of 100 mg/L had no effect on phosphotungstate reduction method. *3393*

Meprobamate At 200 mg/L on reversed phase LC procedure of Zhiri et al. *3942* At acute overdose concentration (20 mg/dL) on Technicon® SMAC® method. *3719* At concentration of 200 mg/L had no effect on phosphotungstate reduction method. *3393*

Meralluride No effect at 0.07 mL/dL on SMA 12/60 method. *2636*

Mercaptomerin At concentration of 58 mg/L had no effect on phosphotungstate reduction method. *3393*

Mesoridazine At acute overdose concentration (20 mg/dL) on Technicon® SMAC® method. *3719*

Metformin At 2 g/L on reversed phase LC procedure of Zhiri et al. *3942*

Methadone At acute overdose concentration (2.5 mg/dL) on Technicon® SMAC® method. *3719* At concentration of 25 mg/L had no effect on phosphotungstate reduction method. *3393*

Methamphetamine At concentration of 2 mg/L had no effect on phosphotungstate reduction method. *3393*

Methanol At acute overdose concentration (20 mg/dL) on Technicon® SMAC® method. *3719*

Methapyrilene At concentration of 13 mg/L had no effect on phosphotungstate reduction method. *3393*

Methaqualone At acute overdose concentration (2.5 mg/dL) on Technicon® SMAC® method. *3719* No effect at therapeutic concentration on Reflotron method. *1984* At concentration of 25 mg/L had no effect on phosphotungstate reduction method. *3393*

Methicillin At 20 mg/dL on Nishi phosphotungstate procedure. *2230* At 5 times upper limit of therapeutic range on SMAC®, Abbott-VP, aca, Ektachem®, Hitachi® 705 and KDA. *2138* At concentration of 900 mg/L had no effect on phosphotungstate reduction method. *3393*

Methohexital At concentration of 50 mg/L had no effect on phosphotungstate reduction method. *3393*

Methotrexate At 5 times upper limit of therapeutic range on methods on SMAC®, Abbott-VP, aca, Cobas-Bio, Ektachem® and KDA. *2138* At concentration of 1 mg/L had no effect on Kageyama-Hantzsch method. *3393* At concentration of 1 mg/L had no effect on catalase-AIDH method. *3393* At concentration of 25 mg/L had no effect on uricase-PAP method. *3393*

Methotrimeprazine At concentration of 1 mg/L had no effect on phosphotungstate reduction method. *3393*

Methsuximide At concentration of 40 mg/L had no effect on phosphotungstate reduction method. *3393*

Methyldopa At 30 mg/L on reversed phase LC procedure of Zhiri et al. *3942* No effect at therapeutic concentration on Reflotron method. *1984* At concentration of 320 mg/L had no effect on catalase-AIDH method. *3393*

Methylphenidate At acute overdose concentration (20 mg/dL) on Technicon® SMAC® method. *3719* At concentration of 200 mg/L had no effect on phosphotungstate reduction method. *3393*

Methylphenobarbital At concentration of 150 mg/L had no effect on phosphotungstate reduction method. *3393*

Methyprylon At acute overdose concentration (10 mg/dL) on Technicon® SMAC® method. *3719* At concentration of 100 mg/L had no effect on phosphotungstate reduction method. *3393*

Metoprolol At concentration of 0.34 mg/L had no effect on phosphotungstate reduction method. *3393* At concentration of 3 mg/L had no effect on uricase-PAP method. *3393*

Morphine At acute overdose concentration (20 mg/dL) on Technicon® SMAC® method. *3719* At concentration of 200 mg/L had no effect on phosphotungstate reduction method. *3393*

Moxalactam At 2.50 mmol/L on method on Eppendorf Epos. *1414* At concentration of 96 mg/L had no effect on phosphotungstate reduction method. *3393*

Nafcillin At concentration of 50 mg/L had no effect on phosphotungstate reduction method. *3393*

Nalidixic Acid At concentration of 50 mg/L had no effect on uricase-PAP method. *3393*

Naproxen At 5 times upper limit of therapeutic range on methods on SMAC®, Abbott-VP, aca, Ektachem® and KDA. *2138*

Niacin No effect at therapeutic concentration on Reflotron method. *1984*

Niacinamide At concentration of 40 mg/L had no effect on Kageyama-Hantzsch method. *3393* At concentration of 40 mg/L had no effect on catalase-AIDH method. *3393*

Nicergoline At 10 mg/L on reversed phase LC procedure of Zhiri et al. *3942*

Niflumic Acid At 200 mg/L on reversed phase LC procedure of Zhiri et al. *3942* At concentration of 200 mg/L had no effect on Kageyama-Hantzsch method. *3393* At concentration of 150 mg/L had no effect on catalase-AIDH method. *3393*

Nitrofurantoin No effect at therapeutic concentration on Reflotron method. *1984* On routine methods in use on SMAC®, Ektachem®, Abbott-VP, Cobas-Bio, aca, Hitachi® 705, KDA at 5 times normal upper therapeutic concentration. *2138* At concentration of 30 mg/L had no effect on Kageyama-Hantzsch method. *3393* At concentration of 30 mg/L had no effect on catalase-AIDH method. *3393* At concentration of 4 mg/L had no effect on phosphotungstate reduction method. *3393* At concentration of 0.5 mg/L had no effect on uricase-PAP method. *3393*

Norfenefrin At concentration of 6 mg/L had no effect on Kageyama-Hantzsch method. *3393* At concentration of 6 mg/L had no effect on catalase-AIDH method. *3393*

Norgestrel At 10 mg/L on reversed phase LC procedure of Zhiri et al. *3942*

Nortriptyline At acute overdose concentration (20 mg/dL) on Technicon® SMAC® method. *3719* At concentration of 200 mg/L had no effect on phosphotungstate reduction method. *3393*

Novaminsulfon At concentration of 800 mg/L had no effect on catalase-AIDH method. *3393*

Orphenadrine At concentration of 9 mg/L had no effect on phosphotungstate reduction method. *3393*

Ouabain At 0.02 mg/dL no effect on SMA 12/60 method. *2636*

Oxacillin At concentration of 40 mg/L had no effect on phosphotungstate reduction method. *3393*

Oxazepam At 10 mg/L on reversed phase LC procedure of Zhiri et al. *3942* No effect at therapeutic concentration on Reflotron method. *1984* At concentration of 1 mg/L had no effect on phosphotungstate reduction method. *3393*

Uric Acid *(continued)*

Serum   No Effect   Analytical *(continued)*

Oxedrine At concentration of 60 mg/L had no effect on uricase-PAP method. *3393*

Oxyphenbutazone At concentration of 120 mg/L had no effect on catalase-AIDH method. *3393*

Oxytetracycline No effect at therapeutic concentration on Reflotron method. *1984*

Papaverine At concentration of 10 mg/L had no effect on phosphotungstate reduction method. *3393*

Paraldehyde At concentration of 2,000 mg/L had no effect on phosphotungstate reduction method. *3393*

Penicillamine At concentration of 480 mg/L had no effect on catalase-AIDH method. *3393*

Penicillin At 1,000 units/mL on Nishi phosphotungstate procedure. *2230*

Penicillin G At 5 times upper limit of therapeutic range on methods on SMAC®, Abbott-VP, Cobas-Bio, aca, Ektachem®, Hitachi® 705 and KDA. *2138* At concentration of 2,000 mg/L had no effect as measured by phosphotungstate reduction method. *3393* At concentration of 90 mg/L had no effect as measured by uricase-PAP method. *3393*

Pentobarbital At acute overdose concentration (20 mg/dL) on Technicon® SMAC® method. *3719* At concentration of 340 mg/L had no effect on phosphotungstate reduction method. *3393*

Perphenazine At acute overdose concentration (20 mg/dL) on Technicon® SMAC® method. *3719* At concentration of 200 mg/L had no effect on phosphotungstate reduction method. *3393*

Phenazopyridine No effect at therapeutic concentration on Reflotron method. *1984*

Phencyclidine At concentration of 6 mg/L had no effect on phosphotungstate reduction method. *3393*

Phenelzine At concentration of 136 mg/L had no effect on phosphotungstate reduction method. *3393*

Phenobarbital At 50 mg/L on reversed phase LC procedure of Zhiri et al. *3942* No effect at therapeutic concentrations on uricase-catalase, aldehyde dehydrogenase, direct UV-test and phosphotungstate methods. *1786* At acute overdose concentration (20 mg/dL) on Technicon® SMAC® method. *3719* No effect at therapeutic concentration on Reflotron method. *1984* At concentration of 80 mg/L had no effect on Kageyama-Hantzsch method. *3393* At concentration of 250 mg/L had no effect on phosphotungstate reduction method. *3393* At concentration of 100 mg/L had no effect on uricase-PAP method. *3393*

Phenothiazines At concentration of 30 mg/L had no effect on Kageyama-Hantzsch method. *3393* At concentration of 30 mg/L had no effect on catalase-AIDH method. *3393*

Phenprocoumon No effect at therapeutic concentration on Reflotron method. *1984* At concentration of 6 mg/L had no effect on Kageyama-Hantzsch method. *3393* At concentration of 6 mg/L had no effect on catalase-AIDH method. *3393*

Phensuximide At concentration of 120 mg/L had no effect on phosphotungstate reduction method. *3393*

Phenylbutazone At concentration of 200 mg/L had no effect on Kageyama-Hantzsch method. *3393* At concentration of 120 mg/L had no effect on catalase-AIDH method. *3393* At concentration of 750 mg/L had no effect on phosphotungstate reduction method. *3393* At concentration of 200 mg/L had no effect on uricase-PAP method. *3393*

Phenylephrine At concentration of 4 mg/L had no effect on phosphotungstate reduction method. *3393*

Phenytoin No effect at 1.8 mg/dL on SMA 12/60 method. *2636* At acute overdose concentration (20 mg/dL) on Technicon® SMAC® method. *3719* No effect at therapeutic concentration on Reflotron method. *1984* At concentration of 240 mg/L had no effect on phosphotungstate reduction method. *3393*

Prednisolone At concentration of 200 mg/L had no effect on Kageyama-Hantzsch method. *3393* At concentration of 200 mg/L had no effect on catalase-AIDH method. *3393* At concentration of 25 mg/L had no effect on uricase-PAP method. *3393*

Probenecid No effect at therapeutic concentration on Reflotron method. *1984* At concentration of 280 mg/L had no effect on Kageyama-Hantzsch method. *3393* At concentration of 200 mg/L had no effect on catalase-AIDH method. *3393* At concentration of 1300 mg/L had no effect on phosphotungstate reduction method. *3393* At concentration of 570 mg/L had no effect on Seralyzer method. *3393*

Procainamide No effect on SMA 12/60 method at 35 mg/dL. *2636* At concentration of 50 mg/L had no effect on phosphotungstate reduction method. *3393*

Procaine No effect at therapeutic concentration on Reflotron method. *1984* At concentration of 2 mg/L had no effect on phosphotungstate reduction method. *3393*

Prochlorperazine At concentration of 1 mg/L had no effect on phosphotungstate reduction method. *3393*

Promethazine At acute overdose concentration (20 mg/dL) on Technicon® SMAC® method. *3719* At concentration of 200 mg/L had no effect on phosphotungstate reduction method. *3393*

Propoxyphene At acute overdose concentration (2.5 mg/dL) on Technicon® SMAC® method. *3719* At concentration of 25 mg/L had no effect on phosphotungstate reduction method. *3393*

Propranolol No effect at 0.1 mg/dL on SMA 12/60 method. *2636* At 100 mg/L on reversed phase LC procedure of Zhiri et al. *3942* At concentration of 0.2 mg/L had no effect on phosphotungstate reduction method. *3393* At concentration of 80 mg/L had no effect on uricase-PAP method. *3393*

Pyribenzamine® At acute overdose concentration (20 mg/dL) on Technicon® SMAC® methods. *3719*

Pyridamole No effect at therapeutic concentration on Reflotron method. *1984*

Pyritinol No effect at therapeutic concentration on Reflotron method. *1984*

Quinidine At 26 mg/dL no effect on SMA 12/60 method. *2636* At acute overdose concentration (20 mg/dL) Technicon® SMAC® method. *3719* No effect at therapeutic concentration on Reflotron method. *1984* At concentration of 210 mg/L had no effect on phosphotungstate reduction method. *3393*

Quinine At acute overdose concentration (1.5 mg/dL) on Technicon® SMAC® method. *3719* At concentration of 30 mg/L had no effect on phosphotungstate reduction method. *3393*

Reserpine No effect at 0.02 mg/dL on SMA 12/60 method. *2636* At concentration of 61 mg/L had no effect on phosphotungstate reduction method. *3393*

Resorcinol At concentration of 50 mg/L had no effect on uricase method on aca. *3393*

Rifampin At 5 times upper limit of therapeutic range on methods on SMAC®, Abbott-VP, aca, Ektachem® and Cobas-Bio. *2138*

Rolitetracycline At concentration of 4 mg/L had no effect on uricase-PAP method. *3393*

Salicylate No effect on Tripyridyl-s-triazine method of Morin. *2486* At 30 mg/dL on copper chelate procedure. *2230* At concentration of 500 mg/L had no effect on phosphotungstate reduction method. *3393* At concentration of 300 mg/L had no effect on Seralyzer method. *3393* At concentration of 300 mg/L had no effect on uricase method on aca. *3393* At concentration of 967 mg/L had no effect on uricase-PAP method. *3393*

Scopolamine Bromide At concentration of 20 mg/L had no effect on uricase-PAP method. *3393*

Secobarbital At acute overdose concentration (20 mg/dL) on Technicon® SMAC® method. *3719* At concentration of 200 mg/L had no effect on phosphotungstate reduction method. *3393*

Sodium Citrate At concentration of 5,000 mg/L had no effect on Kageyama-Hantzsch method. *3393* At concentration of 5,000 mg/L had no effect on catalase-AIDH method. *3393* At concentration of 11764 mg/L had no effect on uricase-PAP method. *3393*

Sodium Fluoride At concentration of 2,000 mg/L had no effect on Kageyama-Hantzsch method. *3393* At concentration of 2,000 mg/L had no effect on catalase-AIDH method. *3393* At concentration of 2,000 mg/L had no effect on uricase method on aca. *3393* At concentration of 15116 mg/L had no effect on uricase-PAP method. *3393*

Sodium Heparin At concentration of 750 mg/L had no effect on Kageyama-Hantzsch method. *3393* At concentration of 750 mg/L had no effect on catalase-AIDH method. *3393* At concentration of 4,000 mg/L had no effect on uricase-PAP method. *3393*

Sodium Oxalate At concentration of 3,000 mg/L had no effect on Kageyama-Hantzsch method. *3393* At concentration of 3,000 mg/L had no effect on catalase-AIDH method. *3393* At concentration of 4620 mg/L had no effect on uricase-PAP method. *3393*

Sodium Salicylate 10 mg/dL no effect on method of Klein. *1949*

Spironolactone At concentration of 20 mg/L had no effect on Kageyama-Hantzsch method. *3393* At concentration of 20 mg/L had no effect on catalase-AIDH method. *3393* At concentration of 1 mg/L had no effect on uricase-PAP method. *3393*

Standing of Sample If on polystyrene beads 24 h at 25°. *2624*

Storage of Sample No effect for 3 d at room temperature, 6 mo frozen. *1563*

Streptomycin At concentration of 58 mg/L had no effect on phosphotungstate reduction method. *3393*

Strychnine At concentration of 12 mg/L had no effect on phosphotungstate reduction method. *3393*

Sulfadiazine At concentration of 1500 mg/L had no effect on phosphotungstate reduction method. *3393*

Sulfaguanidine At concentration of 500 mg/L had no effect on phosphotungstate reduction method. *3393*

Sulfamethoxazole No effect at therapeutic concentration on Reflotron method. *1984*

Sulfamethoxydiazine At concentration of 230 mg/L had no effect on Kageyama-Hantzsch method. *3393* At concentration of 300 mg/L had no effect on catalase-AIDH method. *3393*

Sulfamethoxypyridazine At concentration of 70 mg/L had no effect on uricase-PAP method. *3393*

Sulfanilamide At concentration of 1,000 mg/L had no effect on phosphotungstate reduction method. *3393*

Sulforidazine At concentration of 200 mg/L had no effect on phosphotungstate reduction method. *3393*

Tetracycline At 5 times upper limit of therapeutic range on methods on SMAC®, Abbott-VP, aca, Cobas-Bio, Ektachem®, Hitachi® 705 and KDA. *2138* At concentration of 200 mg/L had no effect on catalase-AIDH method. *3393* At concentration of 100 mg/L had no effect on Seralyzer method. *3393*

Theobromine At concentration of 2,000 mg/L had no effect on phosphotungstate reduction method. *3393*

Theophylline At 1.0 g/L on reversed phase LC procedure of Zhiri et al. *3942* At 5 times upper limit of therapeutic range on methods on SMAC®, Abbott-VP, aca, Cobas-Bio, Ektachem®, Hitachi® 705 and KDA. *2138* No effect at therapeutic concentration on Reflotron method. *1984* At concentration of 2,000 mg/L had no effect on phosphotungstate reduction method. *3393*

Thiamazole At concentration of 114 mg/L had no effect on phosphotungstate reduction method. *3393*

Thiamine At 500 mg/L on reversed phase LC procedure of Zhiri et al. *3942*

Thioguanine At 10 mg/dL on Nishi phosphotungstate procedure At 10 mg/dL on copper chelate procedure. *2230*

Thiopental At acute overdose concentration (20 mg/dL) on Technicon® SMAC® method. *3719* At concentration of 200 mg/L had no effect on phosphotungstate reduction method. *3393*

Timolol At concentration of 0.01 mg/L had no effect on phosphotungstate reduction method. *3393*

Tolbutamide At concentration of 260 mg/L had no effect on Kageyama-Hantzsch method. *3393* At concentration of 400 mg/L had no effect on catalase-AIDH method. *3393* At concentration of 270 mg/L had no effect on phosphotungstate reduction method. *3393* At concentration of 500 mg/L had no effect on uricase-PAP method. *3393*

Transport of Specimen By pneumatic tube if specimen protected. *3217*

Trazodone At concentration of 50 mg/L had no effect on uricase-PAP method. *3393*

Tribromethanol At concentration of 90 mg/L had no effect on phosphotungstate reduction method. *3393*

Trichlorethanol At concentration of 1,000 mg/L had no effect on phosphotungstate reduction method. *3393*

Trifluoperazine At concentration of 1 mg/L had no effect on phosphotungstate reduction method. *3393*

Trimethoprim No effect at therapeutic concentration on Reflotron method. *1984*

Tripelennamine At concentration of 200 mg/L had no effect on phosphotungstate reduction method. *3393*

Troleandomycin At 200 mg/L on reversed phase LC procedure of Zhiri et al. *3942*

Tryptophan At 5 mg/dL on copper chelate procedure. *2230*

Tyrosine At 5 mg/dL on copper chelate procedure. *2230* At 5 mg/dL no effect on method of Klein. *1949*

Valproic Acid At 1.0 g/L on reversed phase LC procedure of Zhiri et al. *3942* At concentration of 100 mg/L had no effect on uricase-PAP method. *3393*

Vitamin B Complex At concentration of 32 mg/L had no effect on Kageyama-Hantzsch method. *3393* At concentration of 12.9 mg/L had no effect on catalase-AIDH method. *3393*

Vitamin Preparations No effect at expected concentration with SMA 12/60 procedure. *2637*

Warfarin At concentration of 26 mg/L had no effect on Kageyama-Hantzsch method. *3393* At concentration of 100 mg/L had no effect on phosphotungstate reduction method. *3393*

Xanthine At 10 mg/dL on copper chelate procedure. At 10 mg/dL on Nishi phosphotungstate procedure. *2230*

Serum   No Effect   Physiological

Acetylsalicylic Acid No effect seen in patients undergoing treatment. *1787*

Amiloride No effect observed. *1487* No effect seen in 13 men treated for 8 weeks. *0656*

Atenolol After 100 mg/d for 5 weeks in 20 hypertensive men. *2135*

Bendrofluazide After 12 mo treatment of 53 previously untreated hypertensives. *0315*

Captopril No significant change in 7,000 hypertensives treated for 3 y. *1407*

Cimetidine No effect in volunteers receiving 1.2 g daily for 12 d. *0300* Unchanged although there may be slight increases in creatinine, not progressive: decreases with continued treatment. *2368*

Diatrizoic acid No significant effect of bilateral renal arteriography in 23 patients with hypertension. *2594*

Diclofenac No effect seen in patients undergoing treatment. *1787*

Glucose No change observed after i.v. infusion. *1160*

Indomethacin No effect seen in patients undergoing treatment. *1787*

Ketoprofen No effect seen in patients undergoing treatment. *1787*

Levonorgestrel No significant difference over 3 y between implant recipients and IUD users. *0763* In a group of women using levonorgestrel covered rods versus controls using copper IUD. *0899*

Meals No effect after standard breakfast. *0579*

Nisoldipine In 14 mild to moderately hypertensive noninsulin dependent diabetic patients. *2653*

Oxprenolol In 20 hypertensive men given 160 mg/d for 5 weeks. *2135*

Pindolol No effect in short-term study of 10 hypertensive men with 15 mg/d. *2135*

Prazosin In 23 healthy hypertensives treated for 8 weeks. *2136*

Propranolol After 12 mo treatment of 53 previously untreated hypertensives. *0315* In 340 patients given drug for 10 weeks to reduce diastolic BP to less than 90 mm Hg. *3707*

Sorbitol Increase by 0.3-0.8 mg/dL after i.v. infusion. *1160*

Spironolactone No effect observed. *1487*

Triamterene Unless predisposition to gouty arthritis. *1680*

Serum   Increase   Analytical

Acetaminophen Reacts as if uric acid with SMA 12/60 method. *3335* Falsely high values with phosphotungstate methods. *1311* But increase only of 0.12 mmol/L at concentrations found in overdosage (40 mg/dL) when measured by phosphotungstate reduction. *3363* 40% increase on SMAC® at 5 times upper limit of therapeutic concentration on phosphotungstate procedures routinely used. 14% increase on KDA at 5 times upper limit of therapeutic concentration on phosphotungstate procedures routinely used. *2138* At concentrations above 50 mg/L (normal therapeutic concentration 20 mg/L) raised concentration as measured by phosphotungstate reduction method. *3393*

Acetylsalicylic Acid Increase by 6% with 8326 μmol/L on uricase-catalase method. *1786*

Uric Acid (continued)

Serum Increase Analytical (continued)

Aminophenol At 1 mmol/L affects SMA 12/60 and Henry methods. *3335* At concentrations above 5 mg/L raised concentration as measured by phosphotungstate reduction method. *3393*

Aminophylline Reacts as if uric acid in method of Bittner. *2752*

Ampicillin 20 mg/dL equivalent to 2.6 mg/dL copper chelate procedure. *2230* At concentrations above 100 mg/L (therapeutic concentration about 320 mg/L) raised concentration as measured by Seralyzer. *3393*

Ascorbic Acid 10 mg/dL equivalent to 9.4 mg/dL copper chelate procedure. *2230* At 5 mg/dL increases by 3.15 mg/dL method of Klein. *1949* Measured as reducing substance. *0652* 10 mg/dL equivalent to 0.6 mg/dL with Nishi procedure. *2230* Each 1 mg/dL of drug increases concentration as measured by phosphotungstate method by 0.05 mg/dL. *3719* At concentrations above 100 mg/L (maximum serum concentration 34 mg/L) raised concentration when measured by phosphotungstate reduction. *3393*

Aspirin Acts as reducing substance with nonspecific methods. *1488*

Bilirubin 6% with phosphotungstate procedure at 8.5 mg/dL. *0904* At 32.0 mg/dL and above on recommended method on RA-1000. At 5.0 mg/dL and above on Hitachi® 705 using Behring reagent. *3407* At 16.0 mg/dL and above on method recommended on I.L. Multistat III. *3407*

Caffeine Reduction of phosphotungstate by metabolites. *0583*

Carbenicillin 40 mg/dL equivalent to 0.9 mg/dL copper chelate procedure. *2230*

Chloral Hydrate Affects phosphotungstate reduction methods. *2220*

Chlorine If standards made in contaminated water. *0581*

Coffee Component of coffee measured (method of Bittner). *2752*

Cysteine 5 mg/dL equivalent to 0.5 mg/dL with Nishi procedure 5 mg/dL equivalent to 2.8 mg/dL copper chelate procedure. *2230* Marked increase at 5 times upper limit of therapeutic range on phosphotungstate methods on SMAC® and KDA. *2138*

Cystine 5 mg/dL equivalent to 0.3 mg/dL with Nishi procedure. *2230* Marked increase at 5 times upper limit of therapeutic range on phosphotungstate methods on SMAC® and KDA. *2138*

Dextran Reduces phosphotungstate. *2220*

Epinephrine At 1 mmol/L affects SMA 12/60, Henry methods. *3335* At concentrations above 100 mg/L raised concentration as measured by phosphotungstate reduction method. *3393*

Ergothionine Falsely high values with phosphotungstate methods. *1311*

Etoposide Values of more than 3 times uricase values with direct phosphotungstate method. *2333*

Formaldehyde If added to standards inhibits uricase. *0582* If added to standards retards dialysis etc. *1950*

Gentisic Acid Falsely high values with phosphotungstate methods. *1311*

Glucose At 1 g/dL increases by 0.2 mg/dL with method of Klein. *1949* Reducing substance reacts with phosphotungstate. *0652*

Glutathione Acts as reducing agent with phosphotungstate. *0582*

Hemoglobin At 200 mg/dL 2%, at 424 mg/dL 11% phosphotungstate procedures. *0904*

Hemolysis False positive with glutathione and ergothionine. *0580*

Hydralazine At 1 mmol/L affects SMA 12/60 and Henry methods. *3335* At concentrations above 6 mg/L (normal therapeutic concentration 2.3mg/L) raised concentration as measured by phosphotungstate reduction method. *3393*

Hypoxanthine Measured as uric acid by reducing methods. *3500*

Isoniazid At concentrations above 14 mg/L (normal therapeutic concentration 10 mg/L) raised concentration as measured by phosphotungstate reduction method. *3393*

Levodopa Falsely high values with phosphotungstate methods. *1311* 10 mg/dL equivalent to 31.0 mg/dL copper chelate procedure 10 mg/dL equivalent to 2.3 mg/dL Nishi procedure. *2230* At concentrations above 20 mg/L raised concentration as measured by phosphotungstate reduction method. *3393*

Mercaptopurine At 1 mmol/L affects SMA 12/60 method. *3335* At concentrations above 14 mg/L raised concentration as measured by phosphotungstate reduction method. *3393*

Methicillin 20 mg/dL equivalent to 1.6 mg/dL copper chelate procedure. *2230*

Methyldopa Interferes with phosphotungstate procedure. *1488* At concentrations above 20 mg/L (normal therapeutic concentration 2 mg/L) raised concentration as measured by phosphotungstate reduction method. *3393*

Naproxen At 5 times upper limit of therapeutic range on method on Hitachi® 705. *2138*

Oxyphenbutazone At concentrations above 200 mg/L (normal therapeutic concentration 20 mg/L) raised concentration as measured by Kageyama-Hantzsch method. *3393*

Oxytetracycline At concentrations above 50 mg/L raised concentration as measured by phosphotungstate reduction method. *3393*

Penicillin 1,000 units/mL equivalent to 2.0 mg/dL copper chelate procedure. *2230*

Phenelzine At 1 mmol/L slightly affects SMA 12/60, Henry methods. *3335*

Potassium Salts Turbidity with phosphotungstate procedure. *0583*

Propylthiouracil At 1 mmol/L affects SMA 12/60, Henry methods. *3335* At concentrations above 150 mg/L raised concentration as measured by phosphotungstate reduction method. *3393*

Protein Incomplete precipitation affects method of Morin. *2486*

Protocatechuic Acid Falsely high values with phosphotungstate methods. *1311*

Resorcinol 5 mg/dL equivalent to 5.4 mg/dL copper chelate procedure 5 mg/dL equivalent to 0.3 mg/dL by Nishi procedure. *2230*

Rifampin Substantial increase at upper limit of therapeutic range on methods on KDA and Hitachi® 705. *2138*

Salicylate 30 mg/dL equivalent to 0.1 mg/dL by Nishi procedure. *2230*

Sulfanilamide At concentrations above 250 mg/L raised concentration as measured by Seralyzer method. *3393*

Tetracycline At concentrations above 30 mg/L (normal therapeutic concentration 8 mg/L) raised concentration as measured by Phosphotungstate reduction method. *3393*

Theophylline Reduction of phosphotungstate by metabolites. *2220*

Thioneine Affects phosphotungstate reduction methods. *3500*

Tryptophan 5 mg/dL equivalent to 0.3 mg/dL by Nishi procedure. *2230*

Tyramine Affects phosphotungstate reduction methods. *3500*

Tyrosine 5 mg/dL equivalent to 0.3 mg/dL by Nishi procedure. *2230*

Xanthine Measured as uric acid by nonspecific methods. *3500*

Serum Increase Physiological

Acetazolamide Decreased urate clearance (?also stimulates synthesis). *1022*

Acetoacetate Inhibits tubular secretion of urate. *1331*

Aging Slight to puberty then big increase in men. *0887*

Aldatense Increase by 10-20% in a group of hypertensives compared with controls. *1611*

Altitude Good correlation with hematocrit or hemoglobin level. *3374*

Aluminum Nicotinate May increase uric acid to produce gout. *0126*

Amiloride Possibly due to decreased clearance. *3745* May occur with prolonged therapy particularly if combined with thiazides. *3711*

Anabolic Steroids Improves nitrogen balance. *2313*

Androgens Augments nitrogen balance. *2313*

Angiotensin Reduced urate clearance if given i.v. *1124*

Antineoplastic Agents Destruction of nucleoprotein. *0127*

Aspirin Low doses reduce renal excretion of uric acid. *3879*

Azathioprine Rapid destruction of tissues and nucleic acid catabolism. *1343*

Azathymine Due to tissue destruction. *2313*

Azauridine Due to tissue destruction. *2313*

Azosemide Slight effect noted in 2 volunteers. *2047*

Bendrofluazide Inhibition of tubular secretion of urate. *1433* In men and women in 3 y study of patients with mild hypertension. *2506* Increase by 2.7 to 4 mg/dL approximately in 5 or 6 hypertensives given 5 mg daily for 6 weeks. *3866*

Benzthiazide May occur as with other thiazides. *1680*

Beryllium Salts Nephropathy associated with decreased excretion per nephron. *1903*

Blindness Significantly higher (average 1.4 mg/dL) than normal. *1637*

Blood Transfusions Catabolism of infused blood cells. *1237*

Body Habitus In males and females with overweight. *3872*

Bopindolol Minimal effect observed in large population of hypertensives over 12 weeks study period. *3779*

Bumetanide Increase by 1 mg/dL over 3 mo. *2666*

Busulfan As a result of tissue destruction. *2313*

Capreomycin Nephrotoxic effect associated with hyperuricemia. *1552*

Chloroform May occur with marked tissue destruction. *0279*

Chlorothiazide Decreased urate clearance. *3505*

Chlorthalidone Decreases urate clearance. *0514* Significantly different in post-menopausal women but still increased in premenopausal given drug 100 mg/d for 6 weeks. *0402* From 1.2 to 1.8 mg/dL at doses from 25 to 200 mg/d in 37 patients over 8 weeks. *3643* By 1.7 mg/dL in 39 mg/dL mildly hypertensives treated for 6 weeks. *1404* Slight and insignificant in 10 hypertensive patients given 25 mg daily for 16 week. *1229* 0.9 mg/dL in drug-treated versus placebo-treated patients. *3367* Increased in hypertensives; most marked in those individuals whose initial values were lower. *2904*

Cimetidine Slight increase in patients with renal failure. *2085*

Cisplatin Probable consequence of drug-induced nephrotoxicity. *2553*

Citrates Antagonism of uricosuric effect. *1237*

Clorexolone Occurs as with thiazide drugs. *2427*

Cyclosporine Mean increase from 5.7 to 7.0 mg/dL in 26 patients with inflammatory ocular disease after 3 to 6 mo treatment. 9 patients had increase greater than 2.0 mg/dL. *2713* In 62 renal allograft recipients from 28 to 90 d after transplant: effect greater than with azathioprine and prednisolone: reversible. *0635* Changed from mean of 5.7 mg/dL pretreatment to 7.0 mg/dL in patients with ocular inflammatory disease, not associated with impaired renal function. *2713*

Cyclothiazide Inhibition of tubular secretion of urate. *1433*

Cytarabine Secondary to rapid lysis of neoplastic cells. *1680*

Dantrolene Isolated observation in one individual. *0659*

Diapamide Impaired excretion. *2256*

Diazoxide Inhibition of tubular uric acid excretion. *1805*

Diuretics Reduced clearance. *0127*

Epinephrine Result of vasoconstriction in kidney. *1488*

Ethacrynic Acid Decreased urate clearance with low doses. *3505*

Ethambutol May occur within 24 h, by decreasing clearance. *2852* In 66% patients with tuberculosis when combined with streptomycin and isoniazid not seen when same drugs given with thioacetazone, during first 60-90 d of treatment. Reverted to normal when ethambutol withdrawn. *1928* Due to decreased renal clearance, usually occurs by third week of treatment. *2852*

Ethanol Seen in acutely intoxicated and in response to diet. *2166* Increase by 14% increase in heavy versus occasional drinker. *3273* Significant increase in heavy drinkers versus non-drinkers or occasional drinker. *0059* Common effect and transient, usually returns to normal after 1 week: effect due to decreased renal excretion. *2428* Significantly correlated with alcohol abuse. *0956*

Ethoxzolamide Decreases urate clearance. *2220*

Flumethiazide Inhibition of tubular secretion of urate. *1433*

Fluorides ?related to lactic acidemia. *0010*

Fructose Following i.v. infusion: dose related effect. *3505*

Furosemide Decreased urate clearance. *3505* In about 0.4% of patients receiving drug. *1391* In 10% of 204 hospitalized patients receiving the drug. *3419* Observed after drug administration. *2047*

Gentamicin Reported effect of i.m. injection. *2832*

Halothane In postoperative period with normal renal function. *0369*

Height Positively correlated in children. *0887*

Hydrochlorothiazide Decreased urate clearance. *3505* Slight effect in hypertensive patients given either 25 or 50 mg/d. *1980* Increase by 2.0 mg/dL in 343 patients with hypertension given drug for 10 weeks. *3707* Increase by 1.37 mg/dL in approximately 175 patients with essential hypertension treated for 1 y. *3708*

Hydroflumethiazide Probably due to impaired clearance. *1680*

Hydroxybutyrate Inhibits tubular secretion of urate. *1331*

Hydroxyurea Probable effect of cell catabolism. *1488*

Ibufenac Side effect similar to that of aspirin. *2313*

Ibuprofen Significant effect in group of patients with rheumatoid arthritis, although most values remained within normal range. *0885*

Indapamide Observed in normal men and those with mild essential hypertension. *3780* Increase by 1.4 mg/dL in 17 patients with essential hypertension treated with 2.5 mg daily for 3 mo. *2010*

Indomethacin Oliguric renal failure developed in one man during treatment: reversible. *1245*

Lactate Inhibits tubular secretion of urate. *3930*

Lead Nephropathy associated with decreased secretion per nephron. *0215*

Levarterenol Result of decreased urate clearance. *1124*

Levodopa Two cases reported, exaggerated by fructose. *0042*

Light Sunlight- probably due to increased nucleic acid catabolism. *0264*

Lipomul® Decreased clearance. *2313*

Mannose Similar action to fructose. *3895*

Meals Affected by meals (most marked in women). *3455*

Mecamylamine ?due to reduced renal blood flow. *0924*

Mechlorethamine Leucocyte destruction, catabolism of nucleic acids. *3505*

Mefruside Inhibits excretion of urate. *3854*

Menopause Change of 0.45 mg/dL from 5th to 6th decade. *3830*

Meralluride May cause occasional hyperuricemia. *1680*

Mercaptopurine Leucocyte destruction, catabolism of nucleic acids. *3505*

Mercurial Diuretics May precipitate attacks of gout. *0464*

Methanol Observed with tissue destruction in fatal poisoning. *0279*

Methicillin Nephrotoxic effect. *2313*

Methotrexate Seen in gouty patients — some decrease in leukemics. *2316*

Methoxyflurane Decreased urate clearance: contraction of extracellular fluid volume. *2355*

Methyclothiazide Reduces clearance. *1353*

Metolazone Diuretic action of drug acting on distal tubules. *2816* Increase by 3-5 mmol/L after 6 weeks treatment with 5 mg daily in 5 or 6 hypertensives. *3866*

Mitomycin Due to nephrotoxicity. *2194*

Morinamide Decreased renal clearance. *2427*

Muzolimine Average 26 μmol/L in 10 hypertensive patients given 30 mg daily for 16 weeks. *1229*

Niacin May rise by up to 1.5 mg/dL if large doses. *2736*

Obesity Proportion of subjects with high values increases. *2123*

Pempidine ?due to reduced renal blood flow. *0924*

Phenothiazines Reported effect. *2754*

Phenylbutazone Antiuricosuric action at low doses. *1433*

Piroxicam In patients with normal concentrations pretreatment. Variable effect in people with initial high concentrations. *3819*

Polythiazide Due to impaired renal clearance. *1680*

Prednisone Promotes nucleic acid catabolism. *3505*

Probucol Effect noted in women only. *0806*

Propranolol In 15 patients with essential hypertension when treated for 1 mo. *3776* Mean increase from 6.1 to 6.6 mg/dL in approximately 120 hypertensives after 16 weeks therapy. *2331* But only by 0.1 mg/dL in approximately 120 patients with essential hypertension treated for 1 y. *3708* Approximately 10% increase in both men and women with treatment for 3 y. *2506* Increase by 10% in 23 hypertensive men aged 47-55 y treated for 8 weeks. *1551* Increase by 10% in 23 hypertensive men given up to 160 mg/d for 8 weeks. *2135*

Uric Acid *(continued)*

Serum Increase Physiological *(continued)*

Pyrazinamide Inhibits tubular secretion of urate. *3441* Decreases tubular secretion, may result in clinical manifestations of gout. *3683*

Quinethazone Increased up to 4 mg/dL, inhibits tubular secretion. *0464*

Radioactive Compounds May occur with tissue destruction. *0279*

Reducing Diet Probably due to tissue destruction. *3217*

Rifampin Temporary renal failure reported. *0944*

Season Significantly higher in summer even with controlled intake. *0264*

Spironolactone Decreased urate clearance. *2313*

Starvation Impaired renal excretion and tissue catabolism. *2522*

Theophylline Significant effect dose related and unrelated to reduced clearance. Slight inhibitory effect on hypoxanthine guanine phosphoribosyl transferase. *2488*

Thiazides Decreased renal excretion. *1237*

Thioguanine Rapid destruction of tissues — nucleic acid catabolism. *1343*

Thiotepa Due to cell destruction, may cause nephropathy. *1680*

Triamterene Significant effect in about 17% cases. *1488*

Trichlormethiazide Inhibition of tubular secretion of urate. *1433*

Trimetrexate Nephrotoxicity reported in some patients following treatment, possibly associated with prior reduced renal function. *1613*

Vincristine Increased nucleic acid catabolism. *3505*

Warfarin Noted in men at all levels of renal function without alteration of renal clearance of uric acid, possibly related to increased production: effect up to 25%. *2411*

Weight Positively correlated in children. *0887*

X-Ray Therapy Due to cellular destruction. *1320*

Xylitol Increases by 1.5 to 2 times normal. *0925*

Yeast Equivalent of 8 g/d produce increase of approximately 8 mg/dL. *0286*

Serum Decrease Analytical

Ascorbic Acid At concentrations above 10 mg/L (maximum serum concentration 34 mg/L) lowered concentration as measured by Seralyzer. *3393*

Bilirubin At concentrations as low as 2.5 mg/dL and higher on BMD-8700 using manufacturer recommended method. At 5.0 mg/dL and above on methods recommended on Hitachi® 705. *3407* At 5.0 mg/dL and above on method recommended on Hitachi® 705 with ferrocyanide. *3407*

Cefotaxime Effect of drug on specimens containing tobramycin on Parallel method. *0201*

Chloramphenicol At concentrations greater than 600 mg/L (normal therapeutic concentration 10 mg/L) lowered concentration as measured by Kageyama-Hantzsch method. *3393*

Chlorine If in water affects phosphotungstate procedure. *0583*

Cyanides Uricase is sensitive to cyanide. *1563*

Hemolysis Calculated error -1.4%/g with phosphotungstate procedure. *0515*

Levodopa At concentrations above 3 mg/L lowered concentration as measured by uricase-PAP method. *3393*

Methotrexate At 5 times upper limit of therapeutic range on enzymatic method on Hitachi® 705. *2138*

Methyldopa At concentrations above 300 mg/L (normal therapeutic concentration 2 mg/L) lowered concentration as measured by Kageyama-Hantzsch method. *3393* At concentrations above 3mg/L (normal therapeutic concentration 2 mg/L) lowered concentration as measured by uricase-PAP method. *3393*

Novaminsulfon At concentrations above 100 mg/L (normal therapeutic concentration 15 mg/L) lowered concentration as measured by Kageyama-Hantzsch method. *3393*

Phenylmercuric Acetate At 280 mg/L on uricase procedure of Kabasak. *1840*

Piperazine Reported observation. *0355*

Plasma Mean 0.17 mg/dL higher in serum. *2228*

Tetracycline At concentrations above 150 mg/L (normal therapeutic concentration 8 mg/L) lowered concentration as measured by Kageyama-Hantzsch method. *3393*

Xanthine Linear decrease of from 3 to 36.5% at concentrations of from 100 to 520 mg/L on Kodak Ektachem®. Note of possible clinical importance in patients responding to chemotherapy. Hypoxanthine had no effect up to 400 mg/L. *2854*

Serum Decrease Physiological

Acetohexamide Mild uricosuric action. *3928*

Alanine Decreases high of fasting but not to normal. *1263*

Allopurinol Inhibition of xanthine oxidase. *1022* Therapeutic effect due to inhibition of xanthine oxidase. *2909*

Amiloride Effect observed after 8 weeks in one study. *3745*

Ascorbic Acid From 1.2 to 3.1 mg/dL in 3 subjects ingesting 8 g/d due to uricosuria. *3446*

Aspirin Mild uricosuric action (large dose effect). *0652* Progressive effect during three days but also noted over longer term study at beginning, eventually reverted towards normal. Effect statistically significant. *3076* Noted in several patients receiving therapeutic amounts of drug. *2909* Significant effect observed in group of patients with rheumatoid arthritis. *0885*

Azapropazone Therapeutic intent. *3565* Mean fall after 4 d was 31%, and 46% on 2400 mg daily. *1589*

Azathioprine In patients with gout decreases concentration. *3395*

Azlocillin Fell from average of 6.4 mg/dL to 2.3 mg/dL in 20 hospitalized patients not observed in controls without azlocillin. *1046*

Benzbromarone Potent uricosuric agent: effect enhanced when oxipurinol co-administered. *0695*

Benziodarone Stimulates secretion of urate. *1363*

Cadmium Generalized impairment of proximal tubular function. *0019*

Cannabis Reported effect. *2220*

Chlorothiazide If given i.v. in large doses has uricosuric effect. *1433*

Chlorpromazine Uricosuric action within 2-3 d. *3505*

Chlorprothixene Uricosuric effect within 2-3 d of start of drug. *3505* Reduction of from 8 to 69% in 30 psychiatric patients. *3268*

Cinchophen Uricosuric action. *2313*

Clofibrate Occurs in about 10% (may occasionally cause gout). *0048* Reduction observed in diabetics after 1 week treatment. *1120*

Corticosteroids Produces negative nitrogen balance. *2313*

Corticotropin Uricosuric action. *1433*

Cortisone Observed in normals after 200 mg daily. *3191*

Coumarin Uricosuric effect. *1237*

Diatrizoic acid Uricosuric effect. *2507*

Dicumarol Uricosuric action. *0111*

Diethylstilbestrol In men on therapy (hormonal effect). *2590*

Diflunisal Dose-dependent effect: at dose of 1 g/d concentration was 30% lower than at baseline, maximum effect within 2 weeks. *2390*

Diurnal Variation Minimum 11 am, maximum 5 pm and again at 8 am. *3428*

Enalapril After enalapril alone (up to 160 mg/daily) but may be increased when combined with hydrochlorothiazide. *2277*

Estrogens In men on therapy (hormonal effect). *2590*

Ethacrynic Acid If given i.v. in large doses has uricosuric effect. *3442*

Ethinyl Estradiol In men on therapy (hormonal effect). *2590*

Ethyl Biscoumacetate Uricosuric action. *1433*

Flufenamic Acid Uricosuric action of drug. *3013*

Glucose Increased urate clearance also seen in diabetics. *3344*

Griseofulvin Effective in treatment of gout ?mechanism. *3754*

Guaifenesin Modest effect, eg 3 mg/dL over 3 d. *2909* Modest hypouricemic effect, with fall of up to 3 mg/dL over 3 d in some patients. *2909*

Halofenate Uricosuric action. *2483* 500 mg/d/48 weeks causes reduction of 35%. *0160*

Hydroxyhexamide Mild uricosuric action. *3928*

Ibuprofen Significantly less in drug treated group than in controls. Same effect noted with initiation of therapy. *1787*

Indomethacin Of some value in treatment of gouty arthritis. *1332*

Iodipamide Uricosuric effect. *2507*

Iodopyracet Uricosuric action. *1433*

Iopanoic Acid Uricosuric effect. *2507*

Ipodate Uricosuric effect. *2507*

Lithium Reported to have uricosuric effect. *0144*

Mannitol Reported to have uricosuric action. *1198*

Meals Significant effect in subjects 20-40 y. *3455*

Mechlorethamine More effective than several uricosuric agents. *3013*

Mefenamic Acid Uricosuric action of drug. *3013*

Mercury Compounds Occurs with Fanconi syndrome. *2108*

Mersalyl Uricosuric action. *1433*

Methotrexate Inhibits synthesis. *3217*

Metiazinic Acid Possible interference with tubular reabsorption. *2255*

Oxyphenbutazone Uricosuric effect. *0710*

Phenindione Uricosuric action. *1433*

Phenolsulfonphthalein Uricosuric action. *1433*

Phenothiazines Reported uricosuric action. *2220*

Phenylbutazone Uricosuric action at high doses. *1433*

Pregnancy Mean decrease of 9.8 mg/dL compared with normal. *2439*

Probenecid Uricosuric action. *1433*

Radiographic Agents Interferes with reabsorption. *2909*

Renal Transplant Significant effect over first 3 mo. *3191*

Saline Diminishes urate reabsorption. *3440*

Seclazone Marked antigout response. *1767*

Smoking Less in smokers who inhale than others. *2123* Lower than in non smokers. *2124*

Spironolactone Average fell from 380 to 342 mmol/L in 15 primary hypertension patients at 6 mo after 100 mg drug daily. *1072*

Sulfamethoxazole Presumed uricosuric effect. *3920*

Sulfinpyrazone Uricosuric action. *1433*

Sulfonamides Presumed uricosuric effect. *3920*

Tiaprofenic Acid In 10 healthy subjects given 300 mg bid. *2239*

Vinblastine Of theoretical value in acute gout. *1332*

Urine No Effect Analytical

Storage of Sample For 3 d usually unless bacterial contamination. *0904*

Urine No Effect Physiological

Acenocoumarol No uricosuric action observed. *1487*

Anisindione No uricosuric action observed. *1487*

Bumetanide No effect from 0-2 h. *0822*

Cimetidine No effect in volunteers receiving 1.2 g daily for 12 d. *0300* Although drug cleared principally by glomerular filtration and partial reabsorption in proximal renal tubules. *2368*

Citrates In 89 patients with hypocitraturic calcium nephrolithiasis treated for up to 4 y. *2708*

Urine Increase Analytical

Acetaminophen Falsely high values with phosphotungstate methods. *1311*

Ascorbic Acid Measured as reducing substance. *1563*

Aspirin Acts as reducing substance with nonspecific methods. *1488*

Caffeine Reduction of phosphotungstate by metabolites. *0580*

Ergothionine Falsely high values with phosphotungstate methods. *1311*

Gentisic Acid Falsely high values with phosphotungstate methods. *1311*

Levodopa Falsely high values with phosphotungstate methods. *1311*

Methyldopa Interferes with phosphotungstate procedure. *1237*

Protocatechuic Acid Falsely high values with phosphotungstate methods. *1311*

Theophylline Reduction of phosphotungstate by metabolites. *0580*

Urine Increase Physiological

Acetohexamide Uricosuric action. *1433*

Alanine Reciprocal change with serum when given to fasting. *1263*

Ampicillin Competes with uric acid for reabsorption and potentiates response to probenecid. Significant uricosuria with ampicillin alone from 0-2 h and 8-24 h after administration. *3386*

Ascorbic Acid Substantial effect with large amounts of compound: effect inhibited by pyrazinamide. *3446*

Asparaginase Marked response if lysis of tissues. *2657*

Aspirin High dose effect (greater than 3 g daily). *1022*

Azapropazone Potent uricosuric agent (2400 mg daily comparable to 1 g probenecid). *1589*

Benzbromarone Potent uricosuric agent: effect enhanced when oxipurinol co-administered. *0695*

Benziodarone Stimulates secretion of urate. *1363*

Cadmium Generalized impairment of proximal tubular function. *0019*

Chlorothiazide If given i.v. in large doses has uricosuric effect. *1433*

Chlorpromazine Uricosuric action reported. *2220*

Chlorprothixene Uricosuric action. *1537* Significant increase during treatment. *3268*

Clofibrate Weak transient uricosuric effect. *0071*

Corticotropin Uricosuric action. *1433*

Cortisone Observed in normals after 200 mg daily. *3191*

Coumarin Uricosuric effect. *1237*

Diatrizoic acid Uricosuric effect. *2507*

Dicumarol Uricosuric action. *1433*

Diethylstilbestrol In men on therapy (hormonal effect). *2590*

Estrogens In men on therapy (hormonal effect). *2590*

Ethacrynic Acid If given i.v. in large doses has uricosuric effect. *3442*

Ethinyl Estradiol In men on therapy (hormonal effect). *2590*

Ethyl Biscoumacetate Uricosuric action. *1433*

Fructose Dose related effect. *3113*

Glucose Infusions have uricosuric action. *3344*

Glycine Increased tubular secretion especially in gouty subjects. *3929*

Halofenate Effect independent of GFR. *2483*

Hydroxyhexamide Mild uricosuric action. *3928*

Iodipamide Uricosuric effect. *2507*

Iodopyracet Uricosuric action. *1433*

Iopanoic Acid Uricosuric effect. *2507*

Ipodate Uricosuric effect. *2507*

Lithium Reported effect of LiCO$_3$ in manic-depressives. *1487*

Mannitol Intravenously produces uricosuria. *1487*

Mannose Great increase after i.v. infusion. *0358*

Mercaptopurine Increased nuclear protein breakdown. *1237*

Mercury Compounds Occurs with Fanconi syndrome. *2108*

Mersalyl Uricosuric action. *1433*

Methotrexate Increased excretion associated with cell destruction. *1320*

Niacinamide By up to 40% at 2 h maximum at 4 h. *1622*

Nifedipine Significant effect following 20 mg sublingual dose. *0650*

Phenindione Uricosuric action. *1433*

Phenolsulfonphthalein Uricosuric action. *1433*

Phenothiazines Reported uricosuric action. *2220*

Phenylbutazone Uricosuric action. *1433*

Phloridzin Uricosuric action in man. *3343*

Probenecid Uricosuric action. *1433*

Protein Increase in proportion to dietary purine. *0987*

Pyruvate Observed after oral feeding. *2422*

Saline Hypertonic solution may have marked effect. *0574*

SC-16102 Minor increased clearance noted. *1804*

Uric Acid *(continued)*

Urine Increase Physiological *(continued)*

Seclazone Uricosuric action. *1767*

Sulfamethoxazole Presumed uricosuric effect. *3920*

Sulfinpyrazone Uricosuric action. *1433*

Sulfonamides Uricosuric effect. *1237*

Thioguanine Due to augmented tissue catabolism. *1343*

Tiaprofenic Acid In 10 healthy subjects given 300 mg bid, but effect occurs early so no increased excretion may be observed later. *2239*

Triamterene Slight uricosuric action. *1237*

X-Ray Therapy Due to cellular destruction. *1320*

Xylitol As result of increased serum concentration (may be doubled). *0925*

Yeast Equivalent of 8 g/d produce increase of approximately 1.2-1.4g/d. *0286*

Urine Decrease Analytical

Formaldehyde Inhibits uricase when added as preservative. *0583*

Storage of Sample At any temperature unless alkali added. *2954*

Urine Decrease Physiological

Acetazolamide Inhibition of tubular secretion of urate. *1022*

Acetoacetate Inhibits tubular secretion of urate. *1331*

Allopurinol Inhibition of xanthine oxidase. *1022*

Angiotensin Decreases urate excretion and renal plasma flow. *1124*

Aspirin Low doses effect. *3879*

Azathioprine In patients with gout reduces excretion. *3395*

Azosemide Slight effect noted on first 2 d. *2047*

Benzbromarone Statistically significant reduction in excretion and in clearance versus creatinine clearance in comparison with no treatment. *3386*

Beryllium Salts Nephropathy associated with decreased excretion per nephron. *1903*

Blindness Uric acid/creatinine ratio lower than normal. *1637*

Bumetanide Significantly reduced from 3-6 h. *0822*

Chlorothiazide Impaired clearance. *2220*

Chlorthalidone Decreased clearance. *2220*

Citrates Antagonism of uricosuric effect. *1237*

Diazoxide Effect over 2 h of 4 mg/kg given i.v. *1805*

Ethacrynic Acid Decreased urate clearance. *1022*

Ethambutol Decreases clearance but mechanism not known. *2852*

Ethanol Impaired excretion during period of intake. *1343*

Ethoxzolamide Decreases urate clearance. *2220*

Furosemide Slight effect observed on first and second day. *2047*

Hydrochlorothiazide Decreased urate clearance. *2220*

Hydroxybutyrate Inhibits tubular secretion of urate. *1331*

Lactate Inhibits tubular secretion of urate. *3930*

Lead Nephropathy associated with decreased secretion per nephron. *0215*

Levarterenol Decreases urate excretion and renal plasma flow. *1124*

Niacin Decreases by approximately 50%. *1253*

Probenecid Small doses depress secretion. *1343*

Pyrazinamide Increased tubular reabsorption of urate. *3193*

Pyrazinoic Acid Inhibition of renal tubular secretion. *0297*

Thiazides Decreased renal excretion (18% in long term). *0269*

CSF Increase Physiological

Ethanol Especially high on withdrawal, remains increased in alcoholics. *0590*

Test Conditions Increase Analytical

Aminoimidazoleacetic Acid Positive spot test with phosphotungstate. *3765*

Dopamine Positive spot test with phosphotungstate. *3765*

Epinephrine Positive spot test with phosphotungstate. *3765*

Ferulic Acid Positive spot test with phosphotungstate. *3765*

Homogentisic Acid Positive spot test with phosphotungstate. *3765*

Homovanillic Acid Positive spot test with phosphotungstate. *3765*

5-Hydroxyindoleacetic Acid Positive spot test with phosphotungstate. *3765*

5-Hydroxytryptamine Positive spot test with phosphotungstate. *3765*

Levarterenol Positive spot test with phosphotungstate. *3765*

Metanephrine Positive spot test with phosphotungstate. *3765*

Methoxytyramine Positive spot test with phosphotungstate. *3765*

Normetanephrine Positive spot test with phosphotungstate. *3765*

Thiamine Positive spot test with phosphotungstate. *3765*

Uric Acid Positive spot test with phosphotungstate. *3765*

Vanillic Acid Positive spot test with phosphotungstate. *3765*

Vanillylmandelic Acid (VMA) Positive spot test with phosphotungstate. *3765*

Uric Acid Clearance

Urine No Effect Physiological

Radiographic Agents No effect of arteriography. *2594*

Urine Increase Physiological

Amphotericin B Varies inversely with GFR (consistent change). *0536*

Diethylstilbestrol In men on therapy (hormonal effect). *2590*

Estrogens In men on therapy (hormonal effect). *2590*

Ethinyl Estradiol In men on therapy (hormonal effect). *2590*

Fructose Dose related effect. *3113*

Phloridzin Probably inhibits tubular reabsorption. *3343*

Sex Difference In women average of 12.67 mL/min, in men 10.38 mL/min. *2590*

Urine Decrease Physiological

Angiotensin 44% reduction with infusion. *1125*

Diazoxide Effect over 2 h of 4 mg/kg given i.v. *1805*

Epinephrine Sharp fall following i.m. injection. *2427*

Levarterenol 20% reduction with infusion. *1125*

Niacin Decreases by 75%, ?due to altered tubular handling. *1253*

Thiazides Decreased renal clearance. *2427*

Triamterene Reduced in proportion to creatinine clearance. *3745*

Urobilin

Urine No Effect Physiological

Cromoglycate No clinical significant change observed. *0484*

Urine Increase Analytical

Acriflavine Produces yellow-green fluorescence. *1563*

Fluorescein Produces yellow-green fluorescence. *1563*

Merbromin Yields pink color and mauve fluorescence. *1563*

Phylloerythrinogen Produces red fluorescence. *1563*

Porphyrins Produce red fluorescence. *1563*

Riboflavin Produces yellow-green fluorescence. *1563*

Tetraiodofluorescein Pink color with mauve fluorescence. *1563*

Urine Increase Physiological

Allopurinol Transient increases reported. *2427*

Ampicillin Necrosis of tubules due to nephrotoxicity. *3542*

Aspirin Occurs with poisoning. *1302*

Capreomycin May cause nephrotoxicity. *1680*

Heroin Occasional microscopic evidence in addicts. *2914*

Ipodate Nephrotoxic effect. *0838*
Iron Salts May exacerbate urinary tract infections. *2427*
Kanamycin Nephrotoxic effect. *1680*
Levodopa Usually minor and unimportant. *0824*
Methicillin Pyuria reported as complication. *1680*
Muscular Exercise Noted after heavy exercise. *1842*

Urobilinogen

Urine   No Effect   Analytical

Bilirubin If p-MDFB procedure and absorption read at 436 nm. *2044*
Indican No effect p-MDFB procedure of Rutter. *2044*
Indole No effect on p-MDFB procedure of Rutter. *2044*
Phenazopyridine No effect in quantitative p-MDFB procedure of Rutter. *2044*
Porphobilinogen Extractable and no effect on p-MDFB procedure. *2044* No effect on Urobilistix® in porphyric patients. *1858*
Skatole No effect on p-MDFB procedure of Rutter. *2044*
Thymol No effect on Waldenström procedure. *2981*
Vanillylmandelic Acid (VMA) No effect on p-MDFB procedure of Rutter. *2044*

Urine   Increase   Analytical

Acetone May produce similar color in reaction. *2313*
Aminosalicylic Acid Reacts with Ehrlich's reagent (yellow color). *0252*
Antipyrine Produces similar color with Ehrlichs. *3879*
Apronalide Produces red color with Ehrlich's reagent. *1563*
Bananas 5-HIAA produced reacts with Ehrlich's reagent. *1563*
Bile Salts Converted to acids, produce turbidity. *1563*
Bilirubin Produces green color due to biliverdin. *1563*
Biliverdin May produce green color. *1563*
Cascara Reacts with Ehrlich's reagent. *1563*
Chlorophyll Metabolite phylloerythrinogen produces red color. *0583*
Chlorpromazine Reacts with Erhlich's aldehyde reagent. *3879*
Coffee Acetyltryptophan produced reacts with Ehrlich's. *1563*
Diatrizoic acid Heavy yellow precipitate with Ehrlich's. *0583*
Epinephrine Positive spot test with Ehrlich's reagent. *3765*
Formaldehyde Interferes with color reaction. *0757*
5-Hydroxyindoleacetic Acid Reacts with Ehrlich's reagent. *1563*
Indican Produces yellow color with Ehrlich's reagent. *1563*
Indole Produces red color with Ehrlich's reagent. *1563*
Indoleacetic Acid Produces red color with Ehrlich's reagent. *1563*
Iopanoic Acid Produce white cloudy precipitate with Ehrlich's. *0583*
Melanin Melanogen reacts with Ehrlich's reagent. *1563*
Methenamine Produces formaldehyde which interferes. *0252*
Phenazopyridine Orange-red with Ehrlich's aldehyde reaction. *3879*
Phenothiazines May react with Ehrlich's aldehyde reagent. *2950*
Phylloerythrinogen Produces red color with Ehrlich's reagent. *1563*
Porphobilinogen But not extractable into chloroform. *0251*
Procaine Interferes with Ehrlich's aldehyde reaction. *3879*
Proteinuria Produces turbidity in direct methods. *1563*
Radiographic Agents Turbidity in acid solutions. *0583*
Skatole Produces blue color with Ehrlich's reagent. *1563*
Smog Oxidation of spots on chromatogram by ozone. *0583*
Sulfadiazine May produce greenish color with Ehrlich's reagent. *0459* Gives greenish-yellow color with Ehrlich's. *0251*
Sulfamethizole Yellow color with Ehrlich's (extracted into $CHCl_3$). *2359*
Sulfamethoxazole May react with Ehrlich's to produce false color. *0085*
Sulfanilamide Gives greenish-yellow color with Ehrlich's. *0251*
Sulfapyridine Yields greenish color with Ehrlichs. *0459*

Sulfathiazole Yields greenish color with Ehrlichs. *0459*
Sulfisoxazole Reacts with Urobilistix®. *1487*
Sulfobromophthalein Color development with Ehrlich's aldehyde reagent. *3879*
Sulfonamides React with Ehrlich's reagent. *1563*
Tetracycline Yellow color extracted into $CHCl_3$ (Ehrlich's procedure). *2359*
Tryptophan Produces an orange color with Ehrlich's reagent. *1563*

Urine   Increase   Physiological

Acetazolamide Excretion increased by alkalinization of urine. *2452*
Amyl Nitrite Less reliable index than fecal if hemolysis occurs. *2426*
Antipyrine Occurs with hemolytic anemia. *0788*
Chloroquine Hemolytic anemia in G-6-PD deficiency. *1680*
Dapsone Associated with hemolytic anemia when occurs as main side effect. *3423*
Diurnal Variation Maximum excretion in afternoon. *0459*
Hemolysis Further metabolism of hemoglobin. *2313*
Imipramine May cause cholestatic jaundice. *1873*
Lead Due to hemolysis of poisoning. *0987*
Methyldopa Autoimmune hemolytic anemia. *0597*
Nalidixic Acid May occur if cholestasis. *2808*
Novobiocin Hemolytic anemia in G-6-PD deficiency. *1343*
Phenazopyridine Gives false positive with Ehrlich's reaction. *0251*
Pipobroman Hemolytic anemia has been described. *0070*
Sodium Bicarbonate Increased clearance when urine alkaline. *2452*
Sulfobromophthalein Presence of BSP in urine and liver disease. *1022*
Sulfonamides Hemolytic anemia with G-6-PD deficiency. *1343*
Tolmetin Observed in single case of 49 y old man who had taken drug as needed for arthritis for 1 y. *3421*
Vitamin K Hemolytic anemia in G-6-PD deficiency. *1343*

Urine   Decrease   Analytical

Formaldehyde Marked effect at 200 mg/dL on p-MDFB procedure. *2044*
Light Oxidation by light occurs. *0582*
Nitrites Slight effect only p-MDFB procedure of Rutter Marked effect with Ehrlich's procedure. *2044*
Urea Produces yellow color with Ehrlich's reagent. *1563*

Urine   Decrease   Physiological

Acetohexamide Theoretical effect if cholestasis. *0071*
Ajmaline Intrahepatic cholestasis. *0268*
Ammonium Salts Acidification of urine decreases excretion. *2452*
Antibiotics Inhibit gastrointestinal tract flora. *1238*
Ascorbic Acid Lowered pH reduces excretion. *2452*
Chloramphenicol Reduction of flora in gastrointestinal tract. *1488*
Chlordiazepoxide Infrequent cholestatic effect. *1713*
Chlorpromazine May be cholestasis. *1713*
Chlorpropamide Cholestasis may occur infrequently. *1713*
Erythromycin May occur with cholestasis. *1713*
Mepazine Cholestatic effect. *1713*
Methandrostenolone Due to cholestasis. *1713*
Methimazole Cholestatic effect. *1713*
Methyltestosterone Cholestatic effect. *1713*
Neomycin Reduces flora in gastrointestinal tract. *1488*
Nitrofurantoin Intrahepatic cholestatic jaundice. *1045*
Norethandrolone Manifestation of cholestasis. *1280*
Oral Contraceptives May induce cholestasis. *1713*
Oxymetholone Cholestatic effect. *1713*
Prochlorperazine Cholestatic effect. *0071*
Promazine Cholestatic effect. *1713*

Urobilinogen *(continued)*

Urine   Decrease   Physiological *(continued)*
Spinal Anesthesia   Related to reduced hepatic blood flow. *1343*
Sulfadimethoxine   Occurs with reversible cholestasis. *1049*
Sulfamethizole   Occurs with reversible cholestasis. *1974*
Sulfamethoxazole   Occurs with reversible cholestasis. *2313*
Sulfamethoxypyridine   Occurs with reversible cholestasis. *2313*
Sulfisoxazole   With reversible cholestasis. *1974*
Sulfonamides   Cholestatic effect. *1713*
Testosterone   May cause cholestasis. *2313*
Tetracycline   Reduces flora in gastrointestinal tract. *1488*
Thiabendazole   Cholestatic effect. *2413*
Thiacetazone   May cause cholestasis. *0248*
Thiazides   Cholestatic effect. *1713*
Tolazamide   Cholestatic effect. *2313*
Tolbutamide   Possible cholestatic effect. *1713*
Trifluoperazine   Cholestatic effect. *1713*

Feces   Increase   Physiological
Amyl Nitrite   If hemolytic anemia occurs. *2426*
Antipyrine   Occurs with hemolytic anemia. *0788*
Methyltestosterone   Pale stools, may cause cholestasis. *1713*

Feces   Decrease   Analytical
Storage of Sample   Decreased up to 46% within 24 h at 4°. *1563*

Feces   Decrease   Physiological
Acetohexamide   Theoretically, if cholestasis occurs. *0071*
Ajmaline   Intrahepatic cholestasis. *0268*
Aminosalicylic Acid   Pale stools may occur if cholestasis. *1713*
Antibiotics   Inhibit gastrointestinal tract flora. *1237*
Chloramphenicol   Reduction of flora in gastrointestinal tract. *1488*
Chlordiazepoxide   Infrequent cholestatic effect. *1713*
Chlorpromazine   Pale stools, due to cholestasis. *1713*
Chlorpropamide   Pale stools due to cholestasis (infrequent). *1713*
Erythromycin   May cause cholestasis (reversible). *1713*
Mepazine   Due to cholestasis (pale stools result). *1713*
Methandrostenolone   Light stools, due to cholestasis. *1713*
Methimazole   Cholestatic effect. *1713*
Nalidixic Acid   May occur with cholestasis. *2808*
Neomycin   Reduces flora in gastrointestinal tract. *1488*
Norethandrolone   Light stools due to cholestasis. *1713*
Oral Contraceptives   Pale stools, due to cholestasis. *1713*
Oxymetholone   Cholestatic effect. *1713*
Prochlorperazine   Pale stools associated with cholestasis. *0071*
Promazine   Pale stools as result of cholestasis. *1713*
Sulfadimethoxine   Pale stools with reversible cholestasis. *1049*
Sulfamethizole   Pale stools with reversible cholestasis. *1974*
Sulfamethoxazole   Pale stools with reversible cholestasis. *2313*
Sulfamethoxypyridine   May cause reversible cholestasis. *2313*
Sulfisoxazole   Pale stools with reversible cholestasis. *1974*
Sulfonamides   Pale stools with reversible cholestasis. *1713*
Tetracycline   Reduces flora in gastrointestinal tract. *1488*
Thiabendazole   Pale stools, cholestatic effect. *2413*
Thiacetazone   Pale stools, occurs with cholestasis. *0248*
Thiazides   May cause cholestasis with pale stools. *1713*
Tolazamide   Pale stools with cholestasis. *2313*
Tolbutamide   Pale stools with cholestasis may occur. *1713*
Trifluoperazine   Pale stools with cholestasis. *1713*

Test Conditions   Increase   Analytical
Acetylhistamine   Positive spot test with Ehrlich's reagent. *3765*
Cresol   Positive spot test with Ehrlich's reagent. *3765*
Dopamine   Positive spot test with Ehrlich's reagent. *3765*

Ergothionine   Positive spot test with Ehrlich's reagent. *3765*
Ferulic Acid   Positive spot test with Ehrlich's reagent. *3765*
Homogentisic Acid   Positive spot test with Ehrlich's reagent. *3765*
Homovanillic Acid   Positive spot test with Ehrlich's reagent. *3765*
Hydroxybenzoic Acid   Positive spot test with Ehrlich's reagent. *3765*
Hydroxycinnamic Acid   Positive spot test with Ehrlich's reagent. *3765*
5-Hydroxyindoleacetic Acid   Positive spot test with Ehrlich's reagent. *3765*
Hydroxykynurenine   Positive spot test with Ehrlich's reagent. *3765*
Hydroxyphenylacetic Acid   Positive spot test with Ehrlich's reagent. *3765*
5-Hydroxytryptamine   Positive spot test with Ehrlich's reagent. *3765*
Levarterenol   Positive spot test with Ehrlich's reagent. *3765*
Metanephrine   Positive spot test with Ehrlich's reagent. *3765*
Methoxytyramine   Positive spot test with Ehrlich's reagent. *3765*
Normetanephrine   Positive spot test with Ehrlich's reagent. *3765*
Phenols   Positive spot test with Ehrlich's reagent. *3765*
Pyridoxal   Positive spot test with Ehrlich's reagent. *3765*
Pyridoxamine   Positive spot test with Ehrlich's reagent. *3765*
Pyridoxine   Positive spot test with Ehrlich's reagent. *3765*
Thiamine   Positive spot test with Ehrlich's reagent. *3765*
Tyramine   Positive spot test with Ehrlich's reagent. *3765*
Tyrosine   Positive spot test with Ehrlich's reagent. *3765*
Vanillic Acid   Positive spot test with Ehrlich's reagent. *3765*
Vanillylmandelic Acid (VMA)   Positive spot test with Ehrlich's reagent. *3765*
Xanthurenic Acid   Positive spot test with Ehrlich's reagent. *3765*

Urocanic Acid

Urine   Increase   Physiological
Methotrexate   With treatment of folic acid/B_{12} deficiency. *3681*

Urocanylglycine

Urine   Increase   Physiological
Methotrexate   Large effect of treatment observed. *3681*

Urokinase

Urine   Increase   Physiological
Proteinuria   If renal infarction or hypoxia. *2897*

Uropepsin

Urine   No Effect   Analytical
Storage of Sample   Several days at room temperature if no bacterial growth. *0877*

Urine   Decrease   Analytical
Storage of Sample   Increase by 10-30% in 4 d even with toluene at 4°. *0877*

Uropepsinogen

Urine   Increase   Physiological
Corticosteroids   Increase renal clearance. *0814*
Pregnancy   Moderate increase observed. *0987*

Uroporphyrin

Urine   No Effect   Physiological
Cimetidine   No measurable change in one patient with acute intermittent porphyria. *1659*

Urine   Increase   Physiological
Anticonvulsants   Drug related effect reported in one child. *2427*
Barbiturates   May precipitate acute porphyria. *1322*
Chloroquine   Transient increase may occur with cutaneous porphyria. *1322*
Chlorpropamide   May precipitate cutaneous porphyria. *1322*
Diethylstilbestrol   May induce porphyria cutanea tarda. *3769*
Estrogens   May induce porphyria cutanea tarda. *3769*
Estrogens, Conjugated   May induce porphyria cutanea tarda. *3821*
Ethanol   May precipitate attack of porphyria. *1322*
Ethinyl Estradiol   May induce porphyria cutanea tarda. *3769*
Oral Contraceptives   May induce porphyria cutanea tarda. *3029*
Sulfomethane   May provoke attack of porphyria. *1322*
Tolbutamide   May precipitate porphyria attack. *1322*

Urine   Decrease   Physiological
Actinomycin   Suppresses induction of ALA synthetase in porphyria. *0938*
Glucose   Suppresses ALA synthetase in acute porphyria. *0938*
Vitamin E   Reduce to normal levels in porphyria. *2549*

Feces   Increase   Physiological
Diethylstilbestrol   May induce porphyria cutanea tarda. *3769*
Estrogens   May induce porphyria cutanea tarda. *3769*
Estrogens, Conjugated   May induce porphyria cutanea tarda. *3821*
Ethinyl Estradiol   May induce porphyria cutanea tarda. *3769*

Valine

Plasma   Increase   Physiological
Obesity   Characteristic finding correlated with insulin. *1106*
Starvation   Marked effect even after 1 d. *0021*

Plasma   Decrease   Physiological
Alanine   Decrease of about 40% with 1 week administration. *1263*
Glucose   Probable muscle uptake. *1403*
Histidine   Probably due to flow into gut after oral load. *1644*
Oral Contraceptives   Hormonal effect (second part of cycle). *0742*

Urine   Increase   Physiological
Starvation   Observed effect due to tissue catabolism. *0021*

Urine   Decrease   Physiological
Ascorbic Acid   In 10 healthy females given 10 g/d. *3624*

Valproic Acid

Serum   Decrease   Physiological
Carbamazepine   Significant reduction of half-life from 10.9 h to 6.4 h due to enzyme induction. *2233*   Concentration reduced to about 66% when carbamazepine coadministered. *2340*
Phenobarbital   Concentration reduced to about 75% when phenobarbital coadministered. *2340*   Significant reduction in half-life from 10.9 h to 8.2 h due to enzyme induction. *2233*
Phenytoin   Significant reduction of half-life from 10.9 h to 6.9 h when drugs coadministered. *2233*   Concentration approximately 50% less when phenytoin co-administered. *2340*

Vancomycin

Serum   No Effect   Analytical
Mezlocillin   Negligible interference at up to 500 mg/L on Abbott TDx. *2417*
Piperacillin   Negligible interference at up to 500 mg/L on Abbott TDx. *2417*

Vanillic Acid

Urine   Increase   Physiological
3,4-Dihydroxybenzoic Acid   Significant increase with glycine conjugate. *2870*

Vanillylamine

Urine   Increase   Physiological
Vanilla   Occurs after ingestion of food with vanilla flavor. *2784*

Vanillylmandelic Acid

Plasma   No Effect   Physiological
Chlorthalidone   In 22 premenopausal and 18 postmenopausal women given 100 mg/d for 6 weeks. *0402*

Urine   No Effect   Analytical
Amphetamine   No effect observed. *0913*
Chlorpromazine   No effect on Gitlow method. *1487*
Dextroamphetamine   Apparently no method interference. *1487*
Digoxin   No effect observed. *0913*
Isoproterenol   No apparent methodological effect. *1487*
Labetalol   On method of Pisano et al. *2430*
Meprobamate   No effect observed. *0913*
Methyldopa   No interference with Pisano procedure. *0583*
Reserpine   No effect on Gitlow method. *1487*
Storage of Sample   Acidified several days room temperature, 1 week at 4°. *3873*
Vanillin   Dietary up to 62 mg/d no effect Pisano procedure. *0583*

Urine   No Effect   Physiological
Amphetamine   No effect observed. *0913*
Angiotensin   Epinephrine decrease compensates for norepinephrine increase. *3967*
Antibiotics   Sterilization of gut has no effect on excretion. *0426*
Barbiturates   No effect with addiction or withdrawal. *3634*
Chlorothiazide   No effect observed. *0913*
Clopamide   Single case of diuretic associated myopathy. *1793*
Dextroamphetamine   No physiological effect. *1487*
Digoxin   No effect observed. *0913*
Ephedrine   No effect observed. *0913*
Hydralazine   No effect observed. *0913*
Isoproterenol   No apparent physiological effect. *1487*
Mecamylamine   No effect observed. *0913*
Meprobamate   No effect observed. *0913*
Molindone   Relatively unaffected over short term. *1635*
Neomycin   No effect on excretion observed. *0426*
Oral Contraceptives   No effect observed with ingestion. *3217*
Phenoxybenzamine   No effect observed. *0913*
Phentolamine   No effect observed. *0913*
Prochlorperazine   No significant effect with up to 30 mg in controls. *1607*
Season   Same in summer and winter in Antarctica. *3217*
Smoking   No effect usual smoking in smokers. *3626*
Storage of Sample   No effect observed for 6 y at -20°. *0426*
Thiothixene   No effect on short term administration. *1635*

Vanillylmandelic Acid *(continued)*

Urine   Increase   Analytical

Aminosalicylic Acid   Reacts with diazo reagent if used. *0086*

Anileridine   Interferes with Pisano procedure. *1022*

Aspirin   Interferes with fluoro-, colorimetric procedures. *0502*

Bananas   Metabolites interfere with analysis. *1237*

Caffeine   Interference with method reported. *1237*

Chocolate   Catechols in chocolate. *1237*

Coffee   Interference with nonspecific methods. *2704*

Disulfiram   ?interference from acetaldehyde from ethanol. *1488*

Guaifenesin   Affects p-nitraniline in initial part of reaction. *0757*

5-Hydroxyindoleacetic Acid   Linear relationship with concentration affect Mahler method. *1149*   In screening method 5-HIAA reacts as if VMA. *0583*

Hydroxymandelic Acid   Unless corrected for, affects method of Sunderman. *2928*

Mephenesin   Purple or red color in initial part of screening test. *0570*

Methenamine   Interferes with fluorometric procedures. *2220*

Methocarbamol   False positive with screening, no effect quant method. *1022*

Methyldopa   Interference by 5-HIAA with nonspecific diazo reaction. *3218*

Nalidixic Acid   Apparent 4 fold with normal regime. *0583*   Affects Pisano procedure. *0121*

Oxytetracycline   Produces interfering color. *3879*

Phenazopyridine   Yields similar color in reaction. *2220*

Phenolsulfonphthalein   May interfere with colorimetric method. *0086*

Stibophen   At neutral or alkaline pH affect Gitlow procedure. *3873*

Sulfobromophthalein   Occurs unless completely extracted. *0086*

Tea   Metabolites interfere with analysis. *3506*

Vanilla   May be measured by colorimetric method. *1237*

Urine   Increase   Physiological

Ajmaline   Release of stored norepinephrine. *2220*

Chlorpromazine   Increased metabolism and decreased organ uptake of norepinephrine. *3879*

Creatinine Clearance   Positive correlation with clearance. *1636*

Diurnal Variation   Significantly higher during day than at night. *1176*

Epinephrine   Measured as endogenous compound. *1488*

Glucagon   Significant effect during i.v. in patients with congestive cardiac failure. *1607*

Guanethidine   Initial pharmacological response (changes later). *0913*

Histamine   Due to release from adrenals. *1176*

Insulin   Increase after insulin shock, none with normal dose. *3503*

Isoproterenol   Pharmacol effect (still seen 4 d after stopping). *0913*

Ketosis   Marked effect ?due to hypovolemia. *0171*

Levarterenol   Normal metabolite, effect slight usually. *3503*

Levodopa   Small increase, larger increase HVA. *0562*

Lithium   Slight increase only, mechanism not clear. *1488*

Methyldopa   Initial physiological response, changes later. *0913*

Muscular Exercise   Significant effect after physical stress. *1176*

Nitroglycerin   Effect greater than that on catecholamines. *3713*

Prochlorperazine   Increased metabolism, decreased organ uptake of norepinephrine. *3879*

Rauwolfia   Release of stored norepinephrine. *2220*

Reserpine   Release of stored norepinephrine. *2220*

Starvation   Similar response to other forms of stress. *2382*

Syrosingopine   Release of stored norepinephrine. *2220*

Urine   Decrease   Analytical

Alkaline Urine   Unstable at room temperature above pH 3. *0913*

Aspirin   At therapeutic physiological concentrations on Pisano procedure. *1097*

Clofibrate   Due to interfering glucuronides of drug. *3109*

Dihydroxyphenylacetic Acid   At therapeutic physiological concentrations on Pisano procedure. *1097*

Gentisic Acid   At therapeutic physiological concentrations on Pisano procedure. *1097*

Homogentisic Acid   At therapeutic physiological concentrations on Pisano procedure. *1097*

Homovanillic Acid   At therapeutic physiological concentrations on Pisano procedure. *1097*

Levodopa   At therapeutic physiological concentrations on Pisano procedure. *1097*

Urine   Decrease   Physiological

Clonidine   Dose related effect. *1627*

Debrisoquin   Mechanism not discussed. *0121*

Disulfiram   Although HMPG excretion increased. *0426*   Probably due to inhibition of dopamine hydroxylase. *3703*

Ethanol   Following ingestion after 0.4 g alcohol/kg in 56 healthy males. *0015*

Guanethidine   30% reduction seen if given therapeutically. *3897*

Hydrazine Derivatives   May potentiate action of drugs on CNS. *2220*

Imipramine   Approximately 30% decrease, block of uptake into cells. *1520*

Isocarboxazid   Inhibition of formation. *1488*

Levodopa   ?dopamine as neurotransmitter suppresses normetanephrine. *1700*

MAO Inhibitors   Inhibition of normetanephrine conversion to VMA. *1488*

Methyldopa   Depletion of tissue stores. *0923*

Morphine   Clinically significant but small effect. *2356*

Nialamide   Effect observed in schizophrenics. *2427*

Phenothiazines   Alters blood concentration. *0127*

Radiographic Agents   Competes for excretion after i.v. pyelography. *0913*

Reserpine   Depletion of catecholamine stores. *2425*

Uremia   Probably due to toxicity and impaired excretion. *2472*

Vasoactive Intestinal Polypeptide

Plasma   No Effect   Physiological

Morphine   No effect observed in 6 healthy volunteers after test meal and drug i.v. *0626*

Naloxone   In 6 healthy volunteers after ingestion of test meal and intravenous administration of drug. *0626*

Viscosity

Blood   No Effect   Physiological

Ethanol   Unaffected by drinking ethanol (33 mmol/kg). *1599*

Blood   Increase   Physiological

Diurnal Variation   Maximum about 8 am, minimum about midnight. *0242*

Ethanol   Significant increase at 15 h after ingestion of 100 mg in 4 healthy male volunteers. *1223*

Vital Capacity

Patient   No Effect   Physiological

Oral Contraceptives   No significant effect observed. *1185*

Vitamin A

Serum   No Effect   Analytical

Carotene   On modified fluorometric procedure of Drujan. *3685*

Hemolysis   No effect of method of Price and Carr. *2734*

Heparin   No effect on method of Price and Carr. *2734*

Phytofluene   No effect on method of Bubb-Murphy. *0517*
Storage of Sample   No effect on method of Price/Carr (4° for 6 d). *2734*

Serum   No Effect   Physiological
Anticonvulsants   In 146 epileptics with long-term treatment. *2007*
Ascorbic Acid   No effect with varying degrees of supplementation over 14 week study period in young men. *1751*
Cholestyramine   Absorption not significantly impaired. *2427*
Clofibrate   Insignificant change in about 26 subjects given 2 g daily. *2875*
Colestipol   No effect with 30 g daily of drug. *2875*
Ethanol   No different in chronic alcoholics and controls. *0367*
Phenobarbital   No difference between treated and untreated cognitively delayed children. *2003*

Serum   Increase   Analytical
Canthoxanthine   During administration and for 10 d after may affect certain analytical methods. *3023*
Carotene   If high in serum will affect UV methods. *3028*
Carotenoid   May interfere with some analytical methods. *3023*
Storage of Sample   After frozen storage and analyzed by method of Price/Carr. *2734*

Serum   Increase   Physiological
Ethinyl Estradiol   Due to alteration in concentration of binding globulin. *1221*
Norethindrone   Due to alteration in concentration of binding globulin. *1221*
Oral Contraceptives   Increased to about 150% of controls in women on a variety of oral contraceptives. *1303*   Values doubled in 6 mo in 8 women taking estrogenic combination but diminished to some extent after 4 y. *0773*   Values almost doubled within 6 mo but diminished somewhat after 4 y. *0773*   Due to influence on binding protein. *3023*
Phenytoin   In comparison with untreated cognitively delayed children 28.8 μg/dL versus 19.4 μg/dL. *2003*
Pregnancy   Increase to above normal in 2nd and 3rd trimester. *1222*
Tomatoes   Significant rise of vitamin A maximum at 90 minutes. *0517*
Vitamin A   Ingested compound (with overdose over 1200 IU/dL). *1343*   Observed with acute intoxication. *3867*

Serum   Decrease   Analytical
Citrates   Dilutional effect when compared against serum or heparin as anticoagulant. *3361*
Light   Destroyed by sunlight and fluorescent lamps. *1205*

Serum   Decrease   Physiological
Aluminum Hydroxide   Reduced absorption due to adsorption and oxidation. *3794*
Cholestyramine   Due to bile acid sequestration, but values still remain within normal range. *3023*
Colestipol   With prolonged treatment due to bile acid sequestration but values still in normal range. *3023*
Diethylstilbestrol   Hormonal effect when given to lactating women. *1221*
Ethanol   With alcoholic hepatitis and cirrhosis, at same liver concentration of vitamin reduced. *2428*
Neomycin   Probably related to reduced absorption bile salts. *1077*   Reduced absorption due to sprue-like syndrome. *3794* Value lowered with bile acid sequestration but probably remains within normal range. *3023*
Organophosphorus Insecticides   Metabolism impaired. *1031*
Pregnancy   Decreased in first trimester but then increasing. *1222*

Vitamin B$_1$

Serum   No Effect   Physiological
Anticonvulsants   In 146 epileptics with long-term treatment. *2007*

Vitamin B$_2$

Serum   No Effect   Physiological
Anticonvulsants   In 146 epileptics with long-term treatment. *2007*

Vitamin B$_{12}$

Serum   No Effect   Analytical
Chloral Hydrate   No effect found on Magic (Ciba-Corning) and Combostat II (Micromedics) assays. *2040*

Serum   No Effect   Physiological
Anticonvulsants   In 146 epileptics with long-term treatment. *2007*
Ascorbic Acid   No effect with varying degrees of supplementation over 14 week study period in young men. *1751*
Cyclosporine   No significant effect observed in 23 patients with inflammatory ocular disease treated for 6 mo. *2713*
Ethanol   In some alcoholics ?decreases myelopoiesis. *2196*
Medroxyprogesterone   No effect seen in Africans at least. *0472*
Nitrous Oxide   With intraoperative exposure for as long as 6 h. *1985*
Oral Contraceptives   No significant difference between oral contraceptive users and control adolescents. *1368*

Serum   Increase   Analytical
Chloral Hydrate   Falsely high values obtained with a radioimmunoassay even at very low drug concentrations. *2614*   Significant increase on Dual Count (Diagnostic Products) methods, both boil and and no-boil. *2040*

Serum   Increase   Physiological
Blood Transfusions   Significant effect in patients with intraoperative transfusion. *1985*
Race   In European women (420 pg/mL) vs African (310 pg/ml). *0472*

Serum   Decrease   Analytical
Chlorpromazine   Possible inhibition effect on some strains *E. gracilis*. *2427*

Serum   Decrease   Physiological
Aminosalicylic Acid   Impairs absorption. *3773*   Reduced absorption noted. *3794*
Anticonvulsants   Significantly lower than normal controls. *2427*
Ascorbic Acid   When ingested with food compound causes destruction of substantial amounts of B$_{12}$. *0233*
Cholestyramine   Reported to cause ileal malabsorption. *3023*
Colchicine   Impairs absorption. *3773*   Reduced absorption due to ileal blockade. *3794*
Ethanol   Impairs absorption. *3773*
Metformin   Due to impaired B$_{12}$ absorption. *3605*   Reported to cause ileal malabsorption. *3023*
Neomycin   Impairs absorption. *3773*   Reduced absorption due to sprue-like syndrome. *3794*
Oral Contraceptives   Probable interference with gastrointestinal absorption. *3802*   Mild effect in oral contraceptive users. *3310*
Potassium Chloride   Reported to cause ileal malabsorption. *3023*
Pregnancy   Decrease by 25%-33% in third trimester. *0472*
Ranitidine   Significant decreased absorption of protein bound compound. *1938*
Smoking   Inverse correlation with urine thiocyanate. *2363*
Vegetarian Diet   Deficiency levels observed in many subjects. *2468*

Urine   Decrease   Physiological
Aminosalicylic Acid   Occurs with vitamin B$_{12}$ deficiency. *3773*
Cimetidine   Excretion reduced because of markedly reduced absorption. *3123*
Colchicine   Occurs with impaired absorption of B$_{12}$. *3773*
Metformin   May cause megaloblastic anemia. *3605*
Neomycin   With impaired absorption. *3773*

Vitamin B₁₂ *(continued)*

CSF   Decrease   Physiological
Anticonvulsants   Significantly lower with long term therapy. *1192*
Phenobarbital   Significantly lower with long term therapy. *1192*
Phenytoin   Significantly lower with long term therapy. *1192*

Red Blood Cells   Increase   Physiological
Folic Acid   During treatment of folate deficient anemia. *1510*
Iron Salts   During treatment of iron-deficient anemia. *1510*
Vitamin B₁₂   During treatment of vitamin B₁₂ deficiency. *1510*

Vitamin B₁₂ (Unsaturated)

Serum   Decrease   Physiological
Ethanol   In some alcoholics ?decreases myelopoiesis. *2196*

Vitamin D Binding Globulin

Serum   No Effect   Physiological
Anticonvulsants   No significant difference in men or women on long-term treatment. *0257*
Phenytoin   In 10 women receiving anticonvulsants mainly phenytoin versus 10 controls, drug taken on average for 15 y. *0257*

Serum   Decrease   Physiological
Stanozolol   From 270 to 230 mg/L after 1 week in 9 healthy men given 10 mg/d for 14 d. *3354*

Vitamin E (Tocopherol)

Serum   No Effect   Physiological
Anticonvulsants   In 146 epileptics with long-term treatment. *2007*

Serum   Decrease   Physiological
Cholestyramine   Decreased but values still in normal range over 1-2 y in children with familial hypercholesterolemia. *3023*
Phenobarbital   Mean of 0.58 mg/dL versus 0.67 mg/dL in untreated control of handicapped children. *1587*
Phenytoin   Significantly reduced values in treated handicapped children. *1587*   Mean of 0.58 mg/dL versus 0.67 mg/dL in untreated control of handicapped children. *1587*

VLDL-Apoprotein B

Serum   Decrease   Physiological
Bezafibrate   Residence time in plasma fell from 3.4 to 1.0 h. *3291*

VLDL-Remnant Apoprotein B

Serum   Increase   Physiological
Bezafibrate   Metabolism little effect but plasma concentration rose 30%. *3291*

Volume

Plasma   No Effect   Physiological
Amiloride   No effect seen in 13 men treated for 8 weeks. *0656*
Chlorthalidone   In 22 premenopausal and 18 postmenopausal women given 100 mg/d for 6 weeks. *0402*
Propranolol   No significant change observed. *1714*

Plasma   Increase   Physiological
Altitude   Common finding after ascent to 5400 metres. *2957*
Aminopyrine   Fluid retention may occur. *2427*

Corticosteroids   Expansion of extracellular fluid due to water retention. *1343*
Corticotropin   Moderate doses can cause salt and water retention. *1680*
Cyclophosphamide   Water retention due to metabolites. *0854*
Desoxycorticosterone   Small increase of extracellular fluid with exogenous DOCA®. *3263*   Associated with fluid retention. *0019*
Dextran   Hemodynamic effect, also increases urine flow. *1343*   Effect of Macrodex®, not Rheomacrodex®. *3346*
Diatrizoic acid   Hypertonicity may occur in children. *0071*
Diazoxide   Causes salt and water retention. *1343*
Dienestrol   Fluid retention with large dose estrogens. *1680*
Estrogens   Retention of salt and water common. *2427*
Fluoxymesterone   With retention of water and electrolytes. *1680*
Methyldopa   Occurs with sodium retention if sole agent. *2706*
Methylprednisolone   May cause fluid retention. *1680*
Oxyphenbutazone   May cause marked salt and water retention. *1343*
Oxytocin   Isolated cases of water intoxication. *1680*
Phenylbutazone   May increase by 50% due to salt and water retention. *1343*
Pregnancy   Increased to maximum of plus 45% by 32nd week. *0987*   Starts in first 3 mo maximal in 2nd and sustained. *2180*   Increased by plus 25 to 55% by 32nd week. *0987*
Testosterone   Salt and water retention, also increased RBC mass. *0330*
Triamcinolone   May cause sodium and fluid retention. *1680*
Verapamil   Due to decreased arterial and venous resistance in 15 patients with uncomplicated essential hypertension. *0867*

Plasma   Decrease   Physiological
β-Adrenergic Drugs   Decreased by approximately 10% shortly after i.v. injection. *1832*
Altitude   Probably related in some way to ADH secretion. *1836*
Angiotensin   Promotes loss of protein-free filtrate to tissues. *1343*
Atropine   Increase by 10-15% after i.v. injection. *1832*
Chlorthalidone   At 3 d but not after 3 mo in 8 patients with mild hypertension given 50 mg/d. *3040*
Dehydration   By approximately 13.6% after heat exposure for 2-11 h. *3254*
Diuretics   May occur if rapid diuretic action. *1343*
Epinephrine   Loss of protein-free fluid to tissues. *1343*
Erect Posture   Decreases by about 12% on standing. *1087*
Ethacrynic Acid   Initial diuretic response may cause hypovolemia. *0105*
Furosemide   Excess diuresis may cause reduction in blood volume. *0019*
Histamine   Due to increased capillary permeation and hemoconcentration. *1176*
Levarterenol   Due to loss of protein-free fluid to tissues. *1343*
Nandrolone   May occur from 10 weeks onwards (anabolic effect). *0330*
Pregnancy   In pre-eclampsia compared with normal pregnancy. *2180*
Propranolol   Decreased by approximately 8% usually, ?mechanism. *3544*
Reducing Diet   May constrict by 30% with 300-600 Cal/d. *3217*
Spironolactone   Dehydration observed in 3.4%. *1393*
Thiazides   Due to diuretic action. *1343*

Blood   No Effect   Physiological
Enalapril   As part of long-term response in 6 responders of 10 treated hypertensives. *1166*

Blood   Decrease   Physiological
Chlorthalidone   At 3 d but not after 3 mo in 8 patients with mild hypertension given 50 mg/d. *3040*
Methylprednisolone   Transient but significant effect observed in patients given i.v. infusion after myocardial infarction in comparison with placebo treated control population. *1560*
Testosterone   Produces polycythemic-like state. *2427*

Urine   No Effect   Physiological

Azosemide   Effect observed in normal male volunteers after 60 mg orally. *2658*

Bopindolol   In 10 hypertensive patients treated for either 1 or 21 d. *1063*

Cimetidine   Although drug cleared principally by glomerular filtration and partial reabsorption in proximal renal tubules. *2368*

Dopamine   Did not change with infusion of up to 3.0 µg/kg/min for 2 h in 6 normal men. *2149*

Levarterenol   Observed after i.v. 0.2-44.0 µg/min. *1176*

Sulindac   In 9 individuals with stable renal insufficiency when treated for 9 d. *2457*   In 10 furosemide treated patients with well controlled congestive heart failure given 400 mg/d for 2 d. *1042*

Urine   Increase   Physiological

Acetazolamide   Initial response noted. *1680*

Altitude   Diuresis usually begins within 4 d. *1836*

Ammonium Salts   Diuretic action of acid forming salt. *1343*

Aspirin   Decreased renal tubular reabsorption. *1343*

Azosemide   Significant effect on first day, but decrease on second. *2047*

Benzthiazide   Therapeutic diuretic intent. *1680*

Bumetanide   After 1 mg maximum within 2 h over by 4 h. *0822*   Marked dose related effect. *2666*

Caffeine   Diuretic action. *1343*

Chlorpromazine   ?depresses ADH secretion or water reabsorption. *1343*

Chlorthalidone   May produce marked diuresis. *2427*

Citrates   Weak diuretic effect. *0071*

Clopamide   Diuretic action. *2900*

Cyclothiazide   Potent diuretic, same as hydrochlorothiazide. *0859*

Dehydrocholic Acid   Mild diuretic action. *0071*

Demeclocycline   Reversible nephrogenic diabetes insipidus may occur. *3333*   Rare induction of reversible nephrogenic diabetes. *1680*

Desoxycorticosterone   Associated with polydipsia also. *1343*

Dextran   Hemodynamic effect. *1343*

Diapamide   Diuretic action. *2256*

Digitalis   Diuretic action in cardiac failure. *1343*

Dopamine   Directly inhibits tubular solute reabsorption. *1591*

Ethacrynic Acid   Therapeutic intent. *3214*

Ethanol   Diuretic action, inhibits pituitary ADH secretion. *1343*

Furosemide   Diuretic action. *1617*   Significant effect, peak between 1 and 3 h after single 40 mg dose. *2658*   Significant increase on first day but decrease on second. *2047*

Glyburide   Normal diuretic response. *2499*

Glycerin   Weak osmotic diuretic action. *2605*

Hydration   In patients with proteinuria. *2818*

Hydrochlorothiazide   Significant effect in 5 h after 150 mg orally. *1189*

Intra-Amniotic Saline   Significant increase over next 24 h. *3892*

Inulin   Massive doses produce marked diuresis. *0071*

Isosorbide   Diuretic action alone, potentiates others. *2605*

Lithium   Rare inhibition of vasopressin action on tubules. *2910*   Occurs in about 60% patients with start of treatment and persists in about 20--25%. *3118*   Large proportion of patients with polyuria with long term treatment. *3705*

Mefruside   Diuretic action, less effective than thiazides. *0180*

Menstruation   Post-menstrual diuresis (decrease premenstrually). *0987*

Mercurial Diuretics   Therapeutic intent. *1343*

Methoxyflurane   Impaired distal renal tubular function. *0749*

Methyclothiazide   Intended diuretic action (maximum in 6 h). *1680*

Metolazone   Diuretic action. *1425*

Muscular Exercise   With mild exercise. *1842*

Nifedipine   After 10 mg sublingually in next 2 h. *2917*

Nitrates   Renal tubular epithelium relatively impermeable. *1343*

Parathyroid Extract   Hormonal action. *1343*

Pregnancy   May increase by up to 25% in last trimester. *0459*

SC-16102   Duration of diuretic action short. *1804*

Sleep   Significant effect over control. *1190*

Starvation   Observed response to stress. *2382*

Streptozocin   Noted ten days after infusion (with polydipsia). *2532*

Tetracycline   May have diuretic or nephrotoxic action. *0491*

Theophylline   4.7 mL/min versus 1.9 mL/min in 11 healthy subjects when given drug intravenously. *1593*

Tolazamide   Normal diuretic response. *2499*

Torasemide   Similar effects to that of furosemide. *0919*

Urea   Acts as osmotic diuretic. *2354*

Xanthine   Diuretic action. *1343*

Xylitol   Increased up to 3 times normal (osmotic diuretic effect). *3212*

Urine   Decrease   Physiological

Acetaminophen   Increase transport of water in diabetes insipidus. *1343*

Acids   May produce oliguria with renal damage. *0279*

β-Adrenergic Drugs   Antidiuretic effect mediated centrally. *3205*

Amitriptyline   May cause urinary retention. *0071*

Anesthetic Agents   Induction of anesthesia produces marked effect. *0916*

Angiotensin   Vasoconstrictive effect in kidney. *1343*

Atropine   Very large doses cause release of ADH. *1343*

Barbiturates   May produce renal shutdown with oliguria. *0279*

Benzene   May produce oliguria with renal failure. *0279*

Biperiden   Decreased flow noted in a few patients. *1680*

Boric Acid   May cause oliguria with renal failure. *0279*

Bufotenine   6-7% antidiuretic activity of serotonin. *1176*

Cantharides   May cause oliguria with renal damage. *0279*

Carbamazepine   Isolated case of dilutional hyponatremia with water intoxication: 7 previous cases reported. *1882*

Cephaloridine   Suggestive of nephrotoxicity. *1019*

Chlorate   May cause oliguria with renal damage. *0279*

Chlorpropamide   In patients with diabetes insipidus. *3214*

Cyclophosphamide   To 60 mL/h from 160 mL/h for d (4-12 h after drug). *0855*

Desipramine   Temporary due to anticholinergic action. *1680*

Diazoxide   Effect over 2 h of 4 mg/kg given i.v. *1805*

Diurnal Variation   Volume reduced at night (max following breakfast). *2222*

Epinephrine   i.m. injection causes sharp decrease. *2427*

Ether   Due to vasoconstriction. *0071*

Ethylene Glycol   May cause oliguria with renal failure. *0279*

Guancydine   Probable action on renal tubules. *3715*

Histamine   After 0.3-0.5 mg subcutaneously in humans. *1176*

Indomethacin   Caused by increased tubular absorption with marked oliguria as a result. *3261*

Isoproterenol   Antidiuretic effect mediated through ADH release. *3205*

Kanamycin   Nephrotoxicity may occur with oliguria, azotemia. *1680*

Levarterenol   Slight fall after i.v. infusion. *2427*

Mepenzolate   Rare urinary retention may occur. *1680*

Meperidine   Causes release of ADH. *1343*

Mercurial Diuretics   May produce oliguria and tubular necrosis. *1188*

Methylserotonin   30-40% antidiuretic activity of 5-HT. *1176*

Morphine   May stimulate release of ADH. *0071*

Muscular Exercise   With heavy exercise. *1842*

Naproxen   Increase by 19% in 10 furosemide treated patients with well controlled congestive heart failure. *1042*

Oxyphenbutazone   May cause marked water retention. *1343*

Pain   Slight changes in response to stress situations. *1990*

Phenylbutazone   Causes significant water retention. *1343*

Phosphorus   Toxicity effect. *1343*

Prolactin   Action for up to 8 h when given i.m. *1662*

Spectinomycin   Mechanism not discussed. *0122*

Volume *(continued)*

Urine   Decrease   Physiological *(continued)*
Spinal Anesthesia   Found postoperatively, but normal renal function. *1343*
Sulfadiazine   Associated with crystalluria and hematuria. *2808*
Trihexyphenidyl   Retention due to parasympatholytic effect. *1680*
Vasopressin   Therapeutic intent. *0071*

Bile   Increase   Physiological
Secretin   Slight effect in normals, marked in cirrhotics. *0371*

Duodenal Contents   Increase   Physiological
Amino Acids   With duodenal infusion, no effect if given i.v. *2918*

Feces   Increase   Physiological
Lactulose   Slight effect observed. *0032*

Gastric Material   Increase   Physiological
Chewing Gum   60% increase observed in normals. *1480*
Histamine   Characteristic normal response. *1176*
Smoking   Effect of 4-6 cigarettes in 1 h. *0295*

Gastric Material   Decrease   Physiological
Burimamide   Marked reduction occurs. *3905*
Diazepam   Presumed central action lasts for 5 h after 10 mg. *0362*
Flucytosine   Effect on both basal and submaximal pentagastrin-stimulated gastric acid secretion. Effect dose related. *0559*
Hexocyclium   Effective anticholinergic action. *1679*
5-Hydroxytryptamine   But mucus secretion increased. *3579*
Insulin   In response to injection or infusion. *0599*
Loperamide   Effect on both basal and submaximal pentagastrin-stimulated gastric acid secretion. Effect dose related. *0559*
Oxmetidine   Significant inhibition following treatment. *2578*
Oxyphencyclimine   Basal output decreased by 60%. *2537*
Ranitidine   Reduced to 10% of normal in healthy volunteers. *3388* Significant inhibition following treatment. *2578*

Pancreatic Juice   No Effect   Physiological
5-Hydroxytryptamine   No effect following injection. *3579*

Pancreatic Juice   Increase   Physiological
Secretin   Normal response over 3.2 mL/kg/80 minutes. *0486*

Pancreatic Juice   Decrease   Physiological
Ethanol   Oral or i.v. causes direct inhibition. *2504*
Smoking   Chronic effect heavy smokers, less in light. *0554*

Saliva   Increase   Physiological
Flurazepam   Excessive salivation may occur. *1679*

Saliva   Decrease   Physiological
Clonidine   Mechanism not discussed. *0393*

Warfarin

Plasma   No Effect   Physiological
Allopurinol   No significant effect observed on plasma concentrations. *2927* Overall no change in steady-state plasma concentration although some individual variation. *2927*
Aluminum Hydroxide   No effect observed. *0079*
Antacids   Neither concentration nor effect affected. *1487*
Chlorothiazide   No effect of 1 g/d for 21 d. *1487*
Disulfiram   No effect on plasma concentrations of either R- or S-warfarin. *2644*
Magnesium Hydroxide   No effect observed. *0079* No effect observed. *3794*
Ranitidine   No effect on plasma concentration or prothrombin time during coadministration. *1938*

Spironolactone   No consistent change in concentration when spironolactone coadministered. *2643*

Plasma   Increase   Analytical
Chlordiazepoxide   Spectrophotometric method shows interference. *3011*

Plasma   Increase   Physiological
Amiodarone   Potentiates action of warfarin due to depression of vitamin-K dependent clotting factors. *2011* Effect potentiated and concentration increased as early as third day. *3950*
Chloral Hydrate   TCA displaces from protein (temporary effect). *1487*
Cimetidine   Clearance reduced: probably by inhibition of drug metabolism with increased half-life. *3256*
Co-Trimoxazole   Effects of compound may be potentiated. *0679*
Disulfiram   If administered together. *2641*
Erythromycin   Clearance decreased by 14% in 12 individuals given 1 g/d for 8 d. *0198*
Metronidazole   May potentiate effects by effect on metabolism. *3057*
Phenylbutazone   Displaces warfarin from protein binding sites. *0528*
Quinidine   Potentiates action. *1487*
Sulfinpyrazone   Increased concentration due to displacement from protein. *2642*
Trimethoprim   May potentiate action by effects on metabolism. *0678*

Plasma   Decrease   Physiological
Carbamazepine   Reported interaction due to alteration of metabolism. *2042*
Cholestyramine   Reduced absorption from gastrointestinal tract due to adsorption. *3794*
Rifampin   Decreased serum concentration due to hepatic enzyme induction. *0014* Catabolism of drug increased due to induction of hepatic microsomal enzymes. *3036*

Wassermann Reaction

Serum   Positive   Physiological
Hydralazine   False positive reaction reported. *2427*

Water

Red Blood Cells   No Effect   Physiological
Race   No difference between whites and blacks. *2523*

Red Blood Cells   Increase   Physiological
Digoxin   Slight effect only due to action on membrane. *1923*

Water Clearance

Urine   Increase   Physiological
Acetohexamide   Normal slight diuretic response. *2499*
Sleep   Significant effect paralleling volume. *1190*

Water Clearance, Free

Urine   No Effect   Physiological
Methoxyflurane   No effect observed with high or low concentrations. *1380*

Urine   Increase   Physiological
Glyburide   Normal diuretic response. *2499*
Tolazamide   Normal diuretic response. *2499*

Urine   Decrease   Physiological
Bumetanide   Marked dose related effect. *2666*

Chlorpropamide   Normal antidiuretic response in normals. *2499*
Metolazone   Diuretic action of drug. *3494*

Xanthine

Serum   Increase   Physiological
Allopurinol   Inhibits xanthine oxidase (slight increase only). *0071*

Urine   Increase   Physiological
Allopurinol   Inhibits xanthine oxidase. *0071*

Xanthine Calculi

Urine   Positive   Physiological
Allopurinol   Rare side effect. *1395*

Xanthochromia

CSF   Positive   Analytical
Blood   Spurious effect 8 h after blood added to clear CSF. *2959*
Cork Stoppers   Contact with cork for 30 minutes may produce color. *0251*

Xanthurenic Acid

Urine   No Effect   Physiological
Isoniazid   Initial response may be decreased later. *3045*

Urine   Increase   Physiological
Deoxypyridoxine   After 2 g tryptophan load. *3050*
Estrogens   Induce tryptophan pyrrolase. *3045*
Glucocorticoids   Induce tryptophan pyrrolase. *3045*
Hydrocortisone   Causes induction of tryptophan pyrrolase. *3047*
Oral Contraceptives   Estrogen component induces tryptophan pyrrolase. *3045*
Penicillamine   Induces pyridoxine antagonism. *1343*
Pregnancy   Causes induction of tryptophan pyrrolase. *3045*
Sleep Deprivation   Increase between second and fifth day. *2032*
Stress Cold Exposure   Possibly due to activation of tryptophan oxygenase. *1169*

Urine   Decrease   Physiological
Anabolic Steroids   Effect of synthetic androgens given to males. *3045*
Androgens   Effect of synthetic androgens given to males. *3045*

Xylose Excretion

Serum   No Effect   Physiological
Neomycin   Single doses did not affect absorption of d-xylose. *2178*

Urine   No Effect   Analytical
Thymol   No effect on analytical method. *2981*

Urine   No Effect   Physiological
Diurnal Variation   No evidence of cycle observed. *0273*

Urine   Increase   Analytical
Galactose   Interferes with bromoaniline procedure if over 2 g/dL. *2981*
Glucose   Interferes with bromoaniline procedure if over 2 g/dL. *2981*
Phenazopyridine   Due to interfering background color. *2945*

Urine   Increase   Physiological
Ethanol   Observed in chronic alcoholics after usual dose. *0589*

Fruit   Increased by much fruit in diet compared with fasting. *0273*
Vegetables   Increase compared with fasting with high intake. *0273*

Urine   Decrease   Physiological
Aminosalicylic Acid   2 times normal dose produced reversible effect on absorption. *3855*
Arsenicals   May produce gastrointestinal irritation, impaired absorption. *0612*
Aspirin   Affects renal elimination. *1913*   Reduced renal excretion reported with tolerance tests. *2866*
Bacteriuria   Theoretical effect (note marked gastrointestinal tract effect). *1914*
Colchicine   Decreased absorption from gut, disturbs epithelial cell function. *2899*
Digitalis   May produce gastrointestinal intolerance, impaired absorption. *0612*
Ethionamide   May cause gastrointestinal irritation, impaired absorption. *0612*
Gold   May produce gastrointestinal irritation, impaired absorption. *0612*
Indomethacin   ?due to increased motility of gut. *1913*
Isocarboxazid   Decreased gastrointestinal tract absorption. *3679*
Kanamycin   Due to impaired gastrointestinal absorption. *1680*
MAO Inhibitors   Decreased gastrointestinal tract absorption. *1237*
Meals   Less absorbed in nonfasting state. *2945*
Metformin   Probably dose related malabsorption. *3605*
Nalidixic Acid   May cause gastrointestinal irritation, impaired absorption. *0612*
Neomycin   Affects intestinal absorption with mucosal damage. *1756*
Opium Alkaloids   Some may produce gastrointestinal irritation impaired absorption. *0612*
Phenelzine   Decreased absorption of xylose from gastrointestinal tract. *3679*

Zinc

Serum   No Effect   Analytical
Acetates   At 50 to 1 with flameless atomic absorption. *2038*
Calcium   At 100 to 1 with flameless atomic absorption. *3336*
Chloride Salts   At 6,000 to 1 with flameless atomic absorption. *2038*
Copper   At 50 to 1 with flameless atomic absorption. *2038*
Iodine   At 50 to 1 with flameless atomic absorption. *2038*
Iron   At 50 to 1 with flameless atomic absorption. *2038*
Lithium   At 50 to 1 with flameless atomic absorption. *2038*
Nitrates   At 150 to 1 with flameless atomic absorption. *2038*
Potassium   At 50 to 1 with flameless atomic absorption. *2038*
Sodium   At 6,000 to 1 with flameless atomic absorption. *2038*
Sulfates   At 50 to 1 with flameless atomic absorption. *2038*
Urea   At 15,000 to 1 with flameless atomic absorption. *2038*

Serum   No Effect   Physiological
Cellulose Phosphate   With decrease of activity product ratio. *2710*
Cisplatin   Insignificant change in 15 patients treated with drug. *3964*
Naproxen   Mean increase of 35% in 10 healthy volunteers receiving 250 mg 3 times per day for 1 week mechanism not known. *1025*
Oral Contraceptives   No difference from normal observed. *0532* In 22 women ingesting combination type contraceptives. *1604*
Phenobarbital   Observed in both hospitalized and home-living patients with long-term administration. *3803*   No significant difference on average between treated and untreated handicapped children, although hypozincemia in 7 of 32 treated children and 0 of 13 controls. *1587*

Zinc (continued)

Serum No Effect Physiological (continued)

Phenytoin Observed in both hospitalized and home-living individuals with long-term administration. *3803* No significant difference between treated and untreated groups of handicapped children. *1587* No significant difference in average values between treated and untreated handicapped children although hypozincemia in 7 of 32 treated children and 0 of 13 Controls. *1587*

Piretanide No significant effect with up to 12 mg/d for 3 mo. *3696*

Serum Increase Analytical

Trichloracetic Acid Volume reduction effect with protein precipitation. *1131* Enhances absorption in atomic absorption methods. *1769*

Serum Increase Physiological

Auranofin In 10 of 12 patients with initially low concentration. *0617*

Chlorthalidone In hypertensive males taking drug for at least 6 mo: probably due to release from tissues but also possibly due to contracted blood volume. *1259* Significantly higher, also increased hair concentration. *1259*

Corticotropin Occurs in most subjects (decreased in others). *3552*

Estrogens Metabolic effect. *0475*

Hydrochlorothiazide Slight effect observed or no effect. *2712*

Penicillamine In rheumatoid arthritis patients when treated for 6 mo. Changes not related to activity of disease. *1794* Related to transient changes in zinc metabolism. *1794*

Zinc Maximal 2 h after ingestion, raised for 5 h. *0439*

Serum Decrease Analytical

Citrates Dilutional effect when compared against serum or heparin as anticoagulant. *3361*

Serum Decrease Physiological

Corticosteroids Rapid sustained drop in burn and surgical patients. *1153*

Oral Contraceptives Duration of use unrelated to Zn concentration. *1467*

Phytate Early effect but increased if continued. *2947*

Prednisone Within 3 d average fell from 12.6 μmol/L to 11.1 μmol/L in 14 adult hospitalized patients given 40-50 mg daily; level rose to above normal then reverted normal in 2 weeks. *3790*

Urine No Effect Analytical

Acetates At 50 to 1 with flameless atomic absorption. *2038*
Calcium At 100 to 1 with flameless atomic absorption. *2038*
Chloride Salts At 6,000 to 1 with flameless atomic absorption. *2038*
Copper At 50 to 1 with flameless atomic absorption. *2038*
Iodine At 50 to 1 with flameless atomic absorption. *2038*
Iron At 50 to 1 with flameless atomic absorption. *2038*
Lithium At 50 to 1 with flameless atomic absorption. *2038*
Nitrates At 150 to 1 with flameless atomic absorption. *2038*
Potassium At 50 to 1 with flameless atomic absorption. *2038*
Sodium At 6,000 to 1 with flameless atomic absorption. *2038*
Sulfates At 50 to 1 with flameless atomic absorption. *2038*
Urea At 15,000 to 1 with flameless atomic absorption. *2038*

Urine No Effect Physiological

Cellulose Phosphate No significant effect observed. *2710*
Furosemide No effect observed on excretion. *2712*
Mercaptomerin No effect observed on excretion. *2712*
Prednisone No consistent change noted in 14 hospitalized adult patients when given 40-50 mg/d. *3790*
Sodium Salts No effect observed with oral doses. *2712*
Triamterene No effect observed. *2712*

Urine Increase Physiological

Bendrofluazide Increase by 60% in 9 hypertensives treated for 2 weeks. *3806*
Bumetanide Increase by 14% in 9 patients with hypertension over 2 weeks. *3806*
Chlorthalidone Increase by 65% in 9 hypertensives treated for 2 weeks. *3806*
Cisplatin Marked effect in 15 patients treated with drug. *3964*
EDTA If poisoning due to zinc. *0071*
Ethacrynic Acid When given i.v. brief effect only. *3439*
Ethanol In absence of effect on serum concentration. *0586*
Furosemide When given i.v. brief effect only. *3439* Increase by 12% in 9 hypertensive patients treated for 2 weeks. *3806*
Hydrochlorothiazide Twice normal reached on fifth day. *2712* Increase by 60% in 9 patients with hypertension over 2 weeks. *3806*
Naproxen Mean increase of 35% in 10 healthy volunteers receiving 250 mg 3 times per day for 1 week mechanism no known. *1025*
Penicillamine If poisoning due to zinc. *1343* In rheumatoid arthritis patients when treated for 6 mo. Changes not related to activity of disease. *1794* Related to transient changes in zinc metabolism. *1794*
Saline Increase by 28% with isotonic saline loading. *3439*
Triamterene Increase by 18% in 9 patients with hypertension for 2 weeks. *3806*
Zinc Slight increase related to dose. *0439*

Urine Decrease Physiological

Diuretics Although other cations increased. *3439*

Feces Increase Physiological

Phytate Significant effect observed with 2.5 g/d. *2947*

Hair No Effect Physiological

Oral Contraceptives No significant effect in young women studied for at least 3 mo. *3720*

Hair Increase Analytical

Hair Treatments Threefold with permanent waving lotion. *1594*

Red Blood Cells No Effect Physiological

Oral Contraceptives In 22 women ingesting combination type contraceptives. *1604*

Red Blood Cells Increase Physiological

Penicillamine In rheumatoid arthritis patients when treated for 6 mo. Changes not related to activity of disease. *1794* But granulocyte concentration decreased related to transient changes in zinc metabolism. *1794*

White Blood Cells No Effect Physiological

Oral Contraceptives In 22 women ingesting combination type contraceptives. *1604*

White Blood Cells Decrease Physiological

Penicillamine In rheumatoid arthritis patients when treated for 6 mo. Changes not related to activity of disease. *1794*

Zinc Sulfate Turbidity

Serum No Effect Analytical

Bilirubin Probably same noninterference as with thymol turbidity. *1563*
Heat Little change with temperature observed. *0579*
Hemoglobin Probably same noninterference as with thymol turbidity. *1563*

Serum Increase Analytical

Heparin Affects physico-chemical properties. *1563*
Lipemia Effect less marked than with thymol turbidity. *1563*

Serum   Increase   Physiological
Rifampin   Hepatotoxic effect (cholestasis). *2820*

Serum   Decrease   Analytical
Storage of Sample   Effect observed after 1 d at 4°. *1563*

4 DRUG LISTINGS

Acebutolol

Antinuclear Antibodies *Serum Positive Physiological* In 3 of 11 patients treated for 48 weeks: readministration after titer became negative resulted in significant rise. *0386* Observed in 18.6% diabetics with drug versus 3.8% in normals and 1.3% in diabetics without drug. *2104* In 20% of 35 men and 44% of 23 women in comparison with 10.2% in men and 12.9% women with other antihypertensive drugs. *0419* Developed in 8 of 9 patients treated for 12 to 24 weeks. *0680*

Apolipoprotein AI *Serum No Effect Physiological* Increase by 1 mg/dL in 11 patients given 400 mg daily for 3 mo. *2550*

Apolipoprotein B *Serum No Effect Physiological* Decrease by -2 mg/dL in 11 patients given 400 mg daily for 3 mo. *2550*

Apolipoprotein CII *Serum No Effect Physiological* No change in 11 patients given 400 mg daily for 3 mo. *2550*

Apolipoprotein CIII *Serum No Effect Physiological* Increase by 1 mg/dL in 11 patients given 400 mg daily for 3 mo. *2550*

Apolipoprotein E *Serum No Effect Physiological* Increase by 1 mg/dL in 11 patients given 400 mg daily for 3 mo. *2550*

Cholesterol *Serum No Effect Physiological* In small numbers of patients when treated for 1-12 mo. *0088* Decrease by -3 mg/dL in 11 patients given 400 mg daily for 3 mo. *2550* Insignificant change on average in one 6 mo long study. *2116* In several studies of about 15 patients treated for 1 to 12 mo. *0089*

Cholesterol, High Density Lipoprotein
Serum No Effect Physiological In small numbers of patients when treated for 1-12 mo. *0088* Decrease by -2 mg/dL in 11 patients given 400 mg daily for 3 mo. *2550* Insignificant change on average in one 6 mo long study. *2116* In several studies of about 15 patients treated for 1 to 12 mo. *0089*
Serum Decrease Physiological Slight effect, but no change in ratio of HDL/total cholesterol. *2116* But insignificantly in 18 patients treated for 6 mo. *2115*

Cholesterol, Low Density Lipoprotein
Serum No Effect Physiological In small numbers of patients when treated for 1-12 mo. *0088* Insignificant change on average in one 6 mo long study. *2116* In several studies of about 15 patients treated for 1 to 12 mo. *0089*

Cholesterol, Very Low Density Lipoprotein
Serum No Effect Physiological In small numbers of patients when treated for 1-12 mo. *0088* In several studies of about 15 patients treated for 1 to 12 mo. *0089*

Creatinine *Serum No Effect Analytical* At 100 mg/L on reversed phase LC procedure of Zhiri et al. *3942*

Fatty Acids, Free (FFA) *Serum Decrease Physiological* Significant effect after one month's treatment and then plateaued. *2116* Significant reduction in 18 patients treated for 6 mo. *2115*

Glucose Tolerance *Serum Decrease Physiological* Increased glucose at 60 and 120 minutes without impaired release of insulin. *2115*

LE Cells *Blood Positive Physiological* Observed in several patients treated for 12 to 24 weeks. *0680*

Triglycerides *Serum No Effect Physiological* In small numbers of patients when treated for 1-12 mo. *0088* Increase by 1 mg/dL in 11 patients given 400 mg daily for 3 mo. *2550* Insignificant change on average in one 6 mo long study. *2116* In several studies of about 15 patients treated for 1 to 12 mo. *0089*

Uric Acid *Serum No Effect Analytical* At 100 mg/L on reversed phase LC procedure of Zhiri et al. *3942*

Acenocoumarol

Diphenylhydantoin *Serum No Effect Physiological* Interactions not been reported (and unlikely to occur due to different routes of elimination). *3556*

Prothrombin Time *Plasma Increase Physiological* Lowers prothrombin of blood: intended action. *0019*

Uric Acid *Urine No Effect Physiological* No uricosuric action observed. *1487*

Acetaldehyde

Ketones *Urine Increase Analytical* Affects alkaline nitroprusside procedure. *0583*

Acetaminophen

Acetaminophen Screening Test *Urine Positive Analytical* Blue with o-cresol at therapeutic concentration. *3326* Red color with 1-naphthol at therapeautic concentration. *3326*
Urine Negative Analytical Negative Bratton-Marshall reaction at therapeutic concentrations. *3326*

Alanine Aminotransferase *Serum No Effect Analytical* At acute overdose concentration (10 mg/dL) on SMAC® method. *3719* On continuous method at 10 times maximal therapeutic concentration. *1785* No effect at therapeutic concentration on Reflotron method. *1984*
Serum Increase Analytical On colorimetric method at 10 times therapeutic concentration. *1785*
Serum Increase Physiological Hepatic necrosis with dose of 10 g reported. *0438* Manifestation usually of single toxic high dose ingestion in suicide attempt. Liver damage usually of centrizonal hemorrhagic hepatic necrosis. *3948*

Albumin *Serum No Effect Analytical* At 5 times therapeutic concentration on BCG methods on SMAC®, Ektachem® 400, Hitachi® 705 and KDA. *2138* At acute overdose concentration (10 mg/dL) on SMAC® method. *3719* At concentration of 1500 mg/L had no effect on BCG method. *3393*
Serum Decrease Physiological In severe cases of poisoning (also in some mild). *3470*

Alkaline Phosphatase *Serum No Effect Analytical* At acute overdose concentration (10 mg/dL) on SMAC® method. *3719*
Serum Increase Analytical On continuous method at 10 times maximal therapeutic concentration. *1785*
Serum Increase Physiological In severe cases of poisoning (also in some mild). *3470* With overdose and centrolobular hep-

Acetaminophen *(continued)*

Alkaline Phosphatase *(continued)*
atic necrosis usually associated with high drug concentration. *1771*

Alkaloids *Urine No Effect Analytical* With iodoplatinate of Frings TLC procedure at 10 mg/dL. With ninhydrin in Frings TLC procedure at 10 mg/dL. *1204*

Amino Acids *Urine Positive Analytical* Metabolite reacting with ninhydrin migrates near dihydroxyphenylalanine with 2-dimensional high voltage electrophoresis. *3304*

Ammonia *Plasma No Effect Analytical* At concentration of 50 mg/L had no effect on Ektachem® method. *3393*

Ammonium Ions *Urine No Effect Physiological* Without overdose not usually any effect. *2866*

Amylase *Serum No Effect Analytical* At 5 times therapeutic concentration on maltotriose method on aca, maltotetrose method on Cobas-Bio and amylopectin method on Ektachem®. *2138*

Aspartate Aminotransferase *Serum No Effect Analytical* At acute overdose concentration (10 mg/dL) on SMAC® method. *3719* At 10 times maximal therapeutic concentration on colorimetric method. *1785* No effect at therapeutic concentration on Reflotron method. *1984*
Serum Increase Analytical Reacts in Morgenstern procedure. *0583* At therapeutic concentration may affect SMA 12/60 method. *3335*
Serum Increase Physiological Manifestation usually of single toxic high dose ingestion in suicide attempt. Liver damage usually of centrizonal hemorrhagic hepatic necrosis. *3948* With overdose and centrolobular hepatic necrosis usually associated with high drug concentration. *1771* Hepatic necrosis with dose of 10 g reported. *0438*
Serum Decrease Analytical On continuous method at 10 times maximal therapeutic concentration. *1785*

Barbiturate *Urine No Effect Analytical* With diphenylcarbazone in Frings TLC procedure at 10 mg/dL. With mercuric SO$_4$ in Frings TLC procedure at 10 mg/dL. With mercurous NO$_3$ in Frings TLC procedure at 1 mg/dL. With vanillin in Frings procedure at 10 mg/dL. *1204*

Bicarbonate *Serum No Effect Analytical* At concentration of 1500 mg/L had no effect on method using phenolphthalein method. *3393*

Bile *Urine Increase Physiological* Hepatic necrosis may occur with dose of 10 g. *0438*

Bile Acids *Serum Increase Physiological* Manifestation usually of single toxic high dose ingestion in suicide attempt. Liver damage usually of centrizonal hemorrhagic hepatic necrosis. *3948* With overdose and centrolobular hepatic necrosis usually associated with high drug concentration. *1771*

Bilirubin *Serum No Effect Analytical* At 5 times therapeutic concentration on routine methods in use on SMAC®, aca, Cobas-Bio, Ektachem®, Hitachi® and KDA. *2138* At acute overdose concentration (10 mg/dL) on SMAC® method. *3719* At concentration of 50 mg/L had no effect on Ektachem® method. *3393* At concentration of 1500 mg/L had no effect on Jendrassik and Grof method. *3393*
Serum No Effect Physiological Insignificant displacement from protein in neonates. *3748*
Serum Increase Physiological Hepatic damage reported with overdose. *3052* Manifestation usually of single toxic high dose ingestion in suicide attempt. Liver damage usually of centrizonal hemorrhagic hepatic necrosis. *3948* With overdose and centrolobular hepatic necrosis usually associated with high drug concentration. *1771*

Bilirubin, Direct *Serum Increase Physiological* Hepatic necrosis with dose of 10 g reported. *0438*

Bromide *Urine No Effect Analytical* No effect on method of Frings. *1204*

Calcium *Serum No Effect Analytical* At 5 times therapeutic concentration on o-CPC methods on SMAC®, Abbott-VP, aca Hitachi® 705 and KDA and arsenazo III method on Ektachem® 400. *2138* At acute overdose concentration (10 mg/dL) on SMAC® method. *3719* At concentration of 1500 mg/L had no effect on cresolphthalein method. *3393*

Casts *Urine Increase Physiological* Renal damage due to hemolysis/anuria. *1343*

Catecholamines *Urine No Effect Physiological* No effect with short term ingestion 2.6 g/d. *0766*

Cells *Urine Increase Physiological* May be marked increase in renal tubular cells. *1343* 3-6 g for 5 d produces increase in tubular cells but less than with Aspirin or Phenacetin. *2866*

Chloramphenicol *Serum No Effect Analytical* At 100 mg/L no effect on coupled enzymatic procedure. *2490*

Chlordiazepoxide *Blood No Effect Analytical* On method of Riddick at 5 mg/dL. *2983*

Chloride *Serum No Effect Analytical* At concentration of 1500 mg/L had no effect on mercurimetric method. *3393*

Cholesterol *Serum No Effect Analytical* At 5 times therapeutic concentration on Liebermann-Burchard method on SMAC® and enzymatic methods on Abbott-VP, Cobas-Bio, Ektachem®, Hitachi® 705 and KDA. *2138* At acute overdose concentration (10 mg/dL) on SMAC® method. *3719* No effect at therapeutic concentration on Reflotron method. *1984* At concentration of 6 mg/L had no effect on CHOD-PAP method. *3393* At concentration of 1500 mg/L had no effect on Liebermann-Burchard method. *3393*

Color *Urine Positive Analytical* Dark brown urine observed in some patients with overdose, probably due to p-Aminophenol. *0669*

Creatine Kinase *Serum No Effect Analytical* At acute overdose concentration (10 mg/dL) on SMAC® method. *3719* On continuous method at 10 times maximal therapeutic concentration. *1785*

Creatinine *Serum No Effect Analytical* At 200 mg/L on reversed phase LC procedure of Zhiri et al. *3942* At 5 times upper limit of therapeutic concentration on routine methods in use on SMAC®, Abbott-VP, aca, Cobas-Bio, Ektachem® 400, Hitachi® 705, KDA. *2138* At acute overdose concentration (10 mg/dL) on SMAC® method. *3719* At concentration of 50 mg/L had no effect on Ektachem® method. *3393* At concentration of 180 mg/L had no effect on creatinine iminohydrolase method. *3393* At concentration of 1500 mg/L had no effect on AutoAnalyzer Jaffé method. *3393*
Serum No Effect Physiological Without overdose not usually any effect. *2866*
Serum Increase Analytical Mild acute renal failure in two patients following therapeutic ingestion of drug. *1219*
Serum Increase Physiological Reversible tubular necrosis reported. *0438*

Creatinine Clearance *Urine No Effect Physiological* Without overdose not usually any effect. *2866*

Epinephrine *Urine No Effect Physiological* No effect with short term ingestion 2.6 g/d. *0766*

Erythrocytes *Blood Decrease Physiological* May cause hemolytic anemia. *2313*
Urine Increase Physiological Renal damage due to hemolysis. *1343*

Ethchlorvynol *Serum No Effect Analytical* At 100 mg/L on GLC procedure of Evenson. *1058*
Urine No Effect Analytical No effect on method of Frings and Cohen at 10 mg/dL. *1205*

FPN Test *Urine No Effect Analytical* No effect at 10 mg/dL on method of Frings. *1204*

Glucose *Serum No Effect Analytical* At 5 times upper limit of therapeutic concentration on routine methods in use on SMAC®, Abbott-VP, aca, Cobas-Bio, Ektachem® 400, Hitachi® 705, KDA. *2138* No effect observed with colorimetric glucose oxidase procedure on AutoAnalyzer using Boehringer-Mannheim kit. No effect on Beckman glucose analyzer with oxygen produced measured by specific electrode. *3018* At acute overdose concentration (10 mg/dL) on SMAC® method. *3719* No effect at therapeutic concentration on Reflotron method. *1984* At concentration of 57 mg/L had no effect on GOD/POD-PAP method. *3393* At concentration of 600 mg/L had no effect on hexokinase/G-6-PDH method. *3393*
Serum Increase Analytical At 1 mmol/L affects SMA 12/60 method. *3335* In YSI glucose analyzer with potentiometric measurement of hydrogen peroxide produced. Effect can be quite marked. *3018*
Serum Decrease Analytical At concentrations above 25 mg/L (normal therapeutic concentration 20 mg/L) lowered concentration as measured by GOD-Perid method. *3393*
Serum Decrease Physiological Reported effect of metabolite. *0817*

γ-Glutamyltransferase (GGT) *Serum No Effect Analytical* No effect at therapeutic concentration on Reflotron method. *1984*
Serum Increase Physiological In severe cases of poisoning (also in some mild). *3470*

Hemoglobin *Blood Decrease Physiological* Anemia/pancytopenia. *0071*

Hydroxybutyrate Dehydrogenase
Serum No Effect Analytical On continuous method at 10 times maximal therapeutic concentration. *1785*

11-Hydroxycorticosteroids *Urine No Effect Physiological* No effect with short term ingestion 2.6 g/d. *0766*

17-Hydroxycorticosteroids *Urine No Effect Physiological* No effect with short term ingestion 2.6 g/d. *0766*

5-Hydroxyindoleacetic Acid (5-HIAA)
Urine Increase Analytical May cause false high colorimetric results. *0913* Affects nitrosonaphthol procedures. *1487*

Icteric Index *Serum Increase Physiological* May cause hepatic toxicity. *2313*

Indomethacin *Serum No Effect Analytical* No effect on HPLC method of Roberts and Smith. *3002*

Iron *Serum No Effect Analytical* At 5 times therapeutic concentration on Ferrozine method on SMAC®. *2138* At acute overdose concentration (10 mg/dL) on SMAC® method. *3719* At concentration of 100 mg/L had no effect on Ferrozine method. *3393*

17-Ketosteroids *Urine No Effect Physiological* No effect with short term ingestion 2.6 g/d. *0766*

Lactate *Plasma No Effect Analytical* At concentration of 600 mg/L had no effect on enzymatic method. *3393*

Lactate Dehydrogenase *Serum No Effect Analytical* At acute overdose concentration (10 mg/dL) on SMAC® method. *3719* On continuous methods using either pyruvate or lactate as substrate at 10 times maximal therapeutic concentration. *1785*
Serum Increase Analytical On colorimetric method at 10 times maximal therapeutic concentration. *1785*

Leukocytes *Blood Decrease Physiological* May affect bone marrow function/pancytopenia. *2198*

Metanephrines, Total *Urine Positive Analytical* Interference with unmodified ion-exchange chromatographic method of Shoup and Kissinger. *3861*

Methaqualone *Serum No Effect Analytical* At 100 mg/L on GLC procedure of Evenson. *1057*

Methemoglobin *Blood Increase Physiological* May rarely cause hemolysis. *1343*
Urine Increase Physiological Renal damage due to hemolysis. *1343*

Neutrophils *Blood Decrease Physiological* May cause neutropenia/pancytopenia. *1343*

Norepinephrine *Urine No Effect Physiological* No effect with short term ingestion 2.6 g/d. *0766*

5'-Nucleotidase *Serum Increase Physiological* In severe cases of poisoning (also in some mild). *3470*

Phosphate *Serum No Effect Analytical* At acute overdose concentration (10 mg/dL) on SMAC® method. *3719* At concentration of 1500 mg/L had no effect on phosphomolybdate method. *3393* At 5 times therapeutic concentration on molybdate procedures on SMAC®, aca, Hitachi® 705 and KDA. *2138*

Platelets *Blood Decrease Physiological* Single case of antibodies to SO_4 conjugate of drug. *1009*

Potassium *Serum No Effect Analytical* At concentration of 6 mg/L had no effect on flame-photometric method. *3393* At concentration of 1500 mg/L had no effect on measurement by ISE with predilution. *3393*

Protein *Serum No Effect Analytical* At 5 times therapeutic concentration on biuret procedures on SMAC®, Abbott-VP, Ektachem® 400, Hitachi® 705 and KDA. *2138* At acute overdose concentration (10 mg/dL) on SMAC® method. *3719* At concentration of 1500 mg/L had no effect on biuret method with blank correction. *3393*
Urine Increase Physiological Nephrotoxic effect of drug. *0438*

Prothrombin Time *Plasma No Effect Physiological* No effect observed in volunteers. *3292*
Plasma Increase Physiological Depresses clotting factor synthesis. *0438* Manifestation usually of single toxic high dose

ingestion in suicide attempt. Liver damage usually of centrizonal hemorrhagic hepatic necrosis. *3948* Significant effect in patients receiving warfarin, possibly due to interference with synthesis of factors II, VII, IX, and X. *0403*

Pseudocholinesterase *Serum Decrease Physiological* In severe cases of poisoning (also in some mild). *3470*

Salicylate *Urine No Effect Analytical* At 10 mg/dL no effect on Trinder method. *1204*

Sodium *Serum No Effect Analytical* At concentration of 6 mg/L had no effect on flame-photometric method. *3393* At concentration of 1500 mg/L had no effect on measurement by ISE with predilution. *3393*

Sulfate *Urine Decrease Physiological* Reduces increased output of rheumatoids. *3102*

Titratable Acidity *Urine No Effect Physiological* Without overdose not usually any effect. *2866*

Triglycerides *Serum No Effect Analytical* At 5 times upper limit of therapeutic concentration on routine methods in use on SMAC®, Abbott-VP, Cobas-Bio, Ektachem® 400, Hitachi® 705, KDA. *2138* At acute overdose concentration (10 mg/dL) on SMAC® method. *3719* No effect at therapeutic concentration on Reflotron method. *1984* At concentration of 300 mg/L had no effect on GPO-PAP method. *3393* At concentration of 1500 mg/L had no effect on lipase/esterase method. *3393*

Urea Nitrogen *Serum No Effect Analytical* as routine methods in use on SMAC®, Abbott-VP, Cobas-Bio, Ektachem® 400, Hitachi® 705, KDA. *2138* At acute overdose concentration (10 mg/dL) on SMAC® method. *3719* No effect at therapeutic concentration on Reflotron method. *1984* At concentration of 1500 mg/L had no effect on diacetylmonoxime method. *3393*
Serum No Effect Physiological Without overdose not usually any effect. *2866*
Serum Increase Physiological Reversible tubular necrosis reported. *0438* Mild acute renal failure in two patients following therapeutic ingestion of drug. *1219*

Uric Acid *Serum No Effect Analytical* At 200 mg/L on reversed phase LC procedure of Zhiri et al. *3942* At 5 times upper limit of therapeutic concentration on uricase procedures on Abbott-VP, aca, Cobas-Bio, Ektachem® 400, Hitachi® 705. *2138* At acute overdose concentration (10 mg/dL) on SMAC® method. *3719* No effect at therapeutic concentration on Reflotron method. *1984* At concentration of 810 mg/L had no effect on uricase-PAP method. *3393*
Serum Increase Analytical Reacts as if uric acid with SMA 12/60 method. *3335* Falsely high values with phosphotungstate methods. *1311* But increase only of 0.12 mmol/L at concentrations found in overdosage (40 mg/L) when measured by phosphotungstate reduction. *3363* 40% increase on SMAC® at 5 times upper limit of therapeutic concentration on phosphotungstate procedures routinely used. 14% increase on KDA at 5 times upper limit of therapeutic concentration on phosphotungstate procedures routinely used. *2138* At concentrations above 50 mg/L (normal therapeutic concentration 20 mg/L) raised concentration as measured by phosphotungstate reduction method. *3393*
Urine Increase Analytical Falsely high values with phosphotungstate methods. *1311*

Volume *Urine Decrease Physiological* Increase transport of water in diabetes insipidus. *1343*

Acetanilid

Bilirubin *Serum Increase Physiological* Hemolysis with G-6-PD deficiency. *2313*

Bilirubin, Direct *Serum Increase Physiological* Hemolysis with G-6-PD deficiency. *2313*

Casts *Urine Increase Physiological* Due to hemolysis and renal damage. *1343*

Color *Blood Increase Physiological* May cause chocolate colored blood. *0279*

Erythrocytes *Blood Decrease Physiological* Hemolytic anemia/agranulocytosis. *2429*
Urine Increase Physiological Due to hemolysis and renal damage. *1343*

Erythrocyte Survival *Blood Decrease Physiological* Due to hemolysis. *1343*

Haptoglobin *Serum Decrease Physiological* Hemolysis with G-6-PD deficiency. *2313*

Acetanilid *(continued)*

Heinz Body Formation *Blood Positive Physiological* Occurs initially prior to overt hemolysis. *1902*

Hematocrit *Blood Decrease Physiological* Hemolytic anemia. *2426*

Hemoglobin *Plasma Increase Physiological* Occurs with marked hemolysis. *1902*
Blood Decrease Physiological Hemolytic anemia. *2426*
Urine Increase Physiological Occurs with hemolysis. *1902*

5-Hydroxyindoleacetic Acid (5-HIAA)
Urine Increase Analytical False positive with nitrosonaphthol, no effect on quantitative test. *2812*

Leukocytes *Blood Decrease Physiological* Leukopenia. *2004*

Methemoglobin *Blood Increase Physiological* Intravascular hemolysis may occur. *2426*
Urine Increase Physiological Due to hemolysis and renal damage. *1343*

Occult Blood *Feces Positive Physiological* Poisoning may cause many gastrointestinal symptoms. *0279*

Protein *Urine Increase Physiological* Hemolysis may cause renal damage. *1343*

Reduced Glutathione
Red Blood Cells Decrease Physiological Sharp fall before overt hemolysis. *1902*

Reticulocytes *Blood Increase Physiological* Occurs during recovery from hemolysis. *1902*

Sugar *Urine Increase Analytical* Acts as reducing substance with Benedict's reagent. *2425* Acts as reducing agent with nonspecific tests. *1237*

Sulfhemoglobin *Blood Increase Physiological* Hemolytic anemia in acute poisoning. *1343*

Acetates

Acetoacetate *Urine Increase Analytical* May interfere with Gerhardt FeCl$_3$ procedure. *0459*

Barbiturate *Serum Increase Analytical* Affect colorimetric method with cobalt acetate. *3054*

Orthophosphate *Test Conditions Decrease Analytical* Interference above 0.7 mol/L on method of Horder. *1658*

pH *Blood Increase Physiological* Alkalinizing action due to rapid metabolism. *1343*

Pyruvate Kinase *Red Blood Cells Decrease Analytical* Observed with 0.3 mmol/L PEP in reaction mixture. *0370*

Zinc *Serum No Effect Analytical* At 50 to 1 with flameless atomic absorption. *2038*
Urine No Effect Analytical At 50 to 1 with flameless atomic absorption. *2038*

Acetazolamide

Albumin *Serum No Effect Analytical* No effect at 12 mg/dL on SMA 12/60 method. *2636* At concentration of 1,000 mg/L had no effect on BCG method. *3393*

Alkaline Phosphatase *Serum No Effect Analytical* No effect at 12 mg/dL on SMA 12/60 method. *2636*

Ammonia *Plasma Increase Physiological* ?diverts NH$_3$ from kidney to general circulation. *1238*
Urine Decrease Physiological Increased alkalinity. *1343*

Amphetamine *Urine No Effect Analytical* At 50 mg/L on fluorescent method of Hayes. *1534*

Aspartate Aminotransferase *Serum No Effect Analytical* No effect at 12 mg/dL on SMA 12/60 method. *2636*

Bicarbonate *Serum No Effect Analytical* At concentration of 1,000 mg/L had no effect on phenolphthalein method. *3393*
Serum Decrease Physiological Carbonic anhydrase inhibition in renal tubules. *0652*
Urine Increase Physiological Inhibition of carbonic anhydrase. *1343*

Bilirubin *Serum No Effect Analytical* No effect at 12 mg/dL on SMA 12/60 method. *2636* At concentration of 1,000 mg/L had no effect on Jendrassik and Grof method. *3393*

Serum No Effect Physiological At pharmacological concentration probably no significant displacement from protein occurs. *3748*
Serum Increase Physiological Single case of cholestatic jaundice reported. *2018*

Calcium *Serum No Effect Analytical* No effect at 12 mg/dL on SMA 12/60 method. *2636* At concentration of 1,000 mg/L had no effect on mercurimetric method. *3393*
Serum Decrease Physiological Defective Calcium and PO$_4$ reabsorption can be induced. *0652*
Urine Increase Physiological Inhibits tubular reabsorption. *2745*
Pancreatic Juice Increase Physiological 60 percent increase with infusion in normals. *0984*

Carbonic Anhydrase
Red Blood Cells Decrease Physiological Inhibitory effect observed. *2893*

Chloride *Serum Increase Physiological* Loss of HCO$_3$ by carbonic anhydrase inhibition. *1022*
Urine Decrease Physiological Significant effect in 5 h after 500 mg orally. *1189*
Pancreatic Juice Increase Physiological 30 percent increase with infusion in normals. *0984*

Cholesterol *Serum No Effect Analytical* No effect at 12 mg/dL on SMA 12/60 method. *2636* At concentration of 1,000 mg/L had no effect on Liebermann-Burchard method. *3393*

Citrate *Urine Decrease Physiological* Alteration of acid base status, diuresis. *2745*

Creatinine *Serum No Effect Analytical* No effect at 12 mg/dL on SMA 12/60 method. *2636* At concentration of 1,000 mg/L had no effect on AutoAnalyzer Jaffé method. *3393*

Crystals *Urine Increase Analytical* Presence of drug. *1022*

Diphenylhydantoin *Serum No Effect Analytical* On GLC procedure of Papadopoulos if added *in vitro*. *2722*

Erythrocytes *Blood Decrease Physiological* May cause pancytopenia/agranulocytosis. *2313*
Urine Increase Physiological Occasionally observed adverse reaction. *1680*

Estriol *Urine No Effect Analytical* No effect observed on acid hydrolysis. *0023*

Estrogens *Urine Increase Physiological* In pregnant but mechanism unknown. *0023*
Urine Decrease Analytical Affects hydrolysis of estrogen conjugates. *0022*

Glucose *Serum No Effect Analytical* No effect at 12 mg/dL on SMA 12/60 method. *2636* At concentration of 200 mg/L had no effect on GOD/POD-PAP method. *3393*
Serum No Effect Physiological Little effect observed in normals. *1487*
Serum Increase Physiological In prediabetics and if hypoglycemic agents used. *2039*
Urine Increase Physiological Occasionally observed adverse reaction. *1680*

Hemoglobin *Blood Decrease Physiological* Anemia due to pancytopenia/agranulocytosis. *2313*

Hydrochloric Acid *Gastric Material Decrease Physiological* In large doses reversible effect. *0814*

17-Hydroxycorticosteroids *Urine Increase Analytical* In vitro interference with Glenn-Nelson method. *1022*

Iodide *Urine Decrease Physiological* Significant effect in 5 h after 500 mg orally. *1189*

^{131}I Uptake *Serum No Effect Physiological* No effect on test. *0583*

17-Ketogenic Steroids *Urine Increase Analytical* Interferes with Zimmermann reaction. *0427*

17-Ketosteroids *Urine No Effect Analytical* Minimal interference with Zimmermann reaction *in vitro*. *0427*

Lactate Dehydrogenase *Serum No Effect Analytical* No effect at 12 mg/dL on SMA 12/60 method. *2636*

LE Cells *Blood Positive Physiological* May produce LE-like syndrome. *3505*

Leukocytes *Blood Decrease Physiological* Leukopenia/agranulocytosis. *2759*

Magnesium *Urine Increase Physiological* Reported to cause minor increase in excretion. *3101*
Urine Decrease Physiological Reported effect. *3745* Following acute i.v. administration of 500 mg. *3101*

Pancreatic Juice *Increase* *Physiological* Approximately 70% increase with infusion in normals. *0984*

Neutrophils *Blood* *Decrease* *Physiological* Occasional agranulocytosis reported. *3717*

Occult Blood *Feces* *Positive* *Physiological* Initial response noted. *1680*

pCO₂ *Blood* *Decrease* *Physiological* Usual effect in bronchitics. *2321*

pH *Blood* *Decrease* *Physiological* Due to inhibition of carbonic anhydrase. *2313*
Urine *Increase* *Physiological* Inhibition of carbonic anhydrase. *1343*

Pheneturide *Serum* *No Effect* *Analytical* On GLC procedure of Papadopoulos if added *in vitro*. *2722*

Phenobarbital *Serum* *No Effect* *Analytical* On GLC procedure of Papadopoulos if added *in vitro*. *2722*

Phosphate *Serum* *No Effect* *Analytical* No effect at 12 mg/dL on SMA 12/60 method. *2636* At concentration of 1,000 mg/L had no effect on phosphomolybdate method. *3393*
Serum *Decrease* *Physiological* Defective Calcium and PO₄ reabsorption can be induced. *0652*
Urine *Increase* *Physiological* Inhibits tubular reabsorption. *0652*

Platelets *Blood* *Decrease* *Physiological* May cause pancytopenia with aplastic anemia. *2074*

Potassium *Serum* *No Effect* *Analytical* At concentration of 1,000 mg/L had no effect on flame-photometric method. *3393*
Serum *Decrease* *Physiological* Diuretic action, carbonic anhydrase inhibition. *0652*
Urine *Increase* *Physiological* Diuretic action. *1343*

Primidone *Serum* *No Effect* *Analytical* On GLC procedure of Papadopoulos when added *in vitro*. *2722*

Protein *Serum* *No Effect* *Analytical* No effect at 12 mg/dL on SMA 12/60 method. *2636* At concentration of 1,000 mg/L had no effect on biuret method with blank correction. *3393*
Urine *Increase* *Analytical* Makes urine highly alkaline, causes false positive. *0111* When measured by Ponceau S dye method in comparison with sulfosalicylic acid or trichloracetic acid methods. *3919*
Pancreatic Juice *Increase* *Physiological* Approximately 40% increase with infusion in normals. *0984*

Protein Bound Iodine (PBI) *Serum* *No Effect* *Physiological* Reported to have no clinical effect. *0830*
Serum *Decrease* *Physiological* Inhibits iodination of tyrosine in thyroxine binding globulin. *0012*

PSP Excretion *Urine* *Decrease* *Physiological* Increased urine flow. *1022*

Quinidine *Serum* *Increase* *Physiological* Alkalinizes urine, increases reabsorption. *1487*

Sodium *Serum* *No Effect* *Analytical* At concentration of 1,000 mg/L had no effect on flame-photometric method. *3393*
Urine *Increase* *Physiological* Diuretic action. *1343*

Titratable Acidity *Urine* *Decrease* *Physiological* Increased alkalinity. *1343*

Triglycerides *Serum* *No Effect* *Analytical* At concentration of 1,000 mg/L had no effect on lipase/esterase method. *3393*

Urea Nitrogen *Serum* *No Effect* *Analytical* At concentration of 1,000 mg/L had no effect on diacetylmonoxime method. *3393*
Serum *Increase* *Physiological* May occur with prolonged treatment. *2220*

Uric Acid *Serum* *No Effect* *Analytical* No effect at 12 mg/dL on SMA 12/60 method. *2636* At concentration of 1,000 mg/L had no effect on phosphotungstate reduction method. *3393*
Serum *Increase* *Physiological* Decreased urate clearance (?also stimulates synthesis). *1022*
Urine *Decrease* *Physiological* Inhibition of tubular secretion of urate. *1022*

Urobilinogen *Urine* *Increase* *Physiological* Excretion increased by alkalinization of urine. *2452*

Volume *Urine* *Increase* *Physiological* Initial response noted. *1680*

Acetic Acid

Casts *Urine* *Increase* *Physiological* May cause nephrotoxicity. *0279*

Urea Nitrogen *Serum* *Increase* *Physiological* May cause nephrotoxicity. *0279*

Acetoacetate

Aspartate Aminotransferase *Serum* *Increase* *Analytical* Interference in diazonium end point reactions. *0077* With non specific diazo procedure on SMA 12/60. *3217*

Creatinine *Serum* *No Effect* *Analytical* No effect on ion-exchange method of Mitchell. *2459* No effect on method of Polar and Metcoff. *1975* No effect on enzymatic procedure on Ektachem®. *1267* No effect even at 20 mmol/L on creatininase/creatinase/sarcosine/peroxidase BM-BCL method. *0765*
Serum *Increase* *Analytical* Slight effect on Lloyd's procedure. Marked effect direct methods. *2459* 1 mmol/L = 0.11 mg/dL method of Heinegard. *1547* Interferes with Grafnetter and AutoAnalyzer methods. *1975* Linear response when added to specimens for analysis by picrate method on Astra™ and SMAC®. Sensitivity of Astra method twice that of SMAC®. Effect less on stored specimens (maybe 5% medical specimens affected). *1267* Can be marked effect on Jaffé based reactions, more marked on Technicon® AutoAnalyzer than Beckman creatinine analyzer. *2470*

Fatty Acids, Free (FFA) *Serum* *Decrease* *Physiological* Effect maximal 40 minutes after load (not marked). *1811*

Ferric Chloride Test *Urine* *Positive* *Analytical* May produce red or red-brown. *0443*

Glucose *Serum* *Decrease* *Physiological* Effect maximal 40 minutes after loading. *1811*
Urine *No Effect* *Analytical* At 5 g/L on Boehringer — Mannheim BM 33071 glucose pad. *3356*
Urine *Decrease* *Analytical* At 5 g/L with Hema-Combistix® caused complete inhibition. At 0.75 g/L cause some inhibition. *3356*

Growth Hormone *Plasma* *Increase* *Physiological* Effect slight after 40 minutes. *1811*

Insulin *Plasma* *Increase* *Physiological* Maximum observed at 40 minutes after loading dose. *1811*

Ketones *Serum* *Increase* *Physiological* Maximum 60 minutes after tolerance load given. *1811*

Phenylketones *Urine* *Positive* *Analytical* red/red-brown with FeCl₃, no effect with Phenistix®. *0775*

Phosphate *Urine* *Increase* *Physiological* Inhibits tubular reabsorption. *0957*

Salicylate *Serum* *Increase* *Analytical* 50 mg/dL give equivalent of 1 mg/dL by Trinder. *2251*

Uric Acid *Serum* *Increase* *Physiological* Inhibits tubular secretion of urate. *1331*
Urine *Decrease* *Physiological* Inhibits tubular secretion of urate. *1331*

Acetohexamide

Alanine Aminotransferase *Serum* *Increase* *Physiological* May cause intrahepatic cholestatic jaundice. *2220*

Alkaline Phosphatase *Serum* *Increase* *Physiological* May cause intrahepatic cholestatic jaundice. *0652*

Aspartate Aminotransferase
Serum *Increase* *Physiological* May cause intrahepatic cholestatic jaundice. *2220*

Bile *Urine* *Increase* *Physiological* May cause intrahepatic cholestatic jaundice. *2313*

Bilirubin *Serum* *Increase* *Physiological* May cause intrahepatic cholestatic jaundice. *0652*
Urine *Increase* *Physiological* May cause intrahepatic cholestatic jaundice. *2313*

Bilirubin, Direct *Serum* *Increase* *Physiological* May cause hemolytic anemia. *1019*

BSP Retention *Serum* *Increase* *Physiological* May cause intrahepatic cholestatic jaundice. *0652*

Cephalin Flocculation *Serum* *Increase* *Physiological* May cause intrahepatic cholestatic jaundice. *2313*

Acetohexamide *(continued)*

Cholesterol *Serum Increase Physiological* May cause intrahepatic cholestatic jaundice. *2220*

Creatinine *Serum Increase Analytical* Significant effect on Jaffé methods on Beckman Astra and Du Pont aca by as much as 2.2 or 3.3 mg/dL respectively at therapeutic concentrations and as little as 0.3 mg/dL on SMA II. *3890*

Erythrocytes *Blood Decrease Physiological* May cause hemolytic anemia. *1019*

Glucose *Serum No Effect Analytical* No effect on Boehringer GOD-PERID method. *3277* At concentration of 200 mg/L had no effect on GOD/POD-PAP method. *3393*
Serum Decrease Physiological Sulfonylurea derivative promotes insulin secretion. *0652*

Haptoglobin *Serum Decrease Physiological* May cause hemolytic anemia. *1019*

Hematocrit *Blood Decrease Physiological* May cause anemia and hemolysis if G-6-PD deficiency. *1019*

Hemoglobin *Blood Decrease Physiological* May cause anemia and hemolysis if G-6-PD deficiency. *1019*

Insulin *Plasma Increase Physiological* Usual effect observed (at 2 mo). *0301*
Plasma Decrease Physiological Effect observed after 3 mo therapy. *0301*

Leukocytes *Blood Decrease Physiological* Leukopenia and aplastic anemia reported. *2313*

5'-Nucleotidase *Serum Increase Physiological* Due to cholestasis. *1778*

Osmolality *Urine Decrease Physiological* Normal slight diuretic response. *2499*

Platelets *Blood Decrease Physiological* Thrombocytopenia and aplastic anemia reported. *2313*

Prothrombin Time *Plasma Increase Physiological* Occurs if failure to excrete bile salts. *0071*

Thymol Turbidity *Serum Increase Physiological* May cause intrahepatic cholestatic jaundice. *2313*

Urea Nitrogen *Serum Increase Analytical* Chemical structure produces reaction with diacetyl. *0652*

Uric Acid *Serum Decrease Physiological* Mild uricosuric action. *3928*
Urine Increase Physiological Uricosuric action. *1433*

Urobilinogen *Urine Decrease Physiological* Theoretical effect if cholestasis. *0071*
Feces Decrease Physiological Theoretically, if cholestasis occurs. *0071*

Water Clearance *Urine Increase Physiological* Normal slight diuretic response. *2499*

Acetohydroxamic Acid

Ammonia *Plasma Decrease Physiological* Potent urease inhibitor, effect seen in 1 patient. *1487*

Urea Nitrogen *Serum Decrease Physiological* Inhibition of urease (limited *in vivo* effect). *3497*

Acetone

Bogen Test *Urine Positive Analytical* Reacts as if ethanol. *1302*

Creatinine *Serum No Effect Analytical* No effect on method of Heinegard and Tiderstrom. *1547*
Serum Increase Analytical Interferes with Jaffé method. *1975*
Urine Increase Analytical 1 mg equivalent to 0.07 mg in automated Jaffé procedure. *2558*

Estriol *Urine No Effect Analytical* No effect observed on hydrolysis. *0406*

17-Hydroxycorticosteroids *Urine Increase Analytical* Reported to interfere with Porter-Silber procedure. *0583*

17-Ketogenic Steroids *Urine Increase Analytical* Brown color with Zimmermann reaction. *0583*

17-Ketosteroids *Urine Increase Analytical* Has absorption at 520 nm in Zimmermann procedure. *0022*

pH *Blood Decrease Physiological* If ingested may cause acidosis. *0279*

Urea Nitrogen *Serum Increase Analytical* Turbidity in Nesslerization procedures. *1563*

Urobilinogen *Urine Increase Analytical* May produce similar color in reaction. *2313*

Acetophenazine

Alanine Aminotransferase *Serum Increase Physiological* Cholestasis with intrahepatic obstruction. *2313*

Alkaline Phosphatase *Serum Increase Physiological* Cholestatic hepatitis with obstruction. *2313*

Aspartate Aminotransferase
Serum Increase Physiological Cholestasis with intrahepatic obstruction. *2313*

Bile *Urine Increase Physiological* Cholestasis with intrahepatic obstruction. *2313*

Bilirubin *Serum Increase Physiological* Cholestasis with intrahepatic obstruction. *1237*
Urine Increase Physiological Cholestasis with intrahepatic obstruction. *2313*

BSP Retention *Serum Increase Physiological* Cholestasis with intrahepatic obstruction. *2313*

Cephalin Flocculation *Serum Increase Physiological* Cholestasis with intrahepatic obstruction. *2313*

Cholesterol *Serum Increase Physiological* Cholestasis with intrahepatic obstruction. *2946*

Eosinophils *Blood Increase Physiological* Allergic manifestation. *1680*

Erythrocytes *Blood Decrease Physiological* Possible aplastic anemia. *2313*

Leukocytes *Blood Decrease Physiological* Transitory leukopenia or agranulocytosis. *0071*

Platelets *Blood Decrease Physiological* Agranulocytosis or aplastic anemia. *0071*

Thymol Turbidity *Serum Increase Physiological* Cholestasis with intrahepatic obstruction. *2313*

Acetophenone

17-Ketogenic Steroids *Urine Increase Analytical* Interferes in measurement by Zimmermann method. *3505*

17-Ketosteroids *Urine Increase Analytical* Interferes in measurement by Zimmermann method. *3505*

Acetrizoate

Aspartate Aminotransferase *Urine Increase Physiological* Sustained increase due to nephrotoxic action. *3538*

Catalase *Urine Increase Physiological* Sustained increase after renal artery injection. *3538*

Creatine Kinase *Urine Increase Physiological* Sustained increase after renal artery injection. *3538*

Glomerular Filtration Rate (GFR)
Urine Decrease Physiological When 70% injected for aortography. *0065*

^{131}I Uptake *Serum Decrease Physiological* Due to iodine component of material. *2220*

Lactate Dehydrogenase *Urine Increase Physiological* Sustained increase due to nephrotoxic action. *3538*

PAH Clearance *Urine Decrease Physiological* When 70% injected for aortography. *0065*

Protein Bound Iodine (PBI) *Serum Increase Analytical* But does not affect T_4 (effect lasts 2 to 4 weeks). *3218*

Acetylacetone

Porphobilinogen *Urine Increase Analytical* Reacts with Ehrlich's reagent unless removed. *3155*

Acetylcholine

Hydrochloric Acid *Gastric Material Increase Physiological* Hypoglycemia is powerful stimulant. *0814*

Pepsin *Gastric Material Increase Physiological* Hypoglycemia is powerful stimulant. *0814*

N-Acetylcysteine

Gold *Urine Increase Physiological* Excretion doubled in response to i.v. drug in patients previously given gold. *1312*

Acetylglucosamine

Protein *Test Conditions Increase Analytical* Slight reaction with Folin-Ciocalteu reagent. *1569*

Acetylhistamine

Urobilinogen *Test Conditions Increase Analytical* Positive spot test with Ehrlich's reagent. *3765*

L-α-Acetylmethadol (LAAM)

Luteinizing Hormone (LH) *Plasma No Effect Physiological* In 9 male heroin addicts maintained on drug and in 2 weeks following abrupt withdrawal. *2409*

Testosterone *Serum No Effect Physiological* In 9 male heroin addicts maintained on drug and in 2 weeks following abrupt withdrawal. *2409*

Acetylpenicillamine

Copper *Serum Decrease Physiological* Elimination of heavy metals in cases of mercury poisoning. *1871*

Mercury *Serum Increase Physiological* Mobilization of mercury in case of poisoning. *1871*
Urine Increase Physiological Elimination of heavy metals in cases of mercury poisoning. *1871*

Acetylphenylhydrazine

Bilirubin *Serum Increase Physiological* May cause hemolytic anemia. *2313*

Bilirubin, Direct *Serum Increase Physiological* May cause hemolytic anemia. *2313*

Erythrocytes *Blood Decrease Physiological* May cause hemolytic anemia. *2313*

Haptoglobin *Serum Decrease Physiological* May cause hemolytic anemia. *2313*

Hematocrit *Blood Decrease Physiological* May cause hemolytic anemia. *2313*

Hemoglobin *Blood Decrease Physiological* May cause hemolytic anemia. *2313*

Acetylsalicylic Acid

Acetaminophen *Serum No Effect Analytical* At 500 mg/L had no effect on HPLC method. *3432*

Alanine Aminotransferase *Serum No Effect Analytical* At 5 times upper limit of therapeutic range on methods on SMAC®, Abbott-VP, aca, Cobas-Bio and KDA. *2138* At 8300 μmol/L on continuous analytical method. *1785* No effect at therapeutic concentration on Reflotron method. *1984*
Serum Decrease Analytical At 8300 μmol/L on colorimetric analytical method. *1785*

Albumin *Serum No Effect Analytical* On bromcresol green method on SMA II at physiological concentration. *1787* At 5 times upper limit of therapeutic range on methods on SMAC®, Ektachem®, Hitachi® 705 and KDA. *2138* At concentration of 2,000 mg/L had no effect on BCG method. *3393*
Serum No Effect Physiological No effect seen in patients undergoing treatment. *1787*

Alkaline Phosphatase *Serum No Effect Analytical* At 5 times upper limit of therapeutic range on methods on SMAC®, Abbott-VP, aca, Cobas-Bio, Hitachi® 705 and KDA. *2138* At 8300 μmol/L on continuous method. *1785*

Ammonia *Plasma No Effect Analytical* At concentration of 300 mg/L had no effect on Ektachem® method. *3393*

Amylase *Serum No Effect Analytical* At 5 times upper limit of therapeutic range on methods on aca, Cobas-Bio, and Ektachem®. *2138*

Angiotensin-Converting Enzyme (ACE)
Serum No Effect Analytical No effect observed even with large amounts. *3038*

Antithrombin III *Plasma No Effect Analytical* At concentration of 3 mg/L had no effect on aca method. *3393*

Aspartate Aminotransferase *Serum No Effect Analytical* At 5 times upper limit of therapeutic range on method on SMAC®, Abbott-VP, Cobas-Bio, aca, and Hitachi® 705. *2138* At 8300 μmol/L on continuous and colorimetric methods. *1785* No effect at therapeutic concentration on Reflotron method. *1984*
Serum Increase Analytical At 5 times upper limit of therapeutic range on method on KDA. *2138*

Bicarbonate *Serum Decrease Analytical* At concentrations greater than 1,000 mg/L (10 times upper limit of therapeutic range) when using a phenolphthalein indicator method. *3393*

Bilirubin *Serum No Effect Analytical* On Jendrassik — Grof, dimethylsulfoxide and spectrophotometric procedures with therapeutic concentrations. *1786* On diazo method on SMA II at physiological concentration. *1787* At 5 times upper limit of therapeutic range on methods on SMAC®, aca, Cobas-Bio, Ektachem®, Hitachi® 705 and KDA. *2138* With 8326 μmol/L on Jendrassik-Grof method. With 8326 μmol/L on method using dimethylsulfoxide. With 8326 μmol/L on direct spectrophotometric method. *1786* With 8326 μmol/L on direct spectrophotometric method. *1786* At concentration of 300 mg/L had no effect on Ektachem® method. *3393* At concentration of 2,000 mg/L had no effect on Jendrassik and Grof method. *3393*
Serum No Effect Physiological No effect seen in patients undergoing treatment. *1787*

Calcium *Serum No Effect Analytical* On cresolphthalein complexone method on SMA II at physiological concentration. *1787* At 5 times upper limit of therapeutic range on method on SMAC®, Abbott-VP, aca, Ektachem®, Hitachi® 705 and KDA. *2138* At concentration of 2,000 mg/L had no effect on cresolphthalein method. *3393*
Serum No Effect Physiological No effect seen in patients undergoing treatment. *1787*

Chloride *Serum No Effect Analytical* At concentration of 2,000 mg/L had no effect on mercurimetric method. *3393*

Cholesterol *Serum No Effect Analytical* At therapeutic concentrations on enzymatic and Liebermann — Burchard methods. *1786* On enzymatic method on SMA II at physiological concentration. *1787* At 5 times upper limit of therapeutic range on method on SMAC®, Abbott-VP, Cobas-Bio, Ektachem®, Hitachi® 705 and KDA. *2138* With 8326 μmol/L on cholesterol esterase/oxidase method. *1786* No effect at therapeutic concentration on Reflotron method. *1984* At concentration of 750 mg/L had no effect on CHOD-Iodide method. *3393* At concentration of 750 mg/L had no effect on CHOD-PAP method. *3393* At concentration of 750 mg/L had no effect on catalase-AIDH method. *3393* At concentration of 750 mg/L had no effect on method using catalase-Hantzsch reaction. *3393* At concentration of 2,000 mg/L had no effect on Liebermann-Burchard method. *3393*
Serum Increase Physiological Gradual increase during course of treatment with drug. *1787*
Serum Decrease Analytical Increase by 9% with 8326 μmol/L on Liebermann-Burchard method. *1786*

Creatine Kinase *Serum No Effect Analytical* At 5 times upper limit of therapeutic range on methods on SMAC®, Abbott-VP, aca, Cobas-Bio, and Hitachi® 705. *2138*
Serum Decrease Analytical With 8300 μmol/L on kinetic method. *1785*

Acetylsalicylic Acid *(continued)*

Creatinine *Serum No Effect Analytical* On alkaline picrate procedures and Slot method at therapeutic concentrations. *1786* On alkaline picrate method on SMA II at physiological concentration. *1787* At 5 times upper limit of therapeutic range on method on SMAC®, Abbott-VP, Cobas-Bio, aca, Ektachem®, Hitachi® 705 and KDA. *2138* With 8326 μmol/L on AutoAnalyzer Jaffé method. With 8326 μmol/L on picrate method with deproteinization. With 8326 μmol/L on Slot procedure. *1786* At concentration of 300 mg/L had no effect on Ektachem® method. *3393* At concentration of 600 mg/L had no effect on creatinine iminohydrolase method. *3393* At concentration of 2,000 mg/L had no effect on AutoAnalyzer Jaffé method. *3393* At concentration of 600 mg/L had no effect on Jaffé-Fading-Fraction method. *3393* At concentration of 600 mg/L had no effect on Jaffé-Fuller's earth method. *3393* At concentration of 500 mg/L had no effect on kinetic Jaffé method on BKA-2. *3393*

Serum No Effect Physiological No effect seen in patients undergoing treatment. *1787*

Fibrinogen *Plasma No Effect Analytical* At concentration of 500 mg/L had no effect on aca method. *3393*

Glucose *Serum No Effect Analytical* At therapeutic concentrations on hexokinase, glucose dehydrogenase 2,4-dichlorophenol, ABTS, and o-toluidine methods. *1786* On hexokinase method on SMA II at physiological concentration. *1787* At 5 times upper limit of therapeutic range on method on SMAC®, Abbott-VP, Cobas-Bio, aca, Ektachem®, Hitachi® 705 and KDA. *2138* With 8326 μmol/L on hexokinase/G-6-PDH method. With 8326 μmol/L on glucose dehydrogenase method. With 8326 μmol/L on o-toluidine method. *1786* No effect at therapeutic concentration on Reflotron method. *1984* At concentration of 100 mg/L had no effect on Ektachem® method. *3393* At concentration of 1400 mg/L had no effect on hexokinase/G-6-PDH method. *3393* At concentration of 500 mg/L had no effect on Seralyzer method. *3393*

Serum No Effect Physiological No effect seen in patients undergoing treatment. *1787*

Serum Increase Analytical Increase by 11% with 8326 μmol/L on glucose-peroxidase method with 2,4-Dichlorophenol. Increase by 3% with 8326 μmol/L on glucose-peroxidase method with ABTS. *1786*

γ-Glutamyltransferase (GGT) *Serum No Effect Analytical* At 5 times upper limit of therapeutic range on methods on SMAC®, Abbott-VP and Hitachi® 705. *2138* No effect at therapeutic concentration on Reflotron method. *1984*

Hydroxybutyrate Dehydrogenase
Serum Decrease Analytical With 8300 μmol/L on kinetic method. *1785*

Iron *Serum No Effect Analytical* On Ramsay and bathophenanthroline methods at therapeutic concentrations. *1786* On Ferrozine method on SMA II at physiological concentration. *1787* At 5 times upper limit of therapeutic range on Ferrozine method of SMAC®. *2138* With 8326 μmol/L on bathophenanthroline method. With 8326 μmol/L on Ramsay method. *1786* At concentration of 1500 mg/L had no effect on Ferrozine method. *3393*

Serum Increase Physiological Significant effect correlated with duration of treatment (17.95 μmol/L vs 14.98 μmol/L). *1787*

Lactate *Serum No Effect Analytical* At concentration of 1400 mg/L had no effect on enzyme method. *3393*

Lactate Dehydrogenase *Serum No Effect Analytical* At 5 times upper limit of therapeutic range on method on SMAC®, Abbott-VP, Cobas-Bio, and Hitachi® 705. *2138* At 8300 μmol/L on continuous method with lactate as substrate. At 8300 μmol/L on colorimetric method. *1785*

Serum Decrease Analytical At 5 times upper limit of therapeutic range on method on KDA. *2138* With 8300 μmol/L on continuous method with pyruvate substrate. *1785*

Phosphate *Serum No Effect Analytical* At 5 times upper limit of therapeutic range on method on SMAC®, aca, Hitachi® 705, and KDA. *2138* At concentration of 2,000 mg/L had no effect on phosphomolybdate method. *3393*

Plasminogen *Plasma No Effect Analytical* At concentration of 300 mg/L had no effect on aca method. *3393*

Potassium *Serum No Effect Analytical* At concentration of 100 mg/L had no effect on flame-photometric method. *3393* At concentration of 3,000 mg/L had no effect on ISE measurement with predilution. *3393* At concentration of 1,000 mg/L had no effect on ISE measurement without predilution. *3393*

Protein *Serum No Effect Analytical* At therapeutic concentrations on biuret and spectrophotometric methods. *1786* On biuret method on SMA II at physiological concentration. *1787* At 5 times upper limit of therapeutic range on methods on SMAC®, Abbott-VP, Hitachi® 705 and KDA. *2138* At concentration of 2,000 mg/L had no effect on biuret method with blank correction. *3393*

Serum No Effect Physiological No effect seen in patients undergoing treatment. *1787*

Serum Increase Analytical Increase by 9% with 8326 μmol/L with direct spectrophotometry at 280 nm. *1786*

Serum Decrease Analytical Increase by 1.5% with 8326 μmol/L on biuret method. *1786*

Protein Electrophoresis *Plasma No Effect Analytical* At concentration of 1,000 mg/L had no effect on automated Olympus-Hite method. *3393*

Sodium *Serum No Effect Analytical* At concentration of 200 mg/L had no effect on flame-photometric method. *3393* At concentration of 3,000 mg/L had no effect on ISE measurement with predilution. *3393* At concentration of 1,000 mg/L had no effect on ISE measurement without predilution. *3393*

Triglycerides *Serum No Effect Analytical* On enzymatic methods at therapeutic concentrations. *1786* On enzymatic method on SMA II at physiological concentration. *1787* At 5 times upper limit of therapeutic range on method on SMAC®, Abbott-VP, Ektachem® and Hitachi® 705. *2138* With 8326 μmol/L on enzymatic procedure with lipase. *1786* No effect at therapeutic concentration on Reflotron method. *1984* At concentration of 2,000 mg/L had no effect on lipase/esterase method. *3393*

Serum Increase Analytical At 5 times upper limit of therapeutic range on method of KDA. *2138*

Serum Increase Physiological Significantly higher in patients receiving drug than in controls (1.69 mmol/L vs 1.47 mmol/L). Also seen in same patients before and after treatment. *1787*

Serum Decrease Analytical At concentrations greater than 75 mg/L (within therapeutic range) lowered analyte concentration as measured by GPO-PAP method. *3393*

Urea Nitrogen *Serum No Effect Analytical* At therapeutic concentrations on glutamate dehydrogenase, phenol-hypochlorite and diacetyl monoxime methods. *1786* On diacetyl monoxime method on SMA II at physiological concentration. *1787* At 5 times upper limit of therapeutic range on method on SMAC®, Abbott-VP, Cobas-Bio, Ektachem®, Hitachi® 705 and KDA. *2138* With 8326 μmol/L on method using glutamate dehydrogenase. With 8326 μmol/L on method using phenol hypochlorite. With 8326 μmol/L on method using diacetylmonoxime. *1786* No effect at therapeutic concentration on Reflotron method. *1984* With 8326 μmol/L on method using diacetylmonoxime. *1786* At concentration of 2,000 mg/L had no effect on diacetylmonoxime method. *3393* At concentration of 100 mg/L had no effect on Ektachem® method. *3393* At concentration of 50 mg/L had no effect on urease-Berthelot method. *3393*

Serum No Effect Physiological No effect seen in patients undergoing treatment. *1787*

Uric Acid *Serum No Effect Analytical* At therapeutic concentrations on uricase-catalase aldehyde dehydrogenase, direct UV-test and phosphotungstate methods. *1786* On phosphotungstate method on SMA II at physiological concentration. *1787* At 5 times upper limit of therapeutic range on method on SMAC®, Abbott-VP, Cobas-Bio, aca, Ektachem®, Hitachi® 705 and KDA. *2138* With 8326 μmol/L on uricase-catalase method with aldehyde dehydrogenase. With 8326 μmol/L on direct UV method with uricase. With 8326 μmol/L on phosphotungstate reduction method. *1786* No effect at therapeutic concentration on Reflotron method. *1984* With 8326 μmol/L on phosphotungstate reduction method. *1786* At concentration of 600 mg/L had no effect on Kageyama-Hantzsch method. *3393* At concentration of 600 mg/L had no effect on catalase-AIDH method. *3393* At concentration of 2,000 mg/L had no effect on phosphotungstate reduction method. *3393* At concentration of 900 mg/L had no effect on uricase-PAP method. *3393*

Serum No Effect Physiological No effect seen in patients undergoing treatment. *1787*

Serum Increase Analytical Increase by 6% with 8326 μmol/L on uricase-catalase method. *1786*

Acetylsulfadiazine

Crystals *Urine Increase Physiological* Wheatsheaves eccentric binding (not in alkaline urine). *0443*

Acetyltryptophan

Tryptophan *Test Conditions Increase Analytical* Measured as same by Spies/Chambers method. *1415*

Acidosis

Urea Nitrogen *Serum Decrease Analytical* Much lower Azostix® than diacetyl value. *1460*

Acids

Casts *Urine Increase Physiological* Renal damage following ingestion. *1302*

Leukocytes *Blood Increase Physiological* Response to ingestion. *1302*

Occult Blood *Feces Positive Physiological* If ingested may cause bleeding and gastrointestinal symptoms. *0279*

Protein *Urine Increase Physiological* Ingestion may cause severe renal damage. *1302*

Volume *Urine Decrease Physiological* May produce oliguria with renal damage. *0279*

Acid Urine

Phenmetrazine *Urine Increase Physiological* With overdose, acidification increases excretion. *1678*

Trimethoprim *Urine Increase Physiological* If co-trimoxazole administered. *0745*

Aconitine

Occult Blood *Feces Positive Physiological* If ingested may cause many gastrointestinal symptoms. *0279*

Acridine

Chromosomes *Test Conditions Abnormal Physiological* Clastogenic in human diploid fibroblasts *in vitro*. *3282*

Acridine Orange

Chromosomes *Test Conditions Abnormal Physiological* Clastogenic in human diploid fibroblasts *in vitro*. *3282*

Acriflavine

Bile *Urine Increase Analytical* Produces yellow color when urine shaken. *0459*

Color *Urine Increase Analytical* Greenish fluorescence. *2448*

Porphyrins *Urine Increase Analytical* Produce fluorescence. *3879*

Urobilin *Urine Increase Analytical* Produces yellow-green fluorescence. *1563*

Actinomycin

Chromosomes *Test Conditions Abnormal Physiological* Clastogenic in human cells. *3282*

Porphobilinogen *Urine Decrease Physiological* Suppressed induction of ALA synthetase in porphyria. *0938*

Uroporphyrin *Urine Decrease Physiological* Suppresses induction of ALA synthetase in porphyria. *0938*

Acyclovir

Creatinine *Serum Increase Physiological* Reversible increase, especially with doses greater than 5 mg/kg i.v. every 8 h. Major adverse effect occurring in as many as 50% patients. *0936*

Crystals *Urine Increase Physiological* Especially with high-dose bolus infusion and with dehydration and pre-existing renal insufficiency. *0936*

Erythrocytes *Blood Decrease Physiological* One case megaloblastic anemia documented. *0936*

Hemoglobin *Blood Decrease Physiological* One case megaloblastic anemia documented. *0936*

Lymphocyte Response *Blood No Effect Physiological* No effect on blastogenic response to mitogens-phytohemagglutin, pokeweed, and concanavilin A *in vitro*. *0147*

Urea Nitrogen *Serum Increase Physiological* Transient rises reported in some patients. *0147*

Adenine

Chromosomes *Test Conditions Abnormal Physiological* Clastogenic in human cells. *3282*

Adenine Deoxyribose

Chromosomes *Test Conditions Abnormal Physiological* Clastogenic in human cells. *3282*

Adenine Diphosphate

Phosphate *Test Conditions Increase Analytical* Affects Fiske-Subbarow unless Seddon procedure used. *3236*

Adenine Monophosphate

Phosphate *Test Conditions Increase Analytical* Affects Fiske-Subbarow unless Seddon procedure used. *3236*

Adenosine

Epinephrine *Plasma Increase Physiological* Increase by 213% after i.v. infusion of 10 to 140 μg/kg/min in 7 normal subjects. *0338*

Norepinephrine *Plasma Increase Physiological* Increase by 44% after i.v. infusion of 10 to 140 μg/kg/min in 7 normal subjects. *0338*

Renin Activity *Plasma No Effect Physiological* After i.v. infusion of 10 to 140 μg/kg/min in 7 normal subjects. *0338*

Adenosine Triphosphate

Phosphate *Serum Decrease Analytical* At concentrations above 0.56 mmol/L had linear decreasing effect on method of Daly and Ertingshausen. *0628*

Adiphenine

Amphetamine *Urine No Effect Analytical* No effect at 100 mg/L on method of Rutter. *3097*

β-Adrenergic Drugs

Glucose *Serum Increase Physiological* Masks spurious hypoglycemia from insulin overdose. *1237*

Hematocrit *Blood Increase Physiological* But less than expected with change in plasma volume. *1832*

Volume *Plasma Decrease Physiological* Decreased by approximately 10% shortly after i.v. injection. *1832*

Urine Decrease Physiological Antidiuretic effect mediated centrally. *3205*

Aflatoxin

Chromosomes *Test Conditions Abnormal Physiological* Probably clastogenic in human cells. *3282*

Aging

α_1-Acid Glycoprotein *Serum Increase Physiological* Some increase with age. *0954*

Acid Phosphatase *Serum Decrease Physiological* Steep fall to about age 20 y then almost steady. *1043*

Alkaline Phosphatase *Serum Increase Physiological* Peaks at puberty then steep fall. *0887*

α_1-Antitrypsin *Serum Increase Physiological* Very slight increase with age. *0954*

Aspartate Aminotransferase
Serum Decrease Physiological From 47 U/L at 6 y to 35 at 20 y (women). From 45 U/L at 6 y to 29 at 20 y (men). *0887*

BSP Retention *Serum No Effect Physiological* No change with increasing age noted. *1976*

Calcitonin *Plasma Decrease Physiological* Generally higher in infants at birth. *3127*

Calcium *Serum No Effect Physiological* No significant change from age 6-20 y. *0887*
Urine Decrease Physiological In women over 50 y more constant than younger. *2481*

Carotene *Serum Decrease Physiological* Significant correlations especially after 40 y in women increase to age 40. *2880*

Catechol-O-Methyl Transferase
Red Blood Cells No Effect Physiological No difference between ages 16 and 89 y. *1401*

Ceruloplasmin *Serum Increase Physiological* Slight effect in men around 46 y. *0954*

Cholesterol *Serum No Effect Physiological* No significant change from age 6-20 y. *0887*

Creatine Kinase *Serum Increase Physiological* 5 U/L in normal adult women age 40 y and above. *2744*

Creatinine *Serum Increase Physiological* From 0.5 mg/dL at 6 y to 0.95 at 20 y (men). From 0.5 mg/dL at 6 y to 0.75 at 16 y (women). *0887*

Dehydroepiandrosterone Sulfate (DHEA-S)
Plasma Decrease Physiological In women from 160 µg/dL at 25 to 60 at 55 y. *0462*

Diphenylhydantoin *Serum Increase Physiological* Concentration significantly lower in children at same dose. *3054*

Dopamine Hydroxylase *Serum Increase Physiological* Low or not detectable in newborns. *3807*

Effective Renal Plasma Flow
Patient Decrease Physiological Approximately 45% linear decrease from age 45 to 85 y. *2617*

Erythrocyte Sedimentation Rate
Blood Increase Physiological Steady to puberty then rise in men, fall in women. *0887*

Estradiol *Urine Increase Physiological* Maximal about 30-35 y, smooth fall thereafter. *1304*

Estriol *Urine Increase Physiological* Maximal about 30-35 y, smooth fall thereafter. *1304*

Estrone *Urine Increase Physiological* Maximal about 30-35 y, smooth fall thereafter. *1304*

Etiocholanolone *Urine Decrease Physiological* In women from 2.7 mg/24 h at 25 to 1.3 at 55 y. *0462*

α-Fucosidase *Serum Increase Physiological* Steep rise to age 25 y then little change. *1043*

β-Galactosidase *Serum Decrease Physiological* Steep fall to age 25 y then less steep. *1043*

Glucose *Serum No Effect Physiological* No significant change from age 6-20 y. *0887*

β-Glucosidase *Serum No Effect Physiological* Little change throughout life. *1043*

β-Glucuronidase *Serum Increase Physiological* Little change to age 45 y then sharp increase. *1043*

β-Glusosaminidase *Serum Increase Physiological* Steady increase from puberty onwards. *1043*

Haptoglobin *Serum Increase Physiological* Mean 120 mg/dL at 20 y, 250 at 70 y. *0954*

Hemoglobin *Blood Increase Physiological* 13.5 g/dL at 6 to 16 y, at 16 y 15.7, at 20 y 16.3 (men). 13.5 g/dL at 6 y to 14.3 at puberty then steady (women). *0887*

5-Hydroxyindoleacetic Acid (5-HIAA)
Urine Decrease Physiological Lower in elderly compared with 20-60 y. *1176*

17-Ketosteroids *Plasma Decrease Physiological* In women from 160 µg/dL at 25 y to 80 µg/dL at 45 y. *0462*
Urine Increase Physiological Progressive increase to puberty, continues to increase after in men. *0887*

Lactate Dehydrogenase *Serum Decrease Physiological* Slow fall to puberty then steep to age 20 y. *0887*

Leukocytes *Blood Decrease Physiological* Fall from age 20 to 80 y in non smokers. *1195*

Lipoprotein Lipase *Serum Decrease Physiological* Mean 0.13 meq/L at age 26 y, 0.06 at 86 y. *0480*

Magnesium *Serum No Effect Physiological* No significant change from age 6-20 y. *0887*

α-Mannosidase *Serum Increase Physiological* Very slight increase after puberty. *1043*

PAH Clearance *Urine Decrease Physiological* Approximately 60% linear decrease from age 40 to 85 y. *2617*

Phosphate *Serum Decrease Physiological* Approximately 4.6 mg/dL 6 y to puberty then fall to 3.5 at 20 y. *0887*

Potassium *Serum No Effect Physiological* No significant change from age 6-20 y. *0887*

Pregnanediol *Urine Increase Physiological* In women progressive increase to 20 y, slight decrease after. In males progressive increase to 20 y, slight decrease after. *1304*

Pregnanetriol *Urine Increase Physiological* In women progressive increase to 20 y, slight decrease after. In men progressive increase to 20 y, slight decrease after. *1304*

Prolactin *Plasma Decrease Physiological* Maximum in premature, falls to 5 ng/mL ages 2-12 y. *1435*

Renin Activity *Plasma Increase Physiological* Much higher in neonates, than children, than adults. *2006*

Rheumatoid Factor *Serum Positive Physiological* In older nonsmokers. *2335*

Ribonuclease *Serum Increase Physiological* In males over 30 y, but not females. *1854*

Sodium *Serum No Effect Physiological* No significant change from age 6-20 y. *0887*

Testosterone *Serum Increase Physiological* In women due to decreased metabolic clearance rate. *1941*

α-Tocopherol *Serum Increase Physiological* From 0.8-1.0 mg/dL at 5 y to 1.2 at 50 y. *2158*

Transferrin *Serum Decrease Physiological* Slight effect with aging. *0954*

Triglycerides *Serum Increase Physiological* Positive correlation in sedentary but not athletes. *1702*

Urea Nitrogen *Serum No Effect Physiological* No significant change from age 6-20 y. *0887*

Uric Acid *Serum Increase Physiological* Slight to puberty then big increase in men. *0887*

Ajmaline

Alanine Aminotransferase *Serum Increase Physiological* Intrahepatic cholestasis. *0268* Hepatotoxicity and intrahepatic cholestasis reported. *3215* In 4 patients with centrilobular cholestasis and mild hepatocytic lesions: recovery with drug withdrawal. *2729*

Alkaline Phosphatase *Serum Increase Physiological* Intrahepatic cholestasis. *0268* Hepatotoxicity and intrahepatic cholestasis reported. *3215* In 4 patients with centrilobular cholestasis and mild hepatocytic lesions: recovery with drug withdrawal. *2729*

Aspartate Aminotransferase
Serum Increase Physiological Intrahepatic cholestasis. *0268* Hepatotoxicity and intrahepatic cholestasis reported. *3215* In 4 patients with centrilobular cholestasis and mild hepatocytic lesions: recovery with drug withdrawal. *2729*

Bile *Urine Increase Physiological* Intrahepatic cholestasis. *0268*

Bilirubin *Serum Increase Physiological* Intrahepatic cholestasis. *0268* Hepatotoxicity and intrahepatic cholestasis reported. *3215* In 4 patients with centrilobular cholestasis and mild hepatocytic lesions: recovery with drug withdrawal. *2729*

Catecholamines *Plasma Increase Physiological* Release of stored norepinephrine. *2220*

Digoxin *Serum No Effect Physiological* No effect on serum concentration reported. *0344*

Eosinophils *Blood Increase Physiological* In 4 patients with centrilobular cholestasis and mild hepatocytic lesions: recovery with drug withdrawal. *2729*

Glucose *Serum No Effect Analytical* At concentration of 3.3 mg/L had no effect on GOD/POD-PAP method. *3393*

HMPG (4-Hydroxy-3-Methoxy-Phenylethylene Glycol) *Urine Increase Physiological* Release of stored norepinephrine. *2220*

Hydrochloric Acid *Gastric Material Increase Physiological* Strong stimulant action. *2220*

Migration Inhibition Factor *Blood Positive Physiological* In 6 patients with as high frequency as with Quinidine, Procainamide and Nifedipine. *3073*

Neutrophils *Blood Decrease Physiological* Isolated reversible cases of agranulocytosis. *3215*

Thymol Turbidity *Serum Increase Physiological* Intrahepatic cholestasis. *0268*

Triglycerides *Serum No Effect Analytical* At concentration of 3.3 mg/L had no effect on GPO-PAP method. *3393*

Uric Acid *Serum No Effect Analytical* At concentration of 3.3 mg/L had no effect on uricase-PAP method. *3393*

Urobilinogen *Urine Decrease Physiological* Intrahepatic cholestasis. *0268*
Feces Decrease Physiological Intrahepatic cholestasis. *0268*

Vanillylmandelic Acid *Urine Increase Physiological* Release of stored norepinephrine. *2220*

Alanine

Ammonia *Plasma Decrease Analytical* With 2500 nmoles produces 8% inhibition indophenol react. *1290*

Creatinine *Serum No Effect Physiological* Unaffected by alanine administration to fasting obese. *1263*
Urine No Effect Physiological Unaffected by alanine administration to fasting obese. *1263*

Glucose *Serum Increase Physiological* Slight effect in normals, greater in diabetics. *3880* After prolonged fast in obese. *1263*
Serum Decrease Physiological After short term fast in obese. *1263*

Growth Hormone *Plasma Increase Physiological* Infusion caused increase by at least 40 pg/mL. *3880*

β-Hydroxybutyrate *Blood Decrease Physiological* Marked fall when given orally to fasting obese. *1263*
Urine Decrease Physiological Falls in parallel with blood with chronic administration. *1263*

Insulin *Plasma Increase Physiological* Marked increase after fasting in obese. *1263*

Isoleucine *Plasma Decrease Physiological* Decrease of about 40% with 1 week administration. *1263*

Leucine *Plasma Decrease Physiological* Decrease of about 40% with 1 week administration. *1263*

Phosphate *Serum Increase Physiological* Slight effect of chronic administration in fasting. *1263*
Urine Increase Physiological Inhibits tubular reabsorption. *0957*
Urine Decrease Physiological Marked effect when given to fasting obese. *1263*

Potassium *Serum Decrease Physiological* Significant effect with chronic administration in fasting. *1263*
Urine Decrease Physiological Marked effect when given to fasting obese. *1263*

Pyruvate Kinase *Red Blood Cells Decrease Analytical* Most marked with larger side chain. *0370*

Tyrosine *Plasma No Effect Analytical* On fluorometric procedure of Ambrose. *0080*

Urea Nitrogen *Serum No Effect Analytical* No effect on Berthelot reaction. *1862*
Serum Increase Physiological Significant effect of chronic administration in fasting. *1263*
Urine Increase Physiological Significant effect of chronic administration in fasting. *1263*

Uric Acid *Serum Decrease Physiological* Decreases high of fasting but not to normal. *1263*
Urine Increase Physiological Reciprocal change with serum when given to fasting. *1263*

Valine *Plasma Decrease Physiological* Decrease of about 40% with 1 week administration. *1263*

β-Alanine

Tyrosine *Plasma No Effect Analytical* On fluorometric procedure of Ambrose. *0080*

Albumin

Alkaline Phosphatase *Serum Increase Physiological* Placental alkaline phosphate if from Pitman-Moore or Parke-Davis. *1488*
Serum Decrease Analytical If prepared from venous blood. *0583*

Calcium *Serum No Effect Analytical* method of Morin at 8 mg/dL. *2485*

Creatinine *Serum Increase Analytical* 50 g/L equivalent to 0.02 mg/dL with method of Heinegard. *1547*

Estriol *Urine No Effect Analytical* No effect observed on hydrolysis. *0406*

Lipase *Serum Increase Analytical* By preventing inactivation of enzyme. *3589*

Protein *CSF Decrease Analytical* Turbidity < globulins with sulfosalicylic acid. *2313*

Protein Bound Iodine (PBI) *Serum Increase Physiological* Causes effect if iodinated. *3505*

Specific Gravity *Urine Increase Analytical* Each 1 g/dL increases S.G. by 0.003. *0251*

Thymol Turbidity *Serum Increase Analytical* If high. *2313*
Serum Decrease Analytical High serum concentration causes decrease. *0355*

Albuterol

Acetaminophen Screening Test *Urine Negative Analytical* No reaction with o-cresol at therapeutic concentrations. *3326*

Aldosterone *Plasma Decrease Physiological* No significant change in 4 healthy men given i.v. infusion of therapeutic amount. *2807*

Calcium *Serum Decrease Physiological* Significant dose-related effect after i.v. infusion of therapeutic doses in 4 healthy male volunteers. *2807*

Cholesterol, High Density Lipoprotein
Serum Increase Physiological After 2 weeks of treatment average increase of 6.9% when receiving 8 mg twice daily. *0640*

Corticosteroids *Plasma Decrease Physiological* Significant dose-related effect after i.v. infusion of therapeutic doses in 4 healthy male volunteers. *2807*

Estriol *Plasma Decrease Physiological* Significant reduction when given to women in premature labor. *1525*

Glucose *Serum Increase Physiological* Significant dose-related effect after i.v. infusion of therapeutic doses in 4 healthy male volunteers. *2807*

β-Hydroxybutyrate *Serum No Effect Physiological* No significant change in 4 healthy men given i.v. infusion of therapeutic amount. *2807*

Insulin *Plasma Increase Physiological* Significant dose-related effect after i.v. infusion of therapeutic doses in 4 healthy male volunteers. *2807*

Ketones *Serum Increase Physiological* Significant dose-related effect after i.v. infusion of therapeutic doses in 4 healthy male volunteers. *2807*

Lactate *Serum Increase Physiological* Significant dose-related effect after i.v. infusion of therapeutic doses in 4 healthy male volunteers. *2807*

Albuterol (continued)

Magnesium *Serum Decrease Physiological* Significant dose-related effect after i.v. infusion of therapeutic doses in 4 healthy male volunteers. *2807*

Phosphate *Serum Decrease Physiological* Significant dose-related effect after i.v. infusion of therapeutic doses in 4 healthy male volunteers. *2807*

Platelets *Blood Decrease Physiological* In 5 healthy volunteers given drug i.v.: significant reduction occurred over 6 minutes. *2045*

Potassium *Serum Decrease Physiological* Significant dose-related effect after i.v. infusion of therapeutic doses in 4 healthy male volunteers. *2807*

Renin Activity *Plasma Increase Physiological* Significant dose-related effect after i.v. infusion of therapeutic doses in 4 healthy male volunteers. *2807*

Albutoin

Protein Bound Iodine (PBI) *Serum Decrease Physiological* Side effects less marked than with other hydantoins. *2449*

Alclofenac

Tryptophan *Plasma Decrease Physiological* (total measured) in rheumatoids on therapy. *0188*

Tryptophan, Free *Plasma Increase Physiological* In rheumatoids on therapy. *0188*

Alclometasone

Cortisol *Plasma No Effect Analytical* In 10 volunteers given 30 g over 80% of body surface twice daily. *3586* With twice daily topical application in 39 children over 3 weeks. *0755*

Cortisol, Free *Urine No Effect Analytical* In 10 volunteers given 30 g over 80% of body surface twice daily. *3586*

17-Hydroxycorticosteroids *Urine No Effect Analytical* In 10 volunteers given 30 g over 80% of body surface twice daily. *3586*

Alcohols

Occult Blood *Feces Positive Physiological* May cause many gastrointestinal symptoms including bleeding. *0279*

Alcuronium

Albumin *Serum Increase Physiological* Sensitivity to drug associated with albumin concentration. *3481*

Alcuronium Chloride

Albumin *Serum Decrease Physiological* Negative correlation between sensitivity and albumin concentration. *3481*

Aldatense

Alkaline Phosphatase *Serum Decrease Physiological* Increase by 80% in a group of hypertensives compared with controls. *1611*

Creatinine *Serum Increase Physiological* Increase by 10-20% in a group of hypertensives compared with controls. *1611*

γ-Glutamyltransferase (GGT) *Serum Decrease Physiological* Increase by 80% in a group of hypertensives compared with controls. *1611*

Phosphate *Serum Increase Physiological* Increase by 10-20% in a group of hypertensives compared with controls. *1611*

Triglycerides *Serum Increase Physiological* Increase by 30% in a group of hypertensives compared with controls. *1611*

Urea Nitrogen *Serum Increase Physiological* Increase by 10-20% in a group of hypertensives compared with controls. *1611*

Uric Acid *Serum Increase Physiological* Increase by 10-20% in a group of hypertensives compared with controls. *1611*

Aldosterone

Bicarbonate *Serum Increase Physiological* Metabolic alkalosis. *1022*

Chloride *Serum Decrease Physiological* Hypokalemic hypochloremic alkalosis may occur. *1343*

Corticosteroids *Plasma Increase Analytical* 10% (cortisol 100%) competitive protein binding method of Ficher. *1128*

Cortisol *Plasma No Effect Analytical* Reactivity of less than 1% possible with RIA. *1081*

Estrogens *Urine No Effect Analytical* At 50 mg/L on fluorescent method of Corns. *0730*

Magnesium *Urine Increase Physiological* Magnesium excretion increased to same extent as Potassium. *3730*

pH *Urine Increase Physiological* H and K ions exchanged for Na ion. *1343*

Potassium *Serum Decrease Physiological* Hypokalemic hypochloremic alkalosis may occur. *1343*
Urine Increase Physiological Conservation of Na with increased K loss. *3730*

Sodium *Urine Decrease Physiological* Due to hormonal action. *3730*

Aldrin

Lactate Dehydrogenase *Serum Decrease Physiological* In vitro 4.5% decrease at 1 x 10⁻⁴ mol/L. *1270*

Pseudocholinesterase *Serum Decrease Physiological* In vitro 6.0% decrease at 1 x 10⁻⁴ mol/L. *1270*

Aletamine

Amphetamine *Urine No Effect Analytical* No effect at 100 mg/L on method of Rutter. *3097*

Alkalies

Occult Blood *Feces Positive Physiological* If ingested may cause many gastrointestinal symptoms. *0279*

Alkaline Antacids

Calcium *Serum Increase Physiological* Theoretical possibility with absorption of Calcium salts. *3369*

Color *Feces Decrease Analytical* White discoloration or speckling. *2313*

Creatinine *Serum Increase Physiological* May cause milk-alkali syndrome. *2220*

Magnesium *Serum Increase Physiological* Theoretical possibility. *3369*

Phosphate *Serum Decrease Physiological* Theoretical possibility. *3369*

Urea Nitrogen *Serum Increase Physiological* Prolonged use may cause nephrotoxicity. *2220*

Alkaline Urine

Calcium *Urine Decrease Analytical* Due to precipitated Calcium salts. *1841*

Hemoglobin *Urine Decrease Analytical* May interfere with hemagglutination test. *0017*

5-Hydroxyindoleacetic Acid (5-HIAA) *Urine Decrease Analytical* Unstable at room temperature at alkaline pH, ok if acidified. *0913*

Myoglobin *Urine Decrease Analytical* May interfere with hemagglutination test. *0017*

Norepinephrine *Urine Decrease Analytical* May be almost total destruction unless kept at pH 4-5. *1117*

Protein *Urine Increase Analytical* Possible false positive with Albustix, Combistix® etc. *1872* Highly buffered urine may affect Combistix®. *2313*
Urine Decrease Analytical Possible false negative with sulfosalicylic acid. Possible false negative with heat and acetic acid. *1872*

Quinidine *Serum Increase Physiological* Marked effect if given when pH over 7. *1487*

Sulfamethoxazole *Urine Increase Physiological* If co-trimoxazole administered. *0745*

Vanillylmandelic Acid *Urine Decrease Analytical* Unstable at room temperature above pH 3. *0913*

Alkalosis

Urea Nitrogen *Serum Increase Analytical* With Azostix® compared to diacetyl. *1460*

Alkoxyglycerols

Leukocytes *Blood Increase Physiological* Protect against decrease caused by X-rays. *0483*

Ornithine Carbamoyltransferase (OCT)
Serum Decrease Physiological Protects against elevation caused by X-rays. *0483*

Platelets *Blood Increase Physiological* Protect against decrease caused by X-rays. *0483*

Allantoin

Ammonia *Plasma No Effect Analytical* On indophenol reaction with 5,000 nmoles. *1290*

Amphetamine *Urine No Effect Analytical* At 50 mg/L on fluorescent method of Hayes. *1534*

Urea Nitrogen *Serum Increase Analytical* Reacts with xanthydrol in method of Fosse. *1563*

Allopurinol

Alanine Aminotransferase *Serum Increase Physiological* Reversible clinical hepatotoxicity reported. *2220* In 6 of 73 patients treated for 6 mo. *3565*

Albumin *Serum No Effect Analytical* At concentration of 5 mg/L had no effect on BCG method. *3393*

Alkaline Phosphatase *Serum Increase Physiological* Reversible clinical hepatotoxicity noted. *0652* Occasional granulomatous hepatitis as well as massive hepatic necrosis and severe hepatitis. *2389* In 1 of 73 patients treated for 6 mo. *3565*

Alloxanthine *Serum Increase Physiological* Accumulates with chronic administration. *1343*

Amphetamine *Urine No Effect Analytical* No effect at 100 mg/L on method of Rutter. *3097*

Antipyrine *Plasma Increase Physiological* Impairs metabolism. *2641*

Aspartate Aminotransferase
Serum Increase Physiological Reversible clinical hepatotoxicity reported. *2220* Occasional granulomatous hepatitis as well as massive hepatic necrosis and severe hepatitis. *2389* In 6 of 73 patients treated for 6 mo. *3565*

Bicarbonate *Serum No Effect Analytical* At concentration of 5 mg/L had no effect on method using phenolphthalein. *3393*

Bile *Urine Increase Physiological* Hepatotoxic effect. *2220*

Bilirubin *Serum No Effect Analytical* At concentration of 136 mg/L had no effect on Jendrassik and Grof method. *3393*
Serum Increase Physiological Reported in renal failure also in other patients. *2165*

Bishydroxycoumarin *Plasma Increase Physiological* Impairs metabolism. *2641* Long term use of drug prolongs half-life. *2927*

BSP Retention *Serum Increase Physiological* Associated with reversible hepatotoxicity. *3226*

Calcium *Serum No Effect Analytical* At concentration of 136 mg/L had no effect on cresolphthalein method. *3393*

Chloride *Serum No Effect Analytical* At concentration of 5 mg/L had no effect on mercurimetric method. *3393*

Cholesterol *Serum No Effect Analytical* At concentration of 250 mg/L had no effect on CHOD-Iodide method. *3393* At concentration of 250 mg/L had no effect on CHOD-PAP method. *3393* At concentration of 250 mg/L had no effect on catalase-AIDH method. *3393* At concentration of 250 mg/L had no effect on method using catalase-Hantzsch reaction. *3393* At concentration of 250 mg/L had no effect on Liebermann-Burchard method. *3393*
Serum Decrease Physiological Hepatotoxic effect. *2220*

Creatinine *Serum No Effect Analytical* at 200 mg/L on reversed phase LC procedure of Zhiri et al. *3942* At concentration of 60 mg/L had no effect on AutoAnalyzer Jaffé method. *3393* At concentration of 80 mg/L had no effect on Jaffé-Fading-Fraction method. *3393* At concentration of 80 mg/L had no effect on Jaffé-Fuller's earth method. *3393* At concentration of 10 mg/L had no effect on kinetic Jaffé method on BKA-2. *3393*

Eosinophils *Blood Increase Physiological* May cause severe sensitivity reaction. *2451* Frequently associated with drug-induced hepatitis when it occurs. *2906*

Erythrocytes *Blood Decrease Physiological* Rare case of anemia may occur. *0071*
Urine Increase Physiological Associated with severe sensitivity reaction. *2451*

Glucose *Serum No Effect Analytical* At concentration of 18 mg/L had no effect on Ektachem® method. *3393* At concentration of 100 mg/L had no effect on GOD/POD-PAP method. *3393*
Serum Decrease Physiological Hepatotoxic effect. *2220*

γ-Glutamyltransferase (GGT)
Serum Increase Physiological In 3 of 73 patients treated for 6 mo. *3565*

Hematocrit *Blood Decrease Physiological* Rare case of anemia reported. *0071*

Hemoglobin *Blood Decrease Physiological* Rare case of anemia reported. *0071*

Hypoxanthine *Serum Increase Physiological* Inhibits xanthine oxidase (slight increase only). *0071*
Urine Increase Physiological Inhibits xanthine oxidase. *0071*

Indomethacin *Serum No Effect Analytical* No effect on HPLC method of Roberts and Smith. *3002*

Iron *Serum Decrease Physiological* 40% reduction in 1 week (accumulates in liver). *1672*

Isocitrate Dehydrogenase *Serum Increase Physiological* Reversible clinical hepatotoxicity reported. *2220*

Leukocytes *Blood Increase Physiological* Hypersensitivity reaction occurs with fever. *2313*
Blood Decrease Physiological Hypersensitivity reaction (often transient). *2313* In 1 of 73 patients treated for 6 mo. *3565* Noted in 3 individuals undergoing therapeutic starvation. *3224*
Urine Increase Physiological Associated with severe sensitivity reaction. *2451*

6-Mercaptopurine *Plasma Increase Physiological* Impairs metabolism. *2641*

Neutrophils *Blood Decrease Physiological* Noted in 3 individuals undergoing therapeutic starvation. *3224* Occasional agranulocytosis reported. *3717*

Phenprocoumon *Plasma Increase Physiological* Accumulates for several weeks when drugs coadministered. *1765*

Phenylbutazone *Serum No Effect Physiological* No significant effect observed on plasma concentrations. *2927*

Phosphate *Serum No Effect Analytical* At concentration of 136 mg/L had no effect on phosphomolybdate reduction method. *3393*

Platelets *Blood Decrease Physiological* Rare potentially dangerous complication. *1332* In 4 of 73 patients treated for 6 mo. *3565*

Potassium *Serum No Effect Analytical* At concentration of 5 mg/L had no effect on flame-photometric method. *3393* At concentration of 180 mg/L had no effect on ISE measurement without predilution. *3393*

Protein *Serum No Effect Analytical* At concentration of 136 mg/L had no effect on biuret method with blank correction. *3393*

Protein Electrophoresis *Serum No Effect Analytical* At concentration of 180 mg/L had no effect on automated Olympus-Hite method but with slight displacement of fractions. *3393*

Allopurinol *(continued)*

Prothrombin Time *Plasma Increase Physiological* Patients on coumarins (unconfirmed clinically). *3704*

Reticulocytes *Blood Increase Physiological* Mild (may be associated with other drugs). *0542*

Sodium *Serum No Effect Analytical* At concentration of 5 mg/L had no effect on flame-photometric method. *3393* At concentration of 180 mg/L had no effect on ISE measurement without predilution. *3393*

Theophylline *Serum Increase Physiological* Increased half-life and decreased clearance probably due to effect on hepatic metabolizing enzymes. *2284*

Triglycerides *Serum No Effect Analytical* At concentration of 100 mg/L had no effect on GPO-PAP method. *3393* At concentration of 5 mg/L had no effect on lipase/esterase method. *3393*

Urea Nitrogen *Serum No Effect Analytical* At concentration of 136 mg/L had no effect on diacetylmonoxime method. *3393* At concentration of 18 mg/L had no effect on Ektachem® method. *3393*
Serum Increase Physiological Azotemia as sensitivity reaction in 2 cases. *2451* In 2 of 73 patients treated for 6 mo. *3565*

Uric Acid *Serum No Effect Analytical* At 200 mg/L on reversed phase LC procedure of Zhiri et al. *3942* At concentration of 250 mg/L had no effect on Kageyama-Hantzsch method. *3393* At concentration of 80 mg/L had no effect on catalase-AIDH method. *3393* At concentration of 5 mg/L had no effect on phosphotungstate reduction method. *3393* At concentration of 100 mg/L had no effect on uricase method on aca. *3393* At concentration of 204 mg/L had no effect on uricase-PAP method. *3393*
Serum Decrease Physiological Inhibition of xanthine oxidase. *1022* Therapeutic effect due to inhibition of xanthine oxidase. *2909*
Urine Decrease Physiological Inhibition of xanthine oxidase. *1022*

Urobilin *Urine Increase Physiological* Transient increases reported. *2427*

Warfarin *Plasma No Effect Physiological* No significant effect observed on plasma concentrations. *2927* Overall no change in steady-state plasma concentration although some individual variation. *2927*

Xanthine *Serum Increase Physiological* Inhibits xanthine oxidase (slight increase only). *0071*
Urine Increase Physiological Inhibits xanthine oxidase. *0071*

Xanthine Calculi *Urine Positive Physiological* Rare side effect. *1395*

Alloxan

Acid Phosphatase *White Blood Cells No Effect Analytical* Glycerophosphate enzyme no effect with 45 mmol/L. *0183*
White Blood Cells Decrease Analytical Phenylphosphate enzyme inhibited by 45 mmol/L. *0183*

Casts *Urine Increase Physiological* May cause nephrotoxicity. *0279*

Urea Nitrogen *Serum Increase Physiological* If ingested may cause nephrotoxicity. *0279*

Allymid

Platelets *Blood Decrease Physiological* Immunologically induced thrombocytopenia. *1877*

Aloin

Casts *Urine Increase Physiological* May cause nephrotoxicity. *0279*

Color *Urine Increase Analytical* red-brown/yellow pink (alkaline),yellow brown (acid). *2313*

Urea Nitrogen *Serum Increase Physiological* May cause nephrotoxicity. *0279*

Alphaprodine

Amphetamine *Urine No Effect Analytical* No effect at 100 mg/L on method of Rutter. *3097*

Alprazolam

Alanine Aminotransferase *Serum Increase Physiological* Isolated case, abnormal test value reverted to normal with withdrawal of drug. *3079*

Aspartate Aminotransferase
Serum Increase Physiological Isolated case, abnormal test value reverted to normal with withdrawal of drug. *3079*

Alprenolol

Platelets *Blood Decrease Physiological* Marked reduction on two occasions in one patient due to increased destruction. *2267* In one woman with essential hypertension after 1.5 y treatment: disappeared with withdrawal of drug. *0613*

Potassium *Serum Increase Physiological* Slight increase in patients treated with moderate doses of drug. *2836*

Althesin

Glucose *Serum Increase Physiological* 30 mg/dL increase observed 1 h after. *0291*

pCO₂ *Blood Increase Physiological* Post-induction measurement in arterial blood. *3160*

pO₂ *Blood Decrease Physiological* Post-induction measurement in arterial blood. *3160*

Altitude

Albumin *Serum Decrease Physiological* After 6 weeks at 5400 metres. *2957*

Aldosterone *Plasma Decrease Physiological* Effect most marked in older subjects. *1836*
Urine Decrease Physiological Effect observed on acute exposure. *1836*

Cortisol *Plasma Increase Physiological* Probably related to hypoxia. *1836*
Urine Increase Physiological Probably related to acute hypoxia. *1836*

Creatinine *Serum Increase Physiological* But less than 0.1 mg/dL. *2957*
Urine Decrease Physiological Associated with decreased creatinine clearance. *2957*

Creatinine Clearance *Urine Decrease Physiological* Possibly due to altered renal blood flow. *2957*

Epinephrine *Plasma Increase Physiological* By third day at 12,000 feet. *2001*
Urine Increase Physiological From first day at 12,000 feet. *2001*

Estriol *Urine Decrease Physiological* At 4200 meters mean 10.7 mg/24 h vs 23.0 mg/24 h at sea level. *0406*

β-Globulin *Serum Increase Physiological* 43% increase after 6 weeks at 5400 metres. *2957*

γ-Globulin *Serum Increase Physiological* 72% increase after 6 weeks at 5400 metres. *2957*

Hematocrit *Blood No Effect Physiological* No effect if plasma volume not contracted. *2957*
Blood Increase Physiological At 4500 feet 7-8% higher than at sea level. *3742*

Hemoglobin *Blood Increase Physiological* At 4500 feet 7-8% higher than at sea level. *3742*

Norepinephrine *Plasma Increase Physiological* By third day at 12,000 feet. *2001*
Urine Increase Physiological On third day at 12,000 feet. *2001*

Osmolality *Serum Decrease Physiological* Significant effect at 3800 m compared with sea level. *3939*

Protein *Serum No Effect Physiological* No effect at 5400 meters unless dehydrated. *2957*
Urine Increase Physiological Associated with decreased creatinine clearance. *2957*

Renin Activity *Plasma Decrease Physiological* At 12,000 feet (both supine and standing). *2001*

Transferrin *Serum Decrease Physiological* After rapid ascent to 5400 meters some depletion. *2957*

Uric Acid *Serum Increase Physiological* Good correlation with hematocrit or hemoglobin level. *3374*

Volume *Plasma Increase Physiological* Common finding after ascent to 5400 metres. *2957*
Plasma Decrease Physiological Probably related in some way to ADH secretion. *1836*
Urine Increase Physiological Diuresis usually begins within 4 d. *1836*

Aluminum Hydroxide

Acetaminophen Screening Test *Urine Negative Analytical* No reaction with o-cresol at therapeutic concentrations. *3326*

Alkaline Phosphatase *Serum Increase Physiological* Mild effect with drug associated osteomalacia. *1054* Observed in one patient secondary to ingestion of large amounts of compound for long time. *1313*

Aluminum *Serum Increase Physiological* In 7 nondialyzed patients with chronic renal failure (increase from mean 1.7 to to 3.6 μg/dL) with 15-18 g/d. *3533* Moderate increase after ingestion of aluminum hydroxide as antacid. *3348*

Bishydroxycoumarin *Plasma No Effect Physiological* No effect if co-administered. *0079*

Calcitonin *Plasma No Effect Physiological* With 15-18 g/d in 7 nondialyzed patients with chronic renal failure. *3533*

Calcium *Serum No Effect Physiological* With 15-18 g/d in 7 nondialyzed patients with chronic renal failure. *3533* Observed in one patient secondary to ingestion of large amounts of compound for long time. *1313*
Serum Increase Physiological Small increase to 2.37-2.40 mmol/L during administration of 100 mL gel/d for for 28 d. *0564* Mild effect with osteomalacia due to inhibition of entry into bone. *1054*
Urine Increase Physiological Observed in one patient secondary to ingestion of large amounts of compound for long time. *1313*

Color *Feces Increase Analytical* White discoloration/speckling if taken orally. *1187*

Cyclic Adenosine Monophosphate
Urine No Effect Physiological Observed in one patient secondary to ingestion of large amounts of compound for long time. *1313*

1,25-Dihydroxy Vitamin D$_3$ *Serum Increase Physiological* Observed in one patient secondary to ingestion of large amounts of compound for long time. *1313*
Serum Decrease Physiological Low compared with normals in 7 patients (nondialyzed) with chronic renal failure. Decreased significantly (19.4 to 11.4 pg/mL) with treatment (15-18 g/d). *3533*

Fluoride *Serum Decrease Physiological* When aluminum hydroxide coadministered to osteoporotic patients receiving 50 mg fluoride/d. *3415*
Feces Increase Physiological In osteoporotic patients receiving fluoride (50 mg/d) with net decreased absorption of fluoride by 57% when aluminum hydroxide co-administered. *3415*

25-Hydroxy Vitamin D$_3$ *Serum No Effect Physiological* Observed in one patient secondary to ingestion of large amounts of compound for long time. *1313*

Parathyroid Hormone *Plasma Decrease Physiological* Fell approximately 40% in 7 nondialyzed patients with chronic renal failure with 15-18 g/d. *3533*

Phosphate *Serum Increase Physiological* In 5 normal subjects given 100 mL/d for 28 d (rose to 1.32 mmol/L from 1.18 mmol/L). *0564* Mild effect with osteomalacia due to inhibition of entry into bone. *1054*
Serum Decrease Physiological In 7 nondialyzed patients with chronic renal failure: fell from average 6.3 mg/dL to 3.7 mg/dL with 15-18 g/d. *3533* Observed in one patient secondary to ingestion of large amounts of compound for long time. *1313*

Pseudoephedrine *Serum Increase Physiological* Increased rate of absorption due to raised gastric pH. *3794*

Salicylate *Serum Decrease Physiological* Significant reduction from 19.8 mg/dL to 15.8 mg/dL with several days antacid, due to increased alkalinity of urine. *1489*

Vitamin A *Serum Decrease Physiological* Reduced absorption due to adsorption and oxidation. *3794*

Warfarin *Plasma No Effect Physiological* No effect observed. *0079*

Aluminum Nicotinate

Alanine Aminotransferase *Serum Increase Physiological* Rare hepatotoxic effects. *1678*

Aspartate Aminotransferase
Serum Increase Physiological Rare hepatotoxic effects. *1678*

Bilirubin *Serum Increase Physiological* Rare hepatotoxic effects. *1678*

Cholesterol *Serum Decrease Physiological* Therapeutic intent through nicotinic acid release. *0126*

Glucose *Serum Increase Physiological* Produces carbohydrate intolerance over long term. *1979*

Glucose Tolerance *Serum Decrease Physiological* Rare hepatotoxic effects. *1678*

Uric Acid *Serum Increase Physiological* May increase uric acid to produce gout. *0126*

Aluminum Salts

Aspartate Aminotransferase *Serum No Effect Analytical* No effect on Karmen procedure. No effect on Babson procedure. *2760*

Color *Feces Decrease Analytical* White discoloration or speckling. *2313*

Diagnex Blue Excretion *Urine Increase Physiological* Heavy metal displacement of diagnex blue. *3879*

Phosphate *Serum Decrease Physiological* Binding of phosphate in gastrointestinal tract. *3879*
Urine Decrease Physiological Decreased absorption and excretion. *0071*

Amantadine

Alkaline Phosphatase *Serum Increase Physiological* Mild transient increase, other liver function tests normal. *3213*

Aspartate Aminotransferase
Serum Increase Physiological Reported effect, no mechanism discussed. *2313*

Dopamine *Urine No Effect Physiological* In normals and Parkinson patients. *1817*

Epinephrine *Urine No Effect Physiological* In normals and Parkinson patients. *1817*

Growth Hormone *Plasma No Effect Physiological* No effect observed. *2329*

Histamine *Urine No Effect Physiological* In normals and Parkinson patients. *1817*

Homovanillic Acid *Urine No Effect Physiological* In normals and Parkinson patients. *1817*

5-Hydroxyindoleacetic Acid (5-HIAA)
Urine No Effect Physiological In normals but increased in Parkinsonism. *1817*

1,4-Methylhistamine *Urine No Effect Physiological* In normals but increased in Parkinsonism. *1817*

Norepinephrine *Urine No Effect Physiological* In normals and Parkinson patients. *1817*

Urea Nitrogen *Serum Increase Physiological* Mild transient elevations reported. *3213*

Ambazone

Amphetamine *Urine No Effect Analytical* No effect at 100 mg/L on method of Rutter. *3097*

Ambenonium

Pseudocholinesterase *Serum Decrease Physiological* Direct action of drug. *1147*

Ambulation

Alkaline Phosphatase *Serum Increase Physiological* Following prolonged bed rest (?increase bone formation). *1688*

Calcium *Serum Decrease Physiological* But not to normal in 4 d from high of bed rest. *1539*
Urine Decrease Physiological Following several weeks of bed rest. *1688*
Feces Decrease Physiological After several weeks bed rest (gradual effect). *1688*

Cholesterol *Serum Increase Physiological* 30 mg/dL higher in outpatients than inpatients. *2978*

Glucose *Serum Decrease Physiological* 1/2 mile walk in normal causes decrease of 5 mg/dL. *1946*

Hydroxyproline *Urine Decrease Physiological* Falls below normal after high of bed rest. *0927*

Ionized Calcium *Serum Decrease Physiological* But not to normal in 4 d from high of bed rest. *1539*

Myoglobin *Serum Increase Physiological* Follows physical exertion. *1657*
Urine Increase Physiological Follows physical exertion. *1657*

Pyrophosphate *Urine Decrease Physiological* Falls below normal after high of bed rest. *0927*

Renin Activity *Plasma Increase Physiological* Activity increased by moderate to severe exercise. *0182*

Thyroxine (T4) *Serum Decrease Physiological* Transient due to increased cellular utilization. *0879*

Amdinocillin

Glucose *Urine No Effect Analytical* On Clinitest® at physiological concentration. On Diastix® at physiological concentration. *2733* On TesTape® at physiological concentration. *0191*

Amikacin

Chloramphenicol *Serum No Effect Analytical* No effect at 100 mg/L on coupled enzymatic method. *2490*

Creatinine *Serum Increase Physiological* Nephrotoxicity observed in 7 of 54 patients whose drug concentrations were monitored. *1250*

Glucose *Urine No Effect Analytical* No effect at up to 2 mg/mL on Clinitest® method and Diastix® and TesTape® procedures. *2250*

Amiloride

Aldosterone *Plasma Increase Physiological* 3 fold increase observed in 6 healthy subjects given 10 mg/d for 8 d. *3737* In 5 normal subjects given 75 mg daily for 7 d. *2440*

Angiotensin II *Plasma Increase Physiological* Change varies with renin activity. *2440*

Bicarbonate *Serum No Effect Physiological* No effect seen in 13 men treated for 8 weeks. *0656*
Serum Decrease Physiological Reduces alkalosis induced by other diuretics. *0142*
Urine Increase Physiological Associated with potassium sparing diuretic action. *2618*

Calcium *Serum No Effect Physiological* No effect seen in 13 men treated for 8 weeks. *0656*
Urine Increase Physiological Observed in normals if given for 7 d. *3745*

Chloride *Serum No Effect Physiological* No effect seen in 13 men treated for 8 weeks. *0656*
Urine Increase Physiological Diuretic action. *1343*

Cholesterol *Serum Decrease Physiological* Significant fall with treatment in 13 men treated with drug: concentration rose when drug discontinued. *0656*

Corticosterone *Plasma No Effect Physiological* In 5 normal subjects given 75 mg daily for 7 d. *2440*

Cortisol *Plasma No Effect Physiological* In 5 normal subjects given 75 mg daily for 7 d. *2440*

Creatinine *Serum No Effect Physiological* No effect seen in 13 men treated for 8 weeks. *0656*
Serum Increase Physiological May occur with prolonged therapy. *3711*
Urine No Effect Physiological No effect seen in 13 men treated for 8 weeks. *0656*

Deoxycorticosterone *Plasma No Effect Physiological* In 5 normal subjects given 75 mg daily for 7 d. *2440*

Digoxin *Serum Increase Analytical* At 50 ng/mL equals 0.6 ng/mL by RIA. *3940*
Serum Decrease Physiological Mean clearance increased due to increased tubular secretion of digoxin. *3737*

Glucose *Serum No Effect Physiological* No effect seen in 13 men treated for 8 weeks. *0656*

Hematocrit *Blood No Effect Physiological* No effect seen in 13 men treated for 8 weeks. *0656*

18-Hydroxycorticosterone *Plasma Increase Physiological* In 5 normal subjects given 75 mg daily for 7 d. *2440*

18-Hydroxydeoxycorticosterone *Plasma Increase Physiological* In 5 normal subjects given 75 mg daily for 7 d. *2440*

Insulin *Plasma Increase Physiological* Marked effect of i.v. injection in rat. *0190*

Inulin Clearance *Urine No Effect Physiological* No effect on glomerular filtration rate or effective renal plasma flow. *0656* Unchanged or may be slight increase. *3711*

Magnesium *Serum Increase Physiological* 20 mg/d produces significant effect sustained for duration of treatment. *3101*
Urine Decrease Physiological 5 or 10 mg when administered alone, and also blocked enhanced excretion caused by hydrochlorothiazide. *3101*

PAH Clearance *Urine No Effect Physiological* No effect on glomerular filtration rate or effective renal plasma flow. *0656*

pH *Urine Increase Physiological* Moderate increase sometimes observed. *1343*

Potassium *Serum Increase Physiological* Also inhibits kaliuresis caused by thiazides. *2969* Significant rise in 3 patients with creatinines from 0.2 to 0.70 mmol/L when given with hydrochlorothiazide. *3816* Increase by 0.9 mmol/L over 7 d in 5 volunteers with 75 mg daily. *2440* Binds reversibly to luminal membrane, blocking reabsorption of filtered sodium and inhibiting passive potassium and hydrogen ion secretion. *2836*
Urine No Effect Physiological No effect seen in 13 men treated for 8 weeks. *0656*
Urine Decrease Physiological May occur even if marked sodium loss. *1343* Decreased significantly with start of treatment. *2440* 20 mg/d produces significant effect sustained for duration of treatment. *3101*

Renin Activity *Plasma Increase Physiological* 13.6 fold increase, maximal after 5 d treatment. *2440*

Sodium *Serum No Effect Physiological* No effect seen in 13 men treated for 8 weeks. *0656*
Serum Increase Physiological Significant effect in 13 men given drug alone: remained high throughout treatment. *0656*
Serum Decrease Physiological In 3 women when drug given with hydrochlorothiazide probably due to direct effect on distal nephrons. *3545* Increase by 10 mmol/L over 7 d in 5 volunteers with 75 mg daily. *2440*
Urine No Effect Physiological No effect seen in 13 men treated for 8 weeks. *0656*
Urine Increase Physiological Diuretic action of drug. *2969* Marked natriuresis observed with start of therapy. *2440*

Urea Nitrogen *Serum No Effect Physiological* No effect seen in 13 men treated for 8 weeks. *0656*
Serum Increase Physiological May occur with excessive saluresis. *2427*

Uric Acid *Serum No Effect Physiological* No effect observed. *1487* No effect seen in 13 men treated for 8 weeks. *0656*
Serum Increase Physiological Possibly due to decreased clearance. *3745* May occur with prolonged therapy particularly if combined with thiazides. *3711*
Serum Decrease Physiological Effect observed after 8 weeks in one study. *3745*

Volume *Plasma No Effect Physiological* No effect seen in 13 men treated for 8 weeks. *0656*

Aminacrine

Amphetamine *Urine No Effect Analytical* No effect at 100 mg/L on method of Rutter. *3097*

Amines

Bilirubin *Serum No Effect Analytical* Spectropolarographic procedure of Grahnen. *1374*

Amino Acids

Adenosine Triphosphate (ATP)
Red Blood Cells Decrease Physiological Associated with low serum phosphate. *3619*

Amino Acids *Serum Increase Physiological* Marked if intravenously, less if intraduodenal. *2918*

Bicarbonate *Duodenal Contents Increase Physiological* With duodenal infusion, no effect if given i.v. *2918*

Bilirubin *Serum Increase Analytical* If given i.v. may affect accuracy of estimation. *0120*
Duodenal Contents Increase Physiological Increase by 800% with duodenal infusion, no effect if given i.v. *2918*

2,3-Diphosphoglycerate
Red Blood Cells Decrease Physiological Associated with low serum phosphate. *3619*

Fatty Acids, Free (FFA) *Serum Decrease Physiological* More marked if intraduodenal than i.v. *2918*

Fructose-6-Phosphate
Red Blood Cells Decrease Physiological Associated with low serum phosphate. *3619*

Gastrin *Serum Increase Physiological* Oral feeding produces up to tenfold increase. *1995*

Glucose *Serum Increase Physiological* Significant if i.v. infusion, not if intraduodenal. *2918*
Serum Decrease Physiological Fall may occur after intraduodenal infusion. *2918*

Glucose-6-Phosphate
Red Blood Cells Decrease Physiological Associated with low serum phosphate. *3619*

Glycerol *Serum Decrease Physiological* More marked if intraduodenal than i.v. *2918*

Growth Hormone *Plasma Increase Physiological* Slight effect, mode of infusion makes no difference. *2918*

Hemoglobin *Blood Increase Analytical* If i.v. reported to affect accuracy of measurement. *0120*

Insulin *Plasma Increase Physiological* ?due to action on beta cells. *1775* Max higher if intraduodenal than if i.v. *2918*

Phosphate *Serum Decrease Physiological* If given i.v. hyperalimentation. *3619*

Phosphoenolpyruvate
Red Blood Cells Decrease Physiological Associated with low serum phosphate. *3619*

2-Phosphoglycerate
Red Blood Cells Decrease Physiological Associated with low serum phosphate. *3619*

3-Phosphoglycerate
Red Blood Cells Decrease Physiological Associated with low serum phosphate. *3619*

Protein *Serum Increase Analytical* If given i.v. may affect accuracy of measurement. *0120*

Sodium *Serum Increase Physiological* If Aminosol given i.v. may cause sodium retention. *0120*

Sugar *Urine Increase Analytical* False positive caused by some with Benedict's. *1022*

Trypsin *Duodenal Contents Increase Physiological* Increase by 200% with duodenal infusion, no effect if given i.v. *2918*

Urea Nitrogen *Serum Increase Analytical* May react with Berthelot if urease not used. *2981*
Serum Increase Physiological As result of i.v. infusion. *1215*

Volume *Duodenal Contents Increase Physiological* With duodenal infusion, no effect if given i.v. *2918*

Aminoantipyrine

Cholesterol *Serum Decrease Analytical* At concentrations above 400 mg/L lowered concentration as measured by CHOD-PAP method. *3393*

Aminobenzoic Acid

Alanine Aminotransferase *Serum Increase Physiological* Hepatotoxicity with centrolobular necrosis. *1948*

Aspartate Aminotransferase
Serum Increase Physiological Toxic hepatitis reported. *1343*

Bilirubin *Serum Increase Physiological* Hepatotoxicity with centrolobular necrosis. *1948*

Erythrocytes *Blood Decrease Physiological* Induces malabsorption and megaloblastic anemia. *2716*

Glucose *Serum Decrease Physiological* Rare case of hypoglycemia observed. *3251*

Hematocrit *Blood Decrease Physiological* Induces malabsorption and megaloblastic anemia. *2716*

Hemoglobin *Blood Decrease Physiological* Induces malabsorption and megaloblastic anemia. *2716*

^{131}I Uptake *Serum Decrease Physiological* May impair uptake. *2220*

MCV *Blood Increase Physiological* Induces malabsorption and megaloblastic anemia. *2716*

Methotrexate *Serum Increase Physiological* Displaces from plasma protein binding. *1487*

PAH Clearance *Urine Increase Analytical* Measured as if PAH. *2981*

pH *Blood Decrease Physiological* Acidosis from use of free acid reported. *1343*

Phenylketones *Urine Positive Analytical* Green with $FeCl_3$. *0775*

Platelets *Blood Decrease Physiological* Induces malabsorption and megaloblastic anemia. *3602*

Protein *CSF Increase Analytical* 1.0 mg = 1.8 mg in Folin-Ciocalteu procedure. *0583*

Protein Bound Iodine (PBI) *Serum Decrease Physiological* Inhibits iodination of tyrosine in thyroxine binding globulin. *0012*

Salicylate *Serum Increase Physiological* By competing for glycine for conjugation. *1343*

Sulfa as Sulfanilamide *Serum Increase Analytical* Measured as if sulfonamide. *1343*

Aminobutyric Acid

Histidine *Plasma Increase Analytical* May affect fluorometric method of Ambrose. *0081*

α-Amino-N-Butyric Acid

Tyrosine *Plasma No Effect Analytical* On fluorometric procedure of Ambrose. *0080*

Aminocaproic Acid

Amino Acids *Urine Increase Analytical* Reacts with ninhydrin, measured as amino acid. *3920*

α_1-Antitrypsin *Serum Increase Physiological* Inhibits activation of plasminogen, plasmin and trypsin. *1988*

Aspartate Aminotransferase
Serum Increase Physiological Rare myopathy six weeks after treatment. *0296*

Bleeding Time *Patient Increase Physiological* Effect noted in majority of patients after 24 g/d after 9 d treatment. *1387*

Creatine Kinase *Serum Increase Physiological* Rare myopathy 6 weeks after treatment. *0296*

Leucine Aminopeptidase *Urine Decrease Physiological* Due to antifibrinolytic action. *2897*

Plasmin Activity *Blood Decrease Physiological* Therapeutic intent. *1680*

Potassium *Serum Increase Physiological* Reported effect especially if renal function impaired. *0130*

Aminoglutethimide

Alanine Aminotransferase *Serum Increase Physiological* Casual association between drug administration and cholestasis in 2 cases. *2779*

Aldosterone *Urine Decrease Physiological* May inhibit aldosterone production. *2775*

Alkaline Phosphatase *Serum Increase Physiological* Casual association between drug administration and cholestasis in 2 cases. *2779*

Apolipoprotein AI *Serum Increase Physiological* In 73 patients with advanced breast cancer receiving 500 mg/d with 40 mg hydrocortisone. *0415*

Aspartate Aminotransferase
Serum Increase Physiological Casual association between drug administration and cholestasis in 2 cases. *2779*

Bilirubin *Serum Increase Physiological* Casual association between drug administration and cholestasis in 2 cases. *2779*

Cholesterol *Serum Increase Physiological* In 73 patients with advanced breast cancer receiving 500 mg/d with 40 mg hydrocortisone. *0415*

Cholesterol, High Density Lipoprotein
Serum Increase Physiological In 73 patients with advanced breast cancer receiving 500 mg/d with 40 mg hydrocortisone. *0415*

Cholesterol, Low Density Lipoprotein
Serum Increase Physiological In 73 patients with advanced breast cancer receiving 500 mg/d with 40 mg hydrocortisone. *0415*

Estradiol *Plasma Decrease Physiological* By approximately 50% in women with advanced breast cancer (Postmenopausal). *0946*

Estradiol, Free *Plasma Decrease Physiological* Marked reduction in small number of postmenopausal women with severe breast cancer although no difference in proportion (%) of total estradiol. *0946*

γ-Glutamyltransferase (GGT)
Serum Increase Physiological Increased in 26 of 45 patients with breast cancer. *2546*

17-Hydroxycorticosteroids *Urine Decrease Physiological* Inhibits steroid biosynthesis. *1252*

131I Uptake *Serum Decrease Physiological* Reported effect. *2220*

17-Ketosteroids *Urine Decrease Physiological* Inhibits steroid biosynthesis. *1252*

Leukocytes *Blood Decrease Physiological* May cause leukopenia. *2313*

Protein Bound Iodine (PBI) *Serum Decrease Physiological* Inhibits iodination of tyrosine in thyroxine binding globulin. *0012*

T_3 Uptake *Serum Increase Physiological* Competes for binding sites. *0583*

Thyroid Stimulating Hormone (TSH)
Serum Increase Physiological Observed in one patient in whom tested in association with reduced T_4. *3159*

Thyroxine (T_4) *Serum Decrease Physiological* Competes for T_4 binding sites. *0583* On average 1.6 μg/dL decrease in concentration after 3 mo treatment. *3159*

Aminoguanidine

Histamine *Urine No Effect Physiological* No effect observed on standard diet. *1381*

Imidazoleacetic Acid *Urine Decrease Physiological* Inhibits diamine oxidase. *1382*

Methylhistamine *Urine Increase Physiological* Methylation increased when diamine oxidase inhibited. *1382*

Methylimidazoleacetic Acid *Urine Increase Physiological* As result of metabolism of increased methyl-histamine. *1382*

Aminohippuric Acid

Creatinine *Serum Increase Analytical* Chromogenicity in color reaction. *3505*

Sulfa as Sulfanilamide *Serum Increase Analytical* Measured as if sulfonamide. *2981*

Aminoimidazoleacetic Acid

Millons Test *Test Conditions Positive Analytical* Positive spot test for phenols. *3765*

Sugar *Test Conditions Increase Analytical* Positive spot test with Benedict's reagent. *3765*

Uric Acid *Test Conditions Increase Analytical* Positive spot test with phosphotungstate. *3765*

β-Amino-Isobutyric Acid

Tyrosine *Plasma No Effect Analytical* On fluorometric procedure of Ambrose. *0080*

Aminomel LX 6

Amylase *Serum No Effect Analytical* At concentration of 16 g/L had no effect on maltotetrose method. *3393* At concentration of 16 g/L had no effect on p-nitrophenylmaltoheptoside method. *3393* At concentration of 16 g/L had no effect on p-nitrophenylmaltopentoside/hexoside method. *3393*

Aminomethylcyclohexane

Leucine Aminopeptidase *Urine Decrease Physiological* Due to antifibrinolytic action. *2897*

Aminophenazone

Alanine Aminotransferase *Serum No Effect Analytical* On continuous method at 10 times maximal therapeutic concentration. *1785*
Serum Increase Analytical On continuous method at 10 times maximal therapeutic concentration. *1785*

Alkaline Phosphatase *Serum No Effect Analytical* On continuous method at 10 times maximal therapeutic concentration. *1785*

Aspartate Aminotransferase *Serum Increase Analytical* On continuous method at 10 times maximal therapeutic concentration. *1785*
Serum Decrease Analytical On continuous method at 10 times maximal therapeutic concentration. *1785*

Bilirubin *Serum No Effect Analytical* At therapeutic concentrations no effect on methods of Jendrassik and Grof and those using spectrophotometry. *1786*
Serum Decrease Analytical With 540 μmol/L on dimethylsulfoxide method. *1785*

Cholesterol *Serum No Effect Analytical* No effect at therapeutic concentration on enzymatic and Liebermann-Burchard methods. *1786*

Creatine Kinase *Serum No Effect Analytical* On continuous method at 10 times maximal therapeutic concentration. *1785*

Creatinine *Serum No Effect Analytical* At therapeutic concentrations no effect on alkaline picrate and Slot methods. *1786*

Glucose *Serum No Effect Analytical* At therapeutic concentrations no effects on hexokinase, glucose dehydrogenase 2.4-dichlorophenol and o-toluidine methods. *1786* With 540 μmol/L on hexokinase method. With 540 μmol/L on glucose dehydrogenase method. With 540 μmol/L on GOD-Perid/2,4-dichlorophenol method. With 540 μmol/L on o-toluidine method. *1785*
Serum Decrease Analytical Slight reduction at therapeutic concentration on ABTS method. *1786* With 540 μmol/L on GOD-Perid/ABTS method. *1785*

Hydroxybutyrate Dehydrogenase
Serum No Effect Analytical On continuous method at 10 times maximal therapeutic concentration. *1785*

Iron *Serum No Effect Analytical* At therapeutic concentration on Ramsay and bathophenanthroline methods. *1786*

Lactate Dehydrogenase *Serum No Effect Analytical* On continuous method with pyruvate as substrate at 10 times maximal therapeutic concentration. On continuous method with lactate as substrate at 10 times maximal therapeutic concentration. On colorimetric method at 10 times maximal therapeutic concentration. *1785*

Protein *Serum No Effect Analytical* No effect on biuret and spectrophotometric methods at therapeutic concentration. *1786*

Triglycerides *Serum No Effect Analytical* At therapeutic concentration on enzymatic methods. *1786*

Urea Nitrogen *Serum No Effect Analytical* At therapeutic concentrations no effects on glutamate dehydrogenase, phenol-hypochlorite, diacetylmonoxime methods. *1786* With 540 µmol/L on glutamate dehydrogenase method. With 540 µmol/L on phenol-hypochlorite method. With 540 µmol/L on diacetyl monoxime method. *1785*

Uric Acid *Serum No Effect Analytical* At therapeutic concentrations no effects on methods using uricase-catalase, aldehyde dehydrogenase, direct UV-test phosphotungstate methods. *1786*

Aminophenol

Acetaminophen Screening Test *Urine Positive Analytical* Blue with o-cresol at therapeutic concentration. Red color with 1-naphthol at therapeutic concentrations. *3326*
Urine Negative Analytical Negative Bratton-Marshall reaction at therapeutic concentrations. *3326*

Amphetamine *Urine No Effect Analytical* No effect at 100 mg/L on method of Rutter. *3097*

Aspartate Aminotransferase *Serum Increase Analytical* At 1 mmol/L affects SMA 12/60 method. *3335*

Bilirubin *Serum No Effect Analytical* At concentration of 109 mg/L had no effect on Jendrassik and Grof method. *3393*
Serum Increase Analytical At 1 mmol/L affects SMA 12/60 method. *3335*

Calcium *Serum Increase Analytical* At 0.1 mmol/L affects SMA 12/60 method. *3335* At concentrations above 10 mg/L raised concentration as measured by cresolphthalein method. *3393*

Chloride *Serum No Effect Analytical* At concentration of 5 mg/L had no effect on mercurimetric method. *3393*

Glucose *Serum No Effect Analytical* At concentration of 50 mg/L had no effect on Ektachem® method. *3393* At concentration of 200 mg/L had no effect on hexokinase/G-6-PDH method. *3393*
Serum Increase Analytical At 1 mmol/L affects SMA 12/60 method. *3335*
Serum Decrease Analytical At concentrations above 25 mg/L lowered concentration as measured by GOD-Perid method. *3393*

Phosphate *Serum No Effect Analytical* At concentration of 109 mg/L had no effect on phosphomolybdate method. *3393*

Protein *Serum No Effect Analytical* At concentration of 109 mg/L had no effect on biuret method with blank correction. *3393*

Urea Nitrogen *Serum No Effect Analytical* At concentration of 109 mg/L had no effect on diacetylmonoxime method. *3393*
Serum Increase Analytical May react in Berthelot procedure. *1862*

Uric Acid *Serum Increase Analytical* At 1 mmol/L affects SMA 12/60 and Henry methods. *3335* At concentrations above 5 mg/L raised concentration as measured by phosphotungstate reduction method. *3393*

Aminophylline

Bilirubin *Serum No Effect Physiological* No significant displacement from protein in neonates. *3748*

Catecholamines *Plasma Increase Physiological* Response to i.v. therapeutic dose. *0178*

Coagulation Time *Blood Decrease Physiological* Reported effect. *1343*

Creatinine *Serum No Effect Analytical* At 400 mg/L on reversed phase LC procedure of Zhiri et al. *3942*

Epinephrine *Urine Increase Physiological* Threefold increase in response to i.v. dose. *0178*

Ethchlorvynol *Serum No Effect Analytical* At 80 mg/L on GLC procedure of Evenson. *1058*

Methaqualone *Serum No Effect Analytical* At 80 mg/L on GLC procedure of Evenson. *1057*

Norepinephrine *Urine Increase Physiological* Twofold increase in response to i.v. dose. *0178*

Occult Blood *Feces Positive Physiological* Hematemesis may occur early with poisoning. *2427*

Protein *Urine Increase Physiological* Increase in renal disease (occurs early with poisoning). *1022*

Uric Acid *Serum No Effect Analytical* At 400 mg/L on reversed phase LC procedure of Zhiri et al. *3942*
Serum Increase Analytical Reacts as if uric acid in method of Bittner. *2752*

Aminopropionitrile

Hydroxyproline *Urine Increase Physiological* Observed in experimental studies. *3644*

Aminopterin

Chromosomes *Test Conditions Abnormal Physiological* Clastogenic in human lymphocytes in culture. *3282*

Folate *Serum Decrease Physiological* Inhibits dihydrofolate reductase by combining irreversibly with it. *2062*
Red Blood Cells Decrease Physiological Inhibits dihydrofolate reductase by combining irreversibly with it. *2062*

Aminopyrimidine

Bilirubin *Serum No Effect Analytical* At concentration of 57 mg/L had no effect on Ektachem® method. *3393*

Aminopyrine

Alanine Aminotransferase *Serum Increase Physiological* Liver damage up to hepatic necrosis seen. *2427*

Albumin *Serum No Effect Analytical* At concentration of 125 mg/L had no effect on BCG method. *3393*

Aminolevulinic Acid *Urine Increase Physiological* May precipitate attack of acute porphyria. *1016*

Aspartate Aminotransferase
Serum Increase Physiological Liver damage up to hepatic necrosis seen. *2427*

Bilirubin *Serum No Effect Analytical* At concentration of 125 mg/L had no effect on Jendrassik and Grof method. *3393*
Serum Increase Physiological Hemolysis in G-6-PD deficiency. *0248*

Bilirubin, Direct *Serum Increase Physiological* May cause hemolytic anemia. *2313*

Calcium *Serum No Effect Analytical* At concentration of 125 mg/L had no effect on cresolphthalein method. *3393*

Cholesterol *Serum No Effect Analytical* At concentration of 125 mg/L had no effect on CHOD-PAP method. *3393*
Serum Increase Analytical Interferes with measurement procedure. *1563*

Color *Urine Increase Analytical* Red brown. *2313*

Creatine *Urine Increase Physiological* Reported effect. *2220*

Creatinine *Serum No Effect Analytical* At concentration of 125 mg/L had no effect on AutoAnalyzer Jaffé method. *3393*

Erythrocytes *Blood Decrease Physiological* Hemolytic anemia in G-6-PD deficient persons. *0788*
Urine Increase Physiological Rare nephrotoxicity reported. *2427*

γ-Globulin *Serum Increase Physiological* Specific antibodies to drug may develop. *2161*

Glucaric Acid *Urine Increase Physiological* As result of hepatic enzyme induction. *2220*

Glucose *Serum No Effect Analytical* At concentration of 125 mg/L had no effect on hexokinase/G-6-PDH method. *3393*

Haptoglobin *Serum Decrease Physiological* May cause hemolytic anemia. *2313*

Heinz Body Formation *Blood Positive Physiological* Occurs initially before overt hemolysis. *1902*

Aminopyrine (continued)

Hematocrit *Blood Decrease Physiological* Hemolytic anemia in G-6-PD deficient persons. 0788

Hemoglobin *Plasma Increase Physiological* May occur with marked hemolysis. 1902

Blood Decrease Physiological Hemolytic anemia in G-6-PD deficient persons. 0788

Urine Increase Physiological May occur with marked hemolysis. 1902

Iron *Serum No Effect Analytical* At concentration of 125 mg/L had no effect on Ferrozine method. 3393

Leukocytes *Blood Decrease Physiological* Agranulocytosis occurs within minutes. 3941

Nitrogen *Urine Increase Physiological* Reported effect. 2220

Occult Blood *Feces Positive Physiological* Possibly due to impaired platelet aggregation. 0605

Phosphate *Serum No Effect Analytical* At concentration of 125 mg/L had no effect on phosphomolybdate method. 3393
Urine Increase Physiological Reported effect. 2220

Platelet Aggregation *Blood Decrease Physiological* Observed *in vitro*, might cause gastrointestinal bleeding etc. 0605

Platelets *Blood Decrease Physiological* May cause hemolytic or aplastic anemia. 3941

Porphyrins *Urine Increase Physiological* Stimulates formation of ALA-synthetase. 1343

Protein *Serum No Effect Analytical* At concentration of 125 mg/L had no effect on biuret method with blank correction. 3393
Urine Increase Physiological Nephrotoxicity reported. 2427

Prothrombin Time *Plasma Increase Physiological* Observed in man and experimental animals. 2313

Reduced Glutathione
Red Blood Cells Decrease Physiological Sharp fall before overt hemolysis. 1902

Reticulocytes *Blood Increase Physiological* May occur during recovery from hemolysis. 1902

Sugar *Urine Increase Analytical* Glucuronide metabolism affects Benedict's, Clinitest®. 0583

Urea Nitrogen *Serum No Effect Analytical* At concentration of 125 mg/L had no effect on diacetylmonoxime method. 3393
Serum Increase Physiological Rare suggestion of nephrotoxic action. 2427

Uric Acid *Serum No Effect Analytical* At concentration of 125 mg/L had no effect on phosphotungstate reduction method. 3393

Volume *Plasma Increase Physiological* Fluid retention may occur. 2427

Aminosalicylic Acid

Alanine Aminotransferase *Serum No Effect Analytical* At 5 times upper limit of therapeutic range on methods on Abbott-VP, aca, and Cobas-Bio. 2138
Serum Increase Analytical At 5 times upper limit of therapeutic range on methods on KDA and SMAC®. 2138
Serum Increase Physiological May cause cytotoxic hepatocellular damage. 0248 Rare side effect. 1292

Albumin *Serum No Effect Analytical* At 5 times upper limit of therapeutic range on methods on SMAC®, Ektachem®, Hitachi® 705 and KDA. 2138 At concentration of 460 mg/L had no effect on BCG method. 3393

Alkaline Phosphatase *Serum No Effect Analytical* At 5 times upper limit of therapeutic range on methods on SMAC®, Abbott-VP, aca, Cobas-Bio, Hitachi® 705 and KDA. 2138
Serum Increase Physiological Reversible cholestasis may occur. 1713

Amino-4-Imidazole-5-Carboxamide Ribotide (AICAR)
Urine Increase Physiological Occurs with vitamin B_{12} deficiency. 3773

Amphetamine *Urine No Effect Analytical* At 50 mg/L on fluorescent method of Hayes. 1534

Amylase *Serum No Effect Analytical* At 5 times upper limit of therapeutic range on methods on aca, Cobas-Bio and Ektachem®. 2138
Serum Increase Physiological May cause acute pancreatitis. 1343

Aspartate Aminotransferase *Serum No Effect Analytical* At 5 times upper limit of therapeutic range on methods on Abbott-VP, aca, Cobas-Bio and Hitachi® 705. 2138
Serum Increase Analytical At 1 mmol/L affects SMA 12/60 method. 3335 With diazonium end point method on AutoAnalyzer. 1310 At 5 times upper limit of therapeutic range on methods on KDA and SMAC®. 2138
Serum Increase Physiological May cause cytotoxic hepatocellular damage. 0248 Rare side effect. 1292

Bicarbonate *Serum No Effect Analytical* At concentration of 460 mg/L had no effect on method using phenolphthalein. 3393

Bile *Urine Increase Analytical* Red color with Fouchet procedure. 0121
Urine Increase Physiological Hepatotoxicity. 2313

Bilirubin *Serum No Effect Analytical* At 5 times upper limit of therapeutic range on methods on SMAC®, Ektachem®, Hitachi® 705 and KDA. 2138 At concentration of 230 mg/L had no effect on Ektachem® method. 3393 At concentration of 460 mg/L had no effect on Jendrassik and Grof method. 3393
Serum No Effect Physiological Probable insignificant displacement from protein in neonates. 3748
Serum Increase Analytical Marked increase at upper limit of therapeutic range on aca method. At 5 times upper limit of therapeutic range on Cobas-Bio method. 2138
Serum Increase Physiological Reversible cholestasis caused by drug. 0277

Bilirubin, Direct *Serum Increase Physiological* Reversible cholestasis caused by drug. 0277

BSP Retention *Serum Increase Physiological* Reversible cholestasis caused by drug. 2707

Calcium *Serum No Effect Analytical* At 5 times upper limit of therapeutic range on methods on SMAC®, Abbott-VP, aca, Ektachem®, Hitachi® 705 and KDA. 2138 At concentration of 460 mg/L had no effect on cresolphthalein method. 3393

Cephalin Flocculation *Serum Increase Physiological* Reversible cholestasis caused by drug. 2707

Chloride *Serum No Effect Analytical* At concentration of 460 mg/L had no effect on mercurimetric method. 3393

Cholesterol *Serum No Effect Analytical* At 5 times upper limit of therapeutic range on methods on SMAC®, Abbott-VP, Cobas-Bio, Ektachem®, Hitachi® 705 and KDA. 2138 At concentration of 460 mg/L had no effect on Liebermann-Burchard method. 3393
Serum Decrease Physiological As effective as neomycin, mechanism obscure. 2148

Color *Urine Increase Analytical* Abnormal color (not distinctive). 0382

Coombs' Test, Direct *Serum Positive Physiological* Immunological response to drug. 1486

Creatine Kinase *Serum No Effect Analytical* At 5 times upper limit of therapeutic range on methods on SMAC®, Abbott-VP, aca, Cobas-Bio and Hitachi® 705. 2138

Creatinine *Serum No Effect Analytical* At 5 times upper limit of therapeutic range on methods on SMAC®, Abbott-VP, aca, Cobas-Bio, Ektachem®, Hitachi® 705 and KDA. 2138 At concentration of 460 mg/L had no effect on AutoAnalyzer method. 3393

Eosinophils *Blood Increase Physiological* Hypersensitivity reaction. 1513

Erythrocytes *Blood Decrease Physiological* Hemolytic anemia/megaloblastic anemia. 0788
Urine Increase Physiological Bleeding caused by drug. 1237

Fat *Feces Increase Physiological* Produces steatorrhea possibly because of bile acid chelation. 3023

Ferric Chloride Test *Urine Positive Analytical* Blue-purple color. 0121

Fibrinolytic Activity *Plasma No Effect Physiological* No effect of 1.5 g/d or after 250 mg i.v. in 6 patients. 3876

FIGLU (N-Formiminoglutamic Acid)
Urine Increase Physiological Occurs with vitamin B_{12} deficiency. *3773*

Folate *Serum Decrease Physiological* May occur with protracted therapy. *2427*

FPN Test *Urine Increase Analytical* Reported false positive in patients with liver dysfunction. *1204*

Glucose *Serum No Effect Analytical* At 1 g/dL on glucose-oxidase methods. At 1 g/dL on p-HBAH procedure of Lever. At 1 g/dL on o-toluidine procedure. *2140* At 5 times upper limit of therapeutic range on methods on SMAC®, Abbott-VP, aca, Cobas-Bio, Ektachem®, Hitachi® 705 and KDA. *2138*
Serum Increase Analytical At 1 g/dL on alkaline ferricyanide=1.7 mmol/L. *2140* At 1 mmol/L affects SMA 12/60 method. *3335* At concentrations above 100 mg/L raised concentration as measured by Ektachem® method. *3393*
Serum Increase Physiological Hyperglycemia reported with protracted therapy. *2495*
Serum Decrease Physiological May cause lowering in diabetics. *2220*
Urine Increase Physiological Glycosuria reported with protracted therapy. *2495* Glycosuria reported with long therapy. *1487*

γ-Glutamyltransferase (GGT) *Serum No Effect Analytical* At 5 times upper limit of therapeutic range on methods on SMAC®, Abbott-VP and Hitachi® 705. *2138*

Guanase *Serum Increase Physiological* May cause cytotoxic hepatocellular damage. *0248*

Haptoglobin *Serum Decrease Physiological* Effect of hemolytic anemia. *0788*

Hematocrit *Blood Decrease Physiological* Hemolytic anemia/megaloblastic anemia. *0788*

Hemoglobin *Blood Decrease Physiological* Hemolytic anemia/megaloblastic anemia. *0788* Rare hemolytic anemia. *1292*
Urine Increase Physiological Actual bleeding caused by drug. *1022*

Iron *Serum No Effect Analytical* At 5 times upper limit of therapeutic range on Ferrozine method on SMAC®. *2138*

Isocitrate Dehydrogenase *Serum Increase Physiological* May cause cytotoxic hepatocellular damage. *0248*

Isoniazid *Serum Increase Physiological* Inhibits acetylation, increases serum concentration. *1487*

131I Uptake *Serum Decrease Physiological* May cause goitrous hypothyroidism. *2808*

Lactate Dehydrogenase *Serum No Effect Analytical* At 5 times upper limit of therapeutic range on methods on SMAC®, Abbott-VP, Cobas-Bio, Hitachi® 705 and KDA. *2138*

LE Cells *Blood Positive Physiological* May produce LE-like syndrome. *2325*

Leukocytes *Blood Increase Physiological* Mainly due to eosinophilia due to hypersensitivity. *2313*
Blood Decrease Physiological Leukopenia but no cases of agranulocytosis. *2429*

Lymphocytes *Blood Increase Physiological* May produce syndrome like infectious mononucleosis. *1343*

MCV *Blood Increase Physiological* If megaloblastic anemia occurs. *3773*

Methemoglobin *Blood Increase Physiological* May cause hemolytic anemia. *0071*

Methylmalonic Acid *Urine Increase Physiological* Occurs with vitamin B_{12} deficiency. *3773*

Neutrophils *Blood Decrease Physiological* If severe megaloblastic anemia occurs. *3773* Occasional case of agranulocytosis reported. *3717*

Occult Blood *Feces Positive Physiological* Reversible gastritis caused by drug. *1343*

Ornithine Carbamoyltransferase (OCT)
Serum Increase Physiological May cause cytotoxic hepatocellular damage. *0248*

PAH Clearance *Urine Increase Analytical* Very slight effect, measured as if PAH. *2981*

Peripheral Smear *Blood Abnormal Physiological* Atypical lymphocytosis with eosinophilia. *2427*

pH *Blood Decrease Physiological* Moderate strong acid, loss of fixed cation = acidosis. *1343*

Phenylketones *Urine Positive Analytical* Red-brown with $FeCl_3$, pink to purple Phenistix®. *0775*

Phosphate *Serum No Effect Analytical* On nonprecipitation method of Peynet. *2799* At 5 times upper limit of therapeutic range on methods on SMAC® and KDA. *2138* At concentration of 460 mg/L had no effect on phosphomolybdate method. *3393*
Serum Increase Analytical At 5 times upper limit of therapeutic range on methods on aca and Hitachi® 705. *2138*

Platelet Aggregation *Blood No Effect Physiological* No effect on aggregation or fibrinolytic activity in 6 patients with chronic inflammatory bowel disease. *3876* No effect on epinephrine or ADP threshold value. *3876*

Platelets *Blood Decrease Physiological* May occur with severe megaloblastic anemia. *3773* Several cases of immune mediated thrombocytopenia reported. *2502* Rare reaction. *1292*

Porphobilinogen *Urine Increase Analytical* Reacts with Ehrlich's reagent. *2425*

Potassium *Serum No Effect Analytical* At concentration of 460 mg/L had no effect on ISE measurement with predilution. *3393*
Serum Decrease Physiological Due to action on renal tubules or vomiting. *2372* Rare side effect. *1292*

Protein *Serum No Effect Analytical* At 5 times upper limit of therapeutic range on methods on SMAC®, Abbott-VP, Ektachem®, Hitachi® 705 and KDA. *2138* At concentration of 460 mg/L had no effect on biuret method with blank correction. *3393*
Urine No Effect Analytical No effect on Labstix®, heat with acetic acid tests. *1487*
Urine Increase Analytical Affects acid turbidimetric procedures. *0583* With SSA and heat with acetic acid tests. *0443* Reacts with Folin-Ciocalteu of Lowry procedure. *1237* May cause false positive with sulfosalicylic acid. *1872*
Urine Increase Physiological May occur as result of nephrotoxicity. *2496* May cause nephrotoxicity. *1237*
CSF Increase Analytical Reacts as phenol if Folin-Ciocalteu reaction used. *3505*

Protein Bound Iodine (PBI) *Serum Decrease Physiological* Inhibits iodination of tyrosine in thyroxine binding globulin. *3244*

Prothrombin Time *Plasma Increase Physiological* Drug suppresses formation of prothrombin. *2916* Rare side effect. *1292*

Rifampin *Serum Decrease Physiological* May impair gastrointestinal absorption. *1487* Delayed and reduced absorption due to adsorption to bentonite in PAS granules. *3794*

Sodium *Serum No Effect Analytical* At concentration of 460 mg/L had no effect on ISE measurement with predilution. *3393*

Sugar *Urine Increase Analytical* Acts as reducing agent with Benedict's reagent. *0110* Acts as reducing agent with nonspecific methods. *1237*

Sulfa as Sulfanilamide *Serum Increase Analytical* Very slight effect, measured as if sulfa. *2981*

T3 Uptake *Serum Increase Physiological* Resin uptake increase due to impaired synthesis of T_4. *1915*

β-Thromboglobulin *Plasma No Effect Physiological* No effect of 1.5 g/d or after 250 mg i.v. in 6 patients. *3876*

Thymol Turbidity *Serum Increase Physiological* Liver damage may occur with 10% mortality. *2808*

Thyroxine (T4) *Serum Decrease Physiological* May cause goitrous hypothyroidism. *2808* Prolonged administration may lead to hypothyroidism. *1292*

Triglycerides *Serum No Effect Analytical* At 5 times upper limit of therapeutic range on methods on SMAC®, Abbott-VP, Ektachem®, Hitachi® 705 and KDA. *2138* At concentration of 460 mg/L had no effect on lipase/esterase method. *3393*

Urea Nitrogen *Serum No Effect Analytical* At 5 times upper limit of therapeutic range on methods on SMAC®, Abbott-VP, Cobas-Bio, Ektachem® and Hitachi® 705. *2138* At concentration of 460 mg/L had no effect on diacetylmonoxime method. *3393*
Serum Increase Analytical Effect marked on DMAB method of Morin. *2487* Reacts as urea with dimethylaminobenzaldehyde. *3505* At upper limit of therapeutic range on o-phthaldehyde methoxyquinoline method of KDA. *2138*

Aminosalicylic Acid (continued)

Uric Acid   *Serum   No Effect   Analytical*   At 5 times upper limit of therapeutic range on methods on SMAC®, Abbott-VP, aca, Cobas-Bio, Ektachem®, Hitachi® 705 and KDA. *2138*   At concentration of 460 mg/L had no effect on phosphotungstate reduction method. *3393*   At concentration of 100 mg/L had no effect on Seralyzer method. *3393*

Urobilinogen   *Urine   Increase   Analytical*   Reacts with Ehrlich's reagent (yellow color). *0252*
Feces   Decrease   Physiological   Pale stools may occur if cholestasis. *1713*

Vanillylmandelic Acid   *Urine   Increase   Analytical*   Reacts with diazo reagent if used. *0086*

Vitamin B$_{12}$ *Serum   Decrease   Physiological*   Impairs absorption. *3773*   Reduced absorption noted. *3794*
Urine   Decrease   Physiological   Occurs with vitamin B$_{12}$ deficiency. *3773*

Xylose Excretion   *Urine   Decrease   Physiological*   2 times normal dose produced reversible effect on absorption. *3855*

Aminothiazole

Bilirubin   *Serum   No Effect   Analytical*   At concentration of 23 mg/L had no effect on Ektachem® method. *3393*
Serum   Increase   Physiological   Jaundice with febrile reactions. *1343*

Protein Bound Iodine (PBI)   *Serum   Decrease   Physiological*   Inhibits iodination of tyrosine in thyroxine binding globulin. *0012*

δ-Amino Valeric Acid

Tyrosine   *Plasma   No Effect   Analytical*   On fluorometric procedure of Ambrose. *0080*

Amiodarone

Acenocoumarol   *Plasma   Increase   Physiological*   Significant potentiation of action so reduced dose of anticoagulant required. *0004*

Alanine Aminotransferase   *Serum   Increase   Physiological*   Up to 1.5 to 3 times normal values in 55% patients with hepatitis occurring in 4%. *2011*   Asymptomatic change in liver function with enzyme activity up to 3 times normal. *2478*   1.5 to 4 fold increase in 15 of 100 patients: effect correlated with drug concentration. *1509*   In 5 of 5 patients with toxic and hypersensitivity liver injury. *2987*

Albumin   *Serum   No Effect   Analytical*   No effect at up to 10.0 mg/L on bromcresol green method on Technicon® SMA II and bromcresol purple on Abbott-VP. Metabolite desethylamiodarone also had no effect. *3318*
Serum   Decrease   Physiological   Approximately 10% reduction compared with controls when patients treated for 6 mo: probable direct effect on synthesis or clearance. *1173*

Alkaline Phosphatase   *Serum   No Effect   Physiological*   No effect observed in 100 patients. *1509*
Serum   Increase   Physiological   Up to 1.5 to 3 times normal values in 55% patients with hepatitis occurring in 4%. *2011*   Asymptomatic change in liver function with enzyme activity up to 3 times normal. *2478*   In 4 of 5 patients with toxic and hypersensitivity liver injury. *2987*

Aminolevulinic Acid   *Serum   Increase   Physiological*   Increased in all of 10 patients by average of 79% over 18 mo. *1509*

Aprindine   *Serum   Increase   Physiological*   Concentration increased so dose can be reduced by 50%. *3950*

Aspartate Aminotransferase
Serum   Increase   Physiological   Up to 1.5 to 3 times normal values in 55% patients with hepatitis occurring in 4%. *2011*   Asymptomatic change in liver function with enzyme activity up to 3 times normal. *2478*   1.5 to 4 fold increase in 15 of 100 patients: effect correlated with drug concentration. *1509*   In 5 of 5 patients with toxic and hypersensitivity liver injury. *2987*

Bilirubin   *Serum   No Effect   Physiological*   No effect observed in 100 patients. *1509*   In 0 of 5 patients with toxic and hypersensitivity liver injury. *2987*
Serum   Increase   Physiological   Hepatitis occurs in 4% patients. *2011*

Cholesterol   *Serum   Increase   Physiological*   Observed effect in 3 individuals but without effect on thyroid function. *2833*
Serum   Decrease   Physiological   From 214 to 194 mg/dL in 24 patients in female patients but no effect in men after 90 d. *3392*

Cholesterol, High Density Lipoprotein
Serum   No Effect   Physiological   In 24 patients given drug for 30-90 d. *3392*

Creatinine   *Serum   No Effect   Analytical*   At 50 mg/L on reversed phase LC procedure of Zhiri et al. *3942*

Digoxin   *Serum   Increase   Physiological*   Approximate doubling of effect due to inhibition of metabolism. *2011*   Average increase of 69% with possible toxicity but mechanism not yet established. *2794*   Serum concentration increased 68 to 800% in presence of normal serum creatinine and urea N, possibly due to inhibited tubular secretion of drug. *1993*   Occurs with concomitant administration: mean increase of 280% in four patients, allowing 50%. *1394*   Concentration increased within 24 h of dosing: magnitude of interaction dose related. *3950*   Following treatment with amiodarone for 1 week in 6 volunteers: serum maximum and area under curve increased. *2293*
Urine   Increase   Physiological   Significant correlation between increases in plasma and urine concentrations. Possible displacement of digoxin from its binding sites. *0942*

Diphenylhydantoin   *Serum   Increase   Physiological*   Pharmacokinetic interaction with clinical significance. *2327*   Up to 3 fold increase when drugs coadministered, due to inhibition of hepatic metabolism. *2366*

Disopyramide   *Serum   Increase   Physiological*   Increased concentration with possible adverse effects on EKG. *2298*

Flecainide   *Serum   Increase   Physiological*   Pharmacokinetic interaction with clinical significance. *2327*

Glucose   *Serum   Increase   Physiological*   Observed effect in 3 individuals but without effect on thyroid function. *2833*

γ-Glutamyltransferase (GGT)
Serum   Increase   Physiological   But effect minimal in 100 treated patients. *1509*

Indomethacin   *Serum   No Effect   Analytical*   No effect on HPLC method of Roberts and Smith. *3002*

Lactate Dehydrogenase   *Serum   Increase   Physiological*   Up to 1.5 to 3 times normal values in 55% patients with hepatitis occurring in 4%. *2011*

Malate Dehydrogenase   *Serum   Increase   Physiological*   In 10-20% of 36 patients. *1509*

Mexiletine   *Serum   Increase   Physiological*   Increased concentration with possible adverse effects on EKG. *2298*

Platelets   *Blood   Decrease   Physiological*   Observed in 2 patients with rechallenge probably due to delayed hypersensitivity reaction: occurs early during administration of drug. *3781*

Prealbumin   *Serum   Decrease   Physiological*   Approximately 20% reduction compared with controls when patients treated for 6 mo: probable direct effect on clearance or synthesis. *1173*

Procainamide   *Serum   Increase   Physiological*   Approximate doubling of effect due to inhibition of metabolism. *2011*   In 11 of 12 treated patients, probably due to an effect on tissue binding or decrease of clearance. *3105*   Average 57% in 12 patients allowing 20% reduction in dosage. *1394*

Propafenone   *Plasma   Increase   Physiological*   Increased concentration with probable adverse effects on EKG. *2298*

Prothrombin Time   *Plasma   No Effect   Physiological*   In 0 of 5 patients with toxic and hypersensitivity liver injury. *2987*
Plasma   Increase   Physiological   May double when warfarin coadministered. *2011*   Probably by lowering concentrations of vitamin K-dependent coagulation factors. *1470*   Prolongation sufficiently great that 43% reduction in dose of Coumadin® required in 11 patients. *1394*

Quinidine   *Serum   Increase   Physiological*   Drug reported to produce 50% increase in concentration. *0548*   Rapid effect possibly due to change in protein or receptor binding. *3546*   Average 32% in 11 patients allowing 37% reduction in dosage. *1394*

T$_3$ Uptake   *Serum   Decrease   Physiological*   From 0.98 to 0.96 in 13 patients treated for average of 17 mo. *0425*

Thyroid Stimulating Hormone (TSH)
Serum   No Effect   Physiological   Insignificant change in 13 patients treated for average of 17 mo. *0425*

Serum Increase Physiological Median TSH values in individuals treated for 6 mo increased compared with controls but values still within normal range, probable anterior pituitary effect. *1173* Initial increase returns to normal in a few months possibly due to inhibition of intracellular binding of T_3 to its nuclear receptor by drug. *2327*

Thyroxine (T₄) *Serum Increase Physiological* Significant increase (mean change from 100 to 155 nmol/L, i.e. 55%) after 2 mo probably impaired peripheral conversion of T_4 to T_3. *1316* 2 of 92 patients developed hyperthyroidism, whereas 11 developed hypothyroidism in study of up to 4 y. *2851* After slight decrease in some patients initially in 10 patients over 18 mo. *1509* Mean concentration of 131 nmol/L in 38 patients receiving mean 1420 mg weekly for more than 9 mo. *3142* From 7.4 to 9.7 μg/dL in 13 patients treated for average of 17 mo. *0425* In 9 men treated for 28 d caused insignificant increase of 1.4 μg/dL. *0534* Slight effect in 15 euthyroid volunteers given 400-600 mg/daily for 2 weeks. *0092*
Serum Decrease Physiological Single case of drug induced myxedema coma and death. *2343*

Thyroxine (T₄) Binding Globulin
Serum No Effect Physiological No significant changes after 2 mo. *1316*

Thyroxine (T₄), Free *Serum No Effect Physiological* In 5 healthy subjects given 600 mg daily for 2 weeks. *1764*
Serum Increase Physiological Change from mean of 22 to 32 pmol/L after 2 mo. *1316* In group treated for 6 mo 9 out of 28 had values above normal range. *1173*

Thyroxine (T₄) Index, Free (FTI)
Serum No Effect Physiological In 5 healthy subjects given 600 mg daily for 2 weeks. *1764*
Serum Increase Physiological From 7.1 to 9.3 in 13 patients treated for average of 17 mo. *0425*

Triglycerides *Serum Increase Physiological* Observed effect in 3 individuals but without effect on thyroid function. *2833*
Serum Decrease Physiological From 140 to 123 mg/dL in 24 patients over 90 d treatment. *3392*

Tri-iodothyronine (T₃) *Serum No Effect Physiological* Nonsignificant reduction after 2 mo to lower normal range. *1316* In 5 healthy subjects given 600 mg daily for 2 weeks. *1764*
Serum Increase Physiological From 120 to 138 ng/dL in 13 patients treated for average of 17 mo. *0425*
Serum Decrease Physiological Reduced peripheral conversion from thyroxine in 5% patients. *2011* Mean concentration of 1.89 nmol/L in 38 patients receiving mean 1420 mg weekly for more than 9 mo. *3142* In 9 men treated for 28 d caused mean decrease of 28 ng/dL. *0534* Suggestive decrease in 15 euthyroid volunteers given 400-600 mg/daily for 2 weeks. *0092*

Tri-iodothyronine (T₃), Free *Serum Decrease Physiological* In group treated for 6 mo 10 out of 28 had values below normal range. *1173* Common but not invariable response. *2327*

Tri-iodothyronine (T₃), Reverse
Serum Increase Physiological Exceeded normal range after 2 mo in all subjects. *1316* Significant rise in mean from 0.59 to 1.47 nmol/L in 5 healthy subjects given 600 mg daily for 2 weeks. *1764* Mean concentration of 0.85 nmol/L in 38 patients receiving mean 1420 mg weekly for more than 9 mo. *3142* In 9 men treated for 28 d caused increase of 83 ng/dL. *0534* Significant effect in 15 euthyroid volunteers given 400-600 mg/daily for 2 weeks. *0092*

TSH Response to TRH *Serum Increase Physiological* Maximal increment during treatment 32 mU/mL compared with 20 mU/mL pretreatment. *0534*
Serum Decrease Physiological In 16 of 44 patients, majority of whom were clinically hyperthyroid: normalized with withdrawal of drug. *3436*

Uric Acid *Serum No Effect Analytical* At 50 mg/L on reversed phase LC procedure of Zhiri et al. *3942*

Warfarin *Plasma Increase Physiological* Potentiates action of warfarin due to depression of vitamin-K dependent clotting factors. *2011* Effect potentiated and concentration increased as early as third day. *3950*

Amisometradine

Amphetamine *Urine No Effect Analytical* At 50 mg/L on fluorescent method of Hayes. *1534*

Amitriptyline

Acetaminophen Screening Test *Urine Negative Analytical* No reaction with o-cresol at therapeutic concentrations. *3326*

Alanine Aminotransferase *Serum No Effect Analytical* At acute overdose concentration (2.5 mg/dL) on SMAC® method. *3719*
Serum Increase Physiological Rare cases of transient cholestasis. *2482* Occasional case of hypersensitivity associated hepatitis. *0098*

Albumin *Serum No Effect Analytical* At acute overdose concentration (2.5 mg/dL) on SMAC® method. *3719* At concentration of 25 mg/L had no effect on BCG method. *3393*

Alkaline Phosphatase *Serum No Effect Analytical* At acute overdose concentration (2.5 mg/dL) on SMAC® method. *3719*
Serum Increase Physiological Rare cases of cholestasis (transient). *2482* Occasional case of hypersensitivity associated hepatitis. *0098*

Alkaloids *Urine No Effect Analytical* With ninhydrin in Frings TLC procedure at 10 mg/dL. *1204*
Urine Increase Analytical Red/purple with iodoplatinate in Frings TLC procedure. *1204*

Aspartate Aminotransferase *Serum No Effect Analytical* At acute overdose concentration (2.5 mg/dL) on SMAC® method. *3719*
Serum Increase Physiological Rare cases of transient cholestasis. *2482* Occasional case of hypersensitivity associated hepatitis. *0098*

Barbiturate *Urine No Effect Analytical* With diphenylcarbazone in Frings TLC procedure at 10 mg/dL. With mercurous NO_3 in Frings TLC procedure at 10 mg/dL. With vanillin on Frings procedure at 10 mg/dL. With mercuric SO_4 in Frings TLC procedure at 10 mg/dL. *1204*

Bicarbonate *Serum No Effect Analytical* At concentration of 20 mg/L had no effect on method using phenolphthalein. *3393*

Bile *Urine Increase Physiological* Due to transient cholestasis. *2313*

Bilirubin *Serum No Effect Analytical* At acute overdose concentration (2.5 mg/dL) on SMAC® method. *3719* At concentration of 25 mg/L had no effect on Jendrassik and Grof method. *3393*
Serum No Effect Physiological Insignificant protein-displacement effect in neonates. *3748*
Serum Increase Physiological Rare cases of transient cholestasis. *2482*

Bromide *Urine No Effect Analytical* No effect on method of Frings. *1204*

BSP Retention *Serum Increase Physiological* Due to transient cholestasis. *2313*

Calcium *Serum No Effect Analytical* At acute overdose concentration (2.5 mg/dL) on SMAC® method. *3719* At concentration of 25 mg/L had no effect on cresolphthalein method. *3393*

Cephalin Flocculation *Serum Increase Physiological* Rare cases of transient cholestasis. *2313*

Chlordiazepoxide *Blood No Effect Analytical* On method of Riddick at 5 mg/dL. *2983*

Chloride *Serum No Effect Analytical* At concentration of 20 mg/L had no effect on mercurimetric method. *3393*

Cholesterol *Serum No Effect Analytical* At acute overdose concentration (2.5 mg/dL) on SMAC® method. *3719* At concentration of 0.1 mg/L had no effect on CHOD-PAP method. *3393* At concentration of 25 mg/L had no effect on Liebermann-Burchard method. *3393*

Color *Urine Increase Analytical* Greenish-blue color. *1022* Green urine associated with ingestion of drug. *2613*

Creatine Kinase *Serum No Effect Analytical* At acute overdose concentration (2.5 mg/dL) on SMAC® method. *3719*

Creatinine *Serum No Effect Analytical* At 10 mg/L on reversed phase LC procedure of Zhiri et al. *3942* At acute overdose concentration (2.5 mg/dL) on SMAC® method. *3719* At concentration of 0.4 mg/L had no effect on creatinine iminohydrolase method. *3393* At concentration of 25 mg/L had no effect on AutoAnalyzer Jaffé method. *3393*

Dapsone *Serum No Effect Analytical* At 10 mg/dL on colorimetric procedure of Higgins. *1590*

Amitriptyline (continued)

Diazepam *Serum No Effect Analytical* On cathode-ray Polarographic method of Berry. *0323*

Eosinophils *Blood Increase Physiological* Occasional allergic response. *1680* Occasional case of hypersensitivity associated hepatitis. *0098*

Erythrocytes *Blood Decrease Physiological* Rare transient agranulocytosis. *0071*

Ethchlorvynol *Urine No Effect Analytical* No effect on method of Frings and Cohen at 10 mg/dL. *1205*

FPN Test *Urine No Effect Analytical* No effect at 10 mg/dL on method of Frings. *1204*

Glucose *Serum No Effect Analytical* At acute overdose concentration (2.5 mg/dL) on SMAC® method. *3719* At concentration of 6.3 mg/L had no effect on Ektachem® method. *3393* At concentration of 40 mg/L had no effect on hexokinase/G-6-PDH method. *3393*
Serum Increase Physiological Reported observation. *1680*
Serum Decrease Physiological Reported observation. *1680*

Growth Hormone *Plasma No Effect Physiological* No effect observed with acute or chronic treatment in depressed patients. *1366*

Iron *Serum No Effect Analytical* At acute overdose concentration (2.5 mg/dL) on SMAC® method. *3719* At concentration of 25 mg/L had no effect on Ferrozine method. *3393*

Lactate Dehydrogenase *Serum No Effect Analytical* At acute overdose concentration (2.5 mg/dL) on SMAC® method. *3719*

Leukocytes *Blood Decrease Physiological* Leukopenia/agranulocytosis may occur. *0187*

Methaqualone *Serum No Effect Analytical* At 100 mg/L on GLC procedure of Evenson. *1057*

Monoamine Oxidase *Platelets Decrease Physiological* Mean decrease of 50% in 8 subjects with primary or secondary depression when given 100 to 300 mg/d. *2963*

Neutrophils *Blood Decrease Physiological* May cause agranulocytosis/neutropenia. *2583*

Phosphate *Serum No Effect Analytical* At acute overdose concentration (2.5 mg/dL) on SMAC® method. *3719* At concentration of 25 mg/L had no effect on phosphomolybdate method. *3393*

Platelets *Blood Decrease Physiological* Five cases reported (probably immune response). *2603*

Potassium *Serum No Effect Analytical* At concentration of 0.1 mg/L had no effect on flame-photometric method. *3393* At concentration of 20 mg/L had no effect on ISE measurement with predilution. *3393*

Prolactin *Plasma Increase Physiological* In depressive patients: on first day of administration in 6 of 11 patients: nonsignificant decrease after 28 d treatment. *1366* Marked increase in men and women psychiatric patients treated for up to 4 weeks. *3630*

Propoxyphene *Serum No Effect Analytical* At 10 mg/L on method of Evenson. *1056*

Protein *Serum No Effect Analytical* At acute overdose concentration (2.5 mg/dL) on SMAC® method. *3719* At concentration of 25 mg/L had no effect on biuret method with blank correction. *3393*

Salicylate *Urine No Effect Analytical* At 10 mg/dL no effect on Trinder method. *1204*

Sodium *Serum No Effect Analytical* At concentration of 0.1 mg/L had no effect on flame-photometric method. *3393* At concentration of 20 mg/L had no effect on ISE measurement with predilution. *3393*

Thymol Turbidity *Serum Increase Physiological* Rare cases of transient cholestasis. *2313*

Thyroid Stimulating Hormone (TSH)
Serum No Effect Physiological No effect observed with acute or chronic treatment in depressed patients. *1366*

Triglycerides *Serum No Effect Analytical* At concentration of 132 mg/L had no effect on GPO-PAP method. *3393* At concentration of 25 mg/L had no effect on lipase/esterase method. *3393* At acute overdose concentration (2.5 mg/dL) on SMAC® method. *3719*

Urea Nitrogen *Serum No Effect Analytical* At acute overdose concentration (2.5 mg/dL) on SMAC® method. *3719* At concentration of 25 mg/L had no effect on diacetylmonoxime method. *3393* At concentration of 6.3 mg/L had no effect on Ektachem® method. *3393*

Uric Acid *Serum No Effect Analytical* At 10 mg/L on reversed phase LC procedure of Zhiri et al. *3942* At acute overdose concentration (2.5 mg/dL) on SMAC® method. *3719* At concentration of 25 mg/L had no effect on phosphotungstate reduction method. *3393* At concentration of 40 mg/L had no effect on uricase-PAP method. *3393*

Volume *Urine Decrease Physiological* May cause urinary retention. *0071*

Ammonia

Alanine Aminotransferase *Serum Increase Physiological* May produce severe liver damage. *0279*

Aminolevulinic Acid *Urine Increase Analytical* If no preliminary separation. *2981*

α-Amino-Nitrogen *Plasma Increase Analytical* Unless removed during analytic procedure. *1563*
Urine Increase Analytical Unless removed during analytic procedure. *1563*

Aspartate Aminotransferase
Serum Increase Physiological May produce severe liver damage. *0279*

Histamine *Plasma Increase Analytical* May affect fluorometric method of Shore. *0081*

Histidine *Plasma No Effect Analytical* No significant effect fluorometric method of Ambrose. *0081*

Ammonium Acetate

Ammonia *Plasma Increase Physiological* Maximum in cirrhotics after oral dose. *1289*

Ammonium Chloride

N-Acetylglucosaminidase *Urine No Effect Analytical* No effect at 50 mmol/L on 2 colorimetric analytical methods. *1354*

Calcium *Urine Increase Physiological* During day of administration and following day. *2092*

Chloride *Serum Increase Physiological* Added chloride and metabolic acidosis. *2313*
Urine Increase Physiological Diuretic action. *1343*

Epinephrine *Urine No Effect Physiological* No significant effect observed. *3616*

Magnesium *Urine Increase Physiological* Positive correlation with calcium excretion. *2092*

Norepinephrine *Urine No Effect Physiological* No significant effect observed. *3616*

pH *Blood Decrease Physiological* Following administration (metabolic acidosis). *2313*
Urine Decrease Physiological To below 5.3 if acidification normal. *0913*

Potassium *Serum Decrease Physiological* Perpetuates potassium deficiency, cation loss. *2313*
Urine Increase Physiological Diuretic action and hyperchloremic acidosis. *1343*

Sodium *Serum Decrease Physiological* Cation loss as excess chloride excreted. *0652*
Urine Increase Physiological Diuretic action. *1343*

Ammonium Ions

Protein *Serum Decrease Analytical* Cupric ammonium complex formed (Cu availability decreased). *1563*

Ammonium Nitrate

Methemoglobin *Blood Increase Physiological* May cause hemolytic anemia. *2429*

Ammonium Oxalate

pH *Blood Decrease Analytical* Alters acid-base balance if used as anticoagulant. *2313*

Ammonium Salts

Ammonia *Plasma Increase Analytical* Increased blood levels. *1238*

Magnesium *Serum No Effect Physiological* No effect observed in spite of urinary change. *2315*
Urine Increase Physiological Diuretic action observed with acidosis. *2315*

Urea Nitrogen *Serum Increase Analytical* If urease used and ammonia measured. *1563*

Urobilinogen *Urine Decrease Physiological* Acidification of urine decreases excretion. *2452*

Volume *Urine Increase Physiological* Diuretic action of acid forming salt. *1343*

Amobarbital

Acetaminophen Screening Test *Urine Negative Analytical* No reaction with o-cresol at therapeutic concentrations. *3326*

Albumin *Serum No Effect Analytical* At concentration of 150 mg/L had no effect on BCG method. *3393*

Alkaloids *Urine No Effect Analytical* With ninhydrin in Frings TLC procedure at 10 mg/dL. With iodoplatinate of Frings TLC procedure at 10 mg/dL. *1204*

Amobarbital *Blood Increase Physiological* 600 mg orally produces concentration of 9.6 mg/L. *2348*

Antipyrine *Plasma Decrease Physiological* Significant effect on half-life. *3466*

Barbiturate *Urine No Effect Analytical* With vanillin on Frings procedure at 10 mg/dL. *1204*
Urine Increase Analytical White spot with mercuric SO_4 in Frings TLC procedure. With mercurous NO_3 reagent in Frings TLC procedure. Purple with diphenylcarbazone in Frings TLC procedure. *1204*

Bicarbonate *Serum No Effect Analytical* At concentration of 150 mg/L had no effect on method using phenolphthalein. *3393*

Bilirubin *Serum No Effect Analytical* At concentration of 150 mg/L had no effect on Jendrassik and Grof method. *3393*

Bromide *Urine No Effect Analytical* No effect on method of Frings. *1204*

BSP Retention *Serum Increase Physiological* Probably nonspecific effect on cells. *3699*

Calcium *Serum No Effect Analytical* At concentration of 150 mg/L had no effect on cresolphthalein method. *3393*

Chlordiazepoxide *Blood No Effect Analytical* On method of Riddick at 5 mg/dL. *2983*

Chloride *Serum No Effect Analytical* At concentration of 150 mg/L had no effect on mercurimetric method. *3393*

Cholesterol *Serum No Effect Analytical* At concentration of 5 mg/L had no effect on CHOD-PAP method. *3393* At concentration of 150 mg/L had no effect on Liebermann-Burchard method. *3393*

Creatinine *Serum No Effect Analytical* At concentration of 150 mg/L had no effect on AutoAnalyzer Jaffé method. *3393*

Dapsone *Serum No Effect Analytical* At 5 mg/dL on colorimetric procedure of Higgins. *1590*

Diphenylhydantoin *Serum No Effect Analytical* On GLC procedure of Papadopoulos when added *in vitro*. *2722*

Estriol *Urine No Effect Analytical* No interference with GLC. *1163*

Ethchlorvynol *Serum No Effect Analytical* At 10 μg/mL on colorimetric method of Wallace. *3753* At 600 mg/L on GLC procedure of Evenson. *1058*
Urine No Effect Analytical No effect on method of Frings and Cohen at 10 mg/dL. *1205*

FPN Test *Urine No Effect Analytical* No effect at 10 mg/dL on method of Frings. *1204*

Glucose *Serum No Effect Analytical* At concentration of 5 mg/L had no effect on GOD/POD-PAP method. *3393*

17-Hydroxycorticosteroids *Urine No Effect Physiological* No significant effect (?small decrease). *3466*

6-β-Hydroxycortisol *Urine Increase Physiological* Significant increase (by approximately 50%). *3466*

¹³¹I Uptake *Serum Decrease Physiological* Amytal®, tuinal contain tetraiodofluorescein. *2652*

Methaqualone *Serum No Effect Analytical* At 20 mg/L on GLC procedure of Evenson. *1057*

Methionine *Urine Increase Physiological* Contained in large amount in infant formula. *1749*

Pheneturide *Serum No Effect Analytical* On GLC procedure of Papadopoulos when added *in vitro*. *2722*

Phenobarbital *Serum No Effect Analytical* On GLC procedure of Papadopoulos when added *in vitro*. *2722*

Phenylbutazone *Serum No Effect Physiological* No significant effect on half-life. *3466*

Phosphate *Serum No Effect Analytical* At concentration of 150 mg/L had no effect on phosphomolybdate method. *3393*

Potassium *Serum No Effect Analytical* At concentration of 5 mg/L had no effect on flame-photometric method. *3393* At concentration of 65 mg/L had no effect on ISE measurement with predilution. *3393*

Primidone *Serum No Effect Analytical* On GLC procedure of Papadopoulos when added *in vitro*. *2722*

Propoxyphene *Serum No Effect Analytical* At 25 mg/L on method of Evenson. *1056*

Protein *Serum No Effect Analytical* At concentration of 150 mg/L had no effect on biuret method with blank correction. *3393*

Prothrombin Time *Plasma Decrease Physiological* Antagonizes action of coumarins (enzyme induction). *3011*

Salicylate *Urine No Effect Analytical* At 10 mg/dL no effect on Trinder method. *1204*

Sodium *Serum No Effect Analytical* At concentration of 5 mg/L had no effect on flame-photometric method. *3393* At concentration of 65 mg/L had no effect on ISE measurement with predilution. *3393*

Triglycerides *Serum No Effect Analytical* At concentration of 5 mg/L had no effect on lipase/esterase method. *3393*

Urea Nitrogen *Serum No Effect Analytical* At concentration of 150 mg/L had no effect on diacetylmonoxime method. *3393*

Uric Acid *Serum No Effect Analytical* At concentration of 150 mg/L had no effect on phosphotungstate reduction method. *3393*

Amodiaquine

Alanine Aminotransferase *Serum Increase Physiological* Reported hepatotoxicity. *2313* Abnormality usually mild but associated with an immunoallergic hepatotoxicity. *2081*

Alkaline Phosphatase *Serum Increase Physiological* Reported hepatotoxicity. *2313* Abnormality usually mild but associated with an immunoallergic hepatotoxicity. *2081*

Aspartate Aminotransferase
Serum Increase Physiological Reported hepatotoxicity. *2313* Abnormality usually mild but associated with an immunoallergic hepatotoxicity. *2081*

Bile *Urine Increase Physiological* Reported hepatotoxicity. *2313*

Bilirubin *Serum Increase Physiological* Reported effect. *2313* Abnormality usually mild but associated with an immunoallergic hepatotoxicity. *2081*

BSP Retention *Serum Increase Physiological* Reported hepatotoxicity. *2313*

Cephalin Flocculation *Serum Increase Physiological* Reported hepatotoxicity. *2313*

Erythrocytes *Blood Decrease Physiological* Occasional aplastic anemia reported. *3717*

γ-Glutamyltransferase (GGT)
Serum Increase Physiological Abnormality usually mild but associated with an immunoallergic hepatotoxicity. *2081*

Isocitrate Dehydrogenase *Serum Increase Physiological* Reported hepatotoxicity. *2313*

Amodiaquine (continued)

Leukocytes *Blood Decrease Physiological* Reported effect (AMA — blood dyscrasia committee). *2313* 20 cases associated with agranulocytosis in Switzerland, 12 of which associated with hepatitis. *1610* Occasional aplastic anemia reported. *3717*

Neutrophils *Blood Decrease Physiological* Occasional case of agranulocytosis reported. *3717*

Platelets *Blood Decrease Physiological* Occasional aplastic anemia reported. *3717*

Prothrombin Time *Plasma Increase Physiological* Abnormality usually mild but associated with an immunoallergic hepatotoxicity. *2081*

Thymol Turbidity *Serum Increase Physiological* Reported hepatotoxicity. *2313*

Amoxapine

Granulocytes *Blood Decrease Physiological* Observed in one 35 year old woman after receiving 18 g over 57 d. Granulocytes reappeared in peripheral blood on 15th day after treatment stopped. *3238*

Platelets *Blood Increase Physiological* Observed in one 35 year old woman after receiving 18 g over 57 d. Marked thrombocytosis on 5th day after cessation of treatment. May be early sign of recovery of bone marrow in drug-associated toxic agranulocytosis. *3238*

Prolactin *Plasma Increase Physiological* Significant effect in both men and women possibly by blocking dopamine receptors. *0723*

Amoxicillin

Chloramphenicol *Serum No Effect Analytical* No effect at 100 mg/L on coupled enzymatic method. *2490*

Neutrophils *Blood Decrease Physiological* Absolute count of less than 1500/uL in 54% of 41 children treated for 10 d. *1104*

Amphenone B

Aldosterone *Urine Decrease Physiological* May inhibit aldosterone production. *2775*

^{131}I Uptake *Serum Decrease Physiological* Reported effect. *0830*

Protein Bound Iodine (PBI) *Serum Decrease Physiological* Inhibits iodination of tyrosine in thyroxine binding globulin. *0012*

Amphetamine

Alanine Aminotransferase *Serum No Effect Analytical* At therapeutic concentration on Reflotron method. *1984*

Alkaloids *Urine No Effect Analytical* With iodoplatinate of Frings TLC procedure at 5 mg/dL. *1204*
Urine Increase Analytical Reacts with ninhydrin on TLC method of Frings. *1204*

Amino Acids *Urine Increase Analytical* Reacts with ninhydrin; extra spot with TLC, high voltage electrophoresis. *2855*

Amphetamine *Urine Increase Analytical* Yields fluorophor with NBD chloride reaction. *2473*

Aspartate Aminotransferase *Serum No Effect Analytical* At therapeutic concentration on Reflotron method. *1984*

Barbiturate *Urine No Effect Analytical* With diphenylcarbazone in Frings TLC procedure at 10 mg/dL. With mercuric SO$_4$ in Frings TLC procedure at 10 mg/dL. With mercurous NO$_3$ in Frings TLC procedure at 10 mg/dL. With vanillin on Frings procedure at 10 mg/dL. *1204*

Basal Metabolic Rate *Patient Increase Physiological* Metabolic effect of drugs. *2220*

Bromide *Urine No Effect Analytical* No effect on method of Frings. *1204*

Catecholamines *Urine No Effect Analytical* No effect observed. *0913*
Urine No Effect Physiological No effect observed. *0913*

Chlordiazepoxide *Blood No Effect Analytical* On method of Riddick at 5 mg/dL. *2983*

Cholesterol *Serum No Effect Analytical* At therapeutic concentration on Reflotron method. *1984*

Dapsone *Serum No Effect Analytical* At 10 mg/dL on colorimetric procedure of Higgins. *1590*

Epinephrine *Urine Increase Physiological* 5.8 fold increase noticed with psychosis. *1825*

Erythrocytes *Blood Decrease Physiological* Hemolytic anemia (depends on prior sensitivity). *0788*

Ethchlorvynol *Serum No Effect Analytical* At 40 mg/L on GLC procedure of Evenson. *1058*
Urine No Effect Analytical No effect on method of Frings and Cohen at 10 mg/dL. *1205*

Fatty Acids, Free (FFA) *Serum Increase Physiological* But does not modify carbohydrate utilization. *1343*

FPN Test *Urine No Effect Analytical* No effect at 10 mg/dL on method of Frings. *1204*

Glucose *Serum No Effect Analytical* At therapeutic concentration on Reflotron method. *1984*
Serum Decrease Physiological Dextroamphetamine may produce slight effect. *1487*

γ-Glutamyltransferase (GGT) *Serum No Effect Analytical* At therapeutic concentration on Reflotron method. *1984*

Hematocrit *Blood Decrease Physiological* Hemolytic anemia (depends on prior sensitivity). *0788*

Hemoglobin *Blood Decrease Physiological* Hemolytic anemia (depends on prior sensitivity). *0788*

Homovanillic Acid *CSF No Effect Physiological* No change in CSF with psychotic dose. *1825*

17-Hydroxycorticosteroids *Urine No Effect Analytical* No effect on Glenn-Nelson procedure. *0427*

5-Hydroxyindoleacetic Acid (5-HIAA)
CSF No Effect Physiological No change in CSF with psychotic dose. *1825*

17-Ketosteroids *Urine No Effect Analytical* No effect on Zimmermann reaction. *0427*

Leukocytes *Blood Increase Physiological* Myeloblastic leukemia. *0325*

Methaqualone *Serum No Effect Analytical* At 5 mg/L on GLC procedure of Evenson. *1057*

Myeloblasts *Blood Increase Physiological* Myeloblastic leukemia. *0325*

Norepinephrine *Urine Increase Physiological* 2.2 fold increase noted with psychosis. *1825*

Occult Blood *Feces Positive Physiological* Reported to cause ulcers of gastrointestinal tract. *0612*

Promyeloblasts *Blood Increase Physiological* Myeloblastic leukemia. *0325*

Propoxyphene *Serum No Effect Analytical* At 25 mg/L on method of Evenson. *1056*
Urine Increase Analytical 1% fluorescence in procedure of Valentour. *3662*

Salicylate *Urine No Effect Analytical* At 10 mg/dL no effect on Trinder method. *1204*

Thyroxine (T$_4$) Index, Free (FTI)
Serum Increase Physiological In 7.5% of patients admitted to a psychiatric hospital with many having ingested amphetamine. *2489*

Triglycerides *Serum No Effect Analytical* At therapeutic concentration on Reflotron method. *1984*

Tri-iodothyronine (T$_3$) *Serum Increase Physiological* In 17% of patients admitted to a psychiatric hospital with many having ingested amphetamine. *2489*

Tri-iodothyronine (T$_3$) Index, Free
Serum Increase Physiological In 14% of patients admitted to a psychiatric hospital with many having ingested amphetamine. *2489*

Urea Nitrogen *Serum No Effect Analytical* At therapeutic concentration on Reflotron method. *1984*

Uric Acid *Serum No Effect Analytical* At therapeutic concentration on Reflotron method. *1984*

Vanillylmandelic Acid *Urine No Effect Analytical* No effect observed. *0913*
Urine No Effect Physiological No effect observed. *0913*

Amphotericin B

Alanine Aminotransferase *Serum Increase Physiological* Hepatotoxicity reported. *2313* Noted in one patient with acute myelogenous leukemia when treated with high dose. Probable idiosyncratic response. *2443*

Albumin *Serum No Effect Analytical* At concentration of 20 mg/L had no effect on BCG method. *3393*
Serum No Effect Physiological No effect seen in 10 patients with 6 weeks treatment. *0240*

Alkaline Phosphatase *Serum Increase Physiological* Due to hepatocellular dysfunction. *2313* Noted in one patient with acute myelogenous leukemia when treated with high dose. Probable idiosyncratic response. *2443*

Aspartate Aminotransferase
Serum Increase Physiological Hepatotoxicity reported. *2313* Noted in one patient with acute myelogenous leukemia when treated with high dose. Probable idiosyncratic response. *2443*

Bicarbonate *Serum No Effect Analytical* At concentration of 20 mg/L had no effect on method using phenolphthalein. *3393*

Bile *Urine Increase Physiological* Reported hepatotoxicity. *2313*

Bilirubin *Serum Increase Analytical* At concentrations above 96 mg/L (therapeutic concentration about 3.7 mg/L) raised concentration as measured by Jendrassik and Grof method. *3393*
Serum Increase Physiological May cause hepatocellular dysfunction. *1343* Noted in one patient with acute myelogenous leukemia when treated with high dose. Probable idiosyncratic response. *2443*

BSP Retention *Serum Increase Physiological* Due to hepatocellular dysfunction. *2313*

Calcium *Serum No Effect Analytical* At concentration of 96 mg/L had no effect on cresolphthalein method. *3393*

Casts *Urine Increase Physiological* Granular and hyaline casts with toxicity. *0071* Cylindruria may occur. *3567*

Cephalin Flocculation *Serum Increase Physiological* Due to hepatocellular dysfunction. *2313*

Chloride *Serum No Effect Analytical* At concentration of 20 mg/L had no effect on mercurimetric method. *3393*

Cholesterol *Serum Increase Analytical* At concentrations above 96 mg/L (therapeutic concentration about 3.7 mg/L) raised concentration as measured by Liebermann-Burchard method. *3393*

Coombs' Test, Direct *Serum Positive Physiological* Single case reported with hemolytic anemia. *1487*

Creatine Kinase *Serum Increase Physiological* Rhabdomyolysis caused by severe hypokalemia. *0958*

Creatinine *Serum No Effect Analytical* At concentration of 20 mg/L had no effect on AutoAnalyzer Jaffé method. *3393*
Serum Increase Physiological Nephrotoxic effect. *2313* Increased to average of 1.4 mg/dL from 1.0 mg/dL after 2 weeks in 10 patients; further increase with time. *0240* Nephrotoxicity usually develops after a few weeks of treatment, usually reversible unless more than total of 4 g drug given. *3567*

Creatinine Clearance *Urine Decrease Physiological* Nephrotoxicity effect (decrease up to 36%). *0551* Fell to 69 mL/min from 94 mL/min in 10 patients after 2 weeks. Remained at this value with continued treatment. *0240*

Effective Renal Plasma Flow
Patient Decrease Physiological Occurs in high percentage of patients. *0071*

Eosinophils *Blood Decrease Physiological* Allergic response. *1680*

Erythrocytes *Blood Decrease Physiological* Bone marrow depression with hemolytic anemia. *3645* Frequently develops after several weeks of therapy due to interference with production of erythrocytes. *3567*
Urine Increase Physiological Nephrotoxicity. *1022*

Glomerular Filtration Rate (GFR)
Urine Decrease Physiological Occurs in high percentage of patients. *0071*

Hematocrit *Blood Decrease Physiological* Bone marrow depression with hemolytic anemia. *3645*

Hemoglobin *Blood Decrease Physiological* Bone marrow depression with hemolytic anemia. *3645* Majority of patients receiving drug have fall of 18-35% in concentration (normocytic normochromic anemia). *2262*
Urine Increase Physiological Nephrotoxicity, bleeding actually caused by drug. *1022*

Isocitrate Dehydrogenase *Serum Increase Physiological* Due to hepatocellular dysfunction. *2313*

Lactate Dehydrogenase *Serum Increase Physiological* Noted in one patient with acute myelogenous leukemia when treated with high dose. Probable idiosyncratic response. *2443*

Leukocytes *Blood Increase Physiological* Leukocytosis occurs. *1680*
Blood Decrease Physiological Leukopenia or agranulocytosis. *1680* Occasional complication of treatment. *3567*

Magnesium *Serum Decrease Physiological* (occasional) associated with toxic effect of drug. *0128* Significant reduction by 2 weeks from 2.35 mg/dL to 2.0 mg/dL and to 1.6 mg/dL by 4 weeks in 10 patients. *0240*
Urine Increase Physiological Following i.v. infusion for 2 h. *2427* Significant effect at 4 weeks, but not earlier. Fractional magnesium excretion increased. *0240*

Myoglobin *Urine Increase Physiological* Caused by rhabdomyolysis. *0958*

Nonprotein Nitrogen *Serum Increase Physiological* Nephrotoxic effect. *2313*

Occult Blood *Feces Positive Physiological* Melena and hemorrhagic gastroenteritis. *0071*

pH *Urine Increase Physiological* In absence of acid load indicates pending decrease of GFR. *0536*

Phosphate *Serum No Effect Analytical* At concentration of 96 mg/L had no effect on phosphomolybdate method. *3393*

Platelets *Blood Decrease Physiological* Bone marrow depression with hemolytic anemia. *3645* Occasional complication of treatment. *3567*

Potassium *Serum Increase Physiological* May occur with renal toxicity. *0551*
Serum Decrease Physiological (frequent) associated with renal damage. *0128* Nephrotoxicity usually develops after a few weeks of treatment, usually reversible unless more than total of 4 g drug given. *3567*

Potassium Clearance *Urine Increase Physiological* Dose related inverse relationship to GFR. *0536*

Protein *Serum No Effect Analytical* At concentration of 96 mg/L had no effect on biuret method with blank correction. *3393*
Serum No Effect Physiological No effect seen in 10 patients with 6 weeks treatment. *0240*
Urine Increase Physiological Nephrotoxic effect. Nephrotoxicity. *1022* Nephrotoxicity usually develops after a few weeks of treatment, usually reversible unless more than total of 4 g drug given. *3567*

Sodium *Serum Decrease Physiological* Significant effect even in normal subjects. *0551*

Thymol Turbidity *Serum Increase Physiological* Due to hepatocellular dysfunction. *2313*

Titratable Acidity *Urine Decrease Physiological* Development of renal tubular acidosis. *0536*

Triglycerides *Serum No Effect Analytical* At concentration of 20 mg/L had no effect on lipase/esterase method. *3393*

Urea Nitrogen *Serum No Effect Analytical* At concentration of 96 mg/L had no effect on diacetylmonoxime method. *3393*
Serum Increase Physiological Nephrotoxic effect. *2313*

Uric Acid *Serum No Effect Analytical* At concentration of 96 mg/L had no effect on phosphotungstate reduction method. *3393*

Uric Acid Clearance *Urine Increase Physiological* Varies inversely with GFR (consistent change). *0536*

Ampicillin

Acetic Acid *Feces Increase Physiological* Mild effect when given orally to 6 healthy volunteers for 6 d: normalized after 6 weeks. *1664*

Alanine Aminotransferase *Serum No Effect Analytical* No effect on Reflotron method at therapeutic concentration. *1984*

Ampicillin *(continued)*

Alanine Aminotransferase *(continued)*
Serum Increase Physiological Probable effect of i.m. injection. *1966*

Albumin *Serum No Effect Analytical* At concentration of 40 mg/L had no effect on BCG method. *3393*
Serum Increase Analytical Produces increase and noisy peaks with automated BCG procedure. *0287*

Amino Acids *Urine Increase Analytical* Presence of drug as additional spot. *2856*
Urine Positive Analytical Produces yellow spots with ninhydrin and thin-layer chromatography. *1278*

Amphetamine *Urine No Effect Analytical* At 50 mg/L on fluorescent method of Hayes. *1534*

Androstenetriol Sulfate *Urine No Effect Analytical* No effect on GLC method. *3622*
Urine Decrease Physiological Increase by 52% after 6 d in pregnant women. *3622*

Aspartate Aminotransferase *Serum No Effect Analytical* No effect on Reflotron method at therapeutic concentration. *1984*
Serum Increase Physiological Probable effect of i.m. injection (especially in infants). *1966*

Bicarbonate *Serum No Effect Analytical* At concentration of 40 mg/L had no effect on method using phenolphthalein. *3393*

Bilirubin *Serum No Effect Analytical* At concentration of 40 mg/L had no effect on Jendrassik and Grof method. *3393*
Serum No Effect Physiological Insignificant displacement from protein in neonates. *3748*

Bleeding Time *Blood Increase Physiological* Dose related prolongation of bleeding time in all 5 volunteers studied. *0492*

Calcium *Serum No Effect Analytical* At concentration of 40 mg/L had no effect on cresolphthalein method. *3393*

Casts *Urine Increase Physiological* Occurs as result of nephrotoxicity. *3542*

Catecholamines *Plasma Increase Analytical* Concentrated solutions cause striking fluorescence. *0596*

Cephalin Flocculation *Serum Increase Analytical* Alters time of flocculation. *1966*

Chloramphenicol *Serum No Effect Analytical* At 100 mg/L no effect on coupled enzymatic procedure. *2490*

Chloride *Serum No Effect Analytical* At concentration of 40 mg/L had no effect on mercurimetric method. *3393*

Cholesterol *Serum No Effect Analytical* No effect on Reflotron method at therapeutic concentration. *1984* At concentration of 900 mg/L had no effect on CHOD-PAP method. *3393* At concentration of 900 mg/L had no effect on catalase-AIDH method. *3393* At concentration of 900 mg/L had no effect on catalase-Hantzsch method. *3393* At concentration of 900 mg/L had no effect on Liebermann-Burchard method. *3393*
Serum Decrease Analytical At concentrations above 100 mg/L (therapeutic concentration about 320 mg/L) lowered concentration as measured by CHOD-Iodide method. *3393*

Clot Retraction *Blood No Effect Physiological* In doses as high as 300 mg/kg/d in volunteers. *0492*

Coombs' Test, Direct *Serum Positive Physiological* Hypersensitivity reaction in 39% of 36 patients receiving drug. *1963*

Creatine Kinase *Serum Increase Physiological* Probable effect of i.m. injection. *1966*

Creatinine *Serum No Effect Analytical* At concentration of 12 mg/L had no effect on creatinine iminohydrolase method. *3393* At concentration of 40 mg/L had no effect on AutoAnalyzer Jaffé method. *3393* At concentration of 600 mg/L had no effect on Jaffé-Fading-Fraction method. *3393* At concentration of 600 mg/L had no effect on Jaffé-Fuller's earth method. *3393* At concentration of 1600 mg/L had no effect on kinetic Jaffé method on BKA-2. *3393*

Crystals *Urine Increase Analytical* Presence of drug maximal at pH 5. *2856*

Dehydroepiandrosterone (DHEA)
Urine No Effect Analytical No effect on GLC method. *3622*
Urine Decrease Physiological Increase by 75% after 6 d administration in pregnant women. *3622*

Digitoxin *Serum No Effect Physiological* Insignificant increase in elimination half-life. *2223*

Eosinophils *Blood Increase Physiological* Allergic reaction. *1513* Hypersensitivity reaction in 39% of 36 patients receiving drug. *1963*

Erythrocytes *Blood Decrease Physiological* Reversible anemia may occur. *1679*
Urine Increase Physiological Necrosis of tubules due to nephrotoxicity. *3542*

Estriol *Plasma Decrease Physiological* May inhibit synthesis by fetoplacental unit. *0406* In pregnant women due to alteration of glut flora. *2885*
Urine Decrease Physiological May diminish synthesis by fetoplacental unit. *0406*

Estrogens *Urine No Effect Analytical* No effect on method of Brown et al. *3622*
Urine Decrease Physiological Increase by 46% after 6 d in pregnant, altered gastrointestinal flora. During luteal phase in normal women. During ovulatory phase in normals. *3622*

Estrogens (Conjugated) *Urine No Effect Analytical* No effect on Ittrich/Kober reaction. *3622*
Urine Decrease Physiological Increase by 50% after 6 d in pregnant women. *3622*

Fibrinogen *Plasma No Effect Physiological* In doses as high as 300 mg/kg/d in volunteers. *0492*

Folate *Serum No Effect Analytical* If chromatographic procedure of Landon used. *2068* Allegedly no effect on autoclave method. *1815*
Serum Decrease Analytical Inhibits growth of *L. Casei. 3504*

Glucose *Serum No Effect Analytical* No effect on hexokinase, glucose-oxidase methods. *2568* No effect on Reflotron method at therapeutic concentration. *1984* At concentration of 180 mg/L had no effect on Ektachem® method. *3393* At concentration of 5 mg/L had no effect on GOD/POD-PAP method. *3393* At concentration of 1200 mg/L had no effect on hexokinase/G-6-PDH method. *3393*
Urine No Effect Analytical No effect observed with Tes-Tape®. *1100* No effect at concentration up to 10 mg/mL on Diastix® and TesTape® methods. *2250* At concentration of 4,000 mg/L had no effect on Diabur-test. *3393*
Urine Decrease Analytical Low with Clinistix®, Diastix®. *1100*
Urine Positive Analytical At drug concentration of 4 and 10 mg/mL on Clinitest® when no glucose present, but some reduction with 0.5% glucose and higher concentrations. *2250*

γ-Glutamyltransferase (GGT) *Serum No Effect Analytical* No effect on Reflotron method at therapeutic concentration. *1984*

Hemoglobin *Blood Decrease Physiological* Reversible hypersensitivity reaction. *1680*
Urine Increase Physiological Occurs as result of nephrotoxicity. *3542*

16-α-Hydroxy-dehydroepiandrosterone
Urine Decrease Physiological Increase by 72% after 6 d administration in pregnant women. *3622*

15-α-Hydroxy-dehydroepiandrosterone Glucuronide
Urine No Effect Analytical No effect on GLC method. *3622*

16-α-Hydroxy-dehydroepiandrosterone Glucuronide
Urine Decrease Physiological Increase by 45% after 6 d in pregnant women. *3622*

6-Hydroxyprogesterone Metabolite
Urine No Effect Analytical No effect on method of James. *3622*
Urine Decrease Physiological Increase by 29% after 6 d in pregnant women. *3622*

131I Uptake *Serum Decrease Physiological* Omnipen® contains tetraiodofluorescein. *2652*

17-Ketogenic Steroids *Urine No Effect Analytical* No effect on method of Gray. *3622*
Urine Increase Physiological After 14 d in 1 postmenopausal woman. *3622*
Urine Decrease Physiological Increase by 41% after 7 d administration in pregnant women. *3622*

17-Ketosteroids *Urine No Effect Analytical* No effect on method of Gray. *3622*
Urine Increase Physiological After 14 d in 1 postmenopausal woman. *3622*
Urine Decrease Physiological Increase by 42% after 7 d administration in pregnant women. *3622*

Lactate *Serum No Effect Analytical* At concentration of 1200 mg/L had no effect on enzymatic method. *3393*

Leukocytes *Blood Increase Physiological* Mainly due to eosinophilia of hypersensitivity. *2313*
Blood Decrease Physiological Leukopenia. *1370*

Monocytes *Blood Increase Physiological* Associated with agranulocytosis. *1370*

Neutrophils *Blood Decrease Physiological* Reported observation. *2583*

Partial Thromboplastin Time
Plasma No Effect Physiological In doses as high as 300 mg/kg/d in volunteers. *0492*

Phosphate *Serum No Effect Analytical* At concentration of 40 mg/L had no effect on phosphomolybdate method. *3393*

Platelet Aggregation *Blood No Effect Physiological* No effect on collagen or epinephrine induced aggregation. *0492*
Blood Decrease Physiological Defective platelet aggregation induced by ADP in 4 of 5 volunteers. *0492*

Platelets *Blood No Effect Physiological* In doses as high as 300 mg/kg/d in volunteers. *0492*
Blood Decrease Physiological May be associated with purpura. *1680* Isolated case of platelet-associated IgG and thrombocytopenia. *1910*

Potassium *Serum No Effect Analytical* At concentration of 5 mg/L had no effect on flame-photometric method. *3393* At concentration of 1800 mg/L had no effect on ISE measurement without predilution. *3393*

Pregnanediol *Urine No Effect Analytical* No effect on GLC method of Vela. *3622*
Urine Decrease Physiological Increase by 53% after 5 d administration in pregnant women. During luteal phase in normal women. *3622*

Pregnenolone *Urine No Effect Analytical* No effect on GLC method of Vela. *3622*
Urine Decrease Physiological Increase by 44% after 6 d administration in pregnant women. *3622*

Progesterone *Plasma No Effect Analytical* No effect on GLC method of Willman. *3622*
Plasma Decrease Physiological Probably decreases synthesis. *3622*

Protein *Serum Increase Analytical* At concentrations above 500 mg/L raised concentration as measured by biuret method with blank correction. *3393*
Urine No Effect Analytical No effect in patients receiving up to 8 g daily on sulfosalicylic acid, trichloracetic acid or Ponceau S dye methods. *3919*
Urine Increase Physiological Necrosis of tubules due to nephrotoxicity. *3542*

Protein Electrophoresis *Serum No Effect Analytical* At concentration of 1800 mg/L had no effect on automated Olympus-Hite method. *3393*

Prothrombin Time *Plasma No Effect Physiological* In doses as high as 300 mg/kg/d in volunteers. *0492*

Short-Chain Fatty Acids *Feces Decrease Physiological* Mild effect when given orally to 6 healthy volunteers for 6 d: normalized after 6 weeks. *1664*

Sodium *Serum No Effect Analytical* At concentration of 5 mg/L had no effect on flame-photometric method. *3393*
Serum Increase Analytical At concentrations above 1800 mg/L (therapeutic concentration about 320 mg/L) raised concentration as measured by ISE without predilution. *3393*

Thrombin Time *Plasma No Effect Physiological* In doses as high as 300 mg/kg/d in volunteers. *0492*

Triglycerides *Serum No Effect Analytical* No effect on Reflotron method at therapeutic concentration. *1984* At concentration of 11 mg/L had no effect on GPO-PAP method. *3393* At concentration of 40 mg/L had no effect on lipase/esterase method. *3393*

Urea Nitrogen *Serum No Effect Analytical* No effect on Reflotron method at therapeutic concentration. *1984* At concentration of 40 mg/L had no effect on diacetylmonoxime method. *3393* At concentration of 180 mg/L had no effect on Ektachem® method. *3393*

Uric Acid *Serum No Effect Analytical* At 20 mg/dL on Nishi phosphotungstate procedure. *2230* No effect on Reflotron method at therapeutic concentration. *1984* At concentration of 600 mg/L had no effect on Kageyama-Hantzsch method. *3393* At concentration of 300 mg/L had no effect on catalase-AlDH method. *3393* At concentration of 40 mg/L had no effect on phosphotungstate reduction method. *3393*
Serum Increase Analytical 20 mg/dL equivalent to 2.6 mg/dL copper chelate procedure. *2230* At concentrations above 100 mg/L (therapeutic concentration about 320 mg/L) raised concentration as measured by Seralyzer. *3393*
Urine Increase Physiological Competes with uric acid for reabsorption and potentiates response to probenecid. Significant uricosuria with ampicillin alone from 0-2 h and 8-24 h after administration. *3386*

Urobilin *Urine Increase Physiological* Necrosis of tubules due to nephrotoxicity. *3542*

Amrinone

Digoxin *Serum No Effect Analytical* No effect on Du Pont aca method. *2777*
Serum Increase Analytical Positive interference with TD_x method. *2777*

Amygdalin

Sugar *Urine Increase Analytical* Measured as reducing agents with Benedict's. *2313*

Amyl Alcohol

Alanine Aminotransferase *Serum Increase Physiological* May cause liver damage if ingested. *1302*

Glucose *Urine Increase Physiological* Associated with renal damage. *1302*

Ionized Calcium *Serum Decrease Analytical* At 0.1 mmol/L to 0.1 mol/L with Calcium specific electrode. *0540*

Methemoglobin *Blood Increase Physiological* May cause hemolysis following ingestion. *1302*

Occult Blood *Feces Positive Physiological* May cause gastrointestinal hemorrhage. *1302*

Amyl Nitrite

Bilirubin *Serum Increase Physiological* With increased hemolysis. *2426*

Casts *Urine Increase Physiological* May cause nephrotoxicity. *0279*

Erythrocytes *Blood Decrease Physiological* Hemolytic anemia (slight or marked effect). *2426*

Hemoglobin *Plasma Increase Physiological* If intravascular hemolysis occurs. *2426*

Methemoglobin *Blood Increase Physiological* Hemolytic anemia. *2426*

Reticulocytes *Blood Increase Physiological* Hemolytic anemia (up to 15%). *2426*

Urea Nitrogen *Serum Increase Physiological* May cause nephrotoxicity. *0279*

Urobilinogen *Urine Increase Physiological* Less reliable index than fecal if hemolysis occurs. *2426*
Feces Increase Physiological If hemolytic anemia occurs. *2426*

Anabolic Steroids

Alanine Aminotransferase *Serum Increase Physiological* Cholestatic syndrome (intrahepatic cholestasis). *2220*

Alkaline Phosphatase *Serum Increase Physiological* Cholestatic syndrome. *1343*

Amylase *Serum Decrease Physiological* Mean decrease from 112 to 89 units in patients with chronic pancreatitis after 3 weeks treatment with combination of anabolic steroids. *3617*

Aspartate Aminotransferase
Serum Increase Physiological Cholestatic syndrome (intrahepatic cholestasis). *2220*

Anabolic Steroids *(continued)*

Bile *Urine Increase Physiological* Intrahepatic cholestatic jaundice. *1343*

Bilirubin *Serum Increase Physiological* Cholestatic syndrome. *1022*

Bilirubin, Direct *Serum Increase Physiological* Cholestatic syndrome. *2313*

BSP Retention *Serum Increase Physiological* Cholestatic syndrome. *1022*

Calcium *Serum Increase Physiological* Positive effect on calcium retention. *0070*

Cephalin Flocculation *Serum Increase Physiological* Due to cholestatic syndrome. *2220*

Cholesterol *Serum Increase Physiological* Cholestatic phenomenon. *2220*

Creatine *Urine Decrease Physiological* Anabolic effect. *1237*

Creatinine *Urine Decrease Physiological* Anabolic effect. *1237*

Erythropoietin *Plasma Increase Physiological* Metabolic effect. *0227*

Factor V *Plasma Increase Physiological* Metabolic effect. *0227*

Factor VII *Plasma Increase Physiological* Metabolic effect. *0227*

Factor X *Plasma Increase Physiological* Metabolic effect. *0227*

Fibrinogen *Plasma Decrease Physiological* Metabolic effect. *0227*

Glucose *Serum Decrease Physiological* Anabolic effect in fasting state. *2220*

β-Glucuronidase *Serum Increase Physiological* Metabolic effect. *0227*

Guanase *Serum Increase Physiological* Due to cholestatic syndrome. *2220*

Haptoglobin *Serum Increase Physiological* Metabolic effect. *0227*

3-Hydroxyanthranilic Acid *Urine Decrease Physiological* Effect of synthetic androgens given to males. *3045*

3-Hydroxykynurenine *Urine Decrease Physiological* Effect of synthetic androgens given to males. *3045*

Isocitrate Dehydrogenase *Serum Increase Physiological* Due to cholestatic syndrome. *2220*

^{131}I Uptake *Serum No Effect Physiological* No effect observed. *1915*

Kynurenine *Urine Decrease Physiological* Effect of synthetic androgens given to males. *3045*

Lactate Dehydrogenase *Serum Increase Physiological* ?part of cholestatic syndrome. *0583*

5'-Nucleotidase *Serum Increase Physiological* Due to cholestasis. *1778*

Phosphate *Serum Increase Physiological* Augments phosphate retention. *0652*

Plasminogen *Plasma Increase Physiological* Metabolic effect. *0227*

Prealbumin *Serum Increase Physiological* Metabolic effect. *0227*

Protein *Serum Increase Physiological* Associated with increased protein synthesis. *0697*

Protein Bound Iodine (PBI) *Serum Decrease Physiological* Reduces concentration of circulating thyroxine binding globulin. *0012*

Prothrombin Time *Plasma Increase Physiological* Exaggerated response to anticoagulants reported. *1680*
Plasma Decrease Physiological Increased prothrombin as metabolic effect. *2526*

Sodium *Serum Increase Physiological* Mineralocorticoid effect with retention. *2313*

T$_3$ Uptake *Serum Increase Physiological* Decrease level of thyroxine binding globulin. *0583*

Thymol Turbidity *Serum Increase Physiological* Due to cholestatic syndrome. *2220*

Thyroxine (T$_4$) *Serum Decrease Physiological* Decrease amount of thyroxine binding globulin. *0583*

Thyroxine (T$_4$) Binding Globulin *Serum Decrease Physiological* Direct effect of drug. *0012*

Thyroxine (T$_4$) (Murphy-Pattee) *Serum Decrease Physiological* Decreases circulating thyroxine binding globulin. *0583*

Transcortin *Serum Increase Physiological* Metabolic effect. *0227*

Trypsin *Serum Decrease Physiological* Mean decrease from 7.5 to 5.3 mU/dL in patients with chronic pancreatitis after weeks treatment with combination of anabolic steroids. *3617*

Trypsin Inhibitor *Serum Increase Physiological* Metabolic effect. *0227* Mean increase from 401 to 526 mU/dL in patients with chronic pancreatitis after 3 weeks treatment with combination of anabolic steroids. *3617*

Urea Nitrogen *Serum Increase Physiological* Improves nitrogen balance. *0652*

Uric Acid *Serum Increase Physiological* Improves nitrogen balance. *2313*

Xanthurenic Acid *Urine Decrease Physiological* Effect of synthetic androgens given to males. *3045*

Analgesics

Creatine Kinase *Serum Increase Physiological* May cause effect if injected i.m. *0249*

Methemoglobin *Blood Increase Physiological* May cause intravascular hemolysis. *3504*

Ancrod

Fibrinogen *Plasma Decrease Physiological* Converts to fibrin shreds. *2823*

Androgens

Acid Phosphatase *Serum Increase Physiological* Effect of hormone in females. *1022*

Alanine Aminotransferase *Serum Increase Physiological* May cause cholestatic syndrome. *2220*

Alkaline Phosphatase *Serum Increase Physiological* May cause cholestatic syndrome. *1022*

Aspartate Aminotransferase *Serum Increase Physiological* May cause cholestatic syndrome. *2220*

Bile *Urine Increase Physiological* Intrahepatic cholestatic jaundice. *1343*

Bilirubin *Serum Increase Physiological* May cause cholestatic syndrome. *1022*

Bilirubin, Direct *Serum Increase Physiological* May cause cholestatic syndrome. *2313*

BSP Retention *Serum Increase Physiological* May cause cholestatic syndrome. *1022*

Calcium *Serum Increase Physiological* Positive effect on calcium retention. *0070*

Cephalin Flocculation *Serum Increase Physiological* Cholestatic phenomenon. *2313*

Ceruloplasmin *Serum Increase Physiological* Hormonal effect. *0248*

Chloride *Serum Increase Physiological* May cause salt and water retention. *1343*

Cholesterol *Serum Increase Physiological* Cholestatic phenomenon. *2220*
Serum Decrease Physiological Decreased synthesis. *3879*

Corticosteroids *Plasma Decrease Physiological* May cause cholestatic syndrome. *1237*

Creatine *Urine Decrease Physiological* Anabolic effect. *1237*

Creatinine *Urine Decrease Physiological* Anabolic effect. *1022*

2,3-Diphosphoglycerate *Red Blood Cells Increase Physiological* Metabolic effect. *1091*

Erythrocytes *Blood Increase Physiological* May produce erythrocytemia (increases well-being). *0071*

Glucose *Serum Increase Physiological* Reported effect. *0117*
Serum Decrease Physiological Anabolic effect in fasting state. *2220*

Guanase *Serum Increase Physiological* Cholestatic phenomenon. *2313*

Hematocrit *Blood Increase Physiological* Associated with increased well being. *0071*

3-Hydroxyanthranilic Acid *Urine Decrease Physiological* Effect of synthetic androgens given to males. *3045*

3-Hydroxykynurenine *Urine Decrease Physiological* Effect of synthetic androgens given to males. *3045*

Isocitrate Dehydrogenase *Serum Increase Physiological* Cholestatic phenomenon. *2313*

^{131}I Uptake *Serum No Effect Physiological* No effect on uptake reported. *2220*

Kynurenine *Urine Decrease Physiological* Effect of synthetic androgens given to males. *3045*

5'-Nucleotidase *Serum Increase Physiological* Due to cholestasis. *1778*

Phosphate *Serum Increase Physiological* Augments phosphate balance. *0652*

Protein *Serum Increase Physiological* Associated with increased protein synthesis. *0697*

Protein Bound Iodine (PBI) *Serum Decrease Physiological* Reduced synthesis of thyroxine binding globulin. *2220*

Prothrombin Time *Plasma Increase Physiological* Mechanism not established. *0132*

Sodium *Serum Increase Physiological* Mineralocorticoid effect with increased retention. *0070*

T$_3$ Uptake *Serum Increase Physiological* Decrease level of thyroxine binding globulin. *0583*

Thymol Turbidity *Serum Increase Physiological* Cholestatic phenomenon. *2313*

Thyro-Binding Index *Serum Decrease Physiological* May occur for up to 1 mo after treatment. *0913*

Thyroxine (T$_4$) *Serum Decrease Physiological* For up to 1 mo after treatment (decreased thyroxine binding globulin). *0913* Decreased binding capacity (diminution of transport proteins). *3798*

Thyroxine (T$_4$) Binding Globulin
Serum Decrease Physiological Direct effect of drugs. *0012*

Thyroxine (T$_4$) Index, Free (FTI)
Serum Decrease Physiological Low normal or decreased for 1 mo after therapy. *0913*

Thyroxine (T$_4$) (Murphy-Pattee)
Serum Decrease Physiological Decreases circulating thyroxine binding globulin. *0583*

Tri-iodothyronine (T$_3$) *Serum Decrease Physiological* Decreased binding capacity (diminution of transport proteins). *3798*

Tyrosine *Plasma Decrease Physiological* In 24 h in boys with delayed puberty. *0843*

Urea Nitrogen *Serum Increase Physiological* Augments nitrogen balance. *0652*

Uric Acid *Serum Increase Physiological* Augments nitrogen balance. *2313*

Xanthurenic Acid *Urine Decrease Physiological* Effect of synthetic androgens given to males. *3045*

Androstenedione

Estrogens *Urine No Effect Analytical* At 50 mg/L on fluorescent method of Corns. *0730*

Androsterone

Aldosterone *Urine No Effect Analytical* No significant effect on RIA procedure of Drewes. *0952*
Urine Increase Analytical 1 mg/L equivalent to 2 µg/L in method of Drewes. *0952*

Cholesterol *Serum Decrease Physiological* Therapeutic effect. *1237*

Estrogens *Urine No Effect Analytical* At 50 mg/L on fluorescent method of Corns. *0730*

Anesthetic Agents

Alanine Aminotransferase *Serum Increase Physiological* Occurs without permanent liver damage. *3661*

Aspartate Aminotransferase
Serum Increase Physiological Occurs without permanent liver damage. *3661*

Creatine Kinase *Serum Increase Physiological* Occurs if combined with suxamethonium. *1724*

Fluoride *Serum Increase Physiological* Slight effect following general anesthetic. *0916*

Glucose *Serum Increase Physiological* Presumed response to stress. *0251*

Isocitrate Dehydrogenase *Serum Increase Physiological* Occurs without permanent liver damage. *3661*

Lactate Dehydrogenase *Serum Increase Physiological* Response occurs even with premedication. *3661*

Ornithine Carbamoyltransferase (OCT)
Serum Increase Physiological Occurs without permanent liver damage. *3661*

Osmolality *Urine Increase Physiological* Induction of anesthesia produces marked effect. *0916*

Osmolar Clearance *Urine Decrease Physiological* Significant effect with general anesthesia. *0916*

Phosphate *Serum Decrease Physiological* Observed after anesthesia. *1237*

Potassium *Urine Decrease Physiological* Significant effect with general anesthesia. *0916*

Prothrombin Time *Plasma Increase Physiological* Observed effect. *1680*

Sodium *Urine Decrease Physiological* Significant effect with general anesthesia. *0916*

Thyroid Stimulating Hormone (TSH)
Serum No Effect Physiological No effect observed. *2700*

Volume *Urine Decrease Physiological* Induction of anesthesia produces marked effect. *0916*

Angiotensin

Aldosterone *Plasma Increase Physiological* Increase of 4 times normal after i.v. for 1 h. *1145*
Urine Increase Physiological Approximate doubling in response to i.v. infusion. *3174*

Calcium *Urine Decrease Physiological* Hemodynamic effect of drug. *2173*

Effective Renal Plasma Flow
Patient Decrease Physiological Marked decrease following administration. *2427*

Epinephrine *Plasma Decrease Physiological* Decreased slightly less than norepinephrine increased. *3967*

Glomerular Filtration Rate (GFR)
Urine Decrease Physiological Vasoconstrictive effect in kidney. *1343*

Magnesium *Urine Decrease Physiological* Hemodynamic effect of drug. *2173*

Norepinephrine *Plasma Increase Physiological* Of same magnitude as if norepinephrine given. *3967*

Prostaglandin E *Urine Increase Physiological* 2.3 fold increase after 3 h in women. *1211*

Prostaglandin F *Urine No Effect Physiological* No effect observed after i.v. infusion. *1211*

Protein *Serum Increase Physiological* Hemoconcentration effect. *1343*

Renin Activity *Plasma Decrease Physiological* Causes decrease in normals when given i.v. *1145*

Sodium *Serum Increase Physiological* Due to salt retaining action of aldosterone. *1343*
Urine Decrease Physiological Due to effect on renal tubules. *1343*

Angiotensin *(continued)*

Tetrahydrodeoxycorticosterone
Urine No Effect Physiological No significant effect when infused in normals. *3174*
Uric Acid *Serum Increase Physiological* Reduced urate clearance if given i.v. *1124*
Urine Decrease Physiological Decreases urate excretion and renal plasma flow. *1124*
Uric Acid Clearance *Urine Decrease Physiological* 44% reduction with infusion. *1125*
Vanillylmandelic Acid *Urine No Effect Physiological* Epinephrine decrease compensates for norepinephrine increase. *3967*
Volume *Plasma Decrease Physiological* Promotes loss of protein-free filtrate to tissues. *1343*
Urine Decrease Physiological Vasoconstrictive effect in kidney. *1343*

Anileridine

Amphetamine *Urine No Effect Analytical* At 50 mg/L on fluorescent method of Hayes. *1534*
Vanillylmandelic Acid *Urine Increase Analytical* Interferes with Pisano procedure. *1022*

Aniline

Casts *Urine Increase Physiological* Nephrotoxic effect (chronic effect). *0279*
Color *Blood Increase Physiological* May produce chocolate color. *0279*
Urine Increase Physiological Brown color due to intravascular hemolysis. *1302* May produce red color. *1187*
Erythrocytes *Blood Decrease Physiological* May cause marked hemolytic anemia. *0279*
Hemoglobin *Blood Decrease Physiological* May cause marked hemolytic anemia. *0279*
Ionized Calcium *Serum Decrease Analytical* At 0.1 mmol/L to 0.1 mol/L with Calcium specific electrode. *0540*
Methemoglobin *Blood Increase Physiological* Occurs as result of intravascular hemolysis. *1302*
Urine Increase Physiological Due to intravascular hemolysis. *1302*
Sulfa as Sulfanilamide *Serum Increase Analytical* Reacts in Bratton-Marshall procedure. *0279*
Urea Nitrogen *Serum Increase Physiological* May cause nephrotoxicity (chronic effect). *0279*

Aniline Dyes

Color *Urine Increase Analytical* Red from foods and candy. *0443*

Anisindione

Bilirubin *Serum Increase Physiological* Observed with other indandiones. *1680*
Color *Urine Increase Analytical* Orange (alkaline), pink-red-brown (acid). *2313*
Hemoglobin *Urine Increase Physiological* Overdose manifestation. *1680*
Leukocytes *Blood Decrease Physiological* Observed with other indandiones. *1680*
Occult Blood *Feces Positive Physiological* May indicate toxicity. *1680*
Prothrombin Time *Plasma Increase Physiological* Intended effect. *1680*
Uric Acid *Urine No Effect Physiological* No uricosuric action observed. *1487*

Antacids

Acetylsalicylic Acid *Serum Increase Physiological* Increased rate of absorption due to faster drug release. *3794*

Calcium *Serum Increase Physiological* May occur if calcium containing preparations. *2427*
Chlorpromazine *Serum Decrease Physiological* Absorption decreased by 10 to 45% due to adsorption by antacid. *3794*
Chlortetracycline *Serum Decrease Physiological* Absorption of tetracyclines decreased by up to 80% due to adsorption by antacids. *3794*
Diazepam *Serum Decrease Physiological* Delayed absorption due to adsorption. *3794*
Digoxin *Serum Decrease Physiological* Reduced absorption due to adsorption and faster gastric emptying. *3794*
Folate *Serum Decrease Physiological* Associated with malabsorption. *3023*
Red Blood Cells Decrease Physiological Associated with malabsorption. *3023*
Isoniazid *Serum Decrease Physiological* Delayed and reduced absorption due to adsorption and first-pass metabolism. *3794*
Levodopa *Serum Increase Physiological* Increased absorption with faster gastric emptying. *3794*
Penicillamine *Serum Decrease Physiological* Reduced absorption due to adsorption and chelation. *3794*
pH *Blood Increase Physiological* May cause occasional metabolic alkalosis. *0071*
Proquazone *Serum Decrease Physiological* Delayed absorption noted. *3794*
Prothrombin Time *Plasma Decrease Physiological* May shorten action of anticoagulants. *1679*
Sulfathiazole *Serum Increase Physiological* Due to faster dissolution rate. *3794*
Warfarin *Plasma No Effect Physiological* Neither concentration nor effect affected. *1487*

Antazoline

Creatinine *Serum No Effect Analytical* At concentration of 160 mg/L had no effect on Jaffé-Fading-Fraction method. *3393* At concentration of 160 mg/L had no effect on Jaffé-Fuller's earth method. *3393*
Leukocytes *Blood Decrease Physiological* Leukopenia. *3602*
Platelets *Blood Decrease Physiological* Thrombocytopenia (immunologically induced). *3602*
Uric Acid *Serum No Effect Analytical* At concentration of 160 mg/L had no effect on Kageyama-Hantzsch method. *3393* At concentration of 160 mg/L had no effect on catalase-AIDH method. *3393*

Anthiolimine

LE Cells *Blood Positive Physiological* Observed effect in some cases. *0933*

Anthranilic Acid

Hydroxyproline *Urine No Effect Analytical* At 53 mg/L on method of Goverde. *1365*

Anthraquinone

Color *Urine Increase Analytical* Pink, red, purple, orange and rust. *2425*
Feces Increase Analytical Brownish staining. *2313*
Estrogens *Urine Increase Analytical* Bright yellow colors with fluorometry. *0023*
PSP Excretion *Urine Increase Analytical* Interference by color of compound. *1999*

Antibiotics

Ascorbic Acid *White Blood Cells Decrease Physiological* Observed effect especially in elderly. *3554*
Calcium *Serum No Effect Analytical* No effect on fluorescence of calcein. *2395*

HMPG (4-Hydroxy-3-Methoxy-Phenylethylene Glycol)
Urine No Effect Physiological Orally administered has no effect on excretion. *0426*

NBT Test *Blood Decrease Physiological* Index drops in 4-6 h if therapy satisfactory. *0943*

Potassium *Serum Decrease Physiological* May occur with multiple regime, ?redistribution. *3549*
Urine Increase Physiological Some increase often with multiple regimes. *3549*

Protein Bound Iodine (PBI) *Serum No Effect Physiological* No effect observed. *0830*

Prothrombin Time *Plasma Increase Physiological* Decreased synthesis of vitamin K by gastrointestinal tract. *3879*

Urobilinogen *Urine Decrease Physiological* Inhibit gastrointestinal tract flora. *1238*
Feces Decrease Physiological Inhibit gastrointestinal tract flora. *1237*

Vanillylmandelic Acid *Urine No Effect Physiological* Sterilization of gut has no effect on excretion. *0426*

Anticoagulants

Coagulation Time *Blood Increase Physiological* Therapeutic intent. *2313*

Color *Feces Increase Physiological* Red to black due to internal bleeding. *2313*

Protein Bound Iodine (PBI) *Serum No Effect Physiological* No effect observed. *0830*

Prothrombin Time *Plasma Increase Physiological* Therapeutic intent. *2313*

Anticonvulsants

α_1-Acid Glycoprotein *Serum Increase Physiological* In 8 epileptic patients receiving phenobarbital with carbamazepine or phenytoin. *2670*

Albumin *Serum Decrease Physiological* In 8 epileptic patients receiving phenobarbital with carbamazepine or phenytoin. *2670*

Alkaline Phosphatase *Serum Increase Physiological* Occurs in 24% children (90% bony origin). *1695* In 8 epileptic patients receiving phenobarbital with carbamazepine or phenytoin. *2670*

Amino-4-Imidazole-5-Carboxamide Ribotide (AICAR)
Urine Increase Physiological Occurs with megaloblastic anemia. *3773*

Aminolevulinic Acid *Serum Increase Physiological* Highly significant increase when multiple drugs given (146 nmol/L versus 99 nmol/L in Controls. *1348*
Urine Increase Physiological Reported in one child. *2427*

Antinuclear Antibodies *Serum Positive Physiological* Related to number of drugs, higher in women. *3797*

α_1-Antitrypsin *Serum No Effect Physiological* In 8 epileptic patients receiving phenobarbital with carbamazepine or phenytoin. *2670*

Ascorbic Acid *Plasma No Effect Physiological* In 146 epileptics with long-term treatment. *2007*

Aspartate Aminotransferase
Serum Increase Physiological Observed with chronic administration. *3319*
Red Blood Cells Decrease Physiological In treated patients compared with controls. *2949*

Biotin *Serum Decrease Physiological* Marked effect in 146 epileptics with long-term treatment. *2007*

Calcium *Serum Decrease Physiological* Found in 30% children on prolonged therapy. *1695*

Carotene *Serum No Effect Physiological* In 146 epileptics with long-term treatment. *2007*

Ceruloplasmin *Serum Increase Physiological* In 8 epileptic patients receiving phenobarbital with carbamazepine or phenytoin. *2670*

Complement C3 *Serum No Effect Physiological* In 8 epileptic patients receiving phenobarbital with carbamazepine or phenytoin. *2670*

Complement C4 *Serum No Effect Physiological* In 8 epileptic patients receiving phenobarbital with carbamazepine or phenytoin. *2670*

Cortisol *Plasma No Effect Physiological* No difference in concentrations in men on long-term treatment. *0257*
Plasma Increase Physiological 400 nmol/L versus 260 nmol/L in women on long-term treatment. *0257*

Erythrocyte Sedimentation Rate
Blood Increase Physiological Observed with SLE-like syndrome. *0933*

Estradiol *Plasma No Effect Physiological* Insignificantly higher in women on long-term therapy. *0257*

Factor II *Plasma Increase Physiological* In 8 epileptic patients receiving phenobarbital with carbamazepine or phenytoin. *2670*

Factor VII *Plasma Increase Physiological* In 8 epileptic patients receiving phenobarbital with carbamazepine or phenytoin. *2670*

Factor X *Plasma Increase Physiological* In 8 epileptic patients receiving phenobarbital with carbamazepine or phenytoin. *2670*

Ferritin *Serum No Effect Physiological* In 8 epileptic patients receiving phenobarbital with carbamazepine or phenytoin. *2670*

FIGLU (N-Formiminoglutamic Acid)
Urine Increase Physiological Occurs with megaloblastic anemia. *3773*

Folate *Serum Decrease Physiological* May cause megaloblastic anemia. *3025* Marked effect in 146 epileptics with long-term treatment. *2007*
CSF Decrease Physiological Occurs in many long-treated epileptics. *2967*
Red Blood Cells Decrease Physiological In 39% epileptics on chronic therapy. *3560*

Follicle Stimulating Hormone (FSH)
Plasma Decrease Physiological In 33 male epileptics taking at least one drug for long time. *3020*

γ-Globulin *Serum Increase Physiological* Associated with SLE-like syndrome. *0933*

Glucaric Acid *Urine Increase Physiological* Occurs in 94% children (hepatic enzyme induction). *1695*

γ-Glutamyltransferase (GGT)
Serum Increase Physiological In 8 epileptic patients receiving phenobarbital with carbamazepine or phenytoin. *2670*

Haptoglobin *Serum Increase Physiological* In 8 epileptic patients receiving phenobarbital with carbamazepine or phenytoin. *2670*

Hematocrit *Blood Decrease Physiological* May cause megaloblastic/aplastic anemia. *3025*

Hemoglobin *Blood Decrease Physiological* May cause megaloblastic/aplastic anemia. *3025*

Hydroxybutyrate Dehydrogenase
Serum Increase Physiological In 4% epileptics on chronic therapy. *3560*
Red Blood Cells Increase Physiological In 21% epileptics on chronic therapy. *3560*

5-Hydroxyindoleacetic Acid (5-HIAA)
CSF Decrease Physiological Reduction to 18.3 ng/mL from 25.1 ng/mL in lumbar CSF of treated epileptics. *3926*

25-Hydroxy Vitamin D_3 *Serum No Effect Physiological* No significant difference, although slightly less, in men and women on long-term treatment. *0257*
Serum Decrease Physiological Probably due to increased metabolism of vitamin D. *1451* Marked effect in 146 epileptics with long-term treatment. *2007*

Immunoglobulin IgA *Serum No Effect Physiological* In 8 epileptic patients receiving phenobarbital with carbamazepine or phenytoin. *2670*

Immunoglobulin IgG *Serum No Effect Physiological* In 8 epileptic patients receiving phenobarbital with carbamazepine or phenytoin. *2670*

Immunoglobulin IgM *Serum No Effect Physiological* In 8 epileptic patients receiving phenobarbital with carbamazepine or phenytoin. *2670*

Indocyanine Green *Serum Decrease Physiological* Mechanism not yet established. *2398*

Anticonvulsants *(continued)*

Indoleacetic Acid *CSF No Effect Physiological* Insignificant reduction to 2.60 ng/mL in lumbar CSF from 3.74 ng/mL in untreated epileptics. *3926*

Lactate Dehydrogenase *Serum No Effect Physiological* No effect of chronic therapy. *3560*
Red Blood Cells Increase Physiological In 38% epileptics on chronic therapy. *3560*

LE Cells *Blood Positive Physiological* May induce lupus like syndrome in some cases. *1574*

Leucine Aminopeptidase Isoenzymes
Serum Increase Physiological Increase slower running components. *3087*

Leukocytes *Blood Decrease Physiological* May occur with severe megaloblastic anemia. *3773*

Luteinizing Hormone (LH) *Plasma Decrease Physiological* In 33 male epileptics taking at least one drug for long time. *3020*

MCV *Blood Increase Physiological* May cause megaloblastic/aplastic anemia. *3773*

Neutrophils *Blood Decrease Physiological* May occur without effect on white cell count. *1680*

Phosphate *Serum Decrease Physiological* Disturbance of vitamin D metabolism or hepatic enzyme induction. *3401*

Platelets *Blood Decrease Physiological* May occur with severe megaloblastic anemia. *3773*

Porphobilinogen *Urine Increase Physiological* Drug related effect reported in one child. *2427*

Prealbumin *Serum No Effect Physiological* In 8 epileptic patients receiving phenobarbital with carbamazepine or phenytoin. *2670*

Prolactin *Plasma Decrease Physiological* In 33 male epileptics taking at least one drug for long time. *3020*

Prothrombin Time *Plasma No Effect Physiological* In 8 epileptic patients receiving phenobarbital with carbamazepine or phenytoin. *2670*

Pyridoxal Phosphate *Serum No Effect Physiological* In 146 epileptics with long-term treatment. *2007*
Serum Decrease Physiological Increase by 20% at 4 weeks, 60% at 12 weeks. *2948*

Retinol Binding Protein *Serum No Effect Physiological* In 146 epileptics with long-term treatment. *2007*
Serum Increase Physiological In 8 epileptic patients receiving phenobarbital with carbamazepine or phenytoin. *2670*

Sex Hormone Binding Globulin
Serum No Effect Physiological In 8 epileptic patients receiving phenobarbital with carbamazepine or phenytoin. *2670*
Serum Increase Physiological In 37 male epileptics mean of 49.0 nmol/L versus 23.3 nmol/L in controls receiving chronic therapy. *0803* 83 nmol/L versus 54 nmol/L in women and 32 nmol/L versus 22 nmol/L in men on long-term treatment. *0257*

Testosterone *Serum No Effect Physiological* In 33 male epileptics taking at least one drug for long time. *3020* Insignificant increase in 37 men versus control population on chronic therapy. *0803* No significant difference in men on long-term therapy. *0257*

Testosterone, Free *Serum Decrease Physiological* (free) 0.35 nmol/L versus 0.55 nmol/L in 37 men on chronic therapy. *0803*

Thyroxine (T₄) *Serum Decrease Physiological* 79 nmol/L versus 99 nmol/L in women and 82 nmol/L versus 100 nmol/L in men on long-term treatment. *0257*

Thyroxine (T₄) Binding Globulin
Serum No Effect Physiological In 8 epileptic patients receiving phenobarbital with carbamazepine or phenytoin. *2670* No difference in men or women on long-term therapy. *0257*

Thyroxine (T₄), Free *Serum Decrease Physiological* 16 pmol/L versus 23 pmol/L in women and 16 pmol/L versus 19 pmol/L in men on long-term treatment. *0257*

Transcortin *Serum Increase Physiological* 470 nmol/L versus 320 nmol/L in women and 420 nmol/L versus 320 nmol/L in men on long-term therapy. *0257* In 8 epileptic patients receiving phenobarbitol with carbmazepine or phenytoin. *2670*

Tri-iodothyronine (T₃) *Serum Decrease Physiological* 1.6 nmol/L versus 1.9 nmol/L in women and 1.7 nmol/L versus 2.0 nmol/L in men on long-term treatment. *0257*

Tryptophan *CSF No Effect Physiological* Insignificant reduction to 325 ng/mL from 379 ng/mL in untreated epileptics. *3926*

Uroporphyrin *Urine Increase Physiological* Drug related effect reported in one child. *2427*

Vitamin A *Serum No Effect Physiological* In 146 epileptics with long-term treatment. *2007*

Vitamin B₁ *Serum No Effect Physiological* In 146 epileptics with long-term treatment. *2007*

Vitamin B₂ *Serum No Effect Physiological* In 146 epileptics with long-term treatment. *2007*

Vitamin B₁₂ *Serum No Effect Physiological* In 146 epileptics with long-term treatment. *2007*
Serum Decrease Physiological Significantly lower than normal controls. *2427*
CSF Decrease Physiological Significantly lower with long term therapy. *1192*

Vitamin D Binding Globulin *Serum No Effect Physiological* No significant difference in men or women on long-term treatment. *0257*

Vitamin E (Tocopherol) *Serum No Effect Physiological* In 146 epileptics with long-term treatment. *2007*

Antifungal Agents

Alanine Aminotransferase *Serum Increase Physiological* Hepatotoxicity may occur. *1237*

Alkaline Phosphatase *Serum Increase Physiological* Hepatotoxicity may occur. *1237*

Aspartate Aminotransferase
Serum Increase Physiological Hepatotoxicity may occur. *1237*

Bile *Urine Increase Physiological* Hepatotoxicity may occur. *1237*

Bilirubin *Serum Increase Physiological* Hepatotoxic effect. *1237*

BSP Retention *Serum Increase Physiological* Hepatotoxic effect may impede clearance. *1237*

Antihistamines

Glucose *Serum Decrease Physiological* May occur in susceptible individuals especially children. *0851*

¹³¹I Uptake *Serum Decrease Physiological* Uncommon reported effect. *2220*

Protein Bound Iodine (PBI) *Serum No Effect Physiological* Probably exert no clinically significant effect. *0830*

Prothrombin Time *Plasma Decrease Physiological* Accelerate metabolism of anticoagulants. *2313*

Antihypertensive Agents

17-Hydroxycorticosteroids *Urine Increase Analytical* Some may be measured as analytes. *0279*

17-Ketosteroids *Urine Increase Analytical* Some may be measured as analytes. *0279*

Antilymphocytic Agents

Platelets *Blood Decrease Physiological* Observed less commonly with other immunosuppressants. *3433*

Antimalarials

Bilirubin *Serum Increase Physiological* May cause hemolytic anemia. *1902*

Bilirubin, Direct *Serum Increase Physiological* May cause hemolytic anemia. *1902*

Erythrocytes *Blood Decrease Physiological* May cause hemolytic anemia. *1902*

Heinz Body Formation *Blood Positive Physiological* Occurs prior to overt hemolysis. *1902*

Hematocrit *Blood Decrease Physiological* May cause hemolytic anemia. *1902*

Hemoglobin *Plasma Increase Physiological* Occurs with marked hemolysis. *1902*
Blood Decrease Physiological May cause hemolytic anemia. *1902*
Urine Increase Physiological May occur with marked hemolysis. *1902*

Methemoglobin *Blood Increase Physiological* May cause hemolytic anemia. *1902*

Reduced Glutathione
Red Blood Cells Decrease Physiological Occurs prior to overt hemolysis. *1902*

Reticulocytes *Blood Increase Physiological* Occurs during recovery from hemolysis. *1902*

Antimony Compounds

Alanine Aminotransferase *Serum Increase Physiological* Hepatotoxic effect. *2220*

Alkaline Phosphatase *Serum Increase Physiological* . Hepatotoxic effect. *2220*

Aspartate Aminotransferase
Serum Increase Physiological Hepatotoxic effect. *2220*

Bile *Urine Increase Physiological* Hepatotoxicity. *2313*

Bilirubin *Serum Increase Physiological* Hepatotoxic effect. *2220*

BSP Retention *Serum Increase Physiological* Hepatotoxic effect. *2220*

Casts *Urine Increase Physiological* Nephrotoxic effect. *2313*

Cephalin Flocculation *Serum Increase Physiological* Hepatotoxic effect. *2220*

Cholesterol *Serum Decrease Physiological* Hepatotoxic effect. *2220*

Erythrocytes *Blood Decrease Physiological* pancytopenia/may cause hemolytic anemia. *2313*

Glucose *Serum Decrease Physiological* Hepatotoxic effect. *2220*

Guanase *Serum Increase Physiological* Hepatotoxic effect. *2220*

Haptoglobin *Serum Decrease Physiological* May cause hemolytic anemia. *0279*

Hemoglobin *Blood Decrease Physiological* May cause hemolytic anemia. *0279*

Isocitrate Dehydrogenase *Serum Increase Physiological* Hepatotoxic effect. *2220*

Platelets *Blood Decrease Physiological* Thrombocytopenia. *2313*

Protein *Urine Increase Physiological* Nephrotoxic effect. *1022*

Thymol Turbidity *Serum Increase Physiological* Hepatotoxic effect. *2220*

Urea Nitrogen *Serum Increase Physiological* Nephrotoxic effect (common at therapeutic doses). *1022*

Antineoplastic Agents

Erythrocytes *Blood Decrease Physiological* May cause aplastic anemia. *2313*

Hydroxyproline *Urine Decrease Physiological* Catabolic action of cytostatics. *3644*

Leukocytes *Blood Decrease Physiological* Often therapeutic intent. *2313*

Platelets *Blood Decrease Physiological* May cause aplastic anemia. *2313*

Uric Acid *Serum Increase Physiological* Destruction of nucleoprotein. *0127*

Antipyretics

Bilirubin *Serum Increase Physiological* Occurs with hemolytic anemia. *1902*

Bilirubin, Direct *Serum Increase Physiological* Occurs with hemolytic anemia. *1902*

Erythrocytes *Blood Decrease Physiological* May cause hemolytic anemia. *1902*

Heinz Body Formation *Blood Positive Physiological* Occurs prior to overt hemolysis. *1902*

Hematocrit *Blood Decrease Physiological* May cause hemolytic anemia. *1902*

Hemoglobin *Plasma Increase Physiological* May occur with marked hemolysis. *1902*
Blood Decrease Physiological May cause hemolytic anemia. *1902*
Urine Increase Physiological Occurs with marked hemolysis. *1902*

Porphyrins *Urine Increase Physiological* Reported effect. *1237*

Reduced Glutathione
Red Blood Cells Decrease Physiological Sharp fall before overt hemolysis. *1902*

Reticulocytes *Blood Increase Physiological* Occurs during recovery from hemolysis. *1902*

Antipyrine

Acetoacetate *Urine Increase Analytical* Interferes with ferric chloride test. *2425*

Barbiturate *Serum Increase Analytical* Extraction properties of free form similar. *3608*

Bilirubin *Serum Increase Physiological* Hemolysis in G-6-PD deficiency. *0248*

Casts *Urine Increase Physiological* May cause nephrotoxicity. *0279*

Cells *Urine Increase Physiological* Renal action of drug. *3824*

Color *Urine Increase Analytical* Red brown (green in reflected light). *2313*

Erythrocytes *Blood Decrease Physiological* Hemolytic anemia in G-6-PD deficient persons. *0788*

Ferric Chloride Test *Urine Positive Analytical* May produce red color. *0443*

Glucaric Acid *Urine Increase Physiological* Reported induction of hepatic enzymes. *1694*

Glucose *Serum Increase Physiological* Severe hyperglycemia unresponsive to insulin. *2427*
Urine Increase Physiological Renal action of drug. *3824*

Heinz Body Formation *Blood Positive Physiological* Occurs initially prior to hemolysis. *1902*

Hematocrit *Blood Decrease Physiological* Hemolytic anemia in G-6-PD deficient persons. *0788*

Hemoglobin *Blood Decrease Physiological* Hemolytic anemia in G-6-PD deficient persons. *0788*
Urine Increase Physiological Occurs with marked hemolysis. *1902*

Leukocytes *Blood Decrease Physiological* Leukopenia/agranulocytosis. *2427*

Methemalbumin *Plasma Increase Physiological* Occurs with intravascular hemolysis. *0788*

Methemoglobin *Blood Increase Physiological* Increase seldom occurs. *1343*

Neutrophils *Blood Decrease Physiological* Myelotoxic action of drugs. *2427*

Phenylketones *Urine Positive Analytical* Red fading with FeCl₃, pink to red with Phenistix®. *0775*

Platelets *Blood Decrease Physiological* Associated with hemolytic anemia. *2427*

Protein *Urine Increase Physiological* Renal action of drug. *3824*

Prothrombin Time *Plasma Increase Physiological* Patients on coumarins. *2427*
Plasma Decrease Physiological Observed in man and experimental animals. *2427*

Reticulocytes *Blood Increase Physiological* Occurs with hemolytic anemia (recovery). *0788*

Sugar *Urine Increase Analytical* Acts as reducing agent. *2425*

Urea Nitrogen *Serum Increase Physiological* May cause nephrotoxicity. *0279*

Antipyrine *(continued)*

Urobilinogen *Urine Increase Analytical* Produces similar color with Ehrlichs. *3879*

Urine Increase Physiological Occurs with hemolytic anemia. *0788*

Feces Increase Physiological Occurs with hemolytic anemia. *0788*

Antiseptics

Protein Bound Iodine (PBI) *Serum Increase Analytical* If contain iodine and used to clean skin. *0830*

Antithrombin III

Fibrinogen *Plasma No Effect Analytical* On clottable protein assay procedure. *3463*

Plasma Decrease Analytical Maximum effect on turbidimetric procedures. *3463*

ANTU

Sugar *Urine Increase Analytical* May reduce Fehling's and Benedict's solutions. *0279*

Apalcillin

Antithrombin III *Plasma Decrease Physiological* In 21 volunteers with doses up to 225 mg/kg. *1262*

Fibrinogen *Plasma No Effect Physiological* In 21 volunteers with doses up to 225 mg/kg. *1262*

Partial Thromboplastin Time

Plasma No Effect Physiological In 21 volunteers with doses up to 225 mg/kg. *1262*

Prothrombin Time *Plasma No Effect Physiological* In 21 volunteers with doses up to 225 mg/kg. *1262*

Thrombin Time *Plasma No Effect Physiological* In 21 volunteers with doses up to 225 mg/kg. *1262*

Apiol

Alanine Aminotransferase *Serum Increase Physiological* May cause severe liver damage. *0279*

Aspartate Aminotransferase

Serum Increase Physiological May cause severe liver damage. *0279*

Casts *Urine Increase Physiological* May cause nephrotoxicity. *0279*

Urea Nitrogen *Serum Increase Physiological* May cause nephrotoxicity. *0279*

Apomorphine

Growth Hormone *Plasma Increase Physiological* Effect greater with 5 μg/kg in 8 depressed individuals than with 1.3 μg/kg of clonidine. *0728* In control healthy subjects i.v. injection increased mean concentration from 1.8 to 28.3 ng/mL, but weak effect in schizophrenic patients only. *3178* 7 fold increase at 30-45 minutes in 6 children after 12 mg/kg subcutaneously. *2330*

Homovanillic Acid *Plasma No Effect Physiological* No effect when given i.v. to either healthy controls or schizophrenics. *3178*

Morphine *Urine No Effect Analytical* Insignificant cross reactivity with RIA procedures. *2514*

pCO$_2$ *Blood Increase Physiological* May depress respiration. *1343*

Prolactin *Plasma Decrease Physiological* In control subjects reduced by 57% when given i.v. *3178*

Thyroid Stimulating Hormone (TSH)

Serum No Effect Physiological In euthyroids but can lower basal and TRH-stimulated values. *3798*

Serum Decrease Physiological In hypothyroidism but can lower basal and TRH-stimulated values. *3798*

Aprindine

Alanine Aminotransferase *Serum Increase Physiological* Hepatitis is manifestation of chronic toxicity. *3215* Hepatitis in 2 patients within 3 weeks of start of therapy resolved with withdrawal of drug. *1570*

Alkaline Phosphatase *Serum Increase Physiological* Hepatitis is manifestation of chronic toxicity. *3215* Hepatitis in 2 patients within 3 weeks of start of therapy resolved with withdrawal of drug. *1570*

Aspartate Aminotransferase

Serum Increase Physiological Hepatitis is manifestation of chronic toxicity. *3215* Hepatitis in 2 patients within 3 weeks of start of therapy resolved with withdrawal of drug. *1570*

Bilirubin *Serum Increase Physiological* Hepatitis is manifestation of chronic toxicity. *3215* Hepatitis in 2 patients within 3 weeks of start of therapy resolved with withdrawal of drug. *1570*

Neutrophils *Blood Decrease Physiological* Agranulocytosis may occur between 4th and 16th week of treatment: may be quite severe: usually reversible. Occurs with frequency of 0.1 to 1.0%. *3215*

Apronalide

Aminolevulinic Acid *Urine Increase Physiological* May precipitate acute porphyria. *1322*

Clot Retraction *Blood Decrease Physiological* Immunological effect. *1436*

Coproporphyrin *Urine Increase Physiological* May precipitate acute porphyria. *1322*

Feces Increase Physiological May precipitate acute porphyria. *1322*

Platelets *Blood Decrease Physiological* Thrombocytopenia (immunologically — induced). *2313* Immunological mechanism. *2427*

Porphobilinogen *Urine Increase Analytical* Produces red color with Ehrlich's reagent. *1563*

Urine Increase Physiological May precipitate acute porphyria. *1322*

Porphyrins *Urine Increase Physiological* May precipitate attack of acute porphyria. *2427*

Protoporphyrin *Feces Increase Physiological* May precipitate acute porphyria. *1322*

Urobilinogen *Urine Increase Analytical* Produces red color with Ehrlich's reagent. *1563*

Aprotinin

Amylase *Serum Increase Physiological* One case allergic pancreatitis reported. *2427*

Glucose *Serum No Effect Analytical* At concentration of 150 kU/L had no effect on GOD/POD-PAP method. *3393*

Triglycerides *Serum No Effect Analytical* At concentration of 150 kU/L had no effect on GPO-PAP method. *3393*

Uric Acid *Serum No Effect Analytical* At concentration of 150 kU/L had no effect on uricase-PAP method. *3393*

Arabinose

Glucuronic Acid *Urine No Effect Analytical* Minimal effect on carbazole procedure. On m-hydroxydiphenyl procedure. *0389*

Urine Increase Analytical Large effect on orcinol procedure. *0389*

Sugar *Urine Increase Analytical* False positive with Benedict's. *1563*

Arginine

Aldosterone *Plasma No Effect Physiological* No significant changes observed in 6 normal and 4 diabetic subjects after i.v. infusion. *2328*

Ammonia *Plasma Decrease Analytical* With 2500 nmoles produces 5% inhibition of indophenol reaction. *1290*

Plasma Decrease Physiological Capable of reacting with ammonia. *1943*

Creatine *Serum Increase Physiological* Incorporated into urea cycle. *0071*

Creatinine *Serum Increase Analytical* If method of Voges-Proskauer used. *1563*
Serum Increase Physiological Incorporated into urea cycle. *0071*

Glucagon *Plasma Increase Physiological* Large effect (by 40 pg/mL) after prednisolone. *2296* Moderate increase after infusion of 30 g. *2676*

Glucose *Serum Increase Physiological* Slight, more in diabetics and if prednisolone. *3408* Positive correlation in diabetics given i.v. infusion of 0.5 g/kg. *2328*

Growth Hormone *Plasma Increase Physiological* Effect also shown by protein meals. *2898*

Guanidinosuccinic Acid *Serum No Effect Analytical* No effect on method of Kamoun, Pleau, Man. *1856*

Histidine *Plasma Increase Analytical* May affect fluorometric method of Ambrose. *0081*

Insulin *Plasma Increase Physiological* Slight effect and metabolism by tissues. *1071* Large effect (by 60 uU/mL) after prednisolone. *2296*

Osmolality *Serum Increase Physiological* Slight but not significant increase (about 8 mosmol/L) after i.v. infusion in diabetics. *2328*

pH *Blood No Effect Physiological* No change following i.v. infusion in diabetics. *2328*
Blood Decrease Physiological Chloride salt tends to cause acidosis. *0071*

Phosphate *Serum Decrease Physiological* Fall from 3.3 to 2.2 mg/dL in normals and 3.6 to 3.2 mg/dL in diabetics after i.v. infusion of 0.5 g/kg in 30 minutes. *2328*

Potassium *Serum Increase Physiological* Up to 5.6-6.5 mmol/L in diabetics (less in normals) after 0.5 g/kg i.v. infusion in 30 minutes (maximum increase 1.4 mmol/L). *2328* Causes shift of potassium to extracellular compartment, concentration correlated with aminoacid concentration. *2836*

Prolactin *Plasma Increase Physiological* 2 fold increase at 30 minutes with infusion. *1435*

Proline *Plasma Increase Analytical* Slight effect on method of Goodwin. *1345*

Protein *Serum Decrease Physiological* Slight transient reduction (by up to 0.7 g/dL) after i.v. infusion in diabetics. *2328*

Renin Activity *Plasma No Effect Physiological* No significant changes observed in 6 normal and 4 diabetic subjects after i.v. infusion. *2328*

Sodium *Serum Decrease Physiological* Slight transient reduction (by 4 mmol/L) after i.v. infusion in diabetics. *2328*

Thyroid Stimulating Hormone (TSH)
Serum No Effect Physiological No effect observed. *2700*

Tyrosine *Plasma No Effect Analytical* On fluorometric procedure of Ambrose. *0080*

Urea Nitrogen *Serum No Effect Analytical* No effect on Berthelot reaction. *1862*
Serum Increase Physiological Incorporated into urea cycle. *0071*

Arsenates

Chromosomes *Test Conditions Abnormal Physiological* Clastogenic in human cells. *3282*

Arsenicals

Alanine Aminotransferase *Serum Increase Physiological* Hepatotoxic effect (cholestasis/cholangiolitis). *2220*

Alkaline Phosphatase *Serum Increase Physiological* Hepatotoxic effect (cholestasis/cholangiolitis). *2220*
Serum Decrease Analytical Arsenates are inhibitors of enzyme in lab procedures. *3588*

Aspartate Aminotransferase
Serum Increase Physiological Hepatotoxic effect (cholestasis/cholangiolitis). *2220*

Bile *Urine Increase Physiological* Hepatotoxicity (cholestasis/cholangiolitis). *2313*

Bilirubin *Serum Increase Physiological* Hepatotoxic effect (cholestasis/cholangiolitis). *2220*

BSP Retention *Serum Increase Physiological* Hepatotoxic effect (cholestasis/cholangiolitis). *2220*

Casts *Urine Increase Physiological* Nephrotoxic effect. *2313*

Cephalin Flocculation *Serum Increase Physiological* Hepatotoxic effect (cholestasis/cholangiolitis). *2220*

Cholesterol *Serum Increase Physiological* Hepatotoxic effect (may be very high). *2220*

Creatinine *Serum Increase Physiological* Nephrotoxicity (common with therapeutic doses). *1237*

Eosinophils *Blood Increase Physiological* Up to 50% observed in one case, others 10-20%. *1478*

Erythrocytes *Blood Decrease Physiological* Megaloblastic anemia/pancytopenia. *0788*
Urine Increase Physiological Nephrotoxicity with tubular necrosis. *1343*

Folate *Serum Decrease Physiological* Megaloblastic anemia after Fowler's solution. *2427*

Glucose *Serum Decrease Physiological* Hepatotoxic effect. *2220*

Guanase *Serum Increase Physiological* Hepatotoxic effect. *2220*

Heinz Body Formation *Blood Positive Physiological* Some may cause hemolytic anemia. *0333*

Hematocrit *Blood Decrease Physiological* Pancytopenia. *0788*

Hemoglobin *Blood Decrease Physiological* Megaloblastic anemia/pancytopenia. *0788*
Urine Increase Physiological May be marked hematuria. *1302*

Isocitrate Dehydrogenase *Serum Increase Physiological* Hepatotoxic effect. *3835*

Leukocytes *Blood Decrease Physiological* Pancytopenia. *0788*

Nonprotein Nitrogen *Serum Increase Physiological* Nephrotoxicity (common with therapeutic doses). *1237*

Occult Blood *Feces Positive Physiological* Observed in several cases with poisoning. *1478*

Peripheral Smear *Blood Abnormal Physiological* May produce stippling of red cells. *0279*

Platelets *Blood Decrease Physiological* Pancytopenia. *0788*

Protein *Urine Increase Physiological* Nephrotoxic effect. *1022*

Reticulocytes *Blood Increase Physiological* Values observed ranging up to 18%. *1478*

Thymol Turbidity *Serum Increase Physiological* Hepatotoxic effect. *2220*

Urea Nitrogen *Serum Increase Physiological* Nephrotoxic effect (common with therapeutic doses). *1343*

Xylose Excretion *Urine Decrease Physiological* May produce gastrointestinal irritation, impaired absorption. *0612*

Arsenobenzenes

Platelets *Blood Decrease Physiological* Thrombocytopenia. *2313*

Ascorbate-2-Sulfate

Ascorbic Acid *Plasma No Effect Analytical* Little conjugate in blood so no effect. *0209*
Urine Increase Analytical Affects DNPH procedures. *0209*

Ascorbic Acid

N-Acetylglucosaminidase *Urine No Effect Analytical* No effect at 100 mmol/L on 2 colorimetric analytical methods. *1354*

Alanine *Urine Increase Physiological* In 10 healthy females given 10 g/d. *3624*

Alanine Aminotransferase *Serum No Effect Analytical* No effect at therapeutic concentration on Reflotron method. *1984*

Ascorbic Acid (continued)

Albumin *Serum No Effect Analytical* At concentration of 500 mg/L had no effect on BCG method. *3393*

Ammoniacal Silver Nitrate
Test Conditions Positive Analytical Positive spot test with Tollen's reagent. *3765*

Amylase *Serum No Effect Analytical* Up to 1 mmol/L on method of Rauscher et al. *2926* At concentration of 1,000 mg/L had no effect on maltotetrose method. *3393*

Ascorbic Acid *Plasma Increase Physiological* Significant effect in 50 volunteers given 2 g/d for 2 mo. *1039*
Blood Increase Physiological Significant effect in 50 volunteers given 2 g/d for 2 mo. *1039*
Urine Increase Physiological Significant effect in 50 volunteers given 2 g/d for 2 mo. *1039*
White Blood Cells Increase Physiological Significant effect in 50 volunteers given 2 g/d for 2 mo. *1039*

Aspartate Aminotransferase *Serum No Effect Analytical* No effect at therapeutic concentration on Reflotron method. *1984*
Serum Increase Analytical At 1 mmol/L affects SMA 12/60 method. *3335*

Aspartic Acid *Urine Increase Physiological* In 10 healthy females given 10 g/d. *3624*

Bicarbonate *Serum No Effect Analytical* At concentration of 500 mg/L had no effect on method using phenolphthalein. *3393*

Bilirubin *Serum No Effect Analytical* On SMAC® method at therapeutic concentration. *3719* At concentration of 40 mg/L had no effect on Ektachem® method. *3393*
Serum Increase Analytical At therapeutic concentration may affect SMA 12/60 method. *3335*
Serum Decrease Analytical At concentrations above 200 mg/L (normal maximum serum concentration 34 mg/L) lowered concentration as measured by Jendrassik and Grof method. *3393*

Calcium *Serum No Effect Analytical* At concentration of 2,000 mg/L had no effect on cresolphthalein method. *3393*

Catecholamines *Plasma Increase Analytical* Concentrated solutions cause striking fluorescence. *0596*

Ceruloplasmin *Serum No Effect Physiological* No effect on protein with 605 mg/d for 3 weeks but up to 21% reduction of oxidase activity. *1752*

Chloride *Serum No Effect Analytical* At concentration of 40 mg/L had no effect on Ektachem® method. *3393* At concentration of 500 mg/L had no effect on mercurimetric method. *3393*

Cholesterol *Serum No Effect Analytical* No effect at therapeutic concentration on Reflotron method. *1984* At concentration of 400 mg/L had no effect on catalase-AIDH method. *3393* At concentration of 1,000 mg/L had no effect on method using catalase-Hantzsch reaction. *3393* At concentration of 500 mg/L had no effect on Liebermann-Burchard method. *3393*
Serum No Effect Physiological No influence on concentration of dietary intake in elderly. *1757*
Serum Increase Physiological When atherosclerotic, ?mobilization from arteries. *3420*
Serum Decrease Analytical Increase by 2.1% at 5 mg/dL on enzymatic procedure. *0057* At concentrations above 50 mg/L (maximum serum concentration 34 mg/L) lowered concentration as measured by CHOD-Iodide method. *3393* At concentrations above 50 mg/L (maximum serum concentration 34 mg/L) lowered concentration as measured by CHOD-PAP method. *3393*
Serum Decrease Physiological Tends to fall in people under 25 when 1 g/d given. *3420* 16% decrease on average when 1 g/d given to healthy approximately 29 year olds within 2 mo. 14% fall in 58 year olds but required 12 mo. Administration abolished normal rise observed in winter. *0917*

Cholesterol, High Density Lipoprotein
Serum Increase Physiological Significant correlation between drug intake and analyte concentration in elderly. *1757*
Serum Decrease Analytical Decreases of from 0.3 to 10% in 6 methods to determine compound with drug concentration of 2.0 mg/dL. *2500*

Copper *Serum No Effect Physiological* Not affected by variations in drug intake up to 605 mg/d for 3 weeks. *1752*

Corticosteroids *Urine Increase Analytical* Butanol extract/no hydrolysis Reddy/Porter-Silber reaction. *0022*

Creatinine *Serum No Effect Analytical* At 360 mg/dL on ion-exchange method of Mitchell. No effect on Lloyd's procedure. *2459* No effect on method of Polar and Metcoff. *1975* At 500 mg/L on reversed phase LC procedure of Zhiri et al. *3942* On SMAC® method at therapeutic concentration. *3719* At concentration of 105 mg/L had no effect on creatinine iminohydrolase method. *3393* At concentration of 400 mg/L had no effect on Jaffé-Fading-Fraction method. *3393* At concentration of 400 mg/L had no effect on Jaffé-Fuller's earth method. *3393* At concentration of 100 mg/L had no effect on Jaffé-Heinegard and Tiderstrom method. *3393* At concentration of 1,000 mg/L had no effect on kinetic Jaffé method on BKA-2. *3393*
Serum Increase Analytical Chromogenicity in color reaction (as reducing agent). *1022* 100 mg/L = 0.01 mg/dL method of Heinegard. *1547* Marked effect direct methods. *2459* At concentrations above 250 mg/L (maximum serum concentration 34 mg/L) raised concentration as measured by AutoAnalyzer Jaffé method. *3393* At concentrations above 25 mg/L (maximum serum concentration 34 mg/L) raised concentration as measured by kinetic Jaffe method. *3393*
Serum Decrease Analytical At concentrations above 1,000 mg/L (maximum serum concentration 34 mg/L) lowered concentration as measured by creatinine amidohydrolase method. *3393*
Urine Increase Analytical Acts as reducing agent. *0127*

Crystals *Urine Increase Physiological* Acidification may precipitate oxalates, urates, cystine. *2063*

Cystine *Urine Decrease Physiological* In 10 healthy females given 10 g/d. *3624*

Dapsone *Serum No Effect Analytical* At 5 mg/dL on colorimetric procedure of Higgins. *1590*

Ferritin *Serum Decrease Physiological* During 14 week study of varying intakes of drug probably attributable to phlebotomy only. *1751*

Fibrinogen *Plasma No Effect Analytical* At concentration of 500 mg/L had no effect on aca method. *3393*

Glucose *Serum No Effect Analytical* On p-HBAH procedure of Lever at 1 g/dL. *2140* No effect on Trinder glucose-oxidase method. *2771* Insignificant effect at 5 mg/dL on MBTH procedure of Neeley. *2569* No effect at therapeutic concentration on Reflotron method. *1984* At concentration of 1000 mg/L had no effect on hexokinase/G-6-PDH method. *3393* At concentration of 250 mg/L had no effect on hexokinase/G-6-PDH method on aca. *3393*
Serum Increase Analytical At 1 mmol/L affects SMA 12/60 method. *3335* Increases sensitivity of o-toluidine procedures 1 g/dL equivalent to 3.3 mmol/L with alkaline ferricyanide. *2140* Affects Neocuproin procedure. *2771* At concentrations above 60 mg/L (maximum serum concentration 34 mg/L) raised concentration as measured by glucose dehydrogenase method. *3393*
Serum Decrease Analytical Negative peaks GOD-PERID procedure. *2771* 17% decrease at 10 mg/dL glucose oxidase dianisidine procedure. *3037* Slight effect with coupled glucose-oxidase method. *1311* At very high concentrations on Glucomatic method. *2771* But reduction insignificant (each 1 mg/dL lowers concentration as measured by 0.65 mg/dL) with glucose oxidase method on SMAC®. *3719* At concentrations above 100 mg/L (maximum serum concentration 34 mg/L) lowered concentration as measured by Ektachem® method. *3393* At concentrations above 125 mg/L (maximum serum concentration 34 mg/L) lowered concentration as measured by GOD-Perid method. *3393* At concentrations above 150 mg/L (maximum serum concentration 34 mg/L) lowered concentration as measured by GOD/POD-PAP method. *3393* At concentrations above 25 mg/L (maximum serum concentration 34 mg/L) lowered concentration as measured by Seralyzer. *3393*
Serum Decrease Physiological Significant reduction in fasting concentration of diabetics treated for 15 d. *3136*
Urine No Effect Analytical With DNSA method at usual concentrations. *0076* No effect observed with TesTape®. *3505*
Urine Increase Analytical Effect on Clinitest® causes normal to be read as trace at low concentrations: less effect at high concentrations. *0786*
Urine Decrease Analytical Impaired color develop of chromogen in glucose oxidase method. *1022* May inhibit TesTape® and Clinistix®. *3505* At 0.4 g/L interfered with BM33071 in 20% urines containing 1.0 g glucose/L but not with urines containing 5.0 g glucose/L. No effect of ascorbic acid at 0.1 to 0.2 g/L. At 0.4 g/L interfered with Hema-Combistix® in 90% urines containing glucose at 5.0 g/L. No effect of ascorbic acid at 0.1 to 0.2 g/L. *3356* But only very slight effect with BM33071 procedure

from Boehringer-Mannheim compared with other dipsticks. *0787* At physiological urine amounts, marked reduction of results with especially Ecur-test, Diabur-test 5,000 and Rapignost basis screen but also TesTape®. *0786* At concentration of about 100 mg/dL made 0.1 g/dL react negatively with Chemstrip® 7, 1-lema-Combistix® and give trace reacftion with Chemstrip® UG. *3968* Significant effect with Redia-test, L-Combur-5-test, Labstix®, Rapignost, Meditest but less effect with BM33.071. *0303*
Urine Negative Analytical At concentration of 100 mg/L (normal concentration in urine up to 1290 mg/L) produced false negative result with Diabur-test. *3393*

γ-Glutamyltransferase (GGT) *Serum No Effect Analytical* No effect at therapeutic concentration on Reflotron method. *1984*

Glycine *Urine Decrease Physiological* In 10 healthy females given 10 g/d. *3624*

Hematocrit *Blood No Effect Physiological* No effect with varying degrees of supplementation over 14 week study period in young men. *1751*

Hemoglobin *Blood No Effect Physiological* No effect with varying degrees of supplementation over 14 week study period in young men. *1751*
Urine Decrease Analytical In large amounts inhibits guaiac test. *1022* At up to 140 mg/dL made Chemstrip® 7 react negatively to 250 mg/L hemoglobin, also with Hema-Combistix® and some reactions were negative at approximately same concentration with Chemstrip® 9. *3968*

Homogentisic Acid *Urine No Effect Analytical* On TLC method of Feldman and Bowman. *1096*
Urine Increase Analytical If method of Briggs used. *1096*

17-Hydroxycorticosteroids *Urine Increase Analytical* Interferes with method of Reddy. *1488*

Hydroxyproline *Urine Decrease Physiological* Slight effect in osteogenesis imperfecta. *3644*

Isoleucine *Urine Increase Physiological* In 10 healthy females given 10 g/d. *3624*

17-Ketosteroids *Urine Increase Analytical* Due to chemical structure affects Zimmermann procedure. *3217*

Lactate *Serum No Effect Analytical* At concentration of 600 mg/L had no effect on enzymatic method. *3393*

Lactate Dehydrogenase *Serum Decrease Analytical* At therapeutic concentration may depress SMA 12/60 value. *3335*

Leucine *Urine Increase Physiological* In 10 healthy females given 10 g/d. *3624*

Leukocytes *Urine Decrease Analytical* At concentrations above 2,000 mg/L (maximum concentration in urine up to 1290 mg/L) caused false negative result with Cytur-Test. *3393*

Lymphocytes *Blood No Effect Physiological* No effect with megadose supplementation. *1347*

Metanephrines, Total *Urine No Effect Analytical* At 5 g/L on modified Pisano procedure. *1428*

3-Methylhistidine *Urine Increase Physiological* In 10 healthy females given 10 g/d. *3624*

Neutrophils *Blood No Effect Physiological* No effect with megadose supplementation. *1347*

Occult Blood *Feces Decrease Analytical* False reduction at physiological amounts (if taking added vitamin C) on Hemoccult® and other tests with pseudoperoxidase principle eg guaiac, benzidine, or other diamino compound as indicator. *1763*
Feces Negative Analytical Interferes with analytic methods. *2313* At concentrations above 30 mg/L produced false negative result as measured by Hemoccult®. *3393*

Oxalate *Urine No Effect Analytical* When boric acid used as diluent for low chromatographic measurement procedures. *3007* Minimal effect on gas-chromatographic procedure with alkalinization of urine. *2344*
Urine No Effect Physiological No effect in 50 volunteers given 2 g/d for 2 mo. *1039*
Urine Increase Analytical With Sigma procedure with diluted or undiluted urine, but pretreatment with ferric chloride causes loss of oxalate. *1305*
Urine Increase Physiological Normal metabolite excreted. *2063*
Urine Decrease Analytical Reduced concentration observed with methods involving oxalate decarboxylase. *3030*

pH *Blood No Effect Physiological* Even when 8 g/m²/d ingested. *0233*

Urine Decrease Physiological Acidifies urine when ingested in large amounts. *2063* Some effect with ingestion of 3-6 g/d. *0233*

Phenylalanine *Plasma Decrease Physiological* Reduces elevated level of premature infants. *1343*
Urine Increase Physiological In 10 healthy females given 10 g/d. *3624*

Phosphate *Serum No Effect Analytical* At concentration of 150 mg/L had no effect as measured by aca method. *3393* At concentration of 2,000 mg/L had no effect as measured by phosphomolybdate method. *3393*
Urine No Effect Analytical At 20 mg/dL 1% increase with method of Jung/Parekh. *1835*

Phosphoethanolamine *Urine Increase Physiological* In 10 healthy females given 10 g/d. *3624*

Phosphoserine *Urine Decrease Physiological* In 10 healthy females given 10 g/d. *3624*

Porphobilinogen *Urine Decrease Analytical* Inhibition of color develop if no prior separation. *1116*

Potassium *Serum No Effect Analytical* At concentration of 60 mg/L had no effect as measured by flame-photometric method. *3393* At concentration of 40,000 mg/L had no effect on ISE measurement with predilution. *3393* At concentration of 800 mg/L had no effect on ISE measurement without predilution. *3393*

Protein *Serum No Effect Analytical* At concentration of 2,000 mg/L had no effect as measured by biuret method with blank correction. *3393*
CSF No Effect Analytical No effect on Folin-Ciocalteu procedure. *3958*
Test Conditions Increase Analytical Reacts with Folin-Ciocalteu of Lowry procedure. *0702*

Protein Electrophoresis *Serum No Effect Analytical* At concentration of 800 mg/L had no effect on automated Olympus-Hite method but with slight displacement of fractions. *3393*

Prothrombin Time *Plasma No Effect Physiological* No effect on Thrombotest with dose of 1 g/d. *1691* No effect seen in most patients. *1487*
Plasma Decrease Physiological May shorten action of anticoagulants. *1679* Case reported when compound given to patient whose time was increased by anticoagulant. *0233*

Pyridoxine *Urine No Effect Physiological* No effect with varying degrees of supplementation over 14 week study period in young men. *1751*

Retinol *Plasma No Effect Physiological* No effect with varying degrees of supplementation over 14 week study period in young men. *1751*

Salicylate *Serum No Effect Physiological* No effect on serum concentration with several days treatment. *1489*

Sarcosine *Urine Decrease Physiological* In 10 healthy females given 10 g/d. *3624*

Serine *Urine Decrease Physiological* In 10 healthy females given 10 g/d. *3624*

Sodium *Serum No Effect Analytical* At concentration of 60 mg/L had no effect as measured by flame-photometric method. *3393* At concentration of 40,000 mg/L had no effect on ISE measurement with predilution. *3393* At concentration of 800 mg/L had no effect on ISE measurement without predilution. *3393*

Sugar *Urine Increase Analytical* False positive with Benedict's and Clinitest®. *1022* Positive with Benedict's, Clinitest® at 50 mg/dL. *0583*

Threonine *Urine Increase Physiological* In 10 healthy females given 10 g/d. *3624*

α-Tocopherol *Serum No Effect Physiological* No effect with varying degrees of supplementation over 14 week study period in young men. *1751*

Triglycerides *Serum No Effect Analytical* No effect at therapeutic concentration on Reflotron method. *1984* At concentration of 500 mg/L had no effect as measured by lipase/esterase method. *3393*
Serum Decrease Analytical At concentrations above 30 mg/L (maximum serum concentration 34 mg/L) lowered concentration as measured by GPO-PAP method. *3393*
Serum Decrease Physiological Effect observed in atherosclerotic patients. *3377*

Ascorbic Acid (continued)

TTC Test *Urine Positive Analytical* Oral ingestion causes reduction of TTC = positive test. *3910*

Tyrosine *Plasma Decrease Physiological* Reduces elevated level in premature infants. *1343*

Urea Nitrogen *Serum No Effect Analytical* No effect at therapeutic concentration on Reflotron method. *1984* At concentration of 2,000 mg/L had no effect on diacetylmonoxime method. *3393*

Serum Decrease Analytical At concentrations above 300 mg/L (maximum serum concentration 34 mg/L) lowered concentration as measured by Ektachem® method. *3393* At concentrations above 20 mg/L (maximum serum concentration 34 mg/L) lowered concentration as measured by Seralyzer. *3393*

Uric Acid *Serum No Effect Analytical* At 200 mg/L on uricase procedure of Kabasak. *1840* No effect on Tripyridyl-s-triazine method of Morin. *2486* At 500 mg/L on reversed phase LC procedure of Zhiri et al. *3942* No effect at therapeutic concentration on Reflotron method. *1984* At concentration of 1,000 mg/L had no effect on Kageyama-Hantzsch method. *3393* At concentration of 5,000 mg/L had no effect on catalase-AlDH method. *3393* At concentration of 100 mg/L had no effect on uricase method on aca. *3393* At concentration of 250 mg/L had no effect on uricase-PAP method. *3393*

Serum Increase Analytical 10 mg/dL equivalent to 9.4 mg/dL copper chelate procedure. *2230* At 5 mg/dL increases by 3.15 mg/dL method of Klein. *1949* Measured as reducing substance. *0652* 10 mg/dL equivalent to 0.6 mg/dL with Nishi procedure. *2230* Each 1 mg/dL of drug increases concentration as measured by phosphotungstate method by 0.05 mg/dL. *3719* At concentrations above 100 mg/L (maximum serum concentration 34 mg/L) raised concentration when measured by phosphotungstate reduction. *3393*

Serum Decrease Analytical At concentrations above 10 mg/L (maximum serum concentration 34 mg/L) lowered concentration as measured by Seralyzer. *3393*

Serum Decrease Physiological From 1.2 to 3.1 mg/dL in 3 subjects ingesting 8 g/d due to uricosuria. *3446*

Urine Increase Analytical Measured as reducing substance. *1563*

Urine Increase Physiological Substantial effect with large amounts of compound: effect inhibited by pyrazinamide. *3446*

Urobilinogen *Urine Decrease Physiological* Lowered pH reduces excretion. *2452*

Valine *Urine Decrease Physiological* In 10 healthy females given 10 g/d. *3624*

Vitamin A *Serum No Effect Physiological* No effect with varying degrees of supplementation over 14 week study period in young men. *1751*

Vitamin B₁₂ *Serum No Effect Physiological* No effect with varying degrees of supplementation over 14 week study period in young men. *1751*

Serum Decrease Physiological When ingested with food compound causes destruction of substantial amounts of B₁₂. *0233*

Asparaginase

Alanine Aminotransferase *Serum Increase Physiological* Hepatotoxicity. *1519*

Albumin *Serum Decrease Physiological* Hepatotoxicity (observed in 80% patients). *2657* In 71% children and 82% adults reported in various studies. *2406* Depressed in up to 70% of patients treated in different studies: effects usually mild. *0558*

Alkaline Phosphatase *Serum Increase Physiological* May cause hepatotoxicity (frequent). *1519* In 31% children and 47% adults reported in various studies. *2406* Increased in 30-35% of patients treated in different studies: effects usually mild. *0558* Hemorrhagic pancreatitis in fewer than 0.5% treated patients. *1398*

Ammonia *Plasma Increase Physiological* May be marked, associated with abnormal liver function. *1597* Hemorrhagic pancreatitis in fewer than 0.5% treated patients. *1398*

Amylase *Serum Increase Physiological* May cause pancreatic toxicity. *1519* Reports vary from incidence of 2.5 to 16% cases of acute pancreatitis: usually mild. *0558* Hemorrhagic pancreatitis in fewer than 0.5% treated patients. *1398*

Antithrombin III *Plasma Decrease Physiological* Mean decrease of 68% (immunological) and 74% (functional assay) in different studies. *0558*

Aspartate Aminotransferase
Serum Increase Physiological Hepatotoxicity. *1519* In 46% children and 63% adults reported in various studies. *2406* Increased in 35-45% of patients treated in different studies: effects usually mild. *0558* Hemorrhagic pancreatitis in fewer than 0.5% treated patients. *1398*

Bilirubin *Serum Increase Physiological* Up to 4 mg/dL (dose related effect). *2657* In 29% children and 51% adults reported in various studies. *2406* Increased in 30-60% of patients treated in different studies: effects usually mild. *0558* Hemorrhagic pancreatitis in fewer than 0.5% treated patients. *1398*

BSP Retention *Serum Increase Physiological* Hepatotoxicity (usually mild). *1519* In 57% children and 84% adults reported in various studies. *2406*

Calcium *Serum Decrease Physiological* Observed in 60% (?due to hypoalbuminemia). *2657*
Urine Increase Physiological Atypical response but observed in some. *2657*

Cephalin Flocculation *Serum Increase Physiological* Hepatotoxicity. *1519*

Ceruloplasmin *Serum Decrease Physiological* Diminished hepatic synthesis. *2657* Possibly due to inhibition of protein synthesis reported in various studies. *2406*

Cholesterol *Serum Increase Physiological* Unusual response in some patients. *2657*
Serum Decrease Physiological Hepatotoxicity (effect marked). *1519* In 82-85% of patients reported in various studies. *2406*

Cholesterol Esters *Serum Decrease Physiological* Marked fall maximal at 4 d after single injection. *3957*

Chylomicrons *Serum Increase Physiological* Observed in unusual hyperlipidemic response. *2657*

Coombs' Test, Indirect *Serum Positive Physiological* In 4 children who demonstrated hemolytic anemia (direct Coombs' test negative). *0558*

Euglobulin Clot Lysis Time *Blood Decrease Physiological* Occasionally observed with toxicity. *2864*

Factor V *Plasma Decrease Physiological* Diminished hepatic synthesis. *0578*

Factor VII *Plasma Decrease Physiological* Diminished hepatic synthesis. *0578*

Factor VIII *Plasma Decrease Physiological* Diminished hepatic synthesis. *0578*

Factor IX *Plasma Decrease Physiological* In 75 to 100% in different studies. *0558*

Factor XI *Plasma Decrease Physiological* In 75 to 100% in different studies. *0558*

Fatty Acids, Free (FFA) *Serum Decrease Physiological* Common hepatotoxic response. *2864*

Fibrinogen *Plasma Decrease Physiological* Marked effect in almost all patients. *2657* In 32 of 33 patients reported in various studies. *2406* In 50 to 100% in different studies. *0558*

α₂-Globulin *Serum Decrease Physiological* Decreased to 70% of control at 2 weeks. *2657*

β-Globulin *Serum Decrease Physiological* Decreased to 70% of control at 2 weeks. *2657*

β₁-α-Globulin *Serum Decrease Physiological* Diminished hepatic synthesis. *2657*

γ-Globulin *Serum Increase Physiological* Increased continuously to 170% of mean at 4 weeks. *2657*

Glucose *Serum Increase Physiological* May be hyperosmotic nonketotic hyperglycemia. *0578* Observed in 9.7% children, some within 1 week of start of treatment, observed most commonly in older children. *2883* Reported in 9.7% children, although hypoglycemia reported occasionally. *0558*

β₂-Glycoprotein *Serum Decrease Physiological* Diminished hepatic synthesis. *2657*

Granulocytes *Blood Decrease Physiological* Slight reduction (not dose dependent). *2657*

Haptoglobin *Serum Decrease Physiological* Possibly due to inhibition of protein synthesis reported in various studies. *2406*

Hematocrit *Blood Decrease Physiological* May cause anemia. *1519*

Hemoglobin *Blood Decrease Physiological* May cause anemia. *1519*

Immunoglobulin IgA *Serum Increase Physiological* Increased hepatic synthesis. *2657*

Immunoglobulin IgG *Serum Increase Physiological* Increased hepatic synthesis. *2657*

Immunoglobulin IgM *Serum Increase Physiological* Increased hepatic synthesis. *2657*

Insulin *Plasma Decrease Physiological* ?due to decreased production with decreased protein synthesis. *0578*

Leukocytes *Blood Decrease Physiological* Mild effect in up to 25% patients. *2657*

Lipase *Serum Increase Physiological* Reports vary from incidence of 2.5 to 16% cases of acute pancreatitis: usually mild. *0558*

Lipids, Total *Serum Increase Physiological* Rare response after initial hypolipidemia. *2657*
Serum Decrease Physiological Parallel decrease in cholesterol. *2657*

Lipoprotein Lipase *Serum Decrease Physiological* Observed in unusual hyperlipidemic response. *2657*

Lipoproteins *Serum Decrease Physiological* Possible depressed synthesis. *1519*

β-Lipoproteins *Serum Decrease Physiological* Observed in unusual hyperlipidemic response. *2657*

Lipoproteins, Pre-β *Serum Increase Physiological* Observed in unusual hyperlipidemic response. *2657*

Lymphocytes *Blood Decrease Physiological* Slight reduction (not dose dependent). *2657*

Nitrogen *Urine Increase Physiological* Effect greatest in responders to therapy. *2657*

5'-Nucleotidase *Serum Increase Physiological* Observed in up to 25% patients. *2657* In 15% children and 26% adults reported in various studies. *2406*

Partial Thromboplastin Time
Plasma Increase Physiological Observed frequently with toxicity. *3358*

Phosphate *Urine Increase Physiological* Response in all treated patients. *2657*

Phospholipids, Total *Serum Increase Physiological* Rare response after initial hypolipidemia. *2657*
Serum Decrease Physiological Parallel decrease in cholesterol. *2657*

Plasminogen *Plasma Decrease Physiological* Mean decrease of 59% (immunological) and 62% (functional assay) in different studies. *0558*

Platelets *Blood Decrease Physiological* May cause bone marrow depression. *2657*

Protein *Urine Increase Physiological* Transient slight effect for few days. *2657* Infrequent transient proteinuria. *0558*

Prothrombin Time *Plasma Increase Physiological* Diminished synthesis of prothrombin by liver. *0578* In 18 patients investigated reported in various studies. *2406*

Pseudocholinesterase *Serum Decrease Physiological* Induces impairment of hepatic synthesis. *0769*

Thrombin Time *Plasma Increase Physiological* Occurs with decreased fibrinogen concentration. *2657*

Thymol Turbidity *Serum Increase Physiological* Hepatotoxicity. *1519*

Thyroid Stimulating Hormone (TSH)
Serum No Effect Physiological No effect during treatment within 3 weeks in 14 children. *1543*

Thyroxine (T4) *Serum Decrease Physiological* From 10.7 to 2.9 μg/dL within 3 weeks in 14 children. *1543*

Thyroxine (T4) Binding Globulin
Serum Decrease Physiological From 29.4 to 8.0 μg/mL within 3 weeks in 14 children. *1543*

Thyroxine (T4), Free *Serum Decrease Physiological* From 1.77 to 0.94 ng/dL within 3 weeks in 14 children. *1543*

Transferrin *Serum Decrease Physiological* Reported effect. *2657* Possibly due to inhibition of protein synthesis reported in various studies. *2406*

Triglycerides *Serum Increase Physiological* Rare response after initial hypolipidemia. *2657*
Serum Decrease Physiological Noted after first week in some patients. *2657*

Tri-iodothyronine (T3) *Serum Decrease Physiological* From 0.99 to 0.35 ng/mL within 3 weeks in 14 children. *1543*

Urea Nitrogen *Serum Increase Physiological* Occurs in 50% subjects (prerenal origin). *1519* Increased in from 7.5 to 50% in different studies: usually mild, possibly due to increased availability of ammonia. *0558*

Uric Acid *Urine Increase Physiological* Marked response if lysis of tissues. *2657*

Asparagine

Ammonia *Plasma No Effect Analytical* On indophenol reaction with 5,000 nmoles. *1290*
Plasma Increase Physiological Potential source of ammonia. *1943*

Creatinine *Urine Increase Analytical* 1 mg equivalent to 0.05 mg creatinine. *2558*

Histidine *Plasma Increase Analytical* May affect fluorometric method of Ambrose. *0081*

Tyrosine *Plasma No Effect Analytical* On fluorometric procedure of Ambrose. *0080*

Urea Nitrogen *Serum Increase Analytical* Blue color with Berthelot's reagent. *1936*

Asparagus

Odor *Urine Increase Physiological* Characteristic after ingestion. *0443* Observed in 43% of 800 volunteers. Reproducible effect inherited as autosomal dominant trait. *2461*

Aspartic Acid

Amino Acids *Test Conditions Increase Analytical* Reacts in spot test with ninhydrin. *3765*

Ammonia *Plasma Decrease Analytical* With 1,000 nmoles produces 12% inhibition of indophenol react. *1290*

Hydroxyproline *Urine No Effect Analytical* At 300 mg/L on method of Seymour. *3262*

Tyrosine *Plasma No Effect Analytical* On fluorometric procedure of Ambrose. *0080*

Aspidium

Alanine Aminotransferase *Serum Increase Physiological* May cause hepatic toxicity. *2313*

Alkaline Phosphatase *Serum Increase Physiological* May cause hepatic toxicity. *2313*

Aspartate Aminotransferase
Serum Increase Physiological May cause hepatic toxicity. *2313*

Bile *Urine Increase Physiological* May cause hepatic toxicity. *2313*

Bilirubin *Serum Increase Physiological* May cause hepatic toxicity. *1022*

Bilirubin, Direct *Serum Increase Physiological* Probable inhibition of uptake of bilirubin by liver. *2427*

BSP Retention *Serum Increase Physiological* Probable inhibition of uptake by liver. *2427*

Casts *Urine Increase Physiological* May cause nephrotoxicity. *0279*

Cephalin Flocculation *Serum Increase Physiological* May cause hepatic toxicity. *2313*

Sugar *Urine Increase Analytical* Acts as reducing agent. *1022*

Thymol Turbidity *Serum Increase Physiological* May cause hepatic toxicity. *2313*

Urea Nitrogen *Serum Increase Physiological* May cause nephrotoxicity. *0279*

Aspirin

Acetaminophen *Serum No Effect Analytical* Has no effect on direct acid/ferric reduction method of Liu and Oka. *2195*
Serum Increase Analytical With Glynn and Kendal technique even with arithmetic corrections, due to metabolites. *0199* Significant positive effect on unmodified Glynn-Kendal technique. *2934*

Acetaminophen Screening Test *Urine Negative Analytical* No reaction with o-cresol at therapeutic concentrations. *3326*

Acetoacetate *Serum Increase Physiological* Due to late metabolic acidosis and renal impairment. *1343*
Urine Increase Analytical Reacts with Gerhardt $FeCl_3$ procedure. *3879*
Urine Increase Physiological Acidotic response especially in children. *1302*

N-Acetylglucosaminidase *Urine Increase Physiological* Marked increase observed even when as little as 3.5 g aspirin ingested daily. *2866*

Acetylsalicylic Acid *Blood Increase Physiological* 1 g orally up to 100 mg/L, 12 g = 400 mg/L. *2348*

Alanine Aminotransferase *Serum Increase Physiological* Prolonged use may cause hepatic toxicity. *1022* Noted in some patients at serum concentrations above 25 mg/dL without signs of hypersensitivity reactions but dose related. *3963* In 15% trials of treatment for juvenile rheumatoid arthritis. *0239*

Albumin *Serum No Effect Analytical* No significant effect on BCG method at 600 mg/L. No significant effect on HABA method at 1 g/L. *2625*
Serum Decrease Analytical Decreased dye binding capacity. *3505*
Serum Decrease Physiological Progressive effect during three days but also noted over longer term study at beginning, eventually reverted towards normal. Effect statistically significant. *3076*

Aldolase *Serum Increase Physiological* Experimental effect seen in rabbits with prolonged use. *3092*

Alkaline Phosphatase *Serum No Effect Physiological* When approximately 3 g/d ingested for several weeks. *3076*
Serum Increase Physiological Prolonged use may cause hepatic toxicity. *2111*
Urine Increase Physiological Due to nephrotoxic effect of drug. *2897* Marked increase observed even when as little as 3.5 g aspirin ingested daily. *2866*

Amino Acids *Urine Increase Physiological* Two fold increase after 1.6 g in normals. *1026*

α-Amino-Nitrogen *Urine Increase Physiological* Inhibition of reabsorption, increased protein catabolism. *1343*

Amylase *Serum No Effect Physiological* When approximately 3 g/d ingested for several weeks. *3076*
Serum Increase Physiological Single case reported. *3879*

Ascorbic Acid *Plasma Decrease Physiological* Uptake into leukocytes decreased also. *2205* In children and adults taking large amount of aspirin since it potentiates excretion of vitamin C. *3023*
Urine Increase Physiological Reported effect. *0805*
White Blood Cells Decrease Physiological Prolonged administration decreases concentration in buffy coat. *1132*

Aspartate Aminotransferase
Serum No Effect Physiological When approximately 3 g/d ingested for several weeks. *3076*
Serum Increase Physiological Prolonged administration may cause hepatic toxicity. *2111* Noted in some patients at serum concentrations above 25 mg/dL without signs of hypersensitivity reactions, but dose related. *3963* In 15% trials of treatment for juvenile rheumatoid arthritis. *0239*

Aspirin Esterase *Serum Increase Physiological* In all patients except cirrhotics. *0150*

Barbiturate *Serum Increase Analytical* May interfere with UV spectrophotometry. *0582*
Serum Decrease Analytical Increased absorbance at acid pH, affects UV methods. *0583*

Basal Metabolic Rate *Patient Increase Physiological* Reported metabolic effect. *2220*

Bicarbonate *Serum No Effect Physiological* When approximately 3 g/d ingested for several weeks. *3076*
Serum Increase Physiological Later alteration of acid base balance. *0652*

Serum Decrease Physiological Initial acidosis with excessive doses. *1022*
Urine Increase Physiological Response to respiratory alkalosis of early toxicity. *1343*

Bile *Urine Increase Analytical* Purple color with Fouchet procedure. *0121*

Bilirubin *Serum Increase Physiological* Competition for albumin binding. *1516* Clinically significant displacement from protein in neonates. *3748*
Serum Decrease Physiological Progressive effect during three days but also noted over longer term study at beginning, eventually reverted towards normal. Effect statistically significant. *3076*

Bilirubin, Direct *Serum Increase Physiological* Occurs with hemolytic anemia. *3505*

Bleeding Time *Blood Increase Physiological* Also inhibits platelet glycolysis. *1132* Effect slight in normal subjects, but technical variables such as direction of incision may have influence. *2434*
Patient Increase Physiological Significant effect on template bleeding time over 6 d of study. *0895*

Calcium *Serum Decrease Analytical* Depresses fluorescence of calcein method. *2395*
Serum Decrease Physiological Progressive effect during three days but also noted over longer term study at beginning, eventually reverted towards normal. Effect statistically significant. *3076*

Casts *Urine Increase Physiological* Occurs with poisoning. *1302*

Catecholamines *Urine No Effect Analytical* No effect on fluorometric Crout procedure. *0766*
Urine No Effect Physiological No effect with short term ingestion 2.6 g/d. *0766*
Urine Increase Analytical Interfering fluorescence in many procedures. *2220*

Cells *Urine Increase Physiological* Tubular epithelial cells increased initially, may persist. *1343* Tubular cells increased: occur even with therapeutic doses; effect can be quite marked. *2866*

Chloride *Serum Increase Physiological* Progressive effect during three days but also noted over longer term study at beginning. Eventually reverted towards normal. Effect statistically significant. *3076*

Chlorpropamide *Serum Increase Physiological* Displaces from plasma proteins. *1487*

Cholesterol *Serum Increase Analytical* Alleged effect (but also no effect at 30 mg/dL). *1563*
Serum Decrease Physiological Doses over 5 g reported to have effect. *1343* Progressive effect during three days but also noted over longer term study at beginning, eventually reverted towards normal. Effect statistically significant. *3076*

Color *Feces Increase Physiological* Red or black due to gastrointestinal bleeding. *2313*

Creatinine *Serum No Effect Analytical* At 1.0 g/L on reversed phase LC procedure of Zhiri et al. *3942* No effect on Jaffé procedure on AutoAnalyzer at concentration of 20 mg/dL. *2433*
Serum No Effect Physiological When approximately 3 g/d ingested for several weeks. *3076*
Serum Increase Physiological Average increase of 38% in patients and healthy individuals. *0545* Anti-inflammatory doses produced substantial effect in 13 of 23 patients with systemic lupus erythematosus. *2866*

Creatinine Clearance *Urine Decrease Physiological* As a consequence of increased serum concentrations. *0545* Significant effect correlated with plasma salicylate concentration. *2535* Observed even with therapeutic doses. *2866*

Diatrizoate Clearance *Urine Decrease Physiological* Observed even with therapeutic doses. *2866*

Diphenylhydantoin *Serum Increase Physiological* May displace from plasma protein. *1487*
Serum Decrease Physiological Statistically significant increase of free fraction (0.13 to 0.16) but lower total concentration. *1178*

Dopa Screening Test *Urine Positive Analytical* Light amber color produced. *3075*

EDTA Clearance *Urine No Effect Physiological* Although creatinine clearance apparently affected. *0545*

Eosinophils *Blood Decrease Physiological* May cause aplastic anemia or pancytopenia. *3585*

Epinephrine *Urine No Effect Physiological* No effect with short term ingestion 2.6 g/d. *0766*

Erythrocytes *Blood Decrease Physiological* Hemolysis/G-6-PD deficient/gastrointestinal hemorrhage/direct bone marrow depression. *3227*
Urine Increase Physiological Initial effect always, may persist. *1343* Observed effect with long term low doses for secondary prevention of coronary heart disease. *2866*

Erythrocyte Sedimentation Rate
Blood Increase Physiological Occurs in some patients (reversible). *1343*
Blood Decrease Physiological If elevated, reduces toward normal value. *1343*

Erythrocyte Survival *Blood Decrease Physiological* Large doses increase destruction. *1343*

Estriol *Urine No Effect Analytical* No effect of 188 mg/L on GLC method. *1163*
Urine Decrease Analytical May affect enzyme hydrolysis. *0023*

Ethchlorvynol *Serum No Effect Analytical* At 10 μg/mL on colorimetric method of Wallace. *3753*
Urine No Effect Analytical No effect on Frings and Cohen method at 150 mg/dL. *1205*

Factor VII *Plasma Decrease Physiological* Acts like bishydroxycoumarin. *1343*

Fatty Acids, Free (FFA) *Serum Decrease Physiological* Increased fatty acid oxidation, decreased lipogenesis. *1343* Considerable reduction in concentration in both normals and diabetics and lesser response to oral glucose. *2433*

Ferric Chloride Test *Urine Positive Analytical* Red-purple color may mask true color. *0583*

Fibrinogen *Plasma Increase Physiological* Associated with increased sedimentation rate. *1343*

Fibrin Split Products (FSP) *Urine Decrease Physiological* Occurs in 2/3 patients with proliferative glomerulonephritis. *0671*

Folate *Serum Decrease Physiological* In 1 study subject brisk, significant but reversible fall in total and bound serum folate. Aspirin *in vitro* also displaced significant amounts of bound serum folate. *2095*
Urine No Effect Physiological Small but insignificant rise due to displacement from binding protein in serum. *2095*

Fouchet Test *Urine Positive Analytical* Produces purple color. *0459*

Glomerular Filtration Rate (GFR)
Urine Decrease Physiological Nephrotoxicity of drug occurring acutely. *0267*

Glucagon *Plasma Increase Physiological* Observed in humans and animals but mechanism for this not known. *2433*

Glucose *Serum No Effect Analytical* At 10 mg/dL no effect on glucose oxidase procedure of Gochman. At 10 mg/dL no effect on alkaline ferricyanide procedure. *1311* At 1 g/dL on p-HBAH procedure of Lever. At 1 g/dL on o-toluidine procedure. *2140*
Serum Increase Physiological Increased absorption and steroid release inhibits TCA cycle. *3505*
Serum Decrease Physiological In diabetics and if toxic doses ingested. *1022* Also decreased response to oral glucose in normal subjects and diabetics. *2433*
Urine No Effect Analytical With DNSA method at usual concentrations. *0076*
Urine Increase Physiological Inhibits liver and muscle glycogen synthesis due to hyperglycemia. *1343*
Urine Decrease Analytical Glucose-oxidase methods inhibited by gentisic acid. *1099*
Urine Decrease Physiological May reduce hyperglycemia, glycosuria in diabetes. *1343*

Glucose Tolerance *Serum No Effect Physiological* No significant effect usually observed. *1487*

β-Glucuronidase *Urine Increase Physiological* Marked increase observed even when as little as 3.5 g aspirin ingested daily. *2866*

Guaiacols Spot Test *Urine Positive Analytical* False reaction with screening test of Rogers. *3031*

Heinz Body Formation *Blood Positive Physiological* Occur initially but disappear with hemolysis. *1902*

Hematocrit *Blood Decrease Physiological* Depresses bone marrow, gastrointestinal bleeding, hemolytic anemia. *0788*

Hemoglobin *Plasma Increase Physiological* Occurs with hemolytic anemia. *1902*
Blood Decrease Physiological Depresses bone marrow, gastrointestinal bleeding, hemolytic anemia. *0788*
Urine Increase Physiological Occurs with severe hemolytic anemia. *1902*

Hemoglobin A$_{1c}$ *Blood Increase Analytical* Acetylation of hemoglobin simulates glycosylation when measured by HPLC or electrophoresis but no increase observed with isoelectric focusing and colorimetric techniques. *2564*

Hippuric Acid *Urine Increase Analytical* Salicyluric acid measured by method of Tomokuni. *3607*

Homogentisic Acid *Urine Increase Analytical* Interferes with measurement procedure. *1238*

Homovanillic Acid *Urine Increase Analytical* Affects colorimetric methods of Sandler, Ruthven. *1620* May produce interfering fluorescence. *0583*
Urine Decrease Analytical High blank in fluorometric method of Sato. *1620*
CSF Increase Analytical Interferes with fluorometric method even if 5 d before. *1801*

11-Hydroxycorticosteroids *Urine No Effect Analytical* No effect on fluorometric Mattingly method. *0766*
Urine No Effect Physiological No effect with short term ingestion 2.6 g/d. *0766*

17-Hydroxycorticosteroids *Plasma Increase Physiological* Large doses stimulate adrenocortical activity. *3365*
Urine No Effect Analytical No effect on Porter-Silber procedure. *0766*
Urine No Effect Physiological No effect with short term ingestion 2.6 g/d. *0766*
Urine Decrease Analytical Conjugate inhibits β-glucuronidase, dose > 4.8 g/d. *0766*

5-Hydroxyindoleacetic Acid (5-HIAA)
Urine Decrease Analytical Affects procedure of Udenfriend (modest effect). *1097* Reported to affect fluorometric method. *1487*

Hydroxyproline *Urine Decrease Physiological* At 100 mg/kg in children has significant effect. *2163*

Indomethacin *Serum No Effect Analytical* No effect on HPLC method of Roberts and Smith. *3002*

Insulin *Plasma Increase Physiological* Increased in response to decreased serum glucose. *2433*

Inulin Clearance *Urine Decrease Physiological* Significant effect correlated with plasma salicylate concentration. *2535* Observed even with therapeutic doses. *2866*

Iron *Serum Decrease Physiological* May be markedly reduced with large doses. *1343*

^{131}I Uptake *Serum No Effect Physiological* No effect observed. *1915*
Serum Decrease Physiological With large doses and chronic administration. *1444*

Ketones *Serum Increase Physiological* Due to induced acidosis. *1302*
Serum Decrease Physiological Increased oxidation of ketone bodies in diabetics. *1343*
Urine Increase Analytical Reddish color with Gerhardt's test. *1022*
Urine Increase Physiological Acidotic response especially in children. *1302*
Urine Decrease Physiological Increased oxidation of ketone bodies in diabetics *1343*

17-Ketosteroids *Urine No Effect Analytical* No effect on Zimmermann procedure. *0766*
Urine No Effect Physiological No effect with short term ingestion 2.6 g/d. *0766*

Lactate *Plasma Increase Physiological* Due to late metabolic acidosis and renal impairment. *1343*

Lactate Dehydrogenase *Serum No Effect Physiological* When approximately 3 g/d ingested for several weeks. *3076*
Serum Increase Physiological Experimental effect seen in rabbits with prolonged use. *3092*
Urine Increase Physiological Renal irritation and desquamation of epithelial cells. *2897* Marked increase observed even when as little as 3.5 g aspirin ingested daily. *2866*

Aspirin (continued)

Leucine Aminopeptidase *Urine Increase Physiological* Marked increase observed even when as little as 3.5 g aspirin ingested daily. *2866*

Urine Decrease Physiological Due to antifibrinolytic action. *2897*

Leukocytes *Blood Decrease Physiological* Depresses leukocytosis of acute rheumatic fever. *3825*

Magnesium *Serum Increase Physiological* Prolonged therapy likely to cause elevation. *0251*

Methemoglobin *Blood Increase Physiological* May cause hemolysis with G-6-PD deficiency. *3581*

Methotrexate *Serum Increase Physiological* Displaces from plasma protein binding, if present. *1487* Clearance by kidneys may be halved by large doses of drug. *2941*

NBT Test *Blood Decrease Physiological* Mechanism not discussed. *0943*

Nitrogen *Urine Increase Physiological* Effect observed in adults. *3365*

Norepinephrine *Urine No Effect Physiological* No effect with short term ingestion 2.6 g/d. *0766*

5'-Nucleotidase *Serum Increase Physiological* Reversible hepatotoxicity with prolonged administration. *3092*

Occult Blood *Feces Positive Physiological* In over 70% patients when more than 3 g/d given. *3761* In 1.5% trials of treatment for juvenile rheumatoid arthritis. *0239*

PAH Clearance *Urine No Effect Physiological* Insignificant reduction observed. *2535*

Urine Decrease Physiological Observed even with therapeutic doses. *2866*

pCO$_2$ *Blood Decrease Physiological* In toxicity with increased respiratory rate and pulmonary ventilation. *1343*

pH *Blood Increase Physiological* Initial respiratory alkalosis. *3687*

Blood Decrease Physiological May cause acidosis later (respiratory and metabolic). *2313* Systemic acidosis common in poisoning, more frequent in chronic situation, in patients with severe manifestations and dehydration. *1251*

Phenylketones *Urine Positive Analytical* Purple with FeCl$_3$, purple with Phenistix®. *1022*

Phosphate *Serum No Effect Analytical* On nonprecipitation method of Peynet. *2799*

Serum No Effect Physiological When approximately 3 g/d ingested for several weeks. *3076*

Urine No Effect Analytical No effect at 20 mg/dL on method of Jung/Parekh. *1835*

Urine Increase Physiological Inhibits tubular reabsorption. *1343*

Phospholipids, Total *Serum Decrease Physiological* Increased fatty acid oxidation, decreased lipogenesis. *1343*

Platelet Aggregation *Blood Decrease Physiological* Inhibits release of ADP from platelets. *1343* Inhibits collagen induced. *3068*

Platelets *Blood Decrease Physiological* Decreased platelet survival time, may be purpura. *3566* Several cases of immune-mediated thrombocytopenia reported. *2502*

Urine Increase Physiological Loss through damaged glomeruli may occur. *3510*

Potassium *Serum No Effect Physiological* When approximately 3 g/d ingested for several weeks. *3076*

Serum Decrease Physiological Diuretic action, respiratory alkalosis. *0652*

Urine Increase Physiological Direct effect on renal tubules. *1343*

Pregnancy Tests *Urine Positive Analytical* Large dose effect on mouse, rabbit tests. *1487*

Prolactin Response to TRH
Plasma No Effect Physiological In up to 35 healthy volunteers given 3.6 g daily for 1 week. *2904*

Propoxyphene *Serum No Effect Analytical* At 25 mg/L on method of Evenson. *1056*

Propylthiouracil *Serum Decrease Analytical* Produces negative interference (procedure of Ratliff). *2922*

Prostaglandin E *Plasma Decrease Physiological* In up to 35 healthy volunteers given 3.6 g daily for 1 week. *2904*

Urine Decrease Physiological Clearance-reduced increased: clearance with sodium restriction. *2535*

Prostaglandin F *Plasma Decrease Physiological* In up to 35 healthy volunteers given 3.6 g daily for 1 week. *2904*

Protein *Serum Decrease Physiological* Progressive effect during three days but also noted over longer term study at beginning, eventually reverted towards normal. Effect statistically significant. *3076*

Urine Increase Analytical Interference with Folin-Ciocalteu reaction. *3879*

Urine Increase Physiological May cause nephrotoxicity. *1488* Observed effect with long term low doses for secondary prevention of coronary heart disease. *2866*

CSF Increase Analytical False positive with Folin-Ciocalteu reagent. *3958*

Protein Bound Iodine (PBI) *Serum Decrease Physiological* Competes for thyroxine binding prealbumin also uncouples phosphorylation. *1022*

Prothrombin Time *Plasma No Effect Physiological* No effect observed in volunteers. *3292*

Plasma Increase Physiological Large dose effect (decreased synthesis of clot factors). *0894*

Plasma Decrease Physiological Small dose effect. *2313*

PSP Excretion *Urine Decrease Physiological* Competes with PSP for excretion. *3879*

Pyruvate *Serum Increase Physiological* Due to late metabolic acidosis and renal impairment. *1343*

Reduced Glutathione
Red Blood Cells Decrease Physiological Occurs initially before overt hemolysis. *1902*

Renin Activity *Plasma No Effect Physiological* In patients in whom it had been increased with sodium restriction. *2535*
Plasma Decrease Physiological Mean reduction from 2.94 ng/mL/h to 1.41 ng/mL/h when treatment started, but rose following treatment before returning to baseline in 10 weeks. *0487*

Reticulocytes *Blood Increase Physiological* Response during recovery from hemolysis. *1902*

Salicylate *Serum Increase Physiological* Due to ingestion of compound. *1302* When approximately 3 g/d ingested for several weeks. *3076*

Sodium *Serum No Effect Physiological* When approximately 3 g/d ingested for several weeks. *3076*
Urine Increase Physiological Response to respiratory alkalosis of early toxicity. *1343*

Sugar *Urine Increase Analytical* False positive with Clinitest® or Benedict's. *1022* Conjugate may react with Benedict's. *1563*

T$_3$ Uptake *Serum No Effect Physiological* When approximately 3 g/d ingested for several weeks. *3076*
Serum Increase Physiological Red cell uptake affected, also affects resin test. *1022*

Taurine *Urine Decrease Physiological* Reduces elevated concentration in rheumatoid patients. *3102*

Thromboxane B$_2$ *Serum Decrease Physiological* Reduction by about 99% over 24 h, gradually reverted towards normal. *2046*

Thyro-Binding Index *Serum Decrease Physiological* For up to 1 week after treatment. *0913*

Thyroid Stimulating Hormone (TSH)
Serum Decrease Physiological Decreased release after administration. *2427*

Thyroxine (T$_4$) *Serum Decrease Physiological* Displaces thyroxine from binding sites (thyroxine binding prealbumin). *3505* Progressive effect during three days but also noted over longer term study at beginning, eventually reverted towards normal. Effect statistically significant. *3076* In up to 35 healthy volunteers given 3.6 g daily for 1 week. *2904*

Thyroxine (T$_4$), Free *Serum Increase Physiological* Interferes with binding to thyroxine binding globulin and thyroxine binding prealbumin. *2083*

Thyroxine (T$_4$) Index, Free (FTI)
Serum Decrease Physiological Low normal or decreased for 1 week after therapy. *0913*

Thyroxine (T$_4$) (Murphy-Pattee)
Serum Decrease Physiological Binds to thyroxine binding prealbumin but not thyroxine binding globulin. *0583*

Tolbutamide *Serum Increase Physiological* Displaces from plasma proteins if present. *1487*

Tri-iodothyronine (T₃) *Serum Decrease Physiological* In up to 35 healthy volunteers given 3.6 g daily for 1 week. *2904*

Tri-iodothyronine (T₃), Free *Serum Increase Physiological* Interferes with binding to thyroxine binding globulin. *2083*

Tryptophan *Plasma Decrease Physiological* (total measured) in rheumatoids on therapy. *0188*

Tryptophan, Free *Serum Increase Physiological* In rheumatoids on therapy. *0188*

TSH Response to TRH *Serum Decrease Physiological* In up to 35 healthy volunteers given 3.6 g daily for 1 week. *2904*

Tyrosine *Plasma Increase Analytical* Interferes with fluorometric method of Scott. *3228*

Urea Nitrogen *Serum No Effect Analytical* No effect on Berthelot procedure. *1862*
Serum No Effect Physiological When approximately 3 g/d ingested for several weeks. *3076* No effect although creatinine affected. *0545*
Serum Increase Physiological May have nephrotoxic effect. *3678* Anti-inflammatory doses produced substantial effect in 13 of 23 patients with systemic lupus erythematosus. Observed effect with long term low doses for secondary prevention of coronary heart disease. *2866*

Uric Acid *Serum No Effect Analytical* No effect on methods using uricase. *2954* At 1 g/L on uricase procedure of Kabasak. *1840* At 1.0 g/L on reversed phase LC procedure of Zhiri et al. *3942*
Serum Increase Analytical Acts as reducing substance with nonspecific methods. *1488*
Serum Increase Physiological Low doses reduce renal excretion of uric acid. *3879*
Serum Decrease Physiological Mild uricosuric action (large dose effect). *0652* Progressive effect during three days but also noted over longer term study at beginning, eventually reverted towards normal. Effect statistically significant. *3076* Noted in several patients receiving therapeutic amounts of drug. *2909* Significant effect observed in group of patients with rheumatoid arthritis. *0885*
Urine Increase Analytical Acts as reducing substance with nonspecific methods. *1488*
Urine Increase Physiological High dose effect (greater than 3 g daily). *1022*
Urine Decrease Physiological Low doses effect. *3879*

Urobilin *Urine Increase Physiological* Occurs with poisoning. *1302*

Vanillylmandelic Acid *Urine Increase Analytical* Interferes with fluoro-, colorimetric procedures. *0502*
Urine Decrease Analytical At therapeutic physiological concentrations on Pisano procedure. *1097*

Volume *Urine Increase Physiological* Decreased renal tubular reabsorption. *1343*

Xylose Excretion *Urine Decrease Physiological* Affects renal elimination. *1913* Reduced renal excretion reported with tolerance tests. *2866*

Aspoxicillin

Alanine Aminotransferase *Serum Increase Physiological* In 2 of 5 patients given up to 112 mg/kg intravenously in 3 or 4 divided doses. *1517*

Aspartate Aminotransferase
Serum Increase Physiological In 2 of 5 patients given up to 112 mg/kg intravenously in 3 or 4 divided doses. *1517*

Eosinophils *Blood Increase Physiological* In 1 patient given up to 112 mg/kg intravenously in 3 or 4 divided doses. *1517*

Atenolol

Apolipoprotein AI *Serum Decrease Physiological* Reduction by 6% reported from one study. *0088*

Apolipoprotein B *Serum Increase Physiological* Increase by 3% reported from one study. *0088*

Cholesterol *Serum No Effect Physiological* Insignificant change after 6 mo treatment of 14 hypertensives. *0836* No significant change in 15 hypertensives treated for 8 mo. *1021* Insignificant change in 3 studies of from 1 to 3 mo. *2116* In 20 patients with 50 mg/d compared with pre-treatment values after 3 mo. *3071* Typically no significant change from several studies. *0088*
Serum Increase Physiological Increase by 49% after 100 mg/d for 5 weeks in 20 hypertensive men. *2135* Slight increase (8%) in one 6 mo study. *2116*

Cholesterol, High Density Lipoprotein
Serum No Effect Physiological Typically no significant change but may be slight reduction. *0088* No significant change in 15 hypertensives treated for 8 mo. *1021* After 100 mg/d for 5 weeks in 20 hypertensive men. *2135* No significant change in one 3 mo long study. *2116* In 20 patients with 50 mg/d compared with pre-treatment values after 3 mo. *3071*
Serum Decrease Physiological Reduction by 7 to 12% in 2 studies of 3 to 6 mo. *2116* From 1.21 to 1.13 mmol/L with 100 mg drug/d. *2192*

Cholesterol, Low Density Lipoprotein
Serum No Effect Physiological No significant change in 4 studies of from 3 to 6 mo. *2116* In 20 patients with 50 mg/d compared with pre-treatment values after 3 mo. *3071* No effect of treatment with 100 mg/d. *2192*
Serum Increase Physiological Slight increase reported from one study. *0088* Increase by 5.9% after 100 mg/d for 5 weeks in 20 hypertensive men. *2135*

Cholesterol, Very Low Density Lipoprotein
Serum No Effect Physiological In 20 patients with 50 mg/d compared with pre-treatment values after 3 mo. *3071*
Serum Increase Physiological Effect observed in 2 studies, but not significant in one other. *0088* From 1.21 to 1.62 mmol/L in 15 hypertensives treated for 8 mo. *1021* From 0.90 to 1.14 mmol/L with 100 mg drug/d. *2192* Increase by 5.9% after 100 mg/d for 5 weeks in 20 hypertensive men. *2135*
Serum Decrease Physiological Increase by 5.9% after 100 mg/d for 5 weeks in 20 hypertensive men. *2135*

C-Peptide *Plasma Decrease Physiological* Reduced activity at 2 and 3 h during glucose tolerance test versus control. *1977*

Disopyramide *Serum No Effect Physiological* Significant reduction of clearance while half-life, volume of distribution remain unchanged. *0413*

Fatty Acids, Free (FFA) *Serum Decrease Physiological* Significant reduction after 3 mo treatment in 14 hypertensives. *0836* Marked effect with 3 mo treatment given 100 mg/d. *0835* In fasting state and after 30 minutes during oral glucose tolerance Test. *1977*

Glucagon *Plasma No Effect Physiological* No change in basal concentration or after glucose in 14 hypertensives. *0836*
Plasma Decrease Physiological In 18 patients with mild essential hypertension treated with chlorothiazide concomitantly over 4 weeks. *1035*

Glucose *Serum No Effect Physiological* No significant change after 3 or 6 mo treatment in 14 hypertensives. *0836* In 18 patients with mild essential hypertension treated with chlorothiazide concomitantly over 4 weeks. *1035* Compared with controls during oral glucose tolerance test. *1977*

Growth Hormone *Plasma No Effect Physiological* No change in basal concentration or after glucose in 14 hypertensives. *0836*

Indomethacin *Serum No Effect Analytical* No effect on HPLC method of Roberts and Smith. *3002*

Insulin *Plasma No Effect Physiological* No change in basal concentration or after glucose in 14 hypertensives. *0836* In 18 patients with mild essential hypertension treated with chlorothiazide concomitantly over 4 weeks. *1035* Compared with controls during glucose tolerance test. *1977*

Triglycerides *Serum No Effect Physiological* In 20 patients with 50 mg/d compared with pre-treatment values after 3 mo. *3071*
Serum Increase Physiological Up to 60% increase in several studies although no significant chance in some. *0088* Moderate increase after 6 mo treatment of 14 hypertensives. *0836* Marked effect in 18 patients with mild essential hypertension treated with chlorothiazide concomitantly over 14 weeks. *1035* From 1.94 to 2.44 mmol/L in 15 hypertensives treated for 8 mo. *1021* Increase by 28% after 100 mg/d for 5 weeks in 20 hyper-

Atenolol *(continued)*

Triglycerides *(continued)*
tensive men. *2135* Increase by 24 to 62% in 4 studies of from 1 to 6 mo. *2116*

Triglycerides, High Density Lipoprotein
Serum No Effect Physiological In 20 patients with 50 mg/d compared with pre-treatment values after 3 mo. *3071*
Serum Increase Physiological 34% increase reported from one study. *0088*

Triglycerides, Low Density Lipoprotein
Serum No Effect Physiological In 20 patients with 50 mg/d compared with pre-treatment values after 3 mo. *3071*
Serum Increase Physiological From 0.46 to 0.51 mmol/L in 15 hypertensives treated for 8 mo. *1021* Slight increase reported from one study. *0088*

Triglycerides, Very Low Density Lipoprotein
Serum No Effect Physiological In 20 patients with 50 mg/d compared with pre-treatment values after 3 mo. *3071*
Serum Increase Physiological 48% effect observed in one study. *0088* From 1.21 to 1.62 mmol/L in 15 hypertensives treated for 8 mo. *1021* From 0.90 to 1.14 mmol/L with 100 mg drug/d. *2192*

Uric Acid *Serum No Effect Physiological* After 100 mg/d for 5 weeks in 20 hypertensive men. *2135*

Atropine

Barbiturate *Serum Increase Analytical* False positive screen test with Hg complex formation. *0581*

Basal Metabolic Rate *Patient Increase Physiological* Temporary effect observed. *0251*

Ethanol *Blood Decrease Physiological* Reduced rate of absorption due to delayed gastric emptying. *3794*

Gastrin *Serum Decrease Physiological* Significant effect if given i.m. *1236*

Glucose *Serum Decrease Physiological* Possible slight fall if given as premedication. *2394*

Hematocrit *Blood Increase Physiological* But not to same extent as plasma volume. *1832*

Histamine *Plasma Increase Physiological* Associated with dose given associated with anesthesia. *2212*

Hydrochloric Acid *Gastric Material Decrease Physiological* Volume also reduced. *1343*

Leukocytes *Blood Increase Physiological* May cause leukocytosis (effect in children). *0071*

Pepsin *Gastric Material Decrease Physiological* Antagonizes cholinergic stimulation. *0814*

Prolactin *Plasma No Effect Physiological* No effect on basal concentration in morning or evening but evening response to hypoglycemia significantly inhibited. *2565*

PSP Excretion *Urine Decrease Physiological* Interferes with secretion by tubules. *2981*

Volume *Plasma Decrease Physiological* Increase by 10-15% after i.v. injection. *1832*
Urine Decrease Physiological Very large doses cause release of ADH. *1343*

Augmentin®

Chloramphenicol *Serum No Effect Analytical* No effect at 100 mg/L on coupled enzymatic method. *2490*

Auranofin

Eosinophils *Blood Increase Physiological* Infrequent reversible side effect (occurring in up to 13% patients). *0617*

Erythrocytes *Blood Decrease Physiological* Infrequent reversible side effect (occurring in up to 13% patients). *0617*

Hematocrit *Blood Decrease Physiological* Infrequent reversible side effect (occurring in up to 13% patients). *0617*

Hemoglobin *Blood Decrease Physiological* Infrequent reversible side effect (occurring in up to 13% patients). *0617*
Urine Increase Physiological Occasional drug-associated hematuria: may be partially responsible for anemia. *0617*

Immunoglobulin IgG *Serum No Effect Physiological* No effect of 6 mg/d for 1 y. *0617*
Serum Increase Physiological 26.5% reduction after 12 weeks treatment in patients with rheumatoid arthritis. *0617*

Leukocytes *Blood Decrease Physiological* Infrequent reversible side effect (occurring in up to 13% patients). *0617*

Platelets *Blood Increase Physiological* Infrequent reversible side effect (occurring in up to 13% patients). *0617*

Protein *Urine Increase Physiological* In 3% of patients receiving drug, but proteinuria did not persist beyond 12 mo: effect less than when gold given. *1886*

Zinc *Serum Increase Physiological* In 10 of 12 patients with initially low concentration. *0617*

Aurothioglucose

Immunoglobulin IgA *Serum Decrease Physiological* Reduced only at 3 mo in 25 patients with rheumatoid arthritis. *3680*

Immunoglobulin IgG *Serum Decrease Physiological* Reduced only at 12 mo in 25 patients with rheumatoid arthritis. *3680*

Immunoglobulin IgM *Serum Decrease Physiological* Substantial reduction at 3 mo in 25 patients with rheumatoid arthritis. *3680*

Aurothiomalate

Platelets *Blood Decrease Physiological* In patient with rheumatoid arthritis with shortened platelet survival and platelet phagocytosis. *2144*

Avocados

5-Hydroxyindoleacetic Acid (5-HIAA)
Urine Increase Physiological May contain large amounts of serotonin. *1487*

Aza Drugs

BSP Retention *Serum Increase Physiological* May cause hepatotoxicity. *1237*

Azapropazone

Alanine Aminotransferase *Serum Increase Physiological* In 9 of 83 patients treated for 6 mo. *3565*

Aspartate Aminotransferase
Serum Increase Physiological In 9 of 83 patients treated for 6 mo. *3565*

Chloride *Serum Increase Physiological* In 1 of 83 patients treated for 6 mo. *3565*

Cholesterol *Serum No Effect Analytical* At concentration of 250 mg/L had no effect as measured by CHOD-Iodide method. *3393* At concentration of 250 mg/L had no effect as measured by CHOD-PAP method. *3393* At concentration of 250 mg/L had no effect as measured by catalase-AlDH method. *3393* At concentration of 250 mg/L had no effect as measured by catalase-Hantzsch method. *3393* At concentration of 250 mg/L had no effect as measured by Liebermann-Burchard method. *3393*

Creatinine *Serum No Effect Analytical* At concentration of 360 mg/L had no effect as measured by Jaffé-Fading-Fraction method. *3393* At concentration of 360 mg/L had no effect as measured by Jaffé-Fuller's earth method. *3393* At concentration of 625 mg/L had no effect as measured by kinetic Jaffé method on BKA-2. *3393*
Serum Increase Physiological Initial 10% increase observed in 83 treated patients but thereafter no further increase *3565*

Diphenylhydantoin *Serum Increase Physiological* Probably due to decreased clearance of phenytoin. *1257*

Eosinophils *Blood Increase Physiological* In 1 of 83 patients treated for 6 mo. *3565*

Erythrocytes *Blood Decrease Physiological* In 1 of 83 patients treated for 6 mo. *3565*

Glucose *Serum No Effect Analytical* At concentration of 36 mg/L had no effect as measured by Ektachem® method. *3393*

Hemoglobin *Urine Increase Physiological* In 1 of 83 patients treated for 6 mo. *3565*

Monocytes *Blood Decrease Physiological* In 1 of 83 patients treated for 6 mo. *3565*

Neutrophils *Blood Increase Physiological* In 1 of 83 patients treated for 6 mo. *3565*

Platelets *Blood Decrease Physiological* In 3 of 83 patients treated for 6 mo. *3565*

Potassium *Serum No Effect Analytical* At concentration of 360 mg/L had no effect on ISE measurement without predilution. *3393*

Protein Electrophoresis *Serum No Effect Analytical* At concentration of 360 mg/L had no effect as measured by automated Olympus-Hite method except for slight displacement of fractions. *3393*

Prothrombin Time *Plasma Increase Physiological* Drug has ulcerogenic potential by itself, but enhanced with warfarin. *0095*

Sodium *Serum No Effect Analytical* At concentration of 360 mg/L had no effect on ISE measurement without predilution. *3393*

Urea Nitrogen *Serum No Effect Analytical* At concentration of 36 mg/L had no effect as measured by Ektachem® method. *3393*
Serum Increase Physiological Initial 10% increase observed in 83 treated patients but thereafter no further increase *3565*

Uric Acid *Serum No Effect Analytical* At concentration of 360 mg/L had no effect as measured by Kageyama-Hantzsch method. *3393* At concentration of 360 mg/L had no effect as measured by catalase-AIDH method. *3393*
Serum Decrease Physiological Therapeutic intent. *3565* Mean fall after 4 d was 31%, and 46% on 2400 mg daily. *1589*
Urine Increase Physiological Potent uricosuric agent (2400 mg daily comparable to 1 g probenecid). *1589*

Azaserine

Alanine Aminotransferase *Serum Increase Physiological* May cause hepatotoxicity. *3835*

Aspartate Aminotransferase
Serum Increase Physiological May cause hepatotoxicity. *3835*

Azathioprine

Alanine Aminotransferase *Serum Increase Physiological* May cause hepatotoxicity. *3433* Minimal cholestasis and reversible portal fibrosis observed in 8 patients: occasional hepatotoxicity. *0036* Canalicular cholestasis and centrilobular ballooning of hepatocytes in one patient. *0846*

Albumin *Serum Decrease Physiological* May cause hepatotoxicity. *3433*

Alkaline Phosphatase *Serum Increase Physiological* May be very high, liver damage or biliary stasis. *1680* Minimal cholestasis and reversible portal fibrosis observed in 8 patients: occasional hepatotoxicity. *0036* Canalicular cholestasis and centrilobular ballooning of hepatocytes in one patient. *0846*
Serum Decrease Physiological Improves biliary excretion in biliary cirrhosis. *3063*

Amylase *Serum Increase Physiological* Unusual side effect but may cause pancreatitis. *1680* 6.2% of 116 patients receiving only drug demonstrated clinical pancreatitis. *2281*

Aspartate Aminotransferase *Serum No Effect Analytical* No effect on Babson procedure. No effect on Karmen procedure. *2760*
Serum Increase Physiological May cause hepatotoxicity. *3433* Minimal cholestasis and reversible portal fibrosis observed in 8 patients: occasional hepatotoxicity. *0036* Canalicular cholestasis and centrilobular ballooning of hepatocytes in one patient. *0846*

Bilirubin *Serum Increase Physiological* May cause hepatotoxicity (not usually very high). *3433* Canalicular cholestasis and centrilobular ballooning of hepatocytes in one patient. *0846* Hypersensitivity reaction without effect on aminotransferases. *0827*

Cholesterol *Serum Decrease Physiological* Improves biliary excretion in biliary cirrhosis. *3063*

Chromosomes *Test Conditions Abnormal Physiological* Clastogenic to bone marrow and leukocytes. *3282*

Erythrocytes *Blood Decrease Physiological* Bone marrow depression with anemia. *0071* Principal manifestation of bone marrow depression. *0036* 18 cases of pancytopenia out of 79 cases with hematological problems out of 328 reports of severe adverse effects. *2097*

Fat *Feces Increase Physiological* May occasionally cause steatorrhea. *1680*

Hematocrit *Blood Decrease Physiological* In 6 cases of anemia out of 79 cases with hematological problems out of 328 reports of severe adverse effects. *2097*

Hemoglobin *Blood Decrease Physiological* Principal manifestation of bone marrow depression. *0036* In 6 cases of anemia out of 79 cases with hematological problems out of 328 reports of severe adverse effects. *2097*

Immunoglobulin IgM *Serum Decrease Physiological* Reduced when biliary cirrhosis treated. *3063*

Leukocytes *Blood Decrease Analytical* Low count by Coulter S, ?due to fragile cells. *2227*
Blood Decrease Physiological Drug related leukopenia (probable effect). *3287* Principal manifestation of bone marrow depression. *0036* In 30 of 79 cases with hematological problem out of 328 reports of severe adverse effects. *2097*

Lipase *Serum Increase Physiological* 6.2% of 116 patients receiving only drug demonstrated clinical pancreatitis. *2281*

MCV *Blood Increase Physiological* Macrocytosis in two-thirds of renal transplant patients. *0036*

Neutrophils *Blood Decrease Physiological* Drug related leukopenia (cured by cessation). *1508* In 6 of 79 cases with hematological problem out of 328 reports of severe adverse effects. *2097*

Platelets *Blood Decrease Physiological* May cause bone marrow depression. *3287* Principal manifestation of bone marrow depression. *0036* In 13 of 79 cases with hematological problem out of 328 reports of severe adverse effects. *2097*

Potassium *Serum Increase Physiological* Observed in only 1 of 13 renal transplant patients after 1 mo treatment also with prednisone. *1154*

Prothrombin Time *Plasma Decrease Physiological* May cause hepatotoxicity. *3433*

Reticulocytes *Blood Decrease Physiological* Gradual reduction observed. *1904*

Sperm Count *Semen Decrease Physiological* Not uncommon in males on drug. *3061*

Uric Acid *Serum Increase Physiological* Rapid destruction of tissues and nucleic acid catabolism. *1343*
Serum Decrease Physiological In patients with gout decreases concentration. *3395*
Urine Decrease Physiological In patients with gout reduces excretion. *3395*

Azathymine

Uric Acid *Serum Increase Physiological* Due to tissue destruction. *2313*

Azauridine

Crystals *Urine Increase Physiological* May occur due to response to tissue destruction. *0278*

Orotic Acid *Urine Increase Physiological* Metabolic effect may cause crystalluria. *0278*

Uric Acid *Serum Increase Physiological* Due to tissue destruction. *2313*

Drug Listings

Azide

Ethanol *Blood No Effect Analytical* Over 16 weeks at 20° at concentration of 0.5% w/v. *3355*

Azidothymidine

Hemoglobin *Blood Decrease Physiological* Profound effect observed in one male homosexual with AIDS unlikely related to other factors. *1157*

Leukocytes *Blood Decrease Physiological* Profound effect observed in one male homosexual with AIDS unlikely related to other factors. *1157*

Platelets *Blood Decrease Physiological* Profound effect observed in one male homosexual with AIDS unlikely related to other factors. *1157*

Aziridine

Chromosomes *Test Conditions Abnormal Physiological* Clastogenic in human cells. *3282*

Azlocillin

Bilirubin *Serum No Effect Physiological* Probably clinically insignificant displacement from protein in neonates. *3748*

Glucose *Urine No Effect Analytical* Concentrations accurately measured by Diastix®. Concentrations accurately measured by TesTape®. *2100*
Urine Increase Analytical Falsely elevated values with Clinitest®. *2100*

Uric Acid *Serum Decrease Physiological* Fell from average of 6.4 mg/dL to 2.3 mg/dL in 20 hospitalized patients not observed in controls without azlocillin. *1046*

Azosemide

Aldosterone *Plasma Increase Physiological* Effect observed in normal male volunteers after 60 mg orally. *2658*
Urine Increase Physiological Effect observed in normal male volunteers after 60 mg orally. *2658*

Antidiuretic Hormone *Plasma Increase Physiological* Effect observed in normal male volunteers after 60 mg orally. *2658*

Calcium *Urine Increase Physiological* Tended to increase but not marked. *2047*

Chloride *Serum Increase Physiological* Delayed effect observed in normal male volunteers after 60 mg orally. *2658*
Serum Decrease Physiological Slight effect noted in 2 volunteers. *2047*
Urine Increase Physiological Effect observed in normal male volunteers after 60 mg orally. *2658* Significant effect then gradually decreased. *2047*

Creatinine *Urine No Effect Physiological* No clear pattern observed. *2047*

Epinephrine *Plasma No Effect Physiological* Effect observed in normal male volunteers after 60 mg orally. *2658*

Norepinephrine *Plasma Increase Physiological* Effect observed in normal male volunteers after 60 mg orally. *2658*
Urine Increase Physiological Effect observed in normal male volunteers after 60 mg orally. *2658*

Osmolality *Serum No Effect Physiological* Effect observed in normal male volunteers after 60 mg orally. *2658*

Phosphate *Urine Increase Physiological* Tended to increase but not marked. *2047*

Potassium *Serum No Effect Physiological* Effect observed in normal male volunteers after 60 mg orally. *2658*
Serum Decrease Physiological Slight effect noted in 2 volunteers. *2047*
Urine Increase Physiological Slight effect observed in normal male volunteers after 60 mg orally. *2658* Tended to increase but not marked. *2047*

Prolactin *Plasma Increase Physiological* Effect observed in normal male volunteers after 60 mg orally. *2658*

Protein *Serum Increase Physiological* Slight and delayed effect observed in normal male volunteers after 60 mg orally. *2658*

Renin Activity *Plasma Increase Physiological* Effect observed in normal male volunteers after 60 mg orally. *2658*

Sodium *Serum No Effect Physiological* Effect observed in normal male volunteers after 60 mg orally. *2658*
Serum Decrease Physiological Slight effect noted in 2 volunteers. *2047*
Urine Increase Physiological Effect observed in normal male volunteers after 60 mg orally. *2658* Significant effect then gradually decreased. *2047*

Urea Nitrogen *Urine No Effect Physiological* No clear pattern observed. *2047*

Uric Acid *Serum Increase Physiological* Slight effect noted in 2 volunteers. *2047*
Urine Decrease Physiological Slight effect noted on first 2 d. *2047*

Volume *Urine No Effect Physiological* Effect observed in normal male volunteers after 60 mg orally. *2658*
Urine Increase Physiological Significant effect on first day, but decrease on second. *2047*

Azostix®

Urea Nitrogen *Serum Decrease Analytical* Result low compared with diacetyl monoxime value. *1460*

Azosulfamide

PAH Clearance *Urine Increase Analytical* Slight effect only, measured as if PAH. *2981*

Sulfa as Sulfanilamide *Serum Increase Analytical* Slight effect, measured as if sulfonamide. *2981*

Aztreonam

Bilirubin *Serum No Effect Physiological* Probably clinically insignificant displacement from protein in neonates. *3748*

Glucose *Urine No Effect Analytical* Dark green-black color with Clinitest®, but no effect or slight reduction with glucose concentrations of 1% and up. Concentrations measured accurately by Diastix®. Concentrations measured accurately by TesTape®. *2100* Concentrations measured accurately by TesTape®. *0086*

Azuresin

Color *Urine Increase Analytical* Blue or green for a few days after test. *1488*

Bacitracin

Casts *Urine Increase Physiological* Nephrotoxic effect (cylindruria may occur). *2313*

Erythrocytes *Urine Increase Physiological* Actual bleeding caused by drug. *1022*

Hemoglobin *Urine Increase Physiological* Actual bleeding caused by drug. *1022*

Protein *Urine Increase Physiological* May cause nephrotoxicity. *1237* Nephrotoxic effect. *1022*

Urea Nitrogen *Serum Increase Physiological* Renal toxicity (especially if given i.v.). *1343*

Bacterial Contamination

Cephalin Flocculation *Serum Increase Analytical* If reagent contaminated turbidity measured. *3879*

Chylomicrons *Serum Increase Analytical* If contaminated with phospholipase producing bacteria. *3223*

α-Lipoproteins *Serum Decrease Analytical* If contaminated with phospholipase producing bacteria. *3223*

β-Lipoproteins *Serum Decrease Analytical* If contaminated with phospholipase producing bacteria. *3223*

Lipoproteins, Pre-β *Serum Decrease Analytical* If contaminated with phospholipase producing bacteria. *3223*

Phospholipids, Total *Serum Decrease Analytical* If contaminated with phospholipase producing bacteria. *3223*

Triglycerides *Serum Increase Analytical* If contaminated with phospholipase producing bacteria. *3223*

Urea Nitrogen *Urine Decrease Analytical* Rapid evolution of NH₃ changes pH, loss of NH₃. *1563*

Bacteriuria

Acetic Acid *Urine Increase Physiological* Experimental studies — all 11 species produced much acid. *1483*

Albumin *Urine No Effect Analytical* No effect on automated BCG method. *2919*

Appearance *Urine Abnormal Physiological* Cloudy, insoluble in dilute acetic acid. *0443*

Benzoic Acid *Urine Increase Physiological* 8 of 11 species hydrolyzed hippuric acid. *1483*

Bile *Urine Decrease Analytical* Negative due to bacterial metabolism of bili diglucuronide. *2308*

Butyric Acid *Urine Increase Physiological* Some organisms can produce in urine. *1483*

Caproic Acid *Urine Increase Physiological* Some organisms can produce in urine. *1483*

Carcinoembryonic Antigen *Urine No Effect Analytical* No effect irrespective of number and type or organisms. *1459*

Catalase *Urine Increase Physiological* Infected urines have activity. *2897*

Color *Urine Increase Physiological* Blue diapers in babies with P. Aeruginosa (pyocyanin). *0524*

Glucose *Urine Decrease Analytical* False negative through metabolic actions. *2308*

β-Glucuronidase *Urine Increase Physiological* High activity possible with bacterial contamination. *2897*

Hemoglobin *Urine Increase Analytical* False positive with Hematest®. *3879*

Isovaleric Acid *Urine Increase Physiological* Some organisms can produce in urine. *1483*

Propionic Acid *Urine Increase Physiological* In experimental studies 8 of 11 species produced acid. *1483*

Protein *Urine Increase Analytical* If turbidimetric methods used. *2919*

Xylose Excretion *Urine Decrease Physiological* Theoretical effect (note marked gastrointestinal tract effect). *1914*

Bamethan

Amphetamine *Urine No Effect Analytical* No effect at 100 mg/L on method of Rutter. *3097*

Bananas

Catecholamines *Urine Increase Analytical* Catechols in bananas. *0759*

4-Hydroxybenzoic Acid *Urine Increase Physiological* Possible effect of bacterial metabolism. *3281*

5-Hydroxyindoleacetic Acid (5-HIAA) *Urine Increase Analytical* Indoles in bananas. *0759*

Urobilinogen *Urine Increase Analytical* 5-HIAA produced reacts with Ehrlich's reagent. *1563*

Vanillylmandelic Acid *Urine Increase Analytical* Metabolites interfere with analysis. *1237*

Barbital

Alanine Aminotransferase *Serum No Effect Analytical* At acute overdose concentration (20 mg/dL) on Technicon® SMAC® method. *3719*

Albumin *Serum No Effect Analytical* At acute overdose concentration (20 mg/dL) on Technicon® SMAC® method. *3719* At concentration of 500 mg/L had no effect as measured by BCG method. *3393*

Alkaline Phosphatase *Serum No Effect Analytical* At acute overdose concentration (20 mg/dL) on Technicon® SMAC® method. *3719*

Aspartate Aminotransferase *Serum No Effect Analytical* At acute overdose concentration (20 mg/dL) on Technicon® SMAC® method. *3719*

Bicarbonate *Serum No Effect Analytical* At concentration of 500 mg/L had no effect on method using phenolphthalein. *3393*

Bilirubin *Serum No Effect Analytical* At acute overdose concentration (20 mg/dL) on Technicon® SMAC® method. *3719* At concentration of 500 mg/L had no effect as measured by Jendrassik and Grof method. *3393*
Serum No Effect Physiological Insignificant protein-displacing effect in neonates. *3748*

Calcium *Serum No Effect Analytical* At acute overdose concentration (20 mg/dL) on Technicon® SMAC® method. *3719* At concentration of 500 mg/L had no effect as measured by cresolphthalein method. *3393*

Chloride *Serum No Effect Analytical* At concentration of 500 mg/L had no effect as measured by mercurimetric method. *3393*

Cholesterol *Serum No Effect Analytical* At acute overdose concentration (20 mg/dL) on Technicon® SMAC® method. *3719* At concentration of 500 mg/L had no effect as measured by Liebermann-Burchard method. *3393*

Creatine Kinase *Serum No Effect Analytical* At acute overdose concentration (20 mg/dL) on Technicon® SMAC® method. *3719*

Creatinine *Serum No Effect Analytical* At acute overdose concentration (20 mg/dL) on Technicon® SMAC® method. *3719* At concentration of 500 mg/L had no effect as measured by AutoAnalyzer Jaffé method. *3393*

Glucose *Serum No Effect Analytical* At acute overdose concentration (20 mg/dL) on Technicon® SMAC® method. *3719*

Iron *Serum No Effect Analytical* At acute overdose concentration (20 mg/dL) on Technicon® SMAC® method. *3719* At concentration of 200 mg/L had no effect as measured by Ferrozine method. *3393*

Lactate Dehydrogenase *Serum No Effect Analytical* At acute overdose concentration (20 mg/dL) on Technicon® SMAC® method. *3719*

Phosphate *Serum No Effect Analytical* At acute overdose concentration (20 mg/dL) on Technicon® SMAC® method. *3719* At concentration of 500 mg/L had no effect as measured by phosphomolybdate method. *3393*

Potassium *Serum No Effect Analytical* At concentration of 500 mg/L had no effect on measurement by ISE with predilution. *3393*

Protein *Serum No Effect Analytical* At acute overdose concentration (20 mg/dL) on Technicon® SMAC® method. *3719* At concentration of 500 mg/L had no effect as measured by biuret method with blank correction. *3393*

Prothrombin Time *Plasma Decrease Physiological* Antagonizes effect of bishydroxycoumarin. *3292*

Sodium *Serum No Effect Analytical* At concentration of 500 mg/L had no effect on measurement by ISE with predilution. *3393*

Triglycerides *Serum No Effect Analytical* At acute overdose concentration (20 mg/dL) on Technicon® SMAC® method. *3719* At concentration of 500 mg/L had no effect as measured by lipase/esterase method. *3393*

Urea Nitrogen *Serum No Effect Analytical* At acute overdose concentration (20 mg/dL) on Technicon® SMAC® method. *3719* At concentration of 500 mg/L had no effect as measured by diacetylmonoxime method. *3393*

Uric Acid *Serum No Effect Analytical* At acute overdose concentration (20 mg/dL) on Technicon® SMAC® method. *3719* At concentration of 500 mg/L had no effect as measured by phosphotungstate reduction method. *3393*

Barbiturates

Alanine Aminotransferase *Serum Increase Physiological* Occurs with poisoning, probable muscle origin. *3901*

Alkaline Phosphatase *Serum Increase Physiological* Rare case of hepatotoxicity. *0357*

Amino-4-Imidazole-5-Carboxamide Ribotide (AICAR) *Urine Increase Physiological* Occurs if megaloblastic anemia. *3773*

Barbiturates *(continued)*

Aminolevulinic Acid *Urine Increase Physiological* May precipitate acute porphyria. *1322*

Amitriptyline *Serum Decrease Physiological* Stimulates metabolism of tricyclic antidepressants. *1487*

Ammonia *Plasma Increase Physiological* Impaired metabolism in dogs. *3601*

Aspartate Aminotransferase
Serum Increase Physiological Occurs with poisoning, probable muscle origin. *3901*

Basal Metabolic Rate *Patient Decrease Physiological* Decreases rate by about 10%. *0251*

Bilirubin *Serum Increase Physiological* Rare cases of jaundice following use. *0357*
Serum Decrease Physiological Induces glucuronyl transferase in newborn infants. *1488*

BSP Retention *Serum Increase Physiological* May increase retention if given within 24 h. *1342*
Serum Decrease Physiological Increases conjugation with glutathione. *1343*

Calcium *Serum No Effect Analytical* No effect on fluorescence of calcein. *2395*

Coproporphyrin *Blood Increase Physiological* May precipitate acute cutaneous porphyria. *1322*
Urine Increase Physiological May precipitate acute porphyria. *1322*
Feces Increase Physiological May precipitate acute porphyria. *1322*

Cortisol *Plasma Decrease Physiological* If used preoperatively lower concentration. *3657*

Creatine Kinase *Serum Increase Physiological* Occurs with poisoning, probable muscle origin. *3901*

Creatinine *Serum Increase Physiological* Shock and renal failure in intoxication. *2425*

Dopamine *Urine No Effect Physiological* No effect with addiction or withdrawal. *3634*

Epinephrine *Urine No Effect Physiological* No effect with addiction or withdrawal. *3634*

Erythrocytes *Blood Decrease Physiological* Aplastic or megaloblastic anemia. *2313*

Estriol *Urine Increase Physiological* Theoretical due to increased hydroxylation. *0023*

FIGLU (N-Formiminoglutamic Acid)
Urine Increase Physiological Occurs if megaloblastic anemia. *3773*

Folate *Serum Decrease Physiological* May cause megaloblastic anemia (impairs absorption). *3773*

Glucaric Acid *Urine Increase Physiological* As result of hepatic enzyme induction. *2220*

Glucose *Serum Decrease Physiological* Reported effect. *0117*

γ-Glutamyltransferase (GGT)
Serum Increase Physiological Possibly due to enzyme induction. *3815*

Hematocrit *Blood Decrease Physiological* May cause megaloblastic anemia. *3773*

Hemoglobin *Blood Decrease Physiological* May cause megaloblastic anemia. *3773*

Homovanillic Acid *Urine No Effect Physiological* No effect with addiction or withdrawal. *3634*

Hydrochloric Acid *Gastric Material Decrease Physiological* Secretion slightly depressed. *1343*

17-Hydroxycorticosteroids *Urine No Effect Analytical* No interference with Porter-Silber procedure. *0766*
Urine No Effect Physiological No effect with short term ingestion. *3879*
Urine Decrease Physiological Chronic ingestion metabolism diverted to 6-β-hydroxy-cortisol. *0766*

5-Hydroxyindoleacetic Acid (5-HIAA)
Urine No Effect Physiological No effect with addiction or withdrawal. *3634*

^{131}I Uptake *Serum Increase Physiological* ?due to enzyme induction. *2220*

17-Ketosteroids *Urine No Effect Analytical* No interference with Zimmermann procedure. *0766*

Urine No Effect Physiological Short term ingestion produces no effect. *0766*

Leukocytes *Blood Decrease Physiological* Leukopenia. *3669*

MCV *Blood Increase Physiological* May cause megaloblastic anemia. *3773*

Myoglobin *Urine Increase Physiological* May be increased in barbiturate poisoning. *0017*

Norepinephrine *Urine No Effect Physiological* No effect with addiction or withdrawal. *3634*

pCO$_2$ *Blood Increase Physiological* Respiratory depressant. *1343*

Pepsin *Gastric Material Decrease Physiological* Secretion slightly depressed. *1343*

Platelets *Blood Decrease Physiological* Thrombocytopenia. *2313*

pO$_2$ *Blood Decrease Physiological* Slight decrease during sleep with hypnotic dose. *1343*

Porphobilinogen *Urine Increase Physiological* May precipitate acute porphyria. *1322*

Porphyrins *Urine Increase Physiological* May precipitate attack of acute porphyria. *1016*

Protein Bound Iodine (PBI) *Serum Decrease Physiological* Competes with T$_4$ for thyroxine binding prealbumin binding sites. *0583*

Prothrombin Time *Plasma Increase Physiological* Theoretically if cholestasis occurs. *2313*
Plasma Decrease Physiological Metabolism of coumarins enhanced (enzyme induction in liver). *1359*

Protoporphyrin *Blood Increase Physiological* May precipitate acute cutaneous porphyria. *1322*
Feces Increase Physiological May precipitate acute porphyria. *1322*

Pseudocholinesterase *Serum Decrease Physiological* May inhibit activity. *1680*

T$_3$ Uptake *Serum Increase Physiological* Competes for thyroxine binding prealbumin sites. *0583*

Testosterone *Serum Increase Physiological* Due to decreased metabolic clearance rate (in women). *1941*

Thyroxine (T$_4$) *Serum Decrease Physiological* Compete with T$_4$ for thyroxine binding prealbumin. *0583*

Uroporphyrin *Urine Increase Physiological* May precipitate acute porphyria. *1322*

Vanillylmandelic Acid *Urine No Effect Physiological* No effect with addiction or withdrawal. *3634*

Volume *Urine Decrease Physiological* May produce renal shutdown with oliguria. *0279*

Barium

Alanine Aminotransferase *Serum Increase Physiological* Severe liver damage if tannic acid in enema. *0071*

Aspartate Aminotransferase
Serum Increase Physiological Severe liver damage if tannic acid in enema. *0071*

Color *Feces Increase Analytical* White discoloration/speckling if taken orally. *1187*

Diagnex Blue Excretion *Urine Increase Physiological* Heavy metal displacement of diagnex blue. *3879*

^{131}I Uptake *Serum Decrease Physiological* Theoretically if contaminated with I$_2$. *0583*

Magnesium *Urine Increase Analytical* Measured by fluorometric method of Schachter. *3573*

Occult Blood *Feces Positive Physiological* May cause severe gastrointestinal tract hemorrhage. *1343*

Protein Bound Iodine (PBI) *Serum Increase Analytical* Occasionally contaminated with inorganic iodine. *0830*

BCG Vaccine

γ-Globulin *Serum Decrease Physiological* Associated with severe reaction. *2427*

Beclamide

Diphenylhydantoin *Serum No Effect Analytical* On GLC procedure of Papadopoulos at *in vivo* concentration. *2722*

Pheneturide *Serum No Effect Analytical* On GLC procedure of Papadopoulos at *in vivo* concentration. *2722*

Phenobarbital *Serum No Effect Analytical* On GLC procedure of Papadopoulos at *in vivo* concentration. *2722*

Primidone *Serum No Effect Analytical* On GLC procedure of Papadopoulos at *in vivo* concentration. *2722*

Beclomethasone

Cortisol *Plasma Decrease Physiological* Significant effect observed in asthmatic children given drug by inhalation reflecting compromizing of pituitary-adrenal axis. *3693*

Bed Rest

Calcium *Serum No Effect Physiological* May be no change in total although ionized increased. *1539*
Serum Increase Physiological Increase of approximately 0.1 meq/L. *1688*
Urine Increase Physiological Up to 50 meq/d in healthy males. *1688*
Feces Increase Physiological Up to 70 meq/d in healthy males. *1688*

Citrate *Urine No Effect Physiological* Not significantly affected by immobilization. *0863*

Estriol *Urine No Effect Physiological* No significant influence during pregnancy. *0966*

Estrogens *Plasma Decrease Physiological* Reported observation. *3217*

Glucose *Serum Decrease Physiological* Observed effect with prolonged inactivity. *0117*

17-Hydroxycorticosteroids *Urine No Effect Physiological* No effect observed on excretion. *3217*

Hydroxyproline *Urine Increase Physiological* Maximal during second month bed rest. *1688*

Ionized Calcium *Serum Increase Physiological* In normals increased up to 2.9 meq/L also in patients. *1539*

Magnesium *Urine Increase Physiological* Slight effect only. *1688*

Nitrogen *Urine Increase Physiological* Negative nitrogen balance (over 1 g/d). *1688*

pH *Urine Increase Physiological* Average increase of 0.1 to 0.2 with immobilization. *0863*

Phosphate *Serum Increase Physiological* Increased up to 0.5 mg/dL. *1688*
Urine Increase Physiological Increase of approximately 200 mg/d. *1688*
Feces Increase Physiological Slight effect only. *1688*

Potassium *Urine Increase Physiological* Increased loss of up to 14 g with immobilization. *0863*

Pyrophosphate *Urine Increase Physiological* Mild elevation with bed rest, fall with reambulation. *0927*

Sodium *Urine Increase Physiological* Slight increase during immobilization. *0863*

Beets

Color *Urine Increase Analytical* Red due to anthocyanins and other pigments. *2313*

Belladonna

Leukocytes *Blood Increase Physiological* May cause leukocytosis. *0071*

Bendrofluazide

Acetaminophen Screening Test *Urine Negative Analytical* No reaction with o-cresol at therapeutic concentrations. *3326*

Angiotensin II *Plasma Increase Physiological* Approximately doubled in 5 or 6 hypertensives given 5 mg daily for 6 weeks. *3866*

Bicarbonate *Serum Increase Physiological* Metabolic alkalosis. *1129* By 2 mmol/L approximately in 5 or 6 hypertensives given 5 mg daily for 6 weeks. *3866*

Calcium *Serum Increase Physiological* Due to increased concentration of protein due to temporary Na depletion. *2727*
Urine Increase Physiological Initial diuretic response. *0105*
Urine Decrease Physiological Impaired excretion may occur with thiazides. *2727*

Cholesterol *Serum No Effect Physiological* No significant change in 15 individuals when given as sole treatment for less than 1 y. *3826* No effect with treatment for several years in many patients. *0089*
Serum Increase Physiological 5% increase in 66 subjects when treated for less than 1 y when used as only drug. *3706* 5% increase in short term studies. *0089*

Cholesterol, High Density Lipoprotein
Serum No Effect Physiological No significant change in 15 individuals when given as sole treatment for less than 1 y. *3826*

Citrate *Urine Decrease Physiological* Up to 30% decrease observed. *2427*

1,25-Dihydroxy Vitamin D$_3$ *Serum Decrease Physiological* In 19 healthy early menopausal women given 5 mg/d with calcium supplement due to primary effect on renal tubules and secondary change on vitamin D metabolism. *2989*

24,25-Dihydroxy Vitamin D *Serum Increase Physiological* In 19 healthy early menopausal women given 5 mg/d with calcium supplement due to primary effect on renal tubules and secondary change on vitamin D metabolism. *2989*

Glucose *Serum No Effect Physiological* After 12 mo treatment of 53 previously untreated hypertensives. *0315*

Glucose Tolerance *Serum Decrease Physiological* In men and women in 3 y study of patients with mild hypertension. *2506*

25-Hydroxy Vitamin D$_3$ *Serum No Effect Physiological* In 19 healthy early menopausal women given 5 mg/d with calcium supplement due to primary effect on renal tubules and secondary change on vitamin D metabolism. *2989*

Insulin *Plasma No Effect Physiological* After 12 mo treatment of 53 previously untreated hypertensives. *0315*

Lithium *Serum Increase Physiological* Substantial increase in one patient receiving lithium to whose regime was subsequently added. *1919*

Magnesium *Serum Decrease Physiological* Significant effect noted with prolonged therapy. *3285*

Osmolality *Serum Decrease Physiological* Due to hyponatremia. *1129*

Potassium *Serum No Effect Physiological* After 12 mo treatment of 53 previously untreated hypertensives. *0315*
Serum Decrease Physiological Marked diuretic response. *1129* In men and women in 3 y study of patients with mild hypertension. *2506* Mean fall from 4.3 to 3.9 mmol/L in 7 essential hypertensives treated for 16 weeks. *0352* From 0.2 to 0.6 mmol/L in 5 or 6 hypertensives given 5 mg daily for 6 weeks. *3866*

Renin Activity *Plasma Increase Physiological* Mean increase from 3.3 to 5.6 mmol/L in 7 essential hypertensives treated for 16 weeks. *0352* Approximately doubled in 5 or 6 hypertensives given 5 mg daily for 6 weeks. *3866*

Sodium *Serum No Effect Physiological* In 7 essential hypertensives treated for 16 weeks. *0352*
Serum Decrease Physiological May cause hyponatremia due to potassium depletion. *1129*

Triglycerides *Serum No Effect Physiological* No significant change in 15 individuals when given as sole treatment for less than 1 y. *3826* After 12 mo treatment of 53 previously untreated hypertensives. *0315*
Serum Increase Physiological 20% increase in 66 subjects when treated for less than 1 y when used as only drug. *3706* 20% increase in short term studies. *0089*

Urea Nitrogen *Serum Increase Physiological* In men and women in 3 y study of patients with mild hypertension. *2506*
Serum Decrease Physiological Mean increase from 5.6 to 6.5 mmol/L in 7 essential hypertensives treated for 16 weeks. *0352*

Bendrofluazide *(continued)*

Uric Acid *Serum No Effect Physiological* After 12 mo treatment of 53 previously untreated hypertensives. *0315*
Serum Increase Physiological Inhibition of tubular secretion of urate. *1433* In men and women in 3 y study of patients with mild hypertension. *2506* Increase by 2.7 to 4 mg/dL approximately in 5 or 6 hypertensives given 5 mg daily for 6 weeks. *3866*

Zinc *Urine Increase Physiological* Increase by 60% in 9 hypertensives treated for 2 weeks. *3806*

Benfluorex

Cholesterol *Serum Decrease Physiological* Increase by 17% in 12 hypertriglyceridemic type 2 diabetic patients given 150 mg tid after 1 mo treatment. *3385*

Cholesterol, High Density Lipoprotein
Serum No Effect Physiological In 12 hypertriglyceridemic type 2 diabetic patients given 150 mg tid after 1 mo treatment. *3385*

Cholesterol, Very Low Density Lipoprotein
Serum Decrease Physiological Increase by 20% in 12 hypertriglyceridemic type 2 diabetic patients given 150 mg tid after 1 mo treatment. *3385*

Hemoglobin A₁c *Blood Decrease Physiological* Significant effect in hypertriglyceridemic type 2 diabetic patients given 150 mg tid after 1 mo treatment. *3385*

Triglycerides *Serum Decrease Physiological* Increase by 33% in 12 hypertriglyceridemic type 2 diabetic patients given 150 mg tid after 1 mo treatment. *3385*

Triglycerides, Very Low Density Lipoprotein
Serum Decrease Physiological Increase by 36% in 12 hypertriglyceridemic type 2 diabetic patients given 150 mg tid after 1 mo treatment. *3385*

Benserazide

Prolactin *Plasma Increase Physiological* Significant increase in 5 patients with primary hypothyroidism; possibly due to reduction in circulating dopamine. *0869*

Thyroid Stimulating Hormone (TSH)
Serum Increase Physiological In hypothyroidism: is a dopamine antagonist. *3798* Significant increase in 5 patients with primary hypothyroidism; possibly due to reduction in circulating dopamine. *0869*

Benzaldehyde

Ionized Calcium *Serum Decrease Analytical* At 0.1 mmol/L to 0.1 mol/L with Calcium specific electrode. *0540*

Benzaprine

Imipramine *Serum Increase Analytical* In liquid-chromatographic method described has similar retention time. *3195*

Benzazepine

Hematocrit *Blood Decrease Physiological* Slight effect observed. *1738*

Hemoglobin *Blood Decrease Physiological* Slight effect observed. *1738*

Leukocytes *Blood Decrease Physiological* Slight effect observed. *1738*

Benzbromarone

Cholesterol *Serum No Effect Analytical* At concentration of 60 mg/L had no effect as measured by CHOD-Iodide method. *3393* At concentration of 60 mg/L had no effect as measured by CHOD-PAP method. *3393* At concentration of 60 mg/L had no effect as measured by catalase-AIDH method. *3393* At concentration of 60 mg/L had no effect as measured by method using catalase-Hantzsch reaction. *3393* At concentration of 60 mg/L had no effect as measured by Liebermann-Burchard method. *3393*

Creatinine *Serum No Effect Analytical* At concentration of 20 mg/L had no effect as measured by Jaffé-Fading-Fraction method. *3393* At concentration of 20 mg/L had no effect as measured by Jaffé-Fuller's earth method. *3393* At concentration of 47 mg/L had no effect as measured by kinetic Jaffé method on BKA-2. *3393*

Glucose *Serum No Effect Analytical* At concentration of 8 mg/L had no effect as measured by Ektachem® method. *3393*

Potassium *Serum No Effect Analytical* At concentration of 80 mg/L had no effect on ISE measurement without predilution. *3393*

Protein Electrophoresis *Serum No Effect Analytical* At concentration of 80 mg/L had no effect as measured by automated Olympus-Hite method. *3393*

Sodium *Serum No Effect Analytical* At concentration of 80 mg/L had no effect on ISE measurement without predilution. *3393*

Urea Nitrogen *Serum No Effect Analytical* At concentration of 8 mg/L had no effect as measured by Ektachem® method. *3393*

Uric Acid *Serum No Effect Analytical* At concentration of 20 mg/L had no effect as measured by Kageyama-Hantzsch method. *3393* At concentration of 20 mg/L had no effect as measured by catalase-AIDH method. *3393*
Serum Decrease Physiological Potent uricosuric agent: effect enhanced when oxipurinol co-administered. *0695*
Urine Increase Physiological Potent uricosuric agent: effect enhanced when oxipurinol co-administered. *0695*
Urine Decrease Physiological Statistically significant reduction in excretion and in clearance versus creatinine clearance in comparison with no treatment. *3386*

Benzene

Alanine Aminotransferase *Serum Increase Physiological* May cause liver damage. *0279*

Aspartate Aminotransferase
Serum Increase Physiological May cause liver damage. *0279*

Color *Urine Increase Analytical* Red brown. *2313*

Erythrocytes *Blood Decrease Physiological* Mild macrocytic anemia/aplastic anemia. *2429*

β₁-α-Globulin *Serum Decrease Physiological* Significant effect in people occupationally exposed. *3370*

Glucose *Serum Decrease Physiological* Toxic effect. *2313*

Haptoglobin *Serum Decrease Physiological* May cause hemolytic anemia. *0279*

Hemoglobin *Blood Decrease Physiological* Due to aplastic anemia. *1343*

Immunoglobulin IgA *Serum Decrease Physiological* Significant effect in people occupationally exposed. *3370*

Immunoglobulin IgG *Serum Decrease Physiological* Significant effect in people occupationally exposed. *3370*

Immunoglobulin IgM *Serum No Effect Physiological* No effect in people occupationally exposed. *3370*

Ionized Calcium *Serum Decrease Analytical* At 0.1 mmol/L to 0.1 mol/L with Calcium specific electrode. *0540*

Lactate Dehydrogenase *Serum Decrease Physiological* In vitro 6.1% decrease at 1×10^{-4} mol/L. *1270*

Leukocytes *Blood Increase Physiological* Lymphocytic or polymorphonuclear early. *1343*
Blood Decrease Physiological Probable myelotoxic effect (usually late). *2429*

Lysozyme *Urine Increase Physiological* With chronic exposure-related to monocytic leukemia. *0040*

Myeloblasts *Blood Increase Physiological* Chronic exposure may cause acute leukemia. *0040*

Occult Blood *Feces Positive Physiological* May cause gastrointestinal bleeding and other symptoms. *0279*

Phenol *Urine Increase Physiological* Relationship between exposure to benzene vapor. *3301*

Platelets *Blood Decrease Physiological* Probable myelotoxic effect (common). *2429*

Protein *Serum Decrease Physiological* Reduced synthesis due to liver damage. *0279*

Pseudocholinesterase *Serum Decrease Physiological* In vitro 7.1% decrease at 1 x 10⁻⁴ mol/L. *1270*

Volume *Urine Decrease Physiological* May produce oliguria with renal failure. *0279*

Benziodarone

Alanine Aminotransferase *Serum Increase Physiological* May cause hepatic toxicity. *2313*

Alkaline Phosphatase *Serum Increase Physiological* May cause hepatic toxicity. *2313*

Aspartate Aminotransferase
Serum Increase Physiological May cause hepatic toxicity. *2313*

Bile *Urine Increase Physiological* May cause hepatic toxicity. *2313*

Bilirubin *Serum Increase Physiological* May cause hepatic toxicity. *2313*

BSP Retention *Serum Increase Physiological* May cause hepatic toxicity. *2313*

Cephalin Flocculation *Serum Increase Physiological* May cause hepatic toxicity. *2313*

¹³¹I Uptake *Serum Decrease Physiological* Due to iodine component of drug. *2220*

Protein Bound Iodine (PBI) *Serum Increase Analytical* Contains 46% iodine. *0012*

Thymol Turbidity *Serum Increase Physiological* May cause hepatic toxicity. *2313*

Thyroxine (T₄) *Serum No Effect Physiological* Relatively constant in 9 normal volunteers receiving 100 mg three times daily for 14 d due to diversion of peripheral metabolism of T_4 to rT_3 rather than T_3. *3907*

Tri-iodothyronine (T₃) *Serum Increase Physiological* Significant effect in 9 normal volunteers receiving 100 mg three times daily for 14 d due to diversion of peripheral metabolism of T_4 to rT_3 rather than T_3. *3907*

Tri-iodothyronine (T₃), Reverse
Serum Increase Physiological Significant effect in 9 normal volunteers receiving 100 mg three times daily for 14 d due to diversion of peripheral metabolism of T_4 to rT_3 rather than T_3. *3907*

Uric Acid *Serum Decrease Physiological* Stimulates secretion of urate. *1363*
Urine Increase Physiological Stimulates secretion of urate. *1363*

Benzocaine

Amphetamine *Urine No Effect Analytical* At 50 mg/L on fluorescent method of Hayes. *1534*

Erythrocytes *Blood Decrease Physiological* May cause hemolysis. *0380*

Hematocrit *Blood Decrease Physiological* May cause hemolysis. *0380*

Hemoglobin *Blood Decrease Physiological* May cause hemolysis. *0380*

Methemoglobin *Blood Increase Physiological* Hemolysis. *0380*

Sulfa as Sulfanilamide *Serum Increase Analytical* Diazotizes and interferes. *0251*

Benzoic Acid

Glucose *Serum No Effect Analytical* If saturated on o-toluidine methods. If saturated on alkaline ferricyanide procedures. If saturated on glucose-oxidase methods. If saturated on p-HBAH procedure of Lever. *2140*

Benzphetamine

Amphetamine *Urine No Effect Analytical* No effect at 100 mg/L on method of Rutter. *3097*

Benzquinamide

Fatty Acids, Free (FFA) *Serum Increase Physiological* 136% increase following single injection i.m. of 300 mg. *1634*

Benzthiazide

Amylase *Serum Increase Physiological* Rare, may occur as with other thiazides. *1680*

Bilirubin *Serum Increase Physiological* Rare, may occur as with other thiazides. *1680*

Chloride *Urine Increase Physiological* Therapeutic diuretic intent. *1680*

Erythrocytes *Blood Decrease Physiological* May occur as with other thiazides. *1680*

Glucose *Serum Increase Physiological* May occur as with other thiazides. *1680*
Urine Increase Physiological Rare, may occur as with other thiazides. *1680*

Leukocytes *Blood Decrease Physiological* May occur as with other thiazides. *1680*

Platelets *Blood Decrease Physiological* May occur as with other thiazides. *1680*

Potassium *Serum Decrease Physiological* May occur as with other thiazides. *1680*

Sodium *Urine Increase Physiological* Therapeutic intent (maximum effect 4-6 h). *1680*

Urea Nitrogen *Serum Increase Physiological* May occur as with other thiazides. *1680*

Uric Acid *Serum Increase Physiological* May occur as with other thiazides. *1680*

Volume *Urine Increase Physiological* Therapeutic diuretic intent. *1680*

Benztropine

Alkaloids *Urine No Effect Analytical* With ninhydrin in Frings TLC procedure at 1 mg/dL. *1204*
Urine Increase Analytical Purple with iodoplatinate in Frings TLC procedure. *1204*

Barbiturate *Urine No Effect Analytical* With diphenylcarbazone in Frings TLC procedure at 1 mg/dL. With mercurous NO_3 in Frings TLC procedure at 1 mg/dL. With mercuric SO_4 in Frings TLC procedure at 1 mg/dL. With vanillin on Frings procedure at 1 mg/dL. *1204*

Bromide *Urine No Effect Analytical* No effect on method of Frings. *1204*

Ethchlorvynol *Urine No Effect Analytical* No effect on method of Frings and Cohen. *1204*

FPN Test *Urine No Effect Analytical* No effect at 1 mg/dL on method of Frings. *1204*

Prolactin *Plasma No Effect Physiological* No significant response to 1.0 mg given i.m. *1416*

Salicylate *Urine No Effect Analytical* At 1 mg/dL no effect on Trinder method. *1204*

Benzyl Alcohol

Cortisol *Plasma Increase Analytical* As contaminant may affect fluorometric procedures. *0583*

Glucose *Serum Increase Physiological* Used as saline preservative, effect in mice ?humans. *0210*

Ionized Calcium *Serum Decrease Analytical* At 0.1 mmol/L to 0.1 mol/L with Calcium specific electrode. *0540*

Beryllium Salts

Alkaline Phosphatase *Serum Decrease Analytical* Inhibitors of enzyme in laboratory procedures. *3588*

Protein *Urine Increase Physiological* May cause nephrotoxicity. *3204*

Urea Nitrogen *Serum Increase Physiological* May cause nephrotoxicity. *3204*

Uric Acid *Serum Increase Physiological* Nephropathy associated with decreased excretion per nephron. *1903*

Beryllium Salts (continued)

Uric Acid (continued)
Urine Decrease Physiological Nephropathy associated with decreased excretion per nephron. *1903*

Betamethasone

Bicarbonate *Serum Increase Physiological* May cause hypokalemic alkalosis. *1680*

Cortisol *Plasma Decrease Physiological* Very low values observed in children in whom cream applied topically. *0777*

17-Hydroxycorticosteroids *Plasma Increase Physiological* Slight effect, absorbed through skin. *0071*
Urine Increase Physiological Slight increase, may be absorbed through skin. *0071*

17-Ketosteroids *Urine Decrease Physiological* Feedback pituitary suppression of ACTH. *1238*

Occult Blood *Feces Positive Physiological* May activate peptic ulcer. *1680*

Potassium *Serum Decrease Physiological* Occurs infrequently. *0071*
Urine Increase Physiological Occurs infrequently. *0182*

Sodium *Serum Increase Physiological* Occurs infrequently with edema. *0071*

Betazole

Amphetamine *Urine No Effect Analytical* No reaction with NBD chloride procedure of Monforte. *2473*

Hydrochloric Acid *Gastric Material Increase Physiological* Used to stimulate secretion. *0071*

Ionized Calcium *Serum Decrease Physiological* Increase by 0.062 mmol/L 1 h after 1.5 mg/kg s.c. *1682*

Propoxyphene *Urine No Effect Analytical* Less than 1% fluorescence in procedure of Valentour. *3662*

Bethanechol

Albumin *Serum No Effect Analytical* No effect at 0.09 mg/dL on SMA 12/60 method. *2636*

Alkaline Phosphatase *Serum No Effect Analytical* No effect at 0.09 mg/dL on SMA 12/60 method. *2636*

Amylase *Serum Increase Physiological* May cause increased secretion, spasm of sphincter of Oddi. *0816*

Aspartate Aminotransferase *Serum No Effect Analytical* No effect at 0.09 mg/dL on SMA 12/60 method. *2636*
Serum Increase Physiological Impaired excretion due to spasm of sphincter. *0652*

Bilirubin *Serum No Effect Analytical* No effect at 0.09 mg/dL on SMA 12/60 method. *2636*
Serum Increase Physiological Impaired excretion due to spasm of sphincter. *0652*

BSP Retention *Serum Increase Physiological* Impaired excretion due to spasm of sphincter. *0652*

Calcium *Serum No Effect Analytical* No effect at 0.09 mg/dL on SMA 12/60 method. *2636*

Cholesterol *Serum No Effect Analytical* No effect at 0.09 mg/dL on SMA 12/60 method. *2636*

Creatinine *Serum No Effect Analytical* No effect at 0.09 mg/dL on SMA 12/60 method. *2636*

Glucose *Serum No Effect Analytical* No effect at 0.09 mg/dL on SMA 12/60 method. *2636*

Lactate Dehydrogenase *Serum No Effect Analytical* No effect at 0.09 mg/dL on SMA 12/60 method. *2636*

Lipase *Serum Increase Physiological* May cause increased secretion, spasm of sphincter of Oddi. *0816*

Phosphate *Serum No Effect Analytical* No effect at 0.09 mg/dL on SMA 12/60 method. *2636*

Protein *Serum No Effect Analytical* No effect at 0.09 mg/dL on SMA 12/60 method. *2636*

Uric Acid *Serum No Effect Analytical* No effect at 0.09 mg/dL on SMA 12/60 method. *2636*

Bethanidine

Amphetamine *Urine No Effect Analytical* No effect at 100 mg/L on method of Rutter. *3097*

Creatinine *Serum Increase Physiological* Effect reported in one patient. *3479*

Norepinephrine *Urine Decrease Physiological* Reported effect. *1488*

Platelets *Blood Decrease Physiological* Mild thrombocytopenia reported. *3856*

Bezafibrate

Alanine Aminotransferase *Serum No Effect Analytical* No effect at therapeutic concentration on Reflotron method. *1984*

Apolipoprotein AI *Serum No Effect Physiological* In 11 hypertriglyceridemic subjects. *3291*

Apolipoprotein AII *Serum No Effect Physiological* In 11 hypertriglyceridemic subjects. *3291*

Aspartate Aminotransferase *Serum No Effect Analytical* No effect at therapeutic concentration on Reflotron method. *1984*

Cholesterol *Serum No Effect Analytical* No effect at therapeutic concentration on Reflotron method. *1984*

Cholesterol, High Density Lipoprotein
Serum Increase Physiological Increase by 21% in 7 hypertriglyceridemic type 2 diabetic patients given 200 mg tid after 1 mo treatment. *3385* Increase by 13% in 11 hypertriglyceridemic subjects. *3291*

Cholesterol, Low Density Lipoprotein
Serum Increase Physiological Increase by 16% in 7 hypertriglyceridemic type 2 diabetic patients given 200 mg tid after 1 mo treatment. *3385*

Creatine Kinase *Serum Increase Physiological* In four patients with poor renal function produced myolysis as result of overdose due to renal dysfunction. *3086*

Creatinine *Serum Increase Physiological* In four patients with poor renal function produced myolysis as result of overdose due to renal dysfunction. *3086*

Glucose *Serum No Effect Analytical* No effect at therapeutic concentration on Reflotron method. *1984*

γ-Glutamyltransferase (GGT) *Serum No Effect Analytical* No effect at therapeutic concentration on Reflotron method. *1984*

Myoglobin *Serum Increase Physiological* In four patients with poor renal function produced myolysis as result of overdose due to renal dysfunction. *3086*
Urine Increase Physiological In four patients with poor renal function produced myolysis as result of overdose due to renal dysfunction. *3086*

Triglycerides *Serum No Effect Analytical* No effect at therapeutic concentration on Reflotron method. *1984*
Serum Decrease Physiological Increase by 44% in 7 hypertriglyceridemic type 2 diabetic patients given 200 mg tid after 1 mo treatment. *3385* Increase by 54% in 7 hypertriglyceridemic type 2 diabetic patients given 200 mg tid after 1 mo treatment. *3385* By average of 58% in 11 hypertriglyceridemic subjects. *3291*

Urea Nitrogen *Serum No Effect Analytical* No effect at therapeutic concentration on Reflotron method. *1984*

Uric Acid *Serum No Effect Analytical* No effect at therapeutic concentration on Reflotron method. *1984*

VLDL-Apoprotein B *Serum Decrease Physiological* Residence time in plasma fell from 3.4 to 1.0 h. *3291*

VLDL-Remnant Apoprotein B
Serum Increase Physiological Metabolism little effect but plasma concentration rose 30%. *3291*

Bicarbonates

Bicarbonate *Serum Increase Physiological* Induces metabolic alkalosis. *0652*

Calcium *Urine Decrease Physiological* Protects skeleton from reabsorption of PO_4. *0243*

Chloride *Serum Decrease Physiological* Induces metabolic alkalosis. *0652*

Lactate *Serum Increase Physiological* Small effect after i.v. Na bicarbonate. *1487*

Phosphate *Urine Increase Physiological* Possible competition for same excretory mechanism. *0957*

Potassium *Serum Decrease Physiological* Induces metabolic alkalosis. *0652*

Protein *Urine Increase Analytical* Highly alkaline urine causes false positive. *1487*

Sodium *Serum Increase Physiological* Induces metabolic alkalosis. *0652*

Bicine

Protein *Test Conditions Decrease Analytical* Lowry procedure ?non linear absorption. *1796*

Bile

Color *Urine Increase Analytical* Green color with oxidized bile on standing. *0459*

PSP Excretion *Urine Increase Analytical* Interfering color. *2313*

Bile Salts

Cholesterol *Serum Increase Physiological* Augment cholesterol pool. *0652*

Urobilinogen *Urine Increase Analytical* Converted to acids, produce turbidity. *1563*

Bilirubin

N-Acetylglucosaminidase *Urine No Effect Analytical* No effect at 5 g/L on 2 colorimetric analytical methods. *1354*

Albumin *Serum No Effect Analytical* No effect if bromocresol green used in method. *2131*
Serum Decrease Analytical If HABA, methyl orange, or biuret methods used. *2131* Albumin-bilirubin binding firm with eosin procedure. *1673*
Test Conditions Decrease Analytical Competes for binding sites ANSA method. *3596*

Alkaline Phosphatase *Serum No Effect Analytical* No effect up to 15 mg/dL method of Proksch. *2878*
Serum Increase Analytical Forms diazonium salt with alpha-naphthyl phosphate. *0251*

Ammonia *Plasma No Effect Analytical* No effect on measurement procedure. *1936*

Amylase *Serum No Effect Analytical* No effect on Phadebas procedure. *1726* At 0.17 mmol/L on method of Rauscher et al. *2926*

Angiotensin-Converting Enzyme (ACE)
Serum No Effect Analytical No effect on colorimetric method of Boomsma and Schalekamp. *2814*

Aspartate Aminotransferase *Serum Increase Analytical* Interference in diazonium end point reactions. *0077*

Bacteria *Urine Negative Analytical* May interfere with color on Microstix. *0129*

Bicarbonate *Serum Increase Analytical* At 300 mg/L increased concentration by as much as 13 mmol/L when Worthington procedure used on Olympus Demand. *1746*

Bile Acids, Conjugated *Serum No Effect Analytical* At up to 300 mg/L had insignificant effect on Becton-Dickinson RIA method. *2715*

Bilirubin *Serum Increase Analytical* Conjugated bilirubin increases polarographic concentration of unconjugated. *1374*

Calcium *Serum No Effect Analytical* With method of Morin at 10 mg/dL. *2485*
Serum Increase Analytical If high may confuse EDTA titration endpoint. *1563*
Serum Decrease Analytical Quenches fluorescence in method of Rushton. *3091* If calcein titration used. *0055*

Ceruloplasmin *Serum No Effect Analytical* No effect observed on Henry PPD procedure. *0877*

Cholesterol *Serum No Effect Analytical* When measured enzymatically with oxidase. *3027* Up to 0.5 mmol/L on Leffler's procedure. *2854* At concentration of 40.0 mg/dL on Ektachem® method. *3407* At 40.0 mg/dL on method recommended on I.L. Multistat III. *3407* At concentration of 40.0 mg/dL on Du Pont aca II. *3407*
Serum Increase Analytical Liebermann-Burchard chromogen 5-9 x cholesterol. Moderate effect with ferric chloride 1 mg equivalent to 4.5 mg/dL with AutoAnalyzer procedure. *3217* Interferes with direct procedures. *1237* Esters increased absorption at 0.2 mmol/L (can be extracted). *2854* At 24.0 mg/dL and above on recommended method on RA-1000. At 5.0 mg/dL and above on methods recommended on BMD-8700 but with ferrocyanide. *3407* At 24.0 mg/dL and above on method recommended on Hitachi® 705 with ferrocyanide. *3407*
Serum Decrease Analytical Increase by 1.0% at 15 mg/dL on enzymatic procedure. *0057* At concentrations as low as 2.5 mg/dL and higher on BMD-8700 using manufacturer recommended method. At 8.0 mg/dL and above on methods recommended on Hitachi® 705. *3407* At 5.0 mg/dL and above on method recommended on Centrifichem. *3407* From less than 2.5 mg/dL on method recommended on Cobas-Bio. *3407*

Color *Urine Increase Analytical* Yellowish foam. *2448*

Creatinine *Serum No Effect Analytical* By up to 200 mg/L on enzymatic procedure. *2524*
Serum Decrease Analytical On kinetic Jaffé methods: related to bilirubin concentration, method and temperature dependent, caused by both conjugated and unconjugated bilirubin, unrelated to creatinine concentration: effects due to oxidation of bilirubin by alkali. *1964*

Crystals *Urine Increase Physiological* Red-brown amorphous needles in acid pH. *0443*

Digoxin *Serum Increase Analytical* May affect radioimmunoassay if beta count. *0615*

Fatty Acids, Free (FFA) *Serum No Effect Analytical* Even at 30 mg/dL on method of Mikac-Devic. *2438* No effect on method of Soloni. *3381* No effect on method of Pinelli. *2819*
Serum Increase Analytical In methods using diethyldithiocarbamate. *2438* Extracted in $CHCl_3$ in procedure of Itaya and Ui. *2079*

Ferric Chloride Test *Urine Positive Analytical* May produce blue-green. *0443*

Gentamicin *Serum No Effect Analytical* No difference between plate-diffusion and tube dilution. *1266*

Glucose *Serum No Effect Analytical* No effect on hexokinase method of Yee. *3912* At 20 mg/dL on MBTH procedure of Neeley. *2569* At 15 mg/dL on hexokinase method of Coburn. *0676*
Serum Increase Analytical May have marked effect on o-toluidine method. *3902*
Serum Decrease Analytical Slight effect on hexokinase procedures. *3217* Effect can be avoided by using Somogyi filtrate. *1563* Very slight effect with automated alkaline ferricyanide. *3217*
Urine Decrease Analytical May produce false negative with Clinistix®. *2544*

Guanase *Serum No Effect Analytical* Up to 150 mg/L on method of Nishikawa. *2602*

Hemoglobin *Plasma No Effect Analytical* No effect on method of de Mendonca. *0672*
Blood Increase Analytical In high concentrations affects oxyhemoglobin methods. *2981*

Indocyanine Green *Serum No Effect Analytical* No effect at high plasma concentrations. *2398*

Iron *Serum No Effect Analytical* No effect on iron measurement. *2981*

Iron-Binding Capacity, Total (TIBC)
Serum No Effect Analytical No effect on measurement. *2981*

Lipase *Serum Increase Analytical* Turbidimetric but not titrimetric methods. *0583*

Oxytetracycline *Serum Increase Physiological* Affects fluorometric method of Murthy. *2533*

Phenylketones *Urine Positive Analytical* Blue-green with $FeCl_3$. *0775*

Phosphate *Serum No Effect Analytical* If free, on method of Peynet. *2799* On enzymatic procedure with centrifugal analyzer. *2790*

Bilirubin (continued)

Phosphate (continued)
Serum Increase Analytical With UV molybdate procedure on centrifugal analyzer. 2789

Protein *Serum No Effect Analytical* On ACA using dual wavelength biuret procedure. 2782
Serum Increase Analytical No effect up to 29 mg/dL. 1563 Above 5 mg/dL (20 mg/dL = 0.2 g/L). 3217
Urine No Effect Analytical At 4 mg/dL on TCA dye method of Pesce and Strande. 2791
CSF No Effect Analytical At 4 mg/dL with TCA dye method of Pesce and Strande. 2791
CSF Increase Analytical May affect turbidimetric procedures. 0583

Salicylate *Serum No Effect Analytical* Up to 20 mg/dL no effect on Trinder procedure. 2251

Thymol Turbidity *Serum No Effect Analytical* No effect up to 14 mg/dL. 1563

Thyroxine (T₄) *Serum No Effect Analytical* At 10 mg/dL on competitive protein binding procedure of Alexander. 0054

Triglycerides *Serum No Effect Analytical* No effect at 17 mg/dL on method of Levy. 2151 At concentration of 40.0 mg/dL on Ektachem® method. At 40.0 mg/dL on recommended methods on Centrifichem. At 40.0 mg/dL on recommended method on I.L. Multistat III. At 40.0 mg/dL on method recommended on Hitachi® 705 with ferrocyanide. 3407 At concentration of 40.0 mg/dL on Du Pont aca II. 3407
Serum Increase Analytical At 24.0 mg/dL and above on methods recommended on BMD-8700 but with ferrocyanide. At 5.0 mg/dL and above on method recommended on Hitachi® 705 with ferrocyanide. From below 2.5 mg/dL and above on recommended methods on Centrifichem. At 40.0 mg/dL on recommended method on I.L. Multistat III. 3407
Serum Decrease Analytical From less than 2.5 mg/dL on recommended method on RA-1000. At concentrations as low as 2.5 mg/dL and higher on BMD-8700 using manufacturer recommended method. 3407 At 8.0 mg/dL and above on method recommended on Hitachi® 705. 3407 From below 2.5 mg/dL on method recommended on Centrifichem. 3407 At 24.0 mg/dL and above on method recommended on Cobas-Bio. 3407

Urea Nitrogen *Serum No Effect Analytical* No effect on Berthelot reaction. 1862 With DMAB procedure of Morin below 3 mg/dL. 2487
Serum Increase Analytical With DMAB method of Morin above 3 mg/dL. 2487

Uric Acid *Serum No Effect Analytical* At 200 mg/L on uricase procedure of Kabasak. 1840 With phosphotungstate procedures up to 6.6 mg/dL. 0904 At concentration of 40.0 mg/dL on Ektachem® method. 3407 At 40.0 mg/dL on method recommended on Centrifichem. 3407 At 40.0 mg/dL on method recommended on Cobas-Bio. 3407 At concentration of 40.0 mg/dL on Du Pont aca II. 3407
Serum Increase Analytical 6% with phosphotungstate procedure at 8.5 mg/dL. 0904 At 32.0 mg/dL and above on recommended method on RA-1000. At 5.0 mg/dL and above on Hitachi® 705 using Behring reagent. 3407 At 16.0 mg/dL and above on method recommended on I.L. Multistat III. 3407
Serum Decrease Analytical At concentrations as low as 2.5 mg/dL and higher on BMD-8700 using manufacturer recommended method. At 5.0 mg/dL and above on methods recommended on Hitachi® 705. 3407 At 5.0 mg/dL and above on method recommended on Hitachi® 705 with ferrocyanide. 3407

Urobilinogen *Urine No Effect Analytical* If p-MDFB procedure and absorption read at 436 nm. 2044
Urine Increase Analytical Produces green color due to biliverdin. 1563

Zinc Sulfate Turbidity *Serum No Effect Analytical* Probably same noninterference as with thymol turbidity. 1563

Bilirubin Glucuronide

Glucose *Serum Decrease Analytical* Inhibits coupled glucose-oxidase reaction. 0583

Biliverdin

Color *Urine Increase Analytical* Greenish foam. 2448

Urobilinogen *Urine Increase Analytical* May produce green color. 1563

Biperiden

Volume *Urine Decrease Physiological* Decreased flow noted in a few patients. 1680

Bisacodyl

Acetaminophen Screening Test *Urine Negative Analytical* No reaction with o-cresol at therapeutic concentrations. 3326

Cholesterol *Serum No Effect Analytical* At concentration of 40 mg/L had no effect as measured by CHOD-Iodide method. 3393 At concentration of 40 mg/L had no effect as measured by CHOD-PAP method. 3393 At concentration of 40 mg/L had no effect as measured by catalase-AIDH method. 3393 At concentration of 40 mg/L had no effect as measured by method using catalase-Hantzsch reaction. 3393 At concentration of 40 mg/L had no effect as measured by Liebermann-Burchard method. 3393

Creatinine *Serum No Effect Analytical* At concentration of 4 mg/L had no effect as measured by Jaffé-Fading-Fraction method. 3393 At concentration of 4 mg/L had no effect as measured by Jaffé-Fuller's earth method. 3393 At concentration of 2 mg/L had no effect as measured by kinetic Jaffé method on BKA-2. 3393

Fat *Feces Increase Physiological* May cause steatorrhea if protracted ingestion. 3855 Observed in laxative abusers. 3023

Glucose *Serum No Effect Analytical* At concentration of 0.2 mg/L had no effect as measured by Ektachem® method. 3393 At concentration of 4,000 mg/L had no effect as measured by hexokinase/G-6-PDH method. 3393
Urine No Effect Analytical No effect observed with Tes-Tape®. 1100
Urine Decrease Analytical Low with Clinistix®, Diastix®. 1100

Lactate *Serum No Effect Analytical* At concentration of 4,000 mg/L had no effect as measured by enzymatic method. 3393

Occult Blood *Feces No Effect Analytical* No effect noted on Hemoquant method. 3671

Potassium *Serum No Effect Analytical* At concentration of 2 mg/L had no effect on ISE measurement without predilution. 3393
Serum Decrease Physiological Associated with steatorrhea if used in excess. 3855

Protein Electrophoresis *Serum No Effect Analytical* At concentration of 2 mg/L had no effect as measured by automated Olympus-Hite method. 3393

Sodium *Serum No Effect Analytical* At concentration of 2 mg/L had no effect on ISE measurement without predilution. 3393

Urea Nitrogen *Serum No Effect Analytical* At concentration of 0.2 mg/L had no effect as measured by Ektachem® method. 3393

Uric Acid *Serum No Effect Analytical* At concentration of 20 mg/L had no effect as measured by Kageyama-Hantzsch method. 3393 At concentration of 4 mg/L had no effect as measured by catalase-AIDH method. 3393

Bismuth Salts

Alanine Aminotransferase *Serum Increase Physiological* Hepatotoxicity. 2313

Alkaline Phosphatase *Serum Increase Physiological* Hepatotoxicity. 2313

Amino Acids *Serum Increase Physiological* Toxicity effect. 1022
Urine Increase Physiological Due to proximal tubular dysfunction (Fanconi syndrome). 2911

Aspartate Aminotransferase
Serum Increase Physiological Possibly of renal origin due to nephrotoxicity. 2911

Bile *Urine Increase Physiological* Hepatotoxicity. 2313

Bilirubin *Serum   Increase   Physiological* Hepatotoxicity. *2313*

Bismuth *Serum   Increase   Physiological* Measurable in poisoning. *2427*
Urine   Increase   Physiological Measurable with poisoning. *2427*

BSP Retention *Serum   Increase   Physiological* Hepatotoxicity. *2313*

Casts *Urine   Increase   Physiological* Mainly granular due to nephrotoxicity. *2911*

Cells *Urine   Increase   Physiological* Mainly tubular due to nephrotoxicity. *2911*

Cephalin Flocculation *Serum   Increase   Physiological* Hepatotoxicity. *2313*

Color *Feces   Increase   Analytical* Blackens or discolors stool with 5 g orally. *2313*

Erythrocytes *Blood   Decrease   Physiological* May cause aplastic anemia. *0279*
Urine   Increase   Physiological May cause severe renal damage. *2427*

Glucose *Urine   Increase   Physiological* Due to proximal tubular dysfunction (Fanconi syndrome). *2911*

Hemoglobin *Urine   Increase   Physiological* May cause severe renal damage. *2427*

Lactate Dehydrogenase Isoenzymes
Serum   Increase   Physiological Especially of LDH 1 and 2 — probably of renal origin. *2911*

Leukocytes *Blood   Decrease   Physiological* May cause aplastic anemia/agranulocytosis. *2313*

Phosphate *Urine   Increase   Physiological* Fanconi syndrome with poisoning. *2427*

Platelets *Blood   Decrease   Physiological* May cause aplastic anemia. *0279*

Protein *Urine   Increase   Physiological* Nephrotoxic effect. *1022* Nephrotoxic effect reported. *3204*

Protein Bound Iodine (PBI) *Serum   Increase   Physiological* Many bismuth salts contain iodine. *0012*

Sugar *Urine   Increase   Analytical* Interferes with Benedict's reaction. *1022*

Thymol Turbidity *Serum   Increase   Physiological* Hepatotoxicity. *2313*

Urea Nitrogen *Serum   Increase   Physiological* Nephrotoxic effect reported, and even renal failure. *3204*

Bismuth Subnitrate

Methemoglobin *Blood   Increase   Physiological* May cause hemolytic anemia. *2429*

Bisoprolol

Cholesterol, High Density Lipoprotein
Serum   Decrease   Physiological From 1.22 to 1.10 mmol/L with 20 mg drug/d. *2192*

Cholesterol, Low Density Lipoprotein
Serum   No Effect   Physiological No effect with 10 or 20 mg/d. *2192*

Triglycerides, Very Low Density Lipoprotein
Serum   Increase   Physiological Significant effect with 10 mg/d (from 1.04 to 1.31 mmol/L). *2192*

Bisulfite

Ethanol *Blood   No Effect   Analytical* Over 16 weeks at 20° at concentration of 0.5% w/v. *3355*

Indocyanine Green *Serum   Decrease   Analytical* As contaminant of heparin may reduce peak. *0071*

Bleomycin

Alkaline Phosphatase *Serum   Increase   Physiological* Rare mild abnormality reported: reversible. *2406*

Aspartate Aminotransferase
Serum   Increase   Physiological Rare mild abnormality reported: reversible. *2406*

Bilirubin *Serum   Increase   Physiological* Rare mild abnormality reported: reversible. *2406*

Platelets *Blood   Decrease   Physiological* May cause bone marrow aplasia. *1436*

Blindness

Bilirubin *Serum   Increase   Physiological* Significantly higher (average 0.17 mg/dL) than normal. *1637*

Calcium *Serum   Increase   Physiological* Slight effect (average 0.03 meq/L) higher. *1637*
Urine   Decrease   Physiological Significantly lower (average 92 meq/d) than normal. *1637*

Chloride *Serum   Decrease   Physiological* Significantly lower (average 7 meq/L) than normal. *1637*
Urine   Decrease   Physiological Significantly lower (average 49 meq/d) than normal. *1637*

Cholesterol *Serum   Increase   Physiological* Significantly higher (average 51 mg/dL) than normal. *1637*

Cortisol *Plasma   Decrease   Physiological* Significantly lower (average 4.4 μg/dL) than normal. *1637*

Creatinine *Serum   Increase   Physiological* Significantly higher (average 0.2 mg/dL) than normal. *1637*

Glucose *Serum   Decrease   Physiological* Significantly lower (average 14 mg/dL) than normal. *1637*

17-Hydroxycorticosteroids *Urine   Decrease   Physiological* Significantly lower (average 5.1 mg/d) than normal. *1637*

17-Ketosteroids *Urine   Decrease   Physiological* Significantly lower (average 3.5 mg/d) than normal. *1637*

Nonprotein Nitrogen *Serum   Increase   Physiological* Significantly higher (average 6 mg/dL) than normal. *1637*

Phosphate *Serum   Decrease   Physiological* Significantly lower (average 0.2 mg/dL) than normal. *1637*
Urine   Decrease   Physiological Significantly lower (average 176 mg/d) than normal. *1637*

Potassium *Serum   Increase   Physiological* Significantly higher (average 0.4 meq/L) than normal. *1637*
Urine   Decrease   Physiological sodium/potassium ratio higher than normal. *1637*

Protein *Serum   Decrease   Physiological* Significantly lower (average 0.57 g/dL) than normal. *1637*

Sodium *Serum   Decrease   Physiological* Significantly lower (average 5 meq/L) than normal. *1637*

Urea Nitrogen *Urine   Decrease   Physiological* Significantly lower (average 4.3 g/d) than normal. *1637*

Uric Acid *Serum   Increase   Physiological* Significantly higher (average 1.4 mg/dL) than normal. *1637*
Urine   Decrease   Physiological Uric acid/creatinine ratio lower than normal. *1637*

Blood

Color *Urine   Increase   Physiological* Smoky red to brown due to hematuria. *2448*

Xanthochromia *CSF   Positive   Analytical* Spurious effect 8 h after blood added to clear CSF. *2959*

Blood Collection Tube

Cortisol *Plasma   No Effect   Analytical* No effect on concentration even with storage for 6 d in gel-barrier tubes. *1592*

Estradiol *Plasma   No Effect   Analytical* No effect on concentration even with storage for 6 d in gel-barrier tubes. *1592*

Meperidine *Serum   Decrease   Analytical* Reduced concentration when blood drawn into some Vacutainers™. *3834*

Progesterone *Plasma   Decrease   Analytical* When gel-barrier tubes used to collect specimen: effect progressively increases with time. *1592*

Propranolol *Serum   Decrease   Analytical* Spuriously low values when specimen drawn into certain Vacutainers™. *0734*

Blood Collection Tube *(continued)*

Quinidine *Serum Decrease Analytical* Observed 4 to 8% reduction in plasma concentration when some Vacutainers™ used compared with glass. *3271*

Thyroxine (T₄) *Serum No Effect Analytical* No effect on concentration even with storage for 6 d in gel-barrier tubes. *1592*

Tri-iodothyronine (T₃) *Serum No Effect Analytical* No effect on concentration even with storage for 6 d in gel-barrier tubes. *1592*

Blood Group

Alkaline Phosphatase *Serum Increase Physiological* O secretor subjects — highest intestinal component. *3743*

Cholesterol *Serum Increase Physiological* Slightly higher level reported in group A males. *0219*

Blood Pressure

Cholesterol *Serum Increase Physiological* Strongly correlated with increased systolic pressure. *2123*

Fatty Acids, Free (FFA) *Serum Increase Physiological* Strongly correlated with increased systolic pressure. *2123*

Glucose *Serum Increase Physiological* Strongly correlated with increased systolic pressure. *2123*

Hematocrit *Blood Increase Physiological* Strongly correlated with increased systolic pressure. *2123*

Protein *Serum Increase Physiological* Strongly correlated with increased systolic pressure. *2123*

Triglycerides *Serum Increase Physiological* Strongly correlated with increased systolic pressure. *2123*

Urea Nitrogen *Serum Decrease Physiological* Strongly correlated with increased systolic pressure. *2123*

Blood Transfusions

Ammonia *Plasma Increase Physiological* Especially if stored previously. *3217*

Citrate *Serum Increase Physiological* If citrated blood used. *2220*

2,3-Diphosphoglycerate
Red Blood Cells Increase Physiological Especially if stored blood used. *2226*

Folate *Serum No Effect Physiological* No effect with intraoperative transfusion. *1985*

Hemoglobin *Urine Increase Physiological* May occur especially with frozen glycerolized blood. *2220*

Iron *Serum Increase Physiological* Siderosis may occur with multiple transfusions. *3742*

pH *Blood Decrease Physiological* If massive transfusions given. *2220*

Phthalates *Plasma Positive Physiological* When blood stored for few days in plastic pack. *2295*

Potassium *Serum Increase Physiological* May occur with release from RBC. *2220* Plasma concentration affected by preservative, storage temperature and length of storage. *2836*

Urea Nitrogen *Serum Increase Physiological* Catabolism of infused protein. *1022*

Uric Acid *Serum Increase Physiological* Catabolism of infused blood cells. *1237*

Vitamin B₁₂ *Serum Increase Physiological* Significant effect in patients with intraoperative transfusion. *1985*

Body Habitus

Uric Acid *Serum Increase Physiological* In males and females with overweight. *3872*

Bombesin

Follicle Stimulating Hormone (FSH)
Plasma Increase Physiological After LHRH after 5 ng/nk/min x 2.5 h in healthy men. *2841*

Luteinizing Hormone (LH) *Plasma Increase Physiological* After LHRH after 5 ng/nk/min x 2.5 h in healthy men. *2841*

Prolactin *Plasma Decrease Physiological* After TRH after 5 ng/nk/min x 2.5 h in healthy men. *2841*

Thyroid Stimulating Hormone (TSH)
Serum Decrease Physiological After TRH after 5 ng/nk/min x 2.5 h in healthy men. *2841*

Bopindolol

Chloride *Serum No Effect Physiological* In 10 hypertensive patients treated for either 1 or 21 d. *1063*
Urine No Effect Physiological In 10 hypertensive patients treated for either 1 or 21 d. *1063*

Cholesterol *Serum No Effect Physiological* In 24 hypertensives treated with drug for 3 mo. *3667*

Cholesterol, High Density Lipoprotein
Serum No Effect Physiological In 24 hypertensives treated with drug for 3 mo. *3667*

Cholesterol, Low Density Lipoprotein
Serum No Effect Physiological In 24 hypertensives treated with drug for 3 mo. *3667*

Creatinine *Serum No Effect Physiological* In 10 hypertensive patients treated for either 1 or 21 d. *1063*
Serum Increase Physiological Minimal effect observed in large population of hypertensives over 12 weeks study period. *3779*

Fractional Sodium Excretion *Urine No Effect Physiological* In 10 hypertensive patients treated for either 1 or 21 d. *1063*

Free Water Clearance *Urine No Effect Physiological* In 10 hypertensive patients treated for either 1 or 21 d. *1063*

Hematocrit *Blood Decrease Physiological* After 1 d but reverted to normal in 10 hypertensives treated for 21 d. *1063*

Osmolality *Serum No Effect Physiological* In 10 hypertensive patients treated for either 1 or 21 d. *1063*
Urine No Effect Physiological In 10 hypertensive patients treated for either 1 or 21 d. *1063*

Potassium *Serum No Effect Physiological* In 10 hypertensive patients treated for either 1 or 21 d. *1063*
Serum Increase Physiological Minimal effect observed in large population of hypertensives over 12 weeks study period. *3779*
Urine No Effect Physiological In 10 hypertensive patients treated for either 1 or 21 d. *1063*

Protein *Serum No Effect Physiological* In 10 hypertensive patients treated for either 1 or 21 d. *1063*
Urine No Effect Physiological In 10 hypertensive patients treated for either 1 or 21 d. *1063*

Renin Activity *Plasma Decrease Physiological* In 19 patients with essential hypertension reduced mean pretreatment concentration of 2.6 µg/L/h to 1.6 µg/L/h with chronic treatment. *1689*

Sodium *Serum No Effect Physiological* In 10 hypertensive patients treated for either 1 or 21 d. *1063*
Urine No Effect Physiological In 10 hypertensive patients treated for either 1 or 21 d. *1063*

Triglycerides *Serum Increase Physiological* After 4-8 weeks by as much as 41% but disappeared after 12 weeks of therapy in 24 hypertensives. *3667*

Uric Acid *Serum Increase Physiological* Minimal effect observed in large population of hypertensives over 12 weeks study period. *3779*

Volume *Urine No Effect Physiological* In 10 hypertensive patients treated for either 1 or 21 d. *1063*

Boric Acid

Alanine Aminotransferase *Serum Increase Physiological* May cause liver damage. *0279*

Albumin *Serum No Effect Analytical* At concentration of 50 mg/L had no effect as measured by BCG method. *3393*

Aspartate Aminotransferase
Serum Increase Physiological May cause liver damage. *0279*

Bilirubin *Serum   No Effect   Analytical* At concentration of 50 mg/L had no effect as measured by Jendrassik and Grof method. *3393*
Serum   Increase   Physiological Toxicity effect with hepatomegaly. *1302*

Calcium *Serum   No Effect   Analytical* At concentration of 50 mg/L had no effect as measured by cresolphthalein method. *3393*

Chloride *Serum   No Effect   Analytical* At concentration of 50 mg/L had no effect as measured by mercurimetric method. *3393*
Serum   Increase   Physiological Toxicity effect (with acute tubular necrosis). *2313*

Cholesterol *Serum   No Effect   Analytical* At concentration of 50 mg/L had no effect as measured by Liebermann-Burchard method. *3393*

Creatinine *Serum   No Effect   Analytical* At concentration of 50 mg/L had no effect as measured by AutoAnalyzer Jaffé method. *3393*

Occult Blood *Feces   Positive   Physiological* Toxicity effect. *2313*

Orthophosphate *Test Conditions   Decrease   Analytical* Slight interference reported only. *1658*

Phosphate *Serum   No Effect   Analytical* At concentration of 50 mg/L had no effect as measured by phosphomolybdate method. *3393*

Potassium *Serum   Increase   Physiological* Toxicity effect with acute tubular necrosis. *0652*

Pregnancy Tests *Urine   No Effect   Analytical* Even in gelatin on Planotest, Pregnosticon®. *1075*
Urine   Positive   Analytical If in gelatin affects Gravindex™. *1075*

Protein *Serum   No Effect   Analytical* At concentration of 50 mg/L had no effect as measured by method using biuret with blank correction. *3393*
Urine   Increase   Physiological Due to renal damage. *1302*

Sodium *Serum   Increase   Physiological* Toxicity effect. *2313*

Triglycerides *Serum   No Effect   Analytical* At concentration of 50 mg/L had no effect as measured by lipase/esterase method. *3393*

Urea Nitrogen *Serum   No Effect   Analytical* At concentration of 50 mg/L had no effect as measured by diacetylmonoxime method. *3393*
Serum   Increase   Physiological Toxicity effect with acute tubular necrosis. *0652*

Uric Acid *Serum   No Effect   Analytical* At concentration of 50 mg/L had no effect as measured by phosphotungstate reduction method. *3393*

Volume *Urine   Decrease   Physiological* May cause oliguria with renal failure. *0279*

Boxidine

Cholesterol *Serum   Decrease   Physiological* Inhibits transformation of 7-dehydrocholesterol. *1220*

7-Dehydrocholesterol *Serum   Increase   Physiological* Inhibits transformation to cholesterol. *1220*

Urea Nitrogen *Serum   Increase   Physiological* Possible relationship to tumor destruction. *1220*

Bran

Alkaline Phosphatase *Serum   No Effect   Physiological* No significant effect with 38 g/d for 5 weeks. *1540*

Calcium *Serum   Decrease   Physiological* 0.15 meq/L with 38 g/d for 5 weeks. *1540*

Cholesterol *Serum   No Effect   Physiological* No significant effect with 38 g/d for 5 weeks. *1540*
Serum   Decrease   Physiological Possible effect with very large amounts. *1540*

Phosphate *Serum   No Effect   Physiological* No significant effect with 38 g/d for 5 weeks. *1540*

Triglycerides *Serum   Decrease   Physiological* Unprocessed bran for 5 weeks caused decreased of 18 mg/dL. *1540*

Bromate

Casts *Urine   Increase   Physiological* Due to renal damage. *1302*

Erythrocytes *Blood   Decrease   Physiological* May cause hemolysis. *1343*
Urine   Increase   Physiological Due to renal damage. *1302*

Hemoglobin *Blood   Decrease   Physiological* May cause hemolysis. *1343*

Methemoglobin *Blood   Increase   Physiological* May cause hemolysis. *1343*

Protein *Urine   Increase   Physiological* Due to renal damage. *1302*

Urea Nitrogen *Serum   Increase   Physiological* Anuria and azotemia may occur within hours. *1302*

Bromazepam

Alanine Aminotransferase *Serum   No Effect   Analytical* At 5 times therapeutic concentration on methods on SMAC®, Abbott-VP, aca, Cobas-Bio and KDA. *2138*

Albumin *Serum   No Effect   Analytical* At 5 times therapeutic concentration on BCG methods on SMAC®, Ektachem® 400, Hitachi® 705 and KDA. *2138*

Alkaline Phosphatase *Serum   No Effect   Analytical* At 5 times therapeutic concentration on methods on SMAC®, Abbott-VP, aca, Cobas-Bio, Hitachi® 705 and KDA. *2138*

Amylase *Serum   No Effect   Analytical* At 5 times therapeutic concentration on methods on aca, Cobas-Bio, Ektachem®. *2138*

Aspartate Aminotransferase *Serum   No Effect   Analytical* At 5 times therapeutic concentration on methods on SMAC®, Abbott-VP, aca, Cobas-Bio, Hitachi® 705 and KDA. *2138*

Bilirubin *Serum   No Effect   Analytical* At 5 times therapeutic concentration on methods on SMAC®, aca, Cobas-Bio, Ektachem®, Hitachi® 705 and KDA. *2138*

Calcium *Serum   No Effect   Analytical* At 5 times therapeutic concentration on methods of SMAC®, Abbott-VP, aca, Ektachem® 400, Hitachi® 705 and KDA. *2138*

Cholesterol *Serum   No Effect   Analytical* At 5 times therapeutic concentration on methods on SMAC®, Abbott-VP, Cobas-Bio, Ektachem®, Hitachi® 705 and KDA. *2138*

Creatine Kinase *Serum   No Effect   Analytical* At 5 times therapeutic concentration on methods on SMAC®, Abbott-VP, aca, Cobas-Bio and and Hitachi® 705. *2138*

Creatinine *Serum   No Effect   Analytical* At 5 times therapeutic concentration on methods of SMAC®, Abbott-VP, aca, Cobas-Bio, Hitachi® 705 and KDA. *2138*

Glucose *Serum   No Effect   Analytical* At 5 times therapeutic concentration on methods of SMAC®, Abbott-VP, aca, Cobas-Bio, Ektachem® 400, KDA and Hitachi® 705. *2138*

γ-Glutamyltransferase (GGT) *Serum   No Effect   Analytical* At 5 times therapeutic concentration on SMAC®, Abbott-VP, and Hitachi® 705. *2138*

Iron *Serum   No Effect   Analytical* At 5 times therapeutic concentration on Ferrozine method on SMAC®. *2138*

Lactate Dehydrogenase *Serum   No Effect   Analytical* At 5 times therapeutic concentration on SMAC®, Abbott-VP, Cobas-Bio, Hitachi®, KDA. *2138*

Metanephrine *Urine   No Effect   Analytical* No effect at 2 mg/L on HPLC method. *0342*

Normetanephrine *Urine   No Effect   Analytical* No effect at 2 mg/L on HPLC method. *0342*

Phosphate *Serum   No Effect   Analytical* At 5 times therapeutic concentration on methods of SMAC®, aca, Hitachi® 705, KDA. *2138*

Protein *Serum   No Effect   Analytical* At 5 times therapeutic concentration on methods of SMAC®, Abbott-VP, Ektachem® 400, Hitachi® 705 and KDA. *2138*

Triglycerides *Serum   No Effect   Analytical* At 5 times therapeutic concentration on methods on SMAC®, Abbott-VP, Ektachem® 400, Hitachi® 705 and KDA. *2138*

Urea Nitrogen *Serum   No Effect   Analytical* At 5 times therapeutic concentration on methods of SMAC®, Abbott-VP, Cobas-Bio, Ektachem® 400, Hitachi® 705 and KDA. *2138*

Bromazepam (continued)

Uric Acid *Serum No Effect Analytical* At 5 times therapeutic concentration on methods of SMAC®, Abbott-VP, aca, Cobas-Bio, Hitachi® 705, Ektachem® 400 and KDA. *2138*

Bromelains

Fibrinogen *Plasma Decrease Physiological* One drug related case reported. *1680*

Platelet Aggregation *Blood Decrease Physiological* Reduced sensitivity to ADP induced aggregation. *1549*

Prothrombin Time *Plasma Increase Physiological* Effect observed over 4 h period in animals. *0071*

Bromides

Barbiturate *Urine No Effect Analytical* With mercuric SO_4 in Frings TLC procedure at 100 mg/dL. *1204*

Bromide *Serum Increase Physiological* Concentration measurable with therapeutic doses. *2427*

Chlordiazepoxide *Blood No Effect Analytical* On method of Riddick at 5 mg/dL. *2983*

Chloride *Serum Increase Analytical* 0.03 meq/L at 20 meq/L method of Fingerhut. *1135* Measured as Cl mercurimetric, electrometric methods. *1563*
Serum Decrease Physiological Reversible halide dysequilibration. *0652*
Urine Increase Analytical Bromide measured as chloride. *1237*

Cholesterol *Serum No Effect Analytical* Insignificant effect on Leffler's procedure. *1743*
Serum Increase Analytical Interferes with Zlatkis-Zak method (up to 10%). *2973*
Serum Decrease Analytical Increase by 0.5% at 10 mg/dL (in serum) on enzymatic procedure. *0057*

Ethchlorvynol *Urine No Effect Analytical* No effect on method of Frings and Cohen at 250 mg/dL. *1205*

Hemoglobin *Urine Increase Analytical* Interferes with guaiac and benzidine tests. *3879*

^{131}I Uptake *Serum Decrease Physiological* Theoretically if contaminated with I_2. *0583*

Occult Blood *Feces Positive Analytical* In vitro reaction, ?high enough concentration in vivo. *0583*

Protein Bound Iodine (PBI) *Serum Increase Analytical* May be contaminated with iodine. *1237*

Pyruvate Kinase *Red Blood Cells Decrease Analytical* Observed with 0.3 mmol/L PEP in reaction mixture. *0370*

Salicylate *Urine No Effect Analytical* At 100 mg/dL no effect on Trinder method. *1204*

Bromisovalum

Bromide *Serum Increase Physiological* Metabolite (concentration may exceed 25 meq/L). *1343*

Chloride *Serum Increase Analytical* Metabolite may be measured as chloride. *1343*

Bromocriptine

Alkaline Phosphatase *Serum Increase Physiological* In 10 of 45 patients with parkinson's disease increased by up to 25% without other laboratory abnormalities. *2167*

Luteinizing Hormone (LH) *Plasma No Effect Physiological* No effect observed. *1736*

Norepinephrine *Plasma Decrease Physiological* Significant effect of 5 mg/d for 5 d in healthy women but not in hyperprolactinemic women. *2283*

Prolactin *Plasma Decrease Physiological* 70% decrease in healthy women and those with prolactin secreting tumors when given 5 mg/d for 5 d. *2283*

Testosterone *Serum Increase Physiological* Causes increase of about 20%. *1736*

Thyroxine (T$_4$) *Serum No Effect Physiological* In euthyroid patients although response to TRH shows no significant change. *3798*

Serum Decrease Physiological In hypothyroid patients although response to TRH shows no significant change. *3798*

Bromodeoxyuridine

Chromosomes *Test Conditions Abnormal Physiological* Clastogenic in human cells. *3282*

Brompheniramine

Amino Acids *Urine Positive Analytical* Orange-brown spot with ninhydrin on thin-layer chromatography. *1278*

Erythrocytes *Blood Decrease Physiological* Occasionally observed with antihistamines. *1679*

Hematocrit *Blood Decrease Physiological* Occasionally observed with antihistamines. *1679*

Hemoglobin *Blood Decrease Physiological* Occasionally observed with antihistamines. *1679*

^{131}I Uptake *Serum Decrease Physiological* Effect observed in some patients. *0830*

Leukocytes *Blood Decrease Physiological* Occasional leukopenia/agranulocytosis. *1679*

Neutrophils *Blood Decrease Physiological* Occasional case of agranulocytosis reported. *3717*

Protein Bound Iodine (PBI) *Serum Decrease Physiological* Inhibits iodination of tyrosine in thyroxine binding globulin. *0012*

Bucindol

Bicarbonate *Blood No Effect Physiological* No effects noted in 8 patients following several weeks on treatment. *1283*

Calcium *Serum No Effect Physiological* No effects noted in 8 patients following several weeks on treatment. *1283*

Chloride *Serum No Effect Physiological* No effects noted in 8 patients following several weeks on treatment. *1283*

Creatine Kinase *Serum Increase Physiological* Effect noted in 3 of 6 patients studied with increased activity originating from skeletal muscle. *1283*

Hemoglobin *Blood No Effect Physiological* No effects noted in 8 patients following several weeks on treatment. *1283*

Leukocytes *Blood No Effect Physiological* No effects noted in 8 patients following several weeks on treatment. *1283*

Platelets *Blood No Effect Physiological* No effects noted in 8 patients following several weeks on treatment. *1283*

Potassium *Serum No Effect Physiological* No effects noted in 8 patients following several weeks on treatment. *1283*

Sodium *Serum No Effect Physiological* No effects noted in 8 patients following several weeks on treatment. *1283*

Urea Nitrogen *Serum No Effect Physiological* No effects noted in 8 patients following several weeks on treatment. *1283*

Buformin

Amphetamine *Urine No Effect Analytical* At 50 mg/L on fluorescent method of Hayes. *1534*

Cholesterol *Serum Decrease Physiological* Probably inhibits synthesis in liver. *0878*

Glucose *Serum Decrease Physiological* Therapeutic intent. *2427*

Lactate *Serum Increase Physiological* Rises but not usually to level of acidosis. *2427*

Bufotenine

Volume *Urine Decrease Physiological* 6-7% antidiuretic activity of serotonin. *1176*

Bumetanide

Bicarbonate *Serum Increase Physiological* By approximately 1 meq/L. *2666*

Bilirubin *Serum No Effect Physiological* Although displacement observed in vitro, insignificant effect likely in vivo at pharmacological concentration. *3748*

Calcium *Urine Increase Physiological* After 1 mg maximum within 2 h over by 4 h. *0822*

Chloride *Serum Decrease Physiological* By approximately 3 meq/L. *2666*
Urine Increase Physiological Marked dose related effect. *2666*

Creatinine *Serum No Effect Physiological* No significant effect observed. *2666*

Leukocytes *Blood Increase Physiological* By up to 6,000/uL over 3 mo. *2666*

Lithium *Serum Increase Physiological* Substantial increase in one patient receiving lithium to whose regime diuretic was subsequently added. *1919*

Magnesium *Serum No Effect Physiological* No significant effect observed. *2666*
Urine Increase Physiological After 1 mg maximum within 2 h over by 4 h. *0822*

Neutrophils *Blood Decrease Physiological* Occasional case of agranulocytosis reported. *3717*

Osmolality *Serum No Effect Physiological* No significant effect observed. *2666*

Osmolar Clearance *Urine Increase Physiological* Marked dose related effect. *2666*

Potassium *Serum Decrease Physiological* By approximately 0.3 meq/L. *2666*
Urine Increase Physiological After 1 mg maximum within 2 h over by 4 h. *0822*

Protein *Serum Increase Physiological* By up to 0.6 g/dL over 3 mo. *2666*

Sodium *Serum No Effect Physiological* No significant effect observed. *2666*
Urine Increase Physiological Marked dose related effect. *2666* After 1 mg maximum within 2 h over by 4 h. *0966*

Uric Acid *Serum Increase Physiological* Increase by 1 mg/dL over 3 mo. *2666*
Urine No Effect Physiological No effect from 0-2 h. *0822*
Urine Decrease Physiological Significantly reduced from 3-6 h. *0822*

Volume *Urine Increase Physiological* After 1 mg maximum within 2 h over by 4 h. *0822* Marked dose related effect. *2666*

Water Clearance, Free *Urine Decrease Physiological* Marked dose related effect. *2666*

Zinc *Urine Increase Physiological* Increase by 14% in 9 patients with hypertension over 2 weeks. *3806*

Bunamiodyl

Bilirubin *Serum Increase Physiological* Competition for hepatic uptake of unconjugated. *0349*

BSP Retention *Serum Increase Physiological* Competes for hepatocellular protein binding sites. *0327*

Creatinine Clearance *Urine Decrease Physiological* At dose of 4.5 g; occurred without liver effect. *3796*

Protein *Urine Increase Analytical* Affects turbidity tests for up to 3 d. *0065*
Urine Increase Physiological May cause nephrotoxicity. *3204*

Protein Bound Iodine (PBI) *Serum Increase Physiological* Effect up to 2 mo (affects T_4 at high concentrations). *0604*

Urea Nitrogen *Serum Increase Physiological* May cause nephrotoxicity. *3204*

Bunitrolol

Apolipoprotein AI *Serum No Effect Physiological* When 30 mg/d given for 12 weeks in normolipidemic patients with mild essential hypertension. *3151*

Apolipoprotein AII *Serum No Effect Physiological* When 30 mg/d given for 12 weeks in normolipidemic patients with mild essential hypertension. *3151*

Apolipoprotein B *Serum No Effect Physiological* When 30 mg/d given for 12 weeks in normolipidemic patients with mild essential hypertension. *3151*

Apolipoprotein CII *Serum No Effect Physiological* When 30 mg/d given for 12 weeks in normolipidemic patients with mild essential hypertension. *3151*

Apolipoprotein CIII *Serum No Effect Physiological* When 30 mg/d given for 12 weeks in normolipidemic patients with mild essential hypertension. *3151*

Apolipoprotein E *Serum No Effect Physiological* When 30 mg/d given for 12 weeks in normolipidemic patients with mild essential hypertension. *3151*

Cholesterol *Serum No Effect Physiological* When 30 mg/d given for 12 weeks in normolipidemic patients with mild essential hypertension. *3151*

Cholesterol, High Density Lipoprotein
Serum Decrease Physiological When 30 mg/d given for 12 weeks in normolipidemic patients with mild essential hypertension. *3151*

Cholesterol, Low Density Lipoprotein
Serum No Effect Physiological When 30 mg/d given for 12 weeks in normolipidemic patients with mild essential hypertension. *3151*

Cholesterol, Very Low Density Lipoprotein
Serum No Effect Physiological When 30 mg/d given for 12 weeks in normolipidemic patients with mild essential hypertension. *3151*

Triglycerides *Serum No Effect Physiological* When 30 mg/d given for 12 weeks in normolipidemic patients with mild essential hypertension. *3151*

Bupropion

Homovanillic Acid *CSF No Effect Physiological* Insignificant effect in approximately 40 patients with depression or Alzheimer's disease after chronic treatment. *2991*

5-Hydroxyindoleacetic Acid (5-HIAA)
CSF No Effect Physiological Insignificant effect in approximately 40 patients with depression or Alzheimer's disease after chronic treatment. *2991*

Burimamide

Hydrochloric Acid *Gastric Material Decrease Physiological* (up to 60%) mainly due to decreased volume. *3905*

Volume *Gastric Material Decrease Physiological* Marked reduction occurs. *3905*

Busulfan

Bilirubin *Serum Increase Physiological* Occurs with hemolytic anemia. *1680*

Erythrocytes *Blood Decrease Physiological* pancytopenia/hemolytic anemia. *1680*

Hematocrit *Blood Decrease Physiological* pancytopenia/hemolytic anemia. *1680*

Hemoglobin *Blood Decrease Physiological* pancytopenia/hemolytic anemia. *1680*

Leukocytes *Blood Decrease Physiological* Pancytopenia/leukopenia. *2427*

Platelets *Blood Decrease Physiological* Pancytopenia/thrombocytopenia. *1680*

Protein Bound Iodine (PBI) *Serum No Effect Physiological* Normal levels observed even over long time. *0583*

Urea Nitrogen *Serum Increase Physiological* Renal damage may occur from urate deposition. *1343*

Uric Acid *Serum Increase Physiological* As a result of tissue destruction. *2313*

Butabarbital

Albumin *Serum No Effect Analytical* At concentration of 100 mg/L had no effect as measured by BCG method. *3393*

Alkaloids *Urine No Effect Analytical* With ninhydrin in Frings TLC procedure at 10 mg/dL. With iodoplatinate of Frings TLC procedure at 10 mg/dL. *1204*

Barbiturate *Urine No Effect Analytical* With vanillin on Frings procedure at 10 mg/dL. *1204*

Butabarbital (continued)

Barbiturate (continued)

Urine Increase Analytical With mercurous NO_3 reagent in Frings TLC procedure. Purple with diphenylcarbazone in Frings TLC procedure. White spot with mercuric SO_4 in Frings TLC procedure. 1204

Bilirubin Serum No Effect Analytical At concentration of 100 mg/L had no effect as measured by Jendrassik and Grof method. 3393

Bromide Urine No Effect Analytical No effect on method of Frings. 1204

Butabarbital Blood Increase Physiological 600 mg orally produces concentration of 14 mg/L. 2348

Calcium Serum No Effect Analytical At concentration of 100 mg/L had no effect as measured by cresolphthalein method. 3393

Chlordiazepoxide Blood No Effect Analytical On method of Riddick at 5 mg/dL. 2983

Chloride Serum No Effect Analytical At concentration of 100 mg/L had no effect as measured by mercurimetric method. 3393

Cholesterol Serum No Effect Analytical At concentration of 100 mg/L had no effect as measured by Liebermann-Burchard method. 3393

Creatinine Serum No Effect Analytical At concentration of 100 mg/L had no effect as measured by AutoAnalyzer Jaffé method. 3393

Estriol Urine No Effect Analytical No effect on GLC method. 1163

Ethchlorvynol Urine No Effect Analytical No effect on method of Frings and Cohen at 10 mg/dL. 1205

Folate Serum No Effect Analytical If chromatographic procedure of Landon used. 2068

FPN Test Urine No Effect Analytical No effect at 10 mg/dL on method of Frings. 1204

Phosphate Serum No Effect Analytical At concentration of 100 mg/L had no effect as measured by phosphomolybdate method. 3393

Protein Serum No Effect Analytical At concentration of 100 mg/L had no effect on method using biuret with blank correction. 3393

Prothrombin Time Plasma Decrease Physiological Decreased response to anticoagulants (enzyme induction). 3379

Salicylate Urine No Effect Analytical At 10 mg/dL no effect on Trinder method. 1204

Triglycerides Serum No Effect Analytical At concentration of 100 mg/L had no effect as measured by lipase/esterase method. 3393

Urea Nitrogen Serum No Effect Analytical At concentration of 100 mg/L had no effect as measured by diacetylmonoxime method. 3393

Uric Acid Serum No Effect Analytical At concentration of 100 mg/L had no effect as measured by phosphotungstate reduction method. 3393

Butacaine

Amphetamine Urine No Effect Analytical At 50 mg/L on fluorescent method of Hayes. 1534

Butanol

Ionized Calcium Serum Decrease Analytical At 0.1 mmol/L to 0.1 mol/L with Calcium specific electrode. 0540

Butaperazine

Alanine Aminotransferase Serum Increase Physiological Cholestatic hepatitis with obstruction. 0071

Aspartate Aminotransferase
Serum Increase Physiological Cholestatic hepatitis with obstruction. 0071

Bilirubin Serum Increase Physiological Cholestatic hepatitis with obstruction. 0071

Eosinophils Blood Increase Physiological Allergic manifestation. 1680

Erythrocytes Blood Decrease Physiological Transitory anemia. 0071

Hematocrit Blood Decrease Physiological Transitory anemia. 0071

Hemoglobin Blood Decrease Physiological Transitory anemia reported. 0071

Leukocytes Blood Decrease Physiological Transitory leukopenia. 0071

Platelets Blood Decrease Physiological purpura/pancytopenia observed. 1680

Pregnancy Tests Urine Positive Physiological May occur with delayed menstruation and ovulation. 1680

Prolactin Plasma Increase Physiological Significant response to 5 mg given orally. 1416

Butethamine

Amphetamine Urine Increase Analytical Yields fluorophor with NBD chloride reaction. 2473

Propoxyphene Urine Increase Analytical 1% fluorescence in procedure of Valentour. 3662

Butizide

Cholesterol Serum No Effect Analytical At concentration of 440 mg/L had no effect as measured by CHOD-Iodide method. 3393 At concentration of 440 mg/L had no effect as measured by CHOD-PAP method. 3393 At concentration of 440 mg/L had no effect as measured by catalase-AIDH method. 3393 At concentration of 440 mg/L had no effect as measured by method using catalase-Hantzsch reaction. 3393 At concentration of 440 mg/L had no effect as measured by Liebermann-Burchard method. 3393

Creatinine Serum No Effect Analytical At concentration of 2.6 mg/L had no effect as measured by Jaffé-Fading-Fraction method. 3393 At concentration of 2.6 mg/L had no effect as measured by Jaffé-Fuller's earth method. 3393 At concentration of 2 mg/L had no effect as measured by kinetic Jaffé method on BKA-2. 3393

Glucose Serum No Effect Analytical At concentration of 0.2 mg/L had no effect as measured by Ektachem® method. 3393

Potassium Serum No Effect Analytical At concentration of 1.98 mg/L had no effect on ISE measurement without predilution. 3393

Protein Electrophoresis Serum No Effect Analytical At concentration of 1.98 mg/L had no effect as measured by automated Olympus-Hite method. 3393

Sodium Serum No Effect Analytical At concentration of 1.98 mg/L had no effect on ISE measurement without predilution. 3393

Urea Nitrogen Serum No Effect Analytical At concentration of 0.2 mg/L had no effect as measured by Ektachem® method. 3393

Uric Acid Serum No Effect Analytical At concentration of 260 mg/L had no effect as measured by Kageyama-Hantzsch method. 3393 At concentration of 2.6 mg/L had no effect as measured by catalase-AIDH method. 3393

Butorphanol

Cortisol Plasma No Effect Physiological No significant effect in 6 healthy male volunteers given 2 mg i.m. 3033

Follicle Stimulating Hormone (FSH)
Plasma No Effect Physiological No significant effect in 6 healthy male volunteers given 2 mg i.m. 3033

Growth Hormone Plasma No Effect Physiological No significant effect in 6 healthy male volunteers given 2 mg i.m. 3033

Luteinizing Hormone (LH) Plasma No Effect Physiological No significant effect in 6 healthy male volunteers given 2 mg i.m. 3033

Prolactin Plasma Increase Physiological Significant rise in 6 healthy male volunteers given 2 mg i.m. 3033

Thyroid Stimulating Hormone (TSH)
Serum No Effect Physiological No significant effect in 6 healthy male volunteers given 2 mg i.m. *3033*

Butylgallate

Catechol-O-Methyl Transferase
Test Conditions Decrease Physiological Observed *in vivo* and *in vitro*. *2913*

Cadmium

α-**Amino-Nitrogen** *Urine Increase Physiological* Occasionally with long exposure, proteinuria. *2217*

Bilirubin *Serum Increase Physiological* May occur even following respiratory exposure. *1302*

Calcium *Urine Increase Physiological* Occasionally observed with proximal tubular dysfunction. *2217*

Glucose *Urine Increase Physiological* Occasionally observed with proteinuria. *2217*

Magnesium *Urine Increase Analytical* Measured in fluorometric method of Schachter. *3573*

Phosphate *Urine Increase Physiological* Occasionally observed with proximal tubular dysfunction. *2217*

Protein *Urine Increase Physiological* Generalized impairment of proximal tubular function. *0019*

Urea Nitrogen *Serum Increase Physiological* May cause nephrotoxicity. *3204*

Uric Acid *Serum Decrease Physiological* Generalized impairment of proximal tubular function. *0019*
Urine Increase Physiological Generalized impairment of proximal tubular function. *0019*

Caffeine

Acetaminophen *Serum No Effect Analytical* At concentration of 10 mg/L had no effect on HPLC method. *3432*

Alanine Aminotransferase *Serum No Effect Analytical* At acute overdose concentration (10 mg/dL) on Technicon® SMAC® method. *3719* On either continuous or colorimetric methods at 10 times maximal therapeutic methods. *1785* At therapeutic concentration on Reflotron method. *1984*

Albumin *Serum No Effect Analytical* At acute overdose concentration (10 mg/dL) on Technicon® SMAC® method. *3719* At concentration of 160 mg/L had no effect on BCG method. *3393*

Alkaline Phosphatase *Serum No Effect Analytical* At acute overdose concentration (10 mg/dL) on Technicon® SMAC® method. *3719* On continuous method at 10 times maximal therapeutic concentration. *1785*

Amphetamine *Urine No Effect Analytical* At 50 mg/L on fluorescent method of Hayes. *1534*

Aspartate Aminotransferase *Serum No Effect Analytical* At acute overdose concentration (10 mg/dL) on Technicon® SMAC® method. *3719* On either continuous or colorimetric methods at 10 times maximal therapeutic concentration. *1785* At therapeutic concentration on Reflotron method. *1984*

Basal Metabolic Rate *Patient Increase Physiological* Metabolic effect of drug (temporary effect). *2220*

Bicarbonate *Serum No Effect Analytical* At concentration of 160 mg/L had no effect on method using phenolphthalein. *3393*

Bilirubin *Serum No Effect Analytical* No effect on Jendrassik-Grof, dimethylsulfoxide and spectrophotometric methods at therapeutic concentrations. *1786* At acute overdose concentration (10 mg/dL) on Technicon® SMAC® method. *3719* At concentration of 194 mg/L had no effect on Jendrassik and Grof method. *3393*
Serum No Effect Physiological No significant displacement from protein in neonates. *3748*
Serum Decrease Analytical Allegedly lower values when used as coupling agent. *1022*

Bilirubin, Direct *Serum Decrease Analytical* Allegedly lower values if used as coupling agent. *1238*

Caffeine *Serum Increase Physiological* 24-hour average concentration correlated poorly with caffeine intake. *2125*

Calcium *Serum No Effect Analytical* At acute overdose concentration (10 mg/dL) on Technicon® SMAC® method. *3719* At concentration of 194 mg/L had no effect on cresolphthalein method. *3393*

Cells *Urine Increase Physiological* May cause marked increase in renal tubular cells. *1343*

Chloride *Serum No Effect Analytical* At concentration of 160 mg/L had no effect on mercurimetric method. *3393*

Cholesterol *Serum No Effect Analytical* No effect on enzymatic and Liebermann-Burchard methods at therapeutic concentrations. *1786* At acute overdose concentration (10 mg/dL) on Technicon® SMAC® method. *3719* At therapeutic concentration on Reflotron method. *1984* At concentration of 194 mg/L had no effect on Liebermann-Burchard method. *3393*
Serum Increase Physiological Positive relationship between coffee consumption and lipid concentration: association possibly more with coffee than caffeine. *1447*

Cholesterol, Low Density Lipoprotein
Serum Increase Physiological Positive relationship between coffee consumption and lipid concentration: association possibly more with coffee than caffeine. *1447*

Chromosomes *Test Conditions Abnormal Physiological* Clastogenic in human cells. *3282*

Cortisol *Plasma No Effect Physiological* No effect of 250 mg in men or women but slight increase with 500 mg after 90 minutes in men. *3418* In men no significant effect of single dose of 500 mg. In women no significant effect of single dose of 500 mg. *3418*

Creatine *Urine Increase Physiological* Acts on intermediary metabolism. *1343*

Creatine Kinase *Serum No Effect Analytical* At acute overdose concentration (10 mg/dL) on Technicon® SMAC® method. *3719* On continuous method at 10 times maximal therapeutic concentration. *1785*

Creatinine *Serum No Effect Analytical* At 500 mg/L on reversed phase LC method of Zhiri et al. *3942* No effect on alkaline picrate and Slot methods at therapeutic concentrations. *1786* At concentration of 160 mg/L had no effect on AutoAnalyzer Jaffé method. *3393* At acute overdose concentration (10 mg/dL) on Technicon SMAC® method. *3719*

Cyclic Adenosine Monophosphate
Plasma Increase Physiological Inhibits phosphodiesterase increases lipolysis. *1529*

Diagnex Blue Excretion *Urine Decrease Physiological* As Na benzoate salt; low gastric acidity. *1022*

Dopamine *Urine Increase Physiological* After 250 mg increased for 2 h then returned to baseline in next 2 h. *2751*

β-**Endorphin** *Plasma Increase Physiological* Caused increase with peak at 60 minutes after 500 mg in both men and women but with no effect when 250 mg given (increase of order of 25 to 50%). *3418* 500 mg increased concentration in both men and women. *3418*

Epinephrine *Urine Increase Physiological* After 250 mg increased for 2 h then returned to baseline in next 2 h. *2751*

Erythrocytes *Urine Increase Physiological* Daily administration causes moderate increase. *2427*

Ethchlorvynol *Serum No Effect Analytical* At 130 mg/L on GLC procedure of Evenson. *1058*

Fatty Acids, Free (FFA) *Serum Increase Physiological* 220 mg raised nonesterified fatty acids by approximately 30%. *1894* Concentration doubled 1 h after 250 mg and remained increased for 4 h. *2751*

Glucose *Serum No Effect Analytical* No effect at therapeutic concentrations on hexokinase, glucose dehydrogenase 2,4-dichlorophenol, ABTS and o-toluidine methods. *1786* At acute overdose concentration (10 mg/dL) on Technicon® SMAC® method. *3719* At therapeutic concentration on Reflotron method. *1984* At concentration of 300 mg/L had no effect on hexokinase/G-6-PDH method. *3393*
Serum Increase Physiological Effect in order of 10 mg/dL only. *0644*

Glucose Tolerance *Serum Increase Physiological* Decrease of glucose at 30 and 60 minutes. *1093*
Serum Decrease Physiological Reduced pancreatic insulin release in normals. *0049*

γ-**Glutamyltransferase (GGT)** *Serum No Effect Analytical* At therapeutic concentration on Reflotron method. *1984*

Caffeine *(continued)*

Growth Hormone   *Plasma   No Effect   Physiological*   No significant effect of single dose of 250 mg in men or women. *3418* In men no significant effect of single dose of 500 mg. *3418*

Hydrochloric Acid   *Gastric Material   Increase   Physiological* Effect of i.v. infusion, also if given orally. *0689*

Hydroxybutyrate Dehydrogenase
Serum   No Effect   Analytical   On continuous method at 10 times maximal therapeutic concentration. *1785*

5-Hydroxyindoleacetic Acid (5-HIAA)
Urine   Increase   Physiological   Alleged effect. *1487*

Indomethacin   *Serum   No Effect   Analytical*   No effect on HPLC method of Roberts and Smith. *3002*

Iron   *Serum   No Effect   Analytical*   No effect of therapeutic concentrations on Ramsay and bathophenanthroline methods. *1786*   At acute overdose concentration (10 mg/dL) on Technicon® SMAC® method. *3719*   At concentration of 150 mg/L had no effect on Ferrozine method. *3393*

Lactate   *Serum   No Effect   Analytical*   At concentration of 300 mg/L had no effect on enzymatic method. *3393*

Lactate Dehydrogenase   *Serum   No Effect   Analytical*   At acute overdose concentration (10 mg/dL) on Technicon® SMAC® method. *3719*   On continuous method with pyruvate as substrate at 10 times maximal therapeutic concentration. On colorimetric method at 10 times maximal therapeutic concentration. *1785*
Serum   Increase   Analytical   On continuous method with lactate as substrate at 10 times maximal therapeutic concentration. *1785*

Linoleic Acid   *Serum   Increase   Physiological*   Significant increase within 1 h, still elevated after 4 h, when 250 mg ingested. *2751*

Methaqualone   *Serum   No Effect   Analytical*   At 130 mg/L on GLC procedure of Evenson. *1057*

Norepinephrine   *Urine   No Effect   Physiological*   Effect not significant over 2 h after ingestion of 250 mg. *2751*

Oleic Acid   *Serum   Increase   Physiological*   Significant increase within 1 h, still elevated after 4 h, when 250 mg ingested. *2751*

Palmitic Acid   *Serum   Increase   Physiological*   Significant increase within 1 h, still elevated after 4 h, when 250 mg ingested. *2751*

Palmitoleic Acid   *Serum   Increase   Physiological*   Significant increase within 1 h, still elevated after 4 h, when 250 mg ingested. *2751*

Pepsin   *Gastric Material   Increase   Physiological*   Effect of i.v. infusion, also if given orally. *0689*

Phosphate   *Serum   No Effect   Analytical*   At acute overdose concentration (10 mg/dL) on Technicon® SMAC® method. *3719* At concentration of 194 mg/L had no effect on phosphomolybdate method. *3393*

Prolactin   *Plasma   No Effect   Physiological*   No significant effect of single dose of 250 mg in men or women. *3418*   In men no significant effect of single dose of 500 mg. In women no significant effect of single dose of 500 mg. *3418*

Protein   *Serum   No Effect   Analytical*   No effect on biuret and spectrophotometric methods at therapeutic concentrations. *1786*   At acute overdose concentration (10 mg/dL) on Technicon® SMAC® method. *3719*   At concentration of 194 mg/L had no effect on biuret method with blank correction. *3393*

Prothrombin Time   *Plasma   Decrease   Physiological* Observed in patients receiving anticoagulants. *2313*

Stearic Acid   *Serum   Increase   Physiological*   Significant increase within 1 h, still elevated after 4 h, when 250 mg ingested. *2751*

Theophylline   *Serum   Increase   Analytical*   When present in tea, coffee or cola may raise apparent concentration in certain fluoroimmunoassays. *1614*

Thyroid Stimulating Hormone (TSH)
Serum   No Effect   Physiological   No significant effect of single dose of 250 mg in men or women. *3418*   In men no significant effect of single dose of 500 mg. *3418*

Triglycerides   *Serum   No Effect   Analytical*   At therapeutic concentration on Reflotron method. *1984*   At concentration of 160 mg/L had no effect on lipase/esterase method. *3393*

Plasma   No Effect   Analytical   No effect on enzymatic methods at therapeutic concentrations. *1786*

Tri-iodothyronine (T₃)   *Serum   No Effect   Physiological*   No significant effect of single dose of 250 mg in men or women. *3418*   In women no significant effect of single dose of 500 mg. *3418*

Urea Nitrogen   *Serum   No Effect   Analytical*   No effect at therapeutic concentrations on glutamate dehydrogenase, phenol-hypochlorite, diacetylmonoxime methods. *1786*   At acute overdose concentration (10 mg/dL) on Technicon® SMAC® method. *3719*   At therapeutic concentration on Reflotron method. *1984*   At concentration of 194 mg/L had no effect on diacetylmonoxime method. *3393*

Uric Acid   *Serum   No Effect   Analytical*   At 500 mg/L on reversed phase LC method of Zhiri et al. *3942*   No effect at therapeutic concentrations on uricase-catalase, aldehyde dehydrogenase direct UV-test, and phosphotungstate methods. *1786*   At acute overdose concentration (10 mg/dL) on Technicon® SMAC® method. *3719*   At therapeutic concentration on Reflotron method. *1984*   At concentration of 194 mg/L had no effect on phosphotungstate reduction method. *3393*
Serum   Increase   Analytical   Reduction of phosphotungstate by metabolites. *0583*
Urine   Increase   Analytical   Reduction of phosphotungstate by metabolites. *0580*

Vanillylmandelic Acid   *Urine   Increase   Analytical*   Interference with method reported. *1237*

Volume   *Urine   Increase   Physiological*   Diuretic action. *1343*

Calcitonin

Calcium   *Serum   Decrease   Physiological*   Decrease of greater than 0.5 meq/L observed. *1439*
Urine   Increase   Physiological   Acts independently of parathyroid. *1439*

Casts   *Urine   Increase   Physiological*   Possible nephrotoxic effect (increased granular casts). *1490*

Cells   *Urine   Increase   Physiological*   Possible nephrotoxic effect (increased epithelial cells). *1490*

C-Peptide   *Plasma   Decrease   Physiological*   Response to glucose significantly reduced. *2738*

Glucagon   *Plasma   Increase   Physiological*   Inhibitory action of oral glucose on glucagon secretion partially prevented in comparison with control. *2738*

Glucose Tolerance   *Serum   Decrease   Physiological*   Rise in plasma glucose exaggerated after oral sugar. *2738*

Hydroxyproline   *Urine   Decrease   Physiological*   Due to anticatabolic action. *3644*

Insulin   *Plasma   Decrease   Physiological*   Response to glucose significantly reduced. *2738*

Magnesium   *Urine   Increase   Physiological*   Acts independently of parathyroid. *1439*

Phosphate   *Serum   Decrease   Physiological*   Due to urinary loss. *1343*
Urine   Increase   Physiological   Acts independently of parathyroid. *1439*

Potassium   *Urine   Increase   Physiological*   Acts independently of parathyroid. *1439*

Prolactin   *Plasma   Decrease   Physiological*   In 9 healthy subjects and 4 patients with hyperprolactinemia after i.v. infusion of salmon preparation. *1729*

Sodium   *Urine   Increase   Physiological*   Acts independently of parathyroid. *0346*

Calcium

Calcium   *Duodenal Contents   Increase   Physiological*   Calcium infusion (10 mg/kg/h) causes increase x 2. *1314*

Chymotrypsin   *Duodenal Contents   Increase   Physiological* Calcium infusion (10 mg/kg/h) causes increase x 2. *1314*

Hydrochloric Acid   *Gastric Material   Increase   Physiological* Strong response to infusion of Calcium gluconate. *2739*

¹³¹I Uptake   *Serum   No Effect   Physiological*   Loading and withdrawal no effect. *0583*

Lead   *Urine   Increase   Analytical*   75 meq/L equivalent to 110 µg/L by atomic absorption. *3240*

Lipase *Serum No Effect Analytical* At 5 mmol/L no effect on method of Tietz, Repique. *3589*
Duodenal Contents Increase Physiological Calcium infusion (10 mg/kg/h) causes doubling. *1314*

Magnesium *Serum No Effect Analytical* At concentration of 2.5 to 40 mg/dL did not significantly interfere with measurement of magnesium on Du Pont aca-III. *0949*

Pepsin *Gastric Material Increase Physiological* Marked increase if serum concentration increased above 6 meq/L. *2689*

Potassium *Serum Increase Analytical* Emission spectrum may interfere. *2313*

Protein Bound Iodine (PBI) *Serum No Effect Physiological* No effect with calcium loading. *0583*

Sodium *Serum Increase Analytical* Emission spectrum may interfere. *2313*

Trypsin *Duodenal Contents Increase Physiological* Calcium infusion (10 mg/kg/h) causes increase x 2. *1314*

Zinc *Serum No Effect Analytical* At 100 to 1 with flameless atomic absorption. *3336*
Urine No Effect Analytical At 100 to 1 with flameless atomic absorption. *2038*

Calcium Bromogalactogluconate

Bromide *Serum Increase Physiological* In patient taking 12 g drug daily for 3 mo serum concentration of 18.7 mmol/L. *2867*

Chloride *Serum Increase Analytical* Relatively small error by coulometric method as used on Beckman Astra but significant increase at clinical concentrations with thiocyanate methods on Technicon® RA-1000 or SMAC®. *2867*

Calcium Carbimide

Oxalate *Urine Decrease Analytical* Inhibits enzymatic procedure for measurement. *3936*

Calcium Carbonate

Calcium *Serum No Effect Physiological* No rise after 4 g observed. *3302*

Crystals *Urine Increase Physiological* Small colorless dumbbells/spheres (not in acid). *0443*

Gastrin *Serum Increase Physiological* 2 fold increase, maximum 30 to 75 minutes after. *3302*

Calcium Chloride

Gastrin *Serum Increase Physiological* If given intraluminally 13% increase over NaCl. *1643*

Hydrochloric Acid *Gastric Material Increase Physiological* Intraluminally (3 x increase over sodium chloride). *1643*

pH *Blood Decrease Physiological* Is an acidifying salt. *2313*

Calcium Gluconate

Calcium *Serum Increase Physiological* Marked effect noted in newborns. *1550*

Glucose *Serum Decrease Physiological* Slight effect observed in newborns. *1550*

11-Hydroxycorticosteroids *Plasma Increase Physiological* Transient effect maximum at 15 minutes after i.v. *1487*

17-Hydroxycorticosteroids *Urine Decrease Physiological* Reduced value reported in a single case. *0531*

Insulin *Plasma Increase Physiological* Marked effect noted in newborns. *1550*

131I Uptake *Serum Decrease Physiological* Wafer of Upjohn contains tetraiodofluorescein. *2652*

Magnesium *Serum No Effect Physiological* No effect on dihydroxyazobenzene method. *3501*
Serum Decrease Analytical False decrease if measured by titan-yellow. *0091*
Urine Decrease Analytical False decrease if measured by titan-yellow. *0091*

Calcium Ions

Lipase *Serum No Effect Analytical* Up to 5 mmol/L on method of Tietz. *3589*
Serum Decrease Analytical Above 5 mmol/L on method of Tietz. *3589*

Calcium Salts

Calcium *Serum Increase Analytical* Possible contamination of distilled water. *3879*
Serum Increase Physiological Direct effect of increased gut absorption. *0071*
Urine Increase Analytical Possible contamination of distilled water. *3879*
Urine Increase Physiological Increased excretion with large doses. *1343*

Diagnex Blue Excretion *Urine Increase Physiological* Heavy metal displacement from resin. *2313*

Hydrochloric Acid *Gastric Material Increase Physiological* Rebound increase at night if CaCO₃ ingested early in day. *3264*

Magnesium *Serum Decrease Physiological* Competes for absorption from gut and tubules. *0091*

Urea Nitrogen *Serum Increase Physiological* Azotemia with milk-alkali syndrome. *1343*

Calomel

Color *Feces Increase Analytical* Green color observed. *1187*

Campesterol

Cholesterol *Serum No Effect Analytical* On GLC procedure of MacGee. *2252*

Cannabis

Catecholamines *Urine No Effect Physiological* Essentially unchanged with moderate dose. *1633*

Chloride *Serum Increase Physiological* Reported effect. *2220*

Cortisol *Plasma No Effect Physiological* Essentially unchanged with moderate dose. *1633*

Creatinine *Serum Decrease Physiological* Reported effect. *2220*

Creatinine Clearance *Urine Decrease Physiological* Temporary decrease noted. *1633*

Fatty Acids, Free (FFA) *Serum No Effect Physiological* No effect observed. *1633*

Glucose *Serum No Effect Physiological* No effect observed. *1633*
Serum Increase Physiological Reported effect. *0117*
Serum Decrease Physiological Hypoglycemic effect approximately 4 h after use. *0912*

Glucose Tolerance *Serum Decrease Physiological* Increased concentrations noted in some subjects at 1/2 to 1 h. *2830*

Insulin *Plasma Increase Physiological* Responsible for hypoglycemia. *0912*

Phosphate Clearance *Urine Decrease Physiological* Temporary decrease noted. *1633*

Potassium *Serum Increase Physiological* Reported effect. *2220*

Sodium *Serum Increase Physiological* Reported effect. *2220*

Urea Nitrogen *Serum Increase Physiological* Reported effect. *2220*

Uric Acid *Serum Decrease Physiological* Reported effect. *2220*

Canrenoate Potassium

Digoxin *Serum Increase Analytical* At normal concentrations in serum if no preincubation. *2804*

Cantharides

Protein *Urine Increase Physiological* May cause nephrotoxicity. *3204*

Urea Nitrogen *Serum Increase Physiological* May cause nephrotoxicity. *3204*

Volume *Urine Decrease Physiological* May cause oliguria with renal damage. *0279*

Canthoxanthine

Carotene *Serum Increase Analytical* During administration and for 10 d after may affect certain analytical methods. *3023*

Vitamin A *Serum Increase Analytical* During administration and for 10 d after may affect certain analytical methods. *3023*

Capreomycin

Alkaline Phosphatase *Serum Increase Physiological* Reported effect on liver. *1705*

Aspartate Aminotransferase
Serum Increase Physiological Reported effect on liver. *1705*

BSP Retention *Serum Increase Physiological* Transient increase in 5% subjects. *2427*

Casts *Urine Increase Physiological* Nephrotoxic effect (usually granular casts). *2313*

Creatinine *Serum Increase Physiological* Nephrotoxic effect. *2313*

Eosinophils *Blood Increase Physiological* Allergic reaction (may be up to 35%). *0505*

Erythrocytes *Urine Increase Physiological* May cause nephrotoxicity. *1680*

Leukocytes *Blood Increase Physiological* Due to eosinophilia of hypersensitivity. *2313*
Blood Decrease Physiological May cause leukopenia or leukocytosis. *1680*

Nonprotein Nitrogen *Serum Increase Physiological* Nephrotoxic effect. *2313*

Potassium *Serum Decrease Physiological* Observed occasionally with therapy. *1679*

Protein *Urine Increase Physiological* Nephrotoxic effect (transient). *1022*

PSP Excretion *Urine Decrease Physiological* Observed with other signs of decreased renal function. *1680*

Urea Nitrogen *Serum Increase Physiological* Nephrotoxic effect (in 6% subjects). *1343*

Uric Acid *Serum Increase Physiological* Nephrotoxic effect associated with hyperuricemia. *1552*

Urobilin *Urine Increase Physiological* May cause nephrotoxicity. *1680*

Caproic Acid

Fatty Acids, Free (FFA) *Serum No Effect Analytical* No effect on method of Pinelli. *2819*

Caproxamine

Albumin *Serum Increase Analytical* At 1.5 mg/dL conventional methods if added to sera. *1758*

γ-Globulin *Serum Decrease Analytical* At 1.5 mg/dL conventional methods when added to sera. *1758*

Protein *Serum Decrease Analytical* At 1.5 mg/dL conventional methods when added to sera. *1758*

Urea Nitrogen *Serum Decrease Analytical* At 1.5 mg/dL conventional methods when added to sera. *1758*

Caprylate

Fatty Acids, Free (FFA) *Serum No Effect Analytical* No effect on method of Pinelli. *2819*

Captopril

Aldosterone *Plasma Decrease Physiological* From 1.08 nmol/L to 0.22 nmol/L after 3 d treatment in 1 patient with Bartter's syndrome. *1558* Parallel decline with sodium in congestive heart failure patients. *2592* Fell in conjunction with fall of systemic arterial pressure. *1716* In 11 patients with resistant heart failure reduced mean concentration from 62 ng/dL to 26 ng/dL. *1165* Gradual reduction probably due to longer half-life than angiotensin II. *0177* With doses up to 800 mg/d for 10 d in 23 hypertensive patients. *2836*
Urine Decrease Physiological Fell in conjunction with fall of systemic arterial pressure. *1716* Sustained effect: extent related to pretreatment plasma renin activity. *0177* With doses up to 800 mg/d for 10 d in 23 hypertensive patients. *2836*

Alkaline Phosphatase *Serum Increase Physiological* Isolated case of cholestasis in patient receiving 25 mg tid for 1 mo. *2732* Characteristic pattern of hepatocellular jaundice in one patient. But with secondary cholestatic elements. *3688*

Angiotensin-Converting Enzyme (ACE)
Serum Decrease Analytical 31% inhibition of method using benzyloxycarbonyl-phenylalanyl-histidyl-leucine as substrate. *3221* Marked inhibition of enzyme by drug. *3038*
Serum Decrease Physiological May be marked reduction in drug-treated patients. *1855*

Angiotensin II *Plasma Decrease Physiological* Significant effect in 1 h after single dose of 25 mg. *0181* Parallel decline with sodium in congestive heart failure patients. *2592* Fell in parallel with reduction of blood pressure. *1716* Prompt and striking reduction following oral administration over 30 minutes. *0177*

Antibodies to DNA *Serum Decrease Physiological* In 3 of 78 patients, of IgM class, treated for mean of 11 mo. *1850*

Antinuclear Antibodies *Serum Increase Physiological* In 13 of 78 patients, mainly of IgM class treated for mean of 11 mo. *1850*

Apolipoprotein AI *Serum No Effect Physiological* Increase by 1 mg/dL in 15 patients given 75 mg daily for 8 weeks. *2550*
Serum Increase Physiological After 12 weeks treatment in 18 patients with mild essential hypertension. *3150*

Apolipoprotein AII *Serum Increase Physiological* After 12 weeks treatment in 18 patients with mild essential hypertension. *3150*

Apolipoprotein B *Serum No Effect Physiological* After 12 weeks treatment in 18 patients with mild essential hypertension. *3150* Decrease by -2 mg/dL in 15 patients given 75 mg daily for 8 weeks. *2550*

Apolipoprotein CII *Serum No Effect Physiological* No change in 15 patients given 75 mg daily for 8 weeks. *2550* After 12 weeks treatment in 18 patients with mild essential hypertension. *3150*

Apolipoprotein CIII *Serum No Effect Physiological* Increase by 1 mg/dL in 15 patients given 75 mg daily for 8 weeks. *2550* After 12 weeks treatment in 18 patients with mild essential hypertension. *3150*

Apolipoprotein E *Serum No Effect Physiological* After 12 weeks treatment in 18 patients with mild essential hypertension. *3150* No change in 15 patients given 75 mg daily for 8 weeks. *2550*

Aspartate Aminotransferase
Serum Increase Physiological Isolated case of cholestasis in patient receiving 25 mg tid for 1 mo. *2732* Characteristic pattern of hepatocellular jaundice in one patient. But with secondary cholestatic elements. *3688*

Bilirubin *Serum Increase Physiological* Isolated case of cholestasis in patient receiving 25 mg tid for 1 mo. *2732* Characteristic pattern of hepatocellular jaundice in one patient. But with secondary cholestatic elements. *3688*

Bilirubin, Direct *Serum Increase Physiological* Characteristic pattern of hepatocellular jaundice in one patient. But with secondary cholestatic elements. *3688*

Catecholamines *Plasma Decrease Physiological* Mean decrease from 695 ng/L to 476 ng/L but not significant change in patients with heart failure. *1165*

Cholesterol *Serum No Effect Physiological* No significant change in patients in 2 studies treated for 2 and 24 mo. *0088* After 12 weeks treatment in 18 patients with mild essential hypertension. *3150* No significant change in 7,000 hypertensives treated for 3 y. *1407* Decrease by -2 mg/dL in 15 patients given 75 mg daily for 8 weeks. *2550*

Cholesterol, High Density Lipoprotein
Serum No Effect Physiological After 12 weeks treatment in 18 patients with mild essential hypertension. *3150* Increase by 1 mg/dL in 15 patients given 75 mg daily for 8 weeks. *2550*

Cholesterol, Low Density Lipoprotein
Serum No Effect Physiological After 12 weeks treatment in 18 patients with mild essential hypertension. *3150*

Cholesterol, Very Low Density Lipoprotein
Serum No Effect Physiological After 12 weeks treatment in 18 patients with mild essential hypertension. *3150*

Creatinine *Serum No Effect Physiological* No significant change in 7,000 hypertensives treated for 3 y. *1407*
Serum Increase Physiological Acute reversible renal failure may occur: transient increases common. *3712* Occasional reversible azotemia, either due to hypotension or direct renal damage. *0283* Severe reversible azotemia in few patients with peripheral vascular disease two probably associated with GFR reduction. *0655* Eosinophilic interstitial nephritis and membranous glomerulopathy reported. Cases of nephrotic syndrome also reported. *0929*

Eosinophils *Blood Increase Physiological* Several cases of rash and eosinophilia reported. *0929*
Blood Decrease Physiological Isolated cases of pancytopenia, usually with pre-existing renal disease. *1255*

Erythrocytes *Blood Decrease Physiological* Isolated case of pancytopenia reported. *1012* Observed in 9 of 12 hypertensive patients on maintenance hemodialysis, maximum effect achieved after about 11 mo. *1608*

Factor XI *Plasma Decrease Physiological* High value of essential hypertension significantly reduced. *2748*

Factor XII *Plasma Decrease Physiological* High value of essential hypertension significantly reduced. *2748*

Glucose *Serum No Effect Physiological* No significant change in 7,000 hypertensives treated for 3 y. *1407*
Urine Increase Physiological Reversible glycosuria reported in one boy with abdominal aortitis and resistant hypertension. *0929*

Glucose Tolerance *Serum No Effect Physiological* No significant deterioration with long-term treatment in diabetic hypertensives. *3307*

γ-Glutamyltransferase (GGT)
Serum Increase Physiological Characteristic pattern of hepatocellular jaundice in one patient. But with secondary cholestatic elements. *3688*

Hematocrit *Blood Decrease Physiological* Observed in 9 of 12 hypertensive patients on maintenance hemodialysis, maximum effect achieved after about 11 mo. *1608*

Hemoglobin *Blood Decrease Physiological* Observed in 9 of 12 hypertensive patients on maintenance hemodialysis, maximum effect achieved after about 11 mo. *1608*

Insulin *Plasma No Effect Physiological* No decrease during treatment nor effect on glucose tolerance. *2902*

Iron *Serum No Effect Physiological* In 9 of 12 hypertensive patients although other hematological effects observed. *1608*

Iron-Binding Capacity, Total (TIBC)
Serum No Effect Physiological In 9 of 12 hypertensive patients although other hematological effects observed. *1608*

Ketones *Urine Positive Analytical* False positive at concentration of 25 mmol/L on Ames Keto-diastix®, also affected Boehringer Combur Test. *1373* Trace to 3 + reactions in 9 patients with both Diastix® and Chemstrip-6®. *3766*

Lactate Dehydrogenase *Serum Increase Physiological* Isolated case of cholestasis in patient receiving 25 mg tid for 1 mo. *2732* Characteristic pattern of hepatocellular jaundice in one patient. But with secondary cholestatic elements. *3688*

Leukocytes *Blood Decrease Physiological* Isolated reports of neutropenia and agranulocytosis when first introduced and given in high doses. *1761* Agranulocytosis observed in several cases. *3712* Isolated case of pancytopenia reported. *1012* Isolated cases of pancytopenia, usually with pre-existing renal disease. *1255*

MCHC *Blood No Effect Physiological* In 9 of 12 hypertensive patients although other hematological effects observed. *1608*

MCV *Blood No Effect Physiological* In 9 of 12 hypertensive patients although other hematological effects observed. *1608*

Neutrophils *Blood Decrease Physiological* In approximately 0.3% patients: develops within first 3 to 12 weeks of treatment associated with myeloid hypoplasia of bone marrow. *3712* Isolated cases of pancytopenia, usually with pre-existing renal disease. *1255* Agranulocytosis reported to occur in 1 of 250 treated patients. *3717*

Norepinephrine *Plasma No Effect Physiological* No effect of drug and concentration responds appropriately to postural changes. *3712*

Platelets *Blood Decrease Physiological* Isolated case of pancytopenia reported. *1012*

Potassium *Serum No Effect Physiological* No significant change in 7,000 hypertensives treated for 3 y. *1407*
Serum Increase Physiological From 2.9 to 3.3 mmol/L after 1 mo in patient with Bartter's syndrome due to inhibited aldosterone production. *1558* Resulting from decreased secretion of aldosterone. *1716* Rise less than 1.0 mmol/L, but greatest in patients with high baseline renin activity. *2836*

Prekallikrein *Plasma Decrease Physiological* Rapid decrease following institution of therapy. *2748*

Protein *Serum No Effect Physiological* In 9 of 12 hypertensive patients although other hematological effects observed. *1608*
Urine Increase Physiological Isolated reports of immune complex glomerulopathy when first introduced and high doses given. *1761* Effect observed in small number of patients with excellent control of hypertension. *0181* Greater than 1.0 g/d occurs in about 1.2% patients may subside with continuing treatment. *3712* Occurs in approximately 1% of 7100 hypertensives most often who had pre-existing renal disease and receiving high doses of drug. *1407* In patients with pre-existing renal dysfunction with proteinuria. *0181* Some patients develop heavy proteinuria during use of drug. Reversible with discontinuation. *2263* Eosinophilic interstitial nephritis and membranous glomerulopathy reported. Cases of nephrotic syndrome also reported. *0929*
Urine Decrease Physiological Significant reduction in patients with advanced diabetic nephropathy. *3528*

Renin Activity *Plasma Increase Physiological* From 10.0 to 34.7 pmol/L/s in patient with Bartter's syndrome after 1 mo. *1558* Significant effect in 1 h after single dose of 25 mg. *0181* Marked effect within 30 minutes on active and total renin with reduction of inactive renin. *2748* Gradually increases over 1 h due to negative feedback on renin activity. *0177*
Plasma Decrease Physiological In 6 nonazotemic patients with cirrhosis and ascites due to increased renal vasodilatation. *2728*

Renin Substrate *Plasma No Effect Physiological* No significant effect although marked changes in PRA and angiotensin II. *0181*

Reticulocytes *Blood No Effect Physiological* In 9 of 12 hypertensive patients although other hematological effects observed. *1608*

Sodium *Serum Decrease Physiological* In five men with congestive heart failure with fall of sodium by 7 mmol/L on 3rd to 4th day. *2592*

Triglycerides *Serum No Effect Physiological* No significant change in patients in 2 studies treated for 2 and 24 mo. *0088* After 12 weeks treatment in 18 patients with mild essential hypertension. *3150* Decrease by -3 mg/dL in 15 patients given 75 mg daily for 8 weeks. *2550*

Urea Nitrogen *Serum Increase Physiological* Occasional reversible azotemia, either due to hypotension or direct renal damage. *0655* Severe reversible azotemia in few patients with peripheral vascular disease two weeks after start of therapy, probably associated with GFR reduction. *0655*

Uric Acid *Serum No Effect Physiological* No significant change in 7,000 hypertensives treated for 3 y. *1407*

Carbacrylamine Resin

Potassium *Serum Increase Physiological* Part of resin is in form of potassium salt. 2313

Sodium *Serum Decrease Physiological* Cation-exchange with reduced gastrointestinal tract absorption. 2313

Carbamazepine

α_1-**Acid Glycoprotein** *Serum Increase Physiological* Significantly higher in children treated with drug compared with Controls. 2994

Alanine Aminotransferase *Serum Increase Physiological* Cholestatic and hepatocellular damage. 2907

Albumin *Serum No Effect Physiological* In about 20 epileptic patients treated for 2 y. 0873 Osteomalacia observed in 3 of 31 patients given average of 758 mg/d for average of 20.5 mo. 2635

Alkaline Phosphatase *Serum No Effect Physiological* Observed 2 times upper limit of normal in none of 36 adult inpatient epileptics (dose and duration of treatment unknown). 1162
Serum Increase Physiological Cholestatic and hepatocellular damage. 2907 4 of 21 patients with epilepsy (19%) treated with drug only for average of 40 mo. 1625 In about 20 epileptic patients treated for 2 y. 0873 Granulomatous hepatitis observed in small proportion of patients treated for less than 1 mo; resolved within 3 d of cessation of treatment. 2153 Osteomalacia observed in 3 of 31 patients given average of 758 mg/d for average of 20.5 mo. 2635

Aminolevulinic Acid *Serum Increase Physiological* Modest but highly significant statistical increase in drug treated population (132 nmol/L versus 99 nmol/L). 1348

Ammonia *Plasma No Effect Physiological* No striking effect when given to epileptic patients. 3697

Androstenedione *Plasma Decrease Physiological* Within 7 d of starting 400 mg/d treatment in 6 healthy males probably due to induction of hepatic monooxygenase activity. 0707

Aspartate Aminotransferase
Serum Increase Physiological Cholestatic and hepatocellular damage. 2907 Granulomatous hepatitis observed in small proportion of patients treated for less than 1 mo; resolved within 3 d of cessation of treatment. 2153

Bilirubin *Serum No Effect Physiological* Insignificant displacement from protein in neonates. 3748
Serum Increase Physiological Cholestatic and hepatocellular damage. 2907 Granulomatous hepatitis observed in small proportion of patients treated for less than 1 mo: resolved within 3 d of cessation of treatment. 2153
Serum Decrease Physiological In about 20 epileptic patients treated for 2 y. 0873 Significant reduction compared with controls in patients treated for average of 20.5 mo due to hepatic microsomal enzyme induction. 2635

Biotin *Serum Decrease Physiological* Dose related effect observed in long term treated epileptic compared with controls. 2008

BSP Retention *Serum Increase Physiological* Cholestatic and hepatocellular damage. 1488

Calcium *Serum Decrease Physiological* Observed in 3 of 21 (14%) patients whose epilepsy was treated only with drug (mean duration of treatment 40 mo). 1625 In about 20 epileptic patients treated for 2 y. 0873 Osteomalacia observed in 3 of 31 patients given average of 758 mg/d for average of 20.5 mo. 2635

Carbamazepine *Serum Decrease Physiological* Plasma concentration may decrease by 20 to 30% after 2 to 3 weeks of treatment. 1085

Chloride *Serum Decrease Physiological* Isolated case of dilutional hyponatremia with water intoxication: 7 previous cases reported. 1882

Copper *Serum Increase Physiological* As result of increased ceruloplasmin synthesis. 1275

Cortisol *Plasma Increase Analytical* When determined by Mattingly method. 2901
Urine Increase Analytical When determined by Mattingly method. 2901

Creatinine *Serum No Effect Analytical* At 100 mg/L on reversed phase LC method of Zhiri et al. 3942

Cyclosporine *Serum Decrease Physiological* Increases hepatic metabolism with rate of hydroxylation and elimination. 3732

Dehydroepiandrosterone Sulfate (DHEA-S)
Plasma Decrease Physiological Significant reduction in women treated with drug compared with control untreated epileptics. Same effect observed when given in combination with phenytoin. 2141 Within 7 d of starting 400 mg/d treatment in 6 healthy males probably due to induction of hepatic monooxygenase activity. 0707

Diphenylhydantoin *Serum No Effect Analytical* On GLC procedure of Papadopoulos at *in vivo* concentration. 2722
Serum Increase Physiological Significant increase (36%) after drug added to regime. 3945
Serum Decrease Physiological Half life reduced from 10.6 to 6.4 h. 2041

Eosinophils *Blood Increase Physiological* Eosinophilia observed occasionally. 0019 May intensify eosinophilia of filarial infection. 1343 Isolated hypersensitivity reaction reported. Associated with fever, rash, lymphadenopathy, hepatosplenomegaly and asthma. 2156

Erythrocytes *Blood Decrease Physiological* Pancytopenia. 0928
Urine Increase Physiological Associated with bleeding tendency. 0115

Erythrocyte Sedimentation Rate
Blood Increase Physiological Associated with rare cases of granulomatous hepatitis. 3375

Folate *Serum No Effect Physiological* In about 20 epileptic patients treated for 2 y. 0873
Red Blood Cells No Effect Physiological In about 20 epileptic patients treated for 2 y. 0873

Glucaric Acid *Urine Increase Physiological* Dose dependent correlation: marked effect. 2788

Glucose *Urine Increase Physiological* Single case of glycosuria reported. 1488

Glutamine *Plasma No Effect Physiological* No striking effect when given to epileptic patients. 3697

γ-**Glutamyltransferase (GGT)**
Serum Increase Physiological Observed 2 times upper limit of normal in 2 of 35 adult inpatient epileptics (dose and duration of treatment unknown). 1162 Associated with rare cases of granulomatous hepatitis. 3375

Glycine *Plasma No Effect Physiological* No striking effect when given to epileptic patients. 3697

Granulocytes *Blood Decrease Physiological* May cause agranulocytosis. 0019

Hematocrit *Blood Decrease Physiological* Pancytopenia. 0928

Hemoglobin *Blood Decrease Physiological* Pancytopenia. 0928 In about 20 epileptic patients treated for 2 y. 0873
Urine Increase Physiological Associated with bleeding tendency. 0115

11-Hydroxycorticosteroids *Plasma No Effect Analytical* No effect on fluorometric Mattingly procedure at 1 mg/ml. 2977
Plasma Increase Analytical When determined by Mattingly method. 2901

17-Hydroxycorticosteroids *Urine Increase Analytical* Purple color so impossible to quantify with Silber and Porter method at physiological amounts. 3943
Urine Decrease Analytical Forms colored compound at 430 nm in Zimmermann procedure. 2977

25-Hydroxy Vitamin D$_3$ *Serum Decrease Physiological* Significantly lower (11.1 ng/mL vs 17.6 ng/mL) in 21 patients treated with drug only for average of 40 mo. 1625

Immunoglobulin IgA *Serum Decrease Physiological* Significant effect within 1 mo, remained low over next 30 mo. 1282

Immunoglobulin IgG *Serum No Effect Physiological* No effect regardless of duration of treatment. 1282

Immunoglobulin IgM *Serum Decrease Physiological* Significant effect within 1 mo, slight rebound with continuation of treatment for 3 mo. 1282

Indomethacin *Serum No Effect Analytical* No effect on HPLC method of Roberts and Smith. 3002

17-Ketogenic Steroids *Urine No Effect Analytical* No effect on method of Normyberski. 2901

Urine   No Effect   Physiological   No physiological effect observed. *2901*

17-Ketosteroids   *Urine   No Effect   Physiological*   No physiological effect observed. *2901*
Urine   Decrease   Analytical   Forms colored compound at 430 nm in Zimmermann procedure. *2977*

Lactate   *Plasma   Increase   Physiological*   Dose related effect observed in long term treated epileptic compared with controls. *2008*

Leukocytes   *Blood   Increase   Physiological*   Often marked leukocytosis maximal on fourth day. *1343*   Leukocytosis occasionally observed. *0019*   Granulomatous hepatitis observed in small proportion of patients treated for less than 1 mo; results within 3 d of cessation of treatment. *2153*
Blood   Decrease   Physiological   Pancytopenia (leukopenia in 15% patients). *0928*   Many patients have drop to 3,000/uL but returns to normal with continued treatment. *1085*

Luteinizing Hormone (LH)   *Plasma   No Effect   Physiological*   Insignificant fall in 6 healthy males at 14 d after 400 mg/d treatment. *0707*

MCV   *Blood   Increase   Physiological*   In about 20 epileptic patients treated for 2 y. *0873*

Neutrophils   *Blood   Decrease   Physiological*   Reported to cause neutropenia/agranulocytosis. *2583*   Neutropenia associated with decrease of WBC in small proportion of patients. *3090*   Occasional case of agranulocytosis reported. *3717*

5'-Nucleotidase   *Serum   No Effect   Physiological*   Observed 2 times upper limit of normal in none of 34 adult inpatient epileptics given drug alone (dose and duration of treatment unknown). *1162*
Serum   Increase   Physiological   Due to cholestasis. *1778*

Organic Acids   *Urine   Increase   Physiological*   Dose related effect observed in long-term treated epileptics compared with controls. *2008*

Ornithine   *Plasma   No Effect   Physiological*   No striking effect when given to epileptic patients. *3697*

Osmolality   *Serum   Decrease   Physiological*   Mean concentration reduced in carbamazepine treated patients, possibly due to stimulation of release of ADH. *2787*   Isolated case of dilutional hyponatremia with water intoxication: 7 previous cases reported. *1882*

Pheneturide   *Serum   No Effect   Analytical*   On GLC procedure of Papadopoulos at *in vivo* concentration. *2722*

Phenobarbital   *Serum   No Effect   Analytical*   On GLC procedure of Papadopoulos at *in vivo* concentration. *2722*

Phosphate   *Serum   No Effect   Physiological*   Osteomalacia observed in 3 of 31 patients given average of 758 mg/d for average of 20.5 mo. *2635*
Serum   Decrease   Physiological   Hypophosphatemia observed in 1 of 21 patients (5%) treated for average of 40 mo. *1625*   In about 20 epileptic patients treated for 2 y. *0873*

Platelets   *Blood   Decrease   Physiological*   Thrombocytopenia may occur after 1 y (immunologic). *0928*   Within 2 weeks of commencement of treatment in a single patient. Associated with petechiae. *3096*   Marked effect in one patient with strongly positive migration inhibition (MIF) test. *3308*   In one case drug-dependent IgG antibodies identified. *1996*

Potassium   *Serum   No Effect   Analytical*   At concentration of 20 mg/L had no effect on ISE measurement with predilution. *3393*
Urine   Decrease   Physiological   Isolated case of dilutional hyponatremia with water intoxication: 7 previous cases reported. *1882*

Pregnancy Tests   *Urine   Decrease   Physiological*   False negative or inconclusive value with Prepurex, Predictor, Gonavislide, Pregnosticon®. *2182*

Primidone   *Serum   No Effect   Analytical*   On GLC procedure of Papadopoulos at *in vivo* concentration. *2722*

Protein   *Serum   Decrease   Physiological*   In about 20 epileptic patients treated for 2 y. *0873*
Urine   Increase   Physiological   Manifestation of renal damage. *0019*

Prothrombin Time   *Plasma   Decrease   Physiological*   Enhances warfarin metabolism by enzyme induction. *1487*

Sex Hormone Binding Globulin
Serum   Increase   Physiological   Within 7 d of starting 400 mg/d treatment in 6 healthy males probably due to induction of hepatic monooxygenase activity. *0707*

Sodium   *Serum   No Effect   Analytical*   At concentration of 20 mg/L had no effect on ISE measurement with predilution. *3393*
Serum   Decrease   Physiological   Although mean concentration in population of epileptics not affected, significant reduction in 5 of 80 patients. *2787*   Isolated case of dilutional hyponatremia with water intoxication: 7 previous cases reported. *1882*   In 28 of 674 epileptic patients most often when drug combined with barbiturates. *1848*
Urine   Decrease   Physiological   Isolated case of dilutional hyponatremia with water intoxication: 7 previous cases reported. *1882*

Somatostatin   *CSF   Decrease   Physiological*   Appears to affect compound whereas other drugs do not. *3081*

Testosterone   *Serum   No Effect   Physiological*   Usually no effect but may be slight increase. *1736*
Serum   Decrease   Physiological   Within 7 d of starting 400 mg/d treatment in 6 healthy males probably due to induction of hepatic monooxygenase activity. *0707*

Testosterone, Free   *Serum   Decrease   Physiological*   Observed effect. *1736*   Within 7 d of starting 400 mg/d treatment in 6 healthy males probably due to induction of hepatic monooxygenase activity. *0707*

Theophylline   *Serum   Decrease   Physiological*   Half-life reduced from 5.25 h to 2.75 h during treatment probably due to enzyme interaction. *3055*

Thyroglobulin   *Serum   No Effect   Physiological*   No significant effect with 400 mg/d for 21 d. *0708*

Thyroid Stimulating Hormone (TSH)
Serum   No Effect   Analytical   At therapeutic and toxic concentrations on RIA methods. *0374*   No effect observed at both therapeutic and toxic concentrations when added to serum *in vitro*. *0374*
Serum   No Effect   Physiological   No effect of 400 mg/d on basal or stimulated TSH. *0708*   Observed in 9 hypothyroid patients given substitution therapy due to increased extra-thyroidal metabolism of thyroid hormones. *0001*   No change in TSH or in its response to TRH. *3798*   No significant change in patients on long-term treatment. *2168*
Serum   Decrease   Physiological   Marked reduction in 26 patients on long-term therapy. *2169*

Thyroxine (T₄)   *Serum   No Effect   Analytical*   At therapeutic and toxic concentrations on an equilibrium dialysis method. *0374*   No effect observed at both therapeutic and toxic concentrations when added to serum *in vitro*. *0374*
Serum   Decrease   Physiological   Fall from mean of 82 to 75 nmol/L with 400 mg/d carbamazepine after 14 d treatment in 10 healthy males, secondary to hepatic enzyme induction with accelerated nondeiodinative hepatic hormone disposal. *0708*   Observed in 9 hypothyroid patients given substitution therapy due to increased extra-thyroidal metabolism of thyroid hormones. *0001*

Thyroxine (T₄) Binding Globulin
Serum   No Effect   Analytical   At therapeutic and toxic concentrations on Corning radioimmunometric kit method. *0374*   No effect observed at both therapeutic and toxic concentrations when added to serum *in vitro*. *0374*
Serum   Increase   Physiological   Observed in 9 hypothyroid patients given substitution therapy due to increased extra-thyroidal metabolism of thyroid hormones. *0001*

Thyroxine (T₄), Free   *Serum   No Effect   Analytical*   At therapeutic and toxic concentrations on RIA method. *0374*   No effect observed at both therapeutic and toxic concentrations when added to serum *in vitro*. *0374*
Serum   Increase   Analytical   As measured by addition of drug to control sera and measured by analog radioimmunoassay. *2169*
Serum   Decrease   Physiological   Fall from mean of 16.0 to 14.2 pmol/L with 400 mg/d in healthy males. *0708*   As measured by equilibrium dialysis and analog-type radioimmunoassays; not due to to displacement from protein. Therapy. *2169*   Observed in 9 hypothyroid patients given substitution therapy due to increased extra-thyroidal metabolism of thyroid hormones. *0001*

Thyroxine (T₄) Index, Free (FTI)
Serum   Decrease   Physiological   No significant change in patients on long-term treatment. *2168*

Carbamazepine (continued)

Tri-iodothyronine (T₃) *Serum No Effect Analytical* At therapeutic and toxic concentrations on RIA methods. *0374* No effect observed at both therapeutic and toxic concentrations when added to serum *in vitro*. *0374*
Serum Decrease Physiological Fall from mean of 1.6 to 1.4 nmol/L with 400 mg/d in healthy males. *0708* Observed in 9 hypothyroid patients given substitution therapy due to increased extra-thyroidal metabolism of thyroid hormones. *0001*

Tri-iodothyronine (T₃), Free *Serum Increase Analytical* As measured by addition of drug to control sera and measured by analog radioimmunoassay. *2169*
Serum Decrease Physiological As measured by equilibrium dialysis and analog-type radioimmunoassays; not due to displacement from protein. *2169* Observed in 9 hypothyroid patients given substitution therapy due to increased extra-thyroidal metabolism of thyroid hormones. *0001*

Tri-iodothyronine (T₃), Reverse
Serum No Effect Physiological No significant effect with 400 mg/d for 21 d. *0708* No significant change in patients on long-term treatment. *2168*

Urea Nitrogen *Serum Increase Physiological* May cause kidney dysfunction. *0071*

Uric Acid *Serum No Effect Analytical* At 100 mg/L on reversed phase LC method of Zhiri et al. *3942*

Valproic Acid *Serum Decrease Physiological* Significant reduction of half-life from 10.9 h to 6.4 h due to enzyme induction. *2233* Concentration reduced to about 66% when carbamazepine coadministered. *2340*

Volume *Urine Decrease Physiological* Isolated case of dilutional hyponatremia with water intoxication: 7 previous cases reported. *1882*

Warfarin *Plasma Decrease Physiological* Reported interaction due to alteration of metabolism. *2042*

Carbarsone

Alanine Aminotransferase *Serum Increase Physiological* May cause hepatitis/liver necrosis. *0071*

Aspartate Aminotransferase
Serum Increase Physiological May cause hepatitis/liver necrosis. *0071*

Bilirubin *Serum Increase Physiological* Cholestasis with cholangiolitis may occur. *1948*

BSP Retention *Serum Increase Physiological* Toxic hepatitis may occur. *2575*

¹³¹I Uptake *Serum Decrease Physiological* Lilly compound contains tetraiodofluorescein. *2652*

Protein *Urine Increase Physiological* Nephrotoxic effect. *1022*

Carbenicillin

Alanine Aminotransferase *Serum No Effect Analytical* At 5 times upper limit of therapeutic range on methods on SMAC®, Abbott-VP, aca, Cobas-Bio, and KDA. *2138*
Serum Increase Physiological Elevation reported, ?due to hepatotoxicity. *1680* In several patients as evidence of drug induced hepatotoxicity. *1371* In 7 of 27 patients given 270 mg/kg i.v. for 6 d. *3275*

Albumin *Serum No Effect Analytical* At 5 times upper limit of therapeutic range on BCG method on SMAC®, Ektachem® 400 Hitachi® 705 and KDA. *2138*
Serum Decrease Analytical At concentrations above 5,000 mg/L lowered the concentration as measured by BCG method. *3393*

Alkaline Phosphatase *Serum No Effect Analytical* At 5 times upper limit of therapeutic range on methods on SMAC®, Abbott-VP, aca, Cobas-Bio, Hitachi® 705 and KDA. *2138*
Serum Increase Physiological Transient elevations reported. *1343* In 4 of 27 patients given 270 mg/kg i.v. for 6 d. *3275*

Amino Acids *Urine Positive Analytical* Unusual ninhydrin positive spot on TLC. *1596*

Amylase *Serum No Effect Analytical* At 5 times upper limit of therapeutic range on methods on aca, Cobas-Bio and Ektachem® 400. *2138*

Aspartate Aminotransferase *Serum No Effect Analytical* At 5 times upper limit of therapeutic range on methods on SMAC®, Abbott-VP, aca, Cobas-Bio, Hitachi® 705 and KDA. *2138*
Serum Increase Physiological Effect of drug ?hepatotoxic. *1237* In 6 of 27 patients given 270 mg/kg i.v. for 6 d. *3275*

Bicarbonate *Serum No Effect Analytical* At concentration of 15,000 mg/L had no effect on method using phenolphthalein. *3393*
Serum Increase Physiological Observed in approximately 8% patients. *1947*

Bilirubin *Serum No Effect Analytical* At 5 times upper limit of therapeutic range on methods on SMAC®, Cobas-Bio, Ektachem® 400, Hitachi® 705 and KDA. *2138* At concentration of 2,000 mg/L had no effect on Jendrassik and Grof method. *3393*
Serum Increase Analytical At 5 times upper limit of therapeutic range on aca method. *2138*

Calcium *Serum No Effect Analytical* At 5 times upper limit of therapeutic range on methods on SMAC®, Abbott-VP, aca, Ektachem®, Hitachi® 705 and KDA. *2138* At concentration of 15,000 mg/L had no effect on cresolphthalein method. *3393*

Chloride *Serum No Effect Analytical* At concentration of 15,000 mg/L had no effect on mercurimetric method. *3393*

Cholesterol *Serum No Effect Analytical* At 5 times upper limit of therapeutic range on methods on SMAC®, Abbott-VP, Cobas-Bio, Ektachem® 400, Hitachi® 705 and KDA. *2138* At concentration of 15,000 mg/L had no effect on Liebermann-Burchard method. *3393*

Clotting Time *Blood Increase Physiological* Reported effect. *3734*

Creatine Kinase *Serum No Effect Analytical* At 5 times upper limit of therapeutic range on methods on SMAC®, Abbott-VP, aca, Cobas-Bio and Hitachi® 705. *2138*
Serum Increase Physiological Probably due to trauma of i.m. injection. *1966*

Creatinine *Serum No Effect Analytical* At 5 times upper limit of therapeutic range on methods on SMAC®, Abbott-VP, aca, Ektachem® 400, Cobas-Bio, Hitachi® 705 and KDA. *2138*

Eosinophils *Blood Increase Physiological* Manifestation of allergic response. *1680* In 2 of 27 patients given 270 mg/kg i.v. for 6 d. *3275*

Erythrocytes *Blood Decrease Physiological* Hemolytic anemia reported. *1680*

Glucose *Serum No Effect Analytical* At 5 times upper limit of therapeutic range on methods on SMAC®, Abbott-VP, aca, Ektachem® 400, Cobas-Bio, Hitachi® 705 and KDA. *2138*
Urine No Effect Analytical No influence of drug at up to 20 mg/mL on Diastix® and TesTape® procedures. *2250*
Urine Positive Analytical At 10 and 20 mg/mL gave false positive with negative urine using Clinitest® but with positive glucose specimens gave falsely low results. *2250*

γ-Glutamyltransferase (GGT) *Serum No Effect Analytical* At 5 times upper limit of therapeutic range on methods on SMAC®, Abbott-VP, and Hitachi® 705. *2138*

Hemoglobin *Blood Decrease Physiological* Hemolytic anemia reported. *1680*

Iron *Serum No Effect Analytical* At 5 times upper limit of therapeutic range on Ferrozine method on SMAC®. *2138*
Serum No Effect Physiological No effect although marrow depressed. *2964*

Iron-Binding Capacity, Total (TIBC)
Serum No Effect Physiological No effect although marrow depressed. *2964*

Lactate Dehydrogenase *Serum No Effect Analytical* At 5 times upper limit of therapeutic range on methods on SMAC®, Abbott-VP, Cobas-Bio, Hitachi® 705 and KDA. *2138*
Serum Increase Physiological Transient elevations reported. *1343*

Leukocytes *Blood Decrease Physiological* Leukopenia may occur infrequently. *1680*

Monocytes *Blood Increase Physiological* In 1 of 27 patients given 270 mg/kg i.v. for 6 d. *3275*

Neutrophils *Blood Decrease Physiological* Neutropenia reported occasionally. *1680*

pH *Blood Increase Physiological* High pH observed in most patients. *1947*

Phosphate *Serum No Effect Analytical* At 5 times upper limit of therapeutic range on methods on SMAC®, aca, Hitachi® 705 and KDA. *2138* At concentration of 15,000 mg/L had no effect on phosphomolybdate method. *3393*

Platelets *Blood Decrease Physiological* Occasionally thrombocytopenia may occur. *1680*

Potassium *Serum Decrease Physiological* Observed effect ?redistribution or increased excretion. *3549* Nonreabsorbable anion increases electrical negativity of lumen of distal nephron with enhanced potassium and hydrogen excretion. *0645* In one child due to drug having impermeant anion effect on renal tubule. *3431*

Urine Increase Physiological Possible increased distal tubular excretion in some patients. *3549*

Protein *Serum No Effect Analytical* At 5 times upper limit of therapeutic range on biuret method on KDA. *2138* At concentrations above 500 mg/L raised the concentration as measured by biuret method with blank correction. *3393*

Serum Increase Analytical At 5 times upper limit of therapeutic range on methods on SMAC®, Abbott-VP, Ektachem® 400 and Hitachi® 705. *2138*

Prothrombin Time *Plasma Increase Physiological* Effect observed especially in uremics. *1680*

Triglycerides *Serum No Effect Analytical* At 5 times upper limit of therapeutic range on methods on SMAC®, Ektachem® 400, Hitachi® 705 and KDA. *2138* At concentration of 15,000 mg/L had no effect on lipase/esterase method. *3393*

Serum Increase Analytical At 5 times upper limit of therapeutic range on enzymatic method on Abbott-VP. *2138*

Urea Nitrogen *Serum No Effect Analytical* At 5 times upper limit of therapeutic range on methods on SMAC®, Abbott-VP, Cobas-Bio, Ektachem® 400, Hitachi® 705 and KDA. *2138* At concentration of 15,000 mg/L had no effect on diacetylmonoxime method. *3393*

Uric Acid *Serum No Effect Analytical* At 40 mg/dL on Nishi phosphotungstate procedure. *2230* At 5 times upper limit of therapeutic range on methods on SMAC®, Abbott-VP, aca, Ektachem® 400, Cobas-Bio, Hitachi® 705 and KDA. *2138* At concentration of 15,000 mg/L had no effect on phosphotungstate reduction method. *3393*

Serum Increase Analytical 40 mg/dL equivalent to 0.9 mg/dL copper chelate procedure. *2230*

Carbenoxolone

Alanine Aminotransferase *Serum Increase Physiological* Associated with myopathy. *2458*

Aldolase *Serum Increase Physiological* Due to hypokalemic myopathy. *0232*

Aspartate Aminotransferase
Serum Increase Physiological Associated with myopathy. *2458*

Bicarbonate *Serum Increase Physiological* Hypokalemic alkalosis. *2458*

Bilirubin *Serum Increase Physiological* Possible hepatotoxic effect of drug. *2458*

BSP Retention *Serum Increase Physiological* Reversible hepatotoxic effect. *2458*

Calcium *Serum Decrease Physiological* Aldosterone like effect of drug. *2458*

Chloride *Serum Decrease Physiological* Average decrease of 3 meq/L, associated with alkalosis. *1179*

Creatine Kinase *Serum Increase Physiological* May cause hypokalemic myopathy. *1179*

Creatinine Clearance *Urine Decrease Physiological* Due to nephropathy. *2458*

Hematocrit *Blood Decrease Physiological* By approximately 4% in adrenalectomized patients. *3801*

11-Hydroxycorticosteroids *Plasma Increase Physiological* Transient effect after 100 mg orally. *1487*

Magnesium *Serum Decrease Physiological* By approximately 0.1 meq/L in adrenalectomized patients. *3801*

Myoglobin *Urine Increase Physiological* Myopathy following hypokalemia. *2458*

5'-Nucleotidase *Serum Increase Physiological* Possible direct hepatotoxicity of drug. *2458*

pH *Blood Increase Physiological* Alkalosis occurs in approximately 15% patients. *2849*
Urine Increase Physiological Impaired acidification with hypokalemia. *2458*

Potassium *Serum Decrease Physiological* By approximately 0.5 meq/L in adrenalectomized patients. *3801* Aldosterone like effect. *1159*
Urine Increase Physiological Aldosterone like effect. *2458*

Protein *Urine Increase Physiological* Due to myoglobinuria and nephropathy. *2458*

Renin Activity *Plasma Decrease Physiological* Significant decrease after adrenalectomy (standing and lying). *3801*

Sodium *Serum Increase Physiological* Significant increase (approximately 5 meq/L) in adrenalectomized patients. *3801*

Specific Gravity *Urine Decrease Physiological* Impaired concentration with hypokalemia Impaired ability to concentrate with low potassium. *2458*

Urea Nitrogen *Serum Increase Physiological* Due to nephropathy with hypokalemia. *0232*

Carbidopa

Cortisol *Plasma No Effect Physiological* In 10 normal volunteers given 300 mg daily for 1 week. *0494*

Dihydroxyphenylalanine *Plasma Increase Physiological* Concentration doubled in 6 men over 5 h after administration, especially if protein also given. *3842*

Dopamine *Urine Decrease Physiological* 70% reduction in excretion noted over 5 h. *3842*

Epinephrine *Urine No Effect Physiological* No effect over 5 h after ingestion. *3842*

Growth Hormone *Plasma No Effect Physiological* In 10 normal volunteers given 300 mg daily for 1 week. *0494*

Hydrazine *Plasma Increase Analytical* On fluorometric method of Vickers. *3710*

Norepinephrine *Urine No Effect Physiological* No effect over 5 h after ingestion. *3842*

Prolactin *Plasma Increase Physiological* In 10 normal volunteers given 300 mg daily for 1 week. *0494*

Tryptamine *Urine Decrease Physiological* Marked effect due to decarboxylase inhibition. *0494*

Carbimazole

Alanine Aminotransferase *Serum Increase Physiological* Isolated case of jaundice conclusively linked to drug. *0911*

Alkaline Phosphatase *Serum Increase Physiological* Isolated case of jaundice conclusively linked to drug. *0911* Reversible hypersensitivity response to drug. *1790*

Apolipoprotein AI *Serum No Effect Physiological* From 2.62 to 2.82 g/L in 12 hyperthyroid women patients treated with 10-30 mg daily. *1201*

Aspartate Aminotransferase
Serum Increase Physiological Isolated case of jaundice conclusively linked to drug. *0911* Reversible hypersensitivity response to drug. *1790*

Bilirubin *Serum Increase Physiological* Isolated case of jaundice conclusively linked to drug. *0911*

Cholesterol *Serum Increase Physiological* From 4.4 to 5.4 mmol/L in 12 hyperthyroid women patients treated with 10-30 mg daily. *1201*

Cholesterol, High Density Lipoprotein
Serum Increase Physiological From 1.3 to 1.6 mmol/L in 12 hyperthyroid women patients treated with 10-30 mg daily. *1201*

Cholesterol, Low Density Lipoprotein
Serum Increase Physiological From 2.7 to 3.5 mmol/L in 12 hyperthyroid women patients treated with 10-30 mg daily. *1201*

Erythrocytes *Blood Decrease Physiological* Occasional aplastic anemia reported. *3717*

γ-Glutamyltransferase (GGT)
Serum Increase Physiological Reversible hypersensitivity response to drug. *1790*

131I Uptake *Serum No Effect Physiological* No effect on uptake by thyroid. *2220*

Carbimazole *(continued)*

Leukocytes *Blood Decrease Physiological* Agranulocytosis. *0541* Occasional aplastic anemia reported. *3717* Two case reports with considerable reduction in white cell count and neutrophil count. *1473*

Lymphocyte Autoantibodies
Blood Decrease Physiological Impaired thyroid microsomal or thyroglobulin antibody secretion due to effect on lymphocytes within thyroid. *2376*

Neutrophils *Blood Decrease Physiological* May cause agranulocytosis/neutropenia. *2583* Occasional agranulocytosis reported. *3717* Two case reports with considerable reduction in white cell count and neutrophil count. *1473* Occasional case of drug-induced neutropenia. *0155*

Platelets *Blood Decrease Physiological* Occasional aplastic anemia reported. *3717*

Protein Bound Iodine (PBI) *Serum Decrease Physiological* Inhibits iodination of tyrosine in thyroxine binding globulin. *0012*

T_3 Uptake *Serum Decrease Physiological* From 1.33 to 0.90 arbitrary units in 12 hyperthyroid women patients treated with 10-30 mg daily. *1201*

Thyroxine (T_4) *Serum Decrease Physiological* From 184 to 101 nmol/L in 12 hyperthyroid women patients treated with 10-30 mg daily. *1201*

Triglycerides *Serum Increase Physiological* From 0.77 to 0.89 mmol/L in 12 hyperthyroid women patients treated with 10-30 mg daily. *1201*

Tri-iodothyronine (T_3) *Serum Decrease Physiological* From 4.55 to 1.68 nmol/L in 12 hyperthyroid women patients treated with 10-30 mg daily. *1201*

Carbochromen

Alanine Aminotransferase *Serum No Effect Analytical* No effect at therapeutic concentration on Reflotron method. *1984*

Aspartate Aminotransferase *Serum No Effect Analytical* No effect at therapeutic concentration on Reflotron method. *1984*

Cholesterol *Serum No Effect Analytical* No effect at therapeutic concentration on Reflotron method. *1984*

Glucose *Serum No Effect Analytical* No effect at therapeutic concentration on Reflotron method. *1984*

γ-Glutamyltransferase (GGT) *Serum No Effect Analytical* No effect at therapeutic concentration on Reflotron method. *1984*

Triglycerides *Serum No Effect Analytical* No effect at therapeutic concentration on Reflotron method. *1984*

Urea Nitrogen *Serum No Effect Analytical* No effect at therapeutic concentration on Reflotron method. *1984*

Uric Acid *Serum No Effect Analytical* No effect at therapeutic concentration on Reflotron method. *1984*

Carbohydrate

Aldosterone *Plasma No Effect Physiological* No effect when starved patients refed. *1242*

Gastrin *Serum No Effect Physiological* No effect on endogenous secretion if given orally. *1236*

Renin Activity *Plasma No Effect Physiological* No effect when starved patients refed. *1242*

Sodium *Urine Decrease Physiological* Marked effect when starved patients refed. *1242*

Carbonates

Appearance *Urine Abnormal Analytical* Cloudy, soluble in dilute acetic acid. *0443*

Carbon Disulfide

Alanine Aminotransferase *Serum Increase Physiological* May cause liver damage. *0279*

Albumin *Serum Decrease Physiological* Effect greater than on total protein. *0818*

Aspartate Aminotransferase
Serum Increase Physiological May cause liver damage. *0279*

Cholesterol *Serum Increase Physiological* With other liver function tests normal on chronic exposure. *0256*
Serum Decrease Physiological In people with exposure to 37 ppm for 5.7 y. *0818*

Cholesterol Esters *Serum Decrease Physiological* While cholesterol was increased. *0256*

Eosinophils *Blood Increase Physiological* Occasional response. *1302*

Erythrocytes *Urine Increase Physiological* Manifestation of renal damage. *0818*

Hematocrit *Blood Decrease Physiological* Hypochromic anemia common. *0818*

Hemoglobin *Blood Decrease Physiological* Hypochromic anemia common. *0818*
Urine Increase Physiological Manifestation of renal damage. *0818*

Hydrochloric Acid *Gastric Material Decrease Physiological* Frequent;also high incidence chronic gastritis. *0818*

17-Hydroxycorticosteroids *Urine Decrease Physiological* Reduction in relation to time of exposure. *0818*

17-Ketosteroids *Urine Decrease Physiological* Reduction in relation to time of exposure. *0818*

Lymphocytes *Blood Increase Physiological* Effect usually slight in poisoning. *1302*

Monocytes *Blood Increase Physiological* Occurs in 25% cases regardless of duration. *0818*

Protein *Serum Decrease Physiological* Effect slight. *0818*
Urine Increase Physiological Manifestation of renal damage. *0818*

Pseudocholinesterase *Serum Decrease Physiological* Not usually severe enough to produce symptoms. *1302*

Urea Nitrogen *Serum Increase Physiological* Renal impairment observed. *0818*

Carbon Monoxide

Aspartate Aminotransferase
Serum Increase Physiological Associated with hepatomegaly. *1302*

Carboxyhemoglobin *Blood Increase Physiological* Due to presence of carbon monoxide. *1302*

Color *Blood Increase Physiological* May cause cherry-red color (carboxyhemoglobin). *0279*

Glomerular Filtration Rate (GFR)
Urine Increase Physiological Occurs 12 to 24 h after exposure. *3454*

Glucose *Serum No Effect Analytical* No effect on o-toluidine procedure. *2717*
Serum Decrease Analytical Saturation may cause decrease of 15% with glucose oxidase methods. *2717*

Myoglobin *Urine Increase Physiological* May be observed in intoxication. *0017*

Protein *Urine Increase Physiological* May cause nephrotoxicity. *3204* Associated with oliguria and renal damage. *1302*

Urea Nitrogen *Serum Increase Physiological* May cause nephrotoxicity. *3204*

Carboplatin

Platinum *Red Blood Cells Increase Physiological* 1 h after bolus injection 2% of drug bound to erythrocytes, stabilizes after 3 h. *3286*

Carbromal

Alanine Aminotransferase *Serum Increase Physiological* Occurs with poisoning, probable muscle origin. *3901*

Aspartate Aminotransferase
Serum Increase Physiological Occurs with poisoning, probable muscle origin. *3901*

Bromide *Serum Increase Physiological* Metabolite concentration may exceed 25 meq/L. *1343*

Chloride *Serum Increase Analytical* Bromide as metabolite may be measured as chloride. *1343*

Coombs' Test, Direct *Serum Positive Physiological* Unusual cause of hemolytic anemia. *2365*

Coombs' Test, Indirect *Serum Positive Physiological* Unusual cause of hemolytic anemia. *2365*

Creatine Kinase *Serum Increase Physiological* Occurs with poisoning, probable muscle origin. *3901*

Ethchlorvynol *Serum No Effect Analytical* At 10 μg/mL on colorimetric method of Wallace. *3753*

Porphyrins *Urine Increase Physiological* May precipitate attack of acute porphyria. *2220*

Carbutamide

Alanine Aminotransferase *Serum Increase Physiological* Hepatotoxicity. *2313*

Alkaline Phosphatase *Serum Increase Physiological* Hepatotoxicity. *2313*

Aspartate Aminotransferase
Serum Increase Physiological Hepatotoxicity. *2313*

Bile *Urine Increase Physiological* Hepatotoxicity. *2313*

Bilirubin *Serum Increase Physiological* Hepatotoxicity. *2313*

BSP Retention *Serum Increase Physiological* Hepatotoxicity. *2313*

Cholesterol *Serum Decrease Physiological* Hepatotoxicity. *2313*

Creatinine *Serum Increase Physiological* Nephrotoxic effect. *2313*

Fatty Acids, Free (FFA) *Serum Increase Physiological* Reported to cause hyperlipemia. *2220*

Glucose *Serum Decrease Physiological* Hepatotoxicity. *2313*

^{131}I Uptake *Serum Decrease Physiological* Substantial effect observed in elderly. *0830*

Leukocytes *Blood Decrease Physiological* Leukopenia/agranulocytosis. *3878*

Nonprotein Nitrogen *Serum Increase Physiological* Nephrotoxic effect. *2313*

Platelets *Blood Decrease Physiological* May cause aplastic anemia or thrombocytopenia. *3878*

Protein *Urine Increase Physiological* Nephrotoxic effect. *3729*

Protein Bound Iodine (PBI) *Serum Decrease Physiological* Inhibits iodination of tyrosine in thyroxine binding globulin. *0012*

Urea Nitrogen *Serum Increase Physiological* Nephrotoxic effect. *2313*

Carfecillin

Chloramphenicol *Serum No Effect Analytical* No effect at 100 mg/L on coupled enzymatic method. *2490*

Carinamide

Protein *Urine Increase Analytical* Produces precipitate with acid tests. *0459*

PSP Excretion *Urine Decrease Physiological* Inhibits secretion. *1022*

Sugar *Urine Increase Analytical* False positive with Benedict's. *1563*

Carisoprodol

Alkaloids *Urine No Effect Analytical* With ninhydrin in Frings TLC procedure at 10 mg/dL. With iodoplatinate of Frings TLC procedure at 10 mg/dL. *1204*

Barbiturate *Urine No Effect Analytical* With mercuric SO$_4$ in Frings TLC procedure at 10 mg/dL. With mercurous NO$_3$ in Frings TLC procedure at 10 mg/dL. With vanillin on Frings procedure at 10 mg/dL. With diphenylcarbazone in Frings TLC procedure at 10 mg/dL. *1204*

Bromide *Urine No Effect Analytical* No effect on method of Frings. *1204*

Eosinophils *Blood Increase Physiological* Allergic manifestation. *1680*

Ethchlorvynol *Urine No Effect Analytical* No effect on method of Frings and Cohen. *1204*

FPN Test *Urine No Effect Analytical* No effect at 10 mg/dL on method of Frings. *1204*

Leukocytes *Blood Decrease Physiological* Possible consequence/not marked. *1680*

Salicylate *Urine No Effect Analytical* At 10 mg/dL no effect on Trinder method. *1204*

Carmine

Color *Feces Increase Analytical* Produces red color. *1187*

Carmustine (BCNU)

Alanine Aminotransferase *Serum Increase Physiological* Usually mild and return to normal over few days. *2406*

Alkaline Phosphatase *Serum Increase Physiological* Usually mild and return to normal over few days. *2406*

Aspartate Aminotransferase
Serum Increase Physiological Usually mild and return to normal over few days. *2406*

Bilirubin *Serum Increase Physiological* Usually mild and return to normal over few days. *2406*

Carotene

Bilirubin *Serum Increase Analytical* Color may interfere with direct methods. *1022*

Icteric Index *Serum Increase Analytical* Interfering background color. *1237*

Vitamin A *Serum No Effect Analytical* On modified fluorometric procedure of Drujan. *3685*
Serum Increase Analytical If high in serum will affect UV methods. *3028*

Carotenoid

Carotene *Serum Increase Analytical* May interfere with some analytical methods. *3023*

Color *Serum Increase Analytical* Color serum orange. *3023*

Vitamin A *Serum Increase Analytical* May interfere with some analytical methods. *3023*

Carphenazine

Alanine Aminotransferase *Serum Increase Physiological* Probable cholestatic effect (reversible). *2313*

Aspartate Aminotransferase
Serum Increase Physiological Probable cholestatic effect (reversible). *2313*

Bile *Urine Increase Physiological* Probable cholestatic effect (reversible). *2313*

Bilirubin *Serum Increase Physiological* Probable cholestatic effect (reversible). *2313*

BSP Retention *Serum Increase Physiological* Probable cholestatic effect (reversible). *2313*

Cephalin Flocculation *Serum Increase Physiological* Probable cholestatic effect (reversible). *2313*

Thymol Turbidity *Serum Increase Physiological* Probable cholestatic effect (reversible). *2313*

Carprazidil

Aldosterone *Plasma No Effect Physiological* No consistent effect in 15 men with mild to moderate essential hypertension treated for up to 16 weeks. *1268*

Cholesterol *Serum No Effect Physiological* No significant effect in 12 subjects treated for 4 mo. *0088* In one study involving 12 patients treated for 4 mo. *0089*

Cholesterol, High Density Lipoprotein
Serum No Effect Physiological In one study involving 12 patients treated for 4 mo. *0089*
Serum Increase Physiological In 15 men with mild to moderate essential hypertension by 26% over 8 weeks. *1268*

Cholesterol, Low Density Lipoprotein
Serum No Effect Physiological No consistent effect in 15 men with mild to moderate essential hypertension treated for up to 16 weeks. *1268* In one study involving 12 patients treated for 4 mo. *0089*

Cholesterol, Very Low Density Lipoprotein
Serum No Effect Physiological No consistent effect in 15 men with mild to moderate essential hypertension treated for up to 16 weeks. *1268* In one study involving 12 patients treated for 4 mo. *0089*

Epinephrine *Plasma No Effect Physiological* No consistent effect in 15 men with mild to moderate essential hypertension treated for up to 16 weeks. *1268*

α-Lipoproteins *Serum Increase Physiological* In 15 men with mild to moderate essential hypertension by 26% over 8 weeks. *1268*

Norepinephrine *Plasma No Effect Physiological* No consistent effect in 15 men with mild to moderate essential hypertension treated for up to 16 weeks. *1268*

Renin Activity *Plasma No Effect Physiological* No consistent effect in 15 men with mild to moderate essential hypertension treated for up to 16 weeks. *1268*

Triglycerides *Serum No Effect Physiological* No consistent effect in 15 men with mild to moderate essential hypertension treated for up to 16 weeks. *1268* In one study involving 12 patients treated for 4 mo. *0089*

Carrots

Bilirubin *Serum Increase Analytical* Color may be misinterpreted. *1238*

Bilirubin, Direct *Serum Increase Analytical* Color may be misinterpreted. *1238*

Color *Urine Increase Analytical* Yellow color soluble in petroleum ether. *2313*

Icteric Index *Serum Increase Analytical* Color may be misinterpreted. *1237*

Methemoglobin *Blood Increase Physiological* With fresh or juice if contaminated with nitrites. *1893*

Carteolol

Creatine Kinase *Serum Increase Physiological* Significant effect in 10 of 15 patients with essential hypertension. MM isoenzyme most affected. *3149*

Cascara

Color *Urine Increase Analytical* Brown (acid),yellow-pink(alkaline)black on standing. *1022*

Estradiol *Urine Decrease Analytical* Affects unmodified Brown/Kober procedure. *0583*

Estrogens *Urine Increase Analytical* Reacts with Kober procedure Reacts in Brown's procedure. *0279*
Urine Decrease Analytical Affects colorimetric/fluorometric procedures. *0022*

131I Uptake *Serum Decrease Physiological* Lilly compound contains tetraiodofluorescein. *2652*

Porphobilinogen *Urine Increase Analytical* Color extractable into chloroform. *1563*

PSP Excretion *Urine Increase Analytical* Converted to anthraquinone-gastrointestinal tract (red color). *0757*

Urobilinogen *Urine Increase Analytical* Reacts with Ehrlich's reagent. *1563*

Castor Oil

Cells *Urine Increase Physiological* Multinucleated giant cells in 50% patients. *2427*

Protein *Urine Increase Physiological* May cause renal damage. *0279*

Urea Nitrogen *Serum Increase Physiological* May cause renal damage. *0279*

Catechol

Glucose *Serum Decrease Analytical* May inhibit glucose-oxidase procedures. *1563*

Guaiacols Spot Test *Urine Positive Analytical* Action on procedure of Rogers. *3031*

Homogentisic Acid *Urine Increase Analytical* If method of Nuberger used. *1096*

Catecholamines

Gastrin *Serum Increase Physiological* High concentration associated with pheochromocytoma and on injection. *1533*

Cathartics

Albumin *Serum Decrease Physiological* Rare protein-losing gastroenteropathy. *1343*

Potassium *Serum Decrease Physiological* Excessive use may cause hypokalemia. *1343*
Urine Increase Physiological Secondary aldosteronism if blood volume reduced. *1343*

Protein *Serum Decrease Physiological* Rare protein losing gastroenteropathy. *1343*

Prothrombin Time *Plasma Increase Physiological* Accelerated passage and decreased absorption of vitamin K. *1487*

Sodium *Serum Decrease Physiological* Excessive use may cause sodium depletion. *1343*

Cefaclor

Platelets *Blood Decrease Physiological* Rare immune-mediated toxicity. *2615*

Theophylline *Serum No Effect Physiological* No effect on steady state kinetics in healthy young adults. *1824*

Cefamandole

Alanine Aminotransferase *Serum Increase Physiological* Occurs in about 4% treated cases. *3134*

Alkaline Phosphatase *Serum Increase Physiological* Fewer than 3 of 53 patients developed reversible hepatotoxicity. *0969*

Aspartate Aminotransferase
Serum Increase Physiological Occurs in about 4% treated cases. *3134* Fewer than 3 of 53 patients developed reversible hepatotoxicity. *0969*

Bilirubin *Serum No Effect Analytical* At 2.50 mmol/L on method on Eppendorf Epos. *1414*

Cholesterol *Serum No Effect Analytical* At 2.50 mmol/L on method on Eppendorf Epos. *1414*

Coombs' Test, Direct *Serum Positive Physiological* In about 1% of treated patients (dose related). *3134*

Creatinine *Serum No Effect Analytical* At concentration of 1500 mg/L had no effect on AutoAnalyzer Jaffé method. *3393* At concentration of 1500 mg/L had no effect on kinetic Jaffé method on aca. *3393* At concentration of 1500 mg/L had no effect on kinetic method on BKA-2. *3393*
Serum Increase Analytical Slow and slight reaction in Jaffé methods. *1414*

Eosinophils *Blood Increase Physiological* In 5% cases. *3134*

Factor VII *Plasma Decrease Physiological* Effect on activity in 30 patients with serious infections. Possibly associated with sulfhydryl group. *0031*

Fibrinogen *Plasma Decrease Physiological* Effect observed in some cases: associated with vitamin K associated hypoprothrombinemia. *3103*

Hemoglobin *Blood Decrease Physiological* Rare hemolytic anemia. *3134*

Leukocytes *Blood Decrease Physiological* In 2 patients given 6-9 g daily intravenously. *1646*

Lymphocytes *Blood No Effect Physiological* In 2 patients given 6-9 g daily intravenously. *1646*

Neutrophils *Blood Decrease Physiological* Occasional count below 500/uL observed. *2615*

Occult Blood *Feces Positive Physiological* Observed in 3 of 37 patients receiving intravenous nutrition. *3134*

Partial Thromboplastin Time
Plasma Increase Physiological In 7 cases developed vitamin K deficient hypoprothrombinemia. *3103*

Platelets *Blood No Effect Physiological* Not known as complication. *3134*
Blood Decrease Physiological Rare immune-mediated toxicity reported. *2615*

Protein *Serum No Effect Analytical* At 2.50 mmol/L on method on Eppendorf Epos. *1414*
Urine No Effect Analytical No effect on sulfosalicylic acid method and Albustix method at concentration of 1 g/L. *2388*

Prothrombin Time *Plasma Increase Physiological* In 3 of 37 patients receiving intravenous nutrition. *3134* In 7 cases developed vitamin K deficient hypoprothrombinemia. *3103* 4 of 31 patients receiving drug for more than 48 h developed response; reversible but significant bleeding occurred in some. *0329*

Urea Nitrogen *Serum No Effect Analytical* At 2.50 mmol/L on method on Eppendorf Epos. *1414*

Uric Acid *Serum No Effect Analytical* At 2.50 mmol/L on method on Eppendorf Epos. *1414*

Cefazedone

Bilirubin *Serum No Effect Analytical* On method on Eppendorf Epos at concentration of 2.50 mmol/L. *1414*

Cholesterol *Serum No Effect Analytical* On method on Eppendorf Epos at concentration of 2.50 mmol/L. *1414*

Creatinine *Serum No Effect Analytical* On Jaffé method at concentration of 2.50 mmol/L. *1414*

Protein *Serum No Effect Analytical* On method on Eppendorf Epos at concentration of 2.50 mmol/L. *1414*

Urea Nitrogen *Serum No Effect Analytical* On method on Eppendorf Epos at concentration of 2.50 mmol/L. *1414*

Uric Acid *Serum No Effect Analytical* On method on Eppendorf Epos at concentration of 2.50 mmol/L. *1414*

Cefazolin

Albumin *Serum No Effect Analytical* At concentration of 117 mg/L had no effect on BCG method. *3393*

Alkaline Phosphatase *Serum Increase Physiological* Reversible hepatotoxicity noted in some patients. *0969* In 4 of 31 treated cases: transient 2 to 3 fold increase. *2273*

Aspartate Aminotransferase
Serum Increase Physiological Reversible hepatotoxicity noted in some patients. *0969* In 4 of 31 treated cases: transient 2 to 3 fold increase. *2273*

Bicarbonate *Serum No Effect Analytical* At concentration of 110 mg/L had no effect on method using phenolphthalein. *3393*

Bilirubin *Serum No Effect Analytical* At 2.50 mmol/L on method on Eppendorf Epos. *1414* At concentration of 117 mg/L had no effect on Jendrassik and Grof method. *3393*
Serum Increase Physiological In 4 of 31 treated cases: transient 2 to 3 fold increase. *2273*

Calcium *Serum No Effect Analytical* At concentration of 110 mg/L had no effect on cresolphthalein method. *3393*

Chloride *Serum No Effect Analytical* At concentration of 110 mg/L had no effect on mercurimetric method. *3393*

Cholesterol *Serum No Effect Analytical* At 2.50 mmol/L on method on Eppendorf Epos. *1414* At concentration of 110 mg/L had no effect on CHOD-PAP method. *3393*

Creatinine *Serum No Effect Analytical* No effect at up to 250 mg/L on Kodak DT-60. *2554* At concentration of 2,000 mg/L had no effect on AutoAnalyzer Jaffé method. *3393* At concentration of 2,000 mg/L had no effect on kinetic Jaffé method on aca. *3393*
Serum Increase Analytical Increase of 10-20 μmol/L for every 20 mg/L of drug on Abbott Vision. *2554* Slow and slight reaction in Jaffé methods. *1414* At concentrations above 2,000 mg/L (normal therapeutic concentration 150 mg/L) kinetic Jaffé method on BKA-2. *3393* Increase of 10-20 μmol/L for every 20 mg/L of drug on Ames Seralyzer. *2554*

Glucose *Serum No Effect Analytical* At concentration of 110 mg/L had no effect on GOD/POD-PAP method. *3393*

Phosphate *Serum No Effect Analytical* At concentration of 110 mg/L had no effect on phosphomolybdate method. *3393*

Platelets *Blood Decrease Physiological* Rare immune-mediated toxicity reported. *2615*

Potassium *Serum No Effect Analytical* At concentration of 110 mg/L had no effect on flame-photometric method. *3393*

Protein *Serum No Effect Analytical* At 2.50 mmol/L on method on Eppendorf Epos. *1414* At concentration of 117 mg/L had no effect on biuret method with blank correction. *3393*
Urine No Effect Analytical No effect in 9 patients receiving up to 4 g daily on sulfosalicylic acid, trichloracetic acid and Ponceau S dye methods. *3919*

Sodium *Serum No Effect Analytical* At concentration of 110 mg/L had no effect on flame-photometric method. *3393*

Triglycerides *Serum No Effect Analytical* At concentration of 110 mg/L had no effect on lipase/esterase method. *3393*

Urea Nitrogen *Serum No Effect Analytical* At 2.50 mmol/L on method on Eppendorf Epos. *1414* At concentration of 117 mg/L had no effect on diacetylmonoxime method. *3393*

Uric Acid *Serum No Effect Analytical* At 2.50 mmol/L on method on Eppendorf Epos. *1414* At concentration of 117 mg/L had no effect on phosphotungstate reduction method. *3393*

Cefodizime

Bilirubin *Serum No Effect Analytical* No effect at 2.50 mmol/L on method on Eppendorf Epos. *1414*

Cholesterol *Serum No Effect Analytical* No effect at 2.50 mmol/L on method on Eppendorf Epos. *1414*

Creatinine *Serum No Effect Analytical* No effect on Jaffé method at concentration up to 2.50 mmol/L. *1414*

Protein *Serum No Effect Analytical* No effect at 2.50 mmol/L on method on Eppendorf Epos. *1414*

Urea Nitrogen *Serum No Effect Analytical* No effect at 2.50 mmol/L on method on Eppendorf Epos. *1414*

Uric Acid *Serum No Effect Analytical* No effect at 2.50 mmol/L on method on Eppendorf Epos. *1414*

Cefoperazone

Alanine Aminotransferase *Serum Increase Physiological* Transient increase in 5% of 450 patients. *0690*

Aspartate Aminotransferase
Serum Increase Physiological Transient increase in 5% of 450 patients. *0690*

Bilirubin *Serum No Effect Analytical* At 2.50 mmol/L on method on Eppendorf Epos. *1414*
Serum Increase Physiological Significant displacement in neonates but possibly only at concentrations above therapeutic. *3748*

Cholesterol *Serum No Effect Analytical* At 2.50 mmol/L on method on Eppendorf Epos. *1414*

Creatinine *Serum No Effect Analytical* No effect on Jaffé methods. *2024* At up to 1,000 μg/mL on Technicon® SMAC® Jaffé method. *3405*
Serum Increase Analytical Slow and slight reaction in Jaffé methods. *1414*

Cefoperazone (continued)

Eosinophils *Blood Increase Physiological* In 8% of 450 patients: effect mild and reversible. *0690*

Glucose *Urine No Effect Analytical* Dark green-black color with Clinitest® but no effect or slight reduction with glucose concentrations of 1% and up. Concentration measured accurately by Diastix®. Concentrations measured accurately by Tes-Tape®. *2100*

Leukocytes *Blood Decrease Physiological* In 2% of 450 patients: resolved on withdrawal of drug. *0690*

Occult Blood *Feces Positive Physiological* Associated with vitamin K correctable hypoprothrombinemia but mechanism unclear. *2685*

Partial Thromboplastin Time
Plasma Increase Physiological Reported in previous study to occur in about 4% treated patients. Mechanism unclear but possibly due to inhibition of prothrombin production. Effect correctable with vitamin K. *2685*

Protein *Serum No Effect Analytical* At 2.50 mmol/L on method on Eppendorf Epos. *1414*
Urine No Effect Analytical No effect at 1 g/L on sulfosalicylic acid and Albustix methods. *2388*

Prothrombin Time *Plasma Increase Physiological* Reported in previous study to occur in about 4% treated patients. Mechanism unclear but possibly due to inhibition of prothrombin production. Effect correctable with vitamin K. *2685* Reported effect in some patients with occasional bleeding. *2615* Associated with vitamin K correctable hypoprothrombinemia but mechanism unclear. *2685* In 4% of 450 patients but correctable with vitamin K. *0690*

Urea Nitrogen *Serum No Effect Analytical* At 2.50 mmol/L on method on Eppendorf Epos. *1414*

Uric Acid *Serum No Effect Analytical* At 2.50 mmol/L on method on Eppendorf Epos. *1414*

Ceforanide

Creatinine *Serum No Effect Analytical* At up to 1,000 μg/mL on Technicon® SMAC® Jaffé method. *3405*

Protein *Urine No Effect Analytical* No effect of 1 g/L on sulfosalicylic acid and Albustix methods. *2388*

Cefotaxime

Albumin *Serum Increase Analytical* Slight effect of drug and metabolite on method on Parallel. *0201*

Alkaline Phosphatase *Serum Increase Analytical* Consistent slight increase on American Monitor Parallel method by metabolite. *0201*

Amylase *Serum Decrease Analytical* General effect of both drug and metabolite on Parallel method. *0201*

Bilirubin *Serum No Effect Analytical* No effect at 2.50 mmol/L on method on Eppendorf Epos. *1414*

Calcium *Serum Increase Analytical* Statistically significant effect of metabolite on Parallel method. *0201*

Chloride *Serum Increase Analytical* Statistically significant effect of drug on Parallel method. *0201*
Serum Decrease Analytical Statistically significant effect of metabolite on Parallel method. *0201*

Cholesterol *Serum No Effect Analytical* No effect at 2.50 mmol/L on method on Eppendorf Epos. *1414*
Serum Increase Analytical Generally significant increase of drug and metabolite on parallel method. *0201*

Creatine Kinase *Serum Increase Analytical* Statistically significant effect of metabolite on Parallel method. *0201*

Creatinine *Serum No Effect Analytical* No effect on Jaffé methods. *2024* No effect at up to 250 mg/L on Kodak DT-60. No effect at up to 250 mg/L on Abbott Vision. No effect at up to 250 mg/L on Ames Seralyzer. *2554* At up to 1,000 μg/mL on Technicon® SMAC® Jaffé method. *3405* No effect observed on Jaffé procedure at up to 2.50 mmol/L. *1414* At concentration of 500 mg/L had no effect on AutoAnalyzer Jaffé method. *3393* At concentration of 500 mg/L had no effect on Jaffé-Fuller's earth method. *3393* At concentration of 500 mg/L had no effect on kinetic Jaffé method on aca. *3393*

Glucose *Serum Increase Analytical* Statistically significant effect of metabolite on Parallel method. *0201*
Urine No Effect Analytical Dark green-black color with Clinitest® but no effect or slight reduction with glucose concentrations of 1% and up. Concentrations measured accurately by Diastix®. Concentrations measured accurately by TesTape®. *2100*

γ-Glutamyltransferase (GGT) *Serum Decrease Analytical* 10% reduction in specimens from patients with liver disease measured on American Monitor Parallel. *0201*

Iron *Serum Increase Analytical* General effect of both drug and metabolite on Parallel method. *0201*

Iron Saturation *Serum Increase Analytical* Slight but significant increase with both drug and metabolite on method on American Monitor Parallel. *0201*

Lactate Dehydrogenase *Serum Decrease Analytical* Generally significant decrease of drug and metabolite on parallel method. *0201*

Leukocytes *Blood Decrease Physiological* Observed in 0.5% treated patients. *2615*

Magnesium *Serum Increase Analytical* Effect of metabolite on method on American Monitor Parallel. *0201*
Serum Decrease Analytical Marked effect of drug and metabolite on specimens containing tobramycin on American Monitor method on Parallel. *0201*

Neutrophils *Blood Decrease Physiological* Occasional count of less than 500/uL observed. *2615*

Phosphate *Serum Increase Analytical* Marked increase with both drug and metabolite on specimens containing gentamicin and tobramycin on American Monitor Parallel method. *0201*
Serum Decrease Analytical Effect of drug on most specimens (also of metabolite) on method on American Monitor Parallel. *0201*

Potassium *Serum Increase Analytical* By up to 0.2 mmol/L with both drug and metabolite on method on American Monitor Parallel. *0201*

Protein *Serum No Effect Analytical* No effect at 2.50 mmol/L on method on Eppendorf Epos. *1414*
Serum Decrease Analytical Effect of drug and metabolite on specimens containing gentamicin and tobramycin on American Monitor Parallel. *0201*
Urine No Effect Analytical No effect of 1 g/L on sulfosalicylic acid and Albustix methods. *2388*

Sodium *Serum Increase Analytical* By up to 5 mmol/L with both drug and metabolite on method on American Monitor Parallel. *0201*

Urea Nitrogen *Serum No Effect Analytical* No effect at 2.50 mmol/L on method on Eppendorf Epos. *1414*
Serum Decrease Analytical Effect of drug on specimens containing tobramycin on Parallel method. *0201*

Uric Acid *Serum No Effect Analytical* No effect at 2.50 mmol/L on method on Eppendorf Epos. *1414*
Serum Decrease Analytical Effect of drug on specimens containing tobramycin on Parallel method. *0201*

Cefotetan

Prothrombin Time *Plasma Increase Physiological* Reported prolongation and occasional bleeding. *2615*

Cefotiam

Bilirubin *Serum No Effect Analytical* At 2.50 mmol/L on method on Eppendorf Epos. *1414*

Cholesterol *Serum No Effect Analytical* At 2.50 mmol/L on method on Eppendorf Epos. *1414*

Creatinine *Serum Increase Analytical* Slow and slight reaction in Jaffé methods. *1414*

Protein *Serum No Effect Analytical* At 2.50 mmol/L on method on Eppendorf Epos. *1414*

Urea Nitrogen *Serum No Effect Analytical* At 2.50 mmol/L on method on Eppendorf Epos. *1414*

Uric Acid *Serum No Effect Analytical* At 2.50 mmol/L on method on Eppendorf Epos. *1414*

Cefoxitin

N-Acetylglucosaminidase *Urine No Effect Analytical* At 2 g/L on 2 colorimetric analytical methods. *1354*

Alanine Aminotransferase *Serum No Effect Analytical* At 5 times upper limit of therapeutic range on methods on SMAC®, Abbott-VP, aca, Cobas-Bio and KDA. *2138*
Serum Increase Physiological Reported incidence of 3%. *3134*

Albumin *Serum No Effect Analytical* At 5 times upper limit of therapeutic range on methods on SMAC®, Ektachem®, Hitachi® 705 and KDA. *2138*

Alkaline Phosphatase *Serum No Effect Analytical* At 5 times upper limit of therapeutic range on methods on SMAC®, Abbott-VP, aca, Cobas-Bio, Hitachi® 705 and KDA. *2138*
Serum Increase Physiological Transient 2 to 3-fold increase in 3 of 31 patients following colorectal surgery. *2273*

Amylase *Serum No Effect Analytical* At 5 times upper limit of therapeutic range on methods on aca, Cobas-Bio and Ektachem®. *2138*

Aspartate Aminotransferase *Serum No Effect Analytical* At 5 times upper limit of therapeutic range on methods on SMAC®, Abbott-VP, aca, Cobas-Bio, Hitachi® 705 and KDA. *2138*
Serum Increase Physiological Reported incidence of 3%. *3134* Transient 2 to 3-fold increase in 3 of 31 patients following colorectal surgery. *2273*

Bilirubin *Serum No Effect Analytical* At 5 times upper limit of therapeutic range on methods on SMAC®, aca, Cobas-Bio, Ektachem®, Hitachi® 705 and KDA. *2138* At concentrations up to 2.50 mmol/L on methods on Eppendorf Epos. *1414*
Serum Increase Physiological Transient 2 to 3-fold increase in 3 of 31 patients following colorectal surgery. *2273*

Calcium *Serum No Effect Analytical* At 5 times upper limit of therapeutic range on methods on SMAC®, Abbott-VP, aca, Ektachem®, Hitachi® 705 and KDA. *2138*

Cholesterol *Serum No Effect Analytical* At 5 times upper limit of therapeutic range on methods on SMAC®, Abbott-VP, Cobas-Bio, Ektachem®, Hitachi® 705, and KDA. *2138* At concentrations up to 2.50 mmol/L on methods on Eppendorf Epos. *1414*

Coombs' Test *Serum Positive Physiological* 6 of 77 patients developed rapidly reversible direct Coombs' test without hemolysis. *2615*

Coombs' Test, Direct *Plasma Positive Physiological* In about 2% of treated patients (dose related). *3134*

Creatine Kinase *Serum No Effect Analytical* At 5 times upper limit of therapeutic range on methods on SMAC®, Abbott-VP, aca, Cobas-Bio, and Hitachi® 705. *2138*

Creatinine *Serum No Effect Analytical* At 5 times upper limit of therapeutic range on enzymatic method on Ektachem® 400. *2138* No effect at up to 250 mg/L on Ektachem® DT-60. *2554* No effect of up to 160 mg/L on Boehringer creatinase enzymatic kit. *3794* At concentration of 160 mg/L had no effect on creatinine amidohydrolase method. *3393* At concentration of 1 mg/L had no effect on creatinine iminohydrolase method. *3393* At concentration of 500 mg/L had no effect as measured by Jaffé-Fuller's earth method. *3393* In patients at 2 h after 2 g given i.v. *0976*
Serum Increase Analytical Profound interference with alkaline picrate based reactions. *1414* At 5 times upper limit of therapeutic range on methods on SMAC®, Abbott-VP, aca, Cobas-Bio, Hitachi® 705 and KDA. *2138* Increase of 50-80 µmol/L for every 100 mg/L of drug on Abbott Vision. *2554* When used with Jaffé method on Greiner selective analyzer: concentration related effect. *0061* Concentration dependent increase with kinetic Jaffé reaction on Beckman Astra or Eppendorf system with Merck reagents. *3449* Effect noted on Technicon® SMAC® less than with other commercial systems due to lesser dialysis of drug than creatinine. *2022* Linear correlation between drug concentration and that of analyte as measured by Jaffé reaction. *1414* At concentrations above 80 mg/L (normal therapeutic concentration 150 mg/L) raised the concentration as measured by AutoAnalyzer Jaffé method. *3393* At concentrations above 70 mg/L (normal therapeutic concentration 150 mg/L) raised the concentration as measured by kinetic Jaffé method. *3393* At concentrations above 25 mg/L (normal therapeutic concentration 150 mg/L) raised the concentration as measured by kinetic Jaffé method on aca. *3393* At concentrations above 50 mg/L (normal therapeutic concentration 150 mg/L) raised the concentration as measured by kinetic Jaffé method on BKA-2. *3393* Increase of 50-80 µmol/L for every 100 mg/L of drug on Ames Seraltzer. *2554*
Urine Increase Analytical Falsely high values with Jaffé methods. *0976* When used with Jaffé method on Greiner selective analyzer: concentration related effect. *0061*

Creatinine Clearance *Urine Increase Analytical* Artifactually increased for more than 4 h after 2 g drug given i.v. *0976*
Urine Decrease Analytical If measured with Jaffé procedure at peak drug concentration, otherwise value will be low, especially near trough concentration. *0061*

Eosinophils *Blood Increase Physiological* Reported incidence of 3%. *3134*

Glucose *Serum No Effect Analytical* At 5 times upper limit of therapeutic range on methods on SMAC®, Abbott-VP, aca, Cobas-Bio, Ektachem®, Hitachi® 705 and KDA. *2138* At concentration of 1,000 mg/L had no effect on GOD/POD-PAP method. *3393*

γ-Glutamyltransferase (GGT) *Serum No Effect Analytical* At 5 times upper limit of therapeutic range on methods on SMAC®, Abbott-VP, and Hitachi® 705. *2138*

Hemoglobin *Blood Decrease Physiological* Rare case of hemolytic anemia reported. *3134*

17-Hydroxycorticosteroids *Urine Increase Analytical* Methodological interference with Porter-Silber reaction not eliminated by sodium bisulfite. *1062* Increased from 3 to 10-fold in urine from patients when Amberlite XAD-2 Clini-Screen used for measurement. *2023* Substantial effect (up to 3 times actual concentration) when Porter-Silber reaction used on specimen from patients (*in vitro* 5 mg/L reacted as if 14-4 mg/L. *1062*

Iron *Serum No Effect Analytical* At 5 times upper limit of therapeutic range on Ferrozine method on SMAC®. *2138*

Lactate Dehydrogenase *Serum No Effect Analytical* At 5 times upper limit of therapeutic range on methods on SMAC®, Abbott-VP, Cobas-Bio, Hitachi® 705 and KDA. *2138*

Leukocytes *Blood Decrease Physiological* Isolated case of drug-induced leukopenia. *1839* Reported in fewer than 0.1% treated patients. *2615*

Neutrophils *Blood Decrease Physiological* Isolated case of drug-induced leukopenia. *1839*

Phosphate *Serum No Effect Analytical* At 5 times upper limit of therapeutic range on methods on SMAC®, aca, Hitachi® 705 and KDA. *2138*

Platelets *Blood No Effect Physiological* No report of thrombocytopenia in response to therapy. *3134*
Blood Decrease Physiological Rare immune-mediated toxicity reported. *2615*

Protein *Serum No Effect Analytical* At 5 times upper limit of therapeutic range on methods on SMAC®, Abbott-VP, Ektachem®, Hitachi® 705 and KDA. *2138* At concentrations up to 2.50 mmol/L on methods on Eppendorf Epos. *1414*
Urine No Effect Analytical No effect of 1 g/L on sulfosalicylic acid and Albustix methods. *2388* No effect in 4 patients receiving up to 8 g daily on trichloracetic acid, sulfosalicylic acid and Ponceau S dye methods. *3919*

Triglycerides *Serum No Effect Analytical* At 5 times upper limit of therapeutic range on methods on SMAC®, Abbott-VP, Ektachem®, Hitachi® 705 and KDA. *2138* At concentration of 1025 mg/L had no effect on GPO-PAP method. *3393*

Urea Nitrogen *Serum No Effect Analytical* At 5 times upper limit of therapeutic range on methods on SMAC®, Abbott-VP, Cobas-Bio, Ektachem®, Hitachi® 705, and KDA. *2138* At concentrations up to 2.50 mmol/L on methods on Eppendorf Epos. *1414*

Uric Acid *Serum No Effect Analytical* At 5 times upper limit of therapeutic range on methods on SMAC®, Abbott-VP, aca, Cobas-Bio, Ektachem®, Hitachi® 705 and KDA. *2138* At concentrations up to 2.50 mmol/L on methods on Eppendorf Epos. *1414* At concentration of 1,000 mg/L had no effect on uricase-PAP method. *3393*

Cefpirome

Bilirubin *Serum No Effect Analytical* At concentration of up to 2.50 mmol/L on methods of Eppendorf Epos. *1414*

Cholesterol *Serum No Effect Analytical* At concentration of up to 2.50 mmol/L on methods of Eppendorf Epos. *1414*

Cefpirome (continued)

Creatinine *Serum Increase Analytical* Profound interference with Jaffé reaction. *1414* Linear correlation between drug concentration and apparent analyte concentration. *1414*

Protein *Serum No Effect Analytical* At concentration of up to 2.50 mmol/L on methods of Eppendorf Epos. *1414*

Urea Nitrogen *Serum No Effect Analytical* At concentration of up to 2.50 mmol/L on methods of Eppendorf Epos. *1414*

Uric Acid *Serum No Effect Analytical* At concentration of up to 2.50 mmol/L on methods of Eppendorf Epos. *1414*

Cefsulodin

Creatinine *Serum No Effect Analytical* At up to 1,000 μg/mL on Technicon® SMAC® Jaffé method. *3405*

Ceftazidime

Alanine Aminopeptidase *Urine Increase Physiological* Significant increase in association with reduced GFR. *0051*

Creatinine *Serum No Effect Analytical* No effect on Jaffé methods. *2024* No effect at concentrations up to 250 mg/L on Kodak DT-60. No effect at concentrations up to 250 mg/L on Ames Seralyzer. *2554* No effect at concentrations up to 250 mg/L on Abbott vision. *2554*
Serum No Effect Physiological Unchanged in most patients with GFR above 30 mL/min. *0051*
Serum Increase Physiological Effect noted in 3 patients given high doses in relation to their renal function. *0050*

Factor VII *Plasma No Effect Physiological* In 30 patients with serious infections. *0031*

Glomerular Filtration Rate (GFR)
Urine Decrease Physiological Decreased by mean of 10 mL/min with 4 g drug/d in 16 patients: initial GFR in patients 30 to 110 mL/min. *0051* Slight but significant reduction after 3 g daily i.v. in 15 patients for 4 to 9 d. *2616*

Glucose *Urine No Effect Analytical* Dark green-black color with Clinitest® but no effect or slight reduction with glucose concentrations of 1% and up. Concentrations measured accurately by Diastix®. Concentrations measured accurately by Tes-Tape®. *2100*

Leukocytes *Blood Decrease Physiological* Observed in 0.6% of patients receiving drug. *2615*

β_2-Microglobulin *Serum No Effect Physiological* Unchanged in most patients with GFR above 30 mL/min. *0051*
Urine No Effect Physiological Unchanged in most patients with GFR above 30 mL/min. *0051* No effect on proximal tubular function in 15 patients given 3 g daily for 4-9 d. *2616*

Protein *Urine No Effect Analytical* No effect of 1 g/L on sulfosalicylic acid and Albustix methods. *2388*

Prothrombin Time *Plasma No Effect Physiological* In 30 patients with serious infections. *0031*

Ceftizoxime

Alkaline Phosphatase *Serum Increase Physiological* In 4 of 110 patients: reversible hepatotoxicity noted. *0969* In 3 of 30 patients following colorectal surgery had transient 2 to 3-fold increase. *2273*

Aspartate Aminotransferase
Serum Increase Physiological In 4 of 110 patients: reversible hepatotoxicity noted. *0969* In 3 of 30 patients following colorectal surgery had transient 2 to 3-fold Increase. *2273*

Bilirubin *Serum Increase Physiological* In 3 of 30 patients following colorectal surgery had transient 2 to 3-fold Increase. *2273*

Glucose *Urine No Effect Analytical* Dark green-black color with Clinitest® but no effect or slight reduction with glucose concentrations of 1% and up. Concentrations measured accurately by Diastix®. Concentrations measured accurately by Tes-Tape®. *2100*

Protein *Urine No Effect Analytical* No effect of 1 g/L on sulfosalicylic acid and Albustix methods. *2388*

Ceftriaxone

Bilirubin *Serum No Effect Analytical* On method on Eppendorf Epos at concentration of 2.50 mmol/L. *1414*

Cholesterol *Serum No Effect Analytical* On method on Eppendorf Epos at concentration of 2.50 mmol/L. *1414*

Creatinine *Serum No Effect Analytical* On Jaffé method at concentration of 2.50 mmol/L. *1414*

Factor VII *Plasma Decrease Physiological* In 30 patients with serious infections possibly due to presence of sulfhydryl group. *0031*

Glucose *Urine No Effect Analytical* Dark green-black color with Clinitest® but no effect or slight reduction with glucose concentrations of 1% and up. Concentrations measured accurately by Diastix®. Concentrations measured accurately by Tes-Tape®. *2100*

Leukocytes *Blood Decrease Physiological* Reported to occur in fewer than 0.1% patients. *2615*

Protein *Serum No Effect Analytical* On method on Eppendorf Epos at concentration of 2.50 mmol/L. *1414*
Urine No Effect Analytical No effect at 1 g/L on sulfosalicylic acid and Albustix methods. *2388*

Prothrombin Time *Plasma Increase Physiological* In 30 patients with serious infections possibly due to presence of sulfhydryl group. *0031*

Urea Nitrogen *Serum No Effect Analytical* On method on Eppendorf Epos at concentration of 2.50 mmol/L. *1414*

Uric Acid *Serum No Effect Analytical* On method on Eppendorf Epos at concentration of 2.50 mmol/L. *1414*

Cefuroxime

Alanine Aminotransferase *Serum Increase Physiological* Observed in 24% of 89 patients in one study, but many had pre-existing liver problems. *2615*

Aspartate Aminotransferase
Serum Increase Physiological Observed in 24% of 89 patients in one study, but many had pre-existing liver problems. *2615*

Bilirubin *Serum No Effect Analytical* On method on Eppendorf Epos at concentration of 2.50 mmol/L. *1414*

Cholesterol *Serum No Effect Analytical* On method on Eppendorf Epos at concentration of 2.50 mmol/L. *1414*

Creatinine *Serum No Effect Analytical* On Jaffé method at concentration of 2.50 mmol/L. *1414*
Serum Increase Physiological Mean change from 112 to 137 μmol/L in 3 patients with chronic osteomyelitis over over 14 d when lysine salt used. *3620*

Leukocytes *Blood Decrease Physiological* In one case who developed drug-dependent neutrophil antibodies. *2529*

β_2-Microglobulin *Serum Increase Physiological* Mean change from 4,000 to 6,000 μg/L in 3 patients with chronic osteomyelitis 14 d when lysine salt used. *3620*

Neutrophils *Blood Decrease Physiological* In one case who developed drug-dependent neutrophil antibodies. *2529*

Protein *Serum No Effect Analytical* On method on Eppendorf Epos at concentration of 2.50 mmol/L. *1414*

Urea Nitrogen *Serum No Effect Analytical* On method on Eppendorf Epos at concentration of 2.50 mmol/L. *1414*

Uric Acid *Serum No Effect Analytical* On method on Eppendorf Epos at concentration of 2.50 mmol/L. *1414*

Celiprolol

Cholesterol *Serum No Effect Physiological* In patients with hyperlipoproteinemia types IIa, IIb or IV with 300 mg drug for 4 weeks. *2873*

Cholesterol, High Density Lipoprotein
Serum Increase Physiological From 42 to 54 mg/dL in patients with hyperlipoproteinemia types IIa, IIb or IV with 300 mg drug for 4 weeks. *2873*

Cholesterol, Low Density Lipoprotein
Serum Increase Physiological Effect slight in patients with hyperlipoproteinemia types IIa, IIb, or IV with 300 mg drug for 4 weeks. *2873*

Triglycerides *Serum Decrease Physiological* In patients with hyperlipoproteinemia types IIa, IIb or IV with 300 mg drug for 4 weeks. *2873*

Cellobiose

Glucose *Test Conditions Increase Analytical* If glucose-oxidase contaminated with glucosidase. *2803*

Cellopentose

Glucose *Test Conditions Increase Analytical* If glucose-oxidase contaminated with glucosidase. *2803*

Cellotriose

Glucose *Test Conditions Increase Analytical* If glucose-oxidase contaminated with glucosidase. *2803*

Cellulose

Cholesterol *Serum Decrease Physiological* Observed in young girls with 100 g/d. *1540*

Cellulose Acetate

Acetone *Urine Increase Analytical* Gives false positive with Rothera's test. *0251*

Thyroxine (T₄) *Serum Increase Analytical* Due to phthalates as plasticizer in containers. *3153*

Tri-iodothyronine (T₃) *Serum Increase Analytical* Inhibits binding of T_4 to thyroxine binding globulin. *3153*

Cellulose Phosphate

Calcium *Serum Increase Physiological* Without changing calcium balance. *2711*
Urine Decrease Physiological By impairing absorption decreases output. *2711* As result of decreased gastrointestinal absorption. *2710*

Citrate *Urine Decrease Physiological* Observed in 3 of 5 cases. *2710*

Copper *Serum No Effect Physiological* No significant effect observed. *2710*
Urine No Effect Physiological No significant effect observed. *2710*

Creatinine Clearance *Urine No Effect Physiological* No significant effect observed. *2710*

Parathyroid Hormone *Plasma No Effect Physiological* In patients with absorptive hypercalciuria. *2711*

Phosphate *Urine Increase Physiological* With decrease of activity product ratio. *2710*

Potassium *Serum No Effect Physiological* No significant effect observed. *2710*

Sodium *Serum No Effect Physiological* With decrease of activity product ratio. *2710*

Zinc *Serum No Effect Physiological* With decrease of activity product ratio. *2710*
Urine No Effect Physiological No significant effect observed. *2710*

Cephalexin

Amino Acids *Urine Increase Analytical* Reacts with ninhydrin; extra spot with TLC, high voltage electrophoresis. *2855*
Urine Positive Analytical Yellow spot with ninhydrin on thin-layer chromatography. *1278*

Bicarbonate *Serum Increase Physiological* If given with gentamicin to leukemics. *3922*

Coombs' Test, Direct *Serum Positive Physiological* Incidence low. *1060*

Eosinophils *Blood Increase Physiological* Allergic response. *2808* In 16% of 74 patients all children given 25-150 mg/kg/d for 5 to 15 d. *2505*

Glucose *Urine No Effect Analytical* No effect on Tes-Tape®. *1705* At concentration of 10,000 mg/L had no effect on Diabur-test. *3393*

Leukocytes *Blood Decrease Physiological* In 12% of 74 patients all children given 25-150 mg/kg/d for 5 to 15 d. *2505*

Neutrophils *Blood No Effect Physiological* In 74 patients all children given 25-150 mg/kg/d for 5 to 15 d. *2505*
Blood Decrease Physiological Agranulocytosis reported to occur occasionally. *3717*

Potassium *Serum Decrease Physiological* If given with gentamicin to leukemics. *3922*

Sugar *Urine Increase Analytical* False positive with Benedict's, Fehlings, Clinitest®. *1705*

Cephaloglycin

Alanine Aminotransferase *Serum Increase Physiological* Slight elevation reported. *1680*

Aspartate Aminotransferase
Serum Increase Physiological Slight elevation reported. *1680*

Coombs' Test *Serum Positive Physiological* Rare observed side effect. *0131*

Eosinophils *Blood Increase Physiological* Allergic response. *1680*

Glucose *Urine No Effect Analytical* No effect on Tes-Tape®. *1680*

Neutrophils *Blood Decrease Physiological* Neutropenia reported. *1680*

Protein *Urine Increase Physiological* Of cephalosporins one of most likely to cause nephrotoxicity. *2615*

Sugar *Urine Increase Analytical* Affects Benedict's, Fehlings, Clinitest®. *1680*

Cephaloridine

Alanine Aminotransferase *Serum Increase Physiological* Transient slight rise ?hepatotoxicity. *1019*

Alkaline Phosphatase *Serum Increase Physiological* Mechanism not listed. *1488*

Aspartate Aminotransferase
Serum Increase Physiological Transient slight rise ?hepatotoxicity. *1019*

Bilirubin *Serum No Effect Analytical* At 2.50 mmol/L on method on Eppendorf Epos. *1414*

Casts *Urine Increase Physiological* Nephrotoxicity (hyaline and granular). *1019*

Cholesterol *Serum No Effect Analytical* At 2.50 mmol/L on method on Eppendorf Epos. *1414*

Coombs' Test, Direct *Serum Positive Physiological* No immunological response (complex binds to cell in 8% cases). *1486*

Creatinine *Serum Increase Analytical* Pronounced interference with alkaline picrate based reactions. *1414* Linear correlation between drug concentration and analyte concentration as measured by Jaffé method. *1414*
Serum Increase Physiological Nephrotoxic especially if combined with diuretic. *2808*

Eosinophils *Blood Increase Physiological* May be up to 10% in 1% people. *1019*

Erythrocytes *Blood Decrease Physiological* May cause anemia. *0071*

Folate *Serum No Effect Analytical* If chromatographic procedure of Landon used. *2068*

Glucose *Urine No Effect Analytical* No effect observed with TesTape®. *1019*

Hematocrit *Blood Decrease Physiological* May cause anemia. *0071*

Hemoglobin *Blood Decrease Physiological* May cause anemia. *0071*

17-Ketogenic Steroids *Urine Increase Analytical* On day of administration — method of Wilson and Lipsett. *1779*

17-Ketosteroids *Urine Increase Analytical* On day of administration — Zimmermann reaction. *1779*

Cephaloridine *(continued)*

Leukocytes *Blood Decrease Physiological* May cause neutropenia/leukopenia. *1019*

Nonprotein Nitrogen *Serum Increase Physiological* Nephrotoxic especially if combined with diuretic. *2808*

Potassium *Serum Increase Physiological* May occur with nephrotoxicity. *2313*

Protein *Serum No Effect Analytical* At 2.50 mmol/L on method on Eppendorf Epos. *1414*

Urine Increase Analytical Affects acid turbidimetric procedures. *0583* Precipitate occurs with sulfosalicylic acid Weak positive with Albustix at high concentrations. *2152*

Urine Increase Physiological May be result of nephrotoxicity. *1605* Of cephalosporins one of most likely to produce nephrotoxicity. *2615*

Prothrombin Time *Plasma Increase Physiological* Reported effect. *0071*

Sugar *Urine Increase Analytical* Abnormal dark color with Benedict's and Clinitest®. *1099*

Urea Nitrogen *Serum No Effect Analytical* At 2.50 mmol/L on method on Eppendorf Epos. *1414*

Serum Increase Physiological High doses, nephrotoxicity reported. *0085*

Uric Acid *Serum No Effect Analytical* At 2.50 mmol/L on method on Eppendorf Epos. *1414*

Volume *Urine Decrease Physiological* Suggestive of nephrotoxicity. *1019*

Cephalosporin

Alanine Aminotransferase *Serum Increase Physiological* Observed with all cephalosporins at frequency of from 1 to 7%. *2615*

Alkaline Phosphatase *Serum Increase Physiological* Observed with all cephalosporins at frequency of from 1 to 7%. *2615*

Antithrombin III *Plasma No Effect Analytical* At concentration of 30 mg/L had no effect on aca method. *3393*

Aspartate Aminotransferase
Serum Increase Physiological Observed with all cephalosporins at frequency of from 1 to 7%. *2615*

Coombs' Test, Direct *Serum Positive Physiological* Approximately 3% incidence, immunologic effect. *1487*

Eosinophils *Blood Increase Physiological* In about 4% patients treated with drugs. *2615*

Gentamicin *Serum No Effect Analytical* If enzymatic procedure of Daigneault used. *0791*

Plasminogen *Plasma No Effect Analytical* At concentration of 30 mg/L had no effect on aca method. *3393*

Potassium *Serum No Effect Analytical* At concentration of 100 mg/L had no effect on ISE measurement with predilution. *3393*

Sodium *Serum No Effect Analytical* At concentration of 100 mg/L had no effect on ISE measurement with predilution. *3393*

Cephalothin

Alanine Aminotrans. .ase *Serum No Effect Analytical* At 5 times upper limit of therapeutic range on methods on SMAC®, Abbott-VP, aca, Cobas-Bio and KDA. *2138*
Serum Increase Physiological Single case reported. *2427*

Albumin *Serum No Effect Analytical* At 5 times upper limit of therapeutic range on methods on SMAC®, Ektachem®, Hitachi® 705 and KDA. *2138* At concentration of 1,000 mg/L had no effect on BCG method. *3393*

Alkaline Phosphatase *Serum No Effect Analytical* At 5 times upper limit of therapeutic range on methods on SMAC®, Abbott-VP, aca, Cobas-Bio, Hitachi® 705 and KDA. *2138*

Amylase *Serum No Effect Analytical* At 5 times upper limit of therapeutic range on methods on aca, Cobas-Bio and Ektachem®. *2138*

Aspartate Aminotransferase *Serum No Effect Analytical* At 5 times upper limit of therapeutic range on methods on SMAC®, Abbott-VP, aca, Cobas-Bio, Hitachi® 705 and KDA. *2138*

Serum Increase Physiological Transient rise reported. *1680*

Bicarbonate *Serum No Effect Analytical* At concentration of 1,000 mg/L had no effect on method using phenolphthalein. *3393*

Bilirubin *Serum No Effect Analytical* At 5 times upper limit of therapeutic range on methods on SMAC®, aca, Cobas-Bio, Ektachem®, Hitachi® 705 and KDA. *2138* No effect a 2.50 mmol/L on method on Eppendorf Epos. *1414* At concentration of 1,000 mg/L had no effect on Jendrassik and Grof method. *3393*

Serum No Effect Physiological Clinically insignificant displacement from protein in neonates. *3748*

Serum Increase Physiological May cause hemolytic anemia. *1377*

Bilirubin, Direct *Serum Increase Physiological* May cause hemolytic anemia. *1377*

Calcium *Serum No Effect Analytical* At 5 times upper limit of therapeutic range on methods on SMAC®, Abbott-VP, aca, Ektachem®, Hitachi® 705 and KDA. *2138* At concentration of 1,000 mg/L had no effect on cresolphthalein method. *3393*

Chloride *Serum No Effect Analytical* At concentration of 1,000 mg/L had no effect on mercurimetric method. *3393*

Cholesterol *Serum No Effect Analytical* At 5 times upper limit of therapeutic range on methods on SMAC®, Abbott-VP, Cobas-Bio, Ektachem®, Hitachi® 705 and KDA. *2138* No effect a 2.50 mmol/L on method on Eppendorf Epos. *1414* At concentration of 1,000 mg/L had no effect on CHOD-PAP method. *3393*

Coombs' Test, Direct *Serum Positive Physiological* Nonimmunologic phenomenon, complex binds to cell. *1486*

Cortisol *Plasma No Effect Physiological* No physiological effect. *1779*

Creatine Kinase *Serum No Effect Analytical* At 5 times upper limit of therapeutic range on methods on SMAC®, Abbott-VP, aca, Cobas-Bio and Hitachi® 705. *2138*

Creatinine *Serum No Effect Analytical* At 5 times upper limit of therapeutic range on methods on aca, Ektachem® and KDA. *2138* At concentration of 1,000 mg/L had no effect on kinetic Jaffé method on aca. *3393*

Serum Increase Analytical Pronounced interference with alkaline picrate based reactions. *1414* At 5 times upper limit of therapeutic range on methods on SMAC®, Abbott-VP, Cobas-Bio and Hitachi® 705. *2138* Dose related interference with Jaffé procedure on Greiner selective analyzer. *0061* Concentration related increase when Jaffé method used. *1414* At concentrations above 500 mg/L (normal therapeutic concentration 17 mg/L) raised concentration as measured by AutoAnalyzer Jaffé method. *3393* At concentrations above 100 mg/L (normal therapeutic concentration 17 mg/L) raised concentration as measured by kinetic Jaffé method on BKA-2. *3393*

Serum Increase Physiological High rate of nephrotoxicity observed with combined with aminoglycosides, more than when other drugs given with aminoglycosides. *3731*

Serum Decrease Analytical As measured by Du Pont aca due to decrease of absorbance of product with picrate with time. *2025* Positive interference with kinetic Jaffé reactions in first 45 s, thereafter negative effect with reduced concentration. *2025*

Urine Increase Analytical Dose related interference with Jaffé procedure on Greiner selective analyzer. *0061*

Creatinine Clearance *Urine Decrease Analytical* If Jaffé procedure used and plasma concentration of drug is near peak; values low if drug concentration low. *0061*

Eosinophils *Blood Increase Physiological* Modest increase with prolonged high dose i.v. administration. *3133* Allergic response. *2416*

Erythrocytes *Blood Decrease Physiological* Hemolytic anemia. *1378*

Erythrocyte Sedimentation Rate
Blood Increase Physiological With high doses and prolonged i.v. administration. *3133*

Glucose *Serum No Effect Analytical* At 5 times upper limit of therapeutic range on methods on SMAC®, Abbott-VP, aca, Cobas-Bio, Ektachem®, Hitachi® 705 and KDA. *2138* At concentration of 1046 mg/L had no effect on glucose dehydrogenase method. *3393* At concentration of 1,000 mg/L had no effect on GOD/POD-PAP method. *3393*

γ-Glutamyltransferase (GGT) *Serum No Effect Analytical* At 5 times upper limit of therapeutic range on methods on SMAC®, Abbott-VP, and Hitachi® 705. *2138*

Hematocrit *Blood Decrease Physiological* Hemolytic anemia. *1378*

Hemoglobin *Blood Decrease Physiological* Coombs' positive hemolytic anemia. *1378*

17-Hydroxycorticosteroids *Urine Increase Analytical* 5 mg/L reacted as if 7.2 mg/L: effect eliminated if Allen correction used. *1062*

Iron *Serum No Effect Analytical* At 5 times upper limit of therapeutic range on methods on Ferrozine method on SMAC®. *2138*

17-Ketogenic Steroids *Urine Increase Analytical* Non specificity of method of Wilson and Lipsett. *1779*

17-Ketosteroids *Plasma Increase Analytical* Marked effect on Zimmermann procedure with 6 g dose. *1932*
Urine Increase Analytical Non specificity of Zimmermann reaction on day of administration. *1779* Marked effect on Zimmermann procedure with 6 g dose. *1932*

Lactate Dehydrogenase *Serum No Effect Analytical* At 5 times upper limit of therapeutic range on methods on SMAC®, Abbott-VP, Cobas-Bio, Hitachi® 705 and KDA. *2138*

Leukocytes *Blood Increase Physiological* Due to eosinophilia of hypersensitivity. *2313*
Blood Decrease Physiological Marked decrease 1 case prolonged high i.v. doses. *3133* Rare response to therapy. *2313* In 2 patients given 12 g daily intravenously. *1646*

Lymphocytes *Blood No Effect Physiological* In 2 patients given 12 g daily intravenously. *1646*

Neutrophils *Blood Decrease Physiological* Neutropenia or leukopenia rare. *0070* Occasional count below 500/uL observed. *2615*

Phosphate *Serum No Effect Analytical* At 5 times upper limit of therapeutic range on methods on SMAC®, aca, Hitachi® 705 and KDA. *2138* At concentration of 1,000 mg/L had no effect on phosphomolybdate method. *3393*

Platelets *Blood Decrease Physiological* Increased resistance to osmotic fragility. *2426*

Potassium *Serum No Effect Analytical* At concentration of 1,000 mg/L had no effect on flame-photometric method. *3393*

Protein *Serum No Effect Analytical* At 5 times upper limit of therapeutic range on methods on SMAC®, Abbott-VP, Ektachem®, Hitachi® 705 and KDA. *2138* No effect a 2.50 mmol/L on method on Eppendorf Epos. *1414*
Serum Increase Analytical At concentrations above 500 mg/L (normal therapeutic concentration 17 mg/L) raised concentration as measured by biuret method with blank correction. *3393*
Urine No Effect Analytical No effect of 1 g/L on sulfosalicylic acid and Albustix methods. *2388*
Urine Increase Analytical Affects acid turbidimetric procedures. *0583* Precipitate occurs with sulfosalicylic acid Weak positive with Albustix at high concentrations. *2152*
Urine Increase Physiological Rare effect following i.v. infusion. *2427*

Sodium *Serum No Effect Analytical* At concentration of 1,000 mg/L had no effect on flame-photometric method. *3393*

Sugar *Urine Increase Analytical* False positive with copper reduction procedures. *3505* Brown-black color with Clinitest® (false positive). *0110*

Thymol Turbidity *Serum Increase Physiological* Reported in alcoholic patient with cirrhosis. *2313*

Triglycerides *Serum No Effect Analytical* At 5 times upper limit of therapeutic range on methods on SMAC®, Abbott-VP, Ektachem®, Hitachi® 705 and KDA. *2138* At concentration of 1,000 mg/L had no effect on lipase/esterase method. *3393*

Urea Nitrogen *Serum No Effect Analytical* At 5 times upper limit of therapeutic range on methods on SMAC®, Abbott-VP, Cobas-Bio, Ektachem®, Hitachi® 705 and KDA. *2138* No effect a 2.50 mmol/L on method on Eppendorf Epos. *1414* At concentration of 1,000 mg/L had no effect on diacetylmonoxime method. *3393*
Serum Increase Physiological Rare elevation noted (possible nephrotoxicity). *1019*

Uric Acid *Serum No Effect Analytical* At 5 times upper limit of therapeutic range on methods on SMAC®, Abbott-VP, aca, Cobas-Bio, Ektachem®, Hitachi® 705 and KDA. *2138* No effect a 2.50 mmol/L on method on Eppendorf Epos. *1414* At concentration of 1,000 mg/L had no effect on phosphotungstate reduction method. *3393*

Cephapirin

Creatinine *Serum No Effect Analytical* At concentration of 1,000 mg/L had no effect on AutoAnalyzer Jaffé method. *3393* At concentration of 1,000 mg/L had no effect on kinetic Jaffé method on aca. *3393* At concentration of 100 mg/L had no effect on kinetic Jaffé method on BKA-2. *3393*

Eosinophils *Blood Increase Physiological* Modest increase with prolonged high dose i.v. administration. *3133*

Erythrocyte Sedimentation Rate
Blood Increase Physiological With high doses and prolonged i.v. administration. *3133*

Leukocytes *Blood Decrease Physiological* Marked decrease 1 case prolonged high i.v. doses. *3133*

Neutrophils *Blood Decrease Physiological* Occasional count of below 500/uL reported. *2615*

Cephradine

Chloramphenicol *Serum No Effect Analytical* No effect at 100 mg/L on coupled enzymatic method. *2490*

Eosinophils *Blood Increase Physiological* In 28% of 86 children given 25-110 mg/kg daily orally for 5 to 15 d. *2505*

Leukoagglutinins *Serum Increase Physiological* In a single patient either due to a toxic effect on bone marrow or immune mediated mechanism, probably immune basis. *2096*

Leukocytes *Blood Decrease Physiological* In 12% of 86 children given 25-110 mg/kg daily orally for 5 to 15 d. *2505*

Neutrophils *Serum Decrease Physiological* In a single patient either due to toxic effect on bone marrow or immune-mediated. *2096*
Blood Decrease Physiological In 1 of 86 children given 25-110 mg/kg daily orally for 5 to 15 d. *2505*

Cerulein

Glucagon *Plasma Increase Physiological* After i.v. infusion in normals. *1076*

Glucose *Serum Increase Physiological* Slight effect after i.v. infusion. *1076*

Cetoxime

Amphetamine *Urine No Effect Analytical* No effect at 100 mg/L on method of Rutter. *3097* At 50 mg/L on fluorescent method of Hayes. *1534*

Charcoal

Acetylsalicylic Acid *Serum Decrease Physiological* Reduced absorption due to adsorption. *3794*

Carbamazepine *Serum Decrease Physiological* Reduced absorption due to adsorption. *3794*

Color *Feces Increase Analytical* Black due to ingested material. *1187*

Digoxin *Serum Decrease Physiological* Reduced absorption due to adsorption. *3794*

Diphenylhydantoin *Serum Decrease Physiological* Reduced absorption due to adsorption. *3794*

Charcoal *(continued)*

Phenobarbital *Serum Decrease Physiological* Reduced absorption and increased elimination rate with adsorption reducing availability. *3794*

Phenylbutazone *Serum Decrease Physiological* Reduced absorption and increased elimination rate with adsorption reducing availability. *3794*

Chenodeoxycholic Acid

Alanine Aminotransferase *Serum Increase Physiological* Slight transient increases may be observed: clinically significant hepatic injury in 3%. *3198* Mild elevation usually less than 3 times normal in up to 30% patients receiving 750 mg/d. *3199*

Alkaline Phosphatase *Serum No Effect Physiological* After 3 mo in patients with gallstones. *0274*
Serum Increase Physiological Mild transient elevation observed. *0808* In 7-2% in 916 patients given up to 750 mg/d for 2 y. *3199*

Apolipoprotein AI *Serum No Effect Physiological* No significant effect of 12 mo treatment in 252 patients. *0046*

Apolipoprotein AII *Serum No Effect Physiological* No significant effect of 12 mo treatment in 252 patients. *0046*

Apolipoprotein B *Serum No Effect Physiological* No significant effect of 12 mo treatment in 252 patients. *0046*

Aspartate Aminotransferase
Serum No Effect Physiological After 3 mo in patients with gallstones. *0274*
Serum Increase Physiological Mild transient elevation observed. *0808* Slight transient increases may be observed: clinically significant hepatic injury in 3%. *3198* Mild elevation usually less than 3 times normal in up to 30% patients receiving 750 mg/d. *3199*

Bile Salts *Bile No Effect Physiological* No change in patients with gallstones. *0274*

Bilirubin *Serum Increase Physiological* In 4-9% in 916 patients given up to 750 mg/d for 2 y. *3199*

BSP Retention *Serum No Effect Physiological* No effect noticed even if ICD increased. *0274*

Cholesterol *Serum No Effect Physiological* No significant change over 3 mo treatment of gallstones. *0274* No significant change with 150 mg four times daily in patients with endogenous hypertriglyceridemias. *0601*
Serum Increase Physiological Elevation of about 10 mg/dL attributable to drug. *3198* Mean increase of 20 mg/dL in 252 individuals treated for 12 mo. *0046* Increase by 10% or more in 82% in 916 patients given up to 750 mg/d for 2 y. *3199*
Bile Decrease Physiological In patients with gallstones, not in normals. *0274*

Cholesterol, High Density Lipoprotein
Serum No Effect Physiological No significant effect of 12 mo treatment in 252 patients. *0046*
Serum Increase Physiological Significant effect with 150 mg four times daily in patients with endogenous hypertriglyceridemias. *0601*
Serum Decrease Physiological Increase by 46% in 8 normolipemic patients receiving 16 g drug daily. *2121*

Cholesterol, Low Density Lipoprotein
Serum No Effect Physiological By 26% in 8 normolipemic patients receiving 16 g drug daily. *2121*
Serum Increase Physiological Increase by 12.2 mg/dL in 916 patients given up to 750 mg/d for 2 y. *3199*

Cholesterol, Very Low Density Lipoprotein
Serum No Effect Physiological By 26% in 8 normolipemic patients receiving 16 g drug daily. *2121*

Creatinine *Serum No Effect Physiological* No significant effect with up to 750 mg/d in 916 patients treated for 2 y. *3199*

Erythrocytes *Blood No Effect Physiological* No significant effect with up to 750 mg/d over 2 y in 916 patients. *3199*
Urine No Effect Physiological No significant effect with up to 750 mg/d in 916 patients treated for 2 y. *3199*

Glucose *Urine No Effect Physiological* No significant effect with up to 750 mg/d in 916 patients treated for 2 y. *3199*

Hematocrit *Blood No Effect Physiological* No significant effect with up to 750 mg/d over 2 y in 916 patients. *3199*

Hemoglobin *Blood No Effect Physiological* No significant effect with up to 750 mg/d over 2 y in 916 patients. *3199*

Urine No Effect Physiological No significant effect with up to 750 mg/d in 916 patients treated for 2 y. *3199*

Isocitrate Dehydrogenase *Serum Increase Physiological* Slight effect in almost 30% cases. *0274*

Leukocytes *Blood No Effect Physiological* No significant effect with up to 750 mg/d over 2 y in 916 patients. *3199*
Urine No Effect Physiological No significant effect with up to 750 mg/d in 916 patients treated for 2 y. *3199*

β-Lipoproteins *Serum No Effect Physiological* No effect observed during treatment of gallstones. *0808*

Phospholipids, Total *Bile No Effect Physiological* No significant effect in patients with gallstones. *0274*

Platelets *Blood No Effect Physiological* No significant effect with up to 750 mg/d in 916 patients treated for 2 y. *3199*

Protein *Urine No Effect Physiological* No significant effect with up to 750 mg/d in 916 patients treated for 2 y. *3199*

Reticulocytes *Blood No Effect Physiological* No significant effect with up to 750 mg/d over 2 y in 916 patients. *3199*

Triglycerides *Serum Decrease Physiological* Mean fall by 20 mg/dL over 3 mo. *0274* Significant effect with 150 mg four times daily in patients with endogenous hypertriglyceridemias. *0601* By 26% in 8 normolipemic patients receiving 16 g drug daily. *2121* By up to 20% in patients given 10 to 25 mg/kg body weight for 6 mo. Increase by 18.6% in 916 patients given up to 750 mg/d for 2 y. *3199*

Triglycerides, High Density Lipoprotein
Serum Decrease Physiological In 8 normolipemic patients receiving 1 g drug daily. *2121*

Triglycerides, Low Density Lipoprotein
Serum Decrease Physiological In 8 normolipemic patients receiving 1 g drug daily. *2121*

Triglycerides, Very Low Density Lipoprotein
Serum No Effect Physiological In 8 normolipemic patients receiving 1 g drug daily. *2121*

Chewing Gum

Hydrochloric Acid *Gastric Material Increase Physiological* But insignificantly in normals. *1480*

Volume *Gastric Material Increase Physiological* 60% increase observed in normals. *1480*

Chiniofon

Alanine Aminotransferase *Serum Increase Physiological* Large doses may occasionally cause hepatic damage. *1343*

Protein Bound Iodine (PBI) *Serum Increase Analytical* Contains 27.5% iodine. *0012*

Chloral Hydrate

Acetaminophen Screening Test *Urine Negative Analytical* No reaction with o-cresol at therapeutic concentrations. *3326*

Alanine Aminotransferase *Serum Increase Physiological* May cause liver damage. *0279*

Alkaloids *Urine No Effect Analytical* With ninhydrin in Frings TLC procedure at 10 mg/dL. With iodoplatinate of Frings TLC procedure at 10 mg/dL. *1204*

Aspartate Aminotransferase
Serum Increase Physiological May cause liver damage. *0279*

Barbiturate *Urine No Effect Analytical* With mercurous NO_3 in Frings TLC procedure at 10 mg/dL. With mercuric SO_4 in Frings TLC procedure at 10 mg/dL. With vanillin on Frings procedure at 10 mg/dL. With diphenylcarbazone in Frings TLC procedure at 10 mg/dL. *1204*

Bilirubin *Serum Increase Physiological* Probably nonspecific effect on cells. *1500*

Bromide *Urine No Effect Analytical* No effect on method of Frings. *1204*

BSP Retention *Serum Increase Physiological* Probably nonspecific effect on cells. *1500*

Catecholamines *Urine Increase Analytical* Interferes with fluorometric procedures. *1488*

Chloral Hydrate   *Blood   No Effect   Physiological*   Not detectable after 1 g orally. *2348*

Chlordiazepoxide   *Blood   No Effect   Analytical*   On method of Riddick at 5 mg/dL. *2983*

Corticosteroids   *Urine   Increase   Analytical*   Butanol extract/no hydrolysis Reddy/Porter-Silber reaction. *0022*

Estriol   *Urine   No Effect   Analytical*   No effect on GLC method. *1163*

Ethchlorvynol   *Urine   No Effect   Analytical*   No effect on method of Frings and Cohen at 10 mg/dL. *1205*

FPN Test   *Urine   No Effect   Analytical*   No effect at 10 mg/dL on method of Frings. *1204*

Glucose   *Urine   No Effect   Analytical*   No effect observed with TesTape®. *1100*
Urine   Decrease   Analytical   Low with Clinistix®, Diastix®. *1100*

17-Hydroxycorticosteroids   *Urine   Increase   Analytical*   Interferes with Porter-Silber reaction. *3505*

Methylmalonic Acid   *Urine   No Effect   Analytical*   Drug has no effect on GC/MS assay which can be used to assess vitamin B_{12} deficiency. *2614*

Porphyrins   *Urine   Increase   Physiological*   May precipitate attack of acute porphyria. *2220*

Potassium   *Serum   No Effect   Analytical*   At concentration of 200 mg/L had no effect on measurement by ISE with predilution. *3393*

Prolactin   *Plasma   No Effect   Physiological*   No significant effect in response to up to 9 g/d. *1416*

Protein   *Urine   Increase   Physiological*   Renal damage with high concentration Parenchymatous renal injury in chronic intoxication. *1343*

Prothrombin Time   *Plasma   No Effect   Physiological*   No significant effect noted in 10 patients on warfarin when receiving 0.5 g daily. *3649*
Plasma   Increase   Physiological   Displaces anticoagulants from albumin. *3379*
Plasma   Decrease   Physiological   Accelerates rate of inactivation of coumarins. *2313*

Salicylate   *Urine   No Effect   Analytical*   At 10 mg/dL no effect on Trinder method. *1204*

Sodium   *Serum   No Effect   Analytical*   At concentration of 200 mg/L had no effect on measurement by ISE with predilution. *3393*

Sugar   *Urine   Increase   Analytical*   Excreted as glucuronide false positive with Benedict's. *1022*

Trichlorethanol   *Blood   Increase   Physiological*   1 g orally produces concentration of 1.5 mg/L. *2348*

Urea Nitrogen   *Serum   Increase   Analytical*   Reacts with Nessler reaction. *3505*

Uric Acid   *Serum   Increase   Analytical*   Affects phosphotungstate reduction methods. *2220*

Vitamin B_{12}   *Serum   No Effect   Analytical*   No effect found on Magic (Ciba-Corning) and Combostat II (Micromedics) assays. *2040*
Serum   Increase   Analytical   Falsely high values obtained with a radioimmunoassay even at very low drug concentrations. *2614* Significant increase on Dual Count (Diagnostic Products) methods, both boil and and no-boil. *2040*

Warfarin   *Plasma   Increase   Physiological*   TCA displaces from protein (temporary effect). *1487*

Chlorambucil

Alanine Aminotransferase   *Serum   Increase   Physiological*   Occasional hepatotoxicity reported. *1343*

Aspartate Aminotransferase
Serum   Increase   Physiological   Hepatotoxicity with centrolobular necrosis. *1948*

Bilirubin   *Serum   Increase   Physiological*   Hepatotoxicity with centrolobular necrosis. *1948*

Chromosomes   *Test Conditions   Abnormal   Physiological*   In lymphocytes with chronic lympholeukemia. *2094*

Erythrocytes   *Blood   Decrease   Physiological*   Dose dependent bone marrow depression. *0071*

Leukocytes   *Blood   Decrease   Physiological*   Neutropenia (dose related). *0543*

Lymphocytes   *Blood   Decrease   Physiological*   Marked decrease may occur. *1680*

Lymphocytic Response to PHA
Blood   Decrease   Physiological   Significant reduction with chemotherapy observed. *0445*

Neutrophils   *Blood   Decrease   Physiological*   Severe neutropenia may be produced. *1680*

Platelets   *Blood   Decrease   Physiological*   Dose dependent bone marrow depression. *0071*

Chloramine

Chromosomes   *Test Conditions   Abnormal   Physiological*   Clastogenic in human lymphocyte cultures. *3282*

Chloramphenicol

Acetaminophen   *Serum   No Effect   Analytical*   No effect at therapeutic concentration on o-cresol reaction based method. *0621*

Alanine Aminotransferase   *Serum   No Effect   Analytical*   No effect at therapeutic concentration on Reflotron method. *1984*
Serum   Increase   Physiological   Hepatotoxic-cholestatic effect. *2220*

Alkaline Phosphatase   *Serum   Increase   Physiological*   ?hepatotoxic-cholestatic effect. *2220*

Antithrombin III   *Plasma   No Effect   Analytical*   At concentration of 25 mg/L had no effect on aca method. *3393*

Aspartate Aminotransferase   *Serum   No Effect   Analytical*   No effect at therapeutic concentration on Reflotron method. *1984*
Serum   Increase   Physiological   Hepatotoxic-cholestatic effect. *2220* Isolated case of hepatitis and pancytopenia reported which resolved with discontinuation of drug. *0602*

Bile   *Urine   Increase   Physiological*   Hepatotoxicity. *2313*

Bilirubin   *Serum   No Effect   Physiological*   Clinically insignificant displacement from protein in neonates. *3748*
Serum   Increase   Physiological   Hepatotoxic-cholestatic effect. *2220*

BSP Retention   *Serum   Increase   Physiological*   Hepatotoxicity. *2220*

Cephalin Flocculation   *Serum   Increase   Physiological*   Hepatotoxicity. *2220*

Chlorpropamide   *Blood   Increase   Physiological*   Inhibits hepatic microsomal metabolizing enzymes thus prolonging half-life. *3864*

Cholesterol   *Serum   No Effect   Analytical*   No effect at therapeutic concentration on Reflotron method. *1984* At concentration of 600 mg/L had no effect on CHOD-Iodide method. *3393* At concentration of 600 mg/L had no effect on CHOD-PAP method. *3393* At concentration of 600 mg/L had no effect on catalase-AIDH method. *3393* At concentration of 600 mg/L had no effect on method using catalase-Hantzsch reaction. *3393* At concentration of 600 mg/L had no effect on Liebermann-Burchard method. *3393*
Serum   No Effect   Physiological   No effect seen even when administration orally. *1487*

Chromosomes   *Test Conditions   No Effect   Physiological*   Not clastogenic in human cells. *3282*

Creatinine   *Serum   No Effect   Analytical*   At concentration of 20 mg/L had no effect on creatinine iminohydrolase method. *3393* At concentration of 600 mg/L had no effect on Jaffé-Fading-Fraction method. *3393* At concentration of 600 mg/L had no effect on Jaffé-Fuller's earth method. *3393* At concentration of 100 mg/L had no effect on kinetic Jaffé method on BKA-2. *3393*

Dicumarol   *Plasma   Increase   Physiological*   Inhibits hepatic microsomal metabolizing enzymes thus prolonging half-life. *3864*

Diphenylhydantoin   *Serum   Increase   Physiological*   May increase from 2.5 to 9 µg/mL (at 2 g/d). *2041* Inhibition of metabolism. *2042* Inhibits hepatic microsomal metabolizing enzymes thus prolonging half-life. *3864*

Eosinophils   *Blood   Increase   Physiological*   Allergic reaction. *1513* Occasional hypersensitivity reaction occurs. *3864*

Chloramphenicol (continued)

Erythrocytes *Blood Decrease Physiological* Dose related response usual/aplastic anemia. *0907* Occurs either as dose-related or idiosyncratic bone marrow suppression: dose- related usually occurs 5-7 d after start of therapy: idiosyncratic is rare occurring 1 case per 40,000 courses of therapy. *3864* Aplastic anemia reported to occur in from 1 to 10,000 or 40,000 cases. *3717*

Folate *Serum Decrease Analytical* Inhibits growth of L. Casei. *3504*

Gentamicin *Serum No Effect Analytical* If enzymatic procedure of Daigneault used. *0791*

Glucose *Serum No Effect Analytical* No effect at therapeutic concentration on Reflotron method. *1984* At concentration of 100 mg/L had no effect on Ektachem® method. *3393* At concentration of 3,000 mg/L had no effect on GOD/POD-PAP method. *3393* At concentration of 400 mg/L had no effect on hexokinase/G-6-PDH method. *3393*
Serum Decrease Physiological Reported effect. *0117*
Urine No Effect Analytical At concentration of 400 mg/L had no effect on Diabur-test. *3393*

γ-Glutamyltransferase (GGT) *Serum No Effect Analytical* No effect at therapeutic concentration on Reflotron method. *1984*
Serum Increase Physiological Isolated case of Hepatitis and Pancytopenia reported, which resolved with discontinuation of drug. *0602*

Hematocrit *Blood Decrease Physiological* Normally slight response/may be pancytopenia. *0907*

Hemoglobin *Blood Decrease Physiological* Normally slight response/may be pancytopenia. *0907* Isolated cases of hepatitis and pancytopenia reported which resolved with discontinuation of drug. *0602*

17-Hydroxycorticosteroids *Urine No Effect Analytical* No effect modified Reddy procedure. *1487*

Iron *Serum No Effect Analytical* At concentration of 5,000 mg/L had no effect on Ferrozine method. *3393*
Serum Increase Physiological Reversible toxic reaction. *2808* Occurs either as dose-related or idiosyncratic bone marrow suppression: dose- related usually occurs 5-7 d after start of therapy: idiosyncratic is rare occurring 1 case per 40,000 courses of therapy. *3864*

Iron-Binding Capacity, Total (TIBC)
Serum Decrease Physiological Decreased uptake by erythroid tissue. *2220*

17-Ketosteroids *Urine Increase Analytical* Affects some methods but not modified Reddy method. *1488*

Lactate *Serum No Effect Analytical* At concentration of 400 mg/L had no effect on enzymatic method. *3393*

Lactate Dehydrogenase *Serum Increase Physiological* Isolated case of hepatitis and pancytopenia reported which resolved with discontinuation of drug. *0602*

Leukocytes *Blood Increase Physiological* Leukemia may occur as sequel to marrow depression. *0071*
Blood Decrease Physiological Pancytopenia/aplastic anemia. *0907* Occurs either as dose-related or idiosyncratic bone marrow suppression: dose- related usually occurs 5-7 d after start of therapy: idiosyncratic is rare occurring 1 case per 40,000 courses of therapy. *3864* Aplastic anemia reported to occur in from 1 to 10,000 or 40,000 cases. *3717* Isolated case of hepatitis and pancytopenia reported which resolved with discontinuation of drug. *0602*

Methemoglobin *Blood Increase Physiological* May cause hemolysis in G-6-PD deficiency. *3581*

Mexiletine *Serum Increase Physiological* Decreases clearance of drug and prolongs half-life. *2011*

Neutrophils *Blood Decrease Physiological* Common toxic reaction/aplastic anemia. *2427* Occurs either as dose-related or idiosyncratic bone marrow suppression: dose- related usually occurs 5-7 d after start of therapy: idiosyncratic is rare occurring 1 case per 40,000 courses of therapy. *3864* Agranulocytosis reported to occur occasionally. *3717*

Nitrofurantoin *Urine No Effect Analytical* No effect on method of Conklin and Hollifield. *0704*

5'-Nucleotidase *Serum Increase Physiological* Due to cholestasis. *2220*

Occult Blood *Feces Positive Physiological* May be gastro-intestinal hemorrhage, with low prothrombin. *0071* Occasional case of pseudomembranous colitis reported with mucus or bloody diarrhea. *0447*

Plasminogen *Plasma No Effect Analytical* At concentration of 25 mg/L had no effect on aca method. *3393*

Platelets *Blood Decrease Physiological* Pancytopenia/aplastic anemia. *0907* Occurs either as dose-related or idiosyncratic bone marrow suppression: dose- related usually occurs 5-7 d after start of therapy: idiosyncratic is rare occurring 1 case per 40,000 courses of therapy. *3864* Aplastic anemia reported to occur in from 1 to 10,000 or 40,000 cases. *3717* Isolated case of hepatitis and pancytopenia reported which resolved with discontinuation of drug. *0602*

Potassium *Serum No Effect Analytical* At concentration of 6,000 mg/L had no effect on measurement by ISE without predilution. *3393*

Protein *Serum Increase Analytical* At concentrations above 500 mg/L (normal therapeutic concentration 10 mg/L) raised concentration as measured by biuret method with blank correction. *3393*
CSF Increase Analytical Slight effect on Folin-Ciocalteu procedure. *3958*

Protein Electrophoresis *Serum No Effect Analytical* At concentration of 6,000 mg/L had no effect on automated Olympus-Hite method but with slight displacement of fractions. *3393*

Prothrombin Time *Plasma Increase Physiological* May cause lowered prothrombin and hemorrhage. *0894*

Reticulocytes *Blood Decrease Physiological* Reversible toxic reaction or pancytopenia. *2426*

Sodium *Serum No Effect Analytical* At concentration of 6,000 mg/L had no effect on measurement by ISE without predilution. *3393*

Sugar *Urine Increase Analytical* False positive with copper reduction procedures. *3505*

Thymol Turbidity *Serum Increase Physiological* Hepatotoxicity. *2220*

Tolbutamide *Serum Increase Physiological* May cause 3 fold increase in half life. *1487* Inhibits hepatic microsomal metabolizing enzymes thus prolonging half-life. *3864*

Triglycerides *Serum No Effect Analytical* No effect at therapeutic concentration on Reflotron method. *1984* At concentration of 3,000 mg/L had no effect on GPO-PAP method. *3393*

Urea Nitrogen *Serum No Effect Analytical* No effect at therapeutic concentration on Reflotron method. *1984* At concentration of 600 mg/L had no effect on Ektachem® method. *3393* At concentration of 2500 mg/L had no effect on urease/Berthelot method. *3393*
Serum Increase Analytical Reacts with Nessler reagent. *3505*
Serum Decrease Analytical Inhibits Berthelot reaction. *3505*

Uric Acid *Serum No Effect Analytical* No effect at therapeutic concentration on Reflotron method. *1984* At concentration of 600 mg/L had no effect on catalase-AIDH method. *3393* At concentration of 3,000 mg/L had no effect on uricase-PAP method. *3393*
Serum Decrease Analytical At concentrations greater than 600 mg/L (normal therapeutic concentration 10 mg/L) lowered concentration as measured by Kageyama-Hantzsch method. *3393*

Urobilinogen *Urine Decrease Physiological* Reduction of flora in gastrointestinal tract. *1488*
Feces Decrease Physiological Reduction of flora in gastrointestinal tract. *1488*

Chlorate

Alanine Aminotransferase *Serum Increase Physiological* May cause liver damage. *0279*

Aspartate Aminotransferase
Serum Increase Physiological May cause liver damage. *0279*

Color *Blood Increase Physiological* May cause chocolate color. *0279*

Erythrocytes *Blood Decrease Physiological* Destroys cells on absorption. *1343*

Haptoglobin *Serum Decrease Physiological* May cause hemolytic anemia. *0279*

Hemoglobin *Blood Decrease Physiological* Causes hemolysis. *1343*

Methemoglobin *Blood Increase Physiological* May cause hemolysis. *1343*

Protein *Urine Increase Physiological* May cause renal damage. *0279*

Protein Bound Iodine (PBI) *Serum Decrease Physiological* Interferes with trapping of iodide by thyroid. *0012*

Urea Nitrogen *Serum Increase Physiological* May cause renal damage. *0279*

Volume *Urine Decrease Physiological* May cause oliguria with renal damage. *0279*

Chlordane

Alanine Aminotransferase *Serum Increase Physiological* Due to hepatotoxicity. *1302*

Bilirubin *Serum Increase Physiological* Due to hepatotoxicity. *1302*

Erythrocytes *Blood Decrease Physiological* Pancytopenia (AMA Blood dyscrasias). *2429*

Leukocytes *Blood Decrease Physiological* Pancytopenia (AMA Blood dyscrasias). *2429*

Platelets *Blood Decrease Physiological* Pancytopenia (AMA Blood dyscrasias). *2429*

Prothrombin Time *Plasma Decrease Physiological* Induces hepatic metabolism of anticoagulants. *1343*

Chlordiazepoxide

Acetaminophen Screening Test *Urine Negative Analytical* No reaction with o-cresol at therapeutic concentrations. *3326*

Alanine Aminotransferase *Serum No Effect Analytical* At 5 times upper limit of therapeutic range on methods on SMAC®, Abbott-VP, aca, and Cobas-Bio. *2138* At acute overdose concentration (5 mg/dL) on Technicon® SMAC® method. *3719* No effect at therapeutic concentration on Reflotron method. *1984*
Serum Increase Analytical At 5 times upper limit of therapeutic range on kinetic methods on KDA. *2138*
Serum Increase Physiological May produce hepatotoxic effect. *0006*

Albumin *Serum No Effect Analytical* At 5 times upper limit of therapeutic range on methods on SMAC®, Ektachem®, Hitachi® 705 and KDA. *2138* At acute overdose concentration (5 mg/dL) on Technicon® SMAC® method. *3719* At concentration of 50 mg/L had no effect on BCG method. *3393*

Alkaline Phosphatase *Serum No Effect Analytical* At 5 times upper limit of therapeutic range on methods on SMAC®, Abbott-VP, aca, Cobas-Bio, Hitachi® 705 and KDA. *2138* At acute overdose concentration (5 mg/dL) on Technicon® SMAC® method. *3719*
Serum Increase Physiological Infrequent cholestatic effect. *2220*

Alkaloids *Urine No Effect Analytical* With ninhydrin in Frings TLC procedure at 10 mg/dL. *1204*
Urine Increase Analytical Red/purple with iodoplatinate in Frings TLC procedure. *1204*

Aminolevulinic Acid *Urine Increase Physiological* May precipitate acute porphyria. *1322*

Amitriptyline *Serum No Effect Physiological* No effect on serum concentration if given together. *3321*

Amylase *Serum No Effect Analytical* At 5 times upper limit of therapeutic range on methods on aca, Cobas-Bio and Ektachem®. *2138*

Aspartate Aminotransferase *Serum No Effect Analytical* At 5 times upper limit of therapeutic range on methods on SMAC®, Abbott-VP, aca, Cobas-Bio and Hitachi® 705. *2138* At acute overdose concentration (5 mg/dL) on Technicon® SMAC® method. *0899* No effect at therapeutic concentration on Reflotron method. *1984*
Serum Increase Analytical At 5 times upper limit of therapeutic range on kinetic methods on KDA. *2138*
Serum Increase Physiological May produce hepatotoxic effect. *0006*

Barbiturate *Serum Increase Analytical* Occurs with toxic doses diphenylcarbazone procedure. *0583*
Urine No Effect Analytical With diphenylcarbazone in Frings TLC procedure at 10 mg/dL. With mercuric SO_4 in Frings TLC procedure at 10 mg/dL. With vanillin on Frings procedure at 10 mg/dL. With mercurous NO_3 in Frings TLC procedure at 10 mg/dL. With diphenylcarbazone in Frings TLC procedure at 10 mg/dL. *1204*

Bicarbonate *Serum No Effect Analytical* At concentration of 5 mg/L had no effect on method using phenolphthalein. *3393*

Bile *Urine Increase Physiological* Hepatotoxic effect. *2313*

Bilirubin *Serum No Effect Analytical* At 5 times upper limit of therapeutic range on methods on SMAC®, aca, Cobas-Bio, Ektachem®, Hitachi® 705 and KDA. *2138* At acute overdose concentration (5 mg/dL) on Technicon® SMAC® method. *3719* At concentration of 50 mg/L had no effect on Jendrassik and Grof method. *3393*
Serum No Effect Physiological Insignificant protein displacement effect in neonates. *3748*
Serum Increase Physiological Infrequent cholestatic effect. *1022*

Bromide *Urine No Effect Analytical* No effect on method of Frings. *1204*

BSP Retention *Serum Increase Physiological* Infrequent cholestatic effect. *1713*

Calcium *Serum No Effect Analytical* At 5 times upper limit of therapeutic range on methods on SMAC®, Abbott-VP, aca, Ektachem®, Hitachi® 705 and KDA. *2138* At acute overdose concentration (5 mg/dL) on Technicon® SMAC® method. *3719* At concentration of 50 mg/L had no effect on cresolphthalein method. *3393*

Catecholamines *Urine No Effect Physiological* No effect of short term ingestion of 30 mg/d. *0766*

Cephalin Flocculation *Serum Increase Physiological* Infrequent cholestatic effect. *2313*

Chlordiazepoxide *Blood Increase Physiological* 150 mg orally conc=10 mg/L, 600 mg/d = 40 mg/L. *2348*

Chloride *Serum No Effect Analytical* At concentration of 5 mg/L had no effect on mercurimetric method. *3393*

Cholesterol *Serum No Effect Analytical* At 5 times upper limit of therapeutic range on methods on SMAC®, Abbott-VP, Cobas-Bio, Ektachem®, Hitachi® 705 and KDA. *2138* At acute overdose concentration (5 mg/dL) on Technicon® SMAC® method. *3719* No effect at therapeutic concentration on Reflotron method. *1984* At concentration of 160 mg/L had no effect on CHOD-Iodide method. *3393* At concentration of 160 mg/L had no effect on CHOD-PAP method. *3393* At concentration of 160 mg/L had no effect on catalase-AIDH method. *3393* At concentration of 160 mg/L had no effect on method using catalase-Hantzsch reaction. *3393* At concentration of 160 mg/L had no effect on Liebermann-Burchard method. *3393*

Copper *Urine No Effect Physiological* No significant effect observed. *3728*

Coproporphyrin *Urine Increase Physiological* May precipitate acute porphyria. *1322*
Feces Increase Physiological May precipitate acute porphyria. *1322*

Corticosteroids *Urine Increase Analytical* Increased absorption at 410 nm in Porter-Silber reaction. *0022*

Creatine Kinase *Serum No Effect Analytical* At 5 times upper limit of therapeutic range on methods on SMAC®, Abbott-VP, aca, Cobas-Bio and Hitachi® 705. *2138* At acute overdose concentration (5 mg/dL) on Technicon® SMAC® method. *3719*

Creatinine *Serum No Effect Analytical* At 16 mg/L on reversed phase LC procedure of Zhiri et al. *3942* At 5 times upper limit of therapeutic range on methods on SMAC®, Abbott-VP, aca, *2138* At acute overdose concentration (5 mg/dL) on Technicon® SMAC® method. *3719* At concentration of 20 mg/L had no effect on creatinine iminohydrolase method. *3393* At concentration of 50 mg/L had no effect on AutoAnalyzer Jaffé method. *3393* At concentration of 20 mg/L had no effect on Jaffé-Fading-Fraction method. *3393* At concentration of 20 mg/L had no effect on Jaffé-Fuller's earth method. *3393* At concentration of 20 mg/L had no effect on kinetic Jaffé method on BKA-2. *3393*

Dapsone *Serum No Effect Analytical* At 5 mg/dL on colorimetric procedure of Higgins. *1590*

Chlordiazepoxide (continued)

Diphenylhydantoin *Serum No Effect Analytical* On GLC procedure of Papadopoulos at *in vivo* concentration. *2722* *Serum Increase Physiological* Inhibition of metabolism (by 60%). *2042*

Epinephrine *Urine No Effect Physiological* No effect of short term ingestion of 30 mg/d. *0766*

Erythrocytes *Blood Decrease Physiological* Aplastic anemia occasionally occurs. *3717*

Estriol *Urine No Effect Analytical* No effect on GLC method. *1163*

Ethchlorvynol *Serum No Effect Analytical* At 10 μg/mL on colorimetric method of Wallace. *3753* At 90 mg/L on GLC procedure of Evenson. *1058* *Urine No Effect Analytical* No effect on method of Frings and Cohen at 10 mg/dL. *1205*

FPN Test *Urine No Effect Analytical* No effect at 10 mg/dL on method of Frings. *1204*

Glucose *Serum No Effect Analytical* At 5 times upper limit of therapeutic range on methods on SMAC®, Abbott-VP, aca, Cobas-Bio, Ektachem®, Hitachi® 705 and KDA. *2138* At acute overdose concentration (5 mg/dL) on Technicon® SMAC® method. *3719* No effect at therapeutic concentration on Reflotron method. *1984* At concentration of 1.8 mg/L had no effect on Ektachem® method. *3393* At concentration of 50 mg/L had no effect on GOD/POD-PAP method. *3393* At concentration of 6 mg/L had no effect on hexokinase/G-6-PDH method. *3393*

γ-Glutamyltransferase (GGT) *Serum No Effect Analytical* At 5 times upper limit of therapeutic range on methods on SMAC®, Abbott-VP, and Hitachi® 705. *2138* No effect at therapeutic concentration on Reflotron method. *1984*

11-Hydroxycorticosteroids *Urine No Effect Physiological* No effect of short term ingestion of 30 mg/d. *0766*

17-Hydroxycorticosteroids *Urine No Effect Analytical* No effect Porter-Silber reaction if added *in vitro*. *0766* *Urine No Effect Physiological* No effect of short term ingestion of 30 mg/d. *0766* *Urine Increase Analytical* Interferes with Porter-Silber reaction (vitro, vivo). *0427*

6-β-Hydroxycortisol *Urine Increase Physiological* Observed in 2 out of 5 patients. *2682*

5-Hydroxyindoleacetic Acid (5-HIAA) *Urine Increase Analytical* Slight effect observed *in vitro* in method of Udenfriend. *0670*

Iron *Serum No Effect Analytical* At 5 times upper limit of therapeutic range on Ferrozine method of SMAC®. *2138* At acute overdose concentration (5 mg/dL) on Technicon® SMAC® method. *3719* At concentration of 50 mg/L had no effect on Ferrozine method. *3393*

Isocitrate Dehydrogenase *Serum Increase Physiological* Infrequent cholestatic effect. *2313*

131I Uptake *Serum Decrease Physiological* ?antithyroid effect. *0234*

17-Ketogenic Steroids *Urine Increase Analytical* Interferes with Zimmermann reaction. *2573* *Urine Decrease Analytical* Interferes with Zimmermann reaction. *3505*

17-Ketosteroids *Urine No Effect Analytical* No effect on Zimmermann reaction if added *in vitro*. *0766* *Urine No Effect Physiological* No effect of short term ingestion of 30 mg/d. *0766* *Urine Decrease Analytical* Interferes with Zimmermann reaction (in vitro, vivo). *0427*

Lactate *Serum No Effect Analytical* At concentration of 6 mg/L had no effect on enzymatic method. *3393*

Lactate Dehydrogenase *Serum No Effect Analytical* At 5 times upper limit of therapeutic range on methods on SMAC®, Abbott-VP, Cobas-Bio, Hitachi® 705 and KDA. *2138* At acute overdose concentration (5 mg/dL) on Technicon® SMAC® method. *3719*

Leukocytes *Blood Decrease Physiological* Agranulocytosis reported (leukopenia more common). *2313* Aplastic anemia occasionally occurs. *3717*

Methaqualone *Serum No Effect Analytical* At 90 mg/L on GLC procedure of Evenson. *1057*

Neutrophils *Blood Decrease Physiological* Occasional case of agranulocytosis reported. *3717*

Norepinephrine *Urine No Effect Physiological* No effect of short term ingestion of 30 mg/d. *0766*

Nortriptyline *Serum No Effect Physiological* No effect on serum concentration if given together. *3321*

Pheneturide *Serum No Effect Analytical* On GLC procedure of Papadopoulos at *in vivo* concentration. *2722*

Phenobarbital *Serum No Effect Analytical* On GLC procedure of Papadopoulos at *in vivo* concentration. *2722*

Phosphate *Serum No Effect Analytical* At 5 times upper limit of therapeutic range on methods on SMAC®, aca, Hitachi® 705 and KDA. *2138* At acute overdose concentration (5 mg/dL) on Technicon® SMAC® method. *3719* At concentration of 50 mg/L had no effect on phosphomolybdate method. *3393*

Platelets *Blood Decrease Physiological* May rarely cause bone marrow depression. *1436* Aplastic anemia occasionally occurs. *3717*

Porphobilinogen *Urine Increase Physiological* May precipitate acute porphyria. *1322*

Porphyrins *Urine Increase Physiological* May precipitate attack of acute porphyria. *2220*

Potassium *Serum No Effect Analytical* At concentration of 5 mg/L had no effect on flame-photometric method. *3393* At concentration of 20 mg/L had no effect on measurement by ISE with predilution. *3393* At concentration of 18 mg/L had no effect on measurement by ISE without predilution. *3393*

Pregnancy Tests *Urine No Effect Analytical* No effect on UCG test. *1487* *Urine Positive Analytical* False positive with Gravindex™. *1487*

Primidone *Serum No Effect Analytical* On GLC procedure of Papadopoulos at *in vivo* concentration. *2722*

Propoxyphene *Serum No Effect Analytical* At 50 mg/L on method of Evenson. *1056*

Protein *Serum No Effect Analytical* At 5 times upper limit of therapeutic range on methods on SMAC®, Abbott-VP, Ektachem®, Hitachi® 705 and KDA Hitachi® 705 and KDA. *2138* At acute overdose concentration (5 mg/dL) on Technicon® SMAC® method. *3719* At concentration of 50 mg/L had no effect on biuret method with blank correction. *3393*

Protein Electrophoresis *Serum No Effect Analytical* At concentration of 18 mg/L had no effect on automated Olympus-Hite method except for slight displacement of fractions. *3393*

Prothrombin Time *Plasma No Effect Physiological* Does not affect action of administered coumarins. *3011* *Plasma Increase Physiological* Associated with failure of excretion of bile salts. *1713* *Plasma Decrease Physiological* May shorten action of anticoagulants. *1679*

Protoporphyrin *Feces Increase Physiological* May precipitate acute porphyria. *1322*

Salicylate *Urine No Effect Analytical* At 10 mg/dL no effect on Trinder method. *1204*

Sodium *Serum No Effect Analytical* At concentration of 5 mg/L had no effect on flame-photometric method. *3393* At concentration of 20 mg/L had no effect on measurement by ISE with predilution. *3393* At concentration of 18 mg/L had no effect on measurement by ISE without predilution. *3393*

T₃ Uptake *Serum Decrease Physiological* ?antithyroid effect (resin test affected). *0234*

Thymol Turbidity *Serum Increase Physiological* Infrequent cholestatic effect. *2313*

Triglycerides *Serum No Effect Analytical* At 5 times upper limit of therapeutic range on methods on SMAC®, Abbott-VP, Ektachem®, Hitachi® 705 and KDA. *2138* At acute overdose concentration (5 mg/dL) on Technicon® SMAC® method. *3719* No effect at therapeutic concentration on Reflotron method. *1984* At concentration of 51 mg/L had no effect on GPO-PAP method. *3393* At concentration of 50 mg/L had no effect on lipase/esterase method. *3393*

Urea Nitrogen *Serum No Effect Analytical* At 5 times upper limit of therapeutic range on methods on SMAC®, Abbott-VP, Cobas-Bio, Ektachem®, Hitachi® 705 and KDA. *2138* At acute overdose concentration (5 mg/dL) on Technicon® SMAC® method. *3719* No effect at therapeutic concentration on Reflotron method. *1984* At concentration of 50 mg/L had no effect on diacetylmonoxime method. *3393* At concentration of 1.8 mg/L had no effect on Ektachem® method. *3393*

Uric Acid *Serum No Effect Analytical* At 16 mg/L on reversed phase LC procedure of Zhiri et al. *3942* At 5 times upper limit of therapeutic range on methods on SMAC®, Abbott-VP, aca, *2138* At acute overdose concentration (5 mg/dL) on Technicon® SMAC® method. *3719* No effect at therapeutic concentration on Reflotron method. *1984* At concentration of 20 mg/L had no effect on Kageyama-Hantzsch method. *3393* At concentration of 20 mg/L had no effect on catalase-AlDH method. *3393* At concentration of 50 mg/L had no effect on phosphotungstate reduction method. *3393* At concentration of 50 mg/L had no effect on uricase-PAP method. *3393*

Urobilinogen *Urine Decrease Physiological* Infrequent cholestatic effect. *1713*

Feces Decrease Physiological Infrequent cholestatic effect. *1713*

Warfarin *Plasma Increase Analytical* Spectrophotometric method shows interference. *3011*

Chlorhexidine

Amphetamine *Urine No Effect Analytical* No effect at 100 mg/L on method of Rutter. *3097*

Protein *Urine Increase Analytical* False positive with Stix alkalinization. *0121*

Chloride Salts

Amylase *Serum Increase Analytical* Chloride enhances enzyme activity. *3295*

Chloride *Serum Increase Physiological* Increased absorption with increased serum levels. *0652*

Orthophosphate *Test Conditions Decrease Analytical* 85% inhibition at 0.4 mol/L on method of Horder. *1658*

Protein *Test Conditions Increase Analytical* If glyoxylic acid used to measure tryptophan. *0093*

Protein Bound Iodine (PBI) *Serum Increase Analytical* Laboratory contamination may cause significant effect. *0830*

Pyruvate Kinase *Red Blood Cells Decrease Analytical* Observed with 0.3 mmol/L PEP in reaction mixture. *0370*

Sugar *Urine Increase Analytical* If anthrone method used. *0093*

Zinc *Serum No Effect Analytical* At 6,000 to 1 with flameless atomic absorption. *2038*

Urine No Effect Analytical At 6,000 to 1 with flameless atomic absorption. *2038*

Chlorinated Insecticides

Alanine Aminotransferase *Serum Increase Physiological* May cause liver damage. *0279*

Aldolase *Serum Increase Physiological* Presumed damage of liver cells. *0405*

Alkaline Phosphatase *Serum Increase Physiological* Presumed damage of liver cells. *0405*

Aspartate Aminotransferase
Serum Increase Physiological May cause liver damage. *0279*

Pseudocholinesterase
Red Blood Cells Increase Physiological Powerful inhibitor of enzyme. *0405*

Chlorine

Uric Acid *Serum Increase Analytical* If standards made in contaminated water. *0581*

Serum Decrease Analytical If in water affects phosphotungstate procedure. *0583*

Chlormadinone

Cholesterol *Serum No Effect Physiological* No significant effect observed. *1309*

Factor VIII *Plasma Increase Physiological* Significant effect observed by 3 mo. *2456*

Factor IX *Plasma Increase Physiological* Significant effect observed by 3 mo. *2456*

Factor X *Plasma Decrease Physiological* Significant effect observed by 3 mo. *2456*

Luteinizing Hormone (LH) *Urine Decrease Physiological* Usual effect in normal women. *3464*

Protein Bound Iodine (PBI) *Serum Increase Physiological* Due to increased binding globulin. *2313*

Prothrombin Time *Plasma Increase Physiological* Significant effect observed by 3 mo. *0798*

Thyroxine (T₄) Binding Globulin
Serum Increase Physiological Therapeutic effect. *2313*

Triglycerides *Serum No Effect Physiological* No significant effect observed. *1309*

Chlormerodrin

17-Hydroxycorticosteroids *Urine Increase Analytical* Possible interference with Porter-Silber reaction. *0427*

17-Ketosteroids *Urine Decrease Analytical* Possible interference with Zimmermann reaction. *0427*

Protein *Urine Increase Physiological* May produce nephrotic syndrome. *1188*

Protein Bound Iodine (PBI) *Serum Decrease Analytical* Formation of mercurial iodide during analysis. *0757*

Urea Nitrogen *Serum Increase Physiological* May produce nephrotic syndrome. *1188*

Chlormethiazole

Indomethacin *Serum No Effect Analytical* No effect on HPLC method of Roberts and Smith. *3002*

Chlormethine

Chromosomes *Test Conditions Abnormal Physiological* Clastogenic in human cells. *3282*

Chlormezanone

Alanine Aminotransferase *Serum Increase Physiological* Occasional reversible cholestasis. *2313*

Albumin *Serum No Effect Analytical* At concentration of 100 mg/L had no effect on BCG method. *3393*

Alkaline Phosphatase *Serum Increase Physiological* Occasional reversible cholestasis. *2313*

Aspartate Aminotransferase
Serum Increase Physiological Occasional reversible cholestasis. *2313*

Bicarbonate *Serum No Effect Analytical* At concentration of 100 mg/L had no effect on method using phenolphthalein. *3393*

Bile *Urine Increase Physiological* Occasional reversible cholestasis. *2313*

Bilirubin *Serum No Effect Analytical* At concentration of 100 mg/L had no effect on Jendrassik and Grof method. *3393*
Serum Increase Physiological Occasional reversible cholestasis. *2313*

Calcium *Serum No Effect Analytical* At concentration of 100 mg/L had no effect on cresolphthalein method. *3393*

Cephalin Flocculation *Serum Increase Physiological* Occasional reversible cholestasis. *2313*

Chloride *Serum No Effect Analytical* At concentration of 100 mg/L had no effect on mercurimetric method. *3393*

Cholesterol *Serum No Effect Analytical* At concentration of 100 mg/L had no effect on Liebermann-Burchard method. *3393*

Creatinine *Serum No Effect Analytical* At concentration of 100 mg/L had no effect on AutoAnalyzer Jaffé method. *3393*

Phosphate *Serum No Effect Analytical* At concentration of 100 mg/L had no effect on phosphotungstate reduction method. *3393*

Chlormezanone (continued)

Potassium *Serum No Effect Analytical* At concentration of 100 mg/L had no effect on measurement by ISE with predilution. *3393*

Protein *Serum No Effect Analytical* At concentration of 100 mg/L had no effect on biuret method with blank correction. *3393*

Sodium *Serum No Effect Analytical* At concentration of 100 mg/L had no effect on measurement by ISE with predilution. *3393*

Thymol Turbidity *Serum Increase Physiological* Occasional reversible cholestasis. *2313*

Triglycerides *Serum No Effect Analytical* At concentration of 100 mg/L had no effect on lipase/esterase method. *3393*

Urea Nitrogen *Serum No Effect Analytical* At concentration of 100 mg/L had no effect on diacetylmonoxime method. *3393*

Uric Acid *Serum No Effect Analytical* At concentration of 100 mg/L had no effect on phosphotungstate reduction method. *3393*

Chlorobutanol

Urea Nitrogen *Serum Increase Analytical* False increase with direct Nesslerization. *1563*

Chloroform

Alanine Aminotransferase *Serum Increase Physiological* Hepatotoxic effect with necrosis. *2313*

Alkaline Phosphatase *Serum Increase Physiological* Hepatotoxic effect with necrosis. *2313*

Aspartate Aminotransferase
Serum Increase Physiological Hepatotoxic effect with necrosis. *2313*

Bile *Urine No Effect Analytical* As preservative has no effect on Stix tests. *2308*

Bilirubin *Serum Increase Physiological* Hepatotoxic effect with necrosis. *2313*

BSP Retention *Serum Increase Physiological* Hepatotoxic effect with necrosis. *2313*

Cephalin Flocculation *Serum Increase Physiological* Hepatotoxic effect with necrosis. *2313*

Cholesterol *Serum Decrease Physiological* Hepatotoxic effect with necrosis. *2313*

Glucose *Serum Decrease Physiological* Hepatotoxic effect with necrosis. *2313*
Urine No Effect Analytical As preservative has no effect on Stix tests. *2308*

Hemoglobin *Urine No Effect Analytical* As preservative has no effect on Stix tests. *2308*

Ketones *Urine No Effect Analytical* As preservative has no effect on Stix tests. *2308*

Leukocytes *Blood Increase Physiological* Normal response to anesthesia. *0251*

Ornithine Carbamoyltransferase (OCT)
Serum Increase Physiological Hepatotoxic effect with necrosis. *3835*

pH *Urine No Effect Analytical* As preservative has no effect on Stix tests. *2308*

Phosphate *Serum Decrease Physiological* Follows most forms of anesthesia. *0279*

Protein *Urine No Effect Analytical* As preservative has no effect on Stix tests. *2308*
Urine Increase Physiological May cause renal damage. *0279*

Sugar *Urine Increase Physiological* Positive with Benedict's, Fehling's solutions. *3879*

Thymol Turbidity *Serum Increase Physiological* Hepatotoxic effect with necrosis. *2313*

Urea Nitrogen *Serum Increase Physiological* May cause renal damage. *0279*

Uric Acid *Serum Increase Physiological* May occur with marked tissue destruction. *0279*

Chloroguanide

Casts *Urine Increase Physiological* Nephrotoxic effect. *2313*

Cells *Urine Increase Physiological* May cause transient appearance of epithelial cells. *1343*

Erythrocytes *Urine Increase Physiological* Drug produces actual bleeding. *1022*

Folate *Serum Decrease Physiological* Antagonizes folic acid. *0071*

Gastrin *Serum Decrease Physiological* Significant effect (basal and histamine stimulated) in rats. *0909*

Hemoglobin *Urine Increase Physiological* Drug produces actual bleeding. *1022*

Myelocytes *Blood Increase Physiological* Up to 10% in patients with overt malaria. *1343*

Chlorophenothane

Alanine Aminotransferase *Serum Increase Physiological* May cause liver damage. *0279*

Aldosterone *Urine Decrease Physiological* May inhibit synthesis of hormone. *2775*

Aspartate Aminotransferase
Serum Increase Physiological May cause liver damage. *0279*

Bilirubin *Serum Decrease Physiological* Effect on congenital non-hemolytic hyperbilirubinemia. *3505*

Cortisol *Plasma No Effect Physiological* Occupational exposure no effect. *0674*

Dexamethasone Suppression
Patient No Effect Physiological Normal in subjects with occupational exposure. *0674*

Diphenylhydantoin *Serum Decrease Physiological* Decrease from 9 to 1 parts per billion (with dose of 300-400 mg). *2041*

Erythrocytes *Blood Decrease Physiological* Pancytopenia (AMA blood dyscrasias). *2429*

Leukocytes *Blood Decrease Physiological* Pancytopenia (AMA blood dyscrasias). *2429*

Occult Blood *Feces Positive Physiological* May cause gastrointestinal tract bleeding and other symptoms. *0279*

Phenobarbital *Serum Decrease Physiological* Reported interaction due to alteration of metabolism. *2042*

Platelets *Blood Decrease Physiological* Pancytopenia (AMA blood dyscrasias). *2429*

Chlorophenylalanine

Eosinophils *Blood Increase Physiological* Observed in isolated case. *3686*

Homovanillic Acid *CSF No Effect Physiological* After 1 g/d for 4 d or after single dose. *3342*

5-Hydroxyindoleacetic Acid (5-HIAA)
Urine Decrease Physiological Inhibits biosynthesis of serotonin. *0754*
CSF No Effect Physiological After 1 g/d for 4 d or after single dose. *3342*

Chlorophyll

Urobilinogen *Urine Increase Analytical* Metabolite phylloerythrinogen produces red color. *0583*

Chloroquine

Alkaloids *Urine No Effect Analytical* With ninhydrin in Frings TLC procedure at 10 mg/dL. *1204*
Urine Increase Analytical Purple with iodoplatinate in Frings TLC procedure. *1204*

Aspartate Aminotransferase
Serum Increase Physiological Reported effect no mechanism cited. *2313*

Barbiturate *Urine No Effect Analytical* With mercuric SO_4 in Frings TLC procedure at 10 mg/dL. With diphenylcarbazone in Frings TLC procedure at 10 mg/dL. With mercurous NO_3 in Frings TLC procedure at 10 mg/dL. With vanillin on Frings procedure at 10 mg/dL. *1204*

Bilirubin *Serum Increase Physiological* Hemolytic anemia in G-6-PD deficiency. *1680*

Bromide *Urine No Effect Analytical* No effect on method of Frings. *1204*

Calcium *Urine Decrease Physiological* In 2 patients with sarcoidosis given 500 mg daily while receiving corticosteroid at same time. *2639*

Cholesterol *Serum No Effect Analytical* At concentration of 60 mg/L had no effect on CHOD-Iodide method. *3393* At concentration of 200 mg/L had no effect on CHOD-PAP method. *3393* At concentration of 60 mg/L had no effect on catalase-AIDH method. *3393* At concentration of 60 mg/L had no effect on method using catalase-Hantzsch reaction. *3393* At concentration of 60 mg/L had no effect on Liebermann-Burchard method. *3393*

Color *Urine Increase Analytical* Brown color. *1022*

Creatinine *Serum No Effect Analytical* At concentration of 0.5 mg/L had no effect on creatinine iminohydrolase method. *3393* At concentration of 100 mg/L had no effect on Jaffé-Fading-Fraction method. *3393* At concentration of 100 mg/L had no effect on Jaffé-Fuller's earth method. *3393* At concentration of 50 mg/L had no effect on kinetic Jaffé method on BKA-2. *3393*

1,25-Dihydroxy Vitamin D₃ *Serum Decrease Physiological* In 2 patients with sarcoidosis given 500 mg daily while receiving corticosteroid at same time. *2639*

Erythrocytes *Blood Decrease Physiological* Hemolytic anemia in G-6-PD deficient persons. *3522* Aplastic anemia reported to occur occasionally. *3717*

Ethchlorvynol *Urine No Effect Analytical* No effect on method of Frings and Cohen. *1204*

FPN Test *Urine No Effect Analytical* No effect at 10 mg/dL on method of Frings. *1204*

Glucose *Serum No Effect Analytical* At concentration of 5 mg/L had no effect on Ektachem® method. *3393* At concentration of 250 mg/L had no effect on GOD/POD-PAP method. *3393*

Hematocrit *Blood Decrease Physiological* G-6-PD hemolytic anemia/pancytopenia. *3522*

Hemoglobin *Blood Decrease Physiological* Hemolytic anemia in G-6-PD deficient persons. *3522*

25-Hydroxy Vitamin D₃ *Serum No Effect Physiological* Possibly inhibits conversion of 25-hydroxyvitamin D to 1,25-dihydroxyvitamin D. *2639*

Leukocytes *Blood Decrease Physiological* Agranulocytosis/pancytopenia. *3522* Aplastic anemia reported to occur occasionally. *3717*

Platelets *Blood Decrease Physiological* Pancytopenia. *3522* Aplastic anemia reported to occur occasionally. *3717*

Potassium *Serum No Effect Analytical* At concentration of 50 mg/L had no effect on measurement by ISE without predilution. *3393*
Serum Decrease Physiological Marked prolonged reduction in two cases of acute massive intoxication. *2203* Marked prolonged reduction in two cases of acute massive intoxication. *2203*

Protein Electrophoresis *Serum No Effect Analytical* At concentration of 50 mg/L had no effect on automated Olympus-Hite method. *3393*

Salicylate *Urine No Effect Analytical* At 10 mg/dL no effect on Trinder method. *1204*

Sodium *Serum No Effect Analytical* At concentration of 50 mg/L had no effect on measurement by ISE without predilution. *3393*

Triglycerides *Serum No Effect Analytical* At concentration of 250 mg/L had no effect on GPO-PAP method. *3393*

Urea Nitrogen *Serum No Effect Analytical* At concentration of 5 mg/L had no effect on Ektachem® method. *3393*

Uric Acid *Serum No Effect Analytical* At concentration of 100 mg/L had no effect on Kageyama-Hantzsch method. *3393* At concentration of 100 mg/L had no effect on catalase-AIDH method. *3393* At concentration of 250 mg/L had no effect on uricase-PAP method. *3393*

Urobilinogen *Urine Increase Physiological* Hemolytic anemia in G-6-PD deficiency. *1680*

Uroporphyrin *Urine Increase Physiological* Transient increase may occur with cutaneous porphyria. *1322*

Chlorothiazide

Alanine Aminotransferase *Serum Increase Physiological* May cause cholestatic jaundice. *2313*

Alkaline Phosphatase *Serum Increase Physiological* May cause cholestatic jaundice. *2313*

Ammonia *Plasma Increase Physiological* Decreased potassium and alkalosis. *1238*

Amylase *Serum Increase Physiological* Infrequent consequence of therapy. *0363* 10 of 20 patients developed 50 to 100% increase in serum amylase shortly after beginning treatment. *2281*

Aspartate Aminotransferase *Serum No Effect Analytical* No effect on Babson procedure. No effect on Karmen procedure. *2760*
Serum Increase Physiological May cause cholestatic jaundice. *2313*

Bile *Urine Increase Physiological* May cause cholestatic jaundice. *2313*

Bilirubin *Serum No Effect Physiological* Displacement from protein observed *in vitro*, but unlikely at pharmacological concentrations *in vivo*. *3748*
Serum Increase Physiological Reported to cause cholestatic jaundice. *2745*

BSP Retention *Serum Increase Physiological* May cause cholestatic jaundice. *2313*

Calcium *Serum Increase Physiological* Impaired excretion with chronic administration. *3505* Due to increased tubular reabsorption but usually associated with mild primary hyperparathyroidism. *1897*
Serum Decrease Physiological Initial response to administration. *0105*
Urine Increase Physiological Initial diuretic response. *0105*
Urine Decrease Physiological Impaired excretion with chronic administration. *3505*
Feces Decrease Physiological Accentuates positive calcium balance. *2427*

Catecholamines *Urine No Effect Analytical* No effect on fluorometric method of Crout. *0758*
Urine No Effect Physiological No effect observed. *0913*

Cephalin Flocculation *Serum Increase Physiological* May cause cholestatic jaundice. *2313*

Chloride *Serum Increase Physiological* With prolonged therapy hyperchloremic alkalosis. *2220*

Citrate *Urine Decrease Physiological* Up to 30% decrease observed. *2427*

Creatinine Clearance *Urine Decrease Physiological* Reported effect. *2220*

Erythrocytes *Blood Decrease Physiological* Pancytopenia/aplastic anemia. *0073* Aplastic anemia reported to occur occasionally. *3717*
Urine Increase Physiological occurred in one case after i.v. administration. *1680*

Estriol *Urine No Effect Analytical* No interference with hydrolysis observed. *3060*

Estrogens *Urine No Effect Analytical* No significant effect Kober reaction and Allen correction. *0023*

Glucose *Serum Increase Physiological* Diabetogenic properties of drug affect glucose tolerance test. *3505*
Urine Increase Physiological Diabetogenic-like action of drug affects glucose tolerance test. *3505*

Glucose Tolerance *Serum Decrease Physiological* Diabetogenic-like action of drug. *2745*

Hematocrit *Blood Decrease Physiological* Pancytopenia. *0073*

Hemoglobin *Blood Decrease Physiological* Pancytopenia. *0073*

17-Hydroxycorticosteroids *Urine Increase Analytical* Interferes with Porter-Silber reaction *in vitro*. *3879*

17-Ketogenic Steroids *Urine No Effect Analytical* Probably minimal interference with Zimmermann reaction. *3505*

Chlorothiazide (continued)

17-Ketosteroids Urine No Effect Analytical Minimal interference with Zimmermann reaction in vitro. 0427
Urine Increase Analytical Possible interference in vitro. 0583

LE Cells Blood Positive Physiological May produce LE-like syndrome. 3504

Leukocytes Blood Decrease Physiological Leukopenia/pancytopenia. 0073 Aplastic anemia reported to occur occasionally. 3717

Magnesium Urine Increase Physiological 1.5 g/d for 2 d produces 33% increase. 1487

Neutrophils Blood Decrease Physiological Occasionally observed. 2427 Occasional case of agranulocytosis reported. 3717

5'-Nucleotidase Serum Increase Physiological May cause cholestatic jaundice. 2313

PAH Clearance Urine Decrease Physiological Competitive inhibition of secretion. 1433

Platelets Blood Decrease Physiological Pancytopenia due to aplastic anemia may occur. 0179 Aplastic anemia reported to occur occasionally. 3717

Potassium Serum Decrease Physiological Diuretic action. 3505

Prothrombin Time Plasma No Effect Physiological Does not affect action of administered coumarins. 3011

Quinidine Serum Increase Physiological Alkalinizes urine, increases reabsorption. 1487

Sodium Serum Decrease Physiological May cause hyponatremia. 2220
Urine Increase Physiological Diuretic action. 2220

Thymol Turbidity Serum Increase Physiological May cause cholestatic jaundice. 2313

Uric Acid Serum Increase Physiological Decreased urate clearance. 3505
Serum Decrease Physiological If given i.v. in large doses has uricosuric effect. 1433
Urine Increase Physiological If given i.v. in large doses has uricosuric effect. 1433
Urine Decrease Physiological Impaired clearance. 2220

Vanillylmandelic Acid Urine No Effect Physiological No effect observed. 0913

Warfarin Plasma No Effect Physiological No effect of 1 g/d for 21 d. 1487

Chlorphenesin

Albumin Serum No Effect Analytical At concentration of 150 mg/L had no effect on BCG method. 3393

Bicarbonate Serum No Effect Analytical At concentration of 150 mg/L had no effect on method using phenolphthalein. 3393

Bilirubin Serum No Effect Analytical At concentration of 150 mg/L had no effect on Jendrassik and Grof method. 3393

Calcium Serum No Effect Analytical At concentration of 150 mg/L had no effect on cresolphthalein method. 3393

Chloride Serum No Effect Analytical At concentration of 150 mg/L had no effect on mercurimetric method. 3393

Cholesterol Serum No Effect Analytical At concentration of 150 mg/L had no effect on Liebermann-Burchard method. 3393

Creatinine Serum No Effect Analytical At concentration of 150 mg/L had no effect on AutoAnalyzer Jaffé method. 3393

Leukocytes Blood Decrease Physiological Rare leukopenia/agranulocytosis/pancytopenia. 1680

Occult Blood Feces Positive Physiological Two cases gastrointestinal bleeding reported. 1680

Phosphate Serum No Effect Analytical At concentration of 150 mg/L had no effect on phosphomolybdate method. 3393

Platelets Blood Decrease Physiological Rare thrombocytopenia/pancytopenia. 1680

Potassium Serum No Effect Analytical At concentration of 150 mg/L had no effect on measurement by ISE with predilution. 3393

Protein Serum No Effect Analytical At concentration of 150 mg/L had no effect on biuret method with blank correction. 3393

Sodium Serum No Effect Analytical At concentration of 150 mg/L had no effect on measurement by ISE with predilution. 3393

Triglycerides Serum No Effect Analytical At concentration of 150 mg/L had no effect on lipase/esterase method. 3393

Urea Nitrogen Serum No Effect Analytical At concentration of 150 mg/L had no effect on diacetylmonoxime method. 3393

Uric Acid Serum No Effect Analytical At concentration of 150 mg/L had no effect on phosphotungstate reduction method. 3393

Chlorpheniramine

Alanine Aminotransferase Serum No Effect Analytical At acute overdose concentration (20 mg/dL) on Technicon® SMAC® method. 3719

Albumin Serum No Effect Analytical At acute overdose concentration (20 mg/dL) on Technicon® SMAC® method. 3719 At concentration of 3 mg/L had no effect on BCG method. 3393

Alkaline Phosphatase Serum No Effect Analytical At acute overdose concentration (20 mg/dL) on Technicon® SMAC® method. 3719

Alkaloids Urine No Effect Analytical With ninhydrin in Frings TLC procedure at 12 mg/dL. 1204
Urine Increase Analytical Purple with iodoplatinate in Frings TLC procedure. 1204

Aspartate Aminotransferase Serum No Effect Analytical At acute overdose concentration (20 mg/dL) on Technicon® SMAC® method. 3719

Barbiturate Serum Increase Analytical At toxic levels (over 4 mg/dL) only on screen test. 0583
Urine No Effect Analytical With mercurous NO_3 in Frings TLC procedure at 12 mg/dL. With diphenylcarbazone in Frings TLC procedure at 12 mg/dL. With vanillin on Frings procedure at 12 mg/dL. With mercuric SO_4 in Frings TLC procedure at 12 mg/dL. 1204

Bicarbonate Serum No Effect Analytical At concentration of 200 mg/L had no effect on method using phenolphthalein. 3393

Bilirubin Serum No Effect Analytical At acute overdose concentration (20 mg/dL) on Technicon® SMAC® method. 3719 At concentration of 200 mg/L had no effect on Jendrassik and Grof method. 3393

Bromide Urine No Effect Analytical No effect on method of Frings. 1204

Calcium Serum No Effect Analytical At acute overdose concentration (20 mg/dL) on Technicon® SMAC® method. 3719 At concentration of 200 mg/L had no effect on cresolphthalein method. 3393

Chlordiazepoxide Blood No Effect Analytical On method of Riddick at 5 mg/dL. 2983
Blood Increase Analytical May interfere method of Jatlow. 1777

Chloride Serum No Effect Analytical At concentration of 2 mg/L had no effect on mercurimetric method. 3393

Cholesterol Serum No Effect Analytical At acute overdose concentration (20 mg/dL) on Technicon® SMAC® method. 3719 At concentration of 200 mg/L had no effect on Liebermann-Burchard method. 3393

Creatine Kinase Serum No Effect Analytical At acute overdose concentration (20 mg/dL) on Technicon® SMAC® method. 3719

Creatinine Serum No Effect Analytical At concentration of 200 mg/L had no effect on AutoAnalyzer Jaffé method. 3393 At acute overdose concentration (20 mg/dL) on Technicon SMAC® method. 3719

Erythrocytes Blood Decrease Physiological Reported to cause hemolytic anemia. 1678

Ethchlorvynol Urine No Effect Analytical No effect on method of Frings and Cohen at 12 mg/dL. 1205

FPN Test Urine No Effect Analytical No effect at 12 mg/dL on method of Frings. 1204

Glucose *Serum No Effect Analytical* At acute overdose concentration (20 mg/dL) on Technicon® SMAC® method. *3719*

Hematocrit *Blood Decrease Physiological* Case of aplastic anemia after 10 y of low dose treatment. *1859*

Hemoglobin *Blood Decrease Physiological* Reported to cause hemolytic anemia. *1678* Case of aplastic anemia after 10 y of low dose treatment. *1859* One case of aplastic anemia after 1 mo therapy, reversible with discontinuation of drug discontinuation of drug. *0886*

Iron *Serum No Effect Analytical* At acute overdose concentration (20 mg/dL) on Technicon® SMAC® method. *3719*

¹³¹I Uptake *Serum No Effect Physiological* 16 mg/d reduced uptake by 48%. *0583*
Serum Decrease Physiological Ornade®, Teldrin contain tetraiodofluorescein. *2652*

Lactate Dehydrogenase *Serum No Effect Analytical* At acute overdose concentration (20 mg/dL) on Technicon® SMAC® method. *3719*

Leukocytes *Blood Decrease Physiological* Leukopenia (AMA Blood dyscrasias). *2429* Case of aplastic anemia after 10 y of low dose treatment. *1859* One case of aplastic anemia after 1 mo therapy, reversible with discontinuation of drug. *0886*

Phosphate *Serum No Effect Analytical* At acute overdose concentration (20 mg/dL) on Technicon® SMAC® method. *3719* At concentration of 200 mg/L had no effect on phosphomolybdate method. *3393*

Platelets *Blood Decrease Physiological* Thrombocytopenia reported. *1680* Case of aplastic anemia after 10 y of low dose treatment. *1859* One case of aplastic anemia after 1 mo therapy, reversible with discontinuation of drug. *0886*

Protein *Serum No Effect Analytical* At acute overdose concentration (20 mg/dL) on Technicon® SMAC® method. *3719* At concentration of 200 mg/L had no effect on biuret method with blank correction. *3393*

Salicylate *Urine No Effect Analytical* At 12 mg/dL no effect on Trinder method. *1204*

Triglycerides *Serum No Effect Analytical* At acute overdose concentration (20 mg/dL) on Technicon® SMAC® method. *3719* At concentration of 200 mg/L had no effect on lipase/esterase method. *3393*

Urea Nitrogen *Serum No Effect Analytical* At acute overdose concentration (20 mg/dL) on Technicon® SMAC® method. *3719* At concentration of 200 mg/L had no effect on diacetylmonoxime method. *3393*

Uric Acid *Serum No Effect Analytical* At acute overdose concentration (20 mg/dL) on Technicon® SMAC® method. *3719* At concentration of 200 mg/L had no effect on phosphotungstate reduction method. *3393*

Chlorphentermine

Amobarbital *Urine No Effect Analytical* No interference using TLC with ethyl acetate: methanol: water: ammonium hydroxide and modified Dragendorff's reagent for detection. *3868*

Amphetamine *Urine No Effect Analytical* At 50 mg/L on fluorescent method of Hayes. *1534* No interference using TLC with ethyl acetate: methanol: water: ammonium hydroxide and modified Dragendorff's reagent for detection. *3868*

Hydromorphone *Urine No Effect Analytical* No interference using TLC with ethyl acetate: methanol: water: ammonium hydroxide and modified Dragendorff's reagent for detection. *3868*

Mescaline *Urine No Effect Analytical* No interference using TLC with ethyl acetate: methanol: water: ammonium hydroxide and modified Dragendorff's reagent for detection. *3868*

Methamphetamine *Urine No Effect Analytical* No interference using TLC with ethyl acetate: methanol: water: ammonium hydroxide and modified Dragendorff's reagent for detection. *3868*

Morphine *Urine No Effect Analytical* No interference using TLC with ethyl acetate: methanol: water: ammonium hydroxide and modified Dragendorff's reagent for detection. *3868*

Pentobarbital *Urine No Effect Analytical* No interference using TLC with ethyl acetate: methanol: water: ammonium hydroxide and modified Dragendorff's reagent for detection. *3868*

Phenmetrazine *Urine Positive Analytical* Similar Rf and color reaction on TLC using ethyl acetate: methanol: water: ammonium hydroxide and modified Dragendorff's reagent. *3868*

Phenobarbital *Urine No Effect Analytical* No interference using TLC with ethyl acetate: methanol: water: ammonium hydroxide and modified Dragendorff's reagent for detection. *3868*

Phentermine *Urine Positive Analytical* Similar Rf and color reaction on TLC using ethyl acetate: methanol: water: ammonium hydroxide and modified Dragendorff's reagent. *3868*

Phenylpropanolamine *Urine No Effect Analytical* No interference using TLC with ethyl acetate: methanol: water: ammonium hydroxide and modified Dragendorff's reagent for detection. *3868*

Quinine *Urine Positive Analytical* Similar Rf and color reaction on TLC using ethyl acetate: methanol: water: ammonium hydroxide and modified Dragendorff's reagent. *3868*

Secobarbital *Urine No Effect Analytical* No interference using TLC with ethyl acetate: methanol: water: ammonium hydroxide and modified Dragendorff's reagent for detection. *3868*

Chlorpromazine

Acetaminophen Screening Test *Urine Negative Analytical* No reaction with o-cresol at therapeutic concentrations. *3326*

Alanine Aminotransferase *Serum Increase Physiological* May be damage of biliary canaliculi. *3835*

Albumin *Serum No Effect Analytical* At concentration of 3 mg/L had no effect on BCG method. *3393*

Alkaline Phosphatase *Serum Increase Physiological* Hepatic sensitivity to drug (in up to 2% patients). *3283* In one patient who developed cholestasis with fever and leukopenia after 260 mg drug. *2840*

Alkaloids *Urine No Effect Analytical* With ninhydrin in Frings TLC procedure at 10 mg/dL. *1204*
Urine Increase Analytical Purple with iodoplatinate in Frings TLC procedure. *1204*

Antibodies to dsDNA *Serum Positive Physiological* In 40% to native-DNA of schizophrenic patients given long-term treatment. *3937*

Antidiuretic Hormone *Plasma No Effect Physiological* No effect of intramuscular injection of drug. *2920*

Antinuclear Antibodies *Serum Positive Physiological* Single case reported associated with rash and fever. *0963* In 63% of schizophrenic patients given long-term treatment. *3937*

Antinucleoprotein Antibodies
Serum Positive Physiological In 58% of schizophrenic patients given long-term treatment. *3937*

Antithrombin III *Plasma No Effect Analytical* At concentration of 3 mg/L had no effect on aca method. *3393*

Aspartate Aminotransferase
Serum Increase Physiological Hepatic sensitivity to drug (in up to 2% patients). *3283* In one patient who developed cholestasis with fever and leukopenia after 260 mg drug. *2840*

Barbiturate *Urine No Effect Analytical* With diphenylcarbazone in Frings TLC procedure at 10 mg/dL. With mercurous NO₃ in Frings TLC procedure at 10 mg/dL. *1204*
Urine Increase Analytical Pink spot with vanillin in Frings TLC procedure. Pink spot with mercuric SO₄ in Frings TLC procedure. *1204*

Bicarbonate *Serum No Effect Analytical* At concentration of 3 mg/L had no effect on method using phenolphthalein. *3393*

Bile *Urine Increase Analytical* Alleged interference with BiliLabstix®. *1488*

Bilirubin *Serum No Effect Analytical* At concentration of 3 mg/L had no effect on Jendrassik and Grof method. *3393*
Serum Increase Physiological Sensitivity reaction (may cause jaundice in infant). *3283* In one patient who developed cholestasis with fever and leukopenia after 260 mg drug. *2840*

Chlorpromazine (continued)

Bilirubin, Direct *Serum Increase Physiological* Sensitivity reaction to drug (in up to 2% patients). *3283* In one patient who developed Cholestasis with fever and Leukopenia after 260 mg drug. *2840*

Bromide *Urine No Effect Analytical* No effect on method of Frings. *1204*

BSP Retention *Serum Increase Physiological* Induces transient cholestatic hepatitis (1 case). *0900*

Calcium *Serum No Effect Analytical* At concentration of 3 mg/L had no effect on cresolphthalein method. *3393*

Catecholamines *Plasma Increase Physiological* Increased metabolism, decreased organ uptake of norepinephrine. *1237*
Urine Increase Analytical Affects fluorescent procedures. *3217*
Urine Decrease Analytical Affect measurement by Pisano procedure. *3135*

Chlordiazepoxide *Blood No Effect Analytical* On method of Riddick at 5 mg/dL. *2983*

Chloride *Serum No Effect Analytical* At concentration of 3 mg/L had no effect on mercurimetric method. *3393*

Cholesterol *Serum No Effect Analytical* At concentration of 3 mg/L had no effect on CHOD-PAP method. *3393*
Serum Increase Analytical 50 μg in reaction mixture produces color. *0583*
Serum Increase Physiological Associated with hepato-canalicular cholestatic jaundice. *0248*

Chromosomes *Test Conditions Abnormal Physiological* Clastogenic in human lymphocytes *in vitro*. *3282*

Coombs' Test *Serum Positive Physiological* Immunological response to drug. *1486*

Corticosteroids *Urine Decrease Analytical* Decreased absorption at 410 nm in Porter-Silber procedure. *0022*

Creatine Kinase *Serum Increase Physiological* May be due to injection only (occurs in 20%). *2403*

Creatinine *Serum No Effect Analytical* At concentration of 3 mg/L had no effect on AutoAnalyzer Jaffé method. *3393*

Diazepam *Serum No Effect Analytical* On cathode-ray Polarographic method of Berry. *0323*

Diphenylhydantoin *Serum No Effect Analytical* No effect on TLC method of Simon, Jatlow. *3323* On GLC procedure of Papadopoulos at *in vivo* concentration. *2722*
Serum Increase Physiological Reported impairment of metabolism. *2042*

Effective Renal Plasma Flow
Patient Increase Physiological Slight increase in renal blood flow. *1343*

Eosinophils *Blood Increase Physiological* Often precursor of jaundice. *2427*

Erythrocytes *Blood Decrease Physiological* Hemolytic anemia. *0725* Occasional aplastic anemia reported. *3717*

Estriol *Urine No Effect Analytical* No effect on GLC method. *1163*

Estrogens *Urine Increase Physiological* Blocks ovulation, maintains decidual reaction. *1343*

Ethchlorvynol *Serum No Effect Analytical* At 100 mg/L on GLC procedure of Evenson. *1058*
Urine No Effect Analytical No effect on method of Frings and Cohen with 20 mg/dL. *1205*

Fatty Acids, Free (FFA) *Serum Increase Physiological* 51% increase following single i.m. injection of 50 mg. *1634*

Ferric Chloride Test *Urine Positive Analytical* Purple color. *0121*

FPN Test *Urine No Effect Analytical* Negative FPN test with 0.4 mg/dL concentration in Frings TLC procedure. *1204*
Urine Increase Analytical Pink color observed in method of Frings. Purple with patients also receiving trifluoperazine. Pink FPN test with 0.5 mg/dL concentration in Frings TLC procedure. *1204*

Glucose *Serum No Effect Analytical* At concentration of 0.1 mg/L had no effect on Ektachem® method. *3393* At concentration of 3 mg/L had no effect on GOD/POD-PAP method. *3393*
Serum Increase Physiological Abnormally high with repeated doses. *1595*
Urine Increase Physiological Due to hyperglycemia. *1680*

Glucose Tolerance *Serum No Effect Physiological* Low dose effect in normal subjects. *3002*
Serum Decrease Physiological Abnormal curves in 40% patients. *2427* Acute high dose effect in normal subjects. *1044*

β-Glucuronidase *Serum Increase Physiological* Result of toxic hepatitis. *3835*

Gonadotropins *Urine Decrease Physiological* Blocks ovulation, maintains decidual reaction. *1343*

Growth Hormone *Plasma No Effect Physiological* No significant difference in patients receiving drug on continuing basis versus controls. *3860*
Plasma Decrease Physiological Probably inhibits secretion of pituitary growth hormone. *3297*

Guaiacols Spot Test *Urine Positive Analytical* False reaction with screening test of Rogers. *3031*

Haptoglobin *Serum Decrease Physiological* Hemolytic anemia. *2313*

Hematocrit *Blood Decrease Physiological* Hemolytic anemia. *0725*

Hemoglobin *Blood Decrease Physiological* Hemolytic anemia. *0725*

17-Hydroxycorticosteroids *Urine Increase Analytical* Interferes with Porter-Silber reaction. *0427*
Urine Decrease Physiological Inhibition of hypothalamus and decreased ACTH secretion. *1252*

5-Hydroxyindoleacetic Acid (5-HIAA)
Urine Decrease Analytical Interferes with method of Goldenberg. *1328* Inhibits color development. *0913*

Immunoglobulin IgM *Serum Increase Physiological* Significant correlation with dose and duration of treatment in schizophrenic patients. *3937*

Isocitrate Dehydrogenase *Serum Increase Physiological* May be hypersensitive reaction. *3835*

^{131}I Uptake *Serum Increase Physiological* With procyclidine decreased renal clearance (45 to 27 mL/min). *0583*

17-Ketogenic Steroids *Urine Increase Analytical* Interferes with Zimmermann reaction. *3505*

17-Ketosteroids *Urine Increase Analytical* Interferes with Zimmermann reaction. *3505*
Urine Increase Physiological Alters steroid metabolism. *0427*
Urine Decrease Physiological Inhibition of hypothalamus and decreased ACTH secretion. *1252*

Lactate Dehydrogenase *Serum Increase Physiological* In one patient who developed cholestasis with fever and leukopenia after 260 mg drug. *2840*

LE Cells *Blood Positive Physiological* Single case reported associated with rash and fever. *0963*

Leukocytes *Blood Decrease Physiological* Agranulocytosis/leukopenia/granulocytopenia. *2238* Occasional aplastic anemia reported. *3717* Isolated case of agranulocytosis reported in elderly woman (general incidence of 1 in 1300 in psychiatric population dose related). *2299* In one patient who developed cholestasis with fever and leukopenia after 260 mg drug. *2840*

Lipids, Total *Serum Increase Physiological* Xanthomatous biliary cirrhosis may occur. *1978*

Lymphocyte T-Cells *Blood Decrease Physiological* In 13 of 41 patients with schizophrenia with long term treatment. *3937*

Malondialdehyde *Serum No Effect Physiological* Following treatment in schizophrenic patients significantly lower values. *0351*

Metanephrines, Total *Urine No Effect Analytical* At 15 mg/L on modified Pisano procedure. *1428*
Urine Increase Analytical Interference in Pisano procedure. *0388*

Methaqualone *Serum No Effect Analytical* At 100 mg/L on GLC procedure of Evenson. *1057*

Metyrapone Test *Patient No Effect Physiological* May interfere with response to test. *0913*

Monocytes *Blood Increase Physiological* Occasionally before agranulocytosis. *2238*

Neutrophils *Blood Decrease Physiological* Reported to cause agranulocytosis/neutropenia. *2583* Occasional case of agranulocytosis reported. *3717* Isolated case of agranulocytosis reported in elderly woman (general incidence of 1 in 1300 in psychiatric population dose related). *2299*

Norepinephrine *Plasma Increase Physiological* Increases metabolism, decreases organ uptake. *1343*

Partial Thromboplastin Time
Plasma Increase Physiological Caused by circulating inhibitor resembling that seen in systemic lupus in schizophrenic patients. *3937*

Pheneturide *Serum No Effect Analytical* On GLC procedure of Papadopoulos at *in vivo* concentration. *2722*

Phenobarbital *Serum No Effect Analytical* On GLC procedure of Papadopoulos at *in vivo* concentration. *2722*

Phenylketones *Urine Positive Analytical* Light purple with FeCl₃, same with Phenistix®. *0775*

Phosphate *Serum No Effect Analytical* At concentration of 3 mg/L had no effect on phosphomolybdate method. *3393*

Phospholipids, Total *Serum Increase Physiological* May cause xanthomatous biliary cirrhosis. *1434*

Plasminogen *Plasma No Effect Analytical* At concentration of 5 mg/L had no effect on aca method. *3393*

Platelets *Blood Decrease Physiological* Associated with purpura and pancytopenia. *0071* Occasional aplastic anemia reported. *3717* Significant reduction, although still in normal range in 17 psychiatric patients treated for more than 1 y. *1642*

Porphobilinogen *Urine Increase Analytical* Reacts with Ehrlich's aldehyde reagent. *3879*

Potassium *Serum No Effect Analytical* At concentration of 3 mg/L had no effect on flame-photometric method. *3393* At concentration of 3 mg/L had no effect on measurement by ISE with predilution. *3393*

Pregnancy Tests *Urine Positive Analytical* Gives false positive with frog, rabbit and immunological test. *3505*

Primidone *Serum No Effect Analytical* On GLC procedure of Papadopoulos at *in vivo* concentration. *2722*

Progestins *Urine Decrease Physiological* Blocks ovulation, maintains decidual reaction. *1343*

Prolactin *Plasma Increase Physiological* Marked increase in normals in 2 h. *3631* Significant increase within 5 minutes of ingestion of 50 mg orally and 3 to 27 times baseline at 2 h. Functions as potent dopamine antagonist in the tuberoinfundibular system. Effect dose related. *3108* Marked increase in male and female psychiatric patients treated for up to 4 weeks. *3630* Normal response to intravenous TRH. *0588*

Protein *Serum No Effect Analytical* At concentration of 3 mg/L had no effect on biuret method with blank correction. *3393*
Urine Increase Analytical Affects turbidity tests for up to 3 d. *0065* In 2 patients receiving therapeutic doses on Ponceau S dye method in comparison with sulfosalicylic acid and trichloracetic acid method. *3919*
CSF Increase Analytical Reacts as if phenol with Folin-Ciocalteu reagent. *3505*

Protein Bound Iodine (PBI) *Serum No Effect Physiological* No effect observed with normal doses. *0830*
Serum Decrease Physiological Mechanism obscure (observed if over 600 mg given). *0012*

Prothrombin Time *Plasma Increase Physiological* Associated with failure of excretion of bile salts. *1713*

Salicylate *Serum Increase Analytical* At 100 mg/L had significant positive interference on colorimetric methods of Heller (and modified version) and Trinder. *1776*
Urine No Effect Analytical At 1 mg/dL no effect on Trinder procedure. *1204*
Urine Increase Analytical At 2 mg/dL produces pink color on Trinder procedure. *1204*

Sodium *Serum No Effect Analytical* At concentration of 3 mg/L had no effect on flame-photometric method. *3393* At concentration of 3 mg/L had no effect on measurement by ISE with predilution. *3393*

Thyroid Stimulating Hormone (TSH)
Serum Increase Physiological In hypothyroidism although no change in euthyroidism, increases TRH-induced TSH response. *3798*

Thyroxine (T₄) *Serum Decrease Physiological* Increased metabolism by hepatic microsomes. *3505*

Triglycerides *Serum No Effect Analytical* At concentration of 3 mg/L had no effect on lipase/esterase method. *3393*

Urea Nitrogen *Serum No Effect Analytical* At concentration of 3 mg/L had no effect on diacetylmonoxime method. *3393* At concentration of 0.1 mg/L had no effect on Ektachem® method. *3393*
Test Conditions Increase Analytical Produces turbidity with Berthelot's reagent. *1936*

Uric Acid *Serum No Effect Analytical* At concentration of 3 mg/L had no effect on phosphotungstate reduction method. *3393*
Serum Decrease Physiological Uricosuric action within 2-3 d. *3505*
Urine Increase Physiological Uricosuric action reported. *2220*

Urobilinogen *Urine Increase Analytical* Reacts with Erhlich's aldehyde reagent. *3879*
Urine Decrease Physiological May be cholestasis. *1713*
Feces Decrease Physiological Pale stools, due to cholestasis. *1713*

Vanillylmandelic Acid *Urine No Effect Analytical* No effect on Gitlow method. *1487*
Urine Increase Physiological Increased metabolism and decreased organ uptake of norepinephrine. *3879*

Vitamin B₁₂ *Serum Decrease Analytical* Possible inhibition effect on some strains *E. gracilis*. *2427*

Volume *Urine Increase Physiological* ?depresses ADH secretion or water reabsorption. *1343*

Chlorpropamide

Alanine Aminotransferase *Serum Increase Physiological* May cause cytotoxic liver damage. *0248*

Albumin *Serum No Effect Analytical* At concentration of 150 mg/L had no effect on BCG method. *3393*
Serum Decrease Physiological In isolated case with diabetes who developed proliferative glomerulonephritis indicating immunologically mediated reaction. *0146*

Alkaline Phosphatase *Serum Increase Physiological* Infrequent intrahepatic cholestasis. *1713* Cholestatic jaundice observed in one patient in association with red cell aplasia (prevalence of jaundice as high as 0.5%). *1284*

Aminolevulinic Acid *Urine Increase Physiological* May precipitate cutaneous porphyria. *1322*

Antidiuretic Hormone *Urine Increase Physiological* Stimulates release from neurohypophysis. *1239*

Aspartate Aminotransferase
Serum Increase Physiological May cause cytotoxic liver damage. *0248* Cholestatic jaundice observed in one patient in association with red cell aplasia (prevalence of jaundice as high as 0.5%). *1284* In isolated case of hypersensitivity reaction producing hepatic granulomas also in bone marrow. *2988*

Bicarbonate *Serum No Effect Analytical* At concentration of 150 mg/L had no effect on method using phenolphthalein. *3393*

Bile *Urine Increase Physiological* Cholestatic jaundice. *0085*

Bilirubin *Serum No Effect Analytical* At concentration of 150 mg/L had no effect on Jendrassik and Grof method. *3393*
Serum Increase Physiological Cholestatic jaundice of allergic nature may occur. *1488* Isolated case of drug-induced hemolytic anemia. *3111* Cholestatic jaundice observed in one patient in association with red cell aplasia (prevalence of jaundice as high as 0.5%). *1284*

BSP Retention *Serum Increase Physiological* Infrequent cholestatic effect. *1713*

Calcium *Serum Increase Analytical* At concentrations above 7.5 mg/L (normal therapeutic concentration 150 mg/L) raised concentration as measured by cresolphthalein method. *3393*

Cephalin Flocculation *Serum Increase Physiological* Hepatocellular damage, affects protein synthesis. *2313*

Chloride *Serum No Effect Analytical* At concentration of 150 mg/L had no effect on mercurimetric method. *3393*
Serum Decrease Physiological Drug-induced syndrome of inappropriate ADH secretion. *2601*

Cholesterol *Serum No Effect Analytical* At concentration of 150 mg/L had no effect on CHOD-PAP method. *3393*

Chlorpropamide *(continued)*

Cholesterol *(continued)*
Serum Increase Physiological Infrequent cholestatic effect. *1713*
Serum Decrease Physiological May inhibit hepatic synthesis (?also absorption). *2220* With 8 weeks treatment of 8 C-peptide negative insulin dependent diabetics. *3260*

Cholesterol, High Density Lipoprotein
Serum Decrease Physiological With 8 weeks treatment of 8 C-peptide negative insulin dependent diabetics. *3260*

Cholesterol, Low Density Lipoprotein
Serum Decrease Physiological With 8 weeks treatment of 8 C-peptide negative insulin dependent diabetics. *3260*

Complement C1q *Serum Increase Physiological* In isolated case with diabetes who developed proliferative glomerulonephritis indicating immunologically mediated reaction. *0146*

Complement CH₅₀ *Serum Increase Physiological* In isolated case with diabetes who developed proliferative glomerulonephritis indicating immunologically mediated reaction. *0146*

Coombs' Test *Serum Positive Physiological* Immunological response to drug. *1486*

Coombs' Test, Direct *Serum Increase Physiological* Isolated case of drug-induced hemolytic anemia. *3111*
Serum Positive Physiological Single case with hemolytic anemia. *1487*

Coproporphyrin *Blood Increase Physiological* May precipitate cutaneous porphyria. *1322*
Feces Increase Physiological May precipitate cutaneous porphyria. *1322*

Creatinine *Serum No Effect Analytical* At 5 times therapeutic concentration on Technicon® SMAC®, Beckman Astra and Du Pont aca methods. *2999* At concentration of 150 mg/L had no effect on AutoAnalyzer Jaffé method. *3393*
Serum Increase Physiological In isolated case with diabetes who developed proliferative glomerulonephritis indicating immunologically mediated reaction. *0146*

Creatinine Clearance *Urine Decrease Physiological* In isolated case with diabetes who developed proliferative glomerulonephritis indicating immunologically mediated reaction. *0146*

Eosinophils *Blood Increase Physiological* Allergic response. *1513* In isolated case with diabetes who developed proliferative glomerulonephritis indicating immunologically mediated reaction. *0146*

Erythrocytes *Blood Decrease Physiological* Mild anemia, rare aplastic anemia. *2801* Isolated case of pure red blood cell aplasia. *2828* Occasional aplastic anemia reported. *3717*

Glucagon *Plasma No Effect Physiological* No effect i.v. if given rapidly or slowly. *2766*

Glucose *Serum No Effect Analytical* No effect on Trinder glucose-oxidase procedure. *2771* No effect glucose oxidase (GOD-PERID) method of Boehringer. *3277* On Warner Glucomatic glucose-oxidase method. *2771* At concentration of 1,000 mg/L had no effect on GOD/POD-PAP method. *3393*
Serum Decrease Analytical Negative peaks with Boehringer GOD-PERID method. *2771* At concentrations above 500 mg/L (normal therapeutic concentration 150 mg/L) lowered concentration as measured by GOD-Perid method. *3393*
Serum Decrease Physiological Sulfonylurea derivative promotes insulin secretion. *2801*

γ-Glutamyltransferase (GGT)
Serum Increase Physiological Cholestatic jaundice observed in one patient in association with red cell aplasia (prevalence of jaundice as high as 0.5%). *1284*

Guanase *Serum Increase Physiological* May cause cytotoxic liver damage. *0248*

HDL₃-Cholesterol *Serum Decrease Physiological* With 8 weeks treatment of 8 C-Peptide negative insulin dependent diabetics. *3260*

Hematocrit *Blood Decrease Physiological* Mild anemia, rare aplastic anemia. *2801* Isolated case of drug-induced hemolytic anemia. *3111*

Hemoglobin *Blood Decrease Physiological* Mild anemia, rare aplastic anemia. *2801* Isolated case of drug-induced hemolytic anemia. *3111*
Urine Increase Physiological Isolated case of drug-induced hemolytic anemia. *3111*

Insulin *Plasma Increase Physiological* Effect observed during tolerance test. *2236* Observed in most patients (especially if low initially). *0301*
Plasma Decrease Physiological Observed when initial level high. *0301*

Lactate Dehydrogenase *Serum Increase Physiological* Isolated case of drug-induced hemolytic anemia. *3111* In isolated case of hypersensitivity reaction producing hepatic granulomas also in bone marrow. *2988*

Leukocytes *Blood Increase Physiological* Due to eosinophilia of hypersensitivity. *2313*
Blood Decrease Physiological Agranulocytosis. *3811* Isolated case of agranulocytosis observed. *3627* Occasional aplastic anemia reported. *3717*

Lymphocytes *Blood Increase Physiological* Mild, of no clinical significance. *1680*

Neutrophils *Blood Decrease Physiological* Isolated case of agranulocytosis observed. *3627* Occasional aplastic anemia reported. *3717*

5'-Nucleotidase *Serum Increase Physiological* Due to cholestasis. *1778*

Occult Blood *Feces Positive Physiological* Associated with severe diarrhea and bleeding. *2801*

Ornithine Carbamoyltransferase (OCT)
Serum Increase Physiological May cause cytotoxic liver damage. *0248*

Osmolality *Urine Increase Physiological* Normal diuretic response also if diabetes insipidus. *2499*

Phosphate *Serum No Effect Analytical* At concentration of 150 mg/L had no effect on phosphomolybdate method. *3393*

Platelets *Blood Decrease Physiological* Thrombocytopenia or aplastic anemia may occur. *0073* Occasional aplastic anemia reported. *3717*

Porphobilinogen *Urine Increase Physiological* May aggravate cutaneous porphyria. *1322*

Porphyrins *Urine Increase Physiological* May precipitate attack of acute porphyria. *1016*

Potassium *Serum No Effect Analytical* At concentration of 150 mg/L had no effect on flame-photometric method. *3393*

Protein *Serum No Effect Analytical* At concentration of 150 mg/L had no effect on biuret method with blank correction. *3393*
Urine Increase Physiological Nephrotoxic effect. *3729* In isolated case with diabetes who developed proliferative glomerulonephritis indicating immunologically mediated reaction. *0146* Isolated case of drug-induced hemolytic anemia. *3111*

Protein Bound Iodine (PBI) *Serum Decrease Physiological* Questionable effect (may be slight in diabetics). *0012*

Prothrombin Time *Plasma Increase Physiological* Associated with failure of excretion of bile salts. *1713*

Protoporphyrin *Blood Increase Physiological* May precipitate cutaneous porphyria. *1322*
Urine Increase Physiological May precipitate cutaneous porphyria. *1322*
Feces Increase Physiological May precipitate cutaneous porphyria. *1322*

Reticulocytes *Blood Decrease Physiological* Isolated case of pure red blood cell aplasia. *2828*

Sodium *Serum No Effect Analytical* At concentration of 150 mg/L had no effect on flame-photometric method. *3393*
Serum Decrease Physiological Induces inappropriate ADH secretion. *3791* Drug-induced syndrome of inappropriate ADH secretion. *2601*

T₃ Uptake *Serum Increase Physiological* Competes for thyroxine binding globulin sites. *0583*

Thymol Turbidity *Serum Increase Physiological* May cause cytotoxic liver damage. *2313*

Thyroxine (T₄) *Serum Decrease Physiological* Displaces thyroxine from thyroxine binding globulin. *3505*

Thyroxine (T₄) (Murphy-Pattee)
Serum Decrease Physiological May be competition for thyroxine binding globulin. *0583*

Triglycerides *Serum No Effect Analytical* At concentration of 150 mg/L had no effect on lipase/esterase method. *3393*

Urea Nitrogen *Serum No Effect Analytical* At concentration of 150 mg/L had no effect on diacetylmonoxime method. *3393*

Uric Acid *Serum No Effect Analytical* At concentration of 150 mg/L had no effect on phosphotungstate reduction method. *3393*

Urobilinogen *Urine Decrease Physiological* Cholestasis may occur infrequently. *1713*

Feces Decrease Physiological Pale stools due to cholestasis (infrequent). *1713*

Uroporphyrin *Urine Increase Physiological* May precipitate cutaneous porphyria. *1322*

Volume *Urine Decrease Physiological* In patients with diabetes insipidus. *3214*

Water Clearance, Free *Urine Decrease Physiological* Normal antidiuretic response in normals. *2499*

Chlorprothixene

Alanine Aminotransferase *Serum Increase Physiological* May cause hepatotoxicity (reversible). *2313*

Albumin *Serum No Effect Analytical* At concentration of 1 mg/L had no effect on BCG method. *3393*

Alkaline Phosphatase *Serum Increase Physiological* Hepatotoxicity (reversible) cholestatic effect. *2313*

Aspartate Aminotransferase
Serum Increase Physiological May cause hepatotoxicity (reversible). *2313*

Bicarbonate *Serum No Effect Analytical* At concentration of 1 mg/L had no effect on method using phenolphthalein. *3393*

Bile *Urine Increase Physiological* May cause hepatotoxicity (reversible). *1237*

Bilirubin *Serum No Effect Analytical* At concentration of 1 mg/L had no effect on Jendrassik and Grof method. *3393*
Serum Increase Physiological Hepatotoxicity (reversible) cholestatic effect. *2313*
Urine Increase Physiological May cause hepatotoxicity (reversible). *2313*

BSP Retention *Serum Increase Physiological* May cause hepatotoxicity (reversible). *2313*

Calcium *Serum No Effect Analytical* At concentration of 1 mg/L had no effect on cresolphthalein method. *3393*

Cephalin Flocculation *Serum Increase Physiological* May cause hepatotoxicity (reversible). *2313*

Chloride *Serum No Effect Analytical* At concentration of 1 mg/L had no effect on mercurimetric method. *3393*

Cholesterol *Serum No Effect Analytical* At concentration of 1 mg/L had no effect on Liebermann-Burchard method. *3393*

Creatinine *Serum No Effect Physiological* Constant concentration regardless of dose of drug co-administered and duration. *3268*

Eosinophils *Blood Increase Physiological* Allergic manifestation. *1680*

Glucose *Serum Increase Physiological* Due to altered endocrine function. *1680*
Urine Increase Physiological Due to altered endocrine function. *1680*

LE Cells *Blood Positive Physiological* May produce LE-like syndrome. *3504*

Leukocytes *Blood Decrease Physiological* Possible hematological disorder. *1343*

Phosphate *Serum No Effect Analytical* At concentration of 1 mg/L had no effect on phosphomolybdate method. *3393*

Platelets *Blood Decrease Physiological* Purpura or pancytopenia may occur. *1680*

Pregnancy Tests *Urine Positive Physiological* Endocrine abnormality. *1680*

Protein *Serum No Effect Analytical* At concentration of 1 mg/L had no effect on biuret method with blank correction. *3393*

Thymol Turbidity *Serum Increase Physiological* May cause hepatotoxicity (reversible). *2313*

Triglycerides *Serum No Effect Analytical* At concentration of 1 mg/L had no effect on lipase/esterase method. *3393*

Urea Nitrogen *Serum No Effect Analytical* At concentration of 1 mg/L had no effect on diacetylmonoxime method. *3393*

Serum No Effect Physiological Constant concentration regardless of dose of drug co-administered and duration. *3268*

Uric Acid *Serum No Effect Analytical* At concentration of 1 mg/L had no effect on phosphotungstate reduction method. *3393*

Serum Decrease Physiological Uricosuric effect within 2-3 d of start of drug. *3505* Reduction of from 8 to 69% in 30 psychiatric patients. *3268*

Urine Increase Physiological Uricosuric action. *1537* Significant increase during treatment. *3268*

Chlortetracycline

Alanine Aminotransferase *Serum Increase Physiological* Hepatotoxic effect with centrolobular necrosis. *1237*

Alkaline Phosphatase *Serum Increase Physiological* Hepatotoxic effect with centrolobular necrosis. *1237*

Aspartate Aminotransferase
Serum Increase Physiological Hepatotoxic effect with centrolobular necrosis. *1237*

Bile *Urine Increase Physiological* Hepatotoxic effect with centrolobular necrosis. *1237*

Bilirubin *Serum No Effect Physiological* Critically insignificant displacement from protein in neonates. *3748*
Serum Increase Physiological Hepatotoxic effect with centrolobular necrosis. *1237*

BSP Retention *Serum Increase Physiological* Hepatotoxic effect with centrolobular necrosis. *2425*

Catecholamines *Urine Increase Analytical* Interference with fluorometric methods. *2570*

Cephalin Flocculation *Serum Increase Physiological* Hepatotoxic effect with centrolobular necrosis. *1237*

Cholesterol *Serum Decrease Physiological* Hepatotoxic effect with centrolobular necrosis. *1237* Not as effective as neomycin reacts with bile acids. *1487*

Eosinophils *Blood Increase Physiological* Probably allergic response. *1680*

Erythrocytes *Blood Decrease Physiological* Pancytopenia (AMA Blood dyscrasias). *2429*

Estrogens *Urine Increase Analytical* Possible effect (interference not defined). *0022*

Folate *Urine Increase Physiological* With dose of 3 g/d. *1343*

Hemoglobin *Blood Decrease Physiological* Hemolytic anemia may occur. *1680*

Leukocytes *Blood Decrease Physiological* Pancytopenia (AMA Blood dyscrasias). *2429*

N-Methylnicotinamide *Urine Increase Physiological* With dose of 3 g/d. *1343*

Nitrogen *Urine Increase Physiological* Effect observed in malnourished. *1343*

Nonprotein Nitrogen *Serum Increase Physiological* Effect observed in malnourished. *1343*

Occult Blood *Feces Positive Physiological* Occasional case of pseudomembranous colitis reported with mucus and bloody diarrhea. *0447*

Platelets *Blood Decrease Physiological* Pancytopenia (AMA Blood dyscrasias). *2429*

Prothrombin Time *Plasma Increase Physiological* Reported effect due to action in gastrointestinal tract. *0019*

Riboflavin *Urine Increase Physiological* With dose of 3 g/d. *1343*

Sugar *Urine Increase Analytical* Acts as reducing substance. *1238*

Thymol Turbidity *Serum Increase Physiological* Hepatotoxic effect with centrolobular necrosis. *1237*

Urea Nitrogen *Serum Increase Physiological* Dose related antianabolic action. *1680*

Chlorthalidone

Alanine Aminotransferase *Serum Increase Physiological* Single case of severe reversible myopathy noted. *1793*

Chlorthalidone (continued)

Aldosterone *Plasma Increase Physiological* At least 50% increase in pre- and post-menopausal women given 100 mg/d for 6 weeks. *0402* At 3 d but not after 3 mo in 8 patients with mild hypertension given 50 mg/d. *3040*

Urine Increase Physiological Single case of severe reversible myopathy noted. *1793*

Ammonia *Plasma Increase Physiological* Partially due to decreased potassium and alkalosis. *0109*

Amylase *Serum Increase Physiological* May precipitate acute pancreatitis. *1819*

Antidiuretic Hormone *Plasma Increase Physiological* Secreted in response to hyponatremia. *1129*

Apolipoprotein AI *Serum No Effect Physiological* During diuretic therapy for 6 weeks in 23 subjects. *1307* In 22 premenopausal and 18 postmenopausal women given 100 mg/d for 6 weeks. *0402* Decrease by -1 mg/dL in 10 patients given 100 mg daily for 1 mo. *2550* No effect of monotherapy for 6 weeks. *0089*

Apolipoprotein AII *Serum No Effect Physiological* During diuretic therapy for 6 weeks in 23 subjects. *1307* In 22 premenopausal and 18 postmenopausal women given 100 mg/d for 6 weeks. *0402* No effect of monotherapy for 6 weeks. *0089*

Apolipoprotein B *Serum No Effect Physiological* During diuretic therapy for 6 weeks in 23 subjects. *1307* No effect of monotherapy for 6 weeks. *0089* In 22 premenopausal women given 100 mg/d for 6 weeks. *0402*

Serum Increase Physiological Increase by 16% in 18 postmenopausal women given 100 mg/d for 6 weeks. *0402* Increase by 11 mg/dL in 10 patients given 100 mg daily for 1 mo. *2550*

Apolipoprotein CII *Serum No Effect Physiological* Increase by 1 mg/dL in 10 patients given 100 mg daily for 1 mo. *2550*

Apolipoprotein CIII *Serum No Effect Physiological* Increase by 2 mg/dL in 10 patients given 100 mg daily for 1 mo. *2550*

Aspartate Aminotransferase

Serum Increase Physiological Single case of severe reversible myopathy noted. *1793*

Bicarbonate *Serum Increase Physiological* Metabolic alkalosis in severe cases. *1129* From 2 to 3 mmol/L at doses from 25 to 200 mg/d in 37 patients over 8 weeks. *3643*

Calcium *Serum Increase Physiological* Occurs in normal and hyperparathyroid. *2122*

Urine Decrease Physiological Decreased in normals and hyperparathyroid. *2122*

Casts *Urine No Effect Physiological* Normally but augments effects of acidifying agents. *1720*

Chloride *Serum Decrease Physiological* Single case of severe reversible myopathy noted. *1793* Progressive effect with dose: from 3 mmol/L with 25 mg/d to 6 mmol/L at 200 mg/d in 37 patients treated for 8 weeks. *3643*

Urine Increase Physiological May produce marked diuresis. *2427*

Cholesterol *Serum No Effect Physiological* In 22 premenopausal women without decrease in blood pressure treated for less than 1 y. *0402* In 22 premenopausal women given 100 mg/d for 6 weeks. *0402*

Serum Increase Physiological 5% change during monotherapy in 302 subjects for 1 y. *1334* 9% change in 39 subjects treated with drug only for less than 1 y. *1404* Increase by 13% in 18 postmenopausal women given 100 mg/d for 6 weeks. *0402* Increase by 10 mg/dL in 1,000 men and women with mild hypertension treated for 1 y. *1334* Increase by 26 mg/dL in 10 patients given 100 mg daily for 1 mo. *2550* 5% in long term, 9% in short term in several studies with monotherapy. *0089*

Serum Decrease Physiological Statistically significant decrease of 3 mg/dL in 12 men receiving drug in conjunction with diet to reduce cholesterol. *0090*

Cholesterol, High Density Lipoprotein

Serum No Effect Physiological Nonsignificant change during monotherapy in 302 subjects for 1 y. *1334* Nonsignificant change in 39 subjects treated with drug only for less than 1 y. *1404* In 22 premenopausal women without decrease in blood pressure treated for less than 1 y. *0402* In 22 premenopausal and 18 postmenopausal women given 100 mg/d for 6 weeks. *0402* In 1,000 men and women with mild hypertension treated for 1 y. *1334* Increase by 1 mg/dL in 10 patients given 100 mg daily for 1 mo. *2550* In long or short term in several studies with monotherapy. *0089*

Cholesterol, Low Density Lipoprotein

Serum No Effect Physiological In 22 premenopausal women without decrease in blood pressure treated for less than 1 y. *0402* In 22 premenopausal and 18 postmenopausal women given 100 mg/d for 6 weeks. *0402*

Serum Increase Physiological 10% change during monotherapy in 302 subjects for 1 y. *1334* 10% change in 39 subjects treated with drug only for less than 1 y. *1404* Increase by 21% in 18 postmenopausal women given 100 mg/d for 6 weeks. *0402* Increase by 13 mg/dL in 1,000 men and women with mild hypertension treated for 1 y. *1334* 10% in long term, 10% in short term in several studies with monotherapy. *0089*

Cholesterol, Very Low Density Lipoprotein

Serum No Effect Physiological In 22 premenopausal women without decrease in blood pressure treated for less than 1 y. *0402* In 22 premenopausal and 18 postmenopausal women given 100 mg/d for 6 weeks. *0402*

Serum Increase Physiological 7% change in 39 subjects treated with drug only for less than 1 y. *1404* 7% in short term in several studies with monotherapy. *0089*

Creatine Kinase *Serum Increase Physiological* Single case of severe reversible myopathy noted. *1793*

Creatinine *Serum Increase Physiological* Nephrotoxic effect. *2220* Effect observed at dose of 100 mg/d and upwards in 37 patients treated for 8 weeks. *3643* Increased in hypertensives; most marked in those individuals whose initial values were lower. *2904*

Creatinine Clearance *Urine Decrease Physiological* Significantly reduced in 10 hypertensive patients given 25 mg daily for 16 weeks. *1229*

Epinephrine *Plasma No Effect Physiological* In 22 premenopausal and 18 postmenopausal women given 100 mg/d for 6 weeks. *0402*

Erythrocytes *Blood Decrease Physiological* Aplastic anemia may occur. *1680*

Estradiol *Plasma No Effect Physiological* In 22 premenopausal and 18 postmenopausal women given 100 mg/d for 6 weeks. *0402*

Fatty Acids, Free (FFA) *Serum No Effect Physiological* In 22 premenopausal and 18 postmenopausal women given 100 mg/d for 6 weeks. *0402*

Glucose *Serum No Effect Physiological* In 22 premenopausal and 18 postmenopausal women given 100 mg/d for 6 weeks. *0402*

Serum Increase Physiological Diabetogenic-like action affects pre- or diabetic. *0102* Progressive effect with dose: by 5 mg/dL with 25 mg/d to 14 mg/dL with 200 mg/d in 37 patients treated for 8 weeks. *3643* Increase by 15 mg/dL in 39 mildly hypertensives treated for 6 weeks. *1404*

Urine Increase Physiological Diabetogenic-like action of drug affects glucose tolerance test. *3505* Diabetogenic-like action of drug affects glucose tolerance test. *3504*

Glucose Tolerance *Serum Decrease Physiological* Diabetogenic-like action of drug. *2745* Significant reduction in 10 hypertensive patients given 25 mg daily for 16 weeks. *1229*

Hematocrit *Blood Increase Physiological* Increase by 1.4% in 39 mildly hypertensives treated for 6 weeks. *1404*

17-Hydroxycorticosteroids *Urine Increase Physiological* Reported effect. *2220*

Insulin *Plasma No Effect Physiological* In 22 premenopausal and 18 postmenopausal women given 100 mg/d for 6 weeks. *0402*

17-Ketosteroids *Urine Increase Physiological* Alleged increased excretion. *1488*

Lactate Dehydrogenase *Serum Increase Physiological* Single case of severe reversible myopathy noted. *1793*

Leukocytes *Blood Decrease Physiological* Agranulocytosis. *1953*

Magnesium *Serum Decrease Physiological* Significant effect with prolonged treatment. *3285*

Myoglobin *Urine Increase Physiological* Single case of severe reversible myopathy noted. *1793*

Neutrophils *Blood Decrease Physiological* Few cases reported. *2427* Occasional case of agranulocytosis reported. *3717*

Norepinephrine *Plasma No Effect Physiological* In 22 premenopausal and 18 postmenopausal women given 100 mg/d for 6 weeks. *0402*
Urine Increase Physiological After 3 d and 3 mo in 8 patients with mild hypertension given 50 mg/d. *3040*

Osmolality *Serum Decrease Physiological* With ADH secretion in response to diuresis. *1129*

Parathyroid Hormone *Plasma No Effect Physiological* But probable enhancement of peripheral action. *2122*

Phospholipids, Total *Serum No Effect Physiological* In 22 premenopausal and 18 postmenopausal women given 100 mg/d for 6 weeks. *0402*

Platelets *Blood Decrease Physiological* Associated with rare agranulocytosis. *2427*

Potassium *Serum Decrease Physiological* Diuretic induced depletion. *1237* 43% patients with hypokalemia given long term 25-100 mg/d. *0314* Significant effect in pre- and postmenopausal women given 100 mg/d for 6 weeks. *0402* Progressive effect with dose: from 0.4 mmol/L with 25 mg/d to 1.0 mmol/L at 200 mg/d in 37 patients treated for 8 weeks. *3643* 43% patients with hypokalemia given long term 25-100 mg/d. *0314* Significant fall from 8 weeks in 10 hypertensive patients given 25 mg daily for for 16 weeks. *1229* 0.5 mmol/L reduction in drug-treated versus placebo-treated patients. *3367* Hyperaldosteronism major causal factor in diuretic induced hypokalemia. *1438* In majority of children during course of treatment. *0197*
Urine Increase Physiological May produce marked diuresis. *2427*
Red Blood Cells Decrease Physiological Not significant but tendency to fall in 10 hypertensive patients given 25 mg for 16 weeks. *3341*

Protein *Urine Increase Physiological* Nephrotoxic effect. *1237*

Prothrombin Time *Plasma Increase Physiological* Patients on coumarins. *2426*

Renin Activity *Plasma No Effect Physiological* Insignificant change in 10 hypertensives patients given 25 mg daily for 16 weeks. *1229*
Plasma Increase Physiological At least 3 fold increase in pre- and post-menopausal women given 100 mg/d for 6 weeks. *0402* At 3 d and after 3 mo in 8 patients with mild hypertension given 50 mg/d. *3040*

Sodium *Serum No Effect Physiological* No significant change in 10 hypertensive patients given 25 mg daily for 16 weeks. *1229*
Serum Increase Physiological Single case of severe reversible myopathy noted. *1793*
Serum Decrease Physiological Severe hyponatremia with potassium depletion. *1129*
Urine No Effect Physiological After 3 d and 3 mo in 8 patients with mild hypertension given 50 mg/d. *3040*
Urine Increase Physiological May produce marked diuresis. *2427*

Testosterone *Serum No Effect Physiological* No significant effect observed although higher incidence of sexual dysfunction than in control population. *1259* Although higher incidence of sexual dysfunction. *1259*

Triglycerides *Serum No Effect Physiological* Nonsignificant change during monotherapy in 302 subjects for 1 y. *1334* In 22 premenopausal women without decrease in blood pressure treated for less than 1 y. *0402* In 22 premenopausal and 18 postmenopausal women given 100 mg/d for 6 weeks. *0402* Increase by 6 mg/dL in 10 patients given 100 mg daily for 1 mo. *2550*
Serum Increase Physiological 15% change in 39 subjects treated with drug only for less than 1 y. *1404* Increase by 10 mg/dL in 1,000 men and women with mild hypertension treated for 1 y. *1334* But insignificant change in 10 hypertensive

patients given 25 mg daily for 16 week. *1229* 15% in short term, no effect in long term in several studies with monotherapy. *0089*
Serum Decrease Physiological Statistically significant decrease of 10 mg/dL in 12 men receiving drug in conjunction with diet to reduce cholesterol. *0090*

Triglycerides, High Density Lipoprotein
Serum No Effect Physiological In 22 premenopausal and 18 postmenopausal women given 100 mg/d for 6 weeks. *0402*

Triglycerides, Low Density Lipoprotein
Serum No Effect Physiological In 22 premenopausal and 18 postmenopausal women given 100 mg/d for 6 weeks. *0402*

Triglycerides, Very Low Density Lipoprotein
Serum No Effect Physiological In 22 premenopausal women, without decrease in blood pressure treated for less than 1 y. *0402* In 22 premenopausal and 18 postmenopausal women given 100 mg/d for 6 weeks. *0402*

Urea Nitrogen *Serum Increase Physiological* Nephrotoxic effect. *0070*

Uric Acid *Serum Increase Physiological* Decreases urate clearance. *0514* Significantly different in post-menopausal women but still increased in premenopausal given drug 100 mg/d for 6 weeks. *0402* From 1.2 to 1.8 mg/dL at doses from 25 to 200 mg/d in 37 patients over 8 weeks. *3643* By 1.7 mg/dL in 39 mg/dL mildly hypertensives treated for 6 weeks. *1404* Slight and insignificant in 10 hypertensive patients given 25 mg daily for 16 week. *1229* 0.9 mg/dL in drug-treated versus placebo-treated patients. *3367* Increased in hypertensives; most marked in those individuals whose initial values were lower. *2904*
Urine Decrease Physiological Decreased clearance. *2220*

Vanillylmandelic Acid *Plasma No Effect Physiological* In 22 premenopausal and 18 postmenopausal women given 100 mg/d for 6 weeks. *0402*

Volume *Plasma No Effect Physiological* In 22 premenopausal and 18 postmenopausal women given 100 mg/d for 6 weeks. *0402*
Plasma Decrease Physiological At 3 d but not after 3 mo in 8 patients with mild hypertension given 50 mg/d. *3040*
Blood Decrease Physiological At 3 d but not after 3 mo in 8 patients with mild hypertension given 50 mg/d. *3040*
Urine Increase Physiological May produce marked diuresis. *2427*

Zinc *Serum Increase Physiological* In hypertensive males taking drug for at least 6 mo: probably due to release from tissues but also possibly due to contracted blood volume. *1259* Significantly higher, also increased hair concentration. *1259*
Urine Increase Physiological Increase by 65% in 9 hypertensives treated for 2 weeks. *3806*

Chlorzoxazone

Alanine Aminotransferase *Serum Increase Physiological* May cause hepatotoxicity. *2313*

Alkaline Phosphatase *Serum Increase Physiological* May cause hepatotoxicity. *2313*

Aspartate Aminotransferase
Serum Increase Physiological May cause hepatotoxicity. *2313*

Bile *Urine Increase Physiological* May cause hepatotoxicity. *1022*

Bilirubin *Serum Increase Physiological* May cause hepatotoxicity. *2313*

BSP Retention *Serum Increase Physiological* May cause hepatotoxicity. *2313*

Cephalin Flocculation *Serum Increase Physiological* May cause hepatotoxicity. *2313*

Color *Urine Increase Analytical* Orange to purple-red. *1022*

Thymol Turbidity *Serum Increase Physiological* May cause hepatotoxicity. *2313*

Chocolate

Vanillylmandelic Acid *Urine Increase Analytical* Catechols in chocolate. *1237*

Cholecystokinin

Hydrochloric Acid *Gastric Material No Effect Physiological* May occasionally cause increase. *2049*

Choleretics

BSP Retention *Serum Increase Physiological* Reported effect (?competition for excretion). *1022*

Cholestanol

Cholesterol *Serum Increase Analytical* Same retention time procedure of MacGee. *2252* Yields yellow color with Zlatkis-Zak method. *1563*

Cholesterol

Crystals *Urine Increase Physiological* Flat colorless plates with notch (acid, neutral pH). *0443*

Erythrocyte Sedimentation Rate *Blood Increase Analytical* Probably by increasing viscosity. *0459*

Fatty Acids, Free (FFA) *Serum No Effect Analytical* At 5.2 mmol/L — method of Noma. *2607*

11-Hydroxycorticosteroids *Plasma Increase Analytical* Due to nonspecific fluorescence. *1369*

Testosterone *Serum Increase Analytical* In amounts of 1-20 μg affected competitive protein binding method. *0888*

Cholesteryl Oleate

Fatty Acids, Free (FFA) *Serum Increase Analytical* At 5.2 mmol/L slight effect on method of Noma. *2607*

Cholestyramine

Alanine Aminotransferase *Serum Increase Physiological* Few cases reported, probably not hepatotoxicity. *1680*

Aspartate Aminotransferase
Serum Increase Physiological Few cases reported, probably not hepatotoxicity. *1680*

Bile Acids *Serum Decrease Physiological* Binds acids in gastrointestinal tract. *1680*
Feces Increase Physiological Due to increased binding in gastrointestinal tract. *1680*

Bilirubin, Conjugated *Serum Increase Physiological* In 19 patients suffering from intrahepatic cholestasis of pregnancy. *1545*

BSP Retention *Serum Increase Physiological* Few cases reported, probably not hepatotoxicity. *1680*

Calcium *Urine Increase Physiological* Binds heavy metals. *1022*

Cephalexin *Serum Decrease Physiological* Reduced absorption due to adsorption or steatorrhea. *3794*

Chenodeoxycholic Acid *Serum Decrease Physiological* In 19 patients suffering from intrahepatic cholestasis of pregnancy. *1545*

Chloride *Serum Increase Physiological* Is chloride salt of basic anion exchange resin. *3879* Severe hyperchloremia up to 128 mmol/L in 2 pediatric patients receiving drug for some weeks. *2872*

Cholesterol *Serum Decrease Physiological* Therapeutic goal (?increased binding of bile salts in gut). *1669*

Cholic Acid *Serum No Effect Physiological* In 19 patients suffering from intrahepatic cholestasis of pregnancy. *1545*

Digitoxin *Serum Decrease Physiological* Reduces bioavailability in gastrointestinal tract due to binding of drug. *2794* Increased elimination rate with interrupted enterohepatic circulation. *3794*

Digoxin *Serum Decrease Physiological* Reduces bioavailability in gastrointestinal tract due to binding of drug. *2794*

Fat *Feces Increase Physiological* Occurs with doses over 15 g/d. *1680* Dose-dependent: forms nonabsorbable complexes with bile salts which are lost in feces. *3023*

Hemoglobin *Blood Decrease Physiological* Impairs absorption of iron. *1965*

Iron *Serum Decrease Physiological* Impairs absorption of iron. *1965*

Leukocytes *Blood Decrease Physiological* Possible relationship to drug administration. *1680*

Lipids, Total *Serum Decrease Physiological* Lowers bile acids by ionic binding. *2313*

β-Lipoproteins *Serum Decrease Physiological* But does not affect very low density lipoproteins. *1343*

Occult Blood *Feces No Effect Analytical* No effect noted on Hemoquant procedure. *3671*

Phospholipids, Total *Serum Decrease Physiological* Therapeutic effect. *2220*

Propranolol *Serum Decrease Physiological* Significant reduction of peak plasma concentration and area under curve due to binding in gastrointestinal tract. *1584*

Prothrombin Time *Plasma Increase Physiological* Combines with bile acids (vitamin K not absorbed). *0894*
Plasma Decrease Physiological May shorten action of anticoagulants. *1679*

Sodium *Serum Increase Physiological* Severe hypernatremia (sodium 175 mmol/L) in 2 pediatric patients receiving drug for some weeks. *2872*

Sulfamethoxazole *Serum Decrease Physiological* Reduced absorption due to adsorption or steatorrhea. *3794*

Thyroxine (T₄) *Serum Decrease Physiological* Decreased intestinal absorption of thyroxine. *3505* Reduced absorption from gastrointestinal tract due to adsorption. *3794*

α-Tocopherol *Serum Decrease Physiological* ?inhibits synthesis of carrier lipoprotein. *3788*

Triglycerides *Serum Increase Physiological* Mechanism obscure (may occur in diabetics). *1822*
Serum Decrease Physiological Therapeutic effect. *2220*

Trimethoprim *Serum Decrease Physiological* Delayed absorption due to adsorption. *3794*

Vitamin A *Serum No Effect Physiological* Absorption not significantly impaired. *2427*
Serum Decrease Physiological Due to bile acid sequestration, but values still remain within normal range. *3023*

Vitamin B₁₂ *Serum Decrease Physiological* Reported to cause ileal malabsorption. *3023*

Vitamin E (Tocopherol) *Serum Decrease Physiological* Decreased but values still in normal range over 1-2 y in children with familial hypercholesterolemia. *3023*

Warfarin *Plasma Decrease Physiological* Reduced absorption from gastrointestinal tract due to adsorption. *3794*

Cholinergics

Amylase *Serum Increase Physiological* Cause spasm of sphincter of Oddi. *0127*

Aspartate Aminotransferase
Serum Increase Physiological Impaired excretion due to spasm of sphincter. *0652*

Bilirubin *Serum Increase Physiological* Impaired excretion due to spasm of sphincter. *0652*

BSP Retention *Serum Increase Physiological* Impaired excretion due to spasm of sphincter. *0652*

Lipase *Serum Increase Physiological* Impaired excretion spasm of sphincter of Oddi. *1343*

Chromates

Alanine Aminotransferase *Serum Increase Physiological* May cause liver damage. *0279*

Aspartate Aminotransferase
Serum Increase Physiological May cause liver damage. *0279*

Protein *Urine Increase Physiological* May cause renal damage. *0279*

Urea Nitrogen *Serum Increase Physiological* May cause renal damage. *0279*

Chromatography

L/S Ratio *Amniotic Fluid Increase Analytical* Higher on TLC if H$_2$SO$_4$ spray. Higher on TLC if silica gel h used. *1258*

Chromium

Nickel *Test Conditions Increase Analytical* Possible interference with atomic absorption. *3499*

Chromonar Hydrochloride

Cholesterol *Serum No Effect Analytical* At concentration of 900 mg/L had no effect on CHOD-Iodide method. *3393* At concentration of 900 mg/L had no effect on CHOD-PAP method. *3393* At concentration of 900 mg/L had no effect on catalase-AIDH method. *3393* At concentration of 900 mg/L had no effect on method using catalase-Hantzsch reaction. *3393* At concentration of 900 mg/L had no effect on Liebermann-Burchard method. *3393*

Creatinine *Serum No Effect Analytical* At concentration of 90 mg/L had no effect on Jaffé-Fading-Fraction method. *3393* At concentration of 90 mg/L had no effect on Jaffé-Fuller's earth method. *3393* At concentration of 180 mg/L had no effect on kinetic Jaffé method on BKA-2. *3393*

Glucose *Serum No Effect Analytical* At concentration of 18 mg/L had no effect on Ektachem® method. *3393* At concentration of 180 mg/L had no effect on hexokinase/G-6-PDH method. *3393*

Lactate *Serum No Effect Analytical* At concentration of 180 mg/L had no effect on enzymatic method. *3393*

Potassium *Serum No Effect Analytical* At concentration of 180 mg/L had no effect on ISE measurement without predilution. *3393*

Protein Electrophoresis *Serum No Effect Analytical* At concentration of 180 mg/L had no effect on automated Olympus-Hite method. *3393*

Sodium *Serum No Effect Analytical* At concentration of 180 mg/L had no effect on ISE measurement without predilution. *3393*

Urea Nitrogen *Serum No Effect Analytical* At concentration of 18 mg/L had no effect on Ektachem® method. *3393*

Uric Acid *Serum No Effect Analytical* At concentration of 200 mg/L had no effect on Kageyama-Hantzsch method. *3393* At concentration of 90 mg/L had no effect on catalase-AIDH method. *3393*

Chrysarobin

Color *Urine Increase Analytical* Oxidation product colors alkaline urine red. *1343*

Protein *Urine Increase Physiological* Kidney irritation by metabolite. *0071*

Chyluria

Appearance *Urine Abnormal Analytical* Milky, soluble in either. *0443*

Chymotrypsin

Prothrombin Time *Plasma Increase Physiological* May prolong action of anticoagulants. *1679*

Cibenzoline

Protein *Urine No Effect Analytical* Negative reaction in 23 of 53 patients when sulfosalicylic acid or acetic acid/sodium acetate heat coagulation used to measure protein. *2002* *Urine Increase Analytical* Positive reaction in 23 of 53 patients when bromphenol reagent strips used to measure protein. *2002*

Cimetidine

Alanine Aminotransferase *Serum Increase Physiological* In isolated case treated for 7 mo with 400 mg/d progressed to bridging hepatic necrosis. *3085*

Albumin *Serum No Effect Analytical* At concentration of 1 mg/L had no effect on BCG method. *3393* *Urine No Effect Physiological* No change observed during treatment of 13 ulcer patients. *0649*

Alkaline Phosphatase *Serum Increase Physiological* Mild increase reversible cholestatic jaundice in a few children. *2172* In isolated case treated for 7 mo with 400 mg/d progressed to bridging hepatic necrosis. *3085*

Aminolevulinic Acid *Urine Decrease Physiological* Marked effect in patient with acute intermittent porphyria possibly due to inhibition of hepatic delta-aminolevulinic acid synthetase. *1659*

Antimitochondrial Antibodies *Serum No Effect Physiological* No effect noted in 12 patients treated for acid-peptic disease. *2368*

Antinuclear Antibodies *Serum No Effect Physiological* No effect noted in 12 patients treated for acid-peptic disease. *2368*

Antipyrine *Plasma Increase Physiological* Clearance reduced: probably by inhibition of drug metabolism with increased half-life. *3256* Plasma clearance reduced from 54 mL/min to 48 mL/min in 7 healthy subjects given 1 g/d. *3004*

Antismooth Muscle Antibodies *Serum No Effect Physiological* No effect noted in 12 patients treated for acid-peptic disease. *2368*

Aspartate Aminotransferase *Serum Increase Physiological* Reversible cholestatic jaundice in a few children. *2172* In isolated case treated for 7 mo with 400 mg/d progressed to bridging hepatic necrosis. *3085*

Bicarbonate *Serum No Effect Analytical* At concentration of 1 mg/L had no effect on method using phenolphthalein. *3393*

Bile Acids *Serum Increase Physiological* 8 to 9 fold increase reversible cholestatic jaundice in a few children. *2172*

Bile Salts *Feces Decrease Physiological* By about 30%; in 17 patients with cystic fibrosis receiving constant concomitant therapy with pancreatic enzymes. *0624*

Bilirubin *Serum No Effect Analytical* At concentration of 1 mg/L had no effect on Jendrassik and Grof method. *3393* *Serum Increase Physiological* Reversible cholestatic jaundice in a few children. *2172* In isolated case treated for 7 mo with 400 mg/d progressed to bridging hepatic necrosis. *3085*

Calcium *Serum No Effect Analytical* At concentration of 1 mg/L had no effect on cresolphthalein method. *3393*

Carbamazepine *Serum No Effect Physiological* No significant effect on carbamazepine concentration when co-administered. *2147* *Serum Increase Physiological* Increased concentration when co-administered, as elimination affected, although urinary excretion increased. *0801*

Chlordiazepoxide *Blood Increase Physiological* Due to impaired clearance and impairment of hydroxylation of desmethyl derivative. *3084*

Chloride *Serum No Effect Analytical* At concentration of 1 mg/L had no effect on mercurimetric method. *3393*

Chloroquine *Serum Increase Physiological* Significant reduction in oral clearance from 0.49 L/d/kg to 0.23 L/d/kg and elimination half-life increased from 3.11 d to 4.62 d in test group. *1052*

Cholesterol *Serum No Effect Analytical* At concentration of 1 mg/L had no effect on CHOD-PAP method. *3393* *Serum No Effect Physiological* In healthy individuals and in subjects with high serum glucose concentration. *1653* In 25 patients pretreatment 174 mg/dL, after 181 mg/dL over 5 weeks. *3568*

Cholesterol, High Density Lipoprotein *Serum No Effect Physiological* In healthy individuals and in subjects with high serum glucose concentration. *1653* *Serum Increase Physiological* Significant increase in 25 patients over 5 weeks from 37 mg/dL to 42 mg/dL. *3568*

Cholesterol, Low Density Lipoprotein *Serum No Effect Physiological* In healthy individuals and in subjects with high serum glucose concentration. *1653*

Cimetidine (continued)

Cholesterol, Very Low Density Lipoprotein
Serum No Effect Physiological In healthy individuals and in subjects with high serum glucose concentration. 1653

Coproporphyrin *Urine No Effect Physiological* No measurable change in one patient with acute intermittent porphyria. 1659

Creatine *Urine Increase Physiological* Significant effect at 3 weeks in 9 patients given 1.6 g daily. 0978

Creatine Kinase *Serum No Effect Physiological* No significant effect observed in 9 patients given 1.6 g daily. 0978

Creatinine *Serum No Effect Analytical* At 300 mg/L on reversed phase LC procedure of Zhiri et al. 3942 At concentration of 4 mg/L had no effect on creatinine iminohydrolase method. 3393 At concentration of 1 mg/L had no effect on AutoAnalyzer Jaffé method. 3393
Serum No Effect Physiological No effect in volunteers receiving 1.2 g daily for 12 d. 0300
Serum Increase Physiological Rises because of competitive inhibition of creatinine secretion following i.v. bolus of 300 mg. 0535 Single case of reversible renal failure probably attributable to drug. 3243 In 13 ulcer patients average increase of 22% which fell on cessation on therapy. 0649 Small but detectable increases compared with placebo, but rarely exceeding 2 mg/dl. 2368 Significant effect from first day of treatment and at 3 weeks but not after 12. 0978
Urine No Effect Physiological No significant effect observed in 9 patients given 1.6 g daily. 0978

Creatinine Clearance *Urine Decrease Physiological* By at least 20% in patients with renal failure maximal after 2 to 3 d. 2085 In 13 ulcer patients average fall of 28 ml/min (26%). 0649 Significant effect within 6 h, normalized after several weeks. 0978

Cyclic Adenosine Monophosphate
Platelets Decrease Physiological Probable activation of endogenous cyclic AMP phosphodiesterase involved in favoring action of platelet aggregating agents. 3938

Cyclosporine *Serum Increase Physiological* Increased concentration presumably due to effect on liver metabolism. 0683

Diazepam *Serum Increase Physiological* 40-50% increase due to reduction of total body clearance. 1962 Impaired clearance, increasing half-life by 40%. 3084

Digitoxin *Serum Increase Physiological* Reported to increase concentration, although effect on metabolism uncertain. 0548

Diphenylhydantoin *Serum Increase Physiological* Significant increase due to inhibition of metabolism probably by reversibly binding to hepatic microsomal cytochrome P-450. 2147 Significant increases when coadministered, due to effect on hepatic metabolism. 3120

EDTA Clearance *Urine Decrease Physiological* Significant effect within 6 h, normalized after several weeks. 0978

Encainide *Serum Increase Physiological* Concomitant administration increases drug concentration by 30 to 40%. 2011

Estradiol *Plasma No Effect Physiological* In 25 men treated for duodenal ulcer or duodenitis. 2368
Plasma Decrease Physiological Significantly reduced only in midproliferative phase of cycle. 0271

Ethanol *Serum Increase Physiological* Prior administration of drug caused increased concentration probably due to enhanced absorption rather than effect on metabolism. 3246

Fat *Feces Decrease Physiological* By about 30%; in 17 patients with cystic fibrosis receiving constant concomitant therapy with pancreatic enzymes. 0624

Flecainide *Serum Increase Physiological* Increases elimination half-life and decreases clearance by 13 to 27%. 2011

Flurbiprofen *Serum Increase Physiological* Although peak concentration not increased area under curve increased. 3496

Follicle Stimulating Hormone (FSH)
Plasma No Effect Physiological In 3 studied men with duodenal ulcer or duodenitis. 2368 In 6 patients treated for 1 mo with 1 g/d. 3491
Plasma Increase Physiological With 1.2 g/d caused significant increase in periovulatory period. 0271 In 11 male subjects with chronic duodenal ulcer given drug 1 g/d for 3 mo. 3759

Gastrin *Serum No Effect Physiological* No effect reported in spite of long-term treatment. 1667 No effect noted after short-term treatment and no effect on nocturnal serum concentration. 2368
Serum Increase Physiological Increase higher in people ingesting drug in response to food. 2368

Glucose *Serum No Effect Analytical* At concentration of 4 mg/L had no effect on GOD/POD-PAP method. 3393
Serum No Effect Physiological Oral drug for 48 h had little effect in normal subjects. 3491
Serum Decrease Physiological After 100 mg/h for 4 h caused decrease of 15% at 150 minutes in normal subjects. Mean value fell from 5.4 to 4.8 mmol/L on average in 6 patients given 1 g/d for 1 mo. 3491

γ-Glutamyltransferase (GGT)
Serum Increase Physiological In isolated case treated for 7 mo with 400 mg/d progressed to bridging hepatic necrosis. 3085

Granulocytes *Blood No Effect Physiological* No decrease in bone marrow granulocyte reserves. 3448

Growth Hormone *Plasma No Effect Physiological* In 6 patients treated for 1 mo with 1 g/d. 3491

HDL$_2$-Cholesterol *Serum Increase Physiological* In 8 individuals with peptic ulcer given 1 g/d for 1 mo. 3858 Increased proportion of HDL-cholesterol when 600 mg given bid to 6 males for 1 week. 2677

HDL$_3$-Cholesterol *Serum No Effect Physiological* In 8 individuals with peptic ulcer given 1 g/d for 1 mo. 3858

Hemoglobin *Blood Decrease Physiological* Reported in one case in which drug given intravenously. 1770

Hippuran Clearance *Urine Decrease Physiological* Significant effect within 6 h, normalized after several weeks. 3199

Imipramine *Serum Increase Physiological* Although peak concentration not different; clearance significantly reduced so higher concentration overall. 1555

Insulin *Plasma Decrease Physiological* After 100 mg/h for 4 h caused decrease of 34% at 150 minutes in normal subjects. 3491

Intrinsic Factor *Gastric Material Decrease Physiological* Marked effect on basal and stimulated concentrations. 3451

Lactate Dehydrogenase *Serum Increase Physiological* In isolated case of drug associated bridging hepatic necrosis. 0864 In isolated case treated for 7 mo with 400 mg/d progressed to bridging hepatic necrosis. 3085

Leukocyte Migration Inhibition
Blood No Effect Physiological No effect noted in 12 patients treated for acid-peptic disease. 2368

Leukocytes *Blood Decrease Physiological* Proved case of drug induced agranulocytopenia. 0587 Reported in one case in which drug given intravenously. 1770 Significant reduction observed in one patient. 0696

Lorazepam *Serum No Effect Physiological* Not subject to N-dealkylation or hydroxylation by cytochrome P-450 so not affected by drug. 3084

Luteinizing Hormone (LH) *Plasma No Effect Physiological* In 6 patients treated for 1 mo with 1 g/d. 3491 In 3 studied men with duodenal ulcer or duodenitis. 2368 No significant effect of drug during menstrual cycle. 0271 In 11 male subjects with chronic duodenal ulcer given drug 1 g/d for 3 mo. 3759

Lymphocyte Transformation *Blood Increase Physiological* Serum from patients treated with drug enhanced lymphocyte response to phytohemagglutinin. 2368

Mexiletine *Serum Increase Physiological* Decreases clearance of drug and prolongs half-life. 2011

β_2-Microglobulin *Urine No Effect Physiological* No change observed during treatment of 13 ulcer patients. 0649

Microscopy *Urine No Effect Physiological* Unchanged although there may be slight increases in creatinine, not progressive; decreases with continued treatment. 2368

Neutrophils *Blood Decrease Physiological* Proved case of drug induced agranulocytopenia. 0587 Occasional case of agranulocytosis reported. 3717

Nitrogen *Feces Decrease Physiological* By about 30%; in 17 patients with cystic fibrosis receiving constant concomitant therapy with pancreatic enzymes. 0624

Nortriptyline *Serum Increase Physiological* Area under curve increased, most noticeably for its 10-hydroxy metabolite. 1555

Occult Blood *Feces No Effect Analytical* No effect noted on Hemoquant method. *3671*
Feces Positive Analytical When added to Hemoccult® test paper as pure chemical applied at pH of gastric juice. *1527* At concentrations above 1500 mg/L on Hemoccult® method. *3393*

Oxazepam *Serum No Effect Physiological* Not subject to N-dealkylation or hydroxylation by cytochrome P-450 so not affected by drug. *3084*

Parathyroid Hormone *Plasma Decrease Physiological* Affect C-terminal component only: observed in normals and patients with renal failure: slight increase of N-terminal component. *1137*

Pepsin *Gastric Material Decrease Physiological* Marked effect on basal and stimulated concentrations. *3451*

pH *Urine Increase Physiological* Mean increase of 0.4 in healthy volunteers. *2076*
Gastric Material Increase Physiological Effectively reduces gastric acidity both before and after meals. *2351*

Phosphate *Serum No Effect Analytical* At concentration of 1 mg/L had no effect on phosphomolybdate method. *3393*

Platelets *Blood Decrease Physiological* Isolated case of drug effect together with psoriasis. *3911* Isolated case of platelet-associated IgG and thrombocytopenia. *1910* Reported in one case in which drug given intravenously. *1770* Significant reduction observed in one patient. *0696* Observed in one patient with cancer in absence of leukopenia: fell again with rechallenge by drug. *2286*

Porphobilinogen *Urine Decrease Physiological* Slight decrease in patient with acute intermittent porphyria. *1659*

Porphyrins *Feces No Effect Physiological* No measurable change in one patient with acute intermittent porphyria. *1659*

Potassium *Serum No Effect Analytical* At concentration of 1 mg/L had no effect on flame-photometric method. *3393*

Procainamide *Serum Increase Physiological* NAPA metabolite also increased not due to effect on hepatic blood flow or liver cytochrome enzymes; interaction occurs at renal tubular excretion level. *1588* Area under curve increased by 43%, renal clearance decreased 36% and decreased ratio of clearance to bioavailability. *3022* Decreases renal secretion adn that of N-acetylprocainamide. *3389*

Progesterone *Plasma No Effect Physiological* No significant effect of drug during menstrual cycle. *0271*

Prolactin *Plasma No Effect Physiological* In 6 patients treated for 1 mo with 1 g/d. *3491* In 2 studied men with duodenal ulcer or duodenitis. *2368*
Plasma Increase Physiological Following either i.v. or oral drug, may be associated with larger doses of drug, possible effect on dopamine receptors in anterior pituitary, or on inhibition of uptake in peripheral tissues. *2368* Sustained effect throughout luteal phase of menstrual cycle. *0271* Significant increase in basal values observed. *2780* In 5 of 11 male subjects with chronic duodenal ulcer given drug 1 g/d for 3 mo. *3759*

Propranolol *Serum Increase Physiological* Reduces oral clearance by as much as 50%. *2942*

Prostaglandin, 6-Keto-F₁ₐ *Plasma Decrease Physiological* Probable activation of endogenous cyclic AMP phosphodiesterase involved in favoring action of platelet aggregating agents. *3938*

Protein *Serum No Effect Analytical* At concentration of 1 mg/L had no effect on biuret method with blank correction. *3393*
Urine No Effect Physiological Although drug cleared principally by glomerular filtration and partial reabsorption in proximal renal tubules. *2368*

Prothrombin Time *Plasma Increase Physiological* In 6 patients mean prolongation of time by 12.6 s in patients receiving warfarin, nicoumalone and phenindone. *3256*

Quinidine *Serum Increase Physiological* 55% increase in mean half-life: protein binding and urinary excretion unchanged effect studied in 6 healthy volunteers. *1493* Increased area under curve and prolonged half-life and decreased total body clearance. *1983*

Salicylate *Serum Increase Physiological* Influence on serum concentration in drug treated group compared with enteric-coated drug administered group (180 μg/mL vs 161 μg/mL). *3848*

Sodium *Serum No Effect Analytical* At concentration of 1 mg/L had no effect on flame-photometric method. *3393*

Urine No Effect Physiological Although drug cleared principally by glomerular filtration and partial reabsorption in proximal renal tubules. *2368*

Sperm Count *Semen Decrease Physiological* In 7 patients treated for 9 weeks with 1200 mg/d compared with their pretreatment values. *2368* In 11 male subjects with chronic duodenal ulcer given drug 1 g/d for 3 mo. *3759*

Sperm Morphology *Semen No Effect Physiological* In 11 male subjects with chronic duodenal ulcer given drug 1 g/d for 3 mo. *3759*

Sperm Motility *Semen No Effect Physiological* In 11 male subjects with chronic duodenal ulcer given drug 1 g/d for 3 mo. *3759*

Testosterone *Serum No Effect Physiological* In 6 patients treated for 1 mo with 1 g/d. *3491* In 25 men treated for duodenal ulcer or duodenitis. *2368*
Serum Increase Physiological In 11 male subjects with chronic duodenal ulcer given drug 1 g/d for 3 mo. *3759* Up to 20% increase on chronic administration and displaces DHT from its binding sites at pituitary and hypothalamic level. *1736*
Serum Decrease Physiological Observed in one 66 y old man, together with increased serum gonadotropin possibly due to reversible defect in 17-beta-hydroxysteroid dehydrogenase. *2080*

Theophylline *Serum Increase Physiological* Plasma clearance reduced from 71 mL/min to 56 mL/min in 7 healthy subjects given 1 g/d. *3004* Half-life increased by 60% with coadministration. *3782* Single case reported in which serum concentration doubled: effect reversible: effect possibly due to direct inhibition of hepatic metabolism. *0571* In group of young adults and elderly with increased area under curve, increased half-life and reduced clearance. *0687*

Thromboxane B₂ *Plasma Increase Physiological* Probable activation of endogenous cyclic AMP phosphodiesterase involved in favoring action of platelet aggregating agents. *3938*

Thyroid Stimulating Hormone (TSH)
Serum No Effect Physiological In 6 patients treated for 1 mo with 1 g/d. *3491* No significant changes caused by drug. *2780* No change after 1 g daily for 28 d, although different findings by other group. *3798*

Thyroxine (T₄) *Serum No Effect Physiological* No significant difference in drug treated healthy volunteers: no difference in response to TRH. *1683* In 6 patients treated for 1 mo with 1 g/d. *3491*

Titratable Acidity *Urine No Effect Physiological* Although drug cleared principally by glomerular filtration and partial reabsorption in proximal renal tubules. *2368*

Triglycerides *Serum No Effect Analytical* At concentration of 4 mg/L had no effect on GPO-PAP method. *3393* At concentration of 1 mg/L had no effect on lipase/esterase method. *3393*
Serum No Effect Physiological In healthy individuals and in subjects with high serum glucose concentration. *1653* In 25 patients pretreatment 139 mg/dL, after 146 mg/dL over 5 weeks. *3568*

Triglycerides, Very Low Density Lipoprotein
Serum No Effect Physiological In healthy individuals and in subjects with high serum glucose concentration. *1653*

Tri-iodothyronine (T₃) *Serum No Effect Physiological* In 6 patients treated for 1 mo with 1 g/d. *3491* No significant changes caused by drug. *2780*
Serum Decrease Physiological From average of 1.49 to 1.25 nmol/L after 1.2 g for 5 d in 8 healthy male volunteers reduces response to TRH. *1683*

Tri-iodothyronine (T₃), Reverse
Serum No Effect Physiological No significant difference in drug treated healthy volunteers: no difference in response to TRH. *1683*
Serum Increase Physiological Basal concentration increased by treatment due to drug effect on hepatic metabolism. *2780*

Urea Nitrogen *Serum No Effect Analytical* At concentration of 1 mg/L had no effect on diacetylmonoxime method. *3393*
Serum No Effect Physiological Unchanged although there may be slight increases in creatinine, not progressive: decreases with continued treatment. *2368* No significant effect observed in 9 patients given 1.6 g daily. *0978*
Serum Increase Physiological Single case of reversible renal failure probably attributable to drug. *3243*

Cimetidine (continued)

Uric Acid *Serum No Effect Analytical* At 300 mg/L on reversed phase LC procedure of Zhiri et al. *3942* At concentration of 1 mg/L had no effect on phosphotungstate reduction method. *3393* At concentration of 4 mg/L had no effect on uricase-PAP method. *3393*

Serum No Effect Physiological No effect in volunteers receiving 1.2 g daily for 12 d. *0300* Unchanged although there may be slight increases in creatinine, not progressive: decreases with continued treatment. *2368*

Serum Increase Physiological Slight increase in patients with renal failure. *2085*

Urine No Effect Physiological No effect in volunteers receiving 1.2 g daily for 12 d. *0300* Although drug cleared principally by glomerular filtration and partial reabsorption in proximal renal tubules. *2368*

Uroporphyrin *Urine No Effect Physiological* No measurable change in one patient with acute intermittent porphyria. *1659*

Vitamin B$_{12}$ *Urine Decrease Physiological* Excretion reduced because of markedly reduced absorption. *3123*

Volume *Urine No Effect Physiological* Although drug cleared principally by glomerular filtration and partial reabsorption in proximal renal tubules. *2368*

Warfarin *Plasma Increase Physiological* Clearance reduced: probably by inhibition of drug metabolism with increased half-life. *3256*

Cinchophen

Alanine Aminotransferase *Serum Increase Physiological* May cause hepatotoxicity (viral-hepatitis like). *2313*

Alkaline Phosphatase *Serum Increase Physiological* May cause intrahepatic cholestasis. *2313*

Aspartate Aminotransferase
Serum Increase Physiological May cause hepatotoxicity (viral-hepatitis like). *2313*

Bile *Urine Increase Physiological* May cause hepatotoxicity (viral-hepatitis like). *2313*

Bilirubin *Serum Increase Physiological* May cause hypersensitive cholestasis. *2313*

BSP Retention *Serum Increase Physiological* May cause hepatotoxicity (viral-hepatitis like). *2313*

Cephalin Flocculation *Serum Increase Physiological* May cause hepatotoxicity (viral-hepatitis like). *2313*

Cholesterol *Serum Increase Physiological* May cause intrahepatic cholestasis. *1434*

Color *Urine Increase Analytical* Red brown. *2313*

Leukocytes *Blood Decrease Physiological* Leukopenia or agranulocytosis may occur. *2429*

Occult Blood *Feces Positive Physiological* May cause severe ulceration of gastrointestinal tract. *0612*

Prothrombin Time *Plasma Increase Physiological* May prolong action of anticoagulants. *1679*

Sugar *Urine Increase Analytical* Acts as a reducing substance. *2425*

Thymol Turbidity *Serum Increase Physiological* May cause hepatotoxicity (viral-hepatitis like). *2313*

Uric Acid *Serum Decrease Physiological* Uricosuric action. *2313*

Ciprofloxacin

Alanine Aminotransferase *Serum Increase Physiological* Rare cases reported during clinical trials. *0151*

Alkaline Phosphatase *Serum Increase Physiological* Rare cases reported during clinical trials. *0151*

Aspartate Aminotransferase
Serum Increase Physiological Rare cases reported during clinical trials. *0151*

Bilirubin *Serum No Effect Physiological* Clinically insignificant displacement from protein in neonates. *3748*

Creatinine *Serum Increase Physiological* Mild and transient cases reported although 1 case each of acute renal failure, interstitial nephritis and nonspecific nephritis reported in population of 2829 patients. *0151*

Crystals *Urine Increase Physiological* Observed in 4 of 2829 patients without change in renal function. *0151*

Eosinophils *Blood Increase Physiological* Observed in 16 of 2829 patients and not serious. *0151*

Lactate Dehydrogenase *Serum Increase Physiological* Rare cases reported during clinical trials. *0151*

Leukocytes *Blood Decrease Physiological* Seen in 7 of 2829 patients: in all cases mild and not clearly drug related. *0151*

Neutrophils *Blood Decrease Physiological* Reversible rare effect, often in association with administration of other drugs. *0189*

Theophylline *Serum Increase Physiological* In one case clearance fell from 2.3 L/h to 0.8 L/h when drug added to treatment regime. *3583* Approximate doubling of concentration in 33 hospitalized patients with respiratory tract infections. In 20 patients increase into toxic range. Effect probably due to inhibition of hepatic microsomal enzymes by ciprofloxacin. *2915*

Urea Nitrogen *Serum Increase Physiological* Mild and transient cases reported although 1 case each of acute renal failure, interstitial nephritis and nonspecific nephritis reported in population of 2829 patients. *0151*

Cisplatin

N-Acetylglucosaminidase *Urine No Effect Analytical* At 1 mg/L had no effect on 2 colorimetric analytical methods. *1354*
Urine Increase Physiological Fivefold increase 12 children given drug 100 mg/m^2 for 6 h. Changes observed before alteration in serum creatinine. *1355* Commonly observed increased excretion as a result of nephrotoxicity. *1354*

Adenosine Deaminase Binding Protein
Urine Increase Physiological Fivefold increase 12 children given drug 100 mg/m^2 for 6 h. Changes observed before alteration in serum creatinine. *1355*

Alanine Aminopeptidase *Urine Increase Physiological* Fivefold increase 12 children given drug 100 mg/m^2 for 6 h. Changes observed before alteration in serum creatinine. *1355*

Alanine Aminotransferase *Serum Increase Physiological* Slight and transient increases in 45 patients receiving drug. *3698*

Amylase *Serum Increase Physiological* 10 cycles in 4 patients associated with increased activity: increase mild and transient up to 2 x normal limit. *3698*

Aspartate Aminotransferase
Serum Increase Physiological Slight and transient increases in 45 patients receiving drug. *3698*

Calcium *Serum No Effect Physiological* No effect noted in 11 children treated for 0.1 to 3.8 y. *3286*
Serum Decrease Physiological Observed in 5.8% of 17 patients receiving drug, associated with low serum magnesium increased urine excretion and decreased intestinal absorption of Mg. *3468* Marked effect in 15 patients treated with drug. *3964* In 3rd and 4th week of treatment of one girl with malignant ovarian germ cell tumor probably related to renal tubular dysfunction. *2064* Calcium regulation impaired by magnesium deficit. *0368*

Coombs' Test, Direct *Serum Positive Physiological* Observed in 2 individuals receiving drug. *1272*

Copper *Serum No Effect Physiological* Insignificant change in 15 patients treated with drug. *3964*
Urine Increase Physiological Marked effect in 15 patients treated with drug. *3964*

Creatinine *Serum Increase Physiological* Nephrotoxicity evidenced by increase in 12 children receiving chemotherapy. *1355* Observed in one patient, but major side effect renal toxicity. *1194* In 5% of 96 cancer patients given drug for 5 d repeated at 4-6 week intervals. *3119*

Creatinine Clearance *Urine Decrease Physiological* Observed in one patient, but major side effect renal toxicity. *1194*

EDTA Clearance *Urine Decrease Physiological* From 108 to 90 mL/min/1.73 m^2; irreversible. *3397*

γ-Glutamyltransferase (GGT)
Serum Increase Physiological Slight and transient increases in 45 patients receiving drug. *3698*

Hemoglobin   *Blood   Decrease   Physiological*   By more than 2 g/dL in 2 of 20 patients. *1384*   Observed in 2 individuals receiving drug. *1272*   In 23% of 74 evaluable cancer patients given drug for 5 d repeated at 4-6 weeks intervals. *3119*

Ionized Calcium   *Serum   No Effect   Physiological*   No effect noted in 11 children treated for 0.1 to 3.8 y. *3286*

Iron   *Serum   Increase   Physiological*   Mean increase from 67 to 128 µmol/L observed in 14 of 20 patients. Normalized few months after therapy. *1384*

Magnesium   *Serum   Decrease   Physiological*   In 76% of 50 patients receiving low dose cisplatin in combination with 4 other drugs every 4 weeks. Lower incidence when chemotherapy less frequently. *0520*   Incidence and severity dose-dependent in patients receiving multiple drugs: extent of effect may be related to interaction with another drug. *0520*   Marked effect in 15 patients treated with drug. *3964*   In 3rd and 4th week of treatment of one girl with malignant ovarian germ cell tumor probably related to renal tubular dysfunction. *2064*   Typical response largely avoided by prophylactic magnesium. *2247*   As result of renal tubular defect induced by drug may be severe enough to cause tetany and grand mal fits. *2247*   In 4 % of 140 cycles in 96 cancer patients given drug for 5 d repeated at 4-6 weeks intervals. *3119*   In 41 of 69 gynecologic oncology patients. *0169*   In 22 of 29 of patients with tumors of testis. *3185*   In 76% of 50 patients receiving 5-drug combination including cisplatin: dose dependent effect. *0520*
   Urine   No Effect   Physiological   In 11 children given drug for 0.1-3.8 y although serum concentration reduced. *3286*
   Urine   Increase   Physiological   Renal loss of magnesium responsible for low serum concentration. *3468*   Urine wasting common manifestation. *0368*   Marked effect in 15 patients treated with drug. *3964*
   Red Blood Cells   No Effect   Physiological   In 11 children given drug for 0.1-3.8 y although serum concentration reduced. *3286*

β₂-Microglobulin   *Urine   Increase   Physiological*   Transient 2 to 5-fold increase during treatment. *3397*

Phosphate   *Serum   No Effect   Physiological*   No effect noted in 11 children treated for 0.1 to 3.8 y. *3286*

Platinum   *Serum   Increase   Physiological*   Marked effect in 15 patients treated with drug. *3964*
   Urine   Increase   Physiological   Marked effect in 15 patients treated with drug. *3964*

Potassium   *Serum   Decrease   Physiological*   Marked effect in 15 patients treated with drug. *3964*   Associated with magnesium deficiency. *0368*

Sodium   *Serum   Decrease   Physiological*   In single case in 3rd and 4th treatment course, identified with renal tubular dysfunction. *2064*   In 3rd and 4th week of treatment of one girl with malignant ovarian germ cell tumor probably related to renal tubular dysfunction. *2064*
   Urine   Increase   Physiological   In single case in 3rd and 4th treatment course, identified with renal tubular dysfunction. *2064*

Urea Nitrogen   *Serum   Increase   Physiological*   Observed in one patient, but major side effect renal toxicity. *1194*   In 11 of 18 cases with doses of less than 50 mg/m² surface area: extent of change equivalent to 50% reduction in renal function. *0368*

Uric Acid   *Serum   Increase   Physiological*   Probable consequence of drug-induced nephrotoxicity. *2553*

Zinc   *Serum   No Effect   Physiological*   Insignificant change in 15 patients treated with drug. *3964*
   Urine   Increase   Physiological   Marked effect in 15 patients treated with drug. *3964*

Citrates

Acid Phosphatase   *White Blood Cells   No Effect   Analytical*   Glycerophosphate enzyme no effect of 0.3 mol/L. *0183*
   White Blood Cells   Increase   Analytical   Phenylphosphate enzyme stimulated by 0.3 mol/L. *0183*

Albumin   *Serum   Decrease   Analytical*   Dilutional effect when compared against serum or heparin as anticoagulant. *3361*

Alkaline Phosphatase   *Serum   Decrease   Analytical*   25 mmol/L inhibits 50% activity. *3217*   Inhibitory effect observed. *2228*   Dilutional effect when compared against serum or heparin as anticoagulant. *3361*

Aluminum   *Serum   Increase   Physiological*   When co-administered with aluminum hydroxide ingested as antacid, probably due to formation and absorption of aluminum citrate complexes. *3348*

Amylase   *Serum   Decrease   Analytical*   Inhibition of enzyme activity. *2364*   Sodium salt inhibits by 16% at 5 g/L method of Rauscher et al. *2926*

Calcium   *Serum   Decrease   Physiological*   Complexes calcium. *2313*
   Urine   No Effect   Physiological   Insignificant change with sodium citrate in 5 patients with uric acid lithiasis. *3116*
   Urine   Decrease   Physiological   From 154 to 99 mg/d on average with potassium citrate in 5 patients with uric acid lithiasis. *3116*   From 154 mg/d to 99 mg/d with 60 meq/d K salt: but not significantly to 139 mg/d from 154 mg/d with sodium salt. *3116*

Carotene   *Serum   Decrease   Analytical*   Dilutional effect when compared against serum or heparin as anticoagulant. *3361*

Carotenoids, Total   *Serum   Decrease   Analytical*   Dilutional effect when compared against serum or heparin as anticoagulant. *3361*

Cholesterol   *Serum   Decrease   Analytical*   If used as anticoagulant — causes water shift. *2319*   Dilutional effect when compared against serum or heparin as anticoagulant. *3361*

Citrate   *Urine   Increase   Physiological*   From 398 to 856 mg/d on average with potassium citrate in 5 patients with uric acid lithiasis. From 398 to 799 mg/d with sodium citrate in 5 patients with uric acid lithiasis. *3116*   Restored to normal levels in 89 patients with hypocitraturic calcium nephrolithiasis treated for up to 4 y. *2708*

Copper   *Serum   Decrease   Analytical*   Dilutional effect when compared against serum or heparin as anticoagulant. *3361*

Creatinine   *Serum   Decrease   Analytical*   At concentrations above 1,000 mg/L lowered concentration as measured by creatinine amidohydrolase method. *3393*

Fatty Acids, Free (FFA)   *Serum   Decrease   Analytical*   As anticoagulant or *in vivo* method of Soloni. *3381*

Glucose   *Serum   No Effect   Analytical*   At 5 g/dL on p-HBAH procedure of Lever. At 5 g/dL on o-toluidine procedure. At 5 g/dL on glucose-oxidase procedures. At 5 g/dL on alkaline ferricyanide procedures. *2140*   At concentration of 2,000 mg/L had no effect on hexokinase/G-6-PDH method. *3393*
   Serum   Increase   Analytical   At concentrations above 7591 mg/L raised concentration as measured by Ektachem® method. *3393*

Hemoglobin   *Plasma   No Effect   Analytical*   No effect of citrated blood on method of de Mendonca. *0672*

Hydroxybutyrate Dehydrogenase
   Serum   No Effect   Analytical   No effect on activity. *2981*

Ionized Calcium   *Serum   Decrease   Physiological*   Complexes calcium (effect of transfusions). *2825*

Lipid Glycerol   *Serum   Decrease   Analytical*   Inhibits phospholipase C method of Horney. *2683*

Magnesium   *Serum   No Effect   Analytical*   Has no effect on atomic absorption procedures. *0556*
   Serum   Decrease   Analytical   Affects fluorometric 8-hydroxyquinoline procedure. Affects titan yellow procedures. *0556*
   Serum   Decrease   Physiological   Complexes Mg (may occur with blood transfusions). *0529*

Osmolality   *Serum   Increase   Analytical*   Significant effect when added as anticoagulant. *3361*

Oxalate   *Urine   No Effect   Physiological*   In 89 patients with hypocitraturic calcium nephrolithiasis treated for up to 4 y. *2708*

pH   *Serum   Decrease   Analytical*   *In vitro* addition of sodium salt to plasma/serum. *0106*
   Blood   Increase   Physiological   Restores bicarbonate reserve, may cause alkalosis. *0071*
   Blood   Decrease   Analytical   Significant effect of sodium citrate as anticoagulant. *0579*
   Blood   Decrease   Physiological   Significant acid shift produced. *1233*
   Urine   Increase   Physiological   From average of 5.35 to 6.68 for potassium citrate and to 6.73 for sodium citrate in 5 patients with uric acid lithiasis. *3116*   In 89 patients with hypocitraturic calcium nephrolithiasis treated for up to 4 y. *2708*   60 meq/d increased pH on average by 1.3 with both sodium and potassium salts. *3116*

Phosphate   *Serum   Decrease   Analytical*   Complexes with molybdate decrease color develop. *0583*

Citrates *(continued)*

Phosphate *(continued)*

Urine *No Effect Analytical* No effect at 100 mg/dL on method of Jung/Parekh. *1835*

Urine *No Effect Physiological* In 89 patients with hypocitraturic calcium nephrolithiasis treated for up to 4 y. *2708*

Potassium *Urine Increase Physiological* In 89 patients with hypocitraturic calcium nephrolithiasis treated for up to 4 y. *2708*

Protein *Serum Decrease Analytical* Dilutional effect when compared against serum or heparin as anticoagulant. *3361*

Retinol Binding Protein *Serum Decrease Analytical* Dilutional effect when compared against serum or heparin as anticoagulant. *3361*

Selenium *Serum Decrease Analytical* Dilutional effect when compared against serum or heparin as anticoagulant. *3361*

Sodium *Urine No Effect Physiological* In 89 patients with hypocitraturic calcium nephrolithiasis treated for up to 4 y. *2708*

α-Tocopherol *Serum Decrease Analytical* Dilutional effect when compared against serum or heparin as anticoagulant. *3361*

Triglycerides *Serum Decrease Analytical* Dilutional effect when compared against serum or heparin as anticoagulant. *3361*

Uric Acid *Serum Increase Physiological* Antagonism of uricosuric effect. *1237*

Urine *No Effect Physiological* In 89 patients with hypocitraturic calcium nephrolithiasis treated for up to 4 y. *2708*

Urine *Decrease Physiological* Antagonism of uricosuric effect. *1237*

Vitamin A *Serum Decrease Analytical* Dilutional effect when compared against serum or heparin as anticoagulant. *3361*

Volume *Urine Increase Physiological* Weak diuretic effect. *0071*

Zinc *Serum Decrease Analytical* Dilutional effect when compared against serum or heparin as anticoagulant. *3361*

Citruplexina

Hydroxyproline *Urine No Effect Analytical* At 53 mg/L on method of Goverde. *1365*

Tyrosine *Plasma No Effect Analytical* On fluorometric procedure of Ambrose. *0080*

Urea Nitrogen *Serum Increase Analytical* Absorbance relative to urea = 1.5 diacetyl procedure. *0583*

Clazolam

LE Cells *Blood Positive Physiological* SLE may occur, usually normalized when stopped. *3059*

Cled Medium

Amino Acids *Urine Increase Analytical* Artifactual aminoaciduria if in contact. *2756*

Lactose *Urine Increase Analytical* Artifactual presence if in contact. *2756*

Clindamycin

Acetic Acid *Feces Increase Physiological* Marked proportional increase in 6 healthy individuals fed orally for 6 d. *1664*

Alanine Aminotransferase *Serum Increase Physiological* Mild transient rises seen. *2806* Uncommon side effect. *3864*

Alkaline Phosphatase *Serum Increase Physiological* Transient abnormality noted. *3653* Uncommon side effect. *3864*

Aspartate Aminotransferase
Serum Increase Physiological Transient abnormality noted. *3653* Uncommon side effect. *3864*

Bilirubin *Serum Increase Physiological* Occurs especially if pre-existing liver disease. *2806* Uncommon side effect. *3864*

BSP Retention *Serum Increase Physiological* Probable hepatotoxic effect. *1083*

Chloramphenicol *Serum No Effect Analytical* No effect at 100 mg/L on coupled enzymatic method. *2490*

Creatine Kinase *Serum Increase Physiological* Probably due to muscle damage (common with i.m.). *0841*

Eosinophils *Blood Increase Physiological* Occasional allergic response. *2806* Common hypersensitivity reaction. *3864*

Isocitrate Dehydrogenase *Serum Increase Physiological* Mild transient increase (transient). *1083*

Lactate Dehydrogenase *Serum Increase Physiological* ?hepatic or muscle origin (transient). *1083*

Leukocytes *Blood Decrease Physiological* May cause agranulocytosis. *2808*

Neutrophils *Blood Increase Physiological* Uncommon side effect. *3864*

Blood Decrease Physiological Transient neutropenia may occur. *1679*

Occult Blood *Feces Positive Physiological* Occasionally bloody mucus, especially in elderly. *1679* Reported frequency of 1 case of pseudomembranous colitis in 10 treated cases associated with mucus or bloody diarrhea. *0447*

Ornithine Carbamoyltransferase (OCT)
Serum Increase Physiological Mild increase noted (transient). *1083*

Platelets *Blood Increase Physiological* Uncommon side effect. *3864*

Blood Decrease Physiological Some cases reported. *0137*

Potassium *Serum No Effect Analytical* At concentration of 150 mg/L had no effect on measurement by ISE with predilution. *3393*

Short-Chain Fatty Acids *Feces Decrease Physiological* Marked effect in 6 healthy individuals fed orally for 6 d. *1664*

Sodium *Serum No Effect Analytical* At concentration of 150 mg/L had no effect on measurement by ISE with predilution. *3393*

Clofibrate

Acid Phosphatase *Serum Increase Physiological* Reported effect (?mechanism). *1022*

Alanine Aminotransferase *Serum Increase Physiological* Hepatotoxic effect. *2313* Single case of granulomatous hepatitis associated with drug administration. *2813*

Alkaline Phosphatase *Serum Increase Physiological* Hepatotoxic effect. *2313* Single case of granulomatous hepatitis associated with drug administration. *2813*
Serum Decrease Physiological Continuous reduction from 82 U/L to 52 U/L in 26 type IIa hyperlipoproteinemic patients = same response seen in type IV patients. *3169* Reduction of bone isoenzyme when activity increased in patients on chronic hemodialysis. *1930* Significant reduction in 27 patients, half of whom had hypertriglyceridemia, after 1 week. *1121*

Apolipoprotein AI *Serum Increase Physiological* Increase by 10 to 20% in 10 hyperlipidemic patients. *2577*

Aspartate Aminotransferase
Serum Increase Physiological Transiently elevated during early therapy. *0318*

Bile *Urine Increase Physiological* Hepatotoxic effect. *2313*

Bile Acids *Feces Decrease Physiological* After 2 weeks treatment significant fall. *2463*

Bilirubin *Serum Increase Physiological* Hepatotoxic effect. *2313* Single case of granulomatous hepatitis associated with drug administration. *2813*

Bilirubin, Direct *Serum Increase Physiological* Single case of granulomatous hepatitis associated with drug administration. *2813*

BSP Retention *Serum Increase Physiological* Intrahepatic cholestasis reported. *1022*

Carotenoids, Total *Serum No Effect Physiological* Insignificant change in about 26 subjects given 2 g daily. *2875*

Cephalin Flocculation *Serum Increase Physiological* Hepatotoxic effect. *2313*

Cholesterol *Serum No Effect Analytical* At concentration of 1,000 mg/L had no effect on CHOD-PAP method. *3393* At concentration of 1400 mg/L had no effect on method using catalase-Hantzsch reaction. *3393*
Serum Increase Physiological Paradoxical effect in patients with biliary cirrhosis. *2857*

Serum Decrease Physiological Therapeutic goal (probably diminished synthesis). *0456* By 40% in first year became normal after this in patients with hyperlipoproteinemia and impaired glucose tolerance given diet plus 2 g drug/d. 14 patients monitored for 5 y. *2925* Marked reduction within 6 weeks in 10 hyperlipidemic patients. *2577* Significant reduction in 27 patients, half of whom had hypertriglyceridemia, after 1 week. *1121* In both normals with hypertriglyceridemia and diabetics. *1120* By 40% in first year but normalized later in 14 patients with primary hyperlipoproteinemia given 2.0 g daily over 5 y. *2925*

Cholesterol, High Density Lipoprotein
Serum Increase Physiological Increase by 10 to 20% in 10 hyperlipidemic patients. *2577*

Cortisol *Plasma No Effect Physiological* No effect of drug after 1 week in healthy individuals with hypertriglyceridemia. *1120*

Creatine Kinase *Serum Increase Physiological* Originates from skeletal muscle (in up to 15%). *3505*

Creatinine *Serum No Effect Analytical* At concentration of 500 mg/L had no effect on creatinine iminohydrolase method. *3393* At concentration of 400 mg/L had no effect on Jaffé-Fading-Fraction method. *3393* At concentration of 400 mg/L had no effect on Jaffé-Fuller's earth method. *3393*
Serum Increase Physiological Possibly derived from muscle damage (in 15%). *2220*

Euglobulin Clot Lysis Time *Blood Decrease Physiological* Reported observation. *2427*

Fatty Acids, Free (FFA) *Serum Decrease Physiological* Displacement from albumin. *1343* Reduction observed in diabetics after 1 week treatment. *1120*

Fibrinogen *Plasma Decrease Physiological* Reported observation in some cases. *0620*

Glucose *Serum No Effect Analytical* At concentration of 500 mg/L had no effect on GOD/POD-PAP method. *3393*
Serum Decrease Physiological If previously abnormal. *1487* After 1 week in both normals and diabetics but more marked in latter. *1120*

Glucose Tolerance *Serum Increase Physiological* Improvement relative to degrees of abnormal triglycerides. *0318* Improved in most patients in 14 patients with primary hyperlipoproteinemia given 2.0 g daily over 5 y. *2925*
Serum Decrease Physiological After 1 week in both normals and diabetics but more marked in latter. *1120*

γ-Glutamyltransferase (GGT)
Serum Decrease Physiological Significant reduction in 27 patients, half of whom had hypertriglyceridemia, after 1 week. *1121*

Growth Hormone *Plasma No Effect Physiological* No effect of drug after 1 week in healthy individuals with hypertriglyceridemia. *1120*

Insulin *Plasma Decrease Physiological* After 1 week in both normals and diabetics but more marked in latter. *1120*

^{131}I Uptake *Serum Decrease Physiological* Up to 2.5 g/d produces effect for up to 4 mo. *0583*

Lactate Dehydrogenase *Serum Increase Physiological* May cause muscle fiber atrophy in many patients. *1883*
Serum Decrease Physiological Possibly derived from muscle damage. *2220*

Leukocytes *Blood Decrease Physiological* May cause leukopenia. *3008*

Lipids, Total *Serum Decrease Physiological* Normal response (may increase in diabetics). *2427*

β-Lipoproteins *Serum Decrease Physiological* Therapeutic effect (mechanism disputed). *0071*

Lipoproteins, Pre-β *Serum Decrease Physiological* Therapeutic intent. *0165*

Neutral Steroids *Feces Decrease Physiological* After 2 weeks treatment fall observed. *2463*

Phospholipids, Total *Serum Decrease Physiological* Effect less marked than with triglycerides. *0456*

Protein *Serum Increase Physiological* Reported effect, mechanism not discussed. *2525*

Protein Bound Iodine (PBI) *Serum Increase Physiological* Up to 2.5 g caused increase of 1.5 μg/dL. *0583*

Prothrombin Time *Plasma Increase Physiological* Displaces anticoagulants from binding protein. *2668*

T$_3$ Uptake *Serum Decrease Physiological* Due to increased thyroxine binding globulin. *2858*

Thymol Turbidity *Serum Increase Physiological* Hepatotoxic effect. *2313*

Thyroid Stimulating Hormone (TSH)
Serum Decrease Physiological In hypothyroidism but no change in euthyroid patients presumed to have direct action at hypothalamic or pituitary level. *3798*

Thyroxine (T$_4$) *Serum Increase Physiological* Insignificant increase due to increased binding capacity (augmentation of transport proteins). *3798*
Serum Decrease Physiological Due to increased binding capacity of drug. *2858*

Thyroxine (T$_4$) Binding Globulin
Serum Increase Physiological Increases concentration (small dose effect). *3505*

Thyroxine (T$_4$), Free *Serum Decrease Physiological* Percentage of free decreased (large doses). *2858*

α-Tocopherol *Serum Decrease Physiological* ?inhibits synthesis of carrier lipoprotein. *3788*

Triglycerides *Serum No Effect Analytical* At concentration of 500 mg/L had no effect on GPO-PAP method. *3393*
Serum Decrease Physiological Inhibits transfer from liver to plasma. *0456* Increase by 80% in first year in patients with hyperlipoproteinemia and impaired glucose tolerance given diet plus 2 g drug/d monitored for 5 y. *2925* Marked reduction within 6 weeks in 10 hyperlipidemic patients. *2577* Significant reduction in 27 patients, half of whom had hypertriglyceridemia, after 1 week. *1121* In both normals with hypertriglyceridemia and diabetics. *1120*

Uric Acid *Serum No Effect Analytical* At concentration of 400 mg/L had no effect on Kageyama-Hantzsch method. *3393* At concentration of 400 mg/L had no effect on catalase-AIDH method. *3393* At concentration of 500 mg/L had no effect on uricase-PAP method. *3393*
Serum Decrease Physiological Occurs in about 10% (may occasionally cause gout). *0048* Reduction observed in diabetics after 1 week treatment. *1120*
Urine Increase Physiological Weak transient uricosuric effect. *0071*

Vanillylmandelic Acid *Urine Decrease Analytical* Due to interfering glucuronides of drug. *3109*

Vitamin A *Serum No Effect Physiological* Insignificant change in about 26 subjects given 2 g daily. *2875*

Clomiphene

Androstenedione *Plasma Increase Physiological* Liberates LH, anti-steroid hormone effect of drug. *0609*

BSP Retention *Serum Increase Physiological* May cause hepatotoxicity. *1237*

Cholesterol *Serum Decrease Physiological* Possible interference with synthesis. *0071*

Dehydroepiandrosterone (DHEA)
Plasma Increase Physiological Liberates LH, anti-steroid hormone effect of drug. *0609*

Dehydroepiandrosterone Sulfate (DHEA-S)
Plasma Increase Physiological Liberates LH, anti-steroid hormone effect of drug. *0609*

Estradiol *Plasma Increase Physiological* Maximum with induced ovulation. *1295* Increase in both follicular and luteal phases over normal. *3005*

Estrogens *Urine Increase Physiological* Due to action on hypothalamic-pituitary axis. *2585*

Follicle Stimulating Hormone (FSH)
Plasma Increase Physiological Maximum increase of 350% of control in males. *3145*
Urine Increase Physiological Due to action on hypothalamic-pituitary axis. *2585*

Luteinizing Hormone (LH) *Plasma Increase Physiological* Up to 700% in normal males for first 21 d. *3145* Maximum with induced ovulation. *1295*
Urine Increase Physiological Due to action on hypothalamic-pituitary axis. *2585*

Progesterone *Plasma Increase Physiological* Increased over normal in luteal phase. *3005*

Clomiphene (continued)

Sperm Count *Semen Increase Physiological* In men with oligozoospermia but pregnancy rate unchanged. *0953*

T₃ Uptake *Serum No Effect Physiological* 100 mg/d no effect on test. *0583*

Testosterone *Serum Increase Physiological* Liberates LH, anti-steroid hormone effect of drug. *0609*
Urine Increase Physiological Liberates LH, anti-steroid hormone effect of drug. *0229*

Thyroid Stimulating Hormone (TSH)
Serum Increase Physiological Small but significant effect with basal and TRH-stimulated value: hypophyseal effect. *3798*

Thyroxine (T₄) *Serum No Effect Physiological* Probably direct thyroid inhibition. *3798*

Thyroxine (T₄) Index, Free (FTI)
Serum Decrease Physiological Probably direct thyroid inhibition. *3798*

Tri-iodothyronine (T₃) *Serum Decrease Physiological* Probably direct thyroid inhibition. *3798*

Clomipramine

Growth Hormone *Plasma Increase Physiological* In depressive patients: increase in 5 of 8 patients with acute treatment: no effect observed after 28 d. *1366*

Neutrophils *Blood Decrease Physiological* Occasional case of agranulocytosis reported. *3717*

Prolactin *Plasma Increase Physiological* In depressive patients: temporary increase during first day with lag after drug peak in 6 out of 11 patients. Significant effect after 28 d. *1366*

Thyroid Stimulating Hormone (TSH)
Serum No Effect Physiological No effect with acute or chronic treatment in depressive patients. *1366*

Clonazepam

Bilirubin *Serum No Effect Physiological* Insignificant displacement from protein in neonates. *3748*

Clonidine

Alanine Aminotransferase *Serum Increase Physiological* Single probable case of toxic hepatitis. *1277*

Aldosterone *Urine Decrease Physiological* Secondary to effect on renin. *1627*

Alkaline Phosphatase *Serum Increase Physiological* Single probable case of toxic hepatitis. *1277*

Aspartate Aminotransferase
Serum Increase Physiological Single probable case of toxic hepatitis. *1277*

Bilirubin *Serum Increase Physiological* Toxic hepatitis reported. *1277*

BSP Retention *Serum Increase Physiological* Single probable case of toxic hepatitis. *1277*

Catecholamines *Plasma Increase Physiological* Increase observed in response to exercise but overall response blunted. *3919*
Urine Decrease Physiological Dose related effect (primary action of drug). *1627*

Cholesterol *Serum Increase Physiological* Increase by 6% in 59 patients with primary hypertension given up to 300 mg daily for 6 mo. *1874*
Serum Decrease Physiological In 16 patients treated for 2 mo (by 8%). *0088*

Cholesterol, High Density Lipoprotein
Serum No Effect Physiological In 59 patients with primary hypertension given up to 300 mg daily for 6 mo. *1874*

Cholesterol, Low Density Lipoprotein
Serum No Effect Physiological In 59 patients with primary hypertension given up to 300 mg daily for 6 mo. *1874*

Cholesterol, Very Low Density Lipoprotein
Serum No Effect Physiological In 59 patients with primary hypertension given up to 300 mg daily for 6 mo. *1874*

Coombs' Test, Indirect *Serum Positive Physiological* Mechanism not discussed. *3923*

Corticotropin *Plasma Decrease Physiological* Reduced in response to single oral dose in all of 6 healthy adults studied. *2070*

Cortisol *Plasma No Effect Physiological* No change greater than with normal diurnal variation. *1028*
Plasma Decrease Physiological Lower level in growth hormone deficient children: did not change with 0.1 mg/m² daily for 60 d. *2821* In normal subjects reduced by 50% with 0.15 mg/d but no effect in opiate addicts. *1279*

Creatine Kinase *Serum Increase Physiological* Temporary effect of unknown significance. *1277*

Creatinine *Serum No Effect Analytical* At 10 mg/L on reversed phase LC procedure of Zhiri et al. *3942*
Serum Increase Physiological Approximately 0.1 mg/dL. *3923*

Effective Renal Plasma Flow
Patient Increase Physiological Sustained rise after initial drop with i.v. injection. *0477*

β-Endorphin *Plasma No Effect Physiological* In normal subjects with 0.15 mg/d but raised to normal values in opiate addicts. *1279*

Epinephrine *Urine Decrease Physiological* Primary action of drug. *1627*

Follicle Stimulating Hormone (FSH)
Plasma No Effect Physiological No effect seen in 12 healthy adults after single oral dose. *2070*

Glomerular Filtration Rate (GFR)
Urine Increase Physiological Sustained rise after initial fall with i.v. injection. *0477*

Glucagon *Plasma No Effect Physiological* No effect when given i.v. to 6 healthy volunteers. *2060* No effect observed on normal response to moderate or heavy exercise. *1797*

Glucose *Serum No Effect Physiological* In children with growth hormone deficiency 0.1 mg/m² daily for 60 d. *2821*
Serum Increase Physiological When given i.v. over 10 minutes increase of about 15 mg/dL preceded other changes in normal volunteers. *2060* Increase observed in response to exercise but overall response blunted. *3919*

Gonadotropins *Plasma No Effect Physiological* In children with growth hormone deficiency 0.1 mg/m² daily for 60 d. *2821*

Growth Hormone *Plasma No Effect Physiological* No effect observed on normal response to moderate or heavy exercise. *1797*
Plasma Increase Physiological In children with growth hormone deficiency 0.1 mg/m² daily for 60 d. *2821* Peak of 6.4 to 30 ng/mL when given i.v. to 6 healthy normals. *2060*

Insulin *Plasma No Effect Physiological* No effect observed on normal response to moderate or heavy exercise. *1797*

Luteinizing Hormone (LH) *Plasma No Effect Physiological* No effect seen in 12 healthy adults after single oral dose. *2070*

Norepinephrine *Urine Decrease Physiological* Dose related effect (primary action of drug). *1627*

PAH Clearance *Urine Increase Physiological* Nonsignificant effect in hypertensives. *0393*

Potassium *Saliva Increase Physiological* Mechanism not discussed. *0393*

Prolactin *Plasma No Effect Physiological* In children with growth hormone deficiency 0.1 mg/m² daily for 60 d. *2821* No effect seen in 12 healthy adults after single oral dose. *2070* No effect observed on normal response to moderate or heavy exercise. *1797*

Renin Activity *Plasma Decrease Physiological* Dose related effect (secondary to action of catecholamines). *1627*

Sodium *Serum Increase Physiological* Probable direct action on renal tubules. *1343*

Somatomedin-C *Plasma Increase Physiological* In children with growth hormone deficiency 0.1 mg/m² daily for 60 d. *2821*

Thyroid Stimulating Hormone (TSH)
Serum No Effect Physiological In children with growth hormone deficiency 0.1 mg/m² daily for 60 d. *2821*

Triglycerides, Low Density Lipoprotein
Serum Decrease Physiological Small effect in 59 patients with primary hypertension given up to 300 mg daily for 6 mo. *1874*

Urea Nitrogen *Serum Increase Physiological* Associated with decreased GFR at start of therapy. *2961*

Uric Acid *Serum No Effect Analytical* At 10 mg/L on reversed phase LC procedure of Zhiri et al. *3942*

Vanillylmandelic Acid *Urine Decrease Physiological* Dose related effect. *1627*

Volume *Saliva Decrease Physiological* Mechanism not discussed. *0393*

Clopamide

Alanine Aminotransferase *Serum Increase Physiological* Single case of diuretic associated myopathy. *1793*

Aldolase *Serum Increase Physiological* Single case of diuretic associated myopathy. *1793*

Ammonia *Plasma No Effect Physiological* No significant effects observed. *1487*

Apolipoprotein AI *Serum Increase Physiological* No significant change in 17 individuals treated for less than 1 y when used as sole treatment. *3184*

Apolipoprotein AII *Serum No Effect Physiological* No significant change in 17 individuals treated for less than 1 y when used as sole treatment. *3184*

Apolipoprotein B *Serum No Effect Physiological* No significant change in 17 individuals treated for less than 1 y when used as sole treatment. *3184*

Aspartate Aminotransferase
Serum Increase Physiological Single case of diuretic associated myopathy. *1793*

Catecholamines *Urine No Effect Physiological* Single case of diuretic associated myopathy. *1793*

Chloride *Serum Decrease Physiological* Single case of diuretic associated myopathy. *1793*
Urine Increase Physiological Diuretic action. *2900*

Cholesterol *Serum No Effect Physiological* No significant change in 17 individuals treated for less than 1 y when used as sole treatment. *3184*

Cholesterol, High Density Lipoprotein
Serum No Effect Physiological No significant change in 17 individuals treated for less than 1 y when used as sole treatment. *3184*

Cholesterol, Low Density Lipoprotein
Serum Increase Physiological 13% change in 17 individuals treated for less than 1 y when used as sole treatment. *3184*

Cholesterol, Very Low Density Lipoprotein
Serum No Effect Physiological No significant change in 17 individuals treated for less than 1 y when used as sole treatment. *3184*

Creatine Kinase *Serum Increase Physiological* Single case of diuretic associated myopathy. *1793*

Glucose *Serum Increase Physiological* Mild effect ?due to induced hypokalemia. *1487*

Glucose Tolerance *Serum Decrease Physiological* Possible effect of hypokalemia. *3881*

17-Hydroxycorticosteroids *Urine No Effect Physiological* Single case of diuretic associated myopathy. *1793*

17-Ketosteroids *Urine No Effect Physiological* Single case of diuretic associated myopathy. *1793*

Lactate Dehydrogenase *Serum Increase Physiological* Single case of diuretic associated myopathy. *1793*

Myoglobin *Urine Increase Physiological* Single case of diuretic associated myopathy. *1793*

Potassium *Serum Decrease Physiological* Result of diuretic action. *3881* Single case of diuretic associated myopathy. *1793*
Urine Increase Physiological Diuretic action. *2900*

Sodium *Serum Increase Physiological* Single case of diuretic associated myopathy. *1793*
Urine Increase Physiological Diuretic action (rapid effect). *2900*

Triglycerides *Serum No Effect Physiological* No significant change in 17 individuals treated for less than 1 y when used as sole treatment. *3184*

Vanillylmandelic Acid *Urine No Effect Physiological* Single case of diuretic associated myopathy. *1793*

Volume *Urine Increase Physiological* Diuretic action. *2900*

Clopenthixol

Glucose *Serum Increase Physiological* Hyperglycemic effect reported. *2427*

Clorazepate

Alkaloids *Urine No Effect Analytical* With ninhydrin in Frings TLC procedure at 10 mg/dL. *1204*
Urine Increase Analytical Purple with iodoplatinate in Frings TLC procedure. *1204*

Barbiturate *Urine No Effect Analytical* With diphenylcarbazone in Frings TLC procedure at 10 mg/dL. With mercurous NO_3 in Frings TLC procedure at 10 mg/dL. With mercuric SO_4 in Frings TLC procedure at 10 mg/dL. *1204*
Urine Increase Analytical Yellow spot with vanillin in Frings TLC procedure. *1204*

Bromide *Urine No Effect Analytical* No effect on method of Frings. *1204*

Ethchlorvynol *Urine No Effect Analytical* No effect on method of Frings and Cohen. *1204*

FPN Test *Urine No Effect Analytical* No effect at 10 mg/dL on method of Frings. *1204*

Salicylate *Urine No Effect Analytical* At 10 mg/dL no effect on Trinder method. *1204*

Thymol Turbidity *Serum Increase Physiological* Mild elevation in 10% patients reported. *1878*

Urea Nitrogen *Serum Increase Physiological* Mild elevation in 10% patients reported. *1878*

Clorexolone

Glucose *Serum Increase Physiological* Diabetogenic action. *2313*

Potassium *Serum Decrease Physiological* Diuretic action. *2427*

Uric Acid *Serum Increase Physiological* Occurs as with thiazide drugs. *2427*

Clorgiline

Homovanillic Acid *CSF Decrease Physiological* Significant reduction in 43 patients with depression or Alzheimer's disease chronically treated with drug. *2991*

5-Hydroxyindoleacetic Acid (5-HIAA)
CSF Decrease Physiological Significant reduction in 43 patients with depression or Alzheimer's disease chronically treated with drug. *2991*

Clorprenaline

Amphetamine *Urine No Effect Analytical* No effect at 100 mg/L on method of Rutter. *3097*

Clotermine

Amobarbital *Urine No Effect Analytical* No interference on TLC using ethyl acetate: methanol: water: ammonium hydroxide and modified Dragendorff's reagent for detection. *3868*

Amphetamine *Urine No Effect Analytical* No interference on TLC using ethyl acetate: methanol: water: ammonium hydroxide and modified Dragendorff's reagent for detection. *3868*

Fenfluramine *Urine Positive Analytical* Similar Rfs and color reaction on TLC using ethyl acetate: methanol: water: ammonium hydroxide and modified Dragendorff's reagent. *3868*

Hydromorphone *Urine Positive Analytical* No interference on TLC using ethyl acetate: methanol: water: ammonium hydroxide and modified Dragendorff's reagent for detection. *3868*

Mescaline *Urine No Effect Analytical* No interference on TLC using ethylacetate: methanol: water: ammonium hydroxide and modified Dragendorff's reagent for detection. *3868*

Clotermine (continued)

Methamphetamine *Urine No Effect Analytical* No interference on TLC using ethyl acetate: methanol: water: ammonium hydroxide and modified Dragendorff's reagent for detection. *3868*

Morphine *Urine No Effect Analytical* No interference on TLC using ethyl acetate: methanol: water: ammonium hydroxide and modified Dragendorff's reagent for detection. *3868*

Nicotine *Urine Positive Analytical* Similar Rfs and color reaction on TLC using ethyl acetate: methanol: water: ammonium hydroxide and modified Dragendorff's reagent. *3868*

Pentobarbital *Urine No Effect Analytical* No interference on TLC using ethyl acetate: methanol: water: ammonium hydroxide and modified Dragendorff's reagent for detection. *3868*

Phenobarbital *Urine No Effect Analytical* No interference on TLC using ethyl acetate: methanol: water: ammonium hydroxide and modified Dragendorff's reagent for detection. *3868*

Phenylpropanolamine *Urine No Effect Analytical* No interference on TLC using ethylacetate: methanol: water: ammonium hydroxide and modified Dragendorff's reagent for detection. *3868*

Secobarbital *Urine No Effect Analytical* No interference on TLC using ethyl acetate: methanol: water: ammonium hydroxide and modified Dragendorff's reagent for detection. *3868*

Clothiapine

Aspartate Aminotransferase *Serum Decrease Analytical* At 9 mg/L when added to serum and conventional method. *1758*

Creatine Kinase *Serum Decrease Analytical* At 9 mg/L when added to serum and conventional method. *1758*

γ-Globulin *Serum Decrease Analytical* At 9 mg/L when added to serum and conventional method. *1758*

Thymol Turbidity *Serum Decrease Analytical* At 9 mg/L when added to serum and conventional method. *1758*

Cloxacillin

Alanine Aminotransferase *Serum Increase Physiological* Increases linked to drug administration observed in some hemodialysis patients. *0309*

Albumin *Serum No Effect Analytical* At concentration of 5 mg/L had no effect on BCG method. *3393*

Amino Acids *Urine Positive Analytical* 2 orange colored spots with ninhydrin on paper or thin-layer chromatography, not present following peroxide oxidation of specimen but new spot appeared. *1596*

Aspartate Aminotransferase
Serum Increase Physiological Few cases reported, ?due to i.m. injection. *0071*

Bicarbonate *Serum No Effect Analytical* At concentration of 5 mg/L had no effect on method using phenolphthalein. *3393*

Bilirubin *Serum No Effect Analytical* At concentration of 5 mg/L had no effect on Jendrassik and Grof method. *3393*

Calcium *Serum No Effect Analytical* At concentration of 5 mg/L had no effect on cresolphthalein method. *3393*

Chloride *Serum No Effect Analytical* At concentration of 5 mg/L had no effect on mercurimetric method. *3393*

Cholesterol *Serum No Effect Analytical* At concentration of 5 mg/L had no effect on CHOD-PAP method. *3393*

Creatinine *Serum No Effect Analytical* At concentration of 5 mg/L had no effect on AutoAnalyzer Jaffé method. *3393*

Eosinophils *Blood Increase Physiological* Hypersensitivity response. *1513*

Folate *Serum No Effect Analytical* If chromatographic procedure of Landon used. *2068*

Glucose *Serum No Effect Analytical* At concentration of 5 mg/L had no effect on GOD/POD-PAP method. *3393*

17-Ketosteroids *Urine Increase Analytical* Interferes with Zimmermann reaction. *3505*

Leukocytes *Blood Increase Physiological* Due to eosinophilia of hypersensitivity. *2313*
Blood Decrease Physiological May cause leukopenia. *2313*

Phosphate *Serum No Effect Analytical* At concentration of 5 mg/L had no effect on phosphomolybdate method. *3393*

Potassium *Serum No Effect Analytical* At concentration of 5 mg/L had no effect on flame-photometric method. *3393*

Protein *Serum No Effect Analytical* At concentration of 5 mg/L had no effect on biuret method with blank correction. *3393*

Sodium *Serum No Effect Analytical* At concentration of 5 mg/L had no effect on flame-photometric method. *3393*

Triglycerides *Serum No Effect Analytical* At concentration of 5 mg/L had no effect on lipase/esterase method. *3393*

Urea Nitrogen *Serum No Effect Analytical* At concentration of 5 mg/L had no effect on diacetylmonoxime method. *3393*

Uric Acid *Serum No Effect Analytical* At concentration of 5 mg/L had no effect on phosphotungstate reduction method. *3393*

Clozapine

Neutrophils *Blood Decrease Physiological* Occasional case of agranulocytosis reported. *3717*

Prolactin *Plasma No Effect Physiological* No response to 12.5 mg given orally. *1416*

Coal Tar

Erythrocytes *Blood Decrease Physiological* May cause hemolytic anemia. *0279*

Haptoglobin *Serum Decrease Physiological* May cause hemolytic anemia. *0279*

Hemoglobin *Blood Decrease Physiological* May cause hemolytic anemia. *0279*

Cobalt

Erythrocytes *Blood Increase Physiological* Reported to cause increased red cell production. *2217*

Hematocrit *Blood Increase Physiological* Polycythemia observed with poisoning. *2217*

131I Uptake *Serum Decrease Physiological* Due to impaired synthesis of thyroxine. *1915*

Lipids, Total *Serum Increase Physiological* Profound hyperlipemia, xanthomatosis in one case. *2217*

Nickel *Test Conditions Increase Analytical* Possible interference with atomic absorption. *3499*

T_3 Uptake *Serum Increase Physiological* Resin uptake increased with impaired synthesis of T_4. *1915*

Thyroxine (T_4) *Serum Decrease Physiological* Impaired synthesis from tyrosine. *1915*

Cobalt Salts

131I Uptake *Serum Decrease Physiological* Reported effect. *2220*

Protein Bound Iodine (PBI) *Serum Decrease Physiological* Cause decrease, possibly to extent of goiter. *1343*

Coca Cola™

Caffeine *Serum Increase Physiological* Maximum (6.3 mg/dL) at 60-120 minutes after 36 fl oz. *2303*

Cocaine

Alanine Aminotransferase *Serum No Effect Analytical* At acute overdose concentration (2.5 mg/dL) on Technicon® SMAC® method. *3719*

Albumin *Serum No Effect Analytical* At acute overdose concentration (2.5 mg/dL) on Technicon® SMAC® method. *3719* At concentration of 25 mg/L had no effect on BCG method. *3393*

Alkaline Phosphatase *Serum No Effect Analytical* At acute overdose concentration (2.5 mg/dL) on Technicon® SMAC® method. *3719*

Alkaloids *Urine No Effect Analytical* With ninhydrin in Frings TLC procedure at 10 mg/dL. *1204*
Urine Increase Analytical Purple with iodoplatinate in Frings TLC procedure. *1204*

Aspartate Aminotransferase *Serum No Effect Analytical* At acute overdose concentration (2.5 mg/dL) on Technicon® SMAC® method. *3719*

Barbiturate *Urine No Effect Analytical* With mercuric SO₄ in Frings TLC procedure at 10 mg/dL. With vanillin on Frings procedure at 10 mg/dL. With mercurous NO₃ in Frings TLC procedure at 10 mg/dL. With diphenylcarbazone in Frings TLC procedure at 10 mg/dL. *1204*

Bicarbonate *Serum No Effect Analytical* At concentration of 3 mg/L had no effect on method using phenolphthalein. *3393*

Bilirubin *Serum No Effect Analytical* At acute overdose concentration (2.5 mg/dL) on Technicon® SMAC® method. *3719* At concentration of 25 mg/L had no effect on Jendrassik and Grof method. *3393*

Bromide *Urine No Effect Analytical* No effect on method of Frings. *1204*

Calcium *Serum No Effect Analytical* At acute overdose concentration (2.5 mg/dL) on Technicon® SMAC® method. *3719* At concentration of 25 mg/L had no effect on cresolphthalein method. *3393*

Chlordiazepoxide *Blood No Effect Analytical* On method of Riddick at 5 mg/dL. *2983*

Chloride *Serum No Effect Analytical* At concentration of 3 mg/L had no effect on mercurimetric method. *3393*

Cholesterol *Serum No Effect Analytical* At acute overdose concentration (2.5 mg/dL) on Technicon® SMAC® method. *3719* At concentration of 25 mg/L had no effect on Liebermann-Burchard method. *3393*

Creatine Kinase *Serum No Effect Analytical* At acute overdose concentration (2.5 mg/dL) on Technicon® SMAC® method. *3719*

Creatinine *Serum No Effect Analytical* At acute overdose concentration (2.5 mg/dL) on Technicon® SMAC® method. *3719* At concentration of 25 mg/L had no effect on AutoAnalyzer method. *3393*

Ethchlorvynol *Serum No Effect Analytical* At 8 mg/L on GLC procedure of Evenson. *1058*
Urine No Effect Analytical No effect on method of Frings and Cohen at 2 mg/dL. *1205*

FPN Test *Urine No Effect Analytical* No effect at 10 mg/dL on method of Frings. *1204*

Glucose *Serum No Effect Analytical* At acute overdose concentration (2.5 mg/dL) on Technicon® SMAC® method. *3719*

Iron *Serum No Effect Analytical* At acute overdose concentration (2.5 mg/dL) on Technicon® SMAC® method. *3719* At concentration of 25 mg/L had no effect on Ferrozine method. *3393*

Lactate Dehydrogenase *Serum No Effect Analytical* At acute overdose concentration (2.5 mg/dL) on Technicon® SMAC® method. *3719*

Phosphate *Serum No Effect Analytical* At acute overdose concentration (2.5 mg/dL) on Technicon® SMAC® method. *3719* At concentration of 25 mg/L had no effect on phosphomolybdate method. *3393*

Protein *Serum No Effect Analytical* At acute overdose concentration (2.5 mg/dL) on Technicon® SMAC® method. *3719* At concentration of 25 mg/L had no effect on biuret method with blank correction. *3393*

Salicylate *Urine No Effect Analytical* At 10 mg/dL no effect on Trinder method. *1204*

Triglycerides *Serum No Effect Analytical* At acute overdose concentration (2.5 mg/dL) on Technicon® SMAC® method. *3719* At concentration of 25 mg/L had no effect on lipase/esterase method. *3393*

Urea Nitrogen *Serum No Effect Analytical* At acute overdose concentration (2.5 mg/dL) on Technicon® SMAC® method. *3719* At concentration of 25 mg/L had no effect on diacetylmonoxime method. *3393*

Uric Acid *Serum No Effect Analytical* At acute overdose concentration (2.5 mg/dL) on Technicon® SMAC® method. *3719* At concentration of 25 mg/L had no effect on phosphotungstate reduction method. *3393*

Cocoa

Catecholamines *Plasma Increase Physiological* Catechols present in cocoa. *0596*

Codeine

Acetaminophen *Serum No Effect Analytical* At concentration of 10 mg/L had no effect on HPLC method. *3432*

Alanine Aminotransferase *Serum No Effect Analytical* At acute overdose concentration (2.0 mg/dL) on Technicon® SMAC® method. *3719*
Serum Increase Physiological Rise in intrabiliary pressure especially if liver disease. *3505*

Albumin *Serum No Effect Analytical* At acute overdose concentration (2.0 mg/dL) on Technicon® SMAC® method. *3719* At concentration of 20 mg/L had no effect on BCG method. *3393*

Alkaline Phosphatase *Plasma No Effect Analytical* At acute overdose concentration (2.0 mg/dL) on Technicon® SMAC® method. *3719*

Alkaloids *Urine No Effect Analytical* With ninhydrin in Frings TLC procedure at 10 mg/dL. *1204*
Urine Increase Analytical Purple with iodoplatinate in Frings TLC procedure. *1204*

Amphetamine *Urine No Effect Analytical* No reaction with NBD chloride procedure of Monforte. *2473*

Amylase *Serum Increase Physiological* May cause spasm of sphincter of Oddi. *1343*

Aspartate Aminotransferase *Serum No Effect Analytical* At acute overdose concentration (2.0 mg/dL) on Technicon® SMAC® method. *3719*
Serum Increase Physiological May cause rise in intrabiliary pressure. *3505*

Barbiturate *Urine No Effect Analytical* With mercurous NO₃ in Frings TLC procedure at 10 mg/dL. With mercuric SO₄ in Frings TLC procedure at 10 mg/dL. With diphenylcarbazone in Frings TLC procedure at 10 mg/dL. *1204*
Urine Increase Analytical Purple spot with vanillin in Frings TLC procedure. *1204*

Bicarbonate *Serum No Effect Analytical* At concentration of 1 mg/L had no effect on method using phenolphthalein. *3393*

Bilirubin *Serum No Effect Analytical* At acute overdose concentration (2.0 mg/dL) on Technicon® SMAC® method. *3719* At concentration of 20 mg/L had no effect on Jendrassik and Grof method. *3393*

Bromide *Urine No Effect Analytical* No effect on method of Frings. *1204*

Calcium *Serum No Effect Analytical* At acute overdose concentration (2.0 mg/dL) on Technicon® SMAC® method. *3719* At concentration of 20 mg/L had no effect on cresolphthalein method. *3393*

Chloride *Serum No Effect Analytical* At concentration of 1 mg/L had no effect on mercurimetric method. *3393*

Cholesterol *Serum No Effect Analytical* At acute overdose concentration (2.0 mg/dL) on Technicon® SMAC® method. *3719* At concentration of 0.1 mg/L had no effect on CHOD-PAP method. *3393* At concentration of 20 mg/L had no effect on Liebermann-Burchard method. *3393*

Creatine Kinase *Serum No Effect Analytical* At acute overdose concentration (2.0 mg/dL) on Technicon® SMAC® method. *3719*

Creatinine *Serum No Effect Analytical* At acute overdose concentration (2 mg/dL) on Technicon® SMAC® method. *3719* At concentration of 20 mg/L had no effect on AutoAnalyzer Jaffé method. *3393*

Ethchlorvynol *Serum No Effect Analytical* At 8 mg/L on GLC procedure of Evenson. *1058*
Urine No Effect Analytical No effect on method of Frings and Cohen at 2 mg/dL. *1205*

FPN Test *Urine No Effect Analytical* No effect at 10 mg/dL on method of Frings. *1204*

Glucose *Serum No Effect Analytical* At acute overdose concentration (2.0 mg/dL) on Technicon® SMAC® method. *3719* At concentration of 0.1 mg/L had no effect on GOD/POD-PAP method. *3393*

Codeine (continued)

Iron *Serum No Effect Analytical* At acute overdose concentration (2.0 mg/dL) on Technicon® SMAC® method. *3719* At concentration of 20 mg/L had no effect on Ferrozine method. *3393*

Lactate Dehydrogenase *Serum No Effect Analytical* At acute overdose concentration (2.0 mg/dL) on Technicon® SMAC® method. *3719*
Serum Increase Physiological May cause rise in intrabiliary pressure. *3505*

Lipase *Serum Increase Physiological* Causes spasm of sphincter of Oddi. *3777*

Methaqualone *Serum No Effect Analytical* At 8 mg/L on GLC procedure of Evenson. *1057*

Morphine *Urine Increase Analytical* Cross react equally (or more) with RIA procedures. Greater reaction than morphine EMIT procedure. Substantial cross reaction hemagglutination inhibition procedure. *2514*

pCO₂ *Blood Increase Physiological* May cause respiratory depression. *1343*

Phosphate *Serum No Effect Analytical* At acute overdose concentration (2.0 mg/dL) on Technicon® SMAC® method. *3719* At concentration of 20 mg/L had no effect on phosphomolybdate method. *3393*

Platelets *Blood Decrease Physiological* Thrombocytopenia reported to occur. *1436*

Potassium *Serum No Effect Analytical* At concentration of 0.1 mg/L had no effect on flame-photometric method. *3393*

Propoxyphene *Urine No Effect Analytical* Less than 1% fluorescence in procedure of Valentour. *3662*

Protein *Serum No Effect Analytical* At acute overdose concentration (2.0 mg/dL) on Technicon® SMAC® method. *3719* At concentration of 20 mg/L had no effect on biuret method with blank correction. *3393*
Urine Increase Physiological May cause nephropathy. *2583*

Salicylate *Urine No Effect Analytical* At 10 mg/dL no effect on Trinder method. *1204*

Sodium *Serum No Effect Analytical* At concentration of 0.1 mg/L had no effect on flame-photometric method. *3393*

Triglycerides *Serum No Effect Analytical* At acute overdose concentration (2.0 mg/dL) on Technicon® SMAC® method. *3719* At concentration of 20 mg/L had no effect on lipase/esterase method. *3393*

Urea Nitrogen *Serum No Effect Analytical* At acute overdose concentration (2.0 mg/dL) on Technicon® SMAC® method. *3719* At concentration of 20 mg/L had no effect on diacetylmonoxime method. *3393*
Serum Increase Physiological May cause nephropathy. *2583*

Uric Acid *Serum No Effect Analytical* At acute overdose concentration (2.0 mg/dL) on Technicon® SMAC® method. *3719* At concentration of 20 mg/L had no effect on phosphotungstate reduction method. *3393*

Coffee

Caffeine *Serum Increase Physiological* Maximum (6.7 mg/dL) at 30 minutes after 2 cups. *2303*

Fatty Acids, Free (FFA) *Serum Increase Physiological* Possibly due to stress effect. *2427*

Gastrin *Serum Increase Physiological* 2-5 fold increase after 350 ml. *3302*

Glucose *Serum Increase Physiological* Metabolic action of stimulant. *0117*

4-Hydroxybenzoic Acid *Urine Increase Physiological* Possible effect of bacterial metabolism. *3281*

Lipids, Total *Serum Increase Physiological* Significant correlation with intake in coronary disease. *2193*

Lipoproteins *Serum Increase Physiological* Significant correlation with intake in coronary disease. *2193*

Uric Acid *Serum Increase Analytical* Component of coffee measured (method of Bittner). *2752*

Urobilinogen *Urine Increase Analytical* Acetyltryptophan produced reacts with Ehrlich's. *1563*

Vanillylmandelic Acid *Urine Increase Analytical* Interference with nonspecific methods. *2704*

Colchicine

Alanine Aminotransferase *Serum Increase Physiological* Possible hepatotoxic effect. *2220*

Alkaline Phosphatase *Serum Increase Physiological* Possible hepatotoxic effect. *2220*

Amino-4-Imidazole-5-Carboxamide Ribotide (AICAR) *Urine Increase Physiological* Megaloblastic anemia with B₁₂ deficiency. *3773*

Aspartate Aminotransferase
Serum Increase Physiological Possible hepatotoxic effect. *2220*

Bile *Urine Increase Physiological* May cause hepatotoxicity. *2313*

Bilirubin *Serum Increase Physiological* May cause hepatic toxicity. *2313*

BSP Retention *Serum Increase Physiological* May cause hepatic toxicity. *2313*

Cephalin Flocculation *Serum Increase Physiological* May have hepatotoxic effect. *2313*

Cholesterol *Serum Decrease Physiological* May have hepatotoxic effect. *3082*

Chromosomes *Test Conditions Abnormal Physiological* Inhibits mitosis at metaphase in human cells. *3282*

Corticosteroids *Urine Increase Analytical* Increased absorption at 410 nm in Porter-Silber reaction. Butanol extract/no hydrolysis Reddy/Porter-Silber reaction. *0022*

Erythrocytes *Blood Decrease Physiological* Megaloblastic anemia (impaired absorption of vitamin B₁₂). *3773*
Urine Increase Physiological May cause bleeding. *2313*

Fat *Feces Increase Physiological* May cause villus damage and impaired regeneration of the epithelial cells of small intestine. *3023*

FIGLU (N-Formiminoglutamic Acid)
Urine Increase Physiological Megaloblastic anemia with B₁₂ deficiency. *3773*

Hematocrit *Blood Decrease Physiological* Megaloblastic anemia (impaired absorption of vitamin B₁₂). *3773*

Hemoglobin *Blood Decrease Physiological* Megaloblastic anemia (impaired absorption of vitamin B₁₂). *3773*
Urine Increase Physiological May cause bleeding. *2313*

17-Hydroxycorticosteroids *Urine Increase Analytical* Interferes with Porter-Silber reaction. *3505*

Leukocytes *Blood Increase Physiological* Leukocytosis follows initial leukopenia. *1343*
Blood Decrease Physiological Leukopenia. *2426*

MCV *Blood Increase Physiological* Megaloblastic anemia with B₁₂ deficiency. *3773*

Methylmalonic Acid *Urine Increase Physiological* Occurs with impaired absorption of B₁₂. *3773*

Neutrophils *Blood Decrease Physiological* Occasional case of agranulocytosis reported. *3717*

Occult Blood *Feces Positive Physiological* Toxicity effect. *2313*

Phosphate *Serum No Effect Analytical* On nonprecipitation method of Peynet. *2799*

Platelets *Blood Decrease Physiological* Selective thrombocytopenia or aplastic anemia. *2426*

Prothrombin Time *Plasma Decrease Physiological* Patients on coumarins. *2426*

Sperm Count *Semen Decrease Physiological* Potent antispermatogenic action observed. *2831*

Thymol Turbidity *Serum Increase Physiological* May have hepatotoxic effect. *2313*

Vitamin B₁₂ *Serum Decrease Physiological* Impairs absorption. *3773* Reduced absorption due to ileal blockade. *3794*
Urine Decrease Physiological Occurs with impaired absorption of B₁₂. *3773*

Xylose Excretion *Urine Decrease Physiological* Decreased absorption from gut, disturbs epithelial cell function. *2899*

Cold Agglutinins

Erythrocytes *Blood Decrease Analytical* Spurious macrocytosis due to agglutination. 2797

Hematocrit *Blood Decrease Analytical* Spurious macrocytosis due to agglutination. 2797

Leukocytes *Blood Increase Physiological* Spurious high count until blood warmed to 37°. 2515

MCHC *Blood Increase Analytical* Spurious macrocytosis due to agglutination. 2797

MCV *Blood Increase Analytical* Spurious macrocytosis due to agglutination. 2797

Cold Exposure

Growth Hormone *Plasma Increase Physiological* Observed effect in normals. 0371

Leukocytes *Blood Increase Physiological* Physiological response to prolonged cold baths. 0251

Colestipol

Carotenoids, Total *Serum Decrease Physiological* Increase by 30% with 30 g daily of drug. 2875

Chlorothiazide *Serum Decrease Physiological* Reduced absorption due to binding. 3794

Cholesterol *Serum Decrease Physiological* Therapeutic intent. 0939

Digitoxin *Serum Decrease Physiological* Reduced bioavailability due to binding in gastrointestinal tract. 2794

Digoxin *Serum Decrease Physiological* In one patient drug given for 2 d reduced half-life of digoxin by approximately 50% (from 4 to 2 d). 0548

Fat *Feces Increase Physiological* Bile acid sequestrant may cause steatorrhea with conventional dosage schedules. 3023

Propranolol *Serum Decrease Physiological* Significant reduction of peak concentration and area under curve due to binding in gastrointestinal tract. 1584

T_3 Uptake *Serum Increase Physiological* Observed in euthyroid patients when drug given together with niacin. 0603

Thyroxine (T_4) *Serum No Effect Physiological* Reduced reabsorption of thyroid hormones from gastrointestinal tract. 3798
Serum Decrease Physiological Observed in euthyroid patients when drug given together with niacin. 0603

Thyroxine (T_4) Binding Globulin
Serum Decrease Physiological Observed in euthyroid patients when drug given together with niacin. 0603

Triglycerides *Serum Increase Physiological* Increase of 7% noted. 2874
Serum Decrease Physiological Therapeutic intent. 0939

Tri-iodothyronine (T_3) *Serum Decrease Physiological* Reduced reabsorption of thyroid hormones from gastrointestinal tract. 3798

Vitamin A *Serum No Effect Physiological* No effect with 30 g daily of drug. 2875
Serum Decrease Physiological With prolonged treatment due to bile acid sequestration but values still in normal range. 3023

Colistimethate

Casts *Urine Increase Physiological* Nephrotoxic effect. 2313

Creatinine *Serum Increase Physiological* Nephrotoxic effect (usually reversible). 2808

Erythrocytes *Urine Increase Physiological* Common nephrotoxic effect. 2808

Hemoglobin *Urine Increase Physiological* Common nephrotoxic effect. 2808

Leukocytes *Blood Decrease Physiological* Neutropenia/leukopenia/granulocytopenia. 0071

Nitrofurantoin *Urine No Effect Analytical* No effect on method of Conklin and Hollifield. 0704

Nonprotein Nitrogen *Serum Increase Physiological* Nephrotoxic effect (usually reversible). 2808

Protein *Urine Increase Physiological* Common nephrotoxic effect. 2808 Nephrotoxic effect. 2313

Urea Nitrogen *Serum Increase Physiological* Nephrotoxic effect (usually reversible). 2313

Colistin

Amino Acids *Urine Increase Analytical* Reacts with ninhydrin; extra spot with TLC, high voltage electrophoresis. 2855

Casts *Urine Increase Physiological* Cylindruria may occur with nephrotoxicity. 0971

Chloramphenicol *Serum No Effect Analytical* No effect at 100 mg/L on coupled enzymatic method. 2490

Creatinine *Serum Increase Physiological* Nephrotoxic effect (reversible renal damage). 2313

Erythrocytes *Urine Increase Physiological* Hematuria may occur with nephrotoxicity. 0971

Leucine Aminopeptidase *Urine Increase Physiological* Probably associated with proximal renal tubular injury. 0078

Leukocytes *Blood Decrease Physiological* May cause leukopenia/granulocytopenia. 2313

Nonprotein Nitrogen *Serum Increase Physiological* Nephrotoxic effect (reversible renal damage). 2313

Potassium *Serum No Effect Analytical* At concentration of 150 mg/L had no effect on measurement by ISE with predilution. 3393

Protein *Urine Increase Physiological* Nephrotoxic effect (reversible renal damage). 2495

Sodium *Serum No Effect Analytical* At concentration of 150 mg/L had no effect on measurement by ISE with predilution. 3393

Specific Gravity *Urine Decrease Physiological* Concentrating ability may be impaired. 0971

Urea Nitrogen *Serum Increase Physiological* Nephrotoxic effect/may occur after large doses. 2313

Collidine

Ionized Calcium *Serum Increase Analytical* At concentrations > 0.1 mmol/L on Calcium specific electrode. 0540

Concentrated Urine

Color *Urine Increase Physiological* Yellow orange with dehydration. 0443

Congo Red

Color *Urine Increase Physiological* Dye not taken up by amyloid. 0071

Contact With Clot

Acid Phosphatase *Serum No Effect Analytical* pH change prevented 5 h at room temperature, 24 at 4°. 3217
Serum Decrease Analytical Labile at room temperature. 3879

Albumin *Serum No Effect Analytical* If plastic disc used to separate for 48 h. 1863

Aldolase *Serum Increase Analytical* Due to release from platelets during clotting. 0067

Alkaline Phosphatase *Serum No Effect Analytical* If plastic disc used to separate for 48 h. 1863
Serum Decrease Analytical Reported effect ?temperature or stability. 3879

α-Amino-Nitrogen *Plasma Increase Analytical* Proteolysis occurs even in refrigerator. 1563

Ammonia *Plasma Increase Analytical* Breakdown of urea if delay in analysis. 1237

Angiotensin II *Plasma Increase Analytical* No effect until delay exceeds 5 h. 0977

Contact With Clot (continued)

Bicarbonate *Serum No Effect Analytical* If plastic disc used to separate for 48 h. *1863*
Serum Decrease Analytical pH shift in drawn blood allows at least 5% drop. *1563*

Calcium *Serum No Effect Analytical* If plastic disc used to separate for 48 h. *1863*
Serum Decrease Analytical If prolonged facilitates passage of H_2O from cells. *1563* Absorption by container. *1237*

Chloride *Serum No Effect Analytical* If plastic disc used to separate for 48 h. *1863*
Serum Decrease Analytical If exposed to air water shift from erythrocytes. *1563*

Cholesterol Esters *Serum Increase Analytical* Free decreases at same time due to esterase action. *1563*

Creatinine *Serum No Effect Analytical* If plastic disc used to separate for 48 h. *1863*

Glucose *Serum No Effect Analytical* If plastic disc used to separate for 48 h. *1863*
Serum Decrease Analytical Glucose metabolism by cells. *0579* At 4° changes 1-3 mg/dL/h. *1563*

Iron *Serum No Effect Analytical* If plastic disc used to separate for 24 h. *1863*

Iron-Binding Capacity, Total (TIBC)
Serum No Effect Analytical If plastic disc used to separate for 24 h. *1863*

Lactate Dehydrogenase *Serum Increase Analytical* High intracellular concentration of LDH activity. *1563*

Lithium *Serum No Effect Analytical* No effect observed over 24 h. *2891*

Magnesium *Serum No Effect Analytical* If plastic disc used to separate for 48 h. *1863*

Phosphate *Serum Increase Analytical* Release from erythrocytes and platelets. *1237*

Potassium *Serum No Effect Analytical* If plastic disc used to separate for 48 h. *1863*
Serum Increase Analytical Release from erythrocytes and platelets. *3879*

Protein *Serum No Effect Analytical* If plastic disc used to separate for 48 h. *1863*

Sodium *Serum No Effect Analytical* If plastic disc used to separate for 48 h. *1863*

Testosterone *Serum Increase Analytical* 10-25% increase in serum of females when allowed to stand in contact with clot for 5 d at room temperature. *1475*

Urea Nitrogen *Serum No Effect Analytical* If plastic disc used to separate for 48 h. *1863*

Uric Acid *Serum No Effect Analytical* No effect 8 h at room temperature if uricase used. *2954*

Copper

Acid Phosphatase *Serum Decrease Analytical* Cupric ions inhibit red cell enzyme. *3588*

Alanine Aminotransferase *Serum Increase Physiological* Toxic effect, sharp rise with hemolytic crisis. *0385*

Aldosterone *Urine Increase Analytical* If in reagents affects RIA procedures. *0952*

Amylase *Serum Increase Physiological* Pancreatitis with dialysis induced toxicity. *1955*

Arginase *Serum Increase Physiological* Increased when hemolytic crisis occurs. *0385*

Aspartate Aminotransferase
Serum Increase Physiological Toxic effect, sharp rise with hemolytic crisis. *0385*

Bilirubin *Serum Increase Physiological* Marked increase with hemolysis of toxicity. *0385*

BSP Retention *Serum Increase Physiological* Increased with hemolytic crisis. *0385*

Calcium *Serum Increase Analytical* Interferes with EDTA titration procedures. *1563*

Cephalin Flocculation *Serum Increase Physiological* Transient effect in poisoning. *2427*

Ceruloplasmin *Serum Increase Physiological* Observed with copper poisoning. *2427*

Color *Urine Increase Analytical* Blue diapers (alkaline urine on copper fastenings). *0526*

Copper *Serum Increase Physiological* May be normal in toxic cases with hemolysis. *0385*
Urine Increase Physiological Observed with poisoning. *2427*
Red Blood Cells Increase Physiological Concentration gradient from erythrocytes to plasma. *1955*

Creatine Kinase *Serum Increase Physiological* Observed in one case of dialysis toxicity. *2217*

Glucose-6-Phosphate Dehydrogenase
Red Blood Cells Decrease Physiological Strongly inhibited (no activity at 100 µmol/L) *in vitro*. *0432*

Haptoglobin *Serum Decrease Physiological* May be hemolysis with poisoning. *2427*

Hematocrit *Blood Decrease Physiological* With hemolysis of copper toxicity. *0385*

Hemoglobin *Plasma Increase Physiological* May occur with hemolysis of toxicity. *0385*
Blood Decrease Physiological May be marked decrease. *2427*
Urine Increase Analytical False positive with guaiac and benzidine tests. *2313*
Urine Increase Physiological May occur with copper toxicity. *0385*

Hexokinase *Red Blood Cells Decrease Physiological* Very sensitive to inhibition by copper *in vitro*. *0432*

Iron *Serum Increase Analytical* 500 µg/dL = 50 µg/dL Ferrozine procedure of White. *3810*

Lactate Dehydrogenase *Serum Increase Physiological* Toxic effect, sharp increase with hemolytic crisis. *0385*

Leukocytes *Blood Increase Physiological* Marked leukocytosis with dialysis induced toxicity. *1955* Leukocytosis observed with poisoning. *2427*

Methemalbumin *Plasma Increase Physiological* Occurs with acute intravascular hemolysis. *0385*

Methemoglobin *Blood Increase Physiological* May occur with copper toxicity. *0385*

Myoglobin *Serum Increase Physiological* Observed in one case of dialysis toxicity. *2217*

Nickel *Test Conditions Increase Analytical* Possible interference with atomic absorption. *3499*

Occult Blood *Feces Positive Physiological* May occur with severe poisoning. *2217*

Phosphofructokinase
Red Blood Cells Decrease Physiological several affected *in vitro*. *0432*

6-Phosphoglycerate Dehydrogenase
Red Blood Cells Decrease Physiological Strongly inhibited by addition of copper. *0432*

Phosphoglycerate Kinase
Red Blood Cells Decrease Physiological Less marked inhibition *in vitro*. *0432*

Protein *Urine Increase Analytical* Affects biuret part of Doetsch procedure. *0921*
Urine Increase Physiological May cause nephrotoxicity. *3204*

Protein Bound Iodine (PBI) *Serum Decrease Analytical* As contaminant of water may affect analysis. *2387*

Pyruvate Kinase *Red Blood Cells Decrease Physiological* several affected *in vitro*. *0432*

Reduced Glutathione
Red Blood Cells Decrease Physiological Marked decrease with hemolysis of toxicity. *0385*

Sodium *Serum Increase Analytical* May interfere with flame photometry. *2313*

Urea Nitrogen *Serum Increase Physiological* May cause nephrotoxicity. *3204*

Zinc *Serum No Effect Analytical* At 50 to 1 with flameless atomic absorption. *2038*
Urine No Effect Analytical At 50 to 1 with flameless atomic absorption. *2038*

Corbadrin

Cholesterol *Serum No Effect Analytical* At concentration of 200 mg/L had no effect on CHOD-PAP method. *3393* At concentration of 200 mg/L had no effect on method using catalase-Hantzsch reaction. *3393* At concentration of 200 mg/L had no effect on Kageyama-Hantzsch method. *3393*

Cork Stoppers

Calcium *Serum Increase Analytical* May be considerable increase with storage. *3217*

Xanthochromia *CSF Positive Analytical* Contact with cork for 30 minutes may produce color. *0251*

Corticosteroids

Amylase *Serum Increase Physiological* Condition like acute idiopathic pancreatitis. *3498*

Bilirubin *Serum Decrease Physiological* When elevated as increases bile flow. *2246*

Calcium *Serum Decrease Physiological* Antagonizes action of vitamin D and parathyroid. *0071*
Urine Increase Physiological Metabolic effect. *2220*

Chloride *Serum Increase Physiological* May cause retention. *2313*
Serum Decrease Physiological Metabolic alkalosis with reduced Cl reabsorption. *2448*
Urine Decrease Physiological Promotes retention (mineralocorticoid effect). *0071*

Cholesterol *Serum Increase Analytical* Many steroids react with $FeCl_3$ reagent. *1563*
Serum Increase Physiological Effect of prolonged hormone action. *0697*

Creatinine *Urine Increase Physiological* Associated with negative nitrogen balance. *2313*

Erythrocytes *Blood Decrease Physiological* May cause gastrointestinal bleeding with low iron stores. *0333*

Glucose *Serum Increase Physiological* Tends to be high (as in Cushing's syndrome). *1343*
Urine Increase Physiological As result of hyperglycemia. *2313*

Growth Hormone *Plasma Decrease Physiological* Suppresses secretion of hormone. *2427*

Hematocrit *Blood Decrease Physiological* May cause gastrointestinal bleeding with low iron stores. *0333*

Hemoglobin *Blood Decrease Physiological* May cause gastrointestinal bleeding with low iron stores. *0333*

17-Hydroxycorticosteroids *Urine Decrease Physiological* If given orally, inhaled or topically. *1487*

Hydroxyproline *Urine Increase Physiological* Stimulate growth in children. *2427*
Urine Decrease Physiological Alleged normal metabolic effect. *3644*

25-Hydroxy Vitamin D₃ *Serum Decrease Physiological* Observed effect in some children. *3023*

¹³¹I Uptake *Serum No Effect Physiological* No effect observed. *1915*
Serum Decrease Physiological Observed effect may last 8 d. *1487*

17-Ketosteroids *Urine Decrease Physiological* If given orally, inhaled or topically. *1487*

LE Cells *Blood Positive Physiological* Implicated as activators of SLE. *0962*

Leukocytes *Blood Decrease Physiological* May cause leukopenia. *2313*

Occult Blood *Feces Positive Physiological* May increase incidence of gastric ulcers. *0333*

Osmolality *Serum Increase Physiological* Associated with sodium and chloride retention. *1343*

Phosphate *Urine Increase Physiological* Metabolic effect. *2220*

Potassium *Serum Decrease Physiological* Mineralocorticoid effect with increase renal excretion. *0070*
Urine Increase Physiological Promotes excretion (mineralocorticoid effect). *0071*

Protein *Serum Increase Physiological* Physiological doses promote protein synthesis. *0697*
Urine Increase Physiological Nephrotoxic especially in children with chronic disease. *1487*

Protein Bound Iodine (PBI) *Serum Decrease Physiological* If chronic administration in small or large doses. *0830* Large dose effect with decreased thyroxine binding globulin. *1915*

Prothrombin Time *Plasma Increase Physiological* Alleged potentiation of effect of anticoagulants. *1680*
Plasma Decrease Physiological Accelerate metabolism of anticoagulants. *2313*

Salicylate *Serum Decrease Physiological* Increases glomerular filtration, decreases tubular reabsorption. *1487* Concomitant administration causes accelerated metabolism, most marked in males. *1830* Levels did not rise above 150 µg/mL in patients given corticosteroids and 4.8 g aspirin/d. *1372*

Sodium *Serum Increase Physiological* May cause retention. *2313*
Urine Decrease Physiological Promotes retention (mineralocorticoid effect). *0071*

T₃ Uptake *Serum Increase Physiological* Resin uptake increased due to decreased thyroxine binding globulin. *1915*

Thymol Turbidity *Serum Increase Physiological* Physiological doses increase protein synthesis. *0697*

Thyro-Binding Index *Serum Decrease Physiological* May occur for up to 1 week after therapy. *0913*

Thyroid Stimulating Hormone (TSH)
Serum Decrease Physiological Usual response in patients on steroids. *2220*

Thyroxine (T₄) *Serum Decrease Physiological* May occur for up to 1 week after therapy. *0913*

Thyroxine (T₄) Binding Globulin
Serum Decrease Physiological Large dose response. *1915*

Thyroxine (T₄) Index, Free (FTI)
Serum Decrease Physiological Low normal or slight decrease for 1 week after treat. *0913*

Uric Acid *Serum Decrease Physiological* Produces negative nitrogen balance. *2313*

Uropepsinogen *Urine Increase Physiological* Increase renal clearance. *0814*

Volume *Plasma Increase Physiological* Expansion of extracellular fluid due to water retention. *1343*

Zinc *Serum Decrease Physiological* Rapid sustained drop in burn and surgical patients. *1153*

Corticosterone

Aldosterone *Urine No Effect Analytical* No significant effect on RIA procedure of Drewes. *0952*
Urine Increase Analytical 1 mg/L equivalent to 11 µg/L in method of Drewes. *0952*

Corticosteroids *Plasma Increase Analytical* 73% (cortisol 100%) competitive protein binding method of Ficher. *1128*

Cortisol *Plasma Increase Analytical* If nonselective extraction and competitive protein binding used. *0512*

Estrogens *Urine No Effect Analytical* At 50 mg/L on fluorescent method of Corns. *0730*

Progesterone *Plasma No Effect Analytical* 1% or less cross reactivity with RIA. *0567*

Corticotropin

Aldolase *Serum Increase Physiological* Probably due to muscle damage at site of injection. *0275*

Aldosterone *Urine Increase Physiological* Effect less marked than on tetrahydro-DOC. *3174*

Amino Acids *Urine Increase Physiological* Catabolism of body tissues. *3879*

α-Amino-Nitrogen *Plasma Increase Physiological* Tissue and protein catabolism. *1237*
Urine Increase Physiological Catabolism of body tissues. *3879*

Amylase *Serum Increase Physiological* Increase may be marked (both short and long term). *0622*

Corticotropin *(continued)*

Androsterone *Plasma Increase Physiological* Hormonal effect. *2435*
Urine Increase Physiological Normal response to 2 d ACTH test. *1474*

Ascorbic Acid *Urine Increase Physiological* Mobilizes vitamin stored in adrenal cortex. *0418*

Basal Metabolic Rate *Patient Decrease Physiological* Metabolic action of hormone. *2220*

Calcium *Urine Increase Physiological* Average or large doses promote excretion. *1680*

Chloride *Serum Decrease Physiological* May cause hypochloremic alkalosis. *2312*

Cholesterol *Serum Increase Physiological* Effect of hormone action after initial fall. *2220*
Serum Decrease Physiological Ester concentration reduced by stimulation of adrenal. *0652*

Cholesterol Esters *Serum Increase Physiological* Initial fall then rise about 10%. *2312*

Corticosteroids *Plasma Increase Physiological* Maximum response seen after 4 h. *0168*

Corticosterone *Urine No Effect Physiological* No effect observed in normals. *1982*

Cortisol *Plasma Increase Physiological* Therapeutic intent. *1680*
Urine Increase Physiological Progressive increase with repeated injection. *1982*

Dehydroepiandrosterone (DHEA)
Plasma Increase Physiological Hormonal effect. *2435*
Urine Increase Physiological Normal response to 2 d ACTH test. *1474*

11-Deoxycortisol *Urine Increase Physiological* Progressive increase with repeated injection. *1982*

Eosinophils *Blood Decrease Physiological* Striking response in normals. *1343*

Epinephrine *Urine No Effect Physiological* No effect observed in normals or rheumatoids. *0279*

Erythrocytes *Blood Increase Physiological* Especially marked if given to anemics. *1343*

Erythrocyte Sedimentation Rate
Blood Decrease Physiological Particularly in rheumatoid patients. *0251*

Estradiol *Urine Decrease Analytical* Large amounts affect method of Brown. *0023*

Estriol *Urine Increase Physiological* In pregnant increased production of adrenal precursors. *0023*
Urine Decrease Analytical Large amounts affect method of Brown. *0023*

Estrogens *Urine Increase Physiological* Hormonal effect in men and women. *2211*
Urine Decrease Analytical Affects colorimetric/fluorometric procedures. *0022*

Etiocholanolone *Urine Increase Physiological* Normal response to 2 d ACTH test. *1474*

Glucose *Serum Increase Physiological* Gluconeogenesis, insulin antagonism. *3879*
Urine Increase Physiological Increased blood glucose, reduced TMG. *3879*

Glucose Tolerance *Serum Decrease Physiological* Corticosteroid impairment of insulin secretion. *0203*

Growth Hormone *Plasma Increase Physiological* Up to 1 mg i.v. has marked effect. *1898* Increase in 50%, moderate effect only. *2176*
Plasma Decrease Physiological Reduces maximum of hormone during sleep. *1301*

Hematocrit *Blood Increase Physiological* Especially in anemics. *0275*

Hydrochloric Acid *Gastric Material Increase Physiological* Effect of protracted therapy. *1680*

11-Hydroxyandrosterone *Urine Increase Physiological* Normal response to 2 d ACTH test. *1474*

11-Hydroxycorticosteroids *Plasma Increase Analytical* Increased cholesterol with nonspecific fluorescent procedure. *1369*

17-Hydroxycorticosteroids *Urine Increase Physiological* Marked response to i.v. infusion. *0361*
Urine Decrease Physiological Reduction even if given orally or topically. *0446*

11-Hydroxyetiocholanolone *Urine Increase Physiological* Normal response to 2 d ACTH test. *1474*

5-Hydroxyindoleacetic Acid (5-HIAA)
Urine Decrease Physiological Mechanism not described. *1488*

Iron *Serum Decrease Physiological* Decrease in iron binding globulin. *3879*

Iron-Binding Capacity, Total (TIBC)
Serum Decrease Physiological Decrease in iron binding globulin. *3879*

Iron-Binding Capacity, Unsaturated (UIBC)
Serum Decrease Physiological Decrease in iron binding globulin. *3879*

^{131}I Uptake *Serum Decrease Physiological* ?effect on TSH — lasts up to 8 d. *1444*

11-Ketoetiocholanolone *Urine Increase Physiological* Normal response to 2 d ACTH test. *1474*

17-Ketosteroids *Urine Increase Physiological* Stimulation of adrenal. *1488*
Urine Decrease Physiological Reduction even if orally or topically given. *0446*

Leukocytes *Blood Increase Physiological* Significant neutrophilic granulocytosis. *2427*

Lymphocytes *Blood Decrease Physiological* Marked drop occurs within 2 h. *1343*

Norepinephrine *Urine Decrease Physiological* Observed effect in normals and rheumatoids. *0279*

pH *Urine Decrease Physiological* Metabolic effect. *1622*

Potassium *Serum Decrease Physiological* Increased urinary excretion (mineralocorticoid effect). *3879*
Urine Increase Physiological Mobilization of potassium from tissues. *1343*

Pregnanediol *Urine Increase Physiological* Hormonal effect. *2211*

Protein *Serum Increase Physiological* Physiological doses promote protein synthesis. *0697*
Urine Increase Physiological May be nephrotoxic in chronic disease. *1583*

Protein Bound Iodine (PBI) *Serum Decrease Physiological* Reduced thyro-binding globulin. *3879*

Prothrombin Time *Plasma Increase Physiological* Prolongs action of anticoagulants. *2313*
Plasma Decrease Physiological Patients on coumarins (may induce enzymes). *2426*

Reticulocytes *Blood Increase Physiological* Especially if given to anemics. *1343*

Sodium *Serum Increase Physiological* Causes retention with edema. *0071*

Testosterone *Urine Increase Physiological* Moderate effect in men. *3062*

Tetrahydrodeoxycorticosterone
Urine Increase Physiological Fivefold increase in response to infusion in normals. *3174*

Thymol Turbidity *Serum Increase Physiological* Physiological dose increase protein synthesis. *0697*

Thyro-Binding Index *Serum Decrease Physiological* Occurs for up to 1 week after treatment. *0913*

Thyroxine (T$_4$) *Serum Decrease Physiological* Decreased thyro-binding globulin (occurs for up to 1 week). *3879*

Thyroxine (T$_4$) Binding Globulin
Serum Decrease Physiological Reduced synthesis. *3879*

Thyroxine (T$_4$) Index, Free (FTI)
Serum Decrease Physiological Low normal or slight decrease 1 week after therapy. *0913*

Uric Acid *Serum Decrease Physiological* Uricosuric action. *1433*
Urine Increase Physiological Uricosuric action. *1433*

Volume *Plasma Increase Physiological* Moderate doses can cause salt and water retention. *1680*

Zinc *Serum Increase Physiological* Occurs in most subjects (decreased in others). *3552*

Cortisone

Alanine Aminotransferase *Serum Increase Physiological* May contribute to long duration of hepatitis. *2220*

Aldolase *Serum Increase Physiological* In experimental animals ?of muscle origin. *3835*

Aldosterone *Urine No Effect Analytical* No significant effect on RIA procedure of Drewes. *0952*
Urine Increase Analytical 1 mg/L equivalent to 8 µg/L in method of Drewes. *0952*

α-Amino-Nitrogen *Urine Increase Physiological* Increased tissue catabolism. *3879*

Aspartate Aminotransferase
Serum Increase Physiological May contribute to long duration of hepatitis. *2220*

Calcium *Serum Decrease Physiological* If elevated due to sarcoidosis or vitamin D. *0071*

Cells *Urine Increase Physiological* Increased number observed with long-term treatment. *2427*

Chloride *Serum Decrease Physiological* May cause hypochloremic alkalosis. *3961*
Urine Decrease Physiological Causes retention. *0071*

Cholesterol *Serum Increase Physiological* Increase about 20% with vigorous treatment. *3961*

Cholesterol Esters *Serum Increase Physiological* Increase about 20% with vigorous treatment. *3961*

Corticosteroids *Plasma Increase Analytical* Cross-react 19% (cortisol 100%) competitive protein binding procedure of Ficher. *1128*

Cortisol *Plasma Increase Physiological* Effect lasts for 24 h at least. *1772*

Eosinophils *Blood Decrease Physiological* Normal physiological response to injection. *2222*

Epinephrine *Urine Decrease Physiological* Decreased output with 200 mg/d. *0279*

Erythrocyte Sedimentation Rate
Blood Decrease Physiological Particularly in rheumatoid patients. *0251*

Estradiol *Urine Decrease Analytical* Affects unmodified Brown/Kober procedure. *0583*

Estriol *Urine Decrease Analytical* Sometimes low values produced method of Brown. *0023*

Estrogens *Urine Increase Analytical* Interfere with Brown's method Interferes with Kober procedure. *0279*

Glomerular Filtration Rate (GFR)
Urine Decrease Physiological Probably does this by decreased secretion of creatinine. *0913*

Glucose *Serum Increase Physiological* Gluconeogenesis. *2220*

Glucose Tolerance *Serum Decrease Physiological* Gluconeogenesis and anti-insulin effects. *2220*

17-Hydroxycorticosteroids *Urine Increase Physiological* Measuring excretory products of cortisone. *3879*

Iron *Serum Decrease Physiological* Reduced synthesis of transferrin. *3879*

Iron-Binding Capacity, Total (TIBC)
Serum Decrease Physiological Reduced synthesis of transferrin. *3879*

^{131}I Uptake *Serum Decrease Physiological* Probably diminishes TSH secretion. *2427*

17-Ketogenic Steroids *Urine Increase Physiological* Metabolic products. *2220*

17-Ketosteroids *Urine Decrease Physiological* Pituitary suppression of ACTH. *3879*

Neutrophils *Blood Increase Physiological* Significant granulocytosis observed. *1343*

Occult Blood *Feces Positive Physiological* May cause hemorrhage or ulceration of gastrointestinal tract. *0612*

Potassium *Serum Decrease Physiological* Mineralocorticoid effect. *3879*
Urine Increase Physiological Promotes excretion. *0071*

Protein Bound Iodine (PBI) *Serum Decrease Physiological* Inhibits iodination of tyrosine, also hemodilution. *0012*

Prothrombin Time *Plasma Decrease Physiological* Patients on coumarins. *2407*

Sodium *Serum Increase Physiological* Mineralocorticoid effect. *3879*
Urine Decrease Physiological Causes retention. *0071*

Thyroxine (T₄) *Serum Decrease Physiological* Decreased synthesis. *3505*

Thyroxine (T₄) Binding Globulin
Serum Decrease Physiological Reduced synthesis. *3879*

Transferrin *Serum Decrease Physiological* Reduced synthesis. *3879*

Uric Acid *Serum Decrease Physiological* Observed in normals after 200 mg daily. *3191*
Urine Increase Physiological Observed in normals after 200 mg daily. *3191*

Cotrifamole

Thyroid Stimulating Hormone (TSH)
Serum No Effect Physiological No change in men or women over 10 d. *0686*

Thyroxine (T₄) *Serum No Effect Physiological* Possible intrathyroidal inhibition of conversion of iodine to organic form. *3798* No change in men or women over 10 d. *0686*

Thyroxine (T₄) Index, Free (FTI)
Serum No Effect Physiological No change in men or women over 10 d. *0686*

Tri-iodothyronine (T₃) *Serum Decrease Physiological* Possible intrathyroidal inhibition of conversion of iodine to organic form. *3798* From 2.50 to 2.17 nmol/L in men and 1.90 to 1.65 nmol/L in women for 10 d. *0686*

Co-Trimoxazole

Acetaminophen Screening Test *Urine Positive Analytical* Red color with 1-naphthol at therapeutic concentrations. *3326*
Urine Negative Analytical No reaction with o-cresol at therapeutic concentrations. *3326*

Chloramphenicol *Serum No Effect Analytical* No effect at 100 mg/L on coupled enzymatic method. *2490* No effect at 100 mg/L on coupled enzymatic method. *2490*

Chlorpropamide *Serum Increase Physiological* Effects of compound may be potentiated. *0679* Effects of compound may be potentiated. *0679*

Creatinine *Serum Increase Physiological* Usually reversible effect (in many patients). *1852* By 0.2 mg/dL in 21 patients: probably due to competitive inhibition of tubular secretion of trimethoprim. *0313* Change of 0.12 mg/dL in 5 volunteers after 7 days. *3078*

Creatinine Clearance *Urine Decrease Physiological* Usually reversible effect (in many patients). *1852*

Crystals *Urine Increase Physiological* Single case reported. *1852* May occur with high doses of drug, particularly in patients with severe renal insufficiency. *0679*

Diphenylhydantoin *Serum Increase Physiological* Effects of compound may be potentiated. *0679*

Folic Acid *Serum No Effect Analytical* Drug has no effect on vitamin concentration if measured by radioimmunoassay. *0247*
Serum Decrease Analytical Due to effect on Lactobacillus Casei if organism used to measure amount of vitamin. *0247*

Hematocrit *Blood Increase Physiological* May cause megaloblastic anemia. *0745*

Hemoglobin *Blood Decrease Physiological* May cause hemolysis if G-6-PD deficiency. *2694*

Leukocytes *Blood Decrease Physiological* May cause leukopenia. *0745* Isolated cases reported. *0665*

Methemoglobin *Blood Increase Physiological* Possible effect of drug or sulfonamides. *1678*

Neutrophils *Blood Decrease Physiological* Also reduced survival of transfused platelets due to drug associated antibodies. *0665*

Phenylalanine *Plasma Increase Physiological* 4 h after 0.1 g/kg orally. *1036*

Platelets *Blood Decrease Physiological* May cause thrombocytopenia. *0745* Highly significant effect in warfarin treated patients but no effect on warfarin half-life. *0665* Effects of compound may be potentiated. *1910*

Co-Trimoxazole (continued)

Protein *Urine No Effect Analytical* No effect of up to 4 tablets daily in 4 patients on sulfosalicylic acid, trichloroacetic acid and Ponceau S Dye Methods. *3919*

Prothrombin Time *Plasma Increase Physiological* Several cases of platelet-associated IgG and thrombocytopenia. *2646*

Thyroid Stimulating Hormone (TSH)
Serum No Effect Physiological In men or women with treatment for 10 days. *0686*

Thyroxine (T₄) *Serum Decrease Physiological* Marked decrease possibly due to intrathyroidal inhibition of conversion of iodine to organic form. *3798* From 97.1 to 83.0 in men and 92.6 to 87.2 nmol/L in women with 10 days treatment. *0686*

Thyroxine (T₄) Index, Free (FTI)
Serum Decrease Physiological Marked decrease possibly due to intrathyroidal inhibition of conversion of iodine to organic form. *3798* From 95.2 to 82.6 in men and 90.1 to 86.4 in women with 10 days treatment. *0686*

Tolbutamide *Serum Increase Physiological* Effects of compound may be potentiated. *0679*

Tri-iodothyronine (T₃) *Serum Decrease Physiological* Marked decrease possibly due to intrathyroidal inhibition of conversion of iodine to organic form. *2022* From 2.33 to 1.96 nmol/L in men and 1.76 to 1.53 nmol/L in women after 10 days treatemtn. *0686*

Urea Nitrogen *Serum Increase Physiological* Usually reversible effect (in many patients). *1852* Uremia occurs if anemia, urine infection G-6-PD deficiency. *2694*

Warfarin *Plasma Increase Physiological* Effects of compound may be potentiated. *0679*

Coughing

Creatine Kinase *Serum Increase Physiological* May be doubling with severe coughing. *3179*

Cough Medicines

Alanine Aminotransferase *Serum Increase Physiological* One case chronic use terpin and codeine. *1069*

Alkaline Phosphatase *Serum Increase Physiological* One case chronic use terpin and codeine. *1069*

Aspartate Aminotransferase
Serum Increase Physiological One case chronic use terpin and codeine. *1069*

Protein Bound Iodine (PBI) *Serum Increase Physiological* Some may contain iodides. *1237*

Coumaric Acid

5-Hydroxyindoleacetic Acid (5-HIAA)
Urine Increase Analytical May interfere with colorimetric methods. *0913*

Coumarin

Alanine Aminotransferase *Serum Increase Physiological* May cause hepatic toxicity. *3835*

Alkaline Phosphatase *Serum Increase Physiological* May cause hepatotoxicity. *3835*

Aspartate Aminotransferase
Serum Increase Physiological May cause hepatic toxicity. *3835*

Bile *Urine Increase Physiological* May cause hepatotoxicity. *3835*

Bilirubin *Serum Increase Physiological* May cause hepatotoxicity. *3835*

BSP Retention *Serum Increase Physiological* May cause hepatotoxicity. *3835*

Cephalin Flocculation *Serum Increase Physiological* May cause hepatotoxicity. *3835*

Erythrocytes *Urine Increase Physiological* Actual bleeding may occur with high doses. *1022*

Hemoglobin *Urine Increase Physiological* Actual bleeding may occur with high doses. *1237*

T₃ Uptake *Serum Increase Physiological* Competes with T₃ for thyroxine binding albumin sites. *0583*

Thymol Turbidity *Serum Increase Physiological* May cause hepatotoxicity. *3835*

Uric Acid *Serum Decrease Physiological* Uricosuric effect. *1237*

Urine Increase Physiological Uricosuric effect. *1237*

Creatine

Ammonia *Plasma Decrease Analytical* With 1,000 nmoles produces 7% inhibition of indophenol react. *1290*

Sugar *Urine Increase Analytical* Yellow with Benedict's. *3507*

Tyrosine *Plasma No Effect Analytical* On fluorometric procedure of Ambrose. *0080*

Urea Nitrogen *Serum Increase Analytical* May react with Berthelot if urease not used. *2981*

Creatinine

α-Amino-Nitrogen *Urine No Effect Analytical* No effect if dinitrofluorobenzene method used. *1346*

Ammonia *Plasma Decrease Analytical* With 500 nmoles produces 14% inhibitor indophenol reaction. *1290*

Barbiturate *Serum Increase Analytical* Affect colorimetric method with cobalt acetate. *3054*

Cholesterol *Serum Increase Analytical* Increase by 0.5% at 11 mg/dL on enzymatic procedure. *0057*

Creatinine *Test Conditions Increase Analytical* Positive spot test with Jaffé reagent. *3765*

Glucose *Serum No Effect Analytical* No effect at 100 mg/dL on glucose oxidase procedure of Gochman. *1311* No effect on Warner Glucomatic method. *2771* At 15 mg/dL on MBTH procedure of Neeley. *2569* No effect on Trinder glucose-oxidase method. *2771*
Serum Increase Analytical Affects Neocuproin procedure. *2771* If measured by alkaline ferricyanide reduction. *2857*
Serum Decrease Analytical Negative peaks GOD-PERID procedure. *2771*
Urine No Effect Analytical With DNSA method at usual concentrations. *0076*

Guanidinosuccinic Acid *Serum No Effect Analytical* No effect on method of Kamoun, Pleau, Man. *1856*

Lead *Urine Increase Analytical* 100 mg/dL equivalent to 50 µg/L by atomic absorption. *3240*

Phosphate *Gastric Material No Effect Analytical* No effect at 20 mg/dL on method of Jung/Parekh. *1835*

Platelet Aggregation *Blood No Effect Physiological* No effect observed with concentrations associated with uremia. *0826*

Sugar *Urine Increase Analytical* Acts like reducing substance with Benedict's. *2313*

Thormählen Test *Urine Positive Analytical* Large amounts give brown color. *3588*

Tyrosine *Plasma No Effect Analytical* On fluorometric procedure of Ambrose. *0080*

Urea Nitrogen *Serum No Effect Analytical* No effect on Berthelot procedure. *1862*
Serum Increase Analytical Increases concentrations affect Nesslerization methods slightly. *1563*

Uric Acid *Serum No Effect Analytical* At 10 mg/dL on copper chelate procedure. At 10 mg/dL on Nishi phosphotungstate procedure. *2230*

Creatinine Clearance

Norepinephrine *Urine Increase Physiological* Positive correlation with clearance. *1636*

Vanillylmandelic Acid *Urine Increase Physiological* Positive correlation with clearance. *1636*

Cremomycin

[131]I Uptake *Serum Decrease Physiological* Merck compound contains tetraiodofluorescein. *2652*

Prothrombin Time *Plasma Increase Physiological* May cause hypoprothrombinemia. *0071*

Urea Nitrogen *Serum Increase Physiological* Dehydration may simulate renal disease. *0071*

Creosote

Color *Urine Increase Analytical* Dark green. *2313*

Cresol

Ammoniacal Silver Nitrate
Test Conditions Positive Analytical Positive spot test with Tollen's reagent. *3765*

Color *Urine Increase Analytical* Dark brown on standing. *2313*

Erythrocytes *Blood Decrease Physiological* May cause hemolytic anemia. *0279*

Hemoglobin *Blood Decrease Physiological* May cause hemolytic anemia. *0279*

5-Hydroxyindoleacetic Acid (5-HIAA)
Urine No Effect Analytical No effect 10 mg/dL method of Goldenberg. *1327*

Millons Test *Test Conditions Positive Analytical* Positive spot test for phenols. *3765*

Protein *Urine Increase Physiological* May cause renal damage. *0279*

Sugar *Urine Increase Analytical* Excreted as glucuronide, reacts with Benedict's. *0629*

Sulfa as Sulfanilamide *Serum Increase Analytical* Reacts in Bratton-Marshall procedure. *0279*

Urea Nitrogen *Serum Increase Physiological* May cause renal damage. *0279*

Urobilinogen *Test Conditions Increase Analytical* Positive spot test with Ehrlich's reagent. *3765*

Cromoglycate

Alanine Aminotransferase *Serum No Effect Physiological* No clinical significant change observed. *0484*

Aspartate Aminotransferase
Serum No Effect Physiological No clinical significant change observed. *0484*

Eosinophils *Blood No Effect Physiological* No clinical significant change observed. *0484*

Erythrocyte Sedimentation Rate
Blood No Effect Physiological No clinical significant change observed. *0484*

Hemoglobin *Blood No Effect Physiological* No clinical significant change observed. *0484*
Urine No Effect Physiological No clinical significant change observed. *0484*

17-Hydroxycorticosteroids *Plasma Decrease Physiological* Significant effect observed if given prior to exercise. *1641*
Urine No Effect Physiological No proportional increase in relation to blood change. *1641*

Leukocytes *Blood No Effect Physiological* No clinical significant change observed. *0484*

Protein *Urine No Effect Physiological* No clinical significant change observed. *0484*

Urea Nitrogen *Serum No Effect Physiological* No clinical significant change observed. *0484*

Urobilin *Urine No Effect Physiological* No clinical significant change observed. *0484*

Croton Oil

Agglutinins *Blood Positive Physiological* May induce formation. *0279*

Cryoglobulin

Leukocytes *Blood Increase Analytical* ?particle formation with Coulter model S. *1032*

Cyanates

Acetoacetate *Urine Increase Analytical* May produce interfering color with Gerhardt FeCl₃. *1022*

Ferric Chloride Test *Urine Positive Analytical* May produce red color. *0443*

Cyanides

Acetoacetate Decarboxylase *Serum No Effect Analytical* No inhibition produced. *3684*

Alkaline Phosphatase *Serum Decrease Analytical* Inhibitors of enzyme in laboratory methods. *3588*

Color *Blood Increase Physiological* May cause cherry-red color. *0279*

Protein *Test Conditions Increase Analytical* Interferes with Folin-Ciocalteu of Lowry method. *2202*

Protein Bound Iodine (PBI) *Serum Decrease Physiological* Inhibit iodination of tyrosine in thyroxine binding globulin. *0012*

Uric Acid *Serum Decrease Analytical* Uricase is sensitive to cyanide. *1563*

Cyclacillin

Glucose *Urine No Effect Analytical* No effect on Diastix® at physiological concentration. No effect on TesTape® at physiological concentration. *2733*
Urine Increase Analytical Falsely elevated value of 0.25% approximately at physiological concentration. *2733*

Cyclamate

Chromosomes *Test Conditions Abnormal Physiological* Potent clastogen in human cells. *3282*

Lincomycin *Serum Decrease Physiological* Inhibits gastrointestinal absorption. *1487*

Cyclazocine

Morphine *Urine No Effect Analytical* Insignificant cross reactivity with RIA procedures Insignificant cross reactivity with EMIT procedure for opiates. *2514*

Cyclic AMP

Cortisol *Plasma Increase Physiological* Hormonal action. *1102*

Glucose *Serum Increase Physiological* Hormonal action. *1102*

Growth Hormone *Plasma Increase Physiological* Hormonal action. *1102*

Insulin *Plasma Increase Physiological* Hormonal action. *1102*

Cyclobenzaprine

Amitriptyline *Serum Increase Analytical* Reported interference with EMIT, TLC, GC, LC and GC/MS: may also interfere with other antidepressants. *2888*

Cyclofenil

Alanine Aminotransferase *Serum Increase Physiological* Slightly and transiently increased in 6 of 19 patients over 4 mo of treatment, especially with high doses. *2786* In 1 case, but total of 30 cases reported up to 1980; Considered to be metabolic idiosyncrasy and reversible in all cases. *2672*

Alkaline Phosphatase *Serum Increase Physiological* In 1 case, but total of 30 cases reported up to 1980; Considered to be metabolic idiosyncrasy and reversible in all cases. *2672*

Cyclofenil (continued)

Apolipoprotein AI *Serum Increase Physiological* Increase by 15% after 2 and 4 mo after therapy in 19 patients given 600 mg daily for 4 mo. *2786*

Aspartate Aminotransferase
Serum Increase Physiological Slightly and transiently increased in 6 of 19 patients over 4 mo of treatment, especially with high doses. *2786* In 1 case, but total of 30 cases reported up to 1980; Considered to be metabolic idiosyncrasy and reversible in all cases. *2672*

Bilirubin *Serum Increase Physiological* In 1 case, but total of 30 cases reported up to 1980; Considered to be metabolic idiosyncrasy and reversible in all cases. *2672*

Cholesterol *Serum Increase Physiological* Slight effect in 19 patients given 600 mg daily for 4 mo. *2786*

Cholesterol, High Density Lipoprotein
Serum Increase Physiological By approximately 15% with 2 and 4 mo of therapy in 19 patients given 600 mg daily for 4 mo. *2786*

Cholesterol, Low Density Lipoprotein
Serum No Effect Physiological Nonsignificant tendency to increase in 19 patients given 600 mg daily for 4 mo. *2786*

Estradiol *Urine No Effect Physiological* Normal amounts observed with GC/MS. *0023*
Urine Decrease Analytical Marked effect on colorimetric procedures. *0023*

Estriol *Urine Decrease Analytical* Less intense effect on colorimetric procedures. *0023*

Estrogens *Urine Decrease Analytical* Affects colorimetric/fluorometric procedures. *0022*

Triglycerides *Serum Increase Physiological* Slight increase over 4 mo in 19 patients given 600 mg daily for therapy. *2786*

Cycloheximide

Leukocytes *Blood Decrease Physiological* Leukopenia (AMA Blood dyscrasias). *2429*

Cyclohexylamine

Chromosomes *Test Conditions Abnormal Physiological* Potent clastogen in human cells. *3282*

Cycloleucine

Arginine *Urine Increase Physiological* Reversible marked aminoaciduria. *0499*

Cystine *Urine Increase Physiological* Reversible marked aminoaciduria. *0499*

Lysine *Urine Increase Physiological* Reversible marked aminoaciduria. *0499*

Cyclopentamine

Amphetamine *Urine No Effect Analytical* No effect at 100 mg/L on method of Rutter. *3097*
Urine Increase Analytical Yields fluorophor with NBD chloride reaction. *2473*

Cyclopenthiazide

Coombs' Test *Serum Positive Physiological* Reported observation. *2583*

Neutrophils *Blood Decrease Physiological* Reported to cause neutropenia/agranulocytosis. *2583*

Platelets *Blood Decrease Physiological* Thrombocytopenia reported. *2583*

Cyclophosphamide

Alanine Aminotransferase *Serum No Effect Physiological* No significant perturbation in 41 patients. *2406*
Serum Increase Physiological May cause hepatotoxicity. *3433*

Albumin *Serum Decrease Physiological* May cause hepatotoxicity. *3433*

Alkaline Phosphatase *Serum No Effect Physiological* No significant perturbation in 41 patients. *2406*
Serum Increase Physiological May cause hepatotoxicity. *2313*

Aspartate Aminotransferase
Serum No Effect Physiological No significant perturbation in 41 patients. *2406*
Serum Increase Physiological May cause hepatotoxicity. *3433*

Bile *Urine Increase Physiological* May cause hepatotoxicity. *2313*

Bilirubin *Serum No Effect Physiological* No significant perturbation in 41 patients. *2406*
Serum Increase Physiological May cause hepatotoxicity. *2313*

BSP Retention *Serum Increase Physiological* May cause hepatotoxicity. *2313*

Cephalin Flocculation *Serum Increase Physiological* May cause hepatotoxicity. *2313*

Cholesterol *Serum No Effect Analytical* At concentration of 240 mg/L had no effect on CHOD-Iodide method. *3393* At concentration of 240 mg/L had no effect on CHOD-PAP method. *3393* At concentration of 240 mg/L had no effect on catalase-AIDH method. *3393* At concentration of 240 mg/L had no effect on method using catalase-Hantzsch reaction. *3393* At concentration of 240 mg/L had no effect on Liebermann-Burchard method. *3393*
Serum Increase Physiological Single case of drug induced myxedema. *0681*

Chromosomes *Test Conditions Abnormal Physiological* Clastogenic in human cells. *3282*

Coombs' Test, Direct *Serum Positive Physiological* May cause hemolytic anemia. *2365*

Creatinine *Serum No Effect Analytical* At concentration of 40 mg/L had no effect on Jaffé-Fading-Fraction method. *3393* At concentration of 40 mg/L had no effect on Jaffé-Fuller's earth method. *3393* At concentration of 452 mg/L had no effect on kinetic Jaffé method on BKA-2. *3393*

Creatinine Clearance *Urine No Effect Physiological* In majority of patients although decreased urine volume. With decreased sodium excretion in one patient. *0854*

Erythrocytes *Blood Decrease Physiological* May cause anemia usually reversible. *1680*
Urine Increase Physiological Actual bleeding may be caused by drug. *1022*

Glomerular Filtration Rate (GFR)
Urine No Effect Physiological In majority of patients although decreased urine volume. *0854*

Glucose *Serum No Effect Analytical* At concentration of 8 mg/L had no effect on Ektachem® method. *3393* At concentration of 40 mg/L had no effect on GOD/POD-PAP method. *3393*

Hemoglobin *Blood Decrease Physiological* May cause megaloblastic anemia. *2583*
Urine Increase Physiological Actual bleeding caused by drug. *1022*

^{131}I Uptake *Serum Decrease Physiological* Single case of drug induced myxedema. *0681*

Kininogen *Plasma Decrease Physiological* Maximum effect with onset of leukopenia. *0779*

Leukocytes *Blood Decrease Physiological* Bone marrow depression (reversible leukopenia). *0072*

Lymphocytic Response to PHA
Blood Decrease Physiological Significant reduction with chemotherapy observed. *0445*

Neutrophils *Blood Decrease Physiological* Reported observation. *2583*

Occult Blood *Feces Positive Physiological* Hemorrhagic colitis reported. *2427*

Osmolality *Serum Decrease Physiological* Average 15 mosm/kg for 1 d (due to metabolites). *0855*
Urine Increase Physiological Impaired water excretion 500 mosm/kg increase. *0855*

Platelets *Blood Decrease Physiological* May cause bone marrow depression. *0072*

Potassium *Serum No Effect Analytical* At concentration of 80 mg/L had no effect on measurement by ISE without predilution. *3393*

Protein Electrophoresis *Serum No Effect Analytical* At concentration of 80 mg/L had no effect on automated Olympus-Hite method. *3393*

Prothrombin Time *Plasma Increase Physiological* May cause hypoprothrombinemia. *3433*

Pseudocholinesterase *Serum Decrease Physiological* Causes inhibition of activity. *3959*

Sodium *Serum No Effect Analytical* At concentration of 80 mg/L had no effect on measurement by ISE without predilution. *3393*

Serum Decrease Physiological Average 8.5 meq/L for 1 (H_2O retention by metabolites). *0855*

Testosterone *Serum Decrease Physiological* Due to induced testicular atrophy. *1736*

Thymol Turbidity *Serum Increase Physiological* May cause hepatotoxicity. *2313*

Triglycerides *Serum No Effect Analytical* At concentration of 40 mg/L had no effect on GPO-PAP method. *3393*

Urea Nitrogen *Serum No Effect Analytical* At concentration of 8 mg/L had no effect on Ektachem® method. *3393*

Uric Acid *Serum No Effect Analytical* At concentration of 60 mg/L had no effect on Kageyama-Hantzsch method. *3393* At concentration of 40 mg/L had no effect on catalase-AIDH method. *3393* At concentration of 40 mg/L had no effect on uricase-PAP method. *3393*

Volume *Plasma Increase Physiological* Water retention due to metabolites. *0854*

Urine Decrease Physiological To 60 mL/h from 160 mL/h for d (4-12 h after drug). *0855*

Cyclopropane

Alanine Aminotransferase *Serum Increase Physiological* May cause hepatotoxicity (lasts several days). *2313*

Alkaline Phosphatase *Serum Increase Physiological* May cause hepatotoxicity (lasts several days). *2313*

Aspartate Aminotransferase
Serum Increase Physiological May cause hepatotoxicity (lasts several days). *2313*

Bile *Urine Increase Physiological* May cause hepatotoxicity (lasts several days). *2313*

Bilirubin *Serum Increase Physiological* May cause hepatotoxicity (lasts several days). *2313*

Bleeding Time *Patient No Effect Physiological* Anesthesia has no effect. *1343*

BSP Retention *Serum Increase Physiological* May cause hepatotoxicity (lasts several days). *2313*

Capillary Fragility *Patient No Effect Physiological* Anesthesia has no effect. *1343*

Catecholamines *Plasma Increase Physiological* Significant increase, may reduce blood to kidneys etc. *0071*
Plasma Decrease Analytical Falsely low with ethylene diamine method. *2285*

Cephalin Flocculation *Serum Increase Physiological* May cause hepatotoxicity (lasts several days). *2313*

Clotting Time *Blood No Effect Physiological* Anesthesia has no effect. *1343*

Fibrinogen *Plasma No Effect Physiological* Anesthesia has no effect. *1343*

Glucose *Serum Increase Physiological* Moderate with depletion of liver glycogen. *0071*

Hematocrit *Blood Increase Physiological* Possible effect due to decreased plasma volume. *1343*

Indocyanine Green *Serum Increase Physiological* Hepatic extraction impaired after injection. *1343*

pH *Blood Decrease Physiological* May cause metabolic acidosis. *1343*

Platelets *Blood No Effect Physiological* Anesthesia has no effect. *1343*

Prothrombin Time *Plasma No Effect Physiological* Anesthesia has no effect. *1343*

Thrombin Time *Plasma No Effect Physiological* Anesthesia has no effect. *1343*

Thymol Turbidity *Serum Increase Physiological* May cause hepatotoxicity (lasts several days). *2313*

Cycloserine

Alanine Aminotransferase *Serum Increase Physiological* May cause hepatotoxicity. *2313*

Alkaline Phosphatase *Serum Increase Physiological* May cause hepatotoxicity. *2313*

Amino-4-Imidazole-5-Carboxamide Ribotide (AICAR)
Urine Increase Physiological Occurs with folic acid deficiency. *3773*

Amphetamine *Urine No Effect Analytical* At 50 mg/L on fluorescent method of Hayes. *1534*

Aspartate Aminotransferase
Serum Increase Physiological May cause hepatotoxicity. *2313*

Bile *Urine Increase Physiological* May cause hepatotoxicity. *2313*

Bilirubin *Serum Increase Physiological* May cause hepatotoxicity. *2313*

BSP Retention *Serum Increase Physiological* May cause hepatotoxicity. *2313*

Cephalin Flocculation *Serum Increase Physiological* May cause hepatotoxicity. *2313*

Erythrocytes *Blood Decrease Physiological* May cause megaloblastic anemia. *3773*

FIGLU (N-Formiminoglutamic Acid)
Urine Increase Physiological Occurs with folic acid deficiency. *3773*

Folate *Serum Decrease Physiological* May cause megaloblastic anemia. *3773* Low serum folate in half patients treated with cycloserine plus isoniazid in contrast to 2 of 55 treated with isoniazid alone, but mechanism unknown. *2062*

Hematocrit *Blood Decrease Physiological* May cause megaloblastic anemia. *3773*

Hemoglobin *Blood Decrease Physiological* May cause megaloblastic anemia. *3773*

[131]I Uptake *Serum Decrease Physiological* Seromycin® contains tetraiodofluorescein. *2652*

Leukocytes *Blood Decrease Physiological* May occur with megaloblastic anemia. *3773*

MCV *Blood Increase Physiological* May cause megaloblastic anemia. *3773*

Platelets *Blood Decrease Physiological* May occur with megaloblastic anemia. *3773*

Thymol Turbidity *Serum Increase Physiological* May cause hepatotoxicity. *2313*

Cyclosporine

Alanine Aminotransferase *Serum No Effect Analytical* At concentration of 41.5 mmol/L (50 µg/L) on methods on Technicon® SMAC® II and Hitachi® 705. *1123*
Serum No Effect Physiological No effect observed in 23 patients with inflammatory ocular disease treated for for 6 mo. *2713*
Serum Increase Physiological Increase to 109 U/L on average as manifestation of hepatotoxicity in 18 bone-marrow transplant recipients. *0174* Both early and late toxicity observed but effects mild in comparison with nephrotoxicity in renal transplant patients. *1959*

Albumin *Serum No Effect Analytical* At concentration of 41.5 mmol/L (50 µg/L) on methods on Technicon® SMAC® II and Hitachi® 705. *1123*

Aldosterone *Plasma Decrease Physiological* Inappropriately low in several patients receiving drug possibly due to concomitant beta-blockers. *2836*

Alkaline Phosphatase *Serum No Effect Analytical* At concentration of 41.5 mmol/L (50 µg/L) on methods on Technicon® SMAC® II and Hitachi® 705. *1123*
Serum Increase Physiological Persistent increase reported in some renal transplant patients occurring within of start of treatment, possibly due to cholestasis. *0683* Increase to 124

Cyclosporine (continued)

Alkaline Phosphatase (continued)
U/L on average as manifestation of hepatotoxicity in 18 bone-marrow transplant recipients. 0174 Persistent increase observed in some renal transplant patients: speculation that it is due to increased sensitivity of bone to action of parathyroid hormone. 2201 Slight increase or normal in association with hepatotoxicity in renal transplant patients. 1959

Amylase Serum No Effect Analytical At concentration of 41.5 mmol/L (50 µg/L) on methods on Technicon® SMAC® II and Hitachi® 705. 1123

Androstenedione Plasma No Effect Physiological In 16 patients who developed hypertrichosis but also given added cortisone. 3190

Aspartate Aminotransferase Serum No Effect Analytical At concentration of 41.5 mmol/L (50 µg/L) on methods on Technicon® SMAC® II and Hitachi® 705. 1123
Serum No Effect Physiological No effect observed in 23 patients with inflammatory ocular disease treated for for 6 mo. 2713
Serum Increase Physiological Both early and late toxicity observed but effects mild in comparison with nephrotoxicity in renal transplant patients. 1959

Bilirubin Serum No Effect Analytical At concentration of 41.5 mmol/L (50 µg/L) on methods on Technicon® SMAC® II and Hitachi® 705. 1123
Serum Increase Physiological Observed in 1 of 23 patients with inflammatory ocular disease treated for 6 mo. 2713 Transient increase observed in 20% of renal transplant patients. 0683 In 18 of 21 patients after bone marrow transplantation. 0174 Both early and late toxicity observed but effects mild in comparison with nephrotoxicity in renal transplant patients. 1959 Association in 4 cadaveric renal transplant patients with cyclosporine trough concentration (reversible). 2091

Bilirubin, Direct Serum No Effect Analytical At concentration of 41.5 mmol/L (50 µg/L) on methods on Technicon® SMAC®-II and Hitachi® 705. 1123

Calcium Serum No Effect Analytical At concentration of 41.5 mmol/L (50 µg/L) on methods on Technicon® SMAC® II and Hitachi® 705. 1123

Casts Urine Increase Physiological Usually cellular or granular characteristic feature of nephrotoxicity in most common form of renal damage to drug. 0175

Chloride Serum No Effect Analytical At concentration of 41.5 mmol/L (50 µg/L) on methods on Technicon® SMAC® II and Hitachi® 705. 1123
Serum Increase Physiological Hyperchloremic acidosis in 7 of 43 renal allograft patients. 0027 Hyperchloremic metabolic acidosis out of proportion to reduction in GFR. 2836

Cholesterol Serum No Effect Analytical At concentration of 41.5 mmol/L (50 µg/L) on methods on Technicon®'SMAC® II and Hitachi® 705. 1123

Cortisol Plasma No Effect Physiological In 16 patients who developed hypertrichosis but also given added cortisone. 3190

Creatine Kinase Serum No Effect Analytical At concentration of 41.5 mmol/L (50 µg/L) on methods on Technicon® SMAC® II and Hitachi® 705. 1123

Creatine Kinase Isoenzymes Serum No Effect Analytical At concentration of 41.5 mmol/L (50 µg/L) on methods on Technicon® SMAC® II and Hitachi® 705. 1123

Creatinine Serum No Effect Analytical At concentration of 41.5 mmol/L (50 µg/L) on methods on Technicon® SMAC® II and Hitachi® 705. 1123
Serum Increase Physiological Mean change from 1.0 to 1.5 mg/dL in 8 of 23 patients with 2 to 6 mo treatment (patients with inflammatory ocular disease). 30% increase in more than half patients. 2713 In 4 patients with primary biliary cirrhosis in whom creatinine and drug concentration rose as bleeding occurred. 3014 In 62 renal allograft recipients from 28 to 90 d after transplant: effect greater than with azathioprine and prednisolone: reversible. 0635 In 8 of 23 patients with ocular inflammatory disease treated for 1 mo (mean change from 1.0 to 1.5 mg/dL). 2713 Characteristic feature of nephrotoxicity in most common form of renal damage to drug. 0175 In patients with rheumatoid arthritis: significant effect unrelated to initial. 0905

Dehydroepiandrosterone Sulfate (DHEA-S)
Plasma No Effect Physiological In 16 patients who developed hypertrichosis but also given added cortisone. 3190

Digoxin Serum Increase Physiological Higher concentration after digoxin administered for 4 d possibly due to diminished renal clearance. 3732

Erythrocytes Blood Decrease Physiological In 6 of 23 patients normochromic normocytic anemia in first 6 mo treatment (patients with inflammatory ocular disease). 2713
Urine No Effect Physiological No effect observed in 23 patients with inflammatory ocular disease treated for 6 mo. 2713

Erythrocyte Sedimentation Rate
Blood Increase Physiological Mean increase from 35 to 52 mm/1h in 16 patients with inflammatory ocular disease with 3 to 6 mo treatment. 2713

Estradiol Plasma No Effect Physiological In 16 patients who developed hypertrichosis but also given added cortisone. 3190

Folate Serum No Effect Physiological No significant effect observed in 23 patients with inflammatory ocular disease treated for 6 mo. 2713

Glomerular Filtration Rate (GFR)
Urine Decrease Physiological 51 versus 93 mL/min in 17 cardiac transplant recipients compared with those on azathioprine. 2536

Glucose Serum No Effect Analytical At concentration of 41.5 mmol/L (50 µg/L) on methods on Technicon® SMAC® II and Hitachi® 705. 1123

γ-Glutamyltransferase (GGT) Serum No Effect Analytical At concentration of 41.5 mmol/L (50 µg/L) on methods on Technicon® SMAC® II and Hitachi® 705. 1123

Hemoglobin Blood Decrease Physiological Reduction of 2- 4 g/dL over 6 mo treatment in 6 of 23 patients with inflammatory ocular disease. 2713 In 4 patients with primary biliary cirrhosis in whom creatinine and drug concentration rose as bleeding occurred. 3014 Increase by 2 to 4 g/dL associated with normochromic normocytic anemia in 6 of 23 patients with ocular inflammatory disease over 6 mo. 2713
Urine Increase Physiological In approximately 5% patients given drug versus 8% given azathioprine in over 200 renal transplant patients. 0572

17-Hydroxyprogesterone Plasma No Effect Physiological In 16 patients who developed hypertrichosis but also given added cortisone. 3190

Iron Serum No Effect Analytical At concentration of 41.5 mmol/L (50 µg/L) on methods on Technicon® SMAC® II and Hitachi® 705. 1123
Serum No Effect Physiological No significant effect observed in 23 patients with inflammatory ocular disease treated for 6 mo. 2713

Lactate Dehydrogenase Serum No Effect Analytical At concentration of 41.5 mmol/L (50 µg/L) on methods on Technicon® SMAC® II and. 1123

Leukocytes Blood Decrease Physiological In approximately 1% patients given drug versus 12% given azathioprine in over 200 renal transplant patients. 0572
Urine No Effect Physiological No effect observed in 23 patients with inflammatory ocular disease treated for 6 mo. 2713

Lymphocyte T-Cells Blood Decrease Physiological In renal transplant patients with normal graft function compared with normal controls, higher values seen in individuals with acute cyclosporine nephrotoxicity. 3288

Lymphocytes Blood Decrease Physiological In renal transplant patients with normal graft function compared with normal controls, higher values seen in individuals with acute cyclosporine nephrotoxicity. 3288

Magnesium Serum Decrease Physiological Mean of 1.06 meq/L versus 1.33 meq/L in methotrexate treated patients after 3 mo treatment of patients with bone marrow transplant. 1833
Urine Increase Physiological Renal loss due to nephrotoxicity observed in many patients with bone marrow transplantation. 1833

Neutrophils Blood No Effect Physiological No significant effect observed in 23 patients with inflammatory ocular disease treated for 6 mo. 2713

Occult Blood Feces Increase Physiological In approximately 3% patients given drug versus 6% given azathioprine in over 200 renal transplant patients. 0572

Platelets *Blood No Effect Physiological* No significant effect observed in 23 patients with inflammatory ocular disease treated for 6 mo. *2713*

Potassium *Serum No Effect Analytical* At concentration of 41.5 mmol/L (50 µg/L) on methods on Technicon® SMAC® II and Hitachi® 705. *1123*
Serum Increase Physiological Probably occurs as manifestation of nephrotoxicity with tubular dysfunction. *0683* Reported in 7 of 43 patients with good renal function following kidney transplantation: single case reported here which responded to fludrocortisone. *2795* Sustained hyperkalemia (6.0 to 7.1 mmol/L) inappropriate for renal function in in 7 of 43 renal allograft patients. *0027* Observed in 13 of 50 renal transplant patients after 1 mo. *1154* Hyperkalemia out of proportion to reduction in GFR. *2836*

Prednisolone *Serum Increase Physiological* Increase of area under curve and half-life with reduction of body clearance by 22%. *3732*

Prolactin *Plasma No Effect Physiological* In 16 patients who developed hypertrichosis but also given added cortisone. *3190*

Prostaglandin, 6-Keto-F$_1\alpha$ *Urine No Effect Physiological* No difference in excretion between azathioprine and cyclosporine renal- transplant patients(6-keto). *0223*

Prostaglandin E2 *Urine No Effect Physiological* No difference in excretion between azathioprine and cyclosporine renal-transplant patients. *0223*

Protein *Serum No Effect Analytical* At concentration of 41.5 mmol/L (50 µg/L) on methods on Technicon® SMAC® II and Hitachi® 705. *1123*
Urine No Effect Physiological No effect observed in 23 patients with inflammatory ocular disease treated for for 6 mo. *2713*
Urine Increase Physiological Characteristic feature of nephrotoxicity in most common form of renal damage to drug. *0175*

Renal Blood Flow (RBF) *Patient Decrease Physiological* 320 mL/min versus 480 mL/min in 17 cardiac transplant recipients compared with those on azathioprine. *2536*

Renin Activity *Plasma Decrease Physiological* Significantly lower than in azathioprine treated patients or in those with renal insufficiency, possibly due to expansion of extracellular fluid. *0223* Inappropriately low in several patients receiving drug possibly due to concomitant beta-blockers concentration. *2836*

Sex Hormone Binding Globulin
Serum No Effect Physiological In 16 patients who developed hypertrichosis but also given added cortisone. *3190*

Sodium *Serum No Effect Analytical* At concentration of 41.5 mmol/L (50 µg/L) on methods on Technicon® SMAC® II and Hitachi® 705. *1123*

Testosterone *Serum No Effect Physiological* In 16 patients who developed hypertrichosis but also given added cortisone. *3190*

Urea Nitrogen *Serum No Effect Analytical* At concentration of 41.5 mmol/L (50 µg/L) on methods on Technicon® SMAC® II and Hitachi® 705. *1123*
Serum Increase Physiological Mean change from 13 to 27 mg/dL in 8 of 23 patients with 2 to 6 mo treatment (patients with inflammatory ocular disease). *2713* Increased from 13 to 27 mg/dL in 8 of 23 patients with ocular inflammatory disease. *2713* Characteristic feature of nephrotoxicity in most common form of renal damage to drug. *0175*

Uric Acid *Serum No Effect Analytical* At concentration of 41.5 mmol/L (50 µg/L) on methods on Technicon® SMAC® II and Hitachi® 705. *1123*
Serum Increase Physiological Mean increase from 5.7 to 7.0 mg/dL in 26 patients with inflammatory ocular disease after 3 to 6 mo treatment. 9 patients had increase greater than 2.0 mg/dL. *2713* In 62 renal allograft recipients from 28 to 90 d after transplant: effect greater than with azathioprine and prednisolone: reversible. *0635* Changed from mean of 5.7 mg/dL pretreatment to 7.0 mg/dL in patients with ocular inflammatory disease, not associated with impaired renal function. *2713*

Vitamin B$_{12}$ *Serum No Effect Physiological* No significant effect observed in 23 patients with inflammatory ocular disease treated for 6 mo. *2713*

Cyclothiazide

Alanine Aminotransferase *Serum Increase Physiological* Occasional intrahepatic cholestatic jaundice. *1680*

Amylase *Serum Increase Physiological* Occasional case of pancreatitis observed. *1680*

Aspartate Aminotransferase
Serum Increase Physiological Occasional intrahepatic cholestatic jaundice. *1680*

Bilirubin *Serum Increase Physiological* Occasional intrahepatic cholestatic jaundice. *1680*

Chloride *Urine Increase Physiological* Intended diuretic action. *1680*

Erythrocytes *Blood Decrease Physiological* Occasional response to thiazides. *1680*

Glucose *Urine Increase Physiological* Diabetogenic action of thiazides. *1680*

^{131}I Uptake *Serum Decrease Physiological* Anhydron contains tetraiodofluorescein. *2652*

Leukocytes *Blood Decrease Physiological* Occasional response to thiazides. *1680*

Osmolality *Serum Decrease Physiological* Syndrome of inappropriate ADH secretion seen. *1661*

Platelets *Blood Decrease Physiological* Occasional response to thiazides. *1680*

Potassium *Serum Decrease Physiological* Same degree of hypokalemia as hydrochlorothiazide. *0859*

Sodium *Serum Decrease Physiological* Syndrome of inappropriate ADH secretion seen. *1661*
Urine Increase Physiological Intended diuretic action. *1680*

Urea Nitrogen *Serum Increase Physiological* May occur with decreased renal flow. *1680*

Uric Acid *Serum Increase Physiological* Inhibition of tubular secretion of urate. *1433*

Volume *Urine Increase Physiological* Potent diuretic, same as hydrochlorothiazide. *0859*

Cyproheptadine

Alanine Aminotransferase *Serum Increase Physiological* Isolated case of jaundice due to cholestasis: reversed on withdrawal of drug. *1562*

Alkaline Phosphatase *Serum Increase Physiological* Isolated case of jaundice due to cholestasis: reversed on withdrawal of drug. *1562*

Amylase *Serum Increase Physiological* Mechanism not listed. *1488*

Aspartate Aminotransferase
Serum Increase Physiological Isolated case of jaundice due to cholestasis: reversed on withdrawal of drug. *1562*

Bilirubin *Serum Increase Physiological* Isolated case of jaundice due to cholestasis: reversed on withdrawal of drug. *1562*

Diphenylhydantoin *Serum No Effect Analytical* On GLC procedure of Papadopoulos at *in vivo* concentration. *2722*

Glucose *Serum No Effect Physiological* Reported to have no effect. *1487*
Serum Decrease Physiological Small decrease (10%), often no change. *1488*

Glucose Tolerance *Serum No Effect Physiological* Reported to have no effect. *1487*

Pheneturide *Serum No Effect Analytical* On GLC procedure of Papadopoulos at *in vivo* concentration. *2722*

Phenobarbital *Serum No Effect Analytical* On GLC procedure of Papadopoulos at *in vivo* concentration. *2722*

Primidone *Serum No Effect Analytical* On GLC procedure of Papadopoulos at *in vivo* concentration. *2722*

Prolactin *Plasma No Effect Physiological* No effect on basal concentration or response to hypoglycemia in morning or evening. *2565*

Cyproterone

Androstenedione *Plasma   No Effect   Physiological* In 27 women treated for total of 194 cycles with 50 µg ethinyl estradiol and 2.0 mg cyproterone acetate. *2280*

Plasma   Increase   Physiological By up to 450% if acetate derivative used. *3394*

Cholesterol *Serum   Increase   Physiological* When given in combination with ethinyl estradiol. *2177*

Serum   Decrease   Physiological In oophorectomized women when given alone. In oophorectomized women (slight effect) when given alone. *2177*

Cholesterol, Free *Serum   No Effect   Physiological* When given in combination with Ethinyl Estradiol. *2177*

Serum   Decrease   Physiological In oophorectomized women when given alone. *2177*

C-Peptide *Plasma   No Effect   Physiological* Unchanged following oral glucose when given with ethinyl estradiol combination causes insulin resistance. *3239*

Dehydroepiandrosterone (DHEA)

Plasma   No Effect   Physiological In 27 women treated for total of 194 cycles with 50 µg ethinyl estradiol and 2.0 mg cyproterone acetate. *2280*

Dehydroepiandrosterone Sulfate (DHEA-S)

Plasma   Decrease   Physiological In 27 women treated for total of 194 cycles with 50 µg ethinyl estradiol and 2.0 mg cyproterone acetate. *2280*

Estradiol *Plasma   Decrease   Physiological* In 27 women treated for total of 194 cycles with 50 µg ethinyl estradiol and 2.0 mg cyproterone acetate. *2280*

Estrone *Plasma   Decrease   Physiological* In 27 women treated for total of 194 cycles with 50 µg ethinyl estradiol and 2.0 mg cyproterone acetate. *2280*

Follicle Stimulating Hormone (FSH)

Plasma   No Effect   Physiological No change following oral administration. *3394*

Glucose *Serum   Decrease   Physiological* When given with ethinyl estradiol: combination causes insulin resistance. *3239*

Glucose Tolerance *Serum   Decrease   Physiological* When given with ethinyl estradiol: combination causes insulin resistance. *3239*

Insulin *Plasma   Increase   Physiological* Fasting concentration when given with ethinyl estradiol: combination causes insulin resistance. *3239*

Luteinizing Hormone (LH) *Plasma   No Effect   Physiological* No change following oral administration. *3394*

Phospholipids, Total *Serum   Increase   Physiological* When given in combination with ethinyl estradiol. *2177*

Serum   Decrease   Physiological In oophorectomized women (slight effect) when given alone. *2177*

Progesterone *Plasma   Decrease   Physiological* In 27 women treated for total of 194 cycles with 50 µg ethinyl estradiol and 2.0 mg cyproterone acetate. *2280*

Prolactin *Plasma   No Effect   Physiological* In 27 women treated for total of 194 cycles with 50 µg ethinyl estradiol and 2.0 mg cyproterone acetate 2.0 mg cyproterone acetate. *2280*

Sperm Count *Semen   Decrease   Physiological* May completely inhibit spermatogenesis with high doses. *0953*

Testosterone *Serum   No Effect   Physiological* In 27 women treated for total of 194 cycles with 50 µg ethinyl estradiol and 2.0 mg cyproterone acetate. *2280*

Serum   Decrease   Physiological Increase by 20-40% (antiandrogenic effect). *3394* Decreases hormone in adult men but does not affect concentration in children with precocious puberty. Decreases concentration in hirsute women. *1736*

Testosterone, Free *Serum   Decrease   Physiological* In 27 women treated for total of 194 cycles with 50 µg ethinyl estradiol and 2.0 mg cyproterone. *2280*

Triglycerides *Serum   No Effect   Physiological* In oophorectomized women when given alone. *2177*

Serum   Increase   Physiological When given in combination with ethinyl estradiol. *2177*

Cysteine

Alanine Aminotransferase *Serum   No Effect   Analytical* At 5 times upper limit of therapeutic range on methods on SMAC®, Abbott-VP, aca, Cobas-Bio, and KDA. *2138*

Albumin *Serum   No Effect   Analytical* At 5 times upper limit of therapeutic range on methods on SMAC®, Ektachem®, Hitachi® 705 and KDA. *2138*

Alkaline Phosphatase *Serum   No Effect   Analytical* At 5 times upper limit of therapeutic range on methods on KDA and Cobas-Bio. *2138*

Serum   Decrease   Analytical At 5 times upper limit of therapeutic range on methods on Abbott-VP, aca, and Hitachi® 705. *2138*

Ammonia *Plasma   No Effect   Analytical* On indophenol reaction with 1,000 nmoles. *1290*

Plasma   Decrease   Analytical With 500 nmoles produces 30% inhibitor indophenol reaction. *1290*

Amylase *Serum   No Effect   Analytical* At 5 times upper limit of therapeutic range on methods on aca, Cobas-Bio, and Ektachem®. *2138*

Aspartate Aminotransferase *Serum   No Effect   Analytical* At 5 times upper limit of therapeutic range on methods on SMAC®, Abbott-VP, aca, Cobas-Bio, Hitachi® 705 and KDA. *2138*

Bilirubin *Serum   No Effect   Analytical* At 5 times upper limit of therapeutic range on methods on SMAC®, aca, Cobas-Bio, Ektachem®, Hitachi® 705 and KDA. *2138*

Calcium *Serum   No Effect   Analytical* At 5 times upper limit of therapeutic range on methods SMAC®, Abbott-VP, aca, Ektachem®, Hitachi® 705 and KDA. *2138*

Cholesterol *Serum   No Effect   Analytical* At 5 times upper limit of therapeutic range on methods SMAC®, Abbott-VP, Cobas-Bio, Ektachem®, Hitachi® 705 and KDA. *2138*

Citrulline *Test Conditions   Decrease   Analytical* Inhibits color development in Archibald procedure. *3424*

Creatine Kinase *Serum   No Effect   Analytical* At 5 times upper limit of therapeutic range on methods on Abbott-VP, Cobas-Bio, and Hitachi® 705. *2138*

Serum   Increase   Analytical Stimulates activity of enzyme especially if been stored. *0718* At 5 times upper limit of therapeutic range on methods SMAC® and aca. *2138*

Creatinine *Serum   No Effect   Analytical* At 5 times upper limit of therapeutic range on methods SMAC®, Abbott-VP, Cobas-Bio, Ektachem®, Hitachi® 705 and KDA. *2138*

Glucose *Serum   No Effect   Analytical* At 10 mg/dL no effect on glucose oxidase procedure of Gochman. *1311* At 5 times upper limit of therapeutic range on methods on SMAC®, Abbott-VP aca, Cobas-Bio, Hitachi® 705 and KDA. *2138*

Serum   Increase   Analytical At 10 mg/dL affects alkaline ferricyanide procedure. *1311*

Serum   Decrease   Analytical May affect some glucose-oxidase procedures. *1563* At 5 times upper limit of therapeutic range on glucose oxidase method on Ektachem®. *2138*

γ-Glutamyltransferase (GGT) *Serum   No Effect   Analytical* At 5 times upper limit of therapeutic range on methods on SMAC®, Abbott-VP, and Hitachi® 705. *2138*

Iron *Serum   No Effect   Analytical* At 5 times upper limit of therapeutic range on Ferrozine method on SMAC®. *2138*

Lactate Dehydrogenase *Serum   No Effect   Analytical* At 5 times upper limit of therapeutic range on methods on SMAC®, Abbott-VP and Hitachi® 705. *2138*

Phosphate *Serum   No Effect   Analytical* At 5 times upper limit of therapeutic range on methods on SMAC®, aca, Hitachi® 705 and KDA. *2138*

Protein *Serum   No Effect   Analytical* At 5 times upper limit of therapeutic range on methods on SMAC®, Abbott-VP, Ektachem®, Hitachi® 705 and KDA. *2138*

Test Conditions   Increase   Analytical Reacts with Folin-Ciocalteu of Lowry procedure. *0702*

Triglycerides *Serum   No Effect   Analytical* At 5 times upper limit of therapeutic range on methods on SMAC®, Abbott-VP, Ektachem® and Hitachi® 705. *2138*

Serum   Increase   Analytical At 5 times upper limit of therapeutic range on enzymatic method on KDA. *2138*

Tyrosine *Plasma   No Effect   Analytical* On fluorometric procedure of Ambrose. *0080*

Urea Nitrogen *Serum No Effect Analytical* At 5 times upper limit of therapeutic range on methods SMAC®, Abbott-VP, Cobas-Bio, Ektachem®, Hitachi® 705 and KDA. *2138*

Uric Acid *Serum No Effect Analytical* At 200 mg/L on uricase procedure of Kabasak. *1840* At 5 times upper limit of therapeutic range on uricase methods on Abbott-VP. aca, Cobas-Bio, Ektachem® and Hitachi® 705. *2138*
Serum Increase Analytical 5 mg/dL equivalent to 0.5 mg/dL with Nishi procedure 5 mg/dL equivalent to 2.8 mg/dL copper chelate procedure. *2230* Marked increase at 5 times upper limit of therapeutic range on phosphotungstate methods on SMAC® and KDA. *2138*

Cystine

Alanine Aminotransferase *Serum No Effect Analytical* At 5 times upper limit of therapeutic range on methods on SMAC®, Abbott-VP, aca, Cobas-Bio and KDA. *2138*

Albumin *Serum No Effect Analytical* At 5 times upper limit of therapeutic range on methods on SMAC®, Ektachem®, 705 and KDA. *2138*

Alkaline Phosphatase *Serum No Effect Analytical* At 5 times upper limit of therapeutic range on methods on KDA and Cobas-Bio. *2138*
Serum Decrease Analytical At 5 times upper limit of therapeutic range on methods on SMAC®, Abbott-VP, aca, and Hitachi® 705. *2138*

Amylase *Serum No Effect Analytical* At 5 times upper limit of therapeutic range on methods on aca, Cobas-Bio and and Ektachem®. *2138*

Aspartate Aminotransferase *Serum No Effect Analytical* At 5 times upper limit of therapeutic range on methods on SMAC®, Abbott-VP, aca, Cobas-Bio, Hitachi® 705 and KDA. *2138*

Bilirubin *Serum No Effect Analytical* At 5 times upper limit of therapeutic range on methods on SMAC®, aca, Cobas-Bio, Ektachem®, Hitachi® 705 and KDA. *2138*

Calcium *Serum No Effect Analytical* At 5 times upper limit of therapeutic range on methods on SMAC®, Abbott-VP, aca, Ektachem®, Hitachi® 705 and KDA. *2138*

Cholesterol *Serum No Effect Analytical* At 5 times upper limit of therapeutic range on methods on SMAC®, Abbott-VP, Cobas-Bio, Ektachem®, Hitachi® 705 and KDA. *2138*

Creatine Kinase *Serum No Effect Analytical* At 5 times upper limit of therapeutic range on methods on Abbott-VP, Cobas-Bio, and Hitachi® 705. *2138*
Serum Increase Analytical At 5 times upper limit of therapeutic range on methods on SMAC® and aca. *2138*

Creatinine *Serum No Effect Analytical* At 5 times upper limit of therapeutic range on methods on SMAC®, Abbott-VP, aca, Cobas-Bio, Ektachem®, Hitachi® 705 and KDA. *2138*

Crystals *Urine Increase Physiological* Colorless, hexagonal, flat (in acid). *0443*

Glucose *Serum No Effect Analytical* At 5 times upper limit of therapeutic range on methods on SMAC®, Abbott-VP, aca, Cobas-Bio, Hitachi® 705 and KDA. *2138*
Serum Decrease Analytical At 5 times upper limit of therapeutic range on glucose oxidase method on Ektachem®. *2138*

γ-Glutamyltransferase (GGT) *Serum No Effect Analytical* At 5 times upper limit of therapeutic range on methods on SMAC®, Abbott-VP, and Hitachi® 705. *2138*

Histidine *Plasma Increase Analytical* May affect fluorometric method of Ambrose. *0081*

Iron *Serum No Effect Analytical* At 5 times upper limit of therapeutic range on methods on Ferrozine method on on SMAC®. *2138*

Lactate Dehydrogenase *Serum No Effect Analytical* At 5 times upper limit of therapeutic range on methods on SMAC®, Abbott-VP, and Hitachi® 705. *2138*

Phosphate *Serum No Effect Analytical* At 5 times upper limit of therapeutic range on methods on SMAC®, aca, Hitachi® 705 and KDA. *2138*

Proline *Plasma No Effect Analytical* No effect on method of Goodwin. *1345*

Protein *Serum No Effect Analytical* At 5 times upper limit of therapeutic range on methods on SMAC®, Abbott-VP, Ektachem®, Hitachi® 705 and KDA. *2138*

Triglycerides *Serum No Effect Analytical* At 5 times upper limit of therapeutic range on methods on SMAC®, Abbott-VP, Ektachem® and Hitachi® 705. *2138*
Serum Increase Analytical At 5 times upper limit of therapeutic range on enzymatic methods on KDA. *2138*

Urea Nitrogen *Serum No Effect Analytical* At 5 times upper limit of therapeutic range on methods on SMAC®, Abbott-VP, Cobas-Bio, Ektachem®, Hitachi® 705 and KDA. *2138*

Uric Acid *Serum No Effect Analytical* At 5 mg/dL on copper chelate procedure. *2230* At 5 times upper limit of therapeutic range on uricase methods on Abbott-VP, aca, Cobas-Bio, Ektachem® and Hitachi® 705. *2138*
Serum Increase Analytical 5 mg/dL equivalent to 0.3 mg/dL with Nishi procedure. *2230* Marked increase at 5 times upper limit of therapeutic range on phosphotungstate methods on SMAC® and KDA. *2138*

Cytarabine

Alanine Aminotransferase *Serum Increase Physiological* Hepatotoxic effect. *1343*

Alkaline Phosphatase *Serum Increase Physiological* In two cases with acute leukemia: unlikely to have been due to other drugs co-administered: effects slight. *2827* At 3 g/m^2 every 12 h for 6 d, some mild and reversible abnormalities in about 75% of 12 cases. *1579* Slight effect and returned to normal while still under treatment. *2406*

Aspartate Aminotransferase
Serum Increase Physiological In two cases with acute leukemia: unlikely to have been due to other drugs co-administered: effect slight. *2827* At 3 g/m^2 every 12 h for 6 d, some mild and reversible abnormalities in about 75% of 12 cases. *1579* Slight effect and returned to normal while still under treatment. *2406*

Bilirubin *Serum Increase Physiological* At 3 g/m^2 every 12 h for 6 d, some mild and reversible abnormalities in about 75% of 12 cases. *1579* Significant elevation in 7 of 42 patients with previously normal liver function. *2406*

Bilirubin, Direct *Serum Increase Physiological* In two cases with acute leukemia: unlikely to have been due to other drugs co-administered: effects slight. *2827*

Chromosomes *Test Conditions Abnormal Physiological* Clastogenic in human cells. *3282*

Erythrocytes *Blood Decrease Physiological* Megaloblastic anemia (relatively infrequent). *2313* Aplasia of bone marrow developed in nearly all of 57 leukemics. *1579*

Hemoglobin *Blood Decrease Physiological* May cause anemia. *2313*

Leukocytes *Blood Decrease Physiological* Agranulocytosis/leukopenia. *2313* Aplasia of bone marrow developed in nearly all of 57 leukemics. *1579*

Occult Blood *Feces Positive Physiological* Gastrointestinal hemorrhage in fewer than 2%. *1680*

Platelets *Blood Decrease Physiological* Due to bone marrow depression. *2313* Aplasia of bone marrow developed in nearly all of 57 leukemics. *1579*

Protein *CSF Decrease Analytical* False low value with Folin-Ciocalteu reagent False low with turbidimetric method on standing. *3969*

Reticulocytes *Blood Decrease Physiological* In fewer than 2%. *1680*

Uric Acid *Serum Increase Physiological* Secondary to rapid lysis of neoplastic cells. *1680*

Dactinomycin

Erythrocytes *Blood Decrease Physiological* Anemia/pancytopenia. *0071*

Hemoglobin *Blood Decrease Physiological* May cause anemia. *0071*

Leukocytes *Blood Decrease Physiological* Leukopenia/pancytopenia. *0071*

Platelets *Blood Decrease Physiological* Thrombocytopenia/pancytopenia. *0071*

Reticulocytes *Blood Decrease Physiological* Reticulocytopenia/pancytopenia. *0071*

Danazol

Alanine Aminotransferase *Serum Increase Physiological* Significant effect at 4,8, 12 and 16 weeks after 600 mg daily in 18 women with endometriosis. *0591*

Albumin *Serum No Effect Physiological* No change during treatment of women with endometriosis. *2599* No change in 18 women with endometriosis given 600 mg/d for 6 mo. *0591*

Alkaline Phosphatase *Serum No Effect Physiological* No abnormalities observed in patients taking drug. *3906*

Androstenedione *Plasma Increase Physiological* Increase by 70% in 7 normal subjects given drug 800 mg daily. *2224*

Aspartate Aminotransferase

Serum Increase Physiological In 1 of 4 men and 0 in 7 women. *3906*

BSP Retention *Serum Increase Physiological* Develops in over one third of patients taking drug. *3906*

C_1-Inhibitor *Serum Increase Physiological* Approximate doubling in 5 patients with hereditary angioneurotic edema treated for up to 10 mo. *3219*

Carbamazepine *Serum Increase Physiological* Concentration increased almost 2 fold when drug added for treatment of fibrocystic breast disease. *3946*

Cholesterol *Serum No Effect Physiological* Inconsistent changes observed in 62 patients with endometriosis treated with 600 mg daily for up to 24 weeks. *3222* Insignificant change in 9 women with endometriosis treated for 6 mo. *0060* No significant change over 24 weeks in 12 women with endometriosis given 600 mg daily for 24 weeks. *1066* No effect after 6 mo of 600 mg/d. *3906*

Cholesterol, High Density Lipoprotein

Serum Decrease Physiological After 2 weeks with 600 mg/d in 12 women with endometriosis reduced by 49% and 59% after 6 weeks; returned to normal in 8 weeks after treatment stopped. *1066* By about 45% during first 2 mo in 62 patients with endometriosis treated with 600 mg daily for up to 24 weeks. *3222* Increase by 50% in 6 patients with endometriosis given 600 mg daily for 3 to 6 mo. *2279* In 9 subjects studied for 6 mo significant reduction: returned to normal at end of therapy. *0060* Increase by 49% after 2 weeks and 59% after 8 weeks in 12 women with endometriosis given 600 mg daily for 24 weeks. *1066*

Cholesterol, Low Density Lipoprotein

Serum Increase Physiological After 2 weeks with 600 mg/d in 12 women with endometriosis increased by 14% and and 34% after 8 weeks, normal in 8 weeks after treatment stopped. *1066* Constant but significant effect in 62 patients with endometriosis treated with 600 mg daily for up to 24 weeks. *3222* Increase by 51% in 6 patients with endometriosis given 600 mg daily for 3 to 6 mo. *2279* Increase by 14% after 2 weeks and 34% after 8 weeks in 12 women with endometriosis given 600 mg daily for 24 weeks. *1066*

Complement C4 *Serum Increase Physiological* Up to 8 fold increase in 5 patients with hereditary angioneurotic edema treated for up to 10 mo. *3219*

Cortisol *Plasma Increase Analytical* Probably invalid results because of high cross-reactivity with protein in competitive binding assays. *1479*

Plasma Decrease Analytical Concentrations low: in case of cortisol and testosterone displacement from plasma proteins occurs and protein binding assays are probably invalid. *1479*

Plasma Decrease Physiological Affect binding in women treated with drug so that proportion of free drug high. *1479*

Cortisol, Free *Plasma Increase Physiological* Proportion high due to displacement of hormone from protein. *1479*

Creatinine *Serum Increase Physiological* Slight effect in 18 women with endometriosis given 600 mg/d for 6 mo. *0591*

Dehydroepiandrosterone (DHEA)

Plasma Decrease Physiological Significant effect at 8,12 weeks after 600 mg daily in 18 women with endometriosis. *0591*

Dehydroepiandrosterone Sulfate (DHEA-S)

Plasma No Effect Physiological No consistent change noted in patients with prostatic cancer. *0693*

Plasma Increase Physiological Significant effect at 2,4 and 8 weeks after 600 mg daily in 18 women with endometriosis. *0591* Increase by 40% in 7 normal subjects given drug 800 mg daily. *2224*

Estradiol *Plasma No Effect Physiological* No consistent change noted in patients with prostatic cancer. *0693* In 7 normal women given 800 mg daily for 2 mo amenorrheic state induced drug. *2224*

Plasma Decrease Physiological But normal mid-follicular values during treatment in 62 patients with endometriosis treated with 600 mg daily for up to 24 weeks. *3222*

Estrone *Plasma No Effect Physiological* In 7 normal women given 800 mg daily for 2 mo amenorrheic state induced by drug. *2224*

Factor VIII *Plasma No Effect Physiological* In 21 hemophiliacs given drug for 2 weeks. *2627*

Factor IX *Plasma No Effect Physiological* In 21 hemophiliacs given drug for 2 weeks. *2627*

Fibrinogen *Plasma Decrease Physiological* Significant effect in 21 hemophiliacs given drug for 2 weeks. *2627*

Follicle Stimulating Hormone (FSH)

Plasma No Effect Physiological In 7 normal women given 800 mg daily for 2 mo amenorrheic state induced by drug. *2224*

Plasma Decrease Physiological Effect noted in 6 orchidectomized patients with prostatic cancer. *0693*

Glucagon *Plasma Increase Physiological* Significant reduction in one woman with systemic lupus erythematosus on withdrawal of drug. *0815* Concentration above 50 pmol/L observed in 7 women treated with up to 600 mg/d for up to 24 weeks: slight fall with glucose tolerance test. *3839*

Glucose Tolerance *Serum Decrease Physiological* In patients receiving drug, although baseline concentration normal. *3839* In response to oral and i.v. glucose tolerance associated with insulin resistance. *3906*

HDL_2-Cholesterol *Serum Decrease Physiological* Increase by 73% after 2 weeks in 12 women with endometriosis given 600 mg daily for 24 weeks. *1066*

HDL_3-Cholesterol *Serum Decrease Physiological* Increase by 29% after 2 weeks in 12 women with endometriosis given 600 mg daily for 24 weeks. *1066*

Insulin *Plasma No Effect Physiological* But rose to higher extent than in controls when given oral glucose tolerance test. *3839*

17-Ketosteroids *Urine Increase Physiological* 80% increase in 7 women given 800 mg daily for 2 mo. *2224*

Luteinizing Hormone (LH) *Plasma No Effect Physiological* No significant change in 5 patients with hereditary angioneurotic edema treated for up to 10 mo. *3219* In 7 normal women given 800 mg daily for 2 mo amenorrheic state induced by drug. *2224*

Plasma Decrease Physiological Effect noted in 6 orchidectomized patients with prostatic cancer. *0693*

Plasminogen *Plasma Increase Physiological* Significant effect in 21 hemophiliacs given drug for 2 weeks. *2627*

Potassium *Serum Increase Physiological* Slight effect in 18 women with endometriosis given 600 mg/d for 6 mo. *0591*

Progesterone *Plasma Decrease Physiological* Often to nondetectable level in 62 patients with endometriosis treated with 600 mg daily for up to 24 weeks. *3222*

Prolactin *Plasma Increase Physiological* Effect noted in 6 orchidectomized patients with prostatic cancer. *0693*

Sex Hormone Binding Globulin

Serum Decrease Physiological Reduction of binding capacity by 80% during treatment of women with endometriosis. *2599* When used for treatment of endometriosis can cause increase of up to 20%. *1736* Reduced up to 5 fold in 5 patients with hereditary angioneurotic edema treated for up to 10 mo. *3219* Marked reduction in women over 3 mo. *3906*

Sodium *Serum No Effect Physiological* No change in 18 women with endometriosis given 600 mg/d for 6 mo. *0591*

Testosterone *Serum No Effect Physiological* No consistent change noted in patients with prostatic cancer. *0693* No change during treatment of women with endometriosis. *2599*

Serum Increase Analytical May be cross-reactivity in certain radioimmunoassays. *0693* Probably invalid results because of high cross-reactivity with protein in competitive binding assays. *1479* Nonspecificity in RIA kits produced by diagnostic products corporation, Farmos Diagnostica and Bio-RIA. *3276*

Serum Increase Physiological Increase by 70% in 7 normal subjects given drug 800 mg daily for 2 mo. *2224*

Serum Decrease Analytical Concentrations low: in case of cortisol and testosterone displacement from plasma proteins occurs and protein binding assays are probably invalid. *1479*

Serum Decrease Physiological Throughout treatment period after 600 mg daily in 18 women with endometriosis. *0591* Affect binding in women treated with drug so that proportion of free drug high. *1479* Marked reduction in some patients in 5 patients with hereditary angioneurotic edema treated for up to 10 mo. *3219*

Testosterone, Free *Serum Increase Physiological* Causes displacement of hormone from snbg and may result in increase of up to 90%. *1736* Marked rise observed in women with endometriosis. Due to displacement from sex hormone binding globulin. *2599* Proportion high due to displacement of hormone from protein. *1479*

Thyroid Stimulating Hormone (TSH)
Serum No Effect Physiological No effect after 2 weeks with 800 mg/d but slight decrease after 4 weeks. *0985*
Serum Decrease Physiological Not initially, but after 4 weeks in 12 healthy female post-menopausal volunteers given up to 800 mg daily. *0985*

Thyroxine (T$_4$) *Serum Decrease Analytical* Concentrations low: in case of cortisol and testosterone displacement from plasma proteins occurs and protein binding assays are probably invalid. *1479*
Serum Decrease Physiological In healthy postmenopausal women after 2 weeks treatment with 400 to 800 mg daily through direct effect on thyroxine binding globulin production at cellular level. *0985* Affect binding in women treated with drug so that proportion of free drug high. *1479* In 12 healthy female post-menopausal volunteers given up to 800 mg daily. *0985* Fell by 26% in 9 individuals due to competition for binding sites or androgen-like effect. *2718*

Thyroxine (T$_4$) Binding Globulin
Serum Decrease Physiological In healthy postmenopausal women after 2 weeks treatment with 400 to 800 mg daily through direct effect on thyroxine binding globulin production at cellular level. *0985* Approximately halved in 5 patients with hereditary angioneurotic edema treated for up to 10 mo. *3219* Direct effect on production in 12 healthy female post-menopausal volunteers given up to 800 mg daily. *0985* Mean fall of 19% in 10 people due to competition for binding sites or androgen-like effect. *2718*

Thyroxine (T$_4$), Free *Serum No Effect Physiological* No effect after 2 weeks with 800 mg/d but slight increase after 4 weeks. *0985*
Serum Increase Physiological Proportion high due to displacement of hormone from protein. *1479* Slight effect after 4 weeks in 12 healthy female post-menopausal volunteers. *0985*

Thyroxine (T$_4$) Index, Free (FTI)
Serum No Effect Physiological No consistent change due to competition for binding sites or androgen-like effect. *2718*

Transcortin *Serum No Effect Physiological* No significant change in 5 patients with hereditary angioneurotic edema treated for up to 10 mo. *3219*

Triglycerides *Serum No Effect Physiological* Inconsistent changes observed in 62 patients with endometriosis treated with 600 mg daily for up to 24 weeks. *3222* No significant effect in 9 women with endometriosis treated for 6 mo. *0060* No effect after 6 mo of 600 mg/d. *3906*
Serum Increase Physiological Increase by 20% after 2 weeks and 14% after 8 weeks in 12 women with endometriosis given 600 mg daily for 24 weeks. *1066* In 12 women with endometriosis treated with 600 mg/d increased by 20% after 2 weeks. *1066*

Tri-iodothyronine (T$_3$) *Serum Decrease Physiological* In healthy postmenopausal women after 2 weeks treatment with 400 to 800 mg daily through direct effect on thyroxine binding globulin production at cellular level. *0985* In 12 healthy female post-menopausal volunteers given up to 800 mg daily. *0985*

Dandruff Medication

Protein Bound Iodine (PBI) *Serum Increase Analytical* Iodine contamination (in some preparations). *1237*

Dantrolene

Alanine Aminotransferase *Serum Increase Physiological* Chronic active hepatitis reported in a few patients. *2260* In 1.8% of 1044 patients, hepatocellular damage with acute or sub-acute hepatic disease or chronic active hepatitis. *3656*

Albumin *Serum Decrease Physiological* Single observation, ?significance. *0659*

Alkaline Phosphatase *Serum Increase Physiological* Observed effect, cause uncertain. *0659* Acute hepatitis as result of drug ingestion in 4 patients returned to normal after drug stopped. *3837* In 1.8% of 1044 patients, hepatocellular damage with acute or subacute hepatic disease or chronic active hepatitis. *3656*

Aspartate Aminotransferase
Serum Increase Physiological Observed effect, cause uncertain. *0659* Chronic active hepatitis reported in a few patients. *2260* Acute hepatitis as result of drug ingestion in 4 patients returned to normal after drug stopped. *3837* In 1.8% of 1044 patients, hepatocellular damage with acute or subacute hepatic disease or chronic active hepatitis. *3656*

Bilirubin *Serum Increase Physiological* Isolated observation in one individual. *0659* Acute hepatitis as result of drug ingestion in 4 patients returned to normal after drug stopped. *3837* In 1.8% of 1044 patients, hepatocellular damage with acute or subacute hepatic disease or chronic active hepatitis. *3656*

Cholesterol *Serum Increase Physiological* Isolated observation in one individual. *0659*

Creatine Kinase *Serum Increase Physiological* Isolated observation, uncertain relevance. *0659*

Eosinophils *Blood Increase Physiological* Isolated instance reported. *0659*

Hematocrit *Blood Decrease Physiological* Rare effect of no consequence. *0659*

Hemoglobin *Blood Decrease Physiological* Rare effect of no consequence. *0659*

Leukocytes *Urine Increase Physiological* Possible effect, perhaps urinary infection. *0659*

Protein *Serum Decrease Physiological* Single observation, ?significance. *0659*
Urine Increase Physiological Possible effect, perhaps urinary infection. *0659*

Prothrombin Time *Plasma Increase Physiological* Acute hepatitis as result of drug ingestion in 4 patients returned to normal after drug stopped. *3837*

Uric Acid *Serum Increase Physiological* Isolated observation in one individual. *0659*

Dapsone

Alanine Aminotransferase *Serum Increase Physiological* Sulfone syndrome in one 16 year old girl given 50 mg daily short-term administration. *3604*

Albumin *Serum Decrease Physiological* In 2 patients given long-term treatment. *1935* Significant effect after 32 mo treatment in 1 woman with dermatitis herpetiformis. *1164* To below 2.0 g/dL in one woman with dermatitis herpetiformis: possibly associated with increased intravascular catabolism of albumin. *0740* Sulfone syndrome in one 16 year old girl given 50 mg daily short-term administration. *3604*

Alkaline Phosphatase *Serum Increase Physiological* In 3 individuals treated for 3 to 36 weeks. *3478* Associated with hemolytic anemia when occurs as main side effect. *3423* Sulfone syndrome in one 16 year old girl given 50 mg daily short-term administration. *3604*

Aspartate Aminotransferase
Serum Increase Physiological Sulfone syndrome in one 16 year old girl given 50 mg daily short-term administration. *3604*
Plasma Increase Physiological Components of typical sulfone or dapsone syndrome. *2026*

Bilirubin *Serum Increase Physiological* In 3 individuals treated for 3 to 36 weeks. *3478* Associated with hemolytic anemia when occurs as main side effect. *3423* Sulfone syndrome in one 16 year old girl given 50 mg daily short-term administration. *3604*
Urine Increase Physiological Associated with hemolytic anemia in sulfone syndrome. *3604*

Bilirubin, Direct *Serum Increase Physiological* Sulfone syndrome in one 16 year old girl given 50 mg daily short-term administration. *3604*

Dapsone (continued)

Chloride *Serum Decrease Physiological* Sulfone syndrome in one 16 year old girl given 50 mg daily short-term administration. *3604*

Cholesterol *Serum Increase Physiological* In 3 individuals treated for 3 to 36 weeks. *3478*

Eosinophils *Blood Increase Physiological* Observed in 66% (2 of 3 patients) treated with drug who had rheumatoid arthritis. *3360* Components of typical sulfone or dapsone syndrome. *2026*

Erythrocytes *Blood Decrease Physiological* May cause hemolytic anemia. *1612*

Haptoglobin *Serum Decrease Physiological* Associated with hemolytic anemia when occurs as main side effect. *3423*

Heinz Body Formation *Blood Positive Physiological* Early stage of hemolytic anemia. *0333*

Hematocrit *Blood Decrease Physiological* Hemolysis (hemolytic anemia). *1612*

Hemoglobin *Blood Decrease Physiological* May cause hemolytic anemia. *1612* Associated with hemolytic anemia when occurs as main side effect. *3423* Anemia developed in 61% of 51 patients with leprosy. *2410*

Lactate Dehydrogenase *Serum Increase Physiological* In 3 individuals treated for 3 to 36 weeks. *3478* Associated with hemolytic anemia when occurs as main side effect. *3423* Sulfone syndrome in one 16 year old girl given 50 mg daily short-term administration. *3604*

Leukocytes *Blood Decrease Physiological* Agranulocytosis. *1612* After 7-9 weeks of treatment with drug plus pyrimethamine in 7 patients taking drug as prophylaxis against malaria. *1202*

Methemoglobin *Blood Increase Physiological* May cause hemolytic anemia. *1612*

Neutrophils *Blood Decrease Physiological* Occasional case of agranulocytosis reported. *3717* Isolated case of agranulocytoisis reported. *1140* After 7-9 weeks of treatment with drug plus pyrimethamine in 7 patients taking drug as prophylaxis against malaria. *1202*

Osmotic Fragility *Red Blood Cells Increase Physiological* In 20 of 51 patients with leprosy, most of whom were receiving 51 to 100 mg drug daily. *2410*

Platelets *Blood No Effect Physiological* After 7-9 weeks of treatment with drug plus pyrimethamine in 7 patients taking drug as prophylaxis against malaria. *1202*

Protein *Serum Decrease Physiological* Sulfone syndrome in one 16 year old girl given 50 mg daily short-term administration. *3604*

Sodium *Serum Decrease Physiological* To 117 meq/L in one woman with dermatitis herpetiformis: possibly associated increased intravascular catabolism of albumin. *0740* Sulfone syndrome in one 16 year old girl given 50 mg daily short-term administration. *3604*

Urobilinogen *Urine Increase Physiological* Associated with hemolytic anemia when occurs as main side effect. *3423*

Daunorubicin

Chromosomes *Test Conditions Abnormal Physiological* Clastogenic in human cells. *3282*

Debrisoquin

Amphetamine *Urine No Effect Analytical* No effect at 100 mg/L on method of Rutter. *3097*

Vanillylmandelic Acid *Urine Decrease Physiological* Mechanism not discussed. *0121*

Decaborane

Catecholamines *Urine Decrease Physiological* Markedly decreases output. *3722*

Decamethonium

Protein Bound Iodine (PBI) *Serum Increase Physiological* Contains 49.5% iodine. *0012*

Deferoxamine

Color *Urine Increase Analytical* Forms iron chelate with reddish color. *1488*

Iron *Serum Decrease Analytical* Marked reduction in AutoAnalyzer and other methods. *3180* At concentrations above 140 mg/L lowered concentration as measured by Ferrozine method. *3393*
Serum Decrease Physiological Therapeutic intent. *3180*
Urine Increase Physiological Primary affinity for trivalent iron. *0071*

Dehydration

Albumin *Serum Increase Physiological* By approximately 11.6% after heat exposure for 2-11 h. *3254*

Creatinine Clearance *Urine Decrease Physiological* Following overnight dehydration. *2865*

Hematocrit *Blood Decrease Physiological* Slight hemodilution with low protein fluid. *3254*

Protein *Serum Increase Physiological* By approximately 15.7% after heat exposure for 2-11 h. *3254*

Volume *Plasma Decrease Physiological* By approximately 13.6% after heat exposure for 2-11 h. *3254*

Dehydrobenzperidole

β-Endorphin *Plasma No Effect Physiological* No reduction of increased values induced by stress when given alone but significant decrease when given with fentanyl also. *2077*

Dehydrocholic Acid

BSP Retention *Serum Increase Physiological* Hepatic uptake or biliary excretion impaired. *1488*

Hemoglobin *Blood Decrease Physiological* Mild hemolytic action if injected. *0071*

Volume *Urine Increase Physiological* Mild diuretic action. *0071*

Dehydroemetine

Aspartate Aminotransferase
Serum Increase Physiological Possibly due to generalized myositis. *2427*

Dehydroepiandrosterone

Androstenedione *Plasma Increase Physiological* Mean increase from 4.3 to 8.6 nmol/L in 5 normal men given 1600 mg/d orally for 28 d. *2578*

Cholesterol *Serum Decrease Physiological* Mean fall from 4.82 to 4.48 mmol/L in 5 men given 1600 mg/d orally for 28 d. *2578*

Cholesterol, High Density Lipoprotein
Serum No Effect Physiological In 5 normal men given 1600 mg/d orally for 28 d. *2578*

Cholesterol, Low Density Lipoprotein
Serum Decrease Physiological Increase by 7.5% in 5 normal men given 1600 mg/d orally for 28 d. *2578*

Cholesterol, Very Low Density Lipoprotein
Serum No Effect Physiological In 5 normal men given 1600 mg/d orally for 28 d. *2578*

Dehydroepiandrosterone Sulfate (DHEA-S)
Plasma Increase Physiological Mean increase of 2.5 to 3.5 fold in 5 normal men given 1600 mg/d orally for 28 d. *2578*

Estradiol *Plasma No Effect Physiological* In 5 men given 1600 mg/d orally for 28 d. *2578*

Estrone *Plasma No Effect Physiological* In 5 men given 1600 mg/d orally for 28 d. *2578*

Sex Hormone Binding Globulin
Serum No Effect Physiological In 5 men given 1600 mg/d orally for 28 d. *2578*

Testosterone *Serum No Effect Physiological* In 5 men given 1600 mg/d orally for 28 d. *2578*

Testosterone, Free *Serum No Effect Physiological* In 5 men given 1600 mg/d orally for 28 d. *2578*

Triglycerides *Serum No Effect Physiological* In 5 normal men given 1600 mg/d orally for 28 d. *2578*

Delalande 69276

Thyroxine (T$_4$) *Serum Increase Analytical* At 37.5 mg/dL usual methods when added to serum. *1758*

Thyroxine (T$_4$) Index, Free (FTI) *Serum Increase Analytical* At 37.5 mg/dL usual methods when added to serum. *1758*

Demecarium

Pseudocholinesterase *Serum Decrease Physiological* Therapeutic intent. *0071*

Demeclocycline

Catecholamines *Urine Increase Analytical* Produces interfering fluorescence with analysis. *1961*

Chloramphenicol *Serum No Effect Analytical* No effect at 100 mg/L on coupled enzymatic method. *2490*

Creatinine *Serum Increase Physiological* Dose related nephrotoxicity. *2106*

Eosinophils *Blood Increase Physiological* Rare, but reported to occur. *1680*

Erythrocytes *Blood Decrease Physiological* Rare, but reported to occur. *1680*

^{131}I Uptake *Serum Decrease Physiological* Declomycin® contains tetraiodofluorescein. *2652*

LE Cells *Blood Positive Physiological* Rare exacerbation may occur. *1680*

Neutrophils *Blood Decrease Physiological* Rare, but reported to occur. *1680*

Osmolality *Urine Decrease Physiological* Maximum osmolality in response to dehydration. *3333*

Platelets *Blood Decrease Physiological* Rare, but reported to occur. *1680*

Protein *Urine Increase Physiological* Nephrotoxic effect. Nephrotoxicity. *2106*

Protein Bound Iodine (PBI) *Serum No Effect Physiological* Although discoloration of thyroid occurs. *1680*

Prothrombin Time *Plasma Increase Physiological* Depresses plasma prothrombin activity. *1680*

Urea Nitrogen *Serum Increase Physiological* Dose related nephrotoxicity. *2106*

Volume *Urine Increase Physiological* Reversible nephrogenic diabetes insipidus may occur. *3333* Rare induction of reversible nephrogenic diabetes. *1680*

Demecolcine

Sperm Count *Semen Decrease Physiological* Potent antispermatogenic action observed. *2831*

Deoxycholate

Lipase *Serum Increase Analytical* Sodium salts prevent inactivation of enzyme. *3589*

11-Deoxycortisol

Corticosteroids *Plasma Increase Analytical* 100% = cortisol competitive protein binding method of Ficher. *1128*

Cortisol *Plasma No Effect Analytical* If cross-reactivity of less than 25%. *1081*
Plasma Increase Analytical If nonselective extraction and competitive protein binding used. *0512*

Metyrapone Test *Plasma No Effect Physiological* Not measured by Clark/Rubin procedure. *0512*

21-Deoxycortisol

Cortisol *Plasma No Effect Analytical* No effect on phenylhydrazine (Porter-Silber) method. *0719*
Plasma Increase Analytical Gives 27% of value fluorometrically Gives 87% of value with competitive prot bind techniques. *0719*

Deoxy-Glucose

Glucose *Serum Decrease Analytical* At 4.0 mmol/L if hexokinase from Leuconostoc Mesent. *0411*

Deoxypyridoxine

3-Hydroxyanthranilic Acid *Urine Increase Physiological* After 2 g tryptophan load. *3050*

3-Hydroxykynurenine *Urine Increase Physiological* After 2 g tryptophan load. *2042*

Quinolinic Acid *Urine Increase Physiological* After 2 g tryptophan load. *3050*

Xanthurenic Acid *Urine Increase Physiological* After 2 g tryptophan load. *3050*

Deprenyl

Dihydroxyphenylacetic Acid
Plasma Decrease Physiological From mean of 730 to 370 ng/mL in 12 depressed or Alzheimer's disease patients given 60 mg drug daily for at least 3 weeks. *1766*

Dihydroxyphenylalanine *Plasma No Effect Physiological* In 12 depressed or Alzheimer's disease patients given 60 mg drug daily for at least 3 weeks. *1006*

Dihydroxyphenylglycol *Plasma Decrease Physiological* From mean of 820 to 240 ng/mL in 12 depressed or Alzheimer's disease patients given 60 mg drug daily for at least 3 weeks. *1006*

Epinephrine *Plasma No Effect Physiological* In 12 depressed or Alzheimer's disease patients given 60 mg drug daily for at least 3 weeks. *1006*

Homovanillic Acid *CSF Decrease Physiological* In 43 patients with depression or Alzheimer's disease chronically treated with drug. *2991*

5-Hydroxyindoleacetic Acid (5-HIAA)
CSF Decrease Physiological In 43 patients with depression or Alzheimer's disease chronically treated with drug. *2991*

Norepinephrine *Plasma No Effect Physiological* In 12 depressed or Alzheimer's disease patients given 60 mg drug daily for at. *1006*

Deserpidine

Platelets *Blood Decrease Physiological* Possible effect as related compounds cause this. *1680*

Desiccated Thyroid

Protein Bound Iodine (PBI) *Serum Increase Physiological* Increases by 1-2 µg/dL/d per grain given. *0354*
Serum Decrease Physiological Observed with some batches (T$_4$ converted to T$_3$). *0757*

Thyroxine (T$_4$) *Serum Increase Physiological* Increases by 1-2 µg/dL/d per grain given. *0354*

Desipramine

Alanine Aminotransferase *Serum No Effect Physiological* In 42 children or adolescents treated for up to 24 mo. *1624*
Serum Increase Physiological Rare transient cholestasis. *2313*

Albumin *Serum No Effect Analytical* At concentration of 8 mg/L had no effect on BCG method. *3393*

Desipramine *(continued)*

Alkaline Phosphatase *Serum No Effect Physiological* In 42 children and adolescents treated for up to 24 mo. *1624* *Serum Increase Physiological* Transient cholestasis (rare). *2313* Significant effect in 46 patients but values did not rise above reference range. *2871*

Amphetamine *Urine No Effect Analytical* No reaction with NBD chloride of Monforte. *2473* No effect at 100 mg/L on method of Rutter. *3097*

Aspartate Aminotransferase
Serum No Effect Physiological Increase observed in 4 of 46 patients but previously increased value decreased in 5 others so probably not dose-dependent toxic effect. *2871* In 42 children and adolescents treated for up to 24 mo. *1624*
Serum Increase Physiological Rare transient cholestasis. *2313* Idiosyncratic response observed in 4 of 46 patients. *2871*

Basophils *Blood Increase Physiological* Mild effect noted in some patients. *2427*

Bicarbonate *Serum No Effect Analytical* At concentration of 8 mg/L had no effect on method using phenolphthalein. *3393*

Bile *Urine Increase Physiological* Rare transient cholestasis. *2313*

Bilirubin *Serum No Effect Analytical* At concentration of 8 mg/L had no effect on Jendrassik and Grof method. *3393* *Serum No Effect Physiological* No effect observed in 46 treated patients. *2871* Insignificant protein-displacement effect in neonates. *3748* In 42 children and adolescents treated for up to 24 mo. *1624*
Serum Increase Physiological Rare transient cholestasis. *1488*

Bilirubin, Direct *Serum Decrease Physiological* Slight effect observed in 46 treated patients. *2871*

BSP Retention *Serum Increase Physiological* Rare transient cholestasis. *2313*

Calcium *Serum No Effect Analytical* At concentration of 8 mg/L had no effect on cresolphthalein method. *3393*

Cephalin Flocculation *Serum Increase Physiological* Rare transient cholestasis. *2313*

Chloride *Serum No Effect Analytical* At concentration of 8 mg/L had no effect on mercurimetric method. *3393*

Cholesterol *Serum No Effect Analytical* At concentration of 8 mg/L had no effect on Liebermann-Burchard method. *3393*

Cortisol *Plasma No Effect Physiological* In group of depressed patients. *0560*

Creatinine *Serum No Effect Analytical* At concentration of 8 mg/L had no effect on AutoAnalyzer Jaffé method. *3393*

Eosinophils *Blood Increase Physiological* Allergic reaction. *3857* In 4 of 46 patients as manifestation of idiosyncratic response but no value exceeded 10% of total white cell count: uncorrelated with drug plasma concentration. *2871*
Blood Decrease Physiological Agranulocytosis. *0748*

Epinephrine *Plasma No Effect Physiological* Observed with long term, high or low dose, treatment under experimental conditions. *3066*

Erythrocytes *Blood Decrease Physiological* Transient agranulocytosis (up to 25% reported). *0071*

Ethchlorvynol *Serum No Effect Analytical* At 100 mg/L on GLC procedure of Evenson. *1058*

Fatty Acids, Free (FFA) *Serum Increase Physiological* 41% increase following single injection i.m. of 50 mg. *1634*

Glucose *Serum Increase Physiological* Mechanism not understood. *1680*
Serum Decrease Physiological Mechanism not understood. *1680*

Growth Hormone *Plasma Increase Physiological* Substantial effect in some patients in group of depressed patients. *0560*

Homovanillic Acid *CSF No Effect Physiological* In 43 patients with depression or Alzheimer's disease chronically treated with drug. *2991*

5-Hydroxyindoleacetic Acid (5-HIAA)
CSF Decrease Physiological In 43 patients with depression or Alzheimer's disease chronically treated with drug. *2991*

Leukocytes *Blood Increase Physiological* Due to eosinophilia of hypersensitivity. *2313*
Blood Decrease Physiological May cause agranulocytosis. *2313*

Luteinizing Hormone (LH) *Plasma No Effect Physiological* In group of depressed patients. *0560*

Methaqualone *Serum No Effect Analytical* At 100 mg/L on GLC procedure of Evenson. *1057*

Neutrophils *Blood Decrease Physiological* Occasional case of agranulocytosis reported. *3717*

Norepinephrine *Plasma Increase Physiological* Observed with long term, high or low dose, treatment under experimental conditions. *3066*

Phenylbutazone *Serum Decrease Physiological* Inhibits gastrointestinal absorption. *1487*

Phosphate *Serum No Effect Analytical* At concentration of 8 mg/L had no effect on phosphomolybdate method. *3393*

Platelets *Blood Decrease Physiological* Immune response after some weeks. *1436*

Prolactin *Plasma No Effect Physiological* No significant effect with 50 mg orally. *1416*
Plasma Increase Physiological Acute response in group of depressed patients. *0560*

Propoxyphene *Urine No Effect Analytical* Less than 1% fluorescence in procedure of Valentour. *3662*

Protein *Serum No Effect Analytical* At concentration of 8 mg/L had no effect on biuret method with blank correction. *3393*

Thymol Turbidity *Serum Increase Physiological* Rare transient cholestasis. *2313*

Triglycerides *Serum No Effect Analytical* At concentration of 8 mg/L had no effect on lipase/esterase method. *3393*

Urea Nitrogen *Serum No Effect Analytical* At concentration of 8 mg/L had no effect on diacetylmonoxime method. *3393*

Uric Acid *Serum No Effect Analytical* At concentration of 8 mg/L had no effect on phosphotungstate reduction method. *3393*

Volume *Urine Decrease Physiological* Temporary due to anticholinergic action. *1680*

Deslanoside

Albumin *Serum No Effect Analytical* No effect on SMA 12/60 method at 0.06 mg/dL. *2636*

Alkaline Phosphatase *Serum No Effect Analytical* No effect on SMA 12/60 method at 0.06 mg/dL. *2636*

Aspartate Aminotransferase *Serum No Effect Analytical* No effect on SMA 12/60 method at 0.06 mg/dL. *2636*

Bilirubin *Serum No Effect Analytical* No effect on SMA 12/60 method at 0.06 mg/dL. *2636*

Calcium *Serum No Effect Analytical* No effect on SMA 12/60 method at 0.06 mg/dL. *2636*

Cholesterol *Serum No Effect Analytical* No effect on SMA 12/60 method at 0.06 mg/dL. *2636*

Creatinine *Serum No Effect Analytical* No effect on SMA 12/60 method at 0.06 mg/dL. *2636*

Glucose *Serum No Effect Analytical* No effect on SMA 12/60 method at 0.06 mg/dL. *2636*

Lactate Dehydrogenase *Serum No Effect Analytical* No effect on SMA 12/60 method at 0.06 mg/dL. *2636*

Phosphate *Serum No Effect Analytical* No effect on SMA 12/60 method at 0.06 mg/dL. *2636*

Protein *Serum No Effect Analytical* No effect on SMA 12/60 method at 0.06 mg/dL. *2636*

Uric Acid *Serum No Effect Analytical* No effect on SMA 12/60 method at 0.06 mg/dL. *2636*

Desmethyldiazepam

Methaqualone *Serum No Effect Analytical* At 50 mg/L on GLC procedure of Evenson. *1057*

Desmosterol

Cholesterol *Serum No Effect Analytical* On GLC procedure of MacGee. *2252*

Desogestrel

Apolipoprotein Al *Serum Increase Physiological* Increase by 20% when 150 μg desogestrel given with 30 μg ethinyl estradiol for 3 mo. Results normal within 2 mo after treatment stopped. *0312*

Apolipoprotein B *Serum No Effect Physiological* (ratio apo B: apo A-l decreased by 17%) when 150 μg desogestrel given with 30 μg ethinyl estradiol for 3 mo. Results normal within 2 mo after treatment stopped. *0312*

Ceruloplasmin *Serum No Effect Physiological* With daily dose of 0.125 mg in 30 healthy female volunteers when given alone. Normal values 30 d after treatment stopped. *3089*

Cholesterol *Serum No Effect Physiological* When 150 μg desogestrel given with 30 μg ethinyl estradiol for 3 mo. Results normal within 2 mo after treatment stopped. *0312*

Cholesterol, High Density Lipoprotein
Serum Increase Physiological Increase by 12% (% HDL-cholesterol increased by 15%) when 150 μg desogestrel given with 30 μg ethinyl estradiol for 3 mo. Results normal within 2 mo after treatment stopped. *0312*
Serum Decrease Physiological With daily dose of 0.125 mg in 30 healthy female volunteers when given alone. Normal values 30 d after treatment stopped. *3089*

Glycosylated Proteins *Serum No Effect Physiological* When 150 μg desogestrel given with 30 μg ethinyl estradiol for 3 mo. Results normal within 2 mo after treatment stopped. *0312*

Sex Hormone Binding Globulin
Serum Decrease Physiological With daily dose of 0.125 mg in 30 healthy female volunteers when given alone. Normal values 30 d after treatment stopped. *3089*

Transcortin *Serum No Effect Physiological* With daily dose of 0.125 mg in 30 healthy female volunteers when given alone. Normal values 30 d after treatment stopped. *3089*

Triglycerides *Serum Increase Physiological* Increase by 35% when 150 μg desogestrel given with 30 μg ethinyl estradiol for 3 mo. Results normal within 2 mo after treatment stopped. *0312*

Desoximetasone

Cortisol *Serum Decrease Physiological* Rapid and sustained suppression in 5 patients with psoriasis and topical application of glucocorticoid. *1240*

Glucose Tolerance *Serum Decrease Physiological* In 2 of 5 patients with psoriasis and topical application of glucocorticoid. *1240*

Insulin *Serum Increase Physiological* 2 to 3 fold increase in 5 patients with psoriasis and topical application of glucocorticoid. *1240*

Leukocytes *Blood Increase Physiological* 2 to 3 fold increase in 5 patients with psoriasis and topical application of glucocorticoid. *1240*

Desoxycorticosterone

Aldosterone *Plasma Decrease Physiological* Although on low sodium diet. *3263*
Urine No Effect Analytical No significant effect in RIA procedure of Drewes. *0952*
Urine Increase Analytical 1 mg/L equivalent to 38 μg/L in method of Drewes. *0952*
Urine Decrease Physiological Although on low sodium diet. *3263*

Cortisol *Plasma Decrease Physiological* Rapid and sustained suppression in 5 patients with psoriasis and topical application of glucocorticoid. *1240*

Glucose Tolerance *Serum Decrease Physiological* In 2 of 5 patients with psoriasis and topical application of glucocorticoid. *1240*

16-α-Hydroxyprogesterone *Plasma Increase Analytical* Up to 25% cross reactivity. *0008*

Insulin *Plasma Increase Physiological* 2 to 3 fold increase in 5 patients with psoriasis and topical application of glucocorticoid. *1240*

Leukocytes *Blood Increase Physiological* In 5 patients with psoriasis and topical application of glucocorticoid. *1240*

Potassium *Serum Decrease Physiological* Occurs as result of increased excretion. *0071*
Urine Increase Physiological Promotes increased urinary elimination. *0071*

Progesterone *Plasma No Effect Analytical* 1% or less cross reactivity with RIA. *0567*

Protein Bound Iodine (PBI) *Serum Decrease Physiological* Inhibits iodination of tyrosine in thyroxine binding globulin. *0012*

Renin Activity *Plasma Decrease Physiological* Although on low sodium diet. *3263*

Sodium *Serum Increase Physiological* May cause retention and edema. *0071*
Urine Decrease Physiological Exchanged with potassium ion. *1343*

Volume *Plasma Increase Physiological* Small increase of extracellular fluid with exogenous DOCA®. *3263* Associated with fluid retention. *0019*
Urine Increase Physiological Associated with polydipsia also. *1343*

Detergents

Acid Phosphatase *Serum Decrease Analytical* 10-37% inhibition at 0.6-0.8 mg/ml. *3217*

Alkaline Phosphatase *Serum Decrease Analytical* 17-24% inhibition at 0.6-0.8 mg/ml. *3217*

Antibody Evaluation *Serum Decrease Analytical* Increased concentrations cause decreased agglutination on AutoAnalyzer. *1424*

Aspartate Aminotransferase *Serum Decrease Analytical* 16% inhibition at 0.6-0.8 mg/ml. *3217*

Glutamate Dehydrogenase *Serum Decrease Analytical* 100% inhibition at 0.6-0.8 mg/ml. *3217*

Lactate Dehydrogenase *Serum Decrease Analytical* 30% inhibition at 0.6-0.8 mg/ml. *3217*

Malate Dehydrogenase *Serum Decrease Analytical* 45% inhibition at 0.6-0.8 mg/ml. *3217*

Phenylketones *Urine Negative Analytical* False negative results (brown color) with FeCl₃ test. *3187*

Phosphate *Serum Increase Analytical* Contaminated glassware etc. *2313*

Sodium *Serum Increase Analytical* Contaminated glassware etc. *2313*

Dexamethasone

Absorbance at 450 nm *Amniotic Fluid Decrease Analytical* Time-dependent decrease when added to bilirubin containing amniotic fluid *in vitro*. *2453*

Amylase *Serum Increase Physiological* May cause pancreatitis as side effect. *1680*

Androstenedione *Plasma No Effect Physiological* No effect of treatment on initially high concentrations in hirsute women. *0776*

Androsterone *Urine Decrease Physiological* Suppression of ACTH. *2420*

Bicarbonate *Serum Increase Physiological* May cause hypokalemic alkalosis. *1680*

Calcium *Urine Increase Physiological* Metabolic effect. *1680*

Corticosteroids *Plasma Decrease Physiological* Effect seen following morning if given in evening. *3164*
Urine Increase Analytical Butanol extract/no hydrolysis Reddy/Porter-Silber reaction. *0022*

Corticotropin *Plasma Decrease Physiological* Effect measured after 9 h. *0168*

Cortisol *Plasma No Effect Analytical* Reactivity of less than 1% possible with RIA. *1081*
Plasma Decrease Physiological Marked effect for 4 d after i.v. injection. *0783* Normal response not always observed in psychiatric patients. *2404*

Dehydroepiandrosterone (DHEA)
Urine Decrease Physiological Suppression of ACTH. *2420*

Dexamethasone *(continued)*

Dehydroepiandrosterone Sulfate (DHEA-S)
Plasma Decrease Physiological Suppressed by treatment in hirsute women from mean of 8.6 to 3.4 µmol/L. *0776*

β-Endorphin *Plasma Decrease Physiological* Postoperative induced secretion of compound reduced by all amounts of drug given but amount of pain increased. *1497*

Estradiol *Plasma No Effect Physiological* No effect of treatment on initially high concentrations in hirsute women. *0776*

Estriol *Plasma No Effect Physiological* No effect of treatment on initially high concentrations in hirsute women. *0776*

Estrogens *Urine Decrease Physiological* Decreased conversion of neutral steroids. *0023*

Etiocholanolone *Urine Decrease Physiological* Suppression of ACTH. *2420*

Glucose *Serum Increase Physiological* Hormonal action. *2525*
Urine Increase Physiological Associated with hyperglycemia. *1680*

Glucose Tolerance *Serum Decrease Physiological* May impair carbohydrate tolerance. *1680* Observed in normal subjects and associated with enhanced insulin, C-peptide and glucagon responses to a test meal and blunted gastric inhibitory polypeptide response. *1411*

11-Hydroxycorticosteroids *Plasma No Effect Analytical* Presumed effect on fluorescent method of Mejer. *2396*

17-Hydroxycorticosteroids *Urine Increase Analytical* Measured as endogenous steroids by Reddy method. *2420*
Urine Decrease Physiological Pituitary feedback with suppression of ACTH. *2164*

17-Ketogenic Steroids *Urine Decrease Physiological* Suppression of ACTH. *2420*

17-Ketosteroids *Urine Increase Analytical* Affects method of Reddy. *1686*

Nitrogen Balance *Patient Negative Physiological* Due to protein catabolism. *1680*

Occult Blood *Feces Positive Physiological* May aggravate peptic ulcer and cause bleeding. *1680*

Potassium *Serum No Effect Analytical* At concentration of 1.4 mg/L had no effect on measurement by ISE with predilution. *3393*
Serum Decrease Physiological May promote increased urinary loss. *1680*
Urine Increase Physiological Metabolic effect. *1680*

Prolactin *Plasma Decrease Physiological* Normal response not always observed in psychiatric patients. *2404*

Sex Hormone Binding Globulin
Serum Increase Physiological Mean increase from 35.2 to 47.7 nmol/L in 14 women hirsute women. *0776*

Sodium *Serum No Effect Analytical* At concentration of 1.4 mg/L had no effect on measurement by ISE with predilution. *3393*
Urine Increase Physiological Mild diuresis with sodium loss may occur initially. *0071*

Testosterone *Serum No Effect Physiological* No effect of treatment on initially high concentrations in hirsute women. *0776*
Serum Decrease Physiological Decreased metabolic clearance and binding in women. *1941*
Urine Decrease Physiological Slight effect in men. *3062*

Thromboxane *Plasma No Effect Physiological* No effect of up to 100 µmol/L on formation during clotting. *2890*

Thyroxine (T₄) *Serum Decrease Physiological* 8 mg/d for 2 d reduced concentration in normal individuals due to altered secretion and peripheral metabolism. *0860*

Tri-iodothyronine (T₃) *Serum Decrease Physiological* 8 mg/d for 2 d reduced concentration in normal and hypothyroid patients. *0860*

Dextran

Albumin *Serum No Effect Analytical* At concentration of 30,000 mg/L had no effect on BCG method. *3393*
Serum Decrease Physiological 30% after Macrodex®, no effect after Rheomacrodex®. *3346*

Ammonia *Plasma No Effect Analytical* At concentration of 10,000 mg/L had no effect on Ektachem® method. *3393*

α₁-Antitrypsin *Serum Increase Physiological* Possibly associated with underlying disease. *3346*

Bilirubin *Serum No Effect Analytical* At concentration of 10,000 mg/L had no effect on Ektachem® method. *3393*
Serum Increase Analytical Causes turbidity with methanol in Evelyn-Malloy method. *3505*

Bilirubin, Direct *Serum Increase Analytical* Turbidity develops with Evelyn-Malloy method. *0652*

Bleeding Time *Blood Increase Physiological* Observed effect but explanation uncertain. *2313*

Complement C4 *Serum Decrease Physiological* Complex formation or increased consumption. *3346*

Creatine *Serum No Effect Analytical* No effect on method of Heinegard and Tiderstrom. *1547*

Creatinine *Serum No Effect Analytical* At concentration of 10,000 mg/L had no effect on Ektachem® method. *3393*
Serum Increase Physiological Blocks tubules causing renal failure. *0647*

Erythrocyte Sedimentation Rate
Blood Increase Physiological Due to cell aggregating properties. *2313*

Euglobulin Clot Lysis Time *Blood Decrease Physiological* Marked decrease noted. *0320*

Factor V *Plasma Decrease Physiological* Slight effect (more than hemodilution). *2427*

Factor VII *Plasma Decrease Physiological* Decrease by 50% in normals. *2427*

Factor VIII *Plasma Decrease Physiological* Marked decrease noted in single case. *0320*

Factor IX *Plasma Decrease Physiological* Slight effect (more than hemodilution). *2427*

Factor XI *Plasma Decrease Physiological* Marked reduction in one case. *0320*

Fibrinogen *Plasma Decrease Physiological* Complex formation or increased consumption. *3346*

β₁-α-Globulin *Serum Decrease Physiological* Complex formation or increased consumption. *3346*

Glucose *Serum No Effect Analytical* At 1 g/dL on MBTH procedure of Neeley. *2569* No effect on hexokinase, glucose oxidase methods. *2568* At concentration of 500 mg/L had no effect on Seralyzer method. *3393*
Serum Increase Analytical 10 g/dL equivalent to 0.3 mmol/L alkaline ferricyanide. 10 g/dL equivalent to 0.7 mmol/L p-HBAH procedure. *2140* Affects o-toluidine procedure: turbidity at acid pH. *3505* 10 g/dL equivalent to 6.7 mmol/L o-toluidine. 10 g/dL equivalent to 0.3 mmol/L glucose oxidase procedures. *2140*

Haptoglobin *Serum Decrease Physiological* Complex formation or increased consumption. *3346*

Hematocrit *Blood Decrease Physiological* Macrodex® increases blood vol, no effect Rheomacrodex®. *3346*

Immunoglobulin IgA *Serum Decrease Physiological* Complex formation or increased consumption. *3346*

Immunoglobulin IgG *Serum Decrease Physiological* Complex formation or increased consumption. *3346*

Immunoglobulin IgM *Serum Decrease Physiological* Complex formation or increased consumption. *3346*

Inulin Clearance *Urine Increase Analytical* Interferes with analytical procedure. *2427*

Iron *Serum No Effect Analytical* At concentration of 10,000 mg/L had no effect on Ferrozine method. *3393*
Serum Increase Analytical Causes turbidity with method of Young and Hicks. *2981*

α₂-Macroglobulin *Serum Decrease Physiological* Complex formation or increased consumption. *3346*

Partial Thromboplastin Time
Plasma Increase Physiological 15% increase noted in one case. *0320*

Plasminogen *Plasma Decrease Physiological* Marked reduction. *0320*

Potassium *Serum No Effect Analytical* At concentration of 30,000 mg/L had no effect on Ektachem® ISE method. *3393*

Protein *Serum No Effect Analytical* If dextranase used to remove. *2567* No effect on modified biuret method of Moore. *2477*

Serum Increase Analytical Turbidity effect with biuret reaction. *3505* At concentrations greater than 5,000 mg/L (normal therapeutic concentration 14500 mg/L) raised concentration as measured by biuret method with blank correction. *3393*
Serum Decrease Physiological 30% after Macrodex®, no effect after Rheomacrodex®. *3346*

Prothrombin Time *Plasma Increase Physiological* May prolong action of anticoagulants. *1679*

Specific Gravity *Urine Increase Analytical* High molecular weight. *1022*

Thrombin Time *Plasma Decrease Physiological* Slight reduction noted in one case. *0320*

Transferrin *Serum Decrease Physiological* 30% after Macrodex®, no effect after Rheomacrodex®. *3346*

Urea Nitrogen *Serum No Effect Analytical* Not precipitated in DMAB procedure of Morin. *2487*
Serum Increase Analytical Reacts as if urea with dimethylaminobenzaldehyde. *3505*
Serum Increase Physiological Blocks tubules causing renal failure. *0647*

Uric Acid *Serum Increase Analytical* Reduces phosphotungstate. *2220*

Volume *Plasma Increase Physiological* Hemodynamic effect, also increases urine flow. *1343* Effect of Macrodex®, not Rheomacrodex®. *3346*
Urine Increase Physiological Hemodynamic effect. *1343*

Dextran 40

Amylase *Serum No Effect Analytical* At concentration of 80,000 mg/L had no effect on maltotetrose method. *3393* At concentration of 80,000 mg/L had no effect on p-nitrophenylmaltoheptoside method. *3393* At concentration of 80,000 mg/L had no effect on p-nitrophenylmaltopentoside/-hexoside method. *3393*

Glucose *Serum No Effect Analytical* At concentration of 2500 mg/L had no effect on hexokinase/G-6-PDH method on aca. *3393*
Serum Increase Analytical At concentrations above 10,000 mg/L raised concentration as measured by Ektachem® method. *3393*

Potassium *Serum No Effect Analytical* At concentration of 9100 mg/L had no effect on measurement by ISE with predilution. *3393*

Protein *Serum No Effect Analytical* Since 1987 no interference up to 120 g/L on American Monitor Parallel method. *2651* At concentrations up to 30 g/L no effect on Roche Cobas-Bio, Ektachem® 400 and Beckman Astra 8. *0231*
Serum Decrease Analytical At concentrations above 30 g/L causes falsely low results with Du Pont aca, although results could be corrected if ethylene glycol previously injected into packs. *0231*

Sodium *Serum No Effect Analytical* At concentration of 9100 mg/L had no effect on measurement by ISE with predilution. *3393*

Urea Nitrogen *Serum No Effect Analytical* At concentration of 10,000 mg/L had no effect on Ektachem® method. *3393*

Dextran 60

Cholesterol *Serum No Effect Analytical* At concentration of 6,000 mg/L had no effect on CHOD-Iodide method. *3393* At concentration of 6,000 mg/L had no effect on CHOD-PAP method. *3393* At concentration of 6,000 mg/L had no effect on catalase-AIDH method. *3393* At concentration of 6,000 mg/L had no effect on method using catalase-Hantzsch reaction. *3393* At concentration of 6,000 mg/L had no effect on Liebermann-Burchard method. *3393*

Creatinine *Serum No Effect Analytical* At concentration of 6,000 mg/L had no effect on Jaffé-Fading-Fraction method. *3393* At concentration of 6,000 mg/L had no effect on Jaffé-Fuller's earth method. *3393* At concentration of 81250 mg/L had no effect on kinetic Jaffé method on BKA-2. *3393*

Glucose *Serum No Effect Analytical* At concentration of 1800 mg/L had no effect on Ektachem® method. *3393* At concentration of 6,000 mg/L had no effect on hexokinase/G-6-PDH method. *3393*

Lactate *Serum No Effect Analytical* At concentration of 6,000 mg/L had no effect on enzymatic method. *3393*

Potassium *Serum No Effect Analytical* At concentration of 18,000 mg/L had no effect on measurement by ISE without predilution. *3393*

Protein Electrophoresis *Serum No Effect Analytical* At concentration of 18,000 mg/L had no effect on automated Olympus-Hite method except for slight displacement of fractions. *3393*

Sodium *Serum No Effect Analytical* At concentration of 18,000 mg/L had no effect on measurement by ISE without predilution. *3393*

Urea Nitrogen *Serum No Effect Analytical* At concentration of 1800 mg/L had no effect on Ektachem® method. *3393*

Uric Acid *Serum No Effect Analytical* At concentration of 6,000 mg/L had no effect on Kageyama-Hantzsch method. *3393* At concentration of 6,000 mg/L had no effect on catalase-AIDH method. *3393*

Dextran 70

Antithrombin III *Plasma No Effect Analytical* At concentration of 1500 mg/L had no effect on aca method. *3393*

Fibrinogen *Plasma No Effect Analytical* At concentration of 1500 mg/L had no effect on aca method. *3393*

Glucose *Serum No Effect Analytical* At concentration of 1,000 mg/L had no effect on Ektachem® method. *3393*

Plasminogen *Plasma No Effect Analytical* At concentration of 1500 mg/L had no effect on aca method. *3393*

Potassium *Serum No Effect Analytical* At concentration of 5400 mg/L had no effect on measurement by ISE with predilution. *3393*

Sodium *Serum No Effect Analytical* At concentration of 5400 mg/L had no effect on measurement by ISE with predilution. *3393*

Uric Acid *Serum No Effect Analytical* At concentration of 1,000 mg/L had no effect on Ektachem® method. *3393*

Dextran 75

Glucose *Serum No Effect Analytical* At concentration of 1500 mg/L had no effect on hexokinase/G-6-PDH method. *3393*

Dextroamphetamine

Amphetamine *Serum No Effect Physiological* Not detected after 30 mg/d. *2348*

Chlordiazepoxide *Blood No Effect Analytical* On method of Riddick at 5 mg/dL. *2983*

Corticosteroids *Plasma Increase Physiological* Effect most marked in evenings. *0331*

Epinephrine *Urine Increase Physiological* For 3 h after administration. *1487*

Glucose *Serum Increase Physiological* Metabolic effect. *2313*
Serum Decrease Physiological Reported to produce decrease of fasting glucose. *1487*
Urine Increase Physiological Will occur if hyperglycemia. *2313*

17-Hydroxycorticosteroids *Urine No Effect Analytical* No effect modified Glenn-Nelson procedure. *1487*

17-Ketosteroids *Urine No Effect Analytical* No effect on modified Holtorff-Koch procedure. *1487*

Norepinephrine *Urine No Effect Physiological* No effect observed. *1487*

Platelets *Blood Decrease Physiological* Thrombocytopenia reported to occur. *1436*

Protein Bound Iodine (PBI) *Serum No Effect Physiological* Appears to have no effect. *0830*

Vanillylmandelic Acid *Urine No Effect Analytical* Apparently no method interference. *1487*
Urine No Effect Physiological No physiological effect. *1487*

Dextromethorphan

Morphine *Urine No Effect Analytical* Insignificant cross reactivity with RIA procedures Insignificant cross react with hemagglutination inhibition Insignificant cross react with EMIT procedure for opiates. *2514*

pCO₂ *Blood Increase Physiological* High doses may produce respiratory depression. *1343*

Platelets *Blood Decrease Physiological* Thrombocytopenia (AMA Blood dyscrasias). *2429*

Dextromoramide

Epinephrine *Plasma Decrease Physiological* Mechanism obscure. *1744*

Norepinephrine *Plasma Decrease Physiological* Mechanism obscure. *1744*

Dextrothyroxine

Basal Metabolic Rate *Patient No Effect Physiological* No significant effect noted in normal subjects. *0019*

Bilirubin *Serum Increase Physiological* One case of possible cholestasis reported. *0019*

Cholesterol *Serum Decrease Physiological* Therapeutic intent (enhances excretion). *1488*

Factor VII *Plasma Decrease Physiological* Reported effect. *0019*

Factor VIII *Plasma Decrease Physiological* Reported effect. *0019*

Factor IX *Plasma Decrease Physiological* Reported effect. *0019*

Glucose *Serum Increase Physiological* Effect seen in diabetics. *1488*
Urine Increase Physiological Occurs due to elevation of blood sugar Occurs due to hyperglycemia. *1488*

¹³¹I Uptake *Serum No Effect Physiological* No effect reported. *2220*
Serum Decrease Physiological Marked effect observed even in normals. *0019*

Lipids, Total *Serum Decrease Physiological* Therapeutic effect. *0019*

β-Lipoproteins *Serum Decrease Physiological* Therapeutic effect. *0071*

Phospholipids, Total *Serum Decrease Physiological* Therapeutic effect. *0071*

Protein Bound Iodine (PBI) *Serum Increase Analytical* Reacts like levo compound. *0583*
Serum Increase Physiological Contains 65% iodine (may cause increase to 10-25 µg/dL). *0012*

Prothrombin Time *Plasma Increase Physiological* Increases receptor site affinity for anticoagulants. *3380*

T₃ Uptake *Serum Increase Physiological* Binds to usual T₄ sites, decreases thyroxine binding globulin. *0583*

Thyroxine (T₄) *Serum Increase Physiological* Mechanism not discussed. *1488*

Thyroxine (T₄) (Murphy-Pattee) *Serum Increase Analytical* Reacts as if levo compound. *0583*

Triglycerides *Serum Decrease Physiological* Therapeutic effect. *0071*

1,2-Diaminopropane

Creatinine Clearance *Urine Decrease Physiological* In 11 patients who showed glomerular injury including minimal change nephrotic syndrome with a membranous pattern of immune complex deposition as well as other patterns of deposition. *1170*

Leukocytes *Blood Decrease Physiological* In 3% of 90 patients with rheumatoid arthritis. *1461*

Platelets *Blood Decrease Physiological* In 3% of 90 patients with rheumatoid arthritis. *1461*

Protein *Urine Increase Physiological* Good correlation between amount of protein and duration of proteinuria moderate doses not preclude reinstitution of therapy once protein has cleared. *2586* In 16% of 90 patients with rheumatoid arthritis. *1461* HLA DRB positive patients had 11 times risk of this side effect than those antigen. *1379*

Pyruvate Kinase *Red Blood Cells Decrease Analytical* Most marked with larger side chain. *0370*

Diapamide

Chloride *Urine Increase Physiological* Diuretic action. *2256*

Glucose *Serum Increase Physiological* Alters glucose tolerance. *2256*
Urine Increase Physiological Result of marked hyperglycemia. *2256*

Glucose Tolerance *Serum Decrease Physiological* Modification of glucose tolerance. *2256*

Potassium *Serum Decrease Physiological* Diuretic action. *2256*
Urine Increase Physiological Diuretic action. *2256*

Sodium *Urine Increase Physiological* Diuretic action. *2256*

Uric Acid *Serum Increase Physiological* Impaired excretion. *2256*

Volume *Urine Increase Physiological* Diuretic action. *2256*

Diaprim

Leukocytes *Blood Decrease Physiological* Leukopenia. *2426*

Diatrizoic acid

N-Acetylglucosaminidase *Urine Increase Physiological* Doubled day of bilateral renal arteriography, almost reverted to normal following day. *2594*

Alanine Aminopeptidase *Urine Increase Physiological* Increased almost 3-fold on day of, and day after, bilateral renal arteriography in patients with hypertension. *2594*

Amino Acids *Serum No Effect Analytical* Concentration too low in serum to affect ninhydrin based procedures. *0379*
Urine Positive Analytical May comigrate with asparagine and glutamine on TLC and with isoleucine on amino acid analyzer. *0379*

Ammonia *Plasma No Effect Analytical* At concentration of 5,000 mg/L had no effect on Ektachem® method. *3393*

Amylase *Serum Increase Physiological* Blockage of pancreatic duct for 6-18 h. *0085*

Aspartate Aminotransferase
Serum Increase Physiological Has caused severe muscle spasm and renal failure. *2312*

Bilirubin *Serum No Effect Analytical* At concentration of 5,000 mg/L had no effect on Ektachem® method. *3393*
Serum Increase Physiological Possible displacement from protein because of high circulating concentration achievable. *3748*

Butanol Extractable Iodine (BEI)
Serum No Effect Analytical No effect reported with Hypaque®. *0830*

Calcium *Serum Decrease Physiological* ?due to small amount of EDTA in medium. *2043*

Cholesterol *Serum No Effect Analytical* At concentration of 2600 mg/L had no effect on CHOD-Iodide method. *3393* At concentration of 2600 mg/L had no effect on CHOD-PAP method. *3393* At concentration of 2600 mg/L had no effect on catalase-AIDH method. *3393* At concentration of 2600 mg/L had no effect on method using catalase-Hantzsch reaction. *3393* At concentration of 2600 mg/L had no effect on Liebermann-Burchard method. *3393*

Creatinine *Serum No Effect Analytical* At concentration of 5,000 mg/L had no effect on Ektachem® method. *3393* At concentration of 2600 mg/L had no effect on Jaffé-Fading-Fraction method. *3393* At concentration of 2600 mg/L had no effect on Jaffé-Fuller's earth method. *3393* At concentration of 6100 mg/L had no effect on kinetic Jaffé method on BKA-2. *3393*

Crystals *Urine Increase Physiological* Colorless, thin, rhombic, may be notch (in acid). *0443*

Erythrocytes *Urine Increase Physiological* Has caused acute renal failure with myelography. *2312*

Glucose *Serum No Effect Analytical* At concentration of 2600 mg/L had no effect on hexokinase/G-6-PDH method. *3393*
Serum Decrease Analytical At concentrations above 5140 mg/L (probably below circulating concentration) lowered concentration as measured by Ektachem® method. *3393*

Immunoglobulin IgG *Urine Increase Physiological* Marked increased excretion (20 to 60 times baseline) day of bilateral renal arteriography in 23 patients with hypertension, reverted to normal on following day. *2594*

^{131}I Uptake *Serum Decrease Physiological* Due to iodine component of material. *2220*

Lactate *Serum No Effect Analytical* At concentration of 2600 mg/L had no effect on enzymatic method. *3393*

Lysozyme *Urine No Effect Physiological* No significant effect observed in 23 patients with hypertension and bilateral renal arteriography performed. *2594*

Magnesium *Serum Decrease Physiological* ?due to small amount of EDTA in medium. *2043*

Metanephrines, Total *Urine No Effect Analytical* No effect if Hypaque® administered within 2 d. *1809*
Urine Decrease Analytical False negative if Renovist or Renografin® within 2 d. *1809*

β$_2$-Microglobulin *Urine No Effect Physiological* No significant effect observed in 23 patients with hypertension and bilateral renal arteriography performed. *2594*

Neutrophils *Blood Decrease Physiological* May occur after angiography. *1680*

PAH Clearance *Serum No Effect Physiological* No significant effect of bilateral renal arteriography in 23 patients with hypertension. *2594*

Phosphate Clearance *Urine Decrease Physiological* Reduction from average of 12.9 mL/min/1.73 m^2 to 7.1 mL/min/1.73 m^2 day of bilateral renal arteriography in patients with hypertension. *2594*

Potassium *Serum No Effect Analytical* At concentration of 6080 mg/L had no effect on ISE measurement without predilution. *3393*

Protein *Urine No Effect Analytical* No effect on Albustix, Labstix® No effect on heat acetic acid test. *1487*
Urine Increase Analytical Interferes with sulfosalicylic, nitric acid tests. *1700* Affects biuret part of Doetsch procedure. *0921* False positive with sulfosalicylic acid. *1487*
Urine Increase Physiological Marked increased excretion (20 to 60 times baseline) day of bilateral renal arteriography in 23 patients with hypertension, reverted to normal on following day. *2594*

Protein Bound Iodine (PBI) *Serum Increase Analytical* But does not affect T$_4$, effect lasts 2-6 d. *3218*

Protein Electrophoresis *Serum No Effect Analytical* At concentration of 6080 mg/L had no effect on automated Olympus-Hite method but with displacement of fractions. *3393*

Sodium *Serum No Effect Analytical* At concentration of 6080 mg/L had no effect on ISE measurement without predilution. *3393*

Specific Gravity *Urine Increase Analytical* Presence of high molecular weight compound. *1700*

Sugar *Urine Increase Analytical* Affects copper reduction techniques due to I$_2$. *2425*

Transferrin *Urine Increase Physiological* Marked increased excretion (20 to 60 times baseline) day of bilateral renal arteriography in 23 patients with hypertension, reverted to normal on following day. *2594*

Trypsin *Feces Decrease Analytical* Spectrophotometric assay -inhibits enzyme and absorbs. *0741*

Urea Nitrogen *Serum No Effect Analytical* At concentration of 1300 mg/L had no effect on Ektachem® method. *3393*
Serum Increase Physiological Has caused acute renal failure with myelography. *2312*

Uric Acid *Serum No Effect Analytical* At concentration of 2600 mg/L had no effect on Kageyama-Hantzsch method. *3393* At concentration of 2600 mg/L had no effect on catalase-AIDH method. *3393*

Serum No Effect Physiological No significant effect of bilateral renal arteriography in 23 patients with hypertension. *2594*
Serum Decrease Physiological Uricosuric effect. *2507*
Urine Increase Physiological Uricosuric effect. *2507*

Urobilinogen *Urine Increase Analytical* Heavy yellow precipitate with Ehrlich's. *0583*

Volume *Plasma Increase Physiological* Hypertonicity may occur in children. *0071*

Diazepam

Acetaminophen *Serum No Effect Analytical* At concentration of 2 mg/L had no effect on HPLC method. *3432*

Acetaminophen Screening Test *Urine Negative Analytical* No reaction with o-cresol at therapeutic concentrations. *3326*

Alanine Aminotransferase *Serum No Effect Analytical* At 5 times upper limit of therapeutic range on methods on SMAC®, Abbott-VP, aca, Cobas-Bio, and KDA. *2138* At acute overdose concentration (2.5 mg/dL) on Technicon® SMAC® method. *3719*
Serum Increase Physiological Isolated case of drug induced hepatitis. *3562*

Albumin *Serum No Effect Analytical* At 5 times upper limit of therapeutic range on methods on SMAC®, Ektachem®, Hitachi® 705 and KDA. *2138* At acute overdose concentration (2.5 mg/dL) on Technicon® SMAC® method. *3719* At concentration of 25 mg/L had no effect on BCG method. *3393*

Alkaline Phosphatase *Serum No Effect Analytical* At 5 times upper limit of therapeutic range on methods on SMAC®, Abbott-VP, aca, Cobas-Bio, Hitachi® 705 and KDA. *2138* At acute overdose concentration (2.5 mg/dL) on Technicon® SMAC® method. *3719*
Serum Increase Physiological Mild effect reported in one patient. *1878* Isolated case of drug induced hepatitis. *3562*

Alkaloids *Urine No Effect Analytical* With ninhydrin on Frings TLC procedure at 10 mg/dL. *1204*
Urine Increase Analytical Purple with iodoplatinate on Frings TLC procedure. *1204*

Aminolevulinic Acid *Urine Increase Physiological* May precipitate attack of acute porphyria. *1016*

Amitriptyline *Serum No Effect Physiological* No effect on serum concentration if given together. *3321*

Amylase *Serum No Effect Analytical* At 5 times upper limit of therapeutic range on methods on aca, Cobas-Bio, and Ektachem®. *2138*

Aspartate Aminotransferase *Serum No Effect Analytical* At 5 times upper limit of therapeutic range on methods on SMAC®, Abbott-VP, aca, Cobas-Bio, Hitachi® 705 and KDA. *2138* At acute overdose concentration (2.5 mg/dL) on Technicon® SMAC® method. *3719*
Serum Increase Physiological Isolated case of drug induced hepatitis. *3562*

Barbiturate *Serum Increase Analytical* At toxic levels on diphenylcarbazone procedure. *0583*
Urine No Effect Analytical With mercurous NO$_3$ on Frings TLC procedure at 10 mg/dL. With diphenylcarbazone on Frings TLC procedure at 10 mg/dL. With vanillin on Frings procedure at 10 mg/dL. With mercuric SO$_4$ on Frings TLC procedure at 10 mg/dL. *1204*

Bicarbonate *Serum No Effect Analytical* At concentration of 1.5 mg/L had no effect on method using phenolphthalein. *3393*

Bilirubin *Serum No Effect Analytical* At 5 times upper limit of therapeutic range on methods on SMAC®, aca, Cobas-Bio, Ektachem®, Hitachi® 705 and KDA. *2138* At concentration of 25 mg/L had no effect on Jendrassik and Grof method. *3393* At acute overdose concentration (2.5 mg/dL) on Technicon SMAC® method. *3719* At 5 times upper limit of therapeutic range on methods on SMAC®, aca, Cobas-Bio, Ektachem®, Hitachi® 705 and KDA. *2138*
Serum No Effect Physiological Insignificant displacement from protein in neonates. *3748*
Serum Increase Physiological Presumed hepatic toxic effect. *3879*

Bromide *Urine No Effect Analytical* No effect on method of Frings. *1204*

Diazepam (continued)

Calcium *Serum No Effect Analytical* At 5 times upper limit of therapeutic range on methods on SMAC®, Abbott-VP, aca, Ektachem®, Hitachi® 705 and KDA. *2138* At acute overdose concentration (2.5 mg/dL) on Technicon® SMAC® method. *3719* At concentration of 25 mg/L had no effect on cresolphthalein method. *3393*

Serum No Effect Physiological No effect when given in normal therapeutic amounts to elderly. *3925*

Catecholamines *Urine No Effect Physiological* No effect with short term ingestion of 15 mg/d. *0766*

Chlordiazepoxide *Blood Increase Analytical* 2 mg/dL equivalent to 1.0 mg/dL by method of Riddick. *2983*

Chloride *Serum No Effect Analytical* At concentration of 1.5 mg/L had no effect on mercurimetric method. *3393*

Cholesterol *Serum No Effect Analytical* At 5 times upper limit of therapeutic range on methods on SMAC®, Abbott-VP, Cobas-Bio, Ektachem®, Hitachi® 705 and KDA. *2138* At acute overdose concentration (2.5 mg/dL) on Technicon® SMAC® method. *3719* At concentration of 1.5 mg/L had no effect on CHOD-PAP method. *3393* At concentration of 25 mg/L had no effect on Liebermann-Burchard method. *3393*

Creatine Kinase *Serum No Effect Analytical* At 5 times upper limit of therapeutic range on methods on SMAC®, Abbott-VP, aca, Cobas-Bio, and Hitachi® 705. *2138* At acute overdose concentration (2.5 mg/dL) on Technicon® SMAC® method. *3719*

Creatinine *Serum No Effect Analytical* At 50 mg/L on reversed phase LC procedure of Zhiri et al. *3942* At 5 times upper limit of therapeutic range on methods on SMAC®, Abbott-VP, aca, Cobas-Bio, Ektachem®, Hitachi® 705 and KDA. *2138* At acute overdose concentration (2.5 mg/dL) on Technicon® SMAC® method. *3719* At concentration of 25 mg/L had no effect on AutoAnalyzer Jaffé method. *3393*

Dapsone *Serum No Effect Analytical* At 10 mg/dL on colorimetric procedure of Higgins. *1590*

Diphenylhydantoin *Serum No Effect Analytical* On GLC procedure of Papadopoulos at *in vivo* concentration. *2722*
Serum Increase Physiological From 20 to 40 µg/mL (unknown dose) inhibits metabolism. *2422*

Dopa Screening Test *Urine Positive Analytical* Very slight purple color produced. *3075*

Epinephrine *Urine No Effect Physiological* No effect with short term ingestion of 15 mg/d. *0766*

Estradiol *Plasma Increase Physiological* Observed in 5 of 5 men with gynecomastia. *0316*

Ethchlorvynol *Serum No Effect Analytical* At 10 µg/mL on colorimetric method of Wallace. *3753* At 600 mg/L on GLC procedure of Evenson. *1058*
Urine No Effect Analytical No effect on method of Frings and Cohen at 10 mg/dL. *1205*

FPN Test *Urine No Effect Analytical* No effect at 10 mg/dL on method of Frings. *1204*

Glucose *Serum No Effect Analytical* At 5 times upper limit of therapeutic range on methods on SMAC®, Abbott-VP, aca, Cobas-Bio, Ektachem®, Hitachi® 705 and KDA. *2138* At acute overdose concentration (2.5 mg/dL) on Technicon® SMAC® method. *3719* At concentration of 4 mg/L had no effect on Ektachem® method. *3393* At concentration of 1.5 mg/L had no effect on GOD/POD-PAP method. *3393*
Urine No Effect Analytical No effect observed on TesTape®. *1100*
Urine Decrease Analytical Low with Clinistix®, Diastix®. *1100*

γ-Glutamyltransferase (GGT) *Serum No Effect Analytical* At 5 times upper limit of therapeutic range on methods on SMAC®, Abbott-VP, and Hitachi® 705. *2138*

Homovanillic Acid *CSF Decrease Physiological* Probably due to decreased turnover of dopamine. *2725*

Hydrochloric Acid *Gastric Material Decrease Physiological* Presumed central action lasts for 5 h after 10 mg. *0362*

11-Hydroxycorticosteroids *Urine No Effect Physiological* No effect with short term ingestion of 15 mg/d. *0766*

17-Hydroxycorticosteroids *Urine No Effect Physiological* No effect with short term ingestion of 15 mg/d. *0766*

5-Hydroxyindoleacetic Acid (5-HIAA) *Urine Increase Analytical* With method of Udenfriend et al. due to reaction of nitrosonaphthol on the reactive fused benzene ring of the benzodiazepine nucleus. Effect unlikely to produce clinical misinterpretation. Effect also seen with N-desmethyldiazepam. *0670*

Iron *Serum No Effect Analytical* At 5 times upper limit of therapeutic range on Ferrozine method on SMAC®. *2138* At acute overdose concentration (2.5 mg/dL) on Technicon® SMAC® method. *3719* At concentration of 25 mg/L had no effect on Ferrozine method. *3393*

^{131}I Uptake *Serum No Effect Physiological* No effect after i.v. administration. *0583*
Serum Decrease Physiological Conflicting reports ?no effect. *1237*

17-Ketosteroids *Urine No Effect Physiological* No effect with short term ingestion of 15 mg/d. *0766*

Lactate Dehydrogenase *Serum No Effect Analytical* At 5 times upper limit of therapeutic range on methods on SMAC®, Abbott-VP, Cobas-Bio, Hitachi® 705 and KDA. *2138* At acute overdose concentration (2.5 mg/dL) on Technicon® SMAC® method. *3719*

Leukocytes *Blood Decrease Physiological* Leukopenia. *3017*

Methaqualone *Serum No Effect Analytical* At 50 mg/L on GLC procedure of Evenson. *1057*

Neutrophils *Blood Decrease Physiological* Transitory neutropenia reported. *0071* Occasional case of agranulocytosis reported. *3717*

Norepinephrine *Urine No Effect Physiological* No effect with short term ingestion of 15 mg/d. *0766*

Nortriptyline *Serum No Effect Physiological* No effect on serum concentration if given together. *3321*

pCO₂ *Blood Increase Physiological* In healthy volunteers but change of short duration. *1041*

Pheneturide *Serum No Effect Analytical* On GLC procedure of Papadopoulos at *in vivo* concentration. *2722*

Phenobarbital *Serum No Effect Analytical* On GLC procedure of Papadopoulos at *in vivo* concentration. *2722*

Phosphate *Serum No Effect Analytical* At 5 times upper limit of therapeutic range on methods on SMAC®, aca, Hitachi® 705 and KDA. *2138* At acute overdose concentration (2.5 mg/dL) on Technicon® SMAC® method. *3719* At concentration of 25 mg/L had no effect on phosphomolybdate method. *3393*

pO₂ *Blood Increase Physiological* In healthy volunteers but changes of short duration. *1041*
Blood Decrease Physiological In healthy volunteers but changes of short duration. *1041*

Porphyrins *Urine Increase Physiological* May precipitate attack of acute porphyria. *1016*

Potassium *Serum No Effect Analytical* At concentration of 1.5 mg/L had no effect on flame-photometric method. *3393* At concentration of 10 mg/L had no effect on measurement by ISE with predilution. *3393*

Primidone *Serum No Effect Analytical* On GLC procedure of Papadopoulos at *in vivo* concentration. *2722*

Prolactin *Plasma No Effect Physiological* No effect with 20 mg given i.m. *1416*

Propoxyphene *Serum No Effect Analytical* At 25 mg/L on method of Evenson. *1056*

Protein *Serum No Effect Analytical* At 5 times upper limit of therapeutic range on methods on SMAC®, Abbott-VP, Ektachem®, Hitachi® 705 and KDA. *2138* At acute overdose concentration (2.5 mg/dL) on Technicon® SMAC® method. *3719* At concentration of 25 mg/L had no effect on biuret method with blank correction. *3393*

Protein Bound Iodine (PBI) *Serum No Effect Physiological* Conflicting reports ?no effect. *1488*

Prothrombin Time *Plasma No Effect Physiological* No effect when given 15 mg/d. *2682*

Salicylate *Urine No Effect Analytical* At 10 mg/dL no effect on Trinder method. *1204*

Sodium *Serum No Effect Analytical* At concentration of 1.5 mg/L had no effect on flame-photometric method. *3393* At concentration of 10 mg/L had no effect on measurement by ISE with predilution. *3393*

T₃ Uptake *Serum No Effect Physiological* No effect after i.v. administration. *0583*
Serum Decrease Physiological Resin test affected. *2220*

Thymol Turbidity *Serum Increase Physiological* Mild effect reported in one patient. *1878*

Thyroxine (T₄) *Serum No Effect Physiological* Conflicting reports ?no effect. *1488*
Serum Decrease Physiological Due to competition for transport proteins. *3798*

Triglycerides *Serum No Effect Analytical* At 5 times upper limit of therapeutic range on methods on SMAC®, Abbott-VP Ektachem®, Hitachi® 705 and KDA. *2138* At acute overdose concentration (2.5 mg/dL) on Technicon® SMAC® method. *3719* At concentration of 25 mg/L had no effect on lipase/esterase method. *3393*

Urea Nitrogen *Serum No Effect Analytical* At 5 times upper limit of therapeutic range on methods on SMAC®, Abbott-VP, Cobas-Bio, Ektachem®, Hitachi® 705 and KDA. *2138* At acute overdose concentration (2.5 mg/dL) on Technicon® SMAC® method. *3719* At concentration of 25 mg/L had no effect on diacetylmonoxime method. *3393* At concentration of 4 mg/L had no effect on Ektachem® method. *3393*
Serum Increase Physiological Slight elevation in one hypertensive. *1878*

Uric Acid *Serum No Effect Analytical* At 50 mg/L on reversed phase LC procedure of Zhiri et al. *3942* At 5 times upper limit of therapeutic range on methods on SMAC®, Abbott-VP, aca, Cobas-Bio, Ektachem®, Hitachi® 705 and KDA. *2138* At acute overdose concentration (2.5 mg/dL) on Technicon® SMAC® method. *3719* At concentration of 25 mg/L had no effect on phosphotungstate reduction method. *3393*

Volume *Gastric Material Decrease Physiological* Presumed central action lasts for 5 h after 10 mg. *0362*

Diazo Dyes

Protein Bound Iodine (PBI) *Serum Decrease Physiological* Compete for binding sites. *0583*

T₃ Uptake *Serum Increase Physiological* Compete for binding sites. *0583*

Thyroxine (T₄) *Serum Decrease Physiological* Compete for binding sites. *0583*

Diazoxide

Bicarbonate *Urine Decrease Physiological* Effect not as marked as on sodium excretion. *0124*

Catecholamines *Plasma Increase Physiological* Increased release from tissue. *0116*

Chloride *Serum Increase Physiological* May cause salt retention. *1343*
Urine Decrease Physiological Effect over 2 h of 4 mg/kg given i.v. or orally. *1805*

Cortisol *Urine Decrease Analytical* Decreased unconjugated by fluorometric procedure. *0022*

Creatinine Clearance *Urine Decrease Physiological* Effect over 2 h of 4 mg/kg given i.v. *1805*

Effective Renal Plasma Flow
Patient Decrease Physiological After i.v. injection immediate reduction noted. *0124*

Fatty Acids, Free (FFA) *Serum Increase Physiological* Significant rise observed after oral or i.v. administration. *0124*

Glomerular Filtration Rate (GFR)
Urine Decrease Physiological Effect over 2 h of 4 mg/kg given i.v. *1805*

Glucose *Serum Increase Physiological* Inhibits insulin release, peripheral glucose utilization. *0636* Predictable effect due to direct inhibition of insulin secretion. *0451*
Urine Increase Physiological If hyperglycemia occurs. *0636*

Immunoglobulin IgG *Serum Decrease Physiological* Effect may be persistent ?mechanism. *0124*

Inulin Clearance *Urine Decrease Physiological* Effect over 2 h of 4 mg/kg given i.v. *1805*

Leukocytes *Blood Decrease Physiological* May cause leukopenia. *3740*

Neutrophils *Blood Decrease Physiological* Occasional case of agranulocytosis reported. *3717*

Osmolar Clearance *Urine Decrease Physiological* Effect over 2 h of 4 mg/kg given i.v. *1805*

PAH Clearance *Urine Decrease Physiological* Effect over 2 h of 4 mg/kg given i.v. *1805*

pH *Urine Decrease Physiological* Due to decreased bicarbonate excretion. *0124*

Platelets *Blood Decrease Physiological* Immunologic response occurs from days to months. *3740*

Potassium *Urine Decrease Physiological* Effect over 2 h of 4 mg/kg given i.v. *1805*

Prothrombin Time *Plasma Increase Physiological* Displaces anticoagulants from albumin. *3248*

Renin Activity *Plasma Increase Physiological* Effect observed in some hypertensives. *1343*

Sodium *Serum Increase Physiological* May cause salt retention. *1343*
Urine Decrease Physiological Effect over 2 h of 4 mg/kg given i.v. *1805*

Uric Acid *Serum Increase Physiological* Inhibition of tubular uric acid excretion. *1805*
Urine Decrease Physiological Effect over 2 h of 4 mg/kg given i.v. *1805*

Uric Acid Clearance *Urine Decrease Physiological* Effect over 2 h of 4 mg/kg given i.v. *1805*

Volume *Plasma Increase Physiological* Causes salt and water retention. *1343*
Urine Decrease Physiological Effect over 2 h of 4 mg/kg given i.v. *1805*

Dichloralphenazone

Aminolevulinic Acid *Urine Increase Physiological* May precipitate acute porphyria. *1322*

Coproporphyrin *Urine Increase Physiological* May precipitate acute porphyria. *1322*
Feces Increase Physiological May precipitate acute porphyria. *3895*

Diazepam *Serum No Effect Analytical* On cathode-ray Polarographic method of Berry. *0323*

Porphobilinogen *Urine Increase Physiological* May precipitate acute porphyria. *1322*

Porphyrins *Urine Increase Physiological* May precipitate attack of acute porphyria. *1016*

Prothrombin Time *Plasma Decrease Physiological* Antagonizes action of administered coumarins. *3011*

Protoporphyrin *Feces Increase Physiological* May precipitate acute porphyria. *1322*

Dichlorobenzene

Protein *Urine Increase Physiological* May cause renal damage. *0279*

Urea Nitrogen *Serum Increase Physiological* May cause renal damage. *0279*

Dichlorphenamide

Leukocytes *Blood Decrease Physiological* May cause leukopenia. *2313*

Platelets *Blood Decrease Physiological* Probable effect (like acetazolamide). *0071*

Potassium *Serum Decrease Physiological* Diuretic action (inhibits carbonic anhydrase). *2313*

Sodium *Serum Decrease Physiological* Cation loss as excess chloride is excreted. *0652*

Diclofenac

Alanine Aminotransferase *Serum No Effect Analytical* On continuous method at 10 times maximal therapeutic concentration On colorimetric method at 10 times maximal therapeutic concentration. *1785*

Diclofenac (continued)

Albumin *Serum No Effect Analytical* On bromcresol green method on SMA II at physiological concentration. *1787*
Serum No Effect Physiological No effect seen in patients undergoing treatment. *1787*

Alkaline Phosphatase *Serum No Effect Analytical* On continuous method at 10 times maximal therapeutic concentration. *1785*

Aspartate Aminotransferase *Serum No Effect Analytical* On colorimetric method at 10 times maximal therapeutic concentration. *1785*
Serum Increase Analytical On continuous method at 10 times maximal therapeutic concentration. *1785*

Bilirubin *Serum No Effect Analytical* No effect at therapeutic concentrations on methods of Jendrassik-Grof and those using spectrophotometry. *1786* On diazo method on SMA II at physiological concentration. *1787*
Serum No Effect Physiological No effect seen in patients undergoing treatment. *1787*

Calcium *Serum No Effect Analytical* On cresolphthalein complex one method on SMA II at physiological concentration. *1787* At concentration of 23 mg/L had no effect on cresolphthalein method. *3393*
Serum No Effect Physiological No effect seen in patients undergoing treatment. *1787*

Cholesterol *Serum No Effect Analytical* No effect at therapeutic concentrations on enzymatic and Liebermann-Burchard methods. *1786* On enzymatic method on SMA II at physiological concentration. *1787* At concentration of 23 mg/L had no effect on CHOD-PAP method. *3393*
Serum Increase Physiological Significantly higher (5.16 vs 4.60 mmol/L) than in controls. *1787*

Creatine Kinase *Serum No Effect Analytical* On continuous method at 10 times maximal therapeutic concentration. *1785*

Creatinine *Serum No Effect Analytical* No effect at therapeutic concentrations on alkaline picrate and Slot methods. *1786* On alkaline picrate method on SMA II at physiological concentration. *1787* At concentration of 23 mg/L had no effect on AutoAnalyzer Jaffé method. *3393*
Serum No Effect Physiological No effect seen in patients undergoing treatment. *1787*

Glucose *Serum No Effect Analytical* No effect a therapeutic concentrations on hexokinase, glucose dehydrogenase, 2,4-dichloroaience, and o-toluidine methods. *1786* On hexokinase method on SMA II at physiological concentration. *1787* At concentration of 23 mg/L had no effect on hexokinase/G-6-PDH method. *3393*
Serum No Effect Physiological No effect seen in patients undergoing treatment. *1787*
Serum Increase Analytical Increase at therapeutic concentration with glucose oxidase/peroxidase method with ABTS. *1786*

Hydroxybutyrate Dehydrogenase
Serum No Effect Analytical On continuous method at 10 times maximal therapeutic concentration. *1785*

Iron *Serum No Effect Analytical* No effect at therapeutic concentrations on Ramsay and bathophenanthroline methods. *1786* On Ferrozine method on SMA II at physiological concentration. *1787* At concentration of 23 mg/L had no effect on Ferrozine method. *3393*
Serum No Effect Physiological No effect seen in patients undergoing treatment. *1787*

Lactate Dehydrogenase *Serum No Effect Analytical* On continuous method with lactate or pyruvate as substrate at 10 times maximal therapeutic concentration. On colorimetric method at 10 times maximal therapeutic concentration. *1785*

Lithium *Serum Increase Physiological* Decreased renal clearance and increased plasma concentration by 26% in 5 normal women. *2943*

Phosphate *Serum No Effect Physiological* No effect seen in patients undergoing treatment. *1787*

Protein *Serum No Effect Analytical* On biuret method on SMA II at physiological concentration. *1787* At concentration of 23 mg/L had no effect on biuret method with blank correction. *3393* No effect at therapeutic concentrations on Biuret and Spectrophotometric methods. *1786*
Serum No Effect Physiological No effect seen in patients undergoing treatment. *1787*

Triglycerides *Serum No Effect Analytical* No effect on enzymatic methods at therapeutic concentrations. *1786* On enzymatic method on SMA II at physiological concentration. *1787*
Serum No Effect Physiological No effect seen in patients undergoing treatment. *1787*

Urea Nitrogen *Serum No Effect Analytical* No effect at therapeutic concentrations with glutamate dehydrogenase, phenol hypochlorite and diacetylmonoxime procedures. *1786* On diacetyl monoxime complexone method on SMA II at physiological concentration. *1787*
Serum Increase Physiological Significantly higher in treated patients (6.87 mmol/L vs 5.32 mmol/L) than in controls. *1787*

Uric Acid *Serum No Effect Analytical* On phosphotungstate method on SMA II at physiological concentration. *1787* At concentration of 23 mg/L had no effect on phosphotungstate reduction method. *3393* No effect at therapeutic concentrations on uricase-catalase, aldehyde dehydrogenase, direct UV-test and phosphotungstate methods. *1786*
Serum No Effect Physiological No effect seen in patients undergoing treatment. *1787*

Dicloxacillin

Aspartate Aminotransferase
Serum Increase Physiological Mild hepatic dysfunction observed. *0118*

Bilirubin *Serum Increase Physiological* Possible displacement from protein especially in critically ill neonates. *3748*

Cephalin Flocculation *Serum Increase Physiological* Mild hepatic dysfunction noted. *0118*

Eosinophils *Blood Increase Physiological* Mild allergic response. *0118*

Dicumarol

Alanine Aminotransferase *Serum Increase Physiological* May be high enough to simulate myocardial infarction. *0248*

Aspartate Aminotransferase
Serum Increase Physiological May simulate myocardial infarct (greater than ALT). *0248*

Chlorpropamide *Serum Increase Physiological* Increases half life, augments hypoglycemia. *1487*

Clotting Time *Blood Increase Physiological* Slight effect in glass, greater in silicone. *1343*

Diphenylhydantoin *Serum Increase Physiological* May increase concentration 5 to 15 µg/mL (at dose for prothrombin time=30%). *2041* Increase by 38 to 250% in six volunteers probably due to inhibition of para-hydroxylation in liver. *3556*

Erythrocytes *Blood Decrease Physiological* May cause anemia (AMA Blood dyscrasias committee). *2429*

Erythrocyte Sedimentation Rate
Blood No Effect Physiological Unaltered by coumarins. *1343*

Factor VII *Plasma Decrease Physiological* Dose related effect. *1343*

Factor IX *Plasma Decrease Physiological* Dose related effect. *1343*

Factor X *Plasma Decrease Physiological* Dose related effect. *1343*

Glucose *Serum Decrease Physiological* Reported effect. *0117*

Lactate Dehydrogenase *Serum Increase Physiological* May be high enough to simulate myocardial infarction. *0248*

Leukocytes *Blood Decrease Physiological* Reported effect (AMA Blood dyscrasias committee). *2429*

Mexiletine *Serum Increase Physiological* Decreases clearance and prolongs half-life. *2011*

Occult Blood *Feces Positive Physiological* May cause intramural hemorrhage even if no ulcer. *3498*

Platelet Adhesiveness *Blood Decrease Physiological* Related to dose. *1343*

Platelet Aggregation *Blood Increase Physiological* In response to ADP. *1343*

Prothrombin Time *Plasma Increase Physiological* Dose related effect. *1343*

Recalcification Time *Blood Increase Physiological* Effect on citrated plasma. *1343*

T₃ Uptake *Serum Increase Physiological* Red cell uptake affected. *0582*

Tolbutamide *Serum Increase Physiological* Increases half life, augments hypoglycemia. *1487*

Uric Acid *Serum Decrease Physiological* Uricosuric action. *0111*

Urine Increase Physiological Uricosuric action. *1433*

Dieldrin

Lactate Dehydrogenase *Serum Decrease Physiological* In vitro 8.0% decrease at 1 x 10⁻⁴ mol/L. *1270*

Leukocytes *Blood Increase Physiological* May occur in acute response. *1302*

Pseudocholinesterase *Serum Decrease Physiological* In vitro 7.2% decrease at 1 x 10⁻⁴ mol/L. *1270*

Dienestrol

Alanine Aminotransferase *Serum Increase Physiological* Reported to cause cholestasis. *1680*

Aspartate Aminotransferase

Serum Increase Physiological Reported to cause cholestasis. *1680*

Bilirubin *Serum Increase Physiological* May cause cholestasis. *1680*

Calcium *Serum Increase Physiological* May occur with extended high dosage. *1680*

Volume *Plasma Increase Physiological* Fluid retention with large dose estrogens. *1680*

Diethazine

Leukocytes *Blood Decrease Physiological* May cause leukopenia. *2313*

Diethylaminoethanol

Albumin *Serum No Effect Analytical* At concentration of 30 mg/L had no effect on BCG method. *3393*

Bicarbonate *Serum No Effect Analytical* At concentration of 30 mg/L had no effect on method using phenolphthalein. *3393*

Bilirubin *Serum No Effect Analytical* At concentration of 30 mg/L had no effect on Jendrassik and Grof method. *3393*

Calcium *Serum No Effect Analytical* At concentration of 30 mg/L had no effect on cresolphthalein method. *3393*

Chloride *Serum No Effect Analytical* At concentration of 30 mg/L had no effect on mercurimetric method. *3393*

Cholesterol *Serum No Effect Analytical* At concentration of 30 mg/L had no effect on Liebermann-Burchard method. *3393*

Creatinine *Serum No Effect Analytical* At concentration of 30 mg/L had no effect on AutoAnalyzer method. *3393*

Phosphate *Serum No Effect Analytical* At concentration of 30 mg/L had no effect on phosphomolybdate method. *3393*

Protein *Serum No Effect Analytical* At concentration of 30 mg/L had no effect on biuret method with blank correction. *3393*

Triglycerides *Serum No Effect Analytical* At concentration of 30 mg/L had no effect on lipase/esterase method. *3393*

Urea Nitrogen *Serum No Effect Analytical* At concentration of 30 mg/L had no effect on diacetylmonoxime method. *3393*

Uric Acid *Serum No Effect Analytical* At concentration of 30 mg/L had no effect on phosphotungstate reduction method. *3393*

Diethylpropion

Alanine Aminotransferase *Serum No Effect Analytical* At acute overdose concentration (10 mg/dL) on Technicon® SMAC® method. *3719*

Albumin *Serum No Effect Analytical* At acute overdose concentration (10 mg/dL) on Technicon® SMAC® method. *3719* At concentration of 100 mg/L had no effect on BCG method. *3393*

Alkaline Phosphatase *Serum No Effect Analytical* At acute overdose concentration (10 mg/dL) on Technicon® SMAC® method. *3719*

Alkaloids *Urine No Effect Analytical* With ninhydrin on Frings TLC procedure at 25 mg/dL. *1204*

Urine Increase Analytical Purple with iodoplatinate on Frings TLC procedure. Reacts with ninhydrin on TLC method of Frings. *1204*

Amobarbital *Urine No Effect Analytical* No interference with TLC using ethyl acetate: methanol: water: ammonium hydroxide and modified Dragendorff's reagent for detection. *3868*

Amphetamine *Urine No Effect Analytical* At 50 mg/L on fluorescent method of Hayes. *1534* No interference with TLC using ethyl acetate: methanol: water: ammonium hydroxide and modified Dragendorff's reagent for detection. *3868*

Aspartate Aminotransferase *Serum No Effect Analytical* At acute overdose concentration (10 mg/dL) on Technicon® SMAC® method. *3719*

Barbiturate *Urine No Effect Analytical* With mercuric SO₄ on Frings TLC procedure at 25 mg/dL. With mercurous NO₃ on Frings TLC procedure at 25 mg/dL. With vanillin on Frings procedure at 25 mg/dL. With diphenylcarbazone on Frings TLC procedure at 25 mg/dL. *1204*

Bilirubin *Serum No Effect Analytical* At acute overdose concentration (10 mg/dL) on Technicon® SMAC® method. *3719* At concentration of 100 mg/L had no effect on Jendrassik and Grof method. *3393*

Bromide *Urine No Effect Analytical* No effect on method of Frings. *1204*

Calcium *Serum No Effect Analytical* At acute overdose concentration (10 mg/dL) on Technicon® SMAC® method. *3719* At concentration of 100 mg/L had no effect on cresolphthalein method. *3393*

Cholesterol *Serum No Effect Analytical* At acute overdose concentration (10 mg/dL) on Technicon® SMAC® method. *3719* At concentration of 100 mg/L had no effect on Liebermann-Burchard method. *3393*

Cocaine *Urine Positive Analytical* Similar R_f and color reaction on TLC using ethyl acetate: methanol: water: ammonium hydroxide and modified Dragendorff's reagent. *3868*

Creatine Kinase *Serum No Effect Analytical* At acute overdose concentration (10 mg/dL) on Technicon® SMAC® method. *3719*

Creatinine *Serum No Effect Analytical* At acute overdose concentration (10 mg/dL) on Technicon® SMAC® method. *3719* At concentration of 100 mg/L had no effect on AutoAnalyzer Jaffé method. *3393*

Diazepam *Urine Positive Analytical* Similar R_f and color reaction on TLC using ethyl acetate: methanol: water: ammonium hydroxide and modified Dragendorff's reagent. *3868*

Erythrocytes *Blood Decrease Physiological* May cause bone marrow depression. *0134*

Ethchlorvynol *Urine No Effect Analytical* No effect on method of Frings and Cohen. *1204*

FPN Test *Urine No Effect Analytical* No effect at 25 mg/dL on method of Frings. *1204*

Glucose *Serum No Effect Analytical* At acute overdose concentration (10 mg/dL) on Technicon® SMAC® method. *3719*

Hydromorphone *Urine No Effect Analytical* No interference with TLC using ethyl acetate: methanol: water: ammonium hydroxide and modified Dragendorff's reagent for detection. *3868*

Iron *Serum No Effect Analytical* At acute overdose concentration (10 mg/dL) on Technicon® SMAC® method. *3719* At concentration of 100 mg/L had no effect on Ferrozine method. *3393*

Lactate Dehydrogenase *Serum No Effect Analytical* At acute overdose concentration (10 mg/dL) on Technicon® SMAC® method. *3719*

Leukocytes *Blood Decrease Physiological* Marrow depression with leukopenia/agranulocytosis. *2415*

Diethylpropion (continued)

Mescaline *Urine No Effect Analytical* No interference with TLC using ethyl acetate: methanol: water: ammonium hydroxide and modified Dragendorff's reagent for detection. *3868*

Methamphetamine *Urine No Effect Analytical* No interference with TLC using ethyl acetate: methanol: water: ammonium hydroxide and modified Dragendorff's reagent for detection. *3868*

Methaqualone *Urine Positive Analytical* Similar R_f and color reaction on TLC using ethyl acetate: methanol: water: ammonium hydroxide and modified Dragendorff's reagent. *3868*

Morphine *Urine No Effect Analytical* No interference with TLC using ethyl acetate: methanol: water: ammonium hydroxide and modified Dragendorff's reagent for detection. *3868*

Pentobarbital *Urine No Effect Analytical* No interference with TLC using ethyl acetate: methanol: water: ammonium hydroxide and modified Dragendorff's reagent for detection. *3868*

Phencyclidine *Urine Positive Analytical* Similar R_f and color reaction on TLC using ethyl acetate: methanol: water: ammonium hydroxide and modified Dragendorff's reagent. *3868*

Phenobarbital *Urine No Effect Analytical* No interference with TLC using ethyl acetate: methanol: water: ammonium hydroxide and modified Dragendorff's reagent for detection. *3868*

Phenylpropanolamine *Urine No Effect Analytical* No interference with TLC using ethyl acetate: methanol: water: ammonium hydroxide and modified Dragendorff's reagent for detection. *3868*

Phosphate *Serum No Effect Analytical* At acute overdose concentration (10 mg/dL) on Technicon® SMAC® method. *3719* At concentration of 100 mg/L had no effect on phosphomolybdate method. *3393*

Platelets *Blood Decrease Physiological* Isolated cases of bone marrow depression. *1679*

Protein *Serum No Effect Analytical* At acute overdose concentration (10 mg/dL) on Technicon® SMAC® method. *3719* At concentration of 100 mg/L had no effect on biuret method with blank correction. *3393*

Salicylate *Urine No Effect Analytical* At 25 mg/dL no effect on Trinder method. *1204*

Secobarbital *Urine No Effect Analytical* No interference with TLC using ethyl acetate: methanol: water: ammonium hydroxide and modified Dragendorff's reagent for detection. *3868*

Triglycerides *Serum No Effect Analytical* At acute overdose concentration (10 mg/dL) on Technicon® SMAC® method. *3719* At concentration of 100 mg/L had no effect on lipase/esterase method. *3393*

Urea Nitrogen *Serum No Effect Analytical* At acute overdose concentration (10 mg/dL) on Technicon® SMAC® method. *3719* At concentration of 100 mg/L had no effect on diacetylmonoxime method. *3393*

Uric Acid *Serum No Effect Analytical* At acute overdose concentration (10 mg/dL) on Technicon® SMAC® method. *3719* At concentration of 100 mg/L had no effect on phosphotungstate reduction method. *3393*

Diethylstilbestrol

Acid Phosphatase *Prostatic Fluid Decrease Physiological* Markedly lower in benign hypertrophy. *1917*

Alanine Aminotransferase *Serum Increase Physiological* Hepatotoxicity with centrolobular necrosis. *1948*

Amino Acids *Serum Decrease Physiological* Specific amino acids affected in men. *0742*

Aspartate Aminotransferase
Serum Increase Physiological Hepatotoxicity with centrolobular necrosis. *1948*

Bilirubin *Serum Increase Physiological* Hepatotoxicity with centrolobular necrosis. *1948*

Calcium *Serum Increase Physiological* Rapid increase in 24 h in patients with breast cancer. *3519*

Coproporphyrin *Urine Increase Physiological* May induce porphyria cutanea tarda. *3769*

Feces Increase Physiological May induce porphyria cutanea tarda. *3769*

Corticosteroid-Binding Globulin
Serum Increase Physiological Can be doubled in males with treatment. *3571*

Estradiol *Urine Decrease Analytical* Affects unmodified Brown/Kober procedure. *0583*

Estriol *Urine Decrease Analytical* Affects unmodified Brown/Kober procedure. *0583*

Estrogens *Urine Increase Analytical* Interferes in Kober procedure. Reacts in Brown's procedure. *0279*
Urine Decrease Analytical Affects colorimetric/fluorometric procedures. *0022*

Estrone *Urine Decrease Analytical* method of Brown if very large doses used. *0023*

Follicle Stimulating Hormone (FSH)
Plasma Decrease Physiological At doses affecting pituitary-gonadal axis. *1917*

Glucose Tolerance *Serum Decrease Physiological* May provoke mild to moderate deterioration. *1849*

17-Hydroxycorticosteroids *Urine Increase Physiological* Can be doubled in males with treatment. *3571*

6-β-Hydroxycortisol *Urine Increase Physiological* Increased conversion of cortisol produced. *0022*

Luteinizing Hormone (LH) *Plasma Decrease Physiological* At doses affecting pituitary-gonadal axis. *1917*

Platelets *Blood Decrease Physiological* Thrombocytopenia (AMA Blood dyscrasias). *2429*

Prolactin *Plasma Increase Physiological* Over 1 week in normal males. *3913*

Protein Bound Iodine (PBI) *Serum Increase Physiological* Increases concentration of circulating thyroxine binding globulin. *0012*

Protoporphyrin *Feces Increase Physiological* May induce porphyria cutanea tarda. *3769*

Sex Hormone Binding Globulin
Serum Increase Physiological Observed when given to men with androgen-dependent prostatic cancer. *1736*

Testosterone *Serum Decrease Physiological* At doses affecting pituitary-gonadal axis. *1917* When given to men with androgen-dependent prostatic cancer reduced concentration to normal female range. *1736*

Testosterone, Free *Serum Decrease Physiological* In men with androgen-dependent prostatic cancer and reduction of total hormone concentration with SHBG. *1736*

Thyroxine (T_4) Binding Globulin
Serum Increase Physiological Direct effect of drug. *0012*

Tyrosine *Plasma Decrease Physiological* Observed with 5 mg daily for 5 d. *0843*

Uric Acid *Serum Decrease Physiological* In men on therapy (hormonal effect). *2590*
Urine Increase Physiological In men on therapy (hormonal effect). *2590*

Uric Acid Clearance *Urine Increase Physiological* In men on therapy (hormonal effect). *2590*

Uroporphyrin *Urine Increase Physiological* May induce porphyria cutanea tarda. *3769*
Feces Increase Physiological May induce porphyria cutanea tarda. *3769*

Vitamin A *Serum Decrease Physiological* Hormonal effect when given to lactating women. *1221*

Diflunisal

Bilirubin *Serum No Effect Physiological* At pharmacological concentrations has little or no protein displacement effect. *3748*

Bleeding Time *Patient Increase Physiological* Moderate but not significant increase at 1,000 mg bid whereas 250 mg and 500 mg had no effect. Effect reverted to normal in 24 h. *1386*

Cortisol *Plasma No Effect Analytical* No effect on FETI methods of Syva® advance. *1687*

Digoxin *Serum No Effect Analytical* No effect on FETI methods of Syva® advance. *1687*

Malondialdehyde *Platelets Decrease Physiological* Production reduced with 1,000 mg bid whereas no effect observed at lesser amounts. Effects observed only for 24 h after administration for 1 d. *1386*

Occult Blood *Feces Positive Physiological* Significant increase when 1,000 mg given bid whereas no effect with 250 mg or 500 mg bid. *1386* Two reported cases of hematemesis attributable to drug, although other drugs also ingested. *2326*

Salicylate *Serum Increase Analytical* At 100 mg/L on colorimetric methods of Keller (and modified version) and Trinder - effect approximately half of that of salicylate. *1776* 2 to 3 times higher concentration than salicylate measured by TDX method. *0799* Half to two thirds concentration of salicylate when measured by Trinder and aca methods. *0799*

T₃ Uptake *Serum Decrease Analytical* Same interference as with thyroxine but to lesser extent. *1687*

Thyroxine (T₄) *Serum Decrease Analytical* Using FETI in Syva® advance system in a patient receiving 250 mg drug twice per day although mechanism not clear. *1687*

Uric Acid *Serum Decrease Physiological* Dose-dependent effect: at dose of 1 g/d concentration was 30% lower than at baseline, maximum effect within 2 weeks. *2390*

Difluorophosphate

Protein Bound Iodine (PBI) *Serum Decrease Physiological* Interferes with trapping of iodide by thyroid. *0012*

Digitalis

Barbiturate *Serum No Effect Analytical* No interference with UV absorption methods. *2981*

Chloride *Urine Increase Physiological* Diuretic action in cardiac failure. *1343*

Eosinophils *Blood Increase Physiological* Allergic response may be large with toxicity. *1513*

Erythrocytes *Blood Decrease Physiological* Aplastic anemia/pancytopenia. *2313*

Glomerular Filtration Rate (GFR) *Urine Increase Physiological* Improvement with relief of edema. *1343*

¹³¹I Uptake *Serum Decrease Physiological* Lilly product contains tetraiodofluorescein. *2652*

LE Cells *Blood Positive Physiological* SLE may occur, usually normalizes when stopped. *3059*

Leukocytes *Blood Increase Physiological* Rare leukocytosis. *2313* *Blood Decrease Physiological* Agranulocytosis/leukopenia. *2313*

Occult Blood *Feces Positive Physiological* May occasionally cause gastrointestinal hemorrhagic necrosis. *0658*

Platelets *Blood Decrease Physiological* (rare) pancytopenia/thrombocytopenia. *3924*

Potassium *Serum Increase Physiological* But only with doses 20 to 40 times therapeutic. *2836*

Protein *Serum Increase Physiological* Improved hepatic function and decreased hypovolemia. *1343*

Protein Bound Iodine (PBI) *Serum No Effect Physiological* No effect observed. *0830*

Prothrombin Time *Plasma Decrease Physiological* Animal experiments suggest increased coagulability. *2313*

Sodium *Urine Increase Physiological* Diuretic action in cardiac failure. *1343*

Volume *Urine Increase Physiological* Diuretic action in cardiac failure. *1343*

Xylose Excretion *Urine Decrease Physiological* May produce gastrointestinal intolerance, impaired absorption. *0612*

Digitonin

Cholesterol *Serum Increase Analytical* If p-toluenesulfonic acid reaction used. *1563* Affects L-B and Zlatkis-Zak procedures. *1834*

Digitoxin

Albumin *Serum No Effect Analytical* No effect at 21 mg/dL on SMA 12/60 method. *2636* At concentration of 6 mg/L had no effect on BCG method. *3393*

Alkaline Phosphatase *Serum No Effect Analytical* No effect at 21 mg/dL on SMA 12/60 method. *2636*

Antithrombin III *Plasma No Effect Analytical* At concentration of 20 mg/L had no effect on aca method. *3393*

Aspartate Aminotransferase *Serum No Effect Analytical* No effect at 21 mg/dL on SMA 12/60 method. *2636*

Bicarbonate *Serum No Effect Analytical* At concentration of 6 mg/L had no effect on method using phenolphthalein. *3393*

Bilirubin *Serum No Effect Analytical* No effect at 21 mg/dL on SMA 12/60 method. *2636* At concentration of 6 mg/L had no effect on Jendrassik and Grof method. *3393*

Calcium *Serum No Effect Analytical* No effect at 21 mg/dL on SMA 12/60 method. *2636* At concentration of 6 mg/L had no effect on cresolphthalein method. *3393*

Catecholamines *Urine No Effect Analytical* No effect on fluorometric method of Crout. *0758*

Chloride *Serum No Effect Analytical* At concentration of 6 mg/L had no effect on mercurimetric method. *3393*

Cholesterol *Serum No Effect Analytical* No effect at 21 mg/dL on SMA 12/60 method. *2636* At concentration of 6 mg/L had no effect on Liebermann-Burchard method. *3393*

Creatinine *Serum No Effect Analytical* No effect at 21 mg/dL on SMA 12/60 method. *2636* At concentration of 6 mg/L had no effect on AutoAnalyzer Jaffé method. *3393*

Digoxin *Serum Increase Analytical* At normal concentrations in serum if no preincubation (RIA). *2804* Due to cross-reactivity (RIA) if given i.m. recently. *2033*

Glucose *Serum No Effect Analytical* No effect at 21 mg/dL on SMA 12/60 method. *2636*

17-Hydroxycorticosteroids *Urine Increase Analytical* Moderate effect with *in vitro* test. *1488*

¹³¹I Uptake *Serum No Effect Physiological* With 0.01 mg/kg/week orally no effect. *0583* *Serum Decrease Physiological* Purodigin, Crystodigin® contain tetraiodofluorescein. *2652*

17-Ketogenic Steroids *Urine Increase Analytical* Interferes with Zimmermann reaction *in vitro*. *0427*

17-Ketosteroids *Urine No Effect Analytical* Minimal interference with Zimmermann reaction. *0427*

Lactate Dehydrogenase *Serum No Effect Analytical* No effect at 21 mg/dL on SMA 12/60 method. *2636*

Phosphate *Serum No Effect Analytical* No effect at 21 mg/dL on SMA 12/60 method. *2636* At concentration of 6 mg/L had no effect on phosphomolybdate method. *3393*

Plasminogen *Plasma No Effect Analytical* At concentration of 20 mg/L had no effect on aca method. *3393*

Platelets *Blood Decrease Physiological* Rare thrombocytopenia (due to immune mechanism). *3924* Several cases of immune-mediated thrombocytopenia reported. *2502*

Potassium *Serum No Effect Analytical* At concentration of 6 mg/L had no effect on measurement by ISE with predilution. *3393*

Protein *Serum No Effect Analytical* No effect at 21 mg/dL on SMA 12/60 method. *2636* At concentration of 6 mg/L had no effect on biuret method with blank correction. *3393*

Protein Bound Iodine (PBI) *Serum No Effect Physiological* With 0.1 mg/kg/week no effect observed. *0583*

Sodium *Serum No Effect Analytical* At concentration of 6 mg/L had no effect on measurement by ISE with predilution. *3393*

Triglycerides *Serum No Effect Analytical* At concentration of 6 mg/L had no effect on lipase/esterase method. *3393*

Urea Nitrogen *Serum No Effect Analytical* At concentration of 6 mg/L had no effect on diacetylmonoxime method. *3393*

Uric Acid *Serum No Effect Analytical* No effect at 21 mg/dL on SMA 12/60 method. *2636* At concentration of 6 mg/L had no effect on phosphotungstate reduction method. *3393*

Digoxin

Albumin *Serum No Effect Analytical* No effect at 0.04 mg/dL on SMA 12/60 method. *2636* At concentration of 9 mg/L had no effect on BCG method. *3393*

Alkaline Phosphatase *Serum No Effect Analytical* No effect at 0.04 mg/dL on SMA 12/60 method. *2636*

Androstenedione *Plasma No Effect Physiological* No significant effect with long-term administration. *2576*

Antithrombin III *Plasma No Effect Analytical* At concentration of 10 mg/L had no effect on aca method. *3393*

Aspartate Aminotransferase *Serum No Effect Analytical* No effect at 0.04 mg/dL on SMA 12/60 method. *2636*

Bicarbonate *Serum No Effect Analytical* At concentration of 9 mg/L had no effect on method using phenolphthalein. *3393*

Bilirubin *Serum No Effect Analytical* No effect at 0.04 mg/dL on SMA 12/60 method. *2636* At concentration of 9 mg/L had no effect on Jendrassik and Grof method. *3393*

Calcium *Serum No Effect Analytical* No effect at 0.04 mg/dL on SMA 12/60 method. *2636* At concentration of 9 mg/L had no effect on cresolphthalein method. *3393*

Catecholamines *Urine No Effect Analytical* No effect observed. *0913*

Chloride *Serum No Effect Analytical* At concentration of 9 mg/L had no effect on mercurimetric method. *3393*

Cholesterol *Serum No Effect Analytical* No effect at 0.04 mg/dL on SMA 12/60 method. *2636* At concentration of 100 mg/L had no effect on CHOD-PAP method. *3393* At concentration of 100 mg/L had no effect on method using catalase-Hantzsch reaction. *3393* At concentration of 9 mg/L had no effect on Liebermann-Burchard method. *3393*

Creatine Kinase *Serum Increase Physiological* 15 to 17 x increase after i.m. injection (increase for 8 d). *1392*

Creatinine *Serum No Effect Analytical* No effect at 0.04 mg/dL on SMA 12/60 method. *2636* At 10 mg/L on reversed phase LC procedure of Zhiri et al. *3942* At concentration of 9 mg/L had no effect on AutoAnalyzer Jaffé method. *3393* At concentration of 0.15 mg/L had no effect on Jaffé-Fading-Fraction method. *3393* At concentration of 0.15 mg/L had no effect on Jaffé-Fuller's earth method. *3393* At concentration of 0.15 mg/L had no effect on kinetic Jaffé method on BKA-2. *3393*

Dehydroepiandrosterone (DHEA) *Plasma No Effect Physiological* No significant effect with long-term administration. *2576*

Estriol *Urine No Effect Analytical* No effect on GLC method. *1163*

Estrogens *Plasma Increase Physiological* 2x in men, 1.5 x in postmenopausal women with chronic administration. *3475*

Estrone *Plasma Increase Physiological* Major estrogen with chronic administration. *3475*

Glucose *Serum No Effect Analytical* No effect at 0.04 mg/dL on SMA 12/60 method. *2636* At concentration of 0.015 mg/L had no effect on Ektachem® method. *3393* At concentration of 0.25 mg/L had no effect on GOD/POD-PAP method. *3393* *Urine No Effect Analytical* No effect observed with Tes-Tape®. *1100* *Urine Decrease Analytical* Low with Clinistix®, Diastix®. *1100*

17-Hydroxycorticosteroids *Urine Increase Analytical* Moderate effect with *in vitro* test. *1488*

Indomethacin *Serum No Effect Analytical* No effect on HPLC method of Roberts and Smith. *3002*

17-Ketosteroids *Urine Decrease Analytical* Slight effect on Zimmermann reaction *in vitro*. *0427*

Lactate Dehydrogenase *Serum No Effect Analytical* No effect at 0.04 mg/dL on SMA l1/60 method. *2636*

Luteinizing Hormone (LH) *Plasma Decrease Physiological* 50% in men, 40% decrease in postmenopausal women. *3475*

Magnesium *Serum Decrease Physiological* Important factor in digitalis toxicity, possibly associated with prior diuretic use. *1151*

Neutrophils *Blood Decrease Physiological* Reported observation. *2583*

Phosphate *Serum No Effect Analytical* No effect at 0.04 mg/dL on SMA 12/60 method. *2636* At concentration of 9 mg/L had no effect on phosphomolybdate method. *3393*

Plasminogen *Plasma No Effect Analytical* At concentration of 2 mg/L had no effect on aca method. *3393*

Potassium *Serum No Effect Analytical* At concentration of 0.002 mg/L had no effect on flame-photometric method. *3393* At concentration of 9 mg/L had no effect on measurement by ISE with predilution. *3393* At concentration of 0.15 mg/L had no effect on measurement by ISE without predilution. *3393* *Red Blood Cells Decrease Physiological* 6% drop within 2 d, affects membrane ATP-ase. *1923*

Protein *Serum No Effect Analytical* No effect at 0.04 mg/dL on SMA 12/60 method. *2636* At concentration of 9 mg/L had no effect on biuret method with blank correction. *3393*

Protein Electrophoresis *Serum No Effect Analytical* At concentration of 0.15 mg/L had no effect on automated Olympus-Hite method. *3393*

Sodium *Serum No Effect Analytical* At concentration of 0.002 mg/L had no effect on flame-photometric method. *3393* At concentration of 9 mg/L had no effect on measurement by ISE with predilution. *3393* At concentration of 0.15 mg/L had no effect on measurement by ISE without predilution. *3393* *Red Blood Cells Increase Physiological* 16% increase, due to effect on membrane ATP-ase. *1923*

Testosterone *Serum Decrease Physiological* 30% decrease in men with chronic administration. *3475*

Triglycerides *Serum No Effect Analytical* At concentration of 0.25 mg/L had no effect on GPO-PAP method. *3393* At concentration of 9 mg/L had no effect on lipase/esterase method. *3393*

Urea Nitrogen *Serum No Effect Analytical* At concentration of 9 mg/L had no effect on diacetylmonoxime method. *3393* At concentration of 0.015 mg/L had no effect on Ektachem® method. *3393*

Uric Acid *Serum No Effect Analytical* No effect at 0.04 mg/dL on SMA 12/60 method. *2636* At 10 mg/L on reversed phase LC procedure of Zhiri et al. *3942* At concentration of 0.15 mg/L had no effect on Kageyama-Hantzsch method. *3393* At concentration of 0.15 mg/L had no effect on catalase-AlDH method. *3393* At concentration of 9 mg/L had no effect on phosphotungstate reduction method. *3393* At concentration of 0.25 mg/L had no effect on uricase-PAP method. *3393*

Vanillylmandelic Acid *Urine No Effect Analytical* No effect observed. *0913* *Urine No Effect Physiological* No effect observed. *0913*

Water *Red Blood Cells Increase Physiological* Slight effect only due to action on membrane. *1923*

Digoxin-Specific Fab

Digoxin *Serum Decrease Physiological* Used to remove drug in overdose situations. *1699*

Dihydralazine

Amphetamine *Urine No Effect Analytical* No effect at 100 mg/L on method of Rutter. *3097*

Dihydroergotamine

Creatinine *Serum No Effect Analytical* At 10 mg/L on reversed phase LC procedure of Zhiri et al. *3942*

Uric Acid *Serum No Effect Analytical* At 10 mg/L on reversed phase LC procedure of Zhiri et al. *3942*

Dihydromorphine

Morphine *Urine Increase Analytical* Cross reactivity equal (or more) with RIA procedures. *2514*

Dihydrostreptomycin

Bilirubin *Serum No Effect Physiological* Clinically insignificant displacement from protein in neonates. *3748*

Dihydrotachysterol

Calcium *Serum Increase Physiological* Weak antirachitic activity. *2313* Delayed elimination of drug from plasma in hypothyroid state. *2061*
Urine Increase Physiological Hypercalciuric effect. *1022*

Hemoglobin *Blood Decrease Physiological* Adverse effect with severe hypercalcemia. *1680*

Phosphate *Urine Increase Physiological* Diuresis almost as great as with vitamin D_3. *0071*

Protein *Urine Increase Analytical* Increased glomerular permeability. *1022*
Urine Increase Physiological May induce increased glomerular permeability. *1022*

Dihydrotestosterone

Estrogens *Urine No Effect Analytical* At 50 mg/L on fluorescent method of Corns. *0730*

Testosterone *Serum Increase Analytical* 5-alpha compound affects competitive protein binding method. *2750*

17,21-Di-Hydroxy-4-Pregnene-3,20-Dione

Aldosterone *Urine Increase Analytical* 1 mg/L equivalent to 6 μg/L by method of Drewes. *0952*

Dihydroxyacetone

Triglycerides *Serum Increase Analytical* Measured as glycerol with enzymatic methods. *2981*

3,4-Dihydroxybenzoic Acid

Isovanillic Acid *Urine Increase Physiological* Significant increase with glycine conjugate. *2870*

Vanillic Acid *Urine Increase Physiological* Significant increase with glycine conjugate. *2870*

Dihydroxycholecalciferol

Calcium *Serum No Effect Physiological* No effect observed. *0469*
Urine Increase Physiological 30-100% increase within 2 d up to 2.7 μg/d. *0469*

Phosphate *Serum No Effect Physiological* No effect observed. *0469*
Urine No Effect Physiological No effect observed. *0469*

Dihydroxymandelic Acid

Epinephrine *Test Conditions No Effect Analytical* On fluorescent procedure of Peyrin. *2800*

Homovanillic Acid *Urine No Effect Analytical* At high concentrations with method of Kahane. *1843*

Norepinephrine *Test Conditions No Effect Analytical* On fluorescent procedure of Peyrin. *2800*

Dihydroxyphenylacetic Acid

Catecholamines *Urine Increase Analytical* Reacts like epinephrine/norepinephrine with ethylenediamine. *2285*

Epinephrine *Test Conditions No Effect Analytical* On fluorescent procedure of Peyrin. *2800*

Homovanillic Acid *Urine No Effect Analytical* At high concentrations with method of Kahane. *1843*

5-Hydroxyindoleacetic Acid (5-HIAA)
Urine Decrease Analytical Significant decrease with procedure of Udenfriend. *1097*

Norepinephrine *Test Conditions No Effect Analytical* On fluorescent procedure of Peyrin. *2800*

Vanillylmandelic Acid *Urine Decrease Analytical* At therapeutic physiological concentrations on Pisano procedure. *1097*

3,4-Dihydroxyphenylacetic Acid

Homogentisic Acid *Urine No Effect Analytical* On TLC method of Feldman and Bowman. *1096*

20-α-Di-Hydroxyprogesterone

16-α-Hydroxyprogesterone *Plasma Increase Analytical* Up to 25% cross reactivity. *0008*

Diiodocaffeine

Protein Bound Iodine (PBI) *Serum Increase Physiological* Contains 66.5% iodine. *0012*

Diiodohydroxyquin

Alanine Aminotransferase *Serum Increase Physiological* Hepatic damage reported in animals. *1343*

Erythrocytes *Blood Decrease Physiological* May cause anemia. *2313*

^{131}I Uptake *Serum Decrease Physiological* Effect lasts for several weeks. *1488*

Leukocytes *Blood Decrease Physiological* May cause leukopenia. *2313*

Protein Bound Iodine (PBI) *Serum Increase Analytical* Contains iodine, duration of effect uncertain. *0354* Contains 65% iodine, effect lasts for some weeks. *0012*

Diiodoquinoline

Protein Bound Iodine (PBI) *Serum Increase Analytical* Contains 67% iodine. *0012*

Diiodotyrosine

Thyroxine (T_4) Binding Globulin
Serum No Effect Analytical On radioimmunoassay procedure of Van Herle. *3676*

Tyrosine *Plasma No Effect Analytical* On fluorometric procedure of Ambrose. *0080*

Diltiazem

Carbamazepine *Serum Increase Physiological* Presumed reduction in elimination of drug: drug dose reduced by 60% to reduce toxicity. *1002*

Cholesterol *Serum No Effect Physiological* No significant change in 31 subjects treated for 6 mo. *0088*

Cholesterol, High Density Lipoprotein
Serum Increase Physiological Increase by 15% in 31 subjects treated for 6 mo. *0088*

Cholesterol, Low Density Lipoprotein
Serum No Effect Physiological No significant change in 31 subjects treated for 6 mo. *0088*

Cholesterol, Very Low Density Lipoprotein
Serum No Effect Physiological No significant change in 31 subjects treated for 6 mo. *0088*

Cyclosporine *Serum Increase Physiological* Probable interference with demethylation and binding to cytochrome P-450. *3732*

Digoxin *Serum Increase Physiological* Increased mean trough concentration from 1.11 ng/mL to 1.54 ng/mL after 3 d coadministration in 11 patients with congestive cardiac failure. *0108* Increased plasma concentration after single oral dose, or for 1 week in 6 healthy subjects. Renal clearance decreased. *3917*

Glucose *Serum Increase Physiological* Mean increase from 98 to 105 mg/dL after 16 weeks in approximately 120 hypertensives. *2331*

Prostaglandin E2 *Plasma Increase Physiological* Increase by 63 pg/mL in 20 patients with essential hypertension with 14 weeks treatment. *3521*

Renin Activity *Plasma Increase Physiological* Increase by 0.8 ng/mL/h in 20 patients with essential hypertension with 14 weeks treatment. *3521*

Diltiazem (continued)

Triglycerides *Serum No Effect Physiological* No significant change in 31 subjects treated for 6 mo. *0088*

Dilute Urine

Color *Urine Decrease Physiological* Due to dilution of pigments. *0443*

Dimenhydrinate

Theophylline *Serum Increase Analytical* When Clinical Assays RIA kit used: almost 3 fold increase observed. *1450*

Dimercaprol

Arsenic *Urine Increase Physiological* If poisoning due to arsenic. *0071*

Bicarbonate *Serum Decrease Physiological* Associated with metabolic acidosis. *1022*

Bilirubin *Serum Increase Physiological* May cause hemolysis with G-6-PD deficiency. *0248*

Calcium *Urine Increase Physiological* Effective but less good than edetic acid. *0071*

Copper *Urine Increase Physiological* If cause of poisoning (penicillamine better). *0071*

Erythrocytes *Blood Decrease Physiological* Hemolytic anemia in G-6-PD deficient persons. *0788*

Glucose *Serum Increase Physiological* Initial response to toxic doses. *2313*
Serum Decrease Physiological After initial increase. *2313*

Gold *Urine Increase Physiological* If poisoning due to gold. *0071*

Haptoglobin *Serum Decrease Physiological* May cause hemolysis. *0248*

Hematocrit *Blood Decrease Physiological* Hemolytic anemia in G-6-PD deficient persons. *0788*

Hemoglobin *Plasma Increase Physiological* Occurs with intravascular hemolysis. *0788*
Blood Decrease Physiological Hemolytic anemia in G-6-PD deficient persons. *0788*

[131]I Uptake *Serum Decrease Physiological* Elemental iodine trapped in thyroid. *2313*

Lactate *Serum Increase Physiological* Associated with metabolic acidosis. *0071*

Lead *Urine Increase Physiological* If poisoning due to lead. *0071*

Mercury *Urine Increase Physiological* If poisoning due to mercury. *0071*

Methemalbumin *Plasma Increase Physiological* Occurs with intravascular hemolysis. *0788*

Methemoglobin *Blood Increase Physiological* May cause hemolysis in G-6-PD deficiency. *3581*

pH *Blood Decrease Physiological* May induce metabolic acidosis. *0071*

Reticulocytes *Blood Increase Physiological* Hemolytic anemia in G-6-PD deficient persons. *0788*

Dimercaptoethane

N-Acetylglucosaminidase *Urine No Effect Analytical* At 60 mmol/L on 2 colorimetric analytical methods. *1354*

Dimethadione

Albumin *Serum No Effect Analytical* At concentration of 2,000 mg/L had no effect on BCG method. *3393*

Bicarbonate *Serum No Effect Analytical* At concentration of 2,000 mg/L had no effect on method using phenolphthalein. *3393*
Serum Decrease Physiological Displacement of bicarbonate by dimethadione. *3882*

Bilirubin *Serum No Effect Analytical* At concentration of 2,000 mg/L had no effect on Jendrassik and Grof method. *3393*

Calcium *Serum No Effect Analytical* At concentration of 2,000 mg/L had no effect on cresolphthalein method. *3393*

Chloride *Serum No Effect Analytical* At concentration of 2,000 mg/L had no effect on mercurimetric method. *3393*

Cholesterol *Serum No Effect Analytical* At concentration of 2,000 mg/L had no effect on Liebermann-Burchard method. *3393*

pH *Blood Decrease Physiological* Extracellular acidosis induced. *3882*

Phosphate *Serum No Effect Analytical* At concentration of 2,000 mg/L had no effect on phosphomolybdate method. *3393*

Protein *Serum No Effect Analytical* At concentration of 2,000 mg/L had no effect on biuret method with blank correction. *3393*

Urea Nitrogen *Serum No Effect Analytical* At concentration of 2,000 mg/L had no effect on diacetylmonoxime method. *3393*

Uric Acid *Serum No Effect Analytical* At concentration of 2,000 mg/L had no effect on phosphotungstate reduction method. *3393*

Dimethindene

Erythrocytes *Blood Decrease Physiological* Rare hemolytic anemia with antihistamines. *1678*

Hemoglobin *Blood Decrease Physiological* Rare hemolytic anemia with antihistamines. *1678*

Leukocytes *Blood Decrease Physiological* Rare reported effect of antihistamines. *1680*

N-Dimethyldiazepam

Albumin *Serum No Effect Analytical* At concentration of 6 mg/L had no effect on BCG method. *3393*

Bicarbonate *Serum No Effect Analytical* At concentration of 6 mg/L had no effect on method using phenolphthalein. *3393*

Bilirubin *Serum No Effect Analytical* At concentration of 6 mg/L had no effect on Jendrassik and Grof method. *3393*

Calcium *Serum No Effect Analytical* At concentration of 6 mg/L had no effect on cresolphthalein method. *3393*

Chloride *Serum No Effect Analytical* At concentration of 6 mg/L had no effect on mercurimetric method. *3393*

Cholesterol *Serum No Effect Analytical* At concentration of 6 mg/L had no effect on Liebermann-Burchard method. *3393*

Creatinine *Serum No Effect Analytical* At concentration of 6 mg/L had no effect on AutoAnalyzer Jaffé method. *3393*

Phosphate *Serum No Effect Analytical* At concentration of 6 mg/L had no effect on phosphomolybdate method. *3393*

Protein *Serum No Effect Analytical* At concentration of 6 mg/L had no effect on biuret method with blank correction. *3393*

Triglycerides *Serum No Effect Analytical* At concentration of 6 mg/L had no effect on lipase/esterase method. *3393*

Urea Nitrogen *Serum No Effect Analytical* At concentration of 6 mg/L had no effect on diacetylmonoxime method. *3393*

Uric Acid *Serum No Effect Analytical* At concentration of 6 mg/L had no effect on phosphotungstate reduction method. *3393*

3,4-Dimethylphenol

5-Hydroxyindoleacetic Acid (5-HIAA)
Urine No Effect Analytical No effect with 5 mg/dL on method of Goldenberg. *1327*

Dimethyltryptamine

Amphetamine *Urine No Effect Analytical* At 50 mg/L on fluorescent method of Hayes. *1534*

Dimethyl Tubocurarine

Protein Bound Iodine (PBI) *Serum Increase Physiological* Contains 35% iodine. *0012*

Dimpylate

Pseudocholinesterase *Serum Decrease Physiological* In vitro 8.2% decrease at 1 x 10⁻⁴ mol/L. *1270*

Dinitrophenol

Alanine Aminotransferase *Serum Increase Physiological* Hepatotoxicity with centrolobular necrosis. *1343*

Aspartate Aminotransferase
Serum Increase Physiological Hepatotoxicity with centrolobular necrosis. *1343*

Bile *Urine Increase Physiological* Due to hepatotoxicity. *1302*

Bilirubin *Serum Increase Physiological* Due to toxic hepatitis. *1302*

Casts *Urine Increase Physiological* Due to toxic nephritis. *1302*

Color *Urine Increase Physiological* Red brown due to hematuria. *2313*

Erythrocytes *Blood Decrease Physiological* May cause anemia/aplastic anemia. *1343*
Urine Increase Physiological Due to toxic nephritis. *1302*

Hemoglobin *Blood Decrease Physiological* May cause anemia/aplastic anemia. *1343*

Leukocytes *Blood Decrease Physiological* May cause agranulocytosis/aplastic anemia. *1343*

Methemoglobin *Blood Increase Physiological* May cause hemolysis/aplastic anemia. *2429*

Protein *Urine Increase Physiological* Due to nephrotoxicity. *1302*

Protein Bound Iodine (PBI) *Serum Decrease Physiological* Competes with thyroxine for thyroxine binding globulin. *0012*

T₃ Uptake *Serum Increase Physiological* Competes for sites on thyroxine binding prealbumin. *0583*

Thyroxine (T₄) *Serum Decrease Physiological* Displaced from binding sites on thyroxine binding globulin. *3505*

Thyroxine (T₄) (Murphy-Pattee)
Serum Decrease Physiological Competes with T₄ for thyroxine binding globulin. *0583*

Urea Nitrogen *Serum Increase Physiological* Reported to cause renal damage. *1343*

Dioxane

Protein *Urine Increase Physiological* May cause renal damage. *0279*

Urea Nitrogen *Serum Increase Physiological* May cause renal damage. *0279*

Diphenhydramine

Alanine Aminotransferase *Serum No Effect Analytical* At acute overdose concentration (20 mg/dL) on Technicon® SMAC® method. *3719*

Albumin *Serum No Effect Analytical* At acute overdose concentration (20 mg/dL) on Technicon® SMAC® method. *3719* At concentration of 200 mg/L had no effect on BCG method. *3393*

Alkaline Phosphatase *Serum No Effect Analytical* At acute overdose concentration (20 mg/dL) on Technicon® SMAC® method. *3719*

Alkaloids *Urine No Effect Analytical* With ninhydrin on Frings TLC procedure at 10 mg/dL. *1204*
Urine Increase Analytical Red/purple with iodoplatinate on Frings TLC procedure. *1204*

Aminosalicylic Acid *Serum Decrease Physiological* Delayed absorption due to delayed gastric emptying. *3794*

Ammonia *Plasma Decrease Physiological* Reported effect in exogenous NH₃ toxicity. *1942*

Antipyrine *Plasma No Effect Physiological* No significant effect on half-life. *3466*

Aspartate Aminotransferase *Serum No Effect Analytical* At acute overdose concentration (20 mg/dL) on Technicon® SMAC® method. *3719*

Barbiturate *Urine No Effect Analytical* With diphenylcarbazone on Frings TLC procedure at 10 mg/dL. With vanillin on Frings procedure at 10 mg/dL. With mercurous NO₃ on Frings TLC procedure at 10 mg/dL. With mercuric SO₄ on Frings TLC procedure at 10 mg/dL. *1204*

Bicarbonate *Serum No Effect Analytical* At concentration of 10 mg/L had no effect on method using phenolphthalein. *3393*

Bilirubin *Serum No Effect Analytical* At acute overdose concentration (20 mg/dL) on Technicon® SMAC® method. *3719* At concentration of 200 mg/L had no effect on Jendrassik and Grof method. *3393*
Serum Increase Physiological May occur with hemolytic anemia. *3602*

Bromide *Urine No Effect Analytical* No effect on method of Frings. *1204*

Calcium *Serum No Effect Analytical* At acute overdose concentration (20 mg/dL) on Technicon® SMAC® method. *3719* At concentration of 200 mg/L had no effect on cresolphthalein method. *3393*

Catecholamines *Urine No Effect Analytical* No effect reported on method of Sandhu and Freed. *3135*
Urine No Effect Physiological No effect with short term ingestion of 150 mg/d. *0766*

Chlordiazepoxide *Blood No Effect Analytical* On method of Riddick at 5 mg/dL. *2983*

Chloride *Serum No Effect Analytical* At concentration of 10 mg/L had no effect on mercurimetric method. *3393*

Cholesterol *Serum No Effect Analytical* At acute overdose concentration (20 mg/dL) on Technicon® SMAC® method. *3719* At concentration of 200 mg/L had no effect on Liebermann-Burchard method. *3393*

Creatine Kinase *Serum No Effect Analytical* At acute overdose concentration (20 mg/dL) on Technicon® SMAC® method. *3719*

Creatinine *Serum No Effect Analytical* At acute overdose concentration (20 mg/dL) on Technicon® SMAC® method. *3719* At concentration of 200 mg/L had no effect on AutoAnalyzer Jaffé method. *3393*

Epinephrine *Urine No Effect Physiological* No effect with short term ingestion of 150 mg/d. *0766*

Erythrocytes *Blood Decrease Physiological* Hemolytic anemia. *3602*

Ethchlorvynol *Serum No Effect Analytical* At 80 mg/L on GLC procedure of Evenson. *1058*
Urine No Effect Analytical No effect on method of Frings, Cohen at 10 mg/dL. *1205*

FPN Test *Urine No Effect Analytical* No effect at 10 mg/dL on method of Frings. *1204*

Glucose *Serum No Effect Analytical* At acute overdose concentration (20 mg/dL) on Technicon® SMAC® method. *3719*

Haptoglobin *Serum Decrease Physiological* Consequence of hemolytic anemia. *3602*

Hematocrit *Blood Decrease Physiological* Hemolytic anemia. *3602*

Hemoglobin *Blood Decrease Physiological* Hemolytic anemia. *3602*

11-Hydroxycorticosteroids *Urine No Effect Physiological* No effect with short term ingestion of 150 mg/d. *0766*

17-Hydroxycorticosteroids *Urine No Effect Physiological* No effect with short term ingestion of 150 mg/d. *0766*

Iron *Serum No Effect Analytical* At acute overdose concentration (20 mg/dL) on Technicon® SMAC® method. *3719* At concentration of 200 mg/L had no effect on Ferrozine method. *3393*

¹³¹I Uptake *Serum Decrease Physiological* Benadryl® contains tetraiodofluorescein. *2652*

17-Ketosteroids *Urine No Effect Physiological* No effect with short term ingestion of 150 mg/d. *0766*

Lactate Dehydrogenase *Serum No Effect Analytical* At acute overdose concentration (20 mg/dL) on Technicon® SMAC® method. *3719*

Methadone *Urine Increase Analytical* At concentrations above 100 mg/L positive results obtained with Syva® EMIT-ASSAY for drugs of abuse. *0039*

Methaqualone *Serum No Effect Analytical* At 80 mg/L on GLC procedure of Evenson. *1057*

Diphenhydramine (continued)

Norepinephrine *Urine   No Effect   Physiological* No effect with short term ingestion of 150 mg/d. *0766*

Phenylbutazone *Serum   No Effect   Physiological* No significant effect on half-life. *3466*

Phosphate *Serum   No Effect   Analytical* At acute overdose concentration (20 mg/dL) on Technicon® SMAC® method. *3719* At concentration of 200 mg/L had no effect on phosphomolybdate method. *3393*

Potassium *Serum   No Effect   Analytical* At concentration of 23 mg/L had no effect on measurement by ISE with predilution. *3393*

Prolactin *Plasma   No Effect   Physiological* No significant effect with 50 mg given i.m. *1416*

Protein *Serum   No Effect   Analytical* At acute overdose concentration (20 mg/dL) on Technicon® SMAC® method. *3719* At concentration of 200 mg/L had no effect on biuret method with blank correction. *3393*

Prothrombin Time *Plasma   No Effect   Physiological* Probably no effect on response to anticoagulant. *1487*

Salicylate *Urine   No Effect   Analytical* At 10 mg/dL no effect on Trinder method. *1204*

Sodium *Serum   No Effect   Analytical* At concentration of 23 mg/L had no effect on measurement by ISE with predilution. *3393*

Tricyclic Antidepressants *Serum   Positive   Analytical* False positive observed with EMIT-ST test for tricyclic antidepressants. *3399*

Triglycerides *Serum   No Effect   Analytical* At concentration of 200 mg/L had no effect on lipase/esterase method. *3393* At acute overdose concentration (20 mg/dL) on Technicon SMAC® method. *3719*

Tri-iodothyronine (T$_3$) *Serum   No Effect   Analytical* At acute overdose concentration (20 mg/dL) on Technicon® SMAC® method. *3719*

Urea Nitrogen *Serum   No Effect   Analytical* At acute overdose concentration (20 mg/dL) on Technicon® SMAC® method. *3719* At concentration of 200 mg/L had no effect on diacetylmonoxime method. *3393*

Uric Acid *Serum   No Effect   Analytical* At acute overdose concentration (20 mg/dL) on Technicon® SMAC® method. *3719* At concentration of 200 mg/L had no effect on phosphotungstate reduction method. *3393*

2,3-Diphosphoglycerate

Pyruvate Kinase *Red Blood Cells   Increase   Analytical* Activating effect on enzyme. *0370*

Diphosphonate

Hydroxyproline *Urine   Decrease   Physiological* Normal response in Paget's disease. *3644*

Diprotrizoate

^{131}I Uptake *Serum   Decrease   Physiological* Due to iodine component of drug. *2220*

Protein Bound Iodine (PBI) *Serum   Increase   Analytical* But does not affect T$_4$ (effect for 1-2 weeks). *3218*

Dipyridamole

Glucose *Serum   No Effect   Analytical* At concentration of 40 mg/L had no effect on hexokinase/G-6-PDH method. *3393*

Lactate *Serum   No Effect   Analytical* At concentration of 40 mg/L had no effect on enzymatic method. *3393*

Platelet Aggregation *Blood   Decrease   Physiological* Weak inhibition of adenosine deaminase. *1132*

Platelets *Blood   Increase   Physiological* Possibly due to alteration of turnover. *1132*

Dipyrone

Bilirubin *Serum   Increase   Physiological* May cause hemolytic anemia. *2365*

Bilirubin, Direct *Serum   Increase   Physiological* May cause hemolytic anemia. *2365*

Coombs' Test *Serum   Positive   Physiological* May produce immune hemolytic anemia. *0359*

Erythrocytes *Blood   Decrease   Physiological* Hemolytic anemia. *2098*

Glucose *Urine   Decrease   Analytical* May cause false negative with enzyme tests. *1563*

Haptoglobin *Serum   Decrease   Physiological* May cause hemolytic anemia. *2365*

Hematocrit *Blood   Decrease   Physiological* Hemolytic anemia. *2098*

Hemoglobin *Blood   Decrease   Physiological* Hemolytic anemia. *2098*

Leukocytes *Blood   Decrease   Physiological* Leukopenia or agranulocytosis. *1684*

Methemalbumin *Plasma   Increase   Physiological* Occurs with hemolytic anemia. *2098*

Neutrophils *Blood   Decrease   Physiological* Rate of agranulocytosis 23.7 times higher than in nonusers. *1105*

Occult Blood *Feces   Positive   Physiological* May cause gastrointestinal bleeding. *0071*

Platelets *Blood   Decrease   Physiological* Thrombocytopenic purpura reported. *1680*

Prothrombin Time *Plasma   Increase   Physiological* May aggravate prothrombin deficiency. *1680*

Urea Nitrogen *Serum   Increase   Physiological* Reported to cause anuria. *0071*

Disopyramide

Alanine Aminotransferase *Serum   Increase   Physiological* Occasional case of cholestatic jaundice reported. *3215*

Albumin *Serum   No Effect   Analytical* At concentration of 4 mg/L had no effect on BCG method. *3393*

Alkaline Phosphatase *Serum   Increase   Physiological* Occasional case of cholestatic jaundice reported. *3215*

Aspartate Aminotransferase
Serum   Increase   Physiological Occasional case of cholestatic jaundice reported. *3215*

Bicarbonate *Serum   No Effect   Analytical* At concentration of 4 mg/L had no effect on method using phenolphthalein. *3393*

Bilirubin *Serum   No Effect   Analytical* At concentration of 4 mg/L had no effect on Jendrassik and Grof method. *3393* *Serum   Increase   Physiological* Occasional case of cholestatic jaundice reported. *3215*

Calcium *Serum   No Effect   Analytical* At concentration of 4 mg/L had no effect on cresolphthalein method. *3393*

Chloride *Serum   No Effect   Analytical* At concentration of 4 mg/L had no effect on mercurimetric method. *3393*

Cholesterol *Serum   No Effect   Analytical* At concentration of 4 mg/L had no effect on CHOD-PAP method. *3393*

Creatinine *Serum   No Effect   Analytical* At 200 mg/L on reversed phase LC procedure of Zhiri et al. *3942* At concentration of 4 mg/L had no effect on AutoAnalyzer Jaffé method. *3393*

Creatinine Clearance *Urine   No Effect   Physiological* No effect when drug given to total of 300 or 600 mg daily. *2992*

Digoxin *Serum   No Effect   Physiological* No significant effect after 3 100 mg doses per day. Slight increase noted when daily dose doubled. *2992* *Serum   Increase   Physiological* Mean change for 1.3 to 1.5 nmol/L but clinically unimportant: disopyramide concentrations above therapeutic range when this effect noted. *2291*

Digoxin Clearance *Urine   No Effect   Physiological* No effect when drug given to total of 300 or 600 mg daily. *2992*

Glucose *Serum   No Effect   Analytical* At concentration of 4 mg/L had no effect on GOD/POD-PAP method. *3393* *Serum   Decrease   Physiological* To below 10 mg/dL observed in 2 patients both receiving other drugs and occurring with hypotension. *2557*

Lactate *Blood Increase Physiological* Marked effect observed in 2 patients both receiving other drugs and occurring with hypotension. *2557*

Neutrophils *Blood Decrease Physiological* Isolated case of agranulocytosis reported. *3215* Occasional case of agranulocytosis reported. *3717*

Phosphate *Serum No Effect Analytical* At concentration of 4 mg/L had no effect on phosphomolybdate method. *3393*

Potassium *Serum No Effect Analytical* At concentration of 4 mg/L had no effect on flame photometric method. *3393*

Protein *Serum No Effect Analytical* At concentration of 4 mg/L had no effect on biuret method with blank correction. *3393*

Quinidine *Serum Decrease Physiological* Small effect noted, but elimination half-life not significantly affected. *0208*

Sodium *Serum No Effect Analytical* At concentration of 4 mg/L had no effect on flame photometric method. *3393*

Triglycerides *Serum No Effect Analytical* At concentration of 4 mg/L had no effect on lipase/esterase method. *3393*

Urea Nitrogen *Serum No Effect Analytical* At concentration of 4 mg/L had no effect on diacetylmonoxime method. *3393*

Uric Acid *Serum No Effect Analytical* At 200 mg/L on reversed phase LC procedure of Zhiri et al. *3942* At concentration of 4 mg/L had no effect on phosphotungstate reduction method. *3393*

Disulfiram

Acetaldehyde *Blood Increase Physiological* Ethanol metabolism diverted (10 times normal concentration). *1343*

Acetaldehyde Oxidase
Test Conditions Decrease Analytical Inhibitory effect observed. *2893*

Acetoacetate *Serum Increase Physiological* In 1 of 6 volunteers rapid and short lasting effect. *3482*

Acetone *Blood Increase Physiological* In all of 6 volunteers rapid and short lasting effect. *3482*

Alanine Aminotransferase *Serum Increase Physiological* One case of questionable cholestasis reported. *1680* Reversible toxic liver damage in nonalcoholic woman. *2017* In 6 patients: proved to be drug associated in one by challenge test. *2912* Observed in one patient and similar results followed challenge test. *2491*

Albumin *Serum No Effect Analytical* At concentration of 120 mg/L had no effect on BCG method. *3393*

Alkaline Phosphatase *Serum Increase Physiological* One possible case of cholestasis reported. *1680* Reversible toxic liver damage in nonalcoholic woman. *2017* In 6 patients: proved to be drug associated in one by challenge test. *2912* Observed in one patient and similar results followed challenge test. *2491*

Antipyrine *Plasma Increase Physiological* Inhibits hydroxylation, prolongs action. *2641*

Aspartate Aminotransferase
Serum Increase Physiological Reversible toxic liver damage in nonalcoholic woman. *2017* Observed in one patient and similar results followed challenge test. *2491*

Bicarbonate *Serum No Effect Analytical* At concentration of 120 mg/L had no effect on method using phenolphthalein. *3393*

Bilirubin *Serum No Effect Analytical* At concentration of 120 mg/L had no effect on Jendrassik and Grof method. *3393* *Serum Increase Physiological* One possible case of cholestasis reported. *1680* In 6 patients proved to be drug associated in one by challenge test. *2912* In 6 patients: proved to be drug associated in one by challenge test. *2491*

Calcium *Serum No Effect Analytical* At concentration of 120 mg/L had no effect on cresolphthalein method. *3393*

Chloride *Serum No Effect Analytical* At concentration of 120 mg/L had no effect on mercurimetric method. *3393*

Cholesterol *Serum No Effect Analytical* At concentration of 120 mg/L had no effect on Liebermann-Burchard method. *3393* *Serum Increase Physiological* 500 mg/d raised mean concentration from 193 mg/dL to 227 mg/dL after 3 weeks and 264 mg/dL after 6 weeks in alcoholic subjects. No fall in cholesterol with abstinence. *2272*

Creatinine *Serum No Effect Analytical* At concentration of 120 mg/L had no effect on AutoAnalyzer Jaffé method. *3393*

Diphenylhydantoin *Serum Increase Physiological* Inhibits hydroxylation, prolongs action. *2641* May increase from 7/15 to 25/39 μg/mL (at 400-800 mg/d). *2041* Metabolism inhibited, with half-life increased and decrease in mean metabolic clearance rate. *3555*

Dopamine Hydroxylase *Serum Decrease Physiological* Observed in animals. *2913*

Epinephrine *Plasma No Effect Physiological* No effect observed after 1 week. *1845*

γ-Glutamyltransferase (GGT)
Serum Increase Physiological Reversible toxic liver damage in nonalcoholic woman. *2017*

HMPG (4-Hydroxy-3-Methoxy-Phenylethylene Glycol)
Urine Increase Physiological Inhibition of aldehyde dehydrogenase. *2220*

Homovanillic Acid *Urine Increase Physiological* Probably due to inhibition of dopamine hydroxylase. *3703*

β-Hydroxybutyrate *Blood Increase Physiological* In 1 of 6 volunteers rapid and short lasting effect. *3482*

[131]I Uptake *Serum Decrease Physiological* Uncommon reported effect. *2220*

Mexiletine *Serum Increase Physiological* Decreases clearance and prolongs half-life. *2011*

Norepinephrine *Plasma Decrease Physiological* Significant decrease after 1 week therapy. *1845*

Phosphate *Serum No Effect Analytical* At concentration of 120 mg/L had no effect on phosphomolybdate method. *3393*

Protein *Serum No Effect Analytical* At concentration of 120 mg/L had no effect on biuret method with blank correction. *3393*

Protein Bound Iodine (PBI) *Serum Decrease Physiological* Uncommon reported effect. *2220*

Prothrombin Time *Plasma Increase Physiological* Inhibits metabolism of coumarins. *3072* Augmented S- warfarin hypoprothrombinemia but not that of R- warfarin. *2644*

Triglycerides *Serum No Effect Analytical* At concentration of 120 mg/L had no effect on lipase/esterase method. *3393*

Urea Nitrogen *Serum No Effect Analytical* At concentration of 120 mg/L had no effect on diacetylmonoxime method. *3393*

Uric Acid *Serum No Effect Analytical* At concentration of 120 mg/L had no effect on phosphotungstate reduction method. *3393*

Vanillylmandelic Acid *Urine Increase Analytical* ?interference from acetaldehyde from ethanol. *1488* *Urine Decrease Physiological* Although HMPG excretion increased. *0426* Probably due to inhibition of dopamine hydroxylase. *3703*

Warfarin *Plasma No Effect Physiological* No effect on plasma concentrations of either R- or S- warfarin. *2644* *Plasma Increase Physiological* If administered together. *2641*

Disulphine Blue

Acetaminophen *Serum No Effect Analytical* Using a modified Glynn and Kendal method. *1464*

Acid Phosphatase *Serum No Effect Analytical* With p-nitrophenylphosphate as substrate. *1464*

Alanine Aminotransferase *Serum No Effect Analytical* Using recommended Scandinavian method. *1464*

Albumin *Serum Increase Analytical* At physiological concentration on bromocresol method for at least 2 d. *1464*

Alkaline Phosphatase *Serum No Effect Analytical* With p-nitrophenylphosphate as substrate. *1464*

Amylase *Serum Increase Analytical* At physiological concentration for more than 2 d on Phadebas Cibachron blue starch-complex reaction. *1464*

Aspartate Aminotransferase *Serum No Effect Analytical* Using recommended Scandinavian method. *1464*

Bicarbonate *Serum No Effect Analytical* By continuous flow cresol-red reaction. *1464*

Disulphine Blue (continued)

Bilirubin *Serum No Effect Analytical* With sulfanilic acid/caffeine-benzoate accelerator. *1464*

Calcium *Serum No Effect Analytical* By continuous flow cresolphthalein complexone reaction. *1464*

Cholesterol *Serum No Effect Analytical* By cholesterol oxidase endpoint method. *1464*

Creatinine *Serum No Effect Analytical* By discrete kinetic alkaline-picrate reaction. *1464*

Glucose *Serum No Effect Analytical* By continuous-flow glucose oxidase/4-aminophenazone reaction. *1464*

γ-Glutamyltransferase (GGT) *Serum No Effect Analytical* Using gamma-glutamyl-3-carboxy-4-nitroanilide as substrate. *1464*

Iron *Serum Increase Analytical* At physiological concentration for more than 2 d on continuous flow Ferrozine reaction. *1464*

Iron-Binding Capacity, Total (TIBC)
Serum Increase Analytical At physiological concentration for more than 2 d on magnesium carbonate/ continuous flow Ferrozine reaction. *1464*

Lactate Dehydrogenase *Serum No Effect Analytical* Using recommended Scandinavian method. *1464*

Magnesium *Serum No Effect Analytical* By colorimetric reaction of Mann and Yoe. *1464*

Phosphate *Serum No Effect Analytical* By continuous-flow phosphomolybdate/stannous chloride hydrazine method. *1464*

Protein *Serum Increase Analytical* At physiological concentration on an endpoint biuret reaction for at least 2 d. *1464*

Salicylate *Serum No Effect Analytical* Using a modified Trinder method. *1464*

Triglycerides *Serum No Effect Analytical* By kinetic lipase/glycerol kinase method. *1464*

Urea Nitrogen *Serum No Effect Analytical* By continuous-flow diacetyl monoxime method. *1464*

Uric Acid *Serum No Effect Analytical* By continuous-flow phosphotungstate reduction. *1464*

Dithiazanine

Color *Urine Increase Analytical* Blue. *2313*
Feces Increase Analytical Green to blue. *2313*

^{131}I Uptake *Serum Decrease Physiological* Due to iodine component of drug. *2220*

Protein *Urine Increase Analytical* Reacts with Folin-Ciocalteu of Lowry procedure. *2425*
Urine Increase Physiological Transient proteinuria may occur. *2427*

Protein Bound Iodine (PBI) *Serum Increase Analytical* Contains 24.5% iodine. *0012*

Dithionite

Ethanol *Blood No Effect Analytical* Over 16 weeks at 20° at concentration of 0.5% w/v. *3355*

Dithiothreitol

Protein *Test Conditions Increase Analytical* Reacts with Folin-Ciocalteu of Lowry procedure. *0702*

Diuretics

Calcium *Serum Decrease Physiological* Excretion enhanced by most diuretics. *0105*
Urine Increase Physiological Excretion enhanced by most diuretics. *0105*

Cell Water *White Blood Cells No Effect Physiological* No change observed in normal individuals. *0997*

Chloride *Serum Decrease Physiological* Diuretic action if excessive. *2313*

Creatine Kinase *Serum Increase Physiological* May occur as result of i.m. injections. *0249*

Creatinine *Serum Increase Physiological* May be associated with acute sodium depletion. *0105*

Epinephrine *Urine No Effect Physiological* No significant effect during diuresis. *3616*

Glomerular Filtration Rate (GFR)
Urine Decrease Physiological Effect on clearances. *2745*

Norepinephrine *Urine No Effect Physiological* No significant effect during diuresis. *3616*

Potassium *Serum Decrease Physiological* Loss in urine. *2313*
Urine Increase Physiological Diuretic action. *2313*
White Blood Cells No Effect Physiological No change observed in normal individuals. *0997*

Prothrombin Time *Plasma Increase Physiological* May prolong action of anticoagulants. *1679*
Plasma Decrease Physiological In patients receiving anticoagulants. *2313*

PSP Excretion *Urine Decrease Physiological* Blocking of secretory mechanism. *1022*

Quinidine *Serum Increase Physiological* If administration concomitantly and alkalinize urine. *1487*

Sodium *Serum No Effect Physiological* May appear normal if hypovolemia occurs. *0105*
Serum Decrease Physiological Diuretic action. *2745*
White Blood Cells No Effect Physiological No change observed in normal individuals. *0997*

Urea Clearance *Urine Increase Physiological* May be observed after use of diuretics. *0459*

Urea Nitrogen *Serum Increase Physiological* Reduced clearance. *2745*

Uric Acid *Serum Increase Physiological* Reduced clearance. *0127*

Volume *Plasma Decrease Physiological* May occur if rapid diuretic action. *1343*

Zinc *Urine Decrease Physiological* Although other cations increased. *3439*

Diurnal Variation

Acid Phosphatase *Serum Decrease Physiological* 25-50% less between 9 am and 3 am. *3217*

Albumin *Serum Increase Physiological* Follows calcium; increases blood volume. *1829*

Aldosterone *Urine Increase Physiological* High 6 am- 3 pm, low at night (postural effect). *3886*

α-Amino-Nitrogen *Urine Decrease Physiological* Progressive fall (except for meals) from 8 am to 5 am. *3571*

Androgens *Plasma No Effect Physiological* More effect of exercise than time of day. *0417*

Antidiuretic Hormone *Plasma Increase Physiological* Secretion increased at night. *2222*
Urine No Effect Physiological No effect observed on hourly measurements. *1190*

Arabinose *Urine No Effect Physiological* No evidence of cycle observed. *0273*

Arylsulfatase A *Urine Increase Physiological* Maximum 6-12 am, minimum late afternoon evening. *2320*

Ascorbic Acid *Blood Increase Physiological* At peak in early morning. *2204*
Urine Increase Physiological At maximum in first am urine specimen. *2204*

Calcium *Serum No Effect Physiological* Slightly higher midday, lowest at night (not significant). *3851*
Serum Increase Physiological Constant during day but maximum 8 pm, minimum 3 am. *1829*
Urine Increase Physiological Excretion maximum at night. *2222*

Carbohydrate, Protein Bound
Serum Increase Physiological Maximum observed about midday. *2222*

Catecholamines *Urine Decrease Physiological* Significantly less at night, greater in afternoon that in am. *3616*

Chloride *Urine Increase Physiological* Maximum around noon 2pm, minimum during night. *3217*

Corticotropin *Plasma Increase Physiological* Maximum in early hours of am low in evening. *0926*

Cortisol *Plasma Increase Physiological* Maximum observed about 6 am in morning. *2222*

Creatine *Serum Increase Physiological* 7 pm value 160% of 7 am concentration. *3217*

Creatine Kinase *Serum Increase Physiological* Slight increase (3U/L) observed in women from 9 am to 5pm. *2402*

Creatinine *Serum Increase Physiological* Maximum about 2 am minimum at 8 am. *3428* 7 pm value 130% of 7 am concentration. *3217*
Urine Increase Physiological Noted in day after meals lowest at night. *0242*

Creatinine Clearance *Urine Decrease Physiological* Significant effect with fall at night. *1190*

Cyclic Adenosine Monophosphate
Urine No Effect Physiological No diurnal rhythm observed. *1992*

Dopamine Hydroxylase *Serum No Effect Physiological* No change, or with activity. *3807*

Eosinophils *Blood Increase Physiological* During night (at midnight) minimum in afternoon. *1638*
Blood Decrease Physiological Low between 12 noon and 9 pm, high about 6 am. *2222*

Epinephrine *Urine Increase Physiological* Maximum excretion about 3 pm. *2222* Significantly higher during day than at night. *1176*
Urine Decrease Physiological Reduction in afternoon compared with morning. *1636*

Erythrocyte Sedimentation Rate
Blood No Effect Physiological No effect observed. *0424*
Blood Increase Physiological Maximum observed in mid afternoon. *2222*

Estriol *Urine No Effect Physiological* Variable changes during day in pregnancy. *0966*

Follicle Stimulating Hormone (FSH)
Plasma No Effect Physiological No effect observed with 1/2 h intervals in men. *2014*
Plasma Increase Physiological Cycling throughout day up to 50%, asynchronous with LH. *2542*

Fucose *Urine No Effect Physiological* No evidence of cycle observed. *0273*

β-Galactosidase *Urine Increase Physiological* Maximum 6-12 am, minimum late pm, not marked. *2320*

Gastrin *Serum Increase Physiological* 10% change only but maximum at midnight, min early am. *1235*

α-Glucosidase *Urine Increase Physiological* Maximum 6-12 am, minimum late pm, not marked. *2320*

Growth Hormone *Plasma Increase Physiological* Maximum 2 h after commencement of sleep. *2743*

Hematocrit *Blood Increase Physiological* Maximum observed in morning minimum during night. *2222*

Hemoglobin *Blood Decrease Physiological* Falls at night minimum about midnight. *3428*

Hexosamine *Serum Increase Physiological* Maximum observed about midday. *2222*

Hydrochloric Acid *Gastric Material Decrease Physiological* Minimal at 2 am (almost alkaline). *0814*

17-Hydroxycorticosteroids *Plasma Increase Physiological* Maximum observed in morning (same as in urine). *2222*
Urine Increase Physiological Maximum excretion in morning up to midday, minimum at night. *2222*

5-Hydroxytryptamine (Serotonin)
Plasma Increase Physiological Maximum observed about noon. *2222*

Iron *Serum Increase Physiological* Maximum at 4 pm, minimum at 4 am. *3865*
Serum Decrease Physiological Towards late afternoon or evening. *3873*

Iron-Binding Capacity, Total (TIBC)
Serum No Effect Physiological As iron and UIBC vary reciprocally. *3217*

17-Ketosteroids *Urine Increase Physiological* Maximum in morning, lowest at night. *2222*

Leukocytes *Blood Increase Physiological* Rises to maximum during day, minimum in early am. *2222*

Luteinizing Hormone (LH) *Plasma No Effect Physiological* With 1/2 hourly measurements over 48 h. *2014*
Plasma Increase Physiological Rapid cycling throughout day of 100-300%. *2542*

Lymphocytes *Blood Decrease Physiological* Fall during day, maximum in early am. *2222*

Magnesium *Urine Increase Physiological* Excretion maximum at night. *2222*

Monocytes *Blood Increase Physiological* Rise during evening fall during early am. *2222*

Neutrophils *Blood Increase Physiological* Maximum between 12 noon and 6 pm, minimum about 6 am. *0242*

Norepinephrine *Urine Increase Physiological* Significantly higher during day than at night. *1176* Maximum excretion about 3 pm. *2222*

Osmolality *Serum Decrease Physiological* Associated with water retention at night. *1190*

Parathyroid Hormone *Plasma Increase Physiological* Rises after 8 pm, maximum 3 am, normal by 8 am. *1829*

pCO₂ *Blood Decrease Physiological* On waking, decreased suppression of respiratory center. *0814*

pH *Urine Decrease Physiological* Minimum after midday in afternoon. *2222*

Phosphate *Serum Increase Physiological* Maximum 2-4 am, minimum 8-10 am. *1829*
Serum Decrease Physiological Minimum 11 am, maximum during night about 8 am. *3428*
Urine Decrease Physiological Lowest excretion just after waking. *2222*

Placental Lactogen *Plasma No Effect Physiological* No significant cyclical effect observed. *3201*

Potassium *Serum Decrease Physiological* Minimum about 10 pm maximum at 8 am. *3428*
Urine Increase Physiological Maximum around midday and afternoon. *2222*
Red Blood Cells Increase Physiological Maximum concentration occurs in evening (7-8 pm). *2222*

Prolactin *Plasma Increase Physiological* Maximum in women 1 am to 5 am, in men at 5 am. *2606*

Protein *Serum Increase Physiological* Maximum observed after lunch (and during night). *2222* Maximum at 4-8 pm, minimum at 8 am slight effect. *3865*
Serum Decrease Physiological 7.0 g/dL at 8:30 am but 6.6 g/dL at 10 am. *2123*

Protein Bound Iodine (PBI) *Serum Increase Physiological* Small statistically significant effect. *0830*

Ribose *Urine No Effect Physiological* No evidence of cycle observed. *0273*

Sialic Acid *Serum Increase Physiological* Maximum observed about midday. *2222*

Sodium *Serum Increase Physiological* Maximum observed at night in early am. *2222*
Urine Increase Physiological Maximum around midday and afternoon. *2222*

Specific Gravity *Urine Increase Physiological* Urine concentration increased at night. *2222*

Testosterone *Serum Increase Physiological* Maximum (160%) at 6 am, minimal (60%) at 10 pm. *3201* Maximum observed about 1 pm. *2222*

Tetrahydrocortisol *Urine Increase Physiological* Maximum excretion in morning (about 10 am). *2222*

Tetrahydrocortisone *Urine Increase Physiological* Maximum excretion in early am (about 7 am). *2222*

Thyroid Stimulating Hormone (TSH)
Serum Increase Physiological Maximum between 4 and 6 am (men), 9 pm to 9 am (women). *0417* Significant increase maximum between 2 am and 4 am, minimum 6 pm to 8 pm. *2743*

Transferrin *Serum Increase Physiological* Maximum at 4-8 pm, minimum at 8 am slight effect. *3865*

Tyrosine *Plasma Increase Physiological* Maximum at about 10 am, minimum about 2-4 am. *2222*

Urea Clearance *Urine Decrease Physiological* Clearance least in morning. *0459*

Urea Nitrogen *Serum Increase Physiological* Maximum after midnight, minimum about 5 pm. *3428*

Diurnal Variation (continued)

Urea Nitrogen (continued)
Urine Decrease Physiological Excretion reduced at night. *2222*

Uric Acid *Serum Decrease Physiological* Minimum 11 am, maximum 5 pm and again at 8 am. *3428*

Urobilinogen *Urine Increase Physiological* Maximum excretion in afternoon. *0459*

Vanillylmandelic Acid *Urine Increase Physiological* Significantly higher during day than at night. *1176*

Viscosity *Blood Increase Physiological* Maximum about 8 am, minimum about midnight. *0242*

Volume *Urine Decrease Physiological* Volume reduced at night (max following breakfast). *2222*

Xylose Excretion *Urine No Effect Physiological* No evidence of cycle observed. *0273*

DL-Methionine

Amino Acids *Urine Positive Analytical* Reddish-pink spot with DL-methionine with ninhydrin on thin-layer chromatography. *1278*

Doans® Pills

Color *Urine Increase Analytical* Greenish blue. *2313*

Dobutamine

Antidiuretic Hormone *Plasma No Effect Physiological* In 10 patients with congestive cardiac failure. *3654*

Norepinephrine *Plasma Decrease Physiological* In 10 patients with congestive cardiac failure. *3654*

Renin Activity *Plasma Increase Physiological* In 10 patients with congestive cardiac failure from 11.3 to 17.8 ng/mL/h on average. *3654*

Docusate Sodium

Acetaminophen Screening Test *Urine Negative Analytical* No reaction with o-cresol at therapeutic concentrations. *3326*

Domperidone

Aldosterone *Plasma No Effect Physiological* After 10 mg intravenously in 8 healthy males. *3406* No effect 15 minutes after 1 mg/kg intravenously in healthy volunteers. *3426*

Corticotropin *Plasma No Effect Physiological* No effect 15 minutes after 1 mg/kg intravenously in healthy volunteers. *3426*

Cortisol *Plasma No Effect Physiological* After 10 mg intravenously in 8 healthy males. *3406* No effect 15 minutes after 1 mg/kg intravenously in healthy volunteers. *3426*

Growth Hormone *Plasma No Effect Physiological* No significant change in concentration after 0.17 mg/kg bolus i.v. in 10 children. *2330*

18-Hydroxycorticosterone *Plasma No Effect Physiological* After 10 mg intravenously in 8 healthy males. *3406*

Prolactin *Plasma Increase Physiological* Quick and marked effect when given i.m. to 12 normal subjects and to a group of patients with subclinical hypothyroidism. *0870* After 10 mg intravenously in 8 healthy males. *3406*

Renin Activity *Plasma No Effect Physiological* After 10 mg intravenously in 8 healthy males. *3406*

Thyroid Stimulating Hormone (TSH)
Serum Increase Physiological Small but significant effect when given i.m. to 12 normal subjects and to a group of patients with subclinical hypothyroidism. *0870* After 10 mg intravenously in 8 healthy males. *3406* In euthyroid and hypothyroid patients: drug is a dopamine antagonist. *3798*

Dopamine

Aldosterone *Plasma No Effect Physiological* Did not change with infusion of up to 3.0 µg/kg/min for 2 h in 6 normal men. *2149*

Amino Acids *Urine Increase Analytical* Reacts with ninhydrin; extra spot TLC, high voltage electrophoresis. *2855*
Urine Positive Analytical Unusual ninhydrin reacting spot on TLC. *1596*

Ammoniacal Silver Nitrate
Test Conditions Positive Analytical Positive spot test with Tollen's reagent. *3765*

Amphetamine *Urine No Effect Analytical* At 50 mg/L on fluorescent method of Hayes. *1534* No effect at 100 mg/L on method of Rutter. *3097*

Catecholamines *Urine Increase Analytical* Reacts with ethylenediamine as epinephrine, norepinephrine. *2285*

Cortisol *Plasma No Effect Physiological* Did not change with infusion of up to 3.0 µg/kg/min for 2 h in 6 normal men. *2149*

Dopamine *Urine Increase Physiological* Effect noted with infusion of as little as 0.03 µg/kg/min in 6 normal men for 2 h. *2149*

Effective Renal Plasma Flow
Patient Increase Physiological Intravenous infusion caused increase from 507 to 798 ml/min. *1326*

Epinephrine *Urine Increase Physiological* Effect noted with infusion of 0.3 µg/kg/min and more in 6 normal men for 2 h. *2149*
Test Conditions No Effect Analytical On fluorescent procedure of Peyrin. *2800*

Follicle Stimulating Hormone (FSH)
Plasma No Effect Physiological Did not change with infusion of up to 3.0 µg/kg/min for 2 h in 6 normal men. *2149*

Glucose *Serum Increase Physiological* Inhibits peripheral utilization, increased glycogenolysis. *0481* 5-7 µg/kg/min i.v. causes small increase. *1176*

Growth Hormone *Plasma Increase Physiological* Stimulates release in normal men. *1101*

Guaiacols Spot Test *Urine Negative Analytical* Action on procedure of Rogers. *3031*

Hydrazine *Plasma No Effect Analytical* On fluorometric method of Vickers at 1 µg/mL. *3710*

Inulin Clearance *Urine Increase Physiological* Intravenous infusion caused increase from 109 to 136 mL/min. *1326*

Luteinizing Hormone (LH) *Plasma Decrease Physiological* Effect noted with infusion of 0.3 µg/kg/min and more in 6 normal men for 2 h. *2149*

Norepinephrine *Plasma Increase Physiological* Effect noted with infusion of 3.0 µg/kg/min in 6 normal men for 2 h. *2149*
Test Conditions Increase Analytical 2.9% fluorescence of norepinephrine (Peyrin procedure). *2800*

Potassium *Urine No Effect Physiological* Did not change with infusion of up to 3.0 µg/kg/min for 2 h in 6 normal men. *2149*
Urine Increase Physiological Directly inhibits tubular solute reabsorption. *1591*

Prolactin *Plasma Decrease Physiological* Effect noted with infusion of as little as 0.03 µg/kg/min in 6 normal men for 2 h. *2149*

Renin Activity *Plasma No Effect Physiological* Did not change with infusion of up to 3.0 µg/kg/min for 2 h in 6 normal men. *2149*

Sodium *Urine Increase Physiological* Intravenous infusion in normals (increase = 171 to 571 µEq/min). *1326* Effect noted with infusion of 3.0 µg/kg/min in 6 normal men for 2 h. *2149* Directly inhibits tubular solute reabsorption. *1591*

Thyroid Stimulating Hormone (TSH)
Serum No Effect Physiological Did not change with infusion of up to 3.0 µg/kg/min for 2 h in 6 normal men. *2149*
Serum Decrease Physiological Antagonizes TRH at hypophyseal level: reduces concentration in both euthyroid and hypothyroid individuals. Diminishes TSH response to TRH. *3798*

Uric Acid *Test Conditions Increase Analytical* Positive spot test with phosphotungstate. *3765*

Urobilinogen *Test Conditions Increase Analytical* Positive spot test with Ehrlich's reagent. *3765*

Volume *Urine No Effect Physiological* Did not change with infusion of up to 3.0 µg/kg/min for 2 h in 6 normal men. *2149*
Urine Increase Physiological Directly inhibits tubular solute reabsorption. *1591*

Doxapram

Erythrocytes *Blood Decrease Physiological* Noted postoperatively in a few patients. *2313*

Hematocrit *Blood Decrease Physiological* Noted postoperatively in a few patients. *0034*

Hemoglobin *Blood Decrease Physiological* Noted postoperatively in a few patients. *0034*

Leukocytes *Blood Decrease Physiological* Further decrease noted in patient with leukopenia. *0034*

Protein *Urine Increase Physiological* May have nephrotoxic effect ?nephrotoxic effect. *1022*

Urea Nitrogen *Serum Increase Physiological* ?nephrotoxic effect. *1022*

Doxazosin

Cholesterol *Serum No Effect Physiological* No significant change in 44 patients treated for 3 mo. *0088* Compared with controls over 10-12 weeks, although 9% increase in HDL/total cholesterol ratio. *2842*
Serum Decrease Physiological In 38 treated hypertensives given 1 to 16 mg/d over 10 weeks. *2562*

Cholesterol, High Density Lipoprotein
Serum No Effect Physiological Compared with controls over 10-12 weeks, although 9% increase in HDL/total cholesterol ratio. *2842*

Cholesterol, Very Low Density Lipoprotein
Serum Decrease Physiological In 38 treated hypertensives given 1 to 16 mg/d over 10 weeks. *2562*

Creatinine Clearance *Urine No Effect Physiological* No effect on clearance or on renal blood flow in 24 patients with 6 weeks treatment. *3852*

Norepinephrine *Plasma Increase Physiological* After 6 weeks in 24 patients treated for 6 weeks. *3852*

Renin Activity *Plasma Increase Physiological* Acute increase in 24 patients treated for 6 weeks. *3852*

Triglycerides *Serum No Effect Physiological* No significant change compared with controls. *2842*
Serum Decrease Physiological Increase by 13% in 44 patients treated for 3 mo. *0088* In 38 treated hypertensives given 1 to 16 mg/d over 10 weeks. *2562*

Doxepin

Alkaloids *Urine No Effect Analytical* With ninhydrin on Frings TLC procedure at 1 mg/dL. *1204*
Urine Increase Analytical Purple with iodoplatinate on Frings TLC procedure. *1204*

Barbiturate *Urine No Effect Analytical* With mercurous NO3 on Frings TLC procedure at 1 mg/dL. With vanillin on Frings procedure at 10 mg/dL. With diphenylcarbazone on Frings TLC procedure at 1 mg/dL. With mercuric SO4 on Frings TLC procedure at 10 mg/dL. *1204*

Bile *Urine Increase Physiological* Theoretical possibility due to class of compound. *0071*

Bilirubin *Serum Increase Physiological* Theoretical possibility due to class of compound. *0071*

Bromide *Urine No Effect Analytical* No effect on method of Frings. *1204*

Cholesterol *Serum No Effect Analytical* At concentration of 150 mg/L had no effect on CHOD-Iodide method. *3393* At concentration of 150 mg/L had no effect on CHOD-PAP method. *3393* At concentration of 150 mg/L had no effect on catalase-AIDH method. *3393* At concentration of 150 mg/L had no effect on method using catalase-Hantzsch reaction. *3393* At concentration of 150 mg/L had no effect on Liebermann-Burchard method. *3393*

Creatinine *Serum No Effect Analytical* At 1.2 g/L on reversed phase LC procedure of Zhiri et al. *3942* At concentration of 0.3 mg/L had no effect on creatinine iminohydrolase method. *3393* At concentration of 30 mg/L had no effect on Jaffé-Fading-Fraction method. *3393* At concentration of 30 mg/L had no effect on Jaffé-Fuller's earth method. *3393* At concentration of 10 mg/L had no effect on kinetic Jaffé method on BKA-2. *3393*

Ethchlorvynol *Urine No Effect Analytical* No effect on method of Frings and Cohen. *1204*

FPN Test *Urine No Effect Analytical* No effect at 10 mg/dL on method of Frings. *1204*

Glucose *Serum No Effect Analytical* At concentration of 6 mg/L had no effect on Ektachem® method. *3393*

Leukocytes *Blood Decrease Physiological* Theoretical possibility due to class of compound. *0071*

Platelets *Blood Decrease Physiological* Single case observed (probably immune response). *2603*

Potassium *Serum No Effect Analytical* At concentration of 60 mg/L had no effect on measurement by ISE without predilution. *3393*

Propoxyphene *Serum No Effect Analytical* At 10 mg/L on method of Evenson. *1056*

Protein Electrophoresis *Serum No Effect Analytical* At concentration of 60 mg/L had no effect on automated Olympus-Hite method. *3393*

Salicylate *Urine No Effect Analytical* At 10 mg/dL no effect on Trinder method. *1204*

Sodium *Serum No Effect Analytical* At concentration of 60 mg/L had no effect on measurement by ISE without predilution. *3393*

Urea Nitrogen *Serum No Effect Analytical* At concentration of 6 mg/L had no effect on Ektachem® method. *3393*

Uric Acid *Serum No Effect Analytical* At 1.2 g/L on reversed phase LC procedure of Zhiri et al. *3942* At concentration of 30 mg/L had no effect on Kageyama-Hantzsch method. *3393* At concentration of 30 mg/L had no effect on catalase-AIDH method. *3393*

Doxorubicin

N-Acetylglucosaminidase *Urine No Effect Analytical* At 50 mg/L on 2 colorimetric analytical methods. *1354*

Adenosine Deaminase Binding Protein
Urine No Effect Physiological No effect in 12 children given drug for 3 d. *1355*

Alanine Aminotransferase *Serum Increase Physiological* May produce idiosyncratic hepatic dysfunction, as observed in 6 patients. *0184*

Alkaline Phosphatase *Serum Increase Physiological* May produce idiosyncratic hepatic dysfunction, as observed in 6 patients. *0184*

Aspartate Aminotransferase
Serum Increase Physiological May produce idiosyncratic hepatic dysfunction, as observed in 6 patients. *0184*

Bilirubin, Direct *Serum Increase Physiological* May produce idiosyncratic hepatic dysfunction, as observed in 6 patients. *0184*

Erythrocytes *Blood Decrease Physiological* Impaired but not as marked as WBC decrease. *0387*

Leukocytes *Blood Decrease Physiological* In 70% patients with 60 mg/sq m for 21 d. *0387*

Platelets *Blood Decrease Physiological* Impaired but not as marked as WBC decrease, possible bone marrow aplasia. *0387*

Doxycycline

Aspartate Aminotransferase
Serum Increase Physiological Reported effect (?hepatic origin). *0071*

Chloramphenicol *Serum No Effect Analytical* No effect at 100 mg/L on coupled enzymatic method. *2490*

Creatinine *Serum Increase Physiological* Nephrotoxic effect. *2802*

Doxycycline *(continued)*

Eosinophils *Blood Increase Physiological* Allergic response reported. *2802*

Hemoglobin *Blood Decrease Physiological* Observed with other tetracyclines. *1680*

LE Cells *Blood Positive Physiological* May produce exacerbation of SLE. *1680*

Leukocytes *Blood Decrease Physiological* Neutropenia reported. *2802*

Neutrophils *Blood Decrease Physiological* Observed with other tetracyclines. *1680* Occasional case of agranulocytosis reported. *3717*

Platelets *Blood Decrease Physiological* Thrombocytopenia reported. *2802*

Protein *Urine Increase Physiological* Nephrotoxic effect. *2802*

Prothrombin Time *Plasma Increase Physiological* Depresses prothrombin activity. *1680*

Urea Nitrogen *Serum Increase Physiological* Nephrotoxic effect. *2802*

Driving

Carboxyhemoglobin *Blood Increase Physiological* Most marked increase in day drivers vs night. *1821*

Dromostanolone

Calcium *Serum Increase Physiological* Usually if osteolytic metastases. *0071*
Serum Decrease Physiological May occur with regression. *0071*

Erythrocytes *Blood Increase Physiological* Response to androgens. *0071*

Hematocrit *Blood Increase Physiological* Response to androgens. *0071*

Hemoglobin *Blood Increase Physiological* Response to androgens. *0071*

Dydrogesterone

Leukocytes *Blood Increase Physiological* Mild leukocytosis reported. *0071*

Earlobe Blood

Hematocrit *Blood Increase Physiological* Increased up to 15% above fingerstick blood. *0638*

Hemoglobin *Blood Increase Physiological* Increased up to 15% above fingerstick blood. *0638*

pH *Blood Decrease Physiological* Significantly less in first sample than in subsequent. *0931*

Echothiophate

Pseudocholinesterase *Serum Decrease Physiological* After few weeks of eyedrop therapy. *1678*
Red Blood Cells Decrease Physiological Direct effect of drug (specific inhibitor of enzyme). *3523*

Ectylurea

Alanine Aminotransferase *Serum Increase Physiological* May cause cholestasis with cholangiolitis. *2313*

Alkaline Phosphatase *Serum Increase Physiological* May cause cholestasis with cholangiolitis. *2313*

Aspartate Aminotransferase
Serum Increase Physiological May cause cholestasis with cholangiolitis. *2313*

Bile *Urine Increase Physiological* May cause cholestasis with cholangiolitis. *2313*

Bilirubin *Serum Increase Physiological* May cause cholestasis with cholangiolitis. *2313*

BSP Retention *Serum Increase Physiological* May cause cholestasis with cholangiolitis. *2313*

Cephalin Flocculation *Serum Increase Physiological* May cause cholestasis with cholangiolitis. *2313*

Thymol Turbidity *Serum Increase Physiological* May cause cholestasis with cholangiolitis. *2313*

Edrophonium

Pseudocholinesterase *Serum Decrease Physiological* Direct action of drug. *1147*

EDTA

N-Acetylglucosaminidase *Urine No Effect Analytical* At 20 mmol/L on 2 colorimetric analytical methods. *1354*

Albumin *Serum Decrease Analytical* At concentrations above 25,000 mg/L lowered concentration as measured by BCG method. *3393*

Alkaline Phosphatase *Serum Decrease Analytical* Inhibitory effect observed. *2228* 50 mmol/L is completely inhibitory. *3217*
White Blood Cells Decrease Analytical Reported in presence of low concentrations of metals. *1896*

Amino Acids *Serum Increase Analytical* May be contaminated with ninhydrin positive constituents. *3230*

Aminolevulinic Acid Dehydrase
Red Blood Cells Decrease Analytical 15% decrease compared with heparin sample. *0400*

Ammonia *Plasma No Effect Analytical* Satisfactory as anticoagulant. *0705* At concentration of 800 mg/L had no effect on Ektachem® method. *3393*

Amylase *Serum No Effect Analytical* At concentration of 5380 mg/L had no effect on maltopentose method on aca. *3393* At concentration of 1170 mg/L had no effect on p-nitrophenylmaltopentoside/hexaoside method. *3393*
Serum Decrease Analytical 15% inhibition at 1 g/L, 51% inhibition at 5 g/L, 73% inhibition at 10 g/L of method of Rauscher et al using nitrophenylmaltoheptaoside as substrate. *2926*

Angiotensin-Converting Enzyme (ACE)
Serum Decrease Analytical 40% inhibition reported with method using benzyloxycarbonyl-phenylalanyl-histidyl-leucine as substrate. *3221*

Antithrombin III *Plasma No Effect Analytical* At concentration of 1170 mg/L had no effect on aca method. *3393*

Aspartate Aminotransferase *Serum No Effect Analytical* No effect on activity. *3165*

Bicarbonate *Serum Decrease Analytical* Significant alteration of pH in drawn sample. *1563*
Serum Decrease Physiological Nephrotoxic effect (especially Calcium EDTA). *1237*

Bilirubin *Serum No Effect Analytical* At concentration of 8,000 mg/L had no effect on Ektachem® method. *3393*

Cadmium *Urine Increase Physiological* If poisoning due to cadmium. *0071*

Calcium *Serum No Effect Analytical* No effect on flame photometric methods. *2220*
Serum Decrease Analytical If determined by oxalate or other precipitation technicians. *3505*
Serum Decrease Physiological Complexes calcium by chelation. *2005*
Urine Increase Physiological Excretion of chelate. *0071*

Casts *Urine Increase Physiological* Nephrotoxic effect (especially Calcium EDTA). *2313*

Cholesterol *Serum No Effect Analytical* At concentration of 1,000 mg/L had no effect on CHOD-Iodide method. *3393* At concentration of 1,000 mg/L had no effect on CHOD-PAP method. *3393* At concentration of 1,000 mg/L had no effect on catalase-AIDH method. *3393* At concentration of 10,000 mg/L had no effect on method using catalase-Hantzsch reaction. *3393* At concentration of 1,000 mg/L had no effect on Liebermann-Burchard method. *3393*
Serum Decrease Analytical If used as anticoagulant value low by 5-15 mg/dL. *2319*
Serum Decrease Physiological Reported to occur if given i.v. *1488*

Chromium *Urine Increase Physiological* If poisoning due to chromium. *0071*

Copper *Urine Increase Physiological* If poisoning due to copper. *0071*

Coproporphyrin *Urine Decrease Physiological* If lead poisoning present. *1679*

Creatine Kinase *Serum Decrease Analytical* Inhibits if used as anticoagulant. *3217*

Creatinine *Serum No Effect Analytical* No effect on analytical methods reported. *1563* At concentration of 3360 mg/L had no effect on creatinine iminohydrolase method. *3393* At concentration of 1,000 mg/L had no effect on Jaffé-Fading-Fraction method. *3393* At concentration of 1,000 mg/L had no effect on Jaffé-Fuller's earth method. *3393* At concentration of 1,000 mg/L had no effect on kinetic Jaffé method on BKA-2. *3393*
Serum Increase Analytical At concentrations above 800 mg/L raised concentration as measured by Ektachem® method. *3393*
Serum Decrease Analytical At concentrations above 935 mg/L lowered concentration as measured by creatinine amidohydrolase method. *3393*

Epinephrine *Urine No Effect Analytical* Stabilizes compounds prevents light destruction. *1868*

Erythrocytes *Blood Decrease Physiological* Transient bone marrow depression. *0071*
Urine Increase Physiological Rare side effect, especially with Calcium edetic acid. *1679*

Erythrocyte Sedimentation Rate
Blood Decrease Analytical Retards rate when compared with heparin. *0251*

Factor V *Plasma Decrease Analytical* Unstable ?affects prothrombin determination. *3217*

Fatty Acids, Free (FFA) *Serum No Effect Analytical* No effect of anticoagulant on method of Soloni. *3381*

Fibrinogen *Plasma Decrease Analytical* At concentrations above 1170 mg/L lowered concentration as measured by aca method. *3393*

Glucose *Serum No Effect Analytical* At anticoagulant dose on hexokinase method of Coburn. *0676* At 1.2 g/dL on glucose oxidase procedures At 1.2 g/dL on o-toluidine procedure. *2140* At 200 mg/dL on MBTH procedure of Neeley. *2569* At 1.2 g/dL on p-HBAH procedure of Lever. *2140* No effect on hexokinase method if anticoagulant. *2982* At 1.2 g/dL on alkaline ferricyanide procedure. *2140* At concentration of 2,000 mg/L had no effect on hexokinase/G-6-PDH method. *3393*
Serum Increase Analytical If greater than 1 mg/mL affects o-toluidine procedures. *0724* At concentrations above 100 mg/L raised concentration as measured by Ektachem® method. *3393*
Urine Increase Physiological Reported effect. ?mechanism May cause tubular damage. *1343*

Gold *Blood No Effect Analytical* Atomic absorption method of Harth et al. *1515*

Hematocrit *Blood Decrease Physiological* May occur with prolonged therapy. *0071*

Hemoglobin *Blood Decrease Physiological* May occur with prolonged therapy. *0071*

Hydroxybutyrate Dehydrogenase
Serum No Effect Analytical No effect on activity. *2981*

Insulin *Plasma No Effect Analytical* Results same as when serum used. *0112*

Ionized Calcium *Serum Decrease Physiological* Chelates calcium. *0071*

Iron *Serum Decrease Analytical* Interferes with method of Young and Hicks. *2981* Although no effect with atomic absorption methods low results obtained with colorimetric methods including Ferrozine and bathophenanthroline. *1991* At concentrations above 100 mg/L lowered concentration as measured by Ferrozine method. *3393*

Lead *Blood Decrease Physiological* If lead poisoning present. *1679*
Urine Increase Physiological If poisoning due to lead. *0071*

Leucine Aminopeptidase *Serum Decrease Analytical* Inhibits if used as anticoagulant. *3217*

Leukocytes *Blood Decrease Physiological* Transient bone marrow depression. *0071*

Lipids, Total *Serum Decrease Physiological* Occurs if elevated lipids and given i.v. *1488*

Manganese *Urine Increase Physiological* If poisoning due to manganese. *0071*

NBT Test *Blood No Effect Analytical* If Ficoll used with method of Gordon. *1349*
Blood Decrease Analytical False low result as requires complement. *3507*

Nickel *Urine Increase Physiological* If poisoning due to nickel. *0071*

Nonprotein Nitrogen *Serum Increase Analytical* Theoretical increase of 7 mg/dL as anticoagulant. *1563*
Serum Increase Physiological Nephrotoxic effect (especially Calcium EDTA). *2313*

pH *Serum Decrease Analytical* In vitro addition of sodium salt to plasma/serum. *0106*
Blood Decrease Analytical Affects acid-base balance if used as anticoagulant. *2313*

Plasminogen *Plasma No Effect Analytical* At concentration of 1170 mg/L had no effect on aca method. *3393*

Platelets *Blood Decrease Physiological* Transient bone marrow depression. *0071*

Potassium *Serum No Effect Analytical* At concentration of 1,000 mg/L had no effect on measurement by ISE without predilution. *3393*
Serum Decrease Analytical Reacts with potassium. *2313*
Urine Increase Physiological Occurs particularly if given i.v. *1488*

Pregnancy-Associated Plasma Protein A
Serum Increase Analytical Values higher than serum if used as anticoagulant and test uses crossed immunoelectrophoresis. *3610*

Protein *Urine Increase Physiological* Nephrotoxic effect (especially Calcium EDTA). *0071* Nephrotoxic effect (especially Calcium EDTA). *1237*
Test Conditions Decrease Analytical Variable but significant effect on Lowry procedure. *1796*

Protein Electrophoresis *Serum No Effect Analytical* At concentration of 1,000 mg/L had no effect on automated Olympus-Hite method. *3393*

Prothrombin Time *Plasma Increase Analytical* When used as anticoagulant. *3962*
Plasma Decrease Physiological Transient fall during administration (within 12 h). *2313*

Pyruvate Kinase *Platelets Increase Analytical* 5 fold increase if added to assay mixture. *3543*
Red Blood Cells Increase Analytical 1.5 fold increase if added to assay mixture. *3543*
White Blood Cells Increase Analytical 15 fold increase if added to assay mixture. *3543*

Sodium *Serum Increase Analytical* At concentrations above 500 mg/L raised concentration as measured by ISE without predilution. *3393*

Sorbitol Dehydrogenase *Serum Decrease Analytical* Inhibitory effect. *2981*

Sugar *Urine Increase Analytical* Reduces Benedict's reagent. *1237*

Triglycerides *Serum No Effect Analytical* At concentration of 2922 mg/L had no effect on GPO-PAP method. *3393*

Urea Nitrogen *Serum No Effect Analytical* No effect on analytical methods reported. *1563* At concentration of 100 mg/L had no effect on Ektachem® method. *3393*
Serum Increase Physiological Nephrotoxicity (especially Calcium EDTA). *1237*

Uric Acid *Serum No Effect Analytical* At concentration of 1,000 mg/L had no effect on Kageyama-Hantzsch method. *3393* At concentration of 1,000 mg/L had no effect on catalase-AIDH method. *3393* At concentration of 4,000 mg/L had no effect on uricase method on aca. *3393* At concentration of 2922 mg/L had no effect on uricase-PAP method. *3393*

Zinc *Urine Increase Physiological* If poisoning due to zinc. *0071*

Eggplant

5-Hydroxyindoleacetic Acid (5-HIAA)
Urine Increase Physiological As result of high content in food. *1488*

Eggplant *(continued)*

Thyroxine (T4) *Serum Increase Physiological* Increased binding capacity (augmentation of transport proteins). *3798*

Tri-iodothyronine (T3) *Serum Increase Physiological* Increased binding capacity (augmentation of transport proteins). *3798*

Electrocautery

Creatine Kinase *Serum Increase Physiological* Effect of muscle damage. *1237*

Electroshock

Cyclic Adenosine Monophosphate
Urine Increase Physiological Mean change of from 4.2 μmol/24 h to 14.2 μmol/24 h. *1468*

Dopamine Hydroxylase *Serum No Effect Physiological* No effect observed. *3807*

Myoglobin *Urine Increase Physiological* May occur with electrical shock. *0017*

Protein *Urine Increase Physiological* May cause nephrotoxicity. *3204*

Urea Nitrogen *Serum Increase Physiological* May cause nephrotoxicity. *3204*

Enalapril

Aldosterone *Plasma No Effect Physiological* No significant effect in patients receiving drug. *1616*
Plasma Decrease Physiological Gradual reduction following effect on angiotensin II due to longer half-life. *0177* Increase by 1 mo as part of long-term response in 6 responders of 10 treated hypertensives. *1166*
Urine Decrease Physiological Sustained effect: extent related to pretreatment plasma renin activity. *0177*

Angiotensin-Converting Enzyme (ACE)
Serum Decrease Physiological Good correlation between serum concentration of drug and inhibition of enzyme in both acute and chronic studies. *1747* Significant fall even on first day of treatment. *1616* Marked effect in 4 d as part of long-term response in 6 responders of 10 treated hypertensives. *1166*

Angiotensin I *Plasma Increase Physiological* Increases as angiotensin II falls after first dose. *1616*

Angiotensin II *Plasma Decrease Physiological* Significant fall after first dose, subsequently remains depressed. *1616* Prompt and striking reduction following i.v. administration over 30 min: more gradual reduction when given orally. *0177*

Cholesterol *Serum No Effect Physiological* In 53 patients given up to 160 mg daily or when combined with hydrochlorothiazide. *2277*

Creatinine *Serum Increase Physiological* Reversible renal insufficiency reported in some patients without evidence of renal artery stenosis. *0929* Associated with selective glomerular efferent arteriolar dilatation and possible interference with autoregulatory capacity of kidney in response to severe renovascular hypertension. *0284*

Creatinine Clearance *Urine No Effect Physiological* In 53 patients given up to 160 mg daily or when combined with hydrochlorothiazide. *2277*

Effective Renal Plasma Flow
Patient Decrease Physiological Associated with selective glomerular efferent arteriolar dilatation and possible interference with autoregulatory capacity of kidney in response to severe renovascular hypertension. *0284*

Eosinophils *Blood Increase Physiological* Reversible and associated with rash in one patient. *0929*

Erythrocytes *Blood No Effect Physiological* In 53 patients given up to 160 mg daily or when combined with hydrochlorothiazide. *2277*

Glomerular Filtration Rate (GFR)
Urine Decrease Physiological Associated with selective glomerular efferent arteriolar dilatation and possible interference with autoregulatory capacity of kidney in response to severe renovascular hypertension. *0284*

Glucose *Serum No Effect Physiological* In 53 patients given up to 160 mg daily or when combined with hydrochlorothiazide. *2277*
Urine Increase Physiological Reported in one patient with mild uncomplicated essential hypertension. *0929*

γ-Glutamyltransferase (GGT)
Urine No Effect Physiological In 53 patients given up to 160 mg daily or when combined with hydrochlorothiazide. *2277*

Hematocrit *Blood No Effect Physiological* In 53 patients given up to 160 mg daily or when combined with hydrochlorothiazide. *2277*

Leukocytes *Blood No Effect Physiological* In 53 patients given up to 160 mg daily or when combined with hydrochlorothiazide. *2277*

Lysozyme *Urine No Effect Physiological* In 53 patients given up to 160 mg daily or when combined with hydrochlorothiazide. *2277*

Na/K ATPase *Red Blood Cells Increase Physiological* Effect of drug on erythrocyte membrane so intracellular sodium reduced and potassium increased. *2277*

Norepinephrine *Plasma No Effect Physiological* Non-significant increase as part of long-term response in 6 responders of 10 treated hypertensives. *1166*

Platelets *Blood No Effect Physiological* In 53 patients given up to 160 mg daily or when combined with hydrochlorothiazide. *2277*

Potassium *Serum Increase Physiological* Produces mild potassium retention and hyperkalemia. *2836*

Protein *Urine Decrease Physiological* Often reduction and rarely increased during treatment. *2361*

Renal Blood Flow (RBF) *Patient Increase Physiological* May increase without increase in GFR in patients with normal renal function, but GFR may increase if initial GFR below 80 mL/min/m². *0823*

Renin Activity *Plasma Increase Physiological* Steady increase from administration of first dose. *1616* Gradually increases over 1 h (due to negative feedback on renin secretion). *0177* Within 4 d as part of long-term response in 6 responders of 10 treated hypertensives. *1166*

Triglycerides *Serum No Effect Physiological* In 53 patients given up to 160 mg daily or when combined with hydrochlorothiazide. *2277*

Uric Acid *Serum Decrease Physiological* After enalapril alone (up to 160 mg/daily) but may be increased when combined with hydrochlorothiazide. *2277*

Volume *Blood No Effect Physiological* As part of long-term response in 6 responders of 10 treated hypertensives. *1166*

β-Endorphin

Cortisol *Plasma No Effect Physiological* No effect with i.v. infusion in depressed patients or methadone treated addicts. *0610*

Growth Hormone *Plasma No Effect Physiological* No effect with i.v. infusion in depressed patients or methadone treated addicts. *0610*

Prolactin *Plasma Increase Physiological* Prompt 2-4 fold increase in 4 depressed psychiatric patients with i.v. infusion. *0610*

Endotoxin

BSP Retention *Serum Increase Physiological* In almost 50% patients with induced fever. *0375*

Fatty Acids, Free (FFA) *Serum Increase Physiological* Initial rise occurs, normal later. *1230*

Lipids, Total *Serum Increase Physiological* Observed response. *1230*

Thyroid Stimulating Hormone (TSH)
Serum No Effect Physiological No effect observed. *2700*

Triglycerides *Serum Increase Physiological* Occurs after free fatty acid rise. *1230*

Endralazine

Renin Activity *Plasma Increase Physiological* Significant effect in 20 hypertensive patients (WHO phase I and II). *1524*

Enflurane

Alanine Aminotransferase *Serum Increase Physiological* In up to 10 % patients given 2 to 3 administrations of anesthetic agent. *1089* Evidence of hepatocellular damage; characteristically centrilobular necrosis: mechanism of injury most probably metabolic idiosyncrasy. *2157*

Alkaline Phosphatase *Serum Increase Physiological* Evidence of hepatocellular damage; characteristically centrilobular necrosis: mechanism of injury most probably metabolic idiosyncrasy. *2157*

Aspartate Aminotransferase
Serum No Effect Physiological In all patients given 2 to 3 administrations of anesthetic agent. *1089*
Serum Increase Physiological Slight effect but decreased in others. *0999* Evidence of hepatocellular damage; characteristically centrilobular necrosis: mechanism of injury most probably metabolic idiosyncrasy. *2157*

Bicarbonate *Serum Increase Physiological* Postoperatively moderately high but still normal. *0999*

Bilirubin *Serum Increase Physiological* Evidence of hepatocellular damage; characteristically centrilobular necrosis: mechanism of injury most probably metabolic idiosyncrasy. *2157*

Calcium *Serum Decrease Physiological* Slight decrease noted postoperatively. *0999*

Cortisol *Plasma No Effect Physiological* No difference between pre-induction and values during anesthesia. *3456*

Creatinine *Serum Increase Physiological* In one patient with initially normal renal function following comparatively long exposure to anesthetic. Normal renal function eventually returned. *1001*

Erythrocytes *Urine Increase Physiological* In one patient with initially normal renal function following comparatively long exposures to anesthetic. Normal renal function eventually returned. *1001*

Fatty Acids, Free (FFA) *Serum Decrease Physiological* Fall during anesthesia, but increases in recovery. *2699*

Glucose *Serum Increase Physiological* Effect of anesthesia and surgery. *0999*

γ-Glutamyltransferase (GGT)
Serum Increase Physiological In about 10% patients given 2 to 3 administrations of anesthetic agent. *1089*

Growth Hormone *Plasma No Effect Physiological* No effect unless given with propranolol. *2329* No effect of anesthesia alone. *2699*

Insulin *Plasma No Effect Physiological* No effect during anesthesia, slight increase after. *2699*

Lactate Dehydrogenase *Serum Increase Physiological* In about 10% patients given 2 to 3 administrations of anesthetic agent. *1089*

Leukocytes *Blood Increase Physiological* May also partly reflect surgery. *0999*

Potassium *Serum Decrease Physiological* Slight decrease noted postoperatively. *0999*

Prolactin *Plasma Increase Physiological* In both men and women when used to induce anesthesia with peak at 30 minutes. *3456*

Protein *Serum No Effect Physiological* No effect observed. *0999*
Urine Increase Physiological In one patient with initially normal renal function following comparatively long exposure to anesthetic. Normal renal function eventually returned. *1001*

Urea Nitrogen *Serum Increase Physiological* In one patient with initially normal renal function following comparatively long exposure to anesthetic. Normal renal function eventually returned. *1001*
Serum Decrease Physiological Partly due to surgery possibly. *0999*

Enoxacin

Bilirubin *Serum No Effect Physiological* Clinically insignificant displacement from protein in neonates. *3748*

Enoximone

Antidiuretic Hormone *Plasma No Effect Physiological* In 10 patients with congestive cardiac failure. *3654*

Norepinephrine *Plasma No Effect Physiological* In 10 patients with congestive cardiac failure. *3654*

Renin Activity *Plasma Increase Physiological* In 10 patients with congestive cardiac failure from 13.6 to 16.6 ng/mL/h on average. *3654*

Enprostil

Propranolol *Serum No Effect Physiological* No effect on propranolol elimination or hepatic metabolism. *2942*

Ephedrine

Alkaloids *Urine No Effect Analytical* With iodoplatinate on Frings TLC procedure at 10 mg/dL. With ninhydrin on Frings TLC procedure at 1 mg/dL. *1204*

Amino Acids *Urine Increase Analytical* Reacts with ninhydrin; extra spot TLC, high voltage electrophoresis. *2855*

Amphetamine *Urine No Effect Analytical* At 50 mg/L on fluorescent method of Hayes. *1534* No effect at 100 mg/L on method of Rutter. *3097*
Urine Increase Analytical Reacts with methyl orange in method of Frings. *2474* Yields fluorophor with NBD chloride reaction. *2473*

Barbiturate *Urine No Effect Analytical* With diphenylcarbazone on Frings TLC procedure at 10 mg/dL. With mercuric SO_4 on Frings TLC procedure at 10 mg/dL. With vanillin on Frings procedure at 10 mg/dL. With mercurous NO_3 on Frings TLC procedure at 10 mg/dL. *1204*

Bromide *Urine No Effect Analytical* No effect on method of Frings. *1204*

Catecholamines *Urine No Effect Analytical* No effect observed. *0913*
Urine No Effect Physiological No effect observed. *0913*

Cortisol *Plasma Decrease Physiological* Accelerated steroid clearance due to increased hepatic blood flow and induction of enzymes in liver. *1830*

Dexamethasone *Serum Decrease Physiological* Metabolic clearance enhanced and conjugated fraction in urine increased. *1830*

Epinephrine *Urine Increase Physiological* For 3 h after administration (slight). *1487*

Ethchlorvynol *Urine No Effect Analytical* No effect on method of Frings and Cohen. *1204*

FPN Test *Urine No Effect Analytical* No effect at 10 mg/dL on method of Frings. *1204*

Glucose *Serum Increase Physiological* Less effective than epinephrine. *1343*
Urine Increase Physiological May cause glycosuria Increased excretion of glucose. *1022*

5-Hydroxyindoleacetic Acid (5-HIAA)
Urine Increase Analytical May cause false increase in color. *0913*

^{131}I Uptake *Serum Decrease Physiological* Lilly, P-D products contain tetraiodofluorescein. *2652*

Metanephrines, Total *Urine No Effect Analytical* At 5 mg/L on modified Pisano procedure. *1428*

Norepinephrine *Urine No Effect Physiological* No effect observed. *1487*

Propoxyphene *Urine Increase Analytical* 1% fluorescence in procedure of Valentour. *3662*

Salicylate *Urine No Effect Analytical* At 10 mg/dL no effect on Trinder method. *1204*

Vanillylmandelic Acid *Urine No Effect Physiological* No effect observed. *0913*

Epicillin

Eosinophils *Blood Increase Physiological* Marked increase up to 18% in some cases. *0066*

Epinephrine

Alanine Aminotransferase *Serum Increase Physiological* Following single injection of compound in oil. *1264*

Aldolase *Serum Increase Physiological* Following single injection of compound in oil. *1264*

Amino Acids *Serum Decrease Physiological* Associated with gluconeogenesis. *0355*
Urine Increase Analytical Reacts with ninhydrin; extra spot TLC, high voltage electrophoresis. *2855*
Urine Decrease Physiological Metabolic effect associated with gluconeogenesis. *2313*
Test Conditions Increase Analytical Positive spot test with ninhydrin. *3765*

α-Amino-Nitrogen *Plasma Increase Physiological* Metabolic effect gluconeogenesis. *0355*

Ammoniacal Silver Nitrate
Test Conditions Positive Analytical Positive Tollen's test for phenols. *3765*

Amphetamine *Urine No Effect Analytical* No reaction with NBD chloride of Monforte. *2473* At 50 mg/L on fluorescent method of Hayes. *1534* No effect at 100 mg/L on method of Rutter. *3097*

Aspartate Aminotransferase *Serum Increase Analytical* At 1 mmol/L affects SMA 12/60 method. *3335*
Serum Increase Physiological Following single injection of compound in oil. *1264*

Basal Metabolic Rate *Patient Increase Physiological* Temporary effect observed after injection. *0251*

Bilirubin *Serum No Effect Analytical* At concentration of 183 mg/L had no effect on Jendrassik and Grof method. *3393*
Serum Increase Analytical At 1 mmol/L affects SMA 12/60 method. *3335*

Calcitonin *Plasma No Effect Physiological* Following i.v. infusion for 1 h to produce concentration up to 945 pg/mL. *0401*

Calcium *Serum No Effect Analytical* At concentration of 183 mg/L had no effect on cresolphthalein method. *3393*

Catecholamines *Plasma Increase Physiological* Presence of epinephrine. *0758*

Chloride *Urine Decrease Physiological* With increased filtration fraction. *1343*

Cholesterol *Serum Increase Physiological* Metabolic effect (indirectly through ACTH stimulation). *1488*

Coagulation Time *Blood Decrease Physiological* Probably due to increased activity of factor V. *2313*

Cyclic Adenosine Monophosphate
Plasma No Effect Physiological No effect unless beta-blocking agent also given. *0216*
Plasma Increase Physiological Normal response after subcutaneous injection. *0321* Response to i.v. infusion in normals. *0216*
Urine Increase Physiological 2 fold increase in clearance produced. *0321* Effect less marked than in blood. *0216*

Effective Renal Plasma Flow
Patient Decrease Physiological May decrease up to 40% but no effect on blood pressure. *1343*

Eosinophils *Blood Decrease Physiological* May cause eosinopenia (direct action). *1513*

Epinephrine *Test Conditions Increase Analytical* 34% of fluorescence method of Waldmeier. *3735*

Erythrocytes *Blood Increase Physiological* Due to hemoconcentration. *1343*

Factor V *Plasma Increase Physiological* Transient effect. *2220*

Factor VIII *Plasma Increase Physiological* Occurs following epinephrine i.v. *0883*

Factor VIII Antigen *Plasma Increase Physiological* Occurs following epinephrine i.v. *0883*

Fatty Acids, Free (FFA) *Serum Increase Physiological* Transient metabolic effect. *1343*

Gastrin *Serum Increase Physiological* High concentrations if pheochromocytoma, also if infused. *3472*

Glomerular Filtration Rate (GFR)
Urine Decrease Physiological Slight fall after i.m. injection. *2427*

Glucose *Serum No Effect Analytical* At 10 mg/dL no effect on glucose oxidase procedure of Gochman. *1311*
Serum Increase Analytical At 1 mmol/L affects SMA 12/60 method. *3335* At 10 mg/dL affects alkaline ferricyanide procedure. *1311*
Serum Increase Physiological Characteristic action of hormone. *1488*
Urine Decrease Analytical Inhibits peroxidase reaction of Clinistix®. *2544*

Growth Hormone *Plasma No Effect Physiological* No effect observed in normals. *0371*

Guaiacols Spot Test *Urine Negative Analytical* Action on procedure of Rogers. *3031*

Hemoglobin *Plasma No Effect Physiological* No effect observed with increased blood flow. *1755*

Ionized Calcium *Serum No Effect Physiological* Following i.v. infusion for 1 h to produce concentration up to 945 pg/mL. *0401*

Iron *Serum Decrease Physiological* 1 mL 1% solution effect lasts for 6 h. *0336*

Lactate *Serum Increase Physiological* Marked effect like beta-adrenergic compounds. *1343*

Leukocytes *Blood Increase Physiological* Initially due to lymphocytosis later neutrophilia. *3438*

Lipoproteins *Serum Increase Physiological* Effect on low density components. *1343*

Metanephrines, Total *Urine No Effect Analytical* At 8 mg/L on modified Pisano procedure. *1428*

Millons Test *Test Conditions Positive Analytical* Positive spot test for phenols. *3765*

Norepinephrine *Test Conditions No Effect Analytical* On automated fluorometric procedure of Peyrin. *2800*

Parathyroid Hormone *Plasma No Effect Physiological* Following i.v. infusion for 1 h to produce concentration up to 945 pg/mL. *0401*

pH *Urine Increase Physiological* Metabolic effect. *1622*

Phenylketones *Urine Positive Analytical* Green in high concentrations with FeCl3. *0775*

Phosphate *Serum No Effect Analytical* At concentration of 183 mg/L had no effect on phosphomolybdate method. *3393*
Serum Decrease Physiological Increased gluconeogenesis. *0085* Dose-dependent response to i.v. infusion: maximum decrease of 0.6 mg/dL. *0401*

Phospholipids, Total *Serum Increase Physiological* Metabolic effect. *1343*

Platelets *Blood Increase Physiological* Normal physiological response to injection. *0251*

Potassium *Serum Increase Physiological* Initial rise accompanies glucose mobilization. *1343* Rise within 1 minute of i.v. injection and fall after 4 minutes, eventually dropping below control values. *2836*
Serum Decrease Physiological Intravenous injection causes sharp rise then fall after 4 minutes. *3064* Profound hypokalemia produced by steady-state infusion at 3-5 nmol/L. *3490*
Urine Decrease Physiological With increased filtration fraction. *1343*

Propoxyphene *Urine No Effect Analytical* Less than 1% fluorescence in procedure of Valentour. *3662*

Protein *Serum No Effect Analytical* At concentration of 183 mg/L had no effect on biuret method with blank correction. *3393*
Serum Increase Physiological Due to hemoconcentration. *1343*
CSF Increase Analytical 1.0 mg = 1.5 mg in Folin-Ciocalteu procedure. *0583*
Test Conditions Increase Analytical Reacts with Folin-Ciocalteu of Lowry procedure. *0702*

Protein Bound Iodine (PBI) *Serum Decrease Physiological* Causes release of PBI from thyroid. *0012*

Sodium *Urine Decrease Physiological* With increased filtration fraction. *1343*

Sugar *Urine Increase Physiological* Positive with Fehling's and Benedict's solutions. *0279*
Test Conditions Increase Analytical Positive spot test with Benedict's reagent. *3765*

Thyroxine (T₄) Binding Globulin
Serum Increase Physiological Binding capacity increased after infusion. *1860*

Tyrosine *Plasma Decrease Physiological* Hormonal effect (?exact mechanism). *0843*

Urea Nitrogen *Serum No Effect Analytical* At concentration of 183 mg/L had no effect on diacetylmonoxime method. *3393*

Uric Acid *Serum Increase Analytical* At 1 mmol/L affects SMA 12/60, Henry methods. *3335* At concentrations above 100 mg/L raised concentration as measured by phosphotungstate reduction method. *3393*
Serum Increase Physiological Result of vasoconstriction in kidney. *1488*
Test Conditions Increase Analytical Positive spot test with phosphotungstate. *3765*

Uric Acid Clearance *Urine Decrease Physiological* Sharp fall following i.m. injection. *2427*

Urobilinogen *Urine Increase Analytical* Positive spot test with Ehrlich's reagent. *3765*

Vanillylmandelic Acid *Urine Increase Physiological* Measured as endogenous compound. *1488*

Volume *Plasma Decrease Physiological* Loss of protein-free fluid to tissues. *1343*
Urine Decrease Physiological i.m. injection causes sharp decrease. *2427*

Epinine

Epinephrine *Test Conditions No Effect Analytical* On fluorescent procedure of Peyrin. *2800*

Norepinephrine *Test Conditions Increase Analytical* 3.8% fluorescence of norepinephrine (Peyrin procedure). *2800*

Erect Posture

Albumin *Serum Increase Physiological* Average 0.42 g/dL 30 minutes after rising. *3217*

Aldosterone *Plasma Increase Physiological* Increases approximately 6 times if normal diet. *0071*
Urine Increase Physiological Increased for about 6 h after standing up. *3886*

Antidiuretic Hormone *Plasma Increase Physiological* In comparison with sitting (3.1 pg/mL versus 1.4 pg/mL). *2650*

Calcium *Serum Increase Physiological* Standing for 15 minutes causes increase of 6.7%. *3476*

Cholesterol *Serum Increase Physiological* As bound to lipoproteins increases while standing. *2319*

Cyclic Adenosine Monophosphate
Urine Decrease Physiological In normals but increased in hypertensives. *1471*

Epinephrine *Plasma No Effect Physiological* No effect seen in normals. *1176*
Urine No Effect Physiological No effect seen in normals. *1176*

Hematocrit *Blood Increase Physiological* Standing for 15 minutes causes increase of 12.9%. *3476*

Hemoglobin *Blood Increase Physiological* Due to redistribution of body water. *3879*

MCHC *Blood Decrease Physiological* Approximately 1% decrease in vertical but not significant. *1087*

Norepinephrine *Plasma No Effect Physiological* No effect seen in normals. *1176*
Urine No Effect Physiological No effect seen in normals. *1176*
Urine Increase Physiological 3 fold increase compared with lying. *3217* 30 ng/min in upright whereas 10 ng/min lying. *3351*

Protein *Serum Increase Physiological* Increase by 0.75 g/dL compared with rest and recumbency. *3873*

Protein Bound Iodine (PBI) *Serum Increase Physiological* Change of up to 0.8 μg/dL observed. *3217*

Renin Activity *Plasma Increase Physiological* Increases approximately 2 times if normal diet. *0071*

Testosterone *Serum Increase Physiological* In women due to decreased metabolic clearance rate. *1941*

Urea Clearance *Urine Decrease Physiological* Decrease observed after rising. *0459*

Volume *Plasma Decrease Physiological* Decreases by about 12% on standing. *1087*

Ergocalciferol

Calcium *Serum Increase Physiological* Enhances absorption from gastrointestinal tract. *0652*
Urine Increase Physiological As result of increased absorption. *0652*

Phosphate *Serum Increase Physiological* Better absorption and utilization. *0652*

Ergothionine

Glucose *Serum No Effect Analytical* At 10 mg/dL no effect on glucose oxidase procedure of Gochman. *1311*
Serum Increase Analytical At 10 mg/dL affects alkaline ferricyanide procedure. *1311*

Uric Acid *Serum Increase Analytical* Falsely high values with phosphotungstate methods. *1311*
Urine Increase Analytical Falsely high values with phosphotungstate methods. *1311*

Urobilinogen *Test Conditions Increase Analytical* Positive spot test with Ehrlich's reagent. *3765*

Ergot Preparations

Aminolevulinic Acid *Urine Increase Physiological* May precipitate attack of acute porphyria. *1016*

Occult Blood *Feces Positive Physiological* May cause gastrointestinal bleeding with overdose. *0279*

Porphyrins *Urine Increase Physiological* May precipitate attack of acute porphyria. *1016*

Protein *Urine Increase Physiological* May cause renal damage with poisoning. *0279*

Urea Nitrogen *Serum Increase Physiological* May cause renal damage with poisoning. *0279*

Erythrocytes

Appearance *Urine Abnormal Physiological* Cloudy, lyse in dilute acetic acid. *0443*

Erythromycin

Alanine Aminotransferase *Serum Increase Analytical* Interferes with colorimetric procedures. *2220*
Serum Increase Physiological May cause hepatic toxicity. *1142* Estolate produces mild hepatotoxicity in about 15% treated individuals. Jaundice in 2% patients on drug for more than 2 weeks. *2792* Observed to some extent in some patients with different erythromycin salts: usually cholestatic hepatitis. *1895*

Alkaline Phosphatase *Serum Increase Physiological* May cause intrahepatic cholestasis (reversible). *1142* Moderate increase with hepatitis mixed in type with both cholestasis and mild necrosis. Ethylsuccinate and propionate derivatives produce similar jaundice. *2792* Observed to some extent in some patients with different erythromycin salts: usually cholestatic hepatitis. *1895*

Amino Acids *Urine Positive Analytical* Yellow spot with ninhydrin on thin-layer chromatography. *1278*

Aspartate Aminotransferase *Serum Increase Analytical* Colorimetric assay if DNPH or diazonium salt used. *3505*
Serum Increase Physiological Causes hepatic toxicity in some cases. *1142* Estolate produces mild hepatotoxicity in about 15% treated individuals. Jaundice in 2% patients on drug for more than 2 weeks. *2792* Observed to some extent in some patients with different erythromycin salts: usually cholestatic hepatitis. *1895*

Erythromycin *(continued)*

Bile *Urine Increase Physiological* Intrahepatic cholestatic jaundice. *0085*

Bilirubin *Serum No Effect Analytical* At concentration of 73 mg/L had no effect on Jendrassik and Grof method. *3393*
Serum Increase Physiological Causes cholestasis in approximately 15% patients. *1142* Estolate produces mild hepatotoxicity in about 15% treated individuals. Jaundice in 2% patients on drug for more than 2 weeks. *2792* Observed to some extent in some patients with different erythromycin salts: usually cholestatic hepatitis. *1895*

Bilirubin, Direct *Serum Increase Physiological* Reported effect. *2313*

BSP Retention *Serum Increase Physiological* Intrahepatic cholestatic jaundice. *0085*

Calcium *Serum No Effect Analytical* At concentration of 73 mg/L had no effect on cresolphthalein method. *3393*

Carbamazepine *Serum Increase Physiological* Clearance reduced from 0.36 to 0.29 L/kg/d with 1 g/d for 5 d, probably due to effect on metabolism. *3893* Interferes with liver microsomal metabolism of drug. *1541*

Catecholamines *Urine Increase Analytical* Interferes with fluorometric methods. *0757*

Cephalin Flocculation *Serum Increase Physiological* Weak effect due to cholestasis. *1713*

Chloramphenicol *Serum No Effect Analytical* No effect at 100 mg/L on coupled enzymatic method. *2490*

Cholesterol *Serum Decrease Physiological* Hepatotoxic effect. *2220*

Cyclosporine *Serum Increase Physiological* One case reported in which marked increase in serum concentration occurred after administration of erythromycin possibly due to inhibition of hepatic clearance. *2881* Because of competition for protein binding sites in serum. *3864*

Digoxin *Serum Increase Physiological* Two fold increase noted in some individuals when antibiotic given orally. In 10% patients but bacteria convert digoxin to cardioinactive reduced metabolites. *2179*

Eosinophils *Blood Increase Physiological* Associated with hypersensitivity reaction. *2324* Hypersensitivity reaction in 45% individuals. *2792*

Folate *Serum No Effect Analytical* If chromatographic procedure of Landon used. *2068*
Serum Decrease Analytical Inhibits growth of L. Casei. *3504*

Glucose *Serum Decrease Physiological* Hepatotoxic effect. *2220*

17-Hydroxycorticosteroids *Urine Increase Analytical* Reported interference with measuring procedure. *1488*

Indomethacin *Serum No Effect Analytical* No effect on HPLC method of Roberts and Smith. *3002*

¹³¹I Uptake *Serum Decrease Physiological* Tetraiodofluorescein in pedimycin. *2652*

17-Ketosteroids *Urine Increase Analytical* Interference with measuring procedure. *1488*

Leukocytes *Blood Increase Physiological* Leukocytosis observed. *2324*
Blood Decrease Physiological Leukopenia or neutropenia may occur. *2324*

Nitrofurantoin *Urine No Effect Analytical* No effect on method of Conklin and Hollifield. *0704*

Phosphate *Serum No Effect Analytical* At concentration of 73 mg/L had no effect on phosphomolybdate method. *3393*

Platelets *Blood Decrease Physiological* Thrombocytopenia reported to occur. *1680*

Potassium *Serum No Effect Analytical* At concentration of 10 mg/L had no effect on measurement by ISE with predilution. *3393*

Protein *Serum No Effect Analytical* At concentration of 73 mg/L had no effect on biuret method with blank correction. *3393*
CSF No Effect Analytical No effect on Folin-Ciocalteu procedure. *3958*

Prothrombin Time *Plasma Increase Physiological* Associated with impaired availability of bile salts. *2220* Increase by 40% in a patient within 3 d of adding drug to a regime stabilized with warfarin: possible effect on metabolism. *1706*

Sodium *Serum No Effect Analytical* At concentration of 10 mg/L had no effect on measurement by ISE with predilution. *3393*

Theophylline *Serum Increase Physiological* Clearance significantly reduced. Plasma concentration increased by 28% in 12 patients. *2952* 40% increase in concentration due to 26% reduction in clearance after 1 week therapy. *2057*

Thymol Turbidity *Serum Increase Physiological* Weak effect due to cholestasis. *1713*

Urea Nitrogen *Serum No Effect Analytical* At concentration of 73 mg/L had no effect on diacetylmonoxime method. *3393*

Uric Acid *Serum No Effect Analytical* At concentration of 73 mg/L had no effect on phosphotungstate reduction method. *3393*

Urobilinogen *Urine Decrease Physiological* May occur with cholestasis. *1713*
Feces Decrease Physiological May cause cholestasis (reversible). *1713*

Warfarin *Plasma Increase Physiological* Clearance decreased by 14% in 12 individuals given 1 g/d for 8 d. *0198*

Erythrosine

Iodide *Serum Increase Physiological* Significant dose related increases in 30 men receiving 20 to 200 mg/d for 15 d. *1241*
Urine Increase Physiological Significant effect of daily doses of 60 mg and higher in 30 men treated for 15 d. *1241*

Protein Bound Iodine (PBI) *Serum Increase Physiological* Significant dose related increases in 30 men receiving 20 to 200 mg/d for 15 d. *1241*

T₃ Uptake *Serum No Effect Physiological* In 30 men receiving up to 200 mg daily for 15 d. *1241*

Thyroid Stimulating Hormone (TSH)
Serum Increase Physiological Significant increase from 1.7 to 2.2 μU/mL after 15 d on up to 200 mg/d in 30 men. *1241*

Thyroxine (T₄) *Serum No Effect Physiological* In 30 men receiving up to 200 mg daily for 15 d. *1241*

Tri-iodothyronine (T₃) *Serum No Effect Physiological* In 30 men receiving up to 200 mg daily for 15 d. *1241*

Tri-iodothyronine (T₃), Reverse
Serum No Effect Physiological In 30 men receiving up to 200 mg daily for 15 d. *1241*

TSH Response to TRH *Serum Increase Physiological* Increase from peak increment of 6.3 to 10.5 μU/mL after 15 d on 200 mg/d in 30 men but no significant effect at lower doses. *1241*

Estradiol

Alanine Aminotransferase *Serum Increase Physiological* Cholestatic effect. *2313*

Aldosterone *Urine No Effect Analytical* No significant effect in RIA procedure of Drewes. *0952*

Alkaline Phosphatase *Serum Increase Physiological* Occurs in up to 50% cases. *2508*

Aspartate Aminotransferase
Serum Increase Physiological Cholestatic effect. *2313*

Bile *Urine Increase Physiological* Cholestatic effect. *2313*

Bilirubin *Serum Increase Physiological* Cholestasis may occur. *2313*

BSP Retention *Serum Increase Physiological* Cholestatic effect in almost all subjects. *2508*

Calcium *Serum Increase Physiological* May occur with high doses. *1680*

Cephalin Flocculation *Serum Increase Physiological* Cholestatic effect. *2313*

Cholesterol *Serum Decrease Physiological* In 8 post-menopausal, post-oophorectomy women after 100 mg implant in subcutaneous tissue. *3274* By about 5% with 2 mg estradiol valerate/d in 20 normolipoproteinemic women over 3 mo. *3593*

Cholesterol, High Density Lipoprotein
Serum No Effect Physiological Insignificant increase with 2 mg estradiol valerate/d in 20 normolipoproteinemic women over 3 mo. *3593*
Serum Increase Physiological In 8 post-menopausal, post-oophorectomy women after 100 mg implant in subcutaneous tissue. *3274* With nonalkylated estrogens, eg valerate salt, in treatment of postmenopausal hormone deficiency. *1037*

Cholesterol, Low Density Lipoprotein
Serum No Effect Physiological Insignificant decrease with 2 mg estradiol valerate/d in 20 normolipoproteinemic women over 3 mo. *3593*
Serum Decrease Physiological In 8 post-menopausal, post-oophorectomy women after 100 mg implant in subcutaneous tissue. *3274* With nonalkylated estrogens, eg valerate salt, in treatment of postmenopausal hormone deficiency. *1037*

Cholesterol, Very Low Density Lipoprotein
Serum No Effect Physiological In 8 post-menopausal, post-oophorectomy women after 100 mg implant in subcutaneous tissue. *3274* With nonalkylated estrogens, eg valerate salt, in treatment of postmenopausal hormone deficiency. *1037*
Serum Decrease Physiological By more than 50% with 2 mg estradiol valerate/d in 20 normolipoproteinemic women over 3 mo. *3593*

Dexamethasone *Serum Increase Analytical* Very slight cross reactivity procedure of Hichens. *1585*

Effective Renal Plasma Flow
Patient No Effect Physiological No effect observed in humans. *2590*

Estradiol *Plasma Increase Physiological* Marked effect in 8 post-menopausal, post-oophorectomy. *3274*

Follicle Stimulating Hormone (FSH)
Plasma Decrease Physiological Relatively slower than LH in response to i.v. *3914* Marked effect in 8 post-menopausal, post-oophorectomy women after 100 mg implant in subcutaneous tissue. *3274*

Glomerular Filtration Rate (GFR)
Urine No Effect Physiological No effect observed in humans. *2590*

Hydroxyproline *Urine Decrease Physiological* Due to anti-catabolic action. *3644*

Luteinizing Hormone (LH) *Plasma Decrease Physiological* In response to i.v. infusion for short duration. *3914*

Melatonin *Plasma Decrease Physiological* In one male given valerate derivative 10 μg/kg daily. Episodic pattern decreased. *2772*

5'-Nucleotidase *Serum Increase Physiological* Cholestatic effect. *2313*

Pituitary Gonadotropin *Urine Decrease Physiological* In oophorectomized or postmenopausal women. *3603*

Postheparin Hepatic Lipase
Plasma No Effect Physiological No significant change after 2 mg estradiol valerate/d in 20 normolipoproteinemic women over 3 mo. *3593*

Postheparin Lipoprotein Lipase
Plasma Decrease Physiological Increase by 25% with 2 mg estradiol valerate/d in 20 normolipoproteinemic women over 3 mo. *3593*

Protein Bound Iodine (PBI) *Serum Increase Physiological* Increases concentration of circulating thyroxine binding globulin. *0012*

Thymol Turbidity *Serum Increase Physiological* Cholestatic effect. *2313*

Thyroxine (T₄) Binding Globulin
Serum Increase Physiological Direct effect of drug. *0012*

Triglycerides *Serum No Effect Physiological* In 8 post-menopausal, post-oophorectomy women after 100 mg implant in subcutaneous tissue. *3274* With nonalkylated estrogens, eg valerate salt, in treatment of postmenopausal hormone deficiency. *1037*
Serum Increase Physiological By about 8% (insignificant) with 2 mg estradiol valerate/d in 20 normolipoproteinemic women over 3 mo. *3593*

Triglycerides, High Density Lipoprotein
Serum Increase Physiological By about 25% with 2 mg estradiol valerate/d in 20 normolipoproteinemic women over 3 mo. *3593*

Triglycerides, Low Density Lipoprotein
Serum No Effect Physiological Insignificant increase with 2 mg estradiol valerate/d in 20 normolipoproteinemic women over 3 mo. *3593*

Triglycerides, Very Low Density Lipoprotein
Serum No Effect Physiological Insignificant increase with 2 mg estradiol valerate/d in 20 normolipoproteinemic women over 3 mo. *3593*

Estradiol-17β

Alkaline Phosphatase *Serum Decrease Physiological* In post-menopausal women given different amounts of drug. *0654*

Cholesterol *Serum Decrease Physiological* Almost 10% reduction in 38 healthy post-menopausal women given 2-4 mg orally or 3 mg cutaneously over 6 mo. *1067*

Cholesterol, High Density Lipoprotein
Serum Increase Physiological In 14 oophorectomized women over 6 mo when 50 mg drug given as implant. *1080* By up to 20% in 38 healthy post-menopausal women given 2-4 mg orally or 3 mg cutaneously over 6 mo. *1067*

Cholesterol, Low Density Lipoprotein
Serum Decrease Physiological In 14 oophorectomized women over 6 mo when 50 mg drug given as implant. *1080* Marked reduction in 38 healthy post-menopausal women given 2-4 mg orally or 3 mg cutaneously over 6 mo. *1067*

Cholesterol, Very Low Density Lipoprotein
Serum No Effect Physiological Insignificant change in 38 healthy post-menopausal women given 2-4 mg orally or 3 mg cutaneously over 6 mo. *1067*

Phospholipids, High Density Lipoprotein
Serum Increase Physiological By about 10% in 38 healthy postmenopausal women given 2-4 mg orally or 3 mg subcutaneously over 6 mo. *1067*

Phospholipids, Low Density Lipoprotein
Serum Decrease Physiological By about 12% in 38 healthy postmenopausal women given 2-4 mg orally or 3 mg subcutaneously over 6 mo. *1067*

Phospholipids, Total *Serum No Effect Physiological* Insignificant change in 38 healthy post-menopausal women given 2-4 mg orally or 3 mg cutaneously over 6 mo. *1067*

Triglycerides *Serum Increase Physiological* Slight increase in 38 healthy post-menopausal women given 2-4 mg orally or 3 mg cutaneously over 6 mo. *1067*

Triglycerides, Very Low Density Lipoprotein
Serum No Effect Physiological Insignificant change in 38 healthy postmenopausal women given 2-4 mg orally or 3 mg cutaneously over 6 mo. *1067*

Estriol

Alkaline Phosphatase *Serum Increase Physiological* Possibly related to impaired hepatic excretion. *1498*

Hydroxyproline *Urine Decrease Physiological* Due to anti-catabolic action. *3644*

Estrogens

α_1-Acid Glycoprotein *Serum Decrease Physiological* Altered metabolism. *0473*

Alanine Aminotransferase *Serum Increase Physiological* May cause cholestasis, rare hepatocellular degeneration. *2220*

Albumin *Serum Decrease Physiological* Metabolic changes in liver synthesis. *2189*

Alkaline Phosphatase *Serum Increase Physiological* Transient increases 2 to 50% reported. *2086*

Amino-4-Imidazole-5-Carboxamide Ribotide (AICAR)
Urine Increase Physiological Occurs with megaloblastic anemia. *3773*

Aminolevulinic Acid *Urine Increase Physiological* May precipitate porphyria attack. *1322*

Angiotensin II *Plasma Increase Physiological* Occurs within 5 d, increase may be 3 times normal. *0557*

Antithrombin III *Plasma Decrease Physiological* ?predisposing cause of thrombosis. *3961*

Estrogens *(continued)*

α₁-Antitrypsin *Serum Increase Physiological* Metabolic effect. *0227*

Ascorbic Acid *Plasma Decrease Physiological* Effect marked at 2 weeks, < effect in platelets. *1847*
Urine Decrease Physiological Significant effect (50%) on comparable diets. *1507*

Aspartate Aminotransferase
Serum Increase Physiological May cause cholestasis, rare hepatocellular degeneration. *2220*

Bile *Urine Increase Physiological* Due to cholestasis. *2313*

Bilirubin *Serum Increase Physiological* May cause cholestasis (may be hepatocellular degeneration). *2313*

BSP Retention *Serum Increase Physiological* Cholestatic effect in most cases (dose related). *2508*

Calcium *Serum Increase Physiological* Positive effect on calcium retention. *1343*

Catechol-O-Methyl Transferase
Red Blood Cells Increase Physiological Observed effect of estrogen contraceptives. *1401*

Cephalin Flocculation *Serum Increase Physiological* May occur with cholestasis. *2220*

Ceruloplasmin *Serum Increase Physiological* But no change in ceruloplasmin activity. *2386*

Cervical Secretion *Patient Increase Physiological* Leukorrhea in 20% women on oral contraceptives. *2349*

Chloride *Serum Increase Physiological* May cause salt and water retention. *1343*

Cholesterol *Serum Decrease Physiological* Reduces by up to 18%: used therapeutically. *2508* Increase by 10-13% in post-menopausal women treated for 3 y. *0653*

Cholesterol, Low Density Lipoprotein
Serum Decrease Physiological In post-menopausal women 11-19% reduction compared with women not taking drug. *3733* In post-menopausal women treated for 3 y. *0653*

Copper *Serum Increase Physiological* Associated with increased ceruloplasmin. *1847*

Coproporphyrin *Blood Increase Physiological* May precipitate porphyria attack. *1322*
Urine Increase Physiological May induce porphyria cutanea tarda. *3769*
Feces Increase Physiological May induce porphyria cutanea tarda. *3769*

Corticosteroid-Binding Globulin
Serum Increase Physiological Metabolic effect. *0227*

Cortisol *Plasma Increase Physiological* Increases concentration of binding globulin. *1488*

Cortisol, Free *Plasma Increase Physiological* Effect like glucocorticoids (?physiological sign). *1849*
Urine No Effect Physiological Unaffected if low dose pills. *0537*
Urine Increase Physiological If high dose for cancer, affects diurnal rhythm. *0537*

Cortisol, Protein Bound *Plasma Increase Physiological* Hormonal action. *1488*

C-Reactive Protein *Serum Increase Physiological* Altered liver metabolism. *3332*

Diphenylhydantoin *Serum Increase Physiological* Reported to cause increased plasma levels. *2041*

Electrophoretic Index *Serum Increase Physiological* Increases binding capacity of thyroxine binding globulin. *3871*

Erythrocytes *Blood Decrease Physiological* Megaloblastic anemia (impairs absorption of folate). *3773*

Estrogen Binding Globulin *Serum Increase Physiological* Altered liver metabolism. *3332*

Factor II *Plasma Increase Physiological* Reported effect. *0429*

Factor VII *Plasma Increase Physiological* Altered protein metabolism. *3332*

Factor IX *Plasma Increase Physiological* Altered protein metabolism. *3332*

Factor X *Plasma Increase Physiological* Slight effect observed. *0585*

Fibrinogen *Plasma Increase Physiological* Altered liver metabolism. *3332*

FIGLU (N-Formiminoglutamic Acid)
Urine Increase Physiological Occurs with megaloblastic anemia. *3773*

Folate *Serum Decrease Physiological* May impair absorption. *3773*

Follicle Stimulating Hormone (FSH)
Plasma Decrease Physiological Hormonal effect. *3237*

FPN Test *Urine Increase Analytical* Reported false positive in patients with liver dysfunction. *1204*

GC Globulin *Serum Increase Physiological* Altered metabolism. *0227*

Glucose *Serum Increase Physiological* Metabolic effect. *1680*

Glucose Tolerance *Serum Decrease Physiological* Effect variable depending on strength. *1488* In post-menopausal women on hormone replacement therapy, statistically significant at 30, 60, 90 minutes, but not when fasting and at 120 minutes abnormal in 19% individuals. *3492*

β-Glucuronidase *Serum Increase Physiological* Metabolic effect. *0227*

γ-Glutamyltransferase (GGT)
Serum No Effect Physiological No metabolic effect. *0227*

Growth Hormone *Plasma Increase Physiological* Modulators of high secretion in normals. *2414*

Haptoglobin *Serum Decrease Physiological* Altered liver metabolism. *3332*

Hematocrit *Blood Decrease Physiological* May impair absorption of folate. *3773*

Hemoglobin *Blood Decrease Physiological* Impairs absorption of folate. *3773*

3-Hydroxyanthranilic Acid *Urine Increase Physiological* Induce tryptophan pyrrolase. *3045*

17-Hydroxycorticosteroids *Plasma Increase Physiological* Causes stimulation of pituitary adrenal axis. *3752*
Urine Decrease Analytical Interferes with Zimmermann reactions. *2425*
Urine Decrease Physiological With chronic ingestion cortisol= 6-beta-hydroxy cortisol. *1884*

6-β-Hydroxycortisol *Urine Increase Physiological* Alteration of steroid excretory pattern (long term). *0766*

3-Hydroxykynurenine *Urine Increase Physiological* Induce tryptophan pyrrolase. *3045*

Immunoglobulins *Serum No Effect Physiological* No metabolic effect. *0227*
Serum Decrease Physiological Altered liver metabolism. *3332*

Iron *Serum Increase Physiological* Usually cause effect. *0121*

Iron-Binding Capacity, Total (TIBC)
Serum Increase Physiological Usual effect due to increased carrier protein. *0121*

¹³¹I Uptake *Serum No Effect Physiological* Have no effect on uptake. *1915*
Serum Increase Physiological Increased uptake reported. *2220*
Serum Decrease Physiological SK-estrogens contain tetraiodofluorescein. *2652*

17-Ketosteroids *Urine Decrease Analytical* Interferes with Zimmermann reactions. *1237*

Kynurenine *Urine Increase Physiological* Induce tryptophan pyrrolase. *3045*

Leucine Aminopeptidase *Serum No Effect Physiological* No effect observed. *1498*
Serum Increase Physiological Altered metabolism. *0227*

Lipoprotein Lipase *Serum Decrease Physiological* Metabolic effect. *0227*

Lipoproteins *Serum Increase Physiological* If very low density. *3332*

α-Lipoproteins *Serum Increase Physiological* Metabolic effect (almost doubled in some studies). *0227*

β-Lipoproteins *Serum Decrease Physiological* Metabolic effect (decrease of 30%). *0227*

Lipoproteins, Pre-β *Serum Increase Physiological* Reported effect. *1343*

α₂-Macroglobulin *Serum Increase Physiological* Metabolic effect. *0227*

MCV *Blood Increase Physiological* Megaloblastic anemia. *3773*

Naphthylamidase Isoenzymes
Serum Increase Physiological Metabolic effect. *0227*

NBT Test *Blood No Effect Physiological* When administered alone no effect observed. *0163*

5'-Nucleotidase *Serum Increase Physiological* May cause cholestasis. *2313*

Phospholipids, Total *Serum Increase Physiological* Altered metabolism. *3332*

Plasminogen *Plasma Increase Physiological* Altered metabolism. *3332*

Platelet Aggregation *Blood Decrease Physiological* Reported to be modified by administration. *0619*

Platelets *Blood Decrease Physiological* Thrombocytopenia may occur after some weeks. *1436*

Porphobilinogen *Urine Increase Physiological* May precipitate porphyria attack. *1322*

Porphyrins *Urine Increase Physiological* May precipitate attack of acute porphyria. *1016*

Prealbumin *Serum Decrease Physiological* Metabolic effect. *0227*

Pregnancy Protein *Serum Increase Physiological* Metabolic effect. *0227*

Prolactin *Plasma Increase Physiological* Chronically in men with carcinoma of prostate. *3913*

Protein *Serum Decrease Physiological* Altered metabolism. *3332*

Protein Bound Iodine (PBI) *Serum Increase Physiological* Increases binding capacity of thyroxine binding globulin. *3871*

Protoporphyrin *Blood Increase Physiological* May precipitate porphyria attack. *1322*
Feces Increase Physiological May induce porphyria cutanea tarda. *3769*

Pseudocholinesterase *Serum Decrease Physiological* Altered liver metabolism. *3332*

Pyruvate *Serum Increase Physiological* Action like glucocorticoids. *1849*

Renin Activity *Plasma Increase Physiological* Due to increased substrate. *0557*
Plasma Decrease Physiological Inhibition of circulating level by angiotensin. *0557*

Renin Substrate *Plasma Increase Physiological* Hormonal action. *0557*

Sex Hormone Binding Globulin
Serum Increase Physiological Metabolic effect. *0227*

Sialic Acid *Serum Decrease Physiological* Altered metabolism. *0227*

Sodium *Serum Increase Physiological* May cause salt and water retention. *1343*

T₃ Uptake *Serum Increase Analytical* As measured by Thyopac method. *0632*
Serum Decrease Physiological Increases binding capacity of thyroxine binding globulin. *3871*

Testosterone *Serum Increase Physiological* Probable effect in women — decreased clearance, increased binding. *1941*

Testosterone Binding Globulin
Serum Increase Physiological Metabolic response. *1498*

Thymol Turbidity *Serum Increase Physiological* Occurs with cholestasis. *2220*

Thyro-Binding Index *Serum Decrease Physiological* Increases binding capacity of thyroxine binding globulin for up to 1 mo. *3871*

Thyroid Stimulating Hormone (TSH)
Serum No Effect Physiological High doses did not augment TRH-mediated response in women: in men slight effect only with high doses without influence on basal TSH. *3798*

Thyroxine (T₄) *Serum Increase Physiological* Increases binding capacity of thyroxine binding globulin for up to 1 mo. *0354* Increased binding capacity (augmentation of transport proteins). *3798*

Urine No Effect Physiological No effect observed but increase on cessation. *0630*

Thyroxine (T₄) Binding Globulin
Serum Increase Physiological Increases binding capacity of thyroxine binding globulin. *3871*

Thyroxine (T₄), Free *Serum Increase Physiological* Increases binding capacity of thyroxine binding globulin. *3871*

Thyroxine (T₄) Index, Free (FTI)
Serum Increase Physiological Increased binding capacity of thyroxine binding globulin for up to 1 mo. *3871*

Thyroxine (T₄) (Murphy-Pattee)
Serum Increase Physiological Increases binding capacity of thyroxine binding globulin. *3871*

Transcortin *Serum Increase Physiological* Metabolic response (dose related effect). *1498*

Transferrin *Serum No Effect Physiological* No effect observed. *1498*
Serum Increase Physiological Altered liver synthesis. *2386*

Triglycerides *Serum No Effect Physiological* In postmenopausal women treated for 3 y. *0653*
Serum Increase Physiological Associated with increased very low density lipoproteins. *1535*

Tri-iodothyronine (T₃) *Serum Increase Physiological* Increased binding capacity (augmentation of transport proteins). *3798*

Trypsin Inhibitor *Serum Increase Physiological* Metabolic effect. *0227*

Uric Acid *Serum Decrease Physiological* In men on therapy (hormonal effect). *2590*
Urine Increase Physiological In men on therapy (hormonal effect). *2590*

Uric Acid Clearance *Urine Increase Physiological* In men on therapy (hormonal effect). *2590*

Uroporphyrin *Urine Increase Physiological* May induce porphyria cutanea tarda. *3769*
Feces Increase Physiological May induce porphyria cutanea tarda. *3769*

Volume *Plasma Increase Physiological* Retention of salt and water common. *2427*

Xanthurenic Acid *Urine Increase Physiological* Induce tryptophan pyrrolase. *3045*

Zinc *Serum Increase Physiological* Metabolic effect. *0475*

Estrogens, Conjugated

Coproporphyrin *Urine Increase Physiological* May induce porphyria cutanea tarda. *3821*
Feces Increase Physiological May induce porphyria cutanea tarda. *3821*

Factor V *Plasma Increase Physiological* Small increases reported. *0083*

Factor IX *Plasma Increase Physiological* Small increases reported. *0083*

Fibrinogen *Plasma Increase Physiological* Effect in normal females after i.v. administration. *0083*
Plasma Decrease Physiological Small but significant effect after i.v. in normal males. *0083*

Follicle Stimulating Hormone (FSH)
Urine Decrease Physiological Slight decrease only even with large doses. *3724*

Glucose Tolerance *Serum Decrease Physiological* May provoke mild to moderate deterioration. *1849*

Luteinizing Hormone (LH) *Urine Increase Physiological* Marked initial effect, falls off with large dose. *3724*

NBT Test *Blood No Effect Physiological* No effect observed. *0163*

Partial Thromboplastin Time
Plasma Decrease Physiological Shortening of time in males after i.v. administration. *0083*

Pituitary Gonadotropin *Urine Decrease Physiological* Inhibits pituitary gonadotrophic function. *3603*

Platelet Aggregation *Blood Increase Physiological* Small increases reported. *0083*

Prothrombin Time *Plasma Increase Physiological* In normal women after i.v. administration. *0083*

Estrogens, Conjugated (continued)

Protoporphyrin *Feces Increase Physiological* May induce porphyria cutanea tarda. 3821

Uroporphyrin *Urine Increase Physiological* May induce porphyria cutanea tarda. 3821

Feces Increase Physiological May induce porphyria cutanea tarda. 3821

Estrone

Alanine Aminotransferase *Serum Increase Physiological* May cause cholestasis. 2313

Alkaline Phosphatase *Serum Increase Physiological* Occurs in up to 50% cases. 2508

Aspartate Aminotransferase

Serum Increase Physiological May cause cholestasis. 2313

Bile *Urine Increase Physiological* May cause cholestasis. 2313

Bilirubin *Serum Increase Physiological* May cause cholestasis. 2313

BSP Retention *Serum Increase Physiological* Occurs in almost all cases. 2508

Cephalin Flocculation *Serum Increase Physiological* May cause cholestasis. 2313

Cholesterol *Serum Decrease Physiological* To almost normal values towards end of treatment in 20 women at perimenopause. 2058

Cholesterol, High Density Lipoprotein

Serum Increase Physiological Continuous increase, became significant at 12 mo in 20 women at perimenopause. 2058

Pituitary Gonadotropin *Urine Decrease Physiological* In oophorectomized or postmenopausal women. 3603

Protein Bound Iodine (PBI) *Serum Increase Physiological* Increases concentration of circulating thyroxine binding globulin. 0012

Thymol Turbidity *Serum Increase Physiological* May cause cholestasis. 2313

Thyroxine (T$_4$) Binding Globulin

Serum Increase Physiological Direct effect of drug. 0012

Triglycerides *Serum Increase Physiological* After 3 mo in 20 women in the perimenopause, then gradual decline to almost. 2058

Ethacrynic Acid

Alanine Aminotransferase *Serum Increase Physiological* Cholestasis with hepatocellular damage. 0812

Albumin *Serum No Effect Analytical* At 2.5 mg/dL no effect on SMA 12/60 method. 2636

Alkaline Phosphatase *Serum No Effect Analytical* At 2.5 mg/dL no effect on SMA 12/60 method. 2636

Ammonia *Plasma Increase Physiological* ?due to hypokalemia and alkalosis. 1488

Amylase *Serum Increase Physiological* Isolated case of acute pancreatitis. 1488

Aspartate Aminotransferase *Serum No Effect Analytical* At 2.5 mg/dL no effect on SMA 12/60 method. 2636

Serum Increase Physiological Cholestasis with hepatocellular damage. 0812

Bicarbonate *Serum Increase Physiological* Associated with hypochloremic alkalosis. 3214

Bilirubin *Serum No Effect Analytical* At 2.5 mg/dL no effect on SMA 12/60 method. 2636

Serum No Effect Physiological Although *in vitro* effect observed no significant effect at pharmacological concentrations. 3748

Serum Increase Physiological Cholestasis or hepatocellular damage. 3214

Calcium *Serum No Effect Analytical* At 2.5 mg/dL no effect on SMA 12/60 method. 2636

Urine Increase Physiological Impaired reabsorption. 2745

Casts *Urine Increase Physiological* Hyaline casts without proteinuria (orosomucoid). 1720

Chloride *Serum Decrease Physiological* Diuretic action (inhibits tubular reabsorption). 2745

Urine Increase Physiological Present as major anion. 1343

Cholesterol *Serum No Effect Analytical* At 2.5 mg/dL no effect on SMA 12/60 method. 2636

Cortisol *Plasma No Effect Physiological* No effect observed with therapy. 2427

Urine Decrease Analytical Decreases unconjugated by fluorometric procedure. 0022

Urine Decrease Physiological ?due to changed secretion or renal handling. 1488

Creatinine *Serum No Effect Analytical* At 2.5 mg/dL no effect on SMA 12/60 method. 2636

Erythrocytes *Urine Increase Physiological* Rare case of hematuria reported. 1680

Glucose *Serum No Effect Analytical* At 2.5 mg/dL no effect on SMA 12/60 method. 2636

Serum Increase Physiological Diabetogenic properties. 3505

Serum Decrease Physiological Symptomatic hypoglycemia reported. 3214

Urine Increase Physiological Diabetogenic properties. 3505

Glucose Tolerance *Serum Decrease Physiological* Diabetogenic-like action of drug. 2745

Insulin *Plasma Decrease Physiological* Reduction in fasting state noted. 2427

Lactate Dehydrogenase *Serum No Effect Analytical* At 2.5 mg/dL no effect on SMA 12/60 method. 2636

Leukocytes *Blood Decrease Physiological* Agranulocytosis. 3747

Magnesium *Urine Increase Physiological* Increase up to seven times reported. 1488

Neutrophils *Blood Decrease Physiological* Occasional neutropenia or agranulocytosis. 2427 Occasional case of agranulocytosis reported. 3717

Occult Blood *Feces Positive Physiological* One case reported (26% if given i.v. possibly). 0363

pH *Blood Increase Physiological* Hypochloremic alkalosis may occur. 3214

Phosphate *Serum No Effect Analytical* At 2.5 mg/dL no effect on SMA 12/60 method. 2636

Urine No Effect Physiological No effect observed on excretion. 3745

Platelets *Blood Decrease Physiological* Thrombocytopenia reported. 2745

Potassium *Serum Decrease Physiological* Diuretic action. 3505

Urine Increase Physiological Marked diuretic response may occur. 1488

Protein *Serum No Effect Analytical* At 2.5 mg/dL no effect on SMA 12/60 method. 2636

Prothrombin Time *Plasma Increase Physiological* Displaces coumarins from albumin. 3248

Sodium *Serum Decrease Physiological* Diuretic action. 2313

Urine Increase Physiological Therapeutic intent. 3214

Urea Nitrogen *Serum Increase Physiological* May cause deterioration if renal function impaired. 0085

Uric Acid *Serum No Effect Analytical* At 2.5 mg/dL no effect on SMA 12/60 method. 2636

Serum Increase Physiological Decreased urate clearance with low doses. 3505

Serum Decrease Physiological If given i.v. in large doses has uricosuric effect. 3442

Urine Increase Physiological If given i.v. in large doses has uricosuric effect. 3442

Urine Decrease Physiological Decreased urate clearance. 1022

Volume *Plasma Decrease Physiological* Initial diuretic response may cause hypovolemia. 0105

Urine Increase Physiological Therapeutic intent. 3214

Zinc *Urine Increase Physiological* When given i.v. brief effect only. 3439

Ethambutol

Alanine Aminotransferase *Serum Increase Physiological* Decreased liver function reported. 0071

Amphetamine   *Urine   No Effect   Analytical*   No effect at 100 mg/L on method of Rutter. *3097*

Aspartate Aminotransferase
Serum   Increase   Physiological   Decreased liver function reported. *0071*

BSP Retention   *Serum   Increase   Physiological*   Few cases reported. *2427*

Creatinine   *Serum   Increase   Physiological*   Rare case of renal damage reported. *0131*

Creatinine Clearance   *Urine   Decrease   Physiological*   Rare case of renal damage reported. *0131*

Phentolamine Test   *Patient   Positive   Physiological*   False positive, mechanism unknown. *1218*

Phosphate   *Serum   No Effect   Analytical*   On nonprecipitation method of Peynet. *2799*

Urea Nitrogen   *Serum   Increase   Physiological*   Rare renal damage reported. *0131*

Uric Acid   *Serum   Increase   Physiological*   May occur within 24 h, by decreasing clearance. *2852*   In 66% patients with tuberculosis when combined with streptomycin and isoniazid not seen when same drugs given with thioacetazone, during first 60-90 d of treatment. Reverted to normal when ethambutol withdrawn. *1928*   Due to decreased renal clearance, usually occurs by third week of treatment. *2852*
Urine   Decrease   Physiological   Decreases clearance but mechanism not known. *2852*

Ethamivan

pCO₂   *Blood   Decrease   Physiological*   Increased depth of respiration and improved pulmonary ventilation. *0071*

Ethanediol

Ionized Calcium   *Serum   Decrease   Analytical*   At 0.1-1.0 mol/L with Calcium specific electrode. *0540*

Ethanol

Acetaldehyde   *Blood   Increase   Physiological*   Marked increase in some individuals after 0.4 g alcohol/kg in 56 healthy males. *0015*

Acetate   *Blood   Increase   Physiological*   Mean concentration much higher in chronic alcoholics and heavy drinkers than in nonalcoholics and occasional drinkers. *1998*

Acetoacetate   *Serum   Increase   Physiological*   Sustained lipolysis with i.v. administration. *0202*

Acetylcholinesterase
Red Blood Cells   Decrease   Physiological   Observed in the blood of 36 alcoholics in comparison with 41 healthy volunteers *in vitro* studies showed effect in proportion to concentration of ethanol. *1442*   Reduced activity observed in alcoholics: immediate effect observed with in vitro incubation. *1442*   Mean activity in cells of alcoholics significantly less than in normals. Same effect observed with *in vitro* studies — inhibition in proportion to ethanol concentration with alcohol concentrations at those found in clinically drunk humans. *1442*

Acid Phosphatase   *Serum   Decrease   Analytical*   Inhibits prostatic component. *3587*

Alanine   *Plasma   Decrease   Physiological*   Causes conversion to lactate. *2013*

Alanine Aminopeptidase   *Serum   No Effect   Physiological*   Alcohol consumption did not correlate with activity in men or women. *2509*

Alanine Aminotransferase   *Serum   No Effect   Analytical*   At acute overdose concentration (20 mg/dL) on Technicon® SMAC® methods. *3719*
Serum   No Effect   Physiological   No effect over 110 h following ingestion of 0.75 g/kg body weight on 3 consecutive evenings. *2133*   No significant difference in specimens from volunteers and values compared with control values for up to 110 h, although component/protein ration decreased in all cases. *2134*
Serum   Increase   Physiological   Increased in normals after 1 dose 3 g/kg body weight. *1621*   In 43 to 53% of male alcoholics in various studies with higher proportion with abnormal AST. *3124*

Albumin   *Serum   No Effect   Analytical*   At acute overdose concentration (20 mg/dL) on Technicon® SMAC® methods. *3719*
Serum   Decrease   Physiological   0.5% increase in heavy versus occasional drinkers. *3273*

Alcohol   *Breath   Increase   Physiological*   Presence of ingested alcohol. *1343*

Aldehyde Dehydrogenase
Red Blood Cells   Increase   Physiological   With acute oral administration in normal subjects but reduction seen in chronic. *1575*
Red Blood Cells   Decrease   Physiological   Mean decrease to 4.98 mIU/mg protein in chronic alcoholics from 8.25 mIU/mg in controls: unrelated to degree of alcohol ingestion or extent of liver damage. *2175*

Aldolase   *Serum   Increase   Physiological*   Dramatic increase of LD-1 and LD-2 observed in acute alcoholic myopathy and less marked increase with chronic myopathy. *3124*

Alkaline Phosphatase   *Serum   No Effect   Analytical*   At acute overdose concentration (20 mg/dL) on Technicon® SMAC® methods. *3719*
Serum   No Effect   Physiological   No significant difference between heavy and occasional drinkers. *3273*   Regardless of amount consumed, acute ingestion had no effect on activity. *0890*
Serum   Increase   Physiological   Ratio of GGT to alkaline phosphatase increased in patients with alcoholic liver disease. *2059*   Increased in alcoholic liver disease. *0625*

Amino-4-Imidazole-5-Carboxamide Ribotide (AICAR)
Urine   Increase   Physiological   May occur with folic acid or B₁₂ deficiency. *3773*

β-Amino-Isobutyric Acid   *Plasma   Increase   Physiological*   Due to increased hepatic production and release into circulation. *2428*

Aminolevulinic Acid   *Urine   Increase   Physiological*   May precipitate attack of porphyria. *1322*

Aminolevulinic Acid Dehydrase
Serum   Decrease   Physiological   Falls and returns to normal inversely with alcohol concentration when taken acutely; tends to be low in chronic alcoholics. *3124*

Ammonia   *Plasma   Increase   Physiological*   Dose related significant effect. *2924*

Amylase   *Serum   Increase   Physiological*   Due to stimulation of pancreatic secretion. *1238*

Androstenedione   *Plasma   No Effect   Physiological*   No significant effect on concentration in individuals receiving alcohol over several days: no effect on pattern of secretion in healthy men. *1352*

Antidiuretic Hormone   *Plasma   Decrease   Physiological*   Initial fall after ingestion of 75 mL although rose later. *1007*

α₁-Antitrypsin   *Serum   Increase   Physiological*   Heterozygotes with alcoholic liver disease have higher concentration than usual mean for healthy PI MZ individuals. *3001*

Apolipoprotein AI   *Serum   Increase   Physiological*   Increase by 6.5% in 78 intemperate drinkers on average. *2882*   Significantly higher in drinkers and falls with abstinence, but increases with resumption of moderate drinking. *0565*   In normal volunteers after 60-70 g/d for 2 weeks but concentration did not exceed normal range. However turnover substantially increased. *2282*   Initially high values at end of alcoholic debauch fell to normal with cessation of drinking. *1011*
Serum   Decrease   Physiological   Significantly lower in alcoholic men than in controls. *1701*

Apolipoprotein AII   *Serum   Increase   Physiological*   Increase by 45% in 78 intemperate drinkers on average. *2882*   Significantly higher in drinkers and falls with abstinence, but increases with resumption of moderate drinking. *0565*   Significantly higher in alcoholic women than in controls. *1701*

Apolipoprotein B   *Serum   Increase   Physiological*   In normal volunteers after 60-70 g/d for 2 weeks but concentration did not exceed normal range. However turnover substantially increased. *2282*

Arylsulfatase A   *Serum   Increase   Physiological*   Activity more than double in acute alcoholism. *1265*

Ascorbic Acid   *White Blood Cells   Decrease   Physiological*   In 35 patients with alcohol related illness. *0207*

Aspartate Aminotransferase   *Serum   No Effect   Analytical*   At acute overdose concentration (20 mg/dL) on Technicon® SMAC® methods. *3719*

Ethanol (continued)

Aspartate Aminotransferase (continued)

Serum No Effect Physiological No effect over 110 h following ingestion of 0.75 g/kg body weight on 3 consecutive evenings. *2133* No significant difference in specimens from volunteers and values compared with control values for up to 110 h, although component/protein ratio decreased in all cases. *2134* Regardless of amount consumed, acute ingestion had no effect on activity. *0890*

Serum Increase Physiological Increased in normals after 1 dose 3 g/kg body weight. *1621* 17% increase in heavy versus occasional drinkers. *3273* Observed in different studies in from 18-100% chronic alcoholics or heavy drinkers. *3124* Sensitive indicator of continued alcohol abuse in individuals with known liver disease. *1040* Increased in alcoholic liver disease. *0625*

Red Blood Cells No Effect Physiological Activation by pyridoxine in 35 patients with alcohol related illness. *0207*

Bicarbonate
Serum Decrease Physiological Causes lactic acidosis. *2667*

Pancreatic Juice Decrease Physiological Oral or i.v. affects concentration and output. *2504*

Bilirubin
Serum No Effect Analytical At acute overdose concentration (20 mg/dL) on Technicon® SMAC® methods. *3719*

Serum No Effect Physiological No significant difference between heavy and occasional drinkers. *3273* Clinically insignificant displacement from protein in neonates. *3748*

Serum Increase Physiological May cause centrolobular necrosis of liver. *1948* Increased in alcoholic liver disease. *0625*

Serum Decrease Physiological Decreases in infant if given to pregnant woman. *2160*

Bleeding Time
Patient Increase Physiological Observed after i.v. ethanol in normals. *1528* Potentiation of aspirin induced increased bleeding time in all patients but to variable extent. *0895*

Calcifediol
Serum No Effect Physiological In 8 men who had abused alcohol for at least 10 y. *0347*

Calcium
Serum No Effect Analytical At acute overdose concentration (20 mg/dL) on Technicon® SMAC® methods. *3719*

Serum Increase Physiological 0.6% increase in heavy versus occasional drinkers. *3273*

Serum Decrease Physiological Secondary hyperparathyroidism. Decreased intestinal absorption and reduced bone mass. *1246* At lower limit of normal in 8 men who had abused alcohol for at least 10 y. *0347* Concentration in chronic alcoholics significantly less than in controls. *0367*

Duodenal Contents Increase Physiological If given orally but decreased if i.v. *2504*

Duodenal Contents Decrease Physiological If given intravenously only. *2504*

Catecholamines
Plasma Increase Physiological Slight increase following moderate doses. *0143*

Urine Increase Physiological Produces stress like response. *0143*

Cholesterol
Serum No Effect Analytical At acute overdose concentration (20 mg/dL) on Technicon® SMAC® methods. *3719*

Serum No Effect Physiological No significant difference between heavy and occasional drinkers. *3273* Usually remain unchanged in response to acute ingestion. *2428* No effect of ingestion of 1.5 g/kg at night on concentration measured next morning. *3547*

Serum Increase Physiological Significant increase in heavy drinkers vs nondrinking or occasional drinkers. *0059* Significantly higher in alcoholic men than in controls. *1701*

Serum Decrease Physiological Occurs when cirrhosis develops in alcoholism. *1621* Low values associated with debauch gradually increased to normal. *1011*

Cholesterol, High Density Lipoprotein
Serum No Effect Physiological Not significantly different in alcoholics from nonalcoholic controls. *1701* No effect of ingestion of 1.5 g/kg at night on concentration measured next morning. morning. *3547*

Serum Increase Physiological Increase by 25% in first 2 weeks of ingestion of 30 g alcohol daily, although gradually reverted to normal after another 2 weeks (effect observed in previously non-alcohol drinking healthy young males). *0328* 17% increase in heavy versus occasional drinkers. *3273* Significant increase in heavy drinkers versus nondrinkers or occasional drinker. *0059* Increase by 21% in 78 intemperate drinkers on average. *2882* Possibly due to induction of microsomal

enzymes. *2428* Concentration doubled immediately after debauch compared with 10 d later or control subjects. *1011*

Cholesterol, Low Density Lipoprotein
Serum No Effect Physiological No effect of ingestion of 1.5 g/kg at night on concentration measured next morning. morning. *3547*

Serum Decrease Physiological Associated with alcohol consumption in fasting subjects. *2428* Low values associated with debauch gradually increased to normal. *1011*

Cholesterol, Very Low Density Lipoprotein
Serum No Effect Physiological No effect of ingestion of 1.5 g/kg at night on concentration measured next morning. *3547*

Chylomicrons
Serum Increase Physiological Due to specific disturbance of the clearing mechanism of triglyceride-rich lipoproteins. *3194*

Chymotrypsin
Pancreatic Juice Decrease Physiological Oral or i.v. cause direct inhibition. *2504*

Color
Urine Decrease Analytical Diuresis induced reduces color. *1488*

Coproporphyrin
Blood Increase Physiological May precipitate attack of porphyria. *1322*

Urine Increase Physiological May precipitate attack of porphyria. *1322*

Feces Increase Physiological May precipitate attack of porphyria. *1322*

Cortisol
Plasma No Effect Physiological No effect in men or women at 100 mg/dL. *3611*

Plasma Increase Physiological Effect seen if high doses given i.v. *1789* 12.1 mg/dL in current abusers, 11.0 mg/dL in previous abusers vs 7.7 mg/dL in controls. *3771*

Creatine Kinase
Serum No Effect Analytical At acute overdose concentration (20 mg/dL) on Technicon® SMAC® methods. *3719*

Serum No Effect Physiological No effect over 110 h following ingestion of 0.75 g/kg body weight on 3 consecutive evenings. *2133* No significant difference in specimens from volunteers and values compared with control values for up to 110 h, although component/protein ratio decreases in all cases. *2134*

Serum Increase Physiological Effect noticed in alcoholics after alcohol. *1651*

Serum Decrease Physiological 6 U/L in alcohol drinking versus other women. *2744*

Creatinine
Serum No Effect Analytical At acute overdose concentration (20 mg/dL) on Technicon® SMAC® method. *1817*

Serum Decrease Physiological 2.6% decrease in heavy versus occasional drinkers. *3273*

Cyclic Adenosine Monophosphate
Plasma Increase Physiological At upper limit of normal in 8 men who had abused alcohol for 10 y. *0347*

3,4-Dihydroxyphenylethylene Glycol
Plasma Decrease Physiological Fell immediately after ingestion and remained low for 6 h. *1671*

1,25-Dihydroxy Vitamin D_3
Serum No Effect Physiological In 8 men who had abused alcohol for at least 10 y. *0347*

Diphenylhydantoin
Serum No Effect Physiological No apparent influence when drug coadministered. *0844*

Serum Increase Physiological Reported interaction due to alteration of metabolism. *2042*

Serum Decrease Physiological Half life reduced from 23.5 to 16.3 h. *2041*

Dopamine
Urine No Effect Physiological No change after ingestion after 0.4 g alcohol/kg in 56 healthy males. *0015*

β-Endorphin
Plasma Increase Physiological With severe alcohol abuse in patients with cirrhosis or pancreatitis to 25 pg/mL from 2.5 pg/mL in controls. *3771*

Epinephrine
Plasma Increase Physiological In individuals with unusual aldehyde dehydrogenase after 0.4 g alcohol/kg in 56 healthy males. *0015*

Urine Increase Physiological Slight increase following ingestion. *0143* Large effect observed. *1176* In individuals with unusual aldehyde dehydrogenase after 0.4 g alcohol/kg in 56 healthy males. *0015*

Erythrocytes
Blood Decrease Physiological May affect folic acid absorption and usage. *3773* Increase by 2% decrease in heavy versus occasional drinker. *3273*

Estradiol
Plasma No Effect Physiological No significant effect on concentration in individuals receiving alcohol over several days: no effect on pattern of secretion in healthy men. *1352*

Ethanol *Blood Increase Physiological* 2 g/L = intoxication, stupor at 3 g/L. *2348* Steady fall from 30 minutes after ingestion after 0.4 g alcohol/kg in 56 healthy males. *0015*
Urine Increase Physiological At equilibrium approximately 130% blood concentration. *1343*
CSF Increase Physiological Usually at lower concentration than in blood. *1343*
Red Blood Cells Increase Physiological But lower in RBC than in plasma. *1343*

Fatty Acids, Free (FFA) *Serum Increase Physiological* Altered metabolism with decreased fatty acid oxidation. *1343* But only after high doses. *2428*

Ferritin *Serum Increase Physiological* Increases and decreases in parallel with GGT in 9 alcoholics during drinking and withdrawal. *2423*

FIGLU (N-Formiminoglutamic Acid)
Urine Increase Physiological May occur with folic acid or B_{12} deficiency. *3773*

Folate *Serum Decrease Physiological* Affects absorption if severe alcoholism. *3773* In some alcoholics ?decreased myelopoiesis. *2196* More than half patients with alcoholic cirrhosis had low liver and serum folate, also acutely accelerates deficiency in people with relatively normal folate stores. *2062*

Follicle Stimulating Hormone (FSH)
Plasma No Effect Physiological No consistent effect in healthy men with ingestion over several days. *1352*

Globulin *Serum Increase Physiological* 4% increase in heavy versus occasional drinkers. *3273*

Glucose *Serum No Effect Analytical* At 1 g/dL on p-HBAH procedure of Lever. At 1 g/dL on glucose oxidase procedures. At 1 g/dL on alkaline ferricyanide procedure. At 1 g/dL on o-toluidine procedure. *2140* At acute overdose concentration (20 mg/dL) on Technicon® SMAC® methods. *3719*
Serum No Effect Physiological No change in fasting or postprandial. *1288*
Serum Increase Physiological Transient hyperglycemia in developing intoxication. *1343* 2.2% increase in heavy versus occasional drinkers. *3273*
Serum Decrease Physiological Reduced to below 50 mg/dL: effect for up to 24 h. *0586* In individuals who have depleted glycogen stores after 72 h of fasting. *2428*

Glucose Tolerance *Serum Decrease Physiological* Possible diminished tissue uptake. *0934* In individuals fasted for 12 h in response to acute dose due to decreased peripheral use of glucose with increased glycogenolysis also. *2428*

β-Glucuronidase *Serum Increase Physiological* Marked effect in acute alcoholism. *1265*

Glutamate Dehydrogenase *Serum Increase Physiological* Reflects alcohol induced liver cell necrosis better than AST, ALT and GGT. GDH also increased by recent alcohol consumption reverts rapidly to normal. *3124*

γ-Glutamyltransferase (GGT)
Serum No Effect Physiological Acute consumption has no effect in healthy volunteers or in patients with alcoholic liver disease. *3124* Regardless of amount consumed, acute ingestion had no effect on activity. *0890*
Serum Increase Physiological Seen with other liver functions tests normal in chronic alcoholism. *3043* Observed also in moderate or heavy drinkers. *3042* 68% increase in heavy drinkers vs occasional. *3273* Significant increase in heavy drinkers versus nondrinkers or occasional drinker. *0059* Ratio of GGT to alkaline phosphatase increased in patients with alcoholic liver disease. *2059* Sensitive indicator of continued alcohol abuse in individuals with known liver disease. *1040* Increased in alcoholic liver disease. *0625*

Glutathione Reductase
Red Blood Cells Decrease Physiological Activation by riboflavin in 35 patients with alcohol related illness. *0207*

Glycerol *Serum Increase Physiological* Sustained lipolysis with i.v. administration. *0202*

Growth Hormone *Plasma No Effect Physiological* No effect in men or women at 100 mg/dL. *3611*
Plasma Increase Physiological Transient increase with hypoglycemia and decreased insulin. *0202*

HDL₂-Cholesterol *Serum Increase Physiological* Most of increased HDL-concentration attributable to this protein. *1011*

Hematocrit *Blood Increase Physiological* Increase by 1.5% increase in heavy versus occasional drinker. *3273*

Blood Decrease Physiological May affect folic acid absorption and usage. *3773*

Hemoglobin *Blood Increase Physiological* 1.6% increase in heavy versus occasional drinkers. *3273*
Blood Decrease Physiological May affect folic acid absorption and usage. *3773*

Hepatic Lipase *Serum No Effect Physiological* With acute administration after overnight fast. *2428*
Serum Decrease Physiological With acute administration to fed individuals. *2428*

Hexosaminidase *Serum Increase Physiological* Observed with chronic alcohol consumption. *3124*

Histamine *Urine Increase Physiological* All gastric stimulants produce effect. *0814*

HMPG (4-Hydroxy-3-Methoxy-Phenylethylene Glycol)
Urine Increase Physiological Following ingestion after 0.4 g alcohol/kg in 56 healthy males. *0015*

Hydrochloric Acid *Gastric Material Increase Physiological* Psychically, reflexly and through histamine/gastrin. *1343*

β-Hydroxybutyrate *Serum Increase Physiological* Associated with increased lactate also. *1532*

11-Hydroxycorticosteroids *Plasma No Effect Physiological* No effect seen in chronic alcoholics. *1487*
Plasma Increase Physiological In normals after 1.5 ml/kg. *1487*

5-Hydroxyindoleacetic Acid (5-HIAA)
Urine Decrease Physiological Serotonin metabolism diverted to 5 hydroxy tryptophol. *0833*

5-Hydroxytryptophol *Urine Increase Physiological* Raises excretion from 2% to 50%. *1176*

25-Hydroxy Vitamin D₃ *Serum Decrease Physiological* Of chronic alcoholics 58% had concentration below normal. *0367*
Plasma Decrease Physiological Associated with reduced exposure to sunshine plus inadequate supply of vitamin D. Compound formed in liver where vitamin D binding protein also synthesized. Finding sometimes reported in chronic alcoholics. Normal concentration observed typically in well nourished alcoholics. *1246*

Immunoglobulin IgA *Serum Increase Physiological* In alcoholics continuing to drink. *2065*
Serum Decrease Physiological In alcoholics after 1 y abstinence. *2065*

Insulin *Plasma No Effect Physiological* No change in fasting or postprandial. *1288*
Plasma Increase Physiological In response to 1.5 g/kg at night significantly increased values measured next morning. *3547*
Plasma Decrease Physiological Intravenously caused 65% decrease for 10 mg/dL glucose dec. *3637* Delayed increase seen in all subjects. *3762*

Ionized Calcium *Serum Decrease Analytical* At 0.1 mmol/L to 0.1 mol/L with Calcium specific electrode. *0540*

Iron *Serum No Effect Analytical* At acute overdose concentration (20 mg/dL) on Technicon® SMAC® methods. *3719*
Serum Increase Physiological Enhances absorption from gastrointestinal tract in alcoholics. *1532* From 16 mmol/L to 30 mmol/L at 15 h after 100 mg alcohol in 4 healthy male volunteers. *1223*

Iron-Binding Capacity, Unsaturated (UIBC)
Serum Decrease Physiological Up to 90% increase in saturation in alcoholics. *1532*

Isocitrate Dehydrogenase *Serum Increase Physiological* Occurs within 4 h in normal subjects. *1323*

Isoleucine *Plasma Increase Physiological* Due to increased hepatic production and release in to circulation. *2428*

Ketones *Serum Increase Physiological* May cause marked ketoacidosis. *1291*
Urine Increase Physiological Mainly due to β-hydroxy-butyrate, little acetoacetic. *1291*

Lactate *Serum Increase Physiological* Causes lactic acidosis. *0586* Produced pyruvate with increased NADH/NAD ratio. *2428*

Lactate Dehydrogenase *Serum No Effect Analytical* At acute overdose concentration (20 mg/dL) on Technicon® SMAC® methods. *3719*
Serum Increase Physiological Increased in normals after 1 dose 3 g/kg body weight. *1621* Dramatic increase observed in

Ethanol (continued)

Lactate Dehydrogenase (continued)
acute alcoholic myopathy and less marked increase with chronic myopathy. *3124* Slight effect at 15 and 110 h after 0.75 g/kg ingestion. *2134*

Lactate Dehydrogenase Isoenzymes
Serum Increase Physiological Dramatic increase of LD-1 and LD-2 observed in acute alcoholic myopathy and less marked increase with chronic myopathy. *3124*

Lead *Blood Increase Physiological* 30% increase in heavy versus occasional drinkers. *3273*

Lecithin *Serum Increase Physiological* Moderate increase compared with controls in alcoholics especially in younger adults. *0430*

Leukocytes *Blood Increase Physiological* 3.9% increase in heavy versus occasional drinkers. *3273*
Blood Decrease Physiological In some alcoholics ?decreases myelopoiesis. *2196*

Lipase *Serum Increase Physiological* Chemical or physical pancreatitis. *1237*
Pancreatic Juice Decrease Physiological Oral or i.v. causes direct inhibition. *2504*

Lipids, Total *Serum Decrease Physiological* Occurs when cirrhosis develops in alcoholism. *1621*

Luteinizing Hormone (LH) *Plasma No Effect Physiological* No effect in men or women at 100 mg/dL. *3611*
Plasma Increase Physiological Inconsistent change with ingestion over several days in healthy men. *1352*
Plasma Decrease Physiological Inconsistent change with ingestion over several days in healthy men. *1352*

Lymphocyte T-Cells *Blood No Effect Physiological* Not changed by previous or current abuse. *3771*

Lymphocyte T-Helper Cells *Blood Increase Physiological* But only in patients with alcoholic liver disease or pancreatitis. *3771*

Lymphocyte T-Suppressor Cells
Blood Decrease Physiological But only in patients with alcoholic cirrhosis or pancreatitis. *3771*

Lysolecithin *Serum Increase Physiological* Moderate increase compares with controls in alcoholics especially in younger adults. *3330*

Lysozyme *Serum No Effect Physiological* In some alcoholics ?decreases myelopoiesis. *2196*

Magnesium *Serum Decrease Physiological* Following ethanol induced urinary excretion. *3495* At lower limit of normal in 8 men who had abused alcohol for at least 10 y. *0347*
Urine Increase Physiological Increased excretion seen in chronic alcoholics. *3495*

MCH *Blood Increase Physiological* 3.5% increase in heavy versus occasional drinkers. *3273*

MCHC *Blood No Effect Physiological* No significant difference between heavy and occasional drinkers. *3273*

MCV *Blood Increase Physiological* 3.5% increase in heavy versus occasional drinkers. *3273* By about 1.7 fL for every 10 g alcohol taken daily. *0995* Sensitive indicator of continued alcohol abuse in individuals with known liver disease. *1040* Increased in alcoholic liver disease. *0625*

Methanol *Blood Increase Physiological* Competitive inhibition of alcohol dehydrogenase. *2271*

Methylmalonic Acid *Urine Increase Physiological* May occur if B_{12} deficiency. *3773*

Monoamine Oxidase B *Platelets Decrease Physiological* Marked reduction in alcoholics versus controls (by approximately 38%). *1285*

Myoglobin *Urine Increase Physiological* May be observed in alcoholism. *0017*

Neutrophils *Blood Decrease Physiological* In some alcoholics ?decreased myelopoiesis. *2196*

Nitrogen *Urine Increase Physiological* Result of chronic ingestion with negative nitrogen balance. *2428*

Norepinephrine *Plasma Increase Physiological* In individuals with unusual aldehyde dehydrogenase after 0.4 g alcohol/kg in 56 healthy males. *0015* Begins about 30 minutes after drinking and lasts for 4 h. *1671*

Urine Increase Physiological Slight increase (but less than epinephrine). *0143* In individuals with unusual aldehyde dehydrogenase after 0.4 g alcohol/kg in 56 healthy males. *0015*

Ornithine Carbamoyltransferase (OCT)
Serum Increase Physiological Occurs at 15 h in normal subjects. *1323* Activity correlates with aminotransferases: may be almost as sensitive as AST in detection of alcoholic liver injury. *3124*

Osmolality *Serum Increase Physiological* Osmotic activity of alcohol-minimum from other sources. *3009* Rise precedes rise in vasopressin following ingestion of 75 mL alcohol. *1007*

Parathyroid Hormone *Plasma Increase Physiological* At upper limit of normal in 8 men who had abused alcohol for 10 y. *0347*

Pepsin *Gastric Material Decrease Physiological* Unless major psychic component. *1343*

pH *Blood Decrease Physiological* Causes lactic acidosis. *0568*

Phosphate *Serum No Effect Analytical* At acute overdose concentration (20 mg/dL) on Technicon® SMAC® methods. *3719*
Serum No Effect Physiological In 8 men who had abused alcohol for at least 10 y. *0347*
Serum Increase Physiological 2.8% increase in heavy versus occasional drinkers. *3273*

Phospholipids, High Density Lipoprotein
Serum Increase Physiological Increase by 16% in 78 intemperate drinkers on average. *2882*

Phospholipids, Total *Serum No Effect Physiological* Usually remain unchanged in response to acute ingestion. *2428*
Serum Increase Physiological Moderate increase compared with controls in alcoholics especially in younger adults. *0430*
Serum Decrease Physiological Occurs when cirrhosis develops in alcoholism. *1621*

Platelet Aggregation *Blood No Effect Physiological* No effect observed when ethanol given alone. *0895*
Blood Decrease Physiological Observed after i.v. ethanol in normals. *1528* Highly significant reduction with ADP in 2 h after 100 mg alcohol in 4 healthy volunteers: maximal at 3 h. *1223*

Platelets *Blood No Effect Physiological* No effect observed when ethanol given alone. *0895*
Blood Decrease Physiological Toxic marrow suppression, due to chronic alcoholism. *0568* Decreased lifespan and production in alcoholics. *0739*

Porphobilinogen *Urine Increase Physiological* May precipitate attack of porphyria. *1322*

Porphyrins *Urine Increase Physiological* Increased synthesis of porphyrins. *3879*

Postheparin Lipoprotein Lipase
Plasma No Effect Physiological With acute administration to fed individuals. *2428*
Plasma Increase Physiological Fell by 40% 5 d after cessation of debauch to normal concentration. *1011*
Plasma Decrease Physiological With acute administration after overnight fast. *2428*

Potassium *Serum Increase Physiological* 3% increase in heavy versus occasional drinkers. *3273*
Serum Decrease Physiological Inappropriate secretion of ADH in beer drinkers. *0875* Increase by 0.2 mmol/L 2 h after 0.75 g/kg ingestion. *2134*

Prolactin *Plasma No Effect Physiological* No effect in men or women at 100 mg/dL. *3611*

Protein *Serum No Effect Analytical* At acute overdose concentration (20 mg/dL) on Technicon® SMAC® methods. *3719*

Prothrombin Time *Plasma Increase Physiological* With large quantities and alcoholism. *2313*
Plasma Decrease Physiological May shorten action of anticoagulants. *1679*

Protoporphyrin *Blood Increase Physiological* May precipitate attack of porphyria. *1322*
Feces Increase Physiological May precipitate attack of porphyria. *1322*

Pyridoxal Phosphate *Serum Decrease Physiological* Progressive decrease observed in alcoholics. *1532* Excess consumption affects formation *in vivo*. *2231* In alcoholic patients with or without liver dysfunction. *2428*

Pyruvate Tolerance *Blood Decrease Physiological* In 35 patients with alcohol related illness. *0207*

Retinol Binding Protein *Serum No Effect Physiological* No different in chronic alcoholics and controls. *0367*

Selenium *Serum Decrease Physiological* Mean reduction to 0.065 μg/mL from 0.100 μg/mL in chronic heavy alcohol ingestion versus controls. *0983*

Sodium *Serum Increase Physiological* Slightly higher after heavy beer drinking. *0922* Increase by 2 mmol/L 2 h after ethanol ingestion. *2134*
Serum Decrease Physiological Inappropriate secretion of ADH in beer drinkers. *0875* 0.3% decrease in heavy versus occasional drinkers. *3273*

Testosterone *Serum No Effect Physiological* No effect in men or women at 100 mg/dL. *3611*
Serum Decrease Physiological Observed in male alcoholics. *0780* Reduced number of episodic bursts of secretion and steady decline in with several days of ingestion in healthy men. *1352*

Toluene *Blood Increase Physiological* During exposure if alcohol ingested significant increase in concentration due to competition for alcohol dehydrogenase. *3738*

Transketolase *Red Blood Cells Decrease Physiological* Activation by thiamine in 35 patients with alcohol related illness. *0207*

Triglycerides *Serum No Effect Analytical* At acute overdose concentration (20 mg/dL) on Technicon® SMAC® methods. *3719*
Serum Increase Physiological In fasting hypertriglyceride patients (not normals). *1288* Infusion causes increased hepatic synthesis. *1766* In post-prandial normal and hypertriglyceride patients. *1288* 6.5% increase in heavy versus occasional drinkers. *3273* Significant increase in heavy drinkers versus nondrinkers or occasional drinker. *0059* Usual consequence, but especially with fatty meal. *2428* Significantly higher in alcoholic men than in controls. *1701* In response to 1.5 g/kg at night significantly increased values measured next morning. *3547*
Serum Decrease Physiological Initially low immediately after alcoholic debauch. *1011*

Triglycerides, High Density Lipoprotein
Serum No Effect Physiological No effect of ingestion of 1.5 g/kg at night on concentration measured next morning. *2545*

Triglycerides, Low Density Lipoprotein
Serum Increase Physiological In response to 1.5 g/kg at night significantly increases values measured next morning. *3547*

Triglycerides, Very Low Density Lipoprotein
Serum No Effect Physiological No effect of ingestion of 1.5 g/kg at night on concentration measured next morning. *3547*
Serum Increase Physiological Main fraction affected by acute ingestion. *2428*
Serum Decrease Physiological Initially low immediately after alcoholic debauch. *1011*

TSH Response to TRH *Serum No Effect Physiological* No effect of acute alcohol intake. *3798*

Urea Nitrogen *Serum No Effect Analytical* At acute overdose concentration (20 mg/dL) on Technicon® SMAC® methods. *3719*
Serum Decrease Physiological Increase by 9% decrease in heavy versus occasional drinker. *3273*

Uric Acid *Serum No Effect Analytical* At acute overdose concentration (20 mg/dL) on Technicon® SMAC® methods. *3719*
Serum Increase Physiological Seen in acutely intoxicated and in response to diet. *2166* Increase by 14% increase in heavy versus occasional drinker. *3273* Significant increase in heavy drinkers versus nondrinkers or occasional drinker. *0059* Common effect and transient, usually returns to normal after 1 week: effect due to decreased renal excretion. *2428* Significantly correlated with alcohol abuse. *0956*
Urine Decrease Physiological Impaired excretion during period of intake. *1343*
CSF Increase Physiological Especially high on withdrawal, remains increased in alcoholics. *0590*

Uroporphyrin *Urine Increase Physiological* May precipitate attack of porphyria. *1322*

Vanillylmandelic Acid *Urine Decrease Physiological* Following ingestion after 0.4 g alcohol/kg in 56 healthy males. *0015*

Viscosity *Blood No Effect Physiological* Unaffected by drinking ethanol (33 mmol/kg). *1599*

Blood Increase Physiological Significant increase at 15 h after ingestion of 100 mg in 4 healthy male volunteers. *1223*

Vitamin A *Serum No Effect Physiological* No different in chronic alcoholics and controls. *0367*
Serum Decrease Physiological With alcoholic hepatitis and cirrhosis, at same liver concentration of vitamin reduced. *2428*

Vitamin B₁₂ *Serum No Effect Physiological* In some alcoholics ?decreases myelopoiesis. *2196*
Serum Decrease Physiological Impairs absorption. *3773*

Vitamin B₁₂ (Unsaturated) *Serum Decrease Physiological* In some alcoholics ?decreases myelopoiesis. *2196*

Volume *Urine Increase Physiological* Diuretic action, inhibits pituitary ADH secretion. *1343*
Pancreatic Juice Decrease Physiological Oral or i.v. causes direct inhibition. *2504*

Xylose Excretion *Urine Increase Physiological* Observed in chronic alcoholics after usual dose. *0589*

Zinc *Urine Increase Physiological* In absence of effect on serum concentration. *0586*

Ethanolamine

Ionized Calcium *Serum Decrease Analytical* At concentrations > 0.1 mmol/L on Calcium specific electrode. *0540*

Ethaverine

Alanine Aminotransferase *Serum No Effect Analytical* No effect at therapeutic concentration on Reflotron method. *1984*

Aspartate Aminotransferase *Serum No Effect Analytical* No effect at therapeutic concentration on Reflotron method. *1984*

Cholesterol *Serum No Effect Analytical* No effect at therapeutic concentration on Reflotron method. *1984*

Glucose *Serum No Effect Analytical* No effect at therapeutic concentration on Reflotron method. *1984* At concentration of 400 mg/L had no effect on hexokinase/G-6-PDH method. *3393*

γ-Glutamyltransferase (GGT) *Serum No Effect Analytical* No effect at therapeutic concentration on Reflotron method. *1984*

Lactate *Serum No Effect Analytical* At concentration of 400 mg/L had no effect on enzymatic method. *3393*

Triglycerides *Serum No Effect Analytical* No effect at therapeutic concentration on Reflotron method. *1984*

Urea Nitrogen *Serum No Effect Analytical* No effect at therapeutic concentration on Reflotron method. *1984*

Uric Acid *Serum No Effect Analytical* No effect at therapeutic concentration on Reflotron method. *1984*

Ethchlorvynol

Alanine Aminotransferase *Serum No Effect Analytical* At acute overdose concentration (20 mg/dL) on Technicon® method. *3719*
Serum Increase Physiological Occurs with poisoning, probable muscle origin. *3901*

Albumin *Serum No Effect Analytical* At acute overdose concentration (20 mg/dL) on Technicon® method. *3719* At concentration of 400 mg/L had no effect on BCG method. *3393*

Alkaline Phosphatase *Serum No Effect Analytical* At acute overdose concentration (20 mg/dL) on Technicon® method. *3719*

Alkaloids *Urine No Effect Analytical* With iodoplatinate on Frings TLC procedure at 10 mg/dL. With ninhydrin on Frings TLC procedure at 10 mg/dL. *1204*

Aspartate Aminotransferase *Serum No Effect Analytical* At acute overdose concentration (20 mg/dL) on Technicon® method. *3719*
Serum Increase Physiological Occurs with poisoning, probable muscle origin. *3901*

Barbiturate *Urine No Effect Analytical* With vanillin on Frings procedure at 10 mg/dL. With mercurous NO₃ on Frings TLC procedure at 10 mg/dL. With mercuric SO₄ on Frings TLC procedure at 10 mg/dL. With diphenylcarbazone on Frings TLC procedure at 10 mg/dL. *1204*

Ethchlorvynol (continued)

Bicarbonate *Serum   No Effect   Analytical* At concentration of 400 mg/L had no effect on method using phenolphthalein. *3393*

Bilirubin *Serum   No Effect   Analytical* At acute overdose concentration (20 mg/dL) on Technicon® method. *3719* At concentration of 400 mg/L had no effect on Jendrassik and Grof method. *3393*

Bromide *Urine   No Effect   Analytical* No effect on method of Frings. *1204*

Calcium *Serum   No Effect   Analytical* At acute overdose concentration (20 mg/dL) on Technicon® method. *3719* At concentration of 400 mg/L had no effect on cresolphthalein method. *3393*

Chlordiazepoxide *Blood   No Effect   Analytical* On method of Riddick at 5 mg/dL. *2983*

Chloride *Serum   No Effect   Analytical* At concentration of 400 mg/L had no effect on mercurimetric method. *3393*

Cholesterol *Serum   No Effect   Analytical* At acute overdose concentration (20 mg/dL) on Technicon® method. *3719* At concentration of 400 mg/L had no effect on Liebermann-Burchard method. *3393*

Creatine Kinase *Serum   No Effect   Analytical* At acute overdose concentration (20 mg/dL) on Technicon® method. *3719*
Serum   Increase   Physiological Occurs with poisoning, probable muscle origin. *3901*

Creatinine *Serum   No Effect   Analytical* At acute overdose concentration (20 mg/dL) on Technicon® method. *3719* At concentration of 400 mg/L had no effect on AutoAnalyzer Jaffé method. *3393*

Dapsone *Serum   No Effect   Analytical* At 10 mg/dL on colorimetric procedure of Higgins. *1590*

Estriol *Urine   No Effect   Analytical* No effect on GLC method. *1163*

Ethchlorvynol *Serum   Increase   Physiological* 200 mg orally may produce concentration of 2 mg/L in blood. *2348*
Urine   No Effect   Analytical No effect on method of Frings and Cohen. *1204*
Urine   Increase   Physiological 200 mg orally may produce concentration of 1 mg/L. *2348*

FPN Test *Urine   No Effect   Analytical* No effect at 10 mg/dL on method of Frings. *1204*

Glucose *Serum   No Effect   Analytical* At acute overdose concentration (20 mg/dL) on Technicon® method. *3719*

Iron *Serum   No Effect   Analytical* At acute overdose concentration (20 mg/dL) on Technicon® method. *3719* At concentration of 400 mg/L had no effect on Ferrozine method. *3393*

Lactate Dehydrogenase *Serum   No Effect   Analytical* At acute overdose concentration (20 mg/dL) on Technicon® method. *3719*

Phosphate *Serum   No Effect   Analytical* At acute overdose concentration (20 mg/dL) on Technicon® method. *3719* At concentration of 400 mg/L had no effect on phosphomolybdate method. *3393*

Potassium *Serum   No Effect   Analytical* At concentration of 400 mg/L had no effect on measurement by ISE with predilution. *3393*

Propoxyphene *Serum   No Effect   Analytical* At 5 mg/L on method of Evenson. *1056*

Protein *Serum   No Effect   Analytical* At acute overdose concentration (20 mg/dL) on Technicon® method. *3719* At concentration of 400 mg/L had no effect on biuret method with blank correction. *3393*

Prothrombin Time *Plasma   Decrease   Physiological* Decreased potency of coumarin anticoagulants (enzyme induction). *1802*

Salicylate *Urine   No Effect   Analytical* At 10 mg/dL no effect on Trinder method. *1204*

Sodium *Serum   No Effect   Analytical* At concentration of 400 mg/L had no effect on measurement by ISE with predilution. *3393*

Triglycerides *Serum   No Effect   Analytical* At acute overdose concentration (20 mg/dL) on Technicon® method. *3719* At concentration of 400 mg/L had no effect on lipase/esterase method. *3393*

Urea Nitrogen *Serum   No Effect   Analytical* At acute overdose concentration (20 mg/dL) on Technicon® method. *3719* At concentration of 400 mg/L had no effect on diacetylmonoxime method. *3393*

Uric Acid *Serum   No Effect   Analytical* At acute overdose concentration (20 mg/dL) on Technicon® method. *3719* At concentration of 400 mg/L had no effect on phosphotungstate reduction method. *3393*

Ether

Alanine Aminotransferase *Serum   Increase   Physiological* Hepatic disturbance (transient). *2313*

Alkaline Phosphatase *Serum   Increase   Physiological* Hepatic disturbance (transient). *3218*

Antidiuretic Hormone *Plasma   Increase   Physiological* Effect in moderate to deep anesthesia. *1343*

Aspartate Aminotransferase
Serum   Increase   Physiological Hepatic disturbance (transient). *3218*

Bile *Urine   Increase   Physiological* Hepatic disturbance (transient). *2313*

Bilirubin *Serum   Increase   Physiological* Hepatic disturbance (transient). *3218*

Bleeding Time *Blood   No Effect   Physiological* Anesthesia has no effect. *1343*

Bogen Test *Urine   Positive   Analytical* Reacts as if ethanol. *1302*

BSP Retention *Serum   Increase   Physiological* Hepatic disturbance (transient). *2313*

Capillary Fragility *Patient   No Effect   Physiological* Anesthesia has no effect. *1343*

Catecholamines *Plasma   Increase   Physiological* Response to stress. *2869*
Plasma   Decrease   Analytical Falsely low with ethylene diamine method. *2285*

Cephalin Flocculation *Serum   Increase   Physiological* Hepatic disturbance (transient). *2313*

Cholesterol *Serum   Increase   Physiological* Reportedly may cause hypercholesterolemia. *1487*

Clotting Time *Blood   No Effect   Physiological* No effect of anesthesia. *1343*

Cortisol *Plasma   Increase   Physiological* Effect observed in moderate to deep anesthesia. *1343*

Effective Renal Plasma Flow
Patient   Decrease   Physiological Probably due to renal vasoconstriction. *1343*

Epinephrine *Plasma   Increase   Physiological* Response to stress. *2869*

Glomerular Filtration Rate (GFR)
Urine   Decrease   Physiological Due to renal vasoconstriction. *1343*

Glucose *Serum   Increase   Physiological* Metabolic effect (transient effect). *3918*
Urine   Increase   Physiological Due to hyperglycemia *2313*

Insulin *Plasma   Decrease   Physiological* Due to release of epinephrine causing inhibition. *3918*

Ketones *Urine   Increase   Physiological* May follow anesthesia. *1487*

Lactate *Serum   Increase   Physiological* Metabolic effect. *3918*

Leukocytes *Blood   Increase   Physiological* Normal response to anesthesia. *0251*

Norepinephrine *Plasma   Increase   Physiological* Response to stress. *2869*

pCO₂ *Blood   Decrease   Physiological* May be slight effect during anesthesia. *1343*

pH *Blood   Decrease   Physiological* Metabolic acidosis especially in children. *0071*

Phosphate *Serum   Decrease   Physiological* Observed after most types of anesthesia. *1343*

Platelets *Blood   No Effect   Physiological* Anesthesia has no effect. *1343*

Protein *Urine   Increase   Physiological* May cause nephrotoxicity. *3204*

Protein Bound Iodine (PBI) *Serum Increase Physiological* Causes mobilization of PBI stores. *0012*

Prothrombin Time *Plasma No Effect Physiological* Anesthesia has no effect. *1343*

Pyruvate *Serum Increase Physiological* Metabolic effect. *3918*

Sugar *Urine Increase Physiological* May cause positive Benedict's and Fehling's tests. *0279*

Thymol Turbidity *Serum Increase Physiological* Hepatic disturbance (transient). *2313*

Thyroxine (T₄) *Serum Increase Physiological* Effect observed in moderate to deep anesthesia. *1343*

Urea Nitrogen *Serum Increase Physiological* May cause nephrotoxicity. *3204*

Volume *Urine Decrease Physiological* Due to vasoconstriction. *0071*

Ethinamate

Alanine Aminotransferase *Serum No Effect Analytical* At acute overdose concentration (20 mg/dL) on Technicon® SMAC® method. *3719*

Albumin *Serum No Effect Analytical* At acute overdose concentration (20 mg/dL) on Technicon® SMAC® method. *3719* At concentration of 200 mg/L had no effect on BCG method. *3393*

Alkaline Phosphatase *Serum No Effect Analytical* At acute overdose concentration (20 mg/dL) on Technicon® SMAC® method. *3719*

Alkaloids *Urine No Effect Analytical* With iodoplatinate on Frings TLC procedure at 10 mg/dL. With ninhydrin on Frings TLC procedure at 10 mg/dL. *1204*

Aspartate Aminotransferase *Serum No Effect Analytical* At acute overdose concentration (20 mg/dL) on Technicon® SMAC® method. *3719*

Barbiturate *Urine No Effect Analytical* With vanillin on Frings procedure at 10 mg/dL. *1204*
Urine Increase Analytical With mercurous NO₃ reagent on Frings TLC procedure. Blue with diphenylcarbazone in Frings TLC procedure. White spot with mercuric SO₄ on Frings TLC procedure. *1204*

Bilirubin *Serum No Effect Analytical* At acute overdose concentration (20 mg/dL) on Technicon® SMAC® method. *3719* At concentration of 200 mg/L had no effect on Jendrassik and Grof method. *3393*

Bromide *Urine No Effect Analytical* No effect on method of Frings. *1204*

Calcium *Serum No Effect Analytical* At acute overdose concentration (20 mg/dL) on Technicon® SMAC® method. *3719* At concentration of 200 mg/L had no effect on cresolphthalein method. *3393*

Cholesterol *Serum No Effect Analytical* At acute overdose concentration (20 mg/dL) on Technicon® SMAC® method. *3719* At concentration of 200 mg/L had no effect on Liebermann-Burchard method. *3393*

Creatine Kinase *Serum No Effect Analytical* At acute overdose concentration (20 mg/dL) on Technicon® SMAC® method. *3719*

Creatinine *Serum No Effect Analytical* At acute overdose concentration (20 mg/dL) on Technicon® SMAC® method. *3719* At concentration of 200 mg/L had no effect on AutoAnalyzer Jaffé method. *3393*

Ethchlorvynol *Serum No Effect Analytical* At 10 μg/mL on colorimetric method of Wallace. *3753*
Urine No Effect Analytical No effect on method of Frings and Cohen. *1204*

FPN Test *Urine No Effect Analytical* No effect at 10 mg/dL on method of Frings. *1204*

Glucose *Serum No Effect Analytical* At acute overdose concentration (20 mg/dL) on Technicon® SMAC® method. *3719*

17-Hydroxycorticosteroids *Urine Increase Analytical* Minimum effect with Glenn-Nelson method. *1488*

Iron *Serum No Effect Analytical* At acute overdose concentration (20 mg/dL) on Technicon® SMAC® method. *3719* At concentration of 200 mg/L had no effect on Ferrozine method. *3393*

17-Ketosteroids *Urine Increase Analytical* Interferes with Zimmermann reaction. *0427*

Lactate Dehydrogenase *Serum No Effect Analytical* At acute overdose concentration (20 mg/dL) on Technicon® SMAC® method. *3719*

Phosphate *Serum No Effect Analytical* At acute overdose concentration (20 mg/dL) on Technicon® SMAC® method. *3719* At concentration of 200 mg/L had no effect on phosphomolybdate method. *3393*

Platelets *Blood Decrease Physiological* Thrombocytopenia (AMA Blood dyscrasias). *2429*

Protein *Serum No Effect Analytical* At acute overdose concentration (20 mg/dL) on Technicon® SMAC® method. *3719* At concentration of 200 mg/L had no effect on biuret method with blank correction. *3393*

Salicylate *Urine No Effect Analytical* At 10 mg/dL no effect on Trinder method. *1204*

Sugar *Urine Increase Analytical* Theoretical effect as excreted as glucuronide. *1343*

Triglycerides *Serum No Effect Analytical* At acute overdose concentration (20 mg/dL) on Technicon® SMAC® method. *3719* At concentration of 200 mg/L had no effect on lipase/esterase method. *3393*

Urea Nitrogen *Serum No Effect Analytical* At acute overdose concentration (20 mg/dL) on Technicon® SMAC® method. *3719* At concentration of 200 mg/L had no effect on diacetylmonoxime method. *3393*

Uric Acid *Serum No Effect Analytical* At acute overdose concentration (20 mg/dL) on Technicon® SMAC® method. *3719* At concentration of 200 mg/L had no effect on phosphotungstate reduction method. *3393*

Ethinyl Estradiol

Albumin *Serum Decrease Physiological* Metabolic effect. *0473* Metabolic effect. *0473* In post-menopausal women with 50 μg/d for 14 d. *3474*

Alkaline Phosphatase *Serum Decrease Physiological* In post-menopausal women with 50 μg/d for 14 d. *3474* Increase by 30% after 2 weeks in 5 patients with primary biliary cirrhosis. *1419* Significant effect in postmenopausal women: less marked in perimenopausal women. *2311*

Angiotensin I *Plasma Increase Physiological* Approximately x 3 in 2 weeks in normals. *2357*

Antithrombin III *Plasma Decrease Physiological* In over 20% — not dose related. *3960*

Aspartate Aminotransferase
Serum Decrease Physiological Increase by 50% after 2 weeks in 5 patients with primary biliary cirrhosis. *1419*

BSP Retention *Serum Increase Physiological* Depresses hepatic secretory transport maximum. *1487*

Calcitonin *Plasma Increase Physiological* Significant effect during and after in 7 healthy postmenopausal women treated for 12 weeks. *3467*

Calcium *Serum No Effect Physiological* No effect during and after in 7 healthy postmenopausal women treated for 12 weeks. *3467*
Serum Decrease Physiological Significant fall, associated with decreased albumin. *3923* In post-menopausal women with 50 μg/d for 14 d. *3474* Significant effect in postmenopausal women: less marked in perimenopausal women. *2311*
Urine Decrease Physiological Associated with decreased serum concentration. *3923* In post-menopausal women with 50 μg/d for 14 d. *3474* Significant effect in postmenopausal women: less marked in perimenopausal women. *2311*
Feces Decrease Physiological Decrease less marked than in urine. *3923*

Carotene *Serum Decrease Physiological* Hormonal influence. *1221*

Cephalin *Serum Increase Physiological* Mirrors changes in lipoprotein pattern. *1431*

Ceruloplasmin *Serum Increase Physiological* Metabolic effect on liver synthesis (estrogen effect). *1498*

Cholesterol *Serum Decrease Physiological* Decreases by up to 50% (decreased low density lipoproteins). *1343*

Ethinyl Estradiol *(continued)*

Coproporphyrin *Feces Increase Physiological* May induce porphyria cutanea tarda. *3769*

Creatinine *Serum No Effect Analytical* At 10 mg/L on reversed phase LC procedure of Zhiri et al. *3942*

Cyclic Adenosine Monophosphate
Urine No Effect Physiological In post-menopausal women with 50 μg/d for 14 d. *3474*

1,25-Dihydroxy Vitamin D₃ *Serum Increase Physiological* In post-menopausal women with 50 μg/d for 14 d. *3474* Small effect during and after in 7 healthy postmenopausal women treated for 12 weeks. *3467*

Estradiol *Urine Increase Analytical* 10% color intensity Kober reaction. Up to 30% if method of Brown used. *0023*

Estrogens *Urine Increase Analytical* Reacts in Brown's procedure. Interferes with Kober procedure. *0279*
Urine Decrease Analytical Affects colorimetric/fluorometric procedures. *0022*

Estrone *Urine Increase Analytical* Effect of hot alkali treatment. *0023*

Fibrinopeptide A *Plasma Increase Physiological* Significant effect when 30 μg of drug combined with 2 different progestins. *2399*

Follicle Stimulating Hormone (FSH)
Plasma Decrease Physiological In hypogonadal women in 1 week. *3913*

Glucose Tolerance *Serum Decrease Physiological* Hormonal effect. *2525* In group of women taking drug as anti-androgen therapy. *3239*

γ-Glutamyltransferase (GGT)
Serum Decrease Physiological Increase by 50% after 2 weeks in 5 patients with primary biliary cirrhosis. *1419*

Haptoglobin *Serum Decrease Physiological* Metabolic effect. *0473*

17-Hydroxycorticosteroids *Urine Increase Physiological* Temporary enhanced excretion on stopping drug. *1685*
Urine Decrease Physiological Inhibits response to metyrapone. Decreased excretion of cortisol metabolites. *0022*

Hydroxyproline *Urine Decrease Physiological* In post-menopausal women with 50 μg/d for 14 d. *3474* significant effect in postmenopausal women: less marked in perimenopausal women. *2311*

25-Hydroxy Vitamin D₃ *Serum Decrease Physiological* Small but insignificant effect during and after in 7 healthy postmenopausal women treated for 12 weeks. *3467*

Insulin *Plasma No Effect Physiological* In group of women taking drug as anti-androgen therapy. *3239*

Ionized Calcium *Serum Decrease Physiological* Significant effect in postmenopausal women: less marked in perimenopausal women. *2311*

Iron-Binding Capacity, Total (TIBC)
Serum No Effect Physiological No effect if administered alone. *0473*

¹³¹I Uptake *Serum Decrease Physiological* Feminone contains tetraiodofluorescein. *2652*

17-Ketosteroids *Urine Increase Physiological* Temporary enhanced excretion on stopping drug. *1685*

Lecithin *Plasma Increase Physiological* Mirrors changes in lipoprotein pattern. *1431*

α-Lipoproteins *Serum Increase Physiological* May almost double in some studies. *1343*

β-Lipoproteins *Serum Decrease Physiological* Decreases by up to 30%. *1343*

Luteinizing Hormone (LH) *Plasma Decrease Physiological* In hypogonadal women in 1 week. *3913*

Lysolecithin *Plasma Decrease Physiological* Mirrors change in lipoprotein pattern. *1431*

5'-Nucleotidase *Serum Increase Physiological* Slight effect?impaired excretion. *1498*

Osteocalcin *Plasma Decrease Physiological* In post-menopausal women with 50 μg/d for 14 d. *3474*

Parathyroid Hormone *Plasma Increase Physiological* Significant increase after in 7 healthy postmenopausal women treated for 12 weeks. *3467*

Plasma Decrease Physiological Average decrease of about 25% with RIAs directed against mid-molecule in 10 post-menopausal women treated for 14 d with 50 μg/d. *3474*

Phosphate *Serum Decrease Physiological* Decrease of approximately 2 mg/dL. *2886* In post-menopausal women with 50 μg/d for 14 d. *3474* Significant effect in postmenopausal women: less marked in perimenopausal women. *2311* Small but insignificant effect during and after in 7 healthy postmenopausal women treated for 12 weeks. *3467*

Phospholipids, Total *Serum Increase Physiological* Due to high and very high density lipoproteins. *1431*

Progesterone *Plasma Decrease Physiological* Luteal phase also shortened. *1800*

Prolactin *Plasma No Effect Physiological* At the low doses (20-30 μg) in combined oral contraceptives. *0825*
Plasma Increase Physiological Increase by 1350% in women if on perphenazine. *0522* In hypogonadal women within 1 week of therapy. *3913*

Protein Bound Iodine (PBI) *Serum Increase Physiological* Increases concentration of circulating thyroxine binding globulin. *0012*

Protoporphyrin *Feces Increase Physiological* May induce porphyria cutanea tarda. *3769*

Renin Activity *Plasma Increase Physiological* Men and women, independent of diet and posture. *2427* Usual but not constant in 2 weeks. *2357*
Plasma Decrease Physiological Usual but not constant in 2 weeks. *2357*

Thyroxine (T₄) *Urine Increase Physiological* On withdrawal only in men after 20 μg/d. *0630*

Thyroxine (T₄) Binding Globulin
Serum Increase Physiological Direct effect of drug. *0012*

Tubular Maximum for Phosphate
Urine No Effect Physiological In postmenopausal women with 50 μg/d for 14 d. *3474*

Tyrosine *Plasma Decrease Physiological* Induction of tyrosine transaminase by 0.1 mg/d steroid. *3046*

Uric Acid *Serum No Effect Analytical* At 10 mg/L on reversed phase LC procedure of Zhiri et al. *3942*
Serum Decrease Physiological In men on therapy (hormonal effect). *2590*
Urine Increase Physiological In men on therapy (hormonal effect). *2590*

Uric Acid Clearance *Urine Increase Physiological* In men on therapy (hormonal effect). *2590*

Uroporphyrin *Urine Increase Physiological* May induce porphyria cutanea tarda. *3769*
Feces Increase Physiological May induce porphyria cutanea tarda. *3769*

Vitamin A *Serum Increase Physiological* Due to alteration in concentration of binding globulin. *1221*

Ethiodized Oil

Protein Bound Iodine (PBI) *Serum Increase Analytical* Contains iodine. *0071*

Ethionamide

Alanine Aminotransferase *Serum Increase Physiological* Intrahepatic bile duct damage, toxic hepatitis. *2034*

Alkaline Phosphatase *Serum Increase Physiological* Hepatotoxic effect. *2313*
Urine Decrease Analytical Interference with determination procedure. *2897*

Amphetamine *Urine Increase Analytical* No effect at 100 mg/L on method of Rutter. *3097*

Aspartate Aminotransferase
Serum Increase Physiological Intrahepatic cholestatic jaundice. *1343*

Bile *Urine Increase Physiological* Hepatotoxic effect. *2313*

Bilirubin *Serum Increase Physiological* Hepatotoxicity in about 2% cases. *1488*

BSP Retention *Serum Increase Physiological* Hepatotoxic effect. *2313*

Cephalin Flocculation *Serum Increase Physiological* Hepatotoxic effect. *2313*

Glucose *Serum Increase Physiological* Hyperglycemia reported. *2220*
Urine Increase Physiological Due to hyperglycemia. *2220*

Lactate Dehydrogenase *Urine Decrease Analytical* Interference with determination procedure. *2897*

Phosphate *Serum No Effect Analytical* On nonprecipitation method of Peynet. *2799*

Protein Bound Iodine (PBI) *Serum Decrease Physiological* Antithyroid effect after several weeks. *1488*

Thymol Turbidity *Serum Increase Physiological* Intrahepatic cholestasis. *2313*

Thyroxine (T₄) *Serum Decrease Physiological* Antithyroid effect after several weeks. *1488*

Xylose Excretion *Urine Decrease Physiological* May cause gastrointestinal irritation, impaired absorption. *0612*

Ethionine

Tyrosine *Plasma No Effect Analytical* On fluorometric procedure of Ambrose. *0080*

Ethosuximide

Albumin *Serum No Effect Analytical* At concentration of 390 mg/L had no effect on BCG method. *3393*

Antinuclear Antibodies *Serum Positive Physiological* Related to number of drugs, higher in women. *3797*

Aspartate Aminotransferase
Serum Increase Physiological Increased in one third cases. *2427*

Bicarbonate *Serum No Effect Analytical* At concentration of 390 mg/L had no effect on method using phenolphthalein. *3393*

Bilirubin *Serum No Effect Analytical* At concentration of 390 mg/L had no effect on Jendrassik and Grof method. *3393*
Serum No Effect Physiological Insignificant displacement of protein in neonates. *3748*
Serum Increase Physiological Rare idiosyncratic hepatitis reported. *0071*

Calcium *Serum No Effect Analytical* At concentration of 390 mg/L had no effect on cresolphthalein method. *3393*

Cephalin Flocculation *Serum Increase Physiological* Rare hepatitis reported. *2313*

Chloride *Serum No Effect Analytical* At concentration of 390 mg/L had no effect on mercurimetric method. *3393*

Cholesterol *Serum No Effect Analytical* At concentration of 390 mg/L had no effect on Liebermann-Burchard method. *3393*

Coombs' Test *Serum Positive Physiological* Mechanism obscure. *1486*

Coombs' Test, Direct *Serum Positive Physiological* 3 cases reported with SLE and anemia. *1487*

Creatinine *Serum No Effect Analytical* At concentration of 390 mg/L had no effect on AutoAnalyzer Jaffé method. *3393*

Diphenylhydantoin *Serum No Effect Analytical* On GLC procedure of Papadopoulos at *in vivo* concentration. *2722*
Serum Increase Physiological May be increased plasma levels (impaired metabolism). *2422*

Eosinophils *Blood Increase Physiological* Rare eosinophilia reported. *0071*

Erythrocytes *Blood Decrease Physiological* Aplastic anemia/pancytopenia. *2313*

Glucaric Acid *Urine Increase Physiological* Induces hepatic enzymes. *1694*

Hematocrit *Blood Decrease Physiological* Aplastic anemia. *0071*

Hemoglobin *Blood Decrease Physiological* Aplastic anemia. *0071*
Urine Increase Physiological Rare side effect observed. *2042*

Indomethacin *Serum No Effect Analytical* No effect on HPLC method of Roberts and Smith. *3002*

LE Cells *Blood Positive Physiological* Rare immune response reported. *0518*

Leukocytes *Blood Decrease Physiological* Aplastic anemia/pancytopenia/leukopenia in 10%. *0071*

Neutrophils *Blood Decrease Physiological* Occasional case of agranulocytosis reported. *3717*

Pheneturide *Serum No Effect Analytical* On GLC procedure of Papadopoulos at *in vivo* concentration. *2722*

Phenobarbital *Serum No Effect Analytical* On GLC procedure of Papadopoulos at *in vivo* concentration. *2722*

Phosphate *Serum No Effect Analytical* At concentration of 390 mg/L had no effect on phosphotungstate reduction method. *3393*

Platelets *Blood Decrease Physiological* Aplastic anemia/thrombocytopenia/pancytopenia. *0071*

Potassium *Serum No Effect Analytical* At concentration of 390 mg/L had no effect on measurement by ISE with predilution. *3393*

Primidone *Serum No Effect Analytical* On GLC procedure of Papadopoulos at *in vivo* concentration. *2722*

Protein *Serum No Effect Analytical* At concentration of 390 mg/L had no effect on biuret method with blank correction. *3393*
Urine Increase Physiological Possible reversible nephropathy. *1022*

Sodium *Serum No Effect Analytical* At concentration of 390 mg/L had no effect on measurement by ISE with predilution. *3393*

Triglycerides *Serum No Effect Analytical* At concentration of 390 mg/L had no effect on lipase/esterase method. *3393*

Urea Nitrogen *Serum No Effect Analytical* At concentration of 390 mg/L had no effect on diacetylmonoxime method. *3393*
Serum Increase Physiological Possible reversible nephropathy. *0071*

Uric Acid *Serum No Effect Analytical* At concentration of 390 mg/L had no effect on phosphotungstate reduction method. *3393*

Ethotoin

Alanine Aminotransferase *Serum Increase Physiological* Hepatotoxicity. *2313*

Alkaline Phosphatase *Serum Increase Physiological* Probable idiosyncratic hepatitis. *2313*

Aspartate Aminotransferase
Serum Increase Physiological Hepatotoxicity. *2313*

Bile *Urine Increase Physiological* Hepatotoxicity. *2313*

Bilirubin *Serum Increase Physiological* Probable idiosyncratic hepatitis. *2313*

BSP Retention *Serum Increase Physiological* Probable idiosyncratic hepatitis. *2313*

Cephalin Flocculation *Serum Increase Physiological* Hepatotoxicity. *2313*

Diphenylhydantoin *Serum No Effect Analytical* No detectable influence at 40 μg/ml. *3591* On GLC procedure of Papadopoulos at *in vivo* concentration. *2722*

Erythrocytes *Blood Decrease Physiological* May cause marrow depress or megaloblastic anemia. *0071*

Leukocytes *Blood Decrease Physiological* Possible bone marrow depression. *0071*

MCV *Blood Increase Physiological* Theoretical effect on folic acid metabolism. *1680*

Pheneturide *Serum No Effect Analytical* On GLC procedure of Papadopoulos at *in vivo* concentration. *2722*

Phenobarbital *Serum No Effect Analytical* On GLC procedure of Papadopoulos at *in vivo* concentration. *2722*

Primidone *Serum No Effect Analytical* On GLC procedure of Papadopoulos at *in vivo* concentration. *2722*

Thymol Turbidity *Serum Increase Physiological* Hepatotoxicity. *2313*

Ethoxazene

Alanine Aminotransferase *Serum Increase Physiological* Hepatotoxic effect. *2313*

Alkaline Phosphatase *Serum Increase Physiological* Hepatotoxic effect. *2313*

Aspartate Aminotransferase
Serum Increase Physiological Hepatotoxic effect. *2313*

Bile *Urine Increase Analytical* Atypical red color with Bili-Labstix® and Ictotest®. *1488*
Urine Increase Physiological Hepatotoxic effect. *2313*

Bilirubin *Serum Increase Analytical* Postulated production of color with Ehrlich's diazo reaction. *0085*
Serum Increase Physiological Hepatotoxic effect. *1022*
Urine Increase Physiological Hepatotoxic effect. *1022*

BSP Retention *Serum Increase Analytical* Increases absorbancy in test, falsely high result. *1488*
Serum Increase Physiological Hepatotoxic effect. *2425*

Cephalin Flocculation *Serum Increase Physiological* Hepatotoxic effect. *2313*

Color *Urine Increase Analytical* red, pink, orange and rust colors. *2425*

Porphyrins *Urine Increase Analytical* False positive with fluorescent methods. *3879*
Urine Increase Physiological Hepatotoxic effect. *1022*

PSP Excretion *Urine Increase Analytical* Produces interfering background color. *1022*

Thymol Turbidity *Serum Increase Physiological* Hepatotoxic effect. *2313*

Ethoxycaffeine

Chromosomes *Test Conditions Abnormal Physiological* Clastogenic in human cells. *3282*

Ethoxyethanol

Protein *Urine Increase Physiological* May cause nephrotoxicity. *3204*

Urea Nitrogen *Serum Increase Physiological* May cause nephrotoxicity. *3204*

Ethoxzolamide

Leukocytes *Blood Decrease Physiological* Leukopenia. *2313*

pH *Blood Decrease Physiological* May cause metabolic acidosis. *2220*

Platelets *Blood Decrease Physiological* Thrombocytopenia. *2426*

Potassium *Serum Decrease Physiological* Probable effect (like acetazolamide). *0071*

Uric Acid *Serum Increase Physiological* Decreases urate clearance. *2220*
Urine Decrease Physiological Decreases urate clearance. *2220*

Ethyl Acetate

Aspartate Aminotransferase *Serum Increase Analytical* Significant effect when added to serum, diazo reaction. *3217*

Ethylamine

Pyruvate Kinase *Red Blood Cells Decrease Analytical* Most marked with larger side chain. *0370*

Ethyl Biscoumacetate

Platelet Aggregation *Blood Decrease Physiological* Inhibits if due to ATP, collagen, epinephrine, thrombin. *0614*

Uric Acid *Serum Decrease Physiological* Uricosuric action. *1433*
Urine Increase Physiological Uricosuric action. *1433*

Ethyl Chloride

Alanine Aminotransferase *Serum Increase Physiological* May cause liver damage. *0071*

Aspartate Aminotransferase
Serum Increase Physiological May cause liver damage. *0071*

Bilirubin *Serum Increase Physiological* May cause liver damage. *0071*

Ethylene

Glucose *Serum Increase Physiological* After prolonged use may be moderate increase. *0071*

Phosphate *Serum Decrease Physiological* Observed after most forms of anesthesia. *0279*

Ethylenediamine

Amino Acids *Urine Increase Analytical* Reacts with ninhydrin; extra spot TLC, high voltage electrophoresis. *2855*

Ethylene Glycol

Calcium *Serum Decrease Physiological* Characteristic finding, often marked, due to deposition of oxalate and interference with normal homeostasis. *3325*

Creatinine *Serum Increase Physiological* Due to tubular necrosis caused by toxic metabolites and possibly also due to deposition of calcium oxalate crystals in kidney. *3325*

Crystals *Urine Increase Physiological* Oxalate crystals observed in poisoning. *3325*

Erythrocytes *Urine Increase Physiological* Due to nephrotoxicity. *1302*

Ethylene Glycol *Urine Increase Physiological* Readily detectable in urine with poisoning. *3325*

Hemoglobin *Urine Increase Physiological* Due to nephrotoxicity. *1302*

Lactate *Serum Increase Physiological* Due to NADH formation from oxidation of ethylene glycol and inhibition of tricarboxylic acid cycle by toxic metabolites. *3325*

Oncotic Pressure *Plasma Increase Physiological* Also observed in patients who ingested several different alcohols. *3325*

Osmolality *Serum Increase Physiological* Greater if measured by freezing point depression than calculated value. *3325*

Oxalate *Urine Increase Physiological* Increased excretion following ingestion. *3325*

Parathyroid Hormone *Plasma No Effect Physiological* Although normal homeostasis interfered with. *3325*

pH *Blood Decrease Physiological* May cause marked acidosis with poisoning. *0279*

Protein *Urine Increase Physiological* May cause nephrotoxicity. *3204* Due to nephrotoxicity. *1302*

Urea Nitrogen *Serum Increase Physiological* May cause nephrotoxicity. *3204* Due to tubular necrosis caused by toxic metabolites and possibly also due to deposition of calcium oxalate crystals in kidney. *3325*

Volume *Urine Decrease Physiological* May cause oliguria with renal failure. *0279*

Ethylestrenol

Haptoglobin *Serum Increase Physiological* Metabolic effect. *0227*

Plasminogen *Plasma Increase Physiological* Metabolic effect. *0227*

Sialic Acid *Serum Increase Physiological* Metabolic effect. *0227*

Ethylmethylketone

Creatinine *Test Conditions Increase Analytical* Positive spot test with Jaffé reagent. *3765*

Ethylnorepinephrine

Amphetamine *Urine No Effect Analytical* No effect at 100 mg/L on method of Rutter. *3097*

Ethylphenacemide

Alkaline Phosphatase *Serum Increase Physiological* Frequently occurs. *1343*

Glucaric Acid *Urine Increase Physiological* Induces hepatic enzymes (more potent than phenobarbital). *1694*

Leukocytes *Blood Decrease Physiological* Leukopenia reported. *1343*

Ethynodiol

Glucose *Serum No Effect Physiological* With doses of 0.25, 0.35, 0.5 mg/d. *1335*

Insulin *Plasma No Effect Physiological* With doses of 0.25, 0.35, 0.5 mg/d. *1335*

Triglycerides *Serum Increase Physiological* Decreased removal from circulation. *3505*

Etidronate

Calcium *Serum Decrease Physiological* In majority of patients with hypercalcemia of malignancy when drug given i.v. for 3-5 d. *1521*

Creatinine *Serum Increase Physiological* Transient mild effect in patients with hypercalcemia of malignancy given drug i.v. for 3-5 d. *1521*

Etintidine

Propranolol *Serum Increase Physiological* Significantly increased area under curve and prolonged elimination half-life: also protracted elimination of 4-hydroxypropranolol, an active metabolite. *1676*

Etiocholanolone

Aldosterone *Urine No Effect Analytical* No significant effect in RIA procedure of Drewes. *0952*
Urine Increase Analytical 1 mg/L equivalent to 2 μg/L in method of Drewes. *0952*

BSP Retention *Serum Increase Physiological* In almost 50% patients with induced fever. *0375*

Estrogens *Urine No Effect Analytical* At 50 mg/L on fluorescent method of Corns. *0730*

Leukocytes *Blood Increase Physiological* Pyrogen — increases number of granulocytes. *2427*

Etodolac

Chloride *Urine Decrease Physiological* Both in normal subjects but also to greater extent in individuals with renal insufficiency. *0453*

Creatinine *Serum No Effect Physiological* No cumulative effect observed over 24 h in individuals with normal renal function or with renal insufficiency when given 500 mg bid for 4 d. *0453*

Creatinine Clearance *Urine No Effect Physiological* No cumulative effect observed over 24 h in individuals with normal renal function or with renal insufficiency when given 500 mg bid for 4 d. *0453*

Inulin Clearance *Urine No Effect Physiological* No effect with acute or chronic treatment in individuals with normal renal function but transient reduction in people with renal insufficiency. *0453*

PAH Clearance *Urine No Effect Physiological* No effect with acute or chronic treatment in individuals with normal renal function but transient reduction in people with renal insufficiency. *0453*

Potassium *Serum No Effect Physiological* No cumulative effect observed over 24 h in individuals with normal renal function or with renal insufficiency when given 500 mg bid for 4 d. *0453*

Prostaglandin, 6-Keto-F$_{1\alpha}$ *Urine Decrease Physiological* 6-keto- less than 30% in people with normal renal function but about 60% in in individuals with renal insufficiency. *0453*

Prostaglandin E2 *Urine Decrease Physiological* About 40% reduction in people with normal renal function but 70% in people with renal insufficiency. *0453*

Sodium *Urine No Effect Physiological* No cumulative effect observed over 24 h in individuals with normal renal function or with renal insufficiency when given 500 mg bid for 4 d. *0453*
Urine Decrease Physiological Both in normal subjects but also to greater extent in individuals with renal insufficiency. *0453*

Urea Nitrogen *Serum No Effect Physiological* Fewer than 2% patients showed pattern of deviant renal function tests. *3272*

Etomidate

Aldosterone *Plasma Decrease Physiological* Clear suppression when induction compared with induction by thiopentone. *0063*

Corticosterone *Plasma Decrease Physiological* Clear suppression when induction compared with induction by thiopentone. *0063*

Corticotropin *Plasma Increase Physiological* Nonsignificant increase vs thiopentone controls in 7 men. *0063* Apparently direct suppressive effect on adrenal cortical function. *0641*

Cortisol *Plasma Decrease Physiological* Clear suppression when induction compared with induction by thiopentone. *0063* Apparently direct suppressive effect on adrenal cortical function. *0641* Similar effect to that of thiopental but reduction with induction of anesthesia. *3235* Attributable to direct antisteroidogenic effects on adrenal gland. *0709*

11-Deoxycorticosterone *Plasma Increase Physiological* Clear effect 3.5 h after induction demonstrating inhibition of 11β-hydroxylation of glucocorticoid and mineralocorticoid intermediates. *0063*

11-Deoxycortisol *Plasma Increase Physiological* Clear effect 3.5 h after induction demonstrating inhibition of 11β-hydroxylation of glucocorticoid and mineralocorticoid intermediates. *0063*

17-Hydroxyprogesterone *Plasma No Effect Physiological* No significant difference in 7 men given drug in induction dose versus thiopentone. *0063*

Progesterone *Plasma No Effect Physiological* No significant difference in 7 men given drug in induction dose versus thiopentone. *0063*

Etoposide

Adenosine Deaminase Binding Protein
Urine No Effect Physiological No effect in 12 children given drug for 2 d. *1355*

Alanine Aminotransferase *Serum Increase Physiological* Hepatic toxicity reported in 3% treated patients. *0141*

Alkaline Phosphatase *Serum Increase Physiological* Hepatic toxicity reported in 3% treated patients. *0141*

Aspartate Aminotransferase
Serum Increase Physiological Hepatic toxicity reported in 3% treated patients. *0141*

Bilirubin *Serum Increase Physiological* Hepatic toxicity reported in 3% treated patients. *0141*

Leukocytes *Blood Decrease Physiological* Dose-limiting noncumulative leukopenia in 60-90% patients. *0141* 12 of 37 patients had transient leukopenia for 3-4 d. *3891* Occurs 7-10 d after therapy started: recovery occurs by day 20 to 24. *3339*

Neutrophils *Blood Decrease Physiological* Nadir at 7 to 14 d after administration. *0141*

Platelets *Blood No Effect Physiological* In all of 37 patients given drug orally. *3891*
Blood Decrease Physiological Nadir at 9 to 16 d after administration. *0141* Occurs 9-13 d after therapy started. *3339*

Etoposide (continued)

Uric Acid   *Serum   No Effect   Analytical*   No effect on uricase method on Du Pont aca at therapeutic concentration. *2333*
Serum   Increase   Analytical   Values of more than 3 times uricase values with direct phosphotungstate method. *2333*

Etretinate

Alanine Aminotransferase   *Serum   Increase   Physiological*   In 2 of 20 patients treated for 1 y. *1866*   In one 74-year old woman treated for severe psoriasis: normalized with withdrawal of therapy and recurred when reinstituted. *3789*

Alkaline Phosphatase   *Serum   Increase   Physiological*   In one 74-year old woman treated for severe psoriasis: normalized with withdrawal of therapy and recurred when reinstituted. *3789*

Apolipoprotein AI   *Serum   No Effect   Physiological*   In 13 patients with hyperkeratotic disorders given 50 mg daily for 2 mo. *3659*

Apolipoprotein AII   *Serum   No Effect   Physiological*   In 13 patients with hyperkeratotic disorders given 50 mg daily for 2 mo. *3659*

Apolipoprotein B   *Serum   Increase   Physiological*   From 133 to 147% in 13 patients with hyperkeratotic disorders given 50 mg daily for 2 mo. *3659*

Aspartate Aminotransferase
Serum   Increase   Physiological   In 2 of 20 patients treated for 1 y. *1866*   In one 74-year old woman treated for severe psoriasis: normalized with withdrawal of therapy and recurred when reinstituted. *3789*

Bilirubin   *Serum   No Effect   Physiological*   In 2 of 20 patients treated for 1 y. *1866*
Serum   Increase   Physiological   No marked effect seen but slight change noted. *3867*

Cholesterol   *Serum   Increase   Physiological*   From 5.75 to 6.05 mmol/L in 13 patients with hyperkeratotic disorders given 50 mg daily for 2 mo. *3659*   In 4 of 11 patients who showed hypertriglyceridemia: effect slight but measured over 1 y therapy. *1866*

Cholesterol, High Density Lipoprotein
Serum   No Effect   Physiological   In 13 patients with hyperkeratotic disorders given 50 mg daily for 2 mo. *3659*

Cholesterol, Low Density Lipoprotein
Serum   Increase   Physiological   From 3.82 to 4.16 mmol/L in 13 patients with hyperkeratotic disorders given 50 mg daily for 2 mo. *3659*

Cholesterol, Very Low Density Lipoprotein
Serum   No Effect   Physiological   In 13 patients with hyperkeratotic disorders given 50 mg daily for 2 mo. *3659*

Eosinophils   *Blood   Increase   Physiological*   In one 74-year old woman treated for severe psoriasis: normalized with withdrawal of therapy and recurred when reinstituted. *3789*

Erythrocytes   *Blood   Decrease   Physiological*   No marked effect seen but slight change noted. *3867*

γ-Glutamyltransferase (GGT)
Serum   Increase   Physiological   In 2 of 20 patients treated for 1 y. *1866*

Lactate Dehydrogenase   *Serum   Increase   Physiological*   In one 74-year old woman treated for severe psoriasis: normalized with withdrawal of therapy and recurred when reinstituted. *3789*

Triglycerides   *Serum   No Effect   Physiological*   In 13 patients with hyperkeratotic disorders given 50 mg daily for 2 mo. *3659*
Serum   Increase   Physiological   In 11 of 20 patients but not to more than 2 times normal over 1 y. *1866*

Triglycerides, High Density Lipoprotein
Serum   No Effect   Physiological   In 13 patients with hyperkeratotic disorders given 50 mg daily for 2 mo. *3659*

Triglycerides, Very Low Density Lipoprotein
Serum   No Effect   Physiological   In 13 patients with hyperkeratotic disorders given 50 mg daily for 2 mo. *3659*

Etryptamine

17-Hydroxycorticosteroids   *Urine   Increase   Analytical*   Glucuronide interferes with Zimmermann reaction. *0427*

17-Ketogenic Steroids   *Urine   Increase   Analytical*   Interferes with Zimmermann reaction. *0427*

17-Ketosteroids   *Urine   Increase   Analytical*   Slight increase reported with *in vivo* studies. *0583*

Evans Blue

Color   *Urine   Increase   Analytical*   Blue color due to presence of dye. *0459*

Protein Bound Iodine (PBI)   *Serum   Decrease   Physiological*   Mechanism obscure. *0012*

Thyroxine (T₄)   *Serum   Decrease   Physiological*   Compete for binding sites. *0583*

Exchange Transfusion

Digoxin   *Serum   No Effect   Physiological*   No effect observed. *0700*

Factor IX Complex, Human

Alkaline Phosphatase   *Serum   Increase   Physiological*   62% of cases developed hepatitis after use. *2663*

Aspartate Aminotransferase
Serum   Increase   Physiological   62% cases developed hepatitis after use. *2663*

Bilirubin   *Serum   Increase   Physiological*   62% cases developed hepatitis after use. *2663*

Famotidine

Alanine Aminotransferase   *Serum   No Effect   Physiological*   In 9 patients with Zollinger-Ellison syndrome when followed for 33 weeks. *1667*

Aspartate Aminotransferase
Serum   No Effect   Physiological   In 9 patients with Zollinger-Ellison syndrome when followed for 33 weeks. *1667*

Creatinine   *Serum   No Effect   Physiological*   No significant change from pretreatment values in 9 patients with Zollinger-Ellison syndrome over 33 weeks. *1667*

Estradiol   *Plasma   No Effect   Physiological*   No effect of short-term or long-term treatment in male patients with duodenal ulcers. *0727*

Follicle Stimulating Hormone (FSH)
Plasma   No Effect   Physiological   No effect of short-term or long-term treatment in male patients with duodenal ulcers. *0727*

Gastrin   *Serum   No Effect   Physiological*   No effect in 9 patients with Zollinger-Ellison syndrome treated for 33 weeks. *1667*

Hydrochloric Acid   *Gastric Material   Increase   Physiological*   Therapeutic effect as drug is long-acting histamine H_2-receptor antagonist. *1667*

Leukocytes   *Blood   No Effect   Physiological*   No significant change from pretreatment values in 9 patients with Zollinger-Ellison syndrome over 33 weeks. *1667*

Luteinizing Hormone (LH)   *Plasma   No Effect   Physiological*   No effect of short-term or long-term treatment in male patients with duodenal ulcers. *0727*

Progesterone   *Plasma   No Effect   Physiological*   No effect of short-term or long-term treatment in male patients with duodenal ulcers. *0727*

Testosterone   *Serum   No Effect   Physiological*   No effect of short-term or long-term treatment in male patients with duodenal ulcers. *0727*

Thyroid Stimulating Hormone (TSH)
Serum   No Effect   Physiological   No effect of short-term or long-term treatment in male patients with duodenal ulcers. *0727*

Thyroxine (T₄)   *Serum   No Effect   Physiological*   No effect of short-term or long-term treatment in male patients with duodenal ulcers. *0727*

Tri-iodothyronine (T₃)   *Serum   No Effect   Physiological*   No effect of short-term or long-term treatment in male patients with duodenal ulcers. *0727*

Fansidar®

Alanine Aminotransferase *Serum Increase Physiological* Biopsy-proven granulomatous hepatitis due to sulfadoxine moiety of drug reported in two individuals receiving drug prophylactically possibly due to sulfonamide hypersensitivity. *2099*

Alkaline Phosphatase *Serum Increase Physiological* Biopsy-proven granulomatous hepatitis due to sulfadoxine moiety of drug reported in two individuals receiving drug prophylactically possibly due to sulfonamide hypersensitivity. *2099*

Aspartate Aminotransferase
Serum Increase Physiological Biopsy-proven granulomatous hepatitis due to sulfadoxine moiety of drug reported in two individuals receiving drug prophylactically possibly due to sulfonamide hypersensitivity. *2099*

Eosinophils *Blood Increase Physiological* Biopsy-proven granulomatous hepatitis due to sulfadoxine moiety of drug reported in two individuals receiving drug prophylactically possibly due to sulfonamide hypersensitivity. *2099*

Fasting

Aminolevulinic Acid *Urine Increase Physiological* In normals, if porphyria may precipitate attack. *0938*

Bilirubin *Serum Increase Physiological* Increase of 1.3 g after 24 h and 2.2x after 48 h in normals. *0237*

Coproporphyrin *Urine Increase Physiological* In normals, if porphyria may precipitate attack. *0938*

Cortisol *Plasma Increase Physiological* Associated also with increased number of secretory episodes and plasma half-life. *1130*

Creatinine *Serum No Effect Analytical* No effect with 96 h fasting when measured by enzymatic method. *2323*
Serum Increase Analytical Observed in 5 healthy volunteers when measured by Jaffé method: probably due to accumulation and measurement of acetoacetate (increase by 0.7 mg/dL after 96 h). *2323*

Growth Hormone *Plasma Increase Physiological* Values low and also blunted response to TRH. *1130*

Prostaglandin E1 *Plasma Decrease Physiological* Mean decrease of 4.5 ng/mL compared with postprandial. *1385*

Renin Activity *Plasma Increase Physiological* During total fasting in obese individuals. *3695*

Thyroid Stimulating Hormone (TSH)
Serum Decrease Physiological Values low and also blunted response to TRH. *1130*

Urea Nitrogen *Serum No Effect Analytical* No effect with 96 h fasting in 5 healthy volunteers. *2323*

Fat Emulsions

Alkaline Phosphatase *Serum Increase Physiological* Prolonged infusion may impair liver function. *0120*

Bilirubin *Serum Increase Analytical* If given i.v. may affect accuracy of measurement. *0120*

BSP Retention *Serum Increase Physiological* Long infusion may impair liver function. *0120*

Hemoglobin *Blood Increase Analytical* If given i.v. may affect accuracy of measurement. *0120*

Protein *Serum Increase Analytical* If given i.v. may affect accuracy of measurement. *0120*

Fatty Acids

Bilirubin *Serum No Effect Analytical* Spectropolarographic procedure of Grahnen. *1374*

Fatty Foods

Alkaline Phosphatase *Serum Increase Physiological* Effect most marked 4 h after meal in O secretors. *3743*

Aminolevulinic Acid *Urine Increase Physiological* Effect of high fat diet on hepatic porphyria. *0938*

Thymol Turbidity *Serum Increase Analytical* Lipemia causes false high values due to turbidity. *1238*

Fava Beans

Alanine Aminotransferase *Serum Increase Physiological* May cause liver damage. *0279*

Aspartate Aminotransferase
Serum Increase Physiological May cause liver damage. *0279*

Bilirubin *Serum Increase Physiological* May cause hemolysis with G-6-PD deficiency. *0248*

Bilirubin, Direct *Serum Increase Physiological* May cause hemolysis with G-6-PD deficiency. *0248*

Erythrocytes *Blood Decrease Physiological* May cause hemolysis with G-6-PD deficiency. *0248*

Hematocrit *Blood Decrease Physiological* May cause hemolysis with G-6-PD deficiency. *0248*

Hemoglobin *Blood Decrease Physiological* May cause hemolysis with G-6-PD deficiency. *0248*

Protein *Urine Increase Physiological* May cause renal damage. *0279*

Urea Nitrogen *Serum Increase Physiological* May cause renal damage. *0279*

Felodipine

Aldosterone *Urine Increase Physiological* With 20 mg daily in 10 men with essential hypertension over 8 weeks. *1887*

Angiotensin II *Plasma No Effect Physiological* With 20 mg daily in 10 men with essential hypertension over 8 weeks. *1887*

Epinephrine *Urine No Effect Physiological* With 20 mg daily in 10 men with essential hypertension over 8 weeks. *1887*

Norepinephrine *Plasma Increase Physiological* When infused in 10 healthy normotensive volunteers. *3352*
Urine Increase Physiological With 20 mg daily in 10 men with essential hypertension over 8 weeks. *1887*

Renin Activity *Plasma Increase Physiological* When infused in 10 healthy normotensive volunteers. *3352* With 20 mg daily in 10 men with essential hypertension over 8 weeks. *1887*

Fencamfamin

Amphetamine *Urine No Effect Analytical* At 50 mg/L on fluorescent method of Hayes. *1534* No effect at 100 mg/L on method of Rutter. *3097*

Fenclofenac

Thyroid Stimulating Hormone (TSH)
Serum No Effect Physiological None of 8 patients receiving 1.2 g/d had altered values. *2971* May be short-term reduction (beginning of therapy) at hypophyseal level. *3798*

Thyroxine (T₄) *Serum Decrease Physiological* In 7 out of 8 individuals receiving 1.2 g/d values below normal range. T₄ displaced from binding protein. *2971* From 173 nmol/L to 70 nmol/L in 4 women with thyrotoxicosis given 1.2 g/d for 7 d. *2758*

Thyroxine (T₄) Binding Globulin
Serum No Effect Physiological In 7 out of 8 patients receiving 1.2 g/d no change observed. *2971*

Thyroxine (T₄), Free *Serum Decrease Physiological* Nonsignificant decrease reported in several studies. *3798* Significant decrease but not to normal in 4 women with thyrotoxicosis given 1.2 g/d for 7 d. *2758*

Thyroxine (T₄) Index, Free (FTI)
Serum Decrease Physiological In 7 out of 8 individuals receiving 1.2 g/d values below normal range. Thyroxine displaced from binding protein. *2971*

Tri-iodothyronine (T₃) *Serum Decrease Physiological* In 7 out of 8 individuals receiving 1.2 g/d values below normal range. Thyroxine displaced from binding protein. *2971* From 6.2 nmol/L to 3.8 nmol/L in 4 women with thyrotoxicosis given 1.2 g/d for 7 d. *2758*

Tri-iodothyronine (T₃), Free *Serum No Effect Physiological* Values at lower end of normal range in individuals receiving 1.2 g/d. *2971*

Fenclofenac *(continued)*

Tri-iodothyronine (T$_3$) Index, Free
Serum Decrease Physiological In 7 out of 8 individuals receiving 1.2 g/d values below normal range. Thyroxine displaced from binding protein. *2971*

Tri-iodothyronine (T$_3$), Reverse
Serum Decrease Physiological From 0.63 nmol/L to 0.52 nmol/L in 4 women with thyrotoxicosis given 1.2 g/d for 7 d. *2758*

Fenfluramine

Amobarbital *Urine No Effect Analytical* No interference on TLC using ethyl acetate: methanol: water: ammonium hydroxide and Dragendorff's reagent for detection. *3868*

Amphetamine *Urine No Effect Analytical* At 50 mg/L on fluorescent method of Hayes. *1534* No effect at 100 mg/L on method of Rutter. *3097* No interference on TLC using ethyl acetate, methanol, water, ammonium hydroxide and modified Dragendorff's reagent for detection. *3868*

Clotermine *Urine Positive Analytical* Similar R$_f$ and color reaction on TLC using ethyl acetate: methanol: water: ammonium hydroxide and modified Dragendorff's reagent. *3868*

Cortisol *Plasma Increase Analytical* Fluoresces with sulfuric acid. *0121*

Erythrocytes *Blood No Effect Physiological* No adverse reaction reported. *0538*

Glucose *Serum Decrease Physiological* Direct effect, increased glucose uptake by muscle. *3638*

Glucose Tolerance *Serum Increase Physiological* Significant improvement (mean decrease of 25 mg/dL). *0377*

Growth Hormone *Plasma Increase Physiological* Action on brain stimulating release. *2160*

Hematocrit *Blood No Effect Physiological* No adverse reaction reported. *0538*

Hydromorphone *Urine No Effect Analytical* No interference on TLC using ethyl acetate: methanol: water: ammonium hydroxide and modified Dragendorff's reagent for detection. *3868*

Insulin *Plasma No Effect Physiological* Variable response (some increased, some decreased). *0377*

Ketones *Serum Increase Physiological* (average increase of 57%). *2753*

β-Lipoproteins *Serum Decrease Physiological* Small but significant reduction. *0377*

Mescaline *Urine No Effect Analytical* No interference on TLC using ethyl acetate: methanol: water: ammonium hydroxide and modified Dragendorff's reagent for detection. *3868*

Methamphetamine *Urine No Effect Analytical* No interference on TLC using ethyl acetate: methanol: water: ammonium hydroxide and modified Dragendorff's reagent for detection. *3868*

Morphine *Urine No Effect Analytical* No interference on TLC using ethyl acetate: methanol: water: ammonium hydroxide and modified Dragendorff's reagent for detection. *3868*

Nicotine *Urine Positive Analytical* Similar R$_f$ and color reaction on TLC using ethyl acetate: methanol: water: ammonium hydroxide and modified Dragendorff's reagent. *3868*

Pentobarbital *Urine No Effect Analytical* No interference on TLC using ethyl acetate: methanol: water: ammonium hydroxide and modified Dragendorff's reagent for detection. *3868*

Phenobarbital *Urine No Effect Analytical* No interference on TLC using ethyl acetate: methanol: water: ammonium hydroxide and modified Dragendorff's reagent for detection. *3868*

Phenylpropanolamine *Urine No Effect Analytical* No interference on TLC using ethyl acetate: methanol: water: ammonium hydroxide and modified Dragendorff's reagent for detection. *3868*

Secobarbital *Urine No Effect Analytical* No interference on TLC using ethyl acetate: methanol: water: ammonium hydroxide and modified Dragendorff's reagent for detection. *3868*

Triglycerides *Serum Decrease Physiological* (not constant finding). *3505*

Fenofibrate

Cholesterol *Serum Decrease Physiological* Increase by 18% over 24 weeks in 92 type IIa given 100 mg 3 times per day with similar responses in type IIb patients when compared with baseline values. *0503*

Cholesterol, High Density Lipoprotein
Serum Increase Physiological Increase by 11% over 24 weeks in 92 type IIa given 100 mg 3 times per day with similar responses in type IIb patients when compared with baseline values. *0503* Increase by 16% in 9 hypertriglyceridemic type 2 diabetic patients given 100 mg tid after 1 mo treatment. *3385*

Cholesterol, Low Density Lipoprotein
Serum Increase Physiological Increase by 22% in 9 hypertriglyceridemic type 2 diabetic patients given 100 mg tid after 1 mo treatment. *3385*
Serum Decrease Physiological Increase by 20% over 24 weeks in 92 type IIa given 100 mg 3 times per day with similar responses in type IIb patients when compared with baseline values. *0503*

Cholesterol, Very Low Density Lipoprotein
Serum Decrease Physiological Increase by 38% over 24 weeks in 92 type IIa given 100 mg 3 times per day with similar responses in type IIb patients when compared with baseline values. *0503*

Triglycerides *Serum Decrease Physiological* Increase by 38% over 24 weeks in 92 type IIa given 100 mg 3 times per day with similar responses in type IIb patients when compared with baseline values. *0503* Increase by 50% in 9 hypertriglyceridemic type 2 diabetic patients given 100 mg tid after 1 mo treatment. *3385*

Triglycerides, Very Low Density Lipoprotein
Serum Decrease Physiological Increase by 60% in 9 hypertriglyceridemic type 2 diabetic patients given 100 mg tid after 1 mo treatment. *3385*

Fenoprofen

Alanine Aminotransferase *Serum No Effect Physiological* In 49 patients with juvenile rheumatoid arthritis when some abnormal results in patients with same therapeutic outcome given aspirin. *0465*

Aspartate Aminotransferase
Serum No Effect Physiological In 49 patients with juvenile rheumatoid arthritis when some abnormal results in patients with same therapeutic outcome given aspirin. *0465*

Bilirubin *Serum No Effect Physiological* At pharmacological concentrations has little or no protein displacement effect. *3748*

Creatinine *Serum Increase Physiological* In 2 isolated cases of patients with arthritis in the absence of hypertension. *0466*

Creatinine Clearance *Urine Decrease Physiological* Rare case of acute tubulointerstitial nephritis with acute renal failure (probable association with drug administration). *3582*

Erythrocytes *Blood Decrease Physiological* In 6.5% in patients treated for juvenile rheumatoid arthritis. *0239*
Urine Increase Physiological Rare case of acute tubulointerstitial nephritis with acute renal failure (probable association with drug administration). *3582*

Hemoglobin *Urine Increase Physiological* In 6.5% in patients treated for juvenile rheumatoid arthritis. *0239* Rare case of acute tubulointerstitial nephritis with acute renal failure (probable association with drug administration). *3582*

Leukocytes *Urine Increase Physiological* Rare case of acute tubulointerstitial nephritis with acute renal failure (probable association with drug administration). *3582*

Neutrophils *Blood Decrease Physiological* Occasional case of agranulocytosis reported. *3717*

Occult Blood *Feces No Effect Physiological* In 49 patients with juvenile rheumatoid arthritis when some abnormal results in patients with same therapeutic outcome given aspirin. *0465*
Feces Positive Physiological Significant effect but less than with aspirin. *2985* In 9% in patients treated for juvenile rheumatoid arthritis. *0239*

Platelets *Blood Decrease Physiological* Single case with approximately 1 g/d for 6 to 8 weeks, although patient also taking niacin. *3330*

Protein *Urine Increase Physiological* In 4% in patients treated for juvenile rheumatoid arthritis. *0239* In 2 isolated cases of patients with arthritis in the absence of hypertension. *0466*

Thyroid Stimulating Hormone (TSH)
Serum No Effect Physiological No effect noted physiologically or on methods. *3594*

Thyroxine (T₄) *Serum No Effect Physiological* No effect noted physiologically or on methods. *3594*

Tri-iodothyronine (T₃) *Serum Increase Analytical* Doubled in 2 volunteers over 2 weeks as also observed in some patients due to metabolite cross-reacting with antisera from Amersham and to lesser extent with that from Corning. *3594*

Tri-iodothyronine (T₃), Free *Serum Increase Analytical* Tripled in 2 volunteers over 2 weeks as also observed in some patients due to metabolite cross-reacting with antisera from Amersham and to lesser extent with that from Corning. *3594*

Fenoterol

Potassium *Serum Decrease Physiological* Reduction of up to 0.9 mmol/L in normal subjects with inhalation of drug. *1437*

Fenpipramide

Amphetamine *Urine No Effect Analytical* No effect at 100 mg/L on method of Rutter. *3097*

Fentanyl

Amylase *Serum Increase Physiological* May cause spasm of sphincter of Oddi. *0071*

Epinephrine *Plasma Increase Physiological* Mechanism obscure. *1744*

Hemoglobin *Plasma Increase Physiological* In a single patient with low haptoglobin following high dose administration dose related effect. *1719*
Urine Increase Physiological In a single patient with low haptoglobin following high dose administration. dose related effect. *1719*

Norepinephrine *Plasma Decrease Physiological* Mechanism obscure. *1744*

Ferricyanide

Hemoglobin *Urine Increase Analytical* Interferes with benzidine test. *1022*

Ferrous Ascorbate

Glucose Tolerance *Serum Decrease Physiological* Mechanism not reported. *1488*

Ferrous Gluconate

Acetaminophen Screening Test *Urine Negative Analytical* No reaction with o-cresol at therapeutic concentrations. *3326*

Occult Blood *Feces Positive Analytical* 50% false positive with Hemoccult®; 65% false positive with Hematest® in 10 male volunteers taking 300 mg 3 times per day for 1 week. *2170*

Ferrous Sulfate

Aspartate Aminotransferase *Serum No Effect Analytical* No effect on Karmen procedure. No effect on Babson procedure. *2760*

Doxycycline *Serum Decrease Physiological* Reduced absorption due to chelation which also affects elimination. *3794*

Glucose *Urine No Effect Analytical* No effect observed with TesTape®. *1100*
Urine Decrease Analytical Low with Clinistix®, Diastix®. *1100*

Methacycline *Serum Decrease Physiological* Reduced absorption due to chelation which also affects elimination. *3794*

Occult Blood *Feces Positive Analytical* 65% false positive with Hemoccult®; 25% false positive with Hematest® in 10 male volunteers taking 300 mg 3 times per day for 1 week. *2170*

Oxytetracycline *Serum Decrease Physiological* Reduced absorption due to chelation which also affects elimination. *3794*

Penicillamine *Serum Decrease Physiological* Reduced absorption due to chelation which also affects elimination. *3794*

Tetracycline *Serum Decrease Physiological* Reduced absorption due to chelation which also affects elimination. *3794*

Ferulic Acid

Ammoniacal Silver Nitrate *Test Conditions Positive Analytical* Positive spot test with Tollen's reagent. *3765*

Millons Test *Test Conditions Positive Analytical* Positive spot test for phenols. *3765*

Uric Acid *Test Conditions Increase Analytical* Positive spot test with phosphotungstate. *3765*

Urobilinogen *Test Conditions Increase Analytical* Positive spot test with Ehrlich's reagent. *3765*

Fever

BSP Retention *Serum Increase Physiological* In almost 50% patients with induced fever. *0375*

Glucose *Serum Decrease Physiological* Observed effect. *0117*

Fibrin Hydrolysate

Ammonia *Plasma Increase Physiological* Due to high ammonia content of solution. *3746*

Hemoglobin *Blood Decrease Physiological* ?due to septicemia or hypophosphatemia. *3746*

Phosphate *Serum Decrease Physiological* Intracellular transfer. *3746*

Fibrinogen

Erythrocyte Sedimentation Rate *Blood Increase Analytical* Increased plasma concentration accelerates rate. *0251*

Fibrinolysin

Aspartate Aminotransferase *Serum Increase Physiological* Hepatitis as late complication of administration. *0071*

Urea Nitrogen *Serum Increase Physiological* Renal failure reported (rare complication). *0071*

Filter Paper

Ammonia *Plasma Increase Analytical* Especially if acid-washed paper (absorbs NH₃ from air). *0579*

Barbiturate *Serum Increase Analytical* Affects diphenylcarbazone procedure if contains tin. *0583* May contain tin, react in Baer's procedure. *2300*
Serum Decrease Analytical If siliconized paper used for method of Broughton. *2935*

Calcium *Serum Increase Analytical* Ordinary paper contains high concentration. *0579*

Hemoglobin *Urine Increase Analytical* Interferes with benzidine test. *0459*

Lipids, Total *Serum Increase Analytical* If low fat paper not used in some methods. *0579*

Nonprotein Nitrogen *Serum Increase Analytical* If contains ammonia affects some methods. *1563*

Fish

Mercury *Serum Increase Physiological* High concentrations found in some fish-eaters. *3270*

Fish *(continued)*

Occult Blood *Feces Positive Analytical* Affect most methods if not withheld for 3 d. *0849*

Protein Bound Iodine (PBI) *Serum Decrease Physiological* May be associated with high Hg content of some fish. *2935*

Flavaspidic Acid

Bilirubin *Serum Increase Physiological* Inhibits UDP-glucuronyl transferase. *2621*

Bilirubin, Direct *Serum Increase Physiological* Inhibits uptake of bilirubin by liver cells. *0348*

BSP Retention *Serum Increase Physiological* Competition for hepatocellular binding sites. *2621*

Flavoxate

Eosinophils *Blood Increase Physiological* Single reversible case reported. *1680*

Leukocytes *Blood Decrease Physiological* Single reversible case reported. *1680*

Flecainide

Alkaline Phosphatase *Serum Increase Physiological* Mild increases in activity reported. *2011*

Digoxin *Serum Increase Physiological* Increases serum concentration. *2011*

Propranolol *Serum Increase Physiological* Increases hypotensive and negative inotropic effects as well as concentration. *2011*

Florantyrone

Alanine Aminotransferase *Serum Increase Physiological* Hepatotoxic effect. *2313*

Alkaline Phosphatase *Serum Increase Physiological* Hepatotoxic effect. *2313*

Aspartate Aminotransferase
Serum Increase Physiological Hepatotoxic effect. *2313*

Bile *Urine Increase Physiological* Hepatotoxic effect. *2313*

BSP Retention *Serum Increase Physiological* Hepatotoxic effect. *1022*

Cephalin Flocculation *Serum Increase Physiological* Hepatotoxic effect. *2313*

Eosinophils *Blood Increase Physiological* In patients with pre-existing liver disease. *2313*

Leukocytes *Blood Increase Physiological* In patients with pre-existing liver disease. *2313*

Thymol Turbidity *Serum Increase Physiological* Hepatotoxic effect. *2313*

Floxuridine

Alanine Aminotransferase *Serum Increase Physiological* Manifestation of toxicity. *1680*

Alkaline Phosphatase *Serum Increase Physiological* Hepatotoxic manifestation. *1680*

Aspartate Aminotransferase
Serum Increase Physiological Manifestation of toxicity. *1680*

Bilirubin *Serum Increase Physiological* Hepatotoxic manifestation. *1680*

BSP Retention *Serum Increase Physiological* Manifestation of toxicity. *1680*

Chromosomes *Test Conditions Abnormal Physiological* Clastogenic in human cells. *3282*

Erythrocytes *Blood Decrease Physiological* Bone marrow depression. *1680*

Erythrocyte Sedimentation Rate
Blood Increase Physiological Manifestation of toxicity. *1680*

Hematocrit *Blood Decrease Physiological* Bone marrow depression. *1680*

Hemoglobin *Blood Decrease Physiological* Bone marrow depression. *1680*

Lactate Dehydrogenase *Serum Increase Physiological* Manifestation of toxicity. *1680*

Leukocytes *Blood Decrease Physiological* Early indication of toxicity on bone marrow. *1680*

Occult Blood *Feces Positive Physiological* May cause gastrointestinal bleeding. *3850*

Platelets *Blood Decrease Physiological* Toxic effect on bone marrow. *3850*

Protein *Serum Decrease Physiological* Manifestation of toxicity. *1680*

Prothrombin Time *Plasma Increase Physiological* Manifestation of toxicity. *1680*

Flubiprofen

Bilirubin *Serum No Effect Physiological* At pharmacological concentrations has little or no effect on displacement from protein. *3748*

Flucloxacillin

Alanine Aminotransferase *Serum Increase Physiological* In a single patient on hemodialysis: abnormal results only mild. *2200*

Alkaline Phosphatase *Serum Increase Physiological* In a single patient on hemodialysis: abnormal results only mild. *2200*

Aspartate Aminotransferase
Serum Increase Physiological In a single patient on hemodialysis: abnormal results only mild. *2200*

Bilirubin *Serum No Effect Physiological* In a single patient on hemodialysis: abnormal results only mild. *2200*

Chloramphenicol *Serum No Effect Analytical* No effect at 100 mg/L on coupled enzymatic method. *2490*

Lactate Dehydrogenase *Serum No Effect Physiological* In a single patient on hemodialysis: abnormal results only mild. *2200*

Flucytosine

Alanine Aminotransferase *Serum Increase Physiological* Reversible hepatotoxicity in 10%. *3443*

Alkaline Phosphatase *Serum Increase Physiological* Reversible hepatotoxicity in 10%. *3443* Occurs in about 25% treated patients: usually mild but necessitates discontinuation of therapy. *3567*

Aspartate Aminotransferase
Serum Increase Physiological Reversible hepatotoxicity in 10%. *3443* Occurs in about 25% treated patients: usually mild but necessitates discontinuation of therapy. *3567*

Bilirubin *Serum Increase Physiological* Reversible hepatotoxicity in 10%. *3443*

Cephalin Flocculation *Serum Increase Physiological* Reversible hepatotoxicity in 10%. *3443*

Creatinine *Serum No Effect Analytical* At 100 µg/mL on Jaffé methods on Technicon® SMAC® and Du Pont aca. *3404* *Serum Increase Analytical* At 10 µg/mL and above on Ektachem® 700 2 slide procedure: change in proportion to concentration. *3404* *Serum Increase Physiological* May have nephrotoxic effect. *1679*

Erythrocytes *Blood Decrease Physiological* May depress bone marrow function. *1679* Usually mild, occurs in 8 to 13% patients: occasionally pancytopenia occurs. *3567*

Hemoglobin *Blood Decrease Physiological* May depress bone marrow function. *1679*

Leukocytes *Blood Decrease Physiological* Reported effect (with agranulocytosis). *3443* Usually mild, occurs in 8 to 13% patients: occasionally pancytopenia occurs. *3567*

Neutrophils *Blood Decrease Physiological* Occasional case of agranualocytosis reported. *3717*

pH *Gastric Material Decrease Physiological* Effect on both basal and submaximal pentagastrin simulated gastric acid secretion. Effect dose related. *0559*

Platelets *Blood Decrease Physiological* Occasional thrombocytopenia observed. *1679* Usually mild, occurs in 8 to 13% patients: occasionally pancytopenia occurs. *3567*

Thymol Turbidity *Serum Increase Physiological* Reversible hepatotoxicity in 10%. *3443*

Urea Nitrogen *Serum Increase Physiological* May have nephrotoxic effect. *1679*

Volume *Gastric Material Decrease Physiological* Effect on both basal and submaximal pentagastrin-stimulated gastric acid secretion. Effect dose related. *0559*

Fludrocortisone

Aldosterone *Urine Decrease Physiological* Marked reduction following administration. *3174*

Amylase *Serum Increase Physiological* May cause hemorrhagic pancreatitis. *1680*

Bicarbonate *Serum Increase Physiological* May cause hypokalemic alkalosis. *1680*

Glucose *Serum Increase Physiological* Endocrine response. *1680*
Urine Increase Physiological Endocrine response. *1680*

Glucose Tolerance *Serum Decrease Physiological* Endocrine response. *1680*

Nitrogen Balance *Patient Negative Physiological* Due to protein catabolism. *1680*

Occult Blood *Feces Positive Physiological* May activate peptic ulcer. *1680*

Potassium *Serum Decrease Physiological* May cause increased urinary excretion. *0071*
Urine Increase Physiological Increases urinary elimination. *0071*

Sodium *Serum Increase Physiological* May cause sodium retention and edema. *0071*

Tetrahydrodeoxycorticosterone
Urine Decrease Physiological Slight effect following administration. *3174*

Flufenamic Acid

Alanine Aminotransferase *Serum Increase Physiological* Transient elevation reported. *2427*

Aspartate Aminotransferase
Serum Increase Physiological Transient elevation reported. *2427*

Bile *Urine Increase Analytical* Probably due to interfering metabolite. *1488*

Uric Acid *Serum Decrease Physiological* Uricosuric action of drug. *3013*

Flumecinolone

Bilirubin *Serum Decrease Physiological* Inhibits hyperbilirubinemia of term or premature newborns due to enzyme induction. *2304*

Flumethiazide

Uric Acid *Serum Increase Physiological* Inhibition of tubular secretion of urate. *1433*

Flunarizine

Follicle Stimulating Hormone (FSH)
Plasma No Effect Physiological After 90 d in 8 women. *2265*

Luteinizing Hormone (LH) *Plasma No Effect Physiological* After 90 d in 8 women. *2265*

Lymphocytic Response to PHA
Blood Increase Physiological Enhanced response suggesting differential sensitivity of lymphocytes to calcium-entry blockers. *0482*

Prolactin *Plasma Increase Physiological* Significant increase of baseline value in men and women. *2265*

Thyroid Stimulating Hormone (TSH)
Serum No Effect Physiological Basal concentration not affected but increase of domperidone blunted in women. *2265*
Serum Increase Physiological Significant increase in men noted. *2265*

Fluocinolone

Corticotropin *Plasma Decrease Physiological* About 23% reduced when 30 g/d drug applied as topical ointment over large area. *0069*

Cortisol *Plasma Decrease Physiological* About 43% suppression in 4 d when 30 g/d drug applied as topical ointment over large area. *0069*

α-Melanocyte Stimulating Hormone
Plasma No Effect Physiological When 30 g/d drug applied as topical ointment over large area. *0069*

Fluorescein

Alkaline Phosphatase *Serum Increase Analytical* Slight effect at 200 mg/L on Ektachem® 700 method. *1722*

Amylase *Serum Increase Analytical* At expected concentration of 1.0 to 100 mg/L interfered with method on Ektachem®-700. *1722*
Urine No Effect Analytical In specimen spiked to 250 mg/L on Ektachem® 700 method. *1722*

Aspartate Aminotransferase *Serum Increase Analytical* Slight effect at 200 mg/L on Ektachem® 700 method. *1722*

Bilirubin *Serum No Effect Analytical* Minimal interference at expected concentration on Ektachem® 700. *1722*
Serum Decrease Analytical By up to 25% at up to 100 mg/L on Beckman Astra method. *1722*

Bilirubin, Conjugated *Serum Increase Analytical* At expected concentration of 1 to 100 mg/L interfered with method on Ektachem® 700. *1722*

Bilirubin, Direct *Serum Increase Analytical* At expected concentration of 1.0 to 100 mg/L interfered with method on Ektachem®-700. *1722*

Bilirubin, Neonatal *Serum Decrease Analytical* Minimal interference at expected concentration on Ektachem® 700. *1722*

Bilirubin, Unconjugated *Serum Decrease Analytical* At expected concentration of 1 to 100 mg/L interfered with method on Ektachem® 700. *1722*

Cholesterol *Serum No Effect Analytical* No effect at 200 mg/L on Ektachem® 700 method. *1722*

Color *Urine Increase Analytical* Intravenous administration may cause yellow-orange color. *1488*

Cortisol *Plasma Negative Analytical* No result obtainable at concentration of 1.0 mg/L on Abbott TD$_x$ and carryover effect into subsequent specimen. *1722*

Digoxin *Serum Negative Analytical* No result obtainable at concentration of 1.0 mg/L on Abbott TD$_x$ and carryover effect into subsequent specimen. *1722*

Glucose *Serum No Effect Analytical* No effect at 200 mg/L on Ektachem® 700 method. *1722*
Urine No Effect Analytical At 250 mg/L in spiked specimen run on Ames Clinitek 200. *1722*

Hemoglobin *Urine No Effect Analytical* At 250 mg/L in spiked specimen run on Ames Clinitek 200. *1722*

Ketones *Urine No Effect Analytical* At 250 mg/L in spiked specimen run on Ames Clinitek 200. *1722*

Leukocyte Esterase *Urine No Effect Analytical* At 250 mg/L in spiked specimen run on Ames Clinitek 200. *1722*

Nitrite *Urine No Effect Analytical* At 250 mg/L in spiked specimen run on Ames Clinitek 200. *1722*

pH *Urine No Effect Analytical* At 250 mg/L in spiked specimen run on Ames Clinitek 200. *1722*

Protein *Serum No Effect Analytical* No effect at 200 mg/L on Ektachem® 700 method. *1722*
Urine No Effect Analytical At 250 mg/L in spiked specimen run on Ames Clinitek 200. *1722*

Triglycerides *Serum No Effect Analytical* No effect at 200 mg/L on Ektachem® 700 method. *1722*

Urobilin *Urine Increase Analytical* Produces yellow-green fluorescence. *1563*

Fluorides

Acid Phosphatase *Serum Decrease Analytical* Enzyme activity inhibited by fluoride. *0011*
Serum Decrease Physiological In vivo inhibition observed. *1168*
White Blood Cells No Effect Analytical Phenylphosphate enzyme no effect of 15 mmol/L. *0183*
White Blood Cells Decrease Analytical Glycerophosphate enzyme inhibited by 15 mmol/L. *0183*

Alkaline Phosphatase *Serum Decrease Analytical* Inhibitory action on enzyme. *1563*
Serum Decrease Physiological In vivo inhibition with doses for fluoridation. *1112*

Ammonia *Plasma Increase Analytical* Cause increase if used as anticoagulants. *0705*

Amylase *Serum Increase Analytical* Effect of fluoride, activates amylase. *3588*
Serum Decrease Analytical Sodium salt inhibits by 4% at 10 g/L method of Rauscher et al. *2926*

Aspartate Aminotransferase *Serum Increase Analytical* Increased activity observed with colorimetric method. *3165*
Serum Increase Physiological Due to anoxia with tissue damage. *0010*
Serum Decrease Physiological In vivo inhibition with low doses. *1168*

Bicarbonate *Serum Decrease Analytical* Alteration of pH with loss of CO_2. *3879*

Calcium *Serum Decrease Analytical* May precipitate calcium (forms insoluble salt). *1563*
Serum Decrease Physiological Combines with calcium, accretion in bone crystals. *0010*
Feces Decrease Physiological Reduces loss in bone disorders. *1343*

Chloride *Serum Decrease Analytical* 0.05 meq/L at 20 meq/L in method of Fingerhut. *1135*

Cholesterol *Serum Decrease Analytical* Value lower by 30-50 mg/dL. *2319*

Eosinophils *Blood Increase Physiological* Eosinophilia reported as allergic reaction. *2427*

Erythrocytes *Blood Decrease Physiological* May cause aplastic anemia. *0279*

Erythrocyte Sedimentation Rate
Blood Decrease Analytical Retards rate when compared with heparin. *0459*

Fluoride *Serum Increase Physiological* Due to absorbed material in poisoning. *0010*

Glucose *Serum No Effect Analytical* No effect on hexokinase procedure. *2982* At 2.5 g/dL on alkaline ferricyanide procedure. At 2.5 g/dL on p-HBAH procedure of Lever. At 2.5 g/dL on o-toluidine procedure. At 2.5 g/dL on glucose oxidase procedures. *2140*
Serum Increase Analytical If greater than 5 mg/mL affects o-toluidine procedure. *0724*

Hemoglobin *Blood Decrease Physiological* May cause aplastic anemia. *0279*

Lactate Dehydrogenase *Serum Increase Physiological* Due to anoxia with tissue damage. *0010*
Serum Decrease Physiological In vivo inhibition with low doses. *1168*

Lactate Dehydrogenase Isoenzymes
Serum Increase Physiological Mainly fraction 4 and 5 in poisoning. *0010*

Leukocytes *Blood Decrease Physiological* May cause aplastic anemia. *0279*

Lipid Glycerol *Serum Decrease Analytical* Inhibits phospholipase C method of Horney. *2683*

Occult Blood *Feces Positive Physiological* Due to hemorrhagic gastroenteritis if ingested. *0010*

pH *Blood Decrease Physiological* Combined respiratory and metabolic acidosis in poisoning. *0010*

Platelets *Blood Decrease Physiological* May cause aplastic anemia. *0279*

Potassium *Serum Increase Analytical* With inhibition of phosphorylation. *1563*

Protein *Urine Increase Physiological* May cause renal damage. *0279*

Protein Bound Iodine (PBI) *Serum No Effect Physiological* At concentrations used to cause fluoridation of water. *0830*
Serum Decrease Physiological Large doses interfere with I_2 trapping by thyroid. *0012*

Prothrombin Time *Plasma Increase Physiological* Mechanism not discussed causes bleeding. *0010*

Pseudocholinesterase *Serum Decrease Analytical* Total inhibition occurs. *3651*

Sodium *Serum Increase Analytical* If salt of fluoride. *1563*

Urea Nitrogen *Serum No Effect Analytical* No effect on urease if concentration less than 2 mg/ml. *0279*
Serum Increase Physiological May cause renal damage. *0279*
Serum Decrease Analytical Urease activity inhibited by fluoride. *0579*

Uric Acid *Serum Increase Physiological* ?related to lactic acidemia. *0010*

Fluoroborate

Protein Bound Iodine (PBI) *Serum Decrease Physiological* Interferes with trapping of iodide by thyroid. *0012*

5-Fluoronicotinic Acid

Fatty Acids, Free (FFA) *Serum Decrease Physiological* Max effect of 50% at 2 h, starts in 15 minutes. *3077*

Glucose *Serum No Effect Physiological* No effect seen in normals. *3077*

Fluorophenylalanine

Chromosomes *Test Conditions Abnormal Physiological* Clastogenic in human lymphocyte cultures. *3282*

Fluoropyrimidine

Protein Bound Iodine (PBI) *Serum Increase Physiological* Mechanism obscure. *0012*

Fluorouracil

Bilirubin *Serum No Effect Analytical* At concentration of 130 mg/L had no effect on Jendrassik and Grof method. *3393*
Serum Increase Physiological Single case reported. *1488*

Calcium *Serum No Effect Analytical* At concentration of 130 mg/L had no effect on cresolphthalein method. *3393*

Erythrocytes *Blood Decrease Physiological* May cause bone marrow depression. *0071*

5-Hydroxyindoleacetic Acid (5-HIAA)
Urine Increase Physiological In patients with carcinoid, due to cell destruction. *1488*

Kininogen *Plasma Decrease Physiological* Maximum effect with onset of leukopenia. *0779*

Leukocytes *Blood Decrease Physiological* May cause bone marrow depression. *0071*

Occult Blood *Feces Positive Physiological* Manifestation of toxicity. *1680*

Phosphate *Serum No Effect Analytical* At concentration of 130 mg/L had no effect on phosphomolybdate method. *3393*

Platelets *Blood Decrease Physiological* May cause bone marrow depression. *0071*

Protein *Serum No Effect Analytical* At concentration of 130 mg/L had no effect on biuret method with blank correction. *3393*

Protein Bound Iodine (PBI) *Serum No Effect Physiological* No effect observed. *0830*

Thyroxine (T₄) *Serum Increase Physiological* Increased binding capacity (augmentation of transport proteins) *3798*

Tri-iodothyronine (T₃) *Serum Increase Physiological* Increased binding capacity (augmentation of transport proteins) *3798*

Urea Nitrogen *Serum No Effect Analytical* At concentration of 130 mg/L had no effect on diacetylmonoxime method. *3393*

Uric Acid *Serum No Effect Analytical* At concentration of 130 mg/L had no effect on phosphotungstate reduction method. *3393*

Fluosol-DA

Alanine Aminotransferase *Serum No Effect Analytical* On SMA IIC at concentration of 50%. *2518*

Serum Decrease Analytical Increase by 86% on Du Pont aca III at concentration of 50%. *2518*

Albumin *Serum No Effect Analytical* On SMA IIC at concentration of 50% On Du Pont aca III at concentration of 50%. *2518*

Serum Increase Analytical Albumin results very variable but apparent effect. *2518*

Alkaline Phosphatase *Serum No Effect Analytical* On SMA IIC at concentration of 50%. On Du Pont aca III at concentration of 50%. *2518*

Aspartate Aminotransferase *Serum No Effect Analytical* On SMA IIC at concentration of 50%. *2518*

Serum Decrease Analytical Increase by 100% on Du Pont aca III at concentration of 50%. *2518*

Bicarbonate *Serum No Effect Analytical* On SMA IIC at concentration of 50%. *2518* By 18% on Kodak Ektachem® 400 at concentration of 50%. *2518*

Bilirubin *Serum No Effect Analytical* On SMA IIC at concentration of 50%. *2518*

Serum Increase Analytical Increase by 500% on Du Pont aca III at concentration of 50%. *2518*

Calcium *Serum No Effect Analytical* On SMA IIC at concentration of 50%. By 18% on Kodak Ektachem® 400 at concentration of 50%. *2518* On Du Pont aca III at concentration of 50%. *2518*

Carbamazepine *Serum No Effect Analytical* As measured on Du Pont aca III at concentration up to 50%. *2518*

Chloride *Serum Increase Analytical* Unpredictable variation in concentrations as measured on SMA IIC, Beckman Astra 8 and Ektachem® 400. *2518*

Cholesterol *Serum No Effect Analytical* On Du Pont aca III at concentration of 50%. *2518*

Serum Increase Analytical Increase by 20% on SMA IIC at concentration of 50%. *2518*

Creatine Kinase Isoenzymes *Serum No Effect Analytical* No effect on electrophoresis although added fluorescent band. *2518*

Creatinine *Serum No Effect Analytical* On SMA IIC at concentration of 50%. *2518*

Digoxin *Serum No Effect Analytical* As measured by Abbott TD$_x$ at concentration up to 50%. *2518*

Diphenylhydantoin *Serum No Effect Analytical* As measured on Du Pont aca III at concentration up to 50%. *2518*

Glucose *Serum No Effect Analytical* On SMA IIC at concentration of 50%. *2518* By 18% on Kodak Ektachem® 400 at concentration of 50%. *2518*

Serum Increase Analytical Increase by 14% on Du Pont aca III at concentration of 50%. *2518*

Lactate Dehydrogenase *Serum Increase Analytical* Increase by 37% on SMA IIC at concentration of 50%. Increase by 18% on Du Pont aca III at concentration of 50%. *2518*

Lipoprotein Electrophoresis *Serum No Effect Analytical* Not affected by presence of compound in specimen. *2518*

Magnesium *Serum No Effect Analytical* No effect at 50% on atomic absorption measurement. *2518*

Phenobarbital *Serum No Effect Analytical* As measured on Du Pont aca III at concentration up to 50%. *2518*

Phosphate *Serum Increase Analytical* Increase by 16% on SMA IIC at concentration of 50%. Increase by 82% on Du Pont aca III at concentration of 50%. *2518*

Potassium *Serum No Effect Analytical* By 18% on Kodak Ektachem® 400 at concentration of 50%. *2518*

Serum Increase Analytical Icrease by 33% on SMA IIC at concentration of 50%. Increase by 23% on SMA IIC at concentration of 50%. *2518*

Protein *Serum No Effect Analytical* On SMA IIC at concentration of 50%. *2518*

Serum Increase Analytical By 18% on Kodak Ektachem® 400 at concentration of 50%. *2518*

Serum Decrease Analytical Increase by 20% on Du Pont aca III at concentration of 50%. *2518*

Protein Electrophoresis *Serum No Effect Analytical* Not affected by presence of compound in specimen. *2518*

Sodium *Serum Increase Analytical* Unpredictable variation in concentrations as measured on SMA IIC, Beckman Astra 8 and Ektachem® 400. *2518*

Theophylline *Serum No Effect Analytical* As measured by Syva® EMIT at concentration up to 50%. *2518*

Triglycerides *Serum Increase Analytical* Directly proportional to Fluosinal concentration since emulsion contains glycerol. *2518*

Urea Nitrogen *Serum No Effect Analytical* On SMA IIC at concentration of 50%. *2518* By 18% on Kodak Ektachem® 400 at concentration of 50%. *2518*

Uric Acid *Serum No Effect Analytical* On SMA IIC at concentration of 50%. On Du Pont aca III at concentration of 50%. *2518*

Fluoxymesterone

Alanine Aminotransferase *Serum Increase Physiological* Intrahepatic cholestatic jaundice. *1343*

Alkaline Phosphatase *Serum Increase Physiological* Intrahepatic cholestatic jaundice. *1343*

Aspartate Aminotransferase
Serum Increase Physiological Intrahepatic cholestatic effect. *0019*

Bile *Urine Increase Physiological* Intrahepatic cholestatic jaundice. *1343*

Bilirubin *Serum Increase Physiological* Intrahepatic cholestatic jaundice. *1343*

BSP Retention *Serum Increase Physiological* ?hepatotoxic, cholestatic effect. *1237*

Calcium *Serum Increase Physiological* May occur in immobilized patients or if cancer. *1678*

Cholesterol *Serum Increase Physiological* Observed effect. *1680*
Serum Decrease Physiological Observed effect. *1680*

Creatine *Urine Increase Physiological* For up to 2 weeks when treatment stopped. *1680*

Creatinine *Urine Increase Physiological* For up to 2 weeks when treatment stopped. *1680*

Erythropoietin *Serum Increase Physiological* Observed if anemia of renal failure. *1048*

Factor II *Plasma Increase Physiological* Metabolic effect. *0019*

Factor V *Plasma Increase Physiological* Metabolic effect. *0019*

Factor VII *Plasma Increase Physiological* Metabolic effect. *0019*

Factor X *Plasma Increase Physiological* Metabolic effect. *0019*

Glucose *Serum Increase Physiological* Endocrine response. *1680*
Serum Decrease Physiological Metabolic action. *0019*

Glucose Tolerance *Serum Increase Physiological* Metabolic effect. *0019*

β-Glucuronidase *Serum Increase Physiological* Metabolic effect. *0227*

Haptoglobin *Serum Increase Physiological* Metabolic effect. *0227*

Hematocrit *Blood Increase Physiological* Observed if anemia of renal failure. *1048*

^{131}I Uptake *Serum Decrease Physiological* Metabolic effect. *0019*

17-Ketosteroids *Urine Decrease Physiological* Endogenous hormone suppression. *1680*

Leukocytes *Blood Decrease Physiological* Leukopenia may occur. *0019*

Metyrapone Test *Patient Positive Physiological* At concentration of 23 mg/L had no effect on Ektachem® method. *0019*

Fluoxymesterone *(continued)*

Plasminogen *Plasma Increase Physiological* Metabolic effect. *0227*

Protein Bound Iodine (PBI) *Serum Decrease Physiological* Due to decreased binding capacity (androgen effect). *1680*

Prothrombin Time *Plasma Increase Physiological* Increased sensitivity to anticoagulant reported. *0019*

Sperm Count *Semen Decrease Physiological* After prolonged administration or excess dosage. *1678*

T₃ Uptake *Serum Increase Physiological* Affects RBC and resin uptake. *0019*

Thyroxine (T₄) Binding Globulin *Serum Decrease Physiological* Metabolic effect. *1680*

Thyroxine (T₄), Free *Serum No Effect Physiological* No change observed. *0019*

Volume *Plasma Increase Physiological* With retention of water and electrolytes. *1680*

Fluphenazine

Alanine Aminotransferase *Serum Increase Physiological* Hypersensitivity response. *2313* Isolated case of hepatotoxicity with centrilobular cholestasis. *3372*

Alkaline Phosphatase *Serum Increase Physiological* Hypersensitivity response. *2313* Isolated case of hepatotoxicity with centrilobular cholestasis. *3372*

Alkaloids *Urine No Effect Analytical* With ninhydrin on Frings TLC procedure at 10 mg/dL. *1204*

Aspartate Aminotransferase
Serum Increase Physiological Hypersensitivity response. *2313* Isolated case of hepatotoxicity with centrilobular cholestasis. *3372*

Barbiturate *Urine No Effect Analytical* With diphenylcarbazone on Frings TLC procedure at 10 mg/dL. With mercurous NO₃ on Frings TLC procedure at 10 mg/dL. *1204*
Urine Increase Analytical Orange spot with vanillin on Frings procedure. Orange spot with mercuric SO₄ in Frings TLC procedure. *1204*

Bile *Urine Increase Physiological* Hypersensitivity response. *2313*

Bilirubin *Serum Increase Physiological* Hypersensitivity response. *2313*
Urine Increase Physiological Hypersensitivity response. *2313*

Bilirubin, Direct *Serum Increase Physiological* Relatively high compared with total. *1343*

Bromide *Urine No Effect Analytical* No effect on method of Frings. *1204*

BSP Retention *Serum Increase Physiological* Hypersensitivity response. *2313*

Cephalin Flocculation *Serum Increase Physiological* Hypersensitivity (maybe normal when other tests not). *2313*

Chlordiazepoxide *Blood Increase Analytical* 2 mg/dL equivalent to 1.0 mg/dL by method of Riddick. *2983*

Eosinophils *Blood Increase Physiological* Allergic response with phenothiazines. *1679*

Erythrocytes *Blood Decrease Physiological* Rare pancytopenia with phenothiazines. *1679*

FPN Test *Urine No Effect Analytical* Negative FPN test with 1 mg/dL concentration in Frings TLC procedure. *1204*
Urine Positive Analytical Pink FPN test with 2 mg/dL concentration in Frings TLC procedure. *1204*

Lactate Dehydrogenase *Serum Increase Physiological* Isolated case of hepatotoxicity with centrilobular cholestasis. *3372*

Leukocytes *Blood Decrease Physiological* Agranulocytosis/leukopenia rarely. *1679*

Neutrophils *Blood Decrease Physiological* Occasional case of agranulocytosis reported. *3717*

Platelets *Blood Decrease Physiological* Rare thrombocytopenia with/without purpura. *1679* Significant reduction observed in 18 psychiatric patients treated for at least 1 y although mean value still in normal range. *1642*

Pregnancy Tests *Urine Positive Analytical* False reactions with phenothiazines. *1679*

Prolactin *Plasma Increase Physiological* Typical dose-related response to i.m. administered drug due to antidopaminergic action. *2072* Marked increase in male and female psychiatric patients treated for up to 4 weeks. *3630*

Salicylate *Urine No Effect Analytical* Up to 10 mg/dL no effect on Trinder procedure. *1204*

Thymol Turbidity *Serum Increase Physiological* Hypersensitivity response. *2313*

Flurazepam

Alanine Aminotransferase *Serum No Effect Analytical* At 5 times upper limit of therapeutic range on methods on SMAC®, Abbott-VP, aca, Cobas-Bio, and KDA. *2138* At acute overdose concentration (2.5 mg/dL) on Technicon® SMAC® method. *3719*
Serum Increase Physiological May cause hepatic toxicity. *3016*

Albumin *Serum No Effect Analytical* At 5 times upper limit of therapeutic range on methods on SMAC®, Ektachem®, Hitachi® 705 and KDA. *2138* At acute overdose concentration (2.5 mg/dL) on Technicon® SMAC® method. *3719* At concentration of 25 mg/L had no effect on BCG method. *3393*

Alkaline Phosphatase *Serum No Effect Analytical* At 5 times upper limit of therapeutic range on methods on SMAC®, Abbott-VP, Cobas-Bio, aca, Hitachi® 705 and KDA. *2138* At acute overdose concentration (2.5 mg/dL) on Technicon® SMAC® method. *3719*
Serum Increase Physiological May cause hepatic toxicity. *3016*

Alkaloids *Urine No Effect Analytical* With ninhydrin on Frings TLC procedure at 1 mg/dL. *1204*
Urine Increase Analytical Reacts with ninhydrin on TLC method of Frings Purple with iodoplatinate on Frings TLC procedure. *1204*

Amylase *Serum No Effect Analytical* At 5 times upper limit of therapeutic range on methods on aca, Cobas-Bio, and Ektachem®. *2138*

Aspartate Aminotransferase *Serum No Effect Analytical* At 5 times upper limit of therapeutic range on methods on SMAC®, Abbott-VP, Cobas-Bio, aca, Hitachi® 705 and KDA. *2138* At acute overdose concentration (2.5 mg/dL) on Technicon® SMAC® method. *3719*
Serum Increase Physiological May cause hepatic toxicity. *3016*

Barbiturate *Urine No Effect Analytical* With mercuric SO₄ on Frings TLC procedure at 10 mg/dL. With mercurous NO₃ on Frings TLC procedure at 10 mg/dL. With vanillin on Frings procedure at 10 mg/dL. With diphenylcarbazone on Frings TLC procedure at 10 mg/dL. *1204*

Bicarbonate *Serum No Effect Analytical* At concentration of 0.1 mg/L had no effect on method using phenolphthalein. *3393*

Bile *Urine Increase Physiological* May cause hepatotoxicity. *3016*

Bilirubin *Serum No Effect Analytical* At 5 times upper limit of therapeutic range on methods on SMAC®, Cobas-Bio, aca, Ektachem®, Hitachi® 705 and KDA. *2138* At acute overdose concentration (2.5 mg/dL) on Technicon® SMAC® method. *3719* At concentration of 25 mg/L had no effect on Jendrassik and Grof method. *3393*
Serum Increase Physiological May cause hepatic toxicity. *3016*

Bilirubin, Direct *Serum Increase Physiological* May cause hepatic toxicity. *3016*

Bromide *Urine No Effect Analytical* No effect on method of Frings. *1204*

BSP Retention *Serum Increase Physiological* May cause hepatotoxicity. *3016*

Calcium *Serum No Effect Analytical* At 5 times upper limit of therapeutic range on methods on SMAC®, Abbott-VP, aca, Ektachem®, Hitachi® 705 and KDA. *2138* At acute overdose concentration (2.5 mg/dL) on Technicon® SMAC® method. *3719* At concentration of 25 mg/L had no effect on cresolphthalein method. *3393*

Cephalin Flocculation *Serum Increase Physiological* May cause hepatotoxicity. *3016*

Chlordiazepoxide *Blood No Effect Analytical* On method of Riddick at 5 mg/dL. *2983*

Chloride *Serum No Effect Analytical* At concentration of 0.1 mg/L had no effect on mercurimetric method. *3393*

Cholesterol *Serum No Effect Analytical* At 5 times upper limit of therapeutic range on methods on SMAC®, Abbott-VP, Cobas-Bio, Ektachem®, Hitachi® 705 and KDA. *2138* At acute overdose concentration (2.5 mg/dL) on Technicon® SMAC® method. *3719* At concentration of 0.1 mg/L had no effect on CHOD-PAP method. *3393* At concentration of 25 mg/L had no effect on Liebermann-Burchard method. *3393*

Creatine Kinase *Serum No Effect Analytical* At 5 times upper limit of therapeutic range on methods on SMAC®, Abbott-VP, aca, Cobas-Bio, and Hitachi® 705. *2138* At acute overdose concentration (2.5 mg/dL) on Technicon® SMAC® method. *3719*

Creatinine *Serum No Effect Analytical* At 5 times upper limit of therapeutic range on methods on SMAC®, Abbott-VP, aca, Cobas-Bio, Ektachem®, Hitachi® 705, and KDA. *2138* At acute overdose concentration (2.5 mg/dL) on Technicon® SMAC® method. *3719* At concentration of 25 mg/L had no effect on AutoAnalyzer Jaffé method. *3393*

Ethchlorvynol *Urine No Effect Analytical* No effect on method of Frings and Cohen at 10 mg/dL. *1205*

FPN Test *Urine No Effect Analytical* No effect at 10 mg/dL on method of Frings. *1204*

Glucose *Serum No Effect Analytical* At 5 times upper limit of therapeutic range on methods on SMAC®, Abbott-VP, aca, Cobas-Bio, Ektachem®, Hitachi® 705, and KDA. *2138* At acute overdose concentration (2.5 mg/dL) on Technicon® SMAC® method. *3719* At concentration of 0.1 mg/L had no effect on GOD/POD-PAP method. *3393*
Urine No Effect Analytical No effect observed with Tes-Tape®. *1100*
Urine Decrease Analytical Low with Clinistix®, Diastix®. *1100*

γ-Glutamyltransferase (GGT) *Serum No Effect Analytical* At 5 times upper limit of therapeutic range on methods on SMAC®, Abbott-VP, and Hitachi® 705. *2138*

5-Hydroxyindoleacetic Acid (5-HIAA) *Urine Increase Analytical* Slight effect when added *in vitro* in method of Udenfriend. *0670*

Iron *Serum No Effect Analytical* At 5 times upper limit of therapeutic range on methods on Ferrozine method on SMAC®. *2138* At acute overdose concentration (2.5 mg/dL) on Technicon® SMAC® method. *3719* At concentration of 25 mg/L had no effect on Ferrozine method. *3393*

Lactate Dehydrogenase *Serum No Effect Analytical* At 5 times upper limit of therapeutic range on methods on SMAC®, Abbott-VP, aca, Cobas-Bio, Hitachi® 705 and KDA. *2138* At acute overdose concentration (2.5 mg/dL) on Technicon® SMAC® method. *3719*

Phosphate *Serum No Effect Analytical* At 5 times upper limit of therapeutic range on methods on SMAC®, aca, Hitachi® 705 and KDA. *2138* At acute overdose concentration (2.5 mg/dL) on Technicon® SMAC® method. *3719* At concentration of 25 mg/L had no effect on phosphomolybdate method. *3393*

Potassium *Serum No Effect Analytical* At concentration of 0.1 mg/L had no effect on flame-photometric method. *3393*

Protein *Serum No Effect Analytical* At 5 times upper limit of therapeutic range on methods on SMAC®, Abbott-VP, Ektachem®, Hitachi® 705 and KDA. *2138* At acute overdose concentration (2.5 mg/dL) on Technicon® SMAC® method. *3719* At concentration of 25 mg/L had no effect on biuret method with blank correction. *3393*

Salicylate *Urine No Effect Analytical* At 10 mg/dL no effect on Trinder method. *1204*

Sodium *Serum No Effect Analytical* At concentration of 0.1 mg/L had no effect on flame-photometric method. *3393*

Sugar *Urine Increase Analytical* Excreted as glucuronide, sulfate — affect Benedict's. *3216*

Thymol Turbidity *Serum Increase Physiological* May cause hepatotoxicity. *3016*

Triglycerides *Serum No Effect Analytical* At 5 times upper limit of therapeutic range on methods on SMAC®, Abbott-VP, Ektachem®, Hitachi® 705 and KDA. *2138* At acute overdose concentration (2.5 mg/dL) on Technicon® SMAC® method. *3719* At concentration of 25 mg/L had no effect on lipase/esterase method. *3393*

Urea Nitrogen *Serum No Effect Analytical* At 5 times upper limit of therapeutic range on methods on SMAC®, Abbott-VP, Cobas-Bio, Ektachem®, Hitachi® 705 and KDA. *2138* At acute overdose concentration (2.5 mg/dL) on Technicon® SMAC® method. *3719* At concentration of 25 mg/L had no effect on diacetylmonoxime method. *3393*

Uric Acid *Serum No Effect Analytical* At 5 times upper limit of therapeutic range on methods on SMAC®, Abbott-VP, aca, Cobas-Bio, Ektachem®, Hitachi® 705, and KDA. *2138* At acute overdose concentration (2.5 mg/dL) on Technicon® SMAC® method. *3719* At concentration of 25 mg/L had no effect on phosphomolybdate reduction method. *3393*

Volume *Saliva Increase Physiological* Excessive salivation may occur. *1679*

Flurazepam Metabolite

Alkaloids *Urine Increase Analytical* Purple with iodoplatinate on Frings TLC procedure. *1204*

Fluroxene

Alanine Aminotransferase *Serum Increase Physiological* Potentially hepatotoxic. *0071*

Aspartate Aminotransferase
Serum Increase Physiological Potentially hepatotoxic. *0071*

Bilirubin *Serum Increase Physiological* Potentially hepatotoxic. *0071*

Bleeding Time *Blood Increase Physiological* Prolonged during anesthesia, normal in 24 h. *0071*

BSP Retention *Serum Increase Physiological* Potentially hepatotoxic. *0071*

Factor V *Plasma No Effect Physiological* No significant effect noted. *1814*

Fibrinogen *Plasma Decrease Physiological* From 283 mg/dL to 257 mg/dL postoperatively. *1814*

Hemoglobin *Urine No Effect Physiological* No significant effect noted. *1814*

Leukocytes *Blood Increase Physiological* Normal response to anesthesia. *3465*

Occult Blood *Feces Positive Physiological* In 1 of 8 patients (slight only). *1814*

Partial Thromboplastin Time
Plasma No Effect Physiological No significant effect noted. *1814*

Platelets *Blood Increase Physiological* Slight from 217 m/cmm to 251 m postoperatively. *1814*

Prothrombin Time *Plasma No Effect Physiological* No significant effect noted. *1814*

Thrombin Time *Plasma No Effect Physiological* No significant effect noted. *1814*

Fluspirilene

Alkaline Phosphatase *Serum Decrease Physiological* Slight but observable trend. *0221*

Aspartate Aminotransferase
Serum Decrease Physiological Slight but observable trend. *0221*

Granulocytes *Blood Decrease Physiological* Decreased values noted (not marked). *0221*

Leukocytes *Blood Increase Physiological* Initial effect observed. *0221*

Folic Acid

Acetaminophen Screening Test *Urine Negative Analytical* No reaction with o-cresol at therapeutic concentrations. *3326*

N-Acetylglucosaminidase *Urine No Effect Analytical* At 50 mmol/L no effect on 2 colorimetric analytical methods. *1354*

Folic Acid (continued)

Diphenylhydantoin *Serum Decrease Physiological* If patients folate deficient, stimulate metabolism. *1487* Occurs with pharmacologic doses; lowers plasma concentration. *3024* Increase by 8 to 48% in 4 male patients associated with increase in drug oxidative metabolism. *0307* Due to increased clearance: with doubling of amount of dose excreted in urine. *0308*

Folate *Serum Increase Analytical* Increase by 20% if chloramphenicol method of O'Broin used. *2631*

131I Uptake *Serum Decrease Physiological* Filibon, Iberet® contain tetraiodofluorescein. *2652*

Lymphocytes *Blood Decrease Physiological* Significant effect with megadose supplementation. *1347*

Neutrophils *Blood No Effect Physiological* No effect with megadose supplementation. *1347*

Phenobarbital *Serum Decrease Physiological* Occurs with pharmacologic doses; lowers plasma concentration. *3024*

Vitamin B₁₂ *Red Blood Cells Increase Physiological* During treatment of folate deficient anemia. *1510*

Food

Acetaminophen *Serum Decrease Physiological* Absorption delayed when taken with food. *3024*

Acetylsalicylic Acid *Serum Decrease Physiological* Absorption reduced when taken with food. *3024*

Amoxicillin *Serum Decrease Physiological* Absorption delayed when taken with food Absorption reduced when taken with food. *3024*

Ampicillin *Serum Decrease Physiological* Absorption reduced when taken with food. *3024*

Cephalexin *Serum Decrease Physiological* Absorption delayed when taken with food. *3024*

Cephradine *Serum Decrease Physiological* Absorption delayed when taken with food. *3024*

Creatinine *Serum Increase Physiological* Average increase of 52% 1.5 to 3.5 h after ingestion of 225 g meat. *2996* *Urine Increase Physiological* Increased during day of meat consumption in 4 of 7 healthy volunteers. *2580* Average increase of 19% during 24 h after ingestion of 225 g meat. *2996*

Creatinine Clearance *Urine No Effect Physiological* No effect in individuals in response to ingestion of a cooked meat meal. *2996*

Demethylchlortetracycline *Serum Decrease Physiological* Absorption reduced when taken with food. *3024*

Digoxin *Serum Decrease Physiological* Absorption delayed when taken with food. *3024*

Doxycycline *Serum Decrease Physiological* Absorption reduced when taken with food. *3024*

Ethanol *Serum Decrease Physiological* Absorption slowed when taken with food. *3024*

Furosemide *Serum Decrease Physiological* Absorption delayed when taken with food. *3024*

Griseofulvin *Serum Increase Physiological* Better absorbed with meal containing fat. *3024*

5-Hydroxyindoleacetic Acid (5-HIAA)
Urine No Effect Physiological No significant change in hourly excretion rate after a meal. *0099*

5-Hydroxytryptamine (Serotonin)
Blood No Effect Physiological No significant increase for 1 h after a meal. *0099*

Isoniazid *Serum Decrease Physiological* Absorption reduced when taken with food. *3024*

Levodopa *Serum Decrease Physiological* Absorption reduced when taken with food. *3024*

Methacycline *Serum Decrease Physiological* Absorption reduced when taken with food. *3024*

3-Methylhistidine *Urine Increase Physiological* Quantitatively excreted in 2 d in 7 healthy people. *2580*

Oxytetracycline *Serum Decrease Physiological* Absorption reduced when taken with food. *3024*

Penicillin G *Serum Decrease Physiological* Absorption reduced when taken with food. *3024*

Penicillin V *Serum Decrease Physiological* Absorption reduced when taken with food. *3024*

Phenethicillin *Serum Decrease Physiological* Absorption reduced when taken with food. *3024*

Phenobarbital *Serum Decrease Physiological* Absorption reduced when taken with food. *3024*

Potassium *Serum Decrease Physiological* Absorption delayed when taken with food. *3024*

Prolactin *Plasma Increase Physiological* Abrupt increase within 45 minutes of starting a meal. *2895*

Propafenone *Plasma Increase Physiological* Maximum plasma drug concentration higher and reached earlier. *0186*

Propantheline *Serum Decrease Physiological* Absorption reduced when taken with food. *3024*

Prostaglandin E1 *Plasma Increase Physiological* Probably due to ingestion of fats. *1385*

Rifampin *Serum Decrease Physiological* Absorption reduced when taken with food. *3024*

Sulfadiazine *Serum Decrease Physiological* Absorption delayed when taken with food. *3024*

Sulfamethoxine *Serum Decrease Physiological* Absorption delayed when taken with food. *3024*

Sulfamethoxypyridazine *Serum Decrease Physiological* Absorption delayed when taken with food. *3024*

Sulfanilamide *Serum Decrease Physiological* Absorption delayed when taken with food. *3024*

Sulfasymazine *Serum Decrease Physiological* Absorption delayed when taken with food. *3024*

Sulfisoxazole *Serum Decrease Physiological* Absorption delayed when taken with food. *3024*

Tetracycline *Serum Decrease Physiological* Absorption reduced when taken with food. *3024*

Tryptophan *Plasma No Effect Physiological* No significant increase for 1 h after a meal. *0099*

Formaldehyde

Acid Phosphatase *Serum Decrease Analytical* Inhibits erythrocytic component. *0078*

Bile *Urine Increase Analytical* May give yellow ring with Gmelin test. *0459*

Bogen Test *Urine Positive Analytical* Reacts as if ethanol. *1302*

Catecholamines *Urine Increase Analytical* Interferes with fluorometric method. *1022* Interferes with procedures for analysis. *0757*

Crystals *Urine Increase Analytical* Urea precipitation. *2313*

Erythrocytes *Urine Increase Physiological* Ingestion may cause hematuria and renal damage. *1302*

Estriol *Urine Decrease Analytical* Interferes with acid hydrolysis stage. *0406*

Glucose *Urine Decrease Analytical* Produces orange color (interferes) with o-toluidine. Inhibits reaction in coupled glucose oxidase procedure. *0583*

Hemoglobin *Urine Increase Analytical* Reaction with benzidine. *2313*

5-Hydroxyindoleacetic Acid (5-HIAA)
Urine Decrease Analytical Inhibits color develop with nitrosonaphthol. *0583*

Indican *Urine Decrease Analytical* Prevents Obermeyer test reaction. *0459*

Obermayer Test *Urine Negative Analytical* Inhibits test for indican. *0459*

Occult Blood *Feces Positive Physiological* Ingestion may cause hematemesis. *1302*

PSP Excretion *Urine Increase Analytical* Produces interfering color reaction. *0757*

Specific Gravity *Urine Increase Analytical* Causes slight increase only. *0251*

Sugar *Urine Increase Analytical* May reduce Fehlings and Clinitest®. *1563*

Urea Nitrogen *Serum Increase Physiological* May cause renal damage if ingested. *0279*

Urine Decrease Analytical Inhibits urease (can not be used as preservative). *1563*

Uric Acid *Serum Increase Analytical* If added to standards inhibits uricase. *0582* If added to standards retards dialysis etc. *1950*
Urine Decrease Analytical Inhibits uricase when added as preservative. *0583*

Urobilinogen *Urine Increase Analytical* Interferes with color reaction. *0757*
Urine Decrease Analytical Marked effect at 200 mg/dL on p-MDFB procedure. *2044*

Freezing

Angiotensin-Converting Enzyme (ACE)
Serum Increase Analytical Freezing with subsequent thawing increases activity by 15%. *3221*

Pseudocholinesterase *Serum No Effect Analytical* Stable for up to 12 mo at -20° C and unaffected by freezing and thawing with benzoyl choline or butyrylthiocholine as substrate. *3635*

Fructose

Alanine Aminotransferase *Serum No Effect Physiological* No effect observed after i.v. infusion. *1160*

Aspartate Aminotransferase
Serum No Effect Physiological No effect observed after i.v. infusion. *1160*

Bicarbonate *Serum Decrease Physiological* Metabolic response to large dose fructose. *3113*

Bilirubin *Serum Increase Physiological* Increase by 0.3-0.8 mg/dL after i.v. infusion. *1160*

Citrate *Serum Increase Physiological* Rises more rapidly than after glucose. *1709*

Citrulline *Test Conditions Increase Analytical* Enhances color development in Archibald procedure. *3424*

Corticosteroids *Urine Increase Analytical* Increased absorption at 410 nm in Porter-Silber reaction. *0022*

Creatinine *Serum Increase Analytical* Interference with Jaffé reaction. *3879*
Urine Increase Analytical 1 mg equivalent to 0.07 mg with automated Jaffé. *2558*

Estriol *Urine Decrease Analytical* Probably destroys estriol during acid hydrolysis. *0023*

Estrogens *Urine Decrease Analytical* Affects hydrolysis of estrogen conjugates. *0022*

Glucose *Serum No Effect Analytical* No effect on GOD-PERID procedure. *2771* No effect if measured by hexokinase, glucose oxidase. *2568* At 200 mg/dL on hexokinase method of Coburn. *0676* No effect on Trinder glucose oxidase procedure No effect on o-toluidine procedure. *2771*
Serum Increase Analytical Mole for mole effect alkaline ferricyanide. Mole for mole effect p-HBAH procedure of Lever. *2140* Affects Neocuproin procedures. *2771* 1 g/dL equivalent to 0.8 mmol/L o-toluidine procedure. *2140* Non specificity of o-toluidine, Neocuproin, FeCN procedures. *2857*
Serum Decrease Analytical At 3.7 mmol/L if hexokinase from Leuconostoc Mesent. *0411*
Serum Decrease Physiological Marked fall may occur 1 h after i.v. fructose. *3113*
Red Blood Cells No Effect Analytical No effect in RBC hemolysates if hexokinase used. *0785*

17-Hydroxycorticosteroids *Urine Increase Analytical* Interferes with Porter-Silber reaction. *3505*

α-Ketoglutarate *Serum Increase Physiological* Rises more rapidly than after glucose. *1709*

Lactate *Serum Increase Physiological* Metabolic response to i.v. or oral fructose. *3113*
Saliva Increase Physiological Occurs at 1 h, as with blood. *1907*

Lactate Dehydrogenase *Serum No Effect Physiological* No effect observed with ingestion. *1907*

Magnesium *Urine Increase Physiological* Observed after test meals. *2422*

pH *Blood Decrease Physiological* Associated with lactic acidosis. *0744*

Phosphate *Serum Decrease Physiological* Phosphorylation occurs after i.v. injection. *1709*

Protein *Test Conditions Increase Analytical* Interferes with Folin-Ciocalteu method of Lowry. *3954*

Pyruvate *Serum Increase Physiological* Rises more rapidly than after glucose. *1709*
Saliva Increase Physiological Occurs at 1 h, as with blood. *1907*

Sugar *Urine Increase Analytical* False positive with Benedict's, Clinitest®. *1563*

Triglycerides *Serum Increase Physiological* In experimental animals, contradictory in humans. *1709*

Uric Acid *Serum Increase Physiological* Following i.v. infusion: dose related effect. *3505*
Urine Increase Physiological Dose related effect. *3113*

Uric Acid Clearance *Urine Increase Physiological* Dose related effect. *3113*

Fructose-1,6-Diphosphate

Pyruvate Kinase *Red Blood Cells Increase Analytical* Activating effect on enzyme. *0370*

Fruit

Arabinose *Urine Increase Physiological* Increased by much fruit in diet compared with fasting. *0273*

Fucose *Urine No Effect Physiological* No effect compared with fasting. *0273*

Ribose *Urine No Effect Physiological* No effect compared with fasting. *0273*

Xylose Excretion *Urine Increase Physiological* Increased by much fruit in diet compared with fasting. *0273*

Fumagillin

Leukocytes *Blood Decrease Physiological* Leukopenia (AMA Blood dyscrasias). *2429*

Furadaltone

Bilirubin *Serum Increase Physiological* May cause hemolysis with G-6-PD deficiency. *0248*

Erythrocytes *Blood Decrease Physiological* May cause hemolysis with G-6-PD deficiency. *0248*

Haptoglobin *Serum Decrease Physiological* Due to hemolysis. *0248*

Heinz Body Formation *Blood Positive Physiological* Early stage of hemolytic anemia. *0333*

Hematocrit *Blood Decrease Physiological* May cause hemolysis with G-6-PD deficiency. *0248*

Hemoglobin *Blood Decrease Physiological* May cause hemolysis with G-6-PD deficiency. *0248*

Methemalbumin *Plasma Increase Physiological* May occur with hemolysis. *0248*

Reticulocytes *Blood Increase Physiological* Occurs with hemolysis. *0248*

Furapromidium

Ammonia *Plasma Increase Physiological* Occasional increase observed. *2427*

Furazolidone

Bilirubin *Serum Increase Physiological* May cause hemolysis with G-6-PD deficiency. *0248*

Color *Urine Increase Analytical* Metabolites may produce brown color. *1488*

Erythrocytes *Blood Decrease Physiological* May cause hemolysis with G-6-PD deficiency. *0248*

Glucose *Urine Increase Analytical* Due to presence of glucose in vaginal powder. *1237*

Furazolidone *(continued)*

Haptoglobin *Serum Decrease Physiological* Due to hemolysis. *0248*

Heinz Body Formation *Blood Positive Physiological* Early stage of hemolytic anemia. *0333*

Hematocrit *Blood Decrease Physiological* May cause hemolysis with G-6-PD deficiency. *0248*

Hemoglobin *Blood Decrease Physiological* May cause hemolysis with G-6-PD deficiency. *0248*

Methemalbumin *Plasma Increase Physiological* May occur with hemolysis. *0248*

Methemoglobin *Blood Increase Physiological* May cause hemolysis with G-6-PD deficiency. *3581*

Reticulocytes *Blood Increase Physiological* Occurs with hemolysis. *0248*

Sugar *Urine Increase Analytical* Metabolites may give false positive with Benedict's. *1488*

Furazolium

Color *Urine Increase Analytical* Red, pink, purple, orange and rust colors. *1237*

Furosemide

Acetaminophen Screening Test *Urine Negative Analytical* No reaction with o-cresol at therapeutic concentrations. *3326*

N-Acetylglucosaminidase *Urine Increase Physiological* Excretion increased by drug induced diuresis. *1492*

Alanine Aminotransferase *Serum No Effect Analytical* No effect at therapeutic concentration on Reflotron method. *1984*

Albumin *Serum No Effect Analytical* No effect at 1.4 mg/dL on SMA 12/60 method. *2636* At concentration of 4 mg/L had no effect on BCG method. *3393*

Aldosterone *Plasma Increase Physiological* Slower response than angiotensin to i.v. injection. *2359*
Plasma Decrease Physiological May be marked reduction in patients with congestive cardiac failure when given drug i.v. if initial concentration high, otherwise usually slight increase. *2513*
Urine Increase Physiological Effect observed after 40 mg orally in normal male volunteers. *2658*

Alkaline Phosphatase *Serum No Effect Analytical* No effect at 1.4 mg/dL on SMA 12/60 method. *2636*

Ammonia *Plasma Increase Physiological* Acts like thiazides causes hypokalemia and alkalosis. *1488*

Amylase *Serum Increase Physiological* May induce mild pancreatitis. *3505* Isolated case of acute hemorrhagic pancreatitis. *1820* Not significant increase in total amylase in 12 hypertensives Increase by 16% on average in 12 hypertensives. *2015*

Amylase, Pancreatic *Serum Increase Physiological* Increase by 17% on average in 12 hypertensives. *2015*

Angiotensin II *Plasma Increase Physiological* Rapid response to i.v. injection. *0977*

Antidiuretic Hormone *Plasma Increase Physiological* Effect observed after 40 mg orally in normal male volunteers. *2658*

Apolipoprotein AI *Serum No Effect Physiological* Decrease by -2 mg/dL in 12 patients given 40 to 80 mg daily for 3 mo. *2550*

Apolipoprotein B *Serum No Effect Physiological* Increase by 2 mg/dL in 12 patients given 40 to 80 mg daily for 3 mo. *2550*

Apolipoprotein CII *Serum No Effect Physiological* Increase by 1 mg/dL in 12 patients given 40 to 80 mg daily for 3 mo. *2550*

Apolipoprotein CIII *Serum No Effect Physiological* Increase of 4 mg/dL in 12 patients given 40 to 80 mg daily for 3 mo. *2550*

Aspartate Aminotransferase *Serum No Effect Analytical* No effect at 1.4 mg/dL on SMA 12/60 method. *2636* No effect on Karmen procedure. *2760* No effect at therapeutic concentration on Reflotron method. *1984*

Bicarbonate *Serum No Effect Analytical* At concentration of 4 mg/L had no effect on method using phenolphthalein. *3393*

Serum Increase Physiological In 24% in 204 hospitalized patients receiving the drug. *3419*

Bilirubin *Serum No Effect Analytical* No effect at 1.4 mg/dL on SMA 12/60 method. *2636* At concentration of 4 mg/L had no effect on Jendrassik and Grof method. *3393*
Serum No Effect Physiological Although *in vitro* effect observed at pharmacological concentrations no significant effect. *3748*

Calcium *Serum No Effect Analytical* No effect at 1.4 mg/dL on SMA 12/60 method. *2636* At concentration of 4 mg/L had no effect on cresolphthalein method. *3393*
Serum No Effect Physiological In 8 normal subjects associated with secondary hyperparathyroidism. *1214*
Serum Decrease Physiological If increase due to dihyrotachysterol, also if normal. *2944* Diuretic action (different effect hydrochlorothiazide). *0468*
Urine Increase Physiological Impaired reabsorption (initial effect only). *2745* In 8 normal subjects associated with secondary hyperparathyroidism. *1214* Slight increase with administration. *2047* In 8 normal subjects associated with secondary hyperparathyroidism. *1214*

Casts *Urine Increase Physiological* Hyaline casts without proteinuria (orosomucoid). *1720*

Chloride *Serum No Effect Analytical* At concentration of 4 mg/L had no effect on mercurimetric method. *3393*
Serum Decrease Physiological Diuretic action (inhibits tubular reabsorption). *0070* In 36% of 204 hospitalized patients receiving the drug. *3419* Observed after drug administration. *2047* Reduced often in association with hyponatremia. *1391*
Urine Increase Physiological Diuretic action. *1617* Marked effect: decreased sharply when administration stopped. *2047*

Cholesterol *Serum No Effect Analytical* No effect at 1.4 mg/dL on SMA 12/60 method. *2636* No effect at therapeutic concentration on Reflotron method. *1984* At concentration of 60 mg/L had no effect on CHOD-Iodide method. *3393* At concentration of 60 mg/L had no effect on CHOD-PAP method. *3393* At concentration of 60 mg/L had no effect on catalase-AIDH method. *3393* At concentration of 60 mg/L had no effect on method using catalase-Hantzsch reaction. *3393* At concentration of 60 mg/L had no effect on Liebermann-Burchard method. *3393*
Serum No Effect Physiological Increase by 2 mg/dL in 12 patients given 40 to 80 mg daily for 3 mo. *2550*
Serum Increase Physiological 5% change in 12 normotensive men when treated with drug alone for less than 1 y. *1826* 6% change in 16 subjects treated for less than 1 y with drug only. *1306*

Cholesterol, High Density Lipoprotein
Serum No Effect Physiological Not significant change in 12 normotensive men when treated with drug alone for less than 1 y. *1826* No significant change in 16 subjects treated for less than 1 y with drug only. *1306* Reduced by -4 mg/dL in 12 patients given 40 to 80 mg daily for 3 mo. *2550*

Cholesterol, Low Density Lipoprotein
Serum No Effect Physiological Not significant change in 12 normotensive men when treated with drugs alone for for less than 1 y. *1826*
Serum Increase Physiological 15% change in 16 subjects treated for less than 1 y with drug only. *1306*

Cholesterol, Very Low Density Lipoprotein
Serum No Effect Physiological No significant change in 16 subjects treated for less than 1 y with drug only. *1306*
Serum Increase Physiological 56% change in 12 normotensive men when treated with drug alone for less than 1 y. *1826* 6% change in 16 subjects treated for less than 1 y with drug only. *1306*

Creatinine *Serum No Effect Analytical* No effect at 1.4 mg/dL on SMA 12/60 method. *2636* At concentration of 5 mg/L had no effect on creatinine iminohydrolase method. *3393* At concentration of 4 mg/L had no effect on AutoAnalyzer Jaffé method. *3393* At concentration of 20 mg/L had no effect on Jaffé-Fading-Fraction method. *3393* At concentration of 20 mg/L had no effect on Jaffé-Fuller's earth method. *3393* At concentration of 30 mg/L had no effect on kinetic Jaffé method on BKA-2. *3393*
Serum No Effect Physiological In 8 normal subjects associated with secondary hyperparathyroidism. *1214*
Urine No Effect Physiological No clear pattern observed with drug administration. *2047*

Creatinine Clearance *Urine Increase Physiological* If given intravenously especially. *2818*

Cyclic Adenosine Monophosphate
Plasma No Effect Physiological In 8 normal subjects associated with secondary hyperparathyroidism. *1214*
Urine Increase Physiological In 8 normal subjects associated with secondary hyperparathyroidism. *1214*

Dopamine *Urine Increase Physiological* Significant effect in 15 minutes following 30 mg intravenously. *1784*

Epinephrine *Plasma No Effect Physiological* Effect observed after 40 mg orally in normal male volunteers. *2658*

Erythrocytes *Blood Decrease Physiological* May cause anemia. *1617*

Glomerular Filtration Rate (GFR)
Urine Decrease Physiological Excessive diuresis may cause effect. *2427*

Glucose *Serum No Effect Analytical* No effect at 1.4 mg/dL on SMA 12/60 method. *2636* No effect at therapeutic concentration on Reflotron method. *1984* At concentration of 40 mg/L had no effect on Ektachem® method. *3393* At concentration of 5 mg/L had no effect on GOD/POD-PAP method. *3393* At concentration of 100 mg/L had no effect on hexokinase/G-6-PDH method. *3393*
Serum Increase Physiological Diabetogenic-like action of drug affects glucose tolerance test. *3505* In 6% in 204 hospitalized patients receiving the drug. *3419*
Urine No Effect Analytical No effect observed with Tes-Tape®. *1100*
Urine Increase Physiological Diabetogenic-like action of drug. *3505* Observed in 0.2% of 2580 medical inpatients. *2218*
Urine Decrease Analytical Low with Clinistix®, Diastix®. *1100*

Glucose Tolerance *Serum Decrease Physiological* Diabetogenic-like action of drug. *2745*

γ-Glutamyltransferase (GGT) *Serum No Effect Analytical* No effect at therapeutic concentration on Reflotron method. *1984*

Hematocrit *Blood Decrease Physiological* May cause anemia. *1617*

Hemoglobin *Blood Decrease Physiological* May cause anemia. *1617*

Indomethacin *Serum No Effect Analytical* No effect on HPLC method of Roberts and Smith. *3002*

Insulin *Plasma Decrease Physiological* Intravenous njection effect, little on blood sugar. *2427*

Ionized Calcium *Serum No Effect Physiological* In 8 normal subjects associated with secondary hyperparathyroidism. *1214*

Lactate *Serum No Effect Analytical* At concentration of 100 mg/L had no effect on enzymatic method. *3393*

Lactate Dehydrogenase *Serum No Effect Analytical* No effect at 1.4 mg/dL on SMA 12/60 method. *2636*

Leukocytes *Blood Decrease Physiological* May cause leukopenia or aplastic anemia. *2313*

Lipase *Serum Increase Physiological* Isolated case of acute hemorrhagic pancreatitis. *1820*

Lithium *Serum No Effect Physiological* No effect observed in normal volunteers given drug over 2 weeks (40 mg/d). *1782*

Lymphocytes *Lymphocytes Decrease Physiological* Significantly reduced concentration in congestive heart failure patients treated with drug. *3101*

Magnesium *Serum Decrease Physiological* Observed in several patients on long term therapy or with short term vigorous treatment. *3285* Significant reduction may occur with prolonged treatment. *3285*
Urine Increase Physiological Diuretic action on divalent cations. *3745* Renal wasting of magnesium reported with loop-blocking diuretics. *3101*
Lymphocytes Decrease Physiological Significantly reduced concentration in congestive heart failure patients treated with drug. *3101*

Neutrophils *Blood Decrease Physiological* Reported to cause neutropenia. *2583*

Norepinephrine *Plasma Increase Physiological* Effect observed after 40 mg orally in normal male volunteers. *2658*
Urine Increase Physiological Effect observed after 40 mg orally in normal male volunteers. *2658*

Parathyroid Hormone *Plasma Increase Physiological* 10% increase in normal individuals after 8 d administration. *1214*

Phosphate *Serum No Effect Analytical* No effect at 1.4 mg/dL on SMA 12/60 method. *2636* At concentration of 4 mg/L had no effect on phosphomolybdate method. *3393*
Serum Increase Physiological Temporary increase when fluid losses continuously replaced. *2944*
Urine Increase Physiological Slight increase with administration. *2047*

Platelet Aggregation *Blood Decrease Physiological* Inhibits primary ADP-induced agglutination. Effect seen *in vitro* but not *in vivo*. *3069*

Platelets *Blood Decrease Physiological* May be associated with purpura. *1617* Observed in 0.2% of 2580 medical inpatient. *2218*

Potassium *Serum No Effect Analytical* At concentration of 4 mg/L had no effect on flame-photometric method. *3393* At concentration of 6,000 mg/L had no effect on measurement by ISE with predilution. *3393* At concentration of 400 mg/L had no effect on measurement by ISE without predilution. *3393*
Serum Decrease Physiological Diuretic action. *3505* Clinically important in about 3.6% patients. *1391* In 25% of 204 hospitalized patients receiving the drug. *3419* 12% patients with hypokalemia given 40-80 mg/d over long term. *0314* Observed after drug administration. *2047* In 12% patients with hypokalemia given 40-80 mg/d over long term. *0314*
Urine Increase Physiological Diuretic action. *1617* Slight increase with administration. *2047*

Prolactin *Plasma Increase Physiological* Effect observed after 40 mg orally in normal male volunteers. *2658*

Protein *Serum No Effect Analytical* No effect at 1.4 mg/dL on SMA 12/60 method. *2636* At concentration of 4 mg/L had no effect on biuret method with blank correction. *3393*
Urine Increase Physiological If pre-existing proteinuria. *2818*

Protein Electrophoresis *Serum No Effect Analytical* At concentration of 400 mg/L had no effect on automated Olympus-Hite method. *3393*

Renin Activity *Plasma Increase Physiological* Most often slight increase in patients with chronic congestive heart failure in response to i.v. drug but not consistent pattern of response. *2513* Effect observed after 40 mg orally in normal male volunteers. *2658*

Rifampin *Serum Increase Physiological* Effect observed after 40 mg orally in normal male volunteers. *2658*

Salicylate *Serum Increase Physiological* If given concomitantly, competes for excretion. *1678*

Sodium *Serum No Effect Analytical* At concentration of 4 mg/L had no effect on flame-photometric method. *3393* At concentration of 6,000 mg/L had no effect on measurement by ISE with predilution. *3393* At concentration of 400 mg/L had no effect on measurement by ISE without predilution. *3393*
Serum Decrease Physiological Diuretic action with sodium depletion. *0070* In 25% of 204 hospitalized patients receiving the drug. *3419* Observed after drug administration. *2047*
Urine Increase Physiological Diuretic action. *1617* Therapeutic intent of drug administration. *1214* Marked effect: decreased sharply when administration stopped. *2047*

T₃ Uptake *Serum Increase Physiological* Diminished protein binding at time of peak drug concentration. *3473* Changes observed in opposite direction to total T₄ concentration. *2584* Transient significant effect seen 2-5 h after ingestion of various amounts of drug chronically in 34 patients with congestive cardiac failure. *2584*

Theophylline *Serum Decrease Physiological* Significantly lower when coadministered with drug. *0594*

Thyroid Stimulating Hormone (TSH)
Serum No Effect Physiological No effect typically observed in spite of changes in concentration of peripheral hormones. *2584*
Serum Increase Physiological Increased after drug treatment stopped. *3473*

Furosemide (continued)

Thyroxine (T₄) *Serum Decrease Physiological* In 4 patients high concentrations of drug inhibit T_4 binding in plasma. *3473* Probably due to enhanced clearance due to displacement from serum protein binding sites. *2584* Associated with high dose i.v. administration at peak drug concentration. *3473* Transient significant effect seen 2-5 h after ingestion of various amounts of drug chronically in 34 patients with congestive cardiac failure. *2584*

Thyroxine (T₄) Binding Globulin
Serum No Effect Physiological At same time as thyroxine concentration reduced. *3473*

Thyroxine (T₄), Free *Serum Increase Physiological* Effect similar to that of fT_4 index with analog tracer assay. *2584* Transient significant effect seen 2-5 h after ingestion of various amounts of drug chronically in 34 patients with congestive cardiac failure. *2584*
Serum Decrease Physiological At same time as thyroxine concentration reduced. *3473*

Thyroxine (T₄) Index, Free (FTI)
Serum Increase Physiological occurring 2-5 h after single morning dose of 80 to 250 mg. *2584* Transient significant effect seen 2-5 h after ingestion of various amounts of drug chronically in 34 patients with congestive cardiac failure. *2584*

Triglycerides *Serum No Effect Analytical* No effect at therapeutic concentration on Reflotron method. *1984* At concentration of 5 mg/L had no effect on GPO-PAP method. *3393* At concentration of 4 mg/L had no effect on lipase/esterase method. *3393*
Serum No Effect Physiological No significant change in 16 subjects treated for less than 1 y with drug only. *1306*
Serum Increase Physiological 37% change in 12 normotensive men when treated with drug alone for less than 1 year. *1826* Increase by 29 mg/dL in 12 patients given 40 to 80 mg daily for 3 mo. *2550*

Triglycerides, Very Low Density Lipoprotein
Serum Increase Physiological 56% change in 12 normotensive men when treated with drug alone for less than 1 y. *1826* 6% change in 16 subjects treated for less than 1 y with drug only. *1306*

Tri-iodothyronine (T₃) *Serum Decrease Physiological* At same time as thyroxine concentration reduced. *3473*

Urea Nitrogen *Serum No Effect Analytical* No effect at therapeutic concentration on Reflotron method. *1984* At concentration of 4 mg/L had no effect on diacetylmonoxime method. *3393* At concentration of 40 mg/L had no effect on Ektachem® method. *3393*
Serum Increase Physiological ?nephrotoxic effect (reversible) usually dehydration. *0085* In 9 patients with chronic failure when treated for 6 d average plasma urea rose from 18.7 mmol/L to 28.8 mmol/L secondary to reduced urea excretion which occurred in spite of increased urea filtration. *0792* Largely as consequence of volume depletion: effect most marked in patients receiving additional diuretics. *1391* In patients with chronic renal failure: associated with decreased renal clearance and increased tubular reabsorption of urea. *0792* In 8% of 204 hospitalized patients receiving the drug. *3419*
Urine No Effect Physiological No clear pattern observed with drug administration. *2047*
Urine Decrease Physiological Occurs when diuretic given to patients with chronic failure, due to increased tubular reabsorption of urea, presumably in distal part of nephron, secondary to extracellular fluid volume depletion. *0792*

Uric Acid *Serum No Effect Analytical* No effect at 1.4 mg/dL on SMA 12/60 method. *2636* No effect at therapeutic concentration on Reflotron method. *1984* At concentration of 20 mg/L had no effect on Kageyama-Hantzsch method. *3393* At concentration of 20 mg/L had no effect on catalase-AIDH method. *3393* At concentration of 4 mg/L had no effect on phosphotungstate reduction method. *3393* At concentration of 5 mg/L had no effect on uricase-PAP method. *3393*
Serum Increase Physiological Decreased urate clearance. *3505* In about 0.4% of patients receiving drug. *1391* In 10% of 204 hospitalized patients receiving the drug. *3419* Observed after drug administration. *2047*
Urine Decrease Physiological Slight effect observed on first and second day. *2047*

Volume *Plasma Decrease Physiological* Excess diuresis may cause reduction in blood volume. *0019*

Urine Increase Physiological Diuretic action. *1617* Significant effect, peak between 1 and 3 h after single 40 mg dose. *2658* Significant increase on first day but decrease on second. *2047*

Zinc *Urine No Effect Physiological* No effect observed on excretion. *2712*
Urine Increase Physiological When given i.v. brief effect only. *3439* Increase by 12% in 9 hypertensive patients treated for 2 weeks. *3806*

Fusaric Acid

Dopamine Hydroxylase *Serum Decrease Physiological* Most effective if given with levodopa. *2913* Total inhibitor *in vivo* possible for up to 122 h after 300 mg. *2545*

Norepinephrine *Urine Increase Physiological* May cause increase release early, later increased metabolism. *2702*

Thyroid Stimulating Hormone (TSH)
Serum Decrease Physiological Inhibition of dopamine degradation in hypothyroidism. *3798*

Fuscin

Color *Urine Increase Analytical* Red from foods and candy. *0443*

Fusidic Acid

Alanine Aminotransferase *Serum Increase Physiological* Reported in 6 patients in UK, although causal relationship could not be established with certainty. *3535*

Alkaline Phosphatase *Serum Increase Physiological* Reported in 6 patients in UK, although causal relationship could not be established with certainty. *3535* Jaundice developed in 34% of 112 patients given drug, highest when drug given intravenously. Jaundice resolved when drug withdrawn. *1690*

Aspartate Aminotransferase
Serum Increase Physiological Reported in 6 patients in UK, although causal relationship could not be established with certainty. *3535* Jaundice developed in 34% of 112 patients given drug, highest when drug given intravenously. Jaundice resolved when drug withdrawn. *1690*

Bile Acids *Serum Increase Analytical* In methods using 3-alpha-hydroxysteroid dehydrogenase for which drug functions as substrate. *1485*

Bilirubin *Serum Increase Physiological* Clinically significant displacement from protein in neonates. *3748* Reported in 6 patients in UK, although causal relationship could not be established with certainty. *3535* Jaundice developed in 34% of 112 patients given drug, highest when drug given intravenously. Jaundice resolved when drug withdrawn. *1690*

Cortisol *Plasma Increase Analytical* Interfering fluorescence. *1772*

11-Hydroxycorticosteroids *Plasma Increase Analytical* When determined by Mattingly method. *2901*

Galactosamine

Protein *Test Conditions Increase Analytical* Reacts with Folin-Ciocalteu of Lowry method. *1569*

Galactose

Estriol *Urine Decrease Analytical* Probably destroys estriol during acid hydrolysis. *0023* Interferes with GLC method. *3186*

Estrogens *Urine No Effect Analytical* If sodium borohydride used with fluorescence procedures. *3899*
Urine Decrease Analytical Affects fluorometric procedure unless removed. *3899*

Glucagon *Plasma Increase Physiological* Much smaller response than after glucose. *3306*

Glucose *Serum No Effect Analytical* No effect on Trinder glucose oxidase procedure. *2771* At 200 mg/dL on hexokinase method of Coburn. *0676* No effect if measured by hexokinase or glucose oxidase. *2568* No effect on GOD-PERID procedure. *2771*

Serum Increase Analytical Affects o-toluidine procedure. Affects Neocuproin procedures. Affects alkaline ferricyanide method. *2771* Mg for mg effect in MBTH procedure of Neeley. *2569*

Insulin *Plasma No Effect Physiological* No effect observed in peripheral blood. *0310*

Sugar *Urine Increase Analytical* Gives positive with Benedict's, Clinitest®. *1563*

Xylose Excretion *Urine Increase Analytical* Interferes with bromoaniline procedure if over 2 g/dL. *2981*

Gallamine

Albumin *Serum Increase Physiological* Sensitivity to drug correlated with concentration. *3481*

Histamine *Plasma Increase Physiological* Observed with injection for anesthesia. *2212*

131I Uptake *Serum Decrease Physiological* Due to iodine component of drug. *2220*

Protein Bound Iodine (PBI) *Serum Increase Analytical* Organic iodine contamination. *2220*

Gallium Nitrate

Calcium *Serum Decrease Physiological* In two patients with hypercalcemia due to parathyroid carcinoma. *3764* *Urine Decrease Physiological* In two patients with hypercalcemia due to parathyroid carcinoma. *3764*

Creatinine Clearance *Urine No Effect Physiological* In two patients with hypercalcemia due to parathyroid carcinoma. *3764*

Hydroxyproline *Urine Decrease Physiological* In two patients with hypercalcemia due to parathyroid carcinoma. *3764*

Ionized Calcium *Serum Decrease Physiological* Change paralleled that of total calcium in two patients with hypercalcemia due to parathyroid carcinoma. *3764*

Parathyroid Hormone *Plasma Decrease Physiological* In two patients with hypercalcemia due to parathyroid carcinoma. *3764*

Gallopamil

Digoxin *Serum Increase Physiological* 16% increase when concomitantly administered with drug. *0283*

Ganglionic Blocking Agents

Effective Renal Plasma Flow
Patient Decrease Physiological Returns to normal usually within 2 h. *1343*

Glomerular Filtration Rate (GFR)
Urine Decrease Physiological Returns to normal usually within 2 h. *1343*

Hydrochloric Acid *Gastric Material Decrease Physiological* Volume and acidity generally reduced. *1343*

Gargles

Protein Bound Iodine (PBI) *Serum Increase Analytical* Some may contain organic and inorganic iodine. *1237*

Garlic

Cholesterol *Serum Decrease Physiological* Significant decrease although given with fatty meal. *0421*

Coagulation Time *Blood Decrease Physiological* Significant decrease although given with fatty meal. *0421*

Fibrinogen *Plasma Decrease Physiological* Significant decrease although given with fatty meal. *0421*

Fibrinolytic Time *Plasma Increase Physiological* Significant effect although given with fatty meal. *0421*

Gasoline

Protein *Urine Increase Physiological* May cause renal damage if ingested. *0279*

Urea Nitrogen *Serum Increase Physiological* May cause renal damage if ingested. *0279*

Gastrin

Calcium *Serum Decrease Physiological* Mean decrease of 0.7 mg/dL after 5 h. *2370*

Cholecystokinin (CCK) *Plasma No Effect Analytical* No significant cross reactivity in RIA procedure of Harvey. *1518*

Gastrin *Serum Increase Physiological* From average fasting level of 87.3 pg/mL to 245 pg/mL/3 h. *2370*

Hydrochloric Acid *Gastric Material Increase Physiological* Hormonal action. *2049*

Ionized Calcium *Serum Decrease Physiological* Increase by 0.052 mmol/L 1 h after 30 μg i.v. *1682*

Gemfibrozil

Apolipoprotein AI *Serum Increase Physiological* Increase by 29% in 6 patients with primary familial endogenous hypertriglyceridemia from baseline values. Synthetic rates of apo A-I and apo A-II increased by 27% and 34% respectively. *3117*

Apolipoprotein AII *Serum Increase Physiological* Increase by 38% in 6 patients with primary familial endogenous hypertriglyceridemia from from baseline values. Synthetic rates of apo A-I and apo A-II increased by 27% and 34% respectively. *3117*

Cholesterol *Serum Decrease Physiological* Increase by 11% in extensive 5 year double-blinded trial with 2,000 men receiving 600 mg drug twice daily. *1193*

Cholesterol, High Density Lipoprotein
Serum Increase Physiological Increase by 36% in 6 patients with primary familial endogenous hypertriglyceridemia from baseline values. Synthetic rates of apo A-I and apo A-II increased by 27% and 34% respectively. *3117* By more than 10% in extensive 5 year double-blinded trial with 2,000 men receiving 600 mg drug twice daily. *1193* Increase by 50% in 18 patients with chronic renal failure treated with 1200 mg/d for 28 weeks. Simultaneous activation of postheparin plasma lipoprotein and hepatic lipases. Effects reversed when drug discontinued. *2741*

Cholesterol, Low Density Lipoprotein
Serum Decrease Physiological Increase by 10% in extensive 5 year double-blinded trial with 2,000 men receiving 600 mg drug twice daily. *1193*

Cholesterol, Very Low Density Lipoprotein
Serum Decrease Physiological Increase by 50% in 18 patients with chronic renal failure treated with 1200 mg/d for 28 weeks. Simultaneous activation of postheparin plasma lipoprotein and hepatic lipases. Effects reversed when drug discontinued. *2741*

Triglycerides *Serum Decrease Physiological* Increase by 54% in 6 patients with primary familial endogenous hypertriglyceridemia from baseline values. Synthetic rates of apo A-I and apo A-II increased by 27% and 34% respectively. *3117* Increase by 43% in extensive 5 year double-blinded trial with 2,000 men receiving 600 mg drug twice daily. *1193*

Triglycerides, Very Low Density Lipoprotein
Serum Decrease Physiological Increase by 50% in 18 patients with chronic renal failure treated with 1200 mg/d for 28 weeks. Simultaneous activation of postheparin plasma lipoprotein and hepatic lipases. Effects reversed when drug discontinued. *2741*

Gentamicin

N-Acetylglucosaminidase *Urine No Effect Analytical* No effect at 50 g/L on 2 colorimetric analytical methods. *1354* *Urine Increase Physiological* Marked effect especially with treatment for more than 12 d. *2593*

Alanine Aminopeptidase *Urine Increase Physiological* Marked effect especially with treatment for more than 12 d. *2593*

Alanine Aminotransferase *Serum Increase Physiological* May cause hepatotoxicity. *2220*

Gentamicin *(continued)*

Albumin *Serum No Effect Analytical* At concentration of 150 mg/L had no effect on BCG method. *3393*

Aldosterone *Plasma No Effect Physiological* Effect observed in 2 patients following larger doses of gentamicin. Associated with massive urinary loss of magnesium and potassium. *0224*

Alkaline Phosphatase *Serum Increase Physiological* Hepatotoxic effect. *1343*

Amino Acids *Urine Increase Analytical* Reacts with ninhydrin; extra spot TLC, high voltage electrophoresis. *2855*

Aspartate Aminotransferase *Serum Increase Analytical* False elevation with Babson procedure. *3107*
Serum Increase Physiological May cause hepatotoxicity (or due to i.m. injection). *2220*

Bicarbonate *Serum No Effect Analytical* At concentration of 150 mg/L had no effect on method using phenolphthalein. *3393*
Serum Increase Physiological If given with cephalexin to leukemics. *3922*

Bilirubin *Serum No Effect Analytical* At concentration of 150 mg/L had no effect on Jendrassik and Grof method. *3393*
Serum No Effect Physiological Clinically insignificant displacement from protein in neonates. *3748*
Serum Increase Physiological Affects liver function. *0071*

Calcium *Serum No Effect Analytical* At concentration of 150 mg/L had no effect on cresolphthalein method. *3393*
Serum Decrease Physiological Effect observed in 2 patients following larger doses of gentamicin. Associated with massive urinary loss of magnesium and potassium. *0224*

Casts *Urine Increase Physiological* Observed in association with drug associated nephrotoxicity. *3182*

Chloramphenicol *Serum No Effect Analytical* No effect at 100 mg/L on coupled enzymatic method. *2490*

Chloride *Serum No Effect Analytical* At concentration of 150 mg/L had no effect on mercurimetric method. *3393*

Cholesterol *Serum No Effect Analytical* At concentration of 6 mg/L had no effect on CHOD-PAP method. *3393* At concentration of 150 mg/L had no effect on Liebermann-Burchard method. *3393*

Creatinine *Serum No Effect Analytical* At concentration of 150 mg/L had no effect on AutoAnalyzer Jaffé method. *3393* At concentration of 6 mg/L had no effect on Jaffé-Fading-Fraction method. *3393* At concentration of 6 mg/L had no effect on Jaffé-Fuller's earth method. *3393*
Serum Increase Physiological Nephrotoxic effect. *2220* 10.2% incidence in 49 patients given drug by McHenry method versus 8% in 50 patients given drug by Sawchuk/Zaske method. *2339* Nephrotoxicity observed in 26% patients treated with drug for sepsis. Mean increase for all population studied 0.4 mg/dL. *3359* 24% incidence of nephrotoxicity but unrelated to initial renal function or prior use of drug, drug concentration, amount given, duration of treatment or concurrent treatment with other drugs. *3182*

Erythrocytes *Blood Decrease Physiological* May cause anemia. *0071*

Folate *Serum No Effect Analytical* If chromatographic procedure of Landon used. *2068*

Glomerular Filtration Rate (GFR)
Urine Decrease Physiological Using chromium labeled EDTA even with subnormal amounts of drug. Noticeable before effect on serum creatinine. *3621*

Glucose *Serum No Effect Analytical* At concentration of 10 mg/L had no effect on GOD/POD-PAP method. *3393*
Urine No Effect Analytical At up to 250 µg/mL had no effect on measured glucose concentrations using Clinitest®, Diastix® and TesTape®. *2250* At concentration of 200 mg/L had no effect on Diabur-test. *3393*

Heparin Sulfate *Urine Increase Physiological* Reacts with heparin to form precipitate. *0897*

Immunoglobulin Light Chains *Urine Increase Physiological* Marked effect especially with treatment for more than 12 d. *2593*

17-Ketosteroids *Urine No Effect Analytical* With normal dose on Zimmermann procedure. *1932*

Lysozyme *Serum No Effect Physiological* Remained in normal range in 26 patients given course of treatment. *2593*

Urine Increase Physiological Marked effect especially with treatment for more than 12 d. *2593*

Magnesium *Serum Decrease Physiological* Effect observed in 2 patients following larger doses of gentamicin. Associated with massive urinary loss of magnesium and potassium. *0224*
Urine Increase Physiological Effect observed in 2 patients following larger doses of gentamicin. Associated with massive urinary loss of magnesium and potassium. *0224*

β₂-Microglobulin *Serum No Effect Physiological* No difference during or following treatment in 26 patients given drug. *2593*
Serum Increase Physiological Significant effect observed in the absence of change in the serum creatinine. *3621*
Urine Increase Physiological Competitively inhibits reabsorption of compound when drug excretion rates exceed 150 mg/min. *3739* Significant effect observed in the absence of change in the serum creatinine. *3621* Marked effect especially with treatment for more than 12 d. *2593*

Neutrophils *Blood Decrease Physiological* Occasional case of agranulocytosis reported. *3717*

Parathyroid Hormone *Plasma Decrease Physiological* Effect observed in 2 patients following larger doses of gentamicin. Associated with massive urinary loss of magnesium and potassium. *0224*

Phosphate *Serum No Effect Analytical* At concentration of 150 mg/L had no effect on phosphomolybdate method. *3393*

Potassium *Serum No Effect Analytical* At concentration of 6 mg/L had no effect on flame-photometric method. *3393* At concentration of 14 mg/L had no effect on measurement by ISE with predilution. *3393*
Serum Decrease Physiological If given with cephalexin to leukemics. *3922* Effect observed in 2 patients following larger doses of gentamicin. Associated with massive urinary loss of magnesium and potassium. *0224*
Urine Increase Physiological Effect observed in 2 patients following larger doses of gentamicin. Associated with massive urinary loss of magnesium and potassium. *0224*

Protein *Serum No Effect Analytical* At concentration of 150 mg/L had no effect on biuret method with blank correction. *3393*
Urine No Effect Analytical No difference between sulfosalicylic acid and trichloracetic acid methods in patients receiving therapeutic doses. *3760*
Urine Increase Analytical On Ponceau S dye method in comparison with sulfosalicylic acid method in 7 patients receiving therapeutic doses. *3760*
Urine Increase Physiological Nephrotoxic effect. *1022* Manifestation of drug-induced nephrotoxicity. *3182*

Renin Activity *Plasma No Effect Physiological* Effect observed in 2 patients following larger doses of gentamicin. Associated with massive urinary loss of magnesium and potassium. *0224*

Retinol Binding Protein *Serum No Effect Physiological* No difference during or following treatment in 26 patients given drug. *2593*
Urine Increase Physiological Marked effect especially with treatment for more than 12 d. *2593*

Sodium *Serum No Effect Analytical* At concentration of 6 mg/L had no effect on flame-photometric method. *3393* At concentration of 14 mg/L had no effect on measurement by ISE with predilution. *3393*

Transferrin *Urine No Effect Physiological* No effect observed in 26 patients given drug. *2593*

Triglycerides *Serum No Effect Analytical* At concentration of 10 mg/L had no effect on GPO-PAP method. *3393* At concentration of 150 mg/L had no effect on lipase/esterase method. *3393*

Urea Nitrogen *Serum No Effect Analytical* At concentration of 150 mg/L had no effect on diacetylmonoxime method. *3393*
Serum Increase Physiological Nephrotoxic effect with large doses. *2220*

Uric Acid *Serum No Effect Analytical* At concentration of 6 mg/L had no effect on Kageyama-Hantzsch method. *3393* At concentration of 6 mg/L had no effect on catalase-AlDH method. *3393* At concentration of 150 mg/L had no effect on phosphotungstate reduction method. *3393* At concentration of 10 mg/L had no effect on uricase-PAP method. *3393*

Serum Increase Physiological Reported effect of i.m. injection. *2832*

Gentisic Acid

Glucose *Serum No Effect Analytical* Has no effect on glucose oxidase method of Gochman. *1311*
Serum Increase Analytical Falsely high with alkaline ferricyanide procedure. *1311*
Serum Decrease Analytical 29% decrease at 10 mg/dL glucose oxidase dianisidine procedures. *3037*
Urine Decrease Analytical Affects Clinistix® at 0.05 mg/ml. *1924*

Homogentisic Acid *Urine No Effect Analytical* On TLC method of Feldman and Bowman. *1096*
Urine Increase Analytical If method of Nuberger used. *1096*

5-Hydroxyindoleacetic Acid (5-HIAA)
Urine Decrease Analytical Significant decrease with procedure of Udenfriend. *1097*

Protein Bound Iodine (PBI) *Serum Decrease Physiological* Competes for binding with thyroxine binding prealbumin. *0012*

Sugar *Urine Increase Analytical* Reducing substance affects Benedict's, Clinitest®. *0583*

Thyroxine (T4) *Serum Decrease Physiological* Competes with T4 for thyroxine binding prealbumin but not thyroxine binding globulin. *0583*

Thyroxine (T4) (Murphy-Pattee)
Serum Decrease Physiological Decreases T4, does not bind to thyroxine binding globulin. *0583*

Uric Acid *Serum Increase Analytical* Falsely high values with phosphotungstate methods. *1311*
Urine Increase Analytical Falsely high values with phosphotungstate methods. *1311*

Vanillylmandelic Acid *Urine Decrease Analytical* At therapeutic physiological concentrations on Pisano procedure. *1097*

Glass Containers

NBT Test *Blood Increase Analytical* When glass used instead of plastic. *2338*

Glassware

Cephalin Flocculation *Serum Increase Analytical* If heavy metals or strong acids present. *3879*

Chloride *Serum Increase Analytical* If chloride present. *3879*
Urine Increase Analytical If chloride present. *3879*

Mercury *Urine Increase Physiological* Risk of exposure if Van Slyke etc apparatus used. *1504*

Triglycerides *Serum Increase Analytical* Vacutainers™ contaminated with glycerol affect enzyme procedure. *3590*

Urea Nitrogen *Serum Decrease Analytical* If mercury present affects Nesslerization. *3879*
Urine Decrease Analytical If mercury present affects Nesslerization. *3879*

Glaucarubin

Leukocytes *Blood Decrease Physiological* Leukopenia. *2313*

Glibenclamide

Alanine Aminotransferase *Serum No Effect Analytical* No effect at therapeutic concentration on Reflotron method. *1984*
Serum Increase Physiological Intrahepatic cholestasis described in 61 y old diabetic. *3894*

Alkaline Phosphatase *Serum Increase Physiological* Intrahepatic cholestasis described in 61 y old diabetic. *3894*

Aspartate Aminotransferase *Serum No Effect Analytical* No effect at therapeutic concentration on Reflotron method. *1984*
Serum Increase Physiological Intrahepatic cholestasis described in 61 y old diabetic. *3894*

Bilirubin *Serum Increase Physiological* Intrahepatic cholestasis described in 61 y old diabetic. *3894*

Cholesterol *Serum No Effect Analytical* No effect at therapeutic concentration on Reflotron method. *1984* At concentration of 32 mg/L had no effect on CHOD-Iodide method. *3393* At concentration of 32 mg/L had no effect on CHOD-PAP method. *3393* At concentration of 32 mg/L had no effect on catalase-AlDH method. *3393* At concentration of 32 mg/L had no effect on method using catalase-Hantzsch reaction. *3393* At concentration of 32 mg/L had no effect on Liebermann-Burchard method. *3393*

Creatinine *Serum No Effect Analytical* At 10 mg/L on reversed phase LC procedure of Zhiri et al. *3942* At concentration of 3 mg/L had no effect on Jaffé-Fading-Fraction method. *3393* At concentration of 3 mg/L had no effect on Jaffé-Fuller's earth method. *3393* At concentration of 3 mg/L had no effect on kinetic Jaffé method on BKA-2. *3393*

Glucose *Serum No Effect Analytical* No effect at therapeutic concentration on Reflotron method. *1984* At concentration of 0.3 mg/L had no effect on Ektachem® method. *3393*

γ-Glutamyltransferase (GGT) *Serum No Effect Analytical* No effect at therapeutic concentration on Reflotron method. *1984*

Indomethacin *Serum No Effect Analytical* No effect on HPLC method of Roberts and Smith. *3002*

Potassium *Serum No Effect Analytical* At concentration of 3 mg/L had no effect on measurement by ISE with predilution. *3393*

Protein *Urine No Effect Analytical* No significant difference in 12 patients receiving up to 15 mg daily with sulfosalicylic acid and trichloracetic methods. *3919*
Urine Increase Analytical Significant effect with Ponceau S dye method in comparison with sulfosalicylic method in 12 patients receiving up to 15 mg daily. *3919*

Protein Electrophoresis *Serum No Effect Analytical* At concentration of 3 mg/L had no effect on automated Olympus-Hite method. *3393*

Sodium *Serum No Effect Analytical* At concentration of 3 mg/L had no effect on measurement by ISE with predilution. *3393*

Triglycerides *Serum No Effect Analytical* No effect at therapeutic concentration on Reflotron method. *1984*

Urea Nitrogen *Serum No Effect Analytical* No effect at therapeutic concentration on Reflotron method. *1984* At concentration of 0.3 mg/L had no effect on Ektachem® method. *3393*

Uric Acid *Serum No Effect Analytical* At 10 mg/L on reversed phase LC procedure of Zhiri et al. *3942* No effect at therapeutic concentration on Reflotron method. *1984* At concentration of 3 mg/L had no effect on Kageyama-Hantzsch method. *3393* At concentration of 3 mg/L had no effect on catalase-AlDH method. *3393*

Glibonuride

Cholesterol *Serum No Effect Physiological* No significant effect observed after 4 weeks. *0301*

Fatty Acids, Free (FFA) *Serum Decrease Physiological* Small immediate rise then fall to 25% at 90 minutes. *1526*

Glucose *Serum Decrease Physiological* Therapeutic intent. *1526*

Glucose Tolerance *Serum Increase Physiological* Reported but not universal effect. *0301*

Glycerol *Serum Decrease Physiological* Response similar to that of free fatty acids. *1526*

Insulin *Plasma Increase Physiological* Immediate sharp increase, lasting for 20 minutes. *1526*

Lipoproteins *Serum No Effect Physiological* No significant effect observed after 4 weeks. *0301*

Triglycerides *Serum No Effect Physiological* No significant effect observed after 4 weeks. *0301*

Glipizide

Glucose *Serum Decrease Physiological* At dose of 2.5-5.0 mg produces rapid decrease. *2735*

Glipizide (continued)

Insulin *Plasma Increase Physiological* Stimulated beta cells in pancreas. *2735*

Glisoxepide

Fatty Acids, Free (FFA) *Serum Decrease Physiological* Small immediate rise then fall to 25% at 90 minutes. *1526*

Glycerol *Serum Decrease Physiological* Response similar to that of free fatty acids. *1526*

Insulin *Plasma Increase Physiological* Immediate sharp increase lasting for 20 minutes. *1526*

Globulin

Cholesterol *Serum Decrease Analytical* Gamma globulin increases viscosity in AutoAnalyzer methods. *1645*

Erythrocyte Sedimentation Rate
Blood Increase Physiological High molecular weight proteins contribute. *2313*

Protein *CSF Increase Analytical* Color equivalent > albumin with Folin-Ciocalteu. *1196*

Glucagon

Calcium *Serum Decrease Physiological* Occurs 10 to 20 minutes after administration if hypoglycemia. *2207*

Cholecystokinin (CCK) *Plasma No Effect Analytical* No significant cross reactivity in RIA procedure of Harvey. *1518*

Cholesterol *Serum Decrease Physiological* Reported effect, mechanism not listed. *0802*

Cyclic Adenosine Monophosphate
Plasma Increase Physiological Normal response after subcutaneous injection. *0321*
Urine Increase Physiological 4 fold increase in clearance produced. *0321*

Dopamine *Urine Decrease Physiological* Marked initial effect after i.v. in patients with congestive cardiac failure. *1607*

Epinephrine *Urine No Effect Physiological* No significant effects with i.v. in patients with congestive cardiac failure. *1607*

Gastrin *Serum Decrease Physiological* Suppresses response to meals. *0263*

Glucose *Serum Increase Physiological* Counteracts hypoglycemia by glycogenolysis. *1019*
Urine Increase Physiological Mobilizes hepatic glycogen, causes glycosuria. *2313* Mobilizes enough glycogen to cause glycosuria. *0071*

Glucose Tolerance *Serum Decrease Physiological* Decreased insulin sensitivity. *1271*

Growth Hormone *Plasma Increase Physiological* Hormonal action (potent stimulant). *1271*

Hydrochloric Acid *Gastric Material Decrease Physiological* Intravenously 0.015 mg/kg/h affects for hours. *0814*

Insulin *Plasma Increase Physiological* Stimulates beta cells of pancreas. *3128*

Lipids, Total *Serum Decrease Physiological* Transfer of lipids to platelets. *0584*

Magnesium *Serum Decrease Physiological* Significant effect at 120 minutes of i.v. infusion. *0715*

Norepinephrine *Urine Increase Physiological* Mainly conjugated, after i.v. in patients with congestive cardiac failure. *1607*

Potassium *Serum Increase Physiological* Immediate increase in dogs then fall to normal or less. *0691*

Prolactin *Plasma No Effect Physiological* On already elevated values in newborn. *1435*

Prothrombin Time *Plasma Increase Physiological* Patients on coumarins. *1971*

Triglycerides *Serum Decrease Physiological* Transfer of lipids to platelets. *0584*

Tyrosine *Plasma Decrease Physiological* Hormonal effect (?exact mechanism). *0843*

Vanillylmandelic Acid *Urine Increase Physiological* Significant effect during i.v. in patients with congestive cardiac failure. *1607*

Glucocorticoids

Aldosterone *Urine Decrease Physiological* May inhibit aldosterone production. *3045*

Amino Acids *Serum Increase Physiological* Due to breakdown of tissue proteins. *0071*

Amylase *Serum Increase Physiological* Well documented early and late pancreatitis. *2427*

Basal Metabolic Rate *Patient Decrease Physiological* Metabolic action of hormones. *2220*

Calcium *Serum Decrease Physiological* Effective if hypercalcemia due to sarcoid, vitamin D. *0071*
Urine Increase Physiological Promote renal excretion. *0071*

Eosinophils *Blood Decrease Physiological* Reduced inflammatory response. *0071* Characteristic finding as a result of redistribution of cells rather than cell lysis. Effect usually transient lasting less than 24 h. *0410*

Erythrocytes *Blood Increase Physiological* Stimulate erythropoiesis. *0071*

Glucagon *Plasma No Effect Physiological* Intravenous injection failed to modify concentration. *2296*
Plasma Increase Physiological After 4 d fasting level increased 24 pg/ml. *2296* Increase basal and stimulated concentrations and promote gluconeogenesis. *0410*

Glucose *Serum Increase Physiological* After i.v. 6 mg/dL at 150 minutes to 15 at 4 h. *2296* Diabetogenic action (increased gluconeogenesis). *2313*
Urine Increase Physiological Due to hyperglycemia. *2313*

Glucose Tolerance *Serum Decrease Physiological* In majority of nondiabetic patients although adaptive response usually occurs with time and glucose concentrations revert to pretreatment values. *0410*

3-Hydroxyanthranilic Acid *Urine Increase Physiological* Induce tryptophan pyrrolase. *3045*

3-Hydroxykynurenine *Urine Increase Physiological* Induce tryptophan pyrrolase. *3045*

Hydroxyproline *Urine Decrease Physiological* In rheumatoids under treatment. *2919*

Immunoglobulin IgA *Serum Decrease Physiological* Less marked effect than with IgG. *0410*

Immunoglobulin IgG *Serum Decrease Physiological* May be reduced by 50% for up to 3 mo after 1 week treatment. *0410*

Immunoglobulin IgM *Serum No Effect Physiological* No effect seen, although other immunoglobulins affected. *0410*

Insulin *Plasma No Effect Physiological* Intravenous injection failed to modify concentration. *2296*

[131]I Uptake *Serum Decrease Physiological* Associated with reduced BMR. *2220*

Kynurenine *Urine Increase Physiological* Induce tryptophan pyrrolase. *3045*

Lymphocytes *Blood Decrease Physiological* Also decrease in lymphoid tissue. *0071* Characteristic finding as a result of redistribution of cells rather than cell lysis. Effect usually transient lasting less than 24 h. *0410*

Monocytes *Blood Decrease Physiological* Characteristic finding as a result of redistribution of cells rather than cell lysis. Effect usually transient lasting less than 24 h. *0410*

Neutrophils *Blood Increase Physiological* Characteristic finding as a result of redistribution of cells rather than cell lysis. Effect usually transient lasting less than 24 h. *0410*

Occult Blood *Feces Positive Physiological* May cause lower intestinal ulceration, perforation and hemorrhage. *0410*

Platelets *Blood Increase Physiological* Stimulate production of platelets. *0071*

Potassium *Serum Decrease Physiological* May cause potassium loss. *0071*

Protein Bound Iodine (PBI) *Serum Decrease Physiological* Reduced amount of thyroxine binding globulin. *1237*

Sodium *Serum Increase Physiological* May cause sodium retention. *0071*

Thyroid Stimulating Hormone (TSH)
Serum Decrease Physiological In euthyroidism and hypo-thyroidism with diminished response to TRH due to hypophyseal inhibition. *3798*

Thyroxine (T₄) *Serum No Effect Physiological* Commonly observed typically with inhibition of conversion. *3798*
Serum Increase Physiological In Hashimoto's disease typi-cally with inhibition of conversion. *3798*
Serum Decrease Physiological Or no change typically with inhibition of conversion. *3798*

Triglycerides *Serum Increase Physiological* Mean 16% increased observed with low dose. *0915*

Tri-iodothyronine (T₃) *Serum Decrease Physiological* Typically with inhibition of conversion. *3798*

Xanthurenic Acid *Urine Increase Physiological* Induce tryptophan pyrrolase. *3045*

Glucosamine

Aminolevulinic Acid *Urine Increase Analytical* If no pre-liminary separation procedure. *2981*

Protein *Test Conditions Increase Analytical* Hexosamines, N-acetyl-derivatives affect Lowry method. *0304* Reacts with Folin-Ciocalteu of Lowry method. *1569*

Sugar *Urine Increase Analytical* Positive with Benedict's, Clinitest®. *0583*

Urea Nitrogen *Serum No Effect Analytical* No effect on Berthelot reaction. *1862*

Glucose

N-Acetylglucosaminidase *Urine No Effect Analytical* No effect at 100 mmol/L on 2 colorimetric analytical methods. *1354*

Alanine *Plasma Increase Physiological* Rise related to gluconeogenesis after load. *1403*

Alanine Aminotransferase *Serum No Effect Physiological* No effect observed after i.v. infusion. *1160*

Amino Acids *Serum Decrease Physiological* (some affected only) deposited in muscle. *1403*

Ammonia *Plasma Increase Physiological* May increase as glucose increases in cirrhotics. *1488*
Urine Decrease Physiological In starved individuals even at 7.5 g/d. *1242*

Amylase *Serum No Effect Analytical* No effect on Phadebas procedure. *1726* At 100 mmol/L on method of Rau-scher et al. *2926*

Apolipoprotein AI *Serum No Effect Physiological* In response to infusion of 1 g/kg body weight in normal volunteers and non-insulin dependent diabetics. *0476*

Apolipoprotein AII *Serum No Effect Physiological* In response to infusion of 1 g/kg body weight in normal volunteers and non-insulin dependent diabetics. *0476*

Apolipoprotein CII *Serum Increase Physiological* Signifi-cant change in normals in response to infusion of 1 g/kg body weight, although not very marked. *0476*

Apolipoprotein CIII *Serum Increase Physiological* Signifi-cant change in normals in response to infusion of 1 g/kg body weight, although not very marked. *0476*

Arginine *Plasma Decrease Physiological* Probable muscle uptake. *1403*

Aspartate Aminotransferase
Serum No Effect Physiological No effect observed after i.v. infusion. *1160*

Bilirubin *Serum Increase Physiological* Increase by 0.3-0.8 mg/dL after rapid i.v. infusion. *1160*

Calcium *Serum Decrease Physiological* May fall by 0.5 meq/L during glucose tolerance test. *2207*
Urine Increase Physiological Nutrient induced augmentation ?mechanism. *2126*

Carcinoembryonic Antigen *Urine No Effect Analytical* No effect observed with glycosuria. *1459*

Catecholamines *Urine No Effect Physiological* No signifi-cant response to hyperglycemia. *3373*

Chloride *Serum Decrease Physiological* Dilutional effect when infused. *0652*

Cholesterol *Serum No Effect Physiological* In response to infusion of 1 g/kg body weight in normal volunteers and non-insulin dependent diabetics. *0476*
Serum Decrease Analytical Increase by 1.0% at 500 mg/dL on enzymatic procedure. *0057*

Cholesterol, High Density Lipoprotein
Serum No Effect Physiological In response to infusion of 1 g/kg body weight in normal volunteers and non-insulin depen-dent diabetics. *0476*

Cortisol *Plasma Decrease Physiological* Progressive fall when serum glucose increased. *3373*
Urine No Effect Analytical On fluorometric method of Ratliff and Hall. *2923*

Creatinine *Serum Increase Analytical* 7 g/L =0.06 mg/dL method of Heinegard. *1547* Interferes with Jaffé reaction. *2313*

Epinephrine *Urine No Effect Physiological* No significant response to hyperglycemia. *3373*

Erythrocyte Sedimentation Rate
Blood Decrease Analytical High blood sugar lowers sedi-mentation rate. *2313*

Estriol *Urine No Effect Analytical* With 80 mg glucose, enzyme hydrolysis and Kober procedure. *0762*
Urine Decrease Analytical Interference with GLC method. *3186* Probably destroys estriol during acid hydrolysis. *0023*

Estrogens *Urine No Effect Analytical* If sodium borohydride used with fluorescence procedures. *3899*
Urine Decrease Analytical 10% decrease if fluorometric method used. *1457* Affects hydrolysis of estrogen conjugates. *0022*

Fatty Acids, Free (FFA) *Serum No Effect Analytical* Up to 480 mg/dL on method of Soloni. *3381*
Serum Increase Physiological Delayed rise 3 h after admin-istration. *1697*
Serum Decrease Physiological Effect greater in normals than in diabetics. *2837*

Fucose *Serum No Effect Physiological* No change observed during glucose tolerance test in all subjects. *1665*

Glucagon *Plasma Increase Physiological* Of gastrointesti-nal tract origin after glucose load. *2676* Slight increase maxi-mum at 30 minutes after oral dose. *3306*

Glucose *Serum Increase Physiological* Normal metabolic response usually maximum at 60 minutes. *3306* Normal response in normals and diabetics to ingestion or infusion. *0476*

Glucuronic Acid *Urine No Effect Analytical* On orcinol pro-cedure. On m-hydroxydiphenyl procedure. *0389*
Urine Increase Analytical Significant effect on carbazole reaction. *0389*

Growth Hormone *Plasma Increase Physiological* Delayed rise after initial slight fall. *1697*

11-Hydroxycorticosteroids *Urine No Effect Physiological* No significant response to hyperglycemia. *3373*

Hydroxyproline *Urine No Effect Analytical* At 10 g/L on method of Goverde. *1365*

Insulin *Plasma Increase Physiological* 50 g orally caused increase from 28 to 39 pmol/L. *2939* Marked rise immediately after oral or i.v. administration. *3909* Marked effect (maximum at 60 minutes) for 2 h. *0310* Rises in parallel with glucose in normals or in noninsulin dependent diabetics. *0476*
Urine Increase Physiological Occurs 3 h after glucose load. *3080*

Ionized Calcium *Serum Increase Analytical* At concentra-tions above 1 g/dL ion specific electrode. *0540*

Isoleucine *Plasma Decrease Physiological* Probable mus-cle uptake. *1403*

17-Ketogenic Steroids *Urine Decrease Analytical* Acts as reducing agent in Zimmermann reaction. *0583* Interferes with Norymberski reaction. *3505*

Ketones *Urine Decrease Physiological* In starved individu-als even at 7.5 g/d. *1242*

17-Ketosteroids *Urine Decrease Analytical* Interferes with Zimmermann reaction. *3505*

Lactate *Serum Increase Physiological* Increased concen-tration observed after i.v. infusion. *1160*

Leucine *Plasma Decrease Physiological* Probable muscle uptake. *1403*

Glucose (continued)

Magnesium *Urine Increase Physiological* Nutrient induced augmentation ?mechanism. 2126

Metanephrines, Total *Urine No Effect Analytical* At 5 g/L on modified Pisano procedure. 1428

3-O-Methyl-D-Glucose *Urine Increase Analytical* Several methods without removal of glucose. 1183

Norepinephrine *Urine No Effect Physiological* No significant response to hyperglycemia. 3373

Orthophosphate *Test Conditions No Effect Analytical* No effect up to 1 mol/L. 1658

Osmolality *Serum Increase Analytical* Osmotically active constituent in samples. 1810
Urine Increase Analytical Osmotically active constituent in samples. 1810

pH *Urine Decrease Physiological* Associated with increased net acid excretion. 2126

Phenylalanine *Plasma Decrease Physiological* Probable muscle uptake. 1403

Phosphate *Serum Decrease Physiological* During glucose tolerance test, less marked and longer than calcium. 2207
Urine No Effect Analytical No effect at 20 mg/dL method of Jung/Parekh. 1835

Porphobilinogen *Urine Decrease Physiological* Suppresses ALA synthetase in acute porphyria. 0938

Potassium *Serum Decrease Physiological* Shift intracellularly when infusion given. 3362
Urine Decrease Physiological After oral glucose and fall in serum concentration. 1623

Protein *Test Conditions Increase Analytical* Interferes with Folin-Ciocalteu method of Lowry. 3636

Pyruvate *Serum Increase Physiological* Increased concentration observed after i.v. infusion. 1160

Salicylate *Serum No Effect Analytical* Up to 1,000 mg/dL no effect on Trinder method. 2251

Sialic Acid *Serum No Effect Physiological* No change observed during glucose tolerance test in all subjects. 1665

Specific Gravity *Urine Increase Analytical* Each 1% increases by 0.004. 0251

Sugar *Urine Increase Analytical* Positive with Clinitest®, Benedict's, Fehling's. 2308

Testosterone *Serum Decrease Physiological* Mean maximum fall of 34% normal in 150 minutes. 3751

Threonine *Plasma Decrease Physiological* Probable muscle uptake. 1403

Triglycerides *Serum No Effect Analytical* No effect at 500 mg/dL on method of Levy. 2151
Serum Increase Physiological Significant change in normals in response to infusion of 1 g/kg body weight, although not very marked. 0476

Tryptophan *Plasma Decrease Physiological* After 75 g nonprotein bound fell by 35% in 3 h. 2188

Tyrosine *Plasma Decrease Physiological* Hormonal effect (?exact mechanism). 0843

Urea Nitrogen *Serum Decrease Physiological* Protein sparing effect of glucose and hemodilution. 3879

Uric Acid *Serum No Effect Analytical* At 5 g/L on uricase procedure of Kabasak. 1840
Serum No Effect Physiological No change observed after i.v. infusion. 1160
Serum Increase Analytical At 1 g/dL increases by 0.2 mg/dL with method of Klein. 1949 Reducing substance reacts with phosphotungstate. 0652
Serum Decrease Physiological Increased urate clearance also seen in diabetics. 3344
Urine Increase Physiological Infusions have uricosuric action. 3344

Uroporphyrin *Urine Decrease Physiological* Suppresses ALA synthetase in acute porphyria. 0938

Valine *Plasma Decrease Physiological* Probable muscle uptake. 1403

Xylose Excretion *Urine Increase Analytical* Interferes with bromoaniline procedure if over 2 g/dL. 2981

Glucose-6-Phosphate

Pyruvate Kinase *Red Blood Cells Increase Analytical* Activating effect on enzyme. 0370

Glucosidase

Glucose *Test Conditions Increase Analytical* May be contaminant of glucose oxidase kit. 2803

Glucosulfone

Erythrocytes *Blood Decrease Physiological* Hemolytic anemia. 2429

Hematocrit *Blood Decrease Physiological* Hemolytic anemia. 0788

Hemoglobin *Blood Decrease Physiological* Hemolytic anemia. 0788

Leukocytes *Blood Decrease Physiological* Leukopenia. 2313

Methemoglobin *Blood Increase Physiological* Hemolytic anemia. 0788

Sulfhemoglobin *Blood Increase Physiological* Hemolytic anemia. 2429

Glucuronic Acid

Glucose *Serum Increase Analytical* Interferes with some methods. 2313
Urine No Effect Analytical No effect on Clinistix®, Labstix® methods. No effect on TesTape®. 3877

Magnesium *Serum No Effect Analytical* Calcium gluconate has no effect on dihydroxyazobenzene procedure. 3501
Serum Decrease Analytical Falsely low with titan yellow method. 3501
Urine Decrease Analytical Falsely low with titan yellow method. 3501

Sugar *Urine Increase Analytical* Interferes with Benedict's, Fehlings, Clinitest®. 3877

Glutamate

Diphenylhydantoin *Serum Decrease Physiological* Occurs with pharmacologic doses: lowers plasma concentrations. 3024

γ-Glutamyltransferase (GGT) *Serum Increase Analytical* Free glutamate effect 100% in normal range. 0412

Lymphocytes *Blood Decrease Physiological* Significant effect with megadose supplementation. 1347

Neutrophils *Blood No Effect Physiological* No effect with megadose supplementation. 1347

Phenobarbital *Serum Decrease Physiological* Occurs with pharmacologic doses: lowers plasma concentrations. 3024

Glutamic Acid

Alanine *Plasma Increase Physiological* Small increase noted after single 6 g dose. 3444

Ammonia *Plasma Decrease Analytical* With 1,000 nmoles produces 9% inhibition of indophenol reaction. 1290
Plasma Decrease Physiological Capable of reacting with ammonia. 1943

Aspartic Acid *Plasma Increase Physiological* Small increase noted after single 6 g dose. 3444

Glutamic Acid *Plasma Increase Physiological* Small increase noted after single 6 g dose. 3444

pH *Blood Increase Physiological* Sodium salt may cause alkalosis. 0071

Tyrosine *Plasma No Effect Analytical* On fluorometric procedure of Ambrose. 0080

Glutarimide

Barbiturate   *Serum   No Effect   Analytical*   No effect on UV absorption methods. *2981*

Glutathione

N-Acetylglucosaminidase   *Urine   No Effect   Analytical*   At 1 mmol/L had no effect on 2 colorimetric analytical methods. *1354*

Citrulline   *Test Conditions   Decrease   Analytical*   Inhibits color development in Archibald procedure. *3424*

Creatinine   *Serum   No Effect   Analytical*   No effect on method of Polar and Metcoff. *1975*

Glucose   *Serum   No Effect   Analytical*   No effect on Warner Glucomatic method. No effect on Trinder glucose oxidase method. *2771*   At 10 mg/dL no effect on glucose oxidase procedure of Gochman. *1311*
Serum   Increase   Analytical   Affects Neocuproin procedure. Affects alkaline ferricyanide method. *2771*
Serum   Decrease   Analytical   May affect some glucose oxidase methods. *1563*   Negative peaks GOD-PERID procedure. *2771*

Pseudocholinesterase   *Serum   Increase   Analytical*   If used as standard reacts with ACH in Levine method. *1024*

Uric Acid   *Serum   No Effect   Analytical*   If blank at 300 mg/L procedure of Kabasak. *1840*   No effect on Tripyridyl-s-triazine method of Morin. *2486*
Serum   Increase   Analytical   Acts as reducing agent with phosphotungstate. *0582*

Glutethimide

Alanine Aminotransferase   *Serum   No Effect   Analytical*   At acute overdose concentration (5 mg/dL) on Technicon® SMAC® method. *3719*

Albumin   *Serum   No Effect   Analytical*   At acute overdose concentration (5 mg/dL) on Technicon® SMAC® method. *3719*   At concentration of 50 mg/L had no effect on BCG method. *3393*

Alkaline Phosphatase   *Serum   No Effect   Analytical*   At acute overdose concentration (5 mg/dL) on Technicon® SMAC® method. *3719*

Alkaloids   *Urine   No Effect   Analytical*   With iodoplatinate on Frings TLC procedure at 10 mg/dL. With ninhydrin on Frings TLC procedure at 10 mg/dL. *1204*

Aminolevulinic Acid   *Urine   Increase   Physiological*   May precipitate acute porphyria. *1322*

Aspartate Aminotransferase   *Serum   No Effect   Analytical*   At acute overdose concentration (5 mg/dL) on Technicon® SMAC® method. *3719*

Barbiturate   *Serum   Increase   Analytical*   At 5 mg/dL false positive with diphenylcarbazone. *0583*
Serum   Decrease   Analytical   Negative interference with colorimetric method. *0293*
Urine   No Effect   Analytical   With vanillin on Frings procedure at 10 mg/dL. With mercuric SO_4 on Frings TLC procedure at 1 mg/dL. *1204*
Urine   Increase   Analytical   With mercurous NO_3 reagent on Frings TLC procedure. Purple with diphenylcarbazone on Frings TLC procedure. *1204*

Bilirubin   *Serum   No Effect   Analytical*   At acute overdose concentration (5 mg/dL) on Technicon® SMAC® method. *3719*   At concentration of 50 mg/L had no effect on Jendrassik and Grof method. *3393*
Serum   No Effect   Physiological   Insignificant protein-displacement effect in neonates. *3748*

Bromide   *Urine   No Effect   Analytical*   No effect on method of Frings. *1204*

Calcium   *Serum   No Effect   Analytical*   At acute overdose concentration (5 mg/dL) on Technicon® SMAC® method. *3719*   At concentration of 50 mg/L had no effect on cresolphthalein method. *3393*

Chlordiazepoxide   *Blood   No Effect   Analytical*   On method of Riddick at 5 mg/dL. *2983*

Cholesterol   *Serum   No Effect   Analytical*   At acute overdose concentration (5 mg/dL) on Technicon® SMAC® method. *3719*   At concentration of 50 mg/L had no effect on Liebermann-Burchard method. *3393*
Serum   Increase   Physiological   Significant effect from about 15 d of treatment with 500 mg drug/d in 6 volunteers. *0407*

Cholesterol, High Density Lipoprotein
Serum   Increase   Physiological   Significant effect from about 15 d of treatment with 500 mg drug/d in 6 volunteers. *0407*

Cholesterol, Low Density Lipoprotein
Serum   Increase   Physiological   Significant effect from about 15 d of treatment with 500 mg drug/d in 6 volunteers. *0407*

Cholesterol, Very Low Density Lipoprotein
Serum   Increase   Physiological   Significant effect from about 15 d of treatment with 500 mg drug/d in 6 volunteers. *0407*

Coproporphyrin   *Urine   Increase   Physiological*   May precipitate acute porphyria. *1322*
Feces   Increase   Physiological   May precipitate acute porphyria. *1322*

Creatine Kinase   *Serum   No Effect   Analytical*   At acute overdose concentration (5 mg/dL) on Technicon® SMAC® method. *3719*

Creatinine   *Serum   No Effect   Analytical*   At acute overdose concentration (5 mg/dL) on Technicon® SMAC® method. *3719*   At concentration of 50 mg/L had no effect on AutoAnalyzer Jaffé method. *3393*

Dapsone   *Serum   No Effect   Analytical*   At 10 mg/dL on colorimetric procedure of Higgins. *1590*

Diazepam   *Serum   No Effect   Analytical*   On cathode-ray Polarographic method of Berry. *0323*

Erythrocytes   *Blood   Decrease   Physiological*   May cause aplastic/megaloblastic anemia. *0071*

Estriol   *Urine   No Effect   Analytical*   No effect on GLC method. *1163*

Ethchlorvynol   *Serum   No Effect   Analytical*   At 600 mg/L on GLC procedure of Evenson. *1058*   At 10 µg/mL on colorimetric method of Wallace. *3753*   At 10 mg/L on GLC procedure of Evenson. *1058*
Urine   No Effect   Analytical   No effect on method of Frings and Cohen at 10 mg/dL. *1205*

Folate   *Serum   Decrease   Physiological*   Characteristic with anticonvulsant intake, possibly due to enhanced catabolism. *3023*
Red Blood Cells   Decrease   Physiological   Characteristic with anticonvulsant intake, possibly due to enhanced catabolism. *3023*

FPN Test   *Urine   No Effect   Analytical*   No effect at 10 mg/dL on method of Frings. *1204*

Glucaric Acid   *Urine   Increase   Physiological*   Manifestation of hepatic enzyme induction when 500 mg daily ingested for 21 d by 6 volunteers. *0407*

Glucose   *Serum   No Effect   Analytical*   At acute overdose concentration (5 mg/dL) on Technicon® SMAC® method. *3719*   At concentration of 17.5 mg/L had no effect on Ektachem® method. *3393*

Glutethimide   *Blood   Increase   Physiological*   1 g orally may produce concentration of 7 mg/L. *2348*

17-Hydroxycorticosteroids   *Urine   Increase   Analytical*   Interferes with moderate Glenn-Nelson procedure *in vitro*. *0427*

Iron   *Serum   No Effect   Analytical*   At acute overdose concentration (5 mg/dL) on Technicon® SMAC® method. *3719*   At concentration of 50 mg/L had no effect on Ferrozine method. *3393*

17-Ketogenic Steroids   *Urine   No Effect   Analytical*   Probably minimum interference with Zimmermann reaction. *3505*

17-Ketosteroids   *Urine   No Effect   Analytical*   Minimal interference with Zimmermann reaction *in vitro*. *0427*

Lactate Dehydrogenase   *Serum   No Effect   Analytical*   At acute overdose concentration (5 mg/dL) on Technicon® SMAC® method. *3719*

Leukocytes   *Blood   Decrease   Physiological*   May cause aplastic anemia. *0071*

MCV   *Blood   Increase   Physiological*   May cause megaloblastic anemia. *1343*

Methaqualone   *Serum   No Effect   Analytical*   At 20 mg/L on GLC procedure of Evenson. *1057*

Methemoglobin   *Blood   Increase   Physiological*   Bleeding associated with overdosage. *2427*

Glutethimide *(continued)*

Phosphate *Serum No Effect Analytical* At acute overdose concentration (5 mg/dL) on Technicon® SMAC® method. *3719* At concentration of 50 mg/L had no effect on phosphomolybdate method. *3393*

Platelets *Blood Decrease Physiological* Thrombocytopenia or aplastic anemia. *0071*

Porphobilinogen *Urine Increase Physiological* May precipitate acute porphyria. *1322*

Porphyrins *Urine Increase Physiological* May precipitate attack of acute porphyria. *2220*

Potassium *Serum No Effect Analytical* At concentration of 10 mg/L had no effect on measurement by ISE with predilution. *3393*

Propoxyphene *Serum No Effect Analytical* At 25 mg/L on method of Evenson. *1056*

Protein *Serum No Effect Analytical* At acute overdose concentration (5 mg/dL) on Technicon® SMAC® method. *3719* At concentration of 50 mg/L had no effect on biuret method with blank correction. *3393*

Prothrombin Time *Plasma Increase Physiological* Effect observed with overdosage. *2427*
Plasma Decrease Physiological Increases rate of degradation of administered coumarins. *3011* In 10 patients receiving 0.5 g/d when on warfarin due to hepatic microsomal enzyme Induction. *3649*

Protoporphyrin *Feces Increase Physiological* May precipitate acute porphyria. *1322*

Salicylate *Urine No Effect Analytical* At 10 mg/dL no effect on Trinder method. *1204*

Sodium *Serum No Effect Analytical* At concentration of 10 mg/L had no effect on measurement by ISE with predilution. *3393*

Triglycerides *Serum No Effect Analytical* At acute overdose concentration (5 mg/dL) on Technicon® SMAC® method. *3719* At concentration of 50 mg/L had no effect on lipase/esterase method. *3393*
Serum No Effect Physiological No significant effect with 21 d of 500 mg daily in 6 volunteers. *0407*

Triglycerides, High Density Lipoprotein
Serum No Effect Physiological No significant effect with 21 d of 500 mg daily in 6 volunteers. *0407*

Triglycerides, Low Density Lipoprotein
Serum No Effect Physiological No significant effect with 21 d of 500 mg daily in 6 volunteers. *0407*

Triglycerides, Very Low Density Lipoprotein
Serum No Effect Physiological No significant effect with 21 d of 500 mg daily in 6 volunteers. *0407*

Urea Nitrogen *Serum No Effect Analytical* At acute overdose concentration (5 mg/dL) on Technicon® SMAC® method. *3719* At concentration of 50 mg/L had no effect on diacetylmonoxime method. *3393*

Uric Acid *Serum No Effect Analytical* At acute overdose concentration (5 mg/dL) on Technicon® SMAC® method. *3719* At concentration of 17.5 mg/L had no effect on Kageyama-Hantzsch method. *3393* At concentration of 50 mg/L had no effect on phosphotungstate reduction method. *3393*

Glyburide

Alanine Aminotransferase *Serum Increase Physiological* Rare and reverts to normal if therapy continued. *0125*

Alkaline Phosphatase *Serum Increase Physiological* Rare and reverts to normal if therapy continued. *0125*

Antidiuretic Hormone *Urine Increase Physiological* Slight (not significant) effect. *2499*

Aspartate Aminotransferase
Serum Increase Physiological Rare and reverts to normal if therapy continued. *0125*

Cholesterol *Serum Decrease Physiological* Fall by over 50 mg/dL in treated diabetics. *0125*

Cortisol *Plasma Increase Physiological* Occasional slightly higher, no significant change after glucose. *0125*

Creatinine *Serum No Effect Analytical* At 5 times expected therapeutic concentration had no effect on Du Pont aca, Beckman Astra and Technicon® SMAC® methods. *2999*

Serum No Effect Physiological No significant effect observed. *0125*

Fatty Acids, Free (FFA) *Serum Decrease Physiological* Protracted fall maximal at 4.5 h. *1526*

Glucose *Serum No Effect Analytical* No effect on glucose oxidase method of Boehringer. *3277*
Serum Decrease Physiological Long acting sulfonylurea = insulin secretion. *2216*

Glucose Tolerance *Serum Increase Physiological* Marked improvement noted. *2216*

Glycerol *Serum Decrease Physiological* Response similar to that of free fatty acids. *1526*

Growth Hormone *Plasma No Effect Physiological* No effect observed. *0125*

Insulin *Plasma Increase Physiological* Significant increase reported. *2216*

Osmolality *Urine Decrease Physiological* Normal diuretic response. *2499*

Platelets *Blood Decrease Physiological* Rare thrombocytopenia reported. *0125*

Proinsulin *Plasma No Effect Physiological* No significant effect after 6 mo. *0965*

Triglycerides *Serum Decrease Physiological* Fall by 50% in treated diabetics. *0125*

Urea Nitrogen *Serum No Effect Physiological* No significant effect observed. *0125*

Volume *Urine Increase Physiological* Normal diuretic response. *2499*

Water Clearance, Free *Urine Increase Physiological* Normal diuretic response. *2499*

Glyceraldehyde

Triglycerides *Serum Increase Analytical* Measured as glycerol with enzymatic procedures. *2981*

Glyceric Acid

Lactate *Plasma Decrease Analytical* High concentrations interfere with enzymatic method. *1150*

Glycerin

Creatinine *Serum Increase Analytical* 100 mg/L = 0.03 mg/dL method of Heinegard. *1547*

Glucose *Serum No Effect Physiological* No effect of ingestion of glycerol. *3615*

Glycerol *Serum Increase Physiological* Rose from normal 0.51 mmol/L to 20 mmol/L. *3615*

Hematocrit *Blood Decrease Physiological* Hemolysis may occur after i.v. administration. *1448* Maximum decrease about 80 minutes after ingestion. *3615*

Hemoglobin *Plasma Increase Physiological* Hemolysis may occur after i.v. administration. *1448*
Blood Decrease Physiological Hemolysis may occur after i.v. administration. *1448*
Urine Increase Physiological Rare transient effect after i.v. infusion. *3615* Isolated cases after 20% solution i.v. *1504*

Ionized Calcium *Serum Decrease Analytical* At 0.1-1.0 mol/L with Calcium specific electrode. *0540*

Osmolality *Serum Increase Physiological* Rose an average of 19 mosm/kg. *3615*

Protein *Urine Increase Physiological* Hemolysis may occur after i.v. administration. *1448*
Test Conditions Increase Analytical Interferes with Lowry and biuret methods. *3954*

Sodium *Serum Decrease Physiological* Maximum effect of less than 5%. *3615*

Triglycerides *Serum Increase Analytical* When enzymatic procedure used is end product of reaction. *3590*

Volume *Urine Increase Physiological* Weak osmotic diuretic action. *2605*

Glycine

Alkaline Phosphatase *Serum Decrease Analytical* If used as buffer in lab procedure as complexes Mg. *3588*

Aminolevulinic Acid *Urine Increase Analytical* If no preliminary separation. *2981*
Urine Increase Physiological In patients with acute porphyria in remission. *0938*

Ammonia *Plasma Decrease Analytical* With 2500 nmoles produces 63% inhibitor indophenol reaction. *1290*
Urine Increase Physiological Same extent in gouty and normal subjects. *3929*

Glycine *Plasma Increase Physiological* Similar marked increase in gouty and normals. *3929*
Urine Increase Physiological Similar marked increase in gouty and normals. *3929*

Megaloblasts *Blood Increase Physiological* May provoke or accentuate megaloblastosis. *3025*

Phosphate *Urine Increase Physiological* Inhibits tubular reabsorption. *0957*

Porphobilinogen *Urine Increase Physiological* In patients with acute porphyria in remission. *0938*

Protein *Test Conditions Decrease Analytical* Lowry procedure due to chelation of chemicals. *1796*

Pyruvate Kinase *Red Blood Cells Decrease Analytical* Most marked with larger side chain. *0370*

Serine *Plasma Increase Physiological* Similar marked increase in gouty and normals. *3929*
Urine Increase Physiological Similar marked increase in gouty and normals. *3929*

Tyrosine *Plasma No Effect Analytical* On fluorometric procedure of Ambrose. *0080*

Uric Acid *Urine Increase Physiological* Increased tubular secretion especially in gouty subjects. *3929*

Glycocholate

Lipase *Serum Increase Analytical* Sodium salts prevent inactivation of enzyme. *3589*

Glycocyamidine

Creatinine *Serum Increase Analytical* Reacts to give false increase with Jaffé reagent. *3505*
Urine Increase Analytical 1 mg equivalent to 0.30 mg with automated Jaffé. *2558*

Glycocyamine

Creatinine *Serum Increase Analytical* If reaction of Voges-Proskauer used. *1563*

Glycopyrrolate

Alanine Aminotransferase *Serum Increase Physiological* Hepatotoxicity. *2313*

Alkaline Phosphatase *Serum Increase Physiological* Hepatotoxicity. *2313*

Alkaloids *Urine No Effect Analytical* With ninhydrin on Frings TLC procedure at 2 mg/dL. With iodoplatinate on Frings TLC procedure at 2 mg/dL. *1204*

Aspartate Aminotransferase
Serum Increase Physiological Hepatotoxicity. *2313*

Barbiturate *Urine No Effect Analytical* With vanillin on Frings procedure at 2 mg/dL. With mercurous NO_3 on Frings TLC procedure at 2 mg/dL. With diphenylcarbazone on Frings TLC procedure at 2 mg/dL. With mercuric SO_4 on Frings TLC procedure at 2 mg/dL. *1204*

Bile *Urine Increase Physiological* Hepatotoxicity. *2313*

Bilirubin *Serum Increase Physiological* Hepatotoxicity. *2313*

Bromide *Urine No Effect Analytical* No effect on method of Frings. *1204*

BSP Retention *Serum Increase Physiological* Hepatotoxicity. *2313*

Cephalin Flocculation *Serum Increase Physiological* Hepatotoxicity. *2313*

Ethchlorvynol *Urine No Effect Analytical* No effect on method of Frings and Cohen. *1204*

FPN Test *Urine No Effect Analytical* No effect at 2 mg/dL on method of Frings. *1204*

Salicylate *Urine No Effect Analytical* At 2 mg/dL no effect on Trinder method. *1204*

Thymol Turbidity *Serum Increase Physiological* Hepatotoxicity. *2313*

Glycylglycine

Protein *Test Conditions Decrease Analytical* Lowry procedure due to chelation of chemicals. *1796*

Glycylproline

Proline *Plasma Increase Analytical* Equivalent effect on method of Goodwin. *1345*

Glycyrrhiza

Aldosterone *Plasma Decrease Physiological* Hormonal like action of drug. *0232*
Urine Decrease Physiological Pseudoaldosteronism effect. *3755*

Estradiol *Urine Increase Analytical* Affects methods of Brown and Beling. *0023*

Estriol *Urine Increase Analytical* Affects methods of Brown and Beling. *0023*

Estrogens *Urine Increase Analytical* Affects colorimetric and fluorometric procedures. *0022*

Myoglobin *Urine Increase Physiological* May follow hypokalemia. *2458*

pH *Blood Increase Physiological* May cause alkalosis. *2220*

Potassium *Serum Decrease Physiological* Aldosterone like action. *2458*
Urine Increase Physiological Aldosterone like action. *2458*

Renin Activity *Plasma Decrease Physiological* Pseudoaldosteronism effect. *3755*

Urea Nitrogen *Serum Increase Physiological* May cause nephropathy. *2458*

Glymidine

Erythrocyte Sedimentation Rate
Blood Increase Physiological Slight rise in mean rate reported. *2427*

Glucose *Serum Decrease Physiological* Plasma half-life approximately 4 h. *2563*

Gold

N-Acetylglucosaminidase *Serum No Effect Physiological* No significant changes of microsomal enzyme in patients treated with parenteral drug. *2933*
Urine Increase Physiological In 77% in 31 patients treated with parenteral gold. *2933*

Alanine Aminopeptidase *Urine Increase Physiological* In 13% (not significant) increase of microsomal enzyme in 31 patients treated with parenteral gold. *2933*

Alanine Aminotransferase *Serum Increase Physiological* Hepatotoxicity with centrolobular necrosis. *2313* Due to toxic effect on drug on liver. *1674*

Alkaline Phosphatase *Serum Increase Physiological* Rare side effect with liver damage (reversible). *0085* Cholestatic jaundice in 3 patients with rheumatoid arthritis given gold sodium thiomalate: all patients recovered spontaneously. *1084* In 3 patients with rheumatoid arthritis during chrysotherapy. *0993*

Aminolevulinic Acid *Urine Increase Physiological* Occurs with panmyelopathy. *1343*

Gold (continued)

Aspartate Aminotransferase
Serum Increase Physiological Hepatotoxicity with centrolobular necrosis. 2313 Cholestatic jaundice in 3 patients with rheumatoid arthritis given gold sodium thiomalate: all patients recovered spontaneously. 1084 In 3 patients with rheumatoid arthritis during chrysotherapy. 0993 Due to toxic effect on drug on liver. 1674

Bile *Urine Increase Physiological* Hepatotoxicity with centrolobular necrosis. 2313

Bilirubin *Serum Increase Physiological* Rare side effect with liver damage. 2427 Cholestatic jaundice in 3 patients with rheumatoid arthritis given gold sodium thiomalate: all patients recovered spontaneously. 1084 In 3 patients with rheumatoid arthritis during chrysotherapy. 0993 Due to toxic effect on drug on liver. 1674

Bilirubin, Direct *Serum Increase Physiological* Cholestatic jaundice in 3 patients with rheumatoid arthritis given gold sodium thiomalate; all patients recovered spontaneously. 1084 In 3 patients with rheumatoid arthritis during chrysotherapy. 0993

BSP Retention *Serum Increase Physiological* Hepatotoxicity with centrolobular necrosis. 2313

Cephalin Flocculation *Serum Increase Physiological* Hepatotoxicity with centrolobular necrosis. 2313

Cholesterol *Serum Increase Physiological* May cause hypersensitive cholestasis. 1434

Coproporphyrin *Urine Increase Physiological* Occurs with panmyelopathy. 1343

Creatinine Clearance *Urine Decrease Physiological* In 11 patients who showed glomerular injury, including minimal change nephrotic syndrome with a membranous pattern of immune complex deposition as well as other patterns of deposition. 1170

Eosinophils *Blood Increase Physiological* Transient allergic response. 0114 Occurred in 21% patients taking gold sodium thiomalate and in 13% taking auranofin. Occurred in 24% and 30% respectively with toxicity. 0992 Observed in 11% of 64 treated patients. 3360

Erythrocytes *Blood Decrease Physiological* Aplastic anemia. 0114 Blood dyscrasias major side effect as reported to uk committee on safety of medicines. 0782 Occasional case of aplastic anemia reported. 3717 7 cases out of 246 treated cases. 0829
Urine Increase Physiological Produces actual bleeding. 1022

Erythrocyte Sedimentation Rate
Blood Decrease Physiological With 200 mg aurothioglucose at 4 weeks intervals mean ESR fell from 46 to 26 mm/h in 30 patients. 2619

γ-Globulin *Serum Decrease Physiological* Single case reported. 2427

γ-Glutamyltransferase (GGT)
Serum No Effect Physiological No significant changes of microsomal enzyme in patients treated with parenteral drug. 2933
Urine Increase Physiological In 6.5% (not significant) in 31 patients treated with parenteral gold. 2933

Gold *Red Blood Cells Increase Physiological* Up to 45% of blood concentration in RBC. 3366

Hematocrit *Blood Decrease Physiological* Aplastic anemia. 0114

Hemoglobin *Blood Decrease Physiological* Aplastic anemia. 0114
Urine Increase Physiological Produces actual bleeding. 1022

Immunoglobulin IgA *Serum Decrease Physiological* Substantial lowering at 3 mo in 25 patients with rheumatoid arthritis treated with gold. 3680

Immunoglobulin IgG *Serum Decrease Physiological* Substantial lowering at 3 mo in 25 patients with rheumatoid arthritis treated with gold. 3680

Immunoglobulin IgM *Serum Decrease Physiological* Substantial lowering at 12 mo in 25 patients with rheumatoid arthritis treated with gold. 3680

[131]I Uptake *Serum No Effect Physiological* No effect observed. 1915

Lactate Dehydrogenase *Serum Increase Physiological* Cholestatic jaundice in 3 patients with rheumatoid arthritis given gold sodium thiomalate: all patients recovered spontaneously. 1084

LE Cells *Blood Positive Physiological* SLE may occur, usually normalizes when stopped. 3059

Leukocytes *Blood Increase Physiological* Due to eosinophilia of hypersensitivity. 2313
Blood Decrease Physiological Aplastic anemia/leukopenia/agranulocytosis. 0114 Blood dyscrasias major side effect as reported to uk committee on safety of medicines. 0782 Occasional case of aplastic anemia reported. 3717 7 cases out of 246 treated cases. 0829 Reported in 6 patients: brief self-limiting process. 1362 In 3% of 90 patients with rheumatoid arthritis. 1461

Lymphocytes *Blood No Effect Physiological* No effect observed with megadose supplementation. 1347

Microsomal Aminopeptidase
Serum No Effect Physiological No significant changes in patients treated with parenteral drug. 2933

Neutrophils *Blood No Effect Physiological* No effect observed with megadose supplementation. 1347
Blood Decrease Physiological Occasional case of sodium aurothiomalate induced neutropenia. 0155 Reported in 6 patients: brief self-limiting process. 1362

Occult Blood *Feces Positive Physiological* Associated with thrombocytopenia. 2427 Rare side effect as reported to uk committee on safety of. 0782

Partial Thromboplastin Time
Plasma Increase Physiological In one 3 y old with rheumatoid arthritis. 1674

Peripheral Smear *Blood Abnormal Physiological* May cause stippling of cells. 1205

Platelets *Blood Decrease Physiological* Aplastic anemia (thrombocytopenia). 0114 Blood dyscrasias major side effect as reported to uk committee on safety of medicines. 0782 Several cases of platelet-associated IgG and thrombocytopenia. 1910 Occasional case of aplastic anemia reported. 3717 Severe thrombocytopenia observed 18 mo after end of gold therapy for rheumatoid arthritis. 3427 7 cases out of 246 treated cases. 0829 In 23 patients treated for 25 y: apparently associated with HLA-DR3 alloantigen. 0675 In 3% of 90 patients with rheumatoid arthritis. 1461

Protein *Urine Increase Physiological* May cause nephrotoxicity. 2427 Nephrotoxic effect (at least in 50% cases). 1022 Observed in 16% of 90 patients with rheumatoid arthritis in one study. 0782 Prevalence of gold nephropathy about 1 in 500. Nephropathy not necessarily related to other side effects. 0829 Observed in 3% of 1283 patients ranging from mild to heavy: did not persist beyond 12 mo. Some patients showed membranous glomerulonephritis. 1886 In 11 patients who showed glomerular injury, including minimal change nephrotic syndrome with a membranous pattern of immune complex deposition as well as other patterns of deposition. 1170 Good correlation between amount of protein and duration of proteinuria. Moderate proteinuria does not preclude reinstitution of therapy once protein has cleared. 2586 In 16% of 90 patients with rheumatoid arthritis. 1461 HLA DR3 positive patients had 3 times risk of this side effect than those without antigen. 1379

Protein Bound Iodine (PBI) *Serum Decrease Analytical* Interferes with chloric acid, Barker methods. 0354

Prothrombin Time *Plasma Increase Physiological* In one 3 year old with rheumatoid arthritis. 1674

T3 Uptake *Serum No Effect Physiological* No effect observed with resin. 1915

Thymol Turbidity *Serum Increase Physiological* Hepatotoxicity with centrolobular necrosis. 2313

Urea Nitrogen *Serum Increase Physiological* May cause nephrotoxicity. 3204

Xylose Excretion *Urine Decrease Physiological* May produce gastrointestinal irritation, impaired absorption. 0612

Gonadotropin

Androsterone *Urine Increase Physiological* 150% increase when given to normal men. 3231

Estrogens *Urine Increase Physiological* Hormonal action. *2211*

Etiocholanolone *Urine Increase Physiological* 150% increase when given to normal men. *3231*

17-Hydroxycorticosteroids *Urine Increase Physiological* Hormonal action. *3225*

17-Ketosteroids *Urine Increase Physiological* Metabolic response. *3225*

Pregnanediol *Urine Increase Physiological* Hormonal action. *2211*

Pregnanetriol *Urine Increase Physiological* Variable change when given to normal men. *3231*

Protein Bound Iodine (PBI) *Serum No Effect Physiological* No significant effect observed with chorionic hormone. *0583*

Testosterone *Serum Increase Physiological* Twofold increase observed in males. *3338*

Urine Increase Physiological Marked effect in normal men. *3062*

GPA-1714

Alanine Aminotransferase *Serum Increase Physiological* 4 of 11 patients increased when dose of approximately 3 g. *3328*

Griseofulvin

Alanine Aminotransferase *Serum Increase Physiological* May be hepatotoxic. *2220*

Alkaline Phosphatase *Serum Increase Physiological* May be hepatotoxic. *2220*

Aminolevulinic Acid *Urine Increase Physiological* May precipitate acute porphyria attack. *1322*

Aspartate Aminotransferase
Serum Increase Physiological May be hepatotoxic. *2220*

Casts *Urine Increase Physiological* Cylindruria may occur without renal insufficiency. *1343*

Coproporphyrin *Urine Increase Physiological* May precipitate acute porphyria attack. *1322*
Feces Increase Physiological May precipitate acute porphyria attack. *1322*

Creatinine *Serum Increase Physiological* Rare renal damage reported. *0131*

Creatinine Clearance *Urine Decrease Physiological* Rare renal damage reported. *0131*

LE Cells *Blood Positive Physiological* May produce LE-like syndrome. *3504*

Leukocytes *Blood Decrease Physiological* May cause leukopenia/neutropenia. *2427*

Lymphocytes *Blood Increase Physiological* (relative lymphocytosis). *2427*

Monocytes *Blood Increase Physiological* (relative monocytosis). *2427*

Neutrophils *Blood Decrease Physiological* May cause decrease by up to 20% with fall in total count. *2353* Occasional case of agranulocytosis reported. *3717*

Porphobilinogen *Urine Increase Physiological* May precipitate acute porphyria attack. *1322*

Porphyrins *Blood Increase Physiological* May precipitate acute porphyria attack. *2808*
Urine Increase Physiological Stimulates formation of ALA-synthetase. *1343*

Protein *Urine Increase Physiological* May cause nephrotoxicity. *0633* ?nephrotoxic effect (transient and reversible). *1022*

Prothrombin Time *Plasma Decrease Physiological* Induces hepatic metabolism of anticoagulants. *0772*

Protoporphyrin *Feces Increase Physiological* May precipitate acute porphyria attack. *1322*

Urea Nitrogen *Serum Increase Physiological* Rare renal damage reported. *0131*

Uric Acid *Serum Decrease Physiological* Effective in treatment of gout ?mechanism. *3754*

Growth Hormone

α-Amino-Nitrogen *Plasma Decrease Physiological* Increased peripheral removal. *2312*

Calcium *Urine Increase Physiological* Normal metabolic response. *2414*

Fatty Acids, Free (FFA) *Serum Increase Physiological* Induces lipolysis (?at adenyl cyclase step). *0819*

Glucose *Urine Increase Physiological* May cause hyperglycemia and glycosuria *2313*

Glucose Tolerance *Serum Decrease Physiological* Reduces disappearance rate after i.v. glucose tolerance test. *0819*

Hydroxyproline *Urine Increase Physiological* Metabolic effect. *1343*

Insulin *Plasma Increase Physiological* Small postabsorptive rise. *0138*

Ketones *Serum Increase Physiological* Small postabsorptive rise. *0138*
Urine Increase Physiological Small postabsorptive rise. *0138*

Phosphate *Serum Increase Physiological* Gradual slight rise — metabolic effect. *2414*

Potassium *Urine Decrease Physiological* Causes positive balance. *0138*

Protein *Serum Increase Physiological* Associated with increased protein synthesis. *0697*

Thymol Turbidity *Serum Increase Physiological* Increased protein synthesis. *0697*

Thyroxine (T₄) *Serum Decrease Physiological* Observed in most patients initially euthyroid given drug possibly due to inhibition of TSH response to TRH. *1181*

TSH Response to TRH *Serum Decrease Physiological* Attributable to T_3 increment through acceleration of T_4 conversion. *3798*

Urea Nitrogen *Serum Decrease Physiological* Metabolic effect. *2414*

Guaiacol

Color *Urine Increase Analytical* May produce green color. *1187*

Guaifenesin

Guaiacols Spot Test *Urine Positive Analytical* False reaction with screening test of Rogers. *3031*

5-Hydroxyindoleacetic Acid (5-HIAA)
Urine No Effect Analytical No effect on TLC method of McGregor. *2367* No effect in vivo dose on method of Goldenberg. *1327*
Urine Increase Analytical Interferes with nitrosonaphthol method. *1237* Affects quantitative method of Udenfriend. *0583*

2-Methoxyphenoxy-Lactic Acid
Urine Increase Physiological Major metabolite (44% of 1 g in 3 h). *3690*

Platelet Aggregation *Blood Decrease Physiological* Reported effect. *3322*

Uric Acid *Serum Decrease Physiological* Modest effect, eg 3 mg/dL over 3 d. *2909* Modest hypouricemic effect, with fall of up to 3 mg/dL over 3 d in some patients. *2909*

Vanillylmandelic Acid *Urine Increase Analytical* Affects p-nitraniline in initial part of reaction. *0757*

Guanabenz

Apolipoprotein AI *Serum No Effect Physiological* Increase by 5 mg/dL in patients given 4 to 8 mg daily for 12 to 16 weeks. *2550*

Apolipoprotein B *Serum No Effect Physiological* Increase by 2 mg/dL in patients given 4 to 8 mg daily for 12 to 16 weeks. *2550*

Apolipoprotein CII *Serum No Effect Physiological* No change in patients given 4 to 8 mg daily for 12 to 16 weeks. *2550*

Guanabenz (continued)

Apolipoprotein CIII *Serum No Effect Physiological* No change in patients given 4 to 8 mg daily for 12 to 16 weeks. *2550*

Apolipoprotein E *Serum Decrease Physiological* No change in patients given 4 to 8 mg daily for 12 to 16 weeks. *2550*

Cholesterol *Serum Decrease Physiological* Mean concentration in 480 patients treated for up to 2 y decreased by 10 mg/dL. Decrease maintained throughout subsequent therapy. *1865* Decrease by -10 mg/dL in patients given 4 to 8 mg daily for 12 to 16 weeks. *2550*

Cholesterol, High Density Lipoprotein
Serum No Effect Physiological No effect in 39 hypertensives treated for 2 y. *1865* Increase by 1 mg/dL in patients given 4 to 8 mg daily for 12 to 16 weeks. *2550*

Cholesterol, Low Density Lipoprotein
Serum Decrease Physiological In 39 patients treated for 2 y: approximately 23 mg/dL decrease. *1865*

Glucagon *Plasma Increase Physiological* In 30 hypertensives: effect most marked in individuals with higher doses. No change observed on withdrawal of drug. *1018*

Growth Hormone *Plasma No Effect Physiological* In 30 patients treated with twice daily doses of from 4 to 32 mg. *1018*

Insulin *Plasma No Effect Physiological* In 30 patients treated with twice daily doses of from 4 to 32 mg. *1018*

Prolactin *Plasma No Effect Physiological* In 30 patients treated with twice daily doses of from 4 to 32 mg. *1018*

Triglycerides *Serum No Effect Physiological* No effect in 39 hypertensives treated for 2 y. *1865* Decrease by -13 mg/dL in patients given 4 to 8 mg daily for 12 to 16 weeks. *2550*

Guanazole

Leukocytes *Blood Decrease Physiological* Observed in approximately 80% patients. *1581*

Platelets *Blood Decrease Physiological* Observed in almost all patients. *1581*

Guancydine

Aspartate Aminotransferase
Serum Increase Physiological Transient increases observed in some cases. *0668*

Chloride *Urine Decrease Physiological* Reduced renal blood flow causes retention. *2132*

Creatinine Clearance *Urine Decrease Physiological* Due to decreased renal blood flow. *2132*

Effective Renal Plasma Flow
Patient Decrease Physiological Due to hypotensive action of drug. *2132*

Sodium *Urine Decrease Physiological* Probable action on renal tubules. *3715*

Volume *Urine Decrease Physiological* Probable action on renal tubules. *3715*

Guanethidine

Alanine Aminotransferase *Serum Increase Physiological* Possible myopathic complication. *2220*

Aspartate Aminotransferase
Serum Increase Physiological Possible myopathic complication. *2220*

Catecholamines *Urine Decrease Physiological* Inhibits release of norepinephrine. *3722*

Chloride *Serum Increase Physiological* Salt retention ?due to tubular effect. *1343*

Epinephrine *Urine Increase Physiological* Slight increased output reported. *3083*

Glucose *Serum Decrease Physiological* Antidiabetic activity (rise once stop therapy). *1426*

Glucose Tolerance *Serum Increase Physiological* Antidiabetic activity (change at end of treatment). *1426*

Norepinephrine *Urine Decrease Physiological* Non significant decrease reported. *3083*

Prothrombin Time *Plasma Increase Physiological* Enhances anticoagulant activity. *1343*

Sodium *Serum Increase Physiological* Salt retention ?due to tubular effect. *1343*

Tyramine Test *Patient Increase Physiological* Increased response to tyramine due to MAO inhibition. *1325*

Urea Nitrogen *Serum Increase Analytical* Due to chemical similarity to urea. *0085*
Serum Increase Physiological Associated with blood pressure reduction, decreased blood flow. *1680*

Vanillylmandelic Acid *Urine Increase Physiological* Initial pharmacological response (changes later). *0913*
Urine Decrease Physiological 30% reduction seen if given therapeutically. *3897*

Guanfacine

Cholesterol *Serum Decrease Physiological* Increase by 14% in 30 patients treated for 2 y. *0088*

Metanephrine *Urine No Effect Analytical* No effect at 2 mg/L on HPLC method. *0342*

Normetanephrine *Urine No Effect Analytical* No effect at 2 mg/L on HPLC method. *0342*

Triglycerides *Serum Decrease Physiological* Increase by 15% in 30 patients treated for 2 y. *0088*

Guanidine

Ammonia *Plasma Decrease Analytical* With 500 nmoles produces 34% inhibitor indophenol reaction. *1290*

Creatinine *Serum Increase Analytical* If reaction of Voges-Proskauer used. *1563*

Urea Nitrogen *Serum No Effect Analytical* No effect on Berthelot reaction. *1862*

Guanidinoacetic Acid

Ammonia *Plasma Decrease Analytical* With 1,000 nmoles produces 22% inhibition of indophenol reaction. *1290*

Arginine *Plasma Increase Analytical* Measured as analyte in method of Bacchus. *0196*

Uric Acid *Serum No Effect Analytical* At 10 mg/dL on Nishi phosphotungstate procedure At 10 mg/dL on copper chelate procedure. *2230*

Guanidinosuccinic Acid

Platelet Aggregation *Blood No Effect Physiological* No effect of concentrations occurring in uremia. *0826*

Uric Acid *Serum No Effect Analytical* At 3 mg/dL on Nishi phosphotungstate procedure At 3 mg/dL on copper chelate procedure. *2230*

Guanoclor

Alanine Aminotransferase *Serum Increase Physiological* Possible effect on liver. *2220*

Amphetamine *Urine No Effect Analytical* No effect at 100 mg/L on method of Rutter. *3097*

Aspartate Aminotransferase
Serum Increase Physiological Possible effect on liver. *2220*

Glucose *Serum Decrease Analytical* Affects glucose oxidase method of Boehringer. *3277*

Urea Nitrogen *Serum Increase Physiological* May decrease renal blood flow. *1022*

Guanoxan

Alanine Aminotransferase *Serum Increase Physiological* May cause hepatic toxicity. *2313*

Alkaline Phosphatase *Serum Increase Physiological* May cause hepatic toxicity. *2313*

Amphetamine *Urine No Effect Analytical* No effect at 100 mg/L on method of Rutter. *3097*

Aspartate Aminotransferase
Serum Increase Physiological May cause hepatic toxicity. *2313*

Bile *Urine Increase Physiological* May cause hepatic toxicity. *2313*

Bilirubin *Serum Increase Physiological* May cause hepatic toxicity. *2313*

BSP Retention *Serum Increase Physiological* May cause hepatic toxicity. *1210*

Cephalin Flocculation *Serum Increase Physiological* May cause hepatic toxicity. *2313*

LE Cells *Blood Positive Physiological* SLE may occur, usually normalizes when stopped. *3059*

Norepinephrine *Urine Decrease Physiological* Decreased output reported. *3083*

Thymol Turbidity *Serum Increase Physiological* May cause hepatic toxicity. *2313*

Urea Nitrogen *Serum Increase Physiological* Due to decreased blood flow. *2220*

Gum Tragacanth

Cholesterol *Serum No Effect Physiological* In 5 male volunteers consuming 9.9 g daily for 21 d. *0990*

Fat *Feces Increase Physiological* In 5 male volunteers consuming 9.9 g daily for 21 d. *0990*

Glucose Tolerance *Serum No Effect Physiological* In 5 male volunteers consuming 9.9 g daily for 21 d. *0990*

Hydrogen *Breath No Effect Physiological* In 5 male volunteers consuming 9.9 g daily for 21 d. *0990*

Methane *Breath No Effect Physiological* In 5 male volunteers consuming 9.9 g daily for 21 d. *0990*

Phospholipids, Total *Serum No Effect Physiological* In 5 male volunteers consuming 9.9 g daily for 21 d. *0990*

Triglycerides *Serum No Effect Physiological* In 5 male volunteers consuming 9.9 g daily for 21 d. *0990*

Haemaccel

Amylase *Serum No Effect Analytical* At concentration of 7,000 mg/L had no effect on maltotetrose method. *3393* At concentration of 7,000 mg/L had no effect on p-nitrophenylmaltoheptoside method. *3393* At concentration of 7,000 mg/L had no effect on p-nitrophenylmaltohepto-side/hexoside method. *3393*

Hair Lacquer

Leukocytes *Blood Decrease Physiological* Some may cause leukopenia (AMA Blood dyscrasias). *2429*

Hair Treatments

Calcium *Hair Increase Analytical* Marked with feminine deodorants. *1594*
Hair Decrease Analytical Waving and setting lotions cause decrease. *1594*

Copper *Hair Decrease Analytical* Lowering seen with hair tonics. *1594*

Magnesium *Hair Increase Analytical* Marked with feminine deodorants. *1594*
Hair Decrease Analytical With setting permanent waving lotions. *1594*

Zinc *Hair Increase Analytical* Threefold with permanent waving lotion. *1594*

Halofenate

Alanine Aminotransferase *Serum Increase Physiological* Mild transient effect observed ?origin. *0161*

Aspartate Aminotransferase
Serum Increase Physiological Mild transient effect observed ?origin. *0161*

Bilirubin *Serum Decrease Physiological* No reversal of early effect with long treatment. *1708*

Cholesterol *Serum No Effect Physiological* No significant effect observed in hyperlipemics. *0161*
Serum Decrease Physiological Irregular effect, mean decrease up to 9%. *0317*

Creatine Kinase *Serum Increase Physiological* Mild transient effect observed ?origin. *0161*

Dialyzable Free Thyroxine *Serum Increase Physiological* Lowers binding to thyroxine binding globulin and albumin. *0831*

Prealbumin *Serum No Effect Physiological* No effect observed after 3 weeks. *2483*

Protein Bound Iodine (PBI) *Serum Decrease Physiological* Inhibits binding of T_4 to thyroxine binding globulin. *2483*

T_3 Uptake *Serum Increase Physiological* Resin uptake increased by 20% after 3 weeks. *2483*

Thyroxine (T_4) *Serum Increase Physiological* Associated with competition for transport proteins. *3798*
Serum Decrease Physiological Interferes with binding of T_4 to thyroxine binding globulin. *2483*

Thyroxine (T_4) Binding Globulin
Serum No Effect Physiological No effect observed after 3 weeks. *2483*

Thyroxine (T_4), Free *Serum Increase Physiological* Interferes with binding of T_4 to thyroxine binding globulin. *2483* Associated with competition for transport proteins. *3798*

Thyroxine (T_4) Index, Free (FTI)
Serum Increase Physiological Associated with competition for transport proteins. *3798*

Thyroxine (T_4) (Murphy-Pattee)
Serum Increase Physiological Significant increase observed. *2483*

Triglycerides *Serum Decrease Physiological* Reduced by up to 50%. *2483*

Tri-iodothyronine (T_3) *Serum Increase Physiological* Associated with competition for transport proteins. *3798*

Uric Acid *Serum Decrease Physiological* Uricosuric action. *2483* 500 mg/d/48 weeks causes reduction of 35%. *0160*
Urine Increase Physiological Effect independent of GFR. *2483*

Haloperidol

Acetaminophen Screening Test *Urine Negative Analytical* No reaction with o-cresol at therapeutic concentrations. *3326*

Alanine Aminotransferase *Serum Increase Physiological* May cause hepatocellular changes. *2313* In isolated case producing cholestatic liver disease: typical frequency of liver disease is 0.2%. *0910*

Alkaline Phosphatase *Serum Increase Physiological* May cause hepatocellular changes. *2313* In isolated case producing cholestatic liver disease: typical frequency of liver disease is 0.2%. *0910*

Alkaloids *Urine No Effect Analytical* With ninhydrin on Frings TLC procedure at 1 mg/dL. *1204*
Urine Increase Analytical Purple with iodoplatinate on Frings TLC procedure. *1204*

Antidiuretic Hormone *Plasma No Effect Physiological* No effect of intravenous injection. *2920*

Aspartate Aminotransferase
Serum Increase Physiological May cause hepatocellular changes. *2313* In isolated case producing cholestatic liver disease: typical frequency of liver disease is 0.2%. *0910*

Barbiturate *Urine No Effect Analytical* With vanillin on Frings procedure at 1 mg/dL. With mercurous NO_3 on Frings TLC procedure at 1 mg/dL. With diphenylcarbazone on Frings TLC procedure at 1 mg/dL. With mercuric SO_4 on Frings TLC procedure at 1 mg/dL. *1204*

Bile *Urine Increase Physiological* May cause hepatocellular changes. *2313*

Bilirubin *Serum Increase Physiological* May cause hepatocellular changes. *2313* In isolated case producing cholestatic liver disease: typical frequency of liver disease is 0.2% In isolated case producing cholestatic liver disease: typical frequently of liver disease is 0.2%. *0910*

Haloperidol (continued)

Bilirubin, Direct *Serum Increase Physiological* In isolated case producing cholestatic liver disease; typical frequency of liver disease is 0.2%. *0910*

Bromide *Urine No Effect Analytical* No effect on method of Frings. *1204*

BSP Retention *Serum Increase Physiological* May cause hepatocellular changes. *2313*

Cephalin Flocculation *Serum Increase Physiological* May cause hepatocellular changes. *2313*

Cholesterol *Serum No Effect Physiological* No detectable effect observed in man. *3327*
Serum Decrease Physiological Inhibits cholesterol biosynthesis. *2313*

Desmosterol *Serum Increase Physiological* Further metabolism inhibited so accumulates. *1343*

Diphenylhydantoin *Serum No Effect Analytical* On GLC procedure of Papadopoulos at *in vivo* concentration. *2722*

Eosinophils *Blood Increase Physiological* In isolated case producing cholestatic liver disease: typical frequency of liver disease is 0.2%. *0910*

Erythrocytes *Blood Decrease Physiological* May cause anemia. *2313*

Ethchlorvynol *Urine No Effect Analytical* No effect on method of Frings and Cohen. *1204*

FPN Test *Urine No Effect Analytical* No effect at 1 mg/dL on method of Frings. *1204*

Glucose *Serum Increase Physiological* Observed endocrinological disorder. *1680*
Serum Decrease Physiological Insulin like action of drug reported in one case. *1488*

γ-Glutamyltransferase (GGT)
Serum Increase Physiological In isolated case producing cholestatic liver disease: typical frequency of liver disease is 0.2%. *0910*

Indocyanine Green *Serum Decrease Physiological* Mechanism not yet established. *2398*

Leukocytes *Blood Increase Physiological* Rarely reported leukocytosis. *0071*
Blood Decrease Physiological May cause anemia. *2313* Several cases reported but none of agranulocytosis. *0250*

Lymphocytes *Blood Increase Physiological* Slight effect may occur. *1680*

Monoamine Oxidase *Platelets Decrease Physiological* Significant reduction in both acute and chronic schizophrenics. Effect seen after 14 d and results did not correlate with response to treatment. *2270*

Monocytes *Blood Increase Physiological* Slight effect may be observed. *1680*

Pheneturide *Serum No Effect Analytical* On GLC procedure of Papadopoulos at *in vivo* concentration. *2722*

Phenobarbital *Serum No Effect Analytical* On GLC procedure of Papadopoulos at *in vivo* concentration. *2722*

Platelets *Blood No Effect Physiological* No significant reduction, in contrast to phenothiazines, in psychiatric patients treated for more than 1 y. *1642*

Primidone *Serum No Effect Analytical* On GLC procedure of Papadopoulos at *in vivo* concentration. *2722*

Prolactin *Plasma Increase Physiological* Repeated administration of 1 mg i.m. to 19 normal men produced Reproducible dose-response curve. *2072*

Prothrombin Time *Plasma Decrease Physiological* May shorten action of anticoagulants. *1679*

Salicylate *Urine No Effect Analytical* At 1 mg/dL no effect on Trinder method. *1204*

Testosterone *Serum No Effect Physiological* Has no effect, unlike some other neuroleptics. *1736*

Thymol Turbidity *Serum Increase Physiological* May cause hepatocellular changes. *2313*

Halothane

Alanine Aminotransferase *Serum No Effect Physiological* No effect seen in patients who demonstrated mild increase of GST activity. *1707*

Serum Increase Physiological Allergic hepatic hypersensitivity. *3294* Liver disease from mild focal necrosis and inflammation to massive necrosis. *2260* Granulomatous hepatitis in single case following second exposure to anesthetic. *3266* In one nurse occupationally exposed to gas. *1900* Low frequency but greater number of abnormalities than with enflurane. Occurred most frequently in obese patients. *1089*

Albumin *Serum No Effect Physiological* In one nurse occupationally exposed to gas. *1900*
Serum Decrease Physiological May induce hepatitis. *0593* Liver disease from mild focal necrosis and inflammation to massive necrosis. *2260*

Alkaline Phosphatase *Serum Increase Physiological* Reversible depression of liver function. *3294* In one case after repeated exposure to anesthetic: evidence of acute hepatitis with bridging necrosis and piecemeal necrosis. *2027* Granulomatous hepatitis in single case following second exposure to anesthetic. *3266* In one nurse occupationally exposed to gas. *1900* Low frequency but greater number of abnormalities than with enflurane. Occurred most frequently in obese patients. *1089*

Antimitochondrial Antibodies
Serum Positive Physiological Liver disease from mild focal necrosis and inflammation to massive necrosis. *2260*

Aspartate Aminotransferase
Serum Increase Physiological Allergic hepatic hypersensitivity. *3294* Liver disease from mild focal necrosis and inflammation to massive necrosis. *2260* In one case after repeated exposure to anesthetic: evidence of acute hepatitis with bridging necrosis and piecemeal necrosis. *2027* Granulomatous hepatitis in single case following second exposure to anesthetic. *3266* Low frequency but greater number of abnormalities than with enflurane. Occurred most frequently in obese patients. *1089*

Bile *Urine Increase Physiological* Allergic hepatic hypersensitivity. *2313*

Bilirubin *Serum No Effect Physiological* In one nurse occupationally exposed to gas. *1900*
Serum Increase Physiological Reversible depression of liver function. *3294* In one case after repeated exposure to anesthetic: evidence of acute hepatitis with bridging necrosis and piecemeal necrosis. *2027* Granulomatous hepatitis in single case following second exposure to anesthetic. *3266*

Bilirubin, Conjugated *Serum Increase Physiological* Granulomatous hepatitis in single case following second exposure to anesthetic. *3266*

Bleeding Time *Patient Increase Physiological* Significant correlation between aggregation *in vitro* and bleeding time *in vivo*. *0800*

Bromide *Serum Increase Analytical* Peak value of from 0.65 mmol/L to 2.25 mmol/L 48 to 72 h after anesthesia in 16 patients. *3597*

BSP Retention *Serum Increase Physiological* Reversible depression of liver function. *3294*

Cephalin Flocculation *Serum Increase Physiological* Reversible depression of liver function. *2313*

Creatine Kinase *Serum Increase Physiological* Marked effect on CK if administered during anesthesia. *2809* Marked effect if given with suxamethonium. *1724*

Diphenylhydantoin *Serum Increase Physiological* Reported impairment of metabolism. *2042*

Effective Renal Plasma Flow
Patient Decrease Physiological Up to 38% decrease reported in normals. *2427*

Eosinophils *Blood No Effect Physiological* No significant change noted even with multiple exposures. *1089*
Blood Increase Physiological Granulomatous hepatitis in single case following second exposure to anesthetic. *3266*

Fibrinogen *Plasma No Effect Physiological* No significant effect noted. *1814*

Fluoride *Serum Increase Physiological* Slight increase following operation. *0916*

Glomerular Filtration Rate (GFR)
Urine Decrease Physiological Up to 19% decrease reported in normals. *2427*

Glucose *Serum Increase Physiological* Response to stress of surgery. *2701*

γ-Glutamyltransferase (GGT)
Serum Increase Physiological Granulomatous hepatitis in single case following second exposure to anesthetic. *3266* Low frequency but greater number of abnormalities than with enflurane. Occurred most frequently in obese patients. *1089*

Glutathione S-Transferase *Serum Increase Physiological* 16 of 20 patients who received halothane anesthesia for minor urological problems showed small transient rise 1 to 3 h after surgery. *1707*

Growth Hormone *Plasma Increase Physiological* ?metabolic response to surgical stress. *2701*

Hemoglobin *Urine No Effect Physiological* No significant effect noted. *1814*

Lactate Dehydrogenase *Serum Increase Physiological* Hepatotoxic in rare cases, ?idiosyncrasy. *3294* Low frequency but greater number of abnormalities than with enflurane. Occurred most frequently in obese patients. *1089*

Leukocytes *Blood Increase Physiological* Allergic hypersensitive response. *1343*

Occult Blood *Feces Positive Physiological* In 3 of 8 patients (slight only). *1814*

Platelet Aggregation *Blood Decrease Physiological* Significant correlation between aggregation *in vitro* and bleeding time *in vivo*. *0800*

Platelets *Blood No Effect Physiological* No significant effect noted. *1814*

Prothrombin Time *Plasma No Effect Physiological* In one nurse occupationally exposed to gas. *1900*
Plasma Increase Physiological May induce hepatitis. *0593* Granulomatous hepatitis in single case following second exposure to anesthetic. *3266*

Pseudocholinesterase *Serum Decrease Physiological* Low frequency but greater number of abnormalities than with enflurane. Occurred most frequently in obese patients. *1089*

Testosterone *Serum Decrease Physiological* Mechanism unclear ?inhibits release of gonadotropin. *2697*

Thymidine Incorporation
Lymphocytes Increase Physiological Liver disease from mild focal necrosis and inflammation to massive necrosis. *2260*

Thymol Turbidity *Serum Increase Physiological* Allergic hepatic hypersensitivity. *2313*

Thyroid Stimulating Hormone (TSH)
Serum No Effect Physiological No appreciable effect observed. *2698*

Thyroxine (T₄) *Serum Increase Physiological* Increase of almost 2 µg/dL observed. *2698* Probably due to release from liver. *3798*

Thyroxine (T₄) Index, Free (FTI)
Serum Increase Physiological Probably due to release from liver. *3798*

Urea Nitrogen *Serum Increase Physiological* Observed in postoperative period with norm renal function. *0369*

Uric Acid *Serum Increase Physiological* In postoperative period with normal renal function. *0369*

Heat

Acetoacetate Decarboxylase *Serum Decrease Analytical* Inactivation of enzyme observed at 80° C. *3684*

Aldosterone *Plasma Increase Physiological* After week of thermal stress increased by 76%. *0206*

Cephalin Flocculation *Serum Increase Analytical* As room temperature is increased. *1237*

Ionized Calcium *Serum Decrease Analytical* 2-3% at 37° compared with room temperature. *2056*

Lactate Dehydrogenase Isoenzymes
Serum Decrease Analytical Inactivates liver components to large extent. *0718*

Lead *Blood Increase Physiological* Increased temperature may cause mobilization of fixed. *0489*

pH *Blood Increase Analytical* Increased 0.015 per degree (may also affect buffers). *0579*

Phenol Turbidity *Serum Increase Analytical* Increases greatly with change of temperature. *0579*

Prothrombin Time *Plasma Increase Physiological* Prolonged hot weather may have effect. *1679*

Renin Activity *Plasma Increase Physiological* After week of thermal stress increased by 174%. *0206*

Thymol Turbidity *Serum Decrease Analytical* As room temperature is increased. *0579*

Zinc Sulfate Turbidity *Serum No Effect Analytical* Little change with temperature observed. *0579*

Height

Alkaline Phosphatase *Serum Increase Physiological* Positively correlated in children. *0887*

Aspartate Aminotransferase
Serum No Effect Physiological No correlation observed in children. *0887*

Hemoglobin *Blood Increase Physiological* Positively correlated in children. *0887*

Lactate Dehydrogenase *Serum No Effect Physiological* No correlation in children. *0887*

Phosphate *Serum Increase Physiological* Positively correlated in children. *0887*

Uric Acid *Serum Increase Physiological* Positively correlated in children. *0887*

Hematin

Aminolevulinic Acid *Serum Decrease Physiological* In a patient with acute intermittent porphyria. *0414*
Urine Decrease Physiological In a patient with elevation due to porphyria. *0414*

Crystals *Urine Increase Physiological* Small, biconcave, whetstone (in acid). *0443*

Porphobilinogen *Serum Decrease Physiological* In a patient with acute intermittent porphyria. *0414*
Urine Decrease Physiological In a patient with elevation due to porphyria. *0414*

Protein *Test Conditions Increase Analytical* Even 1 µg has significant effect on 150 µg protein. *0396*

Hematoxylin

Color *Feces Increase Analytical* Reddish-brown with 1 gram. *0902*

Hemodialysis

Aldosterone *Plasma Increase Physiological* Mean increase of 8.2 ng/dL during dialysis. *2350*

Ascorbic Acid *Blood Decrease Physiological* Low values observed with long term dialysis. *3885*

Aspartate Aminotransferase
Serum Decrease Physiological Possible loss of pyridoxal phosphate. *3885*

Cortisol *Plasma No Effect Physiological* No significant effect observed during dialysis. *2350*

2,3-Diphosphoglycerate
Red Blood Cells No Effect Physiological Chronic hemodialysis has no effect. *2740*

Folate *Serum Decrease Physiological* Repeated may even lead to megaloblastic anemia. *3885*

Glucose *Serum Increase Physiological* Usual response. *2547*

Growth Hormone *Plasma Decrease Physiological* Normal response to increased glucose. *2547*

Haptoglobin *Serum No Effect Physiological* Unless intravascular hemolysis occurs. *3613*

Insulin *Plasma Increase Physiological* Normal physiological response to increased glucose. *2547*

Lactate Dehydrogenase *Serum No Effect Physiological* Unless intravascular hemolysis occurs. *3613*
Serum Increase Physiological Variable response of total LDH. *2990*
Serum Decrease Physiological Variable response of total LDH. *2990*

Hemodialysis *(continued)*

Lactate Dehydrogenase Isoenzymes
Serum Increase Physiological LD-5 of renal origin if acute tubular necrosis. *2990*

Prolactin *Plasma No Effect Physiological* Elevated value not decreased by dialysis. *2547*

Renin Activity *Plasma No Effect Physiological* Not affected by period requiring dialysis. *0750*

Thyroid Stimulating Hormone (TSH)
Serum Increase Physiological Increased to abnormal levels in some subjects. *2547*
Serum Decrease Physiological Probably effect of heparin. *1816*

Thyroxine (T$_4$), Free *Serum Increase Physiological* Probably effect of heparin. *1816*

Tri-iodothyronine (T$_3$) *Serum Increase Physiological* Probably effect of heparin. *1816*

Hemoglobin

N-Acetylglucosaminidase *Urine No Effect Analytical* At 75 mg/L had no effect on 2 colorimetric analytical methods. *1354*

Albumin *Serum Increase Analytical* If no blank with HABA dye, slight effect BCG. *2467* 100 mg/dL equivalent to 100 mg/dL BCG method. *3217*

Alkaline Phosphatase *Serum Increase Analytical* Linear increase of 2 U/L per 0.1 g/dL method of Proksch. *2878*

α-Amino-Nitrogen *Plasma No Effect Analytical* 20 g/dL has no effect on Goodwin procedure. *1712*

Ammonia *Plasma Increase Analytical* Slight interference with color development. *1936*

Amylase *Serum No Effect Analytical* No effect on Phadebas procedure. *1726* With up to 35 μmol/L with method of Rauscher et al. *2926*

Angiotensin-Converting Enzyme (ACE)
Serum No Effect Analytical No effect reported on colorimetric method of Boomsma and Schalekamp. *2814*

Aspartate Aminotransferase *Serum Increase Analytical* If greater than 25 mg/dL. *3165*

Bence-Jones Protein *Urine Positive Analytical* False positive heat test, paraprotein on electrophoresis. *2755*

Bilirubin *Serum Increase Analytical* Percentage of unconjugated decreases with increased Hemoglobin. *2313* At 1 g/L by 1% polarometric procedure of Grahnen. *1374* Cause spurious elevation in direct procedures. *0583*
Serum Decrease Analytical Competes for nitrite with sulfanilic acid. *0583*

Bilirubin, Direct *Serum No Effect Analytical* At concentrations up to 1.5 g/L on bilirubin oxidase method of Doumas et al. *0940*
Serum Increase Analytical Occurs with direct spectrophotometric method. *0454*

Calcium *Serum No Effect Analytical* method of Morin at 500 mg/dL. *2485*
Serum Decrease Analytical Interference with calcein titration. *0055*

Carcinoembryonic Antigen *Urine No Effect Analytical* No effect observed with hematuria. *1459*

Ceruloplasmin *Serum No Effect Analytical* With up to 200 mg/dL on Henry PPD procedure. *0877*

Cholesterol *Serum No Effect Analytical* On Leffler's procedure at 20 g/L. *1743* With up to 2 g/L on enzymatic procedure. *3027*
Serum Increase Analytical Increase by 2.1% at 10 mg/dL on enzymatic procedure. *0057* Interference with Zlatkis-Zak and direct methods. *1237*

Color *Urine Increase Analytical* Clear red to reddish brown. *2448*

Creatine Kinase *Serum No Effect Analytical* If AMP added to inhibit adenylate kinase. *3217*
Serum Increase Analytical Marked effect on UV coupled procedures with 1 g Hemoglobin. *3217*
Serum Decrease Analytical At 2.5 g/dL inhibits by 12%. *3217*

Digoxin *Serum Decrease Analytical* Estimated by radioimmunoassay unless bleached. *2805*

Glucose *Serum No Effect Analytical* At 200 mg/dL on MBTH procedure of Neeley. *2569* No effect on hexokinase method of Yee. *3912*
Serum Increase Analytical May have marked effect with o-toluidine. *3902*
Urine No Effect Analytical At 0.5 g/L on Boehringer-Mannheim BM33071 glucose pads. *3356*

Glucose-6-Phosphate Dehydrogenase
Red Blood Cells Decrease Analytical Minimal quenching of fluorescence may occur. *2264*

Guanase *Serum No Effect Analytical* Up to 5 g/L on method of Nishikawa. *2602*

Haptoglobin *Serum Increase Analytical* If radial immunoassay used. *0455*
Serum Decrease Analytical If peroxidase activation method used. *0455*

Hydroxybutyrate Dehydrogenase
Serum Increase Analytical Large effect on UV procedures per g Hemoglobin. *3217*

Icteric Index *Serum Increase Analytical* Contributes background color. *0582*

Iron *Serum No Effect Analytical* With up to 20 mg/dL method of Megraw. *2391* No effect unless concentration greater than 500 mg/dL. *2981*
Serum Increase Analytical Slight effect of 1 g on Tripyridyl-s-triazine procedure. *3217*

Iron-Binding Capacity, Total (TIBC)
Serum No Effect Analytical No effect on measurement. *2981*

Lactate *Serum Decrease Analytical* At concentrations where no visible hemolysis: effect greater on Du Pont aca method than Abbott TD$_x$. *3154*

Lactate Dehydrogenase *Serum Increase Analytical* Marked effect on UV procedures. *3217*

Lipase *Serum No Effect Analytical* At 5 g/L no effect on method of Tietz, Repique. *3589*
Serum Decrease Analytical Inhibits lipase activity. *1936*

L/S Ratio *Amniotic Fluid Decrease Analytical* Variable effect seen if more than 10%. *0513*

Magnesium *Serum Increase Analytical* Slight effect of 1 g on atomic absorption. *3217*

Phosphate *Serum Increase Analytical* Slight effect phosphomolybdate reductions. *3217*

6-Phosphoglycerate Dehydrogenase
Red Blood Cells Decrease Analytical May cause up to 7% quenching of fluorescence. *2219*

Porphyrins *Urine Increase Analytical* Affects UV absorption screening methods. *2981*

Protein *Serum Increase Analytical* 1 mg measured as equivalent to 1.9 mg protein. *1563*

PSP Excretion *Urine Increase Analytical* Interfering color. *2313*

Thymol Turbidity *Serum No Effect Analytical* No effect up to concentration of 14 mg/dL. *1563*

Thyroxine (T$_4$) *Serum No Effect Analytical* At 1 g/dL on competitive protein binding procedure of Alexander. *0054*

Urea Nitrogen *Serum Increase Analytical* May affect color of urease/Berthelot procedure. *2981*

Uric Acid *Serum Increase Analytical* At 200 mg/dL 2%, at 424 mg/dL 11% phosphotungstate procedures. *0904*

Zinc Sulfate Turbidity *Serum No Effect Analytical* Probably same noninterference as with thymol turbidity. *1563*

Hemolysis

Acid Phosphatase *Serum No Effect Analytical* If moderate no effect with thymolphthalein phosphate. *0877*
Serum Increase Analytical Erythrocytes contain considerable activity. *2981*

Acid Phosphatase, Tartrate Labile
Serum No Effect Analytical No effect with hemoglobin up to 1 g/dL. *3217*

Alanine Aminotransferase *Serum Increase Analytical* Effect minimal as only 3-5 times concentration in RBC as in sera. *0877*
Serum Increase Physiological Released from RBCs and platelets. *2313*

Albumin *Serum Decrease Analytical* Affects measurement by HABA, biuret, methyl orange. *3873*

Aldolase *Serum No Effect Analytical* Slight hemolysis has no effect. *2981*

Serum Increase Analytical High content in erythrocytes and platelets. *0797*

Alkaline Phosphatase *Serum Increase Analytical* RBC concentration 6 x serum so effect minimal. *1563*

Amino Acids *Serum Increase Analytical* Causes release of intracellular constituents. *3230*

α-Amino-Nitrogen *Plasma Increase Analytical* Release of glutathione from erythrocytes. *1563*

Ammonia *Plasma Increase Analytical* Some interference with color development. *1237*

Amylase *Serum Increase Analytical* Reported effect. *0248*

Angiotensin II *Plasma Decrease Analytical* Significant effect observed. *0977*

α₁-Antitrypsin *Serum No Effect Analytical* On AutoAnalyzer immunological method. *2297*

Arginase *Serum Increase Analytical* Ratio of 1,000 to 1 between cells and serum. *0579*

Arginine *Plasma Decrease Analytical* Due to arginase in RBCs. *3230*

Aspartate Aminotransferase *Serum Increase Analytical* Just visible (50 mg/dL) produces 1-2% increase. *0877*

Serum Increase Physiological Released from RBCs and platelets. *2313*

CSF Increase Analytical If more than 400 RBC or 200 WBC per cubic mm. *0877*

Bicarbonate *Serum No Effect Analytical* Calculated error less than 1%/g. *0515*

Bilirubin *Serum Decrease Analytical* Methemoglobin formation affects diazotization. *2362*

Bilirubin, Direct *Serum Decrease Analytical* Interference with direct spectrophotometry. *1563*

Bromide *Serum Increase Analytical* May cause color in protein-free filtrate. *0579*

BSP Retention *Serum Increase Analytical* Absorbance of Hemoglobin varies with pH (affects some procedures). *1237*

Calcium *Serum Decrease Analytical* Interferes with calcein fluorescence methods. *2395*

Carotene *Serum Decrease Analytical* Significant decrease observed with hemolysis. *2734*

Ceruloplasmin *Serum Decrease Analytical* May affect linearity of reaction curve. *2981*

Chloride *Serum Decrease Analytical* Dilutional effect. *2313*

Cholesterol *Serum No Effect Analytical* Calculated error less than 1 %/g after extract. *0515* When measured enzymatically with oxidase. *3027*

Serum Increase Analytical Values high by 10% if extraction not used. *1563*

Cortisol *Plasma Increase Analytical* Produces interference with fluorometric method. *1772*

Creatine *Serum Increase Analytical* Even if slight may increase by 100-200%. *0904*

Creatine Kinase *Serum No Effect Analytical* Slight hemolysis has no effect. *2981*

Serum Increase Analytical 5 U/L in hemolyzed samples. *2744*

Serum Decrease Analytical -3%/g hemoglobin with Oliver procedure. *0515*

Creatinine *Serum No Effect Analytical* Moderate amount no effect on method of Heinegard. *1547* Moderate hemolysis has no effect. *3505*

Serum Increase Analytical At 500 mg/L hemoglobin on enzymatic procedure. *2524* Calculated error 1.8 %/g with alkaline picrate procedure. *0515*

Cystine *Plasma Decrease Analytical* Due to binding with protein or dilution. *3230*

Digoxin *Serum No Effect Analytical* If dioxane based cocktail used in assay. *3823*

Serum Increase Analytical May affect radioimmunoassay with beta count. *0615*

Diphenylhydantoin *Serum No Effect Analytical* When determined by gas chromatographic methods. *3828*

Fatty Acids, Free (FFA) *Serum No Effect Analytical* No effect on method of Soloni. *3381*

Glucose *Serum Decrease Analytical* Affects glucose oxidase procedure, avoidable by Somogyi filtrate. *1563*

Glutamate Dehydrogenase *Serum No Effect Analytical* Slight hemolysis has no effect. *2981*

Glutathione *Plasma Increase Analytical* Reduced and oxidized in RBCs. *3230*

Histidine *Plasma Increase Analytical* Released from red cells. *0081*

Hydroxybutyrate Dehydrogenase
Serum Increase Analytical Erythrocytes contain large amounts. *2981*

Icteric Index *Serum Increase Analytical* Presence of hemoglobin produces red color. *1565*

Immunoglobulin IgA *Serum No Effect Analytical* On AutoAnalyzer immunological method. *2297*

Immunoglobulin IgG *Serum No Effect Analytical* On AutoAnalyzer immunological method. *2297*

Insulin *Plasma Decrease Analytical* Causes destruction of hormone for radioimmunoassay. *0478*

Iron *Serum Increase Analytical* Interferes with colorimetric procedure. *1564*

Iron-Binding Capacity, Total (TIBC)
Serum No Effect Analytical No effect observed with 1 g/dL. *3217*

Isocitrate Dehydrogenase *Serum Increase Analytical* Activity high in erythrocytes. *0279*

Lactate *Plasma Increase Analytical* Due to metabolic activity of cells. *3873*

Lactate Dehydrogenase *Serum Increase Analytical* Released from RBCs and platelets. *1237*

Lipase *Serum Decrease Analytical* Inhibition of lipase activity. *1566*

Magnesium *Serum Increase Analytical* Calculated error of 7.6%/g. *0515*

5'-Nucleotidase *Serum No Effect Analytical* No effect hemoglobin up to 1 g/dL. *3217*

Ornithine *Plasma Increase Analytical* Due to arginase in RBCs. *3230*

Ornithine Carbamoyltransferase (OCT)
Serum No Effect Analytical Slight has no effect, massive produces slight effect. *0877*

Oxytetracycline *Serum Increase Physiological* Affects fluorometric method of Murthy. *2533*

Partial Thromboplastin Time *Plasma Decrease Analytical* Significant effect in normal subjects (but not abnormals). *1244*

Phosphate *Serum Increase Physiological* Released from RBCs and platelets. *2334*

Phosphohexoseisomerase *Serum No Effect Analytical* Small degree has no effect. *3873*

Phospholipids, Total *Serum Decrease Analytical* Organic phosphate hydrolysis to inorganic. *3873*

Potassium *Serum Increase Physiological* Released from RBCs and platelets. *0579*

Propylthiouracil *Serum No Effect Analytical* No effect with 2,6-DQC (procedure of Ratliff). *2922*

Protein *Serum Increase Analytical* Added protein measured. *2313*

Protein Bound Iodine (PBI) *Serum Decrease Analytical* 5% decrease if at 700 nmol/L. *3257*

Serum Decrease Physiological If marked, RBCs contain no I₂ (dilution effect). *0354*

Protein Electrophoresis *Serum Positive Analytical* May augment beta, between alpha₂ and beta. *1563*

Prothrombin Time *Plasma No Effect Analytical* Has negligible effect. *3873*

Pseudocholinesterase *Serum No Effect Analytical* If minor does not affect Acholest method. *1680*

Pyruvate *Serum Increase Analytical* Due to metabolic activity of cells. *3873*

Sodium *Serum Decrease Physiological* Dilution effect. *2313*

Hemolysis (continued)

Sorbitol Dehydrogenase *Serum No Effect Analytical* Slight hemolysis has no effect. *2981*

Thyroxine (T₄) *Serum No Effect Analytical* If minimal on competitive protein binding methods. *3257*
Serum Decrease Analytical 5% decreased if at 700 nmol/L. *3257* When severe, decreased recovery in competitive protein binding procedures. *0054*

Tyrosine *Plasma Increase Analytical* Substantially increased values occur. *0080*

Urea Nitrogen *Serum No Effect Analytical* DMAB procedure of Morin, Hemoglobin below 100 mg/dL. *2487* Calculated error less than 1%/g. *0515*
Serum Increase Analytical If Berthelot procedure, and wavelength less than 600 nm. *3314* With method of Morin with hemoglobin above 100 mg/dL. *2487* Color produced in protein-free filtrates. *0579*
Serum Decrease Analytical Red cell concentration less than serum. *0579*

Uric Acid *Serum No Effect Analytical* No effect if measured by uricase and is moderate. *2954*
Serum Increase Analytical False positive with glutathione and ergothionine. *0580*
Serum Decrease Analytical Calculated error -1.4%/g with phosphotungstate procedure. *0515*

Urobilinogen *Urine Increase Physiological* Further metabolism of hemoglobin. *2313*

Vitamin A *Serum No Effect Analytical* No effect of method of Price and Carr. *2734*

Hemosiderin

Crystals *Urine Increase Physiological* Clumps golden brown granules (in acid and neutral). *0443*

Heparin

N-Acetylglucosaminidase *Urine No Effect Analytical* At 200,000 U/L had no effect on 2 colorimetric analytical methods. *1354*

Acid Phosphatase *Serum Decrease Analytical* Reported observation (significant observation). *0839*
Red Blood Cells Decrease Analytical Reported observation (significant observation). *0839*
White Blood Cells Decrease Analytical More than 1 unit/mL inhibits by 80%. *0839*

Alanine Aminotransferase *Serum Increase Physiological* In 59% patients receiving bovine deprived drug. In 27% patients receiving porcine derived drug. *0968* Mean maximal value 67 U/L in 89% of 46 patients given up to 120,000 IU daily subcutaneously after 8 d. *2454* On 5th and 10th postoperative day (from 27 to 40 U/L) compared with control group. *2595* In 2/3 of patients treated with heparin for deep venous thrombosis. *1064* In patients with cerebrovascular accidents after low-dose heparin treatment. *2671*

Albumin *Serum No Effect Analytical* No effect on binding of spectrum AB₂. *0583* No significant effect on BCG method on Du Pont aca II or Beckman Astra 8 analyzer. *1494*
Serum Increase Analytical Large amounts may cause HABA binding to fibrinogen. *2228* Promotes binding of HABA dye to globulins. *3505*
Serum Decrease Analytical Significant reduction with bromocresol purple method on aca II: concentration effect less at higher pH values. *1494* Negative bias progressively greater with increasing amount of heparin on dye- binding procedures using bromocresol purple but not with bromocresol green. *1494*

Aldosterone *Plasma Decrease Physiological* Reversible toxic effect on glomerulosa cells (Probably). *2638* Possibly due to toxic effect on glomerulosa cells: effect readily reversible with withdrawal of drug. *2638*
Urine Decrease Physiological Suppressed secretion with protracted treatment. *1680*

Alkaline Phosphatase *Serum No Effect Analytical* Little or no effect observed. *2228*
Serum No Effect Physiological No effect of subcutaneous drug after 8 d. *2454* No effect of surgery in this test group or control group. *2595* After low-dose heparin treatment. *2671*

White Blood Cells *No Effect Analytical* No effect observed. *0839*

Amikacin *Serum Decrease Analytical* Interferes with biological, radioenzymatic and homogeneous enzyme immunoassay but not RIA techniques due to inhibition of acetyltransferase enzymes. *2021*

Amino Acids *Serum Increase Analytical* May cause hemolysis release ninhydrin positive constituents. *3230*

Ammonia *Plasma Increase Analytical* Contains variable amounts of ammonium salts. *0579*

Amylase *Serum No Effect Analytical* No effect on amylase activity. *3873* With sodium salt at 750 mg/L on method of Rauscher et al. *2926* At concentration of 400 mg/L had no effect on maltopentose method on aca. *3393* At concentration of 5 mg/L had no effect on maltotetrose method. *3393* At concentration of 28600 mg/L had no effect on p-nitrophenylmaltopentoside/hexoside method. *3393*

Antithrombin III *Plasma No Effect Analytical* At concentration of 2860 mg/L had no effect on aca method. *3393*

Aspartate Aminotransferase *Serum No Effect Analytical* No effect on activity. *3165*
Serum Increase Physiological In 27% patients receiving porcine derived drug. In 59% patients receiving bovine derived drug. *0968* Mean maximal value 40 U/L in 82% of 46 patients given up to 120,000 IU daily subcutaneously after 8 d. *2454* But significantly different from other individuals in control population. *2595* In 2/3 of patients treated with heparin for deep venous thrombosis. *1064* In patients with cerebrovascular accidents after low-dose heparin treatment. *2671*

Bilirubin *Serum Decrease Physiological* In majority of patients after 8 d after low-dose heparin treatment. *2671*

Bleeding Time *Patient Increase Physiological* Doubled after i.v. injection of 100 U/kg in 34 normal subjects: increase unrelated to changes in platelet count. *1544*

BSP Retention *Serum Increase Analytical* Color intensity increased in serum, wavelength shifted. *0721*

Calcium *Serum Increase Analytical* If calcium salt used may affect result. *3873*
Serum Decrease Analytical Interferes with EDTA and fluorometric methods. *3879*

Carcinoembryonic Antigen *Serum No Effect Analytical* No activity detected in any heparin preparation when Abbott RIA used. *3903*
Serum Increase Analytical Activity present apparently in all heparin preparations when tested by Roche RIA. *3903*

Cholesterol *Serum Increase Physiological* Rebound effect of cessation of treatment. *1680*
Serum Decrease Analytical If used as anticoagulant value low by 5-15 mg/dL. *2319*

Clotting Time *Blood Increase Physiological* Concentration related effect. *2649*

Corticosteroids *Plasma Increase Analytical* If contaminated by impurities. *1912*

Corticosterone *Plasma No Effect Physiological* No effect on components of aldosterone biosynthetic pathway. *2638* No effect noted in spite of effect on aldosterone. *2638*

Creatine Kinase *Serum No Effect Physiological* After low-dose heparin treatment. *2671*
Serum Decrease Analytical Reported effect. *0248*

Creatinine *Serum No Effect Analytical* At concentration of 1,000 mg/L had no effect on creatinine amidohydrolase method. *3393*

11-Deoxycorticosterone *Plasma No Effect Physiological* No effect on components of aldosterone biosynthetic pathway. *2638* No effect noted in spite of effect on aldosterone. *2638*

Diphenylhydantoin *Serum Decrease Physiological* Slight effect when high concentrations of nonesterified fatty acids occur due to heparin due to reduced protein binding. *3211* Decreases by 20% with post-heparin increase of free fatty acids. *3210*

Factor V *Plasma Decrease Physiological* Concentration related effect. *1343*

Factor IX *Plasma Decrease Physiological* Reported effect. *1343*

Factor XI *Plasma Decrease Physiological* Concentration related effect. *1343*

Fatty Acids, Free (FFA) *Serum No Effect Analytical* No effect as anticoagulant on method of Soloni. *3381*

Serum Increase Analytical Some interference observed if plasma used. *2438* Marked increase in some patients when drug given i.v. Effect probably *in vitro* artifact as can be abolished by protamine inhibition of lipoprotein lipase. *2408*

Serum Increase Physiological Also occurs with situational stresses. *3873* Mean concentration increased 4-fold in 19 patients. *0253* From mean of 0.55 to 2.20 mmol/L following administration in 19 patients: extent correlated with pre-heparin concentration of triglycerides. *0253*

Fibrinogen *Plasma No Effect Analytical* On potassium mercuric thiocyanate procedure of Roberts. *3003* On clottable protein assay procedure. *3463*

Plasma Increase Analytical At concentrations above 2,000 mg/L (normal therapeutic concentration 20 mg/L) raised concentration as measured by aca method. *3393*

Plasma Decrease Analytical Maximum effect on turbidimetric procedures. *3463*

Fibrinopeptide A *Plasma Decrease Physiological* High values in acute nonlymphocytic leukemia reduced to normal in 14 of 17 patients after intravenous bolus of heparin. *1421*

Fructosamine *Serum Decrease Analytical* Significantly lower concentrations than in serum when measured by Roche kit. *1698*

Gentamicin *Serum Decrease Analytical* If over 50 U/mL affects agar diffusion. *3927* Interferes with biological, radioenzymatic and homogeneous enzyme immunoassay but RIA techniques due to inhibition of acetyltransferase enzymes. *2021*

Glucose *Serum No Effect Analytical* On glucose oxidase procedures. On o-toluidine procedure. On alkaline ferricyanide procedures. On p-HBAH procedure of Lever. *2140* No effect on hexokinase procedure. *2982* At concentration of 15,000 mg/L had no effect on GOD/POD-PAP method. *3393* At concentration of 2,000 mg/L had no effect on hexokinase/G-6-PDH method. *3393*

Serum Increase Physiological Single report of rise of 30 mg/dL. *0721*

Glucose-6-Phosphate Dehydrogenase

Red Blood Cells Increase Analytical Several additional isoenzymes from complexes. *3217*

β-Glucuronidase *White Blood Cells Decrease Analytical* Significant effect above 1 unit/tube. *0839*

γ-Glutamyltransferase (GGT)

Serum Increase Physiological Increase occurred in 37% of 46 patients given up to 120,000 IU daily subcutaneously after 8 d. *2454* But significantly different from other individuals in control population. *2595*

Serum Decrease Analytical Activity lower in heparinized specimen than in serum: effect can be overcome by using preincubation period of more than 5 minutes, or by addition of 50 mmol/L sodium chloride to reaction mixture. *3775*

Gold *Blood No Effect Analytical* Atomic absorption method of Harth et al. *1515*

Growth Hormone *Plasma No Effect Analytical* With double-antibody procedure and 20-80 I.U./mL. *0761*

Hydroxybutyrate Dehydrogenase

Serum Decrease Analytical Significant inactivation. *2981*

5-Hydroxyindoleacetic Acid (5-HIAA)

Urine Decrease Physiological Reduction in single case carcinoid syndrome. *1742*

Insulin *Plasma Increase Analytical* Spuriously high values reported for immunoassay. *0191*

Plasma Decrease Analytical Effect in heparinized plasma and serum. *2684*

Ionized Calcium *Serum Decrease Analytical* Up to 0.03 mmol/L in Vacutainers™. *2056*

Iron *Serum No Effect Analytical* At concentration of 1,000 mg/L had no effect on Ferrozine method. *3393*

Lactate *Serum No Effect Analytical* No effect observed on enzymatic method. *2268*

Lactate Dehydrogenase *Serum No Effect Physiological* No effect of subcutaneous drug after 8 d. *2454*

Serum Increase Physiological Abnormal in 36% patients receiving either porcine or bovine derived drug. *0968* In 1 of 13 patients with cerebrovascular accident after low-dose heparin treatment. *2671*

Lipase *Serum Increase Physiological* Increase of 150% 10 minutes after injection. *3026*

Lipoprotein Electrophoresis *Serum Positive Analytical* Alters electrophoretic pattern. *1951*

Lipoprotein Lipase *Serum Increase Physiological* Release of tissue lipase into plasma. *1343*

Lipoproteins, Pre-β *Serum Decrease Physiological* Non-sustained response to small i.v. injection. *1343*

Lysozyme *Serum Decrease Analytical* Significant inhibition may occur. *3217*

White Blood Cells No Effect Analytical No effect observed. *0839*

Mucopolysaccharides *Urine Increase Analytical* False positive cetrimide and toluene blue tests — given i.v. *0525*

Partial Thromboplastin Time

Plasma Increase Physiological Concentration related effect. *2649*

pCO₂ *Blood Decrease Analytical* Dilution effect if dead space filled with liquid heparin. *0444*

pH *Blood No Effect Analytical* Even if syringe dead space contains liquid heparin. *0444*

Phosphate *Serum Increase Analytical* Phosphate contamination of heparin reported. *0378*

Placental Protein S *Serum Increase Physiological* Up to 40 times increase above basal level: possibly due to direct effect of heparin on placenta. *2405*

Plasminogen *Plasma No Effect Analytical* At concentration of 2860 mg/L had no effect on aca method. *3393*

Platelets *Blood Decrease Physiological* Reported effect following i.v. infusions. *2426* Count fell in 86% patients with cerebral infarction etc. In 15% count fell by 40%. Note significant association between poor outcome and platelet drop. *0535* Several cases of immune-mediated thrombocytopenia reported. *2502* Occurs in about 5% of all patients receiving drug *in vivo*, usually occurs 6-12 d after start of therapy. *1934* Average 60% decrease in 7 of 137 patients treated for 4 to 24 d. Decreased count observed in 118 patients. *2905* Frequency of 8% in patients with bovine heparin in this study versus 0% in patients porcine heparin. *2862* Apparently associated with increased amounts of platelet IgG and C3. *0663* Greater effect with bovine lung preparations than those from bovine intestinal mucosa. *0276* In 5.2% of 211 patients the majority of whom had received beef lung heparin. Plasma shown to have a heparin-sensitive antiplatelet antibody. *0664* In 9 of 37 patients in a coronary care unit, transient and mild. *2574* Two cases reported: mechanisms not clarified. *2932* Mild reduction observed in some patients immediately following i.v. infusion. Possibly due to reversible aggregation, margination and sequestration due to direct effect of heparin. Usually subsides spontaneously. More severe fall to 2,000 to 90,000 IU/L after several days. Resolves only when heparin discontinued, possibly due to heparin-dependent platelet membrane antibody. *1573*

pO₂ *Blood No Effect Analytical* Even if syringe dead space contains liquid heparin. *0444*

Potassium *Serum Increase Physiological* Decreased renal excretion. *1343* Very rare hyperkalemia reported although consistent impairment of adrenocortical synthesis of aldosterone: in cases of hyperkalemia reported selective hypoaldosteronism present. *2836*

Prazosin *Serum Decrease Physiological* Slight effect when high concentrations of nonesterified fatty acids occur due to heparin due to reduced protein binding. *3211*

Pregnancy-Associated Plasma Protein A

Serum Increase Analytical Values higher than serum when measured by crossed immunoelectrophoresis. *3610*

Pregnenolone *Plasma No Effect Physiological* No effect on components of aldosterone biosynthetic pathway. *2638* No effect noted in spite of effect on aldosterone. *2638*

Progesterone *Plasma No Effect Physiological* No effect on components of aldosterone biosynthetic pathway. *2638*

Plasma Decrease Physiological No effect noted in spite of effect on aldosterone. *2638*

Prothrombin Time *Plasma Increase Physiological* Concentration related effect. *2649*

Salicylate *Serum No Effect Analytical* No effect on Trinder procedure. *2251*

Heparin (continued)

Sodium *Serum Increase Analytical* If sodium salt used may affect result. *3873*
Serum Increase Physiological When 1,000 IU/mL added increased concentration when direct ISE methods used. *3516*
Serum Decrease Analytical At concentration of 300 U/L sodium concentration reduced by up to 5 mmol/L with direct measurement on Corning 902 analyzer. *2287*
Serum Decrease Physiological Increased excretion due to aldosterone suppression. *2313* When 5,000 IU/mL added reduction of sodium concentration when measured by direct ISE. *3516*
Urine Increase Physiological Due to aldosterone suppression. *0652*

Superoxide Dismutase *Plasma Increase Physiological* Marked increase when injected i.v. 200 IU/kg body weight. Enzyme probably arises from endothelial cells. *1876*

T_3 Uptake *Serum No Effect Physiological* No effect in fasting people, or when added *in vitro*. *2622*
Serum Increase Physiological Red cell uptake increased, resin unaffected. *3176*

Thrombin Time *Plasma Increase Physiological* Related to concentration of circulating heparin. *2649*

Thromboplastin Generation *Blood Decrease Physiological* Abnormal response (inhibition). *1343*

Thymol Turbidity *Serum Increase Analytical* Affects physico-chemical properties altering turbidity. *1563*
Serum Decrease Analytical Concentration related effect. *0583*

Thyro-Binding Index *Serum Decrease Physiological* Effect observed during anticoagulant therapy. *0251*

Thyroid Stimulating Hormone (TSH)
Serum Decrease Physiological Interferes with thyroxine binding to protein. *1816*

Thyroxine (T_4) *Serum Decrease Physiological* Probable modification of thyroxine binding. *3110* Due to competition for transport proteins. *3798*

Thyroxine (T_4), Free *Serum Increase Analytical* Marked increase in some patients given drug i.v. Effect probably *in vitro* artifact. *2408* Increased concentration in patients given drug i.v. when measured by equilibrium dialysis and ultrafiltration. *3760* Observed with equilibrium dialysis technique after i.v. administration. *2183*
Serum Increase Physiological Probable modification of thyroxine binding. *2424* Mean concentration increased by 50% in 19 patients. *0253* From 23 to 39 pmol/L by equilibrium dialysis and from 17 to 24 pmol/L by Gamma-Coat RIA probably due to activation of lipases by heparin. *0253*
Serum Decrease Analytical Marked decrease with Amersham analog method within 1 h of i.v. administration. *2183*

Tobramycin *Serum Decrease Analytical* Interferes with biological, radioenzymatic and homogeneous enzyme immunoassay but not RIA techniques due to inhibition of acetyltransferase enzymes. *2021*

Triglycerides *Serum Decrease Physiological* Prompt decrease when small dose given i.v. *1343*

Tri-iodothyronine (T_3) *Serum Increase Physiological* Interferes with thyroxine binding to protein. *1816*

Tri-iodothyronine (T_3), Free *Serum Increase Physiological* Probably acts on thyroxine binding globulin to reduce binding affinity. *1577*

Uric Acid *Serum No Effect Analytical* At concentration of 1200 mg/L had no effect on uricase method on aca. *3393*

Vitamin A *Serum No Effect Analytical* No effect on method of Price and Carr. *2734*

Zinc Sulfate Turbidity *Serum Increase Analytical* Affects physico-chemical properties. *1563*

HEPES

Protein *Test Conditions Increase Analytical* Interferes with Folin-Ciocalteu of Lowry. *3636*

Heptabarbital

Prothrombin Time *Plasma Decrease Physiological* Induces hepatic metabolism of anticoagulants. *2220*

Heptachlor

Alanine Aminotransferase *Serum Increase Physiological* Liver damage occurs as late effect of poisoning. *1302*

Aspartate Aminotransferase
Serum Increase Physiological Liver damage occurs as late effect of poisoning. *1302*

Heroin

Alanine Aminotransferase *Serum Increase Physiological* Normal usually unless alcoholism also. *3471*

Albumin *Serum Decrease Physiological* Observed in addicts with renal disease. *2914*

Alkaline Phosphatase *Serum Increase Physiological* Normal usually unless alcoholism also. *3471*

Aspartate Aminotransferase
Serum Increase Physiological Normal usually unless alcoholism also. *3471*

Cholesterol *Serum Increase Physiological* Observed in addicts with renal disease. *2914*

Creatinine Clearance *Urine Decrease Physiological* Observed in some severe cases. *2914*

Erythrocytes *Urine Increase Physiological* Occasional microscopic evidence in addicts. *2914*

Glomerular Filtration Rate (GFR)
Urine Decrease Physiological Observed in some severe cases. *2914*

Indocyanine Green *Serum Decrease Physiological* Observed in small series with normal liver function test. *2398*

Morphine *Urine Increase Analytical* Cross reactivity equal (or more) with RIA procedures. Substantial cross reactivity with hemagglutination inhibition. *2514*

pCO_2 *Blood Increase Physiological* If pulmonary edema occurs with decreased respiration. *3059*

Platelets *Blood Decrease Physiological* Several cases of immune mediated thrombocytopenia reported. *2502*

pO_2 *Blood Decrease Physiological* If pulmonary edema occurs with decreased respiration. *3059*

Potassium *Serum Increase Physiological* Associated with severe rhabdomyolysis. *2836*

Protein *Urine Increase Physiological* Massive proteinuria may occur in addicts. *2914*

Testosterone *Serum No Effect Physiological* In treated/untreated male addicts. *0780*

Thyroid Stimulating Hormone (TSH)
Serum No Effect Physiological No significant effect observed. *0193*

Thyroxine (T_4) *Serum Increase Physiological* Significant increase in addicts observed. *0193* Increased binding capacity (augmentation of transport proteins). *3798*

Thyroxine (T_4), Free *Serum Decrease Physiological* Slightly lowered percent in addicts. *0193*

Tri-iodothyronine (T_3) *Serum Increase Physiological* Slight (not significant) increase in addicts. *0193* Increased binding capacity (augmentation of transport proteins). *3798*

Urobilin *Urine Increase Physiological* Occasional microscopic evidence in addicts. *2914*

Heroin Withdrawal

Thyroid Stimulating Hormone (TSH)
Serum No Effect Physiological No significant effect observed. *0193*

Thyroxine (T_4) *Serum Increase Physiological* Significant increase in addicts observed. *0193*

Thyroxine (T_4), Free *Serum Decrease Physiological* Significant lower percentage. *0193*

Tri-iodothyronine (T_3) *Serum Increase Physiological* Significant increase observed compared with normals. *0193*

Hetacillin

Aspartate Aminotransferase
Serum Increase Physiological Observed with other penicillins (cause?). *1680*

Eosinophils *Blood Increase Physiological* Probably allergic response. *1680*

Erythrocytes *Blood Decrease Physiological* Occasional anemia reported with penicillins. *1680*

Leukocytes *Blood Decrease Physiological* Probably allergic response (rare). *1680*

Occult Blood *Feces Positive Physiological* Caused bleeding with dose of 4 g. *0853*

Platelets *Blood Decrease Physiological* Thrombocytopenia may occur occasionally. *1680*

Hexachlorobenzene

Erythrocytes *Blood Decrease Physiological* Pancytopenia (AMA Blood dyscrasias). *2429*

Leukocytes *Blood Decrease Physiological* Pancytopenia (AMA Blood dyscrasias). *2429*

Platelets *Blood Decrease Physiological* Pancytopenia (AMA Blood dyscrasias). *2429*

Porphyrins *Urine Increase Physiological* Stimulates formation of ALA-synthetase. *1343*

Hexamethylmelamine

Leukocytes *Blood Decrease Physiological* Moderate effect in half of patients. *3863*

Platelets *Blood Decrease Physiological* Moderate effect in a third of patients. *3863*

Hexocyclium

Hydrochloric Acid *Gastric Material Decrease Physiological* Effective anticholinergic action. *1679*

Volume *Gastric Material Decrease Physiological* Effective anticholinergic action. *1679*

Hexoprenaline

Estriol *Plasma Decrease Physiological* Significant effect when given to women in premature labor. *1525*

Hexoses

Cortisol *Urine No Effect Analytical* On fluorometric method of Ratliff and Hall. *2923*

High Carbohydrate Diet

Bile Acids *Feces Decrease Physiological* Marked if given i.v. *0880*

Cholesterol *Serum Decrease Physiological* Increase by 40% but maximum if given i.v. *0880*

Neutral Sterols *Feces Decrease Physiological* Marked if given i.v. *0880*

Triglycerides *Serum Increase Physiological* Increase by 60% if oral no effect if i.v. *0880*

High Sodium Diet

Aldosterone *Plasma Decrease Physiological* Reverse effect with low sodium diet. *3263*
Urine Decrease Physiological Reverse effect with low sodium diet. *3263*

Renin Activity *Plasma Decrease Physiological* Reverse effect with low sodium diet. *3263*

Hippuric Acid

Ammonia *Plasma No Effect Analytical* On indophenol reaction with 5,000 nmoles. *1290*

Creatinine *Urine Increase Analytical* 1 mg equivalent to 0.08 mg with automated Jaffé. *2558*

Crystals *Urine Increase Physiological* Colorless needles, rhombic plates, prisms (all pH). *0443*

Sugar *Urine Increase Analytical* Reduces Benedict's solution. *1022*

Histamine

Amphetamine *Urine No Effect Analytical* No effect at 100 mg/L on method of Rutter. *3097*

Amylase *Serum Increase Physiological* Subcutaneous injection may cause acute pancreatitis. *3206*

Effective Renal Plasma Flow
Patient Decrease Physiological After 0.3-0.5 mg subcutaneously in humans. *1176*

Epinephrine *Plasma Increase Physiological* Due to release from adrenals. *1176*
Urine Increase Physiological Due to release from adrenals. *1176*

Glomerular Filtration Rate (GFR)
Urine Decrease Physiological After 0.3-0.5 mg subcutaneously in humans. *1176*

Growth Hormone *Plasma Increase Physiological* Observed effect in normals. *0371*

Hematocrit *Blood Increase Physiological* Due to increased capillary permeation and hemoconcentration. *1176*

Histamine *Urine Increase Physiological* Threefold increase in 1 d after injection. *1382*

Histidine *Plasma Increase Analytical* May produce marked effect on method of Ambrose. *0081*

Hydrochloric Acid *Gastric Material Increase Physiological* Used diagnostically. *0071*

17-Hydroxycorticosteroids *Urine Increase Physiological* Due to release from adrenals. *1176*

Magnesium *Gastric Material Increase Physiological* Significant increased output although concentration decrease in first 20 minutes. *3252*

Norepinephrine *Plasma Increase Physiological* Due to release from adrenals. *1176*

Occult Blood *Feces Positive Physiological* May cause gastric mucosal bleeding if given subcutaneously. *3498*

Pepsin *Gastric Material Increase Physiological* Effect occurs with secretion of acid. *1343*

Potassium *Serum Increase Physiological* Marked effect of i.v. injection (?involves gastrointestinal tract). *1343*
Gastric Material Increase Physiological Concentration and output increased after subcutaneous injection. *3252*

Sodium *Gastric Material No Effect Physiological* Output unchanged although concentration decreased. *3252*

Tyrosine *Plasma No Effect Analytical* On fluorometric procedure of Ambrose. *0080*

Vanillylmandelic Acid *Urine Increase Physiological* Due to release from adrenals. *1176*

Volume *Plasma Decrease Physiological* Due to increased capillary permeation and hemoconcentration. *1176*
Urine Decrease Physiological After 0.3-0.5 mg subcutaneously in humans. *1176*
Gastric Material Increase Physiological Characteristic normal response. *1176*

Histidine

Alanine *Plasma Increase Physiological* Interferes with clearance after oral load. *1644*

Ammonia *Plasma Decrease Analytical* With 250 nmoles produces 17% inhibitor indophenol reaction. *1290*

Arginine *Plasma Increase Physiological* Interferes with clearance after oral load. *1644*

Bilirubin *Serum Increase Analytical* Reacts with diazo reagent. *1563*

Citrulline *Plasma Increase Physiological* Interferes with clearance after oral load. *1644*

Creatinine *Urine Increase Analytical* 1 mg equivalent to 0.03 mg with automated Jaffé. *2558*

Histidine *(continued)*

Cystine *Urine Increase Physiological* Increased clearance, plasma concentration unchanged. *1644*

Glycine *Plasma Increase Physiological* Interferes with clearance after oral load. *1644*

Hydroxyproline *Urine No Effect Analytical* At 46 mg/L on method of Goverde. *1365*

Isoleucine *Plasma Decrease Physiological* Probably due to flow into gut after oral load. *1644*

Leucine *Plasma Decrease Physiological* Probably due to flow into gut after oral load. *1644*

Lysine *Plasma Increase Physiological* Interferes with clearance after oral load. *1644*

Methylimidazoleacetic Acid *Urine Increase Physiological* Doubling in days after injection. *1382*

Ornithine *Plasma Decrease Physiological* Probably due to flow into gut after oral load. *1644*

Phenylalanine *Plasma Decrease Physiological* Probably due to flow into gut after oral load. *1644*

Phosphate *Urine No Effect Analytical* No effect at 20 mg/dL method of Jung/Parekh. *1835*

Proline *Plasma No Effect Analytical* No effect on method of Goodwin. *0117*

Serine *Plasma Increase Physiological* Interferes with clearance after oral load. *1644*

Sugar *Urine Increase Analytical* High concentrations cause yellowing of Benedict's color. *3507*

Taurine *Plasma Increase Physiological* Interferes with clearance after oral load. *1644*

Threonine *Plasma Increase Physiological* Interferes with clearance after oral load. *1644*

Tyrosine *Plasma No Effect Analytical* On fluorometric procedure of Ambrose. *0080*
Plasma Decrease Physiological Probably due to flow into gut after oral load. *1644*

Urea Nitrogen *Serum No Effect Analytical* No effect on Berthelot reaction. *1862*

Valine *Plasma Decrease Physiological* Probably due to flow into gut after oral load. *1644*

HMPG

Epinephrine *Test Conditions No Effect Analytical* On fluorescent procedure of Peyrin. *2800*

Guaiacols Spot Test *Urine Positive Analytical* Action on procedure of Rogers. *3031*

Metanephrines, Total *Urine Increase Analytical* At 2 mg/L modified Pisano procedure. *1428*

Norepinephrine *Test Conditions No Effect Analytical* On fluorescent procedure of Peyrin. *2800*

Homocitrulline

Tyrosine *Plasma No Effect Analytical* On fluorometric procedure of Ambrose. *0080*

Homocysteic Acid

Tyrosine *Plasma No Effect Analytical* On fluorometric procedure of Ambrose. *0080*

Homocystine

Tyrosine *Plasma No Effect Analytical* On fluorometric procedure of Ambrose. *0080*

Homogentisic Acid

Ammoniacal Silver Nitrate *Urine Positive Analytical* Rapid darkening of solution. *3588*
Test Conditions Positive Analytical Positive spot test with Tollen's reagent. *3765*

Color *Urine Increase Analytical* Black (occurs with alkaptonuria). *2448*

Creatinine *Test Conditions Increase Analytical* Positive spot test with Jaffé reagent. *3765*

Ferric Chloride Test *Urine Positive Analytical* Transient blue color. *3588*

Glucose *Urine Decrease Analytical* Reducing substance affects coupled glucose oxidase procedure. *0583* Inhibits peroxidase reaction of Clinistix®. *2544*

5-Hydroxyindoleacetic Acid (5-HIAA)
Urine Decrease Analytical Significant effect procedure of Udenfriend Significant effect procedure of Mustala. *1097*

Oxalate *Urine Increase Analytical* When urine from alkaptonuric measured by Sigma oxalate oxidase method. *0345*

Phenylketones *Urine No Effect Analytical* No effect with Phenistix®. *0775*
Urine Positive Analytical Blue/green fading with FeCl₃, nil with Phenistix®. *0775*

Sugar *Urine Increase Analytical* Reduces Benedict's solution, may affect Clinitest®. *1022*

Uric Acid *Test Conditions Increase Analytical* Positive spot test with phosphotungstate. *3765*

Urobilinogen *Test Conditions Increase Analytical* Positive spot test with Ehrlich's reagent. *3765*

Vanillylmandelic Acid *Urine Decrease Analytical* At therapeutic physiological concentrations on Pisano procedure. *1097*

Homoserine

Tyrosine *Plasma No Effect Analytical* On fluorometric procedure of Ambrose. *0080*

Homovanillic Acid

Ammoniacal Silver Nitrate
Test Conditions Positive Analytical Positive spot test with Tollen's reagent. *3765*

Epinephrine *Test Conditions No Effect Analytical* On fluorescent procedure of Peyrin. *2800*

Guaiacols Spot Test *Urine Negative Analytical* Action on procedure of Rogers. *3031*

5-Hydroxyindoleacetic Acid (5-HIAA)
Urine No Effect Analytical No effect at 10 mg/dL method of Goldenberg. *1327* No significant decrease with procedure of Udenfriend. *1097*

Millons Test *Test Conditions Positive Analytical* Positive spot test for phenols. *3765*

Norepinephrine *Test Conditions No Effect Analytical* On fluorescent procedure of Peyrin. *2800*

Uric Acid *Test Conditions Increase Analytical* Positive spot test with phosphotungstate. *3765*

Urobilinogen *Test Conditions Increase Analytical* Positive spot test with Ehrlich's reagent. *3765*

Vanillylmandelic Acid *Urine Decrease Analytical* At therapeutic physiological concentrations on Pisano procedure. *1097*

Hospitalization

Aminolevulinic Acid Dehydrase
Red Blood Cells Decrease Physiological About 60% of normal individuals. *0400*

Hycanthone

Alanine Aminotransferase *Serum Increase Physiological* Transient change in liver function. *0071*

Aspartate Aminotransferase
Serum Increase Physiological Transient change in liver function. *0071*

Bilirubin *Serum Increase Physiological* Induces form of acute toxic hepatitis. *1079*

BSP Retention *Serum Increase Physiological* Induces form of acute toxic hepatitis. *1079*

Hydantoin Derivatives

Alanine Aminotransferase *Serum Increase Physiological* Severe delayed hypersensitivity reported. *2427*

Alkaline Phosphatase *Serum Increase Physiological* Severe delayed hypersensitivity reported. *2427*

Aminolevulinic Acid *Urine Increase Physiological* May precipitate attack of acute porphyria. *1016*

Bilirubin *Serum Increase Physiological* Severe delayed hypersensitivity reported. *2427*

γ-Globulin *Serum Increase Physiological* Observed in cases of hypersensitivity. *2427*

LE Cells *Blood Positive Physiological* May cause appearance of LE cells. *2220*

Methemoglobin *Blood Increase Physiological* Reported effect. *2220*

Platelets *Blood Decrease Physiological* Thrombocytopenia (immunologically induced). *2313*

Porphyrins *Urine Increase Physiological* May precipitate attack of acute porphyria. *1016*

Urea Nitrogen *Serum Increase Analytical* Absorbance relative to urea = 0.3 diacetyl procedure. *0583*

Hydralazine

Alanine Aminotransferase *Serum Increase Physiological* Isolated case with moderate hepatomegaly, abnormal findings resolved when drug discontinued. *1161* Centrilobular necrosis observed in 3 patients 2 to 4 months after drug therapy for hypertension. *1740*

Albumin *Serum Decrease Physiological* Centrilobular necrosis observed in 3 patients 2 to 4 months after drug therapy for hypertension. *1740*

Aldosterone *Plasma Increase Physiological* Significant effect following 20 mg i.v. in 30 to 60 minutes. *1609*

Alkaline Phosphatase *Serum Increase Physiological* Isolated case with moderate hepatomegaly, abnormal findings resolved when drug discontinued. *1161* Cholestatic jaundice observed in one patient. *3469* Asymptomatic or symptomatic reversible hepatitis-like reaction or granulomatous hepatitis. *2374* Centrilobular necrosis observed in 3 patients 2 to 4 months after drug therapy for hypertension. *1740*

Amphetamine *Urine No Effect Analytical* At 50 mg/L on fluorescent method of Hayes. *1534*

Antibodies to dsDNA *Serum No Effect Physiological* Observed in one patient with induced systemic lupus erythematosus. *3784*

Antibodies to Histones *Serum Positive Physiological* Observed in 9 patients developing rapidly progressive glomerulonephritis during treatment. *0366*

Antinuclear Antibodies *Serum Positive Physiological* More common in slow acetylators. *2783* Raised in patients who developed symptoms of lupus, unrelated to dose of drug. *0353*

Aspartate Aminotransferase *Serum Increase Analytical* At 1 mmol/L affects SMA 12/60 method. *3335*
Serum Increase Physiological Isolated case with moderate hepatomegaly, abnormal findings resolved when drug discontinued. *1161* Asymptomatic or symptomatic reversible hepatitis-like reaction or granulomatous hepatitis. *2374* Cholestatic jaundice observed in one patient. *3469* Centrilobular necrosis observed in 3 patients 2 to 4 months after drug therapy for hypertension. *1740*

Bicarbonate *Serum No Effect Analytical* At concentration of 6 mg/L had no effect on method using phenolphthalein. *3393*

Bilirubin *Serum No Effect Analytical* At concentration of 160 mg/L had no effect on Jendrassik and Grof method. *3393*
Serum No Effect Physiological Isolated case with moderate hepatomegaly, abnormal findings resolved when drug discontinued. *1161*
Serum Increase Physiological Observed in a case of obstructive jaundice with pancytopenia. *3469* Asymptomatic or symptomatic reversible hepatitis-like reaction or granulomatous hepatitis. *2374* Centrilobular necrosis observed in 3 patients 2 to 4 months after drug therapy for hypertension. *1740*

Bilirubin, Direct *Serum Increase Physiological* Centrilobular necrosis observed in 3 patients 2 to 4 months after drug therapy for hypertension. *1740*

Calcium *Serum Increase Analytical* At 1 mmol/L has slight effect on SMA 12/60 method. *3335* At concentrations above 150 mg/L (normal therapeutic concentration 2.3 mg/L) raised concentration as measured by cresolphthalein method. *3393*

Catecholamines *Urine No Effect Analytical* No effect on fluorometric method of Crout. *0758*
Urine No Effect Physiological No effect observed. *0913*
Urine Increase Analytical Interferes with some fluorometric methods. *2220*

Chloride *Serum No Effect Analytical* At concentration of 6 mg/L had no effect on mercurimetric method. *3393*

Cholesterol *Serum No Effect Analytical* At concentration of 0.5 mg/L had no effect on CHOD-PAP method. *3393* At concentration of 160 mg/L had no effect on Liebermann-Burchard method. *3393*
Serum Decrease Physiological Increase by 12% in 7 individuals treated for 4 mo. *0088* 12% reduction in 7 patients treated for 4 mo. *0089*

Complement C1q *Serum Decrease Physiological* Observed in one patient with induced systemic lupus erythematosus. *3784*

Complement C3 *Serum Decrease Physiological* Observed in one patient with induced systemic lupus erythematosus. *3784*

Complement C4 *Serum No Effect Physiological* Observed in one patient with induced systemic lupus erythematosus. *3784*

Complement CH₅₀ *Serum Decrease Physiological* Observed in one patient with induced systemic lupus erythematosus. *3784*

Complement Factor B *Serum Decrease Physiological* Observed in one patient with induced systemic lupus erythematosus. *3784*

Coombs' Test *Serum Positive Physiological* Mechanism obscure. *1486*

Coombs' Test, Direct *Serum Positive Physiological* May occur if SLE induced. *1487* Observed in one patient receiving low dose of drug for hypertension. Mechanism involved in producing erythrocyte-reacting antibody not known. *2680*

Creatinine *Serum No Effect Analytical* At concentration of 6 mg/L had no effect on AutoAnalyzer Jaffé method. *3393*

Erythrocytes *Blood Decrease Physiological* Pancytopenia may occur. *0044* Observed in 9 patients developing rapidly progressive glomerulonephritis during treatment. *0366*
Urine Increase Physiological Reported finding. *2427*

Erythrocyte Sedimentation Rate
Blood Increase Physiological Observed in 9 patients developing rapidly progressive glomerulonephritis during treatment. *0366*

γ-Globulin *Serum Increase Physiological* Reported finding. *2427*

Glucose *Serum No Effect Analytical* Affects Boehringer GOD-PERID method. *3277* At concentration of 0.5 mg/L had no effect on GOD/POD-PAP method. *3393*
Serum Increase Analytical At 1 mmol/L affects SMA 12/60 method. *3335*
Serum Decrease Analytical Affects glucose oxidase method of Boehringer. *3277* At concentrations above 115 mg/L (normal therapeutic concentration 2.3mg/L) lowered concentration as measured by GOD-Perid method. *3393*

Hematocrit *Blood Decrease Physiological* May cause hemolytic anemia. *0044*

Hemoglobin *Blood Decrease Physiological* May cause hemolytic anemia. *0044* Observed in one patient receiving low dose of drug for hypertension. Mechanism involved in producing erythrocyte-reacting antibody not known. *2680* Observed in a case of obstructive jaundice with pancytopenia. *3469*
Observed in 9 patients developing rapidly progressive glomerulonephritis during treatment. *0366*
Urine Increase Physiological Reported finding. *2427* Microscopic hematuria observed in 9 patients developing rapidly progressive glomerulonephritis during treatment. *0366*

17-Hydroxycorticosteroids *Urine Decrease Analytical* In vitro effect reported modified Glenn-Nelson. *0427*

17-Ketogenic Steroids *Urine Increase Analytical* Glucuronide interferes with Zimmermann reaction. *3505*

17-Ketosteroids *Urine No Effect Analytical* No significant effect with Zimmermann reaction. *0427*

Hydralazine (continued)

Lactate Dehydrogenase *Serum Increase Physiological* Isolated case with moderate hepatomegaly, abnormal findings resolved when drug discontinued. *1161* Asymptomatic or symptomatic reversible hepatitis-like reaction or granulomatous hepatitis. *2374*

LE Cells *Blood Positive Physiological* More common in slow acetylators. *2783* Observed in one patient with induced systemic lupus erythematosus. *3784* 3% patients developed lupus-like syndrome all of whom were slow acetylators and receiving less than 200 mg drug/d. *0353*

Leukocytes *Blood Decrease Physiological* Agranulocytosis/pancytopenia/leukopenia. *0044* Observed in a case of obstructive jaundice with pancytopenia. *3469*

Neutrophils *Blood Decrease Physiological* Occasional case of agranulocytosis reported. *3717*

Occult Blood *Feces Positive Physiological* Rare gastrointestinal hemorrhage. *1343*

Phosphate *Serum No Effect Analytical* At concentration of 6 mg/L had no effect on phosphomolybdate method. *3393*

Platelets *Blood Decrease Physiological* Pancytopenia may occur with purpura. *0044* Rare finding in patients treated for hypertension: also observed in neonates. *3818*

Potassium *Serum No Effect Analytical* At concentration of 0.5 mg/L had no effect on flame-photometric method. *3393* At concentration of 6 mg/L had no effect on measurement by ISE with predilution. *3393*

Protein *Serum No Effect Analytical* At concentration of 6 mg/L had no effect on biuret method with blank correction. *3393*
Urine Increase Physiological May cause nephrotoxicity. *3204*

Renin Activity *Plasma Increase Physiological* 7 fold increase at 1 mg/kg in rats. *2798* Significant effect following 20 mg i.v. in 30 to 120 minutes. *1609*
Plasma Decrease Physiological May be suppressed activity. *2397*

Reticulocytes *Blood Increase Physiological* Observed in one patient receiving low dose of drug for hypertension. Mechanism involved in producing erythrocyte-reacting antibody not known. *2680*

Sodium *Serum No Effect Analytical* At concentration of 0.5 mg/L had no effect on flame-photometric method. *3393* At concentration of 6 mg/L had no effect on measurement by ISE with predilution. *3393*

Triglycerides *Serum No Effect Analytical* At concentration of 6 mg/L had no effect on lipase/esterase method. *3393*

Tyramine Test *Patient Increase Physiological* Increased responsiveness. *1488*

Urea Nitrogen *Serum No Effect Analytical* At concentration of 160 mg/L had no effect on diacetylmonoxime method. *3393*
Serum Increase Physiological May cause nephrotoxicity. *3204*

Uric Acid *Serum Increase Analytical* At 1 mmol/L affects SMA 12/60 and Henry methods. *3335* At concentrations above 6 mg/L (normal therapeutic concentration 2.3mg/L) raised concentration as measured by phosphotungstate reduction method. *3393*

Vanillylmandelic Acid *Urine No Effect Physiological* No effect observed. *0913*

Wassermann Reaction *Serum Positive Physiological* False positive reaction reported. *2427*

Hydration

Creatinine Clearance *Urine Increase Physiological* In patients with proteinuria. *2818*

Protein *Urine Increase Physiological* In patients with proteinuria. *2818*

Volume *Urine Increase Physiological* In patients with proteinuria. *2818*

Hydrazine Derivatives

Alanine Aminotransferase *Serum Increase Physiological* May cause hepatotoxicity. *2313*

Alkaline Phosphatase *Serum Increase Physiological* May cause hepatotoxicity. *2313*

Aspartate Aminotransferase
Serum Increase Physiological May cause hepatotoxicity. *2313*

Bile *Urine Increase Physiological* May cause hepatotoxicity. *2313*

Bilirubin *Serum Increase Physiological* May cause hepatotoxicity. *2313*

BSP Retention *Serum Increase Physiological* May cause hepatotoxicity. *2313*

Cephalin Flocculation *Serum Increase Physiological* May cause hepatotoxicity. *2313*

Erythrocyte Sedimentation Rate
Blood Increase Physiological Augmentation with SLE-like syndrome. *0933*

Glucose *Serum Decrease Physiological* May potentiate action of insulin in diabetics. *2220*

5-Hydroxyindoleacetic Acid (5-HIAA)
Urine Decrease Physiological May potentiate action of drugs on CNS. *2220*

LE Cells *Blood Positive Physiological* May activate lupus erythematosus. *0616*

Lymphocytes *Blood Decrease Physiological* Significant effect observed with megadose supplementation. *1347*

Metanephrines, Total *Urine Increase Physiological* May potentiate action of drugs on CNS. *2220*

Neutrophils *Blood No Effect Physiological* No effect observed with megadose supplementation. *1347*

Normetanephrine *Urine Increase Physiological* May potentiate action of drugs on CNS. *2220*

Thymol Turbidity *Serum Increase Physiological* May cause hepatotoxicity. *2313*

Vanillylmandelic Acid *Urine Decrease Physiological* May potentiate action of drugs on CNS. *2220*

Hydrochloric Acid

Gastrin *Serum Decrease Physiological* Significant effect if given orally. *1236*

Hydrochlorothiazide

Alanine Aminotransferase *Serum Increase Physiological* May cause cholestasis or cholangiolitic hepatitis. *1680*

Amylase *Serum Increase Physiological* Acute pancreatitis may occur. *3572*

Antidiuretic Hormone *Plasma Increase Physiological* Secreted in response to hyponatremia. *1129*

Apolipoprotein AI *Serum No Effect Physiological* Usual observation when thiazides given to patients. *3780*

Apolipoprotein AII *Serum No Effect Physiological* Usual observation when thiazides given to patients. *3780*

Aspartate Aminotransferase
Serum Increase Physiological May cause cholestasis or cholangiolitic hepatitis. *1680*

Bicarbonate *Serum Increase Physiological* May cause metabolic alkalosis. *1129*

Bilirubin *Serum Increase Physiological* Cholestatic jaundice reported effect of thiazides. *2745*

Calcium *Serum Increase Physiological* Impaired excretion (probably also released from bone). *3505* Increase by 0.3 mg/dL in 343 patients with hypertension given drug for 10 weeks. *3707* Increased renal tubular reabsorption, usually associated with mild primary hyperparathyroidism. *1054*
Serum Decrease Physiological Increase by 0.13 mg/dL in approximately 175 patients with essential hypertension treated for 1 y. *3708*
Urine Decrease Physiological Observed effect. *0451*
Feces Decrease Physiological Accentuates positive calcium balance. *2427*

Casts *Urine No Effect Physiological* Normally but augments effects of acidifying agents. *1720*

Chloride *Serum Increase Physiological* Hyperchloremic alkalosis with prolonged therapy. *2220*

Serum Decrease Physiological May cause marked reduction. *1129*

Urine Increase Physiological Significant effect in 5 h after 150 mg orally. *1189*

Cholesterol *Serum No Effect Physiological* No effect with 25 or 50 mg/d in hypertensive patients. *1980* In approximately 175 patients with essential hypertension treated for 1 y. *3708* In long term involving many patients. *0089*
Serum Increase Physiological Observed with 100 mg/d. *3663* 7% change in individuals given drug alone for less than 1 y. *1404* Increase by 7 mg/dL in 343 patients with hypertension given drug for 10 weeks. *3707* 4-7% in short term study. *0089*
Serum Decrease Physiological In MRFIT study when given with chlorthalidone involving many patients. *0089*

Cholesterol, High Density Lipoprotein
Serum No Effect Physiological Not significant change in 39 individuals given drug alone for less than 1 y. *1404* No effect with 25 or 50 mg/d in hypertensive patients. *1980* In long term over 1 y in many patients. *0089*

Cholesterol, Low Density Lipoprotein
Serum No Effect Physiological Not significant change in 39 individuals given drug alone for less than 1 y. *1404* In short term study involving many patients. *0089*
Serum Increase Physiological Observed with 100 mg/d. *3663*

Cholesterol, Very Low Density Lipoprotein
Serum Increase Physiological 13% change in individuals given drug alone for less than 1 y. *1404* 13% in a short term involving many patients. *0089*

Citrate *Urine Decrease Physiological* By up to 30% reported. *3245*

Cortisol *Urine Decrease Analytical* Decreased unconjugated by fluorometric procedure. *0022*
Urine Decrease Physiological Possibly altered cortisol secretion. *1487*

C-Peptide *Urine No Effect Physiological* In 14 hypertensives men with type 2 diabetes treated for 2 weeks with or without propranolol. *0935*

Creatinine *Serum No Effect Analytical* At concentration of 1.4 mg/L had no effect on creatinine iminohydrolase method. *3393*
Serum No Effect Physiological In approximately 175 patients with essential hypertension treated for 1 y. *3708*
Serum Increase Physiological But by less than 0.1 mg/dL in 33 patients with hypertension given drug for 10 weeks to reduce diastolic BP to less than 90 mm Hg. *3707*

Creatinine Clearance *Urine Decrease Physiological* Reported effect. *2220*

Erythrocytes *Blood Decrease Physiological* Aplastic anemia may occur. *1680* Occasional case of aplastic anemia reported. *3717*

Estriol *Urine No Effect Analytical* At 4 x normal concentration with enzyme hydrolysis and Kober procedure. *0762*
Urine Decrease Analytical Interferes in hydrolysis of conjugates stage. *3060* May have slight effect on enzyme hydrolysis. *0023* Destroys estriol during acid hydrolysis. *0583*

Estrogens *Urine Increase Physiological* In pregnant but mechanism unknown. *0023*
Urine Decrease Analytical Affects colorimetric/fluorometric procedures. *0022* Affects Kober reaction if Allen correction. *0023*

Glucagon *Plasma Increase Physiological* Higher in 15 hypertensives during treatment with 50 mg twice daily than before treatment and after withdrawal. *1018*

Glucose *Serum No Effect Analytical* At concentration of 7 mg/L had no effect on GOD/POD-PAP method. *3393*
Serum No Effect Physiological No effect in hypertensives given potassium supplement if demonstrated no effect in response to hydrochlorothiazide alone. *2360*
Serum Increase Physiological Diabetogenic-like action of drug. *3505* In 14 hypertensive men with type 2 diabetes by 31% over 3 weeks: effect augmented by coadministered propranolol. *0935* To extent of coma in 2 diabetics given hydrochlorothiazide with propranolol: exact mechanism not known. *2559* Increase by 6 mg/dL in 343 patients with hypertension given drug for 10 weeks. *3707* Increase by 4.7 mg/dL in approximately 175 patients with essential hypertension treated for 1 y. *3708*

Urine Increase Physiological Diabetogenic-like action of drug due to hyperglycemia if produced. *3505*

Glucose Tolerance *Serum Decrease Physiological* Diabetogenic-like action of drug. *2745*

Growth Hormone *Plasma No Effect Physiological* In 15 hypertensives treated with 50 mg twice daily. *1018*

Hemoglobin A$_{1c}$ *Blood Increase Physiological* Increase by 6% in 14 hypertensive type 2 diabetic men over 3 weeks: effect augmented by coadministered propranolol. *0935*

Insulin *Plasma No Effect Physiological* In 15 hypertensives treated with 50 mg twice daily. *1018*
Plasma Decrease Physiological Intravenous injection has effect with little on sugar. *2427*

Iodide *Urine Increase Physiological* Significant effect in 5 h after 150 mg orally. *1189*

Ionized Calcium *Serum Increase Physiological* Increased for up to 2 weeks after drug withdrawal. *3480* Observed effect. *0451*

^{131}I Uptake *Serum No Effect Physiological* No change observed. *0583*

Leukocytes *Blood Decrease Physiological* May cause leukopenia/agranulocytosis. *0847* Occasional case of aplastic anemia reported. *3717*

Lithium *Serum Increase Physiological* Significant increase with 50 mg/d over 2 weeks possible effect on reabsorption in loop of Henle. *1782* Reported in 2 cases taking lithium: drug also given with triamterene: due to reduced clearance of lithium. *2393*

Magnesium *Serum No Effect Physiological* No significant difference in hypertensive patients with 50 mg/d. *1980*
Serum Decrease Physiological In 10% patients receiving diuretics, but majority had diseases likely also to hypomagnesemia. *2050*
Urine No Effect Physiological No significant effect on clearance noted with i.v. infusion of 50 mg, although long-term oral studies suggest some increased loss and reduction of plasma concentration. *3101*
Urine Increase Physiological 60% increase after 200 mg in 1 dose orally. *1487*

Neutrophils *Blood Decrease Physiological* Occasionally observed. *2427* Occasional case of agranulocytosis reported. *3717*

Osmolality *Serum Decrease Physiological* Due to hyponatremia of diuretic action. *1129*

Parathyroid Hormone *Plasma No Effect Physiological* No effect (?mechanism of effect on calcium). *3480*

Phosphate *Serum Increase Physiological* Altered parathyroid metabolism. *0468*
Serum Decrease Physiological Observed in some cases prolonged treatment. *1680*
Urine Increase Physiological Altered parathyroid metabolism. *0468*

Platelets *Blood Decrease Physiological* Thrombocytopenia reported. *2745* Several cases of immune mediated thrombocytopenia reported. *2502* Occasional case of aplastic anemia reported. *3717*

Potassium *Serum No Effect Physiological* If potassium supplement coadministered with hydrochlorothiazide in hypertensive women. *2360*
Serum Increase Physiological Marked effect when given with amiloride to 3 patients with renal failure (creatinine 0.2-0.7 mmol/L). *3816*
Serum Decrease Physiological Diuretic action. *3505* During 3 mo treatment in hypertensive patients with either 25 or 50 mg/d, but only significant statistically with 50 mg. *1980* Increase by 0.6 mmol/L in 343 patients with hypertension given drug for 10 weeks. *3707* Increase by 0.6 mmol/L in approximately 175 patients with essential hypertension treated for 1 y. *3708* 28% patients with hypokalemia given long term 25-100 mg/d. *0314*
Urine No Effect Physiological In patients who were normokalemic after receiving drug for treatment for uncomplicated systemic hypertension. *2721*
Urine Increase Physiological Significant effect in 5 h after 150 mg orally. *1189* In hypokalemic patients receiving drug over initial period of therapy: Increase of up to 41 mmol/d. *2721*

Prolactin *Plasma No Effect Physiological* In 15 hypertensives treated with 50 mg twice daily. *1018*

Hydrochlorothiazide *(continued)*

Protein Bound Iodine (PBI) *Serum No Effect Physiological* No effect observed in one study. *0583*
Serum Decrease Physiological Mechanism obscure. *0012*

Pyrophosphate *Urine Increase Physiological* Associated with increased orthophosphate also. *2709*

Quinidine *Serum Increase Physiological* Alkalinizes urine, increases reabsorption. *1487*

Sodium *Serum Decrease Physiological* May cause hyponatremia. *1129* Severe hyponatremia observed in several cases. *0170*
Urine Increase Physiological Diuretic action of drug. *2220*

Triglycerides *Serum No Effect Analytical* At concentration of 7 mg/L had no effect on GPO-PAP method. *3393*
Serum No Effect Physiological No effect with 25 or 50 mg/d in hypertensive patients. *1980* Nonsignificant change by 7 mg/dL in 33 patients with hypertension given drug for 10 weeks to reduce diastolic BP to less than 90 mm Hg. *3707* In approximately 175 patients with essential hypertension treated for 1 y. *3708* In short term, and long term involving many patients. *0089*
Serum Increase Physiological Observed with 100 mg/d. *3663* 17% change in 39 individuals given drug alone for less than 1 y. *1404* 17% in short term, and increase in MRFIT study when given with chlorthalidone. *0089*

Triglycerides, Very Low Density Lipoprotein
Serum Increase Physiological Observed with 100 mg/d. *3663*

Urea Nitrogen *Serum Increase Physiological* May occur with prolonged therapy. *2220* Increase by 2.7 mg/dL in 343 patients with hypertension given drug for 10 weeks to reduce diastolic BP to less than 90 mm Hg. *3707* Increase by 2.5 mg/dL in approximately 175 patients with essential hypertension treated for 1 y. *3708*

Uric Acid *Serum No Effect Analytical* At concentration of 7 mg/L had no effect on uricase-PAP method. *3393*
Serum Increase Physiological Decreased urate clearance. *3505* Slight effect in hypertensive patients given either 25 or 50 mg/d. *1980* Increase by 2.0 mg/dL in 343 patients with hypertension given drug for 10 weeks. *3707* Increase by 1.37 mg/dL in approximately 175 patients with essential hypertension treated for 1 y. *3708*
Urine Decrease Physiological Decreased urate clearance. *2220*

Volume *Urine Increase Physiological* Significant effect in 5 h after 150 mg orally. *1189*

Zinc *Serum Increase Physiological* Slight effect observed or no effect. *2712*
Urine Increase Physiological Twice normal reached on fifth day. *2712* Increase by 60% in 9 patients with hypertension over 2 weeks. *3806*

Hydrocortisone

Aldosterone *Urine No Effect Analytical* No significant effect in RIA procedure of Drewes. *0952*
Urine Increase Analytical 1 mg/L equivalent to 4 μg/L in method of Drewes. *0952*

Amino Acids *Urine Increase Physiological* Glucocorticoid hormonal action. *1841*

Bicarbonate *Serum Increase Physiological* May cause hypochloremic hypokalemic alkalosis. *1343*

Calcium *Serum No Effect Analytical* No effect on fluorescence of calcein. *2395*

Chloride *Serum Increase Physiological* May cause retention and edema. *0071*
Serum Decrease Physiological May cause hypochloremic hypokalemic alkalosis. *1343*

Cortisol *Plasma Increase Analytical* If non selective extraction and competitive protein binding used. *0512*
Plasma Increase Physiological Effect lasts for at least 24 h. *1772*

Dexamethasone *Serum Increase Analytical* Very slight cross react procedure of Hichens. *1585*

Eosinophils *Blood Decrease Physiological* Due to hormonal action on adrenals. *0242*

Estrogens *Urine No Effect Analytical* At 50 mg/L on fluorescent method of Corns. *0730*

Growth Hormone *Plasma Decrease Physiological* Marked effect of 100 mg i.v. by 4 h. *2936*

3-Hydroxyanthranilic Acid *Urine Increase Physiological* Causes induction of tryptophan pyrrolase. *3047*

3-Hydroxykynurenine *Urine Increase Physiological* Causes induction of tryptophan pyrrolase. *3047*

Kynurenine *Urine Increase Physiological* Causes induction of tryptophan pyrrolase. *3047*

Lymphocytes *Blood Decrease Physiological* Due to hormonal action on adrenals. *0242*

Neutrophils *Blood Increase Physiological* Due to hormonal action on adrenals. *0242*

Occult Blood *Feces Positive Physiological* May cause hemorrhage or ulceration of gastrointestinal tract. *0612*

PAH Clearance *Urine Increase Physiological* Increases GFR. *1343*

Parathyroid Hormone *Plasma Increase Physiological* If given i.v. stimulates secretion. *3840*

Platelet Aggregation *Blood Decrease Physiological* Inhibits streptokinase induced aggregation. *3104*

Potassium *Serum Decrease Physiological* Promotes urinary elimination. *0071*
Urine Increase Physiological Increases elimination in urine. *0071*

Progesterone *Plasma No Effect Analytical* 1% or less cross reactivity with RIA. *0567*

Protein Bound Iodine (PBI) *Serum Decrease Physiological* Inhibits iodination of tyrosine in thyroxine binding globulin. *0012*

Sodium *Serum Increase Physiological* May cause retention and edema. *0071*
Urine Increase Physiological Enhances excretion especially if prior loading. *1343*

Thromboxane *Plasma No Effect Physiological* No effect of up to 100 μmol/L on formation during clotting. *2890*

Tyrosine *Plasma Decrease Physiological* Hormonal effect observed in 2 weeks. *0843*

Xanthurenic Acid *Urine Increase Physiological* Causes induction of tryptophan pyrrolase. *3047*

Hydrocyanic Acid

Cyanmethemoglobin *Blood Increase Physiological* Reacts with methemoglobin. *1343*

Hydroflumethiazide

Alanine Aminotransferase *Serum Increase Physiological* May cause intrahepatic cholestasis. *1680*

Alkaline Phosphatase *Serum Increase Physiological* May cause intrahepatic cholestasis. *1680*

Ammonia *Plasma Increase Physiological* May occur especially if pre-existing hepatic impairment. *1680*

Amylase *Serum Increase Physiological* Acute pancreatitis may occur with thiazides. *1680*

Aspartate Aminotransferase
Serum Increase Physiological May cause intrahepatic cholestasis. *1680*

Bicarbonate *Serum Increase Physiological* May cause hypochloremic alkalosis. *1680*

Bilirubin *Serum Increase Physiological* May cause intrahepatic cholestatic jaundice. *1680*

BSP Retention *Serum Increase Physiological* May cause intrahepatic cholestasis. *1680*

Chloride *Serum Decrease Physiological* Diuretic action. *1680*
Urine Increase Physiological Diuretic action. *1680*

Erythrocytes *Blood Decrease Physiological* Pancytopenia, aplastic anemia. *3844*

Glucose *Serum Increase Physiological* May occur (similar action to other thiazides). *1680*
Urine Increase Physiological May occur as consequence of hyperglycemia. *1680*

Glucose Tolerance *Serum Decrease Physiological* Diabetogenic like action of drug. *1680*

Hematocrit *Blood Decrease Physiological* Pancytopenia. *3844*

Hemoglobin *Blood Decrease Physiological* Pancytopenia. *3844*

Leukocytes *Blood Decrease Physiological* Pancytopenia, agranulocytosis or aplastic anemia. *3844*

Platelets *Blood Decrease Physiological* Pancytopenia or thrombocytopenia with purpura. *3844*

Potassium *Serum Decrease Physiological* Diuretic action. *1680*

Sodium *Serum Decrease Physiological* Diuretic action. *1680*
Urine Increase Physiological Diuretic action. *1680*

Urea Nitrogen *Serum Increase Physiological* May occur especially if pre-existing renal disease. *1680*

Uric Acid *Serum Increase Physiological* Probably due to impaired clearance. *1680*

Hydrogen Peroxide

Glucose *Urine Increase Analytical* Oxidises chromogen in glucose oxidase test. *1022*

Hydrogen Sulfide

Casts *Urine Increase Physiological* Occurs due to nephrotoxicity. *1302*

Erythrocytes *Urine Increase Physiological* Occurs with nephrotoxicity (usually marked). *1302*

Hemoglobin *Urine Increase Physiological* Occurs due to nephrotoxicity. *1302*

Protein *Urine Increase Physiological* Occurs due to nephrotoxicity. *1302*

Hydromorphone

Alkaloids *Urine No Effect Analytical* With ninhydrin on Frings TLC procedure at 2 mg/dL. *1204*
Urine Increase Analytical Purple with iodoplatinate on Frings TLC procedure. *1204*

Barbiturate *Urine No Effect Analytical* With diphenyl-carbazone on Frings TLC procedure at 2 mg/dL. With mercuric SO$_4$ on Frings TLC procedure at 2 mg/dL. With vanillin on Frings procedure at 2 mg/dL. With mercurous NO$_3$ on Frings TLC procedure at 2 mg/dL. *1204*

Bromide *Urine No Effect Analytical* No effect on method of Frings. *1204*

Chlordiazepoxide *Blood No Effect Analytical* On method of Riddick at 5 mg/dL. *2983*

Ethchlorvynoi *Urine No Effect Analytical* No effect on method of Frings and Cohen at 2 mg/dL. *1205*

FPN Test *Urine No Effect Analytical* No effect at 2 mg/dL on method of Frings. *1204*

Morphine *Urine Increase Analytical* Cross reactivity equal (or more) with RIA procedures. Substantial cross reactivity with hemagglutination inhibition. *2514*

pCO$_2$ *Blood Increase Physiological* 10 times as potent as morphine. *0493*

Salicylate *Urine No Effect Analytical* At 2 mg/dL no effect on Trinder method. *1204*

Hydroquinone

Catecholamines *Urine Increase Analytical* Reacts like epinephrine with Nelson-Shaw test. *0516*

Glucose *Urine Decrease Analytical* Reported to inhibit glucose oxidase stix reactions. *0076* May inhibit peroxidase reaction on Clinistix®. *2544*

Homogentisic Acid *Urine Increase Analytical* If method of Nuberger used. *1096*

17-Hydroxy-4-Pregnene-3,20-Dione

Aldosterone *Urine Increase Analytical* 1 mg/L equivalent to 5 µg/L by method of Drewes. *0952*

Hydroxyacetamide

Alanine Aminotransferase *Serum Increase Physiological* Toxicity effect. *2313*

Alkaline Phosphatase *Serum Increase Physiological* Hepatotoxicity. *1022*

Aspartate Aminotransferase *Serum Increase Physiological* Toxicity effect. *2313*

Bile *Urine Increase Physiological* Hepatotoxicity. *2313*

Bilirubin *Serum Increase Physiological* Hepatotoxicity. *2313*

BSP Retention *Serum Increase Physiological* Hepatotoxicity. *2313*

Cephalin Flocculation *Serum Increase Physiological* Hepatotoxicity. *2313*

Thymol Turbidity *Serum Increase Physiological* Hepatotoxicity. *2313*

4-Hydroxyacetanilide

5-Hydroxyindoleacetic Acid (5-HIAA) *Urine No Effect Analytical* No effect 10 mg/dL method of Goldenberg. *1327*

Hydroxyaminobutyric Acid

Amino Acids *Urine Increase Analytical* False positive ninhydrin reacting spot on TLC Additional spot by high voltage electrophoresis. *3106*

Hydroxyamphetamine

Amphetamine *Urine No Effect Analytical* No effect at 100 mg/L on method of Rutter. *3097* At 50 mg/L on fluorescent method of Hayes. *1534*

Hydroxyanthranilic Acid

Ammoniacal Silver Nitrate *Test Conditions Positive Analytical* Positive spot test with Tollen's reagent. *3765*

Hydroxybenzoic Acid

Millons Test *Test Conditions Positive Analytical* Positive spot test for phenols. *3765*

Urobilinogen *Test Conditions Increase Analytical* Positive spot test with Ehrlich's reagent. *3765*

Hydroxybutyrate

Glucose *Urine No Effect Analytical* Beta-hydroxybutyrate at 10.0 g/L on Boehringer-Mannheim BM33071 glucose pad. *3356*

Uric Acid *Serum Increase Physiological* Inhibits tubular secretion of urate. *1331*
Urine Decrease Physiological Inhibits tubular secretion of urate. *1331*

Hydroxychloroquine

Erythrocytes *Blood Decrease Physiological* Anemia. *2313*

Leukocytes *Blood Decrease Physiological* Leukopenia. *2313*

Neutrophils *Blood Decrease Physiological* Occasional case of agranulocytosis reported. *3717*

Platelets *Blood Decrease Physiological* Thrombocytopenia. *2879*

Hydroxycinnamic Acid

Millons Test *Test Conditions Positive Analytical* Positive spot test for phenols. *3765*

Urobilinogen *Test Conditions Increase Analytical* Positive spot test with Ehrlich's reagent. *3765*

Hydroxydione

Glucose *Serum Increase Physiological* Significant increase 1 h after anesthesia. *0291*

Hydroxyhexamide

Glucose *Serum Decrease Physiological* Mild effect (stimulates release of insulin). *3928*

Uric Acid *Serum Decrease Physiological* Mild uricosuric action. *3928*
  Urine Increase Physiological Mild uricosuric action. *3928*

Hydroxyhippuric Acid

Millons Test *Test Conditions Positive Analytical* Positive spot test for phenols. *3765*

5-Hydroxyindoleacetic Acid

Ammoniacal Silver Nitrate
  Test Conditions Positive Analytical Positive spot test with Tollen's reagent. *3765*

Glucose *Urine Decrease Analytical* Low with Clinistix®, Diastix®. *1100*

Hydroxyproline *Urine No Effect Analytical* At 100 mg/L on method of Seymour. *3262*

Protein *CSF Increase Analytical* 1.0 mg = 3.2 mg in Folin-Ciocalteu procedure. *0583*

Tyrosine *Plasma Increase Analytical* 21% fluorescence in procedure of Ambrose. *0080*

Uric Acid *Test Conditions Increase Analytical* Positive spot test with phosphotungstate. *3765*

Urobilinogen *Urine Increase Analytical* Reacts with Ehrlich's reagent. *1563*
  Test Conditions Increase Analytical Positive spot test with Ehrlich's reagent. *3765*

Vanillylmandelic Acid *Urine Increase Analytical* Linear relationship with concentration affect Mahler method. *1149* In screening method 5-HIAA reacts as if VMA. *0583*

Hydroxykynurenine

Ammoniacal Silver Nitrate
  Test Conditions Positive Analytical Positive spot test with Tollen's reagent. *3765*

Urobilinogen *Test Conditions Increase Analytical* Positive spot test with Ehrlich's reagent. *3765*

Hydroxymandelic Acid

Vanillylmandelic Acid *Urine Increase Analytical* Unless corrected for, affects method of Sunderman. *2928*

5-(p-Hydroxyphenyl)-5-Phenylhydantoin

Diphenylhydantoin *Serum Decrease Analytical* Reacts quantitatively in radioimmunoassay. *3591*

Hydroxyphenylacetic Acid

Ammoniacal Silver Nitrate
  Test Conditions Positive Analytical Positive spot test given by o-hydroderivative. *3765*

Ferric Chloride Test *Urine Positive Analytical* 2-hydroxy derivative: may produce mauve. *0443*

Millons Test *Test Conditions Positive Analytical* Positive spot test for phenols. *3765*

Hydroxyphenylacetic Acid (continued)

Urobilinogen *Test Conditions Increase Analytical* Positive spot test with Ehrlich's reagent. *3765*

2-Hydroxyphenylacetic Acid

Tyrosine *Plasma Increase Analytical* 8% fluorescence in procedure of Ambrose. *0080*

3-hydroxyphenylacetic Acid

Tyrosine *Plasma Increase Analytical* 10% fluorescence in procedure of Ambrose. *0080*

4-Hydroxyphenylacetic Acid

5-Hydroxyindoleacetic Acid (5-HIAA)
  Urine No Effect Analytical No effect 10 mg/dL method of Goldenberg. *1327*

Tyrosine *Plasma Increase Analytical* 170% fluorescence in procedure of Ambrose. *0080*

Hydroxyphenylpyruvic Acid

Ferric Chloride Test *Urine Positive Analytical* 4-hydroxy derivative: green fades in seconds. 2-hydroxy derivative: red-brown to green-blue to mauve. *0443*

Thyroxine (T4) *Serum Decrease Physiological* Decreased binding capacity (diminution of transport proteins). *3798*

4-Hydroxyphenylpyruvic Acid

Homogentisic Acid *Urine Increase Analytical* If method of Briggs used. *1096*

Tyrosine *Plasma Increase Analytical* 20% fluorescence in procedure of Ambrose. *0080*

Hydroxyprogesterone

Estrogens *Urine No Effect Analytical* At 50 mg/L on fluorescent method of Corns. *0730*

Progesterone *Plasma No Effect Analytical* 1% or less cross reactivity with RIA. *0567*
  Plasma Increase Analytical 46% cross reactivity with alpha compound, 16% with beta. *0567*

17-Hydroxyprogesterone

Corticosteroids *Plasma Increase Analytical* 84% (cortisol 100%) competitive protein binding method of Ficher. *1128*

Cortisol *Plasma Increase Analytical* If nonselective extract and competitive protein binding used. *0512*

Hydroxyproline

Ammonia *Plasma Decrease Analytical* With 500 nmoles produces 17% inhibitor indophenol reaction. *1290*

Proline *Plasma Increase Analytical* Slight effect on method of Goodwin. *1345*

Pyrrole-2-Carboxylate *Urine Increase Physiological* Action of alpha-amino acid oxidases (both d- and l-). *1536*

Hydroxyquinoline

Angiotensin-Converting Enzyme (ACE)
  Serum Decrease Analytical 100% inhibition with method using benzyloxycarbonyl-phenylalanyl-histidyl-leucine as substrate with 8 hydroxyquinolone. *3221*

5-Hydroxytryptamine

Ammoniacal Silver Nitrate
  Test Conditions Positive Analytical Positive spot test with Tollen's reagent. *3765*

Amphetamine *Urine No Effect Analytical* No effect at 100 mg/L on method of Rutter. *3097*

Amylase *Pancreatic Juice No Effect Physiological* No effect following injection. *3579*

Bicarbonate *Pancreatic Juice No Effect Physiological* No effect following injection. *3579*

Creatinine *Test Conditions Increase Analytical* Positive spot test with Jaffé reagent. *3765*

Effective Renal Plasma Flow
Patient Decrease Physiological 30% decrease for 45 minutes after 1 mg i.v. *1176*

Glomerular Filtration Rate (GFR)
Urine Decrease Physiological After 10 μg/kg i.v. infusion. *1176*

Growth Hormone *Plasma Increase Physiological* Probable effect. *1101*

Hydrochloric Acid *Gastric Material Decrease Physiological* But mucus secretion increased. *3579*

5-Hydroxyindoleacetic Acid (5-HIAA)
Urine No Effect Analytical No effect with 5 mg/dL on method of Goldenberg. *1327*

Insulin *Plasma Decrease Physiological* Probable action as inhibits secretion. *1101*

Protein *CSF Increase Analytical* 1.0 mg = 3.1 mg in Folin-Ciocalteu procedure. *0583*

Uric Acid *Test Conditions Increase Analytical* Positive spot test with phosphotungstate. *3765*

Urobilinogen *Test Conditions Increase Analytical* Positive spot test with Ehrlich's reagent. *3765*

Volume *Gastric Material Decrease Physiological* But mucus secretion increased. *3579*
Pancreatic Juice No Effect Physiological No effect following injection. *3579*

5-Hydroxytryptophan

Protein *CSF Increase Analytical* 1.0 mg = 3.9 mg in Folin-Ciocalteu procedure. *0583*

Tyrosine *Plasma Increase Analytical* 6% fluorescence in procedure of Ambrose. *0080*

Hydroxyurea

BSP Retention *Serum Increase Physiological* Abnormal retention reported. *1680*

Creatinine *Serum Increase Physiological* ?related to impaired tubular function. *1680*

Erythrocytes *Blood Decrease Physiological* Anemia (may be transient megaloblastic). *0071*

Erythrocyte Survival *Blood No Effect Physiological* Although reduces rate of iron utilization. *1680*

Hemoglobin *Blood Decrease Physiological* Decreases of from 0.5 to 6.0 g/dL common. *2581*

Leukocytes *Blood Decrease Physiological* Leukopenia may occur. *2427*

Neutrophils *Blood Decrease Physiological* May cause agranulocytosis. *2583*

Platelets *Blood Decrease Physiological* Thrombocytopenia. *0071*

Triglycerides *Serum Decrease Analytical* Inhibits action of glycerol oxidase included as part of Technicon® RA-1000 method but no effect on procedure used on Beckman Astra-8. *2384*

Urea Nitrogen *Serum Increase Physiological* Probably associated with tissue destruction. *2581* ?related to impaired tubular function. *1680*

Uric Acid *Serum Increase Physiological* Probable effect of cell catabolism. *1488*

Hydroxyzine

Corticosteroids *Urine Increase Analytical* Increased absorption at 410 nm in Porter-Silber reaction. *0022*

17-Hydroxycorticosteroids *Urine Increase Analytical* Affects modified Glenn-Nelson method. *0427* Interferes with Porter-Silber reaction. *2425*

17-Ketogenic Steroids *Urine Increase Analytical* Interferes with Zimmermann reaction. *2573*

17-Ketosteroids *Urine No Effect Analytical* No effect reported on Zimmermann reaction. *0427*

Protein Bound Iodine (PBI) *Serum No Effect Physiological* No effect with normal doses. *0830*

Prothrombin Time *Plasma No Effect Physiological* Reported to have no effect. *1487*
Plasma Increase Physiological Prolongs action of anticoagulants. *3879*

Hyoscine-N-Butylbromide

Cholesterol *Serum No Effect Analytical* At concentration of 300 mg/L had no effect on CHOD-Iodide method. *3393* At concentration of 300 mg/L had no effect on CHOD-PAP method. *3393* At concentration of 300 mg/L had no effect on catalase-AlDH method. *3393* At concentration of 300 mg/L had no effect on method using catalase-Hantzsch reaction method. *3393* At concentration of 300 mg/L had no effect on Liebermann-Burchard method. *3393*

Creatinine *Serum No Effect Analytical* At concentration of 12 mg/L had no effect on Jaffé-Fading-Fraction method. *3393* At concentration of 12 mg/L had no effect on Jaffé-Fuller's earth method. *3393*

Glucose *Serum No Effect Analytical* At concentration of 2 mg/L had no effect on Ektachem® method. *3393*

Potassium *Serum No Effect Analytical* At concentration of 20 mg/L had no effect on measurement by ISE without predilution. *3393*

Protein Electrophoresis *Serum No Effect Analytical* At concentration of 20 mg/L had no effect on automated Olympus-Hite method. *3393*

Sodium *Serum Decrease Analytical* At concentrations above 20 mg/L lowered concentration as measured by ISE without predilution. *3393*

Urea Nitrogen *Serum No Effect Analytical* At concentration of 2 mg/L had no effect on Ektachem® method. *3393*

Uric Acid *Serum No Effect Analytical* At concentration of 60 mg/L had no effect on Kageyama-Hantzsch method. *3393* At concentration of 12 mg/L had no effect on catalase-AlDH method. *3393*

Hyperventilation

Ionized Calcium *Serum Decrease Physiological* Effect of respiratory alkalosis. *3234*

pH *Blood Increase Physiological* Due to respiratory alkalosis. *3234*

Phosphate *Serum Decrease Physiological* Effect of respiratory alkalosis. *3234*

Hypochlorites

Glucose *Serum Increase Analytical* Causes direct oxidation of glucose oxidase chromogen. *0583*
Urine Increase Analytical Oxidation of chromogen (o-tolidine). *1022* False positive with glucose oxidase procedures. *0443*

Occult Blood *Feces Positive Physiological* Corrosive action on gastrointestinal tract if swallowed. *1302*

Protein Bound Iodine (PBI) *Serum Decrease Physiological* Interferes with trapping of iodide by thyroid. *0012*

Sugar *Urine Increase Analytical* May cause false positive with Clinistix®. *1563*

Hypothermia

Aspartate Aminotransferase
Serum Increase Physiological Due to hypoxia, acid-base imbalance, hypotension. *2258*

Base Deficit *Blood Increase Physiological* Due to hypoxia, acid-base imbalance, hypotension. *2258*

Hypothermia *(continued)*

Bicarbonate *Serum Decrease Physiological* Due to hypoxia, acid-base imbalance, hypotension. *2258*

Creatine Kinase *Serum Increase Physiological* Due to hypoxia, acid-base imbalance, hypotension. *2258*

Fatty Acids, Free (FFA) *Serum Increase Physiological* Due to hypoxia, acid-base imbalance, hypotension. *2258*

Hydroxybutyrate Dehydrogenase
Serum Increase Physiological Due to hypoxia, acid-base imbalance, hypotension. *2258*

pCO₂ *Blood Increase Physiological* Due to hypoxia, acid-base imbalance, hypotension. *2258*

pH *Blood Decrease Physiological* Due to hypoxia, acid-base imbalance, hypotension. *2258*

pO₂ *Blood Decrease Physiological* Due to hypoxia, acid-base imbalance, hypotension. *2258*

Thyroid Stimulating Hormone (TSH)
Serum No Effect Physiological No effect observed. *2700*

Hypoxanthine

Uric Acid *Serum No Effect Analytical* At 10 mg/dL on Nishi phosphotungstate procedure At 10 mg/dL on copper chelate procedure. *2230*
Serum Increase Analytical Measured as uric acid by reducing methods. *3500*

Hypoxia

Lactate Dehydrogenase Isoenzymes
Serum Increase Physiological Increase of fraction 5 associated with shock. *3485*

Succinate *Blood Increase Physiological* (arterial blood) reversal of path to oxaloacetate. *1717*

Ibopamine

Glucagon *Plasma No Effect Physiological* 300 mg drug had no effect over period of 90 minutes. *3398*

Glucose *Serum Increase Physiological* Peaked after 45 minutes in all subjects: normalized in 90 minutes. *3398*

Insulin *Plasma Increase Physiological* Peaked after 45 minutes in all subjects: normalized in 90 minutes. *3398*

Prolactin *Plasma No Effect Physiological* 300 mg drug had no effect over period of 90 minutes. *3398*

Ibufenac

Alanine Aminotransferase *Serum Increase Physiological* Hepatotoxic effect. *2313*

Alkaline Phosphatase *Serum Increase Physiological* Hepatotoxic effect. *2313*

Aspartate Aminotransferase
Serum Increase Physiological Hepatotoxic effect. *2313*

Bile *Urine Increase Physiological* Hepatotoxic effect. *2313*

Bilirubin *Serum Increase Physiological* Probable effect as bilirubin clearance reduced. *1499*

BSP Retention *Serum Increase Physiological* Hepatotoxic effect. *2313*

Cephalin Flocculation *Serum Increase Physiological* Hepatotoxic effect. *2313*

Glucose *Urine Increase Physiological* Augmentation of diabetic glycosuria seen. *2427*

Occult Blood *Feces Positive Physiological* May cause gastrointestinal tract bleeding. *2313*

Platelet Aggregation *Blood Decrease Physiological* Observed *in vitro*, may cause gastrointestinal bleeding etc. *0605*

Thymol Turbidity *Serum Increase Physiological* Hepatotoxic effect. *2313*

Uric Acid *Serum Increase Physiological* Side effect similar to that of aspirin. *2313*

Ibuprofen

Alanine Aminotransferase *Serum No Effect Analytical* At 5 times upper limit of therapeutic range on methods on SMAC®, Abbott-VP, aca, Cobas-Bio and KDA. *2138* On continuous method at 10 times maximal therapeutic concentration. *1785*
Serum Increase Physiological Rare hepatotoxicity reported: associated in one case with Stevens-Johnson syndrome. *3460*
Serum Decrease Analytical On colorimetric method at 10 times maximal therapeutic concentration. *1785*

Albumin *Serum No Effect Analytical* At 5 times upper limit of therapeutic range on methods on SMAC®, Ektachem®, Hitachi® 705 and KDA. *2138* On bromcresol green method on SMA II at physiological concentration. *1787* At concentration of 200 mg/L had no effect on BCG method. *3393*
Serum Decrease Physiological Significantly lower in patients receiving drug than in controls (4.0 g/dL vs 4.2 g/dL); same effect noted before and after treatment Started. *1787*

Alkaline Phosphatase *Serum No Effect Analytical* At 5 times upper limit of therapeutic range on methods on Abbott-VP, aca, Cobas-Bio, Hitachi® 705 and KDA. *2138* On continuous method at 10 times maximal therapeutic concentration. *1785*
Serum Increase Analytical At 5 times upper limit of therapeutic range on method on SMAC®. *2138*
Serum Increase Physiological Rare case of hepatotoxicity: in one case associated with Stevens-Johnson syndrome. *3460*

Amylase *Serum No Effect Analytical* At 5 times upper limit of therapeutic range on methods on aca, Ektachem® and Hitachi® 705. *2138*

Aspartate Aminotransferase *Serum No Effect Analytical* At 5 times upper limit of therapeutic range on methods on SMAC®, Abbott-VP, aca, Cobas-Bio, Hitachi® 705 and KDA. *2138* On colorimetric method at 10 times maximal therapeutic concentration. *1785*
Serum Increase Physiological Rare hepatotoxicity reported: associated in one case with Stevens-Johnson syndrome. *3460*
Serum Decrease Analytical On continuous method at 10 times maximal therapeutic concentration. *1785*

Bilirubin *Serum No Effect Analytical* No effect at therapeutic concentrations on Jendrassik-Grof,dimethylsulfoxide and spectrophotometric methods. *1786* At 5 times upper limit of therapeutic range on methods on SMAC®, aca, Cobas-Bio, Ektachem®, Hitachi® 705 and KDA. *2138* On diazo method on SMA II at physiological concentration. *1787* At concentration of 200 mg/L had no effect on Jendrassik and Grof method. *3393*
Serum No Effect Physiological At pharmacological concentrations has little or no protein displacement effect. *3748* No effect seen in patients undergoing treatment. *1787*
Serum Increase Physiological Rare hepatotoxicity reported: associated in one case with Stevens-Johnson syndrome. *3460*

Bilirubin, Direct *Serum Increase Physiological* Rare hepatotoxicity reported; associated in one case with Stevens-Johnson syndrome. *3460*

Calcium *Serum No Effect Analytical* At 5 times upper limit of therapeutic range on methods on SMAC®, Abbott-VP, aca, Ektachem®, Hitachi® 705 and KDA. *2138* On cresolphthalein complexone method on SMA II at physiological concentration. *1787* At concentration of 200 mg/L had no effect on cresolphthalein method. *3393*
Serum No Effect Physiological No effect seen in patients undergoing treatment. *1787*

Cholesterol *Serum No Effect Analytical* No effect at therapeutic concentrations on Liebermann-Burchard and enzymatic methods. *1786* On enzymatic method on SMA II at physiological concentration. *1787* At 5 times upper limit of therapeutic range on methods on SMAC®, Abbott-VP, Cobas-Bio, Ektachem®, Hitachi® 705 and KDA. *2138* At concentration of 200 mg/L had no effect on CHOD-PAP method. *3393*
Serum Increase Physiological Significantly higher in patients receiving drug than in controls (5.26 mmol/L vs 4.60 mmol/L). *1787*

Color *Urine Increase Analytical* Pink, red, purple and rust color. *1237*

Creatine Kinase *Serum No Effect Analytical* At 5 times upper limit of therapeutic range on methods on SMAC®, Abbott-VP, Cobas-Bio, aca, and Hitachi® 705. *2138* On continuous method at 10 times maximal therapeutic concentration. *1785*

Creatinine *Serum No Effect Analytical* No effect at therapeutic concentrations on alkaline picrate and Slot methods. *1786* On alkaline picrate method on SMA II at physiological concentration. *1787* At 5 times upper limit of therapeutic range on methods on SMAC®, Abbott-VP, aca, Cobas-Bio, Ektachem®, Hitachi® 705 and KDA. *2138* At concentration of 200 mg/L had no effect on AutoAnalyzer method. *3393*
Serum No Effect Physiological No effect during coadministration with digoxin. *2896*
Serum Increase Physiological In 2 patients with systemic lupus erythematosus. *1155* Increase by 40% in patients with chronic glomerular disease. *0660*
Serum Decrease Physiological Significant reduction in patients receiving medication. *1787*

Creatinine Clearance *Urine Decrease Physiological* Increase by 28% in patients with chronic glomerular disease but no effect in healthy people. *0660*

Digoxin *Serum Increase Physiological* Significant increase after 7 d treatment with average effect of 59%. *2896*

Erythrocytes *Blood Decrease Physiological* In 3% of patients with juvenile rheumatoid arthritis. *0239*
Urine Increase Physiological In 2 patients with systemic lupus erythematosus. *1155*

Glucose *Serum No Effect Analytical* No effect at therapeutic concentrations on hexokinase, glucose dehydrogenase ABTS, 2,4-dichlorophenol or o-toluidine methods. *1786* On hexokinase method on SMA II at physiological concentration. *1787* At 5 times upper limit of therapeutic range on methods on SMAC®, Abbott-VP, aca, Cobas-Bio, Ektachem®, Hitachi® 705 and KDA. *2138* At concentration of 200 mg/L had no effect on hexokinase/G-6-PDH method. *3393*
Serum No Effect Physiological No effect seen in patients undergoing treatment. *1787*

γ-Glutamyltransferase (GGT) *Serum No Effect Analytical* At 5 times upper limit of therapeutic range on methods on SMAC®, Abbott-VP, and Hitachi® 705. *2138*
Serum Increase Physiological Rare case of hepatotoxicity: in one case associated with Stevens-Johnson syndrome. *3460*

Granular Casts *Urine Increase Physiological* In 2 patients with systemic lupus erythematosus. *1155*

Hemoglobin *Urine Increase Physiological* In 3% of patients with juvenile rheumatoid arthritis. *0239*

Hydroxybutyrate Dehydrogenase
Serum No Effect Analytical On continuous method at 10 times maximal therapeutic concentration. *1785*

Iron *Serum No Effect Analytical* No effect at therapeutic concentrations on Ramsay and bathophenanthroline methods. *1786* At 5 times upper limit of therapeutic range on Ferrozine method on SMAC®. *2138* On Ferrozine method on SMA II at physiological concentration. *1787* At concentration of 200 mg/L had no effect on Ferrozine method. *3393*
Serum No Effect Physiological No effect seen in patients undergoing treatment. *1787*

Lactate Dehydrogenase *Serum No Effect Analytical* At 5 times upper limit of therapeutic range on methods on SMAC®, Abbott-VP, Cobas-Bio, Hitachi® 705 and KDA. *2138* On continuous method with lactate or pyruvate as substrate at 10 times maximal therapeutic concentration. On colorimetric method at 10 times maximal therapeutic concentration. *1785*
Serum Increase Physiological Rare case of hepatotoxicity: in one case associated with Stevens-Johnson syndrome. *3460*

Leukocytes *Blood Decrease Physiological* Rare adverse reactions reported by physicians to UK committee on safety of medicines. *0782* In one elderly man associated with a complement-dependent IgG antibody: reversible with cessation of treatment. *1054*
CSF Increase Physiological Few cases of aseptic meningitis described. *1276*

Lithium *Serum No Effect Physiological* Inconsistent change when drug added to therapeutic regime in 3 patients. *2903*
Serum Increase Physiological Mean concentration increased by 15% when ibuprofen co-administered with increased RBC to plasma ratio. *2019*
Urine Decrease Physiological Total body and renal clearance significantly reduced during co-administration. *2019*

Lymphocytes *Blood Decrease Physiological* Observed in one child with juvenile rheumatoid arthritis when dose of drug increased. At same time altered liver function occurred. Resolved with cessation of treatment. *3457*

Neutrophils *Blood Decrease Physiological* In one elderly man associated with a complement-dependent IgG antibody: reversible with cessation of treatment. *1054* Occasional case of agranulocytosis reported. *3717*

Occult Blood *Feces Positive Physiological* May produce gastric irritation and activate ulcer. *1630* Rare adverse reactions reported by physicians to UK committee on safety of medicines. *0782*

PAH Clearance *Urine Decrease Physiological* Increase by 35% in patients with chronic glomerular disease. *0660*

Phosphate *Serum No Effect Analytical* At 5 times upper limit of therapeutic range on methods on SMAC®, aca, Hitachi® 705 and KDA. *2138* At concentration of 200 mg/L had no effect on phosphomolybdate method. *3393*
Serum No Effect Physiological No effect seen in patients undergoing treatment. *1787*

Platelets *Blood Decrease Physiological* Rare adverse reactions reported by physicians to UK committee on safety of medicines. *0782*

Prostaglandin, 6-Keto-F$_1\alpha$ *Urine Decrease Physiological* Reduced by 80% in patients with chronic glomerular disease and in healthy people. *0660*

Prostaglandin E2 *Urine Decrease Physiological* Reduced by 80% in patients with chronic glomerular disease and in healthy people. *0660*

Prostaglandin G2 *Urine Decrease Physiological* In 11 healthy volunteers in randomized trial. *0452*

Protein *Serum No Effect Analytical* No effect at therapeutic concentrations on biuret and spectrophotometric methods. *1786* On biuret method on SMA II at physiological concentration. *1787* At 5 times upper limit of therapeutic range on methods on SMAC®, Abbott-VP, Ektachem®, Hitachi® 705 and KDA. *2138* At concentration of 200 mg/L had no effect on biuret method with blank correction. *3393*
Serum No Effect Physiological No effect seen in patients undergoing treatment. *1787*
Urine Increase Physiological In 2 patients with systemic lupus erythematosus. *1155*
CSF Increase Physiological Few cases of aseptic meningitis described. *1276*

Renin Activity *Plasma Decrease Physiological* 59% reduction in patients being treated. *0660*

Thromboxane B$_2$ *Serum Decrease Physiological* Significant reduction in patients during treatment. *0660*
Urine Decrease Physiological In 11 healthy volunteers in randomized trial. *0452*

Triglycerides *Serum No Effect Analytical* No effect at therapeutic concentrations on enzymatic methods. *1786* On enzymatic method on SMA II at physiological concentration. *1787* At 5 times upper limit of therapeutic range on methods on SMAC®, Abbott-VP, Ektachem®, Hitachi® 705 and KDA. *2138*
Serum No Effect Physiological No effect seen in patients undergoing treatment. *1787*

Urea Nitrogen *Serum No Effect Analytical* No effect at therapeutic concentrations on glutamate, dehydrogenase, phenol- hypochlorite, and diacetylmonoxime methods. *1786* On diacetyl monoxime method on SMA II at physiological concentration. *1787* At concentration of 200 mg/L had no effect on diacetylmonoxime method. *3393* No effect at therapeutic concentrations on glutamate dehydrogenase, phenol- *1786* On diacetylmonoxime method on SMA II at physiological concentration. *1787* At concentration of 200 mg/L had no effect on diacetylmonoxime method. *3393*
Serum Increase Physiological Significant effect in group of patients with rheumatoid arthritis, although most values remained within normal range. *0885* In 2 patients with systemic lupus erythematosus. *1155*

Uric Acid *Serum No Effect Analytical* No effect at therapeutic concentrations on uricase-catalase, aldehyde dehydrogenase direct UV-test and phosphotungstate methods. *1786* On phosphotungstate method on SMA II at physiological concentration. *1787* At 5 times upper limit of therapeutic range on methods on SMAC®, Abbott-VP, aca, Cobas-Bio, Ektachem®, Hitachi® 705 and KDA. *2138* At concentration of 200 mg/L had no effect on phosphotungstate reduction method. *3393*

Ibuprofen (continued)

Uric Acid (continued)

Serum Increase Physiological Significant effect in group of patients with rheumatoid arthritis, although most values remained within normal range. *0885*

Serum Decrease Physiological Significantly less in drug treated group than in controls. Same effect noted with initiation of therapy. *1787*

Ice-Eating

Hemoglobin *Blood Decrease Physiological* Low values observed (probably effect not cause). *0501*

Icterogenin

Alanine Aminotransferase *Serum Increase Physiological* Causes intrahepatic cholestasis. *0153*

Alkaline Phosphatase *Serum Increase Physiological* Causes intrahepatic cholestasis. *0153*

Aspartate Aminotransferase
Serum Increase Physiological Causes intrahepatic cholestasis. *0153*

Bile *Urine Increase Physiological* Causes intrahepatic cholestasis. *0153*

Bilirubin *Serum Increase Physiological* Causes intrahepatic cholestasis. *0153*

BSP Retention *Serum Increase Physiological* Causes intrahepatic cholestasis. *0153*

Cephalin Flocculation *Serum Increase Physiological* Causes intrahepatic cholestasis. *0153*

Thymol Turbidity *Serum Increase Physiological* Causes intrahepatic cholestasis. *0153*

Idoxuridine

Alkaline Phosphatase *Serum Increase Physiological* Cholestatic jaundice reported in one case. *0837*

Aspartate Aminotransferase
Serum Increase Physiological Cholestatic jaundice reported in one case. *0837*

Bilirubin *Serum Increase Physiological* Cholestatic jaundice reported in one case. *0837*

Leukocytes *Blood Decrease Physiological* Effect of high concentration only. *2313*

Ifosfamide

N-Acetylglucosaminidase *Urine No Effect Analytical* At 1 g/L had no effect on 2 colorimetric analytical methods. *1354* *Urine Increase Physiological* Manifestation of nephrotoxicity in 12 children given drug 1.6 g/m² for 5 d. *1355* Commonly observed: seen in all patients in one study in spite of concomitant. *1354*

Adenosine Deaminase Binding Protein
Urine Increase Physiological Manifestation of nephrotoxicity in 12 children given drug 1.6g/m² for 5 d. *1355*

Alanine Aminopeptidase *Urine Increase Physiological* Manifestation of nephrotoxicity in 12 children given drug 1.6 g/m² for 5 d. *1355*

Alanine Aminotransferase *Serum Increase Physiological* Manifestation of hepatotoxicity. *3675*

Alkaline Phosphatase *Serum Increase Physiological* Manifestation of hepatotoxicity. *3675*

Aspartate Aminotransferase
Serum Increase Physiological Manifestation of hepatotoxicity. *3675*

Casts *Urine Increase Physiological* Large number of granular casts in all patients. *3675*

Erythrocytes *Urine Increase Physiological* Occurs in one third patients within 2 d. *3675*

Hemoglobin *Blood Decrease Physiological* Occurs in 32% patients after 150 mg/kg. *3675* *Urine Increase Physiological* Occurs in one third patients within 2 d. *3675*

Leukocytes *Blood Decrease Physiological* Occurs in 80% patients after 150 mg/kg. *3675*

Platelets *Blood Decrease Physiological* Occurs in 13% patients after 150 mg/kg. *3675*

Urea Nitrogen *Serum Increase Physiological* Common toxic manifestation. *3675*

Imidazole

Ionized Calcium *Serum Increase Analytical* At low concentrations but decrease at high, Calcium specific electrode. *0540*

Imidazolepyruvic Acid

Ferric Chloride Test *Urine Positive Analytical* Green or blue-green. *0443*

I.M. Injections

Aspartate Aminotransferase
Serum Increase Physiological Muscle damage. *3218*

Creatine Kinase *Serum Increase Physiological* Muscle damage. *3218*

Imipenem

Bilirubin *Serum No Effect Physiological* Probably clinically insignificant displacement from protein in neonates. *3748*

Imipramine

Acetaminophen Screening Test *Urine Negative Analytical* No reaction with o-cresol at therapeutic concentrations. *3326*

Alanine Aminotransferase *Serum Increase Physiological* May cause cholestatic jaundice. *1873* Either due to hepatotoxicity or hypersensitivity in 33 year old woman. *2501*

Alkaline Phosphatase *Serum Increase Physiological* May cause cholestatic jaundice. *1873* Either due to hepatotoxicity or hypersensitivity in 33 year old woman. *2501*

Alkaloids *Urine No Effect Analytical* With ninhydrin on Frings TLC procedure at 10 mg/dL. *1204* *Urine Increase Analytical* Purple with iodoplatinate on Frings TLC procedure. *1204*

Amphetamine *Urine No Effect Analytical* At 50 mg/L on fluorescent method of Hayes. *1534*

Antidiuretic Hormone *Urine Increase Physiological* Mean excretion of 10.6 mU/h in 7 normal and 10 depressive patients when given 75 mg/d compared to control of 2.6 mU/h. *2889*

Aspartate Aminotransferase
Serum Increase Physiological May cause cholestatic jaundice. *1873* Either due to hepatotoxicity or hypersensitivity in 33 year old woman. *2501*

Barbiturate *Urine No Effect Analytical* With diphenylcarbazone on Frings TLC procedure at 10 mg/dL. With mercurous NO₃ on Frings TLC procedure at 10 mg/dL. *1204* *Urine Increase Analytical* Blue/green spot with vanillin in Frings TLC procedure. Blue spot with mercuric SO₄ in Frings TLC procedure. *1204*

Bile *Urine Increase Physiological* May cause cholestatic jaundice. *2313*

Bilirubin *Serum No Effect Physiological* Insignificant protein displacement effect in neonates. *3748* *Serum Increase Physiological* May cause cholestatic jaundice. *1873* *Urine Increase Physiological* May cause cholestatic jaundice. *1873*

Bromide *Urine No Effect Analytical* No effect on method of Frings. *1204*

BSP Retention *Serum Increase Physiological* May cause cholestatic jaundice. *2447*

Chlordiazepoxide *Blood No Effect Analytical* On method of Riddick at 5 mg/dL. *2983*

Cholesterol *Serum Increase Physiological* Possible cholestatic effect. *2946*

Creatinine *Serum Increase Physiological* Observed in one psychiatric case. *3156*

Dapsone *Serum No Effect Analytical* At 10 mg/dL on colorimetric procedure of Higgins. *1590*

Diazepam *Serum No Effect Analytical* On cathode-ray Polarographic method of Berry. *0323*

Digoxin *Serum No Effect Physiological* No effect on serum concentration reported. *0344*

Diphenylhydantoin *Serum No Effect Analytical* On GLC procedure of Papadopoulos at *in vivo* concentration. *2722*

Eosinophils *Blood Increase Physiological* Allergic response (may produce Löffler's syndrome). *3280*

Ethchlorvynol *Serum No Effect Analytical* At 100 mg/L on GLC procedure of Evenson. *1058*
Urine No Effect Analytical No effect on method of Frings and Cohen at 10 mg/dL. *1205*

Glucose *Serum Increase Physiological* Reported effect. *0019*
Serum Decrease Physiological Reported effect. *0019*

Glucose Tolerance *Serum Decrease Physiological* Preliminary observations only reported. *0158*

5-Hydroxyindoleacetic Acid (5-HIAA)
Urine Decrease Analytical May inhibit color development in reaction. *0913*
Urine Decrease Physiological May decrease up to 50%: decrease cell permeability to 5-HT. *1520*

Imipramine *Blood Increase Physiological* After 150-300 mg/d concentration = 0.1-0.6 mg/L. *2348*

131I Uptake *Serum Decrease Physiological* Tofranil® contains tetraiodofluorescein. *2652*

Lactate Dehydrogenase *Serum Increase Physiological* May cause cholestatic jaundice. *1873*

Leukocytes *Blood Increase Physiological* (transient 1-6 h) (possibly Löffler's syndrome). *3280*
Blood Decrease Physiological May cause leukopenia/agranulocytosis. *3280*

Metanephrines, Total *Urine No Effect Analytical* At 15 mg/L on modified Pisano procedure. *1428*
Urine Increase Analytical Interference with Pisano method. *0388*

Methaqualone *Serum No Effect Analytical* At 100 mg/L on GLC procedure of Evenson. *1057*

Neutrophils *Blood Decrease Physiological* Occasional case of agranulocytosis reported. *3717*

5'-Nucleotidase *Serum Increase Physiological* Due to cholestasis. *1778*

Pheneturide *Serum No Effect Analytical* On GLC procedure of Papadopoulos at *in vivo* concentration. *2722*

Phenobarbital *Serum No Effect Analytical* On GLC procedure of Papadopoulos at *in vivo* concentration. *2722*

Platelets *Blood Decrease Physiological* May be marrow depression and purpura. *1680*

Potassium *Serum No Effect Analytical* At concentration of 30 mg/L had no effect on measurement by ISE with predilution. *3393*

Primidone *Serum No Effect Analytical* On GLC procedure of Papadopoulos at *in vivo* concentration. *2722*

Prolactin *Plasma No Effect Physiological* Like other tricyclic antidepressants no effect. *0723*
Plasma Increase Physiological Marked increase in male and female psychiatric patients treated for up to 4 weeks. *3630*

Protein *CSF Increase Analytical* 1.0 mg = 3.5 mg in Folin-Ciocalteu procedure. *0583*

Protein Bound Iodine (PBI) *Serum No Effect Physiological* No effect reported with normal doses. *0830*

Sodium *Serum No Effect Analytical* At concentration of 30 mg/L had no effect on measurement by ISE with predilution. *3393*

Urea Nitrogen *Serum Increase Physiological* Observed in one psychiatric case. *3156*
Test Conditions Increase Analytical Produces turbidity with Berthelot's reagent. *1936*

Urobilinogen *Urine Increase Physiological* May cause cholestatic jaundice. *1873*

Vanillylmandelic Acid *Urine Decrease Physiological* Approximately 30% decrease, block of uptake into cells. *1520*

Immune Serum Globulin

Platelets *Blood Decrease Physiological* Fall observed several days after injection. *2427*

Urea Nitrogen *Serum Increase Physiological* Nephritis may occur with serum sickness. *2427*

Indandione Derivatives

Alanine Aminotransferase *Serum Increase Physiological* Hepatocellular damage with cholestasis. *2313*

Aspartate Aminotransferase
Serum Increase Physiological Hepatocellular damage with cholestasis. *2313*

Bile *Urine Increase Physiological* May cause hepatotoxicity. *2313*

Bilirubin *Serum Increase Physiological* As result of hypersensitivity. *2427*

BSP Retention *Serum Increase Physiological* May cause hepatotoxicity. *2313*

Cephalin Flocculation *Serum Increase Physiological* May cause hepatotoxicity. *2313*

Color *Urine Increase Physiological* May be orange to red in color. *2425*

Eosinophils *Blood Increase Physiological* With other signs of hypersensitivity. *2427*

Erythrocytes *Urine Increase Physiological* May cause hematuria — manifestation of overdose. *1237*

Hemoglobin *Urine Increase Physiological* May cause hematuria — manifestation of overdose. *1237*

Leukocytes *Blood Decrease Physiological* With other evidence of hypersensitivity. *2313*

Thymol Turbidity *Serum Increase Physiological* May cause hepatotoxicity. *2313*

Indapamide

Apolipoprotein AI *Serum No Effect Physiological* In 13 hypertensive patients with diabetes treated for 24 weeks. *2686*

Apolipoprotein B *Serum No Effect Physiological* In 18 patients treated for 6 weeks. *0089*

Apolipoprotein B100 *Serum Decrease Physiological* In 13 hypertensives patients with diabetes treated for 24 weeks. *2686*

Cholesterol *Serum No Effect Physiological* In 27 subjects treated for less than 1 y with this drug only. *0692* In various studies involving more than 30 patients when used as sole drug for less than 1 y. *0089*
Serum Increase Physiological Slight and insignificant in 13 hypertensive diabetics treated for 24 weeks. *2686* Increase by 17 mg/dL in 17 patients with essential hypertension treated with 2.5 mg daily for 3 mo. *2010*

Cholesterol, High Density Lipoprotein
Serum No Effect Physiological In 43 subjects treated with this drug for less than 1 y. *3780* In various studies involving more than 30 patients when used as sole drug for less than 1 y. *0089*
Serum Decrease Physiological Slight effect in 13 hypertensive diabetics treated for 24 weeks. *2686*

Cholesterol, Low Density Lipoprotein
Serum No Effect Physiological In 27 subjects treated for less than 1 y with this drug only. *0692* In various studies involving more than 30 patients when used as sole drug for less than 1 y. *0089* *Serum Increase Physiological* Slight and insignificant in 13 hypertensive diabetics treated for 24 weeks. *2686*

Cholesterol, Very Low Density Lipoprotein
Serum No Effect Physiological In 27 subjects treated for less than 1 y with this drug only. *0692* In various studies involving more than 30 patients when used as sole drug for less than 1 y. *0089*

C-Peptide *Plasma Increase Physiological* Both mean fasting and stimulated in 13 hypertensive diabetics. *2686*

Indapamide (continued)

Creatinine *Serum No Effect Analytical* At 10 mg/L on reversed phase LC procedure of Zhiri et al. *3942*

Glucose *Serum Increase Physiological* And after 75 g load in 13 hypertensive diabetics. *2686* Increase by 7 mg/dL in 17 patients with essential hypertension treated with 2.5 mg daily for 3 mo. *2010*

Hemoglobin A$_{1c}$ *Blood Increase Physiological* After 24 weeks in 13 hypertensive diabetics. *2686*

Potassium *Serum Decrease Physiological* In 3 of 13 hypertensive patients with diabetes treated for 24 weeks. *2686* Increase by 0.4 mmol/L in 17 patients with essential hypertension treated with 2.5 mg daily for 3 mo. *2010* Mean fall from 4.2 to 3.4 mmol/L in 8 hypertensives treated for 16 weeks. *0352*

Renin Activity *Plasma Increase Physiological* But not significant in 8 hypertensives treated for 16 weeks. *0352*

Sodium *Serum No Effect Physiological* In 8 hypertensives treated for 16 weeks. *0352*

Triglycerides *Serum No Effect Physiological* In 27 subjects treated for less than 1 y with this drug only. *0692* In various studies involving more than 30 patients when used as sole drug for less than 1 y. *0089*
Serum Increase Physiological Slight and insignificant in 13 hypertensive diabetics treated for 24 weeks. *2686*

Urea Nitrogen *Serum No Effect Physiological* In 8 hypertensives treated for 16 weeks. *0352*

Uric Acid *Serum No Effect Analytical* At 10 mg/L on reversed phase LC procedure of Zhiri et al. *3942*
Serum Increase Physiological Observed in normal men and those with mild essential hypertension. *3780* Increase by 1.4 mg/dL in 17 patients with essential hypertension treated with 2.5 mg daily for 3 mo. *2010*

Indican

Bilirubin *Serum No Effect Analytical* No effect on Jendrassik-Grof procedures at concentrations up to 0.9 mmol/L. *2844*
Serum Increase Analytical May produce brown diazo color in uremic serum. *1563* Bilirubin A-Gent Abbott 2,4-dichlorophenyl diazonium procedure increases bilirubin by 50 mg/L for each 1 mmol/L of added indican. Bilirubin C-system (Boehringer-Mannheim) 2,5-dichlorophenyl diazonium procedure increased bilirubin by 33 mg/L per mmol/L of added indican. Slight effect on Harleco micro bilirubin reagent set (Malloy-Evelyn procedure). *2844*

Color *Urine Increase Analytical* Occurs if increased and oxidized on standing. *0459*

Hydroxyproline *Urine No Effect Analytical* At 500 mg/L on method of Seymour. *3262*

Obermayer Test *Urine Positive Analytical* Produces blue color (indigo blue)in chloroform. *3588*

Sugar *Urine Increase Analytical* False positive with Benedict's. *2448*

Urobilinogen *Urine No Effect Analytical* No effect p-MDFB procedure of Rutter. *2044*
Urine Increase Analytical Produces yellow color with Ehrlich's reagent. *1563*

Indigo Blue

Color *Urine Increase Analytical* May produce blue color. *1187*

Indigotin

Crystals *Urine Increase Physiological* Amorphous blue or small crystals (all pH). *0443*

Indigotindisulfonate

Color *Urine Increase Analytical* Color used to measure kidney function. *0071*
Urine Increase Physiological Blue-green, decolorized with alkali. *0443*

Indocyanine Green

^{131}I Uptake *Serum Decrease Physiological* Contains iodine, inhibits further uptake. *2220*

Protein Bound Iodine (PBI) *Serum No Effect Physiological* Not usually affected. *1487*
Serum Increase Analytical Organic iodine contamination. *1237*

Indole

Creatinine *Urine Increase Analytical* 1 mg equivalent to 0.1 mg with automated Jaffé. *2558*

Protein *Test Conditions Increase Analytical* Reacts with Folin-Ciocalteu of Lowry method. *1569*

Sulfa as Sulfanilamide *Serum Increase Analytical* Reacts with diazotization. *0251*

Urobilinogen *Urine No Effect Analytical* No effect on p-MDFB procedure of Rutter. *2044*
Urine Increase Analytical Produces red color with Ehrlich's reagent. *1563*

Indoleacetic Acid

Globulin *Serum Increase Analytical* Reacts as tryptophan in method of Goldenberg. *1329*

5-Hydroxyindoleacetic Acid (5-HIAA)
Urine No Effect Analytical No effect *in vivo* dose on method of Goldenberg. *1327*

Tryptophan *Test Conditions Increase Analytical* Measured as same as Spies/Chambers method. *1415*

Urobilinogen *Urine Increase Analytical* Produces red color with Ehrlich's reagent. *1563*

Indolepropionic Acid

Tryptophan *Test Conditions Increase Analytical* Measured as same as Spies/Chambers method. *1415*

Indolepyruvic Acid

Nitrogen Balance *Patient Positive Physiological* Able to replace tryptophan in diet. *2976*

Indomethacin

Alanine Aminotransferase *Serum No Effect Analytical* On continuous method at 10 times maximal therapeutic concentration. On colorimetric method at 10 times maximal therapeutic concentration. *1785* No effect at therapeutic concentration on Reflotron method. *1984*
Serum Increase Physiological Cytotoxic and cholestatic liver damage. *1908*

Albumin *Serum No Effect Analytical* On bromcresol green method on SMA II at physiological concentration. *1787* At concentration of 13 mg/L had no effect on BCG method. *3393*
Serum No Effect Physiological No effect seen in patients undergoing treatment. *1787*

Aldosterone *Plasma No Effect Physiological* Insignificant change with 1 week treatment. *2464*
Plasma Decrease Physiological Marked reversal on withdrawal of drug in one young woman. *3541*
Urine Decrease Physiological Marked reversal on withdrawal of drug in one young woman. *3541* With 1 week treatment fell from 43 to 18 μg/24 h. *2464*

Alkaline Phosphatase *Serum No Effect Analytical* On continuous method. *1785*
Serum Increase Physiological Cytotoxic and cholestatic liver damage. *1908*

Amylase *Serum Increase Physiological* Single case reported (?correct implication). *1420*

Angiotensin-Converting Enzyme (ACE)
Serum No Effect Physiological In 9 uncomplicated essential hypertensives receiving chronic enalapril treatment. *3126*

Angiotensin II *Plasma No Effect Physiological* In 12 patients with coronary artery disease, although systolic blood pressure increased and coronary blood flow decreased. *1158*

Aspartate Aminotransferase *Serum No Effect Analytical* On continuous method at 10 times maximal therapeutic concentration. On colorimetric method at 10 times maximal therapeutic concentration. *1785* No effect at therapeutic concentration on Reflotron method. *1984*
Serum Increase Physiological Cytotoxic and cholestatic liver damage. *1908*

Bicarbonate *Serum No Effect Analytical* At concentration of 13 mg/L had no effect on method using phenolphthalein. *3393*

Bile *Urine Increase Physiological* Cytotoxic and cholestatic liver damage. *0085*

Bilirubin *Serum No Effect Analytical* No effect at therapeutic concentrations on Jendrassik-Grof, dimethylsulfoxide and spectrophotometric methods. *1786* On diazo method on SMA II at physiological concentration. *1787* At concentration of 13 mg/L had no effect on Jendrassik and Grof method. *3393*
Serum No Effect Physiological No effect seen in patients undergoing treatment. *1787* Although tightly bound to protein at pharmacological concentrations has little or no displacement effect. *3748*
Serum Increase Physiological Cytotoxic and cholestatic types of liver damage. *1908*

Bilirubin, Direct *Serum Increase Physiological* Cytotoxic and cholestatic liver damage. *1908*

BSP Retention *Serum Increase Physiological* Cytotoxic and cholestatic liver damage. *1908*

Calcium *Serum No Effect Analytical* On cresolphthalein complexone method on SMA II at physiological concentration. *1787* At concentration of 13 mg/L had no effect on cresolphthalein method. *3393*
Serum No Effect Physiological No effect seen in patients undergoing treatment. *1787*

Casts *Urine Increase Physiological* Granular casts in one patient. *0506*

Catecholamines *Plasma No Effect Physiological* In 12 patients with coronary artery disease, although systolic blood pressure increased and coronary blood flow decreased. *1158*

Cephalin Flocculation *Serum Increase Physiological* Cytotoxic and cholestatic liver damage. *1908*

Chloride *Serum No Effect Analytical* At concentration of 13 mg/L had no effect on mercurimetric method. *3393*

Cholesterol *Serum No Effect Analytical* No effect at therapeutic concentrations on enzymatic and Liebermann-Burchard methods. *1786* On enzymatic method on SMA II at physiological concentration. *1787* No effect at therapeutic concentration on Reflotron method. *1984* At concentration of 30 mg/L had no effect on CHOD-Iodide method. *3393* At concentration of 30 mg/L had no effect on CHOD-PAP method. *3393* At concentration of 30 mg/L had no effect on catalase-AlDH method. *3393* At concentration of 30 mg/L had no effect on method using catalase-Hantzsch reaction. *3393* At concentration of 30 mg/L had no effect on Liebermann-Burchard method. *3393*
Serum No Effect Physiological No effect seen in patients undergoing treatment. *1787*

Color *Urine Increase Physiological* Indirect result of hepatic toxicity, green urine. *1109*

Creatine Kinase *Serum No Effect Analytical* On continuous method. *1785*

Creatinine *Serum No Effect Analytical* No effect at therapeutic concentrations on alkaline picrate and Slot methods. *1786* On alkaline picrate method on SMA II at physiological concentration. *1787* At concentration of 6 mg/L had no effect on creatinine iminohydrolase method. *3393* At concentration of 13 mg/L had no effect on AutoAnalyzer Jaffé method. *3393* At concentration of 30 mg/L had no effect on Jaffé-Fading-Fraction method. *3393* At concentration of 30 mg/L had no effect on Jaffé-Fuller's earth method. *3393* At concentration of 40 mg/L had no effect on kinetic Jaffé method on BKA-2. *3393*
Serum No Effect Physiological No effect seen in patients undergoing treatment. *1787*
Serum Increase Physiological Oliguric renal failure developed in one man during treatment: reversible. *1245* Associated with increasing risk of renal insufficiency in cirrhosis, nephrotic syndrome, decompensated congestive heart failure and chronic renal disease. *0373* Associated with decreased secretion of renin and aldosterone, decreased sodium delivery to distal tubule and reduction of urinary flow rate. *1228*

Creatinine Clearance *Urine Decrease Physiological* Reduced by about 50% in states of diminished circulatory blood volume. *3261*

Eosinophils *Blood Increase Physiological* Oliguric renal failure developed in one man during treatment: reversible. *1245*
Blood Decrease Physiological Occasional effect. *0392*

Erythrocytes *Blood Decrease Physiological* Secondary to gastrointestinal bleed/agranulocytosis/pancytopenia. *0381* Occasional case of aplastic anemia reported. *3717*
Urine Increase Physiological May cause actual bleeding. *1022* Oliguric renal failure developed in one man during treatment: reversible. *1245*

Erythrocyte Sedimentation Rate
Blood Increase Physiological Oliguric renal failure developed in one man during treatment: reversible. *1245*

Fibrin Split Products (FSP) *Urine Decrease Physiological* In 2/3 patients with proliferative glomerulonephritis. *0671*

Glomerular Filtration Rate (GFR)
Urine Decrease Physiological Transient reduction in individuals during sustained diuresis (fell from 114 mL/minutes to 100 mL/min). *2464*

Glucose *Serum No Effect Analytical* No effect at therapeutic concentrations on hexokinase, glucose dehydrogenase 2,4-dichlorophenol, ABTS and o-toluidine methods. *1786* On hexokinase method on SMA II at physiological concentration. *1787* No effect at therapeutic concentration on Reflotron method. *1984* At concentration of 4 mg/L had no effect on Ektachem® method. *3393* At concentration of 25 mg/L had no effect on GOD/POD-PAP method. *3393* At concentration of 40 mg/L had no effect on hexokinase/G-6-PDH method. *3393*
Serum No Effect Physiological No effect seen in patients undergoing treatment. *1787*
Serum Increase Physiological Rare side effect. *1488* In one patient with psoriatic arthritis, reverted to normal with drug withdrawal. *3600*
Urine Increase Physiological Rare side effect. As result of rare hyperglycemia. *1488*

γ-Glutamyltransferase (GGT) *Serum No Effect Analytical* No effect at therapeutic concentration on Reflotron method. *1984*

Haptoglobin *Serum Decrease Physiological* May cause hemolytic anemia. *0071*

Hematocrit *Blood Decrease Physiological* Secondary to gastrointestinal bleed/agranulocytosis/pancytopenia. *0381*

Hemoglobin *Blood Decrease Physiological* Secondary to gastrointestinal bleed/agranulocytosis/pancytopenia. *0381*
Urine Increase Physiological Produces actual bleeding. *1022*

Hydroxybutyrate Dehydrogenase
Serum No Effect Analytical On continuous method at 10 times maximal therapeutic concentration. *1785*

Iron *Serum No Effect Analytical* No effect at therapeutic concentrations on bathophenanthroline and Ramsay methods. *1786* On Ferrozine method on SMA II at physiological concentration. *1787*
Serum No Effect Physiological No effect seen in patients undergoing treatment. *1787*

Lactate *Serum No Effect Analytical* At concentration of 40 mg/L had no effect on enzymatic method. *3393*

Lactate Dehydrogenase *Serum No Effect Analytical* With lactate or pyruvate as substrate on continuous method. *1785* On colorimetric method at 10 times maximal therapeutic concentration. *1785*

Leukocytes *Blood Decrease Physiological* Agranulocytosis/pancytopenia. *3012* Rare side effect reported to UK committee on safety of medicines. *0782* Occasional case of aplastic anemia reported. *3717*
Urine Increase Physiological Oliguric renal failure developed in one man during treatment: reversible. *1245*

Lipase *Serum Increase Physiological* Associated with cholestatic liver damage. *1237*

Lithium *Serum Increase Physiological* Concentration increased from 0.9 to 1.4 meq/L within 6 d of indomethacin administration in 3 patients. *2903*

NBT Test *Blood Increase Physiological* Mechanism not discussed. *0943*

Indomethacin *(continued)*

Neutrophils *Blood Decrease Physiological* Rare neutropenia/may also cause aplastic anemia. *1343* Occasional case of drug-induced neutropenia. *0155* Occasional case of agranulocytosis reported. *3717*

5'-Nucleotidase *Serum Increase Physiological* Due to cholestasis. *1778*

Occult Blood *Feces Positive Physiological* May cause ulceration of stomach, duodenum, gut. *2412* Rare side effect reported to UK committee on safety of medicines. *0782*

Phosphate *Serum No Effect Analytical* On nonprecipitation method of Peynet. *2799* At concentration of 13 mg/L had no effect on phosphomolybdate method. *3393*

Platelet Aggregation *Blood Decrease Physiological* Observed *in vitro*, might cause gastrointestinal bleeding etc. *0605*

Platelets *Blood Decrease Physiological* Rare agranulocytosis/pancytopenia/aplastic anemia. *3012* Rare side effect reported to UK committee on safety of medicines. *0782* Occasional case of aplastic anemia reported. *3717*

Potassium *Serum No Effect Analytical* At concentration of 13 mg/L had no effect on measurement by ISE with predilution. *3393* At concentration of 40 mg/L had no effect on measurement by ISE without predilution. *3393*
Serum Increase Physiological Marked reversal on withdrawal of drug in one young woman. *3541* Associated with increasing risk of renal insufficiency in cirrhosis, nephrotic syndrome, decompensated congestive heart failure and chronic renal disease. *0373* Rise from 4.3 to 4.6 mmol/L with treatment for 1 week. *2464* Associated with decreased secretion of renin and aldosterone, decreased sodium delivery to distal tubule and reduction of urinary flow rate. *1228* Observed in 3 patients with gouty arthritis who developed renal insufficiency. *1133*

Prolactin *Plasma No Effect Physiological* In up to 35 healthy men given 100 mg/d for 1 week in response to TRH stimulation. *2904*

Prostaglandin, 6-Keto-F$_1\alpha$ *Urine Decrease Physiological* In 9 uncomplicated essential hypertensives receiving chronic enalapril treatment. *3126*

Prostaglandin E *Plasma Decrease Physiological* Decrease by 67% in rats at least. *1760* In up to 35 healthy men given 100 mg/d for 1 week. *2904*

Prostaglandin E2 *Urine Decrease Physiological* Marked reversal on withdrawal of drug in one young woman. *3541*

Prostaglandin F *Plasma Decrease Physiological* In up to 35 healthy men given 100 mg/d for 1 week. *2904*

Protein *Serum No Effect Analytical* No effect at therapeutic concentrations on biuret and spectrophotometric methods. *1786* On biuret method on SMA II at physiological concentration. *1787* At concentration of 13 mg/L had no effect on biuret method with blank correction. *3393*
Serum No Effect Physiological No effect seen in patients undergoing treatment. *1787*
Urine Increase Physiological Reported in one patient. *0506* Oliguric renal failure developed in one man during treatment: reversible. *1245*

Protein Electrophoresis *Serum No Effect Analytical* At concentration of 40 mg/L had no effect on automated Olympus-Hite method except for slight displacement of fractions. *3393*

Prothrombin Time *Plasma Increase Physiological* Displaces anticoagulants from binding protein. *2540*

Renal Blood Flow (RBF) *Patient Decrease Physiological* Associated with decreased secretion of renin and aldosterone, decreased sodium delivery to distal tubule and reduction of urinary flow rate. *1228*

Renin Activity *Plasma No Effect Physiological* In 12 patients with coronary artery disease, although systolic blood pressure increased and coronary blood flow decreased. *1158* Insignificant change with 1 week treatment. *2464*
Plasma Decrease Physiological In 9 uncomplicated essential hypertensives receiving chronic enalapril treatment. *3126* Marked reversal on withdrawal of drug in one young woman. *3541*

Sodium *Serum No Effect Analytical* At concentration of 13 mg/L had no effect on measurement by ISE with predilution. *3393* At concentration of 40 mg/L had no effect on measurement by ISE without predilution. *3393*

Serum Decrease Physiological Marked reduction in some neonates (to below 130 mmol/L) within 48 h of administration to close patent ductus arteriosus. *1476*
Urine Decrease Physiological Caused by increased tubular absorption with marked oliguria as a result. *3261*

Thromboxane B$_2$ *Serum Decrease Physiological* In 9 uncomplicated essential hypertensives receiving chronic enalapril treatment. *3126*
Plasma Decrease Physiological Marked fall from 191 to 1.4 ng/mL in 12 patients with coronary artery disease, although systolic blood pressure increased and coronary blood flow decreased. *1158*

Thymol Turbidity *Serum Increase Physiological* Cytotoxic and cholestatic liver damage. *1908*

Thyroid Stimulating Hormone (TSH)
Serum No Effect Physiological No effect on TSH secretion. *3798*

Thyroxine (T$_4$) *Serum No Effect Physiological* In up to 35 healthy men given 100 mg/d for 1 week. *2904*

Triglycerides *Serum No Effect Analytical* No effect on enzymatic methods at therapeutic concentrations. *1786* On enzymatic method on SMA II at physiological concentration. *1787* No effect at therapeutic concentration on Reflotron method. *1984* At concentration of 13 mg/L had no effect on lipase/esterase method. *3393*
Serum No Effect Physiological No effect seen in patients undergoing treatment. *1787*

Tri-iodothyronine (T$_3$) *Serum No Effect Physiological* In up to 35 healthy men given 100 mg/d for 1 week. *2904*

Tryptophan *Plasma Decrease Physiological* (total measured) in rheumatoids on therapy. *0188*

Tryptophan, Free *Serum Increase Physiological* In rheumatoids on therapy. *0188*

TSH Response to TRH *Serum No Effect Physiological* In up to 35 healthy men given 100 mg/d for 1 week. *2904*
Serum Increase Physiological In up to 35 healthy men given 100 mg/d for 1 week. *2904*

Urea Nitrogen *Serum No Effect Analytical* No effect at therapeutic concentrations on glutamate dehydrogenase, phenol-hypochlorite and diacetylmonoxime methods. *1786* On diacetyl monoxime method on SMA II at physiological concentration. *1787* No effect at therapeutic concentration on Reflotron method. *1984* At concentration of 13 mg/L had no effect on diacetylmonoxime method. *3393* At concentration of 4 mg/L had no effect on Ektachem® method. *3393*
Serum No Effect Physiological No effect seen in patients undergoing treatment. *1787*
Serum Increase Physiological Occasional increase usually to upper normal limit. *2432* Oliguric renal failure developed in one man during treatment: reversible. *1245* Associated with increasing risk of renal insufficiency in cirrhosis, nephrotic syndrome, decompensated congestive heart failure and chronic renal disease. *0373* Associated with decreased secretion of renin and aldosterone, decreased sodium delivery to distal tubule and reduction of urinary flow rate. *1228*

Uric Acid *Serum No Effect Analytical* No effect at therapeutic concentrations on uricase-catalase, aldehyde dehydrogenase, direct UV-test, phosphotungstate methods. *1786* On phosphotungstate method on SMA II at physiological concentration. *1787* No effect at therapeutic concentration on Reflotron method. *1984* At concentration of 30 mg/L had no effect on Kageyama-Hantzsch method. *3393* At concentration of 30 mg/L had no effect on catalase-AIDH method. *3393* At concentration of 13 mg/L had no effect on phosphotungstate reduction method. *3393* At concentration of 25 mg/L had no effect on uricase-PAP method. *3393*
Serum No Effect Physiological No effect seen in patients undergoing treatment. *1787*
Serum Increase Physiological Oliguric renal failure developed in one man during treatment: reversible. *1245*
Serum Decrease Physiological Of some value in treatment of gouty arthritis. *1332*

Volume *Urine Decrease Physiological* Caused by increased tubular absorption with marked oliguria as a result. *3261*

Xylose Excretion *Urine Decrease Physiological* ?due to increased motility of gut. *1913*

Inositol

Ketones *Urine Increase Analytical* Possible reported effect. *1841*

Phosphate *Urine No Effect Analytical* No effect at 20 mg/dL method of Jung/Parekh. *1835*

Insecticides

Bilirubin *Serum Increase Physiological* May cause hemolytic anemia. *2365*

Coombs' Test, Direct *Serum Positive Physiological* May cause hemolytic anemia. *2365*

Protein *Urine Increase Physiological* May cause nephrotoxicity. *3204*

Pseudocholinesterase *Serum Decrease Physiological* Direct effect of drug. *3523*

Urea Nitrogen *Serum Increase Physiological* May cause nephrotoxicity. *3204*

Insulin

Albumin *Serum No Effect Analytical* At concentration of 3 mg/L had no effect on BCG method. *3393*

Amino Acids *Serum Decrease Physiological* Metabolic effect. *0355*
Urine Increase Physiological Metabolic effects. *1022*
Urine Decrease Physiological Metabolic effects. *1237*
Test Conditions Increase Analytical Positive spot test with ninhydrin. *3765*

α-Amino-Nitrogen *Plasma Decrease Physiological* Increased uptake by tissues. *1237*

Bicarbonate *Serum No Effect Analytical* At concentration of 3 mg/L had no effect on method using phenolphthalein. *3393*

Bilirubin *Serum No Effect Analytical* At concentration of 3 mg/L had no effect on Jendrassik and Grof method. *3393*

Calcium *Serum No Effect Analytical* At concentration of 3 mg/L had no effect on cresolphthalein method. *3393*
Serum Decrease Physiological Reported effect. *0355*

Chloride *Serum No Effect Analytical* At concentration of 3 mg/L had no effect on mercurimetric method. *3393*

Cholesterol *Serum No Effect Analytical* At concentration of 3 mg/L had no effect on Liebermann-Burchard method. *3393*
Serum Decrease Physiological Therapeutic goal. *3505*

Corticosteroids *Plasma Increase Physiological* Large effect at 60 minutes after i.v. injection. *3435*

Corticotropin *Plasma Increase Physiological* Significant effect after i.v. in normals 45-90 minutes. *3435* Response to stress. *2442*

Cortisol *Plasma Increase Physiological* Marked effect in insulin induced hypoglycemia. *3570*

Creatine Kinase *Serum Increase Physiological* Is an activator of enzyme. *0688*

Creatinine *Serum No Effect Analytical* At concentration of 3 mg/L had no effect on AutoAnalyzer Jaffé method. *3393*

Epinephrine *Plasma Increase Physiological* Stimulation of adrenal medulla, ?by hypoglycemia. *2442* Marked effect in insulin induced hypoglycemia. *3570*
Urine Increase Physiological Hypoglycemia produces up to tenfold increase. *0279*

Estriol *Urine No Effect Analytical* No effect on GLC method. *1163*

Fatty Acids, Free (FFA) *Serum Decrease Physiological* Effect similar in normals and diabetics. *2837*

Gastrin *Serum Increase Physiological* Similar response (fairly marked). *3904*

Glucagon *Plasma Increase Physiological* Marked increase at 45 minutes after injection. *2676*

Glucose *Serum Decrease Physiological* Natural action of hormone. *2313*

Growth Hormone *Plasma Increase Physiological* Occurs only when glucose down to 10 mg/dL. *3859* Increase of 22 ng/mL after 0.1 U/kg. *2176* Marked effect in insulin induced hypoglycemia. *3570* In 12 healthy male volunteers significant increase at 30 to 60 minutes following injection. *2322*

Hydrochloric Acid *Gastric Material Increase Physiological* Hypoglycemia is powerful stimulant. *0814*
Gastric Material Decrease Physiological Absolute amount decreased by injection or infusion. *0599*

Insulin *Plasma Increase Physiological* In 12 healthy male volunteers significant increase at 30 to 60 minutes following injection. *2322*

Ketones *Urine Increase Physiological* Occurs especially if low liver glycogen stores. *1342*

Magnesium *Serum Decrease Physiological* Effect seen in treatment of diabetic coma. *3730* Significant effect observed at 180 and 210 minutes of glucose tolerance test and when incubated with insulin *in vitro* due to shift of magnesium from plasma to erythrocytes. *2720*
Red Blood Cells Increase Physiological Significant effect observed at 180 and 210 minutes of glucose tolerance test and when incubated with insulin *in vitro* due to shift of magnesium from plasma to erythrocytes. *2720*

Norepinephrine *Plasma Increase Physiological* Marked effect in insulin induced hypoglycemia. *3570*
Urine No Effect Physiological Not appreciably affected by hypoglycemia. *0279*

Pepsin *Gastric Material Increase Physiological* Hypoglycemia is powerful stimulant. *0814*

Phosphate *Serum No Effect Analytical* At concentration of 3 mg/L had no effect on phosphomolybdate method. *3393*
Serum Decrease Physiological Increased phosphorylation of glucose. *3879*

Potassium *Serum No Effect Analytical* At concentration of 3 mg/L had no effect on measurement by ISE with predilution. *3393*
Serum Decrease Physiological Therapeutic effect, causes intracellular shift. *1342*

Prolactin *Plasma No Effect Physiological* No significant effect with hypoglycemia. *1435*
Plasma Increase Physiological In post vagotomy pts marked if i.v. or single injection. *3859* Marked effect in insulin induced hypoglycemia. *3570*

Protein *Serum No Effect Analytical* At concentration of 3 mg/L had no effect on biuret method with blank correction. *3393*
Serum Increase Physiological Associated with increased protein synthesis. *0697*

Protein Bound Iodine (PBI) *Serum Increase Physiological* Mechanism obscure. *0012*

Sodium *Serum No Effect Analytical* At concentration of 3 mg/L had no effect on measurement by ISE with predilution. *3393*
Urine Decrease Physiological Antinatriuresis with/without acidosis may be marked. *3131*

Thymol Turbidity *Serum Increase Physiological* Associated with increased protein synthesis. *0697*

Thyroid Stimulating Hormone (TSH)
Serum No Effect Physiological No effect observed. *2700*

Thyroxine (T₄) *Serum No Effect Physiological* No change during insulin induced hypoglycemia. *3570*
Serum Increase Physiological Probable enhanced release from liver. *3798*

Triglycerides *Serum No Effect Analytical* At concentration of 3 mg/L had no effect on lipase/esterase method. *3393*

Tri-iodothyronine (T₃) *Serum Increase Physiological* Mean increase from 1.86 to 2.51 µg/L at 45 minutes after injection. *3570*

Tri-iodothyronine (T₃), Reverse
Serum Decrease Physiological Fell from mean of 0.184 to 0.171 µg/L but not significant after i.v. injection. *3570*

TSH Response to TRH *Serum No Effect Physiological* Hypoglycemia had no effect on TRH-stimulated TSH release. *3798*

Urea Nitrogen *Serum No Effect Analytical* At concentration of 3 mg/L had no effect on diacetylmonoxime method. *3393*

Uric Acid *Serum No Effect Analytical* At concentration of 3 mg/L had no effect on phosphotungstate reduction method. *3393*

Vanillylmandelic Acid *Urine Increase Physiological* Increase after insulin shock, none with normal dose. *3503*

Volume *Gastric Material Decrease Physiological* In response to injection or infusion. *0599*

α-Interferon

Alanine Aminotransferase *Serum   Increase   Physiological* Dose dependent effect in 3 of 81 patients with malignant disease. *3299*

Aspartate Aminotransferase
Serum   Increase   Physiological Dose dependent effect in 3 of 81 patients with malignant disease. *3299*

Leukocytes *Blood   Decrease   Physiological* Dose dependent leukopenia in 3 of 81 patients with malignant disease. *3299*

Neutrophils *Blood   Decrease   Physiological* Dose dependent leukopenia in 3 of 81 patients with malignant disease. *3299*

Platelets *Blood   Decrease   Physiological* Low value occurred in one patient heavily pretreated with nitrosoureas. *3299*

Protein *Urine   Increase   Physiological* 1 to 2 g/24 h in 2 patients receiving relatively large amount of drug: normalized 1 to 2 weeks after treatment stopped. *3299*

Theophylline *Serum   Increase   Physiological* 1 d after i.m. injection in 5 patients with hepatitis and 4 healthy volunteers significantly reduced clearance and significant increase of elimination half- life. *3843*

Intra-Amniotic Saline

Chloride *Serum   Increase   Physiological* Increased by approximately 3 meq/L after 1 h. *3892*
Urine   Increase   Physiological Significant effect for 24 h. *3892*

Potassium *Serum   Increase   Physiological* No change until 8 h after then by 0.6 meq/L. *3892*
Urine   Increase   Physiological Significant effect for 24 h. *3892*

Sodium *Serum   Increase   Physiological* Increased by approximately 4 meq/L after 1 h. *3892*
Urine   Increase   Physiological Significant effect for 24 h. *3892*

Volume *Urine   Increase   Physiological* Significant increase over next 24 h. *3892*

Inulin

Estriol *Urine   Decrease   Analytical* Probably destroys estriol during acid hydrolysis. *0023*

Estrogens *Urine   Decrease   Analytical* Affects hydrolysis of estrogen conjugates. *0022*

Osmolality *Serum   Increase   Physiological* Massive doses have marked effect. *0071*

Volume *Urine   Increase   Physiological* Massive doses produce marked diuresis. *0071*

Iobenzamic Acid

Thyroid Stimulating Hormone (TSH)
Serum   Increase   Physiological Blocks intrahypophyseal conversion of T_4 to T_3. *3798*

Thyroxine (T_4) *Serum   Increase   Physiological* Impaired conversion of T_4 to T_3. *3798*

Tri-iodothyronine (T_3) *Serum   Decrease   Physiological* Impaired conversion of T_4 to T_3. *3798*

Iodates

Cholesterol *Serum   Increase   Analytical* Interference with Zlatkis-Zak reaction. *3505*

Protein Bound Iodine (PBI) *Serum   Decrease   Physiological* Interferes with trapping of iodide by thyroid. *0012*

Iodides

Chloride *Serum   Increase   Analytical* 0.51 meq/L at 20 meq/L in method of Fingerhut. *1135*

Cholesterol *Serum   Increase   Analytical* Interference with Zlatkis-Zak reaction. *3505*

Eosinophils *Blood   Increase   Physiological* Allergic response. *2313*

Hemoglobin *Urine   Increase   Analytical* Interferes with guaiac and benzidine tests. *1237*

17-Hydroxycorticosteroids *Urine   Increase   Analytical* Interferes with Porter-Silber reaction. *3505*

Ionized Calcium *Serum   Increase   Analytical* Observed interference ion specific electrode. *0540*

^{131}I Uptake *Serum   Decrease   Physiological* Massive doses increase pool. *0583*

Leukocytes *Blood   Increase   Physiological* Due to eosinophilia. *2313*

Occult Blood *Feces   Positive   Analytical* In vitro reaction, ?high enough concentration in vivo. *0583* Interferes with benzidine test. *0579*
Feces   Positive   Physiological Chronic poisoning may cause bloody diarrhea. *1343*

Protein Bound Iodine (PBI) *Serum   Increase   Analytical* Iodide salts if more than 1 g/d lasts 28 d. *0012*

Thyroid Stimulating Hormone (TSH)
Serum   Increase   Physiological Also increased response to TRH related to decreased T_4 and T_3 concentrations. *3798*

Thyroxine (T_4) *Serum   Decrease   Physiological* Decrease synthesis if diagnostic or therapeutic I 131. *3505* Associated with inhibition of conversion. *3798*

Tri-iodothyronine (T_3) *Serum   Decrease   Physiological* Associated with inhibition of conversion. *3798*

Iodinated Glycerol

^{131}I Uptake *Serum   Decrease   Physiological* Daily use may give up to 200 mg/d. *0583*

Protein Bound Iodine (PBI) *Serum   Increase   Physiological* Effect lasts up to 2 d. *0354*

Iodine

Thyroxine (T_4) *Serum   Increase   Analytical* Some organic iodine may elute (odd pattern). *0583*

Zinc *Serum   No Effect   Analytical* At 50 to 1 with flameless atomic absorption. *2038*
Urine   No Effect   Analytical At 50 to 1 with flameless atomic absorption. *2038*

Iodine Containing Drugs

Casts *Urine   Increase   Physiological* Hemorrhagic nephritis with toxic doses. *1302*

Erythrocytes *Urine   Increase   Physiological* Hemorrhagic nephritis with toxic doses. *1302*

Iodine, Total *Serum   Increase   Physiological* Iodine contamination. *1237*

^{131}I Uptake *Serum   Decrease   Physiological* Small doses may affect but not PBI. *0583*

Occult Blood *Feces   Positive   Physiological* May occur with toxicological doses. *1302*

Protein *Urine   Increase   Physiological* Result of hemorrhagic nephritis at toxic doses. *1302*

Protein Bound Iodine (PBI) *Serum   Increase   Physiological* Iodine contamination. *1237*

T_3 Uptake *Serum   No Effect   Physiological* No effect on resin test observed. *1915*

Iodipamide

Bilirubin *Serum   Increase   Physiological* Clinically significant displacement from protein potentially in neonates. *3748*

BSP Retention *Serum   Increase   Physiological* Reported without evidence of liver damage. *2220*

^{131}I Uptake *Serum   Decrease   Physiological* Due to iodine component of material. *2220*

17-Ketogenic Steroids *Urine   Decrease   Analytical* Interferes with reaction. *2573*

Protein Bound Iodine (PBI) *Serum   Increase   Analytical* But does not affect T_4 effect lasts 3-4 mo. *3218*

Pseudocholinesterase *Serum Decrease Physiological* Fairly powerful inhibitor. *2088*

Uric Acid *Serum Decrease Physiological* Uricosuric effect. *2507*

Urine Increase Physiological Uricosuric effect. *2507*

Iodized Oil

^{131}I Uptake *Serum Decrease Physiological* Interferes with uptake. *0071*

Protein Bound Iodine (PBI) *Serum Increase Physiological* Does not affect T_4 (effect 1-5 y). *3218*

Iodoacetate

Acetoacetate Decarboxylase *Serum Decrease Analytical* Inhibition of enzyme produced. *3684*

Cysteine *Plasma Increase Analytical* Effect of oxidizing agent (added *in vitro*). *0384*

Glucose *Serum No Effect Analytical* At 0.5 g/dL on o-toluidine procedure. At 0.5 g/dL on alkaline ferricyanide methods. At 0.5 g/dL on p-HBAH procedure of Lever. At 0.5 g/dL on glucose oxidase methods. *2140*

Methionine *Plasma Decrease Analytical* Effect of oxidizing agent (added *in vitro*). *0384*

Methionine Sulfoxide *Plasma Increase Analytical* Effect of oxidizing agent (added *in vitro*). *0384*

Taurine *Plasma Increase Analytical* Effect of oxidizing agent (added *in vitro*). *0384*

Iodoacetic Acid

Glucose *Serum No Effect Analytical* At 0.5 g/dL on alkaline ferricyanide procedure. At 0.5 g/dL on o-toluidine procedure. At 0.5 g/dL on p-HBAH procedure of Lever. At 0.5 g/dL on glucose oxidase procedures. *2140*

Iodoalphionic Acid

BSP Retention *Serum Increase Physiological* Effect reported without abnormal liver function. *2220*

Creatinine Clearance *Urine Decrease Physiological* Reported to cause renal failure. *0065*

^{131}I Uptake *Serum Decrease Physiological* Organic iodine contamination. *1237*

Protein *Urine Increase Analytical* Affects acid precipitation methods. *0065*

Protein Bound Iodine (PBI) *Serum Increase Analytical* Contains iodine (effect lasts 2-12 mo). *1563*

Thyroxine (T_4) *Serum Increase Analytical* Interferes with direct bromination. *0583*

Urea Nitrogen *Serum Increase Physiological* Reported to cause renal failure. *0065*

Iodoamide

Thyroid Stimulating Hormone (TSH)
Serum Decrease Physiological In 17 euthyroid men following arteriography: maximum effect of more than 50% and at 10-12 weeks. *1030*

Thyroxine (T_4) *Serum No Effect Physiological* In 17 euthyroid men following arteriography for up to 14 weeks. *1030*

Tri-iodothyronine (T_3) *Serum No Effect Physiological* In 17 euthyroid men following arteriography for up to 14 weeks, although initial slight reduction. *1030*

Tri-iodothyronine (T_3), Reverse
Serum No Effect Physiological In 17 euthyroid men following arteriography for up to 14 weeks. *1030*

Iodoantipyrine

Protein Bound Iodine (PBI) *Serum Increase Analytical* Contains iodine. *2425*

Iodocasein

Protein Bound Iodine (PBI) *Serum Increase Analytical* Organic iodine contamination. *1237*

Iodochlorhydroxyquin

Butanol Extractable Iodine (BEI)
Serum Increase Physiological Contains organically bound iodine. *2145*

Ferric Chloride Test *Urine Positive Analytical* If on diaper may give false positive. *1680*

^{131}I Uptake *Serum Decrease Physiological* Contains organically bound iodine. *2145*

Neutrophils *Blood Increase Physiological* Toxic effect reported. *3569*

Phenylketones *Urine Positive Analytical* Green with $FeCl_3$. *0775*

Protein Bound Iodine (PBI) *Serum Increase Analytical* Contains iodine (effect lasts 2-3 mo). *3652*

Iodoform

^{131}I Uptake *Serum Decrease Physiological* Organically bound iodine, inhibits further uptake. *2220*

Protein Bound Iodine (PBI) *Serum Increase Analytical* Contains iodine. *2220*

Iodohippurate

Protein Bound Iodine (PBI) *Serum Increase Analytical* But does not affect T_4. *3218*

Thyroxine (T_4) *Serum Increase Analytical* Interferes with measurement by column. *0251*

Iodophor Detergents

Protein Bound Iodine (PBI) *Serum No Effect Physiological* No effect with occupational exposure. *0583*

Iodopyracet

^{131}I Uptake *Serum Decrease Physiological* Due to iodine component of material. *2220*

Protein *Urine Increase Analytical* If acid precipitation methods used. *2958* Gives false positive with turbidity tests. *0757*

Protein Bound Iodine (PBI) *Serum Increase Analytical* Contains iodine effect lasts 2 weeks. *3505*

PSP Excretion *Urine Decrease Physiological* May compete for excretion through renal tubules. *0974*

Uric Acid *Serum Decrease Physiological* Uricosuric action. *1433*

Urine Increase Physiological Uricosuric action. *1433*

Iodoxyl

Protein Bound Iodine (PBI) *Serum Increase Physiological* Contains iodine, effect lasts for 1-2 weeks. *0354*

Iohexol

Creatinine *Serum Increase Physiological* Transient increase at 3rd day after i.v. administration for angiography. *1053*

Ion Exchange Resins

Ammonia *Plasma Increase Physiological* Mechanism not reported ?depends on resin. *1488*

Chloride *Serum Increase Physiological* Mechanism not cited — ?depends on resin. *0652*

Iopamidol

Creatinine *Serum Increase Physiological* Transient increase at 3rd day after i.v. administration for angiography. *1053*

Iopanoic Acid

Alanine Aminotransferase *Serum Increase Physiological* May cause hepatotoxicity. *2313*

Aspartate Aminotransferase
Serum Increase Physiological May cause hepatotoxicity. *2313*

Bilirubin *Serum Increase Physiological* May cause hepatotoxicity. *2313* Clinically significant displacement from protein observable in neonates. *3748*

BSP Retention *Serum Increase Physiological* Competes for hepatocellular protein binding sites. *3200*

Cephalin Flocculation *Serum Increase Physiological* May cause hepatotoxicity. *2313*

Creatinine Clearance *Urine Decrease Physiological* Reported cause of acute renal failure. *0065*

^{131}I Uptake *Serum Decrease Physiological* Due to iodine component of material. *2220*

Platelets *Blood Decrease Physiological* Few cases transient thrombocytopenia noted. *2292* 3 episodes in one patient occurring from 8 to 40 d after drug ingestion, resolution occurred in 4 to 8 d. *1711*

Protein *Urine No Effect Analytical* No effect on heat acetic acid test No effect on Albustix, Labstix®. *1487*
Urine Increase Analytical Gives turbidity if acid precipitation tests used. *2958* Causes false positive with turbidity tests. *0065*

Protein Bound Iodine (PBI) *Serum Increase Analytical* Organic iodine affects test, effect lasts 1-4 mo. *3218*

Pseudocholinesterase *Serum Decrease Physiological* Potent inhibitor at 0.06 mmol/L. *2088*

Thymol Turbidity *Serum Increase Physiological* May cause hepatotoxicity. *2313*

Thyroid Stimulating Hormone (TSH)
Serum Increase Physiological Still increased 1 week after course of agent. *3511* Blocks intrahypophyseal conversion of T_4 to T_3. *3798*

Thyroxine (T_4) *Serum Increase Analytical* Contaminating iodine affects test. *3218*
Serum Increase Physiological Inhibition of conversion of thyroxine to tri-iodothyronine. *3798*

Thyroxine (T_4), Free *Serum Increase Physiological* Inhibition of conversion of thyroxine to tri-iodothyronine. *3798*

Tri-iodothyronine (T_3) *Serum Decrease Physiological* Inhibition of conversion of thyroxine to tri-iodothyronine. *3798*

Uric Acid *Serum Decrease Physiological* Uricosuric effect. *2507*
Urine Increase Physiological Uricosuric effect. *2507*

Urobilinogen *Urine Increase Analytical* Produce white cloudy precipitate with Ehrlich's. *0583*

Iophendylate

Protein Bound Iodine (PBI) *Serum Increase Analytical* Contains iodine effect lasts up to 5 y. *0354*

Iophenoxic Acid

^{131}I Uptake *Serum Decrease Physiological* Due to iodine component of material. *2220*

Protein *Urine Increase Analytical* Affects turbidity tests for up to 3 d. *0065*

Protein Bound Iodine (PBI) *Serum Increase Analytical* Contains iodine effect lasts up to 30 y. *0354*

Iopydone

^{131}I Uptake *Serum Decrease Physiological* Due to iodine component of material. *2220*

Protein Bound Iodine (PBI) *Serum Increase Analytical* Contains iodine. *2220*

Iothalamate

17-Hydroxycorticosteroids *Urine No Effect Physiological* No physiological effect observed. *2573*

^{131}I Uptake *Serum Decrease Physiological* Interferes with uptake. *0071*

17-Ketogenic Steroids *Urine No Effect Physiological* No physiological effect observed. *2573*
Urine Decrease Analytical Acts as reducing substance (Rutherford and Nelson method). *2573*

17-Ketosteroids *Urine No Effect Physiological* No physiological effect observed. *2573*

Protein Bound Iodine (PBI) *Serum Increase Analytical* Contains iodine. *2220*

Iothiouracil

^{131}I Uptake *Serum Decrease Physiological* Drug consists of organically bound iodine. *0830*

Leukocytes *Blood Decrease Physiological* Agranulocytosis. *2313*

Protein Bound Iodine (PBI) *Serum Increase Analytical* Contains organic iodine (effect for several months). *3505*

T_3 Uptake *Serum Decrease Physiological* Depresses thyroid function. *0583*

Thyroxine (T_4) *Serum Increase Analytical* Organic iodine compound (satisfactory with Murphy-Pattee). *0354*
Serum Decrease Analytical Depresses thyroid function. *0583*

Thyroxine (T_4) (Murphy-Pattee)
Serum No Effect Analytical No effect on method. *2220*
Serum Decrease Physiological Depresses thyroid function. *0583*

Ioxitalamic Acid

Histamine *Plasma Increase Physiological* Major increase in arterial and mixed venous blood in 30 s following injection for translumbar arterial aortography in 16 patients. *1364*

Osmolality *Serum Increase Physiological* Major increase in arterial osmolality within 30 s following injection of 77 $\pm$ 16 mL Telebrix 38 for translumbar arterial aortography in 16 patients. *1364*

Protein Bound Iodine (PBI) *Serum Increase Analytical* May take up to 8 d to become normal. *0149*

Thyroxine (T_4) *Serum No Effect Analytical* No effect observed after i.v. injection. *0149*

Ipecac

Lysergic Acid Diethylamide (LSD)
Gastric Material Increase Analytical Fluorescent spectrum may interfere. *2511*

Ipodate

Bilirubin *Serum Increase Physiological* Probably due to competition for excretion. *0652*

Creatinine *Serum Increase Physiological* Nephrotoxic effect. *0838*

Erythrocytes *Urine Increase Physiological* Nephrotoxic effect. *0838*

Hemoglobin *Urine Increase Physiological* Nephrotoxic effect. *0838*

^{131}I Uptake *Serum Decrease Physiological* Interferes with uptake. *0071*

Protein *Urine Increase Physiological* Nephrotoxic effect. *0838*

Protein Bound Iodine (PBI) *Serum Increase Analytical* Contains organic iodine. *1563*

T_3 Uptake *Serum No Effect Physiological* Significant increase unrelated to I_2 content. *0583*

Thyroid Stimulating Hormone (TSH)
Serum Increase Physiological Affects intrahypophyseal conversion of T_4 to T_3. *3798*

Thyroxine (T_4) *Serum Increase Physiological* Inhibition of T_4 to T_3 conversion. *3798*

Tri-iodothyronine (T_3) *Serum Decrease Physiological* Inhibition of T_4 to T_3 conversion. *3798*

Urea Nitrogen *Serum Increase Physiological* Nephrotoxic effect. *0838*

Uric Acid *Serum Decrease Physiological* Uricosuric effect. *2507*
Urine Increase Physiological Uricosuric effect. *2507*

Urobilin *Urine Increase Physiological* Nephrotoxic effect. *0838*

Iprindole

Alanine Aminotransferase *Serum Increase Physiological* Hepatic cholestasis without inflammation. *0038*

Alkaline Phosphatase *Serum Increase Physiological* Mild elevation with cholestasis. *0038*

Aspartate Aminotransferase
Serum Increase Physiological Hepatic toxicity. *3505*

Bile *Urine Increase Physiological* Due to hepatic toxicity. *0038*

Bilirubin *Serum Increase Physiological* Associated with low serum phosphate. *0038*

Bilirubin, Direct *Serum Increase Physiological* Associated with low serum phosphate. *0038*

Eosinophils *Blood Increase Physiological* Allergic response reported. *0038*

Iproniazid

Alanine Aminotransferase *Serum Increase Physiological* Prolonged use may cause hepatotoxicity. *3835*

Albumin *Serum No Effect Analytical* At concentration of 40 mg/L had no effect on BCG method. *3393*

Alkaline Phosphatase *Serum Increase Physiological* May cause cholestatic jaundice. *2313*

Amphetamine *Urine No Effect Analytical* No effect at 100 mg/L on method of Rutter. *3097*

Aspartate Aminotransferase
Serum Increase Physiological Prolonged use may cause hepatotoxicity. *3835*

Bicarbonate *Serum No Effect Analytical* At concentration of 40 mg/L had no effect on method using phenolphthalein. *3393*

Bile *Urine Increase Physiological* Prolonged use may cause hepatotoxicity. *2313*

Bilirubin *Serum No Effect Analytical* At concentration of 40 mg/L had no effect on Jendrassik and Grof method. *3393*
Serum Increase Physiological May cause cholestatic and cytotoxic jaundice. *0248*

BSP Retention *Serum Increase Physiological* May cause cholestatic jaundice. *2313*

Calcium *Serum No Effect Analytical* At concentration of 40 mg/L had no effect on cresolphthalein method. *3393*

Cephalin Flocculation *Serum Increase Physiological* May cause cholestatic jaundice. *2313*

Chloride *Serum No Effect Analytical* At concentration of 40 mg/L had no effect on mercurimetric method. *3393*

Cholesterol *Serum No Effect Analytical* At concentration of 40 mg/L had no effect on Liebermann-Burchard method. *3393*

Creatinine *Serum No Effect Analytical* At concentration of 40 mg/L had no effect on AutoAnalyzer Jaffé method. *3393*

Erythrocytes *Blood Decrease Physiological* Pancytopenia (AMA Blood dyscrasias). *2429*

Glucose *Serum No Effect Analytical* Affects Boehringer GOD-PERID method. *3277*
Serum Decrease Analytical Depresses glucose oxidase method of Boehringer. *3277* At concentrations above 110 mg/L lowered concentration as measured by GOD-Perid method. *3393*

β-Glucuronidase *Serum Increase Physiological* Effect of toxic hepatitis. *3835*

Guanase *Serum Increase Physiological* May cause cholestatic and cytotoxic jaundice. *0248*

Isocitrate Dehydrogenase *Serum Increase Physiological* May cause cholestatic and cytotoxic jaundice. *0248*

Leukocytes *Blood Decrease Physiological* Pancytopenia (AMA Blood dyscrasias). *2429*

Ornithine Carbamoyltransferase (OCT)
Serum Increase Physiological May cause cholestatic and cytotoxic jaundice. *0248*

Phosphate *Serum No Effect Analytical* At concentration of 40 mg/L had no effect on phosphomolybdate method. *3393*

Platelets *Blood Decrease Physiological* Pancytopenia (AMA Blood dyscrasias). *2429*

Potassium *Serum No Effect Analytical* At concentration of 40 mg/L had no effect on measurement by ISE with predilution. *3393*

Protein *Serum No Effect Analytical* At concentration of 40 mg/L had no effect on biuret method with blank correction. *3393*

Pyridoxal *Serum Decrease Physiological* Observed in patient with toxicity. *0738*

Pyridoxine *Serum Decrease Physiological* Observed in patient with toxicity. *0738*

Sodium *Serum No Effect Analytical* At concentration of 40 mg/L had no effect on measurement by ISE with predilution. *3393*

Thymol Turbidity *Serum Increase Physiological* May cause cholestatic jaundice. *2313*

Triglycerides *Serum No Effect Analytical* At concentration of 40 mg/L had no effect on lipase/esterase method. *3393*

Urea Nitrogen *Serum No Effect Analytical* At concentration of 40 mg/L had no effect on diacetylmonoxime method. *3393*

Uric Acid *Serum No Effect Analytical* At concentration of 40 mg/L had no effect on phosphotungstate reduction method. *3393*

Iron

Ferritin *Serum Increase Physiological* With 200 mg iron daily, most patients serum ferritin increased as well as bone marrow stainable iron. *1501*

Nickel *Test Conditions Increase Analytical* If ferrous and atomic absorption used. *3499*

Occult Blood *Feces No Effect Analytical* No effect on Hemoquant procedure. *3671*

Zinc *Serum No Effect Analytical* At 50 to 1 with flameless atomic absorption. *2038*
Urine No Effect Analytical At 50 to 1 with flameless atomic absorption. *2038*

Iron Dextran

Color *Urine Increase Physiological* Black urine on standing reported. *1678*

Creatinine Clearance *Urine No Effect Physiological* No effect if prior normal renal function. *1678*

Erythrocytes *Urine Increase Physiological* Reversible after chronic administration in one patient. *1678*

Ferritin *Serum Increase Physiological* Peak values 7 to 9 d after i.v. infusion, thereafter declined. *0390*

Glucose *Serum Increase Analytical* At 5 g Fe/dL on glucose oxidase (slight) procedure. 5 g Fe/dL equivalent to 15.5 mmol/L o-toluidine procedure. 5 g Fe/dL equivalent to 3.5 mmol/L p-HBAH procedure. Slight effect at 5 g Fe/dL on alkaline ferricyanide procedure. *2140*

Hemoglobin *Blood Increase Physiological* Typically rises after i.v. infusion. *0390*
Urine Increase Physiological Reversible after chronic administration in one patient. *1678*

Iron *Serum Increase Physiological* Increased iron stores. *2313* Rose to exceedingly high values (up to 8,000 μmol/L) immediately after intravenous infusion. *0390*
Urine Increase Physiological Approximately 30% Fe Sorbitex in urine in 24 h. *1678*

Iron Dextran *(continued)*

Iron-Binding Capacity, Unsaturated (UIBC)
Serum Decrease Physiological Due to increased availability of iron. *2313*

Leukocytes *Blood Increase Physiological* Leukemoid reaction reported. *2427*

Protein *Urine Increase Physiological* Reversible after chronic administration in one patient. *1678*

Iron Salts

Alanine Aminotransferase *Serum Increase Physiological* May cause severe liver damage with poisoning. *0279*

Aspartate Aminotransferase
Serum Increase Physiological May cause severe liver damage with poisoning. *0279*

Bilirubin *Serum Increase Physiological* With poisoning 3-4 d after ingestion. *3817*

Bilirubin, Direct *Serum Increase Physiological* With poisoning 3-4 d after ingestion. *3817*

Calcium *Serum Increase Analytical* Interfere with direct EDTA titration. *1563*

Color *Urine Increase Analytical* Iron Sorbitex can cause brown urine (Fe sulfide). *1237*
Feces Increase Analytical Black (gray-black) darkens in air with about 70 mg. *2313*

Diagnex Blue Excretion *Urine Increase Physiological* Heavy metal displacement of diagnex blue. *3879*

Erythrocytes *Urine Increase Physiological* Hematuria reported after chronic administration Fe sorbitol. *1680*

Fibrinogen *Plasma Decrease Physiological* One case reported with poisoning. *3817*

Glucose Tolerance *Serum Decrease Physiological* As Fe ascorbate before glucose tolerance test. *1487*

Hemoglobin *Urine Increase Physiological* Hematuria reported after iron Sorbitex. *2427*

Iron *Serum Increase Physiological* Effect of i.m. iron. *1237*

Iron-Binding Capacity, Total (TIBC)
Serum Increase Physiological Effect of i.m. iron. *1237*

Iron-Binding Capacity, Unsaturated (UIBC)
Serum Decrease Physiological Effect of i.m. iron. *1237*

Occult Blood *Feces No Effect Analytical* If 3,3-dimethylnaphthidine used as chromogen. *0849*
Feces Positive Analytical Interferes with guaiac test (?benzidine). *3584*
Feces Positive Physiological With poisoning may cause gastrointestinal hemorrhage. *3817*

Protein *Urine Increase Physiological* May cause nephrotoxicity. *1488*

Sedoheptulose *Urine Decrease Analytical* Ferric iron inhibits cysteine-H_2SO_4 reaction. *2346*

Urea Nitrogen *Serum Increase Physiological* May cause nephrotoxicity. *3204*

Urobilin *Urine Increase Physiological* May exacerbate urinary tract infections. *2427*

Vitamin B_{12} *Red Blood Cells Increase Physiological* During treatment of iron-deficient anemia. *1510*

Iron Sorbitex

Color *Urine Increase Physiological* Complex with citrate produce black urine. *1187*

Glucose *Serum Increase Analytical* 5 g Fe/dL equivalent to 7.1 mmol/L o-toluidine. 5 g Fe/dL equivalent to 1.9 mmol/L p-HBAH procedure. 5 g Fe/dL equivalent to 1.0 mmol/L alkaline ferricyanide. 5 g Fe/dL equivalent to 1/6 mmol/L glucose oxidase. *2140*

Isocarbazide

Glucose *Serum No Effect Analytical* At concentration of 55 mg/L had no effect on GOD-Perid method. *3393*

Isocarboxamide

Propoxyphene *Urine No Effect Analytical* Less than 1% fluorescence in procedure of Valentour. *3662*

Isocarboxazid

Alanine Aminotransferase *Serum Increase Physiological* Cholestatic effect. *2313*

Alkaline Phosphatase *Serum Increase Physiological* Cholestatic effect. *2313*

Ammonia *Plasma Decrease Physiological* Reportedly effective in reducing NH_3 intoxication. *1488*

Amphetamine *Urine No Effect Analytical* No reaction with NBD chloride of Monforte. *2473*

Aspartate Aminotransferase
Serum Increase Physiological Cholestatic effect. *2313*

Bile *Urine Increase Physiological* Cholestatic effect. *2313*

Bilirubin *Serum Increase Physiological* Cholestatic effect. *2313*

BSP Retention *Serum Increase Physiological* Intrahepatic cholestasis reported. *0553*

Cephalin Flocculation *Serum Increase Physiological* Hepatotoxic effect. *2313*

Erythrocytes *Blood Decrease Physiological* Agranulocytosis/anemia — rare. *2313*

Glucose *Serum No Effect Analytical* Affects Boehringer GOD-PERID method. *3277*
Serum Decrease Analytical Affects glucose oxidase method of Boehringer. *3277*
Serum Decrease Physiological MAO inhibitors have slight effect. *1488*

Hematocrit *Blood Decrease Physiological* May occasionally produce anemia. *2427*

Hemoglobin *Blood Decrease Physiological* May occasionally produce anemia. *2427*

5-Hydroxyindoleacetic Acid (5-HIAA)
Urine Decrease Physiological Due to inhibition of conversion of 5-HT to 5-HIAA. *1488*

Leukocytes *Blood Decrease Physiological* Agranulocytosis/leukopenia — rare. *1612*

Protein Bound Iodine (PBI) *Serum No Effect Physiological* No effect reported with normal doses. *0830*

Thymol Turbidity *Serum Increase Physiological* Cholestatic effect. *2313*

Vanillylmandelic Acid *Urine Decrease Physiological* Inhibition of formation. *1488*

Xylose Excretion *Urine Decrease Physiological* Decreased gastrointestinal tract absorption. *3679*

Isoflurane

Leukocytes *Blood Increase Physiological* Normal response to anesthesia. *3465*

Isoflurophate

Pseudocholinesterase *Serum Decrease Physiological* Therapeutic action of drug. *0071*

Isoleucine

Ammonia *Plasma Decrease Analytical* With 500 nmoles produces 59% inhibitor indophenol reaction. *1290*

Pyruvate Kinase *Red Blood Cells Decrease Analytical* Most marked with larger side chain. *0370*

Tyrosine *Plasma No Effect Analytical* On fluorometric procedure of Ambrose. *0080*

Isometheptene

Amphetamine *Urine No Effect Analytical* No effect at 100 mg/L on method of Rutter. *3097*

Urine Positive Analytical Reacts as if amphetamine in EMIT screening and confirmatory assays (Note compound is a component of Midrin® used to treat migraine). 2146

Isomil®

Amino Acids Urine Positive Analytical Reddish-pink spot with DL-methionine with ninhydrin on thin-layer chromatography. 1278

Isoniazid

Alanine Aminotransferase Serum Increase Physiological Probable intrahepatic cholestatic jaundice. 2220 Mild liver injury occurs in approximately 10% patients taking drug, possibly due to conversion of drug to acetylhydrazine or related hepatotoxic derivatives. 2260 Liver toxicity occurred in 18% patients receiving combined isoniazid and rifampin: effect slight in 14% and severe in 4%. 1409 Risk of hepatitis caused by or exacerbated by drug over 12 mo 5.2 per 1,000 patients; clinically picture resembles viral hepatitis. 1292 15 of 89 patients developed significant liver disease, typically hepatitis, although one developed cholestasis, in patients taking drug for at least 2 mo. 0901 Observed in up to 0.5% patients slightly greater than in placebo treated group. 1725 In 3.3% children also treated with rifampin: all reactions occurring in first 10 weeks. 2630

Albumin Serum No Effect Analytical At concentration of 100 mg/L had no effect on BCG method. 3393

Alkaline Phosphatase Serum Increase Physiological Probable cholestatic effect. 2220 15 of 89 patients developed significant liver disease, typically hepatitis, although one developed cholestasis, in patients taking drug for at least 2 mo. 0901

Amino-4-Imidazole-5-Carboxamide Ribotide (AICAR)
Urine Increase Physiological Occurs if megaloblastic anemia. 3773

Ammonia Plasma No Effect Physiological No effect observed in most studies. 1487
Plasma Increase Physiological Due to metabolism of isonicotinic acid. 3951

Amphetamine Urine No Effect Analytical At 50 mg/L on fluorescent method of Hayes. 1534 No reaction with NBD chloride of Monforte. 2473 No effect at 100 mg/L on method of Rutter. 3097

Amylase Serum Increase Physiological Reported cause of acute pancreatitis, much doubt. 1488

Antinuclear Antibodies Serum Positive Physiological Up to 78% tuberculous patients develop antibodies. 0043

Aspartate Aminotransferase Serum Increase Analytical At therapeutic concentration may affect SMA 12/60 method. 3335
Serum Increase Physiological Probable intrahepatic cholestatic jaundice. 2220 Mild liver injury occurs in approximately 10% patients taking drug, possibly due to conversion of drug to acetylhydrazine or related hepatotoxic derivatives. 2260 In 18.3% versus 6.7% of controls in adult patients receiving drug for prophylaxis. 0555 Liver toxicity occurred in 18% patients receiving combined isoniazid and rifampin: effect slight in 14% and severe in 4%. 1409 Risk of hepatitis caused by or exacerbated by drug over 12 mo 5.2 per 1,000 patients; clinically picture resembles viral hepatitis. 1292 15 of 89 patients developed significant liver disease, typically hepatitis, although one developed cholestasis, in patients taking drug for at least 2 mo. 0901 Occurs in about 20% of treated patients due to acetylhydrazine metabolite: higher incidence in older and in consumers of excess alcohol. 0053 Observed in up to 0.5% patients slightly greater than in placebo treated group. 1725 In 3.3% children also treated with rifampin: all reactions occurring in first 10 weeks. 2630

Bicarbonate Serum No Effect Analytical At concentration of 100 mg/L had no effect on method using phenolphthalein. 3393

Bile Urine Increase Physiological Intrahepatic cholestatic jaundice. 0085

Bile Acids Serum Increase Physiological Reported increases when given alone or with rifampin in the absence of other abnormal liver function tests. 0306 15 of 89 patients developed significant liver disease, typically hepatitis, although one developed cholestasis, in patients taking drug for at least 2 mo. 0901 In 72% of 61 patients studied for 80 d when treatment combined with rifampin but most other liver function tests including bilirubin could be normal. 0305

Bilirubin Serum No Effect Analytical At concentration of 100 mg/L had no effect on Jendrassik and Grof method. 3393
Serum Increase Physiological Intrahepatic cholestasis. 1022 15 of 89 patients developed significant liver disease, typically hepatitis, although one developed cholestasis, in patients taking drug for at least 2 mo. 0901

BSP Retention Serum Increase Physiological Intrahepatic cholestatic jaundice. 0085

Calcium Serum No Effect Analytical At concentration of 100 mg/L had no effect on cresolphthalein method. 3393

Carbamazepine Serum Increase Physiological Due to inhibition of liver enzymes by drug: caused intoxication in 10 of 13 epileptic patients. 3665

Casts Urine Increase Physiological Nephrotoxic effect. 2313

Cephalin Flocculation Serum Increase Physiological Cytotoxic hepatocellular damage. 0248

Chloride Serum No Effect Analytical At concentration of 100 mg/L had no effect on mercurimetric method. 3393

Cholesterol Serum No Effect Analytical At concentration of 100 mg/L had no effect on Liebermann-Burchard method. 3393
Serum Decrease Physiological Probable hepatotoxic effect. 2220

Coombs' Test Serum Positive Physiological Immunological response to drug. 1486

Coombs' Test, Direct Serum Positive Physiological Complement type of positive response. 1487

Creatinine Serum No Effect Analytical At 20 mg/L on reversed phase LC procedure of Zhiri et al. 3942 At concentration of 100 mg/L had no effect on AutoAnalyzer method. 3393

Cyclosporine Serum Decrease Physiological Enhances metabolism by hepatic enzyme induction. 3732

Diphenylhydantoin Serum Increase Physiological Impairs metabolism in approximately 10%. 2042 From 12 to 42 µg/mL (with dose of 300 mg). 2041 Increases blood concentration and toxicity. 3683

Eosinophils Blood Increase Physiological Allergic phenomenon. 1513

Erythrocytes Blood Decrease Physiological Hemolytic anemia (rare complication). 2313

FIGLU (N-Formiminoglutamic Acid)
Urine Increase Physiological Occurs if megaloblastic anemia. 3773

Folate Serum Decrease Physiological Low incidence of impaired absorption. 3773

Glucose Serum No Effect Analytical Affects Boehringer GOD-PERID method. 3277
Serum Increase Physiological Large doses cause hyperglycemia by glycogenolysis. 1488
Serum Decrease Analytical Affects glucose oxidase method of Boehringer. 3277 At concentrations above 55 mg/L (normal therapeutic concentration 10 mg/L) lowered concentration as measured by GOD-Perid method. 3393 At concentrations above 500 mg/L (normal therapeutic concentration 10 mg/L) lowered concentration as measured by Seralyzer method. 3393
Urine Increase Physiological Glycosuria may follow induced hyperglycemia. 1488 Due to hyperglycemia. 2220

Guanase Serum Increase Physiological Cytotoxic hepatocellular damage. 0248

Haptoglobin Serum Decrease Physiological Hemolytic anemia. 2313

Hematocrit Blood Decrease Physiological Hemolytic anemia/rare megaloblastic anemia. 2313

Isoniazid (continued)

Hemoglobin *Blood Decrease Physiological* Hemolytic anemia/rare megaloblastic anemia. *2313* Very rare cases of pure red cell aplasia, other cases of hemolytic anemia pyridoxine-responsive sideroblastic anemia also reported. *0667* Rare hemolytic anemia in patients with glucose-6-phosphate deficiency. *1292*

5-Hydroxyindoleacetic Acid (5-HIAA)
Urine Decrease Physiological Causes decarboxylase inhibition with reduced 5-HT. *0832*

3-Hydroxykynurenine *Urine Increase Physiological* Induces pyridoxal PO$_4$ deficiency. *3045*

INH-Ketoglutarate *Urine Increase Physiological* Major metabolite in urine. *3093*

INH-Pyruvate *Urine Increase Physiological* Major metabolite in urine. *3093*

Isocitrate Dehydrogenase *Serum Increase Physiological* Cytotoxic hepatocellular jaundice. *0248*

^{131}I Uptake *Serum Decrease Physiological* Reduces uptake. *2220*

Ketones *Urine Increase Physiological* Mechanism not listed. *1488*

Kynurenine *Urine Increase Physiological* Induces pyridoxal PO$_4$ deficiency. *3045*

Lactate *Serum Increase Physiological* If overdose with inhibition of NAD activity. *1487*
Plasma No Effect Physiological Probable effect of normal dose and renal function. *1487*

LE Cells *Blood Positive Physiological* More common in slow acetylators. *2325*

Leukocytes *Blood Increase Physiological* Due to eosinophilia. *2313*
Blood Decrease Physiological Agranulocytosis — rare. *0222* Occasional case of agranulocytosis reported. *3683*

MCV *Blood Increase Physiological* If megaloblastic anemia occurs. *2313*

Methemoglobin *Blood Increase Physiological* Reported effect. *1343*

Mexiletine *Serum Increase Physiological* Decreases clearance and prolongs half-life. *2011*

Neutrophils *Blood Decrease Physiological* May cause neutropenia. *2583* Occasional case of agranulocytosis reported. *3717*

Ornithine Carbamoyltransferase (OCT)
Serum Increase Physiological Cytotoxic hepatocellular damage. *0248*

pH *Blood Decrease Physiological* Large doses may produce severe acidosis. *1343*

Phosphate *Serum No Effect Analytical* On nonprecipitation method of Peynet. *2799* At concentration of 100 mg/L had no effect on phosphomolybdate method. *3393*

Platelets *Blood Decrease Physiological* May rarely cause bone marrow aplasia. *0222*

Potassium *Serum Increase Physiological* Reported effect of overdose. *1488*

Protein *Serum No Effect Analytical* At concentration of 100 mg/L had no effect on biuret method with blank correction. *3393*
Urine Increase Physiological May have nephrotoxic effect. *1488* Nephrotoxic effect. *1022*

Protein Bound Iodine (PBI) *Serum Decrease Physiological* Reduces thyroid synthesis. *2220*

Pyridoxine *Serum Decrease Physiological* Observed in 13% of 38 children when measured with protozoan procedure: also reported in 2% of adults after 6 mo. *2769*
Urine Increase Physiological Dose related effect. *3025* Particularly prevalent with use of large doses of drug or in nutritionally deficient persons such as alcoholics. *3683*

Sugar *Urine Increase Analytical* False positive with Benedict's and Clinitest®. *1099*

Triglycerides *Serum No Effect Analytical* At concentration of 100 mg/L had no effect on lipase/esterase method. *3393*

Urea Nitrogen *Serum No Effect Analytical* At concentration of 100 mg/L had no effect on diacetylmonoxime method. *3393*

Uric Acid *Serum No Effect Analytical* At 20 mg/L on reversed phase LC procedure of Zhiri et al. *3942*
Serum Increase Analytical At concentrations above 14 mg/L (normal therapeutic concentration 10 mg/L) raised concentration as measured by phosphotungstate reduction method. *3393*

Xanthurenic Acid *Urine No Effect Physiological* Initial response may be decreased later. *3045*

Isonicotinic Acid

Glucose *Serum No Effect Analytical* At concentration of 5 mg/L had no effect on Ektachem® method. *3393*

Isoprenaline

Amphetamine *Urine No Effect Analytical* No effect at 100 mg/L on method of Rutter. *3097* At 50 mg/L on fluorescent method of Hayes. *1534*

Calcium *Serum No Effect Analytical* At concentration of 211 mg/L had no effect on cresolphthalein method. *3393*

Phosphate *Serum No Effect Analytical* At concentration of 211 mg/L had no effect on phosphomolybdate method. *3393*

Protein *Serum No Effect Analytical* At concentration of 211 mg/L had no effect on biuret method with blank correction. *3393*

Urea Nitrogen *Serum No Effect Analytical* At concentration of 211 mg/L had no effect on diacetylmonoxime method. *3393*

Uric Acid *Serum No Effect Analytical* At concentration of 211 mg/L had no effect on phosphotungstate reduction method. *3393*

Isopropamide

^{131}I Uptake *Serum Decrease Physiological* Contains iodine — reduces further uptake. *1488*

Protein Bound Iodine (PBI) *Serum No Effect Physiological* No definite effect observed. *0583*
Serum Increase Analytical Contains iodine. *0012*

Isopropanol

Acetone *Blood Increase Physiological* Metabolite of isopropanol. *2348*
Urine Increase Physiological Metabolized partially to acetone. *1488*

Alanine Aminotransferase *Serum Increase Physiological* Transient and mild late toxic effect. *1302*

Alcohol *Breath Increase Analytical* Can produce measurable levels. *1238*

Bogen Test *Urine Positive Analytical* Reacts as if ethanol. *1302*

Isopropanol *Blood Increase Physiological* Fatal poisoning at 1 g/L. *2348*

Ketones *Urine Increase Physiological* In intoxication acetone is normal metabolite. *1487*

Occult Blood *Feces Positive Physiological* May cause hematemesis and gastroenteritis. *1302*

Protein *Urine Increase Physiological* Transient and mild late toxic effect. *1302*

Isopropyl Dipyrone

Aminolevulinic Acid *Urine Increase Physiological* May precipitate attack of acute porphyria. *1016*

Porphyrins *Urine Increase Physiological* May precipitate attack of acute porphyria. *1016*

Isopropylepinephrine

Epinephrine *Test Conditions Increase Analytical* 25% of fluorescence method of Waldmeier. *3735*

Isopropylnorepinephrine

Epinephrine *Test Conditions Increase Analytical* 58% fluorescence of epinephrine in procedure of Peyrin. *2800*

Norepinephrine *Test Conditions No Effect Analytical* On fluorescent procedure of Peyrin. *2800*

Isoproterenol

Aspartate Aminotransferase *Serum Increase Analytical* At 1 mmol/L affects SMA 12/60 method. *3335*

Bilirubin *Serum Increase Analytical* At 1 mmol/L affects SMA 12/60 method. *3335*

Catecholamines *Plasma Increase Physiological* Due to inhalation — effect slight. *0758*
Urine Increase Analytical Interferes with resin and fluorometric procedures. *1205* Metabolite produces fluorescence in screening. *0583*
Urine Increase Physiological Observed for up to 4 d after drug stopped. *0913*

Cyclic Adenosine Monophosphate
Plasma Increase Physiological No effect unless beta-blocking agent also given Response to i.v. infusion in normals. *0216*
Urine Increase Physiological Effect less marked than in blood. *0216*
Urine Decrease Physiological In normals but increased in hypertensives. *1471*

Epinephrine *Urine Increase Physiological* Probably small effect with usual doses. *0758*

Fatty Acids, Free (FFA) *Serum Increase Physiological* As effective as epinephrine. *1343*

Glomerular Filtration Rate (GFR)
Urine Decrease Physiological Frequently observed to be diminished. *3205*

Glucose *Serum Increase Analytical* At 1 mmol/L affects SMA 12/60 method. *3335*
Serum Increase Physiological Not as marked as with epinephrine. *1343*

pO$_2$ *Blood Decrease Physiological* By approximately 10 mm Hg in chronic lung disease. *2427*

Vanillylmandelic Acid *Urine No Effect Analytical* No apparent methodological effect. *1487*
Urine No Effect Physiological No apparent physiological effect. *1487*
Urine Increase Physiological Pharmacol effect (still seen 4 d after stopping). *0913*

Volume *Urine Decrease Physiological* Antidiuretic effect mediated through ADH release. *3205*

Isosorbide

Chloride *Urine Increase Physiological* Similar to natriuretic effect. *2605*

Creatinine Clearance *Urine Increase Physiological* ?due to decreased tubular reabsorption. *2605*

Indomethacin *Serum No Effect Physiological* No effect on HPLC method of Roberts and Smith. *3002*

Methemoglobin *Blood Increase Physiological* Significant increase in angina patients but probably not of routine significance, but may be important in anemics or in patients with coronary insufficiency (difference = 1.13 vs 0.99 in controls). *2096* Commonly used nitrates at regular doses capable of causing usually clinically insignificant increases. *0164*

Osmolar Clearance *Urine Increase Physiological* Diuretic action alone, potentiates others. *2605*

Potassium *Urine Increase Physiological* Slight effect only if given alone. *2605*

Sodium *Serum Increase Physiological* Dehydration with overdosage. *3315*
Urine Increase Physiological Diuretic action alone, potentiates others. *2605*

Urea Nitrogen *Serum Increase Physiological* Dehydration with overdosage. *3315*

Volume *Urine Increase Physiological* Diuretic action alone, potentiates others. *2605*

Isotretinoin

Alanine Aminotransferase *Serum No Effect Physiological* In 18 patients with severe acne given 0.8 mg/kg daily for 3 mo: changes reverted to normal after treatment stopped. *2242* Significant effect at 6 weeks in 7 patients with severe rosacea treated with 1 mg/kg/d. *2307*
Serum Increase Physiological In 10 of 523 patients with mean daily dose of 109 mg for 150 d. *3867*

Albumin *Serum No Effect Physiological* In 18 patients with severe acne given 0.8 mg/kg daily for 3 mo: changes reverted to normal after treatment stopped. *2242*

Alkaline Phosphatase *Serum Increase Physiological* In 7 patients with severe rosacea treated with 1 mg/kg/d for 12 weeks. Effects possibly due to induction of hepatic microsomal enzymes. *2307* In 10 of 523 patients with mean daily dose of 109 mg for 150 d. *3867*

Apolipoprotein AI *Serum No Effect Physiological* In 12 patients with hyperkeratotic disorders given 40 mg daily for 2 mo. *3659*

Apolipoprotein AII *Serum No Effect Physiological* In 12 patients with hyperkeratotic disorders given 40 mg daily for 2 mo. *3659*

Apolipoprotein B *Serum Increase Physiological* From 132 to 157% in 12 patients with hyperkeratotic disorders given 40 mg daily for 2 mo. *3659*

Aspartate Aminotransferase
Serum No Effect Physiological In 18 patients with severe acne given 0.8 mg/kg daily for 3 mo: changes reverted to normal after treatment stopped. *2242*
Serum Increase Physiological In 7 patients with severe rosacea treated with 1 mg/kg/d for 12 weeks. Effects possibly due to induction of hepatic microsomal enzymes. *2307* Reversible dose-related effect noted in patients with myelodysplastic syndrome and leukemia at high doses. *1319* Significant effect at 6 weeks in 7 patients with severe rosacea treated with 1 mg/kg/d. *2307*

Bilirubin *Serum No Effect Physiological* In 18 patients with severe acne given 0.8 mg/kg daily for 3 mo: changes reverted to normal after treatment stopped. *2242*
Serum Increase Physiological Reversible dose-related effect noted in patients with myelodysplastic syndrome and leukemia at high doses. *1319*
Serum Decrease Physiological In 7 patients with severe rosacea treated with 1 mg/kg/d for 12 weeks effect possibly due to induction of hepatic microsomal enzymes. *2307* Significant at 12 weeks in 7 patients with severe rosacea treated with 1 mg/kg/d. *2307*

Blast Count *Bone Marrow Decrease Physiological* Therapeutic response is normalization but effect not usually observed until after 2 weeks. *1319*

Cholesterol *Serum Increase Physiological* In 7 patients with severe rosacea treated with 1 mg/kg/d for 12 weeks. Effects possibly due to induction of hepatic microsomal enzymes. *2307* In both men and women with 1 mg/kg/d for 20 weeks when given for nodulocystic acne. *0326* From 0.9 to 2.2 mmol/L in 18 patients with severe acne given 0.8 mg/kg daily for 3 mo: changes reverted to normal after treatment stopped. *2242* From 0.8 to 1.6 mmol/L at 6 weeks in 7 patients with severe rosacea treated with mg/kg/d. *2307* During treatment but not to abnormal levels. *1818* From 5.75 to 6.49 mmol/L in 12 patients with hyperkeratotic disorders given 40 mg daily for 2 mo. *3659* Increase in both men and women with 1 mg/kg/d for 20 weeks when given for nodulocystic acne. *0326*

Cholesterol, High Density Lipoprotein
Serum Decrease Physiological In 7 patients with severe rosacea treated with 1 mg/kg/d for 12 weeks. Effects possibly due to induction of hepatic microsomal enzymes. *2307* In both men and women with 1 mg/kg/d for 20 weeks when given for nodulocystic acne. *0326* From 1.28 to 1.14 mmol/L in 12 patients with hyperkeratotic disorders given 40 mg daily for 2 mo. *3659* From 1.1 to 0.9 mmol/L in 18 patients with severe acne given 0.8 mg/kg daily for 3 mo: changes reverted to normal after treatment stopped. *2242* From 1.30 to 1.04 mmol/L at 6 weeks in 7 patients with severe rosacea treated with 1 mg/kg/d. *2307* Increase in both men and women with 1 mg/kg/d for 20 weeks when given for nodulocystic acne. *0326*

Isotretinoin (continued)

Cholesterol, Low Density Lipoprotein
Serum Increase Physiological In 7 patients with severe rosacea treated with 1 mg/kg/d for 12 weeks. Effects possibly due to induction of hepatic microsomal enzymes. *2307* Increase in both men and women with 1 mg/kg/d for 20 weeks when given for nodulocystic acne. *0326* From 3.92 to 5.33 mmol/L at 6 weeks in 7 patients with severe rosacea treated with 1 mg/kg/d. *2307* From 3.67 to 4.37 mmol/L in 12 patients with hyperkeratotic disorders given 40 mg daily for 2 mo. *3659* Increase in both men and women with 1 mg/kg/d for 20 weeks when given for nodulocystic acne. *0326*

Cholesterol, Very Low Density Lipoprotein
Serum Increase Physiological In 7 patients with severe rosacea treated with 1 mg/kg/d for 12 weeks. Effects possibly due to induction of hepatic microsomal enzymes. *2307* From 0.17 to 0.39 mmol/L at 6 weeks in 7 patients with severe rosacea treated with 1 mg/kg/d. *2307* From 0.69 to 0.93 mmol/L in 12 patients with hyperkeratotic disorders given 40 mg daily for 2 mo. *3659*

Erythrocytes *Blood Decrease Physiological* In 10 of 523 patients with mean daily dose of 109 mg for 150 d. *3867*

Erythrocyte Sedimentation Rate
Blood Increase Physiological In 50 of 523 patients with mean daily dose of 109 mg for 150 d. *3867*

Follicle Stimulating Hormone (FSH)
Plasma No Effect Physiological In 7 patients with severe rosacea treated with 1 mg/kg/d for 12 weeks. Effects possibly due to induction of hepatic microsomal enzymes. *2307* With 12 weeks treatment with 1 mg/kg/d. *2307*

FSH Response to LHRH *Serum No Effect Physiological* In 7 patients with severe rosacea treated with 1 mg/kg/d for 12 weeks. Effects possibly due to induction of hepatic microsomal enzymes. *2307*

γ-Glutamyltransferase (GGT)
Serum Increase Physiological In 7 patients with severe rosacea treated with 1 mg/kg/d for 12 weeks. Effects possibly due to induction of hepatic microsomal enzymes. *2307* From 13.0 to 21.1 U/L in 18 patients with severe acne given 0.8 mg/kg daily for 3 mo: changes reverted to normal after treatment stopped. *2242* Significant effect at 6 weeks in 7 patients with severe rosacea treated with 1 mg/kg/d. *2307* Slight increase with high dose treatment. *1818*

Hemoglobin *Blood Increase Physiological* Therapeutic response is normalization but effect not usually observed until after 3 weeks. *1319*

Hepatic Triglyceride Lipase
Serum No Effect Physiological Increase in both men and women with 1 mg/kg/d for 20 weeks when given for nodulocystic acne. *0326*

Leukocytes *Blood Increase Physiological* Therapeutic response is normalization but effect not usually observed until after 3 weeks. *1319*
Blood Decrease Physiological In 10 of 523 patients with mean daily dose of 109 mg for 150 d. *3867*
Urine Increase Physiological In 10 of 523 patients with mean daily dose of 109 mg for 150 d. *3867*

LH Response to LHRH *Plasma No Effect Physiological* In 7 patients with severe rosacea treated with 1 mg/kg/d for 12 weeks. Effects possibly due to induction of hepatic microsomal enzymes. *2307*

Lipoprotein Lipase *Serum No Effect Physiological* Increase in both men and women with 1 mg/kg/d for 20 weeks when given for nodulocystic acne. *0326* Increase in both men and women with 1 mg/kg/d for 20 weeks when given for nodulocystic acne. *0326*

Luteinizing Hormone (LH) *Plasma No Effect Physiological* In 7 patients with severe rosacea treated with 1 mg/kg/d for 12 weeks. Effects possibly due to induction of hepatic microsomal enzymes. *2307* With 12 weeks treatment with 1 mg/kg/d. *2307*

Platelets *Blood Increase Physiological* Therapeutic response is normalization but effect not usually observed until after 3 weeks. *1319* In 10 of 523 patients with mean daily dose of 109 mg for 150 d. *3867*

Protein *Serum No Effect Physiological* In 18 patients with severe acne given 0.8 mg/kg daily for 3 mo: changes reverted to normal after treatment stopped. *2242*

Thyroid Stimulating Hormone (TSH)
Serum No Effect Physiological In 7 patients with severe rosacea treated with 1 mg/kg/d for 12 weeks. Effects possibly due to induction of hepatic microsomal enzymes. *2307* Or no significant change in 24 healthy men given 1 mg/kg/d for 16 weeks. *2640* In 7 patients given 1 mg/kg/d for 12 weeks. *2307* In 18 patients with severe acne given 0.8 mg/kg daily for 3 mo: changes reverted to normal after treatment stopped. *2242* Significant at 12 weeks in 7 patients with severe rosacea treated with 1 mg/kg/d. *2307*

Thyroxine (T₄) *Serum Decrease Physiological* In 7 patients with severe rosacea treated with 1 mg/kg/d for 12 weeks. Effects possibly due to induction of hepatic microsomal enzymes. *2307* From 98 to 85 nmol/L in 18 patients with severe acne given 0.8 mg/kg daily for 3 mo: changes reverted to normal after treatment stopped. *2242* Significant at 12 weeks in 7 patients with severe rosacea treated with 1 mg/kg/d. *2307*

Thyroxine (T₄), Free *Serum No Effect Physiological* No significant change in 24 healthy men given 1 mg/kg/d for 16 weeks. *2640*
Serum Decrease Physiological In 7 patients given 1 mg/kg/d for 12 weeks. *2307*

Thyroxine (T₄) Index, Free (FTI)
Serum Decrease Physiological From 91 to 78 in 18 patients with severe acne given 0.8 mg/kg daily for 3 mo: changes reverted to normal after treatment stopped. *2242* Significant at 12 weeks in 7 patients with severe rosacea treated with 1 mg/kg/d. *2307*

Triglycerides *Serum No Effect Physiological* From 2.12 to 2.72 mmol/L in 12 patients with hyperkeratotic disorders given 40 mg. *3659*
Serum Increase Physiological In 7 patients with severe rosacea treated with 1 mg/kg/d for 12 weeks. Effects possibly due to induction of hepatic microsomal enzymes. *2307* Increase in both men and women with 1 mg/kg/d for 20 weeks when given for nodulocystic acne. *0326* Reversible dose-related effect noted in patients with myelodysplastic syndrome and leukemia at high doses. *1319* From 5.1 to 6.1 mmol/L in 18 patients with severe acne given 0.8 mg/kg daily for 3 mo: changes reverted to normal after treatment stopped. *2242* From 5.4 to 6.8 mmol/L at 6 weeks in 7 patients with severe rosacea treated with 1 mg/kg/d. *2307* During treatment but not to abnormal levels. *1818* In 25 of 523 patients with mean daily dose of 109 mg for 150 d. *3867* Increase in both men and women with 1 mg/kg/d for 20 weeks when given for nodulocystic acne. *0326*

Triglycerides, High Density Lipoprotein
Serum No Effect Physiological From 0.28 to 0.26 mmol/L in 12 patients with hyperkeratotic disorders given 40 mg daily for 2 mo. *3659*
Serum Increase Physiological From 0.14 to 0.31 mmol/L at 6 weeks in 7 patients with severe rosacea treated with 1 mg/kg/d. *2307* In 7 patients with severe rosacea treated with 1 mg/kg/d for 12 weeks. Effects possibly due to induction of hepatic microsomal enzymes. *2307*

Triglycerides, Low Density Lipoprotein
Serum Increase Physiological From 0.45 to 0.61 mmol/L in 12 patients with hyperkeratotic disorders given 40 mg daily for 2 mo. *3659* In 7 patients with severe rosacea treated with 1 mg/kg/d for 12 weeks. Effects possibly due to induction of hepatic microsomal enzymes. *2307* From 0.26 to 0.56 mmol/L at 6 weeks in 7 patients with severe rosacea treated with 1 mg/kg/d. *2307*

Triglycerides, Very Low Density Lipoprotein
Serum Increase Physiological From 0.43 to 0.77 mmol/L at 6 weeks in 7 patients with severe rosacea treated with 1 mg/kg/d. *2307* From 1.34 to 1.81 mmol/L in 12 patients with hyperkeratotic disorders given 40 mg daily for 2 mo. *3659* In 7 patients with severe rosacea treated with 1 mg/kg/d for 12 weeks. Effects possibly due to induction of hepatic microsomal enzymes. *2307*

Tri-iodothyronine (T₃) *Serum No Effect Physiological* No significant change in 24 healthy men given 1 mg/kg/d for 16 weeks. *2640*
Serum Decrease Physiological In 7 patients with severe rosacea treated with 1 mg/kg/d for 12 weeks. Effects possibly due to induction of hepatic microsomal enzymes. *2307* Significant at 12 weeks in 7 patients with severe rosacea treated with 1 mg/kg/d. *2307*

TSH Response to TRH *Serum No Effect Physiological* In 7 patients with severe rosacea treated with 1 mg/kg/d for 12 weeks. Effects possibly due to induction of hepatic microsomal enzymes. *2307*

Itraconazole

Alanine Aminotransferase *Serum Increase Physiological* Mild reversible increases without serious hepatotoxicity. *3567*

Aspartate Aminotransferase
Serum Increase Physiological Mild reversible increases without serious hepatotoxicity. *3567*

Cyclosporine *Serum No Effect Physiological* No effect reported on metabolism. *3567*

Jaundice

Cholesterol Esters *Serum Decrease Physiological* Obstruction decrease bile salts for esterification. *3947*

Protein Bound Iodine (PBI) *Serum Increase Physiological* If biliary regurgitation as normal excretion in bile. *3947*

Thyroxine (T₄) *Serum Increase Physiological* If biliary obstruction as normal excretion in bile. *3947*

Kanamycin

Alanine Aminotransferase *Serum Increase Physiological* May cause hepatotoxicity. *2313*

Alkaline Phosphatase *Serum Increase Physiological* May cause hepatotoxicity. *2313*
Urine Increase Physiological Due to nephrotoxic effect of drug. *2897*

Amino Acids *Urine Positive Analytical* Reacts with ninhydrin; extra spot TLC, high voltage electrophoresis. *2855* Unusual ninhydrin positive spot observed with TLC. *1596*

Ammonia *Plasma Decrease Physiological* Impairs NH_3 production by gut bacteria. *1487*

Aspartate Aminotransferase
Serum Increase Physiological May cause hepatotoxicity. *2313*

Bilirubin *Serum No Effect Physiological* Clinically insignificant displacement from protein in neonates. *3748*
Serum Increase Physiological May cause hepatotoxicity. *2313*

BSP Retention *Serum Increase Physiological* May cause hepatotoxicity. *1022*

Carotene *Serum Decrease Physiological* May induce malabsorption with diarrhea. *1680*

Casts *Urine Increase Physiological* Nephrotoxic effect (cylindruria and granular casts). *0071*

Cephalin Flocculation *Serum Increase Physiological* May cause hepatotoxicity. *2313*

Cholesterol *Serum Decrease Physiological* Forms salts with bile acids in gut. *1061*

Creatinine *Serum Increase Physiological* Nephrotoxic effect (common but slight). *2220*

Eosinophils *Blood Increase Physiological* Allergic reaction. *1513*

Erythrocytes *Urine Increase Physiological* Actual bleeding may occur. *1022*

Fat *Feces Increase Physiological* May induce malabsorption with diarrhea. *1680* Steatorrhea produced probably by causing mucosal damage. *3023*

Fibrinogen *Plasma Decrease Physiological* May occur at beginning of therapy. *1343*

Hemoglobin *Urine Increase Physiological* Actual bleeding occurs. *1022*

Leucine Aminopeptidase *Urine Increase Physiological* Associated with proximal renal tubular injury. *0078*

Leukocytes *Blood Increase Physiological* Due to eosinophilia. *2313*

Nitrofurantoin *Urine No Effect Analytical* No effect on method of Conklin and Hollifield. *0704*

Nonprotein Nitrogen *Serum Increase Physiological* Nephrotoxic effect. *2313*

Potassium *Serum No Effect Analytical* At concentration of 10 mg/L had no effect on measurement by ISE with predilution. *3393*

Protein *Urine Increase Physiological* Nephrotoxic effect. *1022*

Prothrombin Time *Plasma Increase Physiological* May decrease vitamin K synthesis by gut bacteria. *1487*
Plasma Decrease Physiological Effect noted at beginning of therapy. *2427*

Sodium *Serum No Effect Analytical* At concentration of 10 mg/L had no effect on measurement by ISE with predilution. *3393*

Thymol Turbidity *Serum Increase Physiological* May cause hepatotoxicity. *2313*

Urea Nitrogen *Serum Increase Physiological* Nephrotoxic effect (common slight elevation). *0070*

Urobilin *Urine Increase Physiological* Nephrotoxic effect. *1680*

Volume *Urine Decrease Physiological* Nephrotoxicity may occur with oliguria, azotemia. *1680*

Xylose Excretion *Urine Decrease Physiological* Due to impaired gastrointestinal absorption. *1680*

Kaolin

Clindamycin *Serum Decrease Physiological* With pectin causes delayed absorption. *3794*

Diagnex Blue Excretion *Urine Increase Physiological* Displacement of diagnex blue from resin. *3879*

Lincomycin *Serum Decrease Physiological* Inhibits gastrointestinal absorption. *1487*

Pseudoephedrine *Serum Decrease Physiological* Delayed absorption due to adsorption. *3794*

Kerosene

Alanine Aminotransferase *Serum Increase Physiological* Hepatic damage with large doses. *1343*

Casts *Urine Increase Physiological* With severe toxicity following ingestion. *1302*

Erythrocytes *Urine Increase Physiological* With severe toxicity following ingestion. *1302*

Occult Blood *Feces Positive Physiological* Toxic effect if ingested. *1343*

Protein *Urine Increase Physiological* Renal damage with large doses. *1343*

Urea Nitrogen *Serum Increase Physiological* Renal damage with large doses. *1343*

Ketamine

Alanine Aminotransferase *Serum Increase Physiological* In 14 of 34 individuals who had drug as anesthetic for intermediate operations. *0973*

Alkaline Phosphatase *Serum Increase Physiological* In 14 of 34 individuals who had drug as anesthetic for intermediate operations. *0973*

Aspartate Aminotransferase
Serum Increase Physiological In 14 of 34 individuals who had drug as anesthetic for intermediate operations. *0973*

γ-Glutamyltransferase (GGT)
Serum Increase Physiological In 14 of 34 individuals who had drug as anesthetic for intermediate operations. *0973*

Ketanserin

Cholesterol *Serum No Effect Physiological* In 50 hypertensive patients given 80 mg/daily for 3 mo. *2171*

Cholesterol, High Density Lipoprotein
Serum Increase Physiological In 50 hypertensive patients given 80 mg/daily for 3 mo. *2171*

Cholesterol, Low Density Lipoprotein
Serum Decrease Physiological In 50 hypertensive patients given 80 mg/daily for 3 mo. *2171*

Keto Acids

5-Hydroxyindoleacetic Acid (5-HIAA)
Urine Decrease Analytical Color formation with nitrosonaphthol inhibited. *0583*

α-Ketobutyric Acid

Ferric Chloride Test
Urine Positive Analytical Purple: fades to red brown. *0443*

Ketoconazole

Acid Phosphatase, Prostatic
Serum Decrease Physiological In men with prostatic cancer with 400 mg every 8 h and prolonged treatment. *3390*

Alanine Aminotransferase
Serum Increase Physiological Transient abnormalities of liver function observed in 10% patients but true hepatic injury in only 0.1 to 1%. Probably idiosyncrasy involved but may be immune hypersensitivity in some cases. *3390* Delayed reaction to drug after withdrawal in single case of chronic candidiasis in 61 y old woman. *3513* In 4 of 36 patients treated with 200 mg daily over 8 mo. *3034* Rapidly developing liver failure in 67 year old woman taking 200 mg drug daily for 2 mo. *0960*

Alkaline Phosphatase
Serum No Effect Physiological In 9 healthy men given up to 1200 mg/d for 1 week. *1300*
Serum Increase Physiological Delayed reaction to drug after withdrawal in single case of chronic candidiasis in 61 y old woman. *3513* Rapidly developing liver failure in 67 year old woman taking 200 mg drug daily for 2 mo. *0960*

Androstenedione
Plasma Decrease Physiological Observed in 2 h after single 200 mg dose in normal men. *3390*

Androstenedione, Free
Serum Decrease Physiological Observed in 2 h after single 200 mg dose in normal men. *3390*

Aspartate Aminotransferase
Serum Increase Physiological Transient abnormalities of liver function observed in 10% patients but true hepatic injury in only 0.1 to 1%. Probably idiosyncrasy involved but may be immune hypersensitivity in some cases. *3390* Delayed reaction to drug after withdrawal in single case of chronic candidiasis in 61 y old woman. *3513* In 4 of 36 patients treated with 200 mg daily over 8 mo. *3034* Rapidly developing liver failure in 67 year old woman taking 200 mg drug daily for 2 mo. *0960*

Bilirubin
Serum Increase Physiological Delayed reaction to drug after withdrawal in single case of chronic candidiasis in 61 y old woman. *3513* Rapidly developing liver failure in 67 year old woman taking 200 mg drug daily for 2 mo. *0960*

Bilirubin, Direct
Serum Increase Physiological Rapidly developing liver failure in 67 year old woman taking 200 mg drug daily for 2 mo. *0960*

Calcium
Serum No Effect Physiological In 9 healthy men given up to 1200 mg/d for 1 week. *1300*

Cholesterol
Serum Decrease Physiological After high dose treatment in patients with advanced prostatic cancer; appears to be dose related. *3390*

Cortisol
Plasma No Effect Physiological Observed in 2 h after single 200 mg dose in normal men. *3390*
Plasma Decrease Physiological Significant reduction in patients receiving 800 mg or more daily. *2839*
Urine Decrease Physiological Blocks adrenal response to corticotropin and related to serum concentration of drug. *2839* Significant reduction in 6 patients receiving 1.2 g daily for prostatic cancer, also blunted plasma cortisol response to Synacthen. *3812*

Cortisol, Free
Urine Decrease Physiological Approximate 50% reduction in patients receiving 800 mg daily. *2839*

Cyclosporine
Serum Increase Physiological Prolongs half-life, probably by competing for metabolizing enzymes. *3809*
Blood Increase Physiological Inhibits function of cytochrome P-450 hepatic enzymes inhibiting clearance of cyclosporine. *1113*

1,25-Dihydroxy Vitamin D_3
Serum Decrease Physiological Dose dependent effect after administration to normal volunteers. *3390* In 9 healthy men given up to 1200 mg/d for 1 week. *1300*

Estradiol
Plasma No Effect Physiological In men with prostatic cancer with 400 mg every 8 h and prolonged treatment. *3390*
Plasma Decrease Physiological Mild effect in 4 volunteer males with 600 mg doses. Bound and free ratio unchanged in 5 males receiving high doses for long time effect variable but estradiol testosterone ratio persistently increased. *2838*

Follicle Stimulating Hormone (FSH)
Plasma Increase Physiological Increase by 63% approximately in normal men with dose effect maximal at 900 mg/d. Due to stimulatory effect of dose dependent fall of testosterone. *1299*

17-Hydroxyprogesterone
Plasma Increase Physiological Observed in normal men due to blockade of 17,20-desmolase by drug, with inconsistent effect on serum estradiol. *1299*

25-Hydroxy Vitamin D_3
Serum No Effect Physiological After administration to normal volunteers. *3390* In 9 healthy men given up to 1200 mg/d for 1 week. *1300*

Luteinizing Hormone (LH)
Plasma Increase Physiological Increase by 127% approximately in normal men with dose effect maximal at 900 mg/d. Due to stimulatory effect of dose dependent fall of testosterone. *1299*

Osmolality
Serum Decrease Physiological To 248 mosmol/kg in 73 year old man with prostatic cancer treated with 600 mg daily for 2 1/2 mo. *2817*

Parathyroid Hormone
Plasma No Effect Physiological After administration to normal volunteers. *3390* In 9 healthy men given up to 1200 mg/d for 1 week. *1300*

Phosphate
Serum No Effect Physiological In 9 healthy men given up to 1200 mg/d for 1 week. *1300*

Progesterone
Plasma Increase Physiological In men with prostatic cancer with 400 mg every 8 h and prolonged treatment. *3390*

Prolactin
Plasma No Effect Physiological In men with prostatic cancer with 400 mg every 8 h and prolonged treatment. *3390*

Prothrombin Time
Plasma Increase Physiological Rapidly developing liver failure in 67 year old woman taking 200 mg drug daily for 2 mo. *0960*

Sodium
Serum Decrease Physiological To 121 mmol/L in 73 year old man with prostatic cancer treated with 600 mg daily for 2 1/2 mo. *2817*

Sperm Count
Semen Decrease Physiological With prolonged treatment with 800-1200 mg/d. *3390* Azospermia common in patients receiving drug. *2839*

Testosterone
Serum Decrease Physiological Marked effect in 4 volunteer males with 600 mg doses. With long term high dose treatment same effect observed. *2838* Transiently blocks testosterone synthesis. Doses above 800 mg/d may cause more prolonged blockade. *2839*

Testosterone, Free
Serum Decrease Physiological Observed in 2 h after single 200 mg dose in normal men. *3390* Equally reduced with bound testosterone. *2839*

α-Ketoglutarate

Creatinine
Serum No Effect Analytical No effect ion-exchange method of Mitchell. *2459*
Serum Increase Analytical Slight effect on Lloyd's procedure. Marked effect direct methods. *2459*

Ketones

Cortisol
Urine No Effect Analytical On fluorometric method of Ratliff and Hall. *2923*

Sugar
Urine Increase Analytical False positive with Benedict's. *1563*

Ketoprofen

Alanine Aminotransferase
Serum No Effect Analytical On colorimetric method at 10 times maximal therapeutic concentration. *1785*
Serum Decrease Analytical On continuous method. *1785*

Albumin
Serum No Effect Analytical On bromcresol green method on SMA II at physiological concentration. *1787* At concentration of 60 mg/L had no effect on BCG method. *3393*
Serum No Effect Physiological No effect seen in patients undergoing treatment. *1787*

Alkaline Phosphatase *Serum   No Effect   Analytical* On continuous method at 10 times maximal therapeutic concentration. *1785*

Aspartate Aminotransferase *Serum   No Effect   Analytical* On colorimetric method at 10 times maximal therapeutic concentration. *1785*
Serum   Decrease   Analytical On continuous method. *1785*

Bilirubin *Serum   No Effect   Analytical* No effect at therapeutic concentrations on Jendrassik-Grof, dimethylsulfoxide and spectrophotometric methods. *1786* On diazo method on SMA II at physiological concentration. *1787* At concentration of 60 mg/L had no effect on Jendrassik and Grof method. *3393*
Serum   No Effect   Physiological No effect seen in patients undergoing treatment. *1787*

Bleeding Time *Patient   No Effect   Physiological* In acute studies but prolonged in subacute in 11 patients given drug intravenously. *1234*

Calcium *Serum   No Effect   Analytical* On cresolphthalein complexone method on SMA II at physiological concentration. *1787* At concentration of 60 mg/L had no effect on cresolphthalein method. *3393*
Serum   No Effect   Physiological No effect seen in patients undergoing treatment. *1787*

Cholesterol *Serum   No Effect   Analytical* No effect on Liebermann-Burchard and enzymatic methods at therapeutic concentrations. *1786* On enzymatic method on SMA II at physiological concentration. *1787* At concentration of 60 mg/L had no effect on Liebermann-Burchard method. *3393*
Serum   No Effect   Physiological No effect seen in patients undergoing treatment. *1787*

Creatine Kinase *Serum   No Effect   Analytical* On continuous method at 10 times maximal therapeutic concentration. *1785*

Creatinine *Serum   No Effect   Analytical* No effect at therapeutic concentrations on alkaline picrate and Slot methods. *1786* On alkaline picrate method on SMA II at physiological concentration. *1787* At concentration of 60 mg/L had no effect on AutoAnalyzer Jaffé method. *3393*
Serum   No Effect   Physiological No effect seen in patients undergoing treatment. *1787*

Glucose *Serum   No Effect   Analytical* No effect at therapeutic concentrations on hexokinase, glucose dehydrogenase 2,4-dichlorophenol, ABTS and o-toluidine methods. *1786* On hexokinase method on SMA II at physiological concentration. *1787*
Serum   No Effect   Physiological No effect seen in patients undergoing treatment. *1787*

Hydroxybutyrate Dehydrogenase
Serum   No Effect   Analytical On continuous method at 10 times maximal therapeutic concentration. *1785*

Iron *Serum   No Effect   Analytical* No effect at therapeutic concentrations on Ramsay and bathophenanthroline methods. *1786* On Ferrozine method on SMA II at physiological concentration. *1787* At concentration of 60 mg/L had no effect on Ferrozine method. *3393*
Serum   No Effect   Physiological No effect seen in patients undergoing treatment. *1787*

Lactate Dehydrogenase *Serum   No Effect   Analytical* With pyruvate as substrate on continuous method. *1785*
Serum   Decrease   Analytical With lactate as substrate on continuous method. Marked effect on colorimetric method at 10 times maximal therapeutic concentration. *1785*

Occult Blood *Feces   Positive   Physiological* Gastrointestinal irritation is probably major side effect as reported to UK committee on safety of medicines. *0782*

Partial Thromboplastin Time
Plasma   No Effect   Physiological In 11 patients given drug intravenously. *1234*

Phosphate *Serum   No Effect   Analytical* At concentration of 60 mg/L had no effect on phosphomolybdate method. *3393*

Platelet Aggregation *Blood   Decrease   Physiological* In 11 patients given drug intravenously. *1234*

Protein *Serum   No Effect   Analytical* No effect at therapeutic concentrations on biuret and spectrophotometric methods. *1786* On biuret method on SMA II at physiological concentration. *1787* At concentration of 60 mg/L had no effect on biuret method with blank correction. *3393*
Serum   No Effect   Physiological No effect seen in patients undergoing treatment. *1787*

Prothrombin Time *Plasma   No Effect   Physiological* In 11 patients given drug intravenously. *1234*

Triglycerides *Serum   No Effect   Analytical* No effect at therapeutic concentrations on enzymatic methods. *1786* On enzymatic method on SMA II at physiological concentration. *1787*
Serum   No Effect   Physiological No effect seen in patients undergoing treatment. *1787*

Urea Nitrogen *Serum   No Effect   Analytical* No effect at therapeutic concentrations on glutamate dehydrogenase, phenol- hypochlorite and diacetylmonoxime methods. *1786* On diacetyl monoxime method on SMA II at physiological concentration. *1787* At concentration of 60 mg/L had no effect on diacetylmonoxime method. *3393*
Serum   No Effect   Physiological No effect seen in patients undergoing treatment. *1787*

Uric Acid *Serum   No Effect   Analytical* No effect at therapeutic concentrations on uricase-catalase, aldehyde dehydrogenase, direct UV-test and phosphotungstate methods. *1786* On phosphotungstate method on SMA II at physiological concentration. *1787* At concentration of 60 mg/L had no effect on phosphotungstate reduction method. *3393*
Serum   No Effect   Physiological No effect seen in patients undergoing treatment. *1787*

Ketosis

Aspartate Aminotransferase *Serum   Increase   Analytical* With non specific diazo procedure on SMA 12/60. *3217*

Vanillylmandelic Acid *Urine   Increase   Physiological* Marked effect ?due to hypovolemia. *0171*

Kwashiorkor

Tri-iodothyronine (T₃) *Serum   Decrease   Analytical* Effect on Wien lab test possible related to decrease protein. *3674*

Kynurenine

Amino Acids *Test Conditions   Increase   Analytical* Positive spot test with ninhydrin. *3765*

Histidine *Plasma   Increase   Analytical* May produce slight effect method of Ambrose. *0081*

Tyrosine *Plasma   No Effect   Analytical* On fluorometric procedure of Ambrose. *0080*

Labetalol

Aldosterone *Urine   No Effect   Physiological* With treatment for 4 mo in 15 patients with essential hypertension variable effect observed. *3776*
Urine   Increase   Physiological With treatment for 4 mo in 15 patients with essential hypertension variable effect observed. *3776*
Urine   Decrease   Physiological With treatment for 4 mo in 15 patients with essential hypertension variable effect observed. *3776*

Antinuclear Antibodies *Serum   Increase   Physiological* In one patient previously treated with methyldopa and atenolol. *1400*

Catecholamines *Urine   Increase   Analytical* With fluorometric method of Crout et al. *2430*

Cholesterol *Serum   No Effect   Physiological* No significant change in several studies with patients treated for 1-12 mo. *0088* In 8 patients given 600-1200 mg daily for 4 mo. *1806* In several studies from 1 to 12 mo. *0089*

Cholesterol, High Density Lipoprotein
Serum   No Effect   Physiological No significant change in several studies with patients treated for 1-12 mo. *0088* Although values reported from -12 to +23% in several studies from 1 to 12 mo. *0089*

Cholesterol, Low Density Lipoprotein
Serum   No Effect   Physiological No significant change in several studies with patients treated for 1-12 mo. *0088*

Cholesterol, Very Low Density Lipoprotein
Serum   No Effect   Physiological No significant change in several studies with patients treated for 1-12 mo. *0088*

Labetalol (continued)

C-Peptide Plasma No Effect Physiological In response to i.v. infusion of 100 mg over 10 minutes. 0226

Creatine Kinase Serum Increase Physiological In 3 of 9 patients with essential hypertension. Note MM isoenzyme most affected. 3149

Creatinine Serum No Effect Physiological No significant effect in 15 patients with essential hypertension treated for 1 mo. 3776

Dopamine Urine No Effect Analytical On HPLC method with electrochemical detection. 0434

Epinephrine Plasma No Effect Physiological No significant effect of acute i.v. administration (as measured by HPLC). 2974
Plasma Increase Analytical Elutes simultaneously with epinephrine with HPLC and electrochemical detection detection using method of Krstulovic et al. 0433
Urine Increase Analytical On HPLC method with electrochemical detection. 0434

Fatty Acids, Free (FFA) Serum No Effect Physiological In response to i.v. infusion of 100 mg over 10 minutes. 0226

Glucose Serum Increase Physiological Probably in response to norepinephrine release in response to i.v. infusion of 100 mg over 10 minutes. 0226

Growth Hormone Plasma No Effect Physiological In response to i.v. infusion of 100 mg over 10 minutes. 0226

Leukocytes Blood Decrease Physiological In one patient previously treated with methyldopa and atenolol. 1400

Lipoprotein, High Density Serum Increase Physiological In 15 patients with essential hypertension, also decreased total cholesterol: HDL ratio. 3776

Metanephrines, Total Urine No Effect Analytical On method of Bigelow and Weil-Malherbe. 2430
Urine Increase Analytical With photometric method of Pisano et al and method of Crout et al. 2430 Especially when combined with other antihypertensive drugs and HPLC method unless toluene extraction used. 0751

Norepinephrine Plasma No Effect Physiological No significant effect of acute i.v. administration (as measured by HPLC). 2974
Plasma Increase Physiological In response to i.v. infusion of 100 mg over 10 minutes. 0226
Urine No Effect Analytical On HPLC method with electrochemical detection. 0434

Prolactin Plasma Increase Physiological Marked in women, less so in men: possibly due to antidopaminergic effect of drug in response to i.v. infusion of 100 mg over 10 minutes. 0226

Renin Activity Plasma No Effect Physiological With treatment for 4 mo in 15 patients with essential hypertension variable effect observed. 3776
Plasma Increase Physiological With treatment for 4 mo in 15 patients with essential hypertension variable effect observed. 3776
Plasma Decrease Physiological With treatment for 4 mo in 15 patients with essential hypertension variable effect observed. 3776

Triglycerides Serum No Effect Physiological No significant change in several studies with patients treated for 1-12 mo. 0088 In several studies from 1 to 12 mo. 0089
Serum Increase Physiological Increase by 27% in 8 patients given 600-1200 mg daily for 4 mo. 1806

Urea Nitrogen Serum No Effect Physiological No significant effect in 15 patients with essential hypertension treated for 1 mo. 3776

Vanillylmandelic Acid Urine No Effect Analytical On method of Pisano et al. 2430

Lactate

Creatinine Serum No Effect Analytical No effect on method of Polar and Metcoff. 1975

Fatty Acids, Free (FFA) Serum No Effect Analytical No effect on method of Pinelli. 2819 At 50 mg/dL method of Soloni. 3381

3-Hydroxybutyrate Plasma No Effect Physiological No variability with method of Zivin. 3955

pH Blood Increase Physiological Used in treat of metabolic acidosis. 0071

Uric Acid Serum Increase Physiological Inhibits tubular secretion of urate. 3930
Urine Decrease Physiological Inhibits tubular secretion of urate. 3930

Lactate Dehydrogenase

2,3-Diphosphoglycerate Mutase Blood Increase Analytical Reported to prevent pyruvate inhibition. 3612

Lactation

Albumin Serum Decrease Physiological Remains about 1 g/dL below normal. 0987

Basal Metabolic Rate Patient Increase Physiological Remains elevated from pregnancy. 0987

Ceruloplasmin Serum Increase Physiological Initially high falls to normal. 0987

Copper Serum Increase Physiological Initially high falls to normal. 0987

Estrogens Urine Increase Physiological Excretion 5-10 µg/24 h until normal cycle resumes. 0987

Lactose Urine Increase Physiological Quite common, especially in afternoon. 0987

Prolactin Plasma Increase Physiological Large response ten to forty days postpartum. 3647

Lactobacillus Acidophilus

Ammonia Plasma Decrease Physiological Causes reduction in hepatic encephalopathy. 2249

Lactose

Estriol Urine Decrease Analytical Probably destroys estriol during acid hydrolysis. 0023 Interference with GLC method. 3186

Estrogens Urine No Effect Analytical If sodium borohydride used with fluorescence procedures. 3899
Urine Decrease Analytical Affects fluorometric procedures unless removed. 3899

Galactose Urine Increase Physiological Maximum at 90 minutes in normals after oral load. 2551

Glucose Serum No Effect Analytical No effect on GOD-PERID procedure. 2771 At 200 mg/dL on hexokinase method of Coburn. 0676 No effect on Trinder glucose oxidase procedure. 2771
Serum Increase Analytical 1 g/dL equivalent to 0.3 mmol/L glucose oxidase procedure. 2140 Affects Neocuproin procedures. 2771 1 g/dL equivalent to 3.0 mmol/L with alkaline ferricyanide procedure. 1 g/dL equivalent to 1.3 mmol/L p-HBAH procedure of Lever. 2140 Produces 33 % color of glucose, o-toluidine method. 0583
Urine Increase Physiological Maximum at 90 minutes in normals after oral load. 2551

Lactate Serum Increase Physiological Maximum at 1 h, persists for 2 h. 1907
Saliva Increase Physiological Maximum at 1 h, persists for 2 h. 1907

Lactate Dehydrogenase Serum No Effect Physiological No effect on activity observed. 1907

Lactose Serum No Effect Physiological No change with tolerance test in infants. 2551
Serum Increase Physiological With tolerance test from 0.2 to 0.6 mg/dL. 2551
Urine Increase Physiological Less than 1% of oral load. 2551

Sugar Urine Increase Analytical False positive with Benedict's, Clinitest®. 1563

Lactulose

Ammonia Plasma Decrease Physiological In patients with hepatic encephalopathy. 1446

CSF Decrease Physiological If given as enema has marked effect. *1920*

Feces No Effect Physiological Although increased volume of feces. *0032*

Occult Blood *Feces No Effect Analytical* No effect on Hemoquant procedure. *3671*

pH *Feces Decrease Physiological* When given as retention enema. *1920*

Sodium *Serum Increase Physiological* In 20 of 75 courses of treatment of hepatic failure concentration exceeded 145 mmol/L. *2572*

Volume *Feces Increase Physiological* Slight effect observed. *0032*

Laxatives

Aldosterone *Plasma Increase Physiological* Marked increase in chronic abuser with dehydration. *3682*

Bicarbonate *Serum Increase Physiological* If chronic abuse occurs. *3682*

Calcium *Serum Decrease Physiological* Excessive use may have effect. *2313*

Chloride *Serum Decrease Physiological* If chronic abuse occurs. *3682*

Estriol *Urine Decrease Analytical* If contain phenolphthalein reduce hydrolysis. *0406*

pH *Blood Increase Physiological* If chronic abuse may cause metabolic alkalosis. *3682*

Potassium *Serum Decrease Physiological* Excessive use may have effect. *2313*

Protein *Serum Decrease Physiological* May occur with continued use. *2220*

Prothrombin Time *Plasma Increase Physiological* Accelerated gastrointestinal pass and decrease absorption of vitamin K. *1487*

Renin Activity *Plasma Increase Physiological* Marked increase in chronic abuser with dehydration. *3682*

Sodium *Serum Decrease Physiological* Excessive use may have effect. *2313*

Lead

Alanine *Urine Increase Physiological* Occurs with poisoning. *0987*

Amino Acids *Urine Increase Physiological* Transient observation in poisoning. *0489*

β-Amino-Isobutyric Acid *Urine Increase Physiological* Nephrotoxic effect with lead poisoning. *0987*

Aminolevulinic Acid *Serum Increase Physiological* If over 20 µg/dL indicates poisoning. *0489*
Urine Increase Physiological Observed in poisoning. *1343*

Aminolevulinic Acid Dehydrase
Red Blood Cells Decrease Physiological Negative correlation with blood concentration of lead. *0489*

α-Amino-Nitrogen *Urine Increase Physiological* Occurs with poisoning. *0987*

Basophilic Stippling
Red Blood Cells Abnormal Physiological Occurs in 60% childhood cases. *0489*

Bilirubin *Serum Increase Physiological* May cause hemolytic anemia. *0788*

Casts *Urine Increase Physiological* Nephrotoxicity with poisoning (cylindruria). *1343*

Chromosomes *Test Conditions No Effect Physiological* No effect of chronic occupational exposure. *2648*
Test Conditions Abnormal Physiological Clastogenic in human cells in chronic poisoning. *3282*

Color *Urine Increase Physiological* Red brown (?due to porphyrins and hemoglobin). *2313*

Copper *Red Blood Cells Increase Physiological* Occurs with poisoning. *0987*

Coproporphyrin *Urine Increase Physiological* Observed in poisoning (may occur with 6 µg/m³ in air). *1343*
Red Blood Cells Increase Physiological Occurs with poisoning. *0987*

Cortisol, Free *Urine Decrease Physiological* In some men occupationally exposed to large quantities of lead. *0771*

Erythrocytes *Blood Decrease Physiological* Hemolytic anemia (with basophilic stippling). *0788*
Urine Increase Physiological Nephrotoxicity with poisoning. *1343*

Erythrocyte Survival *Blood Decrease Physiological* Due to hemolysis. *0987*

Follicle Stimulating Hormone (FSH)
Plasma No Effect Physiological In a population of male battery workers compared with control group in cement industry. *0172* In some men occupationally exposed to large quantities of lead. *0771*

Glucose *Urine Increase Physiological* Nephrotoxic effect with lead poisoning. *0987* Nephrotoxicity with poisoning. *1343*

Glycine *Urine Increase Physiological* May be transient increase in poisoning. *0489*

Hematocrit *Blood Decrease Physiological* Hemolytic anemia. *0788*

Hemoglobin *Blood Decrease Physiological* Hemolytic anemia. *0788*
Urine Increase Physiological Occurs with acute hemolytic crisis. *1302*

17-Hydroxycorticosteroids *Urine Decrease Physiological* In some men occupationally exposed to large quantities of lead. *0771*

Iron *Serum Increase Physiological* Reported effect. *0987*

17-Ketosteroids *Urine No Effect Physiological* In a population of male battery workers compared with control group in cement industry. *0172*

Lead *Blood Increase Physiological* May be increased much above normal of 50 µg/dL. *0987* In a population of male battery workers compared with control group in cement industry. *0172*
Urine Increase Physiological Due to increased body load (in poisoning 80 µg/dL). *1302* In a population of male battery workers compared with control group in cement industry. *0172*
Semen Increase Physiological In a population of male battery workers compared with control group in cement industry. *0172*

Leukocytes *Blood Decrease Physiological* Pancytopenia (AMA Blood dyscrasias). *2429*

Luteinizing Hormone (LH) *Plasma No Effect Physiological* In a population of male battery workers compared with control group in cement industry. *0172* In some men occupationally exposed to large quantities of lead. *0771*

MCHC *Blood Decrease Physiological* Hemolytic anemia. *0788*

MCV *Blood Increase Physiological* Rare increase with poisoning. *0987*
Blood Decrease Physiological Hemolytic anemia. *0788*

Occult Blood *Feces Positive Analytical* Lead sulfide may simulate melena. *1302*
Feces Positive Physiological May be bloody diarrhea with poisoning. *1302*

Phosphate *Urine Increase Physiological* Occurs with poisoning. *0987*

Platelets *Blood Decrease Physiological* Pancytopenia (AMA Blood dyscrasias). *2429*

Porphobilinogen *Urine No Effect Physiological* Normal in lead porphyria. *1563*

Prolactin *Plasma No Effect Physiological* In a population of male battery workers compared with control group in cement industry. *0172*

Protein *Urine Increase Physiological* Nephrotoxic effect with poisoning Nephrotoxicity with poisoning. *1343*
CSF Increase Physiological Occurs with lead encephalopathy, encephalitis. *1302*

Protoporphyrin *Urine Increase Physiological* In a population of male battery workers compared with control group in cement industry. *0172*
Red Blood Cells Increase Physiological Occurs with poisoning. *0987*

Reticulocytes *Blood Increase Physiological* Due to stimulation of hemolysis. *1343*

Lead (continued)

Sperm Count *Semen Decrease Physiological* In a population of male battery workers compared with control group in cement industry. *0172* In some men occupationally exposed to large quantities of lead. *0771*

Sperm Motility *Semen Decrease Physiological* In some men occupationally exposed to large quantities of lead. *0771*

Testosterone *Serum No Effect Physiological* In a population of male battery workers compared with control group in cement industry. *0172*

Thyroid Stimulating Hormone (TSH)
Serum No Effect Physiological In some men occupationally exposed to large quantities of lead. *0771*

Thyroxine (T₄) *Serum Decrease Physiological* In some men occupationally exposed to large quantities of lead. *0771*

Thyroxine (T₄), Free *Serum Decrease Physiological* In some men occupationally exposed to large quantities of lead. *0771*

Tri-iodothyronine (T₃) *Serum No Effect Physiological* In some men occupationally exposed to large quantities of lead. *0771*

Urea Nitrogen *Serum Increase Physiological* May cause nephrotoxicity. *3204*

Uric Acid *Serum Increase Physiological* Nephropathy associated with decreased secretion per nephron. *0215*
Urine Decrease Physiological Nephropathy associated with decreased secretion per nephron. *0215*

Urobilinogen *Urine Increase Physiological* Due to hemolysis of poisoning. *0987*

Lecithin

Fatty Acids, Free (FFA) *Serum No Effect Analytical* No effect on method of Pinelli. *2819*
Serum Increase Analytical At 1.6 mmol/L = 100 βEq/L method of Noma. *2607*

Leucine

Ammonia *Plasma Decrease Analytical* With 500 nmoles produces 9% inhibitor indophenol reaction. *1290*

Crystals *Urine Increase Physiological* Yellow spheroids with radial striation (in acid). *0443*

Insulin *Plasma Increase Physiological* Facilitates uptake of amino acids by tissues. *1071*

Pyruvate Kinase *Red Blood Cells Decrease Analytical* Most marked with larger side chain. *0370*

Tyrosine *Plasma No Effect Analytical* On fluorometric procedure of Ambrose. *0080*

Leucovorin

N-Acetylglucosaminidase *Urine No Effect Analytical* Had no effect at 10 g/L on 2 colorimetric analytical methods. *1354*

Leukocytes

Appearance *Urine Abnormal Physiological* Cloudy, insoluble in dilute acetic acid. *0443*

Levamisole

Alanine Aminotransferase *Serum Increase Physiological* In 1 of 11 patients receiving postoperative radiation treatment following breast cancer. *2960*

Aspartate Aminotransferase
Serum Increase Physiological Mild increase in 2 of 11 patients at 2 and 6 mo respectively after start of treatment. Normalized with withdrawal of treatment. *2723* In 1 of 11 patients receiving postoperative radiation treatment following breast cancer. *2960*

Leukocytes *Blood Decrease Physiological* In 2 of 60 patients: sufficiently severe to warrant withdrawal from treatment. *2665* In 16% of 201 patients treated for rheumatoid arthritis had to be withdrawn from study. Occurred after mean treatment time of 7.4 mo. *3709* In 4 of 11 patients with breast cancer and radiation treatment. *2960*

Neutrophils *Blood Decrease Physiological* Occasional case of agranulocytosis reported. *3717* Causally related to presence of autoantibodies in serum. Granulocytoxins found in 6 of 20 patients. *0951* In 16% of 201 patients treated for rheumatoid arthritis had to be withdrawn from study. Occurred after mean treatment time of 7.4 mo. *3709* In 35% of 60 patients treated for rheumatoid arthritis, reversed with withdrawal of drug. *3841* In 4 of 11 patients with breast cancer and radiation treatment. *2960* In one patient with acute lymphoblastic leukemia receiving drug with methotrexate. *3849* Observed in 17 of 174 patients with breast cancer. *3563* In one patient with herpes simplex but also receiving other drugs. *1864*

Platelets *Blood Decrease Physiological* Marked reduction in one patient, recovered on withdrawal of drug, but fell with rechallange. *1014*

Levarterenol

Amino Acids *Serum Increase Physiological* Catabolic effect. *1022*
Urine Increase Analytical Reacts with ninhydrin; extra spot TLC, high voltage electrophoresis. *2855*

Ammoniacal Silver Nitrate
Test Conditions Positive Analytical Positive spot test with Tollen's reagent. *3765*

Amphetamine *Urine No Effect Analytical* No reaction with NBD chloride procedure of Monforte. *2473* No effect at 100 mg/L on method of Rutter. *3097*

Basal Metabolic Rate *Patient Increase Physiological* Normal metabolic response. *1176*

Chloride *Urine Decrease Physiological* Increased tubular resistance and reabsorption. *1176*

Cholesterol *Serum Increase Physiological* Reported effect ?mechanism. *1487*

Creatinine *Test Conditions Increase Analytical* Positive spot test with Jaffé reagent. *3765*

Cyclic Adenosine Monophosphate
Plasma Increase Physiological Response to i.v. infusion in normals No effect unless β-blocking agent also given. *0216*
Urine Increase Physiological Effect less marked than in blood. *0216*

Effective Renal Plasma Flow
Patient Decrease Physiological Blood flow reduced, filtration rate unchanged. *1343* Slight fall after i.v. infusion. *2427*

Epinephrine *Test Conditions Increase Analytical* 3.4% fluorescence of epinephrine in procedure of Peyrin. *2800*

Fatty Acids, Free (FFA) *Serum Increase Physiological* Marked increase observed after i.v. infusion. *3114* Metabolic response. *1176*

Glomerular Filtration Rate (GFR)
Urine No Effect Physiological After i.v. but increased filtration fraction. *1176*
Urine Decrease Physiological Slight fall after i.v. infusion. *2427*

Glucose *Serum No Effect Analytical* At 10 mg/dL no effect on glucose oxidase procedure of Gochman. *1311*
Serum Increase Analytical At 10 mg/dL affects alkaline ferricyanide procedure. *1311*
Serum Increase Physiological Slight increase observed after i.v. infusion. *3114*

Guaiacols Spot Test *Urine Negative Analytical* Action on procedure of Rogers. *3031*

Metanephrines, Total *Urine No Effect Analytical* At 50 mg/L on modified Pisano procedure. *1428*

Norepinephrine *Plasma Increase Physiological* After i.v. infusion. *3967*

Occult Blood *Feces Positive Physiological* Diffuse hemorrhagic enteritis with vasoconstriction. *3498*

Potassium *Urine Decrease Physiological* Increased tubular resistance and reabsorption. *1176*

Propoxyphene *Urine No Effect Analytical* Less than 1% fluorescence in procedure of Valentour. *3662*

Renal Blood Flow (RBF) *Patient Increase Physiological* After i.v. but increased filtration fraction. *1176*

Sodium *Urine Decrease Physiological* Increased tubular resistance and reabsorption. *1176*

Thyroxine (T₄) *Serum Increase Physiological* Metabolic response. *1176*

Urea Nitrogen *Test Conditions Increase Analytical* Brown with Berthelot's reagent. *1936*

Uric Acid *Serum Increase Physiological* Result of decreased urate clearance. *1124*
Urine Decrease Physiological Decreases urate excretion and renal plasma flow. *1124*
Test Conditions Increase Analytical Positive spot test with phosphotungstate. *3765*

Uric Acid Clearance *Urine Decrease Physiological* 20% reduction with infusion. *1125*

Urobilinogen *Test Conditions Increase Analytical* Positive spot test with Ehrlich's reagent. *3765*

Vanillylmandelic Acid *Urine Increase Physiological* Normal metabolite, effect slight usually. *3503*

Volume *Plasma Decrease Physiological* Due to loss of protein-free fluid to tissues. *1343*
Urine No Effect Physiological Observed after i.v. 0.2-44.0 µg/min. *1176*
Urine Decrease Physiological Slight fall after i.v. infusion. *2427*

Levodopa

S-Adenosylmethionine *Blood Decrease Physiological* O-methylation of catecholamines slowed. *2337*

Alanine Aminotransferase *Serum Increase Physiological* Transient effect returns to normal. *1680* Increase by 17% for 2 mo although later normalized, in patients with Parkinsons disease. *1417*

Alkaline Phosphatase *Serum Increase Physiological* Rare elevation reported. *0071*

Amino Acids *Urine Increase Analytical* Reacts with ninhydrin; extra spot TLC, high voltage electrophoresis. *2855*

Ammonia *Plasma Decrease Physiological* Observed in one case ?unrelated. *1487*

Amphetamine *Urine No Effect Analytical* No reaction with NBD chloride procedure of Monforte. *2473*

Amylase *Serum No Effect Analytical* At concentration of 1,000 mg/L had no effect on maltotetrose method. *3393*

Aspartate Aminotransferase *Serum Increase Analytical* At 1 mmol/L affects SMA 12/60 method. *3335*
Serum Increase Physiological Transient effect, normalizes despite continuation. *0071* Increase by 30% for 2 mo although later normalized, in patients with Parkinsons disease. *1417*

Bilirubin *Serum No Effect Analytical* At concentration of 6 mg/L had no effect on Ektachem® method. *3393*
Serum Increase Analytical At 1 mmol/L affects SMA 12/60 method. *3335* Theoretically reacts with diazo reagent. *3220* At concentrations above 80 mg/L raised concentration as measured by Jendrassik and Grof method. *3393*
Serum Increase Physiological Rare elevation reported. *0071*

Biotin *Serum Decrease Physiological* Associated with burning feet syndrome. *0824*

BSP Retention *Serum Increase Physiological* Mild and transient effect. *0824*

Calcium *Serum No Effect Analytical* At concentration of 197 mg/L had no effect on cresolphthalein method. *3393*

Catecholamines *Plasma Increase Analytical* Measured as epinephrine/norepinephrine by ethylenediamine. *2285*

Chloride *Serum No Effect Analytical* At concentration of 80 mg/L had no effect on Ektachem® method. *3393*

Color *Urine Increase Analytical* Red-tinged on voiding, blackens on standing. *3138*
Saliva Increase Analytical Brown reported with treatment. *0798*

Coombs' Test *Serum Positive Physiological* Autoimmune phenomenon (occurs after several months). *1486*

Coombs' Test, Direct *Serum Positive Physiological* Possible dose related without hemolysis. *1487*

Coombs' Test, Indirect *Serum Positive Physiological* Observed in fewer than 1% of patients. *0824*

Corticotropin *Plasma Increase Physiological* Magnitude variable, stress effect. *3324*

Cortisol *Plasma Decrease Physiological* Probably diminished ACTH secretion. *1390*

Creatinine *Serum No Effect Analytical* At concentration of 6 mg/L had no effect on creatinine iminohydrolase method. *3393*
Serum Increase Analytical Acts as reducing agent (probable effect). *0127*
Urine Increase Analytical Probable action as reducing agent. *0127*

Creatinine Clearance *Urine Increase Analytical* Reducing properties affect Jaffé method. *1136*

Dihydroxyphenylalanine
Test Conditions Increase Analytical 85% of fluorescence method of Waldmeier. *3735*

Dopamine *Urine Increase Physiological* Response to therapy in Parkinsonism. *2724*

Eosinophils *Blood Increase Physiological* Occasionally observed without symptoms. *0824*

Epinephrine *Test Conditions No Effect Analytical* On fluorescent procedure of Peyrin. *2800*

Erythrocytes *Blood Decrease Physiological* One case of hemolytic anemia reported. *1680*
Urine Increase Physiological Occasional report of hematuria. *0071*

Estradiol *Urine Increase Analytical* With method of Adlercreutz in one patient. *0023*

Estriol *Urine Increase Analytical* With method of Adlercreutz in one patient. *0023*

Estrogens *Urine Increase Analytical* Affects colorimetric and fluorometric procedures. *0022*

Fatty Acids, Free (FFA) *Serum Increase Physiological* Significant increase if levodopa high in serum (i.v. greater effect). *2997*

Ferric Chloride Test *Urine Positive Analytical* When 1-5 g ingested/d=black/brown. *0583*

Folate *Serum Decrease Physiological* Associated with burning feet syndrome. *0824*

Follicle Stimulating Hormone (FSH)
Plasma Increase Physiological Possible increase over fluctuations in controls. *2778*

Glucose *Serum No Effect Analytical* At 10 mg/dL no effect on glucose oxidase procedure of Gochman. *1311* At 10 mg/dL on MBTH procedure of Neeley. *2569* No effect on hexokinase method. *2568*
Serum No Effect Physiological No effect observed although increased plasma insulin. *1860*
Serum Increase Analytical At 1 mmol/L affects SMA 12/60 method. *3335* At 10 mg/dL affects alkaline ferricyanide procedure. *1311*
Serum Increase Physiological Probably converted to dopamine which acts. *0481*
Serum Decrease Analytical 51% decrease at 10 mg/dL glucose oxidase dianisidine procedure. *3037* May cause marked decrease with glucose oxidase method. *2568* At concentrations above 100 mg/L lowered concentration as measured by Ektachem® method. *3393* At concentrations above 3 mg/L lowered concentration as measured by GOD-Perid method. *3393* At concentrations above 300 mg/L lowered concentration as measured by Seralyzer method. *3393*
Urine No Effect Analytical No effect on TesTape®. *1099*
Urine Decrease Analytical False negative, inhibition of glucose oxidase method False negative if Clinistix® used (no effect on TesTape®). *1099*

Gonadotropins *Plasma Increase Physiological* Reported metabolic effect. *1390*

Growth Hormone *Plasma Increase Physiological* In normals single dose cause increase in 1-2 h. *1860* After i.v. infusion mean rose to 15.5 ng/ml. *1721*

Guaiacols Spot Test *Urine Negative Analytical* Action on procedure of Rogers. *3031*

Hematocrit *Blood Decrease Physiological* Mild not related to hemolysis. *0071*

Levodopa (continued)

Hemoglobin *Blood Decrease Physiological* Mild not related to hemolysis. *0071*
Urine Increase Physiological Occasional report of hematuria. *0071*

Homogentisic Acid *Urine No Effect Analytical* On TLC method of Feldman and Bowman. *1096*
Urine Increase Analytical If method of Briggs used If method of Nuberger used. *1096*

Homovanillic Acid *Urine Increase Physiological* In Parkinsonian patients is response to therapy. *0562*
CSF Increase Physiological ?metabolic response (variable between individuals). *2725*

17-Hydroxycorticosteroids *Urine Decrease Physiological* Possible inhibition of ACTH secretion. *0022*

17-Hydroxycorticosterone *Plasma No Effect Physiological* No effect observed on injection into dogs. *1440*

5-Hydroxyindoleacetic Acid (5-HIAA)
Urine Decrease Analytical Moderate effect on procedure of Udenfriend. *1097*
Urine Decrease Physiological In Parkinson's disease ?increased tryptophan pyrrolase activity. *0509*
CSF Decrease Physiological Inhibition of 5-hydroxy-tryptophan hydroxylase. *2725*

Hydroxy-Methoxymandelic Acid *Urine Increase Analytical* In a patient given Sinemet® (levodopa/carbidopa) using Pisano method. Note high blank. *0698*

Insulin *Plasma Increase Physiological* During therapy of Parkinsonism. *1860*

Inulin Clearance *Urine Increase Physiological* ?secondary to renal vasodilatation or direct action on tubules. *1136*

Isohomovanillic Acid *Urine Increase Physiological* Response to therapy in Parkinson patients. *0703*

^{131}I Uptake *Serum Decrease Physiological* Larodopa® contains tetraiodofluorescein. *2652*

Ketones *Urine Increase Analytical* Affects alkaline nitroprusside procedure. *0583* Intermittent false positive if Ketostix® or Phenistix®. *3884*

Lactate Dehydrogenase *Serum Increase Physiological* Rare instance of elevation, ?origin. *0071*

Leukocytes *Blood Increase Physiological* Unassociated with fever or infection reported. *0824*
Blood Decrease Physiological Transitory depression in a few patients. *0071*

Luteinizing Hormone (LH) *Plasma No Effect Physiological* No effect after 2 weeks in males. *3338*

Lymphocytes *Blood Increase Physiological* Observed with hemolytic anemia. *0824*

Metanephrines, Total *Urine Decrease Physiological* ?dopamine as neurotransmitter suppresses normetanephrine. *1700*

Methionine *CSF Decrease Physiological* In Parkinson patients significant effect after 2 weeks. *1496*

3-Methoxytyrosine *Serum Increase Physiological* Observed after 1 week when given levodopa. *1108*

Monoamine Oxidase *Serum Increase Physiological* Increased activity after 2-3 mo therapy. *3623*

Neutrophils *Blood Decrease Physiological* Occasional case of agranulocytosis reported. *3717*

Norepinephrine *Urine Increase Physiological* No effect on epinephrine excretion. *3146*
Test Conditions Increase Analytical 2.9% fluorescence of norepinephrine (Peyrin procedure). *2800*

Normetanephrine *Urine Decrease Physiological* Sharp decrease in normal after 3 g dose. *2634*

Occult Blood *Feces Positive Physiological* Single case of gastritis with melena. *2984*

PAH Clearance *Urine Increase Physiological* ?secondary to renal vasodilatation or direct action on tubules. *1136*

Phenylalanine *Plasma No Effect Physiological* In patients after 1 week given levodopa. *1108*

Phosphate *Serum No Effect Analytical* At concentration of 197 mg/L had no effect on phosphomolybdate method. *3393*

Platelets *Blood Decrease Physiological* Slight effect observed with hemolytic anemia. *0824*

Potassium *Serum No Effect Analytical* At concentration of 200 mg/L had no effect on measurement by ISE with predilution. *3393*
Serum Decrease Physiological Reduction of up to 0.9 mmol/L in normal subjects with inhalation of drug. *1437*
Urine Increase Physiological ?secondary to renal vasodilatation or direct action on tubules. *1136*

Prolactin *Plasma Decrease Physiological* Fall to 8.7% of baseline after 2 h in normals. *0521* Transient effect in nonpuerperal galactorrhea. *2274* Completely suppressed for 1-4 h after 250 mg orally. *1435*

Proline *Plasma Increase Physiological* In Parkinson patients significant effect after 2 weeks. *1496*

Protein *Serum No Effect Analytical* At concentration of 197 mg/L had no effect on biuret method with blank correction. *3393*

Protein Bound Iodine (PBI) *Serum Increase Physiological* ?effect due to tetraiodofluorescein in capsules. *0420*

PSP Excretion *Urine Increase Physiological* Increased plasma flow. *1136*

Pyridoxine *Serum Decrease Physiological* Associated with burning feet syndrome. *0824*

Reticulocytes *Blood Increase Physiological* Hemolytic anemia when decarboxylase inhibitor also. *0481*

Sodium *Serum No Effect Analytical* At concentration of 200 mg/L had no effect on measurement by ISE with predilution. *3393*
Urine Increase Physiological ?secondary to renal vasodilatation or direct action on tubules. *1136*

Sugar *Urine Increase Analytical* False positive with Clinitest® Produces trace positive if Clinitest® used. *1099*

T$_3$ Uptake *Serum No Effect Physiological* No effect observed in chronic treatment. *1860*

Testosterone *Serum No Effect Physiological* No significant effect observed after 2 weeks in males. *3338*

Thyroid Stimulating Hormone (TSH)
Serum No Effect Physiological No effect observed. *3324*
Serum Decrease Physiological In hypothyroidism but no change in euthyroid — is dopamine precursor. Diminishes response to TRH in hypothyroidism. *3798*

Thyroxine (T$_4$) *Serum No Effect Physiological* No effect observed. *0824*
Serum Increase Physiological During therapy of Parkinsonism. *1860*

Triglycerides *Serum Decrease Analytical* At concentrations above 6 mg/L lowered concentration as measured by GPO-PAP method. *3393*

Trihydroxyphenylacetate *Urine Increase Physiological* Observed in treatment of Parkinsonian patients. *3137*

Tyrosine *Plasma No Effect Physiological* In patients after 1 week given levodopa. *1108*

Urea Nitrogen *Serum No Effect Analytical* At concentration of 197 mg/L had no effect on diacetylmonoxime method. *3393*
Serum Increase Physiological Affects hepatic enzymes, probably not dehydration. *2318*
Serum Decrease Analytical At concentrations above 100 mg/L lowered concentration as measured by Seralyzer method. *3393*

Uric Acid *Serum No Effect Analytical* No increase reported when uricase used. *0071* At concentration of 3 mg/L had no effect on uricase method on aca. *3393*
Serum Increase Analytical Falsely high values with phosphotungstate methods. *1311* 10 mg/dL equivalent to 31.0 mg/dL copper chelate procedure 10 mg/dL equivalent to 2.3 mg/dL Nishi procedure. *2230* At concentrations above 20 mg/L raised concentration as measured by phosphotungstate reduction method. *3393*
Serum Increase Physiological Two cases reported, exaggerated by fructose. *0042*
Serum Decrease Analytical At concentrations above 3 mg/L lowered concentration as measured by uricase-PAP method. *3393*
Urine Increase Analytical Falsely high values with phosphotungstate methods. *1311*

Urobilin *Urine Increase Physiological* Usually minor and unimportant. *0824*

Vanillylmandelic Acid *Urine Increase Physiological* Small increase, larger increase HVA. *0562*
Urine Decrease Analytical At therapeutic physiological concentrations on Pisano procedure. *1097*
Urine Decrease Physiological ?dopamine as neurotransmitter suppresses normetanephrine. *1700*

Levoglutamide

Ammonia *Plasma No Effect Analytical* On indophenol reaction with 5,000 nmoles. *1290*
Plasma Increase Physiological Potential source of additional ammonia. *1943*
Plasma Decrease Analytical Inhibits indophenol color in Berthelot reaction. *2632*

Tyrosine *Plasma No Effect Analytical* On fluorometric procedure of Ambrose. *0080*

Urea Nitrogen *Test Conditions Increase Analytical* Blue color with Berthelot's reagent. *1936*

Levonorgestrel

Albumin *Serum No Effect Physiological* No significant difference over 3 y between implant recipients and IUD users. *0763*
Serum Increase Physiological In a group of women using levonorgestrel covered rods versus controls using copper IUD. *0899*

Alkaline Phosphatase *Serum No Effect Physiological* No significant difference over 3 y between implant recipients and IUD users. *0763* In a group of women using levonorgestrel covered rods versus controls using copper IUD. *0899*

Androstenedione *Plasma Increase Physiological* After 1 mo in 25 female volunteers when given as subdermal implant. *2687*
Plasma Decrease Physiological In 17 women using Norplant implants. *1799*

Apolipoprotein AI *Serum No Effect Physiological* When 150 µg levonorgestrel given with 30 µg ethinyl estradiol for 3 mo. Results normal within 2 mo after treatment stopped. *0312*

Apolipoprotein B *Serum Increase Physiological* Increase by 19% (apo B: apo A-I increased by 18%) when 150 µg levonorgestrel given with 30 µg ethinyl estradiol for 3 mo. Results normal within 2 mo after treatment stopped. *0312*

Aspartate Aminotransferase
Serum No Effect Physiological No significant difference over 3 y between implant recipients and IUD users. *0763* In a group of women using levonorgestrel covered rods versus controls using copper IUD. *0899*

Bilirubin *Serum No Effect Physiological* No significant difference over 3 y between implant recipients and IUD users. *0763* In a group of women using levonorgestrel covered rods versus controls using copper IUD. *0899*

Calcium *Serum No Effect Physiological* No significant difference over 3 y between implant recipients and IUD users. *0763* In a group of women using levonorgestrel covered rods versus controls using copper IUD. *0899*

Ceruloplasmin *Serum Decrease Physiological* Slight effect with daily dose of 0.125 mg in 30 healthy female volunteers when given alone. Values normal within 30 d of end of treatment. *3089*

Cholesterol *Serum No Effect Physiological* When 150 µg levonorgestrel given with 30 µg ethinyl estradiol for 3 mo. Results normal within 2 mo after treatment stopped. *0312* Insignificant decrease in 11 normolipoproteinemic women given 250 µg/d for 2 weeks. *3593* In a group of women using levonorgestrel covered rods versus controls using copper IUD. *0899*
Serum Decrease Physiological Significant reduction in patients over 3 y compared with patients with IUDs for 2 1/2 y. *0763*

Cholesterol, High Density Lipoprotein
Serum No Effect Physiological When 150 µg levonorgestrel given with 30 µg ethinyl estradiol for 3 mo. Results normal within 2 mo after treatment stopped. *0312* No difference between implant recipients and iud users. *0763*
Serum Decrease Physiological With daily dose of 0.125 mg in 30 healthy female volunteers when given alone. Values normal within 30 d of end of treatment. *3089* Increase by 35% in 11 normolipoproteinemic women given 250 µg/d for 2 weeks. *3593*

Cholesterol, Low Density Lipoprotein
Serum No Effect Physiological Insignificant increase in 11 normolipoproteinemic women given 250 µg/d for 2 weeks. *3593*
Serum Decrease Physiological Significant reduction in patients over 3 y compared with patients with IUDs for 2 1/2 y. *0763*

Cholesterol, Very Low Density Lipoprotein
Serum Decrease Physiological By more than 50% in 11 normolipoproteinemic women given 250 µg/d for 2 weeks. *3593*

Cortisol *Plasma No Effect Physiological* No effect observed in group of women 20 and 65 mo after levonorgestrel treated rods inserted in uterus. *0899* In a group of women using levonorgestrel versus control group with copper IUD. *3167*

Estradiol *Plasma No Effect Physiological* No effect observed in group of women 20 and 65 mo after levonorgestrel treated rods inserted in uterus. *0899*
Plasma Decrease Physiological In a group of women using levonorgestrel versus control group with copper IUD. *0899*

Glucose *Serum No Effect Physiological* In a group of women using levonorgestrel covered rods versus controls using copper IUD. *0899*
Serum Increase Physiological In implant users in comparison with iud users (average difference of 5 mg/dL). *0763*

Glycosylated Proteins *Serum No Effect Physiological* When 150 µg levonorgestrel given with 30 µg ethinyl estradiol for 3 mo. Results normal within 2 mo after treatment stopped. *0312*

Lactate Dehydrogenase *Serum No Effect Physiological* No significant difference over 3 y between implant recipients and IUD users. *0763* In a group of women using levonorgestrel covered rods versus controls using copper IUD. *0899*

Phosphate *Serum No Effect Physiological* In a group of women using levonorgestrel covered rods versus controls using copper IUD. *0899*
Serum Decrease Physiological Significant slight reduction in implant recipients compared with IUD users over 3 y. *0763*

Postheparin Hepatic Lipase
Plasma Increase Physiological Increase by 64% in 11 normolipoproteinemic women given 250 µg/d for 2 weeks. *3593*

Postheparin Lipoprotein Lipase
Plasma No Effect Physiological Insignificant increase in 11 normolipoproteinemic women given 250 µg/d for 2 weeks. *3593*

Protein *Serum No Effect Physiological* No significant difference over 3 y between implant recipients and IUD users. *0763*
Serum Increase Physiological In a group of women using levonorgestrel covered rods versus controls using copper IUD. *0899*

Sex Hormone Binding Globulin
Serum Decrease Physiological With daily dose of 0.125 mg in 30 healthy female volunteers when given alone. Values normal within 30 d of end of treatment. *3089* In 17 women using Norplant implants. *1799*

Sperm Count *Semen Decrease Physiological* When combined with testosterone enanthate may cause oligospermia. *0953*

Testosterone *Serum No Effect Physiological* No effect observed in group of women 20 and 65 mo after levonorgestrel treated rods inserted in uterus. *0899*
Serum Increase Physiological Approximately 24% increase after 6 mo use of subdermal implant. *2687*
Serum Decrease Physiological In 17 women using Norplant implants. *1799* In a group of women using levonorgestrel versus control group with copper IUD. *0899*

Testosterone, Free *Serum No Effect Physiological* In 17 women using Norplant implants. *1799*

Thyroid Stimulating Hormone (TSH)
Serum No Effect Physiological No effect observed in group of women 20 and 65 mo after levonorgestrel treated rods inserted in uterus. *0899* In a group of women using levonorgestrel versus control group with copper IUD. *0899*

Thyroxine (T$_4$) *Serum No Effect Physiological* No effect observed in group of women 20 and 65 mo after levonorgestrel treated rods inserted in uterus. *0899* In a group of women using levonorgestrel versus control group with copper IUD. *0899*

Transcortin *Serum No Effect Physiological* With daily dose of 0.125 mg in 30 healthy female volunteers when given alone. Values normal within 30 d of end of treatment. *3089*

Levonorgestrel (continued)

Triglycerides *Serum Increase Physiological* Increase by 48% when 150 μg levonorgestrel given with 30 μg ethinyl estradiol for 3 mo. Results normal within 2 mo after treatment stopped. *0312*
Serum Decrease Physiological Increase by 32 % in 11 normolipoproteinemic women given 250 μg/d for 2 weeks. *3593* Significant reduction in patients over 3 y compared with patients with IUDs for 2 1/2 y. *0763*

Triglycerides, High Density Lipoprotein
Serum No Effect Physiological Insignificant increase in 11 normolipoproteinemic women given 250 μg/d for 2 weeks. *3593*

Triglycerides, Low Density Lipoprotein
Serum Decrease Physiological By about 14 % in 11 normolipoproteinemic women given 250 μg/d for 2 weeks. *3593*

Triglycerides, Very Low Density Lipoprotein
Serum Decrease Physiological Increase by 45% in 11 normolipoproteinemic women given 250 μg/d for 2 weeks. *3593*

Tri-iodothyronine (T$_3$) *Serum No Effect Physiological* No effect observed in group of women 20 and 65 mo after levonorgestrel treated rods inserted in uterus. *0899* In a group of women using levonorgestrel versus control group with copper IUD. *0899*

Urea Nitrogen *Serum No Effect Physiological* No significant difference over 3 y between implant recipients and IUD users. *0763* In a group of women using levonorgestrel covered rods versus controls using copper IUD. *0899*

Uric Acid *Serum No Effect Physiological* No significant difference over 3 y between implant recipients and IUD users. *0763* In a group of women using levonorgestrel covered rods versus controls using copper IUD. *0899*

Levorphanol

Morphine *Urine No Effect Analytical* Insignificant cross react with hemagglutination inhibition. *2514*
Urine Increase Analytical Cross react equally (or more) with RIA procedures. *2514*

Levothyroxine

Basal Metabolic Rate *Patient Increase Physiological* Metabolic effect of hormone (maximum at 1 week). *2220*

Cholesterol *Serum Decrease Physiological* Often therapeutic intent. *3505*

Fatty Acids, Free (FFA) *Serum Decrease Physiological* Correction of hypothyroid state. *3505*

^{131}I Uptake *Serum Decrease Physiological* Due to metabolic effect of drug. *2220*

Neutrophils *Blood Decrease Physiological* May cause neutropenia. *2583*

Phospholipids, Total *Serum Decrease Physiological* Correction of hypothyroid state. *3505*

Protein Bound Iodine (PBI) *Serum Increase Physiological* In patients on thyroid maintenance therapy. *0012*

Prothrombin Time *Plasma Increase Physiological* May potentiate action of anticoagulants. *1680*

Thyroxine (T$_4$) *Serum Increase Physiological* Endogenous hormone suppressed, exogenous measured. *0354*

Thyroxine (T$_4$), Free *Serum Decrease Physiological* Falls with therapy of hypothyroid state. *1680*

Triglycerides *Serum Increase Physiological* Effect observed in hypothyroid patients. *2462*

Lidocaine

Alanine Aminotransferase *Serum No Effect Analytical* No effect SMA 12/60 method with 3.5 mg/dL. *2636*

Albumin *Serum No Effect Analytical* No effect SMA 12/60 method with 3.5 mg/dL. *2636*

Alkaline Phosphatase *Serum No Effect Analytical* No effect SMA 12/60 method with 3.5 mg/dL. *2636*

Bilirubin *Serum No Effect Analytical* No effect SMA 12/60 method with 3.5 mg/dL. *2636*

Calcium *Serum No Effect Analytical* No effect SMA 12/60 method with 3.5 mg/dL. *2636*

Cholesterol *Serum No Effect Analytical* No effect SMA 12/60 method with 3.5 mg/dL. *2636*

Creatinine *Serum No Effect Analytical* No effect SMA 12/60 method with 3.5 mg/dL. *2636*
Serum Increase Analytical Some interference in some specimens with Gen02 slides for Ektachem® system, but rarely more than 3 mg/L. *1358*

Glucose *Serum No Effect Analytical* No effect SMA 12/60 method with 3.5 mg/dL. *2636* At concentration of 3.2 mg/L had no effect on Ektachem® method. *3393*

Lactate Dehydrogenase *Serum No Effect Analytical* No effect SMA 12/60 method with 3.5 mg/dL. *2636*

Phosphate *Serum No Effect Analytical* No effect SMA 12/60 method with 3.5 mg/dL. *2636*

Potassium *Serum No Effect Analytical* At concentration of 0.5 mg/L had no effect on measurement by ISE with predilution. *3393*

Protein *Serum No Effect Analytical* No effect SMA 12/60 method with 3.5 mg/dL. *2636*
CSF Increase Analytical Reacts with Folin-Ciocalteu reagent. *1237*

Sodium *Serum No Effect Analytical* At concentration of 0.5 mg/L had no effect on measurement by ISE with predilution. *3393*

Urea Nitrogen *Serum No Effect Analytical* At concentration of 3.2 mg/L had no effect on Ektachem® method. *3393*

Uric Acid *Serum No Effect Analytical* No effect SMA 12/60 method with 3.5 mg/dL. *2636*

Lidoflazine

Digoxin *Serum No Effect Physiological* Reportedly no effect on serum concentration. *0344*

γ-Glutamyltransferase (GGT)
Serum Increase Physiological Positive correlation with clinical state (normal liver function tests). *0322* ?due to changes in vasculature. *1059*

Light

Alkaline Phosphatase *Serum No Effect Physiological* Sunlight no effect in elderly. *1615*

Aminolevulinic Acid *Urine Decrease Analytical* Unless preserved with acid in dark. *3718*

Bilirubin *Serum Decrease Analytical* Drawn specimens decrease up to 30 %/h in bright daylight. *0251*
Serum Decrease Physiological Breakdown of bilirubin *in vivo* and *in vitro*. *3660*

Bilirubin, Direct *Serum Decrease Physiological* Breakdown of bilirubin (less sensitive than indirect). *1237*

Calcium *Serum Decrease Physiological* In elderly in absence of sun. *1615*

Cephalin Flocculation *Serum Increase Analytical* False positive occur after serum exposed to light. *0244*

Citrulline *Urine Decrease Analytical* Colored complex with diacetylmonoxime unstable. *2776*

Eosinophils *Blood Decrease Physiological* Sharp decrease (up to 30%) in response to bright light. *1638*

Indolylacryloylglycine *Urine Increase Physiological* Probably due to high intensity sunlight. *2302*

Leukocytes *Blood Increase Physiological* Normal response to sun or UV light. *0251*

β-Lipoproteins *Serum Decrease Physiological* Fall of 30% reported on short exposure to UV. *2893*

Phosphate *Serum Decrease Physiological* In elderly in absence of sun. *1615*

Porphyrins *Urine Decrease Analytical* Photosensitive. *2981*

Tryptophan *Plasma Decrease Analytical* Affects fluorometric method of Dencla and Dewey. *2114*
Urine Decrease Analytical Affects fluorometric method of Dencla and Dewey. *2114*

Uric Acid *Serum Increase Physiological* Sunlight- probably due to increased nucleic acid catabolism. *0264*

Urobilinogen *Urine Decrease Analytical* Oxidation by light occurs. *0582*

Vitamin A *Serum Decrease Analytical* Destroyed by sunlight and fluorescent lamps. *1205*

Lincomycin

Alanine Aminotransferase *Serum Increase Physiological* Hepatotoxic-cholestatic effect. *2220*

Alkaline Phosphatase *Serum Increase Physiological* Hepatotoxic-cholestatic effect. *1022*

Aspartate Aminotransferase
Serum Increase Physiological Hepatotoxic-cholestatic effect. *2220*

Bile *Urine Increase Physiological* Occurs with hepatotoxicity. *2313*

Bilirubin *Serum No Effect Physiological* Clinically insignificant displacement from protein in neonates. *3748*
Serum Increase Physiological Hepatotoxic-cholestatic effect. *2220*

Cephalin Flocculation *Serum Increase Physiological* Hepatotoxic-cholestatic effect. *2313*

Chloramphenicol *Serum No Effect Analytical* No effect at 100 mg/L on coupled enzymatic method. *2490*

Cholesterol *Serum Decrease Physiological* Hepatotoxic effect. *2220*

Fat *Feces Increase Physiological* Steatorrhea may result from mucosal damage. *3023*

Folate *Serum No Effect Analytical* If chromatographic procedure of Landon used. *2068*
Serum Decrease Analytical Inhibits growth of L. Casei. *3504*

Glucose *Serum Decrease Physiological* May occur with hepatotoxicity. *2220*

131I Uptake *Serum Decrease Physiological* Lincocin® contains tetraiodofluorescein. *2652*

Leukocytes *Blood Decrease Physiological* Agranulocytosis/leukopenia/neutropenia. *0075*

Neutrophils *Blood Decrease Physiological* Occasional case of agranulocytosis reported. *3717*

5'-Nucleotidase *Serum Increase Physiological* Due to cholestasis. *1778*

Occult Blood *Feces Positive Physiological* May cause severe enterocolitis. *3316* Occasional case of pseudomembranous colitis reported with bloody or mucus diarrhea. *0447*

Platelets *Blood Decrease Physiological* Rare reversible thrombocytopenia. *1343*

Thymol Turbidity *Serum Increase Physiological* Hepatotoxic-cholestatic effect. *2313*

Linoleamide

Cholesterol *Serum Decrease Physiological* Inhibits sterol absorption. *0939*

Liothyronine

Basal Metabolic Rate *Patient Increase Physiological* Metabolic effect of hormone. *2220*

131I Uptake *Serum Decrease Physiological* Except in hyperthyroidism. *2220*

Protein Bound Iodine (PBI) *Serum Decrease Physiological* Remains below normal even with full replacement. *1680*

Thyroxine (T₄) *Serum Decrease Physiological* Depression of endogenous hormone. *2220*

Liotrix

Cholesterol *Serum Decrease Physiological* In hypothyroids falls to within normal range. *1680*

Protein Bound Iodine (PBI) *Serum Decrease Physiological* Therapeutic response. *0071*

Prothrombin Time *Plasma Increase Physiological* May potentiate action of oral anticoagulants. *1680*

T₃ Uptake *Serum Increase Physiological* Increases but remains within normal range. *1680*

Lipemia

Alanine Aminotransferase *Serum Increase Analytical* Turbidity may affect some methods. *0579* With inverse colorimetry: resultant effect of clearing of lipemia and formation of reaction product during SMAC® measurement. *2466*
Serum Increase Physiological Associated with alcoholism. *2313*

Albumin *Serum Increase Analytical* Affects most methods unless blank correction. *3873* Nonblanked BCG procedure in patient specimens measured on SMAC® in comparison with specimens following ultra-centrifugation. *2466*

Alkaline Phosphatase *Serum No Effect Analytical* In patient specimens measured on SMAC® in comparison with specimens following ultra-centrifugation. *2466*

Amylase *Serum No Effect Analytical* Up to 11.4 mmol/L with method of Rauscher et al. *2926*
Serum Increase Analytical Turbidity may affect some methods. *0579*

Angiotensin-Converting Enzyme (ACE)
Serum No Effect Analytical No effect reported on colorimetric method of Boomsma and Schalekamp. *2814*

Aspartate Aminotransferase *Serum Increase Analytical* Turbidity may affect some methods. *0579* With inverse colorimetry: resultant effect of clearing of lipemia and formation of reaction product during SMAC® measurement. *2466*
Serum Increase Physiological Associated with alcoholism. *2313*

Bicarbonate *Serum No Effect Analytical* In patient specimens measured on SMAC® in comparison with specimens following ultra-centrifugation. *2466*

Bilirubin *Serum Increase Analytical* Interferes with direct spectroscopic methods. *0582*

Calcium *Serum No Effect Analytical* In patient specimens measured on SMAC® in comparison with specimens following ultra-centrifugation. *2466*
Serum Increase Analytical Causes turbidity in Ferro-Ham procedure. *1563*

Cephalin Flocculation *Serum Increase Analytical* Reported effect adding to turbidity. *1563*

Chloride *Serum Decrease Analytical* Increase by 1.5% in patient specimens measured on SMAC® in comparison with specimens ultra-centrifugation. *2466*

Cholesterol *Serum Increase Analytical* Turbidity if extraction not used. *1563* To large and variable extent on Technicon® SMAC®. *2466*

Creatinine *Serum No Effect Analytical* Patient specimens measured on SMAC® in comparison with specimens following ultra- centrifugation. *2466*
Serum Increase Analytical At 20 g/L total lipids on enzymatic procedure. *2524*

Erythrocyte Sedimentation Rate
Blood Increase Physiological Large molecular weight components have effect. *2313*

Fibrinogen *Plasma Increase Analytical* Maximum effect on turbidimetric procedure. *3463*

Glucose *Serum No Effect Analytical* No effect on o-toluidine of Feteris or Cooper. *1206*
Serum Increase Analytical Observed effect on "Trucose" o-toluidine method. *1206*
Serum Decrease Analytical Increase by 1.9% patient specimens measured on SMAC® in comparison with specimens following ultra-centrifugation. *2466*

Hemoglobin *Plasma No Effect Physiological* No effect observed with induced lipemia. *1755*
Blood Increase Analytical May cause elevation by up to 3 g/dL. *0579*

Icteric Index *Serum Increase Analytical* Turbidity of serum. *1237*

Iron *Serum No Effect Analytical* If slight or moderate method of Megraw. *2391*

Lipemia *(continued)*

Iron *(continued)*

Serum Decrease Analytical If marked due to high blank (procedure of Megraw). *2391*

Lactate Dehydrogenase *Serum Decrease Analytical* With direct colorimetry: resultant effect of clearing of lipemia and formation of reaction product during SMAC® measurement. *2466*

Methemoglobin *Blood Increase Analytical* Produces turbidity. *1456*

Oxygen Saturation *Blood Increase Analytical* Slight effect on method using CO-Oximeter™. *0842*

Phosphate *Serum No Effect Analytical* In patient specimens measured on SMAC® in comparison with specimens following ultra-centrifugation. *2466*

Potassium *Serum No Effect Analytical* In patient specimens measured on SMAC® in comparison with specimens following ultra-centrifugation. *2466*

Propylthiouracil *Serum No Effect Analytical* No effect with 2,6-DQC (procedure of Ratliff). *2922*

Protein *Serum Increase Analytical* Turbidity of serum. *2313*
Serum Decrease Analytical By less than 1.5% in patient specimens measured on SMAC® in comparison with specimens following ultra-centrifugation. *2466*

Sodium *Serum Decrease Analytical* By less than 1.5% in patient specimens measured on SMAC® in comparison with specimens following ultra-centrifugation. *2466*

Thymol Turbidity *Serum Increase Analytical* Due to inherent turbidity. *2313*

Triglycerides *Serum Increase Analytical* To large and variable extent in patient specimens measured on SMAC® in comparison with specimens following ultra-centrifugation. *2466*

Urea Nitrogen *Serum Increase Analytical* May affect Berthelot reaction. *1862*
Serum Decrease Analytical Increase by 1.6% in patient specimens measured on SMAC® in comparison with specimens following ultra-centrifugation. *2466*

Uric Acid *Serum No Effect Analytical* In patient specimens measured on SMAC® in comparison with specimen following ultra-centrifugation. *2466*

Zinc Sulfate Turbidity *Serum Increase Analytical* Effect less marked than with thymol turbidity. *1563*

Lipochrome

Bilirubin *Serum Increase Analytical* May be measured as bilirubin in some methods. *1237*

Cholesterol *Serum Increase Analytical* Absorb strongly in blue region may affect L-B method. *1563*

Lipomul®

Bicarbonate *Serum Decrease Physiological* Nephrotoxic effect with azotemia. *1022*

BSP Retention *Serum Increase Physiological* Part of fat-overloading syndrome. *2427*

Creatinine *Serum Increase Physiological* Nephrotoxic effect. *2313*

Erythrocytes *Blood Decrease Physiological* May be progressive anemia with excess. *2427*
Urine Increase Physiological May cause actual bleeding. *1022*

Hemoglobin *Blood Decrease Physiological* May be severe hemolytic anemia. *2427*
Urine Increase Physiological Produces actual bleeding. *1022*

Nonprotein Nitrogen *Serum Increase Physiological* Nephrotoxic effect. *2313*

Occult Blood *Feces Positive Physiological* May be severe gastrointestinal tract bleeding. *2427*

Phosphate *Serum Increase Physiological* May occur with azotemia. *2313*

Platelets *Blood Decrease Physiological* Thrombocytopenia with fat-overloading. *2313*

Potassium *Serum Increase Physiological* Nephrotoxic effect. *2313*

Protein *Urine Increase Physiological* Nephrotoxic effect. *1022*

Urea Nitrogen *Serum Increase Physiological* Possible nephrotoxic effect. *0652*

Uric Acid *Serum Increase Physiological* Decreased clearance. *2313*

Liposol

Alanine Aminotransferase *Serum Decrease Analytical* Changed from 41.7 to 8.27 U/L in 26 randomly selected patient sera, before and after addition on Beckman Astra methods. *2054*

Albumin *Serum Increase Analytical* Changed from 34.7 to 39.3 g/L in 26 randomly selected patient sera, before and after addition on Beckman Astra methods. *2054*

Alkaline Phosphatase *Serum No Effect Analytical* Changed from 205.3 to 223.3 U/L in 26 randomly selected patient sera, before and after addition on Beckman Astra methods. *2054*

Aspartate Aminotransferase *Serum Increase Analytical* Changed from 49.3 to 56.6 U/L in 26 randomly selected patient sera, before and after addition on Beckman Astra methods. *2054*

Bicarbonate *Serum Decrease Analytical* Changed from 23.2 to 18.7 mmol/L in 26 randomly selected patient sera, before and after addition on Beckman Astra methods. *2054*

Bilirubin *Serum Decrease Analytical* Changed from 25.5 to 3.1 mg/L in 26 randomly selected patient sera, before and after addition on Beckman Astra methods. *2054*

Calcium *Serum Decrease Analytical* Changed from 90.0 to 84.1 mg/L in 26 randomly selected patient sera, before and after addition on Beckman Astra methods. *2054*

Chloride *Serum Decrease Analytical* Changed from 102.5 to 96.3 mmol/L in 26 randomly selected patient sera, before and after addition on Beckman Astra methods. *2054*

Cholesterol *Serum Decrease Analytical* Changed from 1774 to 0 mg/L in 26 randomly selected patient sera, before and after addition on Beckman Astra methods. *2054*

Creatine Kinase *Serum No Effect Analytical* Changed from 322.6 to 344.3 U/L in 26 randomly selected patient sera, before and after addition on Beckman Astra methods. *2054*

Creatinine *Serum Decrease Analytical* Changed from 22.9 to 21.5 mg/L in 26 randomly selected patient sera, before and after addition on Beckman Astra methods. *2054*

Glucose *Serum No Effect Analytical* Changed from 1295 to 1274 mg/L (insignificant) in 26 randomly selected patient sera, before and after addition on Beckman Astra methods. *2054*

Lactate Dehydrogenase *Serum Decrease Analytical* Changed from 217.2 to 134.4 U/L in 26 randomly selected patient sera, before and after addition on Beckman Astra methods. *2054*

Phosphate *Serum No Effect Analytical* Changed from 40.7 to 41.5 mg/L in 26 randomly selected patient sera, before and after addition on Beckman Astra methods. *2054*

Potassium *Serum Increase Analytical* Changed from 4.45 to 6.59 mmol/L in 26 randomly selected patient sera, before and after addition on Beckman Astra methods. *2054*

Protein *Serum Increase Analytical* Changed from 56.2 to 61.6 g/L in 26 randomly selected patient sera, before and after addition on Beckman Astra methods. *2054*

Sodium *Serum Decrease Analytical* Changed from 138.8 to 131.6 mmol/L in 26 randomly selected patient sera, before and after addition on Beckman Astra methods. *2054*

Triglycerides *Serum Decrease Analytical* Changed from 1752 to 139 mg/L in 26 randomly selected patient sera, before and after addition on Beckman Astra methods. *2054*

Urea Nitrogen *Serum No Effect Analytical* Changed from 262 to 257 mg/L in 26 randomly selected patient sera, before and after addition on Beckman Astra methods. *2054*

Uric Acid *Serum No Effect Analytical* Changed from 43.4 to 45.5 mg/L in 26 randomly selected patient sera, before and after addition on Beckman Astra methods. *2054*

Lipuria

Appearance *Urine Abnormal Physiological* Opalescent, milky soluble in either. *0443*

Liquid Soap

Methadone *Urine Decrease Analytical* Progressively more negative values with increasing amounts of several preparations of liquid soap. *0964*

Lisuride

Thyroid Stimulating Hormone (TSH)
Serum Decrease Physiological Can lower basal and TRH-stimulated values. *3798*

Lithium

Acetaminophen Screening Test *Urine Negative Analytical* No reaction with o-cresol at therapeutic concentrations. *3326*

Adipate *Urine Increase Physiological* May be considerable effect. *2107*

Albumin *Serum No Effect Physiological* No difference between treated manic depressives and controls. *2053*

Aldosterone *Urine Increase Physiological* Occurs after initial fall. *2527*

Alkaline Phosphatase *Serum Increase Physiological* Effect in 20% manic depressives treated for minimum of 20 mo. Markedly increased effect on bone isoenzyme observed in 66% manic depressives treated for minimum of 20 mo. *0488* Significant increase noted in drug treated population and bone isoenzyme increased in 27 of 41 such patients. In 19 patients bone isoenzyme increased although total normal. *0488*

Antidiuretic Hormone *Plasma Increase Physiological* Also acts like vasopressin. *1792* Occurs in majority of patients with polyuria consistent with defect in water balance at level of kidney (lithium administration now most common cause of nephrogenic diabetes Insipidus). *3118*

Bicarbonate *Urine Increase Physiological* Occurs on first day of treatment only. *1792*

Calcium *Serum No Effect Physiological* No significant effect observed with chronic treatment. *0488*
Serum Increase Physiological Mean effect in 130 manic depressives versus controls (about 2% increase). *2053* Increased above normal in about 13% patients, Reversible. *3118* Probably due to unmasking of primary hyperparathyroidism. *1054* Increase observed for next few hours when drug given at 10 p.m. *2401* 8% higher than in controls in 12 patients taking drug for 2 to 13 y. *1175* In 7 of 97 patients not necessarily attributable to drug. *0699*
Urine Decrease Physiological ?affects Calcium dependent catecholamine system. *2275* Picture of primary hyperparathyroidism observed in patients given treatment for long time, but mechanism not fully understood. *3118* Reduced at night in lithium treated patients versus other psychiatric patients and healthy controls. *2401*

Cholesterol *Serum Increase Physiological* Reported to induce myxedema. *2221*

Citrate *Urine No Effect Physiological* No consistent effect observed. *2107* No effect observed after administration. *0409*

Cortisol *Plasma Increase Physiological* Observed in some patients. *2829*
Plasma Decrease Physiological Lower in morning and less diurnal variation. *1466* Significant decrease in a.m. cortisol after 1 y in 48 depressed patients; p.m. values also affected in individuals with greatest change in response to treatment. *3357* Typically reduced in patients treated for 1 y as seen in study of 53 patients. *3357*

Creatine Kinase *Serum Increase Physiological* May be high sustained increase although disease controlled. *1357*

Creatinine *Serum Increase Physiological* Average increase from 0.94 to 1.08 mg/dL in 237 patients on long term treatment. *3705* Observed in a few patients but risk of renal insufficiency is remote even in patients given drug for many years. *3203* In 3 of 97 patients not necessarily attributable to drug. *0699* In lithium treated patients than healthy subjects. *0519*

Creatinine Clearance *Urine Decrease Physiological* Slight decrease: significant negative regression on serum lithium concentration with long-term treatment. *3705*

Fumarate *Urine Increase Physiological* May be considerable effect. *2107*

Glucose *Serum Increase Physiological* Hyperglycemia been reported after use (transient). *3672* In two patients glycosuria and hyperglycemia occurred when lithium started: tests previously normal. *1812*
Urine Increase Physiological Reported effect of lithium therapy in some cases consequence of hyperglycemia. *1488* In two patients glycosuria and hyperglycemia occurred when lithium started: tests previously normal. *1812*

Glucose Tolerance *Serum Decrease Physiological* Associated with hyperglycemia (?mechanism). *1680*

Glutarate *Urine Increase Physiological* Reversible inhibition of renal transport. *0409*

HMPG (4-Hydroxy-3-Methoxy-Phenylethylene Glycol)
CSF Decrease Physiological Significant reduction with treatment of manic patients. Initial high values correlates with severity of disease. *3517*

2-Hydroxy-4-Ketoglutarate *Urine Increase Physiological* May be considerable effect. *2107*

α-Hydroxyglutarate *Urine Increase Physiological* May be considerable effect. *2107*

Ionized Calcium *Serum Increase Physiological* 9% higher than in controls in 12 patients taking drug for 2 to 13 y. *1175*

^{131}I Uptake *Serum Increase Physiological* Mechanism unclear. *0544*
Serum Decrease Physiological Lithionate contains tetraiodofluorescein. *2652*

α-Ketoglutarate *Urine Increase Physiological* Reversible inhibition of renal transport. *0409*

Lactate *Urine No Effect Physiological* No effect observed after administration. *0409*

Leukocytes *Blood Increase Physiological* ?drug associated endocrine effect (may double). *2341* After 4 weeks treatment increase in 21 patients on average from 6.3 to 8.6 thousand/uL. *1798* Mean increase of 2.2 thousand/uL with treatment. *2078*

Lithium *Serum Increase Physiological* Therapeutic level between 0.5 and 1.0 meq/L. *1343*
Urine Increase Physiological Excretion proportional to plasma concentration. *1680*
CSF Increase Physiological Concentration about half in serum. *1343*
Red Blood Cells Increase Physiological But only to max of 50% in plasma. *1441*

Lymphocytes *Blood No Effect Physiological* No change observed although WBC count changed. *2078*
Blood Decrease Physiological ?drug associated endocrine effect. *3313*

Magnesium *Serum Increase Physiological* ?affects membrane transport systems. *3505* Picture of primary hyperparathyroidism observed in patients given treatment for long term, but mechanism not fully understood. *3118* Increased for 24 h after drug administered. *2401*
Urine Increase Physiological Following administration of therapy. *2275* Increased during day in lithium treated patients versus other psychiatric patients and healthy controls. *2401*
CSF Decrease Physiological Significant effect (reverse of serum). *1441*

Malate *Urine Increase Physiological* May be considerable effect. *2107*

Methylmalonic Acid *Urine No Effect Physiological* No consistent effect observed. *2107*

Neutrophils *Blood Increase Physiological* ?drug associated endocrine effect. *3313* Increase largely responsible for increase in WBC count. *2078*

Norepinephrine *Urine Decrease Physiological* Significant reduction with treatment of manic patients. Initial high values correlated with severity of disease. *3517*

Osmolality *Urine Decrease Physiological* Reduced concentrating ability in lithium treated patients than healthy subjects. *0519*

Lithium *(continued)*

Parathyroid Hormone   *Plasma   Increase   Physiological*
Observed in 21% psychiatric patients given compound, positively correlated with serum lithium concentration. *3118*   38% higher than in controls in 12 patients taking drug for 2 to 13 y. *1175*

Phosphate   *Serum   No Effect   Physiological*   No significant effect observed with chronic treatment. *0488*
Serum   Decrease   Physiological   Picture of primary hyperparathyroidism observed in patients given treatment for long time, but mechanism not fully understood. *3118*   When given at 10 p.m. drug caused decrease for next few hours. *2401*
Urine   No Effect   Physiological   No effect observed in lithium treated patients versus other psychiatric patients and healthy controls. *2401*

Pimelate   *Urine   No Effect   Physiological*   No consistent effect observed. *2107*

Platelets   *Blood   Increase   Physiological*   After 4 weeks treatment increase in 21 patients on average from 302 to 342 thousand/uL. *1798*

Potassium   *Serum   No Effect   Physiological*   No significant effect observed with chronic treatment. *0488*   Isolated case of hyperkalemia reported but usually no effect reported. *2836*
Serum   Increase   Physiological   ?by displacing from cells. *2275*
Serum   Decrease   Physiological   Slight decreases observed only. *0530*

Prolactin   *Plasma   No Effect   Physiological*   No significant effect with up to 8 g daily orally. *1416*

Protein   *Urine   Increase   Physiological*   May have slight nephrotoxic effect. *1488*

Protein Bound Iodine (PBI)   *Serum   Decrease   Physiological*   Inhibits iodination of tyrosine in thyroxine binding globulin. *0071*

Pyruvate   *Urine   No Effect   Physiological*   No effect observed after administration. *0409*

Sodium   *Serum   Decrease   Physiological*   Due to initial natriuresis and diuresis. *2275*
Urine   Increase   Physiological   Impairs reabsorption by tubules. *1680*
Urine   Decrease   Physiological   Due to action of aldosterone (initial increase). *2527*

Specific Gravity   *Urine   Decrease   Physiological*   Large proportion of patients with polyuria with long term treatment. *3705*

Suberate   *Urine   No Effect   Physiological*   No consistent effect observed. *2107*

Succinate   *Urine   Increase   Physiological*   May be considerable effect. *2107*

T_3 Uptake   *Serum   No Effect   Physiological*   Observed in one patient who developed thyrotoxicosis without pre-existing goiter. *2418*
Serum   Decrease   Physiological   Reported to induce myxedema. *2221*   Irreversible myxedema observed in 2 patients treated for 1 y. *2773*

Thyroid Antibodies   *Serum   Positive   Physiological*   In two thirds women with hypothyroidism treated for 2 y. *2185*

Thyroid Stimulating Hormone (TSH)
Serum   Increase   Physiological   Observed in long-term therapy in manic depressives. *2184*   Mechanism unclear. *0544*   Compensatory response to reduced peripheral hormone concentrations. Observed in about 15% patients given long-term treatment. TSH response to TRH is increased. *3118*   Hypophyseal influence responsible for augmented TRH-induced values. *3798*   In 10 of 237 patients with smaller increases in others treated for more than 6 mo: normalized in most cases on withdrawal. *0084*   Irreversible myxedema observed in 2 patients treated for 1 y. *2781*
Blood   Increase   Physiological   In 20% of women, but no men among 53 psychiatric patients treated for 2 y. *2185*

Thyroxine (T_4)   *Serum   Increase   Physiological*   Observed in one patient who developed thyrotoxicosis without pre-existing goiter. *2418*
Serum   Decrease   Physiological   Reported to induce myxedema. *2221*   Reduces thyroidal iodine uptake, iodination of tyrosine, release of T_4, to T_3 and hepatic metabolism of T_4 to T_3. *3118*   In 10 of 237 patients with smaller increases in others treated for more than 6 mo: normalized in most cases on withdrawal. *0084*   In 20% of women, but no men among 53 psychiatric patients treated for 2 y. *2185*   Irreversible myxedema

observed in 2 patients treated for 1 y. *2781*   Hypothyroidism observed in 1-2% of 2590 psychiatric patients treated with lithium. *2826*

Thyroxine (T_4), Free   *Serum   Decrease   Physiological*   Mechanism unclear, non toxic goiter may occur. *3312*

Titratable Acidity   *Urine   Decrease   Physiological*   Falls on first day of treatment. *1792*

Tri-iodothyronine (T_3)   *Serum   Decrease   Physiological*   Reduces thyroidal iodine uptake, iodination of tyrosine, release of T_4, to T_3 and hepatic metabolism of T_4 to T_3. *3118*   In 10 of 237 patients with smaller increases in others treated for more than 6 mo: normalized in most cases on withdrawal. *0084*

Uric Acid   *Serum   Decrease   Physiological*   Reported to have uricosuric effect. *0144*
Urine   Increase   Physiological   Reported effect of $LiCO_3$ in manic-depressives. *1487*

Vanillylmandelic Acid   *Urine   Increase   Physiological*   Slight increase only, mechanism not clear. *1488*

Volume   *Urine   Increase   Physiological*   Rare inhibition of vasopressin action on tubules. *2910*   Occurs in about 60% patients with start of treatment and persists in about 20--25%. *3118*   Large proportion of patients with polyuria with long term treatment. *3705*

Zinc   *Serum   No Effect   Analytical*   At 50 to 1 with flameless atomic absorption. *2038*
Urine   No Effect   Analytical   At 50 to 1 with flameless atomic absorption. *2038*

Lithium Heparin

Ammonia   *Plasma   No Effect   Analytical*   At concentration of 500 mg/L had no effect on Ektachem® method. *3393*

Bilirubin   *Serum   No Effect   Analytical*   At concentration of 8,000 mg/L had no effect on Ektachem® method. *3393*

Creatinine   *Serum   No Effect   Analytical*   At concentration of 500 mg/L had no effect on Ektachem® method. *3393*

Potassium   *Serum   No Effect   Analytical*   At concentration of 3,000 mg/L had no effect on Ektachem® method. *3393*

Lithium Lactate

Creatinine   *Serum   Increase   Analytical*   5 mmol/L = 0.01 mg/dL method of Heinegard. *1547*

Local Anesthetics

Erythrocytes   *Blood   Decrease   Physiological*   Bone marrow depression reported. *2427*

Leukocytes   *Blood   Decrease   Physiological*   Bone marrow depression and agranulocytosis reported. *2427*

Methemoglobin   *Blood   Increase   Physiological*   Reported effect. *2427*

Loperamide

Hydrochloric Acid   *Gastric Material   Decrease   Physiological*   Effect on both basal and submaximal pentagastrin-stimulated gastric acid secretion. Effect dose related. *0559*

Volume   *Gastric Material   Decrease   Physiological*   Effect on both basal and submaximal pentagastrin-stimulated gastric acid secretion. Effect dose related. *0559*

Loprazolam

Cortisol   *Urine   Decrease   Physiological*   Observed in 9 poor sleepers given up to 1 mg/d for 3 weeks: overnight urinary cortisol measured. Rebound increase on withdrawal. *0016*

Lorazepam

Albumin   *Serum   No Effect   Analytical*   At concentration of 0.05 mg/L had no effect on BCG method. *3393*

Bicarbonate   *Serum   No Effect   Analytical*   At concentration of 0.05 mg/L had no effect on method using phenolphthalein. *3393*

Bilirubin *Serum No Effect Analytical* At concentration of 0.05 mg/L had no effect on Jendrassik and Grof method. *3393*

Calcium *Serum No Effect Analytical* At concentration of 0.05 mg/L had no effect on cresolphthalein method. *3393*

Chloride *Serum No Effect Analytical* At concentration of 0.05 mg/L had no effect on mercurimetric method. *3393*

Cholesterol *Serum No Effect Analytical* At concentration of 0.05 mg/L had no effect on Liebermann-Burchard method. *3393*

Creatinine *Serum No Effect Analytical* At 10 mg/L on reversed phase LC procedure of Zhiri et al. *3942* At concentration of 0.05 mg/L had no effect on AutoAnalyzer Jaffé method. *3393*

Glucose *Serum No Effect Analytical* At concentration of 0.05 mg/L had no effect on GOD/POD-PAP method. *3393*

Indomethacin *Serum No Effect Analytical* No effect on HPLC method of Roberts and Smith. *3002*

Phosphate *Serum No Effect Analytical* At concentration of 0.05 mg/L had no effect on phosphomolybdate method. *3393*

Potassium *Serum No Effect Analytical* At concentration of 0.05 mg/L had no effect on flame-photometric method. *3393*

Protein *Serum No Effect Analytical* At concentration of 0.05 mg/L had no effect on biuret method with blank correction. *3393*

Sodium *Serum No Effect Analytical* At concentration of 0.05 mg/L had no effect on flame-photometric method. *3393*

Triglycerides *Serum No Effect Analytical* At concentration of 0.05 mg/L had no effect on lipase/esterase method. *3393*

Urea Nitrogen *Serum No Effect Analytical* At concentration of 0.05 mg/L had no effect on diacetylmonoxime method. *3393*

Uric Acid *Serum No Effect Analytical* At 10 mg/L on reversed phase LC procedure of Zhiri et al. *3942* At concentration of 0.05 mg/L had no effect on phosphotungstate reduction method. *3393*

Lorcainide

Osmolality *Serum Decrease Physiological* Significant effect in 16 of 33 patients with organic heart disease and ventricular arrhythmias. Effect observed after single i.v. dose. *3383*

Sodium *Serum Decrease Physiological* Increase by 8 meq/L significant effect in 16 of 33 patients with organic heart disease and ventricular arrhythmias. Effect observed after single i.v. dose. *3383*

Lovastatin

Alanine Aminotransferase *Serum Increase Physiological* Tendency to rise especially with higher doses but not increased above 3 times upper limit of normal. *1530*

Apolipoprotein AI *Serum Increase Physiological* Slight effect in clinical trial of 101 patients with heterozygous familial hypercholesterolemia. *1530*

Apolipoprotein AII *Serum Increase Physiological* Slight effect in clinical trial of 101 patients with heterozygous familial hypercholesterolemia. *1530*

Apolipoprotein B *Serum Decrease Physiological* Increase by 23% at 40 mg bid in clinical trial of 101 patients with heterozygous familial hypercholesterolemia. *1530*

Cholesterol *Serum Decrease Physiological* Increase by 14 to 34% in clinical trial of 101 patients with heterozygous familial hypercholesterolemia. *1530*

Cholesterol, High Density Lipoprotein
Serum Increase Physiological Slight effect in clinical trial of 101 patients with heterozygous familial hypercholesterolemia. *1530*

Cholesterol, Low Density Lipoprotein
Serum Decrease Physiological Increase by 17 to 39% in clinical trial of 101 patients with heterozygous familial hypercholesterolemia. *1530*

Cortisol *Plasma No Effect Physiological* No significant change even in patients with familial hypercholesterolemia receiving 40 mg bid. *1530*

Luteinizing Hormone (LH) *Plasma No Effect Physiological* No significant change even in patients with familial hypercholesterolemia receiving 40 mg bid. *1530*

Testosterone *Serum No Effect Physiological* No significant change even in patients with familial hypercholesterolemia receiving 40 mg bid. *1530*

Triglycerides *Serum Decrease Physiological* Increase by 14% at 40 mg bid in clinical trial of 101 patients with heterozygous familial hypercholesterolemia. *1530*

Loxapine

Prolactin *Plasma Increase Physiological* Significant change in response to 10 mg orally. *1416*

Lucanthone

Alanine Aminotransferase *Serum Increase Physiological* Chronic toxicity may cause liver damage. *1343*

Protein *Urine Increase Physiological* Chronic toxicity may cause renal damage. *1343*

Lugol's Iodine

Bilirubin *Serum Increase Physiological* Occasional hypersensitive response to iodines. *0071*

Eosinophils *Blood Increase Physiological* Rare allergic response to iodines. *0071*

Platelets *Blood Decrease Physiological* Rare possible response with purpura. *0071*

Protein Bound Iodine (PBI) *Serum Increase Analytical* Contains iodine, iodide: effect for up to 3 weeks. *3218*

Thyroxine (T$_4$) *Serum No Effect Analytical* No effect on methods. *3218*

Lutidine

Ionized Calcium *Serum Increase Analytical* At concentrations > 0.1 mmol/L on Calcium specific electrode. *0540*

Lynestrenol

Ceruloplasmin *Serum Increase Physiological* When 5 mg daily given alone to 30 healthy female volunteers. Normal values 30 d after treatment stopped. *3089*

Cholesterol, High Density Lipoprotein
Serum Decrease Physiological When 5 mg daily given alone to 30 healthy female volunteers. Normal values 30 d after treatment stopped. *3089* Icrease by 32% due to progestational activity in 6 women with endometriosis given 5 to 10 mg daily for 6 mo. *2279*

Cholesterol, Low Density Lipoprotein
Serum Increase Physiological Increase by 19% due to progestational activity in 6 women with endometriosis given 5 to 10 mg daily for 6 mo. *2279*

Sex Hormone Binding Globulin
Serum Decrease Physiological When 5 mg daily given alone to 30 healthy female volunteers. Normal values 30 d after treatment stopped. *3089*

Transcortin *Serum Increase Physiological* When 5 mg daily given alone to 30 healthy female volunteers. Normal values 30 d after treatment stopped. *3089*

Lysergide

Chromosomes *Test Conditions Abnormal Physiological* Clastogenic in human cells *in vitro* (?*in vitro*). *3282*

Creatinine Clearance *Urine Decrease Physiological* Temporary effect observed. *1633*

Fatty Acids, Free (FFA) *Serum Increase Physiological* 73% after 1.5 µg/kg orally. *1634*

Glucose *Serum Increase Physiological* Reported effect. *0117*

Phosphate Clearance *Urine Decrease Physiological* Temporary effect observed. *1633*

Lysergide (continued)

Protein Bound Iodine (PBI)   *Serum   Increase   Physiological* Stimulates TSH release, ?effect on I 131 uptake. *0012*

Lysine

Ammonia   *Plasma   Decrease   Analytical* With 500 nmoles produces 6% inhibitor indophenol reaction. *1290*

Insulin   *Plasma   Increase   Physiological* Slight effect, AIDS metabolism of amino acids. *1071*

Proline   *Plasma   No Effect   Analytical* No effect on method of Goodwin. *1345*

Tyrosine   *Plasma   No Effect   Analytical* On fluorometric procedure of Ambrose. *0080*

Urea Nitrogen   *Serum   No Effect   Analytical* No effect on Berthelot reaction. *1862*

Lysol®

Color   *Urine   Increase   Physiological* Brown-black with poisoning. *1187*

Erythrocytes   *Blood   Decrease   Physiological* May cause intravascular hemolysis. *0629*

Heinz Body Formation   *Blood   Positive   Physiological* May cause intravascular hemolysis. *0629*

Hematocrit   *Blood   Decrease   Physiological* May cause intravascular hemolysis. *0629*

Hemoglobin   *Blood   Decrease   Physiological* May cause intravascular hemolysis. *0629*
Urine   Increase   Physiological Occurs with massive hemolysis. *0629*

Methemoglobin   *Blood   Increase   Physiological* May cause intravascular hemolysis. *0629*

Reduced Glutathione
Red Blood Cells   Decrease   Physiological Marked effect with hemolysis. *0629*

Sugar   *Urine   Increase   Analytical* Excreted glucuronide reacts with Benedict's. *0629*

Mafenide

Amino Acids   *Urine   Increase   Analytical* Reacts with ninhydrin; extra spot TLC, high voltage electrophoresis. *2855*

Ammonia   *Urine   Decrease   Physiological* Inhibits carbonic anhydrase if applied topically. *3813*

Bicarbonate   *Urine   Increase   Physiological* Inhibits carbonic anhydrase if applied topically. *3813*

Chloride   *Urine   Decrease   Physiological* Selective retention. *0123*

Leukocytes   *Blood   Decrease   Physiological* Probable effect observed in one child. *0123*

pCO$_2$ *Blood   Decrease   Physiological* Inhibits carbonic anhydrase if applied topically. *3813*

pH   *Blood   Increase   Physiological* Usual finding with respiratory alkalosis. *0123*
Blood   Decrease   Physiological If respiratory impairment as reduced renal buffering. *0123*
Urine   Increase   Physiological Inhibits carbonic anhydrase if applied topically. *3813*

Potassium   *Urine   Increase   Physiological* Inhibits carbonic anhydrase if applied topically. *3813*

Magnesium

Calcium   *Serum   No Effect   Analytical* method of Morin at 10 mg/dL. *2485* On HPE method of American Monitor. *0140*
Serum   Increase   Analytical May be up to 15% of total precipitate. *0140*
Urine   Increase   Analytical Interferes with cresolphthalein procedures. *0140*

Ionized Calcium   *Serum   Decrease   Analytical* As concentration increases. *2056*

Lead   *Urine   Decrease   Analytical* 25 meq/L equivalent to 5 μg/L by atomic absorption. *3240*

Sodium   *Test Conditions   Increase   Analytical* By neutron activation 0.65 μg = 1 mg. *3336*

Magnesium Carbonate

Iron-Binding Capacity, Total (TIBC)
Serum   Increase   Analytical Variable effect method Young/Hicks if not dry. *0280*

Magnesium Hydroxide

Bishydroxycoumarin   *Plasma   Increase   Physiological* 75% increase in peak concentration if coadministered. *0079*

Dicumarol   *Plasma   Increase   Physiological* Increased absorption due to chelation. *3794*

Prothrombin Time   *Plasma   Increase   Physiological* Theoretically if bishydroxycoumarin co-administered. *0079*

Warfarin   *Plasma   No Effect   Physiological* No effect observed. *0079* No effect observed. *3794*

Magnesium Salts

Alkaline Phosphatase   *Serum   Increase   Analytical* Activators of enzyme in laboratory procedures. *3588*

Calcium   *Serum   Increase   Analytical* Measured as Calcium in some EDTA procedures. *1563*
Serum   Decrease   Physiological Competes with calcium for gastrointestinal tract absorption. *2005*

Diagnex Blue Excretion   *Urine   Increase   Physiological* Heavy metal displacement of diagnex blue. *3879*

Magnesium   *Serum   Increase   Physiological* Absorbed from gastrointestinal tract from antacids etc. *1488*

Magnesium Sulfate

Angiotensin-Converting Enzyme (ACE)
Serum   Decrease   Physiological Decreased in 16 women with pregnancy induced hypertension 1-8 h after treatment then plateaued. *1213*

Calcium   *Serum   Increase   Physiological* In one woman given i.v. drug because of Crohn's disease on calcium supplements. *2552*

Orthophosphate   *Test Conditions   No Effect   Analytical* No effect up to 1 mol/L. *1658*

Tetracycline   *Serum   Decrease   Physiological* Reduced absorption noted. *3794*

Magnesium Trisilicate

Nitrofurantoin   *Serum   Decrease   Physiological* Absorption reduced due to adsorption. *3794*

Malic Acid

Lactate   *Serum   Decrease   Analytical* Interferes with enzyme methods. *1150*

Malonic Acid

Methylmalonic Acid   *Urine   Increase   Analytical* Measured as analyte. *1256*

Succinate Dehydrogenase
Test Conditions   Decrease   Physiological Inhibitory effect marked. *2893*

Maltose

Glucose   *Serum   No Effect   Analytical* No effect on o-toluidine procedure. No effect on GOD-PERID procedure. *2771*
Serum   Increase   Analytical Affects Neocuproin procedures. *2771* Produces 5 % color of glucose, o-toluidine method. *0583* Affects alkaline ferricyanide method. Affects Trinder procedure (maltase in Fermcozyme). *2771*
Serum   Increase   Physiological Marked rise observed with oral load. *2551*

Sugar *Urine Increase Analytical* False positive with Benedict's, Clinitest®. *1563*

Mandelic Acid

Erythrocytes *Urine Increase Physiological* Actual bleeding may be caused by drug. *1022*

Estriol *Urine Decrease Analytical* If acid hydrolysis used with hexamine mandelate. *2427*

Hemoglobin *Urine Increase Physiological* Actual bleeding caused by the drug. *1022*

Manganese

Nickel *Test Conditions Increase Analytical* Possible interference with atomic absorption. *3499*

Manganese Dioxide

Color *Feces Increase Analytical* Dark brown to black with 130-140 mg. *0902*

Manganese Salts

Alkaline Phosphatase *Serum Increase Analytical* Activators of enzyme in laboratory procedures. *3588*

Mannitol

Albumin *Serum No Effect Analytical* No effect on SMA 12/60 method at 445 mg/dL. *2636*

Alkaline Phosphatase *Serum No Effect Analytical* No effect on SMA 12/60 method at 445 mg/dL. *2636*

Ammonia *Feces Increase Physiological* Modest rise observed. *0032*

Amylase *Serum No Effect Analytical* At concentration of 40,000 mg/L had no effect on maltotetrose method. *3393* At concentration of 40,000 mg/L had no effect on p-nitrophenylmaltoheptoside method. *3393* At concentration of 40,000 mg/L had no effect on p-nitrophenylmaltopentoside/-hexoside method. *3393*

Aspartate Aminotransferase *Serum No Effect Analytical* No effect on SMA 12/60 method at 445 mg/dL. *2636*

Bilirubin *Serum No Effect Analytical* No effect on SMA 12/60 method at 445 mg/dL. *2636*

Calcium *Serum No Effect Analytical* No effect on SMA 12/60 method at 445 mg/dL. *2636*
Urine Increase Physiological Initial diuretic response. *0105*

Chloride *Serum Decrease Physiological* Effect if marked diuresis. *0071*

Cholesterol *Serum No Effect Analytical* No effect on SMA 12/60 method at 445 mg/dL. *2636*

Creatinine *Serum No Effect Analytical* No effect on SMA 12/60 method at 445 mg/dL. *2636*
Serum Increase Physiological Due to dehydration. *2313*

Glucose *Serum No Effect Analytical* No effect on SMA 12/60 method at 445 mg/dL. *2636*

Hydroxyproline *Urine No Effect Analytical* At 10 g/L on method of Goverde. *1365*

Lactate Dehydrogenase *Serum No Effect Analytical* No effect on SMA 12/60 method at 445 mg/dL. *2636*

Osmolality *Serum Increase Analytical* Concentration related increase: 6% at 3.1 g/L and 12.3% at 6.2 g/L. *2066*
Serum Increase Physiological May cause marked dehydration. *2220*

pH *Feces Decrease Physiological* No effect observed. *0032*

Phosphate *Serum No Effect Analytical* Usually not high enough concentration to interfere. *0583* No effect on SMA 12/60 method at 445 mg/dL. *2636* No effect on method on AutoAnalyzer even at 15.5 g/L. *2066*
Serum Decrease Analytical Inhibition of color development. *0582* 19% reduction at 6.2 g/L on normal phosphate concentration: 23% reduction at 3.1 g/L on increased phosphate specimen on Du Pont aca method affecting also other methods in which

Elon used as reducing agent. *2066* Concentration related reduction, possibly related to nature of reducing agent, on Dade® Paramax and Du Pont aca. *1005* At concentrations above 3100 mg/L lowered concentration as measured by aca method. *3393*
Urine Decrease Analytical Complexes molybdate, decreases color develop. *0583*

Potassium *Serum Increase Physiological* Mechanism not discussed. *2480*

Protein *Serum No Effect Analytical* No effect on SMA 12/60 method at 445 mg/dL. *2636*
Test Conditions Increase Analytical Possible measurement by biuret reaction. *3376*

Sodium *Serum Increase Physiological* May cause marked dehydration. *2220*
Serum Decrease Physiological Effect if marked diuresis. *0071*
Urine Increase Physiological Slight increase occurs only. *2220*

Uric Acid *Serum No Effect Analytical* No effect on SMA 12/60 method at 445 mg/dL. *2636*
Serum Decrease Physiological Reported to have uricosuric action. *1198*
Urine Increase Physiological Intravenously produces uricosuria. *1487*

Mannose

Color *Urine Increase Physiological* Deep orange after i.v. infusion. *0358*

Crystals *Urine Increase Physiological* Massive uric acid crystalluria after infusion. *0358*

Estriol *Urine Decrease Analytical* Interference with GLC method. *3186*

Glucose *Serum No Effect Analytical* No effect on hexokinase, glucose oxidase methods. *2568* At 200 mg/dL on hexokinase method of Coburn. *0676*
Serum Increase Analytical Nonspecificity of FeCN, o-toluidine, Neocuproin. *2857* Mg for mg effect MBTH procedure of Neeley. *2569*
Serum Decrease Analytical Above 0.28 mmol/L if hexokinase from Leuconostoc Mesent. *0411*

Mannose *Serum Increase Physiological* Twice normal level in diabetics (after i.v.). *0358*

Protein *Test Conditions Increase Analytical* Interferes with Folin-Ciocalteu method of Lowry. *3636*

Uric Acid *Serum Increase Physiological* Similar action to fructose. *3895*
Urine Increase Physiological Great increase after i.v. infusion. *0358*

MAO Inhibitors

N-Acetylmetanephrine *Urine Increase Physiological* Due to inhibition of amine oxidase. *0426*

N-Acetylnormetanephrine *Urine Increase Physiological* Due to inhibition of amine oxidase. *0426*

Alanine Aminotransferase *Serum Increase Physiological* Intrahepatic cholestasis. *2220*

Alkaline Phosphatase *Serum Increase Physiological* Cholestatic effect. *2220*

Ammonia *Plasma Decrease Physiological* Reported effect in exogenous NH_3 toxicity. *1942*

Aspartate Aminotransferase
Serum Increase Physiological Intrahepatic cholestasis. *2220*

Bile *Urine Increase Physiological* Hepatotoxic effect. *2313*

Bilirubin *Serum Increase Physiological* Viral hepatitis-like jaundice in some patients. *3265*

BSP Retention *Serum Increase Physiological* Intrahepatic cholestasis with possible cell damage. *1237*

Catecholamines *Plasma Increase Physiological* Prevent deamination but not degradation by catechol-o-methyl transferase. *0127*

Cephalin Flocculation *Serum Increase Physiological* Hepatotoxic effect. *2313*

MAO Inhibitors *(continued)*

Cholesterol *Serum Decrease Physiological* Hepatotoxic effect. *2220*

Dopamine *Plasma Increase Physiological* Effect observed after single large dose. *1343*

Epinephrine *Plasma Increase Physiological* Effect observed after single large dose. *1343*

Erythrocytes *Blood Decrease Physiological* Anemia may occasionally occur. *2313*

Glucose *Serum Decrease Physiological* Mechanism not clear (possible hepatotoxicity). *0025*

Glucose Tolerance *Serum Increase Physiological* Seen in diabetics treated with insulin. *2220*

Guaiacols Spot Test *Urine Positive Analytical* False reaction with screening test of Rogers. *3031*

Hematocrit *Blood Decrease Physiological* Occasional anemia may occur. *2427*

Hemoglobin *Blood Decrease Physiological* Occasional anemia may develop. *2427*

Histamine Test *Patient Increase Physiological* Enhanced responsiveness. *1325*

HMPG (4-Hydroxy-3-Methoxy-Phenylethylene Glycol) *Urine Decrease Physiological* Inhibition of amine oxidase. *0426*

17-Hydroxycorticosteroids *Urine Decrease Physiological* Probably due to depressed central synthesis. *0022*

5-Hydroxyindoleacetic Acid (5-HIAA) *Urine Decrease Physiological* Inhibition of conversion of 5-HT to 5-HIAA. *1937*

5-Hydroxytryptamine Glucuronide *Urine Increase Physiological* Due to inhibition of conversion of 5-HT to 5-HIAA. *1343*

5-Hydroxytryptamine (Serotonin) *Plasma Increase Physiological* Effect observed after single large dose. *1343*

Leukocytes *Blood Decrease Physiological* Occasional leukopenia/agranulocytosis. *2313*

Metanephrines, Total *Urine Increase Physiological* Prevent deamination. *1343*

Norepinephrine *Plasma Increase Physiological* Effect observed after single large dose. *1343*

Normetanephrine *Urine Increase Physiological* Prevent deamination. *1343*

Phentolamine Test *Patient Increase Physiological* Enhanced responsiveness. *1325*

Phenylethylamine *Urine Increase Physiological* Prevent deamination. *1343*

Prothrombin Time *Plasma Increase Physiological* May prolong action of anticoagulants. *1679*

Thymol Turbidity *Serum Increase Physiological* Intrahepatic cholestasis. *2313*

Tryptamine *Urine Increase Physiological* Due to utilization of alternative pathways. *1343*

Tyramine *Urine Increase Physiological* Prevent deamination. *1343*

Tyramine Test *Patient Increase Physiological* Enhanced responsiveness. *1325*

Vanillylmandelic Acid *Urine Decrease Physiological* Inhibition of normetanephrine conversion to VMA. *1488*

Xylose Excretion *Urine Decrease Physiological* Decreased gastrointestinal tract absorption. *1237*

Maprotiline

Alanine Aminotransferase *Serum Increase Physiological* Slight reversible rise in approximately 1% patients. *0214*

Aspartate Aminotransferase *Serum Increase Physiological* Slight reversible rise in approximately 1% patients. *0214*

Bilirubin *Serum No Effect Physiological* Insignificant displacement from protein in neonates. *3748*

Marital Status

Creatine Kinase *Serum Increase Physiological* 4 U in normal married women versus single women. *2744*

Marophen

Eosinophils *Blood Increase Physiological* Low incidence reported. *2427*

Mayo Enema

Bicarbonate *Serum Increase Physiological* May cause retention from bicarbonate in enema. *0726*

pH *Blood Increase Physiological* Due to bicarbonate in enema. *0726*

Sodium *Serum Increase Physiological* High sodium content — may be retained. *0726*

Mazindol

Amobarbital *Urine No Effect Analytical* No interference on TLC using ethyl acetate: methanol: water: ammonium hydroxide and modified Dragendorff's reagent for detection. *3868*

Amphetamine *Urine No Effect Analytical* No interference on TLC using ethyl acetate: methanol: water: ammonium hydroxide and modified Dragendorff's reagent for detection. *3868*

Chlordiazepoxide *Urine Positive Analytical* Same R_f and color reaction on TLC using ethyl acetate: methanol: water: ammonium hydroxide and modified Dragendorff's reagent. *3868*

Flurazepam *Urine Positive Analytical* Same R_f and color reaction on TLC using ethyl acetate: methanol: water: ammonium hydroxide and modified Dragendorff's reagent. *3868*

Hydromorphone *Urine No Effect Analytical* No interference on TLC using ethyl acetate: methanol: water: ammonium hydroxide and modified Dragendorff's reagent for detection. *3868*

Mescaline *Urine No Effect Analytical* No interference on TLC using ethyl acetate: methanol: water: ammonium hydroxide and modified Dragendorff's reagent for detection. *3868*

Methadone *Urine Positive Analytical* Same R_f and color reaction on TLC using ethyl acetate: methanol: water: ammonium hydroxide and modified Dragendorff's reagent. *3868*

Methamphetamine *Urine No Effect Analytical* No interference on TLC using ethyl acetate: methanol: water: ammonium hydroxide and modified Dragendorff's reagent for detection. *3868*

Methapyrilene *Urine Positive Analytical* Same R_f and color reaction on TLC using ethyl acetate: methanol: water: ammonium hydroxide and modified Dragendorff's reagent. *3868*

Methylphenidate *Urine Positive Analytical* Same R_f and color reaction on TLC using ethyl acetate: methanol: water: ammonium hydroxide and modified Dragendorff's reagent. *3868*

Morphine *Urine No Effect Analytical* No interference on TLC using ethyl acetate: methanol: water: ammonium hydroxide and modified Dragendorff's reagent for detection. *3868*

Pentobarbital *Urine No Effect Analytical* No interference on TLC using ethyl acetate: methanol: water: ammonium hydroxide and modified Dragendorff's reagent for detection. *3868*

Phendimetrazine *Urine Positive Analytical* Same R_f and color reaction on TLC using ethyl acetate: methanol: water: ammonium hydroxide and modified Dragendorff's reagent. *3868*

Phenobarbital *Urine No Effect Analytical* No interference on TLC using ethyl acetate: methanol: water: ammonium hydroxide and modified Dragendorff's reagent for detection. *3868*

Phenylpropanolamine *Urine No Effect Analytical* No interference on TLC using ethyl acetate: methanol: water: ammonium hydroxide and modified Dragendorff's reagent for detection. *3868*

Secobarbital *Urine No Effect Analytical* No interference on TLC using ethyl acetate: methanol: water: ammonium hydroxide and modified Dragendorff's reagent for detection. *3868*

Meals

Alanine Aminotransferase *Serum Decrease Physiological* 2 h after standard meal in men. *3455* Observed in women. *3319*

Albumin *Serum No Effect Physiological* No effect after standard breakfast. *0579*
Serum Increase Physiological Affected by meals (most marked in women). *3455*
Serum Decrease Physiological Significant effect in subjects 20-40 y. *3455*

Alkaline Phosphatase *Serum No Effect Analytical* If nitrophenol used as substrate. *3434*
Serum Increase Analytical If phenylphosphate used in substrate. *3434*
Serum Increase Physiological Affected by meals (most marked in women). *3455* Fatty meals especially in O secretors. *3873*
Serum Decrease Physiological Significant effect observed. *3455*

α-Amino-Nitrogen *Plasma Increase Physiological* Small transient rise after protein meals. *2312*
Urine Increase Physiological Significant effect observed after each meal. *3571*

Aspartate Aminotransferase
Serum Increase Physiological Observed in both sexes. *3319* Affected by meals (most marked in women). *3455*
Serum Decrease Physiological Significant effect observed. *3455*

Bicarbonate *Serum No Effect Physiological* No effect after standard breakfast. *0579*

Bilirubin *Serum Increase Physiological* Affected by meals (most marked in women). Observed in both sexes. *3455*

BSP Retention *Serum Increase Physiological* Capacity of liver to eliminate BSP decreased. *0579*

Calcium *Serum No Effect Physiological* Not significantly affected by nonstandardized meals. *3455*
Serum Increase Physiological Affected by meals (most marked in women). *3455* Effect of metabolic alkalosis. *3234*

Chloride *Serum No Effect Physiological* No effect after standard breakfast. *0579*

Cholesterol *Serum No Effect Physiological* Not significantly affected by nonstandardized meals. *3455* No effect after standard breakfast. *0579*
Serum Increase Physiological Reported up to 3% increase after meals. *1563* Variable increase of 40 mg/dL in some people. *3217* Affected by meals (most marked in women). *3455*

Cholesterol Esters *Serum No Effect Physiological* No effect after standard breakfast. *0579*

Creatinine *Serum No Effect Physiological* No effect after standard breakfast. *0579*
Serum Increase Physiological Affected by meals (most marked in women). *3455*
Serum Decrease Physiological Significant effect in subjects 20-40 y. *3455*

Erythrocyte Sedimentation Rate
Blood No Effect Physiological No effect observed. *0424*

Fatty Acids, Free (FFA) *Serum Decrease Physiological* After mixed meal falls for 2 h below fasting. *1230*

Gastrin *Serum Increase Physiological* Similar response (fairly marked). *3904*

Glucose *Serum No Effect Physiological* Not significantly affected by nonstandardized meals. *3455*
Serum Increase Physiological In effect acts as glucose tolerance test. *0579* 2 h after standardized meal in both sexes. *3455*

Ionized Calcium *Serum Decrease Physiological* Increase by 0.053 mmol/L 1 h after steak meal. *1682* Effect more marked than with respiratory alkalosis. *3234*

Lactate Dehydrogenase *Serum Increase Physiological* Significant effect observed Affected by meals (most marked in women). *3455*

Lipoproteins *Serum Decrease Physiological* Affected by meals (most marked in women). *3455*

pCO₂ *Blood Increase Physiological* Slight effect alkaline tide. *0814*

pH *Blood Increase Physiological* Effect of metabolic alkalosis. *3234*

Urine Increase Physiological Due to alkaline tide. *0814*

Phosphate *Serum Increase Physiological* Affected by meals (most marked in women). Observed in men after meals. *3455*
Serum Decrease Physiological Observed in women after meals. *3455* Phosphorylation of glucose and metabolism. *1237*

Potassium *Serum No Effect Physiological* No effect after standard breakfast. *0579*

Protein *Serum No Effect Physiological* No effect after standard breakfast. *0579*
Serum Increase Physiological Affected by meals (most marked in women). *3455*
Serum Decrease Physiological Significant effect in subjects 20-40 y. *3455*

Protein Bound Iodine (PBI) *Serum No Effect Physiological* No effect of meals reported. *0830*

Sodium *Serum No Effect Physiological* Not significantly affected by nonstandardized meals. *3455* No effect after standard breakfast. *0579*

Thymol Turbidity *Serum Increase Analytical* Cause mild lipemia which is measured. *0582*

Triglycerides *Serum Increase Physiological* Increase to maximum at 4 h after, minimum 10 h after. *3162*

Urea Clearance *Urine No Effect Physiological* No significant effect of meals observed. *0459*

Urea Nitrogen *Serum No Effect Physiological* No effect reported after standard breakfast. *0579*
Serum Increase Physiological Affected by meals (most marked in women). *3455*
Serum Decrease Physiological Significant effect in subjects 20-40 y. *3455*

Uric Acid *Serum No Effect Physiological* No effect after standard breakfast. *0579*
Serum Increase Physiological Affected by meals (most marked in women). *3455*
Serum Decrease Physiological Significant effect in subjects 20-40 y. *3455*

Xylose Excretion *Urine Decrease Physiological* Less absorbed in nonfasting state. *2945*

Measles Vaccine

Platelets *Blood Decrease Physiological* May cause thrombocytopenic purpura. *2427*

Meat

Gastrin *Serum Increase Physiological* Maximum effect in normals seen in 30 minutes. *2369*

Histamine *Urine Increase Physiological* All gastric stimulants produce effect. *0814*

Occult Blood *Feces Positive Physiological* Myoglobin in meat, no effect if none for 3 d. *0849*

Meat Extract

Gastrin *Serum Increase Physiological* Similar response (fairly marked). *3904*

Hydrochloric Acid *Gastric Material Increase Physiological* But less than with insulin. *3904*

Mebanazine

Glucose *Serum Decrease Physiological* Appears to potentiate insulin in diabetics. *2427*

Mebhydrolin

Neutrophils *Blood Decrease Physiological* Occasional case of agranulocytoisis reported. *3717*

Mebutamate

Alkaloids *Urine No Effect Analytical* With ninhydrin on Frings TLC procedure at 10 mg/dL With iodoplatinate of Frings TLC procedure at 10 mg/dL. *1204*

Barbiturate *Urine No Effect Analytical* With mercuric SO_4 on Frings TLC procedure at 10 mg/dL. With mercurous NO_3 on Frings TLC procedure at 10 mg/dL. With diphenylcarbazone on Frings TLC procedure at 10 mg/dL. With vanillin on Frings procedure at 10 mg/dL. *1204*

Bromide *Urine No Effect Analytical* No effect on method of Frings. *1204*

Ethchlorvynol *Urine No Effect Analytical* No effect on method of Frings and Cohen. *1204*

FPN Test *Urine No Effect Analytical* No effect at 10 mg/dL on method of Frings. *1204*

Platelets *Blood Decrease Physiological* Rare thrombocytopenic purpura. *1680*

Salicylate *Urine No Effect Analytical* At 10 mg/dL no effect on Trinder method. *1204*

Mecamylamine

Catecholamines *Urine No Effect Analytical* No effect on fluorometric method of Crout. *0758*
Urine No Effect Physiological No effect observed. *0913*

Uric Acid *Serum Increase Physiological* ?due to reduced renal blood flow. *0924*

Vanillylmandelic Acid *Urine No Effect Physiological* No effect observed. *0913*

Mechlorethamine

Alanine Aminotransferase *Serum Increase Physiological* May cause cytotoxic (hepatocellular) damage. *0248*

Alkaline Phosphatase *Serum Increase Physiological* Hepatotoxic effect. *0248*

Aspartate Aminotransferase
Serum Increase Physiological May cause cytotoxic (hepatocellular) damage. *0248*

Bile *Urine Increase Physiological* Hepatotoxic effect. *0248*

BSP Retention *Serum Increase Physiological* Hepatotoxic effect. *0248*

Cephalin Flocculation *Serum Increase Physiological* May cause cytotoxic (hepatocellular) damage. *0248*

Erythrocytes *Blood Decrease Physiological* Mild effect. *0071*

Guanase *Serum Increase Physiological* May cause cytotoxic (hepatocellular) damage. *0248*

Isocitrate Dehydrogenase *Serum Increase Physiological* May cause cytotoxic (hepatocellular) damage. *0248*

Lymphocytes *Blood Decrease Physiological* Occurs within 24 h. *1343*

Ornithine Carbamoyltransferase (OCT)
Serum Increase Physiological May cause cytotoxic (hepatocellular) damage. *0248*

Platelet Aggregation *Blood Decrease Physiological* Observed *in vitro*, might cause gastrointestinal bleeding etc. *0605*

Platelets *Blood Decrease Physiological* Bone marrow depression. *0071*

Pseudocholinesterase *Serum Decrease Physiological* Observed activity *in vitro*, probable *in vivo*. *3959*

Thymol Turbidity *Serum Increase Physiological* Hepatotoxic effect. *0248*

Uric Acid *Serum Increase Physiological* Leucocyte destruction, catabolism of nucleic acids. *3505*
Serum Decrease Physiological More effective than several uricosuric agents. *3013*

Medazepam

5-Hydroxyindoleacetic Acid (5-HIAA)
Urine Increase Analytical Slight effect when added *in vitro* in method of Udenfriend. *0670*

Medroxyprogesterone

Androgens *Urine Decrease Physiological* Inhibition of steroid biosynthesis in adrenals. *3632*

Apolipoprotein AI *Serum Decrease Physiological* Increase by 7% in 11 men with sexual deviation syndrome given approximately 1273 mg over a total of approximately 17 d. *0643*

Apolipoprotein B *Serum Decrease Physiological* Increase by 15% in 11 men with sexual deviation syndrome given approximately 1273 mg over a total of approximately 17 d. *0643*

Calcium *Serum No Effect Physiological* No difference observed compared with controls. *3329*

Cholesterol *Serum Decrease Physiological* Increase by 14% effects observed in 15 postmenopausal women with endometrial cancer after 2 weeks treatment. *2118* Increase by 12% in 11 men with sexual deviation syndrome given approximately 1273 mg over a total of approximately 17 d. *0643*

Cholesterol, High Density Lipoprotein
Serum No Effect Physiological No significant effect in 11 men with sexual deviation syndrome given approximately 1273 mg over a total of approximately 17 d. *0643*
Serum Decrease Physiological Increase by 33% effects observed in 15 postmenopausal women with endometrial cancer after 2 weeks treatment. *2118* Increase by 8% at 2 weeks and more after longer treatment: dose-dependent correlation with results. *1068*

Cholesterol, Low Density Lipoprotein
Serum Decrease Physiological Increase by 13% in 11 men with sexual deviation syndrome given approximately 1273 mg over a total of approximately 17 d. *0643*

Glucose *Serum Increase Physiological* Metabolic effect observed after 1 y. *3414*

Glucose Tolerance *Serum Decrease Physiological* Abnormal in 15% may not return to normal in 1 y. *3414*

Growth Hormone *Plasma No Effect Physiological* No effect observed after 1 y. *3414*

HDL₂-Cholesterol *Serum Decrease Physiological* Increase by 35% effects observed in 15 postmenopausal women with endometrial cancer after 2 weeks treatment. *2118* Increase by 15% at 2 weeks and more after longer treatment: dose-dependent correlation with results. *1068*

HDL₃-Cholesterol *Serum No Effect Physiological* No effect although changes in other fractions. *1068*
Serum Decrease Physiological Increase by 15% effects observed in 15 postmenopausal women with endometrial cancer after 2 weeks treatment. *2118*

17-Hydroxycorticosteroids *Urine Decrease Physiological* Inhibition of steroid biosynthesis in adrenals. *3632*

Insulin *Plasma Increase Physiological* Metabolic effect (?glucocorticoid). *3414*

Lecithin Cholesterol Acyltransferase (LCAT)
Serum Decrease Physiological Significant reduction after treatment. Effects observed in 15 postmenopausal women with endometrial cancer after 2 weeks treatment. *2118*

Magnesium *Serum Increase Physiological* Estrogen type of response. *3329*

Occult Blood *Feces Positive Physiological* Ischemic colitis reported in one case. *1260*

Phosphate *Serum Increase Physiological* Observed effect (unlike oral contraceptives). *3329*

Pregnanediol *Urine Decrease Physiological* Induces anovulatory state. *3632*

Testosterone *Serum Decrease Physiological* Significant reduction after treatment. Effects observed in 15 postmenopausal women with endometrial cancer after 2 weeks treatment. *2118*

Triglycerides *Serum Decrease Physiological* Significant reduction. Effects observed in 15 postmenopausal women with endometrial cancer after 2 weeks treatment. *2118* Increase by 24% in 11 men with sexual deviation syndrome given approximately 1273 mg over a total of approximately 17 d. *0643*

Vitamin B₁₂ *Serum No Effect Physiological* No effect seen in Africans at least. *0472*

Mefenamic Acid

Bile *Urine Increase Analytical* Reported interference with testing procedure. *1881*

Bilirubin *Serum Increase Physiological* May cause autoimmune hemolytic anemia. *1078*

Coombs' Test *Serum Positive Physiological* Autoimmune phenomenon (occurs after several months). *1486*

Coombs' Test, Direct *Serum Positive Physiological* Drug induces autoimmune phenomenon. *1487*

Eosinophils *Blood Increase Physiological* Allergic reactions noted. *1680*

Erythrocytes *Blood Decrease Physiological* Pancytopenia. *2313*
Urine Increase Physiological Actual bleeding caused by drug. *2313*

Glucose Tolerance *Serum Decrease Physiological* Effect noted in a diabetic. *1680*

Hematocrit *Blood Decrease Physiological* May cause autoimmune hemolytic anemia. *1078*

Hemoglobin *Blood Decrease Physiological* May cause autoimmune hemolytic anemia. *1078*
Urine Increase Physiological Actual bleeding caused by the drug. *1022*

Leukocytes *Blood Decrease Physiological* Pancytopenia (temporary depression may occur often). *2313*

MCV *Blood Increase Physiological* Megaloblastic anemia reported. *0071*

Occult Blood *Feces Positive Physiological* Occurs less frequently than with aspirin. *0071*

Platelet Aggregation *Blood Decrease Physiological* Observed *in vitro*, might cause gastrointestinal bleeding etc. *0605*

Platelets *Blood Decrease Physiological* Pancytopenia. *0788*

Protein *Urine Increase Physiological* Nephrotoxic effect. *1237* Nephrotoxic effect. *1022*

Prothrombin Time *Plasma Increase Physiological* Displaces coumarins from albumin. *3248*

Urea Nitrogen *Serum Increase Physiological* Reported in one study with normal volunteers. *0071*

Uric Acid *Serum Decrease Physiological* Uricosuric action of drug. *3013*

Mefruside

Chloride *Urine Increase Physiological* Diuretic action. *0180*

Potassium *Serum Decrease Physiological* Diuretic action less effective than thiazides. *3854*
Urine Increase Physiological Diuretic action, acts up to 20 h. *3854*

Sodium *Urine Increase Physiological* Diuretic action, less effective than thiazides. *3854*

Uric Acid *Serum Increase Physiological* Inhibits excretion of urate. *3854*

Volume *Urine Increase Physiological* Diuretic action, less effective than thiazides. *0180*

Megestrol

11-Deoxycortisol *Plasma Decrease Physiological* In 18 postmenopausal women with breast cancer. *0056*

Estradiol *Plasma Decrease Physiological* In 18 postmenopausal women with breast cancer. *0056*

Follicle Stimulating Hormone (FSH)
Plasma Decrease Physiological In 18 postmenopausal women with breast cancer. *0056*

Glucose *Serum Decrease Physiological* In 18 postmenopausal women with breast cancer. *0056*

Growth Hormone *Plasma No Effect Physiological* In 18 postmenopausal women with breast cancer. *0056*

Insulin *Plasma Increase Physiological* In 18 postmenopausal women with breast cancer. *0056*

Luteinizing Hormone (LH) *Plasma Decrease Physiological* Suppresses LH peak. *3237* In 18 postmenopausal women with breast cancer. *0056*

Megestrol *Serum Increase Physiological* Progressive increase in concentration with time, regardless of dose given. *0056*

NBT Test *Blood Positive Physiological* With 0.5 mg/d 6-9% with positive levels. *0163*

Prolactin *Plasma Increase Physiological* Affects both basal and TRH-stimulated concentration in 18 postmenopausal women with breast cancer. *0056*

Sex Hormone Binding Globulin
Serum Decrease Physiological In 18 postmenopausal women with breast cancer. *0056*

Thyroid Stimulating Hormone (TSH)
Serum No Effect Physiological In 18 postmenopausal women with breast cancer. *0056*

Meglumine

Cholesterol *Serum No Effect Analytical* At concentration of 1200 mg/L had no effect on CHOD-Iodide method. *3393* At concentration of 1200 mg/L had no effect on CHOD-PAP method. *3393* At concentration of 1200 mg/L had no effect on catalase-AlDH method. *3393* At concentration of 1200 mg/L had no effect on method using catalase-Hantzsch reaction. *3393* At concentration of 1200 mg/L had no effect on Liebermann-Burchard method. *3393*

Creatinine *Serum No Effect Analytical* At concentration of 1200 mg/L had no effect on Jaffé-Fading-Fraction method. *3393* At concentration of 1200 mg/L had no effect on Jaffé-Fuller's earth method. *3393* At concentration of 6,000 mg/L had no effect on kinetic Jaffé method on BKA-2. *3393*

Glucose *Serum No Effect Analytical* At concentration of 200 mg/L had no effect on Ektachem® method. *3393* At concentration of 2,000 mg/L had no effect on hexokinase/G-6-PDH method. *3393*

Lactate *Serum No Effect Analytical* At concentration of 1200 mg/L had no effect on enzyme method. *3393*

Metanephrines, Total *Urine Decrease Analytical* Inhibits oxidation to vanillin in Pisano procedure. *0583*

Potassium *Serum No Effect Analytical* At concentration of 2,000 mg/L had no effect on ISE measurement without predilution. *3393*

Protein Electrophoresis *Serum No Effect Analytical* At concentration of 2,000 mg/L had no effect on automated Olympus-Hite method. *3393*

Sodium *Serum No Effect Analytical* At concentration of 1200 mg/L had no effect on ISE measurement without predilution. *3393*

Urea Nitrogen *Serum No Effect Analytical* At concentration of 200 mg/L had no effect on Ektachem® method. *3393*

Uric Acid *Serum No Effect Analytical* At concentration of 2,000 mg/L had no effect on Kageyama-Hantzsch method. *3393* At concentration of 1200 mg/L had no effect on catalase-AlDH method. *3393*

Meglumine Ioglycamate

Bilirubin *Serum Increase Physiological* Possible clinically significant displacement from protein in neonates. *3748*

Melanin

Ammoniacal Silver Nitrate *Urine Positive Analytical* Slow darkening. *3588*

Color *Urine Increase Physiological* Brown to black. *2448*

Ferric Chloride Test *Urine Positive Analytical* Brownish-black color. *3588*

Phenylketones *Urine Positive Analytical* Gray to black with $FeCl_3$. *0775*

Sugar *Urine Increase Analytical* In large quantity may reduce Benedict's. *3588*

Thormählen Test *Urine Positive Analytical* Green-blue to blue-black (normal = olive-brown). *3588*

Melanin *(continued)*

Urobilinogen *Urine Increase Analytical* Melanogen reacts with Ehrlich's reagent. *1563*

Melarsonyl

Alanine Aminotransferase *Serum Increase Physiological* May produce hepatotoxicity. *0071*

Aspartate Aminotransferase
Serum Increase Physiological May produce hepatotoxicity. *0071*

Casts *Urine Increase Physiological* Nephrotoxic effect. *0071*

Erythrocytes *Blood Decrease Physiological* May cause hemolytic anemia in G-6-PD deficiency. *0071*

Hematocrit *Blood Decrease Physiological* May cause hemolytic anemia in G-6-PD deficiency. *0071*

Hemoglobin *Blood Decrease Physiological* May cause hemolytic anemia in G-6-PD deficiency. *0071*

Protein *Urine Increase Physiological* Nephrotoxic effect. *0071*

Melarsoprol

Alanine Aminotransferase *Serum Increase Physiological* May produce hepatotoxicity. *0071*

Aspartate Aminotransferase
Serum Increase Physiological May produce hepatotoxicity. *0071*

Casts *Urine Increase Physiological* Nephrotoxic effect. *2313*

Protein *Urine Increase Physiological* Nephrotoxic effect. *2313*

Melphalan

Bilirubin *Serum Increase Physiological* May cause hemolytic anemia. *2365*

Bilirubin, Direct *Serum Increase Physiological* May cause hemolytic anemia. *2365*

Coombs' Test *Serum Positive Physiological* Immunological response to drug (gamma antibody). *1486*

Erythrocytes *Blood Decrease Physiological* May cause bone marrow depression (dose related). *0543*

Hematocrit *Blood Decrease Physiological* Anemia may occur rarely. *2427*

Hemoglobin *Blood Decrease Physiological* Anemia may occur rarely. *2427*

5-Hydroxyindoleacetic Acid (5-HIAA)
Urine Increase Physiological Probably due to tissue destruction if carcinoid. *2215*

Leukocytes *Blood Increase Physiological* 4 cases of acute leukemia in 474 patients with ovarian carcinoma: all cases in patients who had received at least 300 mg for 3 y. *1003*
Blood Decrease Physiological May cause bone marrow depression (dose related). *0543*

Occult Blood *Feces Positive Physiological* May cause gastrointestinal hemorrhage. *0071*

Platelets *Blood Decrease Physiological* May cause bone marrow depression (dose related). *0543*

Protein Electrophoresis *Serum No Effect Analytical* At concentration of 1.5 mg/L had no effect on automated Olympus-Hite method but with slight displacement of fractions. *3393*

Urea Nitrogen *Serum Increase Physiological* Azotemia may occur. *0071*

Menadione

Corticosteroids *Urine Increase Analytical* Butanol extract/no hydrolysis Reddy/Porter-Silber reaction. *0022*

Menopause

Alkaline Phosphatase *Serum Increase Physiological* Change of 2.1 King-Armstrong units from 5th to 6th decade. *3830*

Calcium *Serum Increase Physiological* Change of 0.06 meq/L from 5th to 6th decade. *3830*

Cholesterol *Serum Increase Physiological* Change of 27 mg/dL from 5th to 6th decade. *3830*

Estrogens *Urine Decrease Physiological* Normal response. *0987*

Gonadotropins *Plasma Increase Physiological* Normal response. *0987*

17-Ketosteroids *Urine Decrease Physiological* Normal response. *0987*

Lipids, Total *Serum Increase Physiological* Change of 0.2 g/dL from 5th to 6th decade. *3830*

Phosphate *Serum Increase Physiological* Change of 0.22 mg/dL from 5th to 6th decade. *3830*

Sodium *Serum Increase Physiological* Change of 0.7 meq/L from 5th to 6th decade. *3830*

Urea Nitrogen *Serum Increase Physiological* Change of 1 mg/dL from 5th to 6th decade. *3830*

Uric Acid *Serum Increase Physiological* Change of 0.45 mg/dL from 5th to 6th decade. *3830*

Menotropins

Cholesterol *Serum Decrease Physiological* Marked fall in type 2 hyperlipoproteinemia. *2235*

Lipids, Total *Serum Decrease Physiological* Marked fall in type 2 hyperlipidemia. *2235*

Phospholipids, Total *Serum Decrease Physiological* Marked fall in type 2 hyperlipidemia. *2235*

Triglycerides *Serum Increase Physiological* May be slight increase initially. *2235*
Serum Decrease Physiological Marked fall in type 2 hyperlipidemia. *2235*

Menstruation

Aldosterone *Plasma Increase Physiological* In 3/4 of subjects peak in mid or late luteal phase. *1885*

Amino Acids *Serum Decrease Physiological* Fall 17-22nd day compared with days 2-7. *0743*

Androstenedione *Plasma Increase Physiological* In follicular phase, highest in mid third of cycle. *1831*

α_1-Antitrypsin *Serum No Effect Physiological* No effect observed during menstrual cycle. *2031*

Ascorbic Acid *Plasma Decrease Physiological* Reverse pattern to dehydroascorbic acid. *2998*
Urine No Effect Physiological No change reported. *2998*

Basal Metabolic Rate *Patient Decrease Physiological* Lowest at time of bleeding, peaks before next period. *0251*

Bicarbonate *Serum Decrease Physiological* Varies in parallel with sodium. *1086*

Chloride *Serum Increase Physiological* Approximately 1.7 meq/L higher premenstrually. *1086*
Urine Increase Physiological Post-menstrual diuresis (decrease premenstrually). *0987*

Cholesterol *Serum Increase Physiological* Increase immediately before menstruation. *1563*

Creatine Kinase *Serum Decrease Physiological* Decrease of 4 U/L from 12 to 26th day. *2744*

Dehydroascorbic Acid *Serum Increase Physiological* At mid cycle, low at ends. *2998*

Estradiol *Plasma Increase Physiological* Observed 1-3 d before LH surge Second peak 30-200 pg/mL in luteal phase. *1295*
Urine Increase Physiological Peak 1 d before LH peak. *1762*

Estriol *Urine Increase Physiological* Peak with ovulation, moderately high thereafter. *1762*

Estrogens *Urine Decrease Physiological* Lowest 2-3 d after onset. *0987*

Estrone *Urine Increase Physiological* Peak 1 d before LH peak. *1762*

Fibrinogen *Plasma Increase Physiological* Normal response. *0987*

Folate *Serum No Effect Physiological* No effect noticed during cycle. *3459*

Follicle Stimulating Hormone (FSH) *Plasma Increase Physiological* Shorter and lower peak than LH peak at mid cycle. *1762*

Hemoglobin *Plasma No Effect Physiological* No effect observed. *1755*

Leukocytes *Blood Increase Physiological* Physiological response. *0251*

Luteinizing Hormone (LH) *Plasma Increase Physiological* Sharp single peak about mid-cycle. *1762* Max with ovulation. *1295*

Magnesium *Serum Increase Physiological* Significant increase with menstruation (by 0.1-0.2 meq/L). *1172*

Methylhistamine *Urine Increase Physiological* Highest value observed at this time. *1381*

Phosphate *Serum Decrease Physiological* Significant decrease with menstruation (by about 0.3 mg/dL). *1172*

Platelets *Blood Increase Physiological* Effect observed with menstruation. *0251*
Blood Decrease Physiological Increase by 50-70 %, rises to normal by 4th day. *0987*

Porphyrins *Urine Increase Physiological* May precipitate episode of hepatic porphyria. *2427*

Pregnanediol *Urine Increase Physiological* Maximum mid-way between ovulation and menstruation. *1762*

Progesterone *Plasma Increase Physiological* Maximum during mid-point luteal phase. *0876*

Prolactin *Plasma Increase Physiological* Significant higher in luteal than in follicular phase. *1435*

Protein *Urine Increase Physiological* Premenstrual urine may contain protein. *0987*

Protein Bound Iodine (PBI) *Serum Decrease Physiological* Slight fall after menstruation observed. *0830*

Pyridoxal *Serum No Effect Analytical* No variation during cycle. *0716*
Urine No Effect Physiological No change observed during cycle. *0716*

Pyridoxal Phosphate *Serum No Effect Physiological* No variation during cycle. *0716*

Pyridoxamine *Serum No Effect Physiological* No variation during cycle. *0716*
Urine No Effect Physiological No change observed during cycle. *0716*

Pyridoxamine Phosphate *Serum No Effect Physiological* No variation during cycle. *0716*

4-Pyridoxic Acid *Serum No Effect Physiological* No variation during cycle. *0716*
Urine No Effect Physiological No change observed during cycle. *0716*

Renin Activity *Plasma Increase Physiological* Follows aldosterone in half cases. *1885*

Renin Substrate *Plasma No Effect Physiological* No change observed. *1885*

Sodium *Serum Decrease Physiological* Decrease by about 2.2 meq/L compared with 16 d after. *1086*
Urine Increase Physiological Post-menstrual diuresis (decreased premenstrually). *0987*

Testosterone *Serum Increase Physiological* Random variations but highest in mid third. *1831*

Thyroxine (T₄) *Urine Increase Physiological* Transient effect during menstruation. *0630*

Volume *Urine Increase Physiological* Post-menstrual diuresis (decrease premenstrually). *0987*

Mepacrine

Erythrocytes *Blood Decrease Physiological* Occasional case of aplastic anemia reported. *3717*

Leukocytes *Blood Decrease Physiological* Occasional case of aplastic anemia reported. *3717*

Platelets *Blood Decrease Physiological* Occasional case of aplastic anemia reported. *3717*

Mepazine

Alanine Aminotransferase *Serum Increase Physiological* Cholestatic effect (up to 6 times normal). *1713*

Alkaline Phosphatase *Serum Increase Physiological* May alter liver function (cholestasis). *2313*

Aspartate Aminotransferase *Serum Increase Physiological* Cholestatic effect (up to 6 times normal). *1713*

Bile *Urine Increase Physiological* May alter liver function (cholestasis). *2313*

Bilirubin *Serum Increase Physiological* Cholestatic effect. *1713*

BSP Retention *Serum Increase Physiological* Cholestatic effect. *1713*

Cephalin Flocculation *Serum Increase Physiological* May alter liver function (cholestasis). *2313*

Cholesterol *Serum Increase Physiological* Cholestatic effect. *1713*

Erythrocytes *Blood Decrease Physiological* Pancytopenia/aplastic anemia. *2313*

Leukocytes *Blood Decrease Physiological* Agranulocytosis. *1138*

Neutrophils *Blood Decrease Physiological* Occasional case of agranulocytosis reported. *3717*

Platelets *Blood Decrease Physiological* Pancytopenia. *2313*

Prothrombin Time *Plasma Increase Physiological* Associated with failure of excretion of bile salts. *1713*

Thymol Turbidity *Serum Increase Physiological* May alter liver function (cholestasis). *2313*

Urobilinogen *Urine Decrease Physiological* Cholestatic effect. *1713*
Feces Decrease Physiological Due to cholestasis (pale stools result). *1713*

Mepenzolate

Volume *Urine Decrease Physiological* Rare urinary retention may occur. *1680*

Meperidine

Alanine Aminotransferase *Serum No Effect Analytical* At acute overdose concentration (5 mg/dL) on Technicon® SMAC® method. *3719*
Serum Increase Physiological May cause rise in intrabiliary pressure. *3505*

Albumin *Serum No Effect Analytical* At acute overdose concentration (5 mg/dL) on Technicon® SMAC® method. *3719* At concentration of 50 mg/L had no effect on BCG method. *3393*

Alkaline Phosphatase *Serum No Effect Analytical* At acute overdose concentration (5 mg/dL) on Technicon® SMAC® method. *3719*

Alkaloids *Urine No Effect Analytical* With ninhydrin on Frings TLC procedure at 10 mg/dL. *1204*
Urine Increase Analytical Red/purple with iodoplatinate on Frings TLC procedure. *1204*

Amphetamine *Urine No Effect Analytical* At 50 mg/L on fluorescent method of Hayes. *1534*

Amylase *Serum Increase Physiological* May cause spasm of sphincter of Oddi. *3316*

Aspartate Aminotransferase *Serum No Effect Analytical* No effect on Karmen procedure. No effect on Babson procedure. *2760* At acute overdose concentration (5 mg/dL) on Technicon® SMAC® method. *3719*
Serum Increase Physiological May cause rise in intrabiliary pressure. *0248*

Barbiturate *Serum Increase Analytical* False positive screen test with Hg complex formation. *0581*
Urine No Effect Analytical With mercurous NO₃ on Frings TLC procedure at 10 mg/dL. With mercuric SO₄ on Frings TLC

Meperidine *(continued)*

Barbiturate *(continued)*
procedure at 10 mg/dL. With vanillin on Frings procedure at 10 mg/dL. With diphenylcarbazone on Frings TLC procedure at 10 mg/dL. *1204*

Bicarbonate *Serum No Effect Analytical* At concentration of 60 mg/L had no effect on method using phenolphthalein. *3393*

Bilirubin *Serum No Effect Analytical* At acute overdose concentration (5 mg/dL) on Technicon® SMAC® method. *3719* At concentration of 60 mg/L had no effect on Jendrassik and Grof method. *3393*

Bromide *Urine No Effect Analytical* No effect on method of Frings. *1204*

BSP Retention *Serum Increase Physiological* Due to spasm of sphincter of Oddi. *0553*

Calcium *Serum No Effect Analytical* At acute overdose concentration (5 mg/dL) on Technicon® SMAC® method. *3719* At concentration of 60 mg/L had no effect on cresolphthalein method. *3393*

Chloride *Serum No Effect Analytical* At concentration of 60 mg/L had no effect on mercurimetric method. *3393*

Cholesterol *Serum No Effect Analytical* At acute overdose concentration (5 mg/dL) on Technicon® SMAC® method. *3719* At concentration of 60 mg/L had no effect on Liebermann-Burchard method. *3393*

Creatine Kinase *Serum No Effect Analytical* At acute overdose concentration (5 mg/dL) on Technicon® SMAC® method. *3719*
Serum No Effect Physiological No reported effect. *2427*
Serum Increase Physiological 2x after single i.m. injection of 50 mg. *3053*

Creatinine *Serum No Effect Analytical* At acute overdose concentration (5 mg/dL) on Technicon® SMAC® method. *3719* At concentration of 60 mg/L had no effect on AutoAnalyzer method. *3393*

Ethchlorvynol *Serum No Effect Analytical* At 80 mg/L on GLC procedure of Evenson. *1058*
Urine No Effect Analytical No effect on method of Frings and Cohen at 10 mg/dL. *1205*

FPN Test *Urine No Effect Analytical* No effect at 10 mg/dL on method of Frings. *1204*

Glucose *Serum No Effect Analytical* At acute overdose concentration (5 mg/dL) on Technicon® SMAC® method. *3719*
Serum Increase Physiological Central effect also involves epinephrine release. *1343*

Histamine *Plasma Increase Physiological* Associated with dose associated with anesthesia. *2212*

Hydroxybutyrate Dehydrogenase
Serum Increase Physiological May cause spasm of sphincter of Oddi. *2427*

17-Hydroxycorticosteroids *Urine Decrease Physiological* Probable effect (inhibits ACTH and PGH release). *1343*

Indocyanine Green *Serum Decrease Physiological* Observed in small series, normal liver function test. *2398*

Iron *Serum No Effect Analytical* At acute overdose concentration (5 mg/dL) on Technicon® SMAC® method. *3719*

17-Ketosteroids *Urine Decrease Physiological* Probable effect (inhibits ACTH and PGH release). *1343*

Lactate Dehydrogenase *Serum No Effect Analytical* At acute overdose concentration (5 mg/dL) on Technicon® SMAC® method. *3719*
Serum Increase Physiological May cause rise in intrabiliary pressure. *0248*

Lipase *Serum Increase Physiological* May cause spasm of sphincter of Oddi. *3316*

Meperidine *Serum Increase Physiological* 100 mg i.m. produces 1 mg/L. *2348*
Urine Positive Physiological Main excretion product in neonates and pregnant. *2427*

Methaqualone *Serum No Effect Analytical* At 80 mg/L on GLC procedure of Evenson. *1057*

Normeperidine *Urine Positive Physiological* Main metabolite in normals. *2427*

pCO₂ *Blood Increase Physiological* Depresses responsiveness of respiratory center to CO_2. *1343*

pH *Blood No Effect Physiological* Insignificant effect observed. *2427*

Phosphate *Serum No Effect Analytical* At acute overdose concentration (5 mg/dL) on Technicon® SMAC® method. *3719* At concentration of 60 mg/L had no effect on phosphomolybdate method. *3393*

pO₂ *Blood Decrease Physiological* Significant reduction in arterial oxygen pressure. *2427*

Protein *Serum No Effect Analytical* At acute overdose concentration (5 mg/dL) on Technicon® SMAC® method. *3719* At concentration of 60 mg/L had no effect on biuret method with blank correction. *3393*

Salicylate *Urine No Effect Analytical* At 10 mg/dL no effect on Trinder method. *1204*

Triglycerides *Serum No Effect Analytical* At acute overdose concentration (5 mg/dL) on Technicon® SMAC® method. *3719* At concentration of 60 mg/L had no effect on lipase/esterase method. *3393*

Urea Nitrogen *Serum No Effect Analytical* At acute overdose concentration (5 mg/dL) on Technicon® SMAC® method. *3719* At concentration of 60 mg/L had no effect on diacetylmonoxime method. *3393*

Uric Acid *Serum No Effect Analytical* At acute overdose concentration (5 mg/dL) on Technicon® SMAC® method. *3719* At concentration of 60 mg/L had no effect on phosphotungstate reduction method. *3393*

Volume *Urine Decrease Physiological* Causes release of ADH. *1343*

Mephenesin

Albumin *Serum No Effect Analytical* At concentration of 100 mg/L had no effect on BCG method. *3393*

Alkaloids *Urine No Effect Analytical* With iodoplatinate of Frings TLC procedure at 10 mg/dL. With ninhydrin on Frings TLC procedure at 10 mg/dL. *1204*

Barbiturate *Urine No Effect Analytical* With diphenylcarbazone on Frings TLC procedure at 10 mg/dL. With mercurous NO₃ on Frings TLC procedure at 10 mg/dL. With mercuric SO₄ on Frings TLC procedure at 10 mg/dL. *1204*
Urine Increase Analytical Pink spot with vanillin in Frings TLC procedure. *1204*

Bicarbonate *Serum No Effect Analytical* At concentration of 100 mg/L had no effect on method using phenolphthalein. *3393*

Bilirubin *Serum No Effect Analytical* At concentration of 100 mg/L had no effect on Jendrassik and Grof method. *3393*

Bromide *Urine No Effect Analytical* No effect on method of Frings. *1204*

Calcium *Serum No Effect Analytical* At concentration of 100 mg/L had no effect on cresolphthalein method. *3393*

Chloride *Serum No Effect Analytical* At concentration of 100 mg/L had no effect on mercurimetric method. *3393*

Cholesterol *Serum No Effect Analytical* At concentration of 100 mg/L had no effect on Liebermann-Burchard method. *3393*

Creatinine *Serum No Effect Analytical* At concentration of 100 mg/L had no effect on AutoAnalyzer Jaffé method. *3393*

Erythrocytes *Urine Increase Physiological* Actual bleeding may be caused by drug. *1022*

FPN Test *Urine No Effect Analytical* No effect at 10 mg/dL on method of Frings. *1204*

Hemoglobin *Urine Increase Physiological* Actual bleeding caused by drug. *1022*

5-Hydroxyindoleacetic Acid (5-HIAA)
Urine No Effect Analytical No effect of *in vivo* dose on method of Goldenberg. *1327*
Urine Increase Analytical Metabolite reacts in quantitative Udenfriend procedure. *0583* Interferes with nitrosonaphthol reaction. *3879*

Leukocytes *Blood Decrease Physiological* May cause leukopenia. *2313*

Phosphate *Serum No Effect Analytical* At concentration of 100 mg/L had no effect on phosphomolybdate method. *3393*

Potassium *Serum No Effect Analytical* At concentration of 100 mg/L had no effect on measurement by ISE with predilution. *3393*

Protein *Serum No Effect Analytical* At concentration of 100 mg/L had no effect on biuret method with blank correction. *3393*

Salicylate *Urine No Effect Analytical* At 10 mg/dL no effect on Trinder method. *1204*

Sodium *Serum No Effect Analytical* At concentration of 100 mg/L had no effect on measurement by ISE with predilution. *3393*

Triglycerides *Serum No Effect Analytical* At concentration of 100 mg/L had no effect on lipase/esterase method. *3393*

Urea Nitrogen *Serum No Effect Analytical* At concentration of 100 mg/L had no effect on diacetylmonoxime method. *3393*
Serum Increase Physiological Fatal anuria with intravascular hemolysis. *1343*

Uric Acid *Serum No Effect Analytical* At concentration of 100 mg/L had no effect on phosphotungstate reduction method. *3393*

Vanillylmandelic Acid *Urine Increase Analytical* Purple or red color in initial part of screening test. *0570*

Mephenesin Carbamate

Alkaloids *Urine No Effect Analytical* With ninhydrin on Frings TLC procedure at 10 mg/dL. *1204*

Barbiturate *Urine Increase Analytical* Pink/orange spot with vanillin in Frings TLC procedure. *1204*

Mephenoxalone

Eosinophils *Blood Increase Physiological* Rare side effect. *0071*

Erythrocytes *Blood Decrease Physiological* May cause anemia. *2313*

Leukocytes *Blood Decrease Physiological* Mild and transitory. *0071*

Mephentermine

Amphetamine *Urine No Effect Analytical* No effect at 100 mg/L on method of Rutter. *3097* At 50 mg/L on fluorescent method of Hayes. *1534*
Urine Increase Analytical Yields fluorophor with NBD chloride reaction. *2473*

Mephenytoin

Alanine Aminotransferase *Serum Increase Physiological* Hepatotoxicity. *2313*

Alkaline Phosphatase *Serum Increase Physiological* Hepatotoxicity. *2313*

Aspartate Aminotransferase
Serum Increase Physiological Hepatotoxicity. *2313*

Bile *Urine Increase Physiological* Hepatotoxicity. *2313*

Bilirubin *Serum Increase Physiological* May cause hemolytic anemia. *2365*

Bilirubin, Direct *Serum Increase Physiological* May cause hemolytic anemia. *2365*

BSP Retention *Serum Increase Physiological* Hepatotoxicity. *2313*

Cephalin Flocculation *Serum Increase Physiological* Hepatotoxicity. *2313*

Coombs' Test, Direct *Serum Positive Physiological* Mechanism obscure. *1486*

Coombs' Test, Indirect *Serum Positive Physiological* Mechanism obscure. *1486*

Diphenylhydantoin *Serum No Effect Analytical* On GLC procedure of Papadopoulos at *in vivo* concentration. *2722*
Serum Decrease Analytical Slight effect only at 4 µg/ml. *3591*

Eosinophils *Blood Increase Physiological* Rare cases described. *1680*

Erythrocytes *Blood Decrease Physiological* Hemolytic anemia. *3008* Occasional case of aplastic anemia reported. *3717*

Haptoglobin *Serum Decrease Physiological* May cause hemolytic anemia. *2365*

Hematocrit *Blood Decrease Physiological* Hemolytic/aplastic/megaloblastic anemia. *3008*

Hemoglobin *Blood Decrease Physiological* Hemolytic/aplastic/megaloblastic anemia. *3008*

^{131}I Uptake *Serum No Effect Physiological* No effect reported. *2220*
Serum Decrease Physiological Mesantoin® contains tetraiodofluorescein. *2652*

LE Cells *Blood Positive Physiological* SLE may occur, usually normalized when stopped. *3059*

Leukocytes *Blood Increase Physiological* Rare cases described. *1680*
Blood Decrease Physiological Agranulocytosis/aplastic anemia. *3008* Occasional case of aplastic anemia reported. *3717*

MCV *Blood Increase Physiological* May cause megaloblastic anemia. *0071*

Monocytes *Blood Increase Physiological* Rare cases described. *1680*

Neutrophils *Blood Decrease Physiological* May occur with/without pancytopenia. *1680* Occasional case of agranulocytosis reported. *3717*

Pheneturide *Serum No Effect Analytical* On GLC procedure of Papadopoulos at *in vivo* concentration. *2722*

Phenobarbital *Serum No Effect Analytical* On GLC procedure of Papadopoulos at *in vivo* concentration. *2722*

Platelets *Blood Decrease Physiological* Secondary to aplastic anemia/pancytopenia. *3008* Occasional case of aplastic anemia reported. *3717*

Potassium *Serum No Effect Analytical* At concentration of 20 mg/L had no effect on measurement by ISE with predilution. *3393*

Primidone *Serum No Effect Analytical* On GLC procedure of Papadopoulos at *in vivo* concentration. *2722*

Protein Bound Iodine (PBI) *Serum Decrease Physiological* Competes with thyroxine for binding sites. *2220*

Sodium *Serum No Effect Analytical* At concentration of 20 mg/L had no effect on measurement by ISE with predilution. *3393*

Thymol Turbidity *Serum Increase Physiological* Hepatotoxicity. *2313*

Mephobarbital

Phenobarbital *Serum No Effect Analytical* On GLC procedure of Papadopoulos when added *in vitro*. *2722*
Serum Increase Physiological Metabolic conversion *in vivo*. *3054*

Meprednisone

Glucose *Serum Increase Physiological* Alteration of carbohydrate metabolism. *1680*
Urine Increase Physiological Alteration of carbohydrate metabolism. *1680*

Hydrochloric Acid *Gastric Material Increase Physiological* Steroid effect. *1680*

Nitrogen Balance *Patient Negative Physiological* Due to protein catabolism. *1680*

Occult Blood *Feces Positive Physiological* Activation or complication of ulcer. *1680*

Meprobamate

Alanine Aminotransferase *Serum No Effect Analytical* At acute overdose concentration (20 mg/dL) on Technicon® SMAC® method. *3719*
Serum Increase Physiological May cause cholestatic (hepatocanalicular) jaundice. *0248*

Meprobamate (continued)

Albumin *Serum No Effect Analytical* At acute overdose concentration (20 mg/dL) on Technicon® SMAC® method. *3719* At concentration of 200 mg/L had no effect on BCG method. *3393*

Alkaline Phosphatase *Serum No Effect Analytical* At acute overdose concentration (20 mg/dL) on Technicon® SMAC® method. *3719*

Serum Increase Physiological May cause cholestatic (hepatocanalicular) jaundice. *0248*

Alkaloids *Urine No Effect Analytical* With ninhydrin on Frings TLC procedure at 10 mg/dL. With iodoplatinate of Frings TLC procedure at 10 mg/dL. *1204*

Aminolevulinic Acid *Urine Increase Physiological* May precipitate acute porphyria. *1322*

Ammonia *Plasma No Effect Analytical* At concentration of 20 mg/L had no effect on Ektachem® method. *3393*

Aspartate Aminotransferase *Serum No Effect Analytical* At acute overdose concentration (20 mg/dL) on Technicon® SMAC® method. *3719*

Serum Increase Physiological May cause cholestatic (hepatocanalicular) jaundice. *0248*

Barbiturate *Urine No Effect Analytical* With diphenyl-carbazone on Frings TLC procedure at 10 mg/dL. With mercuric SO_4 on Frings TLC procedure at 10 mg/dL. With mercurous NO_3 on Frings TLC procedure at 10 mg/dL. *1204*

Urine Increase Analytical Yellow spot with vanillin on Frings TLC procedure. *1204*

Bicarbonate *Serum No Effect Analytical* At concentration of 25 mg/L had no effect on method using phenolphthalein. *3393*

Bilirubin *Serum No Effect Analytical* At acute overdose concentration (20 mg/dL) on Technicon® SMAC® method. *3719* At concentration of 200 mg/L had no effect on Jendrassik and Grof method. *3393*

Serum No Effect Physiological Insignificant displacement from protein in neonates. *3748*

Serum Increase Physiological May cause cholestatic (hepatocanalicular) jaundice. *0248*

Bromide *Urine No Effect Analytical* No effect on method of Frings. *1204*

Calcium *Serum No Effect Analytical* At acute overdose concentration (20 mg/dL) on Technicon® SMAC® method. *3719* At concentration of 200 mg/L had no effect on cresolphthalein method. *3393*

Catecholamines *Urine No Effect Analytical* No effect observed. *0913*

Urine No Effect Physiological No effect observed. *0913*

Chlordiazepoxide *Blood No Effect Analytical* On method of Riddick at 5 mg/dL. *2983*

Chloride *Serum No Effect Analytical* At concentration of 25 mg/L had no effect on mercurimetric method. *3393*

Cholesterol *Serum No Effect Analytical* At acute overdose concentration (20 mg/dL) on Technicon® SMAC® method. *3719* At concentration of 25 mg/L had no effect on CHOD-PAP method. *3393* At concentration of 200 mg/L had no effect on Liebermann-Burchard method. *3393*

Serum Increase Physiological May cause cholestatic (hepatocanalicular) jaundice. *0248*

Chromosomes *Test Conditions Abnormal Physiological* Clastogenic in human lymphocytes *in vitro*. *3282*

Coproporphyrin *Urine Increase Physiological* May precipitate acute porphyria. *1322*

Feces Increase Physiological May precipitate acute porphyria. *1322*

Corticosteroids *Urine No Effect Analytical* If Allen correction and Porter-Silber procedure. *0022*

Creatine Kinase *Serum No Effect Analytical* At acute overdose concentration (20 mg/dL) on Technicon® SMAC® method. *3719*

Creatinine *Serum No Effect Analytical* At 200 mg/L on reversed phase LC procedure of Zhiri et al. *3942* At acute overdose concentration (20 mg/dL) on Technicon® SMAC® method. *3719* At concentration of 20 mg/L had no effect on Ektachem® method. *3393* At concentration of 200 mg/L had no effect on AutoAnalyzer method. *3393*

Dapsone *Serum No Effect Analytical* At 10 mg/dL on colorimetric procedure of Higgins. *1590*

Diazepam *Serum No Effect Analytical* On cathode-ray Polarographic method of Berry. *0323*

Eosinophils *Blood Increase Physiological* Rare allergic manifestation. *1680*

Erythrocytes *Blood Decrease Physiological* Aplastic anemia/erythroid hypoplasia. *2429*

Estriol *Urine No Effect Analytical* No effect on GLC method. *1163* If pretreated with alkali in method of Brown. *0023*

Urine Increase Analytical Affects unmodified Brown/Kober procedure. *0583*

Estrogens *Urine Increase Analytical* May react in Brown's and Kober's procedures. *0279*

Ethchlorvynol *Serum No Effect Analytical* At 10 mg/L on GLC procedure of Evenson. *1058* At 10 µg/mL on colorimetric method of Wallace. *3753*

Urine No Effect Analytical No effect on method of Frings and Cohen at 20 mg/dL. *1205*

FPN Test *Urine No Effect Analytical* No effect at 10 mg/dL on method of Frings. *1204*

Glucose *Serum No Effect Analytical* At acute overdose concentration (20 mg/dL) on Technicon® SMAC® method. *3719* At concentration of 484 mg/L had no effect on Ektachem® method. *3393* At concentration of 25 mg/L had no effect on GOD/POD-PAP method. *3393*

Hematocrit *Blood Decrease Physiological* May cause aplastic anemia. *0333*

Hemoglobin *Blood Increase Physiological* May cause aplastic anemia. *0333*

17-Hydroxycorticosteroids *Urine Increase Analytical* Small effect on modified Glenn-Nelson method. *3879* Glucuronide interferes with Porter-Silber reaction. *2220*

Iron *Serum No Effect Analytical* At acute overdose concentration (20 mg/dL) on Technicon® SMAC® method. *3719* At concentration of 200 mg/L had no effect on Ferrozine method. *3393*

17-Ketogenic Steroids *Urine Increase Analytical* Glucuronide interferes with Zimmermann reaction. *3505*

Urine Decrease Analytical Interferes with Zimmermann reaction (Holtorff-Koch). *3125*

17-Ketosteroids *Urine Increase Analytical* Glucuronide interferes with Zimmermann reaction. *3125*

Urine Decrease Analytical Interferes with Zimmermann react (after Allen correct). *3125*

Lactate Dehydrogenase *Serum No Effect Analytical* At acute overdose concentration (20 mg/dL) on Technicon® SMAC® method. *3719*

Leukocytes *Blood Decrease Physiological* Pancytopenia (may be agranulocytosis). *0072*

Meprobamate *Blood Increase Physiological* 0.4 to 1.2 g orally produces concentrations of 5-15 mg/L. *2348*

Methaqualone *Serum No Effect Analytical* At 20 mg/L on GLC procedure of Evenson. *1057*

Neutrophils *Blood Decrease Physiological* occasional case of agranulocytosis reported. *3717*

Phosphate *Serum No Effect Analytical* At acute overdose concentration (20 mg/dL) on Technicon® SMAC® method. *3719* At concentration of 200 mg/L had no effect on phosphomolybdate method. *3393*

Platelets *Blood Decrease Physiological* Rare bone marrow aplasia, thrombocytopenia. *0072*

Porphobilinogen *Urine Increase Physiological* May precipitate acute porphyria. *1322*

Porphyrins *Urine Increase Physiological* May precipitate attack of acute porphyria. *2220*

Potassium *Serum No Effect Analytical* At concentration of 25 mg/L had no effect on flame-photometric method. *3393* At concentration of 160 mg/L had no effect on measurement by ISE with predilution. *3393*

Propoxyphene *Serum No Effect Analytical* At 25 mg/L on method of Evenson. *1056*

Protein *Serum No Effect Analytical* At acute overdose concentration (20 mg/dL) on Technicon® SMAC® method. *3719* At concentration of 200 mg/L had no effect on biuret method with blank correction. *3393*

Protein Bound Iodine (PBI) *Serum No Effect Physiological* No effect with normal doses. *0830*
Serum Increase Analytical Contains iodine. *3505*

Prothrombin Time *Plasma No Effect Physiological* Probably no significant effect clinically. *1487*
Plasma Decrease Physiological Induces hepatic metabolism of anticoagulants. *1343*

Protoporphyrin *Feces Increase Physiological* May precipitate acute porphyria. *1322*

Salicylate *Urine No Effect Analytical* At 10 mg/dL no effect on Trinder method. *1204*

Sodium *Serum No Effect Analytical* At concentration of 25 mg/L had no effect on flame-photometric method. *3393* At concentration of 160 mg/L had no effect on measurement by ISE with predilution. *3393*

Triglycerides *Serum No Effect Analytical* At acute overdose concentration (20 mg/dL) on Technicon® SMAC® method. *3719* At concentration of 200 mg/L had no effect on lipase/esterase method. *3393*

Urea Nitrogen *Serum No Effect Analytical* At acute overdose concentration (20 mg/dL) on Technicon® SMAC® method. *3719* At concentration of 200 mg/L had no effect on diacetylmonoxime method. *3393* At concentration of 484 mg/L had no effect on Ektachem® method. *3393*

Uric Acid *Serum No Effect Analytical* At 200 mg/L on reversed phase LC procedure of Zhiri et al. *3942* At acute overdose concentration (20 mg/dL) on Technicon® SMAC® method. *3719* At concentration of 200 mg/L had no effect on phosphotungstate reduction method. *3393*

Vanillylmandelic Acid *Urine No Effect Analytical* No effect observed. *0913*
Urine No Effect Physiological No effect observed. *0913*

Meptazinol

Epinephrine *Plasma Increase Physiological* Almost 2 fold increase in 20 minutes with up to 1.4 mg/kg i.v. *2289*

Norepinephrine *Plasma Increase Physiological* Almost 2 fold increase in 20 minutes with up to 1.4 mg/kg i.v. *2289*

Meralluride

Albumin *Serum No Effect Analytical* No effect at 0.07 mL/dL on SMA 12/60 method. *2636*

Alkaline Phosphatase *Serum No Effect Analytical* No effect at 0.07 mL/dL on SMA 12/60 method. *2636*

Aspartate Aminotransferase *Serum No Effect Analytical* No effect at 0.07 mL/dL on SMA 12/60 method. *2636*

Bicarbonate *Serum Increase Physiological* Alkalosis may occur with loss of chloride. *1680*

Bilirubin *Serum No Effect Analytical* No effect at 0.07 mg/dL on SMA 12/60 method. *2636*

Calcium *Serum No Effect Analytical* No effect at 0.07 mL/dL on SMA 12/60 method. *2636*
Urine Increase Physiological Reabsorption impaired. *2745*

Chloride *Serum Decrease Physiological* Diuretic action. *3879*

Cholesterol *Serum No Effect Analytical* No effect at 0.07 mL/dL on SMA 12/60 method. *2636*

Creatinine *Serum No Effect Analytical* No effect at 0.07 mL/dL on SMA 12/60 method. *2636*

Glucose *Serum No Effect Analytical* No effect at 0.07 mL/dL on SMA 12/60 method. *2636*
Urine Decrease Analytical Interferes with glucose oxidase method. *2425*

131I Uptake *Serum No Effect Physiological* No effect on thyroid function. *2220*

Lactate Dehydrogenase *Serum No Effect Analytical* No effect at 0.07 mL/dL on SMA 12/60 method. *2636*

Leukocytes *Blood Decrease Physiological* May cause bone marrow depression. *1680*

Magnesium *Serum Decrease Physiological* Hypomagnesemia especially if NH4Cl also given. *2314*

Neutrophils *Blood Decrease Physiological* May cause bone marrow depression. *1680*

pH *Blood Increase Physiological* Alkalosis may occur with massive diuresis. *1680*

Phosphate *Serum No Effect Analytical* No effect at 0.07 mL/dL on SMA 12/60 method. *2636*

Potassium *Serum Decrease Physiological* Diuretic action. *3505*

Protein *Serum No Effect Analytical* No effect at 0.07 mL/dL on SMA 12/60 method. *2636*

Protein Bound Iodine (PBI) *Serum Decrease Analytical* Interferes with ceric arsenious acid. *3934*

Sodium *Serum Decrease Physiological* Diuretic action. *3879*

Urea Nitrogen *Serum Increase Physiological* May cause transient neutropenia. *1680*

Uric Acid *Serum No Effect Analytical* No effect at 0.07 mL/dL on SMA 12/60 method. *2636*
Serum Increase Physiological May cause occasional hyperuricemia. *1680*

Merbromin

Color *Urine Increase Analytical* Fluorescent pink staining of cells. *0459*

Urobilin *Urine Increase Analytical* Yields pink color and mauve fluorescence. *1563*

Mercaptoethane

N-Acetylglucosaminidase *Urine No Effect Analytical* At 60 mmol/L on 2 colorimetric analytical methods. *1354*

Mercaptoethanol

Chromosomes *Test Conditions Abnormal Physiological* Clastogenic in human cells. *3282*

Mercaptomerin

Bilirubin *Serum No Effect Analytical* At concentration of 58 mg/L had no effect on Jendrassik and Grof method. *3393*

Calcium *Serum No Effect Analytical* At concentration of 58 mg/L had no effect on cresolphthalein method. *3393*
Urine Increase Physiological Reabsorption impaired. *2745*

Magnesium *Urine Increase Physiological* Excretion increased by up to 30%. *3368*

Phosphate *Serum No Effect Analytical* At concentration of 58 mg/L had no effect on phosphomolybdate method. *3393*

Protein *Serum No Effect Analytical* At concentration of 58 mg/L had no effect on biuret method with blank correction. *3393*

Protein Bound Iodine (PBI) *Serum Decrease Analytical* Formation of mercurial iodide during analysis. *0757*

Urea Nitrogen *Serum No Effect Analytical* At concentration of 58 mg/L had no effect on diacetylmonoxime method. *3393*

Uric Acid *Serum No Effect Analytical* At concentration of 58 mg/L had no effect on phosphotungstate reduction method. *3393*

Zinc *Urine No Effect Physiological* No effect observed on excretion. *2712*

Mercaptopurine

Alanine Aminotransferase *Serum Increase Physiological* May cause hepatotoxicity (centrolobular necrosis). *0204* Intrahepatic cholestasis observed in several patients. *2406*

Alkaline Phosphatase *Serum Increase Physiological* Hepatotoxicity (centrolobular necrosis). *2313* Intrahepatic cholestasis observed in several patients. *2406*

Ammonia *Plasma No Effect Analytical* At concentration of 150 mg/L had no effect on Ektachem® method. *3393*

Amylase *Serum Increase Physiological* One case of hemorrhagic pancreatitis. *2427*

Mercaptopurine *(continued)*

Aspartate Aminotransferase

Serum Increase Physiological May cause hepatotoxicity (centrolobular necrosis). *0204* Picture of cholestasis in 10 of 19 leukemic patients, but drug co-administered with Adriamycin® in all and additional drugs in others. *3021* Intrahepatic cholestasis observed in several patients. *2406*

Bile *Urine Increase Physiological* Hepatotoxic effect. *2313*

Bilirubin *Serum No Effect Analytical* At concentration of 152 mg/L had no effect on Jendrassik and Grof method. *3393*

Serum Increase Physiological May cause increase, especially if prior damage. *3293* Picture of cholestasis in 10 of 19 leukemic patients, but drug co-administered with Adriamycin® in all and additional drugs in others. *3021* Reported in up to 53% leukemic patients treated with drug. *2406*

BSP Retention *Serum Increase Physiological* Hepatotoxicity (centrolobular necrosis). *2313*

Calcium *Serum No Effect Analytical* At concentration of 152 mg/L had no effect on cresolphthalein method. *3393*

Cephalin Flocculation *Serum Increase Physiological* Hepatotoxicity (centrolobular necrosis). *2313*

Chloride *Serum No Effect Analytical* At concentration of 15 mg/L had no effect on Ektachem® method. *3393*

Chromosomes *Test Conditions Abnormal Physiological* Clastogenic in human cells. *3282*

Creatinine *Serum No Effect Analytical* At concentration of 150 mg/L had no effect on Ektachem® method. *3393*

Crystals *Urine Increase Physiological* Direct renal damage with doses over 750 mg/sq m. *0980*

Erythrocytes *Blood Decrease Physiological* May occur with bone marrow depression. *2427*

Urine Increase Physiological Direct renal damage with doses over 750 mg/sq m. *0980*

Glucose *Serum No Effect Analytical* At concentration of 20 mg/L had no effect on Ektachem® method. *3393*

Serum Increase Analytical At 1 mmol/L affects SMA 12/60 method. *3335*

Hematocrit *Blood Decrease Physiological* May occur with bone marrow depression. *2427*

Hemoglobin *Blood Decrease Physiological* May occur with bone marrow depression. *2427*

Urine Increase Physiological Direct renal damage with doses over 750 mg/sq m. *0980*

Leukocytes *Blood Decrease Physiological* Agranulocytosis. *2426*

Phosphate *Serum No Effect Analytical* At concentration of 152 mg/L had no effect on phosphomolybdate method. *3393*

Platelets *Blood Decrease Physiological* May cause bone marrow depression. *2426*

Protein *Serum No Effect Analytical* At concentration of 152 mg/L had no effect on biuret method with blank correction. *3393*

Prothrombin Time *Plasma Increase Physiological* Depresses clotting factor synthesis. *3379*

Thymol Turbidity *Serum Increase Physiological* Hepatotoxicity (centrolobular necrosis). *2313*

Urea Nitrogen *Serum No Effect Analytical* At concentration of 152 mg/L had no effect on diacetylmonoxime method. *3393* At concentration of 15 mg/L had no effect on Ektachem® method. *3393*

Uric Acid *Serum Increase Analytical* At 1 mmol/L affects SMA 12/60 method. *3335* At concentrations above 14 mg/L raised concentration as measured by phosphotungstate reduction method. *3393*

Serum Increase Physiological Leucocyte destruction, catabolism of nucleic acids. *3505*

Urine Increase Physiological Increased nuclear protein breakdown. *1237*

Mercaptopyruvate

Chromosomes *Test Conditions Abnormal Physiological* Clastogenic in human cells. *3282*

Mercurial Diuretics

Ammonia *Plasma Increase Physiological* Presumed effect as may precipitate hepatic coma. *1487*

Calcium *Serum Decrease Physiological* Enhances excretion reducing serum concentration. *0105*

Urine Increase Physiological Reabsorption impaired. *2745*

Chloride *Serum Decrease Physiological* May cause hypochloremic alkalosis and diuresis. *2745*

Urine Increase Physiological Therapeutic intent (dominant urinary anion). *2745*

Erythrocytes *Blood Decrease Physiological* Anemia may occur. *2313*

Glucose *Urine Decrease Analytical* May cause false negative results with glucose oxidase methods. *1488* May produce false negative with glucose oxidase based Stix. *2308*

131I Uptake *Serum No Effect Physiological* No effect in euthyroid subjects. *0583*

Leukocytes *Blood Decrease Physiological* neutropenia/agranulocytosis reported. *2000*

Magnesium *Serum Decrease Physiological* Effect most marked if NH_4Cl also given. *2314*

Urine Increase Physiological Excretion increased by up to 30%. *3368*

pH *Blood Increase Physiological* May cause systemic alkalosis especially if hypochloremia. *1343*

Phosphate *Urine Increase Physiological* Reported effect. *3204*

Platelets *Blood Decrease Physiological* Thrombocytopenia reported (sensitization). *1343*

Potassium *Serum Decrease Physiological* May induce hypokalemia in some cases. *1188*

Protein *Urine Increase Physiological* May produce nephrotic syndrome. *1188*

Protein Bound Iodine (PBI) *Serum Increase Analytical* Interferes with digestion technique. *0012*

Serum Decrease Analytical Interfere with acid distill, chloric acid methods. *1488*

Sodium *Serum Decrease Physiological* Diuretic action with sodium depletion. *2313*

Urine Increase Physiological Therapeutic intent. *1343*

Urea Nitrogen *Serum Increase Physiological* May produce renal failure or nephrotic syndrome. *1188*

Uric Acid *Serum Increase Physiological* May precipitate attacks of gout. *0464*

Volume *Urine Increase Physiological* Therapeutic intent. *1343*

Urine Decrease Physiological May produce oliguria and tubular necrosis. *1188*

Mercury Compounds

Acetoacetate Decarboxylase *Serum Decrease Analytical* $HgCl_2$ causes denaturation. *3684*

α-Amino-Nitrogen *Urine Increase Physiological* Fanconi syndrome with laxative overuse. *2108*

Bicarbonate *Serum Decrease Physiological* May be depressed in established poisoning. *0987*

Calcium *Serum Decrease Physiological* Induced with chronic laxative ingestion. *2108*

Casts *Urine Increase Physiological* May cause severe nephritis if absorbed. *0071*

Color *Urine Increase Physiological* Red brown due to hematuria. *2313*

Feces Increase Analytical Green with about 130 mg of calomel. *2313*

Erythrocytes *Urine Increase Physiological* May cause severe nephritis if absorbed. *0071*

Glucose *Urine Increase Physiological* May cause Fanconi syndrome. *2108*

Hemoglobin *Urine Increase Physiological* May cause severe nephritis if absorbed. *0071*

131I Uptake *Serum No Effect Physiological* No effect observed. *1915*

Leukocytes *Blood Increase Physiological* May induce leukocytosis. *0251*

Magnesium *Serum Decrease Physiological* Induced with chronic laxative ingestion. *2108*

Mercury *Urine Increase Physiological* Due to ingestion of compound and if poisoning. *3515*

Occult Blood *Feces Positive Physiological* Bloody diarrhea occurs with poisoning. *1343*

pH *Blood Decrease Physiological* Induced with chronic laxative ingestion. *2108*

Phosphate *Urine Increase Physiological* Occurs with Fanconi syndrome. *2108*

Potassium *Serum Decrease Physiological* Induced with chronic laxative ingestion. *2108*

Protein *Serum Decrease Physiological* Due to albuminuria and starvation. *2313*

Urine Increase Physiological Nephrotoxic effect. *1237*

CSF Increase Physiological May produce Guaillain-Barré like syndrome. *3515*

Protein Bound Iodine (PBI) *Serum No Effect Analytical* No effect Barker dry ash procedure. *0354*

Serum Decrease Analytical Interfere with chloric acid digest procedures. *0012*

Sodium *Serum Decrease Physiological* May occur with established Hg poisoning. *0987*

T₃ Uptake *Serum No Effect Physiological* No effect on resin uptake. *1915*

Urea Nitrogen *Serum Decrease Analytical* Inhibits urease (note present in Nessler's reagent). *1563*

Uric Acid *Serum Decrease Physiological* Occurs with Fanconi syndrome. *2108*

Urine Increase Physiological Occurs with Fanconi syndrome. *2108*

Mersalyl

Calcium *Urine Increase Physiological* Reabsorption impaired. *2745*

Chloride *Serum Decrease Physiological* Consequence of diuretic action. *2745*

Erythrocytes *Urine Increase Physiological* Actual bleeding may be caused by drug. *1022*

Hemoglobin *Urine Increase Physiological* Actual bleeding caused by drug. *1022*

Uric Acid *Serum Decrease Physiological* Uricosuric action. *1433*

Urine Increase Physiological Uricosuric action. *1433*

Mescaline

Alkaloids *Urine No Effect Analytical* With iodoplatinate of Frings TLC procedure at 10 mg/dL. *1204*

Urine Increase Analytical Reacts with ninhydrin on TLC method of Frings. *1204*

Barbiturate *Urine No Effect Analytical* With mercuric SO_4 on Frings TLC procedure at 10 mg/dL. With vanillin on Frings procedure at 10 mg/dL. With diphenylcarbazone on Frings TLC procedure at 10 mg/dL. With mercurous NO_3 on Frings TLC procedure at 10 mg/dL. *1204*

Bromide *Urine No Effect Analytical* No effect on method of Frings. *1204*

Chromosomes *Test Conditions Abnormal Physiological* Teratogenic in experimental animals. *3282*

Ethchlorvynol *Urine No Effect Analytical* No effect on method of Frings and Cohen. *1204*

Fatty Acids, Free (FFA) *Serum Increase Physiological* 115% increase after 5 mg/kg body weight orally. *1634*

FPN Test *Urine No Effect Analytical* No effect at 10 mg/dL on method of Frings. *1204*

Salicylate *Urine No Effect Analytical* At 10 mg/dL no effect on Trinder method. *1204*

Mesna

Ketones *Urine Increase Analytical* Common if not invariable in patients given i.v. mesna. *0577* False-positive with Multistix® and Chemstrip® but red color can be discharged with glacial acetic acid. *0767*

Platelet Aggregation *Blood No Effect Physiological* Not impaired after stimulation with epinephrine, ADP or arachidonic acid. *1567*

Mesoridazine

Alanine Aminotransferase *Serum No Effect Analytical* At acute overdose concentration (20 mg/dL) on Technicon® SMAC® method. *3719*

Albumin *Serum No Effect Analytical* At acute overdose concentration (20 mg/dL) on Technicon® SMAC® method. *3719*

Alkaline Phosphatase *Serum No Effect Analytical* At acute overdose concentration (20 mg/dL) on Technicon® SMAC® method. *3719*

Aspartate Aminotransferase *Serum No Effect Analytical* At acute overdose concentration (20 mg/dL) on Technicon® SMAC® method. *3719*

Serum Increase Physiological Transient effect noted. *0071*

Bilirubin *Serum No Effect Analytical* At acute overdose concentration (20 mg/dL) on Technicon® SMAC® method. *3719*

Serum Increase Physiological Jaundice as manifestation of hepatotoxicity. *1680*

Calcium *Serum No Effect Analytical* At acute overdose concentration (20 mg/dL) on Technicon® SMAC® method. *3719*

Cholesterol *Serum No Effect Analytical* At acute overdose concentration (20 mg/dL) on Technicon® SMAC® method. *3719*

Creatine Kinase *Serum No Effect Analytical* At acute overdose concentration (20 mg/dL) on Technicon® SMAC® method. *3719*

Creatinine *Serum No Effect Analytical* At acute overdose concentration (20 mg/dL) on Technicon® SMAC® method. *3719*

Eosinophils *Blood Increase Physiological* Manifestation of allergic reaction. *1680*

Erythrocytes *Blood Decrease Physiological* Anemia/aplastic anemia/pancytopenia. *1680*

Glucose *Serum No Effect Analytical* At acute overdose concentration (20 mg/dL) on Technicon® SMAC® method. *3719*

Iron *Serum No Effect Analytical* At acute overdose concentration (20 mg/dL) on Technicon® SMAC® method. *3719*

¹³¹I Uptake *Serum Decrease Physiological* Serentil® contains tetraiodofluorescein. *2652*

Lactate Dehydrogenase *Serum No Effect Analytical* At acute overdose concentration (20 mg/dL) on Technicon® SMAC® method. *3719*

Leukocytes *Blood Decrease Physiological* Transient agranulocytosis reported. *0071*

Phosphate *Serum No Effect Analytical* At acute overdose concentration (20 mg/dL) on Technicon® SMAC® method. *3719*

Platelets *Blood Decrease Physiological* Thrombocytopenia may occur. *1680*

Protein *Serum No Effect Analytical* At acute overdose concentration (20 mg/dL) on Technicon® SMAC® method. *3719*

Triglycerides *Serum No Effect Analytical* At acute overdose concentration (20 mg/dL) on Technicon® SMAC® method. *3719*

Urea Nitrogen *Serum No Effect Analytical* At acute overdose concentration (20 mg/dL) on Technicon® SMAC® method. *3719*

Uric Acid *Serum No Effect Analytical* At acute overdose concentration (20 mg/dL) on Technicon® SMAC® method. *3719*

Mesterolone

Tyrosine *Plasma Decrease Physiological* Hormonal effect (?exact mechanism). *0843*

Mestranol

Albumin *Serum Decrease Physiological* Metabolic effect. *3258*

Mestranol (continued)

Antithrombin III *Plasma Decrease Physiological* In over 20% — not dose related. *3960*

BSP Retention *Serum Increase Physiological* Depresses hepatic secretory transport maximum. *1487*

Calcium *Serum Decrease Physiological* Increased sensitivity to calcitonin in postmenopausal women. *0037*
Urine Decrease Physiological Increased sensitivity to calcitonin in postmenopausal women. *0037*

Cortisol, Free *Plasma Increase Physiological* Slight effect if over 0.1 mg. *0537*

Follicle Stimulating Hormone (FSH)
Plasma Decrease Physiological Hormonal effect (inhibitory action of estrogen). *3237*

α_2-**Globulin** *Serum Increase Physiological* Metabolic effect. *3258*

β-**Globulin** *Serum Increase Physiological* Metabolic effect. *3258*

Glucose Tolerance *Serum Decrease Physiological* May provoke mild to moderate deterioration. *1849*

Iron-Binding Capacity, Total (TIBC)
Serum Increase Physiological 20% rise on average. *1660*

Luteinizing Hormone (LH) *Plasma Increase Physiological* Estrogen exerts stimulatory action. *3914*

α_2-**Macroglobulin** *Serum Increase Physiological* Maximum effect 1 week after treatment. *1660*

Phosphate *Serum Decrease Physiological* Increased sensitivity to calcitonin in postmenopausal women. *0037*
Urine Increase Physiological Increased response to calcitonin. *0037*

Prolactin *Plasma Increase Physiological* In 31 of 88 oophorectomized women treated for 3 to 11 y. *0230*

Thyroid Stimulating Hormone (TSH)
Serum No Effect Physiological Insignificant difference in 19 post-menopausal women receiving mean of 24 µg/d versus controls. *0007*

Thyroxine (T$_4$) *Serum Increase Physiological* From 100 to 133 nmol/L in 19 post-menopausal women receiving mean of 24 µg/d versus controls. *0007*

Thyroxine (T$_4$) Binding Globulin
Serum Increase Physiological From 24 to 47 mg/L in 19 post-menopausal women receiving mean of 24 µg/d versus controls. *0007*

Thyroxine (T$_4$), Free *Serum Decrease Physiological* From 19.2 to 15.7 pmol/L in 19 post-menopausal women receiving mean of 24 µg/d versus controls. *0007*

Transferrin *Serum Increase Physiological* Maximum effect 1 week after treatment. *1660*

Triglycerides *Serum Increase Physiological* ?impaired triglyceride removal from circulation. *3505*

Tri-iodothyronine (T$_3$) *Serum Increase Physiological* From 2.22 to 2.72 nmol/L in 19 post-menopausal women receiving mean of 24 µg/d versus controls. *0007*

Tyrosine *Plasma Decrease Physiological* Effect of both estrogen and progestogen. *0843*

Metahexamide

Alanine Aminotransferase *Serum Increase Physiological* Hepatotoxicity (viral hepatitis type). *2313*

Alkaline Phosphatase *Serum Increase Physiological* Hepatotoxicity (viral hepatitis type). *2313*

Aspartate Aminotransferase
Serum Increase Physiological Hepatotoxicity (viral hepatitis type). *2313*

Bile *Urine Increase Physiological* Hepatotoxicity (viral hepatitis type). *2313*

Bilirubin *Serum Increase Physiological* Hepatotoxicity (viral hepatitis type). *2313*

BSP Retention *Serum Increase Physiological* Hepatotoxicity (viral hepatitis type). *2313*

Cephalin Flocculation *Serum Increase Physiological* Hepatotoxicity (viral hepatitis type). *2313*

Protein *Urine Increase Analytical* Interference by drug metabolite. *3729* Interference by drug metabolite. *1022*

Thymol Turbidity *Serum Increase Physiological* Hepatotoxicity (viral hepatitis type). *2313*

Metanephrine

Amino Acids *Urine Increase Analytical* Reacts with ninhydrin; extra spot TLC, high voltage electrophoresis. *2855*

Ammoniacal Silver Nitrate
Test Conditions Positive Analytical Positive spot test with Tollen's reagent. *3765*

Epinephrine *Test Conditions No Effect Analytical* On fluorescent procedure of Peyrin. *2800*

Glucose *Serum Increase Physiological* Intravenous infusion 5 µg/min causes increase of 30%. *1176*

Guaiacols Spot Test *Urine Positive Analytical* Action on procedure of Rogers. *3031*

Norepinephrine *Test Conditions No Effect Analytical* On fluorescent procedure of Peyrin. *2800*

Uric Acid *Test Conditions Increase Analytical* Positive spot test with phosphotungstate. *3765*

Urobilinogen *Test Conditions Increase Analytical* Positive spot test with Ehrlich's reagent. *3765*

Metaphosphoric Acid

Ascorbic Acid *Plasma Decrease Analytical* Increase by 10% if stored at -20° for 21 d. *0441*

Metaproterenol

Sugar *Urine Increase Analytical* Interferes with Benedict's reagent. *1022*

Metaraminol

Amphetamine *Urine No Effect Analytical* No reaction with NBD chloride procedure of Monforte. *2473* No effect at 100 mg/L on method of Rutter. *3097*

Metaxalone

Alanine Aminotransferase *Serum Increase Physiological* Hepatotoxic effect. *2313*

Alkaline Phosphatase *Serum Increase Physiological* Hepatotoxic effect. *2313*

Aspartate Aminotransferase
Serum Increase Physiological Hepatotoxic effect. *2313*

Bile *Urine Increase Physiological* Hepatotoxic effect. *2313*

Bilirubin *Serum Increase Physiological* Hepatotoxic effect. *2313*

BSP Retention *Serum Increase Physiological* Hepatotoxic effect. *1022*

Cephalin Flocculation *Serum Increase Physiological* Hepatotoxic effect. *2313*

Glucose *Urine No Effect Analytical* No effect on glucose oxidase methods. *1487*

Protein *Urine Increase Physiological* May have nephrotoxic effect Nephrotoxic effect. *1022*

Sugar *Urine Increase Analytical* False positive with copper reduction procedures. False positive with Benedict's, Fehling's reactions. *3505*

Thymol Turbidity *Serum Increase Physiological* Hepatotoxic effect. *2313*

Metformin

Amino-4-Imidazole-5-Carboxamide Ribotide (AICAR)
Urine Increase Physiological May cause megaloblastic anemia. *3605*

Bicarbonate *Serum Decrease Physiological* May cause marked acidosis (lactic acidosis). *1022*

Carotene *Serum Decrease Physiological* Probably associated with malabsorption. *3605*

Cholesterol *Serum No Effect Physiological* No effect although phenformin causes decrease. *1487*

Creatinine *Serum No Effect Analytical* At 2 g/L on reversed phase LC procedure of Zhiri et al. *3942*

FIGLU (N-Formiminoglutamic Acid)
Urine Increase Physiological May cause megaloblastic anemia. *3605*

Folate *Serum Increase Physiological* High in patients if B_{12} malabsorption. *3605*
Serum Decrease Physiological Due to decreased absorption of dietary folate. *0302*

Glucose *Serum Decrease Physiological* Mode of action uncertain (occurs with overdose). *1184*

Glucose Tolerance *Serum Increase Physiological* Mode of action uncertain. *1184*

Hematocrit *Blood Decrease Physiological* May be associated with megaloblastic anemia. *3605*

Hemoglobin *Blood Decrease Physiological* Associated with impaired B_{12} absorption. *3605*

Insulin *Plasma Decrease Physiological* Slight, all hyperlipoproteinemias, marked type i.v. *1430*

Iron *Serum Decrease Physiological* Associated with impaired B_{12} absorption. *3605*

Ketones *Urine Increase Physiological* Associated with lactic acidosis. *1022*

Lactate *Serum Increase Physiological* Possibly always with predisposing condition. *2103*

MCV *Blood Increase Physiological* Occurs if megaloblastic anemia. *3605*

Triglycerides *Serum Decrease Physiological* Average of 26% in type i.v. hyperlipoproteinemia. *1430*

Uric Acid *Serum No Effect Analytical* At 2 g/L on reversed phase LC procedure of Zhiri et al. *3942*

Vitamin B_{12} *Serum Decrease Physiological* Due to impaired B_{12} absorption. *3605* Reported to cause ileal malabsorption. *3023*
Urine Decrease Physiological May cause megaloblastic anemia. *3605*

Xylose Excretion *Urine Decrease Physiological* Probably dose related malabsorption. *3605*

Methacholine

Amylase *Serum Increase Physiological* Stimulates pancreatic secretion, constricts ampulla. *1343* Pancreatic stimulation, constriction of ampulla. *1237*

Aspartate Aminotransferase
Serum Increase Physiological Impairs excretion by spasm of sphincter of Oddi. *0652*

Bilirubin *Serum Increase Physiological* Impairs excretion through biliary tract. *0652*

BSP Retention *Serum Increase Physiological* Impairs excretion by spasm of sphincter of Oddi. *0652*

Epinephrine *Urine No Effect Physiological* No effect observed. *0279*

Lipase *Serum Increase Physiological* Constricts sphincter of Oddi. *0652*

Norepinephrine *Urine Increase Physiological* Slight increase observed. *0279*

Methacycline

Alanine Aminotransferase *Serum Increase Physiological* Possible hepatotoxicity. *0071*

Aspartate Aminotransferase
Serum Increase Physiological Possible hepatotoxicity. *0071*

Eosinophils *Blood Increase Physiological* May cause allergic response. *1680*

Erythrocytes *Blood Decrease Physiological* May cause hemolytic anemia. *1680*

Neutrophils *Blood Decrease Physiological* Neutropenia reported. *1680*

Platelets *Blood Decrease Physiological* Thrombocytopenia may occur. *1680*

Urea Nitrogen *Serum Increase Physiological* Possible nephrotoxicity. *0071*

Methadone

Alanine Aminotransferase *Serum No Effect Analytical* At acute overdose concentration (2.5 mg/dL) on Technicon® SMAC® method. *3719*
Serum No Effect Physiological No toxicity if liver function tests normal initially. *2012*

Albumin *Serum No Effect Analytical* At acute overdose concentration (2.5 mg/dL) on Technicon® SMAC® method. *3719* At concentration of 25 mg/L had no effect on BCG method. *3393*

Alkaline Phosphatase *Serum No Effect Analytical* At acute overdose concentration (2.5 mg/dL) on Technicon® SMAC® method. *3719*
Serum No Effect Physiological No toxicity if liver function tests normal initially. *2012*

Alkaloids *Urine No Effect Analytical* With ninhydrin on Frings TLC procedure at 10 mg/dL. *1204*
Urine Increase Analytical Red/purple with iodoplatinate on Frings TLC procedure. *1204*

Aspartate Aminotransferase *Serum No Effect Analytical* At acute overdose concentration (2.5 mg/dL) on Technicon® SMAC® method. *3719*
Serum No Effect Physiological No evidence of toxicity if initially normal liver function test. *2012*

Barbiturate *Urine No Effect Analytical* With vanillin on Frings procedure at 10 mg/dL. With mercurous NO_3 on Frings TLC procedure at 10 mg/dL. With mercuric SO_4 on Frings TLC procedure at 10 mg/dL. With diphenylcarbazone on Frings TLC procedure at 10 mg/dL. *1204*

Bilirubin *Serum No Effect Analytical* At acute overdose concentration (2.5 mg/dL) on Technicon® SMAC® method. *3719* At concentration of 25 mg/L had no effect on Jendrassik and Grof method. *3393*
Serum No Effect Physiological No toxicity if liver function tests normal initially. *2012*

Bromide *Urine No Effect Analytical* No effect on method of Frings. *1204*

BSP Retention *Serum Increase Physiological* Hepatotoxic effect or spasm of sphincter of Oddi. *3879*

Calcium *Serum No Effect Analytical* At acute overdose concentration (2.5 mg/dL) on Technicon® SMAC® method. *3719* At concentration of 25 mg/L had no effect on cresolphthalein method. *3393*

Chlordiazepoxide *Blood No Effect Analytical* On method of Riddick at 5 mg/dL. *2983*

Chloride *Serum No Effect Analytical* At concentration of 7 mg/L had no effect on mercurimetric method. *3393*

Cholesterol *Serum No Effect Analytical* At acute overdose concentration (2.5 mg/dL) on Technicon® SMAC® method. *3719* At concentration of 25 mg/L had no effect on Liebermann-Burchard method. *3393*

Chromosomes *Test Conditions No Effect Physiological* No effect human leucocytes at concentrations 1/6-3x normal. *1073*

Cortisol *Plasma Increase Physiological* Significant response to cold not seen in controls. *2956*

Creatine Kinase *Serum No Effect Analytical* At acute overdose concentration (2.5 mg/dL) on Technicon® SMAC® method. *3719*

Creatinine *Serum No Effect Analytical* At acute overdose concentration (2.5 mg/dL) on Technicon® SMAC® method. *3719* At concentration of 25 mg/L had no effect on AutoAnalyzer Jaffé method. *3393*

Ethchlorvynol *Serum No Effect Analytical* At 8 mg/L on GLC procedure of Evenson. *1058*
Urine No Effect Analytical No effect on method Frings and Cohen at 2 mg/dL. *1205*

FPN Test *Urine No Effect Analytical* No effect at 10 mg/dL on method of Frings. *1204*

Glucose *Serum No Effect Analytical* At acute overdose concentration (2.5 mg/dL) on Technicon® SMAC® method. *3719*

Immunoglobulin IgG *Serum Increase Physiological* Commonly seen in response to treatment. *0781*

Methadone (continued)

Indocyanine Green *Serum Decrease Physiological* Observed in small series, normal liver function test. *2398*

Iron *Serum No Effect Analytical* At acute overdose concentration (2.5 mg/dL) on Technicon® SMAC® method. *3719* At concentration of 25 mg/L had no effect on Ferrozine method. *3393*

Lactate Dehydrogenase *Serum No Effect Analytical* At acute overdose concentration (2.5 mg/dL) on Technicon® SMAC® method. *3719*

Methaqualone *Serum No Effect Analytical* At 8 mg/L on GLC procedure of Evenson. *1057*

Morphine *Urine No Effect Analytical* Insignificant cross reactivity with RIA procedures Insignificant cross reactivity with EMIT procedure for opiates. *2514*

pCO_2 *Blood Increase Physiological* May cause diminished pulmonary ventilation. *1343*

Phosphate *Serum No Effect Analytical* At acute overdose concentration (2.5 mg/dL) on Technicon® SMAC® method. *3719* At concentration of 25 mg/L had no effect on phosphomolybdate method. *3393*

Pregnancy Tests *Urine Positive Analytical* Highest incidence with Gravindex™. *1663*

Propoxyphene *Urine No Effect Analytical* Less than 1% fluorescence in procedure of Valentour. *3662*

Protein *Serum No Effect Analytical* At acute overdose concentration (2.5 mg/dL) on Technicon® SMAC® method. *3719* At concentration of 25 mg/L had no effect on biuret method with blank correction. *3393*

Salicylate *Urine No Effect Analytical* At 10 mg/dL no effect on Trinder method. *1204*

Testosterone *Serum No Effect Physiological* In treated/untreated male addicts. *0780*

Thyroid Stimulating Hormone (TSH)
Serum No Effect Physiological No significant effect observed. *0193*

Thyroxine (T_4) *Serum Increase Physiological* Slight but not significant increase. *0193* Increased binding capacity (augmentation of transport proteins). *3798*

Thyroxine (T_4), Free *Serum Decrease Physiological* Significantly lower percentage. *0193*

Triglycerides *Serum No Effect Analytical* At acute overdose concentration (2.5 mg/dL) on Technicon® SMAC® method. *3719* At concentration of 25 mg/L had no effect on lipase/esterase method. *3393*

Tri-iodothyronine (T_3) *Serum Increase Physiological* Significant increase observed compared with normals. *0193* Increased binding capacity (augmentation of transport proteins). *3798*

Urea Nitrogen *Serum No Effect Analytical* At acute overdose concentration (2.5 mg/dL) on Technicon® SMAC® method. *3719* At concentration of 25 mg/L had no effect on diacetylmonoxime method. *3393*

Uric Acid *Serum No Effect Analytical* At acute overdose concentration (2.5 mg/dL) on Technicon® SMAC® method. *3719* At concentration of 25 mg/L had no effect on phosphotungstate reduction method. *3393*

Methadone Metabolite

Alkaloids *Urine Increase Analytical* Reacts with ninhydrin on TLC method of Frings. *1204*

Methamphetamine

Albumin *Serum No Effect Analytical* At concentration of 2 mg/L had no effect on BCG method. *3393*

Alkaloids *Urine No Effect Analytical* With iodoplatinate of Frings TLC procedure at 10 mg/dL. *1204*
Urine Increase Analytical Reacts with ninhydrin on TLC method of Frings. *1204*

Amino Acids *Urine Increase Analytical* Reacts with ninhydrin; extra spot TLC, high voltage electrophoresis. *2855*

Amphetamine *Urine Increase Analytical* Yields fluorophor with NBD chloride reaction. *2473*

Barbiturate *Urine No Effect Analytical* With mercurous NO_3 on Frings TLC procedure at 10 mg/dL. With mercuric SO_4 on Frings TLC procedure at 10 mg/dL. With diphenylcarbazone on Frings TLC procedure at 10 mg/dL. With vanillin on Frings procedure at 10 mg/dL. *1204*

Bicarbonate *Serum No Effect Analytical* At concentration of 2 mg/L had no effect on method using phenolphthalein. *3393*

Bilirubin *Serum No Effect Analytical* At concentration of 2 mg/L had no effect on Jendrassik and Grof method. *3393*

Bromide *Urine No Effect Analytical* No effect on method of Frings. *1204*

Calcium *Serum No Effect Analytical* At concentration of 2 mg/L had no effect on cresolphthalein method. *3393*

Chloride *Serum No Effect Analytical* At concentration of 2 mg/L had no effect on mercurimetric method. *3393*

Cholesterol *Serum No Effect Analytical* At concentration of 2 mg/L had no effect on CHOD-PAP method. *3393* At concentration of 2 mg/L had no effect on Liebermann-Burchard method. *3393*

Corticosteroids *Plasma Increase Physiological* Effect most marked in am when given i.v. *1488*

Creatinine *Serum No Effect Analytical* At concentration of 2 mg/L had no effect on AutoAnalyzer Jaffé method. *3393*

Epinephrine *Urine Increase Physiological* For 3 h after administration (slight). *1487*

Ethchlorvynol *Serum No Effect Analytical* At 40 mg/L on GLC procedure of Evenson. *1058*
Urine No Effect Analytical No effect on method of Frings and Cohen. *1204*

FPN Test *Urine No Effect Analytical* No effect at 10 mg/dL on method of Frings. *1204*

Growth Hormone *Plasma Increase Physiological* Significant rise. *0331*

5-Hydroxyindoleacetic Acid (5-HIAA)
Urine Increase Physiological Single instance reported. *2142*

Methaqualone *Serum No Effect Analytical* At 5 mg/L on GLC procedure of Evenson. *1057*

Norepinephrine *Urine No Effect Physiological* No effect observed. *1487*

Phosphate *Serum No Effect Analytical* At concentration of 2 mg/L had no effect on phosphomolybdate method. *3393*

Propoxyphene *Serum No Effect Analytical* At 40 mg/L on method of Evenson. *1056*
Urine Increase Analytical 3% fluorescence in procedure of Valentour. *3662*

Protein *Serum No Effect Analytical* At concentration of 2 mg/L had no effect on biuret method with blank correction. *3393*

Salicylate *Urine No Effect Analytical* At 10 mg/dL no effect on Trinder method. *1204*

Triglycerides *Serum No Effect Analytical* At concentration of 2 mg/L had no effect on lipase/esterase method. *3393*

Urea Nitrogen *Serum No Effect Analytical* At concentration of 2 mg/L had no effect on diacetylmonoxime method. *3393*

Uric Acid *Serum No Effect Analytical* At concentration of 2 mg/L had no effect on phosphotungstate reduction method. *3393*

Methandriol

Alanine Aminotransferase *Serum Increase Physiological* Intrahepatic cholestatic jaundice. *1343*

Alkaline Phosphatase *Serum Increase Physiological* Intrahepatic cholestatic jaundice. *1343*

Aspartate Aminotransferase
Serum Increase Physiological Intrahepatic cholestatic jaundice. *1343*

Bile *Urine Increase Physiological* Intrahepatic cholestatic jaundice. *1343*

Bilirubin *Serum Increase Physiological* Intrahepatic cholestatic jaundice. *1343*

BSP Retention *Serum Increase Physiological* Intrahepatic cholestatic jaundice. *1343*

Methandrostenolone

Alanine Aminotransferase *Serum Increase Physiological* Up to 3-6 times normal due to cholestasis. *1713*

Alkaline Phosphatase *Serum Increase Physiological* Intrahepatic cholestatic jaundice. *1343*

Aspartate Aminotransferase
Serum Increase Physiological Up to 3-6 times normal due to cholestasis. *1713* In 2 of 6 body-builders taking up to 20 mg/d intermittently for a year or more. *3290*

Bile *Urine Increase Physiological* Intrahepatic cholestatic jaundice. *1343*

Bilirubin *Serum Increase Physiological* Due to cholestasis. *1713*

BSP Retention *Serum Increase Physiological* Cholestatic phenomenon. *0757* Hepatotoxic effect (common). *2313*

Calcium *Serum Increase Physiological* May occur spontaneously, but especially if breast cancer. *1680*

Cephalin Flocculation *Serum Increase Physiological* Hepatotoxic effect (common). *2313*

Cholesterol *Serum Increase Analytical* Interferes with Zimmermann reaction. *2220*
Serum Increase Physiological Due to cholestasis. *1713*
Serum Decrease Physiological Reported effect (may increase as alternative). *1680*

Creatine *Urine Increase Physiological* May persist up to 2 weeks after treatment. *0019*

Creatinine *Urine Increase Physiological* May persist up to 2 weeks after treatment. *0019*

Factor II *Plasma Increase Physiological* Metabolic effect. *0019*

Factor V *Plasma Increase Physiological* Metabolic effect. *0019*

Factor VII *Plasma Increase Physiological* Metabolic effect. *0019*

Factor X *Plasma Increase Physiological* Metabolic effect. *0019*

Glucose *Serum Decrease Physiological* Anabolic effect. *0019*

Glucose Tolerance *Serum Increase Physiological* Anabolic effect. *0019*
Serum Decrease Physiological Alters curve in diabetic direction. *2427*

β-Glucuronidase *Serum Increase Physiological* Metabolic effect. *0227*

Haptoglobin *Serum Increase Physiological* Metabolic effect. *0227*

17-Hydroxycorticosteroids *Urine Decrease Physiological* Inhibits response to metyrapone. *0022*

^{131}I Uptake *Serum Decrease Physiological* Also modifies binding of thyroid hormones. *2427*

17-Ketosteroids *Urine Decrease Physiological* Metabolic action of drug. *0019*

Luteinizing Hormone (LH) *Plasma Decrease Physiological* In 4 of 6 body-builders taking up to 20 mg/d intermittently for a year or more. *3290*

Metyrapone Test *Patient Positive Physiological* Anabolic effect. *2681*

Oxyphenbutazone *Serum Increase Physiological* Possibly due to inhibition of metabolism. *1487*

Phenylbutazone *Serum No Effect Physiological* Unaffected by concomitant administration. *1487*

Plasminogen *Plasma Increase Physiological* Metabolic effect. *0227*

Protein Bound Iodine (PBI) *Serum Decrease Physiological* Metabolic effect. *0019*

Prothrombin Time *Plasma Increase Physiological* Prolongs action of anticoagulants. *2313* Also seen with other 17-alkyl substituted steroids. *0998*

Sialic Acid *Serum Increase Physiological* Metabolic effect. *0227*

Sodium *Serum Increase Physiological* May be affected but water also retained. *0019*

T$_3$ Uptake *Serum Increase Physiological* Affects RBC and resin uptakes. *0019* In 6 body-builders taking up to 20 mg/d intermittently for a year or more. *3290*

Testosterone *Serum Decrease Physiological* In 4 of 6 body-builders taking up to 20 mg/d intermittently for a year or more. *3290*

Thymol Turbidity *Serum Increase Physiological* Hepatotoxic effect. *2313*

Thyroxine (T$_4$) *Serum No Effect Physiological* In 6 body-builders taking up to 20 mg/d intermittently for a year or more. *3290*

Thyroxine (T$_4$) Binding Globulin
Serum Decrease Physiological Metabolic effect. *0019*

Thyroxine (T$_4$), Free *Serum No Effect Physiological* No change observed. *0019*

Triglycerides *Serum Decrease Physiological* Metabolic (anabolic) effect. *3477*

Urobilinogen *Urine Decrease Physiological* Due to cholestasis. *1713*
Feces Decrease Physiological Light stools, due to cholestasis. *1713*

Methanol

Acetone *Urine Increase Physiological* Slight to moderate effect in poisoning. *1343*

Alanine Aminotransferase *Serum No Effect Analytical* At acute overdose concentration (20 mg/dL) on Technicon® SMAC® method. *3719*

Albumin *Serum No Effect Analytical* At acute overdose concentration (20 mg/dL) on Technicon® SMAC® method. *3719*

Alcohol *Breath Increase Physiological* Presence of ingested alcohol. *1238*

Alkaline Phosphatase *Serum No Effect Analytical* At acute overdose concentration (20 mg/dL) on Technicon® SMAC® method. *3719*

Amylase *Serum Increase Physiological* Elevation due to pancreatitis. *1238*

Aspartate Aminotransferase *Serum No Effect Analytical* At acute overdose concentration (20 mg/dL) on Technicon® SMAC® method. *3719*

Bicarbonate *Serum Decrease Physiological* Metabolic acidosis due to organic acid accumulation. *1343*

Bilirubin *Serum No Effect Analytical* At acute overdose concentration (20 mg/dL) on Technicon® SMAC® method. *3719*
Serum Increase Analytical False value if impure methanol used (Evelyn-Malloy). *0579*
Serum Increase Physiological Manifestation of liver damage. *1238*

Bilirubin, Direct *Serum Increase Physiological* Manifestation of liver damage. *1238*

Bogen Test *Urine Positive Analytical* Reacts as if ethanol. *1302*

Calcium *Serum No Effect Analytical* At acute overdose concentration (20 mg/dL) on Technicon® SMAC® method. *3719*

Cholesterol *Serum No Effect Analytical* At acute overdose concentration (20 mg/dL) on Technicon® SMAC® method. *3719*

Creatine Kinase *Serum No Effect Analytical* At acute overdose concentration (20 mg/dL) on Technicon® SMAC® method. *3719*

Creatinine *Serum No Effect Analytical* At acute overdose concentration (20 mg/dL) on Technicon® SMAC® method. *3719*

Formaldehyde *Urine Increase Physiological* Metabolite of oxidation. *1343*

Formic Acid *Serum Increase Physiological* Metabolite (with formaldehyde) of oxidation. *1343*

Glucose *Serum No Effect Analytical* At acute overdose concentration (20 mg/dL) on Technicon® SMAC® method. *3719*

Ionized Calcium *Serum Decrease Analytical* At 0.1 mmol/L to 0.1 mol/L on Calcium specific electrode. *0540*

Iron *Serum No Effect Analytical* At acute overdose concentration (20 mg/dL) on Technicon® SMAC® method. *3719*

Ketones *Serum Increase Physiological* Moderate effect in comparison with extent of acidosis. *1343*

Methanol *(continued)*

Lactate Dehydrogenase *Serum No Effect Analytical* At acute overdose concentration (20 mg/dL) on Technicon® SMAC® method. *3719*

Methanol *Blood Increase Physiological* Presence of ingested alcohol (fatal at 1 g/L usually). *1343*
CSF Increase Physiological Higher concentration than in blood. *1343*

pH *Blood Decrease Physiological* Causes acidosis. *2313*

Phosphate *Serum No Effect Analytical* At acute overdose concentration (20 mg/dL) on Technicon® SMAC® method. *3719*

Protein *Serum No Effect Analytical* At acute overdose concentration (20 mg/dL) on Technicon® SMAC® method. *3719*
Urine Increase Physiological Nephrotoxic effect with poisoning Occurs with poisoning. *1343*

Triglycerides *Serum No Effect Analytical* At acute overdose concentration (20 mg/dL) on Technicon® SMAC® method. *3719*

Urea Nitrogen *Serum No Effect Analytical* At acute overdose concentration (20 mg/dL) on Technicon® SMAC® method. *3719*
Serum Increase Physiological May cause nephrotoxicity. *3204*

Uric Acid *Serum No Effect Analytical* At acute overdose concentration (20 mg/dL) on Technicon® SMAC® method. *3719*
Serum Increase Physiological Observed with tissue destruction in fatal poisoning. *0279*

Methantheline

[131]I Uptake *Serum Decrease Physiological* Reported to decrease results. *0583*

Methaphentermine

Amphetamine *Urine Increase Analytical* Reacts with methyl orange in method of Frings. *2474*

Methapyrilene

Albumin *Serum No Effect Analytical* At concentration of 13 mg/L had no effect on BCG method. *3393*

Alkaloids *Urine No Effect Analytical* With ninhydrin on Frings TLC procedure at 25 mg/dL. *1204*
Urine Increase Analytical Purple with iodoplatinate on Frings TLC procedure. *1204*

Barbiturate *Urine No Effect Analytical* With mercurous NO_3 on Frings TLC procedure at 25 mg/dL. With mercuric SO_4 on Frings TLC procedure at 25 mg/dL. With diphenylcarbazone on Frings TLC procedure at 25 mg/dL. *1204*
Urine Increase Analytical Purple spot with vanillin on Frings TLC procedure. *1204*

Bicarbonate *Serum No Effect Analytical* At concentration of 13 mg/L had no effect on method using phenolphthalein. *3393*

Bilirubin *Serum No Effect Analytical* At concentration of 13 mg/L had no effect on Jendrassik and Grof method. *3393*

Bromide *Urine No Effect Analytical* No effect on method of Frings. *1204*

Calcium *Serum No Effect Analytical* At concentration of 13 mg/L had no effect on cresolphthalein method. *3393*

Chlordiazepoxide *Blood Increase Analytical* May interfere with method of Jatlow. *1777*

Chloride *Serum No Effect Analytical* At concentration of 13 mg/L had no effect on mercurimetric method. *3393*

Cholesterol *Serum No Effect Analytical* At concentration of 13 mg/L had no effect on Liebermann-Burchard method. *3393*

Creatinine *Serum No Effect Analytical* At concentration of 13 mg/L had no effect on AutoAnalyzer Jaffé method. *3393*

Erythrocytes *Blood Decrease Physiological* Anemia (AMA Blood dyscrasias). *2429*

Ethchlorvynol *Serum No Effect Analytical* At 10 mg/L on GLC procedure of Evenson. *1058*
Urine No Effect Analytical No effect on method of Frings and Cohen. *1204*

Methaqualone *Serum No Effect Analytical* At 10 mg/L on GLC procedure of Evenson. *1057*

Phosphate *Serum No Effect Analytical* At concentration of 13 mg/L had no effect on phosphomolybdate method. *3393*

Potassium *Serum No Effect Analytical* At concentration of 13 mg/L had no effect on measurement by ISE with predilution. *3393*

Protein *Serum No Effect Analytical* At concentration of 13 mg/L had no effect on biuret method with blank correction. *3393*

Sodium *Serum No Effect Analytical* At concentration of 13 mg/L had no effect on measurement by ISE with predilution. *3393*

Triglycerides *Serum No Effect Analytical* At concentration of 13 mg/L had no effect on lipase/esterase method. *3393*

Urea Nitrogen *Serum No Effect Analytical* At concentration of 13 mg/L had no effect on diacetylmonoxime method. *3393*

Uric Acid *Serum No Effect Analytical* At concentration of 13 mg/L had no effect on phosphotungstate reduction method. *3393*

Methaqualone

Alanine Aminotransferase *Serum No Effect Analytical* At acute overdose concentration (2.5 mg/dL) on Technicon® SMAC® method. *3719* No effect at therapeutic concentration on Reflotron method. *1984*

Albumin *Serum No Effect Analytical* At acute overdose concentration (2.5 mg/dL) on Technicon® SMAC® method. *3719* At concentration of 25 mg/L had no effect on BCG method. *3393*

Alkaline Phosphatase *Serum No Effect Analytical* At acute overdose concentration (2.5 mg/dL) on Technicon® SMAC® method. *3719*

Alkaloids *Urine No Effect Analytical* With ninhydrin on Frings TLC procedure at 10 mg/dL. *1204*
Urine Increase Analytical Purple with iodoplatinate on Frings TLC procedure. *1204*

Antipyrine *Plasma No Effect Physiological* No significant effect on half-life. *3466*

Aspartate Aminotransferase *Serum No Effect Analytical* At acute overdose concentration (2.5 mg/dL) on Technicon® SMAC® method. *3719* No effect at therapeutic concentration on Reflotron method. *1984*

Barbiturate *Urine No Effect Analytical* With diphenylcarbazone on Frings TLC procedure at 10 mg/dL. With mercurous NO_3 on Frings TLC procedure at 10 mg/dL. With vanillin on Frings procedure at 10 mg/dL. With mercuric SO_4 on Frings TLC procedure at 10 mg/dL. *1204*

Bilirubin *Serum No Effect Analytical* At acute overdose concentration (2.5 mg/dL) on Technicon® SMAC® method. *3719* At concentration of 25 mg/L had no effect on Jendrassik and Grof method. *3393*

Bromide *Urine No Effect Analytical* No effect on method of Frings. *1204*

Calcium *Serum No Effect Analytical* At acute overdose concentration (2.5 mg/dL) on Technicon® SMAC® method. *3719* At concentration of 25 mg/L had no effect on cresolphthalein method. *3393*

Chlordiazepoxide *Blood No Effect Analytical* On method of Riddick at 5 mg/dL. *2983*

Cholesterol *Serum No Effect Analytical* At acute overdose concentration (2.5 mg/dL) on Technicon® SMAC® method. *3719* No effect at therapeutic concentration on Reflotron method. *1984* At concentration of 25 mg/L had no effect on Liebermann-Burchard method. *3393*

Creatine Kinase *Serum No Effect Analytical* At acute overdose concentration (2.5 mg/dL) on Technicon® SMAC® method. *3719*

Creatinine *Serum No Effect Analytical* At acute overdose concentration (2.5 mg/dL) on Technicon® SMAC® method. *3719* At concentration of 25 mg/L had no effect on AutoAnalyzer Jaffé method. *3393*

Dapsone *Serum No Effect Analytical* At 5 mg/dL on colorimetric procedure of Higgins. *1590*

Diazepam *Serum No Effect Analytical* On cathode-ray Polarographic method of Berry. *0323*

Erythrocytes *Blood Decrease Physiological* One possible case of aplastic anemia reported. *2313*

Estriol *Urine No Effect Analytical* No effect on GLC method. *1163*

Ethchlorvynol *Serum No Effect Analytical* At 10 mg/L on GLC procedure of Evenson. *1058*
Urine No Effect Analytical No effect on method of Frings and Cohen at 10 mg/dL. *1205*

Folate *Serum No Effect Analytical* If chromatographic procedure of Landon used. *2068*

FPN Test *Urine No Effect Analytical* No effect at 10 mg/dL on method of Frings. *1204*

Glucose *Serum No Effect Analytical* At acute overdose concentration (2.5 mg/dL) on Technicon® SMAC® method. *3719* No effect at therapeutic concentration on Reflotron method. *1984* At concentration of 80 mg/L had no effect on hexokinase/G-6-PDH method. *3393*

γ-Glutamyltransferase (GGT) *Serum No Effect Analytical* No effect at therapeutic concentration on Reflotron method. *1984*

Iron *Serum No Effect Analytical* At acute overdose concentration (2.5 mg/dL) on Technicon® SMAC® method. *3719* At concentration of 25 mg/L had no effect on Ferrozine method. *3393*

^{131}I Uptake *Serum Decrease Physiological* Parest® contains tetraiodofluorescein. *2652*

Lactate *Serum No Effect Analytical* At concentration of 25 mg/L had no effect on enzymatic method. *3393*

Lactate Dehydrogenase *Serum No Effect Analytical* At acute overdose concentration (2.5 mg/dL) on Technicon® SMAC® method. *3719*

Leukocytes *Blood Decrease Physiological* 1 case of pancytopenia reported. *1680*

Methaqualone *Serum Increase Physiological* 250 mg orally produced 2 mg/L in 30 minutes. *2348*

Phenylbutazone *Serum No Effect Physiological* No significant effect on half-life. *3466*

Phosphate *Serum No Effect Analytical* At acute overdose concentration (2.5 mg/dL) on Technicon® SMAC® method. *3719* At concentration of 25 mg/L had no effect on phosphomolybdate method. *3393*

Platelets *Blood Decrease Physiological* 1 case of pancytopenia reported. *1680*

Potassium *Serum No Effect Analytical* At concentration of 6,000 mg/L had no effect on measurement by ISE with predilution. *3393*

Protein *Serum No Effect Analytical* At acute overdose concentration (2.5 mg/dL) on Technicon® SMAC® method. *3719*

Prothrombin Time *Plasma Decrease Physiological* In 10 patients receiving 0.3 g/d also on warfarin: effect mild and not significant. Weak inducer of hepatic microsomal enzymes. *3649*

Salicylate *Urine No Effect Analytical* At 10 mg/dL no effect on Trinder method. *1204*

Sodium *Serum No Effect Analytical* At concentration of 6,000 mg/L had no effect on measurement by ISE with predilution. *3393*

Triglycerides *Serum No Effect Analytical* At acute overdose concentration (2.5 mg/dL) on Technicon® SMAC® method. *3719* No effect at therapeutic concentration on Reflotron method. *1984* At concentration of 25 mg/L had no effect on lipase/esterase method. *3393*

Urea Nitrogen *Serum No Effect Analytical* At acute overdose concentration (2.5 mg/dL) on Technicon® SMAC® method. *3719* No effect at therapeutic concentration on Reflotron method. *1984* At concentration of 25 mg/L had no effect on diacetylmonoxime method. *3393*

Uric Acid *Serum No Effect Analytical* At acute overdose concentration (2.5 mg/dL) on Technicon® SMAC® method. *3719* No effect at therapeutic concentration on Reflotron method. *1984* At concentration of 25 mg/L had no effect on phosphotungstate reduction method. *3393*

Metharbital

Barbital *Serum Increase Physiological* Metabolic conversion *in vivo*. *3054*

Prothrombin Time *Plasma Decrease Physiological* Theoretical possibility due to enzyme induction. *1680*

Methazolamide

Hemoglobin *Blood Decrease Physiological* Some cases of aplastic anemia, and leukopenia reported when used in treatment for glaucoma. *3799*

Leukocytes *Blood Decrease Physiological* Probable effect as like acetazolamide. *2313* Some cases of aplastic anemia, and leukopenia reported when used in treatment for glaucoma. *3799*

Neutrophils *Blood Decrease Physiological* Occasional case of agranulocytosis reported. *3717*

Platelets *Blood Decrease Physiological* Probable effect as like acetazolamide. *2313* Some cases of aplastic anemia, and leukopenia reported when used in treatment for glaucoma. *3799*

Potassium *Serum Decrease Physiological* With prolonged use (carbonic anhydrase inhibition). *2313*

Methdilazine

Alanine Aminotransferase *Serum Increase Physiological* Rare case of cholestasis. *1680*

Alkaloids *Urine No Effect Analytical* With ninhydrin on Frings TLC procedure at 10 mg/dL. *1204*
Urine Increase Analytical Purple with iodoplatinate on Frings TLC procedure. *1204*

Aspartate Aminotransferase
Serum Increase Physiological Rare case of cholestasis. *1680*

Barbiturate *Urine No Effect Analytical* With diphenylcarbazone on Frings TLC procedure at 10 mg/dL. With mercurous NO_3 on Frings TLC procedure at 10 mg/dL. *1204*
Urine Increase Analytical Pink spot with vanillin in Frings TLC procedure. Pink/orange with mercuric SO_4 in Frings TLC procedure. *1204*

Bilirubin *Serum Increase Physiological* Rare case of cholestasis. *1680*

Bromide *Urine No Effect Analytical* No effect on method of Frings. *1204*

Ethchlorvynol *Urine No Effect Analytical* No effect on method of Frings and Cohen. *1204* No effect on method Frings and Cohen at 10 mg/dL. *1205*

FPN Test *Urine No Effect Analytical* Negative FPN test with 0.4 mg/dL concentration in Frings TLC procedure. *1204*
Urine Increase Analytical Pink FPN test with 0.5 mg/dL in Frings TLC procedure. *1204*

Salicylate *Urine No Effect Analytical* Up to 5 mg/dL no effect on Trinder procedure. *1204*
Urine Increase Analytical At 10 mg/dL produces pink color on Trinder procedure. *1204*

Methemoglobin

Color *Urine Increase Physiological* May produce red-brown urine. *0443* With oxyhemoglobin produces port-wine color. *1187*

Methenamine

Alanine Aminotransferase *Serum Increase Physiological* Mild transient effect in some cases. *1680*

Ammonia *Urine Increase Physiological* Hydrolyzed in acid urine (to formaldehyde also). *1678*

Aspartate Aminotransferase
Serum Increase Physiological Mild transient effect in some cases. *1680*

Catecholamines *Plasma Increase Analytical* Interference with fluorescence. *0127*

Corticosteroids *Urine No Effect Analytical* If Allen correction and Porter-Silber procedure. *0022*

Crystals *Urine Increase Physiological* Mandelate may occasionally cause crystalluria. *1343*

Methenamine (continued)

Erythrocytes *Urine Increase Physiological* May cause actual bleeding. *1022*

Estriol *Urine No Effect Analytical* At 1 g/dL with enzyme hydrolysis and Oakey procedure. *0762*
Urine Decrease Analytical Interferes with hydrolysis stage of methods. *1038*

Estrogens *Urine Increase Analytical* Interferes in Kober procedure. *0279*
Urine Decrease Analytical Affects hydrolysis of estrogen conjugates. *0022*

Hemoglobin *Urine Increase Physiological* Actual bleeding produced by drug. *1022*

Hippuric Acid *Urine Increase Physiological* If given as hippurate salt. *1678*

17-Hydroxycorticosteroids *Urine Increase Analytical* Affects Porter-Silber and Reddy methods. *0457*

5-Hydroxyindoleacetic Acid (5-HIAA)
Urine Decrease Analytical Slight false negative effect with nitrosonaphthol. *3305*

^{131}I Uptake *Serum Decrease Physiological* Mandelamine® contains tetraiodofluorescein. *2652*

Nitrofurantoin *Urine No Effect Analytical* No effect on method of Conklin and Hollifield. *0704*

pH *Urine Decrease Physiological* Mandelate is an acidifying agent. *1343*

Protein *Urine Increase Physiological* Nephrotoxic in large doses. *1343* Nephrotoxic in large doses. *1022*

PSP Excretion *Urine Increase Analytical* Produces interfering color in urine. *0757*

Sugar *Urine Increase Analytical* False positive with Benedict's reagent. *2150*

Urobilinogen *Urine Increase Analytical* Produces formaldehyde which interferes. *0252*

Vanillylmandelic Acid *Urine Increase Analytical* Interferes with fluorometric procedures. *2220*

Methergoline

Thyroid Stimulating Hormone (TSH)
Serum Decrease Physiological In hypothyroidism — is a serotonin antagonist. *3798*

Methicillin

Alanine Aminotransferase *Serum No Effect Analytical* At 5 times upper limit of therapeutic range on methods on SMAC®, Abbott-VP, Cobas-Bio, aca, and KDA. *2138*

Albumin *Serum No Effect Analytical* At 5 times upper limit of therapeutic range on SMAC®, Abbott-VP, Ektachem®, Hitachi® 705 and KDA. *2138* At concentration of 900 mg/L had no effect on BCG method. *3393*

Alkaline Phosphatase *Serum No Effect Analytical* At 5 times upper limit of therapeutic range on methods on SMAC®, Abbott-VP, aca, Cobas-Bio, Hitachi® 705 and KDA. *2138*

Amylase *Serum No Effect Analytical* At 5 times upper limit of therapeutic range on methods on aca, Cobas-Bio and Ektachem®. *2138*

Aspartate Aminotransferase *Serum No Effect Analytical* At 5 times upper limit of therapeutic range on methods on SMAC®, Abbott-VP, aca, Cobas-Bio, Hitachi® 705 and KDA. *2138*

Bicarbonate *Serum No Effect Analytical* At concentration of 900 mg/L had no effect on method using phenolphthalein. *3393*
Serum Decrease Physiological Nephrotoxicity may cause azotemia. *2313*

Bilirubin *Serum No Effect Analytical* At 5 times upper limit of therapeutic range on SMAC®, aca, Cobas-Bio, Ektachem®, Hitachi® 705 and KDA. *2138* At concentration of 900 mg/L had no effect on Jendrassik and Grof method. *3393*
Serum No Effect Physiological Clinically insignificant displacement from protein in neonates. *3748*

Bleeding Time *Patient No Effect Physiological* Even at 300 mg/kg/d had no effect in volunteers. *0492*

Calcium *Serum No Effect Analytical* At 5 times upper limit of therapeutic range on methods on SMAC®, Abbott-VP, aca, *2138* At concentration of 900 mg/L had no effect on cresolphthalein method. *3393*
Serum Decrease Physiological Reported effect (?mechanism). *2313*

Casts *Urine Increase Physiological* Nephrotoxic effect (cylindruria observed). *2313*

Chloride *Serum No Effect Analytical* At concentration of 900 mg/L had no effect on mercurimetric method. *3393*

Cholesterol *Serum No Effect Analytical* At 5 times upper limit of therapeutic range on methods on SMAC®, Abbott-VP, Cobas-Bio, Ektachem®, Hitachi® 705 and KDA. *2138* At concentration of 20 mg/L had no effect on CHOD-PAP method. *3393* At concentration of 900 mg/L had no effect on Liebermann-Burchard method. *3393*

Clot Retraction *Blood No Effect Physiological* No effect with doses as high as 300 mg/kg/d in volunteers. *0492*

Coombs' Test, Direct *Serum Positive Physiological* In 31% of 45 patients receiving drug (hypersensitivity reaction). *1963*

Creatine Kinase *Serum No Effect Analytical* At 5 times upper limit of therapeutic range on methods on SMAC®, Abbott-VP, Cobas-Bio, aca and Hitachi® 705. *2138*

Creatinine *Serum No Effect Analytical* At 5 times upper limit of therapeutic range on SMAC®, Abbott-VP, Cobas-Bio, aca, Ektachem®, Hitachi® 705 and KDA. *2138* At concentration of 900 mg/L had no effect on AutoAnalyzer Jaffé method. *3393*
Serum Increase Physiological Nephrotoxic effect. *2313*

Eosinophils *Blood Increase Physiological* Hypersensitivity reaction. *0212* In 31% of 45 patients receiving drug (hypersensitivity reaction). *1963* In 3 of 28 children within 5 to 8 d. *2548*

Erythrocytes *Blood Decrease Physiological* May rarely cause bone marrow depression. *2808*
Urine Increase Physiological Hypersensitivity reaction, nephrotoxicity. *0212*

Fibrinogen *Plasma No Effect Physiological* No effect with doses as high as 300 mg/kg/d in volunteers. *0492*

Glucose *Serum No Effect Analytical* At 5 times upper limit of therapeutic range on SMAC®, Abbott-VP, Cobas-Bio, aca, Ektachem®, Hitachi® 705 and KDA. *2138* At concentration of 20 mg/L had no effect on GOD/POD-PAP method. *3393*

γ-Glutamyltransferase (GGT) *Serum No Effect Analytical* At 5 times upper limit of therapeutic range on methods on SMAC®, Abbott-VP and Hitachi® 705. *2138*

Haptoglobin *Serum Decrease Physiological* One doubtful case of hemolytic anemia. *0071*

Hemoglobin *Urine Increase Physiological* Hypersensitivity reaction, nephrotoxicity. *0212*

Iron *Serum No Effect Analytical* At 5 times upper limit of therapeutic range on Ferrozine method on SMAC®. *2138*
Serum Increase Physiological If erythrocyte maturation depressed. *2964*

Iron-Binding Capacity, Unsaturated (UIBC)
Serum Decrease Physiological If erythrocyte maturation depressed. *2964*

17-Ketosteroids *Urine Increase Analytical* Absorption at 520 nm in Zimmermann procedure. *0022*

Lactate Dehydrogenase *Serum No Effect Analytical* At 5 times upper therapeutic range on methods on SMAC®, Abbott-VP, Cobas-Bio, Hitachi® 705 and KDA. *2138*

Leukocytes *Blood Increase Physiological* Due to eosinophilia/leukocytosis. *2313*
Blood Decrease Physiological May cause bone marrow depression. *2313*

Methemoglobin *Blood Increase Physiological* One doubtful case of hemolytic anemia. *0071*

Neutrophils *Blood Decrease Physiological* Neutropenia with granulocytopenia may occur. *1680*

Nonprotein Nitrogen *Serum Increase Physiological* Nephrotoxic effect. *2313*

Partial Thromboplastin Time
Plasma No Effect Physiological No effect with doses as high as 300 mg/kg/d in volunteers. *0492*

Phosphate *Serum No Effect Analytical* At 5 times upper limit of therapeutic range on method on KDA. *2138*

Serum Increase Analytical At 5 times upper limit of therapeutic range on methods on SMAC®, aca, and Hitachi® 705. *2138* At concentrations above 500 mg/L (normal therapeutic concentration 21 mg/L) raised concentration as measured by phosphomolybdate method. *3393*
Serum Increase Physiological Occurs with nephrotoxicity. *2313*

Platelet Aggregation *Blood Decrease Physiological* In response to ADP in 1 of 5 volunteers receiving up to 300 mg/kg/d. *0492*

Platelets *Blood No Effect Physiological* No effect with doses as high as 300 mg/kg/d in volunteers. *0492*
Blood Decrease Physiological May rarely cause bone marrow depression. *2808*

Potassium *Serum No Effect Analytical* At concentration of 20 mg/L had no effect on flame-photometric method. *3393*
Serum Increase Physiological Possible result of nephrotoxicity. *2313*

Protein *Serum No Effect Analytical* At 5 times upper limit of therapeutic range on methods on SMAC®, Abbott-VP, Ektachem®, Hitachi® 705 and KDA. *2138*
Serum Increase Analytical At concentrations above 500 mg/L (normal therapeutic concentration 21 mg/L) raised concentration as measured by biuret method with blank correction. *3393*
Urine Increase Physiological Nephrotoxicity may occur. *0071* Nephrotoxic effect. *2496*

Prothrombin Time *Plasma No Effect Physiological* No effect with doses as high as 300 mg/kg/d in volunteers. *0492*

Sodium *Serum No Effect Analytical* At concentration of 20 mg/L had no effect on flame-photometric method. *3393*

Thrombin Time *Plasma No Effect Physiological* No effect with doses as high as 300 mg/kg/d in volunteers. *0492*

Triglycerides *Serum No Effect Analytical* At 5 times upper limit of therapeutic range on SMAC®, Ektachem®, Hitachi® 705 and KDA. *2138* At concentration of 900 mg/L had no effect on lipase/esterase method. *3393*
Serum Increase Analytical At 5 times upper limit of therapeutic range on enzymatic method on Abbott-VP. *2138*

Urea Nitrogen *Serum No Effect Analytical* At 5 times upper limit of therapeutic range on methods on SMAC®, Abbott-VP, Cobas-Bio, Ektachem®, Hitachi® 705 and KDA. *2138* At concentration of 900 mg/L had no effect on diacetylmonoxime method. *3393*
Serum Increase Physiological May cause azotemia with nephrotoxicity. *0071*

Uric Acid *Serum No Effect Analytical* At 20 mg/dL on Nishi phosphotungstate procedure. *2230* At 5 times upper limit of therapeutic range on SMAC®, Abbott-VP, Cobas-Bio, aca, Ektachem®, Hitachi® 705 and KDA. *2138* At concentration of 900 mg/L had no effect on phosphotungstate reduction method. *3393*
Serum Increase Analytical 20 mg/dL equivalent to 1.6 mg/dL copper chelate procedure. *2230*
Serum Increase Physiological Nephrotoxic effect. *2313*

Urobilin *Urine Increase Physiological* Pyuria reported as complication. *1680*

Methimazole

Alanine Aminotransferase *Serum Increase Physiological* May affect liver function (cholestasis). *2313* Rare case of cholestatic jaundice reported. *1774*

Alkaline Phosphatase *Serum Increase Physiological* May affect liver function (cholestasis). *2313* Rare case of cholestatic jaundice reported. *1774*

Aspartate Aminotransferase
Serum Increase Physiological May affect liver function (cholestasis). *2313* Rare case of cholestatic jaundice reported. *1774*

Bile *Urine Increase Physiological* May affect liver function. *2313*

Bilirubin *Serum Increase Physiological* Toxic effect associated with bone marrow depression. *0262* May affect liver function (cholestasis). *2313* Rare case of cholestatic jaundice reported. *1774*

BSP Retention *Serum Increase Physiological* May affect liver function (cholestasis). *2313*

B-Cell Differentiation Factor
Monocytes No Effect Physiological No effect on production when cells from normal individuals stimulated with mitogens. *3778*

Cephalin Flocculation *Serum Increase Physiological* Hepatotoxic effect. *2313*

Cholesterol *Serum Increase Physiological* Cholestatic effect. *1713*

Erythrocytes *Blood Decrease Physiological* Aplastic anemia. *2313* Occasional case of aplastic anemia reported. *3870*

γ-Globulin *Serum Increase Physiological* Polyclonal hypergammaglobulinemia reported possibly as result of production of nonspecific drug-stimulated polyclonal antibodies. *3870*

Glucose *Serum Increase Analytical* At 1 mmol/L affects SMA 12/60 method. *3335*

Hematocrit *Blood Decrease Physiological* Occasional case of aplastic anemia reported. *3870*

Hemoglobin *Blood Decrease Physiological* Occasional case of aplastic anemia reported. *3870*

IL-2 Receptor Expression
Monocytes No Effect Physiological No effect of the drug on mitogen stimulated mononuclear cells from normal individuals. *3778*

γ-Interferon *Monocytes No Effect Physiological* No effect on production when cells from normal individuals stimulated with mitogens. *3778*

Interleukin-1 *Monocytes No Effect Physiological* No effect on production when cells from normal individuals stimulated with mitogens. *3778*

Interleukin-2 *Monocytes Increase Physiological* Increased activity in culture supernatants when mononuclear cells from normals stimulated with mitogens. Effect apparent between 24 h and 60 h. *3778*

^{131}I Uptake *Serum No Effect Physiological* No effect on uptake by thyroid. *2220*
Serum Decrease Physiological Effect may last from 2 to 8 d. *1444*

LE Cells *Blood Positive Physiological* Lupus-like syndrome reported. *1680*

Leukocytes *Blood Decrease Physiological* Rare agranulocytosis or pancytopenia. *2427* Agranulocytosis. *0072* Occasional case of aplastic anemia reported. *3870*

Neutrophils *Blood Decrease Physiological* Occasional case of drug-induced neutropenia. *0155* Occasional case of drug-induced neutropenia. *0155*

Platelets *Blood Decrease Physiological* Thrombocytopenia. *2426* Occasional case of aplastic anemia reported. *3870*

Protein Bound Iodine (PBI) *Serum Decrease Physiological* Inhibits iodination of tyrosine in thyroxine binding globulin. *1444*

T$_3$ Uptake *Serum Decrease Physiological* Therapeutic result. *1444*

Thymol Turbidity *Serum Increase Physiological* May affect liver function. *2313*

Thyroxine (T$_4$) *Serum Decrease Physiological* Therapeutic intent (stops iodination of tyrosine). *1444*

Thyroxine (T$_4$) Index, Free (FTI)
Serum Decrease Physiological From 23.8 to 17.0 after 4 weeks treatment with 30 mg daily. *0192*

Tri-iodothyronine (T$_3$) Index, Free
Serum Decrease Physiological From 512 to 368 after 4 weeks treatment with 30 mg daily. *0192*

Urobilinogen *Urine Decrease Physiological* Cholestatic effect. *1713*
Feces Decrease Physiological Cholestatic effect. *1713*

Methiodol

Protein Bound Iodine (PBI) *Serum Increase Analytical* But does not affect T$_4$ (effect lasts 1-2 weeks). *3218*

Methionine

Ammonia *Plasma Decrease Analytical* With 250 nmoles produces 7% inhibitor indophenol reaction. *1290*

Methionine *(continued)*

Insulin *Plasma Increase Physiological* Slight effect, AIDS metabolism of amino acids. *1071*

Megaloblasts *Blood Increase Physiological* Can aggravate vitamin B$_{12}$ deficiency. *3025*

pH *Urine Decrease Physiological* Used to acidify urine. *0443*

Tyrosine *Plasma No Effect Analytical* On fluorometric procedure of Ambrose. *0080*

Methocarbamol

Color *Urine Decrease Analytical* Brown, green, blue or black on standing. *0382*
Urine Positive Analytical Green color observed. *2613*

Erythrocytes *Urine Increase Physiological* May cause intravascular hemolysis. *1343*

Hemoglobin *Urine Increase Physiological* May cause intravascular hemolysis. *1343*

5-Hydroxyindoleacetic Acid (5-HIAA)
Urine Increase Analytical Affects quantitative method of Udenfriend. *0583* Metabolite allegedly reacts with nitrosonaphthol. *1649*

Leukocytes *Blood Decrease Physiological* May cause leukopenia. *2313*

Vanillylmandelic Acid *Urine Increase Analytical* False positive with screening, no effect quant method. *1022*

Methohexital

Albumin *Serum No Effect Analytical* At concentration of 50 mg/L had no effect on BCG method. *3393*

Bicarbonate *Serum No Effect Analytical* At concentration of 50 mg/L had no effect on method using phenolphthalein. *3393*

Bilirubin *Serum No Effect Analytical* At concentration of 50 mg/L had no effect on Jendrassik and Grof method. *3393*
Serum No Effect Physiological Insignificant protein-displacing effect in neonates. *3748*

Calcium *Serum No Effect Analytical* At concentration of 50 mg/L had no effect on cresolphthalein method. *3393*

Chloride *Serum No Effect Analytical* At concentration of 50 mg/L had no effect on mercurimetric method. *3393*

Cholesterol *Serum No Effect Analytical* At concentration of 50 mg/L had no effect on Liebermann-Burchard method. *3393*

Creatinine *Serum No Effect Analytical* At concentration of 50 mg/L had no effect on AutoAnalyzer Jaffé method. *3393*

Phosphate *Serum No Effect Analytical* At concentration of 50 mg/L had no effect on phosphomolybdate method. *3393*

Protein *Serum No Effect Analytical* At concentration of 50 mg/L had no effect on biuret method with blank correction. *3393*

Triglycerides *Serum No Effect Analytical* At concentration of 50 mg/L had no effect on lipase/esterase method. *3393*

Urea Nitrogen *Serum No Effect Analytical* At concentration of 50 mg/L had no effect on diacetylmonoxime method. *3393*

Uric Acid *Serum No Effect Analytical* At concentration of 50 mg/L had no effect on phosphotungstate reduction method. *3393*

Methotrexate

N-Acetylglucosaminidase *Urine No Effect Analytical* At 1 mmol/L on 2 colorimetric analytical methods. *1354*
Urine Increase Physiological Manifestation of nephrotoxicity in 12 children receiving chemotherapy: up to 5 fold increase in excretion. *1355*

Adenosine Deaminase Binding Protein
Urine Increase Physiological Manifestation of nephrotoxicity in 12 children receiving chemotherapy; up to 5-fold increase in excretion. *1355*

Alanine Aminopeptidase *Urine Increase Physiological* Manifestation of nephrotoxicity in 12 children receiving chemotherapy: up to 5 fold increase in excretion. *1355*

Alanine Aminotransferase *Serum No Effect Analytical* At 5 times upper limit of therapeutic range on method on Abbott-VP, aca, Cobas-Bio, Hitachi® 705 and KDA. *2138*
Serum No Effect Physiological In 20 patients with psoriasis on long-term therapy. *1556*
Serum Increase Physiological May cause cytotoxic hepatocellular damage. *0248* In 152 of 250 courses of treatment but most regressed. *2406*
Serum Decrease Analytical At 5 times upper limit of therapeutic range on method on SMAC®. *2138*

Albumin *Serum No Effect Analytical* At 5 times upper limit of therapeutic range on method on SMAC®, Ektachem®, Hitachi® 705, and KDA. *2138*

Alkaline Phosphatase *Serum No Effect Analytical* At 5 times upper limit of therapeutic range on method on Abbott-VP, aca, Cobas-Bio, Hitachi® 705 and KDA. *2138*
Serum No Effect Physiological No effect in long-term treatment of 20 patients with psoriasis noted. *1556* In 20 patients with psoriasis on long-term therapy. *1556*
Serum Increase Analytical At 5 times upper limit of therapeutic range on method on SMAC®. *2138*
Serum Increase Physiological Hepatotoxic effect (seen in 5% cases psoriasis). *2608* In psoriatic patients receiving drug in comparison with topically treated and controls. *0364*

Amino-4-Imidazole-5-Carboxamide Ribotide (AICAR)
Urine Increase Physiological Occurs with induced folic acid deficiency. *3773*

Amylase *Serum No Effect Analytical* At 5 times upper limit of therapeutic range on methods on aca, Cobas-Bio and Ektachem®. *2138*

Aspartate Aminotransferase *Serum No Effect Analytical* At 5 times upper limit of therapeutic range on methods on SMAC®, Abbott-VP, aca, Cobas-Bio, Hitachi® 705 and KDA. *2138*
Serum No Effect Physiological No effect in long-term treatment of 20 patients with psoriasis noted. *1556*
Serum Increase Physiological Hepatotoxicity (drug induced cirrhosis). *2313* In 152 of 250 courses of treatment but most regressed. *2406*

Bile *Urine Increase Physiological* Hepatotoxicity. *2313*

Bilirubin *Serum No Effect Analytical* No significant effect at 5 times upper limit of therapeutic range on SMAC®, aca, Cobas-Bio, Hitachi® 705 and KDA. *2138*
Serum Increase Analytical Clinically significant effect at upper limit of therapeutic range on method on Ektachem®. *2138*
Serum Increase Physiological May cause cytotoxic hepatocellular damage. *0248* In 10 of 250 courses of treatment but most regressed. *2406*

Bilirubin, Conjugated *Serum Increase Analytical* At concentration of 1,000 μmol/L as much as 38 mg/L increase in concentration on Kodak Ektachem® 400. *3518*

Bilirubin, Unconjugated *Serum Decrease Analytical* Falsely low values in serum of patients containing large amount of drug following i.v. infusion. *3518*

BSP Retention *Serum Increase Physiological* Hepatotoxicity, may be post-necrotic cirrhosis. *1576*

Calcium *Serum No Effect Analytical* At 5 times upper limit of therapeutic range on method on SMAC®, Abbott-VP, aca, Ektachem®, Hitachi® 705 and KDA. *2138*

Cephalin Flocculation *Serum Increase Physiological* May cause cytotoxic hepatocellular damage. *0248*

Cholesterol *Serum No Effect Analytical* At 5 times upper limit of therapeutic range on methods on SMAC®, Abbott-VP, Cobas-Bio, Ektachem® and Hitachi® 705. *2138* At concentration of 500 mg/L had no effect on CHOD-Iodide method. *3393* At concentration of 500 mg/L had no effect on CHOD-PAP method. *3393* At concentration of 500 mg/L had no effect on catalase-AIDH method. *3393* At concentration of 500 mg/L had no effect on method using catalase-Hantzsch reaction. *3393* At concentration of 500 mg/L had no effect on Liebermann-Burchard method. *3393*
Serum Increase Analytical At 5 times upper limit of therapeutic range on enzymatic method on KDA At 5 times upper limit of therapeutic range on enzymatic method on KDA. *2138*

Chromosomes *Test Conditions Abnormal Physiological* Clastogenic in human lymphocytes in culture. *3282*

Creatine Kinase *Serum No Effect Analytical* At 5 times upper limit of therapeutic range on methods on SMAC®, Abbott-VP, aca, Cobas-Bio, and Hitachi® 705. *2138*

Creatinine *Serum No Effect Analytical* At 5 times upper limit of therapeutic range on method on SMAC®, Abbott-VP, Cobas-Bio, Ektachem®, Hitachi® 705 and KDA. *2138* At concentration of 1 mg/L had no effect on Jaffé-Fading-Fraction method. *3393* At concentration of 1 mg/L had no effect on Jaffé-Fuller's earth method. *3393* At concentration of 55 mg/L had no effect on kinetic Jaffé method on BKA-2. *3393*
Urine Increase Physiological Manifestation of nephrotoxicity. *1355*

Crystals *Urine Increase Analytical* Crystals of unknown identity (probably drug). *0979*

Erythrocytes *Blood Decrease Physiological* May = megaloblastic anemia (folic acid antagonist). *3773*

Fat *Feces Increase Physiological* Probably due to mucosal damage with impaired regeneration of epithelial cells of small intestine. *3023*

FIGLU (N-Formiminoglutamic Acid)
Urine Increase Physiological Large effect of treatment observed. *3681*

Folate *Serum Decrease Physiological* Inhibits folate reductase. *3025* Significantly higher in patients with psoriasis reflecting decreased dihydrofolate reductase activity. Both oxidized forms pteroylglutamate and dihydrofolate affected. *1556*
Red Blood Cells Decrease Physiological Significantly lower in patients with psoriasis with long term treatment reflecting low polyglutamate storage. *1556* In long-term treated group erythrocyte folate significantly reduced reflecting low polyglutamate storage. *1556*

γ-Globulin *Serum Decrease Physiological* Possible immunosuppressive response. *1680*

Glucose *Serum No Effect Analytical* At 5 times upper limit of therapeutic range on method on SMAC®, Abbott-VP, Cobas-Bio, Ektachem®, Hitachi® 705 and KDA. *2138* At concentration of 1200 mg/L had no effect on Ektachem® method. *3393* At concentration of 25 mg/L had no effect on GOD/POD-PAP method. *3393*

γ-Glutamyltransferase (GGT) *Serum No Effect Analytical* At 5 times upper limit of therapeutic range on methods on SMAC®, Abbott-VP, and Hitachi® 705. *2138*
Serum Increase Physiological In psoriatic patients receiving drug in comparison with topically treated and controls. *0364*

Guanase *Serum Increase Physiological* May cause cytotoxic hepatocellular damage. *0248*

Hematocrit *Blood Decrease Physiological* May = megaloblastic anemia (folic acid antagonist). *3773*

Hemoglobin *Blood No Effect Physiological* No effect in long-term treatment of 20 patients with psoriasis noted. *1556* In 20 patients with psoriasis on long-term therapy. *1556*
Blood Decrease Physiological May = megaloblastic anemia (folic acid antagonist). *3773*
Urine Increase Physiological May cause hematuria. *3920*

Hydantoin-5-Propionic Acid *Urine Increase Physiological* Large effect of treatment observed. *3681*

Indocyanine Green Clearance
Serum Decrease Physiological In psoriatic patients receiving drug in comparison with topically treated and controls. *0364*

Iron *Serum No Effect Analytical* At 5 times upper limit of therapeutic range on Ferrozine method on SMAC®. *2138*
Serum Increase Physiological Sharp increase 8-12 h after end of cycle. Maximum value of 295% at 48-60 h; after 108 h had returned to normal in only 50%. *3173*

Isocitrate Dehydrogenase *Serum Increase Physiological* May cause cytotoxic hepatocellular damage. *0248*

Lactate Dehydrogenase *Serum No Effect Analytical* At 5 times upper limit of therapeutic range on method on Abbott-VP, Cobas-Bio, Hitachi® 705 and KDA. *2138*
Serum Increase Physiological Effect in 40% cases of psoriasis (reversible). *2608*
Serum Decrease Analytical At 5 times upper limit of therapeutic range on method on SMAC®. *2138*

Leukocytes *Blood Decrease Physiological* Agranulocytosis/lymphocytopenia. *2426*

MCV *Blood No Effect Physiological* No effect in long-term treatment of 20 patients with psoriasis noted. *1556* In 20 patients with psoriasis on long-term therapy. *1556*

Blood Increase Physiological Occurs with megaloblastic anemia. *3773*

Occult Blood *Feces Positive Physiological* May cause hemorrhagic enteritis. *0136*

Ornithine Carbamoyltransferase (OCT)
Serum Increase Physiological May cause cytotoxic hepatocellular damage. *0248*

Phosphate *Serum No Effect Analytical* At 5 times upper limit of therapeutic range on KDA method. *2138*
Serum Increase Analytical At upper limit of therapeutic range on methods on aca and Hitachi® 705 and at 5 times this on method on SMAC®. *2138*

Platelets *Blood Decrease Physiological* May occur with megaloblastic anemia. *3773*

Potassium *Serum No Effect Analytical* At concentration of 80 mg/L had no effect on measurement by ISE without predilution. *3393*

Protein *Serum No Effect Analytical* At 5 times upper limit of therapeutic range on methods on SMAC®, Abbott-VP, Hitachi® 705 and KDA. *2138*
CSF Increase Analytical High absorbance with turbidimetric methods. Reacts as if phenol with Folin-Ciocalteu. *3969* Marked effect on Du Pont aca method (up to 40 times actual concentration) even at therapeutic concentration; if concentration low enough not to produce yellow color protein concentration not significantly affected. *3701*

Protein Bound Iodine (PBI) *Serum No Effect Physiological* Reported to have no effect. *0830*

Protein Electrophoresis *Serum No Effect Analytical* At concentration of 80 mg/L had no effect on automated Olympus-Hite method. *3393*

Prothrombin Time *Plasma Increase Physiological* May occur with hepatic malfunction. *1487*

Reticulocytes *Blood Decrease Physiological* Affects hematopoiesis. *1343*

Sodium *Serum No Effect Analytical* At concentration of 80 mg/L had no effect on measurement by ISE without predilution. *3393*

Sperm Count *Semen Decrease Physiological* Oligospermia occurs without altering plasma hormone concentrations. *0953*

Thymol Turbidity *Serum Increase Physiological* Hepatotoxic effect. *2313*

Triglycerides *Serum No Effect Analytical* At 5 times upper limit of therapeutic range on methods on Abbott-VP, Ektachem®, Hitachi® 705 and KDA. *2138* At concentration of 25 mg/L had no effect on GPO-PAP method. *3393*
Serum Decrease Analytical At 5 times upper limit of therapeutic range on SMAC® method. *2138*

Urea Nitrogen *Serum No Effect Analytical* At 5 times upper limit of therapeutic range on method on SMAC®, Abbott-VP, Cobas-Bio, Ektachem®, Hitachi® 705 and KDA. *2138* At concentration of 1100 mg/L had no effect on Ektachem® method. *3393*
Serum Increase Physiological May cause severe nephropathy, azotemia. *1680*

Uric Acid *Serum No Effect Analytical* At 5 times upper limit of therapeutic range on methods on SMAC®, Abbott-VP, aca, Cobas-Bio, Ektachem® and KDA. *2138* At concentration of 1 mg/L had no effect on Kageyama-Hantzsch method. *3393* At concentration of 1 mg/L had no effect on catalase-AlDH method. *3393* At concentration of 25 mg/L had no effect on uricase-PAP method. *3393*
Serum Increase Physiological Seen in gouty patients — some decrease in leukemics. *2316*
Serum Decrease Analytical At 5 times upper limit of therapeutic range on enzymatic method on Hitachi® 705. *2138*
Serum Decrease Physiological Inhibits synthesis. *3217*
Urine Increase Physiological Increased excretion associated with cell destruction. *1320*

Urocanic Acid *Urine Increase Physiological* With treatment of folic acid/B12 deficiency. *3681*

Urocanylglycine *Urine Increase Physiological* Large effect of treatment observed. *3681*

Methotrimeprazine

Bicarbonate *Serum No Effect Analytical* At concentration of 1 mg/L had no effect on method using phenolphthalein. *3393*

Bilirubin *Serum No Effect Analytical* At concentration of 1 mg/L had no effect on Jendrassik and Grof method. *3393*
Serum Increase Physiological Three cases reported. *1488*

Calcium *Serum No Effect Analytical* At concentration of 1 mg/L had no effect on cresolphthalein method. *3393*

Chloride *Serum No Effect Analytical* At concentration of 1 mg/L had no effect on mercurimetric method. *3393*

Cholesterol *Serum No Effect Analytical* At concentration of 1 mg/L had no effect on CHOD-PAP method. *3393*

Creatinine *Serum No Effect Analytical* At concentration of 1 mg/L had no effect on AutoAnalyzer Jaffé method. *3393*

Eosinophils *Blood Increase Physiological* Noted with other phenothiazines. *1680*

Ferric Chloride Test *Urine Positive Analytical* Positive if more than 100 mg/d for 6 d. *2427*

Glucose *Serum No Effect Analytical* At concentration of 1 mg/L had no effect on GOD/POD-PAP method. *3393*

Leukocytes *Blood Decrease Physiological* Agranulocytosis with long-term high-dose use. *1680*

Neutrophils *Blood Decrease Physiological* Occasional case of drug-induced neutropenia. *0155*

Phenylketones *Urine Positive Analytical* Phenistix® positive if more than 100 mg/d for 6 d. *2427*

Pheochromocytoma Test *Patient Positive Physiological* May cause false test as produces hypotension. *1488*

Platelets *Blood Decrease Physiological* Noted with other phenothiazines. *1680*

Potassium *Serum No Effect Analytical* At concentration of 1 mg/L had no effect on flame-photometric method. *3393*

Protein *Serum No Effect Analytical* At concentration of 1 mg/L had no effect on biuret method with blank correction. *3393*
CSF Increase Physiological May occur if cerebral edema etc. *1680*

Sodium *Serum No Effect Analytical* At concentration of 1 mg/L had no effect on flame-photometric method. *3393*

Triglycerides *Serum No Effect Analytical* At concentration of 1 mg/L had no effect on lipase/esterase method. *3393*

Urea Nitrogen *Serum No Effect Analytical* At concentration of 1 mg/L had no effect on diacetylmonoxime method. *3393*

Uric Acid *Serum No Effect Analytical* At concentration of 1 mg/L had no effect on phosphotungstate reduction method. *3393*

Methoxamine

Amphetamine *Urine Increase Analytical* Yields fluorophor with NBD chloride reaction. *2473*

Propoxyphene *Urine No Effect Analytical* Less than 1% fluorescence in procedure of Valentour. *3662*

Methoxsalen

Alanine Aminotransferase *Serum Increase Physiological* Hepatotoxic effect. *2313*

Alkaline Phosphatase *Serum Increase Physiological* Hepatotoxic effect. *2313*

Aspartate Aminotransferase
Serum Increase Physiological Hepatotoxic effect. *2313*

Bile *Urine Increase Physiological* Hepatotoxic effect. *2313*

Bilirubin *Serum Increase Physiological* Hepatotoxic effect. *2313*

BSP Retention *Serum Increase Physiological* Hepatotoxic effect. *2313*

Caffeine *Serum No Effect Physiological* Peak concentration and time to reach this not affected although mean elimination greatly increased. *2342*

Cephalin Flocculation *Serum Increase Physiological* Hepatotoxic effect. *2313*

Thymol Turbidity *Serum Increase Physiological* Hepatotoxic effect. *2313*

Methoxyflurane

Alanine Aminotransferase *Serum Increase Physiological* Hepatic toxicity. *3294* In 2 pregnant women when used as analgesia for labor: hepatitis observed. *0872*

Alkaline Phosphatase *Serum Increase Physiological* Hepatic toxicity. *3294* In 2 pregnant women when used as analgesia for labor: hepatitis observed. *0872*

Aspartate Aminotransferase
Serum Increase Physiological Hepatic toxicity. *3294* In 2 pregnant women when used as analgesia for labor: hepatitis observed. *0872*

Bile *Urine Increase Physiological* Hepatotoxic effect. *2313*

Bilirubin *Serum Increase Physiological* Hepatic toxicity. *3294* In 2 pregnant women when used as analgesia for labor: hepatitis observed. *0872*

BSP Retention *Serum Increase Physiological* Hepatotoxicity. *2313*

Calcium *Serum Decrease Physiological* Slight lowering reported in one case. *2876*

Cephalin Flocculation *Serum Increase Physiological* Hepatotoxicity. *2313*

Chloride *Serum Increase Physiological* Toxic nephropathy reported. *2980*

Creatinine *Serum Increase Physiological* Impaired renal tubular function. *2355*

Creatinine Clearance *Urine No Effect Physiological* No effect observed with high or low concentrations. *1380*

Effective Renal Plasma Flow
Patient Decrease Physiological Decrease noted during normal anesthesia. *2427*

Fluoride *Serum Increase Physiological* Metabolic degradation product of anesthetic. *1177*
Urine Increase Physiological Persisting high concentration many days with renal impairment. *3551*

Glomerular Filtration Rate (GFR)
Urine No Effect Physiological No effect observed with high or low concentrations. *1380*
Urine Decrease Physiological Decrease noted during normal anesthesia. *2427*

Osmolality *Serum Increase Physiological* Impaired renal tubular function. *0749*
Urine Decrease Physiological Nephrotoxic effect of drug (dose dependent). *2436*

Oxalate *Serum Increase Physiological* Metabolic degradation product of drug. *2876*
Urine Increase Physiological Metabolic degradation product of drug. *2876*

Proline Hydroxylase *Serum Increase Physiological* Marked elevation observed due to liver cell injury. *3447*

Prothrombin Time *Plasma Increase Physiological* In 2 pregnant women when used as analgesia for labor: hepatitis observed. *0872*

Sodium *Serum Increase Physiological* Impaired renal tubular function. *0749*

Specific Gravity *Urine Decrease Physiological* Impaired renal tubular function (dose dependent) Impaired renal tubular function. *0749*

Thymol Turbidity *Serum Increase Physiological* Hepatotoxicity. *2313*

Thyroid Stimulating Hormone (TSH)
Serum No Effect Physiological No effect observed with anesthesia. *2698*

Thyroxine (T$_4$) *Serum No Effect Physiological* No effect observed with anesthesia. *2698*

Urea Nitrogen *Serum Increase Physiological* Impaired renal tubular function. *0749*

Uric Acid *Serum Increase Physiological* Decreased urate clearance: contraction of extracellular fluid volume. *2355*

Volume *Urine Increase Physiological* Impaired distal renal tubular function. *0749*

Water Clearance, Free *Urine No Effect Physiological* No effect observed with high or low concentrations. *1380*

Methoxyphenamine

Amphetamine *Urine No Effect Analytical* No effect at 100 mg/L on method of Rutter. *3097*
Urine Increase Analytical Yields fluorophor with NBD chloride reaction. *2473*

Propoxyphene *Urine Increase Analytical* 1% fluorescence in procedure of Valentour. *3662*

4-Methoxyphenol

5-Hydroxyindoleacetic Acid (5-HIAA)
Urine No Effect Analytical No effect with 5 mg/dL on method of Goldenberg. *1327*

Methoxytyramine

Amino Acids *Test Conditions Increase Analytical* Positive spot test with ninhydrin. *3765*

Ammoniacal Silver Nitrate
Test Conditions Positive Analytical Positive spot test with Tollen's reagent. *3765*

Dihydroxyphenylalanine
Test Conditions Increase Analytical 24% of fluorescence method of Waldmeier. *3735*

Guaiacols Spot Test *Urine Negative Analytical* Action on procedure of Rogers. *3031*

Millons Test *Test Conditions Positive Analytical* Positive spot test for phenols. *3765*

Uric Acid *Test Conditions Increase Analytical* Positive spot test with phosphotungstate. *3765*

Urobilinogen *Test Conditions Increase Analytical* Positive spot test with Ehrlich's reagent. *3765*

Methsuximide

Acetaminophen Screening Test *Urine Negative Analytical* No reaction with o-cresol at therapeutic concentrations. *3326*

Albumin *Serum No Effect Analytical* At concentration of 40 mg/L had no effect on BCG method. *3393*

Bicarbonate *Serum No Effect Analytical* At concentration of 40 mg/L had no effect on method using phenolphthalein. *3393*

Bilirubin *Serum No Effect Analytical* At concentration of 40 mg/L had no effect on Jendrassik and Grof method. *3393*
Serum Increase Physiological Hepatic damage reported. *0071*

Calcium *Serum No Effect Analytical* At concentration of 40 mg/L had no effect on cresolphthalein method. *3393*

Cephalin Flocculation *Serum Increase Physiological* Hepatic damage reported. *2313*

Chloride *Serum No Effect Analytical* At concentration of 40 mg/L had no effect on Ektachem® method. *3393*

Creatinine *Serum No Effect Analytical* At concentration of 40 mg/L had no effect on Ektachem® method. *3393*

Diphenylhydantoin *Serum No Effect Analytical* On GLC procedure of Papadopoulos at *in vivo* concentration. *2722*

Eosinophils *Blood Increase Physiological* Reported effect. *1680*

Erythrocytes *Blood Decrease Physiological* Rare aplastic anemia reported. *2313*

Glucose *Serum No Effect Analytical* At concentration of 40 mg/L had no effect on Ektachem® method. *3393*

LE Cells *Blood Positive Physiological* Cause of SLE reported. *1680*

Leukocytes *Blood Decrease Physiological* Rare aplastic anemia or reversible leukopenia. *2313*

Monocytes *Blood Increase Physiological* Reported observation. *1680*

Pheneturide *Serum No Effect Analytical* On GLC procedure of Papadopoulos at *in vivo* concentration. *2722*

Phenobarbital *Serum No Effect Analytical* On GLC procedure of Papadopoulos at *in vivo* concentration. *2722*

Phosphate *Serum No Effect Analytical* At concentration of 40 mg/L had no effect on phosphomolybdate method. *3393*

Platelets *Blood Decrease Physiological* May rarely cause bone marrow aplasia. *1436*

Primidone *Serum No Effect Analytical* On GLC procedure of Papadopoulos at *in vivo* concentration. *2722*

Protein *Serum No Effect Analytical* At concentration of 40 mg/L had no effect on biuret method with blank correction. *3393*
Urine Increase Physiological Renal damage reported. *1237*

Triglycerides *Serum No Effect Analytical* At concentration of 40 mg/L had no effect on lipase/esterase method. *3393*

Urea Nitrogen *Serum No Effect Analytical* At concentration of 40 mg/L had no effect on diacetylmonoxime method. *3393*
Serum Increase Physiological Renal damage reported. *0071*

Uric Acid *Serum No Effect Analytical* At concentration of 40 mg/L had no effect on phosphotungstate reduction method. *3393*

Methyclothiazide

Amylase *Serum Increase Physiological* Pancreatitis may occur with thiazide therapy. *1680*

Antidiuretic Hormone *Plasma Increase Physiological* Secreted in response to hyponatremia. *1129*

Bicarbonate *Serum Increase Physiological* Diuretic action (hypochloremic alkalosis). *1353*
Urine Increase Physiological Slight effect only. *1680*

Bilirubin *Serum Increase Physiological* Jaundice may occur with thiazide therapy. *1680*

Calcium *Serum Increase Physiological* ?related to decreased excretion. *1680*
Urine Decrease Physiological Impairs excretion. *1680* Significant reduction compared with placebo in normal volunteers. *2052*

Chloride *Serum Decrease Physiological* Diuretic action (hypochloremic alkalosis). *1353*
Urine Increase Physiological Intended diuretic action (maximum in 6 h). *1680*

Erythrocytes *Blood Decrease Physiological* Aplastic anemia may occur with thiazides. *1680*

Glucose *Serum Increase Physiological* Diabetogenic action of thiazides. *1680*
Urine Increase Physiological May occur as result of hyperglycemia. *1680*

Leukocytes *Blood Decrease Physiological* Agranulocytosis or aplastic anemia may occur. *1680*

Magnesium *Urine Increase Physiological* Small but significant effect in normal volunteers. *2052*

Osmolality *Serum Decrease Physiological* With ADH secretion due to hyponatremia. *1129*

pH *Urine No Effect Physiological* No significant effect usually observed. *1680*

Platelets *Blood Decrease Physiological* Thrombocytopenia with purpura may occur. *1680*

Potassium *Serum Decrease Physiological* Diuretic action. *1353*
Urine Increase Physiological Small but significant effect in normal volunteers. *2052*

Protein Bound Iodine (PBI) *Serum Decrease Physiological* But no signs of thyroid disturbance. *1680*

Renin Activity *Plasma Increase Physiological* Stimulates plasma renin activity. *1353*

Sodium *Serum Decrease Physiological* Diuretic action. *1353*
Urine Increase Physiological Intended diuretic action (maximum in 6 h). *1680*

Urea Nitrogen *Serum Increase Physiological* Slight change only. *1353*

Uric Acid *Serum Increase Physiological* Reduces clearance. *1353*

Volume *Urine Increase Physiological* Intended diuretic action (maximum in 6 h). *1680*

Methylamine

Ionized Calcium *Serum Decrease Analytical* At concentrations > 0.1 mmol/L on Calcium specific electrode. *0540*

Pyruvate Kinase *Red Blood Cells Decrease Analytical* Most marked with larger side chain. *0370*

Methylaminoantipyrine

Cholesterol *Serum Decrease Analytical* At concentrations above 120 mg/L lowered concentration as measured by CHOD-PAP method. *3393*

Methylamphetamine

Amphetamine *Urine No Effect Analytical* No effect at 100 mg/L on method of Rutter. *3097*

Methylbromide

Casts *Urine Increase Physiological* Due to tubular necrosis. *1302*

Erythrocytes *Urine Increase Physiological* Due to tubular necrosis. *1302*

Protein *Urine Increase Physiological* Due to tubular necrosis. *1302*

Urea Nitrogen *Serum Increase Physiological* Late effect with tubular necrosis of poisoning. *1302*

Methyldopa

Alanine Aminotransferase *Serum No Effect Analytical* No effect at therapeutic concentration on Reflotron method. *1984*
Serum Increase Physiological Result of hepatocellular damage or cholestasis. *3879* Disturbances in liver function in 5 to 35% of patients treated for hypertension. Hepatocellular injury may occur after short or long-term exposure. *0162* Chronic active hepatitis associated with immune hemolytic anemia in one case. *3269* Drug induced hepatitis with severe, chronic, aggressive inflammation. *0211*

Albumin *Serum No Effect Analytical* At concentration of 80 mg/L had no effect on BCG method. *3393*
Serum Decrease Physiological Disturbances in liver function in 5 to 35% of patients treated for hypertension. Hepatocellular injury may occur after short- or long-term exposure. *0162*

Alkaline Phosphatase *Serum Increase Physiological* Intrahepatic cholestatic jaundice. *1023* Disturbances in liver function in 5 to 35% of patients treated for hypertension. Hepatocellular injury may occur after short or long-term exposure. *0162* Drug induced hepatitis with severe, chronic, aggressive inflammation. *0211*

Amino Acids *Urine Increase Analytical* Reacts with ninhydrin; extra spot TLC, high voltage electrophoresis. *2855*

Aminolevulinic Acid *Urine Increase Physiological* May precipitate acute porphyria attack. *1322*

Amphetamine *Urine No Effect Analytical* At 50 mg/L on fluorescent method of Hayes. *1534*

Antinuclear Antibodies *Serum No Effect Physiological* In a study of 9 hypertensives. *1909*
Serum Positive Physiological More common in females than males. *1692* Observed in one patient with hepatocellular damage of moderate severity due to sensitization by drug. *0874*

Antismooth Muscle Antibodies
Serum Positive Physiological Disturbances in liver function in 5 to 35% of patients treated for hypertension. Hepatocellular injury may occur after short- or long-term exposure. *0162* Chronic active hepatitis associated with immune hemolytic anemia in one case. *3269*

Apolipoprotein AI *Serum No Effect Physiological* In several patients treated for 2-4 mo. *0088* Decrease by -2 mg/dL in 11 patients given 750 mg daily for 8 weeks. *2550*

Apolipoprotein AII *Serum No Effect Physiological* In several patients treated for 2-4 mo. *0088*

Apolipoprotein B *Serum No Effect Physiological* In several patients treated for 2-4 mo. *0088* Increase by 1 mg/dL in 11 patients given 750 mg daily for 8 weeks. *2550*

Apolipoprotein CII *Serum No Effect Physiological* Increase by 2 mg/dL in 11 patients given 750 mg daily for 8 weeks. *2550*

Apolipoprotein CIII *Serum Increase Physiological* Increase by 3 mg/dL in 11 patients given 750 mg daily for 8 weeks. *2550*

Apolipoprotein E *Serum No Effect Physiological* Increase by 1 mg/dL in 11 patients given 750 mg daily for 8 weeks. *2550*

Aspartate Aminotransferase *Serum No Effect Analytical* No effect on Karmen procedure. *2760* No effect at therapeutic concentration on Reflotron method. *1984*
Serum Increase Analytical Measured as product in AutoAnalyzer Babson method. *2760* At 1 mmol/L affects SMA 12/60 method. *3335*
Serum Increase Physiological Result of hepatocellular damage or cholestasis. *3879* Chronic active hepatitis associated with immune hemolytic anemia in one case. *3269* Disturbances in liver function in 5 to 35% of patients treated for hypertension. Hepatocellular injury may occur after short or long-term exposure. *0162* Chronic active hepatitis associated with immune hemolytic anemia in one case. *3269* Drug induced hepatitis with severe, chronic, aggressive inflammation. *0211*

Bicarbonate *Serum No Effect Analytical* At concentration of 80 mg/L had no effect on method using phenolphthalein. *3393*

Bile *Urine Increase Physiological* Occurs as result of hepatocellular damage. *3646*

Bilirubin *Serum Increase Analytical* At 1 mmol/L affects SMA 12/60 method. *3335* Theoretically reacts with diazo reagent. *3220* At concentrations above 65 mg/L (normal therapeutic concentration 2 mg/L) raised concentration as measured by Jendrassik and Grof method. *3393*
Serum Increase Physiological Mild hepatocellular jaundice in about 1% cases. *1023* Chronic active hepatitis associated with immune hemolytic anemia in one case. *3269* Disturbances in liver function in 5 to 35% of patients treated for hypertension. Hepatocellular injury may occur after short or long-term exposure. *0162*
Urine Increase Physiological Occurs as result of hepatocellular damage. *3646*

Bilirubin, Direct *Serum Increase Physiological* Mild hepatocellular jaundice may occur. *1237*

BSP Retention *Serum Increase Physiological* Result of hepatocellular damage (cholestasis also). *3879*

Calcium *Serum No Effect Analytical* At concentration of 80 mg/L had no effect on cresolphthalein method. *3393*

Catecholamines *Plasma Increase Analytical* Interference with fluorometric methods. *3146* Reacts like catecholamines, may persist for days. *1343*
Plasma Increase Physiological Pronounced increase observed, reduced by barbiturates. *1846*
Urine No Effect Possibly no interference in method Sandhu and Freed. *3135* In volunteers with HPLC methods using electrochemical detection. *2521*
Urine Increase Analytical Has similar fluorescence. *1841* With trihydroxyindole fluorometric procedures. *2521*
Urine Decrease Physiological Depletion of tissue stores. *2220*

Cephalin Flocculation *Serum Increase Physiological* Result of hepatocellular damage. *3879*

Chloride *Serum No Effect Analytical* At concentration of 80 mg/L had no effect on mercurimetric method. *3393*
Serum Increase Physiological May cause salt retention and edema. *1343*

Cholesterol *Serum No Effect Analytical* No effect at therapeutic concentration on Reflotron method. *1984* At concentration of 400 mg/L had no effect on catalase-AIDH method. *3393* At concentration of 400 mg/L had no effect on method using catalase-Hantzsch reaction. *3393* At concentration of 400 mg/L had no effect on Liebermann-Burchard method. *3393*
Serum No Effect Physiological In several patients treated for 2-4 mo. *0088* Increase by 2 mg/dL in 11 patients given 750 mg daily for 8 weeks. *2550* In 3 studies of 7 to 17 patients for up to 3 mo. *0089*
Serum Decrease Analytical At concentrations above 200 mg/L (normal therapeutic concentration 2 mg/L) lowered concentration as measured by CHOD-Iodide method. *3393* At concentrations above 50 mg/L (normal therapeutic concentration 2 mg/L) lowered concentration as measured by CHOD-PAP method. *3393*

Serum Decrease Physiological In 17 hypertensive patients given drug for 3 mo. *0967*

Cholesterol, High Density Lipoprotein
Serum No Effect Physiological In several patients treated for 2-4 mo. *0088* In 17 hypertensive patients given drug for 3 mo. *0967*
Serum Decrease Physiological Increase by 10% in 32 middle-aged hypertensive males treated for 6 weeks. *2129* Decrease by -4 mg/dL in 11 patients given 750 mg daily for 8 weeks. *2550* By up to 15% in 3 studies of 7 to 17 patients for up to 3 mo. *0089*

Cholesterol, Low Density Lipoprotein
Serum No Effect Physiological In several patients treated for 2-4 mo. *0088*
Serum Decrease Physiological In 17 hypertensive patients given drug for 3 mo. *0967*

Cholesterol, Very Low Density Lipoprotein
Serum Decrease Physiological In 17 hypertensive patients given drug for 3 mo. *0967*

Color *Urine Increase Physiological* May produce brown-black on standing. *0443*

Complement C3 *Serum Decrease Physiological* Observed in one patient with hepatocellular damage of moderate severity due to sensitization by drug. *0874*

Complement C4 *Serum Decrease Physiological* Observed in one patient with hepatocellular damage of moderate severity due to sensitization by drug. *0874*

Coombs' Test, Direct *Serum Positive Physiological* Autoimmune phenomenon (occurs with weeks of treatment). *1486* Observed in 5 of 9 patients typically seen in 20% of treated patients. Probably due to impaired F_c-dependent reticuloendothelial function. *1909*

Coombs' Test, Indirect *Serum Positive Physiological* Autoimmune phenomenon (occurs with weeks of treatment). *1486*

Coproporphyrin *Urine Increase Physiological* May precipitate acute porphyria attack. *1322*
Feces Increase Physiological May precipitate acute porphyria attack. *1322*

Creatinine *Serum No Effect Analytical* At 30 mg/L on reversed phase LC procedure of Zhiri et al. *3942* At concentration of 20 mg/L had no effect on creatinine iminohydrolase method. *3393* At concentration of 320 mg/L had no effect on Jaffé-Fading-Fraction method. *3393* At concentrations above 50 mg/L (normal therapeutic concentration 2 mg/L) raised concentration as measured by kinetic Jaffé method on BKA-2. *3393*
Serum Increase Analytical Readily oxidized and affects alkaline picrate method. *2313* Above 2 mg/mL even affects Fuller's earth procedures. *2259* At concentrations above 200 mg/L (normal therapeutic concentration 2 mg/L) raised concentration as measured by AutoAnalyzer Jaffé method. *3393* At concentrations above 2,000 mg/L (normal therapeutic concentration 2 mg/L) raised concentration as measured by Jaffé-Fuller's earth method. *3393*
Urine Increase Analytical Acts as reducing agent with alkaline picrate. *0757*

Dihydroxyphenylalanine
Test Conditions Increase Analytical 11.5% of fluorescence method of Waldmeier. *3735*

Dopamine *Urine No Effect Physiological* No effect when methyldopa administered and HPLC used to measure catecholamines. *2521*

Effective Renal Plasma Flow
Patient Increase Physiological Slight increase or normal in normo- or hypertensives. *1343*

Endoplasmic Reticulum Antibody
Serum Positive Physiological Observed in one patient with chronic active hepatitis and cirrhosis. *0259*

Eosinophils *Blood Increase Physiological* Allergic response. *2313*

Epinephrine *Urine No Effect Physiological* No effect when methyldopa administered and HPLC used to measure catecholamines. *2521*

Erythrocytes *Blood Decrease Physiological* Autoimmune hemolytic anemia/aplastic anemia. *0597* On rare occasion may produce hemolysis. Great majority of patients do not show this. *1909*

Estriol *Urine No Effect Analytical* No effect on GLC method. *1163*

Ferric Chloride Test *Urine Positive Analytical* At 0.2 mg/mL in alkaline urine. *2071*

Follicle Stimulating Hormone (FSH)
Plasma No Effect Physiological No effect in patient on long term treatment who had drug-induced hyperprolactinemia. *0166*

Glomerular Filtration Rate (GFR)
Urine Increase Physiological Slight increase or normal in normo-, hypertensives. *1343*

Glucose *Serum No Effect Analytical* No effect at therapeutic concentration on Reflotron method. *1984* At concentration of 400 mg/L had no effect on hexokinase/G-6-PDH method. *3393*
Serum Increase Analytical At 1 mmol/L affects SMA 12/60 method. *3335*
Serum Decrease Analytical At concentrations above 100 mg/L (normal therapeutic concentration 2 mg/L) lowered concentration as measured by Ektachem® method. *3393* At concentrations above 200 mg/L (normal therapeutic concentration 2 mg/L) lowered concentration as measured by GOD-Perid method. *3393* At concentrations above 7 mg/L (normal therapeutic concentration 2 mg/L) lowered concentration as measured by GOD/POD-PAP method. *3393*
Urine No Effect Analytical No effect on glucose oxidase methods. *1099*

γ-Glutamyltransferase (GGT) *Serum No Effect Analytical* No effect at therapeutic concentration on Reflotron method. *1984*
Serum Increase Physiological Drug induced hepatitis with severe, chronic, aggressive inflammation. *0211*

Growth Hormone *Plasma Decrease Physiological* Slight effect on basal concentration after many mo of treatment for hypertension. Concentration increased to greater extent in response to Insulin in short-term treated individuals than in controls or long-term treated. *3453*

Guaiacols Spot Test *Urine Positive Analytical* False reaction with screening test of Rogers. *3031*

Haptoglobin *Serum Decrease Physiological* May cause hemolytic anemia. *2313*

Hematocrit *Blood Decrease Physiological* Autoimmune hemolytic anemia. *0597*

Hemoglobin *Blood Decrease Physiological* Autoimmune hemolytic anemia. *0597* On rare occasion may produce hemolysis. Great majority of patients do not show this. *1909*

HMPG (4-Hydroxy-3-Methoxy-Phenylethylene Glycol)
Urine Decrease Physiological By more than VMA, ?affects aldehyde reductase Inhibition of Dopa decarboxylase, ?also aldehyde reductase. *0426*

5-Hydroxyindoleacetic Acid (5-HIAA)
Urine Decrease Physiological Inhibition of aromatic amino acid decarboxylation. *0832*

Immune Complexes *Serum No Effect Physiological* In a study of 9 hypertensives. *1909*

Immunoglobulin IgA *Serum Increase Physiological* Observed in one patient with hepatocellular damage of moderate severity due to sensitization by drug. *0874*

Immunoglobulin IgG *Serum Increase Physiological* Observed in one patient with hepatocellular damage of moderate severity due to sensitization by drug. *0874*

Immunoglobulin IgM *Serum Increase Physiological* Observed in one patient with hepatocellular damage of moderate severity due to sensitization by drug. *0874*

Indomethacin *Serum No Effect Analytical* No effect on HPLC method of Roberts and Smith. *3002*

Ketones *Urine Increase Analytical* Affects alkaline nitroprusside procedure. *0583*

Lactate *Serum No Effect Analytical* At concentration of 400 mg/L had no effect on enzymatic method. *3393*

Lactate Dehydrogenase *Serum Increase Physiological* Chronic active hepatitis associated with immune hemolytic anemia in one case. *3269*

LE Cells *Blood Positive Physiological* Autoimmune phenomenon. *3296*

Leukoagglutinins *Serum Positive Physiological* Observed in one patient with hepatocellular damage of moderate severity due to sensitization by drug. *0874*

Methyldopa (continued)

Leukocytes *Blood Increase Physiological* Due to eosinophilia. *2313*

Blood Decrease Physiological Very rare, may cause granulocytopenia. *1940*

Urine No Effect Analytical At concentration of 1,000 mg/L had no effect on Cytur-Test. *3393*

Luteinizing Hormone (LH) *Plasma No Effect Physiological* No effect in patient on long term treatment who had drug-induced hyperprolactinemia. *0166*

Melanogen *Urine Positive Analytical* At 0.2 mg/mL in alkaline urine. *2071*

Metanephrines, Total *Urine Increase Analytical* Questionable interference fluorometric methods. *3879*

Neutrophils *Blood Decrease Physiological* Occasional case of agranulocytosis reported. *3717*

Norepinephrine *Urine No Effect Physiological* No effect when methyldopa administered and HPLC used to measure catecholamines. *2521*

Phosphate *Serum No Effect Analytical* At concentration of 211 mg/L had no effect on phosphomolybdate method. *3393*

Platelet Associated IgG *Serum No Effect Physiological* In a study of 9 hypertensives. *1909*

Platelets *Blood Decrease Physiological* Very rare occurs within days or months. *1940*

Porphobilinogen *Urine Increase Physiological* May precipitate acute porphyria attack. *1322*

Porphyrins *Urine Increase Physiological* May precipitate attack of acute porphyria. *1016*

Potassium *Serum No Effect Analytical* At concentration of 7 mg/L had no effect on flame-photometric method. *3393* At concentration of 100 mg/L had no effect on measurement by ISE with predilution. *3393* At concentration of 800 mg/L had no effect on measurement by ISE without predilution. *3393*

Prolactin *Plasma Increase Physiological* Increased concentration after single doses of drug during long-term treatment. *0166* After single doses of 750 or 1,000 mg peak reached 4 to 6 h after administration. With long term administration 3 to 4 fold increase over normal noted. *3453* Marked effect in male and female hypertensives treated for up to 6 weeks. *3630*

Protein *Serum No Effect Analytical* At concentration of 211 mg/L had no effect on biuret method with blank correction. *3393*

Protein Electrophoresis *Serum No Effect Analytical* At concentration of 800 mg/L had no effect on automated Olympus-Hite method. *3393*

Prothrombin Time *Plasma Increase Physiological* Enhances anticoagulant activity. *3879*

Blood Increase Physiological Decreased synthesis of prothrombin associated with hepatitis in some cases. *2261*

Red cell-associated Immunoglobulin IgG

Blood Positive Physiological Observed in 5 of 9 patients typically seen in 20% of treated patients. Probably due to impaired F_c-dependent reticuloendothelial function. *1909*

Reticulocytes *Blood Increase Physiological* Hemolytic anemia of autoimmune type occurring in fewer than 1% of treated patient. *0461*

Rheumatoid Factor *Serum Positive Physiological* Reported effect. *1343*

Sodium *Serum No Effect Analytical* At concentration of 7 mg/L had no effect on flame-photometric method. *3393* At concentration of 100 mg/L had no effect on measurement by ISE with predilution. *3393* At concentration of 800 mg/L had no effect on measurement by ISE without predilution. *3393*

Serum Increase Physiological May cause salt retention and edema. *2313*

Sugar *Urine No Effect Analytical* No effect reported at 0.2 mg/mL. *2071*

Urine Increase Analytical False positive with Clinitest®, no effect glucose oxidase method. *1099*

Thormählen Test *Urine Positive Analytical* At 0.2 mg/mL in alkaline urine. *2071*

Thymol Turbidity *Serum Increase Physiological* Hepatotoxic effect. *0597*

Triglycerides *Serum No Effect Analytical* No effect at therapeutic concentration on Reflotron method. *1984* At concentration of 80 mg/L had no effect on lipase/esterase method. *3393*

Serum No Effect Physiological In several patients treated for 2-4 mo. *0088* In 3 studies of 7 to 17 patients for up to 3 mo. *0089*

Serum Increase Physiological Increase by 28% in 32 middle-aged hypertensive males treated for 6 weeks. *2129* Increase by 20 mg/dL in 11 patients given 750 mg daily for 8 weeks. *2550*

Serum Decrease Analytical At concentrations above 8 mg/L (normal therapeutic concentration 2 mg/L) lowered concentration as measured by GPO-PAP method. *3393*

Tyramine Test *Patient Increase Physiological* Enhanced responsiveness reported. *1488*

Urea Nitrogen *Serum No Effect Analytical* No effect at therapeutic concentration on Reflotron method. *1984* At concentration of 211 mg/L had no effect on diacetylmonoxime method. *3393* At concentration of 80 mg/L had no effect on Ektachem® method. *3393*

Serum Increase Physiological ?due to decreased renal blood flow. *3879*

Uric Acid *Serum No Effect Analytical* At 30 mg/L on reversed phase LC procedure of Zhiri et al. *3942* No effect at therapeutic concentration on Reflotron method. *1984* At concentration of 320 mg/L had no effect on catalase-AIDH method. *3393*

Serum Increase Analytical Interferes with phosphotungstate procedure. *1488* At concentrations above 20 mg/L (normal therapeutic concentration 2 mg/L) raised concentration as measured by phosphotungstate reduction method. *3393*

Serum Decrease Analytical At concentrations above 300 mg/L (normal therapeutic concentration 2 mg/L) lowered concentration as measured by Kageyama-Hantzsch method. *3393* At concentrations above 3mg/L (normal therapeutic concentration 2 mg/L) lowered concentration as measured by uricase-PAP method. *3393*

Urine Increase Analytical Interferes with phosphotungstate procedure. *1237*

Urobilinogen *Urine Increase Physiological* Autoimmune hemolytic anemia. *0597*

Vanillylmandelic Acid *Urine No Effect Analytical* No interference with Pisano procedure. *0583*

Urine Increase Analytical Interference by 5-HIAA with non-specific diazo reaction. *3218*

Urine Increase Physiological Initial physiological response, changes later. *0913*

Urine Decrease Physiological Depletion of tissue stores. *0923*

Volume *Plasma Increase Physiological* Occurs with sodium retention if sole agent. *2706*

Methyldopa Hydrazine

Dihydroxyphenylalanine *Serum Increase Physiological* 5 fold potentiation with pretreatment. *2306*

Growth Hormone *Plasma Increase Physiological* 2 fold increase when given with Dopa. *2306*

Homovanillic Acid *Plasma Decrease Physiological* 65% reduction when given with Dopa. *2306*

Methyldopamine

Dihydroxyphenylalanine

Test Conditions Increase Analytical 15% of fluorescence method of Waldmeier. *3735*

Methylene Blue

Bacteria *Urine Negative Analytical* May interfere with color on Microstix. *0129*

Bilirubin *Serum Increase Physiological* May cause hemolysis with G-6-PD deficiency. *0248*

Color *Urine Increase Analytical* Blue color. *1342*

Feces Increase Physiological Blue especially on exposure to air with 130-140 mg. *0902*

Diagnex Blue Excretion *Urine Increase Analytical* Detection of methylene blue. *3879*

Erythrocytes *Blood Decrease Physiological* May cause hemolysis with G-6-PD deficiency. *0248*

Haptoglobin *Serum Decrease Physiological* May cause hemolysis. *2313*

Heinz Body Formation *Blood Positive Physiological* Early stage of hemolytic anemia. *0333*

Hematocrit *Blood Decrease Physiological* May cause hemolysis with G-6-PD deficiency. *0248*

Hemoglobin *Blood Decrease Physiological* May cause hemolysis with G-6-PD deficiency. *0248*

Lactate *Plasma Decrease Physiological* Variable response when lactic acidosis. *1487*

Methemoglobin *Blood Increase Physiological* May cause hemolysis (also used as treatment). *0397*

Methylenedioxyamphetamine

Aspartate Aminotransferase
Serum Increase Physiological Reported effect ?of muscle origin. *2975*

Bicarbonate *Serum Decrease Physiological* Respiratory acidosis. *2975*

pCO$_2$ *Blood Increase Physiological* Respiratory acidosis. *2975*

pH *Blood Decrease Physiological* Respiratory acidosis. *2975*

Methylephedrine

Amphetamine *Urine No Effect Analytical* At 50 mg/L on fluorescent method of Hayes. *1534*

Methylethinylestradiol

17-Hydroxycorticosteroids *Urine Decrease Physiological* Decreased excretion of cortisol metabolites. *0022*

Methylmalonic Acid

Propionic Acid *Serum Increase Analytical* May be formed from methylmalonic acid in GLC. *0975*
Urine Increase Analytical May be formed from methylmalonic acid in GLC. *0975*

Methylphenidate

Alanine Aminotransferase *Serum No Effect Analytical* At acute overdose concentration (20 mg/dL) on Technicon® SMAC® method. *3719*

Albumin *Serum No Effect Analytical* At acute overdose concentration (20 mg/dL) on Technicon® SMAC® method. *3719* At concentration of 200 mg/L had no effect on BCG method. *3393*

Alkaline Phosphatase *Serum No Effect Analytical* At acute overdose concentration (20 mg/dL) on Technicon® SMAC® method. *3719*

Alkaloids *Urine No Effect Analytical* With ninhydrin on Frings TLC procedure at 1 mg/dL. *1204*
Urine Increase Analytical Purple with iodoplatinate on Frings TLC procedure. *1204*

Amitriptyline *Serum Increase Physiological* Inhibits metabolism of tricyclic antidepressants. *1487*

Amphetamine *Urine No Effect Analytical* At 50 mg/L on fluorescent method of Hayes. *1534*
Urine Increase Analytical Yields fluorophor with NBD chloride reaction. *2473*

Aspartate Aminotransferase *Serum No Effect Analytical* At acute overdose concentration (20 mg/dL) on Technicon® SMAC® method. *3719*

Barbiturate *Urine No Effect Analytical* With vanillin on Frings procedure at 10 mg/dL. With diphenylcarbazone on Frings TLC procedure at 10 mg/dL. With mercuric SO$_4$ on Frings TLC procedure at 10 mg/dL. With mercurous NO$_3$ on Frings TLC procedure at 10 mg/dL. *1204*

Bilirubin *Serum No Effect Analytical* At acute overdose concentration (20 mg/dL) on Technicon® SMAC® method. *3719* At concentration of 200 mg/L had no effect on Jendrassik and Grof method. *3393*

Bromide *Urine No Effect Analytical* No effect on method of Frings. *1204*

Calcium *Serum No Effect Analytical* At acute overdose concentration (20 mg/dL) on Technicon® SMAC® method. *3719* At concentration of 200 mg/L had no effect on cresolphthalein method. *3393*

Cholesterol *Serum No Effect Analytical* At acute overdose concentration (20 mg/dL) on Technicon® SMAC® method. *3719* At concentration of 200 mg/L had no effect on Liebermann-Burchard method. *3393*

Creatine Kinase *Serum No Effect Analytical* At acute overdose concentration (20 mg/dL) on Technicon® SMAC® method. *3719*

Creatinine *Serum No Effect Analytical* At acute overdose concentration (20 mg/dL) on Technicon® SMAC® method. *3719* At concentration of 200 mg/L had no effect on AutoAnalyzer Jaffé method. *3393*

Diphenylhydantoin *Serum No Effect Physiological* No effect observed in 11 patients in one study. *1487*
Serum Increase Physiological May increase from 9 to 28 μg/mL (at 20-40 mg/d). *2041*

Epinephrine *Urine Increase Physiological* For 3 h after administration (slight). *1487*

Ethchlorvynol *Urine No Effect Analytical* No effect on method of Frings and Cohen. *1204*

FPN Test *Urine No Effect Analytical* No effect at 10 mg/dL on method of Frings. *1204*

Glucose *Serum No Effect Analytical* At acute overdose concentration (20 mg/dL) on Technicon® SMAC® method. *3719*

Iron *Serum No Effect Analytical* At acute overdose concentration (20 mg/dL) on Technicon® SMAC® method. *3719* At concentration of 200 mg/L had no effect on Ferrozine method. *3393*

Lactate Dehydrogenase *Serum No Effect Analytical* At acute overdose concentration (20 mg/dL) on Technicon® SMAC® method. *3719*

Mexiletine *Serum Increase Physiological* Decreases clearance and prolongs half-life. *2011*

Norepinephrine *Urine No Effect Physiological* No effect observed. *1487*

Phosphate *Serum No Effect Analytical* At acute overdose concentration (20 mg/dL) on Technicon® SMAC® method. *3719* At concentration of 200 mg/L had no effect on phosphomolybdate method. *3393*

Platelets *Blood Decrease Physiological* Occasional thrombocytopenic purpura reported. *1678*

Protein *Serum No Effect Analytical* At acute overdose concentration (20 mg/dL) on Technicon® SMAC® method. *3719* At concentration of 200 mg/L had no effect on biuret method with blank correction. *3393*

Prothrombin Time *Plasma Increase Physiological* Inhibits metabolism of coumarins and phytonadione. *1243*

Salicylate *Urine No Effect Analytical* At 10 mg/dL no effect on Trinder method. *1204*

Triglycerides *Serum No Effect Analytical* At acute overdose concentration (20 mg/dL) on Technicon® SMAC® method. *3719* At concentration of 200 mg/L had no effect on lipase/esterase method. *3393*

Urea Nitrogen *Serum No Effect Analytical* At acute overdose concentration (20 mg/dL) on Technicon® SMAC® method. *3719* At concentration of 200 mg/L had no effect on diacetylmonoxime method. *3393*

Uric Acid *Serum No Effect Analytical* At acute overdose concentration (20 mg/dL) on Technicon® SMAC® method. *3719* At concentration of 200 mg/L had no effect on phosphotungstate reduction method. *3393*

Methylphenobarbital

Albumin *Serum No Effect Analytical* At concentration of 150 mg/L had no effect on BCG method. *3393*

Bilirubin *Serum No Effect Analytical* At concentration of 150 mg/L had no effect on Jendrassik and Grof method. *3393*

Calcium *Serum No Effect Analytical* At concentration of 150 mg/L had no effect on cresolphthalein method. *3393*

Chloride *Serum No Effect Analytical* At concentration of 150 mg/L had no effect on mercurimetric method. *3393*

Cholesterol *Serum No Effect Analytical* At concentration of 150 mg/L had no effect on Liebermann-Burchard method. *3393*

Creatinine *Serum No Effect Analytical* At concentration of 150 mg/L had no effect on AutoAnalyzer Jaffé method. *3393*

Diphenylhydantoin *Serum No Effect Analytical* On GLC procedure of Papadopoulos when added *in vitro*. *2722*

Erythrocytes *Blood Decrease Physiological* Megaloblastic anemia. *2313*

Glucose *Serum No Effect Analytical* At concentration of 41 mg/L had no effect on Ektachem® method. *3393*

Hematocrit *Blood Decrease Physiological* May cause megaloblastic anemia. *2313*

Hemoglobin *Blood Decrease Physiological* May cause megaloblastic anemia. *2313*

MCV *Blood Increase Physiological* May cause megaloblastic anemia. *2313*

Pheneturide *Serum No Effect Analytical* On GLC procedure of Papadopoulos when added *in vitro*. *2722*

Phenobarbital *Serum No Effect Analytical* On GLC procedure of Papadopoulos when added *in vitro*. *2722*

Phosphate *Serum No Effect Analytical* At concentration of 150 mg/L had no effect on phosphomolybdate method. *3393*

Primidone *Serum No Effect Analytical* On GLC procedure of Papadopoulos when added *in vitro*. *2722*

Protein *Serum No Effect Analytical* At concentration of 150 mg/L had no effect on biuret method with blank correction. *3393*

Urea Nitrogen *Serum No Effect Analytical* At concentration of 41 mg/L had no effect on Ektachem® method. *3393*

Uric Acid *Serum No Effect Analytical* At concentration of 150 mg/L had no effect on phosphotungstate reduction method. *3393*

Methylprednisolone

Amylase *Serum Increase Physiological* To well above normal in 10 patients given 1 g/d i.v. for 3 d. Maximum effect days 5 to 8. *0804*

Amylase, Pancreatic *Serum Increase Physiological* To well above normal in 10 patients given 1 g/d i.v. for 3 d. Maximum effect days 5 to 8. Possible subclinical damage of pancreatic acinar cell *0804*

Amylase, Salivary *Serum No Effect Physiological* No effect in patients given drug i.v. although pancreatic component significantly affected (Amylase). *0804*

Angiotensin-Converting Enzyme (ACE)
Serum Decrease Analytical Moderate inhibition by doses generally greater than used clinically. *3038* After 96 h incubation *in vitro* at high concentrations but no effect under normal measurement conditions. *3074*

Bicarbonate *Serum Increase Physiological* May cause hypokalemic alkalosis. *1680*

Corticotropin *Plasma Decrease Physiological* Suppression for 3 weeks following single injection in 12 patients. *1754*

Cortisol *Plasma Decrease Physiological* Suppression for 2 weeks following single lumbar extradural injection in 12 patients. *1754*

Creatinine Clearance *Urine Increase Physiological* Varies with inulin clearance. *3774*

Cyclosporine *Serum Increase Physiological* High doses produce effect probably due to effect on hepatic enzymes. *0683*

Digoxin *Serum Increase Analytical* At 50 ng/mL equals 0.2 ng/mL by RIA. *3940*

2,3-Diphosphoglycerate

Red Blood Cells *No Effect Physiological* No significant effect following i.v. infusion of drug in patients with myocardial infarction compared with controls. *1560*

Follicle Stimulating Hormone (FSH)
Plasma No Effect Physiological No apparent effect, unlike that on testosterone, in 67 y old males with chronic pulmonary disease taking drug for more than 1 mo. *2245*

Glucose Tolerance *Serum Decrease Physiological* Decreased carbohydrate tolerance. *1680*

Immunoglobulin IgA *Serum Decrease Physiological* Significant effect in 43% individuals. *0552*

Immunoglobulin IgG *Serum Decrease Physiological* After 96 mg/d for 5 d. *0552*

Immunoglobulin IgM *Serum Decrease Physiological* Noted in 14% individuals. *0552*

Inulin Clearance *Urine Increase Physiological* At 24 h after 1 g i.v. in normals. *3774*

Lactate *Blood Increase Physiological* Significant effect in 10 of 13 patients with myocardial infarction. Effect observed in 1 h, max at 3 h, persisted for 24 h. *1560*

Lipase *Serum Increase Physiological* To well above normal in 10 patients given 1 g/d i.v. for 3 d. Maximum effect days 5 to 8, possible subclinical damage of pancreatic acinar cell. *0804*

Luteinizing Hormone (LH) *Plasma No Effect Physiological* No apparent effect, unlike that on testosterone, in 67 y old males with chronic pulmonary disease taking drug for more than 1 mo. *2245*

NBT Test *Blood Decrease Physiological* False negative in one patient on 1 g/d. *2589*

Nitrogen Balance *Patient Negative Physiological* Due to protein catabolism. *1680*

Osmolality *Serum No Effect Physiological* No significant effect following i.v. infusion of drug in patients with myocardial infarction compared with controls. *1560*

PAH Clearance *Urine Increase Physiological* At 24 h after 1 g i.v. in normals. *3774*

pH *Blood No Effect Physiological* No significant effect following i.v. infusion of drug in patients with myocardial infarction compared with controls. *1560*

pO₂ *Blood Decrease Physiological* Initial reduction in myocardial infarction patients compared with placebo treated controls (112 mm Hg to 88 mm Hg on average). *1560*

Potassium *Serum No Effect Analytical* At concentration of 100 mg/L had no effect on measurement by ISE with predilution. *3393*
Serum Decrease Physiological May cause potassium loss. *1680*

Sodium *Serum No Effect Analytical* At concentration of 100 mg/L had no effect on measurement by ISE with predilution. *3393*

Testosterone *Serum Decrease Physiological* Reduced by more than 50% in 14 of 16 67 y old men with chronic pulmonary disease receiving drug for more than 1 mo. Testosterone concentration inversely correlated with dose of drug. Serum protein-binding unaffected. Effect probably due to suppression of secretion of gonadotropin releasing hormone by hypothalamus. *2245*

Trypsin *Serum Increase Physiological* Increase by 35-109% in 10 patients given 1 g/d i.v. for 3 d. Maximum effect days 5 to 8 possible subclinical damage of pancreatic acinar cell. *0804*

Volume *Plasma Increase Physiological* May cause fluid retention. *1680*
Blood Decrease Physiological Transient but significant effect observed in patients given i.v. infusion after myocardial infarction in comparison with placebo treated control population. *1560*

Methylpromazine

Leukocytes *Blood Decrease Physiological* Leukopenia (AMA Blood dyscrasias). *2429*

Neutrophils *Blood Decrease Physiological* Occasional case of agranulocytosis reported. *3717*

Methylserotonin

Volume *Urine Decrease Physiological* 30-40% antidiuretic activity of 5-HT. *1176*

Methylsulfonal

Aminolevulinic Acid *Urine Increase Physiological* May precipitate acute porphyria. *1322*

Color *Urine Increase Physiological* Red (may provoke porphyria). *2313*

Coproporphyrin *Urine Increase Physiological* May precipitate acute porphyria. *1322*
Feces Increase Physiological May precipitate acute porphyria. *1322*

Methemoglobin *Blood Increase Physiological* May cause hemolytic anemia. *2429*

Porphobilinogen *Urine Increase Physiological* May precipitate acute porphyria. *1322*

Protoporphyrin *Feces Increase Physiological* May precipitate acute porphyria. *1322*

Methyltestosterone

Alanine Aminotransferase *Serum Increase Physiological* Cholestatic effect (moderate elevation). *2313*

Alkaline Phosphatase *Serum Increase Physiological* Cholestatic effect. *1538* In 1 of 60 patients (female transsexuals or impotent males) receiving 150 mg day for long term. *3805*

Aspartate Aminotransferase
Serum Increase Physiological Cholestatic effect (moderate elevation). *1713* In 1 of 60 patients (female transsexuals or impotent males) receiving 150 mg day for long term. *3805*

Bile *Urine Increase Physiological* Cholestatic effect. *1713*

Bilirubin *Serum Increase Physiological* Cholestatic effect. *0153* In 1 of 60 patients (female transsexuals or impotent males) receiving 150 mg day for long term. *3805*

BSP Retention *Serum Increase Physiological* Transport and conjugation of BSP impaired. *0153*

Calcium *Serum Increase Physiological* May occur especially in women on therapy for breast cancer. *1680*

Cephalin Flocculation *Serum Increase Physiological* Cholestatic effect (may not be affected). *1713*

Cholesterol *Serum Increase Physiological* May cause hypersensitive cholestasis. *1434*

Creatine *Urine Increase Physiological* Metabolic effect ?mechanism. *1343*

Cyclosporine *Blood Increase Physiological* One case reported in which administration caused marked increase in blood cyclosporine possibly by inhibition of cytochrome P-450 hepatic enzymes. *2471*

Factor V *Plasma Decrease Physiological* Suppresses activity. *2427*

Factor VII *Plasma Decrease Physiological* Suppresses activity. *2427*

Factor X *Plasma Decrease Physiological* Suppresses activity. *2427*

β-Glucuronidase *Serum Increase Physiological* Metabolic effect. *0227*

Haptoglobin *Serum Increase Physiological* Metabolic effect. *0227*

Lactate Dehydrogenase *Serum Increase Physiological* Infrequent rise observed. *0563*

5'-Nucleotidase *Serum Increase Physiological* Due to cholestasis. *1778*

Plasminogen *Plasma Increase Physiological* Metabolic effect. *0227*

Protein Bound Iodine (PBI) *Serum Decrease Physiological* Lowers concentration of circulating thyroxine binding globulin. *0012*

Prothrombin Time *Plasma Increase Physiological* Associated with failure of excretion of bile salts. *1713*

Sperm Count *Semen Decrease Physiological* May occur after prolonged dose or excess administration. *1678*

Thymol Turbidity *Serum Increase Physiological* Cholestatic effect (may not be affected). *1713*

Thyroxine (T₄) Binding Globulin
Serum Decrease Physiological Direct effect of drug. *0012*

Urobilinogen *Urine Decrease Physiological* Cholestatic effect. *1713*
Feces Increase Physiological Pale stools, may cause cholestasis. *1713*

Methylthiouracil

Alanine Aminotransferase *Serum Increase Physiological* Hepatotoxic effect. *2313*

Alkaline Phosphatase *Serum Increase Physiological* Hepatotoxic effect. *2313*

Aspartate Aminotransferase
Serum Increase Physiological Hepatotoxic effect. *2313*

Bile *Urine Increase Physiological* Hepatotoxic effect. *2313*

Bilirubin *Serum Increase Physiological* Hepatotoxic effect. *2313*

BSP Retention *Serum Increase Physiological* Hepatotoxic effect. *2313*

Cephalin Flocculation *Serum Increase Physiological* Hepatotoxic effect. *2313*

Erythrocytes *Blood Decrease Physiological* Occasional case of aplastic anemia reported. *3717*

¹³¹I Uptake *Serum No Effect Physiological* No effect on uptake by thyroid. *2220*

LE Cells *Blood Positive Physiological* May produce LE-like syndrome. *3505*

Leukocytes *Blood Decrease Physiological* Occasional leukopenia or agranulocytosis. *2313* Occasional case of aplastic anemia reported. *3717*

Neutrophils *Blood Decrease Physiological* Occasional case of agranulocytosis reported. *3717*

Platelets *Blood Decrease Physiological* Occasional case of aplastic anemia reported. *3717*

Protein Bound Iodine (PBI) *Serum Decrease Physiological* Inhibits iodination of tyrosine in thyroxine binding globulin. *0012*

Prothrombin Time *Plasma Increase Physiological* Also exaggerated response to anticoagulants. *1973*

Thymol Turbidity *Serum Increase Physiological* Hepatotoxic effect. *2313*

Thyroxine (T₄) *Serum Decrease Physiological* Inhibits synthesis, stops iodination of tyrosine. *0071*

Methyltryptophan

Tryptophan *Test Conditions Increase Analytical* Measured as same in Spies/Chambers method. *1415*

Methylurea

Urea Nitrogen *Serum Increase Analytical* Absorbance relative to urea = 1.5 diacetyl procedure. *0583*

Methyprylon

Alanine Aminotransferase *Serum No Effect Analytical* At acute overdose concentration (10 mg/dL) on Technicon® SMAC® method. *3719*

Albumin *Serum No Effect Analytical* At acute overdose concentration (10 mg/dL) on Technicon® SMAC® method. *3719* At concentration of 100 mg/L had no effect on BCG method. *3393*

Alkaline Phosphatase *Serum No Effect Analytical* At acute overdose concentration (10 mg/dL) on Technicon® SMAC® method. *3719*

Alkaloids *Urine No Effect Analytical* With ninhydrin on Frings TLC procedure at 10 mg/dL. With iodoplatinate of Frings TLC procedure at 10 mg/dL. *1204*

Aminolevulinic Acid *Urine Increase Physiological* Ala-synthetase stimulated in animals. *1343*

Methyprylon (continued)

Aspartate Aminotransferase *Serum No Effect Analytical* At acute overdose concentration (10 mg/dL) on Technicon® SMAC® method. *3719*

Barbiturate *Serum Increase Analytical* Affects diphenyl-carbazone above 150 mg/dL only. *0583*
Urine No Effect Analytical With mercuric SO_4 on Frings TLC procedure at 10 mg/dL. With diphenylcarbazone on Frings TLC procedure at 10 mg/dL. With mercurous NO_3 on Frings TLC procedure at 10 mg/dL. With vanillin on Frings procedure at 10 mg/dL. *1204*

Bilirubin *Serum No Effect Analytical* At acute overdose concentration (10 mg/dL) on Technicon® SMAC® method. *3719* At concentration of 100 mg/L had no effect on Jendrassik and Grof method. *3393*

Bromide *Urine No Effect Analytical* No effect on method of Frings. *1204*

Calcium *Serum No Effect Analytical* At acute overdose concentration (10 mg/dL) on Technicon® SMAC® method. *3719* At concentration of 100 mg/L had no effect on cresolphthalein method. *3393*

Chlordiazepoxide *Blood No Effect Analytical* On method of Riddick at 5 mg/dL. *2983*

Cholesterol *Serum No Effect Analytical* At acute overdose concentration (10 mg/dL) on Technicon® SMAC® method. *3719* At concentration of 100 mg/L had no effect on Liebermann-Burchard method. *3393*

Creatine Kinase *Serum No Effect Analytical* At acute overdose concentration (10 mg/dL) on Technicon® SMAC® method. *3719*

Creatinine *Serum No Effect Analytical* At acute overdose concentration (10 mg/dL) on Technicon® SMAC® method. *3719* At concentration of 100 mg/L had no effect on AutoAnalyzer Jaffé method. *3393*

Estriol *Urine No Effect Analytical* No effect on GLC method with 125 mg/L. *1163*

Ethchlorvynol *Serum No Effect Analytical* At 10 μg/mL on colorimetric method of Wallace. *3753* At 10 mg/L on GLC procedure of Evenson. *1058*
Urine No Effect Analytical No effect on method Frings and Cohen at 20 mg/ml. *1205*

FPN Test *Urine No Effect Analytical* No effect at 10 mg/dL on method of Frings. *1204*

Glucose *Serum No Effect Analytical* At acute overdose concentration (10 mg/dL) on Technicon® SMAC® method. *3719*

17-Hydroxycorticosteroids *Urine Increase Analytical* Reported effect on Glenn-Nelson procedure. *0427*

Iron *Serum No Effect Analytical* At acute overdose concentration (10 mg/dL) on Technicon® SMAC® method. *3719* At concentration of 100 mg/L had no effect on Ferrozine method. *3393*

17-Ketogenic Steroids *Urine Increase Analytical* Interferes with Zimmermann reaction. *3505*

17-Ketosteroids *Urine Increase Analytical* Reported with Holtorff-Koch modification of Zimmermann. *0427*

Lactate Dehydrogenase *Serum No Effect Analytical* At acute overdose concentration (10 mg/dL) on Technicon® SMAC® method. *3719*

Leukocytes *Blood Decrease Physiological* Due to toxic action of metabolite. *2313*

Methaqualone *Serum No Effect Analytical* At 20 mg/L on GLC procedure of Evenson. *1057*

Phosphate *Serum No Effect Analytical* At acute overdose concentration (10 mg/dL) on Technicon® SMAC® method. *3719* At concentration of 100 mg/L had no effect on phosphomolybdate method. *3393*

Platelets *Blood Decrease Physiological* Isolated reports (?actually responsible). *1680*

Porphyrins *Urine Increase Physiological* May induce porphyria in animals, ?in humans. *1343*

Propoxyphene *Serum No Effect Analytical* At 25 mg/L on method of Evenson. *1056*

Protein *Serum No Effect Analytical* At acute overdose concentration (10 mg/dL) on Technicon® SMAC® method. *3719* At concentration of 100 mg/L had no effect on biuret method with blank correction. *3393*

Salicylate *Urine No Effect Analytical* At 10 mg/dL no effect on Trinder method. *1204*

Sugar *Urine Increase Analytical* Excreted as glucuronide, acts as reducing substance. *1343*

Triglycerides *Serum No Effect Analytical* At acute overdose concentration (10 mg/dL) on Technicon® SMAC® method. *3719* At concentration of 100 mg/L had no effect on lipase/esterase method. *3393*

Urea Nitrogen *Serum No Effect Analytical* At acute overdose concentration (10 mg/dL) on Technicon® SMAC® method. *3719* At concentration of 100 mg/L had no effect on diacetylmonoxime method. *3393*

Uric Acid *Serum No Effect Analytical* At acute overdose concentration (10 mg/dL) on Technicon® SMAC® method. *3719* At concentration of 100 mg/L had no effect on phosphotungstate reduction method. *3393*

Methysergide

Coombs' Test, Direct *Serum Positive Physiological* Single case with retroperitoneal fibrosis. *1487*

Eosinophils *Blood Increase Physiological* Transient effect up to 36 h after i.m. injection. *1274*
Blood Decrease Physiological Starts to occur within 1 h (up to 100% dec). *2427*

Hydrochloric Acid *Gastric Material Increase Physiological* Affect basal juice and after histamine. *2220*

5-Hydroxytryptamine (Serotonin)
Plasma Decrease Physiological Noted in migraine subjects. *0774*

LE Cells *Blood Positive Physiological* Observed effect in some cases. *0933*

Leukocytes *Blood Increase Physiological* Due to neutrophilia. *2313*
Blood Decrease Physiological Neutropenia reported. *2313*

Lymphocytes *Blood Decrease Physiological* Average decrease of 29% noted. *2427*

Neutrophils *Blood Increase Physiological* Average increase of 23% noted. *2427*
Blood Decrease Physiological Reported side effect. *1680*

Protein Bound Iodine (PBI) *Serum Increase Physiological* Mechanism not determined. *0583*

Urea Nitrogen *Serum Increase Physiological* Possible decrease in already impaired renal function. *0085*

Metiamide

Neutrophils *Blood Decrease Physiological* Occasional case of agranulocytosis reported. *3717*

Metiazinic Acid

Uric Acid *Serum Decrease Physiological* Possible interference with tubular reabsorption. *2255*

Metoclopramide

Acetaminophen *Serum Increase Physiological* Increased rate of absorption with faster gastric emptying. *3794*

Aldosterone *Plasma Increase Physiological* Secretion increased but mechanism mediating effect not elucidated but probably through a factor whose concentration is increased in serum. *0569* Observed in normal subjects after 10 mg drug over 2 h; higher response in patients with primary aldosteronism. *3883* Peak 15 minutes after injection: 3 fold higher than control at peak. *0497* Increased by single i.v. bolus of 10 mg: probably related to basal activity of renin angiotensin aldosterone system. *2543* Increase by 99% 15 minutes after 1 mg/kg intravenously in healthy volunteers. *3426*

Androsterone *Urine No Effect Physiological* No significant influence of drug on excretion in menstruating women. *0096*

Bilirubin *Serum No Effect Physiological* No significant displacement from protein observed in neonates. *3748*

Chlorothiazide *Serum Decrease Physiological* Decreased absorption due to absorption window or dissolution. *3794*

Corticosterone *Plasma No Effect Physiological* No change observed in normal individuals or in patients with primary aldosteronism after 10 mg drug. *3883* No effect even though other constituents affected by i.v. bolus of 10 mg. *2543*

Corticotropin *Plasma Increase Physiological* Increase by 55% 15 minutes after 1 mg/kg intravenously in healthy volunteers. *3426*

Cortisol *Plasma No Effect Physiological* No change observed in normal individuals or in patients with primary aldosteronism after 10 mg drug. *3883* No significant change following i.v. injection. *0497* No effect even though other constituents affected by i.v. bolus of 10 mg. *2543*
Plasma Increase Physiological Increase by 75% 15 minutes after 1 mg/kg intravenously in healthy volunteers. *3426*

Cyclosporine *Blood Increase Physiological* Coadministered drug causes increased absorption, area under curve and blood concentration. *3732*

Dehydroepiandrosterone Sulfate (DHEA-S)
Plasma No Effect Physiological No significant differences in either cycling or post-menopausal woman receiving drug. *0096*
Urine No Effect Physiological No significant influence of drug on excretion in menstruating women. *0096*

Deoxycorticosterone *Plasma No Effect Physiological* No effect even though other constituents affected by i.v. bolus of 10 mg. *2543*

Digoxin *Serum Decrease Physiological* Impairs absorption, increases intestinal activity. *2290* With tablets of low bioavailability due to stimulation of bowel motility. *2794* Reduced absorption with limited dissolution. *3794*

Estradiol *Plasma No Effect Physiological* No significant baseline differences in cycling women before and after drug. *0096*

Ethanol *Blood Increase Physiological* Increased rate of absorption with faster gastric emptying. *3794*

Etiocholanolone *Urine No Effect Physiological* No significant influence of drug on excretion in menstruating women. *0096*

Follicle Stimulating Hormone (FSH)
Plasma No Effect Physiological No significant baseline differences in cycling women before and after drug. *0096*

Growth Hormone *Plasma No Effect Physiological* With pharmacological doses in normal men, but increased in hypogonadal men. *0648* In normal subjects after single i.v. injection of 10 mg. *3721*
Plasma Increase Physiological Increased from 30 to 60 minutes after single i.v. injection in 6 normal women. *0648* 5 fold increase at 30-45 minutes after 0.17 mg/kg bolus i.v. in 10 children. *2330* In 5 of 9 cirrhotic male patients after single injection of 10 mg. *3721*

18-Hydroxycorticosterone *Plasma No Effect Physiological* No change observed in normal individuals or in patients with primary aldosteronism after 10 mg drug. *3883*
Plasma Increase Physiological Observed in normal subjects after 10 mg drug over 2 h; higher response in patients with primary aldosteronism. *3883* Increased by single i.v. bolus of 10 mg: probably related to basal activity of renin angiotensin aldosterone system. *2543*

Levodopa *Serum Increase Physiological* Increased rate of absorption with faster gastric emptying. *3794*

Lithium *Serum Increase Physiological* Increased rate of absorption with faster gastric emptying. *3794*

Luteinizing Hormone (LH) *Plasma No Effect Physiological* No significant baseline differences in cycling women before and after drug. *0096*

Occult Blood *Feces No Effect Analytical* No effect on Hemoquant method. *3671*

Potassium *Serum No Effect Physiological* No significant change following i.v. injection. *0497*
Serum Decrease Physiological At dose of 10 mg causes decrease of about 0.3 mmol/L. *2836*

Progesterone *Plasma No Effect Physiological* No significant baseline differences in cycling women before and after drug. *0096*

Prolactin *Plasma Increase Physiological* Significant increase in some menstruating women compared with control cycle. *0096* Observed in normal subjects after 10 mg drug over 2 h; higher response in patients with primary aldosteronism. *3883* 11 fold increase after i.v. injection (0.04 mg/kg). *0497* Increased by single i.v. bolus of 10 mg: probably related to basal activity of renin angiotensin aldosterone system. *2543* Response smaller in thyrotoxic than euthyroid patients with long term treatment. *3520* In both hyperthyroid and euthyroid subjects in response to oral drug. *3163*

Renal Blood Flow (RBF) *Patient Decrease Physiological* In 20 patients given 1 to 2.5 mg/kg (from mean 443 mL/min to 387 mL/min). *1737*

Renin Activity *Plasma No Effect Physiological* No change observed in normal individuals or in patients with primary aldosteronism after 10 mg drug. *3883* No significant change following i.v. injection. *0497* No effect even though other constituents affected by i.v. bolus of 10 mg. *2543*

Thyroid Stimulating Hormone (TSH)
Serum No Effect Physiological Basal values did not change as result of long-term treatment. *3520* In either hyperthyroid or euthyroid subjects in response to oral drug. *3163*
Serum Increase Physiological Following 10 mg orally marked increase within 1 h in euthyroid subjects: effect most marked in patients with primary hypothyroidism. *3168* Increased in euthyroidism and hyperthyroidism; maximum effect 3 to 6 h after administration. *3798*

Metolazone

Ammonia *Urine No Effect Physiological* No effect on carbonic anhydrase. *3358*

Angiotensin II *Plasma Increase Physiological* 2 to 3 fold increase after 6 weeks treatment with 5 mg daily in 5 or 6 hypertensives. *3866*

Bicarbonate *Serum Increase Physiological* Diuretic action of drug acting on distal tubules. *2816* Increase by about 3-5 mmol/L after 6 weeks treatment with 5 mg daily in 5 or 6 hypertensives. *3866*
Urine No Effect Physiological No effect on carbonic anhydrase. *3358*

Calcium *Urine Increase Physiological* Maximum diuretic response of 0.4 µEq/min. *3358*

Chloride *Serum Decrease Physiological* Diuretic action of drug acting on distal tubules. *2816*
Urine Increase Physiological Diuretic action of drug. *1425*

Effective Renal Plasma Flow
Patient No Effect Physiological No effect observed. *3358*

Glomerular Filtration Rate (GFR)
Urine No Effect Physiological No effect observed. *3358*
Urine Decrease Physiological Diuretic action. *3494*

Glucose *Serum Increase Physiological* Diabetic-like action of diuretics. *2816*

Magnesium *Serum Decrease Physiological* In isolated case as result of marked diuresis. *3285*
Urine Increase Physiological Maximum diuretic response of 0.4 µEq/min. *3358*

Osmolality *Urine Increase Physiological* Diuretic action. *3494*

Osmolar Clearance *Urine Increase Physiological* Diuretic action of drug. *3494*

pH *Urine Decrease Physiological* Associated with diuretic response. *3358*

Phosphate *Urine Increase Physiological* Maximum diuretic response up to 8 µEq/min. *3358*

Potassium *Serum Decrease Physiological* Potassium loss in urine. *1425* Significant reduction after 6 weeks treatment of 5 hypertensives given 5 mg daily. *3866* Mean decrease of 0.5-0.6 mmol/L in group of patients as hypertension controlled. *2494*
Urine Increase Physiological Slight diuretic response. *3358*

Renin Activity *Plasma Increase Physiological* Approximately doubled after 6 weeks treatment with 5 mg daily in 5 or 6 hypertensives. *3866*

Sodium *Serum Decrease Physiological* Diuretic action of drug acting on distal tubules. *2816*
Urine Increase Physiological Diuretic action of drug. *1425*

Metolazone (continued)

Titratable Acidity *Urine Increase Physiological* Associated with diuretic response. *3358*

Urea Nitrogen *Serum Increase Physiological* Diuretic action of drug acting on distal tubules. *2816*

Uric Acid *Serum Increase Physiological* Diuretic action of drug acting on distal tubules. *2816* Increase by 3-5 mmol/L after 6 weeks treatment with 5 mg daily in 5 or 6 hypertensives. *3866*

Volume *Urine Increase Physiological* Diuretic action. *1425*

Water Clearance, Free *Urine Decrease Physiological* Diuretic action of drug. *3494*

Metoprolol

Albumin *Serum No Effect Analytical* At concentration of 0.34 mg/L had no effect on BCG method. *3393*

Aldosterone *Urine Decrease Physiological* In 15 patients with essential hypertension treated for 4 weeks, associated with reduction in sympathetic tone and reduced activity of renin-aldosterone system. *1209*

Apolipoprotein B *Serum No Effect Physiological* No significant effect in 15 hypertensive patients given 200 mg/d for 10 weeks. *1118*

Bicarbonate *Serum No Effect Analytical* At concentration of 0.34 mg/L had no effect on method using phenolphthalein. *3393*

Bilirubin *Serum No Effect Analytical* At concentration of 0.34 mg/L had no effect on Jendrassik and Grof method. *3393*

Calcium *Serum No Effect Analytical* At concentration of 0.34 mg/L had no effect on cresolphthalein method. *3393*

Chloride *Serum No Effect Analytical* At concentration of 0.34 mg/L had no effect on mercurimetric method. *3393*

Cholesterol *Serum No Effect Analytical* At concentration of 0.34 mg/L had no effect on CHOD-PAP method. *3393*
Serum No Effect Physiological In several studies for up to 3 mo. *0088* In 20 hypertensive diabetic patients. *3900* No effect in 1 or 3 mo study. *2116* In 20 hypertensives given 200 mg/d for 3 mo. *3071*

Cholesterol, High Density Lipoprotein

Serum No Effect Physiological Typically although slight reduction in 1 study. *0088*
Serum Decrease Physiological Mean decrease from 37 to 31 mg/dL in 15 hypertensives patients given 200 mg/d for 10 weeks. *1118* From 43 to 35 mg/dL in 18 hypertriglyceridemic hypertensives after 12 weeks. *0341* Increase by 6 to 13% in several studies of 2 to 4 mo with up to 400 mg drug/d. *1806* Increase by 13% after 1 mo in one study and 8% (nonsignificant) in another. *2116* Increase by 6 to 13% in several studies of 2 to 4 mo with up to 400 mg drug/d. *1806* Fell from 1.42 to 1.31 mmol/L in 20 hypertensives given 200 mg/d for 3 mo. *3071*

Cholesterol, Low Density Lipoprotein

Serum No Effect Physiological Although slight reduction in one study. *0088* No significant change after 12 weeks in 18 hypertriglyceridemic hypertensives. *0341* Insignificant change after 1 or 3 mo studies. *2116* In 20 hypertensives given 200 mg/d for 3 mo. *3071*
Serum Decrease Physiological Increase by 7 to 8.5% in several studies of 2 to 4 mo with up to 400 mg drug/d. *1806*

Cholesterol, Very Low Density Lipoprotein

Serum No Effect Physiological Nonstatistically significant increase from 61 to 67 mg/dL after 12 weeks in 18 hypertriglyceridemic hypertensives. *0341*
Serum Increase Physiological Increase by 30% in one study. *0088* Rose from 1.00 to 1.29 mmol/L in 20 hypertensives given 200 mg/d for 3 mo. *3071*
Serum Decrease Physiological Increase by 6 to 7% in several studies of 2 to 4 mo with up to 400 mg drug/d. *1806*

Corticosterone *Urine No Effect Physiological* In 15 patients with essential hypertension treated for 4 weeks, associated with reduction in sympathetic tone and reduced activity of renin-aldosterone system. *1209*

Cortisol *Urine No Effect Physiological* In 15 patients with essential hypertension treated for 4 weeks, associated with reduction in sympathetic tone and reduced activity of renin-aldosterone system. *1209*

Creatinine *Serum No Effect Analytical* At concentration of 0.34 mg/L had no effect on AutoAnalyzer Jaffé method. *3393*

Deoxycorticosterone *Urine No Effect Physiological* In 15 patients with essential hypertension treated for 4 weeks, associated with reduction in sympathetic tone and reduced activity of renin-aldosterone system. *1209*

Epinephrine *Plasma No Effect Physiological* No consistent effect but effect generally high when drug concentration high in 11 healthy young men studied under variety of conditions. *0908* No consistent response in individuals after 50 mg orally following different stresses. *0908*

Fatty Acids, Free (FFA) *Serum Decrease Physiological* Observed with pharmacological doses in nondiabetic hypertensives. *1410* Significant effect over 3 mo in 53 patients given 100 mg twice daily. *0835*

Glucagon *Plasma Decrease Physiological* Lower in nondiabetic hypertensives when receiving drug compared with placebo. *1410*

Glucose *Serum No Effect Analytical* At concentration of 3 mg/L had no effect on GOD/POD-PAP method. *3393*
Serum Increase Physiological Increase by 1.0 to 1.5 mmol/L in 20 hypertensive diabetic patients. *3900* Small but significant increase in nondiabetic hypertensives. *1410*

Glucose Tolerance *Serum Decrease Physiological* During early part of i.v. glucose tolerance test. *1410*

18-Hydroxycorticosterone *Urine No Effect Physiological* In 15 patients with essential hypertension treated for 4 weeks, associated with reduction in sympathetic tone and reduced activity of renin-aldosterone system. *1209*

18-Hydroxydeoxycorticosterone
Urine No Effect Physiological In 15 patients with essential hypertension treated for 4 weeks, associated with reduction in sympathetic tone and reduced activity of renin-aldosterone system. *1209*

Insulin *Plasma No Effect Physiological* In 20 hypertensive diabetic patients. *3900* Insignificant change although plasma concentration of glucose changed. *1410*

Kallikrein *Urine Decrease Physiological* In 15 patients with essential hypertension treated for 4 weeks, associated with reduction in sympathetic tone and reduced activity of renin-aldosterone system. *1209*

Lipoprotein Lipase *Serum No Effect Physiological* No significant effect in 15 hypertensive patients given 200 mg/d for 10 weeks. *1118*

Norepinephrine *Plasma Increase Physiological* Significant effect in 11 healthy young men studied under variety of conditions. *0908* General greater increase in subjects after 50 mg drug in response to most stresses. *0908*

Phosphate *Serum No Effect Analytical* At concentration of 0.34 mg/L had no effect on phosphomolybdate method. *3393*

Platelet Aggregation *Blood No Effect Physiological* No effect observed in response ADP compared with placebo in healthy volunteers. *3714*

Platelets *Blood Increase Physiological* When given twice per day for 1 week in healthy volunteers, also after single dose. *3714*

Potassium *Serum No Effect Analytical* At concentration of 0.34 mg/L had no effect on flame-photometric method. *3393*

Protein *Serum No Effect Analytical* At concentration of 0.34 mg/L had no effect on biuret method with blank correction. *3393*

Renin Activity *Plasma Decrease Physiological* In 15 patients with essential hypertension treated for 4 weeks, associated with reduction in sympathetic tone and reduced activity of renin-aldosterone system. *1209*

Sodium *Serum No Effect Analytical* At concentration of 0.34 mg/L had no effect on flame-photometric method. *3393*

Theophylline *Serum Increase Physiological* Clearance reduced by 30-50% in healthy individuals given 200 mg or more per day: possibly due to a metabolite binding to cytochrome P-450. *0712*

Triglycerides *Serum No Effect Analytical* At concentration of 3 mg/L had no effect on GPO-PAP method. *3393* At concentration of 0.34 mg/L had no effect on lipase/esterase method. *3393*
Serum No Effect Physiological Insignificant increase after 12 weeks treatment in hypertriglyceridemic hypertensives. *0341*

Serum Increase Physiological In 15 hypertensive patients given 200 mg/d for 10 weeks caused mean increase from 122 to 142 mg/dL. *1118* Up to 34% in some studies but no effect in almost as many. *0088* Insignificant in 20 hypertensive diabetic patients. *3900* Significant effect of 33% and 14% for 1 mo and 3 mo studies respectively. *2116* Rose from 2.51 to 3.41 mmol/L in 20 hypertensives given 200 mg/d for 3 mo. *3071*

Triglycerides, High Density Lipoprotein
Serum No Effect Physiological In 18 hypertriglyceridemic-hypertensives after 12 weeks. *0341* In 20 hypertensives given 200 mg/d for 3 mo. *3071*

Triglycerides, Low Density Lipoprotein
Serum No Effect Physiological Rose from 1 to 1.29 mmol/L in 20 hypertensives given 200 mg/d for 3 mo. *3071*
Serum Increase Physiological From 24 to 32 mg/dL after 12 wek in 18 hypertriglyceridemic hypertensives. *0341*

Triglycerides, Very Low Density Lipoprotein
Serum No Effect Physiological Non-statistically significant increase from 203 to 232 mg/dL after 12 weeks in 18 hypertriglyceridemic hypertensives. *0341* In 20 hypertensives given 200 mg/d for 3 mo. *3071*
Serum Increase Physiological 36% in one study. *0088*

Urea Nitrogen *Serum No Effect Analytical* At concentration of 0.34 mg/L had no effect on diacetylmonoxime method. *3393*
Serum Increase Physiological Insignificant in 20 hypertensive diabetic patients. *3900*

Uric Acid *Serum No Effect Analytical* At concentration of 0.34 mg/L had no effect on phosphotungstate reduction method. *3393* At concentration of 3 mg/L had no effect on uricase-PAP method. *3393*

Metrifonate

Alanine Aminotransferase *Serum No Effect Physiological* No effect reported on hepatic function. *2427*

Pseudocholinesterase *Serum Decrease Physiological* Powerful inhibitor of enzyme. *2427*

Urea Clearance *Urine No Effect Physiological* No effect reported on renal function. *2427*

Metrizamide

Bilirubin *Serum No Effect Physiological* Probably nonclinically significant protein displacement effect in neonates. *3748*

Metrizoate

Bilirubin *Serum Increase Physiological* Theoretically possible effect in neonates because of high circulating concentration. *3748*

γ-Glutamyltransferase (GGT)
Serum Increase Physiological Effect observed in subjects with liver tumors. *0064*

Metronidazole

Aspartate Aminotransferase *Serum No Effect Analytical* No effect noted on coupled NADH procedure on Technicon® SMA II if appropriate blanking used. *1628* On continuous flow procedure if blank correction incorporated. *1628*
Serum Decrease Analytical Artifactual depression of activity in almost all patients with NADH-coupled analytical methods. *3056* With continuous flow endpoint reaction method because of drug's high absorbance at 340 nm, although no effect on kinetic methods. *0881* On Kessler method with continuous flow AutoAnalyzers with apparent reduction of activity by as much as 50% at concentration of 8 mg/L. *3592*

Chloramphenicol *Serum No Effect Analytical* No effect at 100 mg/L on coupled enzymatic method. *2490*

Color *Urine Increase Analytical* Brown color probably due to metabolite. *1237*

Diphenylhydantoin *Serum Increase Physiological* Increased concentration reported in several patients. *2810*

Glucose *Serum No Effect Analytical* On glucose oxidase method on Beckman Astra at therapeutic concentration. *0937*

Serum Increase Analytical Increased values observed with Technicon® SMAC® and Du Pont aca hexokinase methods. Normal drug concentration increased glucose by about 60 mg/L. *0936* On hexokinase method on Technicon® SMAC® at therapeutic concentration. *0937*
Serum Decrease Analytical At concentrations above 100 mg/L (normal therapeutic concentration 47.5 mg/L) lowered concentration as measured by hexokinase/G-6-PDH method. *3393* At concentrations above 200 mg/L (normal therapeutic concentration 47.5 mg/L) lowered concentration as measured by aca method. *3393*

Indomethacin *Serum No Effect Analytical* No effect on HPLC method of Roberts and Smith. *3002*

131I Uptake *Serum No Effect Physiological* 1200 mg/d for 1 week no effect. *0583*

17-Ketosteroids *Urine Decrease Physiological* If previously elevated, ?depresses adrenal cortex. *1343*

Leukocytes *Blood Decrease Physiological* Leukopenia and reduction of polymorphs. *2313*

Neutrophils *Blood Decrease Physiological* Transient neutropenia may occur. *2427* Occasional case of agranulocytosis reported. *3717*

Partial Thromboplastin Time
Plasma Increase Physiological In one patient stabilized on warfarin to whose therapeutic regime metronidazole was added. *0850*

Protein Bound Iodine (PBI) *Serum No Effect Physiological* 1200 mg/d no effect over 1 week. *0583*

Prothrombin Time *Plasma Increase Physiological* Potentiates effect on warfarin. *1890* In one patient stabilized on warfarin to whose therapeutic regime metronidazole was added. *0850* Significant increase in concentration of warfarin and prolongation of prothrombin time. Effect noted with S (-) — warfarin but not R (+) — warfarin. *2645* Marked effect reported in a single patient previously stabilized on warfarin but to whose regime metronidazole was added. *1890*

Short-Chain Fatty Acids *Feces No Effect Physiological* No significant effect when given orally for 6 d. *1664*

T₃ Uptake *Serum No Effect Physiological* 1200 mg/d for 1 week producing no effect. *0583*

Warfarin *Plasma Increase Physiological* May potentiate effects by effect on metabolism. *3057*

Metyrapone

Aldosterone *Urine Decrease Physiological* May inhibit aldosterone production. *2775*

Androstenedione *Plasma Increase Physiological* Increase in 3-8 h in men, marked increase in women. *2600*

Corticotropin *Plasma Increase Physiological* Response to stress. *2067*

Cortisol *Plasma Decrease Physiological* Normal response to test. *2396*

11-Deoxycortisol *Plasma Increase Physiological* Normal response to test. *0512*

Growth Hormone *Plasma No Effect Physiological* No significant effect with 750 mg orally. *2176*

17-Hydroxycorticosteroids *Plasma Increase Physiological* Indirectly stimulates ACTH production. *2211*
Urine Increase Physiological Normal response to injection is 2-4 times increase. *2211*
Urine Decrease Physiological Direct effect on adrenal steroidogenesis. *0022*

17-Ketogenic Steroids *Urine Increase Physiological* Normal response to injection is doubling of output. *1680*
Urine Decrease Analytical Interferes with Zimmermann reaction. *3505*

17-Ketosteroids *Urine Decrease Analytical* Interferes with Zimmermann reaction. *3505*

Porphyrins *Urine Increase Physiological* Reported to precipitate attack of acute porphyria. *2220*

Testosterone *Serum Decrease Physiological* Fall in 2 h in men, minor increase in women. *2600*

Mexiletine

Alanine Aminotransferase *Serum Increase Physiological* Occurs in fewer than 1% treated patients. *2011*

Antinuclear Antibodies *Serum Positive Physiological* Occurs in fewer than 1% treated patients. *2011*

Aspartate Aminotransferase
Serum Increase Physiological Occurs in fewer than 1% treated patients. *2011*

Digoxin *Serum No Effect Physiological* No significant effect when drug coadministered. *2101*

Leukocytes *Blood Decrease Physiological* Occurs in fewer than 1% treated patients. *2011*

Neutrophils *Blood Decrease Physiological* Occurs in fewer than 1% treated patients. *2011*

Platelets *Blood Decrease Physiological* Occurs in fewer than 1% treated patients. *2011*

Mezlocillin

Glucose *Urine No Effect Analytical* Concentrations measured accurately by Diastix®. Concentration measured accurately by TesTape®. *2100*
Urine Increase Analytical Falsely elevated values with Clinitest®. *2100*

Vancomycin *Serum No Effect Analytical* Negligible interference at up to 500 mg/L on Abbott TD$_x$. *2417*

Mianserin

Leukocytes *Blood Decrease Physiological* Decreased concentration reported in 4 individuals who were also receiving other drugs. *0028*

Neutrophils *Blood Decrease Physiological* Decreased concentration reported in 4 individuals who were also receiving other drugs. *0028*

Miconazole

Creatine Kinase *Serum Increase Physiological* Observed as isolated finding in patients given drug i.v. *3462*

Diphenylhydantoin *Serum Increase Physiological* Isolated case reported with phenytoin concentration markedly increased after miconazole given i.v., probably due to inhibition of hepatic cytochrome P-450. *3032* Inhibits hepatic metabolism thereby increasing concentration. *3032*

Hematocrit *Blood Decrease Physiological* By more than 4% in 44% treated patients. *3462*

Leukocytes *Blood Decrease Physiological* Observed as isolated finding in patients given drug i.v. *3462*

Platelets *Blood Increase Physiological* Occurred in 31% of courses followed. *3462*

Prothrombin Time *Plasma Increase Physiological* In patients receiving warfarin, probably due to displacement from plasma protein. *3770* In patients receiving warfarin whose hepatic metabolism is inhibited. *3461*

Sodium *Serum Decrease Physiological* Occurred in 46% of courses followed, with average decrease of 10 mmol/L. *3462*

Midazolam

pCO$_2$ *Blood Increase Physiological* In healthy volunteers but of short duration. *1041*

pO$_2$ *Blood Decrease Physiological* In healthy volunteers but of short duration. *1041*

Mineral Oil

Carotene *Serum Decrease Physiological* Reduced absorption (may be 50% normal). *1343* In people on diet high in carotene when drug given before meals. *3023*

Prothrombin Time *Plasma Increase Physiological* Inhibits absorption of vitamin K. *2313*

Minocycline

Alkaline Phosphatase *Serum Increase Physiological* Conceivable complication of therapy. *0135*

Aspartate Aminotransferase
Serum Increase Physiological Conceivable complication of therapy. *0135*

Chloramphenicol *Serum No Effect Analytical* No effect at 100 mg/L on coupled enzymatic method. *2490*

Eosinophils *Blood Increase Physiological* Allergic response. *1705*

Hematocrit *Blood Decrease Physiological* May cause hemolytic anemia (rare). *1680*

Hemoglobin *Blood Decrease Physiological* Reported cases of hemolytic anemia. *1680*

LE Cells *Blood Positive Physiological* May cause exacerbation of SLE. *1680*

Leukocytes *Blood Decrease Physiological* Reported as side effects. *1680*

Neutrophils *Blood Decrease Physiological* May cause hypersensitivity reaction. *1705*

pH *Blood Decrease Physiological* May occur if impaired renal function. *1680*

Phosphate *Serum Increase Physiological* If impaired renal function. *1680*

Platelets *Blood Decrease Physiological* Reported as side effects. *1680*

Potassium *Serum No Effect Analytical* At concentration of 6 mg/L had no effect on measurement by ISE with predilution. *3393*

Protein Bound Iodine (PBI) *Serum No Effect Physiological* Although thyroid tissue may be stained. *1680*

Prothrombin Time *Plasma Increase Physiological* May prolong action of anticoagulants. *1705*

Sodium *Serum No Effect Analytical* At concentration of 6 mg/L had no effect on measurement by ISE with predilution. *3393*

Urea Nitrogen *Serum Increase Physiological* Antianabolic action of class of drugs. *1680*

Minoxidil

Cholesterol, High Density Lipoprotein
Serum Increase Physiological Slight effect (order of 10%) after 3 or 6 mo treatment. *1807*

Cholesterol, Low Density Lipoprotein
Serum Decrease Physiological Approximately 10% reduction after 3 mo and 20% after 6 mo. *1807*

Coombs' Test, Direct *Serum Positive Physiological* Occurred without hemolysis in one patient. *2174*

Renin Activity *Plasma No Effect Physiological* No significant effect when given alone or with added diuretics at 3 or 6 mo. *1807*

Mithramycin

Alanine Aminotransferase *Serum Increase Physiological* Hepatocellular damage observed. *0071* Consistent moderate increase in most cases. *2406*

Alkaline Phosphatase *Serum Increase Physiological* Noted in approximately 50% of treated patients. *2406*

Aspartate Aminotransferase
Serum Increase Physiological Hepatocellular damage observed. *0071* Consistent moderate increase in most cases. *2406*

Bleeding Time *Blood Increase Physiological* Reversible effect. *0071*

Calcium *Serum Decrease Physiological* Inhibition of bone resorption of calcium. *3349*
Urine Increase Physiological Inhibition of bone resorption of calcium. *3349*

Clot Retraction *Blood Decrease Physiological* Reversible poor retraction. *0071*

Clotting Time *Blood Increase Physiological* Reversible effect. *0071*

Creatinine *Serum Increase Physiological* Nephrotoxic effect. *0071*

Hemoglobin *Blood Decrease Physiological* Reversible effect. *0071*

Hydroxyproline *Urine Decrease Physiological* Inhibition of bone resorption of calcium. *3349*

Lactate Dehydrogenase *Serum Increase Physiological* Hepatocellular damage observed. *0071*

Leukocytes *Blood Decrease Physiological* Reversible effect. *0071*

Phosphate *Serum Decrease Physiological* Inhibition of bone resorption of calcium. *3349*

Platelets *Blood Decrease Physiological* Thrombocytopenia. *0071*

Potassium *Serum Decrease Physiological* Depression of level reported. *0071*

Protein *Urine Increase Physiological* Nephrotoxic effect. *0071*

Prothrombin Time *Plasma Increase Physiological* Reversible decreased prothrombin level. *0071*

Urea Nitrogen *Serum Increase Physiological* Nephrotoxic effect. *0071*

Mitomycin

Alanine Aminotransferase *Serum Increase Physiological* Centrilobular stasis in 5 of 6 patients and toxic hepatitis in the other. *2406*

Aspartate Aminotransferase
Serum Increase Physiological Centrilobular stasis in 5 of 6 patients and toxic hepatitis in the other. *2406*

Chromosomes *Test Conditions Abnormal Physiological* Clastogenic in human cells. *3282*

Creatinine *Serum Increase Physiological* Nephrotoxic effect. *2194*

Creatinine Clearance *Urine Decrease Physiological* Due to nephrotoxicity. *2194*

Erythrocytes *Blood Decrease Physiological* Pancytopenia. *2194*

Leukocytes *Blood Decrease Physiological* Pancytopenia. *2194*

Platelets *Blood Decrease Physiological* Pancytopenia. *2194*

Protein *Urine Increase Physiological* Nephrotoxic effect. *2194*

Urea Nitrogen *Serum Increase Physiological* Nephrotoxic effect. *2194*

Uric Acid *Serum Increase Physiological* Due to nephrotoxicity. *2194*

Mitotane

Alkaline Phosphatase *Serum Increase Physiological* Possible hepatotoxic effect (reported in dogs,rats). *1680*

Erythrocytes *Urine Increase Physiological* May occasionally produce hematuria. *1680*

Hemoglobin *Urine Increase Physiological* May occasionally produce hematuria. *1680*

17-Hydroxycorticosteroids *Urine Decrease Physiological* Stimulates extra-adrenal hydroxylation of cortisol. *0710*

6-β-Hydroxycortisol *Urine Increase Physiological* Altered cortisol metabolism induced by drug. *2492*

Protein *Urine Increase Physiological* May rarely produce hematuria, renal damage. *1680*

Protein Bound Iodine (PBI) *Serum Decrease Physiological* Rare reported side effect. *1680*

T$_3$ Uptake *Serum Increase Physiological* Competes for sites on thyroxine binding globulin. *0583*

Thyroxine (T$_4$) *Serum Decrease Physiological* Competes with T$_4$ for thyroxine binding globulin. *0583*

Thyroxine (T$_4$) (Murphy-Pattee)
Serum Decrease Physiological Competes with T$_4$ for thyroxine binding globulin. *0583*

Mitoxantrone

Alanine Aminotransferase *Serum Increase Physiological* Mild transient effects in 11 of 26 patients. *2703*

Aspartate Aminotransferase
Serum Increase Physiological Mild transient effects in 11 of 26 patients. *2703*

Bilirubin *Serum Increase Physiological* Mild transient effects in 11 of 26 patients. *2703*

Erythrocytes *Blood Decrease Physiological* Erythrocytes not acutely affected but mild anemia in most patients with successive courses. *3289*

Leukocytes *Blood Decrease Physiological* Granulocytopenia is dose-limiting toxicity. In phase II studies nadir between 8 and 15 d. Less than 5% patients had nadir below 1,000/uL. Fall with each treatment until 5th or 6th course. *3289*

Platelets *Blood Decrease Physiological* Less than 10% patients develop count of less than 100,000/uL. Count usually falls to 5th to 6th course. *3289*

MK-270

Glucose *Serum Decrease Physiological* Slight nonsignificant decrease in fasting value. *3099*

Glucose Tolerance *Serum Increase Physiological* Produces flattening of curve. *3099*

Molindone

Alanine Aminotransferase *Serum Increase Physiological* Single case report (?cause). *1905*

Fatty Acids, Free (FFA) *Serum Increase Physiological* Sustained and significant rise. *1635*

Glucose *Serum No Effect Physiological* Relatively unaffected over short term. *1635*

Prolactin *Plasma Increase Physiological* Significant response to 5 mg given orally. *1416*

Vanillylmandelic Acid *Urine No Effect Physiological* Relatively unaffected over short term. *1635*

Monofluorosulphonate

Protein Bound Iodine (PBI) *Serum Decrease Physiological* Interferes with trapping of iodide by thyroid. *0012*

Monoiodotyrosine

Thyroid Stimulating Hormone (TSH)
Serum Increase Physiological Marked increase within 1 h after 10 mg orally in euthyroid subjects acting through inhibition of tyrosine hydroxylase: effect more marked in hypothyroid subjects. *3168*

Thyroxine (T$_4$) Binding Globulin
Serum No Effect Analytical On radioimmunoassay procedure of Van Herle. *3676*

MOPP

MCV *Blood Increase Physiological* Reflection of bone marrow reaction to cytotoxic therapy in people with malignant disease. *0840*

Morinamide

Bilirubin *Serum Increase Physiological* Jaundice reported as side effect. *2427*

Hydrochloric Acid *Gastric Material Increase Physiological* Hyperacidity reported. *2427*

Uric Acid *Serum Increase Physiological* Decreased renal clearance. *2427*

Morphine

Alanine Aminotransferase *Serum No Effect Analytical* At acute overdose concentration (20 mg/dL) on Technicon® SMAC® method. *3719*
Serum Increase Physiological May cause rise in intrabiliary pressure. *3505*

Albumin *Serum No Effect Analytical* At acute overdose concentration (20 mg/dL) on Technicon® SMAC® method. *3719* At concentration of 200 mg/L had no effect on BCG method. *3393*

Alkaline Phosphatase *Serum No Effect Analytical* At acute overdose concentration (20 mg/dL) on Technicon® SMAC® method. *3719*
Serum Increase Physiological Associated with abnormal liver function. *2313*

Alkaloids *Urine No Effect Analytical* With ninhydrin on Frings TLC procedure at 10 mg/dL. *1204*
Urine Increase Analytical Blue with iodoplatinate in Frings TLC procedure. *1204*

Ammonia *Plasma Increase Physiological* Impairs ability of liver to metabolize NH_3 in dogs. *3601*

Amphetamine *Urine No Effect Analytical* No reaction with NBD chloride procedure of Monforte. *2473*

Amylase *Serum Increase Physiological* Causes spasm of sphincter of Oddi for 48 h. *3505*
Urine Increase Physiological Observed in some patients after administration. *1397*

Aspartate Aminotransferase *Serum No Effect Analytical* At acute overdose concentration (20 mg/dL) on Technicon® SMAC® method. *3719*
Serum Increase Physiological May cause rise in intrabiliary pressure. *3505*

Barbiturate *Serum No Effect Analytical* No interference with UV absorption methods. *2981*
Urine No Effect Analytical With diphenylcarbazone on Frings TLC procedure at 10 mg/dL. With mercuric SO_4 on Frings TLC procedure at 10 mg/dL. With vanillin on Frings procedure at 10 mg/dL. With mercurous NO_3 on Frings TLC procedure at 10 mg/dL. *1204*

Basal Metabolic Rate *Patient Decrease Physiological* Metabolic effect of drug. *2220*

Bicarbonate *Serum No Effect Analytical* At concentration of 1 mg/L had no effect on method using phenolphthalein. *3393*

Bilirubin *Serum No Effect Analytical* At acute overdose concentration (20 mg/dL) on Technicon® SMAC® method. *3719* At concentration of 200 mg/L had no effect on Jendrassik and Grof method. *3393*
Serum Increase Physiological Associated with abnormal liver function. *2313*

Bromide *Urine No Effect Analytical* No effect on method of Frings. *1204*

BSP Retention *Serum Increase Physiological* Abnormal liver function tests reported. *3879*

Calcium *Serum No Effect Analytical* At acute overdose concentration (20 mg/dL) on Technicon® SMAC® method. *3719* At concentration of 200 mg/L had no effect on cresolphthalein method. *3393*

Chlordiazepoxide *Blood No Effect Analytical* On method of Riddick at 5 mg/dL. *2983*

Chloride *Serum No Effect Analytical* At concentration of 1 mg/L had no effect on mercurimetric method. *3393*

Cholesterol *Serum No Effect Analytical* At acute overdose concentration (20 mg/dL) on Technicon® SMAC® method. *3719* At concentration of 200 mg/L had no effect on Liebermann-Burchard method. *3393*

Chromosomes *Test Conditions No Effect Physiological* No effect human leucocytes at concentrations 1/6-3x normal. *1073*

Cortisol *Plasma Decrease Physiological* In 14 volunteer subjects given 5 mg intravenously. *3953*

Creatine Kinase *Serum No Effect Analytical* At acute overdose concentration (20 mg/dL) on Technicon® SMAC® method. *3719*
Serum No Effect Physiological No change, although other enzymes increased. *2427*
Serum Increase Physiological Response to frequent i.m. injections. *0249*

Creatinine *Serum No Effect Analytical* At acute overdose concentration (20 mg/dL) on Technicon® SMAC® method. *3719* At concentration of 200 mg/L had no effect on AutoAnalyzer Jaffé method. *3393*

Enteroglucagon *Plasma Decrease Physiological* Postprandial secretion abolished in 6 volunteers after drug given i.v. *0626*

Epinephrine *Plasma Increase Physiological* Mechanism obscure also involved in glucose release. *1744*

Ethchlorvynol *Serum No Effect Analytical* At 8 mg/L on GLC procedure of Evenson. *1058*
Urine No Effect Analytical No effect on method Frings and Cohen at 2 mg/dL. *1205*

FPN Test *Urine No Effect Analytical* No effect at 10 mg/dL on method of Frings. *1204*

Gastric Inhibitory Polypeptide
Plasma Decrease Physiological Reduction in secretion following test meal and drug in 6 healthy volunteers. *0626*

Gastrin *Serum Increase Physiological* Secretion prolonged following test meal and drug i.v. in 6 healthy volunteers. *0626*

Glucagon, Pancreatic *Plasma No Effect Physiological* No effect observed in 6 healthy volunteers after test meal and drug i.v. *0626*

Glucose *Serum No Effect Analytical* At acute overdose concentration (20 mg/dL) on Technicon® SMAC® method. *3719*
Serum Increase Physiological Minor, clinically insignificant increase. *1343*

Guaiacols Spot Test *Urine Positive Analytical* False reaction with screening test of Rogers. *3031*

Histamine *Plasma Increase Physiological* Observed with injection associated with anesthesia. *2212*

Hydrochloric Acid *Gastric Material Decrease Physiological* Slight decrease in secretion of acid. *1343*

Hydroxybutyrate Dehydrogenase
Serum Increase Physiological Probably due to spasm of sphincter of Oddi. *2427*

17-Hydroxycorticosteroids *Plasma Decrease Physiological* Inhibits ACTH and pituitary gonadotropin release. *1343*
Urine Decrease Physiological Inhibits ACTH and pituitary gonadotropin release. *1343*

Indocyanine Green *Serum Decrease Physiological* Observed in small series, normal liver function test. *2398*

Insulin *Plasma Decrease Physiological* Reduction in secretion following test meal and drug in 6 healthy volunteers. *0626*

Iron *Serum No Effect Analytical* At acute overdose concentration (20 mg/dL) on Technicon® SMAC® method. *3719* At concentration of 200 mg/L had no effect on Ferrozine method. *3393*

17-Ketosteroids *Plasma Decrease Physiological* Inhibits ACTH and pituitary gonadotropin release. *1343*
Urine Increase Analytical Due to chemical structure affects Zimmermann procedure. *3217*
Urine Decrease Physiological Inhibits ACTH and pituitary gonadotropin release. *1343*

Lactate *Serum Decrease Physiological* Intravenous administration of 0.33 mg/kg caused 50% decrease. *1487*

Lactate Dehydrogenase *Serum No Effect Analytical* At acute overdose concentration (20 mg/dL) on Technicon® SMAC® method. *3719*
Serum Increase Physiological May cause rise in intrabiliary pressure. *3505*

Lactate Dehydrogenase Isoenzymes
Serum Increase Physiological Hepatic fraction increase ?due to spasm of sphincter. *2427*

Leucine Aminopeptidase *Serum Increase Physiological* Possibly due to spasm of sphincter of Oddi. *2427*

Lipase *Serum Increase Physiological* Causes spasm of sphincter of Oddi. *3505*

Methaqualone *Serum No Effect Analytical* At 20 mg/L on GLC procedure of Evenson. *1057*

Morphine *Serum Increase Physiological* After 10 mg i.v. concentration is 0.1 mg/L in 1 h. *2348*

Motilin *Plasma Decrease Physiological* Postprandial secretion abolished in 6 volunteers after drug given i.v. *0626*

Neurotensin *Plasma Decrease Physiological* Reduction in secretion following test meal and drug in 6 healthy volunteers. *0626*

Norepinephrine *Plasma Decrease Physiological* Mechanism obscure. *1744*

Pancreatic Polypeptide *Plasma Decrease Physiological* Postprandial secretion abolished in 6 volunteers after drug given i.v. *0626*

pCO₂ *Blood Increase Physiological* Diminishes ventilation, causes hypercapnia. *0071*

Phosphate *Serum No Effect Analytical* At acute overdose concentration (20 mg/dL) on Technicon® SMAC® method. *3719* At concentration of 200 mg/L had no effect on phosphomolybdate method. *3393*

Prolactin *Plasma Increase Physiological* In 14 volunteer subjects given 5 mg intravenously. *3953* 10 mg i.v. produced prompt and significant increase in 7 hypothyroid and 5 healthy individuals. *0891*

Propoxyphene *Urine No Effect Analytical* Less than 1% fluorescence in procedure of Valentour. *3662*

Protein *Serum No Effect Analytical* At acute overdose concentration (20 mg/dL) on Technicon® SMAC® method. *3719* At concentration of 200 mg/L had no effect on biuret method with blank correction. *3393*
Test Conditions Increase Analytical Reacts with Folin-Ciocalteu of Lowry method. *0702*

Salicylate *Urine No Effect Analytical* At 10 mg/dL no effect on Trinder method. *1204*

Somatostatin *Plasma No Effect Physiological* No effect observed in 6 healthy volunteers after test meal and drug i.v. *0626*

Sugar *Urine Increase Analytical* Interferes with copper reduction method. *2425*

Thyroid Stimulating Hormone (TSH)
Serum Increase Physiological 10 mg i.v. produced prompt and significant increase in 7 hypothyroid and 5 healthy individuals. *0891*

Triglycerides *Serum No Effect Analytical* At acute overdose concentration (20 mg/dL) on Technicon® SMAC® method. *3719* At concentration of 200 mg/L had no effect on lipase/esterase method. *3393*

Urea Nitrogen *Serum No Effect Analytical* At acute overdose concentration (20 mg/dL) on Technicon® SMAC® method. *3719* At concentration of 200 mg/L had no effect on diacetylmonoxime method. *3393*

Uric Acid *Serum No Effect Analytical* At acute overdose concentration (20 mg/dL) on Technicon® SMAC® method. *3719* At concentration of 200 mg/L had no effect on phosphotungstate reduction method. *3393*

Vanillylmandelic Acid *Urine Decrease Physiological* Clinically significant but small effect. *2356*

Vasoactive Intestinal Polypeptide
Plasma No Effect Physiological No effect observed in 6 healthy volunteers after test meal and drug i.v. *0626*

Volume *Urine Decrease Physiological* May stimulate release of ADH. *0071*

Morpholine

Ionized Calcium *Serum Decrease Analytical* At concentrations > 0.1 mmol/L on Calcium specific electrode. *0540*

Mouth Washes

Protein Bound Iodine (PBI) *Serum Increase Analytical* Some may contain iodine. *1237*

Moxalactam

Alanine Aminotransferase *Serum Increase Physiological* In about 3% patients as reported from several studies. *0592*

Albumin *Serum No Effect Analytical* At concentration of 96 mg/L had no effect on BCG method. *3393*

Alkaline Phosphatase *Serum Increase Physiological* In about 3% patients as reported from several studies. *0592*

Aspartate Aminotransferase
Serum Increase Physiological In about 3% patients as reported from several studies. *0592*

Bilirubin *Serum No Effect Analytical* At 2.50 mmol/L on method on Eppendorf Epos. *1414* At concentration of 96 mg/L had no effect on Jendrassik and Grof method. *3393*
Serum Increase Physiological Clinically significant displacement from protein in neonates. *3748*

Bleeding Time *Patient Increase Physiological* Reversible bleeding diathesis observed in 5 patients. *3792*

Cholesterol *Serum No Effect Analytical* At 2.50 mmol/L on method on Eppendorf Epos. *1414*

Coombs' Test *Serum Increase Physiological* Observed in some cases with all cephalosporins. *2615*

Coombs' Test, Indirect *Serum Positive Physiological* In up to 0.5% patients as reported from several studies. *0592*

Creatinine *Serum No Effect Analytical* No effect on Jaffé methods. *2024* At up to 1,000 µg/mL on Technicon® SMAC® Jaffé procedure. *3405* At concentration of 400 mg/L had no effect on AutoAnalyzer Jaffé method. *3393* At concentration of 400 mg/L had no effect on kinetic Jaffé method on aca. *3393*
Serum Increase Analytical Slow and slight reaction in Jaffé methods. *1414*
Serum Increase Physiological In about 2% patients as reported from several studies. *0592*

Eosinophils *Blood Increase Physiological* In about 2.5% patients as reported from several studies. *0592*

Erythrocytes *Urine Increase Physiological* In about 2% patients as reported from several studies. *0592*

Factor II *Plasma Decrease Physiological* Depression in 40 preoperative surgical patients: effects mild. *3121*

Factor VII *Plasma Decrease Physiological* Depression in 40 preoperative surgical patients: effects mild. *3121*

Glucose *Urine No Effect Analytical* No effect on copper reduction procedures such as Clinitest®. *2615*

Leukocytes *Blood Increase Physiological* In up to 0.5% patients as reported from several studies. *0592*
Blood Decrease Physiological Reported in fewer than 0.5% treated patients. *2615* In up to 0.5% patients as reported from several studies. *0592*
Urine Increase Physiological In about 2% patients as reported from several studies. *0592*

Neutrophils *Blood Decrease Physiological* Counts below 500/uL reported in some cases. *2615*

Partial Thromboplastin Time
Plasma No Effect Physiological Insignificant change noted in preoperative patients. *3121*

Platelet Aggregation *Blood Decrease Physiological* Reduced in response to ADP in some patients. *2615* Reduced response to ADP demonstrated *in vitro*. *3792*

Platelets *Blood No Effect Physiological* Insignificant changes in preoperative surgical patients. *3121*
Blood Increase Physiological In up to 0.5% patients as reported from several studies. *0592*
Blood Decrease Physiological In up to 0.5% patients as reported from several studies. *0592*

Potassium *Serum Decrease Physiological* Common when coadministered with amikacin. *0592*

Protein *Serum No Effect Analytical* At 2.50 mmol/L on method on Eppendorf Epos. *1414* At concentration of 96 mg/L had no effect on biuret method with blank correction. *3393*
Urine No Effect Analytical No effect of 1 g/L on sulfosalicylic acid and Albustix methods. *2388*
Urine Increase Physiological In about 2% patients as reported from several studies. *0592*

Prothrombin Time *Plasma Increase Physiological* Persistent mild increase of 0.7s in 40 preoperative patients. *3121* Observed in some patients especially malnourished. *2615* Bleeding tendency may occur especially in debilitated patients, also impaired platelet function may occur. *3158* In up to 0.5% patients as reported from several studies. *0592*
Blood No Effect Physiological 4 g intravenously 12 hourly had no effect in 6 subjects. *2629*

Urea Nitrogen *Serum No Effect Analytical* At 2.50 mmol/L on method on Eppendorf Epos. *1414* At concentration of 96 mg/L had no effect on diacetylmonoxime method. *3393*

Moxalactam *(continued)*

Urea Nitrogen *(continued)*
Serum Increase Physiological In about 2% patients as reported from several studies. *0592*

Uric Acid *Serum No Effect Analytical* At 2.50 mmol/L on method on Eppendorf Epos. *1414* At concentration of 96 mg/L had no effect on phosphotungstate reduction method. *3393*

MPCA

Acetoacetate *Serum Decrease Physiological* Anti-lipolytic effect of drug. *1531*

Cholesterol *Serum No Effect Physiological* No significant change after 1 mo treatment (270 mg tid). *1423*

Fatty Acids, Free (FFA) *Serum Decrease Physiological* Anti-lipolytic effect of drug (acts for 4 h). *1531*

β-Hydroxybutyrate *Serum Decrease Physiological* Anti-lipolytic effect of drug. *1531*

Triglycerides *Serum No Effect Physiological* No significant change when given in fed state. *1423*

Mumps Vaccine

Platelets *Blood Decrease Physiological* May cause thrombocytopenic purpura. *1680*

Muscle Massage

Alanine Aminotransferase *Serum Increase Physiological* Significant effect, but not to pathological level 8 h after. *0423*

Creatine Kinase *Serum Increase Physiological* Significant effect but not to pathological level 1 h after. *0423*

Lactate Dehydrogenase *Serum Increase Physiological* Significant effect but not to pathological level 8 h after. *0423*

Myokinase *Serum Increase Physiological* Significant effect 8 h after (may exceed normal limits). *0423*

Muscular Exercise

α$_1$-Acid Glycoprotein *Serum No Effect Physiological* No effect of exercise observed. *2846*
Serum Increase Physiological Occurs within 15 minutes, persists for 1 d. *1491*

Adenylate Kinase *Serum Increase Physiological* Observed after protracted exercise. *0932*

AHF-Like Antigen *Plasma Increase Physiological* Proportional increase in response to extent. *0292*

AHF Procoagulant Activity *Blood Increase Physiological* Proportional increase in response to extent and antigen. *0292*

Alanine Aminotransferase *Serum No Effect Physiological* Insignificant effect of 12 minutes on cycle-ergometer. *1231*
Serum Decrease Physiological Effect of physical training. *2981*

Albumin *Serum Increase Physiological* Significant effect with 12 minutes on cycle-ergometer. *1231* Approximately 10% increase immediately after, delayed fall. *2846*

Aldolase *Serum No Effect Physiological* Physical activity has no effect. *2981*
Serum Increase Physiological Effect of physical training. *2981*

Alkaline Phosphatase *Serum No Effect Physiological* Insignificant effect with 12 minutes on cycle-ergometer. *1231*

Aminoacid Arylpeptidase *Serum Increase Physiological* Increased by exertion. *1491*

Ammonia *Plasma Increase Physiological* Tissue catabolism. *1237*

Amylase *Serum No Effect Physiological* No effect even with strenuous exercise. *2845*

α$_1$-Antitrypsin *Serum No Effect Physiological* No effect of exercise observed. *2846*
Serum Increase Physiological Significant effect at 15 minutes, partial return by 1 d. *1491*

Aspartate Aminotransferase
Serum No Effect Physiological Insignificant effect of 12 minutes on cycle-ergometer. *1231* No effect after 2 h march. *3217*
Serum Increase Physiological Marked after exercise (less in trained individuals). *2981*

Bicarbonate *Serum Decrease Physiological* Vigorous exercise for 30 minutes depresses by 3 meq/L. *1086*

BSP Retention *Serum Increase Physiological* During infusion (related to extent of activity). *1171*

Calcium *Serum Increase Physiological* Observed effect with 12 minutes cycle-ergometer. *1231*

Casts *Urine Increase Physiological* Hyaline, granular both increased with increased exercise. *1842*

Catecholamines *Urine No Effect Physiological* After mild or moderate exercise. *3217*
Urine Increase Physiological Following vigorous exercise (may be increased 7 fold). *0354*

Cells *Urine Increase Physiological* Epithelial cells increase with heavy exercise. *1842*

Ceruloplasmin *Serum Increase Physiological* Occurs within 15 minutes, persists for 1 d. *1491*

Chloride *Serum Increase Physiological* Vigorous exercise for 30 minutes increases by 2 meq/L. *1086*

Cholesterol *Serum No Effect Physiological* Observed effect with 12 minutes cycle-ergometer. *1231* But relative reduction of unsaturated acid. *1702*
Serum Increase Physiological Occasional response to exercise. *0582*

Cortisol *Plasma No Effect Physiological* Usual response in most subjects. *2105*
Plasma Decrease Physiological Slight effect (not significant). *1171*

Creatine Kinase *Serum Increase Physiological* Increase by 7% after 12 minutes on cycle-ergometer. *3319* In women increase of 4 U/L if within past 24 h. *2744* Maximum effect observed following day. *1491* Effect of physical training, increased with exercise. *2981*

Creatinine *Serum No Effect Physiological* Observed effect with 12 minutes cycle-ergometer. *1231*

Creatinine Clearance *Urine Increase Physiological* 40% decrease with severe exercise Mild, walking at 5.6 km/h produces 20% increase. *3217*
Urine Decrease Physiological Decrease with heavy exercise. *1842*

Cyclic Adenosine Monophosphate
Urine Increase Physiological Modest increase in normal people. *0747*

Dopamine Hydroxylase *Serum No Effect Physiological* No effect observed. *3807*

Epinephrine *Plasma Increase Physiological* Significant effect after physical stress. *1176*
Urine Increase Physiological If strenuous may be increased tenfold. *0279*

Erythrocytes *Urine Increase Physiological* Increasing with increasing rates of exercise. *1842*

Euglobulin Clot Lysis Time *Blood Decrease Physiological* Observed effect in exercise. *2847*

Factor II *Plasma Increase Physiological* Observed effect in exercise. *2847*

Factor V *Plasma No Effect Physiological* Observed effect in exercise. *2847*

Factor VIII *Plasma Increase Physiological* Biological activity augmented. *0883* Observed effect in exercise. *2847*

Factor VIII Antigen *Plasma Increase Physiological* Antigen increased following exercise. *0883*

Fatty Acids, Free (FFA) *Serum Increase Physiological* Marked increase after strenuous exercise. *1702*
Serum Decrease Physiological Approximately 15% decrease with bicycle pedalling. *1385*

Fibrinogen *Plasma Decrease Physiological* Observed effect in exercise. *2847*

Fibrinolysin *Plasma Increase Physiological* Activity increased by exercise. *0987*

Fibrinolysis *Plasma Increase Physiological* Observed effect in exercise. *2847*

β₁-α-Globulin *Serum Increase Physiological* Approximately 14% increase immediately after. *2846*

Glomerular Filtration Rate (GFR)
Urine No Effect Physiological If exercise moderate. *1842*
Urine Increase Physiological If exercise mild. *1842*
Urine Decrease Physiological Observed with heavy treadmill exercise. *1842*

Glucagon *Plasma Increase Physiological* May increase 20%, facilitates hepatic glycogenolysis. *1107*

Glucose *Serum No Effect Physiological* Observed effect with 12 minutes cycle-ergometer. *1231*
Serum Increase Physiological Rise due to adrenal activity. *0987*
Serum Decrease Physiological Occurs with strenuous exercise. *0117*

Glutamate Dehydrogenase *Serum No Effect Analytical* Physical activity has no effect. *2981*
Serum Decrease Analytical Observed effect with 12 minutes cycle-ergometer. *1231*

Glycerol *Serum Increase Physiological* Approximately 10-30% increase with bicycle pedalling. *1385*

Growth Hormone *Plasma Increase Physiological* Effect more marked in untrained than trained. *1811*

Haptoglobin *Serum Increase Physiological* Approximately 17% increase immediately after. *2846*
Serum Decrease Physiological Mean effect of 18 mg/dL from pre-exercise. *3217*

Hematocrit *Blood Increase Physiological* 6% increase immediately after exercise normal in 30 minutes. *2846*

Hemoglobin *Plasma No Effect Physiological* No effect with normal activity. *1755*
Plasma Increase Physiological Light activity causes increase x 3-5, heavy increase x 10-30. *2981*
Blood Increase Physiological Mild exercise causes transient decrease in blood volume. *1563*
Urine Increase Physiological May occur after severe exercise. *0987*

Hippuran Retention *Serum Increase Physiological* During infusion (depends on extent of activity). *1171*

α₂-HS Glycoprotein *Serum Increase Physiological* Significant effect at 15 minutes, partial return by 1 d. *1491*

17-Hydroxycorticosteroids *Urine Increase Physiological* Response to stress of exercise. *0582*

Immunoglobulin IgA *Serum No Effect Physiological* No observed effect 15 minutes or 1 d after. *1491*
Serum Increase Physiological Approximately 14% increase immediately after. *2846*

Immunoglobulin IgG *Serum No Effect Physiological* No observed effect 15 minutes or 1 d after. *1491*
Serum Increase Physiological Approximately 10% increase immediately after. *2846*

Immunoglobulin IgM *Serum No Effect Physiological* No observed effect 15 minutes or 1 d after. *1491* No effect of exercise observed. *2846*

Insulin *Plasma No Effect Physiological* No effect if of short or moderate duration. *1171*
Plasma Decrease Physiological If exercise strenuous. *1171*

Iron *Serum Decrease Physiological* Response to stress of exercise. *0582*

Ketones *Serum Increase Physiological* Effect marked in untrained individuals only. *1811*

17-Ketosteroids *Urine Increase Physiological* Effect most marked in well trained individuals. *0185*

Lactate *Serum Increase Physiological* Arterial lactate increased from 5.5 to 20 μmol/ml. *1924* Considerable effect of exercise. *3873*

Lactate Dehydrogenase *Serum Increase Physiological* Maximum effect observed following day. *1491* Marked increase with exercise. *2981*

Lactate Dehydrogenase Isoenzymes
Serum Increase Physiological 3,4,5 increased but no change in 1,2. *3051*

Leukocytes *Blood Increase Physiological* Mainly due to neutrophilia after exercise. *0987*

Linoleate *Serum Increase Physiological* Both absolutely and relatively. *1702*

β-Lipoproteins *Serum No Effect Physiological* No effect of exercise observed. *2846*

Lysozyme *Serum No Effect Physiological* No effect even with strenuous exercise. *2845*
Serum Increase Physiological After protracted exertion. *1491*
Urine Increase Physiological Very high clearance: proximal tubular function affected. *2845*

α₂-Macroglobulin *Serum Increase Physiological* Approximately 5% increase immediately after. *2846* Significant effect at 15 minutes, partial return by 1 d. *1491*

Magnesium *Serum No Effect Physiological* Observed effect with 12 minutes cycle-ergometer. *1231*

Malate Dehydrogenase *Serum Increase Physiological* Significant effect with exercise. *1231* Significant effect after 2 h march. *3217*

Mucoprotein *Urine Increase Physiological* Concentration of Tamm-Horsfall protein increased with decreased volume. *1720*

Norepinephrine *Plasma Increase Physiological* Significant effect after physical stress. *1176*
Urine Increase Physiological May rise up to 200-300 ng/min. *3351* If strenuous may be increased tenfold. *0279*

Ornithine Carbamoyltransferase (OCT)
Serum Increase Physiological Observed to increase following exercise. *1231* Maximum effect observed 7 d after exercise. *1491*

Osmolality *Urine Decrease Physiological* Effect most marked with light exercise. *1842*

Partial Thromboplastin Time
Plasma Decrease Physiological Observed effect in exercise. *2847*

pH *Urine Increase Physiological* Effect noted after mild exercise. *1842*
Urine Decrease Physiological At all rates of exercise (acid metabolites). *1842*
Muscle Decrease Physiological Fell from normal 6.93 to 6.40 after exercise. *1572*

Phosphate *Serum Increase Physiological* Observed effect with 12 minutes cycle-ergometer. *1231* Effect of muscular exercise. *0279*

Phospholipids, Total *Serum No Effect Physiological* But relative reduction of unsaturated acid. *1702*

Plasmin *Plasma Increase Physiological* Observed effect in exercise. *2847*

Plasmin Activity *Urine Increase Physiological* Associated with proteinuria. *1842*

Plasminogen *Plasma Increase Physiological* Observed effect in exercise. *2847*

Platelets *Blood Increase Physiological* Effect of sudden exercise. *0987*

Pregnanetriol *Urine Increase Physiological* Effect most marked in well trained individuals. *0185*

Prostaglandin E1 *Plasma Increase Physiological* Increase of 340% (mean after exercise). *1385*

Protein *Serum Increase Physiological* Significant effect with 12 minutes on cycle-ergometer. *1231* 9% increase immediately after exercise, normal in 30 minutes. *2846*
Urine Increase Physiological More common with heavy exercise than mild. *1842*

Protein Bound Iodine (PBI) *Serum No Effect Physiological* No effect observed. *0830*

Prothrombin Time *Plasma Increase Physiological* Observed effect in exercise. *2847*

Pyruvate *Serum Increase Physiological* After exercise to exhaustion 1.7 times control. *2846*
Serum Decrease Physiological Slight fall within 1 h and then steep drop. *0987*

Sorbitol Dehydrogenase *Serum Increase Physiological* Significant effect after 2 h march. *3217*

Specific Gravity *Urine Decrease Physiological* Apparent reduced ability to concentrate at all rates. *1842*

Standard Bicarbonate *Blood Decrease Physiological* Decreased to 11 meq/L with intermittent exercise. *1924*

Thyroxine (T₄) *Serum Decrease Physiological* Significant effect with strenuous exercise. *3217*

Drug Listings

Muscular Exercise (continued)

Transferrin *Serum Increase Physiological* Raised at 15 minutes, normal within 1 d. *1491* Approximately 10% increase immediately after. *2846*

Trehalase *Serum Increase Physiological* Observed to increase with exercise. *1231*
Urine Increase Physiological Observed to increase with exercise. *1231*

Triglycerides *Serum No Effect Physiological* But relative reduction of unsaturated acid. *1702*

Urea Clearance *Urine Decrease Physiological* Response to stress of exercise. *0582*

Urea Nitrogen *Serum Increase Physiological* Raised at 15 minutes, partial return to normal in 24 h. *1491*
Serum Decrease Physiological Observed effect with 12 minutes cycle-ergometer. *1231*

Urobilin *Urine Increase Physiological* Noted after heavy exercise. *1842*

Vanillylmandelic Acid *Urine Increase Physiological* Significant effect after physical stress. *1176*

Volume *Urine Increase Physiological* With mild exercise. *1842*
Urine Decrease Physiological With heavy exercise. *1842*

Mushroom Poisoning

Agglutinins *Blood Increase Physiological* Result of poisoning. *0279*

Alanine Aminotransferase *Serum Increase Physiological* May produce severe liver damage. *0279*

Aspartate Aminotransferase
Serum Increase Physiological May produce severe liver damage. *0279*

Erythrocytes *Blood Decrease Physiological* May cause severe hemolytic anemia. *0279*

Hemoglobin *Blood Decrease Physiological* May cause severe hemolytic anemia. *0279*

Protein *Urine Increase Physiological* May cause nephrotoxicity. *3204*

Urea Nitrogen *Serum Increase Physiological* May cause nephrotoxicity. *3204*

Mustard

4-Hydroxybenzylamine *Urine Increase Physiological* Excreted after mustard eaten. *2784*

Mustard Gas

Erythrocytes *Blood Decrease Physiological* May cause aplastic anemia. *2429*

Muzolimine

Creatinine Clearance *Urine No Effect Physiological* Not changed in 10 hypertensive patients given 30 mg daily for 16 weeks. *1229*

Glucose Tolerance *Serum Decrease Physiological* Significant reduction in 10 hypertensive patients given 30 mg daily for 16 weeks. *1229*

Potassium *Serum No Effect Physiological* No significant change in 10 hypertensive patients given 30 mg daily for 16 weeks. *1229*
Red Blood Cells No Effect Physiological Not appreciably affected in 10 hypertensive patients given 30 mg daily for 16 weeks. *1229*

Renin Activity *Plasma Increase Physiological* Statistically significant increase in 10 hypertensive patients given 30 mg daily. *1229*

Sodium *Serum No Effect Physiological* No significant change in 10 hypertensive patients given 30 mg daily for 16 weeks. *1229*

Triglycerides *Serum No Effect Physiological* Insignificant change in 10 hypertensive patients given 30 mg daily for 16 weeks. *1229*

Uric Acid *Serum Increase Physiological* Average 26 μmol/L in 10 hypertensive patients given 30 mg daily for 16 weeks. *1229*

Myoglobin

Color *Urine Increase Physiological* Produces red color in urine. *0459*

Nadolol

Cholesterol *Serum No Effect Physiological* In 94 patients treated for 3 mo. *3706* In 13 patients given 50-200 mg daily for 10 weeks. *1806*

Cholesterol, High Density Lipoprotein
Serum Decrease Physiological In hypertensive patients while fasting and during and after a meal and an exercise test. ?secondary to reduction of lipoprotein lipase. *2761* Increase by 3% in 13 patients given 50-200 mg daily for 10 weeks. *1806* Significant reduction in fasting concentration and after breakfast. *2761*

Cholesterol, Low Density Lipoprotein
Serum No Effect Physiological In 13 patients given 50-200 mg daily for 10 weeks. *1806* No difference between treatment with drug and placebo. *2761*

Cholesterol, Very Low Density Lipoprotein
Serum Increase Physiological Increase by 29% in 13 patients given 50-200 mg daily for 10 weeks. *1806*

Thyroid Stimulating Hormone (TSH)
Serum No Effect Physiological No effect over 2 weeks in 10 healthy volunteers. *2938*

Thyroxine (T4) *Serum No Effect Physiological* No effect over 2 weeks in 10 healthy volunteers. *2938*

Triglycerides *Serum Increase Physiological* In hypertensive patients though not significantly, further increased postprandially and after exercise. *2761* Increase by 15% in 13 patients given 50-200 mg daily for 10 weeks. *1806* Higher in fasting state in patients on treatment, but not significantly higher, maybe secondary to reduction in lipoprotein lipase activity. *2761* In 94 patients treated for 3 mo. *3706*

Tri-iodothyronine (T3) *Serum No Effect Physiological* No effect over 2 weeks in 10 healthy volunteers. *2938*

Tri-iodothyronine (T3), Reverse
Serum No Effect Physiological No effect over 2 weeks in 10 healthy volunteers. *2938*

Nafarelin Acetate

Testosterone *Serum Decrease Physiological* In all 9 patients with benign prostatic hypertrophy: due to reduced production in testes by inhibition of pituitary release of gonadotropins. Concentration reversibly reduced from normal to castration level. *2793*

Nafcillin

Alanine Aminotransferase *Serum Increase Physiological* 1 of 32 patients developed abnormal liver function on 3rd day of treatment. *2548*

Albumin *Serum No Effect Analytical* At concentration of 50 mg/L had no effect on BCG method. *3393*

Aspartate Aminotransferase
Serum Increase Physiological Possibly due to trauma of injection. *2313* 1 of 32 patients developed abnormal liver function on 3rd day of treatment. *0052*

Bicarbonate *Serum No Effect Analytical* At concentration of 50 mg/L had no effect on method using phenolphthalein. *3393*

Bilirubin *Serum No Effect Analytical* At concentration of 50 mg/L had no effect on Jendrassik and Grof method. *3393*

Bleeding Time *Patient Increase Physiological* Reportedly 4 times normal in two patients, reverted to normal on withdrawal. *0052*

BSP Retention *Serum Increase Physiological* Reported effect (?hepatotoxicity). *0071*

Calcium *Serum No Effect Analytical* At concentration of 50 mg/L had no effect on cresolphthalein method. *3393*

Chloride *Serum No Effect Analytical* At concentration of 50 mg/L had no effect on mercurimetric method. *3393*

Cholesterol *Serum No Effect Analytical* At concentration of 50 mg/L had no effect on Liebermann-Burchard method. *3393*

Creatinine *Serum No Effect Analytical* At concentration of 50 mg/L had no effect on AutoAnalyzer Jaffé method. *3393*

Eosinophils *Blood Increase Physiological* 3 of 32 children developed eosinophilia within 1-4 d. *2548*

Neutrophils *Blood Decrease Physiological* 2 of 32 children developed neutropenia within 4 to 13 d. *2548*

Phosphate *Serum No Effect Analytical* At concentration of 50 mg/L had no effect on phosphomolybdate method. *3393*

Platelet Aggregation *Blood Decrease Physiological* Reduced response to ADP collagen and epinephrine. *0052*

Platelets *Blood No Effect Physiological* Unchanged in spite of abnormal bleeding time and aggregation. *0052*

Protein *Serum Increase Analytical* At concentrations above 500 mg/L raised concentration as measured by biuret method with blank correction. *3393*

Triglycerides *Serum No Effect Analytical* At concentration of 50 mg/L had no effect on lipase/esterase method. *3393*

Urea Nitrogen *Serum No Effect Analytical* At concentration of 50 mg/L had no effect on diacetylmonoxime method. *3393*

Uric Acid *Serum No Effect Analytical* At concentration of 50 mg/L had no effect on phosphotungstate reduction method. *3393*

Nafenopin

Cholesterol *Serum Decrease Physiological* More effective than clofibrate. *0939*

Triglycerides *Serum Decrease Physiological* More effective than clofibrate. *0939*

Nalidixic Acid

Alanine Aminotransferase *Serum Increase Physiological* May cause cholestatic jaundice. *2808*

Alkaline Phosphatase *Serum Increase Physiological* May cause cholestatic jaundice. *2808*

Aspartate Aminotransferase
Serum Increase Physiological May cause cholestatic jaundice. *2808*

Bilirubin *Serum No Effect Physiological* No clinically significant displacement from protein likely in neonates. *3748*
Serum Increase Physiological May cause cholestatic jaundice. *2808*

Bilirubin, Direct *Serum Increase Physiological* Hemolytic anemia especially if G-6-PD deficiency. *3877*

Cephalin Flocculation *Serum Increase Physiological* May cause cholestatic jaundice. *2808*

Chromosomes *Test Conditions Abnormal Physiological* Clastogenic in human cells. *3282*

Coombs' Test, Direct *Serum Positive Physiological* May cause hemolytic anemia with normal G-6-PD. *1281*

Creatinine *Serum No Effect Analytical* At concentration of 60 mg/L had no effect on creatinine iminohydrolase method. *3393*
Serum Increase Physiological May cause nitrogen retention. *2808*

Eosinophils *Blood Increase Physiological* Allergic response. *1513*

Erythrocytes *Blood Decrease Physiological* Hemolytic anemia with G-6-PD deficiency. *0279*

Glucose *Serum No Effect Analytical* At concentration of 50 mg/L had no effect on GOD/POD-PAP method. *3393*
Serum Increase Analytical Copper reduction methods affected. *3505*
Urine No Effect Analytical No effect with Clinistix®, Tes-Tape®. *3877*

Haptoglobin *Serum Decrease Physiological* May cause hemolytic anemia especially if G-6-PD deficiency. *3877*

Hematocrit *Blood Decrease Physiological* Hemolytic anemia with G-6-PD deficiency. *0279*

Hemoglobin *Blood Decrease Physiological* Hemolytic anemia with G-6-PD deficiency. *0279*

17-Hydroxycorticosteroids *Urine No Effect Analytical* No effect with Porter-Silber reaction. *3877*

Hydroxynalidixic Acid *Urine Increase Physiological* Major metabolites with glucuronide conjugates. *1678*

17-Ketogenic Steroids *Urine Increase Analytical* Interferes with Zimmermann reaction. *2197*

17-Ketosteroids *Urine Increase Analytical* Interferes with Zimmermann reaction. *2197*

Leukocytes *Blood Increase Physiological* Due to eosinophilia. *2313*

Nalidixic Acid *Serum Increase Physiological* After 1 g orally 25-35 µg/mL in 2 h. *1678*
Urine Increase Physiological Major metabolites with glucuronide conjugates. *1678*

Nitrofurantoin *Urine No Effect Analytical* No effect on method of Conklin and Hollifield. *0704*

Nonprotein Nitrogen *Serum Increase Physiological* May cause nitrogen retention. *2808*

5'-Nucleotidase *Serum Increase Physiological* Due to cholestasis. *1778*

Occult Blood *Feces Positive Physiological* May cause bleeding from gastrointestinal tract. *0071*

Platelets *Blood Decrease Physiological* Thrombocytopenia. *3877* Thrombocytopenia occurred in 6 cases in Netherlands characteristically within 10-15 d after 4 g daily. Recovered rapidly once drug stopped but reaction could be severe. *2421*

Potassium *Serum No Effect Analytical* At concentration of 10 mg/L had no effect on measurement by ISE with predilution. *3393*

Prothrombin Time *Plasma Increase Physiological* Displaces coumarins from albumin. *3248*

Sodium *Serum No Effect Analytical* At concentration of 10 mg/L had no effect on measurement by ISE with predilution. *3393*

Sugar *Urine Increase Analytical* False positive with Fehlings, Benedict's, Clinitest®. *3877*

Thymol Turbidity *Serum Increase Physiological* May cause cholestatic jaundice. *2425*

Triglycerides *Serum No Effect Analytical* At concentration of 50 mg/L had no effect on GPO-PAP method. *3393*

Urea Nitrogen *Serum Increase Physiological* May cause nitrogen retention. *2808*

Uric Acid *Serum No Effect Analytical* At concentration of 50 mg/L had no effect on uricase-PAP method. *3393*

Urobilinogen *Urine Increase Physiological* May occur if cholestasis. *2808*
Feces Decrease Physiological May occur with cholestasis. *2808*

Vanillylmandelic Acid *Urine Increase Analytical* Apparent 4 fold with normal regime. *0583* Affects Pisano procedure. *0121*

Xylose Excretion *Urine Decrease Physiological* May cause gastrointestinal irritation, impaired absorption. *0612*

Nalorphine

Chlordiazepoxide *Blood No Effect Analytical* On method of Riddick at 5 mg/dL. *2983*

Ethchlorvynol *Urine No Effect Analytical* No effect on method of Frings and Cohen at 10 mg/dL. *1205*

Morphine *Urine No Effect Analytical* Insignificant cross react with EMIT procedure for opiates Insignificant cross react with hemagglutination inhibition. *2514*

Protein *Test Conditions Increase Analytical* Reacts with Folin-Ciocalteu of Lowry method. *0702*

Naloxone

Antidiuretic Hormone *Plasma No Effect Physiological* No effect of 10 mg on basal values. *0176*

Cortisol *Plasma Increase Physiological* Peak at 60 minutes after start of infusion remained high for duration. *0868*

Naloxone *(continued)*

Follicle Stimulating Hormone (FSH)
Plasma No Effect Physiological Infusion did not affect concentration of hormone at concentrations varying from 0.02 to 0.5 mg/kg. *1970*
Plasma Increase Physiological Significant effect within 60 minutes of start of i.v. infusion. *0868*

Gastric Inhibitory Polypeptide
Plasma No Effect Physiological In 6 healthy volunteers after ingestion of test meal and intravenous administration of drug. *0626*

Gastrin *Serum No Effect Physiological* In 6 healthy volunteers after ingestion of test meal and intravenous administration of drug. *0626*

Glucagon *Plasma No Effect Physiological* With infusion in normal and obese subjects. *0176* In 6 healthy volunteers after ingestion of test meal and intravenous administration of drug. *0626*

Glucagon, Pancreatic *Plasma No Effect Physiological* In 6 healthy volunteers after ingestion of test meal and intravenous administration of drug. *0626*

Glucose *Serum No Effect Physiological* In 6 healthy volunteers after ingestion of test meal and intravenous administration of drug. *0626*

Growth Hormone *Plasma No Effect Physiological* No significant effect with i.v. infusion. *0868*

Hydrochloric Acid *Gastric Material Decrease Physiological* Reduces basal and meal stimulated concentration. *0176*

Insulin *Plasma No Effect Physiological* With infusion in normal and obese subjects. *0176* In 6 healthy volunteers after ingestion of test meal and intravenous administration of drug. *0626*

Luteinizing Hormone (LH) *Plasma Increase Physiological* Significant effect within 30 minutes of start of i.v. infusion. *0868* After 0.08 mg/kg body weight i.v. in girls and boys at most advanced stage of gonadal maturation: ineffective in prepubertal and early pubertal children. *2796* Effect observed regardless of amount infused. *1970*

Morphine *Urine No Effect Analytical* Insignificant cross reactivity with RIA procedures Insignificant cross react with EMIT procedure for opiates. *2514*

Motilin *Plasma No Effect Physiological* In 6 healthy volunteers after ingestion of test meal and intravenous administration of drug. *0626*

Neurotensin *Plasma No Effect Physiological* In 6 healthy volunteers after ingestion of test meal and intravenous administration of drug. *0626*

Oxytocin *Plasma No Effect Physiological* No effect observed for 120 minutes in healthy male volunteers after intravenous injection of 10 mg. *1648*

Pancreatic Polypeptide *Plasma No Effect Physiological* In 6 healthy volunteers after ingestion of test meal and intravenous administration of drug. *0626*

Partial Thromboplastin Time
Plasma Increase Physiological With multiple doses effect observed (no bleeding). *0019*

Prolactin *Plasma No Effect Physiological* No significant effect with i.v. infusion. *0868* Infusion did not affect concentration of hormone at concentrations varying from 0.02 to 0.5 mg/kg. *1970* Basal and stimulated levels not affected. *0176*

Somatostatin *Plasma No Effect Physiological* In 6 healthy volunteers after ingestion of test meal and intravenous administration of drug. *0626*

Thyroid Stimulating Hormone (TSH)
Serum No Effect Physiological No significant effect with i.v. infusion. *0868*

Vasoactive Intestinal Polypeptide
Plasma No Effect Physiological In 6 healthy volunteers after ingestion of test meal and intravenous administration of drug. *0626*

Naltrexone

Cortisol *Plasma No Effect Physiological* No change in unstimulated values observed. *0176*

Estradiol *Plasma No Effect Physiological* With 50 to 100 mg administered daily to obese subjects over 8 weeks. *0176*

Follicle Stimulating Hormone (FSH)
Plasma No Effect Physiological With 50 to 100 mg administered daily to obese subjects over 8 weeks. *0176*

Glucose *Serum No Effect Physiological* No effect of long-term administration to obese individuals. *0176*

Insulin *Plasma No Effect Physiological* No effect of long-term administration to obese individuals. *0176*

Luteinizing Hormone (LH) *Plasma No Effect Physiological* With 50 to 100 mg administered daily to obese subjects over 8 weeks. *0176*

Testosterone *Serum No Effect Physiological* With 50 to 100 mg administered daily to obese subjects over 8 weeks. *0176*

Nandrolone

Alanine Aminotransferase *Serum Increase Physiological* Reported to affect liver function. *2220*

Alkaline Phosphatase *Serum Increase Physiological* Reported to affect liver function. *2220*

Aspartate Aminotransferase
Serum Increase Physiological Reported to affect liver function. *2220*

Calcium *Serum Increase Physiological* May occur in women with neoplasm of breast. *2681*
Urine Increase Physiological Due to hypercalcemia. *2681*

Cholesterol *Serum Increase Physiological* Due to action on liver. *2681*
Serum Decrease Physiological Due to action on liver. *2681*

Creatine *Urine Decrease Physiological* Anabolic effect. *2681*

Creatinine *Urine Decrease Physiological* Anabolic effect. *2681*

Estrogens *Urine Increase Analytical* Normal route of metabolism. *0022*

Factor II *Plasma Increase Physiological* Metabolic effect. *2681*

Factor V *Plasma Increase Physiological* Metabolic effect. *2681*

Factor VII *Plasma Increase Physiological* Metabolic effect. *2681*

Factor X *Plasma Increase Physiological* Metabolic effect. *2681*

Glucose *Serum Decrease Physiological* Anabolic effect. *2681*

Glucose Tolerance *Serum Increase Physiological* Anabolic effect. *2681*

Hematocrit *Blood Increase Physiological* Due to decreased plasma volume. *0330*

^{131}I Uptake *Serum Decrease Physiological* Anabolic effect. *2681*

17-Ketosteroids *Urine Increase Physiological* Anabolic effect. *2681*

Metyrapone Test *Patient Positive Physiological* Anabolic effect. *2681*

Protein Bound Iodine (PBI) *Serum Decrease Physiological* Anabolic effect. *2681*

Prothrombin Time *Plasma Increase Physiological* May increase sensitivity to anticoagulants. *2681*

T_3 Uptake *Serum Increase Physiological* Anabolic effect. *2681*

Thyroxine (T_4) Binding Globulin
Serum Decrease Physiological Anabolic effect. *2681*

Volume *Plasma Decrease Physiological* May occur from 10 weeks onwards (anabolic effect). *0330*

Naphazoline

Amphetamine *Urine No Effect Analytical* No reaction with NBD chloride procedure of Monforte. *2473*

Propoxyphene *Urine No Effect Analytical* Less than 1% fluorescence in procedure of Valentour. *3662*

Naphthalene

Bilirubin *Serum Increase Physiological* May cause hemolysis with G-6-PD deficiency. *0248*

Casts *Urine Increase Physiological* Nephrotoxic effect. *1302*

Color *Urine Increase Physiological* Brown or black due to blood and hemoglobin. *1302*

Erythrocytes *Blood Decrease Physiological* May cause hemolysis with G-6-PD deficiency. *0248*
Urine Increase Physiological Occurs occasionally following inhalation of vapor. *1302*

Haptoglobin *Serum Decrease Physiological* May cause hemolysis. *0248*

Heinz Body Formation *Blood Positive Physiological* Early stage of hemolytic anemia. *0333*

Hematocrit *Blood Decrease Physiological* May cause hemolysis with G-6-PD deficiency. *0248*

Hemoglobin *Blood Decrease Physiological* May cause hemolysis with G-6-PD deficiency. *0248*
Urine Increase Physiological Due to G-6-PD related hemolysis or poisoning. *1302*

Leukocytes *Blood Increase Physiological* Leukocytosis may occur following ingestion. *1302*

Methemoglobin *Blood Increase Physiological* May cause hemolysis with G-6-PD deficiency. *3581*

α-**Naphthol** *Urine Positive Physiological* Present as metabolite. *1302*

Protein *Urine Increase Physiological* Nephrotoxic effect. *1302*

Naphthol

Color *Urine Increase Analytical* Dark color on standing. *2313*

Estrogens *Urine Increase Analytical* If applied topically affects colorimetry. *0023*
Urine Decrease Analytical Affects colorimetric/fluorometric procedures. *0022*

Naphthoxyacetic Acid

Platelets *Blood Decrease Physiological* Reported effect (AMA Blood dyscrasias committee). *2429*

Naproxen

Alanine Aminotransferase *Serum No Effect Analytical* At 5 times upper limit of therapeutic range on methods on SMAC®, Abbott-VP, aca, Cobas-Bio, and KDA. *2138*
Serum Increase Physiological In 6% of patients receiving drug for juvenile rheumatoid arthritis. *0239*

Albumin *Serum No Effect Analytical* At 5 times upper limit of therapeutic range on methods on SMAC®, Ektachem®, Hitachi® 705 and KDA. *2138*

Aldosterone *Plasma No Effect Physiological* In 10 furosemide treated patients with well controlled congestive heart failure. *1042*

Alkaline Phosphatase *Serum No Effect Analytical* At 5 times upper limit of therapeutic range on methods on Abbott-VP, aca, Cobas-Bio, Hitachi® 705 and KDA. *2138*
Serum Increase Analytical At 5 times upper limit of therapeutic range on methods on SMAC®. *2138*

Amylase *Serum No Effect Analytical* At 5 times upper limit of therapeutic range on methods on aca, Cobas-Bio and Ektachem®. *2138*

Aspartate Aminotransferase *Serum No Effect Analytical* At 5 times upper limit of therapeutic range on methods on SMAC®, Abbott-VP, aca, Cobas-Bio, Hitachi® 705 and KDA. *2138*
Serum Increase Physiological In 6% of patients receiving drug for juvenile rheumatoid arthritis. *0239*

Bicarbonate *Serum Increase Analytical* Falsely high value, linearly correlated with serum drug concentration with Technicon® RA-1000 ion-specific electrode. *1512*

Bilirubin *Serum No Effect Analytical* At 5 times upper limit of therapeutic range on methods on SMAC®, aca, Cobas-Bio, Ektachem®, Hitachi® 705, aca, and KDA. *2138*
Serum No Effect Physiological At pharmacological concentration has little or no protein displacement effect. *3748*
Serum Increase Physiological Isolated case of hemolytic anemia: one other case reported previously. *1681*

Calcium *Serum No Effect Analytical* At 5 times upper limit of therapeutic range on methods on SMAC®, Abbott-VP, aca, Ektachem®, Hitachi® 705 and KDA. *2138*

Chloride *Urine Decrease Physiological* Increase by 26% in 10 furosemide treated patients with well controlled congestive heart failure. *1042*

Cholesterol *Serum No Effect Analytical* At 5 times upper limit of therapeutic range on methods on SMAC®, Abbott-VP, Cobas-Bio, Ektachem®, Hitachi® 705 and KDA. *2138*

Creatine Kinase *Serum No Effect Analytical* At 5 times upper limit of therapeutic range on methods on SMAC®, Abbott-VP, aca, Cobas-Bio and Hitachi® 705. *2138*

Creatinine *Serum No Effect Analytical* At 5 times upper limit of therapeutic range on methods on SMAC®, Abbott-VP, aca, Cobas-Bio, Ektachem®, Hitachi® 705 and KDA. *2138*
Serum Increase Physiological In isolated case of patient with arthritis in the absence of hypertension. *0466*

Creatinine Clearance *Urine No Effect Physiological* When given for 14 d to patients with rheumatoid arthritis and heart failure. *3514*

Eosinophils *Blood Increase Physiological* Hypersensitivity reaction in 3 women with pulmonary infiltrates: resolved when drug discontinued. *0547*

Erythrocytes *Blood Decrease Physiological* In 4% of patients receiving drug for juvenile rheumatoid arthritis. *0239*

Glucose *Serum No Effect Analytical* At 5 times upper limit of therapeutic range on methods on SMAC®, Abbott-VP, aca, Cobas-Bio, Ektachem®, Hitachi® 705 and KDA. *2138*

γ-**Glutamyltransferase (GGT)** *Serum No Effect Analytical* At 5 times upper limit of therapeutic range on methods on SMAC®, Abbott-VP, and Hitachi® 705. *2138*

Hemoglobin *Blood Decrease Physiological* Isolated case of hemolytic anemia: one other case reported previously. *1681*
Urine Increase Physiological In 2% of patients receiving drug for juvenile rheumatoid arthritis. *0239*

5-Hydroxyindoleacetic Acid (5-HIAA)
Urine Increase Analytical Due to metabolite desmethylnaproxen on spectrophotometric assays but compound is thermolabile and can be destroyed by heat. *3749*

Iron *Serum No Effect Analytical* At 5 times upper limit of therapeutic range on Ferrozine method of SMAC®. *2138*

Lactate Dehydrogenase *Serum No Effect Analytical* At 5 times upper limit of therapeutic range on methods on SMAC®, Abbott-VP, Cobas-Bio, Hitachi® 705 and KDA. *2138*

Occult Blood *Feces Positive Physiological* Several cases of gastrointestinal bleeding reported to UK committee on safety of medicines. *0782*

Phosphate *Serum No Effect Analytical* At 5 times upper limit of therapeutic range on methods on SMAC® and KDA. *2138*
Serum Increase Analytical At 5 times upper limit of therapeutic range on methods on aca and Hitachi® 705. *2138*

Prostaglandin, 6-Keto-F₁α *Urine Decrease Physiological* Increase by 76% in 10 furosemide treated patients with well controlled congestive heart heart failure. *1042*

Prostaglandin E *Semen Decrease Physiological* Significant effect but returned to prior level 1 week after treatment stopped. *0285*

Prostaglandin E2 *Urine No Effect Physiological* In 11 volunteers in randomized trial. *0452*
Urine Decrease Physiological When given for 14 d days to patients with rheumatoid arthritis and heart failure. Marked effect. *3514*

Prostaglandin E, 19-Hydroxy
Semen Decrease Physiological Significant effect but returned to prior level 1 week after treatment stopped. *0285*

Prostaglandin F *Semen Decrease Physiological* Significant effect but returned to prior level 1 week after treatment stopped. *0285*

Naproxen (continued)

Prostaglandin F2α *Urine Decrease Physiological* When given for 14 d to patients with rheumatoid arthritis and heart failure. Marked effect. *3514*

Prostaglandin F, 19-Hydroxy
Semen Decrease Physiological Significant effect but returned to prior level 1 week after treatment stopped. *0285*

Protein *Serum No Effect Analytical* At 5 times upper limit of therapeutic range on methods on SMAC®, Abbott-VP, Ektachem®, Hitachi® 705 and KDA. *2138*
Urine Increase Physiological In isolated case of patient with arthritis in the absence of hypertension. *0466*

Renin Activity *Plasma No Effect Physiological* In 10 furosemide treated patients with well controlled congestive heart failure. *1042*
Plasma Decrease Physiological When given for 14 d to patients with rheumatoid arthritis and heart failure. Marked effect. *3514*

Reticulocytes *Blood Increase Physiological* Isolated case of hemolytic anemia: one other case reported previously. *1681*

Sodium *Urine Decrease Physiological* Increase by 26% in 10 furosemide treated patients with well controlled congestive heart failure. *1042*

Sperm Count *Semen No Effect Physiological* No change with treatment although prostaglandin concentration reduced. *0285*

Sperm Motility *Semen No Effect Physiological* No change with treatment although prostaglandin concentration reduced. *0285*

Thromboxane B₂ *Urine Decrease Physiological* In 11 volunteers in randomized trial. *0452*

Triglycerides *Serum No Effect Analytical* At 5 times upper limit of therapeutic range on methods on Abbott-VP, Ektachem®, Hitachi® 705 and KDA. *2138*
Serum Decrease Analytical At 5 times upper limit of therapeutic range on enzymatic method on SMAC® At 5 times upper limit of therapeutic range on methods on enzymatic method on SMAC®. *2138*

Urea Nitrogen *Serum No Effect Analytical* At 5 times upper limit of therapeutic range on methods on SMAC®, Abbott-VP, Cobas-Bio, Ektachem®, Hitachi® 705 and KDA. *2138*

Uric Acid *Serum No Effect Analytical* At 5 times upper limit of therapeutic range on methods on SMAC®, Abbott-VP, aca, Ektachem® and KDA. *2138*
Serum Increase Analytical At 5 times upper limit of therapeutic range on method on Hitachi® 705. *2138*

Volume *Urine Decrease Physiological* Increase by 19% in 10 furosemide treated patients with well controlled congestive heart failure. *1042*

Zinc *Serum No Effect Physiological* Mean increase of 35% in 10 healthy volunteers receiving 250 mg 3 times per day for 1 week mechanism not known. *1025*
Urine Increase Physiological Mean increase of 35% in 10 healthy volunteers receiving 250 mg 3 times per day for 1 week mechanism no known. *1025*

Narcotic Antagonists

Prothrombin Time *Plasma Increase Physiological* Reported enhancement of anticoagulant activity. *1487*

Narcotics

Amylase *Serum Increase Physiological* Cause spasm of sphincter of Oddi. *0127*

Aspartate Aminotransferase
Serum Increase Physiological Impaired excretion due to spasm of sphincter. *0652*

Basal Metabolic Rate *Patient Decrease Physiological* Metabolic effect of drugs. *2220*

BSP Retention *Serum Increase Physiological* Impaired excretion — spasm of sphincter of Oddi. *0652*

Creatine Kinase *Serum Increase Physiological* Response to i.m. injections. *0249*

Glucose *Serum Increase Physiological* May be rare insignificant increases. *1487*

Immunoglobulin IgG *Serum Increase Physiological* Often elevated in addicts (?liver problem). *2012*

Immunoglobulin IgM *Serum Increase Physiological* Frequently elevated in addicts (?liver problem). *2012*

Lipase *Serum Increase Physiological* Impaired excretion — spasm of sphincter of Oddi. *0652*

Lymphocytes *Blood Increase Physiological* Frequent absolute and relative increase in addicts. *2012*

Prothrombin Time *Plasma Increase Physiological* Prolonged use may have effect. *1679*

Neoarsphenamine

Hemoglobin *Blood Decrease Physiological* May cause hemolysis. *1902*

Neocinchophen

Sugar *Urine Increase Analytical* Affect Benedict's, Clinitest®. *0583*

Neo-Mull-Soy®

Amino Acids *Urine Positive Analytical* Reddish-pink spot with DL-methionine with ninhydrin on thin-layer chromatography. *1278*

Methionine *Urine Increase Physiological* Contained in large amount in infant formula. *1749*

Neomycin

Amino-4-Imidazole-5-Carboxamide Ribotide (AICAR)
Urine Increase Physiological If megaloblastic anemia develops. *3773*

Amino Acids *Urine Increase Analytical* Reacts with ninhydrin; extra spot TLC, high voltage electrophoresis. *2855*
Urine Positive Analytical Purple spot with ninhydrin on thin-layer chromatography when combined with triamcinolone and nystatin in Kenacomb. *1278*

Ammonia *Plasma Decrease Physiological* Reduces NH₃ producing bacteria in gastrointestinal tract. *3889*
Urine Increase Physiological May occur if hypokalemia induced. *1077*

Bile Acids *Feces Increase Physiological* Precipitates bile salts in gastrointestinal tract. *1077*

Calcium *Urine Decrease Physiological* Occurs when fecal calcium increased. *1077*
Feces Increase Physiological Occurs independent of steatorrhea. *1077*

Carotene *Serum Decrease Physiological* Striking effect even if oral supplements. *1077*

Casts *Urine Increase Physiological* Nephrotoxic effect. *2313*

Chenodeoxycholic Acid *Feces Increase Physiological* Due to altered intestinal flora. *1077*

Chloramphenicol *Serum No Effect Analytical* No effect at 100 mg/L on coupled enzymatic method. *2490*

Cholesterol *Serum Decrease Physiological* Forms salts with bile acids in gut. *3505* In 20 subjects with type II hyperlipoproteinemia given 2 g/d for 9 mo caused 15% decline in cholesterol. *1618* Marked effect over diet in 20 patients with type II hyperlipoproteinemia treated over several months. *1618*

Cholesterol, High Density Lipoprotein
Serum No Effect Physiological No significant effect in 20 type II hyperlipoproteinemic subjects given 2 g/d for 9 mo. *1618* No significant effect in 20 patients with type II hyperlipoproteinemia treated over several months. *1618*

Cholesterol, Low Density Lipoprotein
Serum Decrease Physiological In 20 subjects with type II hyperlipoproteinemia given 2 g/d for 9 mo caused 16% decrease. *1618* Marked effect over diet in 20 patients with type II hyperlipoproteinemia treated over several months. *1618*

Cholesterol, Very Low Density Lipoprotein
Serum No Effect Physiological No significant effect in 20 patients with type II hyperlipoproteinemia treated over several months. *1618*

Cholic Acid *Feces Increase Physiological* Due to altered intestinal flora. *1077*

Creatinine *Serum Increase Physiological* Nephrotoxic effect. *2808*

Deoxycholic Acid *Feces Decrease Physiological* Due to alteration of intestinal flora. *1077*

Digoxin *Serum Decrease Physiological* At doses of 3 g/d reduces bioavailability. *2794* Reduced absorption due to sprue-like syndrome. *3794* At doses of 1-3 g decreased serum concentration and urinary excretion, also prolonged time to peak concentration. *2178*

Erythrocytes *Blood Decrease Physiological* Megaloblastic anemia (impaired B_{12} absorption). *3773*

Estriol *Urine Decrease Physiological* Possibly affects integrity of intestinal microflora. *0023*

Estriol-3-Glucuronide *Urine Decrease Physiological* Possibly affects integrity of intestinal microflora. *0023*

Estrogens *Urine Decrease Physiological* In pregnant women due to alteration of gut flora. *2885*

Fat *Feces Increase Physiological* Alters intestinal villi, inhibits triglyceride hydrolysis. *3855* Steatorrhea produced because of mucosal damage and rendering bile salts less available for fat absorption. *3023*

Fatty Acids, Free (FFA) *Serum Decrease Physiological* Probably related to altered fat absorption. *1077*

FIGLU (N-Formiminoglutamic Acid)
Urine Increase Physiological If megaloblastic anemia develops. *3773*

Hematocrit *Blood Decrease Physiological* Megaloblastic anemia (impaired B_{12} absorption). *3773*

Hemoglobin *Blood Decrease Physiological* Megaloblastic anemia (impaired B_{12} absorption). *3773*

Lactose Tolerance *Serum Decrease Physiological* Significantly lowered glucose response noted. *1077*

Leukocytes *Blood Decrease Physiological* If severe megaloblastic anemia. *3773*

Lithocholic Acid *Feces Decrease Physiological* Due to alteration of intestinal flora. *1077*

Magnesium *Serum Decrease Physiological* May be loss in stools due to steatorrhea. *1343*

MCV *Blood Increase Physiological* If megaloblastic anemia develops. *3773*

Nitrofurantoin *Urine No Effect Analytical* No effect on method of Conklin and Hollifield. *0704*

Nitrogen *Urine Decrease Physiological* Due to reduced absorption of amino acids etc. *1077*
Feces Increase Physiological Alters intestinal villi, inhibits triglyceride hydrolysis. *3855*

Nonprotein Nitrogen *Serum Increase Physiological* Nephrotoxic effect. *2313*

Penicillin *Serum Decrease Physiological* Decrease by 50% when given orally. *1487* Reduced absorption due to sprue-like syndrome. *3794*

Platelets *Blood Decrease Physiological* If severe megaloblastic anemia. *3773*

Potassium *Serum Decrease Physiological* May occur with neomycin induced malabsorption. *1077*
Feces Increase Physiological Induced by steatorrhea. *1077*

Protein *Urine Increase Physiological* Nephrotoxic effect. *1022*

Prothrombin Time *Plasma Increase Physiological* Reduces availability of vitamin K. *3650*

Sodium *Feces Increase Physiological* Induced by steatorrhea. *1077*

Sterols *Feces Increase Physiological* Precipitates bile salts in gastrointestinal tract. *1077*

Thyroglobulin *Serum Decrease Physiological* Mean decrease from 17.3 to 11.7 ng/mL when given 2.0 g/d orally for 7 d. *0959*

Thyroxine (T_4) *Serum No Effect Physiological* When given 2.0 g/d orally for 7 d. *0959*

Triglycerides *Serum No Effect Physiological* No effect in 20 patients with type II hyperlipoproteinemia treated over several months. *1618*

Serum Decrease Physiological Consistent but insignificant decrease in concentration in 20 subjects with type ii hyperlipoproteinemia given 2 g/d for 9 mo. *1618*

Tri-iodothyronine (T_3) *Serum Decrease Physiological* Mean decrease from 104 to 92 ng/dL when given 2.0 g/d for 7 d orally. *0959*

Tri-iodothyronine (T_3), Reverse
Serum No Effect Physiological When given 2.0 g/d orally for 7 d. *0959*

Urea Nitrogen *Serum Increase Physiological* Nephrotoxic effect. *1022*

Urobilinogen *Urine Decrease Physiological* Reduces flora in gastrointestinal tract. *1488*
Feces Decrease Physiological Reduces flora in gastrointestinal tract. *1488*

Vanillylmandelic Acid *Urine No Effect Physiological* No effect on excretion observed. *0426*

Vitamin A *Serum Decrease Physiological* Probably related to reduced absorption bile salts. *1077* Reduced absorption due to sprue-like syndrome. *3794* Value lowered with bile acid sequestration but probably remains within normal range. *3023*

Vitamin B_{12} *Serum Decrease Physiological* Impairs absorption. *3773* Reduced absorption due to sprue-like syndrome. *3794*
Urine Decrease Physiological With impaired absorption. *3773*

Xylose Excretion *Serum No Effect Physiological* Single doses did not affect absorption of d-xylose. *2178*
Urine Decrease Physiological Affects intestinal absorption with mucosal damage. *1756*

Neostigmine

Pseudocholinesterase *Serum Decrease Physiological* Direct effect of drug. *3523*
Test Conditions Decrease Analytical Inhibitory effect observed. *2893*

Nessler's Reagent

Protein Bound Iodine (PBI) *Serum Decrease Analytical* Catalytic effect of iodine inhibited by K+, Hg++. *1563*

Netilmicin

Creatinine *Serum Increase Physiological* 44% increase on average in elderly population with initial clearance of 81 mL/min (increase significant). *1203*

Glucose *Urine No Effect Analytical* On Clinitest® at physiological concentration. On Diastix® at physiological concentration. On TesTape® at physiological concentration. *2733*

Thyroxine (T_4) *Serum No Effect Physiological* No effect when given i.v. with cloxacillin. *0959*

Tri-iodothyronine (T_3) *Serum Decrease Physiological* Mean decrease from 114 ng/dL to 75 ng/dL when combined with cloxacillin given i.v., probably because of increased clearance. Free T_3 proportion increased. *0959*

Tri-iodothyronine (T_3), Reverse
Serum No Effect Physiological No effect when given i.v. with cloxacillin. *0959*

Urea Nitrogen *Serum Increase Physiological* About 2 mg/dL increase in elderly patients with initial poor renal function. *1203*

Neuromuscular Relaxants

pCO$_2$ *Blood Decrease Physiological* Secondary to hyperventilation postoperatively. *1343*

Potassium *Serum Decrease Physiological* If K deficiency potentiates effect. *1343*

Pseudocholinesterase *Serum Decrease Physiological* If enzyme deficiency may cause toxicity. *1343*

Drug Listings

Neutral Red

Chromosomes *Test Conditions Abnormal Physiological* Clastogenic in human cells. *3282*

Niacin

Alanine Aminotransferase *Serum No Effect Analytical* No effect at therapeutic concentration on Reflotron method. *1984*
Serum Increase Physiological Intrahepatic cholestasis observed rarely. *1253*

Albumin *Serum Decrease Physiological* Decreased synthesis due to liver damage. *3196*

Alkaline Phosphatase *Serum Increase Physiological* May cause impairment of hepatic function. *2726*

Apolipoprotein AI *Serum Increase Physiological* Increase by 12% in 34 hypercholesterolemic individuals with 1.5 g/d for 1 mo then 3.0 g/d up to 6 mo. *1967*

Apolipoprotein AII *Serum No Effect Physiological* In 34 hypercholesterolemic individuals with 1.5 g/d for 1 mo, then 3.0 g/d up to 6 mo. *1967*

Aspartate Aminotransferase *Serum No Effect Analytical* No effect at therapeutic concentration on Reflotron method. *1984*
Serum Increase Physiological Intrahepatic cholestasis observed rarely. *1253*

Bile *Urine Increase Physiological* May be impaired hepatic function. *2313*

Bilirubin *Serum Increase Physiological* May cause impairment of hepatic function. *2726*

BSP Retention *Serum No Effect Physiological* Usual unless liver previously damaged. *1622*
Serum Increase Physiological Liver function impairment, ?competition for conjugated. *2726*

Catecholamines *Plasma Increase Analytical* Occurs with large doses, interfering fluorescence. *1488*
Urine Increase Analytical May produce interfering fluorescence. *1487*

Cephalin Flocculation *Serum Increase Physiological* May impair hepatic function. *2313*

Cholesterol *Serum No Effect Analytical* No effect at therapeutic concentration on Reflotron method. *1984*
Serum Decrease Physiological Therapeutic goal (rebound increase when discontinued). *3505*

Cholesterol, High Density Lipoprotein
Serum Increase Physiological Increase by 26% in 34 hypercholesterolemic individuals with 1.5 g/d for 1 mo then 3.0 g/d up to 6 mo. *1967*

Cholesterol, Low Density Lipoprotein
Serum Decrease Physiological Increase by 21% in 34 hypercholesterolemic individuals with 1.5 g/d for 1 mo then 3.0 g/d up to 6 mo. *1967*

Cortisol *Plasma No Effect Physiological* No effect either if fed or starved. *0661*

Diagnex Blue Excretion *Urine Increase Physiological* Displaces diagnex blue from resin. *3879*

Eosinophils *Blood Decrease Physiological* By up to 60% in 2 h, increased after 24 h. *1622*

Fatty Acids, Free (FFA) *Serum No Effect Analytical* No effect on method of Pinelli. *2819*
Serum Decrease Physiological Marked fall then progressive secondary rise. *1727*

Fibrinolytic Time *Plasma Increase Physiological* Significant effect if given parenterally. *0421*

Glucose *Serum No Effect Analytical* No effect at therapeutic concentration on Reflotron method. *1984*
Serum Increase Physiological Mechanism not discussed. *1343*
Urine Increase Physiological As result of hyperglycemia. *1343* Due to hyperglycemia. *1253*

Glucose Tolerance *Serum Increase Physiological* Increases glucose disappearance after i.v. glucose tolerance test. *0819*
Serum Decrease Physiological Reduced tolerance observed in diabetics. *2220*

γ-Glutamyltransferase (GGT) *Serum No Effect Analytical* No effect at therapeutic concentration on Reflotron method. *1984*

Glycine *Urine Decrease Physiological* 2 g causes 30% decrease at 2 h. *1622*

Growth Hormone *Plasma Increase Physiological* Produced by fall in free fatty acids. *1727*

HDL₂-Cholesterol *Serum Increase Physiological* Increase by 36% in 34 hypercholesterolemic individuals with 1.5 g/d for 1 mo then 3.0 g/d up to 6 mo. *1967*

HDL₃-Cholesterol *Serum Increase Physiological* Increase by 35% in 34 hypercholesterolemic individuals with 1.5 g/d for 1 mo then 3.0 g/d up to 6 mo. *1967*

Hemoglobin *Plasma No Effect Physiological* No effect observed with increased blood flow. *1755*

Hydrochloric Acid *Gastric Material Increase Physiological* Increase up to 230% in patients after 0.5 g orally. *0104*

Insulin *Plasma Increase Physiological* ?response to increased glucose output. *1253*

Ketones *Urine Increase Physiological* ?due to hepatic mobilization of ketogenic amino acids. *1253*

Lipids, Total *Serum Decrease Physiological* Prompt and sustained hypolipemic action. *1680*

β-Lipoproteins *Serum Decrease Physiological* Chronic administration has slight effect. *1343*

Lipoproteins, Pre-β *Serum Decrease Physiological* Therapeutic effect. *1343*

Lymphocytes *Blood Decrease Physiological* By up to 20% in 2 h, ?normal at 24 h. *1622* Significant effect with megadose supplementation. *1347*

Neutrophils *Blood No Effect Physiological* No effect with megadose supplementation. *1347*
Blood Increase Physiological By up to 100% in 2 h, high at 24 h. *1622*

pH *Urine Decrease Physiological* Increases acidity at 4 h. *1622*

Phospholipids, Total *Serum Decrease Physiological* Prompt and sustained hypolipemic action. *1680*

Potassium *Urine Decrease Physiological* With 1 g, no change with 2 g. *1622*

Prothrombin Time *Plasma Increase Physiological* Due to hepatocellular and obstructive liver damage. *3196*

Sodium *Urine Increase Physiological* 20% increase at 4 h after 2 g. *1622*

Sugar *Urine Increase Analytical* Interferes with Benedict's reagent. *1022*

Thymol Turbidity *Serum Increase Physiological* Impaired hepatic function. *2313*

Triglycerides *Serum No Effect Analytical* No effect at therapeutic concentration on Reflotron method. *1984*
Serum Decrease Physiological Therapeutic intent. *2736* Increase by 27% in 34 hypercholesterolemic individuals with 1.5 g/d for 1 mo then 3.0 g/d up to 6 mo. *1967*

Urea Nitrogen *Serum No Effect Analytical* No effect at therapeutic concentration on Reflotron method. *1984*

Uric Acid *Serum No Effect Analytical* No effect at therapeutic concentration on Reflotron method. *1984*
Serum Increase Physiological May rise by up to 1.5 mg/dL if large doses. *2736*
Urine Decrease Physiological Decreases by approximately 50%. *1253*

Uric Acid Clearance *Urine Decrease Physiological* Decreases by 75%, ?due to altered tubular handling. *1253*

Niacinamide

Alanine Aminotransferase *Serum Increase Physiological* Rare probable reversible toxic effect. *3875*

Alkaline Phosphatase *Serum Increase Physiological* Rare probable reversible toxic effect. *3875*

Amphetamine *Urine No Effect Analytical* No effect at 100 mg/L on method of Rutter. *3097*

Aspartate Aminotransferase
Serum Increase Physiological Rare probable reversible toxic effect. *3875*

Bilirubin *Serum Increase Physiological* Rare probable reversible toxic effect. *3875*

Bilirubin, Direct *Serum Increase Physiological* Rare probable reversible toxic effect. *3875*

Cholesterol *Serum No Effect Analytical* At concentration of 40 mg/L had no effect on CHOD-Iodide method. *3393* At concentration of 40 mg/L had no effect on CHOD-PAP method. *3393* At concentration of 40 mg/L had no effect on catalase-AIDH method. *3393* At concentration of 400 mg/L had no effect on catalase-Hantzsch reaction. *3393* At concentration of 40 mg/L had no effect on Liebermann-Burchard method. *3393* *Serum No Effect Physiological* No effect observed in normals. *1622*

Creatinine *Serum No Effect Analytical* At concentration of 40 mg/L had no effect on Jaffé-Fading-Fraction method. *3393* At concentration of 40 mg/L had no effect on Jaffé-Fuller's earth method. *3393* At concentration of 120 mg/L had no effect on kinetic Jaffé method on BKA-2. *3393*

Eosinophils *Blood Decrease Physiological* No change after 2 h, marked at 4 h. *1622*

Glucose *Serum No Effect Analytical* At concentration of 12 mg/L had no effect on Ektachem® method. *3393*

Glycine *Urine Increase Physiological* Increased up to 80% with 1 g at 4 h. *1622*

Leukocytes *Blood Increase Physiological* Marked elevation noted at 24 h. *1622*

Lipids, Total *Serum No Effect Physiological* No lowering effect observed. *1343*

Lymphocytes *Blood Increase Physiological* Up by 25% at 4 h 40% at 24 h. *1622*

Neutrophils *Blood Increase Physiological* Up by 40% after 2 g at 4 h. *1622*

pH *Urine Increase Physiological* Reduces acidity at 2-4 h. *1622*

Potassium *Serum No Effect Analytical* At concentration of 120 mg/L had no effect on measurement by ISE without predilution. *3393* *Urine Increase Physiological* 1 g causes marked excretion. *1622*

Protein Electrophoresis *Serum No Effect Analytical* At concentration of 120 mg/L had no effect on automated Olympus-Hite method. *3393*

Prothrombin Time *Plasma Increase Physiological* Rare probable reversible toxic effect. *3875*

Sodium *Serum No Effect Analytical* At concentration of 120 mg/L had no effect on measurement by ISE without predilution. *3393* *Urine Increase Physiological* 1 g causes up to 40% increase at 4 h. *1622*

Urea Nitrogen *Serum No Effect Analytical* At concentration of 12 mg/L had no effect on Ektachem® method. *3393*

Uric Acid *Serum No Effect Analytical* At concentration of 40 mg/L had no effect on Kageyama-Hantzsch method. *3393* At concentration of 40 mg/L had no effect on catalase-AIDH method. *3393* *Urine Increase Physiological* By up to 40% at 2 h maximum at 4 h. *1622*

Niagara Sky Blue

Protein Bound Iodine (PBI) *Serum Decrease Physiological* Mechanism obscure. *0012*

Nialamide

Alkaline Phosphatase *Serum Increase Physiological* Probable hypersensitive hepatitis. *0071*

Amphetamine *Urine No Effect Analytical* No effect at 100 mg/L on method of Rutter. *3097* *Urine Increase Analytical* Yields fluorophor with NBD chloride reaction. *2473*

Aspartate Aminotransferase
Serum Increase Physiological Probable hypersensitive hepatitis. *0071*
Bile Urine Increase Physiological Probable hypersensitive hepatitis. *0071*

Bilirubin *Serum Increase Physiological* Probable hypersensitive hepatitis. *0071*

BSP Retention *Serum Increase Physiological* Hepatotoxic/cholestatic syndromes. *1343*

Glucose *Serum Decrease Physiological* May prolong action of insulin in diabetics. *2427*

Leukocytes *Blood Decrease Physiological* Leukopenia reported. *0071*

Phenylketones *Urine Positive Analytical* Fading green with $FeCl_3$, green with Phenistix®. *0775*

Protein Bound Iodine (PBI) *Serum Increase Physiological* Variable effect on I 131 uptake in rats. *0012*

Urea Nitrogen *Test Conditions Increase Analytical* Bilious yellow-green with Berthelot's reagent. *1936*

Vanillylmandelic Acid *Urine Decrease Physiological* Effect observed in schizophrenics. *2427*

Nicardine

Cyclosporine *Serum Increase Physiological* Doubling of concentration (trough levels) when added to therapeutic regime. *3732*

Nicardipine

Glomerular Filtration Rate (GFR)
Urine Increase Physiological Increase by 35% after 0.5 mg i.v. in 7 patients with mild to moderate hypertension. *0195*

PAH Clearance *Urine Increase Physiological* After 0.5 mg i.v. increased renal blood flow in 7 patients with mild to moderate hypertension by average of 27%. *0195*

Sodium *Urine Increase Physiological* Increase by 56% after 0.5 mg i.v. in 7 patients with mild to moderate hypertension. *0195*

Nicergoline

Creatinine *Serum No Effect Analytical* At 10 mg/L on reversed phase LC procedure of Zhiri et al. *3942*

Uric Acid *Serum No Effect Analytical* At 10 mg/L on reversed phase LC procedure of Zhiri et al. *3942*

Nickel

Leukocytes *Blood Increase Physiological* Occurs with industrial exposure. *2359*

Nicotine

Alkaloids *Urine Increase Analytical* Blue with iodoplatinate in Frings TLC procedure. *1204*

Amphetamine *Urine No Effect Analytical* No effect at 100 mg/L on method of Rutter. *3097* At 50 mg/L on fluorescent method of Hayes. *1534*

Antidiuretic Hormone *Plasma Increase Physiological* In smokers, maximum immediately after smoking. *1703*

Catecholamines *Urine Increase Physiological* Active component of cigarettes etc: same effect. *1921*

Epinephrine *Plasma Increase Physiological* Large effect observed. *1176*

Fatty Acids, Free (FFA) *Serum Increase Physiological* 30% increase during smoking of 3 cigarettes. *1921*

Glucose *Serum Increase Physiological* Due to adrenal response in poisoning. *1302*

11-Hydroxycorticosteroids *Plasma Increase Physiological* Up to 80% increase after heavy smoking. *1922*

5-Hydroxyindoleacetic Acid (5-HIAA)
Urine Increase Physiological Releases serotonin. *1176*

Neurophysin *Plasma Increase Physiological* In smokers, maximum immediately after smoking. *1703*

Norepinephrine *Plasma Increase Physiological* Significant but smaller effect seen. *1176*

Prothrombin Time *Plasma Decrease Physiological* May partially counteract action of heparin. *1679*

Nifedipine

Albumin *Urine Increase Physiological* Increase from 14 to 28 µg/min 40 to 60 minutes after single oral dose of 20 mg sublingually. *0650*

Aldosterone *Plasma No Effect Physiological* Possibly due to drug's calcium antagonizing action since calcium is required to secrete aldosterone. *1609* Insignificant change from 108 to 121 pg/mL in young hypertensives after 3 h. *1609*
Urine Increase Physiological In 25 patients with mild to moderate primary hypertension given up to 80 mg daily. *2659*

Angiotensin-Converting Enzyme (ACE)
Serum No Effect Physiological No significant effect within 3 h after 10 mg orally. *1609*

Angiotensin I *Plasma Increase Physiological* In young but not old people within 3 h of 10 mg orally. *1609* From 1923 to 2669 pg/mL in young hypertensives after 3 h. *1609*

Angiotensin II *Plasma Increase Physiological* In young but not old people within 3 h of 10 mg orally. *1609* From 167 to 215 pg/mL in young hypertensives after 3 h. *1609*

Apolipoprotein AI *Serum No Effect Physiological* Increase by 3 mg/dL in patients given 30 to 60 mg daily for 8 weeks. *2550*
Serum Increase Physiological In 11 patients with 80 mg daily for 6 weeks. *3702* Increase by 5 % in 11 individuals treated for 1.5 mo. *0088*

Apolipoprotein AII *Serum Increase Physiological* In 11 patients with 80 mg daily for 6 weeks. *3702* Increase by 7 % in 11 individuals treated for 1.5 mo. *0088*

Apolipoprotein B *Serum No Effect Physiological* In 11 patients with 80 mg daily for 6 weeks. *3702* Decrease by -6 mg/dL in patients given 30 to 60 mg daily for 8 weeks. *2550*

Apolipoprotein CII *Serum No Effect Physiological* No change in patients given 30 to 60 mg daily for 8 weeks. *2550*

Apolipoprotein CIII *Serum No Effect Physiological* No change in patients given 30 to 60 mg daily for 8 weeks. *2550*

Apolipoprotein E *Serum No Effect Physiological* No change in patients given 30 to 60 mg daily for 8 weeks. *2550*

Atrial Natriuretic Peptide *Plasma Increase Physiological* After 10 mg sublingually increased from mean 19.4 pg/mL to 24.1 pg/mL at 90 minutes. *2917*

Bleeding Time *Blood Increase Physiological* Observed 1 h after ingestion of 20 mg in 20 people. *0796*

Calcium *Serum No Effect Physiological* No effect of 80 mg/d for 2 d. *3694*

Cholesterol *Serum No Effect Physiological* No effect observed in 14 patients with essential hypertension treated over 2 mo with 20 mg twice daily. *2119* Decrease by -5 mg/dL in patients given 30 to 60 mg daily for 8 weeks. *2550*
Serum Decrease Physiological In 23 patients over 60 y old with essential mild to moderate hypertension. *3192*

Cholesterol, High Density Lipoprotein
Serum No Effect Physiological In 11 patients with 80 mg daily for 6 weeks. *3702* No effect observed in 14 patients with essential hypertension treated over 2 mo with 20 mg twice daily. *2119* Increase by 1 mg/dL in patients given 30 to 60 mg daily for 8 weeks. *2550*
Serum Increase Physiological In 100 patients in a double-blind randomized trial. *3965*

Cholesterol, Low Density Lipoprotein
Serum No Effect Physiological No effect observed in 14 patients with essential hypertension treated over 2 mo with 20 mg twice daily. *2119*

Cholesterol, Very Low Density Lipoprotein
Serum No Effect Physiological No effect observed in 14 patients with essential hypertension treated over 2 mo with 20 mg twice daily. *2119*

Cortisol *Plasma Decrease Physiological* Slight but progressive decrease over 3 h after oral administration of 10 mg in young people but not in old. *1609*

Creatinine *Serum Increase Physiological* Acute reversible deterioration of renal function in 4 patients with chronic renal failure. *0898*

Creatinine Clearance *Urine Increase Physiological* After 10 mg sublingually in next 2 h. *2917*

Cyclosporine *Serum No Effect Physiological* No effect on concentration when drug added (20-30 mg/d) to therapeutic regime. *3732*

Digoxin *Serum Increase Physiological* 45% increase in volunteers given both drugs compared with digoxin alone, but mechanism uncertain. *0281* Increase by 15% in patients to whom 20 mg bid were added to a stable digoxin regime. *1957*

1,25-Dihydroxy Vitamin D$_3$ *Serum No Effect Physiological* No effect of 80 mg/d for 2 d. *3694*

Glucagon *Plasma Increase Physiological* Significant effect in normal subjects (0.045 vs 0.034 nmol/L). *0637*

Glucose *Serum No Effect Physiological* In 11 patients with 80 mg daily for 6 weeks. *3702*
Serum Increase Physiological Fasting concentration increased by 10% in normal subjects. *0637*
Serum Decrease Physiological From 102 to 95 mg/dL in 15 hypertensive patients undergoing hemodialysis after 3 weeks treatment. *2986*

Glucose Tolerance *Serum No Effect Physiological* In 11 patients with 80 mg daily for 6 weeks. *3702* In 15 hypertensive patients undergoing hemodialysis after 3 weeks treatment. *2986*
Serum Decrease Physiological Both in normal subjects and those with already impaired glucose tolerance but in normals improved tolerance after 60 minutes. *1296*

Indomethacin *Serum No Effect Analytical* No effect on HPLC method of Roberts and Smith. *3002*

Insulin *Plasma No Effect Physiological* In 11 patients with 80 mg daily for 6 weeks. *3702*
Plasma Decrease Physiological Response to glucose challenge significantly reduced in subjects taking drug. *1296* Basal insulin concentration reduced by 26% in normals. *0637* From 20 to 14 uU/mL in 15 hypertensive patients undergoing hemodialysis after 3 weeks treatment. *2986*

Ketones *Serum Increase Physiological* In 15 hypertensive patients undergoing hemodialysis after 3 weeks treatment. *2986*

Lactate *Serum No Effect Physiological* In 15 hypertensive patients undergoing hemodialysis after 3 weeks treatment. *2986*

β_2-Microglobulin *Urine Increase Physiological* Increase from 0.12 to 0.74 µg/min 40 to 60 minutes after single oral dose of 20 mg sublingually. *0650*

Phosphate *Serum No Effect Physiological* No effect of 80 mg/d for 2 d. *3694*

Platelet Aggregation *Blood Decrease Physiological* Significant reduction of collagen-induced and ADP-induced aggregation probably by inhibiting increase of intracytoplasmic Calcium by blocking Calcium channel through platelet membrane: also inhibits platelet aggregability induced by exercise. *3532* Maximal rate in response to ADP reduced by 20-26% and by 23% in response to collagen. *0796*

Platelets *Blood No Effect Physiological* No effect in 20 people following ingestion of 20 mg. *0796*

Potassium *Serum Increase Physiological* In 23 patients over 60 y old with essential mild to moderate hypertension. *3192* At dose of 40 mg/d caused increase of 0.3 mmol/L when also treated with propranolol. *2836*
Serum Decrease Physiological Single case of drug induced hypokalemia reported. *3599*
Urine No Effect Physiological No effect following single 20 mg sublingual dose. *0650*

Prolactin *Plasma No Effect Physiological* No effect of 80 mg/d for 2 d. *3694*

Protein *Urine Increase Physiological* Acute reversible deterioration of renal function in 4 patients with chronic renal failure. *0898*

Pyruvate *Serum Increase Physiological* In 15 hypertensive patients undergoing hemodialysis after 3 weeks treatment. *2986*

Quinidine *Serum Increase Physiological* Marked effect observed once drug discontinued, probably hemodynamically induced interaction. *1082*
Serum Decrease Physiological Marked decrease when drugs coadministered but mechanism not clearly understood. *1388*

Renin Activity *Plasma Increase Physiological* Observed within 3 h in response to 10 mg drug. *1609* From 2.26 to 3.72 ng/mL/h in young hypertensives after 3 h. *1609* In 25 patients with mild to moderate primary hypertension given up to 80 mg daily. *2659*

Sodium *Serum Decrease Physiological* In 23 patients over 60 y old with essential mild to moderate hypertension. *3192*

Urine _Increase Physiological_ Significant effect following 20 mg sublingually. _0650_ After 10 mg sublingually in next 2 h. _2917_

Triglycerides _Serum No Effect Physiological_ No effect observed in 14 patients with essential hypertension treated over 2 mo with 20 mg twice daily. _2119_ Increase by 4 mg/dL in patients given 30 to 60 mg daily for 8 weeks. _2550_
Serum Decrease Physiological In 23 patients over 60 y old with essential mild to moderate hypertension. _3192_ In 100 patients in a double-blind randomized trial. _3965_

Triglycerides, High Density Lipoprotein
Serum Decrease Physiological In 11 patients with 80 mg daily for 6 weeks. _3702_

Urea Nitrogen _Serum Increase Physiological_ Acute reversible deterioration of renal function in 4 patients with chronic renal failure. _0898_

Uric Acid _Urine Increase Physiological_ Significant effect following 20 mg sublingual dose. _0650_

Volume _Urine Increase Physiological_ After 10 mg sublingually in next 2 h. _2917_

Niflumic Acid

Cholesterol _Serum No Effect Analytical_ At concentration of 200 mg/L had no effect on CHOD-PAP method. _3393_

Creatinine _Serum No Effect Analytical_ At 200 mg/L on reversed phase LC procedure of Zhiri et al. _3942_ At concentration of 150 mg/L had no effect on Jaffé-Fading-Fraction method. _3393_ At concentration of 150 mg/L had no effect on Jaffé-Fuller's earth method. _3393_

Uric Acid _Serum No Effect Analytical_ At 200 mg/L on reversed phase LC procedure of Zhiri et al. _3942_ At concentration of 200 mg/L had no effect on Kageyama-Hantzsch method. _3393_ At concentration of 150 mg/L had no effect on catalase-AIDH method. _3393_

Nifurtimox

Glucose _Serum Decrease Physiological_ May cause decline. _0071_

Nilvadipine

Norepinephrine _Plasma Increase Physiological_ Slight effect only in 10 individuals with mild hypertension. _3530_

Renin Activity _Plasma No Effect Physiological_ In 10 patients with mild essential hypertension. _3530_

Nipagin

Glucose _Serum No Effect Analytical_ At 2 g/dL on glucose oxidase procedure. At 2 g/dL on p-HBAH procedure of Lever. At 2 g/dL on alkaline ferricyanide procedure. At 2 g/dL on o-toluidine procedure. _2140_

Niridazole

Color _Urine Increase Analytical_ Urine becomes dark. _1343_

Eosinophils _Blood Increase Physiological_ Quite common allergic response. _2427_

Erythrocytes _Blood Decrease Physiological_ May cause hemolysis if G-6-PD deficiency. _1343_

Hematocrit _Blood Decrease Physiological_ Occurs with marked hemolysis. _1343_

Hemoglobin _Blood Decrease Physiological_ Occurs with marked hemolysis. _1343_

Methemoglobin _Blood Increase Physiological_ May cause hemolysis if G-6-PD deficiency. _1343_

Nisoldipine

Bilirubin _Serum No Effect Physiological_ In 14 mild to moderately hypertensive noninsulin dependent diabetic patients. _2653_

Calcium _Serum Increase Physiological_ Change from average of 2.4 to 2.5 mmol/L in 14 mild to moderately hypertensive noninsulin dependent diabetic patients. _2653_

Cholesterol _Serum No Effect Physiological_ In 14 mild to moderately hypertensive noninsulin dependent diabetic patients. _2653_

Creatinine _Serum No Effect Physiological_ In 14 mild to moderately hypertensive noninsulin dependent diabetic patients. _2653_

Digoxin _Serum Increase Physiological_ Increased trough values by 15%. _1939_

Glucose _Serum No Effect Physiological_ In 14 mild to moderately hypertensive noninsulin dependent diabetic patients. _2653_

Hemoglobin _Blood Decrease Physiological_ From 14.7 to 14.0 g/dL in 14 mild to moderately hypertensive noninsulin dependent diabetic patients. _2653_

Hemoglobin A$_{1c}$ _Blood Decrease Physiological_ From 14.7 to 14.0 g/dL in 14 mild to moderately hypertensive noninsulin dependent diabetic patients. _2653_

Sodium _Serum Decrease Physiological_ Fell by average of 3 mmol/L in 14 mild to moderately hypertensive noninsulin dependent diabetic patients. _2653_

Triglycerides _Serum Decrease Physiological_ From 1.9 to 1.6 mmol/L in 14 mild to moderately hypertensive noninsulin dependent diabetic patients. _2653_

Urea Nitrogen _Serum No Effect Physiological_ In 14 mild to moderately hypertensive noninsulin dependent diabetic patients. _2653_

Uric Acid _Serum No Effect Physiological_ In 14 mild to moderately hypertensive noninsulin dependent diabetic patients. _2653_

Nitrates

Cholesterol _Serum Decrease Analytical_ Interfere with Zlatkis-Zak reaction. _3505_

Methemoglobin _Blood Increase Physiological_ May cause hemolysis. _2426_

Volume _Urine Increase Physiological_ Renal tubular epithelium relatively impermeable. _1343_

Zinc _Serum No Effect Analytical_ At 150 to 1 with flameless atomic absorption. _2038_
Urine No Effect Analytical At 150 to 1 with flameless atomic absorption. _2038_

Nitrazepam

Acetaminophen _Serum No Effect Analytical_ At 2 mg/L no effect on HPLC method. _3432_

Amitriptyline _Serum No Effect Physiological_ No effect on serum concentration if given together. _3321_

Amphetamine _Urine No Effect Analytical_ No effect at 100 mg/L on method of Rutter. _3097_

Antipyrine _Plasma No Effect Physiological_ No significant effect on half-life. _3466_

Barbiturate _Serum Increase Analytical_ Forms Hg complex with diphenylcarbazone/Baer's method. _0294_

Bilirubin _Serum No Effect Physiological_ Insignificant displacement from protein in neonates. _3748_

Calcium _Serum No Effect Physiological_ No effect when given in normal therapeutic amounts to elderly. _3925_

Diphenylhydantoin _Serum No Effect Analytical_ On GLC procedure of Papadopoulos at _in vivo_ concentration. _2722_
Serum Increase Physiological Probably inhibits metabolism in liver. _1487_

Folate _Serum No Effect Analytical_ If chromatographic procedure of Landon used. _2068_

Glucose _Serum Decrease Analytical_ Slight effect glucose oxidase method of Boehringer. _3277_

17-Hydroxycorticosteroids _Urine No Effect Physiological_ No effect although 6-hydroxycortisol increased. _3466_

6-β-Hydroxycortisol _Urine No Effect Physiological_ No significant effect. _3466_

Nitrazepam *(continued)*

5-Hydroxyindoleacetic Acid (5-HIAA)
Urine Increase Analytical Slight effect when added *in vitro* in method of Udenfriend. *0670*

Nortriptyline *Serum No Effect Physiological* No effect on serum concentration if given together. *3321*

Pheneturide *Serum No Effect Analytical* On GLC procedure of Papadopoulos at *in vivo* concentration. *2722*

Phenobarbital *Serum No Effect Analytical* On GLC procedure of Papadopoulos at *in vivo* concentration. *2722*

Phenylbutazone *Serum No Effect Physiological* No significant effect on half-life. *3466*

Platelets *Blood Decrease Physiological* May cause thrombocytopenia. *2583*

Primidone *Serum No Effect Analytical* On GLC procedure of Papadopoulos at *in vivo* concentration. *2722*

Prothrombin Time *Plasma No Effect Physiological* No effect observed with phenprocoumon. *0340*

Nitric Acid

Hemoglobin *Urine Increase Analytical* Affects benzidine test. *2313*

Nitrites

Cholesterol *Serum Decrease Analytical* 10 μg in reaction mixture decreases by 30%. *0583*

Erythrocytes *Blood Decrease Physiological* May cause hemolytic anemia. *2313*

Ethanol *Blood No Effect Analytical* Over 16 weeks at 20° at concentration of 0.5% w/v. *3355*

Hematocrit *Blood Decrease Physiological* May cause hemolytic anemia. *2313*

Hemoglobin *Blood Decrease Physiological* May cause hemolytic anemia. *2313*

Methemoglobin *Blood Increase Physiological* Effect of organic nitrites less than amyl nitrite. *1343*

Occult Blood *Feces Positive Physiological* May cause bloody diarrhea if ingested. *1302*

Urobilinogen *Urine Decrease Analytical* Slight effect only p-MDFB procedure of Rutter Marked effect with Ehrlich's procedure. *2044*

Nitrobenzene

Color *Urine Increase Analytical* Dark color on standing. *2313*

Erythrocytes *Blood Decrease Physiological* Occurs with hemolysis. *2429*

Hematocrit *Blood Decrease Physiological* Occurs with hemolysis. *2429*

Hemoglobin *Blood Decrease Physiological* Occurs with hemolysis. *2429*

Methemoglobin *Blood Increase Physiological* May cause hemolysis. *2429*

Nitrofurans

Alanine Aminotransferase *Serum Increase Physiological* Hepatotoxicity. *2313*

Alkaline Phosphatase *Serum Increase Physiological* Hepatotoxicity. *2313*

Aspartate Aminotransferase
Serum Increase Physiological Hepatotoxicity. *2313*

Bile *Urine Increase Physiological* Hepatotoxicity. *2313*

Bilirubin *Serum Increase Physiological* Hepatotoxicity or due to hemolytic anemia. *2313*

BSP Retention *Serum Increase Physiological* Hepatotoxicity. *2313*

Cephalin Flocculation *Serum Increase Physiological* Hepatotoxicity. *2313*

Color *Urine Increase Analytical* Brown,green, blue color. *1237*

Creatinine *Urine Increase Analytical* React with color reagent. *2220*

Eosinophils *Blood Increase Physiological* May be serious anaphylactoid reaction. *0071*

Folate *Serum Decrease Physiological* May induce folate deficiency- megaloblastic anemia. *2427*

Heinz Body Formation *Blood Positive Physiological* Occurs initially with hemolysis. *1902*

Hematocrit *Blood Decrease Physiological* May cause hemolytic anemia if G-6-PD deficiency. *0071*

Hemoglobin *Plasma Increase Physiological* Occurs with intravascular hemolysis. *2426*
Blood Decrease Physiological May cause hemolytic anemia if G-6-PD deficiency. *0071*
Urine Increase Physiological May occur with severe hemolytic anemia. *1902*

Leukocytes *Blood Decrease Physiological* May occasionally cause agranulocytosis. *2427*

MCV *Blood Increase Physiological* May cause megaloblastic anemia. *2427*

Methemoglobin *Blood Increase Physiological* May cause hemolytic anemia if G-6-PD deficiency. *0071*

Reduced Glutathione
Red Blood Cells Decrease Physiological Occurs with hemolysis. *1902*

Reticulocytes *Blood Increase Physiological* Occurs during recovery after hemolysis. *1902*

Sugar *Urine Increase Analytical* Metabolites reduce Benedict's reagent. *2220*

Thymol Turbidity *Serum Increase Physiological* Hepatotoxicity. *2313*

Nitrofurantoin

Alanine Aminotransferase *Serum No Effect Analytical* On routine methods in use on SMAC®, Ektachem®, Abbott-VP, Cobas-Bio, aca, Hitachi® 705, KDA at 5 times normal upper therapeutic concentration. *2138* No effect at therapeutic concentration on Reflotron method. *1984*
Serum Increase Physiological May cause cholestatic jaundice. *2808* Moderate increase chronic active hepatitis, much more common in women than men, but still rare. *2444* Moderate increase chronic active hepatitis, much more common in women than men, but still rare. *2444*

Albumin *Serum No Effect Analytical* On routine methods in use on SMAC®, Ektachem®, Abbott-VP, Cobas-Bio, aca, Hitachi® 705, KDA at 5 times normal upper therapeutic concentration. *2138* At concentration of 4 mg/L had no effect on BCG method. *3393*
Serum Decrease Physiological In association with chronic active hepatitis. *3488* In association with chronic active hepatitis. *3488*

Alkaline Phosphatase *Serum No Effect Analytical* On routine methods in use on Abbott-VP, Cobas-Bio, aca, Hitachi® 705, KDA at 5 times normal therapeutic concentration. *2138*
Serum Increase Analytical 19% increase on p-nitrophenyl phosphate method on SMAC® at 5 times normal therapeutic concentration. *2138*
Serum Increase Physiological May cause cholestatic jaundice. *2808* Moderate increase chronic active hepatitis, much more common in women than men, but still rare. *2444* Moderate increase chronic active hepatitis, much more common in women than men, but still rare. *2444* May follow acute hepatitis with or without cholestasis or as chronic active hepatitis or chronic granulomatous reaction. *2374*
Urine Decrease Analytical Interference with determination method. *2620*

Amylase *Serum No Effect Analytical* On routine methods in use on SMAC®, Ektachem®, Abbott-VP, Cobas-Bio, aca, Hitachi® 705, KDA at 5 times normal upper therapeutic concentration. *2138*
Serum Increase Physiological Isolated case confirmed by rechallenge, rapidly resolved on withdrawal of drug. Edema of pancreatic head also caused jaundice. *2571* Isolated case confirmed by rechallenge, rapidly resolved on withdrawal of drug. Edema of pancreatic head also caused jaundice. *2571*

Urine Increase Physiological Isolated case confirmed by rechallenge, rapidly resolved on withdrawal of drug. Edema of pancreatic head also caused jaundice. *2571* Isolated case confirmed by rechallenge, rapidly resolved on withdrawal of drug. Edema of pancreatic head also caused jaundice. *2571*

Antinuclear Antibodies *Serum Increase Physiological* Increased to 1:640 chronic active hepatitis, much more common in women than men, but still rare. *2444* Increased to 1:640 chronic active hepatitis, much more common in women than men, but still rare. *2444*

Antismooth Muscle Antibodies
Serum Increase Physiological Increased to 1:640 chronic active hepatitis, much more common in women than men, but still rare. *2444* Increased to 1:640 chronic active hepatitis, much more common in women than men, but still rare. *2444*

Antithyroglobulin Antibodies
Serum Increase Physiological Mild increase developed lupus-like syndrome associated with pulmonary reaction. *3250*

Aspartate Aminotransferase *Serum No Effect Analytical* On routine methods in use on SMAC®, Abbott-VP, aca, Cobas-Bio, Hitachi® 705 at 5 times normal therapeutic concentration. *2138* No effect at therapeutic concentration on Reflotron method. *1984*
Serum Increase Analytical 36% increase on kinetic MDH procedure on KDA at 5 times normal therapeutic concentration. *2138*
Serum Increase Physiological May cause cholestatic jaundice. *2808* Moderate increase chronic active hepatitis, much more common in women than men, but still rare. *2444* Moderate increase chronic active hepatitis, much more common in women than Men but still rare. *2444* May follow acute hepatitis with or without cholestasis or as chronic active hepatitis or chronic granulomatous reaction. *2374*

Bicarbonate *Serum No Effect Analytical* At concentration of 4 mg/L had no effect on method using phenolphthalein. *3393*
Serum Decrease Physiological Nephrotoxicity may cause azotemia. *1022*

Bile *Urine Increase Physiological* May cause cholestatic jaundice. *2313*

Bilirubin *Serum No Effect Analytical* On routine methods in use on SMAC®, Abbott-VP, Cobas-Bio, aca, Hitachi® 705, KDA at 5 times normal upper therapeutic concentration. *2138* At concentration of 4 mg/L had no effect on Jendrassik and Grof method. *3393*
Serum Increase Analytical Clinically significant effect at upper limit of normal therapeutic range on Ektachem® 400/700 method. *2138* Clinically significant effect at upper limit of normal therapeutic range on Ektachem® 400/700 method. *2138*
Serum Increase Physiological May cause hemolytic anemia or cholestasis. *1974* Chronic active hepatitis much more common in women than men, but still rare. *2444* Chronic active hepatitis much more common in women than men, but still rare. *2444* May follow acute hepatitis with or without cholestasis or as chronic active hepatitis or chronic granulomatous reaction. *2374*

BSP Retention *Serum Increase Physiological* Intrahepatic cholestatic jaundice. *1045*

Calcium *Serum No Effect Analytical* On routine methods in use on SMAC®, Ektachem®, Abbott-VP, aca, Hitachi® 705, KDA at 5 times normal upper therapeutic concentration. *2138* At concentration of 4 mg/L had no effect on cresolphthalein method. *3393*

Cephalin Flocculation *Serum Increase Physiological* May be cholestatic jaundice. *2313*

Chloride *Serum No Effect Analytical* At concentration of 4 mg/L had no effect on mercurimetric method. *3393*

Cholesterol *Serum No Effect Analytical* On routine methods in use on SMAC®, Ektachem®, Abbott-VP, aca, Hitachi® 705, KDA at 5 times normal upper therapeutic concentration. *2138* No effect at therapeutic concentration on Reflotron method. *1984* At concentration of 98 mg/L had no effect on CHOD-Iodide method. *3393* At concentration of 98 mg/L had no effect on CHOD-PAP method. *3393* At concentration of 98 mg/L had no effect on catalase-AIDH method. *3393* At concentration of 98 mg/L had no effect on method using catalase-Hantzsch reaction. *3393* At concentration of 98 mg/L had no effect on Liebermann-Burchard method. *3393*

Color *Urine Increase Analytical* Brown, yellow color. *1022*

Coombs' Test, Direct *Blood Positive Physiological* Developed lupus-like syndrome. Syndrome associated with pulmonary reaction. *3250*

Creatine Kinase *Serum No Effect Analytical* On routine methods in use on SMAC®, Ektachem®, Abbott-VP, aca, Hitachi® 705, KDA at 5 times normal upper therapeutic concentration. *2138*

Creatinine *Serum No Effect Analytical* On routine methods in use on SMAC®, Ektachem®, Abbott-VP, Cobas-Bio, aca, Hitachi® 705, KDA at 5 times normal upper therapeutic concentration. *2138* At concentration of 4 mg/L had no effect on AutoAnalyzer Jaffé method. *3393*
Serum Increase Analytical At concentrations above 5 mg/L (normal therapeutic concentration 5.5 mg/L) raised concentration as measured by Jaffé-Fading-Fraction method. *3393* At concentrations above 5 mg/L (normal therapeutic concentration 5.5 mg/L) raised concentration as measured by Jaffé-Fuller's earth method. *3393* At concentrations above 18 mg/L (normal therapeutic concentration 5.5 mg/L) raised concentration as measured by kinetic Jaffé method on BKA-2. *3393*
Serum Increase Physiological Nephrotoxic effect. *2220* In 8 of 56 acute and 3 of 22 chronic drug induced pulmonary reactions. *1639* In 8 of 56 acute and 3 of 22 chronic drug induced pulmonary reactions. *1639*

Eosinophils *Blood Increase Physiological* Allergic response (greater than 1%). *1974* Mild increase developed lupus-like syndrome associated with pulmonary reaction. *3250* In 158 of 191 acute and 14 of 32 chronic drug induced pulmonary reactions. *1639* Mild increase developed lupus-like syndrome associated with pulmonary reaction. *3250* In 158 of 191 acute and 14 of 32 chronic drug induced pulmonary reactions. *1639*

Erythrocytes *Blood Decrease Physiological* Hypersensitivity (G-6-PD)/megaloblastic anemia. *2358*

Erythrocyte Sedimentation Rate
Blood Increase Physiological Developed lupus-like syndrome associated with pulmonary reaction. *3250* Mild increase developed lupus-like syndrome associated with pulmonary reaction. *3250*

Folate *Serum Decrease Physiological* Inhibits intestinal conjugase. *3058*

γ-Globulin *Serum Increase Physiological* In 3 of 9 acute and 16 of 20 chronic pulmonary reactions. *1640* In 3 of 9 acute and 16 of 20 chronic pulmonary reactions. *1640*

Glucose *Serum No Effect Analytical* No effect at therapeutic concentration on Reflotron method. *1984* On routine methods in use on SMAC®, Ektachem®, Abbott-VP, Cobas-Bio, aca, Hitachi® 705, KDA at 5 times normal upper therapeutic concentration. *2138* At concentration of 5 mg/L had no effect on Ektachem® method. *3393* At concentration of 2.5 mg/L had no effect on GOD/POD-PAP method. *3393* At concentration of 60 mg/L had no effect on hexokinase/G-6-PDH method. *3393*
Urine No Effect Analytical At concentration of 500 mg/L had no effect on Diabur-test. *3393*

Glucose Tolerance *Serum Decrease Physiological* Single case reported. *1463*

γ-Glutamyltransferase (GGT) *Serum No Effect Analytical* On routine methods in use on SMAC®, Ektachem®, Abbott-VP, Cobas-Bio, aca, Hitachi® 705, KDA at 5 times normal upper therapeutic concentration. *2138* No effect at therapeutic concentration on Reflotron method. *1984*

Haptoglobin *Serum Decrease Physiological* Hemolytic anemia. *2313*

Heinz Body Formation *Blood Positive Physiological* May cause hemolytic anemia. *0333*

Hematocrit *Blood Decrease Physiological* Megaloblastic anemia/hypersensitivity (G-6-PD). *2358*

Hemoglobin *Blood Decrease Physiological* Megaloblastic anemia/hypersensitivity (G-6-PD). *2358*

Immunoglobulin IgA *Serum Increase Physiological* Chronic active hepatitis, much more common in women than men, but still rare. *2444* Chronic active hepatitis much more common in women than men, but still rare. *2444*

Immunoglobulin IgG *Serum Increase Physiological* Chronic active hepatitis, much more common in women than men, but still rare. *2444* Chronic active hepatitis, much more common in women than men, but still rare. *2444*

Nitrofurantoin (continued)

Immunoglobulin IgM *Serum Increase Physiological* Chronic active hepatitis, much more common in women than men, but still rare. *2444* Chronic active hepatitis, much more common in women than men, but still rare. *2444*

Indocyanine Green *Serum Decrease Physiological* Mechanism not yet established. *2398*

Iron *Serum No Effect Analytical* On routine methods in use on SMAC®, Ektachem®, Abbott-VP, Cobas-Bio, aca, Hitachi® 705, KDA at 5 times normal upper therapeutic concentration. *2138*

Lactate *Serum No Effect Analytical* At concentration of 60 mg/L had no effect on enzymatic method. *3393*

Lactate Dehydrogenase *Serum No Effect Analytical* On routine methods in use on SMAC®, Ektachem®, Abbott-VP, Cobas-Bio, aca, Hitachi® 705, KDA at 5 times normal upper therapeutic concentration. *2138*
Serum Increase Physiological May cause hemolytic anemia. *1974* In 6 of 17 acute and 12 of 19 chronic drug induced pulmonary reactions. *1639* In 16 of 1756 acute and 12 of 19 chronic drug induced pulmonary reactions. *1639* May follow acute hepatitis with or without cholestasis or as chronic active hepatitis or chronic granulomatous reaction. *2374*
Urine Decrease Analytical Interference with determination method. *2620*

Leukocytes *Blood Increase Physiological* In 80 of 153 acute and 5 of 33 chronic drug induced pulmonary reactions. *1639* In 80 of 153 acute and 5 of 33 chronic drug induced pulmonary reactions. *1639*
Blood Decrease Physiological Leukopenia/agranulocytosis may occur. *0994*

MCV *Blood Decrease Physiological* Megaloblastic anemia/hypersensitivity (G-6-PD). *2358*

Methemoglobin *Blood Increase Physiological* May cause hemolysis with G-6-PD deficiency. *3581*

Neutrophils *Blood Decrease Physiological* Occasional case of agranulocytosis reported. *3717*

Nonprotein Nitrogen *Serum Increase Physiological* Nephrotoxic effect. *2313*

5'-Nucleotidase *Serum Increase Physiological* Due to cholestasis. *1778*

Phosphate *Serum No Effect Analytical* On routine methods in use on SMAC®, Ektachem®, Abbott-VP, Cobas-Bio, aca, Hitachi® 705, KDA at 5 times normal upper therapeutic concentration. *2138* At concentration of 4 mg/L had no effect on phosphomolybdate method. *3393*

Platelet Aggregation *Blood Decrease Physiological* Significant effect on ADP induced aggregation. *3068*

Platelets *Blood Decrease Physiological* Thrombocytopenia. *0994*

Potassium *Serum No Effect Analytical* At concentration of 2.5 mg/L had no effect on flame-photometric method. *3393* At concentration of 14 mg/L had no effect on measurement by ISE with predilution. *3393* At concentration of 50 mg/L had no effect on measurement by ISE without predilution. *3393*

Protein *Serum No Effect Analytical* On routine methods in use on SMAC®, Ektachem®, Abbott-VP, Cobas-Bio, aca, Hitachi® 705, KDA at 5 times normal upper therapeutic concentration. *2138* At concentration of 4 mg/L had no effect on biuret method with blank correction. *3393*

Protein Electrophoresis *Serum No Effect Analytical* At concentration of 50 mg/L had no effect on automated Olympus-Hite method. *3393*

Sodium *Serum No Effect Analytical* At concentration of 2.5 mg/L had no effect on flame-photometric method. *3393* At concentration of 14 mg/L had no effect on method using ISE with predilution. *3393*
Serum Increase Analytical At concentrations above 40 mg/L (normal therapeutic concentration 5.5 mg/L) raised concentration as measured by ISE without predilution. *3393*

Sugar *Urine Increase Analytical* Metabolites may reduce Benedict's, yield false positive. *2220*

Thymol Turbidity *Serum Increase Physiological* May be cholestatic jaundice. *2313*

Triglycerides *Serum No Effect Analytical* On routine methods in use on SMAC®, Ektachem®, Abbott-VP, Cobas-Bio, aca, Hitachi® 705, KDA at 5 times normal upper therapeutic concentration. *2138* No effect at therapeutic concentration on Reflotron method. *1984* At concentration of 0.5 mg/L had no effect on GPO-PAP method. *3393* At concentration of 4 mg/L had no effect on lipase/esterase method. *3393*

Urea Nitrogen *Serum No Effect Analytical* No effect at therapeutic concentration on Reflotron method. *1984* On routine methods in use on SMAC®, Ektachem®, Abbott-VP, Cobas-Bio, aca, Hitachi® 705, KDA at 5 times normal upper therapeutic concentration. *2138* At concentration of 4 mg/L had no effect on diacetylmonoxime method. *3393* At concentration of 5 mg/L had no effect on Ektachem® method. *3393*
Serum Increase Physiological Possible decrease in already impaired renal function. *0085*

Uric Acid *Serum No Effect Analytical* No effect at therapeutic concentration on Reflotron method. *1984* On routine methods in use on SMAC®, Ektachem®, Abbott-VP, Cobas-Bio, aca, Hitachi® 705, KDA at 5 times normal upper therapeutic concentration. *2138* At concentration of 30 mg/L had no effect on Kageyama-Hantzsch method. *3393* At concentration of 30 mg/L had no effect on catalase-AIDH method. *3393* At concentration of 4 mg/L had no effect on phosphotungstate reduction method. *3393* At concentration of 0.5 mg/L had no effect on uricase-PAP method. *3393*

Urobilinogen *Urine Decrease Physiological* Intrahepatic cholestatic jaundice. *1045*

Nitrofurazone

Bilirubin *Serum Increase Physiological* May cause hemolysis with G-6-PD deficiency. *0248*

Creatinine *Urine Increase Analytical* React with color reagent. *1237*

Erythrocytes *Blood Decrease Physiological* May cause hemolysis with G-6-PD deficiency. *0248*

Haptoglobin *Serum Decrease Physiological* May cause hemolysis. *0248*

Heinz Body Formation *Blood Positive Physiological* May occur in early stages of hemolytic anemia. *0333*

Hematocrit *Blood Decrease Physiological* May cause hemolysis with G-6-PD deficiency. *0248*

Hemoglobin *Blood Decrease Physiological* May cause hemolysis with G-6-PD deficiency. *0248*

Methemoglobin *Blood Increase Physiological* May cause hemolytic anemia. *0333*

Sugar *Urine Increase Analytical* Metabolites may reduce Benedict's reagent Reducing action of metabolites. *1237*

Nitrogen Oxides

Methemoglobin *Blood Increase Physiological* Mild elevation may occur after nitrous oxidase inhalation. *1302*

pCO₂ *Blood Increase Physiological* Retention of CO_2 occurs. *1302*

Nitroglycerin

Catecholamines *Plasma Increase Physiological* Effect dosage dependent. *0127*
Urine Increase Physiological 2 fold increase of catecholamines. *1487*

Epinephrine *Urine Increase Physiological* ?due to adrenergic stimulation of hypotension. *3713*

Methemoglobin *Blood Increase Physiological* May cause hemolysis. *2426* Reported occasional complication of i.v. drug. *1598*

Nitroglycerin *Serum Decrease Physiological* Enhanced metabolism when phenobarbital coadministered. *1598*

Norepinephrine *Urine Increase Physiological* ?due to adrenergic stimulation of hypotension. *3713*

Platelets *Blood Decrease Physiological* Immunologic response occurs after 5 mo. *1436*

Triglycerides *Serum No Effect Analytical* No effect on method using glycerol kinase and glycerol-3-phosphate on RA-1000 and Ektachem® DT-60. *2588*
Serum Increase Analytical In method on Technicon® RA-1000 using glycerol oxidase. *2588*

Vanillylmandelic Acid *Urine Increase Physiological* Effect greater than that on catecholamines. *3713*

Nitrophenol

Alanine Aminotransferase *Serum Increase Physiological* Hypersensitive intrahepatic cholestasis. *1434*

Alkaline Phosphatase *Serum Increase Physiological* Hypersensitive intrahepatic cholestasis. *1434*

Aspartate Aminotransferase
Serum Increase Physiological Hypersensitive intrahepatic cholestasis. *1434*

Bilirubin *Serum Increase Physiological* Hypersensitive intrahepatic cholestasis. *1434*

Cholesterol *Serum Increase Physiological* Hypersensitive intrahepatic cholestasis. *1434*

Nitrous Oxide

Erythrocytes *Blood Decrease Physiological* Bone marrow depress (of no significance normally). *1343*

Folate *Serum Increase Physiological* If intraoperative exposure of patient is greater than 6 h, but unchanged if exposure for less than 1 h. *1985*

Vitamin B$_{12}$ *Serum No Effect Physiological* With intraoperative exposure for as long as 6 h. *1985*

Noise

Corticosteroids *Plasma Increase Physiological* Reported max at 10,000 Hz. *3351*

Epinephrine *Urine Increase Physiological* Usual drop in afternoon did not occur. *3351*

11-Hydroxycorticosteroids *Urine No Effect Physiological* Excretion unaffected by 80 decibels for 2 h. *3351*
Urine Increase Physiological Maximum effect at 10,000 Hz. *3351*

17-Ketosteroids *Urine Decrease Physiological* Reported on exposure to 130 decibels. *3351*

Norepinephrine *Urine Increase Physiological* Usual drop in afternoon less marked. *3351*

Noramidopyrine

Glucose *Serum No Effect Analytical* No effect on hexokinase procedures. *3524*
Serum Decrease Analytical A-V decrease 50 mg/L at *in vivo* concentration GOD-ABTS procedure. *3524*

Nordiazepam

Ethchlorvynol *Serum No Effect Analytical* At 600 mg/L on GLC procedure of Evenson. *1058*

Norethandrolone

Alanine Aminotransferase *Serum Increase Physiological* Usually reversible cholestasis. *1280*

Alkaline Phosphatase *Serum Increase Physiological* Cholestasis produced without cholangiolitis. *1538*

Aspartate Aminotransferase
Serum Increase Physiological Cholestasis produced without cholangiolitis. *3172*

Bile *Urine Increase Physiological* Cholestatic effect. *3172*

Bilirubin *Serum Increase Physiological* Intrahepatic cholestasis produced (up to 20%). *1280*
Urine Increase Physiological Cholestasis produced without cholangiolitis. *3172*

Bilirubin, Direct *Serum Increase Physiological* Reversible cholestasis produced. *1280*

BSP Retention *Serum Increase Physiological* Transport and conjugation of BSP impaired. *0153*

Cephalin Flocculation *Serum Increase Physiological* Cholestasis produced without cholangiolitis. *3172*

Cholesterol *Serum Increase Physiological* Due to cholestasis. *1713*

Fibrinogen *Plasma Increase Physiological* Metabolic effect. *0227*

β-Glucuronidase *Serum Increase Physiological* Metabolic effect. *0227*

Haptoglobin *Serum Increase Physiological* Metabolic effect. *0227*

Lactate Dehydrogenase *Serum Increase Physiological* ?part of cholestatic syndrome. *0583*

5'-Nucleotidase *Serum Increase Physiological* Due to cholestasis. *1778*

Plasminogen *Plasma Increase Physiological* Metabolic effect. *0227*

Prealbumin *Serum Increase Physiological* Decreases thyroxine binding globulin however. *0042*

Protein Bound Iodine (PBI) *Serum Decrease Physiological* Concentration of circulating thyroxine binding globulin lowered. *0012*

Prothrombin Time *Plasma Increase Physiological* Also seen with other 17-alkyl substituted steroids. *0998*

Thymol Turbidity *Serum Increase Physiological* Cholestasis produced without cholangiolitis. *3172*

Thyroxine (T$_4$) Binding Globulin
Serum Decrease Physiological Decreases concentration of thyroxine binding globulin, increases thyroxine binding prealbumin. *3505*

Urobilinogen *Urine Decrease Physiological* Manifestation of cholestasis. *1280*
Feces Decrease Physiological Light stools due to cholestasis. *1713*

Norethandrostenolone

Alkaline Phosphatase *Serum Increase Physiological* Intrahepatic cholestatic jaundice. *3197*

Bile *Urine Increase Physiological* Intrahepatic cholestatic jaundice. *3197*

Bilirubin *Serum Increase Physiological* Intrahepatic cholestatic jaundice. *3197*

BSP Retention *Serum Increase Physiological* Impaired hepatic uptake and excretion. *0652*

Norethindrone

α$_1$-Acid Glycoprotein *Serum Decrease Physiological* Metabolic estrogen effect. *0474*

Alanine Aminotransferase *Serum Increase Physiological* Intrahepatic cholestatic jaundice. *2313*

Albumin *Serum No Effect Physiological* No effect observed. *0474*

Alkaline Phosphatase *Serum Increase Physiological* Intrahepatic cholestatic jaundice. *1343*

Aspartate Aminotransferase
Serum Increase Physiological Intrahepatic cholestatic jaundice. *2313*

Bile *Urine Increase Physiological* Intrahepatic cholestatic jaundice. *1343*

Bilirubin *Serum Increase Physiological* Intrahepatic cholestatic jaundice. *1343*

BSP Retention *Serum Increase Physiological* Cholestatic phenomenon (occurs in 20%). *3332*

Carotene *Serum Decrease Physiological* Hormonal influence. *1221*

Cephalin Flocculation *Serum Increase Physiological* Intrahepatic cholestatic jaundice. *2313*

Ceruloplasmin *Serum Increase Physiological* Metabolic estrogen effect. *0474*

Cholesterol *Serum No Effect Physiological* No significant effect with 0.4 mg/d. *1309*

Norethindrone (continued)

Cortisol *Plasma Decrease Physiological* Slight effect compared with controls. *0261*

Urine Decrease Analytical Decreased unconjugated by fluorometric procedure. *0022*

Urine Decrease Physiological When results compared with normal menstrual cycle. *0261*

Glucose Tolerance *Serum Decrease Physiological* After 6 mo treatment compared with control. *3413*

Haptoglobin *Serum Increase Physiological* With high dose, prolonged treatment. *0474*

Serum Decrease Physiological Metabolic estrogen effect. *0474*

Insulin *Plasma Increase Physiological* During glucose tolerance test after 6 mo treat compared with control. *3413*

Luteinizing Hormone (LH) *Plasma Decrease Physiological* Apparent suppression of production or release. *2479*

Protein Bound Iodine (PBI) *Serum Increase Physiological* Reported effect but mechanism unknown. *0757*

Serum Decrease Physiological When treatment results compared with controls. *0260*

T_3 Uptake *Serum No Effect Physiological* When treatment results compared with controls. *0260*

Thymol Turbidity *Serum Increase Physiological* Intrahepatic cholestatic jaundice. *2313*

Thyroxine (T_4) *Serum Decrease Physiological* When treatment results compared with controls. *0260*

Thyroxine (T_4), Free *Serum Decrease Physiological* Slight effect when compared with controls. *0260*

Triglyceride Lipase *Serum Increase Physiological* In normal and hyperlipemic women. *1309*

Triglycerides *Serum Decrease Physiological* In normal and hyperlipemic women. *1309*

Vitamin A *Serum Increase Physiological* Due to alteration in concentration of binding globulin. *1221*

Norethisterone

Alanine Aminotransferase *Serum Increase Physiological* In 5.6% patients being treated for advanced or recurrent breast cancer. *2073*

Alkaline Phosphatase *Serum Increase Physiological* In 5.6% patients being treated for advanced or recurrent breast cancer. *2073*

Antithrombin III *Plasma No Effect Physiological* In 75 women who had received up to 24 intragluteal injections of 200 mg every 56 d. *1666*

Bilirubin *Serum Increase Physiological* In 5.6% patients being treated for advanced or recurrent breast cancer. *2073*

Cholesterol *Serum No Effect Physiological* In 75 women who had received up to 24 intragluteal injections of 200 mg every 56 d. *1666*

Cholesterol, High Density Lipoprotein
Serum Decrease Physiological In 75 women who had received up to 24 intragluteal injections of 200 mg every 56 d. *1666*

Cholesterol, Low Density Lipoprotein
Serum No Effect Physiological In 75 women who had received up to 24 intragluteal injections of 200 mg every 56 d. *1666*

Cholesterol, Very Low Density Lipoprotein
Serum No Effect Physiological In 75 women who had received up to 24 intragluteal injections of 200 mg every 56 d. *1666*

Factor X *Plasma No Effect Physiological* In 75 women who had received up to 24 intragluteal injections of 200 mg every 56 d. *1666*

Glucose Tolerance *Serum Increase Physiological* In 75 women who had received up to 24 intragluteal injections of 200 mg every 56 d. *1666*

Triglycerides *Serum No Effect Physiological* In 75 women who had received up to 24 intragluteal injections of 200 mg every 56 d. *1666*

Norethynodrel

Alanine Aminotransferase *Serum Increase Physiological* Cholestatic effect. *2313*

Alkaline Phosphatase *Serum Increase Physiological* Due to cholestasis. *1778*

Aspartate Aminotransferase
Serum Increase Physiological Cholestatic effect. *2313*

Bile *Urine Increase Physiological* Intrahepatic cholestatic jaundice. *1343*

Bilirubin *Serum Increase Physiological* Intrahepatic cholestatic jaundice. *1343*

BSP Retention *Serum Increase Physiological* Cholestatic phenomenon. *0757*

Cephalin Flocculation *Serum Increase Physiological* Cholestatic effect. *2313*

Glucose Tolerance *Serum Decrease Physiological* Gluconeogenetic effect of steroids. *1237*

17-Hydroxycorticosteroids *Urine Decrease Physiological* Decreased excretion of cortisol metabolites. *0022*

5'-Nucleotidase *Serum Increase Physiological* Due to cholestasis. *1778*

Protein Bound Iodine (PBI) *Serum Increase Physiological* Increases concentration of circulating thyroxine binding globulin. *0012*

Thymol Turbidity *Serum Increase Physiological* Cholestatic effect. *2313*

Thyroxine (T_4) Binding Globulin
Serum Increase Physiological Direct effect of drug. *0012*

Norfenefrin

Cholesterol *Serum No Effect Analytical* At concentration of 2.4 mg/L had no effect on CHOD-Iodide method. *3393* At concentration of 2.4 mg/L had no effect on CHOD-PAP method. *3393* At concentration of 2.4 mg/L had no effect on catalase-AlDH method. *3393* At concentration of 2.4 mg/L had no effect on method using catalase-Hantzsch reaction. *3393* At concentration of 2.4 mg/L had no effect on Liebermann-Burchard method. *3393*

Creatinine *Serum No Effect Analytical* At concentration of 6 mg/L had no effect on Jaffé-Fading-Fraction method. *3393* At concentration of 6 mg/L had no effect on Jaffé-Fuller's earth method. *3393*

Serum Decrease Analytical At concentrations above 4 mg/L (normal therapeutic concentration 0.4 mg/L) lowered concentration as measured by kinetic Jaffé method on BKA-2. *3393*

Glucose *Serum No Effect Analytical* At concentration of 0.48 mg/L had no effect on Ektachem® method. *3393*

Potassium *Serum No Effect Analytical* At concentration of 3 mg/L had no effect on measurement by ISE without predilution. *3393*

Protein Electrophoresis *Serum No Effect Analytical* At concentration of 4.8 mg/L had no effect on automated Olympus-Hite method. *3393*

Sodium *Serum Increase Analytical* At concentrations above 4.8 mg/L (normal therapeutic concentration 0.4 mg/L) raised concentration as measured by ISE without predilution. *3393*

Urea Nitrogen *Serum No Effect Analytical* At concentration of 0.48 mg/L had no effect on Ektachem® method. *3393*

Uric Acid *Serum No Effect Analytical* At concentration of 6 mg/L had no effect on Kageyama-Hantzsch method. *3393* At concentration of 6 mg/L had no effect on catalase-AlDH method. *3393*

Norfloxacin

Alanine Aminotransferase *Serum Increase Physiological* In 2 of 1540 patients in clinical trials. *0731*

Aspartate Aminotransferase
Serum Increase Physiological In 2 of 1540 patients in clinical trials. *0731*

Creatinine *Serum Increase Physiological* Rare complication of treatment. *3750*

Crystals *Urine Increase Physiological* May occur with high doses of drug. *3750*

Eosinophils *Blood Increase Physiological* In 2 of 1540 patients in clinical trials. *0731*

Erythrocytes *Blood Decrease Physiological* Rare complication of treatment. *3750*

Lactate Dehydrogenase *Serum Increase Physiological* In 2 of 1540 patients in clinical trials. *0731*

Leukocytes *Blood Decrease Physiological* Reduction but not to below 1,000/uL in 6 of 1540 patients, usually plateaued between 3,000-4,000/uL. *0731* Occurs in about 1% treated cases probably via an immunologic mechanism. *2747*

Neutrophils *Blood Decrease Physiological* Occurs in about 1% treated cases probably via an immunologic mechanism. *2747*

Urea Nitrogen *Serum Increase Physiological* Rare complication of treatment. *3750*

Norgestrel

Creatinine *Serum No Effect Analytical* At 10 mg/L on reversed phase LC procedure of Zhiri et al. *3942*

Uric Acid *Serum No Effect Analytical* At 10 mg/L on reversed phase LC procedure of Zhiri et al. *3942*

Norleucine

Tyrosine *Plasma No Effect Analytical* On fluorometric procedure of Ambrose. *0080*

Normetanephrine

Amino Acids *Urine Increase Analytical* Reacts with ninhydrin; extra spot TLC, high voltage electrophoresis. *2855*
Test Conditions Increase Analytical Positive spot test with ninhydrin. *3765*

Ammoniacal Silver Nitrate
Test Conditions Positive Analytical Positive spot test with Tollen's reagent. *3765*

Epinephrine *Test Conditions No Effect Analytical* On fluorescent procedure of Peyrin. *2800*
Test Conditions Increase Analytical 11.5% of fluorescence method of Waldmeier. *3735*

Glucose *Serum No Effect Analytical* No effect on glucose oxidase method of Boehringer. *3277*

Guaiacols Spot Test *Urine Positive Analytical* Action on procedure of Rogers. *3031*

HMPG (4-Hydroxy-3-Methoxy-Phenylethylene Glycol)
Urine No Effect Analytical 100 µg = 0-2 µg method of Bigelow. *0343*

^{131}I Uptake *Serum Decrease Physiological* Anhydron contains tetraiodofluorescein. *2652*

Millons Test *Test Conditions Positive Analytical* Positive spot test for phenols. *3765*

Norepinephrine *Test Conditions No Effect Analytical* On fluorescent procedure of Peyrin. *2800*

Uric Acid *Test Conditions Increase Analytical* Positive spot test with phosphotungstate. *3765*

Urobilinogen *Test Conditions Increase Analytical* Positive spot test with Ehrlich's reagent. *3765*

Normorphine

Morphine *Urine No Effect Analytical* Insignificant cross react with hemagglutination inhibition Insignificant cross react with EMIT procedure for opiates. *2514*

Norpropoxyphene

Propoxyphene *Urine Increase Analytical* 5% fluorescence in procedure of Valentour. *3662*

Norpseudoephedrine

Amphetamine *Urine No Effect Analytical* At 50 mg/L on fluorescent method of Hayes. *1534*

Nortriptyline

Alanine Aminotransferase *Serum No Effect Analytical* At acute overdose concentration (20 mg/dL) on Technicon® SMAC® method. *3719*
Serum Increase Physiological May cause cholestasis. *0071*

Albumin *Serum No Effect Analytical* At acute overdose concentration (20 mg/dL) on Technicon® SMAC® method. *3719* At concentration of 200 mg/L had no effect on BCG method. *3393*

Alkaline Phosphatase *Serum No Effect Analytical* At acute overdose concentration (20 mg/dL) on Technicon® SMAC® method. *3719*
Serum Increase Physiological May cause cholestatic jaundice. *0071*

Amphetamine *Urine Increase Analytical* Yields fluorophor with NBD chloride reaction. *2473*

Antipyrine *Plasma Increase Physiological* Impairs metabolism. *2641*

Aspartate Aminotransferase *Serum No Effect Analytical* At acute overdose concentration (20 mg/dL) on Technicon® SMAC® method. *3719*
Serum Increase Physiological May cause cholestasis. *0071*

Bicarbonate *Serum No Effect Analytical* At concentration of 2 mg/L had no effect on method using phenolphthalein. *3393*

Bile *Urine Increase Physiological* May cause cholestatic jaundice. *0071*

Bilirubin *Serum No Effect Analytical* At acute overdose concentration (20 mg/dL) on Technicon® SMAC® method. *3719* At concentration of 200 mg/L had no effect on Jendrassik and Grof method. *3393*
Serum No Effect Physiological Insignificant protein-displacement effect in neonates. *3748*
Serum Increase Physiological May cause cholestatic jaundice. *0071*

Bishydroxycoumarin *Plasma Increase Physiological* Impairs metabolism. *2641*

Calcium *Serum No Effect Analytical* At acute overdose concentration (20 mg/dL) on Technicon® SMAC® method. *3719* At concentration of 200 mg/L had no effect on cresolphthalein method. *3393*

Chloride *Serum No Effect Analytical* At concentration of 2 mg/L had no effect on mercurimetric method. *3393*

Cholesterol *Serum No Effect Analytical* At acute overdose concentration (20 mg/dL) on Technicon® SMAC® method. *3719* At concentration of 200 mg/L had no effect on Liebermann-Burchard method. *3393*

Creatine Kinase *Serum No Effect Analytical* At acute overdose concentration (20 mg/dL) on Technicon® SMAC® method. *3719*

Creatinine *Serum No Effect Analytical* At acute overdose concentration (20 mg/dL) on Technicon® SMAC® method. *3719* At concentration of 200 mg/L had no effect on AutoAnalyzer Jaffé method. *3393*

Eosinophils *Blood Increase Physiological* Presumed allergic response. *1680*

Ethchlorvynol *Serum No Effect Analytical* At 100 mg/L on GLC procedure of Evenson. *1058*

Glucose *Serum No Effect Analytical* At acute overdose concentration (20 mg/dL) on Technicon® SMAC® method. *3719*
Serum Increase Physiological Endocrine response. *1680*
Serum Decrease Physiological Endocrine response. *1680*

5-Hydroxyindoleacetic Acid (5-HIAA)
CSF Decrease Physiological Increase by 4.8 ng/mL in depressed patients. *0167*

Indoleacetic Acid *CSF Decrease Physiological* Increase by 2 ng/mL in depressed patients. *0167*

Iron *Serum No Effect Analytical* At acute overdose concentration (20 mg/dL) on Technicon® SMAC® method. *3719* At concentration of 200 mg/L had no effect on Ferrozine method. *3393*

Nortriptyline *(continued)*

Lactate Dehydrogenase *Serum No Effect Analytical* At acute overdose concentration (20 mg/dL) on Technicon® SMAC® method. *3719*

Leukocytes *Blood Decrease Physiological* May cause agranulocytosis. *0071*

Methaqualone *Serum No Effect Analytical* At 100 mg/L on GLC procedure of Evenson. *1057*

Nortriptyline *Serum Increase Physiological* 30-160 µg/L after 75-225 mg for 4 d. *2348*

Phosphate *Serum No Effect Analytical* At acute overdose concentration (20 mg/dL) on Technicon® SMAC® method. *3719* At concentration of 200 mg/L had no effect on phosphomolybdate method. *3393*

Platelets *Blood Decrease Physiological* Purpura and thrombocytopenia observed. *1680*

Protein *Serum No Effect Analytical* At acute overdose concentration (20 mg/dL) on Technicon® SMAC® method. *3719* At concentration of 200 mg/L had no effect on biuret method with blank correction. *3393*

Prothrombin Time *Plasma Increase Physiological* Patients on coumarins (unconfirmed clinically). *3704*

Triglycerides *Serum No Effect Analytical* At acute overdose concentration (20 mg/dL) on Technicon® SMAC® method. *3719* At concentration of 200 mg/L had no effect on lipase/esterase method. *3393*

Urea Nitrogen *Serum No Effect Analytical* At acute overdose concentration (20 mg/dL) on Technicon® SMAC® method. *3719* At concentration of 200 mg/L had no effect on diacetylmonoxime method. *3393*

Uric Acid *Serum No Effect Analytical* At acute overdose concentration (20 mg/dL) on Technicon® SMAC® method. *3719* At concentration of 200 mg/L had no effect on phosphotungstate reduction method. *3393*

Norvaline

Tyrosine *Plasma No Effect Analytical* On fluorometric procedure of Ambrose. *0080*

Novaminsulfon

Cholesterol *Serum No Effect Analytical* At concentration of 900 mg/L had no effect on CHOD-Iodide method. *3393* At concentration of 900 mg/L had no effect on catalase-AIDH method. *3393* At concentration of 1160 mg/L had no effect on catalase-Hantzsch method. *3393* At concentration of 900 mg/L had no effect on Liebermann-Burchard method. *3393*
Serum Decrease Analytical At concentrations above 150 mg/L (normal therapeutic concentration 15 mg/L) lowered concentration as measured by CHOD-PAP method. *3393*

Creatinine *Serum No Effect Analytical* At concentration of 100 mg/L had no effect on creatinine iminohydrolase method. *3393* At concentration of 8,000 mg/L had no effect on Jaffé-Fading-Fraction method. *3393*
Serum Increase Analytical At concentrations above 8,000 mg/L (normal therapeutic concentration 15 mg/L) raised concentration as measured by Jaffé-Fuller's earth method. *3393*

Glucose *Serum No Effect Analytical* At concentration of 80 mg/L had no effect on Ektachem® method. *3393* At concentration of 800 mg/L had no effect on hexokinase/G-6-PDH method. *3393*
Serum Decrease Analytical At concentrations above 400 mg/L (normal therapeutic concentration 15 mg/L) lowered concentration as measured by GOD-Perid method. *3393* At concentrations above 200 mg/L (normal therapeutic concentration 15 mg/L) lowered concentration as measured by GOD-PAP method. *3393*

Lactate *Serum No Effect Analytical* At concentration of 800 mg/L had no effect on enzymatic method. *3393*

Potassium *Serum No Effect Analytical* At concentration of 800 mg/L had no effect on measurement by ISE without predilution. *3393*

Protein Electrophoresis *Serum No Effect Analytical* At concentration of 800 mg/L had no effect on automated Olympus-Hite method. *3393*

Sodium *Serum No Effect Analytical* At concentration of 800 mg/L had no effect on measurement by ISE without predilution. *3393*

Triglycerides *Serum Decrease Analytical* At concentrations above 100 mg/L (normal therapeutic concentration 15 mg/L) lowered concentration as measured by GPO-PAP method. *3393*

Urea Nitrogen *Serum No Effect Analytical* At concentration of 80 mg/L had no effect on Ektachem® method. *3393*

Uric Acid *Serum No Effect Analytical* At concentration of 800 mg/L had no effect on catalase-AIDH method. *3393*
Serum Decrease Analytical At concentrations above 100 mg/L (normal therapeutic concentration 15 mg/L) lowered concentration as measured by Kageyama-Hantzsch method. *3393*

Novobiocin

Alanine Aminotransferase *Serum Increase Physiological* Intrahepatic cholestasis may occur. *2313*

Alkaline Phosphatase *Serum Increase Physiological* Intrahepatic cholestatic jaundice can occur. *1343*

Aspartate Aminotransferase
Serum Increase Physiological Intrahepatic cholestasis may occur. *2313*

Bile *Urine Increase Physiological* Intrahepatic cholestatic jaundice can occur. *1343*

Bilirubin *Serum Increase Analytical* Interference by metabolite (Evelyn-Malloy). *2313* Yellow metabolite affects direct methods. *0583*
Serum Increase Physiological Especially in newborn: inhibits conjugating mechanism. *0350*

Bilirubin, Direct *Serum Increase Physiological* Competes for conjugation in liver. *0348*

BSP Retention *Serum Increase Physiological* Competition for excretion, may be actual damage. *0350*

Cephalin Flocculation *Serum Increase Physiological* Intrahepatic cholestasis may occur. *2313*

Eosinophils *Blood Increase Physiological* Allergic reaction. *1513*

Erythrocytes *Blood Decrease Physiological* Hemolytic anemia — mild. *0713*

Hematocrit *Blood Decrease Physiological* Hemolytic anemia — mild. *0713*

Hemoglobin *Blood Decrease Physiological* Hemolytic anemia — mild. *0713*

Icteric Index *Serum Increase Physiological* Competition for conjugation mechanism. *2425*

^{131}I Uptake *Serum Decrease Physiological* Albamycin contains tetraiodofluorescein. *2652*

Leukocytes *Blood Increase Physiological* Due to eosinophilia. *2313*
Blood Decrease Physiological May induce blood dyscrasias. *2313*

Occult Blood *Feces Positive Physiological* Intestinal hemorrhage may occur. *1343*

Platelets *Blood Decrease Physiological* Thrombocytopenia (immunologically induced). *1343*

PSP Excretion *Urine Decrease Physiological* ?nephrotoxic effect. *1022*

Thymol Turbidity *Serum Increase Physiological* Intrahepatic cholestasis may occur. *2313*

Urobilinogen *Urine Increase Physiological* Hemolytic anemia in G-6-PD deficiency. *1343*

NSAID (Nonsteroidal Anti-Inflammatory Drugs)

Aldosterone *Plasma Decrease Physiological* Associated with inhibition of prostaglandin synthesis. *1561*

Creatinine *Serum Increase Physiological* Reversible acute renal failure associated with many drugs. *1561*

Eosinophils *Blood Increase Physiological* Observed in 9% of 56 patients treated with these drugs alone for rheumatoid arthritis. *1111*

Potassium *Serum Increase Physiological* Associated with inhibition of prostaglandin synthesis. *1561*

Protein *Urine Increase Physiological* Most of drugs have been associated with reversible clinical syndrome of heavy proteinuria and renal insufficiency. *1561*

Renin Activity *Plasma Decrease Physiological* Associated with inhibition of prostaglandin synthesis. *1561*

Sodium *Serum Decrease Physiological* May be associated with water retention. *1561*

NSD 3004

Bicarbonate *Serum Decrease Physiological* Long acting carbonic anhydrase inhibitor. *2232*

pCO$_2$ *Blood Decrease Physiological* Arterial blood long acting carbonic anhydrase inhibition. *2232*

Standard Bicarbonate *Serum Decrease Physiological* Arterial blood long acting carbonic anhydrase inhibition. *2232*

Nucleoproteins

Sugar *Urine Increase Analytical* Interferes with Benedict's-?reducing sugar. *1022*

Nystatin

Amino Acids *Urine Positive Analytical* Purple spot on thin-layer chromatography with ninhydrin when combined with neomycin and triamcinolone in Kenacomb. *1278*

Eosinophils *Blood Increase Physiological* May cause allergic reaction. *2808*

Platelets *Blood Decrease Physiological* Thrombocytopenia (AMA Blood dyscrasias). *2429*

Obesity

Androsterone *Plasma No Effect Physiological* In females but falls with caloric restriction. *3391*

Dehydroepiandrosterone (DHEA) *Plasma No Effect Physiological* In females but falls with caloric restriction. *3391*

Glucose *Serum No Effect Physiological* Usual unless diabetic. *1106*
Serum Increase Physiological Proportion of subjects with high values increased. *2123*

Glycine *Plasma Decrease Physiological* Characteristic finding correlated with insulin. *1106*

Growth Hormone *Plasma Decrease Physiological* Little secreted at night compared with normals. *2225*

11-Hydroxycorticosteroids *Plasma No Effect Physiological* In females but falls with caloric restriction. *3391*

Insulin *Plasma Increase Physiological* Increase observed in fasting and after glucose. *1015*

Isoleucine *Plasma Increase Physiological* Characteristic finding correlated with insulin. *1106*

Leucine *Plasma Increase Physiological* Characteristic finding correlated with insulin. *1106*

Phenylalanine *Plasma Increase Physiological* Characteristic finding correlated with insulin. *1106*

Thyroxine (T$_4$) *Serum Increase Physiological* Characteristic finding correlated with insulin. *1106*

Triglycerides *Serum Increase Physiological* Proportion of subjects with high values increases. *2123*

Tyrosine *Plasma Increase Physiological* Metabolic effect (?mechanism). *0843*

Uric Acid *Serum Increase Physiological* Proportion of subjects with high values increases. *2123*

Valine *Plasma Increase Physiological* Characteristic finding correlated with insulin. *1106*

Octopamine

Epinephrine *Test Conditions No Effect Analytical* On fluorescent procedure of Peyrin. *2800*

Norepinephrine *Test Conditions No Effect Analytical* On fluorescent procedure of Peyrin. *2800*

Ofloxacin

Alanine Aminotransferase *Serum Increase Physiological* Very infrequent side effect observed during many courses of treatment. *1838*

Aspartate Aminotransferase *Serum Increase Physiological* Very infrequent side effect observed during many courses of treatment. *1838*

Creatinine *Serum Increase Physiological* Very infrequent side effect observed during many courses of treatment. *1838*

Creatinine Clearance *Urine Decrease Physiological* Very infrequent side effect observed during many courses of treatment. *1838*

γ-Glutamyltransferase (GGT) *Serum Increase Physiological* Very infrequent side effect observed during many courses of treatment. *1838*

Ohio 469

Bilirubin *Serum Increase Physiological* Preliminary finding of significant increase post operatively. *1647*

Glucose *Serum Increase Physiological* 10-30% higher in postoperative period. *1647*

Lactate Dehydrogenase *Serum Increase Physiological* Preliminary finding of significant increase postoperatively. *1647*

Leukocytes *Blood Increase Physiological* Significant rise postoperatively especially of segmented neutrophils. *1647*

Oleandomycin

Alanine Aminotransferase *Serum Increase Physiological* May cause hepatotoxicity (cholestatic syndrome). *2313*

Alkaline Phosphatase *Serum Increase Physiological* Cholestatic jaundice reported. *0070*

Aspartate Aminotransferase *Serum Increase Physiological* May cause hepatotoxicity (cholestatic syndrome). *2313*

Bile *Urine Increase Physiological* Hepatotoxic effect. *2313*

Bilirubin *Serum Increase Physiological* May cause intrahepatic cholestatic jaundice. *1237*

BSP Retention *Serum Increase Physiological* May cause intrahepatic cholestasis. *1237*

Cephalin Flocculation *Serum Increase Physiological* Hepatotoxic effect. *2313*

Eosinophils *Blood Increase Physiological* Associated with allergic cholestasis. *1679*

Erythrocytes *Blood Decrease Physiological* Probably due to hemolytic anemia. *2313*

17-Hydroxycorticosteroids *Urine Increase Analytical* Interferes with Porter-Silber reaction. *2628*

^{131}I Uptake *Serum Decrease Physiological* Cyclamycin contains tetraiodofluorescein. *2652*

17-Ketogenic Steroids *Urine Increase Analytical* Interferes with Zimmermann reaction. *3505*

17-Ketosteroids *Urine Increase Analytical* Interferes with Zimmermann reaction. *2628*

Leukocytes *Blood Increase Physiological* Occasional leukocytosis after 2 weeks. *1679*

Thymol Turbidity *Serum Increase Physiological* May cause hepatotoxicity. *2313*

Oleate

Albumin *Test Conditions Decrease Analytical* Competes for binding sites ANSA method. *3596*

Omeprazole

Alanine Aminotransferase *Serum Increase Physiological* Pronounced rise in activity in one patient on eighth day of treatment. *1432*

Cortisol *Plasma No Effect Physiological* No effect observed in 8 volunteers given 30 mg/d for 28 d. *2516*

C-Peptide *Plasma No Effect Physiological* No effect observed in 8 volunteers given 30 mg/d for 28 d. *2516*

Omeprazole *(continued)*

Dehydroepiandrosterone (DHEA)
Plasma No Effect Physiological No effect observed in 8 volunteers given 30 mg/d for 28 d. *2516*

Estradiol *Plasma No Effect Physiological* No effect observed in 8 volunteers given 30 mg/d for 28 d. *2516*

Gastrin *Serum Increase Physiological* Raised to 80.9 pg/mL from 55.5 pg/mL in 8 volunteers after 29 d but reversible. *2516* Significantly increased after 7 and 14 d treatment but not after a single dose. *1126*

Glucagon *Plasma No Effect Physiological* No effect observed in 8 volunteers given 30 mg/d for 28 d. *2516*

Hydrochloric Acid *Gastric Material Decrease Physiological* When stimulated acid output reduced from 27.4 mmol H^+/h to 7.8 mmol H^+/h after 29 d in 8 volunteers but reversible. *2516*

Insulin *Plasma No Effect Physiological* No effect observed in 8 volunteers given 30 mg/d for 28 d. *2516*

Parathyroid Hormone *Plasma No Effect Physiological* No effect observed in 8 volunteers given 30 mg/d for 28 d. *2516*

Pepsinogen I *Serum Increase Physiological* Significant effect after 7 and 14 d therapy (more than doubling concentration). *1126*

Prolactin *Plasma No Effect Physiological* No effect observed in 8 volunteers given 30 mg/d for 28 d. *2516*

Testosterone *Serum No Effect Physiological* No effect observed in 8 volunteers given 30 mg/d for 28 d. *2516*

Thyroid Stimulating Hormone (TSH)
Serum No Effect Physiological No effect observed in 8 volunteers given 30 mg/d for 28 d. *2516*

Thyroxine (T_4) *Serum No Effect Physiological* No effect observed in 8 volunteers given 30 mg/d for 28 d. *2516*

Thyroxine (T_4) Binding Globulin
Serum No Effect Physiological No effect observed in 8 volunteers given 30 mg/d for 28 d. *2516*

Tri-iodothyronine (T_3) *Serum No Effect Physiological* No effect observed in 8 volunteers given 30 mg/d for 28 d. *2516*

Onions

Glucose *Serum Decrease Physiological* As fried onions has effect on diabetes. *1768*

Opiates

Aldosterone *Plasma Increase Physiological* Significant increase within 1 h of i.m. injection of hyoscine, meperidine or omnopon. *1929*

Atrial Natriuretic Peptide *Plasma No Effect Physiological* No effect either with or without anesthesia or surgery. *1929*

Cortisol *Plasma Increase Physiological* Significant increase within 1 h of i.m. injection of hyoscine, meperidine or omnopon. *1929*

Epinephrine *Plasma Increase Physiological* Significant increase within 1 h of i.m. injection of hyoscine, meperidine or omnopon. *1929*

Norepinephrine *Plasma Increase Physiological* Significant increase within 1 h of i.m. injection of hyoscine, meperidine or omnopon. *1929*

Renin Activity *Plasma Increase Physiological* Significant increase within 1 h of i.m. injection of hyoscine, meperidine or omnopon. *1929*

Opipramol

Glucose *Serum No Effect Analytical* No effect on glucose oxidase method of Boehringer. *3277*

Opium Alkaloids

Amylase *Serum Increase Physiological* Impaired excretion spasm of sphincter of Oddi. *0652*

Aspartate Aminotransferase
Serum Increase Physiological Possibly due to spasm of sphincter of Oddi. *0652*

BSP Retention *Serum Increase Physiological* Impaired excretion spasm of sphincter of Oddi. *0652*

Indocyanine Green *Serum Decrease Physiological* Mechanism not yet established. *2398*

Pseudocholinesterase *Serum Decrease Physiological* May cause inhibition of activity. *1680*

Xylose Excretion *Urine Decrease Physiological* Some may produce gastrointestinal irritation impaired absorption. *0612*

Oral Contraceptives

α_1-Acid Glycoprotein *Serum Decrease Physiological* Metabolic changes in liver synthesis (estrogen). *2189*

Alanine *Plasma Decrease Physiological* Hormonal effect (second part of cycle). *0742*

Alanine Aminotransferase *Serum Increase Physiological* Cholestatic-hepatotoxic effect. *2313* Significant increase observed within 3 mo usually continuing for several y. *3170*
Red Blood Cells No Effect Physiological No effect, but stimulated *in vivo* by B_6. *3049* No effect even if treated for 3 y. *3048*
Red Blood Cells Increase Physiological Significantly higher than in women not using oral contraceptives. *0955*
Red Blood Cells Decrease Physiological Possibly associated with vitamin B_6 deficiency. *0563*

Albumin *Serum Decrease Physiological* Not significant after 3 mo cessation, significant after 3 y. *3143* 10% fall if used for several months. *1650*

Aldosterone *Plasma No Effect Physiological* Inconsistent response after 4 weeks. *3785* Normal activity in users in spite of changes in renin substrate concentration. *1333*
Urine Increase Physiological Estrogen effect (in small number of people). *0585*

Alkaline Phosphatase *Serum Increase Physiological* May cause cholestasis, rare hepatocellular degeneration. *1713*
Serum Decrease Physiological Slight decrease reported in one study. *2427* Significant effect observed within 3 mo usually continuing for several y. *3170* But effects less marked with low estrogen preparations. *1568*
White Blood Cells Increase Physiological Observed effect. *1487*

Amino-4-Imidazole-5-Carboxamide Ribotide (AICAR)
Urine Increase Physiological Occurs if megaloblastic anemia develops. *3773*

Aminolevulinic Acid *Urine Increase Physiological* May precipitate porphyria attack. *1322*

α-Amino-Nitrogen *Plasma Decrease Physiological* Anabolic effect of synthetic steroids (progestogen). *0743*

Amylase *Serum Increase Physiological* In approximately 23%: probably of liver origin. *0024* Isolated case of acute pancreatitis reported. *2519*
Urine No Effect Physiological Although high in serum, suggestive of hepatic origin. *0024*

Androsterone *Urine Decrease Physiological* Compared with controls — details not discussed. *0527*

Angiotensin *Plasma Increase Physiological* Twofold increase. *0585*

Angiotensin II *Plasma Increase Physiological* Elevated to 3 times normal during administration. *0557*

Angiotensinogen *Plasma Increase Physiological* Threefold increase with 0.5 mg mestranol. *0585*

Antinuclear Antibodies *Serum Positive Physiological* Slight effect in users. *2335*

Antithrombin III *Plasma No Effect Physiological* No significant changes at various intervals during treatment compared with controls. *1418*
Plasma Decrease Physiological ?predisposing cause of thrombosis. *3961* Slight reduction during treatment with drug low in estrogen content: note effect not as marked as reported when high estrogens content preparations used. *1795*

Antithrombin III Antigen *Plasma Decrease Physiological* By activity unchanged in women given low dose ethinyl estradiol and norethindrone. *2623*

α_1-Antitrypsin *Serum No Effect Physiological* No significant changes at various intervals during treatment compared with controls. *1418*

Serum Increase Physiological Metabolic changes in liver synthesis (estrogen). *2189* In women given low dose ethinyl estradiol and norethindrone. *2623* But effects less marked with low estrogen preparations. *1568*

Apolipoprotein AI *Serum No Effect Physiological* Typically insignificant change over 6 cycles in women with low dose ethinyl estradiol with low dose levonorgestrel or desogestrel. *1249*

Apolipoprotein B *Serum No Effect Physiological* Typically insignificant change over 6 cycles in women with low dose ethinyl estradiol with low dose levonorgestrel or desogestrel. *1249*

Ascorbic Acid *Plasma Decrease Physiological* Maximum effect at 2 weeks, greater effect in platelets. *1847*
Urine Decrease Physiological Significant effect 50% decrease on comparable diets. *1507*
White Blood Cells Decrease Physiological Mean 19 mg/100 g (control 25.7). *2380*

Aspartate Aminotransferase
Serum Increase Physiological Hepatotoxic effect (cholestasis induced). *2086* Significant increase observed within 3 mo usually continuing for several y. *3170*
Red Blood Cells Increase Physiological Compared with normal, B_6 stimulation no effect. *3049* Elevated when treated for 6 mo or longer. *3048*

Bilirubin *Serum Increase Physiological* Interferes with canalicular excretion. *0348*

Bilirubin, Direct *Serum Increase Physiological* Hypersensitivity to estrogen component. *1488*

BSP Retention *Serum Increase Physiological* Occurs in 40%, estrogen depresses secretory mechanism. *2086*

Butanol Extractable Iodine (BEI)
Serum Increase Physiological Due to increased thyroxine binding globulin. *0071*

Calcium *Serum Increase Physiological* Increased ingestion would increase concentration. *0652*
Serum Decrease Physiological Seen in osteoporosis, ?due to fall in albumin. *3923* Isolated case of acute pancreatitis reported. *2519*
Urine Decrease Physiological Occurs with fall in serum concentration. *3923*

Carbonic Anhydrase
Red Blood Cells Increase Physiological Measured as B isoenzyme (hormonal effect). *3181*

Catecholamines *Urine No Effect Physiological* No effect observed with ingestion. *3217*

Catechol-O-Methyl Transferase
Red Blood Cells Increase Physiological When high in estrogen content. *1401*

Cephalin Flocculation *Serum Increase Physiological* Observed in many subjects. *2427*

Cephalin Time *Blood Decrease Physiological* Metabolic effect. *2834*

Ceruloplasmin *Serum Increase Physiological* Estrogen effect on liver, no change in activity. *2189* Increase by 115 to 123% after 3 mo with different preparations. *0311*

Cervical Secretion *Patient No Effect Physiological* Leucorrhea due to estrogen in 20% women. *2349*
Patient Increase Physiological leukorrhea due to estrogen in 20% women. *2349*

Chenodeoxycholic Acid *Serum No Effect Physiological* No effect observed over 12 mo in 29 women given combination of ethinyl estradiol and norgestrel. *1546*

Cholesterol *Serum No Effect Physiological* No effect observed usually. *2427* No difference between treated women and others. *1303* May occur due to opposing effects of estrogen and progestogen components. *0018* In 10 women receiving ethinyl estradiol with norgestrel. *2009* Typically insignificant change over 6 cycles in women with low dose ethinyl estradiol with low dose levonorgestrel or desogestrel. *1249*
Serum Increase Physiological If initially low (no effect if about 200 mg/dL). *0585* Increase by 4 to 15% depending on preparation used. *2860* Extent of effect varies with exact composition of oral contraceptive. *1458* May result with oral contraceptive containing more than 75 µg estrogen. *0018* Slight effect with high-dose combination drugs. *1248* In 10 women receiving ethinyl estradiol with norethindrone. *2009*

Cholesterol Esters *Serum Increase Physiological* Slight effect in primates. *3035*

Cholesterol, High Density Lipoprotein
Serum No Effect Physiological No difference between treated women and others. *1303* With high dose drugs containing norethisterone or ethynodiol. *1248* In 10 women receiving ethinyl estradiol with norgestrel. *2009* Typically insignificant change over 6 cycles in women with low dose ethinyl estradiol with low dose levonorgestrel or desogestrel. *1249*
Serum Increase Physiological Depending on preparation used. Increase by 17 to 81% depending on preparation used. *2860* If oral contraceptive contains more than 80 µg estrogen. *0018* In women taking high estrogen/low progestin combination. *3733* Increase by 35% with high-dose oral contraceptives that are overly estrogenic. *1248* In 10 women receiving ethinyl estradiol with norethindrone. *2009*
Serum Decrease Physiological Depending on preparation used. *2860* If oral contraceptive contains more than 50 µg estrogen. *0018* In women taking estrogen/high or low progestin combination. *3733* Increase by 20% with high dose compound containing 500 µg norgestrel. *1248*

Cholesterol, α-Lipoprotein *Serum Decrease Physiological* Increase by 19 mg/dL in white girl drug users vs controls. *3723*

Cholesterol, Low Density Lipoprotein
Serum No Effect Physiological No difference between treated women and others. *1303* No change with high-dose oral contraceptives that are overly estrogenic. *1248* In 10 women receiving ethinyl estradiol with norethindrone. In 10 women receiving ethinyl estradiol with norgestrel. *2009* Typically insignificant change over 6 cycles in women with low dose ethinyl estradiol with low dose levonorgestrel or desogestrel. *1249*
Serum Increase Physiological Mainly estrogen response. *3525* Increase by 8 to 15% depending on preparation used. *2860* 24% higher median concentration using combination with relatively low estrogen and medium or high progestin. *3733* Increase of 24% when high progestogen compounds ingested. *1248*

Cholesterol, Very Low Density Lipoprotein
Serum No Effect Physiological No difference between treated women and others. *1303*
Serum Increase Physiological Increase by 16 to 40% depending on preparation used. *2860*

Cholic Acid *Serum No Effect Physiological* No effect observed over 12 mo in 29 women given combination of ethinyl estradiol and norgestrel. *1546*

Clotting Time *Blood Decrease Physiological* (silicone clotting time) associated with clot problems. *0760*

Color *Serum Increase Physiological* Ceruloplasmin may be so high blue-green color. *3505*

Complement APH₅₀ *Serum No Effect Physiological* Compared with nonusers in first year of use; largely influence of progestogen. *0866*

Complement C3 *Serum Increase Physiological* Compared with nonusers in first year of use: largely influence of progestogen. *0866*

Complement C4 *Serum Increase Physiological* Compared with nonusers in first year of use: largely influence of progestogen. *0866*

Complement CH₅₀ *Serum Increase Physiological* Compared with nonusers in first year of use: largely influence of progestogen. *0866*

Complement Factor B *Serum Increase Physiological* Compared with nonusers in first year of use: largely influence of progestogen. *0866*

Copper *Serum Increase Physiological* Estrogens increase concentration of binding protein (maybe x 2). *3548* In 22 women ingesting combination type contraceptives. *1604*
Hair No Effect Physiological No significant effect in young women studied for at least 3 mo. *3720*
Red Blood Cells No Effect Physiological In 22 women ingesting combination type contraceptives. *1604*
White Blood Cells No Effect Physiological In 22 women ingesting combination type contraceptives. *1604*

Coproporphyrin *Blood Increase Physiological* May precipitate porphyria attack. *1322*
Urine Increase Physiological May induce porphyria cutanea tarda. *3029*

Oral Contraceptives (continued)

Coproporphyrin (continued)
Feces Increase Physiological May precipitate porphyria attack. 1322

Corticosteroid-Binding Globulin
Serum Increase Physiological Due to estrogenic component. 0071

Corticosteroids *Plasma Increase Physiological* With increased total plasma cortisol. 3525

Corticosterone *Plasma No Effect Physiological* Inconsistent response after 4 weeks. 3785

Cortisol *Plasma Increase Physiological* Decreases cortisol clearance (estrogen effect). 3332 On combined therapy at 9 am compared with female controls and men. 1668
Urine No Effect Physiological No significant difference when referenced to creatinine in women on combined therapy versus controls. 1668
Urine Increase Physiological Significant hormonal effect. 2181
Urine Decrease Physiological When results during therapy compared with normal. 0261
Saliva Increase Physiological Significantly higher on combined therapy versus control women and men. 1668

Cortisol, Free *Plasma Increase Physiological* Significant hormonal effect. 2181 Estrogen effect. 0585
Urine Increase Physiological 50% increase with oral contraceptives. 3217

Cortisol, Protein Bound *Serum Increase Physiological* When results on therapy compared with normal. 0261

C-Reactive Protein *Serum Increase Physiological* Estrogen effect. 3332
Serum Decrease Physiological Progestogen effect. 3505

Creatine Kinase *Serum Increase Physiological* Slight effect only (2 u/L). 2744

Cryofibrinogen *Plasma Increase Physiological* Incidence much higher than in controls. 2385

Dehydroepiandrosterone (DHEA)
Urine Decrease Physiological Compared with controls — details not discussed. 0527

Dehydroepiandrosterone Sulfate (DHEA-S)
Plasma Decrease Physiological Usually reduced in response to most regimens. 2280

Deoxycholic Acid *Serum No Effect Physiological* No effect observed over 12 mo in 29 women given combination of ethinyl estradiol and norgestrel. 3445

Diphenylhydantoin *Serum Increase Physiological* Higher values related to drug dose compared with controls. 0844

Erythrocytes *Blood Decrease Physiological* May cause megaloblastic anemia. 3773 In 46 oral contraceptive users compared with controls studied over at least 2 y. 1182

Erythrocyte Sedimentation Rate
Blood Increase Physiological Associated with increased fibrinogen. 3525

Estradiol *Plasma Decrease Physiological* Inhibits physiological rise. 1944
Urine Decrease Physiological Hormonal effect. 2211

Estradiol Binding Globulin *Serum Increase Physiological* Metabolic changes in liver synthesis. 2189

Estriol *Urine Decrease Physiological* Hormonal effect. 2211

Estrogen Binding Globulin *Serum Increase Physiological* Metabolic effect. 2385

Estrogens *Plasma Increase Physiological* Often related to nausea. 0585
Urine Decrease Physiological Hormonal effect (decreased by 40%). 3217

Etiocholanolone *Urine Decrease Physiological* Compared with controls — details not discussed. 0527

Expiratory Volume *Patient No Effect Physiological* No significant effect observed. 1185

Factor I *Plasma No Effect Physiological* Usually no effect observed. 0339

Factor II *Plasma Increase Physiological* Reported effect of estrogens. 0429

Factor V *Plasma No Effect Physiological* Usually no effect observed. 0339
Plasma Increase Physiological Significant effect of combined oral contraceptives. 2456

Factor VII *Plasma Increase Physiological* Estrogen effect (higher than in pregnancy). 3332
Plasma Decrease Physiological Metabolic effect. 2834

Factor VIII *Plasma Increase Physiological* Slight effect observed. 1462

Factor IX *Plasma Increase Physiological* Estrogen effect. 3332

Factor X *Plasma Increase Physiological* Slight effect observed. 1462
Plasma Decrease Physiological Metabolic effect. 2834

Factor XII *Plasma Increase Physiological* Reported effect. 2385

Fatty Acids, Free (FFA) *Serum Increase Physiological* If given for 3 mo. 3505

Fatty Acids, Total *Serum Increase Physiological* Slight to large effect in primates. 3035

Ferritin *Serum No Effect Physiological* Smaller proportion of contraceptive users had low values than controls. 1368
Serum Increase Physiological Mean of 40 ng/mL vs 25 ng/mL in control population followed for at least 2 y. 1182

Fibrinogen *Plasma Increase Physiological* Metabolic changes in liver synthesis (estrogen). 2189
Plasma Decrease Physiological When combined estrogen and progestogen. 2427

FIGLU (N-Formiminoglutamic Acid)
Urine Increase Physiological Response to histidine tolerance test. 3311 Mild effect in oral contraceptive users. 3310

Folate *Serum Decrease Physiological* Interferes with gastrointestinal absorption. 3802 Greater proportion of contraceptive users had low values than controls. 1368 Mild effect in oral contraceptive users. 3310
Red Blood Cells Decrease Physiological Impaired metabolism due to hormonal factors. 3311 Mild effect in oral contraceptive users. 3310

Follicle Stimulating Hormone (FSH)
Plasma Decrease Physiological Over years depressed to 70% of control values. 1340
Urine Decrease Physiological Marked depression in normal subjects. 3464

α_1-Globulin *Serum No Effect Physiological* No significant effect at 3 mo cessation or 3 y. 3143
Serum Increase Physiological Metabolic change with combined contraceptive. 2312

α_2-Globulin *Serum Increase Physiological* May be increased by as much as 8 times. 0585 After 3 y, not after 3 mo cessation. 3143

β-Globulin *Serum Increase Physiological* After 3 y but not after 3 mo cessation. 3143 Metabolic changes in liver synthesis. 3258

γ-Globulin *Serum Increase Physiological* After 3 y, not after 3 mo cessation. 3143

Glucaric Acid *Urine Increase Physiological* Reported induction of hepatic enzymes. 1694

Glucocorticoids *Plasma Increase Physiological* Expected effect observed. 0820

Glucose *Serum Increase Physiological* Does not affect fasting glucose but alters glucose tolerance test. 3505
Serum Decrease Physiological Estrogen effect of combined oral contraceptive. 0018 Significantly lower in drug users. 1248

Glucose Tolerance *Serum No Effect Physiological* Estrogen effect of combined oral contraceptive. 0018
Serum Decrease Physiological Mainly estrogen effect (reversible in 3 out 4). 1462 Glucose higher by 11 mg/dL at 1 h in drug users. 1248

Glucuronic Acid *Urine Increase Physiological* Extent of effect varies with exact composition of oral contraceptive. 1458

β-Glucuronidase *Serum Increase Physiological* Altered metabolism (estrogen effect). 0227 Significant effect observed within 3 mo usually continuing for several y. 3170

Glutamate Dehydrogenase *Serum Increase Physiological* Significant effect observed within 3 mo usually continuing for several y. 3170

Glutamic Acid *Plasma Decrease Physiological* Hormonal effect (second part of cycle). *0742*

γ-Glutamyltransferase (GGT)
Serum Increase Physiological Significant effect observed within 3 mo usually continuing for several y. *3170* Extent of effect varies with exact composition of oral contraceptive. *1458* Positive association between activity and oral contraceptive use (average 28% Increase). *0157* But effects less marked with low estrogen preparations. *1568*

Glutathione Reductase
Red Blood Cells Decrease Physiological Manifestation of poor riboflavin status in women of low socioeconomic status. *3024*

Glycine *Plasma Decrease Physiological* Hormonal effect (second part of cycle). *0742*

Gonadotropins *Urine Decrease Physiological* Hormonal effect. *2211*

Growth Hormone *Plasma Increase Physiological* During first year of use (may be increased 3 fold). *0071*

Haptoglobin *Serum Decrease Physiological* Metabolic changes in liver synthesis (estrogen). *2189*

HDL$_{2a}$-Cholesterol *Serum No Effect Physiological* In 10 women receiving ethinyl estradiol with norethindrone. In 10 women receiving ethinyl estradiol with norgestrel. *2009*

HDL$_{2b}$-Cholesterol *Serum Increase Physiological* In 10 women receiving ethinyl estradiol with norethindrone. *2009*
Serum Decrease Physiological In 10 women receiving ethinyl estradiol with norethindrone. *2009*

HDL$_3$-Cholesterol *Serum No Effect Physiological* In 10 women receiving ethinyl estradiol with norgestrel. *2009*
Serum Increase Physiological In 10 women receiving ethinyl estradiol with norethindrone. *2009*

Hematocrit *Blood No Effect Physiological* No significant difference between oral contraceptive users and control adolescents. *1368* No significant effect in oral contraceptive users. *3310*
Blood Increase Physiological Probably progestogen effect. *3548*
Blood Decrease Physiological May cause megaloblastic anemia. *3025* In 46 oral contraceptive users compared with controls studied over at least 2 y. *1182*

Hemoglobin *Blood No Effect Physiological* No significant difference between oral contraceptive users and control adolescents. *1368* No significant effect in oral contraceptive users. *3310* In 46 oral contraceptive users compared with controls studied over at least 2 y. *1182*
Blood Increase Physiological Increase after 12 mo with low base combined pill regime. *2996*
Blood Decrease Physiological May cause megaloblastic anemia. *3025*

3-Hydroxyanthranilic Acid *Urine Increase Physiological* Estrogen component induces tryptophan pyrrolase. *3045* After 2 g tryptophan load. *3050*

17-Hydroxycorticosteroids *Urine Decrease Physiological* Probably due to estrogen decreasing cortisol secretion. *2596*

3-Hydroxykynurenine *Urine Increase Physiological* After 2 g tryptophan load. *3050* Estrogen component induces tryptophan pyrrolase. *3045*

3-Hydroxyxanthurenic Acid *Urine Increase Physiological* After 2 g tryptophan load. *3050*

Imidazolepyruvic Acid *Urine Increase Physiological* Mechanism not yet established. *3681*

Immunoglobulin IgA *Serum Decrease Physiological* Estrogen effect. *3505*

Immunoglobulins *Serum Increase Physiological* Metabolic changes in liver synthesis. *2189*
Serum Decrease Physiological Estrogen effect. *3332*

Insulin *Plasma Increase Physiological* In women in whom insulin was initially normal. *3412*

Intermediate density lipoprotein (IDL)
Serum No Effect Physiological In 10 women receiving ethinyl estradiol with norethindrone. In 10 women receiving ethinyl estradiol with norgestrel. *2009*

Iodine, Total *Serum Increase Physiological* Altered metabolism. *1462*

Iron *Serum Increase Physiological* Increase in available binding protein (plus 20% increase). *3548* In 46 oral contraceptive users compared with controls studied over at least 2 y. *1182* Increase after 12 mo with low base combined pill regime. *2996*

Iron-Binding Capacity, Total (TIBC)
Serum Increase Physiological Estrogen or progestogen effect (usually plus 20%). *3548* In 46 oral contraceptive users compared with controls studied over at least 2 y. *1182*

Isoleucine *Plasma Decrease Physiological* Hormonal effect (second part of cycle). *0742*

^{131}I Uptake *Serum No Effect Physiological* Thyroid function unaffected. *0071*

Kaolin-Cephalin Time *Blood Decrease Physiological* Metabolic effect. *2834*

17-Ketogenic Steroids *Urine Decrease Physiological* Probably due to estrogen decreasing cortisol secretion. *2596*

17-Ketosteroids *Urine Decrease Physiological* Probable decrease in cortisol secretion. *2596*

Kynurenine *Urine Increase Physiological* Estrogen component induces tryptophan pyrrolase. *3045*

Lactate *Serum Increase Physiological* Alteration in carbohydrate metabolism. *2385*

Lactate Dehydrogenase *Serum No Effect Physiological* No significant effect observed. *3319*

LE Cells *Blood Positive Physiological* May precipitate or exaggerate LE-like syndrome. *3505*

Leucine *Plasma Decrease Physiological* Hormonal effect (second part of cycle). *0742*

Leucine Aminopeptidase *Serum Increase Physiological* Possible liver damage. *1462*

Leucine Aminopeptidase Isoenzymes
Serum Increase Physiological Increased slow component ?liver involvement. *3087*

Leukocytes *Blood Increase Physiological* ?stimulating effect of steroids on bone marrow. *1141*

Lipase *Serum Increase Physiological* Isolated case of acute pancreatitis reported. *2519*

Lipids, Total *Serum Increase Physiological* 155% higher in pill-users than controls. *2688*
CSF Increase Physiological Altered metabolism. *1601*

Lipoprotein Cholesterol *Serum No Effect Physiological* No difference between treated women and others. *1195*

β-Lipoprotein Cholesterol *Serum Increase Physiological* Icrease by 34 mg/dL in white girl drug users vs controls. *3723*

Lipoprotein, High Density *Serum No Effect Physiological* In 10 women receiving ethinyl estradiol with norgestrel. *2009*
Serum Increase Physiological In 10 women receiving ethinyl estradiol with norethindrone. *2009*

Lipoprotein Lipase *Serum Decrease Physiological* Reduced response to heparin injection. *0585*

Lipoprotein, Low Density *Serum No Effect Physiological* In 10 women receiving ethinyl estradiol with norgestrel. *2009*
Serum Increase Physiological In 10 women receiving ethinyl estradiol with norethindrone. *2009*

Lipoproteins *Serum Increase Physiological* 50% increase after 6 mo use. *3505*

α-Lipoproteins *Serum Increase Physiological* Responsible for increased phospholipids. *0585*

β-Lipoproteins *Serum Increase Physiological* Affects cholesterol (approximately 20% increase after 6 mo). *0585* Extent of effect varies with exact composition of oral contraceptive. *1458*

Lipoproteins, Pre-β *Serum Increase Physiological* Probably due to increased apoprotein synthesis. *2385*

Lipoprotein, Very Low Density
Serum No Effect Physiological In 10 women receiving ethinyl estradiol with norethindrone. *2009*
Serum Increase Physiological In 10 women receiving ethinyl estradiol with norgestrel. *2009*

Luteinizing Hormone (LH) *Plasma Decrease Physiological* Combination type pill lowered value to 20% control. *1340*
Urine Decrease Physiological Marked depression in normal subjects. *3464*

Oral Contraceptives *(continued)*

Lymphocyte Mitotic Index
Test Conditions Decrease Physiological Hormonal action. *1146*

Lymphocytic Response to PHA
Blood Decrease Physiological Hormonal action. *1146*

Lysolecithin *Plasma Decrease Physiological* If administered for long period. *2591*

α₂-Macroglobulin *Serum No Effect Physiological* No significant changes at various intervals during treatment compared with controls. *1418*
Serum Increase Physiological Metabolic changes in liver synthesis. *2189*

Magnesium *Serum No Effect Physiological* No difference from normal observed. *0532*
Serum Decrease Physiological 0.15 meq/L decrease reported (estrogen effect). *1337* Generally significant reduction in all women taking oral contraceptives versus age-matched controls. *3430*
Urine Decrease Physiological Associated with fall in serum concentration. *1337*

MCH *Blood Increase Physiological* Slight increase with continuing use. *1141* In 46 oral contraceptive users compared with controls studied over at least 2 y. *1182*

MCHC *Blood No Effect Physiological* No significant difference between oral contraceptive users and control adolescents. *1368*
Blood Increase Physiological Significant effect in users for less than 5 y. *1141* In 46 oral contraceptive users compared with controls studied over at least 2 y. *1182*

MCV *Blood No Effect Physiological* No significant difference between oral contraceptive users and control adolescents. *1368* In 46 oral contraceptive users compared with controls studied over at least 2 y. *1182*
Blood Increase Physiological Occurs if megaloblastic anemia. *3025*

NBT Test *Blood Positive Physiological* Observed in 4 out of 6 subjects (?mechanism). *2610*

Ornithine Carbamoyltransferase (OCT)
Serum Increase Physiological Often raised during first month of treatment. *1498*

Oxytocin *Plasma Increase Physiological* Significant effect in women taking oral contraceptives. *3320*

Partial Thromboplastin Time
Plasma Decrease Physiological Associated with disordered clotting. *0760* In women given low dose ethinyl estradiol and norethindrone. *2623*

Phosphate *Serum Increase Physiological* 18% increase reported in some patients. *2427*
Serum Decrease Physiological Reduction of approximately 2 mg/dL. *2886*

Phosphatides, Total *Serum Increase Physiological* Altered metabolism. *1601*

Phospholipids, Total *Serum No Effect Physiological* Typically insignificant change over 6 cycles in women with low dose ethinyl estradiol with low dose levonorgestrel or desogestrel. *2216*
Serum Increase Physiological About 20% higher in pill-users after 6 mo. *0585*

Plasmin *Plasma Increase Physiological* Associated also with increased plasminogen (common effect). *0339*

Plasminogen *Plasma Increase Physiological* Metabolic changes in liver synthesis (estrogen). *2189*

Plasminogen Antigen *Plasma Increase Physiological* In women given low dose ethinyl estradiol and norethindrone. *2623*

Platelet Aggregation *Blood Increase Physiological* In response to ADP. *0585*

Platelets *Blood Increase Physiological* Reported effect. *0071* Slight effect during treatment period. *1795*

Porphobilinogen *Urine Increase Physiological* May precipitate porphyria attack. *1322*

Porphyrins *Urine Decrease Physiological* May occur in patients with established disease. *2427*

Postheparin Hepatic Triglyceride Lipase
Plasma Decrease Physiological Reduced by 46% in oral contraceptive treated women. *1303*

Prealbumin *Serum Increase Physiological* Metabolic changes in liver synthesis. *2189*

Pregnanediol *Urine Decrease Physiological* Hormonal effect (may be absent excretion). *2211*

Progesterone *Plasma Decrease Physiological* Less than 100 ng/dL throughout cycle. *0876*

Prolactin *Plasma No Effect Physiological* No significant effect of meals or oral contraceptives. *3320*
Plasma Increase Physiological Mean change from 8.9 ng/mL to 10.2 ng/mL at 3 mo in 120 women with Low estrogen pills. *1710* In 30% women to varying extent, not correlated with dose of estrogen or duration of treatment. *2965*

Proline *Plasma Decrease Physiological* Hormonal effect (second part of cycle). *0742*

Protein *Serum Increase Physiological* Significant increase 3 y after administration, after 3 mo cessation. *3143*
Serum Decrease Physiological Estrogen effect. *3332*

Protein Bound Iodine (PBI) *Serum Increase Physiological* Estrogens increase circulating thyroxine binding globulin. *3548*

Prothrombin Time *Plasma Increase Physiological* Associated with failure of excretion of bile salts. *1713*
Plasma Decrease Physiological Decreases response to oral anticoagulants. *0894*

Protoporphyrin *Blood Increase Physiological* May precipitate porphyria attack. *1322*
Feces Increase Physiological May precipitate porphyria attack. *1322*

Pseudocholinesterase *Serum Decrease Physiological* Estrogen effect. *3505* Significant effect observed within 3 mo usually continuing for several y. *3170* Significant reduction by about 12% in drug users versus controls. *2133*

Pyruvate *Serum Increase Physiological* Greater increase than normal during glucose tolerance test. *0585* Estrogen effect of combined oral contraceptive. *0018*

Renin Activity *Plasma No Effect Physiological* Normal activity in users in spite of changes in renin substrate concentration. *1333*
Plasma Increase Physiological Estrogen effect. *3332* Activity increased although concentration lowered. *0557*
Plasma Decrease Physiological Metabolic changes in liver synthesis. *2189* From 30 ng/ml/h to 13 at 1 week onwards. *0265*

Renin Substrate *Plasma Increase Physiological* Metabolic changes in liver synthesis. *2189* From 1.3 µg/mL to 4 at 1 week, 6 at 3 mo. *0265* Hormonal effect. *0071*
Plasma Decrease Physiological Almost 4 fold increase in oral contraceptive users compared with nonusers probably due to stimulated hepatic synthesis. *1333*

Respiratory Peak Flow *Patient No Effect Physiological* No significant effect observed. *1185*

Retinol Binding Protein *Serum Increase Physiological* Increased to about 150% of controls in women on a variety of oral contraceptives. *1303* Values doubled in 6 mo in 8 women taking estrogenic combination but diminished to some extent after 4 y. *0773* Values almost doubled within 6 mo but diminished somewhat after 4 y. *0773* Due to direct effect on synthesis of protein. *3023*

Retinol Esters *Serum Increase Physiological* Minimally increased fasting concentration in treated women. *1303*

Rheumatoid Factor *Serum Positive Physiological* Slight effect in users. *2335*

Sex Hormone Binding Globulin
Serum Increase Physiological Increase by 80 to 213% after 3 mo with different preparations. *0311* Highly significant increase induced by ethinylestradiol. *2280*

Sialic Acid *Serum Decrease Physiological* Metabolic alteration (estrogen effect). *0227*

Sodium *Serum Increase Physiological* May cause sodium retention. *2313*

T₃ Uptake *Serum Decrease Physiological* Falls from 80-100% to 55-90% (resin test). *0585* Similar values to those of first trimester of pregnancy. *3403*

Testosterone *Serum Increase Physiological* Metabolic changes in liver synthesis. *2189*

Testosterone Binding Globulin
Serum Increase Physiological Metabolic effect. *2385*

Testosterone, Free *Serum Decrease Physiological* As a result of markedly increased sex hormone binding globulin. *2280*

Thromboplastin Generation *Blood Increase Physiological* Significant effect observed. *2456*

Thyro-Binding Index *Serum Increase Physiological* For up to 1 mo after stopping treatment. *0913*

Thyroid Stimulating Hormone (TSH)
Serum No Effect Physiological No significant difference from nonpregnant values. *3403* Hypophyseal mediated effect; results generally contradictory. *3798*

Thyroxine (T₄) *Serum Increase Physiological* Increased binding protein available. *3505* Similar values to those in late first trimester. *3403* Increased binding capacity (augmentation of transport proteins). *3798*

Thyroxine (T₄) Binding Globulin
Serum Increase Physiological Estrogen activation of carrier protein. *0395* In response to amount of circulating estrogen. *0585*

Thyroxine (T₄), Free *Serum No Effect Physiological* No effect observed although PBI may be increased. *2427* No significant difference from nonpregnant values. *3403*
Serum Decrease Physiological When treatment results compared with controls. *0260*

Thyroxine (T₄) Index, Free (FTI)
Serum No Effect Physiological Thyroid function unaffected. *0071*
Serum Increase Physiological May be slightly increased even 1 mo after treatment. *0913*

Thyroxine (T₄) (Murphy-Pattee)
Serum Increase Physiological Increased binding protein available (effect 2-4 weeks). *3175*

Transcortin *Serum Increase Physiological* Metabolic changes in liver synthesis. *2189* Increase by 115 to 140% after 3 mo with different preparations. *0311*

Transferrin *Serum Increase Physiological* Metabolic changes in liver synthesis (estrogen). *2189* In 46 oral contraceptive users compared with controls studied over at least 2 y. *1182*

Triglycerides *Serum No Effect Physiological* Extent of effect varies with exact composition of oral contraceptive. *1458* In 10 women receiving ethinyl estradiol with norethindrone. *2009* Typically insignificant change over 6 cycles in women with low dose ethinyl estradiol with low dose levonorgestrel or desogestrel. *1249*
Serum Increase Physiological If given for 3 mo impaired removal (estrogen). *3505* Increased in response to a variety of different preparations. *1303* Increase by 25 to 62% depending on preparation used. *2860* Increase by 40 mg/dL in white girl drug users vs controls. *3723* Estrogen effect of combined oral contraceptive. *0018* Positive correlation with serum gamma-glutamyltransferase activity suggesting involvement of hepatic enzyme induction. *2317* With high-dose combination drugs. With drug with high progestogen component induce greater than 50% increase on average. *1248* In 10 women receiving ethinyl estradiol with norgestrel. *2009* Increase by 80% with high dose oral contraceptives that are overly estrogenic. *1248*

Triglycerides, High Density Lipoprotein
Serum Increase Physiological Increase by 17 to 81% depending on preparation used. *2860*

Triglycerides, Low Density Lipoprotein
Serum Increase Physiological Increase by 42 to 60% depending on preparation used. *2860*

Triglycerides, Very Low Density Lipoprotein
Serum Increase Physiological Increase by 16 to 52% depending on preparation used. *2860*

Tri-iodothyronine (T₃) *Serum Increase Physiological* Similar values to those in late first trimester. *3403* Increased binding capacity (augmentation of transport proteins). *3798*
Serum Decrease Physiological Due to increase in thyroxine binding globulin. *3175*

Tyrosine *Plasma Decrease Physiological* Hormonal effect (second part of cycle). *0742*

Urobilinogen *Urine Decrease Physiological* May induce cholestasis. *1713*
Feces Decrease Physiological Pale stools, due to cholestasis. *1713*

Uroporphyrin *Urine Increase Physiological* May induce porphyria cutanea tarda. *3029*

Valine *Plasma Decrease Physiological* Hormonal effect (second part of cycle). *0742*

Vanillylmandelic Acid *Urine No Effect Physiological* No effect observed with ingestion. *3217*

Vital Capacity *Patient No Effect Physiological* No significant effect observed. *1185*

Vitamin A *Serum Increase Physiological* Increased to about 150% of controls in women on a variety of oral contraceptives. *1303* Values doubled in 6 mo in 8 women taking estrogenic combination but diminished to some extent after 4 y. *0773* Values almost doubled within 6 mo but diminished somewhat after 4 y. *0773* Due to influence on binding protein. *3023*

Vitamin B₁₂ *Serum No Effect Physiological* No significant difference between oral contraceptive users and control adolescents. *1368*
Serum Decrease Physiological Probable interference with gastrointestinal absorption. *3802* Mild effect in oral contraceptive users. *3310*

Xanthurenic Acid *Urine Increase Physiological* Estrogen component induces tryptophan pyrrolase. *3045*

Zinc *Serum No Effect Physiological* No difference from normal observed. *0532* In 22 women ingesting combination type contraceptives. *1604*
Serum Decrease Physiological Duration of use unrelated to Zn concentration. *1467*
Hair No Effect Physiological No significant effect in young women studied for at least 3 mo. *3720*
Red Blood Cells No Effect Physiological In 22 women ingesting combination type contraceptives. *1604*
White Blood Cells No Effect Physiological In 22 women ingesting combination type contraceptives. *1604*

Oral Resins

Ammonia *Plasma Increase Physiological* Exchanged for other ions in gastrointestinal tract. *1022*

Oranges

Synephrine *Urine Increase Physiological* Due to presence in fruit. *2784*

Organophosphorus Insecticides

Alanine Aminotransferase *Serum Increase Physiological* Hepatotoxic effect. *0405*

Aldolase *Serum Increase Physiological* Hepatotoxic effect. *0405*

Alkaline Phosphatase *Serum Increase Physiological* Hepatotoxic effect. *0405*

Aspartate Aminotransferase
Serum Increase Physiological Hepatotoxic effect. *0405*

Bile *Urine Increase Physiological* Hepatotoxic effect. *0405*

Bilirubin *Serum Increase Physiological* Hepatotoxic effect. *0405*

BSP Retention *Serum Increase Physiological* Hepatotoxic effect. *0405*

Cephalin Flocculation *Serum Increase Physiological* Hepatotoxic effect. *0405*

Pseudocholinesterase *Serum Decrease Physiological* Hepatotoxic effect. *0405*

Thymol Turbidity *Serum Increase Physiological* Hepatotoxic effect. *0405*

Vitamin A *Serum Decrease Physiological* Metabolism impaired. *1031*

Ornithine

Proline *Plasma Increase Analytical* Slight effect on method of Goodwin. *0117*

Tyrosine *Plasma No Effect Analytical* On fluorometric procedure of Ambrose. *0080*

Orotic Acid

Bilirubin *Serum Decrease Physiological* ?induction effect in premature infants only. *3220*

Orphenadrine

Acetaminophen Screening Test *Urine Negative Analytical* No reaction with o-cresol at therapeutic concentrations. *3326*

Albumin *Serum No Effect Analytical* At concentration of 9 mg/L had no effect on BCG method. *3393*

Amphetamine *Urine No Effect Analytical* At 50 mg/L on fluorescent method of Hayes. *1534*

Bicarbonate *Serum No Effect Analytical* At concentration of 9 mg/L had no effect on method using phenolphthalein. *3393*

Bilirubin *Serum No Effect Analytical* At concentration of 9 mg/L had no effect on Jendrassik and Grof method. *3393*

Calcium *Serum No Effect Analytical* At concentration of 9 mg/L had no effect on cresolphthalein method. *3393*

Chloride *Serum No Effect Analytical* At concentration of 9 mg/L had no effect on mercurimetric method. *3393*

Cholesterol *Serum No Effect Analytical* At concentration of 9 mg/L had no effect on Liebermann-Burchard method. *3393*

Creatinine *Serum No Effect Analytical* At concentration of 9 mg/L had no effect on AutoAnalyzer Jaffé method. *3393*

Erythrocytes *Blood Decrease Physiological* 2 cases aplastic anemia reported. *1680*

Leukocytes *Blood Decrease Physiological* 2 cases aplastic anemia reported. *1680*

Phosphate *Serum No Effect Analytical* At concentration of 9 mg/L had no effect on phosphomolybdate method. *3393*

Platelets *Blood Decrease Physiological* 2 cases aplastic anemia reported. *1680*

Potassium *Serum No Effect Analytical* At concentration of 9 mg/L had no effect on measurement by ISE with predilution. *3393*

Protein *Serum No Effect Analytical* At concentration of 9 mg/L had no effect on biuret method with blank correction. *3393*

Prothrombin Time *Plasma Decrease Physiological* May enhance metabolism of anticoagulants. *1679*

Thyroxine (T₄) *Serum Increase Physiological* Reported in two studies. *3798*

Thyroxine (T₄) Index, Free (FTI)
Serum Increase Physiological Reported in two studies. *3798*

Triglycerides *Serum No Effect Analytical* At concentration of 9 mg/L had no effect on lipase/esterase method. *3393*

Tri-iodothyronine (T₃) *Serum No Effect Physiological* Reported in two studies. *3798*

Urea Nitrogen *Serum No Effect Analytical* At concentration of 9 mg/L had no effect on diacetylmonoxime method. *3393*

Uric Acid *Serum No Effect Analytical* At concentration of 9 mg/L had no effect on phosphotungstate reduction method. *3393*

Ouabain

Albumin *Serum No Effect Analytical* At 0.02 mg/dL no effect on SMA 12/60 method. *2636*

Alkaline Phosphatase *Serum No Effect Analytical* At 0.02 mg/dL no effect on SMA 12/60 method. *2636*

Aspartate Aminotransferase *Serum No Effect Analytical* At 0.02 mg/dL no effect on SMA 12/60 method. *2636*

Bilirubin *Serum No Effect Analytical* At 0.02 mg/dL no effect on SMA 12/60 method. *2636*

Calcium *Serum No Effect Analytical* At 0.02 mg/dL no effect on SMA 12/60 method. *2636*

Catecholamines *Urine Decrease Physiological* Marked effect but mechanism unexplained. *3166*

Cholesterol *Serum No Effect Analytical* At 0.02 mg/dL no effect on SMA 12/60 method. *2636*

Creatinine *Serum No Effect Analytical* At 0.02 mg/dL no effect on SMA 12/60 method. *2636*

Digoxin *Serum Increase Analytical* Reported to affect RIA methods. *3940*

Glucose *Serum No Effect Analytical* At 0.02 mg/dL no effect on SMA 12/60 method. *2636*
Serum No Effect Physiological No effect with i.v. infusion for 1 h. *3166*

Insulin *Plasma No Effect Physiological* No effect with i.v. infusion for 1 h. *3166*

Lactate Dehydrogenase *Serum No Effect Analytical* At 0.02 mg/dL no effect on SMA 12/60 method. *2636*

Phosphate *Serum No Effect Analytical* At 0.02 mg/dL no effect on SMA 12/60 method. *2636*

Protein *Serum No Effect Analytical* At 0.02 mg/dL no effect on SMA 12/60 method. *2636*

Uric Acid *Serum No Effect Analytical* At 0.02 mg/dL no effect on SMA 12/60 method. *2636*

Ovulation

Amino Acids *Urine No Effect Physiological* No significant effect following ovulation. *0743*

Ascorbic Acid *Plasma Increase Physiological* Sharp increase most likely at ovulation. *2998*
Urine Increase Physiological Excretion correlated with that of LH. *2206*
Urine Decrease Physiological Sharp decrease reported. *2998*

Cholesterol *Serum Decrease Physiological* Decrease at middle of cycle reported. *2319*

Estradiol *Plasma Increase Physiological* Max of 150-500 pg/mL pre-ovulation. *1295*

Estrogens *Urine Increase Physiological* Rises to peak at ovulation falls after. *0987*

Follicle Stimulating Hormone (FSH)
Plasma Increase Physiological Midcycle peak twice the baseline. *0112*

Gonadotropins *Urine Increase Physiological* May be temporary increase. *0987*

Luteinizing Hormone (LH) *Urine Increase Physiological* Increased by more than 3 times at time of ovulation. *2206*

Platelets *Blood Increase Physiological* Increases from 8th to 14th day of cycle. *0987*

Pregnanediol *Urine Increase Physiological* Slight effect 2 d after ovulation. *0987*

Protein Bound Iodine (PBI) *Serum Increase Physiological* Small increase in ovulatory and luteal phases. *0830*

Tetrahydroaldosterone *Urine Increase Physiological* Higher in luteal phase. *2850*

Oxacillin

Alanine Aminotransferase *Serum Increase Physiological* Possible hepatotoxic effect. *2220* Typically occurring 3 to 14 d after treatment started, may occur without eosinophilia. *2374* Reversible phenomenon observed in 8 patients given drug i.v. *2664* Drug associated hepatitis in patients given high dose drug i.v. *2675* Asymptomatic hepatic dysfunction in 5 cases following high dose treatment: reversible. *2835* Reversible high-dose associated liver injury. *2431* Abnormal liver function in 1 of 8 treated patients on day 15. *2548*

Alkaline Phosphatase *Serum Increase Physiological* Cholestatic jaundice reported in one case. *1022* Asymptomatic hepatic dysfunction in 5 cases following high dose treatment: reversible. *2835* Reversible high-dose associated liver injury. *2431*

Aspartate Aminotransferase
Serum Increase Physiological Possible hepatotoxic effect. *2220* Typically occurring 3 to 14 d after treatment started, may occur without eosinophilia. *2374* Hypersensitivity reaction observed in one patient with Staphylococcus Aureus endocarditis. *0882* Reversible phenomenon observed in 8 patients given drug i.v. *2664* Drug associated hepatitis in patients given high dose drug i.v. *2675* Asymptomatic hepatic dysfunction in 5 cases following high dose treatment: reversible. *2835* Reversible high-dose associated liver injury. *2431* Abnormal liver function in 1 of 8 treated patients on day 15. *2548*

Bile *Urine Increase Physiological* Reversible hepatocellular dysfunction. *2313*

Bilirubin *Serum No Effect Analytical* At concentration of 40 mg/L had no effect on Jendrassik and Grof method. *3393*
Serum No Effect Physiological Asymptomatic hepatic dysfunction in 5 cases following high dose treatment: reversible. *2835*
Serum Increase Physiological Cholestatic jaundice reported in one case. *2737* Reversible high-dose associated liver injury. *2431*

BSP Retention *Serum Increase Physiological* Cholestatic jaundice reported in one case. *1022*

Calcium *Serum No Effect Analytical* At concentration of 40 mg/L had no effect on cresolphthalein method. *3393*

Cephalin Flocculation *Serum Increase Physiological* Hepatotoxic effect. *2313*

Coombs' Test, Direct *Serum No Effect Physiological* 0 of 10 patients demonstrated hypersensitivity reaction. *1963*

Creatinine *Serum Increase Physiological* Transient azotemia with large doses. *0071*

Eosinophils *Blood No Effect Physiological* 0 of 10 patients demonstrated hypersensitivity reaction. *1963*
Blood Increase Physiological Reported effect (?allergic). *0071* Suggestive of hypersensitivity reaction. *2374* Reversible phenomenon observed in 8 patients given drug i.v. *2664*

Erythrocytes *Blood Decrease Physiological* Large doses parenteral penicillin may have effect. *1680*
Urine Increase Physiological Nephrotoxicity with hematuria. *0071*

Hemoglobin *Urine Increase Physiological* Nephrotoxicity with hematuria. *0071*

17-Ketosteroids *Urine Increase Analytical* Absorption at 520 nm in Zimmermann procedure. *0022*

Lactate Dehydrogenase *Serum Increase Physiological* Reversible high-dose associated liver injury. *2431*

Leukocytes *Blood Decrease Physiological* May cause bone marrow depression. *2313* In 2 patients receiving 12-15 g/d intravenously. *1646* Observed in one man, recovery began within 2 d of withdrawal. *1844* In two cases in one of which abnormalities developed within 48 h and in other in 17 d. *0485* Single case of drug-related agranulocytosis in patient with prior history of penicillin sensitivity. *3167*

Lymphocytes *Blood No Effect Physiological* In 2 patients receiving 12-15 g/d intravenously. *1646*

Neutrophils *Blood Decrease Physiological* Observed in 5 patients given high doses intravenously: reversible with cessation of therapy. *0035* Granulocytopenia in 2 cases within 2 d of treatment being started. *1074* Observed in one man, recovery began within 2 d of withdrawal. *1844* Case observed after i.v. drug in 1 y old child: postulated that mechanism due to toxic effect on maturation of cells. *0657* 5 patients developed neutropenia during high-dose i.v. therapy. *0035* In two cases in one of which abnormalities developed within 48 h and in other in 17 d. *0485* Abnormal liver function in 1 of 8 treated patients on day 15. *2548* Single case of drug-related agranulocytosis in patient with prior history of penicillin sensitivity. *3167*

Nitrofurantoin *Urine No Effect Analytical* No effect on method of Conklin and Hollifield. *0704*

5'-Nucleotidase *Serum Increase Physiological* Due to cholestasis. *1778*

Phosphate *Serum No Effect Analytical* At concentration of 40 mg/L had no effect on phosphomolybdate method. *3393*

Platelets *Blood Decrease Physiological* Usually only associated with large parenteral doses. *1680*

Protein *Serum Increase Analytical* At concentrations above 500 mg/L (normal therapeutic concentration 6 mg/L) raised concentration as measured by biuret method with blank correction. *3393*
Urine Increase Physiological Nephrotoxic effect. *0071*

Thymol Turbidity *Serum Increase Physiological* Reversible hepatocellular dysfunction. *2313*

Urea Nitrogen *Serum No Effect Analytical* At concentration of 40 mg/L had no effect on diacetylmonoxime method. *3393*
Serum Increase Physiological Transient azotemia with large doses. *0071*

Uric Acid *Serum No Effect Analytical* At concentration of 40 mg/L had no effect on phosphotungstate reduction method. *3393*

Oxalacetate

Aspartate Aminotransferase *Serum Decrease Analytical* Inhibits colorimetric method. *3835*

Creatinine *Serum No Effect Analytical* No effect on ion-exchange method of Mitchell. *2459*
Serum Increase Analytical Marked effect direct methods. Moderate effect on Lloyd's procedure. *2459*

Oxalate

Acid Phosphatase *Serum Decrease Analytical* Inhibition of enzyme in laboratory procedures. *3588*

Alkaline Phosphatase *Serum No Effect Analytical* Not inhibited when used as anticoagulant. *3217*
Serum Decrease Analytical Inhibition of enzyme in laboratory procedures. *1022*

α-Amino-Nitrogen *Urine Increase Physiological* Oxalic acid may cause renal damage. *0459*

Ammonia *Plasma Increase Analytical* Effect if used as anticoagulants. *0705*

Amylase *Serum Decrease Analytical* Approximately 20% decrease compared with serum/heparinized plasma. *0583* Sodium salt inhibits by 15% at 2 g/L with method of Rauscher et al. *2926*

Aspartate Aminotransferase *Serum No Effect Analytical* No effect on activity. *3165*

Bicarbonate *Serum Decrease Analytical* Significant alteration of pH if used as anticoagulant. *0579*

Calcium *Serum Increase Analytical* May be coprecipitation of Na, K, Mg and protein. *1563*
Serum Decrease Analytical Precipitation of calcium oxalate (may be incomplete). *1563* Unless lanthanum added may decrease atomic absorption. *0583*
Urine Increase Analytical May be coprecipitation of Na, K, Mg. *1563*
Urine Decrease Analytical May be incomplete precipitation. *1563*

Chloride *Serum Decrease Analytical* If used as anticoagulant may be passage into cells. *1563*

Cholesterol *Serum Decrease Analytical* If used as anticoagulant (causes water shift). *1563*

Crystals *Urine Increase Physiological* Oxalate crystals present in urine in poisoning. *1302*

Erythrocytes *Urine Increase Physiological* Nephrotoxic effect if ingested. *1302*

Erythrocyte Sedimentation Rate
Blood Decrease Analytical Retards rate when compared with heparin. *0251*

Factor V *Plasma Decrease Analytical* Unstable ?affects prothrombin determination. *3217*

Fatty Acids, Free (FFA) *Serum No Effect Analytical* No effect as anticoagulant on method of Soloni. *3381*

Glucose *Serum No Effect Analytical* At 2.5 g/dL on p-HBAH procedure of Lever. At 2.5 g/dL on alkaline ferricyanide procedure. At 2.5 g/dL on glucose oxidase procedure. At 2.5 g/dL on o-toluidine procedure. *2140*
Urine No Effect Analytical With DNSA method at usual concentration. *0076*

Gold *Blood No Effect Analytical* Atomic absorption method of Harth et al. *1515*

Hematocrit *Blood Decrease Analytical* Shrinks RBC, plasma volume increased up to 13%. *1563* With K salt may be 8-13% less than with heparin. *0067*

Hydroxybutyrate Dehydrogenase
Serum Decrease Analytical Almost complete inactivation of enzyme. *2981*

Insulin *Plasma Decrease Analytical* Compared with lithium heparin plasma or serum. *3217*

Lactate Dehydrogenase *Serum Decrease Analytical* Inhibition of enzyme activity. *3588*

Lithium *Serum Increase Analytical* If lithium salt of oxalates. *1563*

Nonprotein Nitrogen *Serum Increase Analytical* Balanced oxalate mixture contains NH_4. *1563*
Serum Increase Physiological Acute renal failure with Calcium oxalate deposition. *1343*

Oxalate (continued)

Occult Blood *Feces Positive Physiological* Occurs after ingestion. *1302*

Oxalate *Urine Increase Physiological* Effect greatest with low calcium diet. *2310*

Partial Thromboplastin Time *Plasma Decrease Analytical* If used as anticoagulant compared with citrate. *3382*

pH *Serum Increase Analytical* In vitro addition of K salt to plasma/serum. *0106*
Blood Increase Analytical Occurs with Na or K oxalates as anticoagulants. *0579*
Blood Decrease Analytical Significant effect of NH_4 oxalate as anticoagulant. *0579*

Phosphate *Serum Decrease Analytical* Complexes molybdate, decreases color development. *0583* Excess oxalate interferes with develop of color. *3502*
Urine No Effect Analytical No effect at 50 mg/dL method of Jung/Parekh. *1835*

Potassium *Serum Increase Analytical* If potassium salt of oxalate. *1563*

Protein *Urine Increase Physiological* If ingested may cause nephrotoxicity. *1302*

Salicylate *Serum Increase Analytical* 250 mg/dL give equivalent of 0.3 mg/dL by Trinder. *2251*

Sodium *Serum Increase Analytical* If sodium salt of oxalate. *1563*

Sugar *Urine Increase Analytical* False positive with Benedict's. *1563*

Thymol Turbidity *Serum Decrease Analytical* Decreased turbidity developed. *0583*

Urea Nitrogen *Serum Increase Analytical* Balanced oxalate mixture contains NH_4. *1563*

Oxalate/Fluoride

Alkaline Phosphatase *Serum No Effect Analytical* Little or no effect observed. *2228*

Aminolevulinic Acid Dehydrase
Red Blood Cells Decrease Analytical 15% decrease compared with heparin sample. *0400*

Oxandrolone

BSP Retention *Serum Increase Physiological* Slight increase in one child (other liver function tests normal). *1750*

Cholesterol *Serum Decrease Physiological* Anabolic effect. *1308*

Fibrinogen *Plasma Increase Physiological* Metabolic effect. *0227*

Glucose *Serum Decrease Physiological* Anabolic effect. *1308*

Haptoglobin *Serum Increase Physiological* Metabolic effect. *0227*

Plasminogen *Plasma Increase Physiological* Metabolic effect. *0227*

Sialic Acid *Serum Increase Physiological* Metabolic effect. *0227*

Triglycerides *Serum Decrease Physiological* Increases triglyceride hydrolysis peripherally. *1308*

Oxaprozin

Aldosterone *Plasma No Effect Physiological* No significant change observed in 7 volunteers treated for up to 1 week. *2464*
Urine No Effect Physiological No significant change observed in 7 volunteers treated for up to 1 week. *2464*

Glomerular Filtration Rate (GFR)
Urine No Effect Physiological No significant change observed in 7 volunteers treated for up to 1 week. *2464*

Potassium *Serum No Effect Physiological* No significant change observed in 7 volunteers treated for up to 1 week. *2464*

Renin Activity *Plasma No Effect Physiological* No significant change observed in 7 volunteers treated for up to 1 week. *2464*

Oxazepam

Acetaminophen *Serum No Effect Analytical* At 2 mg/L had no effect on HPLC method. *3432*

Alanine Aminotransferase *Serum No Effect Analytical* No effect at therapeutic concentration on Reflotron method. *1984*
Serum Increase Physiological Possible hepatotoxicity. *2313*

Albumin *Serum No Effect Analytical* At concentration of 1 mg/L had no effect on BCG method. *3393*

Alkaline Phosphatase *Serum Increase Physiological* Possible hepatotoxicity. *2313*

Alkaloids *Urine No Effect Analytical* With ninhydrin on Frings TLC procedure at 10 mg/dL. With iodoplatinate of Frings TLC procedure at 10 mg/dL. *1204*

Amitriptyline *Serum No Effect Physiological* No effect on serum concentration if given together. *3321*

Aspartate Aminotransferase *Serum No Effect Analytical* No effect at therapeutic concentration on Reflotron method. *1984*
Serum Increase Physiological Possible hepatotoxicity. *2313*

Barbiturate *Urine No Effect Analytical* With diphenylcarbazone on Frings TLC procedure at 10 mg/dL. With mercuric SO_4 on Frings TLC procedure at 10 mg/dL. With vanillin on Frings procedure at 10 mg/dL. With mercurous NO_3 on Frings TLC procedure at 10 mg/dL. *1204*

Bicarbonate *Serum No Effect Analytical* At concentration of 1 mg/L had no effect on method using phenolphthalein. *3393*

Bile *Urine Increase Physiological* Possible hepatotoxicity. *2313*

Bilirubin *Serum No Effect Analytical* At concentration of 1 mg/L had no effect on Jendrassik and Grof method. *3393*
Serum No Effect Physiological Insignificant displacement from protein in neonates. *3748*
Serum Increase Physiological Possible hepatotoxicity. *2313*

Bromide *Urine No Effect Analytical* No effect on method of Frings. *1204*

BSP Retention *Serum Increase Physiological* Possible hepatotoxicity. *2313*

Calcium *Serum No Effect Analytical* At concentration of 1 mg/L had no effect on cresolphthalein method. *3393*

Cephalin Flocculation *Serum Increase Physiological* Possible hepatotoxicity. *2313*

Chlordiazepoxide *Blood Increase Analytical* 2 mg/dL equivalent to 3.6 mg/dL by method of Riddick. *2983* Slight effect only on method of Jatlow. *1777*

Chloride *Serum No Effect Analytical* At concentration of 1 mg/L had no effect on mercurimetric method. *3393*

Cholesterol *Serum No Effect Analytical* No effect at therapeutic concentration on Reflotron method. *1984* At concentration of 1 mg/L had no effect on Liebermann-Burchard method. *3393*

Creatinine *Serum No Effect Analytical* At 10 mg/L on reversed phase LC procedure of Zhiri et al. *3942* At concentration of 1 mg/L had no effect on AutoAnalyzer method. *3393*

Eosinophils *Blood Increase Physiological* Rare allergic response. *0071*

Etchlorvynol *Urine No Effect Analytical* No effect on method of Frings and Cohen. *1204*

FPN Test *Urine No Effect Analytical* No effect at 10 mg/dL on method of Frings. *1204*

Glucose *Serum No Effect Analytical* No effect on glucose oxidase method of Boehringer. *3277* No effect at therapeutic concentration on Reflotron method. *1984* At concentration of 30 mg/L had no effect on hexokinase/G-6-PDH method. *3393*
Serum Increase Analytical Filler affects o-toluidine, Neocuproin methods. *2419*

γ-Glutamyltransferase (GGT) *Serum No Effect Analytical* No effect at therapeutic concentration on Reflotron method. *1984*

5-Hydroxyindoleacetic Acid (5-HIAA)
Urine No Effect Analytical Slight effect when added in vitro in method of Udenfriend. *0670*

^{131}I Uptake *Serum Decrease Physiological* Serax® contains tetraiodofluorescein. *2652*

Lactate *Serum No Effect Analytical* At concentration of 30 mg/L had no effect on enzymatic method. *3393*

Leukocytes *Blood Decrease Physiological* Leukopenia occurs rarely. *2313*

Nortriptyline *Serum No Effect Physiological* No effect on serum concentration if given together. *3321*

Phosphate *Serum No Effect Analytical* At concentration of 1 mg/L had no effect on phosphomolybdate method. *3393*

Propoxyphene *Serum No Effect Analytical* At 140 mg/L on method of Evenson. *1056*

Protein *Serum No Effect Analytical* At concentration of 1 mg/L had no effect on biuret method with blank correction. *3393*

Salicylate *Urine No Effect Analytical* At 10 mg/dL no effect on Trinder method. *1204*

Sugar *Urine Increase Analytical* Alleged to reduce copper in Somogyi procedure. *0583*

Thymol Turbidity *Serum Increase Physiological* Possible hepatotoxicity. *2313*

Triglycerides *Serum No Effect Analytical* No effect at therapeutic concentration on Reflotron method. *1984* At concentration of 1 mg/L had no effect on lipase/esterase method. *3393*

Urea Nitrogen *Serum No Effect Analytical* No effect at therapeutic concentration on Reflotron method. *1984* At concentration of 1 mg/L had no effect on diacetylmonoxime method. *3393*

Uric Acid *Serum No Effect Analytical* At 10 mg/L on reversed phase LC procedure of Zhiri et al. *3942* No effect at therapeutic concentration on Reflotron method. *1984* At concentration of 1 mg/L had no effect on phosphotungstate reduction method. *3393*

Oxedrine

Amphetamine *Urine No Effect Analytical* No effect at 100 mg/L on method of Rutter. *3097*

Glucose *Serum No Effect Analytical* At concentration of 60 mg/L had no effect on GOD/POD-PAP method. *3393*

Triglycerides *Serum No Effect Analytical* At concentration of 60 mg/L had no effect on GPO-PAP method. *3393*

Uric Acid *Serum No Effect Analytical* At concentration of 60 mg/L had no effect on uricase-PAP method. *3393*

Oxmetidine

pH *Gastric Material Increase Physiological* pH 5 reached 80 minutes after 200 mg drug. *2578*

Prolactin *Plasma No Effect Physiological* No significant effect during treatment. *2578*

Volume *Gastric Material Decrease Physiological* Significant inhibition following treatment. *2578*

Oxoprogesterone

Progesterone *Plasma Increase Analytical* 22% cross reactivity with RIA method of Cameron. *0567*

Oxprenolol

Cholesterol *Serum No Effect Physiological* Insignificant effect in 12 patients receiving 160 mg/d for 2 weeks. *1806* In 20 hypertensive men given 160 mg/d for 5 weeks. *2135* In several studies treated for more than 1 mo. *0088*
Serum Increase Physiological Mild effect as possessed relatively weak intrinsic sympathomimetic activity. *2116*

Cholesterol, High Density Lipoprotein
Serum No Effect Physiological In several studies of 1-4 mo. *0088* In 20 hypertensive men given 160 mg/d for 5 weeks. *2135*
Serum Decrease Physiological Significant effect in 53 patients given 80 mg/twice daily for 3 mo. *0835* Mild effect as possessed relatively weak intrinsic sympathomimetic activity. *2116*

Cholesterol, Low Density Lipoprotein
Serum No Effect Physiological In several studies of 1-4 mo. *0088*
Serum Increase Physiological Mild effect as possessed relatively weak intrinsic sympathomimetic activity. *2116*

Cholesterol, Very Low Density Lipoprotein
Serum No Effect Physiological In several studies of 1-4 mo. *0088* In 20 hypertensive men given 160 mg/d for 5 weeks. *2135*

Fatty Acids, Free (FFA) *Serum Decrease Physiological* Marked effect in 53 patients given 80 mg/twice daily for 3 mo. *0835*

Glomerular Filtration Rate (GFR)
Urine Decrease Physiological Observed with acute experiments. *1714*

Growth Hormone *Plasma Increase Physiological* Observed with chronic treatment of 5 active hypertensive patients. *1090*

5-Hydroxyindoleacetic Acid (5-HIAA)
Urine Increase Analytical Interferes with screening tests with nitrosonaphthol. *1803*

Metanephrines, Total *Urine Increase Analytical* Especially when combined with other antihypertensive drugs and HPLC method used unless toluene extraction used. *0751*

Platelets *Blood Decrease Physiological* Observed in one patient after slow release drug normalized within 7 d of withdrawal of drug. *0918* Observed with regular preparation in one individual. *1495*

Potassium *Serum Increase Physiological* Slight rise observed in patients treated with moderate doses. *2836*

Renin Activity *Plasma Decrease Physiological* In 10 patients with essential hypertension treated over 5 weeks: significant effect with plasma concentration of about 1,000 ng/mL. *1114*

Terbutaline *Serum Increase Physiological* Decreased clearance and increased area under curve when co-administered. *1823*

Triglycerides *Serum No Effect Physiological* In several studies treated for more than 1 mo. *0088* Insignificant effect in 12 patients receiving 160 mg/d for 2 weeks. *1806*
Serum Increase Physiological Increase by 22-27% in several studies treated for more than 1 mo. *0088* Marked effect in 53 patients given 80 mg/twice daily for 3 mo. *0835* Mild effect as possessed relatively weak intrinsic sympathomimetic activity. *2116* Increase by 22% in 20 hypertensive men given 160 mg/d for 5 weeks. *2135*

Triglycerides, Very Low Density Lipoprotein
Serum Increase Physiological In several studies of 1-4 mo. *0088* Significant effect in 53 patients given 80 mg/twice daily for 3 mo. *0835*

Uric Acid *Serum No Effect Physiological* In 20 hypertensive men given 160 mg/d for 5 weeks. *2135*

Oxycodone

Alkaloids *Urine No Effect Analytical* With ninhydrin on Frings TLC procedure at 5 mg/dL. *1204*
Urine Increase Analytical Purple with iodoplatinate on Frings TLC procedure. *1204*

Barbiturate *Urine No Effect Analytical* With mercurous NO_3 on Frings TLC procedure at 5 mg/dL. With diphenylcarbazone on Frings TLC procedure at 5 mg/dL. With vanillin on Frings procedure at 5 mg/dL. With mercuric SO_4 on Frings TLC procedure at 10 mg/dL. *1204*

Bromide *Urine No Effect Analytical* No effect on method of Frings. *1204*

Ethchlorvynol *Urine No Effect Analytical* No effect on method of Frings and Cohen. *1205*

FPN Test *Urine No Effect Analytical* No effect at 5 mg/dL on method of Frings. *1204*

Salicylate *Urine No Effect Analytical* At 5 mg/dL no effect on Trinder method. *1204*

Oxymetazoline

Alkaloids *Urine No Effect Analytical* With ninhydrin on Frings TLC procedure at 0.5 mg/dL. *1204*
Urine Increase Analytical Gray with iodoplatinate in Frings TLC procedure. *1204*

Oxymetazoline *(continued)*

Barbiturate   *Urine   No Effect   Analytical*   With diphenyl-carbazone on Frings TLC procedure at 0.5 mg/dL. With vanillin on Frings procedure at 0.5 mg/dL. With mercuric SO_4 on Frings TLC procedure at 0.5 mg/dL. With mercurous NO_3 on Frings TLC procedure at 0.5 mg/dL. *1204*

Bromide   *Urine   No Effect   Analytical*   No effect on method of Frings. *1204*

Corticosteroids   *Plasma   Decrease   Physiological*   Clear suppression when induction compared with induction by thiopentone. *0063*

Ethchlorvynol   *Urine   No Effect   Analytical*   No effect on method of Frings and Cohen. *1204*

FPN Test   *Urine   No Effect   Analytical*   No effect at 0.5 mg/dL on method of Frings. *1204*

Salicylate   *Urine   No Effect   Analytical*   At 0.5 mg/dL no effect on Trinder method. *1204*

Oxymetholone

α_1-**Acid Glycoprotein**   *Serum   Increase   Physiological*   Anabolic metabolic effect. *0227*

Alanine Aminotransferase   *Serum   Increase   Physiological*   Cholestatic effect (increased up to 6 times normal). *1713*

Alkaline Phosphatase   *Serum   Increase   Physiological*   Cholestatic effect. *1713*   Cholestatic hepatitis developed in two patients. *3921*   One case of peliosis hepatitis reported. *2539*

α_1-**Antitrypsin**   *Serum   Increase   Physiological*   Metabolic effect. *0227*

Aspartate Aminotransferase
Serum   Increase   Physiological   Cholestatic effect (increased up to 6 times normal). *1713*   Cholestatic hepatitis developed in two patients. *3921*   One case of peliosis hepatitis reported. *2539*

Bilirubin   *Serum   Increase   Physiological*   Cholestatic effect. *1713*   Cholestatic hepatitis developed in two patients. *3921*   One case of peliosis hepatitis reported. *2539*

BSP Retention   *Serum   Increase   Physiological*   Cholestatic effect. *1713*   Cholestatic hepatitis developed in two patients. *3921*

Calcium   *Serum   Increase   Physiological*   May occur spontaneously or if carcinoma of breast. *1679*

Ceruloplasmin   *Serum   No Effect   Physiological*   No metabolic effect. *0227*

Cholesterol   *Serum   Increase   Physiological*   Cholestatic effect. *1713*
Serum   Decrease   Physiological   May cause decrease with therapy. *1679*

Creatine   *Urine   Increase   Physiological*   Anabolic effect (possible for 2 weeks after stop). *1679*

Creatinine   *Urine   Increase   Physiological*   Anabolic effect (possible for 2 weeks after stop). *1679*

Erythropoietin   *Urine   Increase   Physiological*   Enhances production and excretion if anemia. *1679*

Factor II   *Plasma   Increase   Physiological*   Anabolic effect increasing factors. *1679*

Factor V   *Plasma   Increase   Physiological*   Anabolic effect increasing factors. *1679*

Factor VII   *Plasma   Increase   Physiological*   Anabolic effect increasing factors. *1679*

Factor X   *Plasma   Increase   Physiological*   Anabolic effect increasing factors. *1679*

Fibrinogen   *Plasma   Increase   Physiological*   Metabolic effect. *0227*

Glucose   *Serum   Decrease   Physiological*   Anabolic effect (possible for 2 weeks after stop). *1679*

β-**Glucuronidase**   *Serum   Increase   Physiological*   Metabolic effect. *0227*

Haptoglobin   *Serum   Increase   Physiological*   Metabolic effect. *0227*

Iron   *Serum   Decrease   Physiological*   Iron deficiency anemia may occur. *1679*

Iron-Binding Capacity, Unsaturated (UIBC)
Serum   Increase   Physiological   Decreases percentage saturation of transferrin. *1679*

^{131}**I Uptake**   *Serum   Increase   Physiological*   Anabolic effect (possible for 2 weeks after stop). *1679*

17-Ketosteroids   *Urine   Decrease   Physiological*   Anabolic effect (possible for 2 weeks after stop). *1679*

Metyrapone Test   *Patient   Positive   Physiological*   Anabolic effect (possible for 2 weeks after stop). *1679*

Plasminogen   *Plasma   Increase   Physiological*   Metabolic effect. *0227*

Protein Bound Iodine (PBI)   *Serum   Decrease   Physiological*   Reduces concentration of circulating thyroxine binding globulin. *0012*

Prothrombin Time   *Plasma   Increase   Physiological*   Also seen with other 17-alkyl substituted steroids. *0998*   May increase sensitivity to anticoagulants. *1679*

Sialic Acid   *Serum   Increase   Physiological*   Metabolic effect. *0227*

T$_3$ Uptake   *Serum   Increase   Physiological*   Anabolic effect (possible for 2 weeks after stop). *1679*

Thyroxine (T$_4$) Binding Globulin
Serum   Decrease   Physiological   Direct effect of drug. *0012*

Thyroxine (T$_4$), Free   *Serum   No Effect   Physiological*   Anabolic effect (possible for 2 weeks after stop). *1679*

Triglycerides   *Serum   Decrease   Physiological*   Anabolic effect. *0319*

Urobilinogen   *Urine   Decrease   Physiological*   Cholestatic effect. *1713*
Feces   Decrease   Physiological   Cholestatic effect. *1713*

Oxyphenbutazone

Alanine Aminotransferase   *Serum   Increase   Physiological*   Possible hepatotoxicity. *2313*

Alkaline Phosphatase   *Serum   Increase   Physiological*   Possible hepatotoxicity. *2313*

Amylase   *Serum   Increase   Physiological*   Parotitis is rare complication of therapy. *1412*   Occasional case of acute swelling of salivary glands. *0642*

Aspartate Aminotransferase
Serum   Increase   Physiological   Possible hepatotoxicity. *2313*

Bile   *Urine   Increase   Physiological*   Possible hepatotoxicity. *2313*

Bilirubin   *Serum   No Effect   Physiological*   Probably little or no effect on displacement from protein. *3748*
Serum   Increase   Physiological   Possible hepatotoxicity. *2313*

BSP Retention   *Serum   Increase   Physiological*   Possible hepatotoxicity. *2313*

Cephalin Flocculation   *Serum   Increase   Physiological*   Possible hepatotoxicity. *2313*

Chloride   *Serum   Increase   Physiological*   May cause marked salt retention. *2313*

Cholesterol   *Serum   No Effect   Analytical*   At concentration of 600 mg/L had no effect on CHOD-Iodide method. *3393*   At concentration of 600 mg/L had no effect on CHOD-PAP method. *3393*   At concentration of 600 mg/L had no effect on catalase-AIDH method. *3393*   At concentration of 600 mg/L had no effect on method using catalase-Hantzsch reaction. *3393*   At concentration of 600 mg/L had no effect on Liebermann-Burchard method. *3393*

Creatinine   *Serum   No Effect   Analytical*   At concentration of 220 mg/L had no effect on creatinine iminohydrolase method. *3393*   At concentration of 120 mg/L had no effect on Jaffé-Fading-Fraction method. *3393*   At concentration of 120 mg/L had no effect on Jaffé-Fuller's earth method. *3393*   At concentration of 600 mg/L had no effect on kinetic Jaffé method on BKA-2. *3393*
Serum   Increase   Physiological   Numerous reports of kidney damage up to acute renal failure following therapeutic use of drug. *2866*

Creatinine Clearance   *Urine   Decrease   Physiological*   Numerous reports of kidney damage up to acute renal failure following therapeutic use of drug. *2866*

Crystals   *Urine   Increase   Physiological*   May cause crystallization of uric acid. *0019*

Erythrocytes *Blood Decrease Physiological* May cause blood dyscrasias. *2313* Occasional case of aplastic anemia reported. *3717* Aplastic anemia with death in 38 per 100,000 users. *1723*

Urine Increase Physiological May cause actual bleeding. *1022* Numerous reports of kidney damage up to acute renal failure following therapeutic use of drug. *2866*

Glucose *Serum No Effect Analytical* At concentration of 12 mg/L had no effect on Ektachem® method. *3393* At concentration of 200 mg/L had no effect on hexokinase/G-6-PDH method. *3393*

Serum Increase Physiological metabolic/endocrine response. *1680*

Serum Decrease Analytical At concentrations above 25 mg/L (normal therapeutic concentration 20 mg/L) lowered concentration as measured by GOD-Perid method. *3393*

Hematocrit *Blood Decrease Physiological* Dilutional effect of water retention. *1343*

Hemoglobin *Blood Decrease Physiological* Dilutional effect of water retention. *1343*

Urine Increase Physiological Actual bleeding caused by drug. *1022*

¹³¹I Uptake *Serum Decrease Physiological* Impaired synthesis of thyroxine. *1915*

Lactate *Serum No Effect Analytical* At concentration of 120 mg/L had no effect on enzymatic method. *3393*

Leukocytes *Blood Increase Physiological* May cause leukemia type of reaction. *0019*

Blood Decrease Physiological Agranulocytosis/leukopenia. *2313* Rare side effect reported to UK committee on safety of medicines. *0782* Occasional case of aplastic anemia reported. *3717*

Neutrophils *Blood Decrease Physiological* May cause neutropenia. *2583* Occasional case of agranulocytosis reported. *3717* Occasional case of drug-induced neutropenia. *0155*

Occult Blood *Feces Positive Physiological* May cause gastrointestinal bleeding. *0071* Rare side effect reported to UK committee on safety of medicines. *0782*

Platelets *Blood Decrease Physiological* Thrombocytopenia may occur after 1 week therapy. *2313* Rare side effect reported to UK committee on safety of medicines. *0782* Occasional case of aplastic anemia reported. *3717*

Potassium *Serum No Effect Analytical* At concentration of 120 mg/L had no effect on measurement by ISE without predilution. *3393*

Protein *Urine Increase Physiological* May cause renal damage. *0019* May be nephrotoxicity. *1680*

Protein Bound Iodine (PBI) *Serum Decrease Physiological* Inhibits iodination of tyrosine in thyroxine binding globulin. *0012*

Protein Electrophoresis *Serum No Effect Analytical* At concentration of 120 mg/L had no effect on automated Olympus-Hite method. *3393*

Prothrombin Time *Plasma Increase Physiological* Displaces anticoagulants from albumin. *2313*

Sodium *Serum No Effect Analytical* At concentration of 120 mg/L had no effect on measurement by ISE without predilution. *3393*

Serum Increase Physiological May cause marked salt retention. *2313*

T₃ Uptake *Serum Increase Physiological* Impaired synthesis of thyroxine. *1915*

Thymol Turbidity *Serum Increase Physiological* Possible hepatotoxicity. *2313*

Thyroxine (T₄) *Serum Decrease Analytical* Yellow eluate persists in reaction mixture. *0583*

Serum Decrease Physiological Impaired synthesis may occur. *1915*

Triglycerides *Serum No Effect Analytical* At concentration of 120 mg/L had no effect on GPO-PAP method. *3393*

Urea Nitrogen *Serum No Effect Analytical* At concentration of 12 mg/L had no effect on Ektachem® method. *3393*

Serum Increase Physiological May occur with renal damage. *2220* Numerous reports of kidney damage up to acute renal failure following therapeutic use of drug. *2866*

Uric Acid *Serum No Effect Analytical* At concentration of 120 mg/L had no effect on catalase-AIDH method. *3393*

Serum Increase Analytical At concentrations above 200 mg/L (normal therapeutic concentration 20 mg/L) raised concentration as measured by Kageyama-Hantzsch method. *3393*

Serum Decrease Physiological Uricosuric effect. *0710*

Volume *Plasma Increase Physiological* May cause marked salt and water retention. *1343*

Urine Decrease Physiological May cause marked water retention. *1343*

Oxyphencyclimine

Hydrochloric Acid *Gastric Material Decrease Physiological* Absolute decrease by 60%, but not concentration. *2537*

Pepsin *Gastric Material Decrease Physiological* Absolute decrease by 60%, but not concentration. *2537*

Volume *Gastric Material Decrease Physiological* Basal output decreased by 60%. *2537*

Oxyphenisatin

Alanine Aminotransferase *Serum Increase Physiological* May cause hypersensitivity reaction. *2970* Liver damage may present as acute hepatitis or as more advanced chronic disease: usual liver disease is chronic active hepatitis. *2260*

Albumin *Serum Decrease Physiological* May cause hypersensitivity reaction. *2970* Observed with chronic active hepatitis induced by drug. *0903*

Alkaline Phosphatase *Serum Increase Physiological* Hepatic toxicity if over prolonged period. *2757* Observed with chronic active hepatitis induced by drug. *0903*

Antimitochondrial Antibodies *Serum Positive Physiological* Very rare observation with chronic hepatitis. *1297*

Antinuclear Antibodies *Serum Positive Physiological* May cause hypersensitivity reaction. *2970* Liver damage may present as acute hepatitis or as more advanced chronic disease: usual liver disease is chronic active hepatitis. *2260*

Antismooth Muscle Antibodies *Serum Positive Physiological* As component of hypersensitivity response. *1297* Liver damage may present as acute hepatitis or as more advanced chronic disease: usual liver disease is chronic active hepatitis. *2260*

Aspartate Aminotransferase *Serum Increase Physiological* Hepatic toxicity if over prolonged period. *2757* Liver damage may present as acute hepatitis or as more advanced chronic disease: usual liver disease is chronic active hepatitis. *2260* Observed with chronic active hepatitis induced by drug. *0903*

Bile *Urine Increase Physiological* May cause hypersensitivity reaction. *2970*

Bilirubin *Serum Increase Physiological* Hepatic toxicity if over prolonged period. *2757* Liver damage may present as acute hepatitis or as more advanced chronic disease: usual liver disease is chronic active hepatitis. *2260* Observed with chronic active hepatitis induced by drug. *0903*

BSP Retention *Serum Increase Physiological* May cause hypersensitivity reaction. *2970*

Cephalin Flocculation *Serum Increase Physiological* May cause hypersensitivity reaction. *2970*

Coombs' Test *Serum Positive Physiological* May cause hypersensitivity reaction. *2970*

Coombs' Test, Direct *Serum Positive Physiological* Single case with positive LE cell test. *1487*

Globulin *Serum Increase Physiological* May cause hypersensitivity reaction. *2970*

Immunoglobulin IgA *Serum Increase Physiological* As component of hypersensitivity response. *1297*

Immunoglobulin IgG *Serum Increase Physiological* Observed with chronic active hepatitis induced by drug. *0903*

¹³¹I Uptake *Serum Decrease Physiological* Dialose contains tetraiodofluorescein. *2652*

Lactate Dehydrogenase *Serum Increase Physiological* Increase observed with active hepatitis. *0066*

LE Cells *Serum Positive Physiological* Liver damage may present as acute hepatitis or as more advanced chronic disease: usual liver disease is chronic active hepatitis. *2260*

Oxyphenisatin (continued)

LE Cells (continued)
Blood Positive Physiological May cause lupoid hepatitis. *2970*

Ornithine Carbamoyltransferase (OCT)
Serum Increase Physiological Rare increase with chronic active hepatitis. *1297* Observed when combined with iron preparation. *1298*

Prothrombin Time *Plasma Increase Physiological* May cause hypersensitivity reaction. *2970*

Rheumatoid Factor *Serum Positive Physiological* Liver damage may present as acute hepatitis or as more advanced chronic disease: usual liver disease is chronic active hepatitis. *2260*

Thymol Turbidity *Serum Increase Physiological* May cause hypersensitivity reaction. *2970*

Oxyquinoline

Protein Bound Iodine (PBI) *Serum Increase Analytical* In vaginal suppositories (often iodinated). *3505*

11-Oxysteroids

Amino Acids *Serum Increase Physiological* Promote tissue catabolism. *0355*

α-Amino-Nitrogen *Plasma Increase Physiological* Promote tissue catabolism. *1022*
Urine Increase Physiological Promote tissue catabolism. *2313*

Oxytetracycline

Alanine Aminotransferase *Serum No Effect Analytical* No effect at therapeutic concentration on Reflotron method. *1984*

Aspartate Aminotransferase *Serum No Effect Analytical* No effect at therapeutic concentration on Reflotron method. *1984*

Barbiturate *Serum No Effect Analytical* No effect on UV absorption methods. *2981*

Bilirubin *Serum No Effect Physiological* Clinically insignificant displacement from protein in neonates. *3748*
Serum Increase Analytical At concentrations above 50 mg/L raised concentration as measured by Jendrassik and Grof method. *3393*

Calcium *Serum No Effect Analytical* At concentration of 50 mg/L had no effect on cresolphthalein method. *3393*

Catecholamines *Plasma Increase Analytical* Interferes with fluorometric methods. *3879*
Urine Increase Analytical Affects fluorescent procedures. *3217*

Chloramphenicol *Serum No Effect Analytical* No effect at 100 mg/L on coupled enzymatic method. *2490*

Cholesterol *Serum No Effect Analytical* No effect at therapeutic concentration on Reflotron method. *1984*
Serum No Effect Physiological No effect seen even when administered orally. *1487*

Dehydroepiandrosterone Sulfate (DHEA-S)
Plasma No Effect Physiological Despite decreased urine excretion and increased fecal excretion of estrogens due to decreased hydrolysis by beta-glucuronidase in gastrointestinal tract. *1469*

Diphenylhydantoin *Serum No Effect Analytical* On GLC procedure of Papadopoulos when added *in vitro. 2722*

Estradiol *Plasma No Effect Physiological* Despite decreased urine excretion and increased fecal excretion of estrogens due to decreased hydrolysis by beta-glucuronidase in gastrointestinal tract. *1469*

Estriol *Urine Decrease Physiological* Probably due to decreased hydrolysis by beta-glucuronidase of estrogen conjugates in intestinal tract, with increased fecal loss. *1469*

Estriol-3-Glucuronide *Urine Decrease Physiological* Probably due to decreased hydrolysis by beta-glucuronidase of estrogen conjugates in intestinal tract with increased fecal loss. *1469*

Estrone *Plasma No Effect Physiological* Despite decreased urine excretion and increased fecal excretion of estrogens due to decreased hydrolysis by beta-glucuronidase in gastrointestinal tract. *1469*

Folate *Serum No Effect Analytical* If chromatographic procedure of Landon used. *2068*
Urine Increase Physiological With dose of 2.5 g/d. *1343*

Glucose *Serum No Effect Analytical* No effect at therapeutic concentration on Reflotron method. *1984* At concentration of 600 mg/L had no effect on hexokinase/G-6-PDH method. *3393*
Serum Decrease Physiological Mild hypoglycemic effect observed in diabetics. *3253*
Urine Decrease Analytical Affects dipsticks if buffered with ascorbic acid. *1563*

Glucose Tolerance *Serum Decrease Physiological* Effect observed in animals. *0220*

γ-Glutamyltransferase (GGT) *Serum No Effect Analytical* No effect at therapeutic concentration on Reflotron method. *1984*

^{131}I Uptake *Serum Decrease Physiological* Terramycin® contains tetraiodofluorescein. *2652*

Lactate *Serum No Effect Analytical* At concentration of 600 mg/L had no effect on enzymatic method. *3393*

Luteinizing Hormone (LH) *Plasma No Effect Physiological* Despite decreased urine excretion and increased fecal excretion of estrogens due decreased hydrolysis by beta-glucuronidase in gastrointestinal tract. *1469*

Metanephrines, Total *Urine Increase Analytical* Interferes with fluorometric methods. *3879*

N-Methylnicotinamide *Urine Increase Physiological* With dose of 2.5 g/d. *1343*

Nitrofurantoin *Urine No Effect Analytical* At 50 μg on method of Conklin/Hollifield. *0704*

Nitrogen *Urine Increase Physiological* Observed in malnourished. *1343*

Pheneturide *Serum No Effect Analytical* On GLC procedure of Papadopoulos when added *in vitro. 2722*

Phenobarbital *Serum No Effect Analytical* On GLC procedure of Papadopoulos when added *in vitro. 2722*

Phosphate *Serum No Effect Analytical* At concentration of 50 mg/L had no effect on phosphomolybdate method. *3393*

Platelets *Blood Decrease Physiological* Thrombocytopenia reported to occur. *1436*

Porphyrins *Urine Increase Analytical* Produces interfering fluorescence. *1238*

Primidone *Serum No Effect Analytical* On GLC procedure of Papadopoulos when added *in vitro. 2722*

Protein *Serum No Effect Analytical* At concentration of 50 mg/L had no effect on biuret method with blank correction. *3393*
CSF Increase Analytical Reacts as if phenol with Folin-Ciocalteu procedure. *3958*

Riboflavin *Urine Increase Physiological* With dose of 2.5 g/d. *1343*

Sugar *Urine Increase Analytical* Acts as reducing agent. *3879*

Testosterone *Serum No Effect Physiological* Despite decreased urine excretion and increased fecal excretion of estrogens due to decreased hydrolysis by beta-glucuronidase in gastrointestinal tract. *1469*

Testosterone, Free *Serum No Effect Physiological* Despite decreased urine excretion and increased fecal excretion of estrogens due to decreased hydrolysis by beta-glucuronidase in gastrointestinal tract. *1689*

Triglycerides *Serum No Effect Analytical* No effect at therapeutic concentration on Reflotron method. *1984*

Urea Nitrogen *Serum No Effect Analytical* No effect at therapeutic concentration on Reflotron method. *1984* At concentration of 50 mg/L had no effect on diacetylmonoxime method. *3393*

Uric Acid *Serum No Effect Analytical* No effect at therapeutic concentration on Reflotron method. *1984*
Serum Increase Analytical At concentrations above 50 mg/L raised concentration as measured by phosphotungstate reduction method. *3393*

Vanillylmandelic Acid *Urine Increase Analytical* Produces interfering color. *3879*

Oxytocin

Aldosterone *Urine No Effect Physiological* No significant or consistent effect. *1344*

Cortisol *Plasma No Effect Physiological* No significant change observed. *1344*

Fibrinogen *Plasma Decrease Physiological* Single case of possible reduction observed. *1680*

17-Ketogenic Steroids *Urine No Effect Physiological* No significant change observed. *1344*

Renin Activity *Plasma No Effect Physiological* No significant or consistent effect. *1344*

Sodium *Serum Decrease Physiological* In mother during induction of labor due both to oxytocin and glucose in infusion: also observed in infant. *3337*
Urine No Effect Physiological No significant or consistent effect. *1344*

Volume *Plasma Increase Physiological* Isolated cases of water intoxication. *1680*

Ozone

Chromosomes *Test Conditions Abnormal Physiological* Clastogenic in human cell cultures. *3282*

Pain

Antidiuretic Hormone *Urine Increase Physiological* Also seen more markedly with emotional stress. *1990* Response to muscle pain. *3217*

Basal Metabolic Rate *Patient Increase Physiological* Effect observed if severe pain. *0251*

Catecholamines *Urine Increase Physiological* Normal response to stress. *0279*

Epinephrine *Urine Increase Physiological* Also seen more markedly with emotional stress. *1990*

Glucose *Serum Decrease Physiological* Response to physical stress. *0117*

Norepinephrine *Urine No Effect Physiological* Response to muscle pain. *3217*
Urine Increase Physiological Also seen but less markedly with emotional stress. *1990*

Volume *Urine Decrease Physiological* Slight changes in response to stress situations. *1990*

Palmitic Acid

Calcium *Serum No Effect Analytical* On atomic absorption procedures at 4 mmol/L With dialysis procedures at 2 mmol/L. *2673*
Serum Decrease Analytical If titrimetric methods used at concentration over 1.5 mmol/L. *2673*

Pamaquine

Bilirubin *Serum Increase Physiological* May cause hemolysis in G-6-PD deficiency. *0248*

Color *Urine Increase Analytical* Brown color. *1022*

Erythrocytes *Blood Decrease Physiological* May cause hemolytic anemia. *2313* Hemolytic anemia in G-6-PD deficient persons. *0788*

Haptoglobin *Serum Decrease Physiological* May cause hemolytic anemia. *0788*

Heinz Body Formation *Blood Positive Physiological* May occur in early stages of hemolysis. *0333*

Hematocrit *Blood Decrease Physiological* Hemolytic anemia in G-6-PD deficient persons. *0788* May cause hemolytic anemia. *2313*

Hemoglobin *Blood Decrease Physiological* Hemolytic anemia in G-6-PD deficient persons. *0788* May cause hemolytic anemia. *2313*

Methemoglobin *Blood Increase Physiological* May cause hemolysis in G-6-PD deficiency. *3581* May cause hemolytic anemia. *2313*

Pancreozymin

Amylase *Serum Increase Analytical* Preparation contains amylase. *1237*
Serum Increase Physiological Effect seen when pancreatic disorders,?in normals. *3334*

Glucagon *Plasma No Effect Physiological* No effect of 1 unit/kg body weight. *2676*

Glucose *Serum No Effect Physiological* No change after 1 unit/kg i.v. *2676*
Serum Increase Physiological Intravenous infusion causes increase. *1775*

Insulin *Plasma No Effect Physiological* No effect of 1 unit/kg body weight. *2676*
Plasma Increase Physiological Intravenous infusion causes increase. *1775*

Lipase *Serum Increase Analytical* Preparation contains lipase. *1237*

Protein *Pancreatic Juice Increase Physiological* Increase in response to challenge. *0486*

Pancuronium

Epinephrine *Plasma Increase Physiological* During halothane anesthesia if also given. *3410*

Norepinephrine *Plasma Increase Physiological* During halothane anesthesia if also given. *3410*

Pseudocholinesterase *Serum Decrease Physiological* ?in vivo, but in vitro 25 μmol causes 80% inhibition. *3410*

Papaverine

Alanine Aminotransferase *Serum Increase Physiological* Probable hypersensitivity reaction. *3039*

Albumin *Serum No Effect Analytical* At concentration of 10 mg/L had no effect on BCG method. *3393*

Alkaline Phosphatase *Serum Increase Physiological* Reversible hepatotoxic effect. *3039* In 2 patients: fever anorexia and jaundice 4 to 5 weeks after start of treatment due to pericholangitis: reversible. *3371* In 6 of 14 patients abnormal liver function, possibly due to allergic response and to metabolic aberration in host. *2746*

Aspartate Aminotransferase
Serum Increase Physiological Probable hypersensitivity reaction. *3039* In 2 patients: fever anorexia and jaundice 4 to 5 weeks after start of treatment due to pericholangitis: reversible. *3371* In 6 of 14 patients abnormal liver function, possibly due to allergic response and to metabolic aberration in host. *2746*

Bicarbonate *Serum No Effect Analytical* At concentration of 10 mg/L had no effect on method using phenolphthalein. *3393*

Bilirubin *Serum No Effect Analytical* At concentration of 10 mg/L had no effect on Jendrassik and Grof method. *3393*
Serum Increase Physiological Small effect, reversible hypersensitivity reaction. *3267* In 2 patients: fever anorexia and jaundice 4 to 5 weeks after start of treatment due to pericholangitis reversible. *3371* In 6 of 14 patients abnormal liver function, possibly due to allergic response and to metabolic aberration in host. *2746*

Calcium *Serum No Effect Analytical* At concentration of 10 mg/L had no effect on cresolphthalein method. *3393*

Chloride *Serum No Effect Analytical* At concentration of 10 mg/L had no effect on mercurimetric method. *3393*

Cholesterol *Serum No Effect Analytical* At concentration of 10 mg/L had no effect on Liebermann-Burchard method. *3393*

Creatinine *Serum No Effect Analytical* At concentration of 10 mg/L had no effect on AutoAnalyzer Jaffé method. *3393*

Eosinophils *Blood Increase Physiological* Allergic response reported to occur. *1680*

γ-Glutamyltransferase (GGT)
Serum Increase Physiological In 6 of 14 patients abnormal liver function, possibly due to allergic response and to metabolic aberration in host. *2746*

Papaverine *(continued)*

Phosphate *Serum No Effect Analytical* At concentration of 10 mg/L had no effect on phosphomolybdate method. *3393*

Potassium *Serum No Effect Analytical* At concentration of 10 mg/L had no effect on measurement by ISE with predilution. *3393*

Protein *Serum No Effect Analytical* At concentration of 10 mg/L had no effect on biuret method with blank correction. *3393*

Sodium *Serum No Effect Analytical* At concentration of 10 mg/L had no effect on measurement by ISE with predilution. *3393*

Triglycerides *Serum No Effect Analytical* At concentration of 10 mg/L had no effect on lipase/esterase method. *3393*

Urea Nitrogen *Serum No Effect Analytical* At concentration of 10 mg/L had no effect on diacetylmonoxime method. *3393*

Uric Acid *Serum No Effect Analytical* At concentration of 10 mg/L had no effect on phosphotungstate reduction method. *3393*

Paraldehyde

Alanine Aminotransferase *Serum Increase Physiological* Possible hepatotoxicity. *2313*

Albumin *Serum No Effect Analytical* At concentration of 2,000 mg/L had no effect on BCG method. *3393*

Alkaline Phosphatase *Serum Increase Physiological* Possible hepatotoxicity. *2313*

Aspartate Aminotransferase
Serum Increase Physiological Possible hepatotoxicity. *2313*

Bicarbonate *Serum No Effect Analytical* At concentration of 2,000 mg/L had no effect on method using phenolphthalein. *3393*
Serum Decrease Physiological Metabolic acidosis in paraldehyde habitues. *1343*

Bile *Urine Increase Physiological* Possible hepatotoxicity. *2313*

Bilirubin *Serum No Effect Analytical* At concentration of 2,000 mg/L had no effect on Jendrassik and Grof method. *3393*
Serum Increase Physiological Possible hepatotoxicity. *2313*

BSP Retention *Serum Increase Physiological* Possible hepatotoxicity. *2313*

Calcium *Serum No Effect Analytical* At concentration of 2,000 mg/L had no effect on cresolphthalein method. *3393*

Cephalin Flocculation *Serum Increase Physiological* Possible hepatotoxicity. *2313*

Chloride *Serum No Effect Analytical* At concentration of 2,000 mg/L had no effect on mercurimetric method. *3393*

Cholesterol *Serum No Effect Analytical* At concentration of 2,000 mg/L had no effect on Liebermann-Burchard method. *3393*

Corticosteroids *Urine Increase Analytical* Butanol extract/no hydrolysis Reddy/Porter-Silber reaction. Increased absorption at 410 nm in Porter-Silber reaction. *0022*

Creatinine *Serum No Effect Analytical* At concentration of 2,000 mg/L had no effect on AutoAnalyzer Jaffé method. *3393*
Serum Increase Physiological Possible nephrotoxicity. *1443*

Glucose *Serum Increase Physiological* Has caused transient hyperglycemia. *2220*

17-Hydroxycorticosteroids *Urine Increase Analytical* Interferes with Porter-Silber reaction. *3505*

17-Ketogenic Steroids *Urine Increase Analytical* Reported to affect Zimmermann procedure. *0583*

Ketones *Serum Increase Physiological* Transient hyperglycemia and ketosis. *2220*
Urine Increase Analytical Affects alkaline nitroprusside procedure. *0583* False positive when drug combined with ethanol. *1443*

Leukocytes *Blood Increase Physiological* Leukocytosis in severe acute or chronic poisoning. *1343*

Occult Blood *Feces Positive Physiological* Bleeding gastritis in acute or chronic poisoning. *1343*

Paraldehyde *Blood Increase Physiological* Usually fatal at 500 mg/L. *2348*

pH *Blood Decrease Physiological* Acidotic action (decomposes to acetic acid). *2313*

Phosphate *Serum No Effect Analytical* At concentration of 2,000 mg/L had no effect on phosphomolybdate method. *3393*

Protein *Serum No Effect Analytical* At concentration of 2,000 mg/L had no effect on biuret method with blank correction. *3393*
Urine Increase Physiological Nephrotoxic effect (nephrosis with poisoning). *1443*

Prothrombin Time *Plasma Decrease Physiological* May shorten action of anticoagulants. *1679*

Thymol Turbidity *Serum Increase Physiological* Possible hepatotoxicity. *2313*

Triglycerides *Serum No Effect Analytical* At concentration of 2,000 mg/L had no effect on lipase/esterase method. *3393*

Urea Nitrogen *Serum No Effect Analytical* At concentration of 2,000 mg/L had no effect on diacetylmonoxime method. *3393*
Serum Increase Physiological Possible nephrotoxicity in poisoning. *1443*

Uric Acid *Serum No Effect Analytical* At concentration of 2,000 mg/L had no effect on phosphotungstate reduction method. *3393*

Paramethadione

Alanine Aminotransferase *Serum Increase Physiological* Possible hepatotoxicity. *2313*

Alkaline Phosphatase *Serum Increase Physiological* Possible hepatotoxicity. *2313*

Aspartate Aminotransferase
Serum Increase Physiological Possible hepatotoxicity. *2313*

Bile *Urine Increase Physiological* Possible hepatotoxicity. *2313*

Bilirubin *Serum Increase Physiological* Possible hepatotoxicity. *2313*

BSP Retention *Serum Increase Physiological* Possible hepatotoxicity. *2313*

Calcium *Serum Decrease Physiological* Theoretical effect of type of drug. *1452*

Casts *Urine Increase Physiological* May have nephrotoxic effect. *2313*

Cephalin Flocculation *Serum Increase Physiological* Possible hepatotoxicity. *2313*

Cholesterol *Serum Increase Physiological* Possible liver damage. *2313*

Creatinine *Serum Increase Physiological* Possible nephrotoxicity. *2313*

Erythrocytes *Blood Decrease Physiological* Aplastic anemia may occur rarely. *2313*

Glucaric Acid *Urine Increase Physiological* Probable effect as reported to induce hepatic enzymes. *3882*

25-Hydroxy Vitamin D₃ *Serum Decrease Physiological* Theoretical effect of type of drug. *1452*

LE Cells *Blood Positive Physiological* Rare idiosyncratic response. *2042*

Leukocytes *Blood Decrease Physiological* Aplastic anemia/neutropenia. *0071*

Neutrophils *Blood Decrease Physiological* May occur without overall effect on WBC. *1680*

Occult Blood *Feces Positive Physiological* Gastrointestinal bleeding may occur (can affect many organs). *1680*

Platelets *Blood Decrease Physiological* Aplastic anemia may occur rarely. *0071*

Protein *Urine Increase Physiological* May have nephrotoxic effect. *2496*

Thymol Turbidity *Serum Increase Physiological* Possible hepatotoxicity. *2313*

Urea Nitrogen *Serum Increase Physiological* Possible nephrotoxicity. *2313*

Paramethasone

Amylase *Serum Increase Physiological* May cause pancreatitis occasionally. *1680*

Bicarbonate *Serum Increase Physiological* May cause hypokalemic alkalosis. *1680*

Glucose Tolerance *Serum Decrease Physiological* Decreased carbohydrate tolerance. *1680*

Occult Blood *Feces Positive Physiological* May cause peptic ulcer with hemorrhage. *1680*

Potassium *Serum Decrease Physiological* May cause potassium loss. *1680*

Sodium *Urine Increase Physiological* May be excreted although edema may occur. *0071*

Urea Nitrogen *Serum Decrease Physiological* Negative nitrogen balance, protein catabolism. *1680*

Parasympathol

Epinephrine *Test Conditions No Effect Analytical* On fluorescent procedure of Peyrin. *2800*

Norepinephrine *Test Conditions No Effect Analytical* On fluorescent procedure of Peyrin. *2800*

Parathiazine

Leukocytes *Blood Decrease Physiological* May occur with prolonged use. *2313*

Parathion

Leukocytes *Blood Decrease Physiological* Leukopenia (AMA Blood dyscrasias). *2429*

Para-Nitrophenol *Urine Increase Physiological* Roughly increased in proportion to inhibitor of cholinesterase. *3850*

Pseudocholinesterase *Serum Decrease Physiological* Inhibitory action of drug with poisoning. *1302*
Red Blood Cells Decrease Physiological Specific inhibitor of enzyme. *3850*

Parathyroid Extract

Amino Acids *Urine Increase Physiological* Hormonal action. *1343*

Bicarbonate *Urine Increase Physiological* Inhibits tubular exchange of Na for h. *2611*

Calcium *Serum Increase Physiological* Increased calcium mobilization from bone. *0652*
Urine Increase Physiological Due to mobilization from bone. *1841*
Urine Decrease Physiological Increases tubular reabsorption. *0071*

Chloride *Urine Increase Physiological* Hormonal action. *1343*

Citrate *Urine Increase Physiological* Hormonal action. *1343*

Magnesium *Serum Increase Physiological* Due to decreased renal excretion. *1343*
Urine Decrease Physiological Decreased clearance. *1439*

pH *Urine Increase Physiological* Inhibits tubular exchange of Na for h. *2611*

Phosphate *Serum Increase Physiological* Due to increased excretion in urine. *2313*
Urine Increase Physiological Increased clearance. *1439*

Potassium *Urine Increase Physiological* Increased clearance. *1439*

Sodium *Urine Increase Physiological* Increased clearance. *1439*

Sulfate *Urine Increase Physiological* Hormonal action. *1343*

Volume *Urine Increase Physiological* Hormonal action. *1343*

Parathyroid Hormone

Calcitonin *Serum No Effect Physiological* In response to infusion of 150 units in 10 normal subjects. *0467*

Hydroxyproline *Urine Increase Physiological* Due to catabolic action. *3644*

Prolactin *Plasma Increase Physiological* From 5.1 ng/mL to 14.9 ng/mL on average in 10 normal subjects after infusion of 150 units. *0467*

Thyroid Stimulating Hormone (TSH)
Serum No Effect Physiological In response to infusion of 150 units in 10 normal subjects. *0467*

Pargyline

Alanine Aminotransferase *Serum Increase Physiological* Possible hepatotoxicity (reversible). *2313*

Alkaline Phosphatase *Serum Increase Physiological* Possible hepatotoxicity (reversible). *2313*

Amphetamine *Urine No Effect Analytical* At 50 mg/L on fluorescent method of Hayes. *1534*

Aspartate Aminotransferase
Serum Increase Physiological Possible hepatotoxicity (reversible). *2313*

Bile *Urine Increase Physiological* Possible hepatotoxicity (reversible). *2313*

Bilirubin *Serum Increase Physiological* Possible hepatotoxicity (reversible). *2313*

BSP Retention *Serum Increase Physiological* Possible hepatotoxicity (reversible). *2313*

Cephalin Flocculation *Serum Increase Physiological* Possible hepatotoxicity (reversible). *2313*

Glucose *Serum Decrease Physiological* Possible hepatotoxic effect. *2313*

Glucose Tolerance *Serum Increase Physiological* Flat curve may be produced. *2220*

Peripheral Smear *Blood Abnormal Physiological* Poikilocytosis and inisocytosis common. *2427*

Thymol Turbidity *Serum Increase Physiological* Possible hepatotoxicity (reversible). *2313*

Urea Nitrogen *Serum Increase Physiological* Possible decrease in already impaired renal function. *0085*

Parity

Creatine Kinase *Serum Increase Physiological* 5 U/L in parous versus other women. *2744*

Paromomycin

Cholesterol *Serum Decrease Physiological* Reduction up to 18%. ?mechanism. *3129*

Creatinine *Serum Increase Physiological* Frequently observed renal damage. *0131*

Creatinine Clearance *Urine Decrease Physiological* Frequently observed renal damage. *0131*

Protein *Urine Increase Physiological* Potentially nephrotoxic if given parenterally. *0071*

Urea Nitrogen *Serum Increase Physiological* Potentially nephrotoxic if given parenterally. *0071*

Pectin

Cholesterol *Serum Decrease Physiological* 5% decrease observed on adding to diet. *3217*

Pedameth®

Methionine *Urine Increase Analytical* Used for diaper rash in infants. *1749*

Pempidine

Uric Acid *Serum Increase Physiological* ?due to reduced renal blood flow. *0924*

Penethamate

Protein Bound Iodine (PBI)   *Serum   Increase   Analytical*
Contains iodine. *3505*

Penicillamine

Acetylcholine Receptor Antibodies
Serum   Increase   Physiological   Myasthenia gravis may occur as autoimmune syndrome, possibly as result of direct binding of drug to receptor. *1761*

N-Acetylglucosaminidase   *Urine   No Effect   Analytical*   At 20 mmol/L on 2 colorimetric analytical methods. *1354*

α_1-Acid Glycoprotein   *Serum   Decrease   Physiological*   In 21 patients with rheumatoid arthritis treated for 48 weeks given 250 to 750 mg daily. *2345*

Alanine Aminotransferase   *Serum   Increase   Physiological*   Possible hepatotoxicity. *2313*   In 6 of 99 patients treated for rheumatoid arthritis: evidence of toxic liver necrosis observed in two cases. *3887*   Case of acute cholestatic jaundice in patient with systemic lupus erythematosus. *3242*
Serum   Decrease   Physiological   Increase by 45% in 3 patients with Wilson's disease although effect disappeared after 12 mo. *1417*

Albumin   *Serum   No Effect   Physiological*   In 21 patients with rheumatoid arthritis treated for 48 weeks given 250 to 750 mg daily. *2345*

Alkaline Phosphatase   *Serum   No Effect   Physiological*   In rheumatoid arthritis patients when treated for 6 mo. Changes not related to activity of disease. *1794*   But decreased in granulocytes related to transient changes in zinc metabolism. *1794*
Serum   Increase   Physiological   Possible hepatotoxicity. *1022*   In 6 of 99 patients treated for rheumatoid arthritis: evidence of toxic liver necrosis observed in two cases. *3887*   Case of acute cholestatic jaundice in patient with systemic lupus erythematosus. *3242*
White Blood Cells   Decrease   Physiological   In rheumatoid arthritis patients when treated for 6 mo. Changes not related to activity of disease. *1794*

Amino Acids   *Urine   Increase   Analytical*   Reacts with ninhydrin; extra spot TLC, high voltage electrophoresis. *2855*
Urine   Positive   Analytical   2 unusual purple spots with ninhydrin on thin-layer chromatography of urine from patients with Wilson's disease. Spots stain gray with isatin. *1596*

Antibodies to DNA   *Serum   Positive   Physiological*   In 6 patients with rheumatoid arthritis developed systemic lupus erythematosus-like syndrome. *0623*

Antinuclear Antibodies   *Serum   Positive   Physiological*   In 6 patients with rheumatoid arthritis developed systemic lupus erythematosus-like syndrome. *0623*

Aspartate Aminotransferase
Serum   Increase   Physiological   Possible hepatotoxicity. *2313*   In 6 of 99 patients treated for rheumatoid arthritis: evidence of toxic liver necrosis observed in two cases. *3887*   Case of acute cholestatic jaundice in patient with systemic lupus erythematosus. *3242*
Serum   Decrease   Physiological   Increase by 47% in 3 patients with Wilson's disease although effect disappeared after 12 mo. *1417*

Bilirubin   *Serum   Increase   Physiological*   Case of acute cholestatic jaundice in patient with systemic lupus erythematosus. *3242*

Carbonic Anhydrase
Red Blood Cells   Decrease   Physiological   In rheumatoid arthritis patients when treated for 6 mo. Changes not related to activity of disease. *1794*   Related to transient changes in zinc metabolism. *1794*

Cephalin Flocculation   *Serum   Increase   Physiological*   Possible hepatotoxicity. *2313*

Cholesterol   *Serum   No Effect   Analytical*   At concentration of 960 mg/L had no effect on CHOD-PAP method. *3393*   At concentration of 960 mg/L had no effect on catalase-AlDH method. *3393*   At concentration of 960 mg/L had no effect on method using catalase-Hantzsch reaction. *3393*   At concentration of 960 mg/L had no effect on Liebermann-Burchard method. *3393*
Serum   Increase   Physiological   Single case reported. *1488*

Serum   Decrease   Analytical   At concentrations above 10 mg/L (normal therapeutic concentration 11 mg/L) lowered concentration as measured by CHOD-Iodide method. *3393*

Complement C4   *Serum   Decrease   Physiological*   In 6 patients with rheumatoid arthritis developed systemic lupus erythematosus-like syndrome. *0623*

Coombs' Test   *Serum   Positive   Physiological*   In 6 patients with rheumatoid arthritis developed systemic lupus erythematosus-like syndrome. *0623*

Copper   *Urine   Increase   Physiological*   If poisoning due to copper (also in normals). *0071*

Creatine Kinase   *Serum   Increase   Physiological*   Marked increase observed in one patient with rheumatoid arthritis after 1 mo treatment. *0947*

Creatinine   *Serum   No Effect   Analytical*   At concentration of 480 mg/L had no effect on Jaffé-Fading-Fraction method. *3393*   At concentration of 480 mg/L had no effect on Jaffé-Fuller's earth method. *3393*   At concentration of 55 mg/L had no effect on kinetic Jaffé method on BKA-2. *3393*
Serum   Increase   Physiological   Possible nephrotoxicity. *1237*

Cystine   *Urine   Increase   Physiological*   If cystinuria. *0071*

Eosinophils   *Blood   Increase   Physiological*   May occur with rash. *0071*   In 14% patients treated with up to 750 mg daily (63 patients studied who had taken penicillamine). *3360*

Erythrocytes   *Blood   No Effect   Physiological*   In 21 patients with rheumatoid arthritis treated for 48 weeks given 250 to 750 mg daily. *2345*
Blood   Decrease   Physiological   Hypochromic anemia in child, menstruating woman. *0074*   Rare blood dyscrasias as reported to UK committee on safety of medicines. *0782*
Urine   Increase   Physiological   In 1 of 21 patients with rheumatoid arthritis treated for 48 weeks given 250 to 750 mg daily. *2345*

Erythrocyte Sedimentation Rate
Blood   Decrease   Physiological   In 21 patients with rheumatoid arthritis treated for 48 weeks given 250 to 750 mg daily. *2345*

Glucose   *Serum   No Effect   Analytical*   At concentration of 36 mg/L had no effect on Ektachem® method. *3393*
Serum   No Effect   Physiological   In 21 patients with rheumatoid arthritis treated for 48 weeks given 250 to 750 mg daily. *2345*
Urine   No Effect   Physiological   In 21 patients with rheumatoid arthritis treated for 48 weeks given 250 to 750 mg daily. *2345*

Gold   *Blood   No Effect   Physiological*   No effect in patients with rheumatoid arthritis previously treated with gold. *1324*
Urine   Increase   Physiological   Slight effect when given to patients with rheumatoid arthritis previously treated with gold. *1324*

Hematocrit   *Blood   Decrease   Physiological*   Hypochromic anemia in child, menstruating woman. *0074*

Hemoglobin   *Blood   No Effect   Physiological*   In 21 patients with rheumatoid arthritis treated for 48 weeks given 250 to 750 mg daily. *2345*
Blood   Decrease   Physiological   Hypochromic anemia in child, menstruating woman. *0074*

Immunoglobulin IgA   *Serum   Decrease   Physiological*   But not significantly in 21 patients with rheumatoid arthritis treated for 48 weeks given 250 to 750 mg daily. *2345*

Immunoglobulin IgG   *Serum   Decrease   Physiological*   But not significantly in 21 patients with rheumatoid arthritis treated for 48 weeks given 250 to 750 mg daily. *2345*

Immunoglobulin IgM   *Serum   Decrease   Physiological*   But not significantly in 21 patients with rheumatoid arthritis treated for 48 weeks given 250 to 750 mg daily. *2345*

Iron   *Urine   Increase   Physiological*   If poisoning due to iron. *0071*

Kynurenine   *Urine   Increase   Physiological*   Induces pyridoxine antagonism. *1343*

Lactate Dehydrogenase   *Serum   Increase   Physiological*   In 6 of 99 patients treated for rheumatoid arthritis: evidence of toxic liver necrosis observed in two cases. *3887*

Lead   *Urine   Increase   Physiological*   If poisoning due to lead. *0071*

LE Cells *Blood Positive Physiological* Observed effect in some cases. *0933* In 6 patients with rheumatoid arthritis developed systemic lupus erythematosus-like syndrome. *0623*

Leukocytes *Blood No Effect Physiological* In 21 patients with rheumatoid arthritis treated for 48 weeks given 250 to 750 mg daily. *2345*
Blood Decrease Physiological Leukopenia may occur with rash. *0074* Most serious and potentially life-threatening effect of drug. Either idiosyncratic response seen in first year of treatment. Independent of dosage or more commonly dose related with gradual onset. *1761* Rare blood dyscrasias as reported to UK committee on safety of medicines. *0782* In 3% of 90 patients with rheumatoid arthritis previously treated with gold. *1461* Toxicity observed in some patients with rheumatoid disease. *1891*

MCHC *Blood Decrease Physiological* Hypochromic anemia in child, menstruating woman. *0074*

Mercury *Urine Increase Physiological* Related to drug dosage in toxicity cases. *3515*

Neutrophils *Blood Decrease Physiological* Occasional case of drug-induced neutropenia. *0155* In 14 of 84 patients occurring typically in first 6 mo of treatment. *1892*

Occult Blood *Feces Positive Physiological* Occasional side effect as reported to UK committee on safety of medicines. *0782*

Platelets *Blood Decrease Physiological* Thrombocytopenia may occur with rash. *0074* Most serious and potentially life-threatening effect of drug. Either idiosyncratic response seen in first year of treatment. Independent of dosage or more commonly dose related with gradual onset. *1761* Rare blood dyscrasias as reported to UK committee on safety of medicines. *0782* Significant effect from 36th week in 21 patients with rheumatoid arthritis treated for 48 weeks given 250 to 750 mg daily. *2345* In 3% of 90 patients with rheumatoid arthritis previously treated with gold. *1461* In 14 of 84 patients occurring typically in first 6 mo of treatment. *1892* Apparently some bone marrow depression in some patients but normal lifespan of cells. *3574* Toxicity observed in some patients with rheumatoid disease. *1891*

Potassium *Serum No Effect Analytical* At concentration of 360 mg/L had no effect on measurement by ISE without predilution. *3393*

Protein *Urine Increase Physiological* Nephrotoxic effect. *1237* May be minimal and asymptomatic or lead to nephrotic syndrome due to immune complex glomerulitis. *1761* Observed in 16% of 90 patients with rheumatoid arthritis in one study. *1461* In 3 of 21 patients with rheumatoid arthritis treated for 48 weeks given 250 to 750 mg daily. *2345* In 16% of 90 patients with rheumatoid arthritis previously treated with gold. *1461* In 15 of 84 patients occurring most often in 6 to 12 mo after treatment started. *1892* Toxicity observed in some patients with rheumatoid disease. *1891* In 30% patients with rheumatoid arthritis or cystinuria and 4% in patients with Wilson's disease. *0113* In 10 to 20% of patients with rheumatoid arthritis in first year, of which one third may proceed to nephrotic syndrome. *0753*

Protein Electrophoresis *Serum No Effect Analytical* At concentration of 360 mg/L had no effect on automated Olympus-Hite method. *3393*

Rheumatoid Factor *Serum No Effect Physiological* In 21 patients with rheumatoid arthritis treated for 48 weeks given 250 to 750 mg daily. *2345*

Sodium *Serum No Effect Analytical* At concentration of 360 mg/L had no effect on measurement by ISE without predilution. *3393*

Thymol Turbidity *Serum Increase Physiological* Possible hepatotoxicity. *2313*

Thyroxine (T₄) *Serum No Effect Physiological* In 21 patients with rheumatoid arthritis treated for 48 weeks given 250 to 750 mg daily. *2345*

Urea Nitrogen *Serum No Effect Analytical* At concentration of 36 mg/L had no effect on Ektachem® method. *3393*
Serum Increase Physiological Possible nephrotoxicity. *1237*

Uric Acid *Serum No Effect Analytical* At concentration of 480 mg/L had no effect on catalase-AIDH method. *3393*

Xanthurenic Acid *Urine Increase Physiological* Induces pyridoxine antagonism. *1343*

Zinc *Serum Increase Physiological* In rheumatoid arthritis patients when treated for 6 mo. Changes not related to activity of disease. *1794* Related to transient changes in zinc metabolism. *1794*
Urine Increase Physiological If poisoning due to zinc. *1343* In rheumatoid arthritis patients when treated for 6 mo. Changes not related to activity of disease. *1794* Related to transient changes in zinc metabolism. *1794*
Red Blood Cells Increase Physiological In rheumatoid arthritis patients when treated for 6 mo. Changes not related to activity of disease. *1794* But granulocyte concentration decreased related to transient changes in zinc metabolism. *1794*
White Blood Cells Decrease Physiological In rheumatoid arthritis patients when treated for 6 mo. Changes not related to activity of disease. *1794*

Penicillin

Albumin *Serum Decrease Analytical* Competes with HABA for binding (slight effect). *0583* At concentrations above 18,000 mg/L (normal therapeutic concentration about 12 mg/L) lowered concentration as measured by BCG method. *3393*

Aminolevulinic Acid *Urine Increase Analytical* Derivative reacts with Ehrlich's reagent. *1997*

Antithrombin III *Plasma No Effect Analytical* At concentration of 100 mg/L had no effect on aca method. *3393*

Aspartate Aminotransferase
Serum Increase Physiological Nonspecific hepatitis without cholestasis. *3879*

Barbiturate *Serum No Effect Analytical* No effect on UV absorption methods. *2981*

Bilirubin *Serum Increase Physiological* Nonspecific hepatitis without cholestasis. *3879*
Serum Decrease Physiological In newborn combined with sulfisoxazole has effect. *3505*

Casts *Urine Increase Physiological* Renal cell and other types. *1343*

Chromosomes *Test Conditions No Effect Physiological* Not clastogenic in human cells. *3282*

Coombs' Test, Direct *Serum Positive Physiological* Combines to RBC, immunoglobulins develop to drug. *1486*

Coombs' Test, Indirect *Serum Positive Physiological* Combines to RBC, immunoglobulins develop to drug. *1486*

Creatine Kinase *Serum Increase Physiological* Frequent injections may cause increase up to 5 times. *1580*

Creatinine *Serum Increase Physiological* Hypersensitivity reaction or nephropathy. *1343*

Eosinophils *Blood Increase Physiological* Allergic reaction (may be up to 20% of all WBC). *0996*

Erythrocytes *Blood Decrease Physiological* Hemolytic anemia due to binding to erythrocytes. *0996*
Urine Increase Physiological Hypersensitivity reaction, nephrotoxicity. *0212*

Estriol *Plasma Decrease Physiological* In pregnant women due to alteration of gut flora. *2885*

Estrogens *Urine Increase Analytical* Activates enzyme used for hydrolysis. *0022*
Urine Decrease Physiological In pregnant women due to alteration of gut flora. *2885*

Fibrinogen *Plasma No Effect Analytical* At concentration of 100 mg/L had no effect on aca method. *3393*

Folate *Serum No Effect Analytical* If chromatographic procedure of Landon used. *2068* Allegedly no effect on autoclave method. *1815*
Serum Decrease Analytical Inhibits growth of L. Casei. *3505*

Gentamicin *Serum No Effect Analytical* If enzymatic procedure of Daigneault used. *0791*

Glucose *Urine No Effect Analytical* No effect observed with TesTape® No effect on Clinistix® or TesTape®. *2525*

Haptoglobin *Serum Decrease Physiological* May cause hemolytic anemia. *2313*

Hematocrit *Blood Decrease Physiological* Hemolytic anemia due to binding to erythrocytes. *0996*

Hemoglobin *Blood Decrease Physiological* Hemolytic anemia due to binding to erythrocytes. *0996*

Penicillin *(continued)*

Hemoglobin *(continued)*

Urine Increase Physiological Hypersensitivity reaction, nephrotoxicity. *0212*

17-Hydroxycorticosteroids *Urine No Effect Analytical* No effect modified Glenn-Nelson. *1487*
Urine Increase Analytical ?interferes with Porter-Silber reaction. *2220*

^{131}I Uptake *Serum Decrease Physiological* If hydriodide salt given decreases further uptake. *2220* V-cillin contains tetraiodofluorescein. *2652*

17-Ketogenic Steroids *Urine Increase Analytical* Interferes with Zimmermann (Norymberski) reaction. *3505*

17-Ketosteroids *Urine Increase Analytical* Interferes with Zimmermann reaction. *3505*

LE Cells *Blood Positive Physiological* Allergic response with urticaria in one case. *1343*

Leukocytes *Blood Decrease Physiological* Agranulocytosis/leukopenia. *2695*

Mercury *Urine Increase Physiological* Acts as chelating agent in acrodynia at least. *2090*

Neutrophils *Blood Decrease Physiological* Agranulocytosis/leukopenia. *2695*

Nitrofurantoin *Urine No Effect Analytical* No effect on method of Conklin and Hollifield. *0704*

Plasminogen *Plasma No Effect Analytical* At concentration of 100 mg/L had no effect on aca method. *3393*

Platelets *Blood Decrease Physiological* Agranulocytosis/thrombocytopenia. *0996*

Potassium *Serum No Effect Analytical* At concentration of 15,000 mg/L had no effect on measurement by ISE with predilution. *3393*
Serum Increase Physiological May occur if K salt given i.v. also alkalosis. *3505*
Serum Decrease Physiological If i.v. sodium penicillin infused. *3505*
Urine Increase Physiological If i.v. sodium penicillin infused. *3505*

Protein *Urine No Effect Analytical* No effect on Albustix even with massive doses. *0459*
Urine Increase Analytical Causes turbidity if sulfosalicylic acid used. *3505* Massive doses may produce turbidity with acid. *0459* With SSA and acetic acid tests. *0443*
Urine Increase Physiological Nephrotoxicity may occur with large doses. *1343*
CSF Increase Analytical Reacts as if phenol with Folin-Ciocalteu procedure. *3958* Causes turbidity if sulfosalicylic acid used. *3505*
Test Conditions Increase Analytical Massive doses may cause turbidity with acid tests. *2313*

Protein Bound Iodine (PBI) *Serum Increase Analytical* If given as hydriodide salt. *2220*
Serum Decrease Physiological Competes for thyroxine binding prealbumin binding sites. *0583*

Protein Electrophoresis *Serum Positive Analytical* Causes bisalbuminemia. *3505*

Prothrombin Time *Plasma Increase Physiological* Occasional effect. *0019*

PSP Excretion *Urine Decrease Physiological* Interference with PSP excretion. *3879*

Reticulocytes *Blood Increase Physiological* Consequence of hemolytic anemia. *0996*

Sodium *Serum No Effect Analytical* At concentration of 15,000 mg/L had no effect on measurement by ISE with predilution. *3393*

Sugar *Urine Increase Analytical* False positive with copper reduction procedures. *3505* Drug and metabolites act as red substances in high concentrations. *0583*

T$_3$ Uptake *Serum Increase Physiological* Competes for thyroxine binding prealbumin sites. *0583*

Thyroxine (T$_4$) *Serum Decrease Physiological* Competes for thyroxine binding prealbumin binding sites. *0583*

Thyroxine (T$_4$) (Murphy-Pattee)
Serum Decrease Physiological Binds with thyroxine binding prealbumin but not thyroxine binding globulin. *0583*

Urea Nitrogen *Serum Increase Physiological* Rare reaction to large doses given parenterally. *1680*

Uric Acid *Serum No Effect Analytical* At 1,000 units/mL on Nishi phosphotungstate procedure. *2230*
Serum Increase Analytical 1,000 units/mL equivalent to 2.0 mg/dL copper chelate procedure. *2230*

Penicillin G

Alanine Aminotransferase *Serum No Effect Analytical* At 5 times upper limit of therapeutic range on methods on SMAC®, Abbott-VP, Cobas-Bio, aca, and KDA. *2138*

Albumin *Serum No Effect Analytical* At 5 times upper limit of therapeutic range on methods on SMAC®, Ektachem®, Hitachi® 705 and KDA. *2138* At concentration of 2,000 mg/L had no effect as measured by BCG method. *3393*

Alkaline Phosphatase *Serum No Effect Analytical* At 5 times upper limit of therapeutic range on methods on SMAC®, Abbott-VP, Cobas-Bio, aca, Hitachi® 705 and KDA. *2138*

Aminolevulinic Acid *Urine Increase Analytical* In method of Mauzerall and Granick interacts with acetylacetone which reacts with p-dimethylmenine benzaldehyde. *3640*

Amylase *Serum No Effect Analytical* At 5 times upper limit of therapeutic range on methods on aca, Cobas-Bio and Ektachem®. *2138*

Aspartate Aminotransferase *Serum No Effect Analytical* At 5 times upper limit of therapeutic range on methods on SMAC®, Abbott-VP, Cobas-Bio, aca, Hitachi® 705 and KDA. *2138*

Bicarbonate *Serum No Effect Analytical* At concentration of 2,000 mg/L had no effect as measured by method using phenolphthalein. *3393*

Bilirubin *Serum No Effect Analytical* At 5 times upper limit of therapeutic range on methods on SMAC®, aca, Abbott-VP, Ektachem®, Hitachi® 705 and KDA. *2138* At concentration of 2,000 mg/L had no effect as measured by Jendrassik and Grof method. *3393*
Serum No Effect Physiological Insignificant displacement from protein in neonates. *3748*

Bleeding Time *Blood Increase Physiological* Progressive lengthening with dose of 24 million U/d from 3.2 to 6.1 minutes on average, longer with higher dose. *0492*

Calcium *Serum No Effect Analytical* At 5 times upper limit of therapeutic range on methods on SMAC®, Abbott-VP, aca, Ektachem®, Hitachi® 705 and KDA. *2138* At concentration of 2,000 mg/L had no effect as measured by cresolphthalein method. *3393*

Chloride *Serum No Effect Analytical* At concentration of 2,000 mg/L had no effect as measured by mercurimetric method. *3393*

Cholesterol *Serum No Effect Analytical* At 5 times upper limit of therapeutic range on methods on SMAC®, Abbott-VP, Cobas-Bio, Ektachem®, Hitachi® 705 and KDA. *2138* At concentration of 2,000 mg/L had no effect as measured by Liebermann-Burchard method. *3393*

Clot Retraction *Blood No Effect Physiological* No effect with doses as high as 48 million U/d. *0492*

Coombs' Test, Direct *Serum Positive Physiological* 44% of 39 patients demonstrated hypersensitivity reaction. *1963*

Creatine Kinase *Serum No Effect Analytical* At 5 times upper limit of therapeutic range on methods on SMAC®, Abbott-VP, aca, Cobas-Bio and Hitachi® 705. *2138*

Creatinine *Serum No Effect Analytical* At 5 times upper limit of therapeutic range on methods on SMAC®, Abbott-VP, Cobas-Bio, aca, Ektachem®, Hitachi® 705 and KDA. *2138* At concentration of 2,000 mg/L had no effect as measured by AutoAnalyzer Jaffé method. *3393*

Eosinophils *Blood Increase Physiological* 44% of 39 patients demonstrated hypersensitivity reaction. *1963*

Fibrinogen *Plasma No Effect Physiological* No effect with doses as high as 48 million U/d. *0492*

Glucose *Serum No Effect Analytical* At 5 times upper limit of therapeutic range on methods on SMAC®, Abbott-VP, Cobas-Bio, aca, Ektachem®, Hitachi® 705 and KDA. *2138* At concentration of 90 mg/L had no effect as measured by GOD/POD-PAP method. *3393*

Urine   No Effect   Analytical   No effect at up to 16.2 mg/mL on any glucose concentration as measured by Diastix® or Tes-Tape®. *2250*   At concentration of 3.6 mg/L had no effect as measured by Diabur-test. *3393*

Urine   Positive   Analytical   At 16.2 mg/mL gave false positive with negative urine but at higher concentrations gave occasional falsely low value. *2250*

γ-Glutamyltransferase (GGT)   *Serum   No Effect   Analytical*   At 5 times upper limit of therapeutic range on methods on SMAC®, Abbott-VP and Hitachi® 705. *2138*

Immunoglobulin IgE   *Serum   Increase   Physiological*   Present in some patients with drug-induced acute interstitial nephritis. *0145*

Iron   *Serum   No Effect   Analytical*   At 5 times upper limit of therapeutic range on methods on SMAC®. *2138*

Lactate Dehydrogenase   *Serum   No Effect   Analytical*   At 5 times upper limit of therapeutic range on methods on SMAC®, Abbott-VP, Cobas-Bio, Hitachi® 705 and KDA. *2138*

Leukocytes   *Blood   Decrease   Physiological*   Occasional case of drug associated leukopenia. *1839*   In 2 patients receiving 140-150 mg/kg/d intravenously. *1646*

Lymphocytes   *Blood   No Effect   Physiological*   In 2 patients receiving 140-150 mg/kg/d intravenously. *1646*

Neutrophils   *Blood   Decrease   Physiological*   Occasional case of drug associated leukopenia. *1839*

Partial Thromboplastin Time
Plasma   No Effect   Physiological   No effect with doses as high as 48 million U/d. *0492*

Phosphate   *Serum   No Effect   Analytical*   At 5 times upper limit of therapeutic range on methods on SMAC®, aca, Hitachi® 705, and KDA. *2138*   At concentration of 2,000 mg/L had no effect as measured by phosphomolybdate method. *3393*

Platelet Aggregation   *Blood   Decrease   Physiological*   Defective aggregation in response to ADP at 24 million u/d. *0492*

Platelets   *Blood   No Effect   Physiological*   No effect with doses as high as 48 million U/d. *0492*
Blood   Decrease   Physiological   Isolated case of platelet-associated IgG and thrombocytopenia. *1910*

Potassium   *Serum   No Effect   Analytical*   At concentration of 18 mg/L had no effect as measured by flame-photometric method. *3393*
Serum   Decrease   Physiological   Nonreabsorbable anion in distal nephron promoting potassium loss especially with high doses. Effect quite marked in leukemia. *0645*

Protein   *Serum   No Effect   Analytical*   At 5 times upper limit of therapeutic range on methods on SMAC®, Abbott-VP, Ektachem®, Hitachi® 705 and KDA. *2138*
Serum   Increase   Analytical   At concentrations above 500 mg/L (normal therapeutic concentration 12 mg/L) raised concentration as measured by biuret method with blank correction. *3393*
Urine   Increase   Analytical   When measured by Ponceau S dye method in comparison with sulfosalicylic acid or trichloracetic acid methods in 5 patients receiving therapeutic doses. *3919*

Prothrombin Time   *Plasma   No Effect   Physiological*   No effect with doses as high as 48 million U/d. *0492*

Sodium   *Serum   No Effect   Analytical*   At concentration of 18 mg/L had no effect as measured by flame-photometric method. *3393*

Thrombin Time   *Plasma   No Effect   Physiological*   No effect with doses as high as 48 million U/d. *0492*

Triglycerides   *Serum   No Effect   Analytical*   At 5 times upper limit of therapeutic range on methods on SMAC®, Abbott-VP, Ektachem®, Hitachi® 705 and KDA. *2138*   At concentration of 90 mg/L had no effect as measured by GPO-PAP method. *3393*   At concentration of 2,000 mg/L had no effect as measured by lipase/esterase method. *3393*

Urea Nitrogen   *Serum   No Effect   Analytical*   At 5 times upper limit of therapeutic range on methods on SMAC®, Abbott-VP, Cobas-Bio, Ektachem®, Hitachi® 705 and KDA. *2138*   At concentration of 2,000 mg/L had no effect as measured by diacetylmonoxime method. *3393*

Uric Acid   *Serum   No Effect   Analytical*   At 5 times upper limit of therapeutic range on methods on SMAC®, Abbott-VP, Cobas-Bio, aca, Ektachem®, Hitachi® 705 and KDA. *2138*   At concentration of 2,000 mg/L had no effect as measured by phosphotungstate reduction method. *3393*   At concentration of 90 mg/L had no effect as measured by uricase-PAP method. *3393*

Penicillin V

Bilirubin   *Serum   Increase   Physiological*   Possible clinically significant displacement from protein especially in critically ill neonates. *3748*   Isolated case of hemolytic anemia with IgM antibody. *0360*

Chloramphenicol   *Serum   No Effect   Analytical*   No effect at 100 mg/L on coupled enzymatic method. *2490*

Coombs' Test, Direct   *Serum   Positive   Physiological*   Isolated case of hemolytic anemia with IgM antibody. *0360*

Hemoglobin   *Blood   Decrease   Physiological*   Isolated case of hemolytic anemia with IgM antibody. *0360*

Methemalbumin   *Plasma   Positive   Physiological*   Isolated case of hemolytic anemia with IgM antibody. *0360*

Pentaerythritol

Estriol   *Urine   No Effect   Analytical*   No effect on GLC method. *1163*

Pentagastrin

Calcitonin   *Plasma   No Effect   Physiological*   Intravenous infusion had no effect in normals. *0856*
Plasma   Increase   Physiological   Slight effect seen in hypocalcemics if given i.v. *0856*

Calcium   *Serum   No Effect   Physiological*   Intravenous infusion had no effect in normals. *0856*

Growth Hormone   *Plasma   Increase   Physiological*   In 3 of 9 women to 4.87 ng/mL from 1.96 ng/mL, but increase not significant. *3648*

Histamine   *Urine   Increase   Physiological*   All gastric stimulants produce effect. *0814*

Hydrochloric Acid   *Gastric Material   Increase   Physiological*   If infused i.v. *0689*

Pepsin   *Gastric Material   Increase   Physiological*   If infused i.v. *0689*

Prolactin   *Plasma   Increase   Physiological*   Increased in 63% of 9 women: mean concentration increased from 9.67 ng/mL to 16.08 ng/mL in 5 to 10 minutes. *3648*

Pentamidine

Amino-4-Imidazole-5-Carboxamide Ribotide (AICAR)
Urine   Increase   Physiological   If megaloblastic anemia develops. *3773*

Creatinine   *Serum   Increase   Physiological*   Renal toxicity observed in about 25% patients. *3317*   Renal toxicity possibly due to formation of insoluble precipitates of pentamidine with nucleic acids, also associated with hypovolemia. *3445*

Erythrocytes   *Blood   Decrease   Physiological*   Megaloblastic anemia — inhibits dihydrofolate reductase. *3773*
Urine   Increase   Physiological   Gross hematuria observed in a single patient. *3317*

FIGLU (N-Formiminoglutamic Acid)
Urine   Increase   Physiological   If megaloblastic anemia develops. *3773*

Folate   *Serum   Decrease   Physiological*   Inhibits dihydrofolate reductase. *3773*

Glucose   *Serum   Increase   Physiological*   Paradoxical effect observed. *1343*
Serum   Decrease   Physiological   Hypoglycemia — possible effect. *1343*   Hypoglycemia occurred in 10 to 30% patients with pneumocystis carinii pneumonia, usually 5 to 13 d after treatment, but fatal case described after 2 weeks. *3157*   Four patients with pneumocystis carinii developed severe fasting hypoglycemia. *0431*

Pentamidine (continued)

Hematocrit *Blood Decrease Physiological* Megaloblastic anemia — inhibits dihydrofolate reductase. *3773*

Hemoglobin *Blood Decrease Physiological* Megaloblastic anemia — inhibits dihydrofolate reductase. *3773*
Urine Increase Physiological Gross hematuria observed in a single patient. *3317*

MCV *Blood Increase Physiological* Megaloblastic anemia. *3773*

Urea Nitrogen *Serum Increase Physiological* Reversible renal dysfunction reported. *1343* Renal toxicity possibly due to formation of insoluble precipitates of pentamidine with nucleic acids, also associated with hypovolemia. *3445* Renal toxicity observed in about 25% patients. *3317*

Pentaquine

Bilirubin *Serum Increase Physiological* May cause hemolysis with G-6-PD deficiency. *0248*

Erythrocytes *Blood Decrease Physiological* May cause hemolysis with G-6-PD deficiency. *0248*

Haptoglobin *Serum Decrease Physiological* May cause hemolysis. *0248*

Heinz Body Formation *Blood Positive Physiological* May cause hemolysis with G-6-PD deficiency. *0333*

Hematocrit *Blood Decrease Physiological* May cause hemolysis with G-6-PD deficiency. *0248*

Hemoglobin *Blood Decrease Physiological* May cause hemolysis with G-6-PD deficiency. *0248*

Methemoglobin *Blood Increase Physiological* May cause hemolysis in G-6-PD deficiency. *3581*

Pentazocine

Alkaloids *Urine No Effect Analytical* With ninhydrin on Frings TLC procedure at 10 mg/dL. *1204*
Urine Increase Analytical Red/purple with iodoplatinate on Frings TLC procedure. *1204*

Aminolevulinic Acid *Urine Increase Physiological* May precipitate acute porphyria attack. *1322*

Amphetamine *Urine Increase Physiological* Interferes with colorimetric methyl orange method. *1207*

Amylase *Serum Increase Physiological* Causes spasm of sphincter of Oddi. *1606*

Barbiturate *Urine No Effect Analytical* With diphenylcarbazone on Frings TLC procedure at 10 mg/dL. With mercurous NO_3 on Frings TLC procedure at 10 mg/dL. With mercuric SO_4 on Frings TLC procedure at 10 mg/dL. *1204*
Urine Increase Analytical Pale-purple spot with vanillin in Frings TLC procedure. *1204*

Bromide *Urine No Effect Analytical* No effect on method of Frings. *1204*

Coproporphyrin *Urine Increase Physiological* May precipitate acute porphyria attack. *1322*
Feces Increase Physiological May precipitate acute porphyria attack. *1322*

Dapsone *Serum No Effect Analytical* At 10 mg/dL on colorimetric procedure of Higgins. *1590*

Effective Renal Plasma Flow
Patient Decrease Physiological Observed in normal individuals. *1343*

Epinephrine *Plasma Increase Physiological* Significant effect after 0.6 mg/kg i.v. in 20 minutes. *2289*

Estriol *Urine No Effect Analytical* No effect with 12 mg/L on GLC method. *1163*

Ethchlorvynol *Serum No Effect Analytical* At 150 mg/L on GLC procedure of Evenson. *1058*
Urine No Effect Analytical No effect on method of Frings and Cohen at 10 mg/dL. *1205*

FPN Test *Urine No Effect Analytical* No effect at 10 mg/dL on method of Frings. *1204*

11-Hydroxycorticosteroids *Urine Decrease Physiological* ?due to depression of adrenocortical secretion. *0766*

17-Hydroxycorticosteroids *Urine Decrease Physiological* ?due to depression of adrenocortical secretion. *0766*

Leukocytes *Blood Decrease Physiological* Effect observed in 3 patients although taking other drugs, but effect recurred with reinstitution of treatment. *1655*

Lipase *Serum Increase Physiological* Causes spasm of sphincter of Oddi. *2425*

Methaqualone *Serum Decrease Analytical* Insignificant at therapeutic concentration procedure of Evenson. *1057*

Neutrophils *Blood Decrease Physiological* Occasional case of agranulocytosis reported. *3717*

Norepinephrine *Plasma Increase Physiological* After up to 0.6 mg/kg i.v. almost 2-fold increase in 10-20 minutes. *2289*

pCO₂ *Blood Increase Physiological* High doses produce marked respiratory depression. *1343*

Pentazocine *Blood Increase Physiological* Concentration of 150 µg/L achieved. *2348*

pH *Blood Decrease Physiological* Slight fall (average 0.05) 15 minutes after i.v. *2446*

Porphobilinogen *Urine Increase Physiological* May precipitate acute porphyria attack. *1322*

Porphyrins *Urine Increase Physiological* May precipitate attack of acute porphyria. *1016*

Protoporphyrin *Feces Increase Physiological* May precipitate acute porphyria attack. *1322*

Salicylate *Urine No Effect Analytical* At 10 mg/dL no effect on Trinder method. *1204*

Pentobarbital

Alanine Aminotransferase *Serum No Effect Analytical* At acute overdose concentration (20 mg/dL) on Technicon® SMAC® method. *3719*

Albumin *Serum No Effect Analytical* At acute overdose concentration (20 mg/dL) on Technicon® SMAC® method. *3719* At concentration of 340 mg/L had no effect on BCG method. *3393*

Alkaline Phosphatase *Serum No Effect Analytical* At acute overdose concentration (20 mg/dL) on Technicon® SMAC® method. *3719*

Alkaloids *Urine No Effect Analytical* With ninhydrin on Frings TLC procedure at 10 mg/dL. With iodoplatinate of Frings TLC procedure at 10 mg/dL. *1204*

Aspartate Aminotransferase *Serum No Effect Analytical* At acute overdose concentration (20 mg/dL) on Technicon® SMAC® method. *3719*

Barbiturate *Urine No Effect Analytical* With vanillin on Frings procedure at 10 mg/dL. *1204*
Urine Increase Analytical White spot with mercuric SO_4 on Frings TLC procedure. Purple with diphenylcarbazone on Frings TLC procedure. With mercurous NO_3 reagent on Frings TLC procedure. *1204*

Bicarbonate *Serum No Effect Analytical* At concentration of 340 mg/L had no effect on method using phenolphthalein method. *3393*

Bilirubin *Serum No Effect Analytical* At acute overdose concentration (20 mg/dL) on Technicon® SMAC® method. *3719* At concentration of 340 mg/L had no effect on Jendrassik and Grof method. *3393*

Bromide *Urine No Effect Analytical* No effect on method of Frings. *1204*

Calcium *Serum No Effect Analytical* At acute overdose concentration (20 mg/dL) on Technicon® SMAC® method. *3719* At concentration of 340 mg/L had no effect on cresolphthalein method. *3393*

Chloride *Serum No Effect Analytical* At concentration of 340 mg/L had no effect on mercurimetric method. *3393*

Cholesterol *Serum No Effect Analytical* At acute overdose concentration (20 mg/dL) on Technicon® SMAC® method. *3719* At concentration of 340 mg/L had no effect on Liebermann-Burchard method. *3393*

Creatine Kinase *Serum No Effect Analytical* At acute overdose concentration (20 mg/dL) on Technicon® SMAC® method. *3719*

Creatinine *Serum No Effect Analytical* At acute overdose concentration (20 mg/dL) on Technicon® SMAC® method. *3719* At concentration of 340 mg/L had no effect on AutoAnalyzer method. *3393*

Ethchlorvynol *Serum No Effect Analytical* At 600 mg/L on GLC procedure of Evenson. *1058*

Folate *Serum No Effect Analytical* If chromatographic procedure of Landon used. *2068*

FPN Test *Urine No Effect Analytical* No effect at 10 mg/dL on method of Frings. *1204*

Glucose *Serum No Effect Analytical* At acute overdose concentration (20 mg/dL) on Technicon® SMAC® method. *3719*

Iron *Serum No Effect Analytical* At acute overdose concentration (20 mg/dL) on Technicon® SMAC® method. *3719* At concentration of 340 mg/L had no effect on Ferrozine method. *3393*

¹³¹I Uptake *Serum Decrease Physiological* Lilly product contains tetraiodofluorescein. *2652*

Lactate Dehydrogenase *Serum No Effect Analytical* At acute overdose concentration (20 mg/dL) on Technicon® SMAC® method. *3719*

Methaqualone *Serum No Effect Analytical* At 20 mg/L on GLC procedure of Evenson. *1057*

Pentobarbital *Blood Increase Physiological* 600 mg orally produces concentration of 3.3 mg/L. *2348*

Phosphate *Serum No Effect Analytical* At acute overdose concentration (20 mg/dL) on Technicon® SMAC® method. *3719* At concentration of 340 mg/L had no effect on phosphomolybdate method. *3393*

Propoxyphene *Serum No Effect Analytical* At 25 mg/L on method of Evenson. *1056*

Protein *Serum No Effect Analytical* At acute overdose concentration (20 mg/dL) on Technicon® SMAC® method. *3719* At concentration of 340 mg/L had no effect on biuret method with blank correction. *3393*

Salicylate *Urine No Effect Analytical* At 10 mg/dL no effect on Trinder method. *1204*

Triglycerides *Serum No Effect Analytical* At acute overdose concentration (20 mg/dL) on Technicon® SMAC® method. *3719* At concentration of 340 mg/L had no effect on lipase/esterase method. *3393*

Urea Nitrogen *Serum No Effect Analytical* At acute overdose concentration (20 mg/dL) on Technicon® SMAC® method. *3719* At concentration of 340 mg/L had no effect on diacetylmonoxime method. *3393*

Uric Acid *Serum No Effect Analytical* At acute overdose concentration (20 mg/dL) on Technicon® SMAC® method. *3719* At concentration of 340 mg/L had no effect on phosphotungstate reduction method. *3393*

Pentolinium

Catecholamines *Urine No Effect Analytical* No effect on fluorometric method of Crout. *0758*

Pentoses

Glucose *Urine Increase Analytical* Produce orange color with o-toluidine. *0583*

Sugar *Urine Increase Analytical* Positive with Clinitest®, Benedict's Fehling's. *2308*

Pentoxifylline

Hematocrit *Blood Decrease Physiological* In 2 patients receiving other drugs concomitantly but for considerable time previously without ill effects. *2244*

Leukocytes *Blood Decrease Physiological* In 2 patients receiving other drugs concomitantly but for considerable time previously without ill effects. *2244*

Platelets *Blood Decrease Physiological* In 2 patients receiving other drugs concomitantly but for considerable time previously without ill effects. *2244*

Pentylenetetrazole

Cholesterol *Serum Decrease Physiological* Maximal effect seen after 2 weeks. *0332*

Pregnancy Tests *Urine Positive Analytical* In 1 patient affected hCG and Pregslide tests. *1487*

Perchlorate

Aspartate Aminotransferase
Serum Increase Physiological Single case of acute yellow atrophy of liver. *2427*

Bilirubin *Serum Increase Physiological* Single case of acute yellow atrophy of liver. *2427*

Erythrocytes *Blood Decrease Physiological* May cause fatal aplastic anemia. *0071* Occasional case of aplastic anemia reported with potassium salt. *3717*

Ionized Calcium *Serum Increase Analytical* Observed interference ion specific electrode. *0540*

Iron-Binding Capacity, Total (TIBC)
Serum Increase Analytical When Fe salt used to saturate in method Young/Hicks. *1055*

¹³¹I Uptake *Serum No Effect Physiological* No effect on uptake. *2220*

Leukocytes *Blood Decrease Physiological* May cause fatal aplastic anemia. *2313* Occasional case of aplastic anemia reported with potassium salt. *3717*

Neutrophils *Blood Decrease Physiological* Occasional case of agranulocytosis reported with potassium salt. *3717*

Platelets *Blood Decrease Physiological* May cause fatal aplastic anemia. *0071* Occasional case of aplastic anemia reported with potassium salt. *3717*

Protein *Urine Increase Physiological* Reported cases of nephrotic syndrome. *0071*

Protein Bound Iodine (PBI) *Serum Decrease Physiological* Interferes with trapping of iodine by thyroid. *0012*

Urea Nitrogen *Serum Increase Physiological* Reported cases of nephrotic syndrome. *0071*

Pergolide

Prolactin *Plasma Decrease Physiological* Single dose reduced concentration for 24 h in normal subjects, multiple doses reduced concentration by 80%. *2127* Marked reduction in patients with Parkinson's disease to low or undetectable levels. *1956*

Perhexilene

Alkaline Phosphatase *Serum Increase Physiological* Transient elevation observed. *1619*

Aspartate Aminotransferase
Serum Increase Physiological Transient elevation observed. *1619*

Peribedil

Thyroid Stimulating Hormone (TSH)
Serum Decrease Physiological Effect noted in hypothyroid subjects. *3798*

Perindopril

Aldosterone *Plasma Decrease Physiological* After doses of 4 mg twice daily within 4 h of administration. *0550*

Angiotensin-Converting Enzyme (ACE)
Plasma Decrease Physiological With single doses of 8 to 16 mg produced reduction to less than 10% of control in 4 h with lasting effect for 72 h. *0550*

Angiotensin I *Blood Increase Physiological* After doses of 4 mg twice daily within 4 h of administration. *0550*

Angiotensin II *Plasma Decrease Physiological* After doses of 4 mg twice daily within 4 h of administration. *0550*

Renin Activity *Plasma Increase Physiological* After doses of 4 mg twice daily within 4 h of administration. *0550*

Periodate

Protein Bound Iodine (PBI)   *Serum   Decrease   Physiological*
Interferes with trapping of iodine by thyroid. *0012*

Permanganate

Hemoglobin   *Urine   Increase   Analytical*   Interferes with ben-zidine test. *1022*

Peroxide

Glucose   *Urine   Increase   Analytical*   May produce false posi-tive with Clinistix®. *1563*

Perphenazine

Alanine Aminotransferase   *Serum   No Effect   Analytical*   At acute overdose concentration (20 mg/dL) on Technicon® SMAC® method. *3719*

Albumin   *Serum   No Effect   Analytical*   At acute overdose concentration (20 mg/dL) on Technicon® SMAC® method. *3719*   At concentration of 200 mg/L had no effect on BCG method. *3393*

Alkaline Phosphatase   *Serum   No Effect   Analytical*   At acute overdose concentration (20 mg/dL) on Technicon® SMAC® method. *3719*

Alkaloids   *Urine   No Effect   Analytical*   With ninhydrin on Frings TLC procedure at 10 mg/dL. *1204*
Urine   Increase   Analytical   Purple with iodoplatinate on Frings TLC procedure. *1204*

Aspartate Aminotransferase   *Serum   No Effect   Analytical*   At acute overdose concentration (20 mg/dL) on Technicon® SMAC® method. *3719*

Barbiturate   *Urine   No Effect   Analytical*   With mercurous NO_3 on Frings TLC procedure at 10 mg/dL. With diphenyl-carbazone on Frings TLC procedure at 10 mg/dL. *1204*
Urine   Increase   Analytical   Pink spot with vanillin in Frings TLC procedure. Pink spot with mercuric SO_4 in Frings TLC pro-cedure. *1204*

Bicarbonate   *Serum   No Effect   Analytical*   At concentration of 1 mg/L had no effect on measurement by phenolphthalein method. *3393*

Bile   *Urine   Increase   Physiological*   Low incidence of jaun-dice reported. *2313*

Bilirubin   *Serum   No Effect   Analytical*   At acute overdose concentration (20 mg/dL) on Technicon® SMAC® method. *3719*   At concentration of 200 mg/L had no effect on Jendrassik and Grof method. *3393*
Serum   Increase   Physiological   Low incidence of jaundice reported. *0071*
Urine   Increase   Physiological   Low incidence of jaundice reported. *2313*

Bromide   *Urine   No Effect   Analytical*   No effect on method of Frings. *1204*

Calcium   *Serum   No Effect   Analytical*   At acute overdose concentration (20 mg/dL) on Technicon® SMAC® method. *3719*   At concentration of 200 mg/L had no effect on cresolphthalein method. *3393*

Catecholamines   *Plasma   Increase   Physiological*   Increased metabolism, decreased organ uptake of norepinephrine. *1237*

Chloride   *Serum   No Effect   Analytical*   At concentration of 1 mg/L had no effect on mercurimetric method. *3393*

Cholesterol   *Serum   No Effect   Analytical*   At acute overdose concentration (20 mg/dL) on Technicon® SMAC® method. *3719*   At concentration of 1 mg/L had no effect on CHOD-PAP method. *3393*   At concentration of 200 mg/L had no effect on Liebermann-Burchard method. *3393*

Creatine Kinase   *Serum   No Effect   Analytical*   At acute overdose concentration (20 mg/dL) on Technicon® SMAC® method. *3719*

Creatinine   *Serum   No Effect   Analytical*   At acute overdose concentration (20 mg/dL) on Technicon® SMAC® method. *3719*   At concentration of 200 mg/L had no effect on AutoAnalyzer method. *3393*

Ethchlorvynol   *Urine   No Effect   Analytical*   No effect on method of Frings and Cohen at 10 mg/dL. *1205*

FPN Test   *Urine   No Effect   Analytical*   Negative FPN test with 0.1 mg/dL concentration in Frings TLC procedure. *1204*
Urine   Increase   Analytical   Pink FPN test with 0.4 mg/dL con-centration in Frings TLC procedure Pink color observed in method of Frings. *1204*

Glucose   *Serum   No Effect   Analytical*   At acute overdose concentration (20 mg/dL) on Technicon® SMAC® method. *3719*   At concentration of 1 mg/L had no effect on GOD/POD-PAP method. *3393*
Serum   Increase   Physiological   May cause hyperglycemia. *1680*

Glucose Tolerance   *Serum   Decrease   Physiological*   Abnor-mal curves in 35% subjects. *2427*

17-Hydroxycorticosteroids   *Urine   Increase   Analytical*   Abnormal color with Glenn-Nelson procedure. *0427*
Urine   Decrease   Physiological   Acts on hypothalamus to depress ACTH secretion. *0427*

Iron   *Serum   No Effect   Analytical*   At acute overdose con-centration (20 mg/dL) on Technicon® SMAC® method. *3719*   At concentration of 200 mg/L had no effect on Ferrozine method. *3393*

^{131}I Uptake   *Serum   No Effect   Physiological*   No effect observed in euthyroid subjects. *2220*
Serum   Decrease   Physiological   Depresses uptake. *0583*

17-Ketosteroids   *Urine   No Effect   Analytical*   Minimal effect on Zimmermann reaction. *0427*

Lactate Dehydrogenase   *Serum   No Effect   Analytical*   At acute overdose concentration (20 mg/dL) on Technicon® SMAC® method. *3719*

Leukocytes   *Blood   Decrease   Physiological*   Suspected of causing agranulocytosis. *2822*

Neutrophils   *Blood   Decrease   Physiological*   May cause neutropenia/agranulocytosis. *2583*

Phosphate   *Serum   No Effect   Analytical*   At acute overdose concentration (20 mg/dL) on Technicon® SMAC® method. *3719*   At concentration of 200 mg/L had no effect on phosphomolybdate method. *3393*

Potassium   *Serum   No Effect   Analytical*   At concentration of 1 mg/L had no effect on flame-photometric method. *3393*

Prolactin   *Plasma   Increase   Physiological*   By approximately 800% in women after 8 mg. *0522*   Typical dose-related response to i.m. administered drug due to antidopaminergic action. *2072*   Marked effect in male and female psychiatric patients treated for up to 4 weeks. *3630*

Protein   *Serum   No Effect   Analytical*   At acute overdose concentration (20 mg/dL) on Technicon® SMAC® method. *3719*   At concentration of 200 mg/L had no effect on biuret method with blank correction. *3393*
CSF   Increase   Physiological   Altered proteins reported. *1680*

Protein Bound Iodine (PBI)   *Serum   Increase   Physiological*   May contain iodinated contaminants. *1481*

Sodium   *Serum   No Effect   Analytical*   At concentration of 1 mg/L had no effect on flame-photometric method. *3393*

T_3 Uptake   *Serum   Decrease   Physiological*   Occurs with prolonged use. *2313*

Thyroxine (T_4) Binding Globulin
Serum   Increase   Physiological   Direct effect of drug. *0012*

Triglycerides   *Serum   No Effect   Analytical*   At acute over-dose concentration (20 mg/dL) on Technicon® SMAC® method. *3719*   At concentration of 200 mg/L had no effect on lip-ase/esterase method. *3393*

Urea Nitrogen   *Serum   No Effect   Analytical*   At acute over-dose concentration (20 mg/dL) on Technicon® SMAC® method. *3719*   At concentration of 200 mg/L had no effect on diacetylmonoxime method. *3393*

Uric Acid   *Serum   No Effect   Analytical*   At acute overdose concentration (20 mg/dL) on Technicon® SMAC® method. *3719*   At concentration of 200 mg/L had no effect on phosphotung-state reduction method. *3393*

Phenacemide

Alanine Aminotransferase   *Serum   Increase   Physiological*   May affect liver function in about 2% cases. *2313*

Alkaline Phosphatase   *Serum   Increase   Physiological*   May affect liver function in about 2% cases. *2313*

Aspartate Aminotransferase

Serum Increase Physiological May affect liver function in about 2% cases. *2313*

Bile *Urine Increase Physiological* May affect liver function in about 2% cases. *2313*

Bilirubin *Serum Increase Physiological* May affect liver function in about 2% cases. *2313*

BSP Retention *Serum Increase Physiological* May alter liver function in about 2% cases. *2313*

Cephalin Flocculation *Serum Increase Physiological* May affect liver function in about 2% cases. *2313*

Creatinine *Serum Decrease Analytical* Positive interference with kinetic Jaffé reaction within 21 s but negative result thereafter. *2025* As measured by Du Pont aca due to decrease of absorbance of product with picrate with time. *2025*

Erythrocytes *Blood Decrease Physiological* Aplastic anemia/agranulocytosis. *2313*

Ethchlorvynol *Serum No Effect Analytical* At 10 μg/mL on colorimetric method of Wallace. *3753*

Hematocrit *Blood Decrease Physiological* Aplastic anemia. *0071*

Hemoglobin *Blood Decrease Physiological* Aplastic anemia. *0071*

Leukocytes *Blood Decrease Physiological* Aplastic anemia (leukopenia reported most often). *0071*

Platelets *Blood Decrease Physiological* Aplastic anemia may occur rarely. *0071*

Protein *Urine Increase Physiological* Occasional nephropathy. *0071* Occasional nephropathy. *1022*

Thymol Turbidity *Serum Increase Physiological* May affect liver function in about 2% cases. *2313*

Urea Nitrogen *Serum Increase Physiological* Occasional nephropathy. *0071*

Phenaglycodol

Cortisol *Plasma No Effect Analytical* No physiological effect observed. *1487*

Ethchlorvynol *Serum No Effect Analytical* At 10 μg/mL on colorimetric method of Wallace. *3753*

17-Ketogenic Steroids *Urine Increase Analytical* Interferes with Zimmermann reaction. *3505*

17-Ketosteroids *Urine Increase Analytical* Interferes with Zimmermann reaction. *3505*

Phenanthroline

Angiotensin-Converting Enzyme (ACE)
Serum Decrease Analytical 80% inhibition of method using benzyloxycarbonyl-phenylalanyl-histidyl-leucine as substrate. *3221*

Phenazocine

pCO₂ *Blood Increase Physiological* May produce respiratory depression. *1343*

Phenazone

Salicylate *Serum Increase Analytical* At 100 mg/L has slight effect (8%) on method of Keller but 22% on that of Trinder, also observed in one patient who had taken overdose of drug. *1776*

Phenazopyridine

Alanine Aminotransferase *Serum No Effect Analytical* No effect at therapeutic concentration on Reflotron method. *1984*
Serum Increase Physiological Hepatotoxic effect. *2313*

Albumin *Serum Increase Analytical* Contributes to absorption binding procedure. *0583*

Alkaline Phosphatase *Serum Increase Physiological* Hepatotoxic effect. *2313*

Amphetamine *Urine No Effect Analytical* No effect at 100 mg/L on method of Rutter. *3097*

Aspartate Aminotransferase *Serum No Effect Analytical* No effect at therapeutic concentration on Reflotron method. *1984*
Serum Increase Physiological Hepatotoxic effect. *2313*

Bacteria *Urine Negative Analytical* May interfere with color on Microstix. *0129*

Bile *Urine Increase Analytical* False positive with Ictotest®, BiliLabstix®. *2566*
Urine Increase Physiological Hepatotoxic effect. *2313*

Bilirubin *Serum Increase Analytical* Postulated increased color with diazotization. *0652*
Serum Increase Physiological Single report of jaundice (?due to hemolysis). *1652*
Serum Decrease Physiological Atypical color causes interference. *2313*

BSP Retention *Serum Increase Analytical* Increased spectral absorbancy in colorimetric reading. *3879*
Serum Increase Physiological Hepatic damage reported in one case. *1652*
Serum Decrease Analytical High absorbancy in blank if acidified. *0583*

Cephalin Flocculation *Serum Increase Physiological* Hepatotoxic effect. *2313*

Cholesterol *Serum No Effect Analytical* No effect at therapeutic concentration on Reflotron method. *1984*

Color *Urine Increase Analytical* Yellow orange increases with HCl. *0443*
Feces Increase Analytical Orange red. *2313*

Diagnex Blue Excretion *Urine Increase Analytical* Orange color produces interference. *3879*

Erythrocytes *Blood Decrease Physiological* Hemolytic anemia (sensitivity dependent). *2426*

Ethchlorvynol *Serum No Effect Analytical* At 250 mg/L on GLC procedure of Evenson. *1058*
Serum Increase Analytical False increase with method of Frings. *0576*
Urine Increase Analytical Reacts in method of Frings and Cohen. *1205*

Glucose *Serum No Effect Analytical* No effect at therapeutic concentration on Reflotron method. *1984* At concentration of 120 mg/L had no effect on hexokinase/G-6-PDH method. *3393*
Serum Decrease Analytical Delays coupled glucose oxidase reaction. *0583*
Urine Increase Analytical False positive reported with Tes-Tape®. *2566*
Urine Decrease Analytical False negative with glucose oxidase methods. *2566*

γ-Glutamyltransferase (GGT) *Serum No Effect Analytical* No effect at therapeutic concentration on Reflotron method. *1984*

Heinz Body Formation *Blood Positive Physiological* Associated with hemolytic anemia, methemoglobinemia. *2427*

Hematocrit *Blood Decrease Physiological* Hemolytic anemia (sensitivity dependent). *0788*

Hemoglobin *Blood Decrease Physiological* Hemolytic anemia (sensitivity dependent). *0788*

17-Hydroxycorticosteroids *Urine Increase Analytical* Interferes with modified Glenn-Nelson method. *0427*

¹³¹I Uptake *Serum Decrease Physiological* Donnasep contains tetraiodofluorescein. *2652*

17-Ketogenic Steroids *Urine Increase Analytical* Interferes with Zimmermann reaction. *3505*

Ketones *Urine Increase Analytical* False positive with Ketostix® or FeCl₃. *2566*
Urine Decrease Analytical Nitroprusside reaction masked by color. *0583*

17-Ketosteroids *Urine Increase Analytical* Interferes with Zimmermann reaction. *3505*

Lactate *Serum No Effect Analytical* At concentration of 120 mg/L had no effect on enzymatic method. *3393*

Methemoglobin *Blood Increase Physiological* May cause hemolysis. *2426*

Phenazopyridine (continued)

Porphyrins *Urine Increase Analytical* Interference with fluorescence (in screening test). 3879

Potassium *Serum No Effect Analytical* At concentration of 20 mg/L had no effect on measurement by ISE with predilution. 3393

Pregnanediol *Urine Increase Analytical* Mechanism unknown. 2220

Protein *Serum Increase Analytical* Orange-brown color affects absorbance. 0583
Urine No Effect Analytical No effect sulfosalicylic, heat tests. 1487
Urine Increase Analytical False positive with Labstix® etc. 2566

PSP Excretion *Urine Increase Analytical* False color reaction at alkaline pH. 3879

Sodium *Serum No Effect Analytical* At concentration of 20 mg/L had no effect on measurement by ISE with predilution. 3393

Sulfhemoglobin *Blood Increase Physiological* Positive with severe oxidative hemolysis. 1216

Thymol Turbidity *Serum Increase Physiological* Hepatotoxic effect. 2313

Triglycerides *Serum No Effect Analytical* No effect at therapeutic concentration on Reflotron method. 1984

Urea Nitrogen *Serum No Effect Analytical* No effect at therapeutic concentration on Reflotron method. 1984
Serum Increase Physiological Transient acute renal failure reported. 0071

Uric Acid *Serum No Effect Analytical* No effect at therapeutic concentration on Reflotron method. 1984

Urobilinogen *Urine No Effect Analytical* No effect in quantitative p-MDFB procedure of Rutter. 2044
Urine Increase Analytical Orange-red with Ehrlich's aldehyde reaction. 3879
Urine Increase Physiological Gives false positive with Ehrlich's reaction. 0251

Vanillylmandelic Acid *Urine Increase Analytical* Yields similar color in reaction. 2220

Xylose Excretion *Urine Increase Analytical* Due to interfering background color. 2945

Phencyclidine

Albumin *Serum No Effect Analytical* At concentration of 6 mg/L had no effect on BCG method. 3393

Amphetamine *Urine No Effect Analytical* No effect at 100 mg/L on method of Rutter. 3097

Bilirubin *Serum No Effect Analytical* At concentration of 6 mg/L had no effect on Jendrassik and Grof method. 3393

Calcium *Serum No Effect Analytical* At concentration of 6 mg/L had no effect on cresolphthalein method. 3393

Chloride *Serum No Effect Analytical* At concentration of 6 mg/L had no effect on mercurimetric method. 3393

Cholesterol *Serum No Effect Analytical* At concentration of 6 mg/L had no effect on Liebermann-Burchard method. 3393

Creatinine *Serum No Effect Analytical* At concentration of 6 mg/L had no effect on AutoAnalyzer Jaffé method. 3393

Phosphate *Serum No Effect Analytical* At concentration of 6 mg/L had no effect on phosphomolybdate method. 3393

Protein *Serum No Effect Analytical* At concentration of 6 mg/L had no effect on biuret method with blank correction. 3393

Urea Nitrogen *Serum No Effect Analytical* At concentration of 6 mg/L had no effect on diacetylmonoxime method. 3393

Uric Acid *Serum No Effect Analytical* At concentration of 6 mg/L had no effect on phosphotungstate reduction method. 3393

Phendimetrazine

Amobarbital *Urine No Effect Analytical* No interference with TLC using ethyl acetate: methanol: water: ammonium hydroxide and modified Dragendorff's reagent for detection. 3868

Amphetamine *Urine No Effect Analytical* No interference with TLC using ethyl acetate: methanol: water: ammonium hydroxide and modified Dragendorff's reagent for detection. 3868

Chlordiazepoxide *Urine Positive Analytical* Same Rf and color reaction on TLC using ethyl acetate: methanol: water: ammonium hydroxide and modified Dragendorff's reagent. 3868

Hydromorphone *Urine No Effect Analytical* No interference with TLC using ethyl acetate: methanol: water: ammonium hydroxide and modified Dragendorff's reagent for detection. 3868

Mazindol *Urine Positive Analytical* Same Rf and color reaction on TLC using ethyl acetate: methanol: water: ammonium hydroxide and modified Dragendorff's reagent. 3868

Mescaline *Urine No Effect Analytical* No interference with TLC using ethyl acetate: methanol: water: ammonium hydroxide and modified Dragendorff's reagent for detection. 3868

Methadone *Urine Positive Analytical* Same Rf and color reaction on TLC using ethyl acetate: methanol: water: ammonium hydroxide and modified Dragendorff's reagent. 3868

Methamphetamine *Urine No Effect Analytical* No interference with TLC using ethyl acetate: methanol: water: ammonium hydroxide and modified Dragendorff's reagent for detection. 3868

Methapyrilene *Urine Positive Analytical* Same Rf and color reaction on TLC using ethyl acetate: methanol: water: ammonium hydroxide and modified Dragendorff's reagent. 3868

Morphine *Urine No Effect Analytical* No interference with TLC using ethyl acetate: methanol: water: ammonium hydroxide and modified Dragendorff's reagent for detection. 3868

Nicotine *Urine Positive Analytical* Same Rf and color reaction on TLC using ethyl acetate: methanol: water: ammonium hydroxide and modified Dragendorff's reagent. 3868

Pentobarbital *Urine No Effect Analytical* No interference with TLC using ethyl acetate: methanol: water: ammonium hydroxide and modified Dragendorff's reagent for detection. 3868

Phenobarbital *Urine No Effect Analytical* No interference with TLC using ethyl acetate: methanol: water: ammonium hydroxide and modified Dragendorff's reagent for detection. 3868

Phenylpropanolamine *Urine No Effect Analytical* No interference with TLC using ethyl acetate: methanol: water: ammonium hydroxide and modified Dragendorff's reagent for detection. 3868

Secobarbital *Urine No Effect Analytical* No interference with TLC using ethyl acetate: methanol: water: ammonium hydroxide and modified Dragendorff's reagent for detection. 3868

Phenelzine

Alanine Aminotransferase *Serum Increase Physiological* May cause hypersensitive hepatitis. 0071

Amphetamine *Urine No Effect Analytical* At 50 mg/L on fluorescent method of Hayes. 1534

Aspartate Aminotransferase *Serum Increase Analytical* At 1 mmol/L affects SMA 12/60 method. 3335
Serum Increase Physiological May cause hypersensitive hepatitis. 0071

Bile *Urine Increase Physiological* May cause hypersensitive hepatitis. 0071

Bilirubin *Serum No Effect Analytical* At concentration of 136 mg/L had no effect on Jendrassik and Grof method. 3393
Serum Increase Analytical At 1 mmol/L affects SMA 12/60 method slightly. 3335
Serum Increase Physiological May cause hypersensitive hepatitis. 0071

BSP Retention *Serum Increase Physiological* Hepatotoxic/cholestatic syndromes. 1343

Calcium *Serum No Effect Analytical* At concentration of 136 mg/L had no effect on cresolphthalein method. 3393

Glucose *Serum Increase Physiological* Decreases glucose tolerance. 3679

Leukocytes *Blood Decrease Physiological* May cause leukopenia. 0071

Phosphate *Serum No Effect Analytical* At concentration of 136 mg/L had no effect on phosphomolybdate method. *3393*

Protein *Serum No Effect Analytical* At concentration of 136 mg/L had no effect on biuret method with blank correction. *3393*

Pseudocholinesterase *Serum Decrease Physiological* May cause hypersensitive hepatitis. *3505*

Sodium *Serum Increase Physiological* Rare hypernatremia reported. *1680*

Urea Nitrogen *Serum No Effect Analytical* At concentration of 136 mg/L had no effect on diacetylmonoxime method. *3393*

Uric Acid *Serum No Effect Analytical* At concentration of 136 mg/L had no effect on phosphotungstate reduction method. *3393*
Serum Increase Analytical At 1 mmol/L slightly affects SMA 12/60, Henry methods. *3335*

Xylose Excretion *Urine Decrease Physiological* Decreased absorption of xylose from gastrointestinal tract. *3679*

Phenethicillin

Eosinophils *Blood Increase Physiological* Few cases reported only (minor effect). *1680*

Erythrocytes *Blood Decrease Physiological* Theoretically may cause hemolytic anemia. *1680*

Hemoglobin *Blood Decrease Physiological* Theoretical effect of penicillins. *1680*

Leukocytes *Blood Decrease Physiological* Theoretically leukopenia may occur. *1680*

Phenethylamine

Amphetamine *Urine Increase Analytical* False positive fluorescent method of Hayes at 50 mg/L. *1534*

Phenindamine

Dapsone *Serum No Effect Analytical* At 10 mg/dL on colorimetric procedure of Higgins. *1590*

Phenindione

Alanine Aminotransferase *Serum Increase Physiological* May modify liver function (cholestasis). *2313*

Alkaline Phosphatase *Serum Increase Physiological* May modify liver function (cholestasis). *2313*

Aspartate Aminotransferase
Serum Increase Physiological May modify liver function (cholestasis). *2313*

Bile *Urine Increase Physiological* May modify liver function (cholestasis). *2313*

Bilirubin *Serum Increase Physiological* Probable effect as bilirubin clearance reduced. *1499*

Bilirubin, Direct *Serum Increase Physiological* Probable effect as bilirubin conjugation affected. *1499*

BSP Retention *Serum Increase Physiological* May modify liver function (cholestasis). *2313*

Cephalin Flocculation *Serum Increase Physiological* May modify liver function (cholestasis). *2313*

Color *Urine Increase Analytical* Red-orange color produced in alkaline urine. *0720*

Diphenylhydantoin *Serum No Effect Physiological* No effect on metabolism due to different chemical configuration from other coumarins. *3556*

Eosinophils *Blood Increase Physiological* Allergic response after 15 d in one patient. *1574*

Erythrocytes *Urine Increase Physiological* May cause actual bleeding. *1022*

Hemoglobin *Urine Increase Physiological* Actual bleeding caused by drug. *1022*

131I Uptake *Serum Decrease Physiological* Uncommon reported effect. *2220*

Leukocytes *Blood Increase Physiological* Occasional leukocytosis may occur. *1343*
Blood Decrease Physiological Agranulocytosis/leukopenia. *2313*

Protein *Urine Increase Physiological* Nephrotoxicity may occur. *2496*

Protein Bound Iodine (PBI) *Serum Decrease Physiological* Inhibits iodination of tyrosine in thyroxine binding globulin. *0012*

Thymol Turbidity *Serum Increase Physiological* May modify liver function (cholestasis). *2313*

Urea Nitrogen *Serum Increase Physiological* Nephrotoxicity may occur with tubular necrosis. *1343*

Uric Acid *Serum Decrease Physiological* Uricosuric action. *1433*
Urine Increase Physiological Uricosuric action. *1433*

Pheniprazine

Alanine Aminotransferase *Serum Increase Physiological* May cause hepatocellular jaundice. *2313*

Alkaline Phosphatase *Serum Increase Physiological* May cause hepatocellular jaundice. *2313*

Aspartate Aminotransferase
Serum Increase Physiological May cause hepatocellular jaundice. *2313*

Bile *Urine Increase Physiological* May affect liver function. *2313*

Bilirubin *Serum Increase Physiological* Hypersensitive hepatitis. *2313*

BSP Retention *Serum Increase Physiological* May affect liver function. *2313*

Cephalin Flocculation *Serum Increase Physiological* May affect liver function. *2313*

Thymol Turbidity *Serum Increase Physiological* May affect liver function. *2313*

Phenmetrazine

Alkaloids *Urine Increase Analytical* Red/purple with iodoplatinate on Frings TLC procedure. Reacts with ninhydrin on TLC method of Frings. *1204*

Amobarbital *Urine No Effect Analytical* No interference on TLC using ethyl acetate: methanol: water: ammonium hydroxide and modified Dragendorff's reagent for detection. *3868*

Amphetamine *Urine No Effect Analytical* At 50 mg/L on fluorescent method of Hayes. *1534* No effect at 100 mg/L on method of Rutter. *3097* No interference on TLC using ethyl acetate: methanol: water: ammonium hydroxide and modified Dragendorff's reagent for detection. *3868*

Barbiturate *Urine No Effect Analytical* With mercuric SO$_4$ on Frings TLC procedure at 10 mg/dL. With diphenylcarbazone on Frings TLC procedure at 10 mg/dL. *1204*

Bromide *Urine No Effect Analytical* No effect on method of Frings. *1204*

Chlorphentermine *Urine Positive Analytical* Similar R$_f$ and color reaction on TLC using ethyl acetate: methanol: water: ammonium hydroxide and modified Dragendorff's reagent. *3868*

Dapsone *Serum No Effect Analytical* At 10 mg/dL on colorimetric procedure of Higgins. *1590*

Epinephrine *Urine Increase Physiological* For 3 h after administration (slight). *1487*

Ethchlorvynol *Urine No Effect Analytical* No effect on method of Frings and Cohen. *1204*

FPN Test *Urine No Effect Analytical* No effect at 10 mg/dL on method of Frings. *1204*

Hydromorphone *Urine No Effect Analytical* No interference with TLC using ethyl acetate: methanol: water: ammonium and modified Dragendorff's reagent for detection. *3868*

5-Hydroxyindoleacetic Acid (5-HIAA)
Urine Increase Physiological Reported effect. *2142*

131I Uptake *Serum Decrease Physiological* Preludin® contains tetraiodofluorescein. *2652*

Mescaline *Urine No Effect Analytical* No interference with TLC using ethyl acetate: methanol: water: ammonium and modified Dragendorff's reagent for detection. *3868*

Phenmetrazine *(continued)*

Methamphetamine *Urine No Effect Analytical* No interference on TLC using ethyl acetate: methanol: water: ammonium hydroxide and modified Dragendorff's reagent for detection. *3868*

Morphine *Urine No Effect Analytical* No interference on TLC using ethyl acetate: methanol: water: ammonium hydroxide and modified Dragendorff's reagent for detection. *3868*

Norepinephrine *Urine No Effect Physiological* No effect observed. *1487*

Pentobarbital *Urine No Effect Analytical* No interference on TLC using ethyl acetate: methanol: water: ammonium hydroxide and modified Dragendorff's reagent for detection. *3868*

Phenobarbital *Urine No Effect Analytical* No interference on TLC using ethyl acetate: methanol: water: ammonium hydroxide and modified Dragendorff's reagent for detection. *3868*

Phentermine *Urine Positive Analytical* Similar R_f and color reaction on TLC using ethyl acetate: methanol: water: ammonium hydroxide and modified Dragendorff's reagent. *3868*

Phenylpropanolamine *Urine No Effect Analytical* No interference on TLC using ethyl acetate: methanol: water: ammonium hydroxide and modified Dragendorff's reagent for detection. *3868*

Quinine *Urine Positive Analytical* Similar R_f and color reaction on TLC using ethyl acetate: methanol: water: ammonium hydroxide and modified Dragendorff's reagent. *3868*

Salicylate *Urine No Effect Analytical* At 10 mg/dL no effect on Trinder method. *1204*

Secobarbital *Urine No Effect Analytical* No interference on TLC using ethyl acetate: methanol: water: ammonium hydroxide and modified Dragendorff's reagent for detection. *3868*

Phenobarbital

Acetaminophen *Serum No Effect Analytical* At 10 mg/L had no effect on HPLC method. *3432*

Acetaminophen Screening Test *Urine Negative Analytical* No reaction with o-cresol at therapeutic concentrations. *3326*

Alanine Aminotransferase *Serum No Effect Analytical* At acute overdose concentration (20 mg/dL) on Technicon® SMAC® method At acute overdose concentration (20 mg/dL) on Technicon® SMAC® method. *3719* On continuous method at 10 times maximal therapeutic concentration. On colorimetric method at 10 times maximal therapeutic concentration. *1785* No effect at therapeutic concentration on Reflotron method. *1984*
Serum Increase Physiological Hepatotoxicity with centrolobular necrosis. *1948*

Albumin *Serum No Effect Analytical* No significant effect on HABA method at 25 mg/L No significant effect on BCG method at 25 mg/L. *2625* At acute overdose concentration (20 mg/dL) on Technicon® SMAC® method At acute overdose concentration (20 mg/dL) on Technicon® SMAC® method. *3719* At concentration of 250 mg/L had no effect on BCG method. *3393*

Alkaline Phosphatase *Serum No Effect Analytical* At acute overdose concentration (20 mg/dL) on Technicon® SMAC® method At acute overdose concentration (20 mg/dL) on Technicon® SMAC® method. *3719* On continuous method. *1785*
Serum No Effect Physiological No effect observed in 14 adult epileptic inpatients given drug alone (dose and duration of treatment alone). *1162*
Serum Increase Physiological May cause osteomalacia (?also liver effect). *1396* Increased activity occurs early in epileptic children. *2162*

Alkaloids *Urine No Effect Analytical* With iodoplatinate of Frings TLC procedure at 10 mg/dL. With ninhydrin on Frings TLC procedure at 10 mg/dL. *1204*

Amino-4-Imidazole-5-Carboxamide Ribotide (AICAR)
Urine Increase Physiological If megaloblastic anemia develops. *3025*

Amino Acids *Urine Positive Analytical* Unusual ninhydrin positive spot in all systems adjacent to threonine and orange in color. *1596*

Aminolevulinic Acid *Serum Increase Physiological* Slight but not significant increase (112 nmol/L versus 99 nmol/L in controls). *1348*

Aminopyrine *Serum Decrease Physiological* Reported interaction due to alteration of metabolism. *2042*

Ammonia *Plasma No Effect Analytical* At concentration of 30 mg/L had no effect on Ektachem® method. *3393*
Plasma No Effect Physiological No striking abnormality when given to epileptics. *3697*

Aspartate Aminotransferase *Serum No Effect Analytical* At acute overdose concentration (20 mg/dL) on Technicon® SMAC® method. *3719* On continuous method at 10 times maximal therapeutic concentration. On colorimetric method at 10 times maximal therapeutic concentration. *1785* No effect at therapeutic concentration on Reflotron method. *1984*
Serum Increase Physiological Hepatotoxicity with centrolobular necrosis. *1948*

Barbiturate *Urine No Effect Analytical* With vanillin on Frings procedure at 10 mg/dL. *1204*
Urine Increase Analytical White spot with mercuric SO_4 on Frings TLC procedure, Purple with diphenylcarbazone on Frings TLC procedure. With mercurous NO_3 reagent on Frings TLC procedure. *1204*

Bicarbonate *Serum No Effect Analytical* At concentration of 250 mg/L had no effect on method using phenolphthalein. *3393*

Bilirubin *Serum No Effect Analytical* No effect at therapeutic concentrations on Jendrassik-Grof, dimethylsulfoxide and spectrophotometric methods. *1786* At acute overdose concentration (20 mg/dL) on Technicon® SMAC® method. *3719* At concentration of 250 mg/L had no effect on Jendrassik and Grof method. *3393*
Serum No Effect Physiological Insignificant protein-displacement effect in neonates. *3748*
Serum Increase Physiological Hepatotoxicity with centrolobular necrosis. *1948*
Serum Decrease Physiological Induces hepatic microsomal enzymes especially in pregnant women. *3505*

Biotin *Serum Decrease Physiological* Dose related in long-term treated epileptics compared with controls. *2008*

Bromide *Urine No Effect Analytical* No effect on method of Frings. *1204*

BSP Retention *Serum Decrease Physiological* Increased clearance by liver in newborns. *3915*

Calcium *Serum No Effect Analytical* At acute overdose concentration (20 mg/dL) on Technicon® SMAC® method. *3719* At concentration of 250 mg/L had no effect on cresolphthalein method. *3393*
Serum Decrease Physiological Metabolic effect with chronic therapy (osteomalacia). *1452* Slight but significant reduction compared with untreated elderly at doses other than for treatment of epilepsy. *3925*

Catecholamines *Urine No Effect Physiological* No effect short term ingestion of 120 mg/d. *0766*

Ceruloplasmin *Serum Increase Physiological* Observed in both hospitalized and home-living patients with long-term administration. *3803*

Chenodeoxycholic Acid *Serum No Effect Physiological* No effect on increased concentration in patients with intrahepatic cholestasis increased of pregnancy. *1545*

Chlordiazepoxide *Blood No Effect Analytical* On method of Riddick at 5 mg/dL. *2983*

Chloride *Serum No Effect Analytical* At concentration of 250 mg/L had no effect on mercurimetric method. *3393*

Chlorpromazine *Serum Decrease Physiological* Induces hepatic microsomal enzymes. *1487*
Urine Increase Physiological Induces hepatic microsomal enzymes. *1487*

Cholesterol *Serum No Effect Analytical* No effect on enzymatic and Liebermann-Burchard methods at therapeutic concentrations. *1786* At acute overdose concentration (20 mg/dL) on Technicon® SMAC® method. *3719* No effect at therapeutic concentration on Reflotron method. *1984* At concentration of 352 mg/L had no effect on CHOD-Iodide method. *3393* At concentration of 650 mg/L had no effect on CHOD-PAP method. *3393* At concentration of 352 mg/L had no effect on catalase-AIDH method. *3393* At concentration of 520 mg/L had no effect on method using catalase-Hantzsch reaction. *3393* At concentration of 352 mg/L had no effect on Liebermann-Burchard method. *3393*

Cholic Acid *Serum No Effect Physiological* No effect on increased concentration in patients with intrahepatic cholestasis increased of pregnancy. *1545*

Copper *Serum Increase Physiological* Observed in both hospitalized and home-living patients with long-term administration. *3803*

Creatine Kinase *Serum No Effect Analytical* At acute overdose concentration (20 mg/dL) on Technicon® SMAC® method. *3719* On continuous method at 10 times maximal therapeutic concentration. *1785*

Creatinine *Serum No Effect Analytical* At 50 mg/L on reversed phase LC procedure of Zhiri et al. *3942* No effect on alkaline picrate and Slot methods at therapeutic concentrations. *1786* At acute overdose concentration (20 mg/dL) on Technicon® SMAC® method. *3719* At concentration of 30 mg/L had no effect on Ektachem® method. *3393* At concentration of 80 mg/L had no effect on creatinine iminohydrolase method. *3393* At concentration of 250 mg/L had no effect on AutoAnalyzer Jaffé method. *3393* At concentration of 80 mg/L had no effect on Jaffé-Fading-Fraction method. *3393* At concentration of 80 mg/L had no effect on Jaffé-Fuller's earth method. *3393* At concentration of 60 mg/L had no effect on kinetic Jaffé method on BKA-2. *3393*

Cyclosporine *Blood Decrease Physiological* Induces cytochrome P-450 hepatic enzymes thereby increasing clearance of cyclosporine. *0533*

Dapsone *Serum No Effect Analytical* At 10 mg/dL on colorimetric procedure of Higgins. *1590*

Deoxycholic Acid *Serum No Effect Physiological* No effect on increased concentration in patients with intrahepatic cholestasis increased of pregnancy. *1545*

Dexamethasone *Serum Decrease Physiological* 88% increase in metabolic clearance rate with substantial reduction in half-life. *1830*

Digitoxin *Serum Decrease Physiological* Stimulates metabolism, induces hepatic microsomal enzymes. *1487* May depress to subtherapeutic value due to induction of mixed function oxidases. *2794*

Diphenylhydantoin *Serum No Effect Analytical* No effect on fluorometric method of Dill. *0906* No detectable inhibition at 125 µg/ml. *3591*
Serum Increase Analytical Affects titrimetric procedure of Kozelka. *3054*
Serum Increase Physiological Competitive inhibition of metabolism. *1487*

Disopyramide *Serum No Effect Physiological* Concentration unaffected at level used to induce hepatic enzymes. *1861*

Epinephrine *Urine No Effect Physiological* No effect short term ingestion of 120 mg/d. *0766*

Erythrocytes *Blood Decrease Physiological* Megaloblastic anemia secondary to disturbance in folic acid metabolism. *1926*

Estriol *Urine No Effect Analytical* No effect of 50 mg/L on GLC method. *1163*

Ethchlorvynol *Serum No Effect Analytical* At 600 mg/L on GLC procedure of Evenson. *1058* At 10 µg/mL on colorimetric method of Wallace. *3753*
Urine No Effect Analytical No effect on method Frings and Cohen at 20 mg/dL. *1205*

Ferroxidase *Serum Increase Physiological* Significant effect in 40 adult epileptics receiving long term treatment. *3641*

FIGLU (N-Formiminoglutamic Acid)
Urine Increase Physiological If megaloblastic anemia develops. *3025*

Folate *Serum Decrease Physiological* May cause megaloblastic anemia. *3025* Low serum folate in from 27 to 91% of treated epileptics in different studies. *2062*
CSF Decrease Physiological Occurs in many long-treated epileptics. *2967* Low serum folate in from 27 to 91% of treated epileptics in different studies. *2062*
Red Blood Cells Decrease Physiological Inverse correlation with drug concentration. *2966* Low serum folate in from 27 to 91% of treated epileptics in different studies. *2062*
Test Conditions No Effect Analytical No effect on L. casei or S. fecalis. *2427*

FPN Test *Urine No Effect Analytical* No effect at 10 mg/dL on method of Frings. *1204*

Glucaric Acid *Urine Increase Physiological* Due to induction of hepatic enzymes. *2220* In 98% of patients with epilepsy receiving long term treatment. *3641* Dose dependent effect, greater than other anticonvulsants. *2788*

Glucose *Serum No Effect Analytical* No effect at therapeutic concentrations on hexokinase, glucose dehydrogenase 2,4-dichlorophenol, ABTS and o-toluidine methods. *1786* At acute overdose concentration (20 mg/dL) on Technicon® SMAC® method. *3719* No effect at therapeutic concentration on Reflotron method. *1984* At concentration of 30 mg/L had no effect on Ektachem® method. *3393* At concentration of 100 mg/L had no effect on GOD/POD-PAP method. *3393* At concentration of 60 mg/L had no effect on hexokinase/G-6-PDH method. *3393*
Urine No Effect Analytical No effect observed on TesTape®. *1100*
Urine Decrease Analytical Low with Clinistix®, Diastix®. *1100*

Glutamine *Plasma No Effect Physiological* No striking abnormality when given to epileptics. *3697*
Plasma Increase Physiological Significant effect noted in children, possibly due to inhibition of carbamoylphosphate synthase. *3642*

γ-Glutamyltransferase (GGT) *Serum No Effect Analytical* No effect at therapeutic concentration on Reflotron method. *1984*
Serum Increase Physiological ?induction or damage of hepatic microsomes. *3044* Observed 2 times upper limit of normal in 2 of 15 adult epileptic inpatients (dose and duration of treatment unknown). *1162*

Glycine *Plasma No Effect Physiological* No striking abnormality when given to epileptics. *3697*

Griseofulvin *Serum Decrease Physiological* Impairs gastrointestinal absorption. *1487*

Hematocrit *Blood Decrease Physiological* Megaloblastic anemia secondary to disturbance in folic acid metabolism. *2968*

Hemoglobin *Blood Decrease Physiological* Megaloblastic anemia secondary to disturbance in folic acid metabolism. *1926*

Hemopexin *Serum Increase Physiological* Significant effect in 40 adult epileptics receiving long term treatment. *3641*

Hydroxybutyrate Dehydrogenase
Serum Decrease Analytical On continuous method at 10 times maximal therapeutic concentration. *1785*

11-Hydroxycorticosteroids *Plasma No Effect Analytical* No effect on Mattingly procedure at 200 mg/L. *2977*
Urine No Effect Physiological No effect short term ingestion of 120 mg/d. *0766*

17-Hydroxycorticosteroids *Urine No Effect Analytical* No effect on Zimmermann procedure at 200 mg/L. *2977*
Urine No Effect Physiological No effect short term ingestion of 120 mg/d. *0766* No significant change in response to chronic treatment. *3943*
Urine Decrease Physiological With chronic ingestion cortisol= 6-beta-hydroxy cortisol. *0546*

6-β-Hydroxycortisol *Urine Increase Physiological* Alteration of steroid excretory pattern (long term). *0766* Increased conversion of cortisol produced. *0022* Chronic treatment leads to substantially increased excretion compared with controls (approximately 9 fold increase). *3943*

5-Hydroxyindoleacetic Acid (5-HIAA)
Urine Increase Analytical May cause false high colorimetric values. *0913*

Hydroxyproline *Urine Increase Physiological* Increased excretion occurred early in epileptic children without obvious bone changes suggestive of rickets. *2162*

25-Hydroxy Vitamin D$_3$ *Serum Decrease Physiological* Positive correlation with calcium level. *1452* Observed effect in some children. *3023*

Indocyanine Green *Serum Decrease Physiological* Mechanism not yet established. *2398*

Indomethacin *Serum No Effect Analytical* No effect on HPLC method of Roberts and Smith. *3002*

Iron *Serum No Effect Analytical* No effect at therapeutic concentrations on bathophenanthroline and Ramsay methods. *1786* At concentration of 200 mg/L had no effect on Ferrozine method. *3393*

17-Ketosteroids *Urine No Effect Analytical* No effect on Zimmermann procedure at 200 mg/L. *2977*

Phenobarbital *(continued)*

17-Ketosteroids *(continued)*
Urine No Effect Physiological No effect short term ingestion of 120 mg/d. *0766*

Lactate *Serum No Effect Analytical* At concentration of 60 mg/L had no effect on enzymatic method. *3393*
Plasma Increase Physiological Dose related in long-term treated epileptics compared with controls. *2008*

Lactate Dehydrogenase *Serum No Effect Analytical* At acute overdose concentration (20 mg/dL) on Technicon® SMAC® method. *3719* With pyruvate as substrate on continuous method. On colorimetric method at 10 times maximal therapeutic concentration. *1785*
Serum Increase Analytical With lactate as substrate on continuous method. *1785*

LE Cells *Blood Positive Physiological* Rare idiosyncratic response. *2042*

Leucine Aminopeptidase Isoenzymes
Serum No Effect Physiological No effect observed. *3087*

Leukocytes *Blood Decrease Physiological* Pancytopenia (AMA Blood dyscrasias). *2429*

MCV *Blood Increase Physiological* Megaloblastic anemia secondary to disturbance in folic acid metabolism. *1926*

Methaqualone *Serum No Effect Analytical* At 20 mg/L on GLC procedure of Evenson. *1057*

Methylprednisolone *Serum Decrease Physiological* 90% increase in metabolic clearance rate with substantial reduction in half-life. *1830*

Mexiletine *Serum Decrease Physiological* Induces hepatic enzymes: may reduce elimination half-life by 50%. *2011*

Nitroglycerin *Serum Decrease Physiological* When coadministered with nitroglycerin because of enhanced metabolism. *1598*

Norepinephrine *Urine No Effect Physiological* No effect short term ingestion of 120 mg/d. *0766*

5'-Nucleotidase *Serum No Effect Physiological* No effect observed in 12 adult epileptic in patients given drug alone (dose and duration of treatment unknown). *1162*

Organic Acids *Urine Increase Physiological* Dose related in long-term treated epileptics compared with controls. *2008*

Ornithine *Plasma No Effect Physiological* No striking abnormality when given to epileptics. *3697*

Phenobarbital *Serum Increase Physiological* 600 mg orally produces concentration of 23 mg/L in blood. *2348*
CSF Increase Physiological Concentration about half serum level. *3054*

Phenylbutazone *Serum Decrease Physiological* Reported interaction due to alteration of metabolism. *2042*

Phosphate *Serum No Effect Analytical* At acute overdose concentration (20 mg/dL) on Technicon® SMAC® method. *3719* At concentration of 250 mg/L had no effect on phosphomolybdate method. *3393*
Serum Decrease Physiological Increases clearance (secondary hyperparathyroidism). *1396*

Platelets *Blood Decrease Physiological* Thrombocytopenia may occur after some time. *2429*

Potassium *Serum No Effect Analytical* At concentration of 10 mg/L had no effect on flame-photometric method. *3393* At concentration of 250 mg/L had no effect on measurement by ISE with predilution. *3393* At concentration of 60 mg/L had no effect on measurement by ISE without predilution. *3393*

Propoxyphene *Serum No Effect Analytical* At 25 mg/L on method of Evenson. *1056*

Protein *Serum No Effect Analytical* At acute overdose concentration (20 mg/dL) on Technicon® SMAC® method. *3719* No effect at therapeutic concentrations on biuret and spectrophotometric method. *1786* At concentration of 250 mg/L had no effect on biuret method with blank correction. *3393*

Protein Electrophoresis *Serum No Effect Analytical* At concentration of 60 mg/L had no effect on automated Olympus-Hite method. *3393*

Prothrombin Time *Plasma Decrease Physiological* Metabolism enhanced by barbiturates. *0071* With 0.1 g/d in 10 patients receiving warfarin due to induction of hepatic microsomal enzymes. *3649*

Retinol Binding Protein *Serum Increase Physiological* In comparison with untreated cognitively delayed children 4.0 mg/dL on average versus 3.3 mg/dL. *2003*

Salicylate *Urine No Effect Analytical* At 10 mg/dL no effect on Trinder method. *1204*

Sodium *Serum No Effect Analytical* At concentration of 10 mg/L had no effect on flame-photometric method. *3393* At concentration of 250 mg/L had no effect on measurement by ISE with predilution. *3393* At concentration of 60 mg/L had no effect on measurement by ISE without predilution. *3393*

Sulfa as Sulfanilamide *Urine Increase Analytical* Positive reaction but delayed with usual amount present. *0251*

Thyroxine (T₄) *Serum Decrease Physiological* In comparison with untreated cognitively delayed children 7.2 μg/dL on average versus 8.4 μg/dL. *2003* Both significant and nonsignificant changes noted, associated with enzyme induction. *3798*

Thyroxine (T₄) Index, Free (FTI)
Serum Decrease Physiological Both significant and nonsignificant changes noted, associated with enzyme induction. *3798*

Triglycerides *Serum No Effect Analytical* No effect at therapeutic concentrations on enzymatic methods. *1786* At acute overdose concentration (20 mg/dL) on Technicon® SMAC® method. *3719* No effect at therapeutic concentration on Reflotron method. *1984* At concentration of 100 mg/L had no effect on GPO-PAP method. *3393* At concentration of 250 mg/L had no effect on lipase/esterase method. *3393*

Urea Nitrogen *Serum No Effect Analytical* No effect at therapeutic concentrations on glutamate dehydrogenase, phenol- hypochlorite and diacetylmonoxime methods. *1786* At acute overdose concentration (20 mg/dL) on Technicon® SMAC® method. *3719* No effect at therapeutic concentration on Reflotron method. *1984* At concentration of 250 mg/L had no effect on diacetylmonoxime method. *3393* At concentration of 30 mg/L had no effect on Ektachem® method. *3393*

Uric Acid *Serum No Effect Analytical* At 50 mg/L on reversed phase LC procedure of Zhiri et al. *3942* No effect at therapeutic concentrations on uricase-catalase, aldehyde dehydrogenase, direct UV-test and phosphotungstate methods. *1786* At acute overdose concentration (20 mg/dL) on Technicon® SMAC® method. *3719* No effect at therapeutic concentration on Reflotron method. *1984* At concentration of 80 mg/L had no effect on Kageyama-Hantzsch method. *3393* At concentration of 250 mg/L had no effect on phosphotungstate reduction method. *3393* At concentration of 100 mg/L had no effect on uricase-PAP method. *3393*

Valproic Acid *Serum Decrease Physiological* Concentration reduced to about 75% when phenobarbital coadministered. *2340* Significant reduction in half-life from 10.9 h to 8.2 h due to enzyme induction. *2233*

Vitamin A *Serum No Effect Physiological* No difference between treated and untreated cognitively delayed children. *2003*

Vitamin B₁₂ *CSF Decrease Physiological* Significantly lower with long term therapy. *1192*

Vitamin E (Tocopherol) *Serum Decrease Physiological* Mean of 0.58 mg/dL versus 0.67 mg/dL in untreated control of handicapped children. *1587*

Zinc *Serum No Effect Physiological* Observed in both hospitalized and home-living patients with long-term administration. *3803* No significant difference on average between treated and untreated handicapped children, although hypozincemia in 7 of 32 treated children and 0 of 13 controls. *1587*

Phenolphthalein

BSP Retention *Serum Increase Analytical* Color development on alkalinization of sample. *3879*

Color *Urine Increase Analytical* Pink, red, purple (alkaline), orange, rust (acid). *0382*
Feces Increase Analytical Imparts red color. *0071*

Erythrocytes *Urine Increase Physiological* May cause acute nephrosis (K deficiency). *2427*

Estradiol *Urine Increase Analytical* Affects unmodified Brown/Kober procedure. *0583*

Estriol *Urine Decrease Analytical* Interferes with acidic and enzyme hydrolysis. *0406*

Estrogens *Urine Increase Analytical* Affects colorimetric and fluorometric procedures. *0022* Reacts in Brown's procedure. *0279*
Urine Decrease Analytical Competes for enzyme used for hydrolysis. *0022*

Fat *Feces Increase Physiological* Steatorrhea observed in some laxative abusers. *3023*

Glucose *Serum Increase Physiological* Impaired glucose tolerance may occur due to K loss. *0706*

Glucose Tolerance *Serum Decrease Physiological* Result of hypokalemia. *0706*

Hemoglobin *Urine Increase Physiological* May cause acute nephrosis (K deficiency). *2427*

^{131}I Uptake *Serum Decrease Physiological* Phenolax contains tetraiodofluorescein. *2652*

Ketones *Urine Increase Analytical* Presumed false positive from color. *1487* Pink with Rothera, Ketostix® and Acetest® tests. *0121*

LE Cells *Blood Positive Physiological* Associated with hypersensitivity. *2427*

Occult Blood *Feces Positive Physiological* Protracted treatment may cause ulceration. *0612*

Platelets *Blood Decrease Physiological* Immunologically induced thrombocytopenia. *1877*

Potassium *Serum Decrease Physiological* If chronic laxative abuse and aldosteronism. *1148*

Protein *Urine Increase Physiological* May cause acute nephrosis (K deficiency). *2427*

Prothrombin Time *Plasma Increase Physiological* Patients on coumarins. *2427*

PSP Excretion *Urine Increase Analytical* Color development on alkalinization of urine. *3879*

Phenols

Bilirubin *Serum No Effect Analytical* Spectropolarographic procedure of Grahnen. *1374*

Casts *Urine Increase Physiological* Nephrotoxicity with poisoning (usually RBC casts). *1343*

Color *Urine Increase Analytical* Dark green to brownish black on standing. *2313*

Epinephrine *Plasma Increase Analytical* Affects trihydroxyindole method. *1176*

Erythrocytes *Urine Increase Physiological* Occur with renal damage of poisoning. *1302*

Ferric Chloride Test *Urine Positive Analytical* Derivatives may produce violet color. *0443*

Glucose *Serum No Effect Analytical* At 10 mg/dL no effect on glucose oxidase procedure of Gochman. *1311* Minimal effect at 10 mg/dL glucose oxidase dianisidine procedure. *3037* At 10 mg/dL on alkaline ferricyanide procedure. *1311*
Urine No Effect Analytical With DNSA method at usual concentrations. *0076*

Hemoglobin *Urine Increase Physiological* Nephrotoxicity with poisoning. *1343*

Ionized Calcium *Serum Decrease Analytical* At 0.1 mmol/L to 0.1 mol/L on Calcium specific electrode. *0540*

Millons Test *Test Conditions Positive Analytical* Positive spot test for phenols. *3765*

Norepinephrine *Plasma Increase Analytical* Affects trihydroxyindole method. *1176*

Phenol *Urine Increase Physiological* Quantitative relationship between exposure and urine concentration. *2661*

Phenylketones *Urine Positive Analytical* Violet with FeCl$_3$, nil with Phenistix®. *0775*

Protein *Urine Increase Physiological* Nephrotoxicity with poisoning. *1343*

Salicylate *Serum No Effect Analytical* With 25 mg/mL no effect on Trinder procedure. *2251*

Sugar *Urine Increase Analytical* Interferes with Benedict's reagent. *1022*

Sulfa as Sulfanilamide *Serum Increase Analytical* Reacts in Bratton-Marshall procedure. *0279*

Urobilinogen *Test Conditions Increase Analytical* Positive spot test with Ehrlich's reagent. *3765*

Phenolsulfonphthalein

Acetone *Urine Increase Analytical* Chromogenicity in color reaction (Rothera test). *3879*

BSP Retention *Serum Increase Analytical* May increase colorimetric reading. *1488*

Color *Urine Increase Analytical* Purple red or pink in alkaline urine. *1187*

Creatine *Urine Increase Analytical* Chromogenicity in color reaction. *2313*

Creatinine *Serum Increase Analytical* Chromogenicity in color reaction. *3505*
Urine Increase Analytical Interference with Jaffé procedure. *0652*

Ketones *Urine Increase Analytical* Red-purple color with alkaline nitroprusside. *0583* Presumed false positive from color. *1487*

Leucine Aminopeptidase *Urine Increase Physiological* Facilitates permeation of enzyme into tubules. *2897*

Uric Acid *Serum Decrease Physiological* Uricosuric action. *1433*
Urine Increase Physiological Uricosuric action. *1433*

Vanillylmandelic Acid *Urine Increase Analytical* May interfere with colorimetric method. *0086*

Phenothiazines

Acetoacetate *Urine Increase Analytical* Metabolites react with FeCl$_3$. *3879*

Alanine Aminotransferase *Serum Increase Physiological* May cause cholestatic hepatitis (in up to 4% patients). *2313* Liver damage observed after different phenothiazines administered in succession. *1916*
Serum Decrease Physiological Observed in treatment of female schizophrenics. *0289*

Aldolase *Serum Decrease Physiological* In schizophrenics with high initial values. *2427*

Alkaline Phosphatase *Serum Increase Physiological* Intrahepatic cholestatic syndrome. *1563* Liver damage observed after different phenothiazines administered in succession. *1916*

Aspartate Aminotransferase
Serum Increase Physiological May cause cholestatic hepatitis (in up to 4% patients). *2313*

Bile *Urine No Effect Analytical* Ictotest® unaffected. *1487*
Urine Increase Analytical Alleged interference with BiliLabstix®. *1488*
Urine Increase Physiological May cause cholestatic hepatitis. *2313*

Bilirubin *Serum Increase Physiological* Hypersensitivity cholestatic reaction may occur. *1343* Liver damage observed after different phenothiazines administered in succession. *1916*
Urine Increase Physiological May cause cholestatic hepatitis. *2313*

Bilirubin, Direct *Serum Increase Physiological* Relatively large increase compared with total. *1343*

BSP Retention *Serum Increase Physiological* May cause cholestatic hepatitis (in up to 4% patients). *2313*

Catecholamines *Plasma Increase Physiological* Increased metabolism, decreased organ uptake of norepinephrine. *0127*

Cephalin Flocculation *Serum Increase Physiological* May cause cholestatic hepatitis (in up to 4% patients). *2313*

Cholesterol *Serum No Effect Analytical* At concentration of 150 mg/L had no effect on CHOD-Iodide method. *3393* At concentration of 150 mg/L had no effect on CHOD-PAP method. *3393* At concentration of 150 mg/L had no effect on catalase-AIDH method. *3393* At concentration of 150 mg/L had no effect on method using catalase-Hantzsch reaction. *3393* At concentration of 150 mg/L had no effect on Liebermann-Burchard method. *3393*
Serum Increase Physiological Frequently reported effect, ?mechanism. *2946*

Phenothiazines (continued)

Color *Urine Increase Analytical* Pink, red, purple, orange, rust color. 2425

Creatine Kinase *Serum Increase Physiological* Probable effect of i.m. injection. 1488
Serum Decrease Physiological In schizophrenics with high initial values. 2427

Creatinine *Serum No Effect Analytical* At concentration of 30 mg/L had no effect on Jaffé-Fading-Fraction method. 3393 At concentration of 30 mg/L had no effect on Jaffé-Fuller's earth method. 3393 At concentration of 200 mg/L had no effect on kinetic Jaffé method on BKA-2. 3393

Diphenylhydantoin *Serum Decrease Physiological* Decreased by more than 40% with start of drug or increased dose. 1455

Dopa Screening Test *Urine Positive Analytical* May produce false color (usually buff/amber). 3075

Eosinophils *Blood Increase Physiological* Allergic response. 2313

Erythrocytes *Blood Decrease Physiological* Hemolytic anemia (sensitivity dependent). 0788

Estrogens *Urine Decrease Physiological* Block ovulation, inhibit decidual reaction. 1022

Ferric Chloride Test *Urine Positive Analytical* pink/purple color from metabolites. 0583

Follicle Stimulating Hormone (FSH)
Plasma Decrease Physiological Stimulation effect of gonadotropins inhibited. 2220

Glucose *Serum No Effect Analytical* At concentration of 20 mg/L had no effect on Ektachem® method. 3393
Serum Increase Physiological Probable effect, adrenergic response. 1487
Urine Increase Physiological Effect seen especially in long term therapy of diabetics. 2497 Long term effect in some patients. 1487

Glucose Tolerance *Serum Decrease Physiological* May produce diabetic type of curve in normals. 2220

γ-Glutamyltransferase (GGT)
Serum Increase Physiological Liver damage observed after different phenothiazines administered in succession. 1916

Gonadotropins *Urine Decrease Physiological* Associated with other endocrinological changes. 2427

Hematocrit *Blood Decrease Physiological* Hemolytic anemia (sensitivity dependent). 0788

Hemoglobin *Blood Decrease Physiological* Hemolytic anemia (sensitivity dependent). 0788

17-Hydroxycorticosteroids *Urine Increase Analytical* Slight increased absorbance modified Glenn-Nelson method. 0427
Urine Decrease Physiological Inhibit release of steroid hormones. 2220

5-Hydroxyindoleacetic Acid (5-HIAA)
Urine Decrease Analytical False decrease if nitrosonaphthol used. 3879

131I Uptake *Serum No Effect Physiological* No effect observed even with prolonged use. 1915
Serum Increase Physiological Reported effect in hyperthyroidism. 1488

17-Ketogenic Steroids *Urine Increase Analytical* Yield similar color with Zimmermann reaction. 2220

Ketones *Urine Increase Analytical* Pink or purple with Gerhardt's test. 0583

17-Ketosteroids *Urine Increase Analytical* Increased absorbance abnormal color Zimmermann procedure. 0427
Urine Decrease Physiological Inhibit release of steroid hormones. 2220

Leukocytes *Blood Increase Physiological* Due to eosinophilia or generalized leukocytosis. 2313
Blood Decrease Physiological Agranulocytosis/leukopenia especially if low initially. 2113

Luteinizing Hormone (LH) *Plasma Decrease Physiological* Stimulation effect of gonadotropins inhibited. 2220

Metanephrines, Total *Urine Increase Analytical* Interference in Pisano procedure. 0388

5'-Nucleotidase *Serum Increase Physiological* Due to cholestasis. 1778

Phenobarbital *Serum Decrease Physiological* But not to quite same extent as for phenytoin. 1455

Phenothiazine *Urine Positive Physiological* Some detectable up to 18 mo after therapy. 1343

Phenylketones *Urine Positive Analytical* pink/red-purple with FeCl3, same with Phenistix®. 0775

Phosphate *Serum Decrease Analytical* Affect methods using phosphomolybdate. 1013

Platelets *Blood Decrease Physiological* Hemolytic anemia (sensitivity dependent). 0788

Porphobilinogen *Urine Increase Analytical* May react with Ehrlich's aldehyde reagent. 2950

Potassium *Serum No Effect Analytical* At concentration of 200 mg/L had no effect on measurement by ISE without predilution. 3393

Pregnancy Tests *Urine Positive Analytical* False react with frog, rabbit, immunological tests. 2719

Pregnanediol *Urine Decrease Physiological* Reported effect. 2220

Primidone *Serum No Effect Physiological* No significant effect on serum concentration. 1455

Progesterone *Urine Decrease Physiological* Associated with other endocrinological changes. 2427

Protein *CSF Increase Analytical* False positive with Folin-Ciocalteu reagent. 1237

Protein Bound Iodine (PBI) *Serum Increase Physiological* Due to increased thyroxine binding globulin with prolonged use. 1915
Serum Decrease Physiological Antithyroid effect of large doses. 0830

Protein Electrophoresis *Serum No Effect Analytical* At concentration of 200 mg/L had no effect on automated Olympus-Hite method. 3393

Pseudocholinesterase *Serum Decrease Physiological* Also affects erythrocyte enzyme. 1487

Sodium *Serum No Effect Analytical* At concentration of 200 mg/L had no effect on measurement by ISE without predilution. 3393

Sorbitol Dehydrogenase *Serum Decrease Physiological* Observed with therapy of female schizophrenics. 0289

T3 Uptake *Serum Decrease Physiological* Due to increased thyroxine binding globulin with prolonged use. 1915

Thymol Turbidity *Serum Increase Physiological* May cause cholestatic hepatitis (in up to 4% patients). 2313

Thyroxine (T4) *Serum Decrease Physiological* Observed with long-term treatment but cause not clear. 3798

Thyroxine (T4) Binding Globulin
Serum Increase Physiological Occurs with prolonged use. 1915

Urea Nitrogen *Serum No Effect Analytical* At concentration of 20 mg/L had no effect on Ektachem® method. 3393
Serum Decrease Physiological Decreased urea production if hepatic cirrhosis occurs. 0652

Uric Acid *Serum No Effect Analytical* At concentration of 30 mg/L had no effect on Kageyama-Hantzsch method. 3393 At concentration of 30 mg/L had no effect on catalase-AIDH method. 3393
Serum Increase Physiological Reported effect. 2754
Serum Decrease Physiological Reported uricosuric action. 2220
Urine Increase Physiological Reported uricosuric action. 2220

Urobilinogen *Urine Increase Analytical* May react with Ehrlich's aldehyde reagent. 2950

Vanillylmandelic Acid *Urine Decrease Physiological* Alters blood concentration. 0127

Phenoxybenzamine

Catecholamines *Urine No Effect Analytical* No effect on fluorometric method of Crout. 0758
Urine No Effect Physiological No effect observed. 0913

Sodium *Serum Decrease Physiological* Isolated case. Presumed to be due to drug-induced inappropriate release of vasopressin. 0159

Vanillylmandelic Acid *Urine No Effect Physiological* No effect observed. *0913*

Phenoxymethicillin

Bilirubin *Serum Increase Physiological* Possible clinically significant displacement from protein especially in especially in critically ill neonates. *3748*

Phenoxypropazine

Alanine Aminotransferase *Serum Increase Physiological* Produced fatal hepatotoxicity. *2427*

Alkaline Phosphatase *Serum Increase Physiological* Produced fatal hepatotoxicity. *2427*

Aspartate Aminotransferase

Serum Increase Physiological Produced fatal hepatotoxicity. *2427*

Bilirubin *Serum Increase Physiological* Produced fatal hepatotoxicity. *2427*

Phenprocoumon

Alanine Aminotransferase *Serum No Effect Analytical* No effect at therapeutic concentration on Reflotron method. *1984* *Serum Increase Physiological* Mild effect but recurrently in 2 patients having repeated exposure to drug. *3347*

Alkaline Phosphatase *Serum Increase Physiological* Mild effect but recurrently in 2 patients having repeated exposure to drug. *3347*

Aspartate Aminotransferase *Serum No Effect Analytical* No effect at therapeutic concentration on Reflotron method. *1984*
Serum Increase Physiological Mild effect but recurrently in 2 patients having repeated exposure to drug. *3347*

Bilirubin *Serum Increase Physiological* Mild effect but recurrently in 2 patients having repeated exposure to drug. *3347*

Bilirubin, Direct *Serum Increase Physiological* Mild effect but recurrently in 2 patients having repeated exposure to drug. *3347*

BSP Retention *Serum Increase Physiological* Mild effect but recurrently in 2 patients having repeated exposure to drug. *3347*

Cholesterol *Serum No Effect Analytical* No effect at therapeutic concentration on Reflotron method. *1984* At concentration of 80 mg/L had no effect on CHOD-Iodide method. *3393* At concentration of 80 mg/L had no effect on CHOD-PAP method. *3393* At concentration of 80 mg/L had no effect on catalase-AIDH method. *3393* At concentration of 80 mg/L had no effect on method using catalase-Hantzsch reaction. *3393* At concentration of 80 mg/L had no effect on Liebermann-Burchard method. *3393*

Creatinine *Serum No Effect Analytical* At concentration of 6 mg/L had no effect on Jaffé-Fading-Fraction method. *3393* At concentration of 6 mg/L had no effect on Jaffé-Fuller's earth method. *3393* At concentration of 26 mg/L had no effect on kinetic Jaffé method on BKA-2. *3393*

Diphenylhydantoin *Serum Increase Physiological* Half-life increased by 40% in three patients. *3556*

Glucose *Serum No Effect Analytical* No effect at therapeutic concentration on Reflotron method. *1984* At concentration of 0.36 mg/L had no effect on Ektachem® method. *3393* At concentration of 2 mg/L had no effect on hexokinase/G-6-PDH method. *3393*

γ-Glutamyltransferase (GGT) *Serum No Effect Analytical* No effect at therapeutic concentration on Reflotron method. *1984*
Serum Increase Physiological Mild effect but recurrently in 2 patients having repeated exposure to drug. *3347*

Lactate *Serum No Effect Analytical* At concentration of 2 mg/L had no effect on enzymatic method. *3393*

Lactate Dehydrogenase *Serum Increase Physiological* Mild effect but recurrently in 2 patients having repeated exposure to drug. *3347*

Potassium *Serum No Effect Analytical* At concentration of 3.6 mg/L had no effect on measurement by ISE without predilution. *3393*

Protein Electrophoresis *Serum No Effect Analytical* At concentration of 3.6 mg/L had no effect on automated Olympus-Hite method. *3393*

Sodium *Serum No Effect Analytical* At concentration of 3.6 mg/L had no effect on measurement by ISE without predilution. *3393*

Triglycerides *Serum No Effect Analytical* No effect at therapeutic concentration on Reflotron method. *1984*

Urea Nitrogen *Serum No Effect Analytical* No effect at therapeutic concentration on Reflotron method. *1984* At concentration of 0.36 mg/L had no effect on Ektachem® method. *3393*

Uric Acid *Serum No Effect Analytical* No effect at therapeutic concentration on Reflotron method. *1984* At concentration of 6 mg/L had no effect on Kageyama-Hantzsch method. *3393* At concentration of 6 mg/L had no effect on catalase-AIDH method. *3393*

Phenpromethamine

Amphetamine *Urine Increase Analytical* Yields fluorophor with NBD chloride reaction. *2473*

Phensuximide

Albumin *Serum No Effect Analytical* At concentration of 120 mg/L had no effect on BCG method. *3393*

Bicarbonate *Serum No Effect Analytical* At concentration of 120 mg/L had no effect on method using phenolphthalein. *3393*

Bilirubin *Serum No Effect Analytical* At concentration of 120 mg/L had no effect on Jendrassik and Grof method. *3393*

Calcium *Serum No Effect Analytical* At concentration of 120 mg/L had no effect on cresolphthalein method. *3393*

Chloride *Serum No Effect Analytical* At concentration of 120 mg/L had no effect on mercurimetric method. *3393*

Cholesterol *Serum No Effect Analytical* At concentration of 120 mg/L had no effect on Liebermann-Burchard method. *3393*

Color *Urine Increase Analytical* Pink, red, purple, orange and rust color. *2425*

Creatinine *Serum No Effect Analytical* At concentration of 120 mg/L had no effect on AutoAnalyzer Jaffé method. *3393*

Erythrocytes *Urine Increase Physiological* Hematuria and renal damage reported. *1680*

[131]I Uptake *Serum Decrease Physiological* Milontin® contains tetraiodofluorescein. *2652*

Leukocytes *Blood Decrease Physiological* Possible agranulocytosis (very rare). *0071*

Phosphate *Serum No Effect Analytical* At concentration of 120 mg/L had no effect on phosphomolybdate method. *3393*

Potassium *Serum No Effect Analytical* At concentration of 120 mg/L had no effect on measurement by ISE with predilution. *3393*

Protein *Serum No Effect Analytical* At concentration of 120 mg/L had no effect on biuret method with blank correction. *3393*
Urine Increase Physiological Reversible nephropathy (especially in children). *0071*

Sodium *Serum No Effect Analytical* At concentration of 120 mg/L had no effect on measurement by ISE with predilution. *3393*

Triglycerides *Serum No Effect Analytical* At concentration of 120 mg/L had no effect on lipase/esterase method. *3393*

Urea Nitrogen *Serum No Effect Analytical* At concentration of 120 mg/L had no effect on diacetylmonoxime method. *3393*
Serum Increase Physiological Reversible nephropathy (especially in children). *0071*

Uric Acid *Serum No Effect Analytical* At concentration of 120 mg/L had no effect on phosphotungstate reduction method. *3393*

Phentermine

Amobarbital *Urine No Effect Analytical* No interference on TLC using ethylacetate: methanol: water: ammonium hydroxide and modified Dragendorff's reagent for detection. *3868*

Amphetamine *Urine No Effect Analytical* At 50 mg/L on fluorescent method of Hayes. *1534* No interference on TLC using ethylacetate: methanol: water: ammonium hydroxide and modified Dragendorff's reagent for detection. *3868*

Chlorphenmetrazine *Urine Positive Analytical* Similar R_f and color reaction on TLC using ethyl acetate: methanol: water: ammonium hydroxide and modified Dragendorff's reagent. *3868*

Hydromorphone *Urine No Effect Analytical* No interference on TLC using ethylacetate: methanol: water: ammonium hydroxide and modified Dragendorff's reagent for detection. *3868*

Mescaline *Urine No Effect Analytical* No interference on TLC using ethylacetate: methanol: water: ammonium hydroxide and modified Dragendorff's reagent for detection. *3868*

Methamphetamine *Urine No Effect Analytical* No interference on TLC using ethylacetate: methanol: water: ammonium hydroxide and modified Dragendorff's reagent for detection. *3868*

Morphine *Urine No Effect Analytical* No interference on TLC using ethylacetate: methanol: water: ammonium hydroxide and modified Dragendorff's reagent for detection. *3868*

Pentobarbital *Urine No Effect Analytical* No interference on TLC using ethylacetate: methanol: water: ammonium hydroxide and modified Dragendorff's reagent for detection. *3868*

Phenmetrazine *Urine Positive Analytical* Similar R_f and color reaction on TLC using ethyl acetate: methanol: water: ammonium hydroxide and modified Dragendorff's reagent. *3868*

Phenobarbital *Urine No Effect Analytical* No interference on TLC using ethylacetate: methanol: water: ammonium hydroxide and modified Dragendorff's reagent for detection. *3868*

Phenylpropanolamine *Urine No Effect Analytical* No interference on TLC using ethylacetate: methanol: water: ammonium hydroxide and modified Dragendorff's reagent for detection. *3868*

Quinine *Urine Positive Analytical* Similar R_f and color reaction on TLC using ethyl acetate: methanol: water: ammonium hydroxide and modified Dragendorff's reagent. *3868*

Secobarbital *Urine No Effect Analytical* No interference on TLC using ethylacetate: methanol: water: ammonium hydroxide and modified Dragendorff's reagent for detection. *3868*

Phentolamine

Catecholamines *Plasma Increase Physiological* ?due to releasing action or to altered metabolism. *1343*
Urine No Effect Analytical No effect on fluorometric method of Crout. *0758*
Urine No Effect Physiological No effect observed. *0913*

Glucose *Serum Decrease Physiological* Toxic doses over long period of time. *1343*

5-Hydroxyindoleacetic Acid (5-HIAA)
Urine Increase Analytical May cause falsely high colorimetric values. *0913*

Vanillylmandelic Acid *Urine No Effect Physiological* No effect observed. *0913*

Phenylalanine

Amino Acids *Test Conditions Increase Analytical* Positive spot test with ninhydrin. *3765*

Ammonia *Plasma Decrease Analytical* With 250 nmoles produces 14% inhibitor indophenol reaction. *1290*

Insulin *Plasma Increase Physiological* Slight effect, AIDS metabolism of amino acids. *1071*

Probenecid *CSF No Effect Analytical* No effect on method of Korf, Van Praag. *2373*

Proline *Plasma No Effect Analytical* No effect on method of Goodwin. *1345*

Pyruvate Kinase *Red Blood Cells Decrease Analytical* Most marked with larger side chain. *0370*

Tyrosine *Plasma No Effect Analytical* On fluorometric procedure of Ambrose. *0080*

Urea Nitrogen *Serum No Effect Analytical* No effect on Berthelot reaction. *1862*

Phenylbutazone

Alanine Aminotransferase *Serum Increase Physiological* May cause cholestatic and cytotoxic jaundice. *0248* Frequent hypersensitivity reaction but also with hepatotoxic potential: may be hepatocellular injury or systemic vasculitis. *0290*

Albumin *Serum No Effect Analytical* At concentration of 750 mg/L had no effect on BCG method. *3393*

Alkaline Phosphatase *Serum Increase Physiological* Hepatitis may occur as complication of therapy. *2534* Observed in 7 patients with side effects due to drug. *3409* In 2 patients subsided after drug withdrawn associated with granulomas and cholestatic-hepatocellular injury. *1730* Observed in one patient: clear evidence of intrahepatic cholestasis. *3670* Frequent hypersensitivity reaction but also with hepatotoxic potential: may be hepatocellular injury or systemic vasculitis. *0290*

Amylase *Serum Increase Physiological* Parotitis may occur as rare complication. *1412* Sialadenitis reported to occur occasionally but in 5 of 7 patients in this study. *3409*

Aspartate Aminotransferase
Serum Increase Physiological May cause cholestatic and cytotoxic jaundice. *0248* Observed in 7 patients with side effects due to drug. *3409* In 2 patients subsided after drug withdrawn associated with granulomas and cholestatic-hepatocellular injury. *1730* Frequent hypersensitivity reaction but also with hepatotoxic potential: may be hepatocellular injury or systemic vasculitis. *0290*

Bicarbonate *Serum No Effect Analytical* At concentration of 750 mg/L had no effect on method using phenolphthalein. *3393*

Bilirubin *Serum No Effect Analytical* At concentration of 750 mg/L had no effect on Jendrassik and Grof method. *3393*
Serum Increase Physiological Granulomatous reaction in liver. *1338* Clinically significant displacement from protein in neonates. *3748* Observed in 7 patients with side effects due to drug. *3409* In 2 patients subsided after drug withdrawn associated with granulomas and cholestatic-hepatocellular injury. *1730* Observed in one patient: clear evidence of intrahepatic cholestasis. *3670* Frequent hypersensitivity reaction but also with hepatotoxic potential: may be hepatocellular injury or systemic vasculitis. *0290*

Bilirubin, Direct *Serum Increase Physiological* In 2 patients subsided after drug withdrawn, associated with granulomas and cholestatic-hepatocellular injury. *1730* Observed in one patient; clear evidence of intrahepatic cholestasis. *3670*

BSP Retention *Serum Increase Physiological* May cause cholestasis. *2313*

Calcium *Serum No Effect Analytical* At concentration of 750 mg/L had no effect on cresolphthalein method. *3393*

Cephalin Flocculation *Serum Increase Physiological* May cause cholestasis. *2313*

Chloride *Serum No Effect Analytical* At concentration of 750 mg/L had no effect on mercurimetric method. *3393*
Serum Increase Physiological May cause salt retention (?tubular dysfunction). *3687*

Cholesterol *Serum No Effect Analytical* At concentration of 280 mg/L had no effect on CHOD-Iodide method. *3393* At concentration of 280 mg/L had no effect on CHOD-PAP method. *3393* At concentration of 280 mg/L had no effect on catalase-AIDH method. *3393* At concentration of 280 mg/L had no effect on method using catalase-Hantzsch reaction. *3393* At concentration of 750 mg/L had no effect on Liebermann-Burchard method. *3393*
Serum Increase Physiological Observed in one patient: clear evidence of intrahepatic cholestasis. *3670*

Coombs' Test *Serum Positive Physiological* Immunological response to drug. *1486*

Coombs' Test, Direct *Serum Positive Physiological* Single case after 2 weeks therapy. *1487*

Creatinine *Serum No Effect Analytical* At concentration of 750 mg/L had no effect on AutoAnalyzer Jaffé method. *3393* At concentration of 120 mg/L had no effect on Jaffé-Fading-Fraction method. *3393* At concentration of 120 mg/L had no effect on Jaffé-Fuller's earth method. *3393* At concentration of 400 mg/L had no effect on kinetic Jaffé method on BKA-2. *3393*
Serum Increase Physiological May increase especially if coexisting renal damage. *2427*

Digitoxin *Serum Decrease Physiological* Due to induction of mixed function oxidases which may reduce to subtherapeutic level. *2794*

Diphenylhydantoin *Serum Increase Physiological* Reported impairment of metabolism. *2042*

Eosinophils *Blood Increase Physiological* In 2 patients subsided after drug withdrawn associated with granulomas and cholestatic-hepatocellular injury. *1730*

Erythrocytes *Blood Decrease Physiological* Bone marrow depression/secondary to Na and H_2O retention. *0566* Occasional case of aplastic anemia reported. *3717* Fatal aplastic anemia in 2.2 per 100,000. *1723*
Urine Increase Physiological Actual bleeding caused by drug. *1022* Observed effect, may occasionally be marked with oliguria or renal failure. *2866*

Erythrocyte Sedimentation Rate
Blood Increase Physiological Observed in 7 patients with side effects due to drug. *3409*

γ-Globulin *Serum Increase Physiological* Observed in 2 cases (?due to antibodies). *2161*

Glucaric Acid *Urine Increase Physiological* May induce hepatic enzymes. *0002*

Glucose *Serum No Effect Analytical* At concentration of 12 mg/L had no effect on Ektachem® method. *3393* At concentration of 200 mg/L had no effect on GOD-Perid method. *3393* At concentration of 200 mg/L had no effect on GOD/POD-PAP method. *3393* At concentration of 200 mg/L had no effect on hexokinase/G-6-PDH method. *3393*
Serum Increase Physiological metabolic/endocrine response. *1680*
Urine Increase Physiological Observed effect, may occasionally be marked with oliguria or renal failure. *2866*

Granulocytes *Blood Decrease Physiological* Indicative of toxicity developing. *1680*

Guanase *Serum Increase Physiological* May cause cholestatic and cytotoxic jaundice. *0248*

Hematocrit *Blood Decrease Physiological* Bone marrow depression/secondary to Na and H_2O retention. *0566*

Hemoglobin *Blood Decrease Physiological* Bone marrow depression/secondary to Na and H_2O retention. *0566*
Urine Increase Physiological Actual bleeding caused by drug. *1022*

17-Hydroxycorticosteroids *Urine Decrease Physiological* Cortisol metabolism diverted to 6-beta-hydroxy cortisol. *2035*

6-β-Hydroxycortisol *Urine Increase Physiological* Long term change in steroid excretory pattern. *2035*

Indocyanine Green *Serum Decrease Physiological* Mechanism not yet established. *2398*

Isocitrate Dehydrogenase *Serum Increase Physiological* May cause cholestatic and cytotoxic jaundice. *0248*

^{131}I Uptake *Serum Decrease Physiological* May last up to 2 weeks. *3209*

Lactate Dehydrogenase *Serum Increase Physiological* In 2 patients subsided after drug withdrawn associated with granulomas and cholestatic-hepatocellular injury. *1730*

LE Cells *Blood Positive Physiological* May precipitate or exaggerate LE-like syndrome. *3505*

Leukocytes *Blood Increase Physiological* Leukemia drug induced. *2427*
Blood Decrease Physiological Agranulocytosis/aplastic anemia/leukopenia. *2930* Rare side effects reported to UK committee on safety of medicines. *0782* Occasional case of aplastic anemia reported. *3717*

NBT Test *Blood Decrease Physiological* Mechanism not discussed. *0943*

Neutrophils *Blood Decrease Physiological* Occasional case of aplastic anemia reported. *3717*

Occult Blood *Feces Positive Physiological* May cause gastrointestinal tract bleeding. *2313* Rare side effects reported to UK committee on safety of medicines. *0782*

Ornithine Carbamoyltransferase (OCT)
Serum Increase Physiological May cause cholestatic and cytotoxic jaundice. *0248*

pH *Blood Increase Physiological* May cause respiratory or metabolic alkalosis. *1680*

Phenprocoumon *Plasma Increase Physiological* Increased concentration of anticoagulant displaced from protein with increase of free component. *2642*

Phosphate *Serum No Effect Analytical* On nonprecipitation method of Peynet. *2799* At concentration of 750 mg/L had no effect on phosphomolybdate method. *3393*

Platelet Aggregation *Blood Decrease Physiological* Lack second phase ADP agglutination. *0834*

Platelets *Blood Decrease Physiological* May cause aplastic anemia or thrombocytopenia. *0982* Rare side effects reported to UK committee on safety of medicines. *0782* Several cases of immune-mediated thrombocytopenia reported. *2502* Occasional case of aplastic anemia reported. *3717*

Potassium *Serum No Effect Analytical* At concentration of 750 mg/L had no effect on measurement by ISE with predilution. *3393* At concentration of 120 mg/L had no effect on measurement by ISE without predilution. *3393*

Protein *Serum No Effect Analytical* At concentration of 750 mg/L had no effect on biuret method with blank correction. *3393*
Urine Increase Physiological Reported nephrotoxic effect. *1488* Observed effect, may occasionally be marked with oliguria or renal failure. *2866*

Protein Bound Iodine (PBI) *Serum No Effect Physiological* Probably usual effect. *1487*
Serum Decrease Physiological Competes with thyroxine for binding on thyroxine binding globulin. *0830*

Protein Electrophoresis *Serum No Effect Analytical* At concentration of 120 mg/L had no effect on automated Olympus-Hite method. *3393*

Prothrombin Time *Plasma Increase Physiological* Potentiates action of anticoagulants. *0030* Increased concentration of anticoagulant displaced from protein with increase of free component. *2642* Displaces warfarin from protein binding sites. *2929*

Sodium *Serum No Effect Analytical* At concentration of 750 mg/L had no effect on measurement by ISE with predilution. *3393* At concentration of 120 mg/L had no effect on measurement by ISE without predilution. *3393*
Serum Increase Physiological May cause salt retention. *3687*

T$_3$ Uptake *Serum Increase Physiological* Resin and red cell uptake affected. *3209*

Taurine *Urine Decrease Physiological* Reduces increased output in rheumatoids. *3102*

Thymol Turbidity *Serum Increase Physiological* May cause cholestasis. *2313*

Thyro-Binding Index *Serum Decrease Physiological* For up to 3 weeks after therapy. *0913*

Thyroxine (T$_4$) *Serum No Effect Analytical* No effect on competitive protein binding procedure of Alexander. *0054*
Serum Decrease Physiological Due to impaired synthesis, competes for thyroxine binding albumin. *1915* Reported in two studies due to competition for transport proteins. *3798*

Thyroxine (T$_4$), Free *Serum Increase Physiological* Insignificant short-term increase associated with competition for transport proteins. *3798*
Serum Decrease Physiological Long-term effect due to thyroidal inhibition. *3798*

Thyroxine (T$_4$) Index, Free (FTI)
Serum Decrease Physiological Low normal or slight decrease for 3 weeks after therapy. *0913*

Thyroxine (T$_4$) (Murphy-Pattee)
Serum Decrease Physiological Binds to thyroxine binding albumin, has antithyroid activity. *0583*

Tolbutamide *Serum Increase Physiological* Displaces from protein inhibits carboxylation. *1487*

Phenylbutazone (continued)

Triglycerides *Serum No Effect Analytical* At concentration of 200 mg/L had no effect on GPO-PAP method. *3393* At concentration of 750 mg/L had no effect on lipase/esterase method. *3393*

Urea Nitrogen *Serum No Effect Analytical* At concentration of 750 mg/L had no effect on diacetylmonoxime method. *3393* At concentration of 12 mg/L had no effect on Ektachem® method. *3393*
Serum Increase Physiological Reported nephrotoxic effect. *2313*

Uric Acid *Serum No Effect Analytical* At concentration of 200 mg/L had no effect on Kageyama-Hantzsch method. *3393* At concentration of 120 mg/L had no effect on catalase-AIDH method. *3393* At concentration of 750 mg/L had no effect on phosphotungstate reduction method. *3393* At concentration of 200 mg/L had no effect on uricase-PAP method. *3393*
Serum Increase Physiological Antiuricosuric action at low doses. *1433*
Serum Decrease Physiological Uricosuric action at high doses. *1433*
Urine Increase Physiological Uricosuric action. *1433*

Volume *Plasma Increase Physiological* May increase by 50% due to salt and water retention. *1343*
Urine Decrease Physiological Causes significant water retention. *1343*

Warfarin *Plasma Increase Physiological* Displaces warfarin from protein binding sites. *0528*

Phenylenediamine

Methemoglobin *Blood Increase Physiological* May cause hemolysis. *2427*

Phenylephrine

Albumin *Serum No Effect Analytical* At concentration of 4 mg/L had no effect on BCG method. *3393*

Alkaloids *Urine No Effect Analytical* With iodoplatinate of Frings TLC procedure at 10 mg/dL. With ninhydrin on Frings TLC procedure at 1 mg/dL. *1204*

Amino Acids *Urine Increase Analytical* Reacts with ninhydrin; extra spot TLC, high voltage electrophoresis. *2855*
Urine Positive Analytical Orange-brown spot with ninhydrin on thin-layer chromatography. *1278*

Amphetamine *Urine No Effect Analytical* No effect at 100 mg/L on method of Rutter. *3097*
Urine Increase Analytical Yields fluorophor with NBD chloride reaction. *2473*

Barbiturate *Urine No Effect Analytical* With mercuric SO_4 on Frings TLC procedure at 10 mg/dL. With vanillin on Frings procedure at 10 mg/dL. With diphenylcarbazone on Frings TLC procedure at 10 mg/dL. With mercurous NO_3 on Frings TLC procedure at 10 mg/dL. *1204*

Bilirubin *Serum No Effect Analytical* At concentration of 4 mg/L had no effect on Jendrassik and Grof method. *3393*

Bromide *Urine No Effect Analytical* No effect on method of Frings. *1204*

Calcium *Serum No Effect Analytical* At concentration of 4 mg/L had no effect on cresolphthalein method. *3393*

Chloride *Serum No Effect Analytical* At concentration of 4 mg/L had no effect on mercurimetric method. *3393*

Cholesterol *Serum No Effect Analytical* At concentration of 4 mg/L had no effect on Liebermann-Burchard method. *3393*

Creatinine *Serum No Effect Analytical* At concentration of 4 mg/L had no effect on AutoAnalyzer Jaffé method. *3393*

Dapsone *Serum No Effect Analytical* At 10 mg/dL on colorimetric procedure of Higgins. *1590*

Ethchlorvynol *Urine No Effect Analytical* No effect on method of Frings and Cohen. *1204*

FPN Test *Urine No Effect Analytical* No effect at 10 mg/dL on method of Frings. *1204*

Glucose *Serum Increase Physiological* Reported to increase sugar. *2726*

Growth Hormone *Plasma No Effect Physiological* No effect with/without propranolol. *2329*

Metanephrines, Total *Urine No Effect Analytical* At 5 mg/L on modified Pisano procedure. *1428*
Urine Increase Analytical At concentration possibly 10 times higher than would be encountered when measured by Pisano method. *3417*

Occult Blood *Feces Positive Physiological* May = hemorrhagic enteritis with vasoconstriction. *3498*

Phosphate *Serum No Effect Analytical* At concentration of 4 mg/L had no effect on phosphomolybdate method. *3393*

Propoxyphene *Urine Increase Analytical* 1% fluorescence in procedure of Valentour. *3662*

Protein *Serum No Effect Analytical* At concentration of 4 mg/L had no effect on biuret method with blank correction. *3393*

Salicylate *Urine No Effect Analytical* At 10 mg/dL no effect on Trinder method. *1204*

Urea Nitrogen *Serum No Effect Analytical* At concentration of 4 mg/L had no effect on diacetylmonoxime method. *3393*

Uric Acid *Serum No Effect Analytical* At concentration of 4 mg/L had no effect on phosphotungstate reduction method. *3393*

Phenylhydrazine

Bilirubin *Serum Increase Physiological* Causes hemolysis. *0348*

Bilirubin, Direct *Serum Increase Physiological* Causes hemolysis. *0348*

Color *Urine Increase Physiological* May produce dark brown urine. *1187*

Erythrocytes *Blood Decrease Physiological* Hemolytic anemia. *2427*

Haptoglobin *Serum Decrease Physiological* May cause hemolysis (hemolytic anemia). *2427*

Heinz Body Formation *Blood Positive Physiological* May occur in early stages of hemolysis. *0333*

Hematocrit *Blood Decrease Physiological* Hemolytic anemia. *0788*

Hemoglobin *Blood Decrease Physiological* Hemolytic anemia. *0788*

Hydrazine *Plasma No Effect Analytical* On fluorometric method of Vickers at 1 μg/mL. *3710*

Methemoglobin *Blood Increase Physiological* May cause hemolysis. *2427*

Porphyrins *Urine Increase Physiological* May occasionally precipitate attack of porphyria. *3879*

Phenylmercuric Acetate

Uric Acid *Serum Decrease Analytical* At 280 mg/L on uricase procedure of Kabasak. *1840*

Phenylpropanolamine

Alkaloids *Urine No Effect Analytical* With iodoplatinate of Frings TLC procedure at 1 mg/dL. *1204*
Urine Increase Analytical Reacts with ninhydrin on TLC method of Frings. *1204*

Amino Acids *Urine Increase Analytical* Reacts with ninhydrin; extra spot TLC, high voltage electrophoresis. *2855*
Urine Positive Analytical Unusual ninhydrin reacting spot observed on TLC. *1596*

Amphetamine *Urine No Effect Analytical* At 50 mg/L on fluorescent method of Hayes. *1534*
Urine Increase Analytical Yields fluorophor with NBD chloride reaction. *2473* Reacts with methyl orange method of Frings. *2474*

Barbiturate *Urine No Effect Analytical* With mercurous NO_3 on Frings TLC procedure at 10 mg/dL. With mercuric SO_4 on Frings TLC procedure at 10 mg/dL. With diphenylcarbazone on Frings TLC procedure at 10 mg/dL. With vanillin on Frings procedure at 10 mg/dL. *1204*

Bromide *Urine No Effect Analytical* No effect on method of Frings. *1204*

Ethchlorvynol *Urine No Effect Analytical* No effect on method of Frings and Cohen. *1204*

FPN Test *Urine No Effect Analytical* No effect at 10 mg/dL on method of Frings. *1204*

Phentolamine Test
Pancreatic Juice Increase Physiological Single case reported. *0981*

Propoxyphene *Urine No Effect Analytical* Less than 1% fluorescence in procedure of Valentour. *3662*

Salicylate *Urine No Effect Analytical* At 10 mg/dL no effect on Trinder method. *1204*

Phenylpyruvic Acid

Ferric Chloride Test *Urine Positive Analytical* Transient blue-green color formed. *3588*

FPN Test *Urine Increase Analytical* Reported false positive in patients with liver dysfunction. *1204*

Ketones *Urine Increase Analytical* Reacts with alkaline nitroprusside. *0583* False positive with Ketostix®. *2308*

Phenyl Salicylate

Color *Urine Increase Analytical* Dark green. *2313*

Phenylthiourea

Protein *Test Conditions Increase Analytical* Reacts with Folin-Ciocalteu of Lowry method. *0702*

Phenylurea

Urea Nitrogen *Serum Increase Analytical* Absorbance relative to urea = 2.2 diacetyl procedure. *0583*

Phenyramidol

Cholesterol *Serum Decrease Physiological* Probable inhibition of hepatic microsomal enzymes. *3207*

Diphenylhydantoin *Serum Increase Physiological* May increase concentration from 7 to 12 μg/mL (dose of 1.2 g/d). *2041*

Glucose *Serum Decrease Physiological* Reported effect. *0117*

Prothrombin Time *Plasma Increase Physiological* Inhibits metabolism of bishydroxycoumarin. *0600*

Tolbutamide *Serum Increase Physiological* Probably impairs metabolism of drug. *1487*

Phenytoin

Acetaminophen *Urine Increase Physiological* Significantly greater proportion excreted as glucuronide in response to coadministration of drug. *0394*

Acetaminophen Screening Test *Urine Negative Analytical* No reaction with o-cresol at therapeutic concentrations. *3326*

Alanine Aminotransferase *Serum No Effect Analytical* At acute overdose concentration (20 mg/dL) on Technicon® SMAC® method. *3719* No effect at therapeutic concentration on Reflotron method. *1984*
Serum Increase Physiological Intrahepatic cholestatic jaundice. *2313*

Albumin *Serum No Effect Analytical* No effect at 1.8 mg/dL on SMA 12/60 method. *2636* At acute overdose concentration (20 mg/dL) on Technicon® SMAC® method. *3719* At concentration of 240 mg/L had no effect on BCG method. *3393*
Serum No Effect Physiological In about 20 patients treated for 2 y. *0873*

Alkaline Phosphatase *Serum No Effect Analytical* No effect at 1.8 mg/dL on SMA 12/60 method. *2636* At acute overdose concentration (20 mg/dL) on Technicon® SMAC® method. *3719*
Serum Increase Physiological Hepatotoxicity with centrolobular necrosis. *2313* Observed 2 times upper limit of normal in 6 of 112 adult epileptic in patients. *1162* In 42% of 60 epileptics treated for 10 y or more. *2493*

Alkaline Phosphatase Isoenzymes
Serum Increase Physiological Hepatic isoenzyme relatively greater increase than bone fraction in association with increased total enzyme activity (bone still largest single component). *1449*

Alkaloids *Urine No Effect Analytical* With ninhydrin on Frings TLC procedure at 10 mg/dL. With iodoplatinate on Frings TLC procedure at 10 mg/dL. *1204*

Amino-4-Imidazole-5-Carboxamide Ribotide (AICAR)
Urine Increase Physiological Occurs with impaired absorption of folic acid. *3773*

Aminolevulinic Acid *Serum Increase Physiological* Significant increase to 124 nmol/L from 99 nmol/L in controls. *1348*
Urine Increase Physiological May precipitate acute porphyria. *1322*

Ammonia *Plasma No Effect Analytical* At concentration of 20 mg/L had no effect on Ektachem® method. *3393*

Antinuclear Antibodies *Serum No Effect Physiological* Same frequency before and after drug therapy. *0228*
Serum Increase Physiological Occasional increase; possible subclinical collagen-vascular disorder. *1502*
Serum Positive Physiological Elicited in 25% subjects treated. *0616* Related to number of drugs, higher in women. *3797*

Apolipoprotein AI *Serum Increase Physiological* Mean of 2.1 g/L versus 1.8 g/L in controls in 28 patients treated with 200-300 mg/d for 1 to 35 y. *2598*

Apolipoprotein AII *Serum No Effect Physiological* No significant difference between treated epileptics and controls. *2598*

Aspartate Aminotransferase *Serum No Effect Analytical* No effect at 1.8 mg/dL on SMA 12/60 method. *2636* At acute overdose concentration (20 mg/dL) on Technicon® SMAC® method. *3719* No effect at therapeutic concentration on Reflotron method. *1984*
Serum Increase Physiological Intrahepatic cholestatic jaundice. *2313*

Barbiturate *Serum No Effect Analytical* No effect on UV method of Broughton. *1427*
Serum Increase Analytical At 8 mg/dL false positive with diphenylcarbazone. *0583*
Serum Decrease Analytical Negative interference with diphenylcarbazone method. *0293*
Urine No Effect Analytical With vanillin on Frings procedure at 10 mg/dL. *1204*
Urine Increase Analytical With mercurous NO₃ reagent on Frings TLC procedure. White spot with mercuric SO₄ on Frings TLC procedure, Purple with diphenylcarbazone on Frings TLC procedure. *1204*

Bicarbonate *Serum No Effect Analytical* At concentration of 240 mg/L had no effect on method using phenolphthalein. *3393*

Bile *Urine Increase Physiological* Hepatotoxicity. *2313*

Bilirubin *Serum No Effect Analytical* No effect at 1.8 mg/dL on SMA 12/60 method. *2636* At acute overdose concentration (20 mg/dL) on Technicon® SMAC® method. *3719* At concentration of 240 mg/L had no effect on Jendrassik and Grof method. *3393*
Serum No Effect Physiological Insignificant protein displacement effect in neonates. *3748* In about 20 patients treated for 2 y. *0873*
Serum Increase Physiological Rare hypersensitivity reaction. *1488*

Biotin *Serum Decrease Physiological* Dose related in long-term treated epileptics compared with controls. *2008*

Bishydroxycoumarin *Plasma Decrease Physiological* Average decrease from 29-21 μg/mL (DPH dose approximately 1 g/d). *2041*

Bromide *Urine No Effect Analytical* No effect on method of Frings. *1204*

BSP Retention *Serum Increase Physiological* Due to hypersensitivity or intrahepatic cholestasis. *1488*

Calcitonin *Plasma Decrease Physiological* In epileptic children in comparison with controls. *2030*

Calcium *Serum No Effect Analytical* No effect at 1.8 mg/dL on SMA 12/60 method. *2636* At acute overdose concentration (20 mg/dL) on Technicon® SMAC® method. *3719* At concentration of 240 mg/L had no effect on cresolphthalein method. *3393*

Phenytoin *(continued)*

Calcium *(continued)*
Serum Decrease Physiological Metabolic effect with chronic therapy. *1452* With or without osteoporosis in chronic users. *2042* May produce osteomalacia with protracted treatment. *0121* In 7% of 60 epileptics treated for 10 y or more. *2493* In 7% of 60 epileptics treated for 10 y or more. *2493* In about 20 patients treated for 2 y. *0873*
Urine Decrease Physiological In 60 epileptic patients treated for 10 y or more. *2493*

Cephalin Flocculation *Serum Increase Physiological* Hepatotoxicity with centrolobular necrosis. *2313*

Ceruloplasmin *Serum Increase Physiological* Effect on synthesis of protein in liver. *3505* Observed in both hospitalized and home-living individuals with long-term administration. *3803*

Chlordiazepoxide *Blood No Effect Analytical* On method of Riddick at 5 mg/dL. *2983*

Chloride *Serum No Effect Analytical* At concentration of 240 mg/L had no effect on mercurimetric method. *3393*

Cholesterol *Serum No Effect Analytical* No effect at 1.8 mg/dL on SMA 12/60 method. *2636* At acute overdose concentration (20 mg/dL) on Technicon® SMAC® method. *3719* No effect at therapeutic concentration on Reflotron method. *1984* At concentration of 20 mg/L had no effect on CHOD-PAP method. *3393* At concentration of 240 mg/L had no effect on Liebermann-Burchard method. *3393*
Serum No Effect Physiological In women in 27 patients with transient brain ischemia. *1879* No significant difference between epileptics given 200-300 mg/d for 1 to 35 y and controls. *2598*
Serum Increase Physiological Hepatotoxicity with centrolobular necrosis. *3678* In men in 27 patients with transient brain ischemia. *1879* Possibly due to subclinical hypothyroidism caused by drug or due to hepatic synthesis stimulation with increase of pool size of bile acids (increase of 6 to 48% in 11 patients). *2767*

Cholesterol, High Density Lipoprotein
Serum Increase Physiological In both men and women in 27 patients with transient brain ischemia. *1879* In 43% of 28 patients treated with 200-300 mg/d for 1 to 35 y (1.87 mmol/L vs 1.51 mmol/L in controls). vs 1.51 mmol/L in controls). *2598*

Cholesterol, Low Density Lipoprotein
Serum No Effect Physiological In both men and women in 27 patients with transient brain ischemia In 27 patients with transient brain ischemia. *1879*

Cholesterol, Very Low Density Lipoprotein
Serum No Effect Physiological In both men and women in 27 patients with transient brain ischemia In 27 patients with transient brain Ischemia. *1879*

Chromosomes *Test Conditions Abnormal Physiological* One case abnormal y chromosome. *2154*

Color *Urine Increase Analytical* pink, red or red-brown color may occur. *1488*

Complement C3 *Serum No Effect Physiological* In about 118 treated epileptics. *0228*
Serum Decrease Physiological Observed in 50% patients. *0989*

Complement C3c *Serum Decrease Physiological* Observed in 50% patients. *0989*

Complement C4 *Serum No Effect Physiological* In about 118 treated epileptics. *0228*

Complement CH$_{50}$ *Serum No Effect Physiological* No effect of treatment of epileptics. *0228*

Coombs' Test *Serum Positive Physiological* Immunological response to drug. *1486*

Coombs' Test, Direct *Serum Positive Physiological* Few cases with hemolytic anemia. *1487*

Copper *Serum Increase Physiological* Observed in both hospitalized and home-living individuals with long-term administration. *3803*

Coproporphyrin *Urine Increase Physiological* May precipitate acute porphyria. *1322*
Feces Increase Physiological May precipitate acute porphyria. *1322*

Corticosteroids *Plasma Increase Physiological* Alters steroid metabolism. *3700*

Cortisol *Plasma No Effect Physiological* Although increases hepatic turnover but serum concentration remains in normal range due to increased secretion. *1020*
Plasma Decrease Physiological Chronic administration effect in Cushings syndrome. *2220* In 10 women receiving anticonvulsants, mainly phenytoin versus 10 controls, drug taken on average for 15 y. *0257*

Creatine Kinase *Serum No Effect Analytical* At acute overdose concentration (20 mg/dL) on Technicon® SMAC® method. *3719*
Serum Increase Physiological Myopathy observed in one case as part of hypersensitivity reaction. *1502*

Creatine Kinase Isoenzymes
Serum Increase Physiological Myopathy observed in one case as part of hypersensitivity reaction. Increase of M33 isoenzyem. *1502*

Creatinine *Serum No Effect Analytical* No effect at 1.8 mg/dL on SMA 12/60 method. *2636* At acute overdose concentration (20 mg/dL) on Technicon® SMAC® method. *3719* At concentration of 20 mg/L had no effect on Ektachem® method. *3393* At concentration of 240 mg/L had no effect on AutoAnalyzer method. *3393*

Cyclic Adenosine Monophosphate
Urine Increase Physiological In epileptic children in comparison with controls. *2030*

Cyclosporine *Serum Decrease Physiological* Marked lowering of serum or blood concentration, presumably due to hepatic enzyme induction. *0683*
Blood Decrease Physiological Induces cytochrome P-450 hepatic enzymes thereby increasing clearance of cyclosporine. *1186*

Dapsone *Serum No Effect Analytical* At 10 mg/dL on colorimetric procedure of Higgins. *1590*

DDT (Chlorophenothane) *Serum Decrease Physiological* Reported interaction due to alteration of metabolism. *2042*

Dehydroepiandrosterone Sulfate (DHEA-S)
Plasma Decrease Physiological Significant reduction in both men and women compared with untreated epileptics. Same effect observed when combined with carbamazepine. *2141*

Dexamethasone *Serum Decrease Physiological* Due to accelerated hepatic clearance of steroid: may given false impression of Cushing's syndrome. *1830*

Dexamethasone Suppression
Patient Abnormal Physiological Alters steroid metabolism. *1488*

Dicumarol *Plasma Decrease Physiological* With long-term therapy: in one study decreased from 20 to 5 μg/mL. *3556*

Digitoxin *Serum Decrease Physiological* Average decrease from 25 to 10 μg/mL (DPH dose 900 mg/d). *2041* May be reduced to subtherapeutic concentration due to induction of mixed function oxidases. *2794*

Digoxin *Serum Increase Analytical* At 10 mg/mL equals 0.2 ng/mL by RIA. *3940*

Diphenylhydantoin *Serum Increase Physiological* 600 mg orally produces concentration of 10 mg/L. *2348* Therapeutic plasma concentrations obtained within 24 h in most patients: peaking of concentration occurred between 48 and 96 h after loading. *3827*
CSF Increase Physiological Concentration same as unbound concentration in plasma. *3054*

Disopyramide *Serum Increase Physiological* Serum concentration at peak unchanged but area under curve affected suggesting effect on hepatic metabolism. *2597*

Eosinophils *Blood Increase Physiological* Rare hypersensitivity reaction. *1488* In 89% patients who developed hepatotoxicity. *2517* General feature of hepatotoxicity (although this occurs rarely), hypersensitivity usually responsible. *0950* Associated with several cases of hepatotoxicity due to hypersensitivity reaction. *1994*

Erythrocytes *Blood Decrease Physiological* megaloblastic/hemolytic/aplastic anemia/pancytopenia. *3331* Single case of marked aplasia, apparently mediated through an IgG inhibitor requiring the presence of drug to suppress erythroid colony formation *in vitro* inhibitor appears to exert effect on erythroid progenitors at or beyond stage of differentiation of CFU-E. But not on erythroblasts. *0889* Occasional case of aplastic anemia reported. *3717*

Estradiol *Plasma   No Effect   Physiological*   In 10 women receiving anticonvulsants, mainly phenytoin versus 10 controls, drug taken on average for 15 y. *0257*

Estriol   *Urine   No Effect   Analytical*   No effect on GLC method. *1163*

Ethchlorvynol   *Serum   No Effect   Analytical*   At 10 µg/mL on colorimetric method of Wallace. *3753*   At 600 mg/L on GLC procedure of Evenson. *1058*
Urine   No Effect   Analytical   No effect on method of Frings and Cohen at 10 mg/dL. *1205*

Factor VII   *Plasma   Decrease   Physiological*   May cross placenta (vitamin K dependent). *3378*

FIGLU (N-Formiminoglutamic Acid)
Urine   Increase   Physiological   Occurs with impaired absorption of folic acid. *3773*

Folate   *Serum   Decrease   Physiological*   May cause megaloblastic anemia (impairs absorption). *3773*   Low folate in from 27 to 91% of treated epileptics in different studies. *2062*
CSF   Decrease   Physiological   Occurs in many long-treated epileptics. *2967*   Low folate in from 27 to 91% of treated epileptics in different studies. *2062*
Red Blood Cells   Decrease   Physiological   Impaired deconjugation of polyglutamates in gut. *3504*   Low folate in from 27 to 91% of treated epileptics in different studies. *2062*
Test Conditions   Decrease   Analytical   Mild depressant effect on L. casei. *2427*

Follicle Stimulating Hormone (FSH)
Plasma   Increase   Physiological   Mean 3.7 units/L versus 1.9 in controls in approximately 24 male patients given phenytoin alone or with primidone or phenobarbital. *3609*

FPN Test   *Urine   No Effect   Analytical*   No effect at 10 mg/dL on method of Frings. *1204*

α_2-Globulin   *Serum   Increase   Physiological*   Related to duration of therapy. *2427*

β_1-α-Globulin   *Serum   Decrease   Physiological*   Observed in 50% patients. *0989*

β_1-Globulin C/A   *Serum   Decrease   Physiological*   Probable drug induced immunological effect. *1406*

Glucaric Acid   *Urine   Increase   Physiological*   Induces hepatic enzymes. *1694*   Equipotent with phenobarbital. *2089*   Dose dependent effect next to phenobarbital in potency. *2788*

Glucose   *Serum   No Effect   Analytical*   No effect at 1.8 mg/dL on SMA 12/60 method. *2636*   At acute overdose concentration (20 mg/dL) on Technicon® SMAC® method. *3719*   No effect at therapeutic concentration on Reflotron method. *1984*   At concentration of 36 mg/L had no effect on Ektachem® method. *3393*   At concentration of 20 mg/L had no effect on GOD/POD-PAP method. *3393*   At concentration of 160 mg/L had no effect on hexokinase/G-6-PDH method. *3393*
Serum   No Effect   Physiological   No significant difference between drug treated epileptics and controls. *2785*
Serum   Increase   Physiological   Inhibitory effect on insulin secretion. *0789*   Metabolic effect of drug. *0121*   Small number of cases with hyperglycemia; occasional convulsions and Coma reported. Probably due to decreased insulin secretion. *0598*
Urine   Increase   Physiological   Occurs with inhibition of insulin secretion Occurs with hyperglycemia. *0789*

Glucose Tolerance   *Serum   No Effect   Physiological*   No consistent change seen in patients on long-term Treatment. *2767*   No significant difference between drug treated epileptics and controls. *2785*
Serum   Increase   Physiological   Lower in chronic patients than in controls. *0606*
Serum   Decrease   Physiological   Decreases insulin excretion. *2220*

Glucuronic Acid   *Urine   Increase   Physiological*   Due to hepatic enzyme induction in liver. *0848*

Glutamine   *Serum   Increase   Physiological*   Significant increase in children possibly due to inhibition of carbamoylphosphate synthase. *3642*

γ-Glutamyltransferase (GGT)   *Serum   No Effect   Analytical*   No effect at therapeutic concentration on Reflotron method. *1984*
Serum   Increase   Physiological   Possibly due to enzyme induction. *3815*   Observed 2 times upper limit of normal in 58 of 125 adult epileptic inpatients and 5 times upper limit in 9 of 125 (dose and duration of therapy not known). *1162*   Mean 3 fold increase in 90% patients after 6 mo treatment — not influenced

by age or sex or additional anticonvulsant therapy, accentuated by regular consumption of alcohol. *1897*

Hematocrit   *Blood   Decrease   Physiological*   megaloblastic/hemolytic/aplastic anemia,pancytopenia. *3331*

Hemoglobin   *Blood   Decrease   Physiological*   megaloblastic/hemolytic/aplastic anemia,pancytopenia. *3331*

Hydrocortisone   *Serum   Decrease   Physiological*   25% increase in metabolic clearance rate with much reduced half-life. *1830*

11-Hydroxycorticosteroids   *Plasma   No Effect   Analytical*   No effect on Mattingly procedure at 1 mg/ml. *2977*

17-Hydroxycorticosteroids   *Urine   No Effect   Analytical*   No effect on Zimmermann procedure at 200 mg/L. *2977*
Urine   Decrease   Analytical   Inhibit beta-glucuronidase during hydrolysis. *0022*
Urine   Decrease   Physiological   With chronic ingestion cortisol= 6-beta-hydroxy cortisol. *3800*

6-β-Hydroxycortisol   *Urine   Increase   Physiological*   Alters steroid metabolism. *3800*   Marked increase reflecting hepatic enzyme induction. *0394*   Manifestation of drug induced 6-beta-hydroxylase activity. *1020*

Hydroxy-Diphenylhydantoin   *Urine   Increase   Physiological*   Normal metabolite (absolute amount variable in individuals). *1269*

Hydroxyproline   *Urine   Increase   Physiological*   In epileptic children in comparison with controls. *2030*

25-Hydroxy Vitamin D$_3$ *Serum   No Effect   Physiological*   In 10 women receiving anticonvulsants mainly phenytoin versus 10 controls, drug taken on average for 15 y. *0257*
Serum   Decrease   Physiological   Positive correlation with calcium level. *1452*   Reduced by approximately 50% in long term treatment in children: most marked with combination therapy. *1449*   Observed effect in some children. *3023*

Immunoglobulin IgA   *Serum   Decrease   Physiological*   Observed in 21% (mechanism not elucidated). *3400*   Further decrease below low value in epileptics. *0228*   Further decrease from typical low values of epilepsy with drug treatment. *0228*   Further decrease below low value in epileptics. *0228*

Immunoglobulin IgD   *Serum   No Effect   Physiological*   In about 118 treated epileptics. *0228*   No effect of treatment of epileptics. *0228*

Immunoglobulin IgE   *Serum   Decrease   Physiological*   In about 118 treated epileptics. *0228*   Significant decreases in patients with different types of epilepsy. *0228*   In about 118 treated epileptics. *0228*

Immunoglobulin IgG   *Serum   Decrease   Physiological*   Significant effect due to immunosuppressive action. *2257*   Minor decrease noted with treatment of epilepsy. *0228*   Minor effect in about 118 treated epileptics. *0228*

Immunoglobulin IgM   *Serum   Decrease   Physiological*   Minor effect in about 118 treated epileptics. *0228*   Minor decrease noted with treatment of epilepsy. *0228*

Indomethacin   *Serum   No Effect   Analytical*   No effect on HPLC method of Roberts and Smith. *3002*

Insulin   *Plasma   No Effect   Physiological*   No significant difference from controls during glucose tolerance test. *0606*
Plasma   Decrease   Physiological   Reduces insulin response to glucose challenge. *2276*   Lower values in drug treated group than in controls reached significant level at 90 minutes after glucose. *2785*

Iron   *Serum   No Effect   Analytical*   At acute overdose concentration (20 mg/dL) on Technicon® SMAC® method. *3719*   At concentration of 200 mg/L had no effect on Ferrozine method. *3393*

^{131}I Uptake   *Serum   No Effect   Physiological*   No effect observed. *1915*

17-Ketosteroids   *Urine   No Effect   Analytical*   No effect on Zimmermann procedure at 200 mg/L. *2977*
Urine   Decrease   Physiological   Alters steroid metabolism. *1488*

Lactate   *Serum   No Effect   Analytical*   At concentration of 160 mg/L had no effect on enzymatic method. *3393*
Plasma   Increase   Physiological   Dose related in long-term treated epileptics compared with controls. *2008*

Lactate Dehydrogenase   *Serum   No Effect   Analytical*   No effect at 1.8 mg/dL on SMA 12/60 method. *2636*   At acute overdose concentration (20 mg/dL) on Technicon® SMAC® method. *3719*

Phenytoin *(continued)*

LE Cells *Blood Positive Physiological* May activate lupus erythematosus. *0616*

Leucine Aminopeptidase Isoenzymes
Serum Increase Physiological Increased slower running components. *3087*

Leukocytes *Blood Increase Physiological* Due to eosinophilia. *2313* General feature of hepatotoxicity (although this occurs rarely), hypersensitivity usually responsible. *0950*
Blood Decrease Physiological megaloblastic/hemolytic/aplastic anemia/pancytopenia. *0072* Occasional case of aplastic anemia reported. *3717*

Levonorgestrel *Serum Decrease Physiological* Markedly lessened concentrations when drug co-administered due to enhanced metabolism. *2655*

Luteinizing Hormone (LH) *Plasma Increase Physiological* Mean 10.0 units/L versus 4.6 in controls in approximately 24 male patients given phenytoin alone or with primidone or phenobarbital. *3609*

Lymphocytes *Blood Decrease Physiological* Dose related associated with decreased DNA synthesis. *2257*

MCV *Blood Increase Physiological* May occur with megaloblastic anemia. *3331* In about 20 patients treated for 2 y. *0873*

Methaqualone *Serum No Effect Analytical* At 4 mg/L on GLC procedure of Evenson. *1057*

Methemoglobin *Blood Increase Physiological* Occurs with hemolytic anemia. *2220*

Methotrexate *Serum Increase Physiological* Displaces from plasma protein binding. *1487*

Methylprednisolone *Serum Decrease Physiological* 130% increase in metabolic clearance rate with much reduced half-life. *1830*

Metyrapone *Serum Decrease Physiological* Average decrease from 48 to 7 µg/dL (with 3.5 g DPH). *2041*

Mexiletine *Serum Decrease Physiological* Hepatic enzyme induction may reduce elimination half-life by 50%. *2011*

Neutrophils *Blood Decrease Physiological* megaloblastic/hemolytic/aplastic anemia,pancytopenia. *0072* Rare cases of neutropenia reported. *0155* Occasional case of aplastic anemia reported. *3717*

5'-Nucleotidase *Serum Increase Physiological* Observed 2 times upper limit of normal in 10 of 127 adult epileptic inpatients. *1162*

Organic Acids *Urine Increase Physiological* Dose related in long-term treated epileptics compared with controls. *2008*

Parathyroid Hormone *Plasma Increase Physiological* In epileptic children in comparison with controls. *2030*

Partial Thromboplastin Time
Plasma Increase Physiological May cross placenta (vitamin K dependent). *3378*

Phenobarbital *Serum Increase Physiological* Average increase from 22 to 48 µg/mL (DPH dose 900 mg/d). *2041*

Phosphate *Serum No Effect Analytical* No effect at 1.8 mg/dL on SMA 12/60 method. *2636* At acute overdose concentration (20 mg/dL) on Technicon® SMAC® method. *3719* At concentration of 240 mg/L had no effect on phosphomolybdate method. *3393*
Serum No Effect Physiological In 60 epileptic patients treated for 10 y or more. *2493*
Serum Decrease Physiological In about 20 patients treated for 2 y. *0873*

Platelets *Blood Decrease Physiological* megaloblastic/hemolytic/aplastic anemia/pancytopenia. *0072* Several cases of immune-mediated thrombocytopenia reported. *2502* Occasional case of aplastic anemia reported. *3717*

Porphobilinogen *Urine Increase Physiological* May precipitate acute porphyria. *1322*

Potassium *Serum No Effect Analytical* At concentration of 20 mg/L had no effect on flame-photometric method. *3393* At concentration of 240 mg/L had no effect on measurement by ISE with predilution. *3393*

Prednisolone *Serum Decrease Physiological* 77% increase in metabolic clearance rate with much reduced half-life. *1830*

Prolactin *Plasma Increase Physiological* Mean 312 mUnits/L versus 207 in controls in approximately 24 male patients given phenytoin alone or with primidone or phenobarbital. *3609*

Propoxyphene *Serum No Effect Analytical* At 25 mg/L on method of Evenson. *1056*

Protein *Serum No Effect Analytical* No effect at 1.8 mg/dL on SMA 12/60 method. *2636* At acute overdose concentration (20 mg/dL) on Technicon® SMAC® method. *3719* At concentration of 240 mg/L had no effect on biuret method with blank correction. *3393*
Serum No Effect Physiological In about 20 patients treated for 2 y. *0873*

Protein Bound Iodine (PBI) *Serum Decrease Physiological* Competes with thyroxine for binding sites. *1488* Metabolic effect of drug. *0121*

Prothrombin Time *Plasma Increase Physiological* Patients on coumarins which inhibit its metabolism. *2220* Significant increase when drug added to regime of warfarin on which previously stabilized. *2556*

Protoporphyrin *Feces Increase Physiological* May precipitate acute porphyria. *1322*

Pyridoxal Phosphate *Serum Decrease Physiological* Increase by 20% at 4 weeks, 60% at 12 weeks. *2948*

Quinidine *Serum Decrease Physiological* Reduces half-life by more than 50% due to induction of hepatic cytochrome P-450 activity. *2020*

Retinol Binding Protein *Serum Increase Physiological* In comparison with untreated cognitively delayed children 4.0 mg/dL versus 3.3 mg/dL. *2003*

Salicylate *Urine No Effect Analytical* At 10 mg/dL no effect on Trinder method. *1204*

Sex Hormone Binding Globulin
Serum Increase Physiological In 10 women receiving anticonvulsants, mainly phenytoin versus 10 controls, drug taken on average for 15 y. *0257* Mean 8.3 mol/L versus 5.0 in controls in approximately 24 male patients given phenytoin alone or with primidone or phenobarbital. *3609*

Sodium *Serum No Effect Analytical* At concentration of 20 mg/L had no effect on flame-photometric method. *3393* At concentration of 240 mg/L had no effect on measurement by ISE with predilution. *3393*

T₃ Uptake *Serum No Effect Physiological* No effect 300 mg/d for 1 week. *0583* With long-term treatment due to accelerated thyroxine clearance via stimulation of hepatic microsomal enzymes. *2168*
Serum Increase Physiological Affects both resin and red cell uptake. *1488*

Testosterone *Serum No Effect Physiological* Usual effect. *1736*
Serum Increase Physiological Mean 30.7 nmol/L versus 17.8 in controls in approximately 24 male patients given phenytoin alone or with primidone or phenobarbital. *3609*

Testosterone, Free *Serum Decrease Physiological* Observed effect. *1736*

Theophylline *Serum Decrease Physiological* Elimination half-life reduced; total body clearance almost doubled: activity of cytochrome metabolizing enzymes in liver stimulated. *2305*

Thymol Turbidity *Serum Increase Physiological* Intrahepatic cholestatic jaundice. *2313*

Thyro-Binding Index *Serum Decrease Physiological* For up to 3 weeks after therapy. *0913*

Thyroid Stimulating Hormone (TSH)
Serum No Effect Physiological With long-term treatment due to accelerated thyroxine clearance via stimulation of hepatic microsomal enzymes. *2168* No change or in response to TRH: probably central regulation involved. *3798*
Serum Increase Physiological Slight effect observed in 50% patients at dose of 3 mg/kg/d with normal serum concentration. *1293*

Thyroxine (T₄) *Serum No Effect Analytical* No effect on competitive protein binding procedure of Alexander. *0054*
Serum Increase Physiological In 10 women receiving anticonvulsants, mainly phenytoin versus 10 controls, drug taken on average for 15 y. *0257*
Serum Decrease Physiological Displaces thyroxine from binding sites. *3505* Slight effect observed in 50% patients at dose of 3 mg/kg/d with normal serum concentration. *1293* 5.5

µg/dL in treated epileptics vs 8.1 µg/dL in controls. *3512* With long-term treatment due to accelerated thyroxine clearance via stimulation of hepatic microsomal enzymes. *2168* In comparison with untreated cognitively delayed children 7.1 µg/dL versus 8.4 µg/dL. *2003*

Urine Increase Physiological Increases during therapy (as binds to thyroxine binding globulin). *0630*

Thyroxine (T$_4$) Binding Globulin
Serum No Effect Physiological In 10 women receiving anticonvulsants, mainly phenytoin versus 10 controls, drug taken on average for 15 y. *0257*
Serum Decrease Physiological 19.6 µg/mL in treated epileptics vs 23.4 µg/mL in controls. *3512*

Thyroxine (T$_4$), Free *Serum Increase Analytical* As measured by addition of drug to control sera and measured by analog radioimmunoassay. *2169*
Serum Increase Physiological In 10 women receiving anticonvulsants, mainly phenytoin versus 10 controls, drug taken on average for 15 y. *0257*
Serum Decrease Physiological Competes with T$_4$ for thyroxine binding globulin, increased liver degradation. *0012* As measured by equilibrium dialysis and analog-type radioimmunoassays: not due to displacement from binding proteins. *2169* 1.2 ng/dL in treated epileptics vs 1.5 µg/dL in controls. *3512*

Thyroxine (T$_4$) Index, Free (FTI)
Serum Decrease Physiological Low normal or decrease for up to 3 weeks after therapy. *0913* With long-term treatment due to accelerated thyroxine clearance via stimulation of hepatic microsomal enzymes. *2168*

Thyroxine (T$_4$) (Murphy-Pattee) *Serum Increase Analytical* Extractable with ethanol, false increase. *0583*

Transcortin *Serum Increase Physiological* In 10 women receiving anticonvulsants, mainly phenytoin versus 10 controls, drug taken on average for 15 y. *0257*

Triglycerides *Serum No Effect Analytical* At acute overdose concentration (20 mg/dL) on Technicon® SMAC® method. *3719* No effect at therapeutic concentration on Reflotron method. *1984* At concentration of 240 mg/L had no effect on lipase/esterase method. *3393*
Serum No Effect Physiological In both men and women in 27 patients with transient brain ischemia. *1879* No overall significant difference between treated epileptics and controls but higher proportion of males had high concentrations than controls. *2598* No consistent change seen in patients on long-term Treatment. *2767*

Triglycerides, High Density Lipoprotein
Serum Increase Physiological In both men and women in 27 patients with transient brain ischemia. *1879*

Triglycerides, Low Density Lipoprotein
Serum No Effect Physiological In both men and women in 27 patients with transient brain ischemia. *1879*

Triglycerides, Very Low Density Lipoprotein
Serum No Effect Physiological In both men and women in 27 patients with transient brain ischemia. *1879*

Tri-iodothyronine (T$_3$) *Serum Increase Physiological* In 10 women receiving anticonvulsants, mainly phenytoin versus 10 controls, drug taken on average for 15 y. *0257*
Serum Decrease Physiological Slight effect observed in 50% patients at dose of 3 mg/kg/d with normal serum concentration. *1293* With long-term treatment due to accelerated thyroxine clearance via stimulation of hepatic microsomal enzymes. *2168*

Tri-iodothyronine (T$_3$), Free *Serum Increase Analytical* As measured by addition of drug to control sera and measured by analog radioimmunoassay. *2169*
Serum Decrease Physiological As measured by equilibrium dialysis and analog-type radioimmunoassays: not due to displacement from binding proteins. *2169*

Tri-iodothyronine (T$_3$), Reverse
Serum No Effect Physiological With long-term treatment due to accelerated thyroxine clearance via stimulation of hepatic microsomal enzymes. *2168*

TSH Response to TRH *Serum Increase Physiological* Major stimulation when patients given 3 mg/kg/d with normal serum concentration. *1293*

Urea Nitrogen *Serum No Effect Analytical* At acute overdose concentration (20 mg/dL) on Technicon® SMAC® method. *3719* No effect at therapeutic concentration on Reflotron method. *1984* At concentration of 240 mg/L had no effect on diacetylmonoxime method. *3393* At concentration of 36 mg/L had no effect on Ektachem® method. *3393*

Uric Acid *Serum No Effect Analytical* No effect at 1.8 mg/dL on SMA 12/60 method. *2636* At acute overdose concentration (20 mg/dL) on Technicon® SMAC® method. *3719* No effect at therapeutic concentration on Reflotron method. *1984* At concentration of 240 mg/L had no effect on phosphotungstate reduction method. *3393*

Valproic Acid *Serum Decrease Physiological* Significant reduction of half-life from 10.9 h to 6.9 h when drugs coadministered. *2233* Concentration approximately 50% less when phenytoin co-administered. *2340*

Vitamin A *Serum Increase Physiological* In comparison with untreated cognitively delayed children 28.8 µg/dL versus 19.4 µg/dL. *2003*

Vitamin B$_{12}$ *CSF Decrease Physiological* Significantly lower with long term therapy. *1192*

Vitamin D Binding Globulin *Serum No Effect Physiological* In 10 women receiving anticonvulsants mainly phenytoin versus 10 controls, drug taken on average for 15 y. *0257*

Vitamin E (Tocopherol) *Serum Decrease Physiological* Significantly reduced values in treated handicapped children. *1587* Mean of 0.58 mg/dL versus 0.67 mg/dL in untreated control of handicapped children. *1587*

Zinc *Serum No Effect Physiological* Observed in both hospitalized and home-living individuals with long-term administration. *3803* No significant difference between treated and untreated groups of handicapped children. *1587* No significant difference in average values between treated and untreated handicapped children although hypozincemia in 7 of 32 treated children and 0 of 13 Controls. *1587*

Phetharbital

Glucaric Acid *Urine Increase Physiological* Potent enzyme-inducing agent. *1694*

6-β-Hydroxycortisol *Urine Increase Physiological* Increased to over 400 µg/d. *3292*

pHisoHex®

Erythrocytes *Urine Increase Analytical* Red globules that look like red blood cells. *1022*

Phleomycin

Chromosomes *Test Conditions Abnormal Physiological* Clastogenic in human cells. *3282*

Phloridzin

Glucose *Urine Increase Physiological* Decreases glucose reabsorption. *3343* Increases clearance. *2313*

Glucose Clearance *Urine Increase Physiological* Due to decreased reabsorption. *3343*

Inulin Clearance *Urine Decrease Physiological* Decrease of up to 30%. *3343*

Osmolality *Urine Increase Physiological* Due to increased excretion of glucose etc. *3343*

Phosphate *Urine Decrease Physiological* Probably due to increased reabsorption. *3343*

Phosphate Clearance *Urine Decrease Physiological* Due to reduced excretion. *3343*

Uric Acid *Urine Increase Physiological* Uricosuric action in man. *3343*

Uric Acid Clearance *Urine Increase Physiological* Probably inhibits tubular reabsorption. *3343*

Phosgene

Protein *Serum Decrease Physiological* Decreased synthesis due to liver damage. *0279*

Phosphastrate

Alkaline Phosphatase Isoenzymes
Serum Decrease Analytical False low intestinal component (phenylalanine inhibitor). *3845*

Phosphates

Acid Phosphatase *Serum Decrease Analytical* Inhibits reaction if high concentration in substrate. *1022*

Alkaline Phosphatase *Serum Decrease Analytical* Inhibits reaction if high concentration in substrate. *1022*

Ammonia *Plasma Decrease Analytical* Inhibits form of indophenol color in Berthelot reaction. *2632*

Appearance *Urine Abnormal Analytical* Cloudy, soluble in dilute acetic acid. *0443*

Calcium *Serum No Effect Analytical* On HPE method of American Monitor. *0140* method of Morin at 20 mg/dL. *2485*
Serum Decrease Analytical Inhibit emission in some flame methods Compete with EDTA for Calcium (slight effect some methods). *1563*
Serum Decrease Physiological Transient effect due to colloid retention in liver. *0639*
Urine Decrease Analytical Inhibit emission in some flame methods Compete with EDTA for Calcium (more marked than serum). *1563*
Feces Increase Physiological Increased by average of 32 meq/d. *1688*

Chloride *Serum No Effect Analytical* No effect at 20 mg/dL method of Fingerhut. *1135*

Crystals *Urine Increase Physiological* Ammonium magnesium or Calcium hydrogen (soluble in acid). *0443*

2,3-Diphosphoglycerate
Red Blood Cells No Effect Analytical High serum concentration no effect on method of Luisada-Offer. *2226*

Ionized Calcium *Serum Decrease Physiological* Slight effect abolished by Calcium infusion. *2951*

Lipid Glycerol *Serum Decrease Analytical* Inhibits phospholipase with method of Horney. *2683*

Magnesium *Urine Decrease Physiological* Decreases increased excretion of bed rest. *1688*

Parathyroid Hormone *Plasma Increase Physiological* Up to 125% increase at 1 h after 1 g orally. *2951*

Phosphate *Serum Increase Physiological* Increase of greater than 2 mg/dL observed. *1439*
Urine Increase Physiological Doubled excretion compared with bed rest. *1688*
Feces Increase Physiological Doubled excretion compared with bed rest. *1688*

Potassium *Serum Decrease Physiological* Drawn into cells with gluconeogenesis. *3879*

Pyruvate Kinase *Red Blood Cells Increase Analytical* Activating effect on enzyme. *0370*

Phosphoenolpyruvate

Pyruvate Kinase *Red Blood Cells Increase Analytical* Activating effect on enzyme. *0370*

3-Phosphoglyceric Acid

Pyruvate Kinase *Red Blood Cells Increase Analytical* Activating effect on enzyme. *0370*

Phospholipids

Erythrocyte Sedimentation Rate
Blood Decrease Physiological Possibly by increasing viscosity. *0251*

Fatty Acids, Free (FFA) *Serum No Effect Analytical* NaCl used to eliminate in method of Mikac-Devic. *2438*

Phosphorus

Alanine Aminotransferase *Serum Increase Physiological* Hepatotoxic effect with necrosis. *2313*

Alkaline Phosphatase *Serum Increase Physiological* Hepatotoxic effect with necrosis. *2313*

α-Amino-Nitrogen *Urine Increase Physiological* Due to nephrotoxicity. *1302*

Aspartate Aminotransferase
Serum Increase Physiological Hepatotoxic effect with necrosis. *2313*

Bile *Urine Increase Physiological* Hepatotoxic effect with necrosis. *2313*

Bilirubin *Serum Increase Physiological* Hepatotoxic effect with necrosis. *2313*

BSP Retention *Serum Increase Physiological* Hepatotoxic effect with necrosis. *2313*

Casts *Urine Increase Physiological* Nephrotoxic effect. *1302*

Cephalin Flocculation *Serum Increase Physiological* Hepatotoxic effect with necrosis. *2313*

Coagulation Time *Blood Increase Physiological* Due to toxicity. *1343*

Creatinine *Serum Increase Physiological* Nephrotoxic effect with necrosis. *1237*

Erythrocytes *Blood Decrease Physiological* Occurs with chronic poisoning. *1302*
Urine Increase Physiological Nephrotoxic effect. *1022*

Fibrinogen *Plasma Decrease Physiological* Probable hepatotoxic effect. *1302*

Glucose *Serum Decrease Physiological* Toxic effect. *2313*

Hemoglobin *Urine Increase Physiological* Actual bleeding caused by drug. *1022*

Icteric Index *Serum Increase Physiological* Due to nephrotoxicity. *1302*

Leukocytes *Blood Increase Physiological* Reported leukocytosis with poisoning. *1302*
Blood Decrease Physiological Reported effect with poisoning. *1302*

Monocytes *Blood Increase Physiological* Reported effect of poisoning. *1302*

Nonprotein Nitrogen *Serum Increase Physiological* Due to nephrotoxicity. *1302*

Occult Blood *Feces Positive Physiological* Bloody diarrhea may occur with poisoning. *1343*

Platelets *Blood Increase Physiological* Reported effect with poisoning. *1302*

Protein *Serum No Effect Physiological* Decreased synthesis due to liver damage. *0279*
Urine Increase Physiological Renal toxic effect. *1237*

Prothrombin Time *Plasma Increase Physiological* Toxicity effect (hypoprothrombinemia). *2313*

Silicon *Test Conditions Increase Analytical* Up to 60 mg/L affects Jolles/Neurath method. *1773*
Test Conditions Decrease Analytical Above 60 mg/L affects Jolles/Neurath method. *1773*

Thymol Turbidity *Serum Increase Physiological* Hepatotoxic effect with necrosis. *2313*

Urea Nitrogen *Serum Increase Physiological* Nephrotoxic effect. *2313*

Volume *Urine Decrease Physiological* Toxicity effect. *1343*

Phospho-Soda®

Phosphate *Serum Increase Physiological* High concentration of phosphate absorbed from gut. *3505*

Phthalylsulfathiazole

Acetaminophen Screening Test *Urine Positive Analytical* Red color with 1-naphthol at therapeutic concentrations. *3326*
Urine Negative Analytical No reaction with o-cresol at therapeutic concentrations. *3326*

Estriol *Urine Decrease Physiological* Possibly affects integrity of intestinal microflora. *0023*

Estriol-3-Glucuronide *Urine Decrease Physiological* Possibly affects integrity of intestinal microflora. *0023*

Phylloerythrinogen

Urobilin *Urine Increase Analytical* Produces red fluorescence. *1563*

Urobilinogen *Urine Increase Analytical* Produces red color with Ehrlich's reagent. *1563*

Physical Stress

Epinephrine *Plasma Increase Physiological* Significant effect can be observed. *1176*
Urine Increase Physiological Significant effect can be observed. *1176*

Norepinephrine *Plasma Increase Physiological* Significant effect can be observed. *1176*
Urine Increase Physiological Significant effect can be observed. *1176*

Physical Training

Adenosine Triphosphate (ATP)
Muscle Increase Physiological Resting concentration in muscle increased by training. *1875*

Aminoacid Arylpeptidase *Serum Increase Physiological* Higher at rest in trained athletes. *1491*

α₁-Antitrypsin *Serum Increase Physiological* Defense against proteolytic activity in athletes. *1491*

Aspartate Aminotransferase
Serum No Effect Physiological No effect noticed following 10 weeks training. *1696*

Base Excess *Blood No Effect Physiological* No significant difference observed. *1491*

Cholesterol *Serum No Effect Physiological* No difference between trained athletes and others. *1702*

Creatine Kinase *Serum Decrease Physiological* Slight effect noted after 10 weeks training. *1696*

Creatine Phosphate *Muscle No Effect Physiological* No effect of physical training. *1875*

Fatty Acids, Free (FFA) *Serum No Effect Physiological* No difference between trained athletes and others. *1702*

Glutamate Dehydrogenase *Serum Increase Physiological* Significantly higher than in control subjects. *1491*

γ-Glutamyltransferase (GGT)
Serum No Effect Physiological Not systematically affected by training. *1491*

Glycogen Synthetase *Muscle Increase Physiological* Level of activity directly related to fitness. *3553*

Haptoglobin *Serum Increase Physiological* 27% increase after 4 weeks training. *1491*

Hemopexin *Serum Increase Physiological* 6.6 increase after 4 weeks training. *1491*

Ketones *Serum Increase Physiological* Increase in untrained individuals after exercise only. *1811*

Lactate Dehydrogenase *Serum Increase Physiological* Tends to be higher in trained. *1491* Resting level increased, response to exercise dec. *1696*

Lactate Dehydrogenase Isoenzymes
Serum Increase Physiological Heart component increased. *1491*

Malate Dehydrogenase *Serum No Effect Physiological* Not systematically affected by training. *1491*

Phospholipids, Total *Serum No Effect Physiological* No difference between trained athletes and others. *1702*

Standard Bicarbonate *Blood No Effect Physiological* No significant difference observed. *1491*

Transferrin *Serum Increase Physiological* Significantly higher in athletes. *1491*

Triglycerides *Serum Decrease Physiological* 56 mg/dL versus 85 in untrained (on average). *1702*

Physostigmine

Pseudocholinesterase *Serum Decrease Physiological* Direct effect of drug. *3523*

Phytate

Calcium *Serum Decrease Physiological* Falls initially and remains low. *2947*
Urine Decrease Physiological Acts by inhibiting gastrointestinal absorption. *1540*

Iron *Serum Decrease Physiological* Early effect but increased if continued. *2947*

Nitrogen Balance *Patient No Effect Physiological* No effect of 2.5 g/d in healthy. *2947*

Phosphate *Serum Decrease Physiological* Falls initially and remains low. *2947*

Zinc *Serum Decrease Physiological* Early effect but increased if continued. *2947*
Feces Increase Physiological Significant effect observed with 2.5 g/d. *2947*

Phytofluene

Vitamin A *Serum No Effect Analytical* No effect on method of Bubb-Murphy. *0517*

Phytonadione

Bilirubin *Serum Increase Physiological* Large dose effect, or with G-6-PD deficiency. *1022*

Erythrocytes *Blood Decrease Physiological* Hemolysis may occur with G-6-PD deficiency. *2313*
Urine Increase Physiological Actual bleeding caused by drug. *2313*

Hematocrit *Blood Decrease Physiological* Hemolysis may occur with G-6-PD deficiency. *2313*

Hemoglobin *Blood Decrease Physiological* Hemolysis may occur with G-6-PD deficiency. *2313*
Urine Increase Physiological Actual bleeding caused by drug. *1022*

Prothrombin Time *Plasma Decrease Physiological* Stimulates synthesis of clotting factors. *3379*

Picoline

Ionized Calcium *Serum Increase Analytical* At concentrations > 0.1 mmol/L on Calcium specific electrode. *0540*

Picric Acid

Alanine Aminotransferase *Serum Increase Physiological* May cause acute hepatic damage. *1343*

Color *Urine Increase Analytical* Yellow to red brown. *2313*

Erythrocytes *Blood Decrease Physiological* Hemolytic effect. *1343*

Protein *Urine Increase Physiological* Nephrotoxicity. *1343*

PIDH

Glucose *Serum Decrease Physiological* Significant reduction after glucose load. *3100*

Insulin *Plasma Decrease Physiological* Reduces insulin response to glucose load. *3100*

Pilocarpine

Erythrocytes *Blood Increase Physiological* Probably due to contraction of spleen. *1343*

Hydrochloric Acid *Gastric Material Increase Physiological* Produces secretion like that of vagal stimulation. *1343*

Leukocytes *Blood Increase Physiological* Probably due to contraction of spleen. *1343*

Pepsin *Gastric Material Increase Physiological* Produces secretion like that of vagal stimulation. *1343*

Potassium *Saliva Decrease Physiological* Produces secretion like plasma ultrafiltrate. *1343*

Pimozide

Acetaminophen Screening Test *Urine Negative Analytical* No reaction with o-cresol at therapeutic concentrations. *3326*

Follicle Stimulating Hormone (FSH)
Plasma Decrease Physiological Statistically significant decline when given to acutely psychotic males although still within normal range. *3340*

Luteinizing Hormone (LH) *Plasma Decrease Physiological* Statistically significant decline when given to acutely psychotic males although. *3340*

Prolactin *Plasma Increase Physiological* General effect observed. *1736* In acutely psychotic males. *3340*

Testosterone *Serum No Effect Physiological* General effect observed. *1736* In acutely psychotic males. *3340*

Thyroid Stimulating Hormone (TSH)
Serum Decrease Physiological But effect contrary to assumed mechanism of a dopamine blocker. *3798*

Pindolol

Alanine Aminotransferase *Serum Increase Physiological* Minor persistent increases reported in about 7% patients but not progressive. *1208*

Alkaline Phosphatase *Serum Increase Analytical* At 5 mg/L conventional methods when added to serum. *1758*

Apolipoprotein AI *Serum No Effect Physiological* No significant changes after 12 mo treatment. *2116* No significant changes found during 12 mo treatment. *2117*

Apolipoprotein AII *Serum No Effect Physiological* No significant changes after 12 mo treatment. *2116* No significant changes found during 12 mo treatment. *2117*

Aspartate Aminotransferase
Serum Increase Physiological Minor persistent increases reported in about 7% patients but not progressive. *1208*
Serum Decrease Analytical At 5 mg/L conventional methods when added to serum. *1758*

Bilirubin *Serum Decrease Analytical* At 5 mg/L conventional methods when added to serum. *1758*

Calcium *Serum No Effect Physiological* No significant effect even with 6 weeks treatment. *3279*

Cholesterol *Serum No Effect Physiological* No significant change typically seen with treatment for 1 mo or more. *0088* No effect in short-term study of 10 hypertensive men with 15 mg/d. *2135* No significant change in four studies of 1 to 12 mo. *2116*
Serum Increase Physiological Increase by 5.5% in one study of 4 weeks with 15-30 mg/d. *1806*
Serum Decrease Physiological Lower after 6 mo of therapy than after one. *2117*

Cholesterol, High Density Lipoprotein
Serum No Effect Physiological No significant change typically seen with treatment for 1 mo or more. *0088* No effect in short-term study of 10 hypertensive men with 15 mg/d. *2135* No significant change in 2 studies of 2 to 12 mo. *2116*
Serum Increase Physiological Significant increase during first month of therapy. *2117* 20% increase in one 3 mo study. *2116*

Cholesterol, Low Density Lipoprotein
Serum No Effect Physiological No significant change typically seen with treatment for 1 mo or more. *0088* No significant change in 3 studies of 2 to 12 mo. *2116*
Serum Increase Physiological Slight tendency to rise during treatment. *2117*

Cholesterol, Very Low Density Lipoprotein
Serum No Effect Physiological No effect in short-term study of 10 hypertensive men with 15 mg/d. *2135*
Serum Decrease Physiological 35% reduction in concentration after 2 mo in 11 hypertensive patients but after 16 mo treatment had reverted to normal. *1399*

Creatine Kinase *Serum Increase Physiological* In 20 of 25 patients with essential hypertension with increase of 20 to 760% compared with pretreatment values. MM isoenzyme most affected, but 8 of 25 showed slight increase of MB. *3149*
Serum Decrease Analytical At 5 mg/L conventional methods when added to serum. *1758*

Fatty Acids, Free (FFA) *Serum No Effect Physiological* No significant changes after 12 mo treatment. *2116* Remained constant during 12 mo therapy. *2117*

Glucagon *Plasma Decrease Physiological* In 18 patients with mild essential hypertension treated with chlorothiazide concomitantly over 4 weeks. *1035*

Glucose *Serum No Effect Physiological* In 18 patients with mild essential hypertension treated with chlorothiazide concomitantly over 4 weeks. *1035* No significant changes found during 12 mo treatment. *2117*

Glucose Tolerance *Serum Decrease Physiological* Slight effect after 60 minutes, significant effect at 120 minutes during oral glucose tolerance test. *2117*

Insulin *Plasma No Effect Physiological* In 18 patients with mild essential hypertension treated with chlorothiazide. *1035* No significant change in concentration during glucose tolerance test after drug. *2117*

LE Cells *Blood Positive Physiological* Drug induced systemic lupus erythematosus reported in one case. *0299*

Magnesium *Serum No Effect Physiological* No significant effect even with 6 weeks treatment. *3279*

Neutrophils *Blood Decrease Physiological* Occasional case of agranulocytosis reported. *3717*

Parathyroid Hormone *Plasma Decrease Physiological* Observed within 3 h of treatment: significant effect over 6 weeks. *3279*

Phosphate *Serum No Effect Physiological* No significant effect even with 6 weeks treatment. *3279*

Potassium *Serum No Effect Physiological* No significant effect even with 6 weeks treatment. *3279*
Serum Increase Physiological Slight rise in patients treated with moderate doses of drug. *2836*

Sodium *Serum No Effect Physiological* No significant effect even with 6 weeks treatment. *3279*

Triglycerides *Serum No Effect Physiological* No significant change typically seen with treatment for 1 mo or more. *0088* No effect in short-term study of 10 hypertensive men with 15 mg/d. *2135* Insignificant change in 3 studies of 2 to 12 mo. *2116* Remained constant during 12 mo therapy. *2117*
Serum Increase Physiological In 18 patients with mild essential hypertension treated with chlorothiazide concomitantly over 4 weeks. *1035* Increase by 28 % in one study of 4 weeks with 15-30 mg/d. *1806* 28% increase in one 1 mo study. *2116*

Triglycerides, Very Low Density Lipoprotein
Serum Decrease Physiological 35% reduction in concentration after 2 mo in 11 hypertensive patients but after 16 mo treatment had reverted to normal. *1399*

Uric Acid *Serum No Effect Physiological* No effect in short-term study of 10 hypertensive men with 15 mg/d. *2135*

Pineapples

5-Hydroxyindoleacetic Acid (5-HIAA)
Urine Increase Physiological Rich in serotonin. *1167*

Pink Capsules

Protein Bound Iodine (PBI) *Serum Increase Physiological* May contain iodine in dye. *1237*

Pipamazine

Erythrocytes *Blood Decrease Physiological* Pancytopenia (AMA Blood dyscrasias). *2429*

Leukocytes *Blood Decrease Physiological* Pancytopenia (AMA Blood dyscrasias). *2429*

Platelets *Blood Decrease Physiological* Pancytopenia (AMA Blood dyscrasias). *2429*

Pipemidic Acid

N-Acetylglucosaminidase *Urine No Effect Physiological* No significant effect in healthy individuals and those with infections of lower urinary tract but reduction in patients with pyelonephritis when given for 10 d. *0673*

Alanine Aminopeptidase   *Urine   No Effect   Physiological*
No significant effect in healthy individuals and those with infections of lower urinary tract but reduction in patients with pyelonephritis when given for 10 d. *0673*

Piperacetazine

Alanine Aminotransferase   *Serum   Increase   Physiological*
Transient reversible effect. *1227*

Aspartate Aminotransferase
Serum   Increase   Physiological   Transient reversible increase. *1227*

Bilirubin   *Serum   Increase   Physiological*   Rare jaundice (reversible)(like infect hepatitis). *0071*

Eosinophils   *Blood   Increase   Physiological*   Occasional allergic response. *1680*

Erythrocytes   *Blood   Decrease   Physiological*   Rare hemolytic anemia/pancytopenia. *0019*

Glucose   *Serum   Increase   Physiological*   Disordered endocrine response. *0019*
Serum   Decrease   Physiological   Disordered endocrine response. *0019*
Urine   Increase   Physiological   Due to induced hyperglycemia. *0019*

Hematocrit   *Blood   Decrease   Physiological*   Mild transient decrease with hypotension. *1227*

Hemoglobin   *Blood   Decrease   Physiological*   Mild transient decrease with hypotension. *1227*

Leukocytes   *Blood   Decrease   Physiological*   Rare leukopenia/agranulocytosis (reversible). *0071*

Platelets   *Blood   Decrease   Physiological*   Rare thrombocytopenia (reversible). *0071*

Pregnancy Tests   *Urine   Positive   Physiological*   Associated with endocrine disorders. *0019*

Protein   *CSF   Increase   Physiological*   Abnormality produced. *0019*

Piperacillin

Alanine Aminotransferase   *Serum   Increase   Physiological*
In 1 of 29 patients given 181 mg/kg i.v. for 6 d. *3275*

Alkaline Phosphatase   *Serum   Increase   Physiological*   In 1 of 29 patients given 181 mg/kg i.v. for 6 d. *3275*

Aspartate Aminotransferase
Serum   Increase   Physiological   In 1 of 29 patients given 181 mg/kg i.v. for 6 d. *3275*   Transient increase in 1 of 20 treated patients. *3132*   In 1 of 59 patients given drug as sole agent with many patients with severe illness. *3874*

Coombs' Test, Direct   *Serum   Positive   Physiological*   In 1 of 59 patients given drug as sole agent with many patients with severe illness. *3874*

Creatinine   *Serum   Increase   Physiological*   In 1 of 59 patients given drug as sole agent with many patients with severe illness. *3874*

Eosinophils   *Blood   Increase   Physiological*   In 1 of 29 patients given 181 mg/kg i.v. for 6 d. *3275*   In 5 of 59 patients given drug as sole agent with many patients with severe illness. *3874*

Glucose   *Urine   No Effect   Analytical*   Concentrations accurately measured by Diastix®. Concentrations accurately measured by TesTape®. *2100*
Urine   Increase   Analytical   Falsely elevated values with Clinitest®. *2100*

Lactate Dehydrogenase   *Serum   Increase   Physiological*   In 1 of 59 patients given drug as sole agent with many patients with severe illness. *3874*

Leukocytes   *Blood   Decrease   Physiological*   6 of 20 patients had small drop in leukocyte count. *3132*

Monocytes   *Blood   Increase   Physiological*   In 1 of 29 patients given 181 mg/kg i.v. for 6 d. *3275*

Neutrophils   *Blood   Decrease   Physiological*   In 1 of 59 patients given drug as sole agent with many patients with severe illness. *3874*

Vancomycin   *Serum   No Effect   Analytical*   Negligible interference at up to 500 mg/L on Abbott TDx. *2417*

Piperazine

Amphetamine   *Urine   No Effect   Analytical*   No effect at 100 mg/L on method of Rutter. *3097*

Bilirubin   *Serum   Increase   Physiological*   May cause hemolytic anemia. *2365*

Chromosomes   *Test Conditions   Abnormal   Physiological*   Clastogenic to bone marrow and leukocytes. *3282*

Erythrocytes   *Blood   Decrease   Physiological*   May cause hemolytic anemia. *2365*

Hematocrit   *Blood   Decrease   Physiological*   May cause hemolytic anemia. *2365*

Hemoglobin   *Blood   Decrease   Physiological*   May cause hemolytic anemia. *2365*

Uric Acid   *Serum   Decrease   Analytical*   Reported observation. *0355*

Piperidine

17-Hydroxycorticosteroids   *Urine   Increase   Analytical*   Interferes with Porter-Silber reaction *in vitro*. *0427*

Ionized Calcium   *Serum   Decrease   Analytical*   At concentrations > 0.1 mmol/L on Calcium specific electrode. *0540*

17-Ketosteroids   *Urine   Increase   Analytical*   Interference with Zimmermann reaction *in vitro*. *0427*

Pipobroman

Bilirubin   *Serum   Increase   Physiological*   May cause hemolytic anemia. *1022*

Erythrocytes   *Blood   Decrease   Physiological*   May cause anemia (bone marrow depression). *0071*

Hematocrit   *Blood   Decrease   Physiological*   Intended effect when polycythemia present. *1680*

Hemoglobin   *Blood   Decrease   Physiological*   Hemolytic anemia has been described. *0070*

Leukocytes   *Blood   Decrease   Physiological*   Bone marrow depression may occur after 4 weeks. *1680*

Platelets   *Blood   Decrease   Physiological*   May cause bone marrow depression. *1680*

Reticulocytes   *Blood   Increase   Physiological*   Hemolytic anemia has been described. *0070*

Urobilinogen   *Urine   Increase   Physiological*   Hemolytic anemia has been described. *0070*

Pirenzepine

Glucagon   *Plasma   No Effect   Physiological*   Treatment for 1 week had no effect on basal concentrations. *3933*

Insulin   *Plasma   No Effect   Physiological*   Treatment for 1 week had no effect on basal concentrations. *3933*

Leukocytes   *Blood   Decrease   Physiological*   In isolated cases even though other drugs also being ingested. *3486*

Neutrophils   *Blood   Decrease   Physiological*   In isolated cases even though other drugs also being ingested. *3486*

Pancreatic Polypeptide   *Plasma   Decrease   Physiological*   Borderline significant reduction from basal mean value of 37 ng/L to 26.2 ng/L after treatment with 100 mg/d for 7 d. *3933*

Platelets   *Blood   Decrease   Physiological*   In isolated cases even though other drugs also being ingested. *3486*

Piretanide

Cholesterol   *Serum   Increase   Physiological*   Effect observed with 12 mg/d. *3663*

Cholesterol, Low Density Lipoprotein
Serum   Increase   Physiological   Effect observed with 12 mg/d. *3663*

Cobalt   *Serum   No Effect   Physiological*   No significant effect with up to 12 mg/d for 3 mo. *3696*

Copper   *Serum   No Effect   Physiological*   No significant effect with up to 12 mg/d for 3 mo. *3696*

C-Peptide   *Plasma   Increase   Physiological*   61% higher than pretreatment level after 8 weeks in 12 male patients with mild hypertension (6 mg bid). *1503*

Piretanide *(continued)*

Glucose Tolerance *Serum No Effect Physiological* In 12 male patients with mild hypertension (6 mg bid). *1503*

Hemoglobin A₁c *Blood No Effect Physiological* In 12 male patients with mild hypertension (6 mg bid). *1503*

Insulin *Plasma No Effect Physiological* In 12 male patients with mild hypertension (6 mg bid). *1503*

Iron *Serum Decrease Physiological* Slight drop with 12 mg/d for 3 mo. *3696*

Manganese *Serum No Effect Physiological* No significant effect with up to 12 mg/d for 3 mo. *3696*

Triglycerides *Serum Increase Physiological* Effect observed with 12 mg/d. *3663*

Triglycerides, Very Low Density Lipoprotein
Serum Increase Physiological Effect observed with 12 mg/d. *3663*

Zinc *Serum No Effect Physiological* No significant effect with up to 12 mg/d for 3 mo. *3696*

Piroxicam

Alanine Aminotransferase *Serum Increase Physiological* Occasional case of liver damage reported. *2824* Some transient increases reported. *2692*

Alkaline Phosphatase *Serum Increase Physiological* Occasional case of liver damage reported. *2824*

Aspartate Aminotransferase
Serum Increase Physiological Occasional case of liver damage reported. *2824* Some transient increases reported. *2692*

Bilirubin *Serum Increase Physiological* Occasional case of liver damage reported. *2824*

Bleeding Time *Patient Increase Physiological* Prolongs bleeding time and decreases platelet aggregation. *2692*

Creatinine *Serum No Effect Physiological* No effect in spite of effect on urea nitrogen. *2692*
Serum Increase Physiological Occasional drug induced nephrotoxicity, with isolated azotemia, acute interstitial nephritis or nephrotic syndrome. *2824* Possible increase due to drug in some patients. *3819*

Erythrocytes *Urine Increase Physiological* Occasional drug induced nephrotoxicity, with isolated azotemia, acute interstitial nephritis or nephrotic syndrome. *2824*

Granular Casts *Urine Increase Physiological* Few: occasional drug induced nephrotoxicity, with isolated azotemia, acute interstitial nephritis or nephrotic syndrome. *2824*

Hematocrit *Blood Decrease Physiological* Possibly related to drug administration or concomitant therapy in patients with osteoarthrosis. *3564* Reported, unassociated with obvious gastrointestinal bleeding. *2692*

Hemoglobin *Blood Decrease Physiological* Possibly related to drug administration or concomitant therapy in patients with osteoarthrosis. *3564* Reported, unassociated with obvious gastrointestinal bleeding. *2692*
Urine Increase Physiological Occasional drug induced nephrotoxicity, with isolated azotemia, acute interstitial nephritis or nephrotic syndrome. *2824*

Hyaline Casts *Urine Increase Physiological* Many: occasional drug induced nephrotoxicity, with isolated azotemia, acute interstitial nephritis or nephrotic syndrome. *2824*

Occult Blood *Feces Increase Physiological* Gastrointestinal bleeding most common severe problem. *2051*

Protein *Urine Increase Physiological* Occasional drug induced nephrotoxicity, with isolated azotemia, acute acute interstitial nephritis or nephrotic syndrome. *2824*

Urea Nitrogen *Serum Increase Physiological* Possible increase due to drug in some patients. *3819* Occasional increase noted not progressive. *2692*

Uric Acid *Serum Increase Physiological* In patients with normal concentrations pretreatment. Variable effect in people with initial high concentrations. *3819*

Pizotyline

Fatty Acids, Free (FFA) *Serum No Effect Physiological* No effect after 2 mg i.v. *3411*

Glucose *Serum Decrease Physiological* 10% reduction after 3-5 h with 2 mg i.v. *3411*

Insulin *Plasma No Effect Physiological* No effect after 2 mg i.v. *3411*

Plantains

5-Hydroxytryptamine (Serotonin) *Urine Increase Analytical* 5-hydroxy tryptamine contained in plants. *1237*

Plasma

Alanine Aminotransferase *Serum No Effect Analytical* No significant difference as measured on ABA-100. *2228*

Albumin *Serum Decrease Analytical* Increase by 0.12 mg/dL HABA dye on SMA 12/60 method. *2228*

Alkaline Phosphatase *Serum Decrease Analytical* Compared with serum on ABA-100, SMA 12/60 methods. *2228*

Amylase *Serum No Effect Analytical* No significant difference from serum with diamyl procedure. *2228*

Aspartate Aminotransferase *Serum No Effect Analytical* No significant difference from serum on SMA 12/60, LKB 8600. *2228*

Bicarbonate *Serum No Effect Analytical* No significant difference from serum by SMA 12/60. *2228*

Bilirubin *Serum No Effect Analytical* No significant difference from serum on SMA 12/60 method. *2228*

Calcium *Serum No Effect Analytical* No significant difference from serum on SMA 12/60 method. *2228*

Chloride *Serum No Effect Analytical* No significant difference from plasma by chloridometer. *2228*
Serum Decrease Analytical Slightly less in plasma by SMA 6/60 method. *2228*

Cholesterol *Serum Increase Analytical* Compared with serum if unextracted (SMA 12/60). *2228*
Serum Decrease Analytical Compared with serum if extracted (AA2 procedure). *2228*

Creatine Kinase *Serum No Effect Analytical* No significant difference from serum by LKB 8600. *2228*

Creatinine *Serum No Effect Analytical* No significant difference from serum by AutoAnalyzer. *2228*

Folate *Serum Decrease Analytical* Marked effect in some specimens observed. *2455*

Glucose *Serum No Effect Analytical* No significant difference from serum by AutoAnalyzer. *2228*

Lactate Dehydrogenase *Serum No Effect Analytical* When kinetic method on LKB 8600. *2228*
Serum Increase Analytical Plasma high colorimetrically on SMA 12/60. *2228*
Serum Decrease Analytical Probably due to absence of platelet component. *3319*

Phosphate *Serum Decrease Analytical* By average of 0.25 mg/dL. *2228*

Potassium *Serum Decrease Analytical* Mean 0.4 meq/L higher in serum. *2228*

Protein *Serum Increase Analytical* Mean 0.24 g/dL due to fibrinogen. *2228*

Sodium *Serum Decrease Analytical* 2.6 meq/L by 12/60, 0.6 meq/L by I.L. *2228*

Triglycerides *Serum Decrease Analytical* Average 4 mg/dL higher in serum. *2228*

Urea Nitrogen *Serum No Effect Analytical* No significant difference from serum by SMA 12/60. *2228*

Uric Acid *Serum Decrease Analytical* Mean 0.17 mg/dL higher in serum. *2228*

Plastic

Cyclosporine *Serum Increase Analytical* Spurious peak near cyclosporines A and D when HPLC methods used: observed with Sarstedt Monovette tubes. *2269*

Plastic Tubing

Protein Bound Iodine (PBI) *Serum Increase Physiological* Bard Intracath tubing contains I$_2$. *0583*

Plums

5-Hydroxyindoleacetic Acid (5-HIAA)
Urine Increase Physiological Rich in serotonin. *1488*

Poison Ivy

Protein *Urine Increase Physiological* May cause nephrotoxicity. *3204*

Urea Nitrogen *Serum Increase Physiological* May cause nephrotoxicity. *3204*

Poliomyelitis Vaccine

Platelets *Blood Decrease Physiological* May rarely cause thrombocytopenia. *2427*

Protein *CSF Increase Physiological* May rarely cause Guaillain-Barré syndrome. *2427*

Polymagma

^{131}I Uptake *Serum Decrease Physiological* Contains tetraiodofluorescein. *2652*

Polymethylmethacrylate

γ-Glutamyltransferase (GGT)
Serum Increase Physiological 11 of 90 total hip arthroplasty and 7 of 23 knee arthroplasty patients had abnormal GGT increases 5-10 d post-surgery: not observed in controls. *2993*

Polymyxin

Alkaline Phosphatase *Urine Increase Physiological* Due to nephrotoxic effect of drug. *2897*

Amino Acids *Urine Increase Analytical* Reacts with ninhydrin; extra spot TLC, high voltage electrophoresis. *2855*

Casts *Urine Increase Physiological* Cylindruria may occur with daily injection. *1343*

Creatinine *Serum Increase Physiological* Nephrotoxic effect. *2808*

Erythrocytes *Urine Increase Physiological* Actual bleeding may be caused by drug. *1237*

Fat *Feces Increase Physiological* Probably because of mucosal damage. *3023*

Glomerular Filtration Rate (GFR)
Urine Decrease Physiological Nephrotoxicity increases with continued treatment. *1343*

Hemoglobin *Urine Increase Physiological* Actual bleeding caused by drug. *1022*

Leucine Aminopeptidase *Urine Increase Physiological* Facilitates permeation of enzyme into tubules. *2897*

Nitrofurantoin *Urine No Effect Analytical* No effect on method of Conklin and Hollifield. *0704*

Nonprotein Nitrogen *Serum Increase Physiological* Nephrotoxic effect. *2313*

Potassium *Serum Decrease Physiological* Reported to occur in leukemia, maybe steroid effect. *0958* Probably due to direct toxic effect of drug on renal tubular cells. *0645*

Protein *Urine Increase Physiological* Nephrotoxic effect. *1022* Nephrotoxic effect. *1237*

Urea Nitrogen *Serum Increase Physiological* Nephrotoxic effect. *1022*

Polystyrene Sulfonate

Calcium *Serum Increase Physiological* Increased administered calcium (if in Calcium form). *3505*
Serum Decrease Physiological Exchanged for potassium, increased fecal loss. *2313*

Feces Increase Physiological Causes impaired absorption. *1022*

Potassium *Serum Decrease Physiological* Exchanges for sodium (if Na form used). *2313*

Polysucrose

Protein *Test Conditions Increase Analytical* Affects Lowry, Folin-Ciocalteu procedures. *2199* Possible measurement by biuret reaction. *3376*

Polythiazide

Alanine Aminotransferase *Serum Increase Physiological* May affect liver function. *2313*

Alkaline Phosphatase *Serum Increase Physiological* May affect liver function. *2313*

Amylase *Serum Increase Physiological* Rare pancreatitis reported with other thiazides. *1680*

Antidiuretic Hormone *Plasma Increase Physiological* Secreted in response to hyponatremia. *1129*

Aspartate Aminotransferase
Serum Increase Physiological May affect liver function. *2313*

Bicarbonate *Serum Increase Physiological* Metabolic alkalosis with marked diuresis. *1129*
Urine Increase Physiological But effect minimal. *1680*

Bile *Urine Increase Physiological* May affect liver function. *2313*

Bilirubin *Serum Increase Physiological* May affect liver function. *2313*

BSP Retention *Serum Increase Physiological* May affect liver function. *2313*

Calcium *Serum Increase Physiological* May be increased by up to 0.35 meq/L. *2427*
Urine Decrease Physiological Excretion decreased by up to 50%. *2427*
Feces Decrease Physiological Accentuates positive calcium balance. *2427*

Cephalin Flocculation *Serum Increase Physiological* May affect liver function. *2313*

Chloride *Serum Decrease Physiological* Diuretic effect. *1129*

Cholesterol *Serum Increase Physiological* 4% change in 20 people when treated only with drug for less tan 1 y. *1808*

Cholesterol, High Density Lipoprotein
Serum No Effect Physiological Not significant change in 20 people when treated only with drug for less than 1 year. *1808*

Cholesterol, Low Density Lipoprotein
Serum No Effect Physiological Not significant change in 20 people when treated only with drug for less than 1 year. *1808*

Cholesterol, Very Low Density Lipoprotein
Serum No Effect Physiological Not significant change in 20 people when treated only with drug for less than 1 year. *1808*

Citrate *Urine Decrease Physiological* Excretion decreased by up to 30%. *2427*

Glucose *Serum Increase Physiological* Diabetogenic action. *2313*
Urine Increase Physiological Consequence of hyperglycemia. *2313*

Glucose Tolerance *Serum Decrease Physiological* Impaired tolerance may occur. *1680*

Leukocytes *Blood Decrease Physiological* Rare reported side effect. *1680*

Neutrophils *Blood Decrease Physiological* Rare reported side effect. *1680*

Osmolality *Serum Decrease Physiological* ADH secretion with hyponatremia. *1129*

Platelets *Blood Decrease Physiological* with/without purpura may occur. *1680*

Potassium *Serum Decrease Physiological* Diuretic action. *1129*

Protein Bound Iodine (PBI) *Serum Decrease Physiological* Without clinical effect on thyroid function. *1680*

Polythiazide (continued)

Sodium *Serum Decrease Physiological* Diuretic action, with K deficiency. *1129* Severe hyponatremia noted in one patient with mild hypertension. *0170*
Urine Increase Physiological Therapeutic intent (peak effect at 6 h). *1680*

Thymol Turbidity *Serum Increase Physiological* May affect liver function. *2313*

Triglycerides *Serum Increase Physiological* 14% change in 20 people when treated only with drug for less than 1 y. *1808*

Uric Acid *Serum Increase Physiological* Due to impaired renal clearance. *1680*

Porphobilinogen

Aminolevulinic Acid *Urine Increase Analytical* Unless removed by column in method of Vincent. *3718*

Urobilinogen *Urine No Effect Analytical* Extractable and no effect on p-MDFB procedure. *2044* No effect on Urobilistix® in porphyric patients. *1858*
Urine Increase Analytical But not extractable into chloroform. *0251*

Porphyrins

Color *Urine Increase Physiological* Burgundy red, darkens on standing. *2313*

Urobilin *Urine Increase Analytical* Produce red fluorescence. *1563*

Posterior Pituitary

Basal Metabolic Rate *Patient Increase Physiological* Temporary effect after injection. *0251*

Potassium

Aldosterone *Plasma Increase Physiological* If K raised by up to 1 meq/L (orally or i.v.). *0914*

Ammonia *Plasma Decrease Physiological* K repletion in hepatic coma may reduce NH_3. *2498*

Calcium *Serum Increase Analytical* Affects flame photometry if poor instrument. *1563*

Cortisol *Plasma No Effect Physiological* If diurnal effect isolated no change. *0914*

Ionized Calcium *Serum No Effect Analytical* With changes in physiological range. *2056*

Lead *Urine Increase Analytical* 100 meq/L equivalent to 10 μg/L by atomic absorption. *3240*

Renin Activity *Plasma No Effect Physiological* No effect observed with infusion. *0914*

Sodium *Serum Increase Analytical* Affects flame photometry if poor instrument. *1563*

Zinc *Serum No Effect Analytical* At 50 to 1 with flameless atomic absorption. *2038*
Urine No Effect Analytical At 50 to 1 with flameless atomic absorption. *2038*

Potassium Aminobenzoate

Glucose *Serum Decrease Physiological* Reported effect. *2313*

Potassium Chlorate

Methemoglobin *Blood Increase Physiological* May cause hemolytic anemia. *2429*

Potassium Chloride

Chloride *Serum Increase Physiological* Added chloride. *2313*

Glucose *Serum Decrease Physiological* Drawn into cells with potassium. *2313*

Occult Blood *Feces Positive Physiological* May cause gastrointestinal tract ulceration. *0612*

Potassium *Serum Increase Physiological* Over correction of hypokalemia. *0071* Severe hyperkalemia can develop following single oral dose of 30 to 45 meq in patients either with renal potassium excretion or internal disposal disorders. Potassium chloride supplementation was single major cause of fatal drug reactions. *2836*

Vitamin B$_{12}$ *Serum Decrease Physiological* Reported to cause ileal malabsorption. *3023*

Potassium Iodide

Amylase *Serum Increase Physiological* Parotitis reported as result of treatment. *1680*

Corticosteroids *Urine Increase Analytical* Butanol extract/no hydrolysis Reddy/Porter-Silber reaction. *0022*

Eosinophils *Blood Increase Physiological* Allergic response. *1553*

17-Hydroxycorticosteroids *Urine Increase Analytical* Affects Porter-Silber reaction. *3217*

Leukocytes *Blood Increase Physiological* Due to eosinophilia. *2313*

Orthophosphate *Test Conditions Decrease Analytical* Total inhibition at 0.1 mol/L on method of Horder. *1658*

Platelets *Blood Decrease Physiological* Thrombocytopenia reported to occur. *1436*

Protein Bound Iodine (PBI) *Serum Increase Physiological* Small doses elevate in normal range, others above. *0830*

Thyroid Stimulating Hormone (TSH)
Serum Increase Physiological Increase by 2.3 uU/mL over 11 d in normals. *3658*

Thyroxine (T$_4$) *Serum Decrease Physiological* In normals by 1 μg/dL over 11 d. *3658*

Tri-iodothyronine (T$_3$) *Serum Decrease Physiological* In normals by 15 ng/dL over 11 d. *3658*

Potassium Oxalate

Ammonia *Plasma No Effect Analytical* At concentration of 8,000 mg/L had no effect on Ektachem® method. *3393*

Bilirubin *Serum No Effect Analytical* At concentration of 8,000 mg/L had no effect on Ektachem® method. *3393*

Chloride *Serum Decrease Analytical* If used as anticoagulant may be shift into cells. *1563*

Creatinine *Serum Increase Analytical* At concentrations above 8,000 mg/L when combined with sodium fluoride raised concentration as measured by Ektachem® method. *3393*

Glucose *Serum No Effect Analytical* At 200 mg/dL on MBTH procedure of Neeley. *2569*
Serum Decrease Analytical Ineffective as antiglycolytic agent. *2313*

Potassium Salts

Acetaminophen Screening Test *Urine Negative Analytical* No reaction with o-cresol at therapeutic concentrations. *3326*

Diagnex Blue Excretion *Urine Increase Analytical* Displacement of diagnex blue from resin. *3879*

Occult Blood *Feces Positive Physiological* May cause ulceration and hemorrhage in gastrointestinal tract. *1680*

Uric Acid *Serum Increase Analytical* Turbidity with phosphotungstate procedure. *0583*

Povidone-Iodine

Glucose *Serum No Effect Analytical* When used as skin cleansing agent and fingerstick measurement done by Dextrostix®. *2330*
Serum Increase Analytical When povidone-iodine swab used with fingerstick and measured by Chemstrip bG® and Visidex. *1094*

Occult Blood *Feces Positive Analytical* With Hemoccult® reaction probably due to oxidation of alpha-guaiaconic acid impregnated test paper. Same effect with Lugol's iodine. *2678* Occurs with as little as 0.005 mL of a 1 to 1,000 dilution due to iodine component. *0376*

Protein Bound Iodine (PBI) *Serum Increase Physiological* Found by some investigators only. *2892*

T₃ Uptake *Serum No Effect Physiological* No effect observed. *0583*

Practolol

Alkaline Phosphatase *Serum Increase Physiological* After 4 y treatment 2 cases of primary biliary cirrhosis. *0496*

Bilirubin *Serum Increase Physiological* After 4 y treatment 2 cases of primary biliary cirrhosis. *0496*

Cholesterol *Serum No Effect Physiological* Insignificant changes in 2 studies of 2 weeks and 6 mo. *1806*

Triglycerides *Serum Increase Physiological* Increase by 8 to 16% in 2 studies of 2 weeks to 6 mo respectively. *1806*

Prasterone

Alanine Aminotransferase
Red Blood Cells Decrease Physiological Effect observed when given orally. *2210*

Aldosterone *Urine No Effect Analytical* No significant effect in RIA procedure of Drewes. *0952*
Urine Increase Analytical 1 mg/L equivalent to 2 µg/L in method of Drewes. *0952*

Estriol *Urine Increase Physiological* SO₄ comp also affects if i.v. or oral in pregnant women. *0023*

Estrogens *Urine No Effect Analytical* At 50 mg/L on fluorescent method of Corns. *0730*

Prazosin

Aldosterone *Plasma No Effect Physiological* No significant change with monotherapy for 12 mo in 15 patients. *3531*

Apolipoprotein AI *Serum No Effect Physiological* Increase by 1 mg/dL in 16 patients given 1 to 3 mg daily for 8 weeks. *2550*

Apolipoprotein B *Serum No Effect Physiological* Decrease by -8 mg/dL in 16 patients given 1 to 3 mg daily for 8 weeks. *2550* No significant effect in 15 hypertensives given 4 mg/d for 10 weeks. *1118*

Apolipoprotein CII *Serum No Effect Physiological* No change in 16 patients given 1 to 3 mg daily for 8 weeks. *2550*

Apolipoprotein CIII *Serum No Effect Physiological* No change in 16 patients given 1 to 3 mg daily for 8 weeks. *2550*

Cholesterol *Serum No Effect Physiological* In 22 mild/moderate male hypertensives treated for 8 weeks versus controls. Effect not significant though slight decrease. *1693* Decrease by 12 mg/dL in 16 patients given 1 to 3 mg daily for 8 weeks. *2550* No significant change with monotherapy for 12 mo in 15 patients. *3531* Effect observed with doses as low as 1 to 3 mg/d. *3148* In 16 hypertensive patients given 2 mg daily for 3 mo. *2266*
Serum Decrease Physiological In 15 hypertensive patients given 4 mg/d for 10 weeks with mean reduction from 202 to 188 mg/dL. *1118* Increase by 9% in 23 healthy hypertensives treated for 8 weeks. *2136* From 5 to 12% in several studies involving 25 to 50 patients for up to 6 mo. *0089*

Cholesterol, High Density Lipoprotein
Serum No Effect Physiological In 22 mild/moderate male hypertensives treated for 8 weeks versus controls insignificant reduction noted. *1693* In 23 healthy hypertensives treated for 8 weeks. *2136* No change in 16 patients given 1 to 3 mg daily for 8 weeks. *2550* In several studies involving 25 to 50 patients for up to 6 mo. *0089* In 16 hypertensive patients given 2 mg daily for 3 mo. *2266*
Serum Increase Physiological After 6 mg/d for 12 mo average increase of 17% in 15 patients. *3531* Effect observed with doses as low as 1 to 3 mg/d. *3148* Mean increase from 36 to 40.5 mg/dL in 15 hypertensives given 4 mg/d for weeks. *1118*

Cholesterol, Low Density Lipoprotein
Serum No Effect Physiological In 16 hypertensive patients given 2 mg daily for 3 mo. *2266*

Cholesterol, Very Low Density Lipoprotein
Serum No Effect Physiological In 16 hypertensive patients given 2 mg daily for 3 mo. *2266*
Serum Decrease Physiological Increase by 10% in 23 healthy hypertensives treated for 8 weeks. *2136*

Fatty Acids, Free (FFA) *Serum Increase Physiological* With 2 mg in 12 hypertensives (6 with normal and 6 with abnormal glucose tolerance). *0225*

Gastrin *Serum No Effect Physiological* With 2 mg in 12 hypertensives (6 with normal and 6 with abnormal glucose tolerance). *0225*

Glucose *Serum Increase Physiological* With 2 mg in 12 hypertensives (6 with normal and 6 with abnormal glucose tolerance). *0225*

Glucose Tolerance *Serum Decrease Physiological* Decreased glucose response to intravenous glucose tolerance test associated with decreased early insulin response mediated by increased circulating catecholamines. *2191*

Insulin *Plasma Increase Physiological* With 2 mg in 12 hypertensives (6 with normal and 6 with abnormal glucose tolerance). *0225*

Lipoprotein Lipase *Serum Increase Physiological* Mean increase from 28.4 to 37.7 µmol/L per minute given 4 mg/d for 10 weeks. *1118*

Platelet Aggregation *Blood Decrease Physiological* High value in essential hypertension in response to ADP normalized with treatment. *1715*

Prolactin *Plasma No Effect Physiological* With 2 mg in 12 hypertensives (6 with normal and 6 with abnormal glucose tolerance). *0225*

Renin Activity *Plasma No Effect Physiological* No significant change with monotherapy for 12 mo in 15 patients. *3531*

β-Thromboglobulin *Serum Decrease Physiological* 30% higher value in essential hypertensives normalized with treatment. *1715*

Thyroid Stimulating Hormone (TSH)
Serum Increase Physiological In 19 hypertensives treated for 12 weeks caused change from 3.63 uU/mL to 4.83 uU/mL. *0451*

Thyroxine (T₄) *Serum Increase Physiological* In 19 hypertensives treated for 12 weeks caused change from 10.03 µg/dL to 10.85 µg/dL. *0451*

Triglycerides *Serum No Effect Physiological* No significant effect in 15 hypertensives given 4 mg/d for 10 weeks. *1118* No significant change with monotherapy for 12 mo in 15 patients. *3531* In 16 hypertensive patients given 2 mg daily for 3 mo. *2266*
Serum Decrease Physiological In 22 mild/moderate hypertensives treated for 8 weeks versus controls (average effect 9.5%). *1693* Increase by 16% in 23 healthy hypertensives treated for 8 weeks. *2136* Increase by 20 mg/dL in 16 patients given 1 to 3 mg daily for 8 weeks. *2550* Effect observed with doses as low as 1 to 3 mg/d. *3148* From 10 to 22% in several studies involving 25 to 50 patients for up to 6 mo. *0089*

Uric Acid *Serum No Effect Physiological* In 23 healthy hypertensives treated for 8 weeks. *2136*

Prednisolone

Absorbance at 450 nm *Amniotic Fluid Decrease Analytical* Time-dependent reduction in absorbance of bilirubin when added *in vitro*. *2453*

α-Amino-Nitrogen *Plasma Increase Physiological* After 4 d of oral therapy. *2296*

Amylase *Serum Increase Physiological* May rarely cause pancreatitis. *0019*

Angiotensin-Converting Enzyme (ACE)
Serum No Effect Analytical After 96 h incubation *in vitro* at high doses produced marked reduction: unlikely to be factor *in vivo*. *3074*
Serum No Effect Physiological 75 mg drug orally had no effect in 10 patients with sarcoidosis on diurnal variation or enzyme activity. *3580*

Prednisolone *(continued)*

Aspartate Aminotransferase *Serum No Effect Analytical* No effect on Karmen procedure. No effect on Babson procedure. *2760*

Bicarbonate *Serum Increase Physiological* Hypochloremic alkalosis may occur. *0019*

Calcium *Urine Increase Physiological* Effect of all corticosteroids. *0019*

Chloride *Serum Decrease Physiological* Hypochloremic alkalosis may occur. *0019*

Cholesterol *Serum No Effect Analytical* At concentration of 8 mg/L had no effect on CHOD-PAP method. *3393* At concentration of 8 mg/L had no effect on method using catalase-Hantzsch reaction. *3393*
Serum No Effect Physiological In men receiving drug for mean of 3.1 y. *1783*
Serum Increase Physiological Rose from 4.81 to 6.58 mmol/L in women receiving drug for mean of 3.1 y. *1783*

Cholesterol, High Density Lipoprotein
Serum No Effect Physiological In men receiving drug for mean of 3.1 y. *1783*
Serum Decrease Physiological Fell from 1.99 to 1.10 mmol/L in women receiving drug for mean of 3.1 y. *1783*

Cortisol *Plasma Increase Analytical* High and equal cross reactivity with RIA and competitive protein binding. *1081*

Creatinine *Serum No Effect Analytical* At concentration of 0.23 mg/L had no effect on creatinine iminohydrolase method. *3393* At concentration of 200 mg/L had no effect on Jaffé-Fading-Fraction method. *3393* At concentration of 200 mg/L had no effect on Jaffé-Fuller's earth method. *3393*

Digoxin *Serum Increase Analytical* At 50 ng/mL equals 0.5 ng/mL by RIA. *3940* At normal concentrations in serum if no preincubation. *2804*

Glucagon *Plasma No Effect Physiological* Intravenous injection failed to modify concentration. *2296*
Plasma Increase Physiological After 4 d fasting level increased 24 pg/ml. *2296*

Glucose *Serum No Effect Analytical* At concentration of 25 mg/L had no effect on GOD/POD-PAP method. *3393*
Serum Increase Physiological After 4 d of oral therapy After i.v. 6 mg/dL at 150 minutes to 15 at 4 h. *2296*

Glucose Tolerance *Serum Decrease Physiological* Metabolic effect. *0019*

11-Hydroxycorticosteroids *Plasma No Effect Analytical* Presumed effect on fluorometric method of Mejer. *2396*

Insulin *Plasma No Effect Physiological* Intravenous injection failed to modify concentration. *2296*
Plasma Increase Physiological After 4 d of oral therapy. *2296*

Leukocytes *Blood Increase Physiological* leukocytosis observed occasionally. *2313*
Blood Decrease Physiological Leukopenia. *2313*

NBT Test *Blood No Effect Physiological* No effect observed in uremic patients. *3888*

Nitrogen *Urine Increase Physiological* Negative nitrogen balance due to protein catabolism. *0019*

Occult Blood *Feces Positive Physiological* May activate peptic ulcer with hemorrhage. *0019*

Potassium *Serum Decrease Physiological* Slight mineralocorticoid effect. *0071*
Urine Increase Physiological Slight mineralocorticoid effect. *0071*

Protein Bound Iodine (PBI) *Serum Decrease Physiological* Inhibits iodination of tyrosine in thyroxine binding globulin. *0012*

Protein Electrophoresis *Serum No Effect Analytical* At concentration of 7.2 mg/L had no effect on automated Olympus-Hite method except for slight displacement of fractions. *3393*

Sodium *Serum Increase Physiological* Very slight mineralocorticoid effect. *0071*

Thromboxane *Plasma No Effect Physiological* No effect of up to 100 μmol/L on formation during blood clotting. *2890*

Triglycerides *Serum No Effect Analytical* At concentration of 230 mg/L had no effect on GPO-PAP method. *3393*
Serum No Effect Physiological In men receiving drug for mean of 3.1 y. *1783*

Serum Increase Physiological Rose from 0.91 to 1.84 mmol/L in women receiving drug for mean of 3.1 y. *1783*

Uric Acid *Serum No Effect Analytical* At concentration of 200 mg/L had no effect on Kageyama-Hantzsch method. *3393* At concentration of 200 mg/L had no effect on catalase-AIDH method. *3393* At concentration of 25 mg/L had no effect on uricase-PAP method. *3393*

Prednisone

Albumin *Serum Decrease Physiological* Fell within 3 d in adult hospitalized patients given 40-50 mg/d orally. concentration paralleled that of zinc. *3790*

Alkaline Phosphatase *Serum No Effect Physiological* No consistent change noted in 14 hospitalized adult patients when given 40-50 mg/d. *3790*
Serum Decrease Physiological Augmented response in cirrhotics when drug added to therapeutic regime. *0651*

Amylase *Serum Increase Physiological* May cause pancreatitis. *1680*

Angiotensin-Converting Enzyme (ACE)
Serum No Effect Analytical No effect on colorimetric method of Boomsma and Schalekamp. *2814*
Serum Decrease Physiological Marked reduction after 1 week in patients in whom initial value was high, no effect in others. *3038*

Apolipoprotein AI *Serum No Effect Physiological* No change noted after 1 mo but ratio of HDL-C to apo A-I increased, apparent in 48 h. *3949*

Apolipoprotein AII *Serum No Effect Physiological* No change noted after 1 mo. *3949*

Apolipoprotein E *Plasma No Effect Physiological* No change noted after 1 mo. *3949*

Aspartate Aminotransferase
Serum Decrease Physiological Augmented response in cirrhotics when drug added to therapeutic regime. *0651*

Bicarbonate *Serum Increase Physiological* May cause hypokalemic alkalosis. *1680*

Bilirubin *Serum Decrease Physiological* Augmented response in cirrhotics when drug added to therapeutic regime. *0651*

BSP Retention *Serum Decrease Physiological* Augmented response in cirrhotics when drug added to therapeutic regime. *0651*

Calcium *Serum No Effect Physiological* Although given for 1 to 50 mo. *3840*
Serum Decrease Physiological If elevation due to sarcoidosis or vitamin D. *0071*

Cholesterol *Serum Increase Physiological* 17% increase in group of men and women during 1 mo. *3949* Augmented response in cirrhotics when drug added to therapeutic regime. *0651*

Cholesterol, High Density Lipoprotein
Serum Increase Physiological 68% average effect in both men and women. *3949*

Cholesterol, Low Density Lipoprotein
Serum No Effect Physiological Insignificant increase (11%) in both men and women after 1 mo. *3949*

Cortisol *Plasma Increase Analytical* High and equal cross reactivity with RIA and competitive protein binding. *1081*

Creatine Kinase *Serum Decrease Physiological* Low activities below normal range, observed in several patients some of whom were receiving other drugs but not observed in all patients. *1602*

Creatinine *Urine Decrease Physiological* Excretion less than anticipated in comparison with controls. *1656*

Creatinine Clearance *Serum Decrease Physiological* In comparison with controls probably due to decreased muscle mass in drug treated patients in relation to total body weight. *1656*

Digoxin *Serum Increase Analytical* At normal concentrations in serum if no preincubation. *2804* At 50 ng/mL equals 0.2 ng/mL by RIA. *3940*

1,25-Dihydroxy Vitamin D_3 *Serum Decrease Physiological* In children being treated for renal disease: dose dependent. *0646*

Eosinophils *Blood Decrease Physiological* Striking decrease in absolute count with 2 weeks therapy. *3309*

Erythrocyte Sedimentation Rate
Blood Decrease Physiological Augmented response in cirrhotics when drug added to therapeutic regime. *0651*

Fibrin Split Products (FSP) *Urine Decrease Physiological* In 2-3 d in 2/3 patients with proliferative glomerulonephritis. *0671*

Follicle Stimulating Hormone (FSH)
Plasma No Effect Physiological No apparent effect, unlike that on testosterone, in 67 y males with chronic pulmonary disease taking drug for at least 1 mo. *2245*

γ-Globulin *Serum Decrease Physiological* Augmented response in cirrhotics when drug added to therapeutic regime. *0651*

Glucose *Serum Increase Physiological* Glucocorticoid effect. *2525*
Urine No Effect Analytical No effect observed with Tes-Tape®. *1100*
Urine Decrease Analytical Low with Clinistix®, Diastix®. *1100*

Glucose Tolerance *Serum Decrease Physiological* Endocrine action. *1680*

11-Hydroxycorticosteroids *Plasma No Effect Analytical* Presumed effect on fluorometric method of Mejer. *2396*

Immunoglobulin IgG *Serum No Effect Physiological* Shortens half life but increased synthesis (no net change). *1402*

Leukocytes *Blood Increase Physiological* Observed in 2 of 676 cases, no obvious cause. *0428* Significant effect with increase of 6,000/uL by second week. *3309* Augmented response in cirrhotics when drug added to therapeutic regime. *0651*
Blood Decrease Analytical Low count by Coulter S, ?due to fragile cells. *2227*

Luteinizing Hormone (LH) *Plasma No Effect Physiological* No apparent effect, unlike that on testosterone, in 67 y males with chronic pulmonary disease taking drug for at least 1 mo. *2245*

Lymphocyte B-Cells *Blood Decrease Physiological* Reduced but not to same extent as T-Lymphocytes. *3309*

Lymphocyte T-Cells *Blood Decrease Physiological* Proportional greatest reduction with steroids. *3309*

Lymphocytes *Blood Decrease Physiological* Maximal change in third week of therapy due to redistribution of cells out of circulation. *3309*

α₂-Macroglobulin *Serum No Effect Physiological* No consistent change noted in 14 hospitalized adult patients when given 40-50 mg/d. *3790*

Monocytes *Blood Increase Physiological* Changes parallel those of neutrophils. *3309*

Neutrophils *Blood Increase Physiological* Maximum reached in second week of therapy, thereafter falls. *3309*

Nitrogen Balance *Patient Negative Physiological* Due to protein catabolism. *1680*

Occult Blood *Feces Positive Physiological* May cause hemorrhage and ulceration of gastrointestinal tract. *0612*

Parathyroid Hormone *Plasma Increase Physiological* Significant effect although Calcium normal. *3840*

Platelets *Blood Decrease Physiological* Immunologic response occurring in months. *1436*

Potassium *Serum Decrease Physiological* Slight mineralocorticoid effect only. *0071*
Urine Increase Physiological Slight mineralocorticoid effect only. *0071*

Prealbumin *Serum Increase Physiological* Occurs with decreased thyroxine binding globulin. *0042*

Prostaglandin, 6-Keto-F₁α *Plasma No Effect Physiological* No significant change with up to 25 mg drug for up to 14 d: no effect on production during clotting. *2890*

Protein Bound Iodine (PBI) *Serum Decrease Physiological* Inhibits iodination of tyrosine in thyroxine binding globulin. *0012*

Sodium *Serum Increase Physiological* Slight effect only. *0071*

T₃ Binding Capacity *Serum Decrease Physiological* Lowered with 1-2 mg/kg/d for 2-4 weeks in 10 children. *3353*

Testosterone *Serum Decrease Physiological* Reduced by more than 50% in 14 of 16 men aged 67 y with chronic pulmonary disease taking drug for at least 1 mo. Testosterone concentration inversely related to dose of drug. Serum protein binding unaffected. Effect probably due to suppression of secretion of gonadotropin releasing hormone by hypothalamus. *2245*

Thromboxane *Plasma No Effect Physiological* No significant change with up to 25 mg drug for up to 14 d: no effect on production during clotting. *2890*

Thyroid Stimulating Hormone (TSH)
Serum Increase Physiological Increased about twice with 1-2 mg/kg/d for 2-4 weeks in 10 children. *3353* Basal value increased two fold in 10 children given 1-2 mg/kg/h for 2-4 weeks but TSH response to TRH unchanged. *3353*

Thyroxine (T₄) *Serum No Effect Physiological* Insignificant change in 10 children treated for 2-4 weeks. *3353*

Thyroxine (T₄) Binding Globulin
Serum Decrease Physiological Decreases concentration of thyroxine binding globulin, increases thyroxine binding prealbumin. *3505*

Triglycerides *Serum No Effect Physiological* No effect in men during 1 mo. *3949*
Serum Increase Physiological In all 6 women slight effect in 1 mo. *3949*

Tri-iodothyronine (T₃) *Serum Decrease Physiological* Significant decrease with 1-2 mg/kg/d for 2-4 weeks in 10 children. *3353* Significant effect in 10 children given 1-2 mg/kg/h for 2-4 weeks. *3353*

Tri-iodothyronine (T₃), Reverse
Serum Increase Physiological Considerable rise with 1-2 mg/kg/d for 2-4 weeks in 10 children. *3353*
Serum Decrease Physiological Significant effect in 10 children given 1-2 mg/kg/h for 2-4 weeks. *3353*

TSH Response to TRH *Serum No Effect Physiological* No significant effect with 1-2 mg/kg/d for 2-4 weeks in 10 children. *3353*

Uric Acid *Serum Increase Physiological* Promotes nucleic acid catabolism. *3505*

Zinc *Serum Decrease Physiological* Within 3 d average fell from 12.6 μmol/L to 11.1 μmol/L in 14 adult hospitalized patients given 40-50 mg daily; level rose to above normal then reverted normal in 2 weeks. *3790*
Urine No Effect Physiological No consistent change noted in 14 hospitalized adult patients when given 40-50 mg/d. *3790*

Pregnancy

N-Acetylglucosaminidase *Serum Increase Physiological* Three times normal in last trimester. *0987*

AHF-Like Antigen *Plasma Increase Physiological* Progressive increase throughout pregnancy. *0292*

AHF Procoagulant Activity *Blood Increase Physiological* Progressive increase throughout pregnancy. *0292*

Albumin *Serum Decrease Physiological* Falls about 1 g/dL in last 2 trimesters. *0987*

Aldosterone *Plasma Increase Physiological* Possible compensatory mechanism. *2180*
Plasma Decrease Physiological In pre-eclampsia compared with normal pregnancy. *2180*
Urine Increase Physiological Possible compensatory mechanism during last trimester. *2180*

Alkaline Phosphatase *Serum Increase Physiological* 2 to 3 times normal activity in third trimester. *3588*
White Blood Cells Increase Physiological Occurs after third month. *0987*

α-Amino-Nitrogen *Urine Increase Physiological* Renal threshold of amino acids reduced. *0987*

Ammonia *Urine Increase Physiological* Output increased in last trimester. *0987*

Amylase *Serum Increase Physiological* Approximately 70% higher than in normal women. *0024*

Angiotensin II *Plasma Increase Physiological* Increase in 1st trimester due to activity of renin system. *3345*

Antithrombin Titer *Plasma Decrease Physiological* Occurs with pregnancy. *0987*

Pregnancy (continued)

α_1-**Antitrypsin** Serum Increase Physiological In last trimester compared with controls. 2385

Ascorbic Acid Plasma Decrease Physiological Falls significantly after 20 weeks. 2206

Basal Metabolic Rate Patient Increase Physiological Moderately increased especially in last 3 mo. 0987

BSP Retention Serum Increase Physiological May be moderately abnormal in last month. 0987

Calcitonin Plasma Increase Physiological Progressive inc, with decrease of C cell reserves. 1960

Calcium Serum Decrease Physiological Due to decreased albumin (loss of protein bound calcium). 3329

Carbonic Anhydrase
Red Blood Cells Increase Physiological Measured as B isoenzyme — progressive increase. 3181

Carotene Serum No Effect Physiological No significant effect observed. 1222
Serum Increase Physiological Increases with each trimester, maximum at delivery. 0245

Ceruloplasmin Serum Increase Physiological In last trimester compared with controls. 2385

Cholesterol Serum Increase Physiological Increases from 8th week (maximum by 30th). 0987

Cold Agglutinins Blood Positive Physiological Occasional response. 0987

Copper Serum Increase Physiological Metabolic effect maximal in last trimester. 0987

Cortisol Plasma Increase Physiological Higher especially at night than in normals. 1232
Urine Increase Physiological Significant hormonal effect. 2181

Cortisol, Free Plasma Increase Physiological Higher especially at night than in normals (hormonal effect). 1232

Creatine Urine Increase Physiological General effect with pregnancy. 0987
Red Blood Cells Decrease Physiological Decreased at 3 mo, but increased at 8 mo. 3041

Creatine Kinase Serum Increase Physiological Activity increases during last trimester. 2981
Serum Decrease Physiological Significant decrease from 8-20 weeks, max decrease 12-13 weeks. 1933

Cystine Aminopeptidase Serum Increase Physiological Steady increase from 18 to 40 weeks. 0778

Deoxycorticosterone Plasma Increase Physiological Increase from 23 weeks, maximum at term. 0498

Diamine Oxidase Serum Increase Physiological Reaches maximum at 24 weeks, then slight fall only. 0595

2,3-Diphosphoglycerate
Red Blood Cells Increase Physiological Increase by about 30% in normal. 3041

Effective Renal Plasma Flow
Patient Increase Physiological Normal response in first 8 mo. 0987

Erythrocytes Blood Decrease Physiological By up to 10-15% due to expanded plasma volume. 0987

Erythrocyte Sedimentation Rate
Blood Increase Physiological Normal associated with increased fibrinogen after third month. 0987

Estradiol Plasma Increase Physiological Significant increase from 23rd to 41st week. 1143

Estriol Plasma Increase Physiological Average change 2 ng/mL at 20 weeks, 11 at term. 3836
Urine Increase Physiological Progressive increase to 36th weeks, then steady. 3201 Progressive increase to 40th week. 1315

Estrogens Urine Increase Physiological Increase from sixth month until term (up to 100 μg/24 h). 0987

Estrone Plasma Increase Physiological Probably slight increase from 24th to 41st weeks. 1143

Estrone Sulfate Plasma Increase Physiological Progressive increase with time (2x estradiol). 2213

Factor II Plasma Increase Physiological Normal response in late pregnancy. 0987

Factor VII Plasma Increase Physiological Normal response in late pregnancy. 0987

Factor VIII Plasma Increase Physiological Increased 100% above normal in late pregnancy. 0987

Factor IX Plasma Increase Physiological Normal response in late pregnancy. 0459

Factor X Plasma Increase Physiological Normal response in late pregnancy. 0987

Factor XI Plasma Decrease Physiological Normal response in late pregnancy. 0987

α-**Fetoprotein** Serum Increase Physiological Progressive increase until 39 weeks at 200 ng/ml. 1733 Above 250 ng/mL in fetal morbidity or death. 0685

Fibrinogen Plasma Increase Physiological Moderate increase by 16th week (plus 33% by term). 0987

Folate Red Blood Cells Decrease Physiological Maximum decrease in 2nd or 3rd trimester. 1472

GC Globulin Serum Increase Physiological Observed with steroid administration (50% increase in late pregnancy). 3472

α_1-**Globulin** Serum Increase Physiological Marked increase in last 2 trimesters. 0987

α_2-**Globulin** Serum Increase Physiological Occurs in last 2 trimesters. 0987

β-**Globulin** Serum Increase Physiological Slight increase in second trimester. 0987

β_1-α-**Globulin** Serum Increase Physiological Response to steroids (increased in late pregnancy). 3472

γ-**Globulin** Serum Decrease Physiological May fall slightly in last 3 mo. 0987

Glomerular Filtration Rate (GFR)
Urine Increase Physiological Normal during first 8 mo but increases up to 50%. 0987

Glucose Serum Increase Physiological Reported effect. 0279
Serum Decrease Physiological May be low in occasional cases. 0987
Urine Increase Physiological May occur with decreased tolerance. 0987

Glucose-6-Phosphatase Serum No Effect Physiological No significant difference from controls. 1017

Glucose Tolerance Serum Decrease Physiological Decreased tolerance in last trimester. 0987 Intravenous tolerances decreases in 2nd to 3rd trimester. 3509

β-**Glucuronidase** Serum Increase Physiological Slight drop initially then large increase towards term. 1119

Gonadotropins Plasma Increase Physiological Increased early to yield positive pregnancy test. 0459
Urine Increase Physiological HCg increased from 2 to 12 weeks, falls later. 0987

Haptoglobin Serum No Effect Physiological No effect observed. 1498
Serum Decrease Physiological Variable response (may be slight increase in absolute mass). 3472

Hematocrit Blood Decrease Physiological In relation to decreased hemoglobin. 0987

Hemoglobin Blood Decrease Physiological Normally slight reduction (not below 10 mg/dL). 0987

Hemopexin Serum Increase Physiological In effect no change with pregnancy. 3472

Histidine Urine Increase Physiological Progressive increase observed. 3507

Hydrochloric Acid Gastric Material Decrease Physiological May be hyposecretion. 0987

3-Hydroxyanthranilic Acid Urine Increase Physiological Causes induction of tryptophan pyrrolase. 3045

5-Hydroxyindoleacetic Acid (5-HIAA)
Urine Increase Physiological Moderate increase observed. 0987

3-Hydroxykynurenine Urine Increase Physiological Causes induction of tryptophan pyrrolase. 3045

Hydroxyproline Urine Increase Physiological In last trimester, usually maximum 6-8 d postpartum. 2209

Immunoglobulin IgA Serum Decrease Physiological Also absolute decrease in late pregnancy. 3472

Immunoglobulin IgD *Serum Increase Physiological* At labor 0.085 mg/mL versus 0.033 in controls. *1429*

Immunoglobulin IgG *Serum Decrease Physiological* Also absolute decrease in late pregnancy. *3472*

Immunoglobulin IgM *Serum Decrease Physiological* Also absolute decrease in late pregnancy. *3472*

Indocyanine Green *Serum Increase Physiological* Observed in normals greater in pre-eclampsia. *2336*

Iron *Serum Decrease Physiological* Falls from midterm onwards. *0987*

Iron-Binding Capacity, Total (TIBC) *Serum Increase Physiological* With fall of serum iron after midterm. *0987*

Isocitrate Dehydrogenase *Serum Increase Physiological* Placental origin in last trimester. *0987*

17-Ketosteroids *Urine Increase Physiological* Upper limit of normal at term. *0987*

Kynurenine *Urine Increase Physiological* Causes induction of tryptophan pyrrolase. *3045*

Lactate Dehydrogenase *Serum Increase Physiological* Physiological effect observed. *0718*

Lactose *Urine Increase Physiological* May occur in last trimester especially in afternoon. *0987*

Leucine Aminopeptidase *Serum Increase Physiological* Moderately increased throughout pregnancy. *0987*

Leukocytes *Blood Increase Physiological* In late pregnancy and at labor. *0987*

α-Lipoproteins *Serum No Effect Physiological* No significant change observed. *3472*

β-Lipoproteins *Serum Increase Physiological* Increase by 24% (absolute amount greater) in late pregnancy. *3472*

Lymphocytic Response to PHA *Blood Decrease Physiological* Hormonal action. *1146*

α₂-Macroglobulin *Serum Increase Physiological* 20% increase in late pregnancy (other report no change). *3472*

Magnesium *Serum Decrease Physiological* Significant lowering (?related to decreased albumin). *0148*

NBT Test *Blood Positive Physiological* Occurs at all stages (in approximately 30%). *2908*

Neutral Fats *Serum Increase Physiological* Observed throughout pregnancy. *0987*

Nonprotein Nitrogen *Serum Decrease Physiological* Decreased 20-25% (with relatively greater decrease of urea nitrogen). *0459*

Ornithine Carbamoyltransferase (OCT) *Serum Increase Physiological* Physiological and metabolic response. *1498*

Osmotic Fragility *Red Blood Cells Increase Physiological* May be occasional increase. *0987*

Oxytocin *Urine No Effect Physiological* No effect of pregnancy or labor. *0437*

Oxytocinase *Serum Increase Physiological* Rise from tenth week, falls with labor. *0094* In 3rd trimester (55 U/L versus 2.8 normal). *3677*

Pepsin *Gastric Material Decrease Physiological* May be hyposecretion. *0987*

Phosphate *Serum Increase Physiological* Observed effect ?associated with hypocalcemia. *3329*

Phospholipids, Total *Serum Increase Physiological* Increased from 8 weeks. *0987*

Platelets *Blood Increase Physiological* Normal physiological response. *0251*

Polysaccharide, Protein Bound *Serum Increase Physiological* Throughout pregnancy. *0987*

Porphyrins *Urine Increase Physiological* May precipitate episode of hepatic porphyria. *2427*

Prealbumin *Serum Decrease Physiological* Depressed by up to 27%. *3472*

Progesterone *Plasma Increase Physiological* May increase by from 10 to 100 times. *2180*

Prolactin *Plasma Increase Physiological* Rises throughout gestation. *3647*

Prostaglandin F2α *Plasma Decrease Physiological* Lowest values in 2nd and 3rd trimester. *0507*

Protein *Urine Increase Physiological* Moderate proteinuria common Moderate proteinuria may occur commonly. *0987*

Protein Bound Iodine (PBI) *Serum Increase Physiological* Observed throughout pregnancy. *0987*

Pyridoxal Phosphate *Serum Decrease Physiological* Progressive decrease to term. *1472*

Renin Activity *Plasma Increase Physiological* Higher in first half. *3345* Increased in normal pregnancy, decreased with pre-eclampsia. *1981*
Plasma Decrease Physiological If pre-eclampsia compared with normal pregnancy. *2180*

Renin Substrate *Plasma Increase Physiological* Higher in second half. *3345*

Ribonuclease *Serum Increase Physiological* Significant increase, gradual increase with duration. *0365*

Sodium *Red Blood Cells Decrease Physiological* Maternal intracellular Na decrease towards term. *0987*

T₃ Uptake *Serum Increase Analytical* As measured by Thyopac method. *0632*
Serum Decrease Analytical As measured by resin sponge using Triosorb. *0632*
Serum Decrease Physiological Increased level of thyroxine binding globulin. *0583*

Testosterone *Serum Increase Physiological* Possible effect as increased plasma binding. *1941*

Testosterone Binding Globulin *Serum Increase Physiological* Increased during pregnancy, maximum in 3rd trimester. *1498*

Tetrahydroaldosterone *Urine Increase Physiological* Increases with period of gestation. *2850*

Thyro-Binding Index *Serum Increase Physiological* Rises from 3 weeks, normal 8 weeks after birth. *0913*

Thyroxine (T₄) *Serum Increase Physiological* Rises from 3 weeks, normal 8 weeks after birth. *0913*

Thyroxine (T₄) Binding Globulin *Serum Increase Physiological* Increased binding capacity observed. *1498*

Thyroxine (T₄) (Murphy-Pattee) *Serum Increase Physiological* Increases amount of thyroxine binding globulin. *0583*

Transcortin *Serum Increase Physiological* Increased binding capacity observed. *1498*

Transferrin *Serum Increase Physiological* In last trimester compared with controls. *2385*

Triglycerides *Serum Increase Physiological* May increase up to 2 1/2 times in normal. *1968*

Tyrosine *Plasma Decrease Physiological* Lowest values (normally higher in women than men). *0843*

Urea Nitrogen *Serum Decrease Physiological* Marked anabolic state especially in first 6 mo. *3879*
Urine Decrease Physiological Especially in first 6 mo. *0987*

Uric Acid *Serum Decrease Physiological* Mean decrease of 9.8 mg/dL compared with normal. *2439*

Uropepsinogen *Urine Increase Physiological* Moderate increase observed. *0987*

Vitamin A *Serum Increase Physiological* Increase to above normal in 2nd and 3rd trimester. *1222*
Serum Decrease Physiological Decreased in first trimester but then increasing. *1222*

Vitamin B₁₂ *Serum Decrease Physiological* Decrease by 25%-33% in third trimester. *0472*

Volume *Plasma Increase Physiological* Increased to maximum of plus 45% by 32nd week. *0987* Starts in first 3 mo maximal in 2nd and sustained. *2180* Increased by plus 25 to 55% by 32nd week. *0987*
Plasma Decrease Physiological In pre-eclampsia compared with normal pregnancy. *2180*
Urine Increase Physiological May increase by up to 25% in last trimester. *0459*

Xanthurenic Acid *Urine Increase Physiological* Causes induction of tryptophan pyrrolase. *3045*

Pregnanedione

Progesterone *Plasma Increase Analytical* 11-24% cross reactivity RIA procedure of Cameron. *0567*

Pregnanetriol

Estrogens *Urine No Effect Analytical* At 50 mg/L on fluorescent method of Corns. *0730*

Pregnenolone

Progesterone *Plasma No Effect Analytical* 1% or less cross reactivity with RIA. *0567*

Prenalterol

Renin Activity *Plasma No Effect Physiological* Not significantly different from control state in 23 patients with stage III heart disease over 1 mo under controlled conditions. *3768*

Prenylamine

Amphetamine *Urine No Effect Analytical* No effect at 100 mg/L on method of Rutter. *3097*

Prilocaine

Amphetamine *Urine Increase Analytical* Yields fluorophor with NBD chloride reaction. *2473*

Erythrocytes *Blood Decrease Physiological* Hemolysis with doses greater than 400 mg. *0071*

Hematocrit *Blood Decrease Physiological* Hemolysis with doses greater than 400 mg. *0071*

Hemoglobin *Blood Decrease Physiological* Hemolysis with doses greater than 400 mg. *0071*

Methemoglobin *Blood Increase Physiological* O-toluidine produced as metabolite causes hemolysis. *0752*

Primaquine

Alkaloids *Urine Increase Analytical* Pink with iodoplatinate in Frings TLC procedure Reacts with ninhydrin on TLC method of Frings. *1204*

Amphetamine *Urine No Effect Analytical* At 50 mg/L on fluorescent method of Hayes. *1534*

Barbiturate *Urine No Effect Analytical* With mercuric SO_4 on Frings TLC procedure at 10 mg/dL. With diphenylcarbazone on Frings TLC procedure at 10 mg/dL. *1204*
Urine Increase Analytical Orange spot with vanillin in Frings TLC procedure. With mercurous NO_3 reagent on Frings TLC procedure. *1204*

Bilirubin *Serum Increase Physiological* May cause hemolysis with G-6-PD deficiency. *0248*

Bromide *Urine No Effect Analytical* No effect on method of Frings. *1204*

Color *Urine Increase Analytical* Rusty yellow or brown color. *0382*

Erythrocytes *Blood Decrease Physiological* Hemolytic anemia in G-6-PD deficient persons. *0788*

Ethchlorvynol *Urine No Effect Analytical* No effect on method of Frings and Cohen. *1204*

FPN Test *Urine No Effect Analytical* No effect at 10 mg/dL on method of Frings. *1204*

Haptoglobin *Serum Decrease Physiological* May cause hemolytic anemia. *2313*

Heinz Body Formation *Blood Positive Physiological* May occur in early stages of hemolysis. *0333*

Hematocrit *Blood Decrease Physiological* Hemolytic anemia in G-6-PD deficient persons. *0788*

Hemoglobin *Blood Decrease Physiological* Hemolytic anemia in G-6-PD deficient persons. *0788*

Leukocytes *Blood Decrease Physiological* Agranulocytosis/leukopenia. *2313*

Methemoglobin *Blood Increase Physiological* May cause hemolysis in G-6-PD deficiency. *0071*

Reduced Glutathione
Red Blood Cells Decrease Physiological Associated with hemolysis and G-6-PD deficiency. *3171*

Salicylate *Urine No Effect Analytical* At 10 mg/dL no effect on Trinder method. *1204*

Primethamine

Barbiturate *Urine No Effect Analytical* With diphenylcarbazone on Frings TLC procedure at 10 mg/dL. *1204*

Primidone

Acetaminophen Screening Test *Urine Negative Analytical* No reaction with o-cresol at therapeutic concentrations. *3326*

Amino-4-Imidazole-5-Carboxamide Ribotide (AICAR)
Urine Increase Physiological If megaloblastic anemia occurs. *3025*

Amino Acids *Urine Positive Analytical* Unusual orange spot with ninhydrin migrating in all TLC systems close to threonine. *1596*

Barbiturate *Serum Increase Physiological* Phenobarbital is major metabolite. *1225*

Bilirubin *Serum No Effect Physiological* Insignificant displacement from protein in neonates. *3748*

Crystals *Urine Increase Physiological* In acute poisoning case crystals = primidone. *0205*

Diphenylhydantoin *Serum No Effect Analytical* No effect on TLC method of Simon, Jatlow. *3323*

Eosinophils *Blood Increase Physiological* May be allergic type of response. *0416*

Erythrocytes *Blood Decrease Physiological* Megaloblastic anemia secondary to disturbance in folic acid metabolism. *2968*

Ethchlorvynol *Serum No Effect Analytical* At 600 mg/L on GLC procedure of Evenson. *1058*

FIGLU (N-Formiminoglutamic Acid)
Urine Increase Physiological If megaloblastic anemia occurs. *3025*

Folate *Serum Decrease Physiological* May cause megaloblastic anemia (impairs absorption). *3025* Low serum folate in from 27 to 91% of treated epileptics in different studies. *2062*
CSF Decrease Physiological Occurs in many long-treated epileptics. *2967* Low folate in from 27 to 91% of treated epileptics in different studies. *2062*
Red Blood Cells Decrease Physiological Impaired deconjugation of polyglutamates in gut. *0987* Low serum folate in from 27 to 91% of treated epileptics in different studies. *2062*
Test Conditions No Effect Analytical No effect on *L. casei* or *S. fecalis*. *2427*

Glucaric Acid *Urine Increase Physiological* More potent inducer of hepatic enzymes than phenobarbital. *2089* Dose-dependent effect but less potent than carbamazepine. *2788*

Glucose *Serum No Effect Analytical* At concentration of 2.4 mg/L had no effect on Ektachem® method. *3393*

Hematocrit *Blood Decrease Physiological* Megaloblastic anemia secondary to disturbabce in folic acid metabolism. *2968*

Hemoglobin *Blood Decrease Physiological* Megaloblastic anemia secondary to disturbance in folic acid metabolism. *2968*

Indomethacin *Serum No Effect Analytical* No effect on HPLC method of Roberts and Smith. *3002*

LE Cells *Blood Positive Physiological* Less frequent than with many anticonvulsants. *0416*

Leucine Aminopeptidase Isoenzymes
Serum Increase Physiological Increased slower running components. *3087*

Leukocytes *Blood Decrease Physiological* Pancytopenia (AMA Blood dyscrasias). *2429*

MCV *Blood Increase Physiological* Megaloblastic anemia secondary to disturbance in folic acid metabolism. *2968*

Methaqualone *Serum No Effect Analytical* At 3 mg/L on GLC procedure of Evenson. *1057*

Mexiletine *Serum Decrease Physiological* Hepatic enzyme induction may decrease elimination half-life by 50%. *2011*

Neutrophils *Blood Decrease Physiological* Occasional case of agranulocytosis reported. *3717*

Phenobarbital *Serum Increase Physiological* Metabolic conversion *in vivo*. 3054

Platelets *Blood Decrease Physiological* Pancytopenia (AMA Blood dyscrasias). 2429

Potassium *Serum No Effect Analytical* At concentration of 10 mg/L had no effect on measurement by ISE with predilution. 3393

Propoxyphene *Serum No Effect Analytical* At 25 mg/L on method of Evenson. 1056

Prothrombin Time *Plasma Decrease Physiological* May shorten action of anticoagulants. 1679

Sodium *Serum No Effect Analytical* At concentration of 10 mg/L had no effect on measurement by ISE with predilution. 3393

T₃ Uptake *Serum No Effect Physiological* In 5 epileptic patients treated on average for 6 mo. 2168

Testosterone *Serum No Effect Physiological* Usual effect but may be slight increase. 1736

Testosterone, Free *Serum Decrease Physiological* Observed effect. 1736

Thyroid Stimulating Hormone (TSH)
Serum No Effect Physiological In 5 epileptic patients treated on average for 6 mo. 2168

Thyroxine (T₄) *Serum No Effect Physiological* In 5 epileptic patients treated on average for 6 mo. 2168
Serum Decrease Physiological Associated with enzyme induction. 3798

Thyroxine (T₄) Index, Free (FTI)
Serum No Effect Physiological In 5 epileptic patients treated on average for 6 mo. 2168
Serum Decrease Physiological Associated with enzyme induction. 3798

Thyroxine (T₄) (Murphy-Pattee)
Serum No Effect Physiological In 5 epileptic patients treated on average for 6 mo. 2168

Tri-iodothyronine (T₃) *Serum No Effect Physiological* In 5 epileptic patients treated on average for 6 mo. 2168 Associated with enzyme induction. 3798

Tri-iodothyronine (T₃), Reverse
Serum No Effect Physiological In 5 epileptic patients treated on average for 6 mo. 2168

Urea Nitrogen *Serum No Effect Analytical* At concentration of 2.4 mg/L had no effect on Ektachem® method. 3393

Probenecid

Alanine Aminotransferase *Serum No Effect Analytical* No effect at therapeutic concentration on Reflotron method. 1984
Serum Increase Physiological Hepatotoxic effect (centrolobular necrosis). 2313

Albumin *Serum No Effect Analytical* At concentration of 1300 mg/L had no effect on BCG method. 3393

Alkaline Phosphatase *Serum Increase Physiological* Hepatotoxic effect (centrolobular necrosis). 2313

Aminosalicylic Acid *Serum Increase Physiological* Increases by 2-4 fold (inhibits renal excretion). 1487

Aspartate Aminotransferase *Serum No Effect Analytical* No effect at therapeutic concentration on Reflotron method. 1984
Serum Increase Physiological Hepatotoxic effect (centrolobular necrosis). 2313

Bicarbonate *Serum No Effect Analytical* At concentration of 1300 mg/L had no effect on method using phenolphthalein. 3393

Bile *Urine Increase Physiological* Possible hepatotoxicity. 2313

Bilirubin *Serum No Effect Analytical* At concentration of 1300 mg/L had no effect on Jendrassik and Grof method. 3393
Serum Increase Physiological Hepatotoxic effect (centrolobular necrosis). 2313

BSP Retention *Serum Increase Physiological* Hepatotoxic effect (centrolobular necrosis). 3879

Calcium *Serum No Effect Analytical* At concentration of 1300 mg/L had no effect on cresolphthalein method. 3393

Calculi *Urine Increase Physiological* May occur if little fluid drunk or excrete much. 0133

Cefoperazone *Plasma No Effect Physiological* Concomitant administration had no effect on concentration. 0592

Cefsulodin *Plasma No Effect Physiological* Concomitant administration had no effect on concentration. 0592

Ceftazidime *Plasma No Effect Physiological* Concomitant administration had no effect on concentration. 0592

Ceftriaxone *Plasma No Effect Physiological* Concomitant administration had no effect on concentration. 0592

Cephalexin *Serum Increase Physiological* Reduces renal clearance. 1487

Cephalin Flocculation *Serum Increase Physiological* Hepatotoxic effect (centrolobular necrosis). 2313

Cephaloridine *Serum Increase Physiological* Reduces renal clearance. 1487

Cephalothin *Serum Increase Physiological* Reduces renal clearance. 1487

Chloride *Serum No Effect Analytical* At concentration of 1300 mg/L had no effect on mercurimetric method. 3393

Cholesterol *Serum No Effect Analytical* No effect at therapeutic concentration on Reflotron method. 1984 At concentration of 260 mg/L had no effect on CHOD-Iodide method. 3393 At concentration of 280 mg/L had no effect on CHOD-PAP method. 3393 At concentration of 260 mg/L had no effect on catalase-AlDH method. 3393 At concentration of 260 mg/L had no effect on method using catalase-Hantzsch reaction. 3393 At concentration of 1300 mg/L had no effect on Liebermann-Burchard method. 3393

Creatinine *Serum No Effect Analytical* At concentration of 1300 mg/L had no effect on AutoAnalyzer Jaffé method. 3393 At concentration of 200 mg/L had no effect on Jaffé-Fading-Fraction method. 3393 At concentration of 200 mg/L had no effect on Jaffé-Fuller's earth method. 3393 At concentration of 1,000 mg/L had no effect on kinetic Jaffé method on BKA-2. 3393

Cyclic Adenosine Monophosphate
Plasma Increase Physiological In 8 healthy young men. Probable effect on carrier mediated process to clear plasma cyclic AMP. 1318
Urine No Effect Physiological In 8 healthy young men. 1318

Dapsone *Serum Increase Physiological* 50% increase after 4 h, inhibits excretion. 1487

1,25-Dihydroxy Vitamin D₃ *Serum No Effect Physiological* In 8 healthy young men. 1318

Erythrocytes *Blood Decrease Physiological* Hemolytic anemia (sensitivity dependent). 0788
Urine Increase Physiological ?due to sensitivity, or toxicity. 0404

Estradiol *Urine Decrease Physiological* Tubular excretion may be blocked. 0023

Estriol *Urine Decrease Physiological* Tubular excretion may be blocked. 0023

Estrone *Urine Decrease Physiological* Tubular excretion may be blocked. 0023

Glucose *Serum No Effect Analytical* No effect at therapeutic concentration on Reflotron method. 1984 At concentration of 40 mg/L had no effect on Ektachem® method. 3393 At concentration of 200 mg/L had no effect on hexokinase/G-6-PDH method. 3393
Serum Decrease Physiological Reported effect. 0117
Urine No Effect Analytical No effect on glucose oxidase methods. 1487

γ-Glutamyltransferase (GGT) *Serum No Effect Analytical* No effect at therapeutic concentration on Reflotron method. 1984

Hematocrit *Blood Decrease Physiological* Hemolytic anemia (sensitivity dependent). 0788

Hemoglobin *Blood Decrease Physiological* Hemolytic anemia (sensitivity dependent). 0788
Urine Increase Physiological Actual bleeding caused by drug. 1022

HMPG (4-Hydroxy-3-Methoxy-Phenylethylene Glycol)
CSF Increase Physiological 60% increase on average if 100 mg/kg over 18 h. 1351

HMPG Sulfate *CSF Increase Physiological* Small but not significant increase after 9 h. 1351

Probenecid (continued)

Homovanillic Acid *CSF Increase Physiological* Approximately 6 fold at 9 h, 9 fold at 18 h. *1351*

5-Hydroxyindoleacetic Acid (5-HIAA)
CSF Increase Physiological Approximately 4 fold at 9 h, 5 fold at 18 h. *1351*

25-Hydroxy Vitamin D₃ *Serum No Effect Physiological* In 8 healthy young men. *1318*

Indomethacin *Serum Increase Physiological* Inhibits renal tubular secretion. *1487*

17-Ketosteroids *Urine Decrease Physiological* Decrease of up to 50% reported. *1198*

Lactate *Serum No Effect Analytical* At concentration of 200 mg/L had no effect on enzyme method. *3393*

Leukocytes *Blood Decrease Physiological* Occasional aplastic anemia seen. *1679*

Methemoglobin *Blood Increase Physiological* May cause hemolysis with G-6-PD deficiency. *3581*

Moxalactam *Plasma No Effect Physiological* Concomitant administration had no effect on concentration. *0592*

PAH Clearance *Urine Decrease Physiological* Renal clearance impaired. *2375*

Parathyroid Hormone *Plasma No Effect Physiological* In 8 healthy young men. *1318*

Phenylalanine *CSF No Effect Physiological* No effect on concentration observed. *2373*

Phosphate *Serum No Effect Analytical* At concentration of 1300 mg/L had no effect on phosphomolybdate method. *3393*

Platelets *Blood Decrease Physiological* Occasional aplastic anemia seen. *1679*

Potassium *Serum No Effect Analytical* At concentration of 1300 mg/L had no effect on measurement by ISE with predilution. *3393* At concentration of 400 mg/L had no effect on measurement by ISE without predilution. *3393*

Protein *Serum No Effect Analytical* At concentration of 1300 mg/L had no effect on biuret method with blank correction. *3393*
Urine Increase Physiological Nephrotoxic effect. *3879* Nephrotoxic effect. *1488*

Protein Electrophoresis *Serum No Effect Analytical* At concentration of 400 mg/L had no effect on automated Olympus-Hite method. *3393*

Prothrombin Time *Plasma Increase Physiological* Probably displaces anticoagulant from binding protein. *1487*

PSP Excretion *Urine Decrease Physiological* Inhibits renal transport of PSP. *3879*

Rifampin *Serum Increase Physiological* Decreased hepatic uptake, serum concentration increase by 86%. *1918*

Sodium *Serum No Effect Analytical* At concentration of 1300 mg/L had no effect on measurement by ISE with predilution. *3393* At concentration of 400 mg/L had no effect on measurement by ISE without predilution. *3393*

Sugar *Urine Increase Analytical* False positive with Benedict's or Clinitest®. *3505* Reducing substances with Benedict's, Clinitest®. *0583*

Sulfinpyrazone *Serum Increase Physiological* Inhibits renal tubular secretion. *1487*

Thymol Turbidity *Serum Increase Physiological* Hepatotoxic effect (centrolobular necrosis). *2313*

Tolbutamide *Serum No Effect Physiological* Probably has no effect on plasma concentration. *1487*

Triglycerides *Serum No Effect Analytical* No effect at therapeutic concentration on Reflotron method. *1984* At concentration of 1300 mg/L had no effect on lipase/esterase method. *3393*

Urea Nitrogen *Serum No Effect Analytical* No effect at therapeutic concentration on Reflotron method. *1984* At concentration of 1300 mg/L had no effect on diacetylmonoxime method. *3393* At concentration of 40 mg/L had no effect on Ektachem® method. *3393*
Serum Increase Physiological Possible nephrotoxicity. *1488*

Uric Acid *Serum No Effect Analytical* No effect at therapeutic concentration on Reflotron method. *1984* At concentration of 280 mg/L had no effect on Kageyama-Hantzsch method. *3393* At concentration of 200 mg/L had no effect on catalase-AIDH method. *3393* At concentration of 1300 mg/L had no effect on phosphotungstate reduction method. *3393* At concentration of 570 mg/L had no effect on Seralyzer method. *3393*
Serum Decrease Physiological Uricosuric action. *1433*
Urine Increase Physiological Uricosuric action. *1433*
Urine Decrease Physiological Small doses depress secretion. *1343*

Probucol

Aldolase *Serum Decrease Physiological* Mechanism obscure. *0806*

Apolipoprotein AI *Serum Decrease Physiological* In 50 diabetics given 500 mg/d for 16 weeks and reduction greatest in highest cholesterol and triglyceride patients. *1523*

Apolipoprotein CII *Serum Decrease Physiological* In 50 diabetics given 500 mg/d for 16 weeks and reduction greatest in highest cholesterol and triglyceride patients. *1523*

Bicarbonate *Serum Increase Physiological* Mechanism obscure (effect slight). *0806*

Calcium *Serum Decrease Physiological* Mechanism obscure (effect slight). *0806*

Cholesterol *Serum Decrease Physiological* Lowered by more than 20 mg/dL in most patients. *0806* In 50 diabetics given 500 mg/d for 16 weeks and reduction greatest in highest cholesterol and triglyceride patients. *1523*

Cholesterol, High Density Lipoprotein
Serum Decrease Physiological In 50 diabetics given 500 mg/d for 16 weeks and reduction greatest in highest cholesterol and triglyceride patients. *1523*

Eosinophils *Blood Increase Physiological* Hypersensitivity response (less than 10%). *0806*

Growth Hormone *Plasma Decrease Physiological* Mechanism obscure. *0806*

Triglycerides *Serum Decrease Physiological* In 50 diabetics given 500 mg/d for 16 weeks and reduction greatest in highest cholesterol and triglyceride patients. *1523*

Uric Acid *Serum Increase Physiological* Effect noted in women only. *0806*

Procainamide

Alanine Aminotransferase *Serum Increase Physiological* Hepatotoxic effect. *2313*

Albumin *Serum No Effect Analytical* No effect on SMA 12/60 method at 35 mg/dL. *2636* At concentration of 50 mg/L had no effect on BCG method. *3393*

Alkaline Phosphatase *Serum No Effect Analytical* No effect on SMA 12/60 method at 35 mg/dL. *2636*
Serum Increase Physiological Reversible hepatic toxicity reported. *1022*

Alkaloids *Urine No Effect Analytical* With ninhydrin on Frings TLC procedure at 10 mg/dL. With iodoplatinate of Frings TLC procedure at 10 mg/dL. *1204*

Aminobenzoic Acid *Urine Increase Physiological* Up to 10% excreted as this. *1343*

Antibodies to Histones *Serum Positive Physiological* Antibodies to histone induced by drug. *2848*

Antinuclear Antibodies *Serum Positive Physiological* Reported to occur in 50% patients. *3505* 50-80% patients have antibodies after 3 to 6 mo. *1958*

Aspartate Aminotransferase *Serum No Effect Analytical* No effect on SMA 12/60 method at 35 mg/dL. *2636*
Serum Increase Physiological Hepatotoxic effect. *2313*

Barbiturate *Urine No Effect Analytical* With vanillin on Frings procedure at 10 mg/dL. With diphenylcarbazone on Frings TLC procedure at 10 mg/dL. With mercuric SO₄ on Frings TLC procedure at 10 mg/dL. With mercurous NO₃ on Frings TLC procedure at 10 mg/dL. *1204*

Basophils *Blood Decrease Physiological* Occasional severe reduction noted. *1586*

Bicarbonate *Serum No Effect Analytical* At concentration of 50 mg/L had no effect on method using phenolphthalein. *3393*

Bile *Urine Increase Physiological* Hepatotoxic effect. *2313*

Bilirubin *Serum No Effect Analytical* No effect on SMA 12/60 method at 35 mg/dL. *2636* At concentration of 350 mg/L had no effect on Ektachem® method. *3393* At concentration of 50 mg/L had no effect on Jendrassik and Grof method. *3393* *Serum Increase Physiological* Reversible hepatic toxicity reported. *1488* In 3% of all patients receiving drug, but in 14% of patients with direct Coombs' test. *1958*

Bilirubin, Indirect *Serum Increase Physiological* In 3% of all patients receiving drug, but in 14% of patients with direct Coombs' test. *1958*

BSP Retention *Serum Increase Physiological* Reported case of hepatic toxicity. *1488*

Calcium *Serum No Effect Analytical* No effect on SMA 12/60 method at 35 mg/dL. *2636* At concentration of 50 mg/L had no effect on cresolphthalein method. *3393*

Cephalin Flocculation *Serum Increase Physiological* Hepatotoxic effect. *2313*

Chlordiazepoxide *Blood Increase Analytical* 2 mg/dL equivalent to 1.1 mg/dL by method of Riddick. *2983*

Chloride *Serum No Effect Analytical* At concentration of 50 mg/L had no effect on mercurimetric method. *3393*

Cholesterol *Serum No Effect Analytical* No effect on SMA 12/60 method at 35 mg/dL. *2636*

Coombs' Test *Serum Positive Physiological* Mechanism obscure. *1486*

Coombs' Test, Direct *Serum Positive Physiological* Doubled incidence in individuals receiving drug compared with control with production of red cell autoantibody. *1958*

Creatinine *Serum No Effect Analytical* No effect on SMA 12/60 method at 35 mg/dL. *2636* At concentration of 50 mg/L had no effect on AutoAnalyzer Jaffé method. *3393*

Dapsone *Serum Increase Analytical* Develop color in procedure of Higgins. *1590*

Digoxin *Serum No Effect Physiological* No significant effect when drug coadministered. *2101*

Eosinophils *Blood Increase Physiological* Evidence of allergic response to drug. *3215* *Blood Decrease Physiological* Occasional severe reduction noted. *1586*

Erythrocytes *Blood Decrease Physiological* May cause hemolytic anemia. *2583* Pancytopenia may occur with/without lupus-like syndrome. *3215*

Erythrocyte Sedimentation Rate *Blood Increase Physiological* Associated with SLE-like syndrome. *0933*

Estriol *Urine No Effect Analytical* No effect of 12 mg/L on GLC method. *1163*

Ethchlorvynol *Urine No Effect Analytical* No effect on method of Frings and Cohen at 10 mg/dL. *1205*

FPN Test *Urine No Effect Analytical* No effect at 10 mg/dL on method of Frings. *1204*

Glucose *Serum No Effect Analytical* No effect on SMA 12/60 method at 35 mg/dL. *2636*

Haptoglobin *Serum Decrease Physiological* In 3% of all patients receiving drug, but in 14% of patients with direct Coombs' test. *1958*

Hematocrit *Blood Decrease Physiological* Observed with SLE-like syndrome, hemolytic anemia. *0933*

Hemoglobin *Blood Decrease Physiological* Observed with SLE-like syndrome, hemolytic anemia. *0933*

Lactate Dehydrogenase *Serum No Effect Analytical* No effect on SMA 12/60 method at 35 mg/dL. *2636*

Latex Fixation *Serum Positive Physiological* Associated with drug induced lupus. *0930*

LE Cells *Blood Positive Physiological* Reported effect. *2055* 10-20% patients eventually develop drug-induced lupus syndrome. *1958*

Leukocytes *Blood Decrease Physiological* Agranulocytosis. *3537* Significant association observed with therapy with sustained release form of drug at normal therapeutic concentration of drug. *1029* Pancytopenia may occur with/without lupus-like syndrome. *3215*

Lymphocytotoxic Antibodies *Serum Positive Physiological* 50-80% patients have antibodies after 3 to 6 mo. *1958*

Neutrophils *Blood Decrease Physiological* Occasional severe reduction noted. *1586* Occasional case of agranulocytosis reported. *3717*

Phosphate *Serum No Effect Analytical* No effect on SMA 12/60 method at 35 mg/dL. *2636* At concentration of 50 mg/L had no effect on phosphomolybdate method. *3393*

Platelets *Blood Decrease Physiological* Pancytopenia may occur with/without lupus-like syndrome. *3215*

Potassium *Serum Increase Analytical* At concentrations above 8 mg/L (normal therapeutic concentration 10 mg/L) raised concentration as measured by ISE with predilution. *3393*

Protein *Serum No Effect Analytical* No effect on SMA 12/60 method at 35 mg/dL. *2636* At concentration of 50 mg/L had no effect on biuret method with blank correction. *3393*

Pseudocholinesterase *Serum Decrease Physiological* At therapeutic concentration *in vitro* inhibition of enzyme by 15 to 30%. *1853*

Reticulocytes *Blood Increase Physiological* In 3% of all patients receiving drug, but in 14% of patients with direct Coombs' test. *1958*

Salicylate *Urine No Effect Analytical* At 10 mg/dL no effect on Trinder method. *1204*

Sodium *Serum No Effect Analytical* At concentration of 50 mg/L had no effect on measurement by ISE with predilution. *3393*

Thymol Turbidity *Serum Increase Physiological* Hepatotoxic effect. *2313*

Triglycerides *Serum No Effect Analytical* At concentration of 50 mg/L had no effect on lipase/esterase method. *3393*

Urea Nitrogen *Serum No Effect Analytical* At concentration of 50 mg/L had no effect on diacetylmonoxime method. *3393* At concentration of 11.4 mg/L had no effect on Ektachem® method. *3393*

Uric Acid *Serum No Effect Analytical* No effect on SMA 12/60 method at 35 mg/dL. *2636* At concentration of 50 mg/L had no effect on phosphotungstate reduction method. *3393*

Procaine

Alanine Aminotransferase *Serum No Effect Analytical* No effect at therapeutic concentration on Reflotron method. *1984*

Albumin *Serum No Effect Analytical* At concentration of 2 mg/L had no effect on BCG method. *3393*

Aspartate Aminotransferase *Serum No Effect Analytical* No effect at therapeutic concentration on Reflotron method. *1984*

Bicarbonate *Serum No Effect Analytical* At concentration of 2 mg/L had no effect on method using phenolphthalein. *3393*

Bilirubin *Serum No Effect Analytical* At concentration of 2 mg/L had no effect on Jendrassik and Grof method. *3393*

Calcium *Serum No Effect Analytical* At concentration of 2 mg/L had no effect on cresolphthalein method. *3393*

Chloride *Serum No Effect Analytical* At concentration of 2 mg/L had no effect on mercurimetric method. *3393*

Cholesterol *Serum No Effect Analytical* No effect at therapeutic concentration on Reflotron method. *1984* At concentration of 2 mg/L had no effect on Liebermann-Burchard method. *3393*

Creatinine *Serum No Effect Analytical* At concentration of 2 mg/L had no effect on AutoAnalyzer Jaffé method. *3393*

Globulin *CSF Increase Analytical* Gives false positive Pandy test. *0251*

Glucose *Serum No Effect Analytical* No effect at therapeutic concentration on Reflotron method. *1984* At concentration of 40 mg/L had no effect on hexokinase/G-6-PDH method. *3393*

γ-Glutamyltransferase (GGT) *Serum No Effect Analytical* No effect at therapeutic concentration on Reflotron method. *1984*

Lactate *Serum No Effect Analytical* At concentration of 40 mg/L had no effect on enzymatic method. *3393*

Phosphate *Serum No Effect Analytical* At concentration of 2 mg/L had no effect on phosphomolybdate method. *3393*

Procaine (continued)

Porphobilinogen *Urine Increase Analytical* Interferes with Ehrlich's aldehyde reaction. *3879*

Potassium *Serum No Effect Analytical* At concentration of 2 mg/L had no effect on measurement by ISE with predilution. *3393*

Protein *Serum No Effect Analytical* At concentration of 2 mg/L had no effect on biuret method with blank correction. *3393*
CSF Increase Analytical Interferes with Folin-Ciocalteu reagent. *1237*

Sodium *Serum No Effect Analytical* At concentration of 2 mg/L had no effect on measurement by ISE with predilution. *3393*

Sulfa as Sulfanilamide *Serum Increase Analytical* Metabolized to PABA which interferes. *1343*

Triglycerides *Serum No Effect Analytical* No effect at therapeutic concentration on Reflotron method. *1984* At concentration of 2 mg/L had no effect on lipase/esterase method. *3393*

Urea Nitrogen *Serum No Effect Analytical* No effect at therapeutic concentration on Reflotron method. *1984* At concentration of 2 mg/L had no effect on diacetylmonoxime method. *3393*

Uric Acid *Serum No Effect Analytical* No effect at therapeutic concentration on Reflotron method. *1984* At concentration of 2 mg/L had no effect on phosphotungstate reduction method. *3393*

Urobilinogen *Urine Increase Analytical* Interferes with Ehrlich's aldehyde reaction. *3879*

Procarbazine

Bilirubin *Serum Increase Physiological* Reported effect. ?mechanism. *1488*

Erythrocytes *Blood Decrease Physiological* Anemia and bone marrow depression. *0071*

Heinz Body Formation *Blood Positive Physiological* May occur with hemolysis. *1680*

Leukocytes *Blood Decrease Physiological* Leukopenia and bone marrow depression. *0071*

Occult Blood *Feces Positive Physiological* Melena and gastrointestinal tract bleeding. *0071*

Platelets *Blood Decrease Physiological* Thrombocytopenia and bone marrow depression. *0071*

Prochlorperazine

Acetaminophen Screening Test *Urine Negative Analytical* No reaction with o-cresol at therapeutic concentrations. *3326*

Alanine Aminotransferase *Serum Increase Physiological* Cholestatic effect. *0071*

Albumin *Serum No Effect Analytical* At concentration of 1 mg/L had no effect on BCG method. *3393*

Alkaline Phosphatase *Serum Increase Physiological* Cholestatic effect. *0071*

Alkaloids *Urine No Effect Analytical* With ninhydrin on Frings TLC procedure at 10 mg/dL. *1204*
Urine Increase Analytical Purple with iodoplatinate on Frings TLC procedure. *1204*

Aspartate Aminotransferase
Serum Increase Physiological Cholestatic effect. *0071*

Barbiturate *Urine No Effect Analytical* With mercurous NO_3 on Frings TLC procedure at 10 mg/dL. With diphenylcarbazone on Frings TLC procedure at 10 mg/dL. *1204*
Urine Increase Analytical Pink spot with vanillin in Frings TLC procedure. Orange spot with mercuric SO_4 in Frings TLC procedure. *1204*

Bicarbonate *Serum No Effect Analytical* At concentration of 1 mg/L had no effect on method using phenolphthalein. *3393*

Bile *Urine Increase Physiological* Cholestatic effect. *0071*

Bilirubin *Serum No Effect Analytical* At concentration of 1 mg/L had no effect on Jendrassik and Grof method. *3393*
Serum Increase Physiological Cholestatic effect. *0071*

BSP Retention *Serum Increase Physiological* Cholestatic effect. *0071*

Calcium *Serum No Effect Analytical* At concentration of 1 mg/L had no effect on cresolphthalein method. *3393*

Catecholamines *Urine Increase Physiological* Increased metabolism, decreased organ uptake of norepinephrine. *3879*

Cephalin Flocculation *Serum Increase Physiological* Cholestatic effect. *0071*

Chlordiazepoxide *Blood No Effect Analytical* On method of Riddick at 5 mg/dL. *2983*

Chloride *Serum No Effect Analytical* At concentration of 1 mg/L had no effect on mercurimetric method. *3393*

Cholesterol *Serum No Effect Analytical* At concentration of 1 mg/L had no effect on CHOD-PAP method. *3393*
Serum Increase Physiological Cholestatic effect. *0071*

Corticosteroids *Urine Decrease Analytical* Decreased absorption at 410 nm in Porter-Silber procedure. *0022*

Creatinine *Serum No Effect Analytical* At concentration of 1 mg/L had no effect on AutoAnalyzer Jaffé method. *3393*

Diphenylhydantoin *Serum Increase Physiological* Reported impairment of metabolism. *2042*

Dopamine *Urine Increase Physiological* Up to 79% increase after 30 mg in controls. *1607*

Epinephrine *Urine No Effect Physiological* No significant effect with up to 30 mg in controls. *1607*

Estriol *Urine No Effect Analytical* No effect of 25 mg/L on GLC method. *1163*

Ethchlorvynol *Serum No Effect Analytical* At 100 mg/L on GLC procedure of Evenson. *1058*
Urine No Effect Analytical No effect on method of Frings and Cohen at 20 mg/dL. *1205*

FPN Test *Urine No Effect Analytical* Negative FPN test with 0.4 mg/dL concentration in Frings TLC procedure. *1204*
Urine Increase Analytical Pink color observed in method of Frings. Pink FPN test with 0.5 mg/dL concentration in Frings TLC procedure. *1204*

Glucose *Serum No Effect Analytical* At concentration of 1 mg/L had no effect on GOD/POD-PAP method. *3393*

17-Hydroxycorticosteroids *Urine Increase Analytical* Interference with Porter-Silber reaction. *0427*
Urine Decrease Analytical Abnormal yellow-pink color, blank not adequate. *0583*

5-Hydroxyindoleacetic Acid (5-HIAA)
Urine Decrease Analytical Interference with nitrosonaphthol methods. *3065*

17-Ketosteroids *Urine No Effect Analytical* No significant effect with Zimmermann reaction. *0427*

Leukocytes *Blood Decrease Physiological* Agranulocytosis due to interference in development. *0072*

Metanephrines, Total *Urine Increase Physiological* Increased metabolism, decreased organ uptake of norepinephrine. *3879*

Methaqualone *Serum No Effect Analytical* At 100 mg/L on GLC procedure of Evenson. *1057*

Neutrophils *Blood Decrease Physiological* Occasional case of agranulocytosis reported. *3717*

Norepinephrine *Urine No Effect Physiological* No significant effect with up to 30 mg in controls. *1607*

Phenylketones *Urine Positive Analytical* Light purple with $FeCl_3$, also with Phenistix®. *0775*

Phosphate *Serum No Effect Analytical* At concentration of 1 mg/L had no effect on phosphomolybdate method. *3393*

Potassium *Serum No Effect Analytical* At concentration of 1 mg/L had no effect on flame-photometric method. *3393*

Prolactin *Plasma Increase Physiological* Typical dose-related response to i.m. administered drug due to antidopaminergic actions. *2072*

Protein *Serum No Effect Analytical* At concentration of 1 mg/L had no effect on biuret method with blank correction. *3393*

Prothrombin Time *Plasma Increase Physiological* Associated with impaired excretion of bile salts. *0071*

Salicylate *Urine No Effect Analytical* Up to 2 mg/dL no effect or Trinder procedure. *1204*
Urine Increase Analytical 4 mg/dL produces pink color on Trinder procedure. *1204*

Sodium *Serum No Effect Analytical* At concentration of 1 mg/L had no effect on flame-photometric method. *3393*

Thymol Turbidity *Serum Increase Physiological* Cholestatic effect. *0071*

Triglycerides *Serum No Effect Analytical* At concentration of 1 mg/L had no effect on lipase/esterase method. *3393*

Urea Nitrogen *Serum No Effect Analytical* At concentration of 1 mg/L had no effect on diacetylmonoxime method. *3393*

Uric Acid *Serum No Effect Analytical* At concentration of 1 mg/L had no effect on phosphotungstate reduction method. *3393*

Urobilinogen *Urine Decrease Physiological* Cholestatic effect. *0071*
Feces Decrease Physiological Pale stools associated with cholestasis. *0071*

Vanillylmandelic Acid *Urine No Effect Physiological* No significant effect with up to 30 mg in controls. *1607*
Urine Increase Physiological Increased metabolism, decreased organ uptake of norepinephrine. *3879*

Procyclidine

Amylase *Serum Increase Physiological* May cause acute parotitis (theoretical effect). *1680*

Proflavine

Chromosomes *Test Conditions Abnormal Physiological* Clastogenic in human hela cultures. *3282*

Progabide

Alanine Aminotransferase *Serum Increase Physiological* Several fold increase in one patient necessitating stopping treatment. *3188*

Aspartate Aminotransferase
Serum Increase Physiological Several fold increase in one patient necessitating stopping treatment. *3188*

Diphenylhydantoin *Serum Increase Physiological* Increased concentration in patients who had a therapeutic response. *3188*

Phenobarbital *Serum Increase Physiological* Increased concentration in patients who had a therapeutic response. *3188*

Progesterone

Alanine Aminotransferase *Serum Increase Physiological* Hepatotoxicity. *2313*

Albumin *Serum Increase Physiological* Not observed with combined therapy. *0795*

Aldosterone *Urine No Effect Analytical* No significant effect RIA procedure of Drewes. *0952*
Urine Increase Analytical 1 mg/L equivalent to 11 μg/L by method of Drewes. *0952*

Alkaline Phosphatase *Serum Increase Physiological* Hepatotoxicity. *2313*

Amino Acids *Serum Decrease Physiological* Given i.m. to men decreases free and total. *0743* Specific amino acids affected in men. *0742*

α-Amino-Nitrogen *Plasma Decrease Physiological* Catabolic effect. *0743*
Urine No Effect Physiological Almost no effect in normal males. *3571*

Aspartate Aminotransferase
Serum Increase Physiological Hepatotoxicity. *2313*
Serum Decrease Physiological Observed in healthy individuals when only treatment. *0991*

Bile *Urine Increase Physiological* Hepatotoxicity. *2313*

Bilirubin *Serum Increase Physiological* Hepatotoxicity also transient familial increase. *2313*

BSP Retention *Serum Increase Physiological* Hepatic toxicity occasional occurrence. *2498*

Calcium *Serum Increase Physiological* Probable effect with remission of metastases. *2313*

Cephalin Flocculation *Serum Increase Physiological* Hepatotoxicity. *2313*

Ceruloplasmin *Serum No Effect Physiological* No effect reported over several months. *3505*

Cholesterol *Serum Decrease Physiological* Slight effect when only treatment. *0991*

Copper *Serum No Effect Physiological* No effect reported over several months. *3505*

Corticosteroids *Plasma Increase Analytical* 28% (cortisol 100%) competitive protein binding method of Ficher. *1128*

Cortisol *Plasma Increase Analytical* If nonselective extraction and competitive protein binding used. *0512*

Estrogens *Urine No Effect Analytical* At 50 mg/L on fluorescent method of Corns. *0730*

Estrone *Urine Increase Physiological* Not significant change however. *0991*

Globulin *Serum Decrease Physiological* Observed in healthy women when only treatment. *0991*

γ-Globulin *Serum Increase Physiological* Not observed with combined therapy. *0795*

Glucaric Acid *Urine Increase Physiological* Reported induction of hepatic enzymes. *1694*

Glucose *Serum Decrease Physiological* Slight insignificant effect (contrast with estrogens). *0991*

17-Hydroxycorticosteroids *Urine Decrease Physiological* Exact mechanism not known. *0022*

16-α-Hydroxyprogesterone *Plasma Increase Analytical* Up to 25% cross reactivity. *0008*

Luteinizing Hormone (LH) *Plasma Decrease Physiological* Suppresses LH peak. *3237*

Magnesium *Serum Increase Physiological* Significantly higher than in controls. *0794*

Pregnanediol *Urine Decrease Physiological* Significant decrease with treatment (from 3.5 to 2.0 mg/24 h). *0991*

Protein *Serum Increase Physiological* Metabolic effect. *0795*

Protein Bound Iodine (PBI) *Serum Increase Physiological* Increased production of thyroxine binding globulin. *3879*
Serum Decrease Physiological Inhibits iodination of tyrosine in thyroxine binding globulin. *0012*

Sodium *Serum Increase Physiological* May cause sodium retention. *2220*
Urine Increase Physiological May occur with high doses. *2427*

Thymol Turbidity *Serum Increase Physiological* Hepatotoxicity. *2313*

Thyroxine (T$_4$) Binding Globulin
Serum Increase Physiological Increased synthesis. *3879*

Transferrin *Serum No Effect Physiological* No significant effect after 6-9 mo. *3505*

Progestogens

Aminolevulinic Acid *Urine Increase Physiological* May precipitate attack of acute porphyria. *1016*

Cortisol *Plasma No Effect Physiological* No effect with progestogen — only pill. *0537*

Cortisol, Protein Bound *Plasma No Effect Physiological* No effect with progestogen — only pill. *0537*

Leucine Aminopeptidase *Serum No Effect Physiological* No effect observed. *1498*

NBT Test *Blood No Effect Physiological* When administered alone no effect observed. *0163*

Porphyrins *Urine Increase Physiological* May precipitate attack of acute porphyria. *1016*

Transcortin *Serum No Effect Physiological* No effect with progestogen — only pill. *0537*

Proguanil

Amphetamine *Urine No Effect Analytical* No effect at 100 mg/L on method of Rutter. *3097*

Erythrocytes *Urine Increase Physiological* Actual bleeding may be caused by drug. *1022*

Hemoglobin *Urine Increase Physiological* Actual bleeding caused by drug. *1022*

Prolactin

Osmolality *Serum Increase Physiological* Effect not marked. *1662*

pH *Urine Increase Physiological* Individual response — may be marked. *1662*

Potassium *Urine Decrease Physiological* Reduces renal excretion, not as marked as for sodium. *1662*

Sodium *Serum Increase Physiological* Renal retention effect when given i.m. *1662*
Urine Decrease Physiological Reduces renal excretion at tubular level. *1662*

Volume *Urine Decrease Physiological* Action for up to 8 h when given i.m. *1662*

Proline

Ammonia *Plasma Decrease Analytical* With 1,000 nmoles produces 20% inhibition of indophenol reaction. *1290*

Pyrrole-2-Carboxylate *Urine Increase Physiological* Exact mechanism obscure. *1536*

Pyruvate Kinase *Red Blood Cells Decrease Analytical* Most marked with larger side chain. *0370*

Tyrosine *Plasma No Effect Analytical* On fluorometric procedure of Ambrose. *0080*

Promazine

Acetaminophen Screening Test *Urine Negative Analytical* No reaction with o-cresol at therapeutic concentrations. *3326*

Alanine Aminotransferase *Serum Increase Physiological* May cause cholestatic (hepatocanalicular) jaundice. *0248*

Alkaline Phosphatase *Serum Increase Physiological* May cause cholestatic (hepatocanalicular) jaundice. *0248*

Alkaloids *Urine No Effect Analytical* With ninhydrin on Frings TLC procedure at 10 mg/dL. *1204*
Urine Increase Analytical Purple with iodoplatinate on Frings TLC procedure. *1204*

Aspartate Aminotransferase
Serum Increase Physiological May cause cholestatic (hepatocanalicular) jaundice. *0248*

Barbiturate *Urine No Effect Analytical* With diphenylcarbazone on Frings TLC procedure at 10 mg/dL. With mercurous NO_3 on Frings TLC procedure at 10 mg/dL. *1204*
Urine Increase Analytical Orange spot with mercuric SO_4 in Frings TLC procedure. Orange spot with vanillin in Frings TLC procedure. *1204*

Bile *Urine Increase Physiological* May cause cholestasis. *2313*

Bilirubin *Serum Increase Physiological* May cause cholestatic (hepatocanalicular) jaundice. *0248*

Bromide *Urine No Effect Analytical* No effect on method of Frings. *1204*

BSP Retention *Serum Increase Physiological* May cause cholestasis. *2313*

Cephalin Flocculation *Serum Increase Physiological* May cause cholestasis. *2313*

Cholesterol *Serum Increase Physiological* May cause cholestatic (hepatocanalicular) jaundice. *0248*

Estriol *Urine No Effect Analytical* No effect of 25 mg/L on GLC method. *1163*

Ethchlorvynol *Serum No Effect Analytical* At 100 mg/L on GLC procedure of Evenson. *1058*
Urine No Effect Analytical No effect on method of Frings and Cohen at 10 mg/dL. *1205*

Folate *Serum No Effect Analytical* If chromatographic procedure of Landon used. *2068*

FPN Test *Urine No Effect Analytical* Negative FPN test with 0.4 mg/dL concentration in Frings TLC procedure. *1204*
Urine Increase Analytical Orange FPN test with 0.5 mg/dL concentration on Frings procedure Orange color observed in method of Frings. *1204*

17-Hydroxycorticosteroids *Urine Increase Analytical* In vitro effect at least on Glenn-Nelson method. *0427*
Urine Decrease Physiological Acts on hypothalamus to decrease ACTH secretion. *0427*

5-Hydroxyindoleacetic Acid (5-HIAA)

Urine Decrease Analytical Interferes with method of Goldenberg. *1328*

[131]I Uptake *Serum Decrease Physiological* Sparine® contains tetraiodofluorescein. *2652*

17-Ketosteroids *Urine Decrease Analytical* In vitro effect at least on Zimmermann reaction. *0427*

Leukocytes *Blood Decrease Physiological* Agranulocytosis. *1138*

Methaqualone *Serum No Effect Analytical* At 100 mg/L on GLC procedure of Evenson. *1057*

Neutrophils *Blood Decrease Physiological* Occasional case of agranulocytosis reported. *3717*

Platelets *Blood Decrease Physiological* May rarely cause bone marrow aplasia. *1436*

Prolactin *Plasma No Effect Physiological* No significant change in response to 25 mg i.m. *1416*
Plasma Increase Physiological Marked effect in male and female psychiatric patients treated for up to 4 weeks. *3630*

Protein *Urine Increase Analytical* Affects turbidity tests for up to 3 d. *0065*
Urine Increase Physiological Affects turbidity tests for up to 3 d. *0065*

Protein Bound Iodine (PBI) *Serum No Effect Physiological* No effect with normal doses. *0830*

Prothrombin Time *Plasma Increase Physiological* Associated with failure of excretion of bile salts. *1713*

Salicylate *Urine No Effect Analytical* Up to 2 mg/dL no effect on Trinder procedure. *1204*
Urine Increase Analytical 4 mg/dL produces orange color on Trinder procedure. *1204*

Thymol Turbidity *Serum Increase Physiological* May cause cholestasis. *2313*

Urobilinogen *Urine Decrease Physiological* Cholestatic effect. *1713*
Feces Decrease Physiological Pale stools as result of cholestasis. *1713*

Promethazine

Alanine Aminotransferase *Serum No Effect Analytical* At acute overdose concentration (20 mg/dL) on Technicon® SMAC® method. *3719*
Serum Increase Physiological May cause cholestasis. *2313*

Albumin *Serum No Effect Analytical* At acute overdose concentration (20 mg/dL) on Technicon® SMAC® method. *3719* At concentration of 200 mg/L had no effect on BCG method. *3393*

Alkaline Phosphatase *Serum No Effect Analytical* At acute overdose concentration (20 mg/dL) on Technicon® SMAC® method. *3719*
Serum Increase Physiological May cause cholestasis. *2313*

Alkaloids *Urine No Effect Analytical* With ninhydrin on Frings TLC procedure at 10 mg/dL. *1204*
Urine Increase Analytical Purple with iodoplatinate on Frings TLC procedure. *1204*

Aspartate Aminotransferase *Serum No Effect Analytical* At acute overdose concentration (20 mg/dL) on Technicon® SMAC® method. *3719*
Serum Increase Physiological May cause cholestasis. *2313*

Barbiturate *Urine No Effect Analytical* With diphenylcarbazone on Frings TLC procedure at 10 mg/dL. With mercurous NO_3 on Frings TLC procedure at 10 mg/dL. *1204*
Urine Increase Analytical Pink/orange with mercuric SO_4 in Frings TLC procedure. Pink spot with vanillin in Frings TLC procedure. *1204*

Bicarbonate *Serum No Effect Analytical* At concentration of 200 mg/L had no effect on method using phenolphthalein. *3393*

Bile *Urine Increase Physiological* May cause cholestasis. *2313*

Bilirubin *Serum No Effect Analytical* At acute overdose concentration (20 mg/dL) on Technicon® SMAC® method. *3719* At concentration of 200 mg/L had no effect on Jendrassik and Grof method. *3393*
Serum Increase Physiological May cause cholestasis. *2313*

Bromide *Urine No Effect Analytical* No effect on method of Frings. *1204*

BSP Retention *Serum Increase Physiological* May cause cholestasis. *2313*

Calcium *Serum No Effect Analytical* At acute overdose concentration (20 mg/dL) on Technicon® SMAC® method. *3719* At concentration of 200 mg/L had no effect on cresolphthalein method. *3393*

Catecholamines *Plasma Increase Physiological* Increased metabolism, decreased organ uptake of norepinephrine. *0596*

Cephalin Flocculation *Serum Increase Physiological* May cause cholestasis. *2313*

Chlordiazepoxide *Blood No Effect Analytical* On method of Riddick at 5 mg/dL. *2983*

Chloride *Serum No Effect Analytical* At concentration of 1 mg/L had no effect on mercurimetric method. *3393*

Cholesterol *Serum No Effect Analytical* At acute overdose concentration (20 mg/dL) on Technicon® SMAC® method. *3719* At concentration of 1 mg/L had no effect on CHOD-PAP method. *3393* At concentration of 200 mg/L had no effect on Liebermann-Burchard method. *3393*

Corticosteroids *Urine Decrease Analytical* Decreased absorption at 410 nm in Porter-Silber procedure. *0022*

Creatine Kinase *Serum No Effect Analytical* At acute overdose concentration (20 mg/dL) on Technicon® SMAC® method. *3719*

Creatinine *Serum No Effect Analytical* At acute overdose concentration (20 mg/dL) on Technicon® SMAC® method. *3719* At concentration of 200 mg/L had no effect on AutoAnalyzer Jaffé method. *3393*

Ethchlorvynol *Urine No Effect Analytical* No effect on method of Frings and Cohen at 10 mg/dL. *1205*

Folate *Serum No Effect Analytical* If chromatographic procedure of Landon used. *2068*

FPN Test *Urine No Effect Analytical* Negative FPN test with 0.1 mg/dL concentration in Frings TLC procedure. Negative FPN test with 0.5 mg/dL concentration in Frings TLC procedure. *1204*
Urine Increase Analytical Pink FPN test with 1 mg/dL concentration in Frings TLC procedure Pink color observed in method of Frings. *1204*

Glucose *Serum No Effect Analytical* At acute overdose concentration (20 mg/dL) on Technicon® SMAC® method. *3719* At concentration of 1 mg/L had no effect on GOD/POD-PAP method. *3393*
Serum Decrease Physiological If given i.v. or i.m. *0851*

17-Hydroxycorticosteroids *Urine Decrease Analytical* Interference with Porter-Silber reaction. *3505*

5-Hydroxyindoleacetic Acid (5-HIAA)
Urine Decrease Analytical Interference with nitrosonaphthol methods. *3065*

Iron *Serum No Effect Analytical* At acute overdose concentration (20 mg/dL) on Technicon® SMAC® method. *3719* At concentration of 200 mg/L had no effect on Ferrozine method. *3393*

^{131}I Uptake *Serum Decrease Physiological* Phenergan®, Mepergan® contain tetraiodofluorescein. *2652*

Lactate Dehydrogenase *Serum No Effect Analytical* At acute overdose concentration (20 mg/dL) on Technicon® SMAC® method. *3719*

Leukocytes *Blood Decrease Physiological* Agranulocytosis/leukopenia. *3602*

Neutrophils *Blood Decrease Physiological* Occasional case of agranulocytosis reported. *3717*

Phosphate *Serum No Effect Analytical* At acute overdose concentration (20 mg/dL) on Technicon® SMAC® method. *3719* At concentration of 200 mg/L had no effect on phosphomolybdate method. *3393*
Serum Decrease Analytical Turbidity produced, PO$_4$ concentration decrease method of Fiske. *1013*

Potassium *Serum No Effect Analytical* At concentration of 1 mg/L had no effect on flame-photometric method. *3393*

Pregnancy Tests *Urine Positive Analytical* False positive with Gravindex™. *3529*
Urine Negative Analytical False negative with Prepuerin or Pap-test. *3529*

Prolactin *Plasma No Effect Analytical* No significant response to 25 mg i.m. *1416*

Protein *Serum No Effect Analytical* At acute overdose concentration (20 mg/dL) on Technicon® SMAC® method. *3719* At concentration of 200 mg/L had no effect on biuret method with blank correction. *3393*

Salicylate *Urine No Effect Analytical* Up to 5 mg/dL no effect on Trinder procedure. *1204*
Urine Increase Analytical At 10 mg/dL produces pink color on Trinder procedure. *1204*

Sodium *Serum No Effect Analytical* At concentration of 1 mg/L had no effect on flame-photometric method. *3393*

Thymol Turbidity *Serum Increase Physiological* May cause cholestasis. *2313*

Triglycerides *Serum No Effect Analytical* At acute overdose concentration (20 mg/dL) on Technicon® SMAC® method. *3719* At concentration of 200 mg/L had no effect on lipase/esterase method. *3393*

Urea Nitrogen *Serum No Effect Analytical* At acute overdose concentration (20 mg/dL) on Technicon® SMAC® method. *3719* At concentration of 200 mg/L had no effect on diacetylmonoxime method. *3393*

Uric Acid *Serum No Effect Analytical* At acute overdose concentration (20 mg/dL) on Technicon® SMAC® method. *3719* At concentration of 200 mg/L had no effect on phosphotungstate reduction method. *3393*

Propafenone

Digoxin *Serum Increase Physiological* 37% increase when drugs coadministered due to decreased renal clearance. *0283*

Leukocytes *Blood Decrease Physiological* Decreased from mean of 6800 to 5900/uL in 45 patients treated over 1 y. *1477*

Propamidine

Amphetamine *Urine No Effect Analytical* No effect at 100 mg/L on method of Rutter. *3097*

Propanidid

Basophils *Blood Decrease Physiological* Marked fall within 3 minutes of i.v. inject. *2212*

Histamine *Plasma Increase Physiological* Immediate rise after i.v. inject without anaphylaxis. *2212*

Hydrochloric Acid *Gastric Material Increase Physiological* Stimulation paralleled plasma histamine concentration. *2212*

Propanol

Ionized Calcium *Serum Decrease Analytical* At 0.1 mmol/L to 0.1 mol/L on Calcium specific electrode. *0540*

Propantheline

Acetaminophen *Serum Decrease Physiological* Delayed absorption with delayed gastric emptying. *3794*

Chlorothiazide *Serum Increase Physiological* Absorption window increased with absorption window or dissolution. *3794*

Digoxin *Serum Increase Physiological* Improves absorption, decreases gastrointestinal motility activity. *2290* With tablets of low bioavailability due to reduction of bowel motility. *2794* Increased absorption with augmented dissolution. *3794*

Ethanol *Blood Decrease Physiological* Reduced rate of absorption with delayed gastric emptying. *3794*

Hydrochlorothiazide *Serum Increase Physiological* Delayed but increased absorption with delayed gastric emptying. *3794*

Lithium *Serum Decrease Physiological* Delayed absorption with delayed gastric emptying. *3794*

Ranitidine *Serum Increase Physiological* Significantly reduced time to peak concentration but area under curve increased. *1938*

Propericiazine

Acetaminophen Screening Test *Urine   Negative   Analytical*
No reaction with o-cresol at therapeutic concentrations. *3326*

β-Propiolactone

Alkaline Phosphatase *Serum   Decrease   Physiological*
Increase by 8 U/L when plasma from BPL treated blood compared with plasma in 25 specimens. *0217*

Aspartate Aminotransferase
Serum   No Effect   Physiological But inhibition when drug added to plasma negated by increase from hemolysis. *0217*

Bicarbonate *Serum   Decrease   Physiological* Increase by 14 mmol/L when plasma from BPL treated blood compared with plasma in 25 specimens. *0217*

Bilirubin *Serum   Decrease   Physiological* Increase by 1 μmol/L when plasma from BPL treated blood compared with plasma in 25 specimens. *0217*

Calcium *Serum   Increase   Physiological* Increase by 0.1 mmol/L when plasma from BPL treated blood compared with plasma in 25 specimens. *0217*

Chloride *Serum   Decrease   Physiological* Increase by 10 mmol/L when plasma from BPL treated blood compared with plasma in 25 specimens. *0217*

Hemoglobin *Plasma   Increase   Physiological* When plasma from BPL treated blood compared with plasma in 25 specimens. *0217*

Lactate Dehydrogenase *Serum   Increase   Physiological*
Increase by 75 U/L when plasma from BPL treated blood compared with plasma in 25 specimens. *0217*

Potassium *Serum   Increase   Physiological* Significant effect by 0.7 mmol/L when plasma from BPL treated blood compared with plasma in 25 specimens. *0217*

Protein *Serum   Increase   Physiological* Increase by 3.5 g/L when plasma from BPL treated blood compared with plasma in 25 specimens. *0217*

Sodium *Serum   Increase   Physiological* Significant effect by 7 mmol/L when plasma from BPL treated blood compared with plasma in 25 specimens. *0217*

Propionate

Barbiturate *Serum   Increase   Analytical* Affect colorimetric method with cobalt acetate. *3054*

Pyruvate Kinase *Red Blood Cells   Decrease   Analytical*
Most marked with larger side chain. *0370*

Propionic Acid

Methylmalonic Acid *Serum   Decrease   Analytical* MMA may be converted to propionic in GLC. *0975*
Urine   Decrease   Analytical MMA may be converted to propionic in GLC. *0975*

Propoxyphene

Alanine Aminotransferase *Serum   No Effect   Analytical* At acute overdose concentration (2.5 mg/dL) on Technicon® SMAC® method. *3719*
Serum   Increase   Physiological Cholestatic effect. *1954*

Albumin *Serum   No Effect   Analytical* At acute overdose concentration (2.5 mg/dL) on Technicon® SMAC® method. *3719*
At concentration of 25 mg/L had no effect on BCG method. *3393*

Alkaline Phosphatase *Serum   No Effect   Analytical* At acute overdose concentration (2.5 mg/dL) on Technicon® SMAC® method. *3719*
Serum   Increase   Physiological Hepatic toxicity (cholestatic hepatitis). *1954* In 3 patients within 10 d of start of drug treatment. *1156* Hepatotoxic response observed in 2 patients. *2110*

Alkaloids *Urine   No Effect   Analytical* With ninhydrin on Frings TLC procedure at 10 mg/dL. *1204*
Urine   Increase   Analytical Purple with iodoplatinate on Frings TLC procedure. *1204*

Amphetamine *Urine   No Effect   Analytical* At 50 mg/L on fluorescent method of Hayes. *1534*

Aspartate Aminotransferase *Serum   No Effect   Analytical*
At acute overdose concentration (2.5 mg/dL) on Technicon® SMAC® method. *3719*
Serum   Increase   Physiological Cholestatic effect. *1954* In 3 patients within 10 d of start of drug treatment. *1156* Hepatotoxic response observed in 2 patients. *2110*

Barbiturate *Urine   No Effect   Analytical* With mercurous NO₃ on Frings TLC procedure at 10 mg/dL. With vanillin on Frings procedure at 10 mg/dL. With diphenylcarbazone on Frings TLC procedure at 10 mg/dL. With mercuric SO₄ on Frings TLC procedure at 10 mg/dL. *1204*

Bile *Urine   Increase   Physiological* Cholestatic effect. *1954*

Bilirubin *Serum   No Effect   Analytical* At acute overdose concentration (2.5 mg/dL) on Technicon® SMAC® method. *3719*
At concentration of 25 mg/L had no effect on Jendrassik and Grof method. *3393*
Serum   Increase   Physiological Cholestatic effect. *1954* In 3 patients within 10 d of start of drug treatment. *1156* Hepatotoxic response observed in 2 patients. *2110*

Bromide *Urine   No Effect   Analytical* No effect on method of Frings. *1204*

BSP Retention *Serum   Increase   Physiological* Cholestatic effect. *1954*

Calcium *Serum   No Effect   Analytical* At acute overdose concentration (2.5 mg/dL) on Technicon® SMAC® method. *3719*
At concentration of 25 mg/L had no effect on cresolphthalein method. *3393*

Catecholamines *Urine   No Effect   Analytical* No effect on fluorometric Crout procedure. *0766*

Cephalin Flocculation *Serum   Increase   Physiological*
Cholestatic effect. *1954*

Chlordiazepoxide *Blood   No Effect   Analytical* On method of Riddick at 5 mg/dL. *2983*

Cholesterol *Serum   No Effect   Analytical* At acute overdose concentration (2.5 mg/dL) on Technicon® SMAC® method. *3719*
At concentration of 25 mg/L had no effect on Liebermann-Burchard method. *3393*

Creatine Kinase *Serum   No Effect   Analytical* At acute overdose concentration (2.5 mg/dL) on Technicon® SMAC® method. *3719*

Creatinine *Serum   No Effect   Analytical* At acute overdose concentration (2.5 mg/dL) on Technicon® SMAC® method. *3719*
At concentration of 25 mg/L had no effect on AutoAnalyzer Jaffé method. *3393*

Diphenylhydantoin *Serum   Increase   Physiological*
Reported impairment of metabolism, increased plasma concentration. *2042*

Estriol *Urine   No Effect   Analytical* No effect of 12 mg/L on GLC method. *1163*

Ethchlorvynol *Serum   No Effect   Analytical* At 70 mg/L on GLC procedure of Evenson. *1058*
Urine   No Effect   Analytical No effect on method of Frings and Cohen at 10 mg/dL. *1205*

FPN Test *Urine   No Effect   Analytical* No effect at 10 mg/dL on method of Frings. *1204*

Glucose *Serum   No Effect   Analytical* At acute overdose concentration (2.5 mg/dL) on Technicon® SMAC® method. *3719*
At concentration of 1.8 mg/L had no effect on Ektachem® method. *3393*
Serum   Decrease   Physiological Hypoglycemia allegedly occurred in one case. *3820*
Urine   No Effect   Analytical No effect on TesTape®. *1100*
Urine   Decrease   Analytical Low with Clinistix®, Diastix®. *1100*

γ-Glutamyltransferase (GGT)
Serum   Increase   Physiological Hepatotoxic response observed in 2 patients. *2110*

11-Hydroxycorticosteroids *Urine   Decrease   Physiological*
Slight effect only (probably physiological action). *0766*

17-Hydroxycorticosteroids *Urine   No Effect   Analytical*
Added *in vitro* no effect on Porter-Silber procedure. *0766*
Urine   Decrease   Physiological Probable action on hypothalamic pituitary. ACTH secretion. *0766*

Iron *Serum   No Effect   Analytical* At acute overdose concentration (2.5 mg/dL) on Technicon® SMAC® method. *3719* At concentration of 25 mg/L had no effect on Ferrozine method. *3393*

131I Uptake *Serum Decrease Physiological* Darvon® contains tetraiodofluorescein. *2652*

17-Ketosteroids *Urine No Effect Analytical* Added *in vitro* no effect on Zimmermann reaction. *0766*
Urine Decrease Analytical Affects Zimmermann procedure after Allen correction. *0022*
Urine Decrease Physiological Probable action on hypothalamic pituitary. ACTH secretion. *0766*

Lactate Dehydrogenase *Serum No Effect Analytical* At acute overdose concentration (2.5 mg/dL) on Technicon® SMAC® method. *3719*
Serum Increase Physiological Cholestatic effect. *1954*

Methaqualone *Serum No Effect Analytical* At 70 mg/L on GLC procedure of Evenson. *1057*

Morphine *Urine No Effect Analytical* Insignificant cross reactivity with EMIT procedure for opiates. Insignificant cross reactivity with hemagglutination inhibition. Insignificant cross reactivity with RIA procedures. *2514*

5'-Nucleotidase *Serum Increase Physiological* Due to cholestasis. *1778*

pCO₂ *Blood Increase Physiological* Large doses may produce respiratory depression. *1343*

Phosphate *Serum No Effect Analytical* At acute overdose concentration (2.5 mg/dL) on Technicon® SMAC® method. *3719* At concentration of 25 mg/L had no effect on phosphomolybdate method. *3393*

Platelet Aggregation *Blood Decrease Physiological* Observed *in vitro*, may cause gastrointestinal bleeding etc. *0605*

Potassium *Serum No Effect Analytical* At concentration of 5 mg/L had no effect on measurement by ISE with predilution. *3393*

Propoxyphene *Serum Increase Physiological* 0.2 mg/L after 200 mg orally, 0.3 mg/L after 50 mg i.v. *2348*

Protein *Serum No Effect Analytical* At acute overdose concentration (2.5 mg/dL) on Technicon® SMAC® method. *3719* At concentration of 25 mg/L had no effect on biuret method with blank correction. *3393*

Salicylate *Urine No Effect Analytical* At 10 mg/dL no effect on Trinder method. *1204*

Sodium *Serum No Effect Analytical* At concentration of 5 mg/L had no effect on measurement by ISE with predilution. *3393*

Thymol Turbidity *Serum Increase Physiological* Cholestatic effect. *1954*

Triglycerides *Serum No Effect Analytical* At acute overdose concentration (2.5 mg/dL) on Technicon® SMAC® method. *3719* At concentration of 25 mg/L had no effect on lipase/esterase method. *3393*

Urea Nitrogen *Serum No Effect Analytical* At acute overdose concentration (2.5 mg/dL) on Technicon® SMAC® method. *3719* At concentration of 25 mg/L had no effect on diacetylmonoxime method. *3393* At concentration of 1.8 mg/L had no effect on Ektachem® method. *3393*

Uric Acid *Serum No Effect Analytical* At acute overdose concentration (2.5 mg/dL) on Technicon® SMAC® method. *3719* At concentration of 25 mg/L had no effect on phosphotungstate reduction method. *3393*

Propranolol

Adenosine Triphosphate (ATP)
Red Blood Cells Decrease Physiological Significant effect, also inhibits glucose utilization by 60%. *2770*

Alanine Aminotransferase *Serum Increase Physiological* Low incidence drug induced increase. *2427*

Albumin *Serum No Effect Analytical* No effect at 0.1 mg/dL on SMA 12/60 method. *2636* At concentration of 0.2 mg/L had no effect on BCG method. *3393*

Aldosterone *Urine Decrease Physiological* Change less than with renin. *0523*

Alkaline Phosphatase *Serum No Effect Analytical* No effect at 0.1 mg/dL on SMA 12/60 method. *2636*
Serum Increase Physiological Low incidence drug induced increase. *2427*

Apolipoprotein AI *Serum No Effect Physiological* No change observed in 11 patients. *2476* Decrease by -3 mg/dL in 8 patients given 30 to 60 mg daily for 8 weeks. *2550* No change observed in 11 patients. *2476*

Apolipoprotein B *Serum No Effect Physiological* Increase by 5 mg/dL in 8 patients given 30 to 60 mg daily for 8 weeks. *2550* No change observed in 11 patients. *2476*

Apolipoprotein CII *Serum No Effect Physiological* Increase by 2 mg/dL in 8 patients given 30 to 60 mg daily for 8 weeks. *2550*

Apolipoprotein CIII *Serum Increase Physiological* Increase by 4 mg/dL in 8 patients given 30 to 60 mg daily for 8 weeks. *2550*

Apolipoprotein E *Serum No Effect Physiological* Increase by 2 mg/dL in 11 patients given 30 to 60 mg daily for 8 weeks. *2550*

Aspartate Aminotransferase *Serum No Effect Analytical* No effect on Karmen procedure. *2760* No effect at 0.1 mg/dL on SMA 12/60 method. *2636*
Serum Increase Physiological Occasionally seen, probably not due to hepatotoxicity. *2525*

Bicarbonate *Serum No Effect Analytical* At concentration of 0.2 mg/L had no effect on method using phenolphthalein. *3393*

Bilirubin *Serum No Effect Analytical* No effect at 0.1 mg/dL on SMA 12/60 method. *2636* At concentration of 0.2 mg/L had no effect on Jendrassik and Grof method. *3393*
Serum Increase Analytical 4-hydroxypropranolol metabolite increases bilirubin concentration when measured with diazo reaction. Metabolite present as sulfate or glucuronide. *0041*

Bleeding Time *Patient Increase Physiological* Greater than 200% increase observed in 3 of 5 volunteers over 4 day ingestion. *0970*

Calcium *Serum No Effect Analytical* No effect at 0.1 mg/dL on SMA 12/60 method. *2636* At concentration of 0.2 mg/L had no effect on cresolphthalein method. *3393*
Serum No Effect Physiological In 340 patients given drug for 10 weeks to reduce diastolic BP to less than 90 mm Hg. *3707*
Serum Increase Physiological Increase by 0.34 mg/dL in approximately 120 patients with essential hypertension treated for 1 y. *3708*

Chloride *Serum No Effect Analytical* At concentration of 0.2 mg/L had no effect on mercurimetric method. *3393*

Cholesterol *Serum No Effect Analytical* No effect at 0.1 mg/dL on SMA 12/60 method. *2636* At concentration of 0.2 mg/L had no effect on CHOD-PAP method. *3393*
Serum No Effect Physiological In 22 mild/moderate hypertensive males treated for 8 weeks: insignificant increase noted. *1693* General effect observed in multiple studies. *0088* In 23 hypertensive men given up to 160 mg/d for 8 weeks. *2135* Increase by 1.0-1.5 mmol/L in 20 hypertensive diabetic patients. *3900* Insignificant change after 6 mo treatment in 16 hypertensives. *0836* In 53 hypertensives given 80 mg twice daily for 3 mo. *0835* In 340 patients given drug for 10 weeks to reduce diastolic BP to less than 90 mm Hg. *3707* Increase by 2 mg/dL in 8 patients given 30 to 60 mg daily for 8 weeks. *2550* In 23 hypertensive men aged 47-55 y treated for 8 weeks. *1551* In 20 hypertensive diabetic patients. *3900* In 50 volunteers given 160 mg daily for 3 mo. *2171* Range from -2 to 9% change but mostly no change in many studies. *0089*
Serum Increase Physiological Increase by 9 mg/dL in approximately 120 patients with essential hypertension treated for 1 y. *3708* Mean increase from 213 to 222 mg/dL in 16 hypertensives given 80 mg daily for up to 12 mo. *2266*

Cholesterol, High Density Lipoprotein
Serum No Effect Physiological General effect observed in multiple studies. *0088* In 17 patients with hypertension followed for 3 mo. *0967* In 50 volunteers given 160 mg daily for 3 mo. *2171*
Serum Decrease Physiological Approximately 10% decrease noted in 22 mild/moderate hypertensive treated for 8 weeks: inverse relationship to dose given. *1693* Mean decrease from 49 to 45 mg/dL in approximately 120 hypertensives after 16 weeks therapy. *2331* From 42 to 36 mg/dL in 15 hypertriglyceridemic hypertensives after 12 weeks. *0341* Increase by 13% in 23 hypertensive men given up to 160 mg/d for 8 weeks. *2135* Marked effect in 53 hypertensives given 80 mg twice daily for 3 mo. *0835* Decrease by -3 mg/dL in 8 patients given 30 to 60

Propranolol (continued)

Cholesterol, High Density Lipoprotein (continued)

mg daily for 8 weeks. *2550* Increase by 13% in 23 hypertensive men aged 47-55 y treated for 8 weeks. *1551* From no change to 29% reduction in several studies. *0089* Mean decrease from 53 to 48 mg/dL in 16 hypertensives given 80 mg daily for up to 12 mo. *2266*

Cholesterol, Low Density Lipoprotein

Serum No Effect Physiological General effect observed in multiple studies. *0088* Or slight reduction in 53 hypertensives given 80 mg twice daily for 3 mo. *0835* In 17 patients with hypertension followed for 3 mo. *0967* In 50 volunteers given 160 mg daily for 3 mo. *2171* In several studies. *0089*

Serum Increase Physiological Mean increase from 126 to 134 mg/dL in 16 hypertensives given 80 mg daily for up to 12 mo. *2266*

Serum Decrease Physiological Appreciable decrease (130 to 111 mg/dL) after 12 weeks in 15 hypertriglyceridemic hypertensives. *0341*

Cholesterol, Very Low Density Lipoprotein

Serum No Effect Physiological General effect observed in multiple studies. *0088* In 17 patients with hypertension followed for 3 mo. *0967* In 23 hypertensive men given up to 160 mg/d for 8 weeks. *2135* In several studies. *0089*

Serum Increase Physiological Significant increase from 58 to 80 mg/dL after 12 weeks in 15 hypertriglyceridemic hypertensives. *0341* In several studies. *0089* Mean increase from 34 to 41 mg/dL in 16 hypertensives given 80 mg daily for up to 12 mo. *2266*

C-Peptide

Urine No Effect Physiological In 14 hypertensives with type 2 diabetes treated for 3 weeks, with or without added hydrochlorothiazide. *0935*

Creatine Kinase

Serum Increase Physiological In 4 of 27 patients with essential hypertension, effect mainly in MM isoenzyme: effect not very marked. *3149*

Creatinine

Serum No Effect Analytical No effect at 0.1 mg/dL on SMA 12/60 method. *2636* At 100 mg/L on reversed phase LC procedure of Zhiri et al. *3942* At concentration of 0.16 mg/L had no effect on creatinine iminohydrolase method. *3393* At concentration of 0.2 mg/L had no effect on AutoAnalyzer method. *3393*

Serum No Effect Physiological No significant effect on renal function in 15 patients with essential hypertension over 1 mo. *3776* In 340 patients given drug for 10 weeks to reduce diastolic BP to less than 90 mm Hg. *3707* In approximately 120 patients with essential hypertension treated for 1 y. *3708*

2,3-Diphosphoglycerate

Red Blood Cells Decrease Physiological Shifts hemoglobin-O_2 dissociation curve to right. *2770*

Effective Renal Plasma Flow

Patient Decrease Physiological May occur with decreased cardiac output. *3384*

Epinephrine

Plasma Increase Physiological Decreased clearance by 80% in hypertensives and similar effect in normals. *3944*

Fatty Acids, Free (FFA)

Serum No Effect Physiological In 5 hyperthyroid patients given 10 mg every 8 h for 4 d. *0335*

Serum Decrease Physiological Especially during and after exercise. *0062* Significant reduction of basal concentration after 3 mo treatment in 16 hypertensives, but close to normal after 6 mo. *0836* Marked effect in 53 hypertensives given 80 mg twice daily for 3 mo. *0835* But not significant in hyperthyroid patients given 40 mg every 6 h in 6 patients. *0335* Slight effect with pharmacological doses. *1410*

Glomerular Filtration Rate (GFR)

Urine Decrease Physiological 13% decrease on average in hypertensives. *1714*

Glucagon

Plasma No Effect Physiological In hyperthyroid patients given 40 mg every 6 h in 6 patients. *0335* In 5 hyperthyroid patients given 10 mg every 8 h for 4 d. *0335* No change in basal concentration or after glucose in 16 hypertensives. *0836*

Plasma Decrease Physiological In 18 patients with mild essential hypertension treated with chlorothiazide concomitantly over 4 weeks. *1035* Slight effect with pharmacological doses. *1410*

Glucose

Serum No Effect Analytical No effect at 0.1 mg/dL on SMA 12/60 method. *2636* At concentration of 80 mg/L had no effect on GOD/POD-PAP method. *3393*

Serum No Effect Physiological No effect observed with i.v. infusion. *3114* No effect on short term administration. *1635* After 12 mo treatment of 53 previously untreated hypertensives. *0315* Insignificantly reduced after 3 or 6 mo treatment in 16 nondiabetic hypertensives, but lower after glucose ingestion. *0836* In 5 hyperthyroid patients given 10 mg every 8 h for 4 d. *0335* In 18 patients with mild essential hypertension treated with chlorothiazide concomitantly over 4 weeks. *0041*

Serum Increase Physiological Mean increase from 101 to 108 mg/dL after 16 weeks treatment in approximately 120 hypertensives. *2331* Overt diabetes developed or after glucose challenge in 40 hypertensives. *2469* Increase by 1.0-1.5 mmol/L in 20 hypertensive diabetic patients. *3900* To extent of diabetic coma in 2 diabetics when propranolol given with hydrochlorothiazide: exact mechanism not known. *2559* Increase by 2 mg/dL in 340 patients given drug for 10 weeks to reduce diastolic BP to less than 90 mm Hg. *3707* Increase by 8 mg/dL in approximately 120 patients with essential hypertension treated for 1 y. *3708* In 14 hypertensive men with type 2 diabetes when treated for 3 weeks, marked increase when propranolol coadministered. *0935* Slight effect with pharmacological doses. *1410*

Serum Decrease Physiological Has slight effect like that of prolonging insulin. *3505* 6 episodes of hypoglycemia observed in 5 nondiabetic patients on chronic hemodialysis, due to beta-adrenergic blockage. *1376* Rare cases due to inhibition of glycogenolysis in nondiabetics. *2633*

Glucose Tolerance

Serum Decrease Physiological Overt diabetes developed or after glucose challenge in 40 hypertensives. *2469* Raised concentration during later part of i.v. glucose tolerance test. *1410*

Glycerol

Serum No Effect Physiological In 5 hyperthyroid patients given 10 mg every 8 h for 4 d. *0335*

Serum Decrease Physiological In hyperthyroid patients given 40 mg every 6 h in 6 patients. *0335*

Growth Hormone

Plasma No Effect Physiological No effect unless with epinephrine then sharp increase. *2329* No effect on concentration at 3 or 6 mo treatment. *0836*

Plasma Increase Physiological One out of six produce response. *0371* Marked increase observed in 11 hypertensives also treated with diuretics. Also caused marked effect in 3 of 4 acromegalics. *1090*

Hematocrit

Blood Decrease Physiological Associated with altered morphology of red cells. *2770*

Hemoglobin A$_{1C}$

Blood Increase Physiological In 14 hypertensive men with type 2 diabetes when treated for 3 weeks, marked increase when propranolol coadministered. *0935*

Hydrochloric Acid

Gastric Material Decrease Physiological Affects basal and histamine stimulated. *1273*

Hydroxyproline

Urine Decrease Physiological In hyperthyroid patients possible effect due to membrane-stabilizing property of drug. *0334*

Insulin

Plasma No Effect Physiological After 12 mo treatment of 53 previously untreated hypertensives. *0315* In hyperthyroid patients given 40 mg every 6 h in 6 patients. *0335* In 18 patients with mild essential hypertension treated with chlorothiazide concomitantly over 4 weeks. *1035* In 5 hyperthyroid patients given 10 mg every 8 h for 4 d. *0335* No significant effect although drug affects glucose concentration. *1410* In 20 hypertensive diabetic patients. *3900* No change in basal concentrations or after glucose in 16 hypertensives. *0836*

Plasma Decrease Physiological Significantly less in nondiabetic treated hypertensives than in diabetic-treated patients. *2469*

Insulin Tolerance

Plasma Increase Physiological Decreases glucose rebound at end of test. *3292*

[131]I Uptake

Serum No Effect Physiological No effect in euthyroid patients. *0356*

Ketones

Serum No Effect Physiological In 5 hyperthyroid patients given 10 mg every 8 h for 4 d. *0335*

Serum Decrease Physiological In hyperthyroid patients given 40 mg every 6 h in 6 patients. *0335*

Lactate Dehydrogenase

Serum No Effect Analytical No effect at 0.1 mg/dL on SMA 12/60 method. *2636*

Neutrophils

Blood Decrease Physiological Occasional case of agranulocytosis reported. *3717*

Norepinephrine *Plasma Increase Physiological* Decreased clearance by 20% in hypertensives and similar effect in normals. *3944*

Parathyroid Hormone *Plasma No Effect Physiological* In hyperthyroid patients possible effect due to membrane-stabilizing property of drug. *0334*

Phosphate *Serum No Effect Analytical* No effect at 0.1 mg/dL on SMA 12/60 method. *2636* At concentration of 0.2 mg/L had no effect on phosphomolybdate method. *3393*

Platelet Aggregation *Blood No Effect Physiological* No effect on ADP induced aggregation. In healthy volunteers after single dose or after 1 week treatment. *3714*

Platelets *Blood Increase Physiological* Significant increase in healthy volunteers after single dose or after 1 week treatment. *3714*

Blood Decrease Physiological Purpura probably reflects allergic response. *1505*

Potassium *Serum No Effect Analytical* At concentration of 0.2 mg/L had no effect on flame-photometric method. *3393*

Serum No Effect Physiological After 12 mo treatment of 53 previously untreated hypertensives. *0315*

Serum Increase Physiological Usual effect if aldosterone decreased. *0523* In 15 patients with essential hypertension when treated for 1 mo. *3776* Moderate increase observed in several trials due to redistribution from intracellular to extracellular compartments, not due to retention in body. *2234* Increase by 0.2 mmol/L in 340 patients given drug for 10 weeks to reduce diastolic BP to less than 90 mm Hg. *3707* Increase by 0.17 mmol/L in approximately 120 patients with essential hypertension treated for 1 y. *3708* Approximately 5% increase in both men and women with treatment for 3 y. *2506* Pronounced and prolonged increase in patients who had acute myocardial infarction. *2612* Increases of 0.3 to 0.5 mmol/L when doses of 160 to 640 mg/d given, although never to value above 4.5 mmol/L. *2836*

Prealbumin *Serum Increase Physiological* In euthyroid patients due to inhibition of peripheral deiodination of thyroxine and on binding protein metabolism. *1174*

Protein *Serum No Effect Analytical* No effect at 0.1 mg/dL on SMA 12/60 method. *2636* At concentration of 0.2 mg/L had no effect on biuret method with blank correction. *3393*

Renal Blood Flow (RBF) *Patient Decrease Physiological* 13% decrease if continued orally or after i.v. injection. *1714*

Renin Activity *Plasma Decrease Physiological* Observed in all patients, maximum if high initially. *0523* Significant effect over 5 weeks in 10 patients with essential hypertension with mean plasma concentration of drug about 244 ng/mL. *1114*

Sodium *Serum No Effect Analytical* At concentration of 0.2 mg/L had no effect on flame-photometric method. *3393*

Urine Decrease Physiological Impaired in normals and patients. *2427*

Theophylline *Serum Increase Physiological* Clearance markedly reduced in healthy individuals given drug orally 40 mg/6 h, possibly due to a metabolite binding to cytochrome P-450. *0712*

Thyroid Stimulating Hormone (TSH)

Serum No Effect Physiological In 8 healthy volunteers given 80 mg bid and followed serially, in spite of reduction of free T_4. *3833* In 15 euthyroid hypertensives treated for 30 d given 80 to 480 mg/d. *2016* No significant effect in euthyroid patients. *1174* In 10 healthy volunteers treated for 2 weeks. *2938*

Thyroxine (T_4) *Serum Increase Physiological* In 12 hyperthyroid patients given 160 mg/d for 2 weeks due to effect on peripheral metabolism. *1514* Increase by 16% in 15 euthyroid hypertensives treated for 30 d given 80 to 480 mg/d. *2016* In 6 patients on large daily doses of drug (480 ± 155 mg) without clinical evidence of hyperthyroidism, due to drug-induced blockage of iodothyronine deiodination. *0722* Slight tendency in 10 healthy volunteers treated for 2 weeks. *2938*

Thyroxine (T_4) Binding Globulin

Serum Decrease Physiological In 8 healthy volunteers given 80 mg bid and followed serially, fell by average of 1.2 mg/L. *3833* In euthyroid patients due to peripheral inhibition of peripheral deiodination of thyroxine and on binding protein metabolism. *1174*

Thyroxine (T_4), Free *Serum Increase Physiological* In 8 healthy volunteers given 80 mg bid and followed serially, increased by average of 3.3 pmol/L. *3833* Increase by 18% in 15 euthyroid hypertensives treated for 30 d given 80 to 480 mg/d. *2016* In 6 patients on large daily doses of drug (480 ± 155 mg) without clinical evidence of hyperthyroidism, due to drug-induced blockade of iodothyronine deiodination. *0722* In euthyroid patients due to peripheral inhibition of peripheral deiodination of thyroxine and on binding protein metabolism. *1174*

Thyroxine (T_4) Index, Free (FTI)

Serum No Effect Physiological In hyperthyroid patients given 40 mg every 6 h in 6 patients. *0335* In 5 hyperthyroid patients given 10 mg every 8 h for 4 d. *0335*

Triglycerides *Serum No Effect Analytical* At concentration of 92 mg/L had no effect on GPO-PAP method. *3393* At concentration of 0.2 mg/L had no effect on lipase/esterase method. *3393*

Serum No Effect Physiological In 22 mild/moderate hypertensive males treated for 8 weeks: insignificant increase noted. *1693* General effect observed in multiple studies. *0088* After 12 mo treatment of 53 previously untreated hypertensives. *0315*

Serum Increase Physiological Mean increase from 151 to 197 mg/dL after 16 weeks treatment in approximately 120 hypertensives. *2331* From 256 to 369 mg/dL after 12 weeks treatment in 15 hypertriglyceridemic- hypertensive subjects. *0341* Increase by 24% in 23 hypertensive men given up to 160 mg/d for 8 weeks. *2135* In 20 hypertensive diabetic patients. *3900* Marked increase after 6 mo treatment in 16 patients. *0836* Marked effect in 53 hypertensives given 80 mg twice daily for 3 mo. *0835* Increase by 25 mg/dL in 340 patients given drug for 10 weeks to reduce diastolic BP to less than 90 mm Hg. *3707* Increase by 42 mg/dL in approximately 120 patients with essential hypertension treated for 1 y. *3708* Increase by 15 mg/dL in 8 patients given 30 to 60 mg daily for 8 weeks. *2550* Increase by 24% in 23 hypertensive men aged 47-55 y treated for 8 weeks. *1551* From no change to 65% increase in many studies. *0089* Mean increase from 146 to 211 mg/dL in 16 hypertensives given 80 mg daily for up to 12 mo. *2266*

Triglycerides, High Density Lipoprotein

Serum Increase Physiological From 21 to 32 mg/dL in 15 hypertriglyceridemic hypertensives after 12 weeks. *0341*

Triglycerides, Low Density Lipoprotein

Serum No Effect Physiological No change in 15 hypertriglyceridemic hypertensives after 12 weeks. *0341*

Triglycerides, Very Low Density Lipoprotein

Serum Increase Physiological Significant effect in 53 hypertensives given 80 mg twice daily for 3 mo. *0835* Significant increase from 184 mg/dL to 308 mg/dL after 12 weeks in 15 hypertriglyceridemic hypertensives. *0341*

Tri-iodothyronine (T_3) *Serum No Effect Physiological* In 15 euthyroid hypertensives treated for 30 d given 80 to 480 mg/d. *2016* In 6 patients on large daily doses of drug (480 ± 155 mg) without clinical evidence of hyperthyroidism, due to drug-induced blockage of iodothyronine deiodination. *0722* In 5 hyperthyroid patients given 10 mg every 8 h for 4 d. *0335*

Serum Decrease Physiological In hyperthyroid patients possible effect due to membrane-stabilizing property of drug. *0334* In 12 hyperthyroid patients given 160 mg/d for 2 weeks due to effect on peripheral metabolism. *1514* In hyperthyroid patients given 40 mg every 6 h in 6 patients. *0335* In thyrotoxic patients due to effect of drug on peripheral conversion of thyroxine to tri-iodothyronine. *1174* In 10 healthy volunteers treated for 2 weeks. *2938*

Tri-iodothyronine (T_3), Free *Serum No Effect Physiological* In 15 euthyroid hypertensives treated for 30 d given 80 to 480 mg/d. *2016*

Serum Decrease Physiological In 8 healthy volunteers given 80 mg bid and followed serially, fell by average of 1.2 pmol/L. *3833* Slight effect in euthyroid patients due to peripheral inhibition of peripheral deiodination of thyroxine and on binding protein metabolism. *1174*

Propranolol (continued)

Tri-iodothyronine (T₃), Reverse

Serum Increase Physiological In 6 patients on large daily doses of drug (480 ± 155 mg) without clinical evidence of hyperthyroidism, due to drug-induced blockage of iodothyronine deiodination. 0722 In euthyroid patients due to peripheral inhibition of peripheral deiodination of thyroxine and on binding protein metabolism. 1174 In 10 healthy volunteers treated for 2 weeks. 2938

Tri-iodothyronine (T₃), Reverse Free

Serum Increase Physiological In 8 healthy volunteers given 80 mg bid and followed serially increased by average of 0.16 nmol/L. 3833

Urea Nitrogen Serum No Effect Analytical At concentration of 0.2 mg/L had no effect on diacetylmonoxime method. 3393

Serum No Effect Physiological No significant effect on renal function in 15 patients with essential hypertension over 1 mo. 3776 In approximately 120 patients with essential hypertension treated for 1 y. 3708

Serum Increase Physiological May occur secondary to decreased ERPF. 3384 Increase by 1.1 mg/dL in 340 patients given drug for 10 weeks to reduce diastolic BP to less than 90 mm Hg. 3707 Approximately 10% increase in women without significant change in men when treated over 3 y. 2506 Slight but significant in 20 hypertensive diabetic patients. 3900

Uric Acid Serum No Effect Analytical No effect at 0.1 mg/dL on SMA 12/60 method. 2636 At 100 mg/L on reversed phase LC procedure of Zhiri et al. 3942 At concentration of 0.2 mg/L had no effect on phosphotungstate reduction method. 3393 At concentration of 80 mg/L had no effect on uricase-PAP method. 3393

Serum No Effect Physiological After 12 mo treatment of 53 previously untreated hypertensives. 0315 In 340 patients given drug for 10 weeks to reduce diastolic BP to less than 90 mm Hg. 3707

Serum Increase Physiological In 15 patients with essential hypertension when treated for 1 mo. 3776 Mean increase from 6.1 to 6.6 mg/dL in approximately 120 hypertensives after 16 weeks therapy. 2331 But only by 0.1 mg/dL in approximately 120 patients with essential hypertension treated for 1 y. 3708 Approximately 10% increase in both men and women with treatment for 3 y. 2506 Increase by 10% in 23 hypertensive men aged 47-55 y treated for 8 weeks. 1551 Increase by 10% in 23 hypertensive men given up to 160 mg/d for 8 weeks. 2135

Volume Plasma No Effect Physiological No significant change observed. 1714

Plasma Decrease Physiological Decreased by approximately 8% usually, ?mechanism. 3544

Propylene Glycol

Lactate Blood Increase Physiological Statistically significant correlation with serum concentration of drug when used as vehicle for other i.v. drugs. 1906

Triglycerides Serum Increase Analytical Unless blank or Folch extraction. 2128

Propylhexedrine

Amphetamine Urine No Effect Analytical No effect at 100 mg/L on method of Rutter. 3097

Urine Increase Analytical Yields fluorophor with NBD chloride reaction. 2473

Propoxyphene Urine No Effect Analytical Less than 1% fluorescence in procedure of Valentour. 3662

Propyliodone

Protein Bound Iodine (PBI) Serum Increase Analytical Effect lasts 1-5 mo. 3505

Thyroxine (T₄) Serum Increase Analytical At very high concentrations only. 0354

Propylthiouracil

Alanine Aminotransferase Serum Increase Physiological Hepatotoxic effect (centrolobular necrosis). 2313 Single case of chronic active hepatitis reported. 2 others with bridging necrosis. Recovery with drug withdrawal. 2260 In one 24 year old woman who developed fulminant hepatic failure with lymphocyte sensitization. 2437

Alkaline Phosphatase Serum Increase Physiological Hepatotoxic effect (centrolobular necrosis). 2313 Temporary effect associated with drug administration regressed with cessation. 2731 In one 24 year old woman who developed fulminant hepatic failure with lymphocyte sensitization. 2437 Drug induced liver damage: either acute or chronic active hepatitis. 3787

Amylase Serum Decrease Physiological Reported effect. 0248

Antinuclear Antibodies Serum No Effect Physiological Incidence no different from controls. 0616

Serum Increase Physiological Observed in 10 of 53 treated cases without adequate criteria for diagnosis of SLE. 3870

Aspartate Aminotransferase

Serum Increase Physiological Hepatotoxic effect (centrolobular necrosis). 2313 Single case of chronic active hepatitis reported. 2 others with bridging necrosis. Recovery with drug withdrawal. 2260 Temporary effect associated with drug administration regressed with cessation. 2731 In one 24 year old woman who developed fulminant hepatic failure with lymphocyte sensitization. 2437 Drug induced liver damage: either acute or chronic active hepatitis. 3787

Bile Urine Increase Physiological Hepatotoxic effect (centrolobular necrosis). 2313

Bilirubin Serum No Effect Analytical At concentration of 170 mg/L had no effect on Jendrassik and Grof method. 3393

Serum Increase Physiological Rare case of hepatotoxicity reported. 1488 Temporary effect associated with drug administration regressed with cessation. 2731 In one 24 year old woman who developed fulminant hepatic failure with lymphocyte sensitization. 2437 Drug induced liver damage: either acute or chronic active hepatitis. 3787

Bilirubin, Direct Serum Increase Physiological Temporary effect associated with drug administration regressed with cessation of treatment. 2731 In one 24 y old woman who developed fulminant hepatic failure with lymphocyte sensitization. 2437

BSP Retention Serum Increase Physiological Hepatotoxic effect (centrolobular necrosis). 2313

Calcium Serum No Effect Analytical At concentration of 170 mg/L had no effect on cresolphthalein method. 3393

Cephalin Flocculation Serum Increase Physiological Hepatotoxic effect (centrolobular necrosis). 2313

Erythrocytes Blood Decrease Physiological Occasional immunologically associated anemia. 3870 Occasional case of aplastic anemia reported. 3717

Urine Increase Physiological In one 24 y old woman who developed fulminant hepatic failure with lymphocyte sensitization. 2437

Fatty Acids, Free (FFA) Serum Decrease Physiological Marked effect in 5 hyperthyroid individuals given 10 mg every 8 h for 4 d. 0335

γ-Globulin Serum Increase Physiological Possibly related to production of nonspecific drug-stimulated polyclonal antibodies. 3870

Glucagon Plasma No Effect Physiological In 5 hyperthyroid individuals given 10 mg every 8 h for 4 d. 0335

Glucose Serum No Effect Physiological In 5 hyperthyroid individuals given 10 mg every 8 h for 4 d. 0335

Serum Increase Analytical At 1 mmol/L affects SMA 12/60 method. 3335

Glycerol Blood Decrease Physiological Marked effect in 5 hyperthyroid individuals given 10 mg every 8 h for 4 d. 0335

Glycocholic Acid Serum Increase Physiological In one 24 y old woman who developed fulminant hepatic failure with lymphocyte sensitization. 2437

Hematocrit Blood Decrease Physiological Occasional immunologically associated anemia. 3870

Hemoglobin Blood Decrease Physiological Occasional immunologically associated anemia. 3870

Hydroxyproline *Urine No Effect Physiological* In hyperthyroid patients and normal controls. *0334*

Insulin *Plasma No Effect Physiological* In 5 hyperthyroid individuals given 10 mg every 8 h for 4 d. *0335*

^{131}I Uptake *Serum No Effect Physiological* No effect reported. *2220*
Serum Decrease Physiological Effect lasts up to 8 d. *1488*

Ketones *Blood Decrease Physiological* Marked effect in 5 hyperthyroid individuals given 10 mg every 8 h for 4 d. *0335*

Lactate Dehydrogenase *Serum Increase Physiological* Temporary effect associated with drug administration regressed with cessation. *2731* In one 24 year old woman who developed fulminant hepatic failure with lymphocyte sensitization. *2437*

LE Cells *Blood Positive Physiological* May produce LE-like syndrome. *3505*

Leukocytes *Blood Decrease Physiological* Agranulocytosis (incidence 1 in 200). *0701* Occasional case of aplastic anemia reported. *3717* In one 24 year old woman who developed fulminant hepatic failure with lymphocyte sensitization. *2437*

Lymphocytes *Blood Increase Physiological* In one 24 year old woman who developed fulminant hepatic failure with lymphocyte sensitization. *2437*

Monocytes *Blood Increase Physiological* In one 24 y old woman who developed fulminant hepatic failure with lymphocyte sensitization. *2437*

Neutrophils *Blood Decrease Physiological* Occasional reported case of drug-induced neutropenia. *0155* Reported in about 4% of treated individuals. *3870* Occasional case of agranulocytosis reported. *3717*

Occult Blood *Feces Positive Physiological* In one 24 year old woman who developed fulminant hepatic failure with lymphocyte sensitization. *2437*

Parathyroid Hormone *Plasma No Effect Physiological* In hyperthyroid patients and normal controls. *0334*

Phosphate *Serum No Effect Analytical* At concentration of 170 mg/L had no effect on phosphomolybdate method. *3393*

Platelets *Blood Decrease Physiological* Associated with platelet associated IgG. *3870* Occasional case of aplastic anemia reported. *3717* In one 24 year old woman who developed fulminant hepatic failure with lymphocyte sensitization. *2437*

Propylthiouracil *Serum Increase Analytical* Reacts with 2,6-DQC (procedure of Ratliff). *2922*

Protein *Serum No Effect Analytical* At concentration of 170 mg/L had no effect on biuret method with blank correction. *3393*

Protein Bound Iodine (PBI) *Serum Increase Physiological* Reduces metabolism of exogenous thyroxine. *0012*
Serum Decrease Physiological Thyroxine in circulation decreased. *1488*

Prothrombin Time *Plasma Increase Physiological* Deficiency of prothrombin and proconvertin. *0071* Observed occasionally in absence of abnormal liver function. *3870* In one 24 year old woman who developed fulminant hepatic failure with lymphocyte sensitization. *2437*

T_3 Uptake *Serum Decrease Physiological* Action of drug. *1488*

Thymol Turbidity *Serum Increase Physiological* Hepatotoxic effect (centrolobular necrosis). *2313*

Thyroxine (T_4) *Serum Increase Physiological* In one 24 year old woman who developed fulminant hepatic failure with lymphocyte sensitization. *2437*
Serum Decrease Physiological Inhibits synthesis, stops iodination of tyrosine. *1488*

Thyroxine (T_4) Index, Free (FTI)
Serum No Effect Physiological In 5 hyperthyroid individuals given 10 mg every 8 h for 4 d. *0335*

Tri-iodothyronine (T_3) *Serum Decrease Physiological* In hyperthyroid patients and normal controls. *0334* Significant reduction in 5 hyperthyroid individuals given 10 mg every 8 h for 4 d. *0335*

Urea Nitrogen *Serum No Effect Analytical* At concentration of 170 mg/L had no effect on diacetylmonoxime method. *3393*
Serum Increase Physiological Anaphylactic nephritis reported. *0071*

Uric Acid *Serum Increase Analytical* At 1 mmol/L affects SMA 12/60, Henry methods. *3335* At concentrations above 150 mg/L raised concentration as measured by phosphotungstate reduction method. *3393*

Proscillaridin

Digoxin *Serum Increase Analytical* Reported to affect RIA methods. *3940*

ProSobee®

Amino Acids *Urine Positive Analytical* Reddish-pink spot with DL-methionine with ninhydrin on thin-layer chromatography. *1278*

Methionine *Urine Increase Physiological* Contained in large amount in infant formula. *1749*

Prostaglandin-α_1

pH *Gastric Material No Effect Physiological* No consistent effect noted. *0337*

Prostaglandin-α_2

Hydrochloric Acid *Gastric Material Decrease Physiological* Transient effect in all subjects. *0337*

pH *Gastric Material Increase Physiological* Transient inhibition with pH to above 6. *0337*

PSP Excretion *Urine Increase Physiological* If given i.v. as 0.4 μg/min (slight effect). *1739*

Prostaglandin-15-EPI-A_2

pH *Gastric Material No Effect Physiological* No consistent effect noted. *0337*

Prostaglandin F2α

Cortisol *Plasma Increase Physiological* Slight effect at 2 μg/kg/min i.v. (only in men). *0735*

Dopamine Hydroxylase *Serum No Effect Physiological* No effect seen with i.v. infusion in pregnant women. *2510*

Epinephrine *Plasma No Effect Physiological* No effect seen with i.v. infusion in pregnant women. *2510*

Estradiol *Plasma No Effect Physiological* When infused into pregnant women. *0023*

Estriol *Plasma Decrease Physiological* Marked with successful abortion. *3797*

Estrogens *Plasma No Effect Physiological* Little effect in normals if given i.v. *1600*

Estrone *Plasma No Effect Physiological* When infused into pregnant woman. *0023*

Follicle Stimulating Hormone (FSH)
Plasma No Effect Physiological When i.v. up to 2 μg/kg/min in men. *0735*

Growth Hormone *Plasma Increase Physiological* Slight effect at 2 μg/kg/min i.v. (only in men). *0735*

Luteinizing Hormone (LH) *Plasma No Effect Physiological* When i.v. up to 2 μg/kg/min in men. *0735* Little effect in normals when given i.v. *1600*

Norepinephrine *Plasma Increase Physiological* Significant increase if infused i.v. in pregnant women. *2510*

Progesterone *Plasma No Effect Physiological* In early luteal phase with i.v. in normals. *1600*
Plasma Decrease Physiological Marked with successful abortion. *3797* In normals with i.v. in late luteal phase. *1600*

Thyroid Stimulating Hormone (TSH)
Serum No Effect Physiological When i.v. up to 2 μg/kg/min in men. *0735*

Thyroxine (T_4) *Serum Increase Physiological* Peak within 10-48 h up to 2 times original. *1718*

Tri-iodothyronine (T_3) *Serum Increase Physiological* Peak within 10-48 h up to 6 times original. *1718*

Prostaglandins

Sickle Cells *Blood Positive Physiological* Shown to cause sickling *in vitro. 3846*

Thyroid Stimulating Hormone (TSH)
Serum No Effect Physiological No effect on TSH secretion. *3798*

Thyroxine (T$_4$) *Serum Increase Physiological* Probably direct thyroid effect. *3798*

Tri-iodothyronine (T$_3$) *Serum Increase Physiological* Probably direct thyroid effect. *3798*

Prostate Palpation

Acid Phosphatase *Serum Increase Physiological* Release into bloodstream. *1238*

Lactate Dehydrogenase *Serum Increase Physiological* Release into bloodstream. *1731*

Protamine

Antithrombin III *Plasma No Effect Analytical* At concentration of 250 mg/L had no effect on aca method. *3393*

Catecholamines *Plasma Increase Analytical* Concentrated solutions cause striking fluorescence. *0596*

Fibrinogen *Plasma No Effect Analytical* At concentration of 250 mg/L had no effect on aca method. *3393*

Lipase *Serum Decrease Physiological* Inhibition occurs whether heparinized or not. *3026*

Lipids, Total *Serum Increase Physiological* Mechanism not established. *1343*

Lipoprotein Lipase *Serum Decrease Physiological* Inhibition occurs whether heparinized or not. *3026*

Plasminogen *Plasma No Effect Analytical* At concentration of 250 mg/L had no effect on aca method. *3393*

Potassium *Serum No Effect Analytical* At concentration of 10 mg/L had no effect on measurement by ISE with predilution. *3393*

Sodium *Serum No Effect Analytical* At concentration of 10 mg/L had no effect on measurement by ISE with predilution. *3393*

Thromboplastin Generation *Blood Decrease Physiological* Possesses anticoagulant action. *1343*

Protein

α-Amino-Nitrogen *Plasma Increase Physiological* Rises after meals, falls to normal after 4 h. *0987*

Ammonia *Urine Increase Physiological* Due to ingestion of protein. *0987*

Angiotensin I *Plasma Increase Analytical* A protein normally present may interfere. *2705*

Angiotensin II *Plasma Increase Analytical* A protein normally present may interfere. *2705*

Bilirubin *Serum Decrease Analytical* Absence of protein affects diazotization. *2313*

Calcium *Serum Decrease Analytical* With atomic absorption unless added to standards. *0583*

Carotene *Serum Increase Physiological* Rise in parallel with amino acids. *0987*

Chloride *Serum Increase Analytical* Failure to precipitate affects Schales method. *1563*

Cholesterol *Serum No Effect Analytical* When measured enzymatically with oxidase. *3027*
Serum Increase Analytical Tryptophan in protein may react. *2313*

Coproporphyrin *Feces Increase Physiological* Related to the meat content of diet. *0987*

Creatine *Serum Increase Physiological* Especially if meat eaten is raw. *0987*
Urine Increase Physiological Especially if meat eaten is raw. *0987*

Creatinine *Serum Increase Analytical* 5 g/dL produces color equivalent to 2.2 mg/dL. *0583*

Serum Increase Physiological Occurs after large meat intake. *2313*

Dopamine *Urine Increase Physiological* Significant effect of diet on excretion. *1176*

Fucose *Serum Increase Analytical* Interferes with method of Dische and Settles. *1665*

Gastrin *Serum Increase Physiological* Oral feeding may produce up to tenfold increase. *1995*

Glomerular Filtration Rate (GFR)
Urine Increase Physiological Increased in proportion to amount of protein. *0987*

Histidine *Urine Increase Physiological* Occurs after high meat diet. *0987*

Insulin *Plasma Increase Physiological* Stimulation of beta cells by amino acids. *1775*

Magnesium *Serum No Effect Analytical* Serum protein no effect on dihyroxyazobenzene procedure. *3501*

Methylhistidine *Urine Increase Physiological* Occurs with high meat diet. *0987*

Phenylacetylglutamine *Urine Increase Physiological* From bacterial proteinolysis of unabsorbed. *3233*

Potassium *Serum Increase Analytical* Presence raises temperature, increases emission. *2313*

Sodium *Serum Increase Analytical* Presence raises temperature, increases emission. *2313*

Sugar *Urine Increase Analytical* False positive with Benedict's. *2313*

Sulfate *Urine Increase Physiological* Derived from dietary protein. *0948*

Titratable Acidity *Urine Increase Physiological* Due to ingestion of protein. *0987*

Urea Nitrogen *Serum Increase Physiological* 19, 39, 46 mg/dL on 0.5, 1.5, 2.5 g/kg body weight. *0020*
Urine Increase Physiological Increased in proportion to protein intake. *0987*

Uric Acid *Serum Increase Analytical* Incomplete precipitation affects method of Morin. *2486*
Urine Increase Physiological Increase in proportion to dietary purine. *0987*

Protein Deficiency

Ethanol *Blood Decrease Physiological* Causes reduced alcohol dehydrogenase activity. *0398*

Protein Hydrolysate

Alkaline Phosphatase *Serum Increase Physiological* Transient effect hyperalimentation in infants. *3614*

Aspartate Aminotransferase
Serum No Effect Physiological No effect with hyperalimentation in infants. *3614*

Bilirubin *Serum Increase Physiological* Hyperalimentation i.v. infants may cause cholestasis. *3614*

Bilirubin, Direct *Serum Increase Physiological* Hyperalimentation i.v. infants may cause cholestasis. *3614*

Proteinuria

Alanine Aminotransferase *Urine Increase Physiological* Effect much less than with AST. *2897*

Alkaline Phosphatase *Urine Increase Physiological* In renal disease with necrosis or altered GFR. *2897*

Amylase *Urine No Effect Analytical* No effect on Phadebas procedure. *1726*

Aspartate Aminotransferase *Urine Increase Physiological* Associated with infection and tubular damage. *2897*

BSP Retention *Serum Decrease Physiological* False low due to loss of protein bound BSP. *0553*

Cadmium *Urine Increase Physiological* Especially if due to cadmium poisoning. *2155*

Carcinoembryonic Antigen *Urine No Effect Analytical* No effect observed with moderate amount. *1459*

Catalase *Urine Increase Physiological* Renal damage if no bacteriuria. *2897*

Congo Red Test *Serum Decrease Physiological* May cause urinary loss, false indication of amyloid. *0279*

Copper *Urine Increase Physiological* Loss of protein bound material. *1563*

Creatinine *Urine Increase Physiological* Reported effect (but reverse may occur). *0298*

Creatinine Clearance *Urine Increase Physiological* Compared with inulin clearance (but reverse may occur). *0298*

α-Glucosidase *Urine Increase Physiological* Possibly contributed from WBC. *2897*

β-Glucuronidase *Urine Increase Physiological* With all acute or inflammatory renal diseases. *2897*

5-Hydroxyindoleacetic Acid (5-HIAA)
Urine Decrease Analytical If greater than 2 g/L cause low and variable results. *0913*

Lactate Dehydrogenase *Urine Increase Physiological* In large number and variety renal diseases. *2897*

Leucine Aminopeptidase *Urine Increase Physiological* With toxic renal damage. *2897*

Lysozyme *Urine Increase Physiological* Increased in small percentage of cases. *2897*

Peroxidase *Urine Increase Physiological* Indicative of presence of blood cells. *2897*

Pregnancy Tests *Urine Positive Analytical* False positive tests if proteinuria present. *3505*

Protein *Urine Increase Physiological* Actual effect. *2897*
Urine Decrease Analytical Unable to detect paraproteins with BiliLabstix®. *0490*

Sulfatase *Urine Increase Physiological* Increased in renal diseases. *2897*

Transferrin *Serum Decrease Physiological* Fall related to permeability of glomerulus. *1945*
Urine Increase Physiological Clearance related to permeability to protein. *1945*

Urobilinogen *Urine Increase Analytical* Produces turbidity in direct methods. *1563*

Urokinase *Urine Increase Physiological* If renal infarction or hypoxia. *2897*

Protionamide

Alanine Aminotransferase *Serum Increase Physiological* Reported to affect liver function. *2583*

Alkaline Phosphatase *Serum Increase Physiological* Temporary side effect. *2427*

Aspartate Aminotransferase
Serum Increase Physiological Reported to affect liver function. *2583*

Bilirubin *Serum Increase Physiological* Temporary side effect. *2427*

Leukocytes *Blood Decrease Physiological* Single case reported. *2427*

Protocatechuic Acid

Uric Acid *Serum Increase Analytical* Falsely high values with phosphotungstate methods. *1311*
Urine Increase Analytical Falsely high values with phosphotungstate methods. *1311*

Protriptyline

Alanine Aminotransferase *Serum Increase Physiological* Transient reversible cholestasis. *0071*

Alkaline Phosphatase *Serum Increase Physiological* Transient reversible cholestasis. *0071*

Aspartate Aminotransferase
Serum Increase Physiological Transient reversible cholestasis. *0071*

Bile *Urine Increase Physiological* Transient reversible cholestasis. *0071*

Bilirubin *Serum Increase Physiological* Transient reversible cholestasis. *0071*

Eosinophils *Blood Increase Physiological* Rare allergic response. *1678*

Glucose *Serum Increase Physiological* Reported effect. *1678*
Serum Decrease Physiological Reported effect. *1678*

Leukocytes *Blood Decrease Physiological* Transient agranulocytosis or leukopenia. *0071*

Platelets *Blood Decrease Physiological* Rare allergic response. *1678*

Pseudoephedrine

Amino Acids *Urine Positive Analytical* Red spot with ninhydrin on thin-layer chromatography when combined with triprolidine in Actifed®. *1278*

Amphetamine *Urine No Effect Analytical* At 50 mg/L on fluorescent method of Hayes. *1534* No effect at 100 mg/L on method of Rutter. *3097*

Pseudomonas

Color *Urine Increase Physiological* Blue-green color may occur. *0443*

Psyllium Fibre

Occult Blood *Feces No Effect Analytical* No effect on Hemoquant procedure. *3671*

Puromycin

Chromosomes *Test Conditions Abnormal Physiological* Clastogenic in human cells. *3282*

Protein *Urine Increase Physiological* May cause nephrotoxicity. *3204*

Urea Nitrogen *Serum Increase Physiological* May cause nephrotoxicity. *3204*

Putrescine

Amphetamine *Urine No Effect Analytical* No effect at 100 mg/L on method of Rutter. *3097*

Pyrantel

Aspartate Aminotransferase
Serum Increase Physiological Transient increases reported. *2503*

Fibrinogen *Plasma No Effect Analytical* On potassium mercuric thiocyanate procedure of Roberts. *3003*

Pyrazinamide

Alanine Aminotransferase *Serum Increase Physiological* Hepatic toxicity reported (viral-hepatitis like). *3193* Hepatotoxic in 1 to 5% patients (previously reported to be toxic in as many as 25%). *3683* Low toxicity with daily doses of 20 to 30 mg/kg body weight but overall incidence of hepatitis of 0.3%. *0682*

Albumin *Serum Decrease Physiological* Hepatotoxicity (viral-hepatitis like). *0071*

Alkaline Phosphatase *Serum Increase Physiological* Hepatotoxic effect (viral-hepatitis like). *2313* Hepatotoxic in 1 to 5% patients (previously reported to be toxic in as many as 25%). *3683*

Aspartate Aminotransferase
Serum Increase Physiological Hepatic toxicity reported (viral-hepatitis like). *3193* Hepatotoxic in 1 to 5% patients (previously reported to be toxic in as many as 25%). *3683* Low toxicity with daily doses of 20 to 30 mg/kg body weight but overall incidence of hepatitis of 0.3%. *0682*

Bile *Urine Increase Physiological* Hepatotoxic effect (viral-hepatitis like). *2313*

Pyrazinamide (continued)

Bilirubin *Serum Increase Physiological* Hepatic toxicity (approximately 3 %) (dose related). *3193* Hepatotoxic in 1 to 5% patients (previously reported to be toxic in as many as 25%). *3683*

BSP Retention *Serum Increase Physiological* Hepatic toxicity (approximately 4%) (dose related). *3193*

Cephalin Flocculation *Serum Increase Physiological* Hepatotoxic effect (viral-hepatitis like). *2313*

Fibrinogen *Plasma Increase Physiological* Reported effect. *0071*

Globulin *Serum Decrease Physiological* Reported effect (viral-hepatitis like). *0071*

Hematocrit *Blood Decrease Physiological* Rare anemia reported. *2427*

Hemoglobin *Blood Decrease Physiological* Rare anemia reported. *2427*

Ketones *Urine Increase Analytical* Pink-brown with Ketostix®, Acetest® and Rothera. *0121*

17-Ketosteroids *Urine Decrease Physiological* Temperature decreased excretion reported. *1343*

Platelets *Blood Decrease Physiological* Thrombocytopenia reported to occur. *1436*

Protein *Serum Decrease Physiological* Part of hepatotoxicity (viral-hepatitis like). *1343*

Protein Bound Iodine (PBI) *Serum Increase Physiological* Occurs in first week and after 1 mo. *0830*

Prothrombin Time *Plasma Increase Physiological* Reduces concentration of prothrombin. *0071*

Thymol Turbidity *Serum Increase Physiological* Hepatotoxic effect (viral-hepatitis like). *2313*

Uric Acid *Serum Increase Physiological* Inhibits tubular secretion of urate. *3441* Decreases tubular secretion, may result in clinical manifestations of gout. *3683*
Urine Decrease Physiological Increased tubular reabsorption of urate. *3193*

Pyrazinoic Acid

Erythrocytes *Blood Decrease Physiological* Sideroblastic type of anemia may occur. *0333*

Uric Acid *Urine Decrease Physiological* Inhibition of renal tubular secretion. *0297*

Pyrazolones

Erythrocytes *Blood Decrease Physiological* Aplastic/hemolytic anemia. *2313*
Urine Increase Physiological May cause actual bleeding. *1022*

Hematocrit *Blood Decrease Physiological* Aplastic/hemolytic anemia. *2313*

Hemoglobin *Blood Decrease Physiological* Aplastic/hemolytic anemia. *2313*
Urine Increase Physiological Actual bleeding caused by drug. *1022*

Leukocytes *Blood Decrease Physiological* Aplastic/hemolytic anemia (may be agranulocytosis). *2313*

Neutrophils *Blood Decrease Physiological* Myelotoxic effect of drugs. *2427*

Occult Blood *Feces Positive Physiological* May cause gastrointestinal tract bleeding. *2313*

Platelets *Blood Decrease Physiological* Thrombocytopenia. *2313*

Protein *Urine Increase Physiological* Nephrotoxic effect. *1237* Nephrotoxic effect. *1022*

Sugar *Urine Increase Analytical* Interferes with copper reducing methods. *1022*

Urea Nitrogen *Serum Increase Physiological* May cause nephrotoxicity. *1237*

Pyribenzamine®

Alanine Aminotransferase *Serum No Effect Analytical* At acute overdose concentration (20 mg/dL) on Technicon® SMAC® methods. *3719*

Albumin *Serum No Effect Analytical* At acute overdose concentration (20 mg/dL) on Technicon® SMAC® methods. *3719*

Alkaline Phosphatase *Serum No Effect Analytical* At acute overdose concentration (20 mg/dL) on Technicon® SMAC® methods. *3719*

Aspartate Aminotransferase *Serum No Effect Analytical* At acute overdose concentration (20 mg/dL) on Technicon® SMAC® methods. *3719*

Bilirubin *Serum No Effect Analytical* At acute overdose concentration (20 mg/dL) on Technicon® SMAC® methods. *3719*

Calcium *Serum No Effect Analytical* At acute overdose concentration (20 mg/dL) on Technicon® SMAC® methods. *3719*

Cholesterol *Serum No Effect Analytical* At acute overdose concentration (20 mg/dL) on Technicon® SMAC® methods. *3719*

Creatine Kinase *Serum No Effect Analytical* At acute overdose concentration (20 mg/dL) on Technicon® SMAC® methods. *3719*

Creatinine *Serum No Effect Analytical* At acute overdose concentration (20 mg/dL) on Technicon® SMAC® methods. *3719*

Glucose *Serum No Effect Analytical* At acute overdose concentration (20 mg/dL) on Technicon® SMAC® methods. *3719*

Iron *Serum No Effect Analytical* At acute overdose concentration (20 mg/dL) on Technicon® SMAC® methods. *3719*

Lactate Dehydrogenase *Serum No Effect Analytical* At acute overdose concentration (20 mg/dL) on Technicon® SMAC® methods. *3719*

Phosphate *Serum No Effect Analytical* At acute overdose concentration (20 mg/dL) on Technicon® SMAC® methods. *3719*

Protein *Serum No Effect Analytical* At acute overdose concentration (20 mg/dL) on Technicon® SMAC® methods. *3719*

Triglycerides *Serum No Effect Analytical* At acute overdose concentration (20 mg/dL) on Technicon® SMAC® methods. *3719*

Urea Nitrogen *Serum No Effect Analytical* At acute overdose concentration (20 mg/dL) on Technicon® SMAC® methods. *3719*

Uric Acid *Serum No Effect Analytical* At acute overdose concentration (20 mg/dL) on Technicon® SMAC® methods. *3719*

Pyridamole

Alanine Aminotransferase *Serum No Effect Analytical* No effect at therapeutic concentration on Reflotron method. *1984*

Aspartate Aminotransferase *Serum No Effect Analytical* No effect at therapeutic concentration on Reflotron method. *1984*

Cholesterol *Serum No Effect Analytical* No effect at therapeutic concentration on Reflotron method. *1984*

Glucose *Serum No Effect Analytical* No effect at therapeutic concentration on Reflotron method. *1984*

γ-Glutamyltransferase (GGT) *Serum No Effect Analytical* No effect at therapeutic concentration on Reflotron method. *1984*

Triglycerides *Serum No Effect Analytical* No effect at therapeutic concentration on Reflotron method. *1984*

Urea Nitrogen *Serum No Effect Analytical* No effect at therapeutic concentration on Reflotron method. *1984*

Uric Acid *Serum No Effect Analytical* No effect at therapeutic concentration on Reflotron method. *1984*

Pyridine

Ionized Calcium *Serum Increase Analytical* At concentrations > 0.1 mmol/L on Calcium specific electrode. *0540*

Pyridinol Carbamate

Isocitrate Dehydrogenase *Serum Increase Physiological* Noticeable effect, cause not discussed. *2560*

Pyridium®

N-Acetylglucosaminidase *Urine No Effect Analytical* At 40 μmol/L on 2 colorimetric analytical methods. *1354*

Pyridostigmine

Pseudocholinesterase *Serum Decrease Physiological* Direct action of drug. *1147*

Pyridoxal

Aspartate Aminotransferase *Serum Increase Analytical* 25 μmol/L produces average increase of 16%. *2953*

Pyridoxal *Serum Increase Physiological* In response to loading dose. *0716*
Urine Increase Physiological In response to loading dose. *0716*

Pyridoxal Phosphate *Serum Increase Physiological* In response to loading dose. *0716*
Urine Increase Physiological In response to loading dose. *0716*

4-Pyridoxic Acid *Serum Increase Physiological* In response to loading dose. *0716*
Urine Increase Physiological In response to loading dose. *0716*

Urobilinogen *Test Conditions Increase Analytical* Positive spot test with Ehrlich's reagent. *3765*

Pyridoxamine

Urobilinogen *Test Conditions Increase Analytical* Positive spot test with Ehrlich's reagent. *3765*

Pyridoxine

Aspartate Aminotransferase
Serum Increase Physiological Significant increase in elderly after administration. *2427*

Diphenylhydantoin *Serum Decrease Physiological* Occurs with pharmacologic doses: lower plasma concentration. *3024*

Homovanillic Acid *Urine Increase Physiological* When given to patients on L-Dopa. *1675*

Lymphocytes *Blood Decrease Physiological* Significant effect with megadose supplementation. *1347*

Neutrophils *Blood No Effect Physiological* No effect with megadose supplementation. *1347*

Oxalate *Urine Decrease Physiological* In patients with oxalate renal calculi. *1343*

Phenobarbital *Serum Decrease Physiological* Occurs with pharmacologic doses: lower plasma concentration. *3024*

Pyridoxal Phosphate *Serum Increase Physiological* Increased after administration in normal subjects. *2427*

Thyroid Stimulating Hormone (TSH)
Serum Decrease Physiological Decreases concentration in normal and hypothyroid subjects after i.v. injection. *3798*

Urobilinogen *Test Conditions Increase Analytical* Positive spot test with Ehrlich's reagent. *3765*

Pyridyltetrazole

Cholesterol *Serum Decrease Physiological* Effect greater than with nicotinic acid. *2774*

Fatty Acids, Free (FFA) *Serum Decrease Physiological* Reduction of up to 200 βEq/L, then rebound increase. *2774*

Pyrilamine

Alkaloids *Urine No Effect Analytical* With ninhydrin on Frings TLC procedure at 0.5 mg/dL. *1204*
Urine Increase Analytical Purple with iodoplatinate on Frings TLC procedure. *1204*

Barbiturate *Urine No Effect Analytical* With mercuric SO_4 on Frings TLC procedure at 10 mg/dL. With vanillin on Frings procedure at 10 mg/dL. With diphenylcarbazone on Frings TLC procedure at 10 mg/dL. With mercurous NO_3 on Frings TLC procedure at 10 mg/dL. *1204*

Bromide *Urine No Effect Analytical* No effect on method of Frings. *1204*

Ethchlorvynol *Urine No Effect Analytical* No effect on method of Frings and Cohen. *1204*

FPN Test *Urine No Effect Analytical* No effect at 10 mg/dL on method of Frings. *1204*

Salicylate *Urine No Effect Analytical* At 10 mg/dL has no effect on Trinder method. *1204*

Pyrimethamine

Alkaloids *Urine No Effect Analytical* With ninhydrin on Frings TLC procedure at 1 mg/dL. *1204*
Urine Increase Analytical Purple with iodoplatinate on Frings TLC procedure. *1204*

Amino-4-Imidazole-5-Carboxamide Ribotide (AICAR)
Urine Increase Physiological If megaloblastic anemia occurs. *3898*

Barbiturate *Urine No Effect Analytical* With vanillin on Frings procedure at 10 mg/dL. With mercuric SO_4 on Frings TLC procedure at 10 mg/dL. *1204*
Urine Increase Analytical With mercurous NO_3 reagent on Frings TLC procedure. *1204*

Bromide *Urine No Effect Analytical* No effect on method of Frings. *1204*

Erythrocytes *Blood Decrease Physiological* Megaloblastic anemia. *3898*

Ethchlorvynol *Urine No Effect Analytical* No effect on method of Frings and Cohen. *1204*

FIGLU (N-Formiminoglutamic Acid)
Urine Increase Physiological If megaloblastic anemia occurs. *3898*

Folate *Serum Decrease Physiological* Inhibits folate reductase, megaloblastic anemia. *3025* Inhibits dihydrofolate reductase. *2062*
Red Blood Cells Decrease Physiological Inhibits dihydrofolate reductase. *2062*

FPN Test *Urine No Effect Analytical* No effect at 10 mg/dL on method of Frings. *1204*

Hematocrit *Blood Decrease Physiological* Megaloblastic anemia. *3898*

Hemoglobin *Blood Decrease Physiological* Megaloblastic anemia. *3898*

Leukocytes *Blood Decrease Physiological* May occur with severe megaloblastic anemia. *3898*

MCV *Blood Increase Physiological* Megaloblastic anemia. *3898*

Neutrophils *Blood Decrease Physiological* Occasional case of agranulocytosis reported. *3717*

Platelets *Blood Decrease Physiological* May occur with severe megaloblastic anemia. *3898*

Quinine *Serum Increase Physiological* May displace from plasma protein binding. *1487*

Salicylate *Urine No Effect Analytical* At 10 mg/dL no effect on Trinder method. *1204*

Pyrithioxine

Acetylcholine Receptor Antibodies
Serum Increase Physiological With dose of 400-600 mg/d; associated with development of immune complex nephritis and other penicillamine-like toxic reactions. *1761*

Platelets *Blood Decrease Physiological* With dose of 400-600 mg/d: associated with development of immune complex nephritis and other penicillamine-like toxic reactions. *1761*

Protein *Urine Increase Physiological* With dose of 400-600 mg/d: associated with development of immune complex nephritis and other penicillamine-like toxic reactions. *1761*

Pyritinol

Alanine Aminotransferase   *Serum   No Effect   Analytical*   No effect at therapeutic concentration on Reflotron method. *1984*

Aspartate Aminotransferase   *Serum   No Effect   Analytical*   No effect at therapeutic concentration on Reflotron method. *1984*

Cholesterol   *Serum   No Effect   Analytical*   No effect at therapeutic concentration on Reflotron method. *1984*

Glucose   *Serum   No Effect   Analytical*   No effect at therapeutic concentration on Reflotron method. *1984*

γ-Glutamyltransferase (GGT)   *Serum   No Effect   Analytical*   No effect at therapeutic concentration on Reflotron method. *1984*

Triglycerides   *Serum   No Effect   Analytical*   No effect at therapeutic concentration on Reflotron method. *1984*

Urea Nitrogen   *Serum   No Effect   Analytical*   No effect at therapeutic concentration on Reflotron method. *1984*

Uric Acid   *Serum   No Effect   Analytical*   No effect at therapeutic concentration on Reflotron method. *1984*

Pyrogallol

Color   *Urine   Increase   Analytical*   Brown to black, darkens on standing. *2313*

Pyrogens

Corticosteroids   *Plasma   Increase   Physiological*   Effect at 2 h maximum after 4 h after i.v. *3435*

Corticotropin   *Plasma   Increase   Physiological*   Large effect maximum at 2-3 h after i.v. *3435*

Cortisol   *Plasma   Increase   Physiological*   After delay of 1 h following injection. *0254*

Growth Hormone   *Plasma   Increase   Physiological*   After delay of 1 h following injection. *0254*

Pyrophosphate

Orthophosphate   *Test Conditions   No Effect   Analytical*   Due to molybdate catalyzed hydrolysis. *1658*

Phosphate   *Test Conditions   Increase   Analytical*   Affects Fiske-Subbarow unless Seddon procedure used. *3236*

Pyruvate

Acetoacetate Decarboxylase   *Serum   No Effect   Analytical*   No inhibition observed. *3684*

Alanine Aminotransferase   *Serum   Decrease   Analytical*   High concentrations may deplete NADH so even normal enzyme activity gives depletion reaction on Technicon® SMAC®. *1511*

Aspartate Aminotransferase   *Serum   Decrease   Analytical*   High concentrations may deplete NADH so even normal enzyme activity gives depletion reaction on Technicon® SMAC®. *1511*

Creatine   *Serum   Increase   Analytical*   10 mg/L = 0.02 mg/dL by method of Heinegard. *1547*

Creatinine   *Serum   Increase   Analytical*   Interferes - AutoAnalyzer method (200 mg/dL = 7.4 mg/dL). *1975*
Urine   Increase   Analytical   1 mg equivalent to 0.1 mg with automated Jaffé procedure. *2558*

Fatty Acids, Free (FFA)   *Serum   No Effect   Analytical*   At 10 mg/dL on method of Soloni. *3381*

Ferric Chloride Test   *Urine   Positive   Analytical*   Deep gold-yellow or green. *0443*

Lactate   *Serum   Decrease   Analytical*   High concentrations interfere with enzymatic method. *1150*

Phenylketones   *Urine   Positive   Analytical*   Deep yellow/green with FeCl₃, nil Phenistix®. *0775*

Uric Acid   *Urine   Increase   Physiological*   Observed after oral feeding. *2422*

Pyruvate Kinase

Pyruvate   *Serum   Decrease   Analytical*   If LDH used is contaminated with pyruvate kinase. *1557*

Pyuria

Appearance   *Urine   Abnormal   Physiological*   Milky, leukocytes insoluble in dilute acetic acid. *0443*

Quaternary Ammonium Compounds

Protein   *Urine   No Effect   Analytical*   With SSA and acetic acid tests. *0443*
Urine   Increase   Analytical   Gives false positive by changing pH (stix tests). *3505*

Pseudocholinesterase
Test Conditions   Decrease   Analytical   Inhibitory effect observed. *2893*

Quinacrine

Alanine Aminotransferase   *Serum   Increase   Physiological*   Hepatitis reported (centrolobular necrosis). *0071*

Alkaline Phosphatase   *Serum   Increase   Physiological*   Hepatotoxic effect. *2313*

Aspartate Aminotransferase
Serum   Increase   Physiological   Hepatitis reported (centrolobular necrosis). *0071*

Bile   *Urine   Increase   Physiological*   Hepatotoxic effect. *2313*

Bilirubin   *Serum   Increase   Physiological*   Hemolysis may occur with G-6-PD deficiency. *0248*   Hepatotoxic effect. *2313*

Bilirubin, Direct   *Serum   Increase   Physiological*   Hemolysis may occur with G-6-PD deficiency. *1237*

BSP Retention   *Serum   Increase   Physiological*   Hepatotoxic effect. *2313*

Cephalin Flocculation   *Serum   Increase   Physiological*   Hepatotoxic effect. *2313*

Color   *Urine   Increase   Analytical*   Deep yellow color on acidification. *1343*
Urine   Increase   Physiological   Produces yellow coloration. *0443*

Cortisol   *Plasma   Increase   Analytical*   Produces interfering fluorescence. *1772*
Urine   Increase   Analytical   Increased unconjugated by fluorometric procedure. *0022*

Diagnex Blue Excretion   *Urine   Increase   Analytical*   Release of dye from resin. *1238*

Erythrocytes   *Blood   Decrease   Physiological*   May produce aplastic anemia. *2313*   May cause hemolysis with G-6-PD deficiency. *0248*

Hematocrit   *Blood   Decrease   Physiological*   May cause hemolysis with G-6-PD deficiency. *0248*

Hemoglobin   *Blood   Decrease   Physiological*   May cause hemolysis with G-6-PD deficiency. *0248*

11-Hydroxycorticosteroids   *Plasma   Increase   Analytical*   When determined by Mattingly method. *2901*

Icteric Index   *Serum   Increase   Physiological*   Hemolysis may occur with G-6-PD deficiency. *1237*

Leukocytes   *Blood   Decrease   Physiological*   Leukopenia/agranulocytosis/aplastic anemia. *2426*

Methemoglobin   *Blood   Increase   Physiological*   May cause hemolysis with G-6-PD deficiency. *3581*

Platelets   *Blood   Decrease   Physiological*   Thrombocytopenia or aplastic anemia may occur. *2313*

Thymol Turbidity   *Serum   Increase   Physiological*   Hepatotoxic effect. *2313*

Quinethazone

Alanine Aminotransferase   *Serum   Increase   Physiological*   May cause cholestatic jaundice. *2313*

Alkaline Phosphatase   *Serum   Increase   Physiological*   May cause cholestatic jaundice. *2313*

Aspartate Aminotransferase
Serum   Increase   Physiological   May cause cholestatic jaundice. *2313*

Bile   *Urine   Increase   Physiological*   May cause cholestatic jaundice. *2313*

Bilirubin *Serum Increase Physiological* May cause cholestatic jaundice. *2745*

BSP Retention *Serum Increase Physiological* May cause cholestatic jaundice. *2313*

Cephalin Flocculation *Serum Increase Physiological* May cause cholestatic jaundice. *2313*

Eosinophils *Blood Increase Physiological* Isolated case report. *2427*

Erythrocytes *Blood Decrease Physiological* Theoretical bone marrow depression. *1680*

Glucose *Serum Increase Physiological* May precipitate latent diabetes or aggravate exist. *2313*
Urine Increase Physiological Occurs as consequence of hyperglycemia. *2313*

Glucose Tolerance *Serum Decrease Physiological* Similar effect to thiazides. *1680*

LE Cells *Blood Positive Physiological* Theoretical possibility of this type of drug. *1680*

Leukocytes *Blood Decrease Physiological* Theoretical bone marrow depression. *1680*

Platelets *Blood Decrease Physiological* Theoretical bone marrow depression. *1680*

Potassium *Serum Decrease Physiological* Diuretic action. *2427*
Urine Increase Physiological Relatively small increase. *1680*

Sodium *Serum Decrease Physiological* Diuretic action with sodium depletion. *2313*
Urine Increase Physiological Intended effect, high Na/K ratio. *1680*

Sugar *Urine Increase Analytical* False positive with Benedict's. *2313*

Thymol Turbidity *Serum Increase Physiological* May cause cholestatic jaundice. *2313*

Urea Nitrogen *Serum Increase Physiological* Observed effect (?due to dehydration). *2427*

Uric Acid *Serum Increase Physiological* Increased up to 4 mg/dL, inhibits tubular secretion. *0464*

Quingestanol

Cholesterol *Serum No Effect Physiological* No significant effect with 300 μg/d 6 mo. *1309*

Lipoprotein Electrophoresis
Serum No Effect Physiological No significant effect with 300 μg/d 6 mo. *1309*

Monoglyceride Lipase *Serum No Effect Physiological* No significant effect with 300 μg/d 6 mo. *1309*

Postheparin Lipolytic Activity
Plasma No Effect Physiological No significant effect with 300 μg/d 6 mo. *1309*

Triglyceride Lipase *Serum No Effect Physiological* No significant effect with 300 μg/d 6 mo. *1309*

Triglycerides *Serum No Effect Physiological* At 300 μg/d usually no effect. *1309*
Serum Decrease Physiological May be slight effect with 300 μg/d. *1309*

Quinidine

Alanine Aminotransferase *Serum No Effect Analytical* At acute overdose concentration (20 mg/dL) Technicon® SMAC® method. *3719* No effect at therapeutic concentration on Reflotron method. *1984*
Serum Increase Physiological Isolated case of hepatic toxicity (granulomatous hepatitis). *0862* In single case of severe quinidine hypersensitivity. *0858* Hypersensitivity reaction reported in one patient. *3350*

Albumin *Serum No Effect Analytical* At 26 mg/dL no effect on SMA 12/60 method. *2636* At acute overdose concentration (20 mg/dL) Technicon® SMAC® method. *3719* At concentration of 210 mg/L had no effect on BCG method. *3393*
Serum No Effect Physiological Hypersensitivity observed in 32 of 487 patients who had received drug. *1261*
Serum Decrease Physiological In single case of severe quinidine hypersensitivity. *0858*

Aldolase *Serum Increase Physiological* If given intramuscularly, possibly due to increase in intracellular calcium. *3215*

Alkaline Phosphatase *Serum No Effect Analytical* At 26 mg/dL no effect on SMA 12/60 method. *2636* At acute overdose concentration (20 mg/dL) Technicon® SMAC® method. *3719*
Serum Increase Physiological May cause mild hepatic impairment in few patients. *2531* In single case of severe quinidine hypersensitivity. *0858* Hypersensitivity observed in 32 of 487 patients who had received drug. *1261* Isolated case of reversible granulomatous hepatitis from quinidine hypersensitivity and granuloma induction within 3 d of quinidine readministration. *0448*

Alkaloids *Urine No Effect Analytical* With ninhydrin on Frings TLC procedure at 10 mg/dL. *1204*
Urine Increase Analytical Purple with iodoplatinate on Frings TLC procedure. *1204*

Anisindione *Plasma Increase Physiological* Potentiates action. *1487*

Antibodies to dsDNA *Serum No Effect Physiological* Reported in 5 cases of an SLE-like syndrome induced by the drug. *2093*

Antibodies to Histones *Serum No Effect Physiological* Reported in 5 cases of an SLE-like syndrome induced by the drug. *2093*

Antibodies to RNP *Serum No Effect Physiological* Reported in 5 cases of an SLE-like syndrome induced by the drug. *2093*

Antibodies to SCL-70 *Serum No Effect Physiological* Reported in 5 cases of an SLE-like syndrome induced by the drug. *2093*

Antibodies to Sjögren Syndrome A
Serum No Effect Physiological Reported in 5 cases of an SLE-like syndrome induced by the drug. *2093*

Antibodies to Sjögren Syndrome B
Serum No Effect Physiological Reported in 5 cases of an SLE-like syndrome induced by the drug. *2093*

Antibodies to SmAg *Serum No Effect Physiological* Reported in 5 cases of an SLE-like syndrome induced by the drug. *2093*

Antinuclear Antibodies *Serum Increase Physiological* Reported in 5 cases of an SLE-like syndrome induced by the drug. *2093*

Aspartate Aminotransferase *Serum No Effect Analytical* At 26 mg/dL no effect on SMA 12/60 method. *2636* At acute overdose concentration (20 mg/dL) Technicon® SMAC® method. *3719* No effect at therapeutic concentration on Reflotron method. *1984*
Serum Increase Physiological May cause mild hepatic impairment in few patients. *2531* In single case of severe quinidine hypersensitivity. *0858* Hypersensitivity observed in 32 of 487 patients who had received drug. *1261* Hypersensitivity reaction reported in one patient. *3350*

Barbiturate *Urine No Effect Analytical* With mercuric SO₄ on Frings TLC procedure at 10 mg/dL. With vanillin on Frings procedure at 10 mg/dL. With diphenylcarbazone on Frings TLC procedure at 10 mg/dL. With mercurous NO₃ on Frings TLC procedure at 10 mg/dL. *1204*

Bicarbonate *Serum No Effect Analytical* At concentration of 210 mg/L had no effect on method using phenolphthalein. *3393*

Bilirubin *Serum No Effect Analytical* At 26 mg/dL no effect on SMA 12/60 method. *2636* At acute overdose concentration (20 mg/dL) Technicon® SMAC® method. *3719* At concentration of 210 mg/L had no effect on Jendrassik and Grof method. *3393*
Serum Increase Physiological Causes hemolytic anemia. *2365* Hypersensitivity observed in 32 of 487 patients who had received drug. *1261*

Bilirubin, Direct *Serum Increase Physiological* Causes hemolytic anemia. *2365*

Bromide *Urine No Effect Analytical* No effect on method of Frings. *1204*

Calcium *Serum No Effect Analytical* At 26 mg/dL no effect on SMA 12/60 method. *2636* At acute overdose concentration (20 mg/dL) Technicon® SMAC® method. *3719* At concentration of 210 mg/L had no effect on cresolphthalein method. *3393*

Catecholamines *Plasma Increase Analytical* Metabolite causes false positive. *0581*

Quinidine (continued)

Catecholamines (continued)
Urine   No Effect   Analytical   If Hathaway procedure used.
1487
Urine   Increase   Analytical   Affects fluorescent procedures.
3217

Chloride   *Serum   No Effect   Analytical*   At concentration of
210 mg/L had no effect on mercurimetric method. *3393*

Cholesterol   *Serum   No Effect   Analytical*   At 26 mg/dL no
effect on SMA 12/60 method. *2636*   At acute overdose concentration (20 mg/dL) Technicon® SMAC® method. *3719*   No effect
at therapeutic concentration on Reflotron method. *1984*   At concentration of 210 mg/L had no effect on Liebermann-Burchard
method. *3393*

Coombs' Test   *Serum   Positive   Physiological*   Immunological response to drug (gamma antibody). *1486*

Coombs' Test, Direct   *Serum   Positive   Physiological*   Associated with hemolytic anemia. *1487*   Occasional hemolytic anemia and hypersensitivity observed in 32 of 487 patients who had
received drug. *1261*

Corticosteroids   *Urine   Increase   Analytical*   Butanol
extract/no hydrolysis Reddy/Porter-Silber reaction. *0022*

Creatine Kinase   *Serum   No Effect   Analytical*   At acute
overdose concentration (20 mg/dL) Technicon® SMAC® method.
3719
Serum   Increase   Physiological   If given intramuscularly, possibly due to increase in intracellular calcium. *3215*

Creatinine   *Serum   No Effect   Analytical*   At 26 mg/dL no
effect on SMA 12/60 method. *2636*   At acute overdose concentration (20 mg/dL) Technicon® SMAC® method. *3719*   At concentration of 210 mg/L had no effect on AutoAnalyzer Jaffé
method. *3393*

Diagnex Blue Excretion   *Urine   Increase   Analytical*
Release of dye from resin. *1238*

Digoxin   *Serum   Increase   Physiological*   Clearance reduced
by coadministration of quinidine: effect further augmented by
addition of spironolactone to regime. *1110*   Observed in 7 of 9
patients with mean concentration changing from 1.43 to 2.61
nmol/L, probably due to displacement of quinidine from binding
sites in tissues and reduced renal clearance. *2291*   20 to 330%
increase after 3 d of quinidine in 17 patients. *0790*   When
coadministered with quinidine: so effect of 0.2 mg with quinidine
was comparable to that of 0.4 mg without. *0282*   Approximate
2-fold increase due to decreased total body clearance but renal
negligible and not affected by chronic renal failure. *1111*
Absorption rate constant increased, with decreased lag time and
peak time. Systemic availability of digoxin increased from 68% to
79%, but no effect on biotransformation. *2763*   Volume of distribution and nonrenal clearance increased, half-time of elimination
greatly increased but total clearance and renal clearance greatly
decreased. *0344*   2.5 fold increase with more than 50%
decrease in renal clearance. *0920*

Disopyramide   *Serum   Increase   Physiological*   When concurrent quinidine given: effect small but significant no significant
change in elimination half-life. *0208*

Erythrocytes   *Blood   Decrease   Physiological*   Hemolytic
anemia (sensitivity dependent) G-6-PD deficiency. *3673*
Immune mediated hemolytic anemia associated with high titers
IgG antibodies. *3215*   Occasional hemolytic anemia and hypersensitivity observed in 32 of 487 patients who had received
drug. *1261*

Erythrocyte Sedimentation Rate
Blood   Increase   Physiological   Reported in 5 cases of an
SLE-like syndrome induced by the drug. *2093*

Ethchlorvynol   *Serum   No Effect   Analytical*   At 10 mg/L on
GLC procedure of Evenson. *1058*
Urine   No Effect   Analytical   No effect on method of Frings
and Cohen. *1204*

FPN Test   *Urine   No Effect   Analytical*   No effect at 10 mg/dL
on method of Frings. *1204*

Glucose   *Serum   No Effect   Analytical*   At 26 mg/dL no
effect on SMA 12/60 method. *2636*   At acute overdose concentration (20 mg/dL) Technicon® SMAC® method. *3719*   No effect
at therapeutic concentration on Reflotron method. *1984*   At concentration of 5.6 mg/L had no effect on Ektachem® method.
3393

γ-Glutamyltransferase (GGT)   *Serum   No Effect   Analytical*
No effect at therapeutic concentration on Reflotron method.
1984
Serum   Increase   Physiological   Hypersensitivity reaction
reported in one patient. *3350*   Isolated case of reversible granulomatous hepatitis from quinidine hypersensitivity and granuloma
induction within 3 d of quinidine readmission. *0448*

Haptoglobin   *Serum   Decrease   Physiological*   Causes hemolytic anemia. *2365*

Hematocrit   *Blood   Decrease   Physiological*   Hemolytic anemia (sensitivity dependent) G-6-PD deficiency. *3673*

Hemoglobin   *Blood   Decrease   Physiological*   Hemolytic anemia (sensitivity dependent) G-6-PD deficiency. *3673*

17-Hydroxycorticosteroids   *Urine   Increase   Analytical*
Metabolite interferes with Zimmermann reaction. *1238*

Iron   *Serum   No Effect   Analytical*   At acute overdose concentration (20 mg/dL) Technicon® SMAC® method. *3719*   At
concentration of 200 mg/L had no effect on Ferrozine method.
3393

¹³¹I Uptake   *Serum   Decrease   Physiological*   Lilly product
contains tetraiodofluorescein. *2652*

17-Ketosteroids   *Urine   Increase   Analytical*   Metabolite
interferes with Zimmermann reaction. *1022*

Lactate Dehydrogenase   *Serum   No Effect   Analytical*   At
26 mg/dL no effect on SMA 12/60 method. *2636*   At acute overdose concentration (20 mg/dL) Technicon® SMAC® method.
3719
Serum   Increase   Physiological   May cause mild hepatic
impairment in few patients. *2531*   If given intramuscularly, possibly due to increase in intracellular calcium. *3215*   In single case
of severe quinidine hypersensitivity. *0858*   Isolated case of
reversible granulomatous hepatitis from quinidine hypersensitivity and granuloma induction within 3 d of quinidine readministration. *0448*

Leucine Aminopeptidase   *Serum   Increase   Physiological*
Isolated case of hepatic toxicity (granulomatous hepatitis). *0862*

Leukocytes   *Blood   Increase   Physiological*   Marked leukocytosis in 2 patients in association with quinidine fever, normalized after drug discontinued. *0266*
Blood   Decrease   Physiological   Agranulocytosis/leukopenia.
2313   Agranulocytosis after 8 weeks of drug treatment. *1008*

Methaqualone   *Serum   No Effect   Analytical*   At 10 mg/L on
GLC procedure of Evenson. *1057*

Methemoglobin   *Blood   Increase   Physiological*   May cause
hemolysis in G-6-PD deficiency. *3581*

Neutrophils   *Blood   Decrease   Physiological*   Occasional
agranulocytosis reported. *3215*

Phenindione   *Plasma   Increase   Physiological*   Potentiates
action. *1487*

Phosphate   *Serum   No Effect   Analytical*   At 26 mg/dL no
effect on SMA 12/60 method. *2636*   At acute overdose concentration (20 mg/dL) Technicon® SMAC® method. *3719*   At concentration of 210 mg/L had no effect on phosphomolybdate
method. *3393*

Platelets   *Blood   Decrease   Physiological*   Allergic reac-
tion/pancytopenia/purpura. *3673*   Several cases of platelet-associated IgG and thrombocytopenia. *1910*   Isolated case of reversible granulomatous hepatitis from quinidine hypersensitivity and
granuloma induction within 3 d of quinidine readministration.
0448

Potassium   *Serum   No Effect   Analytical*   At concentration of
210 mg/L had no effect on ISE measurement without predilution.
3393

Protein   *Serum   No Effect   Analytical*   At 26 mg/dL no effect
on SMA 12/60 method. *2636*   At acute overdose concentration
(20 mg/dL) Technicon® SMAC® method. *3719*   At concentration
of 210 mg/L had no effect on biuret method with blank correction. *3393*

Protein Bound Iodine (PBI)   *Serum   Decrease   Physiological*
Occasional effect (usually none). *0830*

Prothrombin Time   *Plasma   No Effect   Physiological*   No
effect observed in volunteers. *3292*   Hypersensitivity observed
in 32 of 487 patients who had received drug. *2913*
Plasma   Increase   Physiological   If administration with indandiones, coumarin. *1487*   Depresses prothrombin formation in
liver. *1972*

Reticulocytes *Blood Increase Physiological* Occasional hemolytic anemia and hypersensitivity observed in 32 of 487 patients who had received drug. *1261*

Salicylate *Urine No Effect Analytical* At 10 mg/dL no effect on Trinder method. *1204*

Sodium *Serum No Effect Analytical* At concentration of 210 mg/L had no effect on ISE measurement with predilution. *3393*

Triglycerides *Serum No Effect Analytical* At acute overdose concentration (20 mg/dL) Technicon® SMAC® method. *3719* No effect at therapeutic concentration on Reflotron method. *1984* At concentration of 210 mg/L had no effect on lipase/esterase method. *3393*

Urea Nitrogen *Serum No Effect Analytical* At acute overdose concentration (20 mg/dL) Technicon® SMAC® method. *3719* No effect at therapeutic concentration on Reflotron method. *1984* At concentration of 210 mg/L had no effect on diacetylmonoxime method. *3393* At concentration of 5.6 mg/L had no effect on Ektachem® method. *3393*

Uric Acid *Serum No Effect Analytical* At 26 mg/dL no effect on SMA 12/60 method. *2636* At acute overdose concentration (20 mg/dL) Technicon® SMAC® method. *3719* No effect at therapeutic concentration on Reflotron method. *1984* At concentration of 210 mg/L had no effect on phosphotungstate reduction method. *3393*

Warfarin *Plasma Increase Physiological* Potentiates action. *1487*

Quinine

Alanine Aminotransferase *Serum No Effect Analytical* At acute overdose concentration (1.5 mg/dL) on Technicon® SMAC® method. *3719*

Albumin *Serum No Effect Analytical* At acute overdose concentration (1.5 mg/dL) on Technicon® SMAC® method. *3719* At concentration of 30 mg/L had no effect on BCG method. *3393*

Alkaline Phosphatase *Serum No Effect Analytical* At acute overdose concentration (1.5 mg/dL) on Technicon® SMAC® method. *3719*

Alkaloids *Urine No Effect Analytical* With ninhydrin on Frings TLC procedure at 10 mg/dL. *1204*
Urine Increase Analytical Purple with iodoplatinate on Frings TLC procedure. *1204*

Aspartate Aminotransferase *Serum No Effect Analytical* At acute overdose concentration (1.5 mg/dL) on Technicon® SMAC® method. *3719*

Barbiturate *Urine No Effect Analytical* With mercuric SO$_4$ on Frings TLC procedure at 10 mg/dL. With diphenylcarbazone on Frings TLC procedure at 10 mg/dL. With vanillin on Frings procedure at 10 mg/dL. With mercurous NO$_3$ on Frings TLC procedure at 10 mg/dL. *1204*

Bicarbonate *Serum No Effect Analytical* At concentration of 30 mg/L had no effect on method using phenolphthalein. *3393*

Bilirubin *Serum No Effect Analytical* At acute overdose concentration (1.5 mg/dL) on Technicon® SMAC® method. *3719* At concentration of 30 mg/L had no effect on Jendrassik and Grof method. *3393*
Serum Increase Physiological May cause hemolytic anemia. *2365*

Bilirubin, Direct *Serum Increase Physiological* May cause hemolytic anemia. *2365*

Bromide *Urine No Effect Analytical* No effect on method of Frings. *1204*

Calcium *Serum No Effect Analytical* At acute overdose concentration (1.5 mg/dL) on Technicon® SMAC® method. *3719* At concentration of 30 mg/L had no effect on cresolphthalein method. *3393*

Catecholamines *Plasma Increase Analytical* Interference by metabolite, no effect *in vitro*. *0579*

Chlordiazepoxide *Blood No Effect Analytical* On method of Riddick at 5 mg/dL. *2983*

Chloride *Serum No Effect Analytical* At concentration of 30 mg/L had no effect on mercurimetric method. *3393*

Cholesterol *Serum No Effect Analytical* At acute overdose concentration (1.5 mg/dL) on Technicon® SMAC® method. *3719* At concentration of 30 mg/L had no effect on Liebermann-Burchard method. *3393*

Chromosomes *Test Conditions No Effect Physiological* No effect human leucocytes at concentrations 1/6-3x normal. *1073*

Color *Urine Increase Analytical* Brown color. *1022*

Coombs' Test *Serum Positive Physiological* Immunological response to drug. *1486*

Coombs' Test, Direct *Serum Positive Physiological* Associated with hemolytic anemia. *1487*

Corticosteroids *Urine Increase Analytical* Butanol extract/no hydrolysis Reddy/Porter-Silber reaction. Increased absorption at 410 nm in Porter-Silber reaction. *0022*

Creatine Kinase *Serum No Effect Analytical* At acute overdose concentration (1.5 mg/dL) on Technicon® SMAC® method. *3719*

Creatinine *Serum No Effect Analytical* At acute overdose concentration (1.5 mg/dL) on Technicon® SMAC® method. *3719* At concentration of 30 mg/L had no effect on AutoAnalyzer Jaffé method. *3393*

Diagnex Blue Excretion *Urine Increase Analytical* Release of dye from resin. *2425*

Digoxin *Serum Increase Physiological* Interaction may produce 50% increase in concentration. *0548* Stepwise increase with increasing quinine dose due to impairment of extrarenal digoxin clearance. *2764* Total body clearance reduced with increase of renal elimination half-life. Urine excretion increased. *3758*

Erythrocytes *Blood Decrease Physiological* Hemolytic anemia (sensitivity dependent) G-6-PD deficiency. *0272*

Erythrocyte Sedimentation Rate
Blood Decrease Analytical At therapeutic concentration, maximum at 200 minutes with 2 mg/dL. *3000*

Ethchlorvynol *Urine No Effect Analytical* No effect on method of Frings and Cohen. *1204*

FPN Test *Urine No Effect Analytical* No effect at 10 mg/dL on method of Frings. *1204*

Glucose *Serum No Effect Analytical* At acute overdose concentration (1.5 mg/dL) on Technicon® SMAC® method. *3719*
Serum Decrease Physiological Intravenous drug reduced concentration from 88 to 68 mg/dL in normal volunteers. *3814*

Haptoglobin *Serum Decrease Physiological* May cause hemolytic anemia. *2365*

Hematocrit *Blood Decrease Physiological* Hemolytic anemia (sensitivity dependent) G-6-PD deficiency. *0272*

Hemoglobin *Blood Decrease Physiological* Hemolytic anemia (sensitivity dependent) G-6-PD deficiency. *0272*
Urine Increase Physiological Possible contributing factor. *0071*

Heroin *Urine Increase Analytical* Fluoresces, producing interference. *3869*

17-Hydroxycorticosteroids *Urine No Effect Analytical* No effect modified Porter-Silber reaction. *1487*
Urine Increase Analytical Reddy method affected, not Porter-Silber. *3505*

Insulin *Plasma Increase Physiological* Intravenous drug in normal volunteers increased concentration from 8.9 to 17.1 mU/L. *3814*

Iron *Serum No Effect Analytical* At acute overdose concentration (1.5 mg/dL) on Technicon® SMAC® method. *3719* At concentration of 15 mg/L had no effect on Ferrozine method. *3393*

17-Ketogenic Steroids *Urine Increase Analytical* Reported to affect Zimmermann reaction. *0583*

Lactate Dehydrogenase *Serum No Effect Analytical* At acute overdose concentration (1.5 mg/dL) on Technicon® SMAC® method. *3719*

Leukocytes *Blood Increase Physiological* Primary inc, especially lymphocytes (splenic contractions). *1343*
Blood Decrease Physiological Leukopenia (AMA Blood dyscrasias). *2429*

Methaqualone *Serum No Effect Analytical* At 10 mg/L on GLC procedure of Evenson. *1057*

Methemalbumin *Plasma Increase Physiological* If given concurrently with pamaquine. *1343*

Methemoglobin *Blood Increase Physiological* May cause hemolysis in G-6-PD deficiency. *3581*

Quinine (continued)

Neutrophils *Blood Decrease Physiological* Occasional case of agranulocytosis reported. *3717*

Phosphate *Serum No Effect Analytical* At acute overdose concentration (1.5 mg/dL) on Technicon® SMAC® method. *3719* At concentration of 30 mg/L had no effect on phosphomolybdate method. *3393*

Platelets *Blood Decrease Physiological* Immunological mechanism. *0756* Several cases of immune thrombocytopenia reported. *2502* Quinine dependent antibody caused platelet lysis. *0662*

Potassium *Serum No Effect Analytical* At concentration of 30 mg/L had no effect on measurement by ISE with predilution. *3393*

Protein *Serum No Effect Analytical* At acute overdose concentration (1.5 mg/dL) on Technicon® SMAC® method. *3719* At concentration of 30 mg/L had no effect on biuret method with blank correction. *3393*
Urine Increase Physiological Rare renal damage May rarely cause renal damage. *1343*

Prothrombin Time *Plasma Increase Physiological* Depresses prothrombin formation in liver. *0894*

Salicylate *Urine No Effect Analytical* At 10 mg/dL no effect on Trinder method. *1204*

Sodium *Serum No Effect Analytical* At concentration of 30 mg/L had no effect on measurement by ISE with predilution. *3393*

Triglycerides *Serum No Effect Analytical* At acute overdose concentration (1.5 mg/dL) on Technicon® SMAC® method. *3719* At concentration of 30 mg/L had no effect on lipase/esterase method. *3393*

Urea Nitrogen *Serum No Effect Analytical* At acute overdose concentration (1.5 mg/dL) on Technicon® SMAC® method. *3719* At concentration of 30 mg/L had no effect on diacetylmonoxime method. *3393*
Serum Increase Physiological May cause renal damage (rare). *1343*

Uric Acid *Serum No Effect Analytical* At acute overdose concentration (1.5 mg/dL) on Technicon® SMAC® method. *3719* At concentration of 30 mg/L had no effect on phosphotungstate reduction method. *3393*

Quinine Iodobismuthate

Protein Bound Iodine (PBI) *Serum Increase Physiological* Contains 57% iodine. *0012*

Quinocide

Erythrocytes *Blood Decrease Physiological* May cause hemolysis with G-6-PD deficiency. *1902*

Hematocrit *Blood Decrease Physiological* May cause hemolysis with G-6-PD deficiency. *1902*

Hemoglobin *Blood Decrease Physiological* May cause hemolytic anemia. *1902*

Race

Adenosine Triphosphate (ATP)
Red Blood Cells Decrease Physiological Significantly lower in blacks than caucasians. *2523*

Amino Acids *Hair No Effect Physiological* Same in cephalic hair from black, mongol, caucasian. *1321*

Cholesterol *Serum Decrease Physiological* In blacks significantly < comparable caucasians after age 40. *0286*

Creatine Kinase *Serum Increase Physiological* Significantly higher in blacks than caucasians (for each sex). *2402*

Hematocrit *Blood No Effect Physiological* No difference between blacks and caucasians. *1869*

Hemoglobin *Blood No Effect Physiological* No difference between blacks and caucasians. *1869*

Leukocytes *Blood Decrease Physiological* Significantly lower in blacks (applies to both sexes). *1869*

Lymphocytes *Blood No Effect Physiological* No difference between blacks and caucasians. *1870*

Blood Decrease Physiological In Bantu much below cape colored or whites. *3767*

Neutrophils *Blood Decrease Physiological* Significantly lower in blacks (applies to both sexes). *1870*

Potassium *Red Blood Cells Decrease Physiological* Significantly lower in blacks than caucasians. *2523*

Sodium *Red Blood Cells Increase Physiological* Significantly higher in blacks than caucasians. *2523*

Triglycerides *Serum Decrease Physiological* Significantly less in blacks at all ages. *0286*

Vitamin B$_{12}$ *Serum Increase Physiological* In European women (420 pg/mL) vs African (310 pg/ml). *0472*

Water *Red Blood Cells No Effect Physiological* No difference between whites and blacks. *2523*

Radioactive Compounds

Erythrocytes *Blood Decrease Physiological* Aplastic anemia. *2313*

Leukocytes *Blood Decrease Physiological* May cause marrow depression. *2313*

Prothrombin Time *Plasma Increase Physiological* Also exaggerated response to anticoagulants. *1343*

Uric Acid *Serum Increase Physiological* May occur with tissue destruction. *0279*

Radioactive Iodine

Leukocytes *Blood Increase Physiological* Incidence of acute leukemia higher. *0071*

Protein Bound Iodine (PBI) *Serum Increase Physiological* May cause hypothyroidism (increased by 2-3% per year). *0071*

Radiographic Agents

N-Acetylglucosaminidase *Urine Increase Physiological* Marked increase within 1 d when used for arteriography: effect persisted for many days: effect less when used for urography. *3259*

Alanine Aminopeptidase *Urine Increase Physiological* Marked increase within 1 d when used for arteriography: effect persisted for many days: effect less when used for urography. *3259*

Albumin *Urine Increase Physiological* Mean increase from 34 mg/d to 1873 mg/d in 37 patients day following arteriography. *2594*

Alkaline Phosphatase *Urine Increase Physiological* May occur after i.v. pyelography or aortography. *0078*

Amylase *Serum Increase Physiological* Cholangiography may cause transient increase. *1487*
Pancreatic Juice No Effect Analytical No effect of Gastrografin®. *0741*

Appearance *Urine Abnormal Physiological* Cloudy, in acid urine. *0443*

Aspartate Aminotransferase *Urine Increase Physiological* Transient increase if injected in renal artery. *3538*

Bilirubin *Serum Increase Physiological* Competition for excretion through bile canaliculi. *0127*

BSP Retention *Serum Increase Physiological* Compete for excretory mechanism. *0553*

Casts *Urine Increase Physiological* Nephrotoxic manifestation. *0652*

Catalase *Urine Increase Physiological* Transient increase with slight renal damage if renal artery. *3538*

Catecholamines *Urine Decrease Physiological* Competes for excretion after i.v. pyelography. *0913*

Creatine Kinase *Urine Increase Physiological* Transient increase if injected in renal artery. *3538*

Creatinine *Serum No Effect Physiological* No significant change observed when agents given for either urography or arteriography. *3259*
Serum Increase Physiological Occasional effect following aortography. *0065* Frequency of renal impairment following CT brain scan with infusion 2.1% compared with 1.3% in control group. *0746* In normal people 2%: in patients with renal dys-

function 30% subclinical damage following non-renal angiography. *0784*

Crystals *Urine Increase Physiological* Diatrizoate may produce crystals in acid urine. *0443*

Erythrocytes *Urine Increase Physiological* Nephrotoxic manifestation. *0652*

Glomerular Filtration Rate (GFR)
Urine Decrease Physiological If concentrated solutions used for aortography. *0065*

Hematocrit *Blood Decrease Physiological* Transient fall following rapid i.v. injection. *0611*

Histamine *Plasma Increase Physiological* Observed in some patients if administered i.v. *0449*

Immunoglobulin IgG *Urine Increase Physiological* Mean increase from 6.1 mg/d to 206.3 mg/d in 37 patients day following arteriography. *2594*

17-Ketogenic Steroids *Urine Decrease Analytical* Output halved with Zimmermann procedure. *0583*

Lactate Dehydrogenase *Urine Increase Physiological* Transient increase if injected in renal artery. *3538*

Leucine Aminopeptidase *Urine Increase Physiological* Facilitates permeation of enzyme into tubules. *2897*

Lipase *Pancreatic Juice No Effect Analytical* No effect of Gastrografin®. *0741*

Lysozyme *Urine No Effect Physiological* No change from 3.3 mg/d to 3.3 mg/d in 37 patients day following arteriography. *2594*
Urine Increase Physiological Marked increase within 1 d when used for arteriography: effect persisted for many days: effect less when used for urography. *3259*

β_2-Microglobulin *Urine No Effect Physiological* Mean increase from 2.9 mg/d to 5.4 mg/d in 37 patients day following arteriography. *2594*

PAH Clearance *Urine No Effect Physiological* No effect of arteriography. *2594*
Urine Decrease Physiological If concentrated solutions used for aortography. *0065*

Partial Thromboplastin Time
Plasma Increase Physiological ?transient inactivation of coagulation factors. *1487*

Phosphate Clearance *Urine Decrease Physiological* Mean decrease from 12.9 mL/min/1.73 m^2 to 7.1 mL/min/1.73 m^2 day following arteriography in 37 patients. *2594*

Protein *Serum Increase Analytical* May produce interfering turbidity. *2313*
Urine No Effect Analytical On Combistix®, Urostix, Albustix etc. *1872*
Urine Increase Analytical Turbidity if acid procedures used. *1700* Affects biuret part of Doetsch procedure. *0921* Affects turbidimetric methods for some days. *3505* Affects sulfosalicylic, heat and acetic acid tests. *1872*
Urine Increase Physiological May occur following aortography. *0065* Observed in small number of patients receiving agents for arteriography. *3259* Mean increase from 129 mg/d to 2760 mg/d in 37 patients day following arteriography. *2594*
CSF Increase Analytical Causes turbidity if sulfosalicylic acid used. *3505*

Protein Bound Iodine (PBI) *Serum Increase Analytical* Most radiopaque media have effect. *3217*

Protein Electrophoresis *Serum Positive Analytical* Produces uninterpretable pattern. *3879*

Prothrombin Time *Plasma Increase Physiological* ?transient inactivation of coagulation factors. *1487*

PSP Excretion *Urine Increase Physiological* Affect renal excretion. *1237*

Specific Gravity *Urine Increase Analytical* Presence of high molecular weight substance. *1700* Presence of high molecular weight substance. *3505*

Sugar *Urine Increase Analytical* Green-black reaction with reducing procedures. *0583*

Thyroid Stimulating Hormone (TSH)
Serum Increase Physiological Substantial increase 3 d after oral cholecystography. *2084*

Thyroxine (T₄) *Serum No Effect Analytical* Most agents in low concentration. *3217* No effect on competitive protein binding procedure of Alexander. *0054*
Serum Increase Analytical Orabilix and Dionosil® affect at 1 mg. *3217*
Serum Increase Physiological In first few days following oral cholecystographic agents. *2084*

Transferrin *Urine Increase Physiological* Mean increase from 2.8 mg/d to 147.1 mg/d in 37 patients day following arteriography. *2594*

Tri-iodothyronine (T₃) *Serum Decrease Physiological* In first few days following oral cholecystographic agents. *2084*

Tri-iodothyronine (T₃), Reverse
Serum Increase Physiological In first few days following oral cholecystographic agents. *2084*

Trypsin *Feces Decrease Analytical* Inhibition of esterase by Gastrografin®. *0741*

Urea Nitrogen *Serum Increase Physiological* May produce azotemia or renal failure. *0071*

Uric Acid *Serum Decrease Physiological* Interferes with reabsorption. *2909*

Uric Acid Clearance *Urine No Effect Physiological* No effect of arteriography. *2594*

Urobilinogen *Urine Increase Analytical* Turbidity in acid solutions. *0583*

Vanillylmandelic Acid *Urine Decrease Physiological* Competes for excretion after i.v. pyelography. *0913*

Ragweed

Immunoglobulin IgE *Serum Increase Physiological* In allergic patients increase with pollen season. *3932*

Ramipril

Aldosterone *Plasma Decrease Physiological* In response to 10 mg and 20 mg on successive days in 9 patients with severe chronic congestive heart failure. *0764*

Angiotensin-Converting Enzyme (ACE)
Serum Decrease Physiological In response to 10 mg and 20 mg on successive days in 9 patients with severe chronic congestive heart failure. *0764*

Angiotensin II *Plasma Decrease Physiological* In response to 10 mg and 20 mg on successive days in 9 patients with severe chronic congestive heart failure. *0764*

Antidiuretic Hormone *Plasma No Effect Physiological* In response to 10 mg and 20 mg on successive days in 9 patients with severe chronic congestive heart failure. *0764*

Catecholamines *Plasma No Effect Physiological* In response to 10 mg and 20 mg on successive days in 9 patients with severe chronic congestive heart failure. *0764*

Cortisol *Plasma No Effect Physiological* In response to 10 mg and 20 mg on successive days in 9 patients with severe chronic congestive heart failure. *0764*

Creatinine *Serum Increase Physiological* In response to 10 mg and 20 mg on successive days in 9 patients with severe chronic congestive heart failure. *0764*

Potassium *Serum Increase Physiological* In response to 10 mg and 20 mg on successive days in 9 patients with severe chronic congestive heart failure. *0764*
Urine Decrease Physiological In response to 10 mg and 20 mg on successive days in 9 patients with severe chronic congestive heart failure. *0764*

Renin Activity *Plasma Increase Physiological* In response to 10 mg and 20 mg on successive days in 9 patients with severe chronic congestive heart failure. *0764*

Sodium *Serum Increase Physiological* In response to 10 mg and 20 mg on successive days in 9 patients with severe chronic congestive heart failure. *0764*
Urine Decrease Physiological In response to 10 mg and 20 mg on successive days in 9 patients with severe chronic congestive heart failure. *0764*

Ranitidine

Acetylcholinesterase *Serum   Decrease   Physiological*
Inhibited by therapeutic doses. *1938*

Aldosterone *Plasma   No Effect   Analytical* No effect of up
to 5 μg/mL on method of Sancho and Haber. *3130*
Plasma   No Effect   Physiological Bolus injections have no
effect on basal secretion. *2538*
Plasma   Decrease   Physiological Significant reduction in
plasma aldosterone in both recumbent overnight concentration
and after 2 h ambulation with 3-day oral course. *3130*

Atenolol *Serum   No Effect   Physiological* Minimal effect of
coadministration on atenolol pharmacokinetics. *1938*

Chlormethiazole *Serum   Increase   Physiological* At least in
one patient prolongation of elimination half-life. *1938*

Cholesterol *Serum   No Effect   Physiological* In 25 patients
pretreatment 171 mg/dL, after 175 mg/dL over 5 weeks. *3568*

Cholesterol, High Density Lipoprotein
Serum   No Effect   Physiological In 25 patients pretreatment
40 mg/dL, after 38 mg/dL over 5 weeks. *3568* In 8 ulcer
patients given 300 mg/d for 1 mo. *3858*

Cortisol *Plasma   No Effect   Physiological* No effect with i.v.
bolus of much as 300 mg. *0871*
Plasma   Decrease   Physiological Compared to control state
after 3 d 150 mg/12 h cortisol level significantly lower at rest and
after 30 minutes ambulation. *3130*

Creatinine *Serum   No Effect   Physiological* No effect
reported although other drugs with same overall action blocks
tubular secretion of creatinine. *3389*

Diazepam *Serum   No Effect   Physiological* Steady-state
plasma concentration, clearance and elimination half-life not
affected. *1938*

Diphenylhydantoin *Serum   No Effect   Physiological* No significant effect observed on drug metabolism. *3772*
Serum   Increase   Physiological Mean concentration
increased from 36.1 μmol/L to 39.3 μmol/L after coadministration.
1938

Eosinophils *Blood   Increase   Physiological* Isolated case
reported within 1 mo of treatment being started, reverted to
normal 2 weeks after drug stopped. Probable hypersensitive or
idiosyncratic reaction. *3303*

Estradiol *Plasma   No Effect   Physiological* After 4 weeks or
6 mo treatment (300 mg and 150 mg daily respectively) in male
patients with duodenal ulcer. *0727*

Ethanol *Serum   No Effect   Physiological* No effect on peak
concentration or area under the curve with pretreatment with
drug. *3246* Peak plasma concentration and area under curve
unaffected. *1938*

Follicle Stimulating Hormone (FSH)
Plasma   No Effect   Physiological With up to 450 mg/d in 20
males with chronic duodenal ulcer. *2925* No effect with i.v.
bolus of much as 300 mg. *0871* After 4 weeks or 6 mo treatment (300 mg and 150 mg daily respectively) in male patients
with duodenal ulcer. *0727*

Gastrin *Serum   No Effect   Physiological* No effect reported
in spite of long term treatment. *1667*

Granulocytes *Blood   Decrease   Physiological* Marked
reduction in one elderly woman following 2 weeks of treatment
with 300 mg/ daily: normalized when treatment stopped. *0463*

Growth Hormone *Plasma   No Effect   Physiological* No
effect with i.v. bolus of much as 300 mg. *0871*

HDL₂-Cholesterol *Serum   No Effect   Physiological* In 8
ulcer patients given 300 mg/d for 1 mo. *3858*

HDL₃-Cholesterol *Serum   No Effect   Physiological* In 8
ulcer patients given 300 mg/d for 1 mo. *3858*

Hydrochloric Acid *Gastric Material   Decrease   Physiological*
Reduced to 10% of normal in healthy volunteers. *3388*

Leukocytes *Blood   Decrease   Physiological* Isolated case
reported within 1 mo of treatment being started, reverted to
normal 2 weeks after drug stopped. Probable hypersensitive or
idiosyncratic reaction. *3303* Marked reduction in one elderly
woman following 2 weeks of treatment with 300 mg/ daily: normalized when treatment stopped. *0463*

Lidocaine *Serum   No Effect   Physiological* Insignificant on
drug kinetics when co-administered. *1938*

Lorazepam *Serum   No Effect   Physiological* No affected:
normally conjugated in liver. *1938*

Luteinizing Hormone (LH) *Plasma   No Effect   Physiological*
With up to 450 mg/d in 20 males with chronic duodenal ulcer.
2925 No effect with i.v. bolus of much as 300 mg. *0871* After
4 weeks or 6 mo treatment (300 mg and 150 mg daily respectively) in male patients with duodenal ulcer. *0727*

Metoprolol *Serum   Increase   Physiological* Significantly
increased area under curve and peak plasma concentration.
1938

Midazolam *Serum   Increase   Physiological* Bioavailability
significantly increased. *1938*

Neutrophils *Blood   Decrease   Physiological* Isolated case
reported within 1 mo of treatment being started, reverted to
normal 2 weeks after drug stopped. Probable hypersensitive or
idiosyncratic reaction. *3303*

Nifedipine *Serum   Increase   Physiological* 25% in mean
plasma concentration and area under curve. *1938*

Occult Blood *Feces   No Effect   Analytical* No effect on
Hemoquant method. *3671*

pH *Urine   Increase   Physiological* Mean increase of 0.4 in
healthy volunteers. *3389*
Gastric Material   Increase   Physiological pH 5 reached
approximately 1 h after 50 mg. *2578*

Potassium *Serum   No Effect   Physiological* No effect of 3-d
of 150 mg/12 h. *3130*
Urine   No Effect   Physiological No effect of 3-d of 150 mg/12
h. *3130*

Procainamide *Serum   No Effect   Physiological* No effect on
serum concentration or steady-state pharmacokinetics. *3022*
Serum   Increase   Physiological Significantly increased area
under curve and reduced renal clearance. *1938*
Plasma   Increase   Physiological Renal clearance reduced
without change in half-life. Increased area under curve for both
procainamide and NAPA. Also slightly reduced gastrointestinal
absorption: probably blocks tubular secretion of drug. *3388*

Progesterone *Plasma   No Effect   Physiological* After 4
weeks or 6 mo treatment (300 mg and 150 mg daily respectively) in male patients with duodenal ulcer. *0727*

Prolactin *Plasma   No Effect   Physiological* No difference in
TRH-stimulated concentration compared with control in 10 ulcer
patients or in prestimulation concentration. *1603* No significant
effect during treatment. *2578*
Plasma   Increase   Physiological No effect up to 100 mg i.v.,
but mean increase of 1.3 ng/mL in 5 subjects. *0871* Significant
increase in basal value. *2780*
Plasma   Decrease   Physiological Compared with control
state after 3 d 150 mg/12 h lower during ambulation and at rest.
3130

Propranolol *Serum   No Effect   Physiological* No significant
effect on plasma concentration, area under curve or elimination
half-life. *1938*

Renin Activity *Plasma   No Effect   Physiological* No effect
of 3-d of 150 mg/12 h. *3130*

Sodium *Serum   No Effect   Physiological* No effect of 3-d of
150 mg/12 h. *3130*
Urine   No Effect   Physiological No effect of 3-d of 150 mg/12
h. *3130*

Sperm Count *Semen   No Effect   Physiological* With up to
450 mg/d in 20 males with chronic duodenal ulcer. *2925*

Sperm Morphology *Semen   No Effect   Physiological* With
up to 450 mg/d in 20 males with chronic duodenal ulcer. *2925*

Testosterone *Serum   No Effect   Physiological* With up to
450 mg/d in 20 males with chronic duodenal ulcer. *2925* After 4
weeks or 6 mo treatment (300 mg and 150 mg daily respectively) in male patients with duodenal ulcer. *0727*

Theophylline *Serum   No Effect   Physiological* No significant effect on ranitidine pharmacokinetics. *2859*
Serum   Increase   Physiological Similar effect to cimetidine in
reducing plasma clearance. *1938*

Thyroid Stimulating Hormone (TSH)
Serum   No Effect   Physiological In 10 peptic ulcer patients
no effect of 150 mg twice daily for 28 d. *1603* No effect with
i.v. bolus of much as 300 mg. *0871* No effect of treatment on
values. *2780* After 4 weeks or 6 mo treatment (300 mg and 150
mg daily respectively) in male patients with duodenal ulcer. *0727*

Thyroxine (T₄) *Serum Decrease Physiological* Small reduction noted in 10 ulcer patients after 150 mg twice daily for 28 d ratio: total T₄/total T₃ fell. *1603* Slight reduction after 4 weeks of 300 mg daily, but returned to normal after 6 mo of 150 mg daily in men with duodenal ulcers. *0727*

Thyroxine (T₄) Binding Globulin
Serum No Effect Physiological No difference in thyroid hormone binding protein in 10 ulcer patients after 150 mg twice daily for 28 d. *1603*

Thyroxine (T₄), Free *Serum Decrease Physiological* Similar small effect noted as for total hormone concentration: ratio fT₄/fT₃ fell significantly. *1603*

Triglycerides *Serum No Effect Physiological* In 25 patients pretreatment 121 mg/dL, after 137 mg/dL over 5 weeks. *3568*

Tri-iodothyronine (T₃) *Serum No Effect Physiological* No effect of treatment on values. *2780* After 4 weeks or 6 mo treatment (300 mg and 150 mg daily respectively) in male patients with duodenal ulcer. *0727*
Serum Increase Physiological Small increase noted in ulcer patients after 150 mg twice daily for 28 d. *1603*

Tri-iodothyronine (T₃), Free *Serum Increase Physiological* Similar small effect noted as for total hormone concentration:. *1603*

TSH Response to TRH *Serum Decrease Physiological* In response to i.v. drug: may cause decrease of basal concentration in hypothyroidism. *1938*

Vitamin B₁₂ *Serum Decrease Physiological* Significant decreased absorption of protein bound compound. *1938*

Volume *Gastric Material Decrease Physiological* Reduced to 10% of normal in healthy volunteers. *3388* Significant inhibition following treatment. *2578*

Warfarin *Plasma No Effect Physiological* No effect on plasma concentration or prothrombin time during coadministration. *1938*

Rauwolfia

Catecholamines *Urine Increase Physiological* Release of stored norepinephrine. *2220*

HMPG (4-Hydroxy-3-Methoxy-Phenylethylene Glycol)
Urine Increase Physiological Release of stored norepinephrine. *2220*

Hydrochloric Acid *Gastric Material Increase Physiological* Stimulates gastric secretion. *2220*

17-Hydroxycorticosteroids *Urine Decrease Physiological* Probably due to depressed central synthesis. *0022*

5-Hydroxyindoleacetic Acid (5-HIAA)
Urine Increase Physiological Result of release of 5-HT from brain, tissues. *1343*

Sodium *Serum Increase Physiological* May cause electrolyte retention or edema. *2313*

Vanillylmandelic Acid *Urine Increase Physiological* Release of stored norepinephrine. *2220*

Recumbency

Aldosterone *Plasma Decrease Physiological* One sixth of standing concentration. *0733*

Angiotensin I *Plasma Decrease Physiological* Fell by just over half in 3 h. *0733*

Antidiuretic Hormone *Plasma Decrease Physiological* In comparison with sitting (0.4 pg/mL versus 1.4). *3241*

Calcium *Serum Decrease Physiological* Decreased by about 6.7 % with 15 minutes lying down. *3476*

Cholesterol *Serum Decrease Physiological* 10% drop in 30 minutes in normal. *3540* Decrease of 16% max on lying down (as bound to protein). *3539*

Cyclic Adenosine Monophosphate
Urine No Effect Physiological Posture has no effect. *1992*

Epinephrine *Urine Decrease Physiological* Significantly less than when ambulant. *3616*

Glucose *Serum No Effect Physiological* Change of posture had no effect. *3539*
Serum Increase Physiological Occurs with reduced physical activity. *0279*

Glucose Tolerance *Serum Decrease Physiological* Bed rest associated with increased peripheral insulin resistance. *2187*

Hematocrit *Blood Decrease Physiological* Change of 7% with lying, 4% with sitting. *3539* 6% drop at 20 minutes in normal. *3540*

Hemoglobin *Blood Decrease Physiological* Average maximum decrease of 5% occurs in 20 minutes. *1010*

Insulin *Plasma No Effect Physiological* Change of posture had no effect. *3539*

Norepinephrine *Urine Decrease Physiological* Significantly less than when ambulant. *3616*

Protein *Serum Decrease Physiological* Average decrease of 1.0 g/dL after 40 minutes. *1010*

Renin Activity *Plasma Decrease Physiological* Fell by just over half in 3 h. *0733*

Sodium *Urine Increase Physiological* Metabolic effect. *3578*

Triglycerides *Serum Decrease Physiological* Decrease of 12% maximum on lying down. *3539* 11% drop at 20 minimum in normal. *3540*

Red Dyed Drugs

Protein Bound Iodine (PBI) *Serum Increase Analytical* Presence of erythrocrine dye. *0097*

Red Dyed Food

Protein Bound Iodine (PBI) *Serum Increase Analytical* Presence of erythrocrine dye. *1237*

Reducing Diet

BSP Retention *Serum Increase Physiological* Impaired hepatic perfusion (up to 40). *3217*

Creatinine *Serum Increase Physiological* By up to 2 mg/dL (functional impairment). *3217*

Extracellular Fluid (ECF) *Plasma Decrease Physiological* May constrict by 30% with 300-600 Cal/d. *3217*

Uric Acid *Serum Increase Physiological* Probably due to tissue destruction. *3217*

Volume *Plasma Decrease Physiological* May constrict by 30% with 300-600 Cal/d. *3217*

Renacidin®

Magnesium *Serum Increase Physiological* One case reported of child who developed severe hypermagnesemia. *3853*

Renal Transplant

Uric Acid *Serum Decrease Physiological* Significant effect over first 3 mo. *3191*

Reserpine

Albumin *Serum No Effect Analytical* No effect at 0.02 mg/dL on SMA 12/60 method. *2636*

Alkaline Phosphatase *Serum No Effect Analytical* No effect at 0.02 mg/dL on SMA 12/60 method. *2636*

Aspartate Aminotransferase *Serum No Effect Analytical* No effect at 0.02 mg/dL on SMA 12/60 method. *2636*

Bilirubin *Serum No Effect Analytical* No effect at 0.02 mg/dL on SMA 12/60 method. *2636*
Serum Increase Analytical At concentrations above 61 mg/L raised concentration as measured by Jendrassik and Grof method. *3393*

Calcium *Serum No Effect Analytical* No effect at 0.02 mg/dL on SMA 12/60 method. *2636* At concentration of 61 mg/L had no effect on cresolphthalein method. *3393*

Catecholamines *Plasma Decrease Physiological* Observed normal response to therapy. *2427*

Reserpine *(continued)*

Catecholamines *(continued)*
Urine No Effect Analytical No effect on fluorometric method of Crout. *0758* No effect reported on method of Sandhu and Freed. *1205*
Urine Increase Physiological Release of stored norepinephrine. *2220*
Urine Decrease Physiological Decreased norepinephrine synthesis. *0758*

Cholesterol *Serum No Effect Analytical* No effect at 0.02 mg/dL on SMA 12/60 method. *2636*
Serum No Effect Physiological No significant change in small number of patients treated for up to 2.5 mo. *0088*

Corticosteroids *Urine Increase Analytical* Increased absorption at 410 nm in Porter-Silber reaction. *0022*

Creatinine *Serum No Effect Analytical* No effect at 0.02 mg/dL on SMA 12/60 method. *2636*
Urine Increase Analytical No effect at 0.02 mg/dL on SMA 12/60 method. *2636*

Epinephrine *Urine No Effect Physiological* Not affected (unlike norepinephrine). *2702*

Fatty Acids, Free (FFA) *Serum Increase Physiological* 78% increase after 2.5 mg injected i.m. *1634*

Glucose *Serum No Effect Analytical* No effect at 0.02 mg/dL on SMA 12/60 method. *2636*
Serum Increase Physiological Hyperglycemia may follow administration. *2309*
Urine Increase Physiological Occurs as consequence of hyperglycemia. *2309*

Guaiacols Spot Test *Urine Positive Analytical* False reaction with screening test of Rogers. *3031*

HMPG (4-Hydroxy-3-Methoxy-Phenylethylene Glycol)
Urine Increase Physiological Release of stored norepinephrine. *2220*
Urine Decrease Physiological Long term administration produces decrease. *0426*

Homovanillic Acid *Urine Increase Physiological* Maximum during second day of treatment. *3838*

Hydrochloric Acid *Gastric Material Increase Physiological* Excess secretion may activate peptic ulcers. *1343*

17-Hydroxycorticosteroids *Urine Decrease Analytical* Interference with Porter-Silber reaction. *3505*
Urine Decrease Physiological Probably due to depressed central synthesis. *0022*

5-Hydroxyindoleacetic Acid (5-HIAA)
Urine Increase Physiological Release of 5-HT from brain and tissues. *2425*

5-Hydroxytryptamine (Serotonin)
Plasma Decrease Physiological Observed normal response to therapy. *2427*

17-Ketosteroids *Urine Increase Analytical* Due to chemical structure affects Zimmermann procedure. *3217*
Urine Decrease Analytical Interference with Zimmermann procedure. *0427*

Lactate Dehydrogenase *Serum No Effect Analytical* No effect at 0.02 mg/dL on SMA 12/60 method. *2636*

LE Cells *Blood Positive Physiological* SLE may occur, usually normalizes when stopped. *3059*

Norepinephrine *Urine Decrease Physiological* Contributes to fall of total catecholamines. *2702*

Occult Blood *Feces Positive Physiological* May activate peptic ulcers and cause bleeding. *3856*

Pepsin *Gastric Material Increase Physiological* Greatly augments secretion. *0612*

Phosphate *Serum No Effect Analytical* No effect at 0.02 mg/dL on SMA 12/60 method. *2636* At concentration of 61 mg/L had no effect on phosphomolybdate method. *3393*

Platelets *Blood Decrease Physiological* Thrombocytopenia with purpura may occur. *2429*

Prolactin *Plasma Increase Physiological* Dose-related at doses greater than 0.25 mg/d. *0451* Marked effect in male and female hypertensives treated for up to 6 weeks. *3630*

Protein *Serum No Effect Analytical* No effect at 0.02 mg/dL on SMA 12/60 method. *2636* At concentration of 61 mg/L had no effect on biuret method with blank correction. *3393*

Protein Bound Iodine (PBI) *Serum No Effect Physiological* No effect observed with oral, parenteral administration. *0830*
Serum Increase Physiological Increased metabolism by hepatic microsomes. *1237*
Serum Decrease Physiological Rare depression reported (given orally for 1 mo). *2220*

Prothrombin Time *Plasma Increase Physiological* Long term treatment markedly enhances anticoagulant. *1343*
Plasma Decrease Physiological Short term treatment blocks action of anticoagulant. *1343*

Thyroxine (T$_4$) *Serum Decrease Physiological* Increased metabolism by hepatic microsomes. *3505*

Triglycerides *Serum No Effect Physiological* No significant change in small number of patients treated for up to 2.5 mo. *0088*

Tyramine Test *Patient Decrease Physiological* Inhibits responsiveness (produces false negative). *1325*

Urea Nitrogen *Serum No Effect Analytical* At concentration of 61 mg/L had no effect on diacetylmonoxime method. *3393*

Uric Acid *Serum No Effect Analytical* No effect at 0.02 mg/dL on SMA 12/60 method. *2636* At concentration of 61 mg/L had no effect on phosphotungstate reduction method. *3393*

Vanillylmandelic Acid *Urine No Effect Analytical* No effect on Gitlow method. *1487*
Urine Increase Physiological Release of stored norepinephrine. *2220*
Urine Decrease Physiological Depletion of catecholamine stores. *2425*

Resorcinol

Color *Urine Increase Analytical* Dark green to greenish blue darkens on standing. *2313*

Creatinine *Serum Increase Analytical* Falsely high values if Jaffé reaction used. *1563*

Erythrocytes *Blood Decrease Physiological* Injection or excess absorption may cause hemolysis. *0071*

Haptoglobin *Serum Decrease Physiological* Occurs with hemolytic anemia. *2427*

Hematocrit *Blood Decrease Physiological* Injection or excess absorption may cause hemolysis. *0071*

Hemoglobin *Blood Decrease Physiological* Injection or excess absorption may cause hemolysis. *0071*

^{131}I Uptake *Serum Decrease Physiological* Reported effect of treatment of varicose ulcers. *1343*

Methemoglobin *Blood Increase Physiological* Occurs with hemolytic anemia. *2427* Injection or excess absorption may cause hemolysis. *0071*

Protein Bound Iodine (PBI) *Serum Decrease Physiological* Reported effect of treatment of varicose ulcers. *1343*

Urea Nitrogen *Serum No Effect Analytical* At concentration of 500 mg/L had no effect on Seralyzer method. *3393*

Uric Acid *Serum No Effect Analytical* At concentration of 50 mg/L had no effect on uricase method on aca. *3393*
Serum Increase Analytical 5 mg/dL equivalent to 5.4 mg/dL copper chelate procedure 5 mg/dL equivalent to 0.3 mg/dL by Nishi procedure. *2230*

Rhamnose

Protein *Test Conditions Increase Analytical* Interferes with Folin-Ciocalteu method of Lowry. *3636*

Sugar *Urine Increase Analytical* Reducing substance affects Benedict's, Clinitest®. *0583*

Rhubarb

Color *Urine Increase Analytical* Yellow-brown (acid), yellow-pink (alkaline) darkens. *2313*
Feces Increase Physiological Yellow with 2 ml extract. *0902*

Crystals *Urine Increase Physiological* Calcium oxalate may be deposited. *0152*

Erythrocytes *Urine Increase Physiological* Crystalluria may cause renal damage. *0152*

Protein *Urine Increase Physiological* Crystalluria may cause renal damage. *0152*

Sugar *Urine Increase Analytical* May cause false positive with Benedict's. *2448*

Riboflavin

Acetoacetate Decarboxylase *Serum Increase Analytical* Activation of enzyme produced. *3684*

BSP Retention *Serum Increase Analytical* If given i.v. affects method of Gaebler. *2381*

Catecholamines *Plasma Increase Analytical* Interferes with fluorometric technique. *2425*
Urine Increase Analytical Large doses may produce similar fluorescence. *0583*

Color *Urine Increase Analytical* May produce yellow color with large doses. *1488*

Diagnex Blue Excretion *Urine Increase Analytical* Interfering color. *3879*

Lymphocytes *Blood No Effect Physiological* No effect with megadose supplementation. *1347*

Neutrophils *Blood No Effect Physiological* No effect with megadose supplementation. *1347*

Urobilin *Urine Increase Analytical* Produces yellow-green fluorescence. *1563*

Ribose

Glucose *Serum Increase Analytical* Nonspecificity of ferricyanide, Neocuproin methods. *3902*

Sugar *Urine Increase Analytical* False positive with Benedict's, Clinitest®. *1563*

Rifampin

Acenocoumarol *Plasma Decrease Physiological* Decreased serum concentration due to hepatic enzyme induction. *0014*

Acetaminophen *Urine Increase Physiological* Significantly greater proportion excreted as glucuronide in response to co-administration of drug. *0394*

Alanine Aminotransferase *Serum No Effect Analytical* At 5 times upper limit of therapeutic range on method on Abbott-VP, aca, Cobas-Bio, and KDA. *2138*
Serum Increase Physiological Hepatic toxicity occurs in up to 7% patients. *1553* Minimal abnormalities of liver function are common: severe liver damage in about 0.6% patients. *3683* Observed in one patient receiving drug who developed porphyria cutanea tarda. *2441* In 61% of 18 children receiving drug in combination with isoniazid. *2186*
Serum Decrease Analytical At 5 times upper limit of therapeutic range on method on SMAC®. *2138*

Albumin *Serum No Effect Analytical* At 5 times upper limit of therapeutic range on methods on SMAC®, Ektachem®, Hitachi® 705 and KDA. *2138*

Alkaline Phosphatase *Serum No Effect Analytical* On method involving hydrolysis of p-nitrophenyl phosphate. *0218* At 5 times upper limit of therapeutic range on methods on SMAC®, Cobas-Bio, Abbott-VP, aca, Hitachi® 705 and KDA. *2138*
Serum Increase Physiological Can be serious but reversible liver damage. *2806* Minimal abnormalities of liver function are common: severe liver damage in about 0.6% patients. *3683* Observed in one patient receiving drug who developed porphyria cutanea tarda. *2441*

Amylase *Serum No Effect Analytical* At 5 times upper limit of therapeutic range on methods on aca, Cobas-Bio and Ektachem®. *2138*

Aspartate Aminotransferase *Serum No Effect Analytical* At 5 times upper limit of therapeutic range on methods on SMAC®, Cobas-Bio, Abbott-VP, aca, Hitachi® 705 and KDA. *2138*
Serum Increase Physiological Hepatic toxicity occurs in up to 7% patients. *1553* Minimal abnormalities of liver function are common: severe liver damage in about 0.6% patients. *3683* In 83% of 18 children receiving drug in combination with isoniazid. *2186*

Serum Decrease Analytical At 3.5 mg/L and above decreased activity when methods using absorbance at 340 nm used, since drug absorbs at this wavelength. *0218*

Bile Acids *Serum Increase Physiological* Marked changes may occur when given alone or in combination with isoniazid in the absence of other abnormal liver function tests, possibly due to inhibition of uptake of bile acids into hepatocytes. *0306* Significant increase possibly due to blocking uptake by plasma membrane of hepatocytes. *1224*

Bilirubin *Serum No Effect Analytical* Minimal effect on Jendrassik/Grof methods, although some effect at drug concentration of 15 mg/L. *0218* At 5 times upper limit of therapeutic concentration on methods on SMAC®, aca, Hitachi® 705 and KDA. *2138*
Serum No Effect Physiological Probably no significant displacement from protein in neonates. *3748*
Serum Increase Analytical Due to yellow coloration of drug. *2820* Marked effect when measured by bilirubinometer. *0218* At 5 times upper limit of therapeutic concentration on methods on Ektachem® and Cobas-Bio. *2138*
Serum Increase Physiological Inhibits hepatic excretion. *0013* Minimal abnormalities of liver function are common: severe liver damage in about 0.6% patients. *3683* Observed in one patient receiving drug who developed porphyria cutanea tarda. *2441* Significant increase in all subjects after single dose of 900 mg. *1224*

Bilirubin, Direct *Serum Increase Physiological* Transient effect. *1217*

BSP Retention *Serum Increase Physiological* Inhibits hepatic excretion (probably). *1745* Initial increase of serum concentration due to reduced excretion in bile. *0014*

Calcitonin *Plasma No Effect Physiological* No change observed in 8 healthy men taking drug for 2 weeks. *0479*

Calcium *Serum No Effect Analytical* At 5 times upper limit of therapeutic range on method on SMAC®, Abbott-VP, aca, Ektachem®, Hitachi® 705 and KDA. *2138*

Capillary Fragility *Patient Increase Physiological* Reported in patient with macroglobulinemia. *1050*

Casts *Urine Increase Physiological* Rare reported side effect. *1553*

Cephalin Flocculation *Serum Increase Physiological* Hepatotoxic effect (cholestasis). *2820*

Chenodeoxycholic Acid *Serum Increase Physiological* Significant increase possibly due to blocking uptake by plasma membrane of hepatocytes. *1224*

Chloramphenicol *Serum No Effect Analytical* No effect at 100 mg/L on coupled enzymatic method. *2490*

Cholesterol *Serum No Effect Analytical* At 5 times upper limit of therapeutic concentration on methods on SMAC®, Abbott-VP, aca, Ektachem®, Hitachi® 705 and KDA. *2138*
Serum Decrease Analytical At 5 times upper limit of therapeutic concentration on method on Cobas-Bio. *2138*

Cholic Acid *Serum Increase Physiological* Significant increase possibly due to blocking uptake by plasma membrane of hepatocytes. *1224*

Color *Urine Increase Analytical* Red-orange due to drug and metabolites. *1578* Causes orange-pink color in saliva, tears, urine, sweat. *3683*
Feces Increase Physiological Orange-red color due to drug and metabolites. *0019*

Coombs' Test, Direct *Serum Positive Physiological* Weak response in 8% patients. *2843*

Coombs' Test, Indirect *Serum Positive Physiological* Positive response in 33% patients after 3 mo. *2843*

Corticosteroids *Plasma Decrease Physiological* Decreased serum concentration due to hepatic enzyme induction. *0014*

Cortisone *Serum Decrease Physiological* Metabolism enhanced with reduction of circulating drug concentration: cortisol production rate increased. *1830*

C-Peptide *Plasma Increase Physiological* Increased rate of secretion after oral administration of 100 g glucose. *3534*

Creatine Kinase *Serum No Effect Analytical* At 5 times upper limit of therapeutic range on methods on SMAC®, Abbott-VP, aca, Cobas-Bio, and Hitachi® 705. *2138*

Rifampin *(continued)*

Cyclosporine *Serum Decrease Physiological* Marked reduction of concentration due to hepatic enzyme induction. *0683*

Blood Decrease Physiological Induces cytochrome P-450 hepatic enzymes thereby increasing clearance of cyclosporine. *3668*

Dapsone *Serum Decrease Physiological* Increased plasma clearance, but interaction may not be clinically significant. *0014*

Deoxycholic Acid *Serum Increase Physiological* Significant increase possibly due to blocking uptake by plasma membrane of hepatocytes. *1224*

Digitoxin *Serum Decrease Physiological* May be reduced to subtherapeutic concentration by induction of mixed function oxidases. *2794* Decreased serum concentration due to hepatic enzyme induction. *0014*

Digoxin *Serum Decrease Physiological* Due to induction of detoxifying enzymes. *3683* Observed in patients receiving antituberculous treatment in addition to drug for cardiac irregularity. *0549* Decreased serum concentration due to hepatic enzyme induction. *0014*

1,25-Dihydroxy Vitamin D$_3$ *Serum No Effect Physiological* No change observed in 8 healthy men taking drug for 2 weeks. *0479*

Eosinophils *Blood Increase Physiological* Allergic reaction. *1553* Isolated case of eosinophilia attributable to drug only. *2520* Eosinophilia in one patient clearly linked to drug only since declined once drug withdrawn. *2109*

Erythrocytes *Blood Decrease Physiological* Possible antibody-mediated immune reaction. *3683*

Urine Increase Physiological Rare reported side effect. *1553*

Glucaric Acid *Urine Increase Physiological* Manifestation of hepatic enzyme induction. *0014*

Glucose *Serum No Effect Analytical* At 5 times upper limit of therapeutic concentration on methods on SMAC®, Abbott-VP, Cobas-Bio, Ektachem®, Hitachi® 705 and KDA. *2138* At concentration of 150 mg/L had no effect on hexokinase/G-6-PDH method. *3393*

Serum Increase Analytical At 5 times upper limit of therapeutic concentration on method on aca. *2138*

Serum Increase Physiological Early phase hyperglycemia after oral administration of 100 g glucose. *3534*

γ-Glutamyltransferase (GGT) *Serum No Effect Analytical* On method using hydrolysis of gamma-glutamyl p-nitroanilide with absorbance of product at 410 nm. *0218* At 5 times upper limit of therapeutic range on methods on SMAC®, Abbott-VP, and Hitachi® 705. *2138*

Hemoglobin *Blood Decrease Physiological* Observed effect. *1705* Rarely acute hemolytic anemia occurs associated with circulating drug dependent antibodies in high titer. *1292*

Hydrocortisone *Serum Decrease Physiological* Metabolism enhanced with reduction of circulating drug concentration: cortisol production rate increased. *1830*

6-β-Hydroxycortisol *Urine Increase Physiological* Marked increase reflecting hepatic enzyme induction. *0394*

25-Hydroxy Vitamin D$_3$ *Serum Decrease Physiological* Secondary to induction of hepatic microsomal enzymes: decrease of 70% with short course in 8 healthy men. *0479*

Indocyanine Green *Serum Increase Physiological* Delayed elimination following i.v. injection. *2427*

Insulin *Plasma Increase Physiological* Increased rate of secretion after oral administration of 100 g glucose. *3534*

Iron *Serum No Effect Analytical* At 5 times upper limit of therapeutic range on Ferrozine method on SMAC®. *2138*

Ketoconazole *Serum Decrease Physiological* Plasma concentration reduced by about 33% with concomitant administration. *0450*

Lactate *Serum No Effect Analytical* At concentration of 150 mg/L had no effect on enzymatic method. *3393*

Lactate Dehydrogenase *Serum No Effect Analytical* At 5 times upper limit of therapeutic range on method on SMAC®, Abbott-VP, Cobas-Bio and Hitachi® 705. *2138*

Serum Increase Analytical At 5 times upper limit of therapeutic range on method on KDA. *2138*

Leukocytes *Blood Decrease Physiological* Toxic or allergic response (usually transient). *2820*

Methadone *Serum Decrease Physiological* Decreased serum concentration due to hepatic enzyme induction. *0014*

Mexiletine *Serum Decrease Physiological* Hepatic enzyme induction may reduce elimination half-life by 50%. *2011*

Morphine *Serum Decrease Physiological* Decreased serum concentration due to hepatic enzyme induction. *0014*

Neutrophils *Blood Decrease Physiological* Neutropenia may occur. *3202* Occasional case of agranulocytosis reported. *3717*

Parathyroid Hormone *Plasma No Effect Physiological* No change observed in 8 healthy men taking drug for 2 weeks. *0479*

Phosphate *Serum No Effect Analytical* On nonprecipitation method of Peynet. *2799* At 5 times upper limit of therapeutic concentration on methods on SMAC® and KDA. *2138*

Serum Increase Analytical At 5 times upper limit of therapeutic concentration on methods on aca and Hitachi® 705. *2138*

Platelets *Blood Decrease Physiological* Occurs with antibody production. *1050* Possible antibody-mediated immune reaction. *3683* Several cases of immune mediated thrombocytopenia reported. *2502* Rare thrombocytopenia reported: rapid fall after dose given. *1292* Isolated case of severe reduction in platelet count after one dose in patient receiving treatment 4 mo previously. *1445*

Prednisolone *Serum Decrease Physiological* Significant increase in systemic clearance and plasma concentration and area under curve due to induction of liver enzymes. *2347*

Prolactin *Plasma Decrease Physiological* For at least 12 h in normals. *0865*

Protein *Serum No Effect Analytical* At 5 times upper limit of therapeutic concentration on methods on SMAC®, Abbott-VP, Hitachi® 705 and KDA. *2138*

Serum Increase Analytical At 5 times upper limit of therapeutic concentration on methods on biuret method on Ektachem®. *2138*

Serum Decrease Physiological Due to impaired hepatic metabolism. *2820*

Urine Increase Physiological Attributed to drug in 2 cases. *2427*

Prothrombin Time *Plasma Increase Physiological* May prolong action of anticoagulants. *1679*

Plasma Decrease Physiological Occurs in some patients, especially if on anticoagulants. *1375* Due to induction of enzymes metabolizing warfarin when coadministered. *3683*

Quinidine *Serum Increase Physiological* Concentration of metabolites increased with probable maintenance of therapeutic effects. *0548*

Serum Decrease Physiological Observed in patients receiving antituberculous treatment in addition to drug for cardiac irregularity. *0549* Reduction in peak concentration reported in one patient. *0014*

Testosterone *Serum Increase Physiological* Probably due to increase in microsomal activity and increased biosynthesis. *1736*

Theophylline *Serum Decrease Physiological* 25% reduction in area under curve when coadministered and reduction of half-life from 7.0 to 4.8 h. *0436* Reduced area under concentration curve and increased metabolic clearance and volume of distribution. *2861* Significantly increased mean oral clearance of drug with elimination rate constant increased by mean of 25%. *3484*

Thymol Turbidity *Serum Increase Physiological* Hepatotoxic effect (cholestasis). *2820*

Thyroid Stimulating Hormone (TSH)
Serum Increase Physiological In one patient with primary hypothyroidism receiving constant replacement dose of L-thyroxine: tri-iodothyronine clearance retarded. *1734*

Thyroxine (T$_4$) *Serum Decrease Physiological* In one patient with primary hypothyroidism receiving constant replacement dose of L-thyroxine: tri-iodothyronine clearance retarded. *1734*

Tolbutamide *Serum Decrease Physiological* Decreased serum concentration due to hepatic enzyme induction. *0014*

Triglycerides *Serum No Effect Analytical* At 5 times upper limit of therapeutic concentration on method on SMAC®, Abbott-VP Ektachem® and Hitachi® 705. *2138*

Serum *Decrease Analytical* At 5 times upper limit of therapeutic concentration on method on KDA. *2138*

Tri-iodothyronine (T₃) *Serum Increase Physiological* In one patient with primary hypothyroidism receiving constant replacement dose of L-thyroxine: tri-iodothyronine clearance retarded. *1734*

Urea Nitrogen *Serum No Effect Analytical* At 5 times upper limit of therapeutic range on method on SMAC®, Abbott-VP, Cobas-Bio, Ektachem®, Hitachi® 705 and KDA. *2138*
Serum Increase Physiological Temporary renal failure reported. *0944*

Uric Acid *Serum No Effect Analytical* At 5 times upper limit of therapeutic range on methods on SMAC®, Abbott-VP, aca, Ektachem® and Cobas-Bio. *2138*
Serum Increase Analytical Substantial increase at upper limit of therapeutic range on methods on KDA and Hitachi® 705. *2138*
Serum Increase Physiological Temporary renal failure reported. *0944*

Warfarin *Plasma Decrease Physiological* Decreased serum concentration due to hepatic enzyme induction. *0014* Catabolism of drug increased due to induction of hepatic microsomal enzymes. *3036*

Zinc Sulfate Turbidity *Serum Increase Physiological* Hepatotoxic effect (cholestasis). *2820*

Rimeterol

Aldosterone *Plasma No Effect Physiological* No significant change in 4 healthy men given therapeutic i.v. dose. *2807*

Calcium *Serum Decrease Physiological* Dose related significant change in 4 healthy men given therapeutic i.v. dose. *2807*

Corticosteroids *Plasma Decrease Physiological* Dose related significant change in 4 healthy men given therapeutic i.v. dose. *2807*

Glucose *Serum Increase Physiological* Dose related significant change in 4 healthy men given therapeutic i.v. dose. *2807*

β-Hydroxybutyrate *Serum No Effect Physiological* No significant change in 4 healthy men given therapeutic i.v. dose. *2807*

Insulin *Plasma Increase Physiological* Dose related significant change in 4 healthy men given therapeutic i.v. dose. *2807*

Ketones *Serum Increase Physiological* Dose related significant change in 4 healthy men given therapeutic i.v. dose. *2807*

Lactate *Serum Increase Physiological* Dose related significant change in 4 healthy men given therapeutic i.v. dose. *2807*

Magnesium *Serum Decrease Physiological* Dose related significant change in 4 healthy men given therapeutic i.v. dose. *2807*

Phosphate *Serum Decrease Physiological* Dose related significant change in 4 healthy men given therapeutic i.v. dose. *2807*

Potassium *Serum Decrease Physiological* Dose related significant change in 4 healthy men given therapeutic i.v. dose. *2807*

Renin Activity *Plasma Increase Physiological* Dose related significant change in 4 healthy men given therapeutic i.v. dose. *2807*

Ristocetin

Eosinophils *Blood Increase Physiological* Allergic response. *2313*

Leukocytes *Blood Increase Physiological* Due to eosinophilia. *2313*
Blood Decrease Physiological Leukopenia or agranulocytopenia. *2313*

Platelets *Blood Decrease Physiological* Thrombocytopenia may occur (toxic to platelets). *2313*

Ritodrine

Glucose *Serum Increase Physiological* Moderate effect when given intravenously for premature labor. *0575*

Lactate *Serum Increase Physiological* Moderate effect when given intravenously for premature labor. *0575*

Pyruvate *Serum No Effect Physiological* Little effect noted in patients given drug intravenously for premature labor, with consequent increase in lactate/pyruvate ratio. *0575*

RO4-2137

Urea Nitrogen *Serum Increase Physiological* Slight increase up to 70 mg/dL in some patients. *1925*

Rolitetracycline

Glucose *Serum No Effect Analytical* At concentration of 4 mg/L had no effect on GOD/POD-PAP method. *3393*

Protein *Urine Increase Physiological* May cause nephropathy. *2583*

Triglycerides *Serum No Effect Analytical* At concentration of 4 mg/L had no effect on GPO-PAP method. *3393*

Urea Nitrogen *Serum Increase Physiological* May cause nephropathy. *2583*

Uric Acid *Serum No Effect Analytical* At concentration of 4 mg/L had no effect on uricase-PAP method. *3393*

Room Temperature

α-Amino-Nitrogen *Plasma Increase Analytical* Proteolysis occurs even in refrigerator, satisfactory if frozen. *1563*

Ammonia *Plasma No Effect Analytical* If heparinized and in ice, acid added in 15 minutes. *2877*

Rotenone

Glucose *Serum Decrease Physiological* Severe hypoglycemia reported with poisoning. *1302*

Roxithromycin

Carbamazepine *Serum No Effect Physiological* No effect on pharmacokinetics when added to stable therapeutic regime. *3115*

Theophylline *Serum No Effect Physiological* Little effect on pharmacokinetics when added to stable therapeutic regime. *3115*

Rubella Virus Vaccine

Platelets *Blood Decrease Physiological* Self limiting occurring within 10 d or purpura. *0241*

Saccharated Iron Oxide

Albumin *Serum No Effect Physiological* In 9 individuals given 40 mg/d i.v. daily for up to 42 d. *2662*

Calcium *Serum No Effect Physiological* In 9 individuals given 40 mg/d i.v. daily for up to 42 d. *2662*

Creatinine *Serum No Effect Physiological* In 9 individuals given 40 mg/d i.v. daily for up to 42 d. *2662*

Hematocrit *Blood Increase Physiological* In 9 individuals given 40 mg/d i.v. daily for up to 42 d. *2662*

Hemoglobin *Blood Increase Physiological* In 9 individuals given 40 mg/d i.v. daily for up to 42 d. *2662*

Phosphate *Serum Decrease Physiological* Stepwise significant reduction with 40 mg i.v. daily for up to 42 d. *2662*
Urine No Effect Physiological In 9 individuals given 40 mg/d i.v. daily for up to 42 d. *2662*

Potassium *Serum No Effect Physiological* In 9 individuals given 40 mg/d i.v. daily for up to 42 d. *2662*

Saccharose

Protein *Test Conditions Increase Analytical* Possible measurement by biuret reaction. *3376*

Salicylate

Acetaminophen *Serum No Effect Analytical* At 200 and 500 mg/L had no effect on HPLC method. *3432*
Serum Increase Analytical Increases results with unmodified Glynn and Kendal technique. *0199*

Alanine Aminopeptidase *Urine Decrease Physiological* When 0.5 g given orally to 20 healthy adult volunteers and urine studied over next 3 h. *1828*

Albumin *Serum No Effect Analytical* No significant effect even at 1 g/L on BCG method. *2625* At concentration of 600 mg/L had no effect on BCG method. *3393*
Serum Decrease Analytical At 400 mg/L decreases HABA method by 10%. *2625*

Alkaloids *Urine No Effect Analytical* With iodoplatinate of Frings TLC procedure at 100 mg/dL. With ninhydrin on Frings TLC procedure at 100 mg/dL. *1204*

Aspartate Aminotransferase *Serum Increase Physiological* Concentration related effect in 9 of 17 patients with acute rheumatic fever also given phenoxymethyl penicillin: in patients with low albumin effect most marked. *1294*

Barbiturate *Urine No Effect Analytical* With diphenylcarbazone on Frings TLC procedure at 100 mg/dL. With vanillin on Frings procedure at 100 mg/dL. With mercurous NO_3 on Frings TLC procedure at 10 mg/dL. With mercuric SO_4 on Frings TLC procedure at 100 mg/dL. *1204*

Bicarbonate *Serum No Effect Analytical* At concentration of 500 mg/L had no effect on method using phenolphthalein. *3393*

Bilirubin *Serum No Effect Analytical* At concentration of 350 mg/L had no effect on Ektachem® method. *3393* At concentration of 500 mg/L had no effect on Jendrassik and Grof method. *3393*
Serum Increase Physiological Clinically significant displacement from protein in neonates. *3748*

Bromide *Urine No Effect Analytical* No effect on method of Frings. *1204*

Calcium *Serum No Effect Analytical* At concentration of 500 mg/L had no effect on cresolphthalein method. *3393*

Chloramphenicol *Serum No Effect Analytical* No effect at 100 mg/L on coupled enzymatic method. *2490*

Chlordiazepoxide *Blood No Effect Analytical* On method of Riddick at 5 mg/dL. *2983* On method of Riddick at 5 mg/dL. *2983*

Chloride *Serum No Effect Analytical* At concentration of 500 mg/L had no effect on mercurimetric method. *3393*

Cholesterol *Serum No Effect Analytical* At concentration of 500 mg/L had no effect on method using catalase-Hantzsch reaction. *3393* At concentration of 500 mg/L had no effect on Liebermann-Burchard method. *3393*

Color *Feces Increase Physiological* Pink to red black due to gastrointestinal bleeding. *1187*

Creatinine *Serum No Effect Analytical* At concentration of 150 mg/L had no effect on creatinine iminohydrolase method. *3393* At concentration of 500 mg/L had no effect on AutoAnalyzer Jaffé method. *3393*

Dapsone *Serum No Effect Analytical* At 30 mg/dL on colorimetric procedure of Higgins. *1590*

Diphenylhydantoin *Serum No Effect Analytical* No effect on TLC method of Simon, Jatlow. *3323*

Estrogens *Urine Decrease Analytical* Competes for enzyme used for hydrolysis. *0022*

Ethchlorvynol *Serum No Effect Analytical* At 180 mg/L on GLC procedure of Evenson. *1058* At 180 mg/L on GLC procedure of Evenson. *1058*
Urine No Effect Analytical No effect on method of Frings and Cohen. *1204*

Fatty Acids, Free (FFA) *Serum No Effect Analytical* No effect on method of Pinelli. *2819*

Ferric Chloride Test *Urine Positive Analytical* May produce stable purple color. *0443*

FPN Test *Urine No Effect Analytical* No effect at 100 mg/dL on method of Frings. *1204*

Glucose *Serum No Effect Analytical* At concentration of 133 mg/L had no effect on Ektachem® method. *3393*
Urine No Effect Analytical At concentration of 5,000 mg/L had no effect on Diabur-test. *3393*

γ-Glutamyltransferase (GGT) *Urine Decrease Physiological* When 0.5 g given orally to 20 healthy adult volunteers and urine studied over next 3 h. *1828*

17-Hydroxycorticosteroids *Urine Decrease Analytical* Inhibit beta-glucuronidase during hydrolysis. *0022*

Leucine Aminopeptidase *Urine Decrease Physiological* When 0.5 g given orally to 20 healthy adult volunteers and urine studied over next 3 h. *1828*

Leukocytes *Urine No Effect Analytical* At concentration of 1,000 mg/L had no effect on Cytur-Test. *3393*

Magnesium *Gastric Material Decrease Physiological* When 0.5 g given orally to 20 healthy adult volunteers and urine studied over next 3 h. *1828*

Methaqualone *Serum No Effect Analytical* At 180 mg/L on GLC procedure of Evenson. *1057* At 180 mg/L on GLC procedure of Evenson. *1058*

Phosphate *Serum No Effect Analytical* At concentration of 500 mg/L had no effect on phosphomolybdate method. *3393*

Potassium *Serum No Effect Analytical* At concentration of 100 mg/L had no effect on measurement by ISE with predilution. *3393* At concentration of 350 mg/L had no effect on measurement by ISE on Ektachem®. *3393* At concentration of 30,000 mg/L had no effect on measurement by ISE with predilution. *3393*

Protein *Serum No Effect Analytical* At concentration of 500 mg/L had no effect on biuret method with blank correction. *3393*
Urine No Effect Analytical At 100 mg/dL on TCA dye method of Pesce. *2791*
CSF No Effect Analytical At 100 mg/dL on TCA dye method of Pesce. *2791*

Sodium *Serum No Effect Analytical* At concentration of 100 mg/L had no effect on measurement by ISE with predilution. *3393* At concentration of 30,000 mg/L had no effect on measurement by ISE with predilution. *3393*

Thyroxine (T₄) *Serum Decrease Physiological* Due to competition for transport proteins. *3798*

Triglycerides *Serum No Effect Analytical* At concentration of 594 mg/L had no effect on GPO-PAP method. *3393* At concentration of 500 mg/L had no effect on lipase/esterase method. *3393*

Tri-iodothyronine (T₃) *Serum Decrease Physiological* Due to competition for transport proteins. *3798*

Urea Nitrogen *Serum No Effect Analytical* At concentration of 500 mg/L had no effect on diacetylmonoxime method. *3393* At concentration of 133 mg/L had no effect on Ektachem® method. *3393* At concentration of 300 mg/L had no effect on Seralyzer method. *3393*

Uric Acid *Serum No Effect Analytical* No effect on Tripyridyl-s-triazine method of Morin. *2486* At 30 mg/dL on copper chelate procedure. *2230* At concentration of 500 mg/L had no effect on phosphotungstate reduction method. *3393* At concentration of 300 mg/L had no effect on Seralyzer method. *3393* At concentration of 300 mg/L had no effect on uricase method on aca. *3393* At concentration of 967 mg/L had no effect on uricase-PAP method. *3393*
Serum Increase Analytical 30 mg/dL equivalent to 0.1 mg/dL by Nishi procedure. *2230*

Saline

Aldosterone *Plasma Decrease Physiological* Short term response if given i.v. in hypertensives. *1911*

Calcium *Serum Decrease Physiological* Effect of isotonic solution if hypercalcemia. *0071*
Urine Increase Physiological With isotonic saline loading. *3439* Hypertonic solution has calciuretic effect. *0071*

Chloride *Serum Increase Physiological* Added chloride. *2313*

Cortisol *Urine Increase Physiological* Reported effect. *1000*

Creatine Kinase *Serum Increase Physiological* 40% increase after 1 ml saline i.m. *3053*

Lipase *Serum Decrease Physiological* At molar concentrations (whether heparinized or not). *3026*

Lipoprotein Lipase *Serum Decrease Physiological* At molar concentrations (whether heparinized or not). *3026*

Magnesium *Urine Increase Physiological* With isotonic saline loading. *3439*

Partial Thromboplastin Time *Plasma Increase Analytical* 20% dilution produces significant effect in lab. *1244*

Potassium *Serum Increase Physiological* Effect of hypertonic solution. *2480*

Sodium *Serum Increase Physiological* May occur especially if impaired cardiac/renal function. *0071*
Urine Increase Physiological With isotonic saline loading. *3439*

Uric Acid *Serum Decrease Physiological* Diminishes urate reabsorption. *3440*
Urine Increase Physiological Hypertonic solution may have marked effect. *0574*

Zinc *Urine Increase Physiological* Increase by 28% with isotonic saline loading. *3439*

Saliva

Amylase *Serum Increase Analytical* Saliva contains amylase (may affect pipetting). *1237*

Salol

Color *Urine Increase Analytical* Dark color on standing. *2313*

Salsalate

Albumin *Serum Decrease Physiological* Associated with minimal change nephrotic syndrome in one patient. *3664*

Creatinine *Serum Increase Physiological* Associated with minimal change nephrotic syndrome in one patient. *3664*

Protein *Urine Increase Physiological* Associated with minimal change nephrotic syndrome in one patient. *3664*

Sandfly Fever

Cortisol, Free *Urine Increase Physiological* 3 fold increase over basal with fever. *2929*

Fatty Acids, Free (FFA) *Serum Increase Physiological* Increased 2 fold from baseline. *2929*

Glucose Tolerance *Serum Decrease Physiological* With fever decreased by half. *2929*

Growth Hormone *Plasma Increase Physiological* Increased 8 fold at fasting, 18 fold at 30 minutes. *2929*

Insulin *Plasma Increase Physiological* 2 fold at 30 minutes, 3 fold at 60 minutes. *2929*

Santonin

Color *Urine Increase Analytical* Bright yellow (NaOH changes to pink, scarlet). *2313*
Feces Increase Analytical Deep yellow with 65-70 mg. *2313*

Sugar *Urine Increase Analytical* False positive with Benedict's. *2448*

Sauerkraut

Sugar *Urine Increase Analytical* Metabolite may affect Galatest. *1563*

SC-16102

Bicarbonate *Urine Increase Physiological* Minor increase observed within 2 h. *1804*

Chloride *Urine Increase Physiological* Duration of diuretic action short. *1804*

Creatinine Clearance *Urine No Effect Physiological* No effect noted during study. *1804*

Leukocytes *Blood Increase Physiological* Mechanism not explained. *1804*

Potassium *Serum Decrease Physiological* Duration of diuretic action short. *1804*
Urine Increase Physiological Duration of diuretic action short. *1804*

Sodium *Urine Increase Physiological* Duration of diuretic action short. *1804*

Uric Acid *Urine Increase Physiological* Minor increased clearance noted. *1804*

Volume *Urine Increase Physiological* Duration of diuretic action short. *1804*

Scopolamine

Alkaloids *Urine No Effect Analytical* With ninhydrin on Frings TLC procedure at 10 mg/dL. *1204*
Urine Increase Analytical Purple with iodoplatinate on Frings TLC procedure. *1204*

Barbiturate *Urine No Effect Analytical* With mercuric SO_4 on Frings TLC procedure at 10 mg/dL. With diphenylcarbazone on Frings TLC procedure at 10 mg/dL. With vanillin on Frings procedure at 10 mg/dL. With mercurous NO_3 on Frings TLC procedure at 10 mg/dL. *1204*

Bromide *Urine No Effect Analytical* No effect on method of Frings. *1204*

Chromosomes *Test Conditions Abnormal Physiological* Clastogenic in human cells. *3282*

Ethchlorvynol *Urine No Effect Analytical* No effect on method of Frings and Cohen. *1204*

FPN Test *Urine No Effect Analytical* No effect at 10 mg/dL on method of Frings. *1204*

Salicylate *Urine No Effect Analytical* At 10 mg/dL no effect on Trinder method. *1204*

Scopolamine Bromide

Creatinine *Serum No Effect Analytical* At concentration of 20 mg/L had no effect on creatinine iminohydrolase method. *3393* At concentration of 20 mg/L had no effect on kinetic Jaffé method on BKA-2. *3393*

Glucose *Serum No Effect Analytical* At concentration of 20 mg/L had no effect on GOD/POD-PAP method. *3393*

Triglycerides *Serum No Effect Analytical* At concentration of 20 mg/L had no effect on GPO-PAP method. *3393*

Uric Acid *Serum No Effect Analytical* At concentration of 20 mg/L had no effect on uricase-PAP method. *3393*

Season

Albumin *Serum Increase Physiological* Proportional increase though total protein falls in summer. *2123*

Bicarbonate *Serum Increase Physiological* Significant effect, maximum in April, minimum in summer. *1172*

Calcium *Serum Increase Physiological* Significant effect, sine pattern, maximal in December and May. *1172*
Urine Increase Physiological Up to 50% in July more than January with same diet. *2481*

β-Globulin *Serum Decrease Physiological* Falls during summer. *2123*

γ-Globulin *Serum Increase Physiological* Eskimos- 1.36 g/dL in winter, 2 g/dL in summer. *3742*
Serum Decrease Physiological Falls during summer. *2123*

Glucose *Serum Increase Physiological* Slight increase during summer. *2123*

25-Hydroxy Vitamin D_3 *Plasma Decrease Physiological* Significant decrease both sexes in winter. *3429*

Indolylacryloylglycine *Urine Increase Physiological* Large increase observed with summer. *2302*

Lactate Dehydrogenase *Serum Increase Physiological* High in summer, low in winter. *3872*

Magnesium *Serum Increase Physiological* Significant effect, max in Feb, min in summer. *1172*

Season (continued)

Phosphate *Serum Increase Physiological* Significant effect, max in May and June, min in winter. *1172*

Protein *Serum Increase Physiological* Significant effect maximal in Nov, Dec, minimal in Jan, June. *1172*
Serum Decrease Physiological Slight decrease during summer. *2123*

Sodium *Serum Increase Physiological* Significant effect, maximum in summer. *1172*

Thyroxine (T$_4$) *Serum No Effect Physiological* No seasonal variation observed. *2853*

Tri-iodothyronine (T$_3$) *Serum No Effect Physiological* No seasonal variation observed. *2853*

Urea Nitrogen *Serum Increase Physiological* Slight increase during summer. *2123*

Uric Acid *Serum Increase Physiological* Significantly higher in summer even with controlled intake. *0264*

Vanillylmandelic Acid *Urine No Effect Physiological* Same in summer and winter in Antarctica. *3217*

Seclazone

Uric Acid *Serum Decrease Physiological* Marked antigout response. *1767*
Urine Increase Physiological Uricosuric action. *1767*

Secobarbital

Alanine Aminotransferase *Serum No Effect Analytical* At acute overdose concentration (20 mg/dL) on Technicon® SMAC® method. *3719*

Albumin *Serum No Effect Analytical* At acute overdose concentration (20 mg/dL) on Technicon® SMAC® method. *3719* At concentration of 200 mg/L had no effect on BCG method. *3393*

Alkaline Phosphatase *Serum No Effect Analytical* At acute overdose concentration (20 mg/dL) on Technicon® SMAC® method. *3719*

Alkaloids *Urine No Effect Analytical* With iodoplatinate of Frings TLC procedure at 10 mg/dL. With ninhydrin on Frings TLC procedure at 10 mg/dL. *1204*

Aspartate Aminotransferase *Serum No Effect Analytical* At acute overdose concentration (20 mg/dL) on Technicon® SMAC® method. *3719*

Barbiturate *Urine No Effect Analytical* With vanillin on Frings procedure at 10 mg/dL. *1204*
Urine Increase Analytical White spot with mercuric SO$_4$ on Frings TLC procedure, Purple with diphenylcarbazone on Frings TLC procedure. With mercurous NO$_3$ reagent on Frings TLC procedure. *1204*

Bicarbonate *Serum No Effect Analytical* At concentration of 100 mg/L had no effect on method using phenolphthalein. *3393*

Bilirubin *Serum No Effect Analytical* At acute overdose concentration (20 mg/dL) on Technicon® SMAC® method. *3719* At concentration of 200 mg/L had no effect on Jendrassik and Grof method. *3393*

Bromide *Urine No Effect Analytical* No effect on method of Frings. *1204*

Calcium *Serum No Effect Analytical* At acute overdose concentration (20 mg/dL) on Technicon® SMAC® method. *3719* At concentration of 200 mg/L had no effect on cresolphthalein method. *3393*

Chlordiazepoxide *Blood No Effect Analytical* On method of Riddick at 5 mg/dL. *2983*

Chloride *Serum No Effect Analytical* At concentration of 100 mg/L had no effect on mercurimetric method. *3393*

Cholesterol *Serum No Effect Analytical* At acute overdose concentration (20 mg/dL) on Technicon® SMAC® method. *3719* At concentration of 1 mg/L had no effect on CHOD-PAP method. *3393* At concentration of 200 mg/L had no effect on Liebermann-Burchard method. *3393*

Creatine Kinase *Serum No Effect Analytical* At acute overdose concentration (20 mg/dL) on Technicon® SMAC® method. *3719*

Creatinine *Serum No Effect Analytical* At acute overdose concentration (20 mg/dL) on Technicon® SMAC® method. *3719* At concentration of 200 mg/L had no effect on AutoAnalyzer Jaffé method. *3393*

Diphenylhydantoin *Serum No Effect Analytical* On GLC procedure of Papadopoulos at *in vivo* concentration. *2722*

Estriol *Urine No Effect Analytical* No effect of 50 mg/L on GLC method. *1163*

Ethchlorvynol *Serum No Effect Analytical* At 10 µg/mL on colorimetric method of Wallace. *3753* At 600 mg/L on GLC procedure of Evenson. *1058*
Urine No Effect Analytical No effect on method of Frings and Cohen at 10 mg/dL. *1205*

FPN Test *Urine No Effect Analytical* No effect at 10 mg/dL on method of Frings. *1204*

Glucose *Serum No Effect Analytical* At acute overdose concentration (20 mg/dL) on Technicon® SMAC® method. *3719* At concentration of 1 mg/L had no effect on GOD/POD-PAP method. *3393*
Urine No Effect Analytical No effect observed with Tes-Tape®. *1100*
Urine Decrease Analytical Low with Clinistix®, Diastix®. *1100*

17-Hydroxycorticosteroids *Urine No Effect Analytical* No effect on Glenn-Nelson procedure when added *in vitro*. *0427*

Iron *Serum No Effect Analytical* At acute overdose concentration (20 mg/dL) on Technicon® SMAC® method. *3719* At concentration of 200 mg/L had no effect on Ferrozine method. *3393*

[131]I Uptake *Serum Decrease Physiological* Seconal™, tuinal contain tetraiodofluorescein. *2652*

17-Ketosteroids *Urine Increase Analytical* Metabolite interferes with Zimmermann reaction. *1022*
Urine Decrease Analytical Negative interference on Zimmermann reaction *in vitro*. *0427*

Lactate Dehydrogenase *Serum No Effect Analytical* At acute overdose concentration (20 mg/dL) on Technicon® SMAC® method. *3719*

Methaqualone *Serum No Effect Analytical* At 20 mg/L on GLC procedure of Evenson. *1057*

Pheneturide *Serum No Effect Analytical* On GLC procedure of Papadopoulos at *in vivo* concentration. *2722*

Phenobarbital *Serum No Effect Analytical* On GLC procedure of Papadopoulos at *in vivo* concentration. *2722*

Phosphate *Serum No Effect Analytical* At acute overdose concentration (20 mg/dL) on Technicon® SMAC® method. *3719* At concentration of 200 mg/L had no effect on phosphomolybdate method. *3393*

Potassium *Serum No Effect Analytical* At concentration of 1 mg/L had no effect on flame-photometric method. *3393* At concentration of 100 mg/L had no effect on measurement by ISE with predilution. *3393*

Primidone *Serum No Effect Analytical* On GLC procedure of Papadopoulos at *in vivo* concentration. *2722*

Propoxyphene *Serum No Effect Analytical* At 25 mg/L on method of Evenson. *1056*

Protein *Serum No Effect Analytical* At acute overdose concentration (20 mg/dL) on Technicon® SMAC® method. *3719* At concentration of 200 mg/L had no effect on biuret method with blank correction. *3393*

Prothrombin Time *Plasma Decrease Physiological* Induces metabolism of administered coumarins. *3011* With 0.1 g/d in 10 patients receiving warfarin due to hepatic microsomal enzyme induction. *3649* Increase in plasma clearance of warfarin. *2647*

Salicylate *Urine No Effect Analytical* At 10 mg/dL no effect on Trinder method. *1204*

Secobarbital *Blood Increase Physiological* 600 mg orally produces concentration of 4.8 mg/L. *2348*

Sodium *Serum No Effect Analytical* At concentration of 1 mg/L had no effect on flame-photometric method. *3393* At concentration of 100 mg/L had no effect on measurement by ISE with predilution. *3393*

Triglycerides *Serum No Effect Analytical* At acute overdose concentration (20 mg/dL) on Technicon® SMAC® method. *3719* At concentration of 100 mg/L had no effect on lipase/esterase method. *3393*

Urea Nitrogen *Serum No Effect Analytical* At acute overdose concentration (20 mg/dL) on Technicon® SMAC® method. *3719* At concentration of 200 mg/L had no effect on diacetylmonoxime method. *3393*

Uric Acid *Serum No Effect Analytical* At acute overdose concentration (20 mg/dL) on Technicon® SMAC® method. *3719* At concentration of 200 mg/L had no effect on phosphotungstate reduction method. *3393*

Secretin

Ammonia *Urine Decrease Physiological* Reduced formation with increased alkalinization. *2656*

Amylase *Pancreatic Juice Increase Physiological* Normal above 14.9 U/kg/80 minutes. *0486*

Bicarbonate *Urine Increase Physiological* When i.v. infusion given. *2656*
Bile No Effect Physiological In normals, increased in cirrhotics. *0399*
Pancreatic Juice Increase Physiological Normal above 90 meq/L. *0486*

Bile Acids *Bile Decrease Physiological* Slight in normals, marked in cirrhotics. *0399*

Calcium *Serum Increase Physiological* Increased by 0.6 meq/L in normals, ?cause. *0470*

Chloride *Bile Increase Physiological* Slight in normals, marked decrease in cirrhotics. *0399*

Cholecystokinin (CCK) *Plasma No Effect Analytical* No significant cross reactivity with RIA procedure of Harvey. *1518*

Gastrin *Serum Decrease Physiological* Basal secretion after injection in normals. *1484*

Glucagon *Plasma No Effect Physiological* No effect observed if injected i.v. *2676*

Glucose *Serum No Effect Physiological* No change after 1 unit/kg i.v. *2676*
Serum Increase Physiological Intravenous infusion causes increase. *1775*

Hydrochloric Acid *Gastric Material Decrease Physiological* Pancreatic secretion also at peak. *1989*
Pancreatic Juice Decrease Physiological Reciprocal relationship to bicarbonate. *0194*

Insulin *Plasma Increase Physiological* More than 2 fold increased with 15 U pulse. *2137* Only from 3 to 6 minutes after i.v. injection. *2676*

Lipase *Serum Increase Physiological* May cause spasm of sphincter of Oddi. *3777*

Pepsin *Gastric Material Increase Physiological* Rises to 2.5 times basal, falls in 30 minutes. *1677*

Sodium *Urine Increase Physiological* Interferes with reabsorption, exchange for H+. *2656*

Titratable Acidity *Urine Decrease Physiological* Reduced formation with increased alkalinization. *2656*

Volume *Bile Increase Physiological* Slight effect in normals, marked in cirrhotics. *0371*
Pancreatic Juice Increase Physiological Normal response over 3.2 mL/kg/80 minutes. *0486*

Selenium

Odor *Breath Increase Physiological* Garlic-like odor due to dimethyl selenide. *2217*

Selenium *Urine Increase Physiological* Ingested in diet as well as in poisoning. *2217*

Senna

Acetaminophen Screening Test *Urine Negative Analytical* No reaction with o-cresol at therapeutic concentrations. *3326*

Color *Urine Increase Analytical* Red, orange or rust. *1187*
Feces Increase Analytical Yellow to brown. *2313*

Estradiol *Urine Increase Analytical* Elevated values method of Brown. *0023*

Estrogens *Urine Increase Analytical* Interferes in Kober procedure. *0279*

Estrone *Urine Decrease Analytical* Affects unmodified Brown/Kober procedure. *0583*

Serine

Ammonia *Plasma Decrease Analytical* With 2500 nmoles produces 26% inhibitor indophenol reaction. *1290*

Megaloblasts *Blood Increase Physiological* Can aggravate vitamin B_{12} deficiency. *3025*

Tyrosine *Plasma No Effect Analytical* On fluorometric procedure of Ambrose. *0080*

Serum

Acid Phosphatase *Serum Increase Analytical* Release of enzyme from erythrocytes. *3879*

α-Amino-Nitrogen *Plasma Increase Physiological* Release from RBC with clotting (up to 40% high). *1563*

Cholesterol *Serum No Effect Analytical* If heparin used for plasma. *1563*

Fibronectin *Serum Increase Analytical* Serum concentration 20% higher than all plasma specimens with EDTA heparin, citrate using Boehringer-Mannheim kit and nephelometer. *0770*

Glucose *Serum Increase Physiological* Approximately 13% higher than in whole blood. *3217*
Serum Decrease Analytical By up to 24% if no antiglycolytic agent used. *2690*

Hemoglobin *Plasma Increase Analytical* During clotting may be 100 fold increase. *1755*

Insulin *Plasma Decrease Analytical* Consistently higher in plasma than serum. *0112*

Lactate *Serum Increase Analytical* If no antiglycolytic agent (arises from glucose). *2690*

Lactate Dehydrogenase *Serum Increase Analytical* Up to 40% higher than plasma. *0435*

Lithium *Serum No Effect Analytical* No difference observed. *2891*

pH *Serum No Effect Physiological* Clotting has no effect. *1233*

Phosphate *Serum Increase Analytical* Due to release from cells with clotting. *2229*

Potassium *Serum Increase Physiological* Due to release from erythrocytes with clotting. *3879*

Protein *Serum Decrease Analytical* Due to lack of fibrinogen present in plasma. *2229*

Sex Difference

α₁-Acid Glycoprotein *Serum Increase Physiological* Approximately 10 mg/dL higher in men. *0954*

Alkaline Phosphatase *Serum Increase Physiological* In men up to and including 5th decade. *2130*

Amylase *Serum Increase Physiological* Women approximately 16% higher than men. *0024*

α₁-Antitrypsin *Serum Increase Physiological* Approximately 30 mg/dL higher in women. *0954*

Ascorbic Acid *Plasma Increase Physiological* Postpubertal women higher than men same diet. *2998*

Aspartate Aminotransferase
Serum Increase Physiological In men up to and including 5th decade. *2130*

Bilirubin *Serum Increase Physiological* In males at all ages. *2130*

Catechol-O-Methyl Transferase
Red Blood Cells No Effect Physiological No difference between men and women. *1401*

Ceruloplasmin *Serum Increase Physiological* Slightly higher in women (5 mg/dL). *0954*

Creatinine *Serum Increase Physiological* In males at all ages. *2130*

Glucose *Serum Increase Physiological* In males except in 6th decade. *2130*

Haptoglobin *Serum No Effect Physiological* No effect observed. *0954*

Sex Difference *(continued)*

Ionized Calcium *Serum Increase Physiological* On average 0.025 mmol/L higher in men. *2056*

Norepinephrine *Plasma Increase Physiological* Significant increase in adult men compared with women. *1176*

Prolactin *Plasma Increase Physiological* In woman mean 7.9 ng/mL, in men 5.2 ng/ml. *1678*

Protein Bound Iodine (PBI) *Serum Increase Physiological* Higher in women at all age levels. *2130*

Ribonuclease *Serum Increase Physiological* Higher in nonpregnant women than men. *0365*

Transferrin *Serum Increase Physiological* Approximately 60 mg/dL higher in women. *0954*

Urea Nitrogen *Serum Increase Physiological* In males at all ages except 7th decade. *2130*

Uric Acid Clearance *Urine Increase Physiological* In women average of 12.67 mL/min, in men 10.38 mL/min. *2590*

Sexual Activity

Cortisol *Plasma No Effect Physiological* Probably no significant effect, may be slight decline after. *1735*

Estradiol *Plasma No Effect Physiological* No effect in women. *3437*

Follicle Stimulating Hormone (FSH) *Plasma No Effect Physiological* No effect in either sex. *3437*

Luteinizing Hormone (LH) *Plasma No Effect Physiological* No effect in either sex. *3437*

Progesterone *Plasma No Effect Physiological* No effect in women. *3437*

Prolactin *Plasma Increase Physiological* 8 fold increase in 2 of 6 women (?breast stimulated). *3437*

Testosterone *Serum No Effect Physiological* No effect in men. *3437*
Serum Increase Physiological Reported effect in males. *1735*

Silicones

Prothrombin Time *Plasma No Effect Physiological* If in cooking oil — no effect on anticoagulants absorption. *3536*

Silver

Chloride *Serum Decrease Physiological* Observed after silver nitrate antisepsis. *2427*

[131]I Uptake *Serum No Effect Physiological* No effect reported. *1915*

Occult Blood *Feces Positive Physiological* May cause hemorrhagic gastroenteritis. *1343*

pH *Blood Increase Physiological* Observed after silver nitrate antisepsis. *2427*

Protein *Urine Increase Physiological* May cause nephrotoxicity. *3204*

Protein Bound Iodine (PBI) *Serum Decrease Analytical* Interferes with ashing procedures. *1915*

Sodium *Serum Decrease Physiological* Observed after silver nitrate antisepsis. *2427*

T₃ Uptake *Serum No Effect Physiological* No effect on resin procedures. *1915*

Urea Nitrogen *Serum Increase Physiological* May cause nephrotoxicity. *3204*

Simethicone

Prothrombin Time *Plasma Decrease Physiological* Impairs absorption of oral anticoagulant. *1487*

Sisomicin

N-Acetylglucosaminidase *Urine No Effect Physiological* No significant effect seen in 23 patients given therapeutic amounts for 2 weeks. *2593*

Alanine Aminopeptidase *Urine Increase Physiological* Approximately 8 fold increase in 23 patients given therapeutic amounts of drug for 2 weeks. *2593*

Creatinine *Serum Increase Physiological* 38% increase in elderly patients with average initial creatinine clearance of 66 mL/min. *1203*

Glucose *Urine No Effect Analytical* On TesTape® at physiological concentration. On Diastix® at physiological concentration. On Clinitest® at physiological concentration. *2733*

Immunoglobulin Light Chains *Urine Increase Physiological* Seen in all patients given drug in therapeutic amounts for 2 weeks. *2593*

Lysozyme *Urine No Effect Physiological* No significant effect seen in 23 patients given therapeutic amounts for 2 weeks. *2593*

β₂-Microglobulin *Serum No Effect Physiological* In 23 patients given drug for 2 weeks in therapeutic amounts. *2593*
Urine Increase Physiological Increased up to 2 times in 23 patients given therapeutic amounts for 2 weeks. *2593*

Retinol Binding Protein *Serum No Effect Physiological* In 23 patients given drug for 2 weeks in therapeutic amounts. *2593*
Urine No Effect Physiological No significant effect seen in 23 patients given therapeutic amounts for 2 weeks. *2593*

Transferrin *Urine No Effect Physiological* No significant change in 23 patients given therapeutic amounts of drug for 2 weeks. *2593*

Site of Collection

Cortisol *Plasma No Effect Physiological* Independent of site. *3639*

Epinephrine *Plasma No Effect Physiological* No difference between antecubital and superior vena cava blood. *3639*
Plasma Decrease Physiological Pulmonary artery and right atrium less than arm vein. *3639*

Fatty Acids, Free (FFA) *Serum No Effect Physiological* Independent of site. *3639*

Insulin *Plasma No Effect Physiological* Independent of site. *3639*

Norepinephrine *Plasma No Effect Physiological* Independent of site. *3639*

β-Sitosterol

Cholesterol *Serum No Effect Analytical* On GLC procedure of MacGee. *2252*

Sitosterols

Cholesterol *Serum No Effect Analytical* On GLC procedure of MacGee. *2252*
Serum Decrease Physiological Inhibits absorption of endogenous and exogenous compound. *1336*

β-Lipoproteins *Serum Decrease Physiological* Therapeutic intent. *1680*

Triglycerides *Serum No Effect Physiological* No effect observed. *2112*

Skatole

Tryptophan *Test Conditions Increase Analytical* Measured as same in Spies/Chambers method. *1415*

Urobilinogen *Urine No Effect Analytical* No effect on p-MDFB procedure of Rutter. *2044*
Urine Increase Analytical Produces blue color with Ehrlich's reagent. *1563*

SKF-12185

17-Hydroxycorticosteroids *Urine Decrease Physiological* Direct effect on adrenal steroidogenesis. *0022*

Skin Lightening Cream

Mercury *Urine Increase Physiological* If contain Hg but no obvious nephrotoxicity. *0235*

Protein *Urine Increase Physiological* Trace observed in some cases. *0235*

Protein Electrophoresis *Serum No Effect Physiological* No evidence of nephrotic syndrome. *0235*

Sleep

Angiotensin *Plasma Increase Physiological* Rises during sleep. *1190*

Antidiuretic Hormone *Urine No Effect Physiological* No effect observed on hourly measurements. *1190*

Basal Metabolic Rate *Patient Decrease Physiological* Slight effect observed. *0251*

Bromide *Urine Decrease Physiological* Negligible excretion during night after administration. *3896*

Calcium *Urine Increase Physiological* Related to inactivity: max at night. *1139*

Catecholamines *Urine Decrease Physiological* Significantly less at night. *3616*

Chloride *Urine Decrease Physiological* Reduced excretion compared with day. *1139*

Corticotropin *Plasma Increase Physiological* Maximum secretion at night, especially early am. *3217*

Creatinine Clearance *Urine Increase Physiological* Significant effect observed. *1190*

Fatty Acids, Free (FFA) *Serum Decrease Physiological* Occurs with sleep onset (but in fasting only). *2730*

Glomerular Filtration Rate (GFR) *Urine No Effect Physiological* No effect observed with sleep only. *1190*

Gonadotropins *Plasma Increase Physiological* Maximum at night, especially in early am. *3217*

Growth Hormone *Plasma Increase Physiological* Maximum 2 h after start of sleep (even with naps). *2743*

Magnesium *Urine Increase Physiological* ?related to inactivity: maximum at night. *1139*

Osmolality *Serum Increase Physiological* Reversal with diuresis of sleep. *1190*

Potassium *Urine Decrease Physiological* Reduced excretion compared with day. *1139*

Prolactin *Plasma Increase Physiological* Peak between 5 and 7 am, falls with rising. *3217*

Sodium *Urine Decrease Physiological* Reduced excretion compared with day. *1139* Observed without effect on GFR. *1190*

Tyrosine *Plasma Decrease Physiological* 18% decrease during sleep. *3952*

Volume *Urine Increase Physiological* Significant effect over control. *1190*

Water Clearance *Urine Increase Physiological* Significant effect paralleling volume. *1190*

Sleep Deprivation

Adenosine Triphosphate (ATP) *Red Blood Cells Decrease Physiological* Sleep deprivation for several days causes decrease. *2222*

Anthranilic Acid *Urine Decrease Physiological* Noticed after 5 d (?due to relative B_6 deficiency). *2032*

17-Hydroxycorticosteroids *Urine Increase Physiological* 73% increase between first and second day. *2032*

5-Hydroxyindoleacetic Acid (5-HIAA) *Urine Increase Physiological* Immediate response associated with decreased ATP levels. *2032*

Xanthurenic Acid *Urine Increase Physiological* Increase between second and fifth day. *2032*

Smallpox Vaccine

Platelets *Blood Decrease Physiological* May occasionally cause thrombocytopenic purpura. *2313*

Smog

Urobilinogen *Urine Increase Analytical* Oxidation of spots on chromatogram by ozone. *0583*

Smoking

Acetaminophen *Urine Increase Physiological* Significantly greater amount excreted as glucuronide when ingested in smokers. *0394*

Acetylcholinesterase *Red Blood Cells No Effect Physiological* No significant difference between smokers and nonsmokers. *1548*

α_1-Acid Glycoprotein *Serum No Effect Physiological* No significant difference between smokers and nonsmokers. *0809*

Acid Phosphatase *Serum No Effect Physiological* No significant difference between heavy smokers and nonsmokers. *0961*

Adenosine Triphosphate (ATP) *Red Blood Cells No Effect Physiological* No significant difference between smokers and nonsmokers. *1548* *Red Blood Cells Decrease Physiological* Significantly less in smokers. *3112*

Albumin *Serum Decrease Physiological* Approximately 20% reduction in smokers vs controls. *0809*

Aminolevulinic Acid Dehydrase *Red Blood Cells Decrease Physiological* 13% lower in smokers than nonsmokers. *1548*

Amylase *Serum Increase Physiological* Basal activity 100% higher in heavy smokers than nonsmokers. *0961*

Androstenedione *Plasma Increase Physiological* Significantly higher mean value in male current smokers than in nonsmokers with apparent dose response relationship. *0238*

Anthranilic Acid *Urine No Effect Physiological* No effect of smoking for 1 week. *0068*

α_1-Antichymotrypsin *Serum No Effect Physiological* No significant difference between smokers and nonsmokers. *0809*

Antinuclear Antibodies *Serum Positive Physiological* If smokers all ages both sexes. *2335*

α_1-Antitrypsin *Serum No Effect Physiological* No significant difference between smokers and nonsmokers. *0809*

Apolipoprotein AI *Serum Decrease Physiological* In young nonobese normolipidemic men (same pattern as observed in others with increased risk of atherosclerosis. *0845* Mean 1.09 mg/dL vs 1.29 mg/dL in controls. *0845*

Apolipoprotein AII *Serum No Effect Physiological* Mean 0.38 mg/dL vs 0.36 mg/dL in controls. *0845*

Apolipoprotein B *Serum Increase Physiological* Mean 0.94 mg/dL vs 0.76 mg/dL in controls. *0845*

Apolipoprotein CIII *Serum No Effect Physiological* Mean 0.068 mg/dL vs 0.057 mg/dL in controls. *0845*

Apolipoprotein D *Serum No Effect Physiological* Mean 0.045 mg/dL 0.046 mg/dL in controls. *0845*

Apolipoprotein E *Serum Increase Physiological* In young nonobese normolipidemic men (same pattern as observed in others with increased risk of atherosclerosis). *0845* Mean 0.060 mg/dL vs 0.050 mg/dL in controls. *0845*

Ascorbic Acid *Plasma Decrease Physiological* Lower concentration observed than in nonsmokers. *2768* Significantly lower in smoking adolescent women than in nonsmokers: partly to reduced dietary intake. *1901* *White Blood Cells Decrease Physiological* Lower concentration observed in smokers than nonsmokers. *2768*

Basal Metabolic Rate *Patient No Effect Physiological* No significant constant change. *0251*

Bicarbonate *Pancreatic Juice Decrease Physiological* Chronic effect heavy smokers, less in light. *0554*

Carboxyhemoglobin *Blood Increase Physiological* Maximum after regular, minimum after extra-mild. *3094* Significantly higher in smokers. *3112* No effect in 7 healthy volunteers after 3 cigarettes. *2352* Concentration correlated with extent of smoking. *3255*

Carcinoembryonic Antigen *Serum Positive Physiological* False positives observed in heavy smokers. *0101*

Smoking *(continued)*

Catecholamines *Urine Increase Physiological* Increased by 50% with cigar not inhaled Increased x 2 during 4 h with cigarettes inhaled. *1921*

Cholesterol *Serum No Effect Physiological* No acute short term effects observed. *1927*
Serum Increase Physiological Significantly higher in heavy smokers (?dietary preferences). *1317*

Cholesterol, High Density Lipoprotein
Serum Decrease Physiological Observed in heavy smokers, although rises when smoking given up. *3629*

Cholesterol, α-Lipoprotein *Serum Decrease Physiological* Effect observed in fasting white children aged 8 to 17 y. *3723*

Cholesterol, β-Lipoprotein *Serum No Effect Physiological* Effect observed in fasting white children aged 8 to 17 y. *3723*

Cortisol *Plasma No Effect Physiological* No effect on diurnal rhythm. *3626*
Plasma Increase Physiological Basal levels same in smokers and nonsmokers, but increased significantly on smoking: due to direct effect of nicotine or vasopressin release. *1360*
Urine No Effect Physiological No effect usual smoking in smokers. *3626*

Cotinine *Serum Increase Physiological* Concentration correlated with extent of smoking. *3255*
Urine Positive Physiological Normal nicotine metabolite 5-20 cigarettes/d. *0324*

C-Reactive Protein *Serum Increase Physiological* Up to 11 fold increase in men and 6 fold increase in women. *0809*

Desmethylcotinine *Urine Positive Physiological* Normal nicotine metabolite 5-20 cigarettes/d. *0324*

Diphenylhydantoin *Serum No Effect Physiological* No apparent effect when drug coadministered. *0844*

2,3-Diphosphoglycerate
Red Blood Cells No Effect Physiological No significant difference between smokers and nonsmokers. *1548*

Disopyramide *Serum No Effect Physiological* Similar half-lives in both smokers and nonsmokers. *1861*

Elastase 2 *Serum Increase Physiological* Basal activity of immunoreactive pancreatic enzyme. *0961*

Erythrocytes *Blood No Effect Physiological* No significant difference between smokers and nonsmokers. *1548*
Blood Increase Physiological Significantly higher in smokers. *3112*

Estradiol *Plasma Increase Physiological* Significantly higher mean value in male current smokers than in nonsmokers with apparent dose response relationship. *0238*

Estrone *Plasma Increase Physiological* Significantly higher mean value in male current smokers than in nonsmokers with apparent dose response relationship. *0238*

Fatty Acids, Free (FFA) *Serum No Effect Physiological* If no inhalation with cigarettes With half cigar and no inhalation. *1921*
Serum Increase Physiological 30% increase during smoking of 3 cigarettes. *1921*

Glucose *Serum Increase Physiological* Effect of stimulant. *0117* No effect in 7 healthy volunteers after 3 cigarettes. *2352*

Glucose-6-Phosphate Dehydrogenase
Red Blood Cells No Effect Physiological No significant difference between smokers and nonsmokers. *1548*

β-Glucuronidase *Serum No Effect Physiological* No significant difference between heavy smokers and nonsmokers. *0961*

Glutathione *Red Blood Cells No Effect Physiological* No significant difference between smokers and nonsmokers. *1548*

Glutathione Reductase
Red Blood Cells No Effect Physiological No significant difference between smokers and nonsmokers. *1548*

Granulocytes *Blood Increase Physiological* Approximately 30% — mechanism obscure. *0732*

Growth Hormone *Plasma No Effect Physiological* Mean basal values the same in smokers and nonsmokers. *1360*

Haptoglobin *Serum No Effect Physiological* No significant difference between smokers and nonsmokers. *0809*

Hematocrit *Blood No Effect Physiological* No significant difference between smokers and nonsmokers. *1548*

Blood Increase Physiological Highest in smokers who inhale. *2123* Up to 6% in men (nonsmokers) immediately after 6 cigarettes. *1004*

Hemoglobin *Blood No Effect Physiological* No significant difference between smokers and nonsmokers. *1548*
Blood Increase Physiological Slight effect in all age/class groups. *1195*

Hydrochloric Acid *Gastric Material No Effect Physiological* No effect in 7 healthy volunteers after 3 cigarettes. *2352*
Gastric Material Increase Physiological Effect of 4 to 6 cigarettes in 1 h. *0295* Most subjects, especially in normal initially, show increase. *0139*

3-Hydroxyanthranilic Acid *Urine No Effect Physiological* Probably no significant effect with tryptophan load. *3183*

11-Hydroxycorticosteroids *Plasma Increase Physiological* Up to 77% increase after heavy smoking. *1487*
Urine No Effect Physiological No effect usual smoking in smokers. *3626*

17-Hydroxycorticosteroids *Urine No Effect Physiological* No effect usual smoking in smokers. *3626*

6-β-Hydroxycortisol *Urine Increase Physiological* Greater excretion in response to hepatic enzyme induction. *0394*

Hydroxycotinine *Urine Positive Physiological* Normal nicotine metabolite 5-20 cigarettes/d. *0324*

5-Hydroxyindoleacetic Acid (5-HIAA)
Urine Increase Physiological In habitues and normally nonsmokers. *1176*

3-Hydroxykynurenine *Urine No Effect Physiological* Probably no significant effect with tryptophan load. *3183*

Kynurenine *Urine No Effect Physiological* Probably no significant effect with tryptophan load. *3183* No effect of smoking for 1 week. *0068*

Lead *Blood No Effect Physiological* No difference between smokers and nonsmokers. *0489*
Urine Increase Physiological Non smokers 27.1 μg/L, cigarette 28.6, cigar 29.0 μg/L. *2377*

Leukocytes *Blood Increase Physiological* Higher than in nonsmokers. *2123* Higher in inhalers than other smokers. *1195* Significantly higher in nonsmokers but still in normal range. *1548* Correlated well with carboxyhemoglobin saturation. Average leukocyte count at upper limit of normal in nonsmokers. *3691*

β-Lipoproteins *Serum Increase Physiological* Significantly higher (?due to diet preferences) heavy smokers. *1317*

Lymphocytes *Blood Increase Physiological* Approximately 30% — mechanism obscure. *0732* Unrelated to amount of carboxyhemoglobin. *3691*

MCH *Blood No Effect Physiological* No significant difference between smokers and nonsmokers. *1548*

MCHC *Blood No Effect Physiological* No significant difference between smokers and nonsmokers. *1548*

MCV *Blood No Effect Physiological* No significant difference between smokers and nonsmokers. *1548*

Methemoglobin Reductase
Red Blood Cells No Effect Physiological No significant difference between smokers and nonsmokers. *1548*

Methylimidazoleacetic Acid *Urine Increase Physiological* With standard diet higher values than nonsmokers. *1381*

Methylnicotinamide *Urine No Effect Physiological* No effect of smoking for 1 week. *0068* Probably no significant effect with tryptophan load. *3183*

Monoamine Oxidase *Platelets Decrease Physiological* Low enzyme activity identified in platelets from cigarette smokers. *2679*

Monocytes *Blood Increase Physiological* Approximately 30% — mechanism obscure. *0732*

Narcotics *Urine Positive Analytical* Nicotine spot in TLC method of Berry, Grove. *0324*

Neutrophil Elastase *Plasma No Effect Physiological* Between smokers and nonsmokers but significant rise in smokers after 8 h of abstinence and then intense smoking, probably due to *in vivo* release of neutrophil elastase. *0005*
White Blood Cells Increase Physiological Mean concentration of elastase-derived fibrinopeptide A-alpha-1-21 was 5 fold higher in 10 cigarette smokers than in 20 healthy nonsmokers. Acute effects observed after smoking 3 cigarettes. *3793*

Neutrophils *Blood Increase Physiological* Correlated well with carboxyhemoglobin saturation. Average leukocyte count at upper limit of normal in nonsmokers. *3691*

Nicotine *Plasma Increase Physiological* Sharp rise up to about 40 ng/mL with 1 cigarette. *1728*
Blood Increase Physiological Concentration less than 0.3 mg/L. *2348*
Urine Increase Physiological significantly greater with inhalation than non inhalation. *1921* Normal nicotine metabolite 5-20 cigarettes/d. *0324*

Oxygen-Hemoglobin Affinity
Blood Decrease Physiological Significantly less in smokers. *3112*

Partial Thromboplastin Time
Blood No Effect Physiological No significant difference between smokers and nonsmokers. *1547*

pCO$_2$ *Blood No Effect Physiological* No effect observed. *2028*

pH *Blood No Effect Physiological* No effect observed. *2028*
Gastric Material Decrease Physiological Effect of 4-6 cigarettes in 1 h. *0295*

Platelet Aggregation *Blood Increase Physiological* Following *in vitro* stimulation greater in older smokers than younger, and in those who smoked for longer than shorter time. *2995* Response to thrombin, ADP, collagen and epinephrine increased with cigarettes of higher nicotine content. *2955*

Platelet Factor 4 *Plasma No Effect Physiological* No significant difference between smokers and nonsmokers. *2995*

Platelets *Blood No Effect Physiological* No observed effect. *1004* No significant difference between 5 smokers and nonsmokers. *2995*

pO$_2$ *Blood No Effect Physiological* No effect observed. *2028*

Prolactin *Plasma Decrease Physiological* Significant reduction in women who smoked compared with those who did not. *2609* Mean value less in smokers than nonsmokers but no acute effect of smoking. *1360*

Prostaglandin, 6-Keto-F$_{1}\alpha$ *Urine Decrease Physiological* In heavy smokers following smoking of 4 cigarettes, but no change in nonsmokers. *2541*

Prostaglandin E2 *Gastric Material Increase Physiological* In 7 volunteers after 3 cigarettes: reduced from mean 22.8 ng/15 minutes to 12.2 ng/15 minutes. *2352*

Renin Activity *Plasma No Effect Physiological* No difference between smokers and nonsmokers and not affected by acute smoking. *1361*

Rheumatoid Factor *Serum Positive Physiological* In young smokers. *2335*

Sex Hormone Binding Globulin
Serum No Effect Physiological No difference between male smokers and nonsmokers. *0238*

T$_3$ Uptake *Serum No Effect Physiological* No significant difference between heavy smokers and controls. *3255*

Testosterone *Serum Increase Physiological* Marginal effect in male current smokers compared with nonsmokers. *0238*

Thiocyanate *Serum Increase Physiological* In smokers versus nonsmokers (from cyanide metabolites). *2363* 3-6 fold higher in smokers than nonsmokers. *2278* Concentration correlated with extent of smoking. *3255*
Urine Increase Physiological Doubled excretion in smokers (metabolite of cyanide). *2363* 3-6 fold higher in smokers than nonsmokers. *2278*

β-Thromboglobulin *Plasma No Effect Physiological* No significant difference between smokers and nonsmokers. *2995*

Thyroid Stimulating Hormone (TSH)
Serum No Effect Physiological No significant difference between heavy smokers and controls. *3255*

Thyroxine (T$_4$) *Serum Decrease Physiological* Significant reduction in heavy smokers (6.4 µg/dL versus 7.3 µg/dL in controls). *3255*

Thyroxine (T$_4$) Index, Free (FTI)
Serum No Effect Physiological No significant difference between heavy smokers and controls. *3255*

Transferrin *Serum No Effect Physiological* No significant difference between smokers and nonsmokers. *0809*

Triglycerides *Serum Increase Physiological* Sustained increase with/without inhalation. *1921* Effect observed in fasting white children aged 8 to 17 y. *3723*

Tri-iodothyronine (T$_3$) *Serum Decrease Physiological* Significant reduction in heavy smokers (181.0 ng/dL versus 204.0 ng/dL in Controls). *3255*

Tri-iodothyronine (T$_3$) Index, Free
Serum No Effect Physiological No significant difference between heavy smokers and controls. *3255*

Trypsinogen *Serum No Effect Physiological* No significant difference in basal immunoreactive cationic trypsinogen between smokers and nonsmokers. *0961*

Urea Nitrogen *Serum Decrease Physiological* Less than in nonsmokers, lowest if inhale. *2123*

Uric Acid *Serum Decrease Physiological* Less in smokers who inhale than others. *2123* Lower than in non smokers. *2124*

Vanillylmandelic Acid *Urine No Effect Physiological* No effect usual smoking in smokers. *3626*

Vitamin B$_{12}$ *Serum Decrease Physiological* Inverse correlation with urine thiocyanate. *2363*

Volume *Gastric Material Increase Physiological* Effect of 4-6 cigarettes in 1 h. *0295*
Pancreatic Juice Decrease Physiological Chronic effect heavy smokers, less in light. *0554*

Sodium

Ionized Calcium *Serum Increase Analytical* As concentration changes from 140 to 168 meq/L. *2056*

Lead *Urine Increase Analytical* 200 meq/L equivalent to 50 µg/L by atomic absorption. *3240*

Potassium *Serum Increase Analytical* Variable concentration affects most flame methods. *0582*

Zinc *Serum No Effect Analytical* At 6,000 to 1 with flameless atomic absorption. *2038*
Urine No Effect Analytical At 6,000 to 1 with flameless atomic absorption. *2038*

Sodium Azide

Salicylate *Urine Increase Analytical* Produces orange-red color in Trinders procedure. *2894*

Urea Nitrogen *Serum Decrease Analytical* At 100 mg/dL causes decrease of 60% diacetyl procedure. *0583*

Sodium Bicarbonate

pH *Blood Increase Physiological* Affects acid-base balance *in vivo*. *2313*
Urine Increase Physiological Used to alkalinize urine. *0071*

Potassium *Serum Decrease Physiological* Causes potassium to shift into cells. *0071*

Protein *Urine Increase Analytical* False positive with Labstix® due to high pH. *1488*

Quinidine *Serum Increase Physiological* Alkalinizes urine, increases reabsorption. *1487*

Sodium *Serum Increase Physiological* May cause sodium retention. *0071*

Urobilinogen *Urine Increase Physiological* Increased clearance when urine alkaline. *2452*

Sodium Bisulfite

Protein *Test Conditions Increase Analytical* At concentrations higher than 0.01 % affects Folin-Ciocalteu. *0702*

Sodium Bromide

Alkaloids *Urine No Effect Analytical* With iodoplatinate of Frings TLC procedure at 100 mg/dL. With ninhydrin on Frings TLC procedure at 100 mg/dL. *1204*

Barbiturate *Urine No Effect Analytical* With diphenylcarbazone on Frings TLC procedure at 100 mg/dL With vanillin on Frings procedure at 100 mg/dL. With mercurous NO$_3$ on Frings TLC procedure at 100 mg/dL. *1204*

Sodium Bromide (continued)

Bromide   Urine   No Effect   Analytical   No effect on method of Frings. *1204*

Chloride   Serum   No Effect   Analytical   On Beckman Astra 8 and Corning/EEL 920 chloride meter: chloride meter produced results with 5 mmol/L bromide. *2972*
Serum   Increase   Analytical   As measured by Technicon® C800 instrument in one patient receiving bromide therapy. *2972*

Cortisol   Plasma   No Effect   Physiological   No effect in 10 men, 10 women receiving 1 mg/kg/d during 8 weeks or 2 full menstrual cycles. *3141*

Estradiol   Plasma   No Effect   Physiological   No effect in 10 men, 10 women receiving 1 mg/kg/d during 8 weeks or 2 full menstrual cycles. *3141*

Ethchlorvynol   Urine   No Effect   Analytical   No effect on method of Frings and Cohen. *1204*

Follicle Stimulating Hormone (FSH)
Plasma   No Effect   Physiological   No effect in 10 men, 10 women receiving 1 mg/kg/d during 8 weeks or 2 full menstrual cycles. *3141*

FPN Test   Urine   No Effect   Analytical   No effect at 100 mg/dL on method of Frings. *1204*

Luteinizing Hormone (LH)   Plasma   No Effect   Physiological   No effect in 10 men, 10 women receiving 1 mg/kg/d during 8 weeks or 2 full menstrual cycles. *3141*

Progesterone   Plasma   No Effect   Physiological   No effect in 10 men, 10 women receiving 1 mg/kg/d during 8 weeks or 2 full menstrual cycles. *3141*

Prolactin   Plasma   No Effect   Physiological   No effect in 10 men, 10 women receiving 1 mg/kg/d during 8 weeks or 2 full menstrual cycles. *3141*

Testosterone   Serum   No Effect   Physiological   No effect in 10 men, 10 women receiving 1 mg/kg/d during 8 weeks or 2 full menstrual cycles. *3141*

Thyroid Stimulating Hormone (TSH)
Serum   No Effect   Physiological   No effect in 10 men, 10 women receiving 1 mg/kg/d during 8 weeks or 2 full menstrual cycles. *3141*

Thyroxine (T_4)   Serum   No Effect   Physiological   No effect in 10 men, 10 women receiving 1 mg/kg/d during 8 weeks or 2 full menstrual cycles. *3141*

Thyroxine (T_4) Binding Globulin
Serum   No Effect   Physiological   No effect in 10 men, 10 women receiving 1 mg/kg/d during 8 weeks or 2 Full menstrual cycles. *3141*

Thyroxine (T_4), Free   Serum   No Effect   Physiological   No effect in 10 men, 10 women receiving 1 mg/kg/d during 8 weeks or 2 full menstrual cycles. *3141*

Tri-iodothyronine (T_3)   Serum   No Effect   Physiological   No effect in 10 men, 10 women receiving 1 mg/kg/d during 8 weeks or 2 full menstrual cycles. *3141*

Sodium Chloride

Bilirubin   Serum   Decrease   Analytical   Inhibition of diazo test reported. *1563*

Cortisol   Urine   Increase   Physiological   Significant effect with up to 16 g/d for 10 d. *1487*

Ionized Calcium   Serum   Increase   Analytical   At concentrations above 10 mmol/L on Calcium specific electrode. *0540*

Sodium Citrate

Bilirubin   Serum   No Effect   Analytical   At concentration of 10500 mg/L had no effect on Ektachem® method. *3393*

Cholesterol   Serum   No Effect   Analytical   At concentration of 5,000 mg/L had no effect on CHOD-Iodide method. *3393*   At concentration of 5,000 mg/L had no effect on CHOD-PAP method. *3393*   At concentration of 5,000 mg/L had no effect on catalase-AIDH method. *3393*   At concentration of 30,000 mg/L had no effect on method using catalase-Hantzsch reaction. *3393*   At concentration of 5,000 mg/L had no effect on Liebermann-Burchard method. *3393*

Creatinine   Serum   No Effect   Analytical   At concentration of 5,000 mg/L had no effect on Jaffé-Fading-Fraction method. *3393*   At concentration of 5,000 mg/L had no effect on Jaffé-Fuller's earth method. *3393*   At concentration of 5,000 mg/L had no effect on kinetic Jaffé method on BKA-2. *3393*

Fibrinogen   Plasma   No Effect   Analytical   At concentration of 3794 mg/L had no effect on aca method. *3393*

Glucose   Serum   No Effect   Analytical   At concentration of 500 mg/L had no effect on Ektachem® method. *3393*   At concentration of 20,000 mg/L had no effect on GOD/POD-PAP method. *3393*

Iron   Serum   No Effect   Analytical   At concentration of 10,000 mg/L had no effect on Ferrozine method. *3393*

Orthophosphate   Test Conditions   Decrease   Analytical   Total inhibition at 0.1 mol/L on method of Horder. *1658*

Phosphate   Serum   No Effect   Analytical   At concentration of 5400 mg/L had no effect on aca method. *3393*

Plasminogen   Plasma   No Effect   Analytical   At concentration of 3794 mg/L had no effect on aca method. *3393*

Potassium   Serum   Decrease   Analytical   At concentrations above 2500 mg/L lowered concentration as measured by ISE with predilution. *3393*

Protein   Test Conditions   Decrease   Analytical   Lowry procedure due to chelation of chemicals. *1796*

Protein Electrophoresis   Serum   No Effect   Analytical   At concentration of 5,000 mg/L had no effect on automated Olympus-Hite method. *3393*

Sodium   Serum   Increase   Analytical   At concentrations above 250 mg/L raised concentration as measured by ISE without predilution. *3393*

Triglycerides   Serum   No Effect   Analytical   At concentration of 11764 mg/L had no effect on GPO-PAP method. *3393*

Urea Nitrogen   Serum   No Effect   Analytical   At concentration of 500 mg/L had no effect on Ektachem® method. *3393*

Uric Acid   Serum   No Effect   Analytical   At concentration of 5,000 mg/L had no effect on Kageyama-Hantzsch method. *3393*   At concentration of 5,000 mg/L had no effect on catalase-AIDH method. *3393*   At concentration of 11764 mg/L had no effect on uricase-PAP method. *3393*

Sodium Enibomal

Platelets   Blood   Decrease   Physiological   Immunologically induced thrombocytopenic purpura. *1877*

Sodium Etidronate

Creatinine Clearance   Urine   No Effect   Physiological   No effect of drug alone. *2931*

Cyclic Adenosine Monophosphate
Urine   No Effect   Physiological   No effect of drug alone. *2931*

Diffusible Phosphate   Serum   Increase   Physiological   Increases tubular reabsorption, no effect on PTH action. *2931*

Phosphate   Serum   Increase   Physiological   Increases tubular reabsorption, no effect on PTH action. *2931*
Urine   No Effect   Physiological   Increases tubular reabsorption, no effect on PTH action. *2931*

Phosphate Clearance   Urine   Decrease   Physiological   Increases tubular reabsorption, no effect on PTH action. *2931*

Sodium Fluoride

Ammonia   Plasma   No Effect   Analytical   At concentration of 10,000 mg/L had no effect on Ektachem® method. *3393*

Amylase   Serum   No Effect   Analytical   At concentration of 15 mg/L had no effect on p-nitrophenylmaltopentoside/-hexoside method. *3393*

Antithrombin III   Plasma   No Effect   Analytical   At concentration of 4238 had no effect on aca method. *3393*

Bilirubin   Serum   No Effect   Analytical   At concentration of 10,000 mg/L had no effect on Ektachem® method. *3393*

Cholesterol *Serum No Effect Analytical* At concentration of 2,000 mg/L had no effect on CHOD-Iodide method. *3393* At concentration of 2,000 mg/L had no effect on CHOD-PAP method. *3393* At concentration of 2,000 mg/L had no effect on catalase-AIDH method. *3393* At concentration of 30,000 mg/L had no effect on method using catalase-Hantzsch reaction. *3393* At concentration of 2,000 mg/L had no effect on Liebermann-Burchard method. *3393*

Creatinine *Serum No Effect Analytical* At concentration of 2,000 mg/L had no effect on Jaffé-Fading-Fraction method. *3393* At concentration of 2,000 mg/L had no effect on Jaffé-Fuller's earth method. *3393* At concentration of 2,000 mg/L had no effect on kinetic Jaffé method on BKA-2. *3393*
Serum Increase Analytical At concentrations above 10,000 mg/L when combined with potassium oxalate raised concentration as measured by Ektachem® method. *3393*

Glucose *Serum No Effect Analytical* No effect on alkaline ferricyanide method No effect on o-toluidine method. *2771* At anticoagulant dose on hexokinase method of Coburn. *0676* No effect on GOD-PERID glucose oxidase method No effect on Trinder glucose oxidase method. *2771* At 200 mg/dL on MBTH procedure of Neeley. *2569* No effect on Neocuproin method No effect on Warner Glucomatic method. *2771* At concentration of 200 mg/L had no effect on Ektachem® method. *3393* At concentration of 7,000 mg/L had no effect on GOD-Perid method. *3393* At concentration of 40,000 mg/L had no effect on GOD/POD-PAP method. *3393*

Iron *Serum No Effect Analytical* At concentration of 4,000 mg/L had no effect on Ferrozine method. *3393*

Potassium *Serum No Effect Analytical* At concentration of 2,000 mg/L had no effect on measurement by ISE without predilution. *3393*

Protein Electrophoresis *Serum No Effect Analytical* At concentration of 2,000 mg/L had no effect on automated Olympus-Hite method but with slight displacement of fractions. *3393*

Sodium *Serum Increase Analytical* At concentrations above 250 mg/L raised concentration as measured by ISE without predilution. *3393*

Triglycerides *Serum No Effect Analytical* At concentration of 5879 mg/L had no effect on GPO-PAP method. *3393*

Urea Nitrogen *Serum Increase Analytical* At concentrations above 160 mg/L raised concentration as measured by Ektachem® method. *3393*

Uric Acid *Serum No Effect Analytical* At concentration of 2,000 mg/L had no effect on Kageyama-Hantzsch method. *3393* At concentration of 2,000 mg/L had no effect on catalase-AIDH method. *3393* At concentration of 2,000 mg/L had no effect on uricase method on aca. *3393* At concentration of 15116 mg/L had no effect on uricase-PAP method. *3393*

Sodium Heparin

Cholesterol *Serum No Effect Analytical* At concentration of 750 mg/L had no effect on CHOD-Iodide method. *3393* At concentration of 750 mg/L had no effect on CHOD-PAP method. *3393* At concentration of 750 mg/L had no effect on catalase-AIDH method. *3393* At concentration of 750 mg/L had no effect on Katalase-Hantzsch method. *3393* At concentration of 750 mg/L had no effect on Liebermann-Burchard method. *3393*

Creatinine *Serum No Effect Analytical* At concentration of 750 mg/L had no effect on Jaffé-Fading-fraction method. *3393* At concentration of 750 mg/L had no effect on Jaffé-Fuller's earth method. *3393* At concentration of 750 mg/L had no effect on kinetic Jaffé method on BKA-2. *3393*

Glucose *Serum No Effect Analytical* At concentration of 75 mg/L had no effect on Ektachem® method. *3393*

Phosphate *Serum No Effect Analytical* At concentration of 400 mg/L had no effect on aca method. *3393*

Potassium *Serum No Effect Analytical* At concentration of 750 mg/L had no effect on measurement by ISE without predilution. *3393*

Protein Electrophoresis *Serum No Effect Analytical* At concentration of 750 mg/L had no effect on automated Olympus-Hite method. *3393*

Sodium *Serum No Effect Analytical* At concentration of 750 mg/L had no effect on measurement by ISE without predilution. *3393*

Triglycerides *Serum No Effect Analytical* At concentration of 4,000 mg/L had no effect on GPO-PAP method. *3393*

Urea Nitrogen *Serum No Effect Analytical* At concentration of 75 mg/L had no effect on Ektachem® method. *3393*

Uric Acid *Serum No Effect Analytical* At concentration of 750 mg/L had no effect on Kageyama-Hantzsch method. *3393* At concentration of 750 mg/L had no effect on catalase-AIDH method. *3393* At concentration of 4,000 mg/L had no effect on uricase-PAP method. *3393*

Sodium Iodate

Bilirubin *Serum Increase Physiological* Clinically significant effect with displacement from protein possible in neonate. *3748*

Sodium Lauryl Sulfate

Lipid Glycerol *Serum Decrease Analytical* Inhibits phospholipase C method of Horney. *2683*

Protein *Test Conditions No Effect Analytical* Lowry procedure when referenced with chemical only used. *1796*

Sodium Loading

Prostaglandin A *Plasma Decrease Physiological* 49% to 0.82 ng/mL in normals. *3966*

Prostaglandin E *Plasma No Effect Physiological* No significant effect with variation in intake. *3966*

Prostaglandin F *Plasma No Effect Physiological* No significant effect with variation in intake. *3966*

Sodium Nitroprusside

^{131}I Uptake *Serum Decrease Physiological* Single case reported (?due to thiocyanate). *2626*

Protein Bound Iodine (PBI) *Serum Decrease Physiological* Single case reported (?due to thiocyanate). *2626*

Thyroxine (T$_4$) *Serum Decrease Physiological* Due to thyroidal inhibition especially in patients with some renal impairment. *3798*

Sodium Oxalate

Aminolevulinic Acid Dehydrase
Red Blood Cells No Effect Analytical No effect compared with heparin. *0400*

Ammonia *Plasma No Effect Analytical* At concentration of 1540 mg/L had no effect on Ektachem® method. *3393*

Amylase *Serum No Effect Analytical* At concentration of 3080 mg/L had no effect on p-nitrophenylmaltopentoside/hexoside method. *3393*

Antithrombin III *Plasma No Effect Analytical* At concentration of 17109 mg/L had no effect on aca method. *3393*

Cholesterol *Serum No Effect Analytical* At concentration of 2,000 mg/L had no effect on CHOD-Iodide method. *3393* At concentration of 2,000 mg/L had no effect on CHOD-PAP method. *3393* At concentration of 2,000 mg/L had no effect on catalase-AIDH method. *3393* At concentration of 30,000 mg/L had no effect on method using catalase-Hantzsch reaction. *3393* At concentration of 2,000 mg/L had no effect on Liebermann-Burchard method. *3393*

Creatinine *Serum No Effect Analytical* At concentration of 3,000 mg/L had no effect on Jaffé-Fading-Fraction method. *3393* At concentration of 3,000 mg/L had no effect on Jaffé-Fuller's earth method. *3393*
Serum Decrease Analytical At concentrations above 150 mg/L lowered concentration as measured by kinetic Jaffé method on BKA-2. *3393*

Fibrinogen *Plasma Increase Analytical* At concentrations above 1710 mg/L raised concentration as measured by aca method. *3393*

Glucose *Serum No Effect Analytical* At concentration of 300 mg/L had no effect on Ektachem® method. *3393* At concentration of 8,000 mg/L had no effect on GOD/POD-PAP method. *3393*

Sodium Oxalate *(continued)*

Iron   *Serum   Decrease   Analytical*   At concentrations above 300 mg/L lowered concentration as measured by Ferrozine method. *3393*

Potassium   *Serum   No Effect   Analytical*   At concentration of 3,000 mg/L had no effect on measurement by ISE without predilution. *3393*

Protein Electrophoresis   *Serum   No Effect   Analytical*   At concentration of 3,000 mg/L had no effect on automated Olympus-Hite method but with slight displacement of fractions. *3393*

Sodium   *Serum   Increase   Analytical*   At concentrations above 250 mg/L raised concentration as measured by ISE without predilution. *3393*

Triglycerides   *Serum   No Effect   Analytical*   At concentration of 6160 mg/L had no effect on GPO-PAP method. *3393*

Urea Nitrogen   *Serum   No Effect   Analytical*   At concentration of 300 mg/L had no effect on Ektachem® method. *3393*

Uric Acid   *Serum   No Effect   Analytical*   At concentration of 3,000 mg/L had no effect on Kageyama-Hantzsch method. *3393*   At concentration of 3,000 mg/L had no effect on catalase-AIDH method. *3393*   At concentration of 4620 mg/L had no effect on uricase-PAP method. *3393*

Sodium Phosphate

Protein   *Test Conditions   No Effect   Analytical*   Lowry procedure when referenced with chemical only used. *1796*

Sodium Phosphate ^{32}P

Erythrocytes   *Blood   Decrease   Physiological*   Anemia if excess used. *0071*

Leukocytes   *Blood   Decrease   Physiological*   Leukopenia if excess used. *0071*

Platelets   *Blood   Decrease   Physiological*   Thrombocytopenia if excess used. *0071*

Sodium Phytate

Calcium   *Serum   Decrease   Physiological*   Decreased gastrointestinal tract absorption. *0071*
Urine   Decrease   Physiological   Decreased gastrointestinal tract absorption. *1022*

Sodium Pyruvate

Creatinine   *Serum   No Effect   Analytical*   No effect ion-exchange method of Mitchell. *2459*
Serum   Increase   Analytical   Marked effect direct methods. Moderate effect on Lloyd's procedure. *2459*

Sodium Restriction

Aldosterone   *Plasma   Increase   Physiological*   Trend in 2 d, maximum in 5, with decreased sodium. *0495*

Angiotensin II   *Plasma   Increase   Physiological*   Trend in 2 d, maximum in 5, with decreased sodium. *0495*

Prostaglandin A   *Plasma   Increase   Physiological*   Increase by 36% to 2.14 ng/mL in normals. *3966*

Prostaglandin E   *Plasma   No Effect   Physiological*   No significant effect with variation in intake. *3966*

Prostaglandin F   *Plasma   No Effect   Physiological*   No significant effect with variation in intake. *3966*

Sodium Salicylate

Glucose   *Serum   No Effect   Analytical*   At concentration of 350 mg/L had no effect on Ektachem® method. *3393*   At concentration of 1,000 mg/L had no effect on GOD/POD-PAP method. *3393*

Urea Nitrogen   *Serum   No Effect   Analytical*   At concentration of 350 mg/L had no effect on Ektachem® method. *3393*

Uric Acid   *Serum   No Effect   Analytical*   10 mg/dL no effect on method of Klein. *1949*

Sodium Salts

N-Acetylglucosaminidase   *Urine   No Effect   Analytical*   No effect on 2 colorimetric methods with 1 g/L sodium azide, 100 mmol/L sodium bicarbonate or 10 mmol/L sodium fluoride. *1354*

Aldosterone   *Plasma   Decrease   Physiological*   Comparable decreases observed with chloride and citrate salts in 5 men with essential hypertension. *2037*

Ammonia   *Plasma   Decrease   Physiological*   Reported effect. *2313*

Bicarbonate   *Serum   Increase   Physiological*   Significant but small effect observed with citrate but not chloride salt in 5 men with essential hypertension. *2037*

Cadmium   *Serum   Increase   Analytical*   May interfere with atomic absorption measurements. *2155*

Calcium   *Serum   Increase   Analytical*   Affect flame photometry if poor instrument. *1563*
Urine   Increase   Physiological   Observed with chloride salt supplementation in 5 men with essential hypertension. *2037*
Urine   Decrease   Physiological   Observed with citrate salt supplementation in 5 men with essential hypertension. *2037*

Chloride   *Serum   Decrease   Physiological*   Significant effect with citrate salt not observed with chloride in 5 men with essential hypertension. *2037*

Diagnex Blue Excretion   *Urine   Increase   Analytical*   Displacement of diagnex blue from resin. *3879*

1,25-Dihydroxy Vitamin D$_3$ *Serum   No Effect   Physiological*   No significant effect in 5 men with essential hypertension given either chloride or citrate supplement. *2037*

Ionized Calcium   *Serum   Increase   Physiological*   Slight but not significant increase in 5 men with essential hypertension given citrate salt supplement. *2037*

Parathyroid Hormone   *Plasma   No Effect   Physiological*   No significant effect in 5 men with essential hypertension given either chloride or citrate supplement. *2037*

pH   *Blood   Increase   Physiological*   Significant but small effect observed with citrate but not chloride salt in 5 men with essential hypertension. *2037*

Potassium   *Serum   Decrease   Physiological*   Significant effect with citrate salt not observed with chloride in 5 men with essential hypertension. *2037*

Renin Activity   *Plasma   Decrease   Physiological*   Comparable decreases observed with chloride and citrate salts in 5 men with essential hypertension. *2037*

Sodium   *Serum   Increase   Physiological*   Slight effect with citrate salt compared with chloride in 5 men with essential hypertension. *2037*

Zinc   *Urine   No Effect   Physiological*   No effect observed with oral doses. *2712*

Sodium Sulfate

Ammonia   *Feces   Increase   Physiological*   Modest rise observed. *0032*

Calcium   *Serum   Decrease   Physiological*   If given i.v. may cause hypocalcemia. *0071*
Urine   Increase   Physiological   Promotes excretion. *1343*

Magnesium   *Serum   Decrease   Physiological*   May be excreted combined with sulfate. *0071*

pH   *Feces   Decrease   Physiological*   No effect observed. *0032*

Potassium   *Serum   Decrease   Physiological*   May be excreted combined with sulfate. *3300*

Sodium   *Serum   Increase   Physiological*   If given i.v. may cause fluid retention and coma. *3300*

Sodium Taurocholate

Gentamicin   *Serum   No Effect   Analytical*   No difference between plate-diffusion and tube dilution. *1266*

Soft Water Areas

Sodium   *Urine   Increase   Physiological*   Greater than elsewhere due to added dietary salt. *0813*

Drug Listings

Somatostatin

Sodium *Serum Decrease Physiological* Water intoxication observed in 2 patients given drug i.v., although creatinine remained constant. *1465*

Thyroid Stimulating Hormone (TSH)
Serum Decrease Physiological In hypothyroid and euthyroid patients and reduced response to TRH: drug is inhibitor of TSH release. *3798*

Thyroxine (T4) *Serum Decrease Physiological* With infusion probable inhibition of TSH release. *3798*

Tri-iodothyronine (T3) *Serum Decrease Physiological* With infusion probable inhibition of TSH release. *3798*

Tri-iodothyronine (T3), Free *Serum Decrease Physiological* With infusion probable inhibition of TSH release. *3798*

Somatotropin

Thyroxine (T4) *Serum Decrease Physiological* Or no change associated with accelerated conversion. *3798*

Tri-iodothyronine (T3) *Serum Increase Physiological* Or no change associated with accelerated conversion. *3798*

Sominex®

Ethchlorvynol *Urine No Effect Analytical* No effect on method of Frings and Cohen. *1205*

Sorbitol

Alanine Aminotransferase *Serum No Effect Physiological* Increase by 0.3-0.8 mg/dL after i.v. infusion. *1160*

Ammonia *Feces Increase Physiological* Modest rise observed. *0032*

Aspartate Aminotransferase
Serum No Effect Physiological Increase by 0.3-0.8 mg/dL after i.v. infusion. *1160*

Bilirubin *Serum Increase Physiological* Increase by 0.3-0.8 mg/dL after i.v. infusion. *1160*

Estrogens *Urine No Effect Analytical* On fluorometric procedure at concentration of 1 g/dL. *3899*

Glucose *Serum No Effect Analytical* On glucose oxidase procedure at 1 g/dL On p-HBAH procedure of Lever at 1 g/dL On alkaline ferricyanide procedure at 1 g/dL On o-toluidine procedure at 1 g/dL. *2140*

Lactate *Serum Increase Physiological* In response to i.v. infusion. *1160*

pH *Feces Decrease Physiological* No effect observed. *0032*

Pyruvate *Serum Decrease Physiological* In response to i.v. infusion (effect slight). *1160*

Uric Acid *Serum No Effect Physiological* Increase by 0.3-0.8 mg/dL after i.v. infusion. *1160*

Sorbose

Protein *Test Conditions Increase Analytical* Interferes with Folin-Ciocalteu method of Lowry. *3636*

Sotalol

Cholesterol *Serum No Effect Physiological* General effect in 2 studies but overall progressive deterioration in lipid profile. *0088*
Serum Increase Physiological Significant increase of 16% reported from one long term study (12 mo). *2116* Increased from mean of 5.49 mmol/L to 6.37 mmol/L at 12 mo in group of essential hypertensives. *2120*

Cholesterol, High Density Lipoprotein
Serum Decrease Physiological No change in one study: general effect in 2 studies but overall progressive deterioration in lipid profile. *0088* Significant average decrease of 28% reported from one 12 mo study. *2116* Marked reduction and ratio of HDL-cholesterol to total cholesterol in essential hypertensives after 12 mo. *2120*

Cholesterol, Low Density Lipoprotein
Serum Increase Physiological No change or up to 30%: general effect in 2 studies but overall progressive deterioration in lipid profile. *0088* Significant increase of 32% reported from one 12 mo study. *2116*

Cholesterol, Very Low Density Lipoprotein
Serum No Effect Physiological No change in one study: general effect in 2 studies but overall progressive deterioration in lipid profile. *0088*

Fatty Acids, Free (FFA) *Serum Decrease Physiological* At 1, 3, 6 and 12 mo in group of essential hypertensives given drug orally for 12 mo. *2120*

Metanephrines, Total *Urine Increase Analytical* Cause shift in absorbance peak with pisano method (may be large enough to double apparent concentration). *3161*

Triglycerides *Serum No Effect Physiological* General effect in 2 studies but overall progressive deterioration in lipid profile. *0088*
Serum Increase Physiological Significant increase of 66% reported from one long-term study (12 mo). *2116* Simultaneous increase with other lipid changes; change from 1.14 to 1.89 mmol/L over 12 mo. *2120*

Spartene

Digoxin *Serum No Effect Physiological* 0.8 g/d had no effect on serum concentration. *0282*

Spectinomycin

Alanine Aminotransferase *Serum Increase Physiological* Mechanism not discussed. *0122*

Alkaline Phosphatase *Serum Increase Physiological* Mechanism not discussed. *0122*

Creatinine Clearance *Urine Decrease Physiological* Mechanism not discussed. *0122*

Hematocrit *Blood Decrease Physiological* Mechanism not discussed. *0122*

Hemoglobin *Blood Decrease Physiological* Mechanism not discussed. *0122*

Urea Nitrogen *Serum Increase Physiological* Mechanism not discussed. *0122*

Volume *Urine Decrease Physiological* Mechanism not discussed. *0122*

Spermatozoa

Appearance *Urine Abnormal Physiological* Cloudy, insoluble in dilute acetic acid. *0443*

Spinach

Methemoglobin *Blood Increase Physiological* If contaminated with nitrites. *1893*

Spinal Anesthesia

BSP Retention *Serum Increase Physiological* Seen postoperatively, related to decreased blood flow. *1343*

Chloride *Urine Decrease Physiological* Observed in normal males, pregnant women. *1343*

Effective Renal Plasma Flow
Patient Decrease Physiological In relation to degree of hypotension. *1343*

Glomerular Filtration Rate (GFR)
Urine Decrease Physiological Reported effect, related to degree of hypotension. *1343*

Sodium *Urine Decrease Physiological* Observed in normal males, pregnant women. *1343*

Urobilinogen *Urine Decrease Physiological* Related to reduced hepatic blood flow. *1343*

Volume *Urine Decrease Physiological* Found postoperatively, but normal renal function. *1343*

4-401

Spironolactone

Acetaminophen Screening Test *Urine Negative Analytical* No reaction with o-cresol at therapeutic concentrations. *3326*

Aldosterone *Plasma Increase Physiological* Marked increase in supine concentration on constant diet. *1144* Mean increase of 386 pg/mL in 3 men given drug 100 mg bid for 1 week. *3736* In 5 normal subjects given 300 mg daily for 7 d. *2440* Observed with chronic administration of drug. *2836*

Angiotensin II *Plasma Increase Physiological* Change varies with renin activity. *2440*

Calcium *Urine Increase Physiological* Probably artifact as tablets each contain 40 mg calcium. *2863*

Chloride *Urine Increase Physiological* Diuretic action. *1343*

Cholesterol *Serum No Effect Analytical* At concentration of 20 mg/L had no effect on CHOD-Iodide method. *3393* At concentration of 20 mg/L had no effect on CHOD-PAP method. *3393* At concentration of 20 mg/L had no effect on catalase-AIDH method. *3393* At concentration of 20 mg/L had no effect on method using catalase-Hantzsch reaction. *3393* At concentration of 20 mg/L had no effect on Liebermann-Burchard method. *3393*
Serum No Effect Physiological No significant change in 17 subjects treated with drug for less than 1 y. *1405*
Serum Increase Physiological Maximum average increase of 4% in 3 studies. *0087*
Serum Decrease Physiological Increase by 24 mg/dL in 11 men simultaneously with starting diet to reduce cholesterol. *0090*

Cholesterol, High Density Lipoprotein
Serum No Effect Physiological No significant change in 17 subjects treated with drug for less than 1 y. *1405* No significant change in 3 studies. *0087*
Serum Decrease Physiological Average fell from 1.5 to 1.1 mmol/L at 6 mo, and to 1.0 mmol/L in 15 patient with primary hypertension given 100 mg/d. *1072*

Cholesterol, Low Density Lipoprotein
Serum No Effect Physiological No significant change in 17 subjects treated with drug for less than 1 y. *1405*
Serum Increase Physiological Maximum average increase of 5% in 3 studies. *0087*

Cholesterol, Very Low Density Lipoprotein
Serum No Effect Physiological No significant change in 17 subjects treated with drug for less than 1 y. *1405* No significant change in 3 studies. *0087*

Corticosteroids *Plasma Increase Analytical* Marked effect on fluorometric procedure. *1487*
Urine Increase Analytical Increased absorption at 410 nm in Porter-Silber reaction. *0022*

Corticosterone *Plasma No Effect Physiological* In 5 normal subjects given 300 mg daily for 7 d. *2440*

Cortisol *Plasma No Effect Physiological* In 5 normal subjects given 300 mg daily for 7 d. *2440*
Plasma Increase Analytical Fluorometric methods may be affected. *2237*
Urine Increase Analytical Fluorescence affects method of Ratliff. *2923*

Creatinine *Serum No Effect Analytical* At concentration of 1 mg/L had no effect on creatinine iminohydrolase method. *3393* At concentration of 20 mg/L had no effect on Jaffé-Fading-Fraction method. *3393* At concentration of 20 mg/L had no effect on Jaffé-Fuller's earth method. *3393* At concentration of 50 mg/L had no effect on kinetic Jaffé method on BKA-2. *3393*

Deoxycorticosterone *Plasma Increase Physiological* In 5 normal subjects given 300 mg daily for 7 d. *2440*

Digitoxin *Serum No Effect Physiological* May depress level but not to subtherapeutic value due to induction of mixed function oxidases. *2794*

Digoxin *Serum Increase Analytical* At normal concentrations in serum if no preincubation. *2804* At 250 ng/mL equals 1.4 ng/mL by RIA. *3940* Cross-reactivity possibly with metabolites with antibodies observed with several radioimmunoassay kits. *3577*
Serum Increase Physiological Increased concentration when given with spironolactone than when given alone. *3736* Clearance reduced by coadministration of spironolactone; effect more marked when quinidine also administered: renal tubular secretion of digoxin inhibited. *1110* Drug-induced decrease in renal clearance observed. *3577*

Eosinophils *Blood Increase Physiological* Relative lymphocytosis and 15% eosinophilia observed in a single patient during period of agranulocytosis. *3487*

Estradiol *Plasma No Effect Physiological* No significant effect in males for 2 weeks. *2773*
Urine Increase Physiological In males for 2 weeks from 2.6 to 3.5 µg/24 h. *2773*

Estriol *Plasma No Effect Physiological* No significant effect in males for 2 weeks. *2773*
Urine Increase Physiological In males for 2 weeks from 8.1 to 11.8 µg/24 h. *2773*

Estrone *Plasma No Effect Physiological* No significant effect in males for 2 weeks. *2773*
Urine No Effect Physiological No significant effect in men. *2773*

Glucaric Acid *Urine Increase Physiological* Weak hepatic microsomal enzyme inducer. *3558*

Glucose *Serum No Effect Analytical* At concentration of 8 mg/L had no effect on Ektachem® method. *3393* At concentration of 1 mg/L had no effect on GOD/POD-PAP method. *3393*
Serum No Effect Physiological In 15 patients with primary hypertension treated with 100 mg daily for 1 y In 15 patients with primary hypertension treated with 100 mg daily for 1 y. *1072*

Glucose Tolerance *Serum Decrease Physiological* Transient effect in 15 primary hypertension patients at 6 mo after 100 mg drug daily. *1072*

Granulocytes *Blood Decrease Physiological* Complete agranulocytosis observed in a single patient. *3487*

Growth Hormone *Plasma No Effect Physiological* In 15 patients with primary hypertension treated with 100 mg daily for 1 y. *1072*

Hematocrit *Blood Increase Physiological* When compared with results when patients treated with warfarin alone: augmented effect when spironolactone coadministered probably due to hemoconcentration due to diuresis. *2643*

11-Hydroxycorticosteroids *Plasma Increase Analytical* When determined by Mattingly method. *2901*

17-Hydroxycorticosteroids *Urine Increase Analytical* Metabolite interferes with Porter-Silber reaction. *3505*

18-Hydroxycorticosterone *Plasma Increase Physiological* In 5 normal subjects given 300 mg daily for 7 d. *2440*

18-Hydroxydeoxycorticosterone
Plasma No Effect Physiological In 5 normal subjects given 300 mg daily for 7 d. *2440*

Insulin *Plasma Increase Physiological* Average rose from 16 to 29 mU/L in 15 primary hypertension patients at 6 mo after 100 mg drug daily. *1072*

17-Ketogenic Steroids *Urine Increase Analytical* Metabolite interferes with Zimmermann reaction. *3505*

17-Ketosteroids *Urine Increase Analytical* Metabolite interferes with Zimmermann reaction. *3505*
Urine Decrease Analytical Affects Zimmermann procedure after Allen correction. *0022*
Urine Decrease Physiological Decreased in 5 of 7 men over 2 weeks. *2773*

Leukocytes *Blood Decrease Physiological* Documented single case after 5 weeks of 100 mg/d when no other treatment given in 70 year old woman. Probable immunoallergic mechanism. *3487*

Luteinizing Hormone (LH) *Plasma Increase Physiological* In males for 2 weeks. *2773*

Lymphocytes *Blood Increase Physiological* Relative lymphocytosis and 15% eosinophilia observed in a single patient during period of agranulocytosis. *3487*

Magnesium *Serum No Effect Physiological* 100 mg did not increase serum concentration. *3101*
Urine No Effect Physiological 100 mg/d for 6 mo had sparing properties but may be related to aldosterone status. *3101*

Neutrophils *Blood Decrease Physiological* Occasional case of agranulocytosis reported. *3798*

pH *Blood Decrease Physiological* May cause mild acidosis. *1680*

Platelets *Blood Decrease Physiological* Immunologically induced thrombocytopenia. *1877*

Potassium *Serum No Effect Analytical* At concentration of 80 mg/L had no effect on measurement by ISE without predilution. *3393*

Serum No Effect Physiological In 15 patients with primary hypertension treated with 100 mg daily for 1 y. *1072*

Serum Increase Physiological Inhibits Na/K exchange in renal tubules. *1343* Mean increase of 0.4 mmol/L in 3 men given drug for 1 week. *3736* Increase by 0.4 mmol/L after 7 d in 5 volunteers. *2440* Binds to cytoplasmic hormone receptor on pericapillary side of distal tubular cells, blunting normal response to aldosterone. *2836*

Serum Decrease Physiological Diuretic action (not marked). *3505*

Urine No Effect Physiological No significant change in 5 volunteers over 7 d. *2440*

Protein Electrophoresis *Serum No Effect Analytical* At concentration of 80 mg/L had no effect on automated Olympus-Hite method. *3393*

Prothrombin Time *Plasma Decrease Physiological* When compared with results when patients treated with warfarin alone: augmented effect when spironolactone coadministered probably due to hemoconcentration due to diuresis. *2643*

Renin Activity *Plasma Increase Physiological* Usual response observed. *0684* Marked increase in supine activity on constant diet. *1144* 9.5 fold increase after 5 d in 5 volunteers. *2440*

Sodium *Serum No Effect Analytical* At concentration of 80 mg/L had no effect on measurement by ISE without predilution. *3393*

Serum Decrease Physiological Aldosterone antagonism with consequent diuresis. *2313* Reduction by 4 mmol/L after 5 d in 5 volunteers. *2440*

Urine Increase Physiological Diuretic action. *1343* Marked natriuresis with start of treatment. *2440*

Testosterone *Serum Decrease Physiological* In males for 2 weeks from 729 ng/mL to 634 mg/ml. *2773* Inhibits biosynthesis in the testis. *0953* Also displaces DHT from its cytosolic receptors. *1736*

Thyroid Stimulating Hormone (TSH)
Serum No Effect Physiological Increased response to TRH with action upon hypophyseal T_3 receptors. *3798*

Triglycerides *Serum No Effect Analytical* At concentration of 1.3 mg/L had no effect on GPO-PAP method. *3393*

Serum No Effect Physiological No significant change in 17 subjects treated with drug for less than 1 y. *1405*

Serum Increase Physiological Maximum average increase of 19% in 3 studies. *0087*

Serum Decrease Physiological Increase by 58 mg/dL in 11 men simultaneously with starting diet to reduce cholesterol. *0090* Average fell from 2.4 to 2.0 mmol/L in 15 primary hypertension patients at 6 mo after 100 mg drug daily. *1072*

Triglycerides, Very Low Density Lipoprotein
Serum No Effect Physiological No significant change in 17 subjects treated with drug for less than 1 y. *1405*

Urea Nitrogen *Serum No Effect Analytical* At concentration of 8 mg/L had no effect on Ektachem® method. *3393*

Serum Increase Physiological Especially if increased at start of therapy. *1680*

Uric Acid *Serum No Effect Analytical* At concentration of 20 mg/L had no effect on Kageyama-Hantzsch method. *3393* At concentration of 20 mg/L had no effect on catalase-AIDH method. *3393* At concentration of 1 mg/L had no effect on uricase-PAP method. *3393*

Serum No Effect Physiological No effect observed. *1487*

Serum Increase Physiological Decreased urate clearance. *2313*

Serum Decrease Physiological Average fell from 380 to 342 mmol/L in 15 primary hypertension patients at 6 mo after 100 mg drug daily. *1072*

Volume *Plasma Decrease Physiological* Dehydration observed in 3.4%. *1393*

Warfarin *Plasma No Effect Physiological* No consistent change in concentration when spironolactone coadministered. *2643*

Standing of Sample

Acid Phosphatase *Serum Decrease Analytical* pH increases, inactivates enzyme (by 10% at 4 h). *3217*

Albumin *Serum No Effect Analytical* If on polystyrene beads 24 h at 25°. *2624*

Alkaline Phosphatase *Serum Increase Analytical* Significant if on polystyrene beads 24 h at 25°. *2624*

Amino Acids *Serum Increase Analytical* Reported observation. *0355*

Aspartate Aminotransferase *Serum No Effect Analytical* If on polystyrene beads 24 h at 25°. *2624*

Bicarbonate *Serum Decrease Analytical* By 1.5 mmol/L in open AutoAnalyzer cup in 15 minutes. *3217*

Bilirubin *Serum No Effect Analytical* If on polystyrene beads 24 h at 25°. *2624*

Calcium *Serum No Effect Analytical* If on polystyrene beads 24 h at 25°. *2624*

Serum Increase Analytical Marked top to bottom if frozen and unmixed. *2674*

Urine Increase Analytical Marked top to bottom if frozen and unmixed. *2674*

Urine Decrease Analytical Precipitation of calcium salts unless acidified. *3879*

Cholesterol *Serum No Effect Analytical* If on polystyrene beads 24 h at 25°. *2624*

Copper *Serum Increase Analytical* Marked top to bottom if frozen and unmixed. *2674*

Creatine Kinase *Serum Increase Analytical* 4 U if on cells for 1 h before separation. *2744*

Glucose *Serum Decrease Analytical* Significant if on polystyrene beads 24 h at 25°. *2624*

Homocystine *Plasma Decrease Analytical* Binds to protein with time. *3230*

5-Hydroxyindoleacetic Acid (5-HIAA)
Urine No Effect Analytical Stable by screening test several days at room temperature. *1027*

Lactate *Serum Increase Analytical* May be significantly increased in heparinized sample if over 15 minutes. *2268*

Lactate Dehydrogenase *Serum No Effect Analytical* If on polystyrene beads 24 h at 25°. *2624*

Serum Increase Analytical 25% increase reported if in contact with clot for 1 h. *0877*

Phosphate *Serum No Effect Analytical* If on polystyrene beads 24 h at 25°. *2624*

Serum Increase Analytical May be doubling in 1 h at 37°. *2228*

Potassium *Serum No Effect Analytical* If on polystyrene beads 24 h at 25°. *2624*

Seromucoid *Serum Increase Analytical* Alleged if stands on clot for more than 2 h. *0573*

Sodium *Serum No Effect Analytical* If on polystyrene beads 24 h at 25°. *2624*

Serum Increase Analytical Marked top to bottom if frozen and unmixed. *2674*

Urine Increase Analytical Marked top to bottom if frozen and unmixed. *2674*

Urea Nitrogen *Serum No Effect Analytical* If on polystyrene beads 24 h at 25°. *2624*

Uric Acid *Serum No Effect Analytical* If on polystyrene beads 24 h at 25°. *2624*

Stanozolol

Alanine Aminotransferase *Serum Increase Physiological* May cause intrahepatic cholestasis. *0757*

Alkaline Phosphatase *Serum Increase Physiological* May cause intrahepatic cholestasis. *1680*

Apolipoprotein AI *Serum Decrease Physiological* Increase by 41% in 10 normolipidemic postmenopausal osteoporotic women treated for 6 weeks. *3526*

Apolipoprotein AII *Serum Decrease Physiological* Increase by 24% in 10 normolipidemic postmenopausal osteoporotic women treated for 6 weeks. *3526*

Stanozolol (continued)

Apolipoprotein B *Serum No Effect Physiological* Insignificant increase with 6 weeks treatment reverted to normal by 5 weeks after treatment stopped. *0047*

Apolipoprotein D *Serum Decrease Physiological* Increase by 23% with 6 weeks treatment reverted to normal by 5 weeks after treatment stopped. *0243* By 23% with 6 week's treatement: reverted to normal by 5 weeks after treatment stopped. *0047*

Aspartate Aminotransferase
Serum Increase Physiological May cause intrahepatic cholestasis. *0757*

Bilirubin *Serum Increase Physiological* May cause intrahepatic cholestasis. *1343*

BSP Retention *Serum Increase Physiological* Due to intrahepatic cholestasis. *0757*

Cholesterol *Serum No Effect Physiological* In 10 normolipidemic postmenopausal osteoporotic women treated for 6 weeks. *3526*

Cholesterol, High Density Lipoprotein
Serum Decrease Physiological Increase by 53% in 10 normolipidemic postmenopausal osteoporotic women treated for 6 weeks. *3526*

Cholesterol, Low Density Lipoprotein
Serum Increase Physiological Increase by 21% in 10 normolipidemic postmenopausal osteoporotic women treated for 6 weeks. *3526*

Follicle Stimulating Hormone (FSH)
Plasma Decrease Physiological From 2.7 to 1.8 units/L after 1 week in 9 healthy men given 10 mg/d for 14 d. *3354*

β-Glucuronidase *Serum Increase Physiological* Metabolic effect. *0227*

Haptoglobin *Serum Increase Physiological* Metabolic effect. *0227*

HDL$_2$-Cholesterol *Serum Decrease Physiological* Increase by 85% in 10 normolipidemic postmenopausal osteoporotic women treated for 6 weeks. *3526*

HDL$_3$-Cholesterol *Serum Decrease Physiological* Increase by 35% in 10 normolipidemic postmenopausal osteoporotic women treated for 6 weeks. *3526*

Hemoglobin *Blood Increase Physiological* Effective in some patients with aplastic anemia. *1679*

25-Hydroxy Vitamin D$_3$ *Serum Decrease Physiological* From 64 to 54.5 nmol/L after 1 week in 9 healthy men given 10 mg/d for 14 d. *3354*

Lecithin Cholesterol Acyltransferase (LCAT)
Plasma Decrease Physiological Increase by 30% with 6 weeks treatment: reverted to normal by 5 weeks after treatment stopped. *0047*

Lipoprotein Lp(a) *Serum Decrease Physiological* By average of 65% with 6 weeks treatment reverted to normal by 5 weeks after treatment stopped. *0047*

Luteinizing Hormone (LH) *Plasma Decrease Physiological* From 6.5 to 4.5 units/L after 1 week in 9 healthy men given 10 mg/d for 14 d. *3354*

Plasminogen *Plasma Increase Physiological* Metabolic effect. *0227*

Protein Bound Iodine (PBI) *Serum Decrease Physiological* Observed with other anabolic steroids. *1680*

Sex Hormone Binding Globulin
Serum Decrease Physiological From 20.7 to 12.2 nmol/L after 1 week in 9 healthy men given 10 mg/d for 14 d. *3354*

Sialic Acid *Serum Increase Physiological* Metabolic effect. *0227*

Testosterone *Serum Decrease Physiological* Increase by 55% (from 22.1 to 10.6 nmol/L) after 1 week in 9 healthy men given 10 mg/d for 14 d. *3354*

Testosterone, Free *Serum Decrease Physiological* From 398 to 226 pmol/L after 1 week in 9 healthy men given 10 mg/d for 14 d. *3354*

Thyroid Stimulating Hormone (TSH)
Serum No Effect Physiological 1.95 mU/L before treatment, 2.31 after 2 weeks in 9 healthy men given 10 mg/d for 14 d. *3354*

Thyroxine (T$_4$) *Serum Decrease Physiological* From 106.2 to 90.4 nmol/L after 1 week in 9 healthy men given 10 mg/d for 14 d. *3354*

Thyroxine (T$_4$) Binding Globulin
Serum Decrease Physiological From 22.5 to 17.0 mg/L after 1 week in 9 healthy men given 10 mg/d for 14 d. *3354*

Thyroxine (T$_4$), Free *Serum No Effect Physiological* 18.5 pmol/L before treatment, 19.0 pmol/L after 2 weeks in 9 healthy men given 10 mg/ day for 14 d. *3354*

Triglycerides *Serum No Effect Physiological* In 10 normolipidemic postmenopausal osteoporotic women treated for 6 weeks. *3526*

Tri-iodothyronine (T$_3$) *Serum Decrease Physiological* From 2.21 to 1.52 nmol/L after 1 week in 9 healthy men given 10 mg/d for 14 d. *3354*

Vitamin D Binding Globulin *Serum Decrease Physiological* From 270 to 230 mg/L after 1 week in 9 healthy men given 10 mg/d for 14 d. *3354*

Starvation

Acetoacetate *Serum Increase Physiological* Due to metabolic acidosis. *0021*
Urine Increase Physiological Due to metabolic acidosis. *0021*

Alanine *Plasma Decrease Physiological* Lowered after 1 d lowest on third day. *0021*

Alanine Aminotransferase *Serum Increase Physiological* With absolute fast 15 d ?hepatic focal necrosis. *0422*

Albumin *Serum No Effect Physiological* Usually unaffected if previously well fed. *3783*

Aldosterone *Plasma Increase Physiological* Significant effect in obese after 10 d. *1242*
Urine Increase Physiological Two fold increase by 2nd week of fasting. *3783*

α-Amino-N-Butyric Acid *Serum Increase Physiological* Occurs with starvation even for 1 d. *0021*

α-Amino-Nitrogen *Plasma No Effect Physiological* Usually unaffected by starvation. *3783*

Ammonia *Urine Increase Physiological* Occurs with metabolic acidosis. *3067*

Aspartate Aminotransferase
Serum Increase Physiological With absolute fast 15 d ?hepatic focal necrosis. *0422*

Base Deficit *Blood Decrease Physiological* Marked metabolic acidosis within 5 d. *3067*

Bicarbonate *Serum Decrease Physiological* Due to metabolic acidosis. *2522*

Bilirubin *Serum Increase Physiological* Begins within 8 h, falls within 8 h of refeeding. *2696*

Calcium *Urine Decrease Physiological* Usually lower than controls. *2382*

Chloride *Serum No Effect Physiological* No effect observed usually. *1242*

Creatine *Urine Increase Physiological* Begins to rise from first day. *2382*

Creatine Kinase *Serum Decrease Physiological* Low values observed with malnutrition. *2382*

Creatinine *Serum No Effect Physiological* Little effect over first few days. *2382*

Creatinine Clearance *Urine Decrease Physiological* Falls by 15% in first 4 d. *1242*

Fatty Acids, Free (FFA) *Serum Increase Physiological* Due to release and degradation of fatty acids. *2240*

Glucose *Serum Increase Physiological* Metabolic response (decreased in advanced state). *0279*

Glycerol *Serum Increase Physiological* Occurs soon after food withdrawal. *0021*

Glycine *Plasma No Effect Physiological* No effect over 6 d, but increased if protein-free diet. *0021*

Growth Hormone *Plasma Increase Physiological* Increased fifteen-fold to peak on second day. *0021*

β-Hydroxybutyrate *Serum Increase Physiological* Gradual increase, due to metabolism of fat. *2240*

Urine *Increase Physiological* Due to metabolic acidosis. *0021*

17-Hydroxycorticosteroids *Urine No Effect Physiological* No effect observed in normal or obese. *2382*

Insulin *Plasma Decrease Physiological* 40% decrease occurs after 1 d, then plateaus. *0021*

Isoleucine *Plasma Increase Physiological* Marked effect even after 1 d. *0021*
Urine Increase Physiological Observed effect due to tissue catabolism. *0021*

Ketones *Serum Increase Physiological* Begins increase during 2nd or 3rd day. *3783*
Urine Increase Physiological Occurs within 4 d of commencement of fast. *3783*

Lactate Dehydrogenase *Serum Increase Physiological* With absolute fast 15 d ?hepatic focal necrosis. *0422*

Lactate Dehydrogenase Isoenzymes
Serum Increase Physiological With absolute fast 15 d ?liver fractions increase. *0422*

Leucine *Plasma Increase Physiological* Marked effect even after 1 d. *0021*
Urine Increase Physiological Observed effect due to tissue catabolism. *0021*

Methionine *Plasma Increase Physiological* Slight effect after 4 d. *0021*

Nitrogen *Urine Increase Physiological* Marked increase if previously well fed. *3783*

Osmolality *Urine Decrease Physiological* Observed response to stress. *2382*

pCO$_2$ *Blood Decrease Physiological* Marked metabolic acidosis within 5 d. *3067*

pH *Blood Decrease Physiological* Sharp decrease observed sometimes after 2nd day. *3067*

Phosphate *Urine Increase Physiological* Usually increase in response to stress. *2382*

Potassium *Serum No Effect Physiological* Usually no effect, may fall to low normal. *3783*
Serum Decrease Physiological Probably due to impaired homeostasis. *2522*
Urine Increase Physiological Initially increase then falls to 10-15 meq/d. *3783*

Protein *Serum No Effect Physiological* Usually unaffected if previously well fed. *3783*

Pyridoxine *Urine Decrease Physiological* Falls to level below recommended in 3 d. *0714*

Renin Activity *Plasma Increase Physiological* Significant effect in obese after 10 d. *1242*

Riboflavin *Urine Decrease Physiological* Falls to still acceptable level after 10 d. *0714*

Sodium *Serum No Effect Physiological* No effect observed usually. *1242*
Urine Increase Physiological Initially increase then falls to 1-15 meq/d. *3783*

Specific Gravity *Urine Decrease Physiological* Observed response to stress. *2382*

Thiamine *Urine Decrease Physiological* Falls to level below recommended in 3 d. *0714*

Threonine *Plasma Decrease Physiological* 33% decrease by third day returns to normal at 6th. *0021*

Titratable Acidity *Urine Increase Physiological* Occurs with metabolic acidosis. *3067*

Triglycerides *Serum Increase Physiological* Due to release and degradation of fatty acids. *2240*

Urea Nitrogen *Urine Increase Physiological* Begins to rise from first day. *2382*

Uric Acid *Serum Increase Physiological* Impaired renal excretion and tissue catabolism. *2522*

Valine *Plasma Increase Physiological* Marked effect even after 1 d. *0021*
Urine Increase Physiological Observed effect due to tissue catabolism. *0021*

Vanillylmandelic Acid *Urine Increase Physiological* Similar response to other forms of stress. *2382*

Volume *Urine Increase Physiological* Observed response to stress. *2382*

Stearate

Barbiturate *Serum Increase Analytical* Affect colorimetric method with cobalt acetate. *3054*

Stearic Acid

Calcium *Serum No Effect Analytical* On atomic absorption procedures at 4 mmol/L With dialysis procedures at 2 mmol/L. *2673*
Serum Decrease Analytical If titrimetric methods used at concentration over 1.5 mmol/L. *2673*

Stearylamine

Lipid Glycerol *Serum Increase Analytical* Possible effect in method of Horney. *2683*

Stibophen

Alanine Aminotransferase *Serum Increase Physiological* Hepatitis reported. *0071*

Aspartate Aminotransferase
Serum Increase Physiological Hepatitis reported. *0071*

Bilirubin *Serum Increase Physiological* Produces hemolytic anemia. *2365*

Bilirubin, Direct *Serum Increase Physiological* Produces hemolytic anemia. *2365*

Coombs' Test, Direct *Serum Positive Physiological* Produces hemolytic anemia. *2365*

Erythrocytes *Blood Decrease Physiological* Produces hemolytic anemia. *2365*

Haptoglobin *Serum Decrease Physiological* Produces hemolytic anemia. *2365*

Hematocrit *Blood Decrease Physiological* May produce hemolytic anemia. *2365*

Hemoglobin *Blood Decrease Physiological* May produce hemolytic anemia. *2365*

Platelets *Blood Decrease Physiological* Immunological mechanism (often with purpura). *0756*

Protein *Urine Increase Physiological* Renal irritation reported. *0071*

Vanillylmandelic Acid *Urine Increase Analytical* At neutral or alkaline pH affect Gitlow procedure. *3873*

Stigmasterol

Cholesterol *Serum No Effect Analytical* On GLC procedure of MacGee. *2252*

Storage of Sample

Acid Phosphatase *Serum No Effect Analytical* If sodium citrate added for 7-8 d at 25, 5, -15°. *0877* For 115 d at -20°. *3217* Storage at -20° for over 4 mo. *3873*
Serum Decrease Analytical 28% decrease 6 mo at -10°. *1180* 50% activity lost after 5 h at room temperature. *3873*
Urine Decrease Analytical Destruction of enzyme if frozen (prevented by albumin). *3217*

Adenosine Deaminase *Serum No Effect Analytical* For 30 d at 4-8°, 30 d at -20°. *3217* No effect 1 week at 4°, 1 mo at -20°. *3873*

Alanine Aminotransferase *Serum No Effect Analytical* No effect 1 week at 4°, 1 mo at -20°. *1563* For 3 d at 30°, 1 week refrigerated. *0877*
Serum Decrease Analytical 11% loss in 1 d at -20° 20% decrease in 1 week at -20° if high activity. *3217*

Albumin *Serum No Effect Analytical* No effect for 10 y stored at -10°. *1180*

Aldolase *Serum No Effect Analytical* For 2 d room temperature, 21 d at 4-8°. *3217* 5 h at room temperature, 5 d at 4°, 15 at -15°. *0877* No effect 1 week at 4°, 1 mo at -20°. *3873*

Aldosterone *Urine No Effect Analytical* At room temperature for 45 d if acidified. *1180* 1 mo frozen, 7 d room temperature. *0952*

Storage of Sample (continued)

Alkaline Phosphatase *Serum No Effect Analytical* 8 h room temperature, 7 d 4°, 180 d -20°. *3217* No effect 8 h at room temperature, 1 week at -20°. *3873* For 7 d at 25 or 4°. *0877*

Serum Increase Analytical Up to 2.7% in 1 d room temperature, 5% at 4°, 5% frozen. *2332* May increase 5-10% after 1 d at 4°. *3873*

Serum Decrease Analytical 25% decrease after 6 mo at -10°. *1180*

Urine No Effect Analytical No effect 2 d at 4°, 1 week at -20°. *3873*

Amino Acids *Serum No Effect Analytical* Heparinized plasma stable for 4 d at 30°. *1712*

Urine No Effect Analytical Acidified to pH 3-5 stable for 4 d at 30°. *1712*

Aminolevulinic Acid *Urine No Effect Analytical* If acidified in dark 2% decrease in 3 d. *3217*

Urine Decrease Analytical 50% with light and no preservative in 24 h. *3217*

Aminolevulinic Acid Dehydrase
Red Blood Cells Decrease Analytical 85% 24 h at room temperature, up to 25% at 4°. *0400*

α-Amino-Nitrogen *Plasma No Effect Analytical* For long period if plasma is frozen Several days in protein free filtrate in refrigerator At refrigerator temperature after 48 h. *1712*

Urine No Effect Analytical At room temperature for 4 d without preservative. *1712*

Ammonia *Plasma No Effect Analytical* Stable in iced plasma for several h. *0582* On blood ammonia for 7 d on dry ice. *0904*

Plasma Increase Analytical 3 fold increase in 24 h, 8 fold in 7 d. *3217* Increase by 0.017 μg/mL blood/min at 25°. *0904*

Amylase *Serum No Effect Analytical* 1 week at room temperature, several months at 4°. *0877* 1 week at 4°, 1 mo at -20 at normal concentration. *1563* 7 d at room temperature and 4°, 150 at -20°. *3217*

Serum Decrease Analytical At pancreatitis level loss of 30% 18 h at 4 or -20°. *1670*

Urine No Effect Analytical No effect 2 d at 4°, 1 week at -20°. *3873* 1 week at room temperature, months at 4° if no bacteriuria. *0877*

Antidiuretic Hormone *Plasma No Effect Analytical* No effect probably stored frozen for 45 d. *1180*

Urine No Effect Analytical At room temperature for 45 d if acidified. *1180*

α₁-Antitrypsin *Serum No Effect Analytical* For 7 d at 4 or 30°, indefinitely at -70°. *0573*

Arginase *Serum No Effect Analytical* 1 d at room temperature, 2-3 at 4 d, 60 at -20°. *3217*

Argininosuccinate Lyase *Serum No Effect Analytical* For 6 d at -15°. *0877*

Serum Decrease Analytical 10% loss per day at 4°. *0877*

Ascorbic Acid *Plasma No Effect Analytical* No significant change 21 d at -70°. *0441*

Plasma Decrease Analytical Immediate decrease of 7% on freezing at -20°. *0441*

Blood No Effect Analytical Stable 6 h at 25°, stable 1 mo 4° if + TCA. *0442*

Blood Decrease Analytical Stable after loss of 10-20% in 3 d at -70°. *0442*

Urine Decrease Analytical 50% deterioration per h at 20°. *1506*

Asparagine *Plasma Decrease Analytical* Even with prolonged storage at -20°. *3230*

Aspartate Aminotransferase *Serum No Effect Analytical* 2 d room temperature, 14 at 4°, 30 at -20°. *3217* No effect 1 week at 4°, 1 mo at -20°. *1563*

Serum Decrease Analytical Stable for less time at -20° if high activity. *3217*

Aspartic Acid *Plasma Increase Analytical* Even with prolonged storage at -20° Large amounts in platelets and leukocytes. *3230*

Barbiturate *Serum No Effect Analytical* No effect on UV absorption methods. *2981*

Bicarbonate *Serum No Effect Analytical* No difference with/without oil if tube full. *3217*

Serum No Effect Physiological Insignificant change at 2 h at room temperature. *0931*

Serum Decrease Analytical Substantial decrease after 14 d at 4°. *3862* 6 mmol/L in 2.5-4 h with no oil at room temperature. *3217*

Urine No Effect Analytical If no bacterial growth room temperature for 45 d. *1180*

Bilirubin *Serum No Effect Analytical* No effect for 6 mo at -10°. *1180*

Serum Increase Analytical Standards deteriorate at -23°, stable at -70°. *0941*

BSP Retention *Serum Decrease Analytical* Decreased up to 4% at room temperature for 16 h. *3873*

Calcium *Serum No Effect Analytical* Room temperature 8 h, 4° 1 d, frozen 1 y. *3873*

Serum Increase Analytical Significant increase after 12 d at 4°. *3862*

Urine Decrease Analytical Unless acidified. *3873*

Carotene *Serum No Effect Analytical* No effect at 4° for 6 d. *2734*

Catecholamines *Plasma No Effect Analytical* With NaF-Na₂S₂O₃ 1 mo frozen for ethylenediamine. *2285*

Urine No Effect Analytical No effect at room temperature, several days if acidified. *3873*

Cephalin Flocculation *Serum Increase Analytical* Reported effect after 2 d at 4°. *1563*

Ceruloplasmin *Serum No Effect Analytical* For 2 weeks at 4° or if frozen. *0877*

Serum Decrease Analytical Unstable at room temperature. *3217* Oxidase activity decreased at room temperature, no effect for 2 d at 4°. *3873*

Chloride *Serum No Effect Analytical* No effect after 1 week at room temperature, months at -10. *1563*

Urine No Effect Analytical At room temperature for 45 d, without preservative. *1180*

Cholesterol *Serum No Effect Analytical* No effect 2 d at 4°, 5 y at -20°. *3873*

Cholesterol, Free *Serum No Effect Analytical* At room temperature, 4° for 7 d -20 for 28 d. *2512*

Citrulline *Plasma Decrease Analytical* If picric acid used for deproteinization. *3230*

Complement C3 *Serum Decrease Analytical* Degrades to β₁-globulin, no effect 2 d at room temperature. *0828*

Corticotropin *Plasma No Effect Analytical* No effect probably stored frozen for 45 d. *1180*

Creatine *Urine No Effect Analytical* Stable indefinitely if frozen Without preservative at 30° for 4 h. *0904*

Urine Increase Analytical Significant increase 1 d at room temperature. *1563*

Creatine Kinase *Serum No Effect Analytical* If stored at -20° stable for 2 weeks. *0877* 4 h room temperature, 10 d 4°, 60 at 20°. *3217* 48 h at room temperature, 7 d at 4°, 1 mo at -18. *0877*

Serum Increase Analytical 4 U/L if more than 100 h between venesection and assay. *2744*

Serum Decrease Analytical Possible loss up to 50% with thawing once when frozen. *0718*

Creatinine *Serum No Effect Analytical* No effect for 6 mo at -10°. *1180*

Serum Increase Analytical Significant effect after 15 d at 4°. *3862*

Urine No Effect Analytical Stable indefinitely if frozen. *0904* No effect for 4 to 7 d at room temperature. *1563* For 24 h at room temperature if thymol or toluene added Without preservative at 30° for 4 h. *0904*

Cystine *Urine No Effect Analytical* 1 week room temperature with HCl, indefinite time if frozen With chloroform added, in refrigerator for 3 mo. *1712*

Dehydroepiandrosterone (DHEA)
Plasma No Effect Analytical No effect at -20° for 1 y. *0107*

Diagnex Blue Excretion *Urine No Effect Analytical* Stable 48 h at room temperature, 10 d at 4°. *1679*

Digoxin *Serum Decrease Analytical* 5% 2 d room temperature, 2% 5 d at 4°, ok frozen 1 week. *0615*

Dopamine Hydroxylase *Serum No Effect Analytical* 1 d at 22°, days at -20°. *3807* No effect at -20° for 4 weeks. *1339*

Epinephrine *Urine   No Effect   Analytical* No effect at room temperature, several days if acidified. *3873*

Erythrocytes *Blood   Decrease   Analytical* Lysis produced by freezing But only by 2-3% loss if frozen in liquid N₂. *1180*

Ethanol *Blood   Decrease   Analytical* 0.23 mg/dL/h at 20° if no preservative. *3355*

Fatty Acids, Free (FFA) *Serum   Increase   Analytical* 218% at room temperature for 3 d 43% at 4° for 7 d due to lipoprotein lipase. *2512* Of 45% if few h between drawing and analysis. *2438*

Serum   Decrease   Analytical Significant decrease at -40° for 28 d. *2512*

Fibrinogen *Plasma   No Effect   Analytical* For 7 d at room temperature, 4 weeks at 4°. *0573*

FIGLU (N-Formiminoglutamic Acid)
Urine   No Effect   Analytical If acidified for 2 weeks at 28°, 1 mo at -20. *1712*

Folate *Serum   No Effect   Analytical* At -16 or -76° for 1 week radiometric assay. *2455*
Serum   Decrease   Analytical Room temperature 24 h without preservative procedure of Mincey. *2455*

Galactose-1-Phosphate Uridyl Transferase
Red Blood Cells   No Effect   Analytical 14 d at room temperature, for 4 weeks at 4°. *0877*

β₁-α-Globulin *Serum   Decrease   Analytical* Variable loss of activity after 5 y at -15°. *1180*

γ-Globulin *Serum   Decrease   Analytical* With prolonged storage at -10°. *1180*

Glucose *Serum   Decrease   Analytical* Significant decrease after 36 d in frozen state. *3862* Significant effect after 6 mo at -10°. *1180*

Glucose-6-Phosphate Dehydrogenase
Red Blood Cells   No Effect   Analytical If Alsever's solution added for 3.5 weeks at 30°. *0877*

Glutamate Dehydrogenase *Serum   No Effect   Analytical* 7 d at room temperature and 4°, 14 at -20°. *3217*

Glutamic Acid *Plasma   Increase   Analytical* Large amounts in platelets, leukocytes Even with prolonged storage at -20°. *3230*

Glutamine *Plasma   Decrease   Analytical* Even with prolonged storage at -20°. *3230*

Glutathione *Plasma   No Effect   Analytical* In oxalate reduced form stable for 24 h at room temperature. *1712*
Plasma   Increase   Analytical At refrigerator temperature prone to hemolysis. *1712*

Glutathione Reductase *Serum   No Effect   Analytical* 3 d at room temperature and 4°, 7 at -20°. *3217* No effect 1 week at 4°, 1 mo at -20°. *3873*

Guanase *Serum   No Effect   Analytical* 3 d at room temperature, 2 weeks at 4°, 10 mo at -15. *0877*

Hemoglobin *Blood   No Effect   Analytical* Stable at room temperature for at least 1 week. *1563*
Blood   Decrease   Analytical 2-3 % loss if frozen with liquid nitrogen. *1180*

Hexosamine *Serum   No Effect   Analytical* For 7 d at 30°. *0573*

Histidine *Plasma   No Effect   Analytical* No effect -20° for 6 mo, decreased at room temperature 3 d. *0081*

HMPG (4-Hydroxy-3-Methoxy-Phenylethylene Glycol)
Urine   No Effect   Analytical No effect 6 y at -20° (method of Borud). *0426*

Homocitrulline *Plasma   Decrease   Analytical* If picric acid used for deproteinization. *3230*

Homocystine *Plasma   Decrease   Analytical* Binds to protein with time. *3230*

Homogentisic Acid *Plasma   Decrease   Analytical* 25% loss in frozen plasma in 1 week. *1712*
Urine   No Effect   Analytical If acidified and frozen stable for 1 week. *1712*

Hydroxybutyrate Dehydrogenase
Serum   No Effect   Analytical 5 d at 30 or 4°, at -20° for 10 d. *0877*

17-Hydroxycorticosteroids *Urine   No Effect   Analytical* At room temperature for 45 d if acidified. *1180*

Hydroxyproline *Urine   No Effect   Analytical* At 30° for 5 d if pH reduced to 1-2. *1712* At 0° under toluene. *0561*

5-Hydroxytryptamine (Serotonin)
Urine   No Effect   Analytical At room temperature for 45 d if acidified. *1180*

Immunoglobulin IgA *Serum   No Effect   Analytical* Stable at 4 and 30° for 7 d. *0573*

Immunoglobulin IgD *Serum   No Effect   Analytical* Stable at 4 and 30° for 7 d. *0573*

Immunoglobulin IgG *Serum   No Effect   Analytical* Stable at 4 and 30° for 7 d. *0573*

Immunoglobulin IgM *Serum   No Effect   Analytical* Stable at 4 and 30° for 7 d. *0573*
Serum   Decrease   Analytical Significant effect at -20° for 50 d (though starts immediately). *1367*

Insulin *Plasma   No Effect   Analytical* No change at room temperature for 4 h. *1098*
Plasma   Decrease   Analytical Increase by 75% (serum, plasma) at -20° for 28 mo. *1098*

Ionized Calcium *Serum   No Effect   Analytical* If frozen for up to 3 d. *3493*
Serum   Decrease   Analytical 0.04 meq/L in 2 d, 0.08 meq/L in 7 d (?temperature). *1539*

Iron *Serum   No Effect   Analytical* No effect 4 d room temperature or 7 d at 4°. *1563*

Isocitrate Dehydrogenase *Serum   No Effect   Analytical* No effect 1 week at 4°, 1 mo at -20°. *3873* For 7 d at 30° or if frozen. *0877* 6 h at room temperature, many days at 4°. *3217*
Serum   Decrease   Analytical Freezing reported to cause 10-25% loss of activity. *0877*

Lactate Dehydrogenase *Serum   No Effect   Analytical* 8 h at room temperature, 4 d at 4°, 20 d at -20°. *3217* No effect 1 week at 4°, 1 mo at -20°. *1563*
Serum   Decrease   Analytical Minimal stability observed at 0° 10% loss even with quick freeze and thaw 10% decrease after 24 h at room temperature. *0877*
Urine   Decrease   Analytical Loses activity unless analyzed immediately. *3873*

Lactate Dehydrogenase Isoenzymes
Serum   Decrease   Analytical LD-5 most heat and cold labile, may decrease with storage. *0877*

Leucine Aminopeptidase *Serum   No Effect   Analytical* 7 d at 30°, stable frozen. *0877* No effect 1 week at 4°, 1 mo at -20°. *1563* 1 d room temperature, 21 at 4°, 60 at -20°. *3217*
Urine   No Effect   Analytical No effect 1 d at 4°, 1 week at -20°. *3873*

Leukocytes *Blood   No Effect   Analytical* No lysis produced by freezing with liquid N₂. *1180*

Lipase *Serum   No Effect   Analytical* No effect 1 week at 4°, 1 mo at -20°. *1563* 7 d at room temperature, 4° and -20°. *3217*

Lipoproteins *Serum   No Effect   Analytical* For electrophoresis 3 d room temperature, 14 at -20°. *3217*

β-Lipoproteins *Serum   Decrease   Analytical* Marked with 7 d at -20°, mild 28 d at 0°. *1951*

Lithium *Serum   No Effect   Analytical* No effect 1 week at 4°, frozen for months. *2891*

Lysozyme *Serum   No Effect   Analytical* Stable at -20°. *0877*
Urine   No Effect   Analytical 4 d at room temperature if pH 4.5-6.3. *0877*

Magnesium *Serum   No Effect   Analytical* Room temperature 8 h, 4° 1 d, frozen 1 y. *3873*
Urine   No Effect   Analytical Allegedly satisfactory for 45 d, no preservative. *1180*

Manganese *Urine   No Effect   Analytical* No effect 45 d room temperature, no preservative. *1180*

Methylmalonic Acid *Urine   No Effect   Analytical* Stable 7 d room temperature, 4 mo at -20°. *1712*

Nitrogen *Urine   No Effect   Analytical* No effect 45 d room temperature, no preservative. *1180*

Nonprotein Nitrogen *Serum   No Effect   Analytical* No effect 1 d at room temperature, several at 4°. *1563*
Serum   Decrease   Analytical Significant effect at -10° after 6 mo. *3873*

Norepinephrine *Urine   No Effect   Analytical* No effect at room temperature, several days if acidified. *3873*

5'-Nucleotidase *Serum   No Effect   Analytical* For 4 d at room temperature or 4°, 4.5 mo if frozen. *0877*

Storage of Sample *(continued)*

Ornithine Carbamoyltransferase (OCT)
Serum No Effect Analytical 1 d in blood at room temperature, 1 week at 4° in serum. *0877* 1 y at -20°. *3217*
Serum Decrease Analytical Freezing and thawing once causes 10% decrease. *0877*

Osmolality *Serum No Effect Analytical* No effect for 3 h at room temperature, 10 h at 4°. *1810*

pCO$_2$ *Blood Increase Physiological* Slight increase observed at 2 h at room temperature. *0931*
Blood Decrease Analytical By approximately 0.6 mm Hg per h at 0-4°. *3217*

Pepsinogen *Serum No Effect Analytical* Stable 4 d at room temperature. *0877*

Peripheral Smear *Blood No Effect Analytical* Unstained blood smear usually satisfactory for 45 d. *1180*

pH *Blood Increase Analytical* If sample exposed to air. *3879*
Blood Decrease Analytical By approximately 0.006 per h at 0-4°. *3217* If anaerobic storage at 37° or 2 h at room temperature. *3879*

Phenylalanine *Plasma No Effect Analytical* In diluted serum in refrigerator for 3 d 4 d at room temperature, 7 d if fluoride added Dried blood spots at room temperature for 11 weeks. *1712*

Phosphate *Serum No Effect Analytical* At room temperature 8 h, 4° 1 d, frozen 1 y. *3873*
Serum Increase Analytical From second day at 4 or 37°. *3862*
Urine Decrease Analytical Unless acidified. *3873*

Phosphoglucomutase *Serum No Effect Analytical* 8 h room temperature, 2 d at 4°, 4-7 d at -20°. *3217*
Serum Decrease Analytical After 8 h at room temperature, 2 d at 4°. *3873*

6-Phosphoglycerate Dehydrogenase
Serum Decrease Analytical 50% in 4 d at 4°. *3217*

Phosphohexoseisomerase *Serum No Effect Analytical* 8-12 h at room temperature, 21 d at 4°, 1 y at -20°. *3217* No effect 1 week at 4°, 1 mo at -20°. *3873*

Phospholipids, Total *Serum No Effect Analytical* At room temperature or 4° up to 7 d, 28 d at -20°. *2512*

Platelets *Blood No Effect Analytical* No lysis produced by freezing with liquid N$_2$. *1180*
Blood Decrease Analytical Destroyed by freezing. *1180*

Potassium *Serum No Effect Analytical* No effect for 2 weeks at room temperature or 4°. *1563*
Urine No Effect Analytical No change unprocessed for 45 d at room temperature. *1180*

Prealbumin *Serum No Effect Analytical* No effect for several days at room temperature. *1180*

Protein *Serum No Effect Analytical* No effect for 1 week at room temperature, 1 mo at 4°. *1563*
Serum Decrease Analytical Decrease observed after 26 d at 4°. *3862*

Protein Bound Iodine (PBI) *Serum No Effect Analytical* No effect 5 weeks at room temperature. *1563*

Protein Electrophoresis *Serum No Effect Analytical* No effect 3 d at room temperature or 1 mo at 4°. *1563* No effect at -20° for 6 mo. *0573*

Prothrombin Time *Plasma No Effect Analytical* Stable at room temperature for 18 h in oxalated plasma. *3873*
Plasma Increase Analytical Oxalated plasma loses activity at 4°. *3873* Loss of activity observed after 5 y at -15°. *1180*

Pseudocholinesterase *Serum No Effect Analytical* 2 d room temperature, 14 at 4°, months at -20°. *3217* 4 d room temperature, 6 mo at -10°. *1180* No effect 1 week at 4°, 1 mo at -20°. *3873* For 17 d at 23 or 4°, 3 mo at -20°. *0877*

Pyruvate Kinase *Red Blood Cells No Effect Analytical* If heparin added stable for 8 d at 30° If Alsever's solution added for 3.5 weeks at 30°. *0877*

Seromucoid *Serum No Effect Analytical* For 2-7 d at 30°. *0573*

Sodium *Serum No Effect Analytical* No effect for 2 weeks at room temperature or 4°. *1563*
Urine No Effect Analytical No change unprocessed for 45 d at room temperature. *1180*

Sulfate *Urine No Effect Analytical* No effect at 4° or -20 for 30 d. *2243* Satisfactory for 24 h without preservative at room temperature. *1180*

Taurine *Plasma Increase Analytical* Large amounts in platelets, leukocytes. *3230*

Testosterone *Serum No Effect Analytical* For 2 to 4 mo at -20°. *0888*

Thymol Turbidity *Serum No Effect Analytical* Usually stable at 4° for 1 week. *1563*

Thyroxine (T$_4$) *Serum No Effect Analytical* On competitive protein binding at 4 or -20° for 2 weeks. *3257* No effect for several days at room temperature. *1180*

Titratable Acidity *Urine No Effect Analytical* No significant effect if stored at -20° for 30 d. *0627*

α-Tocopherol *Serum No Effect Analytical* No effect up to 2 weeks frozen. *3692*

Triglycerides *Serum Decrease Analytical* Significant decrease 3 d at room temperature. *2512*

Tri-iodothyronine (T$_3$) *Serum No Effect Analytical* Negligible effect frozen for 2 y. *2853*

Triosephosphate Isomerase
Red Blood Cells No Effect Analytical With Alsever's solution for 3-5 weeks at 30°. *0877*

Trypsin *Feces No Effect Analytical* For 8 d at room temperature, though better stored at 4°. *0877*

Tryptophan *Plasma Decrease Analytical* If picric acid used for deproteinization. *3230*

Tyrosine *Plasma No Effect Analytical* At -20° for 6 mo. *0080* With fluoride added stable 7 d at 30° 4 d at 30°, stable indefinitely frozen. *1712*
Plasma Increase Analytical Above 3°, ?hydrolysis of protein. *0080*

Urea Nitrogen *Serum No Effect Analytical* 3-5 day room temperature, prolonged storage at -10°. *1180*
Urine No Effect Analytical For several days in refrigerator if pH less than 4. *0904*

Uric Acid *Serum No Effect Analytical* No effect for 3 d at room temperature, 6 mo frozen. *1563*
Urine No Effect Analytical For 3 d usually unless bacterial contamination. *0904*
Urine Decrease Analytical At any temperature unless alkali added. *2954*

Urobilinogen *Feces Decrease Analytical* Decreased up to 46% within 24 h at 4°. *1563*

Uropepsin *Urine No Effect Analytical* Several days at room temperature if no bacterial growth. *0877*
Urine Decrease Analytical Increase by 10-30% in 4 d even with toluene at 4°. *0877*

Vanillylmandelic Acid *Urine No Effect Analytical* Acidified several days room temperature, 1 week at 4°. *3873*
Urine No Effect Physiological No effect observed for 6 y at -20°. *0426*

Vitamin A *Serum No Effect Analytical* No effect on method of Price/Carr (4° for 6 d). *2734*
Serum Increase Analytical After frozen storage and analyzed by method of Price/Carr. *2734*

Zinc Sulfate Turbidity *Serum Decrease Analytical* Effect observed after 1 d at 4°. *1563*

STP

Amphetamine *Urine No Effect Analytical* At 50 mg/L on fluorescent method of Hayes. *1534*

Streptokinase

Alanine Aminotransferase *Serum Increase Physiological* Substantial but not as marked as GGT. *3189*

Alkaline Phosphatase *Serum Increase Physiological* Substantial not as marked as GGT. *3189* Transient dysfunction of liver, returning to normal when treatment discontinued. *3122*

α$_1$-Antitrypsin *Serum Increase Physiological* Significant effect after infusion in infarct patients. *1356*

Aspartate Aminotransferase

Serum Increase Physiological Substantial effect not as marked as GGT. *3189* Earlier peak of CK-MB observed in patients on thrombolytic therapy due to reperfusion of ischemic myocardium. *1341* Transient dysfunction of liver, returning to normal when treatment discontinued. *3122*

Bilirubin *Serum No Effect Physiological* Transient dysfunction of liver, returning to normal when treatment discontinued. *3122*

Bilirubin, Direct *Serum No Effect Physiological* Although transient dysfunction of liver observed, which reverts to normal when treatment stopped. *3122*

Bleeding Time *Blood Increase Physiological* Dissolves blood clots. *2313*

BSP Retention *Serum Increase Physiological* Possible acute hypoxic or toxic liver damage. *3189*

Creatine Kinase *Serum No Effect Physiological* Although marked effect on liver. *3189*

Serum Increase Physiological Earlier peak of CK-MB observed in patients on thrombolytic therapy due to reperfusion of ischemic myocardium. *1341*

Creatine Kinase Isoenzymes

Serum Increase Physiological Earlier peak of CK-MB observed in patients on thrombolytic therapy due to reperfusion of ischemic myocardium. *1341*

Creatinine *Serum Increase Physiological* Nephrotoxic effect (with tubular damage). *2313*

Eosinophils *Blood Increase Physiological* Allergic response. *2313*

Euglobulin Clot Lysis Time *Blood Decrease Physiological* Significant effect in patients with myocardial infarct. *1356*

Fibrinogen *Plasma Decrease Physiological* Significant effect in infarct patients. *1356*

Fibrinolytic Time *Plasma Increase Physiological* Significant effect if given parenterally. *0421*

Glutamate Dehydrogenase *Serum Increase Physiological* Disproportionate increase but not as marked as GGT. *3189*

γ-Glutamyltransferase (GGT)

Serum Increase Physiological Occurring in approximately 25% increase by 4x. *3189* Transient dysfunction of liver, returning to normal when treatment discontinued. *3122*

Lactate Dehydrogenase *Serum No Effect Physiological* Although marked effect on liver. *3189*

Serum Increase Physiological Earlier peak of CK-MB observed in patients on thrombolytic therapy due to reperfusion of ischemic myocardium. *1341*

Lactate Dehydrogenase Isoenzymes

Serum Positive Analytical Broadening of band between LD-3 and LD-4 and absent LD-5. Complex formed between LD-3 and IgA. Unusual band disappeared. *3666*

Leucine Aminopeptidase *Urine Increase Physiological* Activates peptidases by plasminogen. *2897*

Leukocytes *Blood Increase Physiological* Due to eosinophilia. *2313*

α₂-Macroglobulin *Serum Decrease Physiological* Significant effect remained low for 2 weeks after infusion. *1356*

Myoglobin *Serum Increase Physiological* Earlier peak of CK-MB observed in patients on thrombolytic therapy due to reperfusion of ischemic myocardium. *1341*

Plasmin *Plasma Increase Physiological* Stimulates conversion of plasminogen to plasmin by forming an activator complex. *3095*

Plasminogen *Plasma Decrease Physiological* Effect observed after myocardial infarct. *1356*

Protein *Urine Increase Physiological* Renal damage reported. *2313*

Pseudocholinesterase *Serum Decrease Physiological* Possible acute hypoxic or toxic liver damage. *3189*

Thrombin Time *Plasma Increase Physiological* Reflects extent of fibrinogen breakdown. *3298*

Urea Nitrogen *Serum Increase Physiological* Renal tubular damage in one case. *0652*

Streptomycin

Acetaminophen *Serum No Effect Analytical* No effect at therapeutic concentration on method using o-cresol. *0621*

Alkaline Phosphatase *Urine Increase Physiological* Due to nephrotoxic effect of drug. *2897*

Bilirubin *Serum No Effect Analytical* At concentration of 58 mg/L had no effect on Jendrassik and Grof method. *3393*

Serum No Effect Physiological Clinically insignificant displacement from protein in neonates. *3748*

Serum Increase Physiological May cause hemolytic anemia. *2313*

Calcium *Serum No Effect Analytical* At concentration of 58 mg/L had no effect on cresolphthalein method. *3393*

Casts *Urine Increase Physiological* Cylindruria may develop. *1343*

Chloramphenicol *Serum No Effect Analytical* No effect at 100 mg/L on coupled enzymatic method. *2490*

Chromosomes *Test Conditions No Effect Physiological* Not clastogenic in human cells. *3282*

Coombs' Test *Serum Positive Physiological* Mechanism obscure. *1486*

Creatinine *Serum Increase Physiological* Nephrotoxicity may occur in 2%. *2808* Occasional nephrotoxicity, although less than with other aminoglycosides. *3683*

Eosinophils *Blood Increase Physiological* Hematopoietic reaction (occurs in 50% cases). *0071*

Erythrocytes *Blood Decrease Physiological* Aplastic/hemolytic anemia (sensitivity dependent). *2313* Occasional case of aplastic anemia reported. *1292*

Glucose *Urine No Effect Analytical* No effect on glucose oxidase methods. *1487* No effect at up to 1 mg/mL on any glucose concentration as measured by Clinitest® Diastix® and TesTape®. *2250*

Granulocytes *Blood Decrease Physiological* Rare agranulocytosis reported. *1292*

Haptoglobin *Serum Decrease Physiological* May cause hemolytic anemia. *2313*

Hematocrit *Blood Decrease Physiological* Hemolytic anemia (sensitivity dependent). *0788*

Hemoglobin *Blood Decrease Physiological* Hemolytic anemia (sensitivity dependent). *0788*

LE Cells *Blood Positive Physiological* May produce LE-like syndrome. *3505*

Leucine Aminopeptidase *Urine Increase Physiological* Facilitates permeation of enzyme into tubules. *2897*

Leukocytes *Blood Decrease Physiological* Agranulocytosis/leukopenia/neutropenia. *2313* Occasional case of aplastic anemia reported. *1292*

Neutrophils *Blood Decrease Physiological* Occasional case of agranulocytosis reported. *3717*

Nitrofurantoin *Urine No Effect Analytical* No effect of 50 μg on method of Conklin/Hollifield. *0704*

Nonprotein Nitrogen *Serum Increase Physiological* Nephrotoxicity may occur in 2%. *2808*

Phosphate *Serum No Effect Analytical* On nonprecipitation method of Peynet. *2799* At concentration of 58 mg/L had no effect on phosphomolybdate method. *3393*

Platelets *Blood Decrease Physiological* Pancytopenia/thrombocytopenia. *0788* Occasional case of aplastic anemia reported. *1292*

Potassium *Serum No Effect Analytical* At concentration of 400 mg/L had no effect on measurement by ISE with predilution. *3393*

Protein *Serum No Effect Analytical* At concentration of 58 mg/L had no effect on biuret method with blank correction. *3393*

Urine Increase Analytical Reacts as phenol if Folin-Ciocalteu reaction used. *3505*

Urine Increase Physiological May cause nephrotoxicity. *1343* Occasional nephrotoxicity, although less than with other aminoglycosides. *3683*

CSF Increase Analytical Reacts as phenol if Folin-Ciocalteu reaction used. *3505*

Prothrombin Time *Plasma Increase Physiological* May decrease vitamin K synthesis by gut bacteria. *1487*

Streptomycin (continued)

Sodium *Serum No Effect Analytical* At concentration of 400 mg/L had no effect on measurement by ISE with predilution. *3393*

Sugar *Urine Increase Analytical* False positive with copper reduction procedures. *3505* Acts as reducing agent affects Benedict's, Galatest. *3879*

Urea Nitrogen *Serum No Effect Analytical* At concentration of 58 mg/L had no effect on diacetylmonoxime method. *3393* At concentration of 1450 mg/L had no effect on urease/Berthelot method. *3393*
Serum Increase Physiological Nephrotoxicity may occur in 2%. *2808* Occasional nephrotoxicity, although less than with other aminoglycosides. *3683*
Serum Decrease Analytical Inhibits Berthelot reaction. *3505*

Uric Acid *Serum No Effect Analytical* At concentration of 58 mg/L had no effect on phosphotungstate reduction method. *3393*

Streptonigrin

Chromosomes *Test Conditions Abnormal Physiological* Clastogenic in human cells *in vitro* for up to 1 mo. *3282*

Streptozocin

Acetoacetate *Serum Increase Physiological* Temporary effect following infusion. *2532*

Bicarbonate *Serum Decrease Physiological* Associated with low potassium and polyuria. *2532*

Calcium *Serum Decrease Physiological* To normal in few cases of following administration. *2087*

Fatty Acids, Free (FFA) *Serum Decrease Physiological* Mechanism not established yet. *2532*

Gastrin *Serum Decrease Physiological* Gastric hypersecretion reduced. *2532*

Glucose *Urine Increase Physiological* Rarely exceeded 25 mg/dL Rarely exceeds 25 mg/dL. *2532*

5-Hydroxyindoleacetic Acid (5-HIAA)
Urine Decrease Physiological If carcinoid treated. *1100*

Insulin *Plasma Increase Physiological* ?due to release from damaged beta cells. *2532*

Lactate *Serum Increase Physiological* Temporary effect following infusion. *2532*

Potassium *Serum Decrease Physiological* Associated with polyuria. *2532*
Urine Increase Physiological In spite of low serum concentration. *2532*

Pyruvate *Serum Increase Physiological* Temporary effect following infusion. *2532*

Volume *Urine Increase Physiological* Noted ten days after infusion (with polydipsia). *2532*

Stress

Antidiuretic Hormone *Urine Increase Physiological* With mental stress increase of 50%. *3217*

Catecholamines *Plasma Increase Physiological* Normal metabolic response. *3527*
Urine Increase Physiological With physical or emotional stress. *3217*

Cholesterol *Serum Increase Physiological* Mental stress such as examinations. *3217*

Cortisol *Plasma Increase Physiological* Marked response to emotional stress. *3217*

Epinephrine *Urine Increase Physiological* With physical or emotional stress. *3217*

Factor VIII *Plasma Increase Physiological* With post operative stress. *0883*

Factor VIII Antigen *Plasma Increase Physiological* Occurs with post operative stress. *0883*

Fatty Acids, Free (FFA) *Serum Increase Physiological* Normal metabolic response. *3527*

Glucose *Serum Increase Physiological* Normal metabolic response. *3527*

Growth Hormone *Plasma Increase Physiological* Observed effect in normals. *0371*

Norepinephrine *Urine Increase Physiological* With physical or emotional stress. *3217*

Protein Bound Iodine (PBI) *Serum Increase Physiological* Large effect with psychological stress. *3217*

Stress Cold Exposure

Antidiuretic Hormone *Urine No Effect Physiological* Response to cold exposure. *3217*

Creatine Kinase *Serum Increase Physiological* Response to cold exposure. *3217*

Epinephrine *Urine No Effect Physiological* Response to cold exposure. *3217*

Kynurenic Acid *Urine Increase Physiological* Possibly due to activation of tryptophan oxygenase. *1169*

Norepinephrine *Urine Increase Physiological* Response to cold exposure. *3217*

Tryptophan *Plasma Decrease Physiological* Although total amino acids unaffected. *1169*

Tyrosine *Plasma Decrease Physiological* Although total amino acids unaffected. *1169*

Xanthurenic Acid *Urine Increase Physiological* Possibly due to activation of tryptophan oxygenase. *1169*

Strontium

Magnesium *Urine Increase Analytical* Measured by fluorometric method of Schachter. *3573*

Strychnine

Albumin *Serum No Effect Analytical* At concentration of 12 mg/L had no effect on BCG method. *3393*

Alkaloids *Urine No Effect Analytical* With ninhydrin on Frings TLC procedure at 10 mg/dL. *1204*
Urine Increase Analytical Purple with iodoplatinate on Frings TLC procedure. *1204*

Barbiturate *Urine No Effect Analytical* With mercuric SO_4 on Frings TLC procedure at 10 mg/dL. With mercurous NO_3 on Frings TLC procedure at 10 mg/dL. With diphenylcarbazone on Frings TLC procedure at 10 mg/dL. With vanillin on Frings procedure at 10 mg/dL. *1204*

Bicarbonate *Serum No Effect Analytical* At concentration of 12 mg/L had no effect on method using phenolphthalein. *3393*

Bilirubin *Serum No Effect Analytical* At concentration of 12 mg/L had no effect on Jendrassik and Grof method. *3393*

Bromide *Urine No Effect Analytical* No effect on method of Frings. *1204*

Calcium *Serum No Effect Analytical* At concentration of 12 mg/L had no effect on cresolphthalein method. *3393*

Chloride *Serum No Effect Analytical* At concentration of 12 mg/L had no effect on mercurimetric method. *3393*

Cholesterol *Serum No Effect Analytical* At concentration of 12 mg/L had no effect on Liebermann-Burchard method. *3393*

Creatinine *Serum No Effect Analytical* At concentration of 12 mg/L had no effect on AutoAnalyzer Jaffé method. *3393*

Ethchlorvynol *Urine No Effect Analytical* No effect on method of Frings and Cohen at 10 mg/dL. *1205*

FPN Test *Urine No Effect Analytical* No effect at 10 mg/dL on method of Frings. *1204*

Leukocytes *Blood Increase Physiological* Probably due to release of epinephrine from adrenal. *1302*

Phosphate *Serum No Effect Analytical* At concentration of 12 mg/L had no effect on phosphomolybdate method. *3393*

Potassium *Serum No Effect Analytical* At concentration of 12 mg/L had no effect on measurement by ISE with predilution. *3393*

Protein *Serum No Effect Analytical* At concentration of 12 mg/L had no effect on biuret method with blank correction. *3393*

Salicylate *Urine No Effect Analytical* At 10 mg/dL no effect on Trinder method. *1204*

Sodium *Serum No Effect Analytical* At concentration of 12 mg/L had no effect on measurement by ISE with predilution. *3393*

Triglycerides *Serum No Effect Analytical* At concentration of 12 mg/L had no effect on lipase/esterase method. *3393*

Urea Nitrogen *Serum No Effect Analytical* At concentration of 12 mg/L had no effect on diacetylmonoxime method. *3393*

Uric Acid *Serum No Effect Analytical* At concentration of 12 mg/L had no effect on phosphotungstate reduction method. *3393*

SU-9055

Aldosterone *Urine Decrease Physiological* May inhibit aldosterone production. *2775*

Succinic Acid

Protein *Test Conditions Decrease Analytical* Lowry procedure due to chelation of chemicals. *1796*

Succinimide

Aminolevulinic Acid *Urine Increase Physiological* May precipitate acute porphyria. *1322*

Coproporphyrin *Urine Increase Physiological* May precipitate acute porphyria. *1322*
Feces Increase Physiological May precipitate acute porphyria. *1322*

Porphobilinogen *Urine Increase Physiological* May precipitate acute porphyria. *1322*

Porphyrins *Urine Increase Physiological* May precipitate attack of acute porphyria. *1016*

Protoporphyrin *Feces Increase Physiological* May precipitate acute porphyria. *1322*

Pyridoxal Phosphate *Serum Decrease Physiological* Increase by 20% at 4 weeks, 60% at 12 weeks. *2948*

Succinylcholine

Creatine Kinase *Serum Increase Physiological* Significant if given with anesthesia (effect of injection). *1724* Occurs if given with halothane. *3628* Marked effect on CK if administered during anesthesia. *2809*

Histamine *Plasma Increase Physiological* Observed with injection for anesthesia. *2212*

Ionized Calcium *Serum Increase Physiological* Rise not marked (i.v. administration). *1047*

Myoglobin *Serum Increase Physiological* Occasional result of i.v. injection in children. *3098*
Urine Increase Physiological Occasional result of i.v. injection in children. *3098*

Potassium *Serum Increase Physiological* If injected i.v. *3505* Caused transient hyperkalemia in patients undergoing general anesthesia. *2836* Due to increased chemosensitivity of muscle membrane due to development of receptor sites in extrajunctional areas. *1408*

Succinylsulfathiazole

Acetaminophen Screening Test *Urine Positive Analytical* Red color with 1-naphthol at therapeutic concentrations. *3326*
Urine Negative Analytical No reaction with o-cresol at therapeutic concentrations. *3326*

Prothrombin Time *Plasma Increase Physiological* May cause vitamin K deficiency. *0071*

Sucralfate

Aluminum *Serum Increase Physiological* Serum concentration in uremic patients comparable to that when patients receiving aluminum hydroxide. *2139*

Calcium *Serum No Effect Physiological* No effect in uremic patients or when previous aluminum hydroxide regime replaced. *2139*

Phosphate *Serum Decrease Physiological* Comparable effect to that of aluminum hydroxide in patients with uremia. *2139*

Sucrose

Calcium *Urine Increase Physiological* Nutrient induced augmentation ?mechanism. *2126*

Cholesterol *Serum Increase Physiological* Effect of sustained high sucrose diet. *3931* Marked increase when sugar substituted for starch. *2036*

Clot Lysis *Blood No Effect Physiological* Effect of sustained high sucrose diet. *3931*

Estriol *Urine Decrease Analytical* Probably destroys estriol during acid hydrolysis. *0023* Affects hydrolysis of conjugates. *0406*

Estrogens *Urine Decrease Analytical* Affects hydrolysis of estrogen conjugates. *0022*

Fatty Acids, Free (FFA) *Serum Decrease Physiological* Mean decrease of 400 mg/dL after 1 g/kg/h orally. *0408*

Fibrinogen *Plasma Decrease Physiological* Effect of sustained high sucrose diet. *3931*

Fructose *Urine Increase Physiological* Maximum at 30 minutes, continues for several h. *2551*

Glucose *Serum No Effect Analytical* At 200 mg/dL on hexokinase method of Coburn. *0676*
Serum Increase Analytical Probably due to acid hydrolysis in o-toluidine method. *0583* Affects GOD-PERID procedure (sucrase in enzyme). *2771*
Serum Increase Physiological Mean increase of 40 mg/dL after 1 g/kg/h orally. *0408* With tolerance test from 80 to 180 mg/dL. *2551*
Urine Increase Physiological Maximum at 60 minutes, continues for several h. *2551*

11-Hydroxycorticosteroids *Plasma Increase Physiological* Increase of 300-400% on high sucrose diet. *3931*

Insulin *Plasma Increase Physiological* By about 30% on high sucrose diet. *3931*

Lactate *Serum Increase Physiological* Increase maximum 1 h after, persists for 2 h. *1907*
Saliva Increase Physiological Increase maximum 1 h after, persists for 2 h. *1907*

Lactate Dehydrogenase *Serum No Effect Physiological* Unaffected by ingestion. *1907*

Magnesium *Urine Increase Physiological* Nutrient induced augmentation ?mechanism. *2126*

pH *Urine Decrease Physiological* Associated with increased acid excretion. *2126*

Phospholipids, Total *Serum Increase Physiological* Marked increase when sugar substituted for starch. *2036*

Platelet Adhesiveness *Blood Decrease Physiological* Effect of sustained high sucrose diet. *3931*

Platelets *Blood Decrease Physiological* Thrombocytopenia may occur in h to days. *1436*

Protein *Test Conditions No Effect Analytical* Lowry procedure when referenced with chemical only used. *1796*

Pyruvate *Serum Increase Physiological* Increase maximum at 1 h, lactate/pyruvate ratio inc. *1907*
Saliva Increase Physiological Increase maximum at 1 h, lactate/pyruvate ratio inc. *1907*

Sucrose *Serum Increase Physiological* With tolerance test from 0.04 to 0.1 mg/dL. *2551*
Urine Increase Physiological But small compared with maltose in tolerance test. *2551*

Triglycerides *Serum Increase Physiological* Marked increase when sugar substituted for starch. *2036* Effect of sustained high sucrose diet. *3931*

Sulfacarbamide

Glucose *Urine No Effect Analytical* At concentration of 1600 mg/L had no effect on Diabur-test. *3393*

Sulfacetamide

Bilirubin *Serum Increase Physiological* May cause hemolysis in G-6-PD deficiency. *0248*

Erythrocytes *Blood Decrease Physiological* May cause hemolysis with G-6-PD deficiency. *1902*

Haptoglobin *Serum Decrease Physiological* May cause hemolysis if G-6-PD deficiency. *0248*

Hematocrit *Blood Decrease Physiological* May cause hemolysis if G-6-PD deficiency. *0248*

Hemoglobin *Blood Decrease Physiological* May cause hemolysis if G-6-PD deficiency. *0248*

Methemoglobin *Blood Increase Physiological* May cause hemolysis if G-6-PD deficiency. *3581*

Sulfachlorpyridazine

Alanine Aminotransferase *Serum Increase Physiological* Occasional hepatitis-like reaction. *1680*

Amylase *Serum Increase Physiological* Occasional cause of pancreatitis. *1680*

Bilirubin *Serum Increase Physiological* Occasional hepatitis-like reaction. *1680*

Erythrocytes *Blood Decrease Physiological* May be aplastic or hemolytic anemia. *1680*

LE Cells *Blood Positive Physiological* Observed occasionally. *1680*

Leukocytes *Blood Decrease Physiological* Leukopenia, aplastic anemia may occur. *1680*

Methemoglobin *Blood Increase Physiological* Due to hemolysis. *1680*

Platelets *Blood Decrease Physiological* Purpura/thrombocytopenia. *1680*

Prothrombin Time *Plasma Increase Physiological* Due to decreased prothrombin. *1680*

Sulfa as Sulfanilamide *Serum Increase Physiological* After 4 g rises to 22 mg/dL at 3 h. *1680*

Sulfadiazine

Acetaminophen Screening Test *Urine Positive Analytical* Red color with 1-naphthol at therapeutic concentrations. *3326*
Urine Negative Analytical No reaction with o-cresol at therapeutic concentrations. *3326*

Alanine Aminotransferase *Serum Increase Physiological* May cause cholestasis with cholangiolitis. *1948*

Albumin *Serum No Effect Analytical* At concentration of 1500 mg/L had no effect on BCG method. *3393*

Alkaline Phosphatase *Serum Increase Physiological* May cause intrahepatic cholestasis. *1434*

Aspartate Aminotransferase
Serum Increase Physiological May cause cholestasis with cholangiolitis. *1948*

Bicarbonate *Serum No Effect Analytical* At concentration of 1500 mg/L had no effect on method using phenolphthalein. *3393*

Bilirubin *Serum No Effect Analytical* At concentration of 150 mg/L had no effect on Ektachem® method. *3393* At concentration of 1500 mg/L had no effect on Jendrassik and Grof method. *3393*
Serum Increase Physiological May cause intrahepatic cholestasis. *1434* Displacement from protein in neonates. *3748*

Calcium *Serum No Effect Analytical* At concentration of 1500 mg/L had no effect on cresolphthalein method. *3393*
Serum Decrease Analytical Depresses fluorescence of calcein method. *2395*

Chlordiazepoxide *Blood Increase Analytical* 5 mg/dL equivalent to 1.0 mg/dL by method of Riddick. *2983*

Chloride *Serum No Effect Analytical* At concentration of 1500 mg/L had no effect on mercurimetric method. *3393*

Cholesterol *Serum No Effect Analytical* At concentration of 1500 mg/L had no effect on Liebermann-Burchard method. *3393*
Serum Increase Physiological May cause intrahepatic cholestasis. *1434*

Creatinine *Serum No Effect Analytical* At concentration of 1500 mg/L had no effect on AutoAnalyzer Jaffé method. *3393*

Crystals *Urine Increase Physiological* Low solubility in acid urine. *2808*

Erythrocytes *Blood Decrease Physiological* May cause hemolytic anemia. *2313*
Urine Increase Physiological Associated with crystalluria and oliguria. *2808*

Glucose *Serum No Effect Analytical* No effect on glucose oxidase method of Boehringer. *3277*

Haptoglobin *Serum Decrease Physiological* May cause hemolytic anemia. *2313*

Hematocrit *Blood Decrease Physiological* May cause hemolytic anemia. *2313*

Hemoglobin *Blood Decrease Physiological* May cause hemolytic anemia. *2313*
Urine Increase Physiological Associated with crystalluria and hematuria. *2808*

131I Uptake *Serum Decrease Physiological* Dulcet contains tetraiodofluorescein. *2652*

Leukocytes *Blood Decrease Physiological* Agranulocytosis/leukopenia. *2313*

Phosphate *Serum No Effect Analytical* At concentration of 1500 mg/L had no effect on phosphomolybdate method. *3393*

Potassium *Serum No Effect Analytical* At concentration of 1500 mg/L had no effect on measurement by ISE with predilution. *3393*

Protein *Serum No Effect Analytical* At concentration of 1500 mg/L had no effect on biuret method with blank correction. *3393*
Urine No Effect Analytical At 100 mg/dL on TCA dye method of Pesce. *2791*
Urine Increase Physiological Due to crystalluria and hematuria Associated with crystalluria and hematuria. *2808*
CSF No Effect Analytical At 100 mg/dL on TCA dye method of Pesce. *2791*
CSF Increase Analytical Reacts as if phenol with Folin-Ciocalteu procedure. *3958*

Sodium *Serum No Effect Analytical* At concentration of 1500 mg/L had no effect on measurement by ISE with predilution. *3393*

Tolbutamide *Serum No Effect Physiological* Appears to have no effect on plasma concentration. *1487*

Triglycerides *Serum No Effect Analytical* At concentration of 1500 mg/L had no effect on lipase/esterase method. *3393*

Urea Nitrogen *Serum No Effect Analytical* No effect on Berthelot procedure. *1862* At concentration of 1500 mg/L had no effect on diacetylmonoxime method. *3393* At concentration of 80 mg/L had no effect on urease/Berthelot method. *3393*

Uric Acid *Serum No Effect Analytical* At concentration of 1500 mg/L had no effect on phosphotungstate reduction method. *3393*

Urobilinogen *Urine Increase Analytical* May produce greenish color with Ehrlich's reagent. *0459* Gives greenish-yellow color with Ehrlich's. *0251*

Volume *Urine Decrease Physiological* Associated with crystalluria and hematuria. *2808*

Sulfadimethoxine

Acetaminophen Screening Test *Urine Positive Analytical* Red color with 1-naphthol at therapeutic concentrations. *3326*
Urine Negative Analytical No reaction with o-cresol at therapeutic concentrations. *3326*

Alanine Aminotransferase *Serum Increase Physiological* Reversible hypersensitive cholestatic response. *1888*

Alkaline Phosphatase *Serum Increase Physiological* Granulomatous reaction in liver. *1049*

Aspartate Aminotransferase
Serum Increase Physiological Granulomatous reaction in liver. *1049*

Bile *Urine Increase Physiological* Granulomatous reaction in liver. *1049*

Bilirubin *Serum Increase Physiological* Granulomatous reaction in liver. *1049*

Urine Increase Physiological Granulomatous reaction in liver. *1049*

Bilirubin, Direct *Serum Increase Physiological* Granulomatous reaction in liver. *1049*

BSP Retention *Serum Increase Physiological* Reversible cholestasis. *1049*

Calcium *Serum Decrease Analytical* Depresses fluorescence of calcein method. *2395*

Cephalin Flocculation *Serum Increase Physiological* Reversible cholestasis. *1049*

LE Cells *Blood Positive Physiological* Implicated as activator of SLE. *0962*

Platelets *Blood Decrease Physiological* Aplastic anemia/thrombocytopenia with sulfonamides. *2313*

Thymol Turbidity *Serum Increase Physiological* Reversible cholestasis. *1049*

Tolbutamide *Serum No Effect Physiological* Appears to have no effect on plasma concentration. *1487*

Urobilinogen *Urine Decrease Physiological* Occurs with reversible cholestasis. *1049*
Feces Decrease Physiological Pale stools with reversible cholestasis. *1049*

Sulfadimidine

Acetaminophen Screening Test *Urine Positive Analytical* Red color with 1-naphthol at therapeutic concentrations. *3326*
Urine Negative Analytical No reaction with o-cresol at therapeutic concentrations. *3326*

Catecholamines *Plasma Increase Analytical* Striking fluorescence. *0596*

Cyclosporine *Serum Decrease Physiological* Reduction when given i.v.; not seen if given orally: mechanism not clear. *3732*

Platelets *Blood Decrease Physiological* Immunologically-induced thrombocytopenic purpura. *1877*

Sulfadoxine

Bilirubin *Serum Increase Physiological* Isolated cases reported when given with pyrimethamine for treatment of malaria. *2669*

Neutrophils *Blood Decrease Physiological* Isolated cases reported when given with pyrimethamine for treatment of malaria. *2669*

Sulfaguanidine

Acetaminophen Screening Test *Urine Positive Analytical* Red color with 1-naphthol at therapeutic concentrations. *3326*
Urine Negative Analytical No reaction with o-cresol at therapeutic concentrations. *3326*

Albumin *Serum No Effect Analytical* At concentration of 500 mg/L had no effect on BCG method. *3393*

Bicarbonate *Serum No Effect Analytical* At concentration of 500 mg/L had no effect on method using phenolphthalein. *3393*

Bilirubin *Serum No Effect Analytical* At concentration of 500 mg/L had no effect on Jendrassik and Grof method. *3393*

Calcium *Serum No Effect Analytical* At concentration of 500 mg/L had no effect on cresolphthalein method. *3393*

Chloride *Serum No Effect Analytical* At concentration of 500 mg/L had no effect on mercurimetric method. *3393*

Cholesterol *Serum No Effect Analytical* At concentration of 500 mg/L had no effect on Liebermann-Burchard method. *3393*

Creatinine *Serum No Effect Analytical* At concentration of 500 mg/L had no effect on AutoAnalyzer Jaffé method. *3393*

Glucose *Serum No Effect Analytical* No effect on Boehringer GOD-PERID glucose oxidase method. *3277*

Phosphate *Serum No Effect Analytical* At concentration of 500 mg/L had no effect on phosphomolybdate method. *3393*

Potassium *Serum No Effect Analytical* At concentration of 500 mg/L had no effect on measurement by ISE with predilution. *3393*

Protein *Serum No Effect Analytical* At concentration of 500 mg/L had no effect on biuret method with blank correction. *3393*
CSF Increase Analytical Reacts as if phenol with Folin-Ciocalteu procedure. *3958*

Sodium *Serum No Effect Analytical* At concentration of 500 mg/L had no effect on measurement by ISE with predilution. *3393*

Triglycerides *Serum No Effect Analytical* At concentration of 500 mg/L had no effect on lipase/esterase method. *3393*

Urea Nitrogen *Serum No Effect Analytical* At concentration of 500 mg/L had no effect on diacetylmonoxime method. *3393*

Uric Acid *Serum No Effect Analytical* At concentration of 500 mg/L had no effect on phosphotungstate reduction method. *3393*

Sulfamerazine

Chlordiazepoxide *Blood Increase Analytical* 5 mg/dL equivalent to 0.9 mg/dL by method of Riddick. *2983*

Corticosteroids *Urine Increase Analytical* Butanol extract/no hydrolysis Reddy/Porter-Silber reaction. *0022*

Crystals *Urine Increase Physiological* Presence of drug. *0071*

17-Hydroxycorticosteroids *Urine Increase Analytical* Alleged effect on method of Reddy. *1488*

Protein *CSF Increase Analytical* Reacts as if phenol with Folin-Ciocalteu procedure. *3958*

Sulfamethizole

Alanine Aminotransferase *Serum Increase Physiological* Reversible cholestasis. *1974*

Alkaline Phosphatase *Serum Increase Physiological* Reversible cholestasis. *1974*

Amylase *Serum Increase Physiological* Case of pancreatitis reported. *0236*

Aspartate Aminotransferase
Serum Increase Physiological Reversible cholestasis. *1974*

Bile *Urine Increase Physiological* Reversible cholestasis. *1974*

Bilirubin *Serum Increase Physiological* May cause cholestatic jaundice. *1974*

BSP Retention *Serum Increase Physiological* Reversible cholestasis. *1974*

Cephalin Flocculation *Serum Increase Physiological* Reversible cholestasis. *1974*

Erythrocytes *Blood Decrease Physiological* Hemolytic anemia/aplastic anemia. *2868*

Glucose *Serum No Effect Analytical* No effect on glucose oxidase method of Boehringer. *3277*

Haptoglobin *Serum Decrease Physiological* May cause hemolytic anemia. *2868*

Hematocrit *Blood Decrease Physiological* Hemolytic anemia. *2868*

Hemoglobin *Blood Decrease Physiological* Hemolytic anemia. *2868*

Leukocytes *Blood Decrease Physiological* Agranulocytosis. *2868*

Methemoglobin *Blood Increase Physiological* May cause hemolytic anemia. *2868*

Platelets *Blood Decrease Physiological* Agranulocytosis/aplastic anemia/thrombocytopenia. *2868*

Prothrombin Time *Plasma Increase Physiological* Reaction may occur with all sulfonamides. *1680*

Sulfhemoglobin *Blood Increase Physiological* May occur with hemolytic anemia. *2868*

Thymol Turbidity *Serum Increase Physiological* Reversible cholestasis. *1974*

Urobilinogen *Urine Increase Analytical* Yellow color with Ehrlich's (extracted into $CHCl_3$). *2359*
Urine Decrease Physiological Occurs with reversible cholestasis. *1974*

Sulfamethizole *(continued)*

Urobilinogen *(continued)*
Feces   Decrease   Physiological   Pale stools with reversible cholestasis. *1974*

Sulfamethoxazole

Alanine Aminotransferase   *Serum   No Effect   Analytical*   No effect at therapeutic concentration on Reflotron method. *1984*
Serum   Increase   Physiological   Cholestatic jaundice may occur. *1974*   Occasional case of cholestasis with or without hepatic necrosis. *3452*   When given with trimethoprim, fever and malaise: successful response to treatment. *1969*

Alkaline Phosphatase   *Serum   Increase   Physiological*   Cholestatic jaundice may occur. *1974*   Occasional case of cholestasis with or without hepatic necrosis. *3452*   When given with trimethoprim, fever and malaise: successful response to treatment. *1969*

α-Amino-Nitrogen   *Plasma   Increase   Analytical*   Measured in naphthoquinone method of Frame. *2981*

Aspartate Aminotransferase   *Serum   No Effect   Analytical*   No effect at therapeutic concentration on Reflotron method. *1984*
Serum   Increase   Physiological   Cholestatic jaundice may occur. *1974*   Occasional case of cholestasis with or without hepatic necrosis. *3452*

Bile   *Urine   Increase   Physiological*   Reversible cholestasis. *2313*

Bilirubin   *Serum   No Effect   Analytical*   At concentration of 60 mg/L had no effect on Ektachem® method. *3393*
Serum   Increase   Physiological   Reversible cholestasis or hemolytic anemia. *2313*   Displacement from protein in neonates. *3748*   Occasional case of cholestasis with or without hepatic necrosis. *3452*

BSP Retention   *Serum   Increase   Physiological*   Reversible cholestasis. *2313*

Cephalin Flocculation   *Serum   Increase   Physiological*   Reversible cholestasis. *2313*

Cholesterol   *Serum   No Effect   Analytical*   No effect at therapeutic concentration on Reflotron method. *1984*

Color   *Urine   Increase   Analytical*   Brown color observed. *1022*

Creatinine   *Serum   No Effect   Analytical*   No effect on alkaline picrate (Jaffé) procedure. *0313*   At concentration of 200 mg/L had no effect on AutoAnalyzer Jaffé method. *3393*
Serum   No Effect   Physiological   Insignificant reduction in 5 volunteers after 7 d. *3078*
Serum   Increase   Analytical   At concentrations above 200 mg/L raised concentration as measured by kinetic Jaffé method. *3393*

Creatinine Clearance   *Urine   Decrease   Physiological*   When given with trimethoprim, fever and malaise: successful response to treatment. *1969*

Crystals   *Urine   Increase   Physiological*   Low solubility in acid urine. *2808*

Cyclosporine   *Serum   Increase   Physiological*   Reversible deterioration in renal function with effect on tubular function and nephrotoxicity. *3732*

Eosinophils   *Blood   Increase   Physiological*   Allergic response. *1974*   When given with trimethoprim, fever and malaise: successful response to treatment. *1969*

Erythrocytes   *Blood   Decrease   Physiological*   May cause hemolytic anemia. *1974*
Urine   Increase   Physiological   May cause hematuria with crystalluria. *2808*

Erythrocyte Sedimentation Rate
Blood   Increase   Physiological   When given with trimethoprim fever and malaise; successful response to treatment. *1969*

Glucose   *Serum   No Effect   Analytical*   No effect at therapeutic concentration on Reflotron method. *1984*   At concentration of 320 mg/L had no effect on hexokinase/G-6-PDH method. *3393*
Urine   No Effect   Analytical   At concentration of 1300 mg/L had no effect on Diabur-test. *3393*

γ-Glutamyltransferase (GGT)   *Serum   No Effect   Analytical*   No effect at therapeutic concentration on Reflotron method. *1984*

Haptoglobin   *Serum   Decrease   Physiological*   May cause hemolytic anemia. *1974*

Hematocrit   *Blood   Decrease   Physiological*   May cause hemolytic anemia. *1974*

Hemoglobin   *Blood   Decrease   Physiological*   May cause hemolytic anemia. *1974*
Urine   Increase   Physiological   May cause hematuria with oliguria. *2808*

Lactate   *Serum   No Effect   Analytical*   At concentration of 320 mg/L had no effect on enzymatic method. *3393*

Lactate Dehydrogenase   *Serum   Increase   Physiological*   May cause hemolytic anemia. *1974*

Leukocytes   *Blood   Increase   Physiological*   When given with trimethoprim, fever and malaise: successful response to treatment. *1969*
Blood   Decrease   Physiological   Toxic reaction to drug. *1974*

Platelets   *Blood   Decrease   Physiological*   Thrombocytopenia. *1974*

Porphyrins   *Urine   Increase   Analytical*   May cause false positive with fluorescent methods. *1022*

Protein   *Urine   Increase   Analytical*   May interfere with sulfosalicylic acid methods. *1022*
Urine   Increase   Physiological   May cause hematuria with crystalluria. *2808*

Protein Bound Iodine (PBI)   *Serum   Decrease   Physiological*   Small effect with therapeutic doses. *1488*

Prothrombin Time   *Plasma   Increase   Physiological*   Possibly due to drug-induced vitamin K deficiency. *1974*   Displaces warfarin from its binding sites on albumin. *3595*

PSP Excretion   *Urine   Increase   Physiological*   Possible large dose effect. *1022*
Urine   Decrease   Physiological   More common effect. *1022*

Sugar   *Urine   Increase   Analytical*   May cause positive with fluorescent methods. *1022*

Thymol Turbidity   *Serum   Increase   Physiological*   Reversible cholestasis. *2313*

Thyroxine (T₄)   *Serum   Increase   Analytical*   Erratic elution from column = false result. *0581*

Triglycerides   *Serum   No Effect   Analytical*   No effect at therapeutic concentration on Reflotron method. *1984*

Urea Nitrogen   *Serum   No Effect   Analytical*   No effect at therapeutic concentration on Reflotron method. *1984*
Serum   Increase   Analytical   Effect marked DMAB method of Morin. *2487*

Uric Acid   *Serum   No Effect   Analytical*   No effect at therapeutic concentration on Reflotron method. *1984*
Serum   Decrease   Physiological   Presumed uricosuric effect. *3920*
Urine   Increase   Physiological   Presumed uricosuric effect. *3920*

Urobilinogen   *Urine   Increase   Analytical*   May react with Ehrlich's to produce false color. *0085*
Urine   Decrease   Physiological   Occurs with reversible cholestasis. *2313*
Feces   Decrease   Physiological   Pale stools with reversible cholestasis. *2313*

Sulfamethoxydiazine

Cholesterol   *Serum   No Effect   Analytical*   At concentration of 231 mg/L had no effect on CHOD-PAP method. *3393*   At concentration of 231 mg/L had no effect on catalase-AlDH method. *3393*   At concentration of 231 mg/L had no effect on method using catalase-Hantzsch reaction. *3393*   At concentration of 231 mg/L had no effect on Liebermann-Burchard method. *3393*

Creatinine   *Serum   No Effect   Analytical*   At concentration of 200 mg/L had no effect on AutoAnalyzer Jaffé method. *3393*   At concentration of 200 mg/L had no effect on Jaffé-Fading-Fraction method. *3393*   At concentration of 200 mg/L had no effect on Jaffé-Fuller's earth method. *3393*   At concentration of 500 mg/L had no effect on kinetic Jaffé method on BKA-2. *3393*

Glucose *Serum No Effect Analytical* At concentration of 20 mg/L had no effect on Ektachem® method. *3393*

Potassium *Serum No Effect Analytical* At concentration of 200 mg/L had no effect on measurement by ISE with predilution. *3393* At concentration of 200 mg/L had no effect on measurement by ISE without predilution. *3393*

Protein Electrophoresis *Serum No Effect Analytical* At concentration of 200 mg/L had no effect on automated Olympus-Hite method. *3393*

Sodium *Serum No Effect Analytical* At concentration of 200 mg/L had no effect on measurement by ISE with predilution. *3393* At concentration of 200 mg/L had no effect on measurement by ISE without predilution. *3393*

Urea Nitrogen *Serum No Effect Analytical* At concentration of 20 mg/L had no effect on Ektachem® method. *3393*

Uric Acid *Serum No Effect Analytical* At concentration of 230 mg/L had no effect on Kageyama-Hantzsch method. *3393* At concentration of 300 mg/L had no effect on catalase-AIDH method. *3393*

Sulfamethoxypyridazine

Creatinine *Serum No Effect Analytical* At concentration of 70 mg/L had no effect on creatinine iminohydrolase method. *3393*

Glucose *Serum No Effect Analytical* At concentration of 70 mg/L had no effect on GOD/POD-PAP method. *3393*

Triglycerides *Serum No Effect Analytical* At concentration of 70 mg/L had no effect on GPO-PAP method. *3393*

Uric Acid *Serum No Effect Analytical* At concentration of 70 mg/L had no effect on uricase-PAP method. *3393*

Sulfamethoxypyridine

Acetaminophen Screening Test *Urine Positive Analytical* Red color with 1-naphthol at therapeutic concentrations. *3326*
Urine Negative Analytical No reaction with o-cresol at therapeutic concentrations. *3326*

Alanine Aminotransferase *Serum Increase Physiological* Reversible cholestasis. *2313*

Alkaline Phosphatase *Serum Increase Physiological* Reversible cholestasis. *2313*

Aspartate Aminotransferase
Serum Increase Physiological Reversible cholestasis. *2313*

Bile *Urine Increase Physiological* Reversible cholestasis. *2313*

Bilirubin *Serum Increase Physiological* May cause hemolysis with G-6-PD deficiency. *0248*

BSP Retention *Serum Increase Physiological* Reversible cholestasis. *2313*

Cephalin Flocculation *Serum Increase Physiological* Reversible cholestasis. *2313*

Eosinophils *Blood Increase Physiological* Observed with hypersensitivity hepatitis. *1987*

Erythrocytes *Blood Decrease Physiological* Anemia (AMA Blood dyscrasias). *2429*

Haptoglobin *Serum Decrease Physiological* May cause hemolytic anemia. *2429*

Hematocrit *Blood Decrease Physiological* Hemolytic/aplastic anemia. *2313*

Hemoglobin *Blood Decrease Physiological* Hemolytic/aplastic anemia. *2313*

Lactate Dehydrogenase *Serum Increase Physiological* May occur with preponderance of LD-5. *1987*

LE Cells *Blood Positive Physiological* Implicated as activator of SLE. *0962*

Leukocytes *Blood Decrease Physiological* Leukopenia (AMA Blood dyscrasias). *2429*

Methemoglobin *Blood Increase Physiological* May cause hemolytic anemia. *2429*

Ornithine Carbamoyltransferase (OCT)
Serum Increase Physiological Reversible cholestasis may occur. *1986*

Platelets *Blood Decrease Physiological* Thrombocytopenia may occur after days to weeks. *2429*

Thymol Turbidity *Serum Increase Physiological* Reversible cholestasis. *2313*

Urobilinogen *Urine Decrease Physiological* Occurs with reversible cholestasis. *2313*
Feces Decrease Physiological May cause reversible cholestasis. *2313*

Sulfanilamide

Alanine Aminotransferase *Serum Increase Physiological* May cause reversible cholestasis. *2220*

Albumin *Serum No Effect Analytical* At concentration of 1,000 mg/L had no effect on BCG method. *3393*

Alkaline Phosphatase *Serum Increase Physiological* May cause reversible cholestasis. *2220*

Alkaloids *Urine No Effect Analytical* With iodoplatinate of Frings TLC procedure at 100 mg/dL. With ninhydrin on Frings TLC procedure at 100 mg/dL. *1204*

Aspartate Aminotransferase
Serum Increase Physiological May cause reversible cholestasis. *2220*

Barbiturate *Urine No Effect Analytical* With diphenylcarbazone on Frings TLC procedure at 100 mg/dL. With mercuric SO_4 on Frings TLC procedure at 100 mg/dL. *1204*
Urine Increase Analytical Yellow spot with vanillin on Frings TLC procedure With mercurous NO_3 reagent on Frings TLC procedure. *1204*

Bicarbonate *Serum No Effect Analytical* At concentration of 1,000 mg/L had no effect on method using phenolphthalein. *3393*

Bilirubin *Serum No Effect Analytical* At concentration of 103 mg/L had no effect on Ektachem® method. *3393* At concentration of 1,000 mg/L had no effect on Jendrassik and Grof method. *3393*
Serum Increase Physiological May cause hemolysis with G-6-PD deficiency. *0248*

Bromide *Urine No Effect Analytical* No effect on method of Frings. *1204*

Calcium *Serum No Effect Analytical* At concentration of 1,000 mg/L had no effect on cresolphthalein method. *3393*

Chloride *Serum No Effect Analytical* At concentration of 1,000 mg/L had no effect on mercurimetric method. *3393*

Cholesterol *Serum No Effect Analytical* At concentration of 1,000 mg/L had no effect on Liebermann-Burchard method. *3393*

Creatinine *Serum No Effect Analytical* At concentration of 1,000 mg/L had no effect on AutoAnalyzer Jaffé method. *3393*

Dapsone *Serum Increase Analytical* Develops color in procedure of Higgins. *1590*

Erythrocytes *Blood Decrease Physiological* May cause hemolytic anemia. *0248*

Estriol *Urine No Effect Analytical* No effect of 125 mg/L on GLC method. *1163*

Ethchlorvynol *Urine No Effect Analytical* No effect on method of Frings and Cohen at 100 mg/dL. *1205*

FPN Test *Urine No Effect Analytical* No effect at 100 mg/dL on method of Frings. *1204*

Glucose *Serum No Effect Analytical* At concentration of 300 mg/L had no effect on Seralyzer method. *3393*

Haptoglobin *Serum Decrease Physiological* May cause hemolytic anemia. *0248*

Heinz Body Formation *Blood Positive Physiological* May occur in early stages of hemolysis. *0333*

Hematocrit *Blood Decrease Physiological* May cause hemolytic anemia. *0248*

Hemoglobin *Blood Decrease Physiological* May cause hemolytic anemia. *0248*

Methemoglobin *Blood Increase Physiological* May cause hemolytic anemia. *2429*

Phosphate *Serum No Effect Analytical* At concentration of 1,000 mg/L had no effect on phosphomolybdate method. *3393*

Potassium *Serum No Effect Analytical* At concentration of 1,000 mg/L had no effect on measurement by ISE with predilution. *3393*

Sulfanilamide (continued)

Protein *Serum No Effect Analytical* At concentration of 1,000 mg/L had no effect on biuret method with blank correction. *3393*

CSF Increase Analytical Reacts as if phenol with Folin-Ciocalteu method. *3958*

Salicylate *Urine No Effect Analytical* At 100 mg/dL no effect on Trinder method. *1204*

Sodium *Serum No Effect Analytical* At concentration of 1,000 mg/L had no effect on measurement by ISE with predilution. *3393*

Sugar *Urine Increase Analytical* False positive with Benedict's. *1563*

Triglycerides *Serum No Effect Analytical* At concentration of 1,000 mg/L had no effect on lipase/esterase method. *3393*

Urea Nitrogen *Serum No Effect Analytical* At concentration of 1,000 mg/L had no effect on diacetylmonoxime method. *3393*

Uric Acid *Serum No Effect Analytical* At concentration of 1,000 mg/L had no effect on phosphotungstate reduction method. *3393*

Serum Increase Analytical At concentrations above 250 mg/L raised concentration as measured by Seralyzer method. *3393*

Urobilinogen *Urine Increase Analytical* Gives greenish-yellow color with Ehrlich's. *0251*

Sulfaphenazole

Acetaminophen Screening Test *Urine Positive Analytical* Red color with 1-naphthol at therapeutic concentrations. *3326*
Urine Negative Analytical No reaction with o-cresol at therapeutic concentrations. *3326*

Diphenylhydantoin *Serum Increase Physiological* Reported impairment of metabolism. *2042*

Glucose *Serum Decrease Physiological* Reported effect. *2313*

Tolbutamide *Serum Increase Physiological* Displaces from protein, inhibits carboxylation. *1487*

Sulfapyridine

Alanine Aminotransferase *Serum Increase Physiological* May cause reversible cholestasis. *2220*

Alkaline Phosphatase *Serum Increase Physiological* May cause reversible cholestasis. *2220*

Aspartate Aminotransferase
Serum Increase Physiological May cause reversible cholestasis. *2220*

Bilirubin *Serum Increase Physiological* May cause hemolysis with G-6-PD deficiency. *0248* Displacement from protein in neonates. *3748*

Chlordiazepoxide *Blood Increase Analytical* 5 mg/dL equivalent to 0.9 mg/dL by method of Riddick. *2983*

Dapsone *Serum No Effect Analytical* At 5 mg/dL on colorimetric procedure of Higgins. *1590*

Erythrocytes *Blood Decrease Physiological* May cause hemolytic anemia. *0248*

Haptoglobin *Serum Decrease Physiological* May cause hemolytic anemia. *0248*

Hematocrit *Blood Decrease Physiological* May cause hemolytic anemia. *0248*

Hemoglobin *Blood Decrease Physiological* May cause hemolytic anemia. *0248*

Leukoagglutinins *Serum Positive Physiological* Reported observation. *0810*

Methemoglobin *Blood Increase Physiological* May cause hemolytic anemia. *2429*

Protein *Serum No Effect Analytical* At concentration of 80 mg/L had no effect on biuret method with blank correction. *3393*

Urobilinogen *Urine Increase Analytical* Yields greenish color with Ehrlichs. *0459*

Sulfarthrol

Platelets *Blood Increase Physiological* Mechanism of action unknown. *1092*

Sulfasalazine

Alanine Aminotransferase *Serum Increase Physiological* Occasional case of drug-induced toxic hepatitis. *1422* Rare hepatotoxicity reported associated with noncaseating granuloma or focal inflammation with or without necrosis. *1127*

Alkaline Phosphatase *Serum Increase Physiological* Occasional case of drug-induced toxic hepatitis. *1422* Rare hepatotoxicity reported associated with noncaseating granuloma or focal inflammation with or without necrosis. *1127* In 2 cases with drug induced hepatotoxicity of hypersensitivity type of reaction: reversed on drug withdrawal. *3364* As part of systemic hypersensitivity reaction in woman with inflammatory bowel disease. *3402*

Aspartate Aminotransferase
Serum Increase Physiological Occasional case of drug-induced toxic hepatitis. *1422* As part of systemic hypersensitivity reaction in woman with inflammatory bowel disease. *3402* Rare hepatotoxicity reported associated with noncaseating granuloma or focal inflammation with or without necrosis. *1127* Hepatotoxic hypersensitivity reaction like sulfonamide hypersensitivity. *2214* In 2 cases with drug induced hepatotoxicity of hypersensitivity type of reaction: reversed on drug withdrawal. *3364*

Bilirubin *Serum No Effect Physiological* Characteristically associated with rare cases of hepatotoxicity. *1127*
Serum Increase Physiological Theoretical displacement from protein in neonates although not observed in practice. *3748* Occasional case of drug-induced toxic hepatitis. *1422* As part of systemic hypersensitivity reaction in woman with inflammatory bowel disease. *3402* Hepatotoxic hypersensitivity reaction like sulfonamide hypersensitivity. *2214* In 2 cases with drug induced hepatotoxicity of hypersensitivity type of reaction: reversed on drug withdrawal. *3364* Occasional case of drug-induced toxic hepatitis. *1422*

Bilirubin, Direct *Serum Increase Physiological* Occasional case of drug-induced toxic hepatitis. *1422*

Creatinine *Serum Increase Analytical* At concentrations above 500 mg/L (normal therapeutic concentration 70 mg/L) raised concentration as measured by kinetic Jaffé method on BKA-2. *3393*

Digoxin *Serum Decrease Physiological* Reduces bioavailability in gastrointestinal tract. *2794*

Eosinophils *Blood Increase Physiological* As part of systemic hypersensitivity reaction in woman with inflammatory bowel disease. *3402* Characteristically associated with rare cases of hepatotoxicity. *1127* Hepatotoxic hypersensitivity reaction like sulfonamide hypersensitivity. *2214* Observed in 11% of 18 patients treated with drug who had rheumatoid arthritis. *3360*

Erythrocytes *Blood Decrease Physiological* Pancytopenia reported plus reductions of individual series of cells. *2887*

Folate *Serum No Effect Physiological* No significant difference from controls in 45 outpatients taking drug orally for ulcerative colitis. *2208*
Serum Decrease Physiological Folate absorption from gastrointestinal tract may be inhibited to a minor degree: may lead to megaloblastic anemia if severe nutritional deficiency or celiac disease. *2062* Due to malabsorption with prolonged use. *3023*
Red Blood Cells Decrease Physiological Folate absorption from gastrointestinal tract may be inhibited to a minor degree: may lead to megaloblastic anemia if severe nutritional deficiency or celiac disease. *2062* Due to malabsorption with prolonged use. *3023* Inversely correlated with drug dosage in 45 outpatients taking drug orally for ulcerative colitis. *2208*

γ-Glutamyltransferase (GGT)
Serum Increase Physiological Occasional case of drug-induced toxic hepatitis. *1422* Rare hepatotoxicity reported associated with noncaseating granuloma or focal inflammation with or without necrosis. *1127*

Hemoglobin *Blood Decrease Physiological* Hemolysis quite common, but frank hemolytic anemia rare. *2887*

Lactate Dehydrogenase *Serum Increase Physiological* Rare hepatotoxicity reported associated with noncaseating granuloma or focal inflammation with or without necrosis. *1127* In 2 cases with drug induced hepatotoxicity of hypersensitivity type of reaction: reversed on drug withdrawal. *3364*

Leukocytes *Blood Increase Physiological* Characteristically associated with rare cases of hepatotoxicity. *1127* Hepatotoxic hypersensitivity reaction like sulfonamide hypersensitivity. *2214*
Blood Decrease Physiological Pancytopenia reported plus reductions of individual series of cells. *2887*

Neutrophils *Blood Decrease Physiological* Rare cases of neutropenia reported. *0155* Hemolysis quite common, but frank hemolytic anemia rare. *2887*

Platelets *Blood Decrease Physiological* Several cases of immune-mediated thrombocytopenia reported. *2502* Pancytopenia reported plus reductions of individual series of cells. *2887*

Potassium *Serum Decrease Analytical* At concentrations above 1,000 mg/L (normal therapeutic concentration 70 mg/L) lowered concentration as measured by ISE without predilution. *3393*

Protein *Serum No Effect Analytical* No effect when Du Pont aca turbidimetric method is used with 100 fold dilution of sperm. *2484* At concentration of 50 mg/L had no effect on biuret method with blank correction. *3393*
Serum Decrease Analytical Spuriously low values with biuret methods on Du Pont aca related to overblanking. *2484*

Protein Electrophoresis *Serum No Effect Analytical* At concentration of 1600 mg/L had no effect on automated Olympus-Hite method except for slight displacement of fractions. *3393*

Sodium *Serum No Effect Analytical* At concentration of 1600 mg/L had no effect on measurement by ISE without predilution. *3393*

Sperm Count *Semen Decrease Physiological* Often with abnormal sperm in high proportion of men: effect reversible. *0953*

Sulfates

Calcium *Serum Decrease Analytical* With atomic absorption. *0583* Theoretical inhibition of emission in flame methods. *1563*
Serum Decrease Physiological Brief hypocalcemic effect if infused i.v. *2313*
Urine Increase Physiological Due to diuresis and CaSO₄ formation. *0071*
Urine Decrease Analytical Inhibit emission in some flame methods. *1563*

Chloride *Serum No Effect Analytical* No effect at 20 mg/dL on method of Fingerhut. *1135*

Glucose-6-Phosphate Dehydrogenase
Red Blood Cells Decrease Physiological Sensitive to sulfate ion at concentration of 5 mmol/L in vitro. *0432*

Potassium *Serum Decrease Physiological* May be excreted combined with sulfate. *2313*
Urine Increase Physiological Increased excretion combines with sulfate. *0071*

Sodium *Serum Decrease Physiological* May be excreted combined with sulfate. *2313*
Urine Increase Physiological Increased excretion combines with sulfate. *0071*

Zinc *Serum No Effect Analytical* At 50 to 1 with flameless atomic absorption. *2038*
Urine No Effect Analytical At 50 to 1 with flameless atomic absorption. *2038*

Sulfathiazole

Ammonia *Plasma No Effect Analytical* At concentration of 60 mg/L had no effect on Ektachem® method. *3393*

Aspartate Aminotransferase *Serum Increase Analytical* At 1 mmol/L affects SMA 12/60 method. *3335*

Bilirubin *Serum No Effect Analytical* At concentration of 60 mg/L had no effect on Ektachem® method. *3393* At concentration of 255 mg/L had no effect on Jendrassik and Grof method. *3393*

Calcium *Serum No Effect Analytical* At concentration of 255 mg/L had no effect on cresolphthalein method. *3393*

Creatinine *Serum No Effect Analytical* At concentration of 60 mg/L had no effect on Ektachem® method. *3393*

Glucose *Serum No Effect Analytical* At concentration of 50 mg/L had no effect on Ektachem® method. *3393*

Phosphate *Serum No Effect Analytical* At concentration of 255 mg/L had no effect on phosphomolybdate method. *3393*

Protein *Serum No Effect Analytical* At concentration of 255 mg/L had no effect on biuret method with blank correction. *3393*

Sugar *Urine Increase Analytical* Yellow-orange with Benedict's. *0583*

Urea Nitrogen *Serum No Effect Analytical* At concentration of 255 mg/L had no effect on diacetylmonoxime method. *3393* At concentration of 50 mg/L had no effect on Ektachem® method. *3393*

Urobilinogen *Urine Increase Analytical* Yields greenish color with Ehrlichs. *0459*

Sulfhydryl Compounds

Alkaline Phosphatase *Serum Decrease Analytical* Inhibit enzyme activity in laboratory methods. *3588*

Sulfinpyrazone

Bilirubin *Serum Increase Physiological* Jaundice reported in cases with overdose. *1680*

Erythrocytes *Blood Decrease Physiological* Rare blood dyscrasia. *0071*

PAH Clearance *Urine Decrease Physiological* Inhibits tubular transport. *0071*

Phenprocoumon *Plasma No Effect Physiological* No effect on plasma concentration. *2642*

Platelet Aggregation *Blood Decrease Physiological* Inhibits collagen induced aggregation. *3068*

Prothrombin Time *Plasma Increase Physiological* May prolong action of anticoagulants. *1679* When free warfarin concentration increased due to displacement from protein. *2642*

PSP Excretion *Urine Decrease Physiological* Competition for excretion. *0974*

Sulfadiazine *Serum Increase Physiological* Displaces from protein binding. *1487*

Sulfisoxazole *Serum Increase Physiological* Displaces from protein binding. *1487*

Uric Acid *Serum Decrease Physiological* Uricosuric action. *1433*
Urine Increase Physiological Uricosuric action. *1433*

Warfarin *Plasma Increase Physiological* Increased concentration due to displacement from protein. *2642*

Sulfisoxazole

Acetaminophen Screening Test *Urine Positive Analytical* Red color with 1-naphthol at therapeutic concentrations. *3326*
Urine Negative Analytical No reaction with o-cresol at therapeutic concentrations. *3326*

Alanine Aminotransferase *Serum Increase Physiological* May cause cholestasis. *1974*

Albumin *Serum No Effect Analytical* No significant effect on BCG method at 500 mg/L No significant effect on HABA method at 500 mg/L. *2625*

Alkaline Phosphatase *Serum Increase Physiological* Reversible cholestasis. *2313*

Amylase *Serum Increase Physiological* May cause pancreatitis occasionally. *1680*

Aspartate Aminotransferase
Serum Increase Physiological Reversible cholestasis. *2313*

Bile *Urine Increase Physiological* May cause reversible cholestasis. *2313*

Sulfisoxazole *(continued)*

Bilirubin *Serum No Effect Analytical* At concentration of 60 mg/L had no effect on Ektachem® method. *3393*
Serum Increase Physiological May cause cholestatic jaundice or hemolysis. *1974* Displacement from protein in neonates. *3748*
Serum Decrease Physiological In newborn combined with penicillin has effect. *3505*

BSP Retention *Serum Increase Physiological* May cause cholestasis. *1974*

Calcium *Serum Decrease Analytical* Depresses fluorescence of calcein method. *2395*

Cephalin Flocculation *Serum Increase Physiological* May cause cholestasis. *1974*

Chlordiazepoxide *Blood Increase Analytical* 5 mg/dL equivalent to 1.0 mg/dL by method of Riddick. *2983*

Crystals *Urine Increase Physiological* May occur particularly in acidic urine. *3832*

Diagnex Blue Excretion *Urine Increase Physiological* Displacement of diagnex blue from resin. *3879*

Diphenylhydantoin *Serum Increase Physiological* Displaces from protein *in vitro. 1487*

Eosinophils *Blood Increase Physiological* Allergic response. *1974*

Erythrocytes *Blood Decrease Physiological* Agranulocytosis/aplastic anemia. *2868* As result of bone marrow depression or hemolysis (in patients with glucose-6-phosphate dehydrogenase deficiency. *3832*

Folate *Serum Decrease Analytical* Affects standard autoclave method. *1815*

Glucose *Serum No Effect Analytical* No effect on Boehringer GOD-PERID glucose oxidase method. *3277*

Haptoglobin *Serum Decrease Physiological* May cause hemolytic anemia. *1974*

Hematocrit *Blood Decrease Physiological* Agranulocytosis/aplastic anemia. *2868*

Hemoglobin *Blood Decrease Physiological* Agranulocytosis/aplastic anemia. *2868*

Lactate Dehydrogenase *Serum Increase Physiological* May cause hemolytic anemia. *1974*

LE Cells *Blood Positive Physiological* May cause LE phenomenon (rare). *1680*

Leukocytes *Blood Decrease Physiological* Agranulocytosis/aplastic anemia. *2868*

Lipase *Serum Increase Physiological* Rare effect on pancreas, salivary glands. *2427*

Methemoglobin *Blood Increase Physiological* May cause hemolytic anemia. *3015*

Methotrexate *Serum Increase Physiological* Displaces from plasma protein binding. *1487*

Nitrofurantoin *Urine No Effect Analytical* No effect with 50 μg method of Conklin/Hollifield. *0704*

Platelets *Blood Decrease Physiological* Agranulocytosis/aplastic anemia (after days). *2868* Isolated case of platelet-associated IgG and thrombocytopenia. *1910*

Potassium *Serum No Effect Analytical* At concentration of 80,000 mg/L had no effect on measurement by ISE with predilution. *3393*

Protein *Urine No Effect Analytical* At 100 mg/dL on TCA dye method of Pesce. *2791* No effect on Albustix. *0459*
Urine Increase Analytical Causes turbidity if sulfosalicylic acid used. *3505* False positive with Exton's reagent No effect with acetic acid and heat. *0443* May produce turbidity with acid methods. *0459*
CSF No Effect Analytical At 100 mg/dL on TCA dye method of Pesce. *2791*
CSF Increase Analytical Causes turbidity if sulfosalicylic acid used. *3505*

Prothrombin Time *Plasma Increase Physiological* Occurs with failure of excretion of bile salts. *0894*

Sodium *Serum No Effect Analytical* At concentration of 80,000 mg/L had no effect on measurement by ISE with predilution. *3393*

Thymol Turbidity *Serum Increase Physiological* Reversible cholestasis. *2313*

Tolbutamide *Serum Increase Physiological* Displaces from protein. *1487*

Urea Nitrogen *Serum Increase Analytical* Effect marked DMAB method of Morin. *2487*

Urobilinogen *Urine Increase Analytical* Reacts with Urobilistix®. *1487*
Urine Decrease Physiological With reversible cholestasis. *1974*
Feces Decrease Physiological Pale stools with reversible cholestasis. *1974*

Sulfobromophthalein

Acetone *Urine Increase Analytical* Development of color at alkaline pH (Rothera test). *3879*

Alkaline Phosphatase *Serum Increase Analytical* If final color developed in alkaline solution. *0583* Interferes with Bessey procedure. *0607*

Calcium *Serum Increase Analytical* Affects methylthymol method of Gindler/King. *1286* May interfere with colorimetric procedures. *2425*

Color *Urine Increase Analytical* Red in alkaline urine. *3362*
Feces Increase Analytical Alkaline stools may appear bloody. *0583*

Creatine *Serum Increase Analytical* Presence of interfering color. *2313*

Creatinine *Serum Increase Analytical* Presence of interfering color. *2313*

¹³¹I Uptake *Serum No Effect Physiological* No effect on uptake reported. *2220*

Ketones *Urine Increase Analytical* Produces purple color with alkaline nitroprusside. *0583* False positive color react with BiliLabstix® in acid urine. *1487*

Occult Blood *Feces No Effect Analytical* No effect on tests for occult blood. *0583*

Protein *Serum Increase Analytical* Color augmentation with alkaline biuret. *0579*

Protein Bound Iodine (PBI) *Serum Increase Analytical* Organic iodide contamination of some BSP solutions. *2815*

PSP Excretion *Urine Increase Analytical* Development of color at alkaline pH. *3879*

Urobilinogen *Urine Increase Analytical* Color development with Ehrlich's aldehyde reagent. *3879*
Urine Increase Physiological Presence of BSP in urine and liver disease. *1022*

Vanillylmandelic Acid *Urine Increase Analytical* Occurs unless completely extracted. *0086*

Sulfomethane

Aminolevulinic Acid *Urine Increase Physiological* May provoke attack of porphyria. *1322*

Color *Urine Increase Physiological* Red-brown (may provoke porphyria). *2313*

Coproporphyrin *Blood Increase Physiological* May provoke attack of porphyria. *1322*
Urine Increase Physiological May provoke attack of porphyria. *1322*
Feces Increase Physiological May provoke attack of porphyria. *1322*

Methemoglobin *Blood Increase Physiological* May cause hemolytic anemia. *2429*

Porphobilinogen *Urine Increase Physiological* May provoke attack of porphyria. *1322*

Porphyrins *Urine Increase Physiological* Stimulates formation of ALA-synthetase. *1343*

Protoporphyrin *Blood Increase Physiological* May provoke attack of porphyria. *1322*
Feces Increase Physiological May provoke attack of porphyria. *1322*

Uroporphyrin *Urine Increase Physiological* May provoke attack of porphyria. *1322*

Sulfonamides

Alanine Aminotransferase *Serum Increase Physiological* May cause cholestasis. *2313* Hepatitis-like reaction in patients with multiple episodes of jaundice after repeated exposure to drug. *1741*

Albumin *Serum Decrease Analytical* Competes with HABA for binding sites. *0583*

Alkaline Phosphatase *Serum Increase Physiological* Hypersensitivity reaction with cholestasis. *1199* Hepatitis-like reaction in patients with multiple episodes of jaundice after repeated exposure to drug. *1741*
Urine Increase Physiological Due to nephrotoxic effect of drug. *2897*

Amino Acids *Serum Increase Analytical* Reported observation. *0355*

Aminolevulinic Acid *Urine Increase Physiological* May precipitate acute porphyria. *1322*

α-Amino-Nitrogen *Plasma Increase Analytical* Reacts with naphthoquinone (method of Frame). *1563*
Urine Increase Analytical Reacts with naphthoquinone (method of Frame). *1563*

Appearance *Urine Abnormal Physiological* Turbid especially, if given with methenamine-due to urates. *1680*

Aspartate Aminotransferase
Serum Increase Physiological May cause cholestasis. *2313*

Barbiturate *Serum Increase Analytical* May interfere with UV absorption method. *2981*

Bile *Urine Increase Physiological* May cause cholestasis. *2313*

Bilirubin *Serum No Effect Analytical* spectropolarographic procedure of Grahnen. *1374*
Serum Increase Physiological Hepatic/cholestatic effect, also affects albumin binding. *0348* Hepatitis-like reaction in patients with multiple episodes of jaundice after repeated exposure to drug. *1741*
Serum Decrease Physiological Displacement from albumin binding sites. *3505*

BSP Retention *Serum Increase Physiological* May cause cholestasis. *2808*

Calcium *Serum Decrease Analytical* Depress fluorescence of calcein method. *2395*

Carbonic Anhydrase
Red Blood Cells Decrease Physiological Inhibitory action observed. *2893*

Casts *Urine Increase Physiological* Hemoglobin casts may cause renal failure. *1343*

Cephalin Flocculation *Serum Increase Physiological* May cause cholestasis. *2808*

Cholesterol *Serum Increase Physiological* Cholestatic effect. *1713*

Color *Urine Increase Analytical* Brown color with some sulfonamides. *1022*

Coombs' Test *Serum Positive Physiological* Immunological response to drug. *1486*

Coproporphyrin *Urine Increase Physiological* May precipitate acute porphyria. *1322*
Feces Increase Physiological May precipitate acute porphyria. *1322*

Crystals *Urine Increase Physiological* Presence of drug. *2313*

Eosinophils *Blood Increase Physiological* Hypersensitivity response. *1513* Associated with hepatitis like attack in patient with prior episodes of jaundice. *1741*

Erythrocytes *Blood Decrease Physiological* May cause Heinz-body hemolytic anemia. *2313* Several cases of aplastic anemia reported. *3717*
Urine Increase Physiological May cause actual bleeding. *1022*

Glucose *Serum Decrease Physiological* Reported effect. *0117*

Haptoglobin *Serum Decrease Physiological* May cause hemolytic anemia if G-6-PD deficiency. *0071*

Heinz Body Formation *Blood Positive Physiological* Occurs with marked hemolytic anemia. *1902*

Hematocrit *Blood Decrease Physiological* May cause Heinz-body hemolytic anemia. *2808*

Hemoglobin *Plasma Increase Physiological* May occur with intravascular hemolysis. *2426*
Blood Decrease Physiological Agranulocytosis/hemolytic anemia/sen. dependent (G-6-PD). *2427*
Urine Increase Physiological Actual bleeding caused by drug (due to hemolysis). *1022*

^{131}I Uptake *Serum Decrease Physiological* Effect lasts about 7 d. *1444*

LE Cells *Blood Positive Physiological* May produce LE-like syndrome. *3505*

Leucine Aminopeptidase *Urine Increase Physiological* Facilitate permeation of enzyme into tubules. *2897*

Leukocytes *Blood Increase Physiological* Associated with hemolysis (especially with long-acting drugs). *2313*
Blood Decrease Physiological Agranulocytosis/aplastic anemia. *2427* Several cases of aplastic anemia reported. *3717*

Methemoglobin *Blood Increase Physiological* Agranulocytosis/aplastic or hemolytic anemia. *2427*

Neutrophils *Blood Decrease Physiological* Several cases of drug induced neutropenia in patients taking only one drug. *0156*

Nonprotein Nitrogen *Serum Increase Physiological* Occasional uremia with/without crystalluria. *2313*

5'-Nucleotidase *Serum Increase Physiological* May cause cholestasis. *1778*

Occult Blood *Feces Positive Physiological* Hematemesis and melena may occur. *2427*

PAH Clearance *Urine Increase Analytical* React as if PAH with Bratton-Marshall method. *2981*

Platelets *Blood Decrease Physiological* Agranulocytosis/aplastic anemia. *2427* Several cases of aplastic anemia reported. *3717* Several cases of immune-mediated thrombocytopenia reported. *2502*

Porphobilinogen *Urine Increase Physiological* May precipitate acute porphyria. *1322*

Porphyrins *Urine Increase Physiological* May precipitate acute porphyria. *2220*

Protein *Urine Increase Analytical* Reacts as phenol with Folin-Ciocalteu reagent. *3505* May cause false positive with sulfosalicylic acid. *1488*
Urine Increase Physiological May cause nephrotoxicity. *1488* May cause nephrotoxicity. *2313*
CSF Increase Analytical Reacts as phenol with Folin-Ciocalteu reagent. *3505*

Protein Bound Iodine (PBI) *Serum Decrease Physiological* Inhibits iodination of tyrosine in thyroxine binding globulin. *0012*

Prothrombin Time *Plasma Increase Physiological* Potentiate action of administered coumarins. *1973*

Protoporphyrin *Feces Increase Physiological* May precipitate acute porphyria. *1322*

PSP Excretion *Urine Decrease Physiological* Reduces renal excretion of PSP. *3879*

Reduced Glutathione
Red Blood Cells Decrease Physiological Occurs early before overt hemolysis. *1902*

Reticulocytes *Blood Increase Physiological* In response to hemolysis occurs with recovery. *1343*
Blood Decrease Physiological Agranulocytosis with aplastic or hemolytic anemia. *1788*

Sugar *Urine Increase Analytical* Affects Benedict's and Clinitest®. *1488*
Urine Decrease Analytical Forms colorless Cu complex with Benedict's. *0583*

Sulfa as Sulfanilamide *Serum Increase Analytical* Reacts in Bratton-Marshall reaction. *0279*

T₃ Uptake *Serum Increase Physiological* Some compete for binding sites. *0583*

Thymol Turbidity *Serum Increase Physiological* May cause cholestasis. *2808*

Thyroxine (T₄) *Serum Decrease Physiological* Acts like thiourea on thyroid gland. *0354*

Urea Nitrogen *Serum Increase Analytical* Reacts like urea with DMAB and Berthelot reactions. *3505*

Sulfonamides *(continued)*

Urea Nitrogen *(continued)*
Serum Increase Physiological Rare uremia may occur with/without crystalluria. *1343*

Uric Acid *Serum Decrease Physiological* Presumed uricosuric effect. *3920*
Urine Increase Physiological Uricosuric effect. *1237*

Urobilinogen *Urine Increase Analytical* React with Ehrlich's reagent. *1563*
Urine Increase Physiological Hemolytic anemia with G-6-PD deficiency. *1343*
Urine Decrease Physiological Cholestatic effect. *1713*
Feces Decrease Physiological Pale stools with reversible cholestasis. *1713*

Sulfones

Alanine Aminotransferase *Serum Increase Physiological* May affect liver function. *2313*

Alkaline Phosphatase *Serum Increase Physiological* May affect liver function. *2313*

Aspartate Aminotransferase
Serum Increase Physiological May affect liver function. *2313*

Bile *Urine Increase Physiological* May affect liver function. *2313*

Bilirubin *Serum Increase Physiological* May cause hemolysis with G-6-PD deficiency. *0248*

BSP Retention *Serum Increase Physiological* May affect liver function. *2313*

Cephalin Flocculation *Serum Increase Physiological* May affect liver function. *2313*

Erythrocytes *Blood Decrease Physiological* Hemolysis may occur with G-6-PD deficiency. *2313*
Urine Increase Physiological May have nephrotoxic effect. *2313*

Haptoglobin *Serum Decrease Physiological* May cause hemolytic anemia. *0248*

Heinz Body Formation *Blood Positive Physiological* Reported effect in first few days. *1343*

Hematocrit *Blood Decrease Physiological* May cause hemolysis. *0071*

Hemoglobin *Plasma Increase Physiological* May occur with hemolysis. *2426*
Blood Decrease Physiological May cause hemolysis. *0071*
Urine Increase Physiological May have nephrotoxic effect. *1022*

Leukocytes *Blood Decrease Physiological* Agranulocytosis/leukopenia. *0071*

Methemoglobin *Blood Increase Physiological* Hemolysis may occur with G-6-PD deficiency. *0071*

Protein *Urine Increase Physiological* May have nephrotoxic effect. *1022*

Reduced Glutathione
Red Blood Cells Decrease Physiological Falls sharply preceding hemolysis. *1902*

Reticulocytes *Blood Increase Physiological* Occurs during recovery phase. *1902*

Thymol Turbidity *Serum Increase Physiological* May affect liver function. *2313*

Sulfonylureas

Acetaldehyde *Blood Increase Physiological* Mechanism not yet elucidated. *1343*

Alanine Aminotransferase *Serum Increase Physiological* Moderate elevation with cholangiolitis. *3171*

Alkaline Phosphatase *Serum Increase Physiological* Elevation with cholangiolitis. *3171*

Aspartate Aminotransferase
Serum Increase Physiological Moderate elevation with cholangiolitis. *3171*

Bilirubin *Serum Increase Physiological* In some cases probably attributable to drug. *2427*

Cholesterol *Serum No Effect Physiological* Usual effect reported. *0301*

Coombs' Test *Serum Positive Physiological* May cause immune hemolytic anemia. *0359*

Eosinophils *Blood Increase Physiological* May occur in association with cholangiolitis. *3171*

Erythrocytes *Blood Decrease Physiological* Pancytopenia. *2313* Occasional case of aplastic anemia reported. *3717*

Glucose *Serum Decrease Physiological* May cause severe hypoglycemia. *2427*

^{131}I Uptake *Serum Decrease Physiological* Due to impaired synthesis of thyroxine. *1915*

Leukocytes *Blood Decrease Physiological* Pancytopenia/agranulocytosis. *2313* Occasional case of aplastic anemia reported. *3717*

Platelets *Blood Decrease Physiological* Pancytopenia. *2313* Occasional case of aplastic anemia reported. *3717*

Protein Bound Iodine (PBI) *Serum No Effect Physiological* Probably no effect in most patients. *1487*
Serum Decrease Physiological Due to impaired synthesis of thyroxine. *1915*

Sodium *Serum Decrease Physiological* Potentiates vasopressin, causes water retention. *1134*

T_3 Uptake *Serum No Effect Physiological* Probably no effect in most patients. *1487*
Serum Increase Physiological Resin uptake increase due to impaired synthesis of T_4. *1915*

Thyroxine (T_4) *Serum Decrease Physiological* Due to impaired synthesis. *1915* Due to competition for transport proteins. *3798*

Triglycerides *Serum Decrease Physiological* Effect reported for some compounds. *0301*

Urea Nitrogen *Serum Increase Analytical* Theoretical effect of ureide group diacetyl method. *0583*

Sulforidazine

Bilirubin *Serum No Effect Analytical* At concentration of 200 mg/L had no effect on Jendrassik and Grof method. *3393*

Calcium *Serum No Effect Analytical* At concentration of 200 mg/L had no effect on cresolphthalein method. *3393*

Cholesterol *Serum No Effect Analytical* At concentration of 200 mg/L had no effect on Liebermann-Burchard method. *3393*

Creatinine *Serum No Effect Analytical* At concentration of 200 mg/L had no effect on AutoAnalyzer Jaffé method. *3393*

Iron *Serum No Effect Analytical* At concentration of 200 mg/L had no effect on Ferrozine method. *3393*

Phosphate *Serum No Effect Analytical* At concentration of 200 mg/L had no effect on phosphomolybdate method. *3393*

Protein *Serum No Effect Analytical* At concentration of 200 mg/L had no effect on biuret method with blank correction. *3393*

Triglycerides *Serum No Effect Analytical* At concentration of 200 mg/L had no effect on lipase/esterase method. *3393*

Urea Nitrogen *Serum No Effect Analytical* At concentration of 200 mg/L had no effect on diacetylmonoxime method. *3393*

Uric Acid *Serum No Effect Analytical* At concentration of 200 mg/L had no effect on phosphotungstate reduction method. *3393*

Sulfoxone

Bilirubin *Serum Increase Physiological* May cause hemolysis in G-6-PD deficiency. *0248*

Erythrocytes *Blood Decrease Physiological* May cause hemolytic anemia. *0248*

Haptoglobin *Serum Decrease Physiological* May cause hemolytic anemia. *0248*

Hematocrit *Blood Decrease Physiological* May cause hemolytic anemia. *0248*

Hemoglobin *Blood Decrease Physiological* May cause hemolytic anemia. *0248*

Methemoglobin *Blood Increase Physiological* May cause hemolysis in G-6-PD deficiency. *3581*

Sulindac

Alanine Aminotransferase *Serum Increase Physiological* Isolated case of cholestatic jaundice 12 d after therapy started. *2371*

Aldosterone *Plasma No Effect Physiological* In 10 furosemide treated patients with well controlled congestive heart failure given 400 mg/d for 2 d. *1042*

Alkaline Phosphatase *Serum Increase Physiological* Isolated case of cholestatic jaundice 12 d after therapy started. *2371*

Aspartate Aminotransferase
Serum Increase Physiological Isolated case of cholestatic jaundice 12 d after therapy started. *2371*

Bilirubin *Serum Increase Physiological* Isolated case of cholestatic jaundice 12 d after therapy started. *2371*

Chloride *Urine No Effect Physiological* In 10 furosemide treated patients with well controlled congestive heart failure given 400 mg/d for 2 d. *1042*

Creatinine *Serum No Effect Physiological* No significant change in patients with chronic glomerular disease. *0660*

Creatinine Clearance *Urine No Effect Physiological* When given for 14 d to patients with rheumatoid arthritis and heart failure. Marked effect. *3514* No significant change in patients with chronic glomerular disease. *0660*
Urine Decrease Physiological Small but significant effect with treatment for 9 d in 9 patients with stable renal insufficiency. *2457*

Effective Renal Plasma Flow
Patient No Effect Physiological In 9 individuals with stable renal insufficiency when treated for 9 d. *2457*

γ-Glutamyltransferase (GGT)
Serum Increase Physiological Isolated case of cholestatic jaundice 12 d after therapy started. *2371*

Lactate Dehydrogenase *Serum Increase Physiological* Observed with other laboratory evidence of hepatitis in 1 child. *1889*

PAH Clearance *Urine No Effect Physiological* No significant change in patients with chronic glomerular disease. *0660*

Platelets *Blood Decrease Physiological* After 3 d of 400 mg/d in one patient reversible but mechanism not understood. No other effects on bone marrow noted. *1867*

Prostaglandin, 6-Keto-$F_1\alpha$ *Urine No Effect Physiological* No significant change in patients with chronic glomerular disease. *0660*

Prostaglandin E2 *Urine No Effect Physiological* No significant change in patients with chronic glomerular disease. *0660*
Urine Decrease Physiological When given for 14 d to patients with rheumatoid arthritis and heart failure. Marked effect. *3514* In 11 normal volunteers contrary to concept of sparing renal but inhibiting systemic prostaglandins. *0452*

Prostaglandin $F_1\alpha$ *Urine No Effect Physiological* In 10 furosemide treated patients with well controlled congestive heart failure. *1042*

Prostaglandin F2α *Urine Decrease Physiological* When given for 14 d to patients with rheumatoid arthritis and heart failure. Marked effect. *3514*

Renin Activity *Plasma No Effect Physiological* In 9 individuals with stable renal insufficiency when treated for 9 d. *2457* No significant change in patients with chronic glomerular disease. *0660* In 10 furosemide treated patients with well controlled congestive heart failure given 400 mg/d for 2 d. *1042*
Plasma Decrease Physiological When given for 14 d to patients with rheumatoid arthritis and heart failure. Marked effect. *3514*

Sodium *Urine No Effect Physiological* In 9 individuals with stable renal insufficiency when treated for 9 d. *2457* In 10 furosemide treated patients with well controlled congestive heart failure given 400 mg/d for 2 d. *1042*

Thromboxane B_2 *Serum Decrease Physiological* 85% reduction in patients with chronic glomerular disease. *0660*
Urine Decrease Physiological In 11 normal volunteers contrary to concept of sparing renal but inhibiting systemic prostaglandins. *0790*

Volume *Urine No Effect Physiological* In 9 individuals with stable renal insufficiency when treated for 9 d. *2457* In 10 furosemide treated patients with well controlled congestive heart failure given 400 mg/d for 2 d. *1042*

Sulpiride

Cholesterol *Serum Decrease Analytical* At 15 mg/dL conventional methods when added to serum. *1758*

Estradiol *Plasma No Effect Physiological* In 11 healthy women between 6 and 9 weeks of pregnancy given 150 mg daily for 2 weeks. *3916*

Glucose *Serum Decrease Analytical* At 15 mg/dL conventional methods when added to serum. *1758*

Metanephrine *Urine No Effect Analytical* At 2 mg/L on HPLC method. *0342*

Normetanephrine *Urine No Effect Analytical* At 2 mg/L on HPLC method. *0342*

Placental Lactogen *Plasma No Effect Physiological* In 11 healthy women between 6 and 9 weeks of pregnancy given 150 mg daily for 2 weeks. *3916*

Progesterone *Plasma No Effect Physiological* In 11 healthy women between 6 and 9 weeks of pregnancy given 150 mg daily for 2 weeks. *3916*

Prolactin *Plasma Increase Physiological* General effect observed. *1736* After 1 and 2 weeks in 11 healthy women between 6 and 9 weeks of pregnancy given 150 mg daily for 2 weeks. *3916*

Testosterone *Serum No Effect Physiological* General effect observed. *1736* In 11 healthy women between 6 and 9 weeks of pregnancy given 150 mg daily for 2 weeks. *3916*

Sulpiridine

Thyroid Stimulating Hormone (TSH)
Serum Increase Physiological As a dopamine blocking substance increases basal and TRH-mediated values. *3798*

Sulthiame

Acetaminophen Screening Test *Urine Negative Analytical* No reaction with o-cresol at therapeutic concentrations. *3326*

Carbonic Anhydrase
Red Blood Cells Decrease Physiological Metabolic action of drug. *1389*

Crystals *Urine Increase Physiological* In poisoning identified as pure drug. *2427*

Diphenylhydantoin *Serum No Effect Analytical* On GLC procedure of Papadopoulos at *in vivo* concentration. *2722*
Serum Increase Physiological Increase from 10 to 20 μg/mL (at dose of 200-800 mg/d). *2041* Probably acts on liver enzymes. *2979*

Leukocytes *Blood Decrease Physiological* leukopenia/also increases half life of diphenylhydantoin. *1343*

Occult Blood *Feces Positive Physiological* One case reported with poisoning. *2427*

Pheneturide *Serum No Effect Analytical* On GLC procedure of Papadopoulos at *in vivo* concentration. *2722*

Phenobarbital *Serum No Effect Analytical* On GLC procedure of Papadopoulos at *in vivo* concentration. *2722*

Primidone *Serum No Effect Analytical* On GLC procedure of Papadopoulos at *in vivo* concentration. *2722*

Sultopride

Cortisol *Plasma No Effect Physiological* When given 300-600 mg/d for 5 weeks in 5 schizophrenic women. *2465*

Estradiol *Plasma No Effect Physiological* When given 300-600 mg/d for 5 weeks in 5 schizophrenic women. *2465*

Follicle Stimulating Hormone (FSH)
Plasma No Effect Physiological When given 300-600 mg/d for 5 weeks in 5 schizophrenic women. *2465*

Growth Hormone *Plasma Decrease Physiological* After 1 week, but normal after 3-5 weeks. *2465*

Sultopride *(continued)*

Insulin *Plasma No Effect Physiological* When given 300-600 mg/d for 5 weeks in 5 schizophrenic women. *2465*

Luteinizing Hormone (LH) *Plasma No Effect Physiological* When given 300-600 mg/d for 5 weeks in 5 schizophrenic women. *2465*

Prolactin *Plasma Increase Physiological* After 1 d, maximum at 1 week increase throughout treatment, probably by blocking pituitary dopamine receptors. *2465*

Thyroid Stimulating Hormone (TSH)
Serum No Effect Physiological When given 300-600 mg/d for 5 weeks in 5 schizophrenic women. *2465*

Suntan Oil

Protein Bound Iodine (PBI) *Serum Increase Analytical* Many preparations contain iodine. *2313*

Suprofen

Creatinine *Serum Increase Physiological* Occasional renal failure observed but mechanism not known. *0009*

Erythrocytes *Urine Increase Physiological* Occasional renal failure observed but mechanism not known. *0009*

Hemoglobin *Urine Increase Physiological* Occasional renal failure observed but mechanism not known. *0009*

Protein *Urine Increase Physiological* Occasional renal failure observed but mechanism not known. *0009*

Urea Nitrogen *Serum Increase Physiological* Occasional renal failure observed but mechanism not known. *0009*

Suramin

Casts *Urine Increase Physiological* Cylindruria reported. *0071*

Erythrocytes *Blood Decrease Physiological* Hemolytic anemia reported. *0071*
Urine Increase Physiological May cause actual bleeding. *1022*

Hematocrit *Blood Decrease Physiological* Hemolytic anemia reported. *0071*

Hemoglobin *Blood Decrease Physiological* Hemolytic anemia reported. *0071*
Urine Increase Physiological Actual bleeding caused by drug. *1237*

Leukocytes *Blood Decrease Physiological* Agranulocytosis reported. *0071*

Protein *Urine Increase Physiological* Usual effect during treatment of acute stage. *0071* Nephrotoxic effect. *1022*

Surgery

Creatine Kinase *Serum Increase Physiological* Significant effect observed 1 d postoperatively. *2809*

Ornithine Carbamoyltransferase (OCT)
Serum No Effect Physiological No significant effect observed. *2809*

Thyroid Stimulating Hormone (TSH)
Serum No Effect Physiological No effect observed. *2700*

Sweat

Amino Acids *Serum Increase Analytical* If contaminated from analyst. *3230*

Sweet Potatoes

Icteric Index *Serum Increase Analytical* Due to color. *1238*

Syringes

pO$_2$ *Blood Decrease Analytical* Occurs through leakage of air in plastic syringes. *3229*

Syrosingopine

Catecholamines *Urine Increase Physiological* Release of stored norepinephrine. *2220*

HMPG (4-Hydroxy-3-Methoxy-Phenylethylene Glycol)
Urine Increase Physiological Release of stored norepinephrine. *2220*

Hydrochloric Acid *Gastric Material Increase Physiological* Excessive amounts may be released. *2220*

Platelets *Blood Decrease Physiological* Purpura due to thrombocytopenia may occur. *0019*

Vanillylmandelic Acid *Urine Increase Physiological* Release of stored norepinephrine. *2220*

Tamoxifen

α_1-Acid Glycoprotein *Serum Decrease Physiological* Significant change due to mild estrogenic effect of drug in breast cancer patients. *3070*

Alkaline Phosphatase *Serum Decrease Physiological* Observed in nonresponders with breast cancer. *1122*

Antithrombin III *Plasma Decrease Physiological* Lowered functional activity in 42% treated patients. *1033*

α_1-Antitrypsin *Serum Increase Physiological* When 10 mg given twice daily to 30 Z homozygous α^1-antitrypsin-deficient subjects for 30 d although change slight. *3808* Significant change due to mild estrogenic effect of drug in breast cancer patients. *3070*

Aspartate Aminotransferase
Serum Increase Physiological Isolated case of drug induced cholestasis in one patient. *0372*

Bilirubin *Serum Increase Physiological* Isolated case of drug induced cholestasis in one patient. *0372*

Calcium *Serum Increase Physiological* Especially in people with pre-existing hypercalcemia and with widespread skeletal metastases. *1542* In patients with previous hypercalcemia and in patients with metastatic bone disease. Probable overall incidence less than 0.1%. *2749* In 2 patients in whom calcium rose to above 18.0 mg/dL: reverted to normal once drug withdrawn. *2069*

Ceruloplasmin *Serum Increase Physiological* Significant change due to mild estrogenic effect of drug in breast cancer patients. *3070*

Cholesterol *Serum Decrease Physiological* Effect on increased concentration in breast cancer patients. *3070*

Cholesterol, Low Density Lipoprotein
Serum Decrease Physiological Observed in one elderly woman treated for breast cancer. Possibly due to reduction of activities of postheparin plasma lipoprotein lipase and hepatic triglyceride lipase. *0508* From 5.11 to 4.10 mmol/L accountable for much of total change. *3070*

Cholesterol, Very Low Density Lipoprotein
Serum Increase Physiological Up to 241 mg/dL observed in one elderly woman treated for breast cancer. Possibly due to reduction of activities of postheparin plasma lipoprotein lipase and hepatic triglyceride lipase. *0508*

Erythrocyte Sedimentation Rate
Blood Decrease Physiological Observed in nonresponders with breast cancer. *1122*

Estradiol *Plasma Increase Physiological* 2 to 8 fold increase in 6 healthy volunteers given 20 mg/d for 5 to 10 d. In postmenopausal breast cancer women who did not respond to drug. *1542*

Follicle Stimulating Hormone (FSH)
Plasma No Effect Physiological After oral administration of 20 mg/d for 5 or 10 d to 6 healthy women during follicular phase of menstrual cycle. In 10 anovulatory women given up to 40 mg/d for 5 d. *1542*
Plasma Decrease Physiological In postmenopausal women with breast cancer treated for 2 weeks. *1542*

Haptoglobin *Serum Decrease Physiological* Significant change due to mild estrogenic effect of drug in breast cancer patients. *3070*

Hemoglobin *Blood Decrease Physiological* In a small proportion of patients with associated reductions of platelet and white cell counts. *1542*

Leukocytes *Blood Decrease Physiological* In up to 20% of treated patients. *1542*

Luteinizing Hormone (LH) *Plasma No Effect Physiological* After oral administration of 20 mg/d for 5 or 10 d to 6 healthy women during follicular phase of menstrual cycle. *1542*
Plasma Increase Physiological In 10 anovulatory women given up to 40 mg/d for 5 d. *1542*
Plasma Decrease Physiological In postmenopausal women with breast cancer treated for 2 weeks. *1542*

Platelets *Blood Decrease Physiological* In up to 20% of treated patients given 400 mg/d for 2 d. *1542*

Progesterone *Plasma No Effect Physiological* After oral administration of 20 mg/d for 5 or 10 d to 6 healthy women during follicular phase of menstrual cycle. *1542*

Prolactin *Plasma Decrease Physiological* Significant effect in 6 healthy volunteers given 20 mg/d for 5 to 10 d. *1542*

Sperm Count *Semen Increase Physiological* Increased in oligozoospermic men if FSH concentration low or normal. *0953*

Thyroxine (T₄) *Serum Increase Physiological* In 10 of 50 postmenopausal patients with breast cancer. Also rose in other patients with start of treatment. Effect not observed in normal volunteers. *1350*

Thyroxine (T₄) Binding Globulin
Serum Increase Physiological Raised in all 6 patients with breast cancer studied. *1350*

Thyroxine (T₄), Free *Serum Increase Physiological* Observed in one patient with persistently high thyroxine of 50 postmenopausal patients with breast cancer. *1350*

Triglycerides *Serum Increase Physiological* Up to 2.8 g/dL observed in one elderly woman treated for breast cancer. Possibly due to reduction of activities of postheparin plasma lipoprotein lipase and hepatic triglyceride lipase. *0508* On slightly higher concentration in breast cancer patients. *3070*

Triglycerides, Low Density Lipoprotein
Serum Increase Physiological From 0.46 to 0.56 mmol/L accountable for much of total change. *3070*

Tri-iodothyronine (T₃) *Serum Increase Physiological* In 2 of 10 postmenopausal patients with breast cancer with increased thyroxine concentration. *1350*

Tartrates

Acid Phosphatase *Serum Decrease Analytical* Strong inhibitors of prostatic enzyme. *3588*
White Blood Cells Increase Analytical Phenylphosphate enzyme activated. *0183*
White Blood Cells Decrease Analytical Glycerophosphate enzyme inhibited by 1.5 mmol/L. *0183*

Phosphate *Serum Decrease Analytical* Complexes molybdate, decreased color develop. *0583*

Protein *Urine Increase Physiological* If absorbed may cause renal damage. *1343*

Urea Nitrogen *Serum Increase Physiological* If absorbed may cause renal damage. *1343*

Taurine

Tyrosine *Plasma No Effect Analytical* On fluorometric procedure of Ambrose. *0080*

Taurocholate

Lipase *Serum Increase Analytical* Sodium salts prevent inactivation of enzyme. *3589*

Tea

Caffeine *Serum Increase Physiological* Maximum (7.5 mg/dL) at 30 minutes after 3 cups. *2303*

Catecholamines *Plasma Increase Analytical* Concentrated solutions cause striking fluorescence. *0596*

3,4-Dimethoxyphenylethylamine (3,4-DMPEA)
Urine Positive Physiological (pink spot) of dietary origin only. *3425*

Vanillylmandelic Acid *Urine Increase Analytical* Metabolites interfere with analysis. *3506*

Temocillin

Prothrombin Time *Plasma No Effect Physiological* 4 g intravenously 12 hourly had no effect in 8 weeks. *2629*

Teprotide

Angiotensin-Converting Enzyme (ACE)
Serum Decrease Physiological 43% inhibition reported with method using benzyloxycarbonyl-phenylalanyl- histidyl-leucine as substrate. *3221*

Bradykinin *Serum Increase Physiological* Transient effect with i.v. or subcutaneous drug in one renal transplant patient. *3689*

Terazosin

Cholesterol *Serum Decrease Physiological* Increase by 5.4 mg/dL with up to 20 mg daily over 4 weeks in patients with moderate hypertension. *0857*

Cholesterol, High Density Lipoprotein
Serum No Effect Physiological No significant effect with up to 20 mg daily over 4 weeks in patients with moderate hypertension. *0857*

LDL + VLDL Cholesterol *Serum Decrease Physiological* Increase by 6.1 mg/dL with up to 20 mg daily over 4 weeks in patients with moderate hypertension. *0857*

Triglycerides *Serum No Effect Physiological* No significant effect with up to 20 mg daily over 4 weeks in patients with moderate hypertension. *0857*

Terbutaline

Albumin *Serum No Effect Physiological* In 6 normal men treated with therapeutic amounts for 2 weeks. *3177*

Cholesterol *Serum No Effect Physiological* No significant effect observed after 2 weeks in 15 subjects. *1654*

Cholesterol, High Density Lipoprotein
Serum Increase Physiological 10% increase with 2 weeks treatment in 15 subjects. *1654*

Cholesterol, Low Density Lipoprotein
Serum No Effect Physiological No significant effect observed after 2 weeks in 15 subjects. *1654*

Epinephrine *Plasma No Effect Physiological* In 6 normal men treated with therapeutic amounts for 2 weeks. *3177*

Fatty Acids, Free (FFA) *Serum No Effect Physiological* In 6 normal men treated with therapeutic amounts for 2 weeks. *3177*

Glucose *Serum No Effect Physiological* In 6 normal men treated with therapeutic amounts for 2 weeks. *3177*
Serum Increase Physiological Observed following infusion of 0.25 mg or 5 mg 3 times/d on first day of treatment. *0288*

Insulin *Plasma No Effect Physiological* In 6 normal men treated with therapeutic amounts for 2 weeks. *3177*
Plasma Increase Physiological Observed following infusion of 0.25 mg or 5 mg 3 times/d on first day of treatment. *0288*

Norepinephrine *Plasma No Effect Physiological* In 6 normal men treated with therapeutic amounts for 2 weeks. *3177*

Potassium *Serum No Effect Physiological* No reduction after oral treatment for 13 d. *0288* In 6 normal men treated with therapeutic amounts for 2 weeks. *3177*
Serum Increase Physiological Due to β₂-adrenergic mediated uptake of potassium in skeletal muscle and other tissues (decrease of about 0.8 mmol/L). *1823*
Serum Decrease Physiological Observed following infusion of 0.25 mg or 5 mg 3 times/d on first day of treatment. *0288*

Prealbumin *Serum No Effect Physiological* In 6 normal men treated with therapeutic amounts for 2 weeks. *3177*

Protein *Serum No Effect Physiological* In 6 normal men treated with therapeutic amounts for 2 weeks. *3177*

Theophylline *Serum Decrease Physiological* 0.075 mg 3 times per day caused average reduction of serum concentration from 13.8 to 10.8 μg/mL when sustained release drug given. *0807*

Thyroid Stimulating Hormone (TSH)
Serum No Effect Physiological In 6 normal men treated with therapeutic amounts for 2 weeks. *3177*

Terbutaline *(continued)*

Thyroxine (T₄) *Serum Decrease Physiological* From 7.2 to 6.7 µg/dL on average in 6 normal men after 2 weeks treatment. *3177* In 6 normal men after 15 mg/d for 2 weeks with non-significant change from 6.7 to 7.2 mg/dL. *3177*

Thyroxine (T₄) Binding Globulin
Serum No Effect Physiological In 6 normal men treated with therapeutic amounts for 2 weeks. *3177*

Thyroxine (T₄), Free *Serum No Effect Physiological* In 6 normal men treated with therapeutic amounts for 2 weeks. *3177*

Triglycerides *Serum No Effect Physiological* No significant effect observed after 2 weeks in 15 subjects. *1654*

Tri-iodothyronine (T₃) *Serum Increase Physiological* From 136 to 160 ng/dL on average in 6 normal men after 2 weeks treatment. *3177* In 6 normal men after 15 mg/d for 2 weeks with average of 160 ng/dL versus 136 ng/dL in controls. *3177*

Tri-iodothyronine (T₃), Free *Serum No Effect Physiological* In 6 normal men treated with therapeutic amounts for 2 weeks. *3177*

Tri-iodothyronine (T₃), Reverse
Serum No Effect Physiological In 6 normal men treated with therapeutic amounts for 2 weeks. *3177*

Testolactone

Calcium *Serum Increase Physiological* Probable effect during remission of metastases. *0071*

17-Ketosteroids *Urine Increase Physiological* Reported effect. *1488*

Testosterone

Alanine Aminotransferase *Serum Increase Physiological* May cause cholestatic and cytotoxic liver damage. *0248*

Aldosterone *Urine No Effect Analytical* No significant effect RIA procedure of Drewes. *0952*
Urine Increase Analytical 1 mg/L equivalent to 8 µg/L by method of Drewes. *0952*

Alkaline Phosphatase *Serum Increase Physiological* May cause cholestasis. *2313*

5-Androstene-3 β, 17 β-Diol
Plasma Decrease Physiological In power athletes with 26 weeks on steroid self administration. *3088*

Androstenedione *Plasma Increase Physiological* In power athletes with 26 weeks on steroid self administration. *3088*

Aspartate Aminotransferase
Serum Increase Physiological May cause cholestatic and cytotoxic liver damage. *0248*

Bile *Urine Increase Physiological* May cause cholestasis. *2313*

Bilirubin *Serum Increase Physiological* May cause cholestatic and cytotoxic liver damage. *0248*

BSP Retention *Serum Increase Physiological* May cause cholestasis. *2313*

Calcium *Serum Increase Physiological* Reported with therapy of breast cancer. *2582*

Cephalin Flocculation *Serum Increase Physiological* May cause cholestasis. *2313*

Cholesterol *Serum Increase Physiological* Cholestatic effect. *1488*

Corticosteroid-Binding Globulin
Serum Decrease Physiological Metabolic effect. *0227*

Corticosteroids *Plasma Increase Analytical* 13% (cortisol 100%) competitive protein binding method of Ficher. *1128*

Dehydroepiandrosterone (DHEA)
Plasma Decrease Physiological In power athletes with 26 weeks on steroid self administration. *3088*

Dexamethasone *Serum Increase Analytical* Very slight cross reactivity procedure of Hichens. *1585*

2,3-Diphosphoglycerate
Red Blood Cells Increase Physiological Increase by 2/3 in patients on hemodialysis for 12 weeks. *2618*

Estradiol *Plasma Increase Physiological* 3 fold increase on days 2 to 7 in 11 hypogonadal men given 200 mg cypionate salt intramuscularly. *2555*

Estrogens *Urine No Effect Analytical* At 50 mg/L on fluorescent method of Corns. *0730*
Urine Increase Physiological Probably due to conversion to estrogen. *3804*

Fibrinogen *Plasma Decrease Physiological* Metabolic effect. *0227*

Guanase *Serum Increase Physiological* May cause cholestatic and cytotoxic liver damage. *0248*

Haptoglobin *Serum Increase Physiological* Metabolic effect (if aqueous solution i.m.). *0227*

17-Hydroxycorticosteroids *Urine Increase Analytical* Common keto group involved in color reaction. *0757*

17-Hydroxypregnenolone *Plasma Decrease Physiological* In power athletes with 26 weeks on steroid self administration. *3088*

17-Hydroxyprogesterone *Plasma Decrease Physiological* In power athletes with 26 weeks on steroid self administration. *3088*

Isocitrate Dehydrogenase *Serum Increase Physiological* May cause cholestatic and cytotoxic liver damage. *0248*

17-Ketosteroids *Urine No Effect Analytical* Measured as a ketosteroid. *1487*

Ornithine Carbamoyltransferase (OCT)
Serum Increase Physiological May cause cholestatic and cytotoxic liver damage. *0248*

Pregnenolone *Plasma Decrease Physiological* In power athletes with 26 weeks on steroid self administration. *3088*

Progesterone *Plasma Decrease Physiological* In power athletes with 26 weeks on steroid self administration. *3088*

Protein Bound Iodine (PBI) *Serum No Effect Physiological* No effect short term administration in men. *0830*
Serum Increase Physiological Androgen effect seen in some patients. *1680*
Serum Decrease Physiological Decreased thyroxine binding globulin synthesis, and affects binding. *3879*

Pseudocholinesterase *Serum Decrease Physiological* Metabolic effect. *0227*

Sex Hormone Binding Globulin
Serum Decrease Physiological Increase by 80-90% in power athletes with 26 weeks on steroid self administration. *3088* Drop below 80% of basal concentration in prepubertal subjects: useful as test of androgen sensitivity. *0270*

Sialic Acid *Serum Increase Physiological* Metabolic effect (if aqueous solution i.m.). *0227*

Testosterone *Serum Increase Physiological* In 11 hypogonadal men given 200 mg cypionate salt intramuscularly, 3 fold rise maximum day 2 to 5. *2555* In power athletes with 26 weeks on steroid self administration. *3088*

Testosterone, Free *Serum Increase Physiological* 4.5 fold increase at days 2-3 in 11 hypogonadal men given 200 mg cypionate salt intramuscularly. *2555*

Thymol Turbidity *Serum Increase Physiological* May cause cholestasis. *2313*

Thyroxine (T₄) Binding Globulin
Serum Decrease Physiological Direct effect of drug. *0012*

Transcortin *Serum No Effect Physiological* In power athletes with 26 weeks on steroid self administration. *3088*

Transferrin *Serum Decrease Physiological* Metabolic effect. *0227*

Tyrosine *Plasma Decrease Physiological* Fall observed in 2 weeks. *0843*

Urobilinogen *Urine Decrease Physiological* May cause cholestasis. *2313*

Volume *Plasma Increase Physiological* Salt and water retention, also increased RBC mass. *0330*
Blood Decrease Physiological Produces polycythemic-like state. *2427*

Tetrabenazine

Homovanillic Acid *CSF Increase Physiological* Significant effect in chorea after oral administration. *2379*

5-Hydroxyindoleacetic Acid (5-HIAA)
CSF No Effect Physiological No significant effect in Huntington's chorea. *2379*

Tetracaine

Protein *CSF Increase Analytical* Interferes with Folin-Cio-calteu reagent. *1237*

Sulfa as Sulfanilamide *Serum Increase Analytical* Yields diazotization reaction. *0251*

Tetrachloroethylene

Protein *Urine Increase Physiological* May cause nephrotoxicity. *3204*

Urea Nitrogen *Serum Increase Physiological* May cause nephrotoxicity. *3204*

Tetrachlorothyronine

Thyroxine (T₄) *Serum Decrease Physiological* Displaces thyroxine from thyroxine binding globulin. *3505*

Tetracosactrin

Alanine *Plasma Increase Physiological* In healthy volunteers given 1 mg intramuscularly for up to 60 h. *1813*

Aldolase *Serum Increase Physiological* Probably due to muscle damage at injection site. *0275*

Corticosteroids *Plasma Increase Physiological* Therapeutic intent has prolonged action. *1899*

Cortisol *Plasma Increase Physiological* Rise of 21 μg/dL in 6 h after i.m. injections. *0275* 3 fold increase in healthy volunteers given 1 mg intramuscularly for up to 60 h. *1813*

Fatty Acids, Free (FFA) *Serum No Effect Physiological* In healthy volunteers given 1 mg intramuscularly for up to 60 h. *1813*

Glucose *Serum Increase Physiological* From 5.2 to 7.2 mmol/L healthy volunteers given 1 mg intramuscularly for up to 60 h. *1813*

Glycerol *Serum No Effect Physiological* In healthy volunteers given 1 mg intramuscularly for up to 60 h. *1813*

Growth Hormone *Plasma Increase Physiological* Associated with change in cortisol. *0275*

Hematocrit *Blood Decrease Physiological* Maximum effect observed after 24 h. *0275*

Insulin *Plasma Increase Physiological* From 5.2 to 13.1 mU/L healthy volunteers given 1 mg intramuscularly for up to 60 h. *1813*

Ketones *Serum No Effect Physiological* In healthy volunteers given 1 mg intramuscularly for up to 60 h. *1813*

Lactate *Serum Increase Physiological* In healthy volunteers given 1 mg intramuscularly for up to 60 h. *1813*

Leukocytes *Blood Increase Physiological* Increase (mainly NPL) of 7,000 seen in 1 d. *3010*

Pyruvate *Serum Increase Physiological* In healthy volunteers given 1 mg intramuscularly for up to 60 h. *1813*

Tetracycline

Acetaminophen *Serum Decrease Analytical* Slight reduction of up to 18% at just above upper therapeutic concentration on on method using o-cresol reaction. *0621*

Acetaminophen Screening Test *Urine Negative Analytical* No reaction with o-cresol at therapeutic concentrations. *3326*

Alanine Aminotransferase *Serum No Effect Analytical* At 5 times upper limit of therapeutic range on methods on SMAC®, Abbott-VP, Cobas-Bio, aca, and KDA. *2138*
Serum Increase Physiological Hepatotoxic especially in pregnant women. *2808*

Albumin *Serum No Effect Analytical* At 5 times upper limit of therapeutic range on methods on SMAC®, Ektachem®, Hitachi® 705 and KDA. *2138* At concentration of 300 mg/L had no effect on BCG method. *3393*

Alkaline Phosphatase *Serum No Effect Analytical* At 5 times upper limit of therapeutic range on methods on SMAC®, Abbott-VP, Cobas-Bio, aca, Hitachi® 705 and KDA. *2138*
Serum Increase Physiological May occur with cholestasis. *1488* 2 cases of drug-associated fatty liver of pregnancy. *3795*

Alkaloids *Urine No Effect Analytical* With iodoplatinate of Frings TLC procedure at 2 mg/dL. With ninhydrin on Frings TLC procedure at 2 mg/dL. *1204*

Amino Acids *Urine Increase Physiological* Nephrotoxic effect with degraded tetracycline. *1237*

Ammonia *Plasma Increase Physiological* If given i.v. in large doses. *1488*
Plasma Decrease Physiological Reduces production by gut bacteria. *1487*

Amylase *Serum No Effect Analytical* At 5 times upper limit of therapeutic range on methods on aca, Cobas-Bio and Ektachem®. *2138*
Serum Increase Physiological Toxic effect especially in pregnant women. *1488*

Antithrombin III *Plasma No Effect Analytical* At concentration of 1,000 mg/L had no effect on aca method. *3393*

Ascorbic Acid *White Blood Cells Decrease Physiological* Other antibiotics also have effect. *3554*

Aspartate Aminotransferase *Serum No Effect Analytical* At 5 times upper limit of therapeutic range on methods on SMAC®, Abbott-VP, Cobas-Bio, aca, Hitachi® 705 and KDA. *2138*
Serum Increase Physiological Hepatotoxic especially in pregnant women. *2808* 2 cases of drug-associated fatty liver of pregnancy. *3795*

Barbiturate *Urine No Effect Analytical* With diphenyl-carbazone on Frings TLC procedure at 2 mg/dL. With mercurous NO₃ on Frings TLC procedure at 2 mg/dL. With mercuric SO₄ on Frings TLC procedure at 2 mg/dL. With vanillin on Frings procedure at 2 mg/dL. *1204*

Bence-Jones Protein *Urine Positive Physiological* Nephrotoxic effect with degraded tetracycline. *1022*

Bicarbonate *Serum No Effect Analytical* At concentration of 4 mg/L had no effect on method using phenolphthalein. *3393*
Serum Decrease Physiological May cause acidosis with nephrotoxicity. *0071*

Bilirubin *Serum No Effect Analytical* At 5 times upper limit of therapeutic range on methods on SMAC®, Cobas-Bio, aca, Ektachem®, Hitachi® 705 and KDA. *2138* At concentration of 10 mg/L had no effect on Ektachem® method. *3393* At concentration of 300 mg/L had no effect on Jendrassik and Grof method. *3393*
Serum Increase Physiological Hepatic injury may occur especially if given i.v. *0348* 2 cases of drug-associated fatty liver of pregnancy. *3795*

Bilirubin, Direct *Serum Increase Physiological* Cholestatic effect. *0945* 2 cases of drug-associated fatty liver of pregnancy. *3795*

Bromide *Urine No Effect Analytical* No effect on method of Frings. *1204*

BSP Retention *Serum Increase Physiological* Cholestatic especially in pregnant women. *2808*

Calcium *Serum No Effect Analytical* At 5 times upper limit of therapeutic range on methods on SMAC®, Abbott-VP, aca, Ektachem®, Hitachi® 705 and KDA. *2138* At concentration of 300 mg/L had no effect on cresolphthalein method. *3393*
Serum Decrease Physiological Observed in pregnant women. *2427*

Catecholamines *Plasma Increase Analytical* Interferes with fluorometric techniques. *0127*
Urine No Effect Analytical No effect reported on method of Sandhu and Freed. *1205*
Urine Increase Analytical Produces interfering fluorescence. *1841*

Cephalin Flocculation *Serum Increase Physiological* May cause cholestasis. *2313*

Chloramphenicol *Serum No Effect Analytical* No effect at 100 mg/L on coupled enzymatic method. *2490*

Chloride *Serum No Effect Analytical* At concentration of 300 mg/L had no effect on mercurimetric method. *3393*

Tetracycline *(continued)*

Cholesterol *Serum No Effect Analytical* At 5 times upper limit of therapeutic range on methods on SMAC®, Abbott-VP, Cobas-Bio, Ektachem®, Hitachi® 705 and KDA. *2138* At concentration of 200 mg/L had no effect on CHOD-PAP method. *3393* At concentration of 200 mg/L had no effect on catalase-AIDH method. *3393* At concentration of 200 mg/L had no effect on method using catalase-Hantzsch reaction. *3393* At concentration of 300 mg/L had no effect on Liebermann-Burchard method. *3393*
Serum No Effect Physiological No effect seen even when administered orally. *1487*
Serum Increase Analytical At concentrations above 50 mg/L (normal therapeutic concentration 8 mg/L) raised concentration as measured by CHOD-Iodide method. *3393*
Serum Decrease Physiological Hepatotoxicity may occur. *2808*

Chromosomes *Test Conditions No Effect Physiological* Not clastogenic in human cells. *3282*

Coagulation Time *Blood Increase Physiological* Delayed coagulation reported. *2313*

Color *Feces Increase Analytical* Red if glucosamine potentiated syrup form. *2313*

Coombs' Test *Serum Positive Physiological* Mechanism obscure. *1486*

Creatine Kinase *Serum No Effect Analytical* At 5 times upper limit of therapeutic range on methods on SMAC®, Abbott-VP, Cobas-Bio, aca and Hitachi® 705. *2138*

Creatinine *Serum No Effect Analytical* At 5 times upper limit of therapeutic range on methods on SMAC®, Abbott-VP, aca, Cobas-Bio, Ektachem®, Hitachi® 705 and KDA. *2138* At concentration of 60 mg/L had no effect on creatinine iminohydrolase method. *3393* At concentration of 300 mg/L had no effect on AutoAnalyzer Jaffé method. *3393* At concentration of 200 mg/L had no effect on Jaffé-Fading-Fraction method. *3393* At concentration of 200 mg/L had no effect on Jaffé-Fuller's earth method. *3393* At concentration of 40 mg/L had no effect on kinetic Jaffé method on BKA-2. *3393*
Serum Increase Physiological Nephrotoxicity may cause Fanconi like syndrome. *2808*

Eosinophils *Blood Increase Physiological* Allergic response. *1513*

Erythrocytes *Blood Decrease Physiological* May cause hemolytic anemia. *2313*

Estriol *Urine No Effect Analytical* No effect on GLC method. *1163*

Estrogens *Urine Increase Analytical* Possible effect (interference not defined). *0022*

Ethchlorvynol *Urine No Effect Analytical* No effect on method of Frings and Cohen. *1204*

FIGLU (N-Formiminoglutamic Acid)
Urine Increase Physiological Isolated case of folic acid deficiency reported. *1815*

Folate *Serum No Effect Analytical* If chromatographic procedure of Landon used. *2068* Allegedly no effect on autoclave method. *1815*
Serum Decrease Analytical Inhibits growth of L. Casei. *3505*
Serum Decrease Physiological Isolated case of impaired absorption. *1815*

FPN Test *Urine No Effect Analytical* No effect at 10 mg/dL on method of Frings. *1204*

Glucose *Serum No Effect Analytical* No effect on glucose oxidase method. *2568* At 5 times upper limit of therapeutic range on methods on SMAC®, Abbott-VP, aca, Cobas-Bio, Ektachem®, Hitachi® 705 and KDA. *2138* At concentration of 20 mg/L had no effect on Ektachem® method. *3393* At concentration of 500 mg/L had no effect on Seralyzer method. *3393*
Serum Increase Analytical Slight effect hexokinase, o-toluidine methods. *2568* Mg for mg (approximately) MBTH procedure of Neeley. *2569*
Serum Decrease Analytical Ascorbic acid in prep may affect glucose oxidase procedure. *0583* At concentrations above 20 mg/L (normal therapeutic concentration 8 mg/L) lowered concentration as measured by GOD/POD-PAP method. *3393*
Urine Increase Physiological Degraded drug may cause Fanconi like syndrome. *2808* Degraded material may cause Fanconi syndrome. *2427*

Urine Decrease Analytical Prevents oxidation of chromogen in glucose oxidase methods. *2425* False negative dipstick test if buffered with ascorbic. *1103*

γ-Glutamyltransferase (GGT) *Serum No Effect Analytical* At 5 times upper limit of therapeutic range on methods on SMAC®, Abbott-VP and Hitachi® 705. *2138*

Hematocrit *Blood Decrease Physiological* May cause hemolytic anemia. *2313*

Hemoglobin *Blood Decrease Physiological* May cause hemolytic anemia. *2313*
Urine Decrease Analytical In large amounts if buffered with ascorbate. *0583*

Histidine *Urine Increase Physiological* Unexplained mechanism. *1343*

Iron *Serum No Effect Analytical* At 5 times upper limit of therapeutic range on Ferrozine method on SMAC®. *2138*

^{131}I Uptake *Serum No Effect Physiological* No effect with 2 g/d for 11 d. *0583*
Serum Decrease Physiological Panmycin® contains tetraiodofluorescein. *2652*

Lactate Dehydrogenase *Serum No Effect Analytical* At 5 times upper limit of therapeutic range on methods on SMAC®, Abbott-VP, Cobas-Bio, Hitachi® 705 and KDA. *2138*
Serum Increase Physiological 2 cases of drug-associated fatty liver of pregnancy. *3795*

LE Cells *Blood Positive Physiological* May produce LE-like syndrome. *3505*

Leukocytes *Blood Increase Physiological* Leukocytosis with atypical lymphocytes. *1343*
Blood Decrease Physiological Leukopenia (AMA Blood dyscrasias). *2429*

Neutrophils *Blood Decrease Physiological* Neutropenia may occur occasionally. *1680*

Nitrofurantoin *Urine No Effect Analytical* No effect on method of Conklin and Hollifield. *0704*

Nitrogen *Urine Increase Physiological* Due to antianabolic effect. *0071*

Nonprotein Nitrogen *Serum Increase Physiological* Due to nephrotoxicity or antianabolic effect. *2313*

Occult Blood *Feces Positive Physiological* May occur with hematemesis and melena. *0945*

pH *Blood Decrease Physiological* May cause acidosis with renal impairment. *2220*

Phosphate *Serum No Effect Analytical* At 5 times upper limit of therapeutic range on methods on SMAC®, aca, Hitachi® 705 and KDA. *2138* At concentration of 300 mg/L had no effect on phosphomolybdate method. *3393*
Serum Increase Physiological May occur with nephrotoxicity. *2313*
Serum Decrease Physiological May occur with Fanconi syndrome. *0071*
Urine Increase Physiological Degraded material may cause Fanconi syndrome. *2427*

Plasminogen *Plasma No Effect Analytical* At concentration of 1,000 mg/L had no effect on aca method. *3393*

Platelets *Blood Decrease Physiological* Thrombocytopenia/pancytopenia. *2313*

Porphyrins *Urine Increase Analytical* Interfering fluorescence. *1238*

Potassium *Serum No Effect Analytical* At concentration of 4 mg/L had no effect on flame-photometric method. *3393* At concentration of 20,000 mg/L had no effect on measurement by ISE with predilution. *3393* At concentration of 400 mg/L had no effect on measurement by ISE without predilution. *3393*
Serum Increase Physiological Occurs with azotemia. *0071*
Serum Decrease Physiological Fanconi like syndrome may occur with degraded comp. *1488*

Protein *Serum No Effect Analytical* At 5 times upper limit of therapeutic range on methods on SMAC®, Abbott-VP, Ektachem®, Hitachi® 705 and KDA. *2138*
Urine Increase Physiological Nephrotoxic effect with degraded tetracycline. *1237*
CSF Increase Analytical Reacts as if phenol with Folin-Ciocalteu procedure. *3958*

Protein Bound Iodine (PBI) *Serum No Effect Physiological* No effect with 2 g/d for 11 d. *0583*

Protein Electrophoresis *Serum No Effect Analytical* At concentration of 400 mg/L had no effect on automated Olympus-Hite method. *3393*

Prothrombin Time *Plasma Increase Physiological* Associated with cholestasis, and reduced activity. *0945*
Plasma Decrease Physiological May partially counteract action of heparin. *1679*

Salicylate *Urine No Effect Analytical* At 2 mg/dL no effect on Trinder method. *1204*

Sex Hormone Binding Globulin
Serum No Effect Physiological In 9 men with acne given compound for 3 d. *2884* No effect reported. *1736*

Sodium *Serum No Effect Analytical* At concentration of 4 mg/L had no effect on flame-photometric method. *3393* At concentration of 20,000 mg/L had no effect on measurement by ISE with predilution. *3393* At concentration of 400 mg/L had no effect on measurement by ISE without predilution. *3393*
Serum Increase Physiological May cause hypernatremia with renal impairment. *2220*
Urine Increase Physiological Natriuresis and diuresis effects of drug. *0491*

Sugar *Urine Increase Analytical* False positive copper reduction procedure. *3505* False positive with Benedict's and Clinitest®. *1488*

T$_3$ Uptake *Serum No Effect Physiological* No effect with 2 g/d for 11 d. *0583*

Testosterone *Serum No Effect Analytical* No effect given at 100 μg/mL on radioimmunoassay. *2884*
Serum Decrease Physiological By about 10-20%. *1736* Fell from 21 to 17 nmol/L in 9 men after 3 d treatment; mechanism not established. *2884*

Testosterone, Free *Serum Decrease Physiological* By about 10-20%. *1736*

Threonine *Urine Increase Physiological* Unexplained mechanism. *1343*

Thromboplastin Generation *Blood Decrease Physiological* Impaired rate of regeneration observed. *1343*

Thymol Turbidity *Serum Increase Physiological* May cause cholestasis. *2313*

Triglycerides *Serum No Effect Analytical* At 5 times upper limit of therapeutic range on methods on SMAC®, Abbott-VP, Ektachem®, Hitachi® 705 and KDA. *2138* At concentration of 60 mg/L had no effect on GPO-PAP method. *3393* At concentration of 300 mg/L had no effect on lipase/esterase method. *3393*

Tryptophan *Urine Increase Physiological* Unexplained mechanism. *1343*

Urea Nitrogen *Serum No Effect Analytical* At 5 times upper limit of therapeutic range on methods on SMAC®, Abbott-VP, Cobas-Bio, Ektachem®, Hitachi® 705 and KDA. *2138* At concentration of 300 mg/L had no effect on diacetylmonoxime method. *3393*
Serum Increase Analytical At concentrations above 40 mg/L (normal therapeutic concentration 8 mg/L) raised concentration as measured by Ektachem® method. *3393*
Serum Increase Physiological Anti-anabolic action, amino acids degraded to urea. *0491* May be dose related renal toxicity. *1680*

Uric Acid *Serum No Effect Analytical* At 5 times upper limit of therapeutic range on methods on SMAC®, Abbott-VP, aca, Cobas-Bio, Ektachem®, Hitachi® 705 and KDA. *2138* At concentration of 200 mg/L had no effect on catalase-AIDH method. *3393* At concentration of 100 mg/L had no effect on Seralyzer method. *3393*
Serum Increase Analytical At concentrations above 30 mg/L (normal therapeutic concentration 8 mg/L) raised concentration as measured by Phosphotungstate reduction method. *3393*
Serum Decrease Analytical At concentrations above 150 mg/L (normal therapeutic concentration 8 mg/L) lowered concentration as measured by Kageyama-Hantzsch method. *3393*

Urobilinogen *Urine Increase Analytical* Yellow color extracted into CHCl$_3$ (Ehrlich's procedure). *2359*
Urine Decrease Physiological Reduces flora in gastrointestinal tract. *1488*
Feces Decrease Physiological Reduces flora in gastrointestinal tract. *1488*

Volume *Urine Increase Physiological* May have diuretic or nephrotoxic action. *0491*

Tetraethyl-Lead

Aminolevulinic Acid Dehydrase
Red Blood Cells Decrease Physiological Powerful inhibiting action. *0258*

Lead *Blood Increase Physiological* Doubled normal concentration observed. *0258*
Urine Increase Physiological Excretion further increased by chelating agents. *0258*

Protoporphyrin *Urine Increase Physiological* Erythrocyte protoporphyrin excretion slight increase. *0258*

Tetragastrin

Glucagon *Plasma No Effect Physiological* No effect observed if given i.v. *2676*

Glucose *Serum No Effect Physiological* No change after 4 μg/kg i.v. *2676*

Hydrochloric Acid *Gastric Material Increase Physiological* Effect similar to pentagastrin. *1482*

Insulin *Plasma Increase Physiological* Slight rise but not significant. *2676*

Tetrahydrocannabinol

Chromosomes *Test Conditions Abnormal Physiological* Clastogenic in human cells. *3282*

Epinephrine *Urine Increase Physiological* Initial marked increase compared with placebo. *3786*

Norepinephrine *Urine Increase Physiological* Possible slight increase initially compared with placebo. *3786*

Tetrahydrocortisol

Aldosterone *Urine No Effect Analytical* No significant effect RIA procedure of Drewes. *0952*
Urine Increase Analytical 1 mg/L equivalent to 2 μg/L by method of Drewes. *0952*

Tetrahydro-DOC

Aldosterone *Urine No Effect Analytical* No significant effect on RIA procedure of Drewes. *0952*
Urine Increase Analytical 1 mg/L equivalent to 28 μg/L by method of Drewes. *0952*

Tetrahydro Reichstein's S

Aldosterone *Urine No Effect Analytical* 1 mg/L equivalent to 2 μg/L by method of Drewes. *0952*

Tetraiodofluorescein

^{131}I Uptake *Serum Decrease Physiological* Constituent of capsules of many pharmaceuticals. *2652*

Protein Bound Iodine (PBI) *Serum Increase Physiological* Red dye used to color many pharmaceuticals. *0012*

Thyroxine (T$_4$) *Serum No Effect Physiological* No effect observed. *3725*

Urobilin *Urine Increase Analytical* Pink color with mauve fluorescence. *1563*

Tetraiodophthalein

BSP Retention *Serum Increase Physiological* Competitive uptake and excretion of dye. *1745*

Precipitable Iodine *Serum Increase Analytical* Iodine contamination. *1237*

Tetralin

Color *Urine Increase Analytical* Greenish blue. *2313*

Tetridaz-B

Protein Bound Iodine (PBI) *Serum Increase Analytical* Iodine contamination. *1237*

Thallium

Alanine Aminotransferase *Serum Increase Physiological* Due to hepatic necrosis. *1302*

Casts *Urine Increase Physiological* Damage to renal tubular epithelium. *1302*

Cells *Urine Increase Physiological* Damage to renal tubular epithelium. *1302*

Occult Blood *Feces Positive Physiological* Due to ingestion of toxic dose. *1302*

Protein *Urine Increase Physiological* Damage to renal tubular epithelium. *1302*

Urea Nitrogen *Serum Increase Physiological* May cause nephrotoxicity. *3204*

Thenalidine

Leukocytes *Blood Decrease Physiological* Leukopenia (AMA Blood dyscrasias). *2429*

Neutrophils *Blood Decrease Physiological* Occasional case of drug-induced neutropenia. *0155*

Theobromine

Acetaminophen *Serum No Effect Analytical* At 10 mg/L had no effect on HPLC method. *3432*

Albumin *Serum No Effect Analytical* At concentration of 2,000 mg/L had no effect on BCG method. *3393*

Bicarbonate *Serum No Effect Analytical* At concentration of 2,000 mg/L had no effect on method using phenolphthalein. *3393*

Bilirubin *Serum No Effect Analytical* At concentration of 2,000 mg/L had no effect on Jendrassik and Grof method. *3393*

Calcium *Serum No Effect Analytical* At concentration of 2,000 mg/L had no effect on cresolphthalein method. *3393*

Chloride *Serum No Effect Analytical* At concentration of 2,000 mg/L had no effect on mercurimetric method. *3393*

Cholesterol *Serum No Effect Analytical* At concentration of 2,000 mg/L had no effect on Liebermann-Burchard method. *3393*

Chromosomes *Test Conditions Abnormal Physiological* Clastogenic in human cells. *3282*

Creatinine *Serum No Effect Analytical* At concentration of 2,000 mg/L had no effect on AutoAnalyzer method. *3393*

Phosphate *Serum No Effect Analytical* At concentration of 2,000 mg/L had no effect on phosphomolybdate method. *3393*

Protein *Serum No Effect Analytical* At concentration of 2,000 mg/L had no effect on biuret method with blank correction. *3393*

Triglycerides *Serum No Effect Analytical* At concentration of 2,000 mg/L had no effect on lipase/esterase method. *3393*

Urea Nitrogen *Serum No Effect Analytical* At concentration of 2,000 mg/L had no effect on diacetylmonoxime method. *3393*

Uric Acid *Serum No Effect Analytical* At concentration of 2,000 mg/L had no effect on phosphotungstate reduction method. *3393*

Theophylline

Acetaminophen *Serum No Effect Analytical* At 20 mg/L had no effect on HPLC method. *3432*

N-Acetylglucosaminidase *Urine No Effect Analytical* At 3 mg/L had no effect on 2 colorimetric analytical methods. *1354*

Alanine Aminotransferase *Serum No Effect Analytical* At 5 times upper limit of therapeutic range on methods on SMAC®, Abbott-VP, aca, Cobas-Bio and KDA. *2138* No effect at therapeutic concentration on Reflotron method. *1984*

Albumin *Serum No Effect Analytical* At 5 times upper limit of therapeutic range on methods on SMAC®, Ektachem®, Hitachi® 705 and KDA. *2138* At concentration of 2,000 mg/L had no effect on BCG method. *3393*

Alkaline Phosphatase *Serum No Effect Analytical* At 5 times upper limit of therapeutic range on methods on Abbott-VP and aca. *2138*
Serum Decrease Analytical At 5 times upper limit of therapeutic range on methods on SMAC®, KDA, Cobas-Bio and Hitachi® 705. *2138* Approximately 10% reduction in activity at drug concentration of 2 mg/dL. *3719*

Amylase *Serum No Effect Analytical* At 5 times upper limit of therapeutic range on aca, Cobas-Bio and Ektachem®. *2138*

Aspartate Aminotransferase *Serum No Effect Analytical* At 5 times upper limit of therapeutic range on methods on SMAC®, Abbott-VP, aca, Cobas-Bio, Hitachi® 705 and KDA. *2138* No effect at therapeutic concentration on Reflotron method. *1984*

Barbiturate *Serum Increase Analytical* Identical ultraviolet Spectra. *3608* In certain gas-chromatographic procedures may produce falsely high values. *0852*

Bicarbonate *Serum No Effect Analytical* At concentration of 2,000 mg/L had no effect on method using phenolphthalein method. *3393*

Bilirubin *Serum No Effect Analytical* At 5 times upper limit of therapeutic range on methods on SMAC®, Cobas-Bio, aca, Ektachem®, Hitachi® 705 and KDA. *2138* At concentration of 2,000 mg/L had no effect on Jendrassik and Grof method. *3393*
Serum Decrease Analytical Causes depression of color formation. *2313*

Calcium *Serum No Effect Analytical* At 5 times upper limit of therapeutic range on methods on SMAC®, Abbott-VP, aca, Ektachem®, Hitachi® 705 and KDA. *2138* At concentration of 120 mg/L had no effect on cresolphthalein method. *3393*
Serum No Effect Physiological No effect in response to i.v. infusion. *3935*
Serum Increase Physiological 11 of 60 patients with theophylline toxicity but reverted to normal when drug withdrawn. Also in normals with 400 mg bid for 5 d. *2383*
Urine No Effect Physiological In normals given 400 mg bid for 5 d. *2383*

Calcium, Ultrafiltrable *Serum Increase Physiological* Apparent log dose response correlation with drug concentration in individuals with toxicity. *2383*

Catecholamines *Urine Increase Physiological* Normal response even when given orally. *0178*

Chloride *Serum No Effect Analytical* At concentration of 2,000 mg/L had no effect on mercurimetric method. *3393*

Cholesterol *Serum No Effect Analytical* At 5 times upper limit of therapeutic range on methods on SMAC®, Abbott-VP, Cobas-Bio, Ektachem®, Hitachi® 705 and KDA. *2138* No effect at therapeutic concentration on Reflotron method. *1984* At concentration of 20 mg/L had no effect on CHOD-PAP method. *3393* At concentration of 2,000 mg/L had no effect on Liebermann-Burchard method. *3393*

Chromosomes *Test Conditions Abnormal Physiological* Clastogenic in human cells. *3282*

Cortisol *Plasma No Effect Physiological* No effect in response to i.v. infusion of aminophylline. *0608* No effect on hepatic metabolism so concentration unaffected. *1830*

Creatine Kinase *Serum No Effect Analytical* At 5 times upper limit of therapeutic range on methods on SMAC®, Abbott-VP, Cobas-Bio, aca and Hitachi® 705. *2138*

Creatinine *Serum No Effect Analytical* At 1.0 g/L on reversed phase LC procedure of Zhiri et al. *3942* At 5 times upper limit of therapeutic range on methods on SMAC®, Abbott-VP, aca, Cobas-Bio, Ektachem®, Hitachi® 705 and KDA. *2138* At concentration of 2,000 mg/L had no effect on AutoAnalyzer Jaffé method. *3393*

Cyclic Adenosine Monophosphate *Urine No Effect Physiological* In normals given 400 mg bid for 5 d. *2383*

Dexamethasone *Serum No Effect Physiological* No effect on hepatic metabolism so concentration unaffected. *1830*

Diphenylhydantoin *Serum Decrease Physiological* Mean value decreased by 21% when theophylline coadministered; rebound when drug discontinued. *3557*

Erythromycin *Serum No Effect Physiological* No effect on area under curve although appeared to be increased elimination. *1593*

Fatty Acids, Free (FFA) *Serum Increase Physiological* Rapid pronounced and prolonged rise associated with aminophylline to produce therapeutic concentration of theophylline. *0608* 123% increase after 4 d of treatment in 10 healthy volunteers. *3757*

Glucagon *Plasma No Effect Physiological* No effect in response to i.v. infusion of aminophylline. *0608*

Glucose *Serum No Effect Analytical* At 5 times upper limit of therapeutic range on methods on SMAC®, Abbott-VP, aca, Cobas-Bio, Ektachem®, Hitachi® 705 and KDA. *2138* No effect at therapeutic concentration on Reflotron method. *1984* At concentration of 23 mg/L had no effect on Ektachem® method. *3393* At concentration of 20 mg/L had no effect on GOD/POD-PAP method. *3393*
Serum Increase Physiological Follows overdose, but decreased glucose tolerance also observed in infants and others. *3422* Effect small in response to i.v. infusion of aminophylline. *0608*
Urine Increase Physiological Follows overdose, but decreased glucose tolerance also observed in infants and others. *3422*

γ-Glutamyltransferase (GGT) *Serum No Effect Analytical* At 5 times upper limit of therapeutic range on methods on SMAC®, Abbott-VP, and Hitachi® 705. *2138* No effect at therapeutic concentration on Reflotron method. *1984*

Growth Hormone *Plasma No Effect Physiological* No effect in response to i.v. infusion of aminophylline. *0608*

Indomethacin *Serum No Effect Analytical* No effect on HPLC method of Roberts and Smith. *3002*

Insulin *Plasma No Effect Physiological* No effect in response to i.v. infusion of aminophylline. *0608*

Iron *Serum No Effect Analytical* At 5 times upper limit of therapeutic range on Ferrozine method on SMAC®. *2138*

Lactate Dehydrogenase *Serum No Effect Analytical* At 5 times upper limit of therapeutic range on methods on SMAC®, Abbott-VP, Cobas-Bio, Hitachi® 705 and KDA. *2138*
Serum Decrease Analytical Effect small and can be ignored at therapeutic concentration. *3719*

Magnesium *Serum No Effect Physiological* No effect in response to i.v. infusion. *3935*

Occult Blood *Feces Positive Physiological* In large doses may cause gastric hemorrhage. *1680*

Parathyroid Hormone *Plasma No Effect Physiological* In patients with drug toxicity. *2383*

pH *Urine No Effect Physiological* In healthy volunteers given drug intravenously. *1593*

Phosphate *Serum No Effect Analytical* At 5 times upper limit of therapeutic range on methods on SMAC®, aca, Hitachi® 705 and KDA. *2138* At concentration of 2,000 mg/L had no effect on phosphomolybdate method. *3393*
Serum Decrease Physiological Response to i.v. infusion, returned to baseline in 4 h. *3935*
Urine No Effect Physiological In normals given 400 mg bid for 5 d. *2383*

Potassium *Serum No Effect Analytical* At concentration of 20 mg/L had no effect on flame-photometric method. *3393* At concentration of 1,000 mg/L had no effect on measurement by ISE with predilution. *3393*
Serum Decrease Physiological Response to i.v. infusion, returned to baseline in 4 h. *3935*

Protein *Serum No Effect Analytical* At 5 times upper limit of therapeutic range on methods on SMAC®, Abbott-VP, Ektachem®, Hitachi® 705 and KDA. *2138* At concentration of 2,000 mg/L had no effect on biuret method with blank correction. *3393*
Urine Increase Physiological When high doses sodium glycinate salt given. *2313*

Sodium *Serum No Effect Analytical* At concentration of 20 mg/L had no effect on flame-photometric method. *3393* At concentration of 1,000 mg/L had no effect on measurement by ISE with predilution. *3393*
Serum Decrease Physiological Significant effect 8 h after i.v. infusion. *3935*

Somatostatin *Plasma Increase Analytical* With a variety of different antibodies for RIA measurement inhibits tracer binding at concentrations normally achieved therapeutically. *3341*
Plasma Decrease Analytical Interferes with binding in radioimmunoassays with 3 different antisera. *3341*

Thyroid Stimulating Hormone (TSH)
Serum No Effect Physiological Increased response to TRH: due to increased intracellular ATP which exerts action on TRH. *3798*

Triglycerides *Serum No Effect Analytical* At 5 times upper limit of therapeutic range on methods on SMAC®, Abbott-VP, Ektachem®, Hitachi® 705 and KDA. *2138* No effect at therapeutic concentration on Reflotron method. *1984* At concentration of 2,000 mg/L had no effect on lipase/esterase method. *3393*

Urea Nitrogen *Serum No Effect Analytical* At 5 times upper limit of therapeutic range on methods on SMAC®, Abbott-VP, Cobas-Bio, Ektachem®, Hitachi® 705 and KDA. *2138* No effect at therapeutic concentration on Reflotron method. *1984* At concentration of 2,000 mg/L had no effect on diacetylmonoxime method. *3393* At concentration of 23 mg/L had no effect on Ektachem® method. *3393*

Uric Acid *Serum No Effect Analytical* At 1.0 g/L on reversed phase LC procedure of Zhiri et al. *3942* At 5 times upper limit of therapeutic range on methods on SMAC®, Abbott-VP, aca, Cobas-Bio, Ektachem®, Hitachi® 705 and KDA. *2138* No effect at therapeutic concentration on Reflotron method. *1984* At concentration of 2,000 mg/L had no effect on phosphotungstate reduction method. *3393*
Serum Increase Analytical Reduction of phosphotungstate by metabolites. *2220*
Serum Increase Physiological Significant effect dose related and unrelated to reduced clearance. Slight inhibitory effect on hypoxanthine guanine phosphoribosyl transferase. *2488*
Urine Increase Analytical Reduction of phosphotungstate by metabolites. *0580*

Volume *Urine Increase Physiological* 4.7 mL/min versus 1.9 mL/min in 11 healthy subjects when given drug intravenously. *1593*

Thiabendazole

Alanine Aminotransferase *Serum Increase Physiological* Rare cholestasis. *2413*

Aldolase *Serum Increase Physiological* Probably of muscle origin although taken orally. *2427*

Alkaline Phosphatase *Serum Increase Physiological* Rare cholestasis. *1778* One case of drug-induced cholestasis reported. *2962*

Aspartate Aminotransferase
Serum Increase Physiological Rare cholestasis and parenchymal liver damage. *2413* One case of drug-induced cholestasis reported. *2962*

Bilirubin *Serum Increase Physiological* Rare cholestasis. *2413* One case of drug-induced cholestasis reported. *2962*

BSP Retention *Serum Increase Physiological* Rare cholestasis. *2413*

Cephalin Flocculation *Serum Increase Physiological* Rare cholestasis and parenchymal liver damage. *2413*

Cholesterol *Serum Increase Physiological* Cholestatic effect. *2413*

Crystals *Urine Increase Analytical* Low solubility of drug, especially at neutral pH. *2413*

Glucose *Serum Increase Physiological* Rare case of hyperglycemia reported. *1488*
Serum Decrease Physiological Asymptomatic lowering observed. *0071*

Leukocytes *Blood Decrease Physiological* Agranulocytosis/leukopenia. *2413*

Myoglobin *Urine No Effect Physiological* Not observed although serum aldolase increased. *2427*

5'-Nucleotidase *Serum Increase Physiological* Rare cholestasis. *1778*

Odor *Urine Increase Analytical* Odor similar to that after ingestion of asparagine. *2413*

Drug Listings

Thiabendazole (continued)

Platelets *Blood Decrease Physiological* Thrombocytopenia. *2413*

Protein *Urine Increase Physiological* May cause nephrotoxicity. *1488*

Thymol Turbidity *Serum Increase Physiological* Rare cholestasis. *2413*

Urobilinogen *Urine Decrease Physiological* Cholestatic effect. *2413*
Feces Decrease Physiological Pale stools, cholestatic effect. *2413*

Thiacetazone

Alanine Aminotransferase *Serum Increase Physiological* May cause cholestatic (hepatocanalicular) jaundice. *0248* May cause hepatitis. *1292*

Alkaline Phosphatase *Serum Increase Physiological* May cause cholestatic (hepatocanalicular) jaundice. *0248*

Aspartate Aminotransferase
Serum Increase Physiological May cause cholestatic (hepatocanalicular) jaundice. *0248* May cause hepatitis. *1292*

Bile *Urine Increase Physiological* May cause cholestasis. *0248*

Bilirubin *Serum Increase Physiological* May cause cholestatic (hepatocanalicular) jaundice. *0248*

BSP Retention *Serum Increase Physiological* May cause cholestasis. *0248*

Cephalin Flocculation *Serum Increase Physiological* May cause cholestasis. *0248*

Cholesterol *Serum Increase Physiological* May cause cholestatic (hepatocanalicular) jaundice. *0248*

Erythrocytes *Blood Decrease Physiological* Rare Stevens-Johnson syndrome induced. *2427*

Hemoglobin *Blood Decrease Physiological* May cause hemolytic anemia; may fall without frank anemia. *1292*

Leukocytes *Blood Decrease Physiological* Drug related agranulocytosis (rare). *2427*

Neutrophils *Blood Decrease Physiological* May cause agranulocytosis. *1292*

Thymol Turbidity *Serum Increase Physiological* May cause cholestasis. *0248*

Urobilinogen *Urine Decrease Physiological* May cause cholestasis. *0248*
Feces Decrease Physiological Pale stools, occurs with cholestasis. *0248*

Thiamazole

Albumin *Serum No Effect Analytical* At concentration of 114 mg/L had no effect on BCG method. *3393*

Bilirubin *Serum No Effect Analytical* At concentration of 114 mg/L had no effect on Jendrassik and Grof method. *3393*

Calcium *Serum No Effect Analytical* At concentration of 114 mg/L had no effect on cresolphthalein method. *3393*

Neutrophils *Blood Decrease Physiological* Occasional case of drug-induced neutropenia. *0155*

Phosphate *Serum No Effect Analytical* At concentration of 114 mg/L had no effect on phosphomolybdate method. *3393*

Protein *Serum No Effect Analytical* At concentration of 114 mg/L had no effect on biuret method with blank correction. *3393*

Triglycerides *Serum Decrease Analytical* At concentrations above 100 mg/L lowered concentration as measured by GPO-PAP method. *3393*

Urea Nitrogen *Serum No Effect Analytical* At concentration of 114 mg/L had no effect on diacetylmonoxime method. *3393*

Uric Acid *Serum No Effect Analytical* At concentration of 114 mg/L had no effect on phosphotungstate reduction method. *3393*

Thiamine

Creatinine *Serum No Effect Analytical* At 500 mg/L on reversed phase LC procedure of Zhiri et al. *3942*

Lymphocytes *Blood Decrease Physiological* Significant effect with megadose supplementation. *1347*

Neutrophils *Blood No Effect Physiological* No effect with megadose supplementation. *1347*

Uric Acid *Serum No Effect Analytical* At 500 mg/L on reversed phase LC procedure of Zhiri et al. *3942*
Test Conditions Increase Analytical Positive spot test with phosphotungstate. *3765*

Urobilinogen *Test Conditions Increase Analytical* Positive spot test with Ehrlich's reagent. *3765*

Thiamylal Sodium

Protein *Test Conditions Increase Analytical* Reacts with Folin-Ciocalteu of Lowry method. *0702*

Protein Bound Iodine (PBI) *Serum No Effect Physiological* No effect on PBI observed. *0583*

T₃ Uptake *Serum No Effect Physiological* No effect on test. *0583*

Thiazides

Alanine Aminotransferase *Serum Increase Physiological* May cause cholestasis. *2313*

Alkaline Phosphatase *Serum Increase Physiological* Cholestatic hepatitis reported. *1343*

Ammonia *Plasma Increase Physiological* Associated with K depletion and alkalosis. *1238*

Amylase *Serum Increase Physiological* May increase up to 200% in a week. *0729*

Aspartate Aminotransferase
Serum Increase Physiological May cause cholestasis. *2313*

Bicarbonate *Serum Increase Physiological* Over 60% have value over 30 meq/L in long term. *0269*

Bilirubin *Serum Increase Physiological* Cholestatic hepatitis reported to occur. *1343*

Bromide *Urine Increase Physiological* In cases of bromide intoxication. *1343*

BSP Retention *Serum Increase Physiological* Reduced plasma volume and hepatic blood flow. *1488*

Calcium *Serum Increase Physiological* May increase up to 0.25 meq/L. *2745*
Urine Decrease Physiological By up to 50% (appear to have extra-renal effect). *2745*

Cephalin Flocculation *Serum Increase Physiological* May cause cholestasis. *2313*

Chloride *Serum Decrease Physiological* Diuretic action (impaired tubular reabsorption). *2313*
Urine Increase Physiological Diuretic action. *1343*

Cholesterol *Serum Increase Physiological* Infrequent cholestatic effect. *1713*

Citrate *Urine Decrease Physiological* By up to 30%. *2745*

Cortisol *Urine Decrease Physiological* ?changed secretion or renal handling. *1000*

Creatine *Urine Decrease Physiological* Clearance tests decreased by 10-20%. *2426*

Creatinine *Serum Increase Physiological* Nephrotoxic effect with large doses. *2220*
Urine Decrease Physiological Clearance tests decreased by 10-20%. *2426*

Creatinine Clearance *Urine Decrease Physiological* May cause decrease by up to 20%. *2426*

Erythrocytes *Blood Decrease Physiological* Rare hypersensitive depression of bone marrow. *2313*
Urine Increase Physiological May cause actual bleeding. *1022*

Glomerular Filtration Rate (GFR)
Urine Decrease Physiological Especially if administered i.v. *1343*

Glucose *Serum Increase Physiological* Decreased glucose tolerance. *0070*

Urine Increase Physiological May occur in prediabetics especially May occur especially in prediabetics. *0460*

Glucose Tolerance *Serum Decrease Physiological* Diabetogenic-like action of drug. *2745*

Hemoglobin *Urine Increase Physiological* Actual bleeding caused by the drug. *1022*

Histamine Test *Patient Decrease Physiological* False negative due to hypotensive effect of drug. *1488*

131I Uptake *Serum No Effect Physiological* No effect observed in most patients. *1487*

LE Cells *Blood Positive Physiological* SLE may occur, usually normalizes when stopped. *3059*

Leukocytes *Blood Decrease Physiological* Reported to cause agranulocytopenia. *2745*

Magnesium *Serum Decrease Physiological* Consequence of diuresis. *1748*
Urine Increase Physiological Increase of 33% reported. *3368*

5'-Nucleotidase *Serum Increase Physiological* Due to cholestasis. *1778*

Phentolamine Test *Patient Decrease Physiological* False negative due to hypotensive effect of drug. *1488*

Platelets *Blood Decrease Physiological* Reported to cause thrombocytopenia. *2745*

Potassium *Serum Decrease Physiological* Consequence of diuresis. *1343*
Urine Increase Physiological Diuretic action. *1343*

Protein Bound Iodine (PBI) *Serum No Effect Physiological* Reported to have no effect. *0830*
Serum Decrease Physiological Increased excretion with prolonged therapy may occur. *1343*

Prothrombin Time *Plasma Increase Physiological* Associated with failure of excretion of bile salts. *1713*

PSP Excretion *Urine Decrease Physiological* Competition for excretion by tubules. *0974*

Quinidine *Serum Increase Physiological* Alkalinize urine, increase reabsorption. *1487*

Sodium *Serum Decrease Physiological* Diuretic action with sodium depletion. *0070*
Urine Increase Physiological Diuretic action. *1343*

Sugar *Urine Increase Analytical* Interferes with Benedict's reagent. *1022*

T₃ Uptake *Serum Decrease Physiological* Slight effect observed only. *2392*

Thymol Turbidity *Serum Increase Physiological* May cause cholestasis. *2313*

Tyramine Test *Patient Decrease Physiological* May attenuate response to tyramine. *1034*

Urea Nitrogen *Serum Increase Physiological* Nephrotoxic effect in large doses. *1022*

Uric Acid *Serum Increase Physiological* Decreased renal excretion. *1237*
Urine Decrease Physiological Decreased renal excretion (18% in long term). *0269*

Uric Acid Clearance *Urine Decrease Physiological* Decreased renal clearance. *2427*

Urobilinogen *Urine Decrease Physiological* Cholestatic effect. *1713*
Feces Decrease Physiological May cause cholestasis with pale stools. *1713*

Volume *Plasma Decrease Physiological* Due to diuretic action. *1343*

Thiazolsulfone

Bilirubin *Serum Increase Physiological* Hemolytic anemia. *2313*

Color *Urine Increase Analytical* Red, pink, purple, orange and rust color. *1237*

Erythrocytes *Blood Decrease Physiological* May cause hemolytic anemia. *2313*

Haptoglobin *Serum Decrease Physiological* Hemolytic anemia. *2313*

Hematocrit *Blood Decrease Physiological* May cause hemolytic anemia. *2313*

Hemoglobin *Blood Decrease Physiological* May cause hemolytic anemia. *2313*

Methemoglobin *Blood Increase Physiological* Hemolytic anemia. *2313*

Thiethylperazine

Alanine Aminotransferase *Serum Increase Physiological* Few cases of cholestasis reported. *1680*

Aspartate Aminotransferase *Serum Increase Physiological* Few cases of cholestasis reported. *1680*

Bilirubin *Serum Increase Physiological* Few cases of cholestasis reported. *1680*

Prolactin *Plasma Increase Physiological* Significant response to drug given i.m. or orally. *1416*

Thimerosal

N-Acetylglucosaminidase *Urine No Effect Analytical* At 20 mg/L had no effect on 2 colorimetric analytical methods. *1354*

Cholesterol *Serum Decrease Analytical* Marked reduction of results with methods using cholesterol oxidase and p-Aminophenazone. *0471*

Glucose *Serum No Effect Analytical* At 2 g/dL on p-HBAH procedure of Lever. At 2 g/dL on alkaline ferricyanide procedures. At 2 g/dL on glucose oxidase procedures. At 2 g/dL on o-toluidine procedure. *2140*

Immunoglobulins *Serum Increase Analytical* As preservative increases immunodiffusion precipitin rings. *0573*
Serum Decrease Analytical Affects immunodiffusion if added as bactericidal. *2579*

Mercury *Urine Increase Physiological* May occur with overdosage. *1680*

Protein Bound Iodine (PBI) *Serum Decrease Analytical* Inhibition of reaction caused by contaminated skin. *3218*

Thiobarbituric Acid

Alkaloids *Urine No Effect Analytical* With ninhydrin on Frings TLC procedure at 10 mg/dL. With iodoplatinate of Frings TLC procedure at 10 mg/dL. *1204*

Barbiturate *Urine No Effect Analytical* With vanillin on Frings procedure at 10 mg/dL. With mercurous NO₃ on Frings TLC procedure at 10 mg/dL. With diphenylcarbazone on Frings TLC procedure at 10 mg/dL. With mercuric SO₄ on Frings TLC procedure at 10 mg/dL. *1204*

Bromide *Urine No Effect Analytical* No effect on method of Frings. *1204*

Ethchlorvynol *Urine No Effect Analytical* No effect on method of Frings and Cohen. *1204*

FPN Test *Urine No Effect Analytical* No effect at 10 mg/dL on method of Frings. *1204*

Salicylate *Urine No Effect Analytical* At 10 mg/dL no effect on Trinder method. *1204*

Thiocarlide

Alanine Aminotransferase *Serum Increase Physiological* Reported effect in 4% subjects. *2427*

Bilirubin *Serum Increase Physiological* Rare observation. *2427*

Glucose *Serum Decrease Physiological* Hypoglycemic reactions observed. *2427*

Leukocytes *Blood Decrease Physiological* Associated with relative lymphocytosis and monocytosis. *2427*

Protein Bound Iodine (PBI) *Serum Decrease Physiological* Myxedema reported in one case. *2427*

Thiocolchicine

Eosinophils *Blood Increase Physiological* May cause eosinophilia. *2143*

Erythrocytes *Blood Decrease Physiological* May cause anemia (rare). *2143*

Thiocolchicine (continued)

Lymphocytes *Blood Increase Physiological* May cause real increase. *2143*

Neutrophils *Blood Decrease Physiological* May be reduced although lymphocytes increased. *2143*

Platelets *Blood Decrease Physiological* May cause thrombocytopenia occasionally. *2143*

Thiocyanate

Alanine Aminotransferase *Serum Increase Physiological* May cause hepatic necrosis. *1343*

Aspartate Aminotransferase
Serum Increase Physiological May cause hepatic necrosis. *1343*

Erythrocytes *Blood Decrease Physiological* May cause anemia. *2313* Several cases of aplastic anemia reported. *3717*

^{131}I Uptake *Serum No Effect Physiological* Does not affect uptake by thyroid. *2220*

Leukocytes *Blood Decrease Physiological* Several cases of aplastic anemia reported. *3717*

Platelets *Blood Decrease Physiological* Several cases of aplastic anemia reported. *3717*

Protein *Urine Increase Physiological* May cause nephrosis. *1343*

Protein Bound Iodine (PBI) *Serum Decrease Physiological* Blockage of thyroxine biosynthesis. *1238*

Urea Nitrogen *Serum Increase Physiological* May cause nephrosis. *1343*

Thioglycolate

Leukocytes *Blood Decrease Physiological* Reported effect. *2429*

Thioguanine

Alanine Aminotransferase *Serum Increase Physiological* May cause cholestasis. *2313*

Alkaline Phosphatase *Serum Increase Physiological* May cause cholestasis. *2313* Transient increase in 3 of 29 patients with polycythemia rubra vera. *2450*

Aspartate Aminotransferase
Serum Increase Physiological May cause cholestasis. *2313* Transient increase in 3 of 29 patients with polycythemia rubra vera. *2450*

Bile *Urine Increase Physiological* May cause cholestasis. *2313*

Bilirubin *Serum Increase Physiological* May cause cholestasis. *2313*

BSP Retention *Serum Increase Physiological* May cause cholestasis. *2313*

Cephalin Flocculation *Serum Increase Physiological* May cause cholestasis. *2313*

Chromosomes *Test Conditions Abnormal Physiological* Clastogenic in human cells. *3282*

Hematocrit *Blood Decrease Physiological* In 29 patients with polycythemia rubra vera; effect usually apparent in 2 weeks. *2450*

Hemoglobin *Blood Decrease Physiological* In 29 patients with polycythemia rubra vera; effect usually apparent in 2 weeks. *2450*

Leukocytes *Blood Decrease Physiological* May cause marrow depression. *0543* In 29 patients with polycythemia rubra vera; effect usually apparent in 2 weeks. *2450* Observed in 7 patients with chronic granulocytic leukemia. *3416*

Neutrophils *Blood Decrease Physiological* Observed in 7 patients with chronic granulocytic leukemia. *3416*

Platelets *Blood Decrease Physiological* Immunologically induced thrombocytopenia. *0543* In 29 patients with polycythemia rubra vera; effect usually apparent in 2 weeks. *2450* Observed in 7 patients with chronic granulocytic leukemia. *3416*

Protein *Test Conditions Increase Analytical* Reacts with Folin-Ciocalteu of Lowry method. *1569*

Thymol Turbidity *Serum Increase Physiological* May cause cholestasis. *2313*

Uric Acid *Serum No Effect Analytical* At 10 mg/dL on Nishi phosphotungstate procedure At 10 mg/dL on copper chelate procedure. *2230*
Serum Increase Physiological Rapid destruction of tissues — nucleic acid catabolism. *1343*
Urine Increase Physiological Due to augmented tissue catabolism. *1343*

Thiols

Creatine Kinase *Serum Increase Analytical* Three to tenfold enhancement. *0249*

Porphobilinogen *Urine Decrease Analytical* Unless prior separation is employed. *1116*

Thioneine

Uric Acid *Serum Increase Analytical* Affects phosphotungstate reduction methods. *3500*

Thiopental

Alanine Aminotransferase *Serum No Effect Analytical* At acute overdose concentration (20 mg/dL) on Technicon® SMAC® method. *3719*
Serum Increase Physiological 1 case out of 24 3-5 d after anesthesia, more longer period after anesthesia pyrexia and jaundice in one case after repeated anesthesia. *0391*

Albumin *Serum No Effect Analytical* At acute overdose concentration (20 mg/dL) on Technicon® SMAC® method. *3719* At concentration of 200 mg/L had no effect on BCG method. *3393*

Alkaline Phosphatase *Serum No Effect Analytical* At acute overdose concentration (20 mg/dL) on Technicon® SMAC® method. *3719*
Serum Increase Physiological 1 case out of 24 3-5 d after anesthesia, more longer period after anesthesia pyrexia and jaundice in one case after reported anesthesia. *0391*

Ammonia *Plasma Increase Physiological* Impaired metabolism in liver of dogs. *1487*

Aspartate Aminotransferase *Serum No Effect Analytical* At acute overdose concentration (20 mg/dL) on Technicon® SMAC® method. *3719*
Serum Increase Physiological 1 case out of 24 3-5 d after anesthesia, more longer period after anesthesia pyrexia and jaundice in one case after repeated anesthesia. *0391*

Basophils *Blood Decrease Physiological* Significant fall within 3 minutes of i.v. injection. *2212*

Bicarbonate *Serum No Effect Analytical* At concentration of 70 mg/L had no effect on method using phenolphthalein method. *3393*

Bilirubin *Serum No Effect Analytical* At acute overdose concentration (20 mg/dL) on Technicon® SMAC® method. *3719* At concentration of 200 mg/L had no effect on Jendrassik and Grof method. *3393*
Serum Increase Physiological 1 case out of 24 3-5 d after anesthesia, more longer period after anesthesia. *0391*

Calcium *Serum No Effect Analytical* At acute overdose concentration (20 mg/dL) on Technicon® SMAC® method. *3719* At concentration of 200 mg/L had no effect on cresolphthalein method. *3393*

Chloride *Serum No Effect Analytical* At concentration of 200 mg/L had no effect on mercurimetric method. *3393*

Cholesterol *Serum No Effect Analytical* At acute overdose concentration (20 mg/dL) on Technicon® SMAC® method. *3719* At concentration of 200 mg/L had no effect on Liebermann-Burchard method. *3393*

Creatine Kinase *Serum No Effect Analytical* At acute overdose concentration (20 mg/dL) on Technicon® SMAC® method. *3719*

Creatinine *Serum No Effect Analytical* At acute overdose concentration (20 mg/dL) on Technicon® SMAC® method. *3719* At concentration of 200 mg/L had no effect on AutoAnalyzer Jaffé method. *3393*

Ethchlorvynol *Serum No Effect Analytical* At 10 μg/mL on colorimetric method of Wallace. *3753*

Glucaric Acid *Urine Increase Physiological* Enhanced excretion with anesthesia. *0003*

Glucose *Serum No Effect Analytical* At acute overdose concentration (20 mg/dL) on Technicon® SMAC® method. *3719* *Serum No Effect Physiological* No effect except during surgery when causes increase. *0972*

γ-Glutamyltransferase (GGT)
Serum Increase Physiological 1 case out of 24 3-5 d after anesthesia, more longer period after anesthesia. *0391*

Guaiacols Spot Test *Urine Positive Analytical* False reaction with screening test of Rogers. *3031*

Histamine *Plasma Increase Physiological* Normal response even up by 350% at 5 minutes. *2212*

Hydrochloric Acid *Gastric Material Increase Physiological* Stimulation of secretion parallels plasma histamine. *2212*

Iron *Serum No Effect Analytical* At acute overdose concentration (20 mg/dL) on Technicon® SMAC® method. *3719* At concentration of 200 mg/L had no effect on Ferrozine method. *3393*

Lactate Dehydrogenase *Serum No Effect Analytical* At acute overdose concentration (20 mg/dL) on Technicon® SMAC® method. *3719* *Serum Increase Physiological* pyrexia and jaundice in one case after repeated anesthesia. *0391*

Phosphate *Serum No Effect Analytical* At acute overdose concentration (20 mg/dL) on Technicon® SMAC® method. *3719* At concentration of 200 mg/L had no effect on phosphomolybdate method. *3393*

Protein *Serum No Effect Analytical* At acute overdose concentration (20 mg/dL) on Technicon® SMAC® method. *3719* At concentration of 200 mg/L had no effect on biuret method with blank correction. *3393*

Protein Bound Iodine (PBI) *Serum No Effect Physiological* No effect on PBI in humans. *0583* *Serum Decrease Physiological* Inhibits iodination of tyrosine in thyroxine binding globulin. *0012*

T₃ Uptake *Serum No Effect Physiological* No effect on test. *0583*

Triglycerides *Serum No Effect Analytical* At acute overdose concentration (20 mg/dL) on Technicon® SMAC® method. *3719* At concentration of 200 mg/L had no effect on lipase/esterase method. *3393*

Urea Nitrogen *Serum No Effect Analytical* At acute overdose concentration (20 mg/dL) on Technicon® SMAC® method. *3719* At concentration of 200 mg/L had no effect on diacetylmonoxime method. *3393*

Uric Acid *Serum No Effect Analytical* At acute overdose concentration (20 mg/dL) on Technicon® SMAC® method. *3719* At concentration of 200 mg/L had no effect on phosphotungstate reduction method. *3393*

Thiopronine

Acetylcholine Receptor Antibodies
Serum No Effect Physiological No occurrence of myasthenia gravis or other autoimmune syndromes reported. *1761*

Leukocytes *Blood No Effect Physiological* Toxicity similar to penicillamine and other sulfhydryl drugs. *1761*

Platelets *Blood Decrease Physiological* Toxicity similar to penicillamine and other sulfhydryl drugs. *1761*

Protein *Urine Increase Physiological* Toxicity similar to penicillamine and other sulfhydryl drugs. *1761*

Thiopropazate

Alkaloids *Urine No Effect Analytical* With ninhydrin on Frings TLC procedure at 10 mg/dL. *1204* *Urine Increase Analytical* Purple with iodoplatinate on Frings TLC procedure. *1204*

Barbiturate *Urine No Effect Analytical* With diphenylcarbazone on Frings TLC procedure at 10 mg/dL. With mercurous NO₃ on Frings TLC procedure at 10 mg/dL. *1204*

Urine Increase Analytical Pink spot with vanillin in Frings TLC procedure. Pink/orange with mercuric SO₄ in Frings TLC procedure. *1204*

Bromide *Urine No Effect Analytical* No effect on method of Frings. *1204*

Ethchlorvynol *Urine No Effect Analytical* No effect on method of Frings and Cohen at 10 mg/dL. *1205*

FPN Test *Urine No Effect Analytical* Negative FPN test with 0.4 mg/dL concentration in Frings TLC procedure. *1204* *Urine Increase Analytical* Pink color observed in method of Frings. Pink FPN test with 0.5 mg/dL concentration in Frings TLC procedure. *1204*

Homovanillic Acid *CSF No Effect Physiological* No significant effect in Huntington's chorea. *2379*

5-Hydroxyindoleacetic Acid (5-HIAA)
CSF No Effect Physiological No significant effect in Huntington's chorea. *2379*

Leukocytes *Blood Decrease Physiological* Rare leukopenia. *0071*

Salicylate *Urine No Effect Analytical* Up to 5 mg/dL has no effect on Trinder procedure. *1204* *Urine Increase Analytical* At 10 mg/dL produces pink color on Trinder procedure. *1204*

Thioridazine

Acetaminophen Screening Test *Urine Negative Analytical* No reaction with o-cresol at therapeutic concentrations. *3326*

Alanine Aminotransferase *Serum Increase Physiological* Hepatotoxicity. *1857*

Alkaline Phosphatase *Serum Increase Physiological* Hypersensitivity reaction. *1857*

Alkaloids *Urine No Effect Analytical* With ninhydrin on Frings TLC procedure at 10 mg/dL. *1204* *Urine Increase Analytical* Purple with iodoplatinate on Frings TLC procedure. *1204*

Amphetamine *Urine No Effect Analytical* No reaction with NBD chloride procedure of Monforte. *2473*

Aspartate Aminotransferase *Serum No Effect Analytical* No effect on Karmen procedure. No effect on Babson procedure. *2760* *Serum Increase Physiological* Hepatotoxicity. *1857*

Barbiturate *Urine No Effect Analytical* With mercurous NO₃ on Frings TLC procedure at 10 mg/dL. With diphenylcarbazone on Frings TLC procedure at 10 mg/dL. *1204* *Urine Increase Analytical* Blue spot with mercuric SO₄ in Frings TLC procedure. Violet spot with vanillin on Frings TLC procedure. *1204*

Bile *Urine Increase Physiological* Hepatotoxicity. *1857*

Bilirubin *Serum Increase Physiological* Hepatotoxicity (questionable). *1857* *Serum Decrease Physiological* ?due to effect on bilirubin metabolism. *1629*

Bromide *Urine No Effect Analytical* No effect on method of Frings. *1204*

BSP Retention *Serum Increase Physiological* Hepatotoxicity. *1857*

Cephalin Flocculation *Serum Increase Physiological* Hepatotoxicity. *1857*

Chlordiazepoxide *Blood No Effect Analytical* On method of Riddick at 5 mg/dL. *2983*

Diazepam *Serum No Effect Analytical* On cathode-ray Polarographic method of Berry. *0323*

Diphenylhydantoin *Serum No Effect Analytical* On GLC procedure of Papadopoulos at *in vivo* concentration. *2722* *Serum Increase Physiological* Two cases reported of increased plasma concentration due to competition for metabolism by cytochrome P-450. *3716* Significant increase in 15% patients with combined therapy, reduction in 7%, No change in others. *3140*

Erythrocytes *Blood Decrease Physiological* Occasionally seen with phenothiazines. *1680*

Estriol *Urine No Effect Analytical* No effect of 62 mg/L on GLC method. *1163*

Ethchlorvynol *Serum No Effect Analytical* At 100 mg/L on GLC procedure of Evenson. *1058*

Thioridazine (continued)

Ethchlorvynol (continued)
Urine No Effect Analytical No effect on method of Frings and Cohen at 10 mg/dL. *1205*

FPN Test *Urine No Effect Analytical* Negative FPN test with 0.4 mg/dL concentration in Frings TLC procedure. *1204*
Urine Increase Analytical Blue color observed in method of Frings Blue FPN test with 0.5 mg/dL concentration in Frings TLC procedure. *1204*

Growth Hormone *Plasma No Effect Physiological* No significant difference in patients receiving drug on continuing basis versus controls. *3860*

Hematocrit *Blood Decrease Physiological* Significant decrease reported. *1629*

Hemoglobin *Blood Decrease Physiological* Significant effect reported. *1629*

Leukocytes *Blood Decrease Physiological* Agranulocytosis due to inhibition of development. *2822*

Luteinizing Hormone (LH) *Plasma Decrease Physiological* Significantly less in 42 male schizophrenics than when they ingested other neuroleptic agents. *0500*

Methaqualone *Serum No Effect Analytical* At 100 mg/L on GLC procedure of Evenson. *1057*

Neutrophils *Blood Decrease Physiological* Occasional case of agranulocytosis reported. *3717*

Pheneturide *Serum No Effect Analytical* On GLC procedure of Papadopoulos at *in vivo* concentration. *2722*

Phenobarbital *Serum No Effect Analytical* On GLC procedure of Papadopoulos at *in vivo* concentration. *2722*

Pituitary Gonadotropin *Urine Decrease Physiological* Total gonadotrophic activity reduced. *3550*

Platelets *Blood Decrease Physiological* Occasionally seen with phenothiazines. *1680*

Pregnancy Tests *Urine Positive Analytical* With Prognosticon and other tests. *1487*

Primidone *Serum No Effect Analytical* On GLC procedure of Papadopoulos at *in vivo* concentration. *2722*

Prolactin *Plasma Increase Physiological* Significant response to drug given orally (50 mg). *1416* Significant increase within 45 minutes of ingestion of 50 mg orally and 8 to 19 times baseline at 2 h. Functions as potent dopamine antagonist in the tuberoinfundibular system. Effect dose related. *3108*

Propoxyphene *Urine No Effect Analytical* Less than 1% fluorescence in procedure of Valentour. *3662*

Protein Bound Iodine (PBI) *Serum No Effect Physiological* No effect on PBI observed. *0583*

Salicylate *Urine No Effect Analytical* Up 1 mg/dL no effect on Trinder procedure. *1204*
Urine Increase Analytical At 2 mg/dL produces green color on Trinder procedure. *1204*

Testosterone *Serum Decrease Physiological* Slight effect in patients on long-term treatment. *1736* Significantly less in 42 male schizophrenics than when they ingested other neuroleptic agents. *0500*

Thymol Turbidity *Serum Increase Physiological* Hepatotoxicity. *1857*

Thiosemicarbazones

Alanine Aminotransferase *Serum Increase Physiological* May affect liver function. *2313*

Alkaline Phosphatase *Serum Increase Physiological* May affect liver function. *2313*

Aspartate Aminotransferase
Serum Increase Physiological May affect liver function. *2313*

Bile *Urine Increase Physiological* May affect liver function. *2313*

Bilirubin *Serum Increase Physiological* May affect liver function. *2313*

BSP Retention *Serum Increase Physiological* May affect liver function. *2313*

Cephalin Flocculation *Serum Increase Physiological* May affect liver function. *2313*

Erythrocytes *Blood Decrease Physiological* Hydrazone complex formed with pyridoxal PO$_4$. *3025*

Hematocrit *Blood Decrease Physiological* Large dose effect. *1343*

Hemoglobin *Blood Decrease Physiological* Large dose effect. *1343*

Leukocytes *Blood Decrease Physiological* leukopenia/agranulocytosis in 0.5%. *1343*

Protein *Urine Increase Physiological* Nephrotoxic effect. *1237* Nephrotoxic effect. *1022*

Thymol Turbidity *Serum Increase Physiological* May affect liver function. *2313*

Thiotepa

Erythrocytes *Blood Decrease Physiological* May cause bone marrow depression. *0071* 10 of 25 patients given drug intravesically had at least one incident of acute myelosuppression; 5 of 29 patients had chronic myelosuppression. *1632*

Hematocrit *Blood Decrease Physiological* May be rapid decrease. *2427*

Hemoglobin *Blood Decrease Physiological* May be rapid decrease. *2427* 10 of 25 patients given drug intravesically had at least one incident of acute myelosuppression; 5 of 29 patients had chronic myelosuppression. *1632*

Leukocytes *Blood Decrease Physiological* May cause bone marrow depression. *0071* 10 of 25 patients given drug intravesically had at least one incident of acute myelosuppression; 5 of 29 patients had chronic myelosuppression. *1632*

Occult Blood *Feces Positive Physiological* May be ulceration of gastrointestinal tract. *2427*

Platelets *Blood Decrease Physiological* May cause bone marrow depression. *0071* 10 of 25 patients given drug intravesically had at least one incident of acute myelosuppression; 5 of 25 patients had chronic myelosuppression. *1632*

Pseudocholinesterase *Serum Decrease Physiological* Mild depressive effects reported. *1487*

Uric Acid *Serum Increase Physiological* Due to cell destruction, may cause nephropathy. *1680*

Thiothixene

Alanine Aminotransferase *Serum Increase Physiological* Hepatotoxic effect (reversible cholestasis). *2313*

Alkaline Phosphatase *Serum Increase Physiological* Hepatotoxic effect (reversible cholestasis). *1022*

Alkaloids *Urine No Effect Analytical* With ninhydrin on Frings TLC procedure at 10 mg/dL. *1204*
Urine Increase Analytical Red/purple with iodoplatinate on Frings TLC procedure. *1204*

Aspartate Aminotransferase
Serum Increase Physiological Hepatotoxic effect (reversible cholestasis). *2313*

Barbiturate *Urine No Effect Analytical* With mercuric SO$_4$ on Frings TLC procedure at 10 mg/dL. With mercurous NO$_3$ on Frings TLC procedure at 10 mg/dL. With diphenylcarbazone on Frings TLC procedure at 10 mg/dL. *1204*
Urine Increase Analytical Orange spot with vanillin in Frings TLC procedure. *1204*

Bile *Urine Increase Physiological* Hepatotoxic effect (reversible cholestasis). *2313*

Bilirubin *Serum Increase Physiological* Hepatotoxic effect (reversible cholestasis). *2313*

Bromide *Urine No Effect Analytical* No effect on method of Frings. *1204*

BSP Retention *Serum Increase Physiological* Hepatotoxic effect (reversible cholestasis). *2313*

Cephalin Flocculation *Serum Increase Physiological* Hepatotoxic effect (reversible cholestasis). *2313*

Eosinophils *Blood Increase Physiological* Reported with other phenothiazines. *1680*

Erythrocytes *Blood Decrease Physiological* Reported with other phenothiazines. *1680*

Ethchlorvynol *Urine No Effect Analytical* No effect on method of Frings and Cohen. *1204*

FPN Test *Urine No Effect Analytical* No effect at 10 mg/dL on method of Frings. *1204*

Glucose *Serum Increase Physiological* Observed with some phenothiazines. *1678*
Serum Decrease Physiological Observed with some phenothiazines. *1678*
Urine Increase Physiological Observed with some phenothiazines. *1678*

Leukocytes *Blood Increase Physiological* Transient leukocytosis may occur. *1678*
Blood Decrease Physiological Transitory leukopenia. *2313*

Platelets *Blood Decrease Physiological* Reported with other phenothiazines. *1680*

Pregnancy Tests *Urine Positive Physiological* Observed with some phenothiazines. *1678*

Prolactin *Plasma Increase Physiological* Typical dose-related response to i.m. administered drug due to antidopaminergic action. *2072*

Prothrombin Time *Plasma Decrease Physiological* Rarely reported side effect. *0071*

Salicylate *Urine No Effect Analytical* At 10 mg/dL no effect on Trinder method. *1204*

Thymol Turbidity *Serum Increase Physiological* Hepatotoxic effect (reversible cholestasis). *2313*

Vanillylmandelic Acid *Urine No Effect Physiological* No effect on short term administration. *1635*

Thiouracil

Alanine Aminotransferase *Serum Increase Physiological* May cause cholestasis with cholangiolitis. *1948*

Alkaline Phosphatase *Serum Increase Physiological* May cause intrahepatic cholestasis. *1434*

Aspartate Aminotransferase
Serum Increase Physiological May cause cholestasis with cholangiolitis. *1948*

Bilirubin *Serum Increase Physiological* May cause intrahepatic cholestasis. *1434*

Cholesterol *Serum Increase Analytical* Interferes with Zlatkis-Zak reaction. *3505*
Serum Increase Physiological May cause intrahepatic cholestasis. *1434*
Serum Decrease Analytical At 13 mg/dL decreased by 30-40 mg/dL. *0583*

Glucose *Serum Decrease Physiological* Reported effect. *0117*

Leukocytes *Blood Decrease Physiological* Agranulocytosis (in about 1% cases). *3618*

Platelets *Blood Decrease Physiological* Thrombocytopenia. *2604*

Protein Bound Iodine (PBI) *Serum Decrease Physiological* Inhibits iodination of tyrosine residues in thyroxine binding globulin. *0012*

Thiourea

Propylthiouracil *Serum Increase Analytical* Reacts with 2,6-DQC (procedure of Ratliff). *2922*

Protein *Test Conditions Increase Analytical* Reacts with Folin-Ciocalteu of Lowry method. *0702*

Urea Nitrogen *Serum Increase Analytical* Reacts with xanthydrol in method of Fosse. *1563*

Thorium Dioxide

Alanine Aminotransferase *Serum Increase Physiological* Increased about 10% after injection. *3147*

Albumin *Serum Decrease Physiological* May produce severe liver damage with years. *3147*

Aldolase *Serum Increase Physiological* Time related liver damage (eventual tumor). *3147*

Alkaline Phosphatase *Serum Increase Physiological* Very common (related to time since injection). *3147*

Aspartate Aminotransferase
Serum Increase Physiological Increased in about 15% after injection. *3147*

Bilirubin *Serum Increase Physiological* Time related liver damage. *3147*

Bilirubin, Direct *Serum Increase Physiological* Time related liver damage. *3147*

BSP Retention *Serum Increase Physiological* May induce liver damage in half cases. *3147*

Cadmium Flocculation *Serum Increase Physiological* May produce severe liver damage after years. *3147*

Chromosomes *Test Conditions Abnormal Physiological* Clastogenic in human cells. *3282*

α_2-Globulin *Serum Increase Physiological* May produce severe liver damage after years. *3147*

β-Globulin *Serum Increase Physiological* May produce severe liver damage after years. *3147*

γ-Globulin *Serum Increase Physiological* May produce severe liver damage after years. *3147*

Leucine Aminopeptidase *Serum Increase Physiological* Common finding (related to time since injection). *3147*

Thymol Turbidity *Serum Increase Physiological* May produce severe liver damage after years. *3147*

Threonine

Ammonia *Plasma Decrease Analytical* With 1,000 nmoles produces 9% inhibition of indophenol reaction. *1290*

Tyrosine *Plasma No Effect Analytical* On fluorometric procedure of Ambrose. *0080*

Thrombin

Fibrinogen *Plasma No Effect Analytical* On potassium mercuric thiocyanate procedure of Roberts. *3003*

Thymol

Acetoacetate *Urine No Effect Analytical* No effect on Gerhardt procedure. *2981*

Acetone *Urine No Effect Analytical* No effect on Rothera procedure. *2981*

Amino Acids *Urine No Effect Analytical* No effect on ninhydrin. *2981*

Ammonia *Urine No Effect Analytical* No effect on Nessler, Berthelot methods. *2981*

Amylase *Urine No Effect Analytical* No effect on starch hydrolysis. *2981*

Bicarbonate *Urine No Effect Analytical* No effect on Van Slyke, Natelson methods. *2981*

Bile *Urine Increase Analytical* Affects determination of bile acids (Hay's test). *2981*

Bilirubin *Urine No Effect Analytical* No effect on Fouchet procedure. *2981*

Calcium *Urine No Effect Analytical* No effect on flame photometric methods. *2981*

Chloride *Urine No Effect Analytical* No effect on Schales and Schales method. *2981*

Color *Urine Increase Analytical* Greenish blue. *2313*

Creatine *Urine No Effect Analytical* No effect on Jaffé, Van Pilsum methods. *2981*

Creatinine *Urine No Effect Analytical* No effect on Jaffé, Van Pilsum methods. *2981*

Estrogens *Urine Increase Physiological* Marked interference with method of Hainsworth. *1457*

Glucose *Serum No Effect Analytical* At 0.7 g/dL o-toluidine procedure At 0.7 g/dL on alkaline ferricyanide procedures At 0.7 g/dL on glucose oxidase procedures. *2140*
Serum Decrease Analytical 1 mg/mL inhibits color = 10 mg/dL by o-toluidine. *0724*
Urine No Effect Analytical No effect on glucose-oxidase methods. *2981*

Indican *Urine No Effect Analytical* No effect on Obermayer and Jaffé methods. *2981*

Thymol *(continued)*

Odor *Urine Increase Physiological* Characteristic after ingestion. *0443*

Phosphate *Urine No Effect Analytical* No effect on Fiske and Subbarow method. *2981*

Porphobilinogen *Urine No Effect Analytical* No effect on Waldenström method. *2981*

Potassium *Urine No Effect Analytical* No effect on flame photometric methods. *2981*

Protein *Urine No Effect Analytical* No effect on biuret, boiling tests. *2981*
Urine Increase Analytical Excess may give false positive with turbidity procedures. *0459*

Protein Bound Iodine (PBI) *Serum Increase Physiological* Affects PBI if given as thymol iodide. *0012*

Sodium *Urine No Effect Analytical* No effect on flame photometric methods. *2981*

Sugar *Urine No Effect Analytical* No effect on reducing methods. *2981*

Urea Nitrogen *Serum Decrease Analytical* Inhibits urease affecting results. *1237*
Urine No Effect Analytical No effect on Berthelot method. *2981*

Urobilinogen *Urine No Effect Analytical* No effect on Waldenström procedure. *2981*

Xylose Excretion *Urine No Effect Analytical* No effect on analytical method. *2981*

Thyroid

Basal Metabolic Rate *Patient Increase Physiological* Metabolic effect of hormone. *2220*

Cholesterol *Serum Decrease Physiological* Physiological effect. *1488*

Glucose *Serum Increase Physiological* Metabolic action of hormone. *0652*

Hydroxyproline *Urine Increase Physiological* Due to catabolic action. *3644*

^{131}I Uptake *Serum Decrease Physiological* Consequence of treatment. *1488*

Protein *Serum Increase Physiological* Physiological effect exerts anabolic effect. *0697*

Protein Bound Iodine (PBI) *Serum Increase Physiological* Increases by 1-2 μg/dL/d/1 grain given. *1488*

Prothrombin Time *Plasma Increase Physiological* Prolongs action of anticoagulants. *2313*

T_3 Uptake *Serum Increase Physiological* Consequence of treatment. *1488*

Thymol Turbidity *Serum Increase Physiological* Increase protein metabolism. *0652*

Thyroxine (T_4) *Serum Increase Physiological* Increased available thyroxine. *2220*

Thyronine

Protein *CSF Increase Analytical* 1.0 mg = 2.9 mg in Folin-Ciocalteu procedure. *0583*

Thyrotropin

^{131}I Uptake *Serum Increase Physiological* Results vary with thyroid status. *2220*

Protein Bound Iodine (PBI) *Serum Increase Physiological* Starts within 15 h, may increase by up to 5.5 μg/dL. *0830*

Thyroxine (T_4) *Serum Increase Physiological* Several h after i.m. injection, less than on T_3. *1631*
Urine Increase Physiological Less marked effect than on T_3. *0518*

Tri-iodothyronine (T_3) *Urine Increase Physiological* Pronounced increase with single i.m. injection. *0631*

Thyrotropin-Releasing Hormone (TRH)

Cortisol *Plasma No Effect Physiological* No effect in normal subjects after i.v. administration. *0100*

Epinephrine *Plasma Increase Physiological* Mean increase of 28% in 2nd to 4th minute after i.v. injection regardless of whether patient was initially hypo- or hyperthyroid. *2921*

Fatty Acids, Free (FFA) *Serum No Effect Physiological* After single dose of 1.0 mg synthetic TRH. *3831*

Follicle Stimulating Hormone (FSH)
Plasma No Effect Physiological No effect in normal subjects after i.v. administration. *0100*

Glucose *Serum No Effect Physiological* After single dose of 1.0 mg synthetic TRH. *3831*

Growth Hormone *Plasma No Effect Physiological* No effect in normal subjects after i.v. administration. *0100*

Insulin *Plasma No Effect Physiological* After single dose of 1.0 mg synthetic TRH. *3831*

Luteinizing Hormone (LH) *Plasma No Effect Physiological* No effect seen with 1 mg i.v. in normals. *3822* No effect in normal subjects. *0100*

Norepinephrine *Plasma Increase Physiological* Mean increase of 21% in 2nd to 4th minute after i.v. injection regardless of whether patient was initially hypo- or hyperthyroid. *2921*

Prolactin *Plasma Increase Physiological* Response within 5 minutes to i.v. injection. *1753* Maximum effect 15-30 minutes after i.v. *1435*

Thyroid Stimulating Hormone (TSH)
Serum Increase Physiological Dose related when given i.v.- effect 20 mg orally. *1065* Threefold increase in 30 minutes. *0100*

Thyroxine (T_4) *Serum Increase Physiological* But no change in % free T_4 (variable response). *1631*

Tri-iodothyronine (T_3) *Serum Increase Physiological* But no change in percent free T_3. *1631*

L-Thyroxine

Acetaminophen Screening Test *Urine Negative Analytical* No reaction with o-cresol at therapeutic concentrations. *3326*

Apolipoprotein A *Serum No Effect Physiological* Insignificant change from 2.7 to 2.8 g/L in 11 hypothyroid women treated with 0.1 to 0.2 mg daily. *1201*

Catechol-O-Methyl Transferase
Test Conditions Decrease Physiological Heavy doses inhibit *in vivo*. *1176*

Cholesterol *Serum Decrease Physiological* From 7.8 to 6.1 mmol/L in 11 hypothyroid women treated with 0.1 to 0.2 mg daily. *1201*

Cholesterol, High Density Lipoprotein
Serum Decrease Physiological From 1.6 to 1.4 mmol/L in 11 hypothyroid women treated with 0.1 to 0.2 mg daily. *1201*

Cholesterol, Low Density Lipoprotein
Serum Decrease Physiological From 5.5 to 4.1 mmol/L in 11 hypothyroid women treated with 0.1 to 0.2 mg daily. *1201*

2,3-Diphosphoglycerate Mutase
Blood No Effect Physiological No effect on amount of 2,3-DPG produced. *3612*

Estriol *Urine Decrease Physiological* Decreased formation due to metabolic action. *0023*

2-Hydroxyestrone *Urine Increase Physiological* Metabolic effect of hormone administration. *0023*

Immunoglobulin IgA *Serum Decrease Physiological* In all 5 children with infantile hypothyroidism soon after start of treatment in 4 concentration returned to normal. *3232*

Immunoglobulin IgG *Serum No Effect Physiological* No significant effect in children with infantile hypothyroidism. *3232*

Immunoglobulin IgM *Serum No Effect Physiological* No significant effect in children with infantile hypothyroidism. *3232*

2-Methoxyestrone *Urine Increase Physiological* Metabolic effect on hormone administration. *0023*

Monoamine Oxidase *Serum Decrease Physiological* Observed *in vitro* and *in vivo*. *1176*

Norepinephrine *Plasma Decrease Physiological* From high pretreatment values to normal range with 30-60 d treatment of 7 hypothyroid women with dry thyroid extract. *2288*

T₃ Uptake *Serum Increase Physiological* From 0.59 to 0.98 arbitrary units in 11 hypothyroid women treated with 0.1 to 0.2 mg daily. *1201*

Thyroid Stimulating Hormone (TSH)
Serum Decrease Physiological In hypothyroidism in response to T₄ analog. *3798*

Thyroxine (T₄) *Serum Increase Physiological* Raised but still less than controls with 30-60 d treatment of 7 hypothyroid women with dry thyroid extract. *2288* From 24 to 124 nmol/L 11 hypothyroid women treated with 0.1 to 0.2 mg daily. *1201*

Thyroxine (T₄) Binding Globulin
Serum No Effect Analytical On radioimmunoassay procedure of Van Herle. *3676*

Thyroxine (T₄), Free *Serum Increase Physiological* Raised but still less than controls with 30-60 d treatment of 7 hypothyroid women with dry thyroid extract. *2288*

Triglycerides *Serum No Effect Physiological* At about 1.3 in 11 hypothyroid women treated with 0.1 to 0.2 mg daily. *1201*

Tri-iodothyronine (T₃) *Serum Increase Analytical* 33% cross reactivity if product from Cyclo. *3247*
Serum Increase Physiological From 0.7 to 17 nmol/L 11 hypothyroid women treated with 0.1 to 0.2 mg daily. *1201*

Tiaprofenic Acid

Uric Acid *Serum Decrease Physiological* In 10 healthy subjects given 300 mg bid. *2239*
Urine Increase Physiological In 10 healthy subjects given 300 mg bid, but effect occurs early so no increased excretion may be observed later. *2239*

Ticarcillin

Glucose *Urine No Effect Analytical* Concentrations measured accurately with Diastix® Concentrations measured accurately with TesTape®. *2100*
Urine Increase Analytical Falsely elevated values with Clinitest®. *2100*

Timegadine

α₁-Acid Glycoprotein *Serum No Effect Physiological* In 23 patients with rheumatoid arthritis given 250 to 750 mg/d for 48 weeks. *2345*

Albumin *Serum No Effect Physiological* In 23 patients with rheumatoid arthritis given 250 to 750 mg/d for 48 weeks. *2345*

Creatinine *Serum No Effect Physiological* In 23 patients with rheumatoid arthritis given 250 to 750 mg/d for 48 weeks. *2345*

Erythrocytes *Blood No Effect Physiological* In 23 patients with rheumatoid arthritis given 250 to 750 mg/d for 48 weeks. *2345*
Urine No Effect Physiological In 23 patients with rheumatoid arthritis given 250 to 750 mg/d for 48 weeks. *2345*

Erythrocyte Sedimentation Rate
Blood No Effect Physiological In 23 patients with rheumatoid arthritis given 250 to 750 mg/d for 48 weeks. *2345*

Glucose *Urine No Effect Physiological* In 23 patients with rheumatoid arthritis given 250 to 750 mg/d for 48 weeks. *2345*

Hemoglobin *Blood No Effect Physiological* In 23 patients with rheumatoid arthritis given 250 to 750 mg/d for 48 weeks. *2345*

Immunoglobulin IgA *Serum Decrease Physiological* But not significantly in 23 patients with rheumatoid arthritis given 250 to 750 mg/d for 48 weeks. *2345*

Immunoglobulin IgG *Serum Decrease Physiological* But not significantly in 23 patients with rheumatoid arthritis given 250 to 750 mg/d for 48 weeks. *2345*

Immunoglobulin IgM *Serum Decrease Physiological* But not significantly in 23 patients with rheumatoid arthritis given 250 to 750 mg/d for 48 weeks. *2345*

Leukocytes *Blood No Effect Physiological* In 23 patients with rheumatoid arthritis given 250 to 750 mg/d for 48 weeks. *2345*

Platelets *Blood No Effect Physiological* In 23 patients with rheumatoid arthritis given 250 to 750 mg/d for 48 weeks. *2345*

Protein *Urine No Effect Physiological* In 23 patients with rheumatoid arthritis given 250 to 750 mg/d for 48 weeks. *2345*

Rheumatoid Factor *Serum No Effect Physiological* In 23 patients with rheumatoid arthritis given 250 to 750 mg/d for 48 weeks. *2345*

Thyroxine (T₄) *Serum No Effect Physiological* In 23 patients with rheumatoid arthritis given 250 to 750 mg/d for 48 weeks. *2345*

Timolol

Albumin *Serum No Effect Analytical* At concentration of 0.01 mg/L had no effect on BCG method. *3393*

Bicarbonate *Serum No Effect Analytical* At concentration of 0.01 mg/L had no effect on method using phenolphthalein. *3393*

Bilirubin *Serum No Effect Analytical* At concentration of 0.01 mg/L had no effect on Jendrassik and Grof method. *3393*

Calcium *Serum No Effect Analytical* At concentration of 0.01 mg/L had no effect on cresolphthalein method. *3393*

Chloride *Serum No Effect Analytical* At concentration of 0.01 mg/L had no effect on mercurimetric method. *3393*

Cholesterol *Serum No Effect Analytical* At concentration of 0.01 mg/L had no effect on CHOD-PAP method. *3393*
Serum No Effect Physiological In 15 patients treated for 1 mo. *3826*

Creatinine *Serum No Effect Analytical* At concentration of 0.01 mg/L had no effect on AutoAnalyzer Jaffé method. *3393*

Fatty Acids, Free (FFA) *Serum No Effect Physiological* In 5 hyperthyroid patients given 10 mg every 8 h for 4 d. *0335*

Glucagon *Plasma No Effect Physiological* In 5 hyperthyroid patients given 10 mg every 8 h for 4 d. *0335*

Glucose *Serum No Effect Analytical* At concentration of 0.01 mg/L had no effect on GOD/POD-PAP method. *3393*
Plasma No Effect Physiological In 5 hyperthyroid patients given 10 mg every 8 h for 4 d. *0335*

Glycerol *Blood No Effect Physiological* In 5 hyperthyroid patients given 10 mg every 8 h for 4 d. *0335*

Hydroxyproline *Urine No Effect Physiological* No effect in hyperthyroid patients. *0334*

Insulin *Plasma No Effect Physiological* In 5 hyperthyroid patients given 10 mg every 8 h for 4 d. *0335*

Ketones *Blood No Effect Physiological* In 5 hyperthyroid patients given 10 mg every 8 h for 4 d. *0335*

Parathyroid Hormone *Plasma No Effect Physiological* No effect in hyperthyroid patients. *0334*

Phosphate *Serum No Effect Analytical* At concentration of 0.01 mg/L had no effect on phosphomolybdate method. *3393*

Potassium *Serum No Effect Analytical* At concentration of 0.01 mg/L had no effect on flame-photometric method. *3393*
Serum Increase Physiological Slight increase in patients treated with moderate doses of drug. *2836* Pronounced and prolonged increase in patients who had acute myocardial infarction. *2612*

Protein *Serum No Effect Analytical* At concentration of 0.01 mg/L had no effect on biuret method with blank correction. *3393*

Sodium *Serum No Effect Analytical* At concentration of 0.01 mg/L had no effect on flame-photometric method. *3393*

Thyroxine (T₄) Index, Free (FTI)
Serum No Effect Physiological In 5 hyperthyroid patients given 10 mg every 8 h for 4 d. *0335*

Triglycerides *Serum No Effect Analytical* At concentration of 0.01 mg/L had no effect on lipase/esterase method. *3393*
Serum No Effect Physiological In 15 patients treated for 1 mo. *3826*

Tri-iodothyronine (T₃) *Serum No Effect Physiological* No effect in hyperthyroid patients. *0334* In 5 hyperthyroid patients given 10 mg every 8 h for 4 d. *0335*

Urea Nitrogen *Serum No Effect Analytical* At concentration of 0.01 mg/L had no effect on diacetylmonoxime method. *3393*

Timolol *(continued)*

Uric Acid *Serum No Effect Analytical* At concentration of 0.01 mg/L had no effect on phosphotungstate reduction method. *3393*

Tin

Barbiturate *Serum Increase Analytical* Affects diphenylcarbazone procedure if in filter paper. *0583*

Magnesium *Urine Increase Analytical* Measured by fluorometric method of Schachter. *3573*

Tobacco Smoke

Potassium *Serum Increase Analytical* Interferes with flame photometry. *1238*

Tobramycin

Antithrombin III *Plasma No Effect Analytical* At concentration of 15 mg/L had no effect on aca method. *3393*

Bilirubin *Serum No Effect Analytical* At concentration of 5 mg/L had no effect on Ektachem® method. *3393*
Serum No Effect Physiological Clinically insignificant displacement from protein in neonates. *3748*

Casts *Urine Increase Physiological* At dose of 4.5 mg/kg/d for 12 d in 90 patients nephrotoxicity observed in up to 39%, reversible in most. *0677* Manifestation of drug induced nephrotoxicity but tend to decrease as serum creatinine begins to climb. *3182*

Chloramphenicol *Serum No Effect Analytical* No effect at 100 mg/L on coupled enzymatic method. *2490*

Creatinine *Serum Increase Physiological* In 12% of patients with sepsis. Mean increase of only 0.1 mg/dL in all patients studied. *3359* 12% incidence of nephrotoxicity but unrelated to initial renal function or prior use of aminoglycosides, drug concentration, amount given duration of treatment or concurrent treatment with other drugs. *3182* 18.4% incidence of nephrotoxicity in 49 patients given drug by McHenry method versus 16.7% in 48 patients given drug by Sawchuk/Zaske method. *2339* In 4 of 59 patients whose drug concentrations were monitored. *1250* At dose of 4.5 mg/kg/d for 12 d in 90 patients nephrotoxicity observed in up to 39%, reversible in most. *0677*

Glucose *Urine No Effect Analytical* No effect at up to 250 μg/mL on any glucose concentration as measured by Clinitest®, Diastix® or TesTape®. *2250*

Hemoglobin *Urine Increase Physiological* At dose of 4.5 mg/kg/d for 12 d in 90 patients nephrotoxicity observed in up to 39%, reversible in most. *0677*

β_2-Microglobulin *Urine No Effect Physiological* Up to i.v. dose of 2 mg/kg does not affect renal excretion. *3726*
Urine Increase Physiological Manifestation of drug induced nephrotoxicity but tend to decrease as serum creatinine begins to climb. *3182*

Plasminogen *Plasma No Effect Analytical* At concentration of 15 mg/L had no effect on aca method. *3393*

Protein *Urine Increase Physiological* Manifestation of drug induced nephrotoxicity but tend to decrease as serum creatinine begins to climb. *3182* At dose of 4.5 mg/kg/d for 12 d in 90 patients nephrotoxicity observed in up to 39%, reversible in most. *0677*

Thyroxine (T₄) *Serum No Effect Physiological* When given i.v. with cloxacillin to 13 patients. *0959*

Tri-iodothyronine (T₃) *Serum No Effect Physiological* When given i.v. with cloxacillin to 13 patients. *0959*

Tri-iodothyronine (T₃), Reverse
Serum No Effect Physiological When given i.v. with cloxacillin to 13 patients. *0959*

Tocainide

Alanine Aminotransferase *Serum Increase Physiological* Occasional abnormality reported associated with reversible hepatitis. *2011*

Antinuclear Antibodies
Serum Positive Physiological Rare finding in fewer than 0.2% patients. *2011*

Aspartate Aminotransferase
Serum Increase Physiological Occasional abnormality reported associated with reversible hepatitis. *2011*

Erythrocytes *Blood Decrease Physiological* Blood dyscrasias occur in fewer than 0.2% patients. *2011*

LE Cells *Blood Positive Physiological* Rare finding in fewer than 0.2% patients. *2011*

Leukocytes *Blood Decrease Physiological* Blood dyscrasias occur in fewer than 0.2% patients. *2011*

Neutrophils *Blood Decrease Physiological* Blood dyscrasias occur in fewer than 0.2% patients. *2011* Estimated incidence of 0.18%. *3019*

Platelets *Blood Decrease Physiological* Blood dyscrasias occur in fewer than 0.2% patients. *2011*

Tolazamide

Alanine Aminotransferase *Serum Increase Physiological* Cholestatic effect. *2313*

Alkaline Phosphatase *Serum Increase Physiological* Intrahepatic cholestatic jaundice. *0085*

Aspartate Aminotransferase
Serum Increase Physiological Cholestasis may occur. *2313*

Bile *Urine Increase Physiological* Cholestatic effect. *2313*

Bilirubin *Serum Increase Physiological* Cholestatic jaundice. *2313*

BSP Retention *Serum Increase Physiological* Cholestasis may occur. *2313*

Cephalin Flocculation *Serum Increase Physiological* Cholestasis may occur. *2313*

Creatinine *Serum No Effect Analytical* At 5 times therapeutic concentration on Du Pont aca, Beckman Astra and Technicon® SMAC® method. *2999*

Erythrocytes *Blood Decrease Physiological* May cause hemolytic anemia. *1343*

γ-Globulin *Serum Increase Physiological* Transient increases reported. *2427*

Glucose *Serum No Effect Analytical* No effect on hexokinase, o-toluidine methods. *2568* If guiacum or phenolaminophenazone with glucose oxidase. *3277* At concentration of 200 mg/L had no effect on GOD/POD-PAP method. *3393*
Serum Decrease Analytical Inhibits oxidation chromogen Boehringer glucose oxidase method. *3277* At concentrations above 10 mg/L lowered concentration as measured by GOD-Perid method. *3393*
Serum Decrease Physiological Sulfonylurea derivative stimulates insulin secretion. *3278*

Insulin *Plasma Increase Physiological* Usual effect observed. *0301*
Plasma Decrease Physiological Return to normal usually if high initially. *0301*

Leukocytes *Blood Decrease Physiological* Leukopenia/agranulocytosis. *1343*

5'-Nucleotidase *Serum Increase Physiological* Due to cholestasis. *1778*

Osmolality *Urine Decrease Physiological* Normal diuretic response. *2499*

Platelets *Blood Decrease Physiological* Thrombocytopenia/pancytopenia. *1343*

Prothrombin Time *Plasma Increase Physiological* Occurs with failure to excrete bile salts. *2313*

Thymol Turbidity *Serum Increase Physiological* Cholestasis may occur. *2313*

Urobilinogen *Urine Decrease Physiological* Cholestatic effect. *2313*
Feces Decrease Physiological Pale stools with cholestasis. *2313*

Volume *Urine Increase Physiological* Normal diuretic response. *2499*

Water Clearance, Free *Urine Increase Physiological* Normal diuretic response. *2499*

Tolazoline

Erythrocytes *Blood Decrease Physiological* Pancytopenia (AMA Blood dyscrasias). *2429*

Hydrochloric Acid *Gastric Material Increase Physiological* Also enhances histamine stimulation. *1343*

Leukocytes *Blood Decrease Physiological* Pancytopenia (AMA Blood dyscrasias). *2429*

Pepsin *Gastric Material Increase Physiological* Also enhances histamine stimulation. *1343*

Platelets *Blood Decrease Physiological* Pancytopenia (AMA Blood dyscrasias). *2429*

Tolbutamide

Alanine Aminotransferase *Serum Increase Physiological* May cause cytotoxic liver damage or cholestasis. *0248*

Albumin *Serum No Effect Analytical* At concentration of 100 mg/L had no effect on BCG method. *3393*

Alkaline Phosphatase *Serum Increase Physiological* May cause intrahepatic cholestatic syndrome. *0248* Bone isoenzyme activity increased: inversely influenced by serum 25-hydroxyvitamin D concentration. *3458*

Aminolevulinic Acid *Urine Increase Physiological* May precipitate porphyria attack. *1322*

Aspartate Aminotransferase *Serum Increase Analytical* At 1 mmol/L affects SMA 12/60 method. *3335*
Serum Increase Physiological May cause cytotoxic liver damage or cholestasis. *0248*

Barbiturate *Serum Increase Analytical* At 10 mg/dL affects diphenylcarbazone procedure. *0583*

Bicarbonate *Serum No Effect Analytical* At concentration of 100 mg/L had no effect on method using phenolphthalein. *3393*

Bile *Urine Increase Physiological* Cholestatic jaundice reported. *1488*

Bilirubin *Serum No Effect Analytical* At concentration of 270 mg/L had no effect on Jendrassik and Grof method. *3393*
Serum Increase Physiological Cholestatic jaundice reported. *1488*

BSP Retention *Serum Increase Physiological* May cause cytotoxic liver damage or cholestasis. *0248*

Calcium *Serum No Effect Analytical* At concentration of 270 mg/L had no effect on cresolphthalein method. *3393*
Serum No Effect Physiological Slight but not significant reduction, associated with increased bone organic matrix turnover. *3458*

Cephalin Flocculation *Serum Increase Physiological* Cholestatic phenomenon. *2313*

Chloride *Serum No Effect Analytical* At concentration of 100 mg/L had no effect on mercurimetric method. *3393*

Cholesterol *Serum No Effect Analytical* At concentration of 480 mg/L had no effect on CHOD-Iodide method. *3393* At concentration of 480 mg/L had no effect on CHOD-PAP method. *3393* At concentration of 480 mg/L had no effect on catalase-AIDH method. *3393* At concentration of 480 mg/L had no effect on method using catalase-Hantzsch reaction. *3393* At concentration of 480 mg/L had no effect on Liebermann-Burchard method. *3393*
Serum Decrease Physiological Inhibits hepatic synthesis (?also absorption). *0878*

Coombs' Test, Direct *Serum Positive Physiological* Probably due to autoantibody complex on RBC surface. *0359* Not definitive proof, hemolytic anemia. *1487*

Coproporphyrin *Blood Increase Physiological* May precipitate cutaneous porphyria. *1322*
Urine Increase Physiological May precipitate porphyria attack. *1322*
Feces Increase Physiological May precipitate cutaneous porphyria. *1322*

Creatinine *Serum No Effect Analytical* At 5 times therapeutic concentration on Du Pont aca, Beckman Astra and Technicon® SMAC® method. *2999* At concentration of 500 mg/L had no effect on creatinine iminohydrolase method. *3393* At concentration of 100 mg/L had no effect on AutoAnalyzer Jaffé method. *3393* At concentration of 400 mg/L had no effect on Jaffé-Fading-Fraction method. *3393* At concentration of 400 mg/L had no effect on Jaffé-Fuller's earth method. *3393* At concentration of 500 mg/L had no effect on kinetic Jaffé method on BKA-2. *3393*

Erythrocytes *Blood Decrease Physiological* Agranulocytosis/aplastic anemia. *0073*

Fatty Acids, Free (FFA) *Serum Increase Physiological* Slight initial rise then fall to 25% at 90 minutes. *1526*

Glucagon *Plasma No Effect Physiological* No effect i.v. if given slowly or rapidly. *2766*

Glucose *Serum No Effect Analytical* No effect on Boehringer GOD-PERID method. *3277* On Warner Glucomatic glucose oxidase method. No effect on Trinder glucose oxidase procedure. *2771* No effect glucose oxidase, o-toluidine, hexokinase procedures. *2568* At concentration of 220 mg/L had no effect on Ektachem® method. *3393* At concentration of 1,000 mg/L had no effect on GOD/POD-PAP method. *3393*
Serum Increase Analytical False increase (little effect at normal concentration) with glucose oxidase. *3278*
Serum Decrease Analytical Negative peaks with Boehringer GOD-PERID method. *2771* At concentrations above 500 mg/L (normal therapeutic concentration 110 mg/L) lowered concentration as measured by GOD-Perid method. *3393*
Serum Decrease Physiological Therapeutic intent (promotes insulin secretion). *1152*

Glycerol *Serum Decrease Physiological* Response similar to that of free fatty acids. *1526*

Hematocrit *Blood Decrease Physiological* Agranulocytosis/aplastic anemia. *0073*

Hemoglobin *Blood Decrease Physiological* Agranulocytosis/aplastic anemia. *0073*

Hydroxyproline *Urine Increase Physiological* Associated with turnover of bone organic matrix. *3458*

25-Hydroxy Vitamin D₃ *Serum Decrease Physiological* Associated with turnover of bone organic matrix. *3458*

Insulin *Plasma Increase Physiological* Marked rise associated with hypoglycemia. *1152* Slight effect max in 15 minutes. *0310*
Plasma Decrease Physiological Effect observed in some patients. *0301*

¹³¹I Uptake *Serum Decrease Physiological* Uncommon reported effect. *2220*

Leukocytes *Blood Decrease Physiological* Agranulocytosis/aplastic anemia. *0634*

Neutrophils *Blood Decrease Physiological* Occasionally observed. *2427* Occasional case of drug-induced neutropenia. *0155* Occasional case of agranulocytosis reported. *3717*

5'-Nucleotidase *Serum Increase Physiological* Due to cholestasis. *1778*

Phosphate *Serum No Effect Analytical* At concentration of 270 mg/L had no effect on phosphomolybdate method. *3393*

Platelets *Blood Decrease Physiological* Agranulocytosis/aplastic anemia. *0634*

Porphobilinogen *Urine Increase Physiological* May precipitate porphyria attack. *1322*

Potassium *Serum No Effect Analytical* At concentration of 100 mg/L had no effect on flame-photometric method. *3393* At concentration of 2,000 mg/L had no effect on measurement by ISE with predilution. *3393* At concentration of 400 mg/L had no effect on measurement by ISE without predilution. *3393*

Protein *Serum No Effect Analytical* At concentration of 270 mg/L had no effect on biuret method with blank correction. *3393*
Urine No Effect Analytical No effect on Albustix. *0459*
Urine Increase Analytical Causes turbidity if sulfosalicylic acid used. *3226* Affects heat and acetic acid test. *1872* Interferes with sulfosalicylic acid method. *3226*
CSF Increase Analytical Causes turbidity if sulfosalicylic acid used. *3505*

Tolbutamide (continued)

Protein Bound Iodine (PBI) *Serum   Decrease   Physiological* Inhibits iodination of tyrosine residues in thyroxine binding globulin. *0012*

Protein Electrophoresis *Serum   No Effect   Analytical* At concentration of 400 mg/L had no effect on automated Olympus-Hite method. *3393*

Prothrombin Time *Plasma   Increase   Physiological* Occurs with failure to excrete bile salts. *0894*

Plasma   Decrease   Physiological Stimulates metabolism of anticoagulants. *1487*

Protoporphyrin *Blood   Increase   Physiological* May precipitate cutaneous porphyria. *1322*

Feces   Increase   Physiological May precipitate cutaneous porphyria. *1322*

Sodium *Serum   No Effect   Analytical* At concentration of 100 mg/L had no effect on flame-photometric method. *3393* At concentration of 2,000 mg/L had no effect on measurement by ISE with predilution. *3393* At concentration of 400 mg/L had no effect on measurement by ISE without predilution. *3393*

T_3 Uptake *Serum   Increase   Physiological* Increase by 5-10% with 1 g i.v. *0583*

Thymol Turbidity *Serum   Increase   Physiological* May cause cytotoxic liver damage or cholestasis. *2313*

Thyroxine (T_4) *Serum   Decrease   Physiological* Displaces from thyroxine binding globulin. *3505*

Triglycerides *Serum   No Effect   Analytical* At concentration of 540 mg/L had no effect on GPO-PAP method. *3393* At concentration of 100 mg/L had no effect on lipase/esterase method. *3393*

Urea Nitrogen *Serum   No Effect   Analytical* At concentration of 270 mg/L had no effect on diacetylmonoxime method. *3393* At concentration of 220 mg/L had no effect on Ektachem® method. *3393*

Uric Acid *Serum   No Effect   Analytical* At concentration of 260 mg/L had no effect on Kageyama-Hantzsch method. *3393* At concentration of 400 mg/L had no effect on catalase-AIDH method. *3393* At concentration of 270 mg/L had no effect on phosphotungstate reduction method. *3393* At concentration of 500 mg/L had no effect on uricase-PAP method. *3393*

Urobilinogen *Urine   Decrease   Physiological* Possible cholestatic effect. *1713*

Feces   Decrease   Physiological Pale stools with cholestasis may occur. *1713*

Uroporphyrin *Urine   Increase   Physiological* May precipitate porphyria attack. *1322*

Tolmetin

Alanine Aminotransferase *Serum   No Effect   Physiological* Observed in single case of 49 y old man who had taken drug as needed for arthritis for 1 y. *3421*

Alkaline Phosphatase *Serum   Increase   Physiological* Observed in single case of 49 y old man who had taken drug as needed for arthritis for 1 yr. *3421*

Bilirubin *Serum   Increase   Physiological* Observed in single case of 49 y old man who had taken drug as needed for arthritis for 1 y. *3421*

Urine   Increase   Physiological Observed in single case of 49 y old man who had taken drug as needed for arthritis for 1 y. *3421*

Coombs' Test, Direct *Serum   Increase   Physiological* Strongly positive: Observed in single case of 49 y old man who had taken drug as needed for arthritis for 1 y. *3421*

Erythrocytes *Blood   Decrease   Physiological* In 2% of patients in trials for treatment of juvenile rheumatoid arthritis. *0239*

γ-Glutamyltransferase (GGT)
Serum   Increase   Physiological Observed in single case of 49 y old man who had taken drug as needed for arthritis for 1 y. *3421*

Haptoglobin *Serum   Decrease   Physiological* Observed in single case of 49 y old man who had taken drug as needed for arthritis for 1 y. *3421*

Hematocrit *Blood   Decrease   Physiological* Observed in single case of 49 y old man who had taken drug as needed for arthritis for 1 y. *3421*

Lactate Dehydrogenase *Serum   Increase   Physiological* Observed in single case of 49 y old man who had taken drug as needed for arthritis for 1 y. *3421*

Partial Thromboplastin Time
Plasma   Increase   Physiological In 1% of patients in trials for treatment of juvenile rheumatoid arthritis. *0239*

Protein *Urine   Increase   Physiological* In 1% of patients in trials for treatment of juvenile rheumatoid arthritis. *0239*

Reticulocytes *Blood   Increase   Physiological* Observed in single case of 49 y old man who had taken drug as needed for arthritis for 1 y. *3421*

Urobilinogen *Urine   Increase   Physiological* Observed in single case of 49 y old man who had taken drug as needed for arthritis for 1 y. *3421*

Tolonium

Color *Urine   Increase   Analytical* Green and blue color. *1022*

Erythrocytes *Blood   Decrease   Physiological* May cause hemolysis with G-6-PD deficiency. *0333*

Heinz Body Formation *Blood   Positive   Physiological* May cause hemolysis with G-6-PD deficiency. *0333*

Hematocrit *Blood   Decrease   Physiological* May cause hemolysis with G-6-PD deficiency. *0333*

Hemoglobin *Blood   Decrease   Physiological* May cause hemolysis with G-6-PD deficiency. *0333*

Toluene

β_1-α-Globulin *Serum   Decrease   Physiological* Significant effect in people occupationally exposed. *3370*

Hippuric Acid *Urine   Increase   Physiological* Metabolites also increased on exposure. *3607*

Immunoglobulin IgA *Serum   Decrease   Physiological* Significant effect in people occupationally exposed. *3370*

Immunoglobulin IgG *Serum   Decrease   Physiological* Significant effect in people occupationally exposed. *3370*

Immunoglobulin IgM *Serum   No Effect   Physiological* No effect in people occupationally exposed. *3370*

Toluene Diamine

Alanine Aminotransferase *Serum   Increase   Physiological* May cause cholestasis with cholangiolitis. *1948*

Aspartate Aminotransferase
Serum   Increase   Physiological May cause cholestasis with cholangiolitis. *1948*

Bilirubin *Serum   Increase   Physiological* May cause cholestasis with cholangiolitis. *1948*

Tomatoes

Phytofluene *Serum   Increase   Physiological* Dramatic rise within 30 minutes of ingestion. *0517*

Vitamin A *Serum   Increase   Physiological* Significant rise of vitamin A maximum at 90 minutes. *0517*

Toothpaste

Protein Bound Iodine (PBI) *Serum   Increase   Physiological* Some may contain enough I_2 to cause effect. *0830*

Torasemide

Calcium *Urine   Increase   Physiological* Similar effects to that of furosemide. *0919*

Chloride *Urine   Increase   Physiological* Similar effects to that of furosemide. *0919*

Creatinine Clearance *Urine   Increase   Physiological* Similar effects to that of furosemide. *0919*

Magnesium *Urine Increase Physiological* Similar effects to that of furosemide. *0919*

Potassium *Urine Increase Physiological* Similar effects to that of furosemide. *0919*

Sodium *Urine Increase Physiological* Similar effects to that of furosemide. *0919*

Volume *Urine Increase Physiological* Similar effects to that of furosemide. *0919*

Tosylate Bretylium

Catecholamines *Urine Decrease Physiological* Inhibits release of norepinephrine. *3722*

Prothrombin Time *Plasma Increase Physiological* Enhances anticoagulant activity. *1343*

Tourniquet

Albumin *Serum Increase Physiological* Affects all protein bound constituents also. *1563*

Calcium *Serum Increase Physiological* Increase by 0.11 meq/L over 3 minutes. *1704* Due to relative increase of plasma proteins. *0884*

Cholesterol *Serum No Effect Physiological* No effect for 1 minute. *3540*
Serum Increase Physiological 5-20% increase with tourniquet on for 5 minutes. *3476*

Erythrocyte Sedimentation Rate
Blood Increase Physiological Hemoconcentration increases rate. *0459*

Hematocrit *Blood No Effect Physiological* No effect for 1 minute. *3540*

Hemoglobin *Plasma No Effect Physiological* No effect observed with local circulatory stasis. *1755*
Blood Increase Physiological Venous stasis may cause significant increase. *3879*

Ionized Calcium *Serum No Effect Physiological* No effect noted for up to 5 minutes. *2056* No effect if used for less than 2 minutes. *3493*

Magnesium *Serum No Effect Physiological* Insignificant over 3 minutes. *1704*

Phosphate *Serum No Effect Physiological* Insignificant over 3 minutes. *1704*

Potassium *Serum Increase Physiological* Especially if forearm muscles exercised also. *1563*

Prostaglandin E1 *Plasma Decrease Physiological* Mean decrease of 7.1 ng/mL after 5 minutes. *1385*

Protein *Serum Increase Physiological* Affects all protein bound constituents also. *1563* Increase by 0.5 g/dL over 3 minutes. *1704*

Protein Bound Iodine (PBI) *Serum Increase Physiological* Occlusion for 5 minutes may cause increase up to 2 µg/dL. *0830*

Triglycerides *Serum No Effect Physiological* No effect for 1 minute. *3540*

Tranquilizers

17-Hydroxycorticosteroids *Urine Increase Analytical* Some artifactually increase value. *0279*

17-Ketosteroids *Urine Increase Analytical* Some artifactually increase value. *0279*

Transport of Specimen

Bicarbonate *Serum No Effect Analytical* By pneumatic tube if specimen protected. *3217*

Bilirubin *Serum No Effect Analytical* By pneumatic tube if specimen protected. *3217*
Amniotic Fluid Decrease Analytical Cardboard container cause decrease of 450 nm peak. *0694*

Calcium *Serum No Effect Analytical* By pneumatic tube if specimen protected. *3217*

Chloride *Serum No Effect Analytical* By pneumatic tube if specimen protected. *3217*

Creatinine *Serum No Effect Analytical* By pneumatic tube if specimen protected. *3217*

Fibrinogen *Plasma No Effect Analytical* By pneumatic tube if specimen protected. *3217*

Glucose *Serum No Effect Analytical* By pneumatic tube if specimen protected. *3217*

Hemoglobin *Plasma Increase Analytical* 7.9 mg/dL in pneumatic tube system. *3217*

Lactate Dehydrogenase *Serum Increase Analytical* Increase by 50 units in pneumatic tube. *3217*

pCO2 *Blood No Effect Analytical* No effect in pneumatic tube system. *3217*

pH *Blood No Effect Analytical* No effect in pneumatic tube system. *3217*

Phosphate *Serum No Effect Analytical* By pneumatic tube if specimen protected. *3217*

pO2 *Blood Increase Analytical* Increase by 0.5 to 4.5 mm Hg in pneumatic tube. *3217*

Potassium *Serum Increase Analytical* 0.1 mmol/L in pneumatic tube system. *3217*

Protein *Serum No Effect Analytical* By pneumatic tube if specimen protected. *3217*

Sodium *Serum No Effect Analytical* By pneumatic tube if specimen protected. *3217*

Urea Nitrogen *Serum No Effect Analytical* By pneumatic tube if specimen protected. *3217*

Uric Acid *Serum No Effect Analytical* By pneumatic tube if specimen protected. *3217*

Tranylcypromine

Alanine Aminotransferase *Serum Increase Physiological* May affect liver function (?hypersensitivity). *2313*

Alkaline Phosphatase *Serum Increase Physiological* May affect liver function (?hypersensitivity). *2313*

α-Amino-Nitrogen *Plasma Increase Physiological* All amino acids increased except those listed below. *0824*

Amphetamine *Urine No Effect Analytical* No reaction with NBD chloride of Monforte. *2473* At 50 mg/L on fluorescent method of Hayes. *1534* No effect at 100 mg/L on method of Rutter. *3097*

Aspartate Aminotransferase
Serum Increase Physiological May affect liver function (?hypersensitivity). *2313*

Aspartic Acid *Plasma Decrease Physiological* Observed effect in normal individuals. *0824*

Bile *Urine Increase Physiological* May affect liver function (?hypersensitivity). *2313*

Bilirubin *Serum Increase Physiological* May affect liver function (?hypersensitivity). *2313*

BSP Retention *Serum Increase Physiological* Hepatotoxic/cholestatic syndromes. *1343*

Cephalin Flocculation *Serum Increase Physiological* May affect liver function (?hypersensitivity). *2313*

Citrulline *Plasma Decrease Physiological* Observed effect in normal individuals. *0824*

Dihydroxyphenylacetic Acid
Plasma Decrease Physiological From mean of 710 to 63 ng/L in 6 patients with depression or Alzheimer's disease treated with up to 40 mg daily for at least 3 weeks. *1006*

Dihydroxyphenylalanine *Plasma No Effect Physiological* In 6 patients with depression or Alzheimer's disease treated with up to 40 mg daily for at least 3 weeks. *1006*

Dihydroxyphenylglycol *Plasma Decrease Physiological* From mean of 850 to 210 ng/L in 6 patients with depression or Alzheimer's disease treated with up to 40 mg daily for at least 3 weeks. *1006*

Epinephrine *Plasma No Effect Physiological* In 6 patients with depression or Alzheimer's disease treated with up to 40 mg daily for at least 3 weeks. *1006*

Glucose *Serum Decrease Physiological* In animals probably stimulates insulin secretion. *1487*

131I Uptake *Serum Decrease Physiological* Parnate contains tetraiodofluorescein. *2652*

Tranylcypromine (continued)

Leukocytes *Blood Decrease Physiological* Theoretical effect of this type of drug. *0071*

Norepinephrine *Plasma No Effect Physiological* In 6 patients with depression or Alzheimer's disease treated with up to 40 mg daily for at least 3 weeks. *1006*

Ornithine *Plasma No Effect Physiological* No effect observed. *0824*

Proline *Plasma No Effect Physiological* No effect observed. *0824*

Taurine *Plasma Decrease Physiological* Observed effect in normal individuals. *0824*

Thymol Turbidity *Serum Increase Physiological* May affect liver function (?hypersensitivity). *2313*

Trazodone

Glucose *Serum No Effect Analytical* At concentration of 50 mg/L had no effect on GOD/POD-PAP method. *3393*

Triglycerides *Serum No Effect Analytical* At concentration of 50 mg/L had no effect on GPO-PAP method. *3393*

Uric Acid *Serum No Effect Analytical* At concentration of 50 mg/L had no effect on uricase-PAP method. *3393*

Tretinoin

Alanine Aminotransferase *Serum Increase Physiological* In 1 of 11 patients given drug orally. *3867*

Alkaline Phosphatase *Serum Increase Physiological* In 1 of 11 patients given drug orally. *3867*

Aspartate Aminotransferase *Serum Increase Physiological* In 1 of 11 patients given drug orally. *3867*

Triamcinolone

Amino Acids *Urine Increase Physiological* Negative nitrogen balance, protein catabolism. *1680*
Urine Positive Analytical Purple spot with ninhydrin on thin-layer chromatography when combined with neomycin and nystatin in Kenacomb. *1278*

Amylase *Serum Increase Physiological* May occasionally cause pancreatitis. *1680*

Calcium *Urine Increase Physiological* Typical action of all corticosteroids. *1680*

Cortisol *Plasma No Effect Analytical* Reactivity of less than 1% possible with RIA. *1081*
Plasma Decrease Physiological Marked decrease in 2 children given injections of compound with suppression of hypothalamic-pituitary axis. *0777*

Glucose *Serum Increase Physiological* Glucocorticoid effect. *2525*
Urine Increase Physiological Consequence of hyperglycemia. *2525*

Glucose Tolerance *Serum Decrease Physiological* Latent diabetes manifestation, endocrine effect. *1680*

Nitrogen Balance *Patient Negative Physiological* Due to protein catabolism. *1680*

Occult Blood *Feces Positive Physiological* May activate peptic ulcer or cause perforation. *1680*

pH *Blood Increase Physiological* May cause hypokalemic alkalosis. *1680*

Potassium *Urine Increase Physiological* Only in exceedingly large doses. *0071*

Sodium *Urine Increase Physiological* Mild diuresis with sodium loss in first few days. *0071*

Volume *Plasma Increase Physiological* May cause sodium and fluid retention. *1680*

Triamterene

Aldosterone *Plasma Increase Physiological* Mean increase of 125 pg/mL in 3 men given drug 50 mg tid for 1 week. *3736*

Amino-4-Imidazole-5-Carboxamide Ribotide (AICAR)

Urine Increase Physiological If megaloblastic anemia occurs. *3773*

Ammonia *Urine No Effect Physiological* Does not cause excessive excretion. *1680*

Bicarbonate *Serum Decrease Physiological* Nephrotoxic effect. *1022*

Calcium *Serum No Effect Physiological* No significant effect observed. *3745*
Urine Increase Physiological Impairs reabsorption. *2745*

Catecholamines *Urine Decrease Analytical* When measured with Bio-Rad laboratories catecholamines column test procedure, fluorescence of drug interferes with high blank production. *0246*

Chloride *Serum Increase Physiological* Nephrotoxic and azotemic effect. *2313*
Serum Decrease Physiological Diuretic action with impaired tubular reabsorption. *0652*
Urine Increase Physiological Diuretic action. *1343*

Color *Urine Increase Analytical* Green, blue with blue fluorescence. *1022*

Creatinine *Serum Increase Physiological* Nephrotoxic effect (causes reduced GFR). *2313*

Creatinine Clearance *Urine Decrease Physiological* Probably reduced renal blood flow. *3745*

Crystals *Urine Increase Physiological* Either free or as part on conglomerations and also in large round brown bodies; common at pH <6.0. *1070*

Diagnex Blue Excretion *Urine Increase Analytical* Increased dye release from resin. *1237*

Digoxin *Serum Increase Analytical* At 500 ng/mL equals 0.3 ng/mL by RIA. *3940*
Serum Increase Physiological Increased concentration when given in association with triamterene than when given alone. *3736*

Eosinophils *Blood Increase Physiological* Allergic response. *1513*

Erythrocytes *Blood Decrease Physiological* May cause megaloblastic anemia (folic acid antagonist). *3773*

FIGLU (N-Formiminoglutamic Acid)
Urine Increase Physiological If megaloblastic anemia occurs. *3773*

Folate *Serum Decrease Physiological* Inhibits dihydrofolate reductase. *3773* Identified as folate antagonist. *3023*

Glomerular Filtration Rate (GFR)
Urine Decrease Physiological Causes reduced glomerular filtration rate. *0652*

Glucose *Serum No Effect Physiological* In normals but increased in diabetics. *3744*
Serum Increase Physiological Effect less common than with thiazides. *2029*

Glucose Tolerance *Serum Decrease Physiological* Increase up to 250 mg/dL seen in normals even. *3744*

Hematocrit *Blood No Effect Physiological* Usually no effect in normals. *3744*
Blood Decrease Physiological May cause megaloblastic anemia (folic acid antagonist). *3773*

Hemoglobin *Blood Decrease Physiological* May cause megaloblastic anemia (folic acid antagonist). *3773*

Lactate Dehydrogenase *Serum No Effect Analytical* At 10 mmol/L on UV methods. *0618*
Serum Increase Analytical At 100 µmol/L on fluorometric procedure. *0618*

Leukocytes *Blood Increase Physiological* Due to eosinophilia. *2313*

Lithium *Serum Increase Physiological* Reported in 2 cases taking lithium: drug also given with hydrochlorothiazide: due to reduced clearance of lithium. *2393*

Magnesium *Serum Increase Physiological* Slight increased effect observed in normals. *3745*
Urine No Effect Physiological Spared renal excretion with 37.5 mg/d. *3101*
Urine Increase Physiological Increased clearance observed. *3745*

MCV *Blood Increase Physiological* Megaloblastic anemia. *3773*

pCO₂ *Blood No Effect Physiological* No effect in normals (decreased in diabetics). *3744*

pH *Blood No Effect Physiological* No effect in normals when given 100 mg bid. *3744*
Urine Increase Physiological Slight alkalinization (mechanism not known). *1343*

Phosphate *Serum No Effect Physiological* No significant effect observed. *3745*

Potassium *Serum Increase Physiological* Potassium sparing action (affects Na/K exchange). *2313* Mean increase of 0.3 mmol/L in 3 men given drug for 1 week. *3736*
Serum Decrease Physiological Diuretic action. *3505*

Quinidine *Serum Increase Analytical* Interfering fluorescence. *1237*

Sodium *Serum Decrease Physiological* Diuretic action. *2313*
Urine Increase Physiological Diuretic action (increased clearance). *1343* Reduces availability of potassium within distal tubular cells for secretion into distal lumen. *2836*

Titratable Acidity *Urine No Effect Physiological* Does not cause excessive excretion. *1680*

Urea Nitrogen *Serum Increase Physiological* Nephrotoxic effect (with excessive diuresis). *1287*

Uric Acid *Serum No Effect Physiological* Unless predisposition to gouty arthritis. *1680*
Serum Increase Physiological Significant effect in about 17% cases. *1488*
Urine Increase Physiological Slight uricosuric action. *1237*

Uric Acid Clearance *Urine Decrease Physiological* Reduced in proportion to creatinine clearance. *3745*

Zinc *Urine No Effect Physiological* No effect observed. *2712*
Urine Increase Physiological Increase by 18% in 9 patients with hypertension for 2 weeks. *3806*

Triaziquone

Chromosomes *Test Conditions Abnormal Physiological* Clastogenic in human cells. *3282*

Triazolam

Cortisol *Urine Decrease Physiological* Observed in 9 poor sleepers given 0.5 mg/d for 3 weeks: overnight urinary cortisol measured. Immediate rebound increase on withdrawal. *0016*

Tribromethanol

Albumin *Serum No Effect Analytical* At concentration of 90 mg/L had no effect on BCG method. *3393*

Bicarbonate *Serum No Effect Analytical* At concentration of 90 mg/L had no effect on method using phenolphthalein. *3393*

Bilirubin *Serum No Effect Analytical* At concentration of 90 mg/L had no effect on Jendrassik and Grof method. *3393*

Calcium *Serum No Effect Analytical* At concentration of 90 mg/L had no effect on cresolphthalein method. *3393*

Chloride *Serum No Effect Analytical* At concentration of 90 mg/L had no effect on mercurimetric method. *3393*

Cholesterol *Serum No Effect Analytical* At concentration of 90 mg/L had no effect on Liebermann-Burchard method. *3393*

Creatinine *Serum No Effect Analytical* At concentration of 90 mg/L had no effect on AutoAnalyzer Jaffé method. *3393*

Phosphate *Serum No Effect Analytical* At concentration of 90 mg/L had no effect on phosphomolybdate method. *3393*

Protein *Serum No Effect Analytical* At concentration of 90 mg/L had no effect on biuret method with blank correction. *3393*

Triglycerides *Serum No Effect Analytical* At concentration of 90 mg/L had no effect on lipase/esterase method. *3393*

Urea Nitrogen *Serum No Effect Analytical* At concentration of 90 mg/L had no effect on diacetylmonoxime method. *3393*

Uric Acid *Serum No Effect Analytical* At concentration of 90 mg/L had no effect on phosphotungstate reduction method. *3393*

Trichloracetic Acid

Ascorbic Acid *Plasma Decrease Analytical* Increase by 10% if stored at -20° for 21 d. *0441*

Iron *Serum Increase Analytical* Volume reduction effect (1.5%) with protein loss. *1131*

Magnesium *Serum Increase Analytical* Enhances absorption in atomic absorption methods. *1769*

Sodium *Serum Increase Analytical* Volume reduction effect with protein precipitation. *1131*

Zinc *Serum Increase Analytical* Volume reduction effect with protein precipitation. *1131* Enhances absorption in atomic absorption methods. *1769*

Trichlorethanol

Albumin *Serum No Effect Analytical* At concentration of 1,000 mg/L had no effect on BCG method. *3393*

Bicarbonate *Serum No Effect Analytical* At concentration of 1,000 mg/L had no effect on method using phenolphthalein. *3393*

Bilirubin *Serum No Effect Analytical* At concentration of 1,000 mg/L had no effect on Jendrassik and Grof method. *3393*

Calcium *Serum No Effect Analytical* At concentration of 1,000 mg/L had no effect on cresolphthalein method. *3393*

Chloride *Serum No Effect Analytical* At concentration of 1,000 mg/L had no effect on mercurimetric method. *3393*

Cholesterol *Serum No Effect Analytical* At concentration of 12 mg/L had no effect on CHOD-PAP method. *3393* At concentration of 1,000 mg/L had no effect on Liebermann-Burchard method. *3393*

Creatinine *Serum No Effect Analytical* At concentration of 1,000 mg/L had no effect on AutoAnalyzer Jaffé method. *3393*

Glucose *Serum No Effect Analytical* At concentration of 12 mg/L had no effect on GOD/POD-PAP method. *3393*

Phosphate *Serum No Effect Analytical* At concentration of 1,000 mg/L had no effect on phosphomolybdate method. *3393*

Potassium *Serum No Effect Analytical* At concentration of 12 mg/L had no effect on flame-photometric method. *3393* At concentration of 1,000 mg/L had no effect on measurement by ISE with predilution. *3393*

Protein *Serum No Effect Analytical* At concentration of 1,000 mg/L had no effect on biuret method with blank correction. *3393*

Sodium *Serum No Effect Analytical* At concentration of 12 mg/L had no effect on flame-photometric method. *3393* At concentration of 1,000 mg/L had no effect on measurement by ISE with predilution. *3393*

Triglycerides *Serum No Effect Analytical* At concentration of 1,000 mg/L had no effect on lipase/esterase method. *3393*

Urea Nitrogen *Serum No Effect Analytical* At concentration of 1,000 mg/L had no effect on diacetylmonoxime method. *3393*

Uric Acid *Serum No Effect Analytical* At concentration of 1,000 mg/L had no effect on phosphotungstate reduction method. *3393*

Trichlormethiazide

Alanine Aminotransferase *Serum Increase Physiological* May rarely cause cholestasis. *1680*

Alkaline Phosphatase *Serum Increase Physiological* May rarely cause cholestasis. *1680*

Amylase *Serum Increase Physiological* May occasionally cause pancreatitis. *1680*

Apolipoprotein AI *Serum No Effect Physiological* Decrease by -5 mg/dL in 15 patients given 4 mg daily for 3 mo but note marked difference between responders and nonresponder populations. *2550*

Apolipoprotein B *Serum No Effect Physiological* Increase by 1 mg/dL in 15 patients given 4 mg daily for 3 mo but note marked difference between responders and nonresponder populations. *2550*

Trichlormethiazide *(continued)*

Apolipoprotein CII *Serum No Effect Physiological* Increase by 2 mg/dL in 15 patients given 4 mg daily for 3 mo but note marked difference between responders and nonresponder populations. *2550*

Apolipoprotein CIII *Serum Increase Physiological* Increase by 6 mg/dL in 15 patients given 4 mg daily for 3 mo but note marked difference between responders and nonresponder populations. *2550*

Apolipoprotein E *Serum No Effect Physiological* Increase by 1 mg/dL in 15 patients given 4 mg daily for 3 mo but note marked difference between responders and nonresponder populations. *2550*

Aspartate Aminotransferase
Serum Increase Physiological May rarely cause cholestasis. *1680*

Bilirubin *Serum Increase Physiological* May rarely cause cholestasis. *1680*

Calcium *Serum Increase Physiological* Impaired excretion. *3505*
Urine Decrease Physiological Impaired excretion. *3505*

Cholesterol *Serum No Effect Physiological* Decrease by -17 mg/dL in 15 patients given 4 mg daily for 3 mo but note marked difference between responders and nonresponder populations. *2550*

Cholesterol, High Density Lipoprotein
Serum Decrease Physiological Decrease by -11 mg/dL in 15 patients given 4 mg daily for 3 mo but note marked difference between responders and nonresponder populations *2550*

Erythrocytes *Blood Decrease Physiological* May cause bone marrow depression. *1680*

Fatty Acids, Free (FFA) *Serum Increase Physiological* Significant increase in fasting state (0.37 vs 0.31 meq/L) and 60 minutes (0.33 vs 0.26 meq/L) after 75 g glucose orally. *2550*

Glucose *Serum Increase Physiological* Diabetogenic like action of drug: affects glucose tolerance test. *3505* At 30 and 60 minutes in 6 patients given orally 75 g glucose values versus controls were 158 and 170 versus 136 and 156 mg/dL respectively. *2550*
Urine Increase Physiological Diabetogenic like action of drug Diabetogenic-like action of drug. *3505*

Insulin *Plasma Increase Physiological* Significant increase versus controls in 6 patients given 75 g glucose orally at 30 minutes (79 vs 54 mU/mL) and 60 minutes (99 vs 76 mU/mL). *2550*

Leukocytes *Blood Decrease Physiological* May cause bone marrow depression. *1680*

Platelets *Blood Decrease Physiological* May occur with/without purpura, bone marrow depression. *1680*

Postheparin Lipase *Plasma Decrease Physiological* Significant effect during 3 mo drug administration (2.3 vs 3.3 μmol free fatty acid/mL/h respectively). *2550*

Potassium *Serum Decrease Physiological* Diuretic action. *3505*

Protein Bound Iodine (PBI) *Serum Decrease Physiological* May occur in absence of thyroid hypofunction. *1680*

Sodium *Urine Increase Physiological* Therapeutic intent — diuretic action. *1680*

Triglycerides *Serum Increase Physiological* Increase by 44 mg/dL in 15 patients given 4 mg daily for 3 mo but note marked difference between responders and nonresponder populations. *2550*

Uric Acid *Serum Increase Physiological* Inhibition of tubular secretion of urate. *1433*

Trichloroethylene

Alanine Aminotransferase *Serum Increase Physiological* Potentially hepatotoxic. *0071*

Aspartate Aminotransferase
Serum Increase Physiological Potentially hepatotoxic. *0071*

Bilirubin *Serum Increase Physiological* Potentially hepatotoxic. *0071*

Tricine

Protein *Test Conditions Decrease Analytical* Lowry procedure, non linear absorption. *1796*

Triethanolamine

Potassium *Serum No Effect Analytical* At concentration of 200 mg/L had no effect on measurement by ISE with predilution. *3393*

Sodium *Serum No Effect Analytical* At concentration of 200 mg/L had no effect on measurement by ISE with predilution. *3393*

Triethylenemelamine

Alanine Aminotransferase *Serum Increase Physiological* Hepatotoxicity (prolonged treatment, large dose). *0071*

Aspartate Aminotransferase
Serum Increase Physiological Hepatotoxicity (prolonged treatment, large dose). *0071*

Bilirubin *Serum Increase Physiological* May cause hemolytic anemia. *2365*

Chromosomes *Test Conditions Abnormal Physiological* Clastogenic in human cells. *3282*

Coombs' Test, Direct *Serum Positive Physiological* May cause hemolytic anemia. *2365*

Crystals *Urine Increase Physiological* Due to crystalization out of drug. *1680*

Erythrocytes *Blood Decrease Physiological* May cause hemolytic anemia. *2313* May cause bone marrow depression. *0071*
Urine Increase Physiological Due to nephrotoxicity. *1680*

Hemoglobin *Blood Decrease Physiological* May cause hemolytic anemia. *2313*
Urine Increase Physiological Due to nephrotoxicity. *1680*

Leukocytes *Blood Decrease Physiological* May cause bone marrow depression. *0071*

Platelets *Blood Decrease Physiological* May cause bone marrow depression. *0071*

Protein *Urine Increase Physiological* Due to nephrotoxicity. *1680*

Pseudocholinesterase *Serum Decrease Physiological* Observed *in vitro*, probable effect *in vivo*. *3959*

Urea Nitrogen *Serum Increase Physiological* Nephrotoxicity with nitrogen retention. *1680*

Triflocin

Ammonia *Urine Increase Physiological* Inhibits carbonic anhydrase. *0033*

Potassium *Urine Increase Physiological* Diuretic action. *0033*

Sodium *Urine Increase Physiological* Diuretic action. *0033*

Titratable Acidity *Urine Increase Physiological* Inhibits carbonic anhydrase. *0033*

Trifluoperazine

Alanine Aminotransferase *Serum Increase Physiological* Cholestatic effect. *1713*

Albumin *Serum No Effect Analytical* At concentration of 1 mg/L had no effect on BCG method. *3393*

Alkaline Phosphatase *Serum Increase Physiological* Cholestatic effect. *1713*
Serum Decrease Physiological Not very marked. *0221*

Alkaloids *Urine No Effect Analytical* With ninhydrin on Frings TLC procedure at 10 mg/dL. *1204*
Urine Increase Analytical Purple with iodoplatinate on Frings TLC procedure. *1204*

Aspartate Aminotransferase
Serum Increase Physiological Cholestatic effect. *1713*
Serum Decrease Physiological Noticeable effect. *0221*

Barbiturate *Urine No Effect Analytical* With diphenylcarbazone on Frings TLC procedure at 10 mg/dL. With mercurous NO_3 on Frings TLC procedure at 10 mg/dL. *1204*
Urine Increase Analytical Orange spot with mercuric SO_4 in Frings TLC procedure. Orange spot with vanillin in Frings TLC procedure. *1204*

Bicarbonate *Serum No Effect Analytical* At concentration of 1 mg/L had no effect on method using phenolphthalein. *3393*

Bilirubin *Serum No Effect Analytical* At concentration of 1 mg/L had no effect on Jendrassik and Grof method. *3393*
Serum Increase Physiological Cholestatic effect. *1713*

Bromide *Urine No Effect Analytical* No effect on method of Frings. *1204*

BSP Retention *Serum Increase Physiological* Cholestatic effect. *1713*

Calcium *Serum No Effect Analytical* At concentration of 1 mg/L had no effect on cresolphthalein method. *3393*

Chlordiazepoxide *Blood No Effect Analytical* On method of Riddick at 5 mg/dL. *2983*

Chloride *Serum No Effect Analytical* At concentration of 1 mg/L had no effect on mercurimetric method. *3393*

Cholesterol *Serum No Effect Analytical* At concentration of 1 mg/L had no effect on Liebermann-Burchard method. *3393*
Serum Increase Physiological Increase of up to 35 mg/dL reported. *2946*

Creatinine *Serum No Effect Analytical* At concentration of 1 mg/L had no effect on AutoAnalyzer method. *3393*

Dapsone *Serum No Effect Analytical* At 10 mg/dL on colorimetric procedure of Higgins. *1590*

Diphenylhydantoin *Serum No Effect Analytical* On GLC procedure of Papadopoulos at *in vivo* concentration. *2722*

Erythrocytes *Blood Decrease Physiological* Pancytopenia (AMA Blood dyscrasias). *2429*

Estriol *Urine No Effect Analytical* No effect of 25 mg/L on GLC method. *1163*

Ethchlorvynol *Serum No Effect Analytical* At 100 mg/L on GLC procedure of Evenson. *1058*
Urine No Effect Analytical No effect on method of Frings and Cohen at 20 mg/dL. *1205*

FPN Test *Urine No Effect Analytical* Negative FPN test with 0.5 mg/dL concentration in Frings TLC procedure. *1204*
Urine Increase Analytical Orange color observed in method of Frings Orange FPN test with 1 mg/dL concentration in Frings TLC procedure Purple with patients also receiving chlorpromazine. *1204*

Glucose *Serum No Effect Analytical* At concentration of 1 mg/L had no effect on GOD/POD-PAP method. *3393*

Granulocytes *Blood Decrease Physiological* Decreased values noted (not marked). *0221*

Leukocytes *Blood Decrease Physiological* Pancytopenia (AMA Blood dyscrasias). *2427*

Methaqualone *Serum No Effect Analytical* At 100 mg/L on GLC procedure of Evenson. *1057*

Pheneturide *Serum No Effect Analytical* On GLC procedure of Papadopoulos at *in vivo* concentration. *2722*

Phenobarbital *Serum No Effect Analytical* On GLC procedure of Papadopoulos at *in vivo* concentration. *2722*

Phosphate *Serum No Effect Analytical* At concentration of 1 mg/L had no effect on phosphomolybdate method. *3393*

Platelets *Blood Decrease Physiological* Pancytopenia (AMA Blood dyscrasias). *2429*

Potassium *Serum No Effect Analytical* At concentration of 1 mg/L had no effect on flame-photometric method. *3393*

Primidone *Serum No Effect Analytical* On GLC procedure of Papadopoulos at *in vivo* concentration. *2722*

Prolactin *Plasma Increase Physiological* Typical response to i.m. administered drug: effect dose related. *2072*

Protein *Serum No Effect Analytical* At concentration of 1 mg/L had no effect on biuret method with blank correction. *3393*
Urine Increase Analytical In 2 patients receiving up to 30 mg daily on Ponceau S dye method in comparison with sulfosalicylic acid and trichloracetic acid methods. *3919*

Protein Bound Iodine (PBI) *Serum No Effect Physiological* No effect with normal doses. *0830*

Prothrombin Time *Plasma Increase Physiological* Associated with failure of excretion of bile salts. *1713*

Salicylate *Urine No Effect Analytical* Up to 10 mg/dL no effect on Trinder procedure. *1204*

Sodium *Serum No Effect Analytical* At concentration of 1 mg/L had no effect on flame-photometric method. *3393*

Triglycerides *Serum No Effect Analytical* At concentration of 1 mg/L had no effect on lipase/esterase method. *3393*

Urea Nitrogen *Serum No Effect Analytical* At concentration of 1 mg/L had no effect on diacetylmonoxime method. *3393*

Uric Acid *Serum No Effect Analytical* At concentration of 1 mg/L had no effect on phosphotungstate reduction method. *3393*

Urobilinogen *Urine Decrease Physiological* Cholestatic effect. *1713*
Feces Decrease Physiological Pale stools with cholestasis. *1713*

Trifluperidol

Alanine Aminotransferase *Serum Increase Physiological* Hepatocellular changes observed. *1343*

Bilirubin *Serum Increase Physiological* Elevated in chronic and acute hepatitis. *1022*

Cholesterol *Serum Decrease Physiological* Inhibits biosynthesis in liver. *1343*

Desmosterol *Serum Increase Physiological* Further metabolism inhibited so accumulates. *1343*

Eosinophils *Blood Increase Physiological* Allergic response. *2313*

Leukocytes *Blood Increase Physiological* Due to eosinophilia. *2313*
Blood Decrease Physiological Rare leukopenia/agranulocytosis. *1343*

Triflupromazine

Alkaloids *Urine No Effect Analytical* With ninhydrin on Frings TLC procedure at 10 mg/dL. *1204*
Urine Increase Analytical Purple with iodoplatinate on Frings TLC procedure. *1204*

Barbiturate *Urine No Effect Analytical* With diphenylcarbazone on Frings TLC procedure at 10 mg/dL. With mercurous NO_3 on Frings TLC procedure at 10 mg/dL. *1204*
Urine Increase Analytical Orange spot with mercuric SO_4 in Frings TLC procedure. Orange spot with vanillin in Frings TLC procedure. *1204*

Bilirubin *Serum Increase Physiological* Liver dysfunction exceptionally rare. *0071*

Bromide *Urine No Effect Analytical* No effect on method of Frings. *1204*

Ethchlorvynol *Urine No Effect Analytical* No effect on method of Frings and Cohen. *1204*

FPN Test *Urine No Effect Analytical* Negative FPN test with 1 mg/dL concentration in Frings TLC procedure. *1204*
Urine Increase Analytical Pink FPN test with 2 mg/dL concentration in Frings TLC procedure Pink color observed in method of Frings. *1204*

Leukocytes *Blood Decrease Physiological* Agranulocytosis due to inhibition of development. *2822*

Trihexyphenidyl

Acetaminophen Screening Test *Urine Negative Analytical* No reaction with o-cresol at therapeutic concentrations. *3326*

Alkaloids *Urine No Effect Analytical* With ninhydrin on Frings TLC procedure at 5 mg/dL. *1204*
Urine Increase Analytical Red/purple with iodoplatinate of Frings TLC procedure. *1204*

Barbiturate *Urine No Effect Analytical* With mercuric SO_4 on Frings TLC procedure at 10 mg/dL. With mercurous NO_3 on Frings TLC procedure at 5 mg/dL. With diphenylcarbazone on Frings TLC procedure at 5 mg/dL. With vanillin on Frings procedure at 5 mg/dL. *1204*

Bromide *Urine No Effect Analytical* No effect on method of Frings. *1204*

Trihexyphenidyl *(continued)*

Ethchlorvynol *Urine No Effect Analytical* No effect on method of Frings and Cohen. *1204*

FPN Test *Urine No Effect Analytical* No effect at 5 mg/dL on method of Frings. *1204*

Salicylate *Urine No Effect Analytical* At 10 mg/dL no effect on Trinder method. *1204*

Volume *Urine Decrease Physiological* Retention due to parasympatholytic effect. *1680*

Tri-iodothyroacetic Acid

Tri-iodothyronine (T$_3$) *Serum Increase Analytical* 29% cross reactivity method of Sedadde. *3247*

Tri-iodothyronine

Angiotensin-Converting Enzyme (ACE)
Serum Increase Physiological In 7 normal women aged 18-27 y given 25 μg/d three times daily for 14 d. Effect observed on endothelium-associated proteins but not on hepatically synthesized proteins. *1383*

Antithrombin III *Plasma No Effect Physiological* In 7 normal women aged 18-27 y given 25 μg/d three times daily for 14 d. Effect observed on endothelium-associated proteins but not on hepatically synthesized proteins. *1383*

α$_1$-Antitrypsin *Serum No Effect Physiological* In 7 normal women aged 18-27 y given 25 μg/d three times daily for 14 d. Effect observed on endothelium-associated proteins but not on hepatically synthesized proteins. *1383*

Ceruloplasmin *Serum No Effect Physiological* In 7 normal women aged 18-27 y given 25 μg/d three times daily for 14 d. Effect observed on endothelium-associated proteins but not on hepatically synthesized proteins. *1383*

Cholesterol *Serum Decrease Physiological* Physiological consequence of hormone. *3505*

2,3-Diphosphoglycerate Mutase
Blood No Effect Physiological No effect on amount of 2,3-DPG produced. *3612*

Factor VIII Antigen *Plasma Increase Physiological* In 7 normal women aged 18-27 y given 25 μg/d three times daily for 14 d. Effect observed on endothelium-associated proteins but not on hepatically synthesized proteins. *1383*

Fibronectin *Serum Increase Physiological* In 7 normal women aged 18-27 y given 25 μg three times daily for 14 d. Effect observed on endothelium-associated proteins but not on hepatically synthesized proteins. *1383*

Haptoglobin *Serum No Effect Physiological* In 7 normal women aged 18-27 y given 25 μg/d three times daily for 14 d. Effect observed on endothelium-associated proteins but not on hepatically synthesized proteins. *1383*

^{131}I Uptake *Serum Decrease Physiological* Marked decrease in obese with 150 μg/d. *0458*

Prealbumin *Serum No Effect Physiological* In 7 normal women aged 18-27 y given 25 μg/d three times daily for 14 d. Effect observed on endothelium-associated proteins but not on hepatically synthesized proteins. *1383*

Protein Bound Iodine (PBI) *Serum Decrease Physiological* Pituitary suppression of TSH. *0012*

T$_3$ Uptake *Serum No Effect Physiological* No significant effect with 150 μg/d in obese. *0458*

Thyroid Stimulating Hormone (TSH)
Serum Decrease Physiological Effect marked in euthyroid, hypothyroid subjects. *3655*

Thyroxine (T$_4$) *Serum No Effect Analytical* No effect on competitive protein binding procedure of Alexander. *0054*
Serum Decrease Physiological Marked decrease in obese with 150 μg/d. *0458* Diminished synthesis. *3505* In 7 normal women aged 18-27 y given 25 μg/d three times daily for 14 d. Effect observed on endothelium-associated proteins but not on hepatically synthesized proteins. *1383*

Thyroxine (T$_4$) Binding Globulin
Serum No Effect Analytical On radioimmunoassay procedure of Van Herle. *3676*
Serum No Effect Physiological In 7 normal women aged 18-27 y given 25 μg/d three times daily for 14 d. Effect observed on endothelium-associated proteins but not on hepatically synthesized proteins. *1383*

Tissue Plasminogen Activator Antigen
Plasma Increase Physiological In 7 normal women aged 18-27 y given 25 μg/d three times daily for 14 d. Effect observed on endothelium-associated proteins but not on hepatically synthesized proteins. *1383*

Transferrin *Serum No Effect Physiological* In 7 normal women aged 18-27 y given 25 μg/d three times daily for 14 d. Effect observed on endothelium-associated proteins but not on hepatically synthesized proteins. *1383*

Tri-iodothyronine (T$_3$) *Serum No Effect Physiological* In 7 normal women aged 18-27 y given 25 μg/d three times daily for 14 d. Effect observed on endothelium-associated proteins but not on hepatically synthesized proteins. *1383*
Serum Increase Physiological Marked increase in obese with 150 μg/d. *0458*

Tyrosine *Plasma Increase Physiological* Slight increase in obese with 150 μg/d. *0458*

Trimazosin

Cholesterol *Serum No Effect Physiological* In 2 studies with 13 and 48 subjects treated for 2.5 to 12 mo. *0088*

Cholesterol, High Density Lipoprotein
Serum No Effect Physiological In 2 studies with 13 and 48 subjects treated for 2.5 to 12 mo. *0088*

Renin Activity *Plasma No Effect Physiological* No significant change observed. *0861*

Sodium *Urine No Effect Physiological* No significant effect noted. *0861*

Triglycerides *Serum No Effect Physiological* In 2 studies with 13 and 48 subjects treated for 2.5 to 12 mo. *0088*

Trimeprazine

Alanine Aminotransferase *Serum Increase Physiological* May cause cholestasis. *0071*

Aspartate Aminotransferase
Serum Increase Physiological May cause cholestasis. *0071*

Bile *Urine Increase Physiological* May cause cholestasis. *0071*

Bilirubin *Serum Increase Physiological* May cause cholestasis. *0071*

Granulocytes *Blood Decrease Physiological* Absence reported in one case. *0440*

Leukocytes *Blood Decrease Physiological* May cause leukopenia or agranulocytosis. *0071*

Neutrophils *Blood Decrease Physiological* Occasional case of agranulocytosis reported. *3717*

Trimethadione

Acetaminophen Screening Test *Urine Negative Analytical* No reaction with o-cresol at therapeutic concentrations. *3326*

Alanine Aminotransferase *Serum Increase Physiological* May cause hepatotoxicity with necrosis. *2313*

Alkaline Phosphatase *Serum Increase Physiological* Hepatotoxicity with centrolobular necrosis. *2313*

Antinuclear Antibodies *Serum Positive Physiological* Related to number of drugs, higher in women. *3797*

Aspartate Aminotransferase
Serum Increase Physiological May cause hepatotoxicity with necrosis. *2313*

Bicarbonate *Serum Decrease Physiological* Converted to dimethadione *in vivo* with same effects. *3882*

Bile *Urine Increase Physiological* May cause hepatotoxicity with necrosis. *2313*

Bilirubin *Serum Increase Physiological* Hepatotoxicity with centrolobular necrosis. *2313*

BSP Retention *Serum Increase Physiological* Hepatotoxicity with centrolobular necrosis. *2313*

Calcium *Serum Decrease Physiological* Theoretical effect of type of drug. *1452*

Casts *Urine Increase Physiological* May have nephrotoxic effect. 2313

Cephalin Flocculation *Serum Increase Physiological* May cause hepatotoxicity with necrosis. 2313

Cholesterol *Serum Increase Physiological* May affect liver function (hepatitis). 2313

Erythrocytes *Blood Decrease Physiological* Aplastic anemia. 2429
Urine Increase Physiological Hematuria may occur especially in children. 2220

Folate *Test Conditions No Effect Analytical* No effect on L. casei or S. fecalis. 2427

Hematocrit *Blood Decrease Physiological* Aplastic anemia. 3008

Hemoglobin *Blood Decrease Physiological* Aplastic anemia. 3008
Urine Increase Physiological Hematuria may occur especially in children. 2220

25-Hydroxy Vitamin D$_3$ *Serum Decrease Physiological* Theoretical effect of type of drug. 1452

LE Cells *Blood Positive Physiological* Rare immune response observed. 1226

Leukocytes *Blood Decrease Physiological* Aplastic anemia. 3008

Neutrophils *Blood Decrease Physiological* Moderate neutropenia quite common. 1343 Occasional agranulocytosis reported. 3717

Occult Blood *Feces Positive Physiological* Gastrointestinal bleeding reported (may affect many organs). 1022

pH *Blood Decrease Physiological* Converted to dimethadione in vivo with same effects. 3882

Platelets *Blood Decrease Physiological* Aplastic anemia may occur rarely. 3008

Protein *Serum Decrease Physiological* Reversible effect due to urinary loss. 2220
Urine Increase Physiological May have nephrotoxic effect. 1488 May have nephrotoxic effect. 1582

Thymol Turbidity *Serum Increase Physiological* May cause hepatotoxicity with necrosis. 2313

Urea Nitrogen *Serum Increase Physiological* Nephropathy reported. 0071

Trimethaphan

Histamine *Plasma Increase Physiological* Observed with injection associated with anesthesia. 2212

Trimethobenzamide

Alkaloids *Urine No Effect Analytical* With ninhydrin on Frings TLC procedure at 25 mg/dL. 1204
Urine Increase Analytical Red/purple with iodoplatinate on Frings TLC procedure. 1204

Barbiturate *Urine No Effect Analytical* With diphenylcarbazone on Frings TLC procedure at 25 mg/dL. With mercuric SO$_4$ on Frings TLC procedure at 25 mg/dL. With mercurous NO$_3$ on Frings TLC procedure at 25 mg/dL. With vanillin on Frings procedure at 25 mg/dL. 1204

Bilirubin *Serum Increase Physiological* Rare cases of jaundice reported. 1680

Bromide *Urine No Effect Analytical* No effect on method of Frings. 1204

Chlordiazepoxide *Blood No Effect Analytical* On method of Riddick at 5 mg/dL. 2983

Ethchlorvynol *Urine No Effect Analytical* No effect on method Frings and Cohen at 25 mg/dL. 1205

FPN Test *Urine No Effect Analytical* No effect at 25 mg/dL on method of Frings. 1204

Salicylate *Urine No Effect Analytical* At 25 mg/dL no effect on Trinder method. 1204

Trimethoprim

Alanine Aminotransferase *Serum No Effect Analytical* No effect at therapeutic concentration on Reflotron method. 1984

Alkaline Phosphatase *Serum Increase Physiological* Transient increase in the absence of abnormal liver function tests in two patients when drug combined with sulfamethoxazole. 0893

Aspartate Aminotransferase *Serum No Effect Analytical* No effect at therapeutic concentration on Reflotron method. 1984

Bilirubin *Serum No Effect Physiological* No clinically significant displacement from protein likely at pharmacological concentrations in neonates. 3748

Chloramphenicol *Serum No Effect Analytical* No effect at 100 mg/L on coupled enzymatic method. 2490

Chlorpropamide *Serum Increase Physiological* May potentiate action by effects on metabolism. 0678

Cholesterol *Serum No Effect Analytical* No effect at therapeutic concentration on Reflotron method. 1984

Creatinine *Serum No Effect Analytical* On Jaffé reaction with alkaline picrate. 0313 At concentration of 5 mg/L had no effect on AutoAnalyzer Jaffé method. 3393 At concentration of 5 mg/L had no effect on kinetic Jaffé method. 3393
Serum Increase Physiological Increase by 0.2 mg/dL in 21 patients when given with sulfamethoxazole; also when given by itself; probably due to competitive inhibition of tubular secretion mechanism. 0313 In elderly increased by more than 50% and by 20% in young males. Tubular secretion affected. 1880 Due to co-trimoxazole component, but effect is slight. 3078

Crystals *Urine Positive Physiological* May occur with high doses especially if severe renal insufficiency. 0678

Cyclosporine *Serum Increase Physiological* Reversible deterioration of renal function with effect on tubular function and nephrotoxicity. 3732
Serum Decrease Physiological Marked reduction when drug given with sulfadimidine i.v. due to effect on hepatic metabolism. 0683

Diphenylhydantoin *Serum Increase Physiological* May potentiate action by effects on metabolism. 0678

EDTA Clearance *Urine No Effect Physiological* True glomerular filtration rate not affected by drug. 1880

Erythrocytes *Blood Decrease Physiological* Folic acid antagonist ?megaloblastic anemia. 3773 Isolated case when drug given alone, but clear evidence of pancytopenia. Megaloblastic anemia may occur with prolonged use. 3284

Erythrocyte Sedimentation Rate
Blood Decrease Physiological With sulfa caused marked decrease in rheumatoids. 1851

Folate *Serum Decrease Physiological* Usually with sulfa — no hematological abnormality. 2808 Reported association with megaloblastic anemia by inhibiting dihydrofolate reductase although other reports dispute drug as cause of megaloblastic anemia. 2062
Red Blood Cells Decrease Physiological Reported association with megaloblastic anemia by inhibiting dihydrofolate reductase although other reports dispute drug as cause of megaloblastic anemia. 2062

Glucose *Serum No Effect Analytical* No effect at therapeutic concentration on Reflotron method. 1984
Urine No Effect Analytical At concentration of 500 mg/L had no effect on Diabur-test. 3393

γ-Glutamyltransferase (GGT) *Serum No Effect Analytical* No effect at therapeutic concentration on Reflotron method. 1984

Hematocrit *Blood Decrease Physiological* Folic acid antagonist ?megaloblastic anemia. 3773

Hemoglobin *Blood Decrease Physiological* Folic acid antagonist ?megaloblastic anemia. 3773 Isolated case when drug given alone, but clear evidence of pancytopenia. Megaloblastic anemia may occur with prolonged use. 3284

Leukocytes *Blood Decrease Physiological* Isolated case when drug given alone, but clear evidence of pancytopenia. Megaloblastic anemia may occur with prolonged use. 3284

MCV *Blood Increase Physiological* Occurs with megaloblastic anemia. 3773

Neutrophils *Blood Decrease Physiological* Occasional cases of neutropenia reported. 0155 Observed at least once in 57% of 49 children given trimethoprim-sulfamethoxazole over 10 day treatment period. 1104

Trimethoprim (continued)

5'-Nucleotidase *Serum Increase Physiological* Transient increase in the absence of abnormal liver function tests in two patients when drug combined with sulfamethoxazole. *0893*

Platelets *Blood Decrease Physiological* Most common serious toxic effect. *2937* Isolated case when drug given alone, but clear evidence of pancytopenia. Megaloblastic anemia may occur with prolonged use. *3284*

Sodium *Serum Decrease Physiological* Impairs free water clearance: effect noted when combined with diuretic. *0986*
Urine Increase Physiological Impairs free water clearance: effect noted when combined with diuretic. *0986*

Tolbutamide *Serum Increase Physiological* May potentiate action by effects on metabolism. *0678*

Triglycerides *Serum No Effect Analytical* No effect at therapeutic concentration on Reflotron method. *1984*

Urea Nitrogen *Serum No Effect Analytical* No effect at therapeutic concentration on Reflotron method. *1984*

Uric Acid *Serum No Effect Analytical* No effect at therapeutic concentration on Reflotron method. *1984*

Warfarin *Plasma Increase Physiological* May potentiate action by effects on metabolism. *0678*

Trimetozine

Glucose *Urine Increase Analytical* False positive with glucose oxidase (Combistix®) reported. *0666*
Urine Increase Physiological Hyperglycemia reported in one patient. *1487*

Sugar *Urine Increase Analytical* Interferes with Benedict's reagent. *1022*

Trimetrexate

Alanine Aminotransferase *Serum Increase Physiological* Transient effect in small proportion of patients although reverted to normal with continued treatment. *0058*

Aspartate Aminotransferase
Serum Increase Physiological Transient effect in small proportion of patients although reverted to normal with continued treatment. *0058*

Bilirubin *Serum Increase Physiological* Transient effect in small proportion of patients although reverted to normal with continued treatment. *0058*

Creatinine *Serum Increase Physiological* Nephrotoxicity reported in some patients following treatment, possibly associated with prior reduced renal function. *1613*

Creatinine Clearance *Urine Decrease Physiological* Nephrotoxicity reported in some patients following treatment, possibly associated with prior reduced renal function. *1613*

Neutrophils *Blood Decrease Physiological* Neutropenia in from 6 to 30% patients given different regimes. *0058*

Urea Nitrogen *Serum Increase Physiological* Nephrotoxicity reported in some patients following treatment, possibly associated with prior reduced renal function. *1613*

Uric Acid *Serum Increase Physiological* Nephrotoxicity reported in some patients following treatment, possibly associated with prior reduced renal function. *1613*

Trinitrotoluene

Color *Urine Increase Physiological* Red brown due to hemoglobin. *2313*

Erythrocytes *Blood Decrease Physiological* May cause hemolysis. *2429* May cause hemolysis with G-6-PD deficiency. *1413*
Urine Increase Physiological May cause hemolysis. *2313*

Hematocrit *Blood Decrease Physiological* May cause hemolysis. *2429*

Hemoglobin *Blood Decrease Physiological* May cause hemolysis with G-6-PD deficiency. *1413*
Urine Increase Physiological May cause hemolysis. *2313*

Methemoglobin *Blood Increase Physiological* Associated with hemolysis. *2429*

Sulfhemoglobin *Blood Increase Physiological* Associated with methemoglobinemia and hemolysis. *2429*

Triolein

Fatty Acids, Free (FFA) *Serum No Effect Analytical* At 2.3 mmol/L — method of Noma. *2607*

Trioxsalen

Alanine Aminotransferase *Serum Increase Physiological* May affect liver function. *2313*

Alkaline Phosphatase *Serum Increase Physiological* May affect liver function. *2313*

Aspartate Aminotransferase
Serum Increase Physiological May affect liver function. *2313*

Bile *Urine Increase Physiological* May affect liver function. *2313*

Bilirubin *Serum Increase Physiological* May affect liver function. *2313*

BSP Retention *Serum Increase Physiological* May affect liver function. *2313*

Cephalin Flocculation *Serum Increase Physiological* May affect liver function. *2313*

Thymol Turbidity *Serum Increase Physiological* May affect liver function. *2313*

Tripamide

Glucose *Serum No Effect Physiological* No effect in hypertensives with or without diabetes. *0711*

Glucose Tolerance *Serum No Effect Physiological* No effect in hypertensives with or without diabetes. *0711*

Insulin *Plasma No Effect Physiological* No effect in hypertensives with or without diabetes. *0711*

Tripelennamine

Albumin *Serum No Effect Analytical* At concentration of 200 mg/L had no effect on BCG method. *3393*

Bilirubin *Serum No Effect Analytical* At concentration of 200 mg/L had no effect on Jendrassik and Grof method. *3393*

Calcium *Serum No Effect Analytical* At concentration of 200 mg/L had no effect on cresolphthalein method. *3393*

Chlordiazepoxide *Blood Increase Analytical* May interfere with method of Jatlow. *1777*

Cholesterol *Serum No Effect Analytical* At concentration of 200 mg/L had no effect on Liebermann-Burchard method. *3393*

Creatinine *Serum No Effect Analytical* At concentration of 200 mg/L had no effect on AutoAnalyzer Jaffé method. *3393*

Erythrocytes *Blood Decrease Physiological* Hemolytic anemia. *2427*

Glucose *Serum Decrease Physiological* Reported effect. *0117*

Haptoglobin *Serum Decrease Physiological* Hemolytic anemia. *0788*

Hematocrit *Blood Decrease Physiological* Hemolytic anemia. *0788*

Hemoglobin *Blood Decrease Physiological* Hemolytic anemia. *0788*

Iron *Serum No Effect Analytical* At concentration of 200 mg/L had no effect on Ferrozine method. *3393*

Leukocytes *Blood Decrease Physiological* Agranulocytosis/leukopenia (rare). *3602*

Phosphate *Serum No Effect Analytical* At concentration of 200 mg/L had no effect on phosphomolybdate method. *3393*

Platelets *Blood Decrease Physiological* Selective thrombocytopenia or aplastic anemia. *1436*

Protein *Serum No Effect Analytical* At concentration of 200 mg/L had no effect on biuret method with blank correction. *3393*

Triglycerides *Serum No Effect Analytical* At concentration of 200 mg/L had no effect on lipase/esterase method. *3393*

Urea Nitrogen *Serum No Effect Analytical* At concentration of 200 mg/L had no effect on diacetylmonoxime method. *3393*

Uric Acid *Serum No Effect Analytical* At concentration of 200 mg/L had no effect on phosphotungstate reduction method. *3393*

Triphosadenine

Phosphate *Test Conditions Increase Analytical* Affects Fiske-Subbarow unless Seddon procedure used. *3236*

Triprolidine

Amino Acids *Urine Positive Analytical* Red spot with ninhydrin on thin-layer chromatography when combined with pseudoephedrine in Actifed®. *1278*

Catecholamines *Urine No Effect Analytical* No effect on method of Sandhu and Freed. *3135*

Tris

Dopamine *Plasma Increase Analytical* Markedly increased concentration when progressively increasingly diluted with tris and radio enzymatic method used for quantitation. *0768*

Epinephrine *Plasma Increase Analytical* Markedly increased concentration when progressively increasingly diluted with tris and radio enzymatic method used for quantitation. *0768*

Norepinephrine *Plasma Increase Analytical* Markedly increased concentration when progressively increasingly diluted with tris and radio enzymatic method used for quantitation. *0768*

Tris(2-Butoxyethyl) Phosphate (TBEP)

Lidocaine *Serum Decrease Analytical* By up to 17% at 12.7 μmol/L when used as plasticizer in evaluated blood tubes. *0892*

Quinidine *Serum Decrease Analytical* By up to 32% at 4.4 μmol/L when used as plasticizer in evaluated blood tubes. *0892*

Triton X-100

Protein *Test Conditions Decrease Analytical* Lowry procedure, non linear absorption. *1796*

Trizma

Creatinine *Serum Decrease Analytical* Present in some quality control materials trizma carbonate interferes with single slide creatinine method on Kodak Ektachem®. *0200*

Trolamine

Ionized Calcium *Serum Decrease Analytical* At concentration more than 0.1 mmol/L on Calcium specific electrode. *0540* Up to 12% if added to standards. *2056*

Protein *Test Conditions Increase Analytical* Possible measurement by biuret reaction. *3376*

Troleandomycin

Alanine Aminotransferase *Serum Increase Physiological* In 30% individuals treated with 1 g/d for 3-4 weeks. Jaundice typically occurs after 2 weeks mixed in type with both cholestasis and mild hepatocytic necrosis. *2792*

Aspartate Aminotransferase
Serum Increase Physiological In 30% individuals treated with 1 g/d for 3-4 weeks. Jaundice typically occurs after 2 weeks mixed in type with both cholestasis and mild hepatocytic necrosis. *2792*

Bilirubin *Serum Increase Physiological* In 4% individuals treated with 1 g/d for 3-4 weeks. Jaundice typically occurs after 2 weeks mixed in type with both cholestasis and mild hepatocytic necrosis. *2792*

BSP Retention *Serum Increase Physiological* In 50% individuals treated with 1 g/d for 3-4 weeks. Jaundice typically occurs after 2 weeks mixed in type with both cholestasis and mild hepatocytic necrosis. *2792*

Corticosteroids *Urine Increase Analytical* Increased absorption at 410 nm in Porter-Silber reaction. *0022*

Creatinine *Serum No Effect Analytical* At 200 mg/L on reversed phase LC procedure of Zhiri et al. *3942*

Uric Acid *Serum No Effect Analytical* At 200 mg/L on reversed phase LC procedure of Zhiri et al. *3942*

Tromethamine

Ammonia *Plasma Decrease Analytical* Indophenol color formation (Berthelot reaction) inhibition. *2632*

Bicarbonate *Serum Increase Physiological* Correction of respiratory acidosis. *3006*

Glucose *Serum Decrease Physiological* Hypoglycemia especially if i.v., also transient if oral. *3006*

Ionized Calcium *Serum Decrease Analytical* At concentrations > 0.1 mmol/L on Calcium specific electrode. *0540*

Orthophosphate *Test Conditions Decrease Analytical* Total inhibition at 0.1 mol/L on method of Horder. *1658*

pCO₂ *Blood Decrease Physiological* Correction of respiratory acidosis. *3006*

pH *Blood Increase Physiological* Can correct metabolic or hypercapnic acidosis. *3006*

Potassium *Serum Increase Physiological* Reported effect. *3006*

Protein *Test Conditions Increase Analytical* Possible measurement by biuret reaction. *3376*
Test Conditions Decrease Analytical Lowry procedure, non linear absorption. *1796*

Trypan Blue

Glucose *Serum Decrease Analytical* Affects glucose oxidase method of Boehringer. *3277*

Protein Bound Iodine (PBI) *Serum Decrease Physiological* Mechanism obscure. *0012*

Tryparsamide

Alanine Aminotransferase *Serum Increase Physiological* May cause liver damage. *0071*

Aspartate Aminotransferase
Serum Increase Physiological May cause liver damage. *0071*

Trypsin

Erythrocytes *Urine Increase Physiological* Occasional reported occurrence. *1022*

Hemoglobin *Urine Increase Physiological* Occasional reported occurrence. *1022*

Ionized Calcium *Serum Decrease Analytical* Up to 2.5% if added to standards. *2056*

Protein *Urine Increase Physiological* Isolated cases reported. *1680*

Tryptamine

Amphetamine *Urine No Effect Analytical* At 50 mg/L on fluorescent method of Hayes. *1534*

Hydrazine *Plasma No Effect Analytical* On fluorometric method of Vickers at 10 μg/mL. *3710*

Pseudocholinesterase *Serum Decrease Physiological* Observed effect in humans. *1176*
Red Blood Cells Decrease Physiological Observed effect in humans. *1176*

Tryptamine *(continued)*

Tryptophan *Test Conditions Increase Analytical* Measured as same in Spies/Chambers method. *1415*

Tyrosine *Plasma No Effect Analytical* On fluorometric procedure of Ambrose. *0080*

Tryptophan

Amino Acids *Test Conditions Increase Analytical* Positive spot test with ninhydrin. *3765*

Ammonia *Plasma Decrease Analytical* With 25 nmoles produces 10% inhibitor indophenol reaction. *1290*

Cholesterol *Serum No Effect Analytical* Insignificant effect on Leffler's procedure. *1743*
Serum Increase Analytical Affects direct reactions using acetic acid. *1563*

Hydroxyproline *Urine No Effect Analytical* At 400 mg/L on method of Seymour. *3262* At 68 mg/L on method of Goverde. *1365*

Indican *Urine Increase Physiological* Direct correlation in normals with amount ingested. *3606*

Phosphate *Urine Increase Physiological* Inhibits tubular reabsorption. *0957*

Protein *CSF Increase Analytical* Interferes with Folin-Ciocalteu reagent. *3958*

Sulfa as Sulfanilamide *Serum Increase Analytical* Yields positive Bratton-Marshall reaction. *0279*

Tyrosine *Plasma Increase Analytical* 2% fluorescence in procedure of Ambrose. *0080*

Uric Acid *Serum No Effect Analytical* At 5 mg/dL on copper chelate procedure. *2230*
Serum Increase Analytical 5 mg/dL equivalent to 0.3 mg/dL by Nishi procedure. *2230*

Urobilinogen *Urine Increase Analytical* Produces an orange color with Ehrlich's reagent. *1563*

Tryptophan-Thiol

Tryptophan *Test Conditions Increase Analytical* Measured as same in Spies/Chambers method. *1415*

Tuaminoheptane

Amphetamine *Urine Increase Analytical* False positive fluorescent method of Hayes at 50 mg/L. *1534*

Tubocurarine

Creatine Kinase *Serum Increase Physiological* Due to i.m. injections or histamine release. *0688*

γ-Globulin *Serum Increase Physiological* Positive correlation between sensitivity and concentration. *3481*

Histamine *Plasma Increase Physiological* Observed with administration for anesthesia. *2212*

pH *Blood Increase Physiological* Respiratory alkalosis with low doses. *2220*
Blood Decrease Physiological Large dose effect with prolonged recovery. *2220*

Pseudocholinesterase
Test Conditions Decrease Analytical Inhibitory effect observed. *2893*

Turpentine

Glucose *Urine Increase Physiological* Manifestation of nephrotoxicity of poisoning Due to ingestion — nephrotoxic effect. *1302*

Hemoglobin *Urine Increase Physiological* Nephrotoxic effect following ingestion. *1302*

Odor *Urine Increase Physiological* Odor resembles that of violets. *1302*

Protein *Urine Increase Physiological* May have nephrotoxic effect. *1022* Nephrotoxic effect. *1237*

Twin Pregnancy

α-Fetoprotein *Urine Increase Physiological* Usually twice as high as in single pregnancy. *1732*

Testosterone Binding *Plasma Increase Physiological* Greater than with single fetus, less than with triplets. *0888*

Tybamate

Alkaloids *Urine No Effect Analytical* With iodoplatinate of Frings TLC procedure at 10 mg/dL. With ninhydrin on Frings TLC procedure at 10 mg/dL. *1204*

Barbiturate *Urine No Effect Analytical* With mercuric SO_4 on Frings TLC procedure at 10 mg/dL. With diphenylcarbazone on Frings TLC procedure at 10 mg/dL. With mercurous NO_3 on Frings TLC procedure at 10 mg/dL. With vanillin on Frings procedure at 10 mg/dL. *1204*

Bromide *Urine No Effect Analytical* No effect on method of Frings. *1204*

Ethchlorvynol *Urine No Effect Analytical* No effect on method of Frings and Cohen. *1204*

FPN Test *Urine No Effect Analytical* No effect at 10 mg/dL on method of Frings. *1204*

Salicylate *Urine No Effect Analytical* At 10 mg/dL no effect on Trinder method. *1204*

Typhoid Vaccine

α₁-Antitrypsin *Serum Increase Physiological* May be considerable rise after single injection. *2031*

NBT Test *Blood Increase Physiological* False positive after typhoid/paratyphoid vaccination. *2589*

Tyramine

Amino Acids *Test Conditions Increase Analytical* Positive spot test with ninhydrin. *3765*

Amphetamine *Urine No Effect Analytical* At 50 mg/L on fluorescent method of Hayes. *1534* No effect at 100 mg/L on method of Rutter. *2173*

Epinephrine *Test Conditions No Effect Analytical* On fluorescent procedure of Peyrin. *2800*

Glucose *Serum Increase Physiological* Observed with high doses. *1176*

Millons Test *Test Conditions Positive Analytical* Positive spot test for phenols. *3765*

Norepinephrine *Test Conditions No Effect Analytical* On fluorescent procedure of Peyrin. *2800*

Protein *CSF Increase Analytical* 1.0 mg = 2.7 mg in Folin-Ciocalteu procedure. *0583*

Tyrosine *Plasma Increase Analytical* 190% fluorescence in procedure of Ambrose. *0080*

Uric Acid *Serum Increase Analytical* Affects phosphotungstate reduction methods. *3500*

Urobilinogen *Test Conditions Increase Analytical* Positive spot test with Ehrlich's reagent. *3765*

Tyropanoic Acid

Thyroid Stimulating Hormone (TSH)
Serum Increase Physiological Intrahypophyseal conversion of T_4 to T_3 involved. *3798*

Thyroxine (T_4) *Serum Increase Physiological* Impaired conversion of T_4 to T_3. *3798*

Tri-iodothyronine (T_3) *Serum Decrease Physiological* Impaired conversion of T_4 to T_3. *3798*

Tyrosine

Ammonia *Plasma Decrease Analytical* With 50 nmoles produces 6% inhibitor indophenol reaction. *1290*

Bilirubin *Serum Increase Analytical* Reacts with diazo reagent. *1563*

Crystals *Urine Increase Physiological* Colorless or yellow needles in sheaves, rosettes. *0443*

Histidine *Plasma Increase Analytical* May produce moderate effect on method of Ambrose. *0081*

Hydrazine *Plasma No Effect Analytical* On fluorometric method of Vickers at 10 µg/mL. *3710*

Hydroxyproline *Urine No Effect Analytical* At 45 mg/L on method of Goverde. *1365*

Millons Test *Test Conditions Positive Analytical* Positive spot test for phenols. *3765*

Protein *CSF Increase Analytical* Reacts as if phenol with Folin-Ciocalteu procedure. *3958*

Uric Acid *Serum No Effect Analytical* At 5 mg/dL on copper chelate procedure. *2230* At 5 mg/dL no effect on method of Klein. *1949*

Serum Increase Analytical 5 mg/dL equivalent to 0.3 mg/dL by Nishi procedure. *2230*

Urobilinogen *Test Conditions Increase Analytical* Positive spot test with Ehrlich's reagent. *3765*

2-Tyrosine

Tyrosine *Plasma Increase Analytical* 10% fluorescence in procedure of Ambrose. *0080*

3-Tyrosine

Tyrosine *Plasma Increase Analytical* 16% fluorescence in procedure of Ambrose. *0080*

Tyrothricin

Bilirubin *Serum Increase Physiological* Hemolysis even when applied topically. *0071*

Erythrocytes *Blood Decrease Physiological* Hemolysis even when applied topically. *0071*

Hematocrit *Blood Decrease Physiological* May cause hemolysis even if applied topically. *0071*

Hemoglobin *Blood Decrease Physiological* May cause hemolysis even if applied topically. *0071*

Ultraviolet Light

Chromosomes *Test Conditions Abnormal Physiological* Clastogenic in human cells. *3282*

Uracil Mustard

Alanine Aminotransferase *Serum Increase Physiological* May affect liver function. *2313*

Alkaline Phosphatase *Serum Increase Physiological* May affect liver function. *2313*

Aspartate Aminotransferase
Serum Increase Physiological May affect liver function. *2313*

Bile *Urine Increase Physiological* May affect liver function. *2313*

Bilirubin *Serum Increase Physiological* May affect liver function. *2313*

BSP Retention *Serum Increase Physiological* May affect liver function. *2313*

Cephalin Flocculation *Serum Increase Physiological* May affect liver function. *2313*

Erythrocytes *Blood Decrease Physiological* May cause bone marrow depression. *0071*

Leukocytes *Blood Decrease Physiological* May cause bone marrow depression. *0071*

Platelets *Blood Decrease Physiological* May cause bone marrow depression. *0071*

Thymol Turbidity *Serum Increase Physiological* May affect liver function. *2313*

Uranium

Protein *Urine Increase Physiological* May cause nephrotoxicity. *3204*

Urea Nitrogen *Serum Increase Physiological* May cause nephrotoxicity. *3204*

Urates

Appearance *Urine Abnormal Analytical* Cloudy: dissolve at 60°. *0443*

Crystals *Urine Increase Physiological* Ca, Mg, K=yellow, NH₄ = brown, sodium=colorless. *0443*

Urea

Acetoacetate Decarboxylase *Serum Decrease Analytical* Inhibition of enzyme produced. *3684*

Albumin *Urine No Effect Analytical* On nephelometric method of Killingsworth. *1931*

α-Amino-Nitrogen *Urine No Effect Analytical* No effect if dinitrofluorobenzene method used. *1346*

Ammonia *Plasma No Effect Analytical* On indophenol reaction with 25,000 nmoles. *1290*
Plasma Increase Analytical Breakdown of urea to ammonia. *1022*

Bilirubin *Serum Decrease Analytical* Brown color in diazo reaction of uremics. *1563*

Cholesterol *Serum Increase Analytical* Increase by 0.5% at 60 mg/dL on enzymatic procedure. *0057*

Creatine *Serum Increase Analytical* 10 g/L=0.03 mg/dL by method of Heinegard. *1547*

Creatinine *Urine Increase Analytical* 1 mg equivalent to 0.05 mg with automated Jaffé procedure. *2558*

Gentamicin *Serum Increase Analytical* Affects measurement by pH method. *1212*

Hemoglobin *Plasma Increase Physiological* Rapid infusion of concentration solution may cause hemolysis. *1022*
Blood Decrease Physiological Rapid infusion of concentration solution may cause hemolysis. *1022*
Urine Increase Physiological Hemoglobinuria may occur especially if hypothermia. *1022*

Hydroxyproline *Urine No Effect Analytical* At 10 g/L on method of Goverde. *1365*

Ionized Calcium *Serum Increase Analytical* At concentrations more than 150 mg/dL ion specific electrode. *0540*

Lactate Dehydrogenase Isoenzymes
Serum Decrease Analytical Inactivates cardiac components (LDH 1, LDH 2). *0718*

Lead *Urine Increase Analytical* 1500 mg/dL equivalent to 60 µg/L by atomic absorption. *3240*

α₂-Macroglobulin *Urine No Effect Analytical* On nephelometric method of Killingsworth. *1931*

Orthophosphate *Test Conditions No Effect Analytical* No effect up to 1 mol/L. *1658*

Osmolality *Serum Increase Analytical* Osmotically active constituent in samples. *1810*
Urine Increase Analytical Osmotically active constituent in samples. *1810*

Platelet Aggregation *Blood Decrease Physiological* At concentrations possibly occurring in uremia. *0826*

Platelets *Blood Decrease Physiological* Reported effect (AMA Blood dyscrasias committee). *2429*

Porphobilinogen *Urine Decrease Analytical* Inhibits color development unless prior separation. *1116*

Potassium *Serum Decrease Physiological* May cause severe depletion with diuresis. *1022*

Protein *Test Conditions No Effect Analytical* Lowry procedure when referenced with chemical only used. *1796*

Salicylate *Serum No Effect Analytical* Up to 1,000 mg/dL no effect on Trinder method. *2251*

Sodium *Serum Decrease Physiological* May cause severe depletion with diuresis. *1022*

Transferrin *Urine No Effect Analytical* On nephelometric method of Killingsworth. *1931*

Urea Nitrogen *Serum Increase Physiological* When massive infusion for sickle cell disease. *2354*

Urea (continued)

Urobilinogen *Urine Decrease Analytical* Produces yellow color with Ehrlich's reagent. *1563*

Volume *Urine Increase Physiological* Acts as osmotic diuretic. *2354*

Zinc *Serum No Effect Analytical* At 15,000 to 1 with flameless atomic absorption. *2038*

Urine No Effect Analytical At 15,000 to 1 with flameless atomic absorption. *2038*

Uremia

Catecholamines *Plasma Increase Analytical* With fluorescence methods due to interfering substances. *2285*

Creatinine *Serum Increase Analytical* Higher concentration of Jaffé chromogens. *2445*

Digoxin *Serum Increase Analytical* May be affected if beta counting used. *0615*

Phentolamine Test *Patient Positive Physiological* Probably related to impaired excretion of drug. *1218*

Vanillylmandelic Acid *Urine Decrease Physiological* Probably due to toxicity and impaired excretion. *2472*

Urethan

Alanine Aminotransferase *Serum Increase Physiological* May cause liver damage and necrosis. *0071*

Aspartate Aminotransferase
Serum Increase Physiological May cause liver damage and necrosis. *0071*

Bilirubin *Serum Increase Physiological* Hepatotoxicity with centrolobular necrosis. *0071*

Erythrocytes *Blood Decrease Physiological* Bone marrow aplasia/pancytopenia. *2313*

Hematocrit *Blood Decrease Physiological* May occur with bone marrow depression. *2427*

Hemoglobin *Blood Decrease Physiological* May occur with bone marrow depression. *2427*

Leukocytes *Blood Decrease Physiological* Bone marrow aplasia/pancytopenia. *0071*

Platelets *Blood Decrease Physiological* Bone marrow aplasia/pancytopenia. *0071*

Uric Acid

Amino Acids *Serum Increase Analytical* Reported observation. *0355*

α-Amino-Nitrogen *Plasma Increase Analytical* Reacts in naphthoquinone method of Frame. *1563*

Ammonia *Plasma Decrease Analytical* With 500 nmoles produces 18% inhibitor indophenol reaction. *1290*

Ammoniacal Silver Nitrate
Test Conditions Positive Analytical Positive spot test with Tollen's reagent. *3765*

Angiotensin-Converting Enzyme (ACE)
Serum Increase Analytical At concentrations above 600 μmol/L on method of Boomsma and Schalekamp. *2814*

Appearance *Urine Abnormal Analytical* Cloudy: dissolve at 60°. *0443*

Cholesterol *Serum Increase Analytical* Increase by 1.6% at 50 mg/dL on enzymatic procedure. *0057*

Creatinine *Serum No Effect Analytical* No effect on method of Polar and Metcoff. *1975*

Crystals *Urine Increase Physiological* Yellow, red-brown, many types in acid. *0443*

Fatty Acids, Free (FFA) *Serum No Effect Analytical* At 25 mg/dL on method of Soloni. *3381*

Glucose *Serum No Effect Analytical* At 10 mg/dL on MBTH procedure of Neeley. *2569* *In vivo* by 10 mg/L per 10 mg/L GOD-ABTS procedure. *3524* No effect on Warner Glucomatic method. *2771* No effect at 100 mg/dL glucose oxidase method of Gochman. *1311* No effect on Trinder glucose oxidase method. *2771*

Serum Increase Analytical Affects Neocuproin procedure. *2771* If measured by alkaline ferricyanide reduction. *2857*
Serum Decrease Analytical If measured by o-toluidine. *2857* Negative peaks GOD-PERID procedure. *2771* If high competes with chromogen in glucose oxidase procedure. *0583*
Urine No Effect Analytical With DNSA method at usual concentrations. *0076*
Urine Decrease Analytical Reported to inhibit glucose oxidase in Stix reactions. *0076*

Homogentisic Acid *Urine Increase Analytical* If method of Walkow, Baumann used. *1096*

Protein *Test Conditions Increase Analytical* Reacts with Folin-Ciocalteu of Lowry method. *1569*

Sugar *Urine Increase Analytical* Acts as reducing agent with Benedict's. *2313*

Urea Nitrogen *Serum Increase Analytical* May react with Berthelot if urease not used. *2981*

Uric Acid *Test Conditions Increase Analytical* Positive spot test with phosphotungstate. *3765*

Urine Flow

Epinephrine *Urine Increase Physiological* Positive correlation between excretion and flow. *1636*

Urine pH

Amphetamine *Urine Increase Physiological* Excretion increased with alkaline urine. *1636*

Epinephrine *Urine Increase Physiological* Excretion increased with alkaline urine. *1636*

Methamphetamine *Urine Increase Physiological* Excretion increased with alkaline urine. *1636*

Norepinephrine *Urine Increase Physiological* Probable effect with alkaline urine. *1636*

Urine Turbidity

Protein *Urine No Effect Analytical* On Stix tests (bromphenol reagent strip). *0443*

Urine Volume

Arylsulfatase A *Urine No Effect Physiological* Excretion not dependent on urine flow. *2320*

Cyclic Adenosine Monophosphate
Urine No Effect Physiological No significant effect with change in volume. *2691*

β-Galactosidase *Urine No Effect Physiological* Excretion not dependent on urine flow. *2320*

α-Glucosidase *Urine No Effect Physiological* Excretion not dependent on urine flow. *2320*

Urobilin

Color *Urine Increase Analytical* Excess may cause yellow to amber color. *2448*

Urokinase

Fibrinolytic Time *Plasma Increase Physiological* Significant effect if given parenterally. *0421*

Plasmin *Plasma Increase Physiological* Acts directly in converting plasminogen to plasmin. *3095*

pO_2 *Blood Increase Physiological* In patients with pulmonary embolism. *3298*

Thrombin Time *Plasma Increase Physiological* In patients with pulmonary embolism. *3298*

Ursodeoxycholic Acid

Cholesterol *Serum No Effect Physiological* No significant effect with 600 mg daily in patients with endogenous hypertriglyceridemia. *0601*

Cholesterol, High Density Lipoprotein
Serum   No Effect   Physiological   No significant effect with 600 mg daily in patients with endogenous hypertriglyceridemia. *0601*   In 8 normolipemic patients receiving 1,000 mg drug daily. *2121*

Cholesterol, Low Density Lipoprotein
Serum   No Effect   Physiological   In 8 normolipemic patients receiving 1,000 mg drug daily. *2121*

Cholesterol, Very Low Density Lipoprotein
Serum   No Effect   Physiological   In 8 normolipemic patients receiving 1,000 mg drug daily. *2121*

Triglycerides   *Serum   No Effect   Physiological*   No significant effect with 600 mg daily in patients with endogenous hypertriglyceridemia. *0601*   In 8 normolipemic patients receiving 1,000 mg drug daily. *2121*

Triglycerides, High Density Lipoprotein
Serum   No Effect   Physiological   In 8 normolipemic patients receiving 1,000 mg drug daily. *2121*

Triglycerides, Low Density Lipoprotein
Serum   No Effect   Physiological   In 8 normolipemic patients receiving 1,000 mg drug daily. *2121*

Triglycerides, Very Low Density Lipoprotein
Serum   No Effect   Physiological   In 8 normolipemic patients receiving 1,000 mg drug daily. *2121*

Vaginal Powder

Glucose   *Urine   Increase   Analytical*   Powders often contain glucose. *2220*   Powders may contain glucose. *1488*

Valine

Ammonia   *Plasma   Decrease   Analytical*   With 500 nmoles produces 47% inhibitor indophenol reaction. *1290*

Insulin   *Plasma   Increase   Physiological*   Slight effect, AIDS metabolism of amino acids. *1071*

Phosphate   *Urine   Increase   Physiological*   Inhibits tubular reabsorption. *0957*

Pyruvate Kinase   *Red Blood Cells   Decrease   Analytical*   Most marked with larger side chain. *0370*

Tyrosine   *Plasma   No Effect   Analytical*   On fluorometric procedure of Ambrose. *0080*

Valproic Acid

Alanine   *Plasma   Increase   Physiological*   When 1 g given orally to fasting individuals. *3633*

Alanine Aminotransferase   *Serum   No Effect   Physiological*   In chronically treated patients with monotherapy only, but increases when combined with other antiepileptics. *1453*
Serum   Increase   Physiological   Observed in 4 of 25 patients treated with drug: reversible with reduction of dose or withdrawal of drug. *3847*   Dose related increase in 44% of treated patients. *0950*   Rise in about 15 to 30% cases transient and usually maximal to 10 to 12 weeks after beginning treatment. *0504*   Transient increase in 44% of patients without change of dosage. *1781*   In 4 of 25 patients, reversed with withdrawal of treatment. *3847*   Fatal case reported in child also given phenytoin. *1626*   Acute hepatic centrilobular necrosis with severe fatty change in small number of patients. *3763*   In 1 of 9 patients with epilepsy poorly controlled. *3508*

Albumin   *Serum   Decrease   Physiological*   Observed in 4 of 25 patients treated with drug: reversible with reduction of dose or withdrawal of drug. *3847*   Reduced to below 35 g/L in 4 of 9 patients. *3508*

Alkaline Phosphatase   *Serum   Increase   Physiological*   Reported in children with coadministration only. *1781*   Acute hepatic centrilobular necrosis with severe fatty change in small number of patients. *3763*   In 1 of 9 patients with epilepsy poorly controlled. *3508*

Aluminum   *Serum   Increase   Physiological*   Reported in one woman on chronic hemodialysis. *2587*

Aminolevulinic Acid   *Serum   Increase   Physiological*   Mean increase to 130 nmol/L from 99 nmol/L in controls. *1348*

Ammonia   *Plasma   No Effect   Physiological*   No striking abnormality when given to epileptics. *3697*   In patients on monotherapy but increases seen with coadministration of other antiepileptics. *1454*
Plasma   Increase   Physiological   Occasional hyperammonia when given alone, more frequent when given with phenytoin when effect also more marked. *2301*   In 29 of 55 patients receiving drug versus none in control population on other anticonvulsants; values especially high when phenytoin also taken. *2528*   About half of patients comedicated with phenobarbital had increase; especially noted when concentrations high. *1454*   Nondose dependent effect in some patients with both normal and abnormal liver function. *1247*   Probably due to inhibition of enzymes involved in glycine clearance in liver analogous to ketotic hyperglycinemias. *1759*   Associated with inhibition of carbamyl phosphate synthetase I and interference with mitochondrial glycine transport. *0736*

Amylase   *Serum   Increase   Physiological*   Almost 20% patients had mild increase. *0213*   Apparently drug related case of acute pancreatitis. *3152*   Associated with dysphagia and epigastric discomfort in one adult patient who subsequently died. *2530*   Reported in one woman on chronic hemodialysis. *2587*   In 1 of 100 epileptic children when given alone or combined with other anticonvulsants. *0737*

Aspartate Aminotransferase
Serum   Increase   Physiological   Observed in 4 of 25 patients treated with drug: reversible with reduction of dose or withdrawal of drug. *3847*   Slight increase in 20% patients on chronic monotherapy: enzyme activity linearly and directly correlated with drug concentration. *1453*   Dose related increase in 44% of treated patients. *0950*   Rise in about 15 to 30% cases transient and usually maximal to 10 to 12 weeks after beginning treatment. *0504*   Transient increase in 44% of patients without change of dosage. *1781*   Observed in 3% of 109 patients: reports of fatalities in some studies. *1780*   Fatal case reported in child also given phenytoin. *1626*   Acute hepatic centrilobular necrosis with severe fatty change in small number of patients. *3763*   In 44% of 100 epileptic children when given alone or combined with other anticonvulsants. *0737*   In 4 of 9 patients with epilepsy poorly controlled. *3508*

Bilirubin   *Serum   Increase   Physiological*   Acute hepatic centrilobular necrosis with severe fatty change in small number of patients. *3763*

Biotin   *Serum   Decrease   Physiological*   Dose related effect in long term treated epileptics compared with controls effect not as marked as with other anticonvulsants. *2008*

Bleeding Time   *Patient   Increase   Physiological*   In 3 of 9 patients with poorly controlled epilepsy. *3508*

Carbamazepine   *Serum   Increase   Physiological*   Reported increase when valproic acid co-administered. *0119*
Serum   Decrease   Physiological   Slight or no effect in 25 patients studied for 5 to 9 mo. *3829*

Carnitine   *Serum   Decrease   Physiological*   Reversibly reduced, inversely related to drug dose and plasma ammonia concentration. *2660*

Clonazepam   *Serum   No Effect   Physiological*   No definite effect in 25 patients studied for 5 to 9 mo. *3829*

Corticotropin   *Plasma   No Effect   Physiological*   Normal concentration in maternal and umbilical cord blood in one pregnant woman. *1522*   In patients with Addison's and Cushing's diseases. *0082*

Cortisol   *Plasma   No Effect   Physiological*   Normal concentrations in maternal and umbilical cord blood in one pregnant woman. *1522*

Creatinine   *Serum   No Effect   Analytical*   At 1.0 g/L on reversed phase LC procedure of Zhiri et al. *3942*   At concentration of 100 mg/L had no effect on creatinine iminohydrolase method. *3393*

Diazepam   *Serum   Increase   Physiological*   Displaces diazepam from plasma protein binding sites and inhibits its metabolism. *0896*

Diphenylhydantoin   *Serum   Increase   Physiological*   Caused increased concentration which led to hepatic damage. *2714*
Serum   Decrease   Physiological   Bound concentration falls, although free concentration unchanged. *2475*   Proportion of free concentration increased from 9.1% to 15.8% as serum concentration fell from 19.7 to 15.3 μg/mL. *1197*   Clearance markedly increased when phenytoin given i.v. *1200*   Due to displacement from albumin in 25 patients studied for 5 to 9 mo. *3829*   Due to

Valproic Acid *(continued)*

Diphenylhydantoin *(continued)*
displacement from protein and reduction in total serum concentration but free concentration unchanged. *3144* Displaces from protein-binding sites and inhibited metabolism in some patients. *0510*

Ethosuximide *Serum No Effect Physiological* No definite effect in 25 patients studied for 5 to 9 mo. *3829*
Serum Increase Physiological Reported increase when valproic acid co-administered. *0119*

Fatty Acids, Free (FFA) *Serum Increase Analytical* With colorimetric method of Duncombe (overestimation by about 40% at plasma concentration. *0045*

Fibrinogen *Plasma Decrease Physiological* Reduced to 0.9 to 1.6 g/L in 9 patients with epilepsy poorly controlled. *3508*

Glucaric Acid *Urine Increase Physiological* Selective induction of hepatic metabolizing enzymes. *0848* Slightly higher but not statistically significant. *2788*

Glucose *Serum No Effect Analytical* At concentration of 100 mg/L had no effect on GOD/POD-PAP method. *3393*

Glutamine *Plasma No Effect Physiological* No striking abnormality when given to epileptics. *3697*

γ-Glutamyltransferase (GGT)
Serum No Effect Physiological In spite of induction of some hepatic enzymes. *0848*
Serum Increase Physiological Transient increase in 44% of patients without change of dosage. *1781*

Glycerol *Serum Increase Physiological* When 1 g given orally to fasting individuals. *3633*

Glycine *Serum Increase Physiological* Associated with inhibition of carbamyl phosphate synthetase I and interference. *0736*
Plasma No Effect Physiological No striking abnormality when given to epileptics. *3697*
Plasma Increase Physiological Probably due to inhibition of enzymes involved in glycine clearance in liver analogous to ketotic hyperglycinemias. *1759*

3-Hydroxybutyrate *Plasma Decrease Physiological* 78% reduction when 1 g given orally to fasting individuals. *3633*

Immunoglobulins *Serum Decrease Physiological* Deficiency occurred in 29% of 41 epileptic patients on anticonvulsant therapy. users of drug in general had lower concentrations than in controls. *1827*

Indomethacin *Serum No Effect Analytical* No effect on HPLC method of Roberts and Smith. *3002*

Ketones *Plasma Decrease Physiological* 60% reduction when 1 g given orally to fasting individuals. *3633*
Urine Increase Analytical Single drug eliminated as ketones may give false positive test. *0504*

Lactate *Plasma Increase Physiological* Dose related effect in long term treated epileptics compared with controls effect not as marked as with other anticonvulsants. *2008* When 1 g given orally to fasting individuals. *3633*

Lactate Dehydrogenase *Serum Increase Physiological* Acute hepatic centrilobular necrosis with severe fatty change in small number of patients. *3763* In 2 of 9 patients with epilepsy poorly controlled. *3508*

Leukocytes *Blood Decrease Physiological* In 27% of 100 epileptic children when given alone or combined with other anticonvulsants. *0737*

Organic Acids *Urine Increase Physiological* Dose related effect in long term treated epileptics compared with controls. Effect not as marked as with other anticonvulsants. *2008*

Ornithine *Plasma No Effect Physiological* No striking abnormality when given to epileptics. *3697*

Phenobarbital *Serum Increase Physiological* In patients taking primidone. *0504* Reported increase when valproic acid coadministered. *0119* In 25 patients studied for 5 to 9 mo. *3829* Increased half-life, relaxed plasma clearance, and other effects suggesting inhibition of metabolism. *2742* In 4 patients decreased conversion to hydroxyphenyl-phenobarbital. *0511*

Platelets *Blood Decrease Physiological* Platelet-bound antibody found in 4 of 31 patients and serum antiplatelet antibody found in 1 patient. *3139* In 1 of 100 epileptic children when given alone or combined with other anticonvulsants. *0737* In 4 of 9 patients with epilepsy poorly controlled. *3508*

Primidone *Serum Increase Physiological* In patients taking primidone. *0504*

Prolactin *Plasma Decrease Physiological* Significant effect in women 30-180 minutes after 400 mg orally. *2400* Significant effect in both normal women and in hyperprolactinemic women. *2400*

Propionic Acid *Serum Increase Physiological* Associated with inhibition of carbamyl phosphate synthetase I and interference with mitochondrial glycine transport. *0736*

Prothrombin Time *Plasma Increase Physiological* Observed in 4 of 25 patients treated with drug: reversible with reduction of dose or withdrawal of drug. *3847* Associated with other effects on coagulation. *0504* In 3 of 9 volunteers with epilepsy poorly controlled. *3508*

Pyruvate *Serum Increase Physiological* When 1 g given orally to fasting individuals. *3633*

T₃ Uptake *Serum No Effect Physiological* In 10 epileptic patients treated average of 8 mo. *2168*

Testosterone *Serum No Effect Physiological* Usual effect although slight increase may occur. *1736*

Testosterone, Free *Serum Decrease Physiological* Observed effect. *1736*

Thyroid Stimulating Hormone (TSH)
Serum No Effect Physiological In 10 epileptic patients treated average of 8 mo. *2168*

Thyroxine (T₄) *Serum No Effect Physiological* In 10 epileptic patients treated average of 8 mo. *2168*

Thyroxine (T₄) Index, Free (FTI)
Serum No Effect Physiological In 10 epileptic patients treated average of 8 mo. *2168*

Triglycerides *Serum No Effect Analytical* At concentration of 1010 mg/L had no effect on GPO-PAP method. *3393*

Tri-iodothyronine (T₃) *Serum No Effect Physiological* In 10 epileptic patients treated average of 8 mo. *2168*

Tri-iodothyronine (T₃), Reverse
Serum No Effect Physiological In 10 epileptic patients treated average of 8 mo. *2168*

Uric Acid *Serum No Effect Analytical* At 1.0 g/L on reversed phase LC procedure of Zhiri et al. *3942* At concentration of 100 mg/L had no effect on uricase-PAP method. *3393*

Vanadate

Phosphate *Test Conditions Decrease Analytical* Interferes with Fiske-Subbarow procedures and thus indirectly with reactions yielding phosphate and measured this way. *1051*

Vanadium

Alanine Aminotransferase *Serum Increase Physiological* Mild hepatic dysfunction observed with poisoning. *2217*

Vancomycin

Bilirubin *Serum No Effect Physiological* Clinically significant displacement from protein in neonates. *3748*

Casts *Urine Increase Physiological* Occasional evidence of mild nephrotoxicity. *1571*

Chloramphenicol *Serum No Effect Analytical* No effect at 100 mg/L on coupled enzymatic method. *2490*

Creatinine *Serum Increase Physiological* Occasional renal damage (usually reversible). *0131* Occasional evidence of mild nephrotoxicity. *1571* Nephrotoxicity in 5% of 60 patients given drug alone but much higher incidence when given with aminoglycosides. *2654*

Creatinine Clearance *Urine Decrease Physiological* Occasional renal damage (usually reversible). *0131*

Eosinophils *Blood Increase Physiological* Allergic response. *1513* Occasional hypersensitivity reaction noted. *1571*

Hemoglobin *Urine Increase Physiological* Occasional evidence of mild nephrotoxicity. *1571*

Leukocytes *Blood Increase Physiological* Due to eosinophilia. *2313*

Blood Decrease Physiological Favorable response of antibiotic-associated colitis to drug. *3561*

Neutrophils *Blood Decrease Physiological* Occasional reversible side effect. *1571* Observed in one patient with renal failure after 3 g drug. *0026*

Protein *Urine Increase Physiological* May cause nephrotoxicity. *3204* Occasional evidence of mild nephrotoxicity. *1571*

Urea Nitrogen *Serum Increase Physiological* Nephrotoxic effect (may even be fatal). *1237* Occasional evidence of mild nephrotoxicity. *1571* Nephrotoxicity in 5% of 60 patients given drug alone but much higher incidence when given with aminoglycosides. *2654*

Vanilla

Vanillylamine *Urine Increase Physiological* Occurs after ingestion of food with vanilla flavor. *2784*

Vanillylmandelic Acid *Urine Increase Analytical* May be measured by colorimetric method. *1237*

Vanillic Acid

Ammoniacal Silver Nitrate
Test Conditions Positive Analytical Positive spot test with Tollen's reagent. *3765*

Millons Test *Test Conditions Positive Analytical* Positive spot test for phenols. *3765*

Phenylketones *Urine Positive Analytical* red-violet/brown with FeCl₃, brown Phenistix®. *0775*

Uric Acid *Test Conditions Increase Analytical* Positive spot test with phosphotungstate. *3765*

Urobilinogen *Test Conditions Increase Analytical* Positive spot test with Ehrlich's reagent. *3765*

Vanillin

Glucose *Serum No Effect Analytical* At 10 mg/dL no effect alkaline ferricyanide procedure At 10 mg/dL no effect on glucose oxidase procedure of Gochman. *1311*

Metanephrines, Total *Urine No Effect Analytical* At 40 mg/L on modified Pisano procedure. *1428*

Vanillylmandelic Acid *Urine No Effect Analytical* Dietary up to 62 mg/d no effect Pisano procedure. *0583*

Vanillylmandelic Acid (VMA)

Ammoniacal Silver Nitrate
Test Conditions Positive Analytical Positive spot test with Tollen's reagent. *3765*

Epinephrine *Test Conditions No Effect Analytical* On fluorescent procedure of Peyrin. *2800*

Glucose *Serum No Effect Analytical* At 10 mg/dL no effect on glucose oxidase procedure of Gochman. At 10 mg/dL no effect on alkaline ferricyanide procedure. *1311*

Guaiacols Spot Test *Urine Positive Analytical* Action on procedure of Rogers. *3031*

HMPG (4-Hydroxy-3-Methoxy-Phenylethylene Glycol)
Urine No Effect Analytical 100 μg = 0-2 μg method of Bigelow. *0343*

Homovanillic Acid *Urine No Effect Analytical* At high concentrations with method of Kahane. *1843*

Metanephrines, Total *Urine No Effect Analytical* At 50 mg/L on modified Pisano procedure. *1428*

Millons Test *Test Conditions Positive Analytical* Positive spot test for phenols. *3765*

Norepinephrine *Test Conditions No Effect Analytical* On fluorescent procedure of Peyrin. *2800*

Uric Acid *Test Conditions Increase Analytical* Positive spot test with phosphotungstate. *3765*

Urobilinogen *Urine No Effect Analytical* No effect on p-MDFB procedure of Rutter. *2044*
Test Conditions Increase Analytical Positive spot test with Ehrlich's reagent. *3765*

Vasopressin

Aldolase *Serum Increase Physiological* Rhabdomyolysis observed in 2 patients following intravenous drug administration. *0029*

Aldosterone *Urine No Effect Physiological* If not overhydrated no effect. *1344*

Aspartate Aminotransferase
Serum Increase Physiological Rhabdomyolysis observed in 2 patients following intravenous drug administration. *0029*

Corticosteroids *Plasma Increase Physiological* Significant effect 30-45 minutes after i.v. in normals. *3435*

Corticotropin *Plasma Increase Physiological* Small but significant effect after i.v. in normals. *3435*

Cortisol *Plasma Increase Physiological* May be rise of up to 6 μg/dL or more. *3625*

Creatine Kinase *Serum Increase Physiological* Rhabdomyolysis observed in 2 patients following intravenous drug administration. *0029*

Creatinine *Serum Increase Physiological* Progressive deterioration in renal function observed in one patient given drug i.v. *0029*

Growth Hormone *Plasma Increase Physiological* Observed effect in normals. *0371*

Lactate Dehydrogenase *Serum Increase Physiological* Rhabdomyolysis observed in 2 patients following intravenous drug administration. *0029*

Leucine Aminopeptidase *Urine Increase Physiological* Activates peptidases by release of plasminogen. *2897*

Myoglobin *Urine Increase Physiological* Rhabdomyolysis observed in 2 patients following intravenous drug administration. *0029*

Protein Bound Iodine (PBI) *Serum Increase Physiological* Stimulates thyroidal I 131 release. *0012*

Renin Activity *Plasma No Effect Physiological* No demonstrable effect if not overhydrated. *1344*

Sodium *Serum Decrease Physiological* May occur with water retention. *0071*

Urea Nitrogen *Serum Increase Physiological* Progressive deterioration in renal function observed in one patient given drug. *0029*

Volume *Urine Decrease Physiological* Therapeutic intent. *0071*

Vegetables

Arabinose *Urine Increase Physiological* Increase compared with fasting with high intake. *0273*

Fucose *Urine No Effect Physiological* No effect with high intake compared with controls. *0273*

Ribose *Urine No Effect Physiological* No effect with high intake compared with controls. *0273*

Xylose Excretion *Urine Increase Physiological* Increase compared with fasting with high intake. *0273*

Vegetarian Diet

Creatine *Urine Increase Physiological* Increased in most vegetarians studied. *3727*

Creatinine *Urine Decrease Physiological* Finding in adults protracted vegetarianism. *3727*

Folate *Serum Increase Physiological* Average of 6.6 ng/mL versus 4.8 (higher in smokers). *0811*

Vitamin B₁₂ *Serum Decrease Physiological* Deficiency levels observed in many subjects. *2468*

Venipuncture

Glutamic Acid *Plasma Decrease Physiological* Effect lasts for up to 1 h. *3230*

Taurine *Plasma Decrease Physiological* Effect lasts for up to 1 h. *3230*

Venoms

Protein *Urine Increase Physiological* May cause nephrotoxicity. *3204*

Urea Nitrogen *Serum Increase Physiological* May cause nephrotoxicity. *3204*

Verapamil

Alanine Aminotransferase *Serum Increase Physiological* Observed in a single case, reverted to normal as soon as drug withdrawn. *2561*

Aldosterone *Plasma No Effect Physiological* No change compared with placebo in 11 patients treated with up to 360 mg daily for 6 weeks. *3396*
Plasma Increase Physiological Concentration initially lower in hypertensives and increased gradually up to fourth month. Exact mechanism still to be elucidated. *3956*
Plasma Decrease Physiological Significant effect in 15 patients with uncomplicated essential hypertension. *0867*

Angiotensin II *Plasma No Effect Physiological* In 15 patients with uncomplicated essential hypertension. *0867* No change compared with placebo in 11 patients treated with up to 360 mg daily for 6 weeks. *3396*

Antidiuretic Hormone *Plasma No Effect Physiological* No change compared with placebo in 11 patients treated with up to 360 mg daily. *3396*

Aspartate Aminotransferase
Serum Increase Physiological Observed in a single case, reverted to normal as soon as drug withdrawn. *2561*

Bilirubin *Serum Increase Physiological* Observed in a single case, reverted to normal as soon as drug withdrawn. *2561*

Bilirubin, Direct *Serum Increase Physiological* Observed in a single case reverted to normal as soon as drug withdrawn. *2561*

Cholesterol *Serum No Effect Physiological* In 64 patients in post-myocardial infarction comparison against placebo. *3489*
Serum Decrease Physiological In 12 patients when angina or hypertension treated for 6 weeks. Where change occurred it was of order of 10%. *3756*

Cholesterol, High Density Lipoprotein
Serum No Effect Physiological In 64 patients in post-myocardial infarction comparison against placebo. *3489*
Serum Decrease Physiological In 12 patients with angina or hypertension treated for 6 weeks. Where change occurred it was of order of 10%. *3756*

Cholesterol, Low Density Lipoprotein
Serum No Effect Physiological In 64 patients in postmyocardial infarction in comparison against placebo. *3489*

Cortisol *Plasma No Effect Physiological* With 80 mg 3-4/d orally no significant effect on resting concentration or after ACTH stimulation in mild hypertensives. *3956*

Digoxin *Serum Increase Physiological* Average increase from 0.96 to 1.63 ng/mL in 41 patients when given 240 mg/d. *1952* Coadministration caused decrease of total body clearance and increased plasma half-life from 33.5 to 41.3 h. *2765*

Epinephrine *Plasma Decrease Physiological* Significant effect in 15 patients with uncomplicated essential hypertension. *0867*

Glomerular Filtration Rate (GFR)
Urine No Effect Physiological No change compared with placebo in 11 patients treated with up to 360 mg daily for 6 weeks. *3396*

Glucagon *Plasma No Effect Physiological* No effect observed even in individuals fasted for 36 h when infused at rate of 5 mg/h or 3 h. *0103*
Plasma Decrease Physiological Significant change during tolerance test when drug and glucose coadministered. *1115*

Glucose *Serum Decrease Physiological* When infused i.v. at rate of 5 mg/h for 3 h in prolonged fasted individuals but not in overnight fasted subjects. *0103*

Glucose Tolerance *Serum Decrease Physiological* Significant impairment when glucose and drug co-administered. *1115*

HMPG (4-Hydroxy-3-Methoxy-Phenylethylene Glycol)
Urine Decrease Physiological In 7 chronically ill schizophrenic patients when administered for 5 weeks. *2811*

Homovanillic Acid *Plasma Increase Physiological* In 7 chronically ill schizophrenic patients when administered for 5 weeks. *2811*
CSF Increase Physiological In 7 chronically ill schizophrenic patients when administered for 5 weeks. *2811*

Insulin *Plasma No Effect Physiological* No effect observed even in individuals fasted for 36 h when infused at rate of 5 mg/h for 3 h. *0103*
Plasma Increase Physiological Significant change during tolerance test when drug and glucose coadministered. *1115*

Lactate Dehydrogenase *Serum Increase Physiological* Observed in a single case, reverted to normal as soon as drug withdrawn. *2561*

Norepinephrine *Plasma Increase Physiological* Insignificant change in 15 patients with uncomplicated essential hypertension. *0867*

Prolactin *Plasma Increase Physiological* In 7 chronically ill schizophrenic patients when administered for 5 weeks. *2811*

Renal Blood Flow (RBF) *Patient No Effect Physiological* No change compared with placebo in 11 patients treated with up to 360 mg daily for 6 weeks. *3396*

Renin Activity *Plasma No Effect Physiological* In 15 patients with uncomplicated essential hypertension. *0867*

Sodium *Urine Increase Physiological* Marked enhancement in 15 patients with uncomplicated essential hypertension. *0867*
Red Blood Cells No Effect Physiological No effect when drug given alone. *2762*
Red Blood Cells Increase Physiological When given with digoxin increased concentration more than with controls with digoxin only. *2762*

Triglycerides *Serum No Effect Physiological* In 12 patients when angina or hypertension treated for 6 weeks. Where change occurred it was of order of 10%. *3756* In 64 patients in post-myocardial infarction comparison against placebo. *3489*

Triglycerides, Very Low Density Lipoprotein
Serum No Effect Physiological In 12 patients with angina or hypertension treated for 6 weeks. Where change occurred it was of order of 10%. *3756*

Volume *Plasma Increase Physiological* Due to decreased arterial and venous resistance in 15 patients with uncomplicated essential hypertension. *0867*

Verdoglobin

Color *Urine Increase Physiological* May cause orange-red-brown color. *1187*

Viloxazine

Metanephrine *Urine No Effect Analytical* No influence on liquid chromatographic measurement as drug elutes at different time. *0342*

Normetanephrine *Urine Increase Analytical* Substantial effect on liquid chromatographic measurement as drug elutes at same time. *0342*

Vinblastine

Erythrocytes *Blood Decrease Physiological* Anemia (secondary to leukopenia). *0071*

Leukocytes *Blood Decrease Physiological* Leukopenia. *0071*

Platelets *Blood Decrease Physiological* May cause bone marrow depression. *0071*

Reticulocytes *Blood Decrease Physiological* Absence from peripheral blood observed. *1343*

Uric Acid *Serum Decrease Physiological* Of theoretical value in acute gout. *1332*

Vincristine

Leukocytes *Blood Decrease Physiological* Usually reversible leukopenia. *0071*

Osmolality *Urine Increase Physiological* May produce clinically impaired water excretion. *0854*

Platelets *Blood Decrease Physiological* May cause bone marrow depression. *1343*

Protein Bound Iodine (PBI) *Serum No Effect Physiological* Reported to have no effect. *0830*

Sodium *Serum Decrease Physiological* May be inappropriate ADH secretion. *1680*
Urine Increase Physiological May be inappropriate ADH secretion. *1680*

Uric Acid *Serum Increase Physiological* Increased nucleic acid catabolism. *3505*

Vinyl Chloride

Chloylglycine *Serum Increase Physiological* Markedly different in individuals with chemical liver injury compared with people with nonchemical liver disease or normals. *2190*

Indocyanine Green Clearance
Serum Decrease Physiological Markedly different in individuals with chemical liver injury compared with people with nonchemical liver disease or normals. *2190*

Vinyl Ether

Alanine Aminotransferase *Serum Increase Physiological* Potentially hepatotoxic. *0071*

Aspartate Aminotransferase
Serum Increase Physiological Potentially hepatotoxic. *0071*

Bilirubin *Serum Increase Physiological* Potentially hepatotoxic. *0071*

Glucose *Serum Increase Physiological* Transient effect (less than with ether). *0071*

Urea Nitrogen *Serum Increase Physiological* Potentially nephrotoxic. *0071*

Viomycin

Bicarbonate *Serum Increase Physiological* Altered electrolyte balance. *1237*

Calcium *Serum Decrease Physiological* May cause tetany with electrolyte imbalance. *0071*
Urine Increase Physiological Promotes urinary loss. *1237*

Casts *Urine Increase Physiological* Nephrotoxicity with cylindruria. *0071*

Chloride *Urine Increase Physiological* Promotes urinary loss. *0071*

Cholesterol *Serum Increase Analytical* Interferes with Zlatkis-Zak reaction. *3505*

Creatinine *Serum Increase Physiological* Nephrotoxic may cause nitrogen retention. *2313*

Creatinine Clearance *Urine Decrease Physiological* Occasional renal damage observed. *0131*

Eosinophils *Blood Increase Physiological* Allergic response. *1513*

Erythrocytes *Urine Increase Physiological* May cause actual bleeding. *1237*

Hemoglobin *Urine Increase Physiological* Actual bleeding caused by drug. *1237*

Leukocytes *Blood Increase Physiological* Due to eosinophilia. *2313*

Potassium *Serum Decrease Physiological* May occur with nephrotoxicity. *2313*
Urine Increase Physiological Promotes urinary loss. *0071*

Protein *Urine Increase Physiological* May have nephrotoxic effect Nephrotoxic effect. *1237*

Urea Nitrogen *Serum Increase Physiological* Frequent complication. *1343*

Vitamin A

Alanine Aminotransferase *Serum Increase Physiological* Observed with acute intoxication. *3867*

Alkaline Phosphatase *Serum Increase Physiological* Observed with acute intoxication. *3867*

Aspartate Aminotransferase
Serum Increase Physiological Observed with acute intoxication. *3867*

Bilirubin *Serum Increase Analytical* Interferes with analysis. *1238*

Bilirubin, Direct *Serum Increase Analytical* Interferes with analysis. *1238*

Calcium *Serum Increase Physiological* Observed with acute intoxication. *3867*

Cholesterol *Serum No Effect Analytical* Unlikely that affects $FeCl_3$ retention. *1487*
Serum Increase Analytical Interferes with Zlatkis-Zak reaction. *3505*

Erythrocytes *Blood Decrease Physiological* With excessive doses and use. *2313*

Erythrocyte Sedimentation Rate
Blood Increase Physiological Observed effect but explanation unknown. *2427*

Hematocrit *Blood Decrease Physiological* Anemia observed. *2427*

Hemoglobin *Blood Decrease Physiological* Anemia observed. *2427*

[131]I Uptake *Serum Decrease Physiological* Significant effect when administered for 3 weeks. *0830*

Leukocytes *Blood Decrease Physiological* Leukopenia with hypoplastic anemia reported. *3908*

Lymphocytes *Blood No Effect Physiological* No apparent effect of megadose supplementation. *1347*

Methemoglobin *Blood Increase Physiological* May cause hemolysis with G-6-PD deficiency. *3581*

Neutrophils *Blood No Effect Physiological* No apparent effect of megadose supplementation. *1347*
Blood Decrease Physiological Leukopenia with hypoplastic anemia reported. *1791*

Phytofluene *Serum No Effect Analytical* No effect on method of Bubb-Murphy. *0517*

Protein Bound Iodine (PBI) *Serum No Effect Physiological* No effect observed when given for 3 weeks. *0830*
Serum Increase Physiological When given in cod liver oil. *2313*
Serum Decrease Physiological Inhibits iodination of tyrosine residues in thyroxine binding globulin. *0012*

Prothrombin Time *Plasma Increase Physiological* Hemorrhagic trend especially if vitamin K restricted. *2313*

Vitamin A *Serum Increase Physiological* Ingested compound (with overdose over 1200 IU/dL). *1343* Observed with acute intoxication. *3867*

Vitamin B₆ Depletion

Alanine Aminotransferase *Serum Decrease Physiological* More affected than AST. *0255*

Aspartate Aminotransferase
Serum Decrease Physiological Effect observed at once, less marked then SGPT. *0255*

Cholesterol *Serum No Effect Physiological* No effect observed with 25 d poor diet. *0255*

Cystathionine *Urine Increase Physiological* Direct correlation between increased excretion and decreased diet. *0148*

Pyridoxal Phosphate *Plasma Decrease Physiological* No change for 15 d then marked fall. *0255*

4-Pyridoxic Acid *Urine Decrease Physiological* Zero detectable after 25 d deprivation. *0255*

Pyridoxine *Serum Decrease Physiological* 20% control value after 5 d, zero after 25 d. *0255*
Urine Decrease Physiological Marked decrease within few days. *0255*

Quinolinic Acid *Urine Increase Physiological* Observed with experimental dietary deficiency. *3045*

Vitamin B₁₂

Ascorbic Acid   *Plasma   No Effect   Physiological*   No effect if individuals saturated with vitamin C. *2468*
White Blood Cells   No Effect   Physiological   No effect if individuals saturated with vitamin C. *2468*

Erythrocytes   *Blood   Increase   Physiological*   Successful treatment may cause mild polycythemia. *2427*

Hematocrit   *Blood   Increase   Physiological*   Successful treat may cause mild polycythemia. *2427*

Lymphocytes   *Blood   No Effect   Physiological*   No effect with megadose supplementation. *1347*

Neutrophils   *Blood   No Effect   Physiological*   No effect with megadose supplementation. *1347*

Vitamin B₁₂   *Red Blood Cells   Increase   Physiological*   During treatment of vitamin B₁₂ deficiency. *1510*

Vitamin B Complex

Catecholamines   *Plasma   Increase   Analytical*   May be interference with fluorescence. *1488*
Urine   Increase   Analytical   Large doses may produce similar fluorescence. *0583*

Cholesterol   *Serum   No Effect   Analytical*   At concentration of 12.9 mg/L had no effect on CHOD-Iodide method. *3393*   At concentration of 12.9 mg/L had no effect on CHOD-PAP method. *3393*   At concentration of 12.9 mg/L had no effect on catalase-AIDH method. *3393*   At concentration of 12.9 mg/L had no effect on method using catalase-Hantzsch reaction. *3393*   At concentration of 12.9 mg/L had no effect on Liebermann-Burchard method. *3393*

Creatinine   *Serum   No Effect   Analytical*   At concentration of 12.9 mg/L had no effect on Jaffé-Fading-Fraction method. *3393*   At concentration of 12.9 mg/L had no effect on Jaffé-Fuller's earth method. *3393*   At concentration of 800 mg/L had no effect on kinetic Jaffé method on BKA-2. *3393*

Diagnex Blue Excretion   *Urine   Increase   Analytical*   Release of dye from resin. *1238*

Estradiol   *Urine   Decrease   Analytical*   Affects method of Beling. *0023*

Estriol   *Urine   Decrease   Analytical*   Affects method of Beling. *0023*

Estrogens   *Urine   Decrease   Analytical*   Affects colorimetric/fluorometric procedures. *0022*

Glucose   *Serum   No Effect   Analytical*   At concentration of 2.3 mg/L had no effect on Ektachem® method. *3393*

Iron   *Serum   No Effect   Analytical*   At concentration of 1.4 mg/L had no effect on Ferrozine method. *3393*

Potassium   *Serum   No Effect   Analytical*   At concentration of 23.3 mg/L had no effect on measurement by ISE without predilution. *3393*

Protein Electrophoresis   *Serum   No Effect   Analytical*   At concentration of 23.3 mg/L had no effect on automated Olympus-Hite method. *3393*

Sodium   *Serum   No Effect   Analytical*   At concentration of 23.3 mg/L had no effect on measurement by ISE without predilution. *3393*

Urea Nitrogen   *Serum   No Effect   Analytical*   At concentration of 2.3 mg/L had no effect on Ektachem® method. *3393*

Uric Acid   *Serum   No Effect   Analytical*   At concentration of 32 mg/L had no effect on Kageyama-Hantzsch method. *3393*   At concentration of 12.9 mg/L had no effect on catalase-AIDH method. *3393*

Vitamin D

Alkaline Phosphatase   *Serum   Increase   Physiological*   May be affected in some cases. *0988*

Calcium   *Serum   Increase   Physiological*   Effect of increased gastrointestinal tract absorption. *3879*
Urine   Increase   Physiological   Associated with hypercalcemia. *1022*
Feces   Decrease   Physiological   Due to excessive absorption. *0987*

Calculi   *Urine   Increase   Physiological*   High incidence in people with self medication. *3559*

Cholesterol   *Serum   Increase   Analytical*   Interferes with Zlatkis-Zak reaction. *3505*
Serum   Increase   Physiological   In men by 25 mg/dL in ages 35-54 y old. *0793*

Creatinine   *Serum   Increase   Physiological*   Manifestation of hypervitaminosis D. *0071*

Hydroxyproline   *Urine   Increase   Physiological*   In vitamin-sensitive rickets. *3644*

Nonprotein Nitrogen   *Serum   Increase   Physiological*   Manifestation of hypervitaminosis D. *2313*

Phosphate   *Serum   Increase   Physiological*   Effect of increased gastrointestinal tract and renal absorption. *3879*
Urine   Increase   Physiological   May be normal in many cases. *3755*
Feces   Decrease   Physiological   Due to excessive absorption. *0987*

Protein   *Urine   Increase   Physiological*   Nephrotoxic effect with hypercalcemia. *1022*

Protein Bound Iodine (PBI)   *Serum   Increase   Physiological*   When given in cod liver oil. *2313*

Urea Nitrogen   *Serum   Increase   Physiological*   Manifestation of hypervitaminosis D. *0071*

Vitamin E

Aminolevulinic Acid   *Urine   Decrease   Physiological*   Possible reduction to normal levels in porphyria. *2549*

Cholesterol   *Serum   No Effect   Physiological*   No significant effect observed. *2158*

Factor VII   *Plasma   Decrease   Physiological*   In 2 patients given 2.3 g/m²/d intravenously for 4 or more days (effect abrogated by prior administration of menadiol sodium diphosphate). *1554*

Factor IX   *Plasma   Decrease   Physiological*   In 2 patients given 2.3 g/m²/d intravenously for 4 or more days (effect abrogated by prior administration of menadiol sodium diphosphate). *1554*

Factor X   *Plasma   Decrease   Physiological*   In 2 patients given 2.3 g/m²/d intravenously for 4 or more days (effect abrogated by prior administration of menadiol sodium diphosphate). *1554*

Lymphocytes   *Blood   Decrease   Physiological*   Significant effect with megadose supplementation. *1347*

Neutrophils   *Blood   No Effect   Physiological*   No effect with megadose supplementation. *1347*

Partial Thromboplastin Time   *Blood   Increase   Physiological*   In 2 patients given 2.3 g/m²/d intravenously for 4 or more days (effect abrogated by prior administration of menadiol sodium diphosphate). *1554*

Porphobilinogen   *Urine   Decrease   Physiological*   Possible reduction to normal levels in porphyria. *2549*

Prothrombin Time   *Blood   Increase   Physiological*   In 2 patients given 2.3 g/m²/d intravenously for 4 or more days (effect abrogated by prior administration of menadiol sodium diphosphate). *1554*

α-Tocopherol   *Serum   Increase   Physiological*   Increase by 50-60% but normal in 4 d. *2158*

Uroporphyrin   *Urine   Decrease   Physiological*   Reduce to normal levels in porphyria. *2549*

Vitamin K

Bilirubin   *Serum   Increase   Physiological*   Large doses in neonates or G-6-PD deficiency. *0248*

Catecholamines   *Plasma   Increase   Analytical*   May react like epinephrine in Shaw test. *0516*

Erythrocytes   *Blood   Decrease   Physiological*   K₃ and K₄ only, especially with G-6-PD deficiency. *2102*

Hematocrit   *Blood   Decrease   Physiological*   K₃ and K₄ only, especially with G-6-PD deficiency. *2102*

Hemoglobin   *Blood   Decrease   Physiological*   K₃ and K₄ only, especially with G-6-PD deficiency. *2102*
Urine   Increase   Physiological   Effect of treatment of hemorrhagic states in children. *0071*

17-Hydroxycorticosteroids   *Urine   Increase   Analytical*
Alleged *in vitro* interference with Reddy method. *1488*

Leukocytes   *Blood   Decrease   Physiological*   Pancytopenia
reported after K$_3$ and K$_4$. *2427*

Platelets   *Blood   Decrease   Physiological*   Pancytopenia
reported after K$_3$ and K$_4$. *2427*

Porphyrins   *Urine   Increase   Physiological*   Reported side
effect. *2427*

Protein   *Urine   Increase   Physiological*   Reported side effect.
2427

Prothrombin Time   *Plasma   Decrease   Physiological*   Therapeutic intent, affects action of warfarin. *2313*

Urobilinogen   *Urine   Increase   Physiological*   Hemolytic anemia in G-6-PD deficiency. *1343*

Vitamin Preparations

Acetaminophen Screening Test   *Urine   Negative   Analytical*
No reaction with o-cresol at therapeutic concentrations. *3326*

Albumin   *Serum   No Effect   Analytical*   No effect at expected
concentration with SMA 12/60 procedure. *2637*

Alkaline Phosphatase   *Serum   No Effect   Analytical*   No
effect at expected concentration with SMA 12/60 procedure.
2637

Aspartate Aminotransferase   *Serum   No Effect   Analytical*
No effect at expected concentration with SMA 12/60 procedure.
2637

BSP Retention   *Serum   Increase   Analytical*   If given i.v. and
contain riboflavin and B vitamins. *2381*

Calcium   *Serum   No Effect   Analytical*   No effect at expected
concentration with SMA 12/60 procedure. *2637*

Cholesterol   *Serum   No Effect   Analytical*   No effect at
expected concentration with SMA 12/60 procedure. *2637*

Creatinine   *Serum   No Effect   Analytical*   No effect at
expected concentration with SMA 12/60 procedure. *2637*

Glucose   *Serum   No Effect   Analytical*   No effect at expected
concentration with SMA 12/60 procedure. *2637*
Urine   No Effect   Analytical   No effect observed with Tes-
Tape®. *1100*
Urine   Decrease   Analytical   Low with Clinistix®, Diastix®.
1100

¹³¹I Uptake   *Serum   Decrease   Physiological*   If preparations
contain iodine. *2220*   Daylets contain tetraiodofluorescein. *2652*

Lactate Dehydrogenase   *Serum   No Effect   Analytical*   No
effect at expected concentration with SMA 12/60 procedure.
2637

Phosphate   *Serum   No Effect   Analytical*   No effect at
expected concentration with SMA 12/60 procedure. *2637*

Protein   *Serum   No Effect   Analytical*   No effect at expected
concentration with SMA 12/60 procedure. *2637*

Protein Bound Iodine (PBI)   *Serum   Increase   Analytical*
Iodine contamination. *1237*

Urea Nitrogen   *Serum   No Effect   Analytical*   No effect at
expected concentration with SMA 12/60 procedure. *2637*

Uric Acid   *Serum   No Effect   Analytical*   No effect at
expected concentration with SMA 12/60 procedure. *2637*

Walnuts

5-Hydroxyindoleacetic Acid (5-HIAA)
Urine   Increase   Physiological   Rich in serotonin. *1488*

Warfarin

Alanine Aminotransferase   *Serum   Increase   Physiological*
Rare cases of intrahepatic cholestasis in patients with previous
history of hypersensitivity reported. *2940*

Albumin   *Serum   No Effect   Analytical*   At concentration of
100 mg/L had no effect on BCG method. *3393*

Alkaline Phosphatase   *Serum   Increase   Physiological*
Rare cases of intrahepatic cholestasis in patients with previous
history of hypersensitivity reported. *2940*

Aspartate Aminotransferase
Serum   Increase   Physiological   Rare cases of intrahepatic
cholestasis in patients with previous history of hypersensitivity
reported. *2940*

Bicarbonate   *Serum   No Effect   Analytical*   At concentration
of 100 mg/L had no effect on method using phenolphthalein.
3393

Bilirubin   *Serum   No Effect   Analytical*   At concentration of
100 mg/L had no effect on Jendrassik and Grof method. *3393*

Bleeding Time   *Blood   Increase   Physiological*   May be prolonged. *1302*

Calcium   *Serum   No Effect   Analytical*   At concentration of
100 mg/L had no effect on cresolphthalein method. *3393*

Chloride   *Serum   No Effect   Analytical*   At concentration of
100 mg/L had no effect on mercurimetric method. *3393*

Cholesterol   *Serum   No Effect   Analytical*   At concentration
of 1.5 mg/L had no effect on CHOD-PAP method. *3393*   At concentration of 100 mg/L had no effect on Liebermann-Burchard
method. *3393*

Clotting Time   *Blood   Increase   Physiological*   May be prolonged. *1302*

Creatinine   *Serum   No Effect   Analytical*   At concentration of
100 mg/L had no effect on AutoAnalyzer Jaffé method. *3393*
Serum   No Effect   Physiological   No effect on renal function
noted with continuing administration. *2411*

Diphenylhydantoin   *Serum   No Effect   Physiological*   With
concurrent therapy typically had no effect. *3556*

Erythrocytes   *Urine   Increase   Physiological*   Excessive
doses may cause hematuria. *0948*

Factor II   *Plasma   Decrease   Physiological*   Therapeutic
action. *1680*

Factor VII   *Plasma   Decrease   Physiological*   Therapeutic
action. *1680*

Factor IX   *Plasma   Decrease   Physiological*   Therapeutic
action. *1680*

Factor X   *Plasma   Decrease   Physiological*   Therapeutic
action. *1680*

Glucose   *Serum   No Effect   Analytical*   At concentration of
1.5 mg/L had no effect on GOD/POD-PAP method. *3393*

γ-Glutamyltransferase (GGT)
Serum   Increase   Physiological   Rare cases of intrahepatic
cholestasis in patients with previous history of. *2940*

Hemoglobin   *Urine   Increase   Physiological*   Due to overdosage in some cases. *1302*

¹³¹I Uptake   *Serum   Decrease   Physiological*   Panwarfin®
contains tetraiodofluorescein. *2652*

MCV   *Blood   Decrease   Physiological*   May be secondary
microcytic hypochromic anemia. *1302*

Neutrophils   *Blood   Decrease   Physiological*   May cause
neutropenia. *2583*

Occult Blood   *Feces   Positive   Physiological*   May cause
intramural hemorrhage even if no ulcer. *3498*

Phenyramidol   *Plasma   Increase   Physiological*   Impairs
metabolism. *2641*

Phosphate   *Serum   No Effect   Analytical*   At concentration of
100 mg/L had no effect on phosphomolybdate method. *3393*

Platelet Aggregation   *Blood   Decrease   Physiological*   If
caused by ADP, thrombin, collagen, epinephrine. *0614*

Potassium   *Serum   No Effect   Analytical*   At concentration of
1.5 mg/L had no effect on flame-photometric method. *3393*   At
concentration of 100 mg/L had no effect on measurement by ISE
with predilution. *3393*

Protein   *Serum   No Effect   Analytical*   At concentration of
100 mg/L had no effect on biuret method with blank correction.
3393

Protein Bound Iodine (PBI)   *Serum   No Effect   Physiological*
No effect with 5 mg/d. *0583*

Prothrombin Time   *Plasma   Increase   Physiological*   Therapeutic intent (inhibits prothrombin formation). *1343*

Sodium   *Serum   No Effect   Analytical*   At concentration of
1.5 mg/L had no effect on flame-photometric method. *3393*   At
concentration of 100 mg/L had no effect on measurement by ISE
with predilution. *3393*

T₃ Uptake   *Serum   No Effect   Physiological*   No effect at
dose of 5 mg/d. *0583*

Warfarin (continued)

Triglycerides Serum No Effect Analytical At concentration of 100 mg/L had no effect on lipase/esterase method. 3393

Urea Nitrogen Serum No Effect Analytical At concentration of 100 mg/L had no effect on diacetylmonoxime method. 3393

Serum No Effect Physiological No effect on renal function noted with continuing administration. 2411

Uric Acid Serum No Effect Analytical At concentration of 26 mg/L had no effect on Kageyama-Hantzsch method. 3393 At concentration of 100 mg/L had no effect on phosphotungstate reduction method. 3393

Serum Increase Physiological Noted in men at all levels of renal function without alteration of renal clearance of uric acid, possibly related to increased production: effect up to 25%. 2411

Warfarin Metabolites

Factor II Plasma Decrease Physiological Slight effect with some compounds only. 2159

Factor VII Plasma Decrease Physiological Usually marked effect all compounds. 2159

Factor IX Plasma Decrease Physiological Usually marked effect all compounds. 2159

Factor X Plasma Decrease Physiological Usually marked effect all compounds. 2159

Water

Calcium Serum No Effect Physiological No discernible difference 2 h after drinking 1 glass. 2123

^{131}I Uptake Serum Decrease Physiological At 1 mg/L in drinking water. 0583

Urea Nitrogen Serum Decrease Physiological 2 h after drinking 1 glass, fell by 1 mg/dL. 2123

Water Load

Prolactin Plasma Decrease Physiological Fall in 2 h to 6.9% of baseline in normals. 0521

Weight

Creatinine Serum Increase Physiological Positively correlated in children. 0887

Hemoglobin Blood Increase Physiological Positively correlated in children. 0887

Uric Acid Serum Increase Physiological Positively correlated in children. 0887

Xanthine

Chloride Urine Increase Physiological Diuretic action. 1343

Crystals Urine Increase Physiological Rare, colorless, rhombic plates (not in alkaline urine). 0443

Factor V Plasma Increase Physiological Effect of methylxanthines. 1343

Fatty Acids, Free (FFA) Serum Increase Physiological Reported effect. 1343

Fibrinogen Plasma Increase Physiological Reported effect. 1343

Guanase Serum No Effect Analytical Up to 1.0 mmol/L on method of Nishikawa. 2602

Potassium Urine Increase Physiological Slight diuretic effect only. 1343

Protein Test Conditions Increase Analytical Reacts with Folin-Ciocalteu of Lowry method. 1569

Prothrombin Time Plasma Decrease Physiological Antagonizes effects of coumarins. 2313

Sodium Urine Increase Physiological Diuretic action. 1343

Uric Acid Serum No Effect Analytical At 10 mg/dL on copper chelate procedure. At 10 mg/dL on Nishi phosphotungstate procedure. 2230

Serum Increase Analytical Measured as uric acid by non-specific methods. 3500

Serum Decrease Analytical Linear decrease of from 3 to 36.5% at concentrations of from 100 to 520 mg/L on Kodak Ektachem®. Note of possible clinical importance in patients responding to chemotherapy. Hypoxanthine had no effect up to 400 mg/L. 2854

Volume Urine Increase Physiological Diuretic action. 1343

Xanthophyll

Bilirubin Serum Increase Analytical May deepen color of serum (?significance). 1022

Xanthurenic Acid

Ammoniacal Silver Nitrate Test Conditions Positive Analytical Positive spot test with Tollen's reagent. 3765

Ferric Chloride Test Urine Positive Analytical Deep green, later brown. 0443

Urobilinogen Test Conditions Increase Analytical Positive spot test with Ehrlich's reagent. 3765

X-Ray Therapy

β-Amino-Isobutyric Acid Urine Increase Physiological Due to tissue destruction. 2378

Amylase Serum Increase Physiological May be up to 20 fold increase at peak with whole body irradiation, mainly of salivary isoenzyme. 1837

Aspartate Aminotransferase Serum Increase Physiological Occurs with local irradiation injury. 0987

Chromosomes Test Conditions Abnormal Physiological Clastogenic in human cells in hela culture. 3282

Creatine Urine Increase Physiological Occurs with tissue destruction. 0987

Deoxycytidine Urine Increase Physiological Due to tissue destruction. 2378

Eosinophils Blood Increase Physiological Occurs after repeated irradiation. 0251

Erythrocytes Blood Decrease Physiological Occurs with onset of aplastic anemia. 0987

Ferritin Serum Increase Physiological Occurs within 2 h of deep x-ray therapy. 0987

Fibrinogen Plasma Increase Physiological Indicative of tissue damage. 0987

α_1-Globulin Serum Increase Physiological Increase due to tissue damage. 0987

α_2-Globulin Serum Increase Physiological Increased rapidly due to tissue damage. 0987

γ-Globulin Serum Increase Physiological May fall in some cases (tissue damage). 0987

Leucine Aminopeptidase Urine Increase Physiological Toxic damage due to released metabolites. 2897

Leukocytes Blood Decrease Physiological Cell destruction in leukemics. 3598

Lipase Serum Increase Physiological May be tripling with total body irradiation. 1837

Lymphocytes Blood Decrease Physiological Due to cell destruction. 0987

Lysozyme Urine Increase Physiological Associated with cell destruction in leukemics Effect observed for more than 45 d after therapy. 3598

Neutrophils Blood Decrease Physiological Also toxic granulation in cells. 0987

Ornithine Carbamoyltransferase (OCT) Serum Increase Physiological Due to breakdown of tissue proteins. 0483

Platelets Blood Decrease Physiological Thrombocytopenia. 0987

Properdin Serum Decrease Physiological Due to tissue destruction. 0987

Protein *Urine Increase Physiological* May cause nephrotoxicity. *3204*

Prothrombin Time *Plasma Increase Physiological* Also exaggerated response to anticoagulants. *1343*

Pseudouridine *Urine Increase Physiological* Due to tissue destruction. *2378*

Taurine *Urine Increase Physiological* Due to tissue destruction. *2378*

Trypsin *Serum Increase Physiological* May be up to 4 fold increase with total body irradiation. *1837*

Urea Nitrogen *Serum Increase Physiological* May cause nephrotoxicity. *3204*

Uric Acid *Serum Increase Physiological* Due to cellular destruction. *1320*

Urine Increase Physiological Due to cellular destruction. *1320*

Xylene

β_1-α-Globulin *Serum Decrease Physiological* Significant effect in people occupationally exposed. *3370*

Hippuric Acid *Urine Increase Analytical* Metabolites measured as hippurate (method of Tomokuni). *3607*

Immunoglobulin IgA *Serum Decrease Physiological* Significant effect in people occupationally exposed. *3370*

Immunoglobulin IgG *Serum Decrease Physiological* Significant effect in people occupationally exposed. *3370*

Immunoglobulin IgM *Serum No Effect Physiological* No effect in people occupationally exposed. *3370*

Xylitol

Alanine Aminotransferase *Serum No Effect Physiological* No effect reported after i.v. infusion. *1160*

Alkaline Phosphatase *Serum Increase Physiological* Dose related hepatocellular damage. *3212*

Aspartate Aminotransferase
Serum No Effect Physiological No effect reported after i.v. infusion. *1160*
Serum Increase Physiological Dose related hepatocellular damage. *3212*

Bicarbonate *Serum Decrease Physiological* May cause azotemia and acidosis. *3576*

Bilirubin *Serum Increase Physiological* Mainly indirect — by 2 to 3 times normal in some. *0925*

Bilirubin, Direct *Serum Increase Physiological* Increased up to 2 mg/dL following loading test. *3212*

Insulin *Plasma No Effect Physiological* Although increased in portal venous blood. *0310*
Plasma Increase Physiological ?same mechanism as glucose to release in dogs. *2048*

Lactate *Serum Increase Physiological* 4 times control after loading test. *3212*

Lactate Dehydrogenase *Serum Increase Physiological* Dose related hepatocellular damage. *3212*

Lactate Dehydrogenase Isoenzymes
Serum Increase Physiological Mainly due to liver component from liver damage. *3576*

pCO_2 *Blood Decrease Physiological* May cause metabolic acidosis. *3576*

pH *Blood Decrease Physiological* Pronounced metabolic acidosis in many patients. *3575*

Phosphate *Serum Increase Physiological* Mechanism not discussed. *3212*

Pyruvate *Serum Decrease Physiological* Slight fall after i.v. infusion reported. *1160*

Urea Nitrogen *Serum Increase Physiological* May cause azotemia. *3576*

Uric Acid *Serum Increase Physiological* Increases by 1.5 to 2 times normal. *0925*
Urine Increase Physiological As result of increased serum concentration (may be doubled). *0925*

Volume *Urine Increase Physiological* Increased up to 3 times normal (osmotic diuretic effect). *3212*

Xylose

N-Acetylglucosaminidase *Urine No Effect Analytical* At 10 mmol/L had no effect on 2 colorimetric analytical methods. *1354*

Glucose *Serum No Effect Analytical* No effect on glucose oxidase, hexokinase methods. *1856* At 200 mg/dL on hexokinase method of Coburn. *0676*
Serum Increase Analytical Nonspecificity of o-toluidine, FeCN, Neocuproin. *3902*

Protein *Test Conditions Increase Analytical* Interferes with Folin-Ciocalteu method of Lowry. *3636*

Sugar *Urine Increase Analytical* False positive with Benedict's, Clinitest®. *1563*

Xylulose

Sugar *Urine Increase Analytical* False positive with Benedict's, Clinitest®. *1563*

Yeast

Appearance *Urine Abnormal Physiological* Cloudy, insoluble in dilute acetic acid. *0443*

Uric Acid *Serum Increase Physiological* Equivalent of 8 g/d produce increase of approximately 8 mg/dL. *0286*
Urine Increase Physiological Equivalent of 8 g/d produce increase of approximately 1.2-1.4g/d. *0286*

Zero Gravity

Norepinephrine *Urine Decrease Physiological* Significant effect observed. *1176*

Zimeldine

Alanine Aminotransferase *Serum Increase Physiological* 7 of 14 patients treated for more than 1 week demonstrated toxic syndrome ? immunological mechanism or related to high initial dose. *2075* Reported in 2 patients in association with headaches, possibly due to fall in blood concentration of 5-hydroxytryptamine. *3387* In 64% of 21 inpatients with endogenous depression given 225 mg daily for 4 weeks. *2082* In several of 147 patients although 16 had initially high values. *3741*

Alkaline Phosphatase *Serum Increase Physiological* 7 of 14 patients treated for more than 1 week demonstrated toxic syndrome ? immunological mechanism or related to high initial dose. *2075* In 45% of 21 inpatients with endogenous depression given 225 mg daily for 4 weeks. *2082* In several of 147 patients although 16 had initially high values. *3741*

Aspartate Aminotransferase
Serum Increase Physiological 7 of 14 patients treated for more than 1 week demonstrated toxic syndrome ? immunological mechanism or related to high initial dose. *2075* Reported in 2 patients in association with headaches, possibly due to fall blood concentration of 5-hydroxytryptamine. *3387* In 64% of 21 inpatients with endogenous depression given 225 mg daily for 4 weeks. *2082* In several of 147 patients although 16 had initially high values. *3741*

Bilirubin *Serum No Effect Physiological* In 1 study involving 147 patients no significant change seen. *3741*

Cortisol *Plasma No Effect Physiological* In group of depressed patients. *0560*

Creatinine *Serum Increase Physiological* 23 of approximately 147 patients. *2082*

Erythrocytes *Blood No Effect Physiological* In 1 study involving 147 patients no significant change seen. *3741*
Urine Increase Physiological 3 of 14 patients treated for more than 1 week presented mild abnormality. *2075*

Erythrocyte Sedimentation Rate
Blood No Effect Physiological In 1 study involving 147 patients no significant change seen. *3741*

γ-Glutamyltransferase (GGT)
Serum Increase Physiological Reported in 2 patients in association with headaches, possibly due to fall in blood concentration of 5-hydroxytryptamine. *3387*

Zimeldine *(continued)*

Homovanillic Acid *CSF Increase Physiological* Insignificant effect in 43 chronically treated patients with depression or Alzheimer's disease. *2991*

5-Hydroxyindoleacetic Acid (5-HIAA)
CSF Decrease Physiological Significant effect in 43 chronically treated patients with depression or Alzheimer's disease. *2991*

Leukocytes *Blood No Effect Physiological* In 1 study involving 147 patients no significant change seen. *3741*
Blood Decrease Physiological 7 of 14 patients treated for more than 1 week demonstrated toxic syndrome ? immunological mechanism or related to high initial dose. *2075*

Luteinizing Hormone (LH) *Plasma No Effect Physiological* In group of depressed patients. *0560*

Platelets *Blood No Effect Physiological* In 1 study involving 147 patients no significant change seen. *3741*
Blood Decrease Physiological 7 of 14 patients treated for more than 1 week demonstrated toxic syndrome ? immunological mechanism or related to high initial dose. *2075*

Prolactin *Plasma Increase Physiological* Response only after chronic pretreatment in group of depressed patients. *0560*

Protein *Urine Increase Physiological* 3 of 14 patients treated for more than 1 week preserved mild abnormality. *2075*

Zinc

Alkaline Phosphatase *Serum Decrease Analytical* Inhibitors of enzyme in laboratory procedures. *3588*

Calcium *Serum Increase Analytical* May affect titration with EDTA procedures. *0583*

Magnesium *Urine Increase Analytical* Measured by fluorometric method of Schachter. *3573*

Nickel *Test Conditions Increase Analytical* Possible interference with atomic absorption. *3499*

Zinc *Serum Increase Physiological* Maximal 2 h after ingestion, raised for 5 h. *0439*
Urine Increase Physiological Slight increase related to dose. *0439*

Zirconium Salts

γ-**Globulin** *Serum Increase Physiological* Observed with zirconium granulomas. *2726*

Zomepirac

Coombs' Test, Direct
Red Blood Cells Increase Physiological Case reported of immune hemolysis. *3208*

Hemoglobin *Blood Decrease Physiological* Case reported of immune hemolysis. *3208*

Zoxazolamine

Alanine Aminotransferase *Serum Increase Physiological* May produce viral-hepatitis like syndrome. *1948*

Aspartate Aminotransferase
Serum Increase Physiological May produce viral-hepatitis like syndrome. *1948*

Bilirubin *Serum Increase Physiological* May produce viral-hepatitis like syndrome. *1948*

Protein *Urine Increase Physiological* Has nephrotoxic effect May have nephrotoxic effect. *1488*

5 REFERENCES

0001 AANDERUD S, MYKING OL, STRANDJORD RE, THE INFLUENCE OF CARBAMAZEPINE ON THYROID HORMONES AND THYROXINE BINDING GLOBULIN IN HYPOTHYROID PATIENTS SUBSTITUTED WITH THYROXINE, *CLIN ENDOCRINOL*, 15, 247-252 (1981)

0002 AARTS EM, DRUG-INDUCED STIMULATION OF THE GLUCURONIC ACID SYSTEM, *PhD THESIS, NIJMEGEN, THE NETHERLANDS* (1968)

0003 AARTS EM, D-GLUCARIC ACID EXCRETION AS A TEST FOR HEPATIC ENZYME INDUCTION, *LANCET*, 1, 859 (1971)

0004 ABBOIX M, FRATI ME, LAPORTE J-R, THE POTENTIATION OF ACENOCOUMAROL ANTICOAGULANT EFFECT BY AMIODARONE, *BR J CLIN PHARMACOL*, 18, 355-360 (1984)

0005 ABBOUD RT, FERA T, JOHAL S, ET AL, EFFECT OF SMOKING ON PLASMA NEUTROPHIL ELASTASE LEVELS, *J LAB CLIN MED*, 108, 294-300 (1986)

0006 ABBRUZZESE A, SWANSON J, JAUNDICE AFTER THERAPY WITH CHLORDIAZEPOXIDE HYDROCHLORIDE, *N ENGL J MED*, 273, 321 (1965)

0007 ABDALLA HI, HART DM, BEASTALL GH, REDUCED SERUM FREE THYROXINE CONCENTRATION IN POSTMENOPAUSAL WOMEN RECEIVING OESTROGEN TREATMENT, *BR MED J [CLIN RES]*, 288, 754-755 (1984)

0008 ABRAHAM GE, SAMOJLIK E, KYLE FW, BUSTER JE, RADIOIMMUNOASSAY OF PLASMA 16 ALPHA-HYDROXYPROGESTERONE, *ANAL LETT*, 6, 675 (1973)

0009 ABREO K, LABARRE J, SUPROFEN, ACUTE RENAL FAILURE, AND HEMATURIA, *ANN INTERN MED*, 105, 799 (1986)

0010 ABUKURAH AR, MOSER AM JR., BAIRD CL, RANDALL RE JR., SETTER JG, BLANKE RV, ACUTE SODIUM FLUORIDE POISONING, *J AM MED ASSOC*, 222, 816 (1972)

0011 ABUL-FADL MAM, KING EJ, PROPERTIES OF THE ACID PHOSPHATASES OF ERYTHROCYTES AND OF THE HUMAN PROSTATE GLAND, *BIOCHEM J*, 45, 51 (1949)

0012 ACLAND JD, THE INTERPRETATION OF THE SERUM PROTEIN BOUND IODINE: A REVIEW, *J CLIN PATHOL*, 24, 187 (1971)

0013 ACOCELLA G, BILLING BH, EFFECT OF DRUGS ON THE HEPATIC TRANSPORT OF BILIRUBIN IN, *THERAPEUTIC AGENTS AND THE LIVER*, N MCINTYRE, S SHERLOCK, EDS. BLACKWELL, OXFORD (1965)

0014 ACOCELLA G, CONTI R, INTERACTION OF RIFAMPICIN WITH OTHER DRUGS, *TUBERCLE*, 61, 171-177 (1980)

0015 ADACHI J, MIZOI Y, ACETALDEHYDE-MEDIATED ALCOHOL SENSITIVITY AND ELEVATION OF PLASMA CATECHOLAMINE IN MAN, *JPN J PHARMACOL*, 33, 531-539 (1983)

0016 ADAM K, OSWALD I, SHAPIRO C, EFFECTS OF LOPRAZOLAM AND TRIAZOLAM ON SLEEP AND OVERNIGHT URINARY CORTISOL, *PSYCHOPHARMACOL*, 82, 389-394 (1984)

0017 ADAMS EC, DIFFERENTIATION OF MYOGLOBIN AND HEMOGLOBIN IN BIOLOGICAL FLUIDS, *ANN CLIN LAB SCI*, 1, 208 (1971)

0018 ADAMS PW, GODSLAND I, MELROSE J, ET AL, THE INFLUENCE OF ORAL CONTRACEPTIVE FORMULATION ON CARBOHYDRATE AND LIPID METABOLISM, *J PHARMACOTHER*, 3, 54-63 (1980)

0019 ADAMS RG, HARRISON JF, SCOTT P, THE DEVELOPMENT OF CADMIUM INDUCED PROTEINURIA, IMPAIRED RENAL FUNCTION AND OSTEOMALACIA, *Q J MED*, 38, 425 (1969)

0020 ADDIS T, BARRETT E, POO LJ, YUEN DW, THE RELATION BETWEEN THE SERUM UREA CONCENTRATION AND THE PROTEIN CONSUMPTION OF NORMAL INDIVIDUALS, *J CLIN INVEST*, 26, 869 (1947)

0021 ADIBI SA, DRASH AL, HORMONE AND AMINO ACID LEVELS IN ALTERED NUTRITIONAL STATES, *J LAB CLIN MED*, 76, 722 (1970)

0022 ADLERCREUTZ H, DRUG INTERFERENCE IN URINARY STEROID HORMONE ASSAY, *FIRST EUROPEAN CONGRESS OF CLINICAL CHEMISTRY, MUNICH*, 1974

0023 ADLERCREUTZ H, DRUG INTERFERENCE IN URINARY ESTROGEN DETERMINATION, *PROC V INTERNATIONAL SYMPOSIUM ON QUALITY CONTROL IN CLINICAL CHEMISTRY, GENEVA, SWITZERLAND*, 1973

0024 ADLERCREUTZ H, SOININEN K, HARKONEN M, ORAL CONTRACEPTIVES AND SERUM AMYLASE, *BR MED J [CLIN RES]*, 3, 529 (1972)

0025 ADNITT PI, HYPOGLYCEMIC ACTION OF MONOAMINE OXIDASE INHIBITORS (MAOI'S), *DIABETES*, 17, 628 (1968)

0026 ADROUNY A, MEGUERDITCHIAN S, KOO CH, ET AL, AGRANULOCYTOSIS RELATED TO VANCOMYCIN THERAPY, *AM J MED*, 81, 1059-1061 (1986)

0027 ADU D, TURNEY J, MICHAEL J, ET AL, HYPERKALAEMIA IN CYCLOSPORIN TREATED RENAL ALLOGRAFT RECIPIENTS, *LANCET*, 2, 370-372 (1983)

0028 ADVERSE DRUG REACTIONS ADVISORY COMMITTEE, MIANSERIN: A POSSIBLE CAUSE OF NEUTROPENIA AND AGRANULOCYTOSIS, *MED J AUST*, 2, 673-674 (1980)

0029 AFFARAH HB, MARS RL, SOMEREN A, ET AL, MYOGLOBINURIA AND ACUTE RENAL FAILURE ASSOCIATED WITH INTRAVENOUS VASOPRESSIN INFUSION, *SOUTH MED J*, 77, 918-921 (1984)

0030 AGGELER PM, O'REILLY RA, LEONG L, POTENTIATION OF ANTICOAGULANT EFFECT OF WARFARIN BY PHENYLBUTAZONE, *N ENGL J MED*, 276, 496 (1967)

0031 AGNELLI G, DEL FAVERO A, PARISE P, ET AL, CEPHALOSPORINS-INDUCED HYPOPROTHROMBINEMIA: IS THE N-METHYL-THIOTETRAZOLE SIDE CHAIN THE CULPRIT?, *ANTIMICROB AGENTS CHEMOTHER*, 29, 1108-1109 (1986)

0032 AGOSTINI L, DOWN PF, MURISON J, WRONG OM, FAECAL AMMONIA AND pH DURING LACTULOSE ADMINISTRATION IN MAN, *GUT*, 13, 859 (1972)

0033 AGUS ZS, GOLDBERG M, RENAL MECHANISMS OF THE NATRIURETIC AND ANTIPHOSPHATURIC EFFECTS OF TRIFLOCIN - A NEW DIURETIC, *J LAB CLIN MED*, 76, 280 (1970)

0034 AH ROBINS, MANUFACTURER'S LITERATURE ON DOPRAM®, *1407 CUMMINGS DR, RICHMOND VA*, 23220

0035 AHEARN MJ, HICKS JE, ANDRIOLE VT, NEUTROPENIA DURING HIGH DOSE INTRAVENOUS OXACILLIN THERAPY, *YALE J BIOL MED*, 49, 351-360 (1976)

0036 AHMED AR, MOY R, AZATHIOPRINE, *INT J DERMATOL*, 20, 461-467 (1981)

0037 AITKEN JM, HART DM, SMITH DA, THE EFFECT OF LONG-TERM MESTRANOL ADMINISTRATION ON CALCIUM AND PHOSPHORUS HOMEOSTASIS IN OOPHORECTOMIZED WOMEN, *CLIN SCI*, 41, 233 (1971)

0038 AJDUKIEWICZ AB, GRAINGER J, SCHEUER PJ, SHERLOCK S, JAUNDICE DUE TO IPRINDOLE, *GUT*, 12, 705 (1971)

0039 AJEL LA, POSITIVE DIPHENHYDRAMINE INTERFERENCE IN THE EMIT-D.A.U. ASSAY, *CLIN CHEM*, 31, 340-341 (1985)

0040 AKSOY M, DINCOL K, ERDEM S, DINCOL G, ACUTE LEUKEMIA DUE TO CHRONIC EXPOSURE TO BENZENE, *AM J MED*, 52, 160 (1972)

0041 AL-DAMLUJI S, MEEK JH, INTERFERENCE OF A PROPRANOLOL METABOLITE WITH SERUM BILIRUBIN ESTIMATION IN CHRONIC RENAL FAILURE, *BR MED J [CLIN RES]*, 2, 1414 (1980)

0042 AL-HUJAJ M, SCHONTHAL H, HYPERURICEMIA AND LEVODOPA, *N ENGL J MED*, 285, 859 (1971)

0043 ALARCON-SEGOVIA D, FISHBEIN E, ALCALA H, ISONIAZID ACETYLATION RATE AND DEVELOPMENT OF ANTINUCLEAR ANTIBODIES UPON ISONIAZID TREATMENT, *ARTHRITIS RHEUM*, 14, 748 (1971)

0044 ALARCON-SEGOVIA D, WAKIM KG, WORTHINGTON JW, WARD LE, CLINICAL AND EXPERIMENTAL STUDIES ON THE HYDRALAZINE SYNDROME, *MEDICINE*, 46, 1 (1967)

0045 ALBANI F, RIVA R, PERUCCA E, ET AL, INTERFERENCE OF VALPROIC ACID IN THE COLORIMETRIC DETERMINATION OF FREE FATTY ACIDS IN PLASMA, *CLIN CHEM*, 28, 1398 (1982)

0046 ALBERS JJ, GRUNDY SM, CLEARY PA, ET AL, NATIONAL COOPERATIVE GALLSTONE STUDY: THE EFFECT OF CHENODEOXYCHOLIC ACID ON LIPOPROTEINS AND APOLIPOPROTEINS, *GASTROENTEROLOGY*, 82, 638-646 (1982)

0047 ALBERS JJ, TAGGART HM, APPELBAUM-BOWDEN D, ET AL, REDUCTION OF LECITHIN CHOLESTEROL ACYLTRANSFERASE, APOLIPOPROTEIN D AND THE Lp (a) LIPOPROTEIN WITH THE ANABOLIC STEROID STANOZOLOL, *BIOCHIM BIOPHYS ACTA*, 795, 293-296 (1984)

0048 ALBERT M, STANSELL MJ, VASCULAR SYMPTOMATIC RELIEF DURING ADMINISTRATION OF ETHYLCHLOROPHENOXYISOBUTYRATE, CLOFIBRATE, *METABOLISM*, 18, 635 (1969)

0049 ALBERTI KGMM, HOCKADAY TDR, WILLIAMSON DM, METABOLIC EFFECTS OF CHRONIC CAFFEINE ADMINISTRATION IN A PATIENT WITH DIABETES MELLITUS, *PROC ROY SOC MED*, 65, 485 (1972)

0050 ALESTIG K, EILARD T, NORRBY R, ET AL, CEFTAZIDIME IN CLINICAL PRACTICE, *J ANTIMICROB CHEMOTHER*, 12, SUPPL A, 111-114 (1983)

0051 ALESTIG K, TROLLFORS B, ANDERSSON R, ET AL, CEFTAZIDINE AND RENAL FUNCTION, *J ANTIMICROB CHEMOTHER*, 13, 177-181 (1984)

0052 ALEXANDER DP, RUSSO ME, FOHRMAN DE, ET AL, NAFCILLIN-INDUCED PLATELET DYSFUNCTION AND BLEEDING, *ANTIMICROB AGENTS CHEMOTHER*, 23, 59-62 (1983)

0053 ALEXANDER MR, LOUIE SG, GUERNSEY BG, ISONIAZID-ASSOCIATED HEPATITIS, *CLIN PHARM*, 1, 148-153 (1982)

0054 ALEXANDER NM, JENNINGS JF, ANALYSIS FOR TOTAL SERUM THYROXINE BY EQUILIBRIUM COMPETITIVE PROTEIN BINDING ON SMALL, RE-USABLE SEPHADEX, *CLIN CHEM*, 20, 553 (1974)

0055 ALEXANDER RL JR., EVALUATION OF AN AUTOMATIC CALCIUM TITRATOR, *CLIN CHEM*, 17, 1171 (1971)

0056 ALEXIEVA-FIGUSCH J, BLANKENSTEIN MA, HOP WCJ, ET AL, TREATMENT OF METASTATIC BREAST CANCER PATIENTS WITH DIFFERENT DOSES OF MEGESTEROL ACETATE: DOSE RELATIONS, METABOLIC AND ENDOCRINE EFFECTS, *EUR J CANCER CLIN ONCOL*, 20, 33-40 (1981)

0057 ALLAIN CC, POON LS, CHAN CSG, RICHMOND W, FU PC, ENZYMATIC DETERMINATION OF TOTAL SERUM CHOLESTEROL, *CLIN CHEM*, 20, 470 (1974)

0058 ALLEGRA CJ, CHABNER BA, TUAZON CU, ET AL, TRIMETREXATE FOR THE TREATMENT OF PNEUMOCYSTIS CARINII PNEUMONIA IN PATIENTS WITH THE ACQUIRED IMMUNODEFICIENCY SYNDROME, *N ENGL J MED*, 317, 978-985 (1987)

0059 ALLEN JK, ADENA MA, THE ASSOCIATION BETWEEN PLASMA CHOLESTEROL, HIGH-DENSITY-LIPOPROTEIN CHOLESTEROL, TRIGLYCERIDES AND URIC ACID IN ETHANOL CONSUMERS, *ANN CLIN BIOCHEM*, 22, 62-66 (1985)

0060 ALLEN JK, FRASER IS, CHOLESTEROL, HIGH DENSITY LIPOPROTEIN AND DANAZOL, *J CLIN ENDOCRINOL METAB*, 53, 149-152 (1981)

0061 ALLEN LC, MICHALKO K, COONS C, MORE ON CEPHALOSPORIN INTERFERENCE WITH CREATININE DETERMINATIONS, *CLIN CHEM*, 28, 555-556 (1982)

0062 ALLISON SP, CHAMBERLAIN MJ, MILLER JE, FERGUSON R, GILLETT AP, BEMAND BV, SAUNDERS RA, EFFECTS OF PROPRANOLOL ON BLOOD SUGAR, INSULIN AND FREE FATTY ACIDS, *DIABETOLOGIA*, 5, 339 (1969)

0063 ALLOLIO B, DORR H, STUTTMAN R, ET AL, EFFECT OF A SINGLE BOLUS OF ETOMIDATE UPON 8 MAJOR CORTICOSTEROID HORMONES AND PLASMA ACTH, *CLIN ENDOCRINOL*, 22, 281-286 (1985)

0064 ALMEN T, A STEERING DEVICE FOR SELECTIVE ANGIOGRAPHY AND SOME VASCULAR AND ENZYMATIC REACTIONS OBSERVED, *ACTA RADIOL SUPPL*, 260, 1 (1966)

0065 ALMEN T, TOXICITY OF RADIO CONTRAST AGENTS IN, *INTERNATIONAL ENCYCLOPEDIA OF PHARMACOLOGY AND THERAPEUTICS* VOL 2, PERGAMON PRESS, NEW YORK, NY (1971) (19)

0066 ALORA BD, ESTRADA FA, LANSANG SL, PARENTERAL SODIUM EPICILLIN IN ACUTE INFECTIONS, *CURR THER RES*, 14, 358 (1972)

0067 ALPER C, SPECIMEN COLLECTION AND PRESERVATION (P 373) IN *CLINICAL CHEMISTRY: PRINCIPLES AND TECHNICS*, RJ HENRY, DC CANNON, JW WINKELMAN, EDS. 2ND ED HARPER AND ROW, HAGERSTOWN, MD (1974)

0068 ALPEROVITCH A, CHABENAT C, FLAMANT R, BOHUON C, ABSENCE OF CHANGE IN URINARY LEVELS OF TRYPTOPHAN METABOLITES DURING SMOKING, *PATHOL BIOL*, 19, 977 (1971)

0069 ALTMEYER P, BUHLES N, HOLZEL C, ET AL, INFLUENCE OF TOPICAL CORTICOSTEROIDS ON HORMONES IN URINE AND PLASMA, *ARZNEIM-FORSCH/DRUG RES*, 36, 993-996 (1986)

0070 AMA COUNCIL ON DRUGS, NEW DRUGS, *1967 EDITION CHICAGO, AMERICAN MEDICAL ASSOCIATION* (1967)

0071 AMA COUNCIL ON DRUGS, AMA DRUG EVALUATIONS, *AMERICAN MEDICAL ASSOCIATION, 535 NORTH DEARBORN ST, CHICAGO, IL, 60610*

0072 AMA COUNCIL ON DRUGS, REGISTRY ON ADVERSE REACTIONS, TABULATION OF REPORTS - PANEL ON HEMATOLOGY, *AMERICAN MEDICAL ASSOCIATION, JULY* (1964)

0073 AMA COUNCIL ON DRUGS, REGISTRY ON ADVERSE REACTIONS, *AMERICAN MEDICAL ASSOCIATION* (1964)

0074 AMA COUNCIL ON DRUGS, A COPPER-CHELATING AGENT: PENICILLAMINE (CUPRIMINE®), *J AM MED ASSOC*, 189, 847 (1964)

0075 AMA COUNCIL ON DRUGS, A NEW ANTIBIOTIC- LINCOMYCIN, *J AM MED ASSOC*, 194, 545 (1965)

0076 AMADOR E, AUTOMATED URINARY GLUCOSE ANALYSES, *AM J CLIN PATHOL*, 59, 735 (1973)

0077 AMADOR E, SALVATORE AC, SERUM GLUTAMIC OXALACETIC TRANSAMINASE ACTIVITY, REVISED MANUAL AND AUTOMATED METHODS USING DIAZONIUM DYES, *AM J CLIN PATHOL*, 55, 686 (1971)

0078 AMADOR E, WACKER WEC, ENZYMES IN GENITOURINARY DISEASE IN, *DIAGNOSTIC ENZYMOLOGY*, EL COODLEY ED. LEA AND FEBIGER, PHILADELPHIA, PA (1970)

0079 AMBRE JJ, FISCHER LJ, EFFECT OF COADMINISTRATION OF ALUMINUM AND MAGNESIUM HYDROXIDES ON ABSORPTION OF ANTICOAGULANTS IN MAN, *CLIN PHARMACOL THER*, 14, 231 (1973)

0080 AMBROSE JA, FLUOROMETRIC MEASUREMENT OF TYROSINE IN SERUM AND PLASMA, *CLIN CHEM*, 20, 505 (1974)

0081 AMBROSE JA, HISTIDINE, *STAND METH CLIN CHEM*, 7, 189 (1972)

0082 AMBROSI B, BOCHICCHIO D, RIVA E, ET AL, EFFECTS OF SODIUM VALPROATE ADMINISTRATION ON PLASMA ACTH LEVELS IN PATIENTS WITH ACTH HYPERSECRETION, *J ENDOCRINOL INVEST*, 6, 305-306 (1983)

0083 AMBRUS JL, SCHIMERT G, LAJOS TZ, AMBRUS CM, MINK IB, LASSMAN HB, MOORE RH, MELZE RJ, EFFECT OF ANTIFIBRINOLYTIC AGENTS AND ESTROGENS ON BLOOD LOSS AND BLOOD COAGULATION FACTORS, *J MED*, 2, 65 (1971)

0084 AMDISEN A, ANDERSEN CJ, LITHIUM TREATMENT AND THYROID FUNCTION. A SURVEY OF 237 PATIENTS IN LONG-TERM LITHIUM TREATMENT, *PHARMACOPSYCHIATRY*, 15, 149-155 (1982)

0085 AMERICAN HOSPITAL FORMULARY SERVICE, AMERICAN HOSPITAL FORMULARY SERVICE,, *AMERICAN SOCIETY OF HOSPITAL PHARMACISTS, WASHINGTON, DC*

0086 AMERY A, CONWAY J, A CRITICAL REVIEW OF DIAGNOSTIC TESTS FOR PHEOCHROMOCYTOMA, *AM HEART J*, 73, 129 (1967)

0087 AMES RP, THE INFLUENCE OF NONBETA-BLOCKING DRUGS ON THE LIPID PROFILE: ARE DIURETICS OUTCLASSED AS INITIAL THERAPY FOR HYPERTENSION?, *AM HEART J*, 114, 998-1006 (1987)

0088 AMES RP, THE EFFECTS OF ANTIHYPERTENSIVE DRUGS ON SERUM LIPIDS AND LIPOPROTEINS II NONDIURETIC DRUGS, *DRUGS*, 32, 335-357 (1986)

0089 AMES RP, THE EFFECT OF ANTIHYPERTENSIVE DRUGS ON SERUM LIPIDS AND LIPOPROTEINS I DIURETICS, *DRUGS*, 32, 260-278 (1986)

0090 AMES RP, PEACOCK PB, SERUM CHOLESTEROL DURING TREATMENT OF HYPERTENSION WITH DIURETIC DRUGS, *ARCH INTERN MED*, 144, 710-714 (1984)

0091 ANAST CS, THE UNRELIABILITY OF THE TITAN YELLOW METHOD FOR THE DETERMINATION OF MAGNESIUM, *CLIN CHEM*, 9, 544 (1963)

0092 ANASTASIOU-NANAM, KOUTRAS DA, LEVIS G, ET AL, THE CORRELATION OF SERUM AMIODARONE LEVELS WITH ABNORMALITIES IN THE METABOLISM OF THYROXINE, *J ENDOCRINOL INVEST*, 7, 405-407 (1984)

0093 ANASTASSIADIS PA, COMMON RH, SOME ASPECTS OF THE RELIABILITY OF CHEMICAL ANALYSES, *ANAL BIOCHEM*, 22, 409 (1968)

0094 ANCES IG, OBSERVATIONS ON THE LEVEL OF BLOOD OXYTOCINASE THROUGHOUT THE COURSE OF LABOR AND PREGNANCY, *AM J OBSTET GYNECOL*, 113, 291 (1972)

0095 ANCILL RJ, ULCEROGENIC ACTION OF AZAPROPAZONE, *BR MED J [CLIN RES]*, 1, 1469-1470 (1977)

0096 ANDERSEN AN, SCHIOLER V, HERTZ J, ET AL, EFFECT OF METOCLOPRAMIDE INDUCED HYPERPROLACTINAEMIA ON THE GONADOTROPHIC RESPONSE TO OESTRADIOL AND LRH, *ACTA ENDOCRINOL*, 100, 1-9 (1982)

0097 ANDERSEN CJ, KEIDING NR, NEILSEN AB, FALSE ELEVATION OF SERUM PROTEIN-BOUND-IODINE CAUSED BY RED COLORED DRUGS OR FOODS, *SCAND J CLIN LAB INVEST*, 16, 249 (1964)

0098 ANDERSON BN, HENRIKSON IR, JAUNDICE AND EOSINOPHILIA ASSOCIATED WITH AMITRIPTYLINE, *J CLIN PSYCHIATRY*, 39, 730-731 (1978)

0099 ANDERSON GM, FEIBEL FC, WETLAUFER LA, ET AL, EFFECT OF A MEAL ON HUMAN WHOLE BLOOD SEROTONIN, *GASTROENTEROLOGY*, 88, 86-89 (1985)

0100 ANDERSON MS, BOWERS CY, KASTIN AJ, ET AL, SYNTHETIC THYROTROPIN-RELEASING HORMONE: A POTENT STIMULATOR OF THYROTROPIN SECRETION IN MAN, *N ENGL J MED*, 285, 1279 (1971)

0101 ANDERSON NG, FALSE POSITIVE GOLD ANTIGEN REACTION IN SMOKERS, *PERSONAL COMMUNICATION*

0102 ANDERSON OO, PERSSON I, CARBOHYDRATE METABOLISM DURING TREATMENT WITH CHLORTHALIDONE AND ETHACRYNIC ACID, *BR MED J [CLIN RES]*, 2, 798 (1968)

0103 ANDERSSON DEH, ROJDMARK S, BLOOD GLUCOSE LOWERING EFFECT OF VERAPAMIL IN FASTED MAN, *HORM METAB RES*, 16, SUPPL 1, 160-163 (1984)

0104 ANDERSSON S, CARLSON LA, ORO L, RICHARDS EA, EFFECT OF NICOTINIC ACID ON GASTRIC SECRETION OF ACID IN HUMAN SUBJECTS AND IN DOGS, *SCAND J GASTROENTEROL*, 6, 693 (1971)

0105 ANDERTON JL, KINCAID-SMITH P, DIURETICS: II CLINICAL CONSIDER-ATIONS, *DRUGS*, 1, 141 (1971)

0106 ANDO Y, FUJII M, FUJITA J, UETE T, EFFECT OF ANTICOAGULANTS ON THE PH, PCO_2 AND PO_2 OF BLOOD, *KITANO BYOIN KIYO*, 15, 14 (1970)

0107 ANDRE CM, JAMES VHT, ASSAY OF PLASMA DEHYDROEPIANDROS-TERONE AND ITS SULFATE BY COMPETITIVE PROTEIN BINDING, *CLIN CHIM ACTA*, 43, 295 (1973)

0108 ANDREJAK M, HARY L, ANDREJAK M-TH, ET AL, DILTIAZEM INCREASES STEADY STATE DIGOXIN SERUM LEVELS IN PATIENTS WITH CARDIAC DISEASE, *J CLIN PHARMACOL*, 27, 967-970 (1987)

0109 ANECKSTEIN AG, WEINGOLD AB, CHLOROTHIAZIDE-INDUCED HEP-ATIC COMA IN PREGNANCY, *AM J OBSTET GYNECOL*, 95, 136 (1966)

0110 ANONYMOUS, FALSE POSITIVE AND FALSE NEGATIVE REACTIONS WITH VARIOUS TESTS FOR GLYCOSURIA, *AMES COMPANY, ELKHART, IN*

0111 ANONYMOUS, FALSE POSITIVE AND FALSE NEGATIVE REACTIONS WITH VARIOUS TESTS FOR PROTEINURIA, *AMES COMPANY, ELKHART, IND*

0112 ANONYMOUS, INSULIN: SERUM OR PLASMA IN, *BIO-SCIENCE REPORTS*, SPRING 1971, BIO-SCIENCE LABORATORIES, VAN NUYS, CA (1971)

0113 ANONYMOUS, PENICILLAMINE NEPHROPATHY, *BR MED J [CLIN RES]*, 1, 761-762 (1981)

0114 ANONYMOUS, TODAY'S DRUGS: DRUGS FOR RHEUMATOID DISOR-DERS, *BR MED J [CLIN RES]*, 1, 545 (1964)

0115 ANONYMOUS, TODAY'S DRUGS: TREATMENT OF TRIGEMINAL NEU-RALGIA, *BR MED J [CLIN RES]*, 2, 583 (1972)

0116 ANONYMOUS, TODAY'S DRUGS: DIAZOXIDE, *BR MED J [CLIN RES]*, 4, 417 (1972)

0117 ANONYMOUS, 52 FACTORS THAT CAN AFFECT BLOOD GLUCOSE LEVELS, *CLIN TOXICOL*, 4, 297 (1971)

0118 ANONYMOUS, FDA APPROVED NEW DRUG LITERATURE-SODIUM DICLOXACILLIN MONOHYDRATE, *DRUG INTELL CLIN PHARM*, 2, 251 (1968)

0119 ANONYMOUS, SODIUM VALPROATE REASSESSED, *DRUG THER BULL*, 19, 93-95 (1981)

0120 ANONYMOUS, INTRAVENOUS FEEDING WITH AMINO ACIDS AND FATS, *DRUG THER BULL*, 10, 49 (1972)

0121 ANONYMOUS, INTERFERENCE OF DRUGS WITH CHEMICAL DIAG-NOSTIC TESTS, *DRUG THER BULL*, 10, 69 (1972)

0122 ANONYMOUS, FDA DRUG EFFICACY STUDY IMPLEMENTATION, *DRUG THERAPY*, 1, 81 (1971)

0123 ANONYMOUS, MAFENIDE ACETATE CREAM: A REVIEW, *DRUGS*, 1, 461 (1971)

0124 ANONYMOUS, DIAZOXIDE: A REVIEW OF ITS PHARMACOLOGICAL PROPERTIES AND THERAPEUTIC USE IN HYPERTENSIVE CRISIS, *DRUGS*, 2, 78 (1971)

0125 ANONYMOUS, GLIBENCLAMIDE: A REVIEW, *DRUGS*, 1, 116 (1971)

0126 ANONYMOUS, ALUMINUM NICOTINATE REDUCES LEVEL OF SERUM CHOLESTEROL, *J AM MED ASSOC*, 209, 353 (1969)

0127 ANONYMOUS, DRUGS AND OTHER FACTORS AFFECTING LABORA-TORY TESTS, *MEDICAL LETTER*, 13, 82 (1971)

0128 ANONYMOUS, DRUGS FOR PARASITIC INFECTIONS, *MEDICAL LET-TER*, 11, 21 (1969)

0129 ANONYMOUS, MICROSTIX AND OTHER OFFICE TESTS FOR DETEC-TION OF URINARY TRACT INFECTION, *MEDICAL LETTER*, 16, 13 (1974)

0130 ANONYMOUS, AMICAR® AND HYPERFIBRINOLYSIS, *MEDICAL LET-TER*, 9, 10 (1967)

0131 ANONYMOUS, HANDBOOK OF ANTIMICROBIAL THERAPY, *MEDICAL LETTER*, 14, 1 (1972)

0132 ANONYMOUS, ADVERSE INTERACTION OF DRUGS, *MEDICAL LET-TER*, 15, 77 (1973)

0133 ANONYMOUS, USE OF PROBENECID IN ANTIMICROBIAL THERAPY, *MEDICAL LETTER*, 13, 85 (1971)

0134 ANONYMOUS, *MEDICAL LETTER*, 13, 101 (1971)

0135 ANONYMOUS, MINOCYCLINE (MINOCIN®), *MEDICAL LETTER*, 14, 9 (1972)

0136 ANONYMOUS, METHOTREXATE IN THE TREATMENT OF PSORIASIS, *MEDICAL LETTER*, 14, 41 (1972)

0137 ANONYMOUS, CLINDAMYCIN, *MEDICAL LETTER*, 15, 25 (1973)

0138 ANONYMOUS, METABOLIC ACTIONS OF GROWTH HORMONE, *NUTR REV*, 30, 79 (1972)

0139 ANONYMOUS, THE HEALTH CONSEQUENCES OF SMOKING, *US DEPT HEW, PHS* (1967)

0140 ANONYMOUS, THE NEW MONITOR HPE CALCIUM, KIT INSTRUC-TIONS, *AMERICAN MONITOR CORPORATION* (1970)

0141 ANONYMOUS, ETOPOSIDE (VP 16-213: VEPESID), *MED LETT DRUGS THER*, 26, 48-49 (1984)

0142 ANTCLIFF AC, BEEVERS DG, HAMILTON M, HARPUR JE, THE USE OF AMILORIDE HYDROCHLORIDE IN THE CORRECTION OF HYPOKALAEMIC ALKALOSIS INDUCED BY DIURETICS, *POST-GRAD MED J*, 47, 644 (1971)

0143 ANTON AH, ETHANOL AND URINARY CATECHOLAMINES IN MAN, *CLIN PHARMACOL THER*, 6, 462 (1965)

0144 ANUMONYE A, READING HW, KNIGHT F, ASHCROFT GW, URIC ACID METABOLISM IN MANIC-DEPRESSIVE ILLNESS AND DURING LITHIUM THERAPY, *LANCET*, 1, 1290 (1968)

0145 APPEL GB, A DECADE OF PENICILLIN RELATED ACUTE INTERSTITIAL NEPHRITIS - MORE QUESTIONS THAN ANSWERS, *CLIN NEPHROL*, 13, 151-154 (1980)

0146 APPEL GB, D'AGATI V, BERGMAN M, ET AL, NEPHROTIC SYNDROME AND IMMUNE COMPLEX GLOMERULONEPHRITIS ASSOCIATED WITH CHLORPROPAMIDE THERAPY, *AM J MED*, 74, 337-342 (1983)

0147 APPELBOOM TM, FLOWERS FP, ACYCLOVIR, *SOUTH MED J*, 76, 905-909 (1983)

0148 APPLEYARD JG, STANLEY DA, THE EVALUATION OF THE VITAMIN B_6 STATUS IN CHILDREN WITH CONVULSIONS, *MED LAB TECHNOL*, 29, 160 (1972)

0149 AQUARON R, PROTEIN-BOUND IODINE AND HORMONAL IODINE AFTER IOXITALAMIC ACID, *CLIN CHIM ACTA*, 41, 175 (1972)

0150 ARAMAKI T, OKUMURA H, ICHIKAWA T, STUDIES ON ASPIRIN INDUCED HEPATIC INJURY, *JPN J PHARMACOL*, 22, 118 (1972)

0151 ARCIERI G, GRIFFITH E, GRUENWALDT G, CIPROFLOXACIN: AN UPDATE ON CLINICAL EXPERIENCE, *AM J MED*, 82 SUPPL 4A, 381-386 (1987)

0152 ARENA JM, RHUBARB HYPERPHAGIA HAZARDS, *J AM MED ASSOC*, 219, 626 (1972)

0153 ARIAS IM, EFFECTS OF PLANT ACID (ICTEROGENIN) AND CERTAIN ANABOLIC STEROIDS ON HEPATIC METABOLISM OF BILIRUBIN AND BSP, *ANN N Y ACAD SCI*, 104, 1014 (1963)

0154 ARKY RA, ABRAMSON EA, INSULIN RESPONSE TO GLUCOSE IN THE PRESENCE OF ORAL HYPOGLYCEMICS, *ANN N Y ACAD SCI*, 148, 768 (1968)

0155 ARNEBORN P, PALMBLAD J, DRUG-INDUCED NEUTROPENIA - A SUR-VEY FOR STOCKHOLM 1973-1978, *ACTA MED SCAND*, 212, 289-292 (1982)

0156 ARNEBORN P, PALMBLAD J, DRUG-INDUCED NEUTROPENIAS IN THE STOCKHOLM REGION 1976-1977, *ACTA MED SCAND*, 206, 241-243 (1979)

0157 ARNESEN E, HUSEBY N-E, BRENN T, ET AL, THE TROMSO HEART STUDY: DISTRIBUTION OF, AND DETERMINANTS FOR, GAMMA-GLUTAMYLTRANSFERASE IN A FREE-LIVING POPULATION, *SCAND J CLIN LAB INVEST*, 46, 63-70 (1986)

0158 ARNESON GA, PHENOTHIAZINE DERIVATIVES AND GLUCOSE METABOLISM, *J NEUROPSYCHIAT*, 5, 181 (1964)

0159 ARON NB, PHENOXYBENZAMINE-INDUCED HYPONATREMIA SIMU-LATING THE SYNDROME OF INAPPROPRIATE ANTIDIURETIC HORMONE SECRETION, *ANN INTERN MED*, 107, 119 (1987)

0160 ARONOW WS, HARDING PR, KHURSHEED M, VANGROW JS, PAPAGEORGIS NP, EFFECT OF HALOFENATE ON SERUM URIC ACID, *CLIN PHARMACOL THER*, 14, 371 (1973)

0161 ARONOW WS, HARDING PR, KHURSHEED M, VANGROW JS, PAPAGEORGIS NP, MAYS J, EFFECT OF HALOFENATE ON SERUM LIPIDS, *CLIN PHARMACOL THER*, 14, 358 (1973)

0162 ARRANTO AJ, SOTANIEMI EA, MORPHOLOGIC ALTERATIONS IN PATIENTS WITH ALPHA-MELTHYLDOPA-INDUCED LIVER DAM-AGE AFTER SHORT- AND LONG-TERM EXPOSURE, *SCAND J GASTROENTEROL*, 16, 853-863 (1981)

0163 ARROWSMITH D, MORIN RJ, ORAL CONTRACEPTIVES AND THE N.B.T. TEST, *LANCET*, 1, 148 (1973)

0164 ARSURA E, LICHSTEIN E, GUADAGNINO V, ET AL, METHEMOGLOBIN LEVELS PRODUCED BY ORGANIC NITRATES IN PATIENTS WITH CORONARY ARTERY DISEASE, *J CLIN PHARMACOL*, 24, 160-164 (1984)

0165 ARTHUR JB, ASHBY DWR, BREMER C, DAVIES DM, TRIAL OF CLOFIBRATE IN THE TREATMENT OF ISCHAEMIC HEART DIS-EASE, *BR MED J [CLIN RES]*, 4, 767 (1971)

0166 ARZE RS, RAMOS JM, RASHID HU, ET AL, AMENORRHOEA, GALACTORRHOEA, AND HYPERPROLACTINAEMIA INDUCED BY METHYLDOPA, *BR MED J [CLIN RES]*, 283, 194 (1981)

0167 ASBERG M, BERTILSSON L, TUCK D, CRONHOLM B, SJOQVIST F, INDOLEAMINE METABOLITES IN THE CEREBROSPINAL FLUID OF DEPRESSED PATIENTS BEFORE AND DURING TREATMENT, *CLIN PHARMACOL THER*, 14, 277 (1973)

0168 ASFELDT VH, PLASMA CORTICOSTEROIDS IN NORMAL INDIVIDU-ALS, *SCAND J CLIN LAB INVEST*, 28, 61 (1971)

0169 ASHRAF M, SCOTCHEL PL, KRALL JM, ET AL, CISPLATINUM-INDUCED HYPOMAGNESEMIA AND PERIPHERAL NEUROPATHY, *GYNECOL ONCOL*, 16, 309-318 (1983)

0170 ASHRAF N, LOCKSLEY R, ARIEFF AI, THIAZIDE-INDUCED HYPONA-TREMIA ASSOCIATED WITH DEATH OR NEUROLOGIC DAMAGE IN OUTPATIENTS, *AM J MED*, 70, 1163-1168 (1981)

0171 ASSAN R, DANCHY F, BOYET F, ELEVATED VANILLYL-MANDELIC ACID EXCRETION IN SEVERE DIABETIC KETOACIDOSIS, *PATHOL BIOL*, 21, 27 (1973)

0172 ASSENNATO G, PACI C, MOLININI R, ET AL, SPERM COUNT SUPPRESSION WITHOUT ENDOCRINE DYSFUNCTION IN LEAD - EXPOSED MEN, *ARCH ENVIRON HEALTH*, 42, 124-127 (1987)

0173 ASURA E, LICHSTEIN E, GUADAGNINI V ET AL, METHEMOGLOBIN LEVELS PRODUCED BY ORGANIC NITRATES IN PATIENTS WITH CORONARY ARTERY DISEASE, *J CLIN PHARMACOL*, 24, 160-164 (1984)

0174 ATKINSON K, BIGGS J, DODDS A, ET AL, CYCLOSPORINE-ASSOCIATED HEPATOTOXICITY AFTER ALLOGENEIC MARROW TRANSPLANTATION IN MAN: DIFFERENTIATION FROM OTHER CAUSES OF POST-TRANSPLANT LIVER DISEASE, *TRANSPLANT PROC*, 15, SUPPL 1, 2761-2767 (1983)

0175 ATKINSON K, BIGGS JC, HAYES J, ET AL, CYCLOSPORIN A ASSOCIATED NEPHROTOXICITY IN THE FIRST 100 DAYS AFTER ALLOGENEIC BONE MARROW TRANSPLANTATION: THREE DISTINCT SYNDROMES, *BR J HAEMATOL*, 54, 59-67 (1983)

0176 ATKINSON RL, ENDOCRINE AND METABOLIC EFFECTS OF OPIATE ANTAGONISTS, *J CLIN PSYCHIATRY*, 45, 20-24 (1984)

0177 ATLAS SA, CASE DB, YU ZY, ET AL, HORMONAL AND METABOLIC EFFECTS OF ANGIOTENSIN CONVERTING ENZYME INHIBITORS: POSSIBLE DIFFERENCES BETWEEN ENALAPRIL AND CAPTOPRIL, *AM J MED*, 77, SUPPL 2A, 13-17 (1984)

0178 ATUK NO, BLAYDES MC, WESTERVELT FB JR., WOOD JE JR., EFFECT OF AMINOPHYLLINE ON URINARY EXCRETION OF EPINEPHRINE AND NOREPINEPHRINE IN MAN, *CIRCULATION*, 35, 745 (1967)

0179 AUBERTIN E, SUDRE J, PURPURA THROMBOCYTOPENIQUE HEMORRHAGIQUE DUE A L'ESIDREX, *J MED BORDEAUX*, 141, 1735 (1964)

0180 AULD WHR, MURDOCH WR, CLINICAL TRIAL OF MEFRUSIDE, A NEW DIURETIC, *BR MED J [CLIN RES]*, 4, 786 (1971)

0181 AURELL M, DELIN K, HERLITZ H, CAPTOPRIL IN TREATMENT-RESISTANT ESSENTIAL AND RENAL HYPERTENSION, *SCAND J UROL NEPHROL*, 16, 243-249 (1982)

0182 AURELL M, VIKGREN P, PLASMA RENIN ACTIVITY IN SUPINE MUSCULAR EXERCISE, *J APPL PHYSIOL*, 31, 839 (1971)

0183 AVILA JL, CONVIT J, HETEROGENEITY OF ACID PHOSPHATASE ACTIVITY IN HUMAN POLYMORPHONUCLEAR LEUKOCYTES, *CLIN CHIM ACTA*, 44, 21 (1973)

0184 AVILES A, HERRERA J, RAMOS E, ET AL, HEPATIC INJURY DURING DOXORUBICIN THERAPY, *ARCH PATHOL LAB MED*, 108, 912-913 (1984)

0185 AVRAMOV R, BRDARIC R, GEREMIC D, CHEN SC, PATRICK JR, EFFECT OF EXERCISE ON THE METABOLISM OF CORTICOSTEROIDS, *CLIN CHEM*, 18, 718 (1972)

0186 AXELSON JE, CHAN GL-Y, KIRSTEN EB, ET AL, FOOD INCREASES THE BIOAVAILABILITY OF PROPAFENONE, *BR J CLIN PHARMACOL*, 23, 735-741 (1987)

0187 AYD FJ, AMITRIPTYLINE REAPPRAISAL AFTER SIX YEARS EXPERIENCE, *DIS NERV SYS*, 26, 719 (1965)

0188 AYLWARD M, MADDOCK J, PLASMA TRYPTOPHAN LEVELS IN DEPRESSION, *LANCET*, 1, 936 (1973)

0189 AYMARD B, AYMARD JP, NETTER P, ET AL, CYTOPENIES SANGUINES ASSOCIEES A LA PRISE DE CIMETIDINE, *THERAPIE*, 39, 545-553 (1984)

0190 AYNSLEY-GREEN A, ALBERTI KGM, THE IMPORTANCE OF THE GUANIDINE GROUP IN THE INSULIN-STIMULATORY EFFECT OF AMILORIDE HYDROCHLORIDE, *HORM METAB RES*, 5, 55 (1973)

0191 AYNSLEY-GREEN A, ALBERTI KGM, SERUM INSULIN OR PLASMA INSULIN?, *LANCET*, 1, 318 (1972)

0192 AZIZI F, ENVIRONMENTAL IODINE INTAKE AFFECTS THE RESPONSE TO METHIMAZOLE IN PATIENTS WITH DIFFUSE TOXIC GOITER, *J CLIN ENDOCRINOL METAB*, 61, 374-377 (1985)

0193 AZIZI F, VAGENAKIS AG, PORTNAY GI, BRAVERMAN LE, INGBAR SH, THYROXINE TRANSPORT AND METABOLISM IN METHADONE AND HEROIN ADDICTS, *ANN INTERN MED*, 80, 194 (1974)

0194 B'HEND P, HADORN B, HALDEMANN B, KLEEB M, LUTHI H, STIMULATION OF PANCREATIC SECRETION IN MAN BY SECRETIN SNUFF, *LANCET*, 1, 509 (1973)

0195 BABA T, BOKU A, ISHIZAKI T, ET AL, RENAL EFFECTS OF NICARDIPINE IN PATIENTS WITH MILD-TO-MODERATE ESSENTIAL HYPERTENSION, *AM HEART J*, 111, 552-557 (1986)

0196 BACCHUS RA, LONDON DR, THE MEASUREMENT OF ARGININE IN PLASMA, *CLIN CHIM ACTA*, 33, 479 (1971)

0197 BACHMANN H, PROPRANOLOL VERSUS CHLORTHALIDONE - A PROSPECTIVE THERAPEUTIC TRIAL IN CHILDREN WITH CHRONIC HYPERTENSION, *HELV PAEDIATR ACTA*, 39, 55-61 (1984)

0198 BACHMANN K, SCHWARTZ JI, FORNEY R JR, ET AL, THE EFFECT OF ERYTHROMYCIN ON THE DISPOSITION KINETICS OF WARFARIN, *PHARMACOLOGIST*, 28, 171-176 (1984)

0199 BADCOCK NR, PENNA AC, EVERETT DS, ET AL, ASPIRIN METABOLITES CAUSING MISINTERPRETATION OF PARACETAMOL RESULTS, *ANN CLIN BIOCHEM*, 21, 527-530 (1984)

0200 BADENOCH JL, O'LEARY TD, EFFECT OF TRIZMA CARBONATE ON RESULTS BY THE KODAK EKTACHEM® SINGLE SLIDE METHOD FOR CREATININE, *CLIN CHEM*, 33, 1476-1477 (1987)

0201 BAER DM, JONES RN, MULLOOLY JP, ET AL, PROTOCOL FOR THE STUDY OF DRUG INTERFERENCES IN LABORATORY TESTS: CEFOTAXIME INTERFERENCE IN 24 CLINICAL TESTS, *CLIN CHEM*, 29, 1736-1740 (1983)

0202 BAGDADE JD, GALE CC, PORTE D JR., HORMONE FUEL INTERRELATIONSHIPS DURING ALCOHOL HYPOGLYCEMIA IN MAN, *PROC SOC EXP BIOL MED*, 141, 540 (1972)

0203 BAGDADE JD, PORTE D JR., BIERMAN EL, STEROID-INDUCED LIPEMIA: A COMPLICATION OF HIGH-DOSAGE CORTICOSTEROID THERAPY, *ARCH INTERN MED*, 125, 129 (1970)

0204 BAILEY CC, GEARY CG, ISRAELS MCG, WHITTAKER JA, BROWN MJ, WEATHERALL DJ, CYTOSINE ARABINOSIDE IN THE TREATMENT OF ACUTE MYELOBLASTIC LEUKEMIA, *LANCET*, 1, 1268 (1971)

0205 BAILEY DN, JATLOW PI, CHEMICAL ANALYSIS OF MASSIVE CRYSTALLURIA FOLLOWING PRIMIDONE OVERDOSE, *AM J CLIN PATHOL*, 58, 583 (1972)

0206 BAILEY RE, BARTOS D, BARTOS F, CASTRO A, DOBSON RL, GRETTIE DP, KRAMER R, ACTIVATION OF ALDOSTERONE AND RENIN SECRETION BY THERMAL STRESS, *EXPERIENTIA*, 28, 159 (1972)

0207 BAINES M, DETECTION AND INCIDENCE OF B AND C VITAMIN DEFICIENCY IN ALCOHOL-RELATED ILLNESS, *ANN CLIN BIOCHEM*, 15, 307-312 (1978)

0208 BAKER BJ, GAMMILL J, MASSENGILL J, ET AL, CONCURRENT USE OF QUINIDINE AND DISOPYRAMIDE: EVALUATION OF SERUM CONCENTRATIONS AND ELECTROCARDIOGRAPHIC EFFECTS, *AM HEART J*, 105, 12-15 (1983)

0209 BAKER EM, HAMMER DC, KENNEDY JE, TOLBERT BM, INTERFERENCE BY ASCORBATE-2-SULFATE IN THE DINITROPHENYLHYDRAZINE ASSAY OF ASCORBIC ACID, *ANAL BIOCHEM*, 55, 641 (1973)

0210 BAKER N, HUEBOTTER RJ, IMMOBILIZING AND HYPERGLYCEMIC EFFECTS OF BENZYL ALCOHOL, *LIFE SCI*, 10, 1193 (1971)

0211 BALAZS M, KOVACH G, CHRONIC AGGRESSIVE HEPATITIS AFTER METHYLDOPA TREATMENT: CASE REPORT WITH ELECTRON MICROSCOPY STUDY, *HEPATOGASTROENTEROLOGY*, 28, 199-202 (1981)

0212 BALDWIN DS, LEVINE BB, MCCLUSKEY RT, GALLO GR, RENAL FAILURE AND INTERSTITIAL NEPHRITIS DUE TO PENICILLIN AND METHICILLIN, *N ENGL J MED*, 279, 1245 (1968)

0213 BALE JF JR, GAY PE, MADSEN JA, MONITORING OF SERUM AMYLASE LEVELS DURING VALPROIC ACID THERAPY, *ANN NEUROL*, 11, 217-218 (1982)

0214 BALESTRIERI A, BENASSI P, CASSANO GB, CASTROGIOVANNI P, CLINICAL COMPARATIVE EVALUATION OF MAPROTILINE, A NEW ANTIDEPRESSANT DRUG, *INTL PHARMACOPSYCHIAT*, 6, 236 (1971)

0215 BALL GV, SORENSEN LB, PATHOGENESIS OF HYPERURICEMIA IN SATURNINE GOUT, *N ENGL J MED*, 280, 1199 (1969)

0216 BALL JH, KAMINSKY NI, HARDMAN JG, BROADUS AE, SUTHERLAND EW, LIDDLE GW, EFFECT OF CATECHOLAMINES AND ADRENERGIC-BLOCKING AGENTS ON PLASMA AND URINARY CYCLIC NUCLEOTIDES IN MAN, *J CLIN INVEST*, 51, 2124 (1972)

0217 BALL MJ, GRIFFITHS D, EFFECT ON CHEMICAL ANALYSES OF BETA-PROPIOLACTONE TREATMENT OF WHOLE BLOOD AND PLASMA, *LANCET*, 1, 1160-1161 (1985)

0218 BALL MJ, PAUL J, KAY JDS, ANALYTICAL INTERFERENCE BY RIFAMPICIN WITH TESTS OF LIVER FUNCTION., *ANN CLIN BIOCHEM*, 24 SUPPL S1, 75-77 (1987)

0219 BANERJEE B, SAHA N, BLOOD-GROUPS AND SERUM-CHOLESTEROL, *LANCET*, 2, 961 (1969)

0220 BANERJEE S, KUMAR KS, BANDYOPADHYAY A, EFFECT OF OXYTETRACYCLINE AND TETRACYCLINE ON GLUCOSE TOLERANCE AND SERUM LIPIDS, *PROC SOC EXP BIOL MED*, 125, 618 (1967)

0221 BANKIER RG, A COMPARISON OF FLUSPIRILENE AND TRIFLUOPERAZINE IN THE TREATMENT OF ACUTE SCHIZOPHRENIC PSYCHOSIS, *J CLIN PHARMACOL*, 13, 44 (1973)

0222 BANSAL OP, AGRANULOCYTOSIS DURING UNITHEBEN THERAPY, *INDIAN PRACTITIONER*, 18, 616 (1965)

0223 BANTLE JP, BOUDREAU RJ, FERRIS TF, SUPPRESSION OF PLASMA RENIN ACTIVITY BY CYCLOSPORINE, *AM J MED*, 83, 59-64 (1987)

0224 BAR RS, WILSON HE, MAZZAFERRI EL, HYPOMAGNESEMIC HYPOCALCEMIA SECONDARY TO RENAL MAGNESIUM WASTING, *ANN INTERN MED*, 82, 646-649 (1975)

0225 BARBIERI C, CALDARA R, FERRARI C, ET AL, METABOLIC EFFECT OF PRAZOSIN, *CLIN PHARMACOL THER*, 27, 313-316 (1980)

0226 BARBIERI C, FERRARI C, CALDARA R, ET AL, ENDOCRINE AND METABOLIC EFFECTS OF LABETALOL IN MAN, *J CARDIOVASC PHARMACOL*, 3, 986-991 (1981)

0227 BARBOSA J, SEAL US, DOE RP, EFFECTS OF ANABOLIC STEROIDS ON HAPTOGLOBIN, OROSOMUCOID, PLASMINOGEN, FIBRINOGEN, TRANSFERRIN, CERULOPLASMIN, *J CLIN ENDOCRINOL METAB*, 33, 388 (1971)

0228 BARDANA EJ JR, GABOUREL JD, DAVIES GH, ET AL, EFFECTS OF PHENYTOIN ON MAN'S IMMUNITY: EVALUATION OF CHANGES IN SERUM IMMUNOGLOBULINS, COMPLEMENT AND ANTINUCLEAR ANTIBODY, *AM J MED*, 74, 289-296 (1984)

0229 BARDIN CW, ROSS GT, LIPSETT MB, SITE OF ACTION OF CLOMIPHENE CITRATE IN MAN: A STUDY OF THE PITUITARY-LEYDIG CELL AXIS, *J CLIN ENDOCRINOL METAB*, 27, 1558 (1967)

0230 BARLOW DH, BEASTALL GH, ABDALLA HI, ET AL, EFFECT OF LONG-TERM HORMONE REPLACEMENT ON PLASMA PROLACTIN CONCENTRATIONS IN WOMEN AFTER OOPHORECTOMY, *BR MED J [CLIN RES]*, 23, 589-591 (1985)

0231 BARNES DB, PIERCE GF, LICHTL D, ET AL, EFFECTS OF DEXTRAN ON FIVE BIURET-BASED PROCEDURES FOR TOTAL PROTEIN IN SERUM, *CLIN CHEM*, 31, 2018-2019 (1985)

0232 BARNES PC, LEONARD JHC, HYPOKALEMIC MYOPATHY AND MYOGLOBINURIA DUE TO CARBENOXOLONE SODIUM, *POSTGRAD MED J*, 47, 813 (1971)

0233 BARNESS LA, SAFETY CONSIDERATIONS WITH HIGH ASCORBIC ACID DOSAGE, *ANN N Y ACAD SCI*, 258, 523-528 (1975)

0234 BARON JM, CHLORDIAZEPOXIDE (LIBRIUM®) AND THYROID FUNCTION TESTS, *BR MED J [CLIN RES]*, 1, 699 (1967)

0235 BARR RD, WOODGER BA, REES PH, LEVELS OF MERCURY IN URINE CORRELATED WITH THE USE OF SKIN LIGHTENING CREAMS, *AM J CLIN PATHOL*, 59, 36 (1973)

0236 BARRETT PV, THIER SO, MENINGITIS AND PANCREATITIS ASSOCIATED WITH SULFAMETHIZOLE, *N ENGL J MED*, 268, 36 (1963)

0237 BARRETT PVD, BILIRUBINEMIA AND FASTING, *N ENGL J MED*, 283, 823 (1970)

0238 BARRETT-CONNOR E, KHAW K-T, CIGARETTE SMOKING AND INCREASED ENDOGENOUS ESTROGEN LEVELS IN MEN., *AM J EPIDEMIOL*, 126, 187-192 (1987)

0239 BARRON KS, PERSON DA, BREWER EJ, THE TOXICITY OF NONSTEROIDAL ANTI-INFLAMMATORY DRUGS IN JUVENILE RHEUMATOID ARTHRITIS, *J RHEUMATOL*, 9, 149-155 (1982)

0240 BARTON CH, PAHL M, VAZIRI ND, ET AL, RENAL MAGNESIUM WASTING ASSOCIATED WITH AMPHOTERICIN B THERAPY, *AM J MED*, 77, 471-474 (1984)

0241 BARTOS HR, THROMBOCYTOPENIA ASSOCIATED WITH RUBELLA VACCINATION, *N Y STATE J MED*, 72, 499 (1972)

0242 BARTTER FC, ELEA CS, HALBERG F, A MAP OF BLOOD AND URINARY CHANGES RELATED TO CIRCADIAN VARIATIONS IN ADRENAL CORTICAL FUNCTION IN NORMAL SUBJECT, *ANN N Y ACAD SCI*, 98, 969 (1962)

0243 BARZEL US, THE ROLE OF BONE IN ACID BASE METABOLISM IN, *OSTEOPOROSIS*, US BARZEL, ED. GRUNE AND STRATTON, NEW YORK, NY (1970)

0244 BASSIR O, HALL J, PHOTO-ACTIVATION AS A SOURCE OF ERROR IN THE CEPHALIN CHOLESTEROL FLOCCULATION TEST, *SCAND J CLIN LAB INVEST*, 7, 274 (1955)

0245 BASU RJ, ARULANANTHAM R, A STUDY OF SERUM PROTEIN AND RETINOL LEVELS IN PREGNANCY AND TOXAEMIA OF PREGNANCY IN WOMAN, *INDIAN J MED RES*, 61, 589 (1973)

0246 BATEH RP, BOWIE LJ, TRIAMTERENE INTERFERES IN URINARY CATECHOLAMINE ANALYSES, *CLIN CHEM*, 29, 1325-1326 (1983)

0247 BATESON MC, HAYES JPLA, PENDHARKAR P, COTRIMOXAZOLE AND FOLATE METABOLISM, *LANCET*, 2, 339-340 (1976)

0248 BATSAKIS JG, BRIERE RO, *INTERPRETIVE ENZYMOLOGY*, CC THOMAS, SPRINGFIELD, IL (1967)

0249 BATSAKIS JG, PRESTON JA, BRIERE RO, GIESEN PC, IATROGENIC ABERRATIONS OF SERUM ENZYME ACTIVITY, *CLIN BIOCHEM*, 2, 125 (1968)

0250 BAUER D, GAERTNER HJ, WIRKUNGEN DER NEUROLEPTIKA AUF DIE LEBERFUNKTION, DAS BLUTBILDENDE SYSTEM, DEN BLUTDRUCK UND DIE TEMPERATUR REGULATION, *PHARMACOPSYCHIATRY*, 16, 23-29 (1983)

0251 BAUER JD, ACKERMANN PG, TORO G, *BRAY'S CLINICAL LABORATORY METHODS* 7TH ED, CV MOSBY, ST LOUIS, MO (1968)

0252 BAUER JD, ACKERMANN PG, TORO G, *BRAY'S CLINICAL LABORATORY METHODS* 6TH ED, CV MOSBY, ST LOUIS, MO (1962)

0253 BAYER MF, EFFECT OF HEPARIN ON SERUM FREE THYROXINE LINKED TO POST-HEPARIN LIPOLYTIC ACTIVITY, *CLIN ENDOCRINOL*, 19, 591-596 (1983)

0254 BAYLIS EM, GREENWOOD FC, JAMES VHT, JENKINS J, LANDON J, MARKS V, SAMOLS E, *GROWTH HORMONE INTERNATIONAL CONGRESS SERIES NO 158*, A PECILE, EE MULLER EDS. EXCERPTA MEDICA FOUNDATION, AMSTERDAM (1968)

0255 BAYSAL A, JOHNSON BA, LINKSWILER H, VITAMIN B₆ DEPLETION IN MAN: BLOOD VITAMIN B₆, PLASMA PYRIDOXAL-PHOSPHATE, SERUM, CHOLESTEROL, SERUM TRANSAMINASE, *J NUTR*, 89, 19 (1966)

0256 BAYSHORE R, HAMILTON A, LEWEY F, ERSKINE L, HAMMOND J, SURVEY OF CARBON DISULFIDE AND HYDROGEN SULFIDE HAZARDS IN THE VISCOSE RAYON INDUSTRY, *OCCUPATIONAL DISEASE PREVENTIVE DIVISION, DEPARTMENT OF LABOR AND INDUSTRY*, BULL, 46 (1938)

0257 BEASTALL GH, COWAN RA, GRAY JMB, ET AL, HORMONE BINDING GLOBULINS AND ANTICONVULSANT THERAPY, *SCOTT MED J*, 30, 101-105 (1985)

0258 BEATTIE AD, MOORE MR, GOLDBERG A, TETRAETHYL-LEAD POISONING, *LANCET*, 2, 12 (1972)

0259 BEAUGRAND M, GAVILLON C, FERRIER J-P, HIGH LEVELS OF ENDOPLASMIC RETICULUM ANTIBODY TITER IN A CASE OF ALPHA-METHYLDOPA-INDUCED CHRONIC ACTIVE HEPATITIS, *GASTROENTEROL CLIN BIOL*, 4, 219-221 (1980)

0260 BECK RP, FAWCETT DM, MORCOS F, THYROID FUNCTION STUDIES IN DIFFERENT PHASES OF THE MENSTRUAL CYCLE AND IN WOMEN RECEIVING NORETHINDRONE, *AM J OBSTET GYNECOL*, 112, 369 (1972)

0261 BECK RP, MORCOS F, FAWCETT D, WATANABE M, ADRENOCORTICAL FUNCTION STUDIES DURING THE NORMAL MENSTRUAL CYCLE AND IN WOMEN RECEIVING NORETHINDRONE, *AM J OBSTET GYNECOL*, 112, 364 (1972)

0262 BECKER CE, GORDEN P, ROBBINS J, HEPATITIS FROM METHIMAZOLE DURING ADRENAL STEROID THERAPY FOR MALIGNANT EXOPHTHALMOS, *J AM MED ASSOC*, 206, 1787 (1968)

0263 BECKER HD, REEDER DD, THOMPSON JC, EFFECT OF GLUCAGON ON CIRCULATING GASTRIN, *GASTROENTEROLOGY*, 62, 720 (1972)

0264 BECKER KL, GOLDSTEIN RA, HYPERURICEMIA IN HEALTHY MEN: INTERMITTENT ELEVATIONS AND THE EFFECT OF SUNLIGHT, *AM J CLIN NUTR*, 25, 453 (1972)

0265 BECKERHOFF R, VETTER W, ARMBRUSTER H, LUETSCHER JA, SIEGENTHALER W, PLASMA-ALDOSTERONE DURING ORAL CONTRACEPTIVE THERAPY, *LANCET*, 1, 1218 (1973)

0266 BEDELL SA, KANG JL, LEUKOCYTOSIS AND LEFT SHIFT ASSOCIATED WITH QUINIDINE FEVER, *AM J MED*, 77, 345-346 (1984)

0267 BEELEY L, KENDALL MJ, EFFECT OF ASPIRIN ON RENAL CLEARANCE OF I-DIATRIZOATE, *BR MED J [CLIN RES]*, 1, 707 (1971)

0268 BEERMAN B, ERICSSON JLE, HELLSTROM K, WENGLE B, WERNER B, TRANSIENT CHOLESTASIS DURING TREATMENT WITH AJMALINE AND CHRONIC XANTHOMATOUS CHOLESTASIS, *ACTA MED SCAND*, 190, 241 (1971)

0269 BEEVERS DG, HAMILTON M, HARPUR JE, THE LONG-TERM TREATMENT OF HYPERTENSION WITH THIAZIDE DIURETICS, *POSTGRAD MED J*, 47, 639 (1971)

0270 BELGOROSKY A, RIVAROLA MA, SEX HORMONE BINDING GLOBULIN RESPONSE TO TESTOSTERONE: AN ANDROGEN SENSITIVITY TEST, *ACTA ENDOCRINOL*, 109, 130-138 (1985)

0271 BELISLE S, PATRY M, TETREAULT L, CIMETIDINE AND PLASMA LEVELS OF GONADOTROPINS, PROLACTIN AND GONADAL STEROIDS IN WOMEN, *CAN MED ASSOC J*, 127, 29-32 (1982)

0272 BELKIN GA, COCKTAIL PURPURA: AN UNUSUAL CASE OF QUININE SENSITIVITY, *ANN INTERN MED*, 66, 583 (1967)

0273 BELL DJ, TALUKDER MQ, RATES OF URINARY EXCRETION OF FREE ALDOPENTOSES AND FUCOSE BY FASTING HEALTHY ADULTS, *CLIN CHIM ACTA*, 40, 13 (1972)

0274 BELL GD, WHITNEY B, DOWLING RH, GALLSTONE DISSOLUTION IN MAN USING CHENODEOXYCHOLIC ACID, *LANCET*, 2, 1213 (1972)

0275 BELL R, JONES J, CORTISOL, GROWTH HORMONE RESPONSE, AND PAIN FOLLOWING TETRACOSACTRIN DEPOT AND ACTH GEL, *S AFR MED J*, 46, 1305 (1972)

0276 BELL WR, ROYALL RM, HEPARIN-ASSOCIATED THROMBOCYTOPENIA: A COMPARISON OF THREE HEPARIN PREPARATIONS, *N ENGL J MED*, 303, 902-907 (1980)

0277 BELLAMY WE JR., MAUCK HP JR., HENNIGAR GR, WIGOD M, JAUNDICE ASSOCIATED WITH THE ADMINISTRATION OF SODIUM P-AMINOSALICYLIC ACID: REVIEW OF THE LITERATURE, *ANN INTERN MED*, 44, 764 (1956)

0278 BELLINGER JF, BUIST NRM, RAPID COLUMN-CHROMATOGRAPHIC MEASUREMENT OF OROTIC ACID, *CLIN CHEM*, 17, 1132 (1971)

0279 BELLON EM, HAEMOLYTIC ANEMIA DUE TO NALIDIXIC ACID, *LANCET*, 2, 691 (1965)

0280 BELTS CA, STUART B, DETERMINATION OF SERUM TOTAL IRON-BINDING CAPACITY, *J CLIN PATHOL*, 26, 457 (1973)

0281 BELZ GG, AUST PE, MUNKES R, DIGOXIN PLASMA CONCENTRATIONS AND NIFEDIPINE, *LANCET*, 1, 844-845 (1981)

0282 BELZ GG, DOERING W, AUST PE, ET AL, QUINIDINE—DIGOXIN INTERACTION: CARDIAC EFFICACY OF ELEVATED SERUM DIGOXIN CONCENTRATION, *CLIN PHARMACOL THER*, 31, 548-554 (1982)

0283 BELZ GG, DOERING W, MUNKES R, ET AL, INTERACTION BETWEEN DIGOXIN AND CALCIUM ANTAGONISTS AND ANTIARRHYTHMIC DRUGS, *CLIN PHARMACOL THER*, 33, 410-417 (1983)

0284 BENDER W, LA FRANCE N, WALKER WG, MECHANISM OF DETERIORATION IN RENAL FUNCTION IN PATIENTS WITH RENOVASCULAR HYPERTENSION TREATED WITH ENALAPRIL, *HYPERTENSION*, 6, SUPPL I, 193-I197 (984)

0285 BENDVOLD E, GOTTLIEB C, SVANBORG K, ET AL, THE EFFECT OF NAPROXEN ON THE CONCENTRATION OF PROSTAGLANDINS IN HUMAN SEMINAL FLUID, *FERTIL STERIL*, 43, 922 (1985)

0286 BENEDEK TG, SUNDER JH, COMPARISONS OF SERUM LIPID AND URIC ACID CONTENT IN WHITE AND NEGRO MEN, *AM J MED SCI*, 260, 331 (1970)

0287 BENG CG, LIM KL, AN IMPROVED AUTOMATED METHOD FOR DETERMINATION OF SERUM ALBUMIN USING BROMCRESOL GREEN, *AM J CLIN PATHOL*, 59, 14 (1973)

0288 BENGTSSON B, PLASMA CONCENTRATION AND SIDE EFFECTS OF TERBUTALINE, *EUR J RESPIR DIS*, 65, SUPPL 134, 231-235 (1984)

0289 BENGZON A, HIPPIUS H, KANIG K, SOME CHANGES IN THE SERUM DURING TREATMENT WITH PSYCHOTROPIC DRUGS, *J NERV MENT DIS*, 143, 369 (1966)

0290 BENJAMIN SB, ISHAK KG, ZIMMERAMAN HJ, ET AL, PHENYLBUTAZONE LIVER INJURY: A CLINICAL PATHOLOGIC SURVEY OF 23 CASES AND REVIEW OF THE LITERATURE, *HEPATOLOGY*, 1, 255-263 (1981)

0291 BENKE A, GOGOLAK G, STUMPF C, TSCHAKALOFF C, ALTHESIN AND HYDROXYDIONE: COMPARATIVE LABORATORY AND CLINICAL INVESTIGATIONS, *POSTGRAD MED* 48, SUPPL, 2, 120 (1972)

0292 BENNETT B, RATNOFF OD, CHANGES IN ANTIHEMOPHILIC FACTOR (AHF, FACTOR VIII) PROCOAGULANT ACTIVITY AND AHF-LIKE ANTIGEN, *J LAB CLIN MED*, 80, 256 (1972)

0293 BENNETT J, HALL RA, BARBITURATE ASSAY, *LANCET*, 1, 1191 (1971)

0294 BENNETT J, HALL RA, BARBITURATE ASSAY, *LANCET*, 1, 298 (1971)

0295 BENNETT JR, SMOKING AND THE GASTROINTESTINAL TRACT, *GUT*, 13, 658 (1972)

0296 BENNETT JR, MYOPATHY FROM ε-AMINOCAPROIC ACID: A SECOND CASE, *POSTGRAD MED J*, 48, 440 (1972)

0297 BENNETT JS, BOND J, SINGER I, GOTTLERB AJ, HYPOURICEMIA IN HODGKIN'S DISEASE, *ANN INTERN MED*, 76, 751 (1972)

0298 BENNETT WM, PORTER GA, ENDOGENOUS CREATININE CLEARANCE AS A CLINICAL MEASURE OF GLOMERULAR FILTRATION RATE, *BR MED J [CLIN RES]*, 4, 84 (1971)

0299 BENSAID J, ALDIGIER J-C, GUALDE N, SYSTEMIC LUPUS ERYTHEMATOSUS SYNDROME INDUCED BY PINDOLOL, *BR MED J [CLIN RES]*, 2, 1603-1604 (1979)

0300 BERARDI RR, HYNECK ML, COHEN IA, EFFECT OF CIMETIDINE ON SERUM URIC ACID CONCENTRATION, *CLIN PHARM*, 3, 56-59 (1984)

0301 BERCHTOLD P, BJORNTORP P, GUSTAFSON A, JONSSON A, FAGERBERG SE, GLUCOSE TOLERANCE, PLASMA INSULIN AND LIPIDS IN DIABETIC SUBJECTS BEFORE AND AFTER TREATMENT, *EUR J CLIN PHARMACOL*, 4, 22 (1971)

0302 BERCHTOLD P, DAHLQUIST A, GUSTAFSON A, ASP NG, EFFECTS OF A BIGUANIDE (METFORMIN) ON VITAMIN B_{12} AND FOLIC ACID ABSORPTION AND INTESTINAL ENZYME ACTIVITIES, *SCAND J GASTROENTEROL*, 6, 751 (1971)

0303 BERG B, ASCORBATE INTERFERENCE IN THE ESTIMATION OF URINARY GLUCOSE BY TEST STRIPS, *J CLIN CHEM CLIN BIOCHEM*, 24, 89-96 (1986)

0304 BERG DH, HEXOSAMINE INTERFERENCE WITH THE DETERMINATION OF PROTEIN BY THE LOWRY PROCEDURE, *ANAL BIOCHEM*, 42, 505 (1971)

0305 BERG JD, PANDOV HI, SAMMONS HG, SERUM TOTAL BILE ACID LEVELS IN PATIENTS RECEIVING RIFAMPICIN AND ISONIAZID, *ANN CLIN BIOCHEM*, 21, 218-222 (1984)

0306 BERG JK, PANDOV HI, SAMMONS HG, SERUM TOTAL BILE ACID LEVELS IN PATIENTS RECEIVING RIFAMPICIN AND ISONIAZID, *ANN CLIN BIOCHEM*, 21, 218-222 (1984)

0307 BERG MJ, FISCHER LJ, RIVEY MP, ET AL, PHENYTOIN AND FOLIC ACID INTERACTION: A PRELIMINARY REPORT, *THER DRUG MONIT*, 5, 389-394 (1983)

0308 BERG MJ, RIVEY MP, VERN BA, ET AL, PHENYTOIN AND FOLIC ACID: INDIVIDUALIZED DRUG-DRUG INTERACTION, *THER DRUG MONIT*, 5, 395-399 (1983)

0309 BERGER M, POTTER DE, PITFALL IN DIAGNOSIS OF VIRAL HEPATITIS IN HAEMODIALYSIS UNIT, *LANCET*, 2, 95-96 (1977)

0310 BERGER W, GOSCHKE H, MOPPERT J, KUNZLI H, INSULIN CONCENTRATIONS IN PORTAL VENOUS AND PERIPHERAL VENOUS BLOOD IN MAN FOLLOWING ADMINISTRATION OF GLUCOSE, *HORM METAB RES*, 5, 4 (1973)

0311 BERGINK EW, HOLMA P, PYORALA T, EFFECT OF ORAL CONTRACEPTIVE COMBINATIONS CONTAINING LEVONORGESTREL OR DESOGESTREL ON SERUM PROTEINS AND ANDROGEN BINDING, *SCAND J CLIN LAB INVEST*, 41, 663-668 (1981)

0312 BERGINK EW, KLOOSTERBOER HJ, LUND L, ET AL, EFFECTS OF LEVONORGESTREL AND DESOGESTREL IN LOW-DOSE ORAL CONTRACEPTIVE COMBINATIONS ON SERUM LIPIDS, APOLIPOPROTEINS A-I AND B AND GLYCOSYLATED PROTEINS, *CONTRACEPTION*, 30, 61-72 (1984)

0313 BERGLUND F, KILLANDER J, POMPEIUS R, EFFECT OF TRIMETHOPRIM-SULFAMETHOXAZOLE ON THE RENAL EXCRETION OF CREATININE IN MAN, *J UROL*, 114, 802-808 (1975)

0314 BERGLUND G, DIURETICS IN LONG-TERM TREATMENT OF HYPERTENSION: A COMPARISON TO β-BLOCKERS, *J CARDIOVASC PHARMACOL*, 6, S256-S259 (1984)

0315 BERGLUND G, ANDERSSON O, LARSSON O, ET AL, ANTIHYPERTENSIVE EFFECT AND SIDE EFFECTS OF BENDROFLUMETHIAZIDE AND PROPRANOLOL, *ACTA MED SCAND*, 199, 499-506 (1976)

0316 BERGMAN D, FUTTERWEIT W, SEGAL R, ET AL, INCREASED OESTRADIOL IN DIAZEPAM RELATED GYNECOMASTIA, *LANCET*, 2, 1225-1226 (1981)

0317 BERKOWITZ D, CLINICAL EXPERIENCES WITH HALOFENATE, A NEW HYPOLIPEMIC AND HYPOURICEMIC AGENT, *CIRCULATION*,, 46/II=256 (197)

0318 BERKOWITZ D, LONG-TERM TREATMENT OF HYPERLIPIDEMIC PATIENTS WITH CLOFIBRATE, *J AM MED ASSOC*, 218, 1002 (1971)

0319 BERKOWITZ D, SPITZER JJ, LIKOFF WP, PRACTICAL SIGNIFICANCE OF SERUM TRIGLYCERIDES AND RADIOACTIVE FAT TOLERANCE: THEIR RELATION TO CURRENT THERAPY, *AM J CARDIOL*, 10, 198 (1962)

0320 BERLINER AD, LACKNER H, HEMORRHAGIC DIATHESIS AFTER PROLONGED INFUSION OF LOW MOLECULAR WEIGHT DEXTRAN, *AM J MED SCI*, 263, 397 (1972)

0321 BERNSTEIN RA, LINARELLI L, FACKTOR MA, FRIDAY GA, DRASH AL, FIREMAN P, DECREASED URINARY ADENOSINE 3'5-MONOPHOSPHATE (CYCLIC AMP) IN ASTHMATICS, *J LAB CLIN MED*, 80, 772 (1972)

0322 BERNSTEIN V, PERETZ DI, LIDOFLAZINE - A NEW DRUG IN THE TREATMENT OF ANGINA PECTORIS, *CURR THER RES*, 14, 483 (1972)

0323 BERRY DJ, THE CATHODE-RAY POLAROGRAPHIC DETERMINATION OF DIAZEPAM IN HUMAN PLASMA, *CLIN CHIM ACTA*, 32, 235 (1971)

0324 BERRY DJ, GROVE J, IMPROVED CHROMATOGRAPHIC TECHNIQUES AND THEIR INTERPRETATION FOR THE SCREENING OF URINE, *J CHROMATOGR*, 61, 111 (1971)

0325 BERRY JN, ACUTE MYELOBLASTIC LEUKEMIA IN A BENZEDRINE ADDICT, *SOUTH MED J*, 59, 1169 (1966)

0326 BERSHAD S, RUBINSTEIN A, PATERNITI JR, ET AL, CHANGES IN PLASMA LIPIDS AND LIPOPROTEINS DURING ISOTRETINOIN THERAPY FOR ACNE, *N ENGL J MED*, 313, 981-985 (1985)

0327 BERTHELOT P, BILLING BH,, EFFECT OF BUNAMIODYL ON HEPATIC UPTAKE OF SULFOBROMOPHTHALEIN IN THE RAT, *AM J PHYSIOL*, 211, 395 (1966)

0328 BERTIERE MC, BETOULLE D, APFELBAUM M, ET AL, TIME-COURSE, MAGNITUDE AND NATURE OF THE CHANGES INDUCED IN HDL BY MODERATE ALCOHOL INTAKE IN YOUNG NONDRINKING MALES., *ATHEROSCLEROSIS*, 61, 7-14 (1986)

0329 BERTINO JS JR, KOZAK AJ, REESE RE, ET AL, HYPOPROTHROMBINEMIA ASSOCIATED WITH CEFAMANDOLE: USE IN A RURAL TEACHING HOSPITAL, *ARCH INTERN MED*, 146, 1125-1128 (1986)

0330 BESA EC, GORSHEIN D, GARDNER FH, BLOOD VOLUME CHANGE IN NORMAL AND VARIOUS ANEMIC STATES IN MAN AFTER ANDROGEN ADMINISTRATION, *ANN INTERN MED*, 76, 869 (1972)

0331 BESSER GM, BUTLER PW, LANDON J, REES L, INFLUENCE OF AMPHETAMINES ON PLASMA CORTICOSTEROID AND GROWTH HORMONE LEVELS IN MAN, *BR MED J [CLIN RES]*, 4, 528 (1969)

0332 BESSMAN AN, CHANDRASEKAR S, EFFECT OF PENTYLENETETRAZOL (METRAZOL) ON BLOOD CHOLESTEROL CONCENTRATION IN MAN, *J AM GERIATR SOC*, 17, 25 (1969)

0333 BEUTLER E, DRUG-INDUCED ANEMIA, *FED PROC*, 31, 141 (1972)

0334 BEYLOT M, BORSON F, DAVID L, ET AL, REDUCTION BY PROPRANOLOL OF URINARY HYDROXYPROLINE EXCRETION IN HUMAN HYPERTHYROIDISM: A BETA-RECEPTOR BLOCKADE OR A MEMBRANE STABILIZING MECHANISM, *METABOLISM*, 33, 124-128 (1984)

0335 BEYLOT M, RIOU JP, BORSON F, ET AL, REDUCTION OF INCREASED LEVELS OF BLOOD GLYCEROL AND KETONE BODIES BY PROPRANOLOL IN HUMAN HYPERTHYROIDISM: ROLE OF THE FALL OF TRIIODOTHYRONINE LEVELS, *METABOLISM*, 33, 1080-1083 (1984)

0336 BEZOS DE, MARGOMENOS D, PAVLIDES P, MARAGOS M, EFFECT OF ADRENALINE ON THE PLASMA IRON LEVEL, *ARCH INST PASTEUR HELLEN*, 15, 55 (1969)

0337 BHANA D, KARIM SM, CARTER DC, GANESAN PA, THE EFFECT OF ORALLY ADMINISTERED PROSTAGLANDINS A1, A2 AND 15-EPI-A2 ON HUMAN GASTRIC SECRETION, *PROSTAGLANDINS*, 3, 307 (1973)

0338 BIAGGIONI I, ONROT J, HOLLISTER AS, ET AL, CARDIOVASCULAR EFFECTS OF ADENOSINE INFUSION IN MAN AND THEIR MODULATION BY DIPYRIDAMOLE, *LIFE SCI*, 39, 2229-2236 (1986)

0339 BICK RL, THOMPSON WB, FIBRINOLYTIC ACTIVITY: CHANGES INDUCED WITH ORAL CONTRACEPTIVES, *OBSTET GYNECOL*, 39, 213 (1972)

0340 BIEGER R, DEJONGE H, LOELIGER EA, INFLUENCE OF NITRAZEPAM ON ORAL ANTICOAGULATION WITH PHENPROCOUMON, *CLIN PHARMACOL THER*, 13, 361 (1972)

0341 BIELMANN P, LEDUC G, JEQUIER J-C, ET AL, CHANGES IN THE LIPOPROTEIN COMPOSITION AFTER CHRONIC ADMINISTRATION OF METOPROLOL AND PROPRANOLOL IN HYPERTRIGLYCERIDEMIC - HYPERTENSIVE SUBJECTS, *CURR THER RES*, 30, 956-967 (1981)

0342 BIEVA C, LADMIRANT IH, SCHEIRS I, ET AL, ADMINISTERED VILOXAZINE INTERFERES IN LIQUID-CHROMATOGRAPHIC ASSAY OF NORMETANEPHRINES, *CLIN CHEM*, 33, 1677-1678 (1987)

0343 BIGELOW LB, NEAL S, WEIL-MALHERBE H, A SPECTROPHOTOMETRIC METHOD FOR THE ESTIMATION OF 3-METHOXY-4-HYDROXYPHENYLGLYCOL IN URINE, *J LAB CLIN MED*, 77, 677 (1971)

0344 BIGGER JT JR, THE QUINIDINE DIGOXIN INTERACTION, *MOD CONCEPT CARDIOVASC DIS*, 51, 73-78 (1982)

0345 BIGGS PA, MIDDLETON JE, WELCH RP, INTERFERENCES IN URINE OXALATE ASSAYS (SIGMA DIAGNOSTICS OXALATE OXIDASE METHOD) FROM HOMOGENTISIC ACID IN ALKAPTONURICS, *CLIN CHEM*, 32, 1598 (1986)

0346 BIJVOET OLM, NATRIURETIC EFFECT OF CALCITONIN IN MAN, *N ENGL J MED*, 284, 681 (1971)

0347 BIKLE DD, GENANT HK, CANN C, ET AL, BONE DISEASE IN ALCOHOL ABUSE, *ANN INTERN MED*, 103, 42-48 (1985)

0348 BILLING BH, BLACK M, THE ACTION OF DRUGS ON BILIRUBIN METABOLISM IN MAN, *ANN N Y ACAD SCI*, 179, 403 (1971)

0349 BILLING BH, MAGGIORE Q, CARTTER MA,, HEPATIC TRANSPORT OF BILIRUBIN, *ANN N Y ACAD SCI*, 111, 319 (1963)

0350 BILLING BH, MAGGIORE QS, GOULIS G,, ACTION OF CHOLECYS-TOGRAPHIC CONTRAST MEDIA AND NOVOBIOCIN ON HEPATIC TRANSPORT OF BILIRUBIN, IN, *BILIARY SYSTEM*, W TAYLOR ED., BLACKWELL, OXFORD (1965)

0351 BINDOLI A, RIGOBELLO MP, CAVALLINI L, ET AL, DECREASE OF SERUM MALONDIALDEHYDE IN PATIENTS TREATED WITH CHLORPROMAZINE, *CLIN CHIM ACTA*, 169, 329-332 (1987)

0352 BING RF, RUSSELL GI, SWALES JD, ET AL, INDAPAMIDE AND BEN-DROFLUAZIDE: A COMPARISON IN THE MANAGEMENT OF ESSENTIAL HYPERTENSION, *BR J CLIN PHARMACOL*, 12, 883-886 (1981)

0353 BING RF, RUSSELL GI, THURSTON H, ET AL, HYDRALLAZINE IN HYPERTENSION: IS THERE A SAFE DOSE?, *BR MED J [CLIN RES]*, 2, 353-354 (1980)

0354 BIO-SCIENCE HANDBOOK, *SPECIALIZED DIAGNOSTIC LABORATORY TESTS* 9TH ED, BIO-SCIENCE LABORATORIES, VAN NUYS, CA (1971)

0355 BIOCHEMEX LABORATORIES, DIAGNOSTIC INTERFERENCES. MATERIALS AFFECTING THE DIAGNOSTIC SPECIFICITY OF LAB-ORATORY TESTS 1271 HEMPSTEAD TURNPIKE, ELMONT, NY 11003, *BIOCHEMEX LABORATORIES INC*

0356 BIRAN S, TAL E, EFFECT OF BETA-BLOCKING DRUG PROPRANOLOL ON ^{131}I UTILIZATION IN EUTHYROID PATIENTS, *J CLIN PHARMACOL*, 12, 105 (1972)

0357 BIRCH CA, JAUNDICE DUE TO PHENOBARBITAL, *LANCET*, 1, 478 (1936)

0358 BIRCH GG, METABOLIC EFFECTS OF RARE FOOD SUGARS, *LAN-CET*, 2, 1419 (1971)

0359 BIRD GW, EELES GH, LITCHFIELD JA, RAHMAN M, WINGHAM J, HAEMOLYTIC ANAEMIA ASSOCIATED WITH ANTIBODIES TO TOLBUTAMIDE AND PHENACETIN, *BR MED J [CLIN RES]*, 1, 718 (1972)

0360 BIRD GWG, MCEVOY MW, WINGHAM J, ACUTE HAEMOLYTIC ANAE-MIA DUE TO IGM PENICILLIN ANTIBODY IN A 3 YEAR OLD CHILD: A SEQUEL TO ORAL PENICILLIN, *J CLIN PATHOL*, 28, 321-323 (1975)

0361 BIRKE G, DICZFALUSY E, PLANTIN LO, ASSESSMENT OF THE FUNC-TIONAL CAPACITY OF THE ADRENAL CORTEX: II CLINICAL APPLICATION OF THE ACTH TEST, *J CLIN ENDOCRINOL METAB*, 20, 593 (1960)

0362 BIRNBAUM D, KARMELI F, TEFERA M, THE EFFECT OF DIAZEPAM ON HUMAN GASTRIC SECRETION, *GUT*, 12, 616 (1971)

0363 BIRNBAUM D, LEVY M, DIURETICS AND ADVERSE GASTROINTESTI-NAL REACTION, *DIGESTION*, 4, 362 (1971)

0364 BIRNIE GG, FITZSIMONS CP, CZARNECKI D, ET AL, HEPATIC META-BOLIC FUNCTION IN PATIENTS RECEIVING LONG-TERM METHO-TREXATE THERAPY: COMPARISON WITH TOPICALLY TREATED PSORIATICS, PATIENT CONTROLS AND CIRRHOTICS, *HEPATO-GASTROENTEROLOGY*, 32, 163-167 (1985)

0365 BISWAS S, HINDOCHA P, SERUM ALKALINE RIBONUCLEASE ACTIV-ITY DURING PREGNANCY, *CLIN CHIM ACTA*, 51, 285 (1974)

0366 BJORCK S, SVALANDER C, WESTBERG G, HYDRALAZINE-ASSOCI-ATED GLOMERULONEPHRITIS, *ACTA MED SCAND*, 218, 261-269 (1985)

0367 BJORNEBOE G-EA, JOHNSEN J, BJORNEBOE A , ET AL, EFFECT OF ALCOHOL CONSUMPTION ON SERUM CONCENTRATION OF 25-HYDROXYVITAMIN D_3, RETINOL, AND RETINOL-BINDING PRO-TEIN, *AM J CLIN NUTR*, 44, 678-682 (1986)

0368 BLACHLEY JD, HILL JB, RENAL AND ELECTROLYTE DISTURBANCES ASSOCIATED WITH CISPLATIN, *ANN INTERN MED*, 95, 628-632 (1981)

0369 BLACK GW, KEILTY SR, RENAL FUNCTION FOLLOWING METHOX-YFLURANE ANESTHESIA, *BR J ANAESTH*, 45, 353 (1973)

0370 BLACK JA, HENDERSON MH, ACTIVATION AND INHIBITION OF HUMAN ERYTHROCYTE PYRUVATE KINASE BY ORGANIC PHOSPHATES, AMINO ACIDS, PEPTIDES, *BIOCHIM BIOPHYS ACTA*, 284, 115 (1972)

0371 BLACKARD WG, CONTROL OF GROWTH HORMONE SECRETION IN MAN, *POSTGRAD MED J* 49, SUPPL, 122 (1973)

0372 BLACKBURN AM, AMIEL SA, MILLIS RR, ET AL, TAMOXIFEN AND LIVER DAMAGE, *BR MED J [CLIN RES]*, 289, 288 (1984)

0373 BLACKSHEAR JL, DAVIDMAN M, STILLMAN MT, IDENTIFICATION OF RISK FOR RENAL INSUFFICIENCY FROM NONSTEROIDAL ANTI-INFLAMMATORY DRUGS, *ARCH INTERN MED*, 143, 1130-1134 (1983)

0374 BLANK DW, JOFFE RT, EFFECT OF CARBAMAZEPINE ON THYROID HORMONE MEASUREMENT IN VITRO, *CLIN CHIM ACTA*, 143, 173-176 (1984)

0375 BLASCHKE TF, ELIN RJ, BERK PD, SONG CS, WOLFF SM, EFFECTS OF INDUCED FEVER ON SULFOBROMOPHTHALEIN KINETICS IN MAN, *ANN INTERN MED*, 78, 221 (1973)

0376 BLEBEA J, MCPHERSON RA, FALSE-POSITIVE GUAIAC TESTING WITH IODINE, *ARCH PATHOL LAB MED*, 109, 437-440 (1985)

0377 BLISS BP, KIRK CJC, NEWALL RG, THE EFFECT OF FENFLURAMINE ON GLUCOSE TOLERANCE, INSULIN, LIPID AND LIPOPROTEIN LEVELS IN PATIENTS, *POSTGRAD MED J*, 48, 409 (1972)

0378 BLITSTEIN I, INFLUENCE DES DIVERS ANTICOAGULANTS SUR LES DOSAGES DE CERTAIN ELEMENTS DU SANG, *REV BELGE SCI MED*, 7, 69 (1935)

0379 BLITZER MG, WILKIE PL, SHAPIRA E, INTERFERENCE OF RADIOOPAQUE DIAGNOSTIC CONTRAST MATERIAL (DIA-TRIZOATE MEGLUMINE) IN ASSAYS OF URINARY AMINOACIDS, *CLIN CHEM*, 32, 559 (1986)

0380 BLOCH A, MORE ON INFANTILE METHEMOGLOBINEMIA DUE TO BENZOCAINE SUPPOSITORY, *J PEDIATR*, 67, 509 (1965)

0381 BLOCH-MICHEL H, KYRIACO C, JUVIN E, NOTRE EXPERIENCE DE L'INDOMETHACINE DANS LES RHEUMATISMES INFLAM-MATOIRES, *THERAPIE*, 22, 45 (1967)

0382 BLOCK LH, LAMY PP, THESE DRUGS DISCOLOR THE FECES OR URINE, *AM PROF PHARM*, 34, 27 (1968)

0383 BLOCK MB, GENANT HK, KIRSNER JB, PANCREATITIS AS AN ADVERSE REACTION TO SALICYLAZOSULFAPYRIDINE, *N ENGL J MED*, 282, 380 (1970)

0384 BLOCK WD, MARKOVS ME, STEELE BF, EFFECT OF IODOACETATE ADDITION TO BLOOD ON THE CONCENTRATION OF PLASMA FREE AMINO ACIDS, *PROC SOC EXP BIOL MED*, 128, 731 (1968)

0385 BLOMFIELD J, DIXON SR, MCCREDIE DA, POTENTIAL HEPATOTOXIC-ITY OF COPPER IN RECURRENT HEMODIALYSIS, *ARCH INTERN MED*, 128, 555 (1971)

0386 BLOOMQUIST JN, LADDU A, ENGLER R, ADVERSE EFFECTS OF ACEBUTOLOL IN CHRONIC STABLE ANGINA: DRUG-INDUCED POSITIVE ANTINUCLEAR ANTIBODY, *J CARDIOVASC PHARMACOL*, 6, 735-738 (1984)

0387 BLUM RH, CARTER SK, ADRIAMYCIN®. A NEW ANTICANCER DRUG WITH SIGNIFICANT CLINICAL ACTIVITY, *ANN INTERN MED*, 80, 249 (1974)

0388 BLUMBERG AG, HEATON AM, VASSILIADES J, THE INTERFERENCE OF CHLORPROMAZINE METABOLITES IN THE ANALYSIS OF URI-NARY METHOXY-CATECHOLAMINES, *CLIN CHEM*, 12, 803 (1966)

0389 BLUMENKRANTZ N, SBOE-HANSEN G, NEW METHOD FOR QUANTI-TATIVE DETERMINATION OF URONIC ACIDS, *ANAL BIOCHEM*, 54, 484 (1973)

0390 BLUNDEN RW, LLOYD JV, RUDZKI Z, ET AL, CHANGES IN SERUM FERRITIN LEVELS AFTER INTRAVENOUS IRON, *ANN CLIN BIOCHEM*, 18, 215-217 (1981)

0391 BLUNNIE WP, ZACHARIAS M, DUNDEE JW, ET AL, LIVER ENZYME STUDIES WITH CONTINUOUS INTRAVENOUS ANESTHESIA, *ANESTHESIA*, 36, 152-156 (1981)

0392 BOARDMAN PL, HART FD, SIDE EFFECTS OF INDOMETHACIN, *ANN RHEUM DIS*, 26, 127 (1967)

0393 BOCK KD, MERGUET P, BRANDT T, MURATA T, EXPERIMENTAL STUDIES WITH CLONIDINE HYDROCHLORIDE IN NORMOTEN-SIVE AND HYPERTENSIVE SUBJECTS IN, *CATAPRES® IN HYPER-TENSION*, ME CONOLLY, ED., BUTTERWORTH, LONDON (1970)

0394 BOCK KW, WILTFANG J, BLUME R, ET AL, PARACETAMOL AS TEST DRUG TO DETERMINE GLUCURONIDE FORMATION IN MAN. EFFECTS OF INDUCERS AND OF SMOKING, *EUR J CLIN PHARMACOL*, 31, 677-683 (1987)

0395 BOCKNER V, ROMAN W, ORAL CONTRACEPTIVES AND THYROID STATUS, *LANCET*, 1, 163 (1967)

0396 BOCTOR FN, HEMATIN INTERFERENCE WITH LOWRY PROTEIN DETERMINATION, *ANAL BIOCHEM*, 50, 500 (1972)

0397 BODANSKY O, METHEMOGLOBINEMIA AND METHEMOGLOBIN-PRO-DUCING COMPOUNDS, *PHARMACOL REV*, 3, 144 (1951)

0398 BODE C, BUCHWALD B, GOEBELL H, INHIBITION OF ETHANOL BREAKDOWN DUE TO PROTEIN DEFICIENCY IN MAN, *GERM MED*, 1, 149 (1971)

0399 BODE C, ZELDER O, GOEBALL H, NEUBERGER HO, CHOLERESIS INDUCED BY SECRETIN: DISTINCTLY INCREASED RESPONSE IN CIRRHOTICS, *SCAND J GASTROENTEROL*, 7, 697 (1972)

0400 BODLAENDER P, ULMER DD, VALLEE BL, AUTOMATED DETERMINA-TION OF DELTA-AMINOLEVULINIC ACID DEHYDRATASE ACTIV-ITY IN HUMAN ERYTHROCYTES, *ANAL BIOCHEM*, 58, 500 (1974)

0401 BODY J-J, CRYER PE, OFFORD KP, ET AL, EPINEPHRINE IS A HYPOPHOSPHATEMIC HORMONE IN MAN: PHYSIOLOGICAL EFFECTS OF CIRCULATING EPINEPHRINE ON PLASMA CAL-CIUM, MAGNESIUM, PHOSPHORUS, PARATHYROID HORMONE, AND CALCITONIN, *J CLIN INVEST*, 71, 572-578 (1983)

0402 BOEHRINGER K, WEIDMANN P, MORDASINI R, ET AL, MENOPAUSE-DEPENDENT PLASMA LIPOPROTEIN ALTERATIONS IN DIURETIC TREATED WOMEN, *ANN INTERN MED*, 97, 206-209 (1982)

0403 BOEIJINGA JJ, BOERSTRA EE, RIS P, ET AL, INTERACTION BETWEEN PARACETAMOL AND COUMARIN ANTICOAGULANTS, *LANCET*, 1, 506 (1982)

0404 BOGER WP, STRICKLAND SC, GOUT, *LANCET*, 1, 420 (1954)

0405 BOGUSZ M, INFLUENCE OF INSECTICIDES ON THE ACTIVITY OF SOME ENZYMES CONTAINED IN HUMAN SERUM, *CLIN CHIM ACTA*, 19, 367 (1968)

0406 BOLOGNESE RJ, CORSON SL, TOUCHSTONE JC, FACTORS AFFECTING THE YIELD OF URINARY ESTRIOL, *OBSTET GYNECOL*, 39, 683 (1972)

0407 BOLTON CH, JACKSON L, ROBERTS CJC, ET AL, ENZYME INDUCTION AND SERUM AND LIPOPROTEIN LIPIDS: A STUDY OF GLUTETHIMIDE IN NORMAL SUBJECTS, *CLIN SCI*, 58, 419-421 (1980)

0408 BOLZANO K, SNADHOFER F, SAILER S, BRAUNSTEINER H, THE EFFECT OF ORAL ADMINISTRATION OF SUCROSE ON THE TURNOVER RATE OF PLASMA FREE FATTY ACIDS, *HORM METAB RES*, 4, 439 (1972)

0409 BOND PA, JENNER FA, LEE CR, LENTON E, POLLITT RJ, SAMPSON GA, THE EFFECT OF LITHIUM SALTS ON THE URINARY EXCRETION OF ALPHA-OXOGLUTARATE IN MAN, *BR J PHARMACOL*, 46, 116 (1972)

0410 BOND WS, TOXIC REACTIONS AND SIDE EFFECTS OF GLUCOCORTICOIDS IN MAN, *AM J HOSP PHARM*, 34, 479-485 (1977)

0411 BONDAR RJL, MEAD DC, EVALUATION OF GLUCOSE-6-PHOSPHATE DEHYDROGENASE FROM LEUCONOSTOC MESENTEROIDES IN THE HEXOKINASE METHOD, *CLIN ACTA*, 20, 586 (1971)

0412 BONDAR RJL, MOSS GA, ENHANCING EFFECT OF GLUTAMATE ON APPARENT SERUM GAMMA-GLUTAMYLTRANSPEPTIDASE ACTIVITY, *CLIN CHEM*, 20, 317 (1974)

0413 BONDE J, BDTKER S, ANGELO HR, ET AL, ATENOLOL INHIBITS THE ELIMINATION OF DISOPYRAMIDE, *EUR J CLIN PHARMACOL*, 28, 41-43 (1985)

0414 BONKOWSKY HL, TSCHUDY DP, COLLINS A, DOHERTY J, BOSSENMAIER I, CARDINAL R, WATSON CJ, REPRESSION OF THE OVERPRODUCTION OF PORPHYRIN PRECURSORS IN ACUTE INTERMITTENT PORPHYRIA BY INTRAVENOUS INFUSION, *PROC NATL ACAD SCI U S A*, 68, 2725 (1971)

0415 BONNETERRE J, NGUYEN M, HECQUET B, ET AL, DYSLIPEMIE INDUITE PAR L'AMINOGLUTETHIMIDE, *BULL CANCER*, 72, 99-103 (1985)

0416 BOOKER HE, PRIMIDONE TOXICITY IN, *ANTIEPILEPTIC DRUGS*, DM WOODBURY, JK PENRY, RP SCHMIDT, EDS. RAVEN PRESS, NEW YORK, NY (1972)

0417 BOON DA, KEENAN RE, VARIABILITY OF MALE PLASMA ANDROGEN CONCENTRATIONS, *ABSTRACTS IV INTL CONGR ENDOCRINOL, WASHINGTON DC, 1972*, 184

0418 BOOTH JB, TODD GB, SUBCLINICAL SCURVY - HYPOVITAMINOSIS C, *GERIATRICS*, 27, 130 (1972)

0419 BOOTH RJ, BULLOCK JY, WILSON JD, ANTINUCLEAR ANTIBODIES IN PATIENTS ON ACEBUTOLOL, *BR J CLIN PHARMACOL*, 9, 515-517 (1980)

0420 BORA SS, RADICHEVICH I, WERNER SC, ARTIFACTUAL ELEVATION OF PBI FROM AN IODINATED DYE USED TO STAIN MEDICINAL CAPSULES PINK, *J CLIN ENDOCRINOL METAB*, 29, 1269 (1969)

0421 BORDIA A, BANSAL HC, LETTER: ESSENTIAL OIL OF GARLIC IN PREVENTION OF ATHEROSCLEROSIS, *LANCET*, 2, 1491 (1973)

0422 BOREL GA, RUEDI B, BRINGOLF M, MAGNENAT P, PERTURBATIONS OF LIVER FUNCTION IN OBESE PATIENTS DURING FAST, *DIGESTION*, 5, 183 (1972)

0423 BORK K, KORTING GW, FAUST G, INCREASE OF CERTAIN SERUM ENZYME LEVELS (GOT, LDH, CPK, MK) AFTER BODY MASSAGE AND ITS SIGNIFICANCE ON DERMATOMYOSITIS, *KLIN WOCHENSCHR*, 50, 332 (1972)

0424 BOROVICZENY KG, BOTTIGER LE, CHATTAS A, DAWSON JB, REFERENCE METHOD FOR THE ERYTHROCYTE SEDIMENTATION RATE (ESR) TEST ON HUMAN BLOOD, *J CLIN PATHOL*, 26, 301 (1973)

0425 BOROWSKI GD, GAROFANO CD, ROSE LI, ET AL, EFFECT OF LONG-TERM AMIODARONE THERAPY ON THYROID HORMONE LEVELS AND THYROID FUNCTION, *AM J MED*, 78, 443-450 (1985)

0426 BORUD O, GJESSING LR, EXCRETION OF VANYLGLYCOL IN HUMAN URINE UNDER DIETARY CONTROL, AND DURING TREATMENT WITH ANTIBIOTICS, DISULFIRAM, *SCAND J CLIN LAB INVEST*, 25, 251 (1970)

0427 BORUSHEK S, GOLD JJ, COMMONLY USED MEDICATIONS THAT INTERFERE WITH ROUTINE ENDOCRINE LABORATORY PROCEDURES, *CLIN CHEM*, 10, 41 (1964)

0428 BOSTON COLLABORATIVE DRUG SURVEILLANCE PROGRAM, ACUTE ADVERSE REACTIONS TO PREDNISONE IN RELATION TO DOSAGE, *CLIN PHARMACOL THER*, 13, 694 (1972)

0429 BOTTIGER LE, WESTERHOLM B, ORAL CONTRACEPTIVES AND THROMBOEMBOLIC DISEASE, *ACTA MED SCAND*, 190, 455 (1971)

0430 BOTTIGER LE,, CARLSON LA, HULTMAN E, ET AL, SERUM LIPIDS IN ALCOHOLICS, *ACTA MED SCAND*, 199, 357-361 (1976)

0431 BOUCHARD P, SAI P, REACH G, ET AL, DIABETES MELLITUS FOLLOWING PENTAMIDINE-INDUCED HYPOGLYCEMIA IN HUMANS, *DIABETES*, 31, 40-45 (1982)

0432 BOULARD M, BLUME KG, BEUTLER E, THE EFFECT OF COPPER ON RED CELL ENZYME ACTIVITIES, *J CLIN INVEST*, 51, 459 (1972)

0433 BOULOUX P-M G, PERRETT D, INTERFERENCE OF LABETALOL METABOLITES IN THE DETERMINATION OF PLASMA CATECHOLAMINES BY HPLC WITH ELECTROCHEMICAL DETERMINATION, *CLIN CHIM ACTA*, 150, 111-117 (1985)

0434 BOULOUX P-M, REES LH, CLEMENT-JONES V, ET AL, ERRONEOUS DIAGNOSIS OF PHAEOCHROMOCYTOMA IN HYPERTENSIVE PATIENT ON LABETALOL, *J ROY SOC MED*, 78, 588-589 (1985)

0435 BOWERS GN JR., LACTIC DEHYDROGENASE, *STAND METH CLIN CHEM*, 4, 163 (1963)

0436 BOYCE EG, DUKES GE, ROLLINS DE, ET AL, THE EFFECT OF RIFAMPIN ON THEOPHYLLINE KINETICS, *J CLIN PHARMACOL*, 26, 696-699 (1986)

0437 BOYD NR, CHARD T, HUMAN URINE OXYTOCIN LEVELS DURING PREGNANCY AND LABOR, *AM J OBSTET GYNECOL*, 115, 827 (1973)

0438 BOYER TD, ROUFF SL, ACETAMINOPHEN-INDUCED HEPATIC NECROSIS AND RENAL FAILURE, *J AM MED ASSOC*, 218, 440 (1971)

0439 BOYETT JD, SULLIVAN JF, THE EFFECT OF ORAL ZINC SULFATE ON PROTEIN-BOUND ZINC IN SERUM, *ALA J MED SCI*, 8, 124 (1971)

0440 BRACHMAN PS, MCCREARY TW, FLORENCE R, AGRANULOCYTOSIS INDUCED BY TRIMEPRAZINE, *N ENGL J MED*, 260, 378 (1959)

0441 BRADLEY DW, EMERY G, MAYNARD JE, VITAMIN C IN PLASMA, *CLIN CHIM ACTA*, 44, 47 (1973)

0442 BRADLEY DW, MAYNARD JE, EMERY G, COMPARISON OF ASCORBIC ACID CONCENTRATIONS IN WHOLE BLOOD OBTAINED BY VENIPUNCTURE OR BY FINGER-PRICK, *CLIN CHEM*, 18, 968 (1972)

0443 BRADLEY GM, BENSON ES, EXAMINATION OF THE URINE IN *TODD-SANFORD CLINICAL DIAGNOSIS BY LABORATORY METHODS*, I DAVIDSOHN, JB HENRY, EDS. SAUNDERS, PHILADELPHIA, PA (1974)

0444 BRADLEY JG, ERRORS IN THE MEASUREMENT OF BLOOD PCO_2 DUE TO DILUTION OF THE SAMPLE WITH HEPARIN SOLUTION, *BR J ANAESTH*, 44, 231 (1972)

0445 BRAEMAN J, PHA RESPONSE AND CYTOTOXIC DRUGS, *LANCET*, 2, 818 (1972)

0446 BRAINERD H, *CURRENT DIAGNOSIS AND TREATMENT-1970*, LANGE MEDICAL PUBLICATIONS, LOS ALTOS, CA (1970)

0447 BRAMBLE MG, RECORD CO, DRUG-INDUCED GASTROINTESTINAL DISEASE, *DRUGS*, 15, 451-463 (1978)

0448 BRAMLET DA, POSALAKY Z, OLSON R, GRANULOMATOUS HEPATITIS AS A MANIFESTATION OF QUINIDINE HYPERSENSITIVITY, *ARCH INTERN MED*, 140, 395-397 (1980)

0449 BRASCH RC, ROCKOFF SC, KUHN C, CHRAPLYVY M, CONTRAST MEDIA AS HISTAMINE LIBERATORS, *INVEST RADIOL*, 6, 510 (1971)

0450 BRASS C, GALGIANI JN, BLASCHKE TF, ET AL, DISPOSITION OF KETOCONAZOLE, AN ORAL ANTIFUNGAL, IN HUMANS, *ANTIMICROB AGENTS CHEMOTHER*, 21, 151-158 (1982)

0451 BRASS EP, EFFECTS OF ANTIHYPERTENSIVE DRUGS ON ENDOCRINE FUNCTION, *DRUGS*, 27, 447-458 (1984)

0452 BRATER DC, ANDERSON S, BAIRD B, ET AL, EFFECTS OF IBUPROFEN, NAPROXEN, AND SULINDAC ON PROSTAGLANDINS IN MEN, *KIDNEY INT*, 27, 66-73 (1985)

0453 BRATER DC, BROWN-CARTWRIGHT D, ANDERSON SA, EFFECT OF HIGH DOSE ETODOLAC ON RENAL FUNCTION, *CLIN PHARMACOL THER*, 42, 283-289 (1987)

0454 BRATLID D, WINSNES A, DETERMINATION OF CONJUGATED AND UNCONJUGATED BILIRUBIN BY METHODS BASED ON DIRECT SPECTROPHOTOMETRY, *SCAND J CLIN LAB INVEST*, 28, 41 (1971)

0455 BRAUN HJ, ALY FW, DIE QUANTITATIVE SERUM HAPTOGLOBINBESTIMMUNG RUIT DER PEROXYDESE AKTIWERUNAPMETHODE, *Z KLIN CHEM KLIN BIOCHEM*, 9, 508 (1971)

0456 BRAUNSTEINER H, HERBST M, SANDHOFER F, LONG-TERM THERAPY OF ADULT CASES OF PRIMARY HYPERTRIGLYCERIDAEMIA WITH CLOFIBRATE, *GERM MED*, 13, 65 (1968)

0457 BRAVERMAN LE, REINSTEIN H, LOAYZA H, POMFRET D, FORMALDEHYDE-FORMING DRUGS AND URINARY 17-HYDROXYCORTICOSTEROIDS, *CLIN CHEM*, 14, 374 (1968)

0458 BRAY GA, MELVIN KEW, CHOPRA IJ, EFFECT OF TRI-IODOTHYRONINE ON SOME METABOLIC RESPONSES OF OBESE PATIENTS, *AM J CLIN NUTR*, 26, 715 (1973)

0459 BRAY WE, *CLINICAL LABORATORY METHODS* 5TH ED, CV MOSBY, ST LOUIS, MO (1957)

0460 BRECKENRIDGE A, WELBORN TA, DOLLERY CT, FRAZER R, GLUCOSE TOLERANCE IN HYPERTENSIVE PATIENTS ON LONG-TERM DIURETIC THERAPY, *LANCET*, 1, 61 (1967)

0461 BRELAND BD, HICKS GS JR., HEPATITIS AND HEMOLYTIC ANEMIA ASSOCIATED WITH METHYLDOPA THERAPY, *DRUG INTELL CLIN PHARM*, 16, 489 (1982)

0462 BRENNAN MJ, BULBROOK RD, DESHPANDE N, WANG DY, HAYWARD JL, URINARY AND PLASMA ANDROGENS IN BENIGN BREAST DISEASE, *LANCET*, 1, 1076 (1973)

0463 BRENNER LO, AGRANULOCYTOSIS AND RANITIDINE, *ANN INTERN MED*, 104, 896 (1986)

0464 BREST AN, HEIDER C, MEHBOD H, DRUG CONTROL OF DIURETIC-INDUCED HYPERURICEMIA, *J AM MED ASSOC*, 195, 42 (1966)

0465 BREWER EJ, GIANNINI EH, BAUM J, ET AL, ASPIRIN AND FENOPROFEN (NALFON®) IN THE TREATMENT OF JUVENILE RHEUMATOID ARTHRITIS, *J RHEUMATOL*, 9, 123-128 (1982)

0466 BREZIN JH, KATZ SM, SCHWARTZ AB, ET AL, REVERSIBLE RENAL FAILURE AND NEPHROTIC SYNDROME ASSOCIATED WITH NON-STEROIDAL ANTI-INFLAMMATORY DRUGS, N ENGL J MED, 301, 1271-1273 (1979)

0467 BRICKMAN AS, CARLSON HE, DEFTOS LJ, PROLACTIN AND CALCITONIN RESPONSES TO PARATHYROID HORMONE INFUSION IN HYPOPARATHYROID, PSEUDOHYPOPARATHYROID AND NORMAL SUBJECTS, J CLIN ENDOCRINOL METAB, 53, 661-664 (1981)

0468 BRICKMAN AS, COBURN JW, KOPPEL M, PEACOCK M, MASSRY SG, EFFECT OF HYDROCHLOROTHIAZIDE ON CALCIUM METABOLISM, J CLIN INVEST, 50, 13 (1971)

0469 BRICKMAN AS, COBURN JW, MASSRY SG, BETHRINE JE, HARRISON HE, NORMAN AW, BIOLOGIC ACTIONS OF 1,25-DIHYDROXYCHOLECALCIFEROL IN MAN, CLIN RES, 21, 250 (1973)

0470 BRICKMAN AS, ISENBERG JI, SECRETIN INCREASES SERUM CALCIUM IN MAN, CLIN RES, 20, 236 (1972)

0471 BRIGGS CJ, ANDERSON D, MUIRHEAD RD, INTERFERENCE BY THIMEROSAL IN CHOLESTEROL ESTIMATION BY THE CHOD-PAP METHOD, ANN CLIN BIOCHEM, 21, 146-147 (1984)

0472 BRIGGS M, BRIGGS M, ENDOCRINE EFFECTS ON SERUM VITAMIN B$_{12}$, LANCET, 2, 1037 (1971)

0473 BRIGGS MH, BRIGGS M, EFFECTS OF ORAL ETHINYLESTRADIOL ON SERUM PROTEINS IN NORMAL WOMEN, CONTRACEPTION, 3, 381 (1971)

0474 BRIGGS MH, BRIGGS M, ANTI-ESTROGENIC EFFECTS OF PROGESTOGENS IN NORMAL WOMEN, LIFE SCI 11, 949 (1972)

0475 BRIGGS MH, BRIGGS M, AUSTIN J, EFFECTS OF STEROID PHARMACEUTICALS ON PLASMA ZINC, NATURE, 232, 480 (1971)

0476 BRIONES ER, PALUMBO PJ, MAO SJT, ET AL, ACUTE EFFECTS OF A GLUCOSE LOAD ON PLASMA APOLIPOPROTEINS A-I, A-II, C-II IN NORMAL AND NONINSULIN DEPENDENT DIABETIC MEN, MAYO CLIN PROC, 59, 399-403 (1984)

0477 BROD J, HORBACH L, JUST H, ROSENTHAL J, NICOLESCU R, ACUTE EFFECTS OF CLONIDINE ON CENTRAL AND PERIPHERAL HAEMODYNAMICS AND PLASMA RENIN ACTIVITY, EUR J CLIN PHARMACOL, 4, 107 (1972)

0478 BRODAL BP, THE INFLUENCE OF HAEMOLYSIS ON THE RADIOIMMUNOASSAY OF INSULIN, SCAND J CLIN LAB INVEST, 28, 287 (1971)

0479 BRODIE MJ, BOOBIS PR, DOLLERY CT, ET AL, RIFAMPICIN AND VITAMIN D METABOLISM, CLIN PHARMACOL THER, 27, 810-814 (1980)

0480 BRODOWS RG, CAMPBELL RG, EFFECT OF AGE ON POST-HEPARIN LIPASE, N ENGL J MED, 287, 969 (1972)

0481 BROGDEN RN, SPEIGHT TM, AVERY GS, LEVODOPA: A REVIEW OF ITS PHARMACOLOGICAL PROPERTIES AND THERAPEUTIC USES, DRUGS, 2, 262 (1971)

0482 BROHEE D, PIRO P, KENNES B, ET AL, IN VITRO EFFECTS OF FLUNARIZINE ON HUMAN LYMPHOCYTES, CYTOBIOS, 45, 139-146 (1986)

0483 BROHULT A, BROHULT J, BROHULT S, EFFECTS OF ALKOXYGLYCEROLS ON THE SERUM ORNITHINE CARBAMOYL TRANSFERASE IN CONNECTION WITH RADIATION TREATMENT, EXPERIENTIA, 28, 146 (1972)

0484 BROMPTON HOSPITAL/MEDICAL RESEARCH COUNCIL COLLABORATIVE TRIAL, LONG-TERM STUDY OF DISODIUM CROMOGLYCATE IN TREATMENT OF SEVERE EXTRINSIC OR INTRINSIC BRONCHIAL ASTHMA IN ADULTS, BR MED J [CLIN RES], 4, 383 (1972)

0485 BROOK I, LEUKOPENIA AND AGRANULOCYTOPENIA AFTER OXACILLIN THERAPY, SOUTH MED J, 70, 565-566 (1977)

0486 BROOKS FP, TESTING PANCREATIC FUNCTION, N ENGL J MED, 286, 300 (1972)

0487 BROOKS PM, COSSUM PA, BOYD GW, REBOUND RISE IN RENIN CONCENTRATIONS AFTER CESSATION OF SALICYLATES, N ENGL J MED, 303, 562-564 (1980)

0488 BROULIK PD, STEPAN JJ, SOUCEK K, ET AL, ALTERATIONS IN HUMAN SERUM ALKALINE PHOSPHATASE AND ITS BONE ISOENZYME BY CHRONIC ADMINISTRATION OF LITHIUM, CLIN CHIM ACTA, 140, 151-155 (1984)

0489 BROWDER AA, JOSELOW MM, LOURIA DB, THE PROBLEM OF LEAD POISONING, MEDICINE, 52, 121 (1973)

0490 BROWES DO, ASSESSMENT OF THE LABSTIX® STRIP TEST: COMMENTS ON DETECTION OF PARAPROTEINS IN URINE, MED LAB TECHNOL, 27, 533 (1970)

0491 BROWN CB, TETRACYCLINE AND RENAL FUNCTION, BR MED J [CLIN RES], 4, 428 (1971)

0492 BROWN CH III, BRADSHAW MW, NATELSON EA, ET AL, DEFECTIVE PLATELET FUNCTION FOLLOWING THE ADMINISTRATION OF PENICILLIN COMPOUNDS, BLOOD, 47, 949-956 (1976)

0493 BROWN CR, FORREST WH JR., HAYDEN J, JAMES KE, RESPIRATORY EFFECTS OF HYDROMORPHONE IN MAN, CLIN PHARMACOL THER, 14, 331 (1973)

0494 BROWN GM, GARFINKEL PE, WARSH JJ, ET AL, EFFECT OF CARBIDOPA ON PROLACTIN, GROWTH HORMONE AND CORTISOL SECRETION IN MAN, J CLIN ENDOCRINOL METAB, 43, 236-239 (1976)

0495 BROWN JJ, FRASER R, LEVER AF, LOVE DR, MORTON JJ, ROBERTSON JIS, RAISED PLASMA ANGIOTENSIN II AND ALDOSTERONE DURING DIETARY SODIUM RESTRICTION IN MAN, LANCET, 2, 1106 (1972)

0496 BROWN PJE, LESNA M, HAMLYN AN, ET AL, PRIMARY BILIARY CIRRHOSIS AFTER LONG-TERM PRACTOLOL ADMINISTRATION, BR MED J [CLIN RES], 2, 1591 (1978)

0497 BROWN RD, BILLMAN GE, KEM DC, ET AL, THE EFFECT OF METOCLOPRAMIDE AND DOPAMINE ON PLASMA ALDOSTERONE CONCENTRATIONS IN NORMAL MAN AND RHESUS MONKEYS (MACACA MULATTA): A NEW MODEL TO STUDY DOPAMINE CONTROL OF ALDOSTERONE SECRETION, J CLIN ENDOCRINOL METAB, 55, 828-832 (1982)

0498 BROWN RD, STROTT CA, LIDDLE GW, PLASMA DEOXYCORTICOSTERONE IN NORMAL AND ABNORMAL HUMAN PREGNANCY, J CLIN ENDOCRINOL METAB, 35, 736 (1972)

0499 BROWN RR, AMINOACIDURIA RESULTING FROM CYCLOLEUCINE ADMINISTRATION IN MAN, SCIENCE, 157, 432 (1967)

0500 BROWN WA, LAUGHREN TP, WILLIAMS B, DIFFERENTIAL EFFECTS OF NEUROLEPTIC AGENTS ON THE PITUITARY GONADAL AXIS IN MEN, ARCH GEN PSYCHIATRY, 38, 1270-1272 (1981)

0501 BROWN WD, DYMENT PG, PAGOPHAGIA AND IRON DEFICIENCY ANEMIA IN ADOLESCENT GIRLS, PEDIATR, 49, 766 (1972)

0502 BROWN WG, OWENS JB JR., BERSON AW, HENRY SK, VANILMANDELIC ACID SCREENING TEST FOR PHEOCHROMOCYTOMA AND NEUROBLASTOMA, AM J CLIN PATHOL, 46, 599 (1966)

0503 BROWN WV, DUJOVNE CA, FARQUHAR JW, ET AL, EFFECTS OF FENOFIBRATE ON PLASMA LIPIDS. DOUBLE-BLIND, MULTICENTER STUDY IN PATIENTS WITH TYPE IIA OR TYPE IIB HYPERLIPIDEMIA, ARTERIOSCLEROSIS, 6, 670-678 (1986)

0504 BROWNE TR, VALPROIC ACID, N ENGL J MED, 302, 661-666 (1980)

0505 BROWNING RH, DONNERBERG RL, CAPREOMYCIN EXPERIENCES IN PATIENT ACCEPTANCE AND TOXICITY, ANN N Y ACAD SCI, 135, 1057 (1966)

0506 BRUCKNER FE, RANDLE AP, THE USE OF INDOMETHACIN IN RHEUMATOID ARTHRITIS, ANN PHYS MED, 8, 100 (1965)

0507 BRUMMER HC, SERUM PGF2 ALPHA LEVELS DURING HUMAN PREGNANCY, PROSTAGLANDINS, 3, 3 (1973)

0508 BRUN LD, GAGNE C, ROUSSEAU C, ET AL, SEVERE LIPEMIA INDUCED BY TAMOXIFEN, CANCER, 57, 2123-2126 (1986)

0509 BRUNE GG, PFLUGHAUPT KW, EFFECTS OF L-DOPA TREATMENT ON INDOLE METABOLISM IN PARKINSON'S DISEASE, EXPERIENTIA, 27, 516 (1971)

0510 BRUNI J, LEE CS, PERCHALSKI RJ, ET AL, INTERACTIONS OF VALPROIC ACID WITH PHENYTOIN, NEUROLOGY, 30, 1233-1236 (1980)

0511 BRUNI J, WILDER BJ, PERCHALSKI RJ, ET AL, VALPROIC ACID AND PLASMA LEVELS OF PHENOBARBITAL, NEUROLOGY, 30, 94-97 (1980)

0512 BRUTON J, LI TK, SMITH GD, COMPARISON OF RESULTS BY 4 DIFFERENT PROCEDURES FOR DETERMINATION OF 17-HYDROXYCORTICOSTEROIDS, CLIN CHEM, 19, 748 (1973)

0513 BRYAN R, VALIDITY OF THE LECITHIN/SPHINGOMYELIN (L/S) RATIO FOR AMNIOTIC FLUID CONTAINING BLOOD, CLIN CHEM, 18, 1551 (1972)

0514 BRYANT JM, YU TF, BERGER L, SCHVARTZ N, HYPERURICEMIA INDUCED BY THE ADMINISTRATION OF CHLORTHALIDONE AND OTHER SULFONAMIDE DIURETICS, AM J MED, 33, 408 (1962)

0515 BRYDON WG, ROBERTS LB, THE EFFECT OF HEMOLYSIS ON THE DETERMINATION OF PLASMA CONSTITUENTS, CLIN CHIM ACTA, 41, 435 (1972)

0516 BRYSON G, BIOGENIC AMINES IN NORMAL AND ABNORMAL BEHAVIORAL STATES, CLIN CHEM, 17, 5 (1971)

0517 BUBB FA, MURPHY GM, DETERMINATION OF SERUM PHYTOFLUENE AND RETINOL, CLIN CHIM ACTA, 48, 329 (1973)

0518 BUCHANAN RA, ETHOSUXIMIDE: TOXICITY, IN, ANTIEPILEPTIC DRUGS, DM WOODBURY, JK PENRY, RP SCHMIDT, EDS. RAVEN PRESS, NEW YORK, NY (1972)

0519 BUCHT G, WAHLIN A, RENAL CONCENTRATING CAPACITY IN LONG-TERM LITHIUM TREATMENT AND AFTER WITHDRAWAL OF LITHIUM, ACTA MED SCAND, 207, 309-314 (1980)

0520 BUCKLEY JE, CLARK VL, MEYER TJ, ET AL, HYPOMAGNESEMIA AFTER CISPLATIN COMBINATION CHEMOTHERAPY, ARCH INTERN MED, 144, 2347-2348 (1984)

0521 BUCKMAN MT, KAMINSKY N, CONWAY M, PEAKE GT, WATER-LOAD - A NEW TEST FOR PROLACTIN SUPPRESSION, CLIN RES, 21, 486 (1973)

0522 BUCKMAN MT, PEAKE GT, ESTROGEN POTENTIATION OF PHENOTHIAZINE-INDUCED PROLACTIN SECRETION IN MAN, J CLIN ENDOCRINOL METAB, 37, 977 (1973)

0523 BUHLER FR, LARAGH JH, BAER L, VAUGHAN ED JR., BRUNNER HR, PROPRANOLOL INHIBITION OF RENIN SECRETION, N ENGL J MED, 287, 1209 (1972)

0524 BUIST NRM, BELLINGER JF, RAMBERG DA, BLUE DIAPERS CAUSED BY PYOCYANIN PRODUCED BY P AERUGINOSA, J PEDIATR, 81, 622 (1972)

0525 BUIST NRM, CURTIS HT, HEPARIN AND FALSE-POSITIVE TESTS FOR MUCOPOLYSACCHARIDURIA, LANCET, 2, 286 (1972)

References

0526 BUIST NRM, RAMBERG DA, STRANDHOLM JJ, FERRY PC, COPPER POPPERS: A BENIGN CAUSE OF BLUE DIAPERS, *ARCH DIS CHILD*, 46, 873 (1971)

0527 BULBROOK RD, HAYWARD JL, EXCRETION OF URINARY 17-HYDROXYCORTICOSTEROIDS AND 11-DEOXY-17-OXOSTEROIDS BY WOMEN, *LANCET*, 2, 1033 (1969)

0528 BULL J, MacKINNON J, PHENYLBUTAZONE AND ANTICOAGULANT CONTROL, *PRACTITIONER*, 215, 767-769 (1975)

0529 BUNKER JP, STETSON JB, COE RC, CITRIC ACID INTOXICATION, *J AM MED ASSOC*, 157, 1361 (1955)

0530 BUNNEY WE JR., GOODWIN FK, DAVIS JM, FAWCETT JA, A BEHAVIORAL-BIOCHEMICAL STUDY OF LITHIUM TREATMENT, *AM J PSYCHIATRY*, 125, 499 (1968)

0531 BURCH PG, MIGEON CJ, SYSTEMIC ABSORPTION OF TOPICAL STEROIDS, *ARCH OPHTHALMOL*, 79, 174 (1968)

0532 BURCH RE, PARKER M, LUBY RJ, SULLIVAN JF, ORAL CONTRACEPTIVES AND TRACE METALS, *CLIN RES*, 20, 774 (1972)

0533 BURCKART GJ, VENKATARAMANAN R, STARZY TE, ET AL, CYCLOSPORINE CLEARANCE IN CHILDREN FOLLOWING ORGAN TRANSPLANTATION, *J CLIN PHARMACOL*, 24, 412 (1984)

0534 BURGER A, DINICHERT D, NICOD P, ET AL, EFFECT OF AMIODARONE ON SERUM TRI-IODOTHYRONINE, REVERSE TRI-IODOTHYRONINE, THYROXINE AND THYROTROPIN: A DRUG INFLUENCING PERIPHERAL METABOLISM OF THYROID HORMONES, *J CLIN INVEST*, 58, 255-259 (1976)

0535 BURGESS E, CUTLER RE, BLAIR AD, CIMETIDINE PHARMACOKINETICS IN HEMODIALYSIS PATIENTS AND INHIBITION OF CREATININE SECRETION IN HEALTHY SUBJECTS, *CLIN PHARMACOL THER*, 27, 247 (1980)

0536 BURGESS JL, BIRCHALL R, NEPHROTOXICITY OF AMPHOTERICIN B, WITH EMPHASIS ON CHANGES IN TUBULAR FUNCTION, *AM J MED*, 53, 77 (1972)

0537 BURKE CW, THE EFFECT OF ORAL CONTRACEPTIVES ON CORTISOL METABOLISM, *J CLIN PATHOL 23 SUPPL*, 3, 11 (1969)

0538 BURLAND WL, FENFLURAMINE AND HAEMOLYTIC ANAEMIA, *BR MED J [CLIN RES]*, 1, 419 (1973)

0539 BURN R, BLOOD GLUCOSE IN PARACETAMOL POISONING, *LANCET*, 1, 728 (1973)

0540 BURR RG, A SOURCE OF ERROR WITH THE CALCIUM-SPECIFIC ION ELECTRODE. SOLVENT EFFECTS IN AQUEOUS SOLUTION, *CLIN CHIM ACTA*, 43, 311 (1973)

0541 BURRELL CD, FATAL MARROW APLASIA AFTER TREATMENT WITH CARBIMAZOLE, *BR MED J [CLIN RES]*, 1, 1456 (1956)

0542 BURROUGHS WELLCOME AND COMPANY, MANUFACTURER'S LITERATURE ON ZYLOPRIM, *1 SCARSDALE RD, TUCKAHOE, NY*, 10707

0543 BURROUGHS WELLCOME AND COMPANY, TREATMENT OF NEOPLASTIC DISEASES - MANUFACTURER'S BROCHURE (1971), *1 SCARSDALE RD, TUCKAHOE, NY*, 10707

0544 BURROW GN, BURKE WR, HIMMELHOCH JM, SPENCER RP, HERSHAM JM, EFFECT OF LITHIUM ON THYROID FUNCTION, *J CLIN ENDOCRINOL METAB*, 32, 647 (1971)

0545 BURRY HC, DIEPPE PA, APPARENT REDUCTION OF ENDOGENOUS CREATININE CLEARANCE BY SALICYLATE TREATMENT, *BR MED J [CLIN RES]*, 2, 16-17 (1976)

0546 BURSTEIN S, KLAIBER EL, PHENOBARBITAL-INDUCED INCREASE IN 6-BETA-HYDROXYCORTISOL EXCRETION: CLUE TO ITS SIGNIFICANCE IN HUMAN URINE, *J CLIN ENDOCRINOL METAB*, 25, 293 (1965)

0547 BUSCAGLIA AJ, COWDEN FE, BRILL H, PULMONARY INFILTRATES ASSOCIATED WITH NAPROXEN, *J AM MED ASSOC*, 25, 65-66 (1984)

0548 BUSSEY HI, UPDATE ON THE INFLUENCE OF QUINIDINE AND OTHER AGENTS ON DIGITALIS GLYCOSIDES, *AM HEART J*, 107, 143-146 (1984)

0549 BUSSEY HI, FARRINGER J, MERRITT GJ, INFLUENCE OF RIFAMPIN ON QUINIDINE AND DIGOXIN, *ARCH INTERN MED*, 144, 1021-1023 (1984)

0550 BUSSIEN JP, d'AMORE TF, PERRET L, ET AL, SINGLE AND REPEATED DOSING OF THE CONVERTING ENZYME INHIBITOR PERINDOPRIL TO NORMAL SUBJECTS, *CLIN PHARMACOL THER*, 39, 554-558 (1986)

0551 BUTLER WT, BENNETT JE, HILL GJ II, SZWED CF, COTLOVE E, ELECTROCARDIOGRAPHIC AND ELECTROLYTE ABNORMALITIES CAUSED BY AMPHOTERICIN B IN DOG AND MAN, *PROC SOC EXP BIOL MED*, 116, 857 (1964)

0552 BUTLER WT, ROSSEN RD, EFFECT OF CORTICOSTEROIDS ON IMMUNITY IN MAN, *J CLIN INVEST*, 52, 2629 (1973)

0553 BUTTNER H, QUESTION AND ANSWER SECTION, *GERM MED*, 13, 247 (1968)

0554 BYNUM TE, SOLOMON TE, JOHNSON LR, JACOBSEN ED, INHIBITION OF PANCREATIC SECRETION IN MAN BY CIGARETTE SMOKING, *GUT*, 13, 361 (1972)

0555 BYRD RB, HORN BR, GRIGGS GA, ET AL, ISONIAZID CHEMOPROPHYLAXIS: ASSOCIATION WITH DETECTION AND INCIDENCE OF LIVER TOXICITY, *ARCH INTERN MED*, 137, 1130-1133 (1977)

0556 CADDELL JL, ERICKSON M, BYRNE PA, INTERFERENCE FROM CITRATE USING THE TITAN YELLOW METHOD AND TWO FLUOROMETRIC METHODS FOR MAGNESIUM DETERMINATION, *CLIN CHIM ACTA*, 50, 9 (1974)

0557 CAIN MD, WALTERS WA, CATT KJ, EFFECTS OF ORAL CONTRACEPTIVE THERAPY ON THE RENIN-ANGIOTENSIN SYSTEM, *J CLIN ENDOCRINOL METAB*, 33, 671 (1971)

0558 CAIRO MS, ADVERSE REACTIONS OF L-ASPARAGINASE, *AM J PEDIATR HEMATOL ONCOL*, 4, 335-339 (1982)

0559 CALDARA R, CAMBIELLI M, MASCI E, ET AL, EFFECT OF LOPERAMIDE AND NALOXONE ON GASTRIC ACID SECRETION IN HEALTHY MAN, *GUT*, 22, 720-723 (1981)

0560 CALIL HM, LESIEUR P, GOLD PW, ET AL, HORMONAL RESPONSES TO ZIMELDINE AND DESIPRAMINE IN DEPRESSED PATIENTS, *PSYCHIATRY RES*, 13, 231-242 (1984)

0561 CALLAHAN PX, SHEPARD JA, ELLIS S, ACCELERATED CHROMATOGRAPHIC METHOD FOR DETERMINATION OF HYDROXYPROLINE, *ANAL BIOCHEM*, 49, 155 (1972)

0562 CALNE DB, KAROUM F, RUTHVEN CRJ, SANDLER M, THE METABOLISM OF ORALLY ADMINISTERED L-DOPA IN PARKINSONISM, *BR J PHARMACOL*, 37, 57 (1969)

0563 CALS MJ, BAILLY M, INTERFERENCE OF DRUGS IN ENZYMOLOGY IN, *REFERENCE VALUES IN HUMAN CHEMISTRY*, G SIEST, ED. KARGER, BASEL, 1973, 292

0564 CAM JM, LUCK VA, EASTWOOD JB, ET AL, THE EFFECT OF ALUMINUM HYDROXIDE ORALLY ON CALCIUM, PHOSPHORUS AND ALUMINUM METABOLISM IN NORMAL SUBJECTS, *CLIN SCI MOL MED*, 51, 407-414 (1976)

0565 CAMARGO CA JR, WILLIAMS PT, VRANIZAN KM, ET AL, THE EFFECT OF MODERATE ALCOHOL INTAKE ON SERUM APOLIPOPROTEIN A-I AND A-II, *J AM MED ASSOC*, 253, 2854-2857 (1985)

0566 CAMERON A, EISEN AA, NIRANJAN LM, APLASTIC ANEMIA DUE TO PHENYLBUTAZONE, *POSTGRAD MED J*, 42, 49 (1966)

0567 CAMERON EHD, SCARISBRICK JJ, RADIOIMMUNOASSAY OF PLASMA PROGESTERONE, *CLIN CHEM*, 19, 1403 (1973)

0568 CAMERON SJ, RIFAMPICIN AND THROMBOCYTOPENIA, *LANCET*, 2, 167 (1971)

0569 CAMPBELL DJ, MENDELSOHN FAO, ADAM WR, ET AL, IS ALDOSTERONE SECRETION UNDER DOPAMINERGIC CONTROL?, *CIRC RES*, 49, 1217-1227 (1981)

0570 CAMPBELL DJ, SHERBANIUK, RIGBY J, FALSE POSITIVE REACTION DUE TO METHOCARBAMOL IN THE SCREENING TEST FOR VANILMANDELIC ACID (VMA), *CLIN CHEM*, 10, 447 (1964)

0571 CAMPBELL MA, PLACHETKA JR, JACKSON JE, ET AL, CIMETIDINE DECREASES THEOPHYLLINE CLEARANCE, *ANN INTERN MED*, 95, 68-69 (1981)

0572 CANADIAN MULTICENTRE TRANSPLANT STUDY GROUP, A RANDOMIZED CLINICAL TRIAL OF CYCLOSPORINE IN CADAVERIC RENAL TRANSPLANTATION, *N ENGL J MED*, 309, 809-815 (1983)

0573 CANNON DC, OLITZKY I, INKPEN JA, PROTEINS (P 405) IN *CLINICAL CHEMISTRY: PRINCIPLES AND TECHNICS*, RJ HENRY, DC CANNON, JW WINKELMAN, EDS. 2ND ED HARPER AND ROW, HAGERSTOWN, MD (1974)

0574 CANNON PJ, SVAHN DS, DEMARTINI FE, INFLUENCE OF HYPERTONIC SALINE INFUSIONS: FRACTIONAL REABSORPTION OF URATE AND OTHER IONS, *CIRCULATION*, 41, 97 (1970)

0575 CANO A, MARTINEZ P, PARILLA JJ, ET AL, EFFECTS OF INTRAVENOUS RITODRINE ON LACTATE AND PYRUVATE LEVELS. ROLE OF GLYCEMIA AND ANAEROBIOSIS, *OBSTET GYNECOL*, 66, 207-210 (1985)

0576 CANTRELL JW, HOPKINS NE, QUEEN CA, FRINGS CS, EFFECT OF PHENAZOPYRIDINE HCL ON DETERMINATION OF ETHCHLORVYNOL, *CLIN CHEM*, 18, 591 (1972)

0577 CANTWELL BMJ, POOLEY J, HARRIS AL, FALSE-POSITIVE KETONURIA DURING IFOSFAMIDE AND MESNA THERAPY, *EUR J CANCER CLIN ONCOL*, 22, 229-230 (1986)

0578 CAPIZZI RL, BERTINO JR, HANDSCHUMACHER RE, L-ASPARAGINASE, *ANN REV MED*, 21, 433 (1970)

0579 CARAWAY WT, CHEMICAL AND DIAGNOSTIC SPECIFICITY OF LABORATORY TESTS, *AM J CLIN PATHOL*, 37, 445 (1962)

0580 CARAWAY WT, NON-URATE CHROMOGENS IN BODY FLUIDS, *CLIN CHEM*, 15, 720 (1969)

0581 CARAWAY WT, ACCURACY IN CLINICAL CHEMISTRY, *CLIN CHEM*, 17, 63 (1971)

0582 CARAWAY WT, SOURCES OF ERROR IN CLINICAL CHEMISTRY, *STAND METH CLIN CHEM*, 5, 19 (1965)

0583 CARAWAY WT, KAMMEYER CW, CHEMICAL INTERFERENCE BY DRUGS AND OTHER SUBSTANCES WITH CLINICAL LABORATORY TEST PROCEDURES, *CLIN CHIM ACTA*, 41, 395 (1972)

0584 CAREN R, CORBO L, DEPRESSION OF PLASMA LIPID FRACTIONS AND INHIBITION OF PLATELET AGGREGATION BY ACTION OF GLUCAGON, *METABOLISM*, 20, 1057 (1971)

0585 CAREY HM, PRINCIPLES OF ORAL CONTRACEPTIVES: 2 SIDE EFFECTS OF ORAL CONTRACEPTIVES, *MED J AUST*, 2, 1242 (1971)

0586 CAREY MA, JONES JD, GASTINEAU CF, EFFECT OF MODERATE ALCOHOL INTAKE ON BLOOD CHEMISTRY VALUES, *J AM MED ASSOC*, 216, 1766 (1971)

0587 CARLOSS HW, TAVASSOLI M, MCMILLAN R, CIMETIDINE-INDUCED GRANULOCYTOPENIA, *ANN INTERN MED*, 93, 57-58 (1980)

0588 CARLSON HE, BRICKMAN AS, BOTTAZZO GF, PROLACTIN DEFICIENCY IN PSEUDOHYPOPARATHYROIDISM, *N ENGL J MED*, 296, 140-144 (1977)

0589 CARLSSON C, DEHLIN O, PANCREATIC EXOCRINE FUNCTION AND D-XYLOSE TESTS IN PATIENTS WITH CHRONIC ALCOHOLISM, *ACTA MED SCAND*, 191, 477 (1972)

0590 CARLSSON C, DENCKER SJ, CEREBROSPINAL URIC ACID IN ALCOHOLICS, *ACTA NEUROL SCAND*, 49, 39 (1973)

0591 CARLSTROM K, DOBERL A, RANNEVIK G, PERIPHERAL ANDROGEN LEVELS IN DANAZOL-TREATED PREMENOPAUSAL WOMEN, *FERTIL STERIL*, 39, 499-504 (1983)

0592 CARMINE AA, BROGDEN RN, HEEL RC, ET AL, MOXALACTAM (LATAMOXEF) A REVIEW OF ITS ANTIBACTERIAL ACTIVITY, PHARMACOKINETIC PROPERTIES AND THERAPEUTIC USE, *DRUGS*, 26, 279-333 (1983)

0593 CARNEY FMT, VAN DYKE RA, HALOTHANE HEPATITIS: A CRITICAL REVIEW, *ANESTH ANALG*, 51, 135 (1972)

0594 CARPENTIERE G, MARINO S, CASTELLO F, FUROSEMIDE AND THEOPHYLLINE, *ANN INTERN MED*, 103, 957 (1985)

0595 CARRINGTON ER, FRISHMUTH GJ, OESTERLING MJ, ADAMS FM, COX SE, GESTATIONAL AND POSTPARTUM PLASMA DIAMINE OXIDASE VALUES, *OBSTET GYNECOL*, 39, 426 (1972)

0596 CARRUTHERS M, TAGGART P, CONWAY N, BATES D, SOMERVILLE W, VALIDITY OF PLASMA CATECHOLAMINE ESTIMATIONS, *LANCET*, 2, 62 (1970)

0597 CARSTAIRS KC, BRECKENRIDGE A, DOLLERY CT, WORLEDGE SM, INCIDENCE OF A POSITIVE DIRECT COOMBS' TEST IN PATIENTS ON ALPHA-METHYLDOPA, *LANCET*, 2, 133 (1966)

0598 CARTER BL, SMALL RE, MANDEL MD, ET AL, PHENYTOIN-INDUCED HYPERGLYCEMIA, *AM J HOSP PHARM*, 38, 1508-1512 (1981)

0599 CARTER DC, DOZOIS RR, KIRKPATRICK JR, INSULIN INFUSION TEST OF GASTRIC ACID SECRETION, *BR MED J [CLIN RES]*, 2, 202 (1972)

0600 CARTER SA, POTENTIATION OF THE EFFECT OF ORALLY ADMINISTERED ANTICOAGULANTS BY PHENYRAMIDOL HYDROCHLORIDE, *N ENGL J MED*, 273, 423 (1965)

0601 CARULLI N, PONZ DE LEON M, PODDA M, ET AL, CHENODEOXYCHOLIC ACID AND URSODEOXYCHOLIC ACID EFFECTS IN ENDOGENOUS HYPERTRIGLYCERIDEMIAS: A CONTROLLED DOUBLE-BLIND TRIAL, *J CLIN PHARMACOL*, 21, 436-442 (1981)

0602 CASALE TB, MACHER AM, FAUCI AS, COMPLETE HEMATOLOGIC AND HEPATIC RECOVERY IN A PATIENT WITH CHLORAMPHENICOL HEPATITIS-PANCYTOPENIA SYNDROME, *J PEDIATR*, 101, 1025-1027 (1982)

0603 CASHIN-HEMPHILL L, SPENCER CA, NICOLOFF JT, ET AL, ALTERATIONS IN SERUM THYROID HORMONAL INDICES WITH COLESTIPOL-NIACIN THERAPY, *ANN INTERN MED*, 107, 324-329 (1987)

0604 CASSIDY CE, THE DURATION OF INCREASED SERUM IODINE CONCENTRATION AFTER INGESTION OF BUNAMIODYL (ORABILEX), *J CLIN ENDOCRINOL METAB*, 20, 1034 (1960)

0605 CASTEELS-VAN DAELE M, GASTROINTESTINAL BLEEDING: A POSSIBLE ASSOCIATION WITH IBUPROFEN, *LANCET*, 1, 1021 (1972)

0606 CASTLEDEN CM, RICHENS A, LETTER: CHRONIC PHENYTOIN THERAPY AND CARBOHYDRATE TOLERANCE, *LANCET*, 2, 966 (1973)

0607 CASTLEMAN B, MCNEELY BU, CASE RECORDS OF THE MASSACHUSETTS GENERAL HOSPITAL, *N ENGL J MED*, 276, 167 (1967)

0608 CATHCART-RAKE WF, KYNER JL, AZARNOFF DL, METABOLIC RESPONSES TO PLASMA CONCENTRATIONS OF THEOPHYLLINE, *CLIN PHARMACOL THER*, 26, 89-95 (1979)

0609 CATHRO DM, SAEZ JM, BERTRAND J, THE EFFECT OF CLOMIPHENE ON THE PLASMA ANDROGENS OF PREPUBERTAL AND PUBERTAL BOYS, *J ENDOCRINOL*, 50, 387 (1971)

0610 CATLIN DH, POLAND RE, GORELICK DA, ET AL, INTRAVENOUS INFUSION OF BETA-ENDORPHIN INCREASES SERUM PROLACTIN, BUT NOT GROWTH HORMONE OR CORTISOL, IN DEPRESSED SUBJECTS AND WITHDRAWING METHADONE ADDICTS, *J CLIN ENDOCRINOL METAB*, 50, 1021-1025 (1980)

0611 CATTELL WR, EXCRETORY PATHWAYS FOR CONTRAST MEDIA, *INVEST RADIOL*, 6, 473 (1971)

0612 CAULIN C, LES TROUBLES DIGESTIFS D'ORIGINE MEDICAMENTEUSE, *PRESSE MED*, 79, 2117 (1971)

0613 CAVIET NL, KLAASSEN CHL, TROMBOCYTOPENIE VEROORZAAKT DOOR ALPRENOLOL, *NED TIJDSCHR GENEESKD*, 123, 18-20 (1979)

0614 CEPELAK V, ROUBAL Z, CEPELAKOVA H, AGGREGATION OF BLOOD PLATELETS AND ITS INHIBITION BY COUMARIN DERIVATIVES, *BLOOD*, 39, 588 (1972)

0615 CERCEO E, ELLOSO CA, FACTORS AFFECTING THE RADIOIMMUNOASSAY OF DIGOXIN, *CLIN CHEM*, 18, 539 (1972)

0616 CETINA JA, FISHBEIN EA, ALARCON-SEGOVIA D, ANTINUCLEAR ANTIBODIES AND PROPYLTHIOURACIL THERAPY, *J AM MED ASSOC*, 220, 1012 (1972)

0617 CHAFFMAN M, BROGDEN RN, HEEL RC, ET AL, AURANOFIN: A PRELIMINARY REVIEW OF ITS PHARMACOLOGICAL PROPERTIES AND THERAPEUTIC USE IN RHEUMATOID ARTHRITIS, *DRUGS*, 27, 378-424 (1984)

0618 CHAINUVATI T, HARINASUTA U, ZIMMERMANN HJ, SPURIOUS ELEVATION OF APPARENT LACTATE DEHYDROGENASE ACTIVITY CAUSED BY TRIAMTERENE, *CLIN CHEM*, 19, 1202 (1973)

0619 CHAKRABARTI R, EVANS JF, FEARNLEY GR, EFFECTS ON PLATELET STICKINESS AND FIBRINOLYSIS OF PHENFORMIN COMBINED WITH ETHYLESTRENOL OR STANOZOLOL, *LANCET*, 1, 591 (1970)

0620 CHAKRABARTI R, FEARNLEY GR, EVANS JF, EFFECTS OF CLOFIBRATE ON FIBRINOLYSIS, PLATELET STICKINESS, PLASMA FIBRINOGEN AND SERUM CHOLESTEROL, *LANCET*, 2, 1007 (1968)

0621 CHAKRABARTY AK, INTERFERENCE BY ANTIBIOTICS IN PLASMA PARACETAMOL DETERMINATION, *ANN CLIN BIOCHEM*, 16, 217 (1979)

0622 CHALLIS TW, REID LC, HINTON JW, STUDY OF SOME FACTORS WHICH INFLUENCE THE LEVEL OF SERUM AMYLASE IN DOGS AND HUMANS, *GASTROENTEROLOGY*, 33, 818 (1957)

0623 CHALMERS A, THOMPSON D, STEIN HE, ET AL, SYSTEMIC LUPUS ERYTHEMATOSUS DURING PENICILLAMINE THERAPY FOR RHEUMATOID ARTHRITIS, *ANN INTERN MED*, 97, 659-663 (1982)

0624 CHALMERS DM, BROWN RC, MILLER MG, ET AL, THE INFLUENCE OF LONG-TERM CIMETIDINE AS AN ADJUVANT TO PANCREATIC ENZYME THERAPY IN CYSTIC FIBROSIS, *ACTA PAEDIATR SCAND*, 74, 114-117 (1985)

0625 CHALMERS DM, RINSLER MG, MacDERMOTT S, ET AL, BIOCHEMICAL AND HAEMATOLOGICAL INDICATORS OF EXCESSIVE ALCOHOL CONSUMPTION, *GUT*, 22, 992-996 (1981)

0626 CHAMPION MC, SULLIVAN SN, BLOOM SR, ET AL, THE EFFECTS OF NALOXONE AND MORPHINE ON POSTPRANDIAL GASTROINTESTINAL HORMONE SECRETION, *AM J GASTROENTEROL*, 77, 617-620 (1982)

0627 CHAN JCM, THE RAPID DETERMINATION OF URINARY TITRATABLE ACID AND AMMONIUM, *CLIN BIOCHEM*, 5, 94 (1972)

0628 CHAN LPE, SWAMINATHAN R, ADENOSINE TRIPHOSPHATE INTERFERES WITH PHOSPHATE DETERMINATION, *CLIN CHEM*, 32, 1981 (1986)

0629 CHAN TK, MAK LW, NG RP, METHEMOGLOBINEMIA, HEINZ BODIES AND ACUTE MASSIVE INTRAVASCULAR HEMOLYSIS IN LYSOL POISONING, *BLOOD*, 38, 739 (1971)

0630 CHAN V, BESSER GM, LANDON J, EFFECTS OF OESTROGENS ON URINARY THYROXINE EXCRETION, *BR MED J [CLIN RES]*, 4, 699 (1972)

0631 CHAN V, BESSER GM, LANDON J, EKINS RP, URINARY TRIIODOTHYRONINE EXCRETION AS INDEX OF THYROID FUNCTION, *LANCET*, 2, 253 (1972)

0632 CHAN V, MCALISTER J, LANDON J, COMPARISON OF TWO TRIIODOTHYRONINE UPTAKE TECHNIQUES FOR THE ASSESSMENT OF THYROID FUNCTION, *J CLIN PATHOL*, 25, 30 (1972)

0633 CHANG TW, COLD URTICARIA AND PHOTOSENSITIVITY DUE TO GRISEOFULVIN, *J AM MED ASSOC*, 193, 848 (1965)

0634 CHAPMAN I, CHEUNG WH, PANCYTOPENIA ASSOCIATED WITH TOLBUTAMIDE THERAPY, *J AM MED ASSOC*, 186, 595 (1963)

0635 CHAPMAN JR, GRIFFITHS D, HARDING NGL, ET AL, REVERSIBILITY OF CYCLOSPORINE NEPHROTOXICITY AFTER THREE MONTHS TREATMENT, *LANCET*, 2, 128-129 (1985)

0636 CHARLES MA, DANFORTH E JR., NONKETOACIDOTIC HYPERGLYCEMIA AND COMA DURING INTRAVENOUS DIAZOXIDE THERAPY IN UREMIA, *DIABETES*, 20, 501 (1971)

0637 CHARLES S, KETELSLEGERS J-M, BUYSSCHAERT M, ET AL, HYPERGLYCAEMIC EFFECT OF NIFEDIPINE, *BR MED J [CLIN RES]*, 283, 19-20 (1981)

0638 CHATTERJEA G, MAASER R, COMPARISON OF HAEMOGLOBIN AND HAEMATOCRIT LEVELS IN WHOLE BLOOD FROM EARLOBE AND FINGER-TIP OF CHILDREN, *GERM MED*, 1, 74 (1971)

0639 CHAUHURI TK, MECHANISM OF HYPOCALCEMIC EFFECT OF INORGANIC PHOSPHATE, *N ENGL J MED*, 285, 691 (1971)

0640 CHAZAN R, DROSZCZ W, BOBILEWICZ D, ET AL, CHANGES IN PLASMA HIGH DENSITY LIPOPROTEINS (HDL) LEVELS AFTER SALBUTAMOL, *INTL J CLIN PHARM THER TOXICOL*, 23, 427-429 (1985)

0641 CHEE HD, BRONSVELD W, LIPS PTAM, ET AL, ADRENOCORTICAL SUPPRESSION IN MULTIPLY INJURED PATIENTS: A COMPLICATION OF ETOMIDATE TREATMENT, *BR MED J [CLIN RES]*, 288, 485 (1984)

0642 CHEN JH, OTTOLENGI P, DISTENFELD A, OXYPHENBUTAZONE-INDUCED SIALADENITIS, *J AM MED ASSOC*, 238, 1399 (1977)

0643 CHEN JJS, BERLIN FS, MARGOLIS S, EFFECT OF LARGE DOSE PROGESTERONE ON PLASMA LEVELS OF LIPIDS, LIPOPROTEINS AND APOLIPOPROTEINS IN MALES., *J ENDOCRINOL INVEST*, 9, 281-285 (1986)

0644 CHERASKIN E, RINGSDORF WM JR., EFFECT OF CAFFEINE VERSUS PLACEBO SUPPLEMENTATION ON BLOOD GLUCOSE CONCENTRATION, *LANCET*, 1, 1299 (1967)

0645 CHESNEY RW, DRUG-INDUCED HYPOKALEMIA, *AM J DIS CHILD*, 130, 1055-1056 (1976)

0646 CHESNEY RW, HAMSTR AJ, MAZEES RB, ET AL, REDUCTION OF SERUM 1,25-DIHYDROXY VITAMIN D IN CHILDREN RECEIVING GLUCOCORTICOIDS, *LANCET*, 2, 1123-1125 (1978)

0647 CHINITZ JL, KIM KE, ONESTI G, SWARTZ C, PATHOPHYSIOLOGY AND PREVENTION OF DEXTR^N-40-INDUCED ANURIA, *J LAB CLIN MED*, 77, 76 (1971)

0648 CHIODERA P, COIRO V, ZANARDI G, ET AL, EFFECT OF METOCLOPRAMIDE ON SERUM GH LEVELS IN NORMAL WOMEN, *HORM METAB RES*, 14, 103-104 (1982)

0649 CHRISTENSEN CK, MORGENSEN CE, HANBERG SORENSEN F, RENAL FUNCTION AND CIMETIDINE. URINARY ALBUMIN AND BETA$_2$ - MICROGLOBULIN EXCRETION AND CREATININE CLEARANCE DURING CIMETIDINE TREATMENT, *SCAND J GASTROENTEROL*, 16, 129-134 (1981)

0650 CHRISTENSEN CK, PEDERSEN OL, MIKKELSEN E, RENAL EFFECTS OF ACUTE CALCIUM BLOCKADE WITH NIFEDIPINE IN HYPERTENSIVE PATIENTS RECEIVING BETA-ADRENOCEPTOR BLOCKING DRUGS, *CLIN PHARMACOL THER*, 32, 572-576 (1982)

0651 CHRISTENSEN E, SCHLICHTING P, FAUERHOLDT L, ET AL, CHANGES OF LABORATORY VARIABLES WITH TIME IN CIRRHOSIS: PROGNOSTIC AND THERAPEUTIC SIGNIFICANCE, *HEPATOLOGY*, 5, 843-853 (1985)

0652 CHRISTIAN DG, DRUG INTERFERENCE WITH LABORATORY BLOOD CHEMISTRY DETERMINATIONS, *AM J CLIN PATHOL*, 54, 118 (1970)

0653 CHRISTIANSEN C, CHRISTENSEN MS, GRANDE P, ET AL, LOW-RISK LIPOPROTEIN PATTERN IN POST-MENOPAUSAL WOMEN ON SEQUENTIAL OESTROGEN/ PROGESTOGEN TREATMENT, *MATURITAS*, 5, 193-199 (1984)

0654 CHRISTIANSEN C, RODBRO P, TJELLSEN L, SERUM ALKALINE PHOSPHATASE DURING HORMONE TREATMENT IN EARLY POSTMENOPAUSAL WOMEN, *ACTA MED SCAND*, 216, 11-17 (1984)

0655 CHRYSANT SG, DUNN M, MARPLES D, ET AL, SEVERE REVERSIBLE AZOTEMIA FROM CAPTOPRIL THERAPY: REPORT OF THREE CASES AND REVIEW OF THE LITERATURE, *ARCH INTERN MED*, 143, 437-441 (1983)

0656 CHRYSANT SG, LUU TM, EFFECTS OF AMILORIDE ON ARTERIAL PRESSURE AND RENAL FUNCTION, *J CLIN PHARMACOL*, 20, 332-337 (1980)

0657 CHU J-Y, O'CONNOR DM, SCHMIDT RR, ET AL, THE MECHANISM OF OXACILLIN-INDUCED NEUTROPENIA, *J PEDIATR*, 90, 668 (1977)

0658 CHUNG EK, DIGITALIS INTOXICATION, *POSTGRAD MED J*, 48, 163 (1972)

0659 CHYATTE SB, BASMAJIAN JV, DANTROLENE SODIUM: LONG-TERM EFFECTS IN SEVERE SPASTICITY, *ARCH PHYS MED REHABIL*, 54, 311 (1973)

0660 CIABATTONI G, CINOTTI GA, PIERUCCI A, ET AL, EFFECTS OF SULINDAC AND IBUPROFEN IN PATIENTS WITH CHRONIC GLOMERULAR DISEASE. EVIDENCE FOR THE DEPENDENCE OF RENAL FUNCTION ON PROSTACYCLIN, *N ENGL J MED*, 310, 279-283 (1984)

0661 CIANCI P, REITZ RE, WERNER HV, WEINSTEIN RL, ROLE OF FREE FATTY ACIDS IN GROWTH HORMONE REGULATION DURING FASTING, *ANN INTERN MED*, 76, 857 (1972)

0662 CIMO PL, DOCUMENTING SUSPECTED DRUG-INDUCED THROMBOCYTOPENIA, *ARCH INTERN MED*, 143, 1117-1118 (1983)

0663 CINES DB, KAYWIN P, BINA M, ET AL, HEPARIN-ASSOCIATED THROMBOCYTOPENIA, *N ENGL J MED*, 303, 788-795 (1980)

0664 CIPOLLE RJ, RODVOLD KA, SEIFERT R, ET AL, HEPARIN-ASSOCIATED THROMBOCYTOPENIA: A PROSPECTIVE EVALUATION OF 211 PATIENTS, *THER DRUG MONIT*, 5, 205-211 (1983)

0665 CLAAS FHJ, VAN DER MEER JWM, LANGERAK J, IMMUNOLOGICAL EFFECT OF CO-TRIMOXAZOLE ON PLATELETS, *BR MED J [CLIN RES]*, 2, 898-899 (1979)

0666 CLAGHORN JL, KINROSE-WRIGHT J, MCISAAC WM, FALSE POSITIVE URINE GLUCOSE TEST IN TWO PATIENTS RECEIVING TRIOXAZINE, *J NEW DRUGS*, 6, 153 (1966)

0667 CLAIBORNE RA, DUTT AK, ISONIAZID-INDUCED PURE RED CELL APLASIA, *AM REV RESPIR DIS*, 131, 947-949 (1985)

0668 CLARK DW, GOLDBERG LI, GUANCYDINE: A NEW ANTIHYPERTENSIVE AGENT, *ANN INTERN MED*, 76, 579 (1972)

0669 CLARK PMS, CLARK JDA, WHEATLEY T, URINE DISCOLORATION AFTER ACETAMINOPHEN OVERDOSE, *CLIN CHEM*, 32, 1777-1778 (1986)

0670 CLARK PMS, KRICKA LJ, INTERFERENCE BY DIAZEPAM IN THE DETERMINATION OF 5-HYDROXYINDOLEACETIC ACID., *ANN CLIN BIOCHEM*, 14, 233-234 (1977)

0671 CLARKSON AR, MACDONALD MK, CASH JD, ROBSON JS, MODIFICATION BY DRUGS OF URINARY FIBRIN/FIBRINOGEN DEGRADATION PRODUCTS IN GLOMERULONEPHRITIS, *BR MED J [CLIN RES]*, 3, 255 (1972)

0672 CLAUVEL DE MENDONCA M, APPEL M, PERNOLLET JC, SCHWARTZ K, A SENSITIVE METHOD FOR AUTOMATED ESTIMATION OF PLASMA HAEMOGLOBIN LEVELS, *CLIN CHIM ACTA*, 39, 149 (1972)

0673 CLEMENZIA G, RUSSO G, GENTILE V, ET AL, REMARKS ON THE BEHAVIOUR OF CERTAIN ENZYMURIAS IN SUBJECTS TREATED WITH PIPEMIDIC ACID, *MINERVA MED*, 77, 621-626 (1986)

0674 CLIFFORD NJ, WEIL J, CORTISOL METABOLISM IN PERSONS OCCUPATIONALLY EXPOSED TO DDT, *ARCH ENVIRON HEALTH*, 24, 145 (1972)

0675 COBLYN JS, WEINBLATT M, HOLDSWORTH D, ET AL, GOLD INDUCED THROMBOCYTOPENIA: A CLINICAL AND IMMUNOGENETIC STUDY OF TWENTY-THREE PATIENTS, *ANN INTERN MED*, 95, 178-181 (1981)

0676 COBURN HJ, CARROLL JJ, IMPROVED MANUAL AND AUTOMATED COLORIMETRIC DETERMINATION OF SERUM GLUCOSE WITH USE OF HEXOKINASE, *CLIN CHEM*, 19, 127 (1973)

0677 COCA A, BLADE J, MARTINEZ A, ET AL, TOBRAMYCIN NEPHROTOXICITY: A PROSPECTIVE CLINICAL STUDY, *POSTGRAD MED J*, 55, 791-796 (1979)

0678 COCKERILL FR III, EDSON RS, TRIMETHOPRIM-SULFAMETHOXAZOLE, *MAYO CLIN PROC*, 58, 147-153 (1983)

0679 COCKERILL FR III, EDSON RS, TRIMETHOPRIM-SULFAMETHOXAZOLE, *MAYO CLIN PROC*, 62, 921-929 (1987)

0680 CODY RJ JR, CALABRESE LH, CLOUGH JD, DEVELOPMENT OF ANTINUCLEAR ANTIBODIES DURING ACEBUTOLOL THERAPY, *CLIN PHARMACOL THER*, 25, 800-805 (1979)

0681 COFFEY VJ, MYXOEDEMA DURING CYCLOPHOSPHAMIDE THERAPY, *BR MED J [CLIN RES]*, 4, 682 (1971)

0682 COHEN CD, SAYED AR, KIRSCH RE, HEPATIC COMPLICATIONS OF ANTITUBERCULOSIS THERAPY REVISITED, *S AFR MED J*, 63, 960-963 (1983)

0683 COHEN DJ, LOERTSCHER R, RUBIN MF, ET AL, CYCLOSPORINE: A NEW IMMUNOSUPPRESSIVE AGENT FOR ORGAN TRANSPLANTION, *ANN INTERN MED*, 101, 667-682 (1984)

0684 COHEN EL, GRIM CE, CONN JW, BLOUGH WM JR., GUYER RB, KEM DC, LUCAS CP, ACCURATE AND RAPID MEASUREMENT OF PLASMA RENIN ACTIVITY BY RADIOIMMUNOASSAY, *J LAB CLIN MED*, 77, 1025 (1971)

0685 COHEN H, GRAHAM H, LAU HL, ALPHA$_1$ FETOPROTEIN IN PREGNANCY, *AM J OBSTET GYNECOL*, 115, 881 (1973)

0686 COHEN HN, BEASTALL GH, RATCLIFFE WA, ET AL, EFFECTS ON HUMAN THYROID FUNCTION OF SULPHONAMIDE AND TRIMETHOPRIM COMBINATION DRUGS, *BR MED J [CLIN RES]*, 281, 646-647 (1980)

0687 COHEN IA, JOHNSON CE, BERARDI RR, ET AL, CIMETIDINE-THEOPHYLLINE INTERACTION: EFFECTS OF AGE AND CIMETIDINE DOSE, *THER DRUG MONIT*, 7, 426-434 (1985)

0688 COHEN LC, CPK TESTS - EFFECT OF INTRAMUSCULAR INJECTION IN MYOCARDIAL INFARCTION, *J AM MED ASSOC*, 219, 625 (1972)

0689 COHEN MM, DEBAS HT, HOLUBITSKY IB, HARRISON RC, CAFFEINE AND PENTAGASTRIN STIMULATION OF HUMAN GASTRIC SECRETION, *GASTROENTEROLOGY*, 61, 440 (1971)

0690 COHEN MS, WASHTON HE, BARRANCO SF, MULTICENTER CLINICAL TRIAL OF CEFOPERAZONE SODIUM IN THE UNITED STATES, *AM J MED*, 77, SUPPL 1B, 35-41 (1984)

0691 COHN KE, AGMON J, GAMBLE OW, THE EFFECT OF GLUCAGON ON ARRHYTHMIAS DUE TO DIGITALIS TOXICITY, *AM J CARDIOL*, 25, 683 (1970)

0692 COL DI G, CEVARO G, LOMBARDO F, ET AL, L'INDAPAMIDE NEI TRAITMENT DELL IPERTENSIONE ARTERIOSA ESSENZIALE DELL ANZIANO: INFLUENZA SUI METABOLISMO GLICO-LIPIDICO, *BOLL SOC ITAL CARDIOL*, 26, 1527-1530 (1981)

0693 COLE RM, RAGHAVAN D, CATERSON I, ET AL, DANAZOL TREATMENT OF ADVANCED PROSTATE CANCER: CLINICAL AND HORMONAL EFFECTS, *PROSTATE*, 9, 15-20 (1986)

0694 COLEMAN RJ, A SPECTROPHOTOMETRIC ARTIFACT IN POSTAL SPECIMENS OF AMNIOTIC FLUID, *N Z MED J*, 76, 272 (1972)

0695 COLIN JN, FARINOTTI R, FREDJ G, ET AL, KINETICS OF ALLOPURINOL AFTER CHRONIC ORAL ADMINISTRATION. INTERACTION WITH BENZBROMARONE, *EUR J CLIN PHARMACOL*, 31, 53-58 (1986)

0696 COLLEN MJ, CIMETIDINE-ASSOCIATED THROMBOCYTOPENIA AND LEUKOPENIA, *WEST J MED*, 132, 257-258 (1980)

0697 COLLINS RD, *ILLUSTRATED MANUAL OF LABORATORY DIAGNOSIS*, JB LIPPINCOTT, PHILADELPHIA, PA (1968)

0698 COLLINSON PO, KIND PRN, SLAVIN B, ET AL, FALSE DIAGNOSIS OF PHAEOCHROMOCYTOMA IN PATIENTS ON SINEMET®, *LANCET*, 1, 14 (1984)

0699 COLT EWD, KIMBRELL D, FIEVE RR, RENAL IMPAIRMENT, HYPERCALCAEMIA, AND LITHIUM THERAPY, *AM J PSYCHIATRY*, 138, 106-108 (1981)

0700 COLTART DJ, WATSON D, HOWARD MR, EFFECT OF EXCHANGE TRANSFUSIONS ON PLASMA DIGOXIN LEVELS, *ARCH DIS CHILD*, 47, 814 (1972)

0701 COLWELL AR JR., SANDO DE, LANG SJ, PROPYLTHIOURACIL-INDUCED AGRANULOCYTOSIS, TOXIC HEPATITIS, AND DEATH, *J AM MED ASSOC*, 148, 639 (1952)

0702 COMBS AB, GIRI SN, PEOPLES SA, A NEW METHOD FOR ANALYSIS OF PHENYLTHIOUREA IN BIOLOGICAL FLUIDS, *ANAL BIOCHEM*, 44, 570 (1971)

0703 COMOY E, BOHOUN C, ISOHOMOVANILLIC ACID DETERMINATION IN HUMAN URINE, *CLIN CHIM ACTA*, 35, 369 (1971)

0704 CONKLIN JD, HOLLIFIELD RD, THE SPECIFICITY OF THE NITROMETHANE-HYAMINE PROCEDURE FOR THE DETERMINATION OF NITROFURANTOIN IN URINE, *CLIN CHEM*, 12, 632 (1966)

0705 CONN HO, BLOOD AMMONIA, *STAND METH CLIN CHEM*, 5, 43 (1965)

0706 CONN JW, HYPERTENSION, THE POTASSIUM ION AND IMPAIRED CARBOHYDRATE TOLERANCE, *N ENGL J MED*, 273, 1135 (1965)

0707 CONNELL JMC, RAPEPORT WG, BEASTALL GH, ET AL, CHANGES IN CIRCULATING ANDROGENS DURING SHORT TERM CARBAMAZEPINE THERAPY, *BR J CLIN PHARMACOL*, 17, 347-351 (1984)

0708 CONNELL JMC, RAPEPORT WG, GORDON S, ET AL, CHANGES IN CIRCULATING THYROID HORMONES DURING SHORT-TERM HEPATIC ENZYME INDUCTION WITH CARBAMAZEPINE, *EUR J CLIN PHARMACOL*, 26, 453-456 (1984)

0709 CONNER CS, ETOMIDATE AND ADRENAL SUPPRESSION, *DRUG INTELL CLIN PHARM*, 18, 393-394 (1984)

0710 CONNEY AH, SPARK RS, ROBINSON SH, HIATT HH, SCHNEIDER P, DRUG METABOLISM AND THERAPEUTICS, *N ENGL J MED*, 280, 653 (1969)

0711 CONRAD KA, FAGAN TC, LEE SM, ET AL, EFFECTS OF TRIPAMIDE ON GLUCOSE TOLERANCE IN PATIENTS WITH HYPERTENSION, *CLIN PHARMACOL THER*, 40, 476-479 (1986)

0712 CONRAD KA, NYMAN DW, EFFECTS OF METOPROLOL AND PROPRONOLOL ON THEOPHYLLINE ELIMINATION, *CLIN PHARMACOL THER*, 28, 463-467 (1980)

0713 CONRAD ME, KNOCHEL JP, CROSBY WH, NOVOBIOCIN JAUNDICE: DEMONSTRATION OF A HEMOLYTIC STATE, *ANTIBIOT MED*, 7, 382 (1960)

0714 CONSOLAZIO CF, JOHNSON HL, KRZYWICKI HJ, DAWS TA, BARNHART RA, THIAMIN, RIBOFLAVIN, AND PYRIDOXINE EXCRETION DURING ACUTE STARVATION AND CALORIE RESTRICTION, *AM J CLIN NUTR*, 24, 1060 (1971)

0715 CONTE N, FEDERSPIL G, FREZZATO S, TRISOTTO A, SCANDELLARI C, PIEMONTE G, GLUCAGON EFFECT ON PLASMA MG CONCENTRATION, *HORM METAB RES*, 4, 48 (1972)

0716 CONTRACTOR SF, SHANE B, ESTIMATION OF VITAMIN B_6 COMPOUNDS IN HUMAN BLOOD AND URINE, *CLIN CHIM ACTA*, 21, 71 (1968)

0717 COODLEY E, DERASSE J, CARVER J, PHENFORMIN AND PANCREATITIS, *ANN INTERN MED*, 78, 307 (1973)

0718 COODLEY EL, ENZYMES IN CARDIAC DISEASE IN, *DIAGNOSTIC ENZYMOLOGY*, EL COODLEY ED. LEA AND FEBIGER, PHILADELPHIA, PA (1970)

0719 COOK DM, ALLEN JP, KENDALL JW, SWANSON R, INTERFERENCE OF 21-DEOXYCORTISOL WITH CORTISOL ASSAY METHODS, *J CLIN ENDOCRINOL METAB*, 36, 608 (1973)

0720 COON WW, WILLIS PW III, SOME ASPECTS OF THE PHARMACOLOGY OF ORAL ANTICOAGULANTS, *CLIN PHARMACOL THER*, 11, 312 (1970)

0721 COON WW, WILLIS PW III, SOME SIDE EFFECTS OF HEPARIN, HEPARINOIDS, AND THEIR ANTAGONISTS, *CLIN PHARMACOL THER*, 7, 379 (1966)

0722 COOPER DS, DANIELS GH, LADENSON PW, ET AL, HYPERTHYROXINEMIA IN PATIENTS TREATED WITH HIGH-DOSE PROPRANOLOL, *AM J MED*, 73, 867-871 (1982)

0723 COOPER DS, GELENBERG AJ, WOJCIK JC, ET AL, THE EFFECT OF AMOXAPINE AND IMIPRAMINE ON SERUM PROLACTIN LEVELS, *ARCH INTERN MED*, 141, 1023-1025 (1981)

0724 COOPER GR, MCDANIEL V, THE DETERMINATION OF GLUCOSE BY THE ORTHO-TOLUIDINE METHOD, *STAND METH CLIN CHEM*, 6, 159 (1970)

0725 COOPERBERG AA, EIDLOW S, HAEMOLYTIC ANEMIA, JAUNDICE AND DIABETES MELLITUS FOLLOWING CHLORPROMAZINE THERAPY, *CAN MED ASSOC J*, 75, 746 (1956)

0726 COPELAND JG, CHANNICK JM, GITTES RF, COMPLICATIONS OF A "MAYO ENEMA", *CALIF MED*, 116, 65 (1972)

0727 CORINALDESI R, PASQUALI R, PATERNICO A, ET AL, EFFECTS OF SHORT- AND LONG-TERM ADMINISTRATIONS OF FAMOTIDINE AND RANITIDINE ON SOME PITUITARY, SEXUAL AND THYROID HORMONES, *DRUGS EXP CLIN RES*, 13, 647-654 (1987)

0728 CORN TH, HALE AS, THOMPSON C, ET AL, A COMPARISON OF THE GROWTH HORMONE RESPONSES TO CLONIDINE AND APOMORPHINE IN THE SAME PATIENT WITH ENDOGENOUS DEPRESSION, *BR J PSYCHIATRY*, 144, 636-639 (1984)

0729 CORNISH AL, MCCLELLAN JT, JOHNSTON DH, EFFECTS OF CHLOROTHIAZIDE ON THE PANCREAS, *N ENGL J MED*, 265, 673 (1961)

0730 CORNS MD, CORNS CM, MILLER AL, STEVENS JF, A COMPARISON OF MANUAL AND A SIMPLE AUTOMATED TECHNIQUE FOR ESTIMATION OF TOTAL ESTROGENS, *CLIN CHIM ACTA*, 48, 335 (1973)

0731 CORRADO ML, STRUBLE WE, CHENNEKATU P, ET AL, NORFLOXACIN: REVIEW OF SAFETY STUDIES, *AM J MED*, 82 SUPPL 6B, 22-26 (1987)

0732 CORRE F, LELLOUCH J, SCHWARTZ D, SMOKING AND LEUKOCYTE COUNTS, *LANCET*, 2, 632 (1971)

0733 CORVOL P, MENARD J, AUZAN C, GONZALES MF, RADIOIMMUNOASSAY OF PLASMA ALDOSTERONE AND PLASMA RENIN ACTIVITY, *ANN ENDOCRINOL*, 33, 285 (1972)

0734 COTHAM RE, SHAND D, SPURIOUSLY LOW PLASMA PROPRANOLOL CONCENTRATION RESULTING FROM BLOOD COLLECTION METHODS, *CLIN PHARMACOL THER*, 18, 535-538 (1975)

0735 COUDERT SP, FAIMAN C, EFFECT OF PROSTAGLANDIN F2 ALPHA ON ANTERIOR PITUITARY FUNCTION IN MAN, *PROSTAGLANDINS*, 3, 89 (1973)

0736 COULTER DL, ALLEN RJ, SECONDARY HYPERAMMONAEMIA: A POSSIBLE MECHANISM FOR VALPROATE ENCEPHALOPATHY, *LANCET*, 1, 1310-1311 (1980)

0737 COULTER DL, WU H, ALLEN RJ, VALPROIC ACID THERAPY IN CHILDHOOD EPILEPSY, *J AM MED ASSOC*, 244, 785-788 (1980)

0738 COURSIN DB, DISCUSSION OF PAPER, *ANN N Y ACAD SCI*, 80, 894 (1959)

0739 COWAN DH, THROMBOKINETIC STUDIES IN ALCOHOL RELATED THROMBOCYTOPENIA, *J LAB CLIN MED*, 81, 64 (1973)

0740 COWAN RE, WRIGHT JT, DAPSONE AND SEVERE HYPOALBUMINAEMIA IN DERMATITIS HERPETIFORMIS, *BR J DERMATOL*, 104, 201-204 (1981)

0741 COWEN AE, MCGEARY HM, CAMPBELL CB, INTERFERENCE BY GASTROGRAFIN® WITH A SPECTROPHOTOMETRIC TRYPSIN ASSAY, *GUT*, 13, 395 (1972)

0742 CRAFT IL, PETERS TJ, QUANTITATIVE CHANGES IN PLASMA AMINO ACIDS INDUCED BY ORAL CONTRACEPTIVES, *CLIN SCI*, 41, 301 (1971)

0743 CRAFT IL, WISE I, PLASMA AMINO ACIDS AND ORAL CONTRACEPTIVES, *LANCET*, 2, 1138 (1969)

0744 CRAIG GM, CRANE CW, LACTIC ACIDOSIS COMPLICATING LIVER FAILURE AFTER INTRAVENOUS FRUCTOSE, *BR MED J [CLIN RES]*, 4, 211 (1971)

0745 CRAIG WA, KUNIN CM, TRIMETHOPRIM-SULFAMETHOXAZOLE: PHARMACODYNAMIC EFFECT OF URINARY PH AND IMPAIRED RENAL FUNCTION, *ANN INTERN MED*, 78, 491 (1973)

0746 CRAMER BC, PARFREY PS, HUTCHINSON TA, ET AL, RENAL FUNCTION FOLLOWING INFUSION OF RADIOLOGIC CONTRAST MATERIAL: A PROSPECTIVE CONTROLLED STUDY, *ARCH INTERN MED*, 145, 87-89 (1985)

0747 CRAMER H, GOODWIN FK, POST RM, BUNNEY WE JR., EFFECTS OF PROBENECID AND EXERCISE ON CEREBROSPINAL FLUID CYCLIC AMP IN AFFECTIVE ILLNESS, *LANCET*, 1, 1346 (1972)

0748 CRAMMER JL, ELKES A, AGRANULOCYTOSIS AFTER DESIPRAMINE, *LANCET*, 1, 105 (1967)

0749 CRANDELL WB, PAPPAS SG, MACDONALD A, NEPHROTOXICITY ASSOCIATED WITH METHOXYFLURANE ANESTHESIA, *ANESTHESIOLOGY*, 27, 591 (1966)

0750 CRASWELL PW, HIRD WM, JUDD PA, VARGHESE Z, MOORHEAD JF, PLASMA RENIN ACTIVITY AND BLOOD PRESSURE IN 89 PATIENTS RECEIVING MAINTENANCE HEMODIALYSIS THERAPY, *BR MED J [CLIN RES]*, 4, 749 (1972)

0751 CRAWFORD GA, GALLERY EDM, GYORY AZ, REMOVAL OF INTERFERENCE BY ANTIHYPERTENSIVE DRUGS IN THE SPECTROPHOTOMETRIC ASSAY OF METANEPHRINES, *CLIN CHIM ACTA*, 169, 117-120 (1987)

0752 CRAWFORD OB, METHEMOGLOBIN IN MAN FOLLOWING THE USE OF PRILOCAINE, *ACTA ANAESTHESIOL SCAND SUPPL*, 16, 183 (1965)

0753 CRAWHALL JC, PROTEINURIA IN D-PENICILLAMINE-TREATED RHEUMATOID ARTHRITIS, *J RHEUMATOL*, 8, SUPPL 7, 161-163 (1981)

0754 CREMATA VY JR., KOE BK, CLINICAL-PHARMACOLOGICAL EVALUATION OF P-CHLOROPHENYLALANINE: NEW SEROTONIN-DEPLETING AGENT, *CLIN PHARMACOL THER*, 7, 768 (1966)

0755 CRESPI HG, TOPICAL CORTICOSTEROID THERAPY FOR CHILDREN: ALCLOMETASONE DIPROPIONATE CREAM 0.05%, *CLIN THER*, 8, 203-210 (1986)

0756 CROFT JD, SWISHER SN, GILLILAND BC, LEDDY JP, WEED RI, COOMBS' - TEST POSITIVITY INDUCED BY DRUGS, *ANN INTERN MED*, 68, 176 (1968)

0757 CROSS FC, CANADA AT JR., DAVIS NM, EFFECTS OF CERTAIN DRUGS ON THE RESULTS OF SOME COMMON LABORATORY DIAGNOSTIC PROCEDURES, *AM J HOSP PHARM*, 23, 235 (1966)

0758 CROUT JR, CATECHOLAMINES IN URINE, *STAND METH CLIN CHEM*, 3, 62 (1961)

0759 CROUT JR, SJOERDSMA A, THE CLINICAL AND LABORATORY SIGNIFICANCE OF SEROTONIN AND CATECHOLAMINES IN BANANAS, *N ENGL J MED*, 261, 23 (1959)

0760 CROWELL EB JR., CLATANOFF DV, KIEKHOFER W, THE EFFECT OF ORAL CONTRACEPTIVES ON FACTOR VIII LEVELS, *J LAB CLIN MED*, 77, 551 (1971)

0761 CROWLEY MF, GARBIEN KJT, ROSSER A, HUMAN GROWTH HORMONE: A COMPARISON OF THE RESULTS OF PLASMA AND SERUM ASSAYS USING DOUBLE ANTIBODY TECHNIQUES, *CLIN CHIM ACTA*, 45, 19 (1973)

0762 CROWLEY MF, ROSSER A, OESTROGEN DETERMINATION IN PREGNANCY URINE USING ENZYMATIC HYDROLYSIS OF OESTROGEN CONJUGATES, *CLIN CHIM ACTA*, 49, 115 (1973)

0763 CROXATTO HB, DIAZ S, ROBERTSON DN, ET AL, CLINICAL CHEMISTRY IN WOMEN WITH LEVONORGESTREL IMPLANTS (NORPLANT ™) OR A TCU 200 IUD, *CONTRACEPTION*, 27, 281-288 (1983)

0764 CROZIER IG, IKRAM H, NICHOLLS MG, ET AL, ACUTE HEMODYNAMIC, HORMONAL AND ELECTROLYTE EFFECTS OF RAMIPRIL IN SEVERE CONGESTIVE HEART FAILURE, *AM J CARDIOL*, 59, 155D-163D (1987)

0765 CRUICKSHANK AM, SHENKIN A, A COMPARISON OF THE EFFECT OF ACETOACETATE CONCENTRATION ON THE MEASUREMENT OF SERUM CREATININE USING TECHNICON® SMAC® II, BECKMAN ASTRA™ AND ENZYMATIC TECHNIQUES, *ANN CLIN BIOCHEM*, 24, 317-319 (1987)

0766 CRYER PE, SODE J, DRUG INTERFERENCE WITH MEASUREMENT OF ADRENAL HORMONES IN URINE:ANALGESICS AND TRANQUIL-IZER-SEDATIVES, *ANN INTERN MED*, 75, 697 (1971)

0767 CSAKO G, FALSE-POSITIVE RESULTS FOR KETONE WITH THE DRUG MESNA AND OTHER FREE SULFHYDRYL COMPOUNDS, *CLIN CHEM*, 33, 289-292 (1987)

0768 CUCHE J-L, SEIZ F, RUGET G, ET AL, DILUTION OF PLASMA WITH TRIS BUFFER INCREASES MEASURED CATECHOLAMINES IN PLASMA, *CLIN CHEM*, 33, 408-411 (1987)

0769 CUCUIANU M, BORNUZ F, MACAVEI I, EFFECT OF L-ASPARAGINASE THERAPY UPON SERUM PSEUDOCHOLINESTERASE AND CERULOPLASMIN LEVELS IN PATIENTS, *CLIN CHIM ACTA*, 38, 97 (1972)

0770 CUIGNIEZ PH, BIESBROUCK M, VAN HOOREN J, A SIMPLE AND RAPID NEPHELOMETRIC PROCEDURE FOR FIBRONECTIN, *ANN CLIN BIOCHEM*, 22, 514-518 (1985)

0771 CULLEN MR, KAYNE RD, ROBINS JM, ENDOCRINE AND REPRODUC-TIVE DYSFUNCTION IN MEN ASSOCIATED WITH OCCUPATIONAL INORGANIC LEAD INTOXICATION, *ARCH ENVIRON HEALTH*, 39, 431-440 (1984)

0772 CULLEN SI, CATALANO PM, GRISEOFULVIN-WARFARIN ANTAGO-NISM, *J AM MED ASSOC*, 199, 582 (1967)

0773 CUMMING FJ, BRIGGS MH, CHANGES IN PLASMA VITAMIN A IN LAC-TATING AND NONLACTATING ORAL CONTRACEPTIVE USERS, *BR J OBSTET GYNAECOL*, 90, 73-77 (1983)

0774 CUMMINGS JN, MIGRAINE - A BIOCHEMICAL DISORDER?, *FOURTH SYMPOSIUM OF THE MIGRAINE TRUST*, LONDON 11 SEPT (1970)

0775 CUNNINGHAM GC, PHENYLKETONURIA TESTING - ITS ROLE IN PEDI-ATRICS AND PUBLIC HEALTH, *CRIT REV CLIN LAB SCI*, 2, 45 (1971)

0776 CUNNINGHAM SK, LOUGHLIN T, CULLITON M, ET AL, THE RELATION-SHIP BETWEEN SEX STEROIDS AND SEX HORMONE-BINDING GLOBULIN IN PLASMA IN PHYSIOLOGICAL AND PATHOLOGICAL CONDITIONS., *ANN CLIN BIOCHEM*, 22, 489-497 (1985)

0777 CURTIS JA, CORMODE E, LASKI B, ET AL, ENDOCRINE COMPLICA-TIONS OF TOPICAL AND INTRALESIONAL CORTICOSTEROID THERAPY, *ARCH DIS CHILD*, 57, 204-207 (1982)

0778 CURZEN P, VARMA R, A COMPARISON OF SERUM CYSTINE AMI-NOPEPTIDASE AND URINARY ESTROGEN EXCRETION AS PLA-CENTAL FUNCTION TESTS, *AM J OBSTET GYNECOL*, 115, 929 (1973)

0779 CUSCHIERI A, ACTIVATION OF THE KININ-FORMING SYSTEM DURING THERAPY WITH CYCLOPHOSPHAMIDE, *CLIN CHEM*, 20, 19 (1974)

0780 CUSHMAN P JR., PLASMA TESTOSTERONE IN NARCOTIC ADDIC-TION, *AM J MED*, 55, 452 (1973)

0781 CUSHMAN P JR., GRIECO MH, HYPERIMMUNOGLOBULINEMIA ASSO-CIATED WITH NARCOTIC ADDICTION: EFFECTS OF METHADONE MAINTENANCE TREATMENT, *AM J MED*, 54, 320 (1973)

0782 CUTHBERT MF, ADVERSE REACTIONS TO NONSTEROIDAL ANTIRHEUMATIC DRUGS, *CURR MED RES OPIN*, 2, 600-610 (1974)

0783 CZERWINSKI AW, CZERWINSKI AB, WHITSETT TL, CLARK ML, EFFECTS OF A SINGLE, LARGE, INTRAVENOUS INJECTION OF DEXAMETHASONE, *CLIN PHARMACOL THER*, 13, 638 (1972)

0784 D'ELIA JA, GLEASON RE, ALDAY M, NEPHROTOXICITY FROM ANGI-OGRAPHIC CONTRAST MATERIAL: A PROSPECTIVE STUDY, *AM J MED*, 72, 719-725 (1982)

0785 DA FONESCA-WOLLHEIM F, ENZYMATIC DETERMINATION OF GLU-COSE IN HEMOLYZED BLOOD SAMPLE WITHOUT INTERFER-ENCE BY FRUCTOSE, *Z KLIN CHEM KLIN BIOCHEM*, 9, 497 (1971)

0786 DAAE LNW, JUELL A, RAPID DIAGNOSTIC TESTS FOR GLUCOSURIA ARE STILL INFLUENCED BY ASCORBIC ACID, *SCAND J CLIN LAB INVEST*, 43, 747-749 (1983)

0787 DAAE LNW, JUELL A, A NEW AND MORE ASCORBIC ACID RESIS-TANT DIPSTICK TEST FOR THE DETECTION OF GLUCOSURIA HAS BEEN INTRODUCED, *SCAND J CLIN LAB INVEST*, 45, 289 (1985)

0788 DACIE JV, *THE HAEMOLYTIC ANEMIAS PART IV, GRUNE AND STRAT-TON INC, NEW YORK, NY* (1967)

0789 DAHL JR, DIPHENYLHYDANTOIN TOXIC PSYCHOSIS WITH ASSOCI-ATED HYPERGLYCEMIA, *CALIF MED*, 107, 345 (1967)

0790 DAHLQUIST R, EJVINSSON G, SCHENCK-GUSTAFSSON K, EFFECT OF QUINIDINE ON PLASMA CONCENTRATION AND RENAL CLEARANCE OF DIGOXIN: A CLINICALLY IMPORTANT DRUG INTERACTION, *BR J CLIN PHARMACOL*, 9, 413-418 (1980)

0791 DAIGNEAULT R, GAGNE M, FOURNIER M, LAZAS A, LEDUC A, COM-PARATIVE STUDY OF GENTAMICIN USING AN ENZYMATIC PRO-CEDURE AND AN AGAR-DIFFUSION TECHNIQUE, *CLIN BIOCHEM*, 6, 326 (1973)

0792 DAL CANTON A, FUIANO G, CONTE G, ET AL, MECHANISM OF INCREASED PLASMA UREA AFTER DIURETIC THERAPY IN URAE-MIC PATIENTS, *CLIN SCI*, 68, 255-261 (1985)

0793 DALDERUP LM, ISCHAEMIC HEART DISEASE AND VITAMIN D, *LAN-CET*, 2, 92 (1973)

0794 DALE E, SIMPSON G, SERUM MAGNESIUM LEVELS OF WOMEN TAK-ING AN ORAL OR LONG-TERM INJECTABLE PROGESTATIONAL CONTRACEPTIVE, *OBSTET GYNECOL*, 39, 115 (1972)

0795 DALE E, SPIVEY SH, SERUM PROTEINS OF WOMEN UTILIZING COM-BINATION ORAL OR LONG-ACTING INJECTABLE PROGESTA-TIONAL CONTRACEPTIVES, *CONTRACEPTION*, 4, 241 (1971)

0796 DALE J, LANDMARK KH, MYHRE E, THE EFFECTS OF NIFEDIPINE, A CALCIUM ANTAGONIST, ON PLATELET FUNCTION, *AM HEART J*, 105, 103-105 (1983)

0797 DALE RA, DEMONSTRATION OF ALDOLASE IN HUMAN PLATELETS: THE RELATION TO PLASMA AND SERUM ALDOLASE, *CLIN CHIM ACTA*, 5, 652 (1960)

0798 DALLOS V, HEATHFIELD K, STONE P, ALLEN F, THE COMPARATIVE VALUE OF AMANTADINE AND LEVODOPA, *POSTGRAD MED J*, 48, 354 (1972)

0799 DALRYMPLE RW, STEARNS FM, DIFLUNISAL INTERFERENCE WITH DETERMINATION OF SALICYLATE BY THE TRINDER, ABBOTT TDX, AND DU PONT ACA METHODS, *CLIN CHEM*, 32, 230 (1986)

0800 DALSGAARD-NIELSEN J, RISBO A, SIMMELKJAER P, ET AL, IMPAIRED PLATELET AGGREGATION AND INCREASED BLEEDING TIME DURING GENERAL ANESTHESIA WITH HALOTHANE, *BR J ANAESTH*, 53, 1039-1041 (1981)

0801 DALTON MJ, POWELL JR, MESSENHEIMER JA JR., THE INFLUENCE OF CIMETIDINE ON SINGLE DOSE CARBAMAZEPINE PHARMACOKINETICS, *EPILEPSIA*, 26, 127-130 (1985)

0802 DAMM HC (ED), *HANDBOOK OF CLINICAL LABORATORY DATA, CRC PRESS, CLEVELAND OH* (1965)

0803 DANA-HAERI J, OXLEY J, RICHENS A, REDUCTION OF FREE TESTOS-TERONE BY ANTIEPILEPTIC DRUGS, *BR MED J [CLIN RES]*, 284, 85-86 (1982)

0804 DANDONA D, JUNGLEE D, KATRAK A, INCREASED SERUM PANCRE-ATIC ENZYMES AFTER TREATMENT WITH METHYL-PREDNISOLONE: POSSIBLE EVIDENCE OF SUBCLINICAL PAN-CREATITIS, *BR MED J [CLIN RES]*, 291, 24 (1985)

0805 DANIELS AL, EVERSON GJ, INFLUENCE OF ACETYLSALICYLIC ACID (ASPIRIN) ON URINARY EXCRETION OF ASCORBIC ACID, *PROC SOC EXP BIOL MED*, 35, 20 (1936)

0806 DANOWSKI TS, VESTER JW, SUNDER JH, GONZALEZ AR, KHURANA RC, JUNG Y, ENDOCRINE AND METABOLIC INDICES DURING ADMINISTRATION OF A LIPOPHILIC BIS-PHENOL, PROBUCOL, *CLIN PHARMACOL THER*, 12, 929 (1971)

0807 DANZIGER Y, GARTY M, VOLWITZ B, ET AL, REDUCTION OF SERUM THEOPHYLLINE LEVELS BY TERBUTALINE IN CHILDREN WITH ASTHMA, *CLIN PHARMACOL THER*, 37, 469-471 (1985)

0808 DANZINGER RG, HOFMANN AF, SCHOENFIELD LJ, THISTLE JL, DIS-SOLUTION OF CHOLESTEROL GALLSTONES BY CHENODEOX-YCHOLIC ACID, *N ENGL J MED*, 286, 1 (1972)

0809 DAS I, RAISED C-REACTIVE PROTEIN LEVELS IN SERUM FROM SMOKERS, *CLIN CHIM ACTA*, 153, 9-13 (1985)

0810 DAS KM, EASTWOOD MA, MCMANUS JPA, SIRCUS W, ADVERSE REACTIONS DURING SALICYLASULFAPYRIDINE THERAPY AND THE RELATION WITH DRUG METABOLISM AND ACETYLATOR, *N ENGL J MED*, 289, 491 (1973)

0811 DASTUR DK, QUADROS EV, WADIA NH, DESAI MM, BHARUCHA EP, EFFECT OF VEGETARIANISM AND SMOKING ON VITAMIN B_{12}, THIOCYANATE, AND FOLATE LEVELS IN THE BLOOD OF NOR-MAL SUBJECTS, *BR MED J [CLIN RES]*, 3, 260 (1972)

0812 DATEY KK, DESHMUKH SN, DALVI CP, PURANDARE NM, HEPATOCELLULAR DAMAGE WITH ETHACRYNIC ACID, *BR MED J [CLIN RES]*, 3, 152 (1967)

0813 DAUNCEY MJ, WIDDOWSON EM, URINARY EXCRETION OF CALCIUM, MAGNESIUM, SODIUM, AND POTASSIUM IN HARD AND SOFT-WATER AREAS, *LANCET*, 1, 711 (1972)

0814 DAVENPORT HW, *PHYSIOLOGY OF THE DIGESTIVE TRACT*, YEAR BOOK MEDICAL PUBLISHERS, CHICAGO, IL (1961)

0815 DAVID J, HYPERGLUCAGONAEMIA AND TREATMENT WITH DANAZOL FOR SYSTEMIC LUPUS ERYTHEMATOSUS, *BR MED J [CLIN RES]*, 291, 1170-1171 (1985)

0816 DAVIDSOHN I, WELLS BB, *CLINICAL DIAGNOSIS BY LABORATORY METHODS* 13TH ED, WB SAUNDERS COMPANY, PHILADELPHIA, PA (1962)

0817 DAVIDSON DGD, EASTHAM WN, ACUTE LIVER NECROSIS FOLLOW-ING OVERDOSE OF PARACETAMOL, *BR MED J [CLIN RES]*, 2, 497 (1966)

0818 DAVIDSON M, FEINLEIB M, CARBON DISULFIDE POISONING: A REVIEW, *AM HEART J*, 83, 100 (1972)

0819 DAVIDSON MB, BERNSTEIN JM, EFFECT OF NICOTINIC ACID ON GROWTH HORMONE INDUCED GLUCOSE INTOLERANCE AND LIPOLYSIS, *CLIN RES*, 20, 237 (1972)

0820 DAVIDSON MB, HOLZMAN GB, ROLE OF GROWTH HORMONE IN THE ALTERATION OF CARBOHYDRATE METABOLISM INDUCED BY ORAL CONTRACEPTIVE AGENTS, *J CLIN ENDOCRINOL METAB*, 36, 246 (1973)

0821 DAVIDSON RJL, PHENACETIN-INDUCED HEMOLYTIC ANEMIA, *J CLIN PATHOL*, 24, 537 (1971)

0822 DAVIES DL, LANT AF, MILLARD NR, SMITH AJ, WARD JW, WILSON GM, SOME ASPECTS OF THE CLINICAL PHARMACOLOGY OF BUMETANIDE, A NEW POTENT ORAL DIURETIC, *BR J PHARMACOL*, 47, 618 P (1973)

0823 DAVIES RO, GOMEZ HJ, IRVIN JD, ET AL, AN OVERVIEW OF THE CLINICAL PHARMACOLOGY OF ENALAPRIL, *BR J CLIN PHARMACOL*, 18 SUPPL 2, 215S-229S (1984)

0824 DAVIS JM, SPAIDE JK, HEIMWICH HE, EFFECTS OF TRANYLCYPROMINE AND L-CYSTEINE ON PLASMA AMINO ACIDS IN CONTROLS AND SCHIZOPHRENIC PATIENTS, *AM J CLIN NUTR*, 25, 302 (1972)

0825 DAVIS JRE, SELBY C, JEFFCOATE WJ, ORAL CONTRACEPTIVE AGENTS DO NOT AFFECT SERUM PROLACTIN IN NORMAL WOMEN, *CLIN ENDOCRINOL*, 20, 427-434 (1984)

0826 DAVIS JW, MCFIELD JR, PHILLIPS PE, GRAHAM BA, EFFECTS OF EXOGENOUS UREA, CREATININE AND GUANIDINOSUCCINIC ACID ON HUMAN PLATELET AGGREGATION, *BLOOD*, 39, 388 (1972)

0827 DAVIS M, EDDLESTON ALWF, WILLIAMS R, HYPERSENSITIVITY AND JAUNDICE DUE TO AZATHIOPRINE, *POSTGRAD MED J*, 56, 274-275 (1980)

0828 DAVIS NC, WEST CD, HO M, EFFECT OF AGING OF SERUM ON QUANTITATION OF COMPLEMENT COMPONENT C3, *CLIN CHEM*, 18, 1485 (1972)

0829 DAVIS P, UNDESIRABLE EFFECTS OF GOLD SALTS, *J RHEUMATOL*, 6, SUPPL 5, 18-24 (1979)

0830 DAVIS PJ, FACTORS AFFECTING THE DETERMINATION OF THE SERUM PROTEIN-BOUND IODINE, *AM J MED*, 40, 918 (1966)

0831 DAVIS PJ, HSU TH, BIANCHINE JR, MORGAN JP, EFFECTS OF A NEW HYPOLIPIDEMIC AGENT, MK-185, ON SERUM THYROXINE BINDING GLOBULIN (TBG), *J CLIN ENDOCRINOL METAB*, 34, 200 (1972)

0832 DAVIS RB, METABOLIC STUDIES IN CARCINOID SYNDROME: OBSERVATIONS ON USE OF ALPHA-METHYLDOPA, ISONICOTINIC ACID HYDRAZIDE, *METABOLISM*, 10, 1035 (1961)

0833 DAVIS VE, BROWN H, HUFF JA, CASHAW JL, THE ALTERATION OF SEROTONIN METABOLISM TO 5-HYDROXYTRYPTOPHOL BY ETHANOL INGESTION IN MAN, *J LAB CLIN MED*, 69, 132 (1967)

0834 DAY HJ, HOLMSEN H, LABORATORY TESTS OF PLATELET FUNCTION, *ANN CLIN LAB SCI*, 2, 63 (1972)

0835 DAY JL, METCALFE J, SIMPSON CN, ADRENERGIC MECHANISMS IN CONTROL OF PLASMA LIPID CONCENTRATIONS, *BR MED J [CLIN RES]*, 284, 1145-1148 (1982)

0836 DAY JL, SIMPSON N, METCALFE J, ET AL, METABOLIC CONSEQUENCES OF ATENOLOL AND PROPRANOLOL IN TREATMENT OF ESSENTIAL HYPERTENSION, *BR MED J [CLIN RES]*, 1, 77-80 (1979)

0837 DAYAN AD, LEWIS PD, IDOXURIDINE AND JAUNDICE, *LANCET*, 2, 1073 (1970)

0838 DAYMOND TJ, CHOLECYSTOGRAPHY AND RENAL FAILURE, *LANCET*, 2, 549 (1971)

0839 DE CHATELET LR, MCCALL CE, COOPER MR, SHIRLEY PS, INHIBITION OF LEUKOCYTE ACID PHOSPHATASE BY HEPARIN, *CLIN CHEM*, 18, 1532 (1972)

0840 DE GRAMONT A, RIOUX E, DROLET Y, ET AL, ERYTHROCYTE MEAN CORPUSCULAR VOLUME DURING CYTOTOXIC THERAPY AND THE RISK OF SECONDARY LEUKEMIA, *CANCER*, 55, 493-495 (1985)

0841 DE HAAN RM, SCHELLENBERG D, VANDEN BOSCH WD, MAILE MH, CLINDAMYCIN PALMITATE IN HEALTHY MEN: GENERAL TOLERANCE AND EFFECT ON STOOLS, *CURR THER RES*, 14, 81 (1972)

0842 DE LA HUERGA J, SHERRICK JC, MEASUREMENT OF OXYGEN SATURATION OF THE BLOOD, *ANN CLIN LAB SCI*, 1, 261 (1971)

0843 DE LANGE WE, VISSER JWE, DOORENBOS H, HORMONAL INFLUENCES ON THE CONCENTRATION OF TYROSINE IN BLOOD, *CLIN CHIM ACTA*, 42, 21 (1972)

0844 DE LEACY EA, McLEAY CD, EADIE MJ, ET AL, EFFECT OF SUBJECTS' SEX, AND INTAKE OF TOBACCO, ALCOHOL AND ORAL CONTRACEPTIVES ON PLASMA PHENYTOIN LEVELS, *BR J CLIN PHARMACOL*, 8, 33-36 (1979)

0845 DE PARSCAU L, FIELDING CJ, ABNORMAL PLASMA CHOLESTEROL METABOLISM IN CIGARETTE SMOKERS, *METABOLISM*, 35, 1070-1073 (1986)

0846 DE PINHO RA, GOLDBERG CS, LEFKOWITCH JH, AZATHIOPRINE AND THE LIVER. EVIDENCE FAVORING IDIOSYNCRATIC, MIXED CHOLESTATIC-HEPATOCELLULAR INJURY IN HUMANS, *GASTROENTEROLOGY*, 86, 162-165 (1984)

0847 DE RITIS L, UN CASO DI NEUTROPENIA DA CHLOROTHIAZIDE, *ARCISPED S ANNA FERRERA*, 16, 985 (1963)

0848 DE WOLFF FA, PETERS ACB, VAN KEMPEN GMJ, VALPROATE INDUCES URINARY D-GLUCARIC ACID EXCRETION, *LANCET*, 1, 843 (1981)

0849 DEADMAN NM, EVALUATION OF CHROMOGENS SUITABLE FOR OCCULT BLOOD IN FAECES, *CLIN CHIM ACTA*, 35, 273 (1971)

0850 DEAN RP, TALBERT RL, BLEEDING ASSOCIATED WITH CONCURRENT WARFARIN AND METRONIDAZOLE THERAPY, *DRUG INTELL CLIN PHARM*, 14, 864-866 (1980)

0851 DEBLASI S, AZIONE DEL FARGAN SULLA GLICEMIA E SULLA CURVA GLICEMIA DE CARICO NELL'UOMO E NEL CANE, *MINERVA ANESTESIOL*, 19, 66 (1953)

0852 DECHTIARUK W, CRAWFORD R, FRYE R, THEOPHYLLINE INTERFERENCE IN PHENOBARBITAL QUANTITATION, *CLIN CHEM*, 25, 2055 (1979)

0853 DEFELICE EA, MEHTA DJ, CAHN MM, LEVY EJ, EVALUATION OF SAFETY, SERUM LEVELS AND URINARY EXCRETION OF HETACILLIN IN NORMAL VOLUNTEERS, *TOXICOL APPL PHARMACOL*, 11, 20 (1967)

0854 DEFRONZO RA, BRAINE H, CALVIN OM, DAVIS PJ, WATER INTOXICATION IN MAN AFTER CYCLOPHOSPHAMIDE THERAPY, *ANN INTERN MED*, 78, 861 (1973)

0855 DEFRONZO RA, COLVIN OM, BRAINE H, DAVIS PJ, CYCLOPHOSPHAMIDE - INDUCED HYPONATREMIA, *CLIN RES*, 20, 891 (1972)

0856 DEFTOS LJ, PARTHEMORE J, ROOS B, EFFECT OF GASTRIN ON CALCITONIN (CT) SECRETION IN MAN, *CLIN RES*, 21, 489 (1973)

0857 DEGER G, EFFECT OF TERAZOSIN ON SERUM LIPIDS, *AM J MED*, 80 SUPPL 5 B, 82-85 (1986)

0858 DEGLIN JM, POLL K, QUINIDINE-INDUCED FEVER AND HEPATIC DYSFUNCTION, *DRUG INTELL CLIN PHARM*, 14, 216-217 (1980)

0859 DEGNBOL B, DORPH S, MARNER T, THE EFFECT OF DIFFERENT DIURETICS ON ELEVATED BLOOD PRESSURE AND SERUM POTASSIUM, *ACTA MED SCAND*, 193, 407 (1973)

0860 DEGROOT LJ, HOYE K, DEXAMETHASONE SUPPRESSION OF SERUM T_3 AND T_4, *J CLIN ENDOCRINOL METAB*, 42, 976-978 (1976)

0861 DEGUIA D, MENDLOWITZ M, VLACHAKIS ND, BERTANI LM, WOLF RL, RUSSO C, GITLOW SE, CP-19,106 - A NEW ORAL ANTIHYPERTENSIVE AGENT, *CLIN RES*, 20, 862 (1972)

0862 DEISSEROTH A, MORGANROTH J, WINOKRU S, QUINIDINE-INDUCED LIVER DISEASE, *ANN INTERN MED*, 77, 595 (1972)

0863 DEITRICK JE, WHEDON GD, SHORR E, EFFECTS OF IMMOBILIZATION UPON VARIOUS METABOLIC AND PHYSIOLOGIC FUNCTIONS OF NORMAL MAN, *AM J MED*, 4, 3 (1948)

0864 DEL ARBOL LR, MOREIRA V, MORENO A, ET AL, BRIDGING HEPATIC NECROSIS ASSOCIATED WITH CIMETIDINE, *AM J GASTROENTEROL*, 74, 267-269 (1980)

0865 DEL POZO E, BRUN DEL RE R, VARGA L, FRIESEN H, THE INHIBITION OF PROLACTIN SECRETION IN MAN BY CB-154 (2-BR-ALPHA-ERGOCRYPTINE), *J CLIN ENDOCRINOL METAB*, 35, 768 (1972)

0866 DELAGE JM, LEHNER-NETSCH G,, BRISSON J, THE CLASSICAL AND ALTERNATE PATHWAYS OF COMPLEMENT IN ORAL CONTRACEPTIVE USERS, *CONTRACEPTION*, 36, 627-632 (1987)

0867 DELEEUW PW, BIRKENHAGER WH, EFFECTS OF VERAPAMIL IN HYPERTENSIVE PATIENTS, *ACTA MED SCAND*, 215, SUPPL 681, 125-128 (1984)

0868 DELITALA G, DEVILLA L, ARATA L, OPIATE RECEPTORS AND ANTERIOR PITUITARY HORMONE SECRETION IN MAN, *ACTA ENDOCRINOL*, 97, 150-156 (1981)

0869 DELITALA G, DEVILLA L, LOTTI G, TSH AND PROLACTIN STIMULATION BY THE DECARBOXYLASE INHIBITOR BENSERAZIDE IN PRIMARY HYPOTHYROIDISM, *CLIN ENDOCRINOL*, 12, 313-316 (1980)

0870 DELITALA G, DEVILLA L, LOTTI G, DOMPERIDONE, AN EXTRACEREBRAL INHIBITOR OF DOPAMINE RECEPTORS, STIMULATES THYROTROPIN AND PROLACTIN RELEASE IN MAN, *J CLIN ENDOCRINOL METAB*, 50, 1127-1130 (1980)

0871 DELITALA G, DEVILLA L, PENDE A, ET AL, EFFECTS OF THE H_2-RECEPTOR ANTAGONIST RANITIDINE ON ANTERIOR HORMONE SECRETION IN MEN, *EUR J CLIN PHARMACOL*, 22, 207-211 (1982)

0872 DELLA JE, MAXSON WS, BREEN JL, METHOXYFLURANE HEPATITIS: TWO CASES FOLLOWING OBSTETRIC ANALGESIA, *INT J GYNAECOL OBSTET*, 21, 89-93 (1983)

0873 DELLAPORTAS DI, SHORVON SD, GALBRAITH AW, ET AL, CHRONIC TOXICITY IN EPILEPTIC PATIENTS RECEIVING SINGLE DRUG TREATMENT, *BR MED J [CLIN RES]*, 285, 409-410 (1982)

0874 DELPRE G, GRINBLAT J, KADISH U, ET AL, IMMUNOLOGICAL STUDIES IN A CASE OF HEPATITIS FOLLOWING METHYLDOPA ADMINISTRATION, *AM J MED SCI*, 277, 207-213 (1979)

0875 DEMANET JC, BONNYNS M, BLEIBERG H, STEVENS-ROCMAN C, COMA DUE TO WATER INTOXICATION IN BEER DRINKERS, *LANCET*, 2, 1115 (1971)

0876 DEMETRIOU JA, AUSTIN FG, A RAPID COMPETITIVE PROTEIN-BINDING ASSAY FOR PLASMA PROGESTERONE, *CLIN CHIM ACTA*, 33, 21 (1971)

0877 DEMETRIOU JA, DREWES PA, GIN JB, ENZYMES (P 815) *IN CLINICAL CHEMISTRY: PRINCIPLES AND TECHNICS*, RJ HENRY, DC CANNON, JW WINKELMAN, EDS. 2ND ED HARPER AND ROW, HAGERSTOWN, MD (1974)

0878 DEMPSEY ME, THE EFFECT OF HYPOGLYCEMIC AGENTS ON CHOLESTEROL BIOSYNTHESIS, *ADV EXP MED BIOL*, 4, 511 (1968)

0879 DENAYER P, MALVAUX P, OSTYN M, SERUM FREE THYROXINE AND BINDING PROTEINS AFTER EXERCISE, *J CLIN ENDOCRINOL METAB*, 28, 714 (1968)

0880 DENBENSTEN L, REYNA RH, CONNOR WE, STEGINK LD, THE DIFFERENT EFFECTS ON THE SERUM LIPIDS AND FECAL STEROIDS OF HIGH CARBOHYDRATES GIVEN ORALLY, *J CLIN INVEST*, 52, 1384 (1973)

0881 DENNIS PM, ERICKSEN CM, INTERFERENCE OF METRONIDAZOLE (FLAGYL®) WITH SERUM ASPARTATE AMINOTRANSFERASE (AST) ASSAYS, *MED J AUST*, 2, 343-344 (1980)

0882 DENNY A, ADAMS J, MILLER TC, OXACILLIN AND THE LIVER, *ANN INTERN MED*, 90, 277 (1979)

0883 DENSON WK, INGRAM GIC, ANTIGEN/BIOLOGICAL - ACTIVITY RATIO FOR FACTOR VIII, *LANCET*, 1, 157 (1973)

0884 DENT CE, SOME PROBLEMS OF HYPERPARATHYROIDISM, *BR MED J [CLIN RES]*, 2, 1419 (1962)

0885 DEQUEKER J, MARDJUARDI A, TREATMENT OF RHEUMATOID ARTHRITIS WITH FLURBIPROFEN: A COMPARISON WITH ENTERIC -COATED ASPIRIN, *CURR MED RES OPIN*, 7, 418-428 (1981)

0886 DERINGER PM, MANIATIS A, CHLORPHENIRAMINE-INDUCED BONE MARROW SUPPRESSION, *LANCET*, 1, 432 (1976)

0887 DESCHAMPS JP, LAHRICHI M, BIOLOGICAL VALUES IN THE CHILD AND ADOLESCENT IN, *REFERENCE VALUES IN HUMAN CHEMISTRY*, G SIEST, ED. KARGER, BASEL 1973, 109

0888 DESSYPRIS A, INFLUENCE OF CHOLESTEROL, HEMOLYSIS AND METALS ON TESTOSTERONE PROTEIN BINDING, *J STEROID BIOCHEM*, 1, 185 (1970)

0889 DESSYPRIS EN, REDLINE S, HARRIS JW, ET AL, DIPHENYLHYDANTOIN-INDUCED PURE RED CELL APLASIA, *BLOOD*, 65, 789-794 (1985)

0890 DEVGUN MS, DUNBAR JA, HAGART J, ET AL, EFFECTS OF ACUTE AND VARYING AMOUNTS OF ALCOHOL CONSUMPTION ON ALKALINE PHOSPHATASE, ASPARTATE TRANSAMINASE AND GAMMA-GLUTAMYLTRANSFERASE, *ALCOHOL CLIN EXP RES*, 9, 235-237 (1985)

0891 DEVILLA L, PENDE A, MORGANO A, ET AL, MORPHINE-INDUCED TSH RELEASE IN NORMAL AND HYPOTHYROID SUBJECTS, *NEUROENDOCRINOLOGY*, 40, 303-308 (1985)

0892 DEVINE JE, DRUG-PROTEIN BINDING INTERFERENCES CAUSED BY THE PLASTICIZER TBEP, *CLIN BIOCHEM*, 17, 345-347 (1984)

0893 DEVITO GA JR., TRANSIENT ELEVATION OF ALKALINE PHOSPHATASE POSSIBLY RELATED TO TRIMETHOPRIM-SULFAMETHOXAZOLE THERAPY, *J PEDIATR*, 100, 998-999 (1982)

0894 DEYKIN D, WARFARIN THERAPY, *N ENGL J MED*, 283, 801 (1970)

0895 DEYKIN D, JANSON P, MCMAHON L, ETHANOL POTENTIATION OF ASPIRIN - INDUCED PROLONGATION OF THE BLEEDING TIME, *N ENGL J MED*, 306, 852-854 (1982)

0896 DHILLON S, RICHENS A, VALPROIC ACID AND DIAZEPAM INTERACTION IN VIVO, *BR J CLIN PHARMACOL*, 13, 553-560 (1982)

0897 DI SANT-AGNESE PA, PERSONAL COMMUNICATION

0898 DIAMOND JR, CHEUNG JY, FANG LST, NIFEDIPINE-INDUCED RENAL DYSFUNCTION. ALTERATIONS IN RENAL HEMODYNAMICS, *AM J MED*, 77, 905-909 (1984)

0899 DIAZ S, CROXATTO HB, PAVEZ M, CLINICAL CHEMISTRY IN WOMEN TREATED WITH SIX LEVONORGESTREL COVERED RODS OR WITH A COPPER IUD, *CONTRACEPTION*, 31, 321-330 (1985)

0900 DICKES R, SCHENKER V, DEUTSCH C, SERIAL LIVER FUNCTION AND BLOOD STUDIES IN PATIENTS RECEIVING CHLORPROMAZINE, *N ENGL J MED*, 256, 1 (1957)

0901 DICKINSON DS, BAILEY WC, HIRSCHOWITZ BI, ET AL, RISK FACTORS FOR ISONIAZID (INH)-INDUCED LIVER DYSFUNCTION, *J CLIN GASTROENTEROL*, 3, 271-279 (1981)

0902 DIEM K, SCIENTIFIC TABLES 6TH EDITION, *DOCUMENTA GEIGY*, GEIGY PHARMACEUTICAL CO, MANCHESTER UK, 1962

0903 DIETRICHSON O, CHRONIC ACTIVE HEPATITIS: AETIOLOGICAL CONSIDERATIONS BASED ON CLINICAL AND SEROLOGICAL STUDIES, *SCAND J GASTROENTEROL*, 10, 617-624 (1975)

0904 DIGIORGIO J, NONPROTEIN NITROGEN CONSTITUENTS (P503) *IN CLINICAL CHEMISTRY: PRINCIPLES AND TECHNICS*, RJ HENRY, DC CANNON, JW WINKELMAN, EDS. 2ND ED HARPER AND ROW, HAGERSTOWN, MD (1974)

0905 DIJKMANS BAC, VAN RIJTHOVEN AWAM, GOEI THE HS, ET AL, EFFECT OF CYCLOSPORINE ON SERUM CREATININE IN PATIENTS WITH RHEUMATOID ARTHRITIS, *EUR J CLIN PHARMACOL*, 31, 541-545 (1987)

0906 DILL WA, GLAZKO AJ, FLUOROMETRIC ASSAY OF DIPHENYLHYDANTOIN IN PLASMA OR WHOLE BLOOD, *CLIN CHEM*, 18, 675 (1972)

0907 DILLON HC JR., BRIDGES RA, NULL WA, BENTLY HP, ERYTHROPOIETIC CHANGES ASSOCIATED WITH CHLORAMPHENICOL THERAPY, *ALA J MED SCI*, 1, 368 (1964)

0908 DIMSDALE JE, HARTLEY LH, RUSKIN J, ET AL, EFFECT OF BETA BLOCKADE ON PLASMA CATECHOLAMINE LEVELS DURING PSYCHOLOGICAL AND EXERCISE STRESS, *AM J CARDIOL*, 54, 182-185 (1984)

0909 DINIZ RS, ABRAHAM GJS, AHMED SS, ANTI-GASTRIC ACID SECRETORY EFFECT OF PHENFORMIN, *EUR J PHARMACOL*, 19, 389 (1972)

0910 DINSCOY HP, SAELINGER DA, HALOPERIDOL-INDUCED CHRONIC CHOLESTATIC LIVER DISEASE, *GASTROENTEROLOGY*, 83, 694-700 (1982)

0911 DINSMORE WW, O'HARA MD, CALLENDER ME, POSTANESTHETIC CARBIMAZOLE JAUNDICE, *N ENGL J MED*, 309, 438 (1983)

0912 DIPPE S, JONES R, MARIJUANA, GLUCOSE AND INSULIN, *CLIN RES*, 20, 237 (1972)

0913 DIVISION OF CLINICAL BIOCHEMISTRY, SPECIAL TESTS INSTRUCTION BOOKLET, *INSTITUTE OF MEDICAL AND VETERINARY SCIENCES, ADELAIDE, AUSTRALIA*

0914 DLUHY RG, AXELROD L, UNDERWOOD RH, WILLIAMS GH, STUDIES OF THE CONTROL OF PLASMA ALDOSTERONE CONCENTRATION IN NORMAL MAN: II EFFECT OF DIETARY POTASSIUM, *J CLIN INVEST*, 51, 1950 (1972)

0915 DOAR JWH, WYNN V, SERUM LIPID LEVELS DURING ORAL CONTRACEPTIVE AND GLUCOCORTICOID ADMINISTRATION, *J CLIN PATHOL*, 55 (1969)

0916 DOBKIN AB, LEVY AA, BLOOD SERUM FLUORIDE LEVELS WITH METHOXYFLURANE ANESTHESIA, *CAN J ANAESTH*, 20, 81 (1973)

0917 DOBSON HM, MUIR MM, HUME R, THE EFFECT OF ASCORBIC ACID ON THE SEASONAL VARIATIONS IN SERUM CHOLESTEROL LEVELS, *SCOTT MED J*, 29, 176-182 (1984)

0918 DODDS WN, DAVIDSON RJL, THROMBOCYTOPENIA DUE TO SLOW RELEASE OXPRENOLOL, *LANCET*, 2, 683 (1978)

0919 DODION L, AMBROES Y, LAMAIRE N, A COMPARISON OF THE PHARMACOKINETICS AND DIURETIC EFFECTS OF TWO LOOP DIURETICS, TORASEMIDE AND FUROSEMIDE, IN NORMAL VOLUNTEERS, *EUR J CLIN PHARMACOL*, 31 SUPPL, 21-27 (1986)

0920 DOERING W, QUINIDINE-DIGOXIN INTERACTION, *N ENGL J MED*, 301, 400-404 (1979)

0921 DOETSCH K, GADSDEN RH, DETERMINATION OF TOTAL URINARY PROTEIN, COMBINING LOWRY SENSITIVITY AND BIURET SPECIFICITY, *CLIN CHEM*, 19, 1170 (1973)

0922 DOIG A, GRAY W, MUNRO JF, STREET HV, WATER INTOXICATION IN BEER DRINKERS, *LANCET*, 2, 1318 (1971)

0923 DOLLERY CT, METHYLDOPA IN HYPERTENSION, *AM HEART J*, 65, 139 (1963)

0924 DOLLERY CT, DUNCAN H, SCHUMER B, HYPERURICAEMIA RELATED TO TREATMENT OF HYPERTENSION, *BR MED J [CLIN RES]*, 2, 832 (1960)

0925 DONAHOE, JF, POWERS RJ, BIOCHEMICAL ABNORMALITIES WITH XYLITOL, *N ENGL J MED*, 282, 690 (1970)

0926 DONALD RA, ESPINER EA, BEAVEN DW, MEASUREMENT OF PITUITARY HORMONES: CLINICAL APPLICATIONS. 2. CORTICOTROPIN, *N Z MED J*, 75, 342 (1972)

0927 DONALDSON CL, HULLEY SB, VOGEL JM, HATTNER RS, BAYERS JH, MCMILLAN DE, EFFECT OF PROLONGED BED REST ON BONE MINERAL, *METABOLISM*, 19, 1071 (1970)

0928 DONALDSON GWK, GRAHAM JG, APLASTIC ANAEMIA FOLLOWING THE ADMINISTRATION OF TEGRETOL®, *BR J CLIN PRACT*, 19, 699 (1965)

0929 DONKER JM, NEPHROTOXICITY OF ANGIOTENSIN CONVERTING ENZYME INHIBITION, *KIDNEY INT*, 31, SUPPL 20, S132-S137 (1987)

0930 DONLAN CJ JR., FORKER AD, CARDIAC TAMPONADE IN PROCAINAMIDE INDUCED LUPUS ERYTHEMATOSUS, *CHEST*, 61, 685 (1972)

0931 DORAN C, KENNY S, LEONARD PJ, CHANGES IN ACID BASE COMPONENTS OF BLOOD ON STORAGE AT ROOM TEMPERATURE, *IR J MED SCI*, 3, 409 (1970)

0932 DORAN GR, WILKINSON JH, SERUM CREATINE KINASE AND ADENYLATE KINASE IN THYROID DISEASE, *CLIN CHIM ACTA*, 35, 115 (1971)

0933 DORFMANN H, KAHN MF, DE SEZE S, IATROGENIC LUPUS: CURRENT STATUS OF THE QUESTION, *NOUV PRESSE MED*, 1, 2907 (1972)

0934 DORNHORST A, OUYANG A, EFFECT OF ALCOHOL ON GLUCOSE TOLERANCE, *LANCET*, 2, 957 (1971)

0935 DORNHORST A, POWELL SH, PENSKY J, AGGRAVATION BY PROPRANOLOL OF HYPERGLYCEMIC EFFECT OF HYDROCHLOROTHIAZIDE IN TYPE II DIABETICS WITHOUT ALTERATION OF INSULIN SECRETION, *LANCET*, 1, 123-126 (1985)

0936 DORSKY DI, CRUMPACKER CS, DRUGS FIVE YEARS LATER: ACYCLOVIR, *ANN INTERN MED*, 107, 859-874 (1987)

0937 DORWART WV, METRONIDAZOLE INTERFERENCE IN HEXOKINASE GLUCOSE DETERMINATIONS, *CLIN CHEM*, 29, 995 (1983)

0938 DOSS M, NAWROCKI P, SCHMIDT A, STROHMEYER G, EGBRING R, SCHIMPFF G, DOLLE W, KOR, THE INFLUENCE OF DIET, GLYCINE AND ALCOHOL ON PORPHYRINURIA IN CHRONIC HEPATIC PORPHYRIA, *GERM MED*, 1, 85 (1971)

0939 DOUGLASS JF, ATHEROSCLEROSIS IN *ANNUAL REPORTS IN MEDICINAL CHEMISTRY 1970*, CK CAIN, ED. ACADEMIC PRESS, NEW YORK, NY (1971)

0940 DOUMAS BT, PERRY B, JENDRZEJCZAK B, ET AL, MEASUREMENT OF DIRECT BILIRUBIN BY USE OF BILIRUBIN OXIDASE, *CLIN CHEM*, 33, 1349-1353 (1987)

0941 DOUMAS BT, PERRY BW, SASSE EA, STRAUMFJORD JV JR., STANDARDIZATION IN BILIRUBIN ASSAYS: EVALUATION OF SELECTED METHODS AND STABILITY OF BILIRUBIN SOLUTIONS, *CLIN CHEM*, 19, 984 (1973)

0942 DOUSTE-BLAZY P, MONTASTRUC JL, BONNET B, ET AL, INFLUENCE OF AMIODARONE ON PLASMA AND URINE DIGOXIN CONCENTRATIONS, *LANCET*, 1, 905 (1984)

0943 DOUWES FR, CLINICAL VALUE OF NBT TEST, *N ENGL J MED*, 287, 822 (1972)

0944 DOW CHEMICAL COMPANY, MANUFACTURER'S LITERATURE (1971), *PO BOX 1656 INDIANAPOLIS IND*, 46206

0945 DOWLING HF, LEPPER MH, HEPATIC REACTIONS TO TETRACY-CLINE, *J AM MED ASSOC*, 188, 307 (1964)

0946 DOWSETT M, MURRAY RML, PITT P, ET AL, BIOCHEMICAL BASIS FOR THE ANTAGONISM BETWEEN AMINOGLUTETHIMIDE AND DANAZOL IN THE ENDOCRINE TREATMENT OF BREAST CANCER, *ANN CLIN BIOCHEM*, 23, 277-284 (1986)

0947 DOYLE DR, McCURLEY TL, SERGENT JS, FATAL POLYMYOSITIS IN D-PENICILLAMINE-TREATED RHEUMATOID ARTHRITIS, *ANN INTERN MED*, 98, 327-330 (1983)

0948 DRASH A, WOLFF FW, DRUG THERAPY IN LEUCINE-SENSITIVE HYPOGLYCEMIA, *METABOLISM*, 13, 487 (1964)

0949 DREHNER DM, WIANS FH JR., STAPLES N, CALCIUM DOES NOT INTERFERE WITH MEASUREMENT OF MAGNESIUM IN THE DU PONT ACA-III, *CLIN CHEM*, 33, 1936 (1987)

0950 DREIFUSS FE, LANGER DH, HEPATIC CONSIDERATIONS IN THE USE OF ANTIEPILEPTIC DRUGS, *EPILEPSIA*, 28, SUPPL 2, S23-S29 (1987)

0951 DREW SI, CARTER BM, NATHANSON DS, ET AL, LEVAMISOLE-ASSOCIATED NEUTROPENIA AND AUTOIMMUNE GRANULOCYTOX-INS, *ANN RHEUM DIS*, 39, 59-63 (1980)

0952 DREWES PA, DEMETRIOU JA, PILEGGI VJ, MEASUREMENT OF URINARY ALDOSTERONE BY A SIMPLIFIED RADIOIMMUNOASSAY PROCEDURE, *CLIN BIOCHEM*, 6, 88 (1973)

0953 DRIFE JO, THE EFFECTS OF DRUGS ON SPERM, *DRUGS*, 53, 610-622 (1987)

0954 DRISCOLL MJ, SPECIFIC PROTEIN DETERMINATIONS USING IMMU-NOELECTROPHORESIS, *ANN CLIN BIOCHEM*, 10, 4 (1973)

0955 DRISKELL JA, GEDERS JM, URBAN MC, VITAMIN B_6 STATUS OF YOUNG MEN, WOMEN, AND WOMEN USING ORAL CONTRACEPTIVES, *J LAB CLIN MED*, 87, 813-821 (1976)

0956 DRUM DE, GOLDMAN PA, JANKOWSKI CB, ET AL, ELEVATION OF SERUM URIC ACID AS A CLUE TO ALCOHOL ABUSE, *ARCH INTERN MED*, 141, 477-479 (1981)

0957 DRUMMOND KN, MICHAEL AF, SPECIFICITY OF THE INHIBITION OF TUBULAR PHOSPHATE REABSORPTION BY CERTAIN AMINO ACIDS, *NATURE*, 201, 1333 (1964)

0958 DRUTZ DJ, FAN JH, TAI TY, CHENG JT, HSIEH WC, HYPOKALEMIC RHABDOMYOLYSIS AND MYOGLOBINURIA FOLLOWING AMPHOTERICIN B THERAPY, *J AM MED ASSOC*, 211, 824 (1970)

0959 DU SOUICH P, PISSON C, PEDNEAULT L, ET AL, EFFECT OF AMINOG-LYCOSIDES ON THE DISPOSITION OF THYROID HORMONES AND THYROGLOBULIN, *CLIN PHARMACOL THER*, 38, 686-691 (1985)

0960 DUARTE PA, CHOW CC, SIMMONS F, RUSKIN J, FATAL HEPATITIS ASSOCIATED WITH KETOCONAZOLE THERAPY, *ARCH INTERN MED*, 144, 1069-1070 (1984)

0961 DUBICK MA, CONTEAS CN, BILLY HT, ET AL, RAISED SERUM CONCENTRATIONS OF PANCREATIC ENZYMES IN CIGARETTE SMOKERS, *GUT*, 28, 330-335 (1987)

0962 DUBOIS EL, CURRENT STATUS OF THE LE CELL TEST, *SEMIN ARTHRITIS RHEUM*, 1, 97 (1971)

0963 DUBOIS EL, TALLMAN E, WONKA RA, CHLORPROMAZINE - INDUCED SYSTEMIC LUPUS ERYTHEMATOSUS, *J AM MED ASSOC*, 221, 595 (1972)

0964 DUC TV, EMIT TESTS FOR DRUGS OF ABUSE: INTERFERENCE BY LIQUID SOAP PREPARATIONS, *CLIN CHEM*, 31, 658-659 (1985)

0965 DUCKWORTH WC, SOLOMON SS, KITABACHI AE, EFFECT OF CHRONIC SULFONYLUREA THERAPY ON PLASMA INSULIN AND PROINSULIN LEVELS, *J CLIN ENDOCRINOL METAB*, 35, 585 (1972)

0966 DUHRING JL, McKEAN HE, GREENE JW JR., DIURNAL VARIATION OF ESTRIOL EXCRETION IN HUMAN PREGNANCY, *AM J OBSTET GYNECOL*, 115, 875 (1973)

0967 DUJOVNE CA, DECOURSEY S, KREHBIEL P, ET AL, SERUM LIPIDS IN NORMO - AND HYPERLIPIDEMICS AFTER METHYLDOPA AND PROPRANOLOL, *CLIN PHARMACOL THER*, 36, 157-162 (1984)

0968 DUKES GE JR., SANDERS SW, RUSSO J JR., ET AL, TRANSAMINASE ELEVATIONS IN PATIENTS RECEIVING BOVINE OR PORCINE HEPARIN, *ANN INTERN MED*, 100, 646-650 (1984)

0969 DUMA RJ, SUMMARY OF COMPARATIVE CLINICAL STUDIES OF CEF-TIZOXIME AND CEFAMANDOLE, CEFAZOLIN AND TOBRAMYCIN, *J ANTIMICROB CHEMOTHER*, 10, SUPPL C, 303-309 (1982)

0970 DUNCAN A, TRACY RP, VLIESTRA RL, INFLUENCE OF PROPRA-NOLOL ON PLATELET FUNCTION, *CLIN CHEM*, 26, 1039 (1980)

0971 DUNCAN DA, COLISTIN TOXICITY: NEUROMUSCULAR AND RENAL MANIFESTATIONS, *MINN MED*, 56, 31 (1973)

0972 DUNDEE JW, EFFECT OF THIOPENTONE ON BLOOD SUGAR AND GLUCOSE TOLERANCE, *BR J PHARMACOL*, 11, 458 (1956)

0973 DUNDEE JW, FEE JPH, MOORE J, ET AL, CHANGES IN SERUM ENZYME LEVELS FOLLOWING KETAMINE INFUSIONS, *ANESTHE-SIA*, 35, 12-16 (1980)

0974 DUNEA G, FREEDMAN P, PHENOLSULFONPHTHALEIN EXCRETION TEST, *J AM MED ASSOC*, 204, 621 (1968)

0975 DURAN M, KETTING D, WADMAN SK, TRIJYBELS JM, BAKKEREN JA, WAELKENS JJ, PROPIONIC ACID, AN ARTIFACT WHICH CAN LEAVE METHYLMALONIC ACIDEMIA UNDISCOVERED, *CLIN CHIM ACTA*, 49, 177 (1973)

0976 DURHAM SR, BIGNELL AHC, WISE R, INTERFERENCE OF CEFOXITIN IN THE CREATININE ESTIMATION AND ITS CLINICAL RELE-VANCE, *J CLIN PATHOL*, 32, 1148-1151 (1979)

0977 DUSTERDIECK G, MCELWEE G, ESTIMATION OF ANGIOTENSIN II CONCENTRATION IN HUMAN PLASMA BY RADIOIMMUNOAS-SAY, *EUR J CLIN INVEST*, 2, 32 (1971)

0978 DUTT MK, MOODY P, NORTHFIELD TC, EFFECT OF CIMETIDINE ON RENAL FUNCTION IN MAN, *BR J CLIN PHARMACOL*, 12, 47-50 (1981)

0979 DUTTERA MJ, *PERSONAL COMMUNICATIONS*

0980 DUTTERA MJ, CAROLLA RL, CALLELLI JF, GULLION DS, KEIM DE, HENDERSON ES, HEMATURIA AND CRYSTALLURIA AFTER HIGH-DOSE 6-MERCAPTOPURINE ADMINISTRATION, *N ENGL J MED*, 287, 292 (1972)

0981 DUVERNOY WFC, POSITIVE PHENTOLAMINE TEST IN HYPERTEN-SION INDUCED BY A NASAL DECONGESTANT, *N ENGL J MED*, 280, 877 (1969)

0982 DVORAK K, BLAZKOVA E, ACUTE THROMBOCYTOPENIC PURPURA AFTER PHENYLBUTAZONE, *VNITR LEK*, 11, 1000 (1965)

0983 DWORKIN BM, ROSENTHAL WS, GORDON GG, ET AL, DIMINISHED BLOOD SELENIUM LEVELS IN ALCOHOLICS, *ALCOHOL CLIN EXP RES*, 8, 535-538 (1984)

0984 DYCK WP, HIGHTOWER NC, JANOWITZ HD, EFFECT OF ACETAZO-LAMIDE ON HUMAN PANCREATIC SECRETION, *GASTROENTER-OLOGY*, 62, 547 (1972)

0985 DYMLING J-F, JEPPSSON S, RANNEVIK G, EFFECT OF DANAZOL ON THYROID FUNCTION IN POST MENOPAUSAL WOMEN, *ACTA OBSTET GYNECOL SCAND*, SUPPL 123, 137-139 (1984)

0986 EASTELL R, EDMONDS CJ, HYPONATRAEMIA ASSOCIATED WITH TRIMETHOPRIM AND A DIURETIC, *BR MED J [CLIN RES]*, 289, 1658-1659 (1984)

0987 EASTHAM RD, *BIOCHEMICAL VALUES IN CLINICAL MEDICINE* 4TH ED, WILLIAMS AND WILKINS, BALTIMORE, MD (1971)

0988 EASTHAM RD, *LAB GUIDE TO CLINICAL DIAGNOSIS*, WILLIAMS AND WILKINS, BALTIMORE, MD (1970)

0989 EASTON JD, POTENTIAL HAZARDS OF HYDANTOIN USE, *ANN INTERN MED*, 77, 998 (1972)

0990 EASTWOOD MA, BRYDON WG, ANDERSON DMW, THE EFFECTS OF DIETARY GUM TRAGACANTH IN MAN, *TOXICOL LETT*, 21, 73-81 (1984)

0991 ECKSTEIN P, WHITBY M, FOTHERBY K, ET AL, CLINICAL AND LABO-RATORY FINDINGS IN A TRIAL OF NORGESTREL, A LOW-DOSE PROGESTOGEN, *BR MED J [CLIN RES]*, 3, 195 (1972)

0992 EDELMAN J, DAVIS P, OWEN ET, PREVALENCE OF EOSINOPHILIA DURING GOLD THERAPY FOR RHEUMATOID ARTHRITIS, *J RHEUMATOL*, 10, 121-123 (1983)

0993 EDELMAN J, DONNELLY R, GRAHAM DN, ET AL, LIVER DYSFUNC-TION ASSOCIATED WITH GOLD THERAPY FOR RHEUMATOID ARTHRITIS, *J RHEUMATOL*, 10, 510-511 (1983)

0994 EDITORIAL, *GENEESK NED T*, 109, 2046 (1965)

0995 EDITORIAL, BLOOD AND ALCOHOL, *LANCET*, 1, 397 (1983)

0996 EDITORIAL, HEMOLYTIC ANEMIA CAUSED BY PENICILLIN, *N ENGL J MED*, 274, 222 (1966)

0997 EDMONDSON RPS, THOMAS R, HILTON PJ, PATRICK J, JONES NF, LEUKOCYTE ELECTROLYTES IN CARDIAC AND NONCARDIAC PATIENTS RECEIVING DIURETICS, *LANCET*, 1, 12 (1974)

0998 EDWARDS MS, CURTIS JR, DECREASED ANTICOAGULANT TOLER-ANCE WITH OXYMETHOLONE, *LANCET*, 2, 221 (1971)

0999 EGILMEZ A, DOBKIN AB, ENFLURANE (ETHRANE®, COMPOUND 347) IN MAN, *ANESTHESIA*, 27, 171 (1972)

1000 EHRLICH EN, RECIPROCAL VARIATIONS IN URINARY CORTISOL AND ALDOSTERONE IN RESPONSE TO INCREASED SALT INTAKE IN HUMANS, *J CLIN ENDOCRINOL METAB*, 26, 1160 (1966)

1001 EICHHORN JH, HEDLEY-WHYTE J, STEINMAN TI, ET AL, RENAL FAIL-URE FOLLOWING ENFLURANE ANESTHESIA, *ANESTHESIOL-OGY*, 45, 557-560 (1976)

1002 EIMER M, CARTER BL, ELEVATED SERUM CARBAMAZEPINE CON-CENTRATIONS FOLLOWING DILTIAZEM INITIATION, *DRUG INTELL CLIN PHARM*, 21, 340-342 (1987)

1003 EINHORN N, ACUTE LEUKEMIA AFTER CHEMOTHERAPY (MELPHALAN), *CANCER*, 41, 444-447 (1978)

1004 EISEN ME, HAMMOND EC, THE EFFECT OF SMOKING ON PACKED CELL VOLUME, RED BLOOD CELL COUNTS, HAEMOGLOBIN AND PLATELET COUNTS, *CAN MED ASSOC J*, 75, 520 (1956)

1005 EISENBERY AB, MATHEW R, KIECHLE FL, MANNITOL INTERFERENCE IN AN AUTOMATED SERUM PHOSPHATE ASSAY, *CLIN CHEM*, 33, 2308-2309 (1987)

1006 EISENHOFER G, GOLDSTEIN DS, STULL R, ET AL, SIMULTANEOUS LIQUID-CHROMATOGRAPHIC DETERMINATION OF 3,4-DIHYDROXYPHENYLGLYCOL, CATECHOLAMINES, AND 3,4-DIHYDROXYPHENYLALANINE IN PLASMA, AND THEIR RESPONSES TO INHIBITION OF MONOAMINE OXIDASE, *CLIN CHEM*, 32, 2030-2033 (1986)

1007 EISENHOFER G, JOHNSON RH, EFFECT OF ETHANOL INGESTION ON PLASMA VASOPRESSIN AND WATER BALANCE IN HUMANS, *AM J PHYSIOL*, 242, R522-R527 (1982)

1008 EISNER EV, CARR RM, MACKINNEY AA, QUINIDINE-INDUCED AGRAN-ULOCYTOSIS, *J AM MED ASSOC*, 238, 884-886 (1977)

1009 EISNER EV, SHAHIDI NT, IMMUNE THROMBOCYTOPENIA DUE TO A DRUG METABOLITE, *N ENGL J MED*, 287, 376 (1972)

1010 EKELUND LG, EKLUND B, KAIJSER L, TIME COURSE FOR THE CHANGE IN HEMOGLOBIN CONCENTRATION WITH CHANGE IN POSTURE, *ACTA MED SCAND*, 190, 335 (1971)

1011 EKMAN R, FEX G, JOHANNSON BG, ET AL, CHANGES IN PLASMA HIGH-DENSITY LIPOPROTEINS AND LIPOLYTIC ENZYMES AFTER LONG-TERM, HEAVY ETHANOL CONSUMPTION, *SCAND J CLIN LAB INVEST*, 41, 709-715 (1981)

1012 EL MATRI A, LARABI MS, KECHRID C, ET AL, FATAL BONE MARROW SUPPRESSION ASSOCIATED WITH CAPTOPRIL, *BR MED J [CLIN RES]*, 283, 277-278 (1981)

1013 EL-DORRY HFA, MEDINA H, BACILA M, INTERFERENCE OF PHENO-THIAZINE COMPOUNDS IN THE COLORIMETRIC DETERMINA-TION OF INORGANIC PHOSPHATE, *ANAL BIOCHEM*, 47, 329 (1972)

1014 EL-GHOBAREY AF, CAPELL HA, LEVAMISOLE-INDUCED THROMBO-CYTOPENIA, *BR MED J [CLIN RES]*, 2, 555-556 (1977)

1015 EL-KHODARY AZ, BALL MF, OWEISS IM, CANARY JJ, INSULIN SECRE-TION AND BODY COMPOSITION IN OBESITY, *METABOLISM*, 21, 641 (1972)

1016 ELDER GH, GRAY CH, NICHOLSON DC, THE PORPHYRIAS: A REVIEW, *J CLIN PATHOL*, 25, 1013 (1972)

1017 ELDER MG, SERUM GLUCOSE-6-PHOSPHATASE LEVELS IN NORMAL PREGNANCY, *J OBSTET GYNAECOL BR COMM*, 80, 109 (1973)

1018 ELDRIDGE JC, STRANDHOY J, BUCKALEW VM JR, ENDOCRINO-LOGIC EFFECTS OF ANTIHYPERTENSIVE THERAPY WITH GUANABENZ OR HYDROCHLOROTHIAZIDE, *J CARDIOVASC PHARMACOL*, 6 SUPPL 5, S776-S780 (1984)

1019 ELI LILLY AND COMPANY, MANUFACTURER'S LITERATURE ON DYMELOR®, LORIDINE, KEFLIN®, GLUCAGON, *307 EAST MCCARTHY, INDIANAPOLIS, IN*, 46206

1020 ELIAS AN, GWINUP G, EFFECTS OF SOME CLINICALLY ENCOUN-TERED DRUGS ON STEROID SYNTHESIS AND DEGRADATION, *METABOLISM*, 29, 582-595 (1986)

1021 ELIASSON K, LINS L-E, ROSSNER S, SERUM LIPOPROTEIN CHANGES DURING ATENOLOL TREATMENT OF ESSENTIAL HYPERTEN-SION, *EUR J CLIN PHARMACOL*, 20, 335-338 (1981)

1022 ELKING MP, KABAT HF, DRUG INDUCED MODIFICATIONS OF LABO-RATORY TEST VALUES, *AM J HOSP PHARM*, 25, 485 (1968)

1023 ELKINGTON SG, HEPATIC INJURY CAUSED BY L-ALPHA-METHYLDOPA, *CIRCULATION*, 40, 589 (1969)

1024 ELLIN RI, GROFF WA, KAMINSKIS A, AN ERROR PRODUCING INTER-ACTION IN AN AUTOMATED METHOD FOR MEASURING CHOLIN-ESTERASE ACTIVITY IN BLOOD, *CLIN CHEM*, 18, 1000 (1972)

1025 ELLING H, KIILERICH, SABRO J, ET AL, INFLUENCE OF A NONSTER-OID ANTIRHEUMATIC DRUG ON SERUM AND URINARY ZINC IN HEALTHY VOLUNTEERS, *SCAND J RHEUMATOL*, 9, 161-163 (1980)

1026 ELLIOTT HC JR., MURDAUGH HV JR., EFFECTS OF ACETYLSALI-CYLIC ACID ON EXCRETION OF ENDOGENOUS METABOLITES BY MAN, *PROC SOC EXP BIOL MED*, 109, 333 (1962)

1027 ELLIOTT HC, CASEY AE, EXPERIENCE WITH A SIMPLE SCREENING TEST FOR SEROTONIN, *SOUTH MED J*, 51, 836 (1958)

1028 ELLIS NF, MACGILLIVRAY MH, VOORHESS ML, EFFECT OF CLONIDINE ON PLASMA CORTISOL CONCENTRATIONS, *CLIN PHARMACOL THER*, 39, 660-663 (1986)

1029 ELLRODT AG, MURATA GH, RIEDINGER MS, ET AL, SEVERE NEU-TROPENIA ASSOCIATED WITH SUSTAINED - RELEASE PRO-CAINAMIDE, *ANN INTERN MED*, 100, 197-201 (1984)

1030 EMANUELE R, ROBUSCHI G, TAGLIAFERRI A, ET AL, IODOTHYRONINE AND THYROTROPIN CONCENTRATIONS AFTER IODOAMIDE ADMINISTRATION FOR ANGIOGRAPHIC STUDIES, *RIC CLIN LAB*, 12, 589-594 (1982)

1031 EMBER M, MINDSZENTY L, RENGEI B, CSASZAR L, CZINA M, SEC-ONDARY VITAMIN A DEFICIENCY IN ORGANOPHOSPHATE FOR-MULATORS AND SPRAY WORKERS, *RES COMMUN CHEM PATHOL PHARMACOL*, 3, 145 (1972)

1032 EMORI HW, BLUESTONE R, GOLDBERG LS, PSEUDOLEUKOCYTOSIS ASSOCIATED WITH CRYOGLOBULINEMIA, *AM J CLIN PATHOL*, 60, 202 (1973)

1033 ENCK RE, RIOS CN, TAMOXIFEN TREATMENT OF METASTATIC BREAST CANCER AND ANTITHROMBIN III LEVELS, *CANCER*, 1984, 2607-2609 (1984)

1034 ENGELMAN K, HORWITZ D, AMBROSE IM, FURTHER EVALUATION OF THE TYRAMINE TEST FOR PHEOCHROMOCYTOMA, *N ENGL J MED*, 278, 705 (1968)

1035 ENGLAND JDF, BETA-ADRENORECEPTOR-BLOCKING DRUGS ONCE DAILY IN ESSENTIAL HYPERTENSION: A COMPARISON OF PROPRANOLOL, PINDOLOL AND ATENOLOL, *AUST N Z J MED*, 11, 35-40 (1981)

1036 ENGLAND JM, COLES M, EFFECT OF CO-TRIMOXAZOLE ON PHE-NYLALANINE METABOLISM IN MAN, *LANCET*, 2, 1341 (1972)

1037 ENK L, CRONA N, SAMSIOE G, ET AL, DOSE AND DURATION EFFECTS OF ESTRADIOL VALERATE ON SERUM AND LIPOPRO-TEIN LIPIDS, *HORM METAB RES*, 18, 551-554 (1986)

1038 ERAZ J AND HAUKNECHT R, DIMINISHED URINARY ESTRIOL DUE TO MANDELAMINE® ADMINISTRATION DURING PREGNANCY, *AM J OBSTET GYNECOL*, 104, 924 (1969)

1039 ERDEN F, HACISALIHOGLU A, KOCER Z, ET AL, EFFECTS OF VITA-MIN C INTAKE ON WHOLE BLOOD PLASMA, LEUCOCYTE AND URINE ASCORBIC ACID AND URINE OXALIC ACID LEVELS, *ACTA VITAMINOL ENZYMOL*, 7, 123-130 (1985)

1040 ERIKSEN J, OLSEN PS, THOMSEN AC, GAMMA-GLUTAMYLTRANS-PEPTIDASE, ASPARTATE AMINOTRANSFERASE, AND ERYTHRO-CYTE MEAN CORPUSCULAR VOLUME AS INDICATORS OF ALCOHOL CONSUMPTION IN LIVER DISEASE, *SCAND J GAS-TROENTEROL*, 19, 813-819 (1984)

1041 ERIKSSON I, BERGGREN L, EFFECT OF REPEATED DOSES OF BENZODIAZEPINES ON ARTERIAL BLOOD GASES AND TRANS-CUTANEOUS PO$_2$, *ACTA ANAESTHESIOL SCAND*, 31, 357-361 (1987)

1042 ERIKSSON L-O, BEERMANN B, KALLNER M, RENAL FUNCTION AND TUBULAR TRANSPORT EFFECT OF SULINDAC AND NAPROXEN IN CHRONIC HEART FAILURE, *CLIN PHARMACOL THER*, 42, 646-654 (1987)

1043 ERIKSSON O, GINSBURG BE, HULTBERG B, OCKERMAN PA, INFLU-ENCE OF AGE AND SEX ON PLASMA ACID HYDROLASES, *CLIN CHIM ACTA*, 40, 181 (1972)

1044 ERLE G, BASSO M, FEDERSPIL G, ET AL, EFFECT OF CHLORPROMA-ZINE ON BLOOD GLUCOSE AND PLASMA INSULIN IN MAN, *EUR J CLIN PHARMACOL*, 11, 15-18 (1977)

1045 ERNAELSTEEN D, WILLIAMS R, JAUNDICE DUE TO NITROFURANTOIN, *GASTROENTEROLOGY*, 41, 590 (1961)

1046 ERNST JA, SY ER, EFFECT OF AZLOCILLIN ON URIC ACID LEVELS IN SERUM, *ANTIMICROB AGENTS CHEMOTHER*, 24, 609-610 (1983)

1047 ERYASA Y, CHANG PM, PITTINGER CB, SERUM IONIC CALCIUM CHANGES IN MAN FOLLOWING SUCCINYLCHOLINE ADMINIS-TRATION, *FED PROC*, 29, 548 (1970)

1048 ESCHBACH JW, ADAMSON JW, IMPROVEMENT IN THE ANEMIA OF CHRONIC RENAL FAILURE WITH FLUOXYMESTERONE, *ANN INTERN MED*, 78, 527 (1973)

1049 ESPIRITU CR, KIM TS, LEVINE RA, GRANULOMATOUS HEPATITIS ASSOCIATED WITH SULFADIMETHOXINE HYPERSENSITIVITY, *J AM MED ASSOC*, 202, 985 (1967)

1050 ESPOSITO R, VITALI D, RIFAMPICIN AND THROMBOCYTOPENIA, *LAN-CET*, 2, 491 (1971)

1051 ESTAE-WAINWRIGHT ES, RODRIGUEZ-SARGENT C, CANGIANO JL, ET AL, HIGH VANADATE INTERFERES WITH THE FISKE-SUB-BAROW DETERMINATION OF INORGANIC PHOSPHATE, *PROC SOC EXP BIOL MED*, 183, 268-272 (1986)

1052 ETTE EI, BROWN-AWALA EA, ESSIEN EE, CHLOROQUINE ELIMINA-TION IN HUMANS: EFFECT OF LOW-DOSE CIMETIDINE, *J CLIN PHARMACOL*, 27, 813-816 (1987)

1053 EVANS JR, SHANKEL SW, CUTLER RE, LOW OSMOLAR CONTRAST AGENTS AND NEPHROTOXICITY, *ANN INTERN MED*, 107, 116 (1987)

1054 EVANS RA, HYPERCALCAEMIA: WHAT DOES IT SIGNIFY?, *DRUGS*, 31, 64-74 (1986)

1055 EVANS RT, HAYNES SP, EFFECT OF PERCHLORATE ON THE MEAS-UREMENT OF SERUM IRON-BINDING CAPACITY, *LANCET*, 2, 387 (1972)

1056 EVENSON MA, KOELLNER S, RAPID METHOD FOR QUANTITATIVE DETERMINATION OF PROPOXYPHENE IN SERUM BY GAS-LIQUID CHROMATOGRAPHY, *CLIN CHEM*, 19, 492 (1973)

1057 EVENSON MA, LENSMEYER GL, QUALITATIVE AND QUANTITATIVE DETERMINATION OF METHAQUALONE IN SERUM BY GAS CHROMATOGRAPHY, *CLIN CHEM*, 20, 249 (1974)

1058 EVENSON MA, POQUETTE MA, RAPID GAS CHROMATOGRAPHIC METHOD FOR QUANTITATION OF ETHCHLORVYNOL (PLACIDYL®) IN SERUM, *CLIN CHEM*, 20, 212 (1974)

1059 EWEN LM, GRIFFITHS J, GAMMA-GLUTAMYL TRANSPEPTIDASE: ELE-VATED ACTIVITIES IN CERTAIN NEUROLOGIC DISEASES, *AM J CLIN PATHOL*, 59, 2 (1973)

1060 EYKYN S, USE AND CONTROL OF CEPHALOSPORINS, *J CLIN PATHOL*, 24, 419 (1971)

1061 EYSSEN H, EVRARD E, VANDERHAEGHE H, CHOLESTEROL LOWER-ING EFFECTS OF N-METHYLATED NEOMYCIN AND BASIC ANTIBIOTICS, *J LAB CLIN MED*, 68, 753 (1966)

1062 FAAS FM, NORMAN J, CARTER WJ, CEFOXITIN INTERFERENCE WITH URINARY 17-HYDROXYCORTICOSTEROID DETERMINATION, *CLIN CHEM*, 29, 1311-1313 (1983)

1063 FABRE J, WINTSCH J, PETER-CONTESSE R, ET AL, EFFECTS OF BOPINDOLOL: ON RENAL FUNCTION, *J CARDIOVASC PHARMACOL*, 8, (SUPPL 6), 545-550 (1986)

1064 FAGHER B, LUNDH B, HEPARIN TREATMENT OF DEEP VENOUS THROMBOSIS, *ACTA MED SCAND*, 210, 357-361 (1981)

1065 FAGLIA G, BECK-PECCOZ P, AMBROSI B, FERRARI C, TRAVAGLINI P, THE EFFECTS OF A SYNTHETIC THYROTROPIN RELEASING HORMONE (TRH) IN NORMAL AND ENDOCRINOPATHIC SUB-JECTS, *ACTA ENDOCRINOL*, 71, 209 (1972)

1066 FAHRAEUS L, LARSSON-COHN U, LJUNGBERG S, ET AL, PLASMA LIPOPROTEINS DURING AND AFTER DANAZOL TREATMENT, *ACTA OBSTET GYNECOL SCAND*, 63, SUPPL 123, 133-135 (1984)

1067 FAHRAEUS L, LARSSON-COHN V, WALLENTIN L, LIPOPROTEINS DURING ORAL AND CUTANEOUS ADMINISTRATION OF OES-TRADIOL-17β TO MENOPAUSAL WOMEN, *ACTA ENDOCRINOL*, 101, 597-602 (1982)

1068 FAHRAEUS L, SYDSJO A, WALLENTIN L, LIPOPROTEIN CHANGES DURING TREATMENT OF PELVIC ENDOMETRIOSIS WITH MEDROXYPROGESTERONE ACETATE, *FERTIL STERIL*, 45, 503-506 (1986)

1069 FAIERMAN D, JACOBS S, LIVER INJURY FROM ELIXIR OF TERPIN HYDRATE WITH CODEINE, *MT SINAI J MED*, 40, 56 (1973)

1070 FAIRLEY KF, BIRCH DF, HAINES I, ABNORMAL URINARY SEDIMENT IN PATIENTS ON TRIAMTERENE, *LANCET*, 1, 421-422 (1983)

1071 FAJANS SS, FLOYD JC JR., KNOPF RF, CONN JW, EFFECT OF AMINO ACIDS AND PROTEINS ON INSULIN SECRETION IN MAN, *RECENT PROG HORM RES*, 23, 617 (1967)

1072 FALCH DK, SCHREINER A, THE EFFECT OF SPIRONOLACTONE ON LIPID, GLUCOSE AND URIC ACID LEVELS IN BLOOD DURING LONG-TERM ADMINISTRATION TO HYPERTENSIVES, *ACTA MED SCAND*, 213, 27-30 (1983)

1073 FALEK A, JORDAN RB, KING BJ, ARNOLD PJ, SKELTON WD, HUMAN CHROMOSOMES AND OPIATES, *ARCH GEN PSYCHIATRY*, 27, 511 (1972)

1074 FALLON JA, TALL AR, JANIS MG, ET AL, OXACILLIN-INDUCED GRANULOCYTOPENIA, *ACTA HAEMATOL*, 59, 167-170 (1978)

1075 FALLON RJ, HOBSON BM, A SOURCE OF ERROR IN PREGNANCY TESTS, *LANCET*, 1, 1243 (1973)

1076 FALLUCCA F, CARRATU R, TAMBURRANO G, JAVICOLI, M MENZINGER G, ANDREANI D, EFFECTS OF CAERULIN AND PANCREOZYMIN ON INSULIN SECRETION IN NORMAL SUBJECTS AND IN PATIENTS WITH INSULINOMA, *FOLIA ENDOCRINOL*, 24, 110 (1971)

1077 FALOON WW, METABOLIC EFFECTS OF NONABSORBABLE ANTIBACTERIAL AGENTS, *AM J CLIN NUTR*, 23, 645 (1970)

1078 FARID NR, JOHNSON RJ, LOW WT, HEMOLYTIC REACTION TO MEFENAMIC ACID, *LANCET*, 2, 382 (1971)

1079 FARID Z, SMITH JH, BASSILY S, SPARKS HA, HEPATOTOXICITY AFTER TREATMENT OF SCHISTOSOMIASIS WITH HYCANTHONE, *BR MED J [CLIN RES]*, 2, 88 (1972)

1080 FARISH E, FLETCHER CD, HART DM, ET AL, THE EFFECTS OF HORMONE IMPLANTS ON SERUM LIPOPROTEINS AND STEROID HORMONES IN BILATERALLY OOPHORECTOMIZED WOMEN, *ACTA ENDOCRINOL*, 106, 116-120 (1984)

1081 FARMER RW, PIERCE CE, PLASMA CORTISOL DETERMINATION: RADIOIMMUNOASSAY AND COMPETITIVE PROTEIN BINDING COMPARED, *CLIN CHEM*, 20, 411 (1974)

1082 FARRINGER JA, GREEN JA, O'ROURKE RA, ET AL, NIFEDIPINE-INDUCED ALTERATIONS IN SERUM QUINIDINE CONCENTRATIONS, *AM HEART J*, 108, 1570-1572 (1984)

1083 FASS RJ, SASLAW S, CLINDAMYCIN: CLINICAL AND LABORATORY EVALUATION OF PARENTERAL THERAPY, *AM J MED SCI*, 263, 368 (1972)

1084 FAVREAU M, TANNENBAUM H, LOUGH J, HEPATIC TOXICITY ASSOCIATED WITH GOLD THERAPY, *ANN INTERN MED*, 87, 717-719 (1977)

1085 FAWCETT J, KRAVITZ HM, THE LONG TERM MANAGEMENT OF BIPOLAR DISORDERS WITH LITHIUM, CARBAMAZEPINE, AND ANTIDEPRESSANTS, *J CLIN PSYCHIATRY*, 46, 58-60 (1985)

1086 FAWCETT JK, WYNN V, VARIATION OF PLASMA ELECTROLYTE AND TOTAL PROTEIN LEVELS IN THE INDIVIDUAL, *BR MED J [CLIN RES]*, 2, 582 (1956)

1087 FAWCETT JK, WYNN V, EFFECTS OF POSTURE ON PLASMA VOLUME AND SOME BLOOD CONSTITUENTS, *J CLIN PATHOL*, 13, 304 (1960)

1088 FEARNLEY GR, CHAKRABARTI R, HOCKING ED, FIBRINOLYTIC EFFECTS OF DIGUANIDES PLUS ETHYLESTRENOL IN OCCLUSIVE VASCULAR DISEASE, *LANCET*, 2, 1008 (1967)

1089 FEE JPH, BLACK GW, DUNDEE JW, ET AL, A PROSPECTIVE STUDY OF LIVER ENZYME AND OTHER CHANGES FOLLOWING REPEAT ADMINISTRATION OF HALOTHANE AND ENFLURANE, *BR J ANAESTH*, 51, 1133-1140 (1979)

1090 FEELY J, BETA-ADRENOCEPTOR-BLOCKING DRUGS, GROWTH HORMONE AND ACROMEGALY, *POSTGRAD MED J*, 56, 236-237 (1980)

1091 FEIG SA, SEGEL GB, ANDROGENS, 2,3-DPG AND SICKLING, *N ENGL J MED*, 287, 1097 (1972)

1092 FEIKS FK, UBER EINE WEITERE INDIKATION ZU PARENTERALER SCHWEFELTHERAPIE, *WIEN MED WOCHENSCHR*, 117, 899 (1967)

1093 FEINBERG LJ, SANDBERG, DECASTRO O, BELLET S, EFFECTS OF COFFEE INGESTION ON ORAL GLUCOSE TOLERANCE CURVES IN NORMAL HUMAN SUBJECTS, *METABOLISM*, 17, 916 (1968)

1094 FEINGOLD KR, SATER B, ENGLE B, IODINE INDUCED ARTIFACTS IN HOME BLOOD GLUCOSE MEASUREMENTS, *DIABETES CARE*, 6, 317-318 (1983)

1095 FEINSTEIN DI, RAPAPORT SI, CHONG MMY, FACTOR V INHIBITOR: REPORT OF A CASE WITH COMMENTS ON A POSSIBLE EFFECT OF STREPTOMYCIN, *ANN INTERN MED*, 78, 385 (1973)

1096 FELDMAN JM, BOWMAN J, URINARY HOMOGENTISIC ACID: DETERMINATION BY THIN LAYER CHROMATOGRAPHY, *CLIN CHEM*, 19, 459 (1973)

1097 FELDMAN JM, BUTLER SS, CHAPMAN BA, INTERFERENCE WITH MEASUREMENT OF 3-METHOXY-4-HYDROXYMANDELIC ACID AND 5-HYDROXYINDOLEACETIC ACID, *CLIN CHEM*, 20, 607 (1974)

1098 FELDMAN JM, CHAPMAN BA, RADIOIMMUNOASSAY OF INSULIN IN SERUM AND PLASMA, *CLIN CHEM*, 19, 1250 (1973)

1099 FELDMAN JM, KELLEY WN, LEBOVITZ HE, INHIBITION OF GLUCOSE OXIDASE PAPER TESTS BY REDUCING METABOLITES, *DIABETES*, 19, 337 (1970)

1100 FELDMAN JM, LEBOVITZ FL, TESTS FOR GLUCOSURIA. AN ANALYSIS OF FACTORS THAT CAUSE MISLEADING RESULTS, *DIABETES*, 22, 115 (1973)

1101 FELDMAN JM, LEBOVITZ HE, CONTROL OF INSULIN AND GROWTH HORMONE SECRETION BY SEROTONIN AND DOPAMINE, *ABSTRACTS IV INTL CONGR ENDOCRINOL, WASHINGTON DC, 1972*, 35

1102 FELDMAN JM, LEBOVITZ HE, ENDOCRINE AND METABOLIC EFFECTS OF GLYBENCLAMIDE, *DIABETES*, 20, 745 (1971)

1103 FELDMAN JM, LEBOVITZ HE, LEVODOPA AND TESTS FOR URINARY GLUCOSE, *N ENGL J MED*, 283, 1053 (1970)

1104 FELDMAN S, DOOLITTLE M, LOTT L, ET AL, SIMILAR HEMATOLOGIC CHANGES IN CHILDREN RECEIVING TRIMETHOPRIM-SULFAMETHOXAZOLE OR AMOXICILLIN FOR OTITIS MEDIA, *J PEDIATR*, 106, 995-1000 (1985)

1105 FELDMANN U, GAUS W, KRETSCHMER F-J, ET AL, ANALGESIC USE, AGRANULOCYTOSIS, AND APLASTIC ANEMIA, *J AM MED ASSOC*, 257, 2590-2591 (1987)

1106 FELIG P, MARLISS E, CAHILL GF JR., PLASMA AMINO ACID LEVELS AND INSULIN SECRETION IN OBESITY, *N ENGL J MED*, 281, 811 (1969)

1107 FELIG P, WAHREN J, HENDLER R, AHLBORG G, PLASMA GLUCAGON LEVELS IN EXERCISING MAN, *N ENGL J MED*, 287, 184 (1972)

1108 FELLMAN JH, JOYCE JR, STRANDHOLM JJ, ANALYSIS IN HUMAN PLASMA OF 3-METHOXYTYROSINE: A METABOLITE OF DOPA, *CLIN CHIM ACTA*, 32, 313 (1971)

1109 FENECH FF, BANNISTER WH, GRECH JL, HEPATITIS WITH BILIVERDINAEMIA IN ASSOCIATION WITH INDOMETHACIN THERAPY, *BR MED J [CLIN RES]*, 3, 155 (1967)

1110 FENSTER PE, HAGER WD, GOODMAN MM, DIGOXIN-QUINIDINE-SPIRONOLACTONE INTERACTION, *CLIN PHARMACOL THER*, 36, 70-73 (1984)

1111 FENSTER PE, HAGER WD, PERRIER D, ET AL, DIGOXIN-QUINIDINE INTERACTION IN PATIENTS WITH CHRONIC RENAL FAILURE, *CIRCULATION*, 66, 1277-1280 (1982)

1112 FERGUSON DB, EFFECTS OF LOW DOSES OF FLUORIDE ON SERUM PROTEINS AND A SERUM ENZYME IN MAN, *NATURE NEW BIOL*, 231, 159 (1971)

1113 FERGUSON RM, SUTHERLAND DER, SIMMONS RL, ET AL, KETOCONAZOLE, CYCLOSPORINE METABOLISM, AND RENAL TRANSPLANTATION, *LANCET*, 2, 882-883 (1982)

1114 FERLINZ J, EASTHOPE JL, HUGHES D, ET AL, RIGHT VENTRICULAR PERFORMANCE IN ESSENTIAL HYPERTENSION AFTER BETA-BLOCKADE, *BR HEART J*, 46, 23-29 (1981)

1115 FERLITO S, MODICA L, ROMANO F, ET AL, EFFECTS OF VERAPAMIL ON GLUCOSE, INSULIN AND GLUCAGON LEVELS AFTER ORAL GLUCOSE LOAD IN NORMAL AND DIABETIC SUBJECTS, *PANMINERVA MED*, 24, 221-226 (1982)

1116 FERNANDEZ AA, JACOBS SL, PORPHYRINS, PORPHOBILINOGEN AND AMINOLEVULINIC ACID IN URINE, *STAND METH CLIN CHEM*, 6, 57 (1970)

1117 FERNARIDO HOI, THE IMPORTANCE OF PROPER COLLECTION OF URINE SAMPLES FOR THE ESTIMATION OF CATECHOLAMINES BY BIOLOGICAL METHODS, *MED LAB TECHNOL*, 29, 188 (1972)

1118 FERRARA LA, MAROTTA T, RUBBA P, ET AL, EFFECTS OF ALPHA-ADRENERGIC AND BETA-ADRENERGIC RECEPTOR BLOCKADE ON LIPID METABOLISM, *AM J MED*, 80, 104-108 (1986)

1119 FERRARA M, MIZRAHY O, SAPOSHINIK A, FEINSTEIN G, SERUM BETA-GLUCURONIDASE ACTIVITY DURING PREGNANCY IN NORMAL AND DIABETIC WOMEN AND ITS RELATION TO ESTROGEN LEVELS, *ISR J MED SCI*, 7, 1214 (1971)

1120 FERRARI C, FREZZATI S, ROMUSSI M, ET AL, EFFECTS OF SHORT TERM CLOFIBRATE ADMINISTRATION ON GLUCOSE TOLERANCE AND INSULIN SECRETION IN PATIENTS WITH CHEMICAL DIABETES OR HYPERTRIGLYCERIDEMIA, *METABOLISM*, 26, 129-139 (1977)

1121 FERRARI C, TESTORI G, SCANNI A, ET AL, REDUCTION OF SERUM ALKALINE PHOSPHATASE AND GAMMA-GLUTAMYL TRANSPEPTIDASE ACTIVITIES BY SHORT-TERM CLOFIBRATE, *N ENGL J MED*, 295, 449 (1976)

1122 FERRAZZI E, CARTEI G, DE BESI P, ET AL, TAMOXIFEN IN DISSEMINATED BREAST CANCER, *TUMORI*, 63, 463-468 (1977)

1123 FERRE C, PANADERO AM, CASTINEIRAS MJ, ET AL, INTERFERENCE BY CYCLOSPORINE ASSESSED, *CLIN CHEM*, 32, 1590-1591 (1986)

1124 FERRIS TF, GORDEN P, EFFECT OF ANGIOTENSIN AND NOREPINEPHRINE UPON URATE CLEARANCE IN MAN, *AM J MED*, 44, 359 (1968)

1125 FESSEL WJ, HYPERURICEMIA IN HEALTH AND DISEASE, *SEMIN ARTHRITIS RHEUM*, 1, 275 (1972)

1126 FESTEN HPM, THIJS JC, LAMERS CBHW, ET AL, EFFECT OF ORAL OMEPRAZOLE ON SERUM GASTRIN AND SERUM PEPSINOGEN LEVELS, *GASTROENTEROLOGY*, 87, 1030-1034 (1984)

1127 FICH A, SCHWARTZ J, BRAVERMAN D, ET AL, SULFASALAZINE HEPATOTOXICITY, *AM J GASTROENTEROL*, 79, 401-402 (1984)

1128 FICHER M, CURTIS GC, GANJAM CK, JOSHLIN L, PERRY S, IMPROVED MEASUREMENT OF CORTICOSTEROIDS IN PLASMA AND URINE BY COMPETITIVE PROTEIN-BINDING RADIOASSAY, *CLIN CHEM*, 19, 511 (1973)

1129 FICHMAN MP, VORHERR H, KLEEMAN CR, TELFER N, DIURETIC - INDUCED HYPONATREMIA, *ANN INTERN MED*, 75, 853 (1971)

1130 FICHTER MM, PIRKE MM, HOLSBOER F, WEIGHT LOSS CAUSES NEUROENDOCRINE DISTURBANCES: EXPERIMENTAL STUDY IN HEALTHY STARVING SUBJECTS, *PSYCHIATRY RES*, 17, 61-72 (1986)

1131 FIELDING J, RYALL RG, SOME CHARACTERISTICS OF TRICHLORACETIC ACID-PRECIPITATED PROTEINS AND THEIR EFFECTS ON BIOCHEMICAL ASSAY, *CLIN CHIM ACTA*, 33, 235 (1971)

1132 FIELDS WS, HASS WK (EDS), *ASPIRIN, PLATELETS AND STROKE: BACKGROUND FOR A CLINICAL TRIAL*, GREEN, ST LOUIS, MO (1971)

1133 FINDLING JW, BACKSTROM D, RAWSTHORNE L, ET AL, INDOMETHACIN-INDUCED HYPERKALEMIA IN THREE PATIENTS WITH GOUTY ARTHRITIS, *J AM MED ASSOC*, 244, 1127-1128 (1980)

1134 FINE D, SHEDROVILZKY H, HYPONATREMIA DUE TO CHLOR-PROPAMIDE, A SYNDROME RESEMBLING INAPPROPRIATE SECRETION OF ANTIDIURETIC HORMONE, *ANN INTERN MED*, 72, 83 (1970)

1135 FINGERHUT B, A NON-MERCURIMETRIC AUTOMATED METHOD FOR SERUM CHLORIDE, *CLIN CHIM ACTA*, 41, 247 (1972)

1136 FINLAY GD, WHITSETT TL, CUCINELL EA, GOLDBERG LI, AUGMEN-TATION OF SODIUM AND POTASSIUM EXCRETION, GLOMERU-LAR FILTRATION RATE AND RENAL PLASMA FLOW BY LEVODOPA, *N ENGL J MED*, 284, 865 (1971)

1137 FIORE CE, MALATINO LS, KANIS JA, EFFECTS OF CIMETIDINE ON PARATHYROID METABOLISM, *LANCET*, 1, 501 (1981)

1138 FIORE JM, NOONAN FM, AGRANULOCYTOSIS DUE TO MEPAZINE (PHENOTHIAZINE), *N ENGL J MED*, 260, 375 (1959)

1139 FIORICA V, BURR MJ, MOSES R, CONTRIBUTION OF ACTIVITY TO THE CIRCADIAN RHYTHM IN EXCRETION OF MAGNESIUM AND CALCIUM, *AEROSPACE MED*, 39, 714 (1968)

1140 FIRKIN FC, MARIANI AF, AGRANULOCYTOSIS DUE TO DAPSONE, *MED J AUST*, 2, 247-251 (1977)

1141 FISCH IR, FREEDMAN SH, ORAL CONTRACEPTIVES AND THE RED BLOOD CELL, *CLIN PHARMACOL THER*, 14, 245 (1973)

1142 FISCHER HW, HOAK JC, MIMICRY OF ACUTE CHOLECYSTITIS BY ERYTHROMYCIN ESTOLATE REACTIONS. REPORT OF TWO CASES, *AM J MED SCI*, 247, 283 (1964)

1143 FISCHER-RUSMUSSEN W, PLASMA OESTRONE (AND OESTRADIOL) IN NORMAL HUMAN PREGNANCY, *J STEROID BIOCHEM*, 2, 371 (1971)

1144 FISHER CE, WOO J, HORTON R, EFFECTS OF SPIRONOLACTONE ON ALDOSTERONE AND RENIN IN MAN, *ABSTRACTS IV INTL CONGR ENDOCRINOL, WASHINGTON DC, 1972*, P, 111

1145 FISHER CE, WOO J, HORTON R, EFFECT OF SPIRONOLACTONE ON ALDOSTERONE AND RENIN IN NORMAL MAN AND IN PRIMARY ALDOSTERONISM, *CLIN RES*, 20, 217 (1972)

1146 FITZGERALD PH, PICKERING AF, FERGUSON DN, DEPRESSED LYM-PHOCYTE RESPONSE TO P.H.A. IN LONG-TERM USERS OF ORAL CONTRACEPTIVES, *LANCET*, 1, 615 (1973)

1147 FLACKE W, DRUG THERAPY. TREATMENT OF MYASTHENIA GRAVIS, *N ENGL J MED*, 288, 27 (1973)

1148 FLEISCHER N, BROWN H, GRAHAM DY, DELENA S, CHRONIC LAXA-TIVE-INDUCED HYPERALDOSTERONISM AND HYPOKALEMIA SIMULATING BARTTER'S SYNDROME, *ANN INTERN MED*, 70, 791 (1969)

1149 FLEISHER M, DOLLINGER MR, SCHWARTZ MK, EFFECT OF 5-HYDROXYINDOLEACETIC ACID IN THE COLORIMETRIC DETER-MINATION OF URINARY VANILMANDELIC ACID, *AM J CLIN PATHOL*, 51, 555 (1969)

1150 FLEISHER WR, ENZYMATIC METHODS FOR LACTIC AND PYRUVIC ACIDS, *STAND METH CLIN CHEM*, 6, 245 (1970)

1151 FLINK EB, HYPOMAGNESEMIA IN THE PATIENT RECEIVING DIGITA-LIS, *ARCH INTERN MED*, 145, 625 (1985)

1152 FLOYD JC JR., FAJANS SS, KNOPF RF, CONN JW, PLASMA INSULIN IN HYPERINSULINISM: COMPARATIVE EFFECTS OF TOLBUTA-MIDE, LEUCINE AND GLUCOSE, *J CLIN ENDOCRINOL METAB*, 24, 747 (1964)

1153 FLYNN A, PORIES WJ, STRAIN WH, HILL OA JR., FRATIANNE RB, RAPID SERUM-ZINC DEPLETION ASSOCIATED WITH CORTICO-STEROID THERAPY, *LANCET*, 2, 1169 (1971)

1154 FOLEY RJ, HAMNER RW, WEINMAN EJ, SERUM POTASSIUM CON-CENTRATIONS IN CYCLOSPORINE AND AZATHIOPRINE-TREATED RENAL TRANSPLANT PATIENTS, *NEPHRON*, 40, 280-285 (1985)

1155 FONG HJ, COHEN AH, IBUPROFEN-INDUCED ACUTE RENAL FAILURE WITH ACUTE TUBULAR NECROSIS, *AM J NEPHROL*, 2, 28-31 (1982)

1156 FORD MJ, KELLETT RJ, BUSUTTIL A, ET AL, DEXTROPROPOX-YPHENE AND JAUNDICE, *BR MED J [CLIN RES]*, 2, 674 (1977)

1157 FORESTER G, PROFOUND CYTOPENIA SECONDARY TO AZIDOTHYMIDINE, *N ENGL J MED*, 317, 772 (1987)

1158 FORMAN MB, UDERMAN H, JACKSON EK, ET AL, EFFECTS OF INDOMETHACIN ON SYSTEMIC AND CORONARY HEMODYNAM-ICS IN PATIENTS WITH CORONARY ARTERY DISEASE, *AM HEART J*, 110, 311-318 (1985)

1159 FORSHAW J, MUSCLE PARESIS AND HYPOKALAEMIA AFTER TREAT-MENT WITH DUOGASTRONE, *BR MED J [CLIN RES]*, 2, 674 (1969)

1160 FORSTER H, SAFETY OF XYLITOL, *N ENGL J MED*, 286, 790 (1972)

1161 FORSTER HW, HEPATITIS FROM HYDRALAZINE, *N ENGL J MED*, 302, 1362 (1980)

1162 FORTMAN CS, WITTE DL, SERUM 5'-NUCLEOTIDASE IN PATIENTS RECEIVING ANTIEPILEPTIC DRUGS, *AM J CLIN PATHOL*, 84, 197-201 (1985)

1163 FOSTER LB, HOCHHOLZER JM, A SINGLE EXTRACTION GAS CHRO-MATOGRAPHIC METHOD FOR THE DETERMINATION OF ESTRIOL IN PREGNANCY URINES, *CLIN CHIM ACTA*, 32, 147 (1971)

1164 FOSTER PN, SWAN CHJ, DAPSONE AND FATAL HYPOALBUMINEMIA, *LANCET*, 2, 806 (1981)

1165 FOUAD FM, EL-TOBGI S, TARAZI RC, ET AL, CAPTOPRIL IN CONGES-TIVE HEART FAILURE RESISTANT TO OTHER VASODILATORS, *EUR HEART J*, 5, 47-54 (1984)

1166 FOUAD FM, TARAZI RC, BRAVO EL, ET AL, HEMODYNAMIC AND ANTIHYPERTENSIVE EFFECTS OF THE NEW ORAL ANGIOTEN-SIN-CONVERTING ENZYME INHIBITOR MK-421 (ENALAPRIL), *HYPERTENSION*, 6, 167-174 (1984)

1167 FOY JM, PARRATT JR, 5-HYDROXYTRYPTAMINE IN PINEAPPLES, *J PHARMACOKINET PHARMACOL*, 13, 382 (1961)

1168 FRAJOLA WJ, *FLUORINE AND DENTAL HEALTH*, JC MUHLER, MK HINE, EDS., STAPLES PRESS, LONDON (1960)

1169 FRANCESCONI RP, BOYD AE III, MAGER M, HUMAN TRYPTOPHAN AND TYROSINE METABOLISM: EFFECTS OF ACUTE EXPOSURE TO COLD STRESS, *J APPL PHYSIOL*, 33, 165 (1972)

1170 FRANCIS KL, JENIS EH, JENSEN GE, ET AL, GOLD-ASSOCIATED NEPHROPATHY, *ARCH PATHOL LAB MED*, 108, 234-238 (1984)

1171 FRANCKSON JRM, VANROUX R, LECLERCQ R, BRUNENGRABER H, OOMS HA, LABELED INSULIN CATABOLISM AND PANCREATIC RESPONSIVENESS DURING LONG TERM EXERCISE IN MAN, *HORM METAB RES*, 3, 366 (1971)

1172 FRANK HA, CARR MH, NORMAL SERUM ELECTROLYTES WITH A NOTE ON SEASONAL AND MENSTRUAL VARIATION, *J LAB CLIN MED*, 49, 246 (1957)

1173 FRANKLYN JA, DAVIS JR, GAMMAGE MD, ET AL, AMIODARONE AND THYROID HORMONE ACTION, *CLIN ENDOCRINOL*, 22, 257-264 (1985)

1174 FRANKLYN JA, WILKINS MR, WILKINSON R, ET AL, THE EFFECT OF PROPRANOLOL ON CIRCULATING THYROID HORMONE MEAS-UREMENTS IN THYROTOXIC AND EUTHYROID SUBJECTS, *ACTA ENDOCRINOL*, 108, 351-355 (1985)

1175 FRANKS RD, DUBOVSKY SL, LIFSHITZ M, ET AL, LONG-TERM LITH-IUM CARBONATE THERAPY CAUSES HYPERPARATHYROIDISM, *ARCH GEN PSYCHIATRY*, 39, 1074-1077 (1982)

1176 FRANZEN F, EYSELL K, *BIOLOGICALLY ACTIVE AMINES FOUND IN MAN*, PERGAMON PRESS, NEW YORK, NY (1969)

1177 FRASCINO JA, EFFECT OF INORGANIC FLUORIDE ON THE RENAL CONCENTRATING MECHANISM, *J LAB CLIN MED*, 79, 192 (1972)

1178 FRASER DG, LUDDEN TM, EVENS RP, ET AL, DISPLACEMENT OF PHENYTOIN FROM PLASMA BINDING SITES BY SALICYLATE, *CLIN PHARMACOL THER*, 27, 165-169 (1980)

1179 FRASER PM, DOLL R, LANGMAN MJS, MISIEWICZ JJ, SHAWDON HH, CLINICAL TRIAL OF A NEW CARBENOXOLONE ANALOGUE (BX24), ZINC SULFATE AND VITAMIN A IN TREATMENT OF GAS-TRIC ULCER, *GUT*, 13, 459 (1972)

1180 FRASER TM, STORAGE OF BIOLOGICAL SAMPLES, *NASA CR-*, 781 (1967)

1181 FRASIER SD, HUMAN PITUITARY GROWTH HORMONE (hGH) THERAPY IN GROWTH HORMONE DEFICIENCY, *ENDOCR REV*, 4, 155-170 (1983)

1182 FRASSINELLI-GUNDERSEN EP, MARGEN S, BROWN JR, IRON STORES IN USERS OF ORAL CONTRACEPTIVE AGENTS, *AM J CLIN NUTR*, 41, 703-712 (1985)

1183 FRAYN KN, ADNITT PI, ESTIMATION OF 3-O-METHYL-D-GLUCOSE IN THE PRESENCE OF GLUCOSE, *J CLIN PATHOL*, 24, 671 (1971)

1184 FRAYN KN, ADNITT PI, TURNER P, THE HYPOGLYCAEMIC ACTION OF METFORMIN, *POSTGRAD MED J*, 47, 777 (1971)

1185 FREEDMAN SH, ANDERSON NE, SPIROMETRY AND ORAL CONTRA-CEPTIVES, *AM J OBSTET GYNECOL*, 116, 682 (1973)

1186 FREEMAN DJ, LAUPACIS A, KEOWN PA, ET AL, EVALUATION OF CYCLOSPORINE-PHENYTOIN INTERACTION WITH OBSERVA-TIONS ON CYCLOSPORINE METABOLITES, *BR J CLIN PHARMACOL*, 18, 887-893 (1984)

1187 FREEMAN JA, BEELER MF, *LABORATORY MEDICINE CLINICAL MICROSCOPY*, LEA AND FEBIGER, PHILADELPHIA, PA (1974)

1188 FREEMAN RB, MAHER JF, SCHREINER GE, MOSTOFI FK, RENAL TUBULAR NECROSIS DUE TO NEPHROTOXICITY OF ORGANIC MERCURIAL DIURETICS, *ANN INTERN MED*, 57, 34 (1962)

1189 FREGLY MJ, MCCARTHY JS, EFFECT OF DIURETICS ON RENAL IODIDE EXCRETION BY HUMANS, *TOXICOL APPL PHARMACOL*, 25, 289 (1973)

1190 FREIBERG M, SABATINO B, DALAKOS TG, MILLER M, MOSES AM, STREETEN DHP, DIURNAL AND SLEEP INFLUENCES ON RENAL EXCRETION, *CLIN RES*, 21, 686 (1973)

1191 FREIDERISZICK FK, TOUSSAINT W, WIRKUNGEN VON PHENACETIN UND NAPAP AUF DAS BLUTBILD DES SAUGLINGS, *MED KLIN*, 61, 304 (1966)

1192 FRENKEL EP, MCCALL MS, SHEEHAN RG, CEREBROSPINAL FLUID FOLATE, AND VITAMIN B_{12} IN ANTICONVULSANT INDUCED MEGALOBLASTOSIS, *J LAB CLIN MED*, 81, 105 (1973)

1193 FRICK MH, ELO O, HAAPA K, ET AL, HELSINKI HEART STUDY: PRIMARY - PREVENTION TRIAL WITH GEMFIBROZIL IN MIDDLE - AGED MEN WITH DYSLIPIDEMIA, *N ENGL J MED*, 317, 1237-1245 (1987)

1194 FRIEDMAN AC, LAUTIN EM, CIS-PLATINUM (II) DIAMINE DICHLORIDE: ANOTHER CAUSE OF BILATERAL SMALL KIDNEYS, *UROLOGY*, 16, 584-586 (1980)

1195 FRIEDMAN GD, SIEGELAUB AB, SELTZER CC, FELDMAN R, COLLEN MF, SMOKING HABITS AND THE LEUKOCYTE COUNT, *ARCH ENVIRON HEALTH*, 26, 137 (1973)

1196 FRIEDMAN HS, TOTAL PROTEINS IN CEREBROSPINAL FLUID (COLORIMETRIC), *STAND METH CLIN CHEM*, 5, 223 (1965)

1197 FRIEL PN, LEAL KW, WILENSKY AJ, VALPROIC ACID-PHENYTOIN INTERACTION, *THER DRUG MONIT*, 1, 243-248 (1979)

1198 FRIEND DG, URICOSURIC DRUGS, *PRACTITIONER*, 200, 153 (1968)

1199 FRIES J, SIRAGANIAN R, SULFONAMIDE HEPATITIS. REPORT OF A CASE DUE TO SULFAMETHOXAZOLE AND SULFISOXAZOLE, *N ENGL J MED*, 274, 95 (1966)

1200 FRIGO GM, LECCHINI S, GATTI G,, ET AL, MODIFICATION OF PHENYTOIN CLEARANCE BY VALPROIC ACID IN NORMAL SUBJECTS, *BR J CLIN PHARMACOL*, 8, 553-556 (1979)

1201 FRIIS T, PEDERSEN LR, SERUM LIPIDS IN HYPER- AND HYPOTHYROIDISM BEFORE AND AFTER TREATMENT, *CLIN CHIM ACTA*, 162, 155-163 (1987)

1202 FRIMAN G, NYSTROM-ROSANDER C, JONSELL G, ET AL, AGRANULOCYTOSIS ASSOCIATED WITH MALARIA PROPHYLAXIS WITH MALOPRIM, *BR MED J [CLIN RES]*, 286, 1244-1245 (1983)

1203 FRIMODT-MOLLER N, MAIGAARD S, MADSEN PO, COMPARATIVE NEPHROTOXICITY AMONG AMINOGLYCOSIDES AND BETA-LACTAM ANTIBIOTICS, *INFECTION*, 6, 283-289 (1980)

1204 FRINGS CS, DRUG SCREENING, *CRIT REV CLIN LAB SCI*, 4, 357 (1974)

1205 FRINGS CS, COHEN PS, ETHCHLORVYNOL (PLACIDYL®) DETERMINATION IN SERUM AND URINE, *STAND METH CLIN CHEM*, 7, 209 (1972)

1206 FRINGS CS, QUEEN C, O-TOLUIDINE METHODS FOR GLUCOSE: EFFECT OF LIPEMIA, *CLIN CHEM*, 18, 488 (1972)

1207 FRINGS CS, QUEEN C, FOSTER LB, IMPROVED COLORIMETRIC METHOD FOR ASSAY OF AMPHETAMINES IN URINE, *CLIN CHEM*, 17, 1016 (1971)

1208 FRISHMAN WH, PINDOLOL: A NEW β-ADRENOCEPTOR ANTAGONIST WITH PARTIAL AGONIST ACTIVITY, *N ENGL J MED*, 308, 940-944 (1983)

1209 FRITSCHKA E, GOTZEN R, KITTLER R, ET AL, EFFECT OF METOPROLOL ON 24 HOUR URINARY EXCRETION OF ADRENAL STEROIDS AND KALLIKREIN IN PATIENTS WITH ESSENTIAL HYPERTENSION, *BR J PHARMACOL*, 81, 245-253 (1984)

1210 FROHLICH ED, DUSTAN HP, PAGE IH, SOME CLINICAL EFFECTS OF GUANOXAN, *CLIN PHARMACOL THER*, 7, 599 (1966)

1211 FROLICH C, WILSON TW, CARR K, NIES AS, OATES JA, URINARY PROSTAGLANDINS: RELEASE FROM THE KIDNEY BY ANGIOTENSIN IN DOG AND MAN, *CLIN RES*, 21, 687 (1973)

1212 FROUD DJR, ESTIMATION OF SERUM GENTAMICIN LEVELS, *LANCET*, 2, 1425 (1971)

1213 FUENTES A, GOLDKRAND JW, ANGIOTENSIN-CONVERTING ENZYME ACTIVITY IN HYPERTENSIVE SUBJECTS AFTER MAGNESIUM SULFATE THERAPY, *AM J OBSTET GYNECOL*, 156, 1375-1379 (1987)

1214 FUJITA T, CHAN JCM, BARTTER FC, EFFECTS OF ORAL FUROSEMIDE AND SALT LOADING ON PARATHYROID FUNCTION IN NORMAL SUBJECTS, *NEPHRON*, 38, 109-114 (1984)

1215 FURST P, GUANIERI G, HULTMAN E, THE EFFECT OF THE ADMINISTRATION OF L-TRYPTOPHAN ON SYNTHESIS OF UREA AND GLUCONEOGENESIS IN MAN, *SCAND J CLIN LAB INVEST*, 27, 183 (1971)

1216 GABOR EP, LETTER: HEMOLYTIC ANEMIA AS ADVERSE REACTION TO SALICYLAZOSULFAPYRIDINE, *N ENGL J MED*, 289, 1372 (1973)

1217 GABRIEL R, RIFAMPICIN JAUNDICE, *BR MED J [CLIN RES]*, 3, 182 (1971)

1218 GABRIEL R, ETHAMBUTOL AND A FALSE-POSITIVE SCREENING TEST FOR PHAEOCHROMOCYTOMA, *BR MED J [CLIN RES]*, 3, 332 (1972)

1219 GABRIEL R, CALDWELL J, HARTLEY RB, ACUTE TUBULAR NECROSIS, CAUSED BY THERAPEUTIC DOSES OF PARACETAMOL?, *CLIN NEPHROL*, 18, 269-271 (1982)

1220 GAILANI S, HOLLAND JF, GLICK A, EFFECTS OF BOXIDINE ON HUMAN SERUM STEROLS AND NEOPLASMS, *CLIN PHARMACOL THER*, 13, 91 (1972)

1221 GAL I, PARKINSON C, CRAFT I, EFFECT OF ORAL CONTRACEPTIVES ON HUMAN PLASMA VITAMIN A LEVELS, *BR MED J [CLIN RES]*, 2, 436 (1971)

1222 GAL I, PARKINSON CE, VARIATIONS IN THE PATTERN OF MATERNAL SERUM VITAMIN A AND CAROTENOIDS DURING HUMAN REPRODUCTION, *INT J VITAM NUTR RES*, 42, 565 (1972)

1223 GALEA G, DAVIDSON RJL, SOME HAEMORHEOLOGICAL AND HAEMATOLOGICAL EFFECTS OF ALCOHOL, *SCAND J HAEMATOL*, 30, 308-310 (1983)

1224 GALEAZZI R, LORENZINI I, ORLANDI F, RIFAMPICIN-INDUCED ELEVATION OF SERUM BILE ACIDS IN MAN, *DIG DIS SCI*, 25, 108-112 (1980)

1225 GALLAGHER B, BAUMEL IP, PRIMIDONE: BIOTRANSFORMATION IN, *ANTIEPILEPTIC DRUGS*, DM WOODBURY, JK PENRY, RP SCHMIDT, EDS. RAVEN PRESS, NEW YORK, NY (1972)

1226 GALLAGHER BB, TRIMETHADIONE AND OTHER OXAGOLIDINEDIONES: TOXICITY, IN, *ANTIEPILEPTIC DRUGS*, DM WOODBURY, JK PENRY, RP SCHMIDT, EDS. RAVEN PRESS, NEW YORK, NY (1972)

1227 GALLANT DM, BISHOP MP, QUIDE VS MELLARIL® IN CHRONIC SCHIZOPHRENIC PATIENTS, *CURR THER RES*, 14, 10 (1972)

1228 GALLER M, FOLKERT VW, SCHLONDORFF D, REVERSIBLE ACUTE RENAL INSUFFICIENCY AND HYPERKALEMIA FOLLOWING INDOMETHACIN THERAPY, *J AM MED ASSOC*, 246, 154-155 (1981)

1229 GALLETTI F, STRAZZULLO P, GAGLIARDI R, ET AL, METABOLIC EFFECTS OF LONG-TERM THERAPY WITH MUZOLIMINE AND CHLORTHALIDONE IN HYPERTENSION, *EUR J CLIN PHARMACOL*, 33, 515-517 (1987)

1230 GALLIN JI, KAYE D, O'LEARY WM, ENDOTOXIN AND LIPIDS, *ANN INTERN MED*, 76, 831 (1972)

1231 GALTEAU MM, SIEST G, VARIATION OF PLASMATIC ENZYMES PRODUCED BY EXERCISE IN, *REFERENCE VALUES IN HUMAN CHEMISTRY*, G SIEST, ED. KARGER, BASEL 1973, 223

1232 GALVAO-TELES A, BURKE CW, CORTISOL LEVELS IN TOXAEMIC AND NORMAL PREGNANCY, *LANCET*, 1, 737 (1973)

1233 GAMBINO SR, PH AND PCO_2, *STAND METH CLIN CHEM*, 5, 169 (1965)

1234 GANDINI R, CUNIETTI E, PAPPALEPORE V, ET AL, EFFECTS OF INTRAVENOUS HIGH DOSES OF KETOPROFEN ON BLOOD CLOTTING, BLEEDING TIME AND PLATELET AGGREGATION, *J INT MED RES*, 11, 243-246 (1983)

1235 GANGULI PC, FORRESTER JM, CIRCADIAN RHYTHM IN PLASMA LEVEL OF GASTRIN, *NATURE NEW BIOL*, 236, 127 (1972)

1236 GANGULI PC, HUNTER WM, RADIOIMMUNOASSAY OF GASTRIN IN HUMAN PLASMA, *J PHYSIOL*, 220, 499 (1972)

1237 GARB S, *CLINICAL GUIDE TO UNDESIRABLE DRUG INTERACTIONS AND INTERFERENCES*, SPRINGER PUB CO, NEW YORK, NY, 10003 (1971)

1238 GARB S, *LABORATORY TESTS IN COMMON USE* 4TH ED, SPRINGER PUB CO, NEW YORK, NY, 10003 (1966)

1239 GARCIA M, MILLER M, MOSES AM, CHLORPROPAMIDE-INDUCED WATER RETENTION IN PATIENTS WITH DIABETES MELLITUS, *ANN INTERN MED*, 75, 549 (1971)

1240 GARDEN JM, FREINKEL RK, SYSTEMIC ABSORPTION OF TOPICAL STEROIDS. METABOLIC EFFECTS AS AN INDEX OF MILD HYPERCORTISOLISM, *ARCH DERMATOL*, 122, 1007-1010 (1986)

1241 GARDNER DF, UTIGER RD, SCHWARTZ SL, ET AL, EFFECTS OF ORAL ERYTHROSINE (2', 4', 5', 7'-TETRAIODOFLUORESCEIN) ON THYROID FUNCTION IN NORMAL MEN, *TOXICOL APPL PHARMACOL*, 91, 299-304 (1987)

1242 GARNETT ES, COHEN H, NAHMIAS C, VIOL G, THE ROLES OF CARBOHYDRATE, RENIN AND ALDOSTERONE IN SODIUM RETENTION DURING AND AFTER TOTAL STARVATION, *METABOLISM*, 22, 867 (1973)

1243 GARRETTSON LK, PEREL JM, DAYTON PG, METHYLPHENIDATE INTERACTION WITH BOTH ANTICONVULSANTS AND ETHYL BISCOUMACETATE: A NEW ACTION OF METHYLPHENIDATE, *J AM MED ASSOC*, 207, 2053 (1969)

1244 GARTON S, LARSEN AE, EFFECT OF HEMOLYSIS ON THE PARTIAL THROMBOPLASTIN TIME, *AM J MED TECHNOL*, 38, 408 (1972)

1245 GARY NE, DODELSON R, EISINGER RP, INDOMETHACIN-ASSOCIATED ACUTE RENAL FAILURE, *AM J MED*, 69, 135-136 (1980)

1246 GASCON-BARRE M, INFLUENCE OF CHRONIC ETHANOL CONSUMPTION ON THE METABOLISM AND ACTION OF VITAMIN D, *J AM COLL NUTR*, 4, 565-574 (1985)

1247 GASKINS JD, HOLT RJ, POSTELNICK M, NONDOSAGE-DEPENDENT VALPROIC ACID INDUCED HYPERAMMONEMIA AND COMA, *CLIN PHARM*, 3, 313-316 (1984)

1248 GASPARD UJ, METABOLIC EFFECTS OF ORAL CONTRACEPTIVES, *AM J OBSTET GYNECOL*, 157, 1029-1041 (1987)

1249 GASPARD UJ, BURET J,, GILLAIN D, ET AL, SERUM LIPID AND LIPO-PROTEIN CHANGES INDUCED BY NEW ORAL CONTRACEPTIVES CONTAINING ETHINYLESTRADIOL PLUS LEVONORGESTREL OR DESOGESTREL, *CONTRACEPTION*, 81, 395-408 (1985)

1250 GATELL JM, SAN MIGUEL JG, ZAMORA L, ET AL, COMPARISON OF THE NEPHROTOXICITY AND AUDITORY TOXICITY OF TOBRAMYCIN AND AMIKACIN, *ANTIMICROB AGENTS CHEMOTHER*, 23, 897-901 (1983)

1251 GAUDREAULT P, TEMPLE AR, LOVEJOY FH JR., THE RELATIVE SEVERITY OF ACUTE VERSUS CHRONIC SALICYLATE POISON-ING IN CHILDREN: A CLINICAL COMPARISON, *PEDIATRICS*, 70, 566-569 (1982)

1252 GAUNT R, STEINETZ BG, CHART JJ, PHARMACOLOGIC ALTERATION OF STEROID HORMONE FUNCTIONS, *CLIN PHARMACOL THER*, 9, 657 (1968)

1253 GAUT ZN, POCELINKO R, SOLOMON HM, THOMAS GB, ORAL GLU-COSE TOLERANCE, PLASMA INSULIN, AND URIC ACID EXCRE-TION IN MAN DURING CHRONIC ADMINISTRATION, *METABO-LISM*, 20, 1031 (1971)

1254 GAUT ZN, TAYLOR WSR, EFFECTS OF LARGE DOSES OF NICOTINYL ALCOHOL ON SERUM LIPID LEVELS AND CARBOHYDRATE, *J CLIN PHARMACOL*, 8, 370 (1968)

1255 GAVRAS I, GRAFF LG, ROSE BD, ET AL, FATAL PANCYTOPENIA ASSOCIATED WITH THE USE OF CAPTOPRIL, *ANN INTERN MED*, 94, 58-59 (1981)

1256 GAWTHORNE JM, WATSON J, STOKSTAD ELR, AUTOMATED METHYLMALONIC ACID ASSAY, *ANAL BIOCHEM*, 42, 555 (1971)

1257 GEANEY DP, CARVER JG, ARONSON JK, ET AL, INTERACTION OF AZAPROPAZONE WITH PHENYTOIN, *BR MED J [CLIN RES]*, 284, 1373 (1982)

1258 GEBHARDT DOE, DUBBELDAM A, RELATION BETWEEN LECI-THIN/SPHINGOMYELIN RATIOS OF AMNIOTIC FLUID AND METHOD OF DETERMINATION, *LANCET*, 1, 726 (1973)

1259 GEISSLER AH, TURNLUND JR, COHEN RD, EFFECT OF CHLORTHALIDONE ON ZINC LEVELS, TESTOSTERONE AND SEXUAL FUNCTION IN MAN, *DRUG NUTR INTERACT*, 4, 275-283 (1986)

1260 GELFAND MD, ISCHEMIC COLITIS ASSOCIATED WITH A DEPOT SYN-THETIC PROGESTOGEN, *AM J DIG DIS*, 17, 275 (1972)

1261 GELTNER D, CHAJEK T, RUBINGER D, ET AL, QUINIDINE HYPERSEN-SITIVITY AND LIVER INVOLVEMENT, *GASTROENTEROLOGY*, 70, 650-652 (1976)

1262 GENTRY LO, WOOD BA, NATELSON EA, EFFECTS OF APALCILLIN ON PLATELET FUNCTION IN NORMAL VOLUNTEERS, *ANTIMICROB AGENTS CHEMOTHER*, 27, 683-684 (1985)

1263 GENUTH SM, EFFECTS OF ORAL ALANINE ADMINISTRATION IN FASTING OBESE SUBJECTS, *METABOLISM*, 22, 927 (1973)

1264 GEOKAS MC, RINDERKNECHT H, PLASMA ARYLSULFATASE AND BETA-GLUCURONIDASE IN ACUTE ALCOHOLISM, *CLIN CHIM ACTA*, 46, 27 (1973)

1265 GEOKAS MC, RINDERKNECHT H, PISES P, OLSEN H, ELEVATED PLASMA ARYLSULFATASE LEVELS IN PLASMA OF ALCOHOLICS, *GASTROENTEROLOGY*, 62, 751 (1972)

1266 GEORGE RH, GENTAMICIN ASSAY IN JAUNDICE, *LANCET*, 1, 838 (1973)

1267 GERARD SK, KHAYAM-BASHI H, CHARACTERIZATION OF CREATI-NINE ERROR IN KETOTIC PATIENTS, *AM J CLIN PATHOL*, 84, 659-664 (1985)

1268 GERBER A, WEIDMANN P, SANER R, ET AL, INCREASED SERUM HIGH DENSITY LIPOPROTEIN CHOLESTEROL IN HYPERTENSIVE MEN TREATED WITH THE POTENT VASODILATOR CAPRAZIDIL, *METABOLISM*, 33, 342-346 (1984)

1269 GERBER N, LYNN R, OATES J, ACUTE INTOXICATION WITH 5,5-DIPHENYLHYDANTOIN (DILANTIN®) ASSOCIATED WITH IMPAIR-MENT OF BIOTRANSFORMATION, *ANN INTERN MED*, 77, 765 (1972)

1270 GERTIG H, NOWACZYK W, GNIADEK M, EFFECT OF ALDRIN, DIEL-DRIN AND LINDANE ON LACTIC ACID DEHYDROGENASE AND CHOLINESTERASE ACTIVITY, *DISS PHARMACEUT PHARMACOL*, 22, 253 (1970)

1271 GESER CA, FELBER JP, BRAND E, SCHULTIS K, UNTERSUCHUNGEN ZUR GLUCAGON-INDUZIERTEN SEKRETION, *KLIN WOCHENSCHR*, 49, 1175 (1971)

1272 GETAZ EP, BECKLEY S, FITZPATRICK J, ET AL, CISPLATIN-INDUCED HEMOLYSIS, *N ENGL J MED*, 302, 334-335 (1980)

1273 GEUMEI A, ISSA I, EL-GENDI M, ABEL-EL-SAMIE Y, INHIBITORY EFFECT OF β-ADRENERGIC-BLOCKING AGENT PROPRANOLOL ON HISTAMINE-STIMULATED GASTRIC ACID SECRETION IN MAN, *AM J DIG DIS*, 17, 55 (1972)

1274 GHANEM MH, FAHMI MH, ABDEL MALEK AT, EOSINOPENIC EFFECT OF ANTISEROTONIN (METHYSERGIDE), *ALEXANDRIA MED J*, 11, 400 (1965)

1275 GHOSE K, TAYLOR A, HYPERCUPRICAEMIA INDUCED BY ANTIEPILEPTIC DRUGS, *HUM TOXICOL*, 3, 519-529 (1983)

1276 GIANSIRACUSA DF, BLUMBERG S, KANTROWITZ FG, ASEPTIC MEN-INGITIS ASSOCIATED WITH IBUPROFEN, *ARCH INTERN MED*, 140, 1553 (1980)

1277 GIFFORD RW JR., CATAPRES® (ST 155) IN THE MANAGEMENT OF HYPERTENSION, IN *CATAPRES® IN HYPERTENSION*, ME CONOLLY,ED., BUTTERWORTH, LONDON (1970)

1278 GIGUERE R, AURAY-BLAIS C, DRAPER P, ET AL, DIET AND MEDICA-TIONS GIVING POSITIVE NINHYDRIN REACTIONS IN TLC IN A NEWBORN URINARY SCREENING PROGRAM, *CLIN BIOCHEM*, 13, 103-105 (1980)

1279 GIL-AD I, BAR-YOSEPH J, SMADJA Y, ET AL, EFFECT OF CLONIDINE ON PLASMA β-ENDORPHIN, CORTISOL AND GROWTH HOR-MONE SECRETION IN OPIATE-ADDICTED SUBJECTS, *ISR J MED SCI*, 21, 601-604 (1985)

1280 GILBERT EF, DASILVA AQ, QUEEN DM, INTRAHEPATIC CHOLES-TASIS WITH FATAL TERMINATION FOLLOWING NORETHAN-DROLONE THERAPY, *J AM MED ASSOC*, 185, 538 (1963)

1281 GILBERTSON C, JONES DR, HAEMOLYTIC ANAEMIA WITH NALIDIXIC ACID, *BR MED J [CLIN RES]*, 4, 493 (1972)

1282 GILHUS NE, STRANDJORD RE, AARLI JA, THE EFFECT OF CARBAM-AZEPINE ON SERUM IMMUNOGLOBULIN CONCENTRATIONS, *ACTA NEUROL SCAND*, 66, 172-179 (1982)

1283 GILL JS, BEEVERS DG, BUCINDOL: EFFECTS ON BLOOD PRESSURE, AIRWAYS RESISTANCE AND SERUM CREATINE PHOSPHOKINASE, *EUR J CLIN PHARMACOL*, 27, 265-268 (1984)

1284 GILL MJ, RATLIFF DA, HARDING LK, HYPOGLYCEMIC COMA, JAUN-DICE, AND PURE RBC APLASIA FOLLOWING CHLORPROPAMIDE THERAPY, *ARCH INTERN MED*, 140, 714-715 (1980)

1285 GILLER E JR., NOCKS J, HALL H, ET AL, PLATELET AND FIBROBLAST MONOAMINE OXIDASE IN ALCOHOLISM, *PSYCHIATRY RES*, 12, 339-347 (1984)

1286 GINDER EM, KING JD, RAPID COLORIMETRIC DETERMINATION OF CALCIUM IN BIOLOGIC FLUIDS WITH METHYLTHYMOL BLUE, *AM J CLIN PATHOL*, 58, 376 (1972)

1287 GINSBERG DJ, SAAD A, GABUZDA GJ, METABOLIC STUDIES WITH THE DIURETIC TRIAMTERENE IN PATIENTS WITH CIRRHOSIS AND ASCITES, *N ENGL J MED*, 271, 1229 (1964)

1288 GINSBERG H, OLEFSKY J, FARQUHAR JW, REAVEN GM, MODERATE ETHANOL INGESTION AND PLASMA TRIGLYCERIDE LEVELS, *ANN INTERN MED*, 80, 143 (1974)

1289 GIPS CH, CURVE PATTERNS AFTER ORAL LOADING WITH AMMO-NIUM ACETATE, *CLIN CHIM ACTA*, 46, 415 (1973)

1290 GIPS CH, REITSEMA A, INFLUENCE OF NONPROTEIN NITROGEN SUBSTANCES ON THE INDOPHENOL REACTION, *CLIN CHIM ACTA*, 33, 257 (1971)

1291 GIRGIS M, DUGA J, LEVY LJ, GORDON EE, NONDIABETIC KETOACIDOSIS: A COMPLICATION OF ALCOHOLISM, *ANN INTERN MED*, 76, 857 (1972)

1292 GIRLING DJ, ADVERSE EFFECTS OF ANTITUBERCULOSIS DRUGS, *DRUGS*, 23, 56-74 (1982)

1293 GIROUD M, FABRE JL, BRUN JM, ET AL, DIPHENYLHYDANTOIN EFFECTS ON THYROID FUNCTION, *THERAPIE*, 40, 119-122 (1985)

1294 GITLIN N, SALICYLATE HEPATOTOXICITY: THE POTENTIAL ROLE OF HYPOALBUMINEMIA, *J CLIN GASTROENTEROL*, 2, 281-285 (1980)

1295 GITSCH E, SPONA J, SERUM ESTRADIOL AND LH DURING NORMAL AND INDUCED MENSTRUAL CYCLES, *ENDOCRINOL EXP*, 7, 53 (1973)

1296 GIUGLIANO D, TORELLA R, CACIAPUOTI F, ET AL, IMPAIRMENT OF INSULIN SECRETION IN MAN BY NIFEDIPINE, *EUR J CLIN PHARMACOL*, 18, 395-398 (1980)

1297 GJONE E, BLOMHOFF JP, RITLAND S, ELGJO K, HUSBY G, LAXA-TIVE-INDUCED CHRONIC LIVER DISEASE, *SCAND J GAS-TROENTEROL*, 7, 395 (1972)

1298 GJONE E, STAVE R, LIVER DISEASE ASSOCIATED WITH A NONCON-STIPATING IRON PREPARATION, *LANCET*, 1, 421 (1973)

1299 GLASS AR, KETOCONAZOLE-INDUCED STIMULATION OF GONADO-TROPIN OUTPUT IN MEN: BASIS FOR A POTENTIAL TEST OF GONADOTROPIN RESERVE., *J CLIN ENDOCRINOL METAB*, 63, 1121-1125 (1986)

1300 GLASS AR, EIL C, KETOCONAZOLE-INDUCED REDUCTION IN SERUM 1, 25-DIHYDROXYVITAMIN D, *J CLIN ENDOCRINOL METAB*, 63, 766-769 (1986)

1301 GLASS D, EVANS JI, DALY JR, SLEEP ASSOCIATED GROWTH HOR-MONE (GH) RELEASE AND THE EFFECT OF CORTICOTROPIN INJECTIONS, *CLIN SCI*, 43, 5P (1972)

1302 GLEASON MN, GOSSLEIN RE, HODGE HC, *CLINICAL TOXICOLOGY OF COMMERCIAL PRODUCTS*, WILLIAMS AND WILKINS, BALTI-MORE, MD (1957)

1303 GLEESON JM, DUKES CS, ELSTAD NL, ET AL, EFFECTS OF ESTROGEN/PROGESTIN AGENTS ON PLASMA RETINOIDS AND CHYLOMICRON REMNANT METABOLISM, *CONTRACEPTION*, 35, 69-78 (1987)

1304 GLEISPACK H, EXCRETION OF ANDROGENS, PREGNANES AND OES-TROGENS DEPENDING ON AGE AND SEX IN, *REFERENCE VAL-UES IN HUMAN CHEMISTRY*, G SIEST, ED. KARGER, BASEL 1973, 68

1305 GLICK JH, DEFICIENCIES OF SIGMA DIAGNOSTICS URINARY OXA-LATE METHOD IN THE PRESENCE OF ASCORBATE, *CLIN CHEM*, 33, 419-420 (1987)

1306 GLUCK Z, BAUMGARTNER G, WEIDMANN P, ET AL, INCREASED RATIO BETWEEN SERUM BETA- AND ALPHA-LIPOPROTEINS DURING DIURETIC THERAPY: AN ADVERSE EFFECT?, *CLIN SCI MOL MED*, 55, 325S-328S (1978)

1307 GLUCK Z, WEIDMANN P, MORDASINI R, ET AL, EINFLUSS EINER DIURETIC-THERAPIE AUF DIE SERUMLIPOPROTEINE: EIN UNERWUNSCHTER EFFECT, *SCHWEIZ MED WOCHENSCHR*, 109, 104-108 (1979)

1308 GLUECK CJ, EFFECTS OF OXANDROLONE ON PLASMA TRIGLYCER-IDES AND POSTHEPARIN LIPOLYTIC ACTIVITY, *METABOLISM*, 20, 691 (1971)

1309 GLUECK CJ, FORD S JR., STEINER P, BUXTON S, FALLAT R, EFFECTS OF A PROGESTATIONAL ORAL CONTRACEPTIVE ON LIPIDS AND LIPASES, *AM J OBSTET GYNECOL*, 116, 689 (1973)

1310 GLYNN KP, CAFARO AF, FOWLER CW, STEAD WW, FALSE ELEVA-TIONS OF SERUM GLUTAMIC-OXALOACETIC TRANSAMINASE DUE TO PARA-AMINOSALICYLIC, *ANN INTERN MED*, 72, 525 (1970)

1311 GOCHMAN N, *PERSONAL COMMUNICATION*

1312 GODFREY NF, PETER A, SIMON TM, ET AL, IV N-ACETYLCYSTEINE TREATMENT OF HEMATOLOGIC REACTIONS TO CHRYSOTHER-APY, *J RHEUMATOL*, 9, 519-526 (1982)

1313 GODSALL JW, BARON R, INSOGNA KL, VITAMIN D METABOLISM AND BONE HISTOMORPHOMETRY IN A PATIENT WITH ANTACID-INDUCED OSTEOMALACIA, *AM J MED*, 77, 747-750 (1984)

1314 GOEBELL H, STEFFEN C, BALTZER G, SCHLOTT KA, BODE C, STIMU-LATION OF ENZYME SECRETION IN THE PANCREAS BY ACUTE HYPERCALCEMIA, *GERM MED*, 2, 16 (1972)

1315 GOEBELSMANN V, THERNEYCROFT IH, NAKAMURA RM, MISHELL DR JR., ESTRIOL IN PREGNANCY 1. RADIOIMMUNOASSAY FOR URINARY ESTRIOL, *AM J OBSTET GYNECOL*, 112, 802 (1972)

1316 GOEDEL-MEINEN L, SCHMIDT G, WIRTZFELD A, ET AL, THE INFLU-ENCE OF AMIODARONE ON THE FUNCTION OF THE THYROID GLAND, *Z KARDIOL*, 73, 399-404 (1984)

1317 GOFMAN JW, LINDGREN FT, STRISOWER B, DELALLA O, GLAZIER F, TAMPLIN A, CIGARETTE SMOKING, SERUM LIPOPROTEINS, AND CORONARY HEART DISEASE, *GERIATRICS*, 10, 349 (1955)

1318 GOGEL E, HALLORAN BP, STREWLER GJ, PROBENECID INHIBITS THE SECRETION OF NEPHROGENOUS ADENOSINE-3'-5'-MONOPHOSPHATE IN NORMAL MAN, *J CLIN ENDOCRINOL METAB*, 57, 689-693 (1983)

1319 GOLD EJ, MERTELSMANN RH, ITRI LM, ET AL, PHASE I CLINICAL TRIAL OF 13 - CIS - RETINOIC ACID IN MYELODYSPLASTIC SYN-DROMES, *CANCER TREAT REP*, 67, 981-986 (1983)

1320 GOLD GL, FRITZ RD, HYPERURICEMIA ASSOCIATED WITH THE TREATMENT OF ACUTE LEUKEMIA, *ANN INTERN MED*, 47, 428 (1957)

1321 GOLD RJM, SCRIVER CR, THE AMINO ACID COMPOSITION OF HAIR FROM DIFFERENT RACIAL ORIGINS, *CLIN CHIM ACTA*, 33, 465 (1971)

1322 GOLDBERG A, PORPHYRINS AND PORPHYRIAS in, *RECENT ADVANCES IN HAEMATOLOGY*, A GOLDBERG, MC BRAIN, EDS., CHURCHILL LIVINGSTONE, EDINBURGH (1971)

1323 GOLDBERG DM, WATTS C, SERUM ENZYME CHANGES AS EVI-DENCE OF LIVER REACTION TO ORAL ALCOHOL, *GASTROEN-TEROLOGY*, 49, 256 (1965)

1324 GOLDBERG IJL, LAWTON K, REDDING JR, ET AL, EFFECT OF PENICILLAMINE ON BLOOD LEVELS AND URINARY EXCRETION OF GOLD, *ANN CLIN BIOCHEM*, 20, 220 (1983)

1325 GOLDBERG LI, MONOAMINE OXIDASE INHIBITORS. ADVERSE REAC-TIONS AND POSSIBLE MECHANISMS, *J AM MED ASSOC*, 190, 456 (1964)

1326 GOLDBERG LI, CARDIOVASCULAR AND RENAL ACTIONS OF DOPAMINE: POTENTIAL CLINICAL APPLICATIONS, *PHARMACOL REV*, 24, 1 (1972)

1327 GOLDENBERG H, SPECIFIC PHOTOMETRIC DETERMINATION OF 5-HYDROXYINDOLEACETIC ACID IN URINE, *CLIN CHEM*, 19, 38 (1973)

1328 GOLDENBERG H, SPECIFIC PHOTOMETRIC DETERMINATION OF 5-HYDROXYINDOLEACETIC ACID IN URINE, *CLIN CHEM*, 13, 697 (1967)

1329 GOLDENBERG H, DREWES PA, DIRECT PHOTOMETRIC DETERMINA-TION OF GLOBULIN IN SERUM, *CLIN CHEM*, 17, 358 (1971)

1330 GOLDFARB S, SINGER EJ, POPPER H, BILIARY DUCTULES AND BILE SECRETION, *J LAB CLIN MED*, 62, 608 (1963)

1331 GOLDFINGER S, KLINENBERG JR, SEEGMILLER JE, RENAL RETEN-TION OF URIC ACID INDUCED BY INFUSION OF BETA-HYDROX-YBUTYRATE AND ACETOACETATE, *N ENGL J MED*, 272, 351 (1965)

1332 GOLDFINGER SE, DRUG THERAPY - TREATMENT OF GOUT, *N ENGL J MED*, 285, 1303 (1971)

1333 GOLDHABER SZ, HENNEKENS CH, SPARK RF, ET AL, PLASMA RENIN SUBSTRATE, RENIN ACTIVITY AND ALDOSTERONE LEVELS IN A SAMPLE OF ORAL CONTRACEPTIVE USERS FROM A COMMU-NITY SURVEY, *AM HEART J*, 107, 119-122 (1984)

1334 GOLDMAN AI, STEELE BW, SCHNAPER HW, ET AL, SERUM LIPOPRO-TEIN LEVELS DURING CHLORTHALIDONE THERAPY, *J AM MED ASSOC*, 244, 1691-1695 (1980)

1335 GOLDMAN JA, ECKERLING B, EFFECT OF A PROGESTOGEN ORAL CONTRACEPTIVE COMPOUND ON CARBOHYDRATE METABO-LISM, *ISR J MED SCI*, 8, 1724 (1972)

1336 GOLDSMITH GA, THERAPY OF HYPERCHOLESTEROLEMIA, *AM J DIG DIS*, 9, 651 (1964)

1337 GOLDSMITH NF, GOLDSMITH JR, EPIDEMIOLOGICAL ASPECTS OF MAGNESIUM AND CALCIUM METABOLISM, *ARCH ENVIRON HEALTH*, 12, 607 (1966)

1338 GOLDSTEIN G, SARCOID REACTION ASSOCIATED WITH PHENYLBUTAZONE HYPERSENSITIVITY, *ANN INTERN MED*, 59, 97 (1963)

1339 GOLDSTEIN M, FREEDMAN LS, BOHUON AC, GUERINOT F, SERUM DOPAMINE-β-HYDROXYLASE ACTIVITY IN NEUROBLASTOMA, *N ENGL J MED*, 286, 1123 (1972)

1340 GOLDZIEHER JW, KLEBER JW, MOSES LE, RATHMACHER RP, A CROSS-SECTIONAL STUDY OF PLASMA FSH AND LH LEVELS IN WOMEN, *CONTRACEPTION*, 2, 225 (1970)

1341 GOLF SW, TEMME H, KEMPF DK, ET AL, SYSTEMIC SHORT-TERM FIBRINOLYSIS WITH HIGH DOSE STREPTOKINASE IN ACUTE MYOCARDIAL INFARCTION: TIME COURSE OF BIOCHEMICAL PARAMETERS, *J CLIN CHEM CLIN BIOCHEM*, 22, 723-729 (1984)

1342 GOODALE RM, *CLINICAL INTERPRETATION OF LABORATORY TESTS* 5TH ED, DAVIS, PHILADELPHIA, PA (1964)

1343 GOODMAN LS, GILMAN A, *THE PHARMACOLOGICAL BASIS OF THERAPEUTICS* 4TH ED, MACMILLAN CO, NEW YORK, NY (1970)

1344 GOODWIN FJ, LEDINGHAM JG, LARAGH JH, THE EFFECTS OF PRO-LONGED ADMINISTRATION OF VASOPRESSIN AND OXYTOCIN ON RENIN, ALDOSTERONE AND SODIUM BALANCE, *CLIN SCI*, 39, 641 (1970)

1345 GOODWIN JF, SPECTROPHOTOMETRY OF PROLINE IN PLASMA AND URINE, *CLIN CHEM*, 18, 449 (1972)

1346 GOODWIN JF, SPECTROPHOTOMETRIC QUANTITATION OF PLASMA AND URINARY NITROGEN WITH FLUORODINITROBENZENE, *STAND METH CLIN CHEM*, 6, 89 (1970)

1347 GOODWIN JS, GARRY PJ, RELATIONSHIP BETWEEN MEGADOSE VITAMIN SUPPLEMENTATION AND IMMUNOLOGICAL FUNCTION IN A HEALTHY ELDERLY POPULATION, *CLIN EXP IMMUNOL*, 51, 647-653 (1983)

1348 GORCHEIN A, WEBBER R, BURNETT D, ET AL, EFFECT OF ANTICON-VULSANT DRUGS ON SERUM DELTA-AMINOLEVULINIC ACID LEVELS IN NON-PORPHYRIC SUBJECTS, *BR J CLIN PHARMACOL*, 24, 847-848 (1987)

1349 GORDON AM, ROWAN RM, BROWN T, CARSON HG, ROUTINE APPLI-CATION OF THE NITROBLUE TETRAZOLIUM TEST IN THE CLINICAL LABORATORY, *J CLIN PATHOL*, 26, 52 (1973)

1350 GORDON D, BEASTALL GH, MCARDLE CS, ET AL, THE EFFECT OF TAMOXIFEN THERAPY ON THYROID FUNCTION TESTS, *CAN-CER*, 58, 1422-1425 (1986)

1351 GORDON EK, OLIVER J, GOODWIN FK, CHASE TN, POST RM, EFFECT OF PROBENECID ON FREE 3-METHOXY-4-HYDROX-YPHENYLETHYLENE GLYCOL (MHPG) AND ITS SULFATE IN HUMAN CSF, *NEUROPHARMACOLOGY*, 12, 391 (1973)

1352 GORDON GG, ALTMAN K, SOUTHERN AL, ET AL, EFFECT OF ALCO-HOL (ETHANOL) ADMINISTRATION ON SEX HORMONE METABO-LISM IN NORMAL MEN, *N ENGL J MED*, 295, 793-797 (1976)

1353 GORDON RD, PAWSEY CGK, O'HALLORAN MW, ABBOTT ML, WIL-SON LL, SILVERSTONE H, USE OF HOME BLOOD PRESSURE MEASUREMENT TO COMPARE THE EFFICACY OF TWO DIURET-ICS, *MED J AUST*, 2, 565 (1971)

1354 GOREN MP, WRIGHT PK, OSBORNE S, TWO AUTOMATED PROCE-DURES FOR N-ACETYL-BETA-GLUCOSAMINIDASE DETERMINA-TION EVALUATED FOR DETECTION OF DRUG-INDUCED TUBU-LAR NEPHROTOXICITY, *CLIN CHEM*, 32, 2052-2055 (1986)

1355 GOREN MP, WRIGHT RK, HOROWITZ ME, ET AL, CANCER CHEMO-THERAPY INDUCED TUBULAR NEPHROTOXICITY EVALUATED BY IMMUNOCHEMICAL DETERMINATION OF URINARY ADENO-SINE BINDING PROTEIN, *AM J CLIN PATHOL*, 86, 780-783 (1986)

1356 GORMSEN J, BIOCHEMICAL EVALUATION OF STANDARD TREAT-MENT WITH STREPTOKINASE IN ACUTE MYOCARDIAL INFARC-TION, *ACTA MED SCAND*, 191, 77 (1972)

1357 GOSLING R, KERRY RJ, OWEN G, CREATINE PHOSPHOKINASE ACTIVITY DURING LITHIUM TREATMENT, *BR MED J [CLIN RES]*, 3, 327 (1972)

1358 GOSNEY K, ADACHI-KIRKLAND J, SCHILLER HS, EVALUATION OF LIDOCAINE INTERFERENCE IN THE KODAK EKTACHEM® 700 ANALYZER SINGLE SLIDE METHOD FOR CREATININE, *CLIN CHEM*, 33, 2311 (1987)

1359 GOSS H, DICKHAUS DW, INCREASED BISHYDROXYCOUMARIN REQUIREMENTS IN PATIENTS RECEIVING PHENOBARBITAL, *N ENGL J MED*, 273, 1094 (1965)

1360 GOSSAIN VV, SHERMA NK, SRIVASTAVA L, ET AL, HORMONAL EFFECTS OF SMOKING. II. EFFECTS ON PLASMA CORTISOL, GROWTH HORMONE, AND PROLACTIN, *AM J MED SCI*, 291, 325-327 (1986)

1361 GOSSAIN VV, SHERMA NK, SRIVASTAVA L, ET AL, HORMONAL EFFECTS OF SMOKING I. EFFECTS ON PLASMA RENIN ACTIVITY, *AM J MED SCI*, 291, 321-324 (1986)

1362 GOTTLIEB NL, BUCHOFF HS, VIDAL AF, ET AL, THE COURSE OF SEVERE GOLD ASSOCIATED GRANULOCYTOPENIA, *CLIN RES*, 30, 659A (1982)

1363 GOUGOUX A, MICHAUD G, VINAY P, THE URICOSURIC ACTION OF BENZIODARONE IN MAN AND DOG, *CLIN RES*, 18, 747 (1970)

1364 GOURSOT G, DERIAZ H, DUFIEUX P, ET AL, HAEMODYNAMICS, PLASMA HISTAMINE AND OSMOLALITY DURING AORTIC ARTE-RIOGRAPHY, *ANN FR ANESTH REANIM*, 3, 90-93 (1984)

1365 GOVERDE BC, VEENKAMP FJN, ROUTINE ASSAY OF TOTAL URI-NARY HYDROXYPROLINE BASED ON RESIN-CATALYZED HYDROLYSIS, *CLIN CHIM ACTA*, 41, 29 (1972)

1366 GOYOT C, DEBRAY Q, HUG R, ET AL, BASAL PLASMA LEVEL EVALU-ATION OF THREE ANTERIOR PITUITARY HORMONES DURING ACUTE OR CHRONIC TREATMENT WITH TRICYCLIC ANTIDEPRESSANTS, *ENCEPHALE*, 11, 45-51 (1985)

1367 GRABNER W, BERGNER D, SAILER D, BERG G, ACCURACY OF QUANTITATIVE IMMUNOGLOBULIN ASSAYS BY SIMPLE RADIAL IMMUNODIFFUSION, *CLIN CHIM ACTA*, 39, 59 (1972)

1368 GRACE E, EMANS SJ, DRUM DE, HEMATOLOGIC ABNORMALITIES IN ADOLESCENTS WHO TAKE ORAL CONTRACEPTIVE PILLS, *J PEDIATR*, 101, 771-774 (1982)

1369 GRAEF V, STAUDINGER H, STUDIES ON SPECIFICITY OF THE FLUO-ROMETRIC DETERMINATION OF PLASMA 11-HYDROXYCORTI-COIDS, *Z KLIN CHEM KLIN BIOCHEM*, 8, 368 (1970)

1370 GRAF M, TARLOV A, AGRANULOCYTOSIS WITH MONOHISTIO-CYTOSIS ASSOCIATED WITH AMPICILLIN THERAPY, *ANN INTERN MED*, 69, 91 (1968)

1371 GRAFT DF, CHESNEY PJ, USE OF TICARCILLIN FOLLOWING CARBENICILLIN-ASSOCIATED HEPATOTOXICITY, *J PEDIATR*, 100, 497-499 (1982)

1372 GRAHAM GG, CHAMPION GD, DAY RO, ET AL, PATTERNS OF PLASMA CONCENTRATIONS AND URINARY EXCRETION OF SAL-ICYLATE IN RHEUMATOID ARTHRITIS, *CLIN PHARMACOL THER*, 22, 410-420 (1977)

1373 GRAHAM P, NAIDOO D, FALSE-POSITIVE KETOSTIX® IN A DIABETIC ON ANTIHYPERTENSIVE THERAPY, *CLIN CHEM*, 33, 1490 (1987)

1374 GRAHNEN A, SJOHOLM I, SPECTROPOLARIMETRIC DETERMINATION OF UNCONJUGATED BILIRUBIN IN HUMAN SERUM, *Z KLIN CHEM KLIN BIOCHEM*, 12, 220 (1974)

1375 GRAISELY B, EMERY JP, HUGUES FC, MARCHE J, RIFAMPICINE ET ICTERE OBSERVATIONS CHEZ L'HOMME, *THERAPIE*, 26, 655 (1971)

1376 GRAJOWER MM, WALTER L, ALBIN J, HYPOGLYCEMIA IN CHRONIC HEMODIALYSIS PATIENTS: ASSOCIATION WITH PROPRANOLOL USE, *NEPHRON*, 26, 126-129 (1980)

1377 GRALNICK HR, MCGINNISS M, ELTON W, MCCURDY P, HEMOLYTIC ANEMIA ASSOCIATED WITH CEPHALOTHIN, *J AM MED ASSOC*, 217, 1193 (1971)

1378 GRALNICK HR, WRIGHT LD JR., MCGINNISS MH, COOMBS' POSITIVE REACTIONS ASSOCIATED WITH SODIUM CEPHALOTHIN THERAPY, *J AM MED ASSOC*, 199, 725 (1967)

1379 GRAN JT, HUSBY G, THORSBY E, HLA DR, ANTIGENS AND GOLD TOXICITY, *ANN RHEUM DIS*, 42, 63-66 (1983)

1380 GRANBERG PO, WAHLIN A, THE EFFECT OF METHOXYFLURANE (PENTHRANE®) ON THE RENAL FUNCTION WITH SPECIAL REF-ERENCE TO TUBULAR REJECTION, *ACTA ANAESTHESIOL SCAND*, 16, 216 (1972)

1381 GRANERUS G, WETTERQUIST H, WHITE T, URINARY EXCRETION OF HISTAMINE, METHYLHISTAMINE AND METHYLIMADAZOLEACETIC ACIDS IN MAN, *SCAND J CLIN LAB INVEST* 22, SUPPL 104, 59 (1968)

1382 GRANERUS G, WETTERQVIST H, WHITE T, HISTAMINE METABOLISM IN HEALTHY SUBJECTS BEFORE AND DURING TREATMENT WITH AMINOGUANIDINE, *SCAND J CLIN LAB INVEST* 22, SUPPL 104, 39 (1968)

1383 GRANINGER W, PIRICH KR, SPEISER W, ET AL, EFFECT OF THYROID HORMONES ON PLASMA PROTEIN CONCENTRATIONS IN MAN, *J CLIN ENDOCRINOL METAB*, 63, 407-411 (1986)

1384 GRAU JJ, ESTAPE J, DANIELS M, ET AL, CISPLATIN AND PLASMA IRON LEVELS, *ANN INTERN MED*, 103, 158-159 (1985)

1385 GREAVES MW, MCDONALD-GIBSON WJ, MCDONALD-GIBSON RG, THE EFFECT OF VENOUS OCCLUSION, STARVATION AND EXER-CISE ON PROSTAGLANDIN ACTIVITY IN WHOLE HUMAN BLOOD, *LIFE SCI 11, PART 2,*, 919 (1972)

1386 GREEN D, DAVIES RO, HOLMES GI, ET AL, EFFECTS OF DIFLUNISAL ON PLATELET FUNCTION AND FECAL BLOOD LOSS, *CLIN PHARMACOL THER*, 30, 378-384 (1981)

1387 GREEN D, TS'AO C-H, CERULLO L, ET AL, CLINICAL AND LABORA-TORY INVESTIGATIONS OF THE EFFECTS OF ε-AMINOCAPROIC ACID ON HEMOSTASIS, *J LAB CLIN MED*, 105, 321-327 (1985)

1388 GREEN JA, CLEMENTI WA, PORTER C, ET AL, NIFEDIPINE-QUINIDINE INTERACTION, *CLIN PHARMACOL*, 2, 461-465 (1983)

1389 GREEN JR, KUPFERBERG HJ, SULFONAMIDES AND DERIVATIVES: SULTHIAME IN, *ANTIEPILEPTIC DRUGS*, DM WOODBURY, JK PENRY, RP SCHMIDT, EDS. RAVEN PRESS, NEW YORK, NY (1972)

1390 GREENBERG SR, MORE ON LEVODOPA - IMPLICATIONS FOR ADRE-NAL FUNCTION, *N ENGL J MED*, 286, 375 (1972)

1391 GREENBLATT DJ, DUHME DW, ALLEN MD, ET AL, CLINICAL TOXICITY OF FUROSEMIDE IN HOSPITALIZED PATIENTS, *AM HEART J*, 94, 6-13 (1977)

1392 GREENBLATT DJ, DUHME DW, KOCH-WESER J, PAIN AND CPK ELE-VATION AFTER INTRAMUSCULAR DIGOXIN, *N ENGL J MED*, 288, 689 (1973)

1393 GREENBLATT DJ, KOCH-WESER J, ADVERSE REACTIONS TO SPIRO-NOLACTONE, *J AM MED ASSOC*, 225, 39 (1973)

1394 GREENE HL, GRAHAM EL, WERNER JA, ET AL, TOXIC AND THERA-PEUTIC EFFECTS OF AMIODARONE IN THE TREATMENT OF CARDIAC ARRHYTHMIAS, *J AM COLL CARDIOL*, 2, 1114-1128 (1983)

1395 GREENE ML, FUJIMOTO WY, SEEGMILLER JE, URINARY XANTHINE STONES, A RARE COMPLICATION OF ALLOPURINOL THERAPY, *N ENGL J MED*, 280, 426 (1969)

1396 GREENLAW R, PRYOR J, WINNACKER J, HAHN T, HADDAD J, ANAST C, OSTEOMALACIA (OM) FROM ANTICONVULSANT DRUGS (ACV): THERAPEUTIC IMPLICATIONS, *CLIN RES*, 20, 56 (1972)

1397 GREENSTEIN AJ, KAYNAN A, SINGER A, DREILING DA, A COMPARA-TIVE STUDY OF PENTAZOCINE AND MEPERIDINE ON THE BILI-ARY PASSAGE PRESSURE, *AM J GASTROENTEROL*, 58, 47 (1972)

1398 GREENSTEIN R, NOGEIRE C, OHNUMA T, ET AL, MANAGEMENT OF ASPARAGINASE INDUCED HEMORRHAGIC PANCREATITIS COM-PLICATED BY PSEUDOCYST, *CANCER*, 43, 718-722 (1979)

1399 GRETZER I, ROSSNER S, LONG-TERM EFFECTS OF PINDOLOL ON SERUM LIPOPROTEINS IN HYPERTENSIVE PATIENTS, *ACTA MED SCAND*, 219, 367-370 (1986)

1400 GRIFFITHS ID, RICHARDSON J, LUPUS-TYPE ILLNESS ASSOCIATED WITH LABETALOL, *BR MED J [CLIN RES]*, 2, 496 (1979)

1401 GRIFFITHS J, LINKLATER H, A RADIOISOTOPE METHOD FOR CAT-ECHOL-O-METHYL TRANSFERASE IN BLOOD, *CLIN CHIM ACTA*, 39, 383 (1972)

1402 GRIGGS RC, CONDEMI JJ, VAUGHAN JH, EFFECT OF THERAPEUTIC DOSAGE OF PREDNISONE ON HUMAN IMMUNOGLOBULIN G METABOLISM, *J ALLERGY CLIN IMMUNOL*, 49, 267 (1972)

1403 GRIMBLE RF, WHITEHEAD RG, THE EFFECT OF AN ORAL GLUCOSE LOAD ON SERUM FREE AMINO ACID CONCENTRATION IN CHIL-DREN BEFORE AND AFTER TREATMENT, *BR J NUTR*, 25, 253 (1971)

1404 GRIMM RH JR., LEON AS, HUNNINGHAKE DB, ET AL, EFFECT OF THIAZIDE DIURETICS ON PLASMA LIPIDS AND LIPOPROTEINS IN MILDLY HYPERTENSIVE PATIENTS, *ANN INTERN MED*, 94, 7-11 (1981)

1405 GRIMM RH JR., LEON AS, HUNNINGHAKE DB, ET AL, DIURETICS AND PLASMA LIPIDS: EFFECTS OF THIAZIDES AND SPIRONO-LACTONE. IN NOSEDA ET AL (EDS) *LIPOPROTEINS AND CORO-NARY ATHEROSCLEROSIS*, ELSEVIER BIOMEDICAL PRESS, AMSTERDAM, 371-376 (1982)

1406 GROB PJ, HEROLD GE, IMMUNOLOGICAL ABNORMALITIES AND HYDANTOINS, *BR MED J [CLIN RES]*, 2, 561 (1972)

1407 GROEL JT, TADROS SS, DRESLINSKI GR, ET AL, LONG-TERM ANTIHYPERTENSIVE THERAPY WITH CAPTOPRIL, *HYPERTEN-SION*, 5 SUPPL 3, 145-151 (1983)

1408 GRONERT GA, THEYE RA, PATHOPHYSIOLOGY OF HYPERKALEMIA INDUCED BY SUCCINYLCHOLINE, *ANESTHESIOLOGY*, 43, 89-99 (1975)

1409 GRONHAGEN-RISKA C, HELLSTROM P-E, FROSETH B, PREDISPOS-ING FACTORS IN HEPATITIS INDUCED BY ISONIAZID-RIFAMPIN TREATMENT OF TUBERCULOSIS, *AM REV RESPIR DIS*, 118, 461-466 (1978)

1410 GROOP L, TOTTERMAN KJ, HARNO K, ET AL, INFLUENCE OF BETA-BLOCKING DRUGS ON GLUCOSE METABOLISM IN HYPERTEN-SIVE, NON-DIABETIC PATIENTS, *ACTA MED SCAND*, 213, 9-14 (1983)

1411 GROOP P-H, GROOP LC, TOTTERMAN KJ, ET AL, THE EFFECT OF DEXAMETHASONE ON THE ENTEROINSULAR AXIS, *SCAND J CLIN LAB INVEST*, 47, 491-495 (1987)

1412 GROSS L, OXYPHENBUTAZONE-INDUCED PAROTITIS, *ANN INTERN MED*, 70, 1229 (1969)

1413 GROSSBARD L, MARKS PA, ENZYMES IN HEMATOLOGIC DISEASE IN, *DIAGNOSTIC ENZYMOLOGY*, EL COODLEY ED. LEA AND FEBIGER, PHILADELPHIA, PA (1970)

1414 GROTSCH H, HAJDU P, INTERFERENCE BY THE NEW ANTIBIOTIC CEFIPIROME AND OTHER CEPHALOSPORINS IN CLINICAL LABO-RATORY TESTS, WITH SPECIAL REGARD TO THE JAFFE REAC-TION, *J CLIN CHEM CLIN BIOCHEM*, 25, 49-52 (1987)

1415 GRUEN LC, RIVETT DE, SPECIFICITY OF SPIES AND CHAMBERS REA-GENT FOR COLORIMETRIC ESTIMATION OF TRYPTOPHAN, *ANAL BIOCHEM*, 44, 519 (1971)

1416 GRUEN PH, SACHAR EJ, LANGER G, ET AL, PROLACTIN RESPONSES TO NEUROLEPTICS IN NORMAL AND SCHIZOPHRENIC SUB-JECTS, *ARCH GEN PSYCHIATRY*, 35, 108-116 (1978)

1417 GRUNDING E, BIRNBAUMER E, KOLLNER U, ET AL, INFLUENCE OF DRUGS ON AMINOTRANSFERASE ACTIVITY IN HUMAN BLOOD SERUM AND CEREBROSPINAL FLUID, *LAB*, 4, 323 (1977)

1418 GUAGNELLINI E, BERTOLINI G, CAPPELLETTI M, ET AL, THROMBIN INHIBITORS IN WOMEN ON ORAL CONTRACEPTIVES, *ACTA HAEMATOL (BASEL)*, 65, 205-210 (1981)

1419 GUATTERY JM, FALOON WW, EFFECT OF ESTRADIOL UPON SERUM ENZYMES IN PRIMARY BILIARY CIRRHOSIS, *HEPATOLOGY*, 7, 737-742 (1987)

1420 GUERRA M, TOXICITY OF INDOMETHACIN REPORT OF A CASE OF ACUTE PANCREATITIS, *J AM MED ASSOC*, 200, 552 (1967)

1421 GUGLIOTTA L, VIGANO S, D'ANGELO A, ET AL, HIGH FIBRINOPEPTIDE A (FPA) LEVELS IN ACUTE NONLYMPHOCYTIC LEUKEMIA ARE REDUCED BY HEPARIN ADMINISTRATION, *THROMB HAEMOST*, 52, 301-304 (1984)

1422 GULLEY RM, MIRZA A, KELLY CE, HEPATOTOXICITY OF SALICYLAZOSULFAPYRIDINE, *AM J GASTROENTEROL*, 72, 561-564 (1979)

1423 GUNDERSEN K, DEMISSIANOS HV, THE EFFECTS OF 5-METHYLPYRAZOLE-3-CARBOXYLIC ACID (V-19425) AND NICOTINIC ACID (NA) ON FREE FATTY ACIDS (FFA), *ADV EXP MED BIOL*, 4, 213 (1968)

1424 GUNSON HH, PHILLIPS PK, STRATTON F, THE EFFECT OF SURFACTANT IN THE AUTOANALYZER, *VOX SANG*, 22, 183 (1972)

1425 GUNSTONE RF, WING AJ, SHANI HGP, NJEMO D, SABUKA EMW, CLINICAL EXPERIENCE WITH METOLAZONE IN FIFTYTWO AFRICAN PATIENTS: SYNERGY WITH FRUSEMIDE, *POSTGRAD MED J*, 47, 789 (1971)

1426 GUPTA KK, GUANETHIDINE AND GLUCOSE TOLERANCE IN DIABETICS, *BR MED J [CLIN RES]*, 3, 679 (1968)

1427 GUPTA RN, KEANE PM, COOPER RG, GREAVES MS, OWEN G, DIPHENYLHYDANTOIN INTERFERENCE WITH SPECTROPHOTOMETRIC BARBITURATE ESTIMATIONS, *CLIN CHEM*, 19, 433 (1973)

1428 GUPTA RN, PRICE D, KEANE PM, MODIFIED PISANO METHOD FOR ESTIMATING URINARY METANEPHRINES, *CLIN CHEM*, 19, 611 (1973)

1429 GUSDON JP JR., PRITCHARD D, IMMUNOGLOBULIN D IN PREGNANCY, *AM J OBSTET GYNECOL*, 112, 867 (1972)

1430 GUSTAFSON A, BJORNTORP P, FAHLEN M, METFORMIN ADMINISTRATION IN HYPERLIPIDEMIC STATES, *ACTA MED SCAND*, 190, 491 (1971)

1431 GUSTAFSON A, SVANBORG A, GONADAL STEROID EFFECTS ON PLASMA LIPOPROTEINS AND INDIVIDUAL PHOSPHOLIPIDS, *J CLIN ENDOCRINOL METAB*, 35, 203 (1972)

1432 GUSTAVSSON S, ADAMI H-O, LOOF L, ET AL, RAPID HEALING OF DUODENAL ULCERS WITH OMEPRAZOLE: DOUBLE-BLIND DOSE COMPARATIVE TRIAL, *LANCET*, 2, 124-125 (1983)

1433 GUTMAN AB, URICOSURIC DRUGS, WITH SPECIAL REFERENCE TO PROBENECID AND SULFINPYRAZONE, *ADV PHARMACOL*, 4, 91 (1966)

1434 GUTMAN AB, DRUG REACTIONS CHARACTERIZED BY CHOLESTASIS ASSOCIATED WITH INTRAHEPATIC BILIARY TRACT OBSTRUCTION, *AM J MED*, 23, 841 (1957)

1435 GUYDA HJ, FRIESEN HG, SERUM PROLACTIN LEVELS IN HUMANS FROM BIRTH TO ADULT LIFE, *PEDIATR RES*, 7, 534 (1973)

1436 GYNN TN, MESSMORE HL, FRIEDMAN IA, DRUG INDUCED THROMBOCYTOPENIA, *MED CLIN NORTH AM*, 56, 65 (1972)

1437 HAALBOOM JRE, DEENSTRA M, STRUYVENBERG A, HYPOKALEMIA INDUCED BY INHALATION OF FENOTEROL, *LANCET*, 1, 1125-1127 (1985)

1438 HAALBOOM JRE, STRUYVENBERG A, THE MECHANISM UNDERLYING CHLORTHALIDONE-INDUCED HYPOKALAEMIA: EFFECTS OF SODIUM RESTRICTION, POTASSIUM SUPPLEMENTATION, SPIRONOLACTONE AND TRIAMTERENE, *NETH J MED*, 25, 184-192 (1982)

1439 HAAS HG, DAMBACHER MA, GUNCAGA J, LAUFFENBERGER T, RENAL EFFECTS OF CALCITONIN AND PARATHYROID EXTRACT IN MAN, *J CLIN INVEST*, 50, 2689 (1971)

1440 HAATAJA M, NIEMINEN L, MOTTONEN M, EFFECT OF LEVODOPA UPON PLASMA LEVELS OF 17-HYDROXYCORTICOSTERONE, *J PHARM SCI*, 61, 481 (1972)

1441 HAAVALDSEN R, INGRALDSEN P, BIOLOGICAL EFFECTS OF LITHIUM SALTS, *LANCET*, 1, 1390 (1973)

1442 HABOUBI NA, THURNHAM DI, EFFECT OF ETHANOL ON ERYTHROCYTE ACETYLCHOLINESTERASE ACTIVITY, *ANN CLIN BIOCHEM*, 23, 458-462 (1986)

1443 HADDEN JW, METZNER RJ, PSEUDOKETOSIS AND HYPERACETALDEHYDEMIA IN PARALDEHYDE ACIDOSIS, *AM J MED*, 47, 642 (1969)

1444 HADEN HT, THYROID FUNCTION TESTS. PHYSIOLOGIC BASIS AND CLINICAL INTERPRETATION, *POSTGRAD MED*, 40, 129 (1966)

1445 HADFIELD JW, RIFAMPICIN-INDUCED THROMBOCYTOPENIA, *POSTGRAD MED J*, 56, 59-60 (1980)

1446 HAEMMERLI UP, BIRCHER J, WRONG IDEA, GOOD RESULTS (THE LACTULOSE STORY), *N ENGL J MED*, 281, 442 (1969)

1447 HAFFNER SM, KNAPP JA, STERN MP, ET AL, COFFEE CONSUMPTION, DIET AND LIPIDS, *AM J EPIDEMIOL*, 122, 1-12 (1985)

1448 HAGNEVIK K, GORDON E, LINS LE, WILHELMSSON S, FORSTER D, GLYCEROL-INDUCED HEMOLYSIS WITH HEMOGLOBINURIA AND ACUTE RENAL FAILURE, *LANCET*, 1, 75 (1974)

1449 HAHN TJ, HENDIN BA, SCHARP CR, ET AL, SERUM 25-HYDROXYCALCIFEROL LEVELS AND BONE MASS IN CHILDREN IN CHRONIC ANTICONVULSANT THERAPY, *N ENGL J MED*, 292, 550-554 (1975)

1450 HAHN E, DIMENHYDRINATE INTERFERES WITH RADIOIMMUNOASSAY OF THEOPHYLLINE, *CLIN CHEM*, 26, 1759-1760 (1980)

1451 HAHN TJ, HADDAD JG, HENDIN BA, EFFECT OF CHRONIC ANTICONVULSANT THERAPY ON SERUM 25 OH VITAMIN D LEVELS, *CLIN RES*, 20, 238 (1972)

1452 HAHN TJ, HENDIN BA, SCHARP CR, HADDAD JG JR., EFFECT OF CHRONIC ANTICONVULSANT THERAPY ON SERUM 25-HYDROXYCALCIFEROL LEVELS IN ADULTS, *N ENGL J MED*, 287, 900 (1972)

1453 HAIDUKEWYCH D, JOHN G, CHRONIC VALPROIC ACID AND COANTILEPTIC DRUG THERAPY AND INCIDENCE OF INCREASES IN SERUM LIVER ENZYMES, *THER DRUG MONIT*, 8, 407-410 (1986)

1454 HAIDUKEWYCH D, JOHN G, ZIELINSKI JJ, ET AL, CHRONIC VALPROIC ACID THERAPY AND INCIDENCE OF INCREASES IN VENOUS PLASMA AMMONIA, *THER DRUG MONIT*, 7, 290-294 (1985)

1455 HAIDUKEWYCH D, RODIN EA, EFFECT OF PHENOTHIAZINES ON SERUM ANTIEPILEPTIC DRUG CONCENTRATIONS IN PSYCHIATRIC PATIENTS WITH SEIZURE DISORDER, *THER DRUG MONIT*, 7, 401-404 (1985)

1456 HAINLINE A, METHEMOGLOBIN, *STAND METH CLIN CHEM*, 5, 143 (1965)

1457 HAINSWORTH IR, HALL PE, A SIMPLE AUTOMATED METHOD FOR THE MEASUREMENT OF OESTROGENS IN THE URINE OF PREGNANT WOMEN, *CLIN CHIM ACTA*, 35, 201 (1971)

1458 HAJOS P, BERLIN I, INTODY Z, ET AL, THE EFFECT OF ORAL CONTRACEPTIVES ON SERUM LIPIDS, GAMMA-GLUTAMYLTRANSPEPTIDASE, AND EXCRETION OF D-GLUCARIC ACID, *INTL J CLIN PHARM THER TOXICOL*, 19, 117-123 (1981)

1459 HALL RR, LAURENCE DJR, DARCY D, STEVENS U, JAMES R, ROBERTS S, NEVILLE AM, CARCINOEMBRYONIC ANTIGEN IN THE URINE OF PATIENTS WITH UROTHELIAL CARCINOMA, *BR MED J [CLIN RES]*, 3, 609 (1972)

1460 HALL SM, PRESTON IW, THE EFFECT OF THE PATIENT'S ACID-BASE BALANCE ON AZOSTIX® STRIPS ESTIMATION OF BLOOD UREA, *ANN CLIN BIOCHEM*, 9, 208 (1972)

1461 HALLA JT, CASSIDY J, HARDIN JG, SEQUENTIAL GOLD AND PENICILLAMINE THERAPY IN RHEUMATOID ARTHRITIS: COMPARATIVE STUDY OF EFFECTIVENESS AND TOXICITY AND REVIEW OF LITERATURE, *AM J MED*, 72, 423-426 (1982)

1462 HALLER J, A REVIEW OF THE LONG TERM EFFECTS OF HORMONAL CONTRACEPTIVES, *CONTRACEPTION*, 1, 233 (1970)

1463 HALLIDAY A, JAWETZ E, SODIUM NITROFURANTOIN ADMINISTERED INTRAVENOUSLY, *N ENGL J MED*, 266, 427 (1962)

1464 HALLORAN SP, TORRENS DJ, EFFECTS OF THE DRUG DISULPHINE BLUE ON ROUTINE BIOCHEMICAL INVESTIGATIONS, *ANN CLIN BIOCHEM*, 20, 317-320 (1983)

1465 HALMA C, JANSEN JBMJ, JANSSENS AR, ET AL, LIFE-THREATENING WATER INTOXICATION DURING SOMATOSTATIN THERAPY, *ANN INTERN MED*, 107, 518-520 (1987)

1466 HALMI KA, NOYES R JC, MILLARD SA, EFFECT OF LITHIUM ON PLASMA CORTISOL AND ADRENAL RESPONSE TO ADRENOCORTICOTROPIN IN MAN, *CLIN PHARMACOL THER*, 13, 699 (1972)

1467 HALSTED JA, HACKLEY BM, SMITH JC, PLASMA ZINC AND COPPER IN PREGNANCY AND AFTER ORAL CONTRACEPTIVES, *LANCET*, 2, 278 (1968)

1468 HAMADAH K, HOLMES H, BARKER GB, HARTMAN GC, PARKE DVW, EFFECT OF ELECTRIC CONVULSION THERAPY ON URINARY EXCRETION OF 3'5'-CYCLIC ADENOSINE MONOPHOSPHATE, *BR MED J [CLIN RES]*, 3, 439 (1972)

1469 HAMALAINEN E, KORPELA JT, ADLERCREUTZ H, EFFECT OF OXYTETRACYCLINE ADMINISTRATION ON INTESTINAL METABOLISM OF OESTROGENS AND ON PLASMA SEX HORMONES IN HEALTHY MEN, *GUT*, 28, 439-445 (1987)

1470 HAMER A, PETER T, MANDEL WJ, ET AL, THE POTENTIATION OF WARFARIN ANTICOAGULATION BY AMIODARONE, *CIRCULATION*, 65, 1025-1029 (1982)

1471 HAMET P, KUCHEL O, GENEST J, EFFECT OF UPRIGHT POSTURE AND ISOPROTERENOL INFUSION ON CYCLIC ADENOSINE MONOPHOSPHATE EXCRETION, *J CLIN ENDOCRINOL METAB*, 36, 218 (1973)

1472 HAMFELT A, TUVEMO T, PYRIDOXAL PHOSPHATE AND FOLIC ACID CONCENTRATION IN BLOOD AND ERYTHROCYTE ASPARTATE AMINOTRANSFERASE, *CLIN CHIM ACTA*, 41, 287 (1972)

1473 HAMILTON DV, SAUNDERS J, CARBIMAZOLE-INDUCED AGRANULOCYTOSIS IN TWO SISTERS, *CURR MED RES OPIN*, 4, 607-608 (1977)

1474 HAMMAN BL, MARTIN MM, SEPARATION OF SIX URINARY 17-KETOSTEROIDS BY TWO-DIMENSIONAL THIN-LAYER CHROMATOGRAPHY: CONTROL VALUES, *J CLIN ENDOCRINOL METAB*, 24, 1195 (1964)

1475 HAMMER EJ, ASTLEY JP, INCREASE IN SERUM TESTOSTERONE FOLLOWING CONTACT WITH BLOOD CELLS., *ANN CLIN BIOCHEM*, 22, 539-540 (1985)

1476 HAMMERMAN C, ZAIA W, WU H-H, SEVERE HYPONATREMIA WITH INDOMETHACIN - A MORE SERIOUS TOXICITY THAN PREVIOUSLY REALIZED?, *DEV PHARMACOL THER*, 8, 260-267 (1985)

1477 HAMMILL SC, SORENSON PB, WOOD DL, ET AL, PROPAFENONE FOR THE TREATMENT OF REFRACTORY COMPLEX VENTRICULAR ECTOPIC ACTIVITY, *MAYO CLIN PROC*, 61, 98-103 (1986)

1478 HANGER FM, GUTMAN AB, POST-ARSPHENAMINE JAUNDICE, *J AM MED ASSOC*, 115, 263 (1940)

1479 HANING RV JR., CARLSON IH, CORTES J, ET AL, DANAZOL AND ITS PRINCIPAL METABOLITES INTERFERE WITH BINDING OF TESTOSTERONE, CORTISOL AND THYROXINE BY PLASMA PROTEINS, *CLIN CHEM*, 28, 696-698 (1982)

1480 HANSEN EA, RUNE SJ, EFFECT OF CHEWING GUM ON GASTRIC ACID SECRETION, *SCAND J GASTROENTEROL*, 7, 733 (1972)

1481 HANSEN JM, SIERSBAECK-NIELSEN K, SERUM PROTEIN-BOUND IODINE AND SERUM THYROXINE DURING PERPHENAZINE THERAPY, *ACTA ENDOCRINOL*, 55, 136 (1967)

1482 HANSEN OH, MADSEN P, THE EFFECT OF TETRAGASTRIN ON GASTRIC ACID SECRETION, *SCAND J GASTROENTEROL*, 7, 171 (1972)

1483 HANSEN S, PERRY TL, LESK D, GIBSON L, URINARY BACTERIA: POTENTIAL SOURCE OF SOME ORGANIC ACIDURIAS, *CLIN CHIM ACTA*, 39, 71 (1972)

1484 HANSKY J, SOVENY C, KORMAN MG, THE EFFECT OF GLUCAGON ON SERUM GASTRIN, I. STUDIES IN NORMAL SUBJECTS, *GUT*, 14, 457 (1973)

1485 HANSSON P, HEDSTROM S-A, HULTBERG B, ET AL, FUSIDIC ACID INTERFERES IN ENZYMATIC DETERMINATION OF BILE ACIDS, *CLIN CHIM ACTA*, 125, 241-243 (1982)

1486 HANSTEN PD, DRUGS IN THE PRODUCTION OF DIRECT COOMBS' TEST POSITIVITY, *AM J HOSP PHARM*, 28, 629 (1971)

1487 HANSTEN PD, *DRUG INTERACTIONS* 2ND ED, LEA AND FEBIGER, PHILADELPHIA, PA (1973)

1488 HANSTEN PD, *DRUG INTERACTIONS*, LEA AND FEBIGER, PHILADELPHIA, PA (1971)

1489 HANSTEN PD, HAYTON WL, EFFECT OF ANTACID AND ASCORBIC ACID ON SERUM SALICYLATE CONCENTRATION, *J CLIN PHARMACOL*, 20, 326-331 (1980)

1490 HANTMAN DA, DONALDSON CL, HULLEY SB, ABNORMAL URINARY SEDIMENT DURING THERAPY WITH SYNTHETIC SODIUM CALCITONIN, *J CLIN ENDOCRINOL METAB*, 33, 564 (1971)

1491 HARALAMBIE G, BIOCHEMICAL CHANGES IN BLOOD (AT REST) INDUCED BY EXERCISE AND TRAINING IN, *REFERENCE VALUES IN HUMAN CHEMISTRY*, G SIEST, ED. KARGER, BASEL 1973, 243

1492 HARDING S, MUNRO AJ, FRUSEMIDE AND RENAL ENZYME EXCRETION, *BR MED J [CLIN RES]*, 2, 1431 (1978)

1493 HARDY BG, ZADOR IT, GOLDEN L, ET AL, EFFECT OF CIMETIDINE ON THE PHARMACOKINETICS AND PHARMACODYNAMICS OF QUINIDINE, *AM J CARDIOL*, 52, 172-175 (1973)

1494 HARDY MJ, BJURSTROM CH, EFFECT OF HEPARIN IN DETERMINATION OF PLASMA ALBUMIN BY BECKMAN ASTRA™ 8 (BCG) AND DU PONT ACA II (BCP) AUTOANALYZERS, *ANN CLIN BIOCHEM*, 21, 387-388 (1984)

1495 HARE DL, HICKS BH, THROMBOCYTOPENIA DUE TO OXPRENOLOL, *MED J AUST*, 2259 (1979)

1496 HARE TA, BEASLEY BL, CHAMBERS RA, BOEHME DH, VOGEL WH, DOPA AND AMINO ACID LEVELS IN PLASMA AND CEREBROSPINAL FLUID OF PATIENTS WITH PARKINSON'S DISEASE, *CLIN CHIM ACTA*, 45, 273 (1973)

1497 HARGREAVES KM, SCHMIDT EA, MUELLER GP, ET AL, DEXAMETHASONE ALTERS PLASMA LEVELS OF BETA-ENDORPHIN AND POSTOPERATIVE PAIN, *CLIN PHARMACOL THER*, 42, 601-607 (1987)

1498 HARGREAVES T, ORAL CONTRACEPTIVES AND LIVER FUNCTION, *J CLIN PATHOL*, 1 (1969)

1499 HARGREAVES T, LATHE GH, DRUGS AFFECTING BILIARY SECRETION IN, *THERAPEUTIC AGENTS AND THE LIVER*, N MCINTYRE, S SHERLOCK, EDS. BLACKWELL, OXFORD (1965)

1500 HARGREAVES T, LATHE GH, INHIBITORY ASPECTS OF BILE SECRETION, *NATURE*, 200, 1172 (1963)

1501 HARJU E, PAKARINEN A, THE EFFECT OF IRON TREATMENT ON SERUM FERRITIN CONCENTRATIONS AND BONE MARROW STAINABLE IRON IN IRON DEFICIENT OUTPATIENTS WITH GASTRITIS, GASTRIC ULCER AND DUODENAL ULCER, *J INT MED RES*, 12, 56-58 (1984)

1502 HARNEY J, GLASBERG MR, MYOPATHY AND HYPERSENSITIVITY TO PHENYTOIN, *NEUROLOGY*, 33, 790-791 (1983)

1503 HARNO K, VALIMAKI M, VERHO M, EFFECTS OF A NEW DIURETIC PIRETANIDE ON GLUCOSE TOLERANCE, INSULIN SECRETION AND ^{125}I-INSULIN BINDING, *EUR J CLIN PHARMACOL*, 27, 697-700 (1985)

1504 HARRINGTON JM, RISK OF MERCURIAL POISONING IN LABORATORIES USING VOLUMETRIC GAS ANALYSIS, *LANCET*, 1, 86 (1974)

1505 HARRIS A, LONG TERM TREATMENT OF PAROXYSMAL CARDIAC ARRHYTHMIAS WITH PROPRANOLOL, *AM J CARDIOL*, 18, 431 (1966)

1506 HARRIS AB, AJOSE D, RAPID ASSAY OF URINARY VITAMIN C, *LANCET*, 1, 671 (1973)

1507 HARRIS AB, HARTLEY J. MOOR A, REDUCED ASCORBIC ACID EXCRETION AND ORAL CONTRACEPTIVES, *LANCET*, 2, 201 (1973)

1508 HARRIS J, JESSOP JD, CHAPUT DE SAINTONGE DM, FURTHER EXPERIENCE WITH AZATHIOPRINE IN RHEUMATOID ARTHRITIS, *BR MED J [CLIN RES]*, 4, 463 (1971)

1509 HARRIS L, MCKENNA WJ, ROWLAND E, ET AL, SIDE EFFECTS OF LONG-TERM AMIODARONE THERAPY, *CIRCULATION*, 67, 45-51 (1983)

1510 HARRISON RJ, VITAMIN B_{12} LEVELS IN ERYTHROCYTES IN HYPOCHROMIC ANAEMIA, *J CLIN PATHOL*, 24, 698 (1971)

1511 HARRISON SP, PYRUVATE INTERFERENCE WITH CONTINUOUS-FLOW (SMAC®) METHODS FOR ASPARTATE AND ALANINE AMINOTRANSFERASES, *CLIN CHEM*, 33, 616-617 (1987)

1512 HARRISON SP, NAPROXEN INTERFERENCE WITH THE ION-SELECTIVE ELECTRODE IN THE RA-1000, *CLIN CHEM*, 33, 421 (1987)

1513 HARRISON TR, *HARRISON'S PRINCIPLES OF INTERNAL MEDICINE* 6TH ED, M WINTROBE, ED., MCGRAW-HILL, NEW YORK, NY (1970)

1514 HARROWER ADB, FYFFE JA, HORN DB, ET AL, THYROXINE AND TRIODOTHYRONINE LEVELS IN HYPERTHYROID PATIENTS DURING TREATMENT WITH PROPRANOLOL, *CLIN ENDOCRINOL*, 7, 41-44 (1977)

1515 HARTH M, HAINES DSM, BONDY DC, A SIMPLE METHOD FOR THE DETERMINATION OF GOLD IN SERUM, BLOOD AND URINE BY ATOMIC ABSORPTION SPECTROSCOPY, *AM J CLIN PATHOL*, 59, 423 (1973)

1516 HARTSHORN EA, DRUG INTERACTIONS, *DRUG INTELL CLIN PHARM*, 2, 174 (1968)

1517 HARUTA T, KUROKI S, OKURA K, ET AL, CLINICAL STUDIES OF ASPOXICILLIN IN PEDIATRICS, *JPN J ANTIBIOT*, 38, 1898-1904 (1985)

1518 HARVEY RF, DOWSETT L, HARTOG M, READ AE, A RADIOIMMUNOASSAY FOR CHOLECYSTOKININ-PANCREOZYMIN, *LANCET*, 2, 826 (1973)

1519 HASKELL CM, CANELLOS GP, LEVENTHAL BG, CARBONE PP, BLOCK JB, SERPICK AA, SELAWRY OS, L-ASPARAGINASE THERAPEUTIC AND TOXIC EFFECTS IN PATIENTS WITH NEOPLASTIC DISEASE, *N ENGL J MED*, 281, 1028 (1969)

1520 HASKOVEC L, RYSANEK K, EXCRETION OF 3-METHOXY-4-HYDROXY-MANDELIC ACID AND 5-HYDROXYINDOLEACETIC ACID IN DEPRESSED PATIENTS, *J PSYCHIATR RES*, 5, 213 (1967)

1521 HASLING C, CHARLES P, MOSEKILDE L, ETIDRONATE SODIUM FOR TREATING HYPERCALCAEMIA OF MALIGNANCY: A DOUBLE BLIND, PLACEBO-CONTROLLED STUDY, *EUR J CLIN INVEST*, 16, 433-437 (1986)

1522 HATJIS CG, ROSE JC, PIPPITT C, ET AL, EFFECT OF TREATMENT WITH SODIUM VALPROATE ON PLASMA ADRENOCORTICOTROPIC HORMONE AND CORTISOL CONCENTRATIONS IN PREGNANCY, *AM J OBSTET GYNECOL*, 152, 315-316 (1985)

1523 HATTORI M, TSUDA K, TAMINATO T, ET AL, EFFECT OF PROBUCOL ON SERUM LIPIDS AND APOPROTEINS IN PATIENTS WITH NONINSULIN-DEPENDENT DIABETES MELLITUS, *CURR THER RES*, 42, 967-973 (1987)

1524 HAUGER-KLEVENE JH, READER C, MAYER E, ET AL, A COMPARATIVE STUDY OF ENDRALAZINE AND CAPTOPRIL IN ESSENTIAL HYPERTENSION: EFFECT ON RENAL LEVELS, PULMONARY FUNCTIONS STUDIES AND LIPID PROFILE, *INT J CLIN PHARMACOL RES*, 6, 275-281 (1986)

1525 HAUKAMAA M, GUMMERUS M, DECREASE OF SERUM OESTRIOL DURING INTRAVENOUS HEXOPRENALINE OR SALBUTAMOL TREATMENT, *BR J OBSTET GYNAECOL*, 89, 917-920 (1982)

1526 HAUPT E, KOBERICH W, BEYER J, SCHOFFLING K, PHARMACODYNAMIC ASPECTS OF TOLBUTAMIDE, GLIBENCLAMIDE, GLIBONURIDE, AND GLISOXEPIDE, *DIABETOLOGIA*, 7, 449 (1971)

1527 HAUSER A, QUIGLEY ML. DRIEVER CW, ET AL, MORE ON FALSE-POSITIVE 'HEMOCCULT®' REACTION WITH CIMETIDINE, *N ENGL J MED*, 304, 847-848 (1981)

1528 HAUT MJ, COWAN DH, THE EFFECT OF ETHANOL ON HEMOSTATIC PROPERTIES OF HUMAN BLOOD PLATELETS, *AM J MED*, 56, 22 (1974)

1529 HAVEL RJ, THE AUTONOMIC NERVOUS SYSTEM AND INTERMEDIARY CARBOHYDRATE AND FAT METABOLISM, *ANESTHESIOLOGY*, 29, 702 (1968)

1530 HAVEL RJ, HUNNINGHAKE DB, ILLINGWORTH DR, ET AL, LOVASTATIN (MEVINOLIN) IN THE TREATMENT OF HETEROZYGOUS FAMILIAL HYPERCHOLESTEROLEMIA, *ANN INTERN MED*, 107, 609-615 (1987)

1531 HAVEL RJ, SEGEL N, BALASSE EO, EFFECT OF 5-METHYLPYRAZOLE-3-CARBOXYLIC ACID (MPCA) ON FAT MOBILIZATION, KETOGENESIS AND GLUCOSE METABOLISM, *ADV EXP MED BIOL*, 4, 105 (1968)

1532 HAWKINS RD, KALANT H, THE METABOLISM OF ETHANOL AND ITS METABOLIC EFFECTS, *PHARMACOL REV*, 24, 67 (1972)

1533 HAYES JR, ARDILL J, KENNEDY TL, SHANKS RG, BUCHANAN KD, STIMULATION OF GASTRIN RELEASE BY CATECHOLAMINES, *LANCET*, 1, 819 (1972)

1534 HAYES TS, AUTOMATED FLUOROMETRIC DETERMINATION OF AMPHETAMINE IN URINE, *CLIN CHEM*, 19, 390 (1973)

1535 HAZZARD WR, SPIGER MJ, BAGDADE JD, BIERMAN EL, STUDIES ON MECHANISM OF INCREASED PLASMA TRIGLYCERIDE LEVELS INDUCED BY ORAL CONTRACEPTIVES, *N ENGL J MED*, 280, 471 (1969)

1536 HEACOCK AM, ADAMS E, SOURCE OF PYRROLE-2-CARBOXYLATE IN MAMMALIAN URINE, *BIOCHEM BIOPHYS RES COMMUN*, 50, 392 (1972)

1537 HEALEY LA, HARRISON M, DECKER JL, URICOSURIC EFFECT OF CHLORPROTHIXENE, *N ENGL J MED*, 272, 526 (1965)

1538 HEANEY RP, WHEDON GD, IMPAIRMENT OF HEPATIC BROMOSULFOPHTHALEIN CLEARANCE BY TWO 17-SUBSTITUTED TESTOSTERONES, *J LAB CLIN MED*, 52, 169 (1955)

1539 HEATH H III, EARLL JM, SCHAAF M, PIECHOCKI JT, LI TK, SERUM IONIZED CALCIUM DURING BED REST IN FRACTURE PATIENTS AND NORMAL MEN, *METABOLISM*, 21, 633 (1972)

1540 HEATON KW, POMARE EW, EFFECT OF BRAN ON BLOOD LIPIDS AND CALCIUM, *LANCET*, 1, 49 (1974)

1541 HEDRICK R, WILLIAMS F, MORIN R, ET AL, CARBAMAZEPINE-ERYTHROMYCIN INTERACTION LEADING TO CARBAMAZEPINE IN FOUR EPILEPTIC CHILDREN, *THER DRUG MONIT*, 5, 405-407 (1983)

1542 HEEL RC, BROGDEN RN, SPEIGHT TM, ET AL, TAMOXIFEN: A REVIEW OF ITS PHARMACOLOGICAL AND THERAPEUTIC USE IN THE TREATMENT OF BREAST CANCER, *DRUGS*, 16, 1-24 (1978)

1543 HEIDEMANN PH, STUBBE P, BECK W, TRANSIENT SECONDARY HYPOTHYROIDISM AND THYROXINE BINDING GLOBULIN DEFICIENCY IN LEUKEMIC CHILDREN DURING POLYCHEMOTHERAPY: AN EFFECT OF L-ASPARAGINASE, *EUR J PEDIATR*, 136, 291-295 (1981)

1544 HEIDEN D, MIELKE CH JR., RODVIEN R, IMPAIRMENT OF HEPARIN OF PRIMARY HAEMOSTASIS AND PLATELET [14c] 5-HYDROXYTRYPTAMINE RELEASE, *BR J HAEMATOL*, 36, 427-436 (1977)

1545 HEIKKINEN J, MAENTAUSTA O, YLOSTALO P, ET AL, SERUM BILE ACID LEVELS IN INTRAHEPATIC CHOLESTASIS OF PREGNANCY DURING TREATMENT WITH PHENOBARBITAL OR CHOLESTYRAMINE, *EUR J OBSTET GYNECOL REPROD BIOL*, 14, 153-162 (1982)

1546 HEIKKINEN J, YLOSTALO P, MAENTAUSTA O, ET AL, SERUM BILE ACIDS DURING BIPHASIC CONTRACEPTIVE TREATMENT WITH ETHINYL ESTRADIOL AND NORGESTREL, *CONTRACEPTION*, 25, 89-95 (1982)

1547 HEINEGARD D, TIDERSTROM G, DETERMINATION OF SERUM CREATININE BY A DIRECT COLORIMETRIC METHOD, *CLIN CHIM ACTA*, 43, 305 (1973)

1548 HEINEMANN G, SCHIEVELBEIN H, EBER S, EFFECT OF CIGARETTE SMOKING ON WHITE BLOOD CELLS AND ERYTHROCYTE ENZYMES, *ARCH ENVIRON HEALTH*, 37, 261-265 (1982)

1549 HEINICKE RM, VAN DER WAL L, YOKOYAMA M, EFFECT OF BROMELAIN (ANANASE) ON HUMAN PLATELET AGGREGATION, *EXPERIENTIA*, 28, 844 (1972)

1550 HEINZE E, FUSSGANGER R, TELLER WM, INFLUENCE OF CALCIUM ON INSULIN SECRETION IN NEWBORNS, *PEDIATR RES*, 7, 100 (1973)

1551 HELGELAND A, THE IMPACT ON SERUM LIPIDS OF COMBINATIONS OF DIURETICS AND β-BLOCKERS AND OF β-BLOCKERS ALONE, *J CARDIOVASC PHARMACOL*, 6, S474-S476 (1984)

1552 HELLSTROM PE, REPO UK, CAPREOMYCIN, ETHAMBUTOL, AND RIFAMPICIN IN APPARENTLY INCURABLE PULMONARY TUBERCULOSIS, *SCAND J RESP DIS* (SUPPL), 69 (1969)

1553 HELLSTROM PE, REPO UK, MATTSON K, NEW DRUGS IN TUBERCULOSIS, *DRUGS*, 1, 349 (1971)

1554 HELSON L, THE EFFECT OF INTRAVENOUS VITAMIN E AND MENADIOL SODIUM DIPHOSPHATE ON VITAMIN K DEPENDENT CLOTTING FACTORS, *THROMB RES*, 35, 11-18 (1984)

1555 HENAUER SA, HOLLISTER LE, CIMETIDINE INTERACTION WITH IMIPRAMINE AND NORTRIPTYLINE, *CLIN PHARMACOL THER*, 35, 183-187 (1984)

1556 HENDEL J, NYFORS A, IMPACT OF METHOTREXATE THERAPY ON THE FOLATE STATUS OF PSORIATIC PATIENTS, *CLIN EXP DERMATOL*, 10, 30-33 (1985)

1557 HENDERSON AR, A SOURCE OF ERROR IN BLOOD PYRUVATE DETERMINATION, *J CLIN PATHOL*, 24, 475 (1971)

1558 HENE RJ, KOOMANS HA, BOER P, ET AL, EFFECT OF CAPTOPRIL IN BARTTER'S SYNDROME, *NEPHRON*, 35, 275 (1983)

1559 HENGSTMANN H, UBER VERBREITUNG UND FOLGEN DES PHENACETIN - ABUSUS AUF DEM LANDE EIN BERICHT UBER 70 FALLE, *MUNCH MED WOCHENSCHR*, 108, 1489 (1966)

1560 HENNING RJ, BECKER H, VINCENT J-L, ET AL, USE OF METHYLPREDNISOLONE IN PATIENTS FOLLOWING ACUTE MYOCARDIAL INFARCTION, *CHEST*, 79, 186-194 (1981)

1561 HENRICH WL, NEPHROTOXICITY OF NONSTEROIDAL ANTI-INFLAMMATORY AGENTS, *AM J KIDNEY DIS*, 11, 478-484 (1983)

1562 HENRY DA, LOWE JM, DONNELLY T, JAUNDICE DURING CYPROHEPTADINE TREATMENT, *BR MED J [CLIN RES]*, 1, 753 (1978)

1563 HENRY RJ, *CLINICAL CHEMISTRY:PRINCIPLES AND TECHNICS*, HOEBER DIVISION, HARPER AND ROW, NEW YORK, NY (1964)

1564 HENRY RJ, BERKMAN S, ABSORBANCE OF VARIOUS PROTEIN-FREE FILTRATES OF SERUM, *CLIN CHEM*, 3, 711 (1957)

1565 HENRY RJ, GOLUB OJ, BERKMAN S, SEGALOVE M, CRITIQUE ON THE ICTERUS INDEX DETERMINATION, *AM J CLIN PATHOL*, 23, 841 (1953)

1566 HENRY RJ, SOBEL C, BERKMAN S, ON THE DETERMINATION OF 'PANCREATIC LIPASE' IN SERUM, *CLIN CHEM*, 3, 77 (1957)

1567 HENRY RL, WASHNOCK MA, TAYLOR GW, MISTABRON EFFECTS ON PLATELETS AND BLOOD COAGULATION, *J INT MED RES*, 12, 277-280 (1984)

1568 HERBETH B, BANGREL A, DALO B, ET AL, INFLUENCE OF ORAL CONTRACEPTIVES OF DIFFERING DOSAGES ON ALPHA₁-ANTITRYPSIN, GAMMA-GLUTAMYLTRANSFERASE AND ALKALINE PHOSPHATASE, *CLIN CHIM ACTA*, 112, 293-299 (1981)

1569 HERD JK, INTERFERENCE OF HEXOSAMINES IN THE LOWRY REACTION, *ANAL BIOCHEM*, 44, 404 (1971)

1570 HERLONG HF, REID PR, BOITNOTT JK, ET AL, APRINDINE HEPATITIS, *ANN INTERN MED*, 89, 359-361 (1978)

1571 HERMANS PE, WILHELM MP, VANCOMYCIN, *MAYO CLIN PROC*, 62, 901-905 (1987)

1572 HERMANSEN L, OSNES JB, BLOOD AND MUSCLE PH AFTER MAXIMAL EXERCISE IN MAN, *J APPL PHYSIOL*, 32, 304 (1972)

1573 HERMELIN LI, HEPARIN-INDUCED THROMBOCYTOPENIA, *AM CLIN PROD REV*, DEC, 15-17 (1984)

1574 HERREMAN F, IATROGENIC PATHOLOGY IN CARDIOLOGY, *NOUV PRESSE MED*, 1, 413 (1972)

1575 HERRING RW, POTTER JJ, MEZEY E, EFFECT OF ACUTE ALCOHOL ADMINISTRATION ON ERYTHROCYTE ALDEHYDE DEHYDROGENASE ACTIVITY IN MAN, *ALCOHOL CLIN EXP RES*, 10, SUPPL, 41-43 (1986)

1576 HERSH EM, WONG VG, HENDERSON ES, HEPATOTOXIC EFFECT OF METHOTREXATE, *CANCER*, 19, 600 (1966)

1577 HERSHMAN JM, JONES CM, BAILEY AL, RECIPROCAL CHANGES IN SERUM THYROTROPIN AND FREE THYROXINE PRODUCED BY HEPARIN, *J CLIN ENDOCRINOL METAB*, 34, 574 (1972)

1578 HERXHEIMER A, RIFAMPICIN, *DRUG THER BULL*, 8, 11 (1970)

1579 HERZIG RH, WOLFF SN, LAZARUS HM, ET AL, HIGH-DOSE CYTOSINE ARABINOSIDE THERAPY FOR REFRACTORY LEUKEMIA, *BLOOD*, 62, 361-369 (1983)

1580 HESS JW, MACDONALD RP, SERUM CREATINE PHOSPHOKINASE ACTIVITY: A NEW DIAGNOSTIC AID IN MYOCARDIAL AND SKELETAL MUSCLE DISEASE, *J MICH MED SOC*, 62, 1095 (1963)

1581 HEWLETT JS, BODEY GP, COLTMAN CA, FREIREICH EJ, HAUT AB, MCCREDIE KB, INTERMITTENT GUANAZOLE THERAPY IN ADULT ACUTE LEUKEMIA, *CLIN PHARMACOL THER*, 14, 271 (1973)

1582 HEYMANN W, NEPHROTIC SYNDROME AFTER USE OF TRIMETHADIONE AND PARAMETHADIONE IN PETIT MAL, *J AM MED ASSOC*, 202, 893 (1967)

1583 HEYMANN W, GRUPE WE, INCREASE IN PROTEINURIA DUE TO STEROID MEDICATION IN CHRONIC RENAL DISEASE, *J PEDIATR*, 74, 356 (1969)

1584 HIBBARD DM, PETERS JR, HUNNINGHAKE DB, EFFECTS OF CHOLESTYRAMINE AND COLESTIPOL ON THE PLASMA CONCENTRATIONS OF PROPRANOLOL, *BR J CLIN PHARMACOL*, 18, 337-342 (1984)

1585 HICHENS M, HOGANS AF, RADIOIMMUNOASSAY FOR DEXAMETHASONE IN PLASMA, *CLIN CHEM*, 20, 266 (1974)

1586 HICKSON B, DAVIDSON RJL, WALKER W, AGRANULOCYTOSIS CAUSED BY PROCAINAMIDE, *SCOTT MED J*, 17, 165 (1972)

1587 HIGASHI A, IKEDA T, MATSUKURA M, ET AL, SERUM ZINC AND VITAMIN E CONCENTRATIONS IN HANDICAPPED CHILDREN TREATED WITH ANTICONVULSANTS, *DEV PHARMACOL THER*, 5, 109-113 (1982)

1588 HIGBEE MD, WOOD JS, MEAD RA, PROCAINAMIDE, CIMETIDINE INTERACTION: A POTENTIAL TOXIC INTERACTION IN THE ELDERLY, *J AM GERIATR SOC*, 32, 162-164 (1984)

1589 HIGGENS CS, SCOTT JT, THE URICOSURIC ACTION OF AZAPROPAZONE: DOSE RESPONSE AND COMPARISON WITH PROBENECID, *BR J CLIN PHARMACOL*, 18, 439-443 (1984)

1590 HIGGINS TN, TAYLOR JD, COLORIMETRIC METHOD FOR THE QUANTITATIVE DETERMINATION OF AVLOSULFON® (DAPSONE) IN SERUM, *CLIN BIOCHEM*, 6, 295 (1973)

1591 HILBERMAN M, MASEDA J, STINSON EB, ET AL, THE DIURETIC PROPERTIES OF DOPAMINE IN PATIENTS AFTER OPEN-HEART OPERATION, *ANESTHESIOLOGY*, 61, 489-494 (1984)

1592 HILBORN S, KRAHN J, EFFECT OF TIME OF EXPOSURE OF SERUM TO GEL-BARRIER TUBES ON RESULTS FOR PROGESTERONE AND SOME OTHER ENDOCRINE TESTS, *CLIN CHEM*, 33, 203-204 (1987)

1593 HILDEBRANDT R, MOLLER H, GUNDERT-REMY U, INFLUENCE OF THEOPHYLLINE ON THE RENAL CLEARANCE OF ERYTHROMYCIN, *INT J CLIN PHARMACOL THER TOXICOL*, 25, 601-604 (1987)

1594 HILDERBRAND DC, WHITE DH, TRACE ELEMENT ANALYSIS IN HAIR: AN EVALUATION, *CLIN CHEM*, 20, 148 (1974)

1595 HILES BW, HYPERGLYCEMIA AND GLYCOSURIA FOLLOWING CHLORPROMAZINE THERAPY, *J AM MED ASSOC*, 162, 1651 (1956)

1596 HILL A, CASEY R, ZALESKI WA, DIFFICULTIES AND PITFALLS IN THE INTERPRETATION OF SCREENING TESTS FOR THE DETECTION OF INBORN ERRORS OF METABOLISM, *CLIN CHIM ACTA*, 72, 1-15 (1976)

1597 HILL JM, ROBERTS J, LOEB E, KAHN A, HILL RW, L-ASPARAGINASE THERAPY FOR LEUKEMIA AND OTHER MALIGNANT NEOPLASMS, *J AM MED ASSOC*, 202, 882 (1967)

1598 HILL NS, ANTMAN EM, GREEN LH, ET AL, INTRAVENOUS NITROGLYCERIN: A REVIEW OF PHARMACOLOGY, INDICATIONS, THERAPEUTIC EFFECTS AND COMPLICATIONS, *CHEST*, 79, 69-76 (1981)

1599 HILLBOM ME, KASTE M, TARSSANEN L, ET AL, EFFECT OF ETHANOL ON BLOOD VISCOSITY AND ERYTHROCYTE FLEXIBILITY IN HEALTHY MEN, *EUR J CLIN INVEST*, 13, 45-48 (1983)

1600 HILLIER K, DUTTON A, CORKER CS, SINGER A, EMBREY MP, PLASMA STEROID AND LUTEINIZING HORMONE LEVELS DURING PROSTAGLANDIN F 2 ALPHA ADMINISTRATION IN LUTEAL PHASE, *BR MED J [CLIN RES]*, 4, 333 (1972)

1601 HILLMER T, FRERICHS H, CREUTZFELDT W, HALLER J, KOLENHYDRAT-UND FETTSTOFFWECHSELVERANDERUNGEN WAHREND DER BEHANDLUNG MIT EINEM OVULATIONS = HEMMER (EUGYNON), *VERH 3 KONGRESS D DTSCH DIABETES-GESELLSCHAFT*, GOTTINGEN (1968)

1602 HINDERKS GJ, FROHLICH J, LOW SERUM CREATINE KINASE VALUES ASSOCIATED WITH ADMINISTRATION OF STEROIDS, *CLIN CHEM*, 25, 2050-2051 (1979)

1603 HINE KR, HARROP JS, HOPTON MR, ET AL, THE EFFECTS OF RANITIDINE ON PITUITARY-THYROID FUNCTION, *BR J CLIN PHARMACOL*, 18, 608-611 (1984)

1604 HINKS LJ, CLAYTON BE, LLOYD RS, ZINC AND COPPER CONCENTRATIONS IN LEUKOCYTES AND ERYTHROCYTES IN HEALTHY ADULTS AND THE EFFECT OF ORAL CONTRACEPTIVES, *J CLIN PATHOL*, 36, 1016-1021 (1983)

1605 HINMAN AR, WOLINSKY E, NEPHROTOXICITY ASSOCIATED WITH THE USE OF CEPHALORIDINE, *J AM MED ASSOC*, 200, 724 (1967)

1606 HINSHAW JR, HOBLER KE, BORJA AR, PENTAZOCINE: A POTENT, NONADDICTING ANALGESIC, *AM J MED SCI*, 251, 57 (1966)

1607 HINTERBERGER H, WILCKEN DE, THE EFFECT ON PROLONGED GLUCAGON INFUSIONS ON THE URINARY EXCRETION OF CATECHOLAMINES, *CLIN CHIM ACTA*, 52, 153 (1974)

1608 HIRAKATA H, ONOYAMA K, ISEKI K, ET AL, WORSENING OF ANEMIA INDUCED BY LONG-TERM USE OF CAPTOPRIL IN HEMODIALYSIS PATIENTS, *AM J NEPHROL*, 4, 355-360 (1984)

1609 HIRAMATSU K, YAMAGISHI F, KUBOTA T, ET AL, ACUTE EFFECTS OF THE CALCIUM ANTAGONIST, NIFEDIPINE, ON BLOOD PRESSURE, PULSE RATE AND RENIN-ALDOSTERONE-ANGIOTENSIN SYSTEM IN PATIENTS WITH ESSENTIAL HYPERTENSION, *AM HEART J*, 104, 1346-1350 (1982)

1610 HIRSCHEL B, AMODIAQUINE AND HEPATITIS, *LANCET*, 1, 467 (1986)

1611 HITZ J, STEINMETZ J, HENNY J, ET AL, EFFETS DE L'ALDATENSE SUR LES EXAMENS DE LABORATOIRE. COMPARAISON AVEC LA POPULATION DE REFERENCE, *ANN BIOL CLIN*, 42, 289-293 (1984)

1612 HJELM M, DEVERDIER CH, BIOCHEMICAL EFFECTS OF AROMATIC AMINES. 1. METHAEMOGLOBINAEMIA, HAEMOLYSIS AND HEINZ-BODY FORMATION, *BIOCHEM PHARMACOL*, 14, 1119 (1965)

1613 HO DHW, COVINGTON WP, LEGHA SS, ET AL, CLINICAL PHARMACOLOGY OF TRIMETREXATE, *CLIN PHARMACOL THER*, 42, 351-356 (1987)

1614 HODGKINSON AJ, SIDKI AM, LANDON J, ET AL, DIRECT DETERMINATION OF THEOPHYLLINE IN SERUM BY FLUOROIMMUNOASSAY USING HIGHLY SPECIFIC ANTIBODIES, *ANN CLIN BIOCHEM*, 22, 519-525 (1985)

1615 HODKINSON HM, STANTON BR, ROUND P, MORGAN C, SUNLIGHT, VITAMIN D, AND OSTEOMALACIA OF THE ELDERLY, *LANCET*, 1, 910 (1973)

1616 HODSMAN GP, BROWN JJ, CUMMING AMM ET AL, ENALAPRIL IN THE TREATMENT OF HYPERTENSION WITH RENAL ARTERY STENOSIS, *BR MED J (CLIN PRACT)*, 287, 1413-1417 (1983)

1617 HOECHST PHARMACEUTICAL COMPANY, MANUFACTURER'S LITERATURE ON LASIX®, *1385 TENNESSEE AVE, CINCINNATI, OHIO*, 45229

1618 HOEG JM, SCHAEFER EJ, ROMANO CA, ET AL, NEOMYCIN AND PLASMA LIPOPROTEINS IN TYPE II HYPERLIPOPROTEINEMIA, *CLIN PHARMACOL THER*, 36, 555-565 (1984)

1619 HOEKENGA MT, BUNDE CA, CHO YW, KUZMA RJ, DOUBLE BLIND MULTICENTER EVALUATION OF A NEW ANTIANGINAL DRUG: PERHEXILINE MALEATE, *CLIN PHARMACOL THER*, 13, 140 (1972)

1620 HOELDTKE R, EFFECT OF ASPIRIN ON THE ASSAY OF HOMOVANILLIC ACID IN URINE, *AM J CLIN PATHOL*, 57, 324 (1972)

1621 HOENSCH H, THE EFFECTS OF ALCOHOL ON THE LIVER, *DIGESTION*, 6, 114 (1972)

1622 HOFFER A, *NIACIN THERAPY IN PSYCHIATRY*, THOMAS, SPRINGFIELD, IL, 1962

1623 HOFFMAN RS, MARTINO JA, WAHL G, ARKY RA, FASTING AND REFEEDING. 3. ANTINATRIURETIC EFFECT OF ORAL AND INTRAVENOUS CARBOHYDRATE, *METABOLISM*, 20, 1065 (1971)

1624 HOGE SK, BIEDERMAN J, LIVER FUNCTION TESTS DURING TREATMENT WITH DESIPRAMINE IN CHILDREN AND ADOLESCENTS, *J CLIN PSYCHOPHARMACOL*, 7, 87-89 (1987)

1625 HOIKKA V, ALHAVA EM, KARJALAINEN P, ET AL, CARBAMAZEPINE AND BONE MINERAL METABOLISM, *ACTA NEUROL SCAND*, 68, 77-80 (1984)

1626 HOJER B, RANE A, FATAL HEPATIC FAILURE IN A CHILD TREATED WITH PHENYTOIN AND VALPROIC ACID, *DEV MED CHILD NEUROL*, 24, 846-849 (1982)

1627 HOKFELT B, HEDELAND H, DYMLING JF, THE INFLUENCE OF CATAPRES® ON CATECHOLAMINES, RENIN AND ALDOSTERONE IN MAN IN, *CATAPRES® IN HYPERTENSION*, ME CONOLLY, ED., BUTTERWORTH, LONDON (1970)

1628 HOKIN DB, BLANK CORRECTION FOR METRONIDAZOLE INTERFERENCE WITH CONTINUOUS FLOW MEASUREMENT OF ASPARTATE AMINOTRANSFERASE, *CLIN CHEM*, 29, 406-407 (1983)

1629 HOLDEN JMC, ITIL TM, LABORATORY CHANGES WITH CHLORDIAZEPOXIDE AND THIORIDAZINE, ALONE AND COMBINED, *CAN J PSYCHIATRY*, 14, 299 (1969)

1630 HOLDSTOCK DJ, GASTROINTESTINAL BLEEDING: A POSSIBLE ASSOCIATION WITH IBUPROFEN, *LANCET*, 1, 541 (1972)

1631 HOLLANDER CS, MITSUMA T, SHENKMAN L, WOOLF P, GERSHENGORN MC, THYROTROPIN-RELEASING HORMONE: EVIDENCE FOR THYROID RESPONSE TO INTRAVENOUS INJECTION IN MAN, *SCIENCE*, 175, 209 (1972)

1632 HOLLISTER D JR., COLEMAN M, HEMATOLOGIC EFFECTS OF INTRAVESICULAR THIOTEPA THERAPY FOR BLADDER CARCINOMA, *J AM MED ASSOC*, 244, 2065-2067 (1980)

1633 HOLLISTER LE, STATUS REPORT ON CLINICAL PHARMACOLOGY OF MARIJUANA, *ANN N Y ACAD SCI*, 191, 132 (1972)

1634 HOLLISTER LE, PREDICTION OF THERAPEUTIC USES OF PSYCHOTHERAPEUTIC DRUGS FROM EXPERIENCES WITH NORMAL VOLUNTEERS, *CLIN PHARMACOL THER*, 13, 803 (1972)

1635 HOLLISTER LE, SOME HUMAN PHARMACOLOGICAL STUDIES OF THREE PSYCHOTROPIC DRUGS: THIOTHIXENE, MOLINDONE, AND W-18677, *J CLIN PHARMACOL*, 8, 95 (1968)

1636 HOLLISTER LE, MOORE FE, FACTORS AFFECTING EXCRETION OF CATECHOLAMINES IN MAN: URINE FLOW, URINE PH AND CREATININE CLEARANCE, *RES COMMUN CHEM PATHOL PHARMACOL*, 1, 193 (1970)

1637 HOLLWICH F, DIECKHUES B, ENDOCRINE SYSTEM AND BLINDNESS, *GERM MED*, 1, 122 (1971)

1638 HOLLWICH F, TILGNER S, CHANGES IN THE EOSINOPHIL COUNT IN RESPONSE TO OCCULAR STIMULATION BY LIGHT, *GERM MED*, 10, 14 (1965)

1639 HOLMBERG L, BOMAN G, PULMONARY REACTIONS TO NITROFURANTOIN, *EUR J RESPIR DIS*, 62, 180-189 (1981)

1640 HOLMBERG L, BOMAN G, BOTTIGER LE, ADVERSE REACTIONS TO NITROFURANTOIN: ANALYSIS OF 921 REPORTS, *AM J MED*, 69, 733-738 (1980)

1641 HOLMES TH, MORGAN BA, WOOLF CR, THE EFFECT OF DISODIUM CROMOGLYCATE ON PLASMA 17-HYDROXYCORTICOID CONCENTRATION DURING EXERCISE, *AM REV RESPIR DIS*, 106, 610 (1972)

1642 HOLT RJ, NEUROLEPTIC DRUG-INDUCED CHANGES IN PLATELET LEVELS, *J CLIN PSYCHOPHARMACOL*, 4, 130-132 (1984)

1643 HOLTERMULLER KH, GO VLW, SIZEMORE GW, ARNAUD CD, GOLDSMITH RS, DISSOCIATION OF CALCITONIN (CT), PARATHYROID HORMONE (PTH) AND INTRALUMINAL CALCIUM EFFECTS IN GASTRIC SECRETION, *CLIN RES*, 20, 732 (1972)

1644 HOLTON JB, THE EFFECT OF A HISTIDINE LOAD ON PLASMA LEVELS AND RENAL CLEARANCES OF OTHER AMINO ACIDS, *CLIN CHIM ACTA*, 21, 241 (1968)

1645 HOLUB WR, GALLI FA, AUTOMATED DIRECT METHOD FOR MEASUREMENT OF SERUM CHOLESTEROL, WITH USE OF PRIMARY STANDARDS AND A STABLE REAGENT, *CLIN CHEM*, 18, 239 (1972)

1646 HOMAYOUNI H, GROSS PA, SETIA U, ET AL, LEUKOPENIA DUE TO PENICILLIN AND CEPHALOSPORIN HOMOLOGUES, *ARCH INTERN MED*, 139, 827-828 (1979)

1647 HOMI J, KONCHIGERI HN, ECKENHOFF JE, LINDE HW, A NEW ANESTHETIC AGENT - FORANE®: PRELIMINARY OBSERVATIONS IN MAN, *ANESTH ANALG*, 51, 439 (1972)

1648 HONER WG, THOMPSON C, LIGHTMAN SC, ET AL, NO EFFECT OF NALOXONE ON PLASMA OXYTOCIN IN NORMAL MEN., *PSYCHONEUROENDOCRINOLOGY*, 11, 245-248 (1986)

1649 HONET JC, FALSE POSITIVE URINARY TEST FOR 5-HIAA DUE TO METHOCARBOMAL AND MEPHENESIN CARBAMATE, *N ENGL J MED*, 261, 188 (1959)

1650 HONGER PE, ROSSING N, ALBUMIN METABOLISM AND ORAL CONTRACEPTION, *CLIN SCI*, 36, 41 (1969)

1651 HONMA Y, NIMURA T, SERUM CREATINE PHOSPHOKINASE ACTIVITY IN CHRONIC ALCOHOLICS, *JPN J STUD ALC*, 4, 87 (1969)

1652 HOOD JW, TOTH WN, JAUNDICE CAUSED BY PHENAZOPYRIDINE HYDROCHLORIDE, *J AM MED ASSOC*, 198, 1366 (1966)

1653 HOOGWERF BJ, GRUND VR, HUNNINGHAKE DB, EFFECT OF CIMETIDINE ON PLASMA LIPIDS AND LIPOPROTEINS, *CLIN PHARMACOL THER*, 36, 217-220 (1984)

1654 HOOPER PL, WOO W, VISCONTI L, ET AL, TERBUTALINE RAISES HIGH-DENSITY-LIPOPROTEIN-CHOLESTEROL LEVELS, *N ENGL J MED*, 305, 1455-1457 (1981)

1655 HOPPIN EC, GREENBERG BR, WALTER RM, AGRANULOCYTOSIS SECONDARY TO PENTAZOCINE THERAPY, *ARCH INTERN MED*, 138, 533-534 (1978)

1656 HORBER FF, SCHEIDEGGER J, FREY FJ, OVERESTIMATION OF RENAL FUNCTION IN GLUCOCORTICOSTEROID TREATED PATIENTS, *EUR J CLIN PHARMACOL*, 28, 537-541 (1985)

1657 HORDER K, HORDER M, PLASMA HAPTOGLOBIN AND PHYSICAL EXERCISE: CHANGES IN HEALTHY INDIVIDUALS CONCOMITANT WITH A STRENUOUS MARCH, *CLIN CHIM ACTA*, 30, 369 (1970)

1658 HORDER M, COLORIMETRIC DETERMINATION OF ORTHOPHOS-PHATE IN THE ASSAY OF INORGANIC PYROPHOSPHATASE ACTIVITY, *ANAL BIOCHEM*, 49, 37 (1972)

1659 HORIE Y, UDAGAWA M, HIRAYAMA C, CLINICAL USEFULNESS OF CIMETIDINE FOR THE TREATMENT OF ACUTE INTERMITTENT PORPHYRIA - A PRELIMINARY REPORT, *CLIN CHIM ACTA*, 167, 267-271 (1987)

1660 HORNE CHW, MALLINSON AC, FERGUSON J, GOUDIE RB, EFFECT OF ESTROGEN AND PROGESTOGEN ON SERUM LEVELS OF ALPHA$_2$ MACROGLOBULIN, TRANSFERRIN AND IgG, *J CLIN PATHOL*, 24, 464 (1971)

1661 HOROWITZ J, KEYNAN A, BEN-ISHAY D, A SYNDROME OF INAPPRO-PRIATE ADH SECRETION INDUCED BY CYCLOTHIAZIDE, *J CLIN PHARMACOL*, 12, 337 (1972)

1662 HORROBIN DF, BURSTYN PG, LLOYD IJ, DURKIN N, LIPTON A, MUIRURI KL, ACTIONS OF PROLACTIN ON HUMAN RENAL FUNC-TION, *LANCET*, 2, 352 (1971)

1663 HORWITZ CA, GARMEZY L, LYON F, HENSLEY M, BURKE MD, A COMPARATIVE STUDY OF FIVE IMMUNOLOGIC PREGNANCY TESTS, *AM J CLIN PATHOL*, 58, 305 (1972)

1664 HOVERSTAD T, CARLSTEDT-DUKE B, LINGAAS E, ET AL, INFLUENCE OF AMPICILLIN, CLINDAMYCIN AND METRONIDAZOLE ON FAE-CAL EXCRETION OF SHORT-CHAIN FATTY ACIDS IN HEALTHY SUBJECTS., *SCAND J GASTROENTEROL*, 21, 621-626 (1986)

1665 HOWARD CB, KELLEHER PC, PLASMA FUCOSE AND SIALIC ACID CONCENTRATIONS DURING ORAL GLUCOSE TOLERANCE TESTS IN NORMAL AND DIABETIC, *CLIN CHIM ACTA*, 31, 75 (1971)

1666 HOWARD G, BLAIR M, FOTHERBY K, ET AL, SOME METABOLIC EFFECTS OF LONG-TERM USE OF THE INJECTABLE CONTRA-CEPTIVE NORETHISTERONE OENANTHATE, *LANCET*, 1, 423-425 (1982)

1667 HOWARD JM, CHREMOS AN, COLLEN MJ, ET AL, FAMOTIDINE, A NEW POTENT, LONG-ACTING HISTAMINE H$_2$-RECEPTOR ANTAG-ONIST: COMPARISON WITH CIMETIDINE AND RANITIDINE IN THE TREATMENT OF ZOLLINGER-ELLISON SYNDROME, *GASTROEN-TEROLOGY*, 88, 1026-1033 (1985)

1668 HOWARD KM, ROELLI AP, WORTH HGJ, SOME EFFECTS OF COM-BINED CONTRACEPTIVE THERAPY ON CORTISOL MEASURE-MENTS, *ANN CLIN BIOCHEM*, 24, SUPPL S1, 191-192 (1987)

1669 HOWARD RP, BRUSCO OJ, FURMAN RH, EFFECT OF CHOLES-TYRAMINE ADMINISTRATION ON SERUM LIPIDS AND ON NITRO-GEN BALANCE IN FAMILIAL HYPERCHOLESTEROLEMIA, *J LAB CLIN MED*, 68, 12 (1966)

1670 HOWE L, ELMSHIE RG, STABILITY OF AMYLASE IN SERUM FROM PATIENTS WITH PANCREATITIS, *AUST J EXP BIOL MED SCI*, 49, 513 (1971)

1671 HOWES LG, REID JL, CHANGES IN PLASMA FREE 3,4-DIHYDROX-YPHENYLETHYLENE GLYCOL AND NORADRENALINE LEVELS AFTER ACUTE ALCOHOL ADMINISTRATION, *CLIN SCI*, 69, 423-428 (1985)

1672 HOWITZ PF, THE EFFECT OF APURIN (ALLOPURINOL) ON LIVER FUNCTION AND SERUM IRON, *DAN MED BULL*, 17, 203 (1970)

1673 HOWORTH PJN, DETERMINATION OF SERUM ALBUMIN IN NEONA-TAL JAUNDICE. THE ALBUMIN SATURATION INDEX, *CLIN CHIM ACTA*, 32, 271 (1971)

1674 HOWRIE DL, GARTNER JC JR., GOLD-INDUCED HEPATOTOXICITY: CASE REPORT AND REVEIW OF THE LITERATURE, *J RHEU-MATOL*, 9, 727-729 (1982)

1675 HSU TH, BIANCHINE JR, MESSIHA FS, EFFECT OF PYRIDOXINE (P) ON METABOLISM OF LEVODOPA (LD) IN HEALTHY VOLUN-TEERS, *CLIN RES*, 20, 863 (1972)

1676 HUANG S-M, WEINTRAUB HS, MARRIOTT TB, ET AL, ETINTIDINE-PROPRANOLOL INTERACTION STUDY IN HUMANS, *J PHARM BIOPHARM*, 15, 557-567 (1987)

1677 HUBEL KA, SECRETIN: A LONG PROGRESS NOTE, *GASTROENTER-OLOGY*, 62, 318 (1972)

1678 HUFF B, BROGELER E, FELKNOR L, MULLER T (EDS), *SUPPLEMENT B. PHYSICIANS' DESK REFERENCE*, MEDICAL ECONOMICS INC, ORADELL, NJ (1973)

1679 HUFF B, BROGELER E, FELKNOR L, MULLER T (EDS), *SUPPLEMENT B. PHYSICIANS' DESK REFERENCE*, MEDICAL ECONOMICS INC, ORADELL, NJ (1972)

1680 HUFF B, BROGELER E, FELKNOR L, MULLER T (EDS), *PHYSICIANS' DESK REFERENCE* 26TH EDITION, MEDICAL ECONOMICS INC, ORADELL, NJ (1972)

1681 HUGHES JA, SUDELL W, HEMOLYTIC ANEMIA ASSOCIATED WITH NAPROXEN, *ARTHRITIS RHEUM*, 26, 1054 (1983)

1682 HUGHES W, SEAMENDS B, COHEN S, ARVAN D, THE RELATION OF SERUM IONIZED CALCIUM TO THE ALKALINE TIDE, *CLIN RES*, 20, 733 (1972)

1683 HUGUES JN, PERRET G, SEBAOUN J, ET AL, EFFECTS OF CIME-TIDINE ON THYROID HORMONES, *CLIN ENDOCRINOL*, 17, 297-302 (1982)

1684 HUGULEY CM JR., AGRANULOCYTOSIS INDUCED BY DIPYRONE, A HAZARDOUS ANTIPYRETIC AND ANALGESIC, *J AM MED ASSOC*, 189, 938 (1964)

1685 HUIS IN'T VELD LG, DRUG INFORMATION SERVICE NOTE NO 5, *IFCC COMMITTEE ON STANDARDS*

1686 HUIS IN'T VELD LG, DRUG INFORMATION SERVICE NOTE NO 6, *IFCC COMMITTEE ON STANDARDS*

1687 HUISMAN JW, INTERFERENCE BY DIFLUNISAL WITH THE FETI METHOD FOR SERUM THYROXINE, *ANN CLIN BIOCHEM*, 23, 223-224 (1986)

1688 HULLEY SB, VOGEL JM, DONALDSON CL, BAYERS JH, FRIEDMAN RJ, ROSEN SN, THE EFFECT OF SUPPLEMENTAL ORAL PHOSPHATE ON THE BONE MINERAL CHANGES DURING PRO-LONGED BED REST, *J CLIN INVEST*, 50, 2506 (1971)

1689 HULTHEN UL, VAN BRUMMELEN P, AMANN FW, ET AL, ANTIHYPERTENSIVE EFFICACY OF THE NEW LONG-ACTING BETA-BLOCKER BOPINDOLOL AS RELATED TO AGE, *J CARDI-OVASC PHARMACOL*, 5, 426-429 (1983)

1690 HUMBLE MW, EYKYN SJ, PHILLIPS I, STAPHYLOCOCCAL BACTER-EMIA, FUSIDIC ACID, AND JAUNDICE, *BR MED J [CLIN RES]*, 1, 1495-1498 (1980)

1691 HUME R, JOHNSTONE JMS, WEYERS E, INTERACTION OF ASCORBIC ACID AND WARFARIN, *J AM MED ASSOC*, 219, 1479 (1972)

1692 HUNTER E, RAIK E, GORDON S, TAYLOR K, INCIDENCE OF POSITIVE COOMBS' TEST, LE CELLS AND ANTINUCLEAR FACTOR IN PATIENTS ON ALPHA-METHYLDOPA (ALDOMET®), *MED J AUST*, 2, 810 (1971)

1693 HUNTER HYPERTENSION GROUP, CHANGES IN SERUM LIPID LEVELS DURING ANTIHYPERTENSIVE THERAPY, *MED J AUST*, 140, 522-524 (1984)

1694 HUNTER J, MAXWELL JD, CARRELLA M, STEWART DA, WILLIAMS R, URINARY D-GLUCARIC ACID EXCRETION AS A TEST FOR HEP-ATIC ENZYME INDUCTION IN MAN, *LANCET*, 1, 572 (1971)

1695 HUNTER J, MAXWELL JD, STEWART DA, PARSONS V, WILLIAMS R, ALTERED CALCIUM METABOLISM IN EPILEPTIC CHILDREN ON ANTICONVULSANTS, *BR MED J [CLIN RES]*, 4, 202 (1971)

1696 HUNTER JB, CRITZ JB, EFFECT OF TRAINING ON PLASMA ENZYME LEVELS IN MAN, *J APPL PHYSIOL*, 31, 20 (1971)

1697 HUNTER WM, *GROWTH AND DEVELOPMENT OF MAMMALS*, GA LODGE, GE LAMMING EDS. BUTTERWORTH, LONDON (1968)

1698 HURST PL, EFFECT OF ANTICOAGULANTS ON FRUCTOSAMINE DETERMINATION, *CLIN CHEM*, 33, 1947 (1987)

1699 HURSTING MJ, RAISYS VA, OPHEIM KE, DRUG SPECIFIC FAB THERAPY IN DRUG OVERDOSE, *ARCH PATHOL LAB MED*, 111, 693-697 (1987)

1700 HURT R, THE EFFECT OF RADIOGRAPHIC CONTRAST MEDIA ON URINALYSIS, *AM J MED TECHNOL*, 26, 122 (1960)

1701 HURT RD, BRIONES ER, OFFORD KP, ET AL, PLASMA LIPIDS AND APOLIPOPROTEIN A-I AND A-II LEVELS IN ALCOHOLIC PATIENTS, *AM J CLIN NUTR*, 43, 521-529 (1986)

1702 HURTER R, SWALE J, PEYMAN MA, BARNETT CWH, SOME IMMEDI-ATE AND LONG-TERM EFFECTS OF EXERCISE ON THE PLASMA LIPIDS, *LANCET*, 2, 671 (1972)

1703 HUSAIN MK, FRANTZ AG, CIAROCHI FF, ROBINSON AG, NICOTINE STIMULATED RELEASE OF NEUROPHYSIN (NP) AND VASOPRES-SIN (VP) IN HUMANS, *CLIN RES*, 21, 494 (1973)

1704 HUSDAN H, RAPOPORT A, LOCKE S, OREPOULOS D, EFFECT OF VENOUS OCCLUSION OF THE ARM ON THE CONCENTRATION OF CALCIUM IN SERUM AND METHODS FOR ITS COMPENSA-TION, *CLIN CHEM*, 20, 529 (1974)

1705 HUSSAR DA, THE NEW DRUGS OF 1971, *AM J PHARM*, 144, 5 (1972)

1706 HUSSERL FE, ERYTHROMYCIN - WARFARIN INTERACTION, *ARCH INTERN MED*, 143, 1831-1832 (1983)

1707 HUSSEY AJ, HOWIE J, ALLAN LG, ET AL, IMPAIRED HEPATOCELLU-LAR INTEGRITY DURING GENERAL ANESTHESIA, AS ASSESSED BY MEASUREMENT OF PLASMA GLUTATHIONE S-TRANSFER-ASE, *CLIN CHIM ACTA*, 161, 19-28 (1986)

1708 HUTCHISON JC, WILKINSON WH, ZEMLIN RD, LONG-TERM EXPER-IENCES WITH HALOFENATE, *CLIN RES*, 21, 469 (1973)

1709 HUTTUNEN JK, FRUCTOSE IN MEDICINE: A REVIEW WITH PARTICU-LAR REFERENCE TO DIABETES MELLITUS, *POSTGRAD MED J*, 47, 654 (1971)

1710 HWANG PLH, NG CSH, CHEONG ST, EFFECT OF ORAL CONTRACEP-TIVES ON SERUM PROLACTIN: LONGITUDINAL STUDY IN 126 NORMAL PREMENOPAUSAL WOMEN, *CLIN ENDOCRINOL*, 24, 127-133 (1986)

1711 HYSELL JK, HYSELL JW, GRAY JM, THROMBOCYTOPENIC PURPURA FOLLOWING IOPANOIC ACID INGESTION, *J AM MED ASSOC*, 237, 361-362 (1977)

1712 IBBOTT FA, AMINO ACIDS AND RELATED SUBSTANCES (P 565) IN *CLINICAL CHEMISTRY: PRINCIPLES AND TECHNICS*, RJ HENRY, DC CANNON, JW WINKELMAN, EDS. 2ND ED HARPER AND ROW, HAGERSTOWN, MD (1974)

1713 IBER FL, CHOLESTASIS, *POSTGRAD MED*, 41, 30 (1967)

1714 IBSEN H, SEDERBERG-OLSEN P, CHANGES IN GLOMERULAR FILTRATION RATE DURING LONG-TERM TREATMENT WITH PROPRANOLOL IN PATIENTS, *CLIN SCI*, 44, 129 (1973)

1715 IKEDA T, NONAKA Y, GOTO A, ET AL, EFFECTS OF PRAZOSIN ON PLATELET AGGREGATION AND PLASMA BETA-THROMBOGLOBULIN IN ESSENTIAL HYPERTENSION, *CLIN PHARMACOL THER*, 37, 601-605 (1985)

1716 IKRAM H, MASLOWSKI AH, NICHOLLS MG, ET AL, HAEMODYNAMIC AND HORMONAL EFFECTS OF CAPTOPRIL IN PRIMARY PULMONARY HYPERTENSION, *BR HEART J*, 48, 541-545 (1982)

1717 ILES RA, BARNETT D, STRUNIN L, STRUNIN JM, SIMPSON BR, COHEN RD, THE EFFECT OF HYPOXIA ON SUCCINATE METABOLISM IN MAN AND THE ISOLATED PERFUSED DOG LIVER, *CLIN SCI*, 42, 35 (1972)

1718 IMAI Y, KATOAKA K, SHENKMAN L, WAN L, HOLLANDER CS, IN VIVO EFFECTS OF PROSTAGLANDIN F2 ALPHA ON THYROID HORMONE RELEASE IN MAN, *CLIN RES*, 21, 494 (1973)

1719 IMAIZUMI H, NAMIKI A, WATANABE A, ET AL, HEMOLYSIS AND HEMOGLOBINURIA FOLLOWING ADMINISTRATION OF HIGH DOSE FENTANYL. A CASE REPORT, *JPN J ANESTHESIOL*, 35, 639-642 (1986)

1720 IMHOF PR, HUSHAK J, SCHUMANN G, DUKOR P, WAGNER J, KELLER HM, EXCRETION OF URINARY CASTS AFTER THE ADMINISTRATION OF DIURETICS, *BR MED J [CLIN RES]*, 2, 199 (1972)

1721 IMURA H, NAKAI Y, MATSUKURA S, MATSUNYAMA H, EFFECT OF INTRAVENOUS INFUSION OF L-DOPA ON PLASMA GROWTH HORMONE LEVELS IN MAN, *HORM METAB RES*, 5, 41 (1973)

1722 INLOES R, CLARK D, DROBNIES A, INTERFERENCE OF FLUORESCEIN, USED IN RETINOL ANGIOGRAPHY, WITH CERTAIN CLINICAL LABORATORY TESTS, *CLIN CHEM*, 33, 2126-2127 (1988)

1723 INMAN WHW, STUDY OF FATAL BONE MARROW DEPRESSION WITH SPECIAL REFERENCE TO PHENYLBUTAZONE AND OXYPHENBUTAZONE, *BR MED J [CLIN RES]*, 1, 1500-1505 (1977)

1724 INNES RKR, STROMME JH, RISE IN SERUM CREATINE PHOSPHOKINASE ASSOCIATED WITH AGENTS USED IN ANESTHESIA, *BR J ANAESTH*, 45, 185 (1973)

1725 INTERNATIONAL UNION AGAINST TUBERCULOSIS COMMITTEE ON PROPHYLAXIS, EFFICACY OF VARIOUS DURATIONS OF ISONIAZID PREVENTIVE THERAPY FOR TUBERCULOSIS: FIVE YEARS OF FOLLOW-UP IN THE IUAT TRIAL, *BULL WHO*, 60, 555-564 (1982)

1726 IRIE A, HUNAKI M, BANDO K, KAWAI K, DETERMINATION OF AMYLASE ACTIVITY IN SERUM AND URINE USING BLUE STARCH SUBSTRATE, *CLIN CHIM ACTA*, 42, 63 (1972)

1727 IRIE M TSUSHIMA T, SAKUMA M, EFFECT OF NICOTINIC ACID ADMINISTRATION ON PLASMA HGH, FFA AND GLUCOSE IN OBESE SUBJECTS, *METABOLISM*, 19, 972 (1971)

1728 ISAAC PF, RAND MJ, CIGARETTE SMOKING AND PLASMA LEVELS OF NICOTINE, *NATURE*, 236, 308 (1972)

1729 ISAAC R, MERCERON R, CAILLENS G, ET AL, EFFECTS OF CALCITONIN ON BASAL AND THYROTROPIN-RELEASING HORMONE STIMULATED PROLACTIN SECRETION IN MAN, *J CLIN ENDOCRINOL METAB*, 50, 1011-1015 (1980)

1730 ISHAK KG, KIRCHNER JP, DHAR JK, GRANULOMAS AND CHOLESTATIC-HEPATOCELLULAR INJURY ASSOCIATED WITH PHENYLBUTAZONE: REPORT OF TWO CASES, *AM J DIG DIS*, 22, 611-617 (1977)

1731 ISHIBE T, ALTERATIONS IN SERUM LACTATE DEHYDROGENASE ACTIVITY AND ITS ISOENZYME PATTERN AFTER MASSAGE OF THE PROSTRATE, *INVEST UROL*, 9, 104 (1971)

1732 ISHIGURO T, ALPHA-FETOPROTEIN IN TWIN PREGNANCY, *LANCET*, 2, 1214 (1973)

1733 ISHIGURO T, ALPHA-FETOPROTEIN IN PREGNANCY, *LANCET*, 1, 1509 (1973)

1734 ISLEY WL, EFFECT OF RIFAMPIN THERAPY ON THYROID FUNCTION TESTS IN A HYPOTHYROID PATIENT ON REPLACEMENT L-THYROXINE, *ANN INTERN MED*, 107, 517-518 (1987)

1735 ISMAIL AA, DAVIDSON DW, LORAINE JA, FOX CA, RELATIONSHIP BETWEEN PLASMA CORTISOL AND HUMAN SEXUAL ACTIVITY, *NATURE*, 237, 288 (1972)

1736 ISMAIL AAA, ASTLEY P, BURR WA, ET AL, THE ROLE OF TESTOSTERONE MEASUREMENT IN THE INVESTIGATION OF ANDROGEN DISORDERS, *ANN CLIN BIOCHEM*, 23, 113-134 (1986)

1737 ISRAEL R, O'MARA V, AUSTIN B, ET AL, METOCLOPRAMIDE DECREASES RENAL PLASMA FLOW, *CLIN PHARMACOL THER*, 39, 261-264 (1986)

1738 ITIL TM, STOCK MJ, DUFFY AD, ESQUENAZI A, SALEUTY B, HAN TH, THERAPEUTIC TRIALS AND EEG INVESTIGATIONS WITH SCH-12,679, *CURR THER RES*, 14, 136 (1972)

1739 ITO H, MITSUHASHI SI, MOMOSE G, PROSTAGLANDINS AND RENAL FUNCTION - THE EFFECT OF PROSTAGLANDIN A2 ON THE PSP EXCRETION, *PROSTAGLANDINS*, 3, 359 (1973)

1740 ITOH S, ICHINOE A, TSUKADA Y, ET AL, HYDRALAZINE-INDUCED HEPATITIS, *HEPATOGASTROENTEROLOGY*, 28, 13-16 (1981)

1741 IWARSON S, LUNDIN P, MULTIPLE ATTACKS OF JAUNDICE ASSOCIATED WITH REPEATED SULFONAMIDE TREATMENT, *ACTA MED SCAND*, 206, 219-222 (1979)

1742 IZAKOVIC V, IZAKOVICOVA A, HEPARIN AND PREDNISONE FOR CARCINOID SYNDROME, *KLIN WOCHENSCHR*, 42, 874 (1964)

1743 IZQUIERDO JM, SOTORRIO P, ROMEO D, QWROS A, EVALUATION OF LEFFLER'S METHOD OF TOTAL CHOLESTEROL, *Z KLIN CHEM KLIN BIOCHEM*, 12, 227 (1974)

1744 JAATELA A, NIKKI P, TAKKI S, TAMMISTO T, EFFECT OF DEXTROMORAMIDE, FENTANYL, AND MORPHINE ON THE PLASMA CATECHOLAMINE LEVELS, *ANN CLIN RES*, 3, 107 (1971)

1745 JABLONSKI P, OWEN JA, THE CLINICAL CHEMISTRY OF BROMOSULFOPHTHALEIN AND OTHER CHOLEPHILIC DYES, *ADV CLIN CHEM*, 12, 209 (1969)

1746 JACKSON B, BALDWIN J, LEIENDECKER-FOSTER C, ET AL, BILIRUBIN INTERFERENCE IN THE ENZYMATIC DETERMINATION OF BICARBONATE IN THE OLYMPUS DEMAND, *CLIN CHEM*, 32, 1233 (1986)

1747 JACKSON B, JOHNSTON CI, ANGIOTENSIN CONVERTING ENZYME DURING ACUTE AND CHRONIC ENALAPRIL THERAPY IN ESSENTIAL HYPERTENSION, *CLIN EXP PHARMACOL PHYSIOL*, 11, 355-359 (1984)

1748 JACKSON CE, MEIER DW, ROUTINE SERUM MAGNESIUM ANALYSIS CORRELATION WITH CLINICAL STATE IN 5,100 PATIENTS, *ANN INTERN MED*, 69, 743 (1968)

1749 JACKSON SH, PROBLEMS IN SCREENING INFANTS FOR DEFECTS OF AMINO ACID METABOLISM, *CLIN BIOCHEM*, 6, 15 (1973)

1750 JACKSON ST, RALLISON ML, BUNTIN WH, JOHNSON SB, FLYNN RR, USE OF OXANDROLONE FOR GROWTH STIMULATION, *AM J DIS CHILD*, 126, 481 (1973)

1751 JACOB RA, OMAYE ST, SKALA JH, ET AL, EXPERIMENTAL VITAMIN C DEPLETION AND SUPPLEMENTATION IN YOUNG MEN. NUTRIENT INTERACTIONS AND DENTAL HEALTH EFFECTS, *ANN N Y ACAD SCI*, 498, 333-346 (1987)

1752 JACOB RA, SKALA JH, OMAYE ST, ET AL, EFFECT OF VARYING ASCORBIC ACID INTAKES ON COPPER ABSORPTION AND CERULOPLASMIN LEVELS OF YOUNG MEN, *J NUTR*, 117, 2109-2115 (1987)

1753 JACOBS LS, SNYDER PJ, WILBER JF, UTIGER RD, DAUGHADAY WH, INCREASED SERUM PROLACTIN AFTER ADMINISTRATION OF SYNTHETIC THYROTROPIN RELEASING HORMONE IN MAN, *J CLIN ENDOCRINOL METAB*, 33, 996 (1971)

1754 JACOBS S, PULLAN PT, POTTER JM, ET AL, ADRENAL SUPPRESSION FOLLOWING EXTRADURAL STEROIDS, *ANESTHESIA*, 38, 953-956 (1983)

1755 JACOBS SL, FERNANDEZ AA, HEMOGLOBIN IN PLASMA, *STAND METH CLIN CHEM*, 6, 107 (1970)

1756 JACOBSON ED, PRIOR JT, FALOON WW, MALABSORPTIVE SYNDROME INDUCED BY NEOMYCIN: MORPHOLOGIC ALTERATIONS IN THE JEJUNAL MUCOSA, *J LAB CLIN MED*, 56, 245 (1960)

1757 JACQUES PF, HARTZ SC, McGANDY RB, ET AL, ASCORBIC ACID, HDL, AND TOTAL PLASMA CHOLESTEROL IN THE ELDERLY, *J AM COLL NUTR*, 6, 169-174 (1987)

1758 JADIN-STARODOUBSKY A, DELWAIDE PA, PENDERS C, COLLARD J, HEUSGHEM, PSYCHOTROPIC DRUG INTERFERENCES WITH CLINICAL CHEMISTRY DETERMINATION IN, *REFERENCE VALUES IN HUMAN CHEMISTRY*, G SIEST, ED. KARGER, BASEL 1973, 299

1759 JAEKEN J, CASAER P, CORBEEL, VALPROATE, HYPERAMMONAEMIA AND HYPERGLYCINAEMIA, *LANCET*, 2, 260 (1980)

1760 JAFFÉ BM, BEHRMAN HR, PARKER CW, RADIOIMMUNOASSAY MEASUREMENT OF PROSTAGLANDINS E, A AND F IN HUMAN PLASMA, *J CLIN INVEST*, 52, 398 (1973)

1761 JAFFE IA, ADVERSE EFFECTS PROFILE OF SULFHYDRYL COMPOUNDS IN MAN, *AM J MED*, 80, 471-476 (1986)

1762 JAFFE RB, HORMONAL PROFILES DURING THE MENSTRUAL CYCLE, *RADIOIMMUNOASSAY SYMPOSIUM, WASHINGTON, DC*, 1974

1763 JAFFE RM, KASTEN B, YOUNG DS, ET AL, FALSE-NEGATIVE STOOL OCCULT BLOOD TEST CAUSED BY INGESTION OF ASCORBIC ACID (VITAMIN C), *ANN INTERN MED*, 83, 824-826 (1975)

1764 JAGGARAO NSV, SHELDON J, GRUNDY EN, ET AL, THE EFFECTS OF AMIODARONE ON THYROID FUNCTION, *POSTGRAD MED J*, 58, 693-696 (1982)

1765 JAHNCHEN E, MEINERTZ T, GILFRICH HJ, INTERACTION OF ALLOPURINOL WITH PHENPROCOUMON IN MAN, *KLIN WOCHENSCHR*, 55, 759-761 (1977)

1766 JAILLARD J, SEZILLE G, SCHERREREEL P, FRUCHART JC, CLINICAL AND EXPERIMENTAL STUDIES OF PLASMA-LIPIDS MODIFICATIONS INDUCED BY ALCOHOL, *NUTR METAB*, 13, 114 (1971)

1767 JAIN A, HOYT H, THE EARLY CLINICAL EVALUATION OF W-2354 IN GOUT, *CLIN PHARMACOL THER*, 13, 141 (1972)

1768 JAIN RC, VYAS CR, MAHATMA OP, HYPOGLYCAEMIC ACTION OF ONION AND GARLIC, *LANCET*, 2, 1491 (1973)

1769 JAMES BE, MACMAHON RA, EFFECT OF TRICHLORACETIC ACID ON THE DETERMINATION OF ZINC BY ATOMIC ABSORPTION SPECTROSCOPY, *CLIN CHIM ACTA*, 32, 307 (1971)

1770 JAMES C, PROUT BJ, MARROW SUSPENSION AND INTRAVENOUS CIMETIDINE, *LANCET*, 1, 987 (1978)

1771 JAMES O, LESNA M, ROBERTS SH, ET AL, LIVER DAMAGE AFTER PARACETAMOL OVERDOSE. COMPARISON OF LIVER FUNCTION TESTS, FASTING SERUM BILE ACIDS AND LIVER HISTOLOGY, *LANCET*, 2, 579-581 (1975)

1772 JAMES VHT, MATTINGLY D, DALY JR, RECOMMENDED METHOD FOR THE DETERMINATION OF PLASMA CORTICOSTEROIDS, *BR MED J [CLIN RES]*, 2, 310 (1971)

1773 JANKOWIAK ME, LEVIER RR, ELIMINATION OF PHOSPHORUS INTERFERENCE IN THE COLORIMETRIC DETERMINATION OF SILICON IN BIOLOGICAL MATERIAL, *ANAL BIOCHEM*, 44, 462 (1971)

1774 JANSEN PLM, FROELING PGAM, SCHADE RWB, ET AL, INTRAHEPATIC CHOLESTASIS IN HYPERTHYROIDISM AND THE EFFECT OF ANTITHYROID AND BETA-BLOCKING DRUGS, *NETH J MED*, 25, 318-324 (1982)

1775 JARRETT RJ, GRAVER HJ, COHEN NM, THE EFFECTS OF INFUSIONS OF AMINOACIDS, SECRETIN AND PANCREOZYMIN UPON LEVELS OF BLOOD SUGAR AND PLASMA INSULIN, *DIABETOLOGIA*, 5, 421 (1969)

1776 JARVIE DR, HEYWORTH R, SIMPSON D, PLASMA SALICYLATE ANALYSIS: A COMPARISON OF COLORIMETRIC, HPLC AND ENZYMATIC TECHNIQUES, *ANN CLIN BIOCHEM*, 24, 364-373 (1987)

1777 JATLOW P, ULTRAVIOLET SPECTROPHOTOMETRIC MEASUREMENT OF CHLORDIAZEPOXIDE IN PLASMA, *CLIN CHEM*, 18, 516 (1972)

1778 JAVITT NB, THE CHOLESTATIC SYNDROME - 1971, *AM J MED*, 51, 637 (1971)

1779 JEANJEAN M, ROUSSEAU M, HARVENGT C, INFLUENCE OF CEPHALOTHIN THERAPY ON THE ESTIMATION OF 17-KETO AND 17-KETOGENIC STEROIDS, *ARCH INTL PHARM THER SUPPLEMENT*, 196, 302 (1972)

1780 JEAVONS PM, SODIUM VALPROATE AND ACUTE HEPATIC FAILURE, *DEV MED CHILD NEUROL*, 22, 547-548 (1980)

1781 JEAVONS PM, NON-DOSE-RELATED SIDE EFFECTS OF VALPROATE, *EPILEPSIA*, 25, SUPPL 1, S50-S55 (1984)

1782 JEFFERSON JW, KALIN NH, SERUM LITHIUM LEVELS AND LONG-TERM DIURETIC USE, *J AM MED ASSOC*, 241, 1134-1136 (1979)

1783 JEFFERYS DB, LESSOF MH, MATTOCK MB, CORTICOSTEROID TREATMENT, SERUM LIPIDS AND CORONARY ARTERY DISEASE, *POSTGRAD MED J*, 56, 491-493 (1980)

1784 JEFFREY RF, MACDONALD TM, RUTTER M, ET AL, THE EFFECT OF INTRAVENOUS FRUSEMIDE ON URINE DOPAMINE IN NORMAL VOLUNTEERS: STUDIES WITH INDOMETHACIN AND CARBIDOPA, *CLIN SCI*, 73, 151-157 (1987)

1785 JELIC Z, MAJKIC-SINGH N, SPASIC S, ET AL, EFFECTS OF ANALGESIC AND ANTIRHEUMATIC DRUGS ON THE ASSAY OF SERUM ENZYMES, *J CLIN CHEM CLIN BIOCHEM*, 22, 559-563 (1984)

1786 JELIC-IVANOVIC Z, MAJKIC-SINGH N, SPASIC S, ET AL, INTERFERENCE BY ANALGESIC AND ANTIRHEUMATIC DRUGS IN 25 COMMON LABORATORY ASSAYS, *J CLIN CHEM CLIN BIOCHEM*, 23, 287-292 (1985)

1787 JELIC-IVANOVIC Z, SPASIC S, MAJKK-SINGH N, ET AL, EFFECTS OF SOME ANTI-INFLAMMATORY DRUGS ON 12 BLOOD CONSTITUENTS: PROTOCOL FOR THE STUDY OF IN VIVO EFFECTS OF DRUGS, *CLIN CHEM*, 31, 1141-1143 (1985)

1788 JENKINS GC, HUGHES DTD, HALL PC, A HAEMATOLOGICAL STUDY OF PATIENTS RECEIVING LONG-TERM TREATMENT WITH TRIMETHOPRIM AND SULPHONAMIDE, *J CLIN PATHOL*, 23, 392 (1970)

1789 JENKINS JS, CONNOLLY J, ADRENOCORTICAL RESPONSE TO ETHANOL IN MAN, *BR MED J [CLIN RES]*, 2, 804 (1968)

1790 JENKINS RM, EVANS DMD, CARBIMAZOLE HYPERSENSITIVITY AND LIVER DAMAGE, *BR J CLIN PRACT*, 35, 415-417 (1981)

1791 JENNEKENS FGI, VAN VEELEN CWM, HYPERVITAMINOSE, *PRESSE MED*, 74, 2925 (1966)

1792 JENNER FA, BIOCHEMICAL STUDIES ON THE EFFECTS OF LITHIUM SALTS, *BIOCHEM SOC TRANS*, 1, 88 (1973)

1793 JENSEN OB, MOSDAL C, RESKE-NIELSEN E, HYPOKALEMIC MYOPATHY DURING TREATMENT WITH DIURETICS, *ACTA NEUROL SCAND*, 55, 465-482 (1977)

1794 JEPSEN LV, PEDERSEN KH, CHANGES IN ZINC AND ZINC-DEPENDENT ENZYMES IN RHEUMATOID PATIENTS DURING PENICILLAMINE TREATMENT, *SCAND J RHEUMATOL*, 13, 282-288 (1984)

1795 JESPERSEN J, INGEBERG S, BACH E, ANTITHROMBIN III AND PLATELETS DURING THE NORMAL MENSTRUAL CYCLE AND IN WOMEN RECEIVING ORAL CONTRACEPTIVES LOW IN OESTROGEN, *GYNECOL OBSTET INVEST*, 15, 153-162 (1983)

1796 JI TH, INTERFERENCE BY DETERGENTS, CHELATING AGENTS AND BUFFERS WITH THE LOWRY PROTEIN DETERMINATION, *ANAL BIOCHEM*, 52, 517 (1973)

1797 JOFFE BI, HAITAS B, EDELSTEIN D, ET AL, CLONIDINE AND THE HORMONAL RESPONSES TO GRADED EXERCISE IN HEALTHY SUBJECTS, *HORM RES*, 23, 136-141 (1986)

1798 JOFFE RT, KELLNER CH, POST RM, ET AL, LITHIUM INCREASES PLATELET COUNT, *N ENGL J MED*, 311, 674 (1984)

1799 JOHANNSON E, ANDROGEN LEVELS IN WOMEN USING NORPLANT IMPLANTS, *CONTRACEPTION*, 34, 157-167 (1986)

1800 JOHANNSON EDB, GEMZELL C, PLASMA LEVELS OF PROGESTERONE DURING THE LUTEAL PHASE IN NORMAL WOMEN TREATED WITH SYNTHETIC ESTROGENS, *ACTA ENDOCRINOL*, 68, 551 (1971)

1801 JOHANSSON B, ROOS BE, 5-HYDROXYINDOLEACETIC AND HOMOVANILLIC ACID LEVELS IN THE CEREBROSPINAL FLUID OF HEALTHY VOLUNTEERS AND PATIENTS, *LIFE SCI*, 6, 1449 (1967)

1802 JOHANSSON SA, APPARENT RESISTANCE TO ORAL ANTICOAGULANT THERAPY AND INFLUENCE OF HYPNOTICS ON SOME COAGULATION FACTORS, *ACTA MED SCAND*, 184, 297 (1968)

1803 JOHN VA, MONK JP, DORHOFER G, INTERFERENCE OF AN OXPRENOLOL METABOLITE WITH SCREENING TESTS FOR 5-HYDROXYINDOLE IN URINE, *CLIN CHEM*, 29, 743-744 (1983)

1804 JOHNSON BF, THE DIURETIC ACTION OF AN AZIDO PYRIMIDINE IN MAN, *CLIN PHARMACOL THER*, 11, 77 (1970)

1805 JOHNSON BF, DIAZOXIDE AND RENAL FUNCTION IN MAN, *CLIN PHARMACOL THER*, 12, 815 (1971)

1806 JOHNSON BF, THE EMERGING PROBLEM OF PLASMA LIPID CHANGES DURING ANTIHYPERTENSIVE THERAPY, *J CARDIOVASC PHARMACOL*, 4 SUPPL 2, S213-S221 (1982)

1807 JOHNSON BF, ERRICHETTI A, URBACH D, ET AL, THE EFFECT OF ONCE DAILY MINOXIDIL ON BLOOD PRESSURE AND PLASMA LIPIDS., *J CLIN PHARMACOL*, 26, 534-538 (1986)

1808 JOHNSON BJ, ROMERO L, JOHNSON J, ET AL, COMPARATIVE EFFECTS OF PROPRANOLOL AND PRAZOSIN UPON SERUM LIPIDS IN THIAZIDE TREATED HYPERTENSIVE PATIENTS, *AM J MED*, 76 (2a), 109-112 (1984)

1809 JOHNSON LR, REESE M, NELSON DH, INTERFERENCE IN PISANO'S URINARY METANEPHRINE ASSAY AFTER USE OF X-RAY CONTRAST MEDIA, *CLIN CHEM*, 18, 209 (1972)

1810 JOHNSON RB JR., OSMOLALITY OF SERUM AND URINE, *STAND METH CLIN CHEM*, 5, 159 (1965)

1811 JOHNSON RH, SULAIMAN WR, WEBSTER MHC, HUMAN GROWTH HORMONE AND KETOSIS IN ATHLETES AND NONATHLETES, *NATURE*, 236, 119 (1972)

1812 JOHNSTON BB, DIABETES MELLITUS IN PATIENTS ON LITHIUM, *LANCET*, 2, 935-936 (1977)

1813 JOHNSTON DG, GILL A, ORSKOV H, ET AL, METABOLIC EFFECTS OF CORTISOL IN MAN - STUDIES WITH SOMATOSTATIN, *METABOLISM*, 31, 312-317 (1982)

1814 JOHNSTON RR, CROMWELL TH, EGER EI II, CULLEN D, STEVENS WC, JOAS T, THE TOXICITY OF FLUROXENE IN ANIMALS AND MAN, *ANESTHESIOLOGY*, 38, 313 (1973)

1815 JONES CC, MEGALOBLASTIC ANEMIA ASSOCIATED WITH LONG-TERM TETRACYCLINE THERAPY, *ANN INTERN MED*, 78, 910 (1973)

1816 JONES CM, ALTERATION OF THYROID FUNCTION BY HEMODIALYSIS - AN EFFECT OF HEPARIN, *ALA J MED SCI*, 8, 265 (1971)

1817 JONES DG, TURNBULL MJ, LENMAN JAR, ROBERTSON MAH, EFFECT OF AMANTADINE ON THE URINARY EXCRETION OF SOME MONOAMINES AND METABOLITES IN NORMAL AND PARKINSONIAN SUBJECTS, *J NEUROL SCI*, 17, 245 (1972)

1818 JONES DH, KING K, MILLER AJ, ET AL, A DOSE-RESPONSE STUDY OF 13-CIS-RETINOIC ACID IN ACNE VULGARIS, *BR J DERMATOL*, 108, 333-343 (1983)

1819 JONES MF, CALDWELL JR, ACUTE HEMORRHAGIC PANCREATITIS ASSOCIATED WITH ADMINISTRATION OF CHLORTHALIDONE, *N ENGL J MED*, 267, 1029 (1962)

1820 JONES PE, OELBAUM MH, FRUSEMIDE-INDUCED PANCREATITIS, *BR MED J [CLIN RES]*, 1, 133-134 (1975)

1821 JONES RD, COMMINS BT, CERNIK AA, BLOOD LEAD AND CARBOXYHEMOGLOBIN LEVELS IN LONDON TAXI DRIVERS, *LANCET*, 2, 302 (1972)

1822 JONES RJ, DOBRILOVIC L, LIPOPROTEIN LIPID ALTERATIONS WITH CHOLESTYRAMINE ADMINISTRATION, *J LAB CLIN MED*, 75, 953 (1970)

1823 JONKERS R, VAN BOXTEL CJ, OOSTERHUIS B, β_2-ADRENOCEPTOR-MEDIATED HYPOKALAEMIA AND ITS ABOLITION BY OXPRENOLOL, *CLIN PHARMACOL THER*, 42, 627-633 (1987)

1824 JONKMAN JM, VAN DER BOON WJ, SCHOENMAKER R, ET AL, CLINICAL PHARMAKINETICS OF THEOPHYLLINE DURING COTREATMENT WITH CEFACLOR, *INT J CLIN PHARMACOL THER TOXICOL*, 24, 88-92 (1986)

1825 JONSSON LE, LEWANDER T, GUNNE LM, AMPHETAMINE PSYCHOSIS: URINARY EXCRETION OF CATECHOLAMINES, *RES COMMUN CHEM PATHOL PHARMACOL*, 2, 355 (1971)

1826 JOOS C, KEWITZ H, REINHOLD-KOURNIATI D, EFFECTS OF DIURETICS ON PLASMA LIPOPROTEINS IN HEALTHY MEN, *EUR J CLIN PHARMACOL*, 17, 251-257 (1980)

1827 JOUBERT PH, AUCAMP AK, POTGIETER GM, ET AL, EPILEPSY AND IgA DEFICIENCY - THE EFFECTS OF SODIUM VALPROATE, *S AFR MED J*, 52, 642-644 (1977)

1828 JOVANOVIC SD, BANIC B, JOVANOVIC E, INFLUENCE OF SALICYLATES ON THE ACTIVITY OF AAP, LAP AND GGT IN URINE, *JUGOSL MED BIOKEM*, 2, 188-191 (1983)

1829 JUBIZ W, CANTERBURY JM, REISS E, TYLER FH, CIRCADIAN RHYTHM IN SERUM PARATHYROID HORMONE CONCENTRATION IN HUMAN SUBJECTS: CORRELATION WITH SERUM CALCIUM, *J CLIN INVEST*, 51, 2040 (1972)

1830 JUBIZ W, MEIKLE AW, ALTERATIONS OF GLUCOCORTICOID ACTIONS BY OTHER DRUGS AND DISEASE STATES, *DRUGS*, 18, 113-121 (1979)

1831 JUDD HL, YEN SSC, SERUM ANDROSTENEDIONE AND TESTOSTERONE LEVELS DURING THE MENSTRUAL CYCLE, *J CLIN ENDOCRINOL METAB*, 36, 475 (1973)

1832 JULIUS S, PASCUAL AV, ABBRECHT PH, LONDON R, EFFECT OF BETA-ADRENERGIC BLOCKAGE ON PLASMA VOLUME IN HUMAN SUBJECTS, *PROC SOC EXP BIOL MED*, 140, 982 (1972)

1833 JUNE CH, THOMPSON CB, KENNEDY MS, ET AL, PROFOUND HYPOMAGNESEMIA AND RENAL MAGNESIUM WASTING ASSOCIATED WITH THE USE OF CYCLOSPORINE FOR MARROW TRANSPLANTATION, *TRANSPLANTATION*, 39, 620-624 (1985)

1834 JUNG DH, PAREKH AC, A NEW COLOR REACTION FOR CHOLESTEROL ASSAY, *CLIN CHIM ACTA*, 35, 73 (1971)

1835 JUNG DH, PAREKH AC, URINARY INORGANIC PHOSPHORUS DETERMINATIONS, *J CLIN PATHOL*, 25, 263 (1972)

1836 JUNG RC, DILL DB, HORTON R, HORVATH SM, EFFECTS OF AGE ON PLASMA ALDOSTERONE LEVELS AND HEMOCONCENTRATION AT ALTITUDE, *J APPL PHYSIOL*, 31, 593 (1971)

1837 JUNGLEE D, KATRAK A, MOHIUDDIN J, ET AL, SALIVARY AMYLASE AND PANCREATIC ENZYMES IN SERUM AFTER TOTAL BODY IRRADIATION, *CLIN CHEM*, 32, 609-610 (1986)

1838 JUNGST G, MOHR R, SIDE EFFECTS OF OFLOXACIN IN CLINICAL TRIALS AND IN POSTMARKETING SURVEILLANCE, *DRUGS*, 34 SUPPL 1, 144-149 (1987)

1839 KAAJA R, VALTONEN VV, LEUCOPENIA ASSOCIATED WITH β-LACTAM ANTIBIOTIC THERAPY, *ACTA MED SCAND*, 216, 531-534 (1984)

1840 KABASAKALIAN P, KALLINEY S, WESTCOTT A, DETERMINATION OF URIC ACID IN SERUM WITH USE OF URICASE AND TRIBROMOPHENOL-AMINOANTIPYRINE CHROMOGEN, *CLIN CHEM*, 19, 522 (1973)

1841 KABAT HF, *CLINICAL PHARMACY HANDBOOK*, LEA AND FEBIGER, PHILADELPHIA, PA (1969)

1842 KACHADORIAN WA, JOHNSON RE, THE EFFECT OF EXERCISE ON SOME CLINICAL MEASURES OF RENAL FUNCTION, *AM HEART J*, 82, 278 (1971)

1843 KAHANE Z, VESTERGAARD P, NON-INTERFERENCE OF COMMON ACID METABOLITES OF CATECHOLAMINES IN A FLUOROMETRIC ASSAY FOR URINARY HOMOVANILLIC ACID, *CLIN CHIM ACTA*, 47, 453 (1973)

1844 KAHN JB, OXACILLIN-INDUCED AGRANULOCYTOSIS, *J AM MED ASSOC*, 240, 2632 (1978)

1845 KALDOR A, DEMECZKY L, JUVANCZ P, THE EFFECT OF DISULFIRAM ON THE BLOOD CATECHOLAMINE LEVEL IN MAN, *INTL ZEITSCHR KLIN PHARMACOL THER TOXICOL*, 5, 284 (1971)

1846 KALDOR A, JUVANCZ P, DEMECZKY M, SEBESTYEN K, PALOTAS J, ENHANCEMENT OF METHYLDOPA METABOLISM WITH BARBITURATE, *BR MED J [CLIN RES]*, 3, 518 (1971)

1847 KALESH DG, MALLIKARJUNESWARA VR, CLEMETSON CAB, EFFECT OF ESTROGEN-CONTAINING ORAL CONTRACEPTIVES ON PLATELET AND PLASMA ASCORBIC ACID CONCENTRATIONS, *CONTRACEPTION*, 4, 183 (1971)

1848 KALFF R, HOUTKOOPER MA, MEYER JWA, ET AL, CARBAMAZEPINE AND SERUM SODIUM LEVELS, *EPILEPSIA*, 25, 390-397 (1984)

1849 KALKHOFF RK, EFFECTS OF ORAL CONTRACEPTIVE AGENTS AND SEX STEROIDS ON CARBOHYDRATE METABOLISM, *ANN REV MED*, 23, 429 (1972)

1850 KALLENBERG CGM, HOORNTJE SJ, SMIT AJ, ET AL, ANTINUCLEAR AND ANTINATIVE DNA ANTIBODIES DURING CAPTOPRIL TREATMENT, *ACTA MED SCAND*, 211, 297-300 (1982)

1851 KALLIOMAKI JL, A THERAPEUTIC TRIAL WITH A COMBINATION OF TRIMETHOPRIM-SULFAMETHOXAZOLE IN RHEUMATOID ARTHRITIS, *CURR THER RES*, 14, 22 (1972)

1852 KALOWSKI S, NANRA RS, MATHEW TH, KINCAID-SMITH P, DETERIORATION IN RENAL FUNCTION IN ASSOCIATION WITH CO-TRIMOXAZOLE THERAPY, *LANCET*, 1, 394 (1973)

1853 KAMBAN JR, NAUKAM RJ, SASTRY BVR, THE EFFECT OF PROCAINAMIDE ON PLASMA CHOLINESTERASE ACTIVITY, *CAN J ANAESTH*, 34, 579-581 (1987)

1854 KAMM RC, SMITH AG, RIBONUCLEASE ACTIVITY IN HUMAN PLASMA, *CLIN BIOCHEM*, 5, 198 (1972)

1855 KAMOUN PP, BARDET JI, DI GIULIO S, ET AL, MEASUREMENTS OF ANGIOTENSIN CONVERTING ENZYME IN CAPTOPRIL-TREATED PATIENTS, *CLIN CHIM ACTA*, 118, 333-336 (1982)

1856 KAMOUN PP, PLEAU JM, MAN NK, SEMIAUTOMATED METHOD FOR MEASUREMENT OF GUANIDINOSUCCINIC ACID IN SERUM, *CLIN CHEM*, 18, 355 (1972)

1857 KANE FJ JR., MOORE LP, HEPATOTOXICITY OCCURRING WITH THIORIDAZINE THERAPY, *SOUTH MED J*, 64, 573 (1971)

1858 KANIS JA, DETECTION OF URINARY PORPHOBILINOGEN, *LANCET*, 1, 1511 (1973)

1859 KANOH T, JINGAMI H, USCHINO H, APLASTIC ANEMIA AFTER PROLONGED TREATMENT WITH CHLORPHENIRAMINE, *LANCET*, 1, 546-547 (1977)

1860 KANSAL PC, BUSE J, TALBERT OR, BUSE MG, THE EFFECT OF L-DOPA ON PLASMA GROWTH HORMONE, INSULIN, AND THYROXINE, *J CLIN ENDOCRINOL METAB*, 34, 99 (1972)

1861 KAPIL RP, AXELSON JE, MANSFIELD IL, ET AL, DISOPYRAMIDE PHARMACOKINETICS AND METABOLISM: EFFECT OF INDUCERS, *BR J CLIN PHARMACOL*, 24, 781-791 (1987)

1862 KAPLAN A, UREA NITROGEN AND URINARY AMMONIA, *STAND METH CLIN CHEM*, 5, 245 (1965)

1863 KAPLAN A, WILLIAMSON L, SUITABILITY OF A PLASTIC DISC FOR USE IN SEPARATING SERUM OR PLASMA, *CLIN CHEM*, 20, 403 (1974)

1864 KAPLAN B, CARDARELLI C, PINNELL SR, LEVAMISOLE AND AGRANULOCYTOSIS, *CUTIS*, 24, 429-430 (1979)

1865 KAPLAN NM, EFFECTS OF GUANABENZ ON PLASMA LIPID LEVELS IN HYPERTENSIVE PATIENTS, *J CARDIOVASC PHARMACOL*, 6 SUPPL 5, S841-S846 (1984)

1866 KAPLAN RP, RUSSELL DH, LOWE NJ, ETRINATE THERAPY FOR PSORIASIS: CLINICAL RESPONSES, REMISSION TIMES, EPIDERMAL DNA, AND POLYAMINE RESPONSES, *J AM ACAD DERMATOL*, 8, 95-102 (1983)

1867 KARACHALIOS GN, PARIGORAKIS JG, THROMBOCYTOPENIA AND SULINDAC, *ANN INTERN MED*, 104, 128 (1986)

1868 KARASAWA T, FUNAKOSHI H, FURUKAWA K, YOSHIDA K, EDTA PREVENTS THE PHOTOCATALYZED DESTRUCTION OF THE PRODUCTS OF CATECHOLAMINE OXIDATION, *ANAL BIOCHEM*, 53, 278 (1973)

1869 KARAYALCIN G, ROSNER F, SAWITSKY A, PSEUDONEUTROPENIA IN AMERICAN NEGROES, *LANCET*, 1, 387 (1972)

1870 KARAYALCIN G, ROSNER F, SAWITSKY A, PSEUDONEUTROPENIA IN NEGROES: A NORMAL PHENOMENON, *N Y STATE J MED*, 72, 1815 (1972)

1871 KARK RAP, POSKANZER DC, BULLOCK JD, BOYLEN G, MERCURY POISONING AND ITS TREATMENT WITH N-ACETYL-DELTA-L-PENICILLAMINE, *N ENGL J MED*, 285, 10 (1971)

1872 KARK RM, PROTEINURIA II: DIAGNOSIS AND MANAGEMENT, *HOSP PRACT*, 6, 59 (1971)

1873 KARKALAS Y, LAL H, JAUNDICE FOLLOWING THERAPY WITH IMIPRAMINE AND CYPROHEPTADINE, *CLIN TOXICOL*, 4, 47 (1971)

1874 KARLBERG BE, LINS LE, ROSSNER S, CLONIDINE IN MILD TO MODERATE HYPERTENSION: EFFECTS ON BLOOD PRESSURE AND SERUM LIPOPROTEINS, *J HYPERTENS*, 3 SUPPL, S69-S71 (1985)

1875 KARLSSON J, NORDESJO LO, JORFELDT L, SALTIN B, MUSCLE LACTATE, ATP AND CPK LEVELS DURING EXERCISE AFTER PHYSICAL TRAINING IN MAN, *J APPL PHYSIOL*, 33, 199 (1972)

1876 KARLSSON K, MARKLUND SL, HEPARIN-INDUCED RELEASE OF EXTRACELLULAR SUPEROXIDE DISMUTASE TO HUMAN BLOOD PLASMA, *BIOCHEM J*, 242, 55-59 (1987)

1877 KARPATKIN S, DRUG-INDUCED THROMBOCYTOPENIA, *AM J MED SCI*, 262, 68 (1971)

1878 KASICH AM, CLORAZEPATE DIPOTASSIUM IN THE TREATMENT OF ANXIETY ASSOCIATED WITH CHRONIC GASTROINTESTINAL DISEASE, *CURR THER RES*, 15, 83 (1973)

1879 KASTE M, MUURONEN A, NIKKILA EA, INCREASE OF LOW SERUM CONCENTRATIONS OF HIGH-DENSITY LIPOPROTEIN (HDL) CHOLESTEROL IN TIA-PATIENTS TREATED WITH PHENYTOIN, *STROKE*, 14, 525-530 (1983)

1880 KASTRUP J, PETERSEN P, BARTRAM R, ET AL, THE EFFECT OF TRIMETHOPRIM ON SERUM CREATININE, *BR J UROL*, 57, 265-268 (1985)

1881 KATER RMH, DOUBLE BLIND EVALUATION OF ANALGESIC AND TOXIC EFFECTS OF FLUFENAMIC ACID AND MEFENAMIC ACID, *MED J AUST*, 1, 848 (1968)

1882 KATO DB, DILUTIONAL HYPONATREMIA AND WATER INTOXICATION DURING CARBAMAZEPINE THERAPY, *DRUG INTELL CLIN PHARM*, 12, 392-396 (1978)

1883 KATSILAMBROS N, BRAATEN J, FERGUSON BD, BRADLEY RF, MUSCULAR SYNDROME AFTER CLOFIBRATE, *N ENGL J MED*, 286, 1110 (1972)

1884 KATZ FH, LIPMAN MM, FRANTZ AG, THE PHYSIOLOGIC SIGNIFICANCE OF 6-BETA-HYDROXYCORTISOL IN HUMAN CORTICOID METABOLISM, *J CLIN ENDOCRINOL METAB*, 22, 71 (1962)

1885 KATZ FH, ROMFH P, PLASMA ALDOSTERONE AND RENIN ACTIVITY DURING THE MENSTRUAL CYCLE, *J CLIN ENDOCRINOL METAB*, 34, 819 (1972)

1886 KATZ WA, BLODGETT RC JR., PIETRUSKO RG, PROTEINURIA IN GOLD-TREATED RHEUMATOID ARTHRITIS, *ANN INTERN MED*, 101, 176-179 (1984)

1887 KATZMAN PL, HULTHEN UL, HOKFELT B, THE EFFECT OF 8 WEEKS TREATMENT WITH THE CALCIUM ANTAGONIST FELODIPINE ON BLOOD PRESSURE, HEART RATE, WORKING CAPACITY, PLASMA RENIN ACTIVITY, PLASMA ANGIOTENSIN II, URINARY CATECHOLAMINES AND ALDOSTERONE IN PATIENTS WITH ESSENTIAL HYPERTENSION, *BR J CLIN PHARMACOL*, 21, 633-640 (1986)

1888 KAUFMAN F, A RARE COMPLICATION OF SULFADIMETHOXINE (MADRIBON) THERAPY, *CALIF MED*, 107, 344 (1967)

1889 KAUL A, REDDY JC, FAGMAN E, ET AL, HEPATITIS ASSOCIATED WITH USE OF SULINDAC IN A CHILD, *J PEDIATR*, 99, 650-651 (1981)

1890 KAZMIER FJ, A SIGNIFICANT INTERACTION BETWEEN METRONIDAZOLE AND WARFARIN, *MAYO CLIN PROC*, 51, 782-784 (1976)

1891 KEAN WF, ANASTASSIADES TP, DWOSH IL, ET AL, EFFICACY AND TOXICITY OF D-PENICILLAMINE FOR RHEUMATOID DISEASE IN THE ELDERLY, *J AM GERIATR SOC*, 30, 94-100 (1982)

1892 KEAN WF, DWOSH IL, ANASTASSIADES TP, ET AL, THE TOXICITY PATTERN OF D-PENICILLAMINE THERAPY: A GUIDE TO ITS USE IN RHEUMATOID ARTHRITIS, *ARTHRITIS RHEUM*, 23, 158-164 (1980)

1893 KEATING JP, LELL ME, STRAUSS AH, ZARKOWSKY H, SMITH GE, INFANTILE METHEMOGLOBINEMIA CAUSED BY CARROT JUICE, *N ENGL J MED*, 288, 824 (1973)

1894 KEDRA M, POLESZAK J, PITERA A, EFFECT OF CAFFEINE ON THE COMPOSITION OF BLOOD LIPIDS, *POL TYG LEK*, 25, 125 (1970)

1895 KEEFE EB, REIS TC, BERLAND JE, HEPATOTOXICITY TO BOTH ERYTHROMYCIN ESTOLATE AND ETHYLSUCCINATE, *DIG DIS SCI*, 27, 701-704 (1982)

1896 KEEFER JH, LAP SPECIMENS, *ASCP SUMMARY REPORT*, 9, 18 (1972)

1897 KEEFFE EB, SUNDERLAND MC, GABOUREL JD, SERUM GAMMA-GLUTAMYL TRANSPEPTIDASE ACTIVITY IN PATIENTS RECEIVING CHRONIC PHENYTOIN THERAPY, *DIG DIS SCI*, 31, 1056-1061 (1986)

1898 KEENAN BS, JOHNSONBAUGH RE, SODE J, DISCORDANT SERUM GROWTH HORMONE RESPONSES TO I.V. AND I.M. ADMINISTRATION OF SYNTHETIC L-24 CORTICOTROPIN (ACTH), *CLIN RES*, 20, 866 (1972)

1899 KEENAN J, THOMPSON JB, CHAMBERLAIN MA, BESSER GM, PROLONGED CORTICOTROPHIC ACTION OF A SYNTHETIC SUBSTITUTED 1-18 ACTH, *BR MED J [CLIN RES]*, 3, 742 (1971)

1900 KEIDING S, DOSSING M, HARDT F, A NURSE WITH LIVER INJURY ASSOCIATED WITH OCCUPATIONAL EXPOSURE TO HALOTHANE IN A RECOVERY UNIT, *DAN MED BULL*, 31, 255-256 (1984)

1901 KEITH RE, MOSSHOLDER SB, ASCORBIC ACID STATUS OF SMOKING AND NONSMOKING ADOLESCENT FEMALES, *INT J VITAM NUTR RES*, 56, 363-366 (1986)

1902 KELLER DF, *G-6-PD DEFICIENCY*, CRC PRESS, CLEVELAND OH, 1971

1903 KELLEY WN, GOLDFINGER SE, HARDY HL, HYPERURICEMIA IN CHRONIC BERYLLIUM DISEASE, *ANN INTERN MED*, 70, 977 (1969)

1904 KELLEY WN, ROSENBLOOM FM, SEEGMILLER JE, THE EFFECTS OF AZATHIOPRINE (IMURAN®) ON PURINE SYNTHESIS IN CLINICAL DISORDERS OF PURINE METABOLISM, *J CLIN INVEST*, 46, 1518 (1967)

1905 KELLNER R, GERVAIS RH, PATHAK D, A PILOT STUDY OF THE SHORT-TERM ANTIANXIETY EFFECTS OF MOLINDONE HCL, *J CLIN PHARMACOL*, 12, 472 (1972)

1906 KELNER MJ, BAILEY DN, PROPYLENE GLYCOL AS A CAUSE OF LACTIC ACIDOSIS, *J ANAL TOXICOL*, 9, 40-42 (1985)

1907 KELSAY JL, BEHALL KM, HOLDEN JM, CRUTCHFIELD HC, PYRUVATE AND LACTATE IN HUMAN BLOOD AND SALIVA IN RESPONSE TO DIFFERENT CARBOHYDRATES, *J NUTR*, 102, 661 (1972)

1908 KELSEY WM, SCHARY JM, FATAL HEPATITIS PROBABLY DUE TO INDOMETHACIN, *J AM MED ASSOC*, 199, 586 (1967)

1909 KELTON JG, IMPAIRED RETICULOENDOTHELIAL FUNCTION IN PATIENTS TREATED WITH METHYLDOPA, *N ENGL J MED*, 313, 596-600 (1985)

1910 KELTON JG, MELTZER D, MOORE J, ET AL, DRUG-INDUCED THROMBOCYTOPENIA IS ASSOCIATED WITH INCREASED BINDING OF IgG TO PLATELETS BOTH IN VIVO AND IN VITRO, *BLOOD*, 58, 524-529 (1981)

1911 KEM DC, WEINBERGER MH, MAYES DM, NUGENT CA, SALINE SUPPRESSION OF PLASMA ALDOSTERONE IN HYPERTENSION, *ARCH INTERN MED*, 128, 380 (1971)

1912 KENDALL JW, EGANS ML, STOTT AK, FLUOROMETRIC DETERMINATION OF CORTICOSTEROIDS: AN INTERFERING SUBSTANCE IN IMPURE DICHLOROMETHANE, *J CLIN ENDOCRINOL METAB*, 28, 1373 (1968)

1913 KENDALL MJ, NUTTER S, HAWKINS CF, XYLOSE TEST: EFFECT OF ASPIRIN AND INDOMETHACIN, *BR MED J [CLIN RES]*, 1, 532 (1971)

1914 KENDALL MJ, NUTTER S, HAWKINS CF, BACTERIA AND THE XYLOSE TEST, *LANCET*, 1, 1017 (1972)

1915 KENDALL-TAYLOR P, HYPERTHYROIDISM, *BR MED J [CLIN RES]*, 2, 337 (1972)

1916 KENNEDY P, LIVER CROSS-SENSITIVITY TO ANTIPSYCHOTIC DRUGS, *BR J PSYCHIATRY*, 143, 312 (1983)

1917 KENT JR, HILL M, PARLOW AF, BISCHOFF AJ, ESTROGENIC SUPPRESSION OF THE PITUITARY-GONADAL AXIS AND PROSTATIC FUNCTION, *CLIN PHARMACOL THER*, 13, 144 (1972)

1918 KENWRIGHT S, LEVI AJ, IMPAIRMENT OF HEPATIC UPTAKE OF RIFAMYCIN ANTIBIOTICS BY PROBENECID AND ITS THERAPEUTIC IMPLICATIONS, *LANCET*, 2, 1401 (1973)

1919 KERRY RJ, LUDLOW JM, OWEN G, DIURETICS ARE DANGEROUS WITH LITHIUM, *BR MED J [CLIN RES]*, 281, 371 (1980)

1920 KERSH ES, RIFKIN H, LACTULOSE ENEMAS, *ANN INTERN MED*, 78, 81 (1973)

1921 KERSHBAUM A, BELLET S, CIGARETTE, CIGAR AND PIPE SMOKING. SOME DIFFERENCES IN BIOCHEMICAL EFFECTS, *GERIATRICS*, 23, 126 (1968)

1922 KERSHBAUM A, PAPPAJOHN DJ, BELLET S, EFFECT OF SMOKING AND NICOTINE ON ADRENOCORTICAL SECRETION, *J AM MED ASSOC*, 203, 275 (1968)

1923 KETTLEWELL M, NOWERS A, WHITE R, EFFECT OF DIGOXIN ON HUMAN RED BLOOD CELL ELECTROLYTES, *BR J PHARMACOL*, 44, 165 (1972)

1924 KEUL J, DOLL E, INTERMITTENT EXERCISE: METABOLITES, PO₂ AND ACID-BASE EQUILIBRIUM IN THE BLOOD, *J APPL PHYSIOL*, 34, 220 (1973)

1925 KEW MC, FIRST RG, A TRIAL OF RO 4-2137 IN THE TREATMENT OF HYPERTENSION, *CURR THER RES*, 14, 343 (1972)

1926 KEW MC, SEFTEL HC, BLOOMBERG BM, PREGNANCY TESTS AND PROTEINURIA, *LANCET*, 1, 902 (1967)

1927 KEYS A, SERUM CHOLESTEROL AND THE QUESTION OF NORMAL IN *MULTIPLE LABORATORY SCREENING*, ES BENSON, PE STRANDJORD, EDS., ACADEMIC PRESS, NEW YORK, NY (1969)

1928 KHANNA BK, GUPTA VP, SINGH MP, ETHAMBUTOL-INDUCED HYPERURICAEMIA, *TUBERCLE*, 65, 195-199 (1984)

1929 KIDD JE, GILCHRIST NL, UTLEY RJ, ET AL, EFFECT OF OPIATE, GENERAL ANESTHESIA AND SURGERY ON PLASMA ATRIAL NATRIURETIC PEPTIDE LEVELS IN MAN, *CLIN EXP PHARMACOL PHYSIOL*, 14, 755-760 (1987)

1930 KIJIMA Y, SASAUKA T, KANAYAMA M, ET AL, CLOFIBRATE EFFECT ON ALKALINE PHOSPHATASE IN RENAL FAILURE, *N ENGL J MED*, 297, 113 (1977)

1931 KILLINGSWORTH LM, SAVORY J, NEPHELOMETRIC METHODS FOR THE DETERMINATION OF URINARY ALBUMIN, TRANSFERRIN AND ALPHA-2 MACROGLOBULIN, *ANN CLIN LAB SCI*, 4, 46 (1974)

1932 KING A, POSITIVE INTERFERENCE BY CEPHALOTHIN WITH ZIMMERMANN REACTION FOR 17-KETOSTEROIDS, *CLIN CHEM*, 20, 401 (1974)

1933 KING B, SPIKESMAN A, EMERY AEH, THE EFFECT OF PREGNANCY ON SERUM LEVELS OF CREATINE KINASE, *CLIN CHIM ACTA*, 36, 267 (1972)

1934 KING DJ, KELTON JG, HEPARIN-ASSOCIATED THROMBOCYTOPENIA, *ANN INTERN MED*, 100, 535-540 (1984)

1935 KINGHAM JGC, SWAIN P, SWARBRICK ET, DAPSONE AND SEVERE HYPOALBUMINEMIA: A REPORT OF TWO CASES, *LANCET*, 2, 662-664 (1979)

1936 KINGSLEY GR, TAGER HS, ION-EXCHANGE METHOD FOR THE DETERMINATION OF PLASMA AMMONIA NITROGEN WITH THE BERTHELOT REACTION, *STAND METH CLIN CHEM*, 6, 115 (1970)

1937 KIRBERGER E, VARIABLE ACTION OF MAO INHIBITING HYDRAZINES ON SEROTONIN METABOLISM, *NATURE*, 197, 1211 (1963)

1938 KIRCH W, HOENSCH H, JANISCH HD, INTERACTIONS AND NON-INTERACTIONS WITH RANITIDINE, *CLIN PHARMACOKINET*, 8, 493-510 (1984)

1939 KIRCH W, STENZEL J, DYLEWICZ P, ET AL, INFLUENCE OF NISOLDIPINE ON HAEMODYNAMIC EFFECTS AND PLASMA LEVELS OF DIGOXIN, *BR J CLIN PHARMACOL*, 22, 155-159 (1986)

1940 KIRCKMAIR H, HUBER H, FIEBERHAFTE REAKTION ('DRUG FEVER') UNTER ALPHA-METHYLDOPA, *WIEN KLIN WOCHENSCHR*, 77, 699 (1965)

1941 KIRSCHNER MA, BARDIN CW, ANDROGEN PRODUCTION AND METABOLISM IN NORMAL AND VIRILIZED WOMEN, *METABOLISM*, 21, 667 (1972)

1942 KIRSH MM, ABRAMS B, COON W, ZUIDEMA G, DIPHENHYDRAMINE (BENADRYL®) HYDROCHLORIDE IN THE TREATMENT OF AMMONIA INTOXICATION, *ARCH SURG*, 91, 466 (1965)

1943 KISFAULDY S, BUKI B, MESZAROS S, EFFECT OF ORALLY ADMINISTERED AMINO ACIDS ON PORTAL BLOOD AMMONIA, *ACTA MED HUNG*, 20, 365 (1964)

1944 KISTNER RW, PRESENT STATUS OF ORAL CONTRACEPTIVES, *DRUG THERAPY*, 1, 14 (1971)

1945 KISTNER S, NORBERG R, TRANSFERRIN EXCRETION IN PATIENTS WITH PROTEINURIA, *ACTA MED SCAND*, 191, 393 (1972)

1946 KLACHKO DM, LIE TH, CHASE GR, BURNS TW, BLOOD GLUCOSE LEVELS DURING WALKING IN NORMAL AND DIABETIC SUBJECTS, *DIABETES*, 21, 89 (1972)

1947 KLASTERSKY J, VANDERKLEN B, DANEAU D, MATHIEW M, CARBENICILLIN AND HYPOKALEMIA, *ANN INTERN MED*, 78, 774 (1973)

1948 KLATSKIN G, TOXIC AND DRUG INDUCED HEPATITIS IN *DISEASES OF THE LIVER* 3RD ED, L SCHIFF, ED., LIPPINCOTT, PHILADELPHIA, PA (1969)

1949 KLEIN B, LUCAS LB, APPLICATION OF FE (11)-5-PYRIDYL BENZODIAZEPIN-2-ONES TO THE MANUAL OR AUTOMATED DETERMINATION OF SERUM URIC ACID, *CLIN CHEM*, 19, 67 (1973)

1950 KLEIN B, SHEEHAN J, FORMALDEHYDE INTERFERENCE IN THE URIC ACID PHOSPHOTUNGSTIC ACID REDOX, *CLIN CHEM*, 19, 531 (1973)

1951 KLEIN GC, COOPER GR, ELECTROPHORETIC DETERMINATION OF SERUM LIPOPROTEINS (PRESTAINING TECHNIQUE), *STAND METH CLIN CHEM*, 6, 127 (1970)

1952 KLEIN HO, LANE R, DI SEGNI E, ET AL, VERAPAMIL-DIGOXIN INTERACTION, *N ENGL J MED*, 303, 160 (1980)

1953 KLEIN M, AGRANULOCYTOSIS SECONDARY TO CHLORTHALIDONE THERAPY (REPORT OF A CASE), *J AM MED ASSOC*, 184, 310 (1963)

1954 KLEIN NC, MAGIDA MG, PROPOXYPHENE (DARVON®) HEPATOTOXICITY, *AM J DIG DIS*, 16, 467 (1971)

1955 KLEIN WJ JR., METZ EN, PRICE AR, ACUTE COPPER INTOXICATION: A HAZARD OF HEMODIALYSIS, *ARCH INTERN MED*, 129, 578 (1972)

1956 KLEINBERG DL, LIEBERMAN A, TODD J, ET AL, PERGOLIDE MESYLATE: A POTENT DAY-LONG INHIBITOR OF PROLACTIN IN RHESUS MONKEYS AND PATIENTS WITH PARKINSON'S DISEASE, *J CLIN ENDOCRINOL METAB*, 51, 152-154 (1980)

1957 KLEINBLOESEM CH, VAN BRUMMELEN P, HILLERS J, ET AL, INTERACTION BETWEEN DIGOXIN AND NIFEDIPINE AT STEADY STATE IN PATIENTS WITH ATRIAL FIBRILLATION, *THER DRUG MONIT*, 7, 372-376 (1985)

1958 KLEINMAN S, NELSON R, SMITH L, ET AL, POSITIVE DIRECT ANTIGLOBULIN TESTS AND IMMUNE HEMOLYTIC ANEMIA IN PATIENTS RECEIVING PROCAINAMIDE, *N ENGL J MED*, 311, 809-812 (1984)

1959 KLINTMALM GBG, IWATSUKI S, STARZL TE, CYCLOSPORINE A HEPATOTOXICITY IN 66 RENAL ALLOGRAFT RECIPIENTS, *TRANSPLANTATION*, 32, 488-489 (1981)

1960 KLOTZ HP, KONOPKA P, DELORME ML, DETERMINATION OF PLASMA CALCITONIN IN PREGNANT WOMEN: CLINICAL DEDUCTIONS, *ANN ENDOCRINOL*, 33, 267 (1972)

1961 KLOTZ MO, RICHTER H, MEUFFELS M, INTERFERENCE BY FORMALDEHYDE FORMING DRUGS IN THE DETERMINATION OF URINARY CATECHOLAMINES, *CLIN CHEM*, 10, 372 (1964)

1962 KLOTZ U, REIMANN I, ELEVATION OF STEADY-STATE DIAZEPAM LEVELS BY CIMETIDINE, *CLIN PHARMACOL THER*, 30, 513-517 (1981)

1963 KLUGE RM, WILKES R, REACTIONS TO ANTIBIOTICS, *J AM MED ASSOC*, 237, 1825 (1977)

1964 KNAPP ML, HADID O, INVESTIGATIONS INTO NEGATIVE INTERFERENCE BY JAUNDICED PLASMA IN KINETIC JAFFE METHODS FOR PLASMA CREATININE DETERMINATION, *ANN CLIN BIOCHEM*, 24, 85-97 (1987)

1965 KNIFFIN JC, NOYES WD, PORTER FS, IRON AND CHOLESTYRAMINE IN ERYTHROPOIETIC PROTOPORPHYRIA, *CLIN RES*, 18, 38 (1970)

1966 KNIRSCH AK, GRALLA EJ, ABNORMAL SERUM TRANSAMINASE LEVELS AFTER PARENTERAL AMPICILLIN AND CARBENICILLIN ADMINISTRATION, *N ENGL J MED*, 282, 1081 (1970)

1967 KNOPP RH, GINSBERG J, ALBERS JJ, ET AL, CONTRASTING EFFECTS OF UNMODIFIED AND TIME-RELEASE FORMS OF NIACIN IN LIPOPROTEINS IN HYPERLIPIDEMIC SUBJECTS: CLUES TO MECHANISM OF ACTION OF NIACIN, *METABOLISM*, 34, 642-650 (1985)

1968 KNOPP RM, BOROUSH M, WORTH MR, PLASMA LIPOPROTEIN COMPOSITION IN PREGNANCY, *CLIN RES*, 20, 882 (1972)

1969 KNUDSEN L, WEISMANN K, PUSTULAR ERUPTION AND DAMAGE TO INTERNAL ORGANS FOLLOWING SULFAMETHOXAZOLE-TRIMETHOPRIM, *UGESKR LAEGER*, 139, 1007-1009 (1977)

1970 KNUTH UA, NIESCHLAG E, EFFECT OF DIFFERENT DOSES OF NALOXONE ON SERUM LEVELS OF PROLACTIN AND GONADOTROPINS IN YOUNG MALE VOLUNTEERS, *J ENDOCRINOL INVEST*, 8, 55-57 (1985)

1971 KOCH-WESER J, POTENTIATION BY GLUCAGON OF THE HYPOPROTHROMBINEMIC ACTION OF WARFARIN, *ANN INTERN MED*, 72, 331 (1970)

1972 KOCH-WESER J, QUINIDINE-INDUCED HYPOPROTHROMBINEMIC HEMORRHAGE IN PATIENTS ON CHRONIC WARFARIN THERAPY, *ANN INTERN MED*, 68, 511 (1968)

1973 KOCH-WESER J, SELLERS EM, COUMARIN ANTICOAGULANTS, *N ENGL J MED*, 285, 555 (1971)

1974 KOCH-WESER J, SIDEL VW, DEXTER M, PARISH C, FINER DC, KANAREK P, ADVERSE REACTIONS TO SULFISOXAZOLE, SULFAMETHOXAZOLE, AND NITROFURANTOIN, *ARCH INTERN MED*, 128, 399 (1971)

1975 KOENEST MH, FRIER EF, AN EVALUATION OF TWO METHODS FOR THE DETERMINATION OF TRUE CREATININE, *AM J MED TECHNOL*, 37, 473 (1971)

1976 KOFF RS, GARVEY AJ, BURNEY SW, BELL B, SULFOBROMOPHTHALEIN (BSP) RETENTION IN HEALTHY MEN: ABSENCE OF AN APE EFFECT, *CLIN RES*, 21, 517 (1973)

1977 KOH H, NAMBU S, TSUSHIMA M, ET AL, A SPECIFIC INHIBITION OF INSULIN SECRETION BY A β_1-SELECTIVE ADRENOCEPTOR BLOCKING DRUG, *ACTA THERAPEUTICA*, 9, 325-332 (1983)

1978 KOHN NN, MYERSON RM, XANTHOMATOUS BILIARY CIRRHOSIS FOLLOWING CHLORPROMAZINE, *AM J MED*, 31, 665 (1961)

1979 KOHN RM, MONTES M, HEPATIC FIBROSIS FOLLOWING LONG ACTING NICOTINIC ACID THERAPY: A CASE REPORT, *AM J MED SCI*, 258, 94 (1969)

1980 KOHVAKKA A, SALO H, GORDIN A, ET AL, ANTIHYPERTENSIVE AND BIOCHEMICAL EFFECTS OF DIFFERENT DOSES OF HYDROCHLOROTHIAZIDE ALONE OR IN COMBINATION WITH TRIAMTERENE, *ACTA MED SCAND*, 219, 381-386 (1986)

1981 KOKOT F, CEKANSKI A, PLASMA RENIN ACTIVITY IN PERIPHERAL AND UTERINE VEIN BLOOD IN PREGNANT AND NONPREGNANT WOMEN, *J OBSTET GYNAECOL BR COMM*, 79, 72 (1972)

1982 KOLANOWSKI J, PIZARRO MA, CRABBE J, CHANGES IN ADRENOCORTICAL RESPONSE TO CORTICOTROPIN (ACTH) ADMINISTERED REPEATEDLY IN MAN, *ABSTRACTS IV INTL CONGR ENDOCRINOL, WASHINGTON DC, 1972*, 115

1983 KOLB KW, GARNETT WR, SMALL RE, ET AL, EFFECT OF CIMETIDINE ON QUINIDINE CLEARANCE, *THER DRUG MONIT*, 6, 306-312 (1984)

1984 KOLLER PU, TRITSCHLER W, CARSTENSEN CA, SYSTEMATIC DRUG INTERFERENCE STUDIES ON REFLOTRON, *CLIN CHEM*, 33, 916 (1987)

1985 KONNO M, NISHIZAWA N, MIYATE Y, ET AL, INTRAOPERATIVE CHANGES IN PLASMA FOLATE AND COBALAMIN LEVELS IN PATIENTS INHALING NITROUS OXIDE, *JPN J ANESTHESIOL*, 34, 1620-1624 (1985)

1986 KONTTINEN A, HEPATOTOXICITY OF SULFAMETHOXYPYRIDAZINE, *BR MED J [CLIN RES]*, 2, 168 (1972)

1987 KONTTINEN A, PERASALO J, EISALO A, SULFONAMIDE HEPATITIS, *ACTA MED SCAND*, 191, 389 (1972)

1988 KONTTINEN YP, EPSILON-AMINOCAPROIC ACID IN TREATMENT OF ACUTE PANCREATITIS, *SCAND J CLIN LAB INVEST*, 27, 41 (1971)

1989 KONTUREK S, GABRYS B, INHIBITION OF GASTRIC SECRETION OF HYDROCHLORIC ACID BY SECRETIN, *POL TYG LEK*, 25, 95 (1970)

1990 KONZETT H, HORNTAGL H, WINKLER H, ON THE URINARY OUTPUT OF VASOPRESSIN, EPINEPHRINE AND NOREPINEPHRINE DURING DIFFERENT STRESS SITUATIONS, *PSYCHOPHARMACOL*, 21, 247 (1971)

1991 KOOPMAN BJ, HINDRIKS FR, LOKERSE WG, ET AL, INJURIOUS EFFECT OF EDTA CONTAMINATION ON COLORIMETRY OF SERUM IRON, *CLIN CHEM*, 31, 2030-2032 (1985)

1992 KOPP LE, LIN T, TUCCI JR, FACTORS AFFECTING CYCLIC AMP (CAMP) EXCRETION IN NORMAL HUMAN SUBJECTS, *CLIN RES*, 20, 866 (1972)

1993 KOREN G, HESSLEIN PS, MACLEOD SM, DIGOXIN TOXICITY ASSOCIATED WITH AMIODARONE THERAPY IN CHILDREN, *J PEDIATR*, 104, 467-470 (1984)

1994 KOREN JF, RANDALL GR,, KINCAID RS, ET AL, PHENYTOIN HYPERSENSITIVITY REACTION: HEPATIC NECROSIS, *DRUG INTELL CLIN PHARM*, 14, 252-257 (1980)

1995 KORMAN MG, SOVENY C, HANSKY J, THE EFFECT OF FOOD ON SERUM GASTRIN, *AUST N Z J MED*, 1, 299 (1971)

1996 KORNBERG A, KOBRIN I, IgG ANTIPLATELET ANTIBODIES DUE TO CARBAMAZEPINE, *ACTA HAEMATOL*, 68, 68-70 (1970)

1997 KORNFELD JM, ULLMANN WW, PENICILLIN INTERFERENCE WITH THE DETERMINATION OF DELTA-AMINOLEVULINIC ACID, *CLIN CHIM ACTA*, 46, 187 (1973)

1998 KORRI U-M, NUUTINEN H, SALASPURO M, INCREASED BLOOD ACETATE: A NEW LABORATORY MARKER OF ALCOHOLISM AND HEAVY DRINKING, *ALCOHOL CLIN EXP RES*, 9, 468-471 (1985)

1999 KOSSOVER MF, BECKHAM ME, THREEFOOT SA, INFLUENCE OF AN ORAL PERISTALTIC STIMULANT, DANTHRON, ON THE PSP EXCRETION TEST, *AM J MED SCI*, 247, 694 (1964)

2000 KOSZEWSKI BJ, HUBBARD TF, IMMUNOLOGIC AGRANULOCYTOSIS DUE TO MERCURIAL DIURETICS, *AM J MED*, 20, 958 (1956)

2001 KOTCHEN TA, HOGAN RP, BOYD AE, LI TK, SING HC, MASON JW, RENIN, NORADRENALINE AND ADRENALINE RESPONSES TO SIMULATED ALTITUDE, *CLIN SCI*, 44, 243 (1973)

2002 KOVACS JL, TUZEL I, AOGAICHI K, ET AL, FALSE-POSITIVE URINE PROTEIN REACTION WITH CIBENZOLINE, A NEW ANTIARRHYTHMIC AGENT, *J CLIN PHARMACOL*, 24, 127-128 (1984)

2003 KOZLOWSKI BW, TAYLOR ML, BAER MT, ET AL, ANTICONVULSANT MEDICATION USE AND CIRCULATING LEVELS OF TOTAL THYROXINE, RETINOL BINDING PROTEIN AND VITAMIN A IN CHILDREN WITH DELAYED COGNITIVE DEVELOPMENT, *AM J CLIN NUTR*, 46, 360-368 (1987)

2004 KRACKE RR, RELATION OF DRUG THERAPY TO NEUTROPENIC STATES, *J AM MED ASSOC*, 111, 1255 (1938)

2005 KRANTZ JC, CARR CJ, *PHARMACOLOGIC PRINCIPLES OF MEDICAL PRACTICE* 6TH ED, WILLIAMS AND WILKINS, BALTIMORE, MD (1965)

2006 KRAUSE DK, SCHILLMOLLER V, HAYDUK K, INCREASED PLASMA RENIN CONCENTRATION IN HEALTHY INFANTS, YOUNG CHILDREN AND SCHOOL-CHILDREN IN COMPARISON TO NORMAL, *GERM MED*, 2, 103 (1972)

2007 KRAUSE K-H, BERLIT P, BONJOUR J-P, ET AL, VITAMIN STATUS IN PATIENTS ON CHRONIC ANTICONVULSANT THERAPY, *INT J VITAM NUTR RES*, 52, 375-385 (1982)

2008 KRAUSE K-H, BONJOUR J-P, BERLIT P, ET AL, BIOTIN STATUS OF EPILEPTICS, *ANN N Y ACAD SCI*, 447, 297-313 (1985)

2009 KRAUSS RM, ROYS S, MISHELL DR JR., ET AL, EFFECTS OF TWO LOW-DOSE ORAL CONTRACEPTIVES ON SERUM LIPIDS AND LIPOPROTEINS: DIFFERENTIAL CHANGES IN HIGH-DENSITY LIPOPROTEIN SUBCLASSES, *AM J OBSTET GYNECOL*, 145, 446-452 (1983)

2010 KREEFT JH, LANGLOIS S, OGILVIE RI, COMPARATIVE TRIAL OF INDAPAMIDE AND HYDROCHLOROTHIAZIDE IN ESSENTIAL HYPERTENSION WITH FOREARM PLETHYSMOGRAPHY, *J CARDIOVASC PHARMACOL*, 6, 622-626 (1984)

2011 KREEGER RW, HAMMILL SC, NEW ANTIARRHYTHMIC DRUGS: TOCAINIDE, MEXILETINE, FLECAINIDE, ENCAINIDE AND AMIODARONE, *MAYO CLIN PROC*, 62, 1033-1050 (1987)

2012 KREEK MJ, DODES L, KANE S, KNOBLER J, MARTIN R, LONG-TERM METHADONE MAINTENANCE THERAPY: EFFECTS ON LIVER FUNCTION, *ANN INTERN MED*, 77, 598 (1972)

2013 KREISBERG RA, SIEGAL AM, STANLEY AW, RACKLEY CE, RUSSELL RO, ALANINE METABOLISM IN MAN: CONVERSION TO ETHANOL, *CLIN RES*, 20, 549 (1972)

2014 KRIEDGER DT, OSSOWSKI R, FOGEL M, ALLEN W, LACK OF CIRCADIAN PERIODICITY OF HUMAN SERUM FSH AND LH LEVELS, *J CLIN ENDOCRINOL METAB*, 35, 619 (1972)

2015 KRISTENSEN BO, SKOV J, PETERSLUND NA, FRUSEMIDE-INDUCED INCREASES IN SERUM AMYLASES, *BR MED J [CLIN RES]*, 281, 978 (1980)

2016 KRISTENSEN BO, WEEKE JO, PROPRANOLOL-INDUCED INCREMENTS IN TOTAL AND FREE SERUM THYROXINE IN PATIENTS WITH ESSENTIAL HYPERTENSION, *CLIN PHARMACOL THER*, 22, 864-867 (1977)

2017 KRISTENSEN ME, TOXIC HEPATITIS INDUCED BY DISULFIRAM IN A NONALCOHOLIC, *ACTA MED SCAND*, 209, 335-336 (1981)

2018 KRISTINSSON A, FATAL REACTION TO ACETAZOLAMIDE, *BR J OPHTHALMOL*, 51, 348 (1967)

2019 KRISTOFF CA, HAYES PE, BARR WH, ET AL, EFFECT OF IBUPROFEN ON LITHIUM PLASMA AND RED BLOOD CELL CONCENTRATIONS, *CLIN PHARM*, 5, 51-55 (1986)

2020 KROBOTH FJ, KROBOTH PD, LOGAN T, PHENYTOIN-THEOPHYLLINE-QUINIDINE INTERACTION, *N ENGL J MED*, 308, 725 (1983)

2021 KROGSTAD DJ, GRANICH GG, MURRAY PR, ET AL, HEPARIN INTERFERES WITH THE RADIOENZYMATIC AND HOMOGENEOUS ENZYME IMMUNOASSAYS FOR AMINOGLYCOSIDES, *CLIN CHEM*, 28, 1517-1521 (1982)

2022 KROLL MH, HAGENGRUBER C, ELIN RJ, EFFECT OF DIALYSIS ON INTERFERENCE BY CEFOXITIN WITH DETERMINATION OF CREATININE, *CLIN CHEM*, 30, 1386-1388 (1984)

2023 KROLL MH, JACKSON AJ, ELIN RJ, CEFOXITIN INTERFERES WITH THE 'CLINI-SKREEN' COLUMN METHOD FOR URINARY 17-HYDROXYCORTICOSTEROIDS, *CLIN CHEM*, 33, 1219-1222 (1987)

2024 KROLL MH, KOCH TR, DRUSANO GL, WARREN JW, LACK OF INTERFERENCE WITH CREATININE ASSAYS BY FOUR CEPHALOSPORIN LIKE ANTIBIOTICS, *AM J CLIN PATHOL*, 82, 214-216 (1984)

2025 KROLL MH, NEALON L, VOGEL MA, ET AL, HOW CERTAIN DRUGS INTERFERE NEGATIVELY WITH THE JAFFE REACTION FOR CREATININE, *CLIN CHEM*, 31, 306-308 (1985)

2026 KROMANN NP, VILHELMSEN R, STAHL D, THE DAPSONE SYNDROME, *ARCH DERMATOL*, 118, 531-532 (1982)

2027 KRONBORG IJ, EVANS DTP, MACKAY IR, ET AL, CHRONIC HEPATITIS AFTER SUCCESSIVE HALOTHANE ANESTHETICS, *DIGESTION*, 27, 123-128 (1983)

2028 KRONE RJ, GOLDBARG AN, BALKOURA M, SCHUESSLER R, RESNEKOV L, EFFECTS OF CIGARETTE SMOKING AT REST AND DURING EXERCISE. II. ROLE OF VENOUS RETURN, *J APPL PHYSIOL*, 32, 745 (1972)

2029 KRUMHOLZ WV, CHIPPS HI, MERLIS S, CLINICAL EFFECTS OF TRIOXAZINE, WITH A CASE REPORT OF HYPERGLYCEMIA AS A SIDE EFFECT, *J CLIN PHARMACOL*, 7, 108 (1967)

2030 KRUSE K, BARTELS H, ZIEGLER R, ET AL, PARATHYROID FUNCTION AND SERUM CALCITONIN IN CHILDREN RECEIVING ANTICONVULSANT DRUGS, *EUR J PEDIATR*, 133, 151-156 (1980)

2031 KUEPPERS F, BRACKERTZ D, CZYGAN PJ, SERUM ALPHA 1-ANTITRYPSIN LEVELS DURING THE OVARIAN CYCLE, *CLIN CHIM ACTA*, 39, 131 (1972)

2032 KUHN E, RYSANEK K, BRODAN V, ALTERATIONS OF TRYPTOPHAN METABOLISM INDUCED BY SLEEP DEPRIVATION, *EXPERIENTIA*, 24, 901 (1968)

2033 KUNO-SAKAI H, SAKAI H, RITZMANN SE, INTERFERENCE OF DIGITOXIN WITH THE RADIOIMMUNOASSAY OF DIGOXIN, *LANCET*, 2, 326 (1972)

2034 KUNTZ E, LIEHR H, PFINGST W, TOXISCHE LEBERSCHADIGUNG DURCH ATHIONAMID, *DTSCH MED WOCHENSCHR*, 92, 1718 (1967)

2035 KUNTZMAN R, JACOBSON M, CONNEY AH, EFFECT OF PHENYLBUTAZONE ON CORTISOL METABOLISM IN MAN, *PHARMACOLOGIST*, 8, 195 (1966)

2036 KUO PT, BASSETT DR, DIETARY SUGAR IN THE PRODUCTION OF HYPERGLYCERIDEMIA, *ANN INTERN MED*, 62, 1199 (1965)

2037 KURTZ TW, AL-BANDER HA, MORRIS RC JR., 'SALT - SENSITIVE ' ESSENTIAL HYPERTENSION IN MEN: IS THE SODIUM ION ALONE IMPORTANT?, *N ENGL J MED*, 317, 1043-1048 (1987)

2038 KURZ D, ROACH J, EYRING EJ, DETERMINATION OF ZINC BY FLAMELESS ATOMIC ABSORPTION SPECTROPHOTOMETRY, *ANAL BIOCHEM*, 53, 586 (1973)

2039 KURZ M, DIAMOX® UND MANIFESTIERUNG VON DIABETES MELLITUS, *WIEN MED WOCHENSCHR*, 118, 239 (1968)

2040 KUTKAITE D, RUDZKI Z, CHLORAL HYDRATE INTERFERES WITH ESTIMATION OF SERUM VITAMIN B_{12} BY SOME RADIOIMMUNOASSAY METHODS, *CLIN CHEM*, 32, 1983 (1986)

2041 KUTT H, DIPHENYLHYDANTOIN: INTERACTION WITH OTHER DRUGS IN MAN IN, *ANTIEPILEPTIC DRUGS*, DM WOODBURY, JK PENRY, RP SCHMIDT, EDS. RAVEN PRESS, NEW YORK, NY (1972)

2042 KUTT H, LOUIS S, ANTICONVULSANT DRUGS: II CLINICAL PHARMACOLOGICAL AND THERAPEUTIC ASPECTS, *DRUGS*, 4, 256 (1972)

2043 KUTT H, MILHORAT TM, MCDOWELL F, THE EFFECT OF IODIZED CONTRAST MEDIA UPON BLOOD PROTEINS, ELECTROLYTES AND RED CELLS, *NEUROLOGY*, 13, 492 (1963)

2044 KUTTER D, HUMBEL R, QUANTITATIVE ASSAY OF URINARY UROBILINOGEN WITH P-METHOXYBENZENE-DIAZONIUM-FLUOBORATE, *CLIN CHIM ACTA*, 45, 61 (1973)

2045 KUTTI J, OLSSON L-B, LUNDBORG P, ET AL, THE PERIPHERAL PLATELET COUNT IN RESPONSE TO INTRAVENOUS INFUSION OF SALBUTAMOL, *ACTA MED SCAND*, 201, 515-517 (1977)

2046 KUTTI J, SAFAI-KUTTI S, SIGVALDASON A, ET AL, THE EFFECT OF ACETYLSALICYLIC ACID IN 3 DIFFERENT FORMULATIONS ON IN VITRO AND IN VIVO PLATELET FUNCTION TESTS, *SCAND J HAEMATOL*, 32, 379-384 (1984)

2047 KUZUYA F, SUGINO N, YOSHIZUMI K, ET AL, CLINICAL INVESTIGATIONS IN THE PHARMACOLOGY OF AZOSEMIDE (SK-110) IN COMPARISON WITH FUROSEMIDE IN HEALTHY VOLUNTEERS, *INT J CLIN PHARMACOL THER TOXICOL*, 22, 291-299 (1984)

2048 KUZUYA T, KANAZAWA Y, STUDIES ON THE MECHANISM OF XYLITOL-INDUCED INSULIN SECRETION IN DOGS, *DIABETOLOGIA*, 5, 248 (1969)

2049 KWONG NK, BROWN BH, WHITTACKER GE, DUTHIE HL, EFFECTS OF GASTRIN 1, SECRETIN AND CHOLECYSTOKININ-PANCREOZYMIN ON THE ELECTRICAL ACTIVITY, MOTOR ACTIVITY, *SCAND J GASTROENTEROL*, 7, 161 (1972)

2050 LAAKE K, HORVEL C, ASPOY B, ET AL, HYPOMAGNESEMIA DURING TREATMENT WITH DIURETICS, *CURR THER RES*, 23, 730-733 (1978)

2051 LAAKE K, KJELDAAS L, BORCHGREVINK CF, SIDE EFFECTS OF PIROXICAM (FELDENE®), *ACTA MED SCAND*, 215, 81-83 (1984)

2052 LABEEUNN M, POZET N, ZECH P, ET AL, MAGNESURIA INDUCED BY THIAZIDES AND THE INFLUENCE OF TRIAMTERENE, *FUNDAM CLIN PHARMACOL*, 1, 225-232 (1987)

2053 LABIB MH, LITHIUM, HYPERCALCAEMIA AND HYPERPARATHYROIDISM, *ANN CLIN BIOCHEM*, 24, SUPPL S1, 147-148 (1987)

2054 LACHER DA, ELSEA AR, EFFECT OF A LIPID-CLARIFYING REAGENT ON RESULTS OF BECKMAN ASTRA™ METHODS, *CLIN CHEM*, 32, 394 (1986)

2055 LADD AT, PROCAINAMIDE-INDUCED LUPUS ERYTHEMATOSUS, *N ENGL J MED*, 267, 1357 (1962)

2056 LADENSON JH, BOWERS GN JR., FREE CALCIUM IN SERUM 1. DETERMINATION WITH THE ION-SPECIFIC ELECTRODE, AND FACTORS AFFECTING THE RESULTS, *CLIN CHEM*, 19, 565 (1973)

2057 LAFORCE CF, MILLER MF, CHAI M, EFFECT OF ERYTHROMYCIN ON THEOPHYLLINE CLEARANCE IN ASTHMATIC CHILDREN, *J PEDIATR*, 99, 153-156 (1981)

2058 LAGRELIUS A, JOHNSON P, LUNELL N-O, ET AL, TREATMENT WITH ORAL ESTRONE SULPHATE IN THE FEMALE CLIMACTERIC, *ACTA OBSTET GYNECOL SCAND*, 60, 27-31 (1981)

2059 LAI CL, NG RP, LOK ASF, THE DIAGNOSTIC VALUE OF THE RATIO OF SERUM GAMMA-GLUTAMYL TRANSPEPTIDASE TO ALKALINE PHOSPHATASE IN ALCOHOLIC LIVER DISEASE, *SCAND J GASTROENTEROL*, 17, 41-47 (1982)

2060 LAL S, TOLIS G, MCDONALD TJ, ET AL, EFFECT OF CLONIDINE ON GROWTH HORMONE AND GLUCAGON SECRETION, *HORM METAB RES*, 13, 648-649 (1981)

2061 LAMBERG B-A, TIKKANEN MJ, HYPERCALCAEMIA DUE TO DIHYDROTACHYSTEROL TREATMENT IN PATIENTS WITH HYPOTHYROIDISM AFTER THYROIDECTOMY, *BR MED J [CLIN RES]*, 283, 461-462 (1981)

2062 LAMBIE DG, JOHNSON RH, DRUGS AND FOLATE METABOLISM, *DRUGS*, 30, 145-155 (1985)

2063 LAMDEN MP, DANGERS OF MASSIVE VITAMIN C INTAKE, *N ENGL J MED*, 284, 336 (1971)

2064 LAMMERS PJ, WHITE L, ETTINGER LJ, CISPLATINUM-INDUCED RENAL SODIUM WASTING, *MED PEDIATR ONCOL*, 12, 343-346 (1984)

2065 LAMY J, ARON E, MARTIN JC, WEILL J, LECLERC M, TITECA C, REVERSIBILITY OF CLINICAL SIGNS OF CIRRHOSIS IN CHRONIC ALCOHOLISM AFTER ABSTENTION, *CLIN CHIM ACTA*, 49, 189 (1973)

2066 LANDESMAN PW, LOTT JA, ZAGER RA, MANNITOL INTERFERES WITH THE DU PONT ACA METHOD FOR INORGANIC PHOSPHORUS, *CLIN CHEM*, 28, 1994-1995 (1982)

2067 LANDON J (1969) CITED BY LORAINE JA, BELL ET, *HORMONE ASSAYS AND THEIR CLINICAL APPLICATION* 3RD ED, WILLIAMS AND WILKINS, BALTIMORE, MD (1971)

2068 LANDON MJ, SEPARATION OF FOLATES FROM ANTIBIOTICS USING TRIETHYLAMINO CELLULOSE, *CLIN CHIM ACTA*, 52, 253 (1974)

2069 LANE SD, BESA EC, JOSEPH RR, TAMOXIFEN AND HYPERCALCEMIA, *ANN INTERN MED*, 91, 414-415 (1979)

2070 LANES R, HERRERA A, PALACIOS A, ET AL, DECREASED SECRETION OF CORTISOL AND ACTH AFTER ORAL CLONIDINE ADMINISTRATION IN NORMAL ADULTS, *METABOLISM*, 32, 568-570 (1983)

2071 LANGELAAN DE, ALDOMET® INTERFERENCE WITH MELANOGEN TESTS, *ACB NEWS SHEET NO 125 SEPT* (1973)

2072 LANGER G, SACHAR EJ, HALPERN FS, ET AL, THE PROLACTIN RESPONSE TO NEUROLEPTIC DRUGS. A TEST OF DOPAMINERGIC BLOCKADE: NEUROENDOCRINE STUDIES IN NORMAL MEN, *J CLIN ENDOCRINOL METAB*, 45, 966-1002 (1977)

2073 LANGLANDS AO, MARTIN WMC, JAUNDICE ASSOCIATED WITH NORETHISTERONE ACETATE TREATMENT OF BREAST CANCER, *LANCET*, 1, 584 (1975)

2074 LANGLOIS J, DIAMOX® ET THROMBOCYTOPENIE, *ARCH OPHTHALMOL*, PARIS, 26, 701 (1966)

2075 LANGLOIS R, COURNOYER G, DE MONTIGNY C, ET AL, HIGH INCIDENCE OF MULTISYSTEMIC REACTIONS TO ZIMELDINE, *EUR J CLIN PHARMACOL*, 28, 67-71 (1985)

2076 LANGMAN S, HENRY DA, BELL GD, ET AL, CIMETIDINE AND RANITIDINE IN DUODENAL ULCER, *BR MED J [CLIN RES]*, 281, 473-474 (1980)

2077 LAORDEN ML, MIRALLES F, FUENTES T, ET AL, EFFECT OF STRESS AND STRESS THERAPY ON PLASMA BETA-ENDORPHIN-LIKE IMMUNOREACTIVITY, *METHODS FUND EXP CLIN PHARMACOL*, 6, 671-674 (1984)

2078 LAPIERRE G, STEWART RB, LITHIUM CARBONATE AND LEUKOCYTOSIS, *AM J HOSP PHARM*, 37, 1525-1528 (1980)

2079 LAPPIN TRJ, THE MEASUREMENT OF NON-ESTERIFIED FATTY ACIDS IN ICTERIC BODY FLUIDS, *CLIN CHIM ACTA*, 33, 153 (1971)

2080 LARDINOIS CK, MAZZAFERRI EL, CIMETIDINE BLOCKS TESTOSTERONE SYNTHESIS, *ARCH INTERN MED*, 145, 920-922 (1985)

2081 LARREY D, CASTOT A, PESSAYRE D, ET AL, AMODIAQUINE-INDUCED HEPATITIS, *ANN INTERN MED*, 104, 801-803 (1986)

2082 LARSEN FW, HANSEN CE, ZIMELDINE VERSUS AMITRIPTYLINE IN ENDOGENOUS DEPRESSION. A DOUBLE-BLIND STUDY, *ACTA PSYCHIATR SCAND*, 69, 343-349 (1984)

2083 LARSEN PR, INHIBITION OF TRI-IODOTHYRONINE (T_3) BINDING TO THYROXINE BINDING GLOBULIN BY SODIUM SALICYLATE, *METABOLISM*, 20, 976 (1971)

2084 LARSEN PR, THYROID—PITUITARY INTERACTION: FEEDBACK REGULATION OF THYROTROPIN SECRETION BY THYROID HORMONES, *N ENGL J MED*, 306, 23-32 (1982)

2085 LARSSON R, BODEMAR G, KAGEDAL B, ET AL, THE EFFECTS OF CIMETIDINE (TAGAMET®) ON RENAL FUNCTION IN PATIENTS WITH RENAL FAILURE, *ACTA MED SCAND*, 209, 27-31 (1980)

2086 LARSSON-COHN U, THE 2-HOUR SULFOBROMOPHTHALEIN RETENTION TEST AND THE TRANSAMINASE ACTIVITY DURING ORAL CONTRACEPTIVE THERAPY, *AM J OBSTET GYNECOL*, 98, 188 (1967)

2087 LARYEA EA, BRODRICK R, HIDVEGI R, HYPERCALCEMIA AND STREPTOZOTOCIN, *ANN INTERN MED*, 80, 276 (1974)

2088 LASSER EC, LANG JH, INHIBITION OF ACETYLCHOLINESTERASE BY SOME ORGANIC CONTRAST MEDIA, *INVEST RADIOL*, 1, 237 (1966)

2089 LATHAM AN, RICHENS A, PHENETURIDE, A MORE POTENT LIVER ENZYME INDUCER IN MAN THAN PHENOBARBITONE?, *BR J PHARMACOL*, 47, 615 (1973)

2090 LAUNAY C, FABIANI P, GRENET P, UN NOUVEAU CAS D'ACRODYNIE AVEC PRESENCE DE MERCURE DANS LES URINES (DISCUSSION THERAPEUTIQUE), *ARCH FR PEDIATR*, 7, 79 (1950)

2091 LAUPACIS A, KEOWN PA, ULAN RA, ET AL, HYPERBILIRUBINAEMIA AND CYCLOSPORIN A LEVELS, *LANCET*, 2, 1426 (1981)

2092 LAVAN, JN, THE EFFECT OF ORAL AMMONIUM CHLORIDE ON THE URINARY EXCRETION OF CALCIUM, MAGNESIUM AND SODIUM, *IR J MED SCI*, 2, 223 (1969)

2093 LAVIE CJ, BIUNDO J, QUINET RJ, ET AL, SYSTEMIC LUPUS ERYTHEMATOSUS (SLE) INDUCED BY QUINIDINE, *ARCH INTERN MED*, 145, 446-448 (1985)

2094 LAWLER SD, LELE KP, CHROMOSOMAL DAMAGE INDUCED BY CHLORAMBUCIL IN CHRONIC LYMPHOCYTIC LEUKAEMIA, *SCAND J HAEMATOL*, 9, 603 (1972)

2095 LAWRENCE VA, LOEWENSTEIN JE, EICHNER ER, ASPIRIN AND FOLATE BINDING: IN VIVO AND IN VITRO STUDIES OF SERUM BINDING AND URINARY EXCRETION OF ENDOGENOUS FOLATE, *J LAB CLIN MED*, 103, 944-948 (1984)

2096 LAWSON AA, MCARDLE T, GHOSH S, CEPHRADINE-ASSOCIATED IMMUNE NEUTROPENIA, *N ENGL J MED*, 312, 651 (1985)

2097 LAWSON DH, LOVATT GE, GURTON CS, ET AL, ADVERSE EFFECTS OF AZATHIOPRINE, *ADV DRUG REACT AC POIS REV*, 3, 161-171 (1984)

2098 LAY WH, DRUG-INDUCED HAEMOLYTIC REACTIONS DUE TO ANTIBODIES AGAINST THE ERYTHROCYTE/DIPYRONE COMPLEX, *VOX SANG*, 11, 601 (1966)

2099 LAZAR HP, MURPHY RL, PHAIR JP, FANSIDAR® AND HEPATIC GRANULOMAS, *ANN INTERN MED*, 102, 722 (1985)

2100 LE BEL M, PAONE RP, LEWIS GP, EFFECT OF TEN NEW β-LACTAM ANTIBIOTICS ON URINE GLUCOSE TEST METHODS, *DRUG INTELL CLIN PHARM*, 18, 617-620 (1984)

2101 LEAHEY EB JR., REIFFEL JA, GIARDINA E-GV, ET AL, THE EFFECT OF QUINIDINE AND OTHER ORAL ANTIARRHYTHMIC DRUGS ON SERUM DIGOXIN, *ANN INTERN MED*, 92, 605-608 (1980)

2102 LEAVELL BS, THORUP OA, *FUNDAMENTALS OF CLINICAL HEMATOLOGY* 3RD ED, WB SAUNDERS COMPANY, PHILADELPHIA, PA (1971)

2103 LEBACQ EG, TIRZMALIS A, METFORMIN AND LACTIC ACIDOSIS, *LANCET*, 1, 314 (1972)

2104 LEBLANC H, LOMBRAIL P, MARRE M, ET AL, PREVALENCE DES ANTICORPS ANTINUCLEAIRES CHEZ LES DIABETIQUES HYPERTENDUS TRAITES PAR L'ACEBUTOLOL, *PRESSE MED*, 13, 2747-2749 (1984)

2105 LECLERCQ R, PORTMANS JR, HORMONAL VARIATIONS DURING MUSCULAR EXERCISE IN MAN, WITH PARTICULAR REFERENCE TO PLASMA CORTISOL IN, *REFERENCE VALUES IN HUMAN CHEMISTRY*, G SIEST, ED. KARGER, BASEL 1973, 264

2106 LEDERLE LABORATORIES DIVISION, PRODUCT INFORMATION (DECLOMYCIN®), AMERICAN CYANAMID COMPANY, PEARL RIVER, NY, 10965

2107 LEE CR, POLLITT RJ, THE EFFECT OF LITHIUM SALTS ON THE URINARY EXCRETION OF SOME DICARBOXYLIC ACIDS, *BIOCHEM SOC TRANS*, 1, 108 (1973)

2108 LEE DB, DRINKARD JP, ROSEN VJ, GONICK HC, THE ADULT FANCONI SYNDROME: OBSERVATIONS ON ETIOLOGY, MORPHOLOGY, RENAL FUNCTION, AND MINERAL METABOLISM, *MEDICINE*, 51, 107 (1972)

2109 LEE M, BERGER HW, EOSINOPHILIA CAUSED BY RIFAMPIN, *CHEST*, 77, 579 (1980)

2110 LEE TH, REES PJ, HEPATOTOXICITY OF DEXTROPROPOXYPHENE, *BR MED J [CLIN RES]*, 2, 296-297 (1977)

2111 LEE-JONES M, SERUM TRANSAMINASES DURING SALICYLATE THERAPY, *BR MED J [CLIN RES]*, 2, 772 (1971)

2112 LEES RS, SCHONFELD G, MYERS GS, GULBRANDSEN C, GEORGE PK, THE EFFICACY OF SITOSTEROLS IN THE TREATMENT OF HYPERCHOLESTEROLEMIA, *CLIN RES*, 20, 409 (1972)

2113 LEFEVRE MJP, INCIDENTS ET ACCIDENTS DES MEDICATIONS NEUROLEPTIQUES ET PSYCHOTROPES, *J MED BORDEAUX*, 144, 139 (1967)

2114 LEHMANN J, LIGHT - A SOURCE OF ERROR IN THE FLUOROMETRIC DETERMINATION OF TRYPTOPHAN, *SCAND J CLIN LAB INVEST*, 28, 49 (1971)

2115 LEHTONEN A, THE EFFECT OF ACEBUTOLOL ON PLASMA LIPIDS, BLOOD GLUCOSE AND SERUM INSULIN LEVELS, *ACTA MED SCAND*, 216, 57-60 (1984)

2116 LEHTONEN A, EFFECT OF BETA BLOCKERS ON BLOOD LIPID PROFILE, *AM HEART J*, 109, 1192-1196 (1985)

2117 LEHTONEN A, LONG-TERM EFFECT OF PINDOLOL ON PLASMA LIPIDS, APOPROTEINS A, BLOOD GLUCOSE AND SERUM INSULIN LEVELS, *INTL J CLIN PHARM THER TOXICOL*, 22, 269-272 (1984)

2118 LEHTONEN A, GRONROOS M, MARNIEMI J, ET AL, EFFECTS OF HIGH DOSE PROGESTIN ON SERUM LIPIDS AND LIPID METABOLIZING ENZYMES IN PATIENTS WITH ENDOMETRIAL CANCER, *HORM METAB RES*, 17, 32-34 (1985)

2119 LEHTONEN A, TANSKANEN A, LEHTO H, ET AL, THE EFFECT OF NIFEDIPINE ON PLASMA LIPIDS IN PATIENTS WITH ESSENTIAL HYPERTENSION, *INTL J CLIN PHARM THER TOXICOL*, 24, 357-358 (1986)

2120 LEHTONEN A, VIIKARI J, LONG-TERM EFFECT OF SOTALOL ON PLASMA LIPIDS, *CLIN SCI*, 57, 405s-407s (1979)

2121 LEISS O, VON BERGMANN K, DIFFERENT EFFECTS OF CHENODEOXYCHOLIC ACID AND URSODEOXYCHOLIC ACID ON SERUM LIPOPROTEIN CONCENTRATIONS IN PATIENTS WITH RADIOLUCENT GALLSTONES, *SCAND J GASTROENTEROL*, 17, 587-592 (1982)

2122 LEKKERKERKER JFF, LIEM CH'ING SZE, DOORENBOS H, THE INFLUENCE OF CHLORTHALIDONE ON CALCIUM ABSORPTION FROM THE GUT IN RELATION TO URINARY CALCIUM EXCRETION, *ABSTRACTS IV INTL CONGR ENDOCRINOL, WASHINGTON DC, 1972*, 239

2123 LELLOUCH J, CLAUDE JR, A STUDY OF SEVERAL BIOLOGICAL PARAMETERS MEASURED IN A LARGE POPULATION OF A SINGLE PROFESSION, *REFERENCE VALUES IN HUMAN CHEMISTRY*, G SIEST, ED. KARGER, BASEL 1973, 100

2124 LELLOUCH J, SCHWARTZ D, TRAN MH, THE RELATIONSHIPS BETWEEN SMOKING AND LEVELS OF SERUM UREA AND URIC ACID, *J CHRON DIS*, 22, 9 (1969)

2125 LELO A, MINERS JO, ROBSON R, ET AL, ASSESSMENT OF CAFFEINE EXPOSURE: CAFFEINE CONTENT OF BEVERAGES, CAFFEINE INTAKE AND PLASMA CONCENTRATIONS OF METHYLXANTHINES, *CLIN PHARMACOL THER*, 39, 54-59 (1986)

2126 LEMANN J JR, PIERING WF, LENNON EJ, POSSIBLE ROLE OF CARBOHYDRATE INDUCED CALCIURIA IN CALCIUM OXALATE KIDNEY STONE FORMATION, *N ENGL J MED*, 280, 232 (1969)

2127 LEMBERGER L, CRABTREE R, CALLAGHAN JT, PERGOLIDE, A POTENT LONG-ACTING DOPAMINE RECEPTOR AGONIST, *CLIN PHARMACOL THER*, 27, 642-651 (1980)

2128 LENZ PG, CARGILL DI, FLEISCHMAN AI, PROPYLENE GLYCOL INTERFERENCE IN DETERMINATION OF SERUM AND LIVER TRIGLYCERIDES, *CLIN CHEM*, 19, 1071 (1973)

2129 LEON AS, AGRE J, MCNALLY C, ET AL, BLOOD LIPID EFFECTS OF ANTIHYPERTENSIVE THERAPY: A DOUBLE-BLIND COMPARISON OF THE EFFECTS OF METHYLDOPA AND PROPRANOLOL, *J CLIN PHARMACOL*, 24, 209-217 (1984)

2130 LEONARD PJ, THE EFFECT OF AGE AND SEX ON BIOCHEMICAL PARAMETERS IN BLOOD OF HEALTHY HUMAN SUBJECTS IN, *REFERENCE VALUES IN HUMAN CHEMISTRY*, G SIEST, ED. KARGER, BASEL 1973, 134

2131 LEONARD PJ, PERSAUD J, MOTWANI R, THE ESTIMATION OF PLASMA ALBUMIN BY BCG DYE BINDING ON THE TECHNICON® SMA 12/60 A, *CLIN CHIM ACTA*, 35, 409 (1971)

2132 LEONETTI G, BONAZZI O, GRAZI S, LIGRESTI A, ROMANO S, ZANCHETTI A, CARDIOVASCULAR AND RENAL EFFECTS OF ACUTE ADMINISTRATION OF A NEW HYPOTENSIVE COMPOUND, GUANCYDINE, *EUR J CLIN PHARMACOL*, 4, 1 (1971)

2133 LEPAGE L, SCHIELE F, GUEGUEN R, ET AL, TOTAL CHOLINESTERASE IN PLASMA: BIOLOGICAL VARIATIONS AND REFERENCE LIMITS, *CLIN CHEM*, 31, 545-550 (1985)

2134 LEPPANEN EA, GRASBECK R, EXPERIMENTAL BASIS OF STANDARDIZED SPECIMEN COLLECTION: THE EFFECT OF MODERATE ETHANOL CONSUMPTION ON SOME SERUM COMPONENTS(K, Na, ASAT, ALT, CK, LD, TOTAL PROTEIN), *SCAND J CLIN LAB INVEST*, 47, 337-343 (1987)

2135 LEREN P, EIDE I, FOSS OP, ET AL, ANTIHYPERTENSIVE DRUGS AND BLOOD LIPIDS: THE OSLO STUDY, *J CARDIOVASC PHARMACOL*, 4, SUPPL 2, S222-S224 (1982)

2136 LEREN P, FOSS PO, HELGELAND A, ET AL, EFFECT OF PROPRANOLOL AND PRAZOSIN ON BLOOD LIPIDS, *LANCET*, II, 4-6 (1980)

2137 LERNER RL, PORTE D JR., STUDIES OF SECRETIN-STIMULATED INSULIN RESPONSES IN MAN, *J CLIN INVEST*, 51, 2205 (1972)

2138 LETELLIER G, DESJARLAIS F, ANALYTICAL INTERFERENCE OF DRUGS IN CLINICAL CHEMISTRY: 1. STUDY OF TWENTY DRUGS ON SEVEN DIFFERENT INSTRUMENTS, *CLIN BIOCHEM*, 18, 345-351 (1985)

2139 LEUNG ACT, HENDERSON IS, HALLS DJ, ET AL, ALUMINUM HYDROXIDE VERSUS SUCRALFATE IS A PHOSPHATE BINDER IN URAEMIA, *BR MED J [CLIN RES]*, 286, 1379-1381 (1983)

2140 LEVER M, POWELL JC, KILLIP M, SMALL CW, A COMPARISON OF 4-HYDROXYBENZOIC ACID HYDRAZIDE (PAHBAH) WITH OTHER REAGENTS FOR THE DETERMINATION OF GLUCOSE, *J LAB CLIN MED*, 82, 649 (1973)

2141 LEVESQUE LA, HERZOG AG, SEIBEL MM, THE EFFECT OF PHENYTOIN AND CARBAMAZEPINE ON SERUM DEHYDROEPIANDROSTERONE SULFATE IN MEN AND WOMEN WHO HAVE PARTIAL SEIZURES WITH TEMPORAL LOBE INVOLVEMENT., *J CLIN ENDOCRINOL METAB*, 63, 243-245 (1986)

2142 LEVI, L, THE EFFECT OF COFFEE ON THE FUNCTION OF THE SYMPATHO-ADRENOMEDULLARY SYSTEM IN MAN, *ACTA MED SCAND*, 181, 431 (1967)

2143 LEVILLAIN R, BOUAZIZ I, CLUZAN R, LEBLAYE O, LEFESVRE A, 15 YEARS OF EXPERIENCE WITH N-DESACETYLTHIOCOLCHICINE (THIOCOLCIRAN) IN CANCEROLOGY, *THERAPIE*, 27, 77 (1972)

2144 LEVIN HA, McMILLAN R, TAVASSOLL M, THROMBOCYTOPENIA ASSOCIATED WITH GOLD THERAPY: OBSERVATIONS ON THE MECHANISM OF PLATELET DESTRUCTION, *AM J MED*, 59, 274-280 (1975)

2145 LEVIN K, JOSEPHSON B, GRUNEWALD G, THE EFFECT OF IODOCHLORO-OXYQUINOLINE AND IOPANOIC ACID ON THE DETERMINATION OF PBI AND BEI, *ACTA ENDOCRINOL*, 52, 627 (1966)

2146 LEVINE BS, CAPLAN YH, ISOMETHEPTENE CROSS REACTS IN THE AMPHETAMINE ASSAY, *CLIN CHEM*, 33, 1264-1265 (1987)

2147 LEVINE M, JONES MW, SHEPPARD I, DIFFERENTIAL EFFECT OF CIMETIDINE ON SERUM CONCENTRATIONS OF CARBAMAZEPINE AND PHENYTOIN, *NEUROLOGY*, 35, 562-565 (1985)

2148 LEVINE RA, STEATORRHEA INDUCED BY PARA-AMINOSALICYLIC ACID, *ANN INTERN MED*, 68, 1265 (1968)

2149 LEVINSON PD, GOLDSTEIN DS, MUNSON PJ, ET AL, ENDOCRINE, RENAL, AND HEMODYNAMIC RESPONSES TO GRADED DOPAMINE INFUSIONS IN NORMAL MEN, *J CLIN ENDOCRINOL METAB*, 60, 821-826 (1985)

2150 LEVINSON SA, MACFATE RP,, *CLINICAL LABORATORY DIAGNOSIS* 6TH ED, LEA AND FEBIGER, PHILADELPHIA, PA (1961)

2151 LEVY AL, MEASUREMENT OF TRIGLYCERIDES USING NONANE EXTRACTION AND COLORIMETRY, *ANN CLIN LAB SCI*, 2, 474 (1972)

2152 LEVY M, ELIAKIM M, URINARY PRECIPITATE DURING CEPHALOTHIN - CEPHALORIDINE TREATMENT, *J AM MED ASSOC*, 219, 908 (1972)

2153 LEVY M, GOODMAN MW, VAN DYNE BJ, ET AL, GRANULOMATOUS HEPATITIS SECONDARY TO CARBAMAZEPINE, *ANN INTERN MED*, 95, 64-65 (1981)

2154 LEWIN PK, PHENYTOIN-ASSOCIATED CONGENITAL DEFECTS WITH Y-CHROMOSOME VARIANT, *LANCET*, 1, 559 (1973)

2155 LEWIS GP, CIGARETTE SMOKING AND CADMIUM ACCUMULATION IN MAN, *LANCET*, 1, 682 (1972)

2156 LEWIS IJ, ROSENBLOOM L, GLANDULAR FEVER-LIKE SYNDROME, PULMONARY EOSINOPHILIA AND ASTHMA ASSOCIATED WITH CARBAMAZEPINE, *POSTGRAD MED J*, 58, 100-101 (1982)

2157 LEWIS JH, ZIMMERMANN HJ, ISHAK KG, ET AL, ENFLURANE HEPATOTOXICITY: A CLINICOPATHOLOGIC STUDY OF 24 CASES, *ANN INTERN MED*, 98, 984-992 (1983)

2158 LEWIS JS, PIAN AK, BAER MT, ACOSTA PB, EMERSON GA, EFFECT OF LONG TERM INGESTION OF POLYUNSATURATED FAT, AGE, PLASMA CHOLESTEROL, DIABETES MELLITUS, *AM J CLIN NUTR*, 26, 136 (1973)

2159 LEWIS RJ, TRAGER WF, ROBINSON AJ, CHEN KK, WARFARIN METABOLITES: THE ANTICOAGULANT ACTIVITY AND PHARMACOLOGY OF WARFARIN ALCOHOLS, *J LAB CLIN MED*, 81, 925 (1973)

2160 LEWIS SA, OSWALD I, DUNLEAVY DLF, CHRONIC FENFLURAMINE ADMINISTRATION: SOME CEREBRAL EFFECTS, *BR MED J [CLIN RES]*, 3, 67 (1971)

2161 LEZA MA, AGRANULOCITOSIS MEDICAMENTOSSAS CON PLASMACITOSIS MEDULAR E HIPERGAMMA GLOBINEMIA, *REV CLIN ESP*, 103, 316 (1966)

2162 LIAKAKOS D, PAPADOPOULOS Z, VLACHOS P, ET AL, SERUM ALKALINE PHOSPHATASE AND URINARY HYDROXYPROLINE VALUES IN CHILDREN RECEIVING PHENOBARBITAL WITH AND WITHOUT VITAMIN D, *J PEDIATR*, 87, 291-296 (1975)

2163 LIAKAKOS D, VALCHOS P, ANOUSSAKIS C, EFFECT OF ACETYLSALICYLIC ACID (ASPIRIN) ON BONE COLLAGEN, *CLIN CHIM ACTA*, 44, 427 (1973)

2164 LIDDLE GW, TESTS OF PITUITARY-ADRENAL SUPRESSIBILITY IN THE DIAGNOSIS OF CUSHING'S SYNDROME, *J CLIN ENDOCRINOL METAB*, 20, 1539 (1960)

2165 LIDSKY MD, SHARP JT, JAUNDICE WITH THE USE OF 4-HYDROXYPYRAZOL-(3,4-D)-PYRIMIDINE (4-HPP), *ARTHRITIS RHEUM*, 10, 294 (1967)

2166 LIEBER CS, HYPERURICEMIA INDUCED BY ALCOHOL, *ARTHRITIS RHEUM*, 8, 786 (1965)

2167 LIEBERMAN AN, KUPERSMITH M, GOPINATHAN G, ET AL, BROMOCRIPTINE IN PARKINSON DISEASE: FURTHER STUDIES, *NEUROLOGY*, 29, 363-369 (1979)

2168 LIEWENDAHL K, MAJURI H, HELENIUS T, THYROID FUNCTION TESTS IN PATIENTS ON LONG-TERM TREATMENT WITH VARIOUS ANTICONVULSANT DRUGS, *CLIN ENDOCRINOL*, 8, 185-191 (1978)

2169 LIEWENDAHL K, TIKANOJA S, HELENIUS T, ET AL, FREE THYROXINE AND FREE TRI-IODOTHYRONINE AS MEASURED BY EQUILIBRIUM DIALYSIS AND ANALOG RADIOIMMUNOASSAY IN SERUM OF PATIENTS TAKING PHENYTOIN AND CARBAMAZEPINE, *CLIN CHEM*, 31, 1993-1996 (1985)

2170 LIFTON LJ, KREISER J, FALSE POSITIVE STOOL OCCULT BLOOD TESTS CAUSED BY IRON PREPARATIONS: A CONTROLLED STUDY AND REVIEW OF LITERATURE, *GASTROENTEROLOGY*, 83, 860-863 (1982)

2171 LIJNEN P, FAGARD R, STAESSEN J, ET AL, SERUM CHOLESTEROL DURING KETANSERIN AND PROPRANOLOL ADMINISTRATION IN HYPERTENSIVE PATIENTS, *J CARDIOVASC PHARMACOL*, 10, 647-650 (1987)

2172 LILLY JR, HITCH DC, JAVITT NB, CIMETIDINE CHOLESTATIC JAUNDICE IN CHILDREN, *J SURG RES*, 24, 384-387 (1978)

2173 LIMA MC, AJZEN HA, RIBEIRO AB, ANDRADE J, RAMOS OL, EFFECT OF ANGIOTENSIN II ON URINARY MAGNESIUM, CALCIUM, AND SODIUM EXCRETION, *AM J MED SCI*, 263, 173 (1972)

2174 LIMAS C, GUIHA NH, FREIS ED, TREATMENT OF SEVERE RESISTANT HYPERTENSION WITH MINOXIDIL, *CLIN PHARMACOL THER*, 13, 145 (1972)

2175 LIN C-C, POTTER JJ, MEZEY E, ERYTHROCYTE ALDEHYDE DEHYDROGENASE ACTIVITY IN ALCOHOLISM, *ALCOHOL CLIN EXP RES*, 8, 539-541 (1984)

2176 LIN T, TUCCI JR, COMPARISON OF PROVOCATIVE TESTS OF GROWTH HORMONE RELEASE, *CLIN RES*, 20, 866 (1972)

2177 LINDBERG U-B, CRONA N, ENK L, ET AL, EFFECTS OF CYPROTERONE ACETATE (CPA) ON SERUM LIPOPROTEINS WHEN ADMINISTERED ALONE AND IN COMBINATION WITH ETHINYL ESTRADIOL (EE), *HORM METAB RES*, 79, 222-225 (1987)

2178 LINDENBAUM J, MAULITZ RM, BUTLER VP JR., INHIBITION OF DIGOXIN ABSORPTION BY NEOMYCIN, *GASTROENTEROLOGY*, 71, 399-404 (1976)

2179 LINDENBAUM J, RUND DG, BUTLER VP JR., ET AL, INACTIVATION OF DIGOXIN BY THE GUT FLORA: REVERSAL BY ANTIBIOTIC THERAPY, *N ENGL J MED*, 305, 789-794 (1981)

2180 LINDHEIMER MD, KATZ AI, SODIUM AND DIURETICS IN PREGNANCY, *N ENGL J MED*, 288, 891 (1973)

2181 LINDHOLM J, SCHULTZ-MOLLER N, PLASMA AND URINARY CORTISOL IN PREGNANCY AND DURING ESTROGEN-GESTAGEN TREATMENT, *SCAND J CLIN LAB INVEST*, 31, 119 (1973)

2182 LINDHOUT D, MEINARDI H, FALSE-NEGATIVE PREGNANCY TEST IN WOMEN TAKING CARBAMAZEPINE, *LANCET*, 2, 505 (1982)

2183 LINDSTEDT G, LUNDBERG P-A, JAGENBURG R, ET AL, EFFECT OF HEPARIN IN VIVO ON THE IN VITRO ASSAY OF FREE THYROXINE, *SCAND J CLIN LAB INVEST*, 43, 643-646 (1983)

2184 LINDSTEDT G, LUNDBERG PA, TOFFT M, AKESSON HO, OHMAN R, SERUM THYROTROPIN AND HYPOTHYROIDISM DURING LITHIUM THERAPY, *CLIN CHIM ACTA*, 48, 127 (1973)

2185 LINDSTEDT G, NILSSON L-A, WALINDER J, ET AL, ON THE PREVALENCE, DIAGNOSIS AND MANAGEMENT OF LITHIUM-INDUCED HYPOTHYROIDISM IN PSYCHIATRIC PATIENTS, *BR J PSYCHIATRY*, 130, 452-458 (1977)

2186 LINNA O, UHARI M, HEPATOTOXICITY OF RIFAMPICIN AND ISONIAZID IN CHILDREN TREATED FOR TUBERCULOSIS, *EUR J PEDIATR*, 134, 227-229 (1980)

2187 LIPMAN RL, RASKIN P, LOVE T, TRIEBWASSER J, LECOCG FR, SCHNURE JJ, GLUCOSE INTOLERANCE DURING DECREASED PHYSICAL ACTIVITY IN MAN, *DIABETES*, 21, 101 (1972)

2188 LIPSETT D, MADRAS BK, WURTMAN RJ, MUNRO HN, SERUM TRYPTOPHAN LEVEL AFTER CARBOHYDRATE INGESTION, *LIFE SCI*, 12, 57 (1973)

2189 LIPSETT MB, COMBS JW, CATT K, SEIGEL DG, PROBLEMS IN CONTRACEPTIVES, *ANN INTERN MED*, 74, 251 (1971)

2190 LISS GM, GREENBERG RA, TAMBURRO CH, USE OF SERUM BILE ACIDS IN THE IDENTIFICATION OF VINYL CHLORIDE HEPATOTOXICITY, *AM J MED*, 78, 68-76 (1985)

2191 LITHELL H, BERNE C, WAERN AU, ET AL, GLUCOSE METABOLISM DURING LONG-TERM TREATMENT WITH PRAZOSIN, *DIABETES RES*, 2, 297-299 (1985)

2192 LITHELL H, WEINER L, SELINUS I, ET AL, A COMPARISON OF THE EFFECTS OF BISOPROLOL AND ATENOLOL ON LIPOPROTEIN CONCENTRATIONS AND BLOOD PRESSURE, *J CARDIOVASC PHARMACOL*, 8, SUPPL, S128-S133 (1986)

2193 LITTLE JA, SHANOFF HM, CSIMA A, YANO R, COFFEE AND SERUM LIPIDS IN CORONARY HEART DISEASE, *LANCET*, 1, 732 (1966)

2194 LIU K, MITTELMAN A, SPROUL EE, ELIAS EG, RENAL TOXICITY IN MAN TREATED WITH MITOMYCIN-C, *CANCER*, 28, 1314 (1971)

2195 LIU TZ, OKA KH, SPECTROPHOTOMETRIC SCREENING METHOD FOR ACETAMINOPHEN IN SERUM AND PLASMA, *CLIN CHEM*, 26, 69-71 (1980)

2196 LIU YK, LEUKOPENIA IN ALCOHOLICS, *ANN INTERN MED*, 79, 605 (1973)

2197 LLERENA O, PEARSON O, INTERFERENCE OF NALIDIXIC ACID IN URINARY 17-KETOSTEROID DETERMINATIONS, *N ENGL J MED*, 279, 983 (1968)

2198 LLOYD TW, AGRANULOCYTOSIS ASSOCIATED WITH PARACETAMOL, *LANCET*, 1, 114 (1961)

2199 LO CH, STELSON H, INTERFERENCE BY POLYSUCROSE IN PROTEIN DETERMINATION BY THE LOWRY METHOD, *ANAL BIOCHEM*, 45, 331 (1971)

2200 LOBATTO S, DIJKMANS BAC, MATTIE H, ET AL, FLUCLOXACILLIN-ASSOCIATED LIVER DAMAGE, *NETH J MED*, 25, 47-48 (1982)

2201 LOERTSCHER R, THIEL G, HARDER F, ET AL, PERSISTENT ELEVATION OF ALKALINE PHOSPHATASE IN CYCLOSPORINE-TREATED RENAL-TRANSPLANT RECIPIENTS, *TRANSPLANTATION*, 36, 115-116 (1983)

2202 LOEWENBERG JR, CYANIDE AND THE DETERMINATION OF PROTEIN WITH THE FOLIN PHENOL REAGENT, *ANAL BIOCHEM*, 19, 95 (1967)

2203 LOFASO F, BAUD FJ, HALNA DU FRETAY X, ET AL, HYPOKALEMIA IN MASSIVE CHLOROQUINE INTOXICATION. TWO CASES, *PRESSE MED*, 16, 22-24 (1987)

2204 LOH HS, SCREENING FOR VITAMIN C STATUS, *LANCET*, 1, 944 (1973)

2205 LOH HS, WATTERS K, WILSON CWM, THE EFFECTS OF ASPIRIN ON THE METABOLIC AVAILABILITY OF ASCORBIC ACID IN HUMANS, *J CLIN PHARMACOL*, 13, 480 (1973)

2206 LOH HS, WILSON CWM, RELATIONSHIP OF HUMAN ASCORBIC ACID METABOLISM TO OVULATION, *LANCET*, 1, 110 (1971)

2207 LONDONO JH, MCGEE JH, SOBEL RE, ROSENBLOOM AC, SAVORY J, VARIATIONS IN SERUM CALCIUM AND PHOSPHORUS CONCENTRATIONS DURING GLUCOSE TOLERANCE TESTS, *CLIN CHEM*, 17, 648 (1971)

2208 LONGSTRETH GF, GREEN R, FOLATE STATUS IN PATIENTS RECEIVING MAINTENANCE DOSES OF SULFASALAZINE, *ARCH INTERN MED*, 143, 902-904 (1983)

2209 LONGSTRETH PL, MALINAK LR, HYDROXYPROLINE AND TRANSIENT OSTEOPOROSIS, *ANN INTERN MED*, 76, 833 (1972)

2210 LOPEZ A, RENE A, EFFECT OF 17 KETOSTEROIDS ON GLUCOSE-6-PHOSPHATE DEHYDROGENASE ACTIVITY ON G6PD ISOENZYMES, *PROC SOC EXP BIOL MED*, 142, 258 (1973)

2211 LORAINE JA, BELL ET, *HORMONE ASSAYS AND THEIR CLINICAL APPLICATION* 3RD ED, WILLIAMS AND WILKINS, BALTIMORE, MD (1971)

2212 LORENZ W, DOENICKE A, MEYER R, REIMANN HJ, KUSCHE J, BARTH H, GEESING H, HUTZEL M, WEISSENBACHER B, HISTAMINE RELEASE IN MAN BY PROPANIDID AND THIOPENTONE: PHARMACOLOGICAL EFFECTS AND CLINICAL CONSEQUENCES, *BR J ANAESTH*, 44, 355 (1972)

2213 LORIAUX DL, RUDER HJ, KNAB DR, LIPSETT MB, ESTRONE SULFATE, ESTRONE, ESTRADIOL AND ESTRIOL PLASMA LEVELS IN HUMAN PREGNANCY, *J CLIN ENDOCRINOL METAB*, 35, 887 (1972)

2214 LOSEK JD, WERLIN SL, SULFASALAZINE HEPATOTOXICITY, *AM J DIS CHILD*, 135, 1070-1072 (1981)

2215 LOTITO CA, MENGEL CE, EFFECT OF MELPHALAN IN THE MALIGNANT CARCINOID SYNDROME, *ARCH INTERN MED*, 124, 36 (1969)

2216 LOUBATIERES AL, CONFRONTATION OF EXPERIMENTAL AND CLINICAL PHARMACOLOGICAL OBSERVATIONS IN THE FIELD OF HYPOGLYCAEMIC, *ARCH INTL PHARM THER (SUPPL)*, 192, 133 (1971)

2217 LOURIA DB, JOSELOW MM, BROWDER AA, THE HUMAN TOXICITY OF CERTAIN TRACE ELEMENTS, *ANN INTERN MED*, 76, 307 (1972)

2218 LOWE J, GRAY J, HENRY DA, ET AL, ADVERSE REACTIONS TO FRUSEMIDE IN HOSPITAL PATIENTS, *BR MED J [CLIN RES]*, 2, 360-362 (1979)

2219 LOWE ML, STELLA AF, MOSHER BS, GIN JB, DEMETRIOU JA, MICROFLUOROMETRY OF GLUCOSE-6-PHOSPHATE DEHYDROGENASE AND 6-PHOSPHOGLUCONATE DEHYDROGENASE IN RED CELLS, *CLIN CHEM*, 18, 440 (1972)

2220 LUBRAN M, THE EFFECTS OF DRUGS ON LABORATORY VALUES, *MED CLIN NORTH AM*, 53, 211 (1969)

2221 LUBY ED, SCHWARTZ D, ROSENBAUM H, LITHIUM CARBONATE-INDUCED MYXEDEMA, *J AM MED ASSOC*, 218, 1298 (1971)

2222 LUCE GG, BIOLOGICAL RHYTHMS IN PSYCHIATRY AND MEDICINE, *PHS PUBLICATION NO 2088*, WASHINGTON DC (1970)

2223 LUCENA MI, MORENO A, FERNANDEZ MC, ET AL, DIGITOXIN ELIMINATION IN HEALTHY SUBJECTS TAKING AMPICILLIN, *INT J CLIN PHARMACOL RES*, 7, 33-37 (1987)

2224 LUCIANO AA, HAUSER KS, CHAPLER FK, ET AL, DANAZOL: ENDOCRINE CONSEQUENCES IN HEALTHY WOMEN, *AM J OBSTET GYNECOL*, 141, 723-727 (1981)

2225 LUCKE C, ADELMAN N, GLICK SM, EFFECT OF PLASMA FREE FATTY ACIDS (FFA) ON SLEEP INDUCED HUMAN GROWTH HORMONE (HGH SECRETION), *ABSTRACTS IV INTL CONGR ENDOCRINOL, WASHINGTON DC, 1972*, 35

2226 LUISADA-OPPER AV, COLORIMETRIC DETERMINATION OF 2,3-DIPHOSPHOGLYCERATE IN WHOLE BLOOD, *CLIN CHEM*, 19, 118 (1973)

2227 LUKE RG, KOEPKE JA, SIEGEL RR, THE EFFECTS OF IMMUNOSUPPRESSIVE DRUGS AND UREMIA ON AUTOMATED LEUKOCYTE COUNTS, *AM J CLIN PATHOL*, 56, 503 (1971)

2228 LUM G, GAMBINO SR, A COMPARISON OF SERUM VERSUS HEPARINIZED PLASMA FOR ROUTINE CHEMISTRY TESTS, *AM J CLIN PATHOL*, 61, 108 (1974)

2229 LUM G, GAMBINO SR, SERUM VERSUS PLASMA DETERMINATIONS IN ROUTINE CHEMISTRY, *CLIN CHEM*, 18, 710 (1972)

2230 LUM G, GAMBINO SR, COMPARISON OF FOUR METHODS FOR MEASURING URIC ACID: COPPER-CHELATE, PHOSPHOTUNG-STATE, MANUAL URICASE, *CLIN CHEM*, 19, 1186 (1973)

2231 LUMENG L, LI TK, INHIBITORY ACTION OF ALCOHOL OXIDATION ON VITAMIN B_6 METABOLISM, *CLIN RES*, 20, 803 (1972)

2232 LUND J, PEDERSEN HE, OLSEN PZ, HVIDBERG EF, EARLY HUMAN STUDIES OF A NEW CARBONIC ANHYDRASE INHIBITOR (NSD 3004) WITH ANTICONVULSANT PROPERTIES, *CLIN PHARMACOL THER*, 12, 902 (1971)

2233 LUNDBERG B, NERGARDH A, BOREUS LO, PLASMA CONCENTRATIONS OF VALPROATE DURING MAINTENANCE THERAPY IN EPILEPTIC CHILDREN, *J NEUROL*, 228, 133-141 (1982)

2234 LUNDBORG P, THE EFFECT OF ADRENERGIC BLOCKADE ON POTASSIUM CONCENTRATIONS IN DIFFERENT CONDITIONS, *ACTA MED SCAND SUPPLEMENT*, 672, 121-125 (1983)

2235 LUPIEN PJ, BRUN D, MORRJANI S, EFFECTIVENESS OF 3-HYDROXY-3-METHYLGLUTARIC ACID IN FAMILIAL HYPERCHOLESTEROL, *LANCET*, 1, 1256 (1973)

2236 LUPOVITCH A,, PHENOTHIAZINE INTERFERENCE AND THE PORTER-SILBER REACTION, *CLIN CHEM*, 14, 179 (1968)

2237 LURIE AO, SPIRONOLACTONE AND STEROID ASSAY, *LANCET*, 2, 326 (1969)

2238 LUTZ EG, MONOCYTOSIS BLOOD DYSCRASIA AND CHLORPROMAZINE TOXICITY, *INTL J NEUROPSYCHIAT*, 1, 76 (1965)

2239 LYFAR VJ, A STUDY OF THE HYPOURICAEMIC EFFECT OF SURGAM., *ANN CLIN BIOCHEM*, 24, SUPPL S1, 32-33 (1987)

2240 LYNGSOE J, BITSCH V, TRAP-JENSEN J, INFLUENCE OF PHENFORMIN ON FAT AND LACTATE METABOLISM AND INSULIN PRODUCTION IN STARVED NORMAL SUBJECTS, *METABOLISM*, 21, 179 (1972)

2241 LYNGSOE J, TRAP-JENSEN J, PHENFORMIN-INDUCED HYPOGLYCAEMIA IN NORMAL SUBJECTS, *BR MED J [CLIN RES]*, 2, 224 (1969)

2242 LYONS F, LAKER MF, MARSDEN JR, ET AL, EFFECT OF ORAL 13 CIS-RETINOIC ACID ON SERUM LIPIDS, *BR J DERMATOL*, 107, 591-595 (1982)

2243 MA RSW, CHAN JCM, ENDOGENOUS SULFURIC ACID PRODUCTION: A METHOD OF MEASUREMENT BY EXTRAPOLATION, *CLIN BIOCHEM*, 6, 82 (1973)

2244 MAAS RD, VENOOK AP, LINKER CA, ET AL, PENTOXIFYLLINE AND APLASTIC ANEMIA, *ANN INTERN MED*, 107, 427-428 (1987)

2245 MACADAMS MR, WHITE RH, CHIPPS BE, REDUCTION OF SERUM TESTOSTERONE LEVELS DURING CHRONIC GLUCOCORTICOID THERAPY, *ANN INTERN MED*, 104, 648-651 (1986)

2246 MACAROL V, MORRIS TQ, BAKER KJ, BRADLEY SE, HYDROCORTISONE CHOLERESIS IN THE DOG, *J CLIN INVEST*, 49, 1714 (1970)

2247 MACAULAY VM, BEGENT RHJ, PHILLIPS ME, ET AL, PROPHYLAXIS AGAINST HYPOMAGNESEMIA INDUCED BY CIS-PLATINUM COMBINATION CHEMOTHERAPY, *CANCER CHEMOTHER PHARMACOL*, 9, 179-181 (1982)

2248 MACBETH WA, BLENNERHASSETT JB, NYE ER, LIVER TOXICITY WITH BETA PYRIDYL CARBINOL, *N Z MED J*, 74, 382 (1971)

2249 MACBETH WA, KASS EH, MCDERMOTT WV, TREATMENT OF HEPATIC ENCEPHALOPATHY BY ALTERATION OF INTESTINAL FLORA WITH LACTOBACILLUS ACIDOPHILUS, *LANCET*, 1, 399 (1965)

2250 MACCARA ME, PARKER WA, IN VITRO EFFECT OF PENICILLINS AND AMINOGLYCOSIDES ON COMMONLY USED TESTS FOR GLYCOSURIA, *AM J HOSP PHARM*, 38, 1340-1345 (1981)

2251 MACDONALD RP, SALICYLATE, *STAND METH CLIN CHEM*, 5, 237 (1965)

2252 MACGEE J, ISHIKAWA T, MILLER W, EVANS G, STEINER P, GLUECK CJ, A MICROMETHOD FOR ANALYSIS OF TOTAL PLASMA CHOLESTEROL USING GAS-LIQUID CHROMATOGRAPHY, *J LAB CLIN MED*, 82, 656 (1973)

2253 MACGIBBON BH, LOUGHRIDGE LW, HOURIHANE DO, BOYD DW, AUTOIMMUNE HAEMOLYTIC ANAEMIA WITH ACUTE RENAL FAILURE DUE TO PHENACETIN AND P-AMINOSALICYLIC ACID, *LANCET*, 1, 7 (1960)

2254 MACGREGOR GA, POOLE-WILSON PA, JONES NF, PHENFORMIN AND METABOLIC ACIDOSIS, *LANCET*, 1, 69 (1972)

2255 MACHTEY I, SERUM URIC ACID LEVELS FOLLOWING ADMINISTRATION OF METIAZINIC ACID, *CLIN CHIM ACTA*, 47, 317 (1973)

2256 MACKAY EV, KHOO SK, CLINICAL AND LABORATORY STUDY OF A NEW DIURETIC AGENT ('VECTREN') IN PREGNANCY, *MED J AUST*, 1, 607 (1969)

2257 MACKINNEY AA, BOOKER HE, DIPHENYLHYDANTOIN EFFECTS ON HUMAN LYMPHOCYTES IN VITRO AND IN VIVO, *ARCH INTERN MED*, 129, 988 (1972)

2258 MACLEAN D, MURISON J, GRIFFITHS PD, SERUM ENZYME ACTIVITIES IN ACCIDENTAL HYPOTHERMIA AND HYPOTHERMIC MYXOEDEMA, *CLIN CHIM ACTA*, 52, 197 (1974)

2259 MADDOCKS J, HANN S, HOPKINS M, COLES GA, EFFECT OF METHYLDOPA ON CREATININE ESTIMATION, *LANCET*, 1, 157 (1973)

2260 MADDREY WC, BOITNOTT JK, DRUG-INDUCED CHRONIC LIVER DISEASE, 72, 1348-1353 (1977)

2261 MADDREY WC, BOITNOTT JK, SEVERE HEPATITIS FROM METHYLDOPA, *GASTROENTEROLOGY*, 68, 351-360 (1975)

2262 MADDUX MS, BARRIERE SL, A REVIEW OF COMPLICATIONS OF AMPHOTERICIN B THERAPY: RECOMMENDATIONS FOR PREVENTION AND MANAGEMENT., *DRUG INTELL CLIN PHARM*, 14, 177-181 (1980)

2263 MADEDDU P, ENA P, DESSI-FULGHERI P, CAPTOPRIL INDUCED PROTEINURIA IN HYPERTENSIVE PSORIATIC PATIENTS, *NEPHRON*, 44, 358-360 (1986)

2264 MADERA-ORSINI FM, VOLINI FI, SCREENING METHODS FOR ENZYME DEFECTS IN ERYTHROCYTES, *ANN CLIN LAB SCI*, 2, 40 (1972)

2265 MAESTRI E, MANZONI GC, MARCHESI G, ET AL, EFFECT OF FLUNARIZINE ON PITUITARY SECRETION BY HEALTHY MEN AND IN WOMEN WITH MIGRAINE, *EUR J CLIN PHARMACOL*, 32, 525-527 (1987)

2266 MAGARIAN EO, DIETZ AJ, FREEMAN DS, ET AL, EFFECT OF PRAZOSIN AND BETA-BLOCKER MONOTHERAPY ON SERUM LIPIDS: A CROSSOVER, PLACEBO-CONTROLLED STUDY, *J CLIN PHARMACOL*, 27, 756-761 (1987)

2267 MAGNUSSON B, RODJER S, ALPRENOLOL-INDUCED THROMBOCYTOPENIA, *ACTA MED SCAND*, 207, 231-233 (1980)

2268 MAGUIRE R, COPELAND BE, A STUDY OF THE EFFECT OF SAMPLE COLLECTION ON THE MEASUREMENT OF LACTATE IN WHOLE BLOOD, *AM J CLIN PATHOL*, 58, 153 (1972)

2269 MAGUIRE S, KYNE F, UA CONAILL D, SPECIMEN COLLECTION IN PLASTIC CONTAINERS GIVES RISE TO INTERFERENTS IN LIQUID-CHROMATOGRAPHIC ASSAY OF CYCLOSPORINE, *CLIN CHEM*, 33, 1493-1494 (1987)

2270 MAJ M, ARIANO MG, PIROZZI R, ET AL, PLATELET MONOAMINE OXIDASE ACTIVITY IN SCHIZOPHRENIA: RELATIONSHIP TO FAMILY HISTORY OF THE ILLNESS AND NEUROLEPTIC TREATMENT, *J PSYCHIAT, RES.*, 18, 131-137 (1984)

2271 MAJCHROWICZ E, MENDELSON JH, BLOOD METHANOL CONCENTRATIONS DURING EXPERIMENTALLY INDUCED ETHANOL INTOXICATION IN ALCOHOLICS, *J PHARMACOL EXP THER*, 179, 293 (1971)

2272 MAJOR LF, GOYER PF, EFFECTS OF DISULFIRAM AND PYRIDOXINE ON SERUM CHOLESTEROL, *ANN INTERN MED*, 88, 53-56 (1978)

2273 MAKI DG, AUGHEY DR, COMPARATIVE STUDY OF CEFAZOLIN, CEFOXITIN AND CEFTIZOXINE FOR SURGICAL PROPHYLAXIS IN COLORECTAL SURGERY, *CHEMOTHERAPY*, 10 SUPPL C, 281-287 (1982)

2274 MALARKEY WB, JACOBS LS, DAUGHADAY WH, LEVODOPA SUPPRESSION OF PROLACTIN IN NONPUERPERAL GALACTORRHEA, *N ENGL J MED*, 285, 1160 (1971)

2275 MALETZKY B, BLACHLY PH, THE USE OF LITHIUM IN PSYCHIATRY, *CRIT REV CLIN LAB SCI*, 2, 279 (1971)

2276 MALHERBE C, BURRILL KC, LEVIN SR, KARAM JH, FORSHAM PH, EFFECT OF DIPHENYLHYDANTOIN ON INSULIN SECRETION IN MAN, *N ENGL J MED*, 286, 339 (1972)

2277 MALINI PL, STROCCHI E, AMBRIOSINI E, COMPARISON OF THE ANTIHYPERTENSIVE, METABOLIC AND CELLULAR EFFECTS OF ENALAPRIL, *J HYPERTENS*, 2, SUPPL, S101-S105 (1984)

2278 MALISZEWSKI TF, BASS DE, 'TRUE AND APPARENT' THIOCYANATE IN BODY FLUIDS OF SMOKERS AND NONSMOKERS, *J APPL PHYSIOL*, 8, 289-291 (1955)

2279 MALKONEN M, MANNINEN V, HIRVONEN E, EFFECTS OF DANAZOL AND LYNESTRENOL ON SERUM LIPOPROTEINS IN ENDOMETRIOSIS, *CLIN PHARMACOL THER*, 28, 602-604 (1980)

2280 MALL-HAEFELI M, DARRAGH A, WERNER-ZODROW I, EFFECTS OF VARIOUS COMBINED ORAL CONTRACEPTIVES ON SEX STEROIDS, GONADOTROPINS AND SHBG, *IR MED J*, 76, 266-272 (1983)

2281 MALLORY A, KERN F JR, DRUG-INDUCED PANCREATITIS: A CRITICAL REVIEW, *GASTROENTEROLOGY*, 78, 813-820 (1980)

2282 MALMENDIER CL, DELCROIX C, EFFECT OF ALCOHOL INTAKE ON HIGH AND LOW DENSITY LIPOPROTEIN METABOLISM IN HEALTHY VOLUNTEERS, *CLIN CHIM ACTA*, 152, 281-288 (1985)

2283 MANCINI AM, BARONTINI M, ARMANDO I, ET AL, EFFECT OF BROMOCRIPTINE ON PLASMA CATECHOLAMINES IN NORMAL SUBJECTS AND PROLACTIN-SECRETING TUMOR PATIENTS, *J ENDOCRINOL INVEST*, 9, 223-226 (1986)

2284 MANFREDI RL, VESSELL ES, INHIBITION OF THEOPHYLLINE METABOLISM BY LONG-TERM ALLOPURINOL ADMINISTRATION, *CLIN PHARMACOL THER*, 29, 224-229 (1981)

2285 MANGER WM, STEINSLAND OS, NAHAS GG, WAKIM KG, DUFTON S, COMPARISON OF IMPROVED FLUOROMETRIC METHODS USED TO QUANTITATE PLASMA CATECHOLAMINES, *CLIN CHEM*, 15, 1101 (1969)

2286 MANN HJ, SCHNEIDER JR, MILLER JB, ET AL, CIMETIDINE-ASSOCIATED THROMBOCYTOPENIA, *DRUG INTELL CLIN PHARM*, 17, 126-128 (1983)

2287 MANN S, GREEN A, MORE ON HEPARIN EFFECT ON SODIUM MEASUREMENT, *CLIN CHEM*, 32, 907 (1986)

2288 MANNELLI M, GHERI RG, DE FEO ML, ET AL, EFFECTS OF THYROID REPLACEMENT THERAPY ON CATECHOLAMINE PLASMA LEVELS, *J ENDOCRINOL INVEST*, 6, 307-309 (1983)

2289 MANNER T, KANTO J, SCHEININ H, ET AL, MEPTAZINOL AND PENTAZOCINE: PLASMA CATECHOLAMINES AND OTHER EFFECTS IN HEALTHY VOLUNTEERS, *BR J CLIN PHARMACOL*, 24, 689-697 (1987)

2290 MANNINEN V, APAJALAHTI A, MELIN J, KARESOJA M, ALTERED ABSORPTION OF DIGOXIN IN PATIENTS GIVEN PROPANTHELINE AND METOCLOPRAMIDE, *LANCET*, 1, 398 (1973)

2291 MANOLAS EG, HUNT D, SLOMAN G, EFFECTS OF QUINIDINE AND DISOPYRAMIDE ON SERUM DIGOXIN CONCENTRATIONS, *AUST N Z J MED*, 10, 426-429 (1980)

2292 MANUFACTURERS LITERATURE ON TELEPAQUE, *MEDICAL DEPARTMENT, WINTHROP LABORATORIES, 90 PARK AVENUE, NEW YORK, NY 10016*

2293 MARAGNO I, SANTOSTASI G, GAION RM, ET AL, INFLUENCE OF AMIODARONE ON ORAL DIGOXIN BIOAVAILABILITY IN HEALTHY VOLUNTEERS, *INT J CLIN PHARMACOL RES*, 4, 149-153 (1984)

2294 MARBLE A, PHARMACOLOGY OF ANTIHYPERGLYCEMIC DRUGS, *N Y STATE J MED*, 72, 2174 (1972)

2295 MARCEL YL, NOEL SP, A PLASTICIZER IN LIPID EXTRACTS OF HUMAN BLOOD, *CHEM PHYS LIPIDS*, 4, 418 (1970)

2296 MARCO J, CALLE C, ROMAN D, DIAZ-FIERROS M, VILLANUEVA ML, VALVERDE I, HYPERGLUCAGONISM INDUCED BY GLUCOCORTICOID TREATMENT IN MAN, *N ENGL J MED*, 288, 128 (1973)

2297 MARCROFT J, NEWLANDS IM, EVALUATION OF A SPECIFIC PROTEIN ANALYZER, *CLIN CHIM ACTA*, 46, 399 (1973)

2298 MARCUS FI, DRUG INTERACTIONS WITH AMIODARONE, *AM HEART J*, 106, 924-930 (1983)

2299 MARCUS J, MULVIHILL FJ, AGRANULOCYTOSIS AND CHLORPROMAZINE, *J CLIN PSYCHIATRY*, 39, 784-786 (1978)

2300 MARCUS M, KLEINBERG S, FILTER PAPER CONTAMINATION: TIN AND CALCIUM, *CLIN CHEM*, 18, 492 (1972)

2301 MARESCAUX C, WARTER JM, BRANDT C, ET AL, ADAPTATION OF HEPATIC AMMONIA METABOLISM AFTER CHRONIC VALPROATE ADMINISTRATION IN EPILEPTICS TREATED WITH PHENYTOIN, *EUR NEUROL*, 24, 191-195 (1985)

2302 MARKLOVA E, HAIS IM, A METHOD FOR THE ESTIMATION OF INDOLYLACRYLOYLGLYCINE IN URINE, *CLIN CHIM ACTA*, 40, 455 (1972)

2303 MARKS V, KELLY JF, ABSORPTION OF CAFFEINE FROM TEA, COFFEE AND COCA COLA, *LANCET*, 1, 827 (1973)

2304 MAROSVARI I, NEUBAUER K, TREATMENT OF NEONATAL HYPERBILIRUBINEMIA WITH FLUMECINOLONE, A NEW ENZYME-INDUCING DRUG, *ACTA PAEDIATR HUNG*, 26, 297-302 (1985)

2305 MARQUIS J-F, CARRUTHER SG, SPENCE JD, ET AL, PHENYTOIN-THEOPHYLLINE INTERACTION, *N ENGL J MED*, 307, 1189-1190 (1982)

2306 MARS H, GENUTH SM, POTENTIATION OF LEVODOPA STIMULATION OF HUMAN GROWTH HORMONE BY SYSTEMIC DECARBOXYLASE INHIBITION, *CLIN PHARMACOL THER*, 14, 390 (1973)

2307 MARSDEN JR, TRINICK TR, LAKER MF, ET AL, EFFECT OF ISOTRETINOIN ON SERUM LIPIDS AND LIPOPROTEINS, LIVER AND THYROID FUNCTION, *CLIN CHIM ACTA*, 143, 243-245 (1984)

2308 MARSH HH, BASIC URINE ANALYSIS: PHYSICO-CHEMICAL TESTS, *AUDIOVISUAL SEMINAR ASCP*, 1973

2309 MARSHALL EF, TAIT AC, TODRICK A, THE EFFECT OF RESERPINE ON THE BLOOD SUGAR LEVEL IN HUMAN SUBJECTS, *BIOCHEM J*, 69, 41 (1958)

2310 MARSHALL RW, COCHRAN M, HODGKINSON A, RELATIONSHIPS BETWEEN CALCIUM AND OXALIC ACID INTAKE IN THE DIET, *CLIN SCI*, 43, 91 (1972)

2311 MARSHALL RW, SELBY PL, CHILVERS DC, ET AL, THE EFFECT OF ETHINYL OESTRADIOL ON CALCIUM AND BONE METABOLISM IN PERI- AND POST MENOPAUSAL WOMEN, *HORM METAB RES*, 16, 97-99 (1984)

2312 MARTIN CM, MYELOGRAPHY WITH SODIUM DIATRIZOATE (HYPAQUE®), *CALIF MED*, 115, 57 (1971)

2313 MARTIN EW, *HAZARDS OF MEDICATION, A MANUAL ON DRUG INTERACTIONS, INCOMPATIBILITIES, CONTRAINDICATIONS AND ADVERSE EFFECTS*, LIPPINCOTT, PHILADELPHIA, PA (1971)

2314 MARTIN HE, CLINICAL STUDIES OF MAGNESIUM METABOLISM, *MED CLIN NORTH AM*, 36, 1157 (1952)

2315 MARTIN HE, JONES R, THE EFFECT OF AMMONIUM CHLORIDE AND SODIUM BICARBONATE ON URINARY EXCRETION OF MAGNESIUM, CALCIUM, AND PHOSPHATE, *AM HEART J*, 62, 206 (1961)

2316 MARTIN JH, GORDON M, WALLACE R, METHOTREXATE IN PSORIASIS. PRECIPITATION OF GOUT, *ARCH DERMATOL*, 96, 431 (1967)

2317 MARTIN JV, MARTIN PJ, GOLDBERG DM, ENZYME INDUCTION AS A POSSIBLE CAUSE OF INCREASED SERUM TRIGLYCERIDES AFTER ORAL CONTRACEPTIVES, *LANCET*, 1, 1107-1108 (1976)

2318 MARTIN WE, ADVERSE REACTIONS DURING TREATMENT OF PARKINSON'S DISEASE WITH LEVODOPA, *J AM MED ASSOC*, 216, 1979 (1971)

2319 MARTINEK RG, IMPROVED SERUM CHOLESTEROL PROCEDURE USING A MODIFIED PEARSON REACTION, *AM J MED TECHNOL*, 34, 32 (1972)

2320 MARUHN D, GIELOW L, STROZYK K, BOCK KD, CIRCADIAN VARIATION OF URINARY ENZYME EXCRETION, *Z KLIN CHEM KLIN BIOCHEM*, 12, 270 (1974)

2321 MARX HH, EMPHYSEMATOUS BRONCHITIS, *GERM MED*, 1, 162 (1971)

2322 MASALA A, ALAGNA S, ROVASIO PP, ET AL, SUPPRESSION OF INSULIN-INDUCED PROLACTIN SECRETION IN MAN BY NOMIFENSINE, *DRUG DEV RES*, 1, 51-54 (1981)

2323 MASCIOLI SR, BANTLE JP, FREIER EF, ET AL, ARTIFACTUAL ELEVATION OF SERUM CREATININE LEVEL DUE TO FASTING, *ARCH INTERN MED*, 144, 1575-1576 (1984)

2324 MASEL MA, ERYTHROMYCIN HEPATO-SENSITIVITY: A PRELIMINARY REPORT OF TWO CASES, *MED J AUST*, 1, 560 (1962)

2325 MASEL MA, A LUPUS-LIKE REACTION TO ANTITUBERCULOSIS DRUGS, *MED J AUST*, 2, 738 (1967)

2326 MASON AM, BLEEDING MASSIVE GASTRIC ULCER ON DIFLUNISAL (DOLOBID®), *BR MED J [CLIN RES]*, 1, 888 (1979)

2327 MASON JW, DRUG THERAPY: AMIODARONE, *N ENGL J MED*, 316, 455-466 (1987)

2328 MASSARA F, CAGLIERO E, BISBOCCI D, ET AL, THE RISK OF PRONOUNCED HYPERKALAEMIA AFTER ARGININE INFUSION IN THE DIABETIC SUBJECT, *DIABETES METAB REV*, 7, 149-153 (1981)

2329 MASSARA F, CAMANNI F, MOLINATTI GM, THE EFFECT OF VARIOUS ADRENERGIC SUBSTANCES ON PLASMA HGH LEVELS, *ABSTRACTS IV INTL CONGR ENDOCRINOL*, WASHINGTON, DC, 30 (1972)

2330 MASSARA F, TANGOLO D, GODANO A, ET AL, EFFECT OF METOCLOPRAMIDE, DOMPERIDONE, AND APOMORPHINE ON GH SECRETION IN CHILDREN AND ADOLESCENTS, *ACTA ENDOCRINOL*, 108, 451-455 (1985)

2331 MASSIE B, MACCARTHY EP, RAMANATHAN KB, ET AL, DILTIAZEM AND PROPRANOLOL IN MILD TO MODERATE ESSENTIAL HYPERTENSION AS MONOTHERAPY OR WITH HYDROCHLOROTHIAZIDE, *ANN INTERN MED*, 107, 150-157 (1987)

2332 MASSION CG, FRANKENFELD JK, ALKALINE PHOSPHATASE: LABILITY IN FRESH AND FROZEN HUMAN SERUM AND IN LYOPHILIZED CONTROL MATERIALS, *CLIN CHEM*, 18, 366 (1972)

2333 MATHEKE ML, KESSLER G, CHAN K-M, INTERFERENCE OF THE CHEMOTHERAPEUTIC AGENT ETOPOSIDE WITH THE DIRECT PHOSPHOTUNGSTIC METHOD FOR URIC ACID, *CLIN CHEM*, 33, 2109-2110 (1987)

2334 MATHER A, MACKIE NR, EFFECTS OF HEMOLYSIS ON SERUM ELECTROLYTES VALUES, *CLIN CHEM*, 6, 223 (1960)

2335 MATHEWS JD, HOOPER BM, WHITTINGHAM S, MACKAY IR, STENHAUSE NS, ASSOCIATION OF AUTOANTIBODIES WITH SMOKING, CARDIOVASCULAR MORBIDITY, AND DEATH IN THE BUSSELTON POPULATION, *LANCET*, 2, 754 (1973)

2336 MATI JKG, KAHUHO S, HORROBIN DF, INDOCYANINE GREEN CLEARANCE TEST FOR SCREENING LIVER FUNCTION IN NORMAL AND PRE-ECLAMPTIC WOMAN, *EAST AFR MED J*, 50, 132 (1973)

2337 MATTHYSSE S, LIPINSKI J, SHIH V, L-DOPA AND S-ADENOSYLMETHIONINE, *CLIN CHIM ACTA*, 35, 253 (1971)

2338 MATULA G, RAAS M, LAUTER CB, EL KHATIB MR, RISING J, ROBIN E, NITROBLUE TETRAZOLIUM DYE TEST, *ANN INTERN MED*, 79, 758 (1973)

2339 MATZKE GR, LUCAROTTI RL, SHAPIRO HS, CONTROLLED COMPARISON OF GENTAMICIN AND TOBRAMYCIN NEPHROTOXICITY, *AM J NEPHROL*, 3, 11-17 (1983)

2340 MAY T, RAMBECK B, SERUM CONCENTRATIONS OF VALPROIC ACID: INFLUENCE OF DOSE AND COMEDICATION, *THER DRUG MONIT*, 7, 387-390 (1985)

2341 MAYFIELD D, BROWN RG, THE CLINICAL LABORATORY AND ELECTROENCEPHALOGRAPHIC EFFECTS OF LITHIUM, *J PSYCHIATR RES*, 4, 207 (1966)

2342 MAYS DC, CAMISA C, CHENEY P, ET AL, METHOXSALEN IS A POTENT INHIBITOR OF THE METABOLISM OF CAFFEINE IN HUMANS, *CLIN PHARMACOL THER*, 42, 621-626 (1987)

2343 MAZONSON PD, WILLIAMS ML, CANTLEY LK, ET AL, MYXEDEMA COMA DURING LONG-TERM AMIODARONE THERAPY, *AM J MED*, 77, 751-754 (1984)

2344 MAZZACHI BC, TEUBNER JK, RYALL RL, FACTORS AFFECTING MEASUREMENT OF URINARY OXALATE, *CLIN CHEM*, 30, 1339-1343 (1984)

2345 MBUYI-MUAMBA JM, DEQUEKER J, A COMPARATIVE TRIAL OF TIMEGADINE AND D-PENICILLAMINE IN RHEUMATOID ARTHRITIS, *CLIN RHEUMATOL*, 2, 369-374 (1983)

2346 MCALLISTER RA, INHIBITION OF THE CYSTEINE-SULFURIC ACID REACTION FOR SEDOHEPTULOSE BY FERRIC IRON, *ANAL BIOCHEM*, 43, 647 (1971)

2347 MCALLISTER WAC, THOMPSON PJ, AL-HABET SM, ET AL, RIFAMPICIN REDUCES EFFECTIVENESS AND BIOAVAILABILITY OF PREDNISOLONE, *BR MED J [CLIN RES]*, 286, 923-925 (1983)

2348 MCBAY AJ, TOXICOLOGICAL FINDINGS IN FATAL POISONINGS, *CLIN CHEM*, 19, 361 (1973)

2349 MCBRIDE WG, ORAL CONTRACEPTIVES, *MED J AUST*, 1, 525 (1965)

2350 MCCAA RE, MCCAA CS, READ DG, BOWER ID, GUYTON AC, INCREASED PLASMA ALDOSTERONE CONCENTRATION IN RESPONSE TO HEMODIALYSIS IN NEPHRECTOMIZED MAN, *CIRC RES*, 31, 473 (1972)

2351 MCCLOY RF, BARON JH, INTRAGASTRIC PH AND CIMETIDINE, FASTING AND AFTER FOOD, *LANCET*, 1, 609-610 (1981)

2352 MCCREADY DR, CLARK L, COHEN MM, CIGARETTE SMOKING REDUCES HUMAN GASTRIC LUMINAL PROSTAGLANDIN E_2, *GUT*, 26, 1192-1196 (1985)

2353 MCCUISTION CH, LAWLIS M, GONZALES BB, HUMAN PHARMACOLOGICAL STUDIES WITH GRISEOFULVIN, *ARCH DERMATOL*, 81, 766 (1960)

2354 MCCURDY PR, MAHMOOD L, INTRAVENOUS UREA TREATMENT OF THE PAINFUL CRISIS OF SICKLE CELL DISEASE: A PRELIMINARY REPORT, *N ENGL J MED*, 285, 992 (1971)

2355 MCDONALD FD, RENAL DYSFUNCTION WITH METHOXYFLURANE, *J AM MED ASSOC*, 217, 79 (1971)

2356 MCDONALD RK, WEISE VK, THE EFFECT OF CERTAIN PSYCHOTROPIC DRUGS ON THE URINARY EXCRETION OF 3-METHOXY-4-HYDROXYMANDELIC ACID IN MAN, *J PHARMACOL EXP THER*, 136, 26 (1962)

2357 MCDONALD WJ, COHEN EL, LUCAS CP, CONN JW, VARIABILITY IN HUMAN PLASMA OF THE RENIN (R) -RENIN SUBSTRATE (RS) REACTION BOTH BEFORE AND AFTER ESTROGEN, *CLIN RES*, 20, 726 (1972)

2358 MCDUFFIE FC, BONE MARROW DEPRESSION AFTER DRUG THERAPY IN PATIENTS WITH SYSTEMIC LUPUS ERYTHEMATOSUS, *ANN RHEUM DIS*, 24, 289 (1965)

2359 MCEWEN J, PATERSON C, DRUGS AND FALSE-POSITIVE SCREENING TESTS FOR PORPHYRIA, *BR MED J [CLIN RES]*, 1, 421 (1972)

2360 MCFARLAND KF, CARR AA, CHANGES IN THE FASTING BLOOD SUGAR AFTER HYDROCHLOROTHIAZIDE AND POTASSIUM SUPPLEMENTATION, *J CLIN PHARMACOL*, 17, 13-17 (1977)

2361 MCFATE SMITH W, KULAGA SF, MONCLOA F, ET AL, OVERALL TOLERANCE AND SAFETY OF ENALAPRIL, *J HYPERTENS*, 2, SUPPL, S113-S117 (1984)

2362 MCGANN CJ, CARTER RE, THE EFFECT OF HEMOLYSIS ON THE VAN DEN BERGH REACTION FOR SERUM BILIRUBIN, *J PEDIATR*, 57, 199 (1960)

2363 MCGARRY JM, ANDREWS J, SMOKING IN PREGNANCY AND VITAMIN B_{12} METABOLISM, *BR MED J [CLIN RES]*, 2, 74 (1972)

2364 MCGEACHIN RL, DAUGHERTY HK, HARGAN LA, POTTER BA, THE EFFECT OF BLOOD ANTICOAGULANTS ON SERUM AND PLASMA AMYLASE ACTIVITIES, *CLIN CHIM ACTA*, 2, 75 (1957)

2365 MCGINNISS M, *PERSONAL COMMUNICATION*

2366 MCGOVERN B, GEER VR, LARAIA PJ, ET AL, POSSIBLE INTERACTION BETWEEN AMIODARONE AND PHENYTOIN, *ANN INTERN MED*, 101, 650-651 (1984)

2367 MCGREGOR RF, PHILLIPS G, ROMSDAHL MM, ANALYSIS OF URINARY 5-HYDROXYINDOLEACETIC ACID BY THIN LAYER CHROMATOGRAPHY: ELIMINATION OF INTERFERENCE, *CLIN CHIM ACTA*, 40, 59 (1972)

2368 MCGUIGAN JE, A CONSIDERATION OF THE ADVERSE EFFECTS OF CIMETIDINE, *GASTROENTEROLOGY*, 80, 181-192 (1981)

2369 MCGUIGAN JE, TRUDEAU WL, DIFFERENCES IN RATES OF GASTRIN RELEASE IN NORMAL PERSONS AND PATIENTS WITH DUODENAL ULCER DISEASE, *N ENGL J MED*, 288, 64 (1973)

2370 MCGUIRE A, COHEN S, BROOKS FP, SERUM GASTRIN AND CALCIUM LEVELS IN MAN DURING THE INFUSION OF SYNTHETIC HUMAN GASTRIN, *CLIN RES*, 20, 734 (1972)

2371 MCINDOE GJ, MENZIES KW, REDDY J, SULINDAC (CLINORIL®) AND CHOLESTATIC JAUNDICE, *N Z MED J*, 2, 430-431 (1981)

2372 MCINTYRE PA, HYPOKALEMIA OCCURRING DURING PARA-AMI-NOSALICYLIC ACID THERAPY, *BULL JOHNS HOPKINS HOSP*, 92, 210 (1953)

2373 MCKEAN CM, THE EFFECTS OF HIGH PHENYLALANINE CONCEN-TRATIONS ON SEROTONIN AND CATECHOLAMINE METABOLISM IN THE HUMAN BRAIN, *BRAIN RES*, 47, 469 (1972)

2374 MCKELVIE GM, BAYLIFF CD, GASKA JA, ET AL, ADVERSE REACTION REVIEWS: DRUG-INDUCED LIVER DISEASES: PART 3; ILLUSTRA-TIVE CASES, *HOSP PHARM*, 17, 562-568 (1982)

2375 MCKINNEY SE, PECK HM, BOCHEY L, BYHAM B, BEYER KH, SCHUCHARDT GS, BENEMID®, P(DI-N-PROPYLSULFAMYL)-BEN-ZOIC ACID, *J PHARMACOL EXP THER*, 102, 208 (1951)

2376 MCLACHLAN S, PEGG CAS, ATHERTON MC, ET AL, THE EFFECT OF CARBIMAZOLE ON THYROID AUTOANTIBODY SYNTHESIS BY THYROID LYMPHOCYTES, *J CLIN ENDOCRINOL METAB*, 60, 1237-1242 (1985)

2377 MCLAUGHLIN M, STOPPS GJ, SMOKING AND LEAD, *ARCH ENVIRON HEALTH*, 26, 131 (1973)

2378 MCLEAN AS, EARLY ADVERSE EFFECTS OF RADIATION, *BR MED BULL*, 29, 69 (1973)

2379 MCLELLAN DL, CHALMERS RJ, JOHNSON RH, A DOUBLE-BLIND TRIAL OF TETRABENAZINE, THIOPROPAZATE AND PLACEBO IN PATIENTS WITH CHOREA, *LANCET*, 1, 104 (1974)

2380 MCLEROY VJ, SCHENDEL HE, INFLUENCE OF ORAL CONTRACEP-TIVES ON ASCORBIC ACID CONCENTRATIONS IN HEALTHY, SEXUALLY MATURE WOMEN, *AM J CLIN NUTR*, 26, 191 (1973)

2381 MCNAIR RD, VITAMIN INTERFERENCE WITH BSP-RETENTION TEST, *CLIN CHEM*, 18, 1039 (1972)

2382 MCNEW JJ, SABBOT IM, HOSHIZAKI T, MANDELL AJ, SPOONER CE, MARCUS I, ADEY WR, URINARY EXCRETION VALUES IN 2-DAY FOOD-DEPRIVED, UNRESTRAINED CHIMPANZEES, *AM J PHYSIOL*, 222, 640 (1970)

2383 MCPHERSON ML, PRINCE SR, ATAMER ER, ET AL, THEOPHYLLINE-INDUCED HYPERCALCEMIA, *ANN INTERN MED*, 105, 52-54 (1986)

2384 MCPHERSON RA, BROWN KD, AGARWAL RP, ET AL, HYDROXYUREA INTERFERES NEGATIVELY WITH TRIGLYCERIDE MEASUREMENT BY A GLYCEROL OXIDASE METHOD, *CLIN CHEM*, 31, 1355-1357 (1985)

2385 MCQUEEN EG, HORMONAL STEROID CONTRACEPTIVES. IV. ADVERSE REACTIONS AND MANAGEMENT OF THE PATIENT, *DRUGS*, 2, 138 (1971)

2386 MEANI A, CARTEI G, VARIATIONS OF TRYPTOPHAN:NICOTINIC ACID METABOLISM AND BLOOD LEVELS OF CERULOPLASMIN AND TRANSFERRIN, *ACTA VITAMINOL ENZYMOL*, 24, 231 (1970)

2387 MEANS JH, DE GROOT LJ, STANBURY JB, *THE THYROID AND ITS DISEASES* 3RD ED, MCGRAW-HILL, NEW YORK, NY (1963)

2388 MEATHERALL RC, GUAY DRP, BAXTER H, CEPHALOSPORINS AND URINARY PROTEIN DETERMINATION, *CLIN CHEM*, 31, 165 (1985)

2389 MEDLINE A, COHEN LB, TOBE BA, ET AL, LIVER GRANULOMAS AND ALLOPURINOL, *BR MED J [CLIN RES]*, 1, 1320-1321 (1978)

2390 MEFFIN PJ, BROOKS PM, BERTOUCH J, ET AL, DIFLUNISAL DISPOSI-TION AND HYPOURICEMIC RESPONSE IN OSTEOARTHRITIS, *CLIN PHARMACOL THER*, 33, 813-821 (1983)

2391 MEGRAW RE, HRITZ AM, BABSON AL, CARROLL JJ, A SINGLE TUBE TECHNIQUE FOR SERUM TOTAL IRON AND TOTAL IRON BIND-ING CAPACITY, *CLIN BIOCHEM*, 6, 266 (1973)

2392 MEHBOD H, SWARTZ CD, BREST AN, THE EFFECT OF PROLONGED THIAZIDE ADMINISTRATION ON THIAZIDE FUNCTION, *ARCH INTERN MED*, 119, 283 (1967)

2393 MEHTA BR, ROBINSON BHB, LITHIUM TOXICITY INDUCED BY TRIAMTERENE HYDROCHLOROTHIAZIDE, *POSTGRAD MED J*, 56, 783-784 (1980)

2394 MEHTA S, THE INFLUENCE OF PREMEDICATION WITH DIAZEPAM ON BLOOD SUGAR LEVEL, *ANESTHESIA*, 26, 468 (1971)

2395 MEITES S, CALCIUM (FLUOROMETRIC), *STAND METH CLIN CHEM*, 6, 207 (1970)

2396 MEJER LE, BLANCHARD RC, FLUOROMETRIC DETERMINATION OF PLASMA 11-HYDROXYCORTICOSTEROIDS, *CLIN CHEM*, 19, 710 (1973)

2397 MELBY J, MINERALOCORTICOID ASSAYS, *RADIOIMMUNOASSAY SYMPOSIUM*, WASHINGTON, DC (1974)

2398 MELIKIAN V, EDDY JD, PATON A, THE STIMULANT EFFECT OF DRUGS ON INDOCYANINE GREEN CLEARANCE BY THE LIVER, *GUT*, 13, 755 (1972)

2399 MELIS GB, FRUZZETTI F, PAOLETTI AM, ET AL, FIBRINOPEPTIDE A PLASMA LEVELS DURING LOW ESTROGEN ORAL CONTRACEP-TIVE TREATMENT, *CONTRACEPTION*, 30, 575-583 (1984)

2400 MELIS GB, PAOLETTI AM, MAIS V, ET AL, THE EFFECTS OF THE GABAERGIC DRUG, SODIUM VALPROATE, ON PROLACTIN SECRETION IN NORMAL AND HYPERPROLACTINEMIC SUB-JECTS, *J CLIN ENDOCRINOL METAB*, 54, 485-489 (1982)

2401 MELLERUP ET, LAURITSEN B, DAM H, ET AL, LITHIUM EFFECTS ON DIURNAL RHYTHM OF CALCIUM, MAGNESIUM, AND PHOSPHATE METABOLISM IN MANIC-MELANCHOLIC DISORDER, *ACTA PSYCHIATR SCAND*, 53, 360-370 (1976)

2402 MELTZER HY, FACTORS AFFECTING SERUM CREATINE PHOSPHOKINASE LEVELS IN THE GENERAL POPULATION. THE ROLE OF RACE, ACTIVITY, *CLIN CHIM ACTA*, 33, 165 (1971)

2403 MELTZER HY, INTRAMUSCULAR CHLORPROMAZINE AND CREATINE KINASE: ACUTE PSYCHOSIS OR LOCAL MUSCLE TRAUMA?, *SCI-ENCE*, 164, 726 (1969)

2404 MELTZER HY, FANG VS, TRICOU BJ, ET AL, EFFECT OF DEX-AMETHASONE ON PLASMA PROLACTIN AND CORTISOL LEVELS IN PSYCHIATRIC PATIENTS, *AM J PSYCHIATRY*, 139, 763-768 (1982)

2405 MENABAWEY M, SILMAN R, RICE A, ET AL, DRAMATIC INCREASE OF PLACENTAL PROTEIN 5 LEVELS FOLLOWING INJECTION OF SMALL DOSES OF HEPARIN, *BR J OBSTET GYNAECOL*, 92, 207-210 (1985)

2406 MENARD DB, GISSELBRECHT C, MARTY M, ET AL, ANTINEOPLASTIC AGENTS AND THE LIVER, *GASTROENTEROLOGY*, 78, 142-164 (1980)

2407 MENCZEL J, DREYFUSS F, EFFECT OF PREDNISONE ON BLOOD COAGULATION TIME IN PATIENTS ON DICUMAROL THERAPY, *J LAB CLIN MED*, 56, 14 (1960)

2408 MENDEL CM, FROST PH, KUNITAKE ST, ET AL, MECHANISM OF THE HEPARIN-INDUCED INCREASE IN THE CONCENTRATION OF FREE THYROXINE IN PLASMA, *J CLIN ENDOCRINOL METAB*, 65, 1259-1264 (1987)

2409 MENDELSON JH, ELLINGBOE J, JODSON BA, ET AL, PLASMA TES-TOSTERONE AND LUTEINIZING HORMONE LEVELS DURING LEVO-ALPHA-ACETYLMETHADOL MAINTENANCE AND WITH-DRAWAL, *CLIN PHARMACOL THER*, 35, 545-547 (1984)

2410 MENEZES S, REGE VIL, SEHGAL VN, DAPSONE HAEMOLYSIS IN LEP-ROSY, *LEPROSY IN INDIA*, 53, 63-69 (1981)

2411 MENON RK, MIKHALIDIS DP, BELL JL, ET AL, WARFARIN ADMINIS-TRATION INCREASES URIC ACID CONCENTRATIONS IN PLASMA, *CLIN CHEM*, 32, 1557-1559 (1986)

2412 MERCK SHARP AND DOHME, MANUFACTURER'S LITERATURE ON INDOMETHACIN, *DIVISION OF MERCK AND CO, WEST POINT, PA 19486*

2413 MERCK SHARP AND DOHME, LITERATURE CONCERNING MINTEZOL®, (THIABENDAZOLE, MSD), *DIVISION OF MERCK AND CO, WEST POINT, PA 19486*

2414 MERIMEE TJ, FINEBERG SE, STUDIES OF THE SEX-BASED VARIA-TION OF HUMAN GROWTH HORMONE SECRETION, *J CLIN ENDOCRINOL METAB*, 33, 896 (1971)

2415 MERRELL-NATIONAL LABORATORIES, DIVISION OF RICHARDSON MERRELL, MANUFACTURER'S LITERATURE ON TENUATE, *CIN-CINNATI, OH 45215*

2416 MERRIEL SL, DAVIS A, SMOLENS B, FINGOLD SM, CEPHALOTHIN IN SERIOUS BACTERIAL INFECTION, *ANN INTERN MED*, 64, 1 (1966)

2417 MERRIT GJ, HUNTER BH, HALL WC, LACK OF MEZLOCILLIN AND PIPERACILLIN INTERFERENCE IN MEASUREMENT OF VANCO-MYCIN IN THE ABBOTT TDx, *CLIN CHEM*, 33, 2304 (1987)

2418 MERRY J, LITHIUM THYROTOXICOSIS, *BR MED J [CLIN RES]*, 2, 765 (1977)

2419 MESTMAN JH, POCOCK DS, KIRCHNER A, LACTIC ACIDOSIS WITH RECOVERY IN DIABETES MELLITUS ON PHENFORMIN THERAPY, *CALIF MED*, 111, 181 (1969)

2420 METCALF MG, RAPID GAS CHROMATOGRAPHIC ASSAY FOR DEHYDROEPIANDROSTERONE IN URINE, *CLIN BIOCHEM*, 4, 241 (1971)

2421 MEYBOOM RHB, THROMBOCYTOPENIA INDUCED BY NALIDIXIC ACID, *BR MED J [CLIN RES]*, 289, 962 (1984)

2422 MEYER FL COOPER K, BOLICK M, NITROGEN AND MINERAL EXCRE-TION AFTER CARBOHYDRATE TEST MEALS, *AM J CLIN NUTR*, 25, 677 (1972)

2423 MEYER TE, KASSIANIDES C, BOTHWELL TH, GREEN A, EFFECTS OF HEAVY ALCOHOL CONSUMPTION ON SERUM FERRITIN CON-CENTRATIONS, *S AFR MED J*, 66, 573-575 (1984)

2424 MEYERS BR, KAPLAN K, WEINSTEIN L, CEPHALEXIN: MICROBIOLOGI-CAL EFFECTS AND PHARMACOLOGIC PARAMETERS IN MAN, *CLIN PHARMACOL THER*, 10, 810 (1969)

2425 MEYERS FH, JAWETZ E, GOLDFIEN A, *REVIEW OF MEDICAL PHAR-MACOLOGY*, LANGE, LOS ALTOS, CALIF (1968)

2426 MEYLER L, HERXHEIMER A (EDS), *SIDE EFFECTS OF DRUGS*, EXCERPTA MEDICA FOUNDATION, MOUTON AND CO, THE HAGUE, 5 (1966)

2427 MEYLER L, HERXHEIMER A (EDS), *SIDE EFFECTS OF DRUGS*, EXCERPTA MEDICA FOUNDATION, MOUTON AND CO, THE HAGUE, 6 (1968)

2428 MEZEY E, METABOLIC EFFECTS OF ALCOHOL, *FED PROC*, 44, 134-138 (1985)

2429 MIALE JB, *LABORATORY MEDICINE - HEMATOLOGY*, 2nd ED, CV MOSBY, ST. LOUIS, MO (1962)

2430 MIANO L, KOLLOCH R, DE QUATTRO V, INCREASED CATECHO-LAMINE EXCRETION AFTER LABETALOL THERAPY: A SPURIOUS EFFECT OF DRUG METABOLITES, *CLIN CHIM ACTA*, 95, 211-217 (1979)

2431 MICHELSON PA, REVERSIBLE HIGH DOSE OXACILLIN-ASSOCIATED LIVER INJURY, *CANAD J HOSP PHARM*, 34, 83-85 (1981)

2432 MICHOTTE LJ, WAUTERS M, CLINICAL TEST OF INDOMETHACIN, *RHEUMATOL SCAND*, 10, 273 (1964)

2433 MICOSSI P, PONTIROLI AE, BARON SH, ET AL, ASPIRIN STIMULATES INSULIN AND GLUCAGON SECRETION AND INCREASES GLUCOSE TOLERANCE IN NORMAL AND DIABETIC SUBJECTS, *DIABETES*, 27, 1196-1204 (1978)

2434 MIELKE CH JR., ASPIRIN PROLONGATION OF THE TEMPLATE BLEEDING TIME: INFLUENCE OF VENOSTASIS AND DIRECTION OF INCISION, *BLOOD*, 60, 1139-1142 (1982)

2435 MIGEON CJ, *CIBA FOUNDATION COLLOQUIA ON ENDOCRINOLOGY*, CHURCHILL, LONDON, 8 (1954)

2436 MIGNAULT G, LABERQUE B, HAMEL S, METHOXYFLURANE AND NEPHROTOXICITY: A STUDY OF RENAL FUNCTION IN 22 PATIENTS ANESTHETIZED WITH METHOXYFLURANE, *CAN J ANAESTH*, 17, 331 (1970)

2437 MIHAS AA, HOLLEY P, KOFF RS, ET AL, FULMINANT HEPATITIS AND LYMPHOCYTE SENSITIZATION DUE TO PROPYLTHIOURACIL, *GASTROENTEROLOGY*, 70, 770-774 (1976)

2438 MIKAC-DEVIC D, STANKOVIC H, BOSKOVIC K, A METHOD FOR THE DETERMINATION OF FREE FATTY ACIDS IN SERUM, *CLIN CHIM ACTA*, 45, 55 (1973)

2439 MIKKELSEN WM, DODGE HJ, VALKENBURG H, THE DISTRIBUTION OF SERUM URIC ACID VALUES IN A POPULATION UNSELECTED AS TO GOUT OR HYPERURICEMIA, *AM J MED*, 39, 242 (1965)

2440 MILLAR JA, FRASER R, MASON P, ET AL, METABOLIC EFFECTS OF HIGH DOSE AMILORIDE AND SPIRONOLACTONE: A COMPARATIVE STUDY IN NORMAL SUBJECTS, *BR J CLIN PHARMACOL*, 18, 369-375 (1984)

2441 MILLAR JW, RIFAMPICIN-INDUCED PORPHYRIA CUTANEA TARDA, *BR J DIS CHEST*, 74, 405-408 (1980)

2442 MILLAR RA, THE FLUORIMETRIC ESTIMATION OF EPINEPHRINE IN PERIPHERAL VENOUS PLASMA DURING INSULIN HYPOGLYCEMIA, *J PHARMACOL EXP THER*, 118, 453 (1956)

2443 MILLER A, REVERSIBLE HEPATOTOXICITY RELATED TO AMPHOTERICIN B, *CAN MED ASSOC J*, 131, 1245-1247 (1984)

2444 MILLER ARO, ADDIS BJ, CLARKE PD, NITROFURANTOIN AND CHRONIC ACTIVE HEPATITIS, *ANN INTERN MED*, 97, 452 (1982)

2445 MILLER BF, DUBOS RJ, DETERMINATION BY A SPECIFIC ENZYMATIC METHOD OF THE CREATININE CONTENT OF BLOOD AND URINE, *J BIOL CHEM*, 121, 457 (1937)

2446 MILLER HC, MCLEOD A, KIRBY BJ, SCOTT DB, JULIAN DG, EFFECT OF PENTAZOCINE ON PULMONARY CIRCULATION, *LANCET*, 2, 1167 (1972)

2447 MILLER M, NEUROPATHY, AGRANULOCYTOSIS AND HEPATOTOXICITY FOLLOWING IMIPRAMINE THERAPY, *AM J PSYCHIATRY*, 120, 185 (1963)

2448 MILLER SE, *A TEXTBOOK OF CLINICAL PATHOLOGY* 6TH ED., WILLIAMS AND WILKINS, BALTIMORE, MD (1960)

2449 MILLICHAP JG, OTHER HYDANTOINS: MEPHENYTOIN, ETHOTOIN ALBUTOIN in, *ANTIEPILEPTIC DRUGS*, DM WOODBURY, JK PENRY, RP SCHMIDT, EDS. RAVEN PRESS, NEW YORK, NY (1972)

2450 MILLIGAN DW, THEIN SL, ROBERTS BE, SECONDARY TREATMENT OF POLYCYTHEMIA RUBRA VERA WITH 6-THIOGUANINE, *CANCER*, 50, 836-839 (1982)

2451 MILLS RM JR., SEVERE HYPERSENSITIVITY REACTIONS ASSOCIATED WITH ALLOPURINOL, *J AM MED ASSOC*, 216, 799 (1971)

2452 MILNE MD, INFLUENCE OF ACID-BASE BALANCE ON EFFICACY AND TOXICITY OF DRUGS, *PROC ROY SOC MED*, 58, 961 (1965)

2453 MILWIDSKY A, YAGEL S, CHAOUAT M, ET AL, GLUCOCORTICOIDS DIRECTLY AFFECT SPECTROPHOTOMETRY OF BILIRUBIN IN AMNIOTIC FLUID, *CLIN CHEM*, 30, 677-680 (1984)

2454 MINAR E, EHRINGER H, HIRSCHL M, ET AL, TRANSAMINASE INCREASE: A LARGELY UNKNOWN SIDE EFFECT OF HEPARIN TREATMENT, *DTSCH MED WOCHENSCHR*, 105, 1713-1717 (1980)

2455 MINCEY EK, WILCOX E, MORRISON RT, ESTIMATION OF SERUM AND RED CELL FOLATE BY A SIMPLE RADIOMETRIC TECHNIQUE, *CLIN BIOCHEM*, 6, 274 (1973)

2456 MINK IB, COUREY NG, MOORE RH, AMBRUS CM, AMBRUS JL, PROGESTATIONAL AGENTS AND BLOOD COAGULATION. IV. CHANGES INDUCED BY PROGESTOGEN ALONE, *AM J OBSTET GYNECOL*, 113, 739 (1972)

2457 MISTRY CD, LOTE CJ, GOKAL R, ET AL, EFFECTS OF SULINDAC ON RENAL FUNCTION AND PROSTAGLANDIN SYNTHESIS IN PATIENTS WITH MODERATE CHRONIC RENAL INSUFFICIENCY, *CLIN SCI*, 70, 501-505 (1986)

2458 MITCHELL ABS, DUOGASTRONE-INDUCED HYPOKALEMIC NEPHROPATHY AND MYOPATHY WITH MYOGLOBINURIA, *POSTGRAD MED J*, 47, 807 (1971)

2459 MITCHELL RJ, IMPROVED METHOD FOR SPECIFIC DETERMINATION OF CREATININE IN SERUM AND URINE, *CLIN CHEM*, 19, 408 (1973)

2460 MITCHELL RT, MARSHALL LH, LEFKOWITZ LB JR, ET AL, FALSELY ELEVATED SERUM CREATININE LEVELS SECONDARY TO THE PRESENCE OF 5-FLUOROCYTOSINE, *AM J CLIN PATHOL*, 84, 251-253 (1985)

2461 MITCHELL SC, WARING RH, LAND D, ET AL, ODOROUS URINE FOLLOWING ASPARAGUS INGESTION IN MAN, *EXPERIENTIA*, 43, 382-383 (1987)

2462 MITCHELL WD, A COMPARISON OF THE EFFECT OF CLOFIBRATE AND THYROXINE ON SERUM LIPIDS IN THREE HYPOTHYROID SUBJECTS, *CLIN CHIM ACTA*, 35, 429 (1971)

2463 MITCHELL WD, MURCHISON LE, THE EFFECT OF CLOFIBRATE ON SERUM AND FAECAL LIPIDS, *CLIN CHIM ACTA*, 36, 153 (1972)

2464 MITNICK PD, GREENBERG A, DEOREO PB, ET AL, EFFECTS OF TWO NONSTEROIDAL ANTI-INFLAMMATORY DRUGS, INDOMETHACIN AND OXAPROZIN, ON THE KIDNEY, *CLIN PHARMACOL THER*, 28, 680-689 (1980)

2465 MIYACHI Y, MIZUCHI A, HAMANO H, ET AL, EFFECTS OF CHRONIC SULTOPRIDE TREATMENT ON ENDOCRINE SYSTEMS IN PSYCHOTIC WOMEN, *PSYCHOPHARMACOL*, 82, 287-290 (1984)

2466 MIYADA D, TIPPER P, JANTSCH D, ET AL, THE EFFECT OF HYPERLIPIDEMIA ON TECHNICON® SMAC® MEASUREMENTS, *CLIN BIOCHEM*, 15, 185-188 (1982)

2467 MIYADA DS, BAYSINGER V, NOTRICA S, NAKAMURA RM, ALBUMIN QUANTITATION BY DYE BINDING AND SALT FRACTIONATION TECHNIQUES, *CLIN CHEM*, 18, 52 (1972)

2468 MOHANRAM M, EFFECT OF VITAMIN B_{12} ADMINISTRATION ON ASCORBIC ACID LEVELS IN PLASMA AND LEUKOCYTES IN HUMAN SUBJECTS, *INT J VITAM NUTR RES*, 42, 50 (1972)

2469 MOHLER H, BRAVO EL, TARAZI RC, GLUCOSE INTOLERANCE DURING CHRONIC β-ADRENERGIC BLOCKADE IN MAN, *CLIN PHARMACOL THER*, 25, 237 (1978)

2470 MOLITCH ME, RODMAN E, HIRSCH CA, ET AL, SPURIOUS SERUM CREATININE ELEVATIONS IN KETOACIDOSIS, *ANN INTERN MED*, 93, 280-281 (1980)

2471 MOLLER BB, EKELUND B, TOXICITY OF CYCLOSPORINE DURING TREATMENT WITH ANDROGENS, *N ENGL J MED*, 313, 1416 (1985)

2472 MOLVALILAR S, KOCAK N, EXCRETION OF VANILLYLMANDELIC ACID IN RENAL INSUFFICIENCY, *LANCET*, 2, 1263 (1972)

2473 MONFORTE J, BATH RJ, SUNSHINE I, FLUOROMETRIC DETERMINATION OF PRIMARY AND SECONDARY AMINES IN BLOOD AND URINE AFTER THIN-LAYER CHROMATOGRAPHY, *CLIN CHEM*, 18, 1329 (1972)

2474 MONFORTE JR, SUNSHINE I, METHYL ORANGE AS A SCREENING REAGENT FOR ORGANIC BASES, *CLIN CHEM*, 18, 593 (1972)

2475 MONKS A, RICHENS A, EFFECT OF SINGLE DOSES OF SODIUM VALPROATE ON SERUM PHENYTOIN LEVELS AND PROTEIN BINDING IN EPILEPTIC PATIENTS, *CLIN PHARMACOL THER*, 27, 89-95 (1980)

2476 MONTANARI G, GIANFRANCESCHI G, SUPPA G, ET AL, PLASMA LIPID AND LIPOPROTEIN CHANGES IN HYPERTENSIVE PATIENTS TREATED WITH PROPRANOLOL AND PRAZOSIN., NOSEDA, ET AL (EDS) *LIPOPROTEINS AND CORONARY ATHEROSCLEROSIS*, AMSTERDAM, ELSEVIER BIOMEDICAL PRESS, 389-398 (1982)

2477 MOORE JJ, SAX SM, ELIMINATION OF DEXTRAN INTERFERENCE IN SERUM PROTEIN DETERMINATIONS, *CLIN CHEM*, 18, 393 (1972)

2478 MORADY F, SAUVE MJ, MALONE P, ET AL, LONG-TERM EFFICACY AND TOXICITY OF HIGH-DOSE AMIODARONE THERAPY FOR VENTRICULAR TACHYCARDIA OR VENTRICULAR FIBRILLATION, *AM J CARDIOL*, 52, 975-979 (1983)

2479 MORCOS F, CROCKFORD PM, BECK RP, THE EFFECT OF NORETHINDRONE WITH AND WITHOUT ESTROGEN ON SERUM IMMUNOREACTIVE LUTEINIZING HORMONE SECRETION, *AM J OBSTET GYNECOL*, 112, 358 (1972)

2480 MORENO M, MURPHY C, GOLDSMITH C, INCREASE IN SERUM POTASSIUM RESULTING FROM THE ADMINISTRATION OF HYPERTONIC MANNITOL OR OTHER SOLUTIONS, *J LAB CLIN MED*, 73, 291 (1969)

2481 MORGAN DB, RIVLIN RS, DAVIS RH, SEASONAL CHANGES IN THE URINARY EXCRETION OF CALCIUM, *AM J CLIN NUTR*, 25, 652 (1972)

2482 MORGAN DH, JAUNDICE ASSOCIATED WITH AMITRIPTYLINE, *BR J PSYCHIATRY*, 115, 105 (1969)

2483 MORGAN JP, BIANCHINE, THE URICOSURIC AND THYROXINE (T_4) DISPLACING EFFECT OF MK-185, A NEW HYPOLIPIDEMIC AGENT, *CLIN RES*, 19, 27 (1971)

2484 MORIARTY AT, MOOREHEAD WR, RYDER KW, ET AL, SULFASALAZINE INTERFERENCE IN TOTAL PROTEIN MEASUREMENTS WITH THE DU PONT ACA, *CLIN CHEM*, 29, 592-593 (1983)

2485 MORIN LG, DIRECT COLORIMETRIC DETERMINATION OF SERUM CALCIUM WITH O-CRESOLPHTHALEIN COMPLEXONE, *AM J CLIN PATHOL*, 61, 114 (1974)

2486 MORIN LG, DETERMINATION OF SERUM URATE BY DIRECT ACID FE 3+ REDUCTION OR BY ABSORBANCE CHANGE (AT 293 NM), *CLIN CHEM*, 20, 51 (1974)

2487 MORIN LG, PROX J, DIMETHYLAMINOBENZALDEHYDE PROCEDURE FOR DETERMINING UREA, *CLIN CHIM ACTA*, 47, 27 (1973)

2488 MORITA Y, NISHIDA Y, KAMATANI N, ET AL, THEOPHYLLINE INCREASES SERUM URIC ACID LEVELS, *J ALLERGY CLIN IMMUNOL*, 74, 707-712 (1984)

2489 MORLEY JE, SHAFER RB, THYROID FUNCTION SCREENING IN NEW PSYCHIATRIC ADMISSIONS, *ARCH INTERN MED*, 142, 591-593 (1982)

2490 MORRIS HC, MILLER J, CAMPBELL RS, ET AL, DEVELOPMENT OF A RAPID ENZYME MEDIATED COLORIMETRIC ASSAY FOR CHLORAMPHENICOL IN SERUM, *ANN CLIN BIOCHEM*, 24, SUPPL S1, 78-80 (1987)

2491 MORRIS SJ, KANNER R, CHIPRUT RO, ET AL, DISULFIRAM HEPATITIS, *GASTROENTEROLOGY*, 75, 100-102 (1978)

2492 MORSELLI PL, MARC V, GARATTINI S, ZACCALA M, METABOLISM OF EXOGENOUS CORTISOL IN HUMANS: INFLUENCE OF PHENOBARBITAL TREATMENT ON PLASMA CORTISOL, *REV EUR ETUDES CLIN BIOL*, 15, 195 (1970)

2493 MOSEKILDE L, MELSEN F, ANTICONVULSANT OSTEOMALACIA DETERMINED BY QUANTITATIVE ANALYSIS OF BONE CHANGES, *ACTA MED SCAND*, 199, 349-355 (1976)

2494 MOSER M, LOW-DOSE DIURETIC THERAPY FOR HYPERTENSION, *CLIN THER*, 8, 554-562 (1986)

2495 MOSER RH, *DISEASES OF MEDICAL PROGRESS: A CONTEMPORARY ANALYSIS OF ILLNESS PRODUCED BY DRUGS* 3RD ED., CC THOMAS, SPRINGFI (1964)

2496 MOSER RH, *DISEASES OF MEDICAL PROGRESS: A STUDY OF IATROGENIC DISEASE* 3RD ED., CC THOMAS, SPRINGFIELD, IL (1969)

2497 MOSER RH, BIBLIOGRAPHIES ON DISEASES OF MEDICAL PROGRESS: REACTIONS TO PHENOTHIAZINE AND RELATED DRUGS, *CLIN PHARMACOL THER*, 7, 683 (1966)

2498 MOSER RH, DISORDERS PRODUCED BY PROGESTATIONAL AGENTS, *CLIN PHARMACOL THER*, 7, 399 (1966)

2499 MOSES, AM, HOWANITZ J, MILLER M, DIURETIC ACTION OF THREE SULFONYLUREA DRUGS, *ANN INTERN MED*, 78, 541 (1973)

2500 MOSHIDES JS, COMPARISON OF ASCORBIC ACID INTERFERENCE IN HDL-CHOLESTEROL ESTIMATION BY SIX PRECIPITATION METHODS, WITH USE OF A SENSITIVE ENZYMIC CHOLESTEROL REAGENT, *CLIN CHEM*, 33, 1467-1468 (1987)

2501 MOSKOVITZ R, DEVANE CL, HARRIS R, ET AL, TOXIC HEPATITIS AND SINGLE DAILY DOSE IMIPRAMINE THERAPY, *J CLIN PSYCHIATRY*, 43, 165-166 (1982)

2502 MOSS RA, DRUG-INDUCED IMMUNE THROMBOCYTOPENIA, *AM J HEMATOL*, 9, 439-446 (1980)

2503 MOST H, TREATMENT OF COMMON PARASITIC INFECTIONS OF MAN ENCOUNTERED IN THE UNITED STATES, *N ENGL J MED*, 287, 495 (1972)

2504 MOTT C, SARLES H, TISCORNIA O, GUILLO L, INHIBITORY ACTION OF ALCOHOL ON HUMAN EXOCRINE PANCREATIC SECRETION, *AM J DIG DIS*, 17, 902 (1972)

2505 MOUALLEM R, COMPARATIVE EFFICACY AND SAFETY OF CEPHRADINE AND CEPHALEXIN IN CHILDREN, *J INT MED RES*, 4, 265-271 (1976)

2506 MRC WORKING PARTY ON MILD TO MODERATE HYPERTENSION, ADVERSE REACTIONS TO BENDROFLUAZIDE AND PROPRANOLOL FOR THE TREATMENT OF MILD HYPERTENSION, *LANCET*, 2, 539-543 (1981)

2507 MUDGE GH, URICOSURIC ACTION OF CHOLECYSTOGRAPHIC AGENTS: A POSSIBLE FACTOR IN NEPHROTOXICITY, *N ENGL J MED*, 284, 929 (1971)

2508 MUELLER MN, KAPPAS A, ESTROGEN PHARMACOLOGY: INFLUENCE OF ESTRADIOL AND ESTRIOL ON HEPATIC DISPOSAL OF SULFOBROMOPHTHALEIN (BSP), *J CLIN INVEST*, 43, 1905 (1964)

2509 MUELLER PW, PHILLIPS DL, STEINBERG KK, ALANINE AMINOPEPTIDASE IN SERUM: AUTOMATED OPTIMIZED ASSAY, AND THE EFFECTS OF AGE, SEX, SMOKING AND ALCOHOL CONSUMPTION IN A SELECTED POPULATION, *CLIN CHEM*, 33, 363-366 (1987)

2510 MUELLER RA, FISHBURNE JI JR., BRENNER WE, BRAAKSMA JJ, STAVROVSKY LG, HOFFER JL, HENDRICKS CH, CHANGES IN HUMAN PLASMA CATECHOLAMINES AND DOPAMINE-β-HYDROXYLASE PRODUCED BY PROSTAGLANDIN F2 ALPHA, *PROSTAGLANDINS*, 2, 219 (1972)

2511 MUELLER RG, LANG GE, FLUORESCENT SPECTRA OF LYSERGIC ACID DIETHYLAMINE, *AM J CLIN PATHOL*, 60, 487 (1973)

2512 MUHLFELLNER O, MUHFELLNER G, ZOFEL P, KAFFARNIK H, THE STABILITY OF PLASMA LIPIDS UNDER VARIOUS STORAGE CONDITIONS, *Z KLIN CHEM KLIN BIOCHEM*, 10, 37 (1972)

2513 MULDER H, SCHOPMAN W JR., VAN DER LELY AJ, ET AL, ACUTE CHANGES IN PLASMA RENIN ACTIVITY, PLASMA ALDOSTERONE CONCENTRATION AND PLASMA ELECTROLYTE CONCENTRATIONS FOLLOWING FUROSEMIDE ADMINISTRATION IN PATIENTS WITH CONGESTIVE HEART FAILURE - INTERRELATIONSHIPS AND DIURETIC RESPONSE, *HORM METAB RES*, 19, 80-83 (1987)

2514 MULE SJ, BASTOS ML, JUKOFSKY D, EVALUATION OF IMMUNOASSAY METHODS FOR DETECTION, IN URINE, OF DRUGS SUBJECT TO ABUSE, *CLIN CHEM*, 20, 243 (1974)

2515 MULKEY DA, CELL COUNTER VERSUS AUTOAGGLUTININS, *ASCP SUMMARY REPORT*, 9, 5 (1972)

2516 MULLER P, SEITZ HK, SIMON B, ET AL, DAY OMEPRAZOLE TREATMENT. GASTRIC ACID SECRETION AND BASAL HORMONE LEVELS IN MAN, *Z GASTROENTEROL*, 22, 236-240 (1984)

2517 MULLICK FG, ISHAK KG, HEPATIC INJURY ASSOCIATED WITH DIPHENYLHYDANTOIN THERAPY, *AM J CLIN PATHOL*, 74, 442-452 (1980)

2518 MULLINS RE, HUTTON PS, CONN RB, EFFECTS OF FLUOSOL-DA (ARTIFICIAL BLOOD) ON CLINICAL CHEMISTRY TESTS AND INSTRUMENTS, *AM J CLIN PATHOL*, 80, 478-483 (1983)

2519 MUNGALL IPF, HAGUE RV, PANCREATITIS AND THE PILL, *POSTGRAD MED J*, 51, 855-857 (1975)

2520 MUNGALL IPF, STANDING VF, EOSINOPHILIA CAUSED BY RIFAMPIN, *CHEST*, 74, 321-322 (1978)

2521 MUNION GL, SEATON JF, HARRISON TS, HPLC FOR URINARY CATECHOLAMINES AND METANEPHRINES WITH ALPHA-METHYLDOPA, *J SURG RES*, 35, 507-514 (1983)

2522 MUNRO JF, DUNCAN LJ, FASTING IN THE TREATMENT OF OBESITY, *PRACTITIONER*, 208, 493 (1972)

2523 MUNRO-FAURE AD, HILL DM, ANDERSON J, ETHNIC DIFFERENCES IN HUMAN BLOOD CELL SODIUM CONCENTRATION, *NATURE*, 231, 457 (1971)

2524 MUNZ E, BERNT E, WAHLEFELD AW, AN EVALUATION OF A FULLY ENZYMATIC METHOD FOR CREATININE DETERMINATION, *Z KLIN CHEM KLIN BIOCHEM*, 12, 259 (1974)

2525 MUNZENBERGER P, EMMANUEL S, THE INCIDENCE OF DRUG-DIAGNOSTIC TEST INTERFERENCES IN OUTPATIENTS, *AM J HOSP PHARM*, 28, 789 (1971)

2526 MURAKAMI M, ODAKE K, MATSUDA T ET AL, EFFECTS OF ANABOLIC STEROIDS ON ANTICOAGULANT REQUIREMENTS, *JPN CIRC J*, 29, 243 (1965)

2527 MURPHY DL, GOODWIN FK, BUNNEY WE, ALDOSTERONE AND SODIUM RESPONSE TO LITHIUM ADMINISTRATION IN MAN, *LANCET*, 2, 458 (1969)

2528 MURPHY JV, MARQUARDT K, ASYMPTOMATIC HYPERAMMONIA IN PATIENTS RECEIVING VALPROIC ACID, *ARCH NEUROL*, 39, 591-592 (1982)

2529 MURPHY MF, METCALFE P, GRINT PCA, ET AL, CEPHALOSPORIN-INDUCED IMMUNE NEUTROPENIA, *BR J HAEMATOL*, 59, 9-14 (1985)

2530 MURPHY MJ, LYON LW, TAYLOR JW, ET AL, VALPROIC ACID ASSOCIATED PANCREATITIS IN AN ADULT, *LANCET*, 1, 41-42 (1981)

2531 MURPHY PJ, RYMER W, QUINIDINE-INDUCED LIVER DISEASE, *ANN INTERN MED*, 78, 785 (1973)

2532 MURRAY-LYON IM, CASSAR J, COULSON R, WILLIAMS R, GANGULI PC, EDWARDS JC, TAYLOR KW, FURTHER STUDIES IN STREPTOZOTOCIN THERAPY FOR A MULTIPLE HORMONE-PRODUCING ISLET CELL CARCINOMA, *GUT*, 12, 717 (1971)

2533 MURTHY VV, GOSWAMI SL, A MODIFIED FLUORIMETRIC PROCEDURE FOR THE RAPID ESTIMATION OF OXYTETRACYCLINE IN BLOOD, *J CLIN PATHOL*, 26, 548 (1973)

2534 MUSCAT-BARON JM, FREEMAN DM, TOXIC HEPATITIS FOLLOWING PHENYLBUTAZONE THERAPY, *BR J CLIN PRACT*, 20, 437 (1966)

2535 MUTHER RS, POTTER DM, BENNETT WM, ASPIRIN-INDUCED DEPRESSION OF GLOMERULAR FILTRATION RATE IN NORMAL HUMANS: ROLE OF SODIUM IMBALANCE, *ANN INTERN MED*, 94, 317-321 (1981)

2536 MYERS MD, ROSS J, NEWTON L, ET AL, CYCLOSPORINE-ASSOCIATED CHRONIC NEPHROPATHY, *N ENGL J MED*, 311, 699-705 (1984)

2537 MYREN J, BERSTAD A, EFFECT OF OXYPHENCYCLIMINE HCL ON BASIC GASTRIC SECRETION OF ACID AND PEPSIN IN MAN, *ACTA HEPATOGASTROENTEROL*, 20, 57 (1973)

2538 McCARTHY DM, RANITIDINE OR CIMETIDINE, *ANN INTERN MED*, 99, 551-553 (1983)

2539 McDONALD EC, SPEICHER CE, PELIOSIS HEPATITIS ASSOCIATED WITH ADMINISTRATION OF OXYMETHOLONE, *J AM MED ASSOC*, 240, 243-244 (1978)

2540 MÜLLER KH, HERRMANN K, VERTRAGT SICH EINE GLEICHZEITIGE BEHANDLUNG MIT ANTIKOAGULANTIEN UND INDOMETHACIN?, *MED WELT*, 17, 1553 (1966)

2541 NADLER JL, VELASCO JS, HORTON R, CIGARETTE SMOKING INHIBITS PROSTACYCLIN FORMATION, *LANCET*, 1, 1248-1250 (1983)

2542 NAFTOLIN F, YEN SS, TSAI CC, RAPID CYCLING OF PLASMA GONADOTROPHINS IN NORMAL MEN AS DEMONSTRATED BY FREQUENT SAMPLING, *NATURE NEW BIOL*, 236, 92 (1972)

2543 NAGAHAMA S, FUJIMAKI M, KAWABE H, ET AL, EFFECT OF METOCLOPRAMIDE ON THE SECRETION OF ALDOSTERONE AND OTHER ADRENOCORTICAL STEROIDS, *CLIN ENDOCRINOL*, 18, 287-293 (1983)

2544 NAGANNA B, RAJAMMA M, RAO KV, ON THE FAILURE OF ENZYME PAPER STRIPS TO DETECT GLUCOSE IN CERTAIN ABNORMAL URINES, *CLIN CHIM ACTA*, 17, 219 (1967)

2545 NAGATSU T, KATO T, KUZUYA H, OKADA T, UMEZAWA H, TAKEUCHI T, INHIBITION OF HUMAN SERUM DOPAMINE-β-HYDROXYLASE AFTER THE ORAL ADMINISTRATION OF FUSARIC ACID, *EXPERIENTIA*, 28, 779 (1972)

2546 NAGEL GA, WANDER HE, BLOSSEY HC, PHASE II STUDY OF AMINOGLUTETHIMIDE AND MEDROXYPROGESTERONE ACETATE IN THE TREATMENT OF PATIENTS WITH ADVANCED BREAST CANCER, *CANCER RES*, 42 SUPPL, 3442-3444 (1982)

2547 NAGEL TC, FREINKEL N, BELL RH, FRIESEN H, WILBER JF, METZGER BE, GYNECOMASTIA, PROLACTIN AND OTHER PEPTIDE HORMONES IN PATIENTS UNDERGOING CHRONIC HEMODIALYSIS, *J CLIN ENDOCRINOL METAB*, 36, 428 (1973)

2548 NAHATA MC, DEBOLT SL, POWELL DA, ADVERSE EFFECTS OF METHICILLIN, NAFCILLIN, AND OXACILLIN IN PEDIATRIC PATIENTS, *DEV PHARMACOL THER*, 4, 117-123 (1982)

2549 NAIR PP, MEZEY E, MURTY H, QUARTNER J, MEDELOFF AL, VITAMIN E AND PORPHYRIN METABOLISM IN MAN, *ARCH INTERN MED*, 128, 411 (1971)

2550 NAKAMURA H, EFFECTS OF ANTIHYPERTENSIVE DRUGS ON PLASMA LIPIDS, *AM J CARDIOL*, 60, 24E-28E (1987)

2551 NAKAMURA H, TAMURA Z, GAS CHROMATOGRAPHIC ANALYSIS OF MONO- AND DISACCHARIDES IN HUMAN BLOOD AND URINE AFTER ORAL ADMINISTRATION, *CLIN CHIM ACTA*, 39, 367 (1972)

2552 NANJI AA, SYMPTOMATIC HYPERCALCAEMIA PRECIPITATED BY MAGNESIUM THERAPY, *POSTGRAD MED J*, 61, 47-48 (1985)

2553 NANJI AA, MIKHAEL NZ, STEWART DJ, INCREASE IN SERUM URIC ACID LEVEL ASSOCIATED WITH CISPLATIN THERAPY, *ARCH INTERN MED*, 145, 2013-2014 (1985)

2554 NANJI AA, POON R, HINBERG I, INTERFERENCE BY CEPHALOSPORINS WITH CREATININE MEASUREMENT BY DESKTOP ANALYZERS, *EUR J CLIN PHARMACOL*, 33, 427-429 (1987)

2555 NANKIN HR, HORMONE KINETICS AFTER INTRAMUSCULAR TESTOSTERONE CYPIONATE, *FERTIL STERIL*, 47, 1004-1009 (1987)

2556 NAPPI JM, WARFARIN AND PHENYTOIN INTERACTION, *ANN INTERN MED*, 90, 852 (1979)

2557 NAPPI JM, DHANANI S, LOVEJOY JR, ET AL, SEVERE HYPOGLYCEMIA ASSOCIATED WITH DISOPYRAMIDE, *WEST J MED*, 138, 95-97 (1983)

2558 NARAYANAN S, APPLETON HD, SPECIFICITY OF ACCEPTED PROCEDURES FOR URINE CREATININE, *CLIN CHEM*, 18, 270 (1972)

2559 NARDONE DA, BOUMA DJ, HYPERGLYCEMIA AND DIABETIC COMA: POSSIBLE RELATIONSHIP TO DIURETIC-PROPRANOLOL THERAPY, *SOUTH MED J*, 72, 1607-1608 (1979)

2560 NASH DT, A STUDY ON THE EFFECTS OF PYRIDINOL CARBAMATE P-23 ON ANGINA PECTORIS, *J CLIN PHARMACOL*, 8, 259 (1968)

2561 NASH DT, FEER TD, HEPATIC INJURY POSSIBLY INDUCED BY VERAPAMIL, *J AM MED ASSOC*, 249, 395-396 (1983)

2562 NASH DT, SCHONFELD G, REEVES RL, ET AL, A DOUBLE BLIND PARALLEL TRIAL TO ASSESS THE EFFICACY OF DOXAZOSIN, ATENOLOL AND PLACEBO IN PATIENTS WITH MILD TO MODERATE SYSTEMIC HYPERTENSION, *AM J CARDIOL*, 59, 87G-90G (1987)

2563 NASH J, CURRENT THERAPEUTICS: CCXLII GLYMIDINE, *PRACTITIONER*, 200, 311 (1968)

2564 NATHAN DM, FRANCIS TB, PALMER JL, EFFECT OF ASPIRIN ON DETERMINATIONS OF GLYCOSYLATED HEMOGLOBIN, *CLIN CHEM*, 29, 466-469 (1983)

2565 NATHAN RS, TABRIZI MA, HALPERN FS, ET AL, EFFECT OF CYPROHEPTADINE AND ATROPINE ON THE DIURNAL PROLACTIN RESPONSES TO INSULIN-INDUCED HYPOGLYCEMIA IN NORMAL MEN, *J CLIN ENDOCRINOL METAB*, 51, 90-92 (1980)

2566 NAUMANN HN, PREVENTION OF PYRIDIUM® INTERFERENCE IN URINALYSIS BY DITHIONITE REDUCTION OR BUTANOL EXTRACTION, *AM J CLIN PATHOL*, 48, 337 (1967)

2567 NAZOR SM, SCHMIDT H, THE USE OF DEXTRANASE TO REMOVE THE INTERFERENCE BY DEXTRAN IN THE DETERMINATION OF SERUM PROTEINS BY THE BIURET, *Z KLIN CHEM KLIN BIOCHEM*, 10, 548 (1973)

2568 NEELEY WE, SIMPLE AUTOMATED DETERMINATION OF SERUM OR PLASMA GLUCOSE BY A HEXOKINASE/G-6-PD METHOD, *CLIN CHEM*, 18, 509 (1972)

2569 NEELEY WE, CUPAS CA, THE USE OF 3-METHYL-2-BENZOTHIAZOLONE HYDRAZONE IN CLINICAL CHEMISTRY, *CLIN BIOCHEM*, 6, 246 (1973)

2570 NEILL DW, CARRE IJ, MCCORRY RL, THOMPSON RH, A POSSIBLE SOURCE OF ERROR IN THE DIAGNOSIS OF PHAEOCHROMOCYTOMA, *J CLIN PATHOL*, 14, 415 (1961)

2571 NELIS GF, NITROFURANTOIN - INDUCED PANCREATITIS: REPORT OF A CASE, *GASTROENTEROLOGY*, 84, 1032-1034 (1983)

2572 NELSON DC, McGREW WRG JR, HOYUMPA AM JR., HYPERNATREMIA AND LACTULOSE THERAPY, *J AM MED ASSOC*, 249, 1295-1298 (1983)

2573 NELSON JC, KRUEGER GG, WILCOX RB, THOMPSON WP, EFFECT OF RADIO-CONTRAST MEDIA ON THE MEASUREMENT OF ADRENAL STEROIDS IN THE URINE, *J CLIN ENDOCRINOL METAB*, 28, 1515 (1968)

2574 NELSON JC, LERNER RG, GOLDSTEIN R, ET AL, HEPARIN-INDUCED THROMBOCYTOPENIA, *ARCH INTERN MED*, 138, 548-552 (1978)

2575 NELSON RB, HEPATITIS DUE TO CARBARSONE, *J AM MED ASSOC*, 160, 764 (1956)

2576 NERI A, ZUKERMAN Z, AYGEN M, ET AL, THE EFFECT OF LONG-TERM ADMINISTRATION OF DIGOXIN ON PLASMA ANDROGENS AND SEXUAL DYSFUNCTION, *J SEX MARITAL THER*, 13, 58-63 (1987)

2577 NESTEL PJ, HUNT D, WAHLQVIST ML, CLOFIBRATE RAISES PLASMA APOPROTEIN A-I AND HDL-CHOLESTEROL CONCENTRATIONS, *ATHEROSCLEROSIS*, 37, 625-629 (1980)

2578 NESTLER JF, BARLASCINI CO, CLORE JN, ET AL, DEHYDROEPIANDROSTERONE REDUCES SERUM LOW DENSITY LIPOPROTEIN LEVELS AND BODY FAT BUT DOES NOT ALTER INSULIN SENSITIVITY IN NORMAL MAN, *J CLIN ENDOCRINOL METAB*, 66, 57-61 (1988)

2579 NEUBERG M, WELTER O, INFLUENCE OF MERTHIOLATE® ON SINGLE RADIAL IMMUNODIFFUSION, *KLIN WOCHENSCHR*, 50, 119 (1972)

2580 NEUHAUSER M, GOTTMANN U, BASSLER KH, UBER DEN EINFLUSS DER NAHRUNGZUSAMMENSETZUNG AUF DIE AUSSCHEIDUNG VON 3-METHYLHISTIDIN UND KREATININ IN HARN, *J CLIN CHEM CLIN BIOCHEM*, 22, 731-734 (1984)

2581 NEVINNY HB, HALL TC, CHEMOTHERAPY WITH HYDROXYUREA (NSC-32065) IN RENAL CELL CARCINOMA, *J CLIN PHARMACOL*, 8, 352 (1968)

2582 NEVINNY HB, HALL TC, HAINES C, KRANT MJ, COMPARISON OF METHOTREXATE (NSC-740) AND TESTOSTERONE PROPIONATE (NSC-9166) IN THE TREATMENT OF BREAST CANCER, *J CLIN PHARMACOL*, 8, 126 (1968)

2583 NEW ZEALAND COMMITTEE ON ADVERSE DRUG REACTIONS, THREE YEAR SURVEY OF REACTIONS 1969-1971, *N Z MED J*, 75, 100 (1972)

2584 NEWNHAM HH, HAMBLIN PS, LONG F, ET AL, EFFECT OF ORAL FRUSEMIDE ON DIAGNOSTIC INDICES OF THYROID FUNCTION, *CLIN ENDOCRINOL*, 26, 423-431 (1987)

2585 NEWTON J, DIXON P, SITE OF ACTION OF CLOMIPHENE AND ITS USE AS A TEST OF PITUITARY FUNCTION, *J OBSTET GYNAECOL BR COMM*, 78, 812 (1971)

2586 NEWTON P, SWINBURN WR, SWINSON DR, PROTEINURIA WITH GOLD THERAPY: WHEN SHOULD GOLD BE PERMANENTLY STOPPED, *BR J RHEUMATOL*, 22, 11-17 (1983)

2587 NG JYK, DISNEY APS, JONES TE, ET AL, ACUTE PANCREATITIS AND SODIUM VALPROATE, *MED J AUST*, 2, 362 (982)

2588 NG RH, GUILMET R, ALTAFFER M, ET AL, FALSELY HIGH RESULTS FOR TRIGLYCERIDES IN PATIENTS RECEIVING INTRAVENOUS NITROGLYCERIN, *CLIN CHEM*, 32, 2098-2099 (1986)

2589 NG RP, CHAN TK, TODD D, NBT TEST - FALSE-NEGATIVE AND FALSE-POSITIVE RESULTS, *LANCET*, 1, 1341 (1972)

2590 NICHOLLS A, SNAITH ML, SCOTT JT, EFFECT OF OESTROGEN THERAPY ON PLASMA AND URINARY LEVELS OF URIC ACID, *BR MED J [CLIN RES]*, 1, 449 (1973)

2591 NICHOLLS DG, HAMPTON JR, MITCHELL JRA, CYCLICAL CHANGES IN PLASMA LYSOLECITHIN INDUCED BY ORAL CONTRACEPTIVES, *LANCET*, 2, 1428 (1971)

2592 NICHOLLS MG, ESPINER EA, IKRAM H, ET AL, HYPONATRAEMIA IN CONGESTIVE HEART FAILURE DURING TREATMENT WITH CAPTOPRIL, *BR MED J [CLIN RES]*, 281, 909 (1980)

2593 NICOT G, MERLE, L, VALETTE J-P, ET AL, GENTAMICIN AND SISOMICIN INDUCED RENAL TUBULAR DAMAGE, *EUR J CLIN PHARMACOL*, 23, 161-166 (1982)

2594 NICOT GS, MEKLE LJ, CHARMES JP, ET AL, TRANSIENT GLOMERULAR PROTEINURIA, ENZYMURIA, AND NEPHROTOXIC REACTIONS INDUCED BY RADIOCONTRAST MEDIA, *J AM MED ASSOC*, 252, 2432-2434 (1984)

2595 NIELSEN HK, HUSTED SE, KOOPMAN HD, ET AL, HEPARIN INDUCED INCREASES IN SERUM LEVELS OF AMINOTRANSFERASES, *ACTA MED SCAND*, 215, 231-233 (1984)

2596 NIELSEN MD, BINDER C, STARUP J, URINARY EXCRETION OF DIFFERENT CORTICOSTEROID METABOLITES IN ORAL CONTRACEPTION AND PREGNANCY, *ACTA ENDOCRINOL*, 60, 473 (1969)

2597 NIGHTINGALE J, NAPPI JM, EFFECT OF PHENYTOIN ON SERUM DISOPYRAMIDE CONCENTRATIONS, *CLIN PHARM*, 6, 46-50 (1987)

2598 NIKKILA EA, KASTE M, EHNHOLM C, ET AL, INCREASE OF SERUM HIGH-DENSITY LIPOPROTEIN IN PHENYTOIN USERS, *BR MED J [CLIN RES]*, 2, 99 (1978)

2599 NILSSON B, SODERGARD R, DAMBER M-G, ET AL, FREE TESTOSTERONE LEVELS DURING DANAZOL THERAPY, *FERTIL STERIL*, 39, 505-509 (1983)

2600 NILSSON KO, HOKFELT B, INFLUENCE OF METYRAPONE ON PLASMA CONCENTRATIONS OF TESTOSTERONE AND ANDROSTEREDIONE IN MAN, *ACTA ENDOCRINOL*, 68, 576 (1971)

2601 NISBET P, CHLORPROPAMIDE-INDUCED HYPONATRAEMIA, *BR MED J [CLIN RES]*, 1, 904 (1977)

2602 NISHIKAWA Y, FUKOMOTO K, WATANABE F, SIMPLE, RAPID DETERMINATION OF SERUM GUANASE ACTIVITY WITH THE HITACHI® 736 AUTOMATED DISCRETE ANALYZER, *CLIN CHEM*, 31, 103-105 (1985)

2603 NIXON DD, THROMBOCYTOPENIA FOLLOWING DOXEPIN TREATMENT, *J AM MED ASSOC*, 220, 418 (1972)

2604 NIZET A, ANTAGONISTIC ACTION OF URACIL, THIOURACIL AND THIOUREA ON RETICULOCYTE RIPENING, *SCIENCE*, 115, 290 (1952)

2605 NODINE JH, MODI KN, RHODES M, PAZ-MARTINEZ V, IBARRA L, SANTOS RJ, PHARMACODYNAMICS AND PHARMACOKINETICS OF ISOSORBIDE IN MAN, *CLIN PHARMACOL THER*, 14, 196 (1973)

2606 NOKIN J, VEKEMANS M, L'HERMITE M, ROBYN C, CIRCADIAN PERIODICITY OF SERUM PROLACTIN CONCENTRATION IN MAN, *BR MED J [CLIN RES]*, 3, 561 (1972)

2607 NOMA A, OKABE H, KITA M, A NEW COLORIMETRIC MICRO-DETERMINATION OF FREE FATTY ACIDS IN SERUM, *CLIN CHIM ACTA*, 43, 317 (1973)

2608 NOOJIN RO, BRADFORD LG, OSMET LS, MONITORING METHOTREX-ATE THERAPY IN PSORIASIS, ARCH DERMATOL, 101, 646 (1970)

2609 NORBURY CG, FRY DE, THE EFFECT OF SMOKING ON THE PROLAC-TIN LEVELS IN SUBFERTILE WOMEN., ANN CLIN BIOCHEM, 24, SUPPL S1, 179-181 (1987)

2610 NORDEN CW, REESE R, ORAL CONTRACEPTIVES AND NBT TEST, N ENGL J MED, 287, 254 (1972)

2611 NORDIN BE, THE EFFECT OF INTRAVENOUS PARATHYROID EXTRACT ON URINARY PH, BICARBONATE AND ELECTROLYTE EXCRETION, CLIN SCI, 19, 311 (1960)

2612 NORDREHAUG JE, JOHANNESSEN K-A, VON DER LIPPE G, ET AL, EFFECT OF TIMOLOL ON CHANGES IN SERUM POTASSIUM CONCENTRATION DURING ACUTE MYOCARDIAL INFARCTION, BR HEART J, 53, 388-393 (1985)

2613 NORFLEET RG, GREEN URINE, J AM MED ASSOC, 247, 29 (1982)

2614 NORMAN EJ, LOGAN D, TERRELL P, FALSELY HIGH SERUM B$_{12}$ LEVELS, AM J CLIN PATHOL, 86, 692 (1986)

2615 NORRBY SR, SIDE EFFECTS OF CEPHALOSPORINS, DRUGS, 34 SUPPL 2, 105-120 (1987)

2616 NORRBY SR, BURMAN LA, LINDERHOLM H, ET AL, CEFTAZIDIME: PHARMACOKINETICS IN PATIENTS AND EFFECTS ON THE RENAL FUNCTION, J ANTIMICROB CHEMOTHER, 10, 199-206 (1982)

2617 NORRIS AH, SHOCK NW, AGING AND VARIABILITY, ANN N Y ACAD SCI, 134, 591 (1966)

2618 NORTH JDK, EXPERIENCE WITH AMILORIDE - A POWERFUL POTAS-SIUM-SPARING DIURETIC, S AFR MED J, 46 SUPPL 9 (1972)

2619 NORTON WL, DONNELLY RJ, HIGH DOSE, LOW FREQUENCY PAREN-TERAL GOLD ADMINISTRATION, J RHEUMATOL, 10, 454-458 (1983)

2620 NOSAL T, KISER WS, ROBITAILLE ML, KING JW, URINARY ALKALINE PHOSPHATASE AND LACTIC DEHYDROGENASE ACTIVITY IN THE DIAGNOSIS OF UROLOGIC NEOPLASMS, CLIN CHEM, 12, 542 (1966)

2621 NOSSLIN B, BROMSULPHALEIN™ RETENTION AND JAUNDICE DUE TO UNCONJUGATED BILIRUBIN FOLLOWING TREATMENT WITH MALE FERN EXTRACT, SCAND J CLIN LAB INVEST SUPPL, 15, 69 (1963)

2622 NOSSLIN B, THORELL JI, EFFECTS OF FAT, MEAT, HEPARIN, AND FATTY ACIDS ON THE TRIIODOTHYRONINE UPTAKE TEST, SCAND J CLIN LAB INVEST, 27, 67 (1971)

2623 NOTELOVITZ M, KITCHENS CS, COONE L, ET AL, LOW-DOSE ORAL CONTRACEPTIVE USAGE AND COAGULATION, AM J OBSTET GYNECOL, 141, 71-75 (1981)

2624 NOTRICA S, KLEIN MW, MIYADA DS, NAKAMURA RM, EFFECT ON CHEMICAL VALUES OF USING POLYSTYRENE BEADS FOR SERUM SEPARATION, CLIN CHEM, 19, 792 (1973)

2625 NOTRICA S, MIYADA DS, BAYSINGER V, NAKAMURA RM, EFFECTS OF VARIOUS MEDICATIONS ON VALUES FROM THE HABA AND BCG METHODS FOR DETERMINING ALBUMIN, CLIN CHEM, 18, 1537 (1972)

2626 NOUROK DS, GLASSOCK RS, SOLOMON DH, HYPOTHYROIDISM FOLLOWING PROLONGED SODIUM NITROPRUSSIDE THERAPY, AM J MED SCI, 248, 129 (1964)

2627 NUGENT DJ, BRAY GL, COUNTS RB, ET AL, DANAZOL FAILS TO INCREASE FACTOR VIII OR IX LEVELS IN A DOUBLE BLIND CROSSOVER STUDY OF PATIENTS WITH HEMOPHILIA A OR B., BR J HAEMATOL, 64, 493-502 (1986)

2628 NUMEROFF M, PERLMUTTER M, SLATER S, FALSELY ELEVATED VALUES FOR URINARY 17 KETOSTEROIDS AND 17 HYDROX-YCORTICOIDS, J CLIN ENDOCRINOL METAB, 19, 1350 (1959)

2629 NUNN B, BAIRD A, CHAMBERLAIN PD, EFFECT OF TEMOCILLIN AND MOXALACTAM ON PLATELET RESPONSIVENESS AND BLEED-ING TIME IN NORMAL VOLUNTEERS, ANTIMICROB AGENTS CHEMOTHER, 27, 858-862 (1985)

2630 O'BRIEN RJ, LONG MW, CROSS FS, ET AL, HEPATOTOXICITY FROM ISONIAZID AND RIFAMPIN AMONG CHILDREN TREATED FOR TUBERCULOSIS, PEDIATRICS, 72, 491-499 (1983)

2631 O'BROIN JD, SCOTT JM, TEMPERLEY IJ, A COMPARISON OF SERUM FOLATE ESTIMATIONS USING TWO DIFFERENT METHODS, J CLIN PATHOL, 26, 80 (1973)

2632 O'DONOVAN DJ, INHIBITION OF THE INDOPHENOL REACTION IN THE SPECTROPHOTOMETRIC DETERMINATION OF AMMONIA, CLIN CHIM ACTA, 32, 59 (1971)

2633 O'DWYER WF, PROPRANOLOL ASSOCIATED HYPOGLYCAEMIA IN NONDIABETICS, J IR MED ASSOC, 73, 173 (1980)

2634 O'GORMAN LP, BORUD O, KHAM IA, GJESSING LR, THE METABO-LISM OF L-3,4-DIHYDROXYPHENYLALANINE IN MAN, CLIN CHIM ACTA, 29, 111 (1970)

2635 O'HARE JA, DUGGAN B, O'DRISCOLL D, ET AL, BIOCHEMICAL EVI-DENCE FOR OSTEOMALACIA WITH CARBAMAZEPINE THERAPY, ACTA NEUROL SCAND, 62, 282-286 (1980)

2636 O'KELL RT, KNEPPER L, MANTZEY L, ELLIOTT JR, EFFECT OF DRUGS ON RESULTS OF LABORATORY TESTS, CLIN CHEM, 18, 1039 (1972)

2637 O'KELL RT, MANTZEY L, KNEPPER L, ELLIOTT JR,, INTRAVENOUS VITAMINS AND CLINICAL LABORATORY TESTS, CLIN CHEM, 18, 403 (1972)

2638 O'KELLY R, McKENNA TJ, EXAMINATION OF THE MECHANISMS WHEREBY HEPARIN IMPAIRS ALDOSTERONE BIOSYNTHESIS, IR J MED SCI, 151, 378-383 (1982)

2639 O'LEARY TJ, JONES G, YIP A, ET AL, THE EFFECTS OF CHLORO-QUINE ON SERUM 1,25-DIHYDROXYVITAMIN D AND CALCIUM METABOLISM IN SARCOIDOSIS, N ENGL J MED, 315, 727-730 (1986)

2640 O'LEARY TJ, SIMO IE, KANIGSBERG ND, ET AL, LACK OF EFFECT OF ISOTRETINOIN ON THYROID FUNCTION TESTS, CLIN CHEM, 32, 913-914 (1986)

2641 O'REILLY RA, INTERACTION OF SODIUM WARFARIN AND DISULFIRAM (ANTABUSE®) IN MAN, ANN INTERN MED, 78, 73 (1973)

2642 O'REILLY RA, PHENYLBUTAZONE AND SULFINPYRAZONE INTERAC-TION WITH ORAL ANTICOAGULANT PHENPROCOUMON, ARCH INTERN MED, 142, 1634-1637 (1982)

2643 O'REILLY RA, SPIRONOLACTONE AND WARFARIN INTERACTION, CLIN PHARMACOL THER, 27, 198-201 (1980)

2644 O'REILLY RA, DYNAMIC INTERACTION BETWEEN DISULFIRAM AND SEPARATED ENANTIOMORPHS OF RACEMIC WARFARIN, CLIN PHARMACOL THER, 29, 332-336 (1981)

2645 O'REILLY RA, THE STEREOSELECTIVE INTERACTION OF WARFARIN AND METRONIDAZOLE IN MAN, N ENGL J MED, 295, 354-357 (1976)

2646 O'REILLY RA, MOTLEY CH, RACEMIC WARFARIN AND TRIMETHOPRIM-SULFAMETHOXAZOLE INTERACTION IN HUMANS, ANN INTERN MED, 91, 34-36 (1979)

2647 O'REILLY RA, TRAGER WF, MOTLEY CH, ET AL, INTERACTION OF SECOBARBITAL WITH WARFARIN PSEUDORACEMATES, CLIN PHARMACOL THER, 28, 187-195 (1980)

2648 O'RIORDAN ML, EVANS HJ, ABSENCE OF SIGNIFICANT CHROMO-SOME DAMAGE IN MALES OCCUPATIONALLY EXPOSED TO LEAD, NATURE, 247, 50 (1974)

2649 O'SHEA MJ, FLUTE PT, PANNELL GM, LABORATORY CONTROL OF HEPARIN THERAPY, J CLIN PATHOL, 24, 542 (1971)

2650 OAKLEY DP, LAUTCH H, HALOPERIDOL AND ANTICOAGULANT TREATMENT, LANCET, 2, 1231 (1963)

2651 OCHS L, SERUM PROTEIN REAGENT FOR USE IN THE PARALLEL ANALYTICAL SYSTEM NOT SUSCEPTIBLE TO INTERFERENCE BY DEXTRANS, CLIN CHEM, 33, 1953 (1987)

2652 ODDIE TH, KOSSLER AW, TETRAIODOFLUORESCEIN, A CONTRIBU-TOR TO VARIATIONS IN RADIOIODINE UPTAKE, AM J HOSP PHARM, 29, 690 (1972)

2653 ODIGNE CE, MCCULLOCH AJ, WILLIAMS DO, ET AL, A TRIAL OF CALCIUM ANTAGONIST NISOLDIPINE IN HYPERTENSIVE NONIN-SULIN DEPENDENT DIABETICS, DIABETIC MED, 3, 463-467 (1986)

2654 ODIO C, McCRACKEN GH JR., NELSON JD, NEPHROTOXICITY ASSO-CIATED WITH VANCOMYCIN-AMINOGLYCOSIDE THERAPY IN FOUR CHILDREN, J PEDIATR, 105, 491-493 (1984)

2655 ODLIND V, OLSSON SE, ENHANCED METABOLISM OF LEVO-NORGESTREL DURING PHENYTOIN TREATMENT IN A WOMAN WITH NORPLANT IMPLANTS, CONTRACEPTION, 33, 257-261 (1986)

2656 OETLIKER O, HADORN B, CHATTAS A, SCHULTZ M, SECRETIN INDUCES RENAL BICARBONATE LOSS IN MAN, BIOL GAS-TROENTEROL BELG, 4, 309 (1971)

2657 OETTGEN HF, STEPHENSON PA, SCHWARTZ MK, LEEPER RD, TAL-LAL L, TAN CC, CLARKSON BD, TOXICITY OF E COLI L-ASPARAGINASE IN MAN, CANCER, 25, 253 (1970)

2658 OGAWA K, HATANO T, YAMAMOTO M, ET AL, HORMONAL RESPONSE TO ACUTE DIURESIS — A COMPARATIVE STUDY OF FUROSEMIDE AND AZOSEMIDE, INTL J CLIN PHARM THER TOX-ICOL, 22, 284-290 (1984)

2659 OHMAN KP, WIENER L, VON SCHENCK H, ET AL, ANTIHYPERTEN-SIVE AND METABOLIC EFFECTS OF NIFEDIPINE AND LABETALOL ALONE AND IN COMBINATION IN PRIMARY HYPER-TENSION, EUR J CLIN PHARMACOL, 29, 149-154 (1985)

2660 OHTANI Y, ENDO F, MATSUDA I, CARNITINE DEFICIENCY AND HYPERAMMONEMIA ASSOCIATED WITH VALPROIC ACID THERAPY, J PEDIATR, 101, 782-785 (1982)

2661 OHTSUJI H, IKEDA M, QUANTITATIVE RELATIONSHIP BETWEEN ATMOSPHERIC PHENOL VAPOUR AND PHENOL IN THE URINE OF WORKERS IN BAKELITE, BR J IND MED, 29, 70 (1972)

2662 OKADA M, IMAMURA K, IIDA M, ET AL, HYPOPHOSPHATEMIA INDUCED BY INTRAVENOUS ADMINISTRATION OF SACCHARATED IRON OXIDE, KLIN WOCHENSCHR, 61, 99-102 (1983)

2663 OKEN MM, HOOTKIN L, DEJAGER RL,, HEPATITIS AFTER KŌNYNE® ADMINISTRATION, AM J DIG DIS, 17, 271 (1972)

2664 OLANS RN, WEINER LB, REVERSIBLE OXACILLIN HEPATOTOXICITY, J PEDIATR, 89, 835-838 (1976)

2665 OLARTE MR, SHAFER SQ, LEVAMISOLE IS INEFFECTIVE IN THE TREATMENT OF AMYOTROPHIC LATERAL SCLEROSIS, NEUROL-OGY, 35, 1063-1066 (1985)

2666 OLESEN KH, SIGURD B, STEINESS E, LETH A, BUMETANIDE, A NEW POTENT DIURETIC, ACTA MED SCAND, 193, 119 (1973)

2667 OLIVA PB, LACTIC ACIDOSIS, AM J MED, 48, 209 (1970)

2668 OLIVER MF, ROBERTS SD, HAYES D, EFFECT OF ATROMID AND ETHYL CHLOROPHENOXYISOBUTYRATE ON ANTICOAGULANT REQUIREMENTS, *LANCET*, 1, 143 (1963)

2669 OLSON VV, LOFT S, CHRISTENSEN KD, SERIOUS REACTIONS DURING MALARIA PROPHYLAXIS WITH PYRIMETHAMINE SULFADOXINE, *LANCET*, 2, 994 (1982)

2670 OLSSON R, HELLNER L, LINDSTEDT G, ET AL, PLASMA PROTEINS IN PATIENTS ON LONG-TERM ANTIEPILEPTIC TREATMENT, *CLIN CHEM*, 29, 728-730 (1984)

2671 OLSSON R, KORSAN-BENGTSEN B-M, KORSAN-BENGTSEN K, ET AL, SERUM AMINOTRANSFERASES AFTER LOW-DOSE HEPARIN TREATMENT, *ACTA MED SCAND*, 204, 229-230 (1978)

2672 OLSSON R, TYLLSTROM J, ZETTERGREN L, HEPATIC REACTIONS TO CYCLOFENIL, *GUT*, 24, 260-263 (1983)

2673 OLTHUIS FM, KRUISINGA K, SOONS JB, INTERFERENCE OF FREE FATTY ACIDS WITH THE DETERMINATION OF CALCIUM IN SERUM, *CLIN CHIM ACTA*, 49, 123 (1973)

2674 OMANG SH, VELLAR OD, ANALYTICAL ERROR DUE TO CONCENTRATION GRADIENTS IN FROZEN AND THAWED SAMPLES, *CLIN CHIM ACTA*, 49, 125 (1973)

2675 ONARATO IM, AXELROD JL, HEPATITIS FROM INTRAVENOUS HIGH-DOSE OXACILLIN THERAPY, *ANN INTERN MED*, 89, 497-500 (1978)

2676 ONEDA A, SATO M, YANBE A, MARUHAMA Y, PLASMA GLUCAGON RESPONSE TO BLOOD GLUCOSE FALL, GASTROINTESTINAL HORMONES AND ARGININE IN MAN, *TOHOKU J EXP MED*, 107, 241 (1973)

2677 OOI TC, PEDEN NR, CHAMPION MC, ET AL, THE EFFECT OF CIMETIDINE AND RANITIDINE ON SERUM HIGH DENSITY LIPOPROTEIN SUBFRACTIONS, *ATHEROSCLEROSIS*, 57, 159-162 (1985)

2678 ORCHARD JL, LAWSONR, FALSE-POSITIVE HEMOCCULT® REACTION CAUSED BY BETADINE®, *N ENGL J MED*, 311, 199 (1984)

2679 ORELAND L, FOWLER CJ, SCHALLING D, LOW PLATELET MONOAMINE OXIDASE ACTIVITY IN CIGARETTE SMOKERS, *LIFE SCI*, 29, 2511-2518 (1981)

2680 ORENSTEIN AA, YAKULIS V, EIPE J, ET AL, IMMUNE HEMOLYSIS DUE TO HYDRALAZINE, *ANN INTERN MED*, 86, 450-451 (1977)

2681 ORGNON INC, MANUFACTURER'S LITERATURE ON DURABOLIN®, *375 MOUNT PLEASANT AVE, WEST ORANGE, NJ*

2682 ORME M, BRECKENRIDGE A, BROOKS RV, INTERACTIONS OF BENZODIAZEPINES WITH WARFARIN, *BR MED J [CLIN RES]*, 3, 611 (1972)

2683 ORNEY DL, AN APPROACH TO THE MEASUREMENT OF TOTAL LIPID GLYCEROL IN SERUM, *CLIN CHEM*, 19, 453 (1973)

2684 OROSZ L, MICHAEL R, ZIEGLER M, SERUM-INSULIN OR PLASMA-INSULIN, *LANCET*, 2, 1149 (1971)

2685 OSBORNE JC, HYPOPROTHROMBINEMIA AND BLEEDING DUE TO CEFOPERAZONE, *ANN INTERN MED*, 102, 721 (1985)

2686 OSEI K, HOLLAND G, FALKO JM, INDAPAMIDE. EFFECT ON APOPROTEIN LIPOPROTEIN, AND GLUCOREGULATION IN AMBULATORY DIABETIC PATIENTS, *ARCH INTERN MED*, 146, 1973-1977 (1986)

2687 OSMAN MI, ABDALLA MI, TOPPOZADA MH, ET AL, SUBDERMAL LEVONORGESTREL IMPLANTS: SERUM ANDROGENS, *CONTRACEPT DELIV SYST*, 4, 127-131 (1983)

2688 OSMAN MM, TOPPOZADA HK, GHANEM MH, GUERGIS FK, THE EFFECT OF AN ORAL CONTRACEPTIVE ON SERUM LIPIDS, *CONTRACEPTION*, 5, 105 (1972)

2689 OTTENJAN R, HYPERCALCEMIA AND GASTRIC SECRETION, *GERM MED*, 1, 62 (1971)

2690 OVERFIELD CV, SAVORY J, HEINTGES MG, GLYCOLYSIS: A RE-EVALUATION OF THE EFFECT OF BLOOD GLUCOSE, *CLIN CHIM ACTA*, 39, 35 (1972)

2691 OWEN P, MOFFAT AC, VARIATION OF CYCLIC-AMP EXCRETION WITH URINE VOLUME, *LANCET*, 2, 1205 (1973)

2692 OWEN RT, PIROXICAM, *DRUGS TODAY*, 16, 115-119 (1980)

2693 OWEN WC, KREISBERG RA, SIEGAL AM, CARBOHYDRATE INDUCED HYPERTRIGLYCERIDEMIA: INHIBITION BY PHENFORMIN, *DIABETES*, 20, 739 (1971)

2694 OWUSU SK, ACUTE HAEMOLYSIS COMPLICATING CO-TRIMOXAZOLE THERAPY FOR TYPHOID FEVER IN A PATIENT WITH G-6-PD DEFICIENCY, *LANCET*, 2, 819 (1972)

2695 OYAKE H, ISONO Y, KUDO M, A CASE OF AGRANULOCYTOSIS POSSIBLY DUE TO AMINOBENZYL PENICILLIN, *IRYO*, 22, 676 (1968)

2696 OYAMA JH, HYPERBILIRUBINEMIA IN HEALTHY MALES ON ACUTELY RESTRICTED DIETARY INTAKES RECEIVING PARENTERAL NUTRITION, *AM J CLIN NUTR*, 25, 459 (1972)

2697 OYAMA T, AOKI N, KUDO T, EFFECT OF HALOTHANE ANESTHESIA AND OF SURGERY ON PLASMA TESTOSTERONE LEVELS IN MEN, *ANESTH ANALG*, 51, 130 (1972)

2698 OYAMA T, MATSUKI A, KUDO T, EFFECT OF HALOTHANE, METHOXYFLURANE ANAESTHESIA AND SURGERY ON PLASMA THYROID-STIMULATING HORMONE, *ANAESTHESIA*, 27, 3 (1972)

2699 OYAMA T, MATSUKI A, KUDO T, EFFECTS OF ENFLURANE (ETHRANE®) ANAESTHESIA AND SURGERY ON CARBOHYDRATE AND FAT METABOLISM IN MAN, *ANAESTHESIA*, 27, 179 (1972)

2700 OYAMA T, MATSUKI A, KUDO T, EFFECT OF ETHER, THIOPENTONE ANESTHESIA AND SURGERY ON PLASMA, *BR J ANAESTH*, 44, 841 (1972)

2701 OYAMA T, TAKAZAWA T, EFFECTS OF HALOTHANE ANESTHESIA AND SURGERY ON HUMAN GROWTH HORMONE AND INSULIN LEVELS IN PLASMA, *BR J ANAESTH*, 43, 573 (1971)

2702 OZAWA N, TOKORO T, HIRAKAWA S, HAYASE S, INFLUENCE OF FUSARIC ACID AND RESERPINE ON THE URINARY EXCRETION OF CATECHOLAMINES, *JPN J PHARMACOL*, 22, 113 (1972)

2703 PACIUCCI PA, SKLARIN NT, MITOXANTRONE AND HEPATIC TOXICITY, *ANN INTERN MED*, 105, 805-806 (1986)

2704 PACKMAN RC, O'NEAL LW, WESSLER S, AVIOLI LV, PHEOCHROMOCYTOMA, *J AM MED ASSOC*, 212, 780 (1970)

2705 PAGE LB, DESSAULLES E, LAGG S, HABER E, INTERFERENCE WITH IMMUNOASSAYS OF ANGIOTENSIN I AND II BY PROTEINS IN HUMAN PLASMA, *CLIN CHIM ACTA*, 34, 55 (1971)

2706 PAGE LB, SIDD JJ, MEDICAL MANAGEMENT OF PRIMARY HYPERTENSION, *N ENGL J MED*, 287, 1018 (1972)

2707 PAINE D, FATAL HEPATIC NECROSIS ASSOCIATED WITH AMINO-SALICYLIC ACID: REVIEW OF LITERATURE AND REPORT OF CASE, *J AM MED ASSOC*, 167, 286 (1958)

2708 PAK CY, FULLER C, SAKHAEE K, ET AL, LONG-TERM TREATMENT OF CALCIUM NEPHROLITHIASIS WITH POTASSIUM CITRATE, *J UROL*, 134, 11-19 (1985)

2709 PAK CYC, HYDROCHLOROTHIAZIDE THERAPY IN NEPHROLITHIASIS, *CLIN PHARMACOL THER*, 14, 209 (1973)

2710 PAK CYC, SODIUM CELLULOSE PHOSPHATE: MECHANISM OF ACTION AND EFFECT ON MINERAL METABOLISM, *J CLIN PHARMACOL*, 13, 15 (1973)

2711 PAK CYC, DELEA CS, BARTTER FC, SUCCESSFUL TREATMENT OF RECURRENT NEPHROLITHIASIS (CALCIUM STONES) WITH CELLULOSE PHOSPHATE, *N ENGL J MED*, 290, 175 (1974)

2712 PAK CYC, RUSKIN B, DILLER E, ENHANCEMENT OF RENAL EXCRETION OF ZINC BY HYDROCHLOROTHIAZIDE, *CLIN CHIM ACTA*, 39, 511 (1972)

2713 PALESTINE AG, NUSSENBLATT RB, CHAN C-C, SIDE EFFECTS OF SYSTEMIC CYCLOSPORINE IN PATIENTS NOT UNDERGOING TRANSPLANTATION, *AM J MED*, 77, 652-656 (1984)

2714 PALM R, SILSETH C, ALVAN G, PHENYTOIN INTOXICATION AS THE FIRST SYMPTOM OF FATAL LIVER DAMAGE INDUCED BY SODIUM VALPROATE, *BR J CLIN PHARMACOL*, 17, 597-598 (1984)

2715 PALOMBO JD, BISTRIAN BR, BLACKBURN GL, ASSESSMENT OF THE EFFECT OF ABOVE NORMAL BILIRUBIN ON RADIOIMMUNOASSAY OF CONJUGATED BILE ACIDS IN SERUM, *CLIN CHEM*, 32, 2204 (1986)

2716 PALVA IP, HEINVAARA O, MATTILA M, DRUG INDUCED MALABSORPTION OF VITAMIN B_{12}. III INTERFERENCE OF PAS AND FOLIC ACID, *ANN MED INTERN FENN*, 54, 37 (1965)

2717 PANKOW D, PANKOW B, PONSOLD W, INTERFERENCE OF THE ENZYMIC GLUCOSE DETERMINATION (GLUCOSE-OXIDASE/PEROXIDASE) BY CARBON MONOXIDE, *ZENTRALBL PHARM PHARMAKROTHER LABORATORIUMSDRAGN*, 111, 165 (1972)

2718 PANNALL PR, MAAS DA, DANAZOL AND THYROID FUNCTION TESTS, *LANCET*, 1, 102-103 (1977)

2719 PAOLETTI F, JUAN A VAZQUEZ J, WOLF PL, POSITIVE PREGNANCY TEST IN AN 82 YEAR-OLD-WOMAN, *AM J MED SCI*, 252, 570 (1966)

2720 PAOLISSO G, SGAMBATO S, PASSARIELLO N, ET AL, INSULIN INDUCES OPPOSITE CHANGES IN PLASMA AND ERYTHROCYTE MAGNESIUM CONCENTRATIONS IN NORMAL MAN, *DIABETOLOGIA*, 29, 644-647 (1986)

2721 PAPADEMETRIOU V, PRICE M, JOHNSON E, ET AL, EARLY CHANGES IN PLASMA AND URINARY POTASSIUM IN DIURETIC-TREATED PATIENTS WITH SYSTEMIC HYPERTENSION, *AM J CARDIOL*, 54, 1015-1019 (1984)

2722 PAPADOPOULOS AS, BYLIS EM, FRY DE, MARKS V, A RAPID MICROMETHOD FOR DETERMINING FOUR ANTICONVULSANT DRUGS BY GAS-LIQUID CHROMATOGRAPHY, *CLIN CHIM ACTA*, 48, 135 (1973)

2723 PAPAGEORGIOU P, KESARWALA HH, ALCID DV, ET AL, LEVAMISOLE IN CHRONIC PYODERMA, *J CLIN LAB IMMUNOL*, 8, 121-127 (1982)

2724 PAPAVASILIOU PS, COTZIAS GC, DUBY SE, STECK AJ, FEHLING C, BELL MA, LEVODOPA IN PARKINSONISM: POTENTIATION OF CENTRAL EFFECTS WITH A PERIPHERAL INHIBITOR, *N ENGL J MED*, 286, 8 (1972)

2725 PAPESCHI R, MOLINA-NEGRO P, SOURKES TL, ERBA G, THE CONCENTRATION OF HOMOVANILLIC AND 5-HYDROXYINDOLEACETIC ACIDS IN VENTRICULAR AND LUMBAR CSF, *NEUROLOGY*, 22, 1151 (1972)

2726 PARDUE WO, SEVERE LIVER DYSFUNCTION DURING NICOTINIC ACID THERAPY, *J AM MED ASSOC*, 175, 137 (1961)

2727 PARFITT AM, CHLOROTHIAZIDE INDUCED HYPERCALCEMIA IN JUVENILE OSTEOPOROSIS AND HYPERPARATHYROIDISM, *N ENGL J MED*, 281, 55 (1969)

2728 PARIENTE EA, BATAILLE C, BERCOFF E, ET AL, ACUTE EFFECTS OF CAPTOPRIL ON SYSTEMIC AND RENAL HEMODYNAMICS AND ON RENAL FUNCTION IN CIRRHOTIC PATIENTS WITH ASCITES, *GASTROENTEROLOGY*, 88, 1255-1259 (1985)

2729 PARIENTE EA, PESSAYRE D, BENTATA-PESSAYRE M, ET AL, HEPATITE A L'AJMALINE. DESCRIPTION DE 4 OBSERVATIONS ET REVUE DE LA LITTERATURE, *GASTROENTEROL CLIN BIOL*, 4, 240-245 (1980)

2730 PARKER DC, ROSSMAN LG, VANDERLAAN EF, PERSISTENCE OF RHYTHMIC HUMAN GROWTH HORMONE RELEASE DURING SLEEP, *METABOLISM*, 21, 241 (1972)

2731 PARKER WA, PROPYLTHIOURACIL-INDUCED HEPATOTOXICITY, *CLIN PHARMACOL*, 1, 471-474 (1982)

2732 PARKER WA, CAPTOPRIL-INDUCED CHOLESTATIC JAUNDICE, *DRUG INTELL CLIN PHARM*, 18, 234-235 (1984)

2733 PARKER WA, MACCARA ME, IN VITRO EFFECT OF NEW PENICILLINS AND AMINOGLYCOSIDES ON TESTS FOR GLYCOSURIA, *AM J HOSP PHARM*, 41, 125-127 (1984)

2734 PARKINSON CE, GAL I, FACTORS AFFECTING THE LABORATORY MANAGEMENT OF HUMAN SERUM AND LIVER VITAMIN A ANALYSIS, *CLIN CHIM ACTA*, 40, 83 (1972)

2735 PARODI FA, PHARMACOLOGICAL AND CLINICAL EVALUATION OF THE ACTIVITY AND TOLERANCE OF NEW ORAL ANTIDIABETIC COMPOUND, *ISR J MED SCI*, 8, 888 (1972)

2736 PARSONS WB JR, EFFECT OF NICOTINIC ACID ON SERUM LEVELS OF CHOLESTEROL, TRIGLYCERIDES, GLUCOSE AND URIC ACID, *CIRCULATION*, SUPPL 6, 154 (1968)

2737 PAS AT, QUINN EL, CHOLESTATIC HEPATITIS FOLLOWING THE ADMINISTRATION OF SODIUM OXACILLIN, *J AM MED ASSOC*, 191, 674 (1965)

2738 PASSARIELLO N, GIUGLIANO D, SGAMBATO S, ET AL, CALCITONIN, A DIABETOGENIC HORMONE, *J CLIN ENDOCRINOL METAB*, 53, 318-323 (1981)

2739 PASSARO E, BASSO N, MENNINI G, LEZOCHE E, SPERANZA V, CALCIUM AND GASTRIC SECRETION IN MAN, *BIOL GASTROENTEROL BELG*, 4, 339 (1971)

2740 PASSO TJ, FARBER MO, APPLEGATE GA, KLEIT SA, SZWEIL JJ, INFLUENCE OF HEMODIALYSIS ON 2,3-DIPHOSPHOGLYCERATE (2,3-DPG) IN CHRONIC UREMICS, *CLIN RES*, 21, 702 (1973)

2741 PASTERNACK A, VANTTINEN T, SOLAKIVI T, ET AL, NORMALIZATION OF LIPOPROTEIN LIPASE AND HEPATIC LIPASE BY GEMFIBROZIL RESULTS IN CORRECTION OF LIPOPROTEIN ABNORMALITIES IN CHRONIC RENAL FAILURE, *CLIN NEPHROL*, 27, 163-168 (1987)

2742 PATEL IH, LEVY RH CUTLER RE, PHENOBARBITAL-VALPROIC ACID INTERACTION, *CLIN PHARMACOL THER*, 27, 515-521 (1980)

2743 PATEL YC, ALFORD FP, BURGER HG, THE 24-HOUR PLASMA THYROTROPHIN PROFILE, *CLIN SCI*, 43, 71 (1972)

2744 PATERSON Y, LAWRENCE EF, FACTORS AFFECTING CREATINE PHOSPHOKINASE LEVELS IN NORMAL ADULT FEMALES, *CLIN CHIM ACTA*, 42, 131 (1972)

2745 PATHY MS, THE USE, ACTION AND SIDE EFFECTS OF DIURETICS, *GERONTOL CLIN*, 13, 261 (1971)

2746 PATHY MS, REYNOLDS AJ, PAPAVERINE AND HEPATOTOXICITY, *POSTGRAD MED J*, 56, 488-490 (1980)

2747 PATOIA L, GUERCIOLINI R, MENICHETTI F, ET AL, NORFLOXACIN AND NEUTROPENIA, *ANN INTERN MED*, 107, 788-789 (1987)

2748 PATRASSI GM, FALLO F, MARTINELLI S, ET AL, THE CONTACT PHASE OF BLOOD COAGULATION AND RENIN ACTIVATION IN ESSENTIAL HYPERTENSION BEFORE AND AFTER CAPTOPRIL, *EUR HEART J*, 5, 561-567 (1984)

2749 PATTERSON JS, FURR BJA, BATTERSBY LA, TAMOXIFEN AND HYPERCALCEMIA, *ANN INTERN MED*, 89, 1013 (1978)

2750 PATTERSON R, LURIE AO, PLASMA TESTOSTERONE: A SIMPLE ASSAY WITHOUT CHROMATOGRAPHY FOR THE CLINICAL LABORATORY, *AM J CLIN PATHOL*, 60, 879 (1973)

2751 PATWARDHAN RV, DESMOND PV, JOHNSON RF, ET AL, EFFECTS OF CAFFEINE ON PLASMA FREE FATTY ACIDS, URINARY CATECHOLAMINES, AND DRUG BINDING, *CLIN PHARMACOL THER*, 28, 398-403 (1980)

2752 PAULUS HE, COUTTS A, CALABRO JJ, KLINENBERG JR, CLINICAL SIGNIFICANCE OF HYPERURICEMIA IN ROUTINELY SCREENED HOSPITALIZED MEN, *J AM MED ASSOC*, 211, 277 (1970)

2753 PAWAN GLS, METABOLIC EFFECTS OF S992 IN MAN, *BR J PHARMACOL*, 41, 416 (1971)

2754 PAYNE R, SERUM URIC ACID AND PHENOTHIAZINES, *LANCET*, 2, 855 (1969)

2755 PAYNE RB, A RED HERRING IN THE DETECTION OF BENCE-JONES PROTEIN, *J CLIN PATHOL*, 25, 183 (1972)

2756 PAYNE RB, JOHNSON KR, BOGUS BRANCHED-CHAIN AMINOACIDURIA, *J CLIN PATHOL*, 26, 313 (1973)

2757 PEARSON AJG, GRAINGER JM, SCHEUER PJ, MCINTYRE N, JAUNDICE DUE TO OXYPHENISATIN, *LANCET*, 1, 994 (1971)

2758 PEARSON DWM, RATCLIFFE WA, THOMSON JA, ET AL, BIOCHEMICAL AND CLINICAL EFFECTS OF FENCLOFENAC IN THYROTOXICOSIS, *CLIN ENDOCRINOL*, 16, 369-373 (1982)

2759 PEARSON JR, BINDER CI, NEBER J, AGRANULOCYTOSIS FOLLOWING DIAMOX® THERAPY, *J AM MED ASSOC*, 157, 339 (1955)

2760 PEARSON M, MEERKIN M, BAIRD C, ALDOMET® INTERFERENCE WITH THE BABSON METHOD OF SGOT ANALYSIS, *MED J AUST*, 2, 84 (1972)

2761 PEDEN NR, DOW RJ, ISLES TE, ET AL, β-ADRENOCEPTOR BLOCKADE AND RESPONSES OF SERUM LIPIDS TO A MEAL AND EXERCISE, *BR MED J [CLIN RES]*, 288, 1788-1790 (1984)

2762 PEDERSEN KE, CHRISTIANSEN BD, KJAER K, ET AL, VERAPAMIL-INDUCED CHANGES IN DIGOXIN KINETICS AND INTRAERYTHROCYTIC SODIUM CONCENTRATIONS, *CLIN PHARMACOL THER*, 34, 8-13 (1983)

2763 PEDERSEN KE, CHRISTIANSEN BD, KLITGAARD NA, ET AL, EFFECT OF QUINIDINE ON DIGOXIN BIOAVAILABILITY, *EUR J CLIN PHARMACOL*, 24, 41-47 (1983)

2764 PEDERSEN KE, MADSEN JL, KLITGAARD NA, ET AL, EFFECT OF QUININE ON PLASMA DIGOXIN CONCENTRATION AND RENAL DIGOXIN CLEARANCE, *ACTA MED SCAND*, 218, 229-232 (1985)

2765 PEDERSEN KE, THAYSSEN P, KLITGAARD NA, ET AL, INFLUENCE OF VERAPAMIL ON THE INOTROPISM AND PHARMACOKINETICS OF DIGOXIN, *EUR J CLIN PHARMACOL*, 25, 199-206 (1983)

2766 PEK S, FAJANS SS, FLOYD JC JR., KNOPF RF, CONN JW, FAILURE OF SULFONYLUREAS TO SUPPRESS PLASMA GLUCAGON IN MAN, *DIABETES*, 21, 216 (1972)

2767 PELKONEN R, FOGELHOLM R, NIKKILA EA, INCREASE IN SERUM CHOLESTEROL DURING PHENYTOIN TREATMENT, *BR MED J [CLIN RES]*, 4, 85 (1975)

2768 PELLETIER O, SMOKING AND VITAMIN C LEVELS IN HUMANS, *AM J CLIN NUTR*, 21, 1359 (1968)

2769 PELLOCK JM, HOWELL J, KENDIG EL JR, PYRIDOXINE DEFICIENCY IN CHILDREN TREATED WITH ISONIAZID, *CHEST*, 87, 658-661 (1985)

2770 PENDLETON RG, NEWMAN DJ, SHERMAN SS, BRANN EG, MAYA WE, EFFECT OF PROPRANOLOL UPON THE HEMOGLOBIN-OXYGEN DISSOCIATION CURVE, *J PHARMACOL EXP THER*, 180, 647 (1972)

2771 PENNOCK CA, MURPHY D, SELLERS J, LANGDON KJ, A COMPARISON OF AUTOANALYZER METHODS FOR THE ESTIMATION OF GLUCOSE IN BLOOD, *CLIN CHIM ACTA*, 48, 193 (1973)

2772 PENNY R, GOEBELSMANN U, EFFECT OF ESTRADIOL ON PLASMA MELATONIN LEVELS, *J ENDOCRINOL INVEST*, 7, 55-57 (1984)

2773 PENTIKAINEN PJ, PENTIKAINEN LA, HUFFMAN DM, AZARNOFF DL, THE EFFECT OF SPIRONOLACTONE ON SEXUAL HORMONES IN MALES, *CLIN RES*, 21, 472 (1973)

2774 PEREIRA JN, HOLLAND GF, HOCHSTEIN F, GILGORE S, DEFELICE S, PINSON R, STUDIES WITH 5-(3-PYRIDYL) TETRAZOLE ON LONG ACTING LIPOLYSIS INHIBITOR, *ADV EXP MED BIOL*, 4, 227 (1968)

2775 PEREZ G, SIEGEL L, SCHREINER GE, SELECTIVE HYPOALDOSTERONISM WITH HYPERKALEMIA, *ANN INTERN MED*, 76, 757 (1972)

2776 PERKINS JR, EFFECTS OF LIGHT ON THE DETERMINATION OF CITRULLINE, *CLIN CHIM ACTA*, 35, 247 (1971)

2777 PERKINS SL, OOI DS, INOCOR® (AMRINONE LACTATE) INTERFERENCES WITH TDx DIGOXIN MEASUREMENT, *CLIN CHEM*, 33, 1944 (1987)

2778 PERLOW MJ, SASSIN JF, BOYAR R, HELLMAN L, WEITZMAN ED, RELEASE OF HUMAN GROWTH HORMONE FOLLICLE STIMULATING HORMONE, LUTEINIZING HORMONE, *DIS NERV SYS*, 33, 804F (1972)

2779 PERRAULT DJ, DOMOVITCH E, AMINOGLUTETHIMIDE AND CHOLESTASIS, *ANN INTERN MED*, 100, 160 (1984)

2780 PERRET G, HUGUES JN, LOUCHAHI M, ET AL, EFFECT OF A SHORT-TERM ORAL ADMINISTRATION OF CIMETIDINE AND RANITIDINE ON THE BASAL AND THYROTROPIN-RELEASING HORMONE-STIMULATED SERUM CONCENTRATIONS OF PROLACTIN, THYROTROPIN AND THYROID HORMONES IN HEALTHY VOLUNTEERS., *PHARMACOLOGY*, 32, 101-108 (1986)

2781 PERRILD H, MADSEN SN, HANSEN JEM, IRREVERSIBLE MYXOEDEMA AFTER LITHIUM CARBONATE, *BR MED J [CLIN RES]*, 1, 1108-1109 (1978)

2782 PERRY BW, HOSTY TA, COKER JG, DUMAS B, STRAUMFJORD JV, A FIELD EVALUATION OF THE DU PONT AUTOMATIC CLINICAL ANALYZER, *UNIVERSITY OF ALABAMA MEDICAL CENTER, BIRMINGHAM, ALABAMA* (1970)

2783 PERRY HM JR., SAKAMOTO A, TAN EM, RELATIONSHIP OF ACETYLATING ENZYME TO HYDRALAZINE TOXICITY, *J LAB CLIN MED*, 70, 1020 (1967)

2784 PERRY TL, HANSEN S, HESTRIN M, MACINTYRE L, EXOGENOUS URINARY AMINES OF PLANT ORIGIN, *CLIN CHIM ACTA*, 11, 24 (1965)

2785 PERRY-KEENE DA, LARKINS RG, HEYMA P, ET AL, THE EFFECT OF LONG-TERM DIPHENYLHYDANTOIN THERAPY ON GLUCOSE TOLERANCE AND INSULIN SECRETION: A CONTROLLED TRIAL, *CLIN ENDOCRINOL*, 12, 575-580 (980)

2786 PERSSON B, FEX G, HDL-INCREASING EFFECT OF CYCLOFENIL, *ACTA MED SCAND*, 208, 205-207 (1980)

2787 PERUCCA E, GARRATT A, HEBDIGE S, ET AL, WATER INTOXICATION IN EPILEPTIC PATIENTS RECEIVING CARBAMAZEPINE., *J NEUROL NEUROSURG PSYCHIATRY*, 41, 713-718 (1978)

2788 PERUCCA E, HEDGES A, MAKKI KA, ET AL, A COMPARATIVE STUDY OF THE RELATIVE ENZYME INDUCING PROPERTIES OF ANTICONVULSANT DRUGS IN EPILEPTIC PATIENTS, *BR J CLIN PHARMACOL*, 18, 401-410 (1984)

2789 PESCE MA, BODOURIAN SH, BILIRUBIN INTERFERENCE WITH ULTRA-VIOLET DETERMINATION OF INORGANIC PHOSPHATE ON THE CENTRIFICHEM, *CLIN CHEM*, 19, 436 (1973)

2790 PESCE MA, BODOURIAN SH, NICHOLSON JF, ENZYMATIC METHOD FOR DETERMINATION OF INORGANIC PHOSPHATE IN SERUM AND URINE WITH A CENTRIFUGAL ANALYZER, *CLIN CHEM*, 20, 332 (1974)

2791 PESCE MA, STRANDE CS, A NEW MICROMETHOD FOR DETERMINATION OF PROTEIN IN CEREBROSPINAL FLUID AND URINE, *CLIN CHEM*, 19, 1265 (1973)

2792 PESSAYRE D, LARREY D, FUNCK-BRENTANO C, ET AL, DRUG INTERACTIONS AND HEPATITIS PRODUCED BY SOME MACROLIDE ANTIBIOTICS, *J ANTIMICROB CHEMOTHER*, 16 SUPPL A, 181-194 (1985)

2793 PETERS CA, WALSH PC, THE EFFECT OF NAFARELIN ACETATE, A LUTEINIZING HORMONE-RELEASING HORMONE AGONIST ON BENIGN PROSTATIC HYPERPLASIA, *N ENGL J MED*, 317, 599-604 (1987)

2794 PETERS U, PHARMACOKINETIC REVIEW OF DIGITALIS GLYCOSIDES, *EUR HEART J*, 3, SUPPL D, 65-78 (1982)

2795 PETERSEN KC, SILBERMAN H, BERNE TV, HYPERKALAEMIA AFTER CYCLOSPORIN THERAPY, *LANCET*, 1, 1470 (1984)

2796 PETRAGLIA F, BERNASCONI S, IUGHETTI L, NALOXONE-INDUCED LUTEINIZING HORMONE SECRETION IN NORMAL, PRECOCIOUS AND DELAYED PUBERTY, *J CLIN ENDOCRINOL METAB*, 63, 1112-1116 (1986)

2797 PETRUCCI JV, DUNNE PA, CHAPMAN CC, SPURIOUS ERYTHROCYTE INDICES AS MEASURED BY THE MODEL S COULTER COUNTER DUE TO COLD AGGLUTININS, *AM J CLIN PATHOL*, 56, 500 (1971)

2798 PETTINGER WA, KEETON K, ALTERED RENIN RELEASE AND SYNERGISTIC ANTIHYPERTENSIVE ACTIVITY OF VASODILATING DRUGS AND PROPRANOLOL, *CLIN RES*, 21, 472 (1973)

2799 PEYNET J, POUILLOT JC, ETUDE CORTIGUE D'UNE TECHNIQUE DE DOSAGE DU PHOSPHORE SANS DEPROTEINISATION IN, *ORGANISATION DES LABORATOIRE BIOLOGIE PROSPECTIVE*, G SIEST (ED), L'EXPANSION SCIENTIFIQUE FRANCAISE (1973)

2800 PEYRIN L, COLTET-EMARD JM, AUTOMATED SPECIFIC FLUORIMETRIC METHODS FOR EPINEPHRINE AND NOREPINEPHRINE ASSAY IN A SINGLE BIOLOGICAL EXTRACT, *ANAL BIOCHEM*, 56, 515 (1973)

2801 PFIZER INC, MANUFACTURER'S LITERATURE ON DIABINASE®, *235 E. 42ND ST, NEW YORK, NY 10017*

2802 PFIZER INC., PRODUCT INFORMATION (VIBRAMYCIN®), *235 E. 42ND ST, NEW YORK, NY 10017*

2803 PHARR DM, DICKINSON DB, β-GLUCOSIDASE CONTAMINANT IN ENZYMATIC REAGENT FOR GLUCOSE DETECTION, *ANAL BIOCHEM*, 51, 315 (1973)

2804 PHILLIPS AP, THE IMPROVEMENT OF SPECIFICITY IN RADIOIMMUNOASSAYS, *CLIN CHIM ACTA*, 44, 333 (1973)

2805 PHILLIPS AP, SOURCES OF ERRORS IN RADIOIMMUNOASSAYS, *LANCET*, 1, 1183 (1972)

2806 PHILLIPS I, CLINICAL USES AND CONTROL OF RIFAMPICIN AND CLINDAMYCIN, *J CLIN PATHOL*, 24, 410 (1971)

2807 PHILLIPS PJ, VEDIG AE, JONES PL, ET AL, METABOLIC AND CARDIOVASCULAR SIDE EFFECTS OF THE β₂-ADRENOCEPTOR AGONISTS SALBUTAMOL AND RIMITEROL, *BR J CLIN PHARMACOL*, 9, 483-491 (1980)

2808 PHILP JR, UNTOWARD EFFECTS OF ANTIMICROBIAL DRUGS: PREVENTION AND CONTROL, *POSTGRAD MED*, 50, 193 (1971)

2809 PHORNPHUTKUL KS, ANURAS S, KOFF RS, SEEFF LB, MAHLER DL, ZIMMERMANN HJ, CAUSES OF INCREASED PLASMA CREATINE KINASE ACTIVITY AFTER SURGERY, *CLIN CHEM*, 20, 340 (1974)

2810 PICARD EH, SIDE EFFECTS OF METRONIDAZOLE, *MAYO CLIN PROC*, 58, 401 (1983)

2811 PICKAR D, WOLKOWITZ OM, DORAN AR, ET AL, CLINICAL AND BIOCHEMICAL EFFECTS OF VERAPAMIL ADMINISTRATION TO SCHIZOPHRENIC PATIENTS, *ARCH GEN PSYCHIATRY*, 44, 113-118 (1987)

2812 PIERCE C, ASSAYS AND IMPORTANCE OF SEROTONIN AND ITS METABOLITES, *AM J CLIN PATHOL*, 30, 230 (1958)

2813 PIERCE EH, CHESLER DL, POSSIBLE ASSOCIATION OF GRANULOMATOUS HEPATITIS WITH CLOFIBRATE THERAPY, *N ENGL J MED*, 299, 314 (1978)

2814 PIETILA K, KOIVULA T, INCREASE OF SERUM ANGIOTENSIN-CONVERTING ENZYME ACTIVITY AFTER FREEZING, *SCAND J CLIN LAB INVEST*, 44, 453-455 (1984)

2815 PILEGGI VJ, SEGAL HA, LANCHANTIN GF, THE EFFECT OF SULFOBROMOPHTHALEIN ON SERUM PBI, BEI AND THYROXINE IODINE, *CLIN CHIM ACTA*, 8, 547 (1963)

2816 PILEWSKI RM, SCHEIB ET, MISAGE JR, KESSLER E, KRIFCHER E, SHAPIRO AP, TECHNIQUE OF CONTROLLED DRUG ASSAY IN HYPERTENSION. V. COMPARISON OF HYDROCHLOROTHIAZIDE WITH A NEW QUINETHAZONE, *CLIN PHARMACOL THER*, 12, 843 (1971)

2817 PILLANS PI, COWAN P, WHITELAW D, HYPONATRAEMIA AND CONFUSION IN A PATIENT TAKING KETOCONAZOLE, *LANCET*, 1, 821-822 (1985)

2818 PILLAY VK, GANDHI VC, SHARMA BK, SMITH EC, DUNEA G, EFFECT OF HYDRATION AND FUROSEMIDE GIVEN INTRAVENOUSLY ON PROTEINURIA, *ARCH INTERN MED*, 130, 90 (1972)

2819 PINELLI A, A NEW COLORIMETRIC METHOD FOR PLASMA FATTY ACID ANALYSIS, *CLIN CHIM ACTA*, 44, 385 (1973)

2820 PINES A, RAAFAT M, SIDDIQUI GM, RIFAMPICIN AND ETHAMBUTOL IN THE TREATMENT OF DRUG RESISTANT AND FAR ADVANCED PULMONARY TUBERCULOSIS, *J IRISH MED ASSOC*, 63, 82 (1970)

2821 PINTOR C, CELLA SE, CORDA R, ET AL, CLONIDINE ACCELERATES GROWTH IN CHILDREN WITH IMPAIRED GROWTH HORMONE SECRETION, *LANCET* 1, 1482-1485 (1985)

2822 PISCIOTTA AV, AGRANULOCYTOSIS INDUCED BY CERTAIN PHENOTHIAZINE DERIVATIVES, *J AM MED ASSOC*, 208, 1862 (1969)

2823 PITNEY WR, OAKLEY CM, GOODWIN JF, THERAPEUTIC DEFIBRINATION WITH ANCRODARVIN, *AM HEART J*, 80, 144 (1970)

2824 PITNICK PD, KLEIN WJ JR., PIROXICAM-INDUCED RENAL DISEASE, *ARCH INTERN MED*, 144, 63-64 (1984)

2825 PITTINGER C, CHANG PM, FAULKNER W, SERUM IONIZED CALCIUM: SOME FACTORS INFLUENCING ITS LEVEL, *SOUTH MED J*, 64, 1211 (1971)

2826 PIZIAK VK, SELLMAN JE, OTHMER E, LITHIUM AND HYPOTHYROIDISM, *J CLIN PSYCHIATRY*, 39, 709-711 (1978)

2827 PIZZUTO J, AVILES A, RAMOS E,, ET AL, CYTOSINE ARABINOSIDE INDUCED LIVER DAMAGE: HISTOPATHOLOGIC DEMONSTRATION, *MED PEDIATR ONCOL*, 11, 287-290 (1983)

2828 PLANAS AT, KRANWINKEL RN, SOLETSKY HB, ET AL, CHLORPROPAMIDE-INDUCED PURE RBC APLASIA, *ARCH INTERN MED*, 140, 707-708 (1980)

2829 PLATMAN SR, FIEVE RR, LITHIUM CARBONATE AND PLASMA CORTISOL RESPONSE IN THE AFFECTIVE DISORDERS, *ARCH GEN PSYCHIATRY*, 18, 591 (1968)

2830 PODOLSKY S, PATTAVINA CG, AMARAL MA, EFFECT OF MARIJUANA ON THE GLUCOSE-TOLERANCE TEST, *ANN N Y ACAD SCI*, 191, 54 (1972)

2831 POFFENBARGER PL, BRINKLEY BR, COLCHICINE FOR FAMILIAL MEDITERRANEAN FEVER: POSSIBLE ADVERSE EFFECTS, *N ENGL J MED*, 290, 56 (1974)

2832 POILLEUX F, LEMERCIER M, USE OF THE NEW ANTIBIOTIC GENTAMYCIN IN SURGERY, *PRESSE MED*, 75, 1611 (1967)

2833 POLITI A, POGGIO G, MARGIOTTA A, CAN AMIODARONE INDUCE HYPERGLYCAEMIA AND HYPERTRIGLYCERIDAEMIA?, *BR MED J [CLIN RES]*, 288, 285 (1984)

2834 POLLER L, THOMSON JM, THOMAS W, ESTROGEN/PROGESTOGEN ORAL CONTRACEPTION AND BLOOD CLOTTING: A LONG-TERM FOLLOW-UP, *BR MED J [CLIN RES]*, 4, 648 (1971)

2835 POLLOCK AA, BERGER SA, SIMBERKOFF MS, ET AL, HEPATITIS ASSOCIATED WITH HIGH-DOSE OXACILLIN THERAPY, *ARCH INTERN MED*, 138, 915-917 (1978)

2836 PONCE SP, JENNINGS AE, MADIAS NE, ET AL, DRUG-INDUCED HYPERKALEMIA, *MEDICINE*, 64, 357-370 (1985)

2837 PONIKOWSKA I, BLOOD PATTERNS OF FREE FATTY ACIDS IN HEALTHY SUBJECTS AND DIABETICS UNDER INSULIN, GLUCOSE AND TOLBUTAMIDE, *POL ARCH MED WEWN*, 45, 71 (1970)

2838 PONT A, GOLDMAN ES, SUGAR EM, ET AL, KETOCONAZOLE-INDUCED INCREASE IN ESTRADIOL - TESTOSTERONE RATIO. PROBABLE EXPLANATION FOR GYNECOMASTIA, *ARCH INTERN MED*, 145, 1429-1431 (1985)

2839 PONT A, GRAYBILL JR, CRAVEN PC, ET AL, HIGH-DOSE KETOCONAZOLE THERAPY AND ADRENAL AND TESTICULAR FUNCTION IN HUMANS, *ARCH INTERN MED*, 144, 2150-2153 (1984)

2840 PONTE CD, DECKER EL, LEUKOPENIA AND HEPATOTOXICITY AS A POSSIBLE CONSEQUENCE OF CHLORPROMAZINE ADMINISTRATION, *DRUG INTELL CLIN PHARM*, 76, 562-565 (1976)

2841 PONTIROLI AE, SCARPIGNATO C, EFFECT OF BOMBESIN ON BASAL AND STIMULATED SECRETION OF SOME PITUITARY HORMONES IN HUMANS, *HORM RES*, 23, 129-135 (1986)

2842 POOL JL, PLASMA LIPID LOWERING EFFECT OF DOXAZOSIN, A NEW SELECTIVE ALPHA₁, ADRENERGIC INHIBITOR FOR SYSTEMIC HYPERTENSION, *AM J CARDIOL*, 59, 46G-50G (1987)

2843 POOLE G, STRADLING P, WORLLEDGE S, POTENTIALLY SERIOUS SIDE-EFFECTS OF HIGH-DOSE TWICE-WEEKLY RIFAMPICIN, *POSTGRAD MED J*, 47, 742 (1971)

2844 POON R, HINBERG IH, INDICAN INTERFERENCE WITH SIX COMMERCIAL PROCEDURES FOR MEASURING TOTAL BILIRUBIN, *CLIN CHEM*, 31, 92-94 (1985)

2845 POORTMANS JR, EFFECT OF EXERCISE ON THE RENAL CLEARANCE OF AMYLASE AND LYSOZYME IN HUMANS, *CLIN SCI*, 43, 115 (1972)

2846 POORTMANS JR, SERUM PROTEIN DETERMINATION DURING SHORT EXHAUSTIVE PHYSICAL ACTIVITY, *J APPL PHYSIOL*, 30, 190 (1971)

2847 POORTMANS JR, BODY FLUIDS FLUCTUATIONS INDUCED BY PHYSICAL ACTIVITIES IN, *REFERENCE VALUES IN HUMAN CHEMISTRY*, G SIEST (ED.), KARGER, BASEL, 255 (1973)

2848 PORTANOVA JP, RUBIN RL, JOSLIN FG, REACTIVITY OF ANTI-HISTONE ANTIBODIES INDUCED BY PROCAINAMIDE AND HYDRALAZINE, *CLIN IMMUNOL IMMUNOPATHOL*, 25, 67-79 (1982)

2849 PORTER GA, STUDIES CONCERNING THE MECHANISM OF SODIUM RETENTION ASSOCIATED WITH CARBENOXOLONE ADMINISTRATION, *GASTROENTEROLOGY*, 62, 795 (1972)

2850 POSADAS C, BLAIR AJ JR, MCCANN DS, EXCRETION OF TETRAHYDROALDOSTERONE IN NORMAL PREGNANCY, *CLIN CHIM ACTA*, 45, 299 (1973)

2851 POSNER J, SOBEL RJ, GLICK S, EFFECT OF AMIODARONE ON THYROID HORMONE ECONOMY, *ISR J MED SCI*, 20, 113-117 (1984)

2852 POSTLETHWAITE AE, BARTEL AG, KELLEY WN, HYPERURICEMIA DUE TO ETHAMBUTOL, *N ENGL J MED*, 286, 761-762 (1972)

2853 POSTMES TJ, HOUT JC, SAAT G, WILLEMS P, COENEGRACHT J, A RADIOIMMUNOASSAY STUDY AND COMPARISON OF SEASONAL VARIATION IN PLASMA TRIIODOTHYRONINE AND THYROXINE, *CLIN CHIM ACTA*, 50, 189 (1974)

2854 POTTER JL, XANTHINE INTERFERENCE IN THE KODAK EKTACHEM® DETERMINATION OF URIC ACID, *CLIN CHEM*, 33, 1265 (1987)

2855 POTTER JL, FURTHER OBSERVATIONS ON NINHYDRIN REACTING COMPOUNDS IN URINE, *J PEDIATR*, 84, 250 (1974)

2856 POTTER JL, WEINBERG AG, WEST R, AMPICILLINURIA AND AMPICILLIN CRYSTALLURIA, *PEDIATR*, 48, 636 (1971)

2857 POWELL JB, DJUH YY, A COMPARISON OF AUTOMATED METHODS FOR GLUCOSE ANALYSIS IN PATIENTS WITH UREMIA BEFORE AND AFTER DIALYSIS, *AM J CLIN PATHOL*, 56, 8 (1971)

2858 POWELL JB, KNESEL EA, HOFFNER PR, IRVIN CW JR., NEWELL JE, EFFECT OF CLOFIBRATE (ATROMID S) ON T_4-BY-COLUMN MEASUREMENTS, *AM J CLIN PATHOL*, 59, 764 (1973)

2859 POWELL JR, ROGERS JF, WARGIN WA, ET AL, INHIBITION OF THEOPHYLLINE CLEARANCE BY CIMETIDINE BUT NOT RANITIDINE, *ARCH INTERN MED*, 144, 484-486 (1984)

2860 POWELL MG, HEDLIN AM, CERSKUS I, ET AL, EFFECTS OF ORAL CONTRACEPTIVES ON LIPOPROTEIN LIPIDS: A PROSPECTIVE STUDY, *OBSTET GYNECOL*, 63, 764-770 (1984)

2861 POWELL-JACKSON PR, JAMIESON AP, GRAY BJ, ET AL, EFFECT OF RIFAMPICIN ADMINISTRATION ON THEOPHYLLINE PHARMACOKINETICS IN HUMANS, *AM REV RESPIR DIS*, 131, 939-940 (1985)

2862 POWERS PJ, KELTON JG, CARTER CJ, STUDIES ON THE FREQUENCY OF HEPARIN-ASSOCIATED THROMBOCYTOPENIA, *THROMB RES*, 33, 439-443 (1984)

2863 PRATI RC, ALFREY AC, HULL AR, SPIRONOLACTONE-INDUCED HYPERCALCIURIA, *J LAB CLIN MED*, 80, 224 (1972)

2864 PRATT CB, VERZOSA M, COMPARISON OF CRYSTALLINE AND AMORPHOUS ASPARAGINASE IN TREATMENT OF ACUTE LEUKEMIA IN CHILDREN, *CLIN PHARMACOL THER*, 13, 343 (1972)

2865 PREECE MJ, RICHARDSON JA, THE EFFECT OF MILD DEHYDRATION ON ONE-HOUR CREATININE CLEARANCE RATES, *NEPHRON*, 9, 106 (1972)

2866 PRESCOTT LF, ANALGESIC NEPHROPATHY: A REASSESSMENT OF THE ROLE OF PHENACETIN AND OTHER ANALGESICS, *DRUGS*, 23, 75-149 (1982)

2867 PRESSAC M, JARDEL C, DURAND D, ET AL, INTERFERENCE OF BROMIDE IN DETERMINATION OF SERUM CHLORIDE, *CLIN CHEM*, 33, 415-416 (1987)

2868 PRETTY HM, GOSSELIN G, COLPRON G, LONG LA, AGRANULOCYTOSIS: A REPORT OF 30 CASES, *CAN MED ASSOC J*, 93, 1058 (1965)

2869 PRICE HL, CIRCULATING ADRENALINE AND NORADRENALINE DURING DIETHYL-ETHER ANAESTHESIA IN MAN, *CLIN SCI*, 16, 377 (1957)

2870 PRICE J, METHYLATION IN SCHIZOPHRENICS: A PHARMACOGENETIC STUDY, *J PSYCHIATR RES*, 9, 345 (1972)

2871 PRICE LH, NELSON JC, JATLOW PI, EFFECT OF DESIPRAMINE ON CLINICAL LIVER FUNCTION TESTS, *AM J PSYCHIATRY*, 141, 798-800 (1984)

2872 PRIMACK WA, GARTNER LM, MCGURK HE, ET AL, HYPERNATREMIA ASSOCIATED WITH CHOLESTYRAMINE THERAPY, *J PEDIATR*, 90, 1024-1025 (1977)

2873 PRISTAUTZ H, STRADNER F, EFFECT OF CELIPROLOL AND METOPROLOL ON SERUM LIPIDS IN PATIENTS WITH DIFFERENT TYPES OF HYPERLIPOPROTEINEMIA, *WIEN MED WOCHENSCHR*, 136, 443-448 (1986)

2874 PROBSTFIELD JL, LEE G, CAMPION B, HUMMINGHAKE DB, COLESTIPOL, AN INVESTIGATIONAL BILE-ACID SEQUESTERING AGENT FOR LOWERING SERUM CHOLESTEROL, *CLIN RES*, 20, 412 (1972)

2875 PROBSTFIELD JL, LIN T-L, PETERS J, ET AL, CAROTENOIDS AND VITAMIN A: THE EFFECT OF HYPOCHOLESTEROLEMIC AGENTS ON SERUM LEVELS, *METABOLISM*, 34, 88-91 (1985)

2876 PROCTOR EA, BARTON FL, POLYURIC ACUTE RENAL FAILURE AFTER METHOXYFLURANE AND TETRACYCLINE, *BR MED J [CLIN RES]*, 4, 661 (1971)

2877 PROELSS HF, WRIGHT BW, RAPID DETERMINATION OF AMMONIA IN A PERCHLORIC ACID SUPERNATE FROM BLOOD, BY USE OF AN AMMONIA-SPECIFIC ELECTRODE, *CLIN CHEM*, 19, 1162 (1973)

2878 PROKSCH GJ, BONDERMAN DP, GRIEP JA, AUTOANALYZER ASSAY FOR SERUM ALKALINE PHOSPHATASE ACTIVITY WITH SODIUM THYMOLPHTHALEIN MONOPHOSPHATE AS SUBSTRATE, *CLIN CHEM*, 19, 103 (1973)

2879 PROPP RP, STILLMAN JS, AGRANULOCYTOSIS AND HYDROXYCHLOROQUINE, *N ENGL J MED*, 277, 492 (1967)

2880 PRYCE JD, WOOTTON IDP, POPULATION STUDIES IN CLINICAL BIOCHEMISTRY, *PROC ASS CLIN BIOCHEM*, 3, 62 (1965)

2881 PTACHCINSKI RJ, CARPENTER BJ, BURCKART GJ, ET AL, EFFECT OF ERYTHROMYCIN ON CYCLOSPORINE LEVELS, *N ENGL J MED*, 313, 1416-1417 (1985)

2882 PUCHOIS P, FONTAN M, GENTILINI J-L, ET AL, SERUM APOLIPOPROTEIN A-II, A BIOCHEMICAL INDICATOR OF ALCOHOL ABUSE, *CLIN CHIM ACTA*, 185, 185-189 (1984)

2883 PUI C-H, BURGHEN GA, BOWMAN WP, ET AL, RISK FACTORS FOR HYPERGLYCEMIA IN CHILDREN WITH LEUKEMIA RECEIVING L-ASPARAGINASE AND PREDNISONE, *J PEDIATR*, 99, 46-50 (1981)

2884 PULKKINEN MO, MAENPAA J, DECREASE IN SERUM TESTOSTERONE CONCENTRATION DURING TREATMENT WITH TETRACYCLINE, *ACTA ENDOCRINOL*, 103, 269-272 (1983)

2885 PULKKINEN MO, WILLMAN K, REDUCTION OF MATERNAL ESTROGEN EXCRETION BY NEOMYCIN, *AM J OBSTET GYNECOL*, 115, 1153 (1973)

2886 PULKKINEN MO, WILLMAN K, SERUM INORGANIC PHOSPHATE DURING ORAL CONTRACEPTIVE THERAPY, *ANN CHIR GYNAECOL*, 57, 172 (1968)

2887 PULLAR T, CAPELL HA, SULFASALAZINE: A 'NEW' ANTIRHEUMATIC DRUG, *BR J RHEUMATOL*, 23, 26-34 (1984)

2888 PUOPOLO PR, FLOOD JG, DETECTION OF INTERFERENCE BY CYCLOBENZAPINE IN LIQUID-CHROMATOGRAPHIC ASSAYS OF TRICYCLIC ANTIDEPRESSANTS, *CLIN CHEM*, 33, 819-820 (1987)

2889 PURI VN, INCREASED URINARY ANTIDIURETIC HORMONE EXCRETION BY IMIPRAMINE, *EXP CLIN ENDOCRINOL*, 88, 112-114 (1986)

2890 PUUSTINEN T, DAHL M-L, UOTILA P, ET AL, GLUCOCORTICOIDS DO NOT DECREASE THROMBOXANE AND PROSTACYCLIN LEVELS IN HUMAN BLOOD, *PROSTAGLANDINS LEUKOT MED*, 15, 409-410 (1984)

2891 PYBUS J, BOWERS GN JR., SERUM LITHIUM DETERMINATION BY ATOMIC ABSORPTION SPECTROSCOPY, *STAND METH CLIN CHEM*, 6, 189 (1970)

2892 QUAGLIANA JM, EFFECT OF TOPICAL POVIDONE-IODINE (BETADINE®) ON SERUM PROTEIN-BOUND IODINE, *J CLIN ENDOCRINOL METAB*, 23, 395 (1963)

2893 QUASTEL JH, THE ACTION OF DRUGS ON ENZYME SYSTEMS IN, *ENZYMES IN HEALTH AND DISEASE*, DM GREENBERG, HA HARPER, EDS. THOMAS, SPRINGFIELD, IL (1960)

2894 QUEEN, CA, FRINGS CS, EFFECT OF SODIUM AZIDE ON TRINDERS METHOD FOR THE DETERMINATION OF SALICYLATE, *CLIN CHIM ACTA*, 45, 307 (1973)

2895 QUIGLEY ME, ROPERT JF, YEN SS, ACUTE PROLACTIN RELEASE TRIGGERED BY FEEDING, *J CLIN ENDOCRINOL METAB*, 52, 1043-1045 (1981)

2896 QUOTTROCCHI FP, ROBINSON JD, CURRY RW JR., ET AL, THE EFFECT OF IBUPROFEN ON SERUM DIGOXIN CONCENTRATIONS, *DRUG INTELL CLIN PHARM*, 17, 286-288 (1983)

2897 RAAB WP, THE DIAGNOSTIC VALUE OF URINARY ENZYME DETERMINATIONS, *CLIN CHEM*, 18, 5 (1972)

2898 RABINOWITZ D, MERIMEE TJ, NELSON JK, SCHULTZ RB, BURGESS JA, *GROWTH HORMONE INTERNATIONAL CONGRESS SERIES NO 158*, EXCERPTA MEDICA, AMSTERDAM (1968)

2899 RACE TF, PAES IC, FALOON WW, INTESTINAL MALABSORPTION INDUCED BY ORAL COLCHICINE: COMPARISON WITH NEOMYCIN AND CATHARTIC AGENTS, *AM J MED SCI*, 259, 32 (1970)

2900 RADCLIFF FJ, WILTON NM, DONNELLY GL, CLOPAMIDE (BRINALDIX), A NEW DIURETIC AGENT: DURATION OF ACTION AND DOSAGE RESPONSE, *CURR THER RES*, 10, 103 (1968)

2901 RADO JP, FALSELY HIGH FLUORESCENCE IN CORTISOL DETERMINATIONS DUE TO THE CARBAMAZEPINE, *HORM METAB RES*, 5, 63 (1973)

2902 RADO JP, BANOS C, GERCSAK G, ET AL, GLUCOSE-INDUCED HYPERKALEMIA DEVELOPING IN THE UPRIGHT POSITION IN CAPTOPRIL-TREATED HYPERTENSIVES, *RES COMMUN CHEM PATHOL PHARMACOL*, 38, 161-164 (1982)

2903 RAGHEB M, BAN TA, BUCHANAN D, ET AL, INTERACTION OF INDOMETHACIN AND IBUPROFEN WITH LITHIUM IN MANIC PATIENTS UNDER A STEADY-STATE LITHIUM LEVEL, *J CLIN PSYCHIATRY*, 41, 397-398 (1980)

2904 RAMEY JN, BURROW GN, SPAULDING SW, ET AL, THE EFFECT OF ASPIRIN AND INDOMETHACIN ON THE TRH RESPONSE IN MAN, *J CLIN ENDOCRINOL METAB*, 43, 107-114 (1976)

2905 RAMIREZ-LASSEPAS M, CIPOLLE RJ, RODVOLD KA, ET AL, HEPARIN-INDUCED THROMBOCYTOPENIA IN PATIENTS WITH CEREBROVASCULAR ISCHEMIC DISEASE, *NEUROLOGY*, 34, 736-740 (1984)

2906 RAMOND M-J, NOVEL O, DEGOTT C, ET AL, HEPATITE A L'ALLOPURINOL, *GASTROENTEROL CLIN BIOL*, 6, 138-142 (1982)

2907 RAMSAY ID, CARBAMAZEPINE-INDUCED JAUNDICE, *BR MED J [CLIN RES]*, 4, 155 (1967)

2908 RAMSDALE EH, MOWBRAY JF, POSTIVE NBT TESTS IN PREGNANCY, *LANCET*, 1, 1246 (1973)

2909 RAMSDELL CM, KELLEY WN, THE CLINICAL SIGNIFICANCE OF HYPOURICEMIA, *ANN INTERN MED*, 78, 239-242 (1973)

2910 RAMSEY TA, MENDELS J, STOKES JW, FITZGERALD RG, LITHIUM CARBONATE AND KIDNEY FUNCTION: A FAILURE IN RENAL CONCENTRATING ABILITY, *J AM MED ASSOC*, 219, 1446 (1972)

2911 RANDALL RE JR., OSHEROFF RJ, BAKERMAN S, SETTER JG, BISMUTH NEPHROTOXICITY, *ANN INTERN MED*, 77, 481 (1972)

2912 RANEK L, ANDREASEN PB, DISULFIRAM HEPATOTOXICITY, *BR MED J [CLIN RES]*, 2, 94-96 (1977)

2913 RAO SD, VAKIL SK, CALNE DB, HILSON A, AUGMENTING THE ACTION OF LEVODOPA, *POSTGRAD MED J*, 48, 653 (1972)

2914 RAO TK, NICASTRI AD, FRIEDMAN EA, NATURAL HISTORY OF HEROIN-ASSOCIATED NEPHROPATHY, *N ENGL J MED*, 290, 19 (1974)

2915 RAOOF S, WOLLSCHLAGER C, KHAN FA, CIPROFLOXACIN INCREASES SERUM LEVELS OF THEOPHYLLINE, *AM J MED*, 82 SUPPL 4A, 115-118 (1987)

2916 RAPOPORT S, WING M, GUEST GM, HYPOPROTHROMBINEMIA AFTER SALICYLATE ADMINISTRATION IN MAN AND RABBITS, *PROC SOC EXP BIOL MED*, 53, 40 (1943)

2917 RAPPELLI A, DESSI-FULGHERI P, BANDIERA F, ET AL, CHANGES IN PLASMA ATRIAL NATRIURETIC PEPTIDE LEVELS AFTER A SINGLE SUBLINGUAL DOSE OF NIFEDIPINE IN HYPERTENSIVE PATIENTS, *MED SCI RES*, 15, 1503-1504 (1987)

2918 RAPTIS S, DOLLINGER HC, SCHRODER KE, SCHLEYER M, ROITHENBUCHNER G, PFEIFFER EF, DIFFERENCE IN INSULIN GROWTH HORMONE AND PANCREATIC ENZYME SECRETION, *N ENGL J MED*, 288, 1199 (1973)

2919 RASANAYAGAM LJ, LIM KL, BENG CG, LAU KS, MEASUREMENT OF URINE ALBUMIN USING BROMOCRESOL GREEN, *CLIN CHIM ACTA*, 44, 53 (1973)

2920 RASKIND MA, COURTNEY N, MURBERY MM, ET AL, ANTIPSYCHOTIC DRUGS AND PLASMA VASOPRESSIN IN NORMALS AND ACUTE SCHIZOPHRENIC PATIENTS, *BIOL PSYCHIATRY*, 22, 53-462 (987)

2921 RATGE D, BARTHELS U, WISSER H, ET AL, NEBENWIRKUNGEN UND VERHALTEN VON NORADRENALIN UND ADRENALIN IN PLASMA BEIM INTRAVENOSEN THYROLIBERIN - TEST BEI PERSONEN MIT NORMALER UND GESTORTER SCHILDDRUSENFUNKTION, *J CLIN CHEM CLIN BIOCHEM*, 25, 393-400 (1987)

2922 RATLIFF CR, GILLILAND PF, HALL FF, SERUM PROPYLTHIOURACIL: DETERMINATION BY A DIRECT COLORIMETRIC PROCEDURE, *CLIN CHEM*, 18, 1373 (1972)

2923 RATLIFF CR, HALL FF, FLUOROMETRIC MEASUREMENT OF URINARY FREE HYDROCORTISONE (CORTISOL), *CLIN CHEM*, 19, 1400 (1973)

2924 RATNAIKE RN, RINGWOOD DIS, HISLOP LG, THE EFFECT OF ALCOHOL ON AMMONIA METABOLISM, *AUST N Z J MED*, 2, 322 (1972)

2925 RATZMANN M-L, RJASANOWSKI I, BRUNS W, ET AL, EFFECTS OF CLOFIBRATE THERAPY ON GLUCOSE TOLERANCE, INSULIN SECRETION AND SERUM LIPIDS IN SUBJECTS WITH HYPERLIPOPROTEINEMIA AND IMPAIRED GLUCOSE TOLERANCE. A FOLLOW-UP STUDY OVER A 5 YEAR PERIOD., *EXP CLIN ENDOCRINOL*, 82, 216-221 (1983)

2926 RAUSCHER E, NEUMANN U, SCHAICH E, ET AL, OPTIMIZED CONDITIONS FOR DETERMINING ACTIVITY CONCENTRATION OF ALPHA-AMYLASE IN SERUM, WITH 1, 4-α-D-4-NITROPHENYLMALTOHEPTAOSIDE AS SUBSTRATE, *CLIN CHEM*, 31, 14-19 (1985)

2927 RAWLINS MD, SMITH SE, INFLUENCE OF ALLOPURINOL ON DRUG METABOLISM IN MAN, *BR J PHARMACOL*, 48, 693-698 (1973)

2928 RAYFIELD EJ, CAIN JP, CASEY MP, WILLIAMS GH, SULLIVAN, JM, INFLUENCE OF DIET ON URINARY VMA EXCRETION, *J AM MED ASSOC*, 221, 704 (1972)

2929 RAYFIELD EJ, CURNOW RT, BEISEL WR, ACUTE GLUCOSE INTOLERANCE DURING SANDFLY FEVER IN MAN: METABOLIC INTERRELATIONSHIPS, *AM J CLIN NUTR*, 26, 463 (1973)

2930 RECHENBERG HK, *PHENYLBUTAZONE*, EDWARD ARNOLD LTD LONDON (1962)

2931 RECKER RR, HASSING GS, LAU JR, SAVILLE PD, THE HYPERPHOSPHATEMIC EFFECT OF DISODIUM ETHANE-1HYDROXY-1, 1-DIPHOSPHONATE (EHDP), *J LAB CLIN MED*, 81, 258 (1973)

2932 RECTOR TS, CIPOLLE RJ, SEIFERT RD, ET AL, CHARACTERISTICS OF HEPARIN-ASSOCIATED THROMBOCYTOPENIA, *AM J HOSP PHARM*, 36, 1561-1565 (1979)

2933 REDONDO FL, BERGON E, TINTURE T, ET AL, URINARY ENZYME ACTIVITIES IN PATIENTS TREATED WITH GOLD AND OTHER ANTIRHEUMATIC DRUGS, *CLIN BIOCHEM*, 20, 343-347 (1987)

2934 REED RG, GUINEY WB, COLLIER SA, SALICYLATE INTERFERENCE WITH MEASUREMENT OF ACETAMINOPHEN, *CLIN CHEM*, 28, 2178-2179 (1982)

2935 REED V, WHITE R, BLOOD BARBITURATE LEVELS, *LANCET*, 1, 634 (1972)

2936 REES LH, RATCLIFFE JG, BESSER GM, KRAMER R, LANDON JR, CHAYEN J, COMPARISON OF THE REDOX ASSAY FOR ACTH WITH PREVIOUS ASSAYS, *NATURE NEW BIOL*, 241, 84 (1973)

2937 REEVES DS, SULFAMETHOXAZOLE/TRIMETHOPRIM: THE FIRST TWO YEARS, *J CLIN PATHOL*, 24, 430 (1971)

2938 REEVES, RA, FROM GLA, PAUL W, ET AL, NADOLOL, PROPRANOLOL AND THYROID HORMONES: EVIDENCE FOR A MEMBRANE STABILIZING ACTION OF PROPRANOLOL, *CLIN PHARMACOL THER*, 37, 157-161 (1985)

2939 REHFELD JF, STADIL F, THE EFFECT OF GASTRIN ON BASAL- AND GLUCOSE-STIMULATED INSULIN SECRETION IN MAN, *J CLIN INVEST*, 52, 1415 (1973)

2940 REHNQVIST N, INTRAHEPATIC JAUNDICE DUE TO WARFARIN THERAPY, *ACTA MED SCAND*, 204, 335-336 (1978)

2941 REIDENBERG MM, KIDNEY FUNCTION AND DRUG ACTION, *N ENGL J MED*, 313, 816-818 (1985)

2942 REILLY CS, BIOALLAZ J, KOSHAKJI RP ET AL, ENPROSTIL, IN CONTRAST TO CIMETIDINE, DOES NOT INHIBIT PROPRANOLOL METABOLISM, *CLIN PHARMACOL THER*, 40, 37-41 (1986)

2943 REIMANN IW, FROLICH JC, EFFECTS OF DICLOFENAC ON LITHIUM KINETICS, *CLIN PHARMACOL THER*, 30, 348-352 (1981)

2944 REIMOLD EW, THE EFFECT OF FUROSEMIDE ON HYPERCALCEMIA DUE TO DIHYDROTACHYSTEROL, *METABOLISM*, 21, 593 (1972)

2945 REINER M, CHEUNG HL, XYLOSE, *STAND METH CLIN CHEM*, 5, 257 (1965)

2946 REINHARDT DJ III, TAUSIG T, ALVAREZ R, SERUM CHOLESTEROL ELEVATION WITH TRIFLUORPERAZINE (STELAZINE®) THERAPY, *DEL MED J*, 34, 318 (1962)

2947 REINHOLD JG, NASR K, LAHIMGARZADEH A, HEDAYATI H, EFFECTS OF PURIFIED PHYTATE AND PHYTATE-RICH BREAD UPON METABOLISM OF ZINC, CALCIUM, PHOSPHORUS AND NITROGEN, *LANCET*, 1, 283 (1973)

2948 REINKEN L, THE EFFECT OF HYDANTOIN AND SUCCINIMIDE ON VITAMIN B_6 METABOLISM, *CLIN CHIM ACTA*, 48, 435 (1973)

2949 REINKEN L, HOHENAUER L, ZIEGLER EE, ACTIVITY OF RED CELL GLUTAMIC OXALOACETIC TRANSAMINASE IN EPILEPTIC CHILDREN UNDER ANTIEPILEPTIC TREATMENT, *CLIN CHIM ACTA*, 36, 270 (1972)

2950 REIO L, WETTERBERG L, FALSE PORPHOBILINOGEN REACTIONS IN THE URINE OF MENTAL PATIENTS, *J AM MED ASSOC*, 207, 148 (1969)

2951 REISS E, CANTERBURY JM, BERCOVITZ MA, KAPLAN EL, ROLE OF PHOSPHATE IN THE SECRETION OF PARATHYROID HORMONE IN MAN, *J CLIN INVEST*, 49, 2146 (1970)

2952 REISZ G, PINGLETON SK, MELETHIL S, ET AL, THE EFFECT OF ERYTHROMYCIN ON THEOPHYLLINE PHARMACOKINETICS IN CHRONIC BRONCHITIS, *AM REV RESPIR DIS*, 127, 581-584 (1983)

2953 REJ R, FASCE CF JR, VANDERLINDE RE, INCREASED ASPARTATE AMINOTRANSFERASE ACTIVITY OF SERUM AFTER IN VITRO SUPPLEMENTATION WITH PYRIDOXAL PHOSPHATE, *CLIN CHEM*, 19, 92 (1973)

2954 REMP DG, URIC ACID (URICASE), *STAND METH CLIN CHEM*, 6, 1 (1970)

2955 RENAUD S, BLACHE D, DUMONT E, ET AL, PLATELET FUNCTION AFTER CIGARETTE SMOKING IN RELATION TO NICOTINE AND CARBON MONOXIDE, *CLIN PHARMACOL THER*, 36, 389-395 (1984)

2956 RENAULT PF, SCHUSTER CR, HEINRICH RL, VAN DER KOLK B, ALTERED PLASMA CORTISOL RESPONSE IN PATIENTS ON METHADONE MAINTENANCE, *CLIN PHARMACOL THER*, 13, 269 (1972)

2957 RENNIE D, FRAYSER R, GRAY G, HOUSTON C, URINE AND PLASMA PROTEINS IN MEN AT 5400 M, *J APPL PHYSIOL*, 32, 369 (1972)

2958 RENNIE ID, KEEN H, EVALUATION OF CLINICAL METHODS FOR DETECTING PROTEINURIA, *LANCET*, 2, 489 (1967)

2959 RESURRECCION EC, ROSENBLUM JA, COMMON CAUSES OF SPURIOUS XANTHOCHROMIA IN CEREBROSPINAL FLUID, *ANGIOLOGY*, 23, 105 (1972)

2960 RETSAS S, PHILLIPS RH, HANHAM IWF, ET AL, AGRANULOCYTOSIS IN BREAST CANCER PATIENTS TREATED WITH LEVAMISOLE, *LANCET*, 2, 324 (1978)

2961 REUBI FC, VORBUGER C, BUTIKOFER E, A COMPARISON OF THE SHORT-TERM AND LONG-TERM HAEMODYNAMIC EFFECT OF ANTIHYPERTENSIVE DRUG THERAPY IN, *CATAPRES® IN HYPERTENSION*, ME CONOLLY (ED.) BUTTERWORTH, LONDON (1970)

2962 REX D, LUMENG L, EBLE J, ET AL, INTRAHEPATIC CHOLESTASIS AND SICCA COMPLEX AFTER THIABENDAZOLE, *GASTROENTEROLOGY*, 85, 718-721 (1983)

2963 REYES E, LISANSKY J, EFFECTS OF TRICYCLIC ANTIDEPRESSANTS ON PLATELET MONOAMINE OXIDASE ACTIVITY, *CLIN PHARMACOL THER*, 35, 531-534 (1984)

2964 REYES MP, PALUTKE M, LERNER AM, GRANULOCYTOPENIA ASSOCIATED WITH CARBENICILLIN, *AM J MED*, 54, 413 (1973)

2965 REYNIAK JV, WENOF M, AUBERT JM, ET AL, INCIDENCE OF HYPERPROLACTINEMIA DURING ORAL CONTRACEPTIVE THERAPY, *OBSTET GYNECOL*, 55, 8-11 (1980)

2966 REYNOLDS EH, ANTICONVULSANTS, FOLIC ACID, AND EPILEPSY, *LANCET*, 1, 1376 (1973)

2967 REYNOLDS EH, GALLAGHER BB, MATTSON RH, BOWERS M, JOHNSON AL, RELATIONSHIP BETWEEN SERUM AND CEREBROSPINAL FLUID FOLATE, *NATURE*, 2540, 155 (1972)

2968 REYNOLDS EH, MILNER G, MATHEWS DM, CHANARIN I, ANTICONVULSANT THERAPY, MEGALOBLASTIC HAEMOPOIESIS AND FOLIC ACID METABOLISM, *Q J MED*, 35, 521 (1966)

2969 REYNOLDS TB, PELLE HC, EFFECTS OF A NEW DIURETIC AMIPRAMIDINE (MK 870) IN PATIENTS WITH CIRRHOSIS AND ASCITES, *CLIN RES*, 14, 184 (1966)

2970 REYNOLDS TB, PETERS RL, YAMADA S, CHRONIC ACTIVE AND LUPOID HEPATITIS CAUSED BY A LAXATIVE, OXYPHENISATIN, *N ENGL J MED*, 285, 813 (1971)

2971 RHYS J, KADURY S, WOODHEAD JS, ET AL, FENCLOFENAC AND THYROID FUNCTION TESTS, *ANN CLIN BIOCHEM*, 20, 381-382 (1983)

2972 RICCI N, TOMA PM, PAZZI P, ET AL, BROMIDE CAN INTERFERE WITH CHLORIDE ESTIMATION, *LANCET*, 2, 100 (1982)

2973 RICE EW, LUKASIEWICZ DB, INTERFERENCE OF BROMIDE IN ZAK FECL$_3$ SULFURIC ACID CHOLESTEROL METHOD AND MEANS OF ELIMINATING THIS INTERFERENCE, *CLIN CHEM*, 3, 160 (1957)

2974 RICHARDS DA, HARRIS DM, MARTIN LE, LABETALOL AND URINARY CATECHOLAMINES, *BR MED J [CLIN RES]*, 1, 685 (1979)

2975 RICHARDS KC, BORGSTEDT HH, NEAR FATAL REACTION TO INGESTION OF THE HALLUCINOGENIC DRUG MDA, *J AM MED ASSOC*, 218, 1826 (1971)

2976 RICHARDS P, BROWN CL, LOWE SM, SYNTHESIS OF TRYPTOPHAN FROM 3-INDOLEPYRUVIC ACID BY A HEALTHY WOMAN, *J NUTR*, 102, 1547 (1972)

2977 RICHARDSON RA, INTERFERENCE IN THE ZIMMERMANN TEST FOR STEROIDS, *CLIN CHIM ACTA*, 50, 151 (1974)

2978 RICHARDSON RW, SETCHELL KDR, WOODMAN DD, ESTIMATION OF THE NORMAL RANGE FOR SERUM CHOLESTEROL BY DIFFERENT METHODS, *CLIN CHIM ACTA*, 37, 305 (1972)

2979 RICHENS A, HOUGHTON GW, LETTER: PHENYTOIN INTOXICATION CAUSED BY SULTHIAME, *LANCET*, 2, 1442 (1973)

2980 RICHEY JE, SMITH RB, RENAL FAILURE AFTER METHOXYFLURANE ANAESTHESIA, *ANAESTHESIA*, 27, 9 (1972)

2981 RICHTERICH R, *CLINICAL CHEMISTRY THEORY AND PRACTICE*, KARGER, BASEL (1969)

2982 RICHTERICH R, DAUWAULDER H, DETERMINATION OF BLOOD GLUCOSE CONCENTRATION BY THE HEXOKINASE/GLUCOSE-6-PHOSPHATE DEHYDROGENASE METHOD, *SCHWEIZ MED WOCHENSCHR*, 101, 615 (1971)

2983 RIDDICK JH JR, AN ULTRAVIOLET AND VISIBLE SPECTROPHOTOMETRIC ASSAY METHOD FOR CHLORDIAZEPOXIDE, *CLIN BIOCHEM*, 6, 189 (1973)

2984 RIDDOCH D, GASTRITIS AND L-DOPA, *BR MED J [CLIN RES]*, 1, 53 (1972)

2985 RIDOLFO AS, RUBIN A, CRABTREE RE, GRUBER CM JR, EFFECTS OF FENOPROFEN AND ASPIRIN ON GASTROINTESTINAL MICROBLEEDING, *CLIN PHARMACOL THER*, 14, 226 (1973)

2986 RIEGEL W, HORL WH, HEIDLAND A, LONG-TERM EFFECTS OF NIFEDIPINE ON CARBOHYDRATE AND LIPID METABOLISM IN HYPERTENSIVE HEMODIALYZED PATIENTS, *KLIN WOCHENSCHR*, 64, 1124-1130 (1986)

2987 RIGAS B, ROSENFELD LE, BARWICK KW, AMIODARONE HEPATOTOXICITY: A CLINICOPATHOLOGIC STUDY OF 5 PATIENTS, *ANN INTERN MED*, 104, 348-351 (1986)

2988 RIGBERG LA, ROBINSON MJ, ESPIRITU CR, CHLORPROPAMIDE - INDUCED GRANULOMAS: A PROBABLE HYPERSENSITIVITY REACTION IN LIVER AND BONE MARROW, *J AM MED ASSOC*, 235, 409-410 (1976)

2989 RIIS B, CHRISTIANSEN C, ACTIONS OF THIAZIDE IN VITAMIN D METABOLISM A CONTROLLED THERAPEUTIC TRIAL IN NORMAL WOMEN EARLY IN THE MENOPAUSE, *METABOLISM*, 34, 421-424 (1985)

2990 RINGOIR S, WIEME RJ, SERUM LDH IN HEMODIALYSIS, *CLIN CHIM ACTA*, 42, 315 (1972)

2991 RISBY ED, HSIAO JK, SUNDERLAND T, ET AL, THE EFFECTS OF ANTIDEPRESSANTS ON THE CEREBROSPINAL FLUID HOMOVANILLIC ACID/ 5-HYDROXYINDOLEACETIC ACID RATIO, *CLIN PHARMACOL THER*, 42, 547-554 (1987)

2992 RISLER T, BURK M, PETERS U, ET AL, ON THE INTERACTION BETWEEN DIGOXIN AND DISOPYRAMIDE, *CLIN PHARMACOL THER*, 34, 176-180 (1983)

2993 RITTER MA, GIOE TJ. SIEBER JM, SYSTEMIC EFFECTS OF POLYMETHYLMETHACRYLATE. INCREASED SERUM LEVELS OF GAMMA-GLUTAMYLTRANSPEPTIDASE FOLLOWING ARTHROPLASTY, *ACTA ORTHOP SCAND*, 55, 411-413 (1984)

2994 RIVA R, CONTIN M, ALBANI F, ET AL, EPILEPTIC CHILDREN BEING TREATED WITH CARBAMAZEPINE, *CLIN CHEM*, 31, 150-151 (1985)

2995 RIVAL J, RIDDLE JM, STEIN PD, EFFECTS OF CHRONIC SMOKING ON PLATELET FUNCTION, *THROMB RES*, 45, 75-85 (1987)

2996 RIVERA R, ALMONTE H, ARREOLA M, ET AL, THE EFFECTS OF THREE DIFFERENT REGIMENS OF ORAL CONTRACEPTIVES AND THREE DIFFERENT INTRAUTERINE DEVICES ON THE LEVELS OF HEMOGLOBIN, SERUM IRON AND IRON-BINDING CAPACITY IN ANEMIC WOMEN, *CONTRACEPTION*, 27, 311-327 (1983)

2997 RIVERA-CALIMLIM L, BIANCHINE JR, EFFECT OF L-DOPA ON PLASMA FREE FATTY ACIDS AND PLASMA GLUCOSE, *METABOLISM*, 21, 611 (1972)

2998 RIVERS JM, DEVINE MM, PLASMA ASCORBIC ACID CONCENTRATIONS AND ORAL CONTRACEPTIVES, *AM J CLIN NUTR*, 25, 684 (1972)

2999 ROACH NA, KROLL MH, ELIN RJ, INTERFERENCE BY SULFONYLUREA DRUGS WITH THE JAFFE METHOD FOR CREATININE, *CLIN CHIM ACTA*, 151, 301-305 (1985)

3000 ROBBINS RC, HARBIN JE JR, THE IN VITRO SENSITIVITY OF ERYTHROCYTE AGGREGATION TO QUININE, *CLIN CHEM*, 17, 31 (1971)

3001 ROBERTS EA, COX DW, MEDLINE A, ET AL, OCCURRENCE OF ALPHA$_1$-ANTITRYPSIN DEFICIENCY IN 155 PATIENTS WITH ALCOHOLIC LIVER DISEASE, *AM J CLIN PATHOL*, 82, 424-427 (1984)

3002 ROBERTS I, SMITH IM, A HIGH PERFORMANCE LIQUID CHROMATOGRAPHY METHOD FOR THE ANALYSIS OF TOTAL AND FREE INDOMETHACIN IN SERUM, *ANN CLIN BIOCHEM*, 24, 167-171 (1987)

3003 ROBERTS PS, REGELSON W, KINGBURY B, DETERMINATION OF FIBRINOGEN IN NORMAL HUMAN PLASMA IN THE PRESENCE AND ABSENCE OF ANTICOAGULANTS, *J LAB CLIN MED*, 82, 822 (1973)

3004 ROBERTS RK, GRICE J, WOOD L, ET AL, CIMETIDINE IMPAIRS THE ELIMINATION OF THEOPHYLLINE AND ANTIPYRINE, *GASTROENTEROLOGY*, 81, 19-21 (1981)

3005 ROBERTSON DM, MESTER J, KELLIE AE, THE MEASUREMENT OF ESTRADIOL AND PROGESTERONE IN PLASMA FROM NORMAL, INFERTILE AND CLOMIPHENE TREATED WOMEN, *ACTA ENDOCRINOL*, 68, 523 (1971)

3006 ROBERTSON NRC, APNEA AFTER THAM ADMINISTRATION IN THE NEWBORN, *ARCH DIS CHILD*, 45, 206 (1970)

3007 ROBERTSON WG, SCURR DS, PREVENTION OF ASCORBIC ACID INTERFERENCE IN THE MEASUREMENT OF OXALIC ACID IN URINE BY ION-CHROMATOGRAPHY, *CLIN CHIM ACTA*, 140, 97-99 (1984)

3008 ROBINS MM, APLASTIC ANEMIA SECONDARY TO ANTICONVULSANTS, *AM J DIS CHILD*, 104, 614 (1962)

3009 ROBINSON AG, LOEB JN, ETHANOL INGESTION, COMMONEST CAUSE OF ELEVATED PLASMA OSMOLALITY?, *N ENGL J MED*, 284, 1253 (1971)

3010 ROBINSON AR, LEUKOCYTOSIS AFTER TETRACOSACTRIN, *BR MED J [CLIN RES]*, 4, 178 (1972)

3011 ROBINSON DS, SYLVESTER D, INTERACTION OF COMMONLY PRESCRIBED DRUGS AND WARFARIN, *ANN INTERN MED*, 72, 853 (1970)

3012 ROBINSON RG, INDOMETHACIN IN RHEUMATIC DISEASE - A REASSESSMENT, *MED J AUST*, 1, 971 (1966)

3013 ROBINSON RG, RADCLIFF FJ, THE EFFECT OF MECLOFENAMIC ACID ON PLASMA URIC ACID LEVELS, *MED J AUST*, 1, 1079 (1972)

3014 ROBSON S, NEUBERGER J, ALEXANDER G, ET AL, CYCLOSPORIN A NEPHROTOXICITY RELATED TO CHANGES IN HAEMOGLOBIN CONCENTRATION, *BR MED J [CLIN RES]*, 288, 1417-1418 (1984)

3015 ROCHE LABORATORIES, MANUFACTURER'S LITERATURE ON GANTRISIN®, *DIVISION OF HOFFMAN LA ROCHE INC, NUTLEY, NJ 07110*

3016 ROCHE LABORATORIES, ROCHE LABORATORIES ADVERTISEMENT: DATA ON FILE, *MEDICAL DEPARTMENT, HOFFMAN LA ROCHE INC, NUTLEY, NJ* (1971)

3017 ROCHE LABORATORIES, *NUTLEY NJ 07110*, DRUG INFORMATION CIRCULAR ON VALIUM® (DIAZEPAM)

3018 RODDIS MJ, PARACETAMOL INTERFERENCE WITH GLUCOSE ANALYSIS, *LANCET*, 2, 634-635 (1981)

3019 RODEN DM, WOOSLEY RL, TOCAINIDE, *N ENGL J MED*, 315, 41-45 (1986)

3020 RODIN E, SUBRAMANIAN MG, GILROY J, INVESTIGATION OF SEX HORMONES IN MALE EPILEPTIC PATIENTS, *EPILEPSIA*, 25, 690-694 (1984)

3021 RODRIGUEZ V, BODEY GP, McCREDIE KB, ET AL, COMBINATION 6-MERCAPTOPURINE, ADRIAMYCIN® IN REFRACTORY ADULT ACUTE LEUKEMIA, *CLIN PHARMACOL THER*, 18, 462-466 (1988)

3022 RODVOLD KA, PALOUCEK FP, JUNG D, ET AL, INTERACTION OF STEADY STATE PROCAINAMIDE WITH H$_2$ RECEPTOR ANTAGONISTS CIMETIDINE AND RANITIDINE, *THER DRUG MONIT*, 9, 378-383 (1987)

3023 ROE DA, DRUG INTERFERENCE WITH THE ASSESSMENT OF NUTRITIONAL STATUS, *CLIN LAB MED*, 1, 647-664 (1981)

3024 ROE DA, INTERACTIONS BETWEEN DRUGS AND NUTRIENTS, *MED CLIN NORTH AM*, 63, 985-1007 (1979)

3025 ROE DA, DRUG-INDUCED DEFICIENCY OF B VITAMINS, *N Y STATE J MED*, 71, 2770 (1971)

3026 ROE JH, TICKTIN HE, SCHNEIDER M, DETERMINATION AND CLINICAL SIGNIFICANCE OF SERUM LIPASE, *ENZYME BIOL CLIN*, 7, 73 (1966)

3027 ROECHLAN P, BERNT E, GRUBER W, ENZYMATIC DETERMINATION OF TOTAL CHOLESTEROL IN SERUM, *Z KLIN CHEM KLIN BIOCHEM*, 12, 226 (1974)

3028 ROELS OA, TROUT M, VITAMIN A AND CAROTENE, *STAND METH CLIN CHEM*, 7, 215 (1972)

3029 ROENIGK HH, GOTTLOB ME, ESTROGEN-INDUCED PORPHYRIA CUTANEA TARDA - REPORT OF 3 CASES, *ARCH DERMATOL*, 102, 260 (1970)

3030 ROFE AM, PHOLENZ SM, BAIS R, ET AL, INHIBITORY EFFECT OF ASCORBIC ACID IN OXALATE ASSAYS INVOLVING OXALATE DECARBOXYLASE, *CLIN CHEM*, 31, 1574-1575 (1985)

3031 ROGERS LE, LYON GM JR, PORTER FS, SPOT TEST FOR VANILLYLMANDELIC ACID AND OTHER GUAIACOLS IN URINE OF PATIENTS WITH NEUROBLASTOMA, *AM J CLIN PATHOL*, 58, 383 (1972)

3032 ROLAN PE, SOMOGYI AA, DREW MJR, ET AL, PHENYTOIN INTOXICATION DURING TREATMENT WITH PARENTERAL MICONAZOLE, *BR MED J [CLIN RES]*, 287, 1760 (1983)

3033 ROLANDI E, FRANCESCHINI R, MARABINI A, ET AL, SERUM CONCENTRATIONS OF PRL, GH, LH, FSH, TSH AND CORTISOL AFTER SINGLE ADMINISTRATION TO MAN OF A NEW SYNTHETIC NARCOTIC ANALGESIC BUTORPHANOL, *EUR J CLIN PHARMACOL*, 26, 563-565 (1984)

3034 ROLLMAN O, LOOF L, HEPATIC TOXICITY OF KETOCONAZOLE, *BR J DERMATOL*, 108, 376-378 (1983)

3035 ROMAN DE ALVAREZ R, JAHED FM, SPITALNY KG, ELKIN H, JAUNAKAIS I, THE INFLUENCE OF ORAL CONTRACEPTIVE STEROIDS ON SERUM LIPIDS, *AM J OBSTET GYNECOL*, 116, 727 (1973)

3036 ROMANKIEWICZ JA, EHRMAN M, RIFAMPIN AND WARFARIN: A DRUG INTERACTION, *ANN INTERN MED*, 82, 224-225 (1975)

3037 ROMANO AT, AUTOMATED GLUCOSE METHODS: EVALUATION OF A GLUCOSE OXIDASE-PEROXIDASE PROCEDURE, *CLIN CHEM*, 19, 1152 (1973)

3038 ROMER FK, JACOBSEN F, THE INFLUENCE OF PREDNISONE ON SERUM ANGIOTENSIN-CONVERTING ENZYME ACTIVITY IN PATIENTS WITH AND WITHOUT SARCOIDOSIS, *SCAND J CLIN LAB INVEST*, 42, 377-382 (1982)

3039 RONNOV-JESSEN V, TJERNLUND A, HEPATOTOXICITY DUE TO TREATMENT WITH PAPAVERINE REPORT OF FOUR CASES, *N ENGL J MED*, 281, 1333 (1969)

3040 ROOS JC, BOER P, KOOMANS HA, ET AL, HAEMODYNAMIC AND HORMONAL CHANGES DURING ACUTE AND CHRONIC DIURETIC TREATMENT IN ESSENTIAL HYPERTENSION, *EUR J CLIN PHARMACOL*, 19, 107-112 (1981)

3041 RORTH M, BILLE BRAHE NE, 2,3-DIPHOSPHOGLYCERATE AND CREATINE IN THE RED CELL DURING HUMAN PREGNANCY, *SCAND J CLIN LAB INVEST*, 28, 271 (1971)

3042 ROSALKI SB, RAU D, SERUM GAMMA GLUTAMYL TRANSPEPTIDASE ACTIVITY IN ALCOHOLISM, *CLIN CHIM ACTA*, 39, 41 (1972)

3043 ROSALKI SB, RAU D, LEHMANN D, PRENTICE M, DETERMINATION OF SERUM GAMMA-GLUTAMYL TRANSPEPTIDASE ACTIVITY AND ITS CLINICAL APPLICATIONS, *ANN CLIN BIOCHEM*, 7, 143 (1970)

3044 ROSALKI SB, TARLOW D, RAU D, PLASMA GAMMA-GLUTAMYL TRANSPEPTIDASE ELEVATION IN PATIENTS RECEIVING ENZYME-INDUCING DRUGS, *LANCET*, 2, 376 (1971)

3045 ROSE DP, ASPECTS OF TRYPTOPHAN METABOLISM IN HEALTH AND DISEASE: A REVIEW, *J CLIN PATHOL*, 25, 17 (1972)

3046 ROSE DP, CRAMP DG, REDUCTION OF PLASMA TYROSINE BY ORAL CONTRACEPTIVES AND ESTROGENS, *CLIN CHIM ACTA*, 29, 49 (1970)

3047 ROSE DP, MCGINTY F, THE INFLUENCE OF ADRENOCORTICAL HORMONES AND VITAMINS UPON TRYPTOPHAN METABOLISM IN MAN, *CLIN SCI*, 35, 1 (1968)

3048 ROSE DP, STRONG R, ADAMS PW, HARDING PE, EXPERIMENTAL VITAMIN B_6 DEFICIENCY AND THE EFFECT OF OESTROGEN-CONTAINING ORAL CONTRACEPTIVES, *CLIN SCI*, 42, 465 (1972)

3049 ROSE DP, STRONG R, FOLKHARD J, ADAMS PW, ERYTHROCYTE AMINOTRANSFERASE ACTIVITIES IN WOMEN USING ORAL CONTRACEPTIVES AND THE EFFECT OF VITAMIN B_6, *AM J CLIN NUTR*, 26, 48 (1973)

3050 ROSE DP, TOSELAND PA, URINARY EXCRETION OF QUINOLINIC ACID AND OTHER TRYPTOPHAN METABOLITES AFTER DEOXYPYRIDOXINE, *METABOLISM*, 22, 165 (1973)

3051 ROSE LI, BOUSSER JE, COOPER KH, SERUM ENZYMES AFTER MARATHON RUNNING, *J APPL PHYSIOL*, 29, 355 (1970)

3052 ROSE PG, PARACETAMOL OVERDOSE AND LIVER DAMAGE, *BR MED J [CLIN RES]*, 1, 381 (1969)

3053 ROSE S, PRICE PG, EFFECT OF INTRAMUSCULAR INJECTIONS ON SERUM CREATINE PHOSPHOKINASE, *J AM MED ASSOC*, 225, 417 (1973)

3054 ROSE SW, SMITH LD, PENRY JK, BLOOD LEVEL DETERMINATIONS OF ANTIEPILEPTIC DRUGS: CLINICAL VALUE AND METHODS, *US DEPT HEW, NIH* (1971)

3055 ROSENBERRY KR, DEFUSCO CJ, MANSMANN HC JR., ET AL, REDUCED THEOPHYLLINE HALF-LIFE INDUCED BY CARBAMAZEPINE THERAPY, *J PEDIATR*, 102, 472-474 (1983)

3056 ROSENBLATT JE, EDSON RS, METRONIDAZOLE, *MAYO CLIN PROC*, 62, 1013-1017 (1987)

3057 ROSENBLATT JE, EDSON RS, METRONIDAZOLE, *MAYO CLIN PROC*, 58, 154-157 (1983)

3058 ROSENOER VM, GILL GM, DRUG INTERACTIONS IN CLINICAL MEDICINE, *MED CLIN NORTH AM*, 56, 585 (1972)

3059 ROSENOW EC III, THE SPECTRUM OF DRUG-INDUCED PULMONARY DISEASE, *ANN INTERN MED*, 77, 977 (1972)

3060 ROSENTHAL AF, TOMSON MR, HYDROCHLOROTHIAZIDE INTERFERENCE WITH URINARY ESTRIOL DETERMINATION, *CLIN CHEM*, 18, 471 (1972)

3061 ROSMAN M, BERTINO JR, AZATHIOPRINE, *ANN INTERN MED*, 79, 694 (1973)

3062 ROSNER JM, CONTE NF, EVALUATION OF TESTICULAR FUNCTION BY MEASUREMENT OF URINARY EXCRETION OF TESTOSTERONE, *J CLIN ENDOCRINOL METAB*, 26, 735 (1966)

3063 ROSS AR, SHERLOCK S, A CONTROLLED TRIAL OF AZATHIOPRINE IN PRIMARY BILIARY CIRRHOSIS, *GUT*, 12, 770 (1971)

3064 ROSS EJ, FUNCTIONAL RELATIONSHIP BETWEEN ADRENAL MEDULLARY AND CORTICAL HORMONES IN MAN, *Q J MED*, 30, 285 (1961)

3065 ROSS G, WEINSTEIN I, KABAKOW B, THE INFLUENCE OF PHENOTHIAZINE AND SOME OF ITS DERIVATIVES ON THE DETERMINATION OF 5-HYDROXYINDOLEACETIC ACID, *CLIN CHEM*, 4, 66 (1958)

3066 ROSS RJ, SAVADIL III AP, CALIL HM, ET AL, EFFECTS OF DESMETHYLIMIPRAMINE ON PLASMA NOREPINEPHRINE, PULSE AND BLOOD PRESSURE, *CLIN PHARMACOL THER*, 33, 429-437 (1983)

3067 ROSS SK, DALE REG, THOMSON WST, IRELAND JT, BLOOD ACID/BASE STUDIES AND URINARY ACID EXCRETION IN OBESE FASTING PATIENTS, *POSTGRAD MED J*, JUNE SUPPL, 447 (1971)

3068 ROSSI EC, LEVIN NW, INHIBITION OF PRIMARY ADP-INDUCED PLATELET AGGREGATION IN NORMAL SUBJECTS, *J CLIN INVEST*, 52, 2457 (1973)

3069 ROSSI EC, LEVIN NW, INHIBITION OF ADP-INDUCED PLATELET AGGREGATION BY FUROSEMIDE, *J LAB CLIN MED*, 81, 140 (1973)

3070 ROSSNER S, WALLGREN A, SERUM LIPOPROTEINS AND PROTEINS AFTER BREAST CANCER SURGERY AND EFFECTS OF TAMOXIFEN, *ATHEROSCLEROSIS*, 52, 339-341 (1984)

3071 ROSSNER S, WEINER L, ATENOLOL AND METOPROLOL: COMPARISON OF EFFECTS ON BLOOD PRESSURE AND SERUM LIPOPROTEINS AND SIDE EFFECTS, *EUR J CLIN PHARMACOL*, 24, 573-577 (1983)

3072 ROTHSTEIN E, WARFARIN EFFECT ENHANCED BY DISULFIRAM, *J AM MED ASSOC*, 206, 1574 (1968)

3073 ROTMENSCH HH, WEINTRAUB M, SOFFERMAN G, ET AL, EXPERIENCE WITH IMMUNOLOGICAL TESTS IN DRUG-INDUCED HEPATITIS, *Z GASTROENTEROL*, 19, 691-697 (1981)

3074 ROULSTON JE, GALLOWAY PJ, IN VITRO INHIBITION OF ANGIOTENSIN-CONVERTING ENZYME BY PREDNISOLONE AND METHYL-PREDNISOLONE, *CLIN CHEM*, 32, 697-698 (1986)

3075 ROUTH JI, BANNOW RE, DEVELOPMENT OF A SCREENING TEST FOR L-DOPA AND ITS METABOLITES IN URINE, *CLIN CHEM*, 17, 872 (1971)

3076 ROUTH JI, PAUL WD, ASSESSMENT OF INTERFERENCE BY ASPIRIN WITH SOME ASSAYS COMMONLY DONE IN THE CLINICAL LABORATORY, *CLIN CHEM*, 22, 837-842 (1976)

3077 ROWE MJ, DOLDER MA, KIRBY BJ, OLIVER MF, EFFECT OF A NICOTINIC ACID ANALOGUE ON RAISED PLASMA FREE FATTY ACIDS AFTER ACUTE MYOCARDIAL INFARCTION, *LANCET*, 2, 814 (1973)

3078 ROY MT, FIRST MR, MYRE SA, ET AL, EFFECT OF CO-TRIMOXAZOLE AND SULFAMETHOXAZOLE ON SERUM CREATININE IN NORMAL SUBJECTS, *THER DRUG MONIT*, 4, 77-79 (1982)

3079 ROY-BYRNE P, VITTONE BJ, UHDE TW, ALPRAZOLAM-RELATED HEPATOTOXICITY, *LANCET*, 2, 786-787 (1983)

3080 RUBENSTEIN AH, LOWY C, WELBORN TA, FRASER TR, URINE INSULIN IN NORMAL SUBJECTS, *METABOLISM*, 16, 234 (1967)

3081 RUBINOW DR, POST RM, GOLD PW, ET AL, EFFECTS OF CARBAMAZEPINE ON CEREBROSPINAL FLUID SOMATOSTATIN, *PSYCHOPHARMACOL*, 85, 210-213 (1985)

3082 RUBULIS A, RUBERT M, FALOON WW, CHOLESTEROL LOWERING FECAL BILE ACID, AND STEROL CHANGES DURING NEOMYCIN AND COLCHICINE, *AM J CLIN NUTR*, 23, 1251 (1970)

3083 RUEDY J, DAVIES RO, A COMPARATIVE CLINICAL TRIAL OF GUANOXAN AND GUANETHIDINE IN ESSENTIAL HYPERTENSION, *CLIN PHARMACOL THER*, 8, 38 (1967)

3084 RUFFALO RL, THOMPSON JF, SEGAL JL, DIAZEPAM-CIMETIDINE DRUG INTERACTION: A CLINICALLY SIGNIFICANT EFFECT, *SOUTH MED J*, 74, 1075-1078 (1081)

3085 RUIZ DEL ARBOL L, MOREIRA V, MORENO A, ET AL, BRIDGING HEPATIC NECROSIS ASSOCIATED WITH CIMETIDINE, *AM J GASTROENTEROL*, 74, 267-269 (1980)

3086 RUMPF KW, BARTH M, BLECH M, ET AL, BEZAFIBRAT-INDUZIERTE MYOLYSE UND MYOGLOBINURIE BEI PATIENTEN MIT EINGESCHRANKTER NIERENFUNKTION, *KLIN WOCHENSCHR*, 62, 346-348 (1984)

3087 RUNDLY AT, SNDELL B, LEUCINE AMINOPEPTIDASE ISOENZYME CHANGES AFTER TREATMENT WITH ANTICONVULSANT DRUGS, *CLIN CHIM ACTA*, 44, 377 (1973)

3088 RUOKONEN A, ALEN M, BOLTON N, ET AL, RESPONSE OF SERUM TESTOSTERONE AND ITS PRECURSOR STEROIDS, SHBG AND CBG TO ANABOLIC STEROID AND TESTOSTERONE SELF-ADMINISTRATION IN MAN, *J STEROID BIOCHEM*, 23, 33-38 (1985)

3089 RUOKONEN A, KAAR K, EFFECTS OF DESOGESTREL, LEVONORGESTREL AND LYNESTRENOL ON SERUM SEX HORMONE BINDING GLOBULIN, CORTISOL BINDING GLOBULIN, CERULOPLASMIN AND HDL-CHOLESTEROL, *EUR J OBSTET GYNECOL REPROD BIOL*, 20, 13-18 (1985)

3090 RUSH JA, BERAN RG, LEUCOPENIA AS AN ADVERSE REACTION TO CARBAMAZEPINE THERAPY, *MED J AUST*, 140, 426-428 (1984)

3091 RUSHTON ML, SAMMONS HG, ROBINSON BHB, A STUDY OF CALCIUM ABSORPTION USING AN AUTOMATED FLUORIMETRIC ASSAY PROCEDURE, *CLIN CHIM ACTA*, 35, 5 (1971)

3092 RUSSEL AS, STURGE RA, SMITH MA, SERUM TRANSAMINASES DURING SALICYLATE THERAPY, *BR MED J [CLIN RES]*, 2, 428 (1971)

3093 RUSSELL DW, LOW CIRCULATING LEVELS OF ACID-LABILE HYDRAZONES AFTER ORAL ADMINISTRATION OF ISONICOTINIC ACID HYDRAZIDE, *CLIN CHIM ACTA*, 41, 163 (1972)

3094 RUSSELL MA, WILSON C, COLE PV, IDLE M, FEYERABEND C, COMPARISON OF INCREASES IN CARBOXYHAEMOGLOBIN AFTER SMOKING 'EXTRA MILD' AND 'NON-MILD' CIGARETTES, *LANCET*, 2, 687 (1973)

3095 RUTKOWSKI DM, BURKLE WS, ADVANCES IN THROMBOLYTIC THERAPY, *DRUG INTELL CLIN PHARM*, 16, 115-121 (1982)

3096 RUTMAN JY, EFFECT OF CARBAMAZEPINE ON BLOOD ELEMENTS, *ANN NEUROL*, 3, 373 (1978)

3097 RUTTER ER, AUTOMATED METHOD FOR SCREENING URINE FOR AMPHETAMINE AND SOME RELATED PRIMARY AMINES, *CLIN CHEM*, 18, 616 (1972)

3098 RYAN JF, KAGEN LJ, HYMAN AI, MYOGLOBINEMIA AFTER A SINGLE DOSE OF SUCCINYLCHOLINE, *N ENGL J MED*, 285, 824 (1971)

3099 RYAN JR, GOMEZ-PEREZ F, JAIN A, MAHA G, MCMAHON FG, METABOLIC AND EFFICACY STUDIES OF MK-270, A NEW ORAL HYPOGLYCEMIC AGENT, *CLIN PHARMACOL THER*, 13, 151 (1972)

3100 RYAN JR, MAHA GE, MCMAHON FG, KYRIAKOPOULOS A, EFFECTS OF A NEW ORAL HYPOGLYCEMIC AGENT ON GLUCOSE AND INSULIN AFTER ORAL GLUCOSE LOADING, *DIABETES*, 20, 734 (1971)

3101 RYAN MP, MAGNESIUM AND POTASSIUM-SPARING DIURETICS, *MAGNESIUM*, 5, 282-292 (1986)

3102 RYLANCE HJ, MYHAL DR, TAURINE EXCRETION AND THE INFLUENCE OF DRUGS, *CLIN CHIM ACTA*, 35, 159 (1971)

3103 RYMER W, GREENLAW CW, HYPOPROTHROMBINEMIA ASSOCIATED WITH CEFAMANDOLE, *DRUG INTELL CLIN PHARM*, 14, 780-783 (1980)

3104 RYSANEK K, KONIG J, SPANKOVA H, MLEJNKOVA M, INFLUENCE OF HYDROCORTISONE AND AN ALLERGEN ON AGGREGATION OF THROMBOCYTES WITH STREPTOKINASE, *BLOOD*, 39, 588 (1972)

3105 SAAL AK, WERNER JA, GREENE HL, ET AL, EFFECT OF AMIODARONE ON SERUM QUINIDINE AND PROCAINAMIDE LEVELS, *AM J CARDIOL*, 53, 1264-1267 (1984)

3106 SABATER J, FERRE C, MAYA A, MISLEADING FINDINGS IN URINE ELECTROPHORESIS, *LANCET*, 2, 316 (1971)

3107 SABATH LD, GERSTEIN DA, FINLAND M, SERUM GLUTAMIC OXALOACETIC TRANSAMINASE, FALSE ELEVATIONS DURING ADMINISTRATION OF ERYTHROMYCIN, *N ENGL J MED*, 279, 1137 (1968)

3108 SACHAR EJ, GRUEN PH, KARASU TB, ET AL, THIORIDAZINE STIMULATES PROLACTIN SECRETION IN MAN, *ARCH GEN PSYCHIATRY*, 32, 885-886 (1975)

3109 SACRE MH, WALKER G, EFFECT OF CLOFIBRATE ADMINISTRATION ON THE URINARY VANILMANDELIC ACID METHOD OF PISANO, CROUT, ABRAHAM, *CLIN CHIM ACTA*, 30, 215 (1971)

3110 SAEED-UZ-ZAFAR M, MILLER JM, BRENEMAN GM, MANSOUR J, OBSERVATIONS ON THE EFFECT OF HEPARINS ON FREE AND TOTAL THYROXINE, *J CLIN ENDOCRINOL METAB*, 32, 633 (1971)

3111 SAFFOURI B, CHO JH, FELBER N, CHLORPROPAMIDE-INDUCED HEMOLYTIC ANEMIA, *POSTGRAD MED J*, 57, 44-45 (1981)

3112 SAGONE AL, LAWRENCE T, BALCERZAK SP, SMOKING, A CAUSE OF 'SPURIOUS' POLYCYTHEMIA, *BLOOD*, 38, 826 (1971)

3113 SAHEBJAMI H, SCALETTAR R, EFFECTS OF FRUCTOSE INFUSION ON LACTATE AND URIC ACID METABOLISM, *LANCET*, 1, 366 (1971)

3114 SAILER S, SANDHOFER F, BOLZANO K, BRAUNSTEINER H, ACTION OF NOREPINEPHRINE AND PROPRANOLOL ON THE TURNOVER RATE OF FREE FATTY ACIDS AND THE ESTERIFICATION RATE, *ADV EXP MED BIOL*, 4, 135 (1968)

3115 SAINT-SALVI B, TREMBLAY D, SURJUS A, ET AL, A STUDY OF THE INTERACTION OF ROXITHROMYCIN WITH THEOPHYLLINE AND CARBAMAZEPINE, *J ANTIMICROB CHEMOTHER*, 20 SUPPL B, 121-129 (1987)

3116 SAKHAEE K, NICAR M, HILL K, ET AL, CONTRASTING EFFECTS OF POTASSIUM CITRATE AND SODIUM CITRATE THERAPIES ON URINARY CHEMISTRIES AND CRYSTALLIZATION OF STONE-FORMING SALTS, *KIDNEY INT*, 24, 348-352 (1983)

3117 SAKU K, GARTSIDE PS, HYND BA, ET AL, MECHANISM OF ACTION OF GEMFIBROZIL ON LIPOPROTEIN METABOLISM, *J CLIN INVEST*, 75, 1702-1712 (1985)

3118 SALATA R, KLEIN I, EFFECTS OF LITHIUM ON THE ENDOCRINE SYSTEM: A REVIEW, *J LAB CLIN MED*, 110, 130-136 (1987)

3119 SALEM P, KHALYL M, JABBOURY K, ET AL, CIS-DIAMINEDICHLOR-PLATINUM (II) 5-DAY CONTINUOUS INFUSION: A NEW DOSE SCHEDULE WITH MINIMAL TOXICITY, *CANCER*, 53, 837-840 (1984)

3120 SALEM RB, BRELAND BD, MISHRA SK, ET AL, EFFECT OF CIMETIDINE ON PHENYTOIN SERUM LEVELS, *EPILEPSIA*, 24, 284-288 (1983)

3121 SALEM RR, MCINDOE A, MATKIN JA, ET AL, THE HEMATOLOGIC EFFECTS OF LATAMOXEF SODIUM WHEN USED AS A PROPHYLAXIS DURING SURGICAL TREATMENT, *SURG GYNECOL OBSTET*, 164, 525-529 (1987)

3122 SALLEN MK, EFRUSY ME, KNIAZ JL, ET AL, STREPTOKINASE-INDUCED HEPATIC DYSFUNCTION, *AM J GASTROENTEROL*, 78, 523-524 (1983)

3123 SALOM IL, SILVIS SE, DOSCHERHOLMEN A, EFFECT OF CIMETIDINE ON THE ABSORPTION OF VITAMIN B_{12}, *SCAND J GASTROENTEROL*, 17, 129-131 (1982)

3124 SALUSPORO M, USE OF ENZYMES FOR THE DIAGNOSIS OF ALCOHOL-RELATED ORGAN DAMAGE, *ENZYME*, 37, 87-107 (1987)

3125 SALVESEN S, NISSEN-MEYER R, INFLUENCE OF MEPROBAMATE THERAPY ON THE ESTIMATION OF 17-KETOSTEROIDS AND 17-KETOGENIC STEROIDS, *J CLIN ENDOCRINOL METAB*, 17, 914 (1957)

3126 SALVETTI A, ABDEL-HAQ B, MAGAGNA A, ET AL, INDOMETHACIN REDUCES THE ANTIHYPERTENSIVE ACTION OF ENALAPRIL, *CLIN EXP HYPERTENS*, 9, 559-567 (1987)

3127 SAMAAN NA, HILL CS JR, BECEIRO JR, SCHULTZ PN, IMMUNOREACTIVE CALCITONIN IN MEDULLARY CARCINOMA OF THE THYROID AND IN MATERNAL AND CORD SERUM, *J LAB CLIN MED*, 81, 671 (1973)

3128 SAMOLS E, MARRI G, MARKS V, PROMOTION OF INSULIN SECRETION BY GLUCAGON, *LANCET*, 2, 415 (1965)

3129 SAMUEL P, SHALCHI OB, HOLTZMAN CM, REDUCTION OF SERUM CHOLESTEROL CONCENTRATIONS BY PAROMOMYCIN IN PATIENTS WITH ARTERIOSCLEROSIS, *PROC SOC EXP BIOL MED*, 115, 718 (1964)

3130 SANCHO JM, ROBLES RG, MANCHENO E, ET AL, INTERFERENCE BY RANITIDINE WITH ALDOSTERONE SECRETION IN VIVO, *EUR J CLIN PHARMACOL*, 27, 495-497 (1984)

3131 SANDEK CD, BOULTER PR, ARKY RA, THE SODIUM RETENTION OF INSULIN THERAPY, *CLIN RES*, 20, 883 (1972)

3132 SANDER S, BERGAN T, FOSSBERG E, PIPERACILLIN IN THE TREATMENT OF URINARY TRACT INFECTIONS, *CHEMOTHERAPY*, 26, 141-144 (1980)

3133 SANDER WE JR, JOHNSON JE III, TAGGART JG, ADVERSE REACTIONS TO CEPHALOTHIN AND CEPHAPIRIN, *N ENGL J MED*, 290, 424 (1974)

3134 SANDERS CV, GREENBERG RN, MARIER RL, CEFAMANDOLE AND CEFOXITIN, *ANN INTERN MED*, 103, 70-78 (1985)

3135 SANDHU RS, FREED RM, CATECHOLAMINES AND ASSOCIATED METABOLITES IN HUMAN URINE, *STAND METH CLIN CHEM*, 7, 231 (1972)

3136 SANDHYA P, DAS UN, VITAMIN C THERAPY FOR MATURITY ONSET DIABETES MELLITUS: RELEVANCE TO PROSTAGLANDIN INVOLVEMENT, *IRCS MED SCI: CLIN BIOCHEM*, 9, 618 (1981)

3137 SANDLER M, JOHNSON RD, RUTHVEN CRJ, REID JL, CALNE DB, TRANSAMINATION IS A MAJOR PATHWAY OF L-DOPA METABOLISM FOLLOWING PERIPHERAL DECARBOXYLASE INHIBITION, *NATURE*, 247, 364 (1974)

3138 SANDLER M, KAROUM F, RUTHVEN CRJ, METABOLISM OF DOPA IN PARKINSONISM, *N ENGL J MED*, 281, 1429 (1969)

3139 SANDLER RM, BEVAN PC, ROBERTS GE, ET AL, INTERACTION BETWEEN SODIUM VALPROATE AND PLATELETS: A FURTHER STUDY, *BR MED J [CLIN RES]*, 2, 1476 (1979)

3140 SANDS CD, ROBINSON JD, SALEM RB, ET AL, EFFECT OF THIORIDAZINE ON PHENYTOIN SERUM CONCENTRATION: A RETROSPECTIVE STUDY, *DRUG INTELL CLIN PHARM*, 21, 267-272 (1987)

3141 SANGSTER B, KRAJNC EJ, LOEBER JG, ET AL, STUDY OF SODIUM BROMIDE IN HUMAN VOLUNTEERS, WITH SPECIAL EMPHASIS ON THE ENDOCRINE SYSTEM, *HUM TOXICOL*, 1, 393-402 (1984)

3142 SANMARTI A, PERMANYER-MIRALDA G, CASTELLANOS JM, ET AL, CHRONIC ADMINISTRATION OF AMIODARONE AND THYROID FUNCTION: A FOLLOW-UP STUDY, *AM HEART J*, 108, 1262-1268 (1984)

3143 SANNOUR MB, IBRAHIM WN, SHAKER A, A THREE-YEAR STUDY OF LIVER FUNCTION DURING LONG-ACTING PROGESTOGEN-ESTROGEN ADMINISTRATION, *CONTRACEPTION*, 7, 403 (1973)

3144 SANSOM LN, BERAN RC, SCHAPEL GJ, INTERACTION BETWEEN PHENYTOIN AND VALPROATE, *MED J AUST*, 2, 212 (1980)

3145 SANTEN RJ, LEONARD JM, SHERINS RJ, GANDY HM, PAULSEN CA, SHORT AND LONG-TERM EFFECTS OF CLOMIPHENE CITRATE ON THE PITUITARY-TESTICULAR AXIS, *J CLIN ENDOCRINOL METAB*, 33, 970 (1971)

3146 SAPIRA JD, KLANIECKI T, RATKIN G, 'NON-PHEOCHROMOCYTOMA', *J AM MED ASSOC*, 212, 2243 (1970)

3147 SARAGOCA A, TAVARES, MH, BARROS FB, HORTA J DA S, SOME CLINICAL AND LABORATORY FINDINGS IN PATIENTS INJECTED WITH THORIUM DIOXIDE, *AM J GASTROENTEROL*, 57, 301 (1972)

3148 SARUTA T, STUDIES ON THE EFFECT OF PRAZOSIN ON BLOOD PRESSURE AND SERUM LIPIDS IN JAPANESE HYPERTENSIVE PATIENTS, *AM J MED*, 76, SUPPL 2A, 117-120 (1984)

3149 SARUTA T, SUZUKI H, KAWAMURA M, ET AL, SERUM CREATINE PHOSPHOKINASE LEVELS DURING TREATMENT WITH β-ADRENORECEPTOR BLOCKING AGENTS, *J CARDIOVASC PHARMACOL*, 7, 805-808 (1985)

3150 SASAKI J, ARAKAWA K, EFFECT OF CAPTOPRIL ON SERUM LIPIDS, LIPOPROTEINS AND APOLIPOPROTEINS IN PATIENTS WITH MILD ESSENTIAL HYPERTENSION, *CURR THER RES*, 40, 898-902 (1986)

3151 SASAKI J, ARAKAWA K, EFFECT OF BUNITROLOL ADMINISTRATION ON SERUM LIPIDS, LIPOPROTEINS, AND APOLIPOPROTEINS IN PATIENTS WITH ESSENTIAL HYPERTENSION, *CURR THER RES*, 40, 903-907 (1986)

3152 SASAKI M, TONODA S, AOKI Y, ET AL, PANCREATITIS DUE TO VALPROIC ACID, *LANCET*, 1, 1196 (1980)

3153 SASAKIN, SHUMURA N, HARIGUCHI T, NAKAJIMA H, EFFECT OF CELLULOSE ACETATE RESIN USED AS A CONTAINER IN TRIOSORB AND TETRASORB TESTS, *HORUMAN TO RINSHO*, 18, 666 (1970)

3154 SASSE EA, EDWARDS JB, LACTATE MEASUREMENT IN THE TDx AND ACA: HEMOGLOBIN INTERFERENCE ASSESSED, *CLIN CHEM*, 33, 189-190 (1987)

3155 SATGUNASINGAM N, BUTTERY JE, DE WITT GF, A SIMPLE AND RAPID ONE-TUBE METHOD FOR THE DETERMINATION OF URINARY DELTA-AMINOLEVULINIC ACID, *J CLIN PATHOL*, 26, 800 (1973)

3156 SATHANANTHAN S, GERSHAN S, RENAL DAMAGE DUE TO IMIPRAMINE, *LANCET*, 1, 833 (1973)

3157 SATTLER FR, WASKIN H, PENTAMIDINE AND FATAL HYPOGLYCEMIA, *ANN INTERN MED*, 197, 789-790 (1987)

3158 SATTLER FR, WEITEKAMP MR,, BALLARD JO, POTENTIAL FOR BLEEDING WITH THE NEW BETA-LACTAM ANTIBIOTICS, *ANN INTERN MED*, 105, 924-931 (1986)

3159 SAVARAJ N, TRONER MB, AMINOGLUTETHIMIDE IN THE MANAGEMENT OF METASTATIC BREAST CANCER, *MED PEDIATR ONCOL*, 8, 251-263 (1980)

3160 SAVEGE TM, FOLEY EI, COULTAS RJ, WALTON B, STRUNIN L, SIMPSON BR, SCOTT DF, CT 1341: SOME EFFECTS IN MAN, *ANAESTHESIA*, 26, 402 (1971)

3161 SAVORY DJ, INTERFERENCE BY SOTALOL WITH THE PISANO METHOD FOR URINARY METANEPHRINES, *ANN CLIN BIOCHEM*, 21, 446 (1984)

3162 SAVORY J, SOBEL RE, PHYSIOLOGICAL VARIATIONS IN SERUM TRIGLYCERIDES, *ANN CLIN LAB SCI*, 2, 126 (1972)

3163 SAWERS JS, KELLETT HA, BROWN NS, ET AL, PROLACTIN RESPONSE TO METOCLOPRAMIDE IN HYPERTHYROIDISM, *J CLIN ENDOCRINOL METAB*, 55, 175-177 (1982)

3164 SAWIN CT, MEASUREMENT OF PLASMA CORTISOL IN THE DIAGNOSIS OF CUSHING'S SYNDROME, *ANN INTERN MED*, 68, 624 (1968)

3165 SAX SM, MOORE JJ, GLUTAMIC OXALOACETIC TRANSAMINASE (COLORIMETRIC), *STAND METH CLIN CHEM*, 6, 149 (1970)

3166 SAXTON C, MAJID PA, CLOUGH G, TAYLOR SH, EFFECT OF OUABAIN ON INSULIN SECRETION IN MAN, *CLIN SCI*, 42, 57 (1972)

3167 SCALLEY RD, ROARK RD, OXACILLIN-INDUCED AGRANULOCYTOSIS, *DRUG INTELL CLIN PHARM*, 11, 420-423 (1977)

3168 SCANLON MF, RODRIGUEZ-ARNAO MD, POURMAND M, ET AL, CATECHOLAMINERGIC INTERACTIONS IN THE REGULATION OF THYROTROPIN (TSH) SECRETION IN MAN, *J ENDOCRINOL INVEST*, 3, 125-129 (1980)

3169 SCHADE RWB, DEMACKER PNM, LAAR AV'T, REDUCTION OF SERUM ALKALINE PHOSPHATASE BY CLOFIBRATE, *LANCET*, 1, 862-863 (1975)

3170 SCHADEL JM, REHFELD N, DER EINFLUSS HORMONALER KONTRAZEPTIVA AUF DIE ENZYME AKTIVITAT VERSCHIEDENERIN BLUTSERUM, *Z MED LABOR DIAGN*, 26, 77-85 (1985)

3171 SCHAFFNER F, IATROGENIC JAUNDICE, *J AM MED ASSOC*, 174, 1690 (1960)

3172 SCHAFFNER F, POPPER H, CHESROW E, CHOLESTASIS PRODUCED BY ADMINISTRATION OF NORETHANDROLONE, *AM J MED*, 26, 249 (1959)

3173 SCHALHORN A, WILMANNS W, KOCZOREK CE, THE INFLUENCE OF HIGH DOSE METHOTREXATE THERAPY ON SERUM IRON, *KLIN WOCHENSCHR*, 64, 475-480 (1986)

3174 SCHAMBELAN M, BIGLIERI EG, DEOXYCORTICOSTERONE PRODUCTION AND REGULATION IN MAN, *J CLIN ENDOCRINOL METAB*, 34, 695 (1972)

3175 SCHATZ DL, PALTER HC, RUSSELL CS, EFFECTS OF ORAL CONTRACEPTIVES AND PREGNANCY ON THYROID FUNCTION, *CAN MED ASSOC J*, 99, 882 (1968)

3176 SCHATZ DL, SHEPPARD RH, STEINER G, CHANDARLAPATY CS, DEVEBER GA, INFLUENCE OF HEPARIN ON SERUM FREE THYROXINE, *J CLIN ENDOCRINOL METAB*, 29, 1015 (1969)

3177 SCHEIDEGGER K, O'CONNELL M, ROBBINS DC, ET AL, EFFECTS OF CHRONIC β-RECEPTOR STIMULATION ON SYMPATHETIC NERVOUS SYSTEM ACTIVITY, ENERGY EXPENDITURE AND THYROID HORMONES, *J CLIN ENDOCRINOL METAB*, 58, 895-903 (1984)

3178 SCHEININ M, ERKKA KG, SYVALAHTI EKG, ET AL, EFFECTS OF APOMORPHINE ON BLOOD LEVELS OF HOMOVANILLIC ACID, GROWTH HORMONE AND PROLACTIN IN MEDICATED SCHIZOPHRENICS AND HEALTHY CONTROL SUBJECTS, *PROG NEUROPSYCHOPHARMACOL BIOL PSYCHIATRY*, 9, 441-449 (1985)

3179 SCHEN RJ, ZURKOWSKI S, INCREASED SERUM CREATINE PHOSPHOKINASE ACTIVITY WITH VIOLENT COUGHING, *N ENGL J MED*, 289, 328 (1973)

3180 SCHENKEN JR, GRASS L, DEFEROXAMINE AND AUTOASSAY OF SERUM IRON, *CLIN TOXICOL*, 4, 641 (1971)

3181 SCHENKER JG, BEN-YOSEPH Y, SHAPIRA E, ERYTHROCYTE CARBONIC ANHYDRASE B LEVELS DURING PREGNANCY AND USE OF ORAL CONTRACEPTIVES, *OBSTET GYNECOL*, 39, 237 (1972)

3182 SCHENTAG JJ, CERRA FB, PLAUT ME, CLINICAL AND PHARMACOKINETIC CHARACTERISTICS OF AMINOGLYCOSIDE NEPHROTOXICITY IN 201 CRITICALLY ILL PATIENTS, *ANTIMICROB AGENTS CHEMOTHER*, 21, 721-726 (1982)

3183 SCHEVELBEIN H, LOSCHENKOHL K, KUNTZE I, TRYPTOPHAN METABOLISM, CANCER OF THE URINARY BLADDER AND SMOKING HABITS, *Z KLIN CHEM KLIN BIOCHEM*, 10, 445 (1972)

3184 SCHIFFL H, WEIDMANN P, MORDASINI R, ET AL, REVERSAL OF DIURETIC INDUCED INCREASES IN SERUM LOW DENSITY LIPOPROTEIN CHOLESTEROL BY THE BETA BLOCKER PINDOLOL., *METABOLISM*, 31, 411-415 (1982)

3185 SCHILSKY RL, BARLOCK A, OZOIS RF, PERSISTENT HYPOMAGNESEMIA FOLLOWING CISPLATIN THERAPY FOR TESTICULAR CANCER, *CANCER TREAT REP*, 66, 1767-1769 (1982)

3186 SCHINDLER AE, RATANASOPA V, HERRMANN WL, ACID HYDROLYSIS OF URINARY ESTRIOL, *CLIN CHEM*, 13, 186 (1967)

3187 SCHLENZKA K, STRECKENBACH K, EFFECT OF WASHING AGENTS ON THE FERRIC CHLORIDE TEST FOR PHENYLKETONURIA, *DEUT GESUNDHEITSW*, 25, 1497 (1970)

3188 SCHMIDT D, UTECH K, PROGABIDE FOR REFRACTORY PARTIAL EPILEPSY: A CONTROLLED ADD-ON TRIAL, *NEUROLOGY*, 36, 217-221 (1986)

3189 SCHMIDT E, POLIWODA H, BUHL V, ALEXANDER K, SCHMIDT FW, OBSERVATIONS OF ENZYME ELEVATIONS IN THE SERUM DURING STREPTOKINASE TREATMENT, *J CLIN PATHOL*, 25, 650 (1972)

3190 SCHMIDT JB, GEBHART W, KOPSA H, ET AL, DOES CYCLOSPORINE A INFLUENCE SEX HORMONE LEVEL?, *EXP CLIN ENDOCRINOL*, 88, 207-211 (1986)

3191 SCHMIDT P, ZAZGORNIK J, KOPSA H, LETTER: HYPOURICEMIA AFTER RENAL TRANSPLANTATION, *N ENGL J MED*, 289, 1373 (1973)

3192 SCHNAPP P, HERMANN H, CERNAK P, ET AL, NIFEDIPINE MONOTHERAPY IN THE HYPERTENSIVE ELDERLY: A PLACEBO CONTROLLED CLINICAL TRIAL, *CURR MED RES OPIN*, 10, 407-413 (1987)

3193 SCHNEEWEISS J, POOLE GW, HYPERURICAEMIA DUE TO PYRAZINAMIDE, *BR MED J [CLIN RES]*, 2, 830 (1960)

3194 SCHNEIDER J, PANNE E, BRAUN H, ET AL, ETHANOL-INDUCED HYPERLIPOPROTEINEMIA, *J LAB CLIN MED*, 101, 114-122 (1983)

3195 SCHNEIDER M, GIARDINA E-GV, INTERFERENCE BY FLEXERIL®, A TRICYCLIC MUSCLE RELAXANT, WITH LIQUID CHROMATOGRAPHIC DETERMINATION OF IMIPRAMINE, *CLIN CHEM*, 32, 1599 (1986)

3196 SCHOCH HK, THE US VETERANS ADMINISTRATION CARDIOLOGY DRUG-LIPID STUDY: AN INTERIM REPORT, *ADV EXP BIOL MED*, 4, 405 (1968)

3197 SCHOENFIELD LJ, SULFOBROMOPHTHALEIN TRANSPORT AND METABOLISM, *GASTROENTEROLOGY*, 48, 530 (1965)

3198 SCHOENFIELD LJ, GRUNDY SM, HOFMAN AF, ET AL, THE NATIONAL COOPERATIVE GALLSTONE STUDY VIEWED BY ITS INVESTIGATORS, *GASTROENTEROLOGY*, 84, 644-647 (1983)

3199 SCHOENFIELD LJ, LACHIN JM, ET AL, CHENODIOL (CHENODEOXYCHOLIC ACID) FOR DISSOLUTION OF GALLSTONES: THE NATIONAL COOPERATIVE GALLSTONE STUDY. A CONTROLLED TRIAL OF EFFICACY AND SAFETY, *ANN INTERN MED*, 95, 257-282 (1981)

3200 SCHOENFIELD LJ, MCGILL DB, FOULK WT, STUDIES OF BSP METABOLISM IN MAN, DEMONSTRATION OF A TRANSPORT MAXIMUM FOR BILIARY EXCRETION OF BSP, *J CLIN INVEST*, 43, 1424 (1964)

3201 SCHOLLER R, ROGER M, AVIGDOR R, INTER- AND INTRA-INDIVIDUAL VARIATIONS IN REFERENCE VALUES FOR HORMONE ASSAYS IN, *REFERENCE VALUES IN HUMAN CHEMISTRY*, G SIEST (ED.), KARGER, BASEL, 47 (1973)

3202 SCHONELL M, DORKEN E, GRZYBOWSKI S, RIFAMPIN, *CAN MED ASSOC J*, 106, 783 (1972)

3203 SCHOU M, VESTERGAARD P, LITHIUM AND THE KIDNEY SCARE, *PSYCHOSOMATICS*, 22, 92-94 (1981)

3204 SCHREINER GE, MAHER JF, TOXIC NEPHROPATHY: ADVERSE RENAL EFFECTS CAUSED BY DRUGS AND CHEMICALS, *J AM MED ASSOC*, 191, 849 (1965)

3205 SCHRIER RW, LIEBERMAN R, UFFERMAN RC, MECHANISM OF ANTIDIURETIC EFFECT OF BETA ADRENERGIC STIMULATION, *J CLIN INVEST*, 51, 97 (1972)

3206 SCHROGIE JJ, HOLT P, HARTLEY RC, BARTHOLOMEW G, 'HISTAMINE INDUCED' PANCREATITIS, *GASTROENTEROLOGY*, 49, 672 (1965)

3207 SCHROGIE JJ, SOLOMON HM, THE HYPOCHOLESTEROLEMIC EFFECT OF PHENYRAMIDOL, *CLIN PHARMACOL THER*, 7, 723 (1966)

3208 SCHULENBURG BJ, BECK ML, PIERCE SR, ET AL, IMMUNE HEMOLYSIS ASSOCIATED WITH ZOMAX, *TRANSFUSION*, 23, 409 (1983)

3209 SCHULTZ AL, INFLUENCE OF DRUGS ON THYROID I 131 UPTAKE AND SERUM CHEMICAL PROTEIN-BOUND IODINE, *POSTGRAD MED*, 31, A-34 (1962)

3210 SCHULZ P, ABANG A, GIACOMINI JC, ET AL, EFFECT OF HEPARIN ON THE RED BLOOD CELL TO PLASMA CONCENTRATION RATIO OF DIPHENYLHYDANTOIN AND PRAZOSIN, *THER DRUG MONIT*, 5, 497-499 (1983)

3211 SCHULZ P, GIACOMINI KM, LUTTRELL S, ET AL, EFFECT OF LOW DOSES OF HEPARIN ON THE PLASMA BINDING OF PHENYTOIN AND PRAZOSIN IN NORMAL MAN, *EUR J CLIN PHARMACOL*, 25, 211-214 (1983)

3212 SCHUMER W, ADVERSE EFFECTS OF XYLITOL IN PARENTERAL ALIMENTATION, *METABOLISM*, 20, 345 (1971)

3213 SCHWAB RS, POSKANZER DC, ENGLAND AC JR, YOUNG RR, AMANTADINE IN PARKINSON'S DISEASE, *J AM MED ASSOC*, 222, 792 (1972)

3214 SCHWARTZ FD, PILLAY VKG, KARK RM, ETHACRYNIC ACID: ITS USEFULNESS AND UNTOWARD EFFECTS, *AM HEART J*, 79, 427 (1970)

3215 SCHWARTZ JB, KEEFE D, HARRISON DC, ADVERSE EFFECTS OF ANTIARRHYTHMIC DRUGS, *DRUGS*, 21, 23-45 (1981)

3216 SCHWARTZ MA, POSTMA E, METABOLISM OF FLURAZEPAM, A BENZODIAZEPINE IN MAN AND DOG, *J PHARM SCI*, 59, 1800 (1970)

3217 SCHWARTZ MK, INTERFERENCES IN DIAGNOSTIC BIOCHEMICAL PROCEDURES, *ADV CLIN CHEM*, 16, 1 (1973)

3218 SCHWARTZ MK, INTERFERENCES IN BIOCHEMICAL TESTS, *MEM SLOAN-KETTERING CLIN BULL*, 1, 11 (1971)

3219 SCHWARZ S, TAPPEINER G, HINTNER H, HORMONE BINDING GLOBULIN LEVELS IN PATIENTS WITH HEREDITARY ANGIOEDEMA DURING TREATMENT WITH DANAZOL, *CLIN ENDOCRINOL*, 14, 563-570 (1981)

3220 SCHWARZE R, KINTZEL HW, HINKEL GK, THE INFLUENCE OF OROTIC ACID ON THE SERUM BILIRUBIN LEVEL OF MATURE NEWBORN, *ACTA PAEDIATR SCAND*, 60, 705 (1971)

3221 SCHWEISFURTH H, SCHIOBERG-SCHIEGNITZ S, ASSAY AND BIOCHEMICAL CHARACTERIZATION OF ANGIOTENSIN-CONVERTING ENZYME IN CEREBROSPINAL FLUID, *ENZYME*, 32, 12-19 (1984)

3222 SCHWEPPE K-W, ASSMAN G, CHANGES OF PLASMA LIPIDS AND LIPOPROTEIN LEVELS DURING DANAZOL TREATMENT FOR ENDOMETRIOSIS, *HORM METAB RES*, 16, 593-597 (1984)

3223 SCHWERTNER HA, FRIEDMAN HS, CHANGES IN LIPID VALUES AND LIPOPROTEIN PATTERNS OF SERUM SAMPLES CONTAMINATED WITH BACTERIA, *AM J CLIN PATHOL*, 59, 829 (1973)

3224 SCOBIE IN, MACCUISH AC, KESSON CM, ET AL, NEUTROPENIA DURING ALLOPURINOL TREATMENT IN TOTAL THERAPEUTIC STARVATION, *BR MED J [CLIN RES]*, 1, 1163 (1980)

3225 SCOMMEGNA A, LASH SR, OVARIAN OVERSTIMULATION, MASSIVE ASCITES, AND SINGLETON PREGNANCY AFTER CLOMIPHENE, *J AM MED ASSOC*, 207, 753 (1969)

3226 SCOTT JT, HALL AP, GRAHAME R, ALLOPURINOL IN TREATMENT OF GOUT, *BR MED J [CLIN RES]*, 2, 321 (1966)

3227 SCOTT JT, PORTER IH, LEWIS SM, DIXON AS, STUDIES OF GASTROINTESTINAL BLEEDING CAUSED BY CORTICOSTEROIDS, SALICYLATES AND OTHER ANALGESICS, *Q J MED*, 30, 167 (1961)

3228 SCOTT PH, A FLUOROMETRIC METHOD FOR THE DETERMINATION OF PLASMA TYROSINE (CONCENTRATIONS), *CLIN CHIM ACTA*, 35, 17 (1971)

3229 SCOTT PV, HORTON JN, MAPLESON WW, LEAKAGE OF OXYGEN FROM BLOOD AND WATER SAMPLES STORED IN PLASTIC AND GLASS SYRINGES, *BR MED J [CLIN RES]*, 3, 512 (1971)

3230 SCRIVER CR, CLOW CL, LAMM P, ON THE SCREENING, DIAGNOSIS AND INVESTIGATION OF HEREDITARY AMINOACIDOPATHIES, *CLIN BIOCHEM*, 6, 142 (1973)

3231 SCURRY MT, BRUTON J, BARRY KG, THE EFFECT OF CHORIONIC GONADOTROPIN ON STEROID EXCRETION, *ARCH INTERN MED*, 128, 561 (1971)

3232 SEAGER J, IGA DEFICIENCY DURING TREATMENT OF INFANTILE HYPOTHYROIDISM WITH THYROXINE, *BR MED J [CLIN RES]*, 288, 1562-1563 (1984)

3233 SEAKINS JWT, THE DETERMINATION OF URINARY PHENYLACETYLGLUTAMINE AS PHENYLACETIC ACID, *CLIN CHIM ACTA*, 35, 121 (1971)

3234 SEAMONDS B, TOWFIGHI J, ARVAN DA, DETERMINATION OF IONIZED CALCIUM IN SERUM BY USE OF AN ION-SELECTIVE ELECTRODE, *CLIN CHEM*, 18, 155 (1972)

3235 SEBEL PS, VERGHESE C, MAKIN HLJ, EFFECT ON PLASMA CORTISOL CONCENTRATIONS OF A SINGLE INDUCTION DOSE OF ETOMIDATE ON THIOPENTONE, *LANCET*, 2, 625 (1983)

3236 SEDDON B, FYNN GH, ORTHOPHOSPHATE ANALYSIS BY THE FISKE-SUBBAROW METHOD AND INTERFERENCE BY ADENOSINE PHOSPHATES, *ANAL BIOCHEM*, 56, 566 (1973)

3237 SEDDON RJ, HORMONAL STEROID CONTRACEPTIVES, 1. PHYSIOLOGICAL AND PHARMACOLOGICAL CONSIDERATIONS, *DRUGS*, 1, 399 (1971)

3238 SEDLACEK SM, RUDOLF PM, KAEHNY WD, AMOXAPINE-ASSOCIATED AGRANULOCYTOSIS WITH THROMBOCYTOSIS OCCURRING EARLY DURING RECOVERY, *AM J MED*, 80, 533-536 (1986)

3239 SEED M, GODSLAND IF, WYNN V, ET AL, THE EFFECTS OF CYPROTERONE ACETATE AND ETHINYL OESTRADIOL ON CARBOHYDRATE METABOLISM, *CLIN ENDOCRINOL*, 21, 689-699 (1984)

3240 SEGAL RJ, NONSPECIFICITY OF URINARY LEAD MEASUREMENTS BY ATOMIC ABSORPTION SPECTROSCOPY, *CLIN CHEM*, 15, 1124 (1969)

3241 SEGAR WE, MOORE WW, THE REGULATION OF ANTIDIURETIC HORMONE RELEASE IN MAN I. EFFECTS OF CHANGE IN POSITIONS AND AMBIENT TEMPERATURE, *J CLIN INVEST*, 47, 2143 (1968)

3242 SEIBOLD JR, LYNCH CJ, MEDSGER JA JR, CHOLESTASIS ASSOCIATED WITH D-PENICILLAMINE THERAPY: CASE REPORT AND REVIEW OF THE LITERATURE, *ARTHRITIS RHEUM*, 24, 554-556 (1981)

3243 SEIDELIN R, CIMETIDINE AND RENAL FAILURE, *POSTGRAD MED J*, 56, 440-441 (1980)

3244 SEINFELD E, STARR P, STUDIES OF THYROID FUNCTION IN PATIENTS TREATED WITH PARA-AMINOSALICYLIC ACID AND ISONIAZID, *AM REV RESPIR DIS*, 80, 845 (1959)

3245 SEITZ H, JAWORSKI ZF, EFFECT OF HYDROCHLOROTHIAZIDE ON SERUM AND URINARY CALCIUM AND URINARY CITRATE, *CAN MED ASSOC J*, 90, 414 (1964)

3246 SEITZ HK, BOSCHE J, CZYGAN P, ET AL, INCREASED BLOOD ETHANOL LEVELS FOLLOWING CIMETIDINE BUT NOT RANITIDINE, *LANCET*, 1, 760 (1983)

3247 SEKADDE CB, SLAUNWHITE WR JR, ACETO T JR, RAPID RADIOIMMUNOASSAY OF TRIIODOTHYRONINE, *CLIN CHEM*, 19, 1016 (1973)

3248 SELLERS EM, KOCH-WESER J, DISPLACEMENT OF WARFARIN FROM HUMAN ALBUMIN BY DIAZOXIDE AND ETHACRYNIC, MEFENAMIC AND NALIDIXIC ACIDS, *CLIN PHARMACOL THER*, 11, 524 (1970)

3249 SELLERS EM, LANG M, KOCH-WESER J, POTENTIATION BY TRICHLOROETHYLPHOSPHATE OF WARFARIN INDUCED HYPOPROTHROMBINEMIA, *CLIN PHARMACOL THER*, 13, 152 (1972)

3250 SELROOS O, EDGREN J, LUPUS-LIKE SYNDROME ASSOCIATED WITH PULMONARY REACTION TO NITROFURANTOIN, *ACTA MED SCAND*, 197, 125-129 (1975)

3251 SELTZER HS, DRUG INDUCED HYPOGLYCEMIA, *DIABETES*, 21, 955 (1972)

3252 SEMB LS, THE RELATIONSHIP BETWEEN ACID, PEPSIN, AND ELECTROLYTES IN GASTRIC JUICE BEFORE AND AFTER SUBCUTANEOUS INJECTION, *SCAND J GASTROENTEROL*, 7, 17 (1972)

3253 SEN S, MUKERJEE AB, HYPOGLYCAEMIC ACTION OF OXYTETRACYCLINE, *J INDIAN MED ASSOC*, 52, 366 (1969)

3254 SENAY LC JR, CHRISTENSEN ML, CHANGES IN BLOOD PLASMA DURING PROGRESSIVE DEHYDRATION, *J APPL PHYSIOL*, 20, 1136 (1965)

3255 SEPKOVIC DW, HALEY NJ, WYNDER EL, THYROID ACTIVITY IN CIGARETTE SMOKERS, *ARCH INTERN MED*, 144, 501-503 (1984)

3256 SERLIN MJ, SIBEON RG, MOSSMAN S, ET AL, CIMETIDINE: INTERACTIONS WITH ORAL ANTICOAGULANTS IN MAN, *LANCET*, 2, 317-319 (1979)

3257 SETH J, WORK SIMPLIFICATION OF A COMPETITIVE PROTEIN BINDING ANALYSIS OF SERUM THYROXINE, *CLIN CHIM ACTA*, 46, 431 (1973)

3258 SETTLAGE DF, NAKAMURA RM, DAVAJAN V, KHARMA K, MISHELL DR JR, A QUANTITATIVE ANALYSIS OF SERUM PROTEINS DURING TREATMENT WITH ORAL CONTRACEPTIVE STEROIDS, *CONTRACEPTION*, 1, 101 (1970)

3259 SEVERINI G, ALIBERTI LM, VARIATION OF URINARY ENZYMES N-ACETYL-BETA-GLUCOSAMINIDASE, ALANINE AMINOPEPTIDASE, AND LYSOZYME IN PATIENTS RECEIVING RADIOCONTRAST AGENTS, *CLIN BIOCHEM*, 20, 339-341 (1987)

3260 SEVIOUR PW, TEAL TK, RICHMOND W, ET AL, CHLORPROPAMIDE LOWERS SERUM AND LIPOPROTEIN CHOLESTEROL IN INSULIN DEPENDENT DIABETICS, *DIABETIC MED*, 3, 152-154 (1986)

3261 SEYBERTH HW, THE SIGNIFICANCE OF RENAL PROSTAGLANDINS FOR KIDNEY FUNCTION IN EARLY CHILDHOOD, *MONATSSCHR KINDERHEILKD*, 135, 178-185 (1987)

3262 SEYMOUR GC, JACKSON MJ, SPECIFIC AUTOMATED METHOD FOR MEASUREMENT OF URINARY HYDROXYPROLINE, *CLIN CHEM*, 20, 544 (1974)

3263 SHADE RE, GRIM CE, RESPONSE OF RENIN-ALDOSTERONE SYSTEM IN MAN TO SMALL AMOUNTS OF DESOXYCORTICOSTERONE, *CLIN RES*, 20, 726 (1972)

3264 SHAFER N, CLINICAL PEARLS, *MED COUNTERPOINT*, 4, 45 (1972)

3265 SHAFFNER F, DIAGNOSIS OF DRUG-INDUCED HEPATIC DAMAGE, *J AM MED ASSOC*, 191, 466 (1965)

3266 SHAH IA, BRANDT H, HALOTHANE-ASSOCIATED GRANULOMATOUS HEPATITIS, *DIGESTION*, 28, 245-249 (1983)

3267 SHAH S, HEPATOTOXICITY DUE TO PAPAVERINE HYDROCHLORIDE, *N ENGL J MED*, 282, 1271 (1970)

3268 SHALEV A, HERMESH H, MUNITZ H, THE HYPOURICEMIC EFFECT OF CHLORPROTHIXENE, *CLIN PHARMACOL THER*, 42. 562-566 (1987)

3269 SHALEV O, MOSSERI M, ARIEL I, ET AL, METHYLDOPA-INDUCED IMMUNE HEMOLYTIC ANEMIA AND CHRONIC ACTIVE HEPATITIS, *ARCH INTERN MED*, 143, 592-593 (1983)

3270 SHAMBERGER RJ, RUKOVENA E, TYTKO SA, WILLIS CE, MERCURY, TROPICAL FISH AND PBIs, *CLIN CHEM*, 18, 719 (1972)

3271 SHAND DG, REDUCED BINDING OF QUINIDINE IN PLASMA FROM VACUTAINERS™, *CLIN PHARMACOL THER*, 20, 120 (1976)

3272 SHAND DG, EPSTEIN C, KINBERG-CALHOUN J, ET AL, THE EFFECT OF ETODOLAC ADMINISTRATION ON RENAL FUNCTION IN PATIENTS WITH ARTHRITIS, *J CLIN PHARMACOL*, 26, 269-274 (1986)

3273 SHAPER AG, POCOCK SJ, ASHBY D, ET AL, BIOCHEMICAL AND HAEMATOLOGICAL RESPONSE TO ALCOHOL INTAKE., *ANN CLIN BIOCHEM*, 22, 50-61 (1985)

3274 SHARF M, OETTINGER M, LANIR A, ET AL, LIPID AND LIPOPROTEIN LEVELS FOLLOWING PURE ESTRADIOL IMPLANTATION IN POST-MENOPAUSAL WOMEN, *GYNECOL OBSTET INVEST*, 19, 207-212 (1985)

3275 SHARIFI R, LEE M, OJEDA L, ET AL, PRELIMINARY REPORT COMPARING PIPERACILLIN AND CARBENICILLIN FOR COMPLICATED URINARY TRACT INFECTIONS, *J UROL*, 128, 755-758 (1982)

3276 SHARP AM, FRASER IS, CATERSON ID, FURTHER STUDIES ON DANAZOL INTERFERENCE IN TESTOSTERONE RADIOIMMUNOASSAYS, *CLIN CHEM*, 29, 141-143 (1983)

3277 SHARP P, INTERFERENCE IN GLUCOSE OXIDASE-PEROXIDASE BLOOD GLUCOSE METHODS, *CLIN CHIM ACTA*, 40, 115 (1972)

3278 SHARP P, RILEY C, COOK JGH, PINK PJF, EFFECT OF TWO SULFONYLUREAS ON GLUCOSE DETERMINATIONS BY ENZYMIC METHODS, *CLIN CHIM ACTA*, 36, 93 (1972)

3279 SHASHA SM, SHILLER M, BEN ARYEH H, ET AL, EFFECT OF PINDOLOL ON BLOOD PARATHYROID HORMONE CONCENTRATION, *ISR J MED SCI*, 17, 1189-1190 (1981)

3280 SHAW D M, ANTI-DEPRESSANT DRUGS, *PRACTITIONER*, 192, 23 (1964)

3281 SHAW KNF, TREVARTHEN J, EXOGENOUS SOURCES OF URINARY PHENOL AND INDOLE ACIDS, *NATURE*, 182, 797 (1958)

3282 SHAW MW, HUMAN CHROMOSOME DAMAGE BY CHEMICAL AGENTS, *ANN REV MED*, 21, 409 (1970)

3283 SHAY H, SIPLET H, STUDY OF CHLORPROMAZINE JAUNDICE: MECHANISM AND PREVENTION, *GASTROENTEROLOGY*, 32, 571 (1957)

3284 SHEEHAN J, TRIMETHOPRIM-ASSOCIATED MARROW TOXICITY, *LANCET*, 2, 692 (1981)

3285 SHEEHAN J, WHITE A, DIURETIC-ASSOCIATED HYPOMAGNESAEMIA, *BR MED J [CLIN RES]*, 285, 1157-1159 (1982)

3286 SHELDON W, WELCH RJ, BONHAM JR, ET AL, HYPOMAGNESAEMIA FOLLOWING TREATMENT OF CHILDHOOD CANCER WITH CIS-PLATINUM., *ANN CLIN BIOCHEM*, 24, SUPPL S1, 85-86 (1987)

3287 SHELP WD, BLOODWORTH JMB JR., RIESELBACH RE, EFFECT OF AZATHIOPRINE ON RENAL HISTOLOGY AND FUNCTION IN LUPUS NEPHRITIS, *ARCH INTERN MED*, 128, 566 (1971)

3288 SHEN S, WEIR M, DAGHER F, ET AL, EFFECT OF CYCLOSPORINE ON TOTAL LYMPHOCYTE AND T-CELL COUNTS IN RENAL TRANSPLANT RECIPIENTS, *N ENGL J MED*, 314, 447-448 (1986)

3289 SHENKENBERG TD, VON HOFF DD, MITOXANTRONE: A NEW ANTI-CANCER DRUG WITH SIGNIFICANT CLINICAL ACTIVITY, *ANN INTERN MED*, 105, 67-81 (1986)

3290 SHEPHARD RJ, KILLINGER D, FRIED T, RESPONSES TO SUSTAINED USE OF ANABOLIC STEROID, *BR J SPORTS MED*, 11, 170-173 (1977)

3291 SHEPHERD J, PACKARD CJ, STEWART JM, ET AL, APOLIPOPROTEIN A AND B (Sf 100-400) METABOLISM DURING BEZAFIBRATE THERAPY IN HYPERTRIGLYCERIDEMIC SUBJECTS, *J CLIN INVEST*, 74, 2164-2177 (1984)

3292 SHER SP, DRUG ENZYME INDUCTION AND DRUG INTERACTIONS: LITERATURE TABULATION, *TOXICOL APPL PHARMACOL*, 18, 780 (1971)

3293 SHERLOCK S, EFFECTS OF DRUGS ON THE LIVER, *ANN CLIN BIOCHEM*, 7, 75 (1970)

3294 SHERLOCK S, HALOTHANE HEPATITIS, *GUT*, 12, 324 (1971)

3295 SHERMAN HC, CALDWELL ML, ADAMS M, A QUANTITATIVE COMPARISON OF THE INFLUENCE OF NEUTRAL SALTS ON THE ACTIVITY OF PANCREATIC AMYLASE, *J AM CHEM SOC*, 50, 2538 (1928)

3296 SHERMAN JD, LOVE DE, HARRINGTON JF, ANEMIA-POSITIVE LUPUS AND RHEUMATOID FACTORS WITH METHYLDOPA A REPORT OF THREE CASES, *ARCH INTERN MED*, 120, 321 (1967)

3297 SHERMAN L, KIM S, BENJAMIN F, KOLODNY HD, EFFECT OF CHLORPROMAZINE ON SERUM GROWTH HORMONE CONCENTRATION IN MAN, *N ENGL J MED*, 284, 72 (1971)

3298 SHERRY S, PROSPECTS IN ANTITHROMBOTIC THERAPY, *AM J CARDIOL*, 29, 81 (1972)

3299 SHERWIN SA, KNOST JA, FEIN S, ET AL, A MULTIPLE DOSE PHASE I TRIAL OF RECOMBINANT LEUKOCYTE A INTERFERON IN CANCER PATIENTS, *J AM MED ASSOC*, 248, 2461-2466 (1982)

3300 SHERWOOD LM, HYPERNATREMIA DURING SODIUM SULFATE THERAPY, *N ENGL J MED*, 277, 314 (1967)

3301 SHERWOOD RJ, EVALUATION OF EXPOSURE TO BENZENE VAPOR DURING LOADING OF PETROL, *BR J IND MED*, 29, 65 (1972)

3302 SHIELDS JE JR, PRAISSMAN M, BERKOWITZ JM, PLASMA GASTRIN RESPONSE TO ORAL COFFEE AND CALCIUM CARBONATE, *CLIN RES*, 21, 524 (1973)

3303 SHIELDS LI, FILES JA, DOLL DC, ET AL, RANITIDINE AND AGRANULO-CYTOSIS, *ANN INTERN MED*, 104, 128 (1986)

3304 SHIH VE, NIKIFOROV V, CARNEY MM, ACETAMINOPHEN METABOLITE INTERFERES IN ANALYSIS FOR AMINO ACIDS, *CLIN CHEM*, 31, 148 (1985)

3305 SHILS ME, INTERFERENCE BY FORMALDEHYDOGENIC DRUGS IN THE QUANTITATIVE DETERMINATION OF URINARY HYDROXY-INDOLACETIC ACID, *CLIN CHEM*, 13, 397 (1967)

3306 SHIMA K, KURODA K, MATSUYAMA T, TARUI S, NISHIKAWA M, PLASMA GLUCAGON AND INSULIN RESPONSES TO VARIOUS SUGARS IN GASTRECTOMIZED AND NORMAL SUBJECTS, *PROC SOC EXP BIOL MED*, 139, 1042 (1972)

3307 SHIONOIRI H, IINO S, INOUE S, GLUCOSE METABOLISM DURING CAPTOPRIL MONOTHERAPY AND COMBINATION THERAPY IN DIABETIC HYPERTENSIVE PATIENTS: A MULTICLINIC TRIAL, *CLIN EXP HYPERTENS*, 9, 671-674 (1987)

3308 SHOENFELD Y, BARUCH NB, LIVNI E, ET AL, CARBAMAZEPINE (TEGRETOL®)-INDUCED THROMBOCYTOPENIA, *ACTA HAEMATOL*, 68, 74 (1982)

3309 SHOENFELD Y, GUREWICH Y, GALLANT LA, ET AL, PREDNISONE-INDUCED LEUKOCYTOSIS, *AM J MED*, 71, 773-778 (1981)

3310 SHOJANIA AM, ORAL CONTRACEPTIVES: EFFECTS ON FOLATE AND VITAMIN B_{12} METABOLISM, *CAN MED ASSOC J*, 126, 244-247 (1982)

3311 SHOJANIA AM, HORNADY GJ, BARNES PH, THE EFFECT OF ORAL CONTRACEPTIVES ON FOLATE METABOLISM, *AM J OBSTET GYNECOL*, 111, 782 (1971)

3312 SHOPSIN B, EFFECTS OF LITHIUM ON THYROID FUNCTION: A REVIEW, *DIS NERV SYS*, 31, 237 (1970)

3313 SHOPSIN B, FRIEDMANN R, GERSHON S, LITHIUM AND LEUKOCYTO-SIS, *CLIN PHARMACOL THER*, 12, 923 (1971)

3314 SHULL B, CHENG CS, RAHILL WJ, EFFECT OF HEMOLYSIS ON VALUES OBTAINED FOR UREA NITROGEN IN PLASMA, *CLIN CHEM*, 19, 1226 (1973)

3315 SHURTLEFF DB, HAYDEN PW, THE TREATMENT OF HYDROCEPHALUS WITH ISOSORBIDE, AN ORAL HYPEROSMOTIC AGENT, *J CLIN PHARMACOL*, 12, 108 (1972)

3316 SHUSTER F, NAPIER EA, HENLEY KS, SERUM TRANSAMINASE ACTIVITY FOLLOWING MORPHINE, MEPERIDINE, AND CODEINE IN NORMALS, *AM J MED SCI*, 246, 714 (1963)

3317 SHUSTER M, DUNN M, PENTAMIDINE AND HEMATURIA, *ANN INTERN MED*, 105, 146 (1986)

3318 SIEBERS RWL, MALING TJB, DINEEN HJ, ET AL, NO INTERFERENCE BY AMIODARONE ON ALBUMIN ESTIMATION BY DYE-BINDING METHODS, *CLIN CHEM*, 32, 1990 (1986)

3319 SIEST G, BAGREL A, PANEK E, GALTEAU MM, BATT AM, SCHIELE F, PLASMA ENZYMES - PHYSIOLOGICAL AND ENVIRONMENTAL VARIATIONS IN, *REFERENCE VALUES IN HUMAN CHEMISTRY*, G SIEST (ED), KARGER, BASEL, 28 (1973)

3320 SILBER M, ALMKVIST O, LARSSON B, ET AL, THE EFFECT OF ORAL CONTRACEPTIVE PILLS ON LEVELS OF OXYTOCIN IN PLASMA AND ON COGNITIVE FUNCTION, *CONTRACEPTION*, 36, 641-650 (1987)

3321 SILVERMAN G, BRAITHWAITE RA, BENZODIAZEPINES AND TRICYCLIC ANTIDEPRESSANT PLASMA LEVELS, *BR MED J [CLIN RES]*, 3, 18 (1973)

3322 SILVERMAN JL, WURZEL HA, THE EFFECT OF GLYCERYL GUAIACOLATE ON PLATELET FUNCTION AND OTHER COAGULATION FACTORS IN VIVO, *AM J CLIN PATHOL*, 51, 35 (1969)

3323 SIMON GE, JATLOW PI, SELIGSON HT, SELIGSON D, MEASUREMENT OF 5,5-DIPHENYLHYDANTOIN IN BLOOD USING THIN LAYER CHROMATOGRAPHY, *AM J CLIN PATHOL*, 55, 145 (1971)

3324 SIMONIN R, ROUX H, OLIVER C, JAQUET P, ARGEMI B, VAGUE P, EFFECT OF ORAL ADMINISTRATION OF 0.500 G L-DOPA ON THE PLASMATIC LEVEL OF GH, ACTH, TSH, *SEMAINE DES HOS-PITAUX*, 49, 617 (1973)

3325 SIMPSON E, SOME ASPECTS OF CALCIUM METABOLISM IN A FATAL CASE OF ETHYLENE GLYCOL POISONING, *ANN CLIN BIOCHEM*, 22, 90-93 (1985)

3326 SIMPSON E, STEWART MJ, SCREENING FOR PARACETAMOL POISONING, *ANN CLIN BIOCHEM*, 10, 171 (1973)

3327 SIMPSON GM, COOPER TB, BRAUN GA, FURTHER STUDIES ON THE EFFECTS OF BUTYROPHENONES ON CHOLESTEROL SYNTHESIS IN HUMANS, *CURR THER RES*, 9, 413 (1967)

3328 SIMPSON GM, VARGA V, AN INVESTIGATION OF THE CLINICAL EFFECT OF GPA-1714, A CATECHOL-O-METHYL TRANSFERASE INHIBITOR, *J CLIN PHARMACOL*, 12, 417 (1972)

3329 SIMPSON GR, DALE E, SERUM LEVELS OF PHOSPHORUS, MAGNESIUM, AND CALCIUM IN WOMEN UTILIZING COMBINATION ORAL OR LONG-ACTING INJECTABLE, *FERTIL STERIL*, 23, 326 (1972)

3330 SIMPSON RE III, GOLDSTEIN DJ, HJELTE GS, ET AL, ACUTE THROMBOCYTOPENIA ASSOCIATED WITH FENOPROFEN, *N ENGL J MED*, 298, 629-630 (1978)

3331 SIMPSON W, SEVERE MEGALOBLASTIC ANEMIA INDUCED BY PHENYTOIN SODIUM, *ORAL SURG*, 22, 302 (1966)

3332 SIMPSON WM JR, ORAL CONTRACEPTIVES AND UNTOWARD EFFECTS COINCIDENT WITH THEIR USE, REVIEW, *SOUTH MED J*, 64, 1184 (1971)

3333 SINGER I ROTENBERG D, DEMECLOCYCLINE-INDUCED NEPHROGENIC DIABETES INSIPIDUS, *ANN INTERN MED*, 79, 679 (1973)

3334 SINGH GB, CLINICAL SIGNIFICANCE AND EXPERIMENTAL STUDIES ON AMYLASE. A REVIEW, *INDIAN J MED SCI*, 20, 181 (1966)

3335 SINGH HP, HERBERT MA, GAULT MH, EFFECT OF SOME DRUGS ON CLINICAL LABORATORY VALUES AS DETERMINED BY THE TECHNICON® SMA 12/60, *CLIN CHEM*, 18, 137 (1972)

3336 SINGH J, DIECKERT JW, MAGNESIUM INTERFERENCE IN THE DETERMINATION OF SODIUM IN BIOLOGICAL MATERIAL BY NEUTRON ACTIVATION ANALYSIS, *ANAL BIOCHEM*, 53, 470 (1973)

3337 SINGHI S, CHOOKANG E, HALL J STE, ET AL, IATROGENIC NEONATAL AND MATERNAL HYPONATREMIA FOLLOWING OXYTOCIN AND AQUEOUS GLUCOSE INFUSION DURING LABOUR, *BR J OBSTET GYNAECOL*, 92, 356-363 (1985)

3338 SINHAMAHAPATRA SB, KIRSCHNER MA, EFFECT OF L-DOPA ON TESTOSTERONE AND LUTEINIZING HORMONE PRODUCTION, *J CLIN ENDOCRINOL METAB*, 34, 756 (1972)

3339 SINKULE JA, ETOPOSIDE: A SEMISYNTHETIC EPIPODOPHYLLOTOXIN, *PHARMACOTHERAPY*, 4, 61-73 (1984)

3340 SIRIS SG, SIRIS ES, VAN KAMMEN DP, ET AL, EFFECTS OF DOPAMINE BLOCKADE ON GONADOTROPINS AND TESTOSTERONE IN MEN, *AM J PSYCHIATRY*, 137, 211-214 (1980)

3341 SIVITZ WI, CHERTOW BS, BARANETSKY NG, ET AL, THEOPHYLLINE INTERFERES WITH LIGAND BINDING IN RADIOIMMUNOASSAYS FOR SOMATOSTATIN, *J IMMUNOASSAY*, 3, 145-154 (1982)

3342 SJOSTROM R, ABSENCE OF EFFECT OF PARA-CHLOROPHENYLALANINE ON 5-HYDROXYINDOLEACETIC ACID IN CEREBROSPINAL FLUID IN MAN, *PSYCHOPHARMACOL*, 27, 393 (1972)

3343 SKEITH MD, HEALEY LA, CUTLER RE, EFFECT OF PHLORIDZIN ON URIC ACID EXCRETION IN MAN, *AM J PHYSIOL*, 219, 1080 (1970)

3344 SKEITH MD, HEALEY LA, CUTLER RE, URATE EXCRETION DURING MANNITOL AND GLUCOSE DIURESIS, *J LAB CLIN MED*, 70, 213 (1967)

3345 SKINNER SL, LUMBERS ER, SYMONDS EM, ANALYSIS OF CHANGES IN THE RENIN-ANGIOTENSIN SYSTEM DURING PREGNANCY, *CLIN SCI*, 42, 479 (1972)

3346 SKREDE S, RO JS, MJOLNEROD O, EFFECTS OF DEXTRANS ON THE PLASMA PROTEIN CHANGES DURING THE POSTOPERATIVE PERIOD, *CLIN CHIM ACTA*, 48, 143 (1973)

3347 SLAGBOOM G, LOELIGER EA, COUMARIN-ASSOCIATED HEPATITIS: REPORT OF TWO CASES, *ARCH INTERN MED*, 140, 1028-1029 (1980)

3348 SLANINA P, FRECH W, EKSTROM L-G, ET AL, DIETARY CITRIC ACID ENHANCES ABSORPTION OF ALUMINUM IN ANTACIDS, *CLIN CHEM*, 32, 539-541 (1986)

3349 SLAYTON RE, SHNIDER BI, ELIAS E, HORTON J, PERLIA CP, NEW APPROACH TO THE TREATMENT OF HYPERCALCEMIA: THE EFFECT OF SHORT-TERM TREATMENT WITH MITHRAMYCIN, *CLIN PHARMACOL THER*, 12, 833 (1971)

3350 SLEZAK P, QUINIDINE HEPATOTOXICITY, *MED J AUST*, 1, 139 (1981)

3351 SLOB A, WINK A, RADDER JJ, THE EFFECT OF ACUTE NOISE EXPOSURE ON THE EXCRETION OF CORTICOSTEROIDS, ADRENALIN AND NORADRENALINE IN MAN, *INTL ARCH ARBEITS MED*, 31, 225 (1973)

3352 SLUITER HE, HUYSMANS FT, THIEN TA, ET AL, HAEMODYNAMIC, HORMONAL, AND DIURETIC EFFECTS OF FELODIPINE IN HEALTHY NORMOTENSIVE VOLUNTEERS, *DRUGS*, 29 SUPPL 2, 26-35 (1985)

3353 SLUSZKIEWICZ E, EFFECT OF PREDNISONE THERAPY ON SERUM LEVELS OF THYROXINE (T_4), TRIIODOTHYRONINE (T_3), REVERSE TRI-IODOTHYRONINE (T_3), T_3-BINDING CAPACITY, BASAL TSH LEVEL, AND TSH RESPONSE TO THYREOLIBERIN (TRH) IN CHILDREN., *EXP CLIN ENDOCRINOL*, 85, 191-198 (1985)

3354 SMALL M, EASTALL GH, SEMPLE CG, ET AL, ALTERATION OF HORMONE LEVELS IN NORMAL MALES GIVEN THE ANABOLIC STEROID STANOZOLOL, *CLIN ENDOCRINOL*, 21, 49-55 (1984)

3355 SMALLDON KW, ETHANOL OXIDATION BY HUMAN ERYTHROCYTES, *NATURE*, 245, 266 (1973)

3356 SMALLEY DL, BRADLEY ME, NEW TEST FOR GLUCOSE (BM33071) EVALUATED, *CLIN CHEM*, 31, 90-92 (1985)

3357 SMIGAN L, PERRIS C, CORTISOL CHANGES IN LONG-TERM LITHIUM THERAPY, *NEUROPSYCHOBIOLOGY*, 11, 219-223 (1984)

3358 SMILEY JW, ONESTI G, SWARTZ C, THE ACUTE EFFECTS OF METOLAZONE ON ELECTROLYTE AND ACID EXCRETION IN MAN, *CLIN PHARMACOL THER*, 13, 336 (1972)

3359 SMITH CR, LIPSKY JJ, LASKIN OL, ET AL, DOUBLE - BLIND COMPARISON ON THE NEPHROTOXICITY AND AUDITORY TOXICITY OF GENTAMICIN AND TOBRAMYCIN, *N ENGL J MED*, 302, 1106-1109 (1980)

3360 SMITH DH, SCOTT DL, ZAPHIROPOULOS GC, EOSINOPHILIA IN D-PENICILLAMINE THERAPY, *ANN RHEUM DIS*, 42, 408-410 (1983)

3361 SMITH JC JR., LEWIS S, HOLBROOK J, ET AL, EFFECT OF HEPARIN AND CITRATE IN MEASURED CONCENTRATIONS OF VARIOUS ANALYSES ON PLASMA, *CLIN CHEM*, 33, 814-816 (1987)

3362 SMITH JW (ED), *MANUAL OF MEDICAL THERAPEUTICS*, LITTLE BROWN AND CO, BOSTON MASS, 19TH ED (1969)

3363 SMITH M, PAYNE RB, REEXAMINATON OF EFFECT OF PARACETAMOL ON SERUM URIC ACID MEASURED BY PHOSPHOTUNGSTIC ACID REDUCTION, *ANN CLIN BIOCHEM*, 16, 96-99 (1979)

3364 SMITH MD, GIBSON GE, ROWLAND R, COMBINED HEPATOTOXICITY AND NEUROTOXICITY FOLLOWING SULFOSALAZINE ADMINISTRATION, *AUST N Z J MED*, 12, 76-80 (1982)

3365 SMITH MJH, SMITH PK, *THE SALICYLATES: A CRITICAL BIBLIOGRAPHIC REVIEW, INTERSCIENCE*, NEW YORK, NY (1966)

3366 SMITH PM, SMITH EM, GOTTLIEB NL, GOLD DISTRIBUTION IN WHOLE BLOOD DURING CHRYSOTHERAPY, *J LAB CLIN MED*, 82, 930 (1973)

3367 SMITH WM, FEIGAL DW, FURBERG CD, ET AL, USE OF DIURETICS IN THE TREATMENT OF HYPERTENSION IN THE ELDERLY, *DRUGS*, 31 SUPPL 4, 154-164 (1986)

3368 SMITH WO, KYRIAKOPOULOS AA, HAMMARSTEIN JF, MAGNESIUM DEPLETION INDUCED BY VARIOUS DIURETICS, *J OKLA STATE MED ASSOC*, 55, 248 (1962)

3369 SMITH, KLINE AND FRENCH, MANUFACTURER'S LITERATURE ON DUCON,® *1500 SPRING GARDEN ST, PHILADELPHIA, PA 19101*

3370 SMOLIK R, GRZYBEK-HRYNCEWICZ K, LANGE A, ZATONSKI W, SERUM COMPLEMENT LEVEL IN WORKERS EXPOSED TO BENZENE, TOLUENE AND XYLENE, *INTL ARCH ARBEITS MED*, 31, 243 (1973)

3371 SNIDER GB, GOGATE SA, CLINICAL OBSERVATIONS FOLLOWING PAPAVERINE THERAPY, *OHIO MED*, 74, 571-573 (1978)

3372 SNYDER S, FLUPHENAZINE JAUNDICE, *AM J GASTROENTEROL*, 73, 336-340 (1980)

3373 SOBER AJ, SODE J, RUDER HJ, PLASMA CORTISOL AND URINARY CATECHOLAMINE RESPONSE TO ORAL GLUCOSE LOADING IN MAN, *CLIN RES*, 20, 866 (1972)

3374 SOBREVILLA LA, SALAZAR F, HIGH ALTITUDE HYPERURICEMIA, *PROC SOC EXP BIOL MED*, 129, 890 (1968)

3375 SOFFER EE, TAYLOR RJ, BERTRAM PD, ET AL, CARBAMAZEPINE-INDUCED LIVER INJURY, *SOUTH MED J*, 76, 681-683 (1983)

3376 SOKOLAWSKI A, JUNG K, EGGER E, FEHLERMOGLICHKEITEN BEI DER ENWEISSBESTIMMUNG MIT DEM BIURET REAGENZ, *Z KLIN CHEM KLIN BIOCHEM*, 10, 531 (1972)

3377 SOKOLOFF B, MICHITERU H, SAELHOF CC, WRZOLEK T, HORI M, AGING: ATHEROSCLEROSIS AND ASCORBIC ACID METABOLISM, *J AM GERIATR SOC*, 14, 1239 (1966)

3378 SOLOMON GE, HILGARTNER MW, KUTT H, COAGULATION DEFECTS CAUSED BY DIPHENYLHYDANTOIN, *NEUROLOGY*, 22, 1165 (1972)

3379 SOLOMON HM, BARAKAT MJ, ASHLEY CJ, MECHANISMS OF DRUG INTERACTION, *J AM MED ASSOC*, 216, 1997 (1971)

3380 SOLOMON HM, SCHROGIE JJ, CHANGE IN RECEPTOR SITE AFFINITY: A PROPOSED EXPLANATION FOR THE POTENTIATING EFFECT OF D-THYROXINE, *CLIN PHARMACOL THER*, 8, 797 (1967)

3381 SOLONI FG, SARDINA LC, COLORIMETRIC MICRODETERMINATION OF FREE FATTY ACIDS, *CLIN CHEM*, 19, 419 (1973)

3382 SOLOWAY HB, COX SP, DONAHOO JW, SENSITIVITY OF THE ACTIVATED PARTIAL THROMBOPLASTIN TIME TO HEPARIN, *AM J CLIN PATHOL*, 59, 760 (1973)

3383 SOMANI P, TEMESY-ARMOS PN, LEIGHTON RF, ET AL, HYPONATREMIA IN PATIENTS TREATED WITH LORCAINIDE, A NEW ANTIARRHYTHMIC DRUG, *AM HEART J*, 108, 1443-1448 (1984)

3384 SOMANI P, WANG RIH, ALPHA AND BETA ADRENERGIC-RECEPTOR BLOCKING DRUGS, *DRUG THERAPY* (MARCH 197)

3385 SOMMARIVA D, BRANCHI A, TIRRITO M, ET AL, DIFFERENTIAL EFFECTS OF BENFLUOREX AND TWO FIBRATE DERIVATIVES IN SERUM LIPOPROTEIN PATTERNS IN HYPERTRIGLYCERIDEMIC TYPE 2 DIABETIC PATIENTS, *CURR THER RES*, 40, 859-870 (1986)

3386 SOMMERS DK, SCHOEMAN HS, DRUG INTERACTIONS WITH URATE EXCRETION IN MAN, *EUR J CLIN PHARMACOL*, 32, 499-502 (1987)

3387 SOMMERVILLE JM, MCLAREN EH, CAMPBELL LM, ET AL, SEVERE HEADACHE AND DISTURBED LIVER FUNCTION DURING TREATMENT WITH ZIMELDINE, *BR MED J [CLIN RES]*, 285, 1009 (1982)

3388 SOMOGYI A, BOCHNER F, DOSE AND CONCENTRATION DEPENDENT EFFECT OF RANITIDINE ON PROCAINAMIDE DISPOSITION AND RENAL CLEARANCE IN MAN, *BR J CLIN PHARMACOL*, 18, 175-181 (1984)

3389 SOMOGYI A, MCLEAN A, HEINZOW B, CIMETIDINE-PROCAINAMIDE PHARMACOKINETICS INTERACTION IN MAN: EVIDENCE OF COMPETITION FOR TUBULAR SECRETION OF BASIC DRUGS, *EUR J CLIN PHARMACOL*, 25, 339-345 (1983)

3390 SONINO N, THE USE OF KETOCONAZOLE AS AN INHIBITOR OF STEROID PRODUCTION, *N ENGL J MED*, 317, 812-818 (1987)

3391 SONKA J, GREGOROVA I, TOMSOVA Z, PAVLOVA A, ZHIRKOVA A, RATH R, URBANEK J, JOSIFKO M, PLASMA ANDROSTERONE, DEHYDROEPIANDROSTERONE AND 11-HYDROXYCORTICOIDS IN OBESITY, *STEROIDS LIPIDS RES*, 3, 65 (1972)

3392 SONNENBLICK M, GOTTLIEB S, GOLDSTEIN R, ET AL, EFFECT OF AMIODARONE ON BLOOD LIPIDS, *CARDIOLOGY*, 73, 147-150 (1986)

3393 SONNTAG O, ARZNEIMITTEL-INTERFERENZEN, GEORG THEME VERLAG, STUTTGART (1985)

3394 SORCINI G, SCIARRA F, DISILVERIO F, FRAIOLI F, FURTHER STUDIES ON PLASMA ANDROGENS AND GONADOTROPINS AFTER CYPROTERONE ACETATE (SH 714), FOLIA ENDOCRINOL, 24, 196 (1971)

3395 SORENSEN LB, SUPPRESSION OF THE SHUNT PATHWAY IN PRIMARY GOUT BY AZATHIOPRINE, PROC NATL ACAD SCI U S A, 55, 571 (1966)

3396 SORENSEN SS, THOMSEN OO, DANIELSEN H, ET AL, EFFECT OF VERAPAMIL ON RENAL PLASMA FLOW, GLOMERULAR FILTRATION RATE AND PLASMA ANGIOTENSIN II, ALDOSTERONE AND ARGININE VASOPRESSIN IN ESSENTIAL HYPERTENSION, EUR J CLIN PHARMACOL, 29, 257-261 (1985)

3397 SORENSON PG, NISSEN MH, GROTH S, ET AL, β_2-MICROGLOBULIN EXCRETION: AN INDICATOR OF LONG TERM NEPHROTOXICITY DURING CIS-PLATINUM TREATMENT?, CANCER CHEMOTHER PHARMACOL, 14, 247-249 (1985)

3398 SORETH JT, DUBB JW, ALLISON NL, ET AL, EFFECT ON THE ENDOCRINE SYSTEM OF A NEW DOPAMINERGIC AGENT, IBOPAMINE, CLIN PHARMACOL THER, 41, 627-632 (1987)

3399 SORISKY A, WATSON DC, POSITIVE DIPHENHYDRAMINE INTERFERENCE IN THE EMIT-ST™ ASSAY FOR TRICYCLIC ANTIDEPRESSANTS IN SERUM, CLIN CHEM, 32, 715 (1986)

3400 SORRELL TC, FORBES IJ, BURNESS FR, RISCHBIETH RHC, DEPRESSION OF IMMUNOLOGICAL FUNCTION IN PATIENTS TREATED WITH PHENYTOIN SODIUM (SODIUM DIPHENYLHYDANTOIN), LANCET, 2, 1233 (1971)

3401 SOTANIEMI EA, HAKKARAINEN HK, PURANEN JA, LAHTI RO, RADIOLOGIC BONE CHANGES AND HYPOCALCEMIA WITH ANTICONVULSANT THERAPY IN EPILEPSY, ANN INTERN MED, 77, 389 (1972)

3402 SOTOLONGO RP, NEEFE LI, RUDZKI C, ET AL, HYPERSENSITIVITY REACTION TO SULFASALAZINE WITH SEVERE HEPATOTOXICITY, GASTROENTEROLOGY, 75, 95-99 (1978)

3403 SOUMA JA, GREEN PJ, COPPAGE AT, ET AL, CHANGES IN THYROID FUNCTION IN PREGNANCY AND WITH ORAL CONTRACEPTIVE USE, SOUTH MED J, 74, 684-687 (1981)

3404 SOUNEY PF, MARIANI G, EFFECT OF VARIOUS CONCENTRATIONS OF FLUCYTOSINE ON THE ACCURACY OF SERUM CREATININE DETERMINATIONS, AM J HOSP PHARM, 42, 621-622 (1985)

3405 SOUNEY PF, MENARD C, CHANG JT, ET AL, EFFECT OF CEPHEN ANTIBIOTICS ON CREATININE ASSAY, AM J HOSP PHARM, 40, 1152 (1983)

3406 SOWERS JR, SHARP B, MCCALLUM RW, EFFECT OF DOMPERIDONE, AN EXTRACEREBRAL INHIBITOR OF DOPAMINE RECEPTORS, ON THYROTROPIN, PROLACTIN, RENIN, ALDOSTERONE, AND 18-HYDROXYCORTICOSTERONE SECRETION IN MAN, J CLIN ENDOCRINOL METAB, 54, 869-871 (1982)

3407 SPAIN MA, WU AHB, BILIRUBIN INTERFERENCE WITH DETERMINATIONS OF URIC ACID, CHOLESTEROL, AND TRIGLYCERIDES IN COMMERCIAL PEROXIDASE-COUPLED ASSAYS, AND THE EFFECT OF FERROCYANIDE, CLIN CHEM, 32, 518-521 (1986)

3408 SPARKS LL, COPINSCHI G, DONALDSON CL, HYPERGLYCEMIC RESPONSE TO INTRAVENOUS ARGININE IN PREMATURITY-ONSET DIABETES, DIABETES, 16, 268 (1967)

3409 SPEED BR, SPELMAN DW, SIALADENITIS AND SYSTEMIC REACTION ASSOCIATED WITH PHENYLBUTAZONE, AUST N Z J MED, 12, 261-264 (1982)

3410 SPEIGHT TM, AVERY GS, PANCURONIUM BROMIDE: A REVIEW, DRUGS, 4, 163 (1972)

3411 SPEIGHT TM, AVERY GS, PIZOTIFEN (BC-105): A REVIEW OF ITS PHARMACOLOGICAL PROPERTIES AND THERAPEUTIC EFFICACY IN VASCULAR HEADACHES, DRUGS, 3, 159 (1972)

3412 SPELLACY WN, BUHI WC, BIRK SA, MCCREARY SA, STUDIES OF ETHYNODIOL DIACETATE AND MESTRANOL ON BLOOD GLUCOSE AND PLASMA INSULIN, CONTRACEPTION, 3, 185 (1971)

3413 SPELLACY WN, BUHI WC, BIRK SA, MCCREARY SA, METABOLIC STUDIES IN WOMAN TAKING NORETHINDRONE FOR 6 MONTHS TIME, FERTIL STERIL, 24, 419 (1973)

3414 SPELLACY WN, MCLEOD AG, BUHI WC, BIRK SA, THE EFFECT OF MEDROXYPROGESTERONE ACETATE ON CARBOHYDRATE METABOLISM: MEASUREMENTS OF GLUCOSE, INSULIN, FERTIL STERIL, 23, 239 (1972)

3415 SPENCER H, KRAMER L, OSIS D, ET AL, EFFECTS OF ALUMINUM HYDROXIDE ON FLUORIDE AND CALCIUM METABOLISM, J ENVIRON PATHOL TOXICOL ONCOL, 6, 33-41 (1985)

3416 SPIERS ASD, GALTON DAG, KAUR J, ET AL, THIOGUANINE AS PRIMARY TREATMENT FOR CHRONIC GRANULOCYTIC LEUKAEMIA, LANCET, 1, 829-832 (1975)

3417 SPILKER B, WATSON BS, WOODS JW, DRUG INTERFERENCE WITH MEASUREMENT OF METANEPHRINES IN URINE, ANN CLIN LAB SCI, 13, 16-19 (1983)

3418 SPINDEL ER, WURTMAN RJ, MCCALL A, ET AL, NEUROENDOCRINE EFFECTS OF CAFFEINE IN NORMAL SUBJECTS, CLIN PHARMACOL THER, 36, 402-407 (1984)

3419 SPINO M, SELLERS EM, KAPLAN HL, ET AL, ADVERSE BIOCHEMICAL AND CLINICAL CONSEQUENCES OF FUROSEMIDE ADMINISTRATION, CAN MED ASSOC J, 118, 1513-1518 (1978)

3420 SPITTLE CR, ATHEROSCLEROSIS AND VITAMIN C, LANCET, 2, 1280 (1971)

3421 SQUIRES JE, MINTZ PD, CLARK S, TOLMETIN-INDUCED HEMOLYSIS, TRANSFUSION, 25, 410-413 (1985)

3422 SRINIVASAN G, SINGH J, CATTAMANCHI G, ET AL, PLASMA GLUCOSE CHANGES IN PRETERM INFANTS DURING ORAL THEOPHYLLINE THERAPY, J PEDIATR, 103, 473-475 (1983)

3423 SRITHARAN V, BHARADWAJ VP, VENKATESAN K, ET AL, DAPSONE INDUCED HYPOHAPTOGLOBINEMIA IN LEPROMATOUS LEPROSY PATIENTS, INT J LEPR OTHER MYCOBACT DIS, 49, 307-310 (1981)

3424 SRIVASTA SK, DHOLAKIA DM, NAIK B, EFFECT OF FRUCTOSE ON CITRULLINE ESTIMATION BY THE ARCHIBALD METHOD, INDIAN J BIOCHEM BIOPHYS, 8, 117 (1971)

3425 STABENAU JR, CREVELING CR, DALY J, THE 'PINK SPOT' 3,4-DIMETHOXYPHENYLETHYLAMINE, COMMON TEA, AND SCHIZOPHRENIA, AM J PSYCHIATRY, 127, 611 (1970)

3426 STAESSEN J, FIOCCHI R, BOUILLON R, ET AL, DIFFERENTIAL RESPONSES OF PLASMA ALDOSTERONE, CORTISOL AND ADRENOCORTICOTROPIN TO TWO DOPAMINE RECEPTOR ANTAGONISTS, METHODS FUND EXP CLIN PHARMACOL, 7, 523-527 (1985)

3427 STAFFORD BT, CROSBY WH, LATE ONSET OF GOLD-INDUCED THROMBOCYTOPENIA: WITH A PRACTICAL NOTE ON THE INJECTIONS OF DIMERCAPROL, J AM MED ASSOC, 239, 50-51 (1978)

3428 STAMM D, TAGESSCHWANKUNGEN DER NORMALBEREICHE DIAGNOSTISCH WICHTIGER BLUBESTANDTEILE, VERH DTSCH GES INN MED, 73, 982 (1967)

3429 STAMP TCB, ROUND JM, SEASONAL CHANGES IN HUMAN PLASMA LEVELS OF 25-HYDROXYVITAMIN D, NATURE, 247, 563 (1974)

3430 STANTON MF, LOWENSTEIN FW, SERUM MAGNESIUM IN WOMEN DURING PREGNANCY, WHILE TAKING ORAL CONTRACEPTIVES AND AFTER MENOPAUSE, J AM COLL NUTR, 6, 313-319 (1987)

3431 STAPLETON FB, NELSON B, VATS TS, ET AL, HYPOKALEMIA ASSOCIATED WITH ANTIBIOTIC TREATMENT: EVIDENCE IN CHILDREN WITH MALIGNANT NEOPLASMS, AM J DIS CHILD, 130, 1104-1108 (1976)

3432 STARKEY BJ, LOSCOMBE SM, SMITH JM, PARACETAMOL (ACETAMINOPHEN) ANALYSIS BY HIGH PERFORMANCE LIQUID CHROMATOGRAPHY: INTERFERENCE STUDIES AND COMPARISON WITH AN ENZYMATIC PROCEDURE, THER DRUG MONIT, 8, 78-84 (1986)

3433 STARZL TE, PUTNAM CW, HALGRIMSON CG, SCHROTER GT, MARTINEAU G, LAUNOIS B, CORMAN JL, CYCLOPHOSPHAMIDE AND WHOLE ORGAN TRANSPLANTATION IN HUMAN BEINGS, SURG GYNECOL OBSTET, 133, 981 (1971)

3434 STATLAND BE, WINKEL P, BOKELUND H, SERUM ALKALINE PHOSPHATASE AFTER FATTY MEALS: THE EFFECT OF SUBSTRATE ON THE ASSAY PROCEDURE, CLIN CHIM ACTA, 49, 299 (1973)

3435 STAUB JJ, JENKINS JS, RATCLIFFE JG, LANDON J, COMPARISON OF CORTICOTROPHIN AND CORTICOSTEROID RESPONSE TO LYSINE VASOPRESSIN, INSULIN AND PYROGEN IN MAN, BR MED J [CLIN RES], 1, 267 (1973)

3436 STAUBLI M, STUDER H, AMIODARONE-TREATED PATIENTS WITH SUPPRESSED TSH TEST AT RISK OF THYROTOXICOSIS, KLIN WOCHENSCHR, 63, 168-175 (1985)

3437 STEARNS E, WINTER JS, FAIMAN C, THE EFFECT OF COITUS ON GONADOTROPIN, PROLACTIN AND SEX STEROID LEVELS IN MAN, CLIN RES, 20, 923 (1972)

3438 STEEL CM, FRENCH EB, AITCHISON WRC, STUDIES ON ADRENALINE INDUCED LEUKOCYTOSIS IN NORMAL MAN, BR J HAEMATOL, 21, 413 (1971)

3439 STEELE TH, DISSOCIATION OF ZINC EXCRETION FROM OTHER CATIONS IN MAN, J LAB CLIN MED, 81, 205 (1973)

3440 STEELE TH, EVIDENCE FOR ALTERED RENAL URATE REABSORPTION DURING CHANGES IN VOLUME OF THE EXTRACELLULAR FLUID, J LAB CLIN MED, 74, 288 (1969)

3441 STEELE TH, CONTROL OF URIC ACID EXCRETION, N ENGL J MED, 284, 1193 (1971)

3442 STEELE TH, OPPENHEIMER S, FACTORS AFFECTING URATE EXCRETION FOLLOWING DIURETIC ADMINISTRATION IN MAN, AM J MED, 47, 564 (1969)

3443 STEER PL, MARKS MI, KLITE PD, EICKHOFF TC, 5-FLUOROCYTOSINE: AN ORAL ANTIFUNGAL AGENT, ANN INTERN MED, 76, 15 (1972)

3444 STEGINK LD, FILER LJ JR, BAKER GL, MONOSODIUM GLUTAMATE: EFFECT ON PLASMA AND BREAST MILK AMINO ACIDS IN LACTATING WOMAN, PROC SOC EXP BIOL MED, 140, 836 (1972)

3445 STEHR-GREEN JK, HELMICK CG, PENTAMIDINE AND RENAL TOXICITY, N ENGL J MED, 313, 694-695 (1985)

3446 STEIN HB, HASAN A, FOX IH, ASCORBIC ACID-INDUCED URICOSURIA, ANN INTERN MED, 84, 385-388 (1976)

3447 STEIN HD, KEISER HR, SJOERDSMA A, PROLINE HYDROXYLASE ACTIVITY IN HUMAN BLOOD, LANCET, 1, 106 (1970)

3448 STEIN RS, HOWARD CL, CLINICAL ASSESSMENT OF CIMETIDINE MYELOTOXICITY, SOUTH MED J, 73, 293-297 (1981)

3449 STEINBACH G, PFLIEGER H, MAIER V, FALSELY INCREASED VALUES FOR SERUM CREATININE DURING THERAPY WITH CEFOXITIN, *CLIN CHEM*, 29, 1700-1701 (1983)

3450 STEINBERG M, LEIFHEIT HC, EFFECT OF METRECAL ON SERUM-PRECIPITABLE IODINE VALUES, *TEX REP BIOL MED*, 23, 122 (1965)

3451 STEINBERG WM, KING CE, TOSKES PP, MALABSORPTION OF PROTEIN-BOUND COBALAMIN BUT NOT UNBOUND COBALAMIN DURING CIMETIDINE ADMINISTRATION, *DIG DIS SCI*, 25, 188-192 (1980)

3452 STEINBRECHER UP, MISHKIN S, SULFAMETHOXAZOLE-INDUCED HEPATIC INJURY, *DIG DIS SCI*, 26, 756-759 (1981)

3453 STEINER J, CASSAR J, MASHITER K, ET AL, EFFECTS OF METHYLDOPA ON PROLACTIN AND GROWTH HORMONE, *BR MED J [CLIN RES]*, 1, 1186-1188 (1976)

3454 STEINER S, LARSEN JK, DONATH A, PAULI HG, RENAL FUNCTION AND PROTEIN ELIMINATION OF HUMAN SUBJECTS DURING CARBON MONOXIDE EXPOSURE, *HELV MED ACTA*, 36, 39 (1972)

3455 STEINMETZ J, PANEK E, SOURIEAU F, SIEST G, INFLUENCE OF FOOD INTAKE ON BIOLOGICAL PARAMETERS IN, *REFERENCE VALUES IN HUMAN CHEMISTRY*, G SIEST (ED), KARGER, BASEL, 193 (1973)

3456 STELLA L, CRESCENTI A, PONTIROLI AE, ET AL, ENFLURANE ANESTHESIA AFFECTS SERUM PROLACTIN LEVELS IN MAN, *IRCS MED SCI*, 12, 572 (1984)

3457 STEMPEL DA, MILLER JJ III, LYMPHOPENIA AND HEPATIC TOXICITY WITH IBUPROFEN, *J PEDIATR*, 90, 657-658 (1977)

3458 STEPAN J, WILCZEK H, JUSTOVA V, ET AL, PLASMA 25-HYDROXYCHOLECALCIFEROL IN ORAL SULFONYLUREA TREATED DIABETES MELLITUS, *HORM METAB RES*, 14, 98-100 (1982)

3459 STEPHENS ME, CRAFT J, PETERS TJ, HOFFBRAND AV, ORAL CONTRACEPTIVES AND FOLATE METABOLISM, *CLIN SCI*, 42, 405 (1972)

3460 STERNLIEB P, ROBINSON RM, STEVENS-JOHNSON SYNDROME PLUS TOXIC HEPATITIS DUE TO IBUPROFEN, *N Y STATE J MED*, JULY, 1239-1243 (1978)

3461 STEVENS DA, MICONAZOLE IN THE TREATMENT OF SYSTEMIC FUNGAL INFECTIONS, *AM REV RESPIR DIS*, 116, 801-806 (1977)

3462 STEVENS DA, MICONAZOLE IN THE TREATMENT OF COCCIDIOIDOMYCOSIS, *DRUGS*, 26, 347-354 (1983)

3463 STEVENS DJ, SANFELIPPO MJ, EVALUATION OF THREE METHODS FOR PLASMA FIBRINOGEN DETERMINATION, *AM J CLIN PATHOL*, 60, 182 (1973)

3464 STEVENS VC, GOLDZIEHER JW, VORYS N, EFFECT OF MESTRANOL AND CHLORMADINONE ACETATE ON URINARY EXCRETION OF FSH AND LH, *AM J OBSTET GYNECOL*, 102, 95 (1968)

3465 STEVENS WC, EGER EL, JOAS TA, CROMWELL TH, WHITE A, DOLAN WM, COMPARATIVE TOXICITY OF ISOFLURANE, HALOTHANE, FLUROXENE, AND DIETHYL ETHER IN HUMAN VOLUNTEERS, *CAN J ANAESTH*, 20, 357 (1973)

3466 STEVENSON IH, BROWNING M, CROOKS J, O'MALLEY K, CHANGES IN HUMAN DRUG METABOLISM AFTER LONG-TERM EXPOSURE TO HYPNOTICS, *BR MED J [CLIN RES]*, 4, 322 (1972)

3467 STEVENSON JC, ABEYASEKERA G, HILLYARD CJ, ET AL, CALCITONIN AND THE CALCIUM REGULATING HORMONES IN POSTMENOPAUSAL WOMEN: EFFECT OF OESTROGENS, *LANCET*, 1, 693-695 (1981)

3468 STEWART AF, KEATING T, SCHWARTZ PE, MAGNESIUM HOMEOSTASIS FOLLOWING CHEMOTHERAPY WITH CISPLATIN: A PROSPECTIVE STUDY, *AM J OBSTET GYNECOL*, 153, 660-665 (1985)

3469 STEWART GW, PEART WS, BOYLSTON AW, OBSTRUCTIVE JAUNDICE, PANCYTOPENIA AND HYDRALAZINE, *LANCET*, 1, 1207 (1981)

3470 STEWART MJ, SIMPSON E, PROGNOSIS IN PARACETAMOL SELF-POISONING: THE USE OF PLASMA PARACETAMOL CONCENTRATION IN A REGION, *ANN CLIN BIOCHEM*, 10, 173 (1973)

3471 STIMMEL B, VERNACE S, TOBIAS H, HEPATIC DYSFUNCTION IN HEROIN ADDICTS. THE ROLE OF ALCOHOL, *J AM MED ASSOC*, 222, 811 (1972)

3472 STIMSON WH, STUDIES ON THE CHANGES IN THE CONCENTRATION AND TOTAL MASS OF INDIVIDUAL SERUM PROTEINS DURING LATE PREGNANCY, *CLIN BIOCHEM*, 5, 3 (1972)

3473 STOCIGT JR, LIM C-F, BARLOW JW, ET AL, HIGH CONCENTRATIONS OF FUROSEMIDE INHIBIT SERUM BINDING OF THYROXINE, *J CLIN ENDOCRINOL METAB*, 59, 62-66 (1984)

3474 STOCK JL, CODERRE JA, MALLETTE LE, EFFECTS OF A SHORT COURSE OF ESTROGEN ON MINERAL METABOLISM IN POSTMENOPAUSAL WOMEN, *J CLIN ENDOCRINOL METAB*, 61, 595-600 (1985)

3475 STOFFER SS, HYNES, KM, JIANG NS, RYON RJ, THE EFFECTS OF CHRONIC DIGOXIN ADMINISTRATION OF SERUM ESTROGEN, SERUM LUTEINIZING HORMONE, *CLIN RES*, 20, 720 (1972)

3476 STOKER DJ, WYNN V, ROBERTSON G, EFFECT OF POSTURE ON THE PLASMA CHOLESTEROL LEVEL, *BR MED J [CLIN RES]*, 1, 336 (1966)

3477 STONE MC, IDIOPATHIC HYPERGLYCERIDAEMIA TREATED WITH METHYLTESTOSTERONE AND METHANDIENONE, *LANCET*, 1, 477 (1963)

3478 STONE SP, GOODWIN RM, DAPSONE-INDUCED JAUNDICE, *ARCH DERMATOL*, 114, 947 (1978)

3479 STORM-MATHISEN H, DISCUSSION, IN *CATAPRES® IN HYPERTENSION*, ME CONOLLY (ED), BUTTERWORTH, LONDON (1970)

3480 STOTE RM, SMITH LH, WILSON DM, DUBE WJ, GOLDSMITH RS, ARNAUD CD, HYDROCHLOROTHIAZIDE EFFECTS ON SERUM CALCIUM AND IMMUNO-ACTIVE PARATHYROID HORMONE CONCENTRATIONS, *ANN INTERN MED*, 77, 587 (1972)

3481 STOVNER J, THEODORSEN L, BJELKE E, SENSITIVITY TO DIMETHYL-TUBOCURARINE AND TOXIFERINE WITH SPECIAL REFERENCE TO SERUM PROTEINS, *BR J ANAESTH*, 44, 374 (1972)

3482 STOWELL A, JOHNSEN J, RIPEL A, ET AL, DISULFIRAM-INDUCED ACETONEMIA, *LANCET*, 1, 882-883 (1983)

3483 STOWERS JM, BREWSTER PD, STUDIES ON THE MECHANISM OF WEIGHT REDUCTION BY PHENFORMIN, *POSTGRAD MED J*, MAY SUPPL, 13 (1969)

3484 STRAUGHN AB, HENDERSON RP, LIEBERMAN PL, ET AL, EFFECT OF RIFAMPIN ON THEOPHYLLINE DISPOSITION, *THER DRUG MONIT*, 6, 153-156 (1984)

3485 STRAUMFJORD JV JR, NOTO TA, HOSTY TA, ISOENZYMES IN DIAGNOSIS IN, *DIAGNOSTIC ENZYMOLOGY*, EL COODLEY (ED), LEA AND FEBIGER, PHILADELPHIA, PA (1970)

3486 STRICKER BHC, MEYBOOM RHB, BLEEKER PA, BLOOD DISORDERS ASSOCIATED WITH PIRENZEPINE, *BR MED J [CLIN RES]*, 293, 1074 (1986)

3487 STRICKER BHC, OEI TT, AGRANULOCYTOSIS CAUSED BY SPIRONOLACTONE, *BR MED J [CLIN RES]*, 289, 731 (1984)

3488 STROMBERG A, WENGLE B, CHRONIC ACTIVE HEPATITIS INDUCED BY NITROFURANTOIN, *BR MED J [CLIN RES]*, 2, 174 (1976)

3489 STRUNGE P, ENGBY B, SCHMIDT E, ET AL, VARIATION OF SERUM LIPIDS IN POSTMYOCARDIAL INFARCTION PATIENTS TREATED WITH VERAPAMIL OR PLACEBO, *ACTA MED SCAND*, 215 SUPPL 681, 53-57 (1984)

3490 STRUTHERS AD, REID JL, THE ROLE OF ADRENAL MEDULLARY CATECHOLAMINES IN POTASSIUM HOMEOSTASIS, *CLIN SCI*, 66, 377-382 (1984)

3491 STUBBS WA, DELITALA G, BESSER GM, ET AL, THE ENDOCRINE AND METABOLIC EFFECTS OF CIMETIDINE, *CLIN ENDOCRINOL*, 18, 167-178 (1983)

3492 STURDEE DW, GUSTAFSON RC, MOORE B, GLUCOSE TOLERANCE AND HORMONE REPLACEMENT THERAPY: A PRELIMINARY STUDY, *POSTGRAD MED J*, 52 (SUPPL 6), 52-54 (1976)

3493 SUBRYAN VL, POPOVTZER MM, PARKS SD, REEVE EB, MEASUREMENT OF SERUM IONIZED CALCIUM WITH THE ION-EXCHANGE ELECTRODE, *CLIN CHEM*, 18, 1458 (1972)

3494 SUKI WN, DAWOUD F, EKNOYAN G, MARTINEZ-MALDONADO M, EFFECTS OF METOLAZONE ON RENAL FUNCTION IN NORMAL MAN, *J CLIN PHARMACOL EXP THER*, 180, 6 (1972)

3495 SULLIVAN JF, LANKFORD HG, SCHWARTZ MJ, FARRELL C, MAGNESIUM METABOLISM IN ALCOHOLISM, *AM J CLIN NUTR*, 13, 297 (1963)

3496 SULLIVAN KM, SMALL RE, ROCK WL, ET AL, EFFECTS OF CIMETIDINE OR RANITIDINE ON THE PHARMACOKINETICS OF FLURBIPROFEN, *CLIN PHARMACOL*, 5, 586-589 (1986)

3497 SUMMERSKILL WHJ, THORSELL F, FEINBERG JH, ALDRETTE JS, EFFECTS OF UREASE INHIBITION IN HYPERAMMONEMIA: CLINICAL AND EXPERIMENTAL STUDIES, *GASTROENTEROLOGY*, 54, 20 (1968)

3498 SUN DCH, IATROGENIC GASTROINTESTINAL DISEASES IN THE AGED, *GERIATRICS*, 27, 89 (1972)

3499 SUNDBERG L, INTERFERENCES IN NICKEL DETERMINATIONS BY ATOMIC ABSORPTION SPECTROMETRY, *ANAL CHEM*, 45, 1460 (1973)

3500 SUNDERMAN FW, URIC ACID IN BIOLOGICAL FLUIDS, *PROFICIENCY TEST SERVICE*, 2 (MAY 1972)

3501 SUNDERMAN FW, THE CHEMICAL MEASUREMENT OF MAGNESIUM IN BIOLOGICAL FLUIDS, *PROFICIENCY TEST SERVICE*, 6 (1972)

3502 SUNDERMAN FW, SERUM MAGNESIUM AND INORGANIC PHOSPHORUS, *PROFICIENCY TEST SERVICE*, 2 (FEB 1972)

3503 SUNDERMAN FW JR, MEASUREMENTS OF VANILMANDELIC ACID FOR THE DIAGNOSIS OF PHEOCHROMOCYTOMA AND NEUROBLASTOMA, *AM J CLIN PATHOL*, 42, 481 (1964)

3504 SUNDERMAN FW JR, EFFECTS OF DRUGS UPON HEMATOLOGICAL TESTS, *ANN CLIN LAB SCI*, 2, 1 (1972)

3505 SUNDERMAN FW JR, DRUG INTERFERENCE IN CLINICAL BIOCHEMISTRY, *CRIT REV CLIN LAB SCI*, 1, 427 (1970)

3506 SUNDERMAN FW JR, COLORIMETRIC DETERMINATION OF VANILMANDELIC ACID IN URINE, *STAND METH CLIN CHEM*, 6, 99 (1970)

3507 SUR BK, SHUKLA RK, AGASHE VS, THE ROLE OF CREATININE AND HISTIDINE IN BENEDICT'S QUALITATIVE TEST FOR REDUCING SUGAR IN URINE, *J CLIN PATHOL*, 25. 892 (1972)

3508 SUSSMAN NM, McLAIN LW JR., A DIRECT HEPATOTOXIC EFFECT OF VALPROIC ACID, *J AM MED ASSOC*, 242, 1173-1174 (1979)

3509 SUTHERLAND HW, STOWERS JM, CORMACK JD, BEWSHER PD, EVALUATION OF CHLORPROPAMIDE IN CHEMICAL DIABETES DIAGNOSED DURING PREGNANCY, *BR MED J [CLIN RES]*, 3, 9 (1973)

3510 SUTOR AH, THROMBOCYTURIA AFTER ASPIRIN, *N ENGL J MED*, 288, 794 (1973)

3511 SUZUKI H, NOGUCHI K, NAKAHATA M, ET AL, EFFECT OF IOPANOIC ACID ON THE PITUITARY-THYROID AXIS: TIME SEQUENCE OF CHANGES IN SERUM IODOTHYRONINES, THYROTROPIN, AND PROLACTIN CONCENTRATIONS AND RESPONSES TO THYROID HORMONES, *J CLIN ENDOCRINOL METAB*, 53, 779-783 (1981)

3512 SUZUKI H, YAMAZAKI N, SUZUKI Y, ET AL, LOWERING EFFECT OF DIPHENYLHYDANTOIN ON SERUM FREE THYROXINE AND THYROXINE BINDING GLOBULIN (TBG), *ACTA ENDOCRINOL*, 105, 477-481 (1984)

3513 SVEDHEM A, TOXIC HEPATITIS FOLLOWING KETOCONAZOLE TREATMENT, *SCAND J INFECT DIS*, 16, 123-125 (1984)

3514 SVENDSEN UG, GERSTOFT J, HANSEN TM, ET AL, THE RENAL EXCRETION OF PROSTAGLANDINS AND CHANGES IN PLASMA RENIN DURING TREATMENT WITH SULINDAC OR NAPROXEN IN PATIENTS WITH RHEUMATOID ARTHRITIS AND THIAZIDE TREATED HEART FAILURE, *J RHEUMATOL*, 11, 779-782 (1984)

3515 SWAIMAN KF, FLAGLER DG, MERCURY POISONING WITH CENTRAL AND PERIPHERAL NERVOUS SYSTEM INVOLVEMENT TREATED WITH PENICILLAMINE, *PEDIATR*, 48, 639 (1971)

3516 SWAMINATHAN R, INTERFERENCE FROM HEPARIN IN THE MEASUREMENT OF PLASMA SODIUM ON ION SELECTIVE ELECTRODES, *ANN CLIN BIOCHEM*, 24, 116-117 (1987)

3517 SWANN AC, KOSLOW SH, KATZ MM, ET AL, LITHIUM CARBONATE TREATMENT OF MANIA. CEREBROSPINAL FLUID AND URINARY MONOAMINE METABOLITES AND TREATMENT OUTCOME, *ARCH GEN PSYCHIATRY*, 44, 345-354 (1987)

3518 SWANSON S, MIFFLIN TE, BOYD JC, METHOTREXATE INTERFERES WITH DETERMINATIONS OF CONJUGATED BILIRUBIN WITH THE KODAK EKTACHEM® 400, *CLIN CHEM*, 32, 863-864 (1986)

3519 SWAROOP S, KRANT MJ, RAPID ESTROGEN-INDUCED HYPERCALCEMIA, *J AM MED ASSOC*, 223, 913 (1973)

3520 SWART S, O'MALLEY BP, VORA J, ET AL, THE EFFECT OF DOPAMINERGIC BLOCKADE ON SERUM TSH AND PROLACTIN LEVELS IN THYROTOXICOSIS, *ACTA ENDOCRINOL*, 106, 330-335 (1984)

3521 SWARTZ SL, ENDOCRINE AND VASCULAR RESPONSES IN HYPERTENSIVE PATIENTS TO LONG-TERM TREATMENT WITH DILTIAZEM, *J CARDIOVASC PHARMACOL*, 9, 391-395 (1987)

3522 SWEENEY GD, SAUNDERS SJ, DOWDLE EB, EALES L, THE EFFECTS OF CHLOROQUINE ON PATIENTS WITH CUTANEOUS PORPHYRIA OF THE 'SYMPTOMATIC' TYPE, *BR MED J [CLIN RES]*, 1, 1281 (1965)

3523 SWIDLER G, *HANDBOOK OF DRUG INTERACTIONS*, WILEY-INTERSCIENCE, NEW YORK, NY (1971)

3524 SZASZ G,, HUTH K, BUSCH EW, KOLLER PU, STAHLER RF, VOLLMAR J, IN VIVO DRUG INTERFERENCE WITH VARIOUS GLUCOSE DETERMINATIONS COMPARED WITH IN VITRO RESULTS, *Z KLIN CHEM KLIN BIOCHEM*, 12, 256 (1974)

3525 TAGATZ GE, MCHUGH RB, ORAL CONTRACEPTIVES, A CONTINUING REAPPRAISAL, *POSTGRAD MED*, 50, 121 (1971)

3526 TAGGART HM, APPLEBAUM-BOWDEN D, HAFFNER S, ET AL, REDUCTION IN HIGH-DENSITY LIPOPROTEINS BY ANABOLIC STEROID (STANOZOLOL) THERAPY FOR POSTMENOPAUSAL OSTEOPOROSIS, *METABOLISM*, 31, 1147-1152 (1982)

3527 TAGGART P, CARRUTHERS M, SUPPRESSION BY OXPRENOLOL OF ADRENERGIC RESPONSE TO STRESS, *LANCET*, 2, 256 (1972)

3528 TAGUMA Y, KITAMOTO Y, FUTAKI G, ET AL, EFFECT OF CAPTOPRIL ON HEAVY PROTEINURIA IN AZOTEMIC DIABETICS, *N ENGL J MED*, 313, 1617-1620 (1985)

3529 TAIT B, INTERFERENCE IN IMMUNOLOGICAL METHODS OF PREGNANCY TESTING BY PROMETHAZINE, *MED J AUST*, 2, 126 (1971)

3530 TAKABATAKE T, YAMAMOTO Y, NAKAMURA S, ET AL, EFFECT OF THE CALCIUM ANTAGONIST NILVADIPINE ON HAEMODYNAMICS AT REST AND DURING COLD STIMULATION IN ESSENTIAL HYPERTENSION, *EUR J CLIN PHARMACOL*, 33, 215-219 (1987)

3531 TAKABATKE T, OHTA H, MAEKAWA M, ET AL, EFFECTS OF LONG-TERM PRAZOSIN THERAPY ON LIPOPROTEIN METABOLISM IN HYPERTENSIVE PATIENTS, *AM J MED*, 76, SUPPL 2A, 113-116 (1984)

3532 TAKAHARA K, KUROIWA A, MATSUSHIMA T, ET AL, EFFECTS OF NIFEDIPINE ON PLATELET FUNCTION, *AM HEART J*, 109, 4-8 (1985)

3533 TAKAMOTO S, ONISHI T, MORIMOTO S, ET AL, SERUM PHOSPHATE, PARATHYROID HORMONE AND VITAMIN D METABOLITES IN PATIENTS WITH CHRONIC RENAL FAILURE: EFFECT OF ALUMINUM HYDROXIDE ADMINISTRATION, *NEPHRON*, 40, 286-291 (1985)

3534 TAKASU N, YAMADA T, MIURA H, ET AL, RIFAMPICIN-INDUCED EARLY PHASE HYPERGLYCEMIA IN HUMANS, *AM REV RESPIR DIS*, 125, 23-27 (1982)

3535 TALBOT J, BEELEY L, FUSIDIC ACID AND JAUNDICE, *BR MED J [CLIN RES]*, 2, 308 (1980)

3536 TALBOT JM, MEADE BW, EFFECT OF SILICONES ON THE ABSORPTION OF ANTICOAGULANT DRUGS, *LANCET*, 1, 1292 (1972)

3537 TALMERS FN, TELMOS AJ, PROCAINE AMIDE HYDROCHLORIDE (PRONESTYL®) INDUCED AGRANULOCYTOSIS, *MICH MED*, 64, 655 (1965)

3538 TALNER LB, RUSHMER HN, COEL MN, THE EFFECT OF RENAL ARTERY INJECTION OF CONTRAST MATERIAL ON URINARY ENZYME EXCRETION, *INVEST RADIOL*, 7, 311 (1972)

3539 TAN MH, WILMSHURST EG, GLEASON RE, SOELDNER JS, EFFECT OF POSTURE ON SERUM LIPID LEVELS, *CLIN RES*, 20, 884 (1972)

3540 TAN MH, WILMSHURST EG, GLEASON RE, SOELDNER JS, EFFECT OF POSTURE ON SERUM LIPIDS, *N ENGL J MED*, 289, 416 (1973)

3541 TAN SY SHAPIRO R, FRANCO R, INDOMETHACIN-INDUCED PROSTAGLANDIN INHIBITION WITH HYPERKALEMIA, *ANN INTERN MED*, 90, 783-785 (1979)

3542 TANNENBERG AM, WICHER KJ, ROSE NR, AMPICILLIN NEPHROPATHY, *J AM MED ASSOC*, 218, 449 (1971)

3543 TANPHAICHITR VS, VAN EYS J, THE ASSAY OF PYRUVATE KINASE ACTIVITY IN BLOOD CELLS, *CLIN CHIM ACTA*, 41, 41 (1972)

3544 TARAZI RC, FROHLICH ED, DUSTAN HP, PLASMA VOLUME CHANGES WITH LONG-TERM BETA-ADRENERGIC BLOCKADE, *AM HEART J*, 82, 770 (1971)

3545 TARSSANEN L, HUIKKO M, ROSSI M, AMILORIDE-INDUCED HYPONATREMIA, *ACTA MED SCAND*, 208, 491-494 (1980)

3546 TARTINI R, KAPPENBERGER L, STEINBRUNN W, ET AL, DANGEROUS INTERACTION BETWEEN AMIODARONE AND QUINIDINE, *LANCET*, 1, 1327-1329 (1982)

3547 TASKINEN M-R, NIKKILA EA, NOCTURNAL HYPERTRIGLYCERIDEMIA AND HYPERINSULINEMIA FOLLOWING MODERATE EVENING INTAKE OF ALCOHOL, *ACTA MED SCAND*, 202, 173-177 (1977)

3548 TATRO DS, EFFECT OF ORAL CONTRACEPTIVES ON PLASMA MINERALS AND PROTEIN, *HOSP FORM MANAGEMENT*, 6, 14 (1971)

3549 TATTERSAL MHN, BATTERSHY G, SPIERS ASD, ANTIBIOTICS AND HYPOKALEMIA, *LANCET*, 1, 630 (1972)

3550 TAUBERT HD, HASKINS AL, MOSZKOWSKI EF, THE INFLUENCE OF THIORIDAZINE UPON URINARY GONADOTROPIN EXCRETION, *SOUTH MED J*, 59, 1301 (1966)

3551 TAVES DR, FRY BW, FREEMAN RB, GILLIES AJ, TOXICITY FOLLOWING METHOXYFLURANE ANESTHESIA: FLUORIDE CONCENTRATION IN NEPHROTOXICITY, *J AM MED ASSOC*, 214, 91 (1970)

3552 TAWODZERA PBP, BELL RMS, JONES JJ, PLASMA ZINC LEVELS AND CORTICOTROPIN GEL, *LANCET*, 1, 1072 (1972)

3553 TAYLOR AW, THAYER R, RAO S, HUMAN SKELETAL MUSCLE GLYCOGEN SYNTHETASE ACTIVITIES WITH EXERCISE AND TRAINING, *CAN J PHYSIOL PHARMACOL*, 50, 411 (1972)

3554 TAYLOR G, VITAMIN C DEFICIENCY, *LANCET*, 2, 1363 (1972)

3555 TAYLOR JW, ALEXANDER B, LYON LW, MATHEMATICAL ANALYSIS OF A PHENYTOIN-DISULFIRAM INTERACTION, *AM J HOSP PHARM*, 38, 93-95 (1981)

3556 TAYLOR JW, ALEXANDER B, LYON LW, A COMPARATIVE EVALUATION OF ORAL ANTICOAGULANT PHENYTOIN INTERACTIONS, *DRUG INTELL CLIN PHARM*, 14, 669-673 (1980)

3557 TAYLOR JW, HENDELES L, WEINBERGER M, THE INTERACTION OF PHENYTOIN AND THEOPHYLLINE, *DRUG INTELL CLIN PHARM*, 14, 638 (1980)

3558 TAYLOR SA, RAWLINS MD, SMITH SE, SPIRONOLACTONE - A WEAK ENZYME INDUCER IN MAN, *J PHARMACOKINET PHARMACOL*, 24, 578 (1972)

3559 TAYLOR WH, RENAL CALCULI AND SELF-MEDICATION WITH MULTIVITAMIN PREPARATIONS CONTAINING VITAMIN D, *CLIN SCI*, 42, 515 (1972)

3560 TEASDALE PR, PEARCE J, COMPARATIVE AND SERIAL ASSAYS OF FOLATE METABOLISM IN ANTICONVULSANT TREATED EPILEPTICS, *J CLIN PATHOL*, 25, 721 (1972)

3561 TEDESCO F, MARKHAM R, GURWITH M, ET AL, ORAL VANCOMYCIN FOR ANTIBIOTIC-ASSOCIATED PSEUDOMEMBRANOUS COLITIS, *LANCET*, 2, 226-228 (1978)

3562 TEDESCO FJ, MILLS LR, DIAZEPAM (VALIUM®) HEPATITIS, *DIG DIS SCI*, 27, 470-471 (1982)

3563 TEERENHOVI L, HEINONEN E, GROHN P, ET AL, HIGH FREQUENCY OF AGRANULOCYTOSIS IN BREAST CANCER PATIENTS TREATED WITH LEVAMISOLE, *LANCET*, 2, 151-152 (1978)

3564 TELHAG H, SAFETY AND EFFICACY OF PIROXICAM IN THE TREATMENT OF OSTEOARTHROSIS, *EUR J RHEUMATOL INFLAMM*, 1, 352-355 (1978)

3565 TEMPLETON JS, AZAPROPAZONE OR ALLOPURINOL IN THE TREATMENT OF CHRONIC GOUT AND/OR HYPERURICEMIA. A PRELIMINARY REPORT, *BR J CLIN PRACT*, 36, 353-358 (1982)

3566 TERRAGNA A, SPIRITO L, PORPORA TROMBOCITOPENIC IN LATTANTE DOPO SOMMINISTRAZIONE DI ACIDO ACETILSALICILICO ALLA NUTRICE, *MINERVA PEDIATR*, 19, 613 (1967)

3567 TERRELL CL, HERMANS PE, ANTIFUNGAL AGENTS USED FOR DEEP-SEATED MYCOTIC INFECTIONS, *MAYO CLIN PROC*, 62, 1116-1128 (1987)

3568 TERRUZZI V, MINOLI G,, TADEO G, ET AL, THE INFLUENCE OF CIMETIDINE AND RANITIDINE OF THE PLASMA LIPID PATTERN, *BR J CLIN PHARMACOL*, 19, 846-848 (1985)

3569 TERRY SI, TRANSIENT DYSAESTHESIAE AND PERSISTENT LEUCOCYTOSIS AFTER CLIOQUINOL THERAPY, *BR MED J [CLIN RES]*, 3, 745 (1971)

3570 TEVAARWERK GJM, HURST CJ, UKSIK P, ET AL, EFFECT OF INSULIN INDUCED HYPOGLYCEMIA ON THE SERUM CONCENTRATIONS OF THYROXINE, TRI-IODOTHYRONINE AND REVERSE TRI-IODOTHYRONINE, *CAN MED ASSOC J*, 20, 1090-1093 (1979)

3571 TEWKSBURY DA, LOHRENZ FN, CIRCADIAN RHYTHM OF HUMAN URINARY AMINO ACID EXCRETION IN FED AND FASTED STATES, *METABOLISM*, 19, 363 (1970)

3572 THENOT A, HYDROCHLOROTHIAZIDE CONTRE PANCREAS, *PRESSE MED*, 71, 572 (1963)

3573 THIERS RE, MAGNESIUM (FLUOROMETRIC), *STAND METH CLIN CHEM*, 5, 131 (1965)

3574 THOMAS D, GALLUS AS, BROOKS PM, ET AL, THROMBOKINETICS IN PATIENTS WITH RHEUMATOID ARTHRITIS TREATED WITH D-PENICILLAMINE, *ANN RHEUM DIS*, 43, 402-406 (1984)

3575 THOMAS DW, EDWARDS JB, EDWARDS RG, EXAMINATION OF XYLITOL, *N ENGL J MED*, 283, 437 (1970)

3576 THOMAS DW, EDWARDS JB, GILLILGAN JE, LAWRENCE JR, EDWARDS RG, COMPLICATIONS FOLLOWING INTRAVENOUS ADMINISTRATION OF SOLUTIONS CONTAINING XYLITOL, *MED J AUST*, 1, 1238 (1972)

3577 THOMAS RW, MADDOX RR, THE INTERACTION OF SPIRONO-LACTONE AND DIGOXIN: A REVIEW AND EVALUATION, *THER DRUG MONIT*, 3, 117-120 (1981)

3578 THOMAS S, EFFECTS OF CHANGE OF POSTURE ON THE DIURNAL RENAL EXCRETORY RHYTHM, *J PHYSIOL*, 148, 489 (1959)

3579 THOMPSON JH, SEROTONIN AND THE ALIMENTARY TRACT, *RES COMMUN CHEM PATHOL PHARMACOL*, 2, 687 (1971)

3580 THOMPSON PJ, KEMP MW, MCALLISTER WC, ET AL, ANGIOTENSIN-CONVERTING ENZYME: INVESTIGATION OF DIURNAL VARIA-TION, THE EFFECT OF A LARGE DOSE OF PREDNISOLONE, AND PREDNISOLONE PHARMACOKINETICS IN PATIENTS WITH SAR-COIDOSIS, *AM REV RESPIR DIS*, 134, 1075-1077 (1986)

3581 THOMPSON RHS, WOOTTON IDP, *BIOCHEMICAL DISORDERS IN HUMAN DISEASE*, 3RD ED, ACADEMIC PRESS, NEW YORK, NY (1971)

3582 THOMSEN BS, FROM A, JACOBSEN IA, ET AL, ACUTE RENAL FAIL-URE POSSIBLY ASSOCIATED WITH FENOPROFEN THERAPY, *ARTHRITIS RHEUM*, 26, 234-235 (1983)

3583 THOMSON AH, THOMSON GD, HEPBURN M, ET AL, A CLINICALLY SIGNIFICANT INTERACTION BETWEEN CIPROFLOXACIN AND THEOPHYLLINE, *EUR J CLIN PHARMACOL*, 33, 435-436 (1987)

3584 THORNTON GHM, ILLINGWORTH DG, AN EVALUATION OF THE BEN-ZIDINE TEST FOR OCCULT BLOOD IN THE FECES, *GASTROEN-TEROLOGY*, 28, 593 (1955)

3585 THRIFT CB, ACUTE SALICYLATE INTOXICATION: EFFECT ON THE DIRECT EOSINOPHIL COUNT, *ILLINOIS MED J*, 128, 39 (1965)

3586 THRONFELDT C, CORNELL RC, STOUGHTON RB, THE EFFECT OF ALCLOMETASONE PROPIONATE CREAM 0.05% ON THE HYPO-THALAMIC-PITUITARY-ADRENAL AXIS OF NORMAL VOLUN-TEERS, *J INT MED RES*, 13, 276-280 (1985)

3587 TICKTIN HE, TRUJILLO NP, ENZYMES IN NEOPLASTIC AND SURGI-CAL DISEASES IN, *DIAGNOSTIC ENZYMOLOGY*, EL COODLEY ED. LEA AND FEBIGER, PHILADELPHIA, PA (1970)

3588 TIETZ NW ED., *FUNDAMENTALS OF CLINICAL CHEMISTRY*, WB SAUNDERS COMPANY, PHILADELPHIA, PA (1970)

3589 TIETZ NW, REPIQUE EV, PROPOSED STANDARD METHOD FOR MEASURING LIPASE ACTIVITY IN SERUM BY A CONTINUOUS SAMPLING TECHNIQUE, *CLIN CHEM*, 19, 1268 (1973)

3590 TIFFANY TO, MORTON JM, HALL EM, GARRETT AS JR, CLINICAL EVALUATION OF KINETIC ENZYMATIC FIXED TIME AND INTE-GRAL ANALYSIS OF SERUM TRIGLYCERIDES, *CLIN CHEM*, 20, 476 (1974)

3591 TIGELAAR RE, RAPPORT RL II, INMAN JK, KUPFERBERG HJ, A RADIO-IMMUNOASSAY FOR DIPHENYLHYDANTOIN, *CLIN CHIM ACTA*, 43, 231 (1973)

3592 TIGHE P, JONES B, METRONIDAZOLE INTERFERENCE WITH CONTIN-UOUS-FLOW SPECTROPHOTOMETRY OF ASPARTATE AMINO-TRANSFERASE, *CLIN CHEM*, 25, 2057-2058 (1979)

3593 TIKKANEN MJ, NIKKILA EA, KUUSI T, ET AL, EFFECTS OF OES-TRADIOL AND LEVONORGESTREL ON LIPOPROTEIN LIPIDS AND POSTHEPARIN PLASMA LIPASE ACTIVITIES IN NORMOLI-POPROTEINAEMIC WOMEN, *ACTA ENDOCRINOL*, 99, 630-635 (1982)

3594 TILLMAN J, LEASK JTS, CAMPBELL J, ET AL, EFFECT OF FENOPROFEN ON THYROID FUNCTION TESTS, *CLIN CHIM ACTA*, 161, 233-238 (1986)

3595 TILSTONE WJ, GRAY JMB, NIMMO-SMITH RH, ET AL, INTERACTION BETWEEN WARFARIN AND SULFAMETHOXAZOLE, *POSTGRAD MED J*, 53, 388-390 (1977)

3596 TILSTONE WJ, KHAN SN, THE EFFECT OF FREE FATTY ACID AND BILIRUBIN ON A DYE-BINDING ASSAY FOR ALBUMIN, *CLIN BIOCHEM*, 6, 5 (1973)

3597 TINKER JH, GANDOLFI AJ, VAN DYKE RA, ELEVATION OF PLASMA BROMIDE LEVELS IN PATIENTS FOLLOWING HALOTHANE ANES-THESIA: TIME CORRELATION WITH TOTAL HALOTHANE DOS-AGE, *ANESTHESIOLOGY*, 44, 194-196 (1976)

3598 TISCHENDORF FW, LEDDEROSE G, MULLER D, WILMANNS W, HEAVY LYSOZYMURIA AFTER X-IRRADIATION OF THE SPLEEN IN HUMAN CHRONIC MYELOCYTIC LEUKAEMIA, *NATURE*, 235, 274 (1972)

3599 TISHLER M, ARMON S, NIFEDIPINE-INDUCED HYPOKALEMIA, *DRUG INTELL CLIN PHARM*, 20, 370-371 (1986)

3600 TKACH JR, INDOMETHACIN-INDUCED HYPERGLYCEMIA IN PSORI-ATIC ARTHRITIS, *J AM ACAD DERMATOL*, 7, 802 (1982)

3601 TOBE BA, GOLDMAN BS, THE METABOLISM OF THE VOLATILE AMINES: THE ROLE OF DRUGS IN THE PATHOGENESIS OF HEP-ATIC ENCEPHALOPATHY, *CAN MED ASSOC J*, 89, 874 (1963)

3602 TODAY'S DRUGS, ANTIHISTAMINES, MECHANISM OF ACTION, *BR MED J [CLIN RES]*, 642 (1962)

3603 TOKUYAMA I, LEACH RB, SHEINFELD S, MADDOCK WO, DEPRES-SION OF GONADOTROPIN EXCRETION AS A METHOD FOR ASSAY OF ESTROGENS IN HUMAN SUBJECTS, *J CLIN ENDOCRI-NOL METAB*, 14, 509 (1954)

3604 TOMECKI KJ, CATALANO CJ, DAPSONE HYPERSENSITIVITY: THE SULFONE SYNDROME REVISITED, *ARCH DERMATOL*, 117, 38-39 (1981)

3605 TOMKIN GH, HADDEN DR, WEAVER JA, MONTGOMERY DAD, VITA-MIN B_{12} STATUS OF PATIENTS ON LONG TERM METFORMIN THERAPY, *BR MED J [CLIN RES]*, 2, 685 (1971)

3606 TOMKIN GH, WEIR DG, INDICANURIA AFTER GASTRIC SURGERY, *Q J MED*, 41, 191 (1972)

3607 TOMOKUNI K, OGATA M, DIRECT COLORIMETRIC DETERMINATION OF HIPPURIC ACID IN URINE, *CLIN CHEM*, 18, 349 (1972)

3608 TOMPSETT SL, INTERFERENCE FROM THE PRESENCE OF OTHER SUBSTANCES IN DETECTING AND DETERMINING BARBITU-RATES, *J CLIN PATHOL*, 22, 291 (1969)

3609 TOONE BK, WHEELER M, FENWICK PBC, SEX HORMONE CHANGES IN MALE EPILEPTICS, *CLIN ENDOCRINOL*, 12, 391-395 (1980)

3610 TOOP KM, KLOPPER A, EFFECTS OF ANTICOAGULANTS ON THE MEASUREMENT OF PREGNANCY-ASSOCIATED PLASMA PRO-TEIN A (PAPP-A), *BR J OBSTET GYNAECOL*, 90, 150-155 (1983)

3611 TORO G, KOLODNY RD, JACOBS LS, MASTERS WH, DAUGHADAY WH, FAILURE OF ALCOHOL TO ALTER PITUITARY AND TARGET ORGAN HORMONE LEVELS, *CLIN RES*, 21, 505 (1973)

3612 TORRANCE JK, DIPHOSPHOGLYCERATE MUTASE ASSAY: THE EFFECT OF PYRUVATE, LACTATE DEHYDROGENASE AND THY-ROID HORMONE ON THE ASSAY, *CLIN CHIM ACTA*, 50, 103 (1974)

3613 TOSHNER D, KRASNER N, MACDOUGALL AL, SERUM HAPTOGLOBIN AND LACTIC DEHYDROGENASE DURING HAEMODIALYSIS, *NEPHRON*, 9, 235 (1972)

3614 TOULOUKIAN RJ, DOWNING SE, CHOLESTASIS ASSOCIATED WITH LONG-TERM PARENTERAL HYPERALIMENTATION, *ARCH SURG*, 106, 58 (1973)

3615 TOURTELLOTTE WW, REINGLASS JL, NEWKIRK TA, CEREBRAL DEHYDRATION ACTION OF GLYCEROL, *CLIN PHARMACOL THER*, 13, 159 (1972)

3616 TOWNSHEND MM, SMITH AJ, FACTORS INFLUENCING THE URINARY EXCRETION OF FREE CATECHOLAMINES IN MAN, *CLIN SCI*, 44, 253 (1973)

3617 TOZHILLIN SA, GONDA M, CARBONELL G, ET AL, SERUM AMYLASES AND THEIR INHIBITORS: 2, CLINICAL AND EXPERIMENTAL OBSERVATIONS -DIET AND STEROID EFFECTS, *AM J GAS-TROENTEROL*, 77, 26-28 (1982)

3618 TRASOFF A, WOHL MG, MINTZ SS, FATAL AGRANULOCYTOSIS WITH AUTOPSY FOLLOWING THE USE OF THIOURACIL IN A CASE OF THYROTOXICOSIS, *AM J MED SCI*, 211, 62 (1946)

3619 TRAVIS SF, SUGERMAN HJ, RUBERG RL, DUDRICK SJ, DELIVORIA-PAPADOPOULOUS M, MILLER LD, OSKI FA, ALTERATIONS OF RED-CELL GLYCOLYTIC INTERMEDIATES AND OXYGEN TRANS-PORT AS A CONSEQUENCE OF HYPOPHOSPHATEMIA, *N ENGL J MED*, 285, 763 (1971)

3620 TROLLFORS B, ALESTIG K, CARLSTEN C, ET AL, UNEXPECTED SIDE EFFECTS OF CEFUROXIME LYSINE, NEW CEFUROXINE SALT, *J ANTIMICROB CHEMOTHER*, 6, 558 (1983)

3621 TROLLFORS B, ALESTIG K, KRANTZ I, ET AL, QUANTITATIVE NEPHROTOXICITY OF GENTAMICIN IN NONTOXIC DOSES, *J INFECT DIS*, 141, 306-309 (1980)

3622 TRYBUCHOWSKI H, EFFECT OF AMPICILLIN ON THE URINARY OUT-PUT OF STEROIDAL HORMONES IN PREGNANT AND NON-PREGNANT WOMEN, *CLIN CHIM ACTA*, 45, 9 (1973)

3623 TRYDING N, TUFVESSON G, NILSSON S, SERUM MONOAMINE-OXI-DASE LEVELS DURING LEVODOPA THERAPY, *LANCET*, 1, 859 (1971)

3624 TSAO CS, MIYASHITA K, EFFECT OF LARGE INTAKE OF ASCORBIC ACID ON THE URINARY EXCRETION OF AMINO ACIDS AND RELATED COMPOUNDS, *IRCS MED SCI*, 13, 855-856 (1985)

3625 TUCCI JR, ESPINER EA, JAGGER PI, LAULER DP, THORN GW, VASO-PRESSIN IN THE EVALUATION OF PITUITARY-ADRENAL FUNC-TION, *ANN INTERN MED*, 69, 191 (1968)

3626 TUCCI JR, SODE J, CHRONIC CIGARETTE SMOKING: EFFECT ON ADRENOCORTICAL AND SYMPATHO-MEDULLARY ACTIVITY IN MAN, *J AM MED ASSOC*, 221, 282 (1972)

3627 TUCKER SC, LYNCH JP, ANSELL BF JR, CHLORPROPAMIDE - INDUCED AGRANULOCYTOSIS, *J AM MED ASSOC*, 238, 422 (1977)

3628 TUMMISTO T, AIRAKSINEN, MM, INCREASE OF CREATINE KINASE ACTIVITY IN SERUM CAUSED BY INTERMITTENTLY ADMINIS-TERED SUXAMETHONIUM, *BR J ANAESTH*, 38, 510 (1966)

3629 TUOMILEHTO J, TANSKANEN A, SALONEN JT, ET AL, EFFECT OF SMOKING ON SERUM HIGH-DENSITY LIPOPROTEIN CHOLESTEROL LEVELS IN A REPRESENTATIVE POPULATION SAMPLE., *PREV MED*, 15, 35-45 (1986)

3630 TURKINGTON RW, PROLACTIN SECRETION IN PATIENTD TREATED WITH VARIOUS DRUGS: PHENOTHIAZINES, TRICYCLIC ANTIDEPRESSANTS, RESERPINE AND METHYLDOPA, *ARCH INTERN MED*, 130, 349-354 (1972)

3631 TURKINGTON RW, PHENOTHIAZINE STIMULATION TEST FOR PROLACTIN RESERVE: THE SYNDROME OF ISOLATED PROLACTIN DEFICIENCY, *J CLIN ENDOCRINOL METAB*, 34, 246 (1972)

3632 TURKINGTON VE, NIXON JC, CAMPBELL JS, HURTEAU,GD, EFFECT OF LONG-ACTING STEROID CONTRACEPTIVE (MEDROXYPROGESTERONE ACETATE) ON HUMAN FEMALE SUBJECTS, *CLIN CHEM*, 17, 667 (1971)

3633 TURNBULL DM, DICK DJ, WILSON L, ET AL, VALPROATE CAUSES METABOLIC DISTURBANCE IN NORMAL MAN, *J NEUROL NEUROSURG PSYCHIATRY*, 49, 405-410 (1986)

3634 TURNBULL MJ, BALLINGER BR, URINARY EXCRETION OF MONOAMINES AND METABOLITES IN PATIENTS DEPENDENT ON AND WITHDRAWN FROM BARBITURATES, *PSYCHOPHARMACOL*, 30, 103 (1973)

3635 TURNER JM, HALL RA, WHITTAKER M, ET AL, EFFECTS OF STORAGE AND REPEATED FREEZING AND THAWING ON PLASMA CHOLINESTERASE ACTIVITY, *ANN CLIN BIOCHEM*, 21, 363-365 (1984)

3636 TURNER LV, MANCHESTER KL, INTERFERENCE OF HEPES WITH THE LOWRY METHOD, *SCIENCE*, 170, 649 (1970)

3637 TURNER RC, OAKLEY NW, NABARRO JD, CHANGES IN PLASMA INSULIN DURING ETHANOL-INDUCED HYPOGLYCEMIA, *METABOLISM*, 22, 111 (1973)

3638 TURTLE JT, BURGESS JA, BAUCKHAN S, THE METABOLIC EFFECTS OF FENFLURAMINE, *S AFR MED J* 44, SUPPL,, 23 (1971)

3639 TURTON MB, DEEGAN T, CENTRAL AND PERIPHERAL LEVELS OF PLASMA CATECHOLAMINES, CORTISOL, INSULIN AND NON-ESTERIFIED FATTY ACIDS, *CLIN CHIM ACTA*, 48, 347 (1973)

3640 TUTOR JC, MECANISMO DE INTERFERENCIA DE LA PENICILINA EN LA DETERMINACION DEL ACIDO δ-AMINOLEVULINICO URINARIO, *LABORATORIO (GRANADA)*, 70, 501-508 (1980)

3641 TUTOR JC, FERNANDEZ MP, PAZ JM, SERUM COPPER CONCENTRATIONS AND HEPATIC ENZYME INDUCTION DURING LONG-TERM THERAPY WITH ANTICONVULSANTS, *CLIN CHEM*, 28, 1367-1370 (1982)

3642 TUTOR JC, PAZ JM, RON R, ET AL, NIVELES PLASMATICAS DE GLUTAMINA Y AMINO-ACIDOS DEL CICLO DE LA UREA EN NINOS TRATADOS CON DIFENILHIDANTOINA Y FENOBARBITAL, *QUIM CLIN*, 4, 115-118 (1985)

3643 TWEEDDALE MG, OGILVIE RI, RUEDY J, ANTIHYPERTENSIVE AND BIOCHEMICAL EFFECTS OF CHLORTHALIDONE, *CLIN PHARMACOL THER*, 22, 519-527 (1977)

3644 TWEK J, GOVERDE BC, *THE SIGNIFICANCE OF HYDROXYPROLINE ASSAY IN URINE*, ORGANON TEKNIKA-OSS, HOLLAND (1972)

3645 TYNES BS, THE CLINICAL USE OF AMPHOTERICIN B, *GP*, 32, 97 (1965)

3646 TYSELLJE JR, HEPATITIS INDUCED BY METHYLDOPA (ALDOMET®) REPORT OF A CASE AND REVIEW OF THE LITERATURE, *AM J DIG DIS*, 16, 849 (1971)

3647 TYSON JE, HWANG P, GUYDA H, FRIESEN HG, STUDIES OF PROLACTIN SECRETION IN HUMAN PREGNANCY, *AM J OBSTET GYNECOL*, 113, 14 (1972)

3648 UBERTI ECD, TRASFORINI G, MARGUTTI AR, ET AL, INTRAVENOUS ADMINISTRATION OF PENTAGASTRIN INCREASES PLASMA PROLACTIN BUT NOT GROWTH HORMONE IN NORMAL WOMEN, *MIN ENDOCR*, 7, 249-252 (1982)

3649 UDALL JA, CLINICAL IMPLICATIONS OF WARFARIN INTERACTIONS WITH FIVE SEDATIVES, *AM J CARDIOL*, 35, 67-71 (1975)

3650 UDALL JA, DRUG INTERFERENCE WITH WARFARIN THERAPY, *CLIN MED*, 77, 20 (1970)

3651 UETE T, MIYAMATO Y, ONISHI M, SHIMANO N, SPECTROPHOTOMETRIC MICROMETHOD FOR MEASURING CHOLINESTERASE ACTIVITY IN SERUM OR PLASMA, *CLIN CHEM*, 18, 454 (1972)

3652 UPJOHN AC, GALBRAITH HJ, SOLOMONS B, RAISED SERUM PROTEIN BOUND IODINE AFTER TOPICAL CLIOQUINOL, *POSTGRAD MED J*, 47, 515 (1971)

3653 UPJOHN COMPANY, LITERATURE ON CLEOCIN®, *7171 PORTAGE RD, KALAMAZOO, MI 49001*

3654 URETSKY BF, GENERALOVICH T, VERBALIS JG, ET AL, COMPARATIVE HEMODYNAMIC AND HORMONAL RESPONSE OF ENOXIMONE AND DOBUTAMINE IN SEVERE CONGESTIVE HEART FAILURE, *AM J CARDIOL*, 58, 110-116 (1986)

3655 UTIGER RD, *CURRENT TOPICS ON THYROID RESEARCH*, C CASSANO, M ANDREOLI, EDS. ACADEMIC PRESS, NEW YORK, NY (1965)

3656 UTILI R, BOITNOTT JK, ZIMMERMANN HJ, DANTROLENE-ASSOCIATED HEPATIC INJURY: INCIDENCE AND CHARACTER, *GASTROENTEROLOGY*, 72, 610-616 (1977)

3657 UTTING JE, WHITFORD JH, ASSESSMENT OF PREMEDICANT DRUGS USING MEASUREMENTS OF PLASMA CORTISOL, *BR J ANAESTH*, 44, 43 (1972)

3658 VAGENAKIS AG, DOWNS P, BRAVERMAN LE, BURGER A, INGBAR SM, CONTROL OF THYROID HORMONE SECRETION IN NORMAL SUBJECTS RECEIVING IODIDES, *J CLIN INVEST*, 52, 528 (1973)

3659 VAHLQUIST C, MICHAELSSON G, VAHLQUIST A, ET AL, A SEQUENTIAL COMPARISON OF ETRETINATE (TIGASON) AND ISOTRETINOIN (ROACCUTANE) WITH SPECIAL REGARD TO THEIR EFFECTS ON SERUM LIPOPROTEINS, *BR J DERMATOL*, 112, 69-76 (1985)

3660 VALDES OS, MAURER HM, SHUMWAY CN, DRAPER DA, HOSSAINI AA, CONTROLLED CLINICAL TRIAL OF PHENOBARBITAL AND/OR LIGHT IN REDUCING NEONATAL HYPERBILIRUBINEMIA, *J PEDIATR*, 79, 1015 (1971)

3661 VALDIGUIE P, LARENG L, DIRAT MF, VALDIGUIE JP, BERTRAND JC, ACTION OF SEVERAL ANESTHETIC DRUGS ON SERUM LACTATE DEHYDROGENASE, *PATHOL BIOL*, 20, 131 (1972)

3662 VALENTOUR JC, MONFORTE JR, SUNSHINE I, FLUOROMETRIC DETERMINATION OF PROPOXYPHENE, *CLIN CHEM*, 20, 275 (1974)

3663 VALIMAKI M, HARNO K, NIKKILA EA, SERUM LIPOPROTEINS AND INDICES OF GLUCOSE TOLERANCE DURING DIURETIC THERAPY. A COMPARISON BETWEEN HYDROCHLOROTHIAZIDE AND PIRETANIDE, *J CARDIOVASC PHARMACOL*, 5, 525-530 (1985)

3664 VALLES M, TOVAR JL, SALSALATE AND MINIMAL CHANGE NEPHROTIC SYNDROME, *ANN INTERN MED*, 107, 116 (1987)

3665 VALSALAN VC, COOPER GL, CARBAMAZEPINE INTOXICATION CAUSED BY INTERACTION WITH ISONIAZID, *BR MED J [CLIN RES]*, 285, 261-262 (1982)

3666 VAN ACKER K, VERBRAEKEN K, BALUWE R, ABNORMAL PATTERNS OF LACTATE DEHYDROGENASE ISOENZYMES AFTER STREPTOKINASE THERAPY, *CLIN CHEM*, 32, 2210 (1986)

3667 VAN BRUMMELEN P, BUHLER FR, AMANN FW, ET AL, EFFECTS OF A NEW LONG-ACTING BETA-BLOCKER BOPINDOLOL (LT 31-200) ON BLOOD PRESSURE, PLASMA CATECHOLAMINES, RENIN AND CHOLESTEROL IN PATIENTS WITH ARTERIAL HYPERTENSION, *EUR J CLIN PHARMACOL*, 22, 491-493 (1982)

3668 VAN BUREN D, WIDEMAN CA, REID M, ET AL, THE ANTAGONISTIC EFFECT OF RIFAMPIN UPON CYCLOSPORINE BIOAVAILABILITY, *TRANSPLANT PROC*, 16, 1642-1645 (1984)

3669 VAN DAM FE, OVERKAMP M, HAANEN C, THE INTERACTION OF DRUGS, *LANCET*, 2, 1027 (1966)

3670 VAN DE MERWE JP, VAN BLANKENSTEIN M, WILSON JHP, INTRAHEPATIC CHOLESTASIS INDUCED BY PHENYLBUTAZONE, *DIGESTION*, 22, 317-320 (1981)

3671 VAN DEN BERG JWO, EDIXHOVEN-BOSDIJK A, KOOLE-LESUIS R, ET AL, FAECAL HAEM ASSAY - SOME PRACTICAL MODIFICATIONS OF THE HEMOQUANT ASSAY FOR HAEMOGLOBIN IN FAECES, *CLIN CHIM ACTA*, 169, 319-322 (1987)

3672 VAN DER VELDE CD, GORDON MW, MANIC-DEPRESSIVE ILLNESS, DIABETES MELLITUS, AND LITHIUM CARBONATE, *ARCH GEN PSYCHIATRY*, 21, 478 (1969)

3673 VAN DER WEERDT CM, THROMBOCYTOPENIA DUE TO QUINIDINE OR QUININE, *VOX SANG*, 12, 265 (1967)

3674 VAN DER WESTHUYZEN JM, LETTER: PLASMA-T_3 ASSAY IN KWASHIORKOR, *LANCET*, 2, 965 (1973)

3675 VAN DYK JJ, FALKSON HC, VAN DER MERWE AM, FALKSON G, UNEXPECTED TOXICITY IN PATIENTS TREATED WITH IPHOSPHAMIDE, *CANCER RES*, 32, 921 (1972)

3676 VAN HERLE AJ, ULLER RP, MATTHEWS NL, BROWN J, RADIOIMMUNOASSAY FOR MEASUREMENT OF THYROGLOBULIN IN HUMAN SERUM, *J CLIN INVEST*, 52, 1320 (1973)

3677 VAN OUDHEUSDEN APM, KINETIC DETERMINATION OF SERUM OXYTOCINASE USING A NEW SUBSTRATE, *Z KLIN CHEM KLIN BIOCHEM*, 10, 345 (1972)

3678 VAN PEENEN MJ, FILES JB, THE EFFECT OF MEDICATION ON LABORATORY TEST RESULTS, *AM J CLIN PATHOL*, 52, 666 (1969)

3679 VAN PRAAG HM, LEIJNSE B, THE XYLOSE METABOLISM IN DEPRESSED PATIENTS AND ITS ALTERATION UNDER THE INFLUENCE OF ANTIDEPRESSIVE HYDRAZINES, *CLIN CHIM ACTA*, 11, 13 (1965)

3680 VAN RIEL PLCM, VAN DE PUTTE LBA, GRIBNAU FWJ, ET AL, SERUM IgA AND GOLD-INDUCED TOXIC EFFECTS IN PATIENTS WITH RHEUMATOID ARTHRITIS, *ARCH INTERN MED*, 144, 1401-1403 (1984)

3681 VAN ROON-DJORDJEVIC B, CERFONTAIN-VAN STAALEN J, URINARY EXCRETION OF HISTIDINE METABOLITES AS AN INDICATION FOR FOLIC ACID AND VITAMIN B_{12} DEFICIENCY, *CLIN CHIM ACTA*, 41, 55 (1972)

3682 VAN ROOYEN RJ, ZIADY F, HYPOKALEMIC ALKALOSIS FOLLOWING THE ABUSE OF PURGATIVES. CASE REPORT, *S AFR MED J*, 46, 998 (1972)

3683 VAN SCOY RE, WILKOWSKE CJ, ANTITUBERCULOUS AGENTS, *MAYO CLIN PROC*, 62, 1129-1136 (1987)

3684 VAN STEKELENBURG GJ, KOOREVAAR G, EVIDENCE FOR THE EXISTENCE OF MAMMALIAN ACETOACETATE DECARBOXYLASE, *CLIN CHIM ACTA*, 39, 191 (1972)

3685 VAN STEVENINCK J, DE GOEIJ AFP, DETERMINATION OF VITAMIN A IN BLOOD PLASMA OF PATIENTS WITH CAROTENEMIA, *CLIN CHIM ACTA*, 49, 61 (1973)

3686 VAN WOERT MH, AMBANI LM, LEVINE RJ, CLINICAL EFFECTS OF PARA-CHLOROPHENYLALANINE IN PARKINSON'S DISEASE, *DIS NERV SYS*, 33, 777 (1972)

3687 VANDAM LD, ANALGESIC DRUGS - THE MILD ANALGESICS, *N ENGL J MED*, 286, 20 (1972)

3688 VANDENBURG M, PARFREY P, WRIGHT P, ET AL, HEPATITIS ASSOCIATED WITH CAPTOPRIL TREATMENT, *BR J CLIN PHARMACOL*, 11, 105-106 (1981)

3689 VANDENBURG MJ, SHARMAN VL, MORTON JJ, ET AL, HORMONAL AND BLOOD PRESSURE CHANGES DURING CONVERTING ENZYME INHIBITION BY TEPROTIDE, *POSTGRAD MED J*, 57, 283-288 (1981)

3690 VANDENHEUVEL WJA, SMITH JL, SILBER RH, B-(2-METHOXYPHE-NOXY)LACTIC ACID, THE MAJOR URINARY METABOLITE OF GLYCERYL GUAIACOLATE IN MAN, *J PHARM SCI*, 61, 1997 (1972)

3691 VANUXEM D, SAMPOL J, WEILLER PJ, ET AL, INFLUENCE DU TABAGISME CHRONIQUE SUR LES LEUCOCYTES, *RESPIRATION*, 46, 258-264 (1984)

3692 VATASSERY GT, MORTENSON GA, MANUAL SPECTROPHOTOMETRIC AND FLUOROMETRIC DETERMINATION OF TOCOPHEROL IN CEREBROSPINAL FLUID, *CLIN CHEM*, 18, 1475 (1972)

3693 VAZ R, SENIOR B, MORRIS M, ET AL, ADRENAL EFFECTS OF BECLOMETHASONE INHALATION THERAPY IN ASTHMATIC CHILDREN, *J PEDIATR*, 100, 660-662 (1982)

3694 VERBEELEN D, VANHAELST L, FUSS M, ET AL, EFFECT OF 1, 25-DIHYDROXYVITAMIN D_3 AND NIFEDIPINE ON PROLACTIN RELEASE IN NORMAL MAN, *J ENDOCRINOL INVEST*, 8, 103-106 (1985)

3695 VERDY M, MARC-AURELE J, FASTING IN OBESE FEMALES: PLASMA RENIN AFTER GLUCOSE REFEEDING, *HORM METAB RES*, 5, 59 (1973)

3696 VERHO M, BOSSALLER W, HEINEN B, SERUM TRACE ELEMENT LEVELS IN PIRETANIDE-TREATED HYPERTENSIVES: A DOUBLE-BLIND TRIAL AGAINST HYDROCHLOROTHIAZIDE PLUS AMILORIDE, *INT J CLIN PHARMACOL RES*, 7, 5-11 (1987)

3697 VERITY CM, APPLEGARTH DA, FARRELL K, ET AL, THE INFLUENCE OF ANTICONVULSANTS ON FASTING PLASMA AMMONIA AND AMINO ACID LEVELS, *CLIN BIOCHEM*, 16, 344-345 (1983)

3698 VERMORKEN JB, PINEDO HM, GASTROINTESTINAL TOXICITY OF CIS-DIAMINEDICHLORO-PLATINUM (II), *NETH J MED*, 25, 270-274 (1982)

3699 VERSCHURE JCM, CLINICAL USE OF MEASUREMENTS OF CLEARANCE AND MAXIMUM CAPACITY OF THE LIVER, *ACTA MED SCAND*, 142, 409 (1952)

3700 VERSTER FDEB, GAROUTTE J, ICHINOSA H, GUERRERO-FIQUEROA R, MODE OF ACTION OF DIPHENYLHYDANTOIN, *FED PROC*, 24, 390 (1965)

3701 VERWEIJ WM, OSSENKOPPELE GJ, WIJERMANS P, INTERFERENCE OF METHOTREXATE WITH THE DETERMINATION OF THE PROTEIN CONTENT OF CEREBROSPINAL FLUID., *ANN CLIN BIOCHEM*, 23, 612-613 (1986)

3702 VESSBY B, ABELIN J, FINNSON M, ET AL, EFFECTS OF NIFEDIPINE TREATMENT ON CARBOHYDRATE AND LIPOPROTEIN METABOLISM, *CURR THER RES*, 33, 1075-1081 (1983)

3703 VESSELL ES, IMPAIRMENT OF DRUG METABOLISM BY DISULFIRAM IN MAN, *CLIN PHARMACOL THER*, 12, 785 (1971)

3704 VESSELL ES, PASSANANTI GT, GREENE FE, IMPAIRMENT OF DRUG METABOLISM IN MAN BY ALLOPURINOL AND NORTRIPTYLINE, *N ENGL J MED*, 283, 1484-1488 (1970)

3705 VESTERGAARD P, AMDISEN A, HANSEN HE, ET AL, LITHIUM TREATMENT AND KIDNEY FUNCTION, A SURVEY OF 237 PATIENTS IN LONG-TERM TREATMENT, *ACTA PSYCHIATR SCAND*, 60, 504-520 (1979)

3706 VETERANS ADMINISTRATION COOPERATIVE STUDY GROUP ON ANTIHYPERTENSIVE AGENTS, EFFICACY OF NADOLOL ALONE AND COMBINED WITH BENDROFLUMETHIAZIDE AND HYDRALAZINE FOR SYSTEMIC HYPERTENSION, *AM J CARDIOL*, 52, 1230-1237 (1983)

3707 VETERANS ADMINISTRATION COOPERATIVE STUDY GROUP ON ANTIHYPERTENSIVE AGENTS, COMPARISON OF PROPRANOLOL AND HYDROCHLOROTHIAZIDE FOR THE INITIAL TREATMENT OF HYPERTENSION: I. RESULTS OF SHORT-TERM TITRATION WITH EMPHASIS ON RACIAL DIFFERENCES IN RESPONSE., *J AM MED ASSOC*, 248, 1996-2003 (1982)

3708 VETERANS ADMINISTRATION COOPERATIVE STUDY GROUP ON ANTIHYPERTENSIVE AGENTS, COMPARISON OF PROPRANOLOL AND HYDROCHLOROTHIAZIDE FOR THE INITIAL TREATMENT OF HYPERTENSION: II. RESULTS OF LONG-TERM THERAPY, *J AM MED ASSOC*, 248, 2004-2011 (1982)

3709 VEYS EM, MIELANTS H, VERBRUGGEN G, ET AL, LEVAMISOLE AS BASIC TREATMENT OF RHEUMATOID ARTHRITIS: LONG TERM EVALUATION, *J RHEUMATOL DIS*, 8, 45-56 (1981)

3710 VICKERS S, STUART EK, SIMPLE, SENSITIVE SPECTROPHOTOFLUOROMETRIC METHOD FOR HYDRAZINE IN PLASMA, *ANAL CHEM*, 46, 138 (1974)

3711 VIDT DG, MECHANISM OF ACTION, PHARMACOKINETICS, ADVERSE EFFECTS AND THERAPEUTIC USES OF AMILORIDE HYDROCHLORIDE, A NEW POTASSIUM-SPARING DIURETIC, *PHARMACOTHERAPY*, 1, 179-187 (1981)

3712 VIDT DG, BRAVO EL, FOUAD FM, CAPTOPRIL, *N ENGL J MED*, 306, 214-219 (1982)

3713 VIGLIANI EC, CAVAGNA G, LOCATIG, FOA V, BIOLOGICAL EFFECTS OF NITROGLYCOL ON THE METABOLISM OF CATECHOLAMINES, *ARCH ENVIRON HEALTH*, 16, 477 (1968)

3714 VILEN L, KUTTI J, FREDEN K, ET AL, THE PERIPHERAL PLATELET COUNT AND ADP-INDUCED PLATELET AGGREGATION IN RESPONSE TO METOPROLOL AND PROPRANOLOL AS STUDIED IN YOUNG HEALTHY MALE VOLUNTEERS, *SCAND J HAEMATOL*, 31, 440-446 (1983)

3715 VILLARREAL H, ARCILA H, RAMIREZMA, SIERRA P, EFFECT OF GUANCYDINE ON SYSTEMIC AND RENAL HEMODYNAMICS IN ARTERIAL HYPERTENSION, *CLIN PHARMACOL THER*, 12, 838 (1971)

3716 VINCENT FM, PHENOTHIAZINE-INDUCED PHENYTOIN INTOXICATION, *ANN INTERN MED*, 93, 56-57 (1980)

3717 VINCENT PC, DRUG-INDUCED APLASTIC ANEMIA AND AGRANULOCYTOSIS: INCIDENCE AND MECHANISMS, *DRUGS*, 31, 52-63 (1986)

3718 VINCENT WF, ULLMANN WW, THE MEASUREMENT OF DELTA-AMINOLEVULINIC ACID IN URINE, *ANN CLIN LAB SCI*, 2, 31 (1972)

3719 VINET B, LETELLIER G, THE IN VITRO EFFECT OF DRUGS ON BIOCHEMICAL PARAMETERS DETERMINED BY A SMAC®SYSTEM, *CLIN BIOCHEM*, 10, 47-51 (1977)

3720 VIR SC, LOVE AHG, ZINC AND COPPER NUTRITURE OF WOMEN TAKING ORAL CONTRACEPTIVE AGENTS, *AM J CLIN NUTR*, 34, 1479-1483 (1981)

3721 VOLPI R, COIRO V, SALVI M, ET AL, EFFECT OF METOCLOPRAMIDE ON SERUM GROWTH HORMONE LEVELS IN CIRRHOTIC MEN, *J ENDOCRINOL INVEST*, 6, 101-105 (1983)

3722 VON EULER US, PATHOPHYSIOLOGICAL ASPECTS OF CATECHOLAMINE PRODUCTION, *CLIN CHEM*, 18, 1445 (1972)

3723 VOORS AW, SRINIVASAN SR, HUNTER SM, ET AL, SMOKING, ORAL CONTRACEPTIVES, AND SERUM LIPID AND LIPOPROTEIN LEVELS IN YOUTHS, *PREV MED*, 11, 1-12 (1982)

3724 VORYS N, ULLERY JC, STEVENS V, THE EFFECTS OF SEX STEROIDS ON GONADOTROPINS, *AM J OBSTET GYNECOL*, 93, 641 (1965)

3725 VOUGHT RL, BROWN FA, WOLFF J, ERYTHROSINE: AN ADVENTITIOUS SOURCE OF IODIDE, *J CLIN ENDOCRINOL METAB*, 34, 747 (1972)

3726 VREE TB, ZWEENS K, HUIGE PJC, ET AL, INTERACTIONS BETWEEN THE RENAL EXCRETION RATES OF β_2-MICROGLOBULIN AND TOBRAMYCIN IN MAN, *CLIN CHIM ACTA*, 138, 49-57 (1984)

3727 VY T, NUTRITIONAL VALUE OF A VEGETARIAN DIET, *AM J CLIN NUTR*, 25, 647 (1972)

3728 VYDEN JK, CURNOW DH, BECK AB, BOUNDY CA, FAILURE OF CHLORDIAZEPOXIDE TO INFLUENCE URINARY COPPER EXCRETION, *LANCET*, 2, 1090 (1972)

3729 WACHTER JP, SMEBY RR, FREE AH, URINALYSIS AND ORAL HYPOGLYCEMIC AGENTS, *AM J MED TECHNOL*, 26, 125 (1960)

3730 WACKER WEC, PARISI AF, MAGNESIUM METABOLISM, *N ENGL J MED*, 278, 712 (1968)

3731 WADE JC, SCHIMPFF SC, WIERNIK PH, ANTIBIOTIC COMBINATION-ASSOCIATED NEPHROTOXICITY IN GRANULOCYTOPENIA PATIENTS WITH CANCER, *ARCH INTERN MED*, 141, 1789-1793 (1981)

3732 WADHWA NF, SCHROEDER TJ, PESCE AJ, ET AL, CYCLOSPORINE DRUG INTERACTIONS: A REVIEW, *THER DRUG MONIT*, 9, 399-406 (1987)

3733 WAHL P, WALDEN C, KNOPP R, ET AL, EFFECT OF ESTROGEN/PROGESTIN POTENCY ON LIPID/LIPOPROTEIN CHOLESTEROL, *N ENGL J MED*, 308, 862-867 (1983)

3734 WAISBREN BA, EVANI SV, ZIEBERT AP, CARBENICILLIN AND BLEEDING, *J AM MED ASSOC*, 217, 1243 (1971)

3735 WALDMEIER P, DE HERDT P, MAITRE L, SIMULTANEOUS AUTOMATED ESTIMATION OF NORADRENALINE AND DOPAMINE IN BRAIN TISSUES, *CLIN CHEM*, 20, 81 (1974)

3736 WALDORFF S, HANSEN PB, EGEBLAD H, ET AL, INTERACTIONS BETWEEN DIGOXIN AND POTASSIUM-SPARING DIURETICS, *CLIN PHARMACOL THER*, 33, 418-423 (1983)

3737 WALDORFF S, HANSEN PB, KJAERGARD H, ET AL, AMILORIDE-INDUCED CHANGES IN DIGOXIN DYNAMICS AND KINETICS: ABOLITION OF DIGOXIN INDUCED INOTROPISM WITH AMILORIDE, *CLIN PHARMACOL THER*, 30, 172-176 (1981)

3738 WALDRON HA, CHERRY N, JOHNSTON JD, THE EFFECTS OF ETHANOL ON BLOOD TOLEUENE CONCENTRATIONS, *INT ARCH OCCUP ENVIRON HEALTH*, 51, 365-369 (1983)

3739 WALENKAMP GHIM, VREE TB, GUELEN PJM, ET AL, INTERACTION BETWEEN THE RENAL EXCRETION RATES OF β_2-MICROGLOBULIN AND GENTAMICIN IN MAN, *CLIN CHIM ACTA*, 127, 229-238 (1983)

3740 WALES JK, WOLFF F, HEMATOLOGICAL SIDE EFFECTS OF DIAZOXIDE, *LANCET*, 1, 53 (1967)

3741 WALINDER J, ARBERG-WISTEDT A, JOZWIAK H, ET AL, THE SAFETY OF ZIMELDINE IN LONG-TERM USE IN DEPRESSIVE ILLNESS, *ACTA PSYCHIATR SCAND*, 68, SUPPL 308, 147-160 (1983)

3742 WALKER ARP, INTERPRETATION OF BIOLOGICAL DATA ON ONE ETHNIC OR REGIONAL GROUP MAY NOT BE EQUALLY APPLICABLE TO OTHER GROUPS, *AM J CLIN NUTR*, 20, 1025 (1967)

3743 WALKER BA, EZE LC, TWEEDIE MCK, EVANS DA, THE INFLUENCE OF ABO BLOOD GROUPS SECRETOR STATUS AND FAT INGESTION ON SERUM ALKALINE PHOSPHATASE, *CLIN CHIM ACTA*, 35, 433 (1971)

3744 WALKER BR, CAPUZZI DM, ALEXANDER F, HYPERKALEMIA AFTER TRIAMTERENE IN DIABETIC PATIENTS, *J CLIN PHARMACOL*, 13, 643 (1972)

3745 WALKER BR, HOPPE RC, ALEXANDER F, EFFECT OF TRIAMTERENE ON THE RENAL CLEARANCE OF CALCIUM, MAGNESIUM, PHOSPHATE AND URIC ACID IN MAN, *CLIN PHARMACOL THER*, 13, 245 (1972)

3746 WALKER FA, AMMONIA IN FIBRIN HYDROLYSATES, *N ENGL J MED*, 285, 1324 (1971)

3747 WALKER JG, FATAL AGRANULOCYTOSIS COMPLICATING TREATMENT WITH ETHACRYNIC ACID, *ANN INTERN MED*, 64, 1303 (1966)

3748 WALKER PC, NEONATAL BILIRUBIN TOXICITY. A REVIEW OF KERNICTERUS AND THE IMPLICATIONS OF DRUG-INDUCED BILIRUBIN DISPLACEMENT, *CLIN PHARMACOKINET*, 13, 26-50 (1987)

3749 WALKER PL, PETTIT BR, SANDLER M, INTERFERENCE BY NAPROXEN IN THE URINARY 5-HYDROXYINDOLEACETIC ACID ASSAY IS DUE TO A METABOLITE DESMETHYLNAPROXEN, *ANN CLIN BIOCHEM*, 24, 177-181 (1987)

3750 WALKER RC, WRIGHT AJ, THE QUINOLONES, *MAYO CLIN PROC*, 62, 1007-1012 (1987)

3751 WALL JR, ZIMMET PZ, JARRETT RJ, BAILES M, RAMAGE CM, FALL IN PLASMA-TESTOSTERONE LEVELS IN NORMAL MALE SUBJECTS IN RESPONSE TO AN ORAL GLUCOSE LOAD, *LANCET*, 1, 967 (1973)

3752 WALLACE EZ, SILVERBERG HI, CARTER AC, EFFECT OF ETHINYL ESTRADIOL ON PLASMA 17-OH CORTICOSTEROIDS, ACTH RESPONSIVENESS AND HYDROCORTISONE CLEARANCE, *PROC SOC EXP BIOL MED*, 95, 805 (1957)

3753 WALLACE JE, HAMILTON HE, RILOFF JA, BLUM K, SPECTROPHOTOMETRIC DETERMINATION OF ETHCHLORVYNOL IN BIOLOGIC SPECIMENS, *CLIN CHEM*, 20, 159 (1974)

3754 WALLACE SL, NISSEN AW, GRISEOFULVIN IN ACUTE GOUT, *N ENGL J MED*, 266, 1099 (1962)

3755 WALLACH EJ, *INTERPRETATION OF DIAGNOSTIC TESTS*, LITTLE BROWN AND CO, BOSTON MA (1970)

3756 WALLDIUS G, EFFECT OF VERAPAMIL ON SERUM LIPOPROTEINS IN PATIENTS WITH ANGINA PECTORIS, *ACTA MED SCAND*, 215 SUPPL 681, 43-48 (1984)

3757 WALSTAD RA, MYHRE KI, WIRUM E, ET AL, THE INFLUENCE OF SUSTAINED RELEASE THEOPHYLLINE THERAPY ON FREE FATTY ACIDS IN SERUM, *ACTA PHARMACOL TOXICOL*, 56, 199-203 (1985)

3758 WANDELL M, POWELL JR, HAGER WD, ET AL, EFFECT OF QUININE ON DIGOXIN KINETICS, *CLIN PHARMACOL THER*, 28, 425-430 (1980)

3759 WANG C, LAI CL, LAM KC, ET AL, EFFECT OF CIMETIDINE ON GONADAL FUNCTION IN MAN, *BR J CLIN PHARMACOL*, 13, 791-794 (1982)

3760 WANG Y-S, HERSHMAN JM, SMITH V, ET AL, EFFECT OF HEPARIN ON FREE THYROXINE AS MEASURED BY EQUILIBRIUM DIALYSIS AND ULTRAFILTRATION, *CLIN CHEM*, 32, 700 (1986)

3761 WANKA J, JONES LI, WOOD PH, DIXON ASJ, INDOMETHACIN IN RHEUMATIC DISEASES, *ANN RHEUM DIS*, 23, 218 (1964)

3762 WAPNICK S, JONES JJ, ALCOHOL AND GLUCOSE TOLERANCE, *LANCET*, 2, 180 (1972)

3763 WARE S, MILLWARD-SADLER GH, ACUTE LIVER DISEASE ASSOCIATED WITH SODIUM VALPROATE, *LANCET*, 2, 1110-1113 (1980)

3764 WARRELL RP JR, ISAACS M, ALCOCK NW, ET AL, GALLIUM NITRATE FOR TREATMENT OF RETRACTORY HYPERCALCEMIA FROM PARATHYROID CARCINOMA, *ANN INTERN MED*, 107, 683-686 (1987)

3765 WARREN KS, SCOTT CD, INTERFERING ULTRAVIOLET-ABSORBING COMPOUNDS AS SOURCES OF ERROR IN COMMON CLINICAL CHEMISTRY TESTS, *CLIN CHEM*, 15, 1146 (1969)

3766 WARREN SE, FALSE-POSITIVE URINE KETONE TEST WITH CAPTOPRIL, *N ENGL J MED*, 303, 1003-1004 (1980)

3767 WASSERMANN HP, LEUKOCYTE COUNTS IN ETHNIC GROUPS, *LANCET*, 1, 852 (1972)

3768 WATHEN CG, MACKAY IG, GLOVER DR, ET AL, EFFECT OF THE PARTIAL BETA-AGONIST PRENALTEROL ON PLASMA RENIN ACTIVITY IN PATIENTS WITH LEFT VENTRICULAR FAILURE, *CLIN CHIM ACTA*, 162, 97-100 (1987)

3769 WATSON CJ, THE PROBLEM OF PORPHYRIA - SOME FACTS AND QUESTIONS, *N ENGL J MED*, 263, 1205 (1960)

3770 WATSON PG, LOCMAN RG, REDDING VJ, DRUG INTERACTION WITH COUMARIN DERIVATIVE ANTICOAGULANTS, *BR MED J [CLIN RES]*, 285, 1045-1046 (1982)

3771 WATSON RR, JACKSON JC, HARTMANN B, ET AL, CELLULAR IMMUNE FUNCTIONS ENDORPHINS, AND ALCOHOL CONSUMPTION IN MALES, *ALCOHOL CLIN EXP RES*, 9, 248-254 (1985)

3772 WATTS RW, HETZEL DJ, BOCHNER F, ET AL, LACK OF INTERACTION BETWEEN RANITIDINE AND PHENYTOIN, *BR J CLIN PHARMACOL*, 15, 499-500 (1983)

3773 WAXMAN S, CORCINO JJ, HERBERT V, DRUGS, TOXINS AND DIETARY AMINO ACIDS AFFECTING VITAMIN B_{12} OR FOLIC ACID ABSORPTION OR UTILIZATION, *AM J MED*, 48, 599 (1970)

3774 WEBEL ML, DONADIO JV JR., EFFECTS OF A LARGE DOSE OF METHYLPREDNISOLONE ON RENAL FUNCTION, *J LAB CLIN MED*, 80, 765 (1972)

3775 WEBER JA, VAN ZANTEN AP, ELIMINATION OF HEPARIN INTERFERENCE IN THE DETERMINATION OF GAMMA-GT, *CLIN CHIM ACTA*, 169, 345-346 (1987)

3776 WEBER MA, DRAYER JIM, KAUFMAN CA, THE COMBINED ALPHA- AND BETA- ADRENERGIC BLOCKER LABETALOL AND PROPRANOLOL IN THE TREATMENT OF HIGH BLOOD PRESSURE: SIMILARITIES AND DIFFERENCES, *J CLIN PHARMACOL*, 24, 103-112 (1984)

3777 WEBSTER PD, ZIEVE L, ALTERATIONS IN SERUM CONTENT OF PANCREATIC ENZYMES, *N ENGL J MED*, 267, 654 (1962)

3778 WEETMAN AP, EFFECT OF THE ANTITHYROID DRUG METHIMAZOLE ON INTERLEUKIN-1 AND INTERLEUKIN-2 IN VITRO, *CLIN ENDOCRINOL*, 25, 133-142 (1986)

3779 WEIDMANN P, BOUSQUET J-C, ET AL, ANTIHYPERTENSIVE TREATMENT WITH BOPINDOLOL ALONE OR COMBINED WITH A DIURETIC IN GENERAL PRACTICE, *J CARDIOVASC PHARMACOL*, 8 (SUPPL 6), 580-587 (1986)

3780 WEIDMANN P, GERBER A, EFFECTS OF TREATMENT WITH DIURETICS ON SERUM LIPOPROTEINS, *J CARDIOVASC PHARMACOL*, 6, S260-S268 (1984)

3781 WEINBERGER I, ROTENBERG Z, FUCHS J, ET AL, AMIODARONE INDUCED THROMBOCYTOPENIA, *ARCH INTERN MED*, 147, 735-736 (1987)

3782 WEINBERGER MM, SMITH G, MILAVETZ G, ET AL, DECREASED THEOPHYLLINE CLEARANCE DUE TO CIMETIDINE, *N ENGL J MED*, 304, 672 (1981)

3783 WEINSIER RL, FASTING - A REVIEW WITH EMPHASIS ON THE ELECTROLYTES, *AM J MED*, 50, 233 (1971)

3784 WEINSTEIN J, HYPOCOMPLEMENTEMIA IN HYDRALAZINE - ASSOCIATED SYSTEMIC LUPUS ERYTHEMATOSUS, *AM J MED*, 65, 553-556 (1978)

3785 WEIR RJ, TREE M, FRASER R, EFFECT OF ORAL CONTRACEPTIVES ON BLOOD PRESSURE AND ON PLASMA RENIN, RENIN SUBSTRATE AND CORTICOSTEROIDS, *J CLIN PATHOL* 23, SUPPL 3, 49 (1969)

3786 WEISS JL, WATANABE AM, LEMBERGER L, TAMARKIN NR, CARDON PV, CARDIOVASCULAR EFFECTS OF DELTA-9-TETRAHYDROCANNABINOL IN MAN, *CLIN PHARMACOL THER*, 13, 671 (1972)

3787 WEISS M, HASSIN D, BANK H, PROPYLTHIOURACIL-INDUCED HEPATIC DAMAGE, *ARCH INTERN MED*, 140, 1184-1185 (1980)

3788 WEISS P, BIANCHINE JR, THE EFFECT ON SERUM TOCOPHEROL LEVELS ON DRUG-INDUCED DECREASE IN SERUM LIPIDS, *AM J MED SCI*, 258, 275 (1969)

3789 WEISS VC, WEST DP, ACKERMAN R, ET AL, HEPATOTOXIC REACTIONS IN A PATIENT TREATED WITH ETRINATE, *ARCH DERMATOL*, 120, 104-106 (1984)

3790 WEISSMAN K, HOYER H, SERUM ZINC LEVELS DURING ORAL GLUCOCORTICOID THERAPY, *J INVEST DERMATOL*, 86, 715-716 (1986)

3791 WEISSMAN PN, SHENKMAN L, GREGERMAN RI, CHLORPROPAMIDE HYPONATREMIA, *N ENGL J MED*, 284, 65 (1971)

3792 WEITEKAMP MR, ABER RC, PROLONGED BLEEDING TIMES AND BLEEDING DIATHESIS ASSOCIATED WITH MOXALACTAM ADMINISTRATION, *J AM MED ASSOC*, 249, 69-71 (1983)

3793 WEITZ JI, CROWLEY KA, LANDMAN SL, ET AL, INCREASED NEUTROPHIL ELASTASE ACTIVITY IN CIGARETTE SMOKERS, *ANN INTERN MED*, 107, 680-682 (1987)

3794 WELLING PG, INTERACTIONS AFFECTING DRUG ABSORPTION, *CLIN PHARMACOKINET*, 9, 404-434 (1984)

3795 WENK RE, GEBHARDT FC, BHAGAVAN BS, ET AL, TETRACYCLINE-ASSOCIATED FATTY LIVER OF PREGNANCY, INCLUDING POSSIBLE PREGNANCY RISK AFTER CHRONIC DERMATOLOGIC USE OF TETRACYCLINE, *J REPROD MED*, 26, 135-141 (1981)

3796 WENNBERG JE, OKUM R, HINMAN EJ, NORTHCUTT RC, GRIEP RJ, WALKER WG, RENAL TOXICITY OF ORAL CHOLECYSTOGRAPHIC MEDIA, *J AM MED ASSOC*, 186, 461 (1963)

3797 WENTZ AC, JONES GS, GRAEBER J, EFFECT OF INFUSED PROSTAGLANDIN F2A ON HORMONAL LEVELS DURING EARLY PREGNANCY, *AM J OBSTET GYNECOL*, 114, 908 (1972)

3798 WENZEL KW, PHARMACOLOGICAL INTERFERENCE WITH IN VITRO TESTS OF THYROID FUNCTION, *METABOLISM*, 30, 717-732 (1981)

3799 WERBLIN TP, POLLACK IP, LISS RA, BLOOD DYSCRASIAS IN PATIENTS USING METHAZOLAMIDE (NEPTAZANE®) FOR GLAUCOMA, *OPHTHALMOLOGY*, 87, 350-354 (1980)

3800 WERK EE JR., MACGEE J, SHOLITON LJ, EFFECT OF DIPHENYLHYDANTOIN ON CORTISOL METABOLISM IN MAN, *J CLIN INVEST*, 43, 1824 (1964)

3801 WERNING C, BAYER JM, FISCHER N, SCHWEIKERT HU, SIEGENTHALER W, THE EFFECT OF CARBENOXOLONE SODIUM ON BLOOD PRESSURE, *GERM MED*, 2, 9 (1972)

3802 WERTALIK LF, METZ EN, LOBUGLIO AF, BALCERZAK SP, DECREASED SERUM B_{12} LEVELS SECONDARY TO ORAL CONTRACEPTIVE AGENTS, *AM J CLIN NUTR*, 24, 603 (1971)

3803 WERTMER CA, CLOUD H, OHTAKE M, ET AL, EFFECT OF LONG-TERM ADMINISTRATION OF ANTICONVULSANTS ON COPPER, ZINC AND CERULOPLASMIN LEVELS, *DRUG NUTR INTERACT*, 4, 269-270 (1986)

3804 WEST CD, DAMAST BL, SARRO SK, PEARSN OM, CONVERSION OF TESTOSTERONE TO ESTROGENS IN CASTRATED ADRENALECTOMIZED HUMAN FEMALES, *J BIOL CHEM*, 218, 409 (1956)

3805 WESTABY D, OGLE SJ, PARADINAS FJ, ET AL, LIVER DAMAGE FROM LONG-TERM METHYLTESTOSTERONE, *LANCET*, 2, 261-263 (1977)

3806 WESTER PO, URINARY ZINC EXCRETION DURING TREATMENT WITH DIFFERENT DIURETICS, *ACTA MED SCAND*, 208, 209-212 (1980)

3807 WETTSBERG L, ABERG H, ROSS SB, FRODEN O, PLASMA DOPAMINE-β-HYDROXYLASE ACTIVITY IN HYPERTENSION AND VARIOUS NEUROPSYCHIATRIC DISORDERS, *SCAND J CLIN LAB INVEST*, 30, 283 (1972)

3808 WEWERS MD, BRANTLY ML, CASOLARO MA, ET AL, EFFECT OF TAMOXIFEN AS A THERAPY TO AUGMENT ALPHA$_1$-ANTITRYPSIN CONCENTRATIONS IN Z HOMOZYGOUS ALPHA$_1$-ANTITRYPSIN-DEFICIENT SUBJECTS., *AM REV RESPIR DIS*, 135, 401-402 (1987)

3809 WHITE DJG, BLATCHFORD NR, CAUWENBERGH G, CYCLOSPORINE AND KETOCONAZOLE, *TRANSPLANTATION*, 37, 214-215 (1984)

3810 WHITE JM, FLASHKA HA, AN AUTOMATED PROCEDURE WITH USE OF FERROZINE FOR ASSAY OF SERUM IRON AND TOTAL IRON-BINDING CAPACITY, *CLIN CHEM*, 19, 526 (1973)

3811 WHITE LL, FATAL MARROW APLASIA DURING CHLORPROPAMIDE THERAPY, *BR MED J [CLIN RES]*, 1, 691 (1962)

3812 WHITE MC, KENDALL-TAYLOR P, ADRENAL HYPOFUNCTION IN PATIENTS TAKING KETOCONAZOLE, *LANCET*, 1, 44 (1985)

3813 WHITE MG, ASCH MJ, ACID-BASE EFFECTS OF TOPICAL MAFENIDE ACETATE IN THE BURNED PATIENT, *N ENGL J MED*, 284, 1281 (1971)

3814 WHITE NJ, WARRELL DJ, CHANTAVANICH P, ET AL, SEVERE HYPO-GLYCEMIA AND HYPERINSULINEMIA IN FALCIPARUM MALARIA, *N ENGL J MED*, 309, 61-66 (1983)

3815 WHITFIELD JB, POUNDER RE, NEALE G, MOSS DW, SERUM ALPHA-GLUTAMYL TRANSPEPTIDASE ACTIVITY IN LIVER DISEASE, *GUT*, 13, 702 (1972)

3816 WHITING GFM, MCLARAN CJ, BOCHNER F, SEVERE HYPERKALEMIA WITH MODURETIC, *MED J AUST*, 1, 409 (1979)

3817 WHITTEN CF, BROUGH AJ, THE PATHOPHYSIOLOGY OF ACUTE IRON POISONING, *CLIN TOXICOL*, 4, 585 (1971)

3818 WIDERLOV E, KARLMAN I, STORSATER J, HYDRALAZINE-INDUCED NEONATAL THROMBOCYTOPENIA, *N ENGL J MED*, 303, 1235 (1980)

3819 WIDMARK P, SAFETY AND EFFICACY OF PIROXICAM IN THE TREAT-MENT OF ACUTE GOUT, *EUR J RHEUMATOL INFLAMM*, 1, 346-355 (1978)

3820 WIEDERHOLT IC, GENCO M, FOLEY JM, RECURRENT EPISODES OF HYPOGLYCEMIA INDUCED BY PROPOXYPHENE, *NEUROLOGY*, 17, 703 (1967)

3821 WIEGAND SE, COPEMAN PWM, PERRY HO, METABOLIC ALKALINIZA-TION IN PORPHYRIA CUTANEA TARDA, *ARCH DERMATOL*, 100, 544 (1969)

3822 WIEGELMANN W, WILDMEISTER W, HORSTER FA, SOLBACH HG, RADIOIMMUNOASSAY OF HGH, ICSH, AND FSH IN PLASMA AFTER I.V. ADMINISTRATION OF THYROTROPIN RELEASING HORMONE (TRH), *HORM METAB RES*, 4, 482 (1972)

3823 WIEN G, KUMAR R, PREVENTION OF HEMOLYSIS - INDUCED INTER-FERENCE WITH DIGOXIN ASSAYS, *CLIN CHEM*, 18, 1443 (1972)

3824 WIGLELY RD, ASPIRIN AND THE KIDNEY, *N Z MED J*, 74, 301 (1971)

3825 WIJNJA LL, SNIJDER JA, NIEWEG HO, ACETYLSALICYLIC ACID AS A CAUSE OF PANCYTOPENIA FROM BONE MARROW DAMAGE, *LANCET*, 2, 768 (1966)

3826 WILCOX RG, RANDOMIZED STUDY OF SIX BETA-BLOCKERS AND A THIAZIDE DIURETIC IN ESSENTIAL HYPERTENSION, *BR MED J [CLIN RES]*, 2, 383-385 (1978)

3827 WILDER BJ, SERRANO EE, RAMSAY RE, PLASMA DIPHENYLHY-DANTOIN LEVELS AFTER LOADING AND MAINTENANCE DOSES, *CLIN PHARMACOL THER*, 14, 797-801 (1973)

3828 WILDER BJ, STREIFF RR, HAMMER RH, DIPHENYLHYDANTOIN: ABSORPTION DISTRIBUTION AND EXCRETION: CLINICAL STUD-IES IN, *ANTIEPILEPTIC DRUGS*, DM WOODBURY, JK PENRY, RP SCHMIDT, EDS. RAVEN PRESS, NEW YORK, NY (1972)

3829 WILDER BJ, WILLMORE LJ, BRUNI J, ET AL, VALPROIC ACID: INTER-ACTION WITH OTHER ANTICONVULSANT DRUGS, *NEUROLOGY*, 28, 892-896 (1978)

3830 WILDING P, CONCEPTS OF NORMALITY IN HEALTH SCREENING, IN *REFERENCE VALUES IN HUMAN CHEMISTRY*, G SIEST, ED. KARGER, BASEL, 168 (1973)

3831 WILDMEISTER W, DAWEKE H, GRIES FA, GRUNEKLEE D, HESSING J, HORSTER FA, INFLUENCE OF SYNTHETIC TRH ON GLUCOSE, FREE FATTY ACIDS AND INSULIN IN PLASMA OF HEALTHY PER-SONS, *HORM METAB RES*, 4, 368 (1972)

3832 WILHELM MP, EDSON RS, ANTIMICROBIAL AGENTS IN URINARY TRACT INFECTIONS, *MAYO CLIN PROC*, 62, 1025-1031 (1987)

3833 WILKINS MR, FRANKLYN JA, WOODS KL, ET AL, EFFECT OF PROPRANOLOL ON THYROID HOMEOSTASIS OF HEALTHY VOL-UNTEERS, *POSTGRAD MED J*, 61, 391-394 (1985)

3834 WILKINSON GR,, SCHENKER S, LETTER TO THE EDITOR, *CLIN PHARMACOL THER*, 19, 486-488 (1976)

3835 WILKINSON JH, AN INTRODUCTION TO DIAGNOSTIC ENZYMOLOGY, *ARNOLD, LONDON, ENGLAND* (1962)

3836 WILKINSON M, EFFER SB, YOUNGLAI EV, GUPTA K, FREE ESTRIOL IN HUMAN PREGNANCY PLASMA, *AM J OBSTET GYNECOL*, 114, 867 (1972)

3837 WILKINSON SP, PORTMANN B, WILLIAMS R, HEPATITIS FROM DAN-TROLENE SODIUM, *GUT*, 20, 33-136 (1979)

3838 WILLIAMS CM, THE EFFECT OF RESERPINE ON DOPAMINE METABO-LISM IN HUMANS, *J NEUROCHEM*, 9, 335 (1962)

3839 WILLIAMS G, LOFTS F, FUESSL H, ET AL, TREATMENT WITH DANAZOL AND PLASMA GLUCAGON CONCENTRATION, *BR MED J [CLIN RES]*, 291, 1155-1156 (1985)

3840 WILLIAMS GA, BOWSER EN, HARGIS GK, HENDERSON WJ, MARTI-NEZ NJ, FUGK R, KUKREYA S, EFFECT OF GLUCOCORTICOIDS ON FUNCTION OF THE PARATHYROID GLANDS IN MAN, *CLIN RES*, 20, 780 (1972)

3841 WILLIAMS GT, JOHNSON SAN, DIEPPE PA, ET AL, NEUTROPENIA DURING TREATMENT OF RHEUMATOID ARTHRITIS, *ANN RHEUM DIS*, 37, 366-369 (1978)

3842 WILLIAMS M, YOUNG JB, ROSA RM, ET AL, EFFECT OF PROTEIN ON URINARY DOPAMINE EXCRETION, *J CLIN INVEST*, 78, 1687-1693 (1986)

3843 WILLIAMS SJ, BAIRD-LAMBERT JA, FARRELL GC, INHIBITION OF THE-OPHYLLINE METABOLISM BY INTERFERON, *LANCET*, 2, 939-941 (1987)

3844 WILLIAMSON JM, GIBBS WN, HYPOPLASTIC ANAEMIA AND HYDROFLUMETHIAZIDE, *SCOTT MED J*, 11, 19 (1966)

3845 WILLIAMSON T, A COMPARISON BETWEEN THE PHOSPHASTRATE AND PHENYL PHOSPHATE METHODS OF ALKALINE PHOSPHA-TASE ASSAY, *MED LAB TECHNOL*, 29, 182 (1972)

3846 WILLIS AL, JOHNSON M, RABINOWITZ I, WOLF PL, PROSTAGLANDIN F2 MAY INDUCE SICKLE CELL CRISIS, *N ENGL J MED*, 286, 783 (1972)

3847 WILLMORE LJ, WILDER BJ, BRUNI J, ET AL, EFFECT OF VALPROIC ACID ON HEPATIC FUNCTION, *NEUROLOGY*, 28, 961-964 (1978)

3848 WILLOUGHBY JS, PATON TW, WALKER SE, ET AL, THE EFFECT OF CIMETIDINE ON ENTERIC-COATED ASA DISPOSITION, *CLIN PHARMACOL THER*, 33, 268 (1983)

3849 WILLOUGHBY MLN, BAIRD GM, CAMPBELL AM, LEVAMISOLE AND NEUTROPENIA, *LANCET*, 1, 657 (1977)

3850 WILLS JH, THE MEASUREMENT AND SIGNIFICANCE OF CHANGES IN THE CHOLINESTERASE ACTIVITIES OF ERYTHROCYTES AND PLASMA IN MAN, *CRIT REV TOXICOL*, 1, 153 (1972)

3851 WILLS MR, THE EFFECT OF DIURNAL VARIATION ON TOTAL PLASMA CALCIUM CONCENTRATION IN NORMAL SUBJECTS, *J CLIN PATHOL*, 23, 772 (1970)

3852 WILNER KD, ZIEGLER MG, EFFECTS OF ALPHA$_1$,-INHIBITION ON RENAL BLOOD FLOW AND SYMPATHETIC NERVOUS ACTIVITY IN SYSTEMIC HYPERTENSION, *AM J CARDIOL*, 59, 82G-86G (1987)

3853 WILSON C, AZMY AF, BEATTIE TJ, ET AL, HYPERMAGNESEMIA AND PROGRESSION OF RENAL FAILURE ASSOCIATED WITH RENACIDIN®THERAPY, *CLIN NEPHROL*, 25, 266-267 (1986)

3854 WILSON CB, KIRKENDALL WM, THE ACUTE EFFECTS OF MEFRUSIDE IN MAN: A NEW DIURETIC COMPOUND, *PHARMACOL EXP THER*, 173, 288 (1970)

3855 WILSON FA, DIETSCHY JM, DIFFERENTIAL DIAGNOSTIC APPROACH TO CLINICAL PROBLEMS OF MALABSORPTION, *GASTROENTER-OLOGY*, 61, 911 (1971)

3856 WILSON GM, TOXICITY OF HYPOTENSIVE DRUGS, *PRACTITIONER*, 194, 51 (1965)

3857 WILSON IC, GAMBILL JM, SANDIFER MG K JR., A DOUBLE BLIND STUDY COMPARING IMIPRAMINE (TOFRANIL®) WITH DESMETHYLIMIPRAMINE (PERTOFRANE®), *PSYCHOSOMATICS*, 5, 88 (1964)

3858 WILSON JA, CRAIG IF, EFFECTS OF CIMETIDINE AND RANITIDINE ON HIGH DENSITY LIPOPROTEIN CHOLESTEROL CONCENTRA-TIONS, *BR MED J [CLIN RES]*, 290, 807-808 (1985)

3859 WILSON RG SINGHAL VK, PERRY-ROBB I, FORREST AP, COLE EN, BOYNS AR, GRIFFITHS K, RESPONSE OF PLASMA PROLACTIN AND GROWTH HORMONE TO INSULIN HYPOGLYCAEMIA, *LAN-CET*, 2, 1283 (1972)

3860 WILSON RG, HAMILTON JR, BOYD WD, ET AL, THE EFFECT OF LONG TERM PHENOTHIAZINE THERAPY ON PLASMA PROLACTIN, *BR J PSYCHIATRY*, 127, 71-74 (1975)

3861 WILSON SP, KAMIN DL, FELDMAN JM, ACETAMINOPHEN ADMINIS-TRATION INTERFERES WITH URINARY METANEPHRINE (AND CATECHOLAMINE) DETERMINATIONS, *CLIN CHEM*, 31, 1093-1094 (1985)

3862 WILSON SS, GUILLAN RA, HOCKER EV, STUDIES OF THE STABILITY OF 18 CHEMICAL CONSTITUENTS OF HUMAN SERUM, *CLIN CHEM*, 18, 1498 (1972)

3863 WILSON WL, BISEL HF, COLE D, ROCHLIN D, RAMIREZ G, MADDEN R, PROLONGED LOW DOSAGE ADMINISTRATION OF HEX-AMETHYLMELAMINE (NC 13875), *CANCER*, 25, 568 (1970)

3864 WILSON WR, COCKERIL FR III, TETRACYCLINES, CHLORAMPHENI-COL, ERYTHROMYCIN AND CLINDAMYCIN, *MAYO CLIN PROC*, 62, 906-915 (1987)

3865 WILTINK WF, KRUITHOF J, MOL C, BOS G, VAN EIJK HG, DIURNAL AND NOCTURNAL VARIATIONS OF THE SERUM IRON IN NOR-MAL SUBJECTS, *CLIN CHIM ACTA*, 49, 99 (1973)

3866 WINCHESTER JF, KELLETT RJ, BODDY K, ET AL, METOLAZONE AND BENDROFLUMETHIAZIDE IN HYPERTENSION: PHYSIOLOGIC AND METABOLIC OBSERVATIONS, *CLIN PHARMACOL THER*, 28, 611-618 (1980)

3867 WINDHORST DB, NIGRA T, GENERAL CLINICAL TOXICOLOGY OF ORAL RETINOIDS, *J AM ACAD DERMATOL*, 6, 675-682 (1982)

3868 WINEK CL, KUHLMAN JJ JR, SHANOR SP, DETECTION AND INTER-FERENCE OF SOME CENTRAL NERVOUS SYSTEM STIMULANTS IN URINE DRUG SCREENING PROCEDURES, *CLIN TOXICOL*, 17, 337-351 (1980)

3869 WINEK CL, SCHWEIGHARDT FK, FOCHTMAN FW, COLLOM WD, QUI-NINE IN URINALYSIS FOR HEROIN, *J AM MED ASSOC*, 217, 1243 (1971)

3870 WING SS, FANTUS IG, ADVERSE IMMUNOLOGIC EFFECTS OF ANTI-THYROID DRUGS, *CAN MED ASSOC J*, 136, 121-127 (1987)

3871 WINIDOFF D, ORAL CONTRACEPTIVES AND THYROID FUNCTION TESTS, THE DIAGNOSIS OF THYROID DISEASE, *MED J AUST*, 1, 1059 (1971)

3872 WINKELMAN JW, CANNON DC, PILEGGI CJ, REED AH, ESTIMATION OF NORMS FROM CONTROLLED SAMPLE SURVEY, *CLIN CHEM*, 19, 488 (1973)

3873 WINSTEN S, COLLECTION AND PRESERVATION OF SPECIMENS, *STAND METH CLIN CHEM*, 5, 1 (1965)

3874 WINSTON DJ, MURPHY W, YOUNG LS, ET AL, PIPERACILLIN THERAPY FOR SERIOUS BACTERIAL INFECTIONS, *AM J MED*, 69, 255-261 (1980)

3875 WINTER SL, BOYER JL, HEPATIC TOXICITY FROM LARGE DOSES OF VITAMIN B$_3$ (NICOTINAMIDE), *N ENGL J MED*, 289, 1180 (1973)

3876 WINTHER K, BONDESEN S, HONORE HANSEN S, ET AL, LACK OF EFFECT OF 5-AMINOSALICYLIC ACID ON PLATELET AGGREGA-TION AND FIBRINOLYTIC ACTIVITY IN VIVO AND IN VITRO, *EUR J CLIN PHARMACOL*, 33, 419-422 (1987)

3877 WINTHROP LABORATORIES, MANUFACTURER'S LITERATURE ON NEGGRAM®, *90 PARK AVENUE, NEW YORK, NY 10016*

3878 WINTROBE MM, *CLINICAL HEMATOLOGY*, LEA AND FEBIGER, PHILA-DELPHIA, PA, 6TH ED (1967)

3879 WIRTH WA, THOMPSON RL, THE EFFECT OF VARIOUS CONDITIONS AND SUBSTANCES ON THE RESULTS OF LABORATORY PROCE-DURES, *AM J CLIN PATHOL*, 43, 579 (1965)

3880 WISE JK, HENDLER R, FELIG P, THE GLYCEMIC RESPONSE TO ALA-NINE: INDEX OF GLUCAGON SECRETION IN MAN, *CLIN RES*, 20, 561 (1972)

3881 WISE PH, CLOPAMIDE AND CARBOHYDRATE METABOLISM, *MED J AUST*, 2, 795 (1969)

3882 WITHROW CD, WOODBURY DM, TRIMETHADIONE AND OTHER OXAZOLIDINEDIONES: INTERACTIONS WITH OTHER DRUGS IN, *ANTIEPILEPTIC DRUGS*, DM WOODBURY, JK PENRY, RP SCHMIDT, EDS. RAVEN PRESS, NEW YORK, NY (1972)

3883 WITZGALL H, V WERDER K, WEBER PC, MINERALOCORTICOID AND PROLACTIN RESPONSE TO THE DOPAMINE ANTAGONIST METOCLOPRAMIDE IN PATIENTS WITH PRIMARY ALDOSTERON-ISM, *J STEROID BIOCHEM*, 19, 1671-1676 (1983)

3884 WOLCOTT GJ, HACKETT TN JR., LEVODOPA AND TESTS FOR KETONURIA, *N ENGL J MED*, 2183, 1522 (1970)

3885 WOLF PL, WILLIAMS D, COPLON N, COULSON AS, LOW ASPARTATE TRANSAMINASE ACTIVITY IN SERUM OF PATIENTS UNDERGO-ING CHRONIC HEMODIALYSIS, *CLIN CHEM*, 18, 567 (1972)

3886 WOLFE LK, GORDON RD, ISLAND DP, LIDDLE GW, AN ANALYSIS OF FACTORS DETERMINING THE CIRCADIAN PATTERN OF ALDOS-TERONE EXCRETION, *J CLIN ENDOCRINOL METAB*, 26, 1261 (1966)

3887 WOLLHEIM FA, LINDSTROM CG, LIVER ABNORMALITIES IN PENICIL-LAMINE TREATED RHEUMATOID ARTHRITIS, *SCAND J RHEU-MATOL*, SUPPL 28, 100-107 (1979)

3888 WOLLMAN MR, DAVID DS, BRENNAN BL, LEWY JE, STENZEL KH, RUBIN AL, MILLER DR, THE NITROBLUE-TETRAZOLIUM TEST, *LANCET*, 2, 289 (1972)

3889 WOLPERT E, PHILLIPS SF, SUMMERSKILL WH, AMMONIA PRODUC-TION IN THE HUMAN COLON: EFFECTS OF CLEANSING NEOMY-CIN AND ACETOHYDROXAMIC ACID (AHA), *N ENGL J MED*, 283, 159 (1970)

3890 WONG L-G, REILLY EB, EFFECT OF ACETOHEXAMIDE ON SERUM CREATININE MEASUREMENT, *CLIN CHEM*, 28, 1651 (1982)

3891 WONG LC, CHOO YC, MA HK, USE OF ORAL VP 16-213 AS PRIMARY CHEMOTHERAPEUTIC AGENT IN TREATMENT OF GESTATIONAL TROPHOBLASTIC DISEASE, *AM J OBSTET GYNECOL*, 15, 924-927 (1984)

3892 WONG TC, PALAV AB, KIN BJ, TRICOMI V, CHANGES IN SERUM AND URINARY ELECTROLYTES FOLLOWING INTRA-AMNIOTIC INJEC-TION OF HYPERTONIC SALINE, *N Y STATE J MED*, 72, 564 (1972)

3893 WONG YY, LUDDEN TM, BELL RD, EFFECT OF ERYTHROMYCIN ON CARBAMAZEPINE KINETICS, *CLIN PHARMACOL THER*, 33, 460-464 (1983)

3894 WONGPAITOON V, MILLS PR, RUSSELL RI, ET AL, INTRAHEPATIC CHOLESTASIS AND CUTANEOUS BULLAE ASSOCIATED WITH GLIBENCLAMIDE THERAPY, *POSTGRAD MED J*, 57, 244-246 (1981)

3895 WOOD FC, CAHILL GF, MANNOSE UTILIZATION IN MAN, *J CLIN INVEST*, 42, 1300 (1963)

3896 WOODBURY DM, OTHER ANTIEPILEPTIC DRUGS: BROMIDES IN, *ANTIEPILEPTIC DRUGS*, DM WOODBURY, JK PENRY, RP SCHMIDT, EDS. RAVEN PRESS, NEW YORK, NY (1972)

3897 WOODS JW, AJZEN H, EFFECT OF RESERPINE AND GUANETHIDINE ON EXCRETION OF 3-METHOXY-4-HYDROXYMANDELIC ACID IN MAN, *PROC SOC EXP BIOL MED*, 114, 107 (1963)

3898 WORLD HEALTH ORGANIZATION, EXPERT COMMITTEE ON MALARIA REPORT, *WORLD HEALTH ORG TECHN REP SER*, 324, 38 (1966)

3899 WORTH HG, THE EFFECT OF SUGARS ON AUTOMATED URINARY OESTROGENS ESTIMATIONS IN PREGNANCY, *CLIN CHIM ACTA*, 49, 53 (1973)

3900 WRIGHT AD, BARBER SG, KENDALL MJ, ET AL, BETA-ADRE-NOCEPTOR-BLOCKING DRUGS AND BLOOD SUGAR CONTROL IN DIABETES MELLITUS, *BR MED J [CLIN RES]*, 1, 159-161 (1979)

3901 WRIGHT N, CLARKSON AR, BROWN SS, FUSTER V, EFFECTS OF POISONING ON SERUM ENZYME ACTIVITIES, *BR MED J [CLIN RES]*, 3, 347 (1971)

3902 WRIGHT WR, RAINWATER JC, TOLLE LD, GLUCOSE ASSAY SYS-TEMS: EVALUATION OF A COLORIMETRIC HEXOKINASE PROCE-DURE, *CLIN CHEM*, 17, 1010 (1971)

3903 WU JT, INTERFERENCE OF HEPARIN IN CARCINOEMBRYONIC ANTI-GEN RADIOIMMUNOASSAYS, *CLIN CHIM ACTA*, 130, 47-54 (1983)

3904 WYLLIE JH, BOULOS PB, LEWIN MR, STAGG BH, CLARK CG, PLASMA GASTRIN AND ACID SECRETION IN MAN FOLLOWING STIMULA-TION BY FOOD, MEAT EXTRACT AND INSULIN, *GUT*, 13, 887 (1972)

3905 WYLLIE JH, HESSELBO T, BLACK JW, EFFECTS IN MAN OF HISTA-MINE H$_2$-RECEPTOR BLOCKADE BY BURIMAMIDE, *LANCET*, 2, 1117 (1972)

3906 WYNN V, METABOLIC EFFECTS OF DANAZOL, *J INT MED RES*, 5 SUPPL 3, 25-35 (1977)

3907 XANTHOPOULOS B, KOUTRAS DA, BOUKIS MA, ET AL, THE EFFECT OF BENZIODARONE ON THE THYROID HORMONE LEVELS AND THE PITUITARY-THYROID AXIS, *J ENDOCRINOL INVEST*, 9, 337-339 (1986)

3908 YAFFE SJ, FILER LJ JR, THE USE AND ABUSE OF VITAMIN A, *PEDIATR*, 48, 655 (1971)

3909 YALOW RS, BERSON SA, IMMUNOASSAY OF ENDOGENOUS PLASMA INSULIN IN MAN, *J CLIN INVEST*, 39, 1157 (1960)

3910 YAMAGISHI T, YOSHIMOTO R, MATSUJARU F, EFFECT OF VITAMIN C ON THE TEST REAGENT TTC (TRIPHENYLTETRAZOLIUM CHLO-RIDE) FOR BACTERIURIA, *RINSHO BYORI*, 19, 489 (1971)

3911 YATES VM, KERR REI, CIMETIDINE AND THROMBOCYTOPENIA, *BR MED J [CLIN RES]*, 280, 1453 (1980)

3912 YEE HY, AUTOMATED HEXOKINASE PROCEDURE FOR ASSAYING GLUCOSE IN URINE, SERUM OR PLASMA, *CLIN CHEM*, 18, 1416 (1972)

3913 YEN SSC, EHARA Y, SILER TM, AUGMENTATION OF PROLACTIN SECRETION BY ESTROGEN IN HYPOGONADAL WOMEN, *J CLIN INVEST*, 53, 652 (1974)

3914 YEN SSC, TSAI CC, THE BIPHASIC PATTERN IN THE FEEDBACK ACTION OF ETHINYL ESTRADIOL ON THE RELEASE OF PITUI-TARY FSH AND LH, *J CLIN ENDOCRINOL METAB*, 33, 882 (1971)

3915 YEUNG CY, YU VYH, PHENOBARBITONE ENHANCEMENT OF BROM-SULPHALEIN™ CLEARANCE IN NEONATAL HYPERBILIRUBINEMIA, *PEDIATR*, 48, 556 (1971)

3916 YLIKORKALA O, KIVINEN S, RONNBERG L, ET AL, SULPIRIDE TREAT-MENT DURING EARLY HUMAN PREGNANCY: EFFECT ON THE LEVELS OF PROLACTIN, SIX STEROIDS, AND PLACENTAL LACTOGEN, *J CLIN ENDOCRINOL METAB*, 51, 155-157 (1980)

3917 YOSHIDA A, FUJITA M, KUROSAWA N, ET AL, EFFECTS OF DIL-TIAZEM ON PLASMA LEVEL AND URINARY EXCRETION OF DIGOXIN IN HEALTHY SUBJECTS, *CLIN PHARMACOL THER*, 35, 681-685 (1984)

3918 YOSHIMURA N, KODAMA K, YOSHITAKE J, CARBOHYDRATE METAB-OLISM AND INSULIN RELEASE DURING ETHER AND HALOTHANE ANESTHESIA, *BR J ANAESTH*, 43, 1022 (1971)

3919 YOSSELSON-SUPERSTINE S, SINAI Y, DRUG INTERFERENCE WITH URINE PROTEIN DETERMINATION, *J CLIN CHEM CLIN BIOCHEM*, 24, 103-106 (1986)

3920 YOUNG DS, *PERSONAL OBSERVATION*

3921 YOUNG GP, BHATHAL PS, SULLIVAN JR, ET AL, FATAL HEPATIC COMA COMPLICATING OXYMETHOLONE THERAPY IN MULTIPLE MYELOMA, *AUST N Z J MED*, 7, 47-51 (1977)

3922 YOUNG GP, SULLIVAN J, HURLEY T, LETTER: HYPOKALAEMIA DUE TO GENTAMICIN/CEPHALEXIN IN LEUKEMIA, *LANCET*, 2, 855 (1973)

3923 YOUNG MM, JASANI C, SMITH DA, NORDIN BEC, SOME EFFECTS OF ETHINYL OESTRADIOL ON CALCIUM AND PHOSPHORUS METABOLISM IN OSTEOPOROSIS, *CLIN SCI*, 34, 411 (1968)

3924 YOUNG RC, NACHMAN RL, HOROWITZ HI, THROMBOCYTOPENIA DUE TO DIGITOXIN, *AM J MED*, 41, 605 (1966)

3925 YOUNG RE, RAMSAY LE, MURRAY TS, BARBITURATES AND SERUM CALCIUM IN THE ELDERLY, *POSTGRAD MED J*, 53, 212-215 (1977)

3926 YOUNG SN, GAUTHIER S, ANDERSON GM, ET AL, TRYPTOPHAN, 5-HYDROXYINDOLEACETIC ACID AND INDOLEACETIC ACID IN HUMAN CEREBROSPINAL FLUID: INTERRELATIONSHIPS AND THE INFLUENCE OF AGE, SEX, EPILEPSY AND ANTICONVULSANT DRUGS, *J NEUROL NEUROSURG PSYCHIATRY*, 43, 438-445 (1980)

3927 YOURASSOWSKY E, DEBROE ME, WIEME RJ, EFFECT OF HEPARIN ON GENTAMICIN CONCENTRATION IN BLOOD, *CLIN CHIM ACTA*, 42, 189 (1972)

3928 YU TF, BERGER L, GUTMAN AB, HYPOGLYCEMIC AND URICOSURIC PROPERTIES OF ACETOHEXAMIDE AND HYDROXYHEXAMIDE, *METABOLISM*, 17, 309 (1968)

3929 YU TF, KAUNG C, GUTMAN AB, EFFECT OF GLYCINE LOADING ON PLASMA AND URINARY URIC ACID AND AMINOACIDS IN NORMAL AND GOUTY SUBJECTS, *AM J MED*, 49, 352 (1970)

3930 YU TF, SIROTA JM, BERGER L, EFFECT OF SODIUM LACTATE INFUSION ON URATE CLEARANCE IN MAN, *PROC SOC EXP BIOL MED*, 96, 809 (1957)

3931 YUDKIN J, SZANTO S, INCREASED LEVELS OF PLASMA INSULIN AND 11 HYDROXYCORTICOSTEROID INDUCED BY SUCROSE AND THEIR REDUCTION, *HORM METAB RES*, 4, 417 (1972)

3932 YUNGINGER JW, GLEICH GJ, SEASONAL CHANGES IN IGE ANTIBODIES AND THEIR RELATIONSHIP TO IGG ANTIBODIES DURING IMMUNOTHERAPY FOR RAGWEED, *J CLIN INVEST*, 52, 1268 (1973)

3933 ZACCARIA M, GIORDANO G, PASQUALI C, ET AL, EFFECTS OF PIRENZEPINE ON PLASMA INSULIN, GLUCAGON AND PANCREATIC POLYPEPTIDE LEVELS IN NORMAL MAN, *EUR J CLIN PHARMACOL*, 27, 701-705 (1985)

3934 ZAK B, WILLARD HH, MYERS GB, BOYLE AJ, CHLORIC ACID METHOD FOR DETERMINATION OF PROTEIN-BOUND IODINE, *ANAL CHEM*, 24, 1345 (1952)

3935 ZANTVOORT FA, DERKX FHM, BOOMSMA F, ET AL, THEOPHYLLINE AND SERUM ELECTROLYTES, *ANN INTERN MED*, 104, 134 (1986)

3936 ZAREMBSKI PM HODGKINSON A, COCHRAN M, TREATMENT OF PRIMARY HYPEROXALURIA WITH CALCIUM CARBIMIDE, *N ENGL J MED*, 277, 1000 (1967)

3937 ZARRABI MH, ZUCKER S, MILLER F, ET AL, IMMUNOLOGIC AND COAGULATION DISORDERS IN CHLORPROMAZINE-TREATED PATIENTS, *ANN INTERN MED*, 91, 194-199 (1979)

3938 ZAVAGLI G, TADDEO U, BOLELLI G, ET AL, PLATELET AGGREGATION AND EVALUATION OF THE RATIO THROMBOXANE B2/6-KETO-PROSTAGLANDIN F1 ALPHA IN THE PLASMA OF PATIENTS ON LONG-TERM CIMETIDINE TREATMENT, *PROSTAGLANDINS LEUKOT MED*, 19, 241-250 (1985)

3939 ZEBALLOS J, GALDOS B, QUINTANILLA A, PLASMA OSMOLALITY IN SUBJECT ACCLIMATISED AT HIGH ALTITUDE, *LANCET*, 1, 230 (1973)

3940 ZEEGERS JJ, MAAS AH, WILLEBRANDS AF, KRUYSWIJK HH, JAMBROES G, THE RADIOIMMUNOASSAY OF PLASMA DIGOXIN, *CLIN CHIM ACTA*, 44, 109 (1973)

3941 ZERNECHEL CF, AMINOPYRINE AND AGRANULOCYTOSIS: REVIEW AND REPORT OF SIX CASES, *N C MED J*, 28, 91 (1967)

3942 ZHIRI A, HOUOT O, WELLMAN-BEDNAWSKA M, ET AL, SIMULTANEOUS DETERMINATION OF URIC ACID AND CREATININE IN PLASMA BY REVERSED-PHASE LIQUID CHROMATOGRAPHY, *CLIN CHEM*, 31, 109-112 (1985)

3943 ZHIRI A, WELLMAN-BEDNAWSKA M, SIEST G, ELISA, 6 BETA-HYDROXYCORTISOL IN HUMAN URINE: DIURNAL VARIATIONS AND EFFECTS OF ANTIEPILEPTIC THERAPY, *CLIN CHIM ACTA*, 157, 267-276 (1986)

3944 ZIEGLER MG, CHERNOW B, WOODSON LC, ET AL, THE EFFECT OF PROPRANOLOL ON CATECHOLAMINE CLEARANCE, *CLIN PHARMACOL THER*, 40, 116-119 (1986)

3945 ZIELINSKI JJ, HAIDUKEWYCH D, LEHATA BJ, CARBAMAZEPINE-PHENYTOIN INTERACTION: ELEVATION OF PLASMA PHENYTOIN CONCENTRATIONS DUE TO CARBAMAZEPINE COMEDICATION, *THER DRUG MONIT*, 7, 51-53 (1985)

3946 ZIELINSKI JJ, LICHTEN EM, HAIDUKEWYCH D, CLINICALLY SIGNIFICANT DANAZOL-CARBAMAZEPINE INTERACTION, *THER DRUG MONIT*, 9, 24-27 (1987)

3947 ZIEVE L, MISINTERPRETATION AND ABUSE OF LABORATORY TESTS BY CLINICIANS, *ANN N Y ACAD SCI*, 134, 563 (1966)

3948 ZIMMERMANN HJ, EFFECTS OF ASPIRIN AND ACETAMINOPHEN ON THE LIVER, *ARCH INTERN MED*, 141, 333-342 (1981)

3949 ZIMMERMANN J, FAINARU M, EISENBERG S, THE EFFECTS OF PREDNISONE THERAPY ON PLASMA LIPOPROTEINS AND APOLIPOPROTEINS: A PROSPECTIVE STUDY, *METABOLISM*, 33, 521-526 (1984)

3950 ZIPES DP, PRYSTOWSKY EN, HEGER JJ, AMIODARONE: ELECTROPHYSIOLOGIC ACTIONS, PHARMACOKINETICS AND CLINICAL EFFECTS, *J AM COLL CARDIOL*, 3, 1057-1071 (1984)

3951 ZIPORIN ZZ, CHAMBERS JS, TAYLOR RR, WIER JA, THE EFFECT OF ISONIAZID ADMINISTRATION ON THE BLOOD AMMONIA IN TUBERCULOUS PATIENTS, *AM REV RESPIR DIS*, 86, 21 (1962)

3952 ZIR LM, PARKER DC, SMITH RA, ROSSMAN LG, THE RELATIONSHIP OF HUMAN GROWTH HORMONE AND PLASMA TYROSINE DURING SLEEP, *J CLIN ENDOCRINOL METAB*, 34, 1 (1972)

3953 ZIS AP, HASKETT RF, ARLAV ALBALA A, ET AL, MORPHINE INHIBITS CORTISOL AND STIMULATES PROLACTIN SECRETION IN MAN, *PSYCHONEUROENDOCRINOLOGY*, 9, 423-427 (1984)

3954 ZISHKA MK, NISHIMURA JS, EFFECT OF GLYCEROL ON LOWRY AND BIURET METHODS OF PROTEIN DETERMINATION, *ANAL BIOCHEM*, 34, 291 (1970)

3955 ZIVIN JA, SNARR JF, AN AUTOMATED COLORIMETRIC METHOD FOR THE MEASUREMENT OF 3-HYDROXYBUTYRATE CONCENTRATION, *ANAL BIOCHEM*, 52, 456 (1973)

3956 ZOFKOVA I, THE EFFECT OF LONG-TERM VERAPAMIL TREATMENT ON THE SECRETION OF CORTISOL AND ALDOSTERONE IN SUBJECTS WITH NORMAL AND HIGH BLOOD PRESSURE, *EXP CLIN ENDOCRINOL*, 85, 217-222 (1985)

3957 ZOLLNER N, HEIMSTADT P, EFFECTS OF A SINGLE ADMINISTRATION OF L-ASPARAGINASE ON SERUM LIPIDS, *NUTR METAB*, 13, 344 (1971)

3958 ZONDAG HA, VAN BOETZELAER GL, DETERMINATION OF PROTEIN IN CEREBROSPINAL FLUID: SOURCES OF ERROR IN THE LOWRY METHOD, *CLIN CHIM ACTA*, 5, 155 (1960)

3959 ZSIGMOND EK, ROBINS G, THE EFFECT OF A SERIES OF ANTI-CANCER DRUGS ON PLASMA CHOLINESTERASE ACTIVITY, *CAN J ANAESTH*, 19, 75 (1972)

3960 ZUCK TF, BERGIN JJ, PERKINS RP, ANTITHROMBIN III AND OESTROGEN CONTENT OF ORAL CONTRACEPTIVES, *LANCET*, 1, 831 (1973)

3961 ZUCK TF, BERGIN JJ, RAYMOND JM, DWYRE WR, IMPLICATIONS OF DEPRESSED ANTITHROMBIN III ACTIVITY ASSOCIATED WITH ORAL CONTRACEPTIVES, *SURG GYNECOL OBSTET*, 133, 609 (1971)

3962 ZUCKER MB, SOME EFFECTS OF DISODIUM ETHYLENEDIAMINE TETRAACETATE ON BLOOD COAGULATION, *AM J CLIN PATHOL*, 24, 39 (1954)

3963 ZUCKER P, DAUM F, COHEN MI, ASPIRIN HEPATITIS, *AM J DIS CHILD*, 129, 1433-1434 (1975)

3964 ZUMKLEY H, BERTRAM HP, PREUSSER P, ET AL, RENAL EXCRETION OF MAGNESIUM AND TRACE ELEMENTS DURING CISPLATIN TREATMENT, *CLIN NEPHROL*, 17, 254-257 (1982)

3965 ZUSMAN R, CHRISTENSEN D, FEDERMAN E, ET AL, COMPARISON OF NIFEDIPINE AND PROPRANOLOL USED IN COMBINATION WITH DIURETICS FOR THE TREATMENT OF HYPERTENSION, *AM J MED*, 82, 37-41 (1987)

3966 ZUSMAN RM, SPECTOR D, CALDWELL BV, SPEROFF L, SCHNEIDER G, MULBOW PJ, THE EFFECT OF CHRONIC SODIUM LOADING AND SODIUM RESTRICTION ON PLASMA PROSTAGLANDIN A, E AND F CONCENTRATIONS, *J CLIN INVEST*, 52, 1093 (1973)

3967 ZUSPAN FP, TALLEDO OE, CHESLEY LC, ABBOTT M, ANGIOTENSIN AND NOREPINEPHRINE INFUSIONS DURING PREGNANCY: ALTERATIONS IN PLASMA EPINEPHRINE AND NOREPINEPHRINE, *J CLIN ENDOCRINOL METAB*, 33, 929 (1971)

3968 ZWEIG MH, JACKSON A, ASCORBIC ACID INTERFERENCE IN REAGENT STRIP REACTIONS FOR ASSAY OF URINARY GLUCOSE AND HEMOGLOBIN, *CLIN CHEM*, 32, 674-677 (1986)

3969 ZWEIG MH, NICHOLS HC, INTERFERENCE WITH SPINAL FLUID PROTEIN RESULTS, *PERSONAL COMMUNICATION*